Table II

Provisional Cases of Selected Notifiable Diseases Preventable by Vaccination (United States, week ending January 3, 1998)

	H. influenzae, invasive	Hepatitis (Viral), by type		Measles (Rubeola)			Meningococcal Disease	Mumps	Pertussis	Rubella
		A	B	Indigenous	Imported[†]	Total				
Reporting Area	Cum. 1997*	Cum. 1997	Cum. 1997	Cum. 1997	Cum. 1997	Cum. 1997	Cum. 1997	Cum. 1997	Cum. 1997	Cum. 1997
UNITED STATES	1,056	27,799	8,749	78	57	135	3,117	612	5,519	161
NEW ENGLAND	65	632	153	11	8	19	205	12	990	2
Maine	5	66	6	—	1	1	18	—	11	—
N.H.	12	35	18	1	—	1	18	1	143	—
Vt.	3	15	10	—	—	—	4	—	262	—
Mass.	40	241	56	10	6	16	102	4	522	1
R.I.	3	130	18	—	—	—	21	6	17	—
Conn.	2	145	45	—	1	1	42	1	35	1
MID. ATLANTIC	147	1,917	1,315	19	8	27	325	59	409	32
Upstate N.Y.	41	373	331	2	3	5	75	13	161	5
N.Y. City	35	705	434	9	2	11	46	3	62	27
N.J.	51	287	222	3	—	3	73	7	11	—
Pa.	20	552	328	5	3	8	131	36	175	—
E.N. CENTRAL	161	2,858	916	6	3	9	461	80	544	5
Ohio	86	332	94	—	—	—	164	35	165	—
Ind.	19	322	93	—	—	—	58	14	85	—
Ill.	38	706	227	6	1	7	148	13	126	2
Mich.	15	1,342	456	—	2	2	53	15	62	—
Wis.	3	156	46	—	—	—	38	3	106	3
W.N. CENTRAL	65	2,201	472	13	5	18	235	18	599	2
Minn.	44	196	44	3	5	8	34	6	384	—
Iowa	7	492	47	1	—	1	48	10	113	—
Mo.	10	1,109	326	1	—	1	109	—	68	2
N. Dak.	—	11	5	—	—	—	2	—	2	—
S. Dak.	2	27	1	8	—	8	6	—	5	—
Nebr.	1	109	16	—	—	—	14	2	14	—
Kans.	1	257	33	—	—	—	22	—	13	—
S. ATLANTIC	174	2,154	1,302	2	15	17	558	85	438	83
Del.	—	31	6	—	—	—	5	—	1	—
Md.	58	215	195	—	2	2	42	10	125	—
D.C.	—	36	30	—	3	3	9	—	3	1
Va.	13	233	128	—	1	1	58	19	56	1
W. Va.	4	12	16	—	—	—	18	—	6	—
N.C.	21	211	265	—	2	2	97	12	118	59
S.C.	4	110	98	—	1	1	61	11	30	19
Ga.	42	657	148	—	1	1	106	10	14	—
Fla.	32	649	416	2	5	7	162	23	85	3
E.S. CENTRAL	48	637	692	—	—	—	237	28	144	—
Ky.	6	76	40	—	—	—	48	3	61	—
Tenn.	27	402	449	—	—	—	85	7	40	—
Ala.	15	90	80	—	—	—	85	9	35	—
Miss.	—	69	123	—	—	—	19	9	8	—
W.S. CENTRAL	54	5,515	1,188	3	5	8	282	75	296	4
Ark.	1	221	63	—	—	—	34	1	60	—
La.	14	239	171	—	—	—	48	16	20	—
Okla.	34	1,424	52	—	1	1	45	—	49	—
Tex.	5	3,631	902	3	4	7	155	58	167	4
MOUNTAIN	99	4,369	898	6	2	8	190	59	1,252	6
Mont.	—	72	12	—	—	—	9	—	19	—
Idaho	1	150	54	—	—	—	15	6	570	1
Wyo.	4	41	40	—	—	—	4	1	7	—
Colo.	21	407	154	—	—	—	51	3	348	—
N. Mex.	10	359	263	—	—	—	30	N	198	—
Ariz.	36	2,376	206	5	—	5	46	34	45	5
Utah	3	547	92	—	1	1	16	8	26	—
Nev.	24	417	77	1	1	2	19	7	39	—
PACIFIC	243	7,516	1,813	18	11	29	624	196	847	27
Wash.	6	673	80	1	1	2	92	21	406	5
Oreg.	35	378	109	—	—	—	126	N	10	—
Calif.	188	6,296	1,592	15	8	23	396	147	403	14
Alaska	7	34	21	—	—	—	3	4	14	—
Hawaii	7	135	11	2	2	4	7	24	14	8
Guam	—	—	3	—	—	—	1	1	—	—
P.R.	—	257	1,376	—	—	—	10	7	2	—
V.I.	—	—	—	—	—	—	—	—	—	—
Amer. Samoa	—	—	—	—	—	—	—	—	—	—
C.N.M.I.	6	1	34	1	—	1	—	4	—	—

Source: Data from *Morbidity and Mortality Weekly Report,* Vol. 46, nos. 52 and 53 (January 9, 1998).

N: Not notifiable —: no reported cases C.N.M.I.: Commonwealth of Northern Mariana Islands

*Of 242 cases among children aged < 5 years, serotype was reported for 126 and of those, 47 were type b.

[†]For imported measles, cases include only those resulting from importation from other countries.

third edition

FOUNDATIONS IN MICROBIOLOGY

Kathleen Park Talaro
Pasadena City College

Arthur Talaro

Boston Burr Ridge, IL Dubuque, IA Madison, WI New York San Francisco St. Louis
Bangkok Bogotá Caracas Lisbon London Madrid
Mexico City Milan New Delhi Seoul Singapore Sydney Taipei Toronto

WCB/McGraw-Hill

A Division of The **McGraw-Hill** *Companies*

FOUNDATIONS IN MICROBIOLOGY, THIRD EDITION

This book is printed on recycled, acid-free paper containing 10% postconsumer waste.

1 2 3 4 5 6 7 8 9 0 QPH/QPH 9 3 2 1 0 9 8

ISBN 0–697–35452–0

Vice president and editorial director: *Kevin T. Kane*
Publisher: *James M. Smith*
Developmental editors: *Terrance Stanton/Jean Sims Fornango*
Marketing manager: *Martin J. Lange*
Senior project manager: *Gloria G. Schiesl*
Senior production supervisor: *Mary E. Hass*
Freelance design coordinator: *Mary L. Christianson*
Senior photo research coordinator: *Carrie K. Burger*
Art editor: *Brenda A. Ernzen*
Supplement coordinator: *Stacy A. Patch*
Compositor: *York Graphic Services, Inc.*
Typeface: *10/12 Times*
Printer: *Quebecor Printing Book Group/Hawkins, TN*

Freelance designer: *Jamie O'Neal*
Cover photograph: *© SuperStock*
Cover: *A colorized electron microscope photograph of* Helicobacter pylori, *a curved motile rod that has established a niche for itself in the stomach of humans and other animals.*

The credits section for this book begins on page C-1 and is considered an extension of the copyright page.

Library of Congress Cataloging-in-Publication Data

Talaro, Kathleen P.
Foundations in microbiology / Kathleen Talaro, Arthur Talaro. — 3rd ed.
p. cm.
Includes index.
ISBN 0–697–35452–0
1. Microbiology. 2. Medical microbiology. I. Talaro, Arthur. II. Title.
QR41.2.T35 1999
579—dc21 98–19122
CIP

When ordering this title, use ISBN 0–07–115580–5.

www.mhhe.com

We dedicate this edition to our parents, Donald and Grace Park of Blackfoot, Idaho, and Arsenio and Florentina Talaro of Honokaa, Hawaii, and acknowledge them for the gifts they gave us. By nurture and example, they taught us the love of knowledge, the benefits of meaningful work, and the joys of creativity and curiosity. Though our fathers are no longer with us, our memories of them are still a daily source of encouragement and inspiration.

Brief Contents

● Denotes chapters (1–17) that appear in the paperback *Basic Principles* version of this text.

CONTENTS

● Denotes chapters (1–17) that appear in the paperback *Basic Principles* version of this text.

Chapter 7

ELEMENTS OF MICROBIAL NUTRITION, ECOLOGY, AND GROWTH 190

Chapter 8

MICROBIAL METABOLISM: THE CHEMICAL CROSSROADS OF LIFE 221

Chapter 9

MICROBIAL GENETICS 256

Chapter 

Chapter 

Chapter 

Drugs, Microbes, Host—The Elements of Chemotherapy 359

Chapter 

Microbe–Human Interactions: Infection and Disease 391

Chapter 

The Nature of Host Defenses 431

Introduction to Identification Techniques in Medical Microbiology 555

Chapter 18

The Cocci of Medical Importance 563

Chapter 19

The Gram-Positive Bacilli of Medical Importance 593

Chapter 20

The Gram–Negative Bacilli of Medical Importance 623

Chapter 21

Miscellaneous Bacterial Agents of Disease 649

Chapter 22

The Fungi of Medical Importance 688

Chapter 23

THE PARASITES OF MEDICAL IMPORTANCE 719

Chapter 24

INTRODUCTION TO THE VIRUSES OF MEDICAL IMPORTANCE: THE DNA VIRUSES 756

Chapter 25

THE RNA VIRUSES OF MEDICAL IMPORTANCE 782

Preface

Microbiology for the Millennium and Beyond

In the three short years since the previous edition of this book was published, events have continued to validate the importance and impact of microorganisms. As we reflect on the recent past and enter the twenty-first century, we appear to be teetering on the brink of what promises to be the "era of microbiology." The science is growing at an astounding rate, and it continues to branch out into new realms. Discoveries keep coming to light that dazzle even seasoned microbiologists. We now know that complex microbial communities living deep in the earth's crust probably outnumber those on the earth's surface and are intimately involved in the geologic processes of the earth. Who would have expected that an insidious type of protein (prion) could be transmitted as an infectious agent and cause serious problems for agriculture and food handling? We watch from the sidelines in surprise and alarm at reports that biological warfare agents such as anthrax are being amassed as instruments of death. And more than ever before, we are made aware (sometimes uncomfortably) of the microbial contents of the food we eat, the water we drink, and even the air we breathe. For the past twenty years, the central role of microorganisms has continued to dominate strides in molecular biology and genetic engineering, and it influences most of the new technologies in industry. As Dr. Samuel Kaplan, chairman of the Department of Microbiology and Molecular Genetics at the University of Texas Medical School, said:

> The science of microbiology is alive and well; it flourishes because of the intrinsic wealth of its subject matter. Its study has spawned the Human Genome Project, plant molecular biology, neurobiology, and all of the rest.

Even a casual observer would have to admit that becoming knowledgeable about microorganisms is no longer a convenience; it is a necessity, regardless of what your future holds. Studying microbiology will add immensely to your personal development, to your future career, and to your ability to make informed, perceptive decisions about the momentous new era of technology. Clearly, the more you learn about the subject of microbiology, the more you will understand about yourself and the world you live in.

Mission and Organization of the Textbook

The driving force of this textbook has always been to provide students with a solid background in the science of microbiology that prepares the student for future coursework and practical applications. As do all technical subjects, microbiology contains a wide array of facts and ideas to be incorporated into your growing store of knowledge. One of the ways to remain grounded in the subject is to concentrate on understanding concepts—important fundamental ideas or themes that explain a process or form a framework for ideas and words. Most of these concepts are laid out like links in a chain of information that lead you to the next level. As you continue to progress through the book, you can branch out into new areas, refine your knowledge, make important connections, and develop sophistication with the subject. Most chapters are structured with two levels of coverage: one that can be used to learn general concepts and one that encourages learning of specific ideas.

The order and style of our presentation are similar to those in the previous edition, but we have incorporated extensive changes. Our primary goals with this edition were to update, simplify, and improve illustrations. Although changes in written text are difficult to detect by a cursory look, every sentence has been carefully evaluated for currency, accuracy, and clarity. We have extensively updated figures and statistics and have introduced pertinent events and discoveries occurring since 1996. We have added about 10 new figures and 30 new photographs and have revised 200 figures. Although the basic design is intact, the use of color in tables and boxes has been altered to improve their readability, and the formats for chapter openers have a new look.

Despite an overwhelming amount of new information being generated every year, we have aimed to present a balanced coverage of traditional and new developments in microbiology without adding to the length of coverage.

Major Areas of Change

We have emphasized the changes in taxonomy that have resulted from ribosomal RNA analysis, and we have updated information on emerging diseases. The presentation of media has been altered, and new photographs have replaced older illustrations on staining techniques. We have added a new illustrated table on bacterial shapes and arrangements, have supplemented information on molecular techniques in identification, and have added an expanded section on the archaea. Tables and figures in the virus chapter have been revised for simplicity. Major changes in the presentation of metabolism and genetics include new and revamped figures, refinements in tables, and fine-tuning of the text to fit figures. Several examples used for biotechnology have been updated, including a box on the Human Genome Project and use of mitochondrial DNA. Improvements have been made to figures in microbial control, including a new box on antibacterial products, new material on the use of UV radiation, and updated information on drugs and drug resistance. To reflect the changing scope of immunology, we have modified coverage on cytokines and have added a flowchart to summarize host defenses. Introduction of new vaccines and vaccine development are significant additions. The coverage of specific diseases has been researched for accuracy, and certain chapters have been retitled for clarity. Chapters 18 through 25 have been supplemented with the latest information on new bacteria, viruses, and parasites. We have also moved the actinomycetes from the fungus chapter to the gram-positive bacteria chapter. The section on AIDS has been amended to include new information on virus structure, infection, drugs, and vaccines. Improvements have been made to the coverage of photosynthesis and some elements of food microbiology. We have added more questions to most chapters and a section suggesting Internet search topics.

The heart and soul of this book join to promote interest in this fascinating subject and to share some of our sense of excitement and awe for it. We hope that our involvement in the subject, our love of language, and our fun with analogies, models, and figures are so contagious that they stimulate your interest and get you caught up in "microfever."

Features of the Third Edition

Chapter Outline and Content

We have retained most elements of the original format from the previous editions. The Brief Contents provides chapter titles and order. Detailed chapter headings, with page numbers are given in the Contents and appear again at the beginning of each chapter.

Organization and Learning Tools

The master plan for chapter organization is as follows:

Opening Page

Every chapter begins with a brief opening statement or description followed by an outline of major chapter headings with page references. Framed in the center is a visual image that focuses attention on an interesting snippet related to the chapter.

Headings

The chapter is organized into sections by means of headings. Primary and secondary headings appear in large type and introduce general sections with a word or description. Lower level headings lead into subsections and help to partition general topics into smaller, more specific areas of discussion.

Illustration and Photography Program

Written text, especially that of a complex nature, is greatly enhanced by illustrations and photographs. A concrete visual image can help the reader picture abstract processes and ideas, and it can make even the most routine concept memorable. In addition to hand-tailoring figures to correspond to the text, we have thoroughly reviewed the original art program for accuracy, use of color, labeling, legends, and placement. Most of the original illustrations have been retained, many have been revised, and several have been completely reworked to increase sophistication and to reflect new information.

Multimedia-Supported Illustrations

Throughout the text the reader will find illustrations of microbiological concepts and processes that can be supplemented with full-color video, animations, or interactive screens from the new second edition of *Microbes in Motion* (0-072-29262-8), an interactive CD-ROM available from WCB/McGraw-Hill. The reader will be able to easily recognize these figures, because the relevant figure legends are preceded by a CD icon. Figure 25.12, reproduced here, is one example of such an illustration.

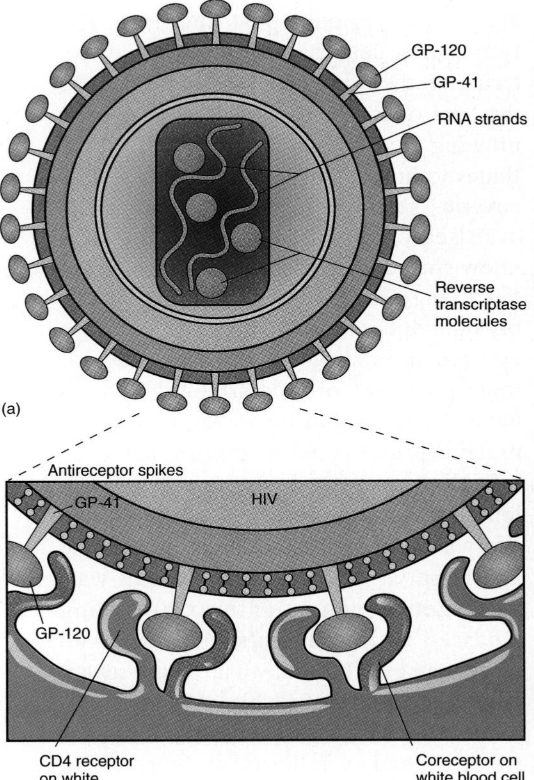

Figure 25.12
(*a*) A cutaway model of HIV. The envelope contains two types of glycoprotein (GP) spikes, two identical RNA strands, and several molecules of reverse transcriptase encased in a protein coating. (*b*) The snug attachment of HIV glycoprotein antireceptors (GP-41 and 120) to their specific receptors on a human cell membrane. These receptors are CD4 and a coreceptor called CCR-5 (fusin) that permit docking with the host cell and fusion with the cell membrane.

Correlation Guide to *Microbes in Motion*

Finding the corresponding information on the *Microbes in Motion* CD-ROM for the multimedia-supported illustrations just described requires a correlation guide, which is available to instructors. The *Microbes in Motion* CD-ROM is organized into 17 topical "books"; the books are divided up into "chapters," and the chapters have numbered "pages." For each multimedia-supported illustration, the correlation guide directs the reader to the book, chapter, and page on the CD-ROM where corresponding material can be found. The correlation guide entry is shown here for multimedia-supported illustration.

Microfiles

Every chapter contains separate boxed-off features called Microfiles. Some of them contain essential information that requires special emphasis; others provide enrichment details on newsworthy, practical, or historical topics that are related to the main text. Your instructor may assign these to you as additional reading. The Microfiles are denoted by icons in the following categories:

History

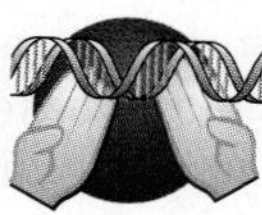
Biotechnology

Genetics

Medical Applications

Immunology

Spotlight

General Topics

Tables and Charts

The text contains about 125 tables and charts. Tables contain data, illustrations, or both, and are used to summarize additional reference and support information. Some figures contain lists of characteristics highlighted by bullets. Flowcharts, also called separation outlines, serve to overview, compare and contrast, and show relationships between various topics.

Chapter Checkpoints

Sometimes, the amount of factual information in a chapter can make it difficult to "see the forest for the trees." A beneficial strategy at such times can be to pause and review some important points before continuing on to the next topic. We have included in each chapter three to six brief summaries called Chapter Checkpoints that concisely state the most important ideas under a major heading and provide the reader with a quick recap of what has been covered to that point. Many instructors assign these as a guide for study and review.

Vocabulary

The study of microbiology will immerse the student in a rich new language. No one expects the beginner to learn all the new terms, but some ability to understand, speak, and write this new language will certainly be essential. To assist you in building vocabulary, we have highlighted principal terms in boldface or italics and have defined them at their first appearance

Chapter Checkpoints

Magnification, resolving power, lens quality, and illumination source all influence the clarity of specimens viewed through the optical microscope.

The maximum resolving power of the optical microscope is 200 nm, or 0.2 μm. This is sufficient to see the internal structures of eucaryotes and the morphology of most bacteria.

There are five types of optical microscopes. Four types use visible light for illumination: bright-field, dark-field, phase-contrast, and interference microscopes. The fifth type, the fluorescence microscope, uses UV light for illumination, but it has the same resolving power as the other optical microscopes.

Electron microscopes (EM) use electrons, not light waves, as an illumination source to provide high magnification (5,000×–1,000,000×) and high resolution (0.5 nm). Electron microscopes can visualize cell ultrastructure (TEM) and three-dimensional images of cell and virus surface features (SEM).

Specimens viewed through optical microscopes can be either alive or dead, depending on the type of specimen preparation, but all EM specimens are dead because they must be treated with metals for effective viewing.

Stains are important diagnostic tools in microbiology because they can be designed to differentiate cell shape, structure, and biochemical composition of the specimens being viewed.

in the text. Terms marked by an asterisk or by a numbered footnote also have pronunciation and word origins given. As a rule, speaking a word will help you to spell it and learn its meaning. Because many scientific terms are derived from Latin and Greek, a grounding in these word roots will help you understand the meanings of other related words. A guide to phonetic pronunciation with examples as used in this text follows:

GENERAL GUIDE TO PRONUNCIATION AS USED IN THE TEXTBOOK

Letter	Pronounced	As In	Example in Text
A	ah	fat	bacteria (1st a)
	ay	day	lipase
	uh	along	schistosoma, bacteria (2nd a)
AE	ee		archaebacteria, salmonellae
C	see	cease	cellulase
c	kay	cat	anaerobic, procaryote
cc	kay, *then* see		vaccine, cocci
Ch	kay	chorus	chemotherapy, spirochete
	ch	chest	chickenpox
	sh	chevrolet	chancre
E	ee	discrete	nucleus, gangrene
	eh	bet	cestode, mesosome
Er	ur	finger	thermophile
Eu	oo	leukemia	*Pseudomonas,* pneumonia
	yoo		eubacteria
G	j	geologic	appendage
	guh	growth	glycolysis
I	eye	bite	halophile, spirochete, lipase
	ih	bit	facultative, aerobic
	ee	nutrient	*Mycobacterium*
O	aw	body	optical
	oh	cold	saprotroph
OY	oh-ee	oil	ameboid
oe	ee		*Entamoeba*
oo	oo	mood	zoonosis
ox	ocks	pox	toxemia
Ph	eff	phase	staphylococci
Ps	sick	Psalm	*Pseudomonas*
S	ess	sense	spirochete
	zee	pleasure	*Blastomyces,* plasmid
tion	shun	mention	sporulation
ture	chur	furniture	denature
U	yoo	uniform	papule
	uh	cup	coccus
UR	yur	cure	purine
Y	y	why	mycosis
	ee	wary	leprosy
	ih	synthetic	Chlamydiales

Chapter Capsules with Key Terms

The major content of each chapter is condensed into short summaries, called Chapter Capsules, in the form of a free-flowing outline. Key terms are placed in the outline to keep them in context with their associated topics. Capsules can be used as both a quick review and an overview of the chapter.

Question Section

Each chapter concludes with an extensive question section intended to guide and supplement your study and self-testing. The number and types of questions are diverse to allow your instructor to assign questions for desired focus and emphasis. Questions with several parts are differentiated by letters so that all or part of a question may be selected. Because of space constraints, this book contains answers only to multiple-choice questions (in Appendix E). A guide for answers to all other questions is available to instructors.

Multiple-Choice (MC) Questions This type of objective question is commonly used in class testing and standardized exams and is a quick way to assess your grasp of chapter content. These questions are compiled randomly from selected vocabulary and factual information. It is assumed that if you answer these accurately, you have a good understanding of other information in the chapter. In general, MC questions have only one correct answer, and this answer may be surmised by a process of eliminating the incorrect answers and narrowing the choices down to one.

Matching Questions Many chapters contain questions that list words and related descriptions. In single-matching questions, only one description matches a given word; in multiple-matching questions, several can match and you should choose all possible descriptions that fit the words in the list.

Concept Questions Education in a subject is not really complete until you can demonstrate understanding by writing in depth about the subject. The concept questions guide your review of the chapter by asking you to compose complete answers that cover essential ideas and correct terminology. This form of expression will require determining what is most important and rewriting it in your own words in concise, logical terms. Writing is an excellent way to learn vocabulary and usage, and working with the concepts will also help you commit them to memory. All information and terminology to answer these questions can be found in the main body of the chapter and in figures, tables, and Microfile readings. Most of these questions will require several sentences or paragraphs to answer fully.

Critical-Thinking Questions
Thought questions challenge you to use scientific thinking, analysis, and problem solving. They assume a substantial grasp of factual information from the chapter in which they appear and earlier chapters, and they require you to reason, find relationships, suggest plausible explanations, and apply concepts to real-world situations. Questions may involve case studies, research findings, and everyday household settings. Some assign demonstrations, models, or ideas for class discussion; some even call for an opinion. By their nature, most of these questions allow more than one interpretation and do not have a predetermined "correct" answer.

Internet Search Topics
One of the most significant and far-reaching sources of information in a rapidly advancing science such as microbiology is the Internet. There are thousands of Web sites covering millions of topics related to this subject. Rather than suggest specific Web sites, we prefer to direct your searches using a search engine such as Yahoo, Webcrawler, or AltaVista. Using a search engine will help you become acquainted with the wide array of coverage available, and you will not encounter inactive Web sites and dead ends. Most chapters suggest one or two topics or terminology to search, but nearly any subject or concept in the book can be accessed by simple word search. We strongly encourage you to use this fantastic and exciting resource!

Appendices
The appendices in this edition cover the following topics:

- Appendix A: Greek letters and units and prefixes of measurement (new)
- Appendix B: exponents and logarithms
- Appendix C: methods for testing sterilization and disinfection
- Appendix D: universal blood and body-fluid precautions and biosafety levels
- Appendix E: answers to multiple-choice questions
- Appendix F: a cross-reference for the major bacterial, fungal, parasitic, and viral agents and diseases, by system affected and mode of transmission

Glossary and Index
The glossary includes about 675 entries that define most of the boldfaced and italicized terms used in the text. Do not overlook the value of the index, which is extensive enough to include the vast majority of major and minor terms used in the text and can be an excellent tool for finding information.

Endpapers
The inside covers of the book depict printouts from the Centers for Disease Control and Prevention on the latest summary data for major reportable infectious diseases in the United States, listed by state and region. Summary data from 1998 were not available at press time, so these data from 1997 are the most recent information available. The back endpapers (hardcover edition only) present a summary table of disease incidence in the United States from 1989 to 1996 and other statistics on diseases such as AIDS.

Study Tips for the Beginning Microbiology Student

Most of you are taking this course in preparation for a career in allied health or some area of the biological sciences. You will probably already have had some experience in studying science. By their nature of being information-intensive, these courses require a significant input of time. They involve a certain amount of pure memorization, of, at the very least, terminology, and they require consistent study habits. The rewards for your commitment will be the gifts of knowledge and awareness that are crucial to your careers and lives. No one will ever be able to take those gifts away from you, and you will probably use them in some way every single day.

Teachers know that for many students the bottom line is their grade, but there are simply no shortcuts to a good grade. It is one of life's little truths that the more time you spend in serious study, the more you will learn, and an improvement in your grade will automatically follow. Just make sure the time you spend is used to maximum benefit. Many students highlight key portions in a chapter as they read, but such passive activity may exercise your hand more than your mind, and it can use up valuable time and energy.

You will retain far more information if you engage your mind in active thought about the words and ideas. Write marginal notes to yourself; question yourself on understanding; and highlight and reread only the most significant points. To add some fun to the process and speed it along, try making up mnemonic devices: little slogans that help you recall a series of ideas. One of our students developed mental pictures that helped her visualize processes. For example, she thought of DNA as the king in a castle, sending out messengers to do his bidding, or the TCA cycle as a weird Ferris wheel that takes on passengers at the top and lets them off at the bottom.

Another active learning process is to write questions and answers on index cards to quiz yourself periodically. Repetition is the secret to recalling information. Spending an hour every day with flash cards is a far more effective way of learning than trying to absorb three chapters of material in a single marathon. Biologists studying brain chemistry have recently determined the reason for that oft-repeated warning "Do not cram because it doesn't work!" It turns out that a certain receptor in the part of the brain that regulates memory must be regenerated about every 30 minutes. Any studying done when the receptors are exhausted will not be placed into memory. This means that you really should study in short bursts with frequent breaks. Even if you have to study over a longer stretch, try to relax for a few moments, take a walk, or involve yourself in something that does not require intense thought. Other worthwhile endeavors include studying with other students or a tutor in small groups, working with the study guide or CD-ROM, and surfing the Internet to find Web sites that have study guides or sample exams. And don't forget to ask your instructor or teaching assistant for help if you are having a difficult time. They have a serious interest in your learning and success.

The subject matter in this text is fundamental, but it is not merely a review of information you have had before. Even if you have already completed a biology and chemistry course, much of the material it contains will be new to you. Microbiology is, after all, a specialized area of biology with its own orientation and emphasis. There is more information presented here than can be covered in a single course, so be guided by your instructor's reading assignments and study guide, because he or she knows what is most important for your course.

Supplemental Materials for Students and Instructors

Learning Aids for Students

A Student Study Guide

Prepared by Jackie Butler of Grayson County Community College, the study guide contains study objectives, activities, and test-taking strategies, and a newly expanded section of questions. Answers to the objective questions are included.

A Multimedia Tie-in to Foundations in Microbiology

An interactive CD-ROM, *Microbes in Motion II,* was developed especially with this textbook. It contains full-colored animations and video programs to supplement the illustration program. Figures that are tied to the CD-ROM program are indicated by a small disc icon. This new version is Mac and Windows compatible.

Web Site

The publisher is launching a Web site for this text with links to more than 100 related microbiology sites. You can find it at http://www.mhhe.com.

Teaching Supplements

Instructor's Manual with Test Item File

Developed by Louis Giacinti, Milwaukee Area Technical College, the instructor's manual is available as both a softbound manual and a computer diskette and contains an extended lecture outline with the complete headings of each chapter down to the fourth-level heads and Microfiles. Instructors can use this to organize their coverage and lectures. It also has a test item file of true/false, multiple-choice, matching, and short-answer questions to be used in compiling exams. A key to all questions is included.

Overhead Transparencies

A set of 200 color transparencies of the most requested figures are offered to all instructors who adopt the text. A Visual Resource Library CD-ROM with key words contains these images plus 200 more. This valuable product can be used for developing lectures and power point presentations.

Microtest Service

A computerized test file containing about 1,000 questions is available to all adopters. The program is compatible with Macintosh and Windows systems.

Acknowledgments

This third edition has been guided by an entirely new and dedicated team from the McGraw-Hill and Wm. C. Brown Publishers merger. Each member has provided valuable support and suggestions for the revision effort. We appreciate the leadership and insights of Ron Worthington, our editor. We are especially grateful for the important input of our developmental editor, Terry Stanton, who was always there to offer calm and patient understanding and advice when the crunch of teaching and authoring overwhelmed us. We also enjoyed collaborating with the able production team, including Gloria Schiesl and Ann Morgan. Other hardworking group members with valuable contributions are art editor Brenda Ernzen and photo editor Carrie Burger. We discovered two true gems during the revision process. One is our copy editor, Margo Quinto, who helped to simplify language and added an elegant, experienced touch to the text. The other is our photo researcher, Connie Mueller, a delightful lady who doggedly and good-naturedly pursued just the right photograph to the ends of the earth, if necessary!

Significant support and feedback have come from our colleagues and students at Pasadena City College, especially Barry Chess, whose keen eye caught several errors and who graciously agreed to proofread a number of new illustrations and made meaningful suggestions for improvements. A heartfelt commendation also goes to a gifted student Garri Tsibel, who read the entire textbook, made detailed marginal notes, and demonstrated a profound love for the subject. The many other fine students of Pasadena City College have been a wonderful source of ideas and inspiration, and they continue to teach us something new every semester.

We have solicited suggestions for changes and improvements from textbook users and student readers and have been able to include most if not all of the ideas submitted. We also recognize the contributions of our team of reviewers. Their critical reading and suggestions for additions and deletions have helped us to keep the appropriate focus and perspective during revision. We owe a debt of gratitude for their able stewardship. Thank you all for being part of this continuing book partnership.

Textbooks go through an extraordinary proofreading process. They are pored over numerous times by authors, reviewers, editors, and proofreaders. Even after all of that attention, it is impossible to produce a work completely free of errors. Please feel free to contact us with any errors you find or other constructive comments for changes we can make in future editions. The e-mail address is ktalaro@aol.com.

Reviewers, First Edition

Shirley M. Bishel
Rio Hondo College

Dale DesLauriers
Chaffey College

Warren R. Erhardt
Daytona Beach Community College

Louis Giacinti
Milwaukee Area Technical College

John Lennox
Penn State, Altoona Campus

Glendon R. Miller
Wichita State University

Joel Ostroff
Brevard Community College

Nancy D. Rapoport
Springfield Technical Community College

Mary Lee Richeson
Indiana University, Purdue University at Fort Wayne

Donald H. Roush
University of North Alabama

Pat Starr
Mt. Hood Community College

Pamela Tabery
Northampton Community College

Reviewers, Second/Third Editions

Rodney P. Anderson
Ohio Northern University

Robert W. Bauman, Jr., Ph.D.
Amarillo College

Leon Benefield
Abraham Baldwin Agricultural College

Lois M. Bergquist
Los Angeles Valley College

L. I. Best
Palm Beach Community College–Central Campus

Bruce Bleakley
South Dakota State University

Kathleen A. Bobbitt
Wagner College

Jackie Butler
Grayson County College

R. David Bynum
SUNY at Stony Brook

David Campbell
St. Louis Community College–Meramec

Joan S. Carter
Durham Technical Community College

Barry Chess
Pasadena City College

John C. Clausz
Carroll College

Margaret Elaine Cox, Ph.D.
Bossier Parish Community College

Kimberlee K. Crum
Mesabi Community College

Paul A. DeLange
Kettering College of Medical Arts

Michael W. Dennis
Montana State University–Billings

William G. Dolak
Rock Valley College

Bob F. Drake
State Technical Institute at Memphis

Mark F. Frana
Salisbury State University

Elizabeth B. Gargus
Jefferson State Community College

Larry Guillou, Ph.D.
Armstrong State College

Safawo Gullo
Abraham Baldwin Agricultural College

Christine Hagelin
Los Medanos College

Geraldine C. Hall
Elmira College

Heather L. Hall
Charles County Community College

Theresa Hornstein
Lake Superior College

Anne C. Jayne
University of San Francisco

Patricia Hilliard Johnson
Palm Beach Community College

Patricia Klopfenstein
Edison Community College

Jacob W. Lam
University of Massachusetts Lowell

James W. Lamb
El Paso Community College

Hubert Ling
County College of Morris

Andrew D. Lloyd
Delaware State University

Marlene McCall
Community College of Allegheny County

Joan H. McCune
Idaho State University

Gordon A. McFeters, Ph.D.
Montana State University

Karen Mock
Yavapai College

Jacquelyn Murray
Garden City Community College

Robert A. Pollack
Nassau Community College

Judith A. Prask
Montgomery College

Leda Raptis
Queen's University

Carol Ann Rush
La Roche College

Andrew M. Scala, Ph.D.
Dutchess Community College

Caren Shapiro
D'Youville College

Linda M. Sherwood, Ph.D.
Montana State University

Lisa A. Shimeld
Crafton Hills College

Cynthia V. Sommer
University of Wisconsin–Milwaukee

Donald P. Stahly
University of Iowa

Terrence Trivett
Pacific Union College

Garri Tsibel
Pasadena City College

Leslie S. Uhazy
Antelope Valley College

Valerie Vander Vliet
Lewis University

Frank V. Veselovsky
South Puget Sound Community College

Katherine Whelchel
Anoka-Ramsey Community College

Vernon L. Wranosky
Colby Community College

Dorothy M. Wrigley
Mankato State University

chapter 1

The Main Themes of Microbiology

These are exciting times in the history of microbiology. We receive nearly daily reminders about the importance of microscopic creatures in our lives. Newspapers and magazines report on newly emerging diseases and keep us abreast of new methods of diagnosis and treatments. This negative view of microbes is balanced with constant reminders of the numerous beneficial roles that microbes play in our everyday lives and in the function of the earth itself. These paradoxical roles are a never-ending source of fascination.

The world continues to experience an increase in new and emerging infectious agents such as the human immunodeficiency virus (HIV) and other retroviruses, and several older diseases such as tuberculosis regularly appear on the increase or continue to wreak their devastating effects (microfile 1.1). Microbiologists turn up intriguing connections between microorganisms and diseases whose causes have been unknown. Surprising correlations exist between type I diabetes and coxsackievirus infection, between schizophrenia and a virus called the borna agent, and between coronary artery disease and a common herpesvirus. These intriguing findings are the focus of serious ongoing research studies. In addition to their invasion of living organisms, microbes are also constantly invading and even thriving in new habitats created by commercial materials including computer chips, jet fuel, paints, concrete, metal, plastic, and paper.

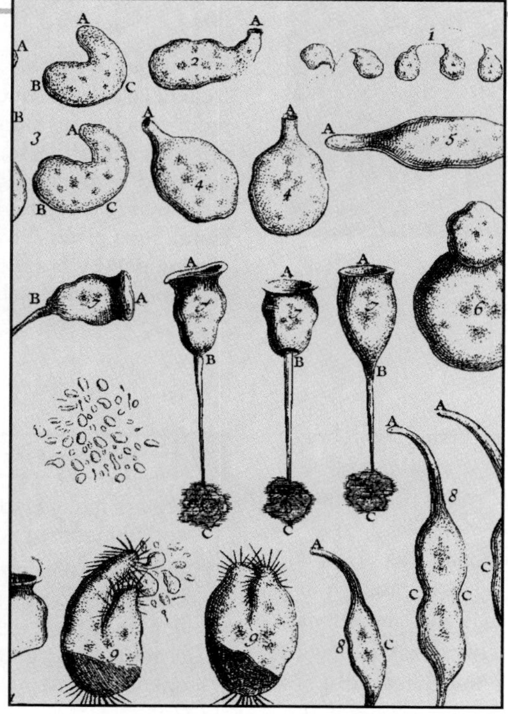

On the plus side, microorganisms contribute to our lives in many ways. They are often called upon to solve various environmental, agricultural, and medical problems. Consider, for instance, the science of bioremediation, defined as the introduction of microbes to restore stability to disturbed or polluted environments (figure 1.1*a*). Increasingly, authorities are using microorganisms to clean up oil slicks or to remove pollutants from

One of the earliest glimpses into the microbial world emerged from the microscopes of the French naturalist Louis Jablot. He made these intricate drawings of protozoa growing in straw infusions in 1711. Despite the fanciful names he gave them, such as "little fish," "caterpillars," and "swans," the accuracy of his observations makes it possible for microbiologists to readily identify them even today.

(a)

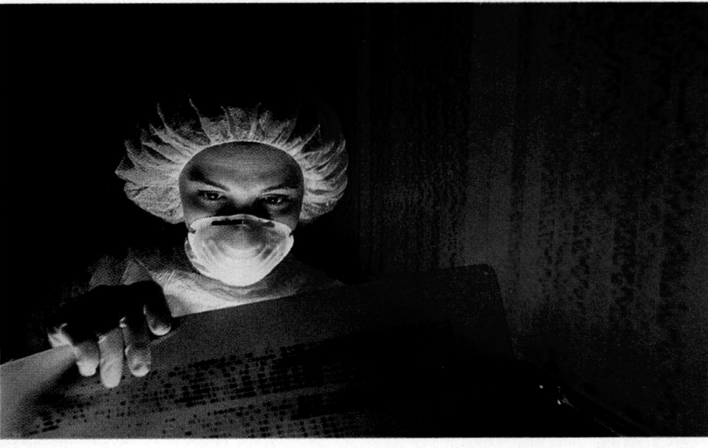

(b)

Figure 1.1
Beneficial roles of microorganisms. (*a*) Polluted water and sediments can be bioremediated by placing submersible platforms in rivers and other waterways. Here, metal cylinders mix bacteria and mud with air to speed up the degradation of polychlorinated biphenyls (PCBs), highly toxic industrial chemicals. (*b*) A technician handles gel plates showing sequences of microbial DNA. She is surrounded by hundreds of these strips, just a tiny fraction of those required to completely unravel the genetics of just one bacterium.

lakes and water supplies. We rely on naturally occurring microbes to degrade the "biodegradable" materials poured into landfills as well as to clean up sewage, extract minerals, and generally "recycle" nutrients in all ecosystems on the earth.

The search for "friendly" microbes is a serious focus of genetic engineering, a field that manipulates genetic material to produce new types of biological materials, microbes, and even plants and animals. Some of the greatest hopes of humankind—a vaccine for AIDS (figure 1.1*b*), self-fertilizing plants, miracle drugs, cures for genetic diseases, elimination of pollution, and solutions to world hunger—seem within the grasp of this "new microbiology."

Clearly, microorganisms pervade our lives in both an everyday, mundane sense and in a far wider view. We wash our clothes with detergents containing microbe-produced enzymes, eat food that derives flavor from microbial action, and, in many cases, even eat microorganisms themselves. We are vaccinated with altered microbes to prevent diseases (that are caused by those very same microbes!); we treat various medical conditions with drugs produced by microbes; we dust our plants with insecticides of microbial origin; and we use microorganisms as tiny factories to churn out various industrial chemicals and plastics.

No one can emerge from a microbiology course without a changed view of the world and of themselves. There is no denying that humans are greatly affected by microbes that act as **pathogens,*** but we also depend upon microbes for many facets of life—one might say even for life itself.

Chapter Checkpoints

There are many kinds of relationships between microorganisms and humans; most are beneficial, but some are harmful.

In the last 110 years, microbiologists have identified the causative agents for most of the infectious diseases. In addition, they have discovered distinct connections between microorganisms and diseases whose causes were previously unknown.

Microorganisms: We have to learn to live with them because we cannot live without them.

THE SCOPE OF MICROBIOLOGY

Microbiology is a specialized area of **biology*** that deals with living things ordinarily too small to be seen without magnification. Such **microscopic*** organisms are collectively referred to as **microorganisms,*** **microbes,*** or several other terms, depending upon the purpose. Some people call them germs or bugs in refer-

*pathogens (patho'-oh-jenz) Gr. *pathos*, disease, and *gennan*, to produce. Disease-causing agents.

*biology Gr. *bios*, life, and *logos*, to study. The study of organisms.

*microscopic (my''-kroh-skaw'-pik) Gr. *mikros*, small, and *scopein*, to see.

*microorganism (my-kroh''-or'-gun-izm)

*microbe (my'-krohb) Gr. *mikros*, small, and *bios*, life.

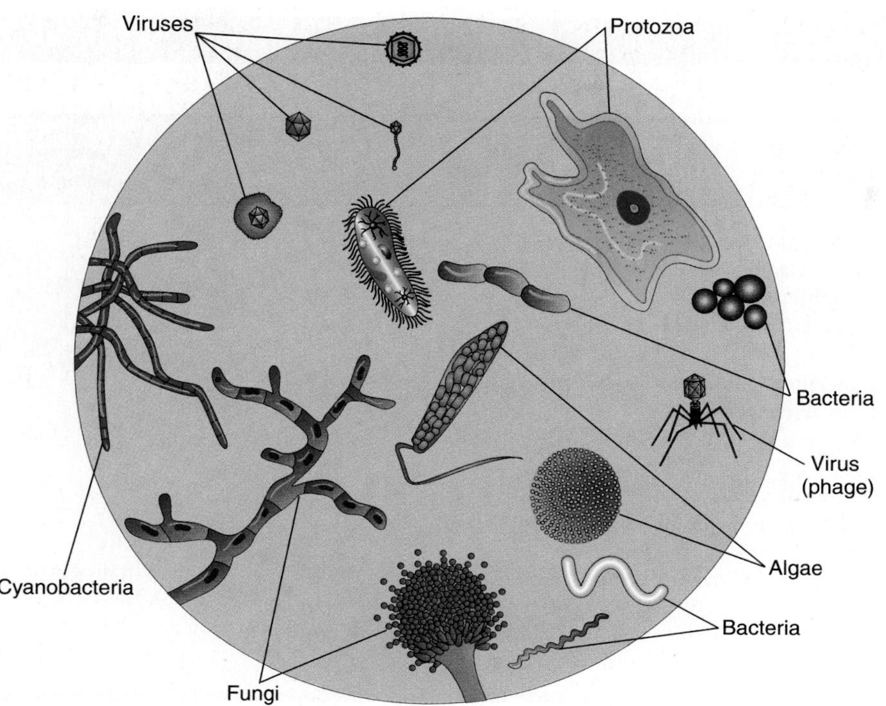

Figure 1.2
The Diversity of the Microbial World
(Not shown to scale)

ence to their role in infection and disease, but those terms have other biological meanings and perhaps place undue emphasis on the disagreeable reputation of microorganisms. Other terms that are encountered in our study are **bacteria, viruses, fungi, protozoa, algae,** and **helminths;** these microorganisms are the major biological groups that microbiologists study (figure 1.2). The very nature of microorganisms makes them ideal subjects for study. They often are more accessible than **macroscopic*** organisms because of their relative simplicity, rapid reproduction, and adaptability, which is the capacity of a living thing to change its structure or function in order to adjust to its environment.

Microbiology is one of the largest and most complex of the biological sciences because it deals with many diverse biological disciplines. In addition to studying the natural history of microbes, it also deals with every aspect of microbe-human and microbe-environmental interactions. These interactions include genetics, metabolism, infection, disease, drug therapy, immunology, genetic engineering, industry, agriculture, and ecology. The subordinate branches that come under the large and expanding umbrella of microbiology are presented in table 1.1.

Microbiology has numerous practical uses in industry and medicine. Some prominent areas that are heavily based on applications in microbiology are as follows:

Immunology studies the system of body defenses that protects against infection. It includes *serology,* a discipline that looks for the products of immune reactions in the blood and tissues and aids in diagnosis of infectious diseases by that means, and *allergy,* the study of hypersensitive responses to ordinary, harmless materials (see chapters 14, 15, 16, and 17).

Public health microbiology and **epidemiology** aim to monitor and control the spread of diseases in communities. The principal U.S. and global institutions involved in this concern are the United States Public Health Service (USPHS) with its main agency, the Centers for Disease Control and Prevention (CDC) located in Atlanta, Georgia, and the World Health Organization (WHO), the medical limb of the United Nations (see chapter 11). The CDC collects information on disease from around the United States and publishes it in a weekly newsletter called the *Morbidity and Mortality Weekly Report* (see data in end papers).

Food microbiology, dairy microbiology, and **aquatic microbiology** examine the ecological and practical roles of microbes in food and water (see chapter 22).

Agricultural microbiology is concerned with the relationships between microbes and crops, with an emphasis on improving yields and combating plant diseases.

Biotechnology includes any process in which humans use the metabolism of living things to arrive at a desired product, ranging from bread making to gene therapy (see chapters 10 and 26).

Industrial microbiology is concerned with the uses of microbes to produce or harvest large quantities of substances such as beer, vitamins, amino acids, drugs, and enzymes (see chapters 7 and 22).

Genetic engineering and **recombinant DNA technology** involve techniques that deliberately alter the genetic makeup of organisms to mass produce human hormones and other drugs, create totally novel substances, and develop organisms with unique methods of synthesis and adaptation. This is the most powerful and rapidly growing area in modern microbiology (see chapter 10).

*macroscopic (mak''-roh-skaw'-pik) Gr. *macros,* large, and *scopein,* to see. Visible with the naked eye.

TABLE 1.1

BRANCHES OF MICROBIOLOGY

Science	Area of Study	Chapter Reference
Bacteriology	The bacteria—the smallest, simplest single-celled organisms	4
Mycology	The fungi, a group of organisms that includes both microscopic forms (molds and yeasts) and larger forms (mushrooms, puffballs)	5, 22
Protozoology	The protozoa—animal-like and mostly single-celled organisms	5, 23
Virology	Viruses—minute, noncellular particles that parasitize living things	6, 24, 25
Parasitology	Parasitism and parasitic organisms—traditionally including pathogenic protozoa, helminth worms, and certain insects	5, 23
Phycology or algology	Simple aquatic organisms called algae, ranging from single-celled forms to large seaweeds	5
Microbial morphology	The detailed structure of microorganisms	4, 5, 6
Microbial physiology	Microbial function (metabolism) at the cellular and molecular levels	7, 8
Microbial taxonomy	Classification, naming, and identification of microorganisms	1, 4, 5
Microbial genetics, molecular biology	The function of genetic material and the biochemical reactions of cells involved in metabolism and growth	9, 10
Microbial ecology	Interrelationships between microbes and the environment; the roles of microorganisms in the nutrient cycles of soil, water, and other natural communities	7, 26

Each of the major disciplines in microbiology contains numerous subdivisions or specialties that in turn deal with a specific subject area or field. In fact, many areas of this science have become so specialized that it is not uncommon for a microbiologist to spend his or her whole life concentrating on a single group or type of microbe, biochemical process, or disease. On the other hand, rarely is one person a single type of microbiologist, and most can be classified in several ways. There are, for instance, bacterial physiologists who study industrial processes, molecular biologists who focus on the genetics of viruses, fungal taxonomists interested in agricultural pests, epidemiologists who are also nurses, and dentists who specialize in the microbiology of gum disease.

Studies in microbiology have led to greater understanding of many theoretical biological principles. For example, the study of microorganisms established universal concepts concerning the chemistry of life (see chapters 2 and 8), systems of inheritance (see chapter 9), and the global cycles of nutrients, minerals, and gases (see chapter 26). Microbiology is often *serendipitous,* meaning that basic research discoveries may later, quite by accident, lead to some new drug, therapy, food, or industrial process. For example, penicillin was discovered purely by a quirk of fate (see microfile 12.1).

INFECTIOUS DISEASES AND THE HUMAN CONDITION

Infectious diseases still devastate human populations worldwide despite wonderful strides made in understanding and treating them (see microfile 1.1). The World Health Organization (WHO), a medical agency of the United Nations, estimates that more than 17 million people die each year worldwide from preventable **infectious diseases** defined as the disruption to the body caused by the action of microbes or their products (figures 1.3 and 1.4). Many of these diseases occur in developing countries where medical care is unavailable or substandard. Adding to this toll are a growing list of emerging diseases discussed in microfile 1.1. To complicate matters even further, increased numbers of patients with severe compromised immune defenses due to AIDS and cancer are being kept alive for extended periods. These individuals are subject to infections from common environmental microorganisms that do not affect healthy people. Even with miracle medical technology, microbes still appear to have "the last word," to quote Louis Pasteur.

Chapter Checkpoints

Microorganisms are defined as "living organisms too small to be seen with the naked eye" with three exceptions: viruses, which are not really alive, but are microscopic; helminth worms, which are not always microscopic; and certain insects, which, although visible, carry microscopic agents of disease.

The scope of microbiology is incredibly diverse. It includes basic microbial research, research on infectious diseases, study of prevention and treatment of disease, environmental functions of microorganisms, and industrial use of microorganisms for commercial, agricultural, and medical purposes.

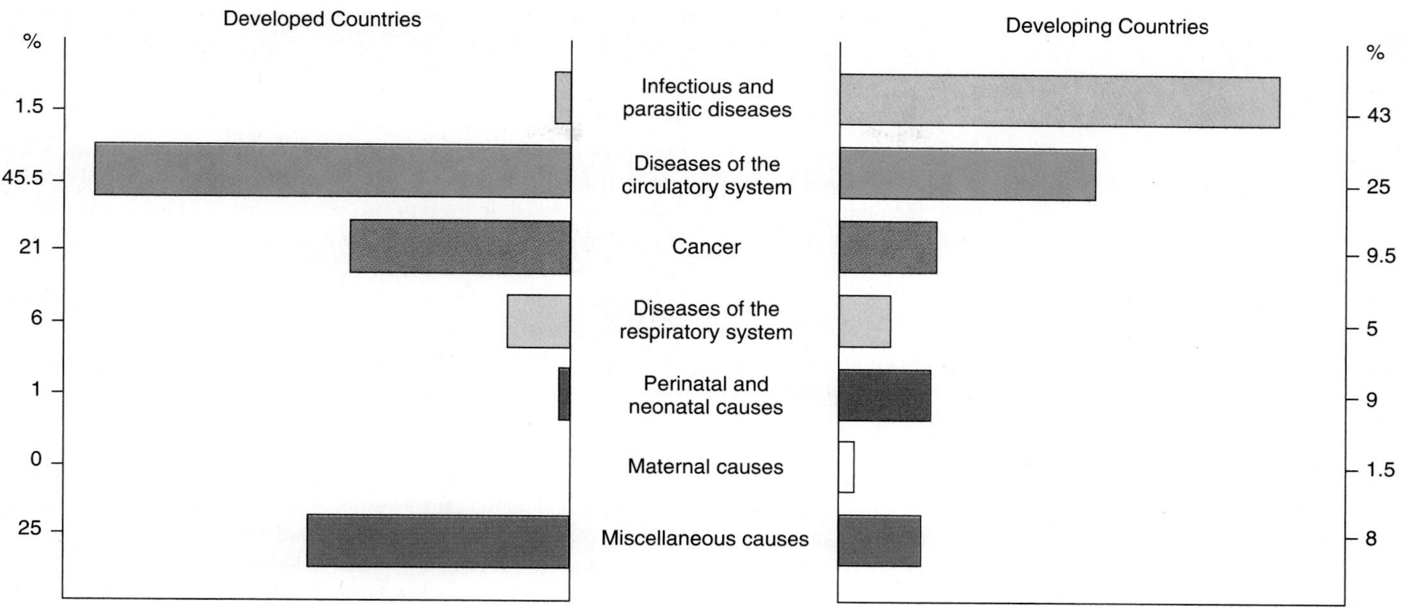

Figure 1.3

Graphs contrast the seven overall causes of death worldwide in developed, industrialized countries and in developing countries. Although mortality from infectious diseases has waned in countries such as the United States, it still has a major impact in countries with less development and access to medical care. Miscellaneous causes of death include accidents, suicides, war, and malnutrition.

Source: Data from World Health Organization 1996 reports.

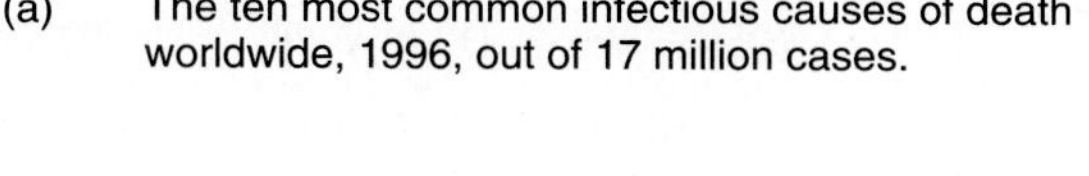

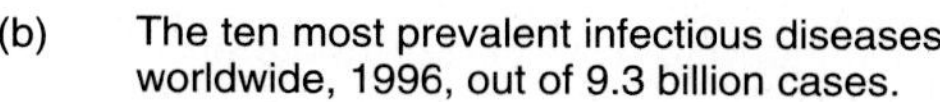

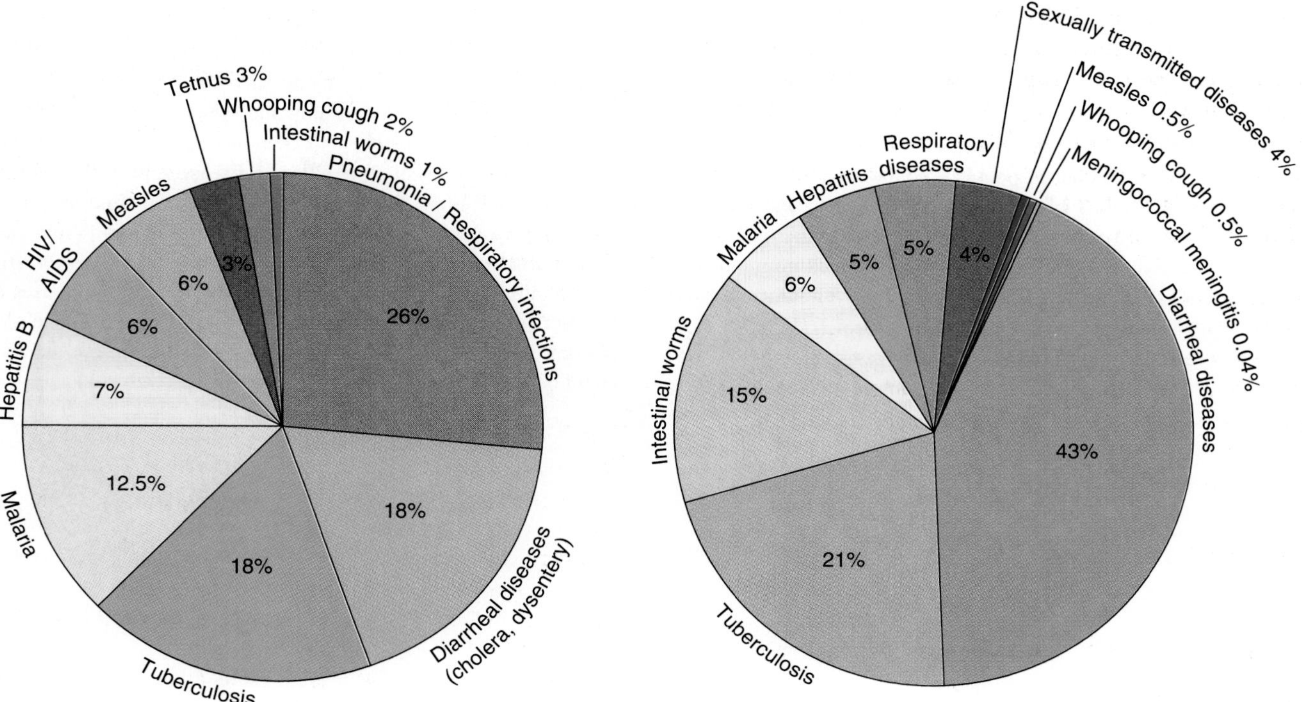

Figure 1.4

Infectious disease statistics rank the major causes of (*a*) mortality and (*b*) morbidity (rate of disease) on a global scale. A large number of diseases can be treated with drugs or prevented altogether with vaccination and improvements in health care and sanitation.

Source: World Health Organization, 1996; most recent data available.

MICROFILE 1.1 INFECTIOUS DISEASES IN THE GLOBAL VILLAGE

Eradicating infectious diseases and arresting their spread have long been goals of medical science. There is no doubt that advances in detection, treatment, and prevention have gradually reduced the numbers of such diseases, but most of these decreases have occurred in more developed countries (see figure 1.3). In less developed countries, however, infections still account for over 40% of deaths. In both groups, infectious diseases are the most common cause of illness, with an estimated 9.3 billion infections a year, amounting to nearly two infections per person (see figure 1.4). In fact, from the standpoint of infectious diseases, the earth serves as a giant incubator for old and new diseases.

An **emerging infectious disease** is defined as one that is becoming more prominent. Most emerging diseases, such as tuberculosis, have already been described and have existed for thousands of years but are presently experiencing a resurgence. Others, such as AIDS, involve entirely new pathogens that have arisen from a changing world scene. Many factors play a part in emergence, but fundamental to all emerging diseases is the formidable capacity of microorganisms to respond and adapt to alterations in the individual, community, and environment. Just when a disease such as polio has been eliminated or greatly reduced through vaccination or chemotherapy, other diseases move in to fill the vacancy. In the past two decades, more than 30 new diseases have been documented.

What events in the world cause emerging diseases? Dr. David Satcher, director of the Centers for Disease Control and Prevention (CDC), sums it up succinctly by stating that "organisms changed and people changed." Among the most profound influences are disruptions in the human population, such as crowding or immigration. For an excellent example of this effect, we have only to look at AIDS, which began as a focus of infection in remote African villages and was transported out of the region through immigration and tourism. When these were combined with changes in sexual mores and behavior, the disease spread rapidly worldwide, and in less than 20 years it has become the sixth cause of death worldwide (see figure 1.4). Another population factor is an increase in the number of people who are susceptible to infections. Some countries have a high percentage of elderly and immunocompromised people at greater risk for pneumonia and other respiratory infections. In other countries, the youngest members lack immunization or are malnourished, both of which increase their susceptibility to common childhood and diarrheal diseases.

A number of prominent emerging diseases are associated with changing methods in agriculture and technology. The mass production and packing of food increases the opportunity for large outbreaks, especially if foods are grown in fecally contaminated soils or are eaten raw or poorly cooked. In the past two years, dozens of food-borne outbreaks caused by emerging pathogens have occurred. Epidemics have been associated with the bacterium *Escherichia coli* 0157:H7 in fresh vegetables, fruits, and meats; outbreaks of the protozoan infection *Cyclospora* were associated with contaminated raspberries; thousands of people were exposed to the hepatitis A virus in strawberries and to *Salmonella* bacteria in eggs and milk. Even municipal water supplies have spread water-borne protozoa such as *Cryptosporidium* and *Giardia* that slipped past the usual water treatment systems.

Many pathogens are emerging because they have developed drug resistance and are no longer responding to traditional medications. This is one reason that both tuberculosis and malaria are increasing worldwide. Some microbes have mutated to become more virulent or have developed the ability to switch hosts. In China, an emerging strain of influenza virus that ordinarily infects chickens was transmitted to humans. In England, an unusual infectious agent called a prion that causes bovine spongiform encephalopathy (mad cow disease) was found to be the origin of an outbreak in humans.

Other influences on emergent diseases are fluctuations in ecology and climate, animal migration, and human travel. Warming in some regions has increased the spread of mosquitos that carry dengue fever or encephalitis viruses. Overgrowth of dinoflagellates (a type of alga) in Northeast rivers has caused massive fish kills and toxic symptoms in fishermen. Evidence is mounting that run-off pollution from nearby farms was responsible for the appearance of this pathogen. The encroachment of humans into wild habitats has opened the way for increased exposure to mammalian pathogens. Monkeys are known to harbor agents such as Ebola virus, and mice are a source of hantavirus.

Increased personal freedom and opportunities for travel favor the rapid dispersal of microbes. A person may become infected and be home for several days before signs of disease appear. Pathogens can literally be transmitted around the globe in a short time. Because of this potential, we can no longer separate the world into "them" and "us." Health authorities from every country must be constantly vigilant to prevent another crisis like AIDS and to keep common diseases in check through vaccination and medication. A conservative estimate of the cost of controlling these diseases is at least $120 billion, but the actual cost could be double that figure.

THE GENERAL CHARACTERISTICS OF MICROORGANISMS

CELLULAR ORGANIZATION

Two basic cell lines have appeared during evolutionary history. These lines, termed **procaryotic cells*** and **eucaryotic cells,*** differ primarily in the complexity of their cell structure (figure 1.5*a*).

In general, procaryotic cells are smaller than eucaryotic cells, and they lack special structures such as a nucleus and **organelles.*** Organelles are small membrane-bound cell structures that perform specific functions in eucaryotic cells. These two cell types and the organisms that possess them (called procaryotes and eucaryotes) are covered in more detail in chapters 2, 4, and 5.

All procaryotes are microorganisms, but only some eucaryotes are microorganisms. The bodies of most microorganisms consist of either a single cell or just a few cells (figure 1.5*c,d,e*). Because of their role in disease, certain animals such as helminth worms and insects, many of which can be seen with the naked eye, are, also considered in the study of microorganisms (see figure 1.4*f*). Even in its seeming simplicity, the microscopic world is every bit as complex and diverse as the macroscopic one. There is no doubt that microorganisms also outnumber macroscopic organisms by a factor of several million.

A NOTE ON VIRUSES

Viruses are subject to intense study by microbiologists. They are small particles that exist at a level of complexity somewhere between large molecules and cells (figure 1.5*b*). Viruses are much simpler than cells; they are composed essentially of a small amount of hereditary material wrapped up in a protein covering. Some biologists refer to viruses as parasitic particles; others consider them to be very primitive organisms. One thing is certain—they are highly dependent on a host cell's machinery for their activities.

MICROBIAL DIMENSIONS: HOW SMALL IS SMALL?

When we say that microbes are too small to be seen with the unaided eye, what sorts of dimensions are we talking about? This concept is best visualized by comparing microbial groups with the larger organisms of the macroscopic world and also with the molecules and atoms of the molecular world (figure 1.6). Whereas the dimensions of macroscopic organisms are usually given in centimeters (cm) and meters (m), those of most microorganisms fall within the range of micrometers (*u*m) and to a lesser extent, nanometers (nm) and millimeters (mm). The size range of most microbes extends from the smallest viruses, measuring around 20 nm and actually not much bigger than a large molecule, to protozoans measuring 3–4 mm and visible with the naked eye.

LIFE-STYLES OF MICROORGANISMS

The majority of microorganisms live a free existence in habitats such as soil and water, where they are relatively harmless and often beneficial. A free-living organism can derive all required foods and other factors directly from the nonliving environment. Some microorganisms require interaction with other organisms. One such group, termed **parasites,** are harbored and nourished by other living organisms, called **hosts.** A parasite's actions cause damage to its host through infection and disease. Most microbial parasites are some type of bacterium, fungus, protozoan, worm, or virus. Although parasites cause important diseases, they make up only a small proportion of microbes. As we shall see later, a few microorganisms can exist on either free-living or parasitic levels.

Chapter Checkpoints

Excluding the viruses, there are two types of microorganisms: procaryotes, which are small and lack a nucleus and organelles, and eucaryotes, which are larger and have both a nucleus and organelles.

Viruses are not cellular and are therefore called particles rather than organisms. They are included in microbiology because of their small size and close relationship with cells.

Most microorganisms are measured in micrometers, with two exceptions. The helminths are measured in millimeters, and the viruses are measured in nanometers.

Contrary to popular belief, most microorganisms are harmless, free-living species that perform vital functions in both the environment and larger organisms. Comparatively few species are agents of disease.

THE HISTORICAL FOUNDATIONS OF MICROBIOLOGY

If not for the extensive interest, curiosity, and devotion of thousands of microbiologists over the last 300 years, we would know little about the microscopic realm that surrounds us. Many of the discoveries in this science have resulted from the prior work of men and women who toiled long hours in dimly lit laboratories with the crudest of tools. Each additional insight, whether large or small, has added to our current knowledge of living things and processes. This treatment of the early history of microbiology will

*procaryotic (proh''-kar-ee-ah'-tik) Gr. *pro,* before, and *karyon,* nucleus.

*eucaryotic (yoo''-kar-ee-ah'-tik) Gr. *eu,* true or good, and *karyon,* nucleus. Sometimes spelled prokaryotic and eukaryotic.

*organelle (or'-gan-el'') L., little organ.

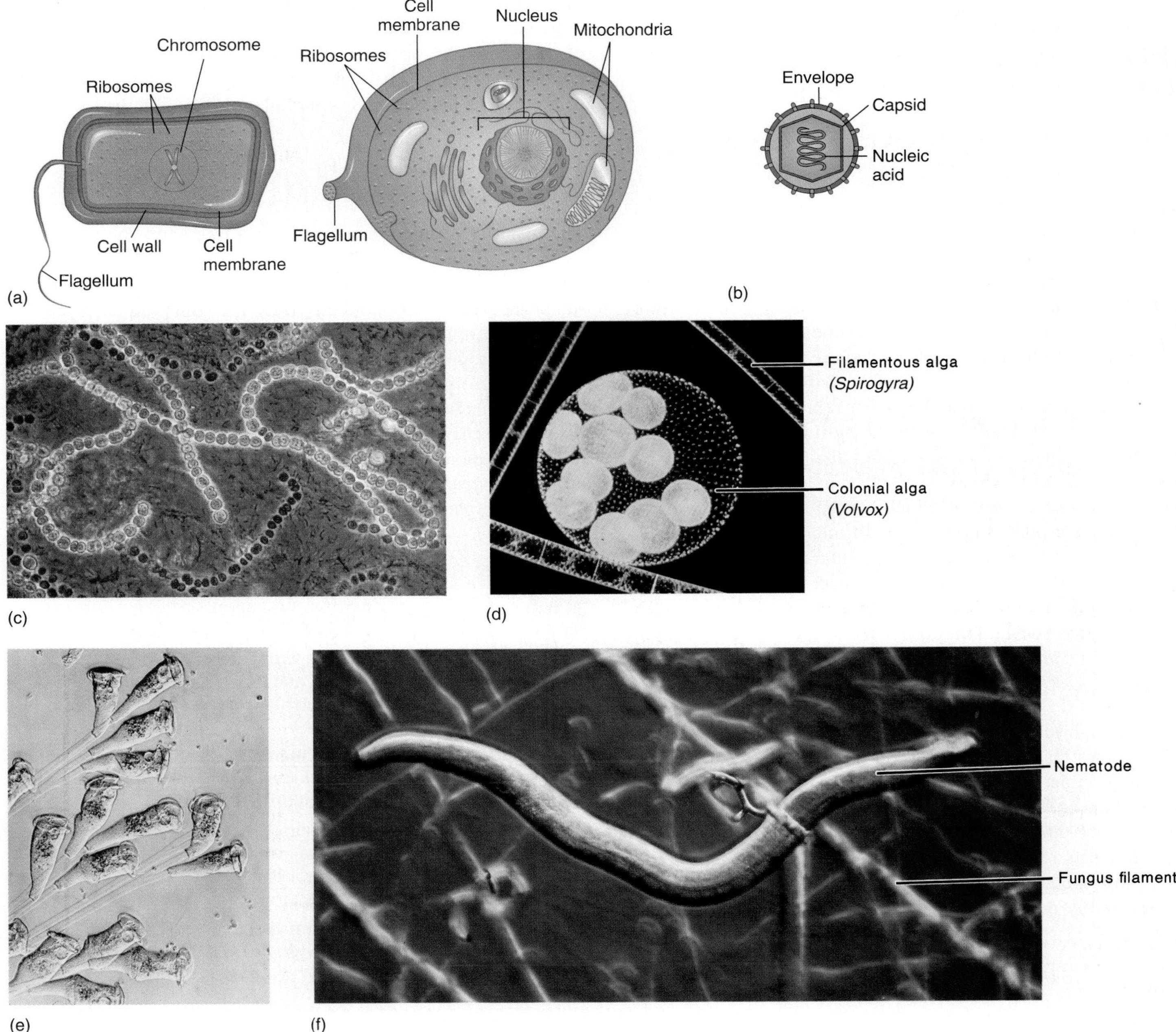

Figure 1.5

The organization of living things and viruses. (*a*) Microbial cells are of the small, relatively simple procaryotic variety (left) or the larger, more complex eucaryotic type (right). (*b*) Viruses are tiny particles, not cells, that consist of genetic material surrounded by a protective covering. (*c*) Example of a prokaryotic organism (the cyanobacterium, *Nostoc*) with cells arranged like a chain of beads. (*d*) Two eucaryotic organisms (algae) organized into more complex cell groupings. *Volvox* is a large, spherical colony composed of smaller colonies (green spheres) and cells (small green dots). *Spirogyra* is a filamentous alga composed of elongate cells joined end to end. (*e*) The stalked protozoan *Vorticella* is shown in feeding mode. These free-living eucaryotes are common in pond water. (*f*) A nematode worm, a multicellular, microscopic animal, is shown being captured in the network of a killer fungus that preys upon it.

summarize the prominent discoveries made in the past 300 years: microscopy, the rise of the scientific method, and the development of medical microbiology, including the germ theory and the origins of modern microbiological techniques. Table 1.2 summarizes some of the pivotal events in microbiology, from its earliest beginnings to the present. Additional historical vignettes are integrated throughout this text to serve as background.

The Development of the Microscope: "Seeing Is Believing"

It is likely that from the very earliest history, humans noticed that when certain foods spoiled they became inedible or caused illness, and yet other spoiled foods did no harm and even had enhanced

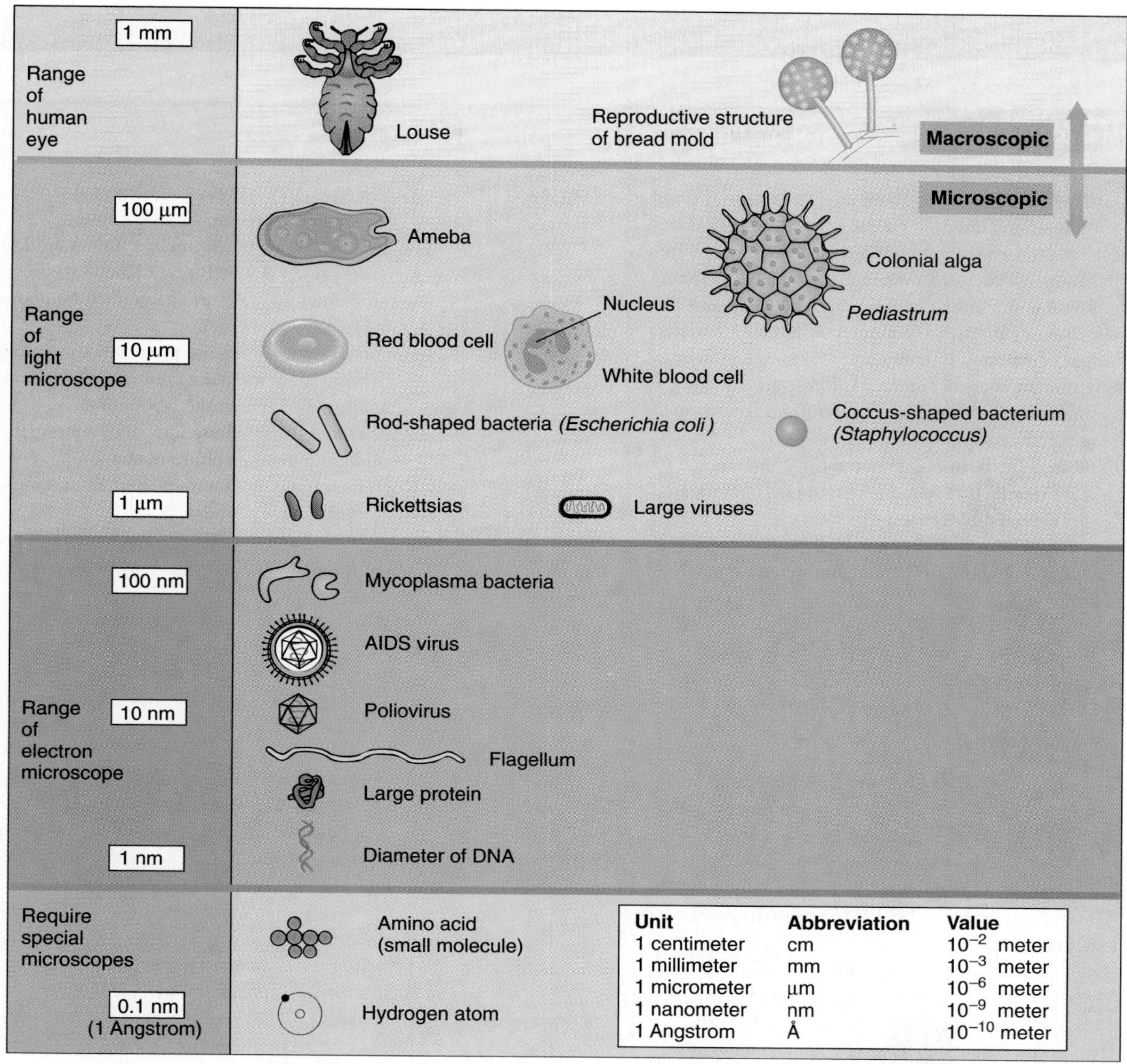

Unit	Abbreviation	Value
1 centimeter	cm	10^{-2} meter
1 millimeter	mm	10^{-3} meter
1 micrometer	µm	10^{-6} meter
1 nanometer	nm	10^{-9} meter
1 Angstrom	Å	10^{-10} meter

Figure 1.6

The size of things. Common measurements encountered in microbiology and a scale of comparison from the macroscopic to the microscopic, molecular, and atomic. Most microbes encountered in our studies will fall between 100 μm and 10 nm in overall dimensions. The microbes shown are more or less to scale within size zone but not between size zones.

flavor. Indeed, several centuries ago there was already a sense that diseases such as the black plague and smallpox were caused by some sort of transmissible matter. But the causes of such phenomena were vague and obscure because the technology to study them was lacking. Consequently, they remained cloaked in mystery and regarded with superstition—a trend that led even well-educated scientists to believe in spontaneous generation (microfile 1.2). True awareness of the widespread distribution of microorganisms and some of their characteristics was finally made possible by the development of the first microscopes. These devices revealed microbes as discrete entities sharing many of the characteristics of larger, visible plants and animals. Several early scientists fashioned magnifying lenses, but their microscopes lacked the optical clarity needed for examining bacteria and other small, single-celled organisms. The first careful and exacting observations awaited the clever single-lens microscope hand-fashioned by Antonie van Leeuwenhoek, a Dutch linen merchant and self-made microbiologist (figure 1.7). The original purpose of the microscopes was to examine cloth for flaws, but Leeuwenhoek turned them to other uses as well.

Leeuwenhoek's wide-ranging investigations included observations of tiny organisms he called *animalcules* (little animals), blood, and other human tissues (including his own tooth scrapings), insects, minerals, and plant materials. He constructed more than 250 small, powerful microscopes that could magnify up to 300 times (figure 1.8). Considering that he had no formal training in science and that he was the first person ever to faithfully record this strange new world, his descriptions of bacteria and protozoa

TABLE 1.2

SIGNIFICANT EVENTS IN MICROBIOLOGY

Date	Event and Its Importance
1361–1380	First widespread use of quarantine to control the spread of epidemic bubonic plague.
1481–1499	Elemental mercury given as a treatment for syphilis.
1546	Italian physician Girolamo Fracastoro suggests that invisible organisms may be involved in disease.
1590	Zaccharias Janssen, a Dutch spectacle maker, invents the first compound microscope.
1660	Englishman Robert Hooke explores various living and nonliving matter with a compound microscope that uses reflected light.
1668	Francesco Redi, an Italian naturalist, conducts experiments that demonstrate the fallacies in the spontaneous generation theory.
1676	Antonie van Leeuwenhoek, a Dutch linen merchant, uses a simple microscope of his own design to observe bacteria and protozoa.
1776	An Italian anatomist, Lazzaro Spallanzani, conducts further convincing experiments that dispute spontaneous generation.
1796	English surgeon Edward Jenner introduces a vaccination for smallpox.
1838	Phillipe Ricord, a French physician, inoculates 2,500 human subjects to demonstrate that syphilis and gonorrhea are two separate diseases.
1839	Theodor Schwann, a German zoologist, and Matthias Schleiden, a botanist, formalize the theory that all living things are composed of cells.
1847–1850	The Hungarian physician Ignaz Semmelweiss substantiates his theory that childbed fever is a contagious disease transmitted to women by their physicians during childbirth; he institutes the first use of antiseptics to reduce hand-borne disease.
1853–1854	John Snow, a London physician, demonstrates the epidemic spread of cholera through a water supply contaminated with human sewage.
1857	French bacteriologist Louis Pasteur shows that fermentations are due to microorganisms and originates the process now known as pasteurization.
1858	Rudolf Virchow, a German pathologist, introduces the concept that all cells originate from preexisting cells.
1861	Louis Pasteur completes the definitive experiments that finally lay to rest the theory of spontaneous generation.
1867	The English surgeon Joseph Lister publishes the first work on antiseptic surgery, beginning the trend toward modern aseptic techniques in medicine.
1869	Johann Miescher, a Swiss pathologist, discovers in the cell nucleus the presence of complex acids, which he terms nuclein (DNA, RNA).
1876–1877	German bacteriologist Robert Koch* studies anthrax in cattle and implicates the bacterium *Bacillus anthracis* as its causative agent.
1881	Pasteur develops a vaccine for anthrax in animals.
	Koch introduces the use of pure culture techniques for handling bacteria in the laboratory.
	Walther and Fanny Hesse introduce agar-agar as a solidifying gel for culture media.
1882	Koch identifies the causative agent of tuberculosis.
1884	Koch outlines his postulates.
	Elie Metchnikoff,* a Russian zoologist, lays groundwork for the science of immunology by discovering phagocytic cells.
	The Danish physician Hans Christian Gram devises the Gram stain technique for differentiating bacteria.
1885	Pasteur develops a special vaccine for rabies.
1887	Julius Petri, a German bacteriologist, adapts two plates to form a container for holding media and culturing microbes.
1890	A German, Emil von Behring,* and a Japanese, Shibasaburo Kitasato, demonstrate the presence of antibodies in serum that neutralize the toxins of diphtheria and tetanus.
1892	A Russian, D. Ivanovski, is the first to isolate a virus (the tobacco mosaic virus) and show that it could be transmitted in a cell-free filtrate.
1895	Jules Bordet,* a Belgian bacteriologist, discovers the antimicrobial powers of complement
1898	R. Ross* and G. Grassi demonstrate that malaria is transmitted by the bite of female mosquitos.
	Germans Friedrich Loeffler and P. Frosch discover that "filterable viruses" cause foot-and-mouth disease in animals.
1899	Dutch microbiologist Martinus Beijerinck further elucidates the viral agent of tobacco mosaic disease and postulates that viruses have many of the properties of living cells and that they reproduce within cells.
1900	The American physician Walter Reed and his colleagues clarify the role of mosquitos in transmitting yellow fever.
	An Austrian pathologist, Karl Landsteiner,* discovers the ABO blood groups.

*These persons were awarded Nobel prizes for their contributions to the field.

Date	Event and Its Importance
1903	American pathologist James Wright and others demonstrate the presence of antibodies in the blood of immunized animals.
1905	Syphilis is shown to be caused by *Treponema pallidum,* through the work of German bacteriologists Fritz Schaudinn and E. Hoffman.
1906	August Wasserman, a German bacteriologist, develops the first serologic test for syphilis.
	Howard Ricketts, an American pathologist, links the transmission of Rocky Mountain spotted fever to ticks.
1908	The German Paul Ehrlich* becomes the pioneer of modern chemotherapy by developing salvarsan, an arsenic-based drug, to treat syphilis.
1910	An American pathologist, Francis Rous,* discovers viruses that can induce cancer.
1915–1917	British scientist F. Twort and French scientist F. D'Herelle independently discover bacterial viruses.
1928	Frederick Griffith lays the foundation for modern molecular genetics by his discovery of transformation in bacteria.
1929	A Scottish bacteriologist, Alexander Fleming,* discovers and describes the properties of the first antibiotic, penicillin.
1933–1938	Germans Ernst Ruska* and von Borries develop the first electron microscope.
1935	Gerhard Domagk,* a German physician, discovers the first sulfa drug and paves the way for the era of antimicrobic chemotherapy.
	Wendell Stanley* is successful in inducing tobacco mosaic viruses to form crystals that still retain their infectiousness.
1941	Australian Howard Florey* and Englishman Ernst Chain* develop commercial methods for producing and purifying penicillin; this first antibiotic is tested and put into widespread use.
1944	Oswald Avery, Colin MacLeod, and Maclyn McCarty show that DNA is the genetic material.
	Joshua Lederberg* and E. L. Tatum* discover conjugation in bacteria.
	The Russian Selman Waksman* and his colleagues discover the antibiotic streptomycin.
1953	James Watson,* Francis Crick,* Rosalind Franklin, and Maurice Wilkins* determine the structure of DNA.
1954	Jonas Salk develops the first polio vaccine.
1957	Alick Isaacs and Lindenmann discover the natural antiviral substance interferon.
	D. Carleton Gajdusek* discovers the underlying cause of slow virus diseases.
1959–1960	Gerald Edelman* and Rodney Porter* determine the structure of antibodies.
1972	Paul Berg* develops the first recombinant DNA in a test tube.
1973	Herb Boyer and Stanley Cohen clone the first DNA using plasmids.
1975	A technique for making monoclonal antibodies is developed by Cesar Milstein, Georges Kohler, and Niels Kai Jerne.
1979	Genetically engineered insulin is first synthesized by bacteria.
1982	Development of first hepatitis B vaccine from virus isolated from human blood.
1983	Isolation and characterization of human immunodeficiency virus (HIV) by Luc Montagnier of France and Robert Gallo of the United States.
	The polymerase chain reaction is invented by Kerry Mullis.*
1987	The molecular genetics of antibody genes is worked out by Susumu Tonegawa.*
	First release of recombinant strain of *Pseudomonas* to prevent frost formation on strawberry plants.
1989	Cancer-causing genes called oncogenes are characterized by J. Michael Bishop, Robert Huber, Hartmut Michel, and Harold Varmus.
1990	First clinical trials in gene therapy testing.
	Vaccine for *Haemophilus influenzae,* a cause of meningitis, is introduced.
1991	Development of transgenic animals to synthesize human hemoglobin.
1994	Human breast cancer gene isolated.
1995	First bacterial genome fully sequenced, for *Haemophilus influenzae.*
1996	Dr. David Ho develops a "cocktail" of drugs to treat AIDS.
1997	Medical researchers at Case Western University construct human artificial chromosomes (HACs).
	Scottish researchers clone first mammal (a sheep) from adult nuclei.

*These persons were awarded Nobel prizes for their contributions to the field.

MICROFILE 1.2 SPONTANEOUS GENERATION AND THE BATTLE OF THE HYPOTHESES

For thousands of years, it was the consensus that certain living things arose intact from vital forces inherent in nonliving or decomposing matter. This ancient belief in **spontaneous generation**—abiogenesis—was continually reinforced as people observed that meat left out in the open soon "produced" maggots, that mushrooms appeared on rotting wood, that rats and mice emerged from piles of litter, and other similar phenomena. On the surface, there is nothing incorrect about the observations themselves; it is the conclusions drawn from them that do not hold up to critical reasoning.

Even the discovery of single-celled organisms during the latter part of the mid-1600s did not eradicate the idea of spontaneous generation, because some scientists assumed that microscopic beings were simply one stage in a process in which larger organisms formed from smaller ones. Though some of these early ideas seem quaint and ridiculous in light of modern knowledge, we must remember that, at the time, mysteries in life were accepted, and the scientific method was not widely practiced. The invalidation of the theory of abiogenesis was a very gradual process.

Two hypotheses attempted to explain the origin of the "simpler forms" of life: (1) They arose spontaneously by the massing of vital forces within nonliving matter **(abiogenesis),*** or (2) they arose only from other living things of their same kind **(biogenesis).** If an investigator was able to verify one, his work automatically falsified the other, because both cannot be true. Throughout history, serious proponents existed on both sides, but determining the correct hypothesis required 200 years of combined inquiries by microbiologists from several countries, each of whom dealt with a particular disputed part of the hypotheses.

Among the important variables to be considered in challenging the hypotheses were the effects of nutrients, air, and heat and the presence of preexisting life forms in the environment. One of the first people to doubt the spontaneous generation theory was Francesco Redi of Italy. He conducted a simple test in which he placed meat in a jar and covered it with fine gauze. Flies gathering at the jar were blocked from entering and thus laid their eggs on the outside of the gauze. The maggots subsequently developed without access to the meat, indicating that maggots were the offspring of flies and did not arise from some "vital force" in the meat. This and related experiments laid to rest the idea that more-complex animals such as insects and mice developed through abiogenesis, but it did not convince many scientists of the day that simpler organisms could not arise in that way. Frenchman Louis Jablot reasoned that even microscopic organisms must have parents, and his experiments with infusions (dried hay steeped in water) supported that hypothesis. He divided an infusion that had been boiled to destroy any living things into two containers: a heated container that was closed to the air and a heated container that was freely open to the air. Only the open vessel developed microorganisms, which he presumed had entered in air laden with dust. Regrettably, the validation of biogenesis was temporarily set back by John Needham, an Englishman who did similar experiments using mutton gravy. His results were in conflict with Jablot's because both his heated and unheated test containers teemed with microbes. Unfortunately, his experiments were done before the concepts of heat-resistant endospores and true methods of sterility were understood and widely known.

Unable to allow Needham's conclusion to remain untested, an Italian naturalist, Lazzaro Spallanzani, took more extensive and sophisticated measures, which included long-term heating of the test vessels and infusions to destroy any lingering resistant forms. He found that a variety of liquid nutrients (broths, urine) thus treated and kept sealed in flasks from air would not develop growth. When he was accused of destroying the vegetative force of the nutrients by overheating, he showed that the heated nutrients could still grow microbes when exposed to air. Undaunted by this argument, other critics objected that Spallanzani's flasks were sealed from oxygen, a gas known to be required in the respiration of animals. Somehow, Leeuwenhoek's earlier discovery that free oxygen was not necessary for microbial growth was disregarded at the time.

Additional experiments further defended biogenesis. Franz Shultze and Theodor Schwann of Germany felt sure that air was the source of microbes and sought to prove this by passing air through strong chemicals or hot glass tubes into heat-treated infusions in flasks. When the infusions again remained devoid of living things, the supporters of abiogenesis claimed that the treatment of the air had made it harmful to the spontaneous development of life. Georg Schroeder and Theodor Van Dusch followed up these studies. They did not treat the air with heat or chemicals but passed it through cotton wool to filter out microscopic organisms. Again, no microbes grew in the infusions. Although all these experiments should have finally laid to rest the arguments for spontaneous generation, they did not.

Then, in the mid-1800s, the acclaimed microbiologist Louis Pasteur entered the arena. He had recently been studying the roles of microorganisms in the fermentation of beer and wine, and it was clear to him that these processes were brought about by the activities of microbes introduced into the beverage from air, fruits, and grains. The methods he used to discount abiogenesis were simple yet brilliant. He repeated the experiments using cotton filters to trap dust from air and observed tiny objects (probably spores) in the filters. He also observed that these same filters would initiate growth in previously sterile broths. To further clarify that air was the source of microbes, he filled flasks with broth and fashioned their openings into elongate, swan neck–shaped tubes. The flasks' openings were freely open to the air but were curved so that gravity would cause any airborne dust particles to deposit in the lower part of the necks. He heated the flasks to sterilize the broth and then incubated them. As long as the flask remained intact, the broth remained sterile, but if the neck was broken off so that dust fell directly down into the container, microbial growth immediately commenced. Some of these ingenious little flasks are still on display at the Institute Pasteur in Paris in their original sterile form. Pasteur summed up his findings, "For I have kept from them, and am still keeping from them, that one thing which is above the power of man to make; I have kept from them the germs that float in the air, I have kept from them life."

*abiogenesis *(ah-bee″-oh-jen-uh-sis) Gr.* a, *not,* bios, *living, and* gennan, *to produce.*

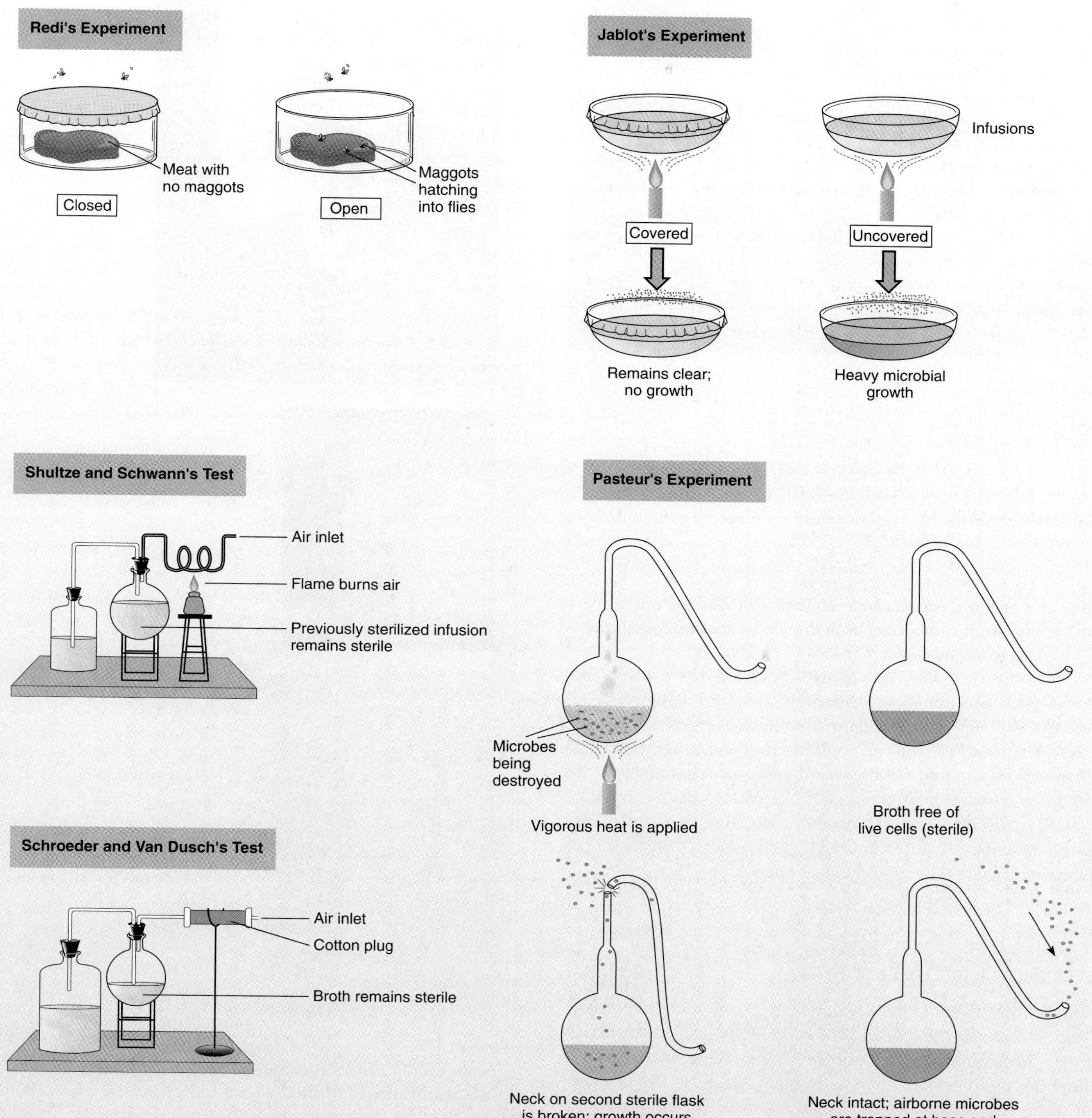
Redi's Experiment
Meat with no maggots
Closed
Open
Maggots hatching into flies
Jablot's Experiment
Infusions
Covered
Uncovered
Remains clear; no growth
Heavy microbial growth
Shultze and Schwann's Test
Air inlet
Flame burns air
Previously sterilized infusion remains sterile
Pasteur's Experiment
Microbes being destroyed
Vigorous heat is applied
Broth free of live cells (sterile)
Schroeder and Van Dusch's Test
Air inlet
Cotton plug
Broth remains sterile
Neck on second sterile flask is broken; growth occurs
Neck intact; airborne microbes are trapped at base and broth is sterile

Figure 1.7
An oil painting of Antonie van Leeuwenhoek (1632–1723) sitting in his laboratory. J. R. Porter and C. Dobell have commented on the unique qualities Leeuwenhoek brought to his craft: "He was one of the most original and curious men who ever lived. It is difficult to compare him with anybody because he belonged to a genus of which he was the type and only species, and when he died his line became extinct."

were astute and precise. Because of Leeuwenhoek's extraordinary contributions to microbiology, he is known as the father of bacteriology and protozoology.

From the time of Leeuwenhoek, microscopes evolved into more complex and improved instruments with the addition of refined lenses, a condenser, finer focusing devices, and built-in light sources. The prototype of the modern compound microscope, in use from about the mid-1800s, was capable of magnifications of 1,000 times or more. Even our modern laboratory microscopes are not greatly different in basic structure and function from those early microscopes. The technical characteristics of microscopes and microscopy are the major focus of chapter 3.

The Rise of the Scientific Method

A serious impediment to the development of true scientific reasoning and testing was the tendency of early scientists to explain natural phenomena by a mixture of belief, superstition, and argument (see the discussion of abiogenesis in microfile 1.2). The development of an experimental system that answered questions objectively and was not based on prejudice marked the beginning of true scientific thinking. These ideas gradually crept into the consciousness of the scientific community during the 1600s. The general approach taken by scientists to explain a certain natural phenomenon is called the **scientific method.** A primary aim of this method is to formulate a **hypothesis,** a tentative explanation to ac-

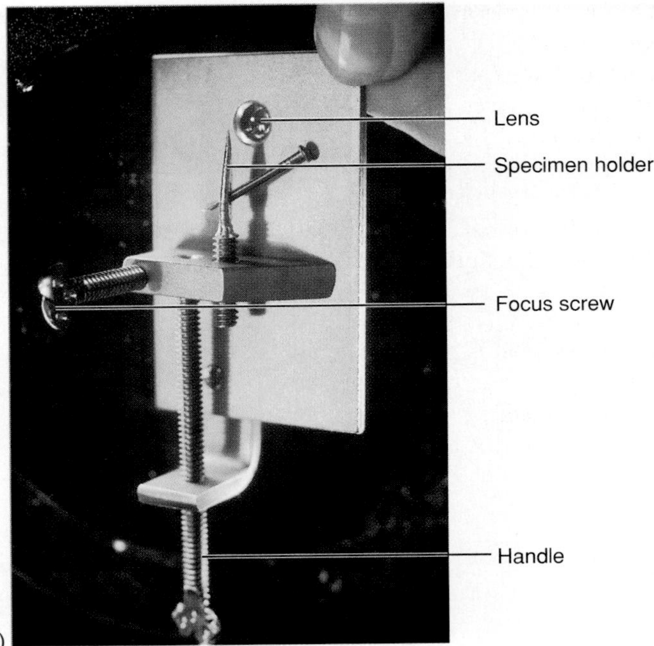

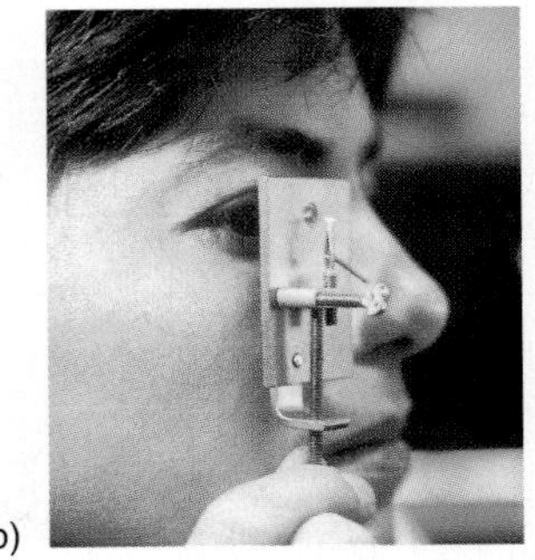

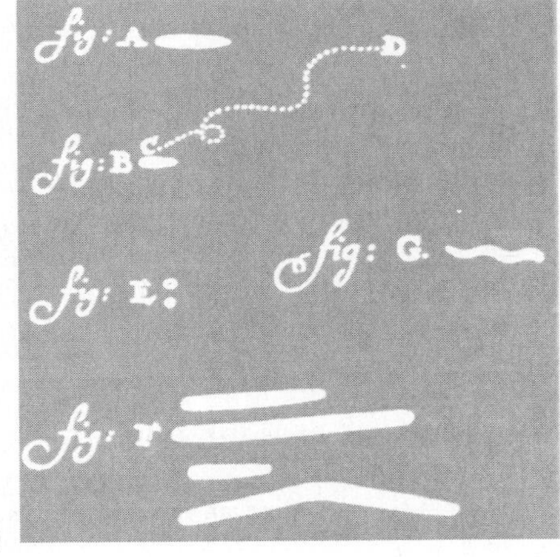

Figure 1.8
(*a*) A brass replica of a Leeuwenhoek microscope made by Dr. William Walter of Montana State University. (*b*) If you can get the lighting just right, it works. (*c*) Examples of bacteria drawn by Leeuwenhoek. He keenly observed, "I discovered living creatures in rain water which had stood but a few days in a new earthen pot. This invited me to view this water with great attention, especially those little animals appearing to me ten thousand times less than those which may be perceived in the water with the naked eye." This is probably the first observation of bacteria.

count for what has been observed or measured. A good hypothesis must be capable of being either supported or discredited by careful, systematic observation or experimentation. For example, the statement that "bees make honey from pollen" can be experimentally determined by the tools of science, but the statement that "bees buzz because they are happy" cannot.

The two types of reasoning that are commonly applied separately or in combination to develop and support hypotheses are **induction** and **deduction** (figure 1.9). In the **inductive approach,** a scientist first accumulates specific data or facts and then formulates a general hypothesis that accounts for those facts. The inductive approach asks, "Are various observed events best explained by this hypothesis or by another one?" In the **deductive approach,** a scientist constructs a hypothesis, tests its validity by outlining particular events that are predicted by the hypothesis, and then performs experiments to test for those events. The deductive process states: "If the hypothesis is valid, then certain specific events can be expected to occur."

Natural processes have numerous physical, chemical, and biological factors, or *variables,* that can hypothetically affect their outcome. To account for these variables, scientists design experiments: (1) to thoroughly test for or measure the consequences of each possible variable and (2) to accompany each variable with one or more *control groups.* A control group is designed exactly as the test group but omits only that variable being tested; thus, it may serve as a basis of comparison for the test group. The reasoning is that, if a certain experimental finding occurs only in the test group and not in the control group, the finding must be due to the variable being tested and not to some uncontrolled, untested factor that is not part of the hypothesis (figure 1.10).

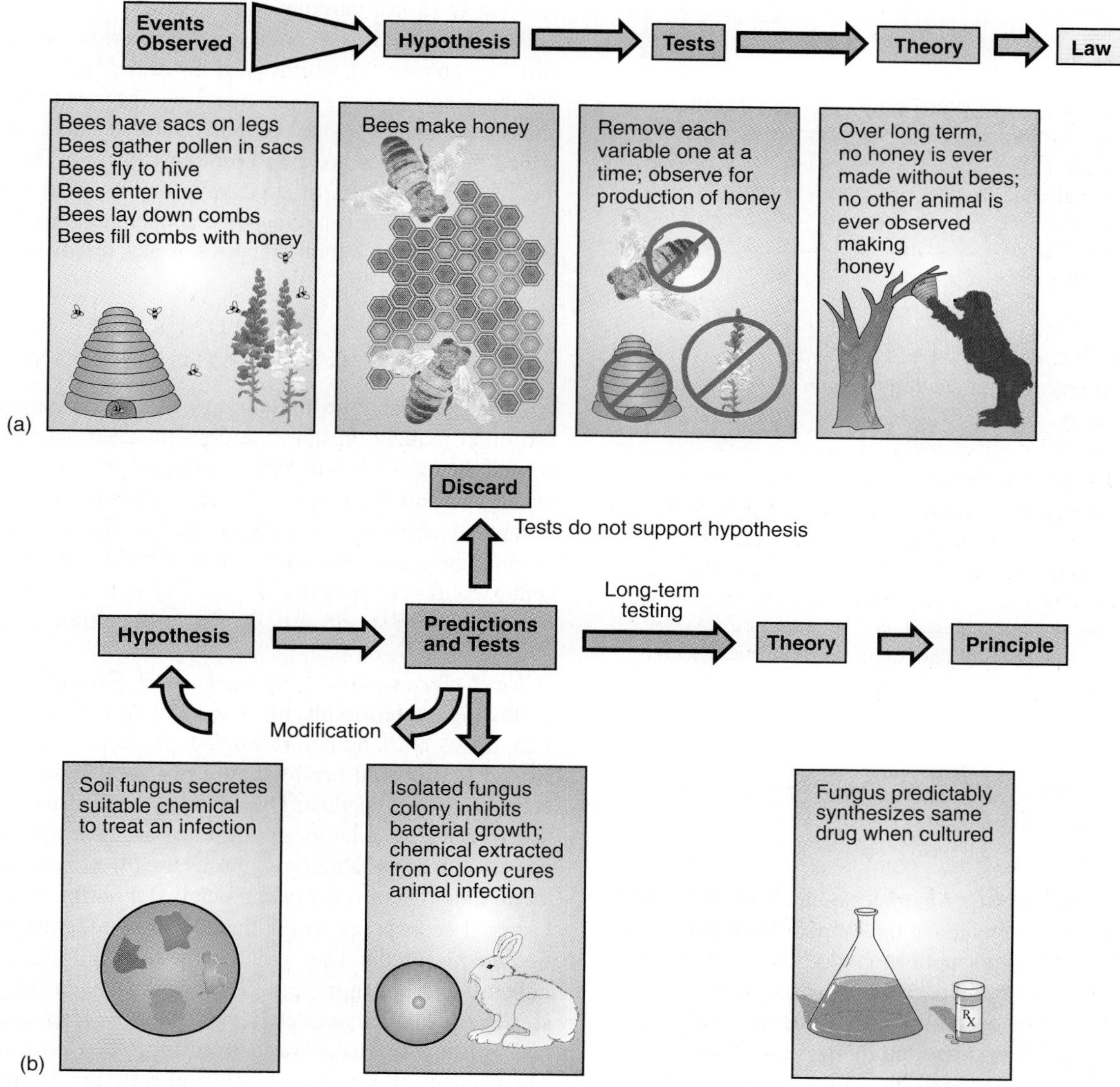

Figure 1.9
Comparing the inductive and deductive approaches to the scientific method. (*a*) The inductive process proceeds from specific observations to general hypothesis. (*b*) The deductive process starts with a general hypothesis that predicts specific expectations.

Subject: Testing the factors responsible for dental caries

Hypothesis: Dental caries (cavities) involve dietary sugar or microbial action or both.

Variables:

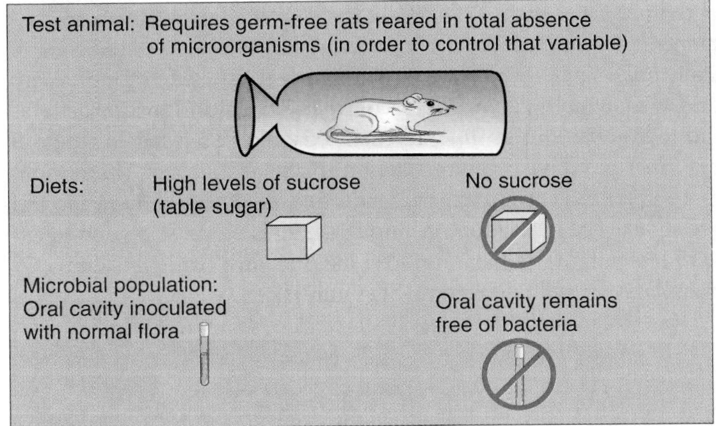

Experimental Protocol:

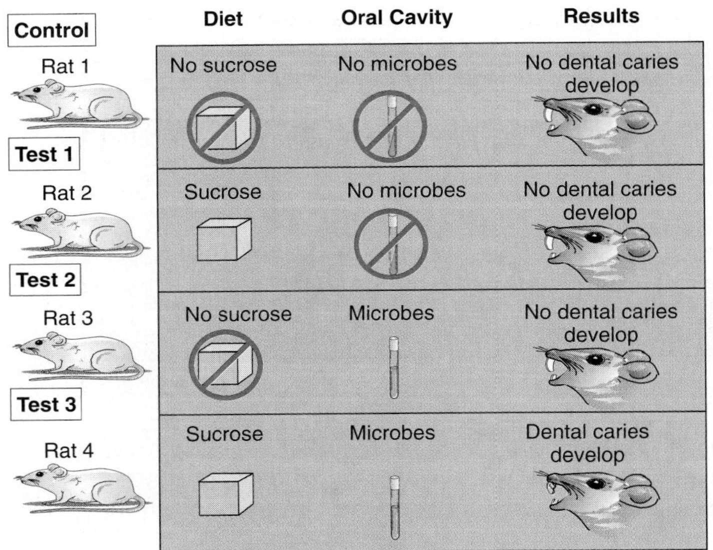

Conclusion: Dental caries will not develop unless both sucrose and microbial action are present. What other variables were not controlled?

Figure 1.10
Any factor that can affect the experimental outcome is called a variable, and each combination of variables must be controlled while the hypothesis is being tested.

A lengthy process of experimentation, analysis, and testing eventually leads to conclusions that either support or refute the hypothesis. If experiments do not uphold the hypothesis—that is, if it is found to be flawed—the hypothesis or some part of it is rejected; it is either discarded or modified to fit the results of the experiment. If the hypothesis is supported by the results from the experiment, it is not (or should not be) immediately accepted as fact. It then must be tested and retested. Indeed, this is an important guideline in the acceptance of a hypothesis. The results of the experiment must be published and then repeated by other investigators. The history of science is littered with the carcasses of apparently ingenious hypotheses that simply could not be repeated and were thus ultimately refuted.

In time, as each hypothesis is supported by a growing body of data and survives rigorous scrutiny, it moves to the next level of acceptance—the **theory.** A theory is a collection of statements, propositions, or concepts that explains or accounts for a natural event. A theory is not the result of a single experiment repeated over and over again but is an entire body of ideas that expresses or explains many aspects of a phenomenon. It is not a fuzzy or weak speculation, as is sometimes the popular notion, but a viable declaration that has stood the test of time and has yet to be disproved by serious scientific endeavors. Often, theories develop and progress through decades of research and are added to and modified by new findings. At some point, evidence of the accuracy and predictability of a theory is so compelling that the next level of confidence is reached and the theory becomes a **law,** or principle. For example, although we still refer to the germ *theory* of disease, so little question remains that microbes can cause disease that it has clearly passed into the realm of law.

Science and its hypotheses and theories must progress along with technology. As advances in instrumentation allow new, more detailed views of living phenomena, old theories may be reexamined and altered and new ones proposed. But scientists do not take the stance that theories are ever absolutely proved. The characteristics that make scientists most effective in their work are curiosity, open-mindedness, skepticism, creativity, cooperation, and readiness to revise their views of natural processes as new discoveries are made.

The Development of Medical Microbiology

Early experiments on the sources of microorganisms led to the profound realization that microbes are everywhere: not only are air and dust full of them, but also the entire surface of the earth, its waters, and all objects are exposed to them. This discovery led to immediate applications in medicine, thus the seeds of medical microbiology were sown in the mid to latter half of the nineteenth century with the introduction of the germ theory of disease and the resulting use of sterile, aseptic, and pure culture techniques.

The Discovery of Spores and Sterilization

Following Pasteur's inventive work with infusions (see microfile 1.2), it was not long before English physicist John Tyndall demonstrated that heated broths would not spoil if stored in chambers completely free of dust. His studies provided the initial evidence that some of the microbes in dust and air have very high heat resistance and that particularly vigorous treatment is required to destroy them. Later, the discovery and detailed description of heat-resistant bacterial endospores by Ferdinand Cohn, a German botanist, clarified the reason that heat would sometimes fail to completely eliminate all microorganisms. The modern meaning of the word **sterile,*** meaning completely free of all life forms including spores and viruses, was established from that point on (see chapter 11). The capacity to sterilize objects and materials is an absolutely essential part of microbiology, medicine, dentistry, and industry.

*sterile (stair′-il) Gr. *steira,* barren.

Figure 1.11
Louis Pasteur (1822–1895), one of the founders of microbiology, is pictured here viewing a sample. Few microbiologists can match the scope and impact of his contributions to the science of microbiology.

The Development of Aseptic Techniques

From earliest history, humans experienced a vague sense that "unseen forces" or "poisonous vapors" emanating from decomposing matter could cause disease. As the study of microbiology became more scientific and the invisible was made visible, the fear of such mysterious vapors was replaced by the knowledge and sometimes even the fear of "germs." About 110 years ago, the first studies by Robert Koch clearly linked a microscopic organism with a specific disease. Since that time, microbiologists have conducted a continuous search for disease-causing agents.

At the same time that abiogenesis was being hotly debated, a few budding microbiologists began to suspect that microorganisms could cause not only spoilage and decay but also infectious diseases. It occurred to these rugged individualists that even the human body itself was a source of infection. Dr. Oliver Wendell Holmes, an American physician, observed that mothers who gave birth at home experienced fewer infections than did mothers who gave birth in the hospital, and the Hungarian Dr. Ignaz Semmelweis showed quite clearly that women became infected in the maternity ward after examinations by physicians coming directly from the autopsy room. The English surgeon Joseph Lister took notice of these observations and was first to introduce **aseptic*** **techniques** aimed at reducing microbes in a medical setting and preventing wound infections. Lister's concept of asepsis was much more limited than our modern precautions. It mainly involved disinfecting the hands and the air with strong antiseptic chemicals, such as phenol, prior to surgery. These techniques and the application of heat for sterilization became the bases for microbial control by physical and chemical methods, which are still in use today.

The Discovery of Pathogens and the Germ Theory of Disease

Two great founders of microbiology, Louis Pasteur of France (figure 1.11) and Robert Koch of Germany (figure 1.12), also introduced techniques that are still used today. Pasteur made enormous contributions to our understanding of the microbial role in wine and beer formation. He invented pasteurization and completed some of the first studies showing that diseases could arise from infection. These studies, supported by the work of other scientists, became known as the **germ theory of disease.** Pasteur's contemporary, Koch, established *Koch's postulates,* a series of proofs that verified the germ theory and could establish whether an organism was pathogenic and which disease it caused (see figure 13.24). About 1875, Koch used this experimental system to show that anthrax was caused by a bacterium called *Bacillus anthracis.* So useful were his postulates that the causative agents of 20 other diseases were discovered between 1875 and 1900, and even today, they are the standard for identifying pathogens.

Numerous exciting technologies emerged from Koch's prolific and probing laboratory work. During this golden age of the 1880s, he realized that study of the microbial world would require

Figure 1.12
Robert Koch (1843–1910), intent at his laboratory workbench and surrounded by the new implements of his trade: Petri plates, tubes, and flasks filled with media; smears of bacteria; and bottles of stains.

*aseptic (ay-sep′-tik) Gr. *a,* no, and *sepsis,* decay or infection. These techniques are aimed at reducing pathogens and do not necessarily sterilize.

separating microbes from each other and growing them in culture. It is not an overstatement to say that he and his colleagues invented most of the techniques that are described in chapter 3: inoculation, isolation, media, maintenance of pure cultures, and preparation of specimens for microscopic examination. Other highlights in this era of discovery are presented in later chapters on microbial control (see chapter 11) and vaccination (see chapter 16).

Chapter Checkpoints

Our current understanding of microbiology is the cumulative work of thousands of microbiologists, many of whom literally gave their lives to advance knowledge in this field.

The microscope made it possible to see microorganisms and thus to identify their widespread presence, particularly as agents of disease.

Antonie van Leeuwenhoek is considered the father of bacteriology and protozoology because he was the first person to produce precise, correct descriptions of these organisms using microscopes he made himself.

The theory of spontaneous generation of living organisms from "vital forces" in the air was disproved once and for all by Louis Pasteur.

The scientific method is a process by which scientists seek to explain natural phenomena. It is characterized by specific procedures that either support or discredit an initial hypothesis.

Knowledge acquired through the scientific method is rigorously tested by repeated experiments by many scientists to verify its validity. A collection of valid hypotheses is called a theory. A theory supported by much data collected over time is called a law.

Scientific truth changes through time as new research brings new information. Scientists must be able and willing to change theory in response to new data.

Medical microbiologists developed the germ theory of disease and introduced the critically important concept of aseptic technique to control the spread of disease agents.

Koch's postulates are the cornerstone of the germ theory of disease. They are still used today to pinpoint the causative agent of a specific disease.

Louis Pasteur and Robert Koch were the leading microbiologists during the golden age of microbiology (1875–1900). Each had his own research institute.

TAXONOMY: ORGANIZING, CLASSIFYING, AND NAMING MICROORGANISMS

Students just beginning their microbiology studies are often dismayed by the seemingly endless array of new, unusual, and sometimes confusing names for groups and specific types of microorganisms. Learning microbial **nomenclature*** is very much like learning a new language, and occasionally its demands may be a bit overwhelming. But paying attention to proper microbial names is just like following a baseball game or a movie plot: You cannot tell the players apart without a program! Your understanding and appreciation of microorganisms will be greatly improved by learning a few general rules about how they are named.

The formal system for organizing, classifying, and naming living things is **taxonomy.*** This science originated more than 250 years ago when Carl von Linné (Linnaeus; 1707–1778), a Swedish botanist, laid down the basic rules for taxonomic categories, or **taxa.*** Von Linné realized early on that a system for recognizing and defining the properties of living things would prevent chaos in scientific studies by providing each organism with a unique name and an exact "slot" in which to catalogue it. This classification would then serve as a means for future identification of that same organism and permit workers in many biological fields to know if they were indeed discussing the same organism. The von Linné system has served well in categorizing the 2 million or more different types of organisms that have been discovered since that time.

The primary concerns of taxonomy are classification, nomenclature, and identification. These three areas are interrelated and play a vital role in keeping a dynamic inventory of the extensive array of living things. **Classification** is the orderly arrangement of organisms into groups, preferably in a format that shows evolutionary relationships. **Nomenclature** is the process of assigning names to the various taxonomic rankings of each microbial species. **Identification** is the process of discovering and recording the traits of organisms so that they may be placed in an overall taxonomic scheme. A survey of some general methods of identification appears in chapter 3.

THE LEVELS OF CLASSIFICATION

The main taxa, or groups, in a classification scheme are organized into several descending ranks beginning with **domain,** which is a giant, all-inclusive category based on a unique cell type, and ending with a **species,*** the smallest and most specific. All the members of a domain share only one or a few general characteristics, whereas members of a species are essentially the same kind of organism—that is, they share the majority of their characteristics. The taxa between the top and bottom levels are, in descending order: **kingdom, phylum*** or **division,**[1] **class, order, family,** and **genus.*** Thus, each domain can be subdivided into a series of kingdoms, each kingdom is made up of several phyla, each phylum contains several classes, and so on. Because taxonomic

1 The term *phylum* is used for protozoa and animals; the term *division* is used for bacteria, algae, plants, and fungi.

*nomenclature (noh'-men-klay''chur) L. *nomen,* name, and *clare,* to call. A system of naming.

*taxonomy (tacks-on''-uh-mee) Gr. *taxis,* arrangement, and *nomos,* name.

*taxa (tacks'-uh) sing. taxon.

*species (spee'-sheez) L. *specere,* kind. In biology, this term is always in the plural form.

*phylum (fy'-lum) pl. phyla (fye'-luh) Gr. *phylon,* race.

*genus (jee'-nus) pl. genera (jen'-er-uh) L. birth, kind.

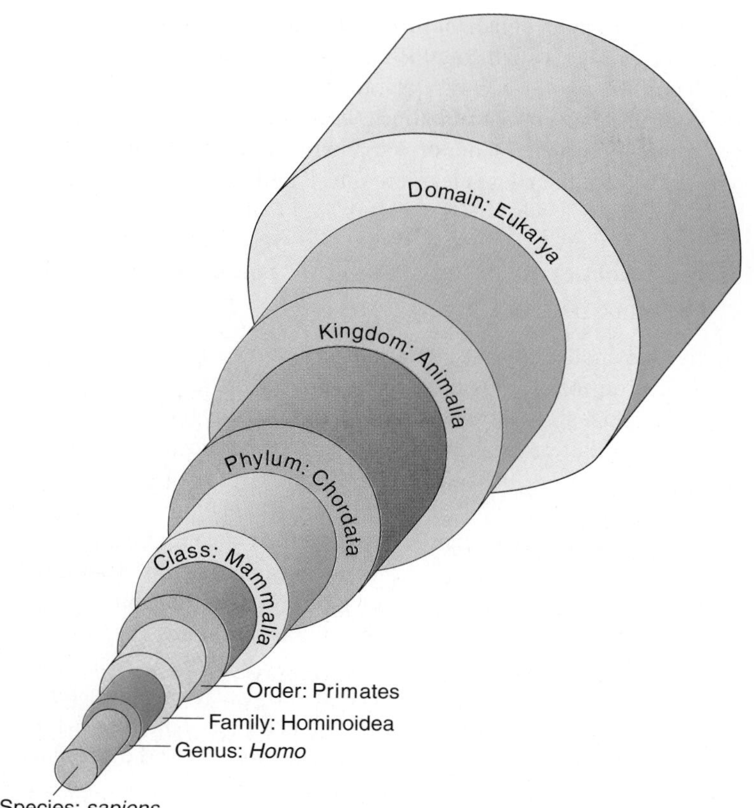

Figure 1.13
The levels in classification from domain to species operate like a set of nesting boxes. Humans are the example here.

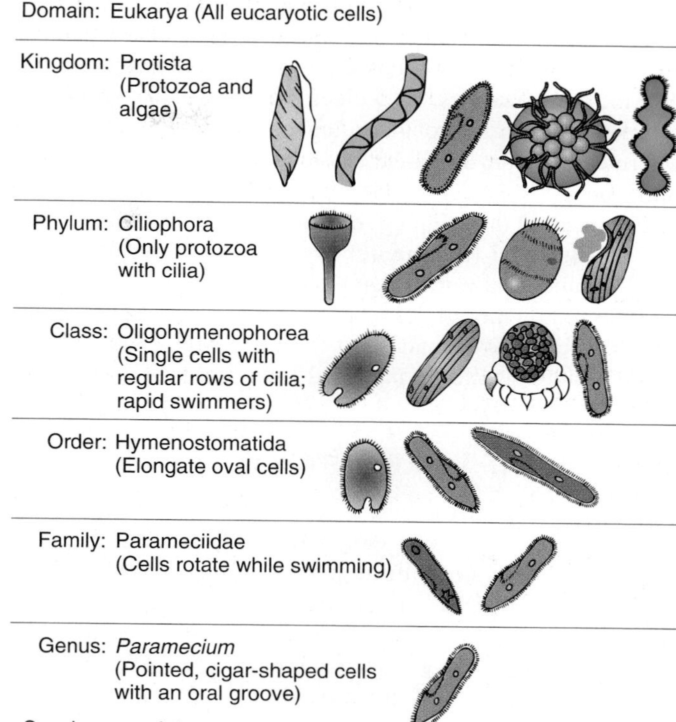

Figure 1.14
A common species of protozoon, *Paramecium caudatum,* traced through its taxonomic series. Note the gradual narrowing of the members, proceeding from general to specific levels.

schemes are to some extent artificial, certain groups of organisms do not exactly fit into the eight taxa. In that case, additional levels can be imposed immediately above (super) or below (sub) a taxon, giving us such categories as superphylum and subclass.

To illustrate the fine points of this system, we compare the taxonomic breakdowns of a human (figure 1.13) and a protozoan (figure 1.14). Humans and protozoa belong to the same domain (Eukarya) but are placed in different kingdoms. To emphasize just how broad the category kingdom is, ponder the fact that we belong to the same kingdom as sponges. Of the several phyla within this kingdom, humans belong to the Phylum Chordata (notochord-bearing animals), but even a phylum is rather all-inclusive, considering that humans share it with other vertebrates as well as creatures called sea squirts. The next level, Class Mammalia, narrows the field considerably by grouping only those vertebrates that have hair and suckle their young. The order that humans belong to is the Primates, a group having flexible hands and feet and five digits that also includes apes, monkeys, and lemurs. Next comes the Family Hominoidea, containing only humans and apes. The final levels are our genus, *Homo* (all races of modern and ancient humans), and our species, *sapiens* (meaning wise). Notice that for both the human and the protozoan, the categories become less inclusive and the individual members more closely related. Other examples of classification schemes are provided in sections of chapters 4 and 5 and in several later chapters.

It would be well to remember that all taxonomic **hierarchies*** are based on the judgment of scientists with certain expertise in a particular group of organisms and that not all other experts may agree with the system being used. Consequently, no taxa are permanent to any degree; they are constantly being revised and refined as new information becomes available or new viewpoints become prevalent. The detailed taxonomy of many groups is subject to lively discussion, but, because this text does not aim to emphasize details of taxonomy, we will usually be concerned with only the most general (kingdom, phylum) and specific (genus, species) levels.

ASSIGNING SPECIFIC NAMES

Many larger organisms are known by a common name suggested by certain dominant features. For example, a bird species might be called a red-headed blackbird or a flowering species a black-eyed susan. Some species of microorganisms (especially pathogens) are also called by informal names such as the gonococcus *(Neisseria gonorrhoeae)* or the leprosy bacillus *(Mycobacterium leprae),* but this is not the usual practice. If we were to adopt common names such as the "little yellow coccus" or the "club-shaped diphtheria

*hierarchy (hy′-ur-ar-kee) L. *hierarchia,* levels of power. Things arranged in the order of rank.

bacterium," the terminology would become even more cumbersome and challenging than scientific names. Even worse, common names are notorious for varying from region to region, even within the same country. A decided advantage of standardized nomenclature is that it provides a universal language, thereby enabling scientists from all countries on the earth to freely exchange information.

The method of assigning the **scientific,** or **specific, name** is called the **binomial (two-name) system of nomenclature.** The scientific name is always a combination of the generic (genus) name followed by the species name. The generic part of the scientific name is capitalized, and the species part begins with a lowercase letter. Both should be italicized (or underlined if italics are not available), as follows:

Campylobacter jejuni

Because other taxonomic levels are not italicized and consist of only one word, one can always recognize a scientific name. An organism's scientific name is sometimes abbreviated to save space, as in *C. jejuni,* but only if the genus name has already been stated. The source for nomenclature is usually Latin or Greek. If other languages such as English or French are used, the endings of these words are revised to have Latin endings. In general, the name first applied to a species will be the one that takes precedence over all others. An international group oversees the naming of every new organism discovered, making sure that standard procedures have been followed and that there is not already an earlier name for the organism or another organism with that same name. The inspiration for names is extremely varied and often rather imaginative. Some species have been named in honor of a microbiologist who originally discovered the microbe or who has made outstanding contributions to the field. Other names may designate a characteristic of the microbe (shape, color), a location where it was found, or a disease it causes. Some examples of specific names, their pronunciations, and their origins are:

1. *Homo sapiens* (hoh′-moh say′-pee-enz) Gr. *homo,* man, and *sapiens,* wise.
2. *Micrococcus luteus* (my′′-kroh-kok′-us loo′-tee-us) Gr. *micros,* small, and *kokkus,* berry; L. *luteus,* yellow. "The little yellow berry."
3. *Pseudomonas tomato* (soo′′-doh-mon′-us toh-may′-toh) Gr. *pseudo,* false, *monas,* unit, and *tomato,* the fruit. A bacterium that infects the common garden tomato.
4. *Corynebacterium diphtheriae* (kor-eye′′-nee-bak-ter′-ee-yum dif′-theer-ee-eye) Gr. *coryne,* club, *bacterion,* little rod, and *diphtheriae,* the causative agent of the disease diphtheria. "The club-shaped diphtheria bacterium."
5. *Campylobacter jejuni* (cam-pee′-loh-bak-ter jee-joo′-neye) Gr. *kampylos,* curved, *bakterion,* little rod, and *jejunum,* a section of intestine. One of the most important causes of intestinal infection worldwide.
6. *Lactobacillus sanfrancisco* (lak′′-toh-bass-ill′-us san-fran-siss′-koh) L. *lacto,* milk, and *bacillus,* little rod. A bacterial species used to make sourdough bread.
7. *Vampirovibrio chlorellavorus* (vam-py′-roh-vib′-ree-oh klor-ell-ah′-vor-us) F. *vampire;* L. *vibrio,* curved cell; *Chlorella,* a genus of green algae; and *vorus,* to devour. A small, curved bacterium that sucks out the cell juices of *Chlorella.*
8. *Giardia lamblia* (jee-ar′-dee-uh lam′-blee-uh) for Alfred Giard, a French microbiologist, and Vilem Lambl, a Bohemian physician, both of whom worked on the organism, a protozoan that causes a severe intestinal infection.

THE ORIGIN AND EVOLUTION OF MICROORGANISMS

Earlier we indicated that taxonomists perfer to use a system of classification that shows the degree of relatedness of organisms, one that places closely related organisms into the same categories. This pattern of organization, called a *natural* or *phylogenetic system,* often uses selected observable traits to form the categories.

A phylogenetic system is based on the concept of evolutionary relationships among types of organisms. **Evolution*** is an important theme that underlies all of biology, including microbiology. From its simplest standpoint, evolution states that living things change gradually through hundreds of millions of years and that these evolvements are expressed in various types of structural and functional changes through many generations. The process of evolution is selective: Those changes that most favor the survival of a particular organism or group of organisms tend to be retained and those that are less beneficial to survival tend to be lost. Space does not permit a detailed analysis of evolutionary theories, but the occurrence of evolution is supported by a tremendous amount of evidence from the fossil record and from the study of **morphology,*** **physiology,*** and **genetics** (inheritance). Evolution accounts for the millions of different species on the earth and their adaptation to its many and diverse habitats.

Evolution is founded on two preconceptions: (1) that all new species originate from preexisting species and (2) that closely related organisms have similar features because they evolved from common ancestral forms. Usually, evolution progresses toward greater complexity, and evolutionary stages range from simple, primitive forms that are close to an ancestral organism to more complex, advanced forms. Although we use the terms *primitive* and *advanced* to denote the degree of change from the original set of ancestral traits, it is very important to realize that all species presently residing on the earth are modern, but some have arisen more recently in evolutionary history than others.

The evolutionary patterns of organisms are often drawn as a family tree, with the trunk representing the main ancestral lines and the branches showing offshoots into specialized groups of organisms. This sort of arrangement places the more ancient groups at the bottom and the more recent ones at the top. The branches may also indicate origins, how closely related various organisms are, and an approximate time scale for evolutionary history (figures 1.15 and 1.16).

*evolution (ev-oh-loo′-shun) L. *evolutio,* to roll out.

*morphology (mor-fol′-oh-jee) Gr. *morphos,* form, and *logos,* to study. The study of organismic structure.

*physiology (fiz′′-ee-ol′-oh-jee) Gr. *physis,* nature. The study of the function of organisms.

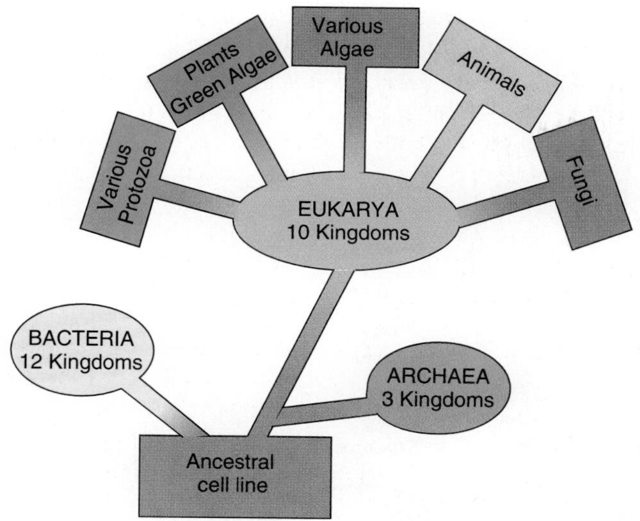

Figure 1.15
A system for representing the origins of cell lines and major taxonomic groups as proposed by Carl Woese and colleagues. They propose three distinct cell lines placed in superkingdoms called domains. The first primitive cells, called progenotes, were ancestors of both lines of procaryotes (Domains Bacteria and Archaea), and the Archaea emerged from the same cell line as eucaryotes (Domain Eukarya). Some of the traditional kingdoms are still present with this system (see Figure 1.16). Protozoa and some algal groups (called various algae here) are lumped into general categories.

Systems of Presenting a Universal Tree of Life

The first phylogenetic trees of life were constructed on the basis of just two kingdoms (plants and animals). In time, it became clear that certain organisms did not truly fit either of those categories, so a third kingdom for simpler organisms that lacked tissue differentiation (protists) was recognized. Eventually, when significant differences became evident even among the protists, Robert Whittaker proposed a fourth kingdom for the bacteria and a fifth one for the fungi.

Although biologists have found the system of five kingdoms and two basic cell types to be a valuable method of classification, recent studies in molecular biology have provided a more accurate view of the relationships and origins of cells. It has been determined that certain types of molecules in cells, called small ribosomal ribonucleic acid (rRNA), provide a "living record" of the evolutionary history of an organism. Analysis of this molecule in procaryotic and eucaryotic cells indicates that certain unusual cells called archaea (originally archaebacteria), are so different from the other two groups that they should be included in a separate superkingdom. Those same studies have also revealed that the cells of archaea though procaryotic in nature, are actually more closely related to eucaryotic cells than to bacterial cells (see table 4.7). To reflect these relationships, Carl Woese and George Fox have proposed a system that assigns all organisms to one of three domains, each described by a different type of cell (figure 1.15). The procaryotic cell types are placed in the Domains **Archaea** and **Bacteria.** Eucaryotes are all placed in the Domain **Eukarya.** It is believed that these three superkingdoms arose from an ancestor most similar to the archaea. This new system is still undergoing analysis and somewhat complicates the presentation of organisms in that it disposes of some traditional groups, although many of the traditional kingdoms still work within this framework (animals, plants, and fungi). The original Kingdom Protista is now a collection of protozoa and algae that exist in several separate kingdoms (see chapter 5). This new scheme will not greatly affect our presentation of most microbes, because we will be discussing them at the genus or species level. It is also an important truism that our methods of classification reflect our current understanding and are constantly changing as new information is uncovered.

In the interest of balance, we will also present the traditional Whittaker system of classification (figure 1.16). This system places all living things in one of five basic kingdoms: (1) the Procaryotae or Monera, (2) the Protista, (3) the Myceteae or Fungi, (4) the Plantae, and (5) the Animalia (figure 1.16). The simple, single-celled organisms at the base of the family tree are in the **Kingdom Procaryotae** (also called **Monera**). Because only those organisms with procaryotic cells are placed in it, the nature of cell structure is the main defining characteristic for this kingdom. It includes all of the microorganisms commonly known as **eubacteria,*** cells with typical procaryotic cell structure, and the **archaebacteria,*** cells with atypical cell structure that live in extreme environments (high salt and temperatures). Primitive procaryotes were the earliest cells to appear on the earth (see figure 4.30) and were the original ancestors of both more advanced bacteria and eucaryotic organisms. The Kingdom Procaryotae is a large and complex group that is surveyed in more detail in chapter 4.

The other four kingdoms contain organisms composed of eucaryotic cells. Their probable origin from procaryotic cells is discussed in microfile 5.1. The **Kingdom Protista*** contains mostly single-celled microbes that lack more complex levels of organization, such as tissues. Its members include both the microscopic algae, defined as independent photosynthetic cells with rigid walls; and the protozoans, animal-like creatures that feed upon other live or dead organisms and lack cell walls. More information on this group's taxonomy is given in chapter 5. The **Kingdom Myceteae*** contains the fungi, single- or multicelled eucaryotes that are encased in cell walls and absorb nutrients from other organisms (see chapter 5). With the exception of certain infectious worms and arthropods, the final two kingdoms, **Animalia** and **Plantae,** are generally not included in the realm of microbiology because most are large, multicellular organisms with tissues, organs, and organ systems. In general, animals move freely and feed on other organisms, whereas plants grow in an attached state and exhibit a nutritional scheme based on photosynthesis. It is possible to integrate the two systems as shown in figure 1.16.

Please note that viruses are *not* included in any of the classification or evolutionary schemes, because they are not cells and their position cannot be given with any confidence. Their special taxonomy is discussed in chapter 6.

---

*eubacteria (yoo''-bak-ter'-ee-uh) Gr. *eu,* true, and *bakterion,* little rod. All bacteria besides the archaebacteria.

*archaebacteria (ark''-ee-uh-bak-ter'-ee-uh) Gr. *archaios,* ancient. See above.

*Protista (pro-tiss'-tah) Gr. *protos,* the first.

*Myceteae (my-cee'-tee-eye) Gr. *mycos,* the fungi.

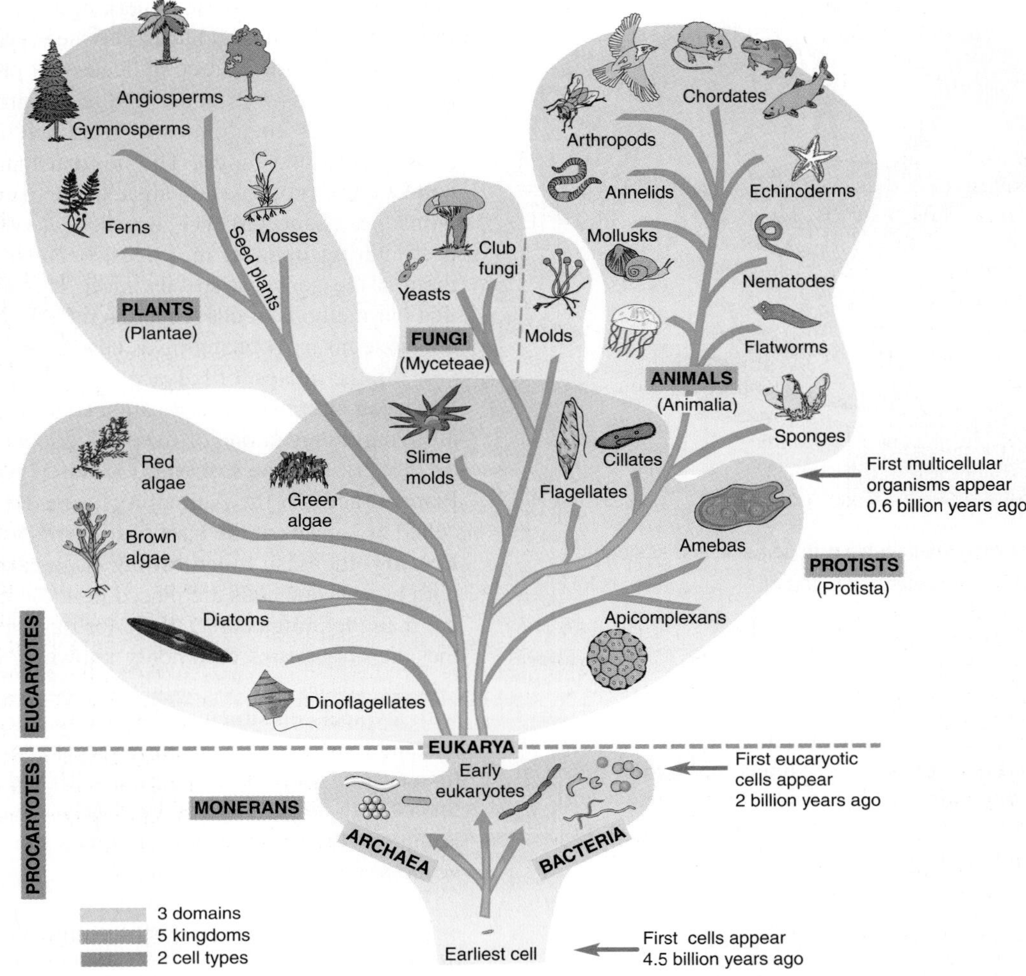

Figure 1.16

Traditional Whittaker system of classification. Kingdoms are based on cell structure and type, the nature of body organization, and nutritional type. Bacteria and Archaea (monerans) have procaryotic cells and are unicellular. Protists have eucaryotic cells and are mostly unicellular. They can be photosynthetic (algae), or they can feed on other organisms (protozoa). Fungi have eucaryotic cells and are unicellular or multicellular; they have cell walls and are not photosynthetic. Plants have eucaryotic cells, are multicellular, have cell walls, and are photosynthetic. Animals have eucaryotic cells, are multicellular, do not have cell walls, and derive nutrients from other organisms.
After Dolphin Biology Lab Manual 4th ed. Fig. 14.1, pg 177 McGraw Hill/Wm. C Brown.

Chapter Checkpoints

Taxonomy is the formal filling system scientists use to classify living organisms. It puts every organism in its place and makes a place for every living organism.

The taxonomic system has three primary functions: classification, nomenclature, and identification of species.

The eight major taxa, or groups, in the taxonomic system are (in descending order): domain, kingdom, phylum or division, class, order, family, genus, and species.

The binomial system of nomenclature describes each living organism by two names: genus and species.

Taxonomy groups organisms by phylogenetic similarity, which in turn is based on evolutionary similarities in morphology, physiology, and genetics.

Evolutionary patterns show a treelike branching from simple, primitive life forms to complex, advanced life forms.

The Woese-Fox classification system places all eucaryotes in the Domain (Superkingdom) Eukarya and subdivides the procaryotes into the two Domains Archaea and Bacteria.

The Whittaker five-kingdom classification system places all bacteria in the Kingdom Procaryotae and subdivides the eucaryotes into Kingdoms Protista, Myceteae, Animalia, and Plantae.

CHAPTER CAPSULE WITH KEY TERMS

Microbiology is the study of **bacteria, viruses, fungi, protozoa,** and **algae,** which are collectively called **microorganisms** or **microbes.** In general, microorganisms are **microscopic** and, unlike **macroscopic** organisms, which are readily visible, they require magnification to be seen.

Microbes can cause disease, but they are also an integral part of nature and are indispensable for normal, balanced life as we know it. Microbes are responsible for spoilage, decay, and disease, but they are also used in the manufacture of foods, beverages, and industrial chemicals, **bioremediation,** drug synthesis, and **genetic engineering.**

The unicellular simplicity, growth rate, and **adaptability** of microbes are some of the reasons that microbiology is so diverse and has branched out into many subsciences and applications. Early microbiology blossomed with the conceptual developments of **sterilization, aseptic techniques,** and the **germ theory of disease.**

CHARACTERISTICS OF MICROORGANISMS

Organisms can be described according to their composition or way of life. For instance, the cells of **eucaryotic** organisms contain a nucleus, but those of **procaryotic** organisms do not. Most organisms are free-living, but a few are **parasites;** some parasites are **infectious.**

Microbes can be measured in micrometers, dimensions too small to be seen without a microscope. With his simple microscope, Leeuwenhoek discovered organisms he called animalcules. As a consequence of his findings and the rise of the **scientific method,** the notion of **spontaneous generation,** or **abiogenesis,** was eventually abandoned for **biogenesis.** The scientific method applies **inductive** and **deductive** reasoning to develop rational **hypotheses** and **theories** that can be tested. Principles that withstand repeated scrutiny become **law** in time.

CLASSIFICATION OF MICROORGANISMS

Taxonomy is a hierarchy scheme for the **classification, identification,** and **nomenclature** of organisms, which are grouped in categories called **taxa,** based on features ranging from general to specific. Starting with the broadest category, the taxa are **domain, kingdom, phylum** (or **division**), **class, order, family, genus,** and **species.** Organisms are assigned **binomial scientific names** consisting of their genus and species names. The latest classification scheme for living things is based on the genetic structure of their ribosomes. The Woese-Fox system recognizes three domains: **Archaea,** simple procaryotes that live in extremes; **Bacteria** typical procaryotes; and **Eukarya,** all type, of eucaryotic organisms.

An alternative classification scheme uses a simpler five-kingdom organization: **Kingdom Procaryotae (Monera),** containing the **eubacteria** and the **archaebacteria; Kingdom Protista,** containing primitive unicellular microbes such as algae and protozoa; **Kingdom Myceteae,** containing the fungi; **Kingdom Animalia,** containing animals; and **Kingdom Plantae,** containing plants. Inspection of the **morphology, physiology,** and **genetics** of organisms reveals an ancestral **evolutionary** relationship among these kingdoms.

MULTIPLE-CHOICE QUESTIONS

Select the correct answer from the answers provided, using strategies suggested in the preface. For questions with blanks, choose the combination of answers that most accurately completes the statement.

1. Which of the following is not considered a microorganism?
   a. alga
   b. bacterium
   c. protozoan
   d. mushroom
2. A prominent difference between procaryotic and eucaryotic cells is the
   a. larger size of procaryotes
   b. lack of pigmentation in eucaryotes
   c. presence of a nucleus in eucaryotes
   d. presence of a cell wall in procaryotes
3. Which of the following parts was absent from Leeuwenhoek's microscopes?
   a. focusing screw
   b. lens
   c. specimen holder
   d. condenser
4. Abiogenesis refers to the
   a. spontaneous generation of organisms from nonliving matter
   b. development of life forms from preexisting life forms
   c. development of aseptic techniques
   d. germ theory of disease
5. A hypothesis can be defined as
   a. a belief based on knowledge
   b. knowledge based on belief
   c. a scientific explanation that is subject to testing
   d. a theory that has been thoroughly tested
6. Which early microbiologist was most responsible for developing sterile laboratory techniques?
   a. Louis Pasteur
   b. Robert Koch
   c. Carl von Linné
   d. John Tyndall
7. Which scientist is most responsible for finally laying the theory of spontaneous generation to rest?
   a. Joseph Lister
   b. Robert Koch
   c. Francesco Redi
   d. Louis Pasteur
8. The process of observing an event and then constructing a hypothesis to explain it involves
   a. inductive reasoning
   b. deductive reasoning
   c. a controlled experiment
   d. guesswork
9. When a hypothesis has been thoroughly supported by long-term study and data, it is considered
   a. a law
   b. a speculation
   c. a theory
   d. proved
10. Which is the correct order of the taxonomic categories, going from most specific to most general?
    a. domain, kingdom, phylum, class, order, family, genus, species
    b. division, domain, kingdom, class, family, genus, species
    c. species, genus, family, order, class, phylum, kingdom, domain
    d. species, family, class, order, phylum, kingdom

11. By definition, organisms in the same ____ are more closely related than are those in the same ____ .
    a. order, family
    b. class, phylum
    c. family, genus
    d. phylum, division

12. Order the following items by size, using numbers: 1 = smallest and 8 = largest.
    ____ AIDS virus
    ____ ameba
    ____ rickettsia
    ____ protein
    ____ worm
    ____ coccus bacterium
    ____ white blood cell
    ____ atom

CONCEPT QUESTIONS

These questions are suggested as a *writing-to-learn* experience. For each question, compose a one- or two-paragraph answer that includes the factual information needed to completely address the question. Discuss the concepts in a sequence that allows you to present the subject using clear logic and correct terminology.

1. Identify the groups of microorganisms included in the scope of microbiology, and explain the criteria for including these groups in the field.
2. Briefly identify the subdivisions of microbiology and tell what is studied in each. What do the following microbiologists study: algologist, epidemiologist, biotechnologist, ecologist, virologist, and immunologist?
3. Why was the abandonment of the spontaneous generation theory so significant? Using the scientific method, describe the steps you would take to test the theory of spontaneous generation.
4. a. Explain how inductive reasoning and deductive reasoning are similar and different.
   b. What are variables and controls?
   c. Look at figure 1.10 and answer the question at the bottom of the figure.
5. a. Differentiate between a hypothesis and a theory.
   b. Is the germ theory of disease really a law, and why?
6. a. Differentiate between taxonomy, classification, and nomenclature.
   b. What is the basis for a phylogenetic system of classification?
   c. What is a binomial system of nomenclature, and why is it used?
   d. Give the correct order of taxa, going from most general to most specific. A mnemonic (memory) device for recalling the order is Darling King Phillip Came Over For Good Spaghetti.
7. a. Construct a table that compares cell types and places them into domains and kingdoms. In which kingdoms do we find microorganisms?
   b. Compare the new domain system with the five-kingdom system. Does the newer system change the basic idea of procaryotes and eucaryotes? What is the third cell type?

CRITICAL-THINKING QUESTIONS

Critical thinking is the ability to reason and solve problems using facts and concepts. It requires you to apply information to new or different circumstances, to integrate several ideas to arrive at a solution, and to perform practical demonstrations as part of your analysis. These questions can be approached from a number of angles, and, in most cases, they do not have a single correct answer.

1. What do you suppose the world would be like if there were cures for all infectious diseases and a means to destroy all microbes? What characteristics of microbes will prevent this from ever happening?
2. a. Where do you suppose the "new" infectious diseases come from?
   b. Name some factors that could cause older diseases to show an increase in the number of cases.
   c. Comment on the sensational ways that some tabloid media portray infectious diseases to the public.
3. a. Add up the numbers of deaths worldwide from infectious diseases (figure 1.4). Look up each disease in the index and see which ones could be prevented by vaccines or treated with drugs. How many do you think could have been prevented by modern medicine?
   b. Think of several reasons for the differences in the major causes of death in developing vs developed nations (figure 1.3)?
4. What events, discoveries, or inventions were probably the most significant in the development of microbiology?
5. List the major variables in abiogenesis outlined in microfile 1.1, and explain how each was tested and controlled by the scientific method.
6. Can you develop a scientific hypothesis and means of testing the cause of stomach ulcers? (Is it caused by an infection? By too much acid? A genetic disorder?)
7. Construct the scientific name of a newly discovered species of bacterium, using your name, a pet's name, a place, or a unique characteristic. Be sure to use proper notation and endings.

INTERNET SEARCH TOPIC

Access a search engine on the World Wide Web under the heading *emerging diseases*. Adding terms like WHO and CDC will refine your search and take you to several appropriate web sites. List the top 10 emerging diseases in the United States and worldwide.

chapter 2

FROM ATOMS TO CELLS: A Chemical Connection

In laboratories all over the world, sophisticated technology is being developed for a wide variety of scientific applications. Refinements in molecular biology techniques now make it possible to routinely identify microorganisms, detect genetic disease, diagnose cancer, sequence the genes of organisms, break down toxic wastes, synthesize drugs and industrial products, and genetically engineer microorganisms, plants, and animals. A common thread that runs through new technologies and hundreds of traditional techniques is that, at some point, they involve chemicals and chemical reactions. In fact, if nearly any biological event is traced out to its ultimate explanation, it will invariably involve atoms, molecules, reactions, and bonding.

It is this relationship between the sciences that makes a background in chemistry necessary to biologists and microbiologists. Students with a basic chemistry background will enhance their understanding of and insight into microbial structure and function, metabolism, genetics, drug therapy, immune reactions, and infectious disease. This chapter has been organized to promote a working knowledge of atoms, molecules, bonding, solutions, pH, and biochemistry and to build foundations to later chapters. It concludes with an introduction to cells and a general comparison of procaryotic and eucaryotic cells as a preparation for chapters 4 and 5.

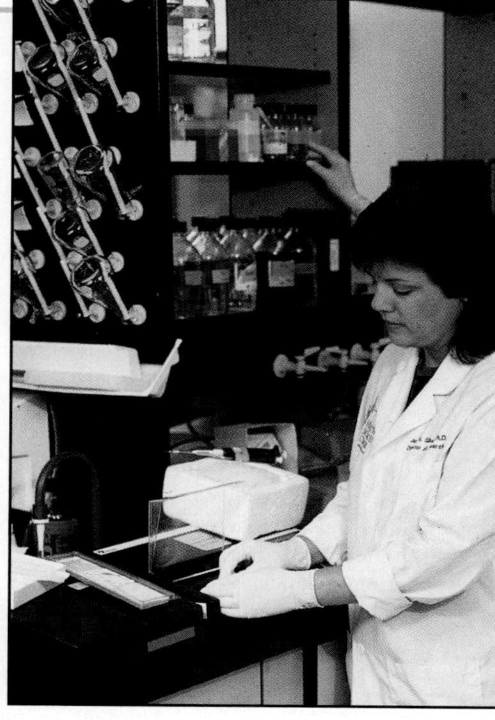

Molecular biologists use sophisticated chemical techniques and equipment to extract and analyze the DNA from human cells. Here Dr. Jane Gibson of Orlando Regional Healthcare System is testing patients' samples to detect mutations in genes for ovarian and breast cancer.

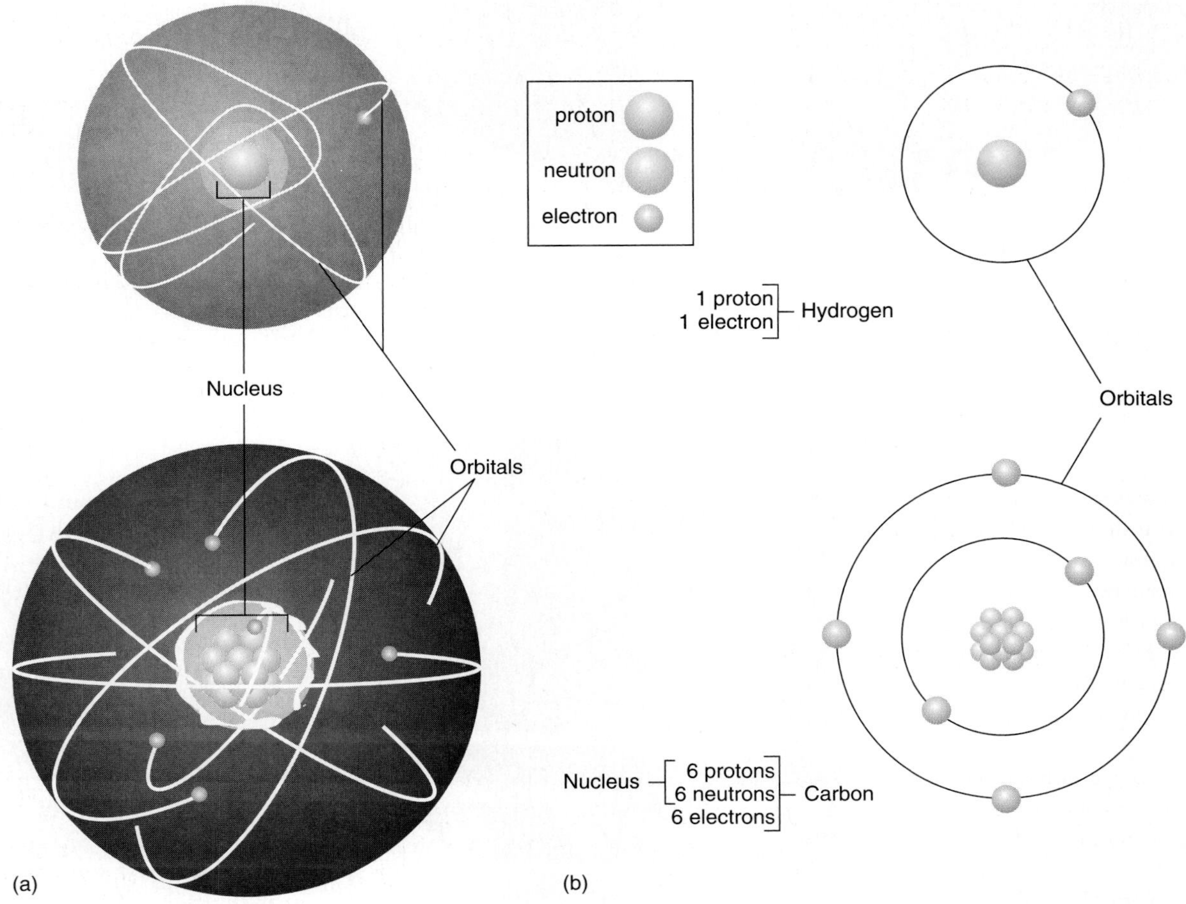

Figure 2.1
Diagrams of atoms. (*a*) Models of a hydrogen atom and a carbon atom depict the nucleus surrounded by a dense cloud, indicating potential position of electrons in orbitals and shells. (*b*) Working models of these same two atoms make it easier to visualize the numbers and arrangements of electrons, with orbitals shown as concentric circles.

ATOMS, BONDS, AND MOLECULES: FUNDAMENTAL BUILDING BLOCKS

The universe is composed of an infinite variety of substances existing in the gaseous, liquid, and solid states. All such tangible materials that occupy space and have mass are called **matter.** The organization of matter—whether air, rocks, or bacteria—begins with individual building blocks called atoms. An **atom*** is defined as a tiny particle that has special structural properties. In general, it cannot be subdivided into smaller substances without losing those properties. Even in a science dealing with very small things, an atom's minute size is striking; for example, an oxygen atom is only 0.0000000013 mm (0.013 Å) in diameter and one million of them in a cluster would barely be visible to the naked eye.

Although scientists have not yet observed the detailed structure of an atom even under the highest power microscopes, the exact composition of atoms has been well established by extensive physical analysis using sophisticated instruments. In general, an atom derives its properties from a combination of subatomic particles called **protons** (p^+), which are positively charged, **neutrons** (n^0), which have no charge (are neutral), and **electrons** (e^-), which are negatively charged. The relatively larger protons and neutrons make up a central core, or **nucleus,** that is surrounded by one or more constantly moving electrons (figure 2.1). The nucleus makes up the larger mass (weight) of the atom, whereas the electron region accounts for the greater volume. To get a perspective on proportions, consider this: If an atom were the size of a football stadium, the nucleus would be about the size of a marble! The stability of atomic structure is largely maintained by: (1) the mutual attraction of the protons and electrons (opposite charges attract each other) and (2) the exact balance of proton number and electron number, which causes the opposing charges to cancel each other out. At least in theory then, isolated, intact atoms do not carry a charge.

*atom (at´-um) Gr. *atomos,* not cut.

TABLE 2.1

THE MAJOR ELEMENTS OF LIFE AND THEIR PRIMARY CHARACTERISTICS

Element	Atomic Symbol*	Atomic Number	Atomic Weight	Ionized Form**	Significance in Microbiology
Calcium	Ca	20	40.1	Ca^{++}	Part of outer covering of certain shelled amebas; stored within bacterial spores
Carbon	C	6	12.0	—	Principal structural component of biological molecules
		6	14.0	—	Isotope used in dating fossils
Chlorine	Cl	17	35.5	Cl^{-}	Component of disinfectants; used in water purification
Cobalt	Co	27	58.9	Co^{++}, Co^{+++}	Trace element needed by some bacteria to synthesize vitamins
		27	60	—	An emitter of gamma rays; used in food sterilization; used to treat cancer
Copper	Cu	29	63.5	Cu^{+}, Cu^{++}	Necessary to the function of some enzymes; Cu salts are used to treat fungal and worm infections
Hydrogen	H	1	1	H^{+}	Necessary component of water and many organic molecules; H_2 gas released by bacterial metabolism
		1	3	—	Tritium has 2 neutrons; radioactive; used in clinical laboratory procedures
Iodine	I	53	126.9	I^{-}	A component of antiseptics and disinfectants; contained in a reagent of the Gram stain
		53	131, 125		Radioactive isotopes for diagnosis and treatment of cancers
Iron	Fe	26	55.8	Fe^{++}, Fe^{+++}	Necessary component of respiratory enzymes; some microbes require it to produce toxin
Magnesium	Mg	12	24.3	Mg^{++}	A trace element needed for some enzymes; component of chlorophyll pigment
Manganese	Mn	25	54.9	Mn^{++}, Mn^{+++}	Trace elements for certain respiratory enzymes
Nitrogen	N	7	14.0	—	Component of all proteins and nucleic acids; the major atmospheric gas
Oxygen	O	8	16.0	—	An essential component of many organic molecules; molecule used in metabolism by many organisms
Phosphorus	P	15	31	—	A component of ATP, nucleic acids, cell membranes; stored in granules in cells
		15	32	—	Radioactive isotope used as a diagnostic and therapeutic agent
Potassium	K	19	39.1	K^{+}	Required for normal ribosome function and protein synthesis; essential for cell membrane permeability
Sodium	Na	11	23.0	Na^{+}	Necessary for transport; maintains osmotic pressure; used in food preservation
Sulfur	S	16	32.1	—	Important component of proteins; makes disulfide bonds; storage element in many bacteria
Zinc	Zn	30	65.4	Zn^{++}	An enzyme cofactor; required for protein synthesis and cell division; important in regulating DNA

*Based on the Latin name of the element. The first letter is always capitalized; if there is a second letter, it is always lowercased.
**A dash indicates that there is no ionized form that is commonly active in living systems.

DIFFERENT TYPES OF ATOMS: ELEMENTS AND THEIR PROPERTIES

All atoms share the same fundamental structure. All protons are identical, all neutrons are identical, and all electrons are identical. But when these subatomic particles come together in specific, varied combinations, unique types of atoms called **elements** result. Each element has a characteristic atomic structure and predictable chemical behavior. To date, 92 naturally occurring elements have been described, and 18 have been produced artificially by physicists. By convention, an element is assigned a distinctive name with an abbreviated shorthand symbol. Table 2.1 lists some of the elements common to biological systems, their atomic characteristics, and some of the natural and applied roles they play.

THE MAJOR ELEMENTS OF LIFE AND THEIR PRIMARY CHARACTERISTICS

The unique properties of each element result from the numbers of protons, neutrons, and electrons it contains, and each element can be identified by certain physical measurements. The **atomic weight,** or **mass,**[1] is equal to the sum of the number of protons plus the number of neutrons. Atomic weight is not only a measure of weight but also a relative scale for comparing atomic size, because larger, heavier elements have higher atomic weights. An element is also assigned an **atomic number (AN),** based on the number of protons alone. By knowing the atomic number, we also automatically know the number of electrons carried by each element because all atoms have an equal number of protons and electrons. Hydrogen is a unique element because it has only a single proton, a single electron, and no neutron, making it the only element with the same atomic weight and number.

Table 2.1 also indicates that certain elements exist in more than one form. Closer inspection reveals that these forms have the same atomic number but different atomic weights (see iodine). These variant forms of an element, called **isotopes,** differ only in the number of neutrons in the nucleus. The nuclei of some isotopes are unstable and spontaneously break down, releasing energy in the form of radiation. Such *radioactive isotopes* are widely used in research, in treatment of disease, as tracers in certain types of immunologic testing, and in some forms of sterilization (see chapter 11 on ionizing radiation).

Electron Orbitals and Shells

Atomic models such as that in figure 2.1*b* make it appear as though electrons are small bodies orbiting the nucleus like planets around a sun. It is important to realize, however, that these diagrams do not show exact structure but rather provide a type of atomic map that economically depicts the major properties of elements. A model of an actual atom is difficult to represent on paper because the electrons move in a three-dimensional cloud and not in perfect concentric circles (figure 2.1*a*). By convention, the overall pathway of an electron's rotational movement is termed its **orbital,** and its energy level falls within a space called a **shell** (figure 2.1). The term *orbital* is derived from the idea that electrons exist in rotating patterns around the nucleus; the shell is named for the notion that electron energy levels form concentric strata like the layers of an onion. Each orbital has a finite number of electrons, and each shell contains one or more orbitals. The electrons and their orbitals vary in energy; the lowest energy electrons lie closest to the nucleus, and the highest energy electrons lie farthest away.

Electrons fill the orbitals *in pairs* according to a predictable pattern (figure 2.2). The first orbital can contain one pair (2 electrons), the second orbital can contain up to four pairs (8 electrons), the third can hold a maximum of nine pairs (18), and the fourth can contain up to sixteen pairs (32). The number of orbitals and how completely they are occupied depends on the number of electrons contained by a given element. Orbitals fill up from inside out, and those elements with higher atomic numbers will have more electrons, and thus more orbitals. For example, helium (AN = 2) has a single filled first orbital of 2 e^-; oxygen (AN = 8) has a filled first orbital (2 e^-) and a partially filled second one (6 e^-); and magnesium (AN = 24) has a filled first orbital (2 e^-), a filled second one (8 e^-), and a nearly empty third orbital (2 e^-). As we will see, the chemical properties of an element are largely controlled by the distribution of electrons in the outermost shell. Selected elements and their electron maps are depicted in figure 2.2.

Chapter Checkpoints

- **Protons (p^+) and neutrons(n^o) make up the nucleus of an atom. Electrons (e^-) orbit the nucleus.**
- **All elements are composed of atoms but differ in the numbers of protons, neutrons, and electrons they possess.**
- **Elements are identified by *atomic weight,* or *mass,* or by *atomic number.***
- **Isotopes are varieties of one element that contain the same number of protons but different numbers of neutrons.**
- **The number of electrons in an element's outermost orbital (compared with the total number possible) determines its chemical nature. As the outer orbital fills with electrons, its reactivity decreases.**

BONDS AND MOLECULES

Most elements do not exist naturally in pure, uncombined form but are bound together as molecules and compounds. A **molecule*** is a distinct chemical substance that results from the combination of two or more atoms. A molecule consists of atoms of the same element, as in oxygen (O_2) and nitrogen gas (N_2), or atoms of dissimilar elements, as is the case with biological molecules (sugars, fats, proteins). When a molecule is a combination of two or more *different* elements in a certain fixed ratio, it is called a **compound.** Water (H_2O), for example, has twice as much hydrogen as oxygen. When atoms bind together in molecules, they lose the properties of the atom and take on the properties of the combined substance. In the same way that an atom has an atomic weight, a molecule has a molecular weight (MW), which is calculated from the sum of all of the atomic weights of the atoms it contains.

The **chemical bonds** of molecules and compounds result when two or more atoms share, donate (lose), or accept (gain) electrons (figure 2.3). The number of electrons in the outermost shell of an element is known as its **valence.*** The valence deter-

1. Atomic weight is measured in atomic mass units, also known as daltons.

*molecule (mol´-ih-kyool) L. *molecula,* little mass.

*valence (vay´-lents) L. *valentia,* strength. A measure of atomic binding capacity.

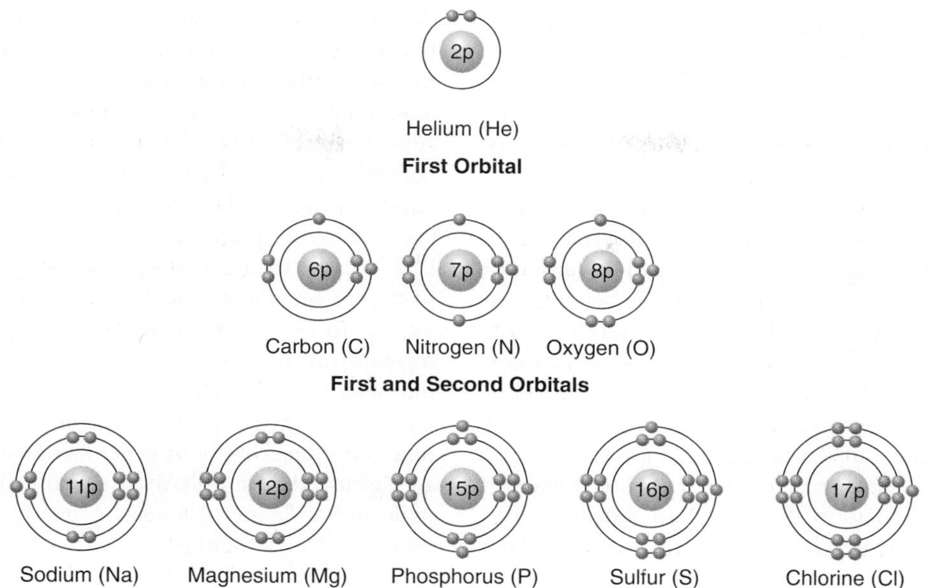

Figure 2.2
Models of several elements show how the shells are filled by electrons as the atomic numbers increase (numbers noted inside nuclei). Electrons tend to appear in pairs, but certain elements have incompletely filled outer shells.

mines the degree of reactivity and the types of bonds an element can make. Elements with a filled outer orbital are relatively stable because they have no extra electrons to share with or donate to other atoms. Elements with partially filled outer orbitals are less stable and are more apt to form some sort of bond. For example, helium has one filled shell, with no tendency either to give up electrons or to take them from other elements, making it a stable, inert (nonreactive) gas. By comparison, an oxygen atom has 6 electrons in its outer shell, which is 2 electrons less than full. This extra capacity confers a quality of reactivity in which oxygen can share or accept electrons. Many bonding interactions are based on the tendency of atoms with unfilled outer shells to gain greater stability by achieving, or at least approximating, a filled outer shell. For example, an atom such as oxygen that can accept 2 additional electrons will bond readily with atoms (such as hydrogen) that can share or donate electrons. We explore some additional examples of the basic types of bonding in the following section.

In addition to reactivity, the number of electrons in the outer shell also dictates the number of chemical bonds an atom can make. For instance, hydrogen can bind with one other atom, oxygen can bind with up to two other atoms, and carbon can bind with four (see figure 2.13).

Covalent Bonds and Polarity: Molecules with Shared Electrons

Covalent (cooperative valence) **bonds** form between atoms with valences that suit them to sharing electrons rather than to donating or receiving them. A simple example is hydrogen gas (H_2), which

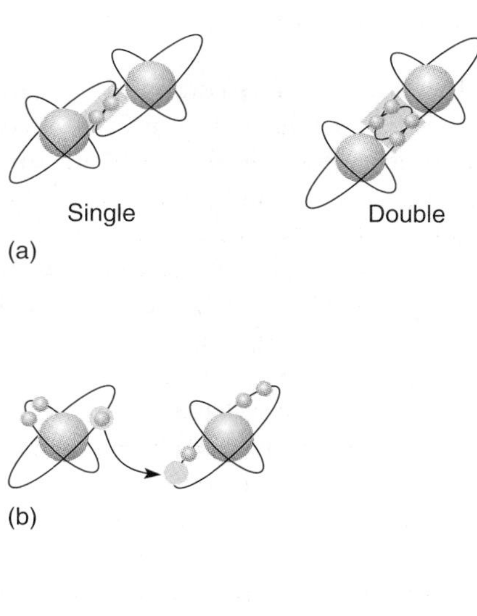

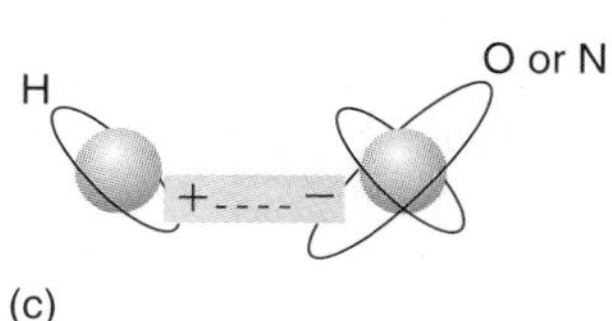

Figure 2.3
General representation of three types of bonding. (*a*) Covalent bonds, both single and double. (*b*) An ionic bond. (*c*) Hydrogen bonds. Note that hydrogen bonds are represented in models and formulas by dotted lines, as shown in (c).

consists of two hydrogen atoms. A hydrogen atom has only a single electron, but when two of them combine, each will bring its electron to orbit about both nuclei, thereby approaching a filled orbital (2 electrons) for both atoms and thus creating a **single covalent bond** (figure 2.4*a*). Covalent bonding also occurs in oxygen gas (O_2), but with a difference. Because each atom has 2 electrons to share in this molecule, the combination creates two pairs of shared electrons, also known as a **double covalent bond** (figure 2.4*b*). The majority of the molecules associated with living things are composed of single and double covalent bonds between the most common biological elements (carbon, hydrogen, oxygen, nitrogen, sulfur, and phosphorus), which are discussed in more depth in chapters 7 and 8. A slightly more complex pattern of covalent bonding is shown for methane gas (CH_4) in figure 2.4*c*.

When covalent bonds are formed between atoms of the same or similar atomic weight, the electrons are shared equally between the two atoms. Because of this balanced distribution no part of the molecule has a greater attraction for the electrons. This sort of electrically neutral molecule is termed **nonpolar.** When two atoms of different atomic weight bind covalently, the electrons are not shared equally and may be pulled more toward one atom than another. This pull causes one end of a molecule to assume a partial negative charge and the other end to assume a partial positive charge. A molecule with such an asymmetrical distribution of charges is termed **polar** and has positive and negative poles. Observe the water molecule shown in figure 2.5 and note that, because the oxygen atom is larger and has more protons than the hydrogen atoms, it will tend to draw the shared electrons with greater force toward its nucleus. This unequal force causes the oxygen part of the molecule to express a negative charge (due to the electrons' being attracted there) and the hydrogens to express a positive charge (due to the protons). The polar nature of water plays an extensive role in a number of biological reactions, which are discussed later. Polarity is a significant property of many large molecules in living systems and greatly influences both their reactivity and their structure.

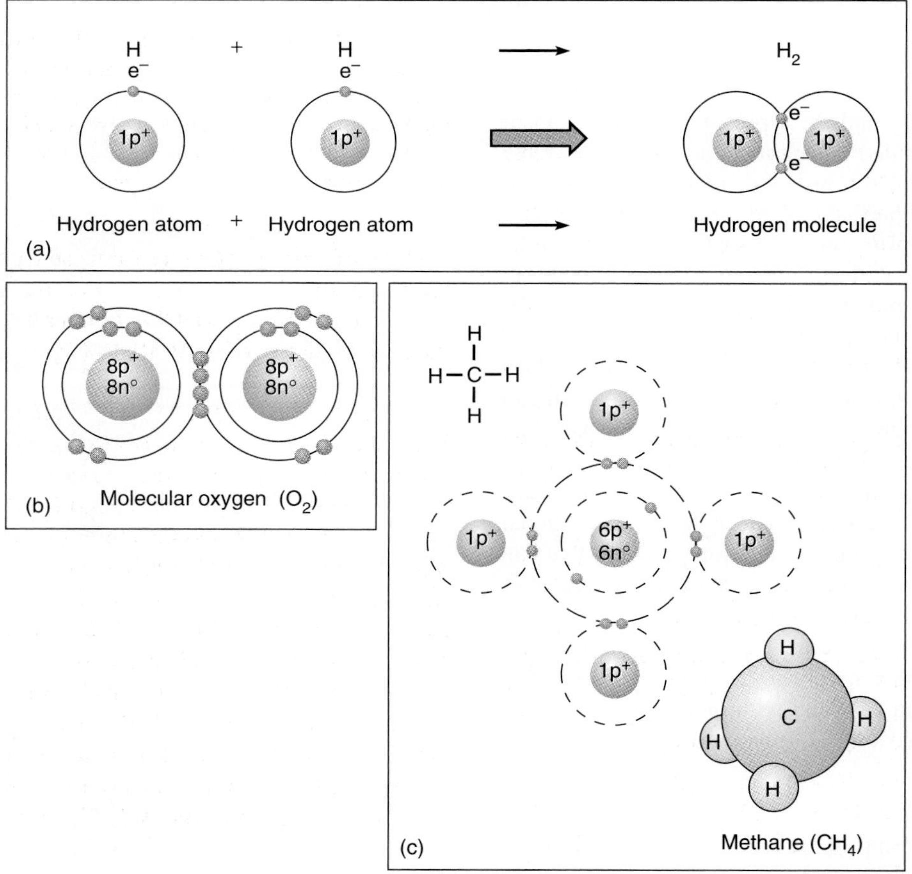

Figure 2.4
Examples of molecules with covalent bonding. (*a*) A hydrogen molecule is formed when two hydrogen atoms share their electrons and form a single bond. (*b*) In a double bond, the outer orbitals of two oxygen atoms overlap and permit the sharing of 4 electrons (one pair from each) and the saturation of the outer orbital for both. (*c*) Simple, working, and three-dimensional models of methane. Note that carbon has 4 electrons to share and hydrogens each have one, thereby completing the shells for all atoms in the compound.

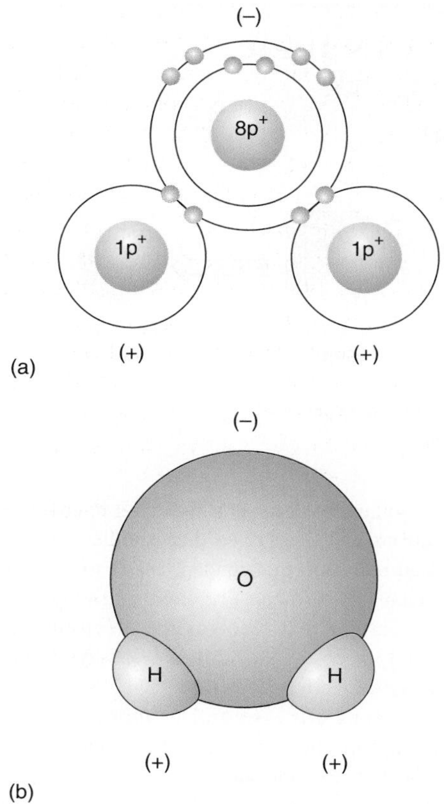

Figure 2.5
(*a*) A simple model and (*b*) a three-dimensional model of a water molecule indicate the polarity, or unequal distribution, of electrical charge, which is caused by the pull of the shared electrons toward the oxygen side of the molecule.

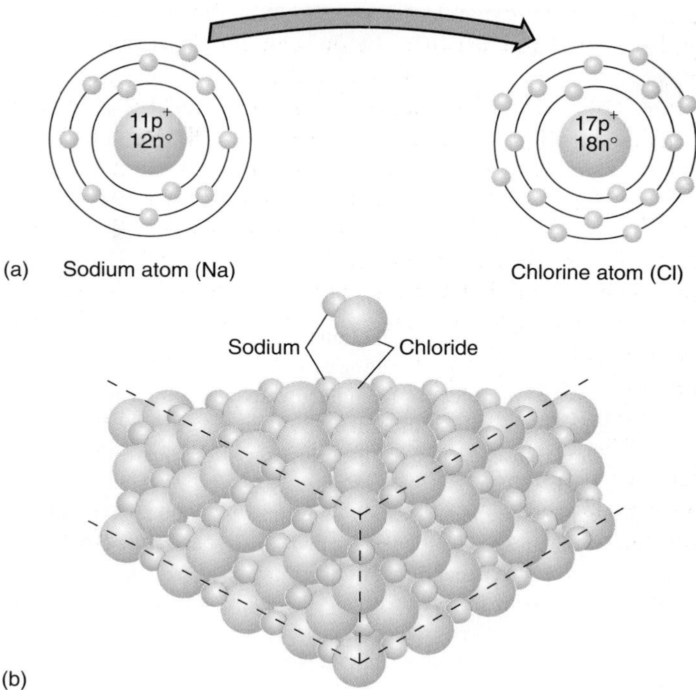

Figure 2.6
Ionic bonding between sodium and chlorine. (*a*) When the two elements are placed together, sodium loses its single outer orbital electron to chlorine, thereby saturating chlorine's outer orbital. (*b*) This reaction produces sodium and chloride ions that form large complexes, or crystals, in which the two atoms alternate in a definite, regular, geometric pattern.

Ionic Bonds: Electron Transfer Among Atoms

In reactions that form **ionic bonds,** electrons are transferred completely from one atom to another and are not shared. These reactions invariably occur between atoms with valences that complement each other, meaning that one atom has an unfilled shell that will readily accept electrons and the other atom has an unfilled shell that will readily lose electrons. A striking example is the reaction that occurs between sodium (Na) and chlorine (Cl) (figure 2.6). Elemental sodium is a soft, lustrous metal so reactive that it can burn flesh, and molecular chlorine is a very poisonous yellow gas. But when the two are combined, they form sodium chloride[2] (NaCl)—the familiar nontoxic table salt—a compound with properties quite different from either parent element.

How does this transformation occur? Sodium has 11 electrons (2 in shell one, 8 in shell two, and only 1 in shell three), so it is 7 short of having a complete outer shell. Chlorine has 17 electrons (2 in shell one, 8 in shell two, and 7 in shell three), making it 1 short of a complete outer shell. These two atoms are very reactive with one another, because a sodium atom will readily donate its single electron and a chlorine atom will avidly receive it. (The reaction is slightly more involved than a single sodium atom's combining with a single chloride atom (microfile 2.1), but this complexity does not detract from the fundamental reaction as described here.) The outcome of this reaction is not many single, isolated molecules of NaCl but rather a solid crystal complex that interlinks millions of these atoms (figure 2.6*b*).

Ionization: Formation of Charged Particles

Molecules with intact ionic bonds are electrically neutral, but they can produce charged particles when dissolved in a liquid called a solvent. This phenomenon, called **ionization,** occurs when the ionic bond is broken and the atoms dissociate (separate) into unattached, charged particles called **ions*** (figure 2.7). To illustrate what imparts a charge to ions, let us look again at the reaction between sodium and chlorine. When a sodium atom reacts with chlorine and loses one electron, the sodium is left with one more proton than electrons. This imbalance produces a positively charged sodium ion (Na^+). Chlorine, on the other hand, has gained one electron and now has one more electron than protons, producing a negatively charged ion (Cl^-). Positively charged ions are termed **cations,*** and negatively charged ions are termed **anions.***

2. In general, when a salt is formed, the ending of the name of the negatively charged ion is changed to *-ide*.

*ion (eye´-on) Gr. *ion,* going.
*cation (kat´-eye-on) An ion that migrates toward the negative pole, or cathode, of an electrical field.
*anion (an´-eye-on) An ion that migrates toward the positive pole, or anode.

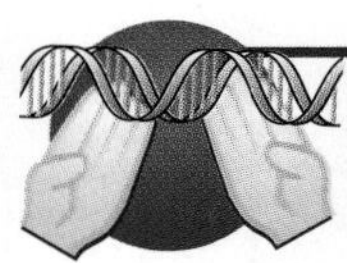

MICROFILE 2.1 REDOX-ELECTRON TRANSFER AND OXIDATION-REDUCTION REACTIONS

The metabolic work of cells, such as synthesis, movement, and digestion, revolves around energy exchanges and transfers. The management of energy in cells is almost exclusively dependent on chemical rather than physical reactions because most cells are far too delicate to operate with heat, radiation, and other more potent forms of energy. The outer orbital electrons are readily portable and easily manipulated sources of energy. It is in fact the movement of electrons from molecule to molecule that accounts for most energy exchanges in cells. Fundamentally then, a cell must have a supply of atoms that can gain or lose electrons if they are to carry out life processes.

The phenomenon in which electrons are transferred from one atom or molecule to another is termed an **oxidation** and **reduction** (shortened to **redox**) **reaction.** Although the term *oxidation* was originally adopted for reactions involving the addition of oxygen, the term oxidation can include any reaction causing electron release, regardless of the involvement of oxygen. By comparison, reduction is any reaction that causes an atom to receive electrons. All redox reactions occur in pairs. To analyze the phenomenon, let us again review the production of NaCl, but from a different standpoint. Although it is true that these atoms form ionic bonds, the chemical combination of the two is also a type of redox reaction.

When these two atoms react to form sodium chloride, a sodium atom gives up an electron to a chlorine atom. During this reaction, sodium is oxidized because it loses an electron, and chlorine is reduced because it gains an electron. To take this definition further, an atom or molecule, such as sodium, that can donate electrons and

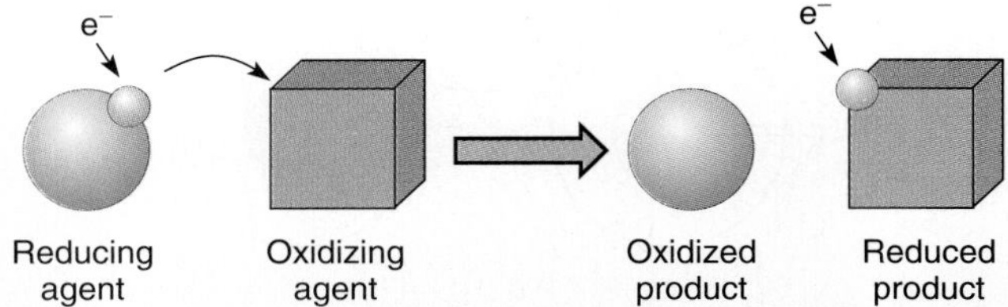

Simplified diagram of the exchange of electrons during an oxidation-reduction reaction.

thereby reduce another molecule is a **reducing agent;** one that can receive extra electrons and thereby oxidize another molecule is an **oxidizing agent.** You may find this concept easier to keep straight if you think of redox agents as partners: The one that gives its electrons away is oxidized; the partner that receives the electrons is reduced. (A mnemonic device to keep track of this is *LEO* says *GER*. "Lose Electrons Oxidized; Gain Electrons Reduced."

Redox reactions are essential to many of the biochemical processes discussed in chapter 8. In cellular metabolism, electrons alone can be transferred from one molecule to another as described here, but sometimes oxidation and reduction occur with the transfer of hydrogen atoms (which are a proton and an electron) from one compound to another. As we will see, many types of electron transfers in cells involve large carrier molecules, such as NAD (nicotinamide adenine dinucleotide), specialized for picking up electrons, hydrogens, or both from one compound and donating them to another.

(A good mnemonic device is to think of the "t" in cation as a plus (+) sign and the first "n" in anion as a negative (–) sign.) Substances such as salts, acids, and bases (discussed later) that release ions when dissolved in water are termed **electrolytes** because their charges enable them to conduct an electrical current. Owing to the general rule that particles of like charge repel each other and those of opposite charge attract each other, we can expect ions to interact electrostatically with other ions and polar molecules. Such interactions are important in many cellular chemical reactions, in the formation of solutions, and in the reactions microorganisms have with dyes. Because electrons are a source of energy, their transferral from one molecule to another constitutes a significant mechanism by which biological systems store and release energy (see microfile 2.1).

Hydrogen Bonding Some types of bonding involve neither sharing, losing, nor gaining electrons but instead are due to attractive forces between nearby molecules or atoms. One such bond is a **hydrogen bond,** a weak type of bond that forms between a hydrogen covalently bonded to one molecule and an oxygen or nitrogen atom on the same molecule or on a different molecule. Because hydrogen in a covalent bond tends to be positively charged, it will attract a nearby negatively charged atom and form an easily disrupted bridge with it. This type of bonding is usually represented in molecular models with a dotted line. A simple example of hydrogen bonding occurs between water molecules (figure 2.8). More extensive hydrogen bonding is partly responsible for the structure and stability of proteins and nucleic acids (see figures 2.22*b* and 2.25*a*).

Chemical Shorthand: Formulas, Models, and Equations

The atomic content of molecules can be represented by a few convenient **formulas.** We have already been exposed to the molecular formula, which concisely gives the atomic symbols and the number of the elements involved in subscript (CO_2, H_2O). More complex molecules such as glucose ($C_6H_{12}O_6$) can also be symbolized this way, but this formula is not unique, since fructose and galac-

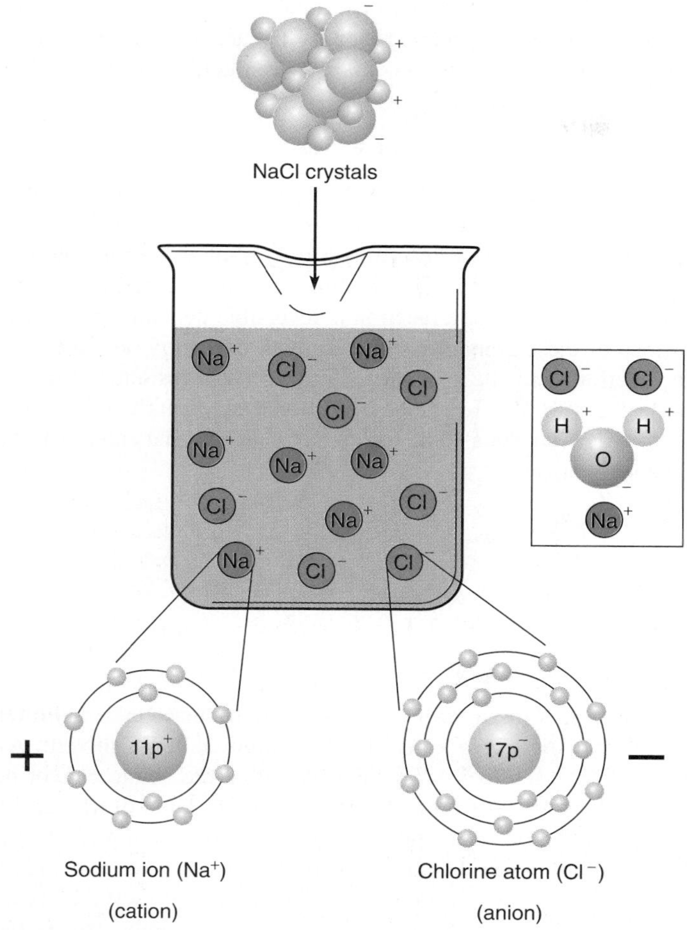

Figure 2.7
When NaCl in the crystalline form is added to water, the ions are released from the crystal as separate charged particles (cations and anions) into solution. (See also figure 2.11.)

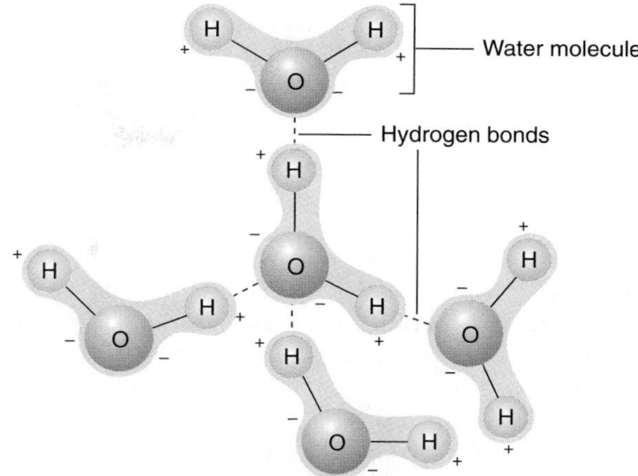

Figure 2.8
Hydrogen bonding in water. Because of the polarity of water molecules, the negatively charged oxygen end of one water molecule is weakly attracted to the positively charged hydrogen end of an adjacent water molecule.

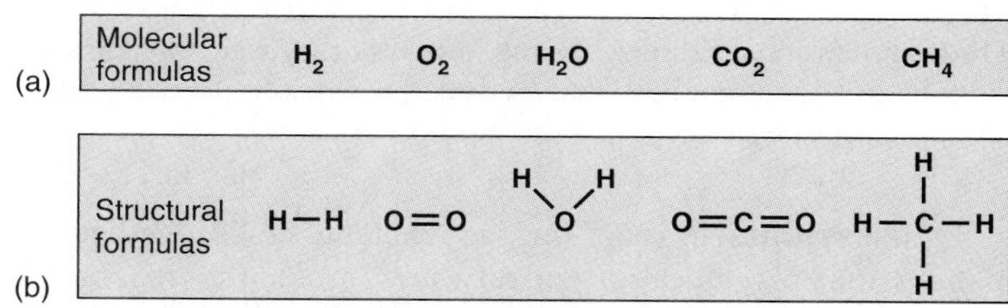

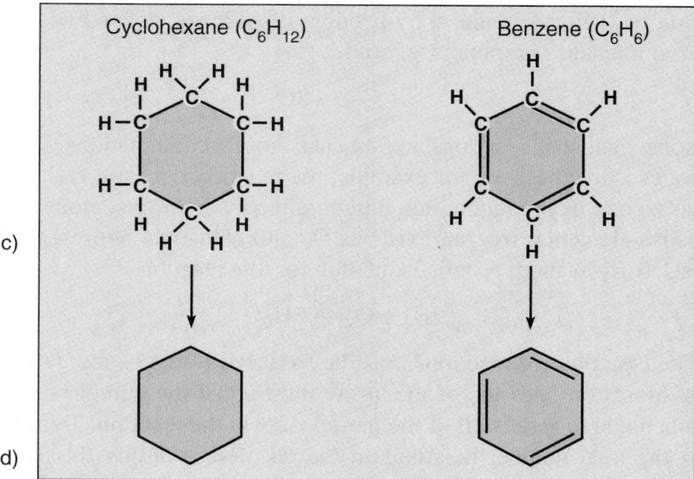

Figure 2.9
Comparison of molecular and structural formulas. (*a*) Molecular formulas provide a brief summary of the elements in a compound. (*b*) Structural formulas clarify the exact relationships of the atoms in the molecule, depicting single bonds by a single line and double bonds by two lines. (*c*) In structural formulas of organic compounds, cyclic or ringed compounds may be completely labeled, or (*d*) they may be presented in a shorthand form in which carbons are assumed to be at the angles and attached to hydrogens. See figure 2.14 for structural formulas of three sugars with the same molecular formula, $C_6H_{12}O_6$.

tose also share it. Molecular formulas are useful, but they only summarize the atoms in a compound; they do not show the position of bonds between atoms. For this purpose, chemists use structural formulas illustrating the relationships of the atoms and the number and types of bonds (figure 2.9). Other structural models present the three-dimensional appearance of a molecule, illustrating the orientation of atoms (differentiated by color codes) and the molecule's overall shape (figure 2.10).

The printed page tends to make molecules appear static, but this picture is far from correct, because molecules are constantly changing through chemical reactions. For ease in tracing chemical exchanges between atoms or molecules, and to derive some sense of the dynamic character of reactions, chemists use shorthand **equations** containing symbols, numbers, and arrows to simplify or summarize the major characteristics of a reaction. Molecules entering or starting a reaction are called **reactants,** and substances left by a reaction are called **products.** In most instances, summary chemical reactions do not give the details of the exchange, in order to keep the expression simple and to save space.

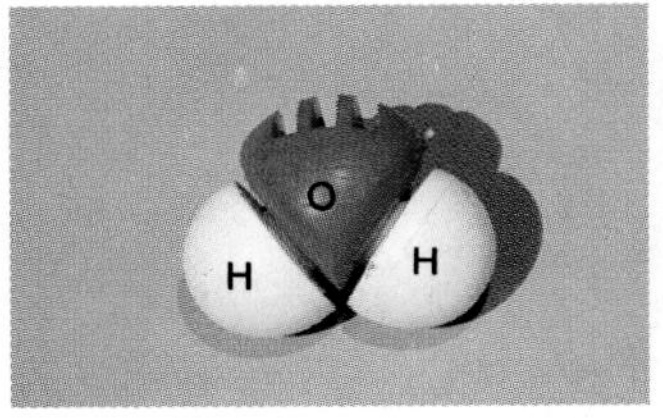

(a)

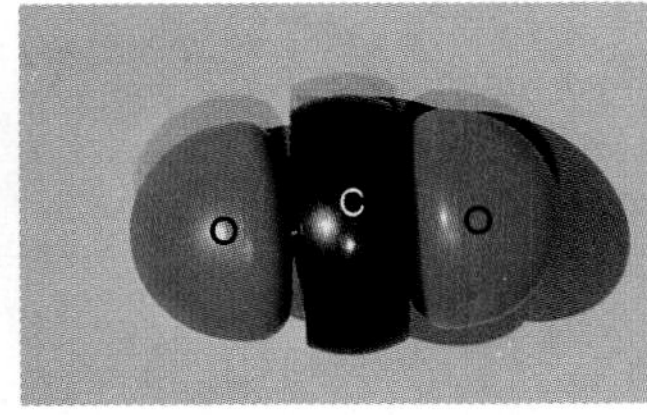

(b)

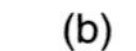

(c)

Figure 2.10
Three-dimensional, or space-filling, models of (*a*) water, (*b*) carbon dioxide, and (*c*) glucose. By convention, the red atoms are oxygen, the white ones hydrogen, and the black ones carbon.

In a **synthesis* reaction,** the reactants bond together in a manner that produces an entirely new molecule (reactant A plus reactant B yields product AB). An example is the production of sulfur dioxide, a by-product of burning sulfur fuels and an important component of smog:

$$S + O_2 \rightarrow SO_2$$

Some synthesis reactions are not such simple combinations. When water is synthesized, for example, the reaction does not really involve one oxygen atom combining with two hydrogen atoms, because elemental oxygen exists as O_2 and elemental hydrogen exists as H_2. A more accurate equation for this reaction is:

$$2H_2 + O_2 \rightarrow 2H_2O$$

The equation for reactions must be balanced—that is, the number of atoms on one side of the arrow must equal the number on the other side to reflect all of the participants in the reaction. To arrive at the total number of atoms in the reaction, multiply the prefix number by the subscript number; if no number is given, it is assumed to be 1.

In **decomposition reactions,** the bonds on a single reactant molecule are permanently broken to release two or more product molecules. One example is the resulting molecules when large nutrient molecules are digested into smaller units; a simpler example can be shown for the common chemical hydrogen peroxide:

$$2H_2O_2 \rightarrow 2H_2O + O_2$$

*synthesis (sin´-thuh-sis) Gr. *synthesis,* putting together.

During **exchange reactions,** the reactants trade portions between each other and release products that are combinations of the two. This type of reaction occurs between acids and bases when they form water and a salt:

$$AB + XY \rightleftharpoons AX + BY$$

The reactions in biological systems can be **reversible,** meaning that reactants and products can be converted back and forth. These reversible reactions are symbolized with a double arrow, each pointing in opposite directions, as in the exchange reaction above. Whether a reaction is reversible depends on the proportions of these compounds, the amount of energy necessary for the reaction, and the presence of catalysts (substances that increase the rate of a reaction). Additional reactants coming from another reaction can also be indicated by arrows that enter or leave at the main arrow:

$$X + Y \xrightarrow{CD \;\curvearrowright\; C} XYD$$

SOLUTIONS: HOMOGENEOUS MIXTURES OF MOLECULES

A **solution** is a mixture of one or more substances called **solutes** uniformly dispersed in a dissolving medium called a **solvent.** An important characteristic of a solution is that the solute cannot be separated by filtration or ordinary settling. The solute can be gaseous, liquid, or solid, and the solvent is usually a liquid. Examples of solutions are salt or sugar dissolved in water and iodine dissolved in alcohol. In general, a solvent will dissolve a solute only if it has similar electrical characteristics as indicated by the rule of solubility, expressed simply as "like dissolves like." For example, water is a polar molecule and will readily dissolve an ionic solute such as NaCl, yet a nonpolar solvent such as benzene will not dissolve NaCl.

Water, the most common solvent in natural systems, has several characteristics that suit it to this role. The polarity of the water molecule causes it to form hydrogen bonds with other water molecules, but it can also interact readily with charged or polar molecules. When an ionic solute such as NaCl crystals is added to water, it is **dissolved**, thereby releasing Na^+ and Cl^- into solution. Dissolution occurs because Na^+ is attracted to the negative pole of the water molecule and Cl^- is attracted to the positive pole; in this way, they are drawn away from the crystal separately into solution. As it leaves, each ion becomes **hydrated,** which means that it is surrounded by a sphere of water molecules (figure 2.11). Molecules such as salt or sugar that attract water to their surface are termed **hydrophilic.*** Nonpolar molecules, such as benzene, that repel water are considered **hydrophobic.*** A third class of molecules, such as the phospholipids in cell membranes, are considered *amphipathic** because they have both hydrophilic and hydrophobic properties.

*hydrophilic (hy-droh-fil´-ik) Gr. *hydros,* water, and *philos,* to love.
*hydrophobic (hy-droh-fob´-ik) Gr. *phobos,* fear,
**amphi;* Gr., both.

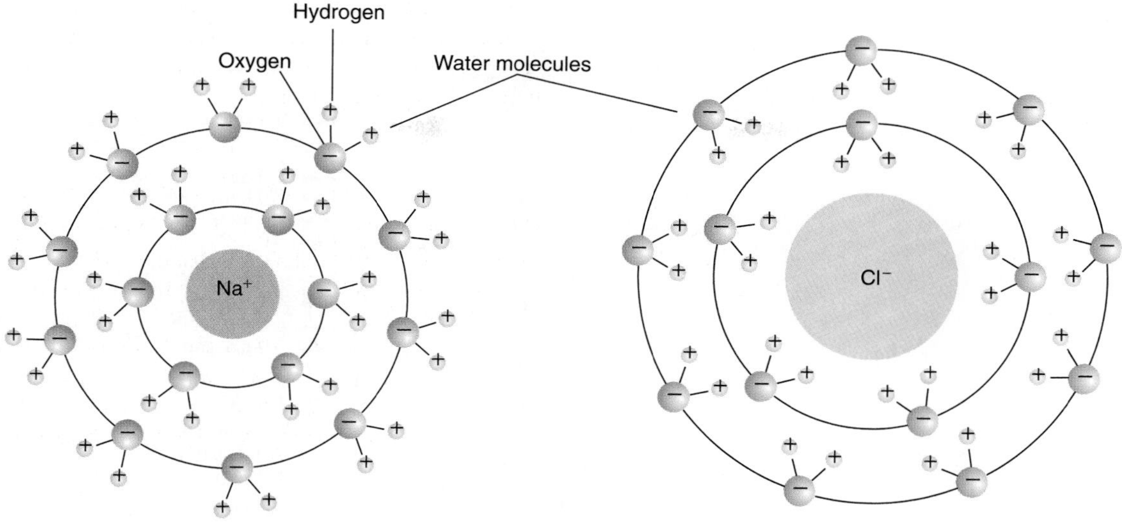

Figure 2.11
Hydration spheres formed around ions in solution. In this example, a sodium cation attracts the negatively charged region of water molecules, and a chloride anion attracts the positively charged region of water molecules. In both cases, the ions become covered with concentric layers of water molecules.

Because most biological activities take place in aqueous (water-based) solutions, the concentration of these solutions can be very important (see chapter 7). The **concentration** of a solution expresses the amount of solute dissolved in a certain amount of solvent. It can be calculated by weight, volume, or percentage. A common way to calculate percentage of concentration is to use the weight of the solute, measured in grams (g), dissolved in a specified volume of solvent, measured in milliliters (ml). For example, dissolving 3 g of NaCl in 100 ml of water produces a 3% solution; dissolving 30 g in 100 ml produces a 30% solution; and dissolving 3 g in 1,000 ml (1 liter) produces a 0.3% solution. A solution with a small amount of solute and a relatively greater amount of solvent (0.3%) is considered dilute or weak. On the other hand, a solution containing significant percentages of solute (30%) is considered concentrated or strong.

A common way to express concentration of biological solutions is by its molar concentration, or *molarity* (M). A standard molar solution is obtained by dissolving one mole, defined as the molecular weight of the compound in grams, in 1 L (1,000 ml) of solution. To make a 1 M solution of sodium chloride, we would dissolve 58 g of NaCl in 1 L of water; a 0.1 M solution would require 5.8 g of NaCl in 1 L.

ACIDITY, ALKALINITY, AND THE pH SCALE

Another factor with far-reaching impact on living things is the concentration of acidic or basic solutions in their environment. To understand how solutions develop acidity or basicity, we must look again at the behavior of water molecules. Hydrogens and oxygen tend to remain bonded by covalent bonds, but in certain instances, a single hydrogen can break away as the ionic form (H^+), leaving the remainder of the molecule in the form of an OH^- ion. The H^+ ion is positively charged because it is essentially a proton that has lost its electron; the OH^- is negatively charged because it remains in possession of that electron. Ionization of water is constantly occurring, but in pure water containing no other ions, H^+ and OH^- are produced in equal amounts, and the solution remains neutral. By one definition, a solution is considered **acidic** when a component dissolved in water (acid) releases excess hydrogen ions[3] (H^+); a solution is **basic** when a component releases excess hydroxyl ions (OH^-), so that there is no longer a balance between the two ions.

To measure the acid and base concentrations of solutions, scientists use the **pH scale,** a graduated numerical scale that ranges from 0 (the most acidic) to 14 (the most basic). This scale is a useful standard for rating relative acidity and basicity; use figure 2.12 to familiarize yourself with pH readings of some common substances. It is not an arbitrary scale but actually a mathematical derivation based on the negative logarithm (reviewed in appendix B) of the concentration of H^+ ions in grams per liter (symbolized as $[H^+]$) in a solution, represented as:

$$pH = -\log[H^+]$$

Acids have a greater concentration of H^+ than OH^-, starting with pH 0, which contains 1.0 g H^+/l. Each of the subsequent whole-number readings in the scale changes in $[H^+]$ by a tenfold reduction, so that pH 1 contains [0.1 g H^+/l], pH 2 contains [0.01 g H^+/l], and so on, continuing in the same manner up to pH 14, which contains [0.00000000000001 g H^+/l]. These same concentrations can be represented more manageably by exponents: pH 2 has a $[H^+]$ of 10^{-2} and pH 14 has a $[H^+]$ of 10^{-14}

3. Actually, it forms a hydronium ion (H_3O^+), but for simplicity's sake, we will use the notation of H^+.

Figure 2.12

The pH scale, showing the relative degrees of acidity and basicity and the approximate pH readings for various substances.

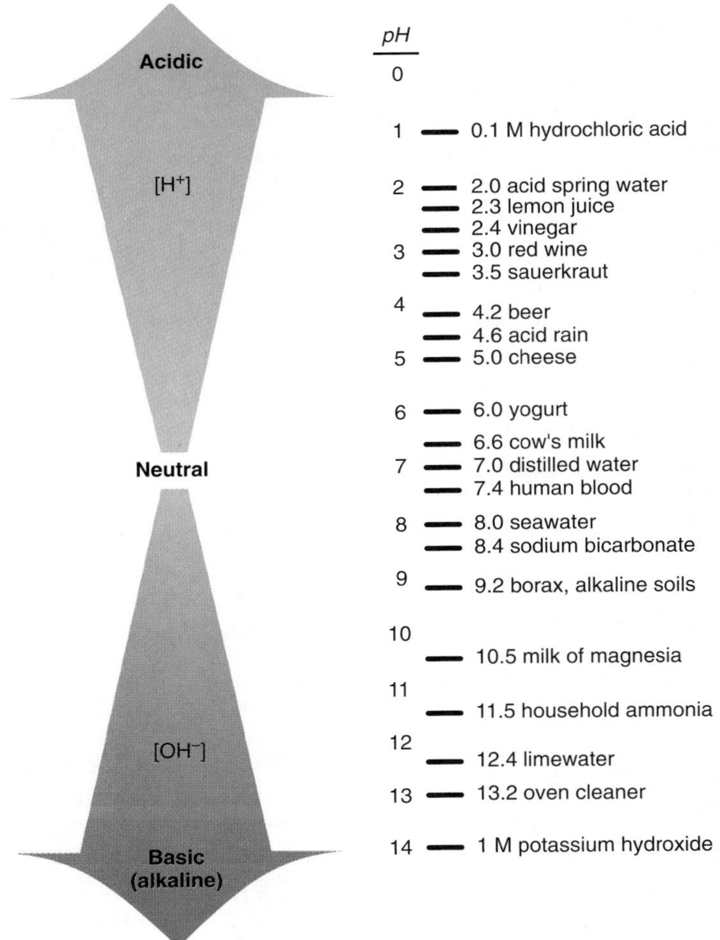

(table 2.2). It is evident that the pH units are derived from the exponent itself. Even though the basis for the pH scale is $[H^+]$, it is important to note that, as the $[H^+]$ in a solution decreases, the $[OH^-]$ increases in direct proportion. At midpoint—pH 7, or neutrality—the concentrations are exactly equal and neither predominates, this being the pH of pure water previously mentioned.

In summary, the pH scale can be used to rate or determine the degree of acidity or basicity (also called alkalinity) of a solution. On this scale, a pH below 7 is acidic, and the lower the pH, the greater the acidity; a pH above 7 is basic, and the higher the pH, the greater the basicity. Incidentally, although pHs are given here in even whole numbers, more often, a pH reading exists in decimal form; for example, pH 4.5 or 6.8 (acidic) and pH 7.4 or 10.2 (basic). Because of the damaging effects of very concentrated acids or bases, most cells operate best under neutral, weakly acidic, or weakly basic conditions (see chapter 7).

Aqueous solutions containing both acids and bases may be involved in **neutralization** reactions, which give rise to water and other neutral by-products. For example, when equal molar solutions of hydrochloric acid (HCl) and sodium hydroxide (NaOH, a base) are mixed, the reaction proceeds as follows:

$$HCl + NaOH \rightarrow H_2O + NaCl$$

TABLE 2.2

HYDROGEN ION AND HYDROXYL ION CONCENTRATIONS AT A GIVEN pH

g/L of Hydrogen Ions	Logarithm	pH	g/L of OH^-
1.0	10^{-0}	0	10^{-14}
0.1	10^{-1}	1	10^{-13}
0.01	10^{-2}	2	10^{-12}
0.001	10^{-3}	3	10^{-11}
0.0001	10^{-4}	4	10^{-10}
0.00001	10^{-5}	5	10^{-9}
0.000001	10^{-6}	6	10^{-8}
0.0000001	10^{-7}	7	10^{-7}
0.00000001	10^{-8}	8	10^{-6}
0.000000001	10^{-9}	9	10^{-5}
0.0000000001	10^{-10}	10	10^{-4}
0.00000000001	10^{-11}	11	10^{-3}
0.000000000001	10^{-12}	12	10^{-2}
0.0000000000001	10^{-13}	13	10^{-1}
0.00000000000001	10^{-14}	14	10^{-0}

Here the acid and base ionize to H^+ and OH^- ions, which form water, and other ions, Na^+ and Cl^-, which form sodium chloride. Any neutral product other than water that arises when acids and bases react is called a **salt.** Many of the organic acids (such as lactic and succinic acids) that function in **metabolism*** alternate between the acid and the salt form (such as lactate, succinate), depending on the conditions in the cell (see chapter 8).

The Chemistry of Carbon and Organic Compounds

So far, our main focus has been on the characteristics of atoms, ions, and small, simple molecules that play diverse roles in the structure and function of living things. These substances are often lumped together in a category called **inorganic* chemicals.** Examples of inorganic substances include NaCl (sodium chloride), $MgPO_4$ (magnesium phosphate), $CaCO_3$ (calcium carbonate), and CO_2 (carbon dioxide). In reality however, most of the chemical reactions and structures of living things occur at the level of more complex molecules, termed **organic* chemicals.** These are defined as compounds that contain a basic framework of the elements carbon and hydrogen. Organic molecules vary in complexity from the simplest, methane (CH_4; figure 2.4*c*), which has a molecular weight of 16, to certain antibody molecules (produced by an immune reaction) that have a molecular weight of nearly 1,000,000 and are among the most complex molecules on the earth.

The role of carbon as the fundamental element of life can best be understood if we look at its chemistry and bonding patterns. The valence of carbon makes it an ideal atomic building block to form the backbone of organic compounds; it has 4 electrons in its outer orbital to be shared with other atoms (including other carbons) through covalent bonding. As a result, it can form stable chains containing thousands of carbon atoms and still has bonding sites available for forming covalent bonds with numerous other atoms. The bonds that carbon forms are linear, branched, or ringed, and it can form four single bonds, two double bonds, or one triple bond (figure 2.13). The atoms with which carbon is most often associated in organic compounds are hydrogen, oxygen, nitrogen, sulfur, and phosphorus.

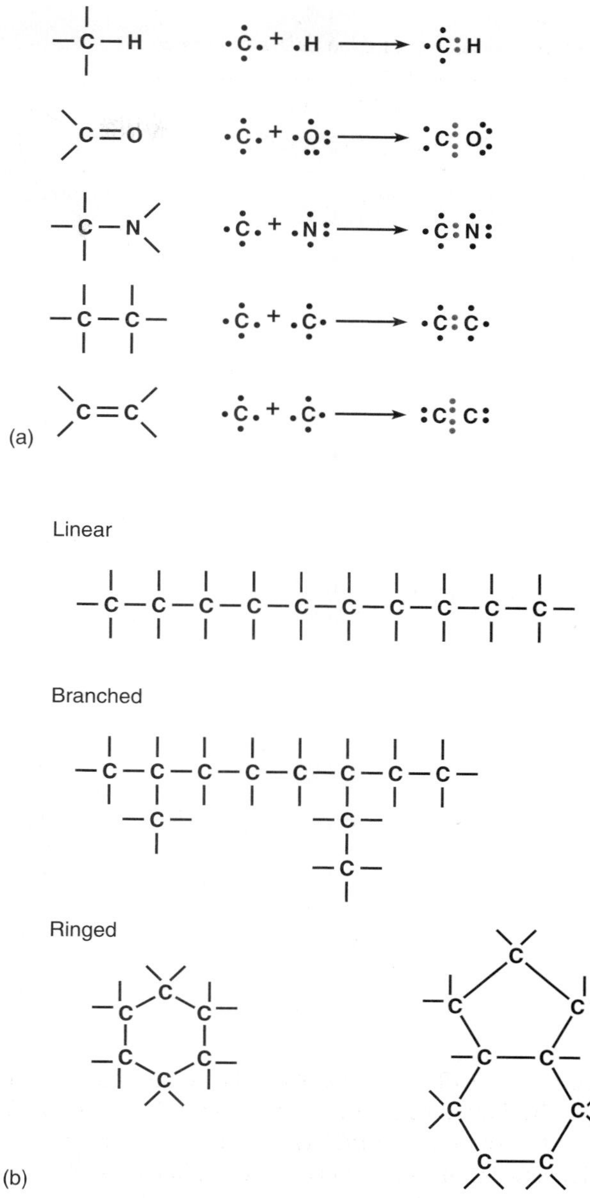

Figure 2.13
The versatility of bonding in carbon. In most compounds, each carbon makes a total of four bonds. (*a*) Both single and double bonds can be made with other carbons, oxygen, and nitrogen; single bonds are made with hydrogen. Simple electron models show how the electrons are shared in these bonds. (*b*) Multiple bonding of carbons can give rise to long chains, branched compounds, and ringed compounds, many of which are extraordinarily large and complex.

Functional Groups of Organic Compounds

One important advantage of carbon's serving as the molecular skeleton for living things is that it is free to bind with an unending array of other molecules. These special molecular groups or accessory molecules that bind to organic compounds are called **functional groups.** Functional groups help define the chemical class of certain groups of organic compounds and confer unique reactive properties on the whole molecule (table 2.3). Because each type of functional group behaves in a distinctive manner, reactions of an organic compound can be predicted by knowing the kind of functional group or groups it carries. Many synthesis, decomposition, and transfer reactions rely upon functional groups such as R—OH or R—NH_2. The **—R** designation on a molecule is shorthand for residue, and its placement in a formula indicates that the group attached at that site varies from one compound to another.

*metabolism (muh-tab´-oh-lizm) A general term referring to the totality of chemical and physical processes occurring in the cell.

*inorganic (in-or-gan´-ik) Any chemical substances that do not contain both carbon and hydrogen.

*organic (or-gan´-ik) Gr. *organikos,* instrumental.

TABLE 2.3

REPRESENTATIVE FUNCTIONAL GROUPS AND CLASSES OF ORGANIC COMPOUNDS

Formula of Functional Group	Name	Class of Compounds
R^*-O-H	Hydroxyl	Alcohols, polysaccharides
$R-C(=O)-OH$	Carboxyl	Fatty acids, proteins, organic acids
$R-CH_2-NH_2$ (H above and below C)	Amino	Proteins, nucleic acids
$R-C(=O)-O-R$	Ester	Lipids
$R-CH_2-SH$ (H above and below C)	Sulfhydryl	Cysteine (amino acid), proteins
$R-C(=O)-H$	Carbonyl, terminal end	Aldehydes, polysaccharides
$R-C(=O)-C-$	Carbonyl, internal	Ketones, polysaccharides
$R-O-P(=O)(OH)-OH$	Phosphate	DNA, RNA, ATP

*The R designation on a molecule is shorthand for residue, and its placement in a formula indicates that what is attached at that site varies from one compound to another.

Chapter Checkpoints

Covalent bonds are chemical bonds in which electrons are shared between atoms. Equally distributed electrons form nonpolar covalent bonds, whereas unequally distributed electrons form polar covalent bonds.

Ionic bonds are chemical bonds in which the outer electron orbital either donates or receives electrons from another atom so that the outer orbital of each atom is completely filled.

Hydrogen bonds are weak chemical bonds that form between covalently bonded hydrogens and either oxygens or nitrogens on different molecules.

Chemical energy is generated by the movement of electrons from one atom or molecule to another. Chemical equations express movement of energy in chemical reactions such as synthesis or decomposition reactions.

Solutions are mixtures of solutes and solvents that cannot be separated by filtration or settling.

The pH, ranging from a highly *acidic* solution to a highly *basic* solution, refers to the concentration of hydrogen ions. It is expressed as a number from 0 to 14.

Biologists define organic molecules as those containing carbon and hydrogen together.

Carbon is the backbone of biological compounds because of its ability to form single, double, or triple covalent bonds with many different elements.

Functional (R) groups are specific arrangements of organic molecules that confer distinct properties including chemical reactivity, to organic compounds.

MACROMOLECULES: SUPERSTRUCTURES OF LIFE

The compounds of life fall into the realm of **biochemistry.** Biochemicals are organic compounds produced by (or components of) living things, and they include four main families: carbohydrates, lipids, proteins, and nucleic acids (table 2.4). The compounds in these groups are assembled from smaller molecular subunits, or building blocks, and because they are often very large compounds, they are termed **macromolecules.** All macromolecules except lipids are formed by **polymerization,** a process in which repeating subunits termed **monomers*** are bound into chains of various lengths termed **polymers.*** For example, proteins (polymers) are composed of a chain of amino acids (monomers) (see figure 2.22*a*). The large size and complex, three-dimensional shape of macromolecules enables them to function as structural components, molecular messengers, energy sources, enzymes (biochemical catalysts), nutrient stores, and sources of genetic information. In the next section and in subsequent chapters, we will consider numerous concepts relating to the roles of macromolecules in cells.

*monomer (mahn′-oh-mur) Gr. *mono,* one, and *meros,* part.

*polymer (pahl′-ee-mur) Gr. *poly,* many; also the root for polysaccharide and polypeptide.

Table 2.4

Macromolecules and Their Functions

Macromolecule	Description	Examples/Functions
Carbohydrates		
Monosaccharides	3–7 carbon sugars	Glucose; fructose/Sugars involved in metabolic reactions; building block of disaccharides and polysaccharides
Disaccharides	Two monosaccharides	Maltose (malt sugar)/Composed of two glucoses; an important breakdown product of starch Lactose (milk sugar)/Composed of glucose and galactose Sucrose (table sugar)/Composed of glucose and fructose
Polysaccharides	Chains of monosaccharides	Starch; cellulose; glycogen/Cell wall, food storage
Lipids		
Triglycerides	Fatty acids + glycerol	Fats; oils/Major component of cell membranes; storage
Phospholipids	Fatty acids + glycerol + phosphate	Membranes
Waxes	Fatty acids, alcohols	Mycolic acid/Cell wall of mycobacteria
Steroids	Ringed structure (not a polymer)	Cholesterol; ergosterol/Membranes of eucaryotes and some bacteria
Proteins	Amino acids	Enzymes; part of cell membrane, cell wall, ribosomes, antibodies/Metabolic reactions; structural components
Nucleic acids	Pentose sugar + phosphate + nitrogenous base Purines: adenine, guanine Pyrimidines: cytosine, thymine, uracil	
Deoxyribonucleic acid (DNA)	Contains deoxyribose sugar and thymine, not uracil	Chromosomes; genetic material of viruses/Inheritance
Ribonucleic acid (RNA)	Contains ribose sugar and uracil, not thymine	Ribosomes; mRNA, tRNA/Expression of genetic traits

Carbohydrates: Sugars and Polysaccharides

The term **carbohydrate** originates from the way that most members of this chemical class resemble combinations of carbon and water. Although carbohydrates can be generally represented by the formula $(CH_2O)_n$, in which *n* indicates the number of units of this combination of atoms, some carbohydrates contain additional atoms of sulfur or nitrogen. In molecular configuration, the carbons form chains or rings with two or more hydroxyl groups and either an aldehyde or a ketone group, giving them the technical designation of *polyhydroxy aldehydes* or *ketones* (figure 2.14).

Carbohydrates exist in a great variety of configurations. The common term **sugar** (*saccharide*)* refers to a simple carbohydrate such as a monosaccharide or a disaccharide that has a sweet taste. A **monosaccharide** is a simple polyhydroxy aldehyde or ketone molecule containing from 3 to 7 carbons; a **disaccharide** is a combination of two monosaccharides; and a **polysaccharide** is a polymer of five or more monosaccharides bound in linear or branched chain patterns (figure 2.14). Monosaccharides and disaccharides are specified by combining a prefix that describes some characteristic of the sugar with the suffix **-ose.** For example, **hexoses** are composed of 6 carbons, and **pentoses** contain 5 carbons. **Glucose** (Gr. sweet) is the most common and universally important hexose; **fructose** is named for fruit (one of its sources); and xylose, a pentose, derives its name from the Greek word for wood. Disaccharides are named similarly: **lactose** (L. milk) is an important component of milk; **maltose** means malt sugar; and **sucrose** (Fr. sugar) is common table sugar or cane sugar.

The Nature of Carbohydrate Bonds

The subunits of disaccharides and polysaccharides are linked by means of **glycosidic bonds,** in which carbons (each is assigned a number) on adjacent sugar units are bonded to the same oxygen atom like links in a chain (figure 2.15). For example, maltose is formed when the number 1 carbon on a glucose bonds to the number 4 carbon on a second glucose; sucrose is formed when glucose and fructose bind between their number 1 and number 2 carbons; and lactose is formed when glucose and galactose connect by their number 1 and number 4 carbons. In order to form this bond, one carbon gives up its OH group and the other (the one contributing the oxygen to the bond) loses the H from its OH group. Because a water molecule is produced, this reaction is known as **dehydration synthesis,** a process common to most polymerization reactions (see proteins in a later section of this chapter). Three polysaccharides (starch, cellulose, and glycogen) are structurally and biochemically distinct, even though all are polymers of the same monosaccharide—glucose. The basis for their differences lies

*saccharide (sak´-uh-ryd) Gr. *sakcharon,* sweet.

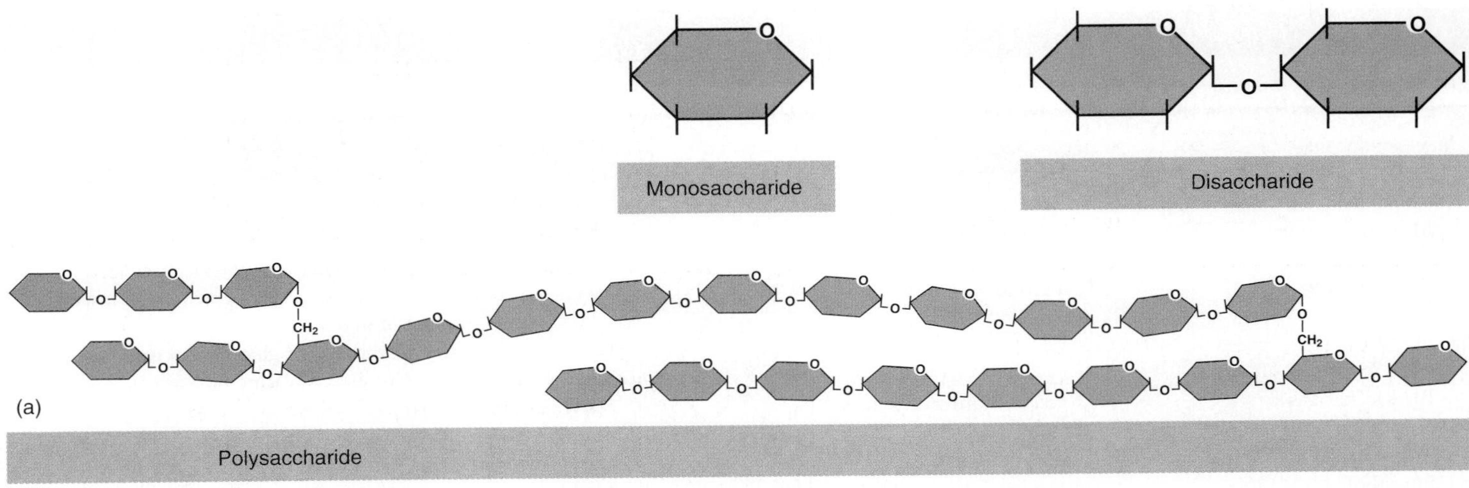

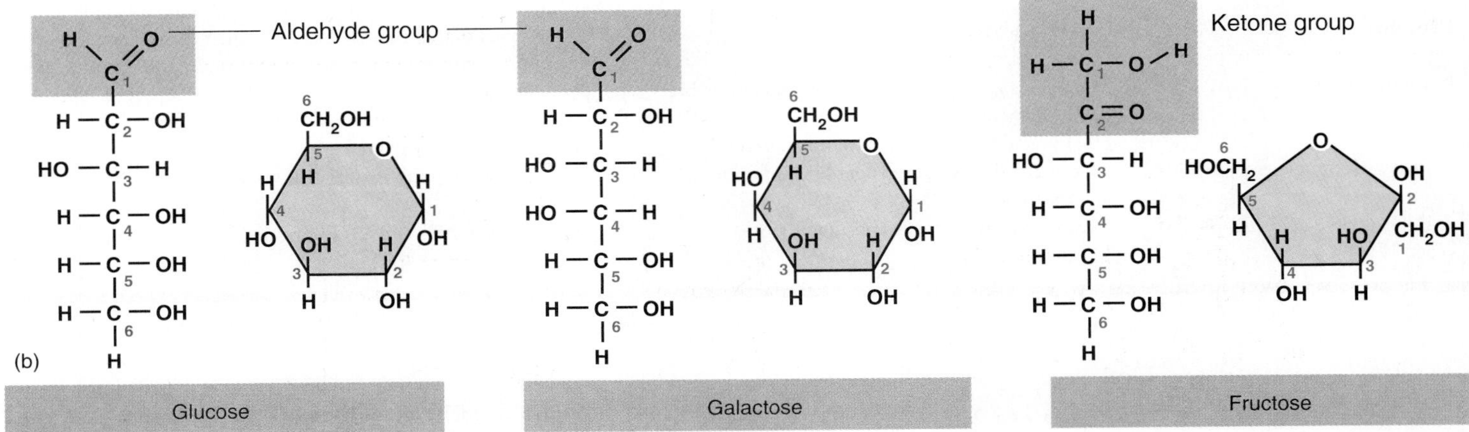

Figure 2.14
Common classes of carbohydrates. (*a*) Major saccharide groups, named for the number of sugar units each contains. (*b*) Three hexoses with the same molecular formula and different structural formulas. Both linear and ring models are given. The linear form indicates aldehyde and ketone groups, although in solution the sugars exist in the ring form. Note that the carbons are numbered so as to keep track of reactions within and between monosaccharides.

primarily in the exact way the glucoses are bound together, which greatly affects the characteristics of the end product (figure 2.16). The synthesis and breakage of each type of bond requires a specialized catalyst called an enzyme (see chapter 8).

The Functions of Polysaccharides

Polysaccharides typically contribute to structural support and protection and serve as nutrient and energy stores. The cell walls in plants and many microscopic algae derive their strength and rigidity from **cellulose,** a long, fibrous polymer (figure 2.16*a*). Because of this role, cellulose is probably one of the most common organic substances on the earth, yet it is digestible only by certain bacteria, fungi, and protozoa. These microbes, called decomposers, play an essential role in breaking down and recycling plant materials (see figure 7.3). Some bacteria secrete slime layers of a glucose polymer called *dextran.* This substance causes a sticky layer to develop in teeth that leads to plaque (see figure 4.12).

Other structural polysaccharides can be conjugated (chemically bonded) to amino acids, nitrogen bases, lipids, or proteins. **Agar,** an indispensable polysaccharide in preparing solid culture media, is a natural component of certain seaweeds. It is a complex polymer of galactose and sulfur-containing carbohydrates. The exoskeletons of certain fungi contain **chitin,** a polymer of glucosamine (a sugar with an amino functional group). **Peptidoglycan*** is one special class of compounds in which polysaccharides (glycans) are linked to peptide fragments (a short chain of amino acids). This molecule provides the main source of structural sup-

*peptidoglycan (pep-tih-doh-gly´-kan).

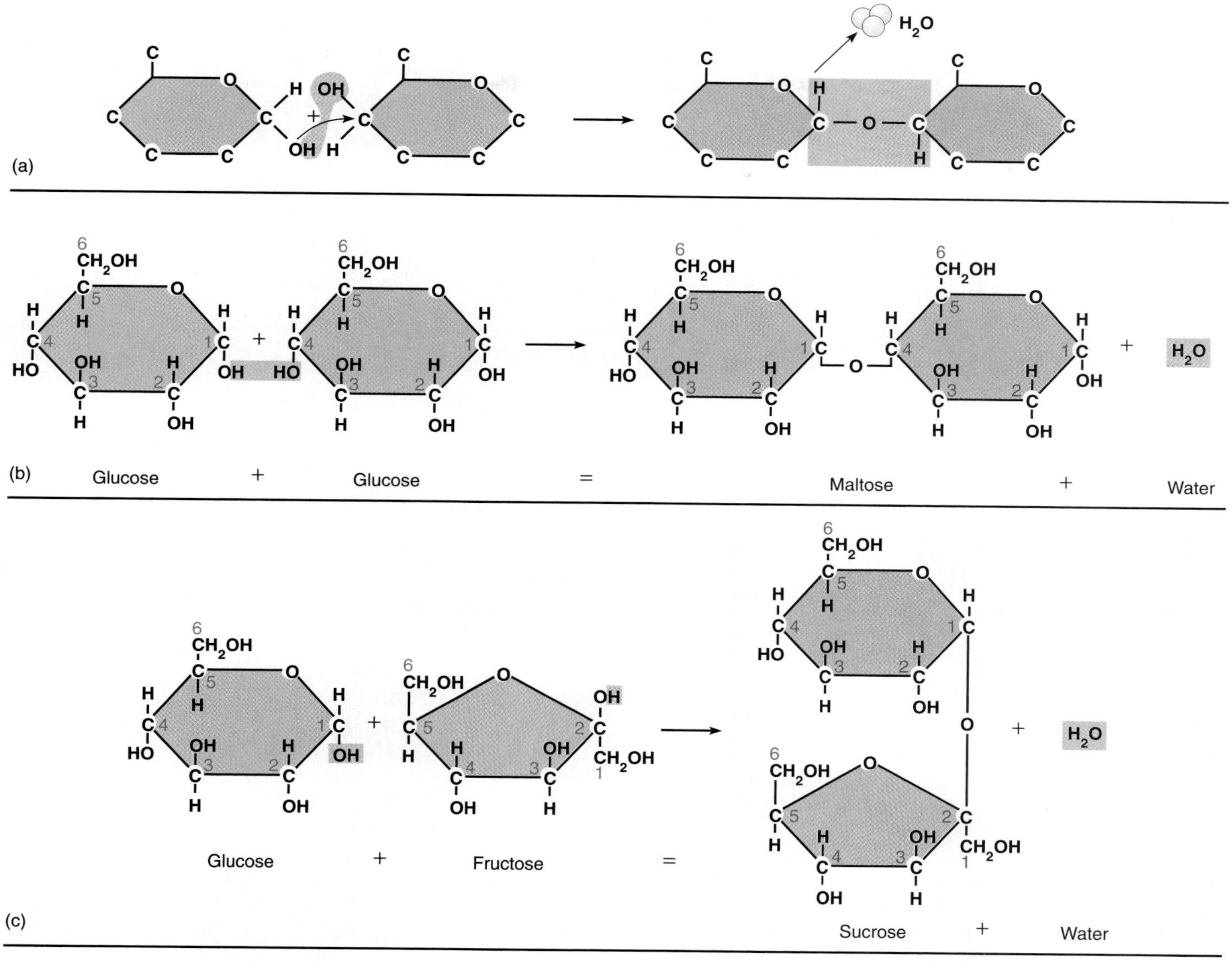

(a)

(b) Glucose + Glucose = Maltose + Water

(c) Glucose + Fructose = Sucrose + Water

Figure 2.15
(*a*) General scheme in the formation of a glycosidic bond by dehydration synthesis. (*b*) Formation of the 1,4 bond between two α glucoses to produce maltose and water. (*c*) Formation of the 1,2 bond between glucose and fructose to produce sucrose and water.

port to the bacterial cell wall. The cell wall of gram-negative bacteria also contains **lipopolysaccharide,** a complex of lipid and polysaccharide responsible for symptoms such as fever and shock (see chapters 4 and 13).

The outer surface of many cells has a delicate "sugar coating" composed of polysaccharides bound in various ways to proteins (the combination is called mucoprotein or glycoprotein). This structure, called the **glycocalyx,*** functions in attachment to other cells or as a site for *receptors*—surface molecules that receive and respond to external stimuli. Small sugar molecules account for the differences in human blood types, and carbohydrates are a component of large protein molecules called antibodies. Some viruses have glycoproteins on their surface with which they adhere to and invade their host cells.

Polysaccharides are usually stored by cells in the form of glucose polymers such as **starch** (figure 2.16*b*) or **glycogen,** but only organisms with the appropriate digestive enzymes can break them down and use them as a nutrient source. Because a water molecule is required for breaking the bond between two glucose molecules, digestion is also termed **hydrolysis.*** Starch is the

*glycocalyx (gly´´-koh-kay´-lix) Gr. *glycos,* sweet, and *calyx,* covering.

*hydrolysis (hy-drol´-uh-sis) Gr. *hydros,* water, and *lyein,* to dissolve.

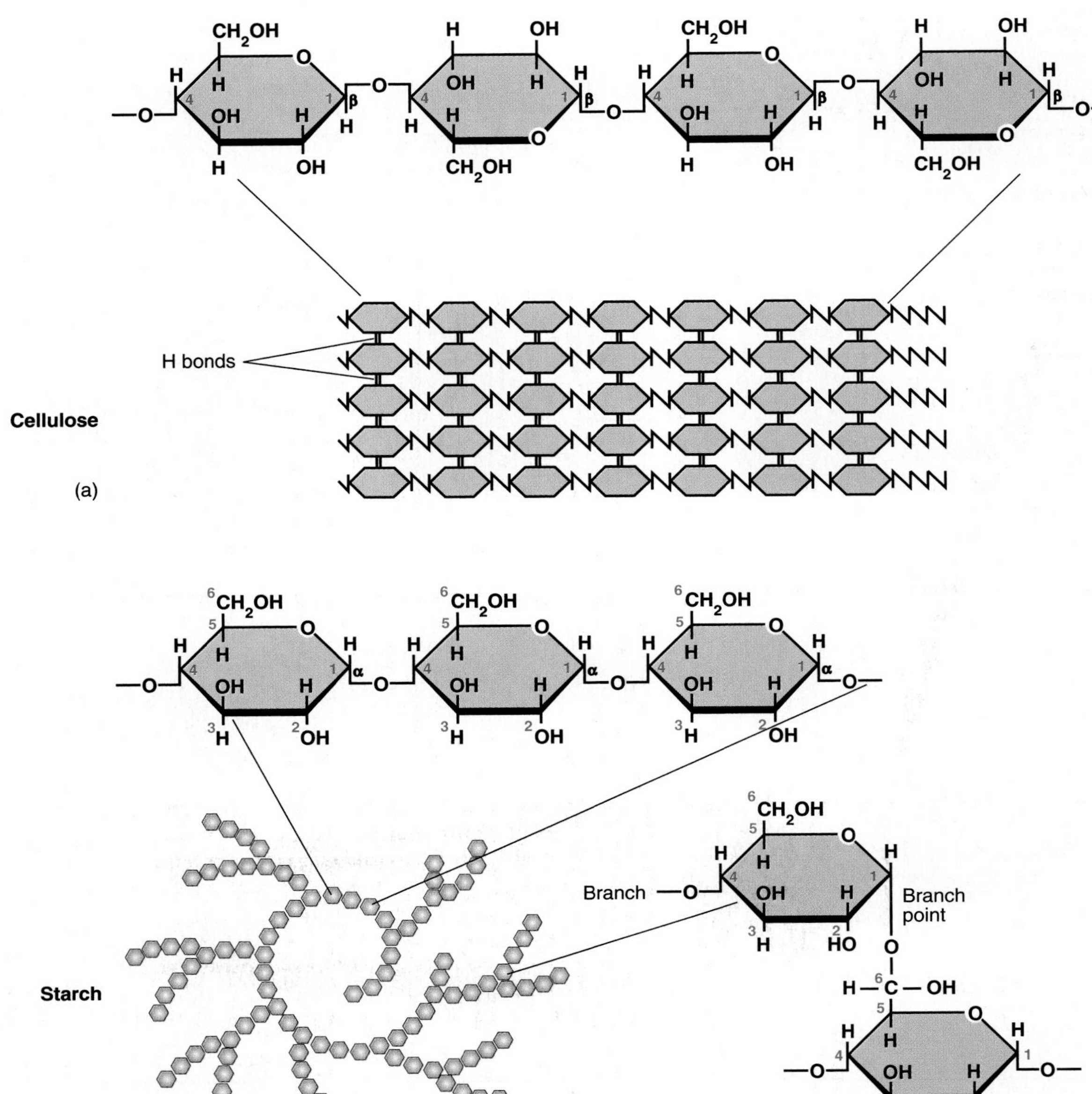

Figure 2.16
Polysaccharides. (*a*) Cellulose is composed of β glucose bonded in 1,4 bonds that produce linear, lengthy chains of polysaccharides that are H-bonded along their length. This is the typical structure of wood and cotton fibers. (*b*) Starch is also *composed of glucose polymers,* in this case α glucose. The main structure is amylose bonded in a 1,4 pattern, with side branches of amylopectin bonded by 1,6 bonds. The entire molecule is compact and granular.

primary storage food of green plants, microscopic algae, and some fungi; glycogen (animal starch) is a stored carbohydrate for animals and certain groups of bacteria.

Lipids: Fats, Phospholipids, and Waxes

The term **lipid,** derived from the Greek word *lipos,* meaning fat, is not a chemical designation, but an operational term for a variety of substances that are not soluble in polar solvents such as water (recall that oil and water do not mix) but will dissolve in nonpolar solvents such as benzene and chloroform. This property occurs because the substances we call lipids contain relatively long or complex C—H (hydrocarbon) chains that are nonpolar and thus hydrophobic. The main groups of compounds classified as lipids are triglycerides, phospholipids, steroids, and waxes.

One important group of storage lipids is the **triglycerides,** a category that includes fats and oils. Triglycerides are composed of a single molecule of glycerol bound to three fatty acids (figure 2.17). **Glycerol** is a 3-carbon alcohol with three OH

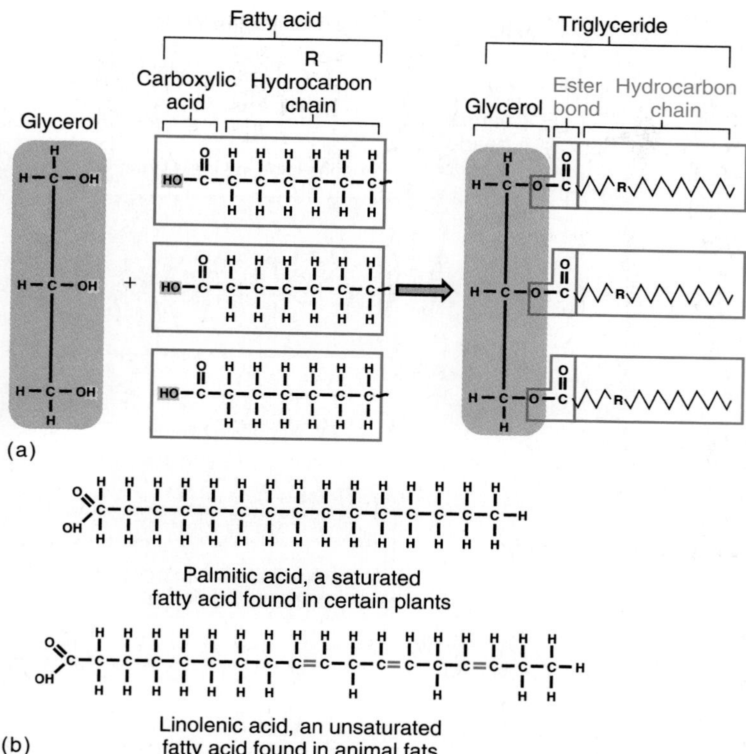

Figure 2.17
(*a*) Synthesis and structure of a triglyceride. Because a water molecule is released at each ester bond, this is another form of dehydration synthesis. The jagged lines and R symbol represent the hydrocarbon chains of the fatty acids, which are commonly very long. (*b*) Structural formulas for saturated and unsaturated fatty acids.

groups that serve as binding sites, and **fatty acids** are long-chain hydrocarbon molecules with a carboxyl group (COOH) at one end that is free to bind to the glycerol. The bond that forms between the —OH group and the —COOH is defined as an **ester bond.** The hydrocarbon portion of a fatty acid can vary in length from 4 to 24 carbons and, depending on the fat, it may be *saturated* or *unsaturated.* If all carbons in the chain are single-bonded to 2 other carbons and 2 hydrogens, the fat is saturated; if there is at least one C=C double bond in the chain, it is unsaturated. The structure of fatty acids is what gives fats and oils (liquid fats) their greasy, insoluble nature. In general, solid fats (such as beef tallow) are more saturated, and oils (or liquid fats) are more unsaturated. In most cells, triglycerides are stored in long-term concentrated form as droplets or globules. When the ester linkage is acted on by digestive enzymes called lipases, the fatty acids and glycerol are freed to be used in metabolism. Fatty acids are a superior source of energy, yielding twice as much per gram as other storage molecules (starch). Soaps are K^+ or Na^+ salts of fatty acids whose qualities make them excellent grease removers and cleaners (see chapter 11).

Membrane Lipids

A class of lipids that serves as a major structural component of cell membranes is the **phospholipids.** Although phospholipids also contain glycerol and fatty acids, they have some significant differences from triglycerides. Phospholipids contain only two fatty acids attached to the glycerol, and the third glycerol binding site holds a phosphate group. The phosphate is in turn bonded to an alcohol,[4] which varies from one phospholipid to another (figure 2.18*a*). These lipids have a hydrophilic region from the charge on the phosphoric acid–alcohol "head" of the molecule and a hydrophobic region that corresponds to the long uncharged "tail" (formed by the fatty acids). When exposed to an aqueous solution, the charged heads are attracted to the water phase and the nonpolar tails are repelled from the water phase (figure 2.18*b*). This property causes lipids to naturally assume single and double layers (bilayers) which contribute to their biological significance in membranes. When two single layers of polar lipids come together to form a double layer, the outer hydrophilic face of each single layer will orient itself toward the solution and the hydrophobic portions will become immersed in the core of the bilayer. The structure of lipid bilayers confers characteristics on membranes such as selective permeability and fluid nature (see microfile 2.3).

Miscellaneous Lipids

Steroids are complex ringed compounds commonly found in cell membranes and animal hormones. The best known of these is the sterol (meaning a steroid with an OH group) called **cholesterol** (figure 2.19). Cholesterol reinforces the structure of the cell membrane in animal cells and an unusual group of cell-wall-deficient bacteria called the mycoplasmas (see chapter 4). The cell membranes of fungi also contain a sterol, called ergosterol. *Prostaglandins* are fatty acid derivatives found in trace amounts that function in inflammatory and allergic reactions, blood clotting, and smooth muscle contraction. Chemically, a *wax* is an ester formed between a long chain alcohol and a saturated fatty acid. The resulting material is typically pliable and soft when warmed but hard and water-resistant when cold (paraffin, for example). Among living things, fur, feathers, fruits, leaves, human skin, and insect exoskeletons are naturally waterproofed with a coating of wax. Bacteria that cause tuberculosis and leprosy produce a wax (wax D) that repels ordinary stains and contributes to their pathogenicity.

PROTEINS: SHAPERS OF LIFE

The predominant organic molecules in cells are **proteins,** a fitting term adopted from the Greek word *proteios,* meaning first or prime. To a large extent, the structure, behavior, and unique qualities of each living thing are a consequence of the proteins they contain. To best explain the origin of the special properties and versatility of proteins, we must examine their general structure. The building blocks of proteins are **amino acids,** which exist in 20 different naturally occurring forms (table 2.5). Various combinations of these amino acids account for the nearly infinite variety of proteins. All amino acids have a basic skeleton consisting of a carbon (called the α carbon) linked to an amino group (NH_2), a

4. Alcohols are hydrocarbons containing OH groups.

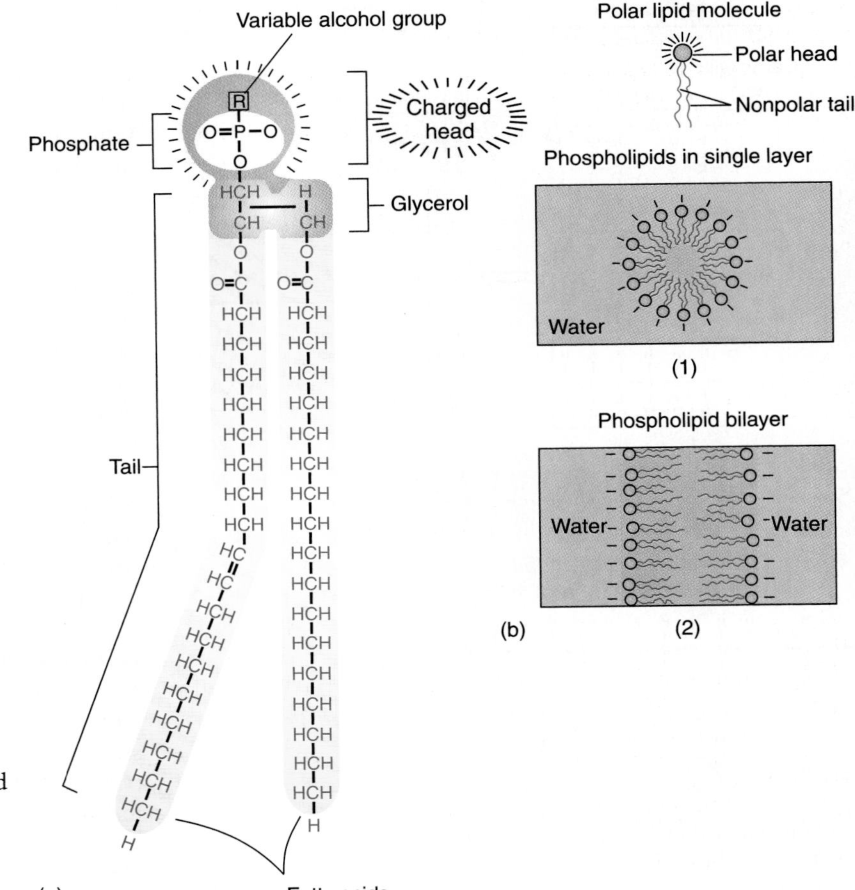

Figure 2.18
Phospholipids—membrane molecules. (*a*) A complex model of a single molecule of a phospholipid. The phosphate-alcohol head lends a charge to one end of the molecule; its long, trailing hydrocarbon chain is uncharged. (*b*) The behavior of phospholipids in water-based solutions causes them to become arranged (*1*) in single layers called micelles, with the charged head oriented toward the water phase and the hydrophobic nonpolar tail buried away from the water phase, or (*2*) in double-layered phospholipid systems with the hydrophobic tails sandwiched between two hydrophilic layers.

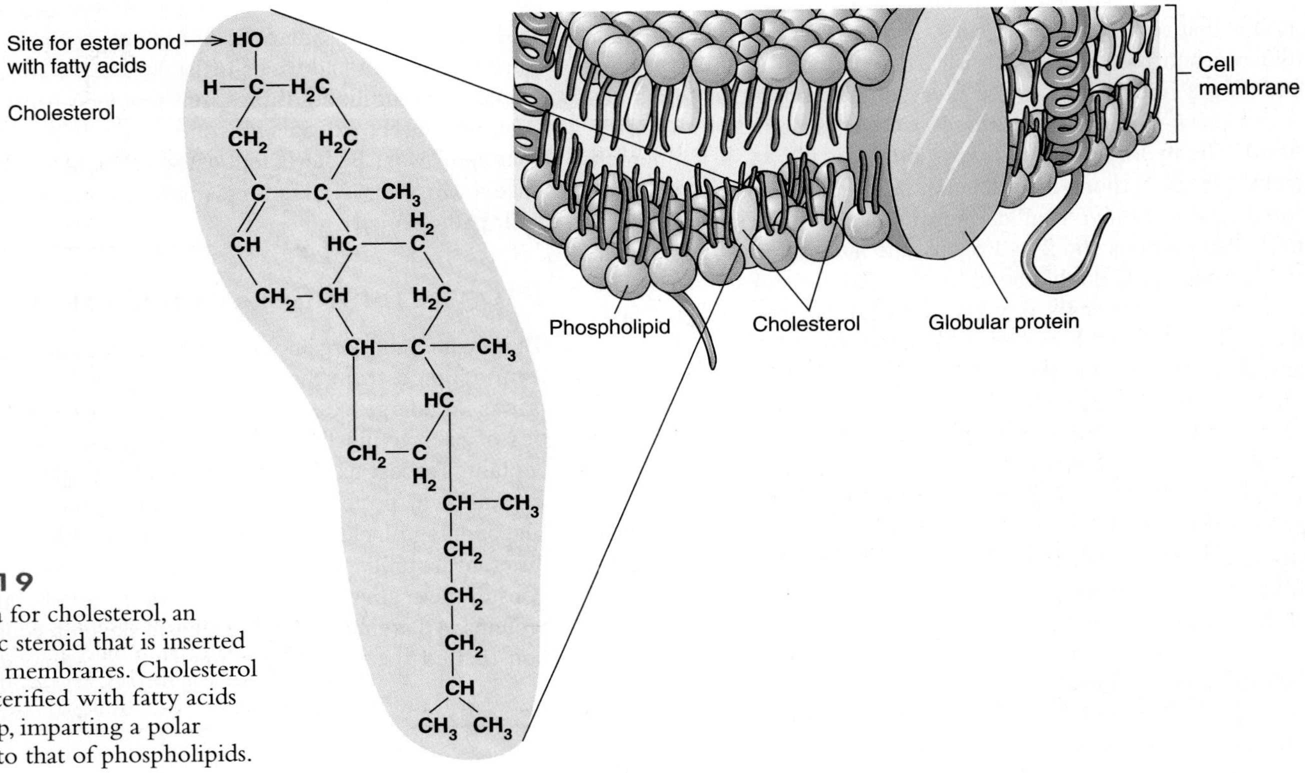

Figure 2.19
Formula for cholesterol, an alcoholic steroid that is inserted in some membranes. Cholesterol can become esterified with fatty acids at its OH group, imparting a polar quality similar to that of phospholipids.

TABLE 2.5

THE 20 AMINO ACIDS AND THEIR ABBREVIATIONS

Acid	Abbreviation	Characteristic of R Groups*
Alanine	Ala	NP
Arginine	Arg	+
Asparagine	Asn	P
Aspartic acid	Asp	–
Cysteine	Cys	P
Glutamic acid	Glu	–
Glutamine	Gln	P
Glycine	Gly	P
Histidine	His	+
Isoleucine	Ile	NP
Leucine	Leu	NP
Lysine	Lys	+
Methionine	Met	NP
Phenylalanine	Phe	NP
Proline	Pro	NP
Serine	Ser	P
Threonine	Thr	P
Tryptophan	Trp	NP
Tyrosine	Tyr	P
Valine	Val	NP

*NP, nonpolar; P, polar; +, positively charged; –, negatively charged.

carboxyl group (COOH), and a hydrogen atom (H). The variations among the amino acids occur at the R group, which is different in each amino acid and imparts the unique characteristics to the molecule and to the proteins that contain it (figure 2.20). A covalent bond called a **peptide bond** forms between the amino group on one amino acid and the carboxyl group on another amino acid. As a result of peptide bond formation, it is possible to produce molecules varying in length from two amino acids to chains containing thousands of them.

Various terms are used to denote the nature of compounds containing peptide bonds. **Peptide*** usually refers to a molecule composed of short chains of amino acids, such as a dipeptide (two amino acids), a tripeptide (three), and a tetrapeptide (four) (figure 2.21). A **polypeptide** contains an unspecified number of amino acids, but usually has more than 20, and is often a smaller subunit of a protein. A protein is the largest of this class of compounds and usually contains a minimum of 50 amino acids. It is common for the terms *polypeptide* and *protein* to be used interchangeably, though not all polypeptides are large enough to be considered proteins. In chapter 8 we see that protein synthesis is not just a random connection of amino acids; it is directed by information provided in DNA.

*peptide (pep´-tyd) Gr. *pepsis,* digestion.

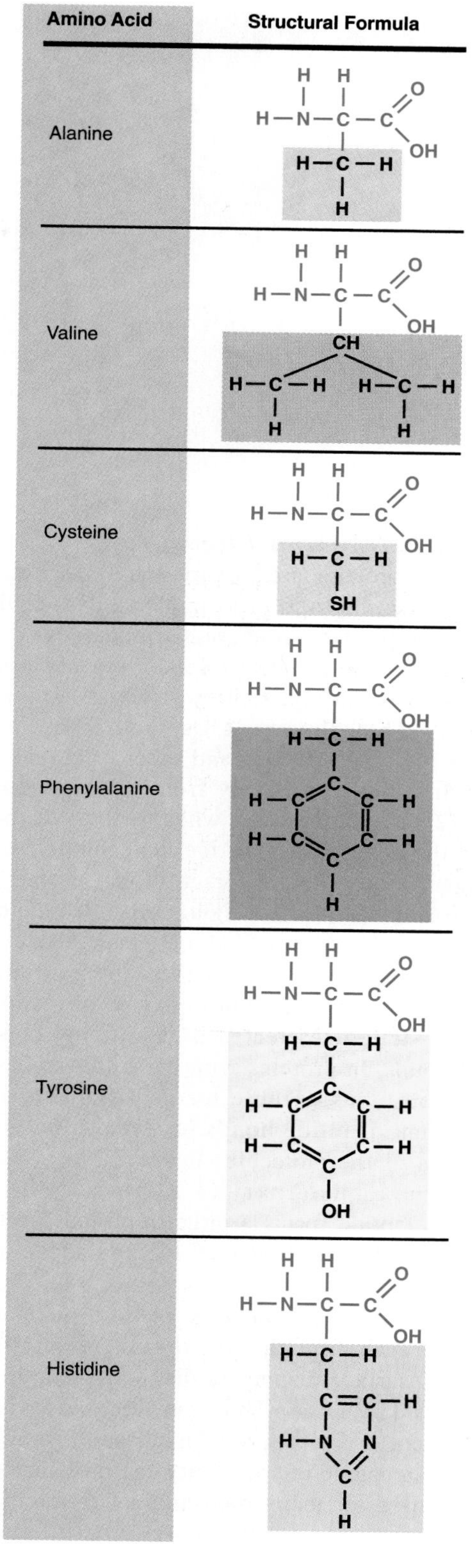

Amino Acid	Structural Formula
Alanine	
Valine	
Cysteine	
Phenylalanine	
Tyrosine	
Histidine	

Figure 2.20

Structural formulas of selected amino acids. The basic structure common to all amino acids is shown in blue, and the variable group, or R group, varies from a methyl (CH_3) group (as in alanine) to more complex ringed compounds (as in tyrosine).

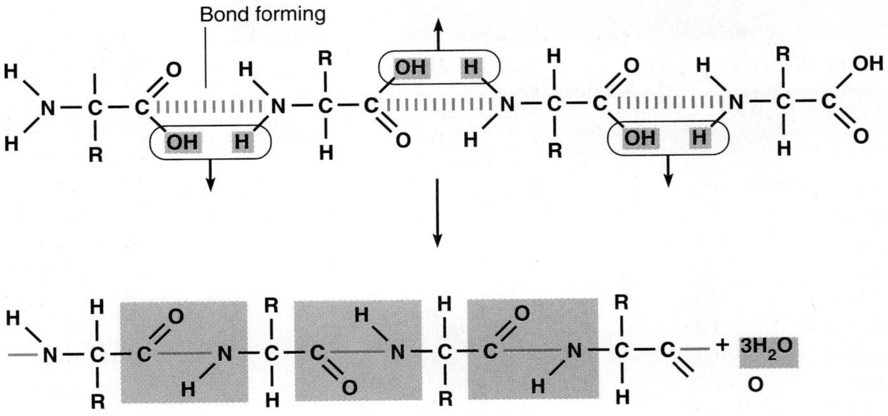

Figure 2.21
The formation of peptide bonds in a tetrapeptide.

Protein Structure and Diversity

The reason that proteins are so varied and specific is that they do not function in the form of a simple straight chain of amino acids (called the primary structure). A protein has a natural tendency to assume more complex levels of organization called the secondary, tertiary, and quaternary structures (figure 2.22 and microfile 2.2). The **primary (1°) structure** is more correctly described as the type, number, and order of amino acids in the chain, which varies extensively from protein to protein. The **secondary (2°) structure** arises when various functional groups exposed on the outer surface of the molecule interact by forming hydrogen bonds. This interaction causes the amino acid chain to twist into a coiled configuration called the α *helix* or to fold into accordian pattern called a *β-pleated sheet.* Some proteins contain both types of secondary configurations. Proteins at the secondary level undergo a third degree of torsion called the **tertiary (3°) structure** created by additional bonds between functional groups. In proteins with the sulfur-containing amino acid cysteine,* considerable tertiary stability is achieved through covalent **disulfide bonds** between sulfur atoms on two different parts of the molecule (figure 2.22*c*). Some complex proteins assume a **quaternary (4°) structure,** in which more than one polypeptide forms a large, multiunit protein. This is typical of antibodies (see chapter 15) and some enzymes that act in cell synthesis.

The most important outcome of intrachain[5] bonding and folding is that each different type of protein develops a unique shape, and its surface displays a distinctive pattern of pockets and bulges. As a result, a protein can react only with molecules that complement or fit its particular surface features like a lock and key. Such a degree of specificity can provide the functional diversity required for many thousands of different cellular activities. **Enzymes** serve as the catalysts for all chemical reactions in cells, and nearly every reaction requires a different enzyme (see chapter 8). **Antibodies** are complex glycoproteins with specific regions of attachment for bacteria, viruses, and other microorganisms; certain bacterial toxins (poisonous products) react with only one specific organ or tissue; and proteins embedded in the cell membrane have reactive sites restricted to a certain nutrient. Some proteins function as *receptors* to receive stimuli from the environment. The functional three-dimensional form of a protein is termed the *native state,* and if it is disrupted by some means, the protein is said to be *denatured.* Such agents as heat, acid, alcohol, and some disinfectants disrupt the stabilizing intrachain bonds and cause the molecule to become nonfunctional (see figure 11.4).

THE NUCLEIC ACIDS: A CELL COMPUTER AND ITS PROGRAMS

The nucleic acids, **deoxyribonucleic acid* (DNA)** and **ribonucleic acid* (RNA),** were originally isolated from the cell nucleus. Shortly thereafter, they were also found in other parts of nucleated cells, in cells with no nuclei (bacteria), and in viruses. The universal occurrence of nucleic acids in all known cells and viruses emphasizes their important roles as informational molecules. DNA, the master computer of cells, contains a special coded genetic program with detailed and specific instructions for each organism's heredity. It transfers the details of its program to RNA, operator molecules responsible for carrying out DNA's instructions and translating the DNA program into proteins that can perform life functions. For now, let us briefly consider the structure and some functions of DNA, RNA, and a close relative, adenosine triphosphate (ATP).

Both nucleic acids are polymers of repeating units called **nucleotides,*** each of which is composed of three smaller units: a *nitrogen base,* a *pentose* (5-carbon) sugar, and a *phosphate* (figure 2.23*a*). The nitrogen base is a cyclic compound that comes in two forms: *purines* (two rings) and *pyrimidines* (one ring). There are two types of purines—**adenine** (A) and **guanine** (G)—and three types of pyrimidines—**thymine** (T), **cytosine** (C), and **uracil** (U) (figure 2.24). A characteristic that differentiates DNA from RNA is that DNA contains all of the nitrogen bases except

5. Meaning within the chain; **inter**chain would be between two chains.

*cysteine (sis´-tuh-yeen) Gr. *kystis,* sac. An amino acid first found in urine stones.

*deoxyribonucleic (dee-ox´´-ee-ry´´-boh-noo-klay´-ik),
*ribonucleic (ry´´-boh-noo-klay´-ik) It is easy to see why the abbreviations are used!
*nucleotide (noo´-klee-oh-tyd) from nucleus and acid.

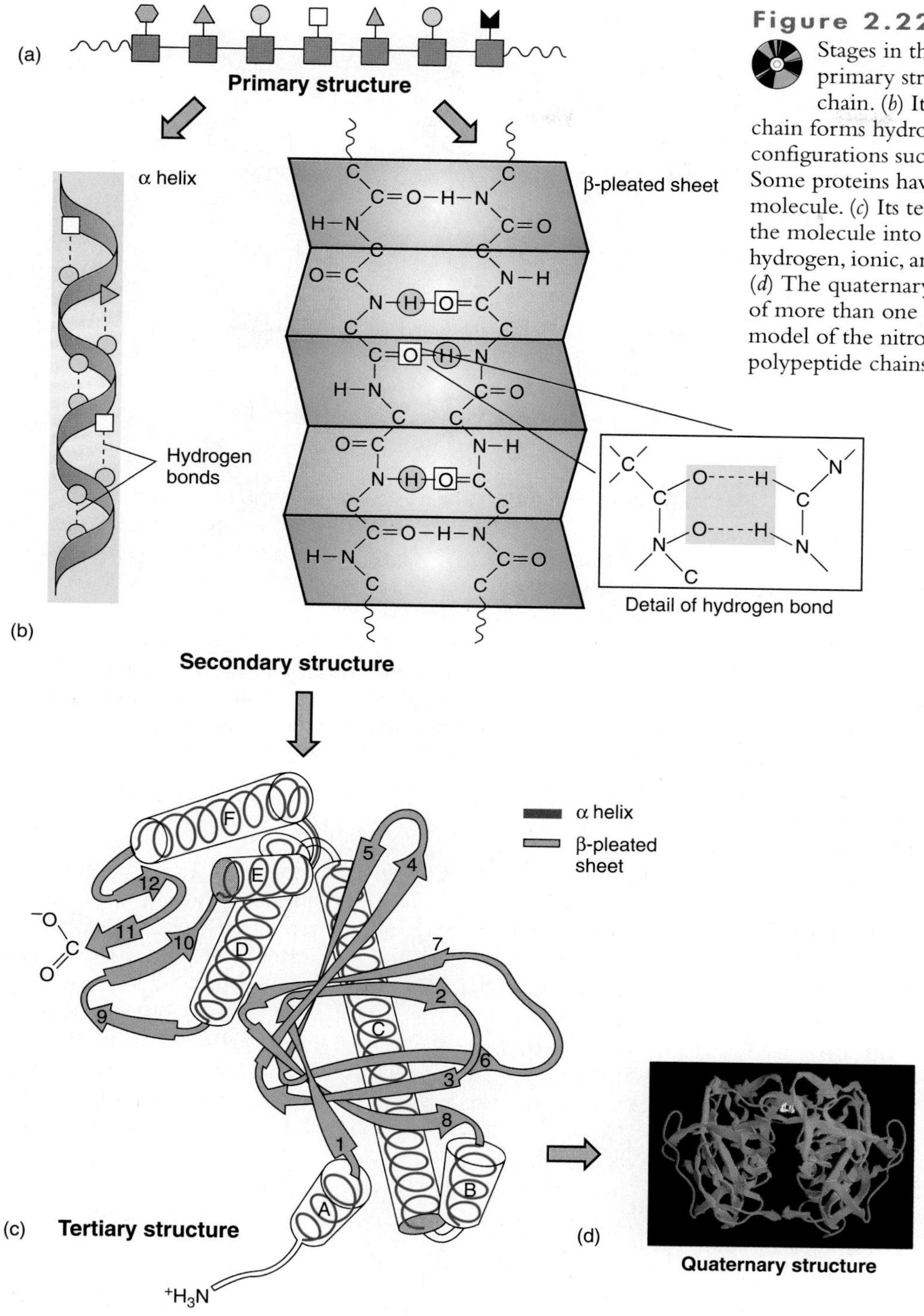

(a) Primary structure

(b) Secondary structure — α helix; β-pleated sheet; Hydrogen bonds; Detail of hydrogen bond

(c) Tertiary structure

(d) Quaternary structure

Figure 2.22

Stages in the formation of a functioning protein: (*a*) Its primary structure is a series of amino acids bound in a chain. (*b*) Its secondary structure develops when the chain forms hydrogen bonds that fold it into one of several configurations such as an alpha helix or beta-pleated sheet. Some proteins have several configurations in the same molecule. (*c*) Its tertiary structure is due to further folding of the molecule into a three-dimensional mass that is stabilized by hydrogen, ionic, and disulfide bonds between functional groups. (*d*) The quaternary structure exists only in proteins that consist of more than one polypeptide chain. Shown here is a computer model of the nitrogenase iron protein, with the two polypeptide chains arranged symmetrically like butterfly wings.

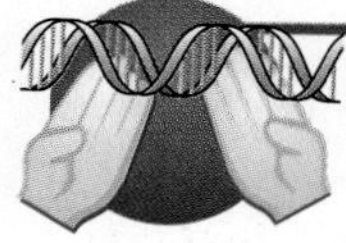

MICROFILE 2.2 PROTEINS AND PAPER AIRPLANES

The levels of protein structure have steps in folding analogous to those of a handmade paper airplane. The flat sheet of paper one starts with is equivalent to the primary state. Initial folds made for wings, body, and tail produce a two-dimensional secondary configuration. When these folds are correctly pulled out and inserted, they yield a three-dimensional tertiary structure that is the airplane. If a more elaborate airplane were constructed, other parts such as wing and tail fins could be added, thus creating a quaternary level of structure. This analogy is more than just structural, it is also functional. The initial shape and size of the paper (1°) will dictate the nature of the folds (2°); these folds will then determine the shape and size of the airplane (3°); and finally, the airplane will fly (function) only if the proper 3° or 4° state is formed.

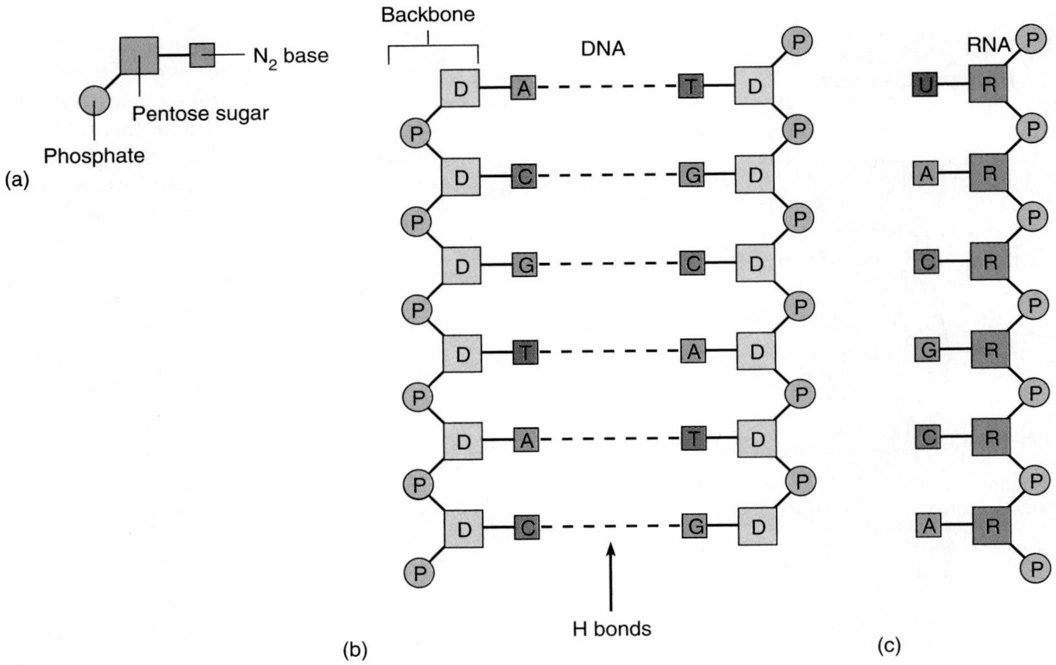

Figure 2.23
The general structure of nucleic acids. (*a*) A nucleotide, composed of a phosphate, a pentose sugar, and a nitrogen base, is the monomer of both DNA and RNA. (*b*) In DNA, the polymer is composed of alternating deoxyribose (D) and phosphate (P) with nitrogen bases (A, T, C, G) attached to the deoxyribose. Two of these polynucleotide strands are oriented so that the bases are paired across the central axis of the molecule. (*c*) In RNA, the polymer is composed of alternating ribose (R) and phosphate (P) attached to nitrogen bases (A, U, C, G), but it is only a single strand.

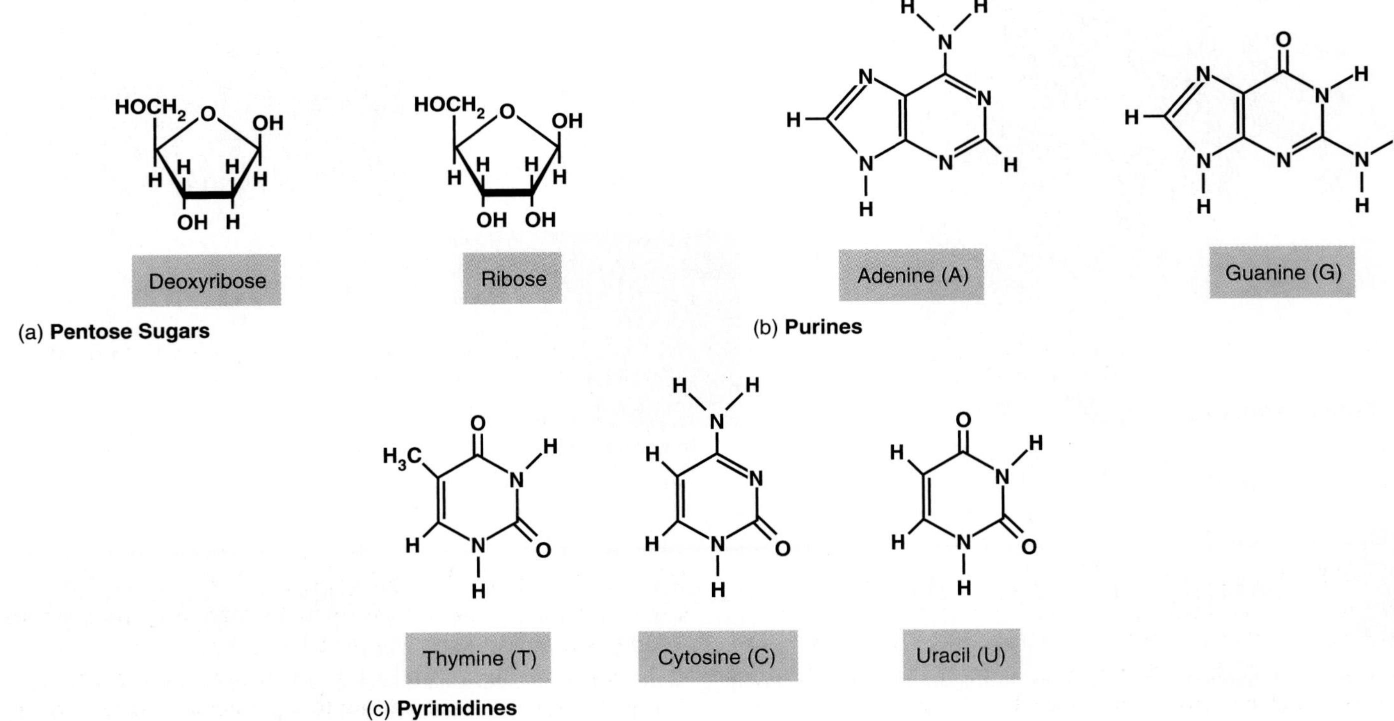

Figure 2.24
The sugars and nitrogen bases that make up DNA and RNA. (*a*) DNA contains deoxyribose, and RNA contains ribose. (*b*) A and G purines are found in both DNA and RNA. (*c*) C pyrimidine is found in both DNA and RNA, but T is found only in DNA, and U is found only in RNA.

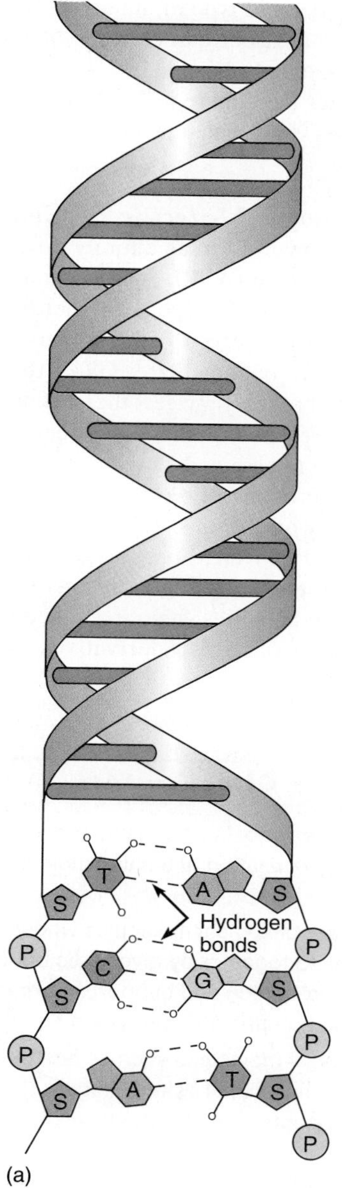

Figure 2.25
(*a*) A structural representation of the double helix of DNA shows the details of hydrogen bonds between the nitrogen bases of the two strands. (*b*) A space-filling model indicates the actual configuration of a small section of a DNA molecule.

uracil, and RNA contains all of the nitrogen bases except thymine. The nitrogen base is covalently bonded to the sugar *ribose* in RNA and *deoxyribose* (because it has one less oxygen than ribose) in DNA. Phosphate (PO_4), a derivative of phosphoric acid (H_3PO_4), provides the final covalent bridge that connects sugars in series. Thus, the backbone of a nucleic acid strand is a chain of alternating phosphate-sugar-phosphate-sugar molecules, and the nitrogen bases branch off the side of this backbone (see figure 2.23*b,c*).

The Double Helix of DNA

DNA is a huge molecule formed by two very long polynucleotide strands linked along their length by hydrogen bonds between complementary pairs of nitrogen bases. The pairing of the nitrogen bases occurs according to a predictable pattern: adenine ordinarily pairs with thymine, and cytosine with guanine. The bases are attracted in this way because each pair shares oxygen, nitrogen, and hydrogen atoms exactly positioned to align perfectly for hydrogen bonds (figure 2.25*a*).

For ease in understanding the structure of DNA, it is sometimes compared to a ladder, with the sugar-phosphate backbone representing the rails and the paired nitrogen bases representing the steps. Owing to the manner of nucleotide pairing and stacking of the bases, the actual configuration of DNA is a *double helix* that looks somewhat like a spiral staircase (figure 2.25). As is true of protein, the structure of DNA is intimately related to its function. DNA molecules are usually extremely long, a feature that satisfies a requirement for storing genetic information in the sequence of base pairs the molecule contains. The hydrogen bonds between pairs can be disrupted when DNA is being copied, and the fixed complementary base pairing is essential to maintain the genetic code.

Making New DNA: Passing on the Genetic Message

The behavior and metabolism of cells and viruses is ultimately programmed by some form of nucleic acid. In all cells and many viruses, these activities are directed by DNA master codes; some viruses contain only RNA as their primary genetic material. Regardless of the type of nucleic acid, cells and viruses will persist only if they are capable of duplicating this genetic material and passing it on to the next generation. Every time a cell divides, it must duplicate its chromosomes so that the new cells will have the information to synthesize proteins and regulate their heredity (figure 2.26). Consequently, two copies of DNA must be synthesized using the original one. Just how is this accomplished?

While reexamining the simplified structure of DNA shown in figure 2.23, you might observe what was first noted in the early 1950s by James Watson and Francis Crick, the two scientists who first deciphered its structure. They saw that the secret to the duplication, or **replication,*** of DNA lay in the double-stranded nature of the molecule and the precise pairing of the bases. The molecule can be opened up by severing the hydrogen bonds between the two halves of the helix. Severing the bonds exposes the nitrogen bases that make up the genetic code and converts each half strand into a template or pattern for synthesizing the new DNA strands. Free nucleotides in the region are subsequently brought in to match up with the corresponding bases on the template strands, conforming to the required base pairing of A-T and C-G. The end result is two separate strands with the same type and order of bases that were in the original molecule (figure 2.26).

RNA: Organizers of Protein Synthesis

Like DNA, RNA consists of a long chain of nucleotides. However, RNA is a single strand containing ribose sugar and uracil instead of thymine (see figure 2.23). Several functional types of RNA are formed using the DNA template through a replication-like process. The three major types of RNA are important for protein synthesis. Messenger RNA (mRNA) is a copy of a gene from DNA giving the order and type of amino acids in a protein; transfer RNA (tRNA) is a carrier that delivers the correct amino acids for protein assembly; and ribosomal RNA (rRNA) is a major component of ribosomes (see figure 4.19). More information on these important processes is presented in chapter 9.

ATP: The Energy Molecule of Cells

A relative of RNA involved in an entirely different cell activity is **adenosine triphosphate (ATP).** ATP is a nucleotide containing adenine, ribose, and three phosphates rather than just one (figure 2.27). It belongs to a category of high-energy compounds (also including guanosine triphosphate, GTP) that give off energy when the bond is broken between the second and third (outermost) PO_4. The presence of these high-energy bonds makes it possible for ATP to release and store energy for cellular chemical reactions. Breakage of the third phosphate not only releases energy to do cellular work but also generates adenosine diphosphate (ADP). ADP can be converted back to ATP when the third phosphate is restored, thereby serving as an energy depot. Carriers for oxidation-reduction activities (nicotinamide adenine dinucleotide [NAD], for instance) are also derivatives of nucleotides (see chapter 8).

Chapter Checkpoints

Macromolecules are very large organic molecules (polymers) built up by polymerization of smaller molecular subunits (monomers).

Carbohydrates are biological molecules whose polymers are monomers linked together by glycosidic bonds. Their main functions are protection and support (in organisms with cell walls) and also nutrient and energy stores.

Lipids are biological molecules such as fats that are insoluble in water and contain special ester linkages. Their main functions are cell components, cell secretions, and nutrient and energy stores.

Proteins are biological molecules whose polymers are chains of amino acid monomers linked together by peptide bonds.

Proteins are called the "shapers of life" because of the many biological roles they play in cell structure and cell metabolism.

Protein shape determines protein function. Shape is dictated by amino acid composition and by the pH and temperature of the protein's immediate environment.

Nucleic acids are biological molecules whose polymers are chains of nucleotide monomers linked together by phosphate–pentose sugar covalent bonds. Double-stranded nucleic acids are linked together by hydrogen bonds. Nucleic acids are information molecules that direct cell metabolism and reproduction. Nucleotides such as ATP also serve as energy transfer molecules in cells.

*replication (reh´´-plih-kay´-shun) A process of making an exact copy of something.

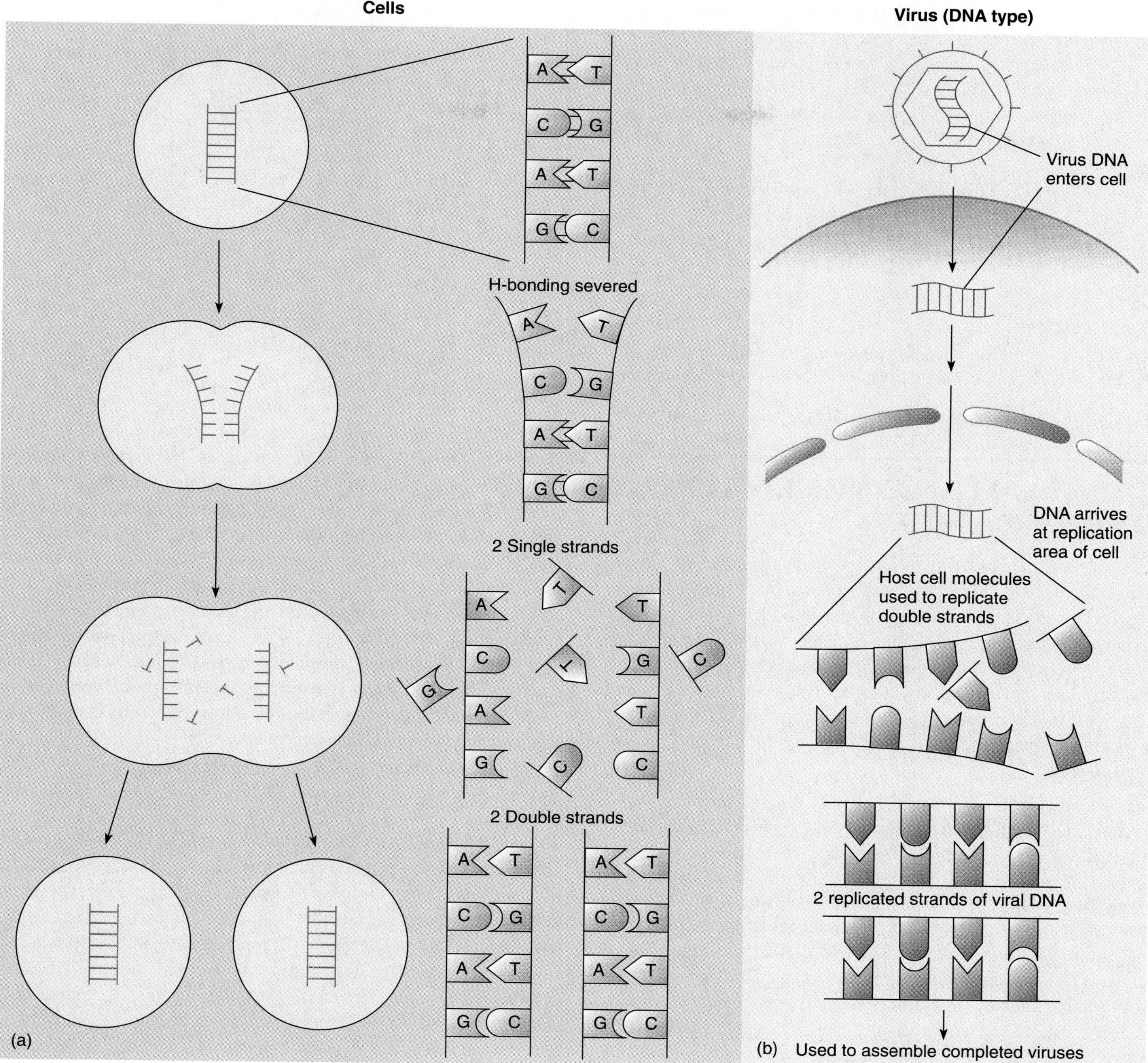

Figure 2.26

Simplified view of DNA replication in cells and viruses. (*a*) The DNA in the cell's chromosome must be duplicated as the cell is dividing. This duplication is accomplished through the separation of the double DNA strand into two single strands. New strands are then synthesized using the original strands as guides to assemble the correct complementary new bases. (*b*) In DNA viruses, the DNA is inserted into its host cell and enters the replication area. Here it is separated into two strands and copied by the machinery of the host cell. Thousands of copies of viral nucleic acid will become part of the completed viruses later released by the infected host cell.

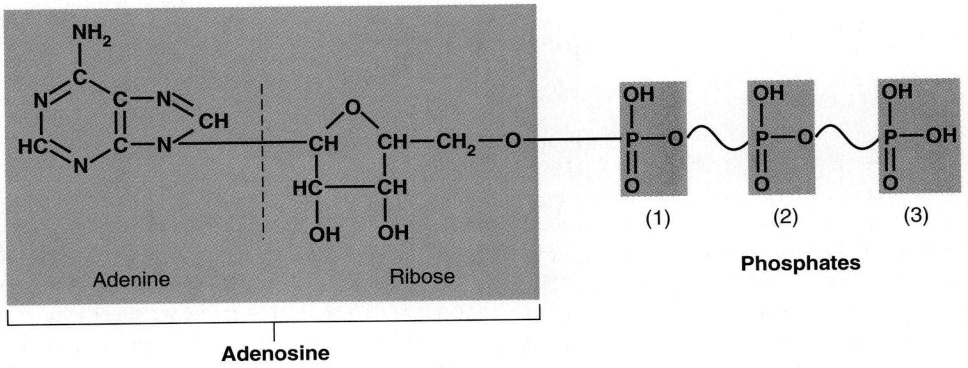

Figure 2.27
The structural formula of an ATP molecule, the chemical form of energy transfer in cells. The wavy lines that connect the phosphates represent bonds that release large amounts of energy when broken.

CELLS: WHERE CHEMICALS COME TO LIFE

As we proceed in this chemical survey from the level of simple molecules to increasingly complex levels of macromolecules, at some point we cross a line from the realm of lifeless molecules and arrive at the fundamental unit of life called a **cell.**[6] A cell is indeed a huge aggregate of carbon, hydrogen, oxygen, nitrogen, and many other atoms, and it follows the basic laws of chemistry and physics, but it is much more. The combination of these atoms produces characteristics, reactions, and products that can only be described as **living.**

FUNDAMENTAL CHARACTERISTICS OF CELLS

The bodies of living things such as bacteria and protozoa consist of only a single cell, whereas those of animals and plants contain trillions of cells. Regardless of the organism, all cells have a few common characteristics. They tend to be spherical, polygonal, cubical, or cylindrical, and their protoplasm (internal cell contents) is encased in a cell or cytoplasmic membrane (microfile 2.3). They have chromosomes containing DNA and ribosomes for protein synthesis and they are exceedingly complex in function. Aside from these few similarities, most cell types fall into one of two fundamentally different lines (discussed in chapter 1): the small, seemingly simple procaryotic cells and the larger, structurally more complicated eucaryotic cells.

Eucaryotic cells are found in animals, plants, fungi, and protists. They contain a number of complex internal parts called organelles that perform useful functions for the cell involving growth, nutrition, or metabolism. By convention, organelles are defined as cell components that perform specific functions and are enclosed by membranes (see microfile 2.3). Organelles also partition the eucaryotic cell into smaller compartments. The most visible organelle is the nucleus, a roughly ball-shaped mass surrounded by a double membrane that contains the DNA of the cell. Other organelles include the Golgi apparatus, endoplasmic reticulum, vacuoles, and mitochondria (table 2.6).

Procaryotic cells are generally single cells found in the bacteria and archaea. Sometimes it may seem that procaryotes are the microbial "have-nots" because, for the sake of comparison, they are described by what they lack. They have no nucleus or other organelles. This apparent simplicity is misleading, because the fine structure of procaryotes is complex. Overall, procaryotic cells can engage in nearly every activity that eucaryotic cells can, and many can function in ways that eucaryotes cannot.

PROCESSES THAT DEFINE LIFE

To lay the groundwork for a detailed coverage of cells in chapters 4 and 5, this section provides an overview of cell structure and function and introduces the primary characteristics of life. The biological activities or properties that help define and characterize cells as living entities are: (1) growth, (2) reproduction and heredity, (3) metabolism, including cell synthesis and the release of energy, (4) movement and/or irritability, (5) the capacity to transport substances into and out of the cell, and (6) cell support and protection and storage mechanisms. Although eucaryotic cells have specific organelles to perform these functions, procaryotic cells must rely on a few simple, multipurpose cell components. As indicated in chapter 1, viruses are not cells, are not generally considered living things, and show certain signs of life only when they invade a host cell. Table 2.6 indicates their relative simplicity compared with cells.

Reproduction: Bearing Offspring

A cell's **genome,*** its complete set of genetic material, is composed of elongate strands of DNA. The DNA is packed into discrete bodies called **chromosomes.*** In eucaryotic cells, the

6. The word was originally coined from an Old English term meaning "small room" because of the way plant cells looked to early microscopists.

*genome (jee´-nohm) A combination of the words *gene* and *chromosome.* Refers to an organism's entire set of hereditary factors.

*chromosome (kro´-moh-sohm) Gr. *chroma,* colored, and *soma,* body. The name is derived from the fact that chromosomes stain readily with dyes.

MICROFILE 2.3 MEMBRANES: CELLULAR SKINS

The word **membrane** appears frequently in descriptions of cells in this chapter and in chapters 4 and 5. The word itself describes any lining or covering, including such multicellular structures as the mucous membranes of the body. From the perspective of a single cell, however, a membrane is a thin, double-layered sheet composed of lipids such as phospholipids and sterols (averaging about 40% of membrane content) and protein molecules (averaging about 60%). The primary role of membranes is as a **cell membrane** that completely encases the cytoplasm. Membranes are also components of eucaryotic organelles such as nuclei, mitochondria, and chloroplasts, and they appear in internal pockets of certain procaryotic cells. Even some viruses, which are not cells at all, can have a membranous protective covering.

Cell membranes are so thin—on the average, just 0.0070 μm (7 nm) thick—that they cannot actually be seen with an optical microscope. Even at magnifications made possible by electron microscopy (500,000×), very little of the precise architecture can be visualized, and a cross-sectional view has the appearance of railroad tracks. Following detailed microscopic and chemical analysis, S. J. Singer and C. K. Nicholson proposed a simple and elegant theory for membrane structure called the **fluid mosaic* model.** According to this theory, a membrane is a continuous bilayer formed by lipids that are oriented with the polar lipid heads toward the outside and the nonpolar heads toward the center of the membrane. Embedded at numerous sites in this bilayer are various-sized globular proteins. Some proteins are situated only at the surface; others extend fully through the entire membrane. The configuration of the inner and outer sides of the membrane can be quite different because of the variations in protein shape and position.

Membranes are dynamic and constantly changing because the lipid phase is in motion and many proteins can migrate freely about, somewhat as icebergs do in the ocean. This fluidity is essential to such activities as engulfment of food and discharge or secretion by cells. The structure of the lipid phase provides an impenetrable barrier to many substances. This property accounts for the **selective permeability** and capacity to regulate transport of molecules. It also serves to segregate activities within the cell's cytoplasm. Membrane proteins function in receiving molecular signals (receptors), in binding and transporting nutrients, and in acting as enzymes (see chapter 8).

*mosaic (moh-zay´-ik) An intricate design made up of many small fragments.

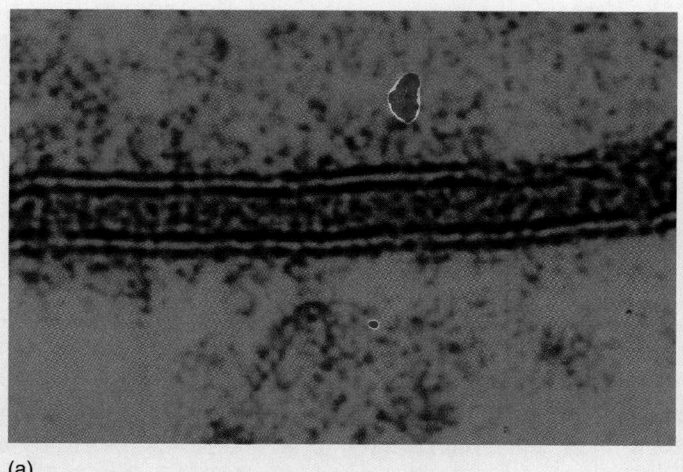

(a)

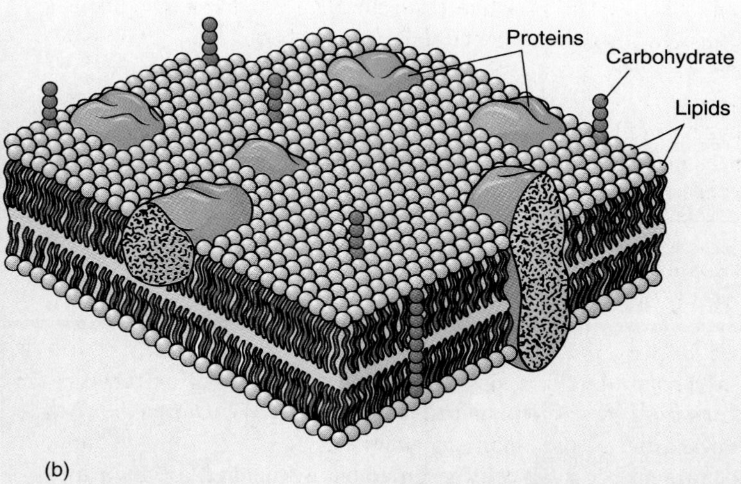

(b)

Extreme magnification of (a) a cross section of a cell membrane, which appears as double tracks. (b) A generalized version of the fluid mosaic model of a cell membrane indicates a bilayer of lipids with globular proteins embedded to some degree in the lipid matrix. This structure explains many characteristics of membranes, including flexibility, solubility, permeability, and transport.

chromosomes are located entirely within a nuclear membrane. Procaryotic DNA occurs in a special type of circular chromosome that is not enclosed by a membrane of any sort.

Living things devote a portion of their life cycle to producing offspring that will carry on their particular genetic line for many generations. In **sexual reproduction,** offspring are produced through the union of sex cells from two parents. In **asexual**[7] **reproduction,** offspring originate through the division of a single parent cell into two daughter cells. Sexual reproduction, often of a complex nature, occurs in most eucaryotes, and eucaryotic cells also reproduce asexually by several processes. One is a type of cell division called *binary fission,* a simple process in which the cell splits equally in two. Many eucaryotic cells engage in **mitosis,*** an orderly division of chromosomes that usually accompanies cell division (see figure 5.6). In contrast, procaryotic cells reproduce primarily by binary fission. They have no mitotic apparatus, nor do they reproduce by typical sexual means.

7. *Asexual* refers to the absence of sexual union. (The prefix *a* or *an* means not or without.)

*mitosis (my-toh´-sis) Gr. *mitos,* thread, and *osis,* a condition. Often the term *mitosis* is used synonymously with eucaryotic cell division.

TABLE 2.6

A GENERAL COMPARISON OF PROCARYOTIC AND EUCARYOTIC CELLS AND VIRUSES*

Function or Structure	Characteristic	Procaryotic Cells	Eucaryotic Cells	Viruses**
Genetics	Nucleic acids	+	+	+
	Chromosomes	+	+	–
	True nucleus	–	+	–
	Nuclear envelope	–	+	–
Reproduction	Mitosis	–	+	–
	Production of sex cells	+/–	+	–
	Binary fission	+	+	–
Biosynthesis	Independent	+	+	–
	Golgi apparatus	–	+	–
	Endoplasmic reticulum	–	+	–
	Ribosomes	+***	+	–
Respiration	Enzymes	+	+	–
	Mitochondria	–	+	–
Photosynthesis	Pigments	+/–	+/–	–
	Chloroplasts	–	+/–	–
Motility/locomotor structures	Flagella	+/–***	+/–	–
	Cilia	–	+/–	–
Shape/protection	Cell wall	+***	+/–	–
	Capsule	+/–	+/–	–
	Spores	+/–	+/–	–
Complexity of function		+	+	+/–
Size (in general)		*0.5–3 μm*	*2–100 μm*	*<0.2 μm*

*+ means most members of the group exhibit this characteristic; – means most lack it; +/– means only some members have it.
**Viruses cannot participate in metabolic or genetic activity outside their host cells.
***The procaryotic type is functionally similar to the eucaryotic, but structurally unique.

Metabolism: Chemical and Physical Life Processes

Protein synthesis in all cells is carried out by hundreds of tiny particles called **ribosomes.** In eucaryotes, they are dispersed throughout the cell or inserted into membranous sacs known as the **endoplasmic reticulum** (see figure 5.8). Procaryotes have smaller ribosomes scattered throughout the protoplasm, since they lack an endoplasmic reticulum. Most energy in the form of ATP is generated by chemical reactions in the **mitochondria** of eucaryotes and in the cell membrane of procaryotes. Photosynthetic microorganisms (algae and some bacteria) trap solar energy by means of pigments and convert it to chemical energy in the cell. Algae (eucaryotes) have compact, membranous bundles called **chloroplasts,** which contain the pigment and perform the photosynthetic reactions. Photosynthetic reactions and pigments in photosynthetic bacteria do not occur in chloroplasts, but in specialized areas of the cell membrane.

Irritability or Motility

All cells have the capacity to respond to chemical, mechanical, or light stimuli. This quality, called **irritability,** can be necessary for adapting to the environment and obtaining nutrition. Although not present in all cells, true **motility,** or self-propulsion, is clearly a sign of life. Eucaryotic cells move by one of the following locomotor organelles: cilia, which are short, hairlike appendages; flagella, which are longer, whiplike appendages; or pseudopods, fingerlike extensions of the cell membrane. Motile procaryotes move by means of unusual, propeller-like flagella unique to bacteria or by special fibrils that produce a gliding form of motility. They have no cilia or pseudopods.

Protection and Storage

Many cells are supported and protected by rigid **cell walls,** which prevent them from rupturing while also providing support and shape. Among eucaryotes, cell walls occur in plants, microscopic algae, and fungi, but not in animals or protozoa. The majority of procaryotes have cell walls, but they differ in composition from the eucaryotic varieties. As protection against depleted nutrient sources, many microbes store nutrients intracellularly. Eucaryotes store nutrients in membranous sacs called vacuoles, and procaryotes concentrate it in crystals called granules or inclusions.

Transport: Movement of Nutrients and Wastes

To survive, a cell needs both to draw nutrients from its external environment and to expel waste and other metabolic products from its internal environment. This two-directional transport is accomplished in both eucaryotes and procaryotes by the cell membrane. This membrane, described in microfile 2.3, has a very similar structure in both eucaryotic and procaryotic cells. Eucaryotes have an additional organelle, the **Golgi apparatus,** that assists in sorting and packaging molecules for transport and removal from the cell.

Table 2.6 summarizes the differences and similarities among procaryotes, eucaryotes, and viruses. You will probably want to refer to this table again after you have completed chapters 4 and 5.

Chapter Checkpoints

As the atom is the fundamental unit of matter, so is the cell the fundamental unit of life.

All true cells contain biological molecules that carry out the processes that define life: metabolism and reproduction. The functions of irritability and motility, protection, storage, and transport support these two basic processes.

The cell membrane is of critical importance to all cells because it controls the interchange between the cell and its environment.

Chapter Capsule with Key Terms

BASIC PROPERTIES OF ATOMS AND ELEMENTS

All **matter** in the universe is composed of minute particles called **atoms**—the simplest form of matter not divisible into a simpler substance by chemical means. Atoms are composed of smaller particles called **protons** (p^+), positive in charge; **neutrons** (n^o), uncharged; and **electrons** (e^-), negative in charge. Protons and neutrons constitute the **nucleus** of an atom, and electrons move about the nucleus in **orbitals** that occupy energy levels called **shells.**

Atoms that differ in numbers of protons, neutrons, and electrons are **elements.** Elements can be described by **atomic weight** (the total number of protons and neutrons in the nucleus) and **atomic number** (the number of protons alone), and each is known by a distinct name and symbol. An unreacted atom is neutral because electron charge cancels out proton charge. Elements exist in variant forms called **isotopes,** which differ in the number of neutrons (atomic weight).

Electrons fill the orbitals in pairs: the first orbital (shell) holds 2 e^-, the second shell can hold 8 e^- (4 pairs), and the third shell can hold 18 e^- (9 pairs). The electron number of each element dictates orbital filling; the outermost orbital becomes the focus of reactivity and bonding. In general, atoms with a filled outermost orbital are less reactive than are those with unfilled outer orbitals.

Atomic Bonds and Molecules

Outer orbitals of atoms interact to form **chemical bonds** and **molecules.** Member atoms of molecules can be the same element or different elements; if the member elements are different, the substance is a **compound.**

The type of bond is dictated by the electron makeup **(valence)** of the outer orbitals of the atoms. **Covalent bonds** occur when the electrons are shared and orbit within the entire molecule. A molecule containing similar elements is **nonpolar;** one with dissimilar elements is **polar.**

Ionic bonds occur when an atom with a low number of valence electrons loses them to an atom that has a nearly filled outer orbital. When ionic bonds are dissolved the atoms **ionize** into charged particles called **ions.** Ions that have lost electrons and have a positive charge are **cations** (Na^+), and those that have gained electrons and have a negative charge are **anions** (Cl^-). Ions of the opposite charge are attracted to each other, and those with the same charges repel each other. **Hydrogen bonds** are caused by weak attractive forces between covalently bonded hydrogen and polarized atoms, such as oxygen, on the same or nearby molecules that bear a negative charge.

Redox Reactions

Chemicals can participate in a transfer of electrons, called an **oxidation-reduction (redox) reaction,** between pairs of atoms or molecules. **Oxidation** is a reaction in which electrons are released, and **reduction** is a reaction in which those same electrons are received. Any atom or molecule that donates electrons to another atom or molecule is a **reducing agent,** and one that picks up electrons is an **oxidizing agent.**

Chemical Formulas, Models, and Equations

Chemical formulas can be presented as simple molecular summaries of the atoms or they can be expressed as structural formulas that give the details of bonding.

A chemical equation summarizes a chemical reaction by showing the reactants (starting chemicals) and the products (resulting chemicals) and can indicate **synthesis reactions, decomposition reactions, exchange reactions,** and **reversible reactions.**

Solutions

A **solution** consists of a solid, liquid, or gaseous chemical termed a **solute** dissolved in a liquid medium called a **solvent.** Water-soluble solutes are either charged (ionic salts) or polar (sugars). The dissolved solute becomes **hydrated** because of electrostatic attraction. Chemicals that attract water are **hydrophilic;** those that repel it are **hydrophobic.** The **concentration** of a solution is defined as the amount of solute dissolved in a given volume of solvent.

Acids, Bases, and pH

Acidity and basicity of solutions are represented by the **pH scale,** based on a standardized range of hydrogen ion **concentration** [H^+]. An **acid** is a solution that contains a [H^+] greater than 0.0000001 g/l (10^{-7} Molar), and a **base** is a solution that contains a [H^+] below that amount. When a solution contains exactly that [H^+], it is considered neutral. pH ranges from the most acidic reading of 0 to the most basic (alkaline) reading of 14; in the exact middle is pH 7, the reading of neutrality. Acidic and basic ions in aqueous solutions can interact to form water and a salt through **neutralization.**

ORGANIC CHEMISTRY

Organic compounds contain carbon and hydrogens (and usually some others). Organic compounds constitute the most prominent molecules in the structure and function of cells.

Carbon forms covalent bonds with other carbons and with hydrogen, oxygen, nitrogen, and phosphorus to create the backbone of biological molecules. **Inorganic compounds** are composed of some combination of atoms other than carbon and hydrogen. **Functional groups** are special accessory molecules that bind to carbon and provide the diversity and reactivity seen in organic compounds.

BIOCHEMISTRY AND MACROMOLECULES

Biochemicals are large organic compounds termed **macromolecules.** Many are assembled from individual smaller building blocks called **monomers** into multiunit chains called **polymers** in a process known as **polymerization.**

Carbohydrates are compounds composed of carbon, hydrogen, and oxygen (CH_2O). **Monosaccharides (glucose)** are the simplest sugars. When two sugars are joined by the —OH group and carbon through **dehydration synthesis,** a **glycosidic bond** occurs. **Disaccharides (lactose, maltose, sucrose)** are composed of two monosaccharides. **Polysaccharides** are chains of five or more monosaccharides; **cellulose** is a long-chain glucose polymer that is a major component of the cell wall in plants and algae; **peptidoglycan** is a combination of glycans and peptides that reinforces the bacterial cell wall; **starch** and **glycogen** are compact storage forms of glucose. Polysaccharides are digested by specific enzymes that break the bond through **hydrolysis.**

Lipids are organic compounds that are not soluble in water and other polar solvents because of their nonpolar, hydrophobic, hydrocarbon chains. **Triglycerides,** including fats and oils, consist of a **glycerol** molecule bound to three **fatty acid** molecules. Triglycerides come in saturated or unsaturated varieties and are important storage lipids. **Phospholipids** are composed of a glycerol bound to two fatty acids and a phosphoric acid–alcohol group; the molecule has a charged (hydrophilic) head and long, uncharged (hydrophobic) fatty acid tails; they form single or double lipid layers in the presence of water and are important constituents of cell membranes.

Proteins are highly complex biochemicals assembled from 20 different subunits called **amino acids** (aa). Amino acids are combined in a certain order by **peptide bonds:**

$$\textbf{peptide bonds: } \mathrm{C{-}\underset{\displaystyle H}{\underset{|}{N}}{-}\overset{\displaystyle O}{\overset{\|}{C}}{-}C.}$$

A **peptide** is a short chain, such as a dipeptide with two aa's or a tripeptide with three; a **polypeptide** is usually 20–50 aa's in a chain; a protein contains more than 50 aa's. Larger proteins predominate in cells.

The chain of amino acids is a polypeptide's **primary (1°) structure.** Functional groups on the amino acids cause proteins to form additional levels of structure. Hydrogen and other weak bonds within the chain twist it first into a helix or sheet called the **secondary (2°) structure.** This folds again, forging stronger **disulfide bonds** on nearby cysteines and producing a three-dimensional **tertiary (3°) structure.** Proteins composed of two or more polypeptides exist in a **quaternary (4°) structure.** The surface configuration of proteins provides their specificity and gives rise to the diversity in **enzymes, antibodies,** and cell receptors.

Nucleic acids—**deoxyribonucleic acid (DNA)** and **ribonucleic acid (RNA)**—are very complex molecules that carry, express, and pass on the genetic information of all cells and viruses. Their basic building block is a nucleotide, composed of a **nitrogen base,** a **pentose sugar,** and a **phosphate.** Nitrogen bases are the ringed compounds: **adenine** (A), **cytosine** (C), **thymine** (T), **guanine** (G), and **uracil** (U); pentose sugars are deoxyribose or ribose. The basic design is a polynucleotide, with the sugars linking up in an alternating series with phosphates to make a backbone and the bases branching off the sugars.

DNA contains **deoxyribose** sugar, has all of the bases except uracil, and occurs as a double-stranded helix with the bases hydrogen-bonded in pairs; the pairs mate according to the pattern A-T and C-G. DNA, the master code for a cell's life processes, can be **replicated** by division of the single strands and synthesis of two new double strands using them as templates. RNA contains **ribose** sugar, has all of the bases except thymine, and is a single-stranded molecule. It helps interpret the DNA code into proteins.

Adenosine triphosphate (ATP) is a nucleotide involved in the transfer and storage of energy in cells. It contains adenine, ribose, and three phosphates in a series. Splitting off the last phosphate in the triphosphate releases a packet of energy that is used to do cell work.

INTRODUCTION TO CELL STRUCTURE

Cells are huge aggregates of macromolecules organized to carry out complex processes described as **living.** All organisms consist of cells, which fall into one of two types: **procaryotic cells,** which are small, structurally simple bacterial cells that lack a **nucleus** and other organelles, and **eucaryotic cells,** which are larger, contain a nucleus and organelles, and are found in the cells of plants, animals, fungi, and protozoa. Viruses are not cells and are not generally considered living because they cannot function independently.

An organism must participate in certain processes in order to be considered alive. Growth and **reproduction** involve producing offspring asexually (with one parent) or sexually (with two parents); **metabolism** refers to the chemical reactions in cells, including the synthesis of proteins on ribosomes and the release of energy (ATP); **motility** originate from special locomotor structures such as flagella and cilia; irritability is the capacity to respond to external stimuli; **protective** external structures include capsules and **cell walls;** nutrient **storage** takes place in compact intracellular masses; and **transport** involves conducting nutrients into the cell and wastes out of the cell.

Important structures that surround the protoplasm of all cells and that are also found internally are **membranes.** Membranes are continuous, ultrathin bilayers of lipids studded with proteins. Lipids provide a flowing network that dictates cell **permeability;** proteins serve as channels of transport and as sites of recognition and chemical reactions.

MULTIPLE-CHOICE QUESTIONS

1. The smallest unit of matter with unique characteristics is
   a. an electron c. an atom
   b. a molecule d. the nucleus
2. The ____ charge of a proton is exactly balanced by the ____ charge of a (an) ____.
   a. negative, positive, electron
   b. positive, neutral, neutron
   c. positive, negative, electron
   d. neutral, negative, electron
3. Electrons move around the nucleus of an atom in pathways called
   a. shells c. circles
   b. orbitals d. rings
4. The atomic weight of an element is about ____ as large as its atomic number.
   a. 4 times b. 3 times c. 2 times
   d. It is not larger, it is the same.
5. The number of electrons of an atom is automatically known if one knows the
   a. atomic number
   b. atomic weight
   c. number of orbitals
   d. valence

6. Elements
   a. are pure substances
   b. have distinctively different atomic structures
   c. vary in atomic weight
   d. a and b are correct
   e. a, b, and c are correct
7. If a substance contains two or more elements of the same or different types, it is considered
   a. a compound
   b. a monomer
   c. a molecule
   d. organic
8. Bonds in which atoms share electrons are defined as _____ bonds.
   a. hydrogen
   b. ionic
   c. double
   d. covalent
9. What kind of bond would you expect potassium to form with chlorine?
   a. ionic
   b. covalent
   c. polar
   d. nonpolar
10. When a compound carries a positive charge on one end and a negative charge on the other end it is said to be
    a. ionized
    b. hydrophilic
    c. polar
    d. oxidized
11. Hydrogen bonds can form between _____ adjacent to each other.
    a. two hydrogen atoms
    b. two oxygen atoms
    c. a hydrogen atom and an oxygen atom
    d. negative charges
12. Ions with the same charge will be _____ each other, and ions with opposite charges will be _____ each other.
    a. repelled by, attracted to
    b. attracted to, repelled by
    c. hydrated by, dissolved by
    d. dissolved by, hydrated by
13. An atom that can donate electrons during a reaction is called
    a. an oxidizing agent
    b. a reducing agent
    c. an ionic agent
    d. an electrolyte
14. In a solution of NaCl and water, NaCl is the _____ and water is the _____.
    a. acid, base
    b. base, acid
    c. solute, solvent
    d. solvent, solute
15. A substance that releases H^+ into a solution
    a. is a base
    b. is ionized
    c. is an acid
    d. has a high pH
16. A solution with a pH of 2 _____ than a solution with a pH of 8.
    a. has fewer H^+
    b. has more H^+
    c. is more concentrated
    d. is less concentrated
17. Fructose is a type of
    a. disaccharide
    b. monosaccharide
    c. polysaccharide
    d. amino acid
18. Bond formation in polysaccharides and polypeptides is accompanied by the removal of a
    a. hydrogen atom
    b. hydroxyl ion
    c. carbon atom
    d. water molecule
19. The monomer unit of polysaccharides such as starch and cellulose is
    a. fructose
    b. glucose
    c. ribose
    d. lactose
20. A phospholipid contains
    a. three fatty acids bound to glycerol
    b. three fatty acids, a glycerol, and a phosphate
    c. two fatty acids and a phosphate bound to glycerol
    d. three cholesterol molecules bound to glycerol
21. Proteins are synthesized by linking amino acids with _____ bonds.
    a. disulfide
    b. glycosidic
    c. peptide
    d. ester
22. The amino acid that accounts for disulfide bonds in the tertiary structure of proteins is
    a. tyrosine
    b. glycine
    c. cysteine
    d. serine
23. DNA is a hereditary molecule that is composed of
    a. deoxyribose, phosphate, and nitrogen bases
    b. deoxyribose, a pentose, and nucleic acids
    c. sugar, proteins, and thymine
    d. adenine, phosphate, and ribose
24. What is meant by DNA replication?
    a. duplication of the sugar-phosphate backbone
    b. matching of base pairs
    c. formation of the double helix
    d. the exact copying of the DNA code into two new molecules

CONCEPT QUESTIONS

1. How are the concepts of an atom and an element related? What causes elements to differ?
2. a. How are atomic weight and atomic number derived?
   b. Using data in table 2.1, give the electron number of nitrogen, sulfur, calcium, phosphorus, and iron.
   c. What is distinctive about isotopes of elements, and why are they important?
3. a. How is the concept of molecules and compounds related?
   b. Compute the molecular weight of oxygen and methane.
4. a. Why is an isolated atom neutral?
   b. Describe the concept of the atomic nucleus, electron orbitals, and shells.
   c. What causes atoms to form chemical bonds?
   d. Why do some elements not bond readily?
   e. Draw the atomic map for magnesium and predict what kinds of bonds it will make.
5. Distinguish between the general reactions in covalent, ionic, and hydrogen bonds.
6. a. Which kinds of elements tend to make covalent bonds?
   b. Distinguish between a single and a double bond.
   c. What is polarity?
   d. Why are some covalent molecules polar and others nonpolar?
   e. What is an important consequence of the polarity of water?
7. a. Which kinds of elements tend to make ionic bonds?
   b. Exactly what causes the charges to form on atoms in ionic bonds?
   c. Verify the proton and electron numbers for Na^+ and Cl^-.
   d. Differentiate between an anion and a cation.
   e. What kind of ion would you expect magnesium to make, on the basis of its valence?
8. Differentiate between an oxidizing agent and a reducing agent.
9. Why are hydrogen bonds relatively weak?
10. a. Compare the three basic types of chemical formulas.
    b. Review the types of chemical reactions and the general ways they can be expressed in equations.
11. a. Define solution, solvent, and solute.
    b. What properties of water make it an effective biological solvent, and how does a molecule like NaCl become dissolved in it?
    c. How is the concentration of a solution determined?
    d. What is molarity? Tell how to make a 1 M solution of $MgPO_4$ and a 0.1 M solution of $CaSO_4$.

12. a. What determines whether a substance is an acid or a base?
   b. Briefly outline the pH scale.
   c. How can a neutral salt be formed from acids and bases?
13. a. What atoms must be present in a molecule for it to be considered organic?
   b. What characteristics of carbon make it ideal for the formation of organic compounds?
   c. What are functional groups?
   d. Differentiate between a monomer and a polymer.
   e. How are polymers formed?
   f. Name several inorganic compounds.
14. a. What characterizes the carbohydrates?
   b. Differentiate between mono-, di-, and polysaccharides, and give examples of each.
   c. What is a glycosidic bond, and what is dehydration synthesis?
   d. What are some of the functions of polysaccharides in cells?
15. a. Draw simple structural molecules of triglycerides and phospholipids to compare their differences and similarities.
   b. What is an ester bond?
   c. How are saturated and unsaturated fatty acids different?
   d. What characteristic of phospholipids makes them essential components of cell membranes?
   e. Why is the hydrophilic end of phospholipids attracted to water?
16. a. Describe the basic structure of an amino acid.
   b. What makes the amino acids distinctive, and how many of them are there?
   c. What is a peptide bond?
   d. Differentiate between a peptide, a polypeptide, and a protein.
   e. Explain what causes the various levels of structure of a protein molecule.
   f. What functions do proteins perform in a cell?
17. a. Describe a nucleotide and a polynucleotide, and compare and contrast the general structure of DNA and RNA.
   b. Name the two purines and the three pyrimidines.
   c. Why is DNA called a double helix?
   d. What is the function of RNA?
   e. What is ATP, and what is its function in cells?
18. a. Outline the general structure of a cell, and describe the characteristics of cells that qualify them as living.
   b. Why are viruses not considered living?
   c. Compare the general characteristics of procaryotic and eucaryotic cells.
   d. What are cellular membranes, and what are their functions?
   e. Explain the fluid mosaic model of a membrane.

CRITICAL-THINKING QUESTIONS

1. The "octet rule" in chemistry helps predict the tendency of atoms to acquire or donate electrons from the outer shell. It says that those with fewer than 4 tend to donate electrons and those with more than 4 tend to accept additional electrons; those with exactly 4 can do both. Using this rule, determine what category each of the following elements falls into: N, S, C, P, O, H, Ca, Fe, and Mg. (You will need to work out the valence of the atoms.)
2. Predict the kinds of bonds that occur in ammonium (NH_3), phosphate (PO_4), disulfide (S-S), and magnesium chloride ($MgCl_2$). (Use simple models such as those in figure 2.3.)
3. Work out the following problems:
   a. What is the number of protons in helium?
   b. Will an H bond form between $H_3C{-}CH{=}O$ and H_2O? Why or why not?
   c. Determine which of the following compounds are polar: Cl_2, NH_3, CH_4, glucose, leucine.
   d. What is the pH of a solution with a concentration of 0.00001 g/ml (M) of H^+?
   e. What is the pH of a solution with a concentration of 0.00001 g/ml (M) of OH^-?
4. a. Describe how hydration spheres are formed around cations and anions.
   b. Which substances will be expected to be hydrophilic and hydrophobic, and what makes them so?
   c. Distinguish between polar and ionic compounds, using your own words.
5. In what way are carbon-based compounds like children's Tinker Toys or Lego blocks?
6. Is galactose an aldehyde or a ketone sugar?
7. a. How many water molecules are released for each triglyceride that is formed?
   b. How many peptide bonds are in a tetrapeptide?
8. a. Use pipe cleaners to help understand the formation of the 2° and 3° structures of proteins.
   b. Note the various ways that your pipe cleaner structure can be folded and the diversity of shapes that can be formed.
9. a. Looking at figure 2.25, can you see why adenine forms hydrogen bonds with thymine and why cytosine forms them with guanine?
   b. Show on paper the steps in replication of the following segment of DNA:

   A T G T T C C C G A T C G G C
   | | | | | | | | | | | | | | |
   T A C A A G G G C T A G C C G

10. A useful mnemonic (memory) device for recalling the major characteristics of life is: *G*iant *R*ats *H*ave *M*any *C*olored *T*eeth. Can you list some of the characteristics for which each letter stands?

chapter 3

Tools of the Laboratory: The Methods for Studying Microorganisms

Every year in the United States hundreds of outbreaks of food-borne illness are reported to public health authorities. Because such episodes could potentially travel through the population, epidemiologists are under pressure to determine, as rapidly as possible, the agent involved, the source of contaminated food, and how the illness was acquired. To gather information on microorganisms that may be involved, infectious disease specialists have developed some remarkable techniques and tools: special media for isolating the microbe, microscopes for observing it, and numerous biochemical and genetic tests. Such refined technology is so sensitive that it can uncover a pathogen lying hidden in samples that often contain large numbers of bacteria that are not involved in disease. It has made it possible to detect *Salmonella* bacteria in ice cream and egg nog, hepatitis A virus in strawberries, *Escherichia coli* 0157:H7 bacteria in meat and produce (see chapter opener), and a veritable "cafeteria" of other common food-borne agents.

The concerns surrounding food-borne disease have led the Department of Agriculture to create new food preparation guidelines. All fresh meats and poultry must include instructions for safe handling, including thorough cooking and clean kitchen techniques (See microfile 20.3). The same agency is also reevaluating the techniques used in the slaughterhouse for determining the safety of food. A final precautionary note on microbes and food emphasizes that their small size and invisibility are serious impediments to their control. It is impossible to judge food fitness and safety on the basis of macroscopic appearance alone.

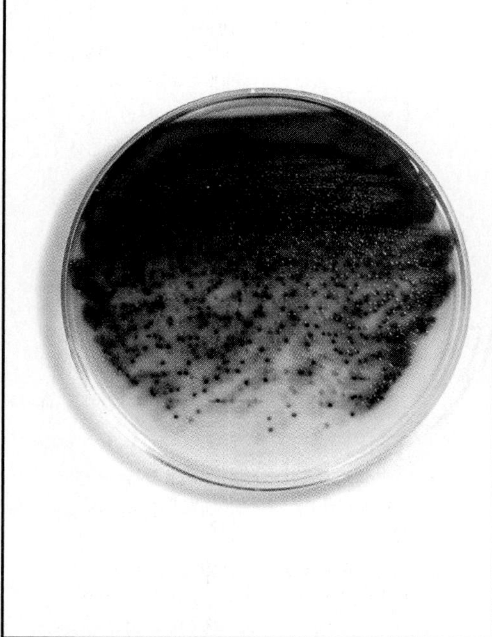

An innovation in media allows rapid isolation and identification of *Escherichia coli* 0157: H7, an emerging food-borne pathogen. On Rainbow™ agar developed by Biolog, it produces black colonies that differentiate it from other strains of *E. coli* and intestinal pathogens.

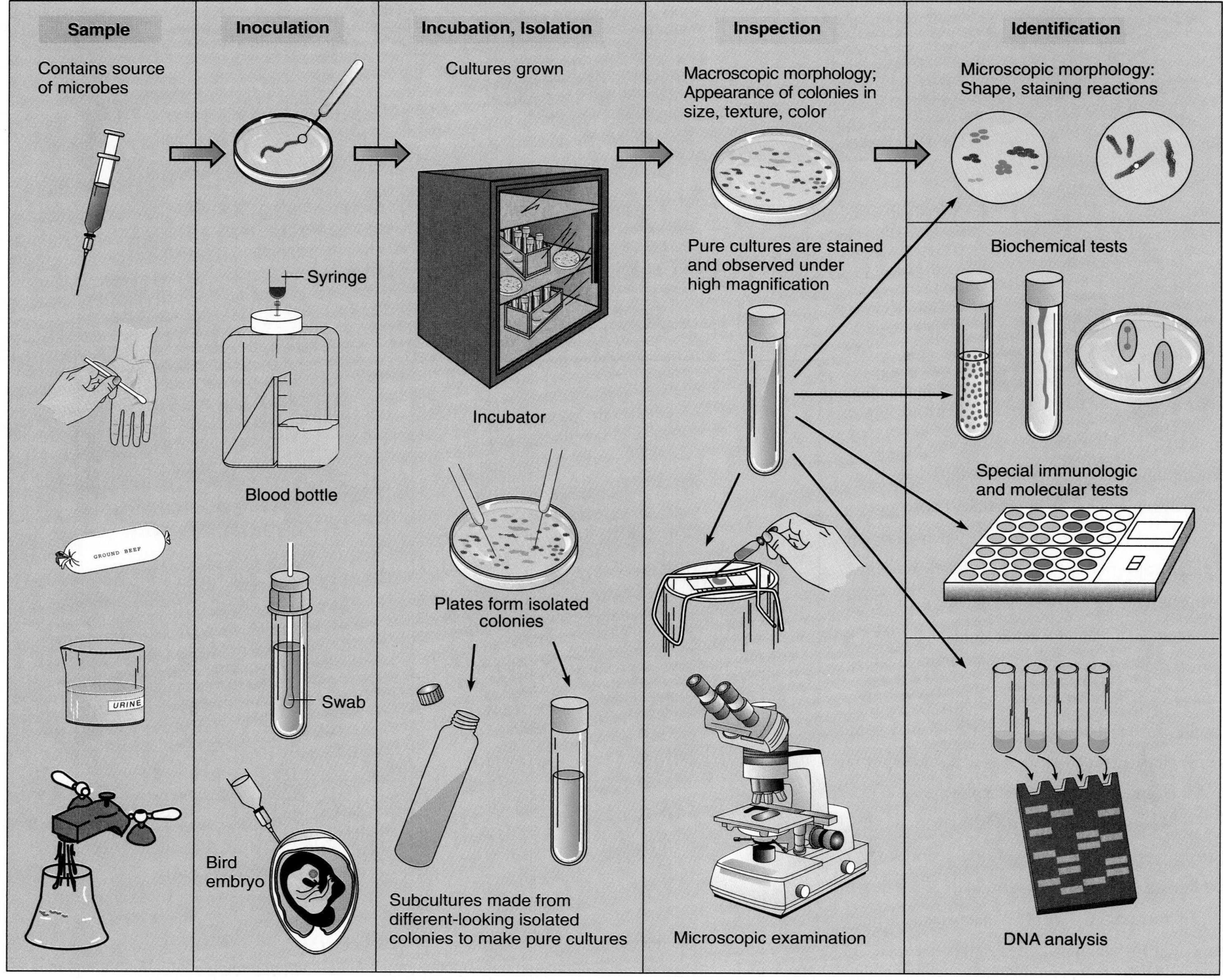

Figure 3.1
A summary of the general laboratory techniques carried out by microbiologists. It is not necessary to perform all the steps shown or to perform them exactly in this order, but all microbiologists participate in at least some of these activities. In some cases, one may proceed right from the sample to inspection, and in others, only inoculation and incubation on special media are required.

METHODS OF CULTURING MICROORGANISMS

Biologists studying large organisms such as animals and plants can, for the most part, immediately see and differentiate their experimental subjects from the surrounding environment and from one another. In fact, they can use their senses of sight, smell, hearing, and even touch to detect and evaluate identifying characteristics and to keep track of growth and developmental changes. Because microbiologists cannot rely as much as other scientists on senses other than sight, they are confronted by some unique problems. First, most habitats (such as the soil and the human mouth) harbor microbes in complex associations, so it is often necessary to separate the species from one another. In addition, to maintain and keep track of such small research subjects, microbiologists usually have to grow them under artificial conditions. A third difficulty in working with microbes is that they are invisible and widely distributed, and undesirable ones can be introduced into an experiment and cause misleading results. These impediments were largely responsible for the development of sterile, pure culture, and aseptic techniques.

Microbiologists have five basic techniques to manipulate, grow, examine, and characterize microorganisms: inoculation, incubation, isolation, inspection, and identification (the Five I's; figure 3.1). Some or all of these procedures are performed by

MICROFILE 3.1 ANIMAL INOCULATION: "LIVING MEDIA"

A great deal of attention has been focused on the uses of animals in biology and medicine. Animal rights activists are vocal about practically any experimentation with animals and have expressed their outrage even through sabotage. Certain kinds of animal testing may seem trivial and unnecessary. But many areas of microbiology and medicine depend on the availability of laboratory animals such as guinea pigs, mice, rats, rabbits, and hamsters specially bred for experimental purposes. These animals, and to a lesser extent primates, monkeys, dogs, cats, chickens, and even armadillos, can be an indispensable aid for studying, growing, and identifying microorganisms. One special use of animals involves inoculation of the early life stages (embryos) of birds. Vaccines for influenza are currently produced in duck embryos. The following is a summary of the major rationales for live animal inoculation:

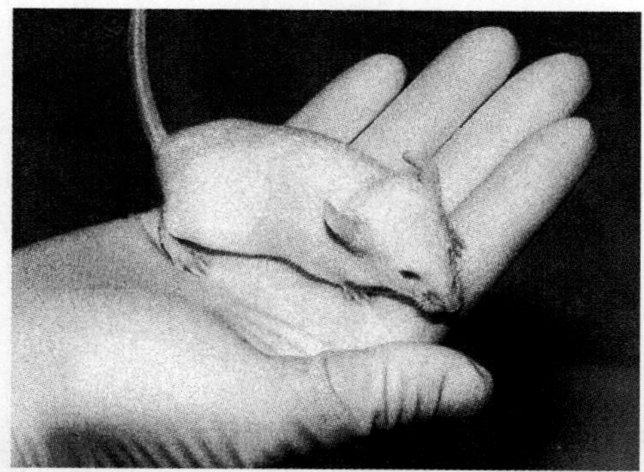

This apparently normal laboratory mouse is an example of a "designer organism" called the HuM-Ab-Mouse. It has been engineered to carry a key piece of the human immune system. When properly stimulated, it will produce protective molecules called antibodies that are identical to those a human would produce. This is just one example of the hundreds of new strains of animals with unusual traits created by researchers to study human diseases.

1. Some of microbes will not grow on artificial media but will grow in a suitable animal and can be recovered in a more or less pure form. These include animal viruses, the spirochete of syphilis, and the leprosy bacillus (grown in armadillos).
2. Researchers can develop animal models for evaluating new diseases or for studying the cause or process of a disease. Koch's postulates is a series of proofs to determine the causative agent of a disease and requires a controlled experiment with an animal that can develop a typical case of the disease. So far, there has not been a completely successful model for HIV infection and human AIDS, but a similar disease in monkeys called simian AIDS has clarified some aspects of human AIDS and is serving as a model for vaccine development.
3. Animals are an important source of antibodies, antisera, antitoxins, and other immune products that can be used in therapy or testing.
4. Animals are sometimes required to determine the pathogenicity or toxicity of certain bacteria. One such test is the mouse neutralization test for the presence of botulism toxin in food. This test can help identify even very tiny amounts of toxin and thereby can avert epidemics of this disease. Occasionally, it is necessary to inoculate an animal to distinguish between pathogenic or nonpathogenic strains of *Listeria* or *Candida* (a yeast).
5. Animal inoculation is an essential step in testing the effects of drugs and the effectiveness of vaccines before they are administered to humans. It makes progress toward prevention, treatment, and cure possible without risking the lives of humans.

microbiologists, whether the beginning laboratory student, the researcher attempting to isolate drug-producing bacteria from soil, or the clinical microbiologist working with a specimen from a patient's infection. These procedures make it possible to handle and maintain microorganisms as discrete entities whose detailed biology can be studied and recorded.

INOCULATION AND ISOLATION

To cultivate, or **culture,** microorganisms, one introduces a tiny sample (the **inoculum**) into a container of nutrient **medium*** (pl. media), which provides an environment in which they multiply. This process is called **inoculation.*** The observable growth that appears in or on the medium is known as a **culture.*** The nature of the sample being cultured depends on the objectives of the analysis. Clinical specimens for determining the cause of an infectious disease are obtained from body fluids (blood, cerebrospinal fluid), discharges (sputum, urine, feces), or diseased tissue. Other samples subject to microbiological analysis are soil, water, sewage, foods, air, and inanimate objects. Procedures for proper specimen collection are discussed on page 556.

Culture media are contained in test tubes, flasks, or Petri plates, and they are inoculated by such tools as loops, needles, pipettes, and swabs. Media are extremely varied in nutrient content and consistency and can be specially formulated for a particular purpose. Culturing obligate parasites (including viruses and certain bacteria) that cannot grow on artificial media requires live cell cultures or host animals (microfile 3.1). For an experiment to be properly controlled, sterile technique is necessary. This means that the inoculation must start with a sterile medium (figure 3.2*a*), and inoculating tools with sterile tips must be used. Measures must be taken to prevent introduction of nonsterile materials such as room air and fingers directly into the culture.

* medium (mee´-dee-um) pl. media; L., middle.
* inoculation (in-ok´´-yoo-lay´-shun) L. *in,* and *oculus,* bud.
* culture (kul´-chur) Gr. *cultus,* to tend or cultivate.

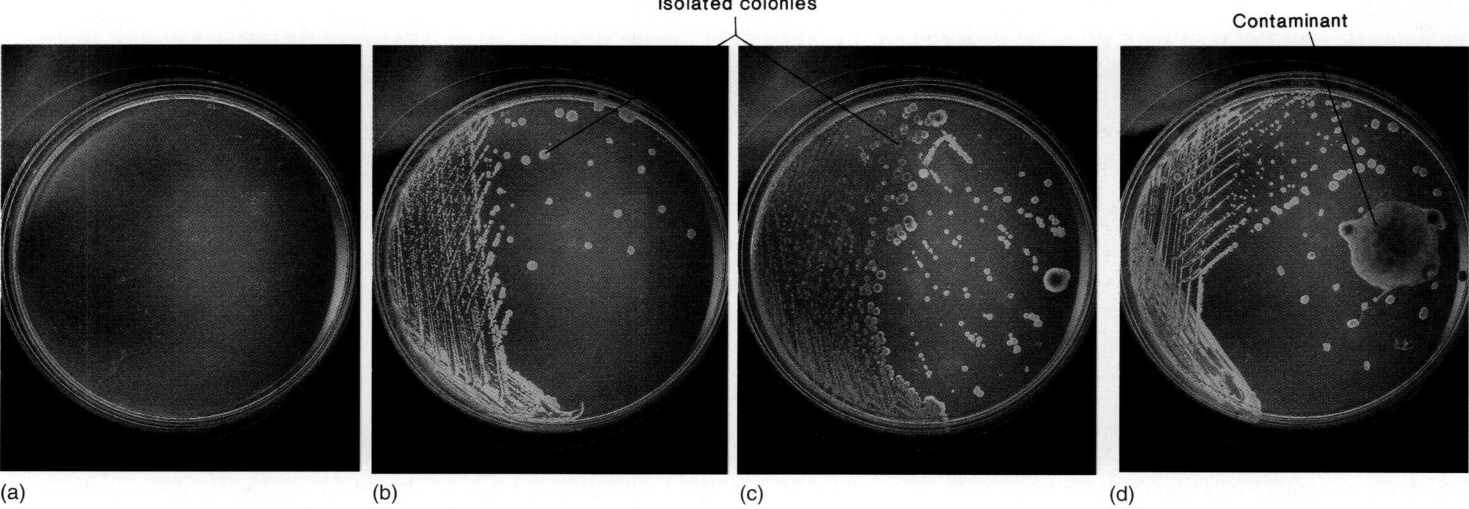

Figure 3.2
Various conditions of media and cultures. The plate in (*a*) contains sterile solid media, shows no visible growth, and is not yet a culture. All procedures in the culturing of microorganisms require sterile media. (*b*) A pure culture of *Micrococcus luteus.* It is evident that only this species is growing in the plate. (*c*) A mixed culture of *M. luteus* and *Serratia marcescens* species, which are readily differentiated by their contrasting pigments. (*d*) This plate was overexposed to room air, and it has developed large, fuzzy colonies of molds. Because these intruders are not desirable and are not identified, the culture is now contaminated.

In some ways, culturing microbes is analogous to gardening on a microscopic scale. Cultures can be compared to carefully separated plots of tiny plants in which extreme care is taken to exclude weeds. A **pure culture** is a container of medium that grows only a single known species or type of microorganism (figure 3.2*b*). This type of culture is most frequently used for laboratory study, because it allows the systematic examination and control of one microorganism by itself. Instead of the term *pure culture* some microbiologists prefer the term **axenic,*** meaning that the culture is free of other living things except for the one being studied. A **mixed culture** (figure 3.2*c*) is a container that holds two or more *identified,* easily differentiated species of microorganisms, not unlike a garden plot containing both carrots and onions. A **contaminated*** **culture** (figure 3.2*d*) was once pure or mixed (and thus a known entity) but has since had **contaminants** (unwanted microbes of uncertain identity) introduced into it, like weeds into a garden. Because contaminants have the potential for causing disruption, constant vigilance is required to exclude them from microbiology laboratories, as you will no doubt witness from your own experience.

Techniques for Isolating Cells

Certain isolation techniques are based on the concept that if an individual bacterial cell is separated from other cells and provided adequate space on a nutrient surface, it will grow into a discrete mound of cells called a **colony** (figure 3.3). Because it was formed from a single cell, a colony consists of just that single species and no other. Proper isolation requires that a small number of cells be inoculated into a relatively large volume or over an expansive area of medium. It generally requires the following materials: a medium that has a relatively firm surface (see agar in "Physical States of Media," page 65), a **Petri plate** (a clear, flat dish with a cover), and an inoculating loop. In the **streak plate method,** a small droplet of culture or sample is spread over the surface of the medium according to a pattern that gradually thins out the sample and separates the cells spatially over several sections of the plate (figure 3.4*a,b*). Because of its effectiveness and economy of materials, the streak plate is the method of choice for most applications.

In the **loop dilution,** or **pour plate, technique,** the sample is inoculated serially into a series of cooled but still liquid agar tubes so as to dilute the number of cells in each successive tube in the series (figure 3.4*c,d*). Inoculated tubes are then plated out (poured) into sterile Petri plates and are allowed to solidify (harden). The end result (usually in the second or third plate) is that the number of cells per volume is so decreased that cells have ample space to grow into separate colonies. One difference between this and the streak plate method is that in this technique some of the colonies will develop in the medium itself and not just on the surface.

MEDIA: PROVIDING NUTRIENTS IN THE LABORATORY

A major stimulus to the rise of microbiology 100 years ago was the development of techniques for growing microbes out of their natural habitats and in pure form in the laboratory. This milestone enabled the close examination of a microbe and its morphology, physiology, and genetics. It was evident from the very first that for successful cultivation, each microorganism had to be provided with all of its required nutrients in an artificial medium.

* axenic (ak-zee′-nik) Gr. *a,* no, and *xenos,* stranger.
* contaminated (kon-tam′-ih-nay-tid) Gr. *con,* together, and L. *tangere,* to touch.

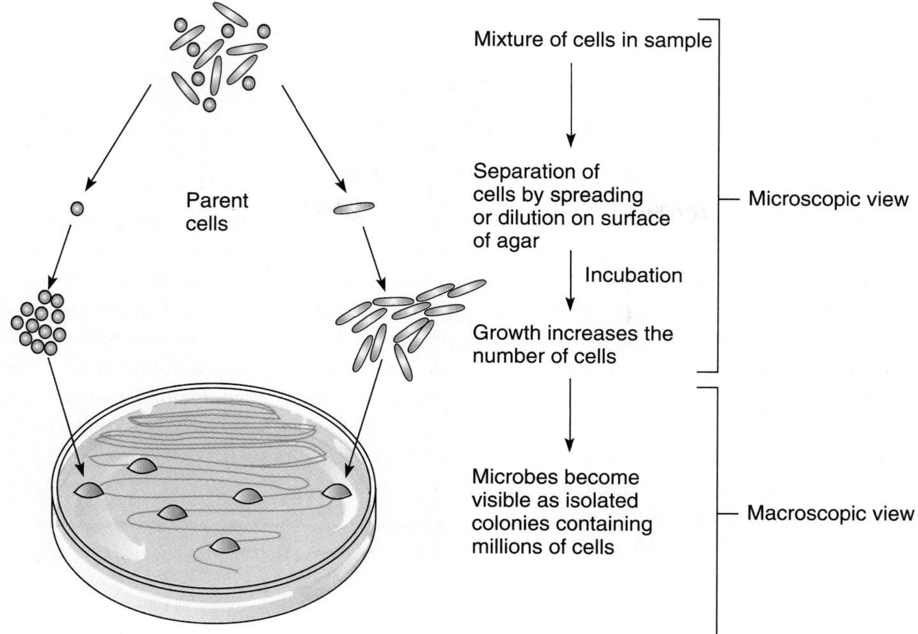

Figure 3.3
Stages in the formation of an isolated colony, showing the microscopic events and the macroscopic result. Separation techniques such as streaking can be used to isolate single cells. After numerous cell divisions, a macroscopic, clinging mound of cells, or a colony, will be formed. This is a relatively simple yet successful way to separate different types of bacteria in a mixed sample.

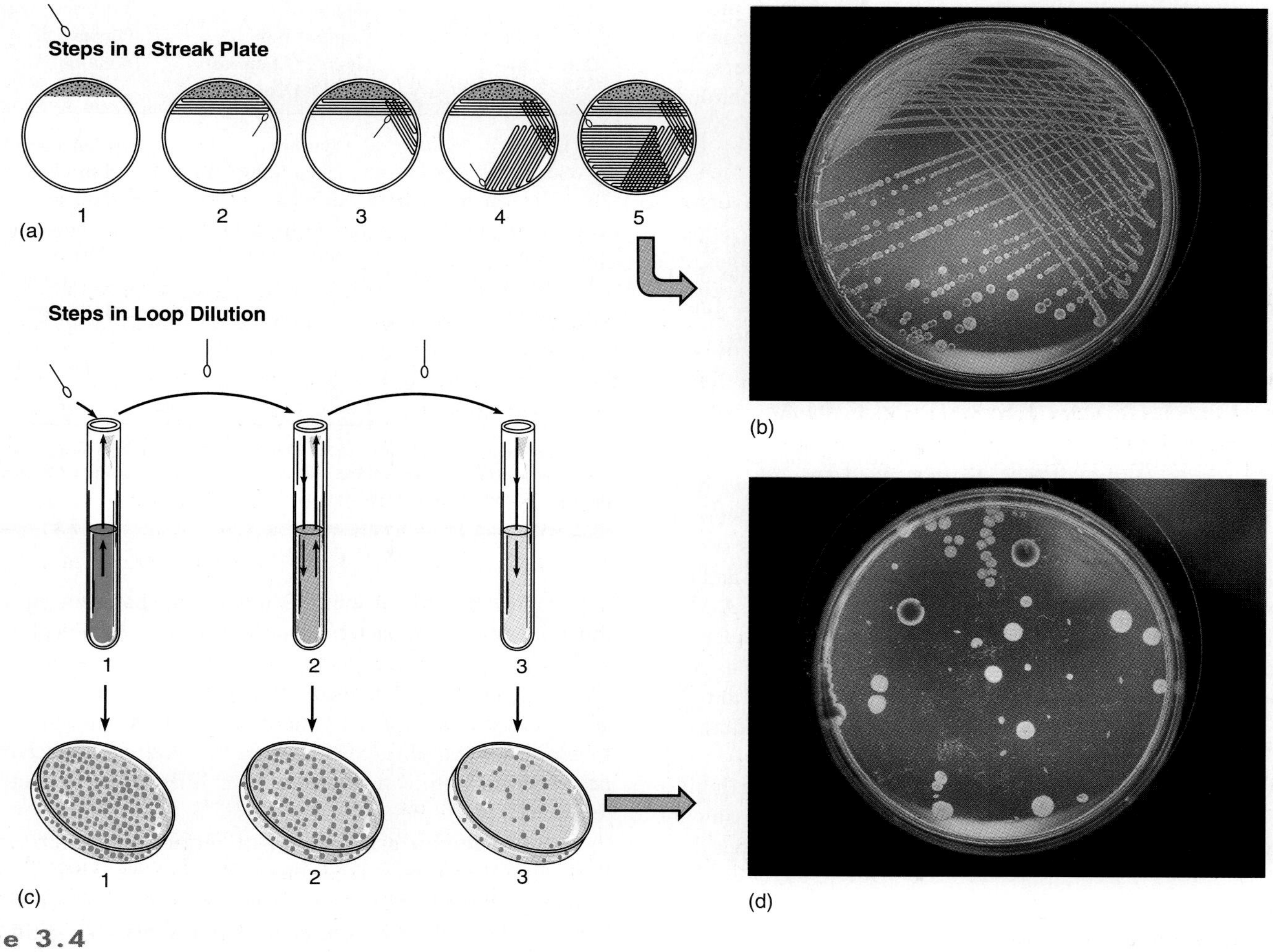

Steps in a Streak Plate

1 2 3 4 5

(a)

(b)

Steps in Loop Dilution

1 2 3

1 2 3

(c)

(d)

Figure 3.4
Methods for isolating bacteria. (*a*) Steps in a quadrant streak plate. (*b*) Resulting isolated colonies of bacteria. (*c*) Steps in the loop dilution method and (*d*) the end result.

MICROFILE 3.2 MEDIA MILESTONES

With the ready availability of commercial products, media preparation nowadays is quite straightforward, but it has not always been so. Many years ago, a media preparation room might have looked like a cross between a kitchen and an alchemist's lab. A standard book reviewing culture media up to 1930* compiled 7,000 different "recipes." Thumbing through this book, one gets a clear idea of the ingenuity and care that media formulators and preparers brought to their task. A fantastic assortment of raw materials were (and occasionally still are) added to preparations, and some formulas were extraordinarily complex. It was not unusual for a single medium to require 25 steps in preparation. The media compositions seem to have been limited only by the imagination of the "cook." Numerous unconventional animal products appear among the recipes: bone, cartilage, placenta, lung, lard, chopped beef, testicles, sperm, pus, feces, and urine. Plant materials, also a good source of nutrients, were ground, mashed, chopped, pureed, filtered, and cooked. A partial list reads like mulligan stew: potatoes, carrots, spinach, onions, peas, fruits (bananas, prunes), grasses (straw, hay), and dozens of different seeds. Foods such as tomato juice, macaroni, coffee, spices, butter, and beer had their functions in certain concoctions. No conceivable source was overlooked, including such curious ingredients as soil, manure, brick, coal, sawdust, wood ashes, lint, rusty nails, asbestos, arsenic, and cyanide!

*A Compilation of Culture Media for the Cultivation of Microorganisms *(1930), Levine and Shoenlein.*

Nutritional requirements of microbes vary from a few very simple inorganic compounds to a complex list of specific inorganic and organic compounds. This tremendous diversity is evident in the types of media that can be prepared. The *Difco Manual*[1] lists approximately 500 entries for media used in modern microbiology laboratories. Most modern media are of the instant variety, prepared from dehydrated commercial powders, but media preparation during the pioneering days of microbiology was very different (microfile 3.2).

TYPES OF MEDIA

Media can be classified on three primary levels: (1) physical form, (2) chemical composition, and (3) functional type (table 3.1). Most media discussed here are designed for bacteria and fungi, though algae and some protozoa can be propagated in media.

TABLE 3.1

THREE CATEGORIES OF MEDIA CLASSIFICATION

Physical State (Medium's Normal Consistency)	Chemical Composition (Type of Chemicals Medium Contains)	Functional Type (Purpose of Medium)*
1. Liquid 2. Semisolid 3. Solid (can be converted to liquid) 4. Solid (cannot be liquefied)	1. Synthetic (chemically defined) 2. Nonsynthetic (not chemically defined)	1. General purpose 2. Enriched 3. Selective 4. Differential 5. Anaerobic growth 6. Specimen transport 7. Assay 8. Enumeration

*Some media can serve more than one function. For example, a medium such as brain-heart infusion is general purpose and enriched; mannitol salt agar is both selective and differential; and blood agar is both enriched and differential.

PHYSICAL STATES OF MEDIA

Liquid media are defined as water-based solutions that do not solidify at temperatures above freezing and that tend to flow freely when the container is tilted (figure 3.5). These media, termed broths, milks, or infusions, are made by dissolving various solutes in distilled water. Growth occurs throughout the container and can then present a dispersed cloudy or particulate appearance. A common laboratory medium, nutrient broth, contains beef extract and peptone dissolved in water. Methylene blue milk and litmus milk are opaque liquids containing whole milk and dyes. Fluid thioglycollate is a slightly viscous broth used for determining patterns of growth in oxygen.

At ordinary room temperature, **semisolid media** exhibit a clotlike consistency (figure 3.6) because they contain an amount of solidifying agent (agar or gelatin) that thickens them but does not produce a firm substrate. Semisolid media are used to determine the motility of bacteria and to localize a reaction at a specific site. Both motility test medium and SIM contain a small amount (0.3–0.5%) of agar. The medium is stabbed carefully in the center and later observed for the pattern of growth around the stab line. In addition to motility, SIM can test for physiological characteristics used in identification (hydrogen sulfide production and indole reaction).

Solid media provide a firm surface on which cells can form discrete colonies (see figure 3.2) and are advantageous for isolating and subculturing bacteria and fungi. They come in two forms: liquefiable and nonliquefiable. **Liquefiable solid media,** sometimes called reversible solid media, contain a solidifying

1. Difco, the Digestive Ferments Company, started in 1895 as a supplier of prepared artificial media.

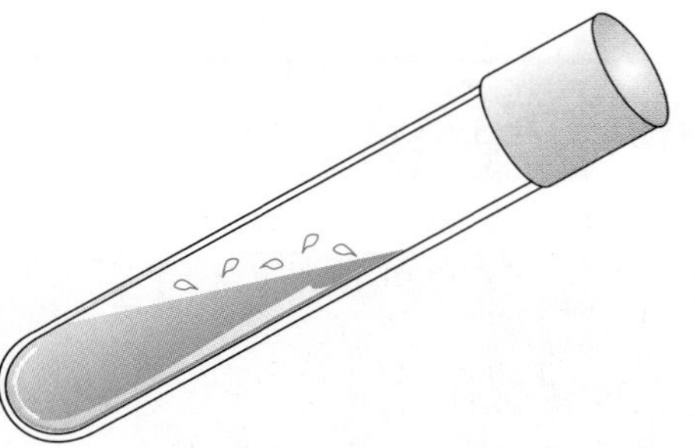

(a)

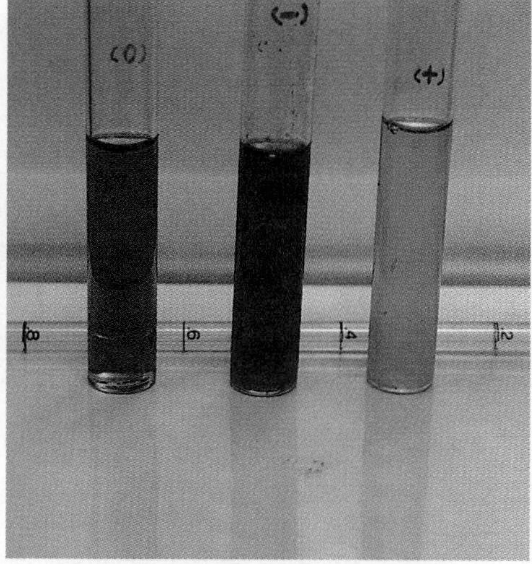

Uninoculated Negative Postive

(b)

Figure 3.5

(*a*) Liquid media tend to flow freely when the container is tilted. (*b*) *Enterococcus faecalis* broth, a selective medium for identifying this species. On the left (O) is a clear, uninoculated broth. The one in the center is growing without a color change (–), and on the right is a broth with growth and color change (+).

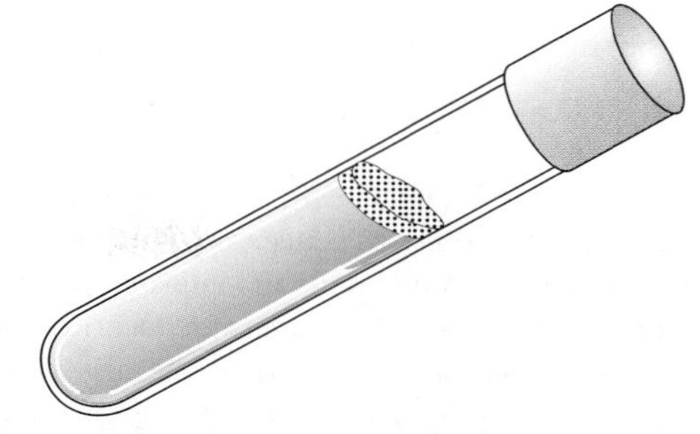

(a)

(b)

Figure 3.6

(*a*) Semisolid media have more body than liquid media but less body than solid media. They do not flow freely and have a soft, clotlike consistency. (*b*) Sulfur indole motility (SIM) medium. The growth patterns that develop in this medium can be used to determine various characteristics. The tube on the left shows no motility; the tube in the center shows motility; the tube on the right shows both motility and production of hydrogen sulfide gas (black precipitate).

agent that is thermoplastic: Its physical properties change in response to temperature. By far the most widely used and effective of these agents is **agar,*** a complex polysaccharide isolated from the red alga *Gelidium.* The benefits of agar are numerous. It is solid at room temperature and most incubation temperatures, and it melts (liquefies) at the boiling temperature of water (100°C). Once liquefied, agar does not resolidify until it cools to 42°C, so it can be inoculated and poured in liquid form at temperatures (45°–50°C) that will not harm the microbes or the handler. Agar is flexible and moldable, and it provides a basic framework to hold moisture and nutrients, though it is not itself a digestible nutrient for the vast majority of microorganisms.

Any medium containing 1% to 5% agar usually has the word *agar* in its name. Nutrient agar is a common one. Like nutrient broth, it contains beef extract and peptone, as well as 1.5% agar by weight. Many of the examples covered in the section on functional categories of media contain agar. Although **gelatin** is not nearly as satisfactory as agar, it will create a reasonably solid surface in concentrations of 10% to 15% (but it probably will not remain solid; microfile 3.3). Agar and gelatin media are illustrated in figure 3.7.

Nonliquefiable solid media have less versatile applications than agar media because they are not thermoplastic. They include materials such as rice grains (used to grow fungi), cooked meat media (good for anaerobes), and potato slices; all of these media start out solid and remain solid after heat sterilization. Other solid media containing egg and serum start out liquid and are permanently coagulated or hardened by moist heat.

CHEMICAL CONTENT OF MEDIA

Media whose compositions are chemically defined are termed **synthetic.** Such media contain pure organic and inorganic compounds that vary little from one source to another and have a molecular content specified by means of an exact formula. Synthetic media come in many forms. Some media, such as minimal media

* agar (ah´-gur) The Malay word for this seaweed product is *agar-agar.*

MICROFILE 3.3 MICROBIOLOGY DISCOVERS AGAR

Early bacteriologists recognized the advantages of growing bacteria in discrete colonies rather than solely in broths and infusions. For this, they needed a surface that was relatively solid and provided nutrients. Two old standbys were sterilized potato slices and gelatin, but they left much to be desired. Potatoes are limited in nutrient content, and gelatin loses its solidity in warm temperatures and can be digested by many microorganisms. About 120 years ago, this problem was solved by a fortunate cultural exchange (of the human kind). The German physician and avid bacteriologist Walther Hesse had been studying airborne microbes and was frustrated by the messy outcome of gelatin cultures. Understanding his need for a gelatin substitute, his wife Fanny suggested that he try a substance her mother had discovered through Dutch friends from Java. This material, called agar-agar, had traditionally been used by Malaysians to stabilize jellies and thicken soup, and it was a nearly perfect solidifying agent. Agar-agar could be liquefied and sterilized without adverse effect. Also, it was solid at room and incubation temperatures, very resistant to digestion, and relatively translucent. Because of Hesse's success with agar-agar, it quickly caught on among bacteriologists, but the credit for its discovery belongs to Fanny Hesse.

Frau Fanny Hesse and her husband Dr. Walther Hesse.

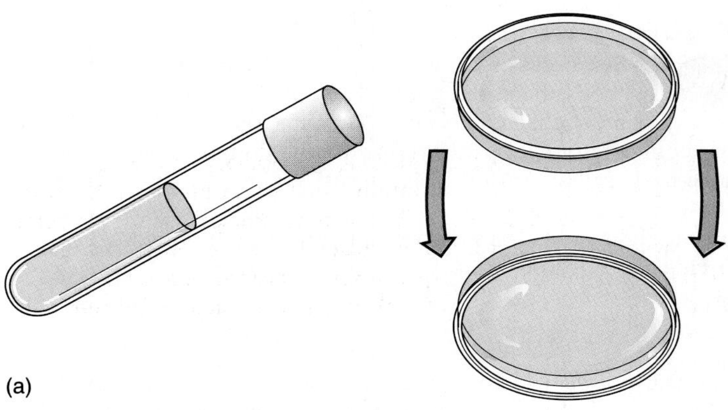

(a)

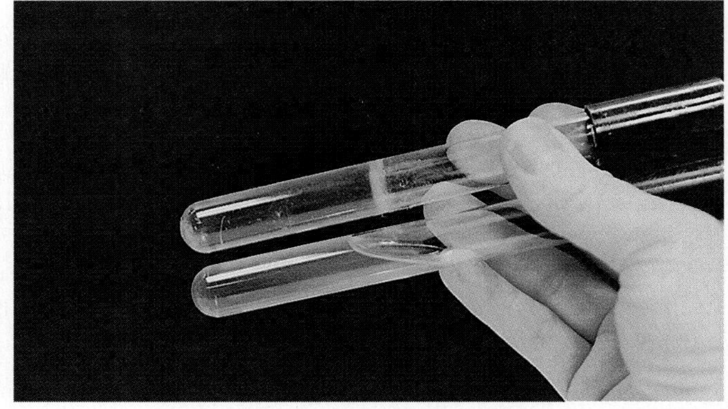

(b)

Figure 3.7

Solid media that are reversible to liquids. (*a*) Media containing 1–5% agar-agar are solid enough to remain in place when containers are tilted or inverted. They are liquefiable by heat but generally not by bacterial enzymes. (*b*) Nutrient gelatin contains enough gelatin (12%) to take on a solid consistency. The top tube shows it as a solid. The bottom tube indicates a result when microbial enzymes digest the gelatin and liquefy it.

for fungi, contain nothing more than a few essential compounds such as salts and amino acids dissolved in water. Others contain a variety of defined organic and inorganic chemicals (table 3.2). Such standardized and reproducible media are most useful in research and cell culture when the exact nutritional needs of the test organisms are known. If even one component of a given medium is not chemically definable, the medium belongs in the next category.

Complex, or **nonsynthetic, media** contain at least one ingredient that is *not* chemically definable—not a simple, pure compound and not representable by an exact chemical formula. Most of these substances are extracts of animals, plants, or yeasts, including such materials as ground-up cells, tissues, and secretions. Examples are blood, serum, and meat extracts or infusions. Infusions are high in vitamins, minerals, proteins, and other organic nutrients. Other nonsynthetic ingredients are milk, yeast extract, soybean digests, and peptone. Peptone is a partially digested protein, rich in amino acids, that is often used as a carbon and nitrogen source. Nutrient broth, blood agar, and MacConkey agar, though different in function and appearance, are all nonsynthetic media. They present a rich mixture of nutrients for microbes that have complex nutritional needs.

TABLE 3.2

MEDIUM FOR THE GROWTH AND MAINTENANCE OF THE GREEN ALGA EUGLENA

Glutamic acid (aa)	6 g
Aspartic acid (aa)	4 g
Glycine (aa)	5 g
Sucrose (c)	30 g
Malic acid (oa)	2 g
Succinic acid (oa)	1.04 g
Boric acid	1.14 mg
Thiamine hydrochloride (v)	12 mg
Monopotassium phosphate	0.6 g
Magnesium sulfate	0.8 g
Calcium carbonate	0.16 g
Ammonium carbonate	0.72 g
Ferric chloride	60 mg
Zinc sulfate	40 mg
Manganese sulfate	6 mg
Copper sulfate	0.62 mg
Cobalt sulfate	5 mg
Ammonium molybdate	1.34 mg

Note: These ingredients are dissolved in 1,000 ml of water. aa, amino acid; c, carbohydrate; oa, organic acid; v, vitamin; g, gram; mg, milligram.

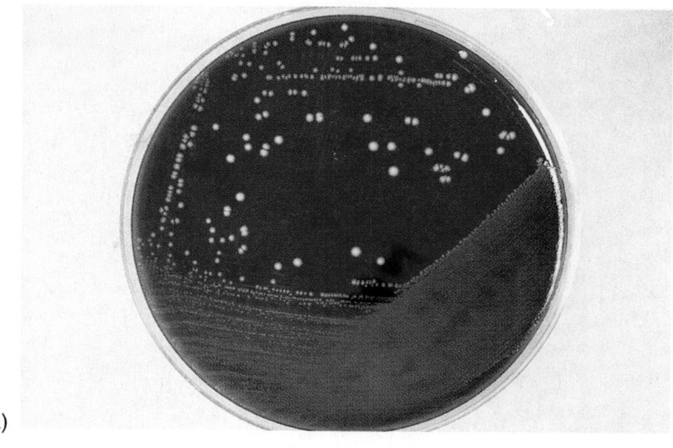

(a)

(b)

Figure 3.8

Examples of enriched media. (*a*) Blood agar growing *Enterococcus faecalis,* a common fecal bacterium. (*b*) Chocolate agar (lower half of plate), a medium that gets its brown color from heated blood, not from chocolate. It is commonly used to culture the fastidious gonococcus *Neisseria gonorrhoeae.* (The upper plate chamber contains MacConkey agar with lactose-fermenting bacteria.)

A specific example can be used to compare what differentiates a synthetic medium from a nonsynthetic one. Both synthetic *Euglena* medium (table 3.2) and nonsynthetic nutrient broth contain amino acids. But *Euglena* medium has three known amino acids in known amounts, whereas nutrient broth contains amino acids (in peptone) in variable types and amounts. Pure inorganic salts and organic acids are added in precise quantities for *Euglena,* whereas those components are provided by undefined beef extract in nutrient broth.

MEDIA TO SUIT EVERY FUNCTION

Microbiologists have many types of media at their disposal, with new ones being devised all the time. Depending upon what is added, a microbiologist can fine-tune a medium for nearly any purpose. As a result, only a few species of bacteria or fungi cannot yet be cultivated artificially. Media are used for primary isolation, to maintain cultures in the lab, to determine biochemical and growth characteristics, and for numerous other functions.

General purpose media are designed to grow as broad a spectrum of microbes as possible. As a rule, they are nonsynthetic and contain a mixture of nutrients that could support the growth of pathogens and nonpathogens alike. Examples include nutrient agar and broth, brain-heart infusion, and trypticase soy agar (TSA). TSA contains partially digested milk protein (casein), soybean digest, NaCl, and agar.

An **enriched medium** contains complex organic substances such as blood, serum, hemoglobin, or special **growth factors** (vitamins, amino acids) that certain species must have in order to grow. Bacteria that require growth factors and complex nutrients are termed **fastidious.*** Blood agar, which is made by adding sterile sheep, horse, or rabbit blood to a sterile agar base (figure 3.8*a*) is widely employed to grow Fastidious streptococci and other pathogens. Pathogenic *Neisseria* (one species causes gonorrhea) are grown on Thayer-Martin medium or chocolate agar, which is made by heating blood agar (figure 3.8*b*).

Selective and Differential Media

Some of the cleverest and most inventive media belong to the categories of selective and differential media (figure 3.9). These media are designed for special microbial groups, and they have extensive applications in isolation and identification. They can permit, in a single step, the preliminary identification of a genus or even a species.

* fastidious (fass-tid´-ee-us) L. *fastidium,* loathing or disgust.

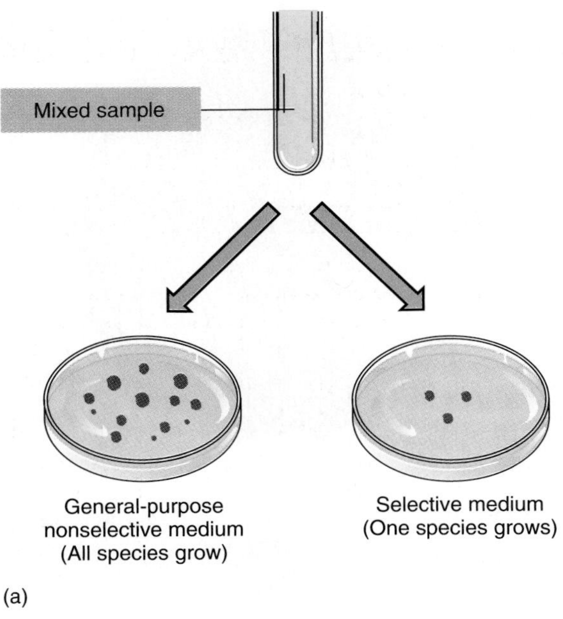

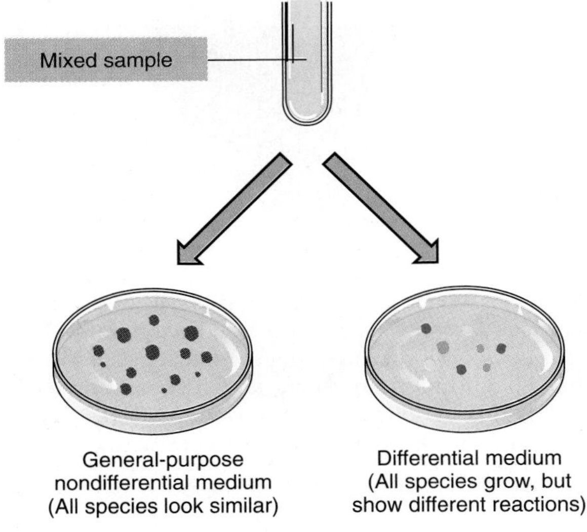

Figure 3.9

Comparison of selective and differential media with general-purpose media. (*a*) The same mixed sample containing five different species is streaked onto plates of general-purpose nonselective medium and selective medium. Note the results. (*b*) Another mixed sample containing three different species is streaked onto plates of general-purpose nondifferential medium and differential medium. Note the results.

A **selective medium** (table 3.3) contains one or more agents that inhibit the growth of a certain microbe or microbes (A, B, C) but not others (D) and thereby encourages, or *selects,* microbe D and allows it to grow. Selective media are very important in primary isolation of a specific type of microorganism from samples containing dozens of different species—for example, feces, saliva, skin, water, and soil. They hasten isolation by suppressing the unwanted background organisms and favoring growth of the desired ones.

TABLE 3.3

SELECTIVE MEDIA, AGENTS, AND FUNCTIONS

Medium	Selective Agent	Used For
Mueller tellurite	Potassium tellurite	Isolation of *Corynebacterium diphtheriae*
Enterococcus faecalis broth	Sodium azide Tetrazolium	Isolation of fecal enterococci
Phenylethanol agar	Phenylethanol chloride	Isolation of staphylococci and streptococci
Tomato juice agar	Tomato juice, acid	Isolation of lactobacilli from saliva
MacConkey agar	Bile, crystal violet	Isolation of gram-negative enterics
Salmonella/Shigella (SS) agar	Bile, citrate, brilliant green	Isolation of *Salmonella* and *Shigella*
Lowenstein-Jensen	Malachite green dye	Isolation and maintenance of *Mycobacteria*
Sabouraud's agar	pH of 5.6 (acid)	Isolation of fungi—inhibits bacteria

Mannitol salt agar (MSA) (figure 3.10*a*) contains a concentration of NaCl (7.5%) that is quite inhibitory to most human pathogens. One exception is the genus *Staphylococcus,* which grows well in this medium and consequently can be amplified in very mixed samples. Bile salts, a component of feces, inhibit most gram-positive bacteria while permitting many gram-negative rods to grow. Media for isolating intestinal pathogens (MacConkey agar, eosin-methylene blue [EMB] agar) contain bile salts as a selective agent (figure 3.10*b*). Dyes such as methylene blue and crystal violet also inhibit certain gram-positive bacteria. Other agents that have selective properties are antimicrobic drugs and acid. Some selective media contain strongly inhibitory agents to favor the growth of a pathogen that would otherwise be overlooked because of its low numbers in a specimen. Selenite and brilliant green dye are used in media to isolate *Salmonella* from feces, and sodium azide is used to isolate enterococci from water and food (see EF broth, figure 10.5*b*).

Differential media grow several types of microorganisms and are designed to display visible differences among those microorganisms. Differentiation shows up as variations in colony size and color, in media color changes, and in the formation of gas bubbles and precipitates (table 3.4). These variations come from the type of chemicals these media contain and the ways that microbes react to them. For example, when microbe X metabolizes a certain substance not used by organism Y, then X will cause a visible change in the medium and Y will not. The

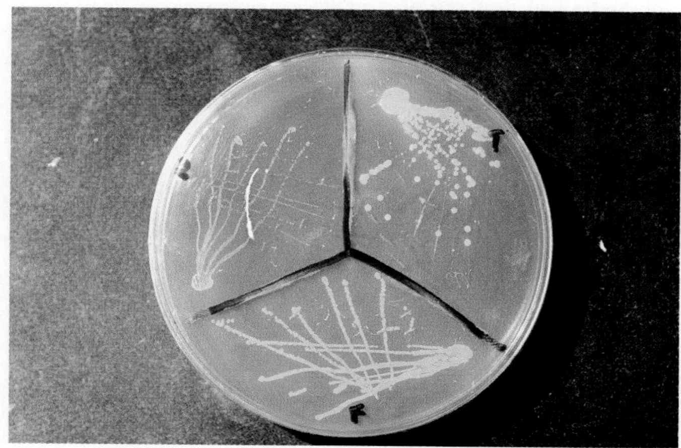

(a)

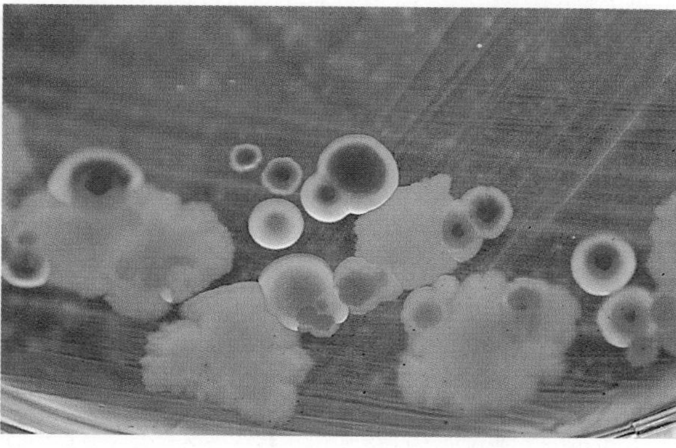

(b)

Figure 3.10

Example of selective media. (*a*) Mannitol salt agar is used to isolate members of the genus *Staphylococcus.* It is selective for this genus because its members can grow in the presence of 7.5% sodium chloride, whereas many other species are inhibited by this high concentration. It contains a dye that also pinpoints those species of *Staphylococcus* that produce acid from mannitol and turn the phenol red dye to a bright yellow.
(*b*) MacConkey agar differentiates between lactose-fermenting bacteria (indicated by a pink-red reaction in the center of the colony) and lactose-negative bacteria (indicated by an off-white colony with no dye reaction).

TABLE 3.4

DIFFERENTIAL MEDIA

Medium	Substances That Facilitate Differentiation	Differentiates Between
Blood agar	Intact red blood cells	Types of hemolysis
Mannitol salt agar	Mannitol, phenol red, and 7.5% NaCl	Species of *Staphylococcus* NaCl also inhibits the salt-sensitive species
Eosin-methylene blue (EMB) agar	Eosin, methylene blue, lactose, and bile	*Escherichia coli,* other lactose fermenters, and lactose nonfermenters Dyes and bile also inhibit gram-positive bacteria
Spirit blue agar	Spirit blue dye and oil	Bacteria that use fats from those that do not
Urea broth	Urea, phenol red	Bacteria that hydrolyze urea to ammonia
Sulfur indole motility (SIM)	Thiosulfate, iron	H_2S gas producers from nonproducers
Triple-sugar iron agar (TSIA)	Triple sugars, iron, and phenol red dye	Fermentation of sugars, H_2S production
XLD agar	Lysine, xylose, iron, thiosulfate, phenol red	*Enterobacter, Escherichia, Proteus, Providencia, Salmonella,* and *Shigella*
Birdseed agar	Seeds from thistle plant	*Cryptococcus neoformans* and other fungi

simplest differential media show two reaction types such as the use or nonuse of a particular nutrient or a color change in some colonies but not in others. Some media are sufficiently complex to show three or four different reactions (TSIA, figure 3.11*a*).

Dyes can be used as differential agents because many of them are pH indicators that change color in response to the production of an acid or a base. For example, MacConkey agar contains neutral red, a dye that is yellow when neutral and pink or red when acidic. A common intestinal bacterium such as *Escherichia coli* that gives off acid when it metabolizes the lactose in the medium develops red to pink colonies, and one like *Salmonella* that does not give off acid remains its natural color (off-white). Spirit blue agar is used to detect the hydrolysis (digestion) of fats by lipase enzyme. Positive hydrolysis is indicated by the dark blue color that develops in colonies (figure 3.11*b*).

Miscellaneous Media

A **reducing medium** contains a substance (thioglycollic acid or cystine) that absorbs oxygen or slows the penetration of oxygen in a medium, thus reducing its availability. Reducing media are important for growing anaerobic bacteria or determining oxygen requirements (see figure 7.13). **Carbohydrate fermentation media** contain sugars that can be fermented (converted to acids) and a pH indicator to show this reaction (figures 3.10*a*, 3.12). Media for other biochemical reactions that provide the basis for identifying bacteria and fungi are presented in the second half of this book.

Transport media are used to maintain and preserve specimens that have to be held for a period of time before clinical analysis or to sustain delicate species that die rapidly if not held

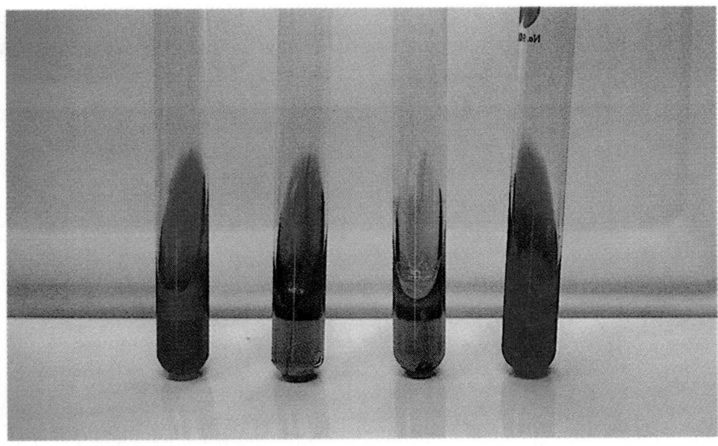

(a)

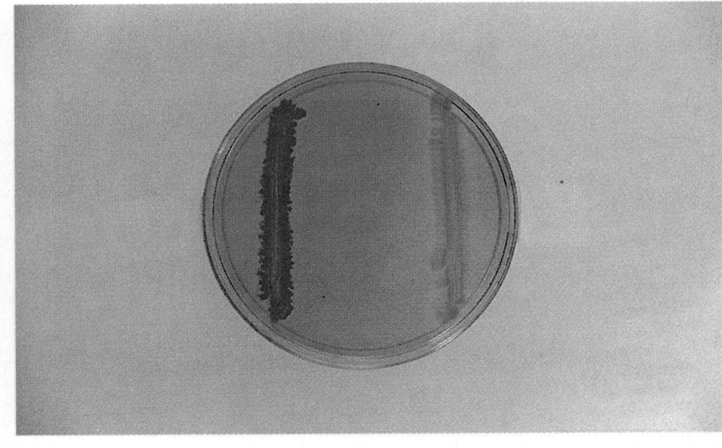

(b)

Figure 3.11

Media that differentiate characteristics. (*a*) Triple-sugar iron agar (TSIA) in a slant. This medium contains three fermentable carbohydrates, phenol red to indicate pH changes, and a chemical (iron) that indicates H_2S gas production. Reactions (from left to right) are: a tube in which no acid was produced (red); a tube in which acid production (yellow) has occurred in the bottom (butt) of the slant; a tube in which acid has been produced throughout the slant; and a tube in which H_2S has been produced, forming a black precipitate. (*b*) A plate of spirit blue agar helps differentiate bacteria that can digest fat (lipase producers) from those that cannot. A positive reaction (left) causes the blue indicator dye to develop a dark blue color on the colony; a negative result (right) is pale yellow or white.

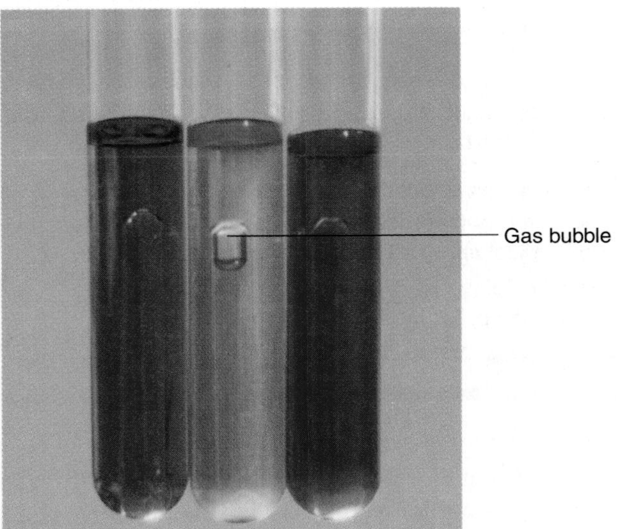

Figure 3.12

Carbohydrate fermentation in broths. This medium is designed to show fermentation (acid production) and gas formation by means of a small, inverted Durham tube for collecting gas bubbles. The tube on the left is an uninoculated negative control; the center tube is positive for acid (yellow) and gas (open space); the tube on the right shows growth but neither acid nor gas.

under stable conditions. Stuart's and Amies transport media contain salts, buffers, and absorbants to prevent cell destruction by enzymes, pH changes, and toxic substances, but will not support growth. **Assay media** are used by technologists to test the effectiveness of antimicrobial drugs (see chapter 2) and by drug manufacturers to assess the effect of disinfectants, antiseptics, cosmetics, and preservatives on the growth of microorganisms. **Enumeration media** are used by industrial and environmental microbiologists to count the numbers of organisms in milk, water, food, soil, and other samples.

Chapter Checkpoints

The Five I's—inoculation, incubation, isolation, inspection, and identification—summarize the kinds of laboratory procedures used in microbiology.

A **pure culture** contains only one species or type of microorganism. A **mixed culture** contains two or more known species. A **contaminated culture** contains both known and unknown (unwanted) microorganisms.

Individual microbial colonies originate from single microbial cells.

Most microorganisms can be cultured on artificial media, but some can be cultured only in living tissue.

Artificial media are classified by their **physical state** as either liquid, semisolid, liquefiable solid, or nonliquefiable solid.

Artificial media are classified by their **chemical content** as either *synthetic* or *nonsynthetic,* depending on whether the exact chemical composition is known.

Artificial media are classified by their **function** as either general purpose media or media with one or more specific purposes. Enriched, selective, differential, transport, assay, and enumerating media are all examples of media designed for specific purposes.

INCUBATION, INSPECTION, AND IDENTIFICATION

Once a container of medium has been inoculated, it is **incubated.*** This means it is placed in a temperature-controlled chamber or incubator to encourage multiplication and produce a culture. Although microbes have adapted to growth at temperatures ranging from freezing to boiling, the usual temperatures used in laboratory propagation fall between 20° and 40°C. Incubators can also control the content of atmospheric gases such as oxygen and carbon dioxide that may be required for the growth of certain microbes. During the incubation period (ranging from a day to several weeks), the microbe multiplies and produces some visible manifestation of growth.

It is common during various stages of incubation to evaluate a culture macroscopically.[2] This is possible because large masses of microorganisms are readily visible even though the individual microbes are too small to be seen individually. Microbial growth in a liquid medium materializes as cloudiness, sediment, scum, or color. A common manifestation of growth on solid media is the appearance of colonies, especially in bacteria and fungi. Colonies are actually large masses of clinging cells with distinctions in size, shape, color, and texture (see figure 4.12*a,b*). Once isolation has occurred, it is standard practice to make a second-level culture, called a **subculture,** by removing a tiny sample from one well-isolated colony and transferring it into a separate container of media. Because a colony consists of only one species of bacterium, this technique can yield a pure culture for further testing and identification.

Cultural inspection is inevitably tied to an even more direct visual method—microscopic inspection. By using a microscope, we are able to see firsthand the tiny cells that appear in cultures. It gives tangible evidence of many characteristics of cell morphology, including size, shape, and details of internal and external structure. So essential is the microscope to accurate cell study that we include the general principles of microscopy in a subsequent section of this chapter.

How does one determine what sorts of microorganisms have been isolated in cultures? Certainly the combination of microscopic and macroscopic appearance can be valuable in differentiating the smaller, simpler procaryotic cells from the larger, more complex eucaryotic cells. Appearance can be especially useful in identifying eucaryotic microorganisms to the level of genus or species because of their distinctive morphological features; however, bacteria are generally not identifiable by these methods because very different species may appear quite similar. For them, we must include techniques that characterize their cellular metabolism. These methods, called biochemical tests, can determine fundamental chemical characteristics such as nutrient requirements, products given off during growth, temperature and gas requirements, and mechanisms for deriving energy.

2. For all intents and purposes, the macroscopic level is synonymous with the cultural level.

* incubate (in´-kyoo-bayt) Gr. *incubatus,* to lie in or upon.

Several modern analytical and diagnostic tools that focus on genetic and molecular characteristics can detect the exact nature of microbial DNA. In the case of certain pathogens, further information on a microbe is obtained by inoculating a suitable laboratory animal. A profile is prepared by compiling physiological testing results with both macroscopic and microscopic traits. The profile then becomes the raw material used in final identification. In several subsequent chapters, we present more detailed examples of identification methods.

MAINTENANCE AND DISPOSAL OF CULTURES

The fate of a laboratory culture depends upon the original purpose of isolation and identification. In some cases, the cultures and specimens constitute a potential hazard and will require immediate and proper disposal. Both steam sterilizing (see autoclave, chapter 11) and incineration (burning) are effective methods of destroying microorganisms. On the other hand, many teaching and research laboratories require a line of stock cultures, continuously maintained species that represent "living catalogues." The largest culture collection can be found at the American Type Culture Collection in Rockville, Maryland, which maintains a voluminous array of frozen and freeze-dried fungal, bacterial, viral, and algal cultures.

Chapter Checkpoints

Microorganisms are identified in terms of their macroscopic, or colony, morphology; their microscopic morphology; and their biochemical reactions and genetic characteristics.

Microbial cultures are disposed of in two ways: steam sterilization or incineration.

THE MICROSCOPE: WINDOW ON AN INVISIBLE REALM

Imagine Leeuwenhoek's excitement and wonder when he first viewed a drop of rainwater and glimpsed an amazing microscopic world teeming with unearthly creatures. Beginning microbiology students still experience this sensation, and even experienced microbiologists never forget their first view. The microbial existence is indeed another world, but it would remain largely uncharted without an essential tool: the microscope. Your efforts in exploring microbes will be more meaningful if you understand some essentials of **microscopy*** and specimen preparation.

* **microscopy** (mye-kraw´-skuh-pee) The science that studies microscope techniques.

Magnification and Microscope Design

The two key characteristics of a reliable microscope are **magnification,** or the ability to enlarge objects, and **resolving power,** or the ability to show detail.

A discovery by early microscopists that spurred the advancement of microbiology was that clear, glass spheres could act as a lens to **magnify*** small objects. Magnification in most microscopes results from a complex interaction between visible light waves and the curvature of the lens. When a beam or ray of light transmitted through air strikes and passes through the convex surface of glass, it experiences some degree of **refraction,*** defined as the bending or change in the angle of the light ray as it passes into the lens (see figure 3.15). The greater the difference in the composition of the two substances the light passes between, the more pronounced is the refraction. When an object is placed a certain distance from the spherical lens and **illuminated*** with light, an optical replica, or **image,** of it is formed by the refracted light. Depending upon the size and curvature of the lens, the image appears enlarged to a particular degree, which is called its power of magnification and is usually identified with a number combined with × (read "times"). This behavior of light is evident if one looks through an everyday object such as a glass ball or a magnifying glass (figure 3.13). It is basic to the function of all **optical,** or **light, microscopes,** though many of them have additional features that define, refine, and increase the size of the image.

The first microscopes were **simple,** meaning they contained just a single magnifying lens and a few working parts. Examples of this type of microscope are a magnifying glass, a hand lens, and Leeuwenhoek's basic little tool shown in figure 1.8*a*. Among the refinements that led to the development of today's **compound microscope** were the addition of a second magnifying lens system, a lamp in the base to give off visible light and **illuminate** the specimen, and a special lens called the **condenser** that converges or focuses the rays of light to a single point on the object. The fundamental parts of a modern compound light microscope are illustrated in figure 3.14.

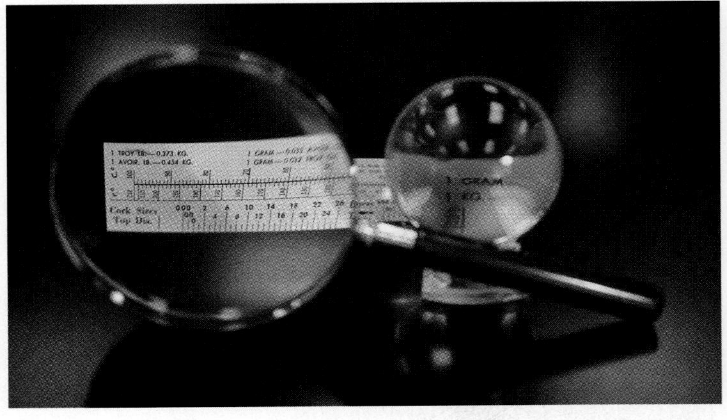

(a)

(b)

Figure 3.13
Demonstration of the magnification and image-forming capacity of clear glass "lenses," given a proper source of illumination. This magnifying glass and crystal ball magnify a ruler two to three times.

Principles of Light Microscopy

To be most effective, a microscope should provide adequate magnification, resolution, and clarity of image. Magnification of the object or specimen by a compound microscope occurs in two phases. The first lens in this system (the one closest to the specimen) is the **objective lens,** and the second (the one closest to the eye) is the **ocular lens,** or eyepiece (figure 3.15). The objective forms the initial image of the specimen, called the **real image.** When this image is projected up through the microscope body to the plane of the eyepiece, the ocular lens forms a second image, the **virtual image.** The virtual image is the one that will be received by the eye and converted to a retinal and visual image. The magnifying power of the objectives alone usually ranges from 4× to 100×, and the power of the ocular alone ranges from 10× to 20×. The **total power of magnification** of the final image formed by the combined lenses is a product of the separate powers of the two lenses:

Power of Objective	X	*Power of Ocular*	=	*Total Magnification*
10× low power objective		10×	=	100×
40× high dry objective		10×	=	400×
100× oil immersion objective		10×	=	1,000×

Microscopes are equipped with a nosepiece holding three or more objectives that can be rotated into position as needed. The power of the ocular usually remains constant for a given microscope. Depending on the power of the ocular, the total magnification of

* **magnify** (mag´-nih-fye) L. *magnus,* great, and *ficere,* to make.
* **refract, refraction** (ree-frakt´, ree-frak´-shun) L. *refringere,* to break apart.
* **illuminate** (ill-oo´-mih-nayt) L. *illuminatus,* to light up.

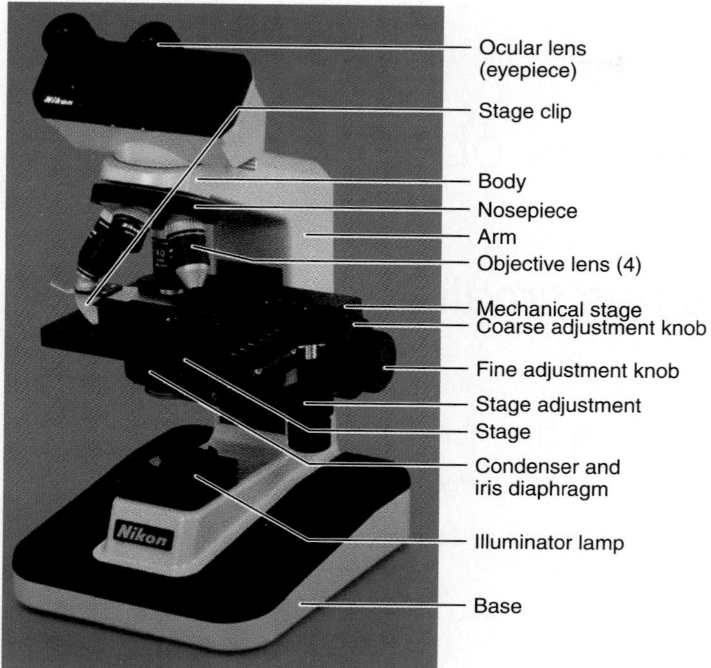

Figure 3.14
The parts of a student laboratory microscope. This microscope is a compound light microscope with two oculars (called binocular). It has four objective lenses, a mechanical stage to move the specimen, a condenser, an iris diaphragm, and a built-in lamp.

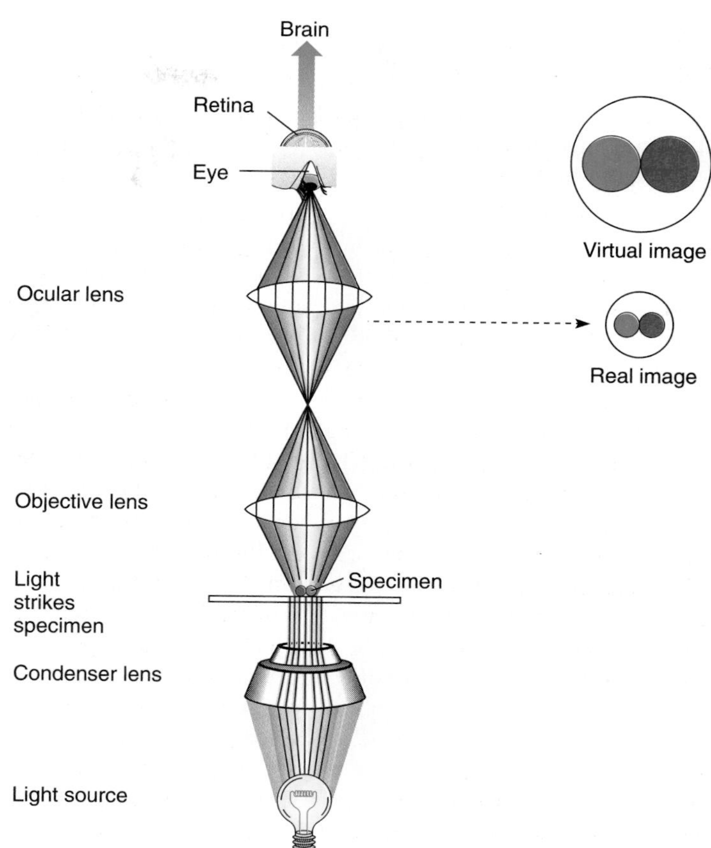

Figure 3.15
The pathway of light and the two stages in magnification of a compound microscope. As light passes through the condenser, it forms a solid beam that is focused on the specimen. Light leaving the specimen that enters the objective lens is refracted so that an enlarged primary image, the real image, is formed. One does not see this image, but its degree of magnification is represented by the lower circle. The real image is projected through the ocular, and a second image, the virtual image, is formed by a similar process. The virtual image is the final magnified image that is received by the retina and perceived by the brain. Notice that the lens systems cause the image to be reversed.

standard light microscopes can vary from 40× with the lowest power objective (called the scanning objective) to 2,000× with the highest power objective (the oil immersion objective).

Resolution: Distinguishing Magnified Objects Clearly

Despite the importance of magnification in visualizing tiny objects or cells, an additional optical property is essential for seeing clearly. That property is **resolution,** or **resolving power.** Resolution is the capacity of an optical system to distinguish or separate two adjacent objects or points from one another. For example, at a certain fixed distance (about 10 inches), the lens in the human eye can resolve two small objects as separate points just as long as the two objects are no closer than 0.2 mm apart (examine figure 3.16 to demonstrate this requirement for yourself). The eye examination given by optometrists is in fact a test of the resolving power of the human eye for various-sized letters read at a distance of 20 feet. Because microorganisms are extremely small and usually very close together, they will not be seen with clarity or any degree of detail unless the microscope's lenses can resolve them.

A simple equation in the form of a fraction expresses the main determining factors in resolution:

$$\text{Resolving power (R.P.)} = \frac{\text{Wavelength of light in nm}}{2 \times \text{Numerical aperture of objective lens}}$$

From this equation it is evident that the resolving power is a function of the wavelength of light that forms the image, along with certain characteristics of the objective. The light source for optical microscopes consists of a band of colored wavelengths in the visible spectrum (see figure 11.7). The shortest visible wavelengths are in the violet-blue portion of the spectrum (400 nm), and the longest are in the red portion (750 nm). Because the wavelength must pass between the objects that are being resolved, shorter wavelengths (in the 400–500 nm range) will provide better resolution (figure 3.17). Some microscopes have a special blue filter placed over the lamp to limit the longer wavelengths of light from entering the specimen.

The other factor influencing resolution is the **numerical aperture,** a mathematical constant that describes the relative efficiency of a lens in bending light rays. Without going into the

Figure 3.16
This is a test of your living optical system's resolving power. Prop your book against a wall about 20 inches away and determine the line that is no longer resolvable by your eye.
Source: Poem by Jonathan Swift.

So, Naturalists observe,

a flea has smaller

fleas that on him prey;

and these have smaller still

to bite 'em; and so proceed,

ad infinitum.

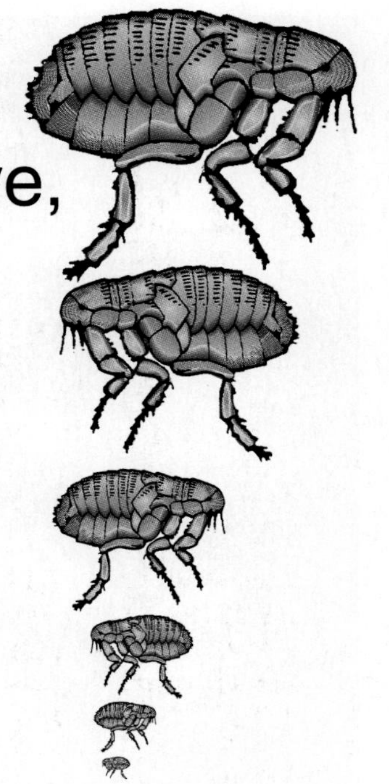

mathematical derivation of this constant, it is sufficient to say that each objective has a fixed numerical aperture reading that is determined by the microscope design and ranges from 0.1 in the lowest power lens to approximately 1.25 in the highest power (oil immersion) lens. The most important thing to remember is that a higher numerical aperture number will provide better resolution. In order for the oil immersion lens to arrive at its maximum resolving capacity, a drop of oil must be inserted between the tip of the lens and the specimen on the glass slide. Because oil has the same optical qualities as glass, it prevents refractive loss that normally occurs as peripheral light passes from the slide into the air; this property effectively increases the numerical aperture (figure 3.18). There is an absolute limitation to resolution in optical microscopes, which can be demonstrated by calculating the resolution of the oil immersion lens using a blue-green wavelength of light:

$$\text{R. P.} = \frac{500 \text{ nm}}{2 \times 1.25}$$
$$= 200 \text{ nm (or } 0.2\ \mu\text{m)}$$

In practical terms, this means that the oil immersion lens can resolve any cell or cell part as long as it is at least 0.2 μm in diameter, and that it can resolve two adjacent objects as long as they are at least 0.2 μm apart (figure 3.19). In general, organisms that are 0.5 μm or more in diameter are readily seen. This includes fungi and protozoa and some of their internal structures and most bacteria. However, a few bacteria and most viruses are far too small to be resolved by the optical microscope and require electron microscopy (discussed later). In summary then, the factor that most limits the clarity of a microscope's image is its re-

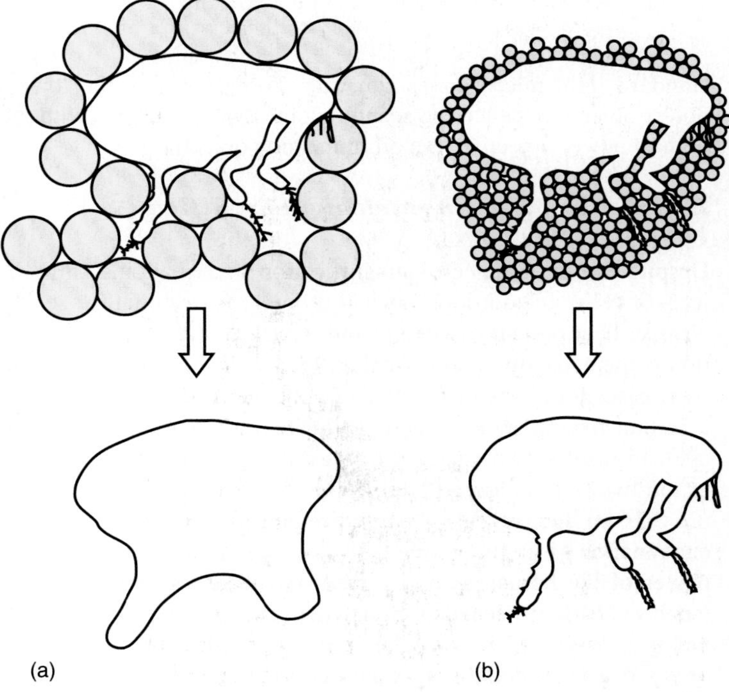

Figure 3.17
A simple model demonstrates how the wavelength influences the resolving power of a microscope. Here an outline of a flea represents the object being illuminated, and two different-sized circles represent the wavelengths of light. In (*a*), the longer waves are too large to penetrate between the finer spaces and produce a fuzzy, undetailed image. In (*b*), shorter waves are small enough to enter small spaces and produce a much more detailed image that is recognizable as a flea.

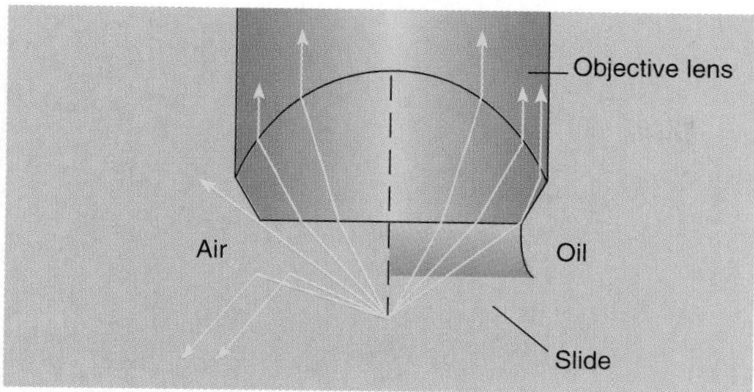

Figure 3.18
To maximize its resolving power, an oil immersion lens (the one with highest magnification) must have a drop of oil placed at its tip. This forms a continuous medium to transmit a beam of light from the condenser to the objective and effectively increase the numerical aperture. Without oil, some of the peripheral light that passes through the specimen is scattered into the air or onto the glass slide; this scattering decreases resolution.

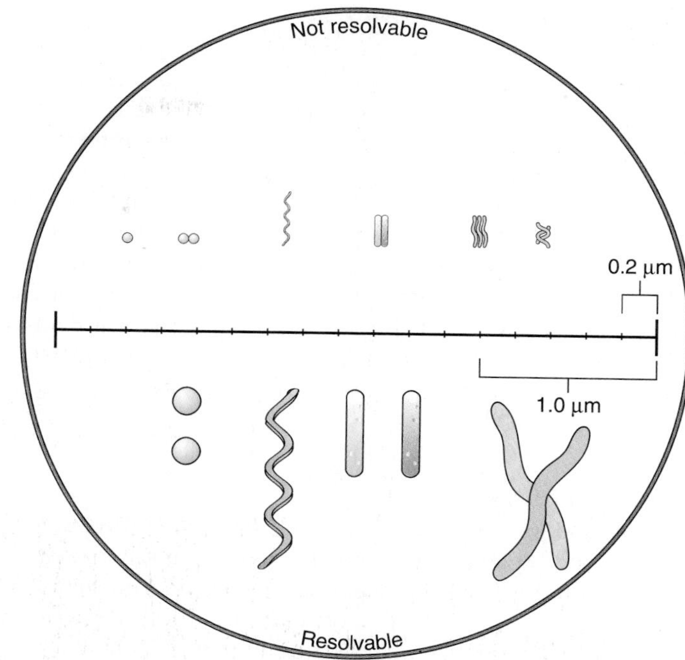

Figure 3.19
Comparison of cells that would not be resolvable versus those that would be resolvable under oil immersion at 1,000× magnification. Note that in addition to differentiating two adjacent things, good resolution also means being able to see an object clearly.

solving power. Even if a light microscope were designed to magnify several thousand times, its resolving power could not be increased, and the image it produced would simply be enlarged and fuzzy.

Other constraints to the formation of a clear image are the quality of the lens and light source and the lack of contrast in the specimen. No matter how carefully a lens is constructed, flaws remain. A typical problem is spherical aberration, a distortion in the image caused by irregularities in the lens, which creates a curved, rather than flat, image (see figure 3.13*a*). Another is chromatic aberration, a blurry, rainbowlike image that is caused by the lens's acting as a prism and separating visible light into its colored bands. To correct for these, the manufacturer incorporates several compensating lenses in the objective and ocular. Brightness and direction of illumination also affect image formation. Because too much light can reduce contrast and burn out the image, an iris diaphragm on most microscopes controls the amount of light entering the condenser. The lack of contrast in cell components is compensated for by using special lenses (the phase-contrast microscope) and by adding dyes.

VARIATIONS ON THE OPTICAL MICROSCOPE

Optical microscopes that use visible light can be described by the nature of their **field,** meaning the circular area viewed through the ocular lens. With special adaptations in lenses, condenser, and light sources, four special types of microscopes can be described: bright-field, dark-field, phase-contrast, and interference. A fifth type of optical microscope, the fluorescence microscope, uses ultraviolet radiation as the illuminating source. Each of these microscopes is adapted for viewing specimens in a particular way, as described in the next sections and summarized in parts of table 3.5.

TABLE 3.5

COMPARISON OF LIGHT MICROSCOPES AND ELECTRON MICROSCOPES

Characteristic	Light or Optical	Electron (Transmission)
Useful magnification	2,000×	1,000,000 or more
Maximum resolution	200 nm	0.5 nm
Image produced by	Visible light rays	Electron beam
Image focused by	Glass objective lens	Electromagnetic objective lenses
Image viewed through	Glass ocular lens	Fluorescent screen
Specimen placed on	Glass slide	Copper mesh
Specimen may be alive	Yes	No
Specimen requires special stains or treatment	Not always	Yes
Colored images produced	Yes	No

Bright-Field Microscopy

The **bright-field microscope** is the most widely used type of light microscope. Although we ordinarily view objects like the words on this page with light reflected off the surface, a bright-field microscope forms its image when light is transmitted through the specimen. The specimen, being denser and more opaque than its surroundings, absorbs some of this light, and the rest of the light is transmitted directly up through the ocular into the field. As a result, the specimen will produce an image that is darker than the surrounding brightly illuminated field. The bright-field microscope is a multipurpose instrument that can be used for both live, unstained material and preserved, stained material. The bright-field image is compared with that of other microscopes in figure 3.20.

Dark-Field Microscopy

A bright-field microscope can be adapted as a **dark-field microscope** by adding a special disc called a *stop* to the condenser. The stop blocks all light from entering the objective lens except peripheral light that is reflected off the sides of the specimen itself. The resulting image is a particularly striking one: brightly illuminated specimens surrounded by a dark (black) field (figure 3.20*b*). Some of Leeuwenhoek's more successful microscopes probably operated with dark-field illumination. The most effective use of dark-field microscopy is to visualize living cells that would be distorted by drying or heat or cannot be stained with the usual methods. It can outline the organism's shape and permit rapid recognition of swimming cells that might appear in dental and other infections (figure 3.21), but it does not reveal fine internal details.

Phase-Contrast and Interference Microscopy

If similar objects made of clear glass, ice, cellophane, or plastic were immersed in the same container of water, an observer would have difficulty telling them apart because they have similar optical properties. Internal components of a live, unstained cell also lack contrast and can be difficult to distinguish. But cell structures do differ slightly in density, enough that they can alter the light that passes through them in subtle ways. The **phase-contrast microscope** has been constructed to take advantage of this characteristic. This microscope contains devices that transform the subtle changes in light waves passing through the specimen into differences in light intensity. For example, denser cell parts such as organelles alter the pathway of light more than less dense regions (the cytoplasm). Light patterns coming from these regions will vary in contrast. The amount of internal detail visible by this method is greater than by either bright-field or dark-field methods. The phase-contrast microscope is most useful for observing intracellular structures such as bacterial spores, granules, and organelles, as well as the locomotor structures of eucaryotic cells (figures 3.20*c* and 3.22*a*).

Like the phase-contrast microscope, the **differential interference contrast DIC) microscope** provides a detailed view of unstained, live specimens by manipulating the light. But this

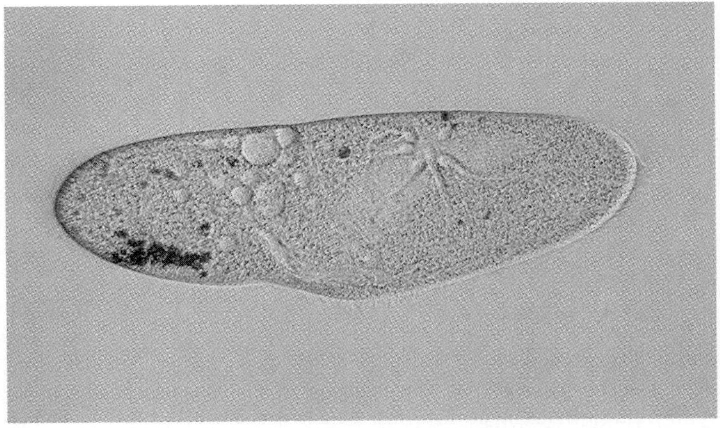

(a)

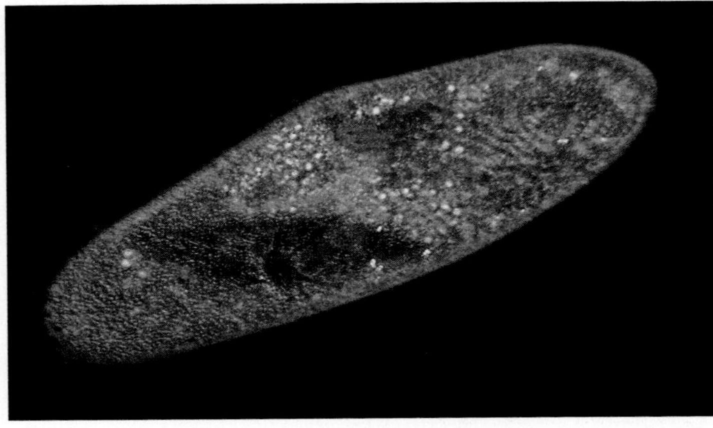

(b)

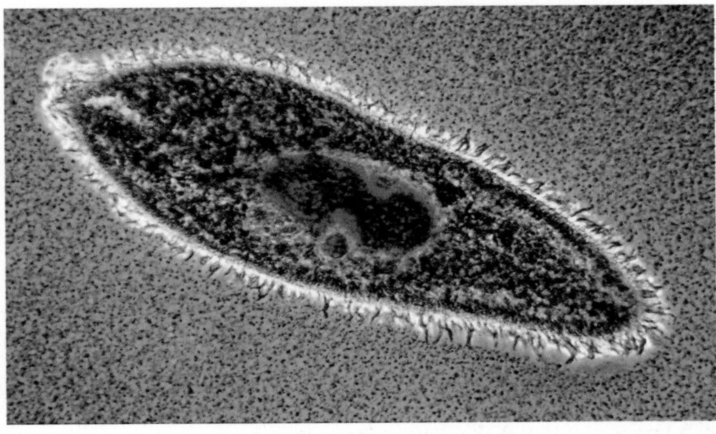

(c)

Figure 3.20
A live cell of *Paramecium* viewed with (*a*) bright-field (400×), (*b*) dark-field (400×), and (*c*) phase-contrast microscopy (400×). Note the difference in the appearance of the field and the degree of detail shown by each method of microscopy. Only in phase-contrast are the cilia (fine hairs) on the cells noticeable. Can you see the nucleus? The oral groove?

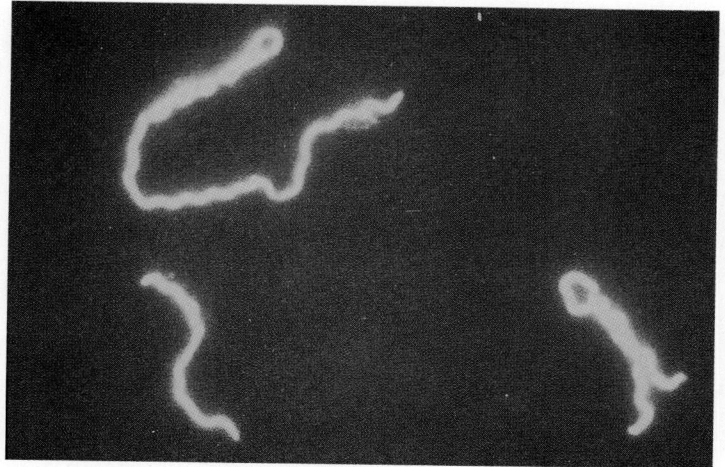

Figure 3.21
Dark-field photomicrograph of a spiral-shaped oral bacterium called *Treponema vincenti* (1,100×). This species lives in the space between the gums and the teeth and may be involved in a deteriorating infection called necrotizing gingivitis.

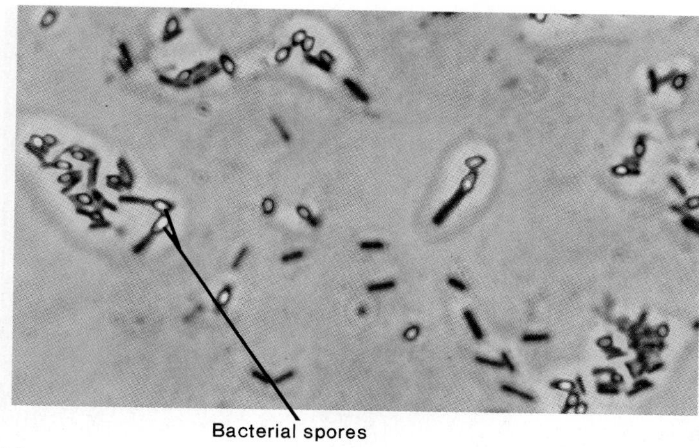

(a)

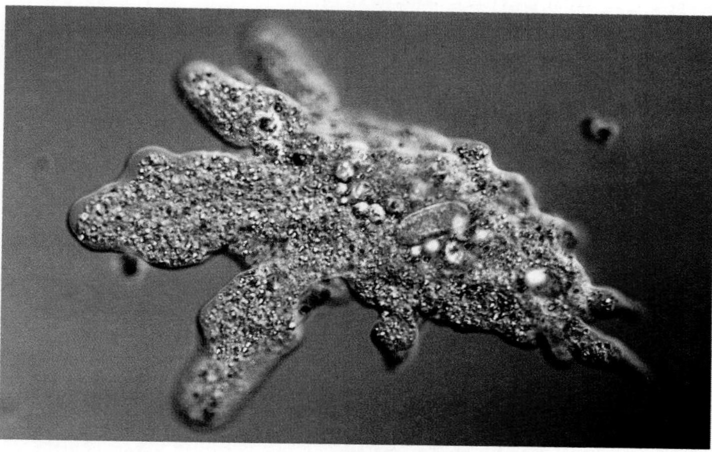

(b)

Figure 3.22
(*a*) Phase-contrast micrograph of a bacterium containing spores. The relative density of the spores causes them to appear as bright, shiny objects against the darker cell parts (600×). (*b*) Differential interference micrograph of *Amoeba proteus,* a common protozoan. Note the outstanding internal detail, the depth of field, and the bright colors, which are not natural (160×).

microscope has additional refinements, including two prisms that add contrasting colors to the image and two beams of light rather than a single one. DIC microscopes produce extremely well-defined images that are vividly colored and appear three-dimensional (figure 3.22*b*).

Fluorescence Microscopy

The **fluorescence microscope** is a specially modified compound microscope furnished with an ultraviolet (UV) radiation source and a filter that protects the viewer's eye from injury by these dangerous rays. The name of this type of microscopy originates from the use of certain dyes (acridine, fluorescein) and minerals that show **fluorescence.** This means that the dyes emit visible light when bombarded by shorter ultraviolet rays. For an image to be formed, the specimen must first be coated or placed in contact with a source of fluorescence. Subsequent illumination by ultraviolet radiation causes the specimen to give off light that will form its own image, usually an intense yellow, orange, or red against a black field.

Fluorescence microscopy has its widest applications in diagnosing infections caused by certain bacteria, protozoans, and viruses. A staining technique with fluorescent dyes is commonly used to detect *Mycobacterium tuberculosis* (the agent of tuberculosis) in patients' specimens (see figure 19.20). In a number of diagnostic procedures, fluorescent dyes are affixed to specific antibodies. These *fluorescent antibodies* can be used to detect the causative agents in such diseases as syphilis, chlamydiosis, trichomoniasis, herpes, and influenza (see figure 21.24*a*). A newer technology using fluorescent nucleic acid stains can differentiate between live and dead cells in mixtures (figure 3.23). A fluorescence microscope can be handy for locating microbes in complex mixtures because only those cells targeted by the technique will fluoresce.

ELECTRON MICROSCOPY

If conventional light microscopes are our windows on the microscopic world, the electron microscope (EM) is our window on the tiniest details of that world. Although this microscope was originally conceived and developed for studying nonbiological materials such as metals and small electronics parts, biologists immediately recognized the importance of the tool and began to use it in the early 1930s. One of the most impressive features of the electron microscope is the resolution it provides.

Unlike the light microscope, the electron microscope forms an image with a beam of electrons that can be made to travel in wavelike patterns when accelerated to high speeds. These waves are 100,000 times shorter than the waves of visible light. Because resolving power is a function of wavelength,

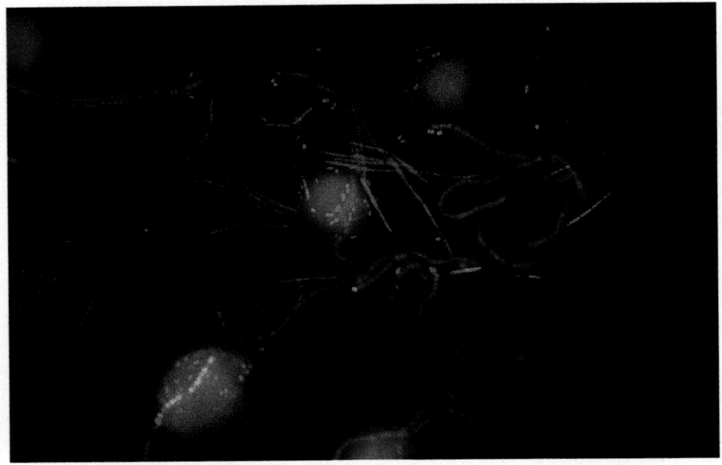

Figure 3.23
Fluorescent staining on fresh sample of cheek scrapings from the oral cavity. Cheek epithelial cells are the larger unfocused red or green cells. Bacteria appearing here are streptococci (tiny spheres in long chains) and filamentous rods. This technique also indicates whether cells are alive or dead; live cells fluoresce green, and dead cells fluoresce red.

electrons have tremendous power to resolve minute structures. Indeed, it is possible to resolve atoms with an electron microscope, though the practical resolution for biological applications is approximately 0.5 nm. Because the resolution is so substantial, it follows that magnification can also be extremely high—usually between 5,000× and 1,000,000× for biological specimens and up to 5,000,000× in some applications. Its capacity for magnification and resolution makes the EM an invaluable tool for seeing the finest structure—or **ultrastructure**—of cells and viruses. If not for electron microscopes, our understanding of biological structure and function would still be in its early theoretical stages.

In fundamental ways, the electron microscope is a derivative of the compound microscope. It employs components analogous to, but not necessarily the same as, those in light microscopy (figure 3.24). For instance, it magnifies in stages by means of two lens systems, and it has a condensing lens, a specimen holder, and focusing apparatus. Otherwise, the two types have numerous differences (table 3.6). An electron gun aims its beam through a vacuum to ring-shaped electromagnets that focus this beam on the specimen. Specimens must be pretreated with chemicals or dyes to increase contrast and cannot be observed in

TABLE 3.6

COMPARISONS OF TYPES OF MICROSCOPY

Microscope	Maximum Practical Magnification	Resolution	Important Features
Visible light as source of illumination			
Bright-field	2,000×	0.2 μm (200 nm)	Common multipurpose microscope for live and preserved stained specimens; specimen is dark, field is white; provides fair cellular detail
Dark-field	2,000×	0.2 μm	Best for observing live, unstained specimens; specimen is bright, field is black; provides outline of specimen with reduced internal cellular detail
Phase-contrast	2,000×	0.2 μm	Used for live specimens; specimen is contrasted against gray background; excellent for internal cellular detail
Differential interference	2,000×	0.2 μm	Provides brightly colored, highly contrasting, three-dimensional images of live specimens
Ultraviolet rays as source of illumination			
Fluorescent	2,000×	0.2 μm	Specimens stained with fluorescent dyes or combined with fluorescent antibodies emit visible light; specificity makes this microscope an excellent diagnostic tool
Electron beam forms image of specimen			
Transmission electron microscope (TEM)	1,000,000×	0.5 nm	Sections of specimen are viewed under very high magnification; finest detailed structure of cells and viruses is shown; used only on preserved material
Scanning electron microscope (SEM)	100,000 ×	10 nm	Scans and magnifies external surface of specimen; produces striking three-dimensional image

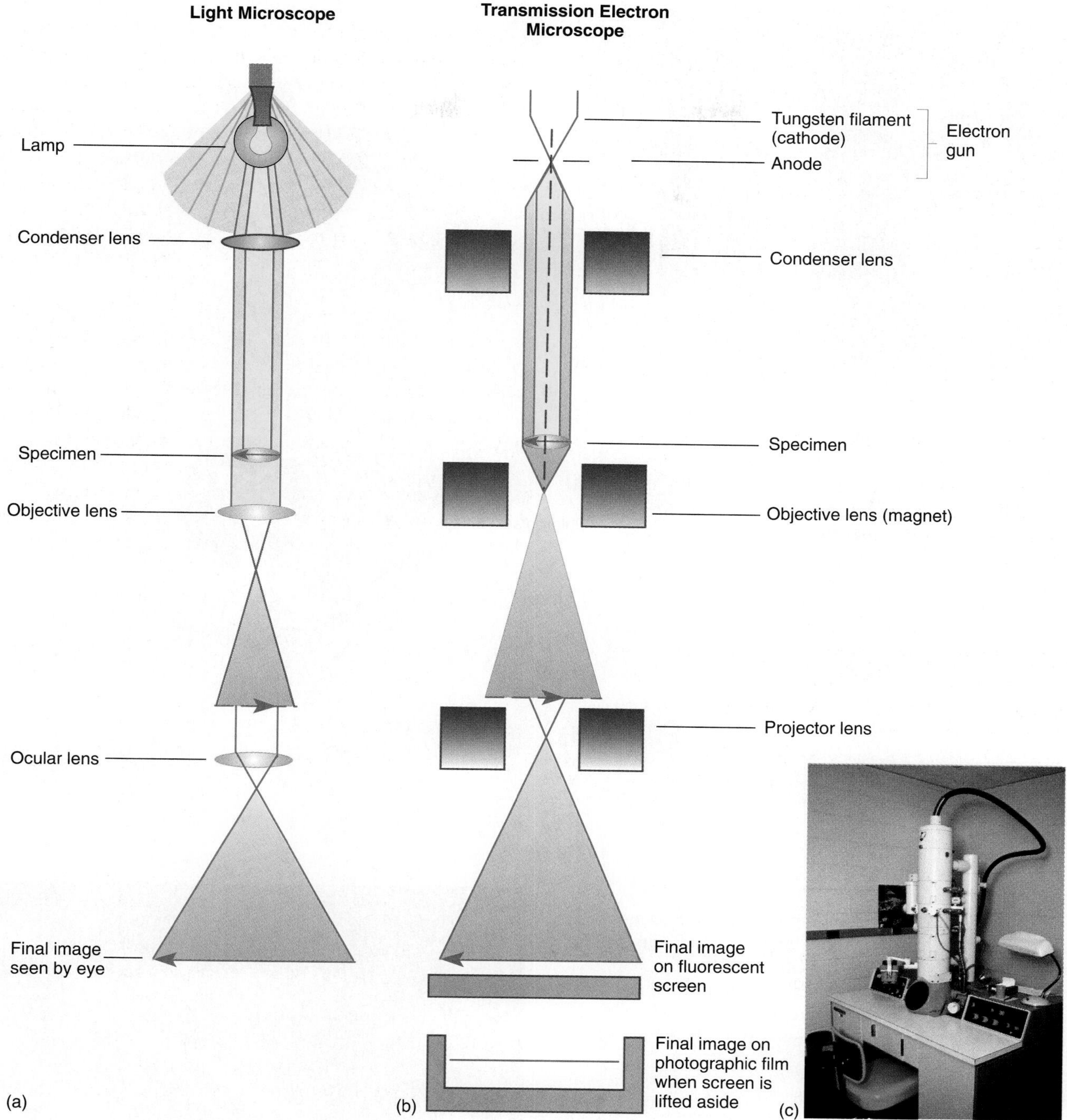

Figure 3.24
Comparison of (*a*) the light microscope and (*b*) one type of electron microscope (EM; transmission type). These diagrams are highly simplified, especially for the electron microscope, to indicate the common components. Note that the EM's image pathway is actually upside-down compared with that of a light microscope. (*c*) The EM is a larger machine with far more complicated working parts than most light microscopes.

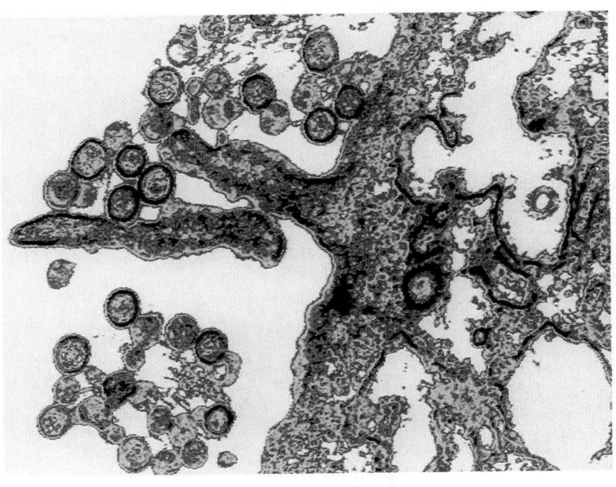

(a)

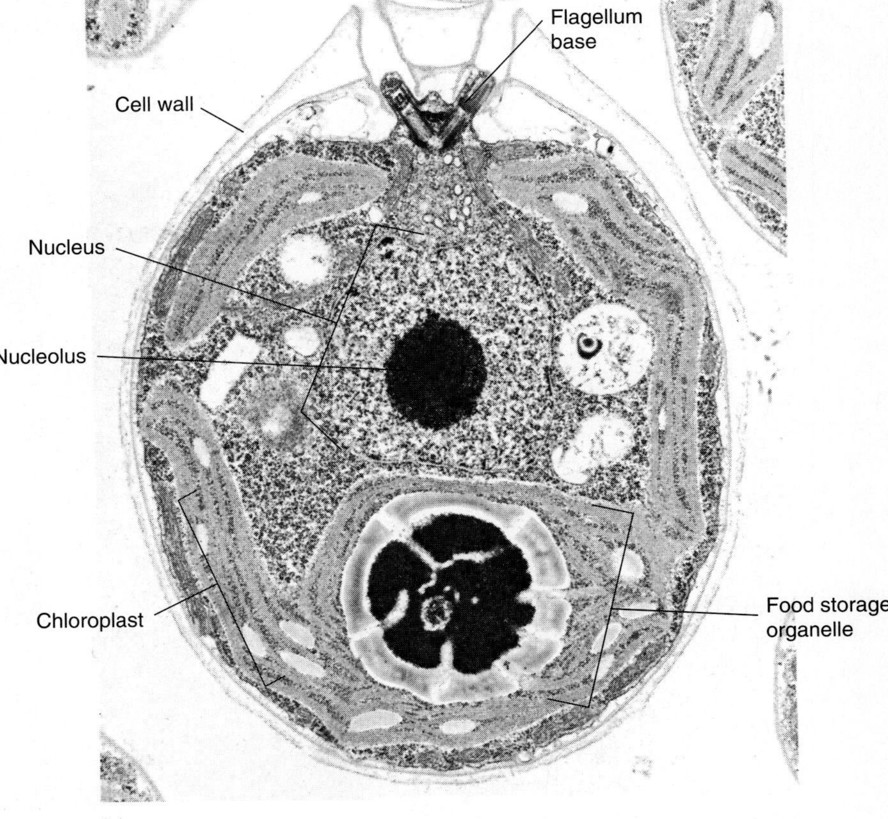

(b)

Figure 3.25
(*a*) Hantaviruses isolated from a patient with hantavirus respiratory syndrome are seen budding off the surface of a cell. The virus particles have been colorized with bright red nucleic acid strands and blue-green capsids. (*b*) TEM section through a cell of *Chlamydomonas,* a motile alga that lives in fresh water. At this magnification, the ultrastructure of several organelles is evident (30,000×).

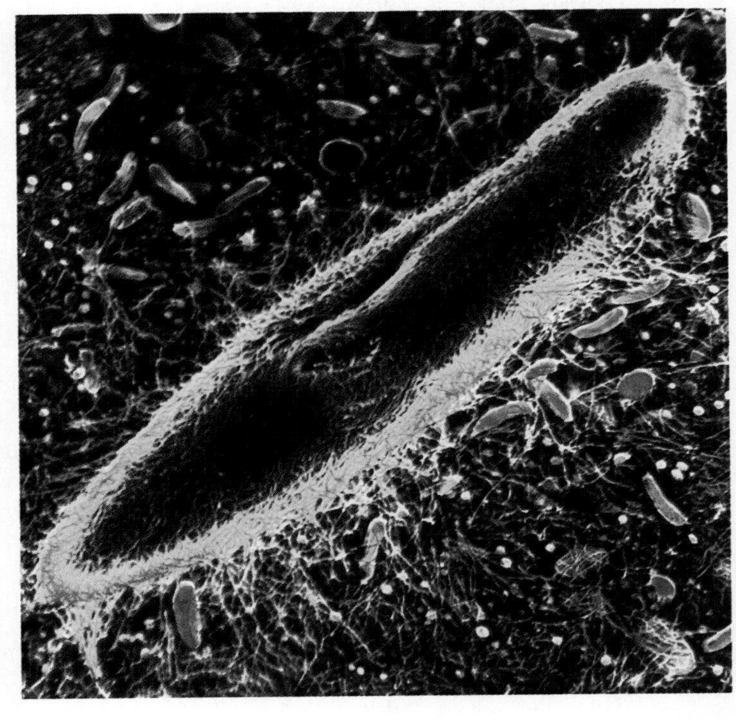

Figure 3.26
A false-color scanning electron micrograph (SEM) of *Paramecium,* covered in masses of fine hairs (100×). These are actually its locomotor and feeding structures—the cilia. Cells in the surrounding medium are bacteria that serve as the protozoan's "movable feast." Compare this with figure 3.20 to appreciate the outstanding three-dimensional detail shown by an SEM.

a live state. The enlarged image is displayed on a viewing screen or photographed for further study rather than being observed directly through an eyepiece. Because images produced by electrons lack color, electron micrographs (a micrograph is a photograph of a microscopic object) are always shades of black, gray, and white. The color-enhanced micrographs used in this and other textbooks have computer-added color (yes, they are colorized!).

Two general forms of EM are the transmission electron microscope (TEM) and the scanning electron microscope (SEM) (see table 3.6). **Transmission electron microscopes** are the method of choice for viewing the detailed structure of cells and viruses. This microscope produces its image by transmitting electrons through the specimen. Because electrons cannot

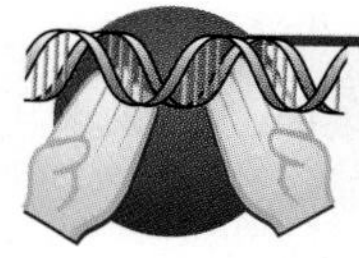

MICROFILE 3.4 A REVOLUTION IN RESOLUTION: PROBING MICROSCOPES

In the past, chemists, physicists, and biologists had to rely on indirect methods to provide information on the fine structures of the smallest molecules. But technological advances in the last decade have created a new generation of microscopes that "see" atomic structure by actually feeling it. *Scanning probe microscopes* operate with a minute needle tapered to a tip that can be as narrow as a single atom! This probe scans over the exposed surface of a material and records an image of its outer texture. These revolutionary microscopes have such profound resolution that they have the potential to image single atoms (but not subatomic structure yet) and to magnify 100 million times. The scanning tunneling microscope (STM) was the first of these microscopes. It uses a tungsten probe that hovers near the surface of an object and follows its topography while simultaneously giving off an electrical signal of its pathway, which is then imaged on a screen. The STM is used primarily for detecting defects on the surfaces of electrical conductors and computer chips composed of silicon, but it recently provided the first incredible close-up views of DNA, the genetic material (see microfile 9.1).

Another exciting new variant is the atomic force microscope (AFM), which gently forces a diamond and metal probe down onto the surface of a specimen like a needle on a record. As it moves along the surface, any deflection of the metal probe is detected by a sensitive device that relays the information to an imager. The AFM is very useful in viewing the detailed patterns of biological molecules.

These powerful new microscopes, along with tools that can move and position atoms, have spawned a field called *nanotechnology*—the science of the "small." Scientists in this area use physics, chemistry, biology, and engineering to explore and manipulate small molecules and atoms. Working at these dimensions, they hope to create tiny molecular tools to miniaturize computers and other electronic devices. In the future, it may be possible to use microstructures to deliver drugs, analyze DNA, and treat disease.

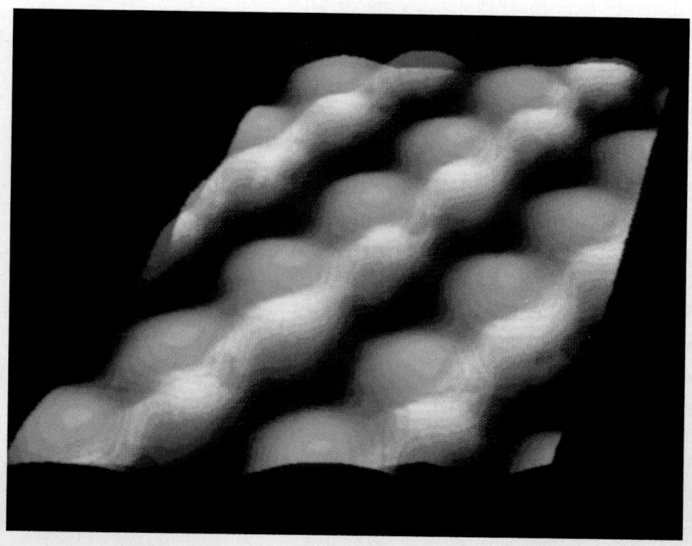

These chains of colored beads are actually atoms of the element gallium (blue) and arsenic (red) as imaged by a scanning tunneling microscope.

readily penetrate thick preparations, the specimen must be sectioned into extremely thin slices (20–100 nm thick) and stained or coated with metals that will increase image contrast. The darkest areas of TEM micrographs represent the thicker (denser) parts, and the lighter areas indicate the more transparent and less dense parts (figure 3.25). The TEM can also be used to produce negative images and shadow casts of whole microbes (see figure 6.1).

The **scanning electron microscope** provides some of the most dramatic and realistic images in existence. This instrument is designed to create an extremely detailed three-dimensional view of anything from a fly's eye to AIDS viruses escaping their host cells. To produce this image, the SEM does not transmit electrons; it bombards the surface of a whole, metal-coated specimen with electrons while scanning back and forth over it. A shower of electrons deflected from the surface is picked up with great fidelity by a sophisticated detector, and the electron pattern is displayed as an image on a television screen. The contours of the specimens resolved with scanning electron micrography are very revealing and often surprising. Areas that look smooth and flat with the light microscope display intriguing surface features with the SEM (figure 3.26). Improved technology has continued to refine electron microscopes and to develop variations on the basic plan. One of the most inventive relatives of the EM is the scanning probe microscope (microfile 3.4).

PREPARING SPECIMENS FOR OPTICAL MICROSCOPES

A specimen for optical microscopy is generally prepared by mounting a sample on a suitable glass slide that sits on the stage between the condenser and the objective lens. The manner in which a slide specimen, or **mount,** is prepared depends upon: (1) the condition of the specimen, either in a living or preserved state; (2) the aims of the examiner, whether to observe overall structure, identify the microorganisms, or see movement; and (3) the type of microscopy available, whether it is bright-field, dark-field, phase-contrast, or fluorescence.

Fresh, Living Preparations

Live samples of microorganisms are placed in **wet mounts** or in **hanging drop mounts** so that they can be observed as near to their natural state as possible. The cells are suspended in a suitable fluid (water, broth, saline) that temporarily maintains viability and provides space and a medium for locomotion. A wet mount consists of a drop or two of the culture placed on a slide and overlaid with a cover glass. Although this type of mount is quick and easy to prepare, it has certain disadvantages. The cover glass can damage larger cells, and the slide is very susceptible to drying and can contaminate the handler's fingers. A more satisfactory alternative is the hanging drop preparation made with a special concave (depression) slide, a Vaseline adhesive or sealant, and a coverslip from which a tiny drop of sample is suspended (see figure 4.4). These types of short-term mounts provide a true assessment of the size, shape, arrangement, color, and motility of cells. Greater cellular detail can be observed with phase-contrast or interference microscopy.

Fixed, Stained Smears

A more permanent mount for long-term study can be obtained by preparing fixed, stained specimens. The **smear** technique, developed by Robert Koch more than 100 years ago, consists of spreading a thin film made from a liquid suspension of cells on a slide and air-drying it. Next, the air-dried smear is usually heated gently by a process called **heat fixation** that simultaneously kills the specimen and secures it to the slide. Another important action of fixation is to preserve various cellular components in a natural state with minimal distortion. Fixation of some microbial cells is performed with chemicals such as alcohol and formalin

Like images on undeveloped photographic film, the unstained cells of a fixed smear are quite indistinct, no matter how great the magnification or how fine the resolving power of the microscope. The process of "developing" a smear to create contrast and make inconspicuous features stand out requires staining techniques. **Staining** is any procedure that applies colored chemicals called **dyes** to specimens. Dyes impart a color to cells or cell parts by becoming affixed to them through a chemical reaction. In general, they are classified as **basic (cationic) dyes,** which have a positive charge, or **acidic (anionic) dyes,** which have a negative charge. Because chemicals of opposite charge are attracted to each other, cell parts that are negatively charged will attract basic dyes and those that are positively charged will attract acidic dyes (table 3.7). Many cells, especially those of bacteria, have numerous negatively charged acidic substances and thus stain more readily with basic dyes. Acidic dyes, on the other hand, tend to be repelled by cells, so they are good for negative staining (discussed in the next section). The chemistry of dyes and their staining reactions is covered in more detail in microfile 3.5.

TABLE 3.7

CATEGORIES OF COMMON MICROBIOLOGICAL STAINS

	Positive Staining	Negative Staining
Appearance of cell	Colored by dye	Clear and colorless
Background	Not stained (generally white)	Stained (dark gray or black)
Dyes employed	Basic dyes: Crystal violet Methylene blue Safranin Malachite green	Acidic dyes: Nigrosin India ink
Subtypes of stains	Several types: Simple stain Differential stains Gram stain Acid-fast stain Spore stain Special stains Capsule Flagella Spore Granules Nucleic acid	Few types: Capsule Spore (Dorner)

Negative Versus Positive Staining Two basic types of staining technique are used, depending upon how a dye reacts with the specimen (summarized in table 3.7). Most procedures involve a **positive stain,** in which the dye actually sticks to the specimen and gives it color. A **negative stain,** on the other hand, is just the reverse (like a photographic negative). The dye does not stick to the specimen but settles around its outer boundary, forming a silhouette. In a sense, negative staining "stains" the glass slide to produce a dark background around the cells. Nigrosin (blue-black) and India ink (a black suspension of carbon particles) are the dyes most commonly used for negative staining. The cells themselves do not stain because these dyes are negatively charged and are repelled by the negatively charged surface of the cells. The value of negative staining is its relative simplicity and the reduced shrinkage or distortion of cells, as the smear is not heat-fixed. A quick assessment can thus be made regarding cellular size, shape, and arrangement. Negative staining is also used to accentuate the capsule that surrounds certain bacteria and yeasts.

Simple Versus Differential Staining Positive staining methods are classified as simple, differential, or special (figure 3.27). Whereas **simple stains** require only a single dye and an uncomplicated procedure, **differential stains** use two different-colored dyes, called the *primary dye* and the *counterstain,* to distinguish between cell types or parts. These staining techniques tend to be more complex and sometimes require additional chemical reagents to produce the desired reaction.

Most simple staining techniques take advantage of the ready binding of bacterial cells to dyes like malachite green, crystal violet, basic fuchsin, and safranin. Simple stains cause all cells in a smear to appear more or less the same color, regardless of type, but they reveal such bacterial characteristics as shape, size, and

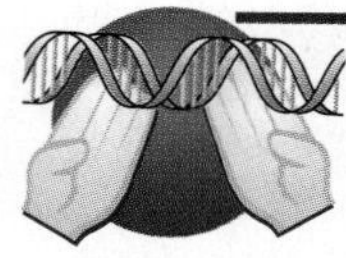

MICROFILE 3.5 THE CHEMISTRY OF DYES AND STAINING

Because many microbial cells lack contrast, it is necessary to use dyes to observe their detailed structure and identify them. Dyes are colored compounds related to or derived from the common organic solvent benzene. When certain double-bonded groups (C=O, C=N, N=N) are attached to complex ringed molecules, the resultant compound gives off a specific color. Most dyes are in the form of a sodium or chloride salt of an acidic or basic compound that ionizes when dissolved in a compatible solvent. The color-bearing ion, termed a **chromophore,** is charged and has an affinity for certain cell parts that are of the opposite charge.

Dyes that ionize into a negatively charged chromophore and a positively charged mineral ion are acidic. An example is sodium eosinate, a bright red dye that dissociates into eosin$^-$ and Na^+. Acidic dyes are attracted to the positively charged molecules of cells such as the granules of some types of white blood cells. In contrast, bacterial cells have numerous acidic substances and carry a slightly negative charge on their surface, so they do not stain well with acidic dyes.

Basic dyes ionize into a positively charged chromophore and a negative mineral ion. For example, basic fuchsin gives off fuchsin$^+$ and Cl^-. Because these dyes are positively charged, they are attracted to negatively charged cell components such as nucleic acids and proteins. Bacteria have a preponderance of negative ions, so they stain readily with basic dyes, such as crystal violet, methylene blue, malachite green, and safranin.

Acidic Dye

Sodium eosinate → Eosin, Na(+); COO(–) reacts with (+) cell

Basic Dye

Basic fuchsin chloride → Fuchsin, Cl(–); NH_2(+) reacts with (–) cell

Examples of the two major groups of dyes and their reactions.

arrangement. A simple stain with Loeffler's methylene blue is distinctive. The blue cells stand out against a relatively unstained background, so that size, shape, and grouping show up easily. This method is also significant because it reveals the internal granules of *Corynebacterium diphtheriae,* a bacterium that is responsible for diphtheria (see chapter 4).

Types of Differential Stains A satisfactory differential stain uses differently colored dyes to clearly contrast two cell types or cell parts. Common combinations are red and purple, red and green, or pink and blue. Differential stains can also pinpoint other characteristics such as the size, shape, and arrangement of cells. Typical examples include Gram, acid-fast, and endospore stains. Some staining techniques (spore, capsule) fall into more than one category.

Gram staining, a century-old method named for its developer, Hans Christian Gram, remains the most universal diagnostic staining technique for bacteria. It permits ready differentiation of major significant categories based upon the color reaction of the cells: **gram positive,** which stain purple, and **gram negative,** which stain pink (red). The Gram stain is the basis of several important bacteriological topics, including bacterial taxonomy, cell wall structure, and identification and diagnosis of infection; in some cases, it even guides the selection of the correct drug for an infection. Gram staining is discussed in greater detail in microfile 4.1.

The **acid-fast stain,** like the Gram stain, is an important diagnostic stain that differentiates acid-fast bacteria (pink) from non-acid-fast bacteria (blue). This stain originated as a specific method to detect *Mycobacterium tuberculosis* in specimens. It was determined that these bacterial cells have a particularly impervious outer wall that holds fast (tightly or tenaciously) to the dye (carbol fuchsin) even when washed with a solution containing acid or acid alcohol. This stain is used for other medically important mycobacteria such as the leprosy bacillus and for *Nocardia,* an agent of lung or skin infections.

The **endospore stain** (spore stain) is similar to the acid-fast method in that a dye is forced by heat into resistant bodies called spores or endospores (their formation and significance are discussed in chapter 4). This stain is designed to distinguish between spores and the cells that they come from (so-called **vegetative cells**). Of significance in medical microbiology are the gram-positive, spore-forming members of the genus *Bacillus* (the cause of anthrax) and *Clostridium* (the cause of botulism and tetanus)—dramatic diseases of universal fascination that we consider later.

Special stains are used to emphasize certain cell parts that are not revealed by conventional staining methods. **Capsule staining** is a method of observing the microbial capsule, an unstructured protective layer surrounding the cells of some bacteria and fungi. Because the capsule does not react with most stains, it is often negatively stained with India ink, or it may be demonstrated by special positive stains. The fact that not all microbes exhibit capsules is a useful feature for identifying pathogens. One example is *Cryptococcus,* which causes a serious fungal infection in AIDS patients.

Flagellar staining is a method of revealing flagella, the tiny, slender filaments used by bacteria for locomotion (see chapter 4). Because the width of bacterial flagella lies beyond the resolving power of the light microscope, in order to be seen, they must be enlarged by depositing a coating on the outside of the filament and then staining it. This stain works best with fresh, young cultures, because flagella are delicate and can be lost or damaged on older cells. Their presence, number, and arrangement on a cell are taxonomically useful.

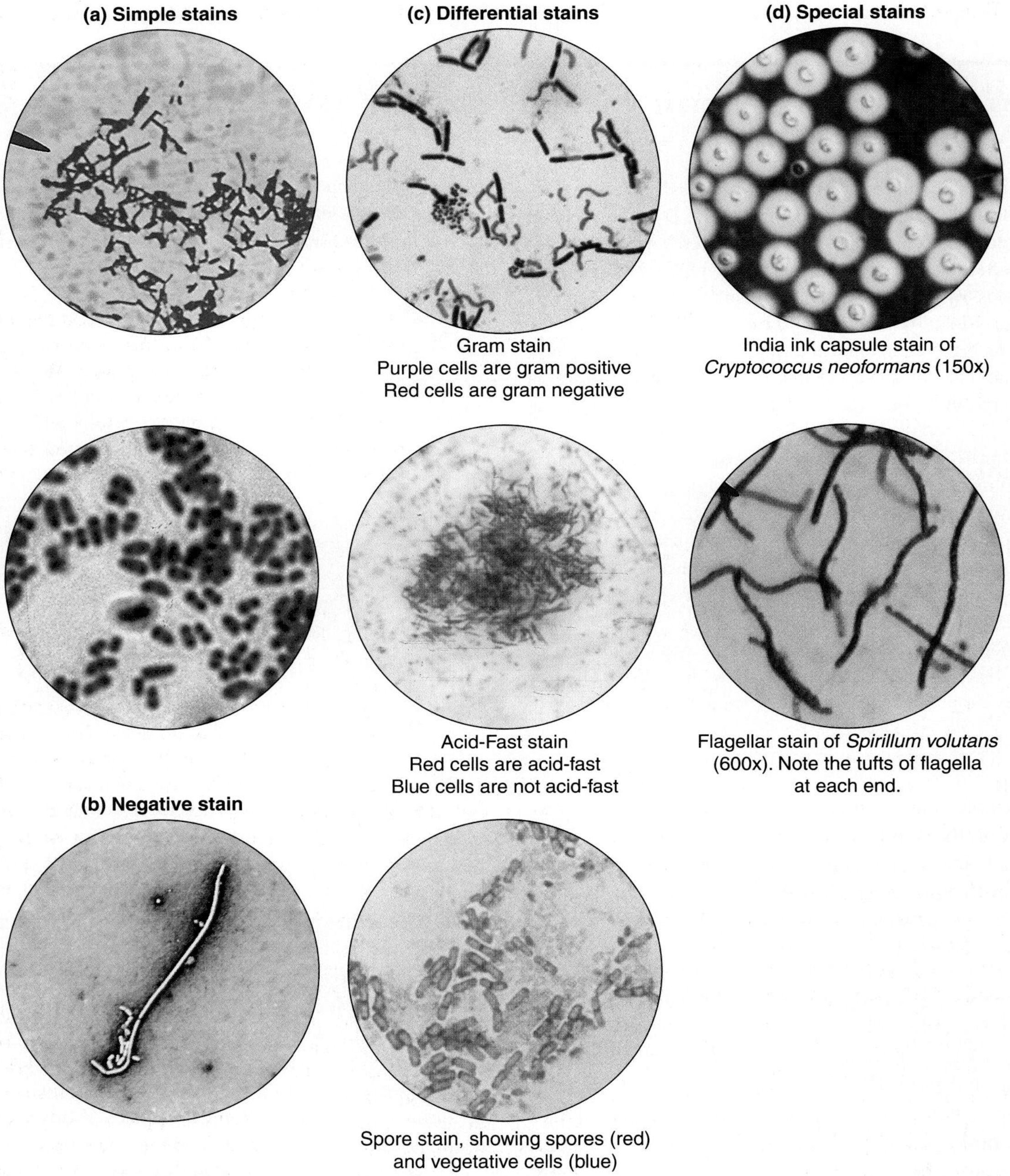

(a) Simple stains

(c) Differential stains

(d) Special stains

Gram stain
Purple cells are gram positive
Red cells are gram negative

India ink capsule stain of *Cryptococcus neoformans* (150x)

Acid-Fast stain
Red cells are acid-fast
Blue cells are not acid-fast

Flagellar stain of *Spirillum volutans* (600x). Note the tufts of flagella at each end.

(b) Negative stain

Spore stain, showing spores (red) and vegetative cells (blue)

Figure 3.27
Types of microbiological stains. (*a*) Simple stains. (*b*) Negative stain. (*c*) Differential stains: Gram, acid-fast, and spore. (*d*) Special stains: capsule and flagellar.

Chapter Checkpoints

Magnification, resolving power, lens quality, and illumination source all influence the clarity of specimens viewed through the optical microscope.

The maximum resolving power of the optical microscope is 200 nm, or 0.2 μm. This is sufficient to see the internal structures of eucaryotes and the morphology of most bacteria.

There are five types of optical microscopes. Four types use visible light for illumination: bright-field, dark-field, phase-contrast, and interference microscopes. The fifth type, the fluorescence microscope, uses UV light for illumination, but it has the same resolving power as the other optical microscopes.

Electron microscopes (EM) use electrons, not light waves, as an illumination source to provide high magnification (5,000×–1,000,000×) and high resolution (0.5 nm). Electron microscopes can visualize cell ultrastructure (TEM) and three-dimensional images of cell and virus surface features (SEM).

Specimens viewed through optical microscopes can be either alive or dead, depending on the type of specimen preparation, but all EM specimens are dead because they must be treated with metals for effective viewing.

Stains are important diagnostic tools in microbiology because they can be designed to differentiate cell shape, structure, and biochemical composition of the specimens being viewed.

Chapter Capsule with Key Terms

CULTURE TECHNIQUES

Microbiology as a science is very dependent on a number of specialized laboratory techniques. Laboratory steps routinely employed in microbiology are inoculation, incubation, isolation, inspection, and identification. To prepare a **culture** (visible growth specimen), a **medium** (nutrient substrate) is **inoculated** (implanted, seeded) and **incubated.** Cultures differ: A **pure culture** consists of a single type of microbe; a **mixed culture** deliberately contains more than one type; and a **contaminated culture** is tainted with some intruding microbe. Methods of isolation: Cells are spread over a large area **(streak plate)** or are diluted in a large volume **(pour plate)** so that individual cells are completely separated and can grow into **colonies.** The cultures are incubated, **subcultured,** observed macroscopically and microscopically, and identified by means of morphological, physiological, genetic, and serological methods.

ARTIFICIAL MEDIA

Artificial nutrient media vary according to their physical form, chemical characteristics, and purpose. They can be **liquid** (broth, milk), **semisolid,** or **solid,** depending upon the absence or the quantity of solidifying agent (usually **agar** or **gelatin**). A **synthetic medium** is any preparation that is chemically defined, but a medium containing a poorly identified component is **nonsynthetic.** A **general purpose medium** is used to grow a wide assortment of microbial types; when supplemented with blood or tissue infusion to culture **fastidious** species, it is called an **enriched medium.** A **selective medium** permits preferential growth of certain organisms in a mixture and inhibits others. Highly selective media are used expressly to favor the growth of an organism that is ordinarily sparse. A **differential medium** distinguishes among different microbes by bringing out their variations in a particular reaction. A **reducing medium** limits oxygen availability and is useful for cultivating anaerobes. The ability to utilize sugars can be determined with **carbohydrate fermentation media.** To convey fragile microbes, special **transport media** are needed to stabilize viability. **Assay media** are used to evaluate the effectiveness of antimicrobial agents. Environmental and industrial surveillance routinely call for **enumeration media** to determine the number of microbes in food, drinking water, soil, sewage, and other sources. In certain instances, microorganisms have to be grown in animals and bird embryos.

MICROSCOPY

Optical, or **light, microscopy** depends upon lenses that **refract** light rays, drawing them to a focus to produce a magnified **image.** A **simple microscope** consists of a single magnifying lens, whereas a **compound microscope** relies on two lenses: the **ocular lens** and the **objective lens.** The objective lens is responsible for the **real image,** and the ocular lens forms the **virtual image.** The **total power of magnification** is calculated from the product of the ocular and objective magnifying powers.

Resolution, or the **resolving power,** is related not only to magnification but also to optical clarity. Resolution is improved with shorter wavelengths of illumination and with a higher **numerical aperture** of the lens. Modifications in the lighting or the lens increase the versatility of the compound microscope system. Such variations give rise to the **bright-field, dark-field, phase-contrast, interference,** and **fluorescence microscopes.** Magnification and resolution in the electron microscope obey the same laws as in optical microscopes. Fundamental substitutions include electron emission for light, electromagnets for focusing the lens, and the proper apparatus to provide the necessary voltage and vacuum. Compared with optical microscopes and imagery, the **transmission electron microscope** (TEM) is analogous to the bright-field microscope, and the **scanning electron microscope** (SEM) corresponds to the dark-field microscope.

TECHNIQUES IN SPECIMEN PREPARATION AND STAINING

Specimen preparation of optical microscopy is governed by the condition of the specimen, the purpose of the inspection, and the type of microscope being used. **Wet mounts** and **hanging drop mounts** permit examination of live organisms, but the resolution is poorer than in **dyed** preparations. A **stained smear** provides higher resolution by virtue of color contrasts, but the dyes used are often toxic. Moreover, a smear must be **heat-fixed** in order to adhere to the slide.

Staining can be controlled by choosing the dye. A **basic (cationic) dye** carries a positive charge, and an **acidic (anionic) dye** bears a negative charge, thus they selectively bind to oppositely charged surfaces. The surfaces of microbes are usually negatively charged and attract basic dyes. This is the basis of **positive staining.** In **negative staining,** the microbe repels the dye and it stains the background.

Staining methods can be distinguished from one another. A **simple stain** uses an uncomplicated procedure and one dye, such as methylene blue, malachite green, crystal violet, basic fuchsin, or safranin. A **differential stain** requires a primary dye and a contrasting counterstain; consequently, the procedure is more involved. Classic examples of differential stains are the **Gram stain, acid-fast stain,** and **endospore stain. Special stains** are designed to bring out distinctive characteristics, as with the spore stain, capsule stain and flagellar stain. Stains have practical diagnostic significance. The ability of the Gram stain to distinguish **gram-positive** bacteria and **gram-negative** bacteria is useful in deciding upon drug therapy. Finding internal granules by means of a simple stain is valuable in diphtheria diagnosis.

The **acid-fast stain** is useful in diagnosing tuberculosis and leprosy. The **spore stain** is helpful in anthrax, botulism, and tetanus diagnosis. The **capsule stain** can assist in diagnosing respiratory infections.

MULTIPLE-CHOICE QUESTIONS

1. The term *culture* refers to the ____ growth of microorganisms in ____.
   a. rapid, an incubator
   b. macroscopic, media
   c. microscopic, the body
   d. artificial, colonies
2. A mixed culture is
   a. the same as a contaminated culture
   b. one that has been adequately stirred
   c. one that contains two or more known species
   d. a pond sample containing algae and protozoa
3. Agar is superior to gelatin as a solidifying agent because agar
   a. does not melt at room temperature
   b. solidifies at 75°C
   c. is not usually decomposed by microorganisms
   d. both a and c
4. The process that most accounts for magnification is
   a. a curved glass surface
   b. refraction of light
   c. illumination
   d. resolution
5. A subculture is a
   a. colony growing beneath the media surface
   b. culture made from a contaminant
   c. culture made in an embryo
   d. culture made from an isolated colony
6. Resolution is ____ with a longer wavelength of light.
   a. improved c. not changed
   b. worsened d. not possible
7. A real image is produced by the
   a. ocular c. condenser
   b. objective d. eye
8. A microscope that has a total magnification of 1,500× with the oil immersion lens has an ocular of what power?
   a. 150× c. 15×
   b. 1.5× d. 30×
9. The specimen for an electron microscope is always
   a. stained with dyes
   b. sliced into thin sections
   c. killed
   d. viewed directly
10. Motility is best seen with a
    a. hanging drop preparation
    b. negative stain
    c. streak plate
    d. flagellar stain
11. Bacteria tend to stain more readily with cationic (positively charged) dyes because bacteria
    a. contain large amounts of alkaline substances
    b. contain large amounts of acidic substances
    c. are neutral
    d. have thick cell walls
12. Multiple Matching: For each type of medium select all descriptions that fit. For media that fit more than one description, briefly explain why this is the case.
    ___ Mannitol salt agar
    ___ Chocolate agar
    ___ MacConkey agar
    ___ Nutrient broth
    ___ Sabouraud's agar
    ___ Triple-sugar iron agar
    ___ *Euglena* agar
    ___ SIM medium
    ___ Stuart's medium
    a. Selective medium
    b. Differential medium
    c. Chemically defined (synthetic) medium
    d. Enriched medium
    e. General purpose medium
    f. Complex medium
    g. Transport medium

CONCEPT QUESTIONS

1. a. Describe briefly what is involved in the Five I's.
   b. Name three basic differences between inoculation and contamination.
2. a. Name two ways that pure, mixed, and contaminated cultures are similar and two ways that they differ from each other.
   b. What must be done to avoid contamination?
3. a. Explain what is involved in isolating microorganisms and why it is necessary to do this.
   b. Compare and contrast two common laboratory techniques for separating bacteria in a mixed sample.
   c. Describe how an isolated colony forms.
   d. Explain why an isolated colony and a pure culture are not the same thing.
4. a. Explain the two principal functions of dyes in media.
   b. Differentiate among the ingredients and functions of enriched, selective, and differential media.
5. Differentiate between microscopic and macroscopic methods of observing microorganisms, citing a specific example of each method.
6. a. Contrast the concepts of magnification, refraction, and resolution.
   b. Briefly explain how an image is made and magnified.
7. a. On the basis of the formula for resolving power, explain why a smaller R. P. value is preferred to a larger one.
   b. What does it mean in practical terms if the resolving power is 1.0 μm?
   c. What does it mean if the value is greater than 1.0 μm?
   d. What if it is less than 1.0 μm?
   e. What can be done to improve resolution?
8. a. Compare bright-field, dark-field, phase-contrast, and fluorescence microscopy as to field appearance, specimen appearance, light source, and uses.
   b. Of what benefit is the dark-field microscope if the field is black?
9. a. Compare and contrast the optical compound microscope with the electron microscope.
   b. Why is the resolution so superior in the electron microscope?
   c. What will you never see in an unretouched electron micrograph?
   d. Compare the way that the image is formed in the TEM and SEM.
10. Evaluate the following preparations in terms of showing microbial size, shape, motility, and differentiation: spore stain, negative stain, simple stain, hanging drop slide, and Gram stain.
11. a. Itemize the various staining methods, and briefly characterize each.
    b. For a stain to be considered a differential stain, what must it do?
    c. Explain what happens in positive staining to cause the reaction in the cell.
    d. Explain what happens in negative staining that causes the final result.

CRITICAL-THINKING QUESTIONS

1. Describe the steps you would take to isolate, cultivate, and identify a microbial pathogen from a urine sample. (Hint: Look at the Five I's.)
2. A certain medium has the following composition:

Glucose	15 g
Yeast extract	5 g
Peptone	5 g
KH_2PO_4	2 g

   Distilled water 1,000 ml
   a. To what chemical category does this medium belong?
   b. How could you convert *Euglena* agar (table 3.2) into a nonsynthetic medium?
3. a. Name four categories that blood agar fits into.
   b. Name four differential reactions that TSIA shows.
   c. Can you tell what functional kind of medium *Enterococcus faecalis* medium is?
4. a. What kind of medium might you make to selectively grow a bacterium that lives in the ocean?
   b. One that lives in the human stomach?
   c. What characteristic of dyes makes them useful in differential media?
   d. Why are intestinal bacteria able to grow on media containing bile?
5. a. When buying a microscope, what features are most important to check for?
   b. What is probably true of a $20 microscope that claims to magnify 1,000×?
6. How can one obtain 2,000× magnification with a 100× objective?
7. a. In what ways are dark-field microscopy and negative staining alike?
   b. How is the dark-field microscope like the scanning electron microscope?
8. Biotechnology companies have engineered hundreds of different types of mice, rats, pigs, goats, cattle, and rabbits to have genetic diseases similar to diseases of humans or to synthesize drugs and other biochemical products. They have patented these animals and sell them to researchers for study and experimentation.
   a. What do you think of creating new life forms just for experimentation?
   b. Comment or start a class discussion on the benefits, safety, and humanity of this trend.

4 chapter

Procaryotic Profiles: The Bacteria and Archaea

Small and deceptively simple, procaryotes are among nature's most abundant and ubiquitous microorganisms. If it were somehow possible to eradicate all bacteria in the world, humans would notice the effects immediately and, for a while, might find it a favorable change. We would not have to be as careful about preparing and refrigerating foods; plaque would no longer develop on our teeth; and there would be fewer cleaning chores around the house. Quite suddenly, the medical community's goal of eradicating certain infectious diseases such as tuberculosis, cholera, syphilis, and tetanus could be a reality. But there are other considerations to take into account. We would also have to do without certain foods, like sauerkraut, yogurt, Swiss cheese, and sourdough bread. At first, this may seem a small price to pay, but are these slight inconveniences the only sacrifices to be expected? Within a few days, industrial processes that produce vitamins, drugs, and solvents would lie silent and useless, and most molecular biology research labs and biotechnology companies would have to be abandoned. In a few months and years, humus containing dead animal and plant matter would build up and trap the very elements needed to sustain the living world. Clearly, bacteria play vital roles in all aspects of our existence. We literally can't live without them!

In order to explore the roles of bacteria in nature we must first understand several aspects of the structure and behavior of procaryotic cells. This understanding will make later studies in nutrition, genetics, drug therapy, infection, and microbial ecology more meaningful. The primary topics to be covered in this chapter are elements of microscopic anatomy, physiology, identification, and classification and a survey of selected bacterial groups.

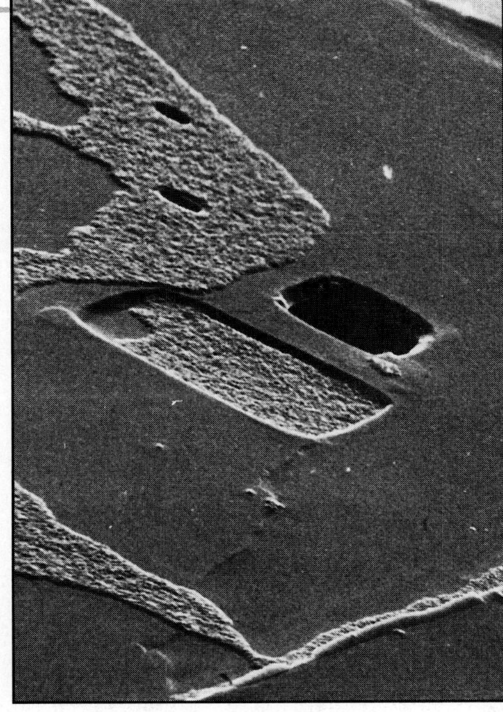

Cells of *Halobacterium salinarium* growing in rectangular pockets of a pure salt crystal. These salt-loving procaryotes are common inhabitants of salt lakes and salt mines. They belong to the superkingdom Archaea, a novel group of cells with spectacular adaptation to extremes.

PROCARYOTIC FORM AND FUNCTION: EXTERNAL STRUCTURE

The evolutionary history of procaryotic cells extends back at least 3.5 billion years (see figure 4.30). It is now generally thought that the very first cells to appear on the earth were a type of archaea possibly related to modern forms that live on sulfur compounds in geothermal ocean vents. The fact that these organisms have endured for so long in such a variety of habitats indicates a cellular structure and function that are amazingly versatile and adaptable. The general cellular organization of a procaryotic cell can be represented with the following flowchart:

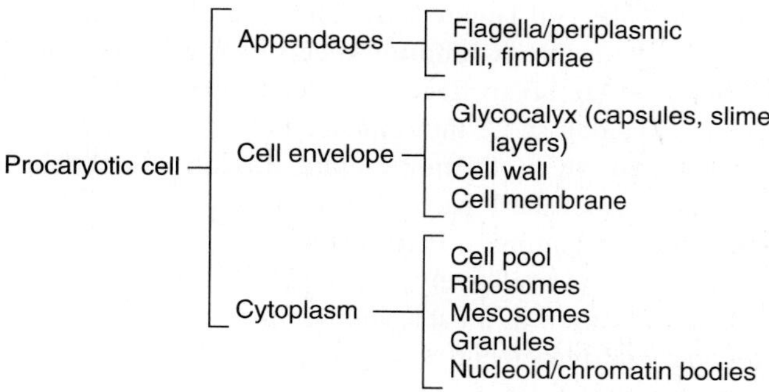

All bacterial cells invariably have a cell membrane, cytoplasm, ribosomes, and chromatin bodies; the majority have a cell wall and some form of surface coating or glycocalyx. Specific structures that are found in some, but not all, bacteria are flagella, pili, fimbriae, capsules, slime layers, and granules.

THE STRUCTURE OF A GENERALIZED PROCARYOTIC CELL

Bacterial cells appear featureless and two-dimensional when viewed with an ordinary microscope. Not until they are subjected to the scrutiny of the electron microscope and biochemical studies does their intricate and functionally complex nature become evident. The descriptions of bacterial structure, except where otherwise noted, refer to the **bacteria,*** a category of procaryotes with peptidoglycan in their cell walls. Figure 4.1 contrasts a three-dimensional anatomical view of a generalized (rod-shaped) bacterial cell with an electron micrograph of an actual cell. As we survey the principal anatomical features of this cell, we will perform a microscopic dissection of sorts, following a course that begins with the outer cell structures and proceeds to the internal contents.

*Formerly known as eubacteria (yoo′-bak-ter-ee-uh) Gr. *eu,* true, and *bakterion,* rod. Includes all procaryotes besides the archaea.

Figure 4.1

(*a*) Cutaway view of a typical rod-shaped bacterium, showing major structural features. Note that not all components are found in all cells. (*b*) Transmission electron micrograph of *Bacillus coagulans* (88,000×).

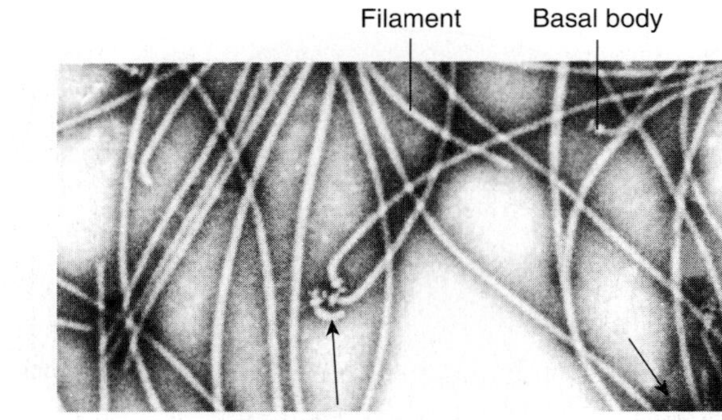

(a)

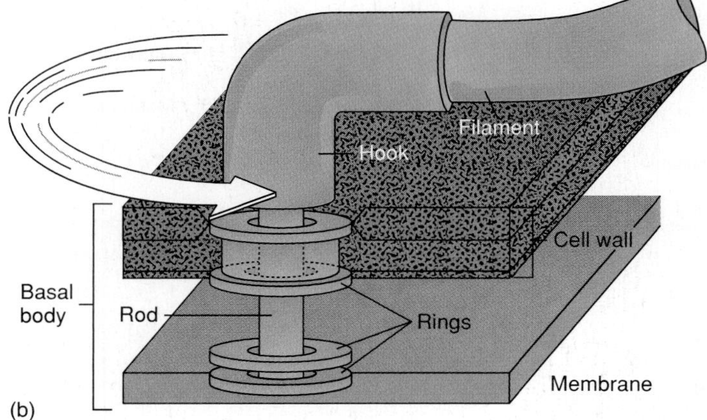

(b)

Figure 4.2

(*a*) Electron micrograph of basal bodies and filaments (at arrows) of bacterial flagella (66,000×). (*b*) Details of the basal body in a gram-negative cell. The hook, rings, and rod function together as a tiny device that rotates the filament 360°.

APPENDAGES: CELL EXTENSIONS

Several discrete types of accessory structures sprout from the surface of bacteria. These elongate **appendages*** are common but are not present on all species. Appendages can be divided into two major groups: those that provide motility (flagella and axial filaments) and those that provide attachments (fimbriae and pili).

Flagella—Bacterial Propellers

The procaryotic **flagellum*** is an appendage of truly amazing construction, certainly unique in the biological world. The primary function of flagella is to confer **motility,** or self-propulsion—that is, the capacity of a cell to swim freely through an aqueous habitat. The extreme thinness of a bacterial flagellum necessitates high magnification to reveal its special architecture, which occurs in three distinct parts: the filament, the hook (sheath), and the basal body (figure 4.2). The **filament,** a helical structure composed of proteins, is approximately 20 nm in diameter and varies from 1 to 70 nm in length. It is inserted into a curved, tubular hook. The hook is anchored to the cell by the basal body, a stack of rings firmly anchored through the cell wall, to the cell membrane. This arrangement permits the hook with its filament to rotate 360°, rather than undulating back and forth like a whip as was once thought. As the flagellum rotates in a counterclockwise direction it causes the cell body to swim in a direct forward path (see figure 4.5).

One can generalize that all spirilla, about half of the bacilli, and a small number of cocci are flagellated. Flagella vary both in number and arrangement according to two general patterns: (1) In a *polar* arrangement, the flagella are attached at one or both ends of the cell. Three subtypes of this pattern are: **monotrichous,*** with a single flagellum; **lophotrichous,*** with small bunches or tufts of flagella emerging from the same site; and **amphitrichous,*** with flagella at both poles of the cell. (2) In a **peritrichous*** arrangement, flagella are dispersed randomly over the surface of the cell (figure 4.3). The type of arrangement has some bearing on the swimming speed of the bacterium. The speediest forms are polar, flagellated cells such as *Thiospirillum,* which can zip along at 5.2 mm/minute, and *Pseudomonas aeruginosa,* which can swim 4.4 mm/minute. Taking into account the small dimensions of these bacteria, such speeds are comparable to some protozoa and animals. Peritrichous rods such as *Escherichia coli* tend to swim at a relatively slower pace (1 mm/minute).

If identification of a bacterium requires detection of the actual number and placement of flagella, special stains or electron microscope preparations are required, since flagella are too minute to be seen in unstained live preparations with an ordinary light microscope. Often it is sufficient to know simply whether a bacterial species is motile. One way to detect motility is to place a tiny mass of cells into a soft (semisolid) medium. Growth spreading rapidly through the entire medium is indicative of motility (see figure 3.6). Alternatively, cells can be observed microscopically with a hanging drop slide (figure 4.4). A truly motile cell will flit, dart, or wobble around the field, making some progress, whereas one that is nonmotile jiggles about in one place and makes no progress at all.

Fine Points of Flagellar Function Flagellated bacterial can perform some rather sophisticated feats. They can detect and move in response to chemical signals—a type of behavior called **chemotaxis.*** Positive chemotaxis is movement of a cell in the direction of a favorable chemical stimulus (usually a nutrient); negative chemotaxis is movement away from a repellent (potentially harmful) compound.

The flagellum is effective in guiding bacteria through the environment primarily because the system for detecting chemicals is linked to the mechanisms that drive the flagellum. Located in the cell membrane are clusters of receptors[1] that bind specific molecules coming from the immediate environment. The attachment of sufficient numbers of these molecules transmits signals to

*appendage (uh-pen′-dij) Any external projection of a body.
*flagellum (flah-jel′-em) pl. flagella; L., a whip.

1. Cell-surface molecules that bind specifically with other molecules.

*monotrichous (mah″-noh-trik′-us) Gr. *mono,* one, and *tricho,* hair.
*lophotrichous (lo″-foh-trik′-us) Gr. *lopho,* tuft or ridge.
*amphitrichous (am″-fee-trik′-us) Gr. *amphi,* on both sides.
*peritrichous (per″-ee-trik′-us) Gr. *peri,* around.
*chemotaxis (ke″-moh-tak′-sis) pl. chemotaxes; Gr. *chemo,* chemicals, and *taxis,* an ordering or arrangement.

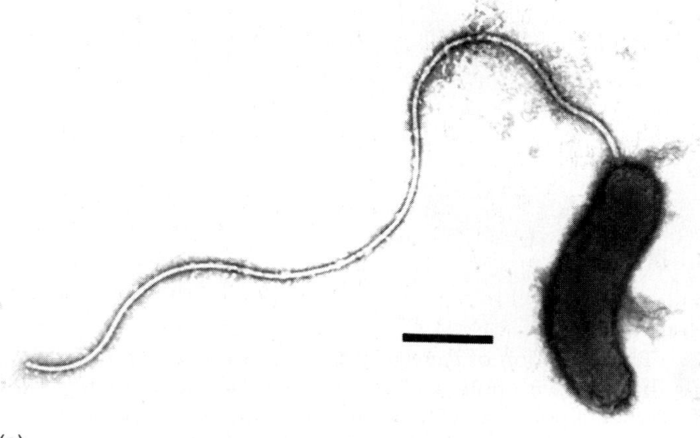

(a)

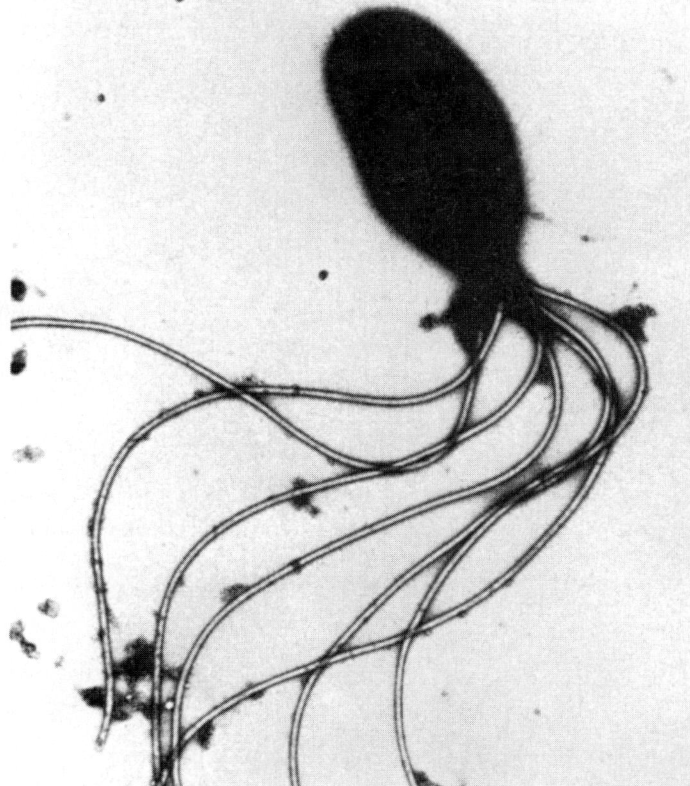

(b)

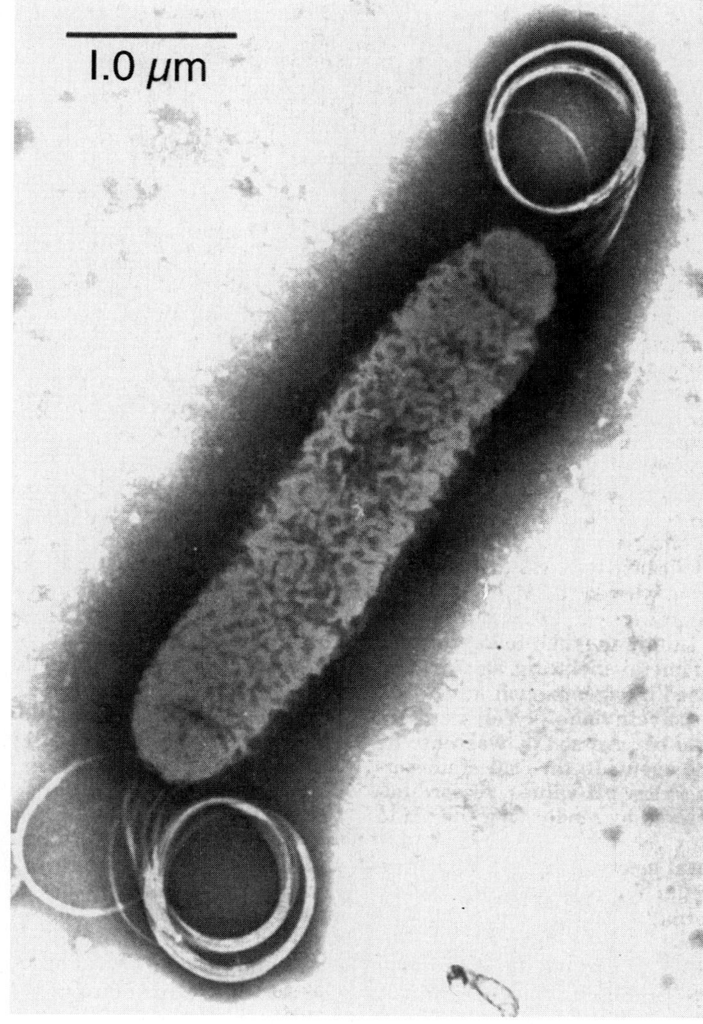

(c)

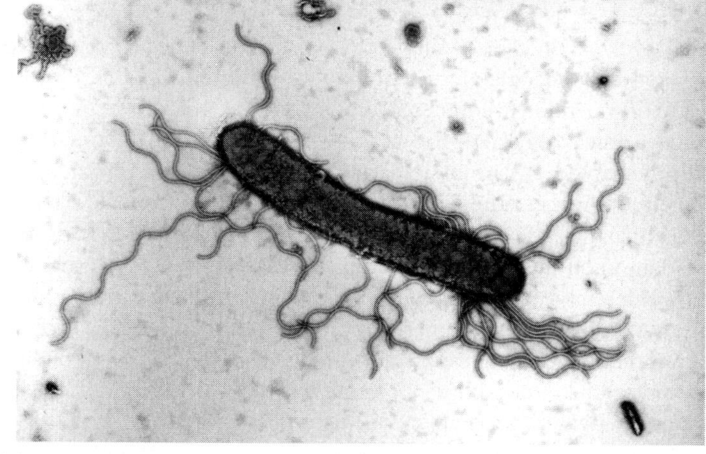

(d)

Figure 4.3

Electron micrographs depicting types of flagellar arrangements. (*a*) Monotrichous flagellum on the predatory bacterium *Bdellovibrio.* (*b*) Lophotrichous flagella on *Vibrio fischeri,* a common marine bacterium (23,000×). (*c*) Unusual flagella on *Aquaspirillum* are amphitrichous (and lophotrichous) in arrangement and coil up into tight loops. (*d*) An unidentified bacterium discovered inside the cells of *Paramecium* exhibits peritrichous flagella.

(b) From Reichelt and Baumann, Arch. Microbiol. *94:283–330. © Springer-Verlag, 1973.*

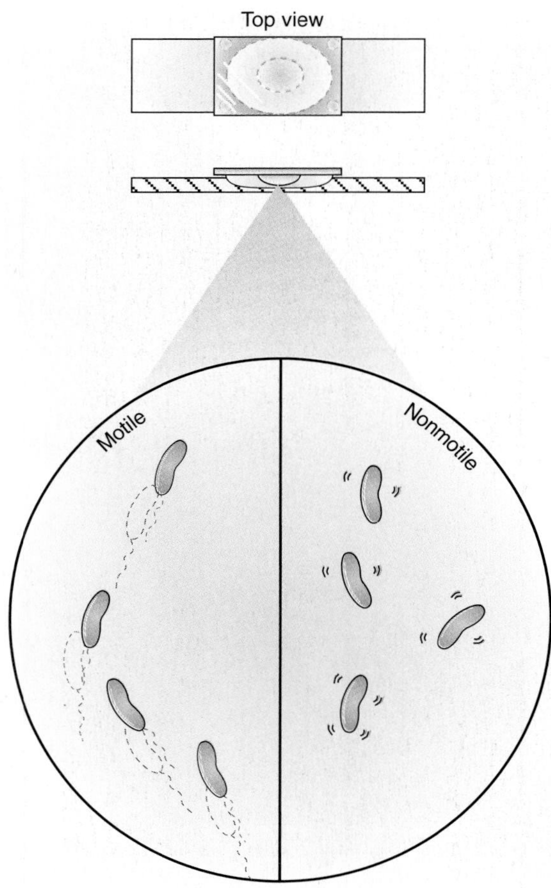

Figure 4.4

A hanging drop slide can be used to detect motility by observing the microscopic behavior of the cell. In true motility, the cell swims and progresses from one point to another. It is assumed that a motile cell has one or more flagella. Nonmotile cells oscillate in the same relative space because of bombardment by molecules, a physical process called Brownian movement.

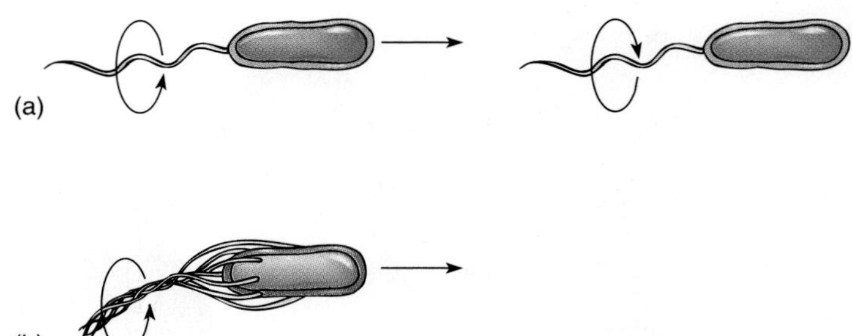

Figure 4.5

The operation of flagella and the mode of locomotion in bacteria with polar and peritrichous flagella. (*a*) In general, when a polar flagellum rotates in a counterclockwise direction, the cell swims forward. When the flagellum reverses direction and rotates clockwise the cell stops. (*b*) In peritrichous forms, all flagella sweep toward one end of the cell and rotate as a single group.

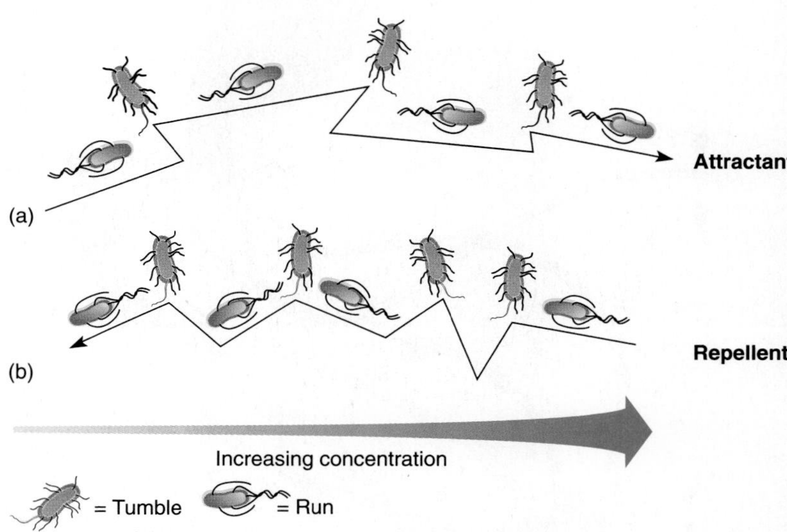

Figure 4.6

Chemotaxis in bacteria. The cell shows a primitive mechanism for progressing (*a*) toward positive stimuli and (*b*) away from irritants by swimming in straight runs or by tumbling. Bacterial runs allow straight, undisturbed progress toward the stimulus, whereas tumbles interrupt progress to allow the bacterium to redirect itself away from the stimulus after sampling the environment.

the flagellum and sets it into rotary motion. If several flagella are present, they become aligned and rotate as a group (figure 4.5). As a flagellum rotates counterclockwise, the cell itself swims in a smooth linear direction toward the stimulus, called a **run.** Runs are interrupted at various intervals by **tumbles,** during which the flagellum reverses direction and causes the cell to stop and change its course. The type of chemotaxis shown at any one time is dependent upon the number of tumbles. It is believed that attractant molecules inhibit tumbles and permit progress toward the stimulus. Repellents cause numerous tumbles, allowing the bacterium to redirect itself away from the stimulus (figure 4.6). Some photosynthetic bacteria exhibit *phototaxis,* a type of movement in response to light rather than chemicals.

Periplasmic Flagella: Internal Flagella

Corkscrew-shaped bacteria called **spirochetes*** show an unusual, wriggly mode of locomotion caused by two or more long, coiled threads, the **periplasmic flagella** or *axial filaments.* A periplasmic flagellum is a type of modified flagellum that consists of a long, thin microfibril inserted into a hook. Unlike regular flagella, however, the entire structure is enclosed in the space between the cell wall and the cell membrane, and for this reason it is also called an endoflagellum (figure 4.7). The filaments curl closely around the spirochete coils yet are free to contract and impart a twisting or flexing motion to the cell. This form of locomotion must be seen in live cells such as the spirochete of syphilis to be truly appreciated.

*spirochete (spy′-roh-keet) Gr. *speira,* coil, and *chaite,* hair.

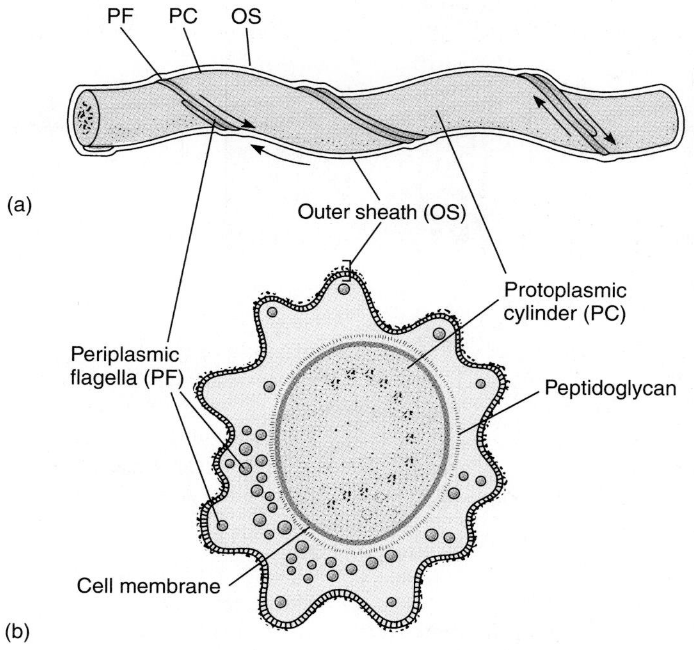

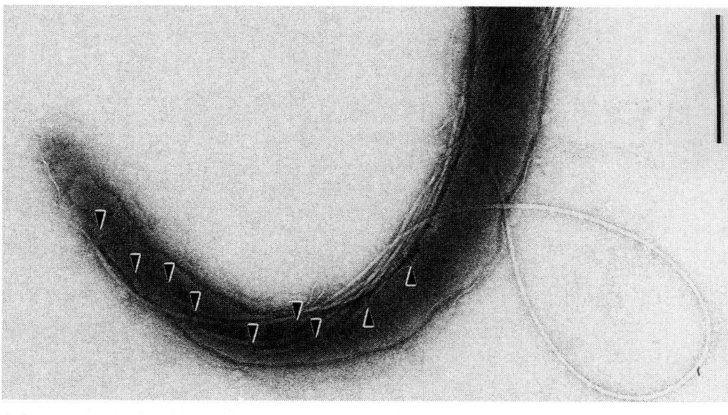

Figure 4.7

The orientation of periplasmic flagella on the spirochete cell. (*a*) Longitudinal section. (*b*) Cross section. Contraction of the filaments imparts a spinning and undulating pattern of locomotion. (*c*) Electron micrograph captures the details of periplasmic flagella and their insertion points (arrows) in *Borrelia burgdorferi*. Bar = 0.2 μm.

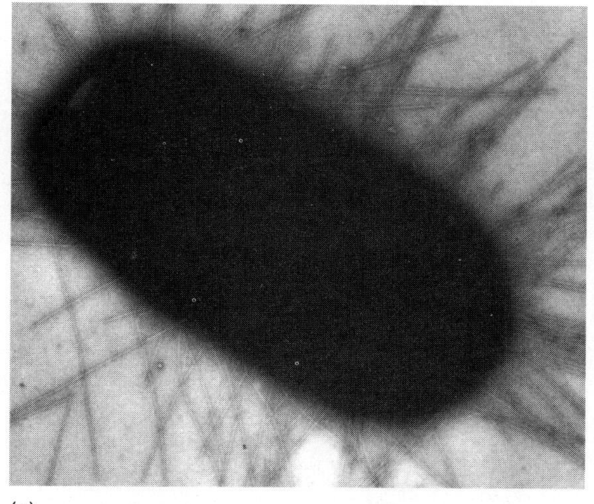

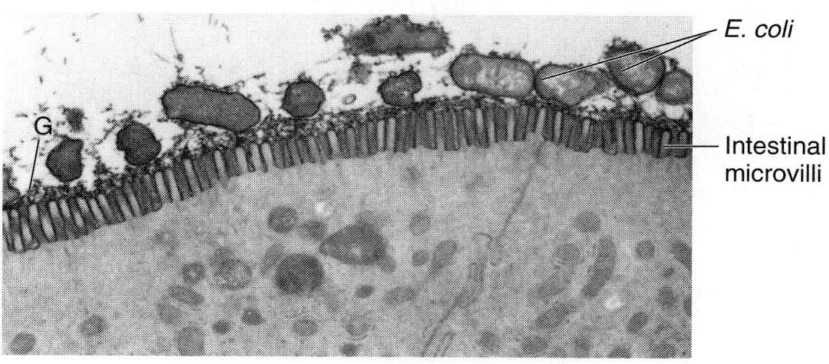

Figure 4.8

Form and function of bacterial fimbriae. (*a*) Single cell of pathogenic *Escherichia coli* is bristling with numerous stiff fibers called fimbriae (40,000×). (*b*) A row of *E. coli* cells tightly adheres by their fimbriae to the surface of intestinal cells (12,000×). This is how the bacterium clings and gains access to the body during an infection. (G = glycocalyx)

Appendages for Attachment and Mating

Customarily, the terms **pilus*** and **fimbria*** have been used interchangeably to indicate any bacterial surface appendage not involved in motility. Because a distinction is helpful, we will use the term *fimbriae* to refer to the shorter, numerous strands and the term *pili* to refer to the longer, sparser appendages.

Fimbriae are small, bristle-like fibers sprouting off the surface of many bacterial cells (figure 4.8). Their exact composition varies, but most of them contain protein. Fimbriae have an inherent tendency to stick to each other and to surfaces. For example, they are often responsible for the mutual clinging of cells that leads to films and other thick aggregates of cells on the surface of liquids and for the microbial colonization of inanimate solids such as rocks and glass (called biofilms). Some pathogens can colonize and infect host tissues because of a tight adhesion between their fimbriae and epithelial cells (figure 4.8*b*). For example, the gonococcus (agent of gonorrhea) invades the genitourinary tract, and *Escherichia coli* invades the intestine by this means. Mutant forms of these pathogens that lack fimbriae are unable to cause infections.

A pilus (also called a *sex pilus*) is an elongate, rigid tubular structure made of a special protein, *pilin*. So far, true pili have been found only on gram-negative bacteria, where they are involved primarily in a mating process between cells called **conjugation,*** which involves partial transfer of DNA from one cell to another (figure 4.9). A pilus from the donor cell unites with a

*pilus (py′-lus) pl. pili; L. hair.
*fimbria (fim′-bree-ah) pl. fimbriae; L. a fringe.

*conjugation (kon-joo-gay′-shun) L. *conjugatus,* linked together.

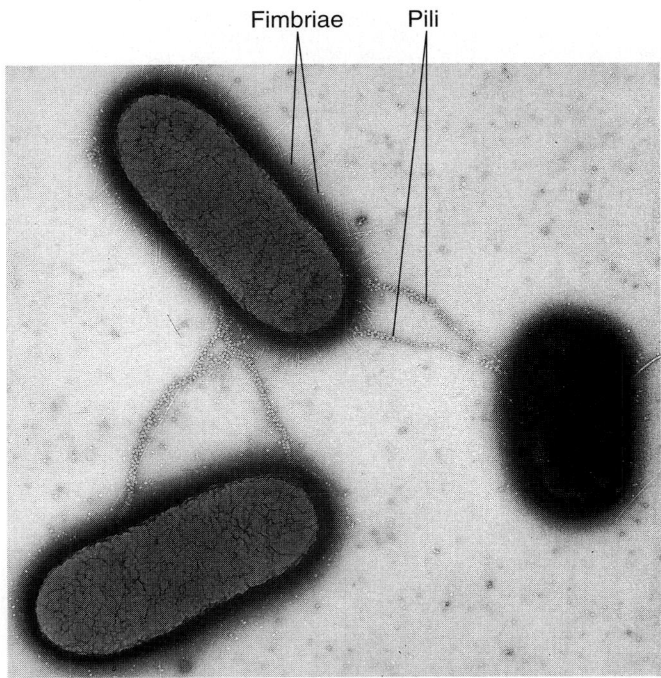

Figure 4.9
Three bacteria in the process of conjugating. Clearly evident are the sex pili forming mutual conjugation bridges between a donor (upper cell) and two recipients (two lower cells). Note the fimbriae on the donor cell and compare their size with the pili.

recipient cell, thereby providing as cytoplasmic connection for making the transfer. Production of pili is controlled genetically, and conjugation takes place only between compatible gram-negative cells. The roles of pili and conjugation are further explored in chapter 9.

The Cell Envelope: The Outer Wrapping of Bacteria

The majority of bacteria have a chemically complex external covering that, for a long time, was simply termed the cell wall. More detailed study eventually revealed that the cell wall may not be the outermost covering and that most cells have surface coatings of various types. The precise anatomy of the bacterial surface and wall is so diverse that we will collectively refer to the complex of layers external to the cell protoplasm as the **cell envelope.** The layers of the envelope are stacked one upon another and are often tightly bonded together like the outer husk and casings of a coconut. The three basic layers that can be identified in electron micrographs are the glycocalyx, the cell wall, and the cell membrane (figure 4.10). Although each envelope layer performs a distinct function, together they act as a single protective unit. The envelope is extensive and can account for one-tenth to one-half of a cell's volume.

The Bacterial Surface Coating, or Glycocalyx

The bacterial cell surface is frequently exposed to severe environmental conditions. The **glycocalyx** develops as a coating of macromolecules to protect the cell and, in some cases, help it adhere to its environment. Glycocalyces differ among bacteria in thickness, organization, and chemical composition. Some bacteria are covered with a loose, soluble shield called a **slime layer** that evidently protects them from loss of water and nutrients (figure 4.11*a*). Other bacteria produce **capsules** of repeating

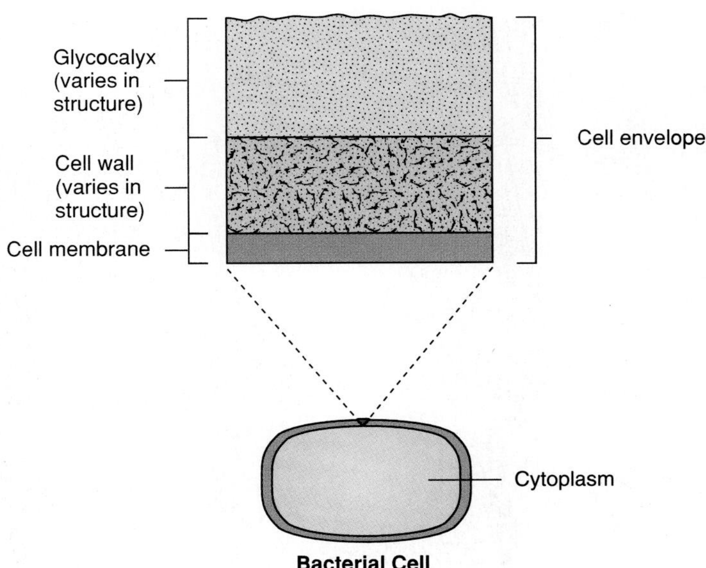

Figure 4.10
The relationship of the three layers of the cell envelope.

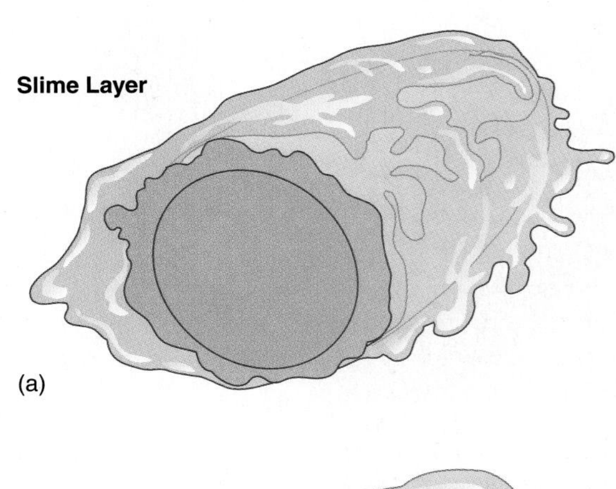

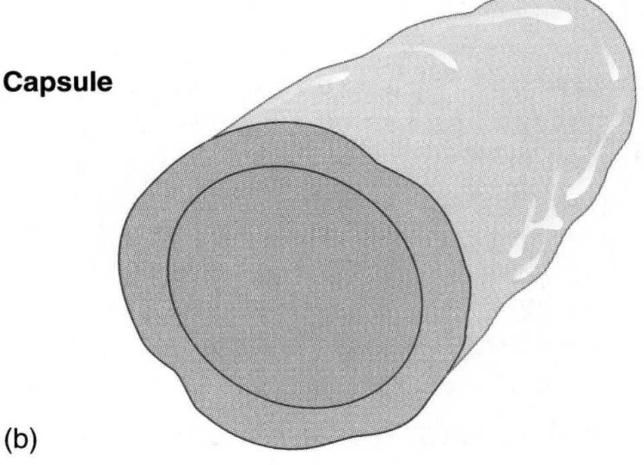

Figure 4.11
Bacterial cells sectioned to show the types of glycocalyces. (*a*) The slime layer is a loose structure that is easily washed off. (*b*) The capsule is a thick, structured layer that is not readily removed.

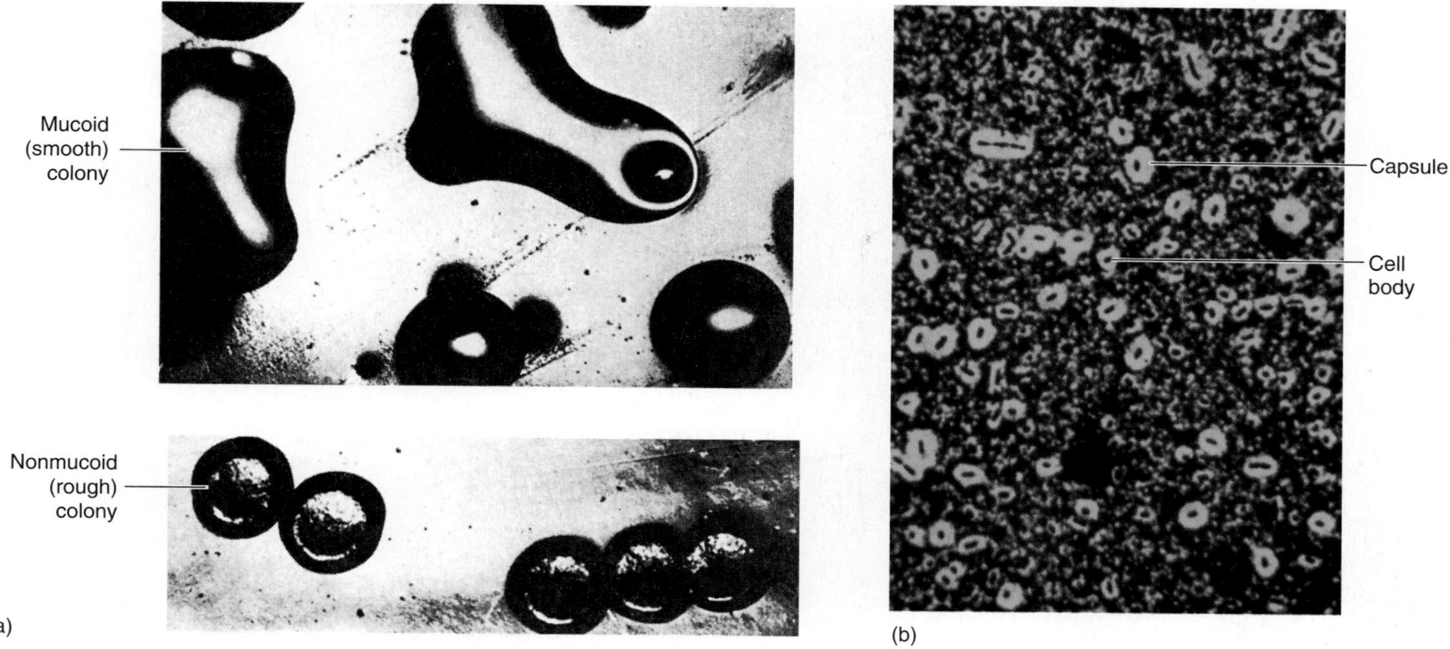

Figure 4.12

(*a*) The appearance of colonies composed of encapsulated cells (mucoid) compared with those lacking capsules (nonmucoid). Even at the macroscopic level, the slippery, gel-like character of the capsule is evident. (*b*) Staining reveals the microscopic appearance of a large, well-developed capsule.

polysaccharide units, of protein, or of both (figure 4.11*b*). A capsule is bound more tightly to the cell than a slime layer is, and it has a thicker, gummy consistency that gives a prominently sticky (mucoid) character to the colonies of most encapsulated bacteria (figure 4.12).

Functions of the Glycocalyx Capsules are formed by several pathogenic species, such as *Streptococcus pneumoniae* (a cause of pneumonia, an infection of the lung), *Haemophilus influenzae* (one cause of meningitis), and *Bacillus anthracis* (the cause of anthrax). Encapsulated bacterial cells generally have greater pathogenicity because capsules protect the bacteria against white blood cells called phagocytes. Phagocytes are a natural body defense that can engulf and destroy foreign cells through phagocytosis, thus preventing infection. A capsular coating makes the bacteria too slippery for the phagocyte to capture or digest. By escaping phagocytosis, the bacteria are free to multiply and infect body tissues. Encapsulated bacteria that mutate to nonencapsulated forms usually lose their pathogenicity.

Other types of glycocalyces can be important in colonization. The thick, white film that forms on teeth (plaque) is formed in part by the surface slimes produced by certain streptococci in the oral cavity. This slime initially allows them to adhere to the teeth and provides a niche for other oral bacteria that, in time, can lead to dental disease. The glycocalyx of some bacteria is so highly adherent that it is responsible for persistent colonization of nonliving materials such as plastic catheters, intrauterine devices, and metal pacemakers that are in common medical use (figure 4.13). Other functions of the glycocalyx are to protect cells from drying and to receive signals from the environment.

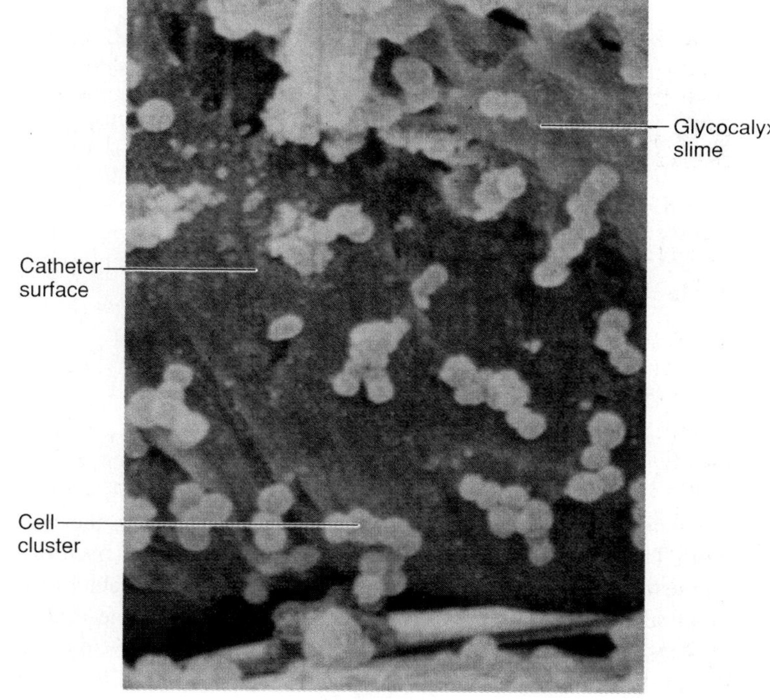

Figure 4.13

Scanning electron micrograph of *Staphylococcus aureus* attached to a catheter.

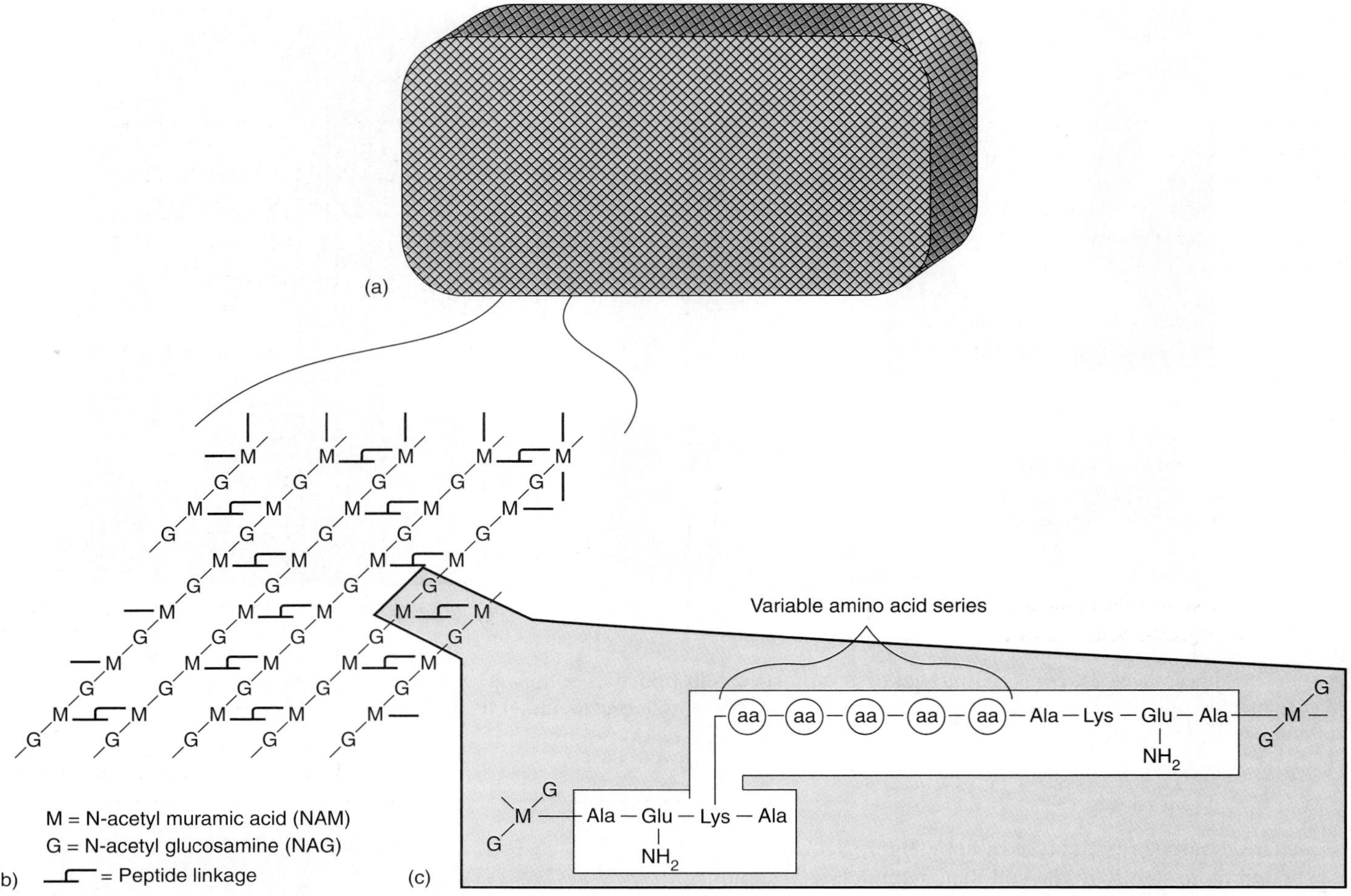

Figure 4.14

General structure of the peptidoglycan component of the cell wall. (*a*) An artist's interpretation of the peptidoglycan meshwork (not to scale). Such is the union of the molecules that the wall is actually one continuous and sometimes many-layered molecule that surrounds the cell like a finely woven basket. (*b*) The structure at the molecular level consists of a backbone of alternating glycan molecules (N-acetyl muramic acid [NAM] and N-acetyl glucosamine [NAG]). These chains are bound together at adjacent NAMs into a meshwork by means of a segment consisting of a short chain of amino acids (peptide). This sort of interlinking pattern confers rigidity yet flexibility. (*c*) The general molecular appearance of a peptide cross-bridge. Its precise structure varies among various groups.

THE CELL WALL: MULTIPURPOSE FRAMEWORK OF THE CELL

Immediately below the glycocalyx lies a second layer, the **cell wall.** This structure accounts for a number of important bacterial characteristics. In general, it determines the shape of a bacterium, and it also provides the kind of strong structural support necessary to keep a bacterium from bursting or collapsing because of changes in osmotic pressure. In this way, the cell wall functions like a bicycle that maintains the necessary shape and prevents the more delicate inner tube from bursting when it is expanded.

The cell walls of most bacteria gain this relatively rigid quality from a unique macromolecule called **peptidoglycan** (PG). This compound is composed of a repeating framework of long *glycan* chains cross-linked by short peptide fragments to provide a strong but flexible support framework (figure 4.14). Peptidoglycan is only one of several materials found in cell walls, and its amount and exact composition vary among the major bacterial groups.

Because many bacteria live in aqueous habitats with a low solute concentration, they are constantly absorbing excess water by osmosis. Were it not for the strength and relative rigidity of the peptidoglycan in the cell wall, they would rupture from internal pressure. Understanding this function of the cell wall has been a tremendous boon to the drug industry. Several types of drugs used to treat infection (penicillin, cephalosporins) are effective because they can destroy the peptidoglycan layer of bacteria. With their cell walls incomplete or missing, such cells have very little protec-

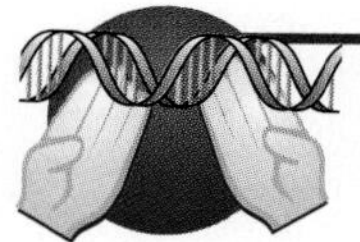

MICROFILE 4.1 THE GRAM STAIN: A GRAND STAIN

In 1884, Hans Christian Gram discovered a staining technique that could be used to make bacteria in infectious specimens more visible. His technique consisted of a timed, sequential application of crystal violet (the primary dye), Gram's iodine (IKI, the mordant), an alcohol rinse (decolorizer), and safranin (the counterstain). In the finished product, bacteria that stained purple were called gram positive and those that stained red were called gram negative.

Although these staining reactions involve an attraction of the cell to a charged dye (see chapter 3), it is important to note that the terms *gram positive* and *gram negative* are not used to indicate the electrical charge of cells or dyes but whether or not a cell retains the primary dye-iodine complex after decolorization. There is nothing specific in the reaction of gram-positive cells to the primary dye or in the reaction of gram-negative cells to the counterstain. The different results in the Gram stain are due to differences in the structure of the cell wall and how it reacts to the series of reagents applied to the cells.

In the first step, crystal violet is attracted to the cells in a smear and stains them all the same purple color. The second and key differentiating step is the addition of the mordant (intensifier)—Gram's iodine. It causes the dye to form large crystals in the peptidoglycan meshwork of the cell wall. Because the peptidoglycan layer in gram-positive cells is thicker, the entrapment of the dye is far more extensive in them than in gram-negative cells. Application of alcohol in the third step acts to reveal this difference by removing the dye from gram-negative cells. It does so by dissolving lipids in the outer membrane and removing the dye from the peptidoglycan layer and the cell itself. By contrast, the crystals of dye tightly embedded in the peptidoglycan of gram-positive bacteria are relatively inaccessible and resistant to removal. Because gram-negative cells are colorless after decolorization, their presence is demonstrated by applying the counterstain safranin in the final step.

	Microscopic Appearance of Cell		Chemical Reaction in Cell Wall (very magnified view)	
Step	Gram (+)	Gram (–)	Gram (+)	Gram (–)
1. Crystal violet			Both cell walls affix the dye	
2. Gram's iodine			Dye crystals trapped in wall	No effect of iodine
3. Alcohol			Crystals remain in cell wall	Cell wall partially dissolved, loses dye
4. Safranin (red dye)			Red dye has no effect	Red dye stains the colorless cell

Gram stain technique and theory.

This century-old staining method remains the universal basis for bacterial classification and identification. It permits differentiation of four major categories based upon color reaction and shape: gram-positive rods, gram-positive cocci, gram-negative rods, gram-negative cocci (see table 4.6). The Gram stain can also be a practical aid in diagnosing infection and in guiding drug treatment. For example, gram staining a fresh urine or throat specimen can help pinpoint the possible cause of infection, and in some cases it is possible to begin drug therapy on the basis of this stain. Even in this day of elaborate and expensive medical technology, the Gram stain remains an important and unbeatable first tool in diagnosis.

tion from **lysis.***[*] Some disinfectants (alcohol, detergents) also kill bacterial cells by damaging the cell wall. Lysozyme, an enzyme contained in tears and saliva, provides a natural defense against certain bacteria by hydrolyzing peptidoglycan.

Differences in Cell Wall Structure

More than a hundred years ago, long before the detailed anatomy of bacteria was even remotely known, a Danish physician named Hans Christian Gram developed a staining technique, the **Gram stain,** that delineates two generally different groups of bacteria (microfile 4.1). We now know that the contrasting staining reactions that occur with this stain are due entirely to some very fundamental differences in the structure of bacterial cell walls. The two major groups shown by this technique are the gram-positive bacteria and the gram-negative bacteria. Because the Gram stain does not actually reveal the nature of these physical differences, we must turn to the electron microscope and to biochemical analysis.

The extent of the differences between gram-positive and gram-negative bacteria is evident in the physical appearance of

[*] lysis (ly'-sis) Gr., to loosen. A process of cell destruction, as occurs in bursting.

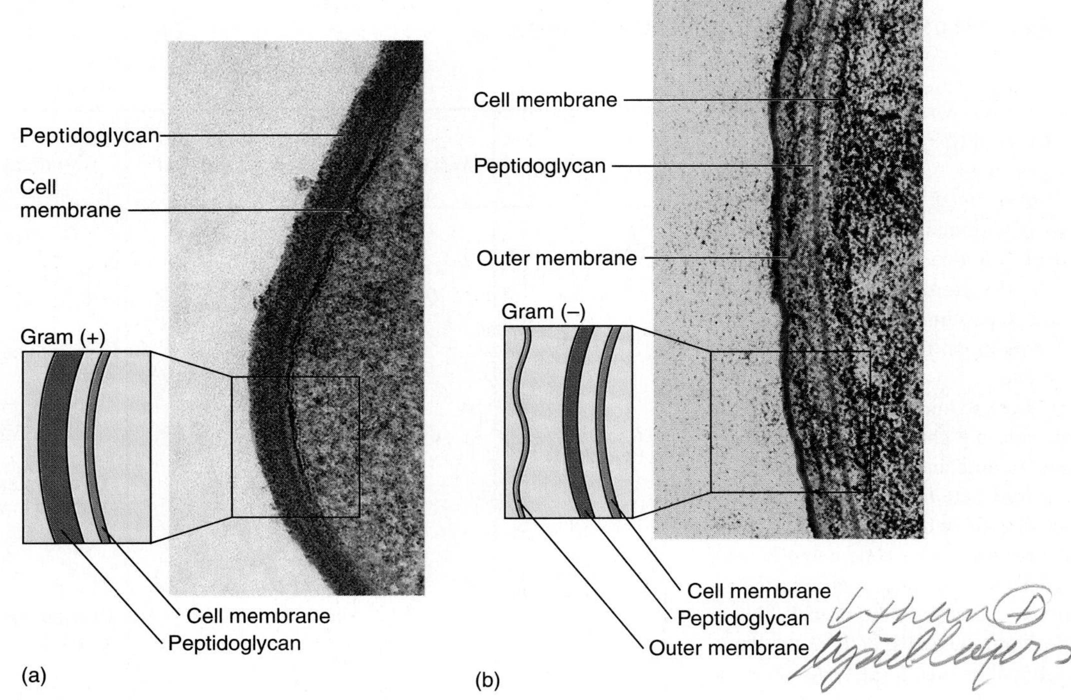

Figure 4.15
A comparison of the envelopes of gram-positive and gram-negative cells. (*a*) A photomicrograph of a gram-positive cell wall/membrane and an artist's interpretations of its open-faced sandwich–style layering with two layers. (*b*) A photomicrograph of a gram-negative cell wall/membrane and an artist's interpretation of its complete sandwich–style layering with three distinct layers.

their cell envelopes (figure 4.15). In gram-positive cells, a microscopic section resembles an open-faced sandwich with two layers: the thick outer cell wall, composed primarily of peptidoglycan, and the cell membrane. A similar section of a gram-negative cell envelope shows a complete sandwich with three layers; the cell wall, composed of an outer membrane and a thin layer of peptidoglycan, and the cell membrane. See table 4.1 for a further comparison of cell wall types.

The Gram-Positive Cell Wall The bulk of the gram-positive cell wall is a thick, homogeneous sheath of peptidoglycan ranging from 20 to 80 nm in thickness. It also contains tightly bound acidic polysaccharides, including techoic acid and lipotechoic acid (figure 4.16). Techoic acid is a polymer of ribitol or glycerol and phosphate embedded in the peptidoglycan sheath. Lipotechoic acid is similar in structure but is attached to the lipids in the plasma membrane. These molecules appear to function in cell wall maintenance and enlargement during cell division, and they also contribute to the acidic charge on the cell surface. In some cases, the cell wall of gram-positive bacteria is pressed tightly against the cell membrane with very little space between them, but in other cells, a thin **periplasmic* space** is evident between the cell membrane and cell wall.

The Gram-Negative Cell Wall The gram-negative cell wall is more complex in morphology because it contains an **outer membrane** (OM), has a thinner shell of peptidoglycan, and has an extensive space surrounding the peptidoglycan (figure 4.16). The outer membrane is somewhat similar in construction to the cell membrane, except that it contains specialized types of polysaccharides and proteins. The uppermost layer of the OM contains *lipopolysaccharide* (LPS), which is essentially a polysaccharide fragment integrated into the membrane lipids. The innermost layer of the OM is another lipid layer anchored by means of proteins to

TABLE 4.1

COMPARISON OF GRAM-POSITIVE AND GRAM-NEGATIVE CELL WALLS

Characteristic	Gram-Positive	Gram-Negative
Number of major layers	1	2
Chemical composition	Peptidoglycan Techoic acid Lipotechoic acid	Lipopolysaccharide Lipoprotein Peptidoglycan
Overall thickness	Thicker (20–80 nm)	Thinner (8–11 nm)
Outer membrane	No	Yes
Periplasmic space	Narrow	Extensive
Porin proteins	No	Yes
Permeability to molecules	More penetrable	Less penetrable

*periplasmic (per''-ih-plaz'-mik) Gr. *peri,* around, and *plastos,* formed.

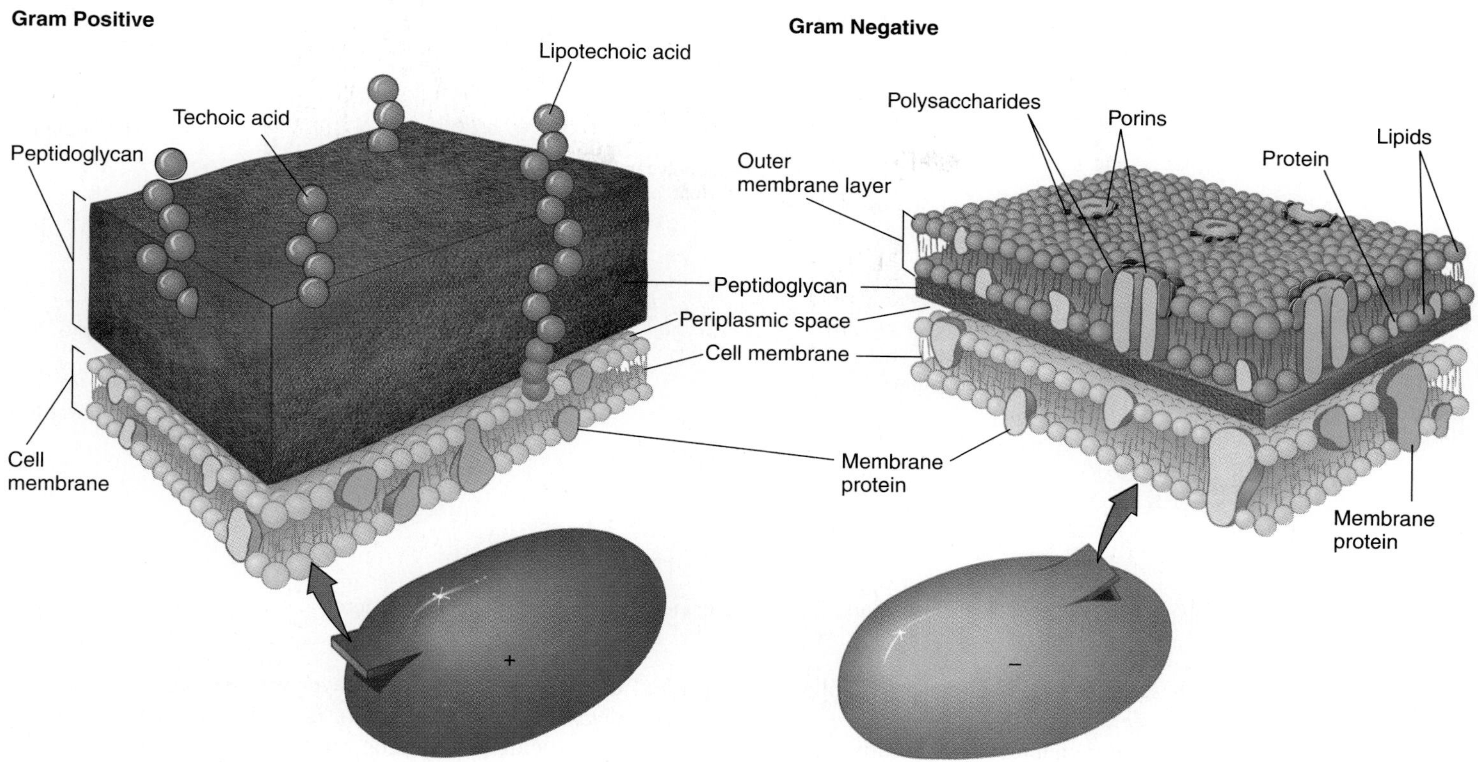

Figure 4.16
A comparison of the detailed structure of gram-positive and gram-negative cell walls.

the peptidoglycan layer below. The outer membrane serves as a partial chemical sieve by allowing only relatively small molecules to penetrate. Access is provided by special membrane channels formed by *porin proteins* that completely span the outer membrane. The size of these porins can be altered so as to block the entrance of harmful chemicals, making them one defense of gram-negative bacteria against certain antibiotics (see chapter 12).

The bottom layer of the gram-negative wall is a single, thin (1–3 nm) sheet of peptidoglycan. Although it acts as a somewhat rigid protective structure as previously described, its thinness gives gram-negative bacteria a relatively greater flexibility and sensitivity to lysis. There is a well-developed *periplasmic space* surrounding the peptidoglycan. This space is an important reaction site for a large and varied pool of substances that enter and leave the cell.

Practical Considerations of Differences in Cell Wall Structure Variations in cell wall anatomy contribute to several other differences between the two cell types. The outer membrane contributes an extra barrier in gram-negative bacteria that makes them more impervious to some antimicrobic chemicals such as dyes and disinfectants, so they are generally more difficult to inhibit or kill than are gram-positive bacteria. One exception is for alcohol-based compounds, which can dissolve the lipids in the outer membrane and disturb its integrity. Treating infections caused by gram-negative bacteria often requires different drugs from gram-positive infections, especially drugs that can cross the outer membrane.

The cell wall or its parts can interact with human tissues and contribute to disease. Gram-negative lipopolysaccharides called *endotoxins* are a complicating factor, causing fever and shock in meningitis and typhoid fever (see microfile 20.1). Proteins attached to the outer portion of the cell wall of several gram-positive species, including *Corynebacterium diphtheriae* (the agent of diphtheria) and *Streptococcus pyogenes* (the cause of strep throat), also have toxic properties. The lipids in the cell walls of certain *Mycobacterium* species are harmful to human cells as well. Because most macromolecules in the cell walls are foreign to humans, they stimulate antibody production by the immune system (see chapter 15).

Exceptions in the Cell Wall

Several bacterial groups lack the cell wall structure of gram-positive or gram-negative bacteria, and some bacteria have no cell wall at all. Although these exceptional forms can stain positive or negative in the Gram stain, examination of their fine structure and chemistry shows that they do not really fit the descriptions for typical gram-negative or -positive cells. For example, the cells of *Mycobacterium* and *Nocardia* contain peptidoglycan and stain gram positive, but the bulk of their cell wall is composed of unique types of lipids. One of these is a very-long-chain fatty acid called *mycolic acid,* or cord factor, that contributes to the pathogenicity of this group (see chapter 19). The thick, waxy nature imparted to the cell wall by these lipids is also responsible for a high degree of resistance to certain chemicals and dyes. Such resistance is the basis for the **acid-fast stain** used to diagnose tuberculosis and leprosy. In this stain, hot carbol fuchsin dye becomes tenaciously

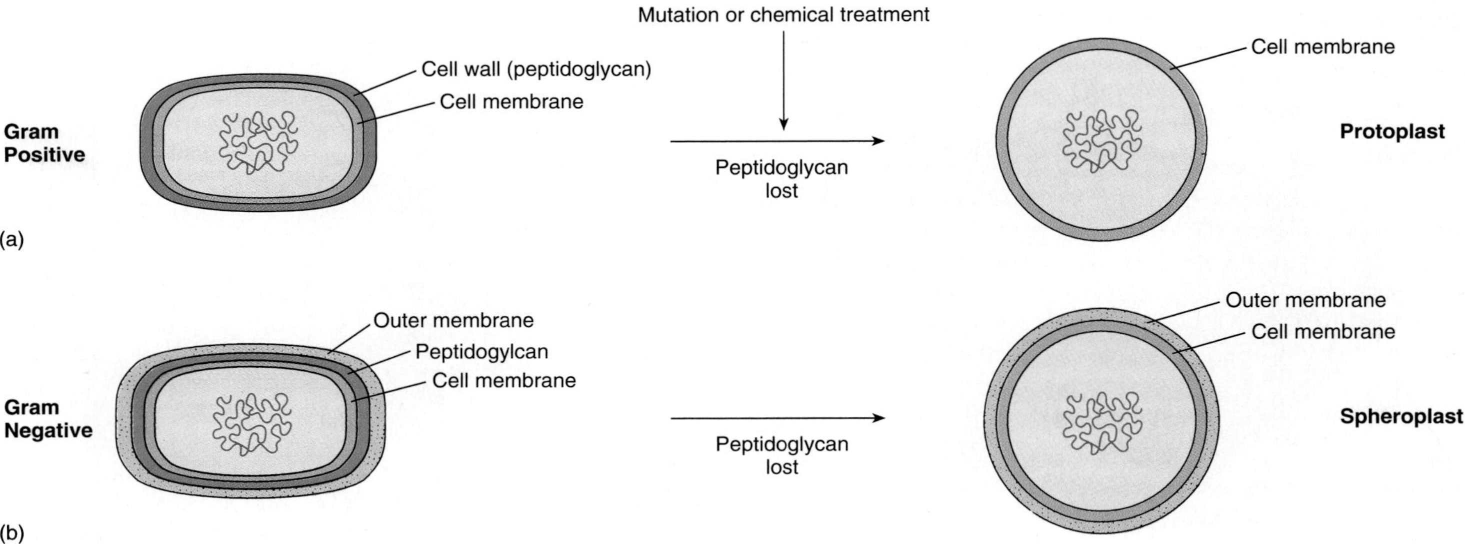

Figure 4.17
The conversion of walled bacterial cells to L forms in (*a*) gram-positive bacteria and (*b*) gram-negative bacteria.

attached (is held fast) to these cells so that an acid-alcohol solution will not remove the dye (see chapter 3).

Because they are from a more ancient and primitive line of procaryotes, the archaea exhibit unusual and chemically distinct cell walls. In some, the walls are composed almost entirely of polysaccharides, and in others, the walls are pure protein; but as a group, they all lack the true peptidoglycan structure described previously. Since a few archaea and all mycoplasmas (discussed in a later section of this chapter) lack a cell wall entirely, their cell membrane must serve the dual functions of protection as well as transport.

Some bacteria that ordinarily have a cell wall can lose it during part of their life cycle. These wall-deficient forms are referred to as **L forms** or L-phase variants (for the Lister Institute, where they were discovered). L forms arise naturally from a mutation in the wall-forming genes, or they can be induced artificially by treatment with a chemical such as lysozyme or penicillin that disrupts the cell wall. When a gram-positive cell is exposed to either of these two chemicals, it will lose the cell wall completely and become a **protoplast,*** a fragile cell bounded only by a membrane that is highly subject to lysis (figure 4.17*a*). A gram-negative cell exposed to these same substances loses its peptidoglycan but retains its outer membrane, leaving a less fragile but nevertheless weakened **spheroplast*** (figure 4.17*b*). Evidence points to a role for L forms in certain infections (see chapter 21).

The Cell Membrane: The Multipurpose Integument

Appearing just beneath the cell wall is the **cell,** or **cytoplasmic, membrane,** a very thin (5–10 nm), flexible sheet molded completely around the cytoplasm. Its general composition was described in chapter 2 as a lipid bilayer with proteins embedded to varying degrees (see microfile 2.3). Bacterial cell membranes have this typical structure, containing primarily phospholipids (making up about 30–40% of the membrane mass) and proteins (contributing 60–70%). Major exceptions to this description are the membranes of mycoplasmas, which contain high amounts of sterols—rigid lipids that stabilize and reinforce the membrane—and the membranes of archaea, which contain unique branched hydrocarbons rather than fatty acids.

In some locations, the cell membrane extends inwardly into coiled passages, or sacs, in the cytoplasm called **mesosomes*** (see figure 4.1). These are prominent in gram-positive bacteria but are harder to see in gram-negative bacteria because of their relatively small size. Mesosomes presumably increase the internal surface area available for membrane activities, although whether they are artifacts of fixation or true functioning structures continues to be a controversial topic. One of the convincing observations that indicates they are real and not artifacts is that specialized procaryotes such as cyanobacteria contain dense stacks of internal membranes (see figure 4.34). Some of the proposed functions of mesosomes in nonphotosynthetic bacteria are to participate in cell wall synthesis and to guide the duplicated chromatin bodies (bacterial chromosomes) into the two daughter cells during cell division (see figure 7.15).

Functions of the Cell Membrane

Since bacteria have none of the eucaryotic organelles, the cell membrane provides a site for functions such as energy reactions, nutrient processing, and synthesis. A major action of the cell membrane is to regulate *transport,* that is, the passage of nutrients into the cell and the discharge of wastes. Although water and small uncharged molecules can diffuse across the membrane unaided, the membrane is a *selectively permeable* structure with spe-

*protoplast (proh′-toh-plast) Gr. *proto,* first, and *plastos,* formed.
*spheroplast (sfer′-oh-plast) Gr. *sphaira,* sphere.

*mesosome (mes′-oh-sohm) Gr. *mesos,* middle, and *soma,* body.

cial carrier mechanisms for passage of most molecules (see chapter 7). The glycocalyx and cell wall can bar the passage of large molecules, but they are not the primary transport apparatus. The cell membrane is also involved in *secretion,* or the discharge of a metabolic product into the extracellular environment.

The membranes of procaryotes are an important site for a number of metabolic activities. It is here that most enzymes of respiration and other energy-processing activities are located (see chapter 8). Enzyme systems located in the cell membrane also help synthesize structural macromolecules to be incorporated into the cell envelope and appendages. Other products (enzymes and toxins) are secreted by the membrane into the extracellular environment.

Chapter Checkpoints

Bacteria are the oldest form of cellular life. They are also the most widely dispersed, occupying every conceivable microclimate on the planet.

The appendages of bacteria provide motility (flagella), attachment (pili and fimbriae), and, in some bacteria, a means of DNA transfer (sex pili).

Flagella vary in number and arrangement as well as in the type and rate of motion they produce.

The cell envelope is the outermost covering of bacteria. It consists of three basic layers: (1) the glycocalyx, (2) the cell wall, and (3) the cell membrane.

The composition of the procaryotic cell wall is used to classify bacteria into four major divisions: gram-positive bacteria, gram-negative bacteria, bacteria with no cell walls, and bacteria with chemically unique cell walls.

Gram-positive bacteria retain the crystal violet and stain purple. Gram-negative bacteria lose the crystal violet and stain red from the safranin counterstain.

Gram-positive bacteria have thick cell walls of peptidoglycan and acidic polysaccharides such as techoic acid, and they have a thin periplasmic space. The cell walls of gram-negative bacteria are thinner but contain an additional outer membrane and have a wide periplasmic space.

The bacterial cell membrane is typically composed of phospholipids and proteins, and it performs many metabolic functions.

The mesosome is an extension of the cell membrane. It is thought to participate in cell wall synthesis and cell division.

BACTERIAL FORM AND FUNCTION: INTERNAL STRUCTURE

CONTENTS OF THE CELL CYTOPLASM

Encased by the cell membrane is a dense, gelatinous solution referred to as **cytoplasm,** which is another prominent site for many of the cell's biochemical and synthetic activities. Its major component is water (70–80%), which serves as a solvent for the **cell**

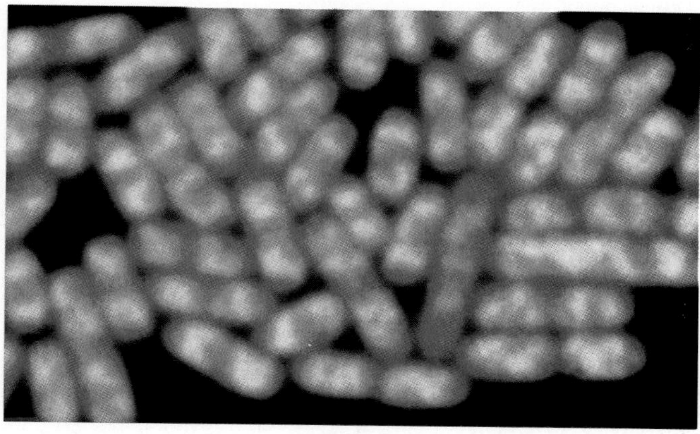

Figure 4.18

Fluorescent staining highlights the chromatin bodies (chromosomes) of the bacterial pathogen *Salmonella enteriditis.* The cytoplasm is orange, and the chromatin body fluoresces bright yellow. Some bacteria appear to have more than one chromatin body because they are in the process of dividing.

pool, a complex mixture of nutrients including sugars, amino acids, and salts. The components of this pool serve as building blocks for cell synthesis or as sources of energy. The cytoplasm also contains larger, discrete cell masses such as the chromatin body, ribosomes, mesosomes, and granules.

Chromatin Bodies and Plasmids: The Sources of Genetic Information

The hereditary material of bacteria exists in the form of a single circular strand of DNA designated as the **chromatin* body,** or bacterial chromosome (see figure 4.1). By definition, bacteria do not have a nucleus; that is, their DNA is not enclosed by a nuclear membrane but instead is aggregated in a dense area of the cell called the **nucleoid.*** The chromatin body is actually an extremely long molecule of DNA that is tightly coiled around special basic protein molecules so as to fit inside the cell compartment (see microfile 9.3). Arranged along its length are genetic units (genes) that carry information required for bacterial maintenance and growth. When exposed to special stains or observed with an electron microscope, chromatin bodies have a granular of fibrous appearance (figure 4.18). Because bacteria have a single chromosome, they are haploid.

Although the chromatin body is the minimal genetic requirement for bacterial survival, many bacteria contain other, nonessential pieces of DNA called **plasmids.*** These tiny, circular extra-chromosomal strands can be free or integrated into the chromosome; they are duplicated and passed on to offspring. They are not essential to bacterial growth and metabolism, but they often confer protective traits such as resisting drugs and producing

*chromatin (kroh′-mah-tin) Gr. *chromato,* color, and *in,* fiber.
*nucleoid (noo′-klee-oid) L. *nucis,* nut, and *oid,* form or like.
*plasmid (plaz′-mid) Gr. *plasma,* a form, and *id,* belonging to.

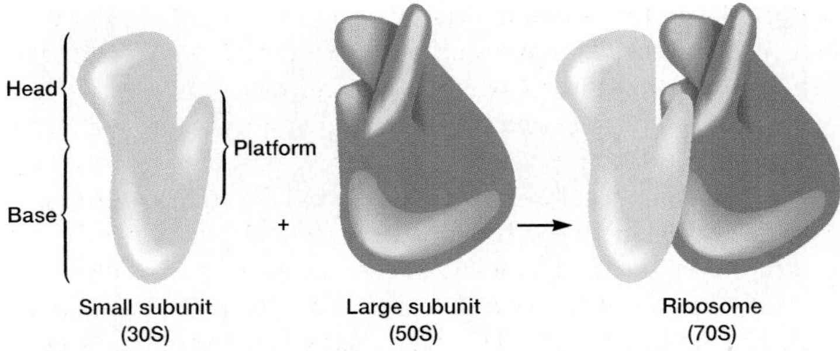

Figure 4.19

A model of a procaryotic ribosome, showing the small (30S) and large (50S) subunits, both separate and joined.

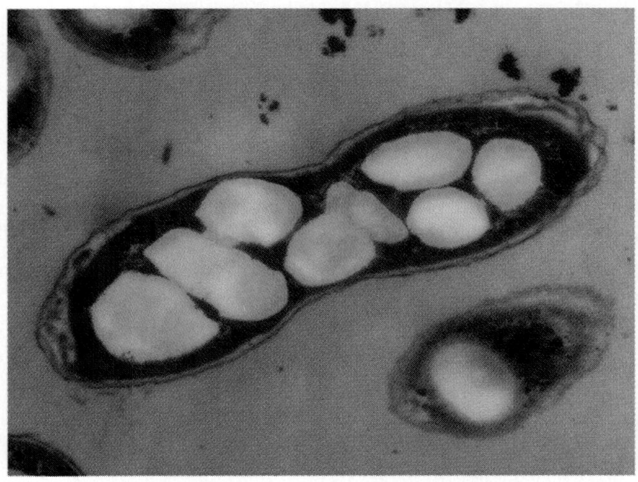

Figure 4.20

An example of a storage inclusion in a bacterial cell (32,500×). Substances such as polyhydroxybutyrate can be stored in an insoluble, concentrated form that provides an ample, long-term supply of that nutrient.

toxins and enzymes (see chapter 9). Because they can be readily manipulated in the laboratory and transferred from one bacterial cell to another, plasmids are an important agent in modern genetic engineering techniques.

Ribosomes: Sites of Protein Synthesis

A bacterial cell contains thousands of tiny, discrete units called **ribosomes.*** When viewed even by very high magnification, ribosomes show up as fine, spherical specks dispersed throughout the cytoplasm that often occur in chains (polysomes). They are also attached to the cell membrane and to mesosomes. Chemically, a ribosome is a combination of a special type of RNA called ribosomal RNA, or rRNA (about 60%), and protein (40%). One method of characterizing ribosomes is by S, or Svedberg,[2] units, which rate the molecular sizes of various cell parts that have been spun down and separated by molecular weight and shape in a centrifuge. Heavier, more compact structures sediment faster and are assigned a higher S rating. Combining this method of analysis with high-resolution electron micrography has revealed that the procaryotic ribosome, which has an overall rating of 70S, is actually composed of two smaller subunits. The 30S unit looks something like a heart; the 50S unit bears a resemblance to a crown (figure 4.19). They fit together to form a miniature platform upon which protein synthesis is performed. We examine the more detailed functions of ribosomes in chapter 9.

Inclusions, or Granules: Storage Bodies

Most bacteria are exposed to severe shifts in the availability of food. During periods of nutrient abundance, they compensate by laying down nutrients intracellularly in **inclusion bodies,** or **inclusions,*** of varying size, number, and content. As the environmental source of these nutrients becomes depleted, the bacterial cell can mobilize its own storehouse as required. Some inclusion bodies enclose condensed, energy-rich organic substances, including glycogen and poly β-hydroxy-butyrate (PHB), within special single-layered membranes (figure 4.20). A unique type of inclusion found in some aquatic bacteria are gas vesicles that provide buoyancy and flotation. Other inclusions, also called granules, contain crystals of inorganic compounds and are not enclosed by membranes. Sulfur granules of photosynthetic bacteria (see figure 4.35*b*) and polyphosphate granules of *Corynebacterium* (see figure 4.22) and *Mycobacterium* are of this type. The latter represent an important source of building blocks for nucleic acid and ATP synthesis. They have been termed **metachromatic* granules** because they stain a contrasting color (red, purple) in the presence of methylene blue dye.

BACTERIAL ENDOSPORES: RESISTANCE IN THE EXTREME

Ample evidence indicates that the anatomy of bacteria helps them adjust rather well to adverse habitats. But of all microbial structures, nothing can compare to the bacterial **endospore*** (or simply spore) for withstanding hostile conditions and facilitating survival. The word *spore* can have more than one usage in microbiology. It is a generic term that refers to any tiny compact cells that are produced by vegetative or reproductive structures of microorganisms. Spores can be quite variable in origin, form, and function. The bacterial type discussed here is called an endospore because it is produced inside a cell. It functions in *survival,* not in reproduction, because no increase in cell numbers is involved in its formation. In contrast, the fungi produce many different types of spores for both survival and reproduction (see chapter 5).

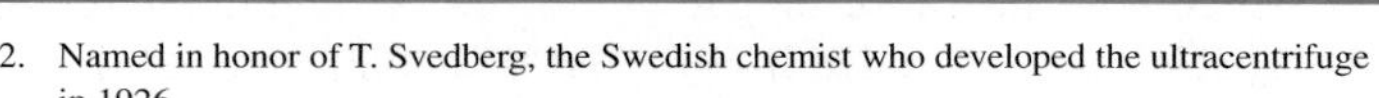

2. Named in honor of T. Svedberg, the Swedish chemist who developed the ultracentrifuge in 1926.

*ribosome (reye'-boh-sohm) Gr. *ribose,* a pentose sugar, and *some,* body.

*inclusion (in-kloo'-zhun). Any substance held in the cell cytoplasm in an insoluble state.

*metachromatic (met''-uh-kroh-mah'-tik) Gr. *meta,* other, and *chromo,* color.

*endospore (en'-doh-spor) Gr. *endo,* inside, and *sporos,* seed.

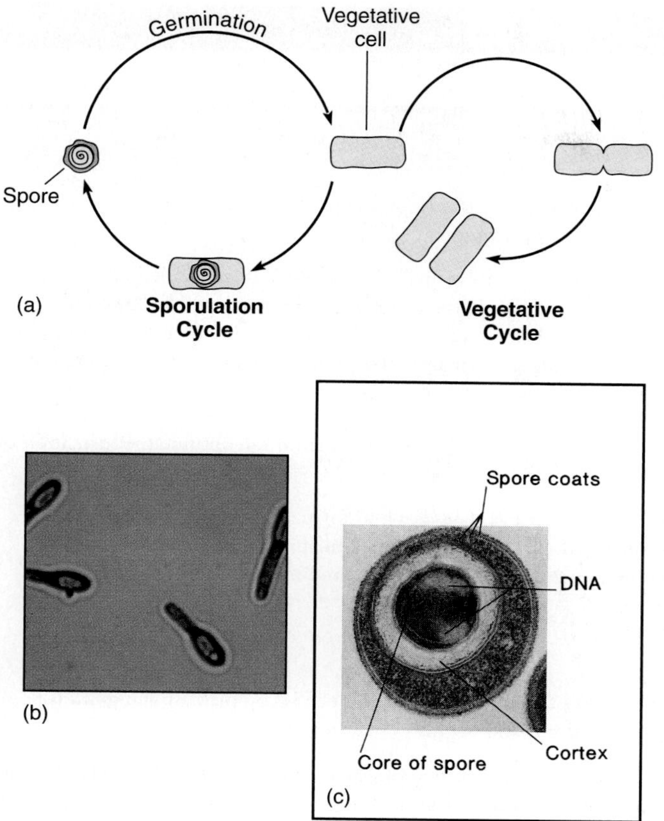

Figure 4.21
(*a*) The general life cycle of a spore-forming bacterium. Sporeformers can exist in an active, growing, vegetative state, or they can enter a dormant survival state through sporulation. Their actual fate depends upon the availability of nutrients. (*b*) A Gram stain of *Clostridium* shows the distinctive appearance of spores swelling up the end of the sporangium. (*c*) A cross section of a single spore, revealing numerous layers.

Endospores are dormant bodies produced by the gram-positive genera *Bacillus, Clostridium,* and *Sporosarcina.* These bacteria have a two-phase life cycle—a vegetative cell and an endospore (figure 4.21*a*). The vegetative cell is a metabolically active and growing entity that can be induced by genetic signals to undergo spore formation, or **sporulation.** Once formed, the spore exists in an inert, resting condition that shows up prominently in a spore or Gram stain (figure 4.21*b*). Spores also stand out in a fresh hanging drop preparation because of their *refractility** (see figure 3.22*a*). Features of spores, including size, shape, and position in the vegetative cell, are somewhat useful in identifying some species.

Endospore Formation and Resistance

The depletion of nutrients, especially an adequate carbon or nitrogen source, is the stimulus for a vegetative cell to begin spore formation. Once this stimulus has been received by the vegetative cell, it undergoes a conversion to a committed sporulating cell called a **sporangium.*** Complete transformation of a vegetative cell into a sporangium and then into a spore requires 6 to 8 hours in most spore-forming species. Table 4.2 illustrates some major physical and chemical events in this process. Bacterial endospores are the hardiest of all life forms, capable of withstanding extremes in heat, drying, freezing, radiation, and chemicals that would readily kill vegetative cells. Their survival under such harsh conditions is due to several factors. The heat resistance of spores has been linked to their high content of calcium and *dipicolinic acid,* although the exact role of these chemicals is not yet clear. We know, for instance, that heat destroys cells by inactivating proteins and DNA and that this process requires a certain amount of water in the protoplasm. Because the deposition of calcium dipicolinate in the spore removes water and leaves the spore very dehydrated, it is less vulnerable to the effects of heat. Moreover, since the spore is already very dehydrated, it is also metabolically inactive and highly resistant to damage from further drying. The thick, impervious cortex and spore coats also protect against radiation and chemicals (figure 4.21*c*). The longevity of bacterial spores verges on immortality. One record describes the isolation of viable spores from a 3,000-year-old archeological specimen, and more recently a California researcher claims to have isolated living spores from a 25-million-year-old fossilized bee! Initial analysis of this ancient microbe indicates it is a species of *Bacillus* that is genetically unique.

The Germination of Endospores

After lying in a state of inactivity for an indefinite time, spores can be revitalized when favorable conditions arise. The breaking of dormancy, or germination, happens in the presence of water and a specific chemical or environmental stimulus (germination agent). Once initiated, it proceeds to completion quite rapidly ($1\frac{1}{2}$ hours). Although the specific germination agent varies among species, it is generally a small organic molecule such as an amino acid or an inorganic salt. This agent stimulates the formation of hydrolytic (digestive) enzymes by the spore membranes. These enzymes digest the cortex and expose the core to water. As the core rehydrates and takes up nutrients, it begins to grow out of the spore coats. In time, it reverts to a fully active vegetative cell, resuming the vegetative cycle.

Practical Significance of Bacterial Spores

Although the majority of spore-forming bacteria are relatively harmless, several bacterial pathogens are sporeformers. In fact, some aspects of the diseases they cause are related to the persistence and resistance of their spores. *Bacillus anthracis* is the agent of anthrax, a skin and lung infection of domestic animals transmissible to humans. The genus *Clostridium* includes even more pathogens, including *C. tetani,* the cause of tetanus (lockjaw), and *C. perfringens,* the cause of gas gangrene. When the spores of these species are embedded in a wound that contains dead tissue,

*refractility (ree-frak-til′-ih-tee) A property of causing light rays to bend, thus creating differences in contrast.

*sporangium (spor-anj′-yum) L *sporos,* and Gr. *angeion,* vessel.

TABLE 4.2

GENERAL STAGES IN ENDOSPORE FORMATION

Stage		State of Cell	Process/Event
1		Vegetative cell	Cell in early stage of binary fission doubles chromatin body.
2		Vegetative cell becomes **sporangium** in preparation for sporulation	One chromatin body and a small bit of cytoplasm are walled off as a protoplast at one end of the cell. This **core** contains the minimum structures and chemicals necessary for guiding life processes. During this time, the sporangium remains active in synthesizing compounds required for spore formation.
3		Sporangium	The protoplast is engulfed by the sporangium to continue the formation of various protective layers around it.
4		Sporangium with **prospore**	Special peptidoglycan is laid down to form a cortex around the spore protoplast, now called the prospore; calcium and dipicolinic acid are deposited; core becomes dehydrated and metabolically inactive.
5		Sporangium with prospore	Three heavy and impervious protein spore coats are added.
6		Mature spore	Spore becomes thicker, and heat resistance is complete; sporangium is no longer functional and begins to deteriorate.
7		Free spore	Complete lysis of sporangium frees spore; it can remain dormant yet viable for thousands of years.
8		Germination	Addition of nutrients and water reverses the dormancy.
9		Vegetative cell	The spore then swells and liberates a young vegetative cell.

they can germinate, grow, and release potent toxins. Another toxin-forming species, *C. botulinum,* is the agent of botulism, a deadly form of food poisoning.

Because they inhabit the soil and dust, spores are a constant intruder where sterility and cleanliness are important. They can resist ordinary cleaning methods that use boiling water, soaps, and disinfectants, and they present problems in the microbiology lab as frequent contaminants of cultures and media. Hospitals and clinics must take precautions to guard against the potential harmful effects of spores in wounds. Spore destruction is a particular concern of the food-canning industry. Several endospore-forming species cause food spoilage or poisoning. Ordinary boiling (100°C) will usually not destroy such spores, so canning is carried out in pressurized steam at 120°C for 20–30 minutes. Such rigorous conditions will ensure that the food is sterile and free from viable bacteria.

Chapter Checkpoints

The cytoplasm of bacterial cells serves as a solvent for materials (the cell pool) used in all cell functions.

The genetic material of bacteria is DNA. Genes are arranged in a circular chromosome. Additional genes are carried on plasmids.

Bacterial ribosomes are dispersed in the cytoplasm in chains (polysomes) and are also embedded in both the cell membrane and the mesosome.

Bacteria store nutrients in their cytoplasm in structures called inclusions. Inclusions vary in structure and the materials that are stored.

A few families of bacteria produce dormant bodies called endospores, which are the hardiest of all life forms, surviving for centuries.

The genera *Bacillus* and *Clostridium* are sporeformers, and both contain deadly pathogens.

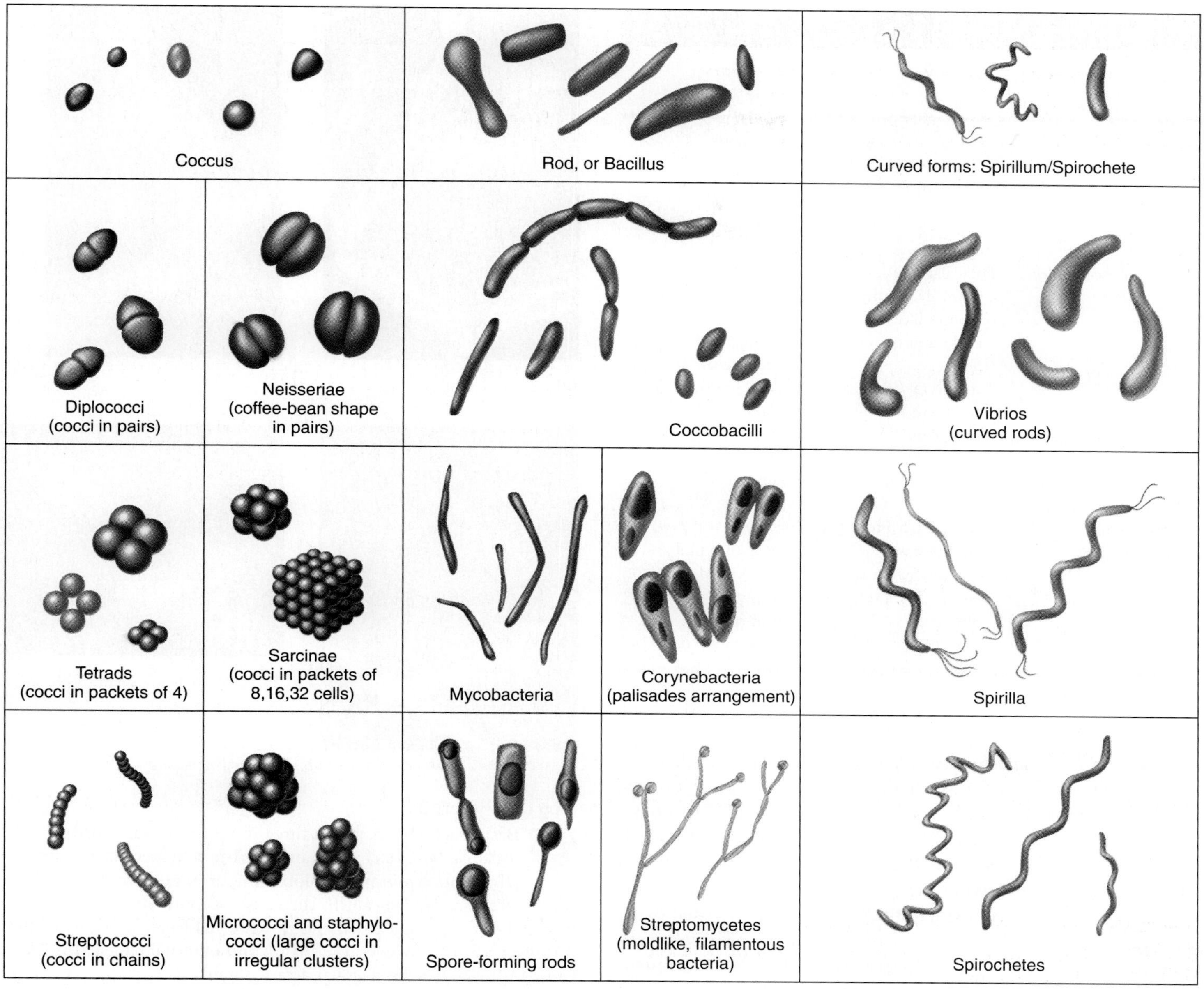

Coccus
Rod, or Bacillus
Curved forms: Spirillum/Spirochete
Diplococci (cocci in pairs)
Neisseriae (coffee-bean shape in pairs)
Coccobacilli
Vibrios (curved rods)
Tetrads (cocci in packets of 4)
Sarcinae (cocci in packets of 8,16,32 cells)
Mycobacteria
Corynebacteria (palisades arrangement)
Spirilla
Streptococci (cocci in chains)
Micrococci and staphylococci (large cocci in irregular clusters)
Spore-forming rods
Streptomycetes (moldlike, filamentous bacteria)
Spirochetes

Figure 4.22

Bacterial shapes and arrangements. May not be shown to exact scale.

BACTERIAL SHAPES, ARRANGEMENTS, AND SIZES

For the most part, bacteria function as independent single-celled, or unicellular, organisms. Although it is true that an individual bacterial cell can live attached to others in colonies or other such groupings, each one is fully capable of carrying out all necessary life activities, such as reproduction, metabolism, and nutrient processing (unlike the more specialized cells of a multicellular organism).

Bacteria exhibit considerable variety in shape, size, and colonial arrangement. It is convenient to describe most bacteria by one of three general shapes as dictated by the configuration of the cell wall (figure 4.22). If the cell is spherical or ball-shaped, the bacterium is described as a **coccus.*** Cocci can be perfect spheres, but they also can exist as oval, bean-shaped, or even pointed variants. A cell that is cylindrical (longer than wide) is termed a **rod,** or **bacillus.*** There is also a genus named *Bacillus* (see microfile

*coccus (kok′-us) pl. cocci (kok′-seye) Gr. *kokkos,* berry.

*bacillus (bah-sil′-lus) pl. bacilli (bah-sil′-eye) L. *bacill,* small staff or rod.

TABLE 4.3

COMPARISON OF THE TWO SPIRAL-SHAPED BACTERIA

	Spirilla	Spirochetes
Overall appearance	Rigid helix 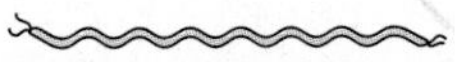	Flexible helix 
Mode of locomotion	Polar flagella; cells swim by rotating around like corkscrews; do not flex 1 to several flagella; can be in tufts	Periplasmic flagella within sheath; cells flex; can swim by rotation or by creeping on surfaces 2 to 100 periplasmic flagella
Number of helical turns	Varies from 1/2 to 20	Varies from 3 to 70
Gram reaction (cell wall type)	Gram negative	Gram negative
Examples of important types	Most are harmless saprobes; one species, *Spirillum minor,* causes rat bite fever	*Treponema pallidum,* cause of syphilis; *Borrelia* and *Leptospira,* important pathogens

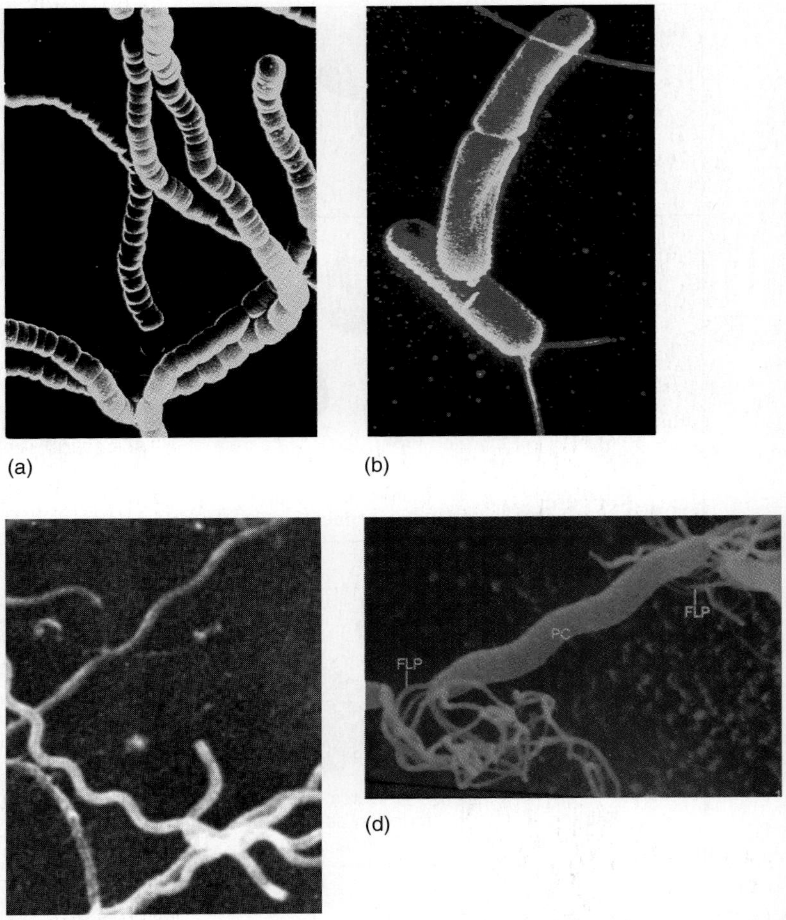

4.2). As might be expected, rods are also quite varied in their actual form. Depending on the bacterial species, they can be blocky, spindle-shaped, round-ended, long and threadlike (filamentous), or even clubbed or drumstick-shaped. When a rod is short and plump, it is called a **coccobacillus;** if it is gently curved, it is a **vibrio.*** A bacterium having the shape of a curviform or spiral-shaped cylinder is called a **spirillum,*** a rigid helix, twisted twice or more along its axis (like a corkscrew). Another spiral cell mentioned earlier in conjunction with periplasmic flagella is the **spirochete,** a more flexible form that resembles a spring or stretched Slinky. Refer to table 4.3 for a comparison of other features of the two helical bacterial forms. Because bacterial cells look rather two-dimensional and flat with traditional staining and microscope techniques, they are seen to best advantage with a scanning electron microscope that emphasizes their striking three-dimensional forms (figure 4.23).

It is rather common for cells of the same species to vary to some extent in shape and size. This phenomenon, called **pleomorphism*** (figure 4.24a), is due to individual variations in cell wall structure caused by nutritional or slight hereditary differences. For example, although the cells of *Corynebacterium diphtheriae* are generally considered rod-shaped, in culture they display variations such as club-shaped, swollen, curved, filamentous, and coccoid.

(a) (b) (c) (d)

Figure 4.23

Basic bacterial shapes in three dimensions with surface features. (*a*) Cocci in chains. (*b*) A rod-shaped bacterium (*Esch-erichia coli*) in a diplobacillus arrangement. (*c*) A spirochete (*Borrelia burgdorferi,* the cause of Lyme disease) is a long, thin cell with irregular coils and no external flagella. (*d*) A spirillum is thicker with a few even coils (PC) and external flagella (FLP). Can you tell what the flagellar arrangement is?

Pleomorphism reaches an extreme in the mycoplasmas, which entirely lack cell walls and thus display extreme variations in shape (see figure 4.33).

The cells of bacteria can also be categorized according to **arrangement,** or style of grouping (see figure 4.22). The main factors influencing the arrangement of a particular cell type are its pattern of division and how the cells remain attached afterward. The greatest variety in arrangement occurs in cocci, which can be single, in pairs **(diplococci),*** in **tetrads** (groups of four), in irregular clusters (both **staphylococci*** and **micrococci),*** or in

*vibrio (vib′-ree-oh) L. *vibrare,* to shake.
*spirillum (spy-ril′-em) pl. spirilla; L. *spira,* a coil.
*pelomorphism (plee″-oh-mor′-fizm) Gr. *pleon,* more, and *morph,* form or shape.

*diplococci; Gr. *diplo,* double.
*staphylococci (staf″-ih-loh-kok′-seye) Gr. *staphyle,* a bunch of grapes.
*micrococci; Gr. *mikros,* small.

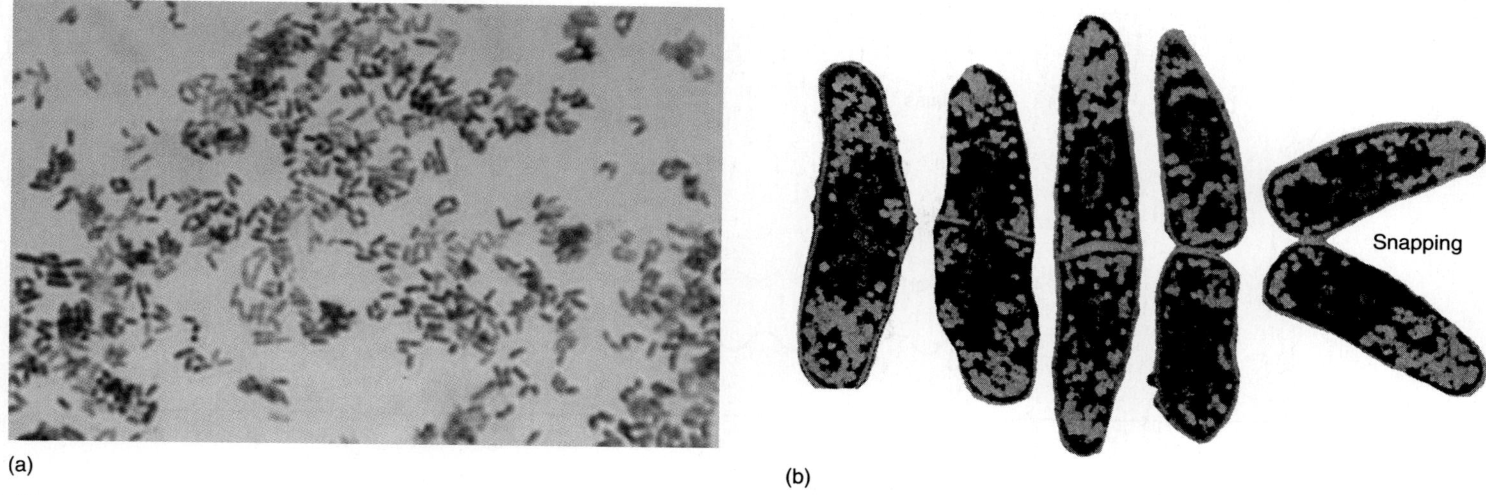

Figure 4.24

(*a*) Pleomorphism in *Corynebacterium*. Cells occur in a great variety of shapes and sizes. (*b*) This genus typically exhibits an unusual formation called a palisades arrangement that is caused by snapping. Close examination will also reveal darkly stained granules inside the cells.

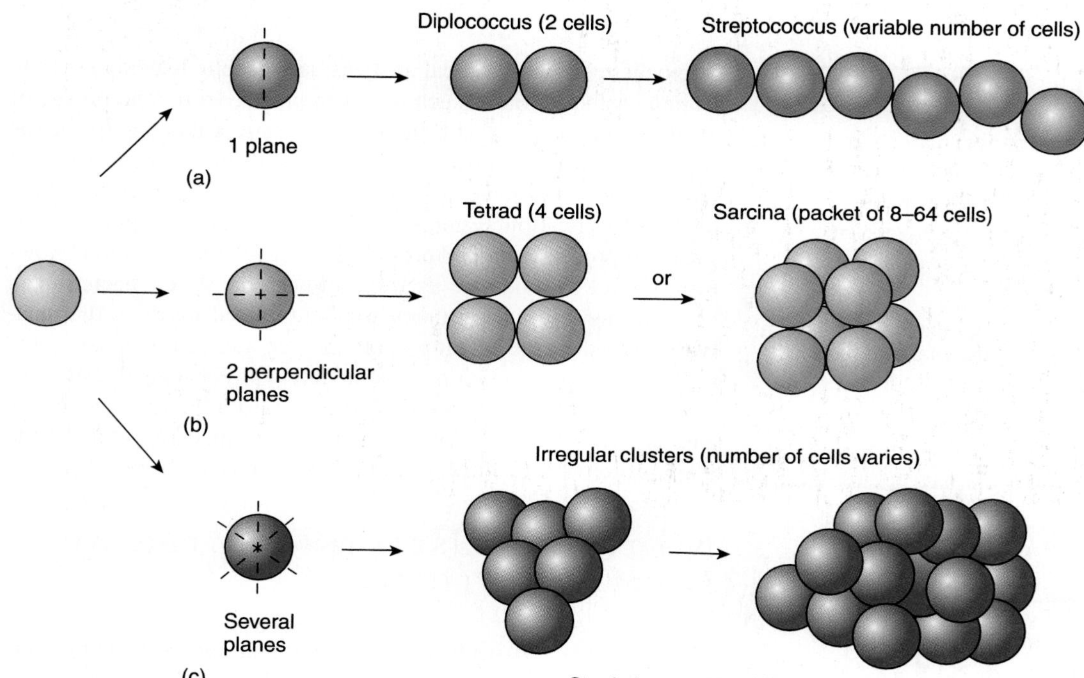

Figure 4.25

Arrangements of cocci resulting from different planes of cell division. (*a*) Division in one plane produces diplococci and streptococci. (*b*) Division in two planes at right angles produces tetrads and packets. (*c*) Division in several planes produces irregular clusters.

chains of a few to hundreds of cells **(streptococci).*** An even more complex grouping is a cubical packet of eight, sixteen, or more cells called a **sarcina** (sar'-sih-nah). These different coccal groupings are the result of the division of a coccus in a single plane, in two perpendicular planes, or in several intersecting planes; after division, the resultant daughter cells remain attached (figure 4.25).

Bacilli are less varied in arrangement because they divide only in the transverse plane (perpendicular to the axis). They occur either as single cells, as a pair of cells with their ends attached (diplobacilli), or as a chain of several cells (streptobacilli). A **palisades*** arrangement, typical of the corynebacteria, is formed when the cells of a chain remain partially attached by a small hinge region at the ends. The cells tend to fold (snap) back upon each other, forming a row of cells oriented side by side (see figure 4.24*b*). The reaction can be compared to the behavior of boxcars on a jackknifed train, and the result looks superficially like an irregular picket fence. Spirilla are occasionally found in short chains, but spirochetes rarely remain attached after division. Comparative sizes of typical cells are presented in figure 4.26.

*streptococci (strep''-toh-kok'-seye) Gr. *streptos,* twisted.

*palisades (pal'-ih-saydz) L. *pale,* a stake. A fence made of a row of stakes.

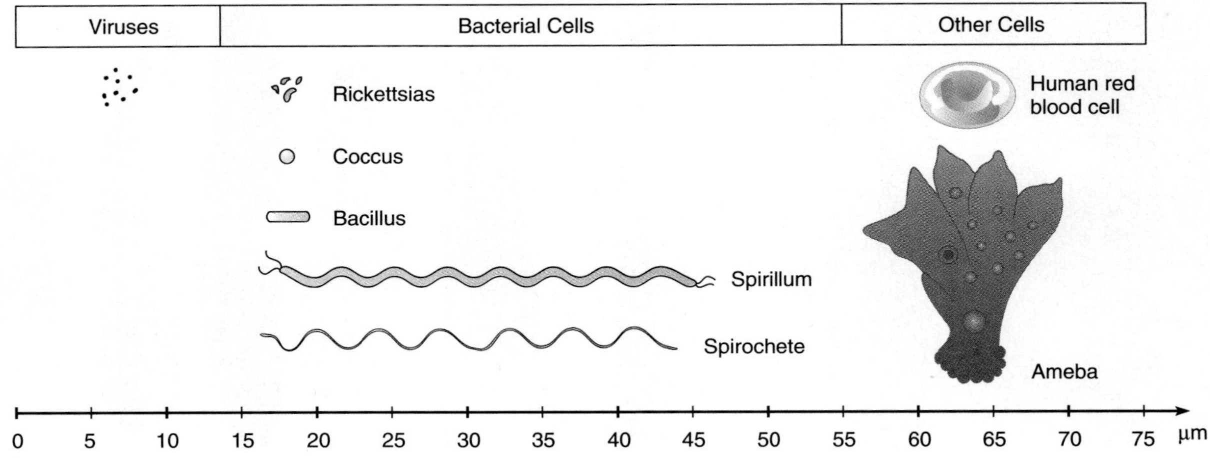

Figure 4.26
The dimensions of bacteria range from those just barely visible with light microscopy (0.2 μm) to those measuring a thousand times that size. Cocci measure anywhere from 0.5 to 3.0 μm in diameter; bacilli range from 0.2 to 2.0 μm in diameter and from 0.5 to 20 μm in length; vibrios and spirilla vary from 0.2 to 2.0 μm in diameter and from 0.5 to 100 μm in length. Spirochetes range from 0.1 to 3.0 μm in diameter and from 0.5 to 250 μm in length. Note the range of sizes as compared with eucaryotic cells and viruses. Comparisons are given as average sizes.

Chapter Checkpoints

Most bacteria have one of three general shapes: coccus (round), bacillus (rod), or spiral, based on the configuration of the cell wall. Two types of spiral cells are spirochetes and spirilla.

Shape and arrangement of cells are key means of describing bacteria. Arrangements of cells are based on the number of planes in which a given species divides.

Cocci can divide in many planes to form pairs, chains, packets, or clumps. Bacilli divide only in the transverse plane. If they remain attached, they form chains or palisades.

Bacteria range in size from the smallest rickettsias to the largest spiral forms.

BACTERIAL IDENTIFICATION AND CLASSIFICATION SYSTEMS

Every speck of soil, dab of saliva, and droplet of pond water is teeming with a rich variety of bacteria. Although study of such mixed populations can be useful, in most laboratory studies, one must prepare pure cultures of the individual members of the population (see chapter 3). This is especially critical in medical labs, which are charged with gathering the characteristics of an infectious agent and identifying it so that proper treatment can be given. Isolation and laboratory growth are also necessary for discovery and verification of new species.

The methods that a microbiologist uses to identify bacteria to the level of genus and species fall into the main categories of morphology (microscopic and macroscopic), bacterial physiology or biochemistry, serological analysis, and genetic techniques. Data from a cross section of such tests can produce a unique profile of each bacterium. Final differentiation of any unknown species is accomplished by comparing its profile with the characteristics of known bacteria in tables, charts, and keys (see figure 20.8). Many of the identification systems are automated and incorporate computers to process data and provide a "best fit" identification. However, not all methods are used on all bacteria. A few bacteria can be identified by placing them in an automated machine that analyzes only the kind of fatty acids they contain; in contrast, some are identifiable by a Gram stain and a few physiological tests; others may require a diverse spectrum of morphological, biochemical, and genetic tests. The following list summarizes some of the general areas of bacteriological testing. (See also figure 3.1.)

METHODS USED IN BACTERIAL IDENTIFICATION

Microscopic Morphology Traits that can be valuable aids to identification are combinations of cell shape and size, Gram stain reaction, acid-fast reaction, and special structures, including endospores, granules, and capsules. Electron microscope studies can pinpoint additional structural features (such as the cell wall, flagella, pili, and fimbriae).

Macroscopic Morphology Appearance of colonies, including texture, size, shape, pigment; speed of growth and patterns of growth in broth and gelatin media.

Physiological/Biochemical Characteristics These have been the traditional mainstay of bacterial identification. Enzymes and other biochemical properties of bacteria are fairly reliable and stable expressions of the chemical identity of each species. Dozens of diagnostic tests exist for determining the presence of

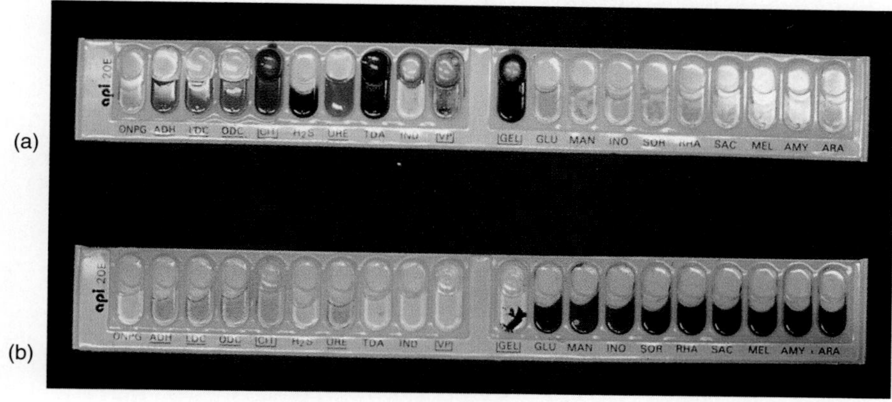

Figure 4.27
Rapid tests. The API 20E manual biochemical system for microbial identification. (*a*) Positive and (*b*) negative results.

specific enzymes and to assess nutritional and metabolic activities. Examples include tests for fermentation of sugars; capacity to digest or metabolize complex polymers such as proteins and polysaccharides; production of gas; presence of enzymes such as catalase, oxidase, and decarboxylases; and sensitivity to antimicrobic drugs. Special rapid identification test systems that record the major biochemical reactions of a culture have streamlined data collection (figure 4.27).

Chemical Analysis Analyzing the types of specific structural substances that the bacterium contains, such as the chemical composition of peptides in the cell wall and lipids in membranes.

Serological Analysis Bacteria have surface and other molecules called antigens that are recognized by the immune system. One immune response to antigens is the production of molecules called antibodies that are designed to bind tightly to the antigens. This response is so specific that antibodies can be used as a means of identifying bacteria in specimens and cultures. Laboratory kits based on this technique are available for immediate identification of a number of pathogens.

Genetic and Molecular Analysis Examining the genetic material itself has revolutionized the identification and classification of bacteria.

G + C base competition. The overall percentage of guanine and cytosine (the G + C content as compared with A + T content) in DNA is a general indicator of relatedness because it is a trait that does not change rapidly. Bacteria with a significant difference in G + C percentage are likely to be genetically distinct species or genera. For example, although superficially similar in Gram reaction, shape, and other morphological characteristics, *Escherichia* has a G + C base composition of 48–52% and *Pseudomonas* has a composition of 58–70%, indicating that they probably are not closely related. This technique is most applicable for clarifying the taxonomic position of a bacterium, but it is too nonspecific to be applicable as a precise identification tool.

DNA analysis using genetic probes. The exact order and arrangement of the DNA code is unique to each organism (see figure 2.26). With a technique called *hybridization,* it is possible to identify a bacterial species by analyzing segments of its DNA (figure 4.28). This requires small fragments of single-stranded DNA (or RNA) called **probes** that are known to be complementary to the specific sequences of DNA from a particular microbe. The test is conducted by extracting unknown test DNA from cells in specimens or cultures and binding it to special blotter paper. After several different probes have been added to the blotter, it is observed for visible signs that the probes have become fixed (hybridized) to the test DNA. The binding of probes onto several areas of the test DNA indicates close correspondence and makes positive identification possible (also discussed in chapter 10).

Nucleic acid sequencing and rRNA analysis. One of the most viable indicators of evolutionary relatedness and affiliation is comparison of the sequence of nitrogen bases in ribosomal RNA, a major component of ribosomes (figure 4.29). Ribosomes have the same function (protein synthesis) in all cells, and they tend to remain more or less stable in their nucleic acid content over long periods. Thus, any major differences in the sequence, or "signature," of the rRNA is likely to indicate some distance in ancestry. This technique is powerful at two levels: It is effective for differentiating general group differences (it was used to separate the three superkingdoms of life discussed in chapter 1), and it can be fine-tuned to identify at the species level (for example in *Mycobacterium* and *Legionella*). Elements of these and other identification methods are presented in more detail in chapters 10, 16, 20, and 21.

CLASSIFICATION SYSTEMS IN THE PROCARYOTAE

Classification systems serve both practical and academic purposes. They aid in differentiating and identifying unknown species in medical and applied microbiology. They are also useful in organizing bacteria and as a means of studying their relationships and origins. Since the classification was started around 200 years ago, several thousand species of bacteria and archaea have been identified, named, and catalogued (microfile 4.2).

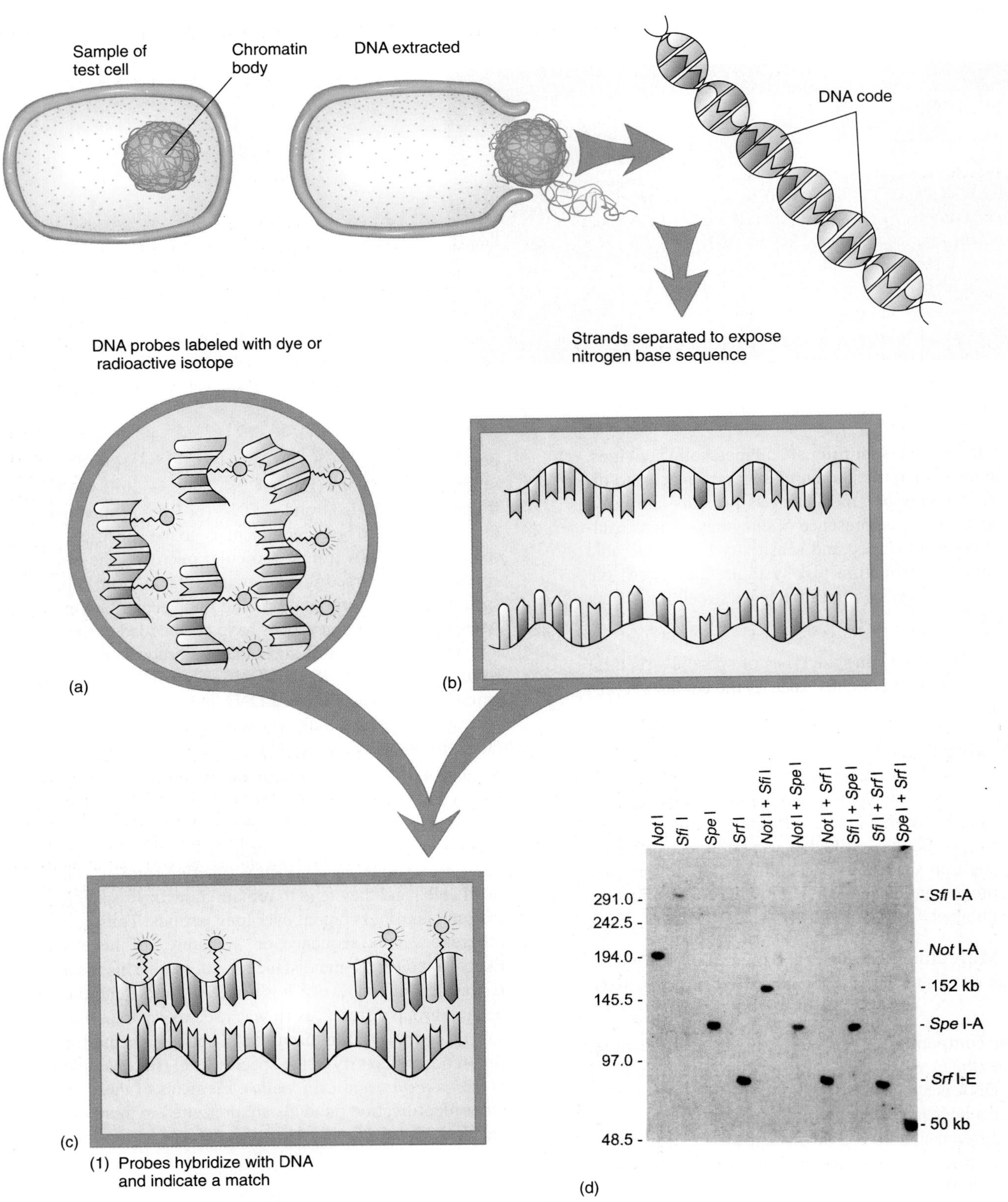

Figure 4.28

DNA hybridization using probes (scale is exaggerated for visibility). (*a*) Probes of known base sequence are labeled with a substance that can be observed when it reacts with DNA. (*b*) Test DNA is blotted onto a special substrate so it will stick. (*c*) When probes contact DNA that has a complementary base sequence, they become affixed through base-pairing and will cause a visible reaction in a localized area of the test DNA (*1*). Probes that are not of the correct specificity will remain free and will not show up as a visible reaction (*2*). (*d*) A DNA hybridization result from the syphilis spirochete, *Treponema pallidum*. Dark bands indicate areas where a specific probe has reacted with *Treponema* DNA.

MICROFILE 4.2 THE NAME GAME

The scope of the bacterial world once seemed much simpler than it does now. During the mid to late 1800s, most known bacteria could be neatly covered by a few generic names. Spherical bacteria were placed in one of three genera: *Micrococcus, Streptococcus,* or *Staphylococcus.* Rod-shaped bacteria were assigned to either the genus *Bacterium* or the genus *Bacillus.* If bacteria had the shape of a curved rod, they were put in the genus *Vibrio,* and if they were spiral, they were placed in either the genus *Spirillum* or *Spirochaeta.* Because very little was known then about the biochemical characteristics of these bacteria, and the Gram stain had not yet been applied as a general classification tool, their morphology was the primary method of identification. As a result, bacteria that we now know are very different often had the same generic name. Consider that *Escherichia coli* was once called *Bacterium coli; Pseudomonas aeruginosa* was *Bacterium aeruginosa,* and *Streptococcus lactis* was *Bacterium lactis,* even though the first two rods and the last is a coccus.

Present classification schemes include over 1,000 genera, and bacterial identification is based on hundreds of characteristics. Although more than 5,000 species are presently known, new ones are discovered every year. The recent increase in new species is probably due in part to an improvement in identification techniques. Most bacteriologists feel that we have only scratched the surface and that myriad bacteria remain to be discovered.

You will notice that certain generic names appear both capitalized and uncapitalized. For example, the word *Staphylococcus* as shown here refers to a particular genus of coccus-shaped bacteria, but when the word appears uncapitalized, unitalicized, or in the plural (staphylococci), it is a general way of designating the members of that genus or of describing an arrangement or cell type. This same usage is true of the use of streptococci, micrococci, mycobacteria, pseudomonads, and clostridia. Likewise, *Bacillus* and *Spirillum* refer to genera, and bacillus and spirillum to shapes.

It is well worth stating that it makes little difference to bacteria what we call them. The plague bacillus, whether named *Yersinia pestis* or *Pasteurella pestis* (its former name), is still quite capable of causing bubonic plague. In the medical microbiology laboratory, the public health laboratory, or the hospital, a scheme of classification becomes something much more than an orderly cataloguing of tongue-twisting names. It presents a foundation of characteristics to be used in identification. With it, not only can technicians identify *Y. pestis* and scores of other pathogens, but they can also distinguish these pathogens from their less injurious relatives.

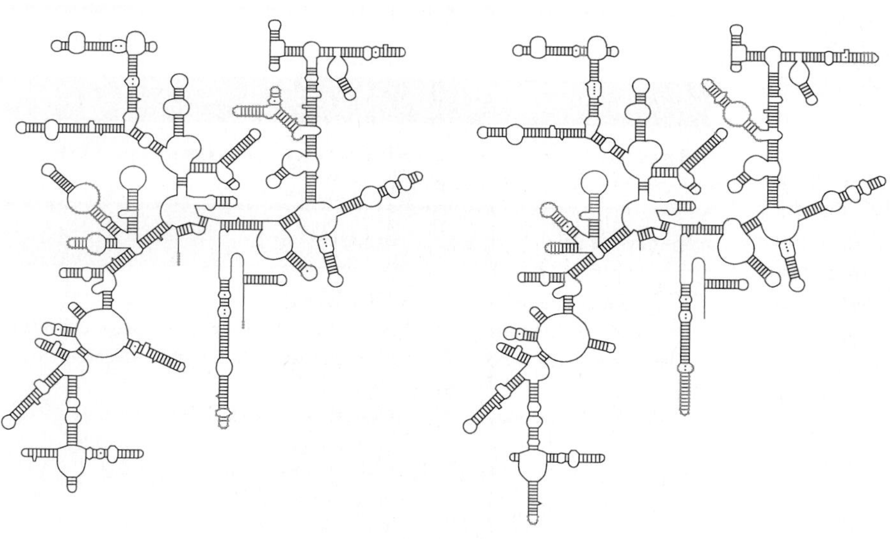

Figure 4.29
A modern molecular technique that identifies the cell type or subtype by analyzing the nitrogen base sequence of its ribosomal RNA. The segments of rRNA shown here are from the small subunit of *Escherichia coli*, a common intestinal bacterium, and *Methanococcus vannielii,* an archaea that inhabits the deep sediments of marshes. The shaded regions indicate areas of rRNA that differ significantly between bacteria *(Escherichia)* and archeae *(Methanococcus)*.

For years there has been intense interest in tracing the origins of and evolutionary relationships among bacteria, but doing so has not been an easy task. As a rule, tiny, relatively soft organisms do not form fossils very readily. Several times since the 1960s, however, scientists have discovered microscopic fossils of procaryotes that look very much like modern bacteria. Some of the rocks that contain these fossils have been dated back billions of years (figure 4.30). One of the questions that has plagued taxonomists is, What characteristics are the most indicative of closeness in ancestry? Early bacteriologists found it convenient to classify bacteria according to shape, variations in arrangement, growth characteristics, and habitat. However, as more species were

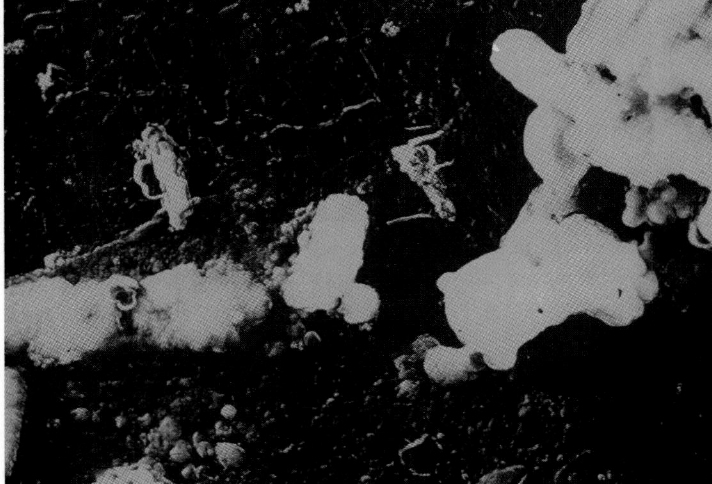

Figure 4.30
Electron micrograph of fossilized rod-shaped bacterial cells preserved in rock sediment. This specimen is more than 2 billion years old (24,000×).

discovered and as techniques for studying their biochemistry were developed, it soon became clear that similarities in cell shape, arrangement, and staining reactions do not automatically indicate relatedness. Even though the gram-negative rods look alike, there are hundreds of different species, with highly significant differences in biochemistry and genetics. If we attempted to classify them on the basis of Gram stain and shape alone, we could not assign them to a more specific level than class. Increasingly, classification of schemes are turning to genetic and molecular traits that cannot be visualized under a microscope or in culture.

The most functional classification schemes make use of current knowledge to show natural relationships. Not only must they be flexible enough to add newly discovered species, but they must also complement a number of microbiology disciplines. In general, schemes for organizing bacteria can be phylogenetic, based on evolutionary relationships, or phenetic, based on their phenotypic traits related to morphology or biochemistry. One system of classification has not been permanent or universally accepted to date; indeed, most systems are in a state of flux as new information and methods of analysis become available. Of the current proposed systems, we present three: (1) the overall scheme of classification from the ninth edition of *Bergey's Manual of Systematic Bacteriology,* a manual of bacterial descriptions and classification published continuously since 1923 (table 4.4); (2) a method that groups bacteria by comparing rRNA sequence; and (3) a practical system that uses a few morphological and physiological traits to categorize the major, medically important bacterial families (see table 4.6).

The ninth edition of *Bergey's Manual* organizes the Kingdom Procaryotae into four major divisions. These somewhat natural divisions are based upon the nature of the cell wall. The **Gracilicutes*** have gram-negative cell walls and thus are thin-skinned; the **Firmicutes*** have gram-positive cell walls that are thick and strong; the **Tenericutes*** lack a cell wall and thus are soft; and the **Mendosicutes*** are the archaea (also called archaebacteria), primitive bacteria with unusual cell walls and nutritional habits. The first two divisions contain the greatest number of species. The 200 or so species that cause human and animal diseases can be found

TABLE 4.4

MAJOR TAXONOMIC GROUPS OF BACTERIA PER BERGEY'S MANUAL

Division I. Gracilicutes: Gram-Negative Bacteria

Class I. Scotobacteria: Gram-negative non-photosynthetic bacteria (examples in table 4.5)
Class II. Anoxyphotobacteria: Gram-negative photosynthetic bacteria that do not produce oxygen (purple and green bacteria)
Class III. Oxyphotobacteria: Gram-negative photosynthetic bacteria that evolve oxygen (cyanobacteria)

Division II. Firmicutes: Gram-Positive Bacteria

Class I. Firmibacteria: Gram-positive rods or cocci (examples in table 4.6)
Class II. Thallobacteria: Gram-positive branching cells (the actinomycetes)

Division III. Tenericutes

Class I. Mollicutes: Bacteria lacking a cell wall (the mycoplasmas)

Division IV. Mendosicutes

Class I. Archaebacteria: Bacteria with atypical compounds in the cell wall and membranes

Source: Data from *Bergey's Manual of Systematic Bacteriology.* 9th ed. Williams & Wilkins. Company, Baltimore. 1984.

TABLE 4.5

BERGEY'S MANUAL TAXONOMIC RANKINGS FOR BORRELIA BURGDORFERI

Taxonomic Rank	Includes
Kingdom: Monera	All bacteria
Division: Gracilicutes	All bacteria with gram-negative cell wall
Class: Scotobacteria	Non-photosynthetic gram-negative bacteria
Order: Spirochaetales	Helical, flexible, motile bacteria (spirochetes)
Family: Spirochaetaceae	Helical, motile bacteria lacking hooked ends
Genus: *Borrelia*	Tiny, loose, irregularly coiled spirochetes
Species: *burgdorferi*	Causative agent of Lyme disease

*Gracilicutes (gras''-ih-lik'-yoo-teez) L. *gracilus,* thin, and *cutis,* skin.
*Firmicutes (fer-mik'-yoo-teez) L. *firmus,* strong.
*Tenericutes (ten''-er-ik'-yoo-teez) L. *tener,* soft.
*Mendosicutes (men-doh-sik'-yoo-teez) L. *mendosus,* to be false.

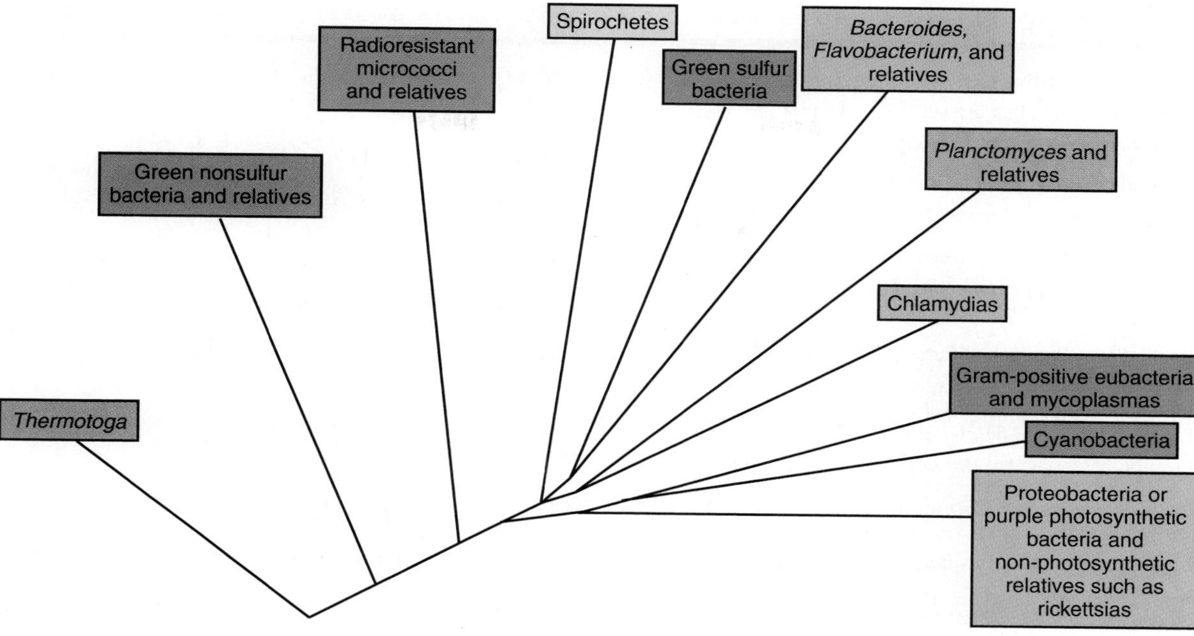

Figure 4.31
Separation chart for the eubacteria proposes phylogenetic relationships based on rRNA sequences. This scheme divides them into 11 genetically discrete groups. The branches suggest evolutionary origins, and branches that have greater similarities appear closer to each other.
Source: Data from C.R. Woese, Microbiol. Rev. *51(1987):221–71. American Society for Microbiology, Washington, D.C.*

in four classes: the Scotobacteria, Firmibacteria, Thallobacteria, and Mollicutes. The system used in *Bergey's Manual* further organizes bacteria into subcategories such as classes, orders, and families, but these are not available for all groups. An example of the entire classification of one bacterial species is shown in table 4.5.

The classification scheme developed by analyzing the similarities in base sequence of rRNA revealed 11 distinct branches on the bacterial "tree" (figure 4.31). It just so happens that some groupings are similar to those used in Bergey's system (gram-positive bacteria), but others are genetically different enough to be placed into their own separate categories (spirochetes).

1. Gram-positive eubacteria: selected representatives are *Bacillus, Clostridium, Mycobacterium, Staphylococcus, Actinomyces,* and the cell-wall-free mycoplasmas.
2. Gram-negative eubacteria (Proteobacteria) includes purple photosynthetic bacteria (*Chromatium*) and non-photosynthetic relatives represented by *Pseudomonas, Vibrio, Neisseria,* and the rickettsias.
3. Cyanobacteria: photosynthetic bacteria with chlorophyll *a* that evolve (give off) oxygen; includes *Oscillatoria* and *Spirulina.*
4. Spirochetes; flexible helical cells with periplasmic flagella such as *Treponema* and *Borrelia.*
5. Walled, budding bacteria that lack peptidoglycan in their cell walls: includes *Planctomyces.*
6. The *Bacteroides, Flavobacterium, Fusobacterium,* and *Cytophaga:* a mixed group morphologically and physiologically.
7. Chlamydias: unusual obligate parasites of vertebrates; lack ability to complete metabolism independently; lack peptidoglycan; one genus—*Chlamydia.*
8. Green sulfur bacteria: anaerobic bacteria that contain bacteriochlorophyll and use sulfur in metabolism; do not give off oxygen during photosynthesis; includes *Chlorobium.*
9. Green nonsulfur bacteria: filamentous, gliding, thermophilic, photosynthetic bacteria that contain bacteriochlorophyll, do not evolve oxygen; includes *Chloroflexus.*
10. Unique bacteria with extreme resistance to electromagnetic radiation: *Deinococcus,* gram-positive cocci, and *Thermus,* thermophilic rods.
11. Unusual thermophilic bacteria inhabiting hot oceanic vents: *Thermotoga.*

Ribosomal RNA analysis of the archaea detected three well-defined groups:

1. Representatives that synthesize methane from simple inorganic compounds (methanogens)
2. Extremely halophilic (salt-loving) archaea
3. Extremely thermophilic (heat-loving) archaea that use sulfur in their metabolism

The last section of this chapter covers features of archaeal form, function, and ecology.

Many medical microbiologists prefer an informal working system that outlines the major families and genera (table 4.6). This system is more applicable for their purposes because it is restricted to bacterial disease agents, depends less on nomenclature, and is based on readily accessible morphological and physiological tests rather than on phylogenetic relationships. It also divides the bacteria into gram positive, gram negative, and those without cell walls and then subgroups them according to cell shape,

Table 4.6

Medically Important Families and Genera of Bacteria, with Notes on Some Diseases*

I. Bacteria with Gram-Positive Cell Wall Structure

Cocci in clusters or packets that are aerobic or facultative

Family Micrococcaceae: *Staphylococcus* (members cause boils, skin infections)

Cocci in pairs and chains that are facultative

Family Streptococcaceae: *Streptococcus* (species cause strep throat, dental caries)

Anaerobic cocci in pairs, tetrads, irregular clusters

Family Peptococcaceae: *Peptococcus, Peptostreptococcus* (involved in wound infections)

Spore-forming rods

Family Bacillaceae: *Bacillus* (anthrax), *Clostridium* (tetanus, gas gangrene, botulism)

Non-spore-forming rods

Family Lactobacillaceae: *Lactobacillus, Listeria* (milk-borne disease), *Erysipelothrix* (erysipeloid)

Family Propionibacteriaceae: *Propionibacterium* (involved in acne)

Family Corynebacteriaceae: *Cornyebacterium* (diphtheria)

Family Mycobacteriaceae: *Mycobacterium* (tuberculosis, leprosy)

Family Nocardiaceae: *Nocardia* (lung abscesses)

Family Actinomycetaceae: *Actinomyces* (lumpy jaw), *Bifidobacterium*

Family Streptomycetaceae: *Streptomyces* (important source of antibiotics)

II. Bacteria with Gram-Negative Cell Wall Structure

Family Neisseriaceae

Aerobic cocci

Neisseria (gonorrhea, meningitis), *Branhamella*

Aerobic coccobacilli

Moraxella, Acinetobacter

Anaerobic cocci

Family Veillonellaceae

Veillonella (dental disease)

Miscellaneous rods

Brucella (undulant fever), *Bordetella* (whooping cough), *Francisella* (tularemia)

Aerobic rods

Family Pseudomonadaceae: *Pseudomonas* (pneumonia, burn infections)

Miscellaneous: *Legionella* (Legionnaires' disease)

Facultative or anaerobic rods and vibrios

Family Enterobacteriaceae: *Escherichia, Edwardsiella, Citrobacter, Salmonella* (typhoid fever), *Shigella* (dysentery), *Klebsiella, Enterobacter, Serratia, Proteus, Yersinia* (one species causes plague)

Family Vibronaceae: *Vibrio* (cholera, food infection), *Campylobacter, Aeromonas*

Miscellaneous genera: *Chromobacterium, Flavobacterium, Haemophilus* (meningitis), *Pasteurella, Cardiobacterium, Streptobacillus*

Anaerobic rods

Family Bacteroidaceae: *Bacteroides, Fusobacterium* (anaerobic wound and dental infections)

Helical and curviform bacteria

Family Spirochetaceae: *Treponema* (syphilis), *Borerelia* (Lyme disease), *Leptospira* (kidney infection)

Obligate intracellular bacteria

Family Rickettsiaceae: *Rickettsia* (Rocky Mountain spotted fever), *Coxiella* (Q fever)

Family Bartonellaceae: *Bartonella* (trench fever, cat scratch disease)

Family Chlamydiaceae: *Chlamydia* (sexually transmitted infection)

III. Bacteria with No Cell Walls

Family Mycoplasmataceae: *Mycoplasma* (pneumonia), *Ureaplasma* (urinary infection)

*Details of pathogens and diseases in chapters 18, 19, 20, and 21.

arrangement, and certain physiological traits such as oxygen usage: *Aerobic* bacteria use oxygen in metabolism; *anaerobic* bacteria do not use oxygen in metabolism; and facultative bacteria may or may not use oxygen. Further tests not listed on the table would be required to separate closely related genera and species. Many of these are included in later chapters on specific diseases.

Species and Subspecies in Bacteria

Among most organisms, the species level is a distinct, readily defined, and natural taxonomic category. In animals, for instance, a species is a distinct type of organism that can produce viable offspring only when it mates with others of its own kind. This definition does not work for bacteria primarily because they do not exhibit a typical mode of sexual reproduction. They can accept genetic information from unrelated forms, and they can also alter their genetic makeup by a variety of mechanisms. Thus, it is necessary to hedge a bit when we define a bacterial species. Theoretically, it is a collection of bacterial cells, all of which share an overall similar pattern of traits, in contrast to other groups whose pattern differs significantly. Although the boundaries that separate two closely related species in a genus are in some cases very arbitrary, this definition still serves as a method to separate the bacteria into various kinds that can be cultured and studied. As additional information on bacterial genomes is discovered, it may be possible to define species according to specific combinations of genetic codes found only in a particular isolated culture.

Since the individual members of given species can show variations, we must also define levels within species (subspecies) called **strains** and **types.** A strain or variety of bacteria is a culture derived from a single parent that differs in structure or metabolism from other cultures of that species (also called biovars or morphovars). For example, there are pigmented and nonpigmented strains of *Serratia marcescens* and flagellated and nonflagellated strains of *Pseudomonas fluorescens.* A type is a subspecies that can show differences in antigenic makeup (serotype or serovar), in susceptibility to bacterial viruses (phage type), and in pathogenicity (pathotype).

Chapter Checkpoints

A bacterial species can be properly identified only if it is grown in pure culture—that is, in isolation from all other forms of life.

Key traits that are used to identify a bacterial species include (1) morphology, (2) gram stain or other stain characteristics, (3) presence of specialized structures, (4) macroscopic appearance of colonies, (5) biochemical reactions, and (6) nucleotide composition of both DNA and rRNA.

Bacteria are formally classified by phylogenetic relationships and phenotypic characteristics.

Medical identification of pathogens uses a more informal system of classification based on Gram stain, morphology, biochemical reactions, and metabolic requirements.

A bacterial species is loosely defined as a collection of bacterial cells that shares an overall similar pattern of traits different from other groups of bacteria.

Variant forms within a species are termed subspecies.

SURVEY OF PROCARYOTIC GROUPS WITH UNUSUAL CHARACTERISTICS

The bacterial world is so diverse that we cannot do complete justice to it in this introductory chapter. This variety extends into all areas of bacterial biology, including nutrition, mode of life, and behavior. Certain types of bacteria exhibit such unusual qualities that they deserve special mention. In this minisurvey, we will consider some medically important groups and some more remarkable representatives of bacteria living free in the environment that are ecologically important. Many of the bacteria mentioned here do not have the morphology typical of bacteria discussed previously, and in a few cases, they are vividly different (microfile 4.3).

UNUSUAL FORMS OF MEDICALLY SIGNIFICANT BACTERIA

Most bacteria are free-living or parasitic forms that can metabolize and reproduce by independent means. Two groups of bacteria—the rickettsias and chlamydias—have adapted to life inside their host cells, where they are considered **obligate intracellular parasites.**

Rickettsias Rickettsias[3] are distinctive, very tiny gram-negative bacteria (figure 4.32). Although they have a somewhat typical bacterial morphology, they are atypical in their life cycle and other adaptations. Most are pathogens that alternate between a mammalian host and blood-sucking arthropods,[4] such as fleas, lice, or ticks (see chapter 21). Rickettsias cannot survive or multiply outside a host cell and cannot carry out metabolism completely on their own, so they are closely attached to their hosts. Several important human diseases are caused by rickettsias. Among these are Rocky Mountain spotted fever, caused by *Rickettsia rickettsii* (transmitted by ticks), and epidemic typhus, caused by *Rickettsia prowazekii* (transmitted by lice). An exceptionally resistant rickettsia, *Coxiella burnetti* (the cause of Q fever), is transmitted in air and dust by arthropods.

Chlamydias Bacteria of the genus *Chlamydia* are similar to the rickettsias in that they require host cells for growth and metabolism, but they are not closely related and are not transmitted by arthropods. Because of their tiny size and obligately parasitic lifestyle, they were at one time considered a type of virus. Later studies indicated that their structure was that of a gram-negative cell and that their mode of cell division (binary fission) was clearly procaryotic. Two species that carry the greatest medical impact are *Chlamydia trachomatis,* the cause of both a severe eye infection (trachoma) that can lead to blindness and one of the most common sexually transmitted diseases (see chapter 21 on chlamydiosis), and *Chlamydia psittaci,* the agent of ornithosis, or parrot fever, a disease of birds that can be transmitted to humans.

3. Name for Howard Ricketts, a physician who first worked with these organisms and later lost his life to typhus.
4. An invertebrate with jointed legs, such as an insect, tick, or spider.

MICROFILE 4.3 BULKY BACTERIUM BAFFLES BIOLOGISTS

Another strange but wondrous member of the bacterial world is a giant creature known as *Epulopiscium fishelsoni.** This unusual procaryote can be over half a millimeter long and actually visible with the naked eye! First discovered in 1985 by a team of researchers studying the intestinal contents of surgeonfish, the bacteria were initially thought to be some form of protozoan. Not only were they very large, but under the light microscope they also appeared to have eucaryotic features such as a nucleus and a fine covering of cilia. Several laboratories took on the challenge of clarifying the exact nature of these mysterious cells. The electron microscope revealed that they did not truly have a nuclear membrane and that the "cilia" were more similar to procaryotic flagella in structure.

When attempts at culturing the bacteria outside of the host's body failed, several researchers turned to molecular analysis of the giant cells. Knowing that certain types of ribosomal RNA are valuable indicators of the relatedness of bacteria and other microbes, they made a detailed analysis of the genetic codes associated with rRNA. The researchers found that the sequence of its rRNA was consistent with that of bacteria and that it most closely resembled the genetic makeup of the gram-positive bacterial group to which *Clostridium* belongs. Other tests showed that the huge cells contained more ribosomes and large amounts of DNA. How a procaryote with such a large volume and no organelles could solve the problems of supplying itself with nutrients and getting rid of wastes is just one of the intriguing questions being explored by the researchers.

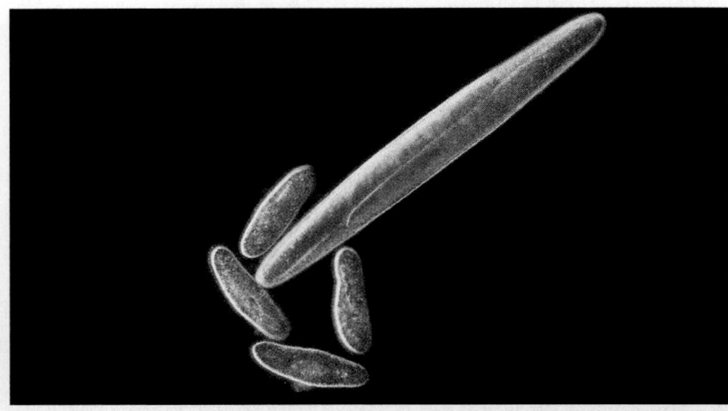

Giant bacterium is flanked by a school of paramecia that look tiny by comparison.

*(ee-pyoo'-loh-pish'-ee-um fish'-ul-son"-eye) Literally means "guest at a banquet of fish."

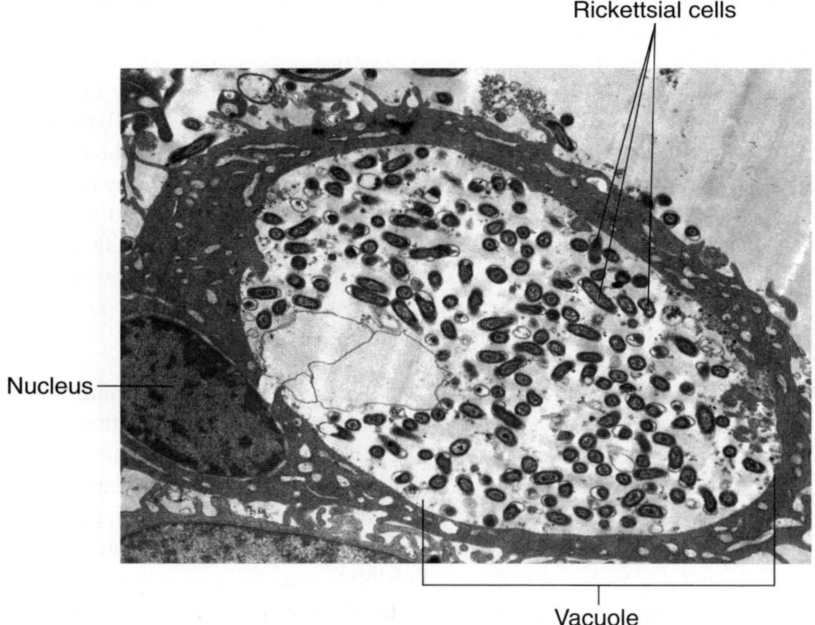

Figure 4.32
Transmission electron micrograph of the rickettsia *Coxiella burnetii,* the cause of Q fever. Its mass growth inside a host cell has filled a vacuole and displaced the nucleus to one side.

Mycoplasmas and Other Cell-Wall-Deficient Bacteria

Mycoplasmas are bacteria that naturally lack a cell wall. Although other bacteria require an intact cell wall to protect the bursting of the cell, the mycoplasmal cell membrane is stabilized by sterols and is resistant to lysis. These extremely tiny, pleomorphic cells have a minimum size of 0.1 μm. They range in shape from filamentous to coccus or doughnut-shaped. They are *not* obligate parasites and can be grown on artificial media, although added sterols are required for the cell membranes of some species. Mycoplasmas are found in many habitats, including plants, soil, and animals. The most important medical species is *Mycoplasma pneumoniae* (figure 4.33), which adheres to the epithelial cells in the lung and causes an atypical form of pneumonia in humans (see chapter 21).

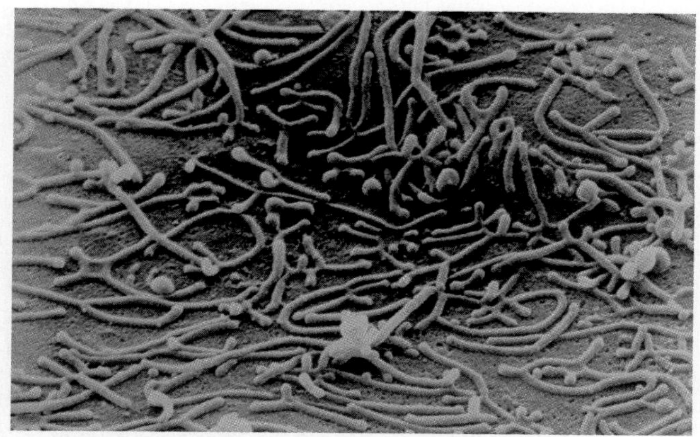

Figure 4.33

Scanning electron micrograph of *Mycoplasma pneumoniae* (62,000×). Cells like these that naturally lack a cell wall exhibit extreme pleomorphism.

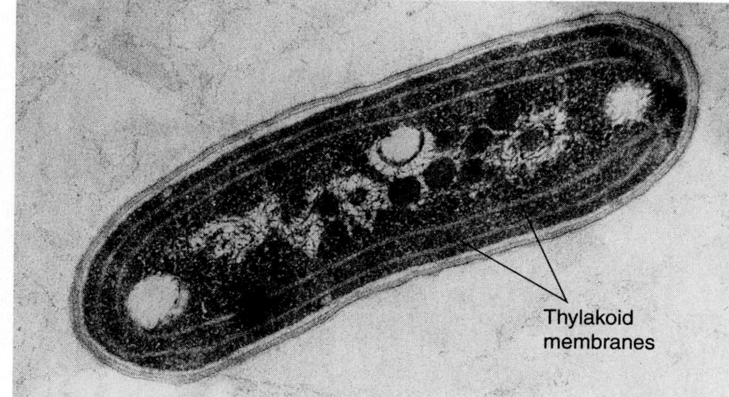

(a)

(b)

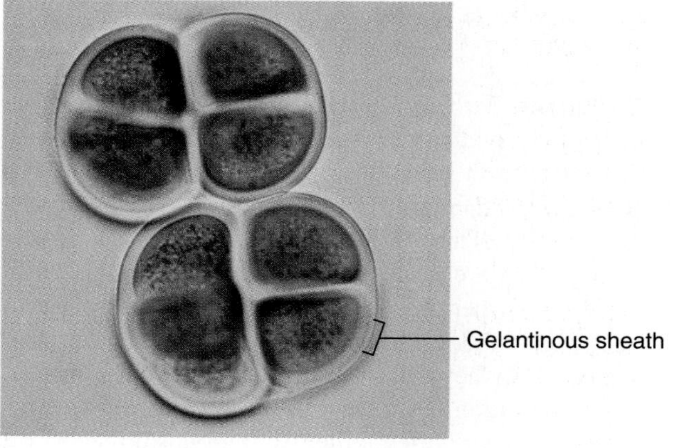

(c)

Figure 4.34

Structure and examples of cyanobacteria. (*a*) Electron micrograph of a cyanobacterial cell (80,000×) reveals folded stacks of membranes that contain the photosynthetic pigments and increase surface area for photosynthesis. (*b*) Two species of *Oscillatoria,* a gliding, filamentous form (100×). (*c*) *Chroococcus,* a colonial form surrounded by a gelatinous sheath (600×).

FREE-LIVING NONPATHOGENIC BACTERIA

Photosynthetic Bacteria

The nutrition of most bacteria is heterotrophic, meaning that they derive their nutrients from other organisms. Photosynthetic bacteria, however, are independent cells that contain special light-trapping pigments and can use the energy of sunlight to synthesize all required nutrients from simple inorganic compounds. The two general types of photosynthetic bacteria are those that produce oxygen during photosynthesis and those that produce some other substance, such as sulfur granules or sulfates.

Cyanobacteria: Blue-Green Bacteria The cyanobacteria were called blue-green algae for many years and were grouped with the eucaryotic algae. However, further study verified that they are indeed bacteria with a gram-negative cell wall (Gracilicutes) and general procaryotic structure. These bacteria range in size from 1 μm to 10 μm, and they can be unicellular or can occur in colonial or filamentous groupings (figure 4.34*b,c*). Some species occur in packets surrounded by a gelatinous sheath (figure 4.34*c*). A specialized adaptation of cyanobacteria are extensive internal membranes called **thylakoids,*** which contain granules of chlorophyll *a* and other photosynthetic pigments (figure 4.34*a*). They also have gas inclusions that permit them to float on the water surface and increase their light exposure and cysts that convert gaseous nitrogen (N_2) into a form usable by plants. This group is sometimes called the blue-green bacteria in reference to their content of **phycocyanin*** pigment that tints some members a shade of blue, although other members are colored yellow and orange. Some representatives glide or sway gently in the water from the action of filaments in the cell envelope that cause wavelike contractions.

---

*thylakoid (theye′-luh-koid) Gr. *thylakon,* and *eidos,* form.

*phycocyanin (feye-koh-seye′-an-in) Gr. *phykos,* seaweed, and *cyan,* blue. A bluish pigment unique to this group.

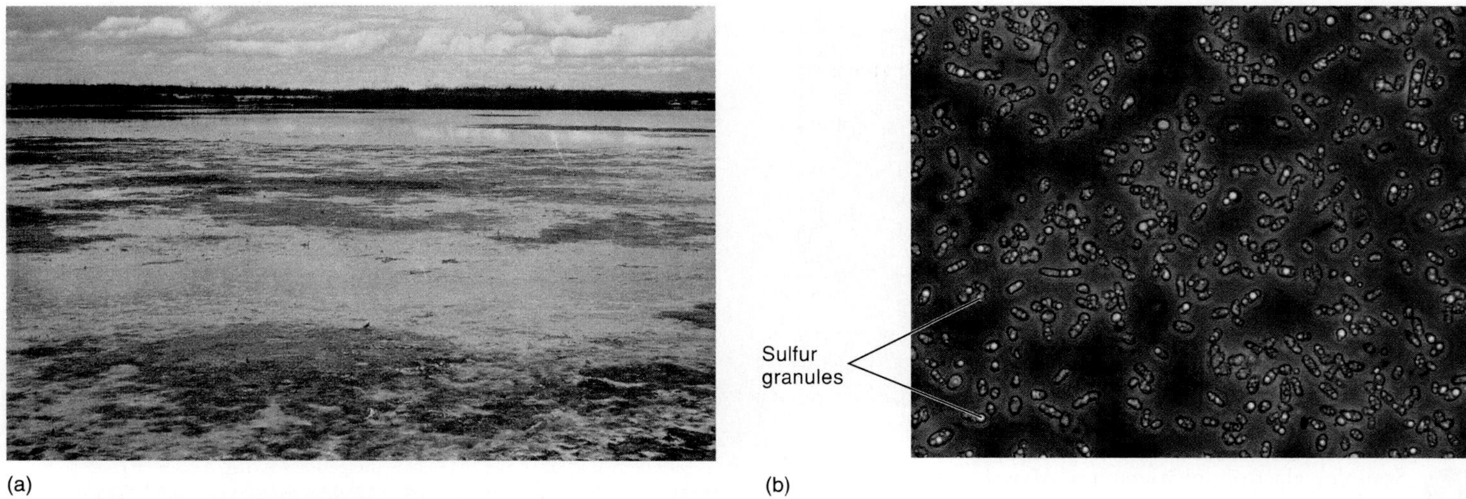

Figure 4.35
(*a*) Floating purple mats are huge masses of purple sulfur bacterial blooming in the Baltic Sea. Photosynthetic bacteria can have significant effects on the ecology of certain habitats. (*b*) Microscopic view of purple sulfur bacteria *(Chromatium vinosum)* carrying large yellow sulfur granules internally.

Cyanobacteria are very widely distributed in nature. They grow profusely in fresh water and seawater and are thought to be responsible for periodic blooms that kill off fish by depleting the available oxygen. Some members are so pollution-resistant that they serve as biological indicators of polluted water. Cyanobacteria inhabit and flourish in hot springs (see microfile 7.1) and have even exploited a niche in dry desert soils and rock surfaces. Some types of lichens are symbiotic associations between fungi and cyanobacteria that assist in the breakdown of rock and soil formation (see chapter 26).

Green and Purple Sulfur Bacteria The green and purple bacteria are also photosynthetic and contain pigments. They differ from the cyanobacteria in having a different type of chlorophyll called *bacteriochlorophyll* and by not giving off oxygen as a product of photosynthesis. They live in sulfur springs, freshwater lakes, and swamps that are deep enough for the anaerobic conditions they require yet where their pigment can still absorb wavelengths of light (figure 4.35*a*). These bacteria are named for their predominant colors, but they can also develop brown, pink, purple, blue, and orange coloration. They exist as single cells of many different shapes and frequently are motile. Both groups utilize sulfur compounds (H_2S, S) in their metabolism, and some can deposit intracellular granules of sulfur or sulfates (figure 4.35*b*).

Gliding, Fruiting Bacteria

The **gliding bacteria** are a mixed collection of gram-negative bacteria that live in water and soil. The name is derived from the tendency of members to glide over moist surfaces. The gliding property evidently involves rotation of filaments or fibers just under the outer membrane of the cell wall. They do not have flagella. There are several morphological forms, including slender rods, long filaments, cocci, and some miniature, tree-shaped fruiting bodies. Probably the most intriguing and exceptional members of this group are the slime bacteria, or **myxobacteria** (figure 4.36). What sets the myxobacteria apart from other bacteria are the complexity and advancement of their life cycle. During this cycle, the vegetative cells swarm together and differentiate into a many-celled, colored structure called the fruiting body. The fruiting body is a survival structure that makes spores by a method very similar to that of certain fungi. These fruiting structures are often large enough to be seen with the unaided eye on tree bark and plant debris.

Appendaged Bacteria

The appendaged bacteria are quite varied in their structure and life cycles, but all of them produce an extended process of the cell wall in the form of a bud, a stalk, or a long thread (figure 4.37). The stalked bacteria live attached to the surface of objects in aquatic environments. One type can even grow in distilled water or tap water. The stalks evidently help them trap minute amounts of organic materials present in the water. Budding bacteria reproduce entirely by budding; that is, they form a tiny bulb (bud) at the end of a thread. The bud then breaks off, enlarges, develops a flagellum, and swarms to another area to start its own cycle. These bacteria can also grow in very low-nutrient habitats.

ARCHAEA: THE OTHER PROCARYOTES

The discovery and characterization of novel procaryotic cells that have unusual anatomy, physiology, and genetics changed our views of microbial taxonomy and classification (see chapter 1). These single-celled, simple organisms, called **archaea*** are now considered a third cell type in a separate superkingdom (the Do-

*Archaea (ark′-ee-uh) Gr. *arche,* beginning.

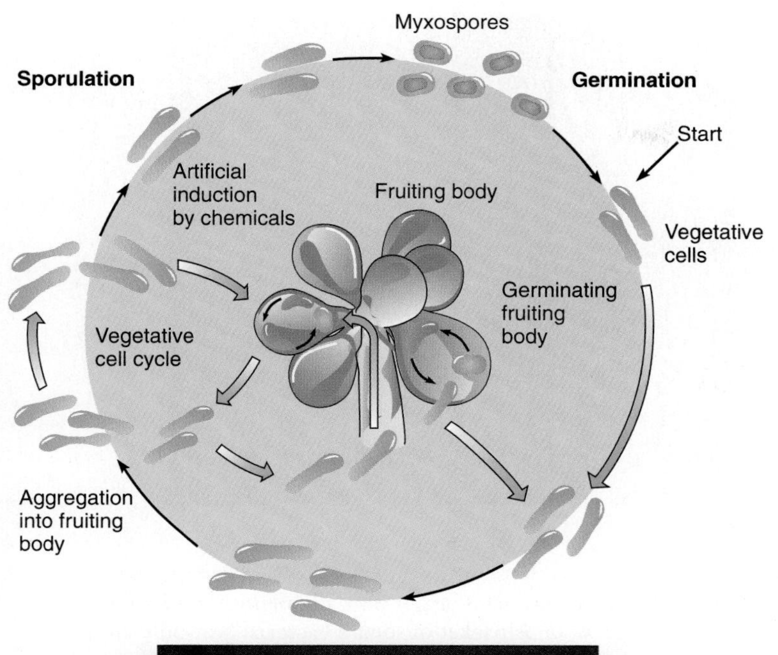

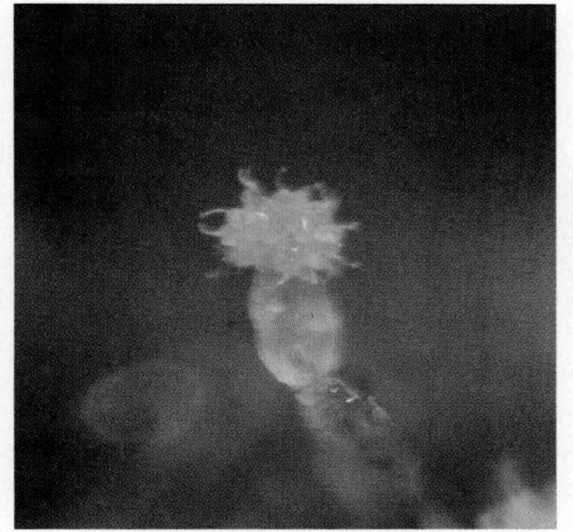

Figure 4.36
The life cycle of a myxobacterium *(Chondromyces),* with a photograph of an actual mature fruiting body.
Source: From ASM News *63(August 1997):425. Photographer David Graham.*

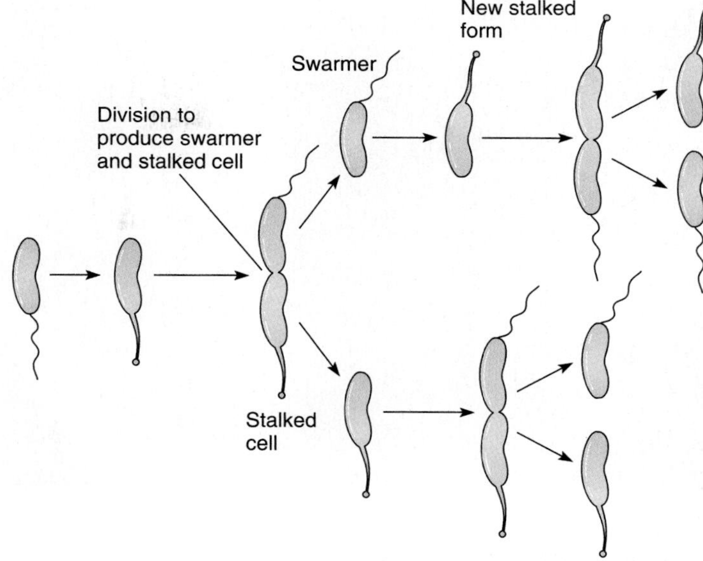

Figure 4.37
Budding bacteria. The life cycle of *Caulobacter,* showing the development of the flagellated swarm cell and stalked phases.

main Archaea). We include them in this chapter because they are procaryotic in general structure and they do share many bacterial characteristics. Evidence is accumulating that they are actually more closely related to Eukarya than to bacteria. For example, archaea and eukarya share a number of ribosomal RNA sequences that are not found in bacteria, and their protein synthesis and ribosomal subunit structure are similar. Table 4.7 outlines selected points of comparison of the three domains.

Among the ways that the archaea differ significantly from other cell types are that certain genetic sequences (CACACACCG) are found only in their rRNA (see figure 4.29) and that they have unique membrane lipids and cell wall construction. It is clear that the archaea are the most primitive of all life forms and are most closely related to the first cells that originated on the earth 4 billion years ago. The early earth is thought to have contained a hot, anaerobic "soup" with sulfuric gases and salts in abundance. The modern archaea still live in the remaining habitats

TABLE 4.7

COMPARISON OF THREE CELLULAR DOMAINS

Characteristic	Bacteria	Archaea	Eukarya
Cell type	Procaryotic	Procaryotic	Eucaryotic
Chromosomes	Single, circular	Single, circular	Several, linear
Type of ribosomes	70S	70S but structure is more similar to 80S	80S
Contains unique ribosomal RNA signature sequences	+	+	+
Number of sequences shared with Eukarya	1	3	
Protein synthesis similar to Eukarya	−	+	
Presence of peptidoglycan in cell wall	+	−	−
Cell membrane lipids	Fatty acids with ester linkages	Long-chain, branched hydrocarbons with ester linkages	Fatty acids with ester linkages
Sterols in membrane	− (some exceptions)	−	+

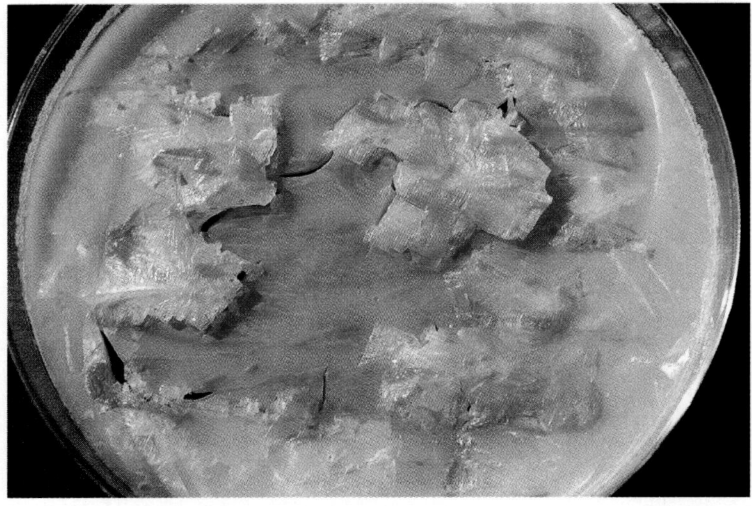
(a)

(b)

Figure 4.38
(*a*) A plate of milk salt agar growing the obligate halophile *Halobacterium salinarium*. This archaea has a bright pink pigment that increases upon exposure to light, and it has a requirement for at least 13% salt concentration to grow. (*b*) A packet of square bacteria. Not only are these unusual salt-loving cells square to rectangular in shape, but also they are extremely flat, giving the appearance of tiles.

on the earth that have these same ancient conditions—the most extreme habitats in nature. It is for this reason that they are often called extremophiles, meaning that they "love" extreme conditions in the environment.

Metabolically, the archaea exhibit nearly incredible adaptations to what would be deadly conditions for other organisms. These hardy microbes have adapted to multiple combinations of heat, salt, acid, pH, pressure, and atmosphere. Included in this group are methane producers, hyperthermophiles, extreme halophiles, and sulfur reducers.

Members of the group called *methanogens* can convert CO_2 and H_2 into methane gas (CH_4) through unusual and complex pathways. These archaea are common inhabitants of anaerobic swamp mud, the bottom sediments of lakes and oceans and even the digestive system of animals. The gas they produce collects in swamps and may become a source of fuel. Methane may also contribute to the "greenhouse effect," which maintains the earth's temperature and can contribute to global warming (see chapter 26).

Other types of archaea—the extreme halophiles—require salt to grow and may have such a high salt adaptation that they can multiply in sodium chloride solutions (36% NaCl) that would destroy most cells (chapter opener and figure 4.38*a*). They exist in the saltiest places on the earth—inland seas, salt lakes, salt mines, and salted fish. They are not particularly common in the ocean because the salt content is not high enough. Many of the "halobacteria" use a red pigment to synthesize ATP in the presence of light. These pigments are responsible for "red herrings," the color of the Red Sea, and the red color of salt ponds. An interesting type of halophile are the square bacteria—tiny packets or sheets (figure 4.38*b*).

Archaea adapted to growth at very high temperatures are defined as hyperthermophilic (they love high temperatures). These organisms flourish at temperatures between 80° and 105°C and cannot grow at 50°C. They live in volcanic waters and soils and submarine vents and are also often salt- and acid-tolerant as well. One member, *Thermoplasma,* lives in hot, acidic habitats in the waste piles around coal mines that regularly sustain a pH of 1 and a temperature of nearly 60°C. Researchers sampling sulfur vents in the deep ocean discovered thermophilic archaea flourishing at temperatures up to 250°C—150° above the temperature of boiling water! Not only were these archaea growing prolifically at this high temperature, but they were also living at 265 atmospheres of pressure. (On the earth's surface, pressure is about one atmosphere.) For additional discussion of the unusual adaptations of archaea, see chapter 7.

Chapter Checkpoints

The rickettsias are a group of bacteria that are intracellular parasites, dependent on their eucaryote host for energy and nutrients. Most are pathogens that alternate between arthropods and mammalian hosts.

The chlamydias are also small, intracellular parasites that infect humans, mammals, and birds. They do not require arthropod vectors.

The mycoplasmas are bacteria that lack cell walls and therefore lack a definite cell shape. Although they are not obligate parasites, many require a eucaryote host and have specialized growth requirements when grown in culture.

Many bacteria are free-living, rather than parasitic. The photosynthetic bacteria, gliding bacteria, and appendaged bacteria encompass many subgroups that colonize specialized habitats, not other living organisms.

Archaea are an unusual type of procaryotic cell that constitute the third domain of life. Members are adapted to extreme habitats with high temperature, salt, pressure, or acid.

Chapter Capsule with Key Terms

Names for Procaryotes: Bacteria, Archaea, Monera, Archaebacteria, Eubacteria, Schizomycetes (the last three are old).

OVERALL MORPHOLOGY

Procaryotic cells lack a nucleus, mitochondria, chloroplasts, or other membranous organelles. The outer covering is a **cell envelope** composed of **glycocalyx (slime layer, capsule), cell walls** (in most), and **cell membrane** with **mesosomes. Gram stain** differentiates two types of cells on the basis of wall structure: **gram-positive** cells have a thick layer of **peptidoglycan; gram-negative** cells have an outer membrane and a thin layer of peptidoglycan. The diameter of most bacteria is 1–5 μm. Special **appendages:** bacterial **flagella** that provide motility are common; **fimbriae** are important in cell adhesion, and **pili** in genetic recombination. Genetic material is in **chromatin bodies** and **plasmids; ribosomes** synthesize proteins. Other special structures include **endospores** that are highly resistant survival cells and **inclusions** and **granules** for storing nutrients.

Most bacteria are unicellular and form colonies on solid nutrients. Multicellularity is rare. Their shapes are distinct, including **cocci, bacilli, spirilla,** and **spirochetes.** Certain groups can assume long filaments; some show variations in shape, or **pleomorphism.** Various **arrangements** based upon mode of cell division are termed **diplo-, strepto-, staphylo-, palisade.**

NUTRITIONAL/HABITAT REQUIREMENTS

A large number of bacteria are heterotrophs that require an organic carbon source derived from dead organisms by absorption or from a live host (parasites). Several groups are capable of synthesizing nutrients using sunlight (photosynthetic). Bacteria can be aerobic, facultative, or anaerobic with respect to oxygen, and most exist between 10° and 40°C. Some species live in very cold habitats, even at subfreezing temperatures, and some at high temperatures (up to 250°C).

REPRODUCTION AND LIFE CYCLES

Bacteria generally have a single chromosome; reproduce primarily by asexual mans, including budding and simple binary fission; and have no mitosis. Genetic exchange can occur through conjugation. The life cycle is complex in spore-formers, gliding bacteria, and appendaged bacteria.

IDENTIFICATION

Bacteria are identified by means of microscopic morphology (Gram stain, shape, special structures), macroscopic morphology, and biochemical, molecular, and genetic characteristics.

CLASSIFICATION OF MAJOR GROUPS

One system divides the Kingdom Procaryotae into four divisions, based on the type of cell wall.

Gracilicutes: The largest group; contains bacteria with gram-negative cell walls; includes several medically significant microbes, such as *Salmonella, Shigella,* and other intestinal pathogens, the **Rickettsias,** and the agents of gonorrhea and syphilis. The photosynthetic, gliding, sheathed, and appendaged bacteria are also in this group.

Firmicutes: Gram-positive cells; examples of pathogens are in the genera *Streptococcus, Staphylococcus, Mycobacterium* (tuberculosis), and *Clostridium* (tetanus).

Tenericutes: Cells without cell walls; primarily in the genus **Mycoplasma** (one types causes pneumonia).

Mendosicutes (also Archaea or archaebacteria): Cells with very atypical cell walls; most species are free-living in extreme environments; includes the methanogens and the extreme heat- and salt-tolerant species. No known medically important species.

DOMAIN ARCHAEA

Unusual procaryotes different from bacteria in genetics and some structural features. Are adapted to extremes of habitat including salt, temperature, and nutrients. Halophiles live in salt concentrations of 10–30%; hyperthermophiles can live at temperatures of 60°–110°C or more. Methanogens live without oxygen and produce methane gas.

ECOLOGICAL IMPORTANCE

Because prokaryotes are highly adaptable as a group, they are found nearly everywhere: water, soil, air, dust, food, deep sea, plants, animals, swamps, hot springs, North and South Poles. They are important as decomposers of organic matter and play a role in the cycles of nitrogen, phosphorus, sulfur, and carbon. The cyanobacteria contribute oxygen to the atmosphere through photosynthesis.

ECONOMIC IMPORTANCE

Bacteria are used in a number of industries, including food, drugs (production of antibiotics, vaccines, and other medicines); and biotechnology (manipulation of bacteria for making large quantities of hormones, enzymes, and other proteins). Bacteria are responsible for spoilage of food and vegetables. Many plant pathogens adversely affect the agricultural industry.

MEDICAL IMPORTANCE

Bacteria are very common pathogens of humans. Approximately 200 species are known to cause disease in humans, and many normally inhabit the bodies of humans and other animals. Infections can be treated with antibiotics.

UNIQUE FEATURES

Bacteria and Archaea are the smallest cells in existence. They have procaryotic structure, unique shapes and arrangements, and lack traditional sexual reproduction and mitosis. Many are pleomorphic. The cell wall of most contains peptidoglycan, and the cytoplasm of some contains mesosomes. Other unique features are the single chromosome and 70S ribosomes.

MULTIPLE-CHOICE QUESTIONS

1. Which of the following is not found in all bacterial cells?
   a. cell membrane c. ribosomes
   b. a nucleoid d. capsule
2. The major locomotor structures in bacteria are
   a. flagella c. fimbriae
   b. pili d. cilia
3. Pili are tubular shafts in ____ bacteria that serve as a means of ____.
   a. gram-positive, genetic exchange
   b. gram-positive, attachment
   c. gram-negative, genetic exchange
   d. gram-negative, protection
4. An example of a glycocalyx is
   a. a capsule c. fimbriae
   b. pili d. a cell wall
5. Which of the following is a primary bacterial cell wall function?
   a. transport c. support
   b. motility d. adhesion
6. Which of the following is present in both gram-positive and gram-negative cell walls?
   a. an outer membrane
   b. peptidoglycan
   c. techoic acid
   d. lipopolysaccharides
7. Mesosomes are internal extensions of the
   a. cell wall c. cell membrane
   b. chromatin body d. capsule
8. Metachromatic granules are concentrated crystals of ____ that are found in ____.
   a. fat, *Mycobacterium*
   b. dipicolinic acid, *Bacillus*
   c. sulfur, *Thiobacillus*
   d. PO_4, *Corynebacterium*
9. Bacterial endospores function in
   a. reproduction c. protein synthesis
   b. survival d. storage
10. A bacterial arrangement in packets of eight cells is described as a ____.
    a. micrococcus c. tetrad
    b. diplococcus d. sarcina
11. The major difference between a spirochete and a spirillum is
    a. presence of flagella
    b. a cell with coils
    c. the nature of motility
    d. size
12. Genetic analysis of bacteria would include
    a. fermentation testing
    b. ability to digest complex nutrients
    c. presence of oxidase
    d. G + C content
13. Which division of bacteria has a gram-positive cell wall?
    a. Gracilicutes c. Firmicutes
    b. Archaea d. Tenericutes
14. To which division of bacteria do cyanobacteria belong?
    a. Tenericutes c. Firmicutes
    b. Gracilicutes d. Mendosicutes

CONCEPT QUESTIONS

1. a. Name several general characteristics that could be used to define the procaryotes.
   b. Do any other microbial groups besides bacteria have procaryotic cells?
   c. What does it mean to say that bacteria are ubiquitous? In what habitats are they found? Give some general means by which bacteria derive nutrients.
2. a. Describe the structure of a flagellum and how it operates. What are the four main types of flagellar arrangement?
   b. How does the flagellum dictate the behavior of a motile bacterium? Differentiate between flagella and periplasmic flagella.
   c. List some direct and indirect ways that one can determine bacterial motility.
3. a. List the components of the cell envelope.
   b. Explain the position of the glycocalyx.
   c. What are the functions of slime layers and capsules?
   d. How is the presence of a slime layer evident even at the level of a colony?
4. a. Differentiate between pili and fimbriae.
   b. How do their structures differ?
   c. How do their functions differ?
5. a. Compare the cell envelopes of gram-positive and gram-negative bacteria.
   b. What function does peptidoglycan serve?
   c. To which part of the cell envelope does it belong?
   d. Give a simple description of its structure.
   e. What happens to a cell that has its peptidoglycan disrupted or removed?
   f. What functions does the LPS layer serve?
6. a. What is the Gram stain?
   b. What is there in the structure of bacteria that causes some to stain purple and others to stain red?
   c. How does the precise structure of the cell walls differ in gram-positive and gram-negative bacteria?
   d. What other properties besides staining are different in gram-positive and gram-negative bacteria?
   e. What is the periplasmic space, and how does it function?
   f. What characteristics does the outer membrane confer on gram-negative bacteria?
7. a. List five functions that the cell membrane performs in bacteria.
   b. What are mesosomes and some of their possible functions?
8. a. Compare the composition of the chromatin body and plasmids.
   b. What are the functions of each?
9. a. What is unique about the structure of bacterial ribosomes?
   b. How do they function?
   c. Where are they located?
10. a. Compare and contrast the structure and function of inclusions and granules.
    b. What are metachromatic granules, and what do they contain?
11. a. Describe the vegetative stage of a bacterial cell.
    b. Describe the structure of an endospore, and explain its function.
    c. Describe the endospore-forming cycle.
    d. Explain why an endospore is not considered a reproductive body.
    e. Why are endospores so difficult to destroy?
12. a. Draw the three bacterial shapes.
    b. How are spirochetes and spirilla different?
    c. What is a vibrio?
    d. A coccobacillus?
    e. What is pleomorphism?
    f. What is the difference between the use of the term bacillus and the name *Bacillus*? *Staphylococcus* and staphylococcus?

13. a. Rank the size ranges in bacteria according to shape.
    b. Rank the bacteria in relationship to viruses and eucaryotic cell size.
14. a. What characteristics are used to classify bacteria?
    b. What are the most useful characteristics for categorizing bacteria into families?
    c. In what ways is ribosomal RNA an important method for differentiating bacteria and groups of organisms?
15. a. How is the species level in bacteria defined?
    b. Name at least three ways bacteria are grouped below the species level.
    c. In what ways are they important?
16. a. Describe at least 2 circumstances that give rise to L forms.
    b. How do L forms survive?
    c. In what ways are they important?
17. Name several ways in which bacteria are medically and ecologically important.
18. a. Explain the characteristics of Archaea that indicate it is a unique domain of living things that is neither a bacterium nor a eucaryote.
    b. What leads microbiologists to believe the Archaea are more closely related to Eukarya than to Bacteria?
    c. What is meant by the term *extremophile?* Describe some archaeal adaptations to extreme habitats.

CRITICAL-THINKING QUESTIONS

1. What would happen if one stained a gram-positive cell with safranin? A gram-negative one with crystal violet? What would happen to the two types if the mordant were omitted? Speculate on what the gram reaction of your body cells would be.
2. What is required to kill endospores? How do you suppose archaeologists were able to date some spores as being 3,000 years old?
3. Using clay, demonstrate how cocci can divide in several planes and show the outcome of this division. Show how the arrangements of bacilli occur, including palisades.
4. Using a corkscrew and a spring to compare the flexibility and locomotion of spirilla and spirochetes, explain which group is represented by each of them.
5. Under the microscope you see a rod-shaped cell that is swimming rapidly forward.
   a. What do you automatically know about that bacterium's structure?
   b. How would a bacterium use its flagellum for phototaxis?
   c. Can you think of another function of flagella besides locomotion?
6. Can you determine which probes in figure 4.28 successfully hybridized and which did not? Account for the reasons that some probes did not attach. What are the implications of the probe's attaching to the test DNA?
7. a. Name an unusual type of bacterium that lives in extreme habitats.
   b. What adaptations enable it to survive there?
8. a. Name a bacterium that has no cell walls.
   b. How is it protected from osmotic destruction?
9. a. Name a bacterium that is aerobic, gram positive, and spore-forming.
   b. What habitat would you expect this species to occupy?
10. Name a bacterium that is pleomorphic and has a palisades arrangement and metachromatic granules.
11. a. Name an acid-fast bacterium.
    b. What characteristics make this bacterium different from other gram-positive bacteria?
12. a. Name two main groups of obligate intracellular parasitic bacteria.
    b. Why can't these groups live independently?
13. Name a bacterium that lives in extremes of heat and salt.
14. Name a coccus that is gram positive and in chains.
15. Name a gram-negative diplococcus.
16. a. Name a bacterium that contains sulfur granules.
    b. What is the advantage in storing these granules?
17. a. Name a bacterium that uses chlorophyll to photosynthesize.
    b. Describe the two major groups of photosynthetic bacteria.
    c. How are they similar?
    d. How are they different?
18. What are some possible adaptations that the giant bacterium *Epulopiscium* has had to make because of its large size?
19. Propose a hypothesis to explain how Bacteria and Archaea could have, together, given rise to the Eukarya.

INTERNET SEARCH TOPIC

Go to a search engine (Yahoo, Alta Vista) and type in "Martian Microbes." Look for papers and information that support or reject the idea that fossil structures discovered in an ancient meteor from Mars could be bacteria. What are two of the main reasons that microbiologists are skeptical of this possiblity?

5 chapter

Eucaryotic Cells and Microorganisms

The eucaryotic cell is a complex, compartmentalized unit that differs from the procaryotic cell by containing a nucleus and several other specialized structures called organelles. Although exact cell structures differ somewhat among the several groups of eucaryotic organisms, the eucaryotic cell is the typical cell of certain microbial groups (fungi, algae, protozoa, and helminth worms) as well as all animals and plants. In this chapter, we examine the overall structure and function of eucaryotic cells in preparation for later chapters that deal with related microbiological concepts, including multiplication of viruses, metabolism, genetics, nutrition, drug therapy, immunology, and disease. Because of the tremendous variety of eucaryotic microorganisms and their practical importance in medicine, industry, and agriculture, this chapter will also cover the major characteristics of each group of eucaryotic microorganisms.

Zoospore of the dinoflagellate *Pfiesteria piscicida,* a protist that displays both algal and protozoan characteristics. Although it is free-living, it is known to parasitize fish and release potent toxins that kill fish and sicken humans.

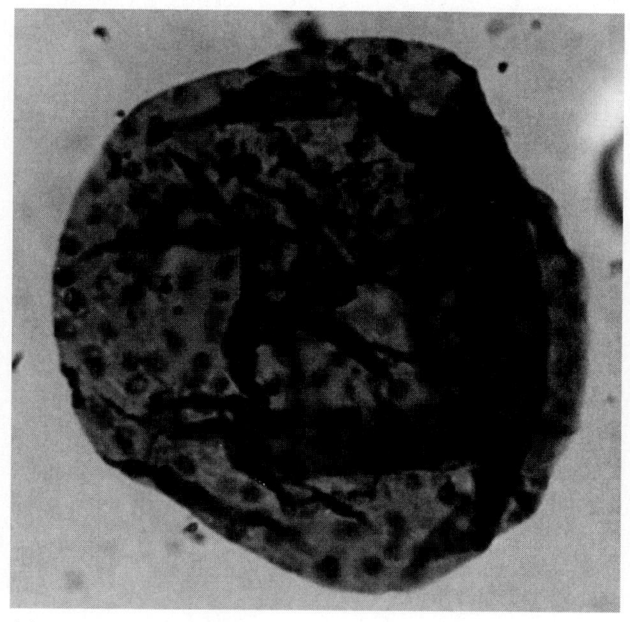

(a)

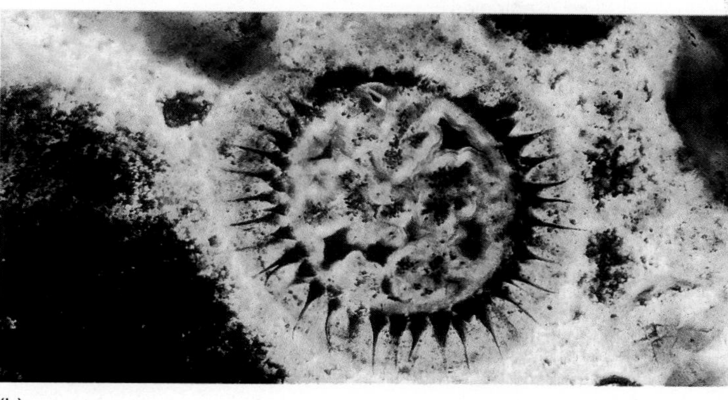

(b)

Figure 5.1

Ancient eucaryotic protists caught up in fossilized rocks. Both forms are called acritarchs. (*a*) An ornamented algalike cell found in the shale of Siberia and dated from 850 million to 950 million years old. (*b*) A large, disclike form with a cell wall bearing a crown of spines from Chinese rock from 590 million to 610 million years ago.

THE NATURE OF EUCARYOTES

Evidence from paleontology indicates that the first eucaryotic cells appeared on the earth approximately 2 billion years ago. Some fossilized cells that look remarkably like algae or protozoa appear in shale sediments from China and Australia that date from 800 million to 900 million years ago (figure 5.1). Biologists have discovered profound evidence to suggest that the eucaryotic cell evolved from procaryotic organisms by a process of intracellular *symbiosis** (microfile 5.1). It now seems clear that some of the **organelles** that distinguish eucaryotic cells originated from procaryotic cells that became trapped inside them. The structure of these first eucaryotic cells was so versatile that eucaryotic microorganisms soon spread out into available habitats and adopted greatly diverse styles of living.

The first primitive eucaryotes were probably single-celled and independent, but, over time, some forms began to aggregate, forming colonies and filaments. With further evolution, some of the cells within colonies became *specialized,* or adapted to perform a particular function advantageous to the whole colony, such as locomotion, feeding, or reproduction. Complex multicellular organisms evolved as individual cells in the organism lost the ability to survive apart from the intact colony. Although a multicellular organism is composed of many cells, it is more than just a disorganized assemblage of cells like a colony. Rather, it is composed of distinct groups of cells that cannot exist independently of the rest of the body. The cell groupings of multicellular organisms that have a specific function are termed *tissues,* and groups of tissues make up *organs.*

Looking at modern eucaryotic organisms, we find examples of many levels of cellular organization. All protozoa, as well as numerous algae and fungi, live a unicellular or colonial existence. Truly multicellular organisms are found only among plants and animals and some of the fungi and algae. Only certain eucaryotes are small enough to fall into the realm of microbiology: namely, the protozoa and some of the algae, fungi, and animal parasites.

✓ Chapter Checkpoints

Eucaryotes are cells with organelles and a nucleus compartmentalized by membranes. They might have originated from procaryote ancestors about 2 billion years ago. Eucaryotic cell structure enabled eucaryotes to diversify from single cells into a huge variety of complex multicellular forms.

FORM AND FUNCTION OF THE EUCARYOTIC CELL: EXTERNAL STRUCTURES

The cells of eucaryotic organisms are so varied that no one member can serve as a typical example. Figure 5.2 presents the generalized structure of typical algal, fungal, and protozoan cells. The outline at the bottom of page 126 shows the organization of a eucaryotic cell. (Only motile algae contain all categories of eucaryotic organelles.)

In general, eucaryotic microbial cells have a cytoplasmic membrane, nucleus, mitochondria, endoplasmic reticulum, Golgi apparatus, vacuoles, and cytoskeleton. A cell wall, locomotor appendages, chloroplasts, and glycocalyx are found only in some groups. In the following sections, we cover the microscopic structure and functions of the eucaryotic cell. As with the procaryotes, we begin on the outside and proceed inward through the cell.

*symbiosis (sim-beye-oh´sis) Gr. *sym,* together, and *bios,* to live.

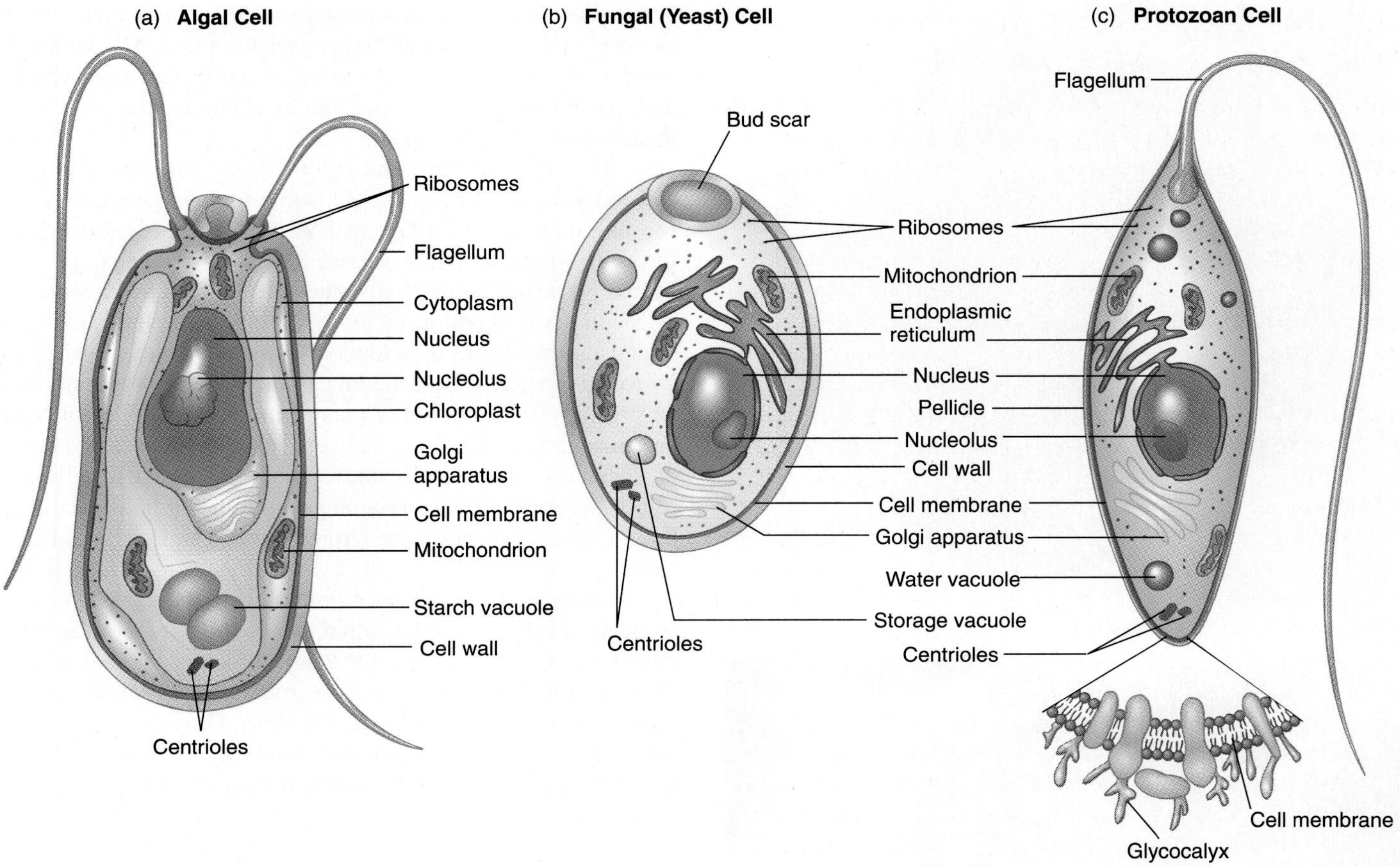

Figure 5.2
The structure of three representative eucaryotic cells. (*a*) *Chlamydomonas,* a unicellular motile alga. (*b*) A yeast cell (fungus). (*c*) *Peranema,* a flagellated protozoan.

Structure Flowchart

- Eucaryotic cell
  - Appendages
    - Flagella
    - Cilia
  - Surface structures
    - Glycocalyx
      - Capsules
      - Slimes
  - Cell wall
  - Cytoplasmic membrane
  - Nucleus
    - Nuclear envelope
    - Nucleolus
    - Chromosomes
  - Cytoplasm
    - Organelles
      - Endoplasmic reticulum
      - Golgi complex
        - Lysosomes
        - Vacuoles
      - Mitochondria
      - Chloroplasts
  - Cytoskeleton
    - Microtubules
    - Microfilaments
  - Ribosomes

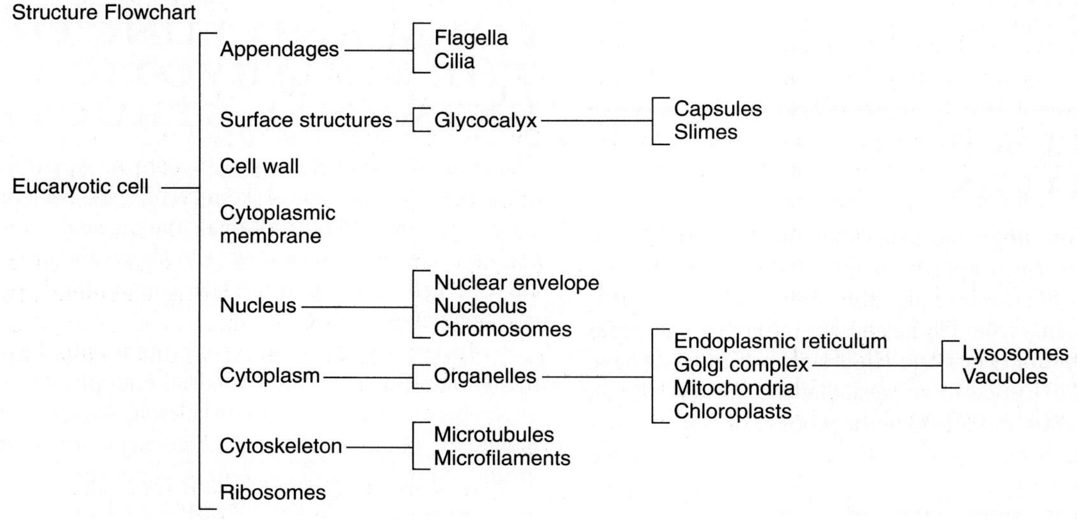

LOCOMOTOR APPENDAGES: CILIA AND FLAGELLA

Motility allows a microorganism to locate life-sustaining nutrients and to migrate toward positive stimuli such as sunlight; it also permits avoidance of harmful substances and stimuli. Locomotion by means of flagella or cilia is a common property of protozoa, many algae, and a few fungal and animal cells.

Although they share the same name, the structure and operation of eucaryotic **flagella** are much different from those of procaryotes. The eucaryotic flagellum is thicker (by a factor of 10), structurally more complex, and covered by an extension of the cell membrane. A single flagellum is a long, sheathed cylinder containing regularly spaced hollow tubules—**microtubules**—that extend along its entire length (figure 5.3*a*). A cross section reveals nine pairs of closely attached microtubules surrounding a single central pair. This scheme, called the 9 + 2 arrangement, is a universal pattern of flagella and cilia. During locomotion, the adjacent microtubules slide past each other, whipping the flagellum back and forth. Although details of this process are too complex to discuss here, it involves expenditure of energy and a coordinating mechanism in the cell membrane. Flagella can move the cell by pushing it forward like a fishtail or by pulling it by a lashing or twirling motion (figure 5.3*b*). The placement and number of flagella can be useful in identifying flagellated protozoa and certain algae.

Cilia* are very similar in overall architecture to flagella, but they are shorter and more numerous (some cells have several thousand). They are found only on a single group of protozoa and certain animal cells. In the ciliated protozoa, the cilia occur in rows over the cell surface, where they beat back and forth in regular oarlike strokes (see figure 5.31). Such protozoa are among the fastest of all motile cells. The fastest ciliated protozoon can swim up to 2,500 μm/s—a meter and a half per minute! On some cells, cilia also function as feeding and filtering structures. A notable example of the function of animal cilia occurs in the epithelial cells in the respiratory tract (see figure 14.3).

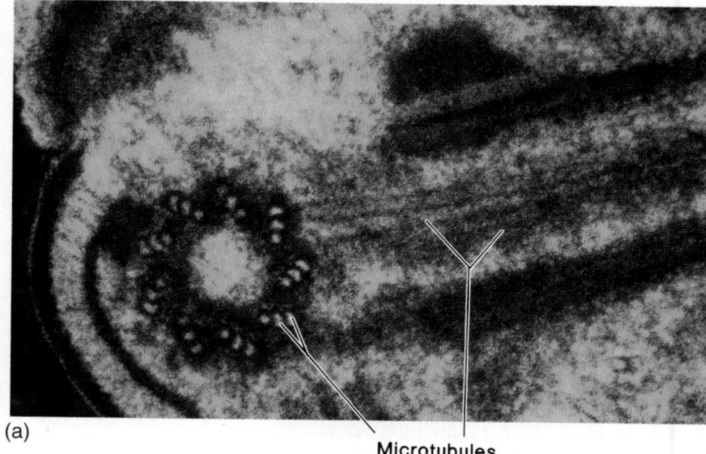

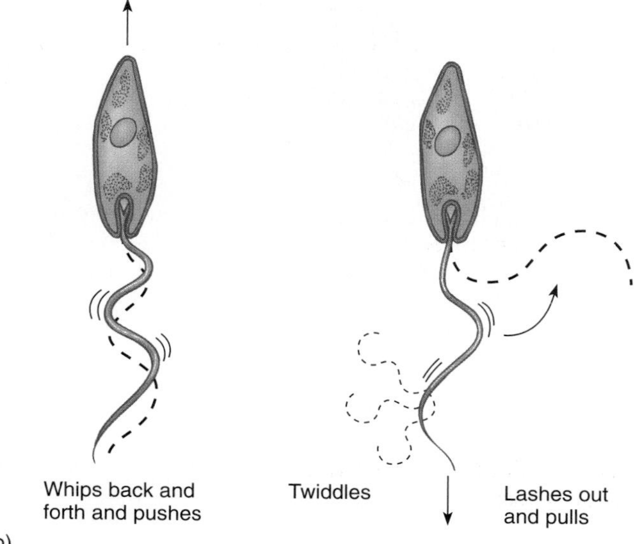

Figure 5.3
(*a*) Longitudinal section through a flagellum, showing microtubules. On the left is a cross section (circular area) that reveals the typical 9 + 2 arrangement found in both flagella and cilia. (*b*) Locomotor patterns seen in flagellates.

SURFACE STRUCTURES: THE GLYCOCALYX

Most eucaryotic cells have a **glycocalyx,** an outermost boundary that comes into direct contact with the environment (see figure 5.2). This structure is usually composed of polysaccharides and appears as a network of fibers, a slime layer, or a capsule much like the glycocalyx of procaryotes. Because of its positioning, the glycocalyx contributes to protection, adherence of cells to surfaces, and reception of signals from other cells and from the environment. We will explore the function of receptors that are part of the glycocalyx in chapters 14 and 15 (host defenses and immune interactions). The nature of the layer supporting the glycocalyx varies among the several eucaryotic groups. Fungi and most algae have a thick, rigid cell wall, whereas protozoa, a few algae, and all animal cells lack a cell wall and have only a cell membrane beneath.

THE CELL WALL

The cell walls of algal and fungal cells are rigid and provide structural support and shape, but they are different in chemical composition from procaryotic cell walls. Fungal cell walls have a thick, inner layer of polysaccharide fibers composed of chitin or cellulose and a thin outer layer of mixed glycans (figure 5.4). The cell walls of algae are quite varied in chemical composition. Substances commonly found among various algal groups are cellulose, pectin,[1] mannans,[2] and minerals such as silicon dioxide and calcium carbonate.

*cilia (sil´-ee-uh) sing. cilium; L. *siliam,* eyelid.

1. A polysaccharide composed of galacturonic acid subunits.
2. A polymer of the sugar known as mannose.

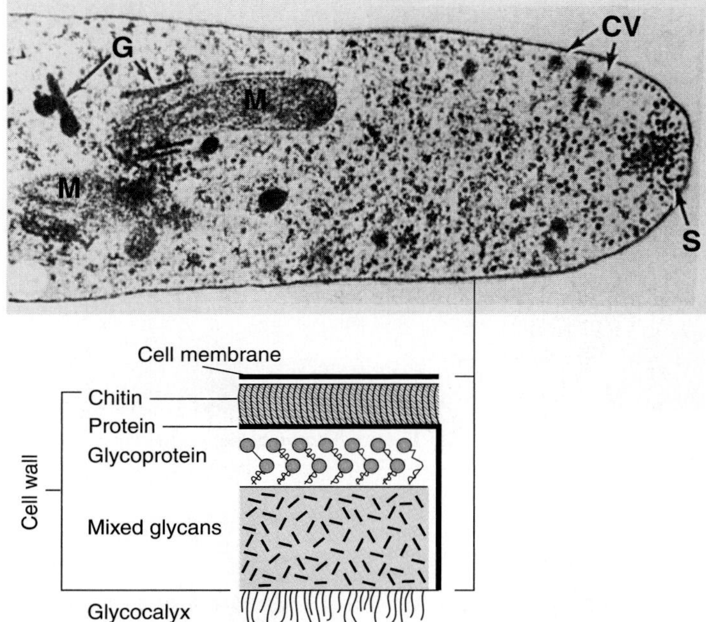

Figure 5.4
Cross section through the tip of a fungal cell to show the general structure of the cell wall and other features. (S, growing tip; CV, coated vesicles; G, Golgi apparatus; M, mitochondrion). The cell wall is a thick, rigid structure composed of complex layers of polysaccharides and proteins.

Chapter Checkpoints

The cell structures common to most eucaryotes are the cell membrane, a membrane-enclosed nucleus with nucleolus inside, vacuoles, mitochondria, endoplasmic reticulum, Golgi apparatus, and a cytoskeleton. Cell walls, chloroplasts, and locomotor organs are present in some eucaryote groups.

Microscopic eucaryotes use locomotor organs such as flagella or cilia for moving themselves or their food.

The glycocalyx is the outmost boundary of most eucaryotic cells. Its functions are protection, adherence, and reception of chemical signals from the environment or from other organisms. The glycocalyx is supported by either a cell wall or a cell membrane.

The cytoplasmic membrane of eucaryotes is similar in function to that of procaryotes, but it differs in composition, possessing sterols as additional stabilizing agents.

Eucaryotic cells have membrane-bound organelles, which are never present in procaryotes. Such compartmentalization allows a wide variety of cellular functions to occur in specialized areas of the cell.

The Cell Membrane

The cell membrane of eucaryotic cells is a typical bilayer of lipids in which protein molecules are embedded. In addition to phospholipids, eucaryotic membranes also contain *sterols* of various kinds. Sterols are different from phospholipids in both structure and behavior (see figure 2.19). Their relative rigidity confers stability on eucaryotic membranes. This strengthening feature is extremely important in cells that lack a cell wall. Cytoplasmic membranes of eucaryotes are functionally similar to those of procaryotes, serving as selectively permeable barriers in transport (see chapter 7). Unlike procaryotes, eucaryotic cells also contain a number of individual membrane-bound organelles that are extensive enough to account for 60% to 80% of their volume.

Survey of Major Organelles

Eucaryotic cells accomplish their cellular work with unique intracellular membrane-wrapped organelles. Each type of organelle carries out one or two functions, and it can partition the cell into many small compartments, thereby serving to organize and separate metabolic events. Functions such as synthesis and secretion often proceed in an assembly-line fashion, with each organelle in the series affecting the outcome of the final product or process.

FORM AND FUNCTION OF THE EUCARYOTIC CELL: INTERNAL STRUCTURES

The Nucleus: The Cell Control Center

The **nucleus** is a compact sphere that is the most prominent organelle of eucaryotic cells. It is separated from the cell cytoplasm by an external boundary called a **nuclear envelope.** The envelope has a unique architecture. It is composed of two parallel membranes separated by a narrow space, and it is perforated with small, regularly spaced openings, or pores, formed at sites where the two membranes unite (figure 5.5). The nuclear pores are passageways through which macromolecules migrate from the nucleus to the cytoplasm and vice versa. The nucleus contains a matrix called the **nucleoplasm** and a granular mass, the **nucleolus,*** that can stain more intensely than the immediate surroundings because of its RNA content. The nucleolus is the site for ribosomal RNA synthesis and a collection area for ribosomal subunits. The subunits are transported through the nuclear pores into the cytoplasm for final assembly into ribosomes.

A prominent feature of the nucleoplasm in stained preparations is a network of dark fibers known as **chromatin*** because of its attraction for dyes. Analysis has shown that chromatin actually

*nucleolus (noo-klee´-oh-lus) pl. nucleoli; L. diminutive of nucleus.

*chromatin (kroh´-muh-tin) Gr. *chromos,* color.

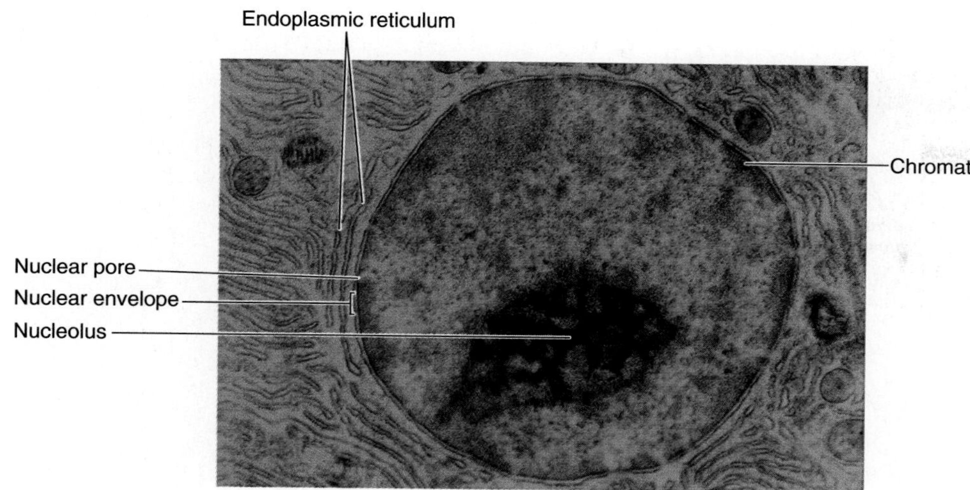

Figure 5.5

Electron micrograph section of an interphase nucleus, showing its most prominent features.

comprises the eucaryotic **chromosomes,** large units of genetic information in the cell. The chromosomes in the nucleus of most cells are not readily visible because they are long, linear DNA molecules bound in varying degrees to *histone*[3] proteins, and they are far too fine to be resolved as distinct structures without extremely high magnification. During **mitosis,*** however, when the duplicated chromosomes are separated equally into daughter cells, the chromosomes themselves finally become readily visible as discrete, X- or Y-shaped structures (figure 5.6). This appearance arises when the DNA becomes highly condensed by forming coils and supercoils around the histones to prevent the chromosomes from tangling as they are separated into new cells.

The Relationship of Chromosome Number to Reproductive Mode

To ensure its continued existence, a eucaryotic cell maintains a specified number of chromosomes that is typical for its species. The set of chromosomes exist in the **haploid*** (single, unpaired) state or in the **diploid*** (matched pair) state. Most fungi, many algae, and some protozoa are examples of organisms whose cells occur in the haploid state during most of the life cycle. For example, the chromosome number of one type of ameba is 25, and one species of the fungus *Penicillium* possesses 5 chromosomes. Regardless of the chromosome state of the adult organism, all gametes (sex cells) are haploid.

The cells of animals and plants, and of some protozoa, fungi, and algae, spend the greater part of their life cycle in the diploid state. This state comes about when the female haploid gamete (egg) fuses with the male haploid gamete (sperm) and produces a double (diploid) set of chromosomes in the offspring during sexual reproduction. Examples are humans, with 46 chromosomes (23 pairs), a parasitic blood fluke, with 16 chromosomes (8 pairs), and a redwood tree, with 22 chromosomes (11 pairs).

During normal cell division (asexual reproduction) of all eucaryotic organisms, the chromosome number, whether haploid or diploid, is maintained by **mitosis** (figures 5.6 and 5.7). When it comes to sexual reproduction in diploid organisms, however, the diploid chromosome number must be reduced to haploid, because gametes must be haploid for proper fertilization. This reduction is accomplished through reduction division, or **meiosis*** (figure 5.7*b*). During this process, diploid cells in the sex organs undergo two sets of divisions that results in haploid sex cells. Formation of a diploid zygote then restores the chromosome number characteristic of that organism. Another example of the function of meiosis occurs in the sexual reproductive cycles of haploid fungi. Normal mitosis alone can produce their gametes, but after these gametes unite, they form a diploid zygote. In order to restore the fungus to the haploid state, the diploid nucleus undergoes meiosis and produces haploid offspring (figure 5.7*a*, step 2, and figure 5.20).

ENDOPLASMIC RETICULUM: A MAZE OF MEMBRANES

The **endoplasmic reticulum* (ER)** can be likened to a labyrinth (an intricate passageway) on a microscopic scale. In sections, it appears as a parallel series of thin, flat pouches bounded by membranes. Two kinds of endoplasmic reticulum are the **rough endoplasmic reticulum (RER)** (figure 5.8) and the **smooth endoplasmic reticulum (SER).** Electron micrographs show that the RER originates from the outer membrane of the nuclear envelope and extends in a continuous network through the cytoplasm, even out to the cell membrane. This architecture permits the spaces in the

3. A simple type of protein containing a high proportion of lysine and arginine.
*mitosis (my-toh´-sis) Gr. *mitos,* thread, and *osis,* a condition of.
*haploid (hap´-loyd) Gr. *haplos,* simple, and *eidos,* form.
*diploid (dip´-loyd) Gr. *di,* two, and *eidos,* form.
*meiosis (my-oh´-sis) Gr. *meiosis,* diminution. A process that permits paired chromosome sets to be reduced to single sets.
*endoplasmic (en´´-doh-plas´-mik) Gr. *endo,* within, and *plasm,* something formed;
*reticulum (rih-tik´-yoo-lum) L. *rete,* net or network.

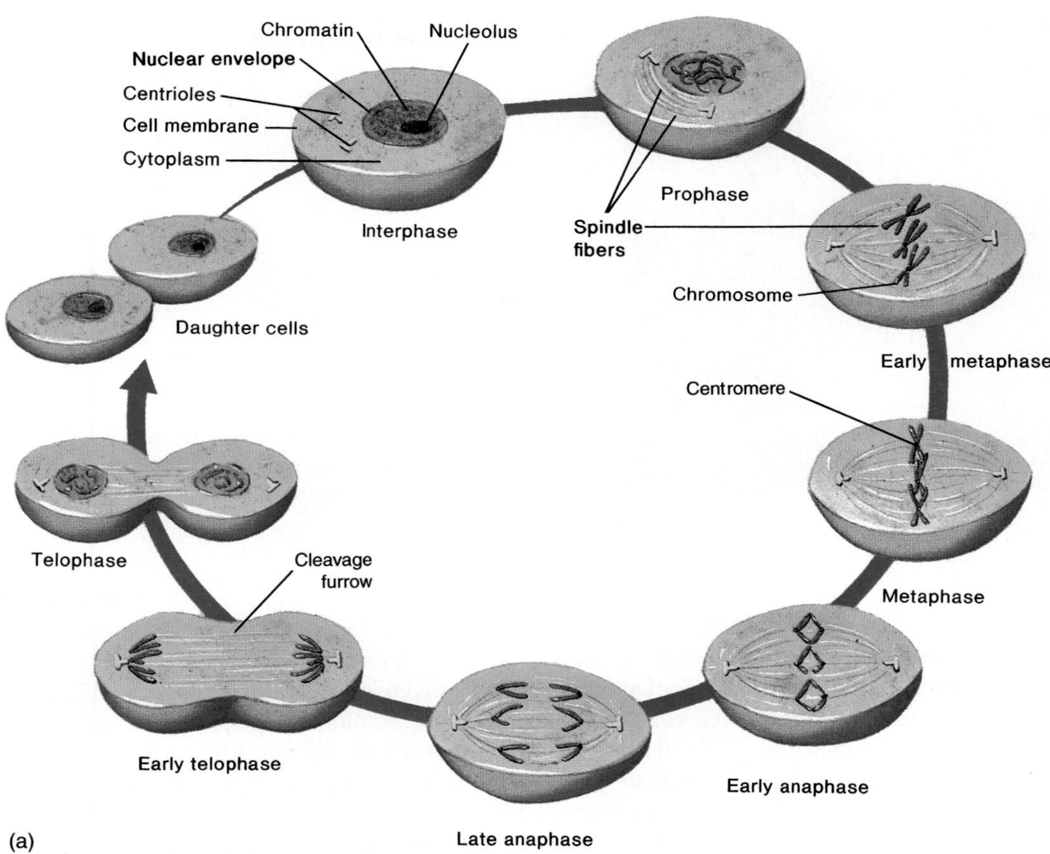

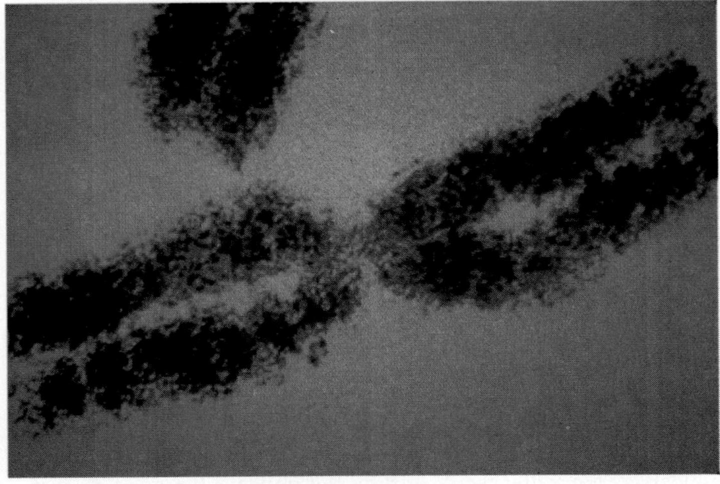

Figure 5.6
Changes in the cell and nucleus that accompany mitosis in a protozoon. (*a*) Before mitosis (at interphase), chromosomes are visible only as chromatin. As mitosis proceeds (early prophase), chromosomes take on a fine, threadlike appearance as they condense, and the nuclear membrane and nucleolus are temporarily disrupted. (*b*) By metaphase, the chromosomes are fully visible as X-shaped structures. The shape is due to duplicated chromosomes attached at a central point, the centromere. Spindle fibers attach to these and facilitate the separation of individual chromosomes during metaphase. Later phases serve in the completion of chromosomal separation and division of the cell proper into daughter cells.

RER, or **cisternae,** to transport materials from the nucleus to the cytoplasm and ultimately to the cell's exterior. The RER appears rough because of large numbers of ribosomes partly attached to its membrane to act as secretory factories. Proteins synthesized on the ribosomes are shunted into the cavity of the reticulum and held there for later packaging and transport. The SER is a closed tubular network without ribosomes that functions in nutrient processing and synthesis and in storage of nonprotein macromolecules such as lipids.

Golgi Apparatus: A Packaging Machine

The **Golgi*** complex, also called the Golgi body or **apparatus,** is a discrete organelle consisting of a stack of several flattened, disc-shaped sacs, also called cisternae. These sacs have outer limiting

*Golgi (gol´-jee) Name for C. Golgi, an Italian histologist who first described the apparatus in 1898.

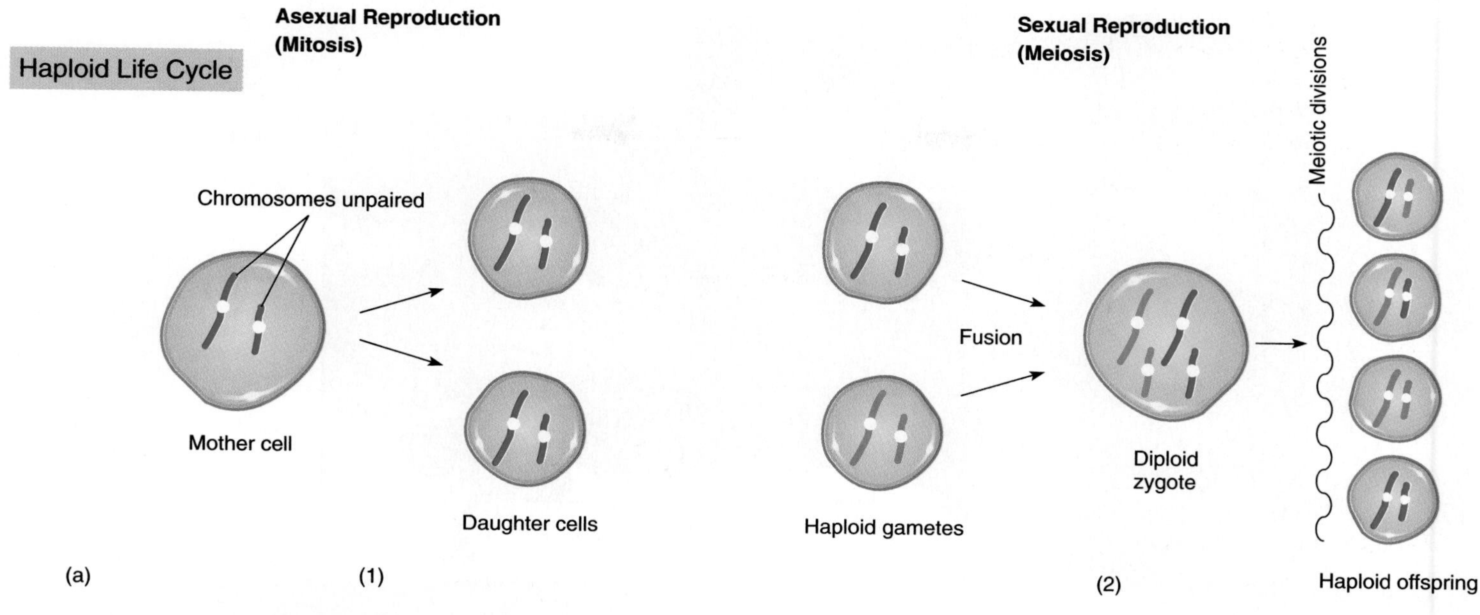

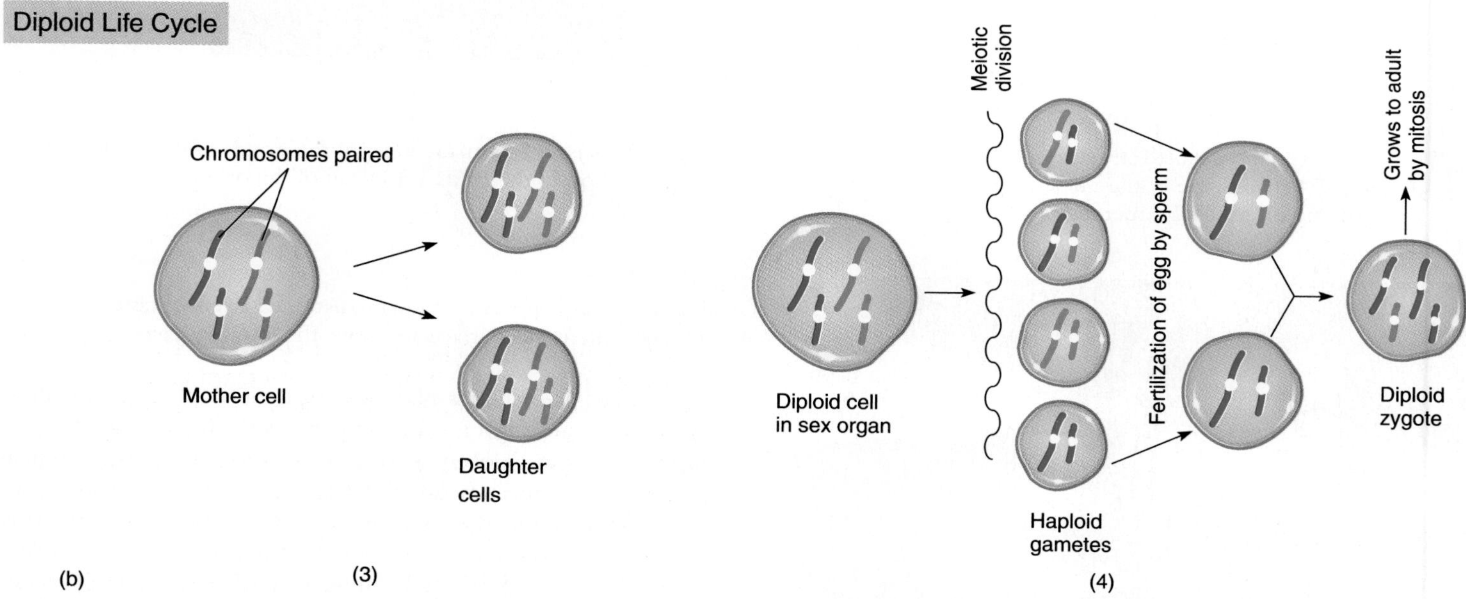

Figure 5.7

Schematic of haploid versus diploid life cycles, showing the roles of mitosis and meiosis in reproduction and maintaining an organism's chromosome number. (*a*) Cells with haploid life cycle (algae, fungi, some protozoa). Chromosomes are unpaired. (*1*) Asexual reproduction is accomplished through mitosis, which maintains the chromosome number in the daughter cells. (*2*) Sexual reproduction involves the fusion of specialized sex cells (already haploid), forming a diploid zygote. The haploid state in the offspring is restored by meiosis of the zygote. Note that reassortment of chromosomes can produce haploid offspring unlike either original gamete. (*b*) Cells with a diploid life cycle (animals, plants, some fungi, algae, and protozoa). Chromosomes occur in pairs. (*3*) Asexual reproduction through mitosis maintains the chromosome number and content. (*4*) In order to reproduce sexually, the normally diploid cells must produce haploid gametes through meiotic division. During fertilization, these haploid gametes (egg, sperm) form a zygote that restores the diploid number. Depending upon which gametes fuse, combinations of chromosomes unlike the original parents are possible in offspring.

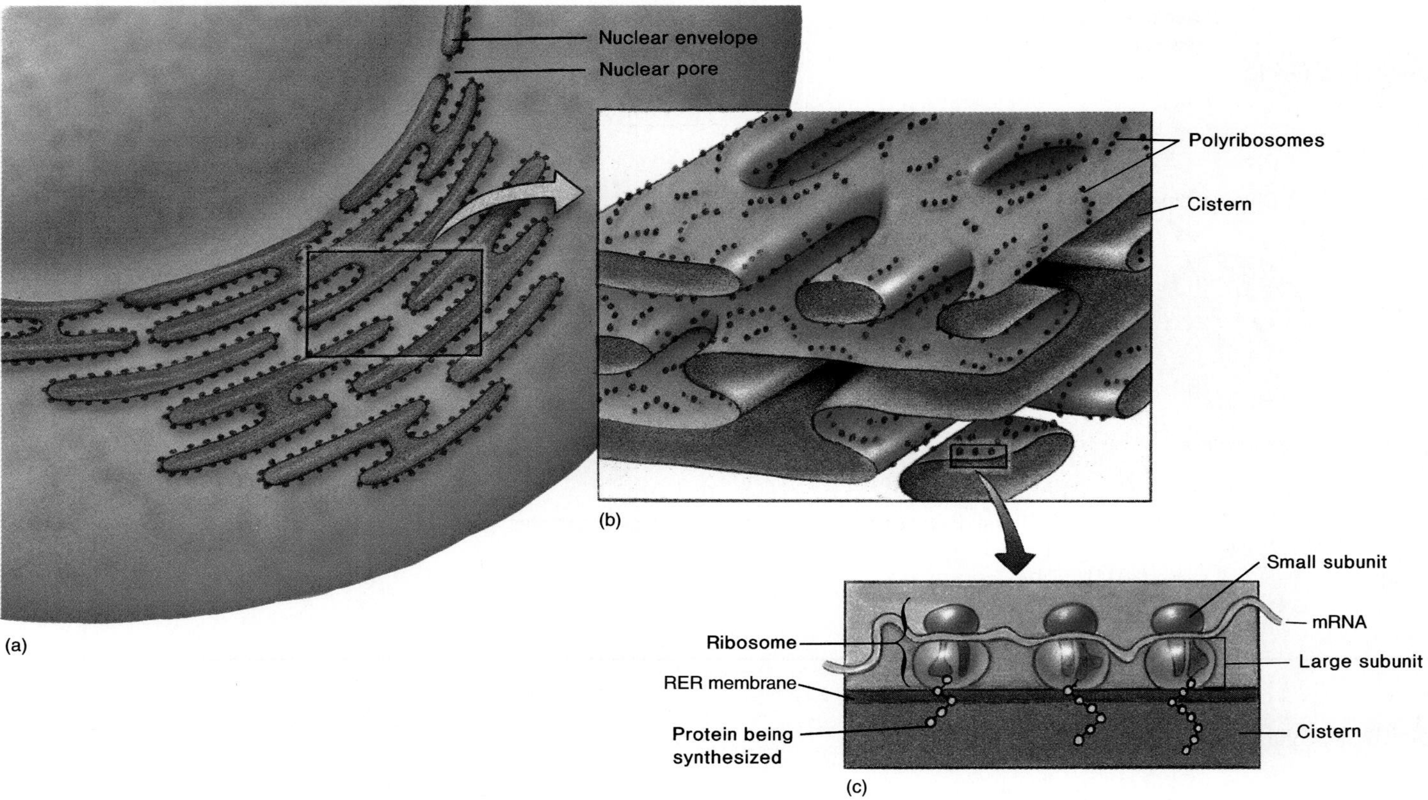

Figure 5.8

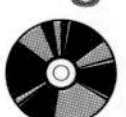

The origin and detailed structure of the rough endoplasmic reticulum (RER). (*a*) Schematic view of the origin of the RER from the outer membrane of the nuclear envelope. (*b*) Three-dimensional projection of the RER. (*c*) Detail of the orientation of a ribosome on the RER membrane.

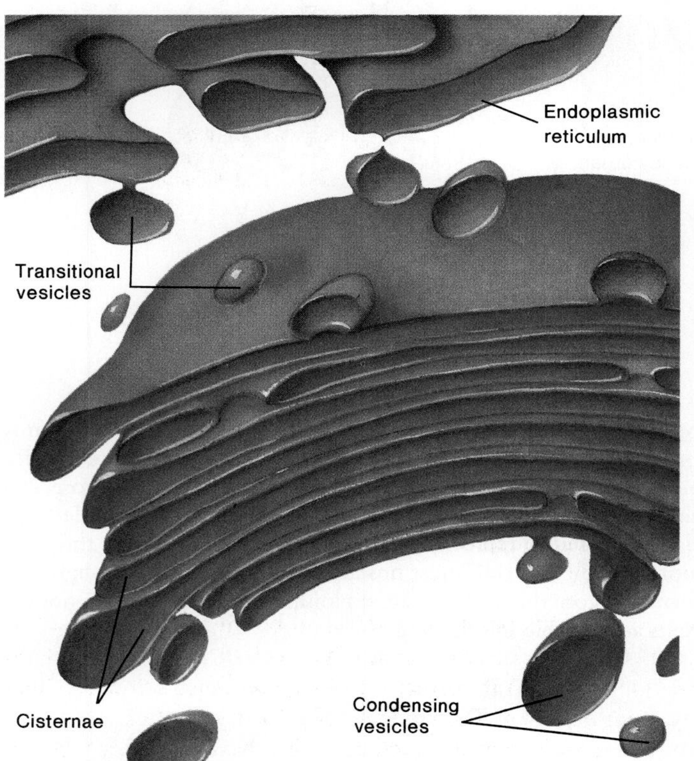

Figure 5.9

Detail of the Golgi apparatus, showing flattened layers of cisternae. Vesicles enter the upper surface and leave the lower surface.

membranes and cavities like those of the endoplasmic reticulum, but they do not form a continuous network (figure 5.9). This organelle is always closely associated with the endoplasmic reticulum both in its location and function. At a site where it meets the Golgi apparatus, the endoplasmic reticulum buds off tiny membrane-bound packets of protein called *transitional vesicles** that are picked up by the forming face of the Golgi apparatus. Once in the complex itself, the proteins are often modified by the addition of polysaccharides and lipids. The final action of this apparatus is to pinch off finished *condensing vesicles* that will be conveyed to organelles such as lysosomes or transported outside the cell as secretory vesicles (figure 5.10). The production of antibodies and receptors in human white blood cells involves this packaging and transport system (see chapter 15).

Nucleus, Endoplasmic Reticulum, and Golgi Apparatus: Nature's Assembly Line

As the keeper of the eucaryotic genetic code, the nucleus ultimately governs and regulates all cell activities. But, because the nucleus re-

*vesicle (ves´-ik-l) L. *vesios,* bladder. A small sac containing fluid.

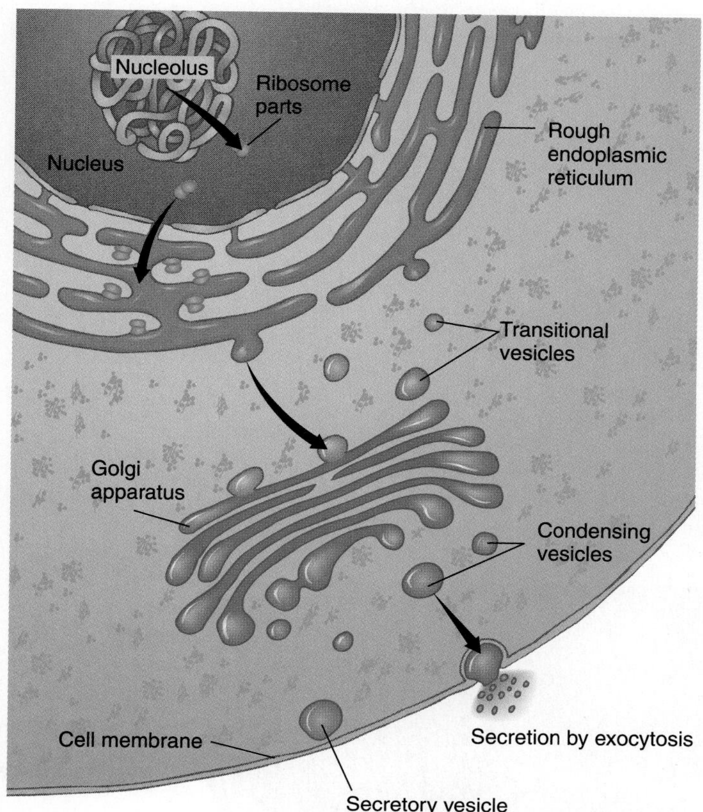

Figure 5.10

The cooperation of organelles in protein synthesis: nucleus → RER → Golgi apparatus → vesicles.

mains fixed in a specific cellular site, it must direct these activities through a structural and chemical network (figure 5.10). This network includes ribosomes, which originate in the nucleus, and the rough endoplasmic reticulum, which is continuously connected with the nuclear envelope. Initially, a segment of the genetic code of DNA (gene) is copied into RNA and passed out through the nuclear pores directly to the ribosomes on the endoplasmic reticulum. Here, specific proteins are synthesized from the RNA code and deposited in the lumen (space) of the endoplasmic reticulum. After being transported by a transitional vesicle to the Golgi apparatus, the protein products are chemically modified and packaged into vesicles that can be used by the cell in a variety of ways. Some of the vesicles contain enzymes to digest food inside the cell; other vesicles are secreted to digest materials outside the cell, and yet others are important in the enlargement and repair of the cell wall and membrane.

A **lysosome*** is one type of vesicle originating from the Golgi apparatus that contains a variety of enzymes. Lysosomes are involved in intracellular digestion of food particles and in protection against invading microorganisms. They also participate in digestion and removal of cell debris in damaged tissue. Other types of vesicles include **vacuoles,*** which are membrane-bound sacs

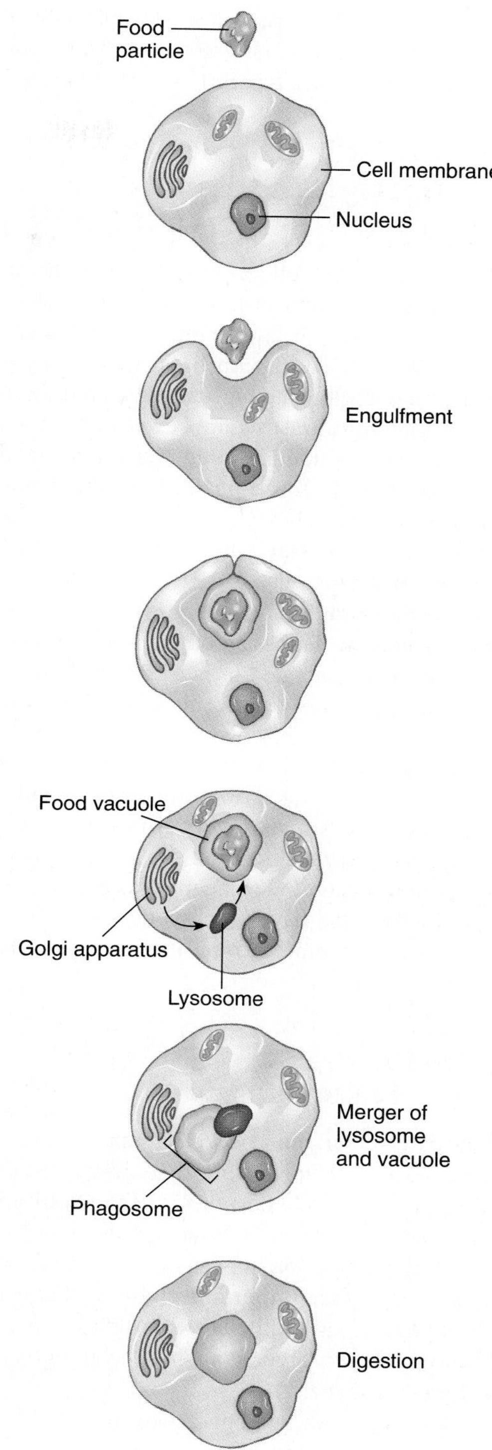

Figure 5.11
The origin and action of lysosomes in phagocytosis.

containing fluids or solid particles to be digested, excreted, or stored. They are formed in phagocytic cells (certain white blood cells and protozoa) in response to food and other substances that have been engulfed. The contents of a food vacuole are digested through the merger of the vacuole with a lysosome (figure 5.11). Other types of vacuoles are used in storing reserve food such as

*lysosome (ly´-soh-sohm) Gr. *lysis,* dissolution, and *soma,* body.

*vacuole (vak´-yoo-ohl) L. *vacuus,* empty. Any membranous space in the cytoplasm.

fats and glycogen. Protozoa living in freshwater habitats regulate osmotic pressure by means of contractile vacuoles, which regularly expel excess water that has diffused into the cell (see figure 5.30).

MITOCHONDRIA: ENERGY GENERATORS OF THE CELL

Although the nucleus is the cell's control center, none of the cellular activities it oversees could proceed without a constant supply of energy, the bulk of which is generated in most eucaryotes by **mitochondria.*** When viewed with light microscopy, mitochondria appear as round to elongate particles scattered throughout the cytoplasm. The internal ultrastructure reveals that a single mitochondrion consists of a smooth, continuous outer membrane that forms the external contour, and an inner, folded membrane nestled neatly within the outer membrane (figure 5.12*a*). The folds on the inner membrane, called **cristae,*** may be tubular, like fingers, or folded into shelf-like bands.

The cristae membranes hold the enzymes and electron carriers of aerobic respiration. This is an oxygen-using process that extracts chemical energy contained in nutrient molecules and stores it in the form of high-energy molecules, or ATP. More detailed functions of mitochondria are covered in chapter 8. The spaces around the cristae are filled with a chemically complex fluid called the **matrix,*** which holds ribosomes, DNA, and the pool of enzymes and other compounds involved in the metabolic cycle. Mitochondria (along with chloroplasts) are unique among organelles in that they divide independently of the cell, contain circular strands of DNA, and have procaryotic-sized 70S ribosomes. These findings have prompted some intriguing speculations on their evolutionary origins (microfile 5.1).

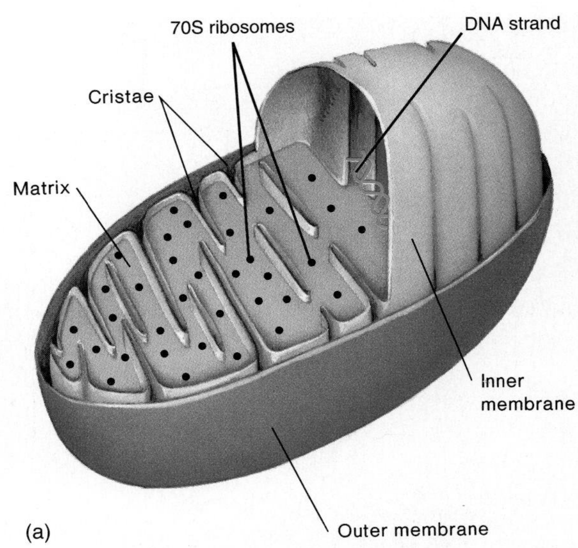

(a)

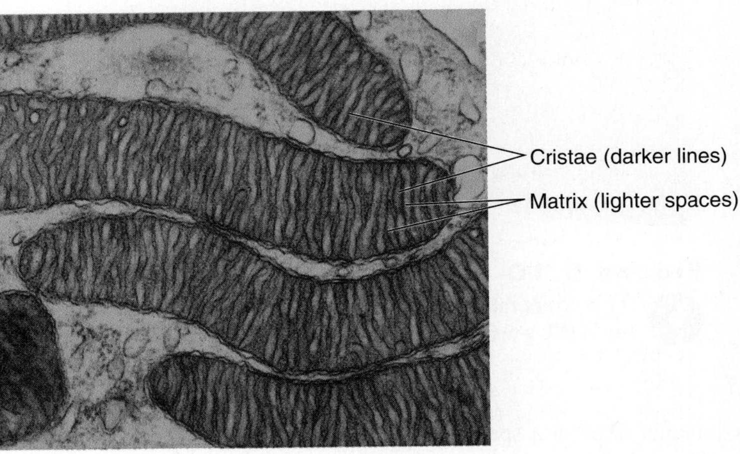

(b)

Figure 5.12

General structure of a composite mitochondrion as seen (*a*) in three-dimensional projection and (*b*) by electron microscopy. In most cells, mitochondria are elliptical or spherical, though in certain fungi, algae, and protozoa, they are long and filament-like.

CHLOROPLASTS: PHOTOSYNTHESIS MACHINES

Chloroplasts* are remarkable packets found in algae and plant cells that are capable of converting the energy of sunlight into chemical energy through photosynthesis. The photosynthetic role of chloroplasts makes them the primary producers of organic nutrients upon which all other organisms (except certain bacteria) ultimately depend. Another important photosynthetic product of chloroplasts is oxygen gas. Although chloroplasts resemble mitochondria, chloroplasts are larger, contain special pigments, and are much more varied in shape.

There are differences among various algal chloroplasts, but most are generally composed of two membranes, one enclosing the other. The even, outer membrane completely covers an inner membrane folded into small, disclike sacs called *thylakoids** that are stacked upon one another into *grana.* These structures carry the green pigment chlorophyll and sometimes additional pigments as well. Surrounding the thylakoids is a ground substance called the *stroma** (figure 5.13). The role of the photosynthetic pigments is to absorb and transform solar energy into chemical energy, which is then converted by reactions in the stroma to carbohydrates. We further explore some important aspects of photosynthesis in chapters 7 and 26.

THE CYTOSKELETON: A SUPPORT NETWORK

The cytoplasm of a eucaryotic cell is much more than a flabby, unstructured gel. It is penetrated by a flexible framework of

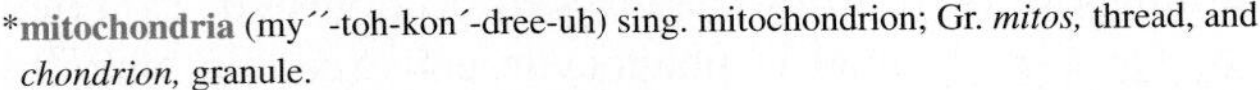

*__mitochondria__ (my″-toh-kon′-dree-uh) sing. mitochondrion; Gr. *mitos,* thread, and *chondrion,* granule.

*__cristae__ (kris′-te) sing. crista; L. *crista,* a comb.

*__matrix__ (may′-triks) L. *mater,* mother or origin.

*__chloroplast__ (klor′-oh-plast) Gr. *chloros,* green, and *plastos,* to form.

*__thylakoid__ (thy′-lah-koid) Gr. *thylakon,* a small sac, and *eidos,* form.

*__stroma__ (stroh′-mah) Gr. *stroma,* mattress or bed.

MICROFILE 5.1 THE EXTRAORDINARY EMERGENCE OF EUCARYOTIC CELLS

For years, biologists have grappled with the problem of how a cell as complex as the eucaryotic cell originated. One of the most fascinating explanations is that of **endosymbiosis,** which proposes that eucaryotic cells arose when a much larger procaryotic cell engulfed smaller bacterial cells that began to live and reproduce there rather than being destroyed. As the smaller cells took up permanent residence, they came to perform specialized functions for the larger cell, such as food synthesis and oxygen utilization that enhanced the cell's versatility and survival. Over time, when the cells evolved into a single functioning entity, the relationship became obligatory. Although the theory of endosymbiosis has been greeted with some controversy, this phenomenon has been demonstrated in the laboratory with amebas infected with bacteria that gradually became dependent upon the bacteria for survival.

The biologist most responsible for validation of the theory of endosymbiosis is Lynn Margulis. Using modern molecular techniques, she has accumulated convincing evidence of the relationships between the organelles of modern eucaryotic cells and the structure of bacteria. For example, the mitochondrion of eucaryotic cells is something like a tiny cell within a cell. The mitochondrion is capable of independent division, contains a circular chromosome that has bacterial DNA sequences, and has ribosomes that are clearly procaryotic. This link is so well established that recent phylogenetic trees of bacteria show "mitochondria" as one of the bacterial branches.

The same holds true for chloroplasts, which could have arisen when endosymbiotic cyanobacteria provided their host cells with a built-in feeding mechanism. Evidence is seen in a modern flagellated protozoan that harbors specialized chloroplasts with cyanobacterial chlorophyll and thylakoids. Margulis also has convincing evidence that eucaryotic cilia and flagella are the consequence of endosymbiosis between spiral bacteria and the cell membrane of early eucaryotic cells.

It is tempting to envision the development of the eucaryotic cell through a fusion of archaeal and bacterial cells. Archaea would have served as a source of ribosomes and certain aspects of protein synthesis, and bacteria would have given rise to mitochondria and chloroplasts. The first eukarya probably resembled modern protozoa such as *Giardia* and *Cyclospora* (see figure 23.16).

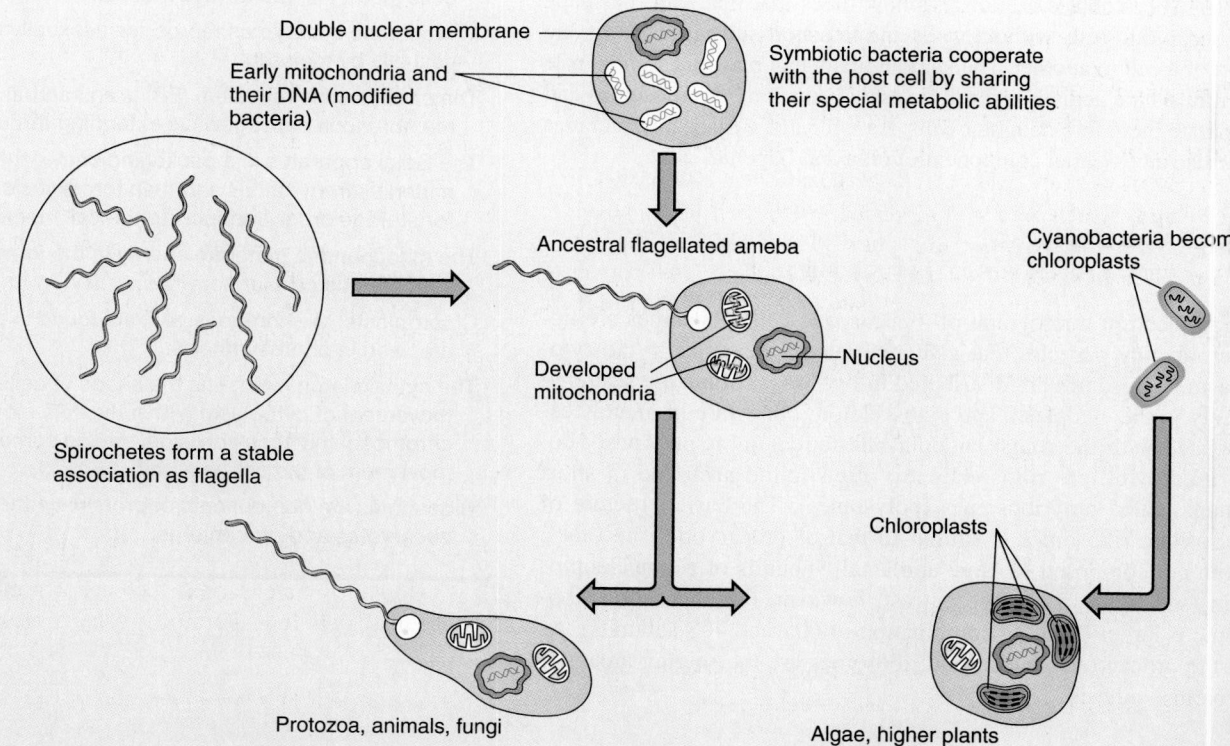

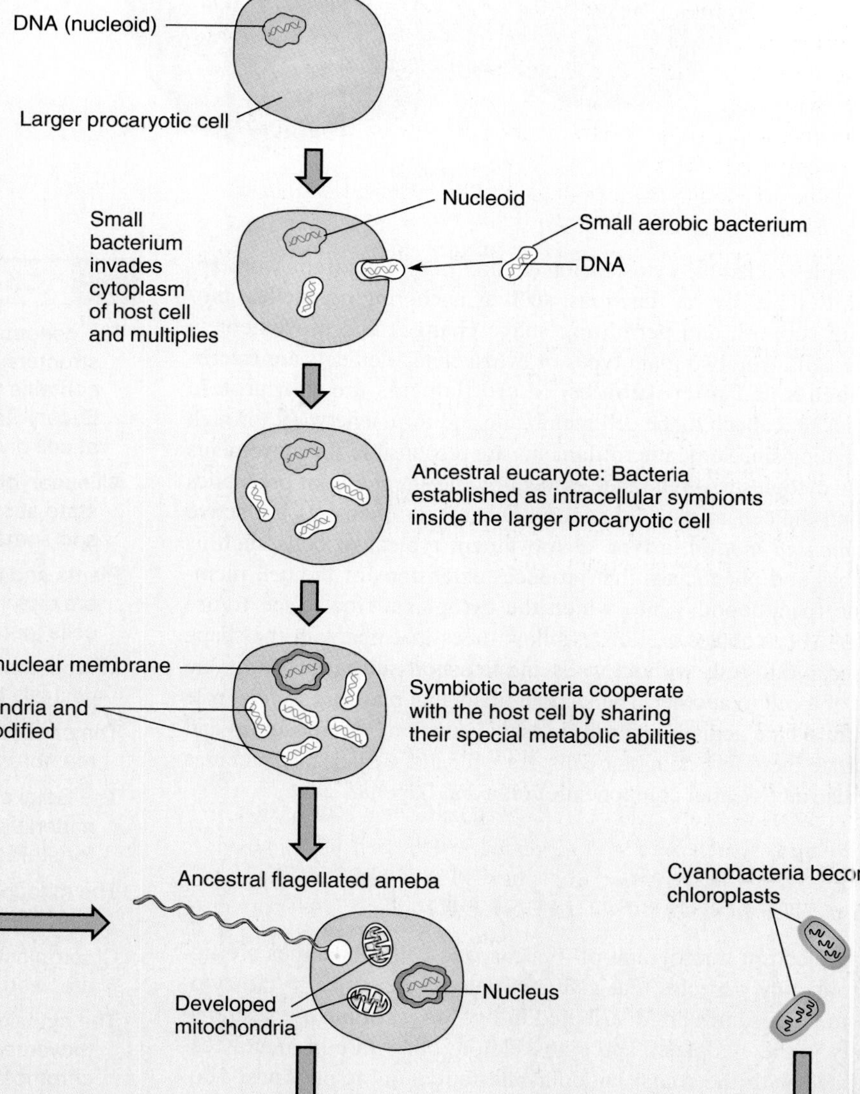

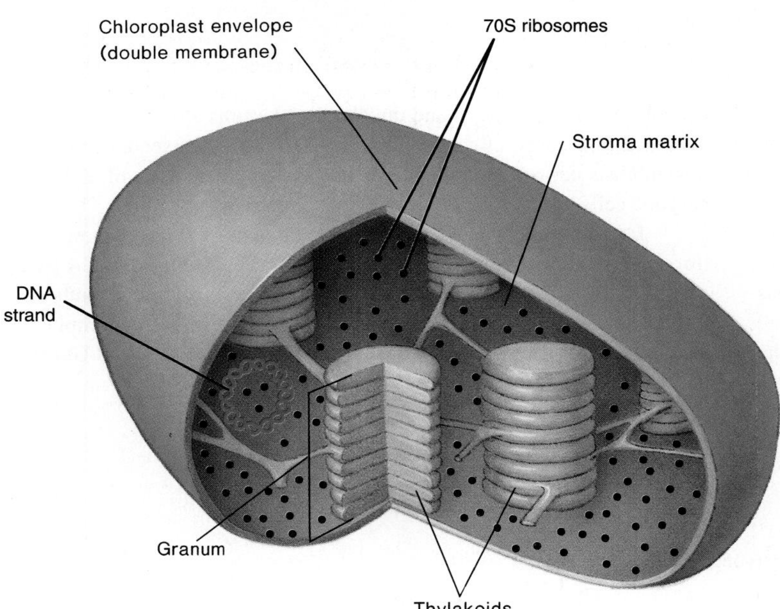

Figure 5.13
Detail of an algal chloroplast.

molecules called the cytoskeleton (figure 5.14). This framework appears to have several functions, such as anchoring organelles, providing support, and permitting shape changes and movement in some cells. The two main types of cytoskeletal elements are **microfilaments** and **microtubules.** Microfilaments are thin protein strands that attach to the cell membrane and form a network through the cytoplasm. Some microfilaments are responsible for movements of the cytoplasm, often made evident by the streaming of organelles around the cell in a cyclic pattern. Other microfilaments are active in *ameboid motion,* a type of movement typical of cells such as amebas and phagocytes that produces extensions of the cell membrane (pseudopods) into which the cytoplasm flows (see figure 5.30). Microtubules are long, hollow tubes that maintain the shape of eucaryotic cells without walls and transport substances from one part of a cell to another. The spindle fibers that play an essential role in mitosis are actually microtubules that attach to chromosomes and separate them into daughter cells. As indicated earlier, microtubules are also an essential component of cilia and flagella.

Ribosomes: Protein Synthesizers

In an electron micrograph of a eucaryotic cell, ribosomes are numerous, tiny particles that give a stippled appearance to the cytoplasm. Ribosomes are distributed in two ways: Some are scattered freely in the cytoplasm and cytoskeleton; others are intimately associated with the rough endoplasmic reticulum as previously described. Multiple ribosomes are often found arranged in short chains called polyribosomes (polysomes). The basic structure of eucaryotic ribosomes is similar to that of procaryotic ribosomes. Both are composed of large and small subunits of ribonucleoprotein (see figure 5.8). By contrast, however, the eucaryotic ribosome is larger (80S—a combination of 60S and 40S subunits). As in the procaryotes, eucaryotic ribosomes are the staging areas for protein synthesis.

Chapter Checkpoints

The genome of eucaryotes is located in the nucleus, a spherical structure surrounded by a double membrane. The nucleus contains the nucleolus, the site of ribosome synthesis. Eucaryote DNA, or chromatin, is organized into chromosomes at cell division.

All eucaryotes alternate between a haploid state and a diploid state at some point in their life cycle. Most fungi, many algae, and some protozoa are haploid except in their zygote stage.

Plants and animals, as well as some protozoa, fungi, and algae, are diploid for most of their life cycles, but their reproductive cells (gametes) are always haploid.

Microscopic eucaryotes reproduce asexually by mitosis and sexually by meiosis.

The endoplasmic reticulum (ER) is an internal network of membranous passageways extending throughout the cell.

The Golgi apparatus is a packaging center that receives materials from the ER and then forms vesicles around them for storage or for transport to the cell membrane for secretion.

The mitochondria generate energy in the form of ATP to be used in numerous cellular activities.

Chloroplasts, membranous packets found in plants and algae, are used in photosynthesis.

The cytoskeleton maintains the shape of cells and produces movement of cytoplasm within the cell, movement of chromosomes at cell division, and, in some groups, movement of the cell as a unit.

Ribosomes are components of protein synthesis present in both eucaryotes and procaryotes.

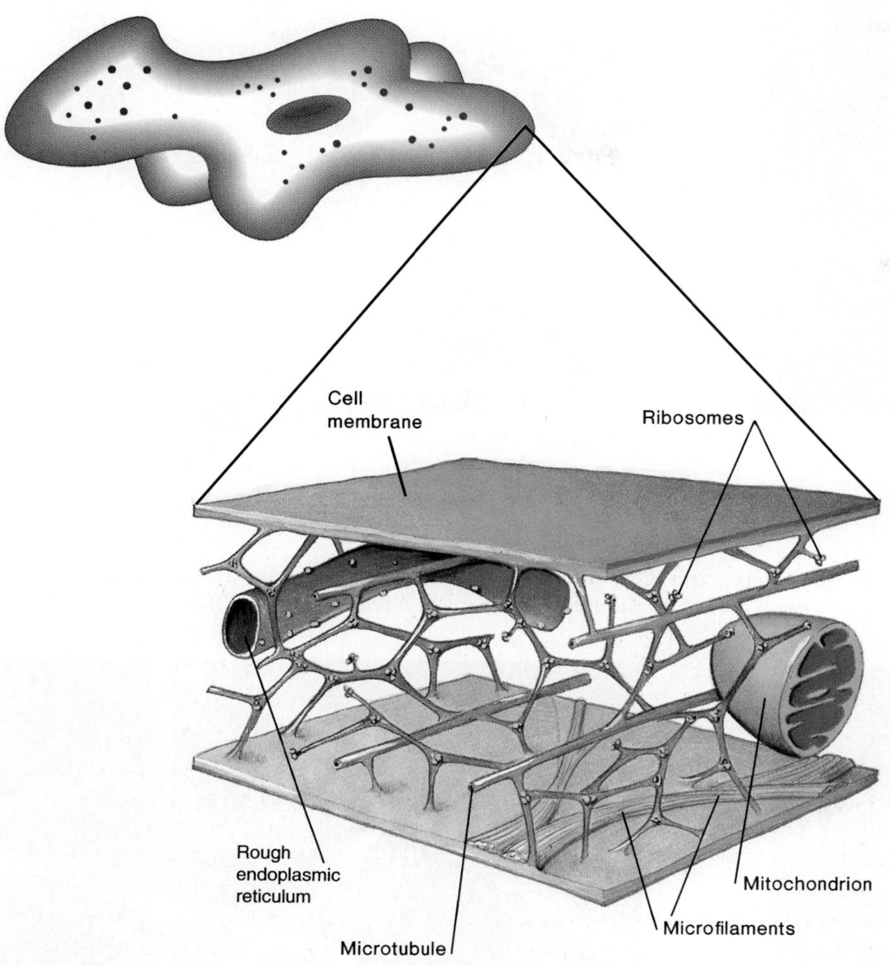

Figure 5.14
A model of the cytoskeleton, depicting the relationship between microtubules, microfilaments, and organelles.

SURVEY OF EUCARYOTIC MICROORGANISMS

With the general structure of the eucaryotic cell in mind, let us next examine the amazingly wide range of adaptations that this cell type has undergone. The following sections contain a general survey of the principal eucaryotic microorganisms—fungi, algae, protozoa, and parasitic worms—while simultaneously introducing elements of their structure, life history, classification, identification, and importance.

THE KINGDOM OF THE FUNGI

The position of the **fungi*** in the biological world has been debated for many years. Although they were originally classified with the green plants (along the algae and bacteria), they were later separated from plants and placed in a group with algae and protozoa (the Protista). Even at that time, however, many *mycologists** were struck by several unique qualities of fungi that warranted their being placed into their own separate kingdom, and eventually they were.

The **Kingdom Fungi,** or Myceteae, is large and filled with forms of great variety and complexity. For practical purposes, the approximately 100,000 species of fungi can be divided into two groups: the *macroscopic fungi* (mushrooms, puffballs, gill fungi) and the *microscopic fungi* (molds, yeasts). Although the majority of fungi are either unicellular or colonial, a few complex forms such as mushrooms and puffballs do have a limited form of cellular specialization. Cells of the microscopic fungi exist in two basic morphological types: yeasts and hyphae. A **yeast** cell is distinguished by its round to oval shape and by its mode of asexual reproduction. It grows swellings on its surface called **buds.** Some species from a **pseudohypha,*** a chain of yeasts formed when buds remain attached in a row (figure 5.15). **Hyphae*** are long, threadlike cells found in the bodies of filamentous fungi, or **molds** (figure 5.16). Some fungal cells exist only in a yeast form; others

*fungi (fun´-jy) sing. fungus; Gr. *fungos,* mushroom.

*mycologist (my-kol´-uh-jist) Gr. *mykes,* standard root for fungus. Scientists who study fungi.

*pseudohypha (soo´´-doh-hy´-fuh) pl. pseudohyphae; Gr. *pseudo,* false, and *hyphe.*

*hypha (hy´-fuh) pl. hyphae (hy´-fee) Gr. *hyphe,* a web.

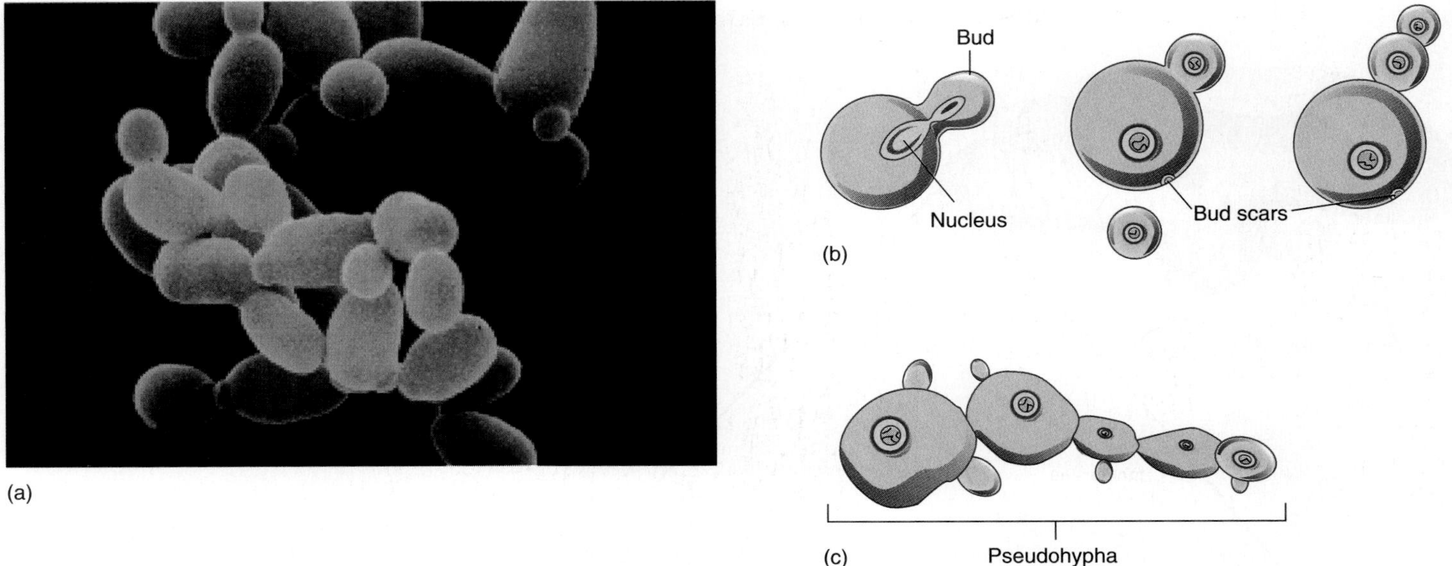

Figure 5.15

Microscopic morphology of yeasts. (*a*) Scanning electron micrograph of the brewer's, or baker's, yeast *Saccharomyces cerevisiae* (21,000×). (*b*) Formation and release of yeast buds. (*c*) Formation of pseudohypha (a chain of budding yeast cells).

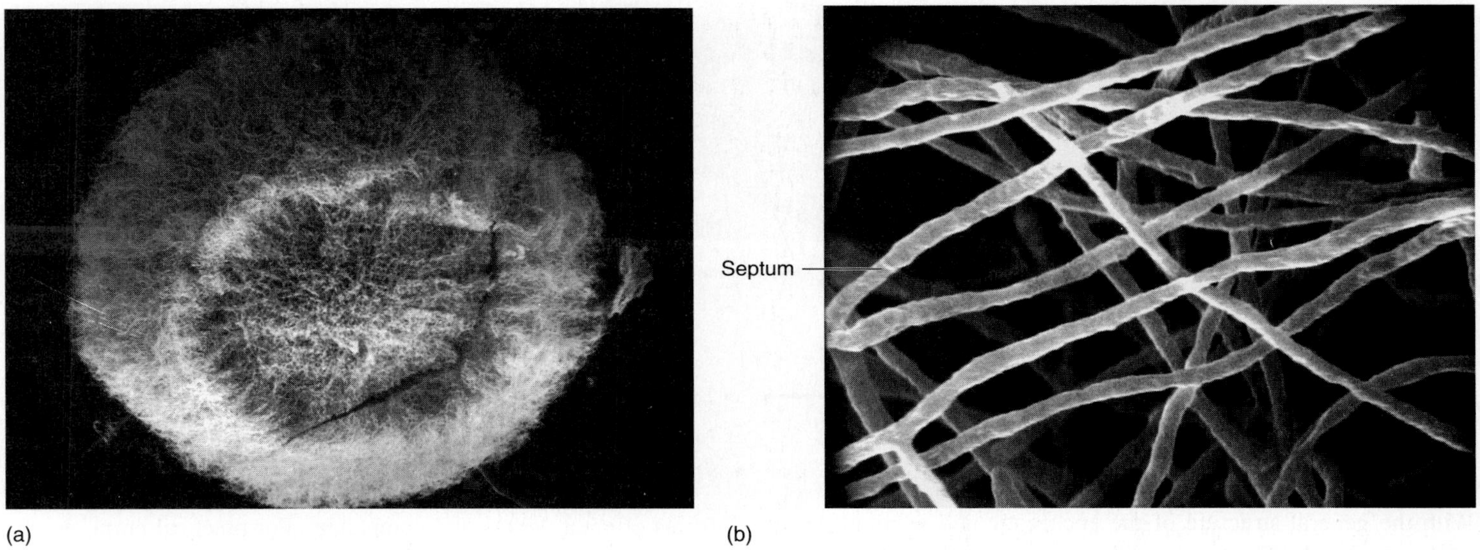

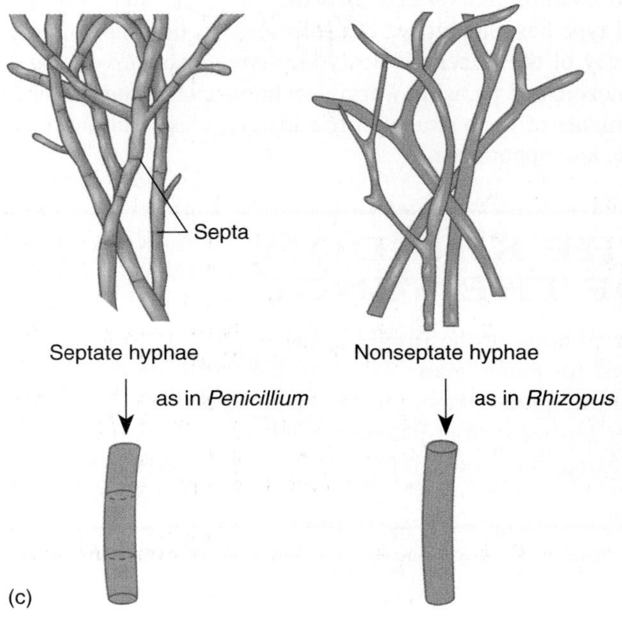

(a)

Septum

(b)

Septa

Septate hyphae

as in *Penicillium*

Nonseptate hyphae

as in *Rhizopus*

(c)

Figure 5.16

Scanning electron micrograph of *Diplodia maydis,* a pathogenic fungus of corn plants. (*a*) Colony or mycelium (24×). (*b*) Close-up of hyphal structure (1,200×). (*c*) Basic structural types of hyphae.

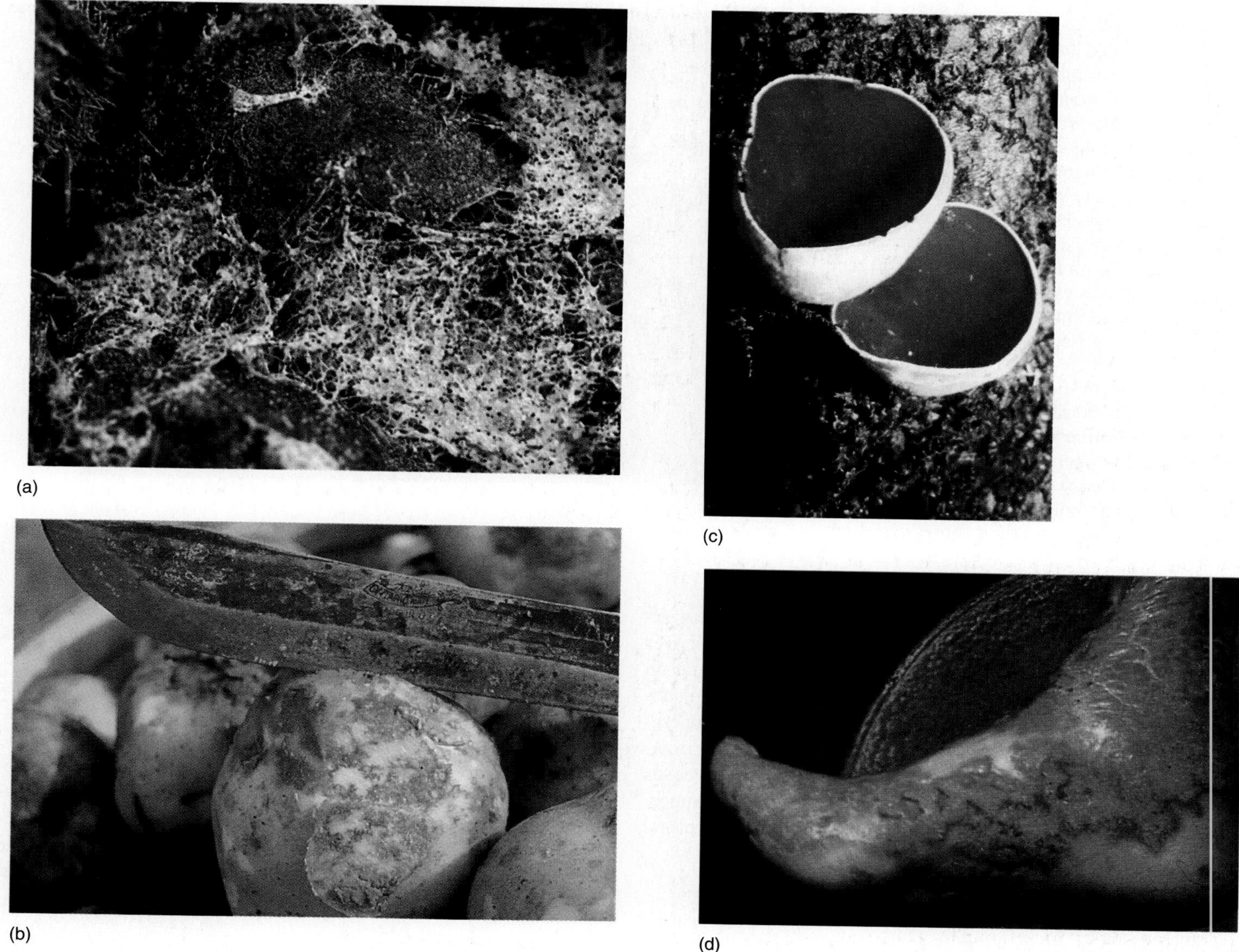

Figure 5.17
A small assortment of nutritional sources (substrates) of fungi. (*a*) A fungal mycelium growing in soil and forest litter (black spheres are spores). (*b*) The fungus *Phytopthora infestans* infecting a potato. This was a cause of the Irish potato famine in the 1840s and is still a problem for potato farmers in many parts of the world. (*c*) Cup fungus on tree bark. The cup is a fruiting body that holds the sexual spores. (*d*) The skin of the foot infected by a soil fungus, *Fonsecaea pedrosoi*.

occur primarily as hyphae; and a few, called *dimorphic,** can take either form, depending upon growth conditions, especially changing temperature. This variability in growth form is particularly characteristic of some pathogenic molds (see chapter 22).

FUNGAL NUTRITION

All fungi are **heterotrophic.*** They acquire nutrients from a wide variety of organic materials called **substrates** (figure 5.17). Most fungi are **saprobes,*** meaning that they obtain these substrates from the remnants of dead plants and animals in soil, fresh water, or salt water. Parasitic fungi can use the bodies of living animals or plants as a substrate, although very few fungi absolutely require a living host. In general, the fungus penetrates the substrate and secretes enzymes that reduce it to small molecules that can be absorbed by the cells. Fungi have enzymes for digesting an incredible array of substances, including feathers, hair, cellulose, petroleum products, wood, rubber. It has been said that every naturally occurring organic material on the earth can be attacked by some

*dimorphic (dy-mor´-fik) Gr. *di,* two, and *morphe,* form.

*heterotrophic (het-ur-oh-tro´-fik) Gr. *hetero,* other, and *troph,* to feed. A type of nutrition that relies on an organic nutrient source.

*saprobe (sap´-rohb) Gr. *sapros,* rotten, and *bios,* to live. Also called saprotroph or sarprophyte.

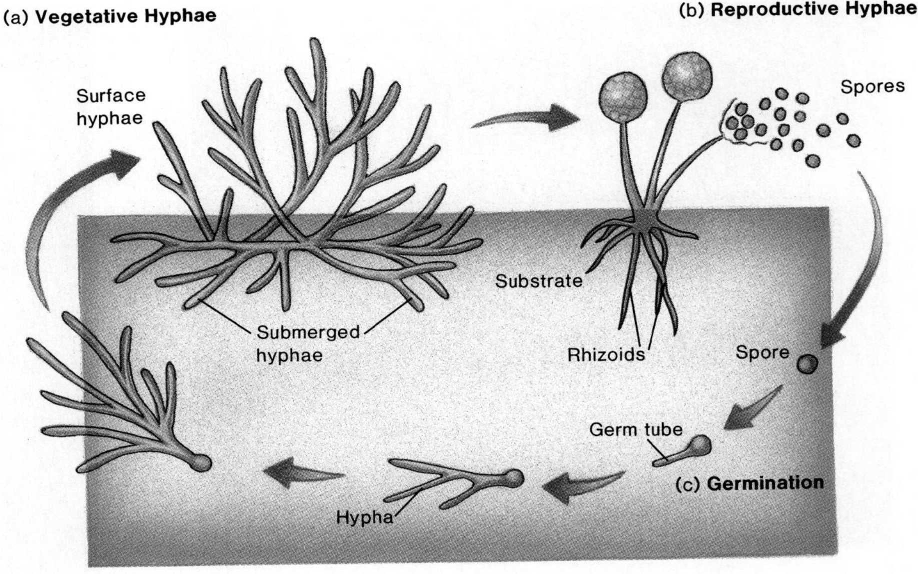

Figure 5.18
Functional types of hyphae using the mold *Rhizopus* as an example. (*a*) Vegetative hyphae are those surface and submerged filaments that digest, absorb, and distribute nutrients from the substrate. This species also has special anchoring structures called rhizoids. (*b*) Later, as the mold matures, it sprouts reproductive hyphae that produce asexual spores (as here) or sexual spores. (*c*) During the asexual life cycle, the dispersed mold spores settle on a usable substrate and send out germ tubes that elongate into hyphae. Through continued growth and branching, an extensive mycelium is produced. So prolific are the fungi that a single colony of mold can easily contain 5,000 spore-bearing structures. If each of these released 2,000 single spores and if every spore released from every sporangium on every mold were able to germinate, we would soon find ourselves in a sea of mycelia. Most spores do not germinate, but enough are successful to keep the numbers of fungi and their spores very high in most habitats.

type of fungus. Fungi are often found in nutritionally poor or adverse environments. In our laboratory, we occasionally discover fungi in mineral and weak acid solutions ("bottle imps") and on specimens preserved in formaldehyde. Various fungi thrive in substrates with high salt or sugar content, at relatively high temperatures, and even in snow and glaciers. Their medical and agricultural impact is extensive. A number of species cause **mycoses** (fungal infections) in animals, and thousands of species are important plant pathogens. Although fungi often have bright colors, these are not a result of photosynthetic pigments.

Organization of Microscopic Fungi

The cells of most microscopic fungi grow in discrete and distinctive colonies. The colonies of yeasts are much like those of bacteria in that they have a soft, uniform texture and appearance. The colonies of filamentous fungi are noted for the striking cottony, hairy, or velvety textures that arise from their microscopic organization and morphology. The woven, intertwining mass of hyphae that makes up the body or colony of a mold is called a **mycelium*** (see figure 5.16).

Although hyphae contain the usual eucaryotic organelles, they also have some unique organizational features. **Nonseptate** hyphae consist of one long, continuous cell not divided into individual compartments by **septa,*** or cross walls (see figure 5.16*c*). With this construction, the cytoplasm and organelles move freely from one region to another, and each hyphal element can have several nuclei. In most fungi, the hyphae are divided into segments by cross walls, a condition called **septate.** The nature of the septa varies from solid partitions with no communication between the compartments to partial walls with small pores that allow the flow of organelles and nutrients between adjacent compartments.

Hyphae can also be classified according to their particular function. **Vegetative hyphae** (mycelia) are responsible for the visible mass of growth that appears on the surface of a substrate and penetrates it to digest and absorb nutrients. During the development of a fungal colony, the vegetative hyphae give rise to structures called **reproductive,** or **aerial, hyphae,** which orient vertically from the vegetative mycelium. These hyphae are responsible for the production of fungal reproductive bodies called **spores.** Other specializations of hyphae are illustrated in figure 5.18.

Reproductive Strategies and Spore Formation

Fungi have many complex and successful reproductive strategies. Most can propagate by the simple outward growth of existing hyphae or by fragmentation, in which a separated piece of mycelium can generate a whole new colony. But the primary reproductive mode of fungi involves the production of various types of spores. Do not confuse fungal spores with the more resistant, nonreproductive bacterial spores. Fungal spores are responsible not only for multiplication but also for survival, producing genetic variation, and dissemination. Because of their compactness and relatively light weight, spores are dispersed widely through the environment by air, water, and living things. Outdoor air commonly contains up to 10,000 fungal spores per cubic meter. Upon encountering a favorable substrate, a spore will germinate and produce a new fungus colony in a very short time (figure 5.18).

The fungi exhibit such a marked diversity in spores that they are largely classified and identified by their spores and spore-forming structures. Although there are some elaborate systems for naming and classifying spores, we will present a basic overview of the

*mycelium (my-see´-lee-um) pl. mycelia; Gr. *mykes,* fungus, and *helos,* nail.

*septa (sep´-tuh) sing. septum; L. *saepio,* to fence or wall.

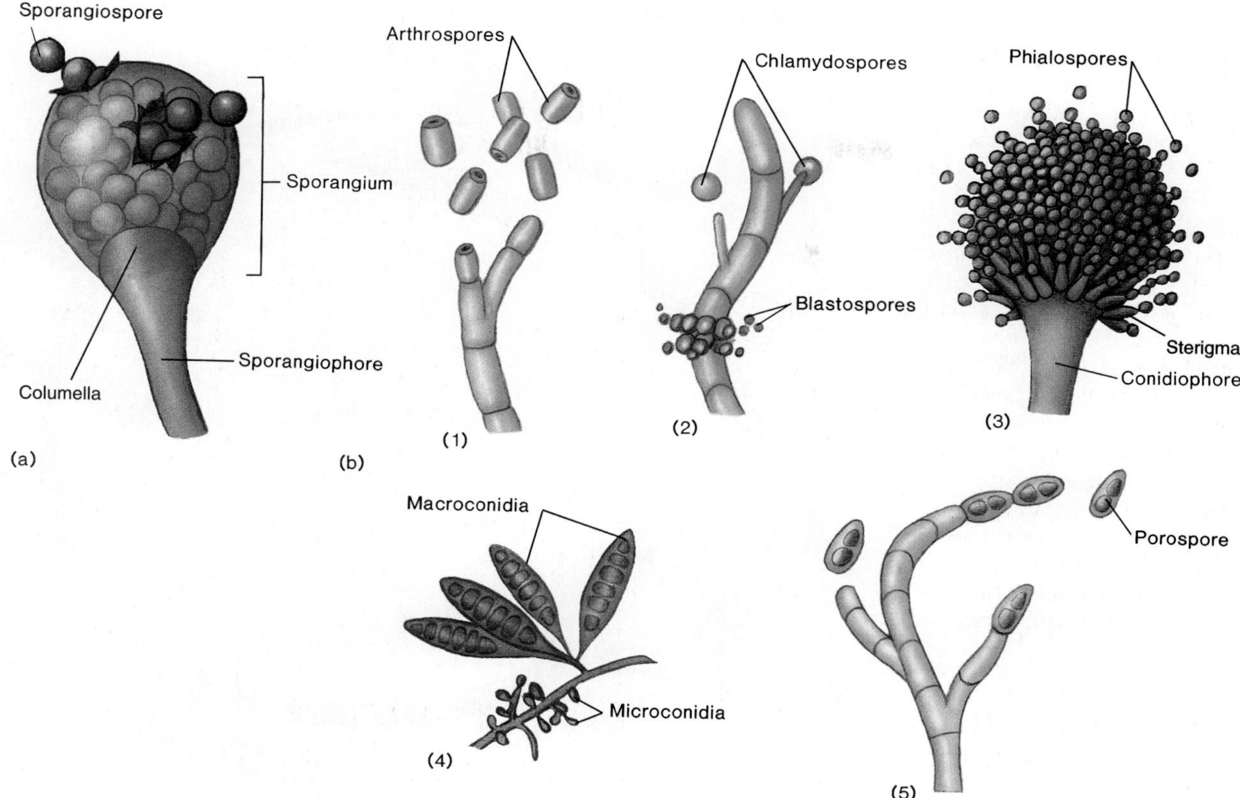

Figure 5.19

Types of asexual mold spores. (*a*) Sporangiospores (e.g., *Absidia*). (*b*) Conidia: (*1*) arthrospores (e.g., *Coccidioides*), (*2*) chlamydospores and blastospores (e.g., *Candida albicans*), (*3*) phialospores (e.g., *Aspergillus*), (*4*) macroconidia and microconidia (e.g., *Microsporum*), and (*5*) porospores (e.g., *Alternaria*).

principal types. The most general subdivision is based on the way the spores arise. **Asexual spores** are the products of mitotic division of a single parent cell, and **sexual spores** are formed through a process involving the fusing of two parental nuclei followed by meiosis.

Asexual Spore Formation

On the basis of the appearance of the fertile hypha and the origin of spores, there are two subtypes of asexual spore (figure 5.19):

1. **Sporangiospores** (figure 5.19*a*) are formed by successive cleavages within a saclike head called a **sporangium,*** which is attached to a stalk, the sporangiophore. These spores are initially enclosed but are released when the sporangium ruptures.
2. **Conidia*** (conidiospores) are free spores not enclosed by a spore-bearing sac (figure 5.19*b*). They develop either by the pinching off of the tip of a special fertile hypha or by the segmentation of a preexisting vegetative hypha. Conidia are the most common asexual spores, and they occur in the following forms:

arthrospore (ar´-thor-spor) Gr. *arthron,* joint. A rectangular spore formed when a septate hypha fragments at the cross walls.

chlamydospore (klams-ih´-doh-spor) Gr. *chlamys,* cloak. A spherical conidium formed by the thickening of a hyphal cell. It is released when the surrounding hypha fractures, and it serves as a survival or resting cell.

blastospore. A spore produced by budding from a parent cell that is a yeast or another conidium; also called a bud.

phialospore (fy´-ah-lo-spor) Gr. *phialos,* a vessel. A conidium that is budded from the mouth of a vase-shaped spore-bearing cell called a phialide or *sterigma,* leaving a small collar.

microconidium and *macroconidium.* The smaller and larger conidia formed by the same fungus under varying conditions. Microconidia are one-celled, and macroconidia have two or more cells.

porospore A conidium that grows out through small pores in the spore-bearing cell; some are composed of several cells.

Sexual Spore Formation

If fungi can propagate themselves successfully with millions of asexual spores, what function does sexual reproduction fulfill? The answer lies in an important interchange that occurs when fungi of different genetic makeup combine their genetic material. Just as in plants and animals, this linking of genes from two parents creates offspring with combinations of genes different from that of either parent. The offspring from such a union can have

*sporangium (spo-ran´-jee-um) pl. sporangia; Gr. *sporos,* seed, and *angeion,* vessel.

*conidia (koh-nid´-ee-uh) sing. conidium; Gr. *konidion,* a particle of dust.

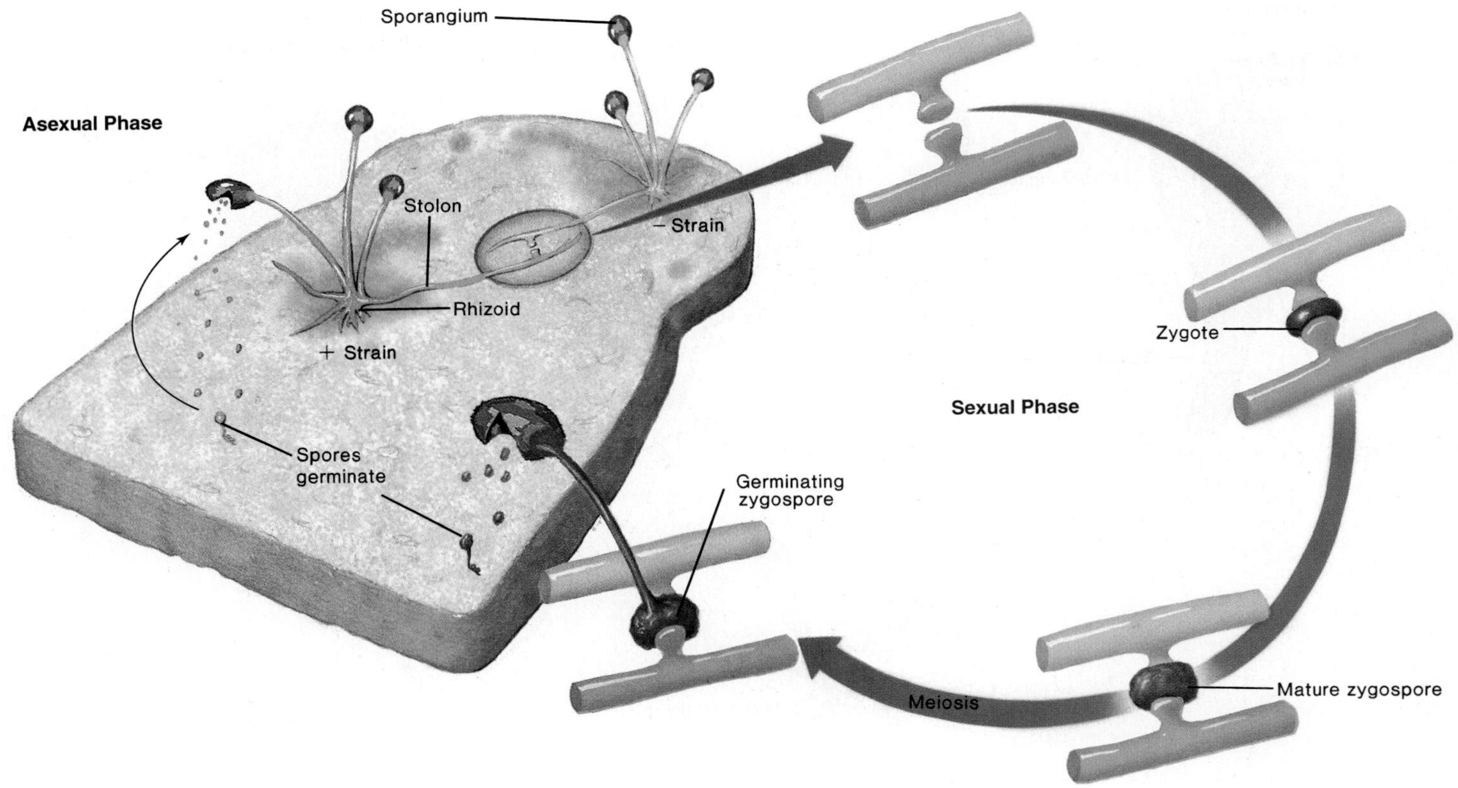

Figure 5.20

Formation of zygospores in *Rhizopus stolonifer.* Sexual reproduction occurs when two mating strains of hyphae grow together, fuse, and form a mature zygospore. Germination of the zygospore involves meiotic division and production of a haploid sporangium that looks just like the asexual one.

slight variations in form and function that are potentially advantageous in the adaptation and survival of their species. Because fungal cells spend most of their life cycle in the haploid state, their cells already have a chromosome number compatible for sexual union. In general, during sexual reproduction, the haploid nuclei from two compatible parental cells fuse, forming a diploid nucleus that subsequently participates in sexual spore development (see figure 5.7*a*, step 2).

The majority of fungi produce sexual spores at some point. The nature of this process varies from the simple fusion of fertile hyphae of two different strains to a complex union of differentiated male and female structures and the development of special fruiting structures. Four different types of sexual spores have been identified, but we will consider the three most common: zygospores, ascospores, and basidiospores. These spore types provide an important basis for classifying the major fungal divisions.

Zygospores* are sturdy diploid spores formed when hyphae of two opposite strains (called the plus and minus strains) fuse and create a diploid zygote that swells and becomes covered by strong, spiny walls (figure 5.20). When its wall is disrupted, and moisture and nutrient conditions are suitable, the zygospore germinates and forms a sporangium. Meiosis of diploid cells of the sporangium results in haploid nuclei that develop into sporangiospores. Both the sporangia and the sporangiospores that arise from sexual processes are outwardly identical to the asexual type, but because the spores arose from the union of two separate fungal parents, they are not genetically identical.

In general, haploid spores called **ascospores*** are created inside a special fungal sac, or **ascus** (pl. asci) (figure 5.21). Although details can vary among types of fungi, the ascus and ascospores are formed when two different strains or sexes join together to produce offspring. In many species, the male sexual organ fuses with the female sexual organ. The end result is a number of terminal cells, each containing a diploid nucleus. Through differentiation, each of these cells enlarges to form an ascus, and its diploid nucleus undergoes meiosis (often followed by mitosis) to form four to eight haploid nuclei that will mature into ascospores. A rise ascus breaks open and releases the ascospores. Some species form an elaborate fruiting body to hold the asci, such as the cup fungus in figure 5.17*c*.

Basidiospores* are haploid sexual spores formed on the outside of a club-shaped cell called a **basidium** (figure 5.22). In general, spore formation follows the same pattern of two mating

*zygospores (zy´-goh-sporz) Gr. *zygon,* yoke, to join.

*ascosphores (as´-koh-sporz) Gr. *ascos,* a sac.

*basidiospores (bah-sid´-ee-oh-sporz) Gr. *basidi,* a pedestal.

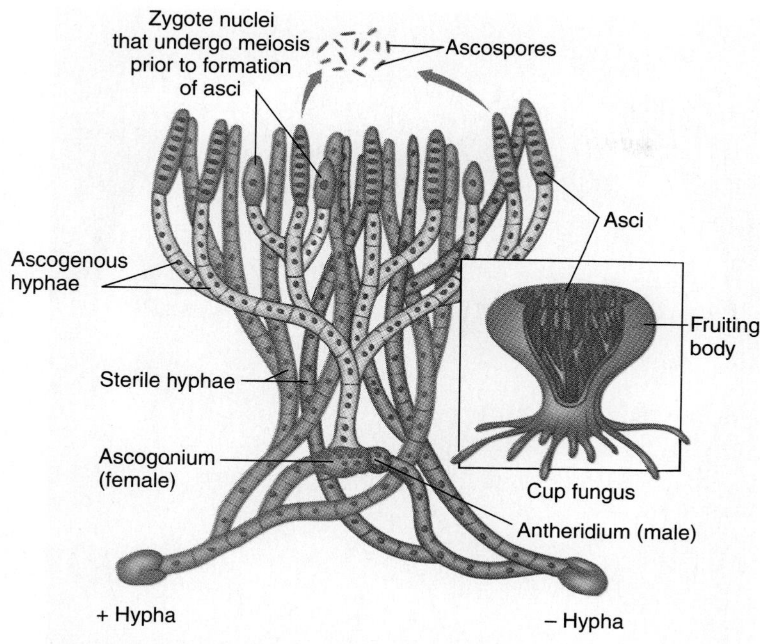

Figure 5.21
Production of ascospores in a cup fungus.

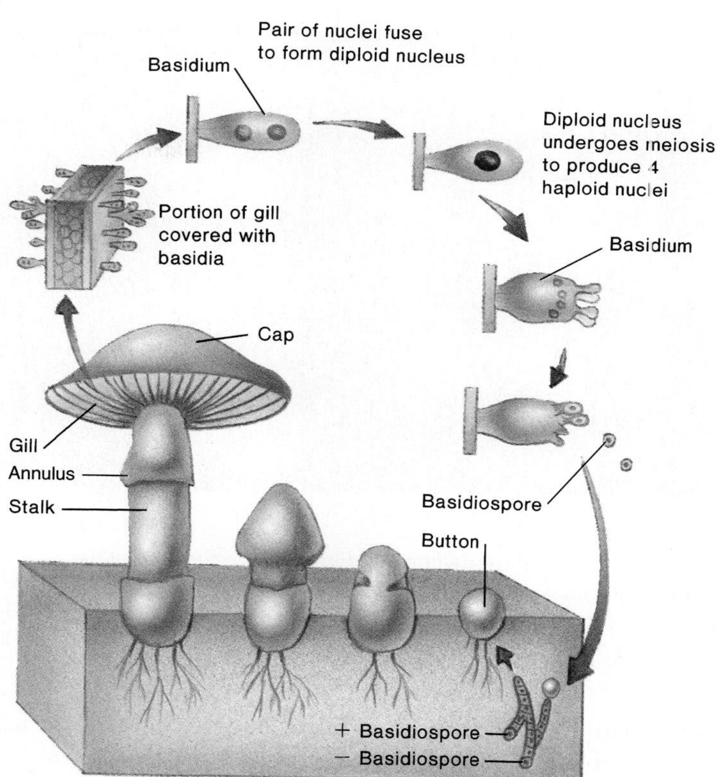

Figure 5.22
Formation of basidiospores in a mushroom.

types coming together, fusing, and forming terminal cells with diploid nuclei. Each of these cells becomes a basidium, and its nucleus produces, through meiosis, four haploid nuclei. These nuclei are extruded through the top of the basidium, where they develop into basidiospores. Notice the location of the basidia along the gills in mushrooms, which are often dark from the spores they contain. It may be a surprise to discover that the fleshy part of a mushroom is actually a fruiting body designed to protect and help disseminate its sexual spores.

Fungal Classification

It is often difficult for microbiologists to assign logical and useful classification schemes to microorganisms while at the same time preserving their relationships. This difficulty is due to the fact that the organisms do not always perfectly fit the neat categories made for them, and even experts cannot always agree on the nature of the categories. The fungi are no exception, and there are several ways to classify them. For our purposes, we will adopt a classification scheme with a medical mycology emphasis, in which the Kingdom Fungi is subdivided into two subkingdoms, the Amastigomycota and the Mastigomycota. The Amastigomycota are common inhabitants of terrestrial habitats. Some of the characteristics used to separate them into subgroups are the type of sexual reproduction and their hyphal structure. The Mastigomycota are primitive filamentous fungi that live primarily in water and may cause disease in potatoes and grapes.

The Myxomycota and the Acrasiomycota are fungus-like groups that originally were considered specialized fungi. Modern genetic analysis has recently revised our view. The myxomycota, or acellular slime molds, are now classified with animals, and the acrasiomycota, or cellular slime molds, are placed in their own separate kingdom. These groups will not be covered here.

From the beginnings of fungal classification, any fungus that lacked a sexual state was called "imperfect" and was placed in a catchall category, the Fungi Imperfecti, or Deuteromycota. A species would remain classified in that category until its sexual state was described (if ever). Gradually, many species of Fungi Imperfecti were found to make sexual spores, and they were assigned to the taxonomic grouping that best fit those spores. This system created a quandary, because several very important species, especially pathogens, had to be regrouped and renamed. In many cases, the older names were so well entrenched in the literature that it was easier to retain them and assign a new generic name for their sexual stage. Consequently, *Blastomyces* and *Histoplasma* are also known as *Ajellomyces* (sexual phase).

The following section outlines the four divisions of Amastigomycota, including major characteristics and important members.

Amastigomycota: Fungi That Produce Sexual and Asexual Spores (Perfect)

Division I—Zygomycota (also Phycomycetes)

Sexual spores: zygospores; asexual spores: mostly sporangiospores, some conidia. Hyphae are usually nonseptate. If septate, the septa are complete. Most species are free-living saprobes; some are animal parasites. Can be obnoxious contaminants in the laboratory and on food and vegetables. Examples are mostly molds: *Rhizopus,* a black bread mold; *Mucor; Syncephalastrum; Circinella* (figure 5.23, asexual phase only).

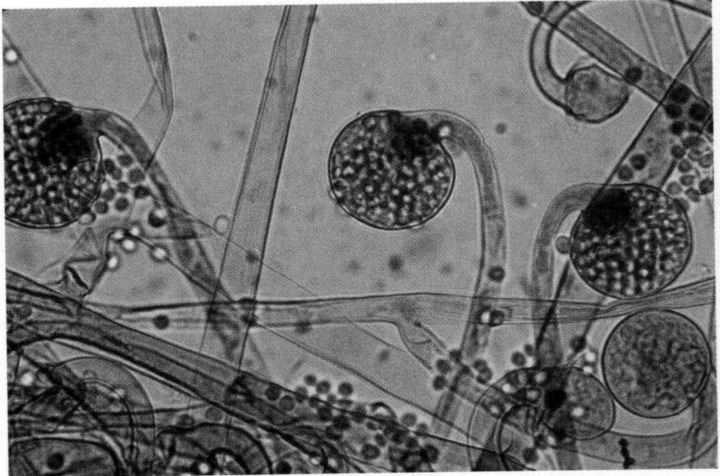

Figure 5.23
A representative Zygomycota, *Circinella.* Note the sporangia, sporangiospores, and nonseptate hyphae.

Division II—Ascomycota (also Ascomycetes) Sexual spores: produce ascospores in asci, asexual spores: many types of conidia, formed at the tips of conidiophores. Hyphae with porous septa. Many important species. Examples: *Histoplasma,* the cause of Ohio Valley fever; *Microsporum,* one cause of ringworm (a common name for certain fungal skin infections that often grow in a ringed pattern); *Penicillium,* one source of antibiotics (figure 5.24, asexual phase only); and *Saccharomyces,* a yeast used in making bread and beer. Most of the species in this division are either molds or yeasts; it also includes many human and plant pathogens, such as *Pneumocystis carinii,* a pathogen of AIDS patients.

Division III—Basidiomycota (also Basidiomycetes) Sexual reproduction by means of basidia and basidospores; asexual spores: conidia. Incompletely septate hyphae. Some plant parasites and one human pathogen. Fleshy fruiting bodies are common. Examples: mushrooms, puffballs, bracket fungi, and plant pathogens called rusts and smuts. The one human pathogen, the yeast *Cryptocococcus neoformans,* causes an invasive systemic infection in several organs, including the skin, brain, and lungs (see figure 22.26).

Amastigomycota: Fungi That Produce Asexual Spores Only (Imperfect)

Division IV—Deuteromycota* Asexual spores: conidia of various types. Hyphae septate. Majority are yeasts or molds, some dimorphic. Saprobes and a few animal and plant parasites. Examples: Several human pathogens were originally placed in this group, especially the imperfect states of *Blastomyces* and *Microsporum.*

*Deuteromycota (doo´ter-oh-my-koh´´-tuh) Gr. *deutero,* second, and *mykes,* fungus.

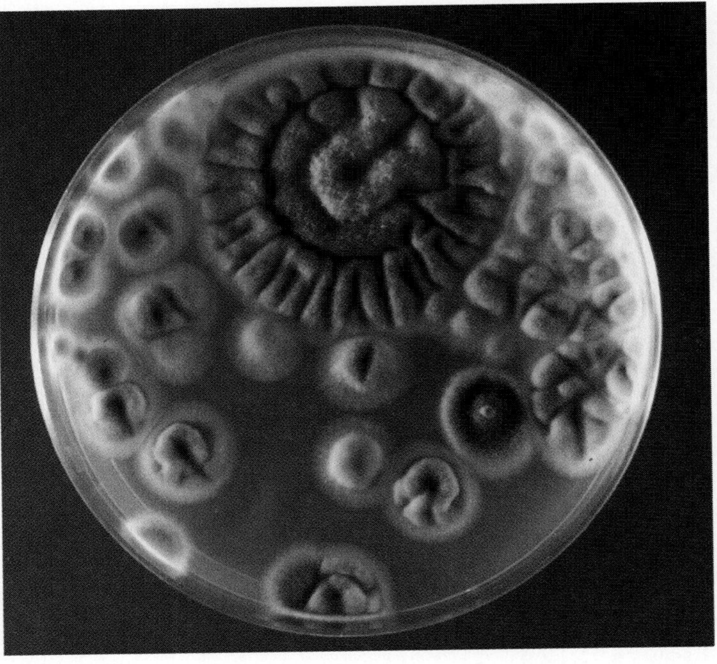

(a)

(b)

Figure 5.24
A common Ascomycota, *Penicillium.* (*a*) Macroscopic view of a typical blue-green colony. (*b*) Microscopic view shows the brush arrangement of phialospores (220×).

Other species are *Coccidioides immitis,* the cause of valley fever; *Candida albicans,* the cause of various yeast infections; and *Cladosporium,* a common mildew fungus (figure 5.25).

FUNGAL IDENTIFICATION AND CULTIVATION

Fungi are identified in medical specimens by first being isolated on special types of media and then being observed macroscopically and microscopically. Examples of media for cultivating fungi

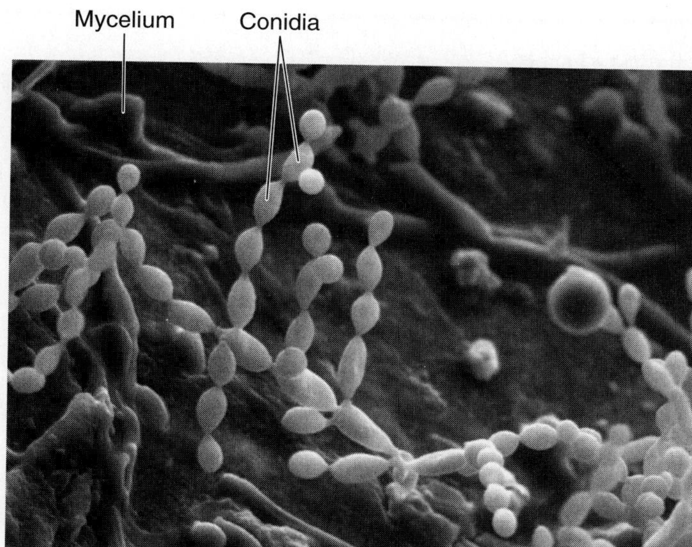

Figure 5.25
Mycelium and spores of a representative of Deuteromycota called *Cladosporium*. This black mold is one of the most common fungi in the world and a ubiquitous air contaminant (see microfile 5.2).

are cornmeal, blood, and Sabouraud's agar. The latter medium is useful in isolating fungi from mixed samples because of its low pH, which inhibits the growth of bacteria but not of most fungi. Because the fungi are classified into general groups by the presence and type of sexual spores, it would seem logical to identify them in the same way, but that is not how it is done. Because sexual spores are rarely if ever demonstrated in the laboratory setting, the sexual spore-forming structures and spores are usually used to identify organisms to the level of genus and species. Other characteristics that contribute to identification are hyphal type, colony texture and pigmentation, and physiological characteristics, and genetic relationships (chapter 22).

The Roles of Fungi in Nature and Industry

Nearly all fungi are free-living and do not require a host to complete their life cycles. Even among those fungi that are pathogenic, most human infection occurs through accidental contact from an environmental source such as soil, water, or dust. Humans are generally quite resistant to fungal infection, except for two main types of fungal pathogens: the true pathogens, which can infect even healthy persons, and the opportunistic pathogens, which attack persons who are already weakened in some way. Chapter 22 is devoted to fungal infections.

Mycoses vary in the way the agent enters the body and the degree of tissue involvement (table 5.1). The list of opportunistic fungal pathogens has been increasing in the past few years because of newer medical techniques that keep compromised patients alive. Even so-called harmless species found in the air and dust around us may be able to cause opportunistic infections in patients who already have AIDS, cancer, or diabetes.

Table 5.1
Major Fungal Infections of Humans

Degree of Tissue Involvement and Area Affected	Name of Infection	Name of Causative Fungus
Superficial (not deeply invasive)		
Hair	Black piedra	*Piedraia hortae*
Outer epidermis	Pityriasis versicolor	*Malassezia furfur*
Epidermis, hair, and dermis can be attacked	Dermatophytosis, also called tinea or ringworm of the scalp, body, feet (athlete's foot), toenails, and beard, depending on the area afflicted	*Microsporum, Trichophyton,* and *Epidermophyton*
Mucous membranes, skin, nails	Candidiasis, or yeast infection	*Candida albicans*
Subcutaneous (invades just beneath the skin)		
Nodules beneath the skin; can invade lymphatics	Sporotrichosis	*Sporothrix schenckii*
Large fungal tumors of limbs and extremities	Mycetoma, madura foot	*Pseudoallescheria boydii, Madurella mycetomatis*
Systemic (deep; organism enters lungs; can invade other organs)		
Lung	Coccidioidomycosis (San Joaquin Valley fever)	*Coccidioides immitis*
	North American blastomycosis (Chicago disease)	*Blastomyces dermatitidis*
	Histoplasmosis (Ohio Valley fever)	*Histoplasma capsulatum*
	Cryptococcosis (torulosis)	*Cryptococcus neoformans*
Lung, skin	Paracoccidioidomycosis (South American blastomycosis)	*Paracoccidioides brasiliensis*

Fungi are involved in other medical conditions besides infections. Fungal cell walls give off chemical substances that can cause allergies. The toxins produced by poisonous mushrooms can induce neurological disturbances and even death. The mold *Aspergillus flavus* (see figure 22.27) synthesizes a potentially lethal poison called aflatoxin,[4] which is the cause of a disease in domestic animals that have eaten grain infested with the mold and is also a cause of liver cancer in humans.

4. From **a**spergillus, **fla**vus, **toxin.**

MICROFILE 5.2 FASCINATING FUNGI

HUMONGOUS FUNGUS AMONG US

Far from being microscopic, one of the largest organisms in the world is a massive basidiomycete growing in a forest near Seattle, Washington. The mycelium of a single colony of *Armillaria ostoye* covers an area of $2\frac{1}{2}$ square miles and could easily weigh 100 tons. Most of the fungus lies hidden beneath the ground, where it periodically fruits into edible mushrooms. Experts with the forest service took random samples over a large area and found that the mycelium is from genetically identical stock. They believe this fungus was started from a single mating pair of sexual hyphae around 1,000 years ago. Another colossal species from Michigan, *A. bulbosa,* is estimated to weigh 250,000 pounds and is perhaps 1,500 years old. Both fungi are able to grow on the stumps and roots of trees and can become pests. For now, the world still awaits the discovery of even larger fungi.

CANCER CURE FROM FUNGI

In yet another forest, a group of researchers has isolated a parasitic fungus, *Taxomyces andreanae,* from the inner bark of the Pacific yew tree. This fungus synthesizes *taxol,* a substance that has shown great promise in treating breast and ovarian cancers. What makes this discovery of particular importance is that the original source of taxol was the Pacific yew itself, yet this tree is somewhat uncommon and slow-growing. Environmentalists feared that the few stands of trees would have to be wiped out to obtain the drug and even then, supplies would not meet the demands for therapy. But it is unlikely that this will happen. The most recent development in the search for a source of taxol is a plan to use genetic engineering for mass production, negating the need to destroy trees.

FUNNY FUNGUS FLOWERS IN THE FOREST

Forests in Colorado have yielded a pathogenic fungus, *Arabis holboellii,* that interacts with its plant host in a most amazing way. When the fungus invades the tissues of the rock cress, it induces the plant to produce a tall spike of leaves that looks like a buttercup flower. The surface of this "flower" becomes a substrate for the fungus to give off yellow reproductive structures and a sweet saplike substance that attracts bees and flies. The fake flower is so attractive to these insects that they prefer it to real flowers. This adaptation not only improves the transmission of spores to other host plants but also helps the fungus complete the sexual phase of its life cycle.

MILDEW FUNGUS A CAUSE OF "SICK BUILDING SYNDROME"?

Hidden away in warm, moist corners in the home are harmless-looking creatures that dominate the fungal world: *Cladosporium* (see figure 5.25) and *Aureobasidium.* These tenacious fungi are to blame for the black mildew spots that appear on walls, shower grouting, and other hard surfaces. The molds are also known to inhabit air vents and internal walls of air conditioners and coolers. They grow tightly against these surfaces, deriving miniscule amounts of nutrients and absorbing periodic infusions of moisture. The dozens of cleaners on the market attest to the difficulty in suppressing and removing these fungi. Researchers of "sick building syndrome" are beginning to collect data on the possible role of airborne fungi in the odd collection of symptoms—headache, rash, fatigue—that some people experience in certain buildings. When the spores released in microscopic clouds are inhaled, they can cause vague discomfort, especially in persons with allergies or high sensitivities to them.

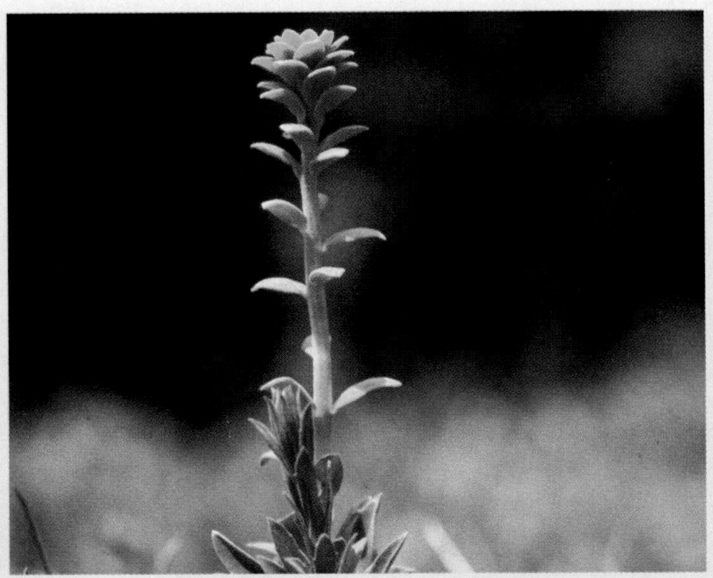

A case of mimicry. This delicate yellow "buttercup" is not a flower at all but a fungus. The phony flower develops as a parasite on a real plant to attract bees and flies that will help it complete its life cycle.

Fungi pose an ever-present economic hindrance to the agricultural industry. A number of species are pathogenic to field plants such as corn and grain, and fungi also rot harvested fresh produce during shipping and storage. It has been estimated that as much as 40% of the yearly fruit crop is consumed not by humans but by fungi. On the beneficial side, however, fungi play an essential role in decomposing organic matter and returning essential minerals to the soil. Industry has tapped the biochemical potential of fungi to produce large quantities of antibiotics, alcohol, organic acids, and vitamins. Some fungi are eaten or used to impart flavorings to food. The yeast *Saccharomyces* produces the alcohol in beer and wine and the gas that causes bread to rise. Blue cheese, soy sauce, and cured meats derive their unique flavors from the actions of fungi (see chapter 26).

THE PROTISTS

The algae and protozoa have been traditionally combined into the Kingdom Protista. The two major taxonomic categories of this kingdom are Subkingdom Algae and Subkingdom Protozoa. Though these general types of microbes are now known to oc-

TABLE 5.2

SUMMARY OF ALGAL CHARACTERISTICS

Group	Organization	Primary Habitat	Cell Wall	Pigmentation	Ecology/ Importance	Example(s)
Euglenophyta (euglenids)	Mainly unicellular; motile by flagella	Fresh water	None; pellicle instead	Chlorophyll, carotenoids, xanthophyll	Some are heterotrophic	*Euglena*
Pyrrophyta (dinoflagellates)	Unicellular, dual flagella	Marine plankton	Cellulose or atypical wall	Chlorophyll, carotenoids	Cause of "red tide"	*Gonyaulax*
Chrysophyta (diatoms or golden-brown algae)	Mainly unicellular, some filamentous forms unusual form of motility	Fresh water and marine	Silicon dioxide	Chlorophyll, fucoxanthin	Diatomaceous earth, major component of plankton	*Navicula,* other diatoms
Phaeophyta (brown algae—kelps)	Multicellular, vascular system, holdfasts	Marine, subtidal forests	Cellulose, alginic acid	Chlorophyll, carotenoids, fucoxanthin	Source of an emulsifier, alginate	*Fucus, Sargassum*
Rhodophyta (red seaweeds)	Multicellular	Marine, intertidal forests	Cellulose	Chlorophyll, carotenoids, xanthophyll, phycobilin	Source of agar and carrageenan, a food additive	*Gelidium*
Chlorophyta (green algae, grouped with plants)	Varies from unicellular, colonial, filamentous, to multicellular	Fresh water and salt water	Cellulose	Chlorophyll, carotenoids, xanthophyll	Precursor of higher plants	*Chlamydomonas, Spirogyra, Volvox*

cupy several kingdoms, it is still useful to retain the concept of a protist as any unicellular or colonial organism that lacks true tissues. This concept is especially helpful in our survey of major groups of microorganisms, because many algae are multicellular and macroscopic, and we cannot present an exhaustive coverage of all forms.

THE ALGAE: PHOTOSYNTHETIC PROTISTS

The **algae*** are a group of photosynthetic organisms usually recognized by their larger members, such as seaweeds and kelps. In addition to being beautifully colored and diverse in appearance, they vary in length from a few micrometers to 100 m. Algae occur in unicellular, colonial, and filamentous forms, and the larger forms can possess tissues but will not be included in our survey. Table 5.2 lists the major characteristics of the various subgroups of algae.

Algal cells as a group contain all of the eucaryotic organelles. The most noticeable of these are the chloroplasts, which contain, in addition to the green pigment chlorophyll, a number of other pigments that create the yellow, red, and brown coloration of some groups (table 5.2). The chloroplasts can be of such unique character that they are used in identification (for example, *Spirogyra* in figure 5.26*b*). Most algal cells are enclosed by an intricately patterned cell wall, which accounts for the distinctive appearance of unicellular members such as diatoms (figure 5.26*a*) and dinoflagellates. The outermost structure in a group called the euglenids (for example, *Euglena* in figure 5.26*a*) is a thick, flexible membrane called a **pellicle.** Motility by flagella or gliding is common among the algae, and many members contain tiny light-sensitive areas (eyespots) that coordinate with the flagella to guide the cell toward the light it requires for photosynthesis.

Algae are widespread inhabitants of fresh and marine waters. They are one of the main components of the large floating community of microscopic organisms called **plankton.** In this capacity, they play an essential role in the aquatic food web and produce most of the earth's oxygen. Other algal habitats include the surface of soil, rocks, and plants, and several species are even hardy enough to live in hot springs or snowbanks (see figure 7.11).

Algae are often described by common names such as green algae, brown algae, golden brown algae, and red algae in reference to their predominant color. A more technical system divides the

*algae (al´-jee) sing. alga; L. seaweeds.

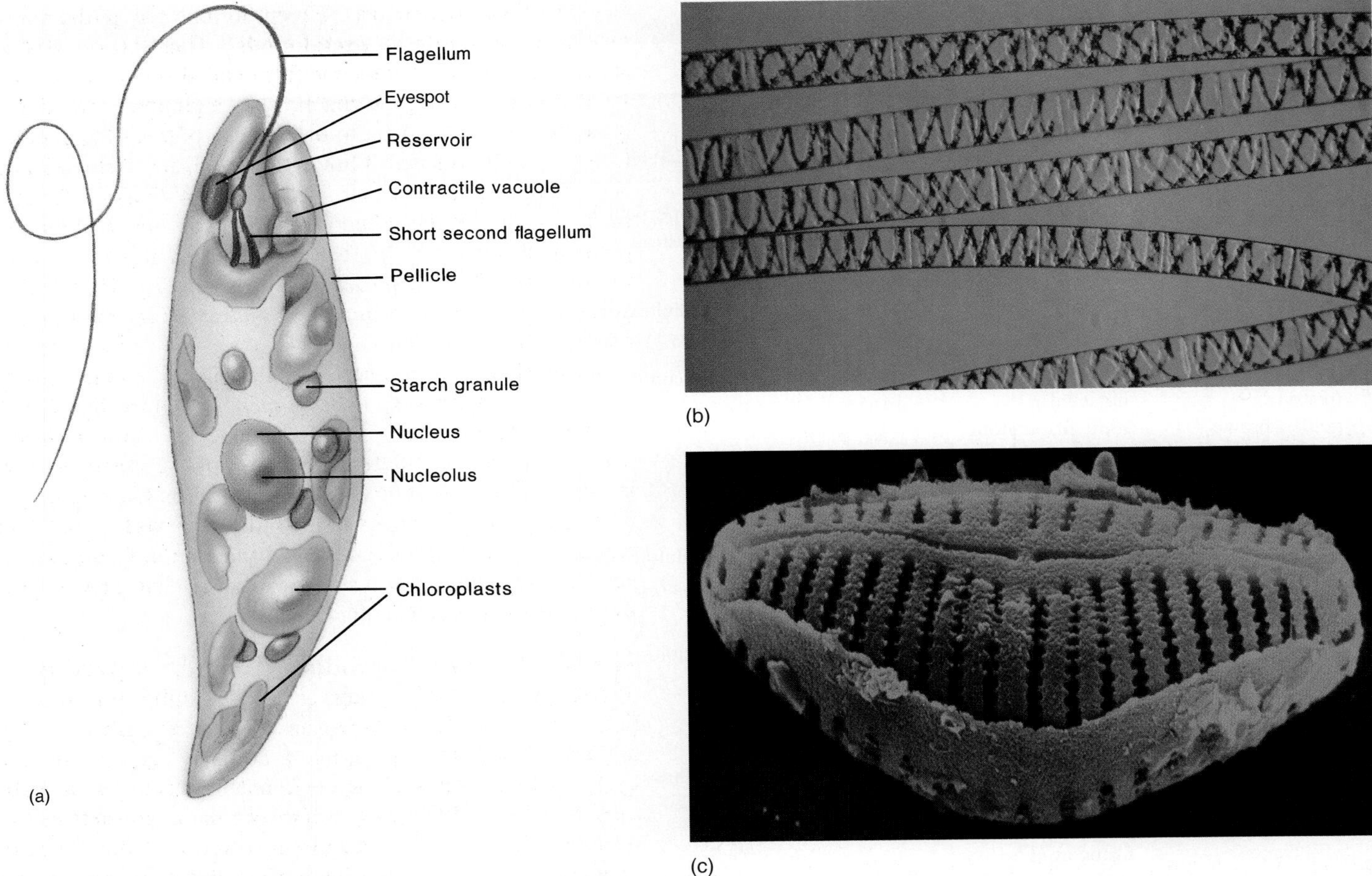

Figure 5.26
Representative microscopic algae. (*a*) *Euglena,* a unicellular motile example (150×). (*b*) *Spirogyra,* a colonial filamentous form with spiral chloroplasts. (*c*) The diatom *Trinacria regina,* a unicellular alga with finely detailed silica cell walls.

microscopic algae into divisions or kingdoms based on the types of chlorophyll and other pigments, the type of cell covering, the nature of their stored foods, and genetic factors. The common names for these groups are: (1) Euglenophyta, or euglenids; (2) Pyorophyta, or dinoflagellates; (3) Chrysophyta, or diatoms; (4) Phaeophyta, or brown algae; (5) Rhodophyta, or red seaweeds; and (6) Chlorophyta, or green algae.

Algae reproduces asexually through fragmentation, binary fission, and mitosis, and some produce motile spores. Their sexual reproductive cycles can be highly complex, with stages similar to those of fungi. The capacity of algae to photosynthesize is extremely important to the earth. They form the basis of aquatic food webs and play an essential role in the earth's oxygen and carbon dioxide balance. Some products of algae have industrial applications. For example, fossilized marine diatoms yield an abrasive powder called diatomaceous earth that is used in polishes, bricks, and filters, and certain seaweeds are a source of agar and algin used in microbiology, dentistry, and the food and cosmetics industries. We consider the role of algae in microbial ecology in chapter 26.

Animal tissues would be rather inhospitable to algae, so they are rarely infectious. One exception, however, is *Prototheca,* an unusual non-photosynthetic alga, which has been associated with skin and subcutaneous infections in humans and animals.

The primary medical threat from algae is due to a type of food poisoning caused by the toxins of certain marine **dinoflagellates.** During particular seasons of the year, the overgrowth of these motile algae imparts a brilliant red color to the water, which is referred to as a "red tide." When intertidal animals feed, their bodies accumulate toxins given off by the dinoflagellates that can persist for several months (see figure 26.20).

Paralytic shellfish poisoning is caused by eating exposed clams or other invertebrates. It is marked by severe neurological symptoms and can be fatal. Recently, an East Coast outbreak of fish disease that spread to humans was traced to a remarkable dinoflagellate called *Pfiesteria piscicida.* This newly identified species occurs in several forms, such as cyst and ameba (see chapter opener) that can release potent toxins. Both fishes and humans develop neurological symptoms and bloody skin lesions. The cause of the epidemic is still a mystery. Ciguatera is a serious intoxication caused by dinoflagellate toxins that have accumulated in fish such as bass and mackerel. Cooking does not destroy the toxin, and there is no antidote.

Biology of the Protozoa

If a poll were taken to choose the most engrossing and vivid group of microorganisms, many people would choose the protozoa. Although their name comes from the Greek for "first animals," they are far from being simple, primitive organisms. The protozoa constitute a very large group (about 65,000 species) of creatures that although single-celled, have startling properties when it comes to movement, feeding, and behavior. Although most members of this group are harmless, free-living inhabitants of water and soil, a few species are parasites collectively responsible for hundreds of millions of infections of humans each year. Before we consider a few examples of important pathogens, let us examine some general aspects of protozoan biology.

Protozoan Form and Function

Most protozoan cells are single cells containing the major eucaryotic organelles except chloroplasts (see figure 5.2*c*). Their organelles can be highly specialized for feeding, reproduction, and locomotion. The cytoplasm is usually divided into a clear outer layer called the *ectoplasm* and a granular inner region called the *endoplasm.* Ectoplasm is involved in locomotion, feeding, and protection. Endoplasm houses the nucleus, mitochondria, and food and contractile vacuoles. Some ciliates and flagellates[5] even have organelles that work somewhat like a primitive nervous system to coordinate movement. Because protozoa lack a cell well, they have a certain amount of flexibility. Their outer boundary is a cell membrane that regulates the movement of food, wastes, and secretions. Cell shape can remain constant (as in most ciliates) or can change constantly (as in amebas). Certain amebas (foraminiferans) encase themselves in hard shells made of calcium carbonate. The size of most protozoan cells falls within the range of 3 to 300 μm. Some notable exceptions are giant amebas and ciliates that are large enough (3–4 mm in length) to be seen swimming in pond water.

Nutritional and Habitat Range Protozoa are heterotrophic and usually require their food in a complex organic form. Free-living species scavenge dead plant or animal debris and even graze on live cells of bacteria and algae. Some species have special feeding structures such as oral grooves, which sweep food particles into a passageway or gullet that packages the captured food into vacuoles for digestion. A remarkable feeding adaptation can be seen in the ciliate *Didinium,* which can easily devour another ciliate that is nearly its size (see figure 5.31*c*). Some protozoa absorb food directly through the cell membrane. Parasitic species live on the fluids of their host, such as plasma and digestive juices, or they can actively feed on tissues.

Although protozoa have adapted to a wide range of habitats, their main limiting factor is the availability of moisture. Their predominant habitats are fresh and marine water, soil, plants, and animals. Even extremes in temperature and pH are not a barrier to their existence; hardy species are found in hot springs, ice, and habitats with low or high pH. Many protozoa can convert to a resistant, dormant stage called a cyst.

Styles of Locomotion Except for one group (the Sporozoa), protozoa are motile by **pseudopods,* flagella,** or **cilia.** A few species have both pseudopods (also called pseudopodia) and flagella. Some unusual protozoa move by a gliding or twisting movement that does not appear to involve any of these locomotor structures. Pseudopods are blunt, branched, or long and pointed, depending on the particular species. As previously described, the flowing action of the pseudopods results in ameboid motion, and pseudopods also serve as feeding structures in many amebas (see figure 5.30). The structure and behavior of flagella and cilia were discussed in the first section of this chapter. Flagella vary in number from one to several, and in certain species they are attached along the length of the cell by an extension of the cytoplasmic membrane called the *undulating membrane* (see figure 5.29). In most ciliates, the cilia are distributed over the entire surface of the cell in characteristic patterns. Because of the tremendous variety in ciliary arrangements and functions, ciliates are among the most diverse and awesome cells in the biological world. In certain protozoa, cilia line the oral groove and function in feeding; in others, they fuse together to form stiff props (cirri) that serve as primitive rows of walking legs (see figure 5.31).

Life Cycles and Reproduction Most protozoa are recognized by a motile feeding stage called the **trophozoite*** that requires ample food and moisture to remain active. A large number of species are also capable of entering into a dormant, resting stage called a **cyst** when conditions in the environment become unfavorable for growth and feeding. During *encystment,* the trophozoite cell rounds up into a sphere, and its ectoplasm secretes a tough, thick cuticle around the cell membrane (figure 5.27). Because cysts are more resistant than ordinary cells to heat, drying, and chemicals, they can survive adverse periods. They can be dispersed by air currents and may even be an important factor in the spread of diseases such as amebic dysentery. If provided with moisture and nutrients, a cyst breaks open and releases the active trophozoite.

The life cycles of protozoans vary from simple to complex. Several protozoan groups exist at all times in the trophozoite state. Many alternate between a trophozoite and a cyst stage, depending on the conditions of the habitat. The life cycle of a parasitic protozoon dictates its mode of transmission to other hosts. For example, the flagellate *Trichomonas vaginalis* causes a common sexually transmitted disease. Because it does not form cysts, it is usually transmitted by intimate contact between sexual partners. In contrast, intestinal pathogens such as *Entamoeba histolytica* and *Giardia lamblia* form cysts and are readily transmitted in contaminated water and foods.

All protozoa reproduce by relatively simple, asexual methods, usually mitotic cell division (figure 5.28). Flagellates tend to divide longitudinally, and ciliates transversely. Several parasitic species, including the agents of malaria and toxoplasmosis, reproduce asexually inside a host cell by multiple fission, or schizogony (figure 5.28*d*). Sexual reproduction also occurs during the life cycle of most protozoa, although several groups of amebas and flagellates appear

5. The terms *ciliate* and *flatellate* are common names of protozoan groups that move by means of flagella and cilia.

*pseudopod (soo´-doh-pod) pl. pseudopods or pseudopodia; Gr. *pseudo,* false, and *pous,* feet.

*trophozoite (trof´´-oh-zoh´yte) Gr. *trophonikos,* to nourish, and *zoon,* animal.

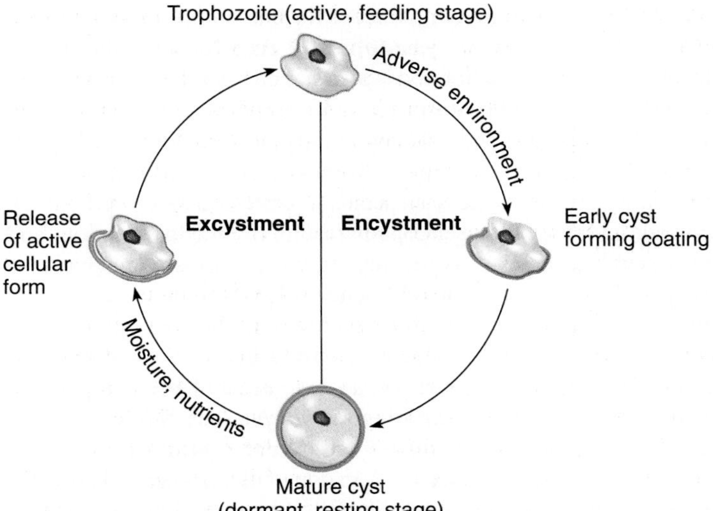

Figure 5.27

The general life cycle exhibited by many protozoa. (All protozoa have a trophozoite, but not all produce cysts.)

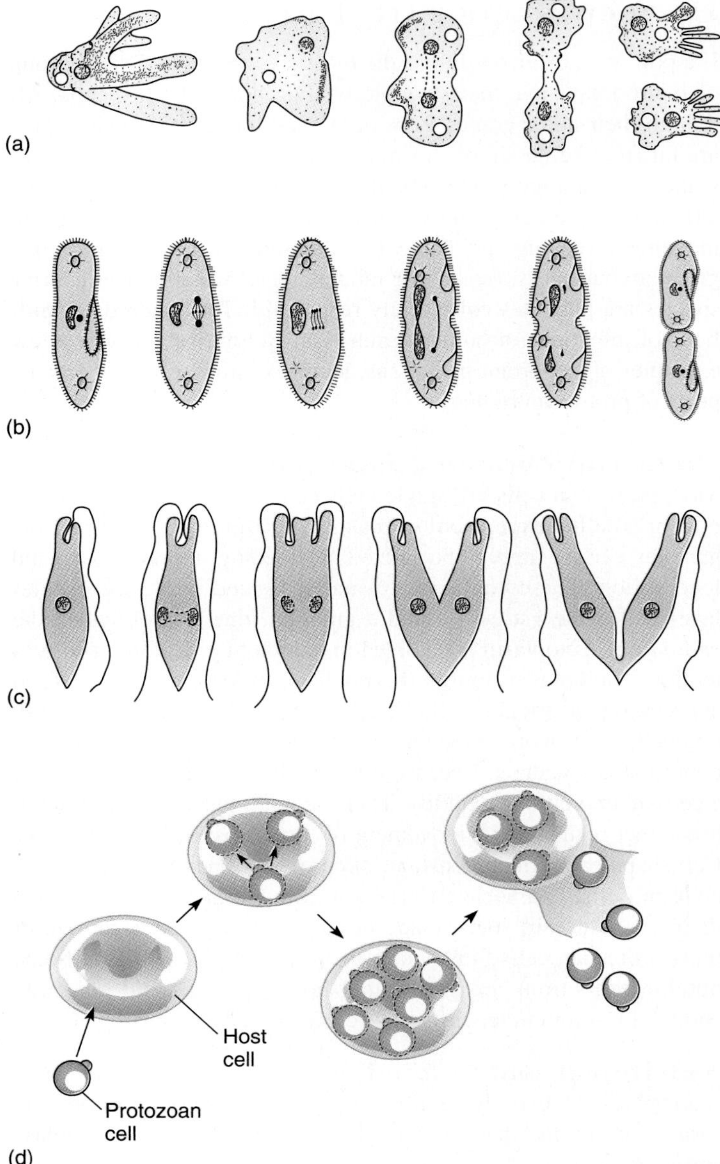

Figure 5.28

Asexual reproductive strategies in protozoa; (*a*) ameba, (*b*) ciliate, (*c*) flagellate. Note that in (*a–c*), the nucleus divides mitotically, followed by fission of the cell into two parts. (*d*) Multiple fission (schizogony) in sporozoa.

to reproduce only asexually. The simplest sexual mode is *syngamy,** in which two haploid motile gametes unite to form a diploid zygote. This zygote subsequently undergoes meiotic division to produce a number of haploid trophozoites. Ciliates participate in *conjugation,* a form of genetic exchange in which members of two different mating types fuse temporarily and exchange micronuclei. A micronucleus is the smaller of the two nuclei of ciliates that functions primarily in sexual recombination which yields new and different genetic combinations that can be advantageous in evolution.

Classification of Selected Medically Important Protozoa

Taxonomists have not escaped problems classifying protozoa. They, too, are very diverse and frequently frustrate attempts to generalize or place them in neat groupings. The most recent system places them in Five Kingdoms on the Eukarya tree (see figure 1.16), but this method may be more complex than is necessary for our survey. We will simplify this system presenting four groups, based on method of motility, mode of reproduction, and stages in the life cycle, summarized as follows:

The Mastigophora (Flagellata)[6] Motility is primarily by flagella alone or both flagella and ameboid motion. Single nucleus. Sexual reproduction, when present, by syngamy; division by longitudinal fission. Several parasitic forms lack mitochondria and Golgi apparatus. Most species form cysts and are free-living; the group also includes several parasites. Some species are found in loose aggregates or colonies, but most are solitary. Examples: *Trypanosoma* and *Leishmania,* important blood pathogens spread by insect vectors: *Giardia,* an intestinal parasite spread in water contaminated with feces; *Trichomonas,* a parasite of the reproductive tract of humans spread by sexual contact (figure 5.29).

The Sarcodina[6] Cell form is primarily an ameba (figure 5.30). Major locomotor organelles are pseudopods, although some species have flagella during reproductive states. Asexual reproduction by fission. Two groups have an external shell; mostly uninucleate; usually encyst. Examples: Most amebas are free-living and not infectious; *Entamoeba* is a pathogen or parasite of humans; shelled amebas called foraminifera and radiolarians are responsible for chalk deposits in the ocean.

*syngamy (sing´-ah-mee) Gr. *syn,* with, and *gamy,* marriage.

6. Some biologists prefer to combine phyla Mastigophora and Sarcodine into the phylum Sarcomastigophora. Because the algal group Euglenophyta has Flagella, it may also be included in the Mastigophora.

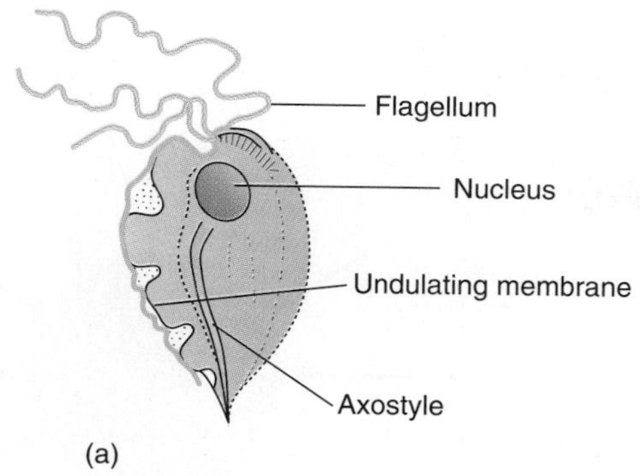

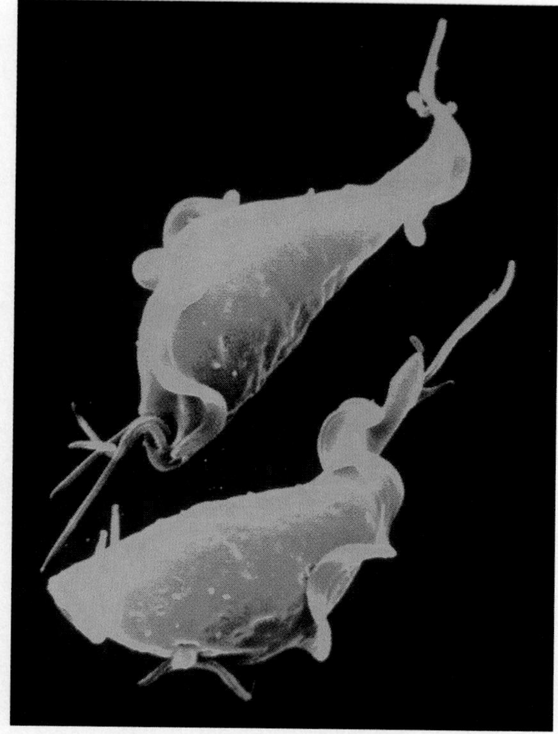

Figure 5.29

The structure of a typical mastigophoran, *Trichomonas vaginalis,* a genital tract pathogen, as shown in (*a*) a drawing, and (*b*) a scanning electron micrograph.

The Ciliophora (Ciliata) Trophozoites are motile by cilia; some have cilia in tufts for feeding and attachment (figure 5.31); most develop cysts; have both macronuclei and micronuclei; division by transverse fission; most have a definite mouth and feeding organelle; show relatively complicated and advanced behavior. Example: The majority of ciliates are free-living and harmless. One important pathogen, *Balantidium coli,* lives in vertebrate intestines and can infect humans.

The Sporozoa, or Apicomplexa Motility is absent in most cells except male gametes. Life cycles are complex, with well-developed asexual and sexual stages. Sporozoa produce special sporelike cells called **sporozoites*** (figure 5.32) following sexual reproduction, which are important in transmission of infections; most form thick-walled zygotes called oocysts; entire group is parasitic. Examples: *Plasmodium,* the most prevalent protozoan parasite, causes 100 million to 300 million cases of malaria each year worldwide. It is an intracellular parasite with a complex cycle alternating between humans and mosquitos (see figure 23.11). *Toxoplasma gondii* causes an acute infection (toxoplasmosis) in humans, which is transmitted primarily by cats.

Protozoan Identification and Cultivation

The unique appearance of most protozoa make it possible for a knowledgeable person to identify them to the level of genus and often species by microscopic morphology alone. Characteristics to consider in identification include the shape and size of the cell; the type, number, and distribution of locomotor structures; the presence of special organelles or cysts; and the number of nuclei. Medical specimens taken from blood, sputum, cerebrospinal fluid, feces, or the vagina are smeared directly onto a slide and observed with or without special stains. Occasionally, protozoa are cultivated on artificial media or in laboratory animals for further identification or study.

Important Protozoan Parasites

Although protozoan infections are very common, they are actually caused by only a small number of species often restricted geographically to the tropics and subtropics. In this survey, we first introduce some concepts of the host-protozoan relationship and then look at examples that illustrate some of the main features of protozoan diseases. Details of several other medically important protozoa are given in chapter 23.

Parasitic protozoa are traditionally studied along with the helminths in the science of parasitology. Although a parasite is generally defined as an organism that obtains food and other needs at the expense of a host, the range of host-parasite relationships can be very broad. At one extreme are the so-called good parasites, which occupy their host with little harm. An example is certain amebas that live in the human intestine and feed off organic matter there. At the other extreme are parasites *(Plasmodium, Trypanosoma)* that multiply in host tissues such as the blood or brain, causing severe damage and disease. Between these two extremes are parasites of varying pathogenicity, depending on their particular adaptations (see chapters 7 and 13).

Most human parasites go through three general stages: (1) The microbe is transmitted to the human host from a source such as soil, water, food, other humans, or animals. (2) The microbe invades and multiples in the host, producing more parasites that can infect other suitable hosts. (3) The microbe leaves the host in large numbers by a specific means and, to survive, must find and enter a new host. There are numerous variations on this simple theme. For instance, the microbe can invade more than one host species (an alternate host) and undergo several changes as it cycles through these hosts, such as sexual reproduction or encystment. Some microbes are spread from human to human by means of **vectors,*** defined as animals such as insects that carry diseases. Others can be spread through bodily fluids and feces.

*sporozoite (spor´´-oh-zoh´-yte) Gr. *sporos,* seed, and *zoon,* animal.

*vector (vek´-tur) L. *vectur,* one who carries.

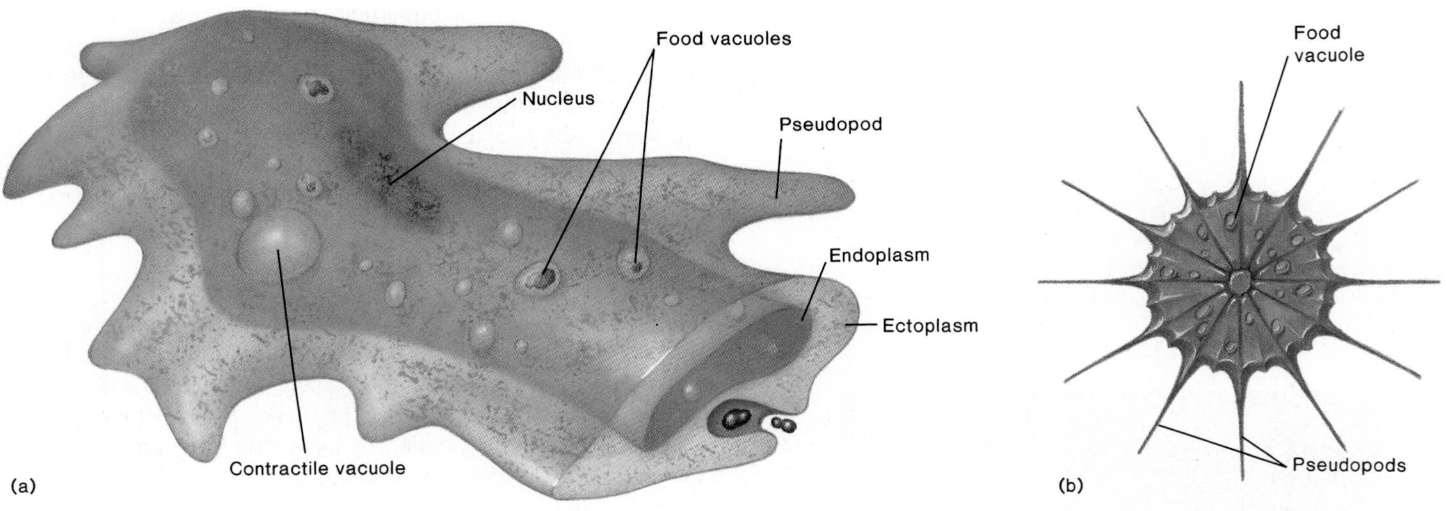

Figure 5.30

Examples of sarcodinians. (*a*) The structure of an ameba. (*b*) Radiolarian, a shelled ameba with long, pointed pseudopods. Pseudopods are used in both movement and feeding.

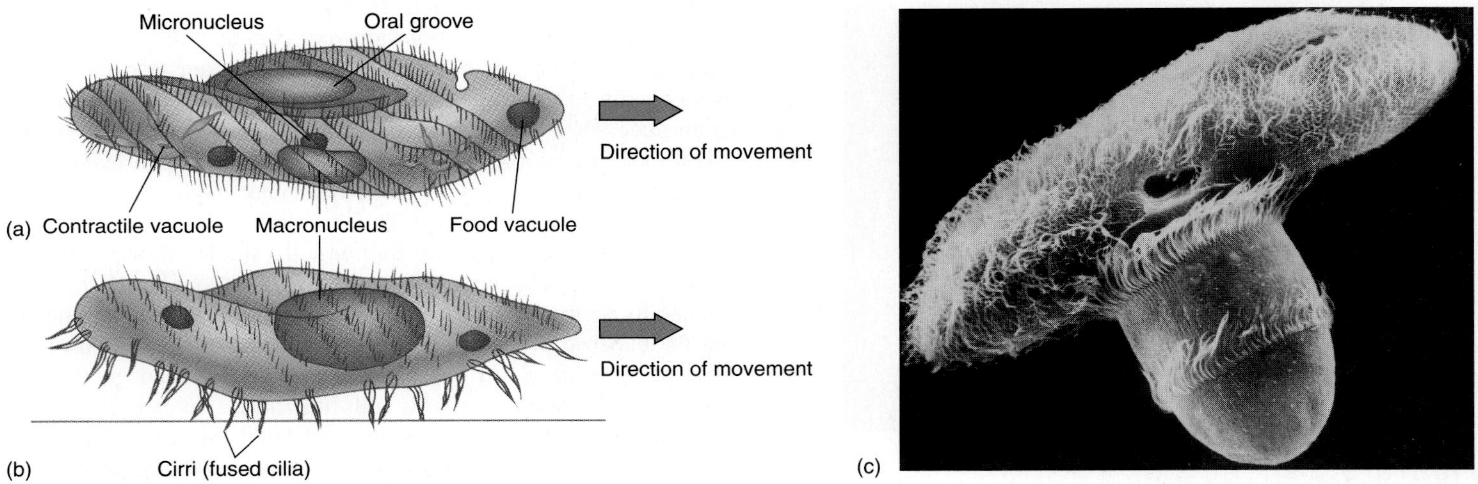

Figure 5.31

Selected ciliate representatives. (*a*) The structure of a typical representative, *Paramecium*. Cilia beat in coordinated waves, driving the cell forward and backward. Cilia are also used to sweep food particles into the gullet to form food vacuoles. (*b*) Fused cilia (cirri) of *Stylonychia* are used for walking over surfaces. (*c*) Scanning electron micrograph of a *Didinium* cell eating a *Paramecium* (1,400×). Note the number and distribution of cilia in each protozoon.

Pathogenic Flagellates: Trypanosomes Trypanosomes are parasites belonging to the genus *Trypanosoma.** The two most important representatives are *T. brucei* and *T. cruzi,* species that are closely related but geographically restricted. *Trypanosoma brucei* occurs in Africa, where it causes approximately 10,000 new cases of sleeping sickness each year. *Trypanosoma cruzi,* the cause of Chagas' disease,[7] is endemic to South and Central America, where it infects approximately 10 million to 20 million people a year and causes 50,000 deaths. Both species have long, crescent-shaped cells with a single flagellum that is sometimes attached to the cell body by an undulating membrane (figure 5.33). Both occur in the blood during infection and are transmitted by blood-sucking vectors. We will use *T. cruzi* to illustrate the phases of a trypanosomal life cycle and to demonstrate the complexity of parasitic relationships.

The trypanosome of Chagas' disease relies on the close relationship of a warm-blooded mammal and an insect that feeds on mammalian blood. The mammalian hosts are numerous, including dogs, cats, oppossums, armadillos, and foxes. The vector is the *reduviid* bug,* an insect that is sometimes called the "kissing bug" because of its habit of biting its host at the corner of the mouth.

7. Named for Carlos Chagas, the discoverer of *T. cruzi.*

**Trypanosoma* (try´´-pan-oh-soh´-mah) Gr. *rypanon,* borer, and *soma,* body.

***reduviid** (ree-doo´-vee-id) A member of a large family of flying insects with sucking, beaklike mouths.

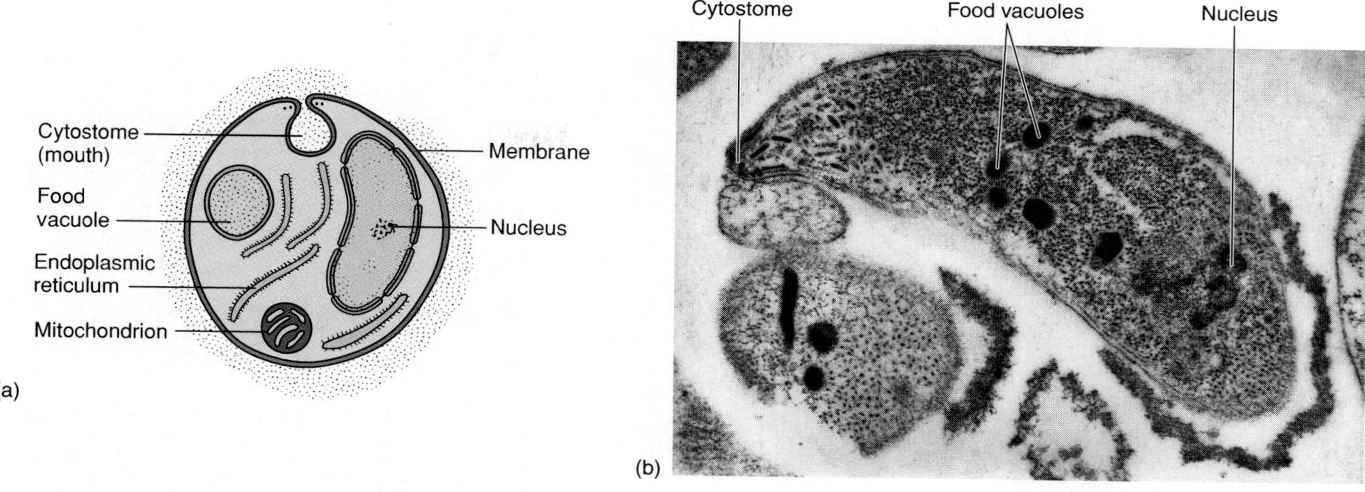

Figure 5.32
Sporozoan parasites. (*a*) General cell structure. Note the lack of specialized locomotor organelles. (*b*) Scanning electron micrograph of the sporozoite of *Cryptosporidium*, an intestinal parasite of humans and other mammals.

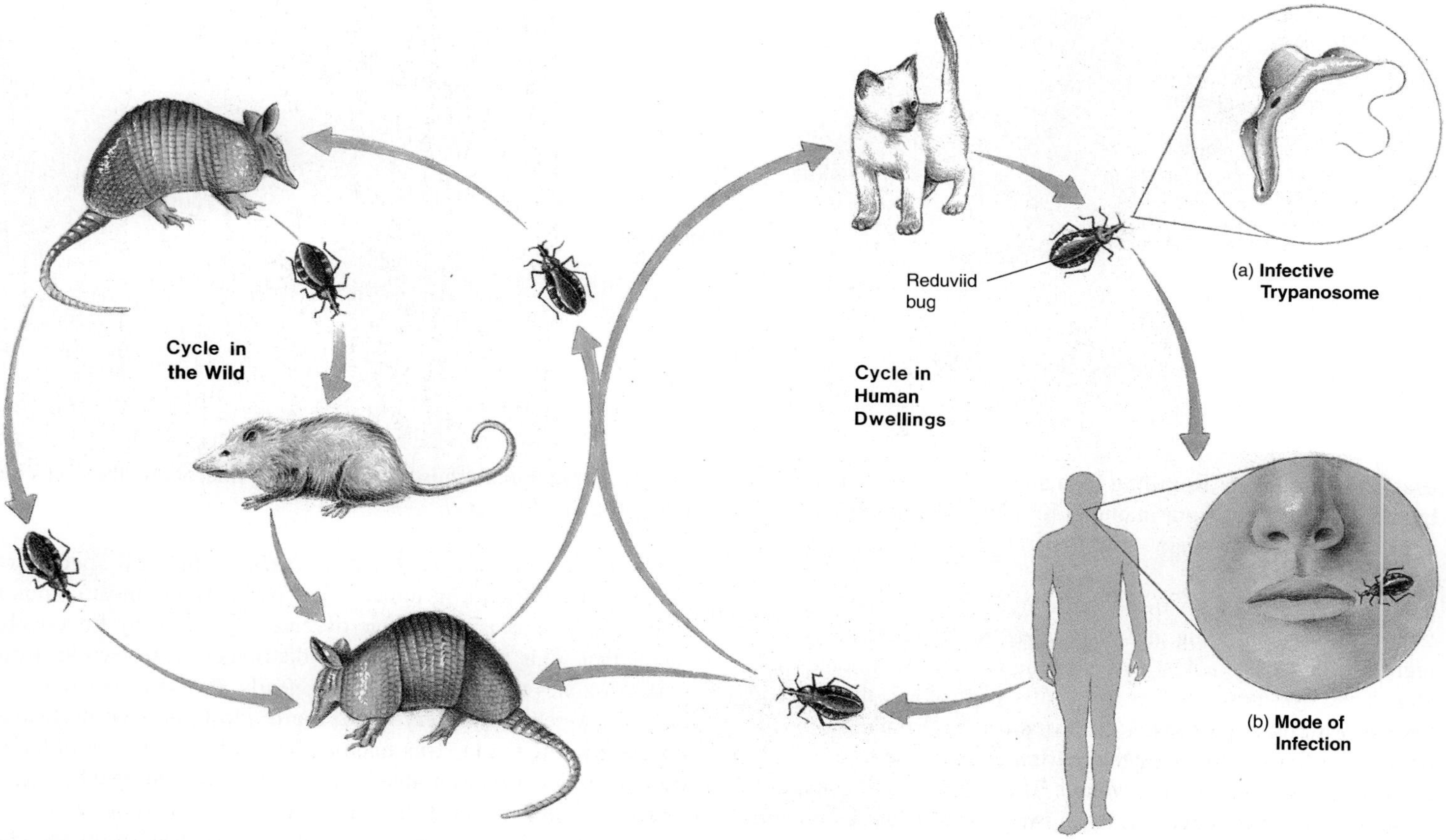

Cycle in the Wild

Reduviid bug

(a) **Infective Trypanosome**

Cycle in Human Dwellings

(b) **Mode of Infection**

Figure 5.33

Cycle of transmission in Chagas' disease. Trypanosomes (inset *a*) are transmitted among mammalian hosts and human hosts by means of a bite from the kissing bug (inset *b*).

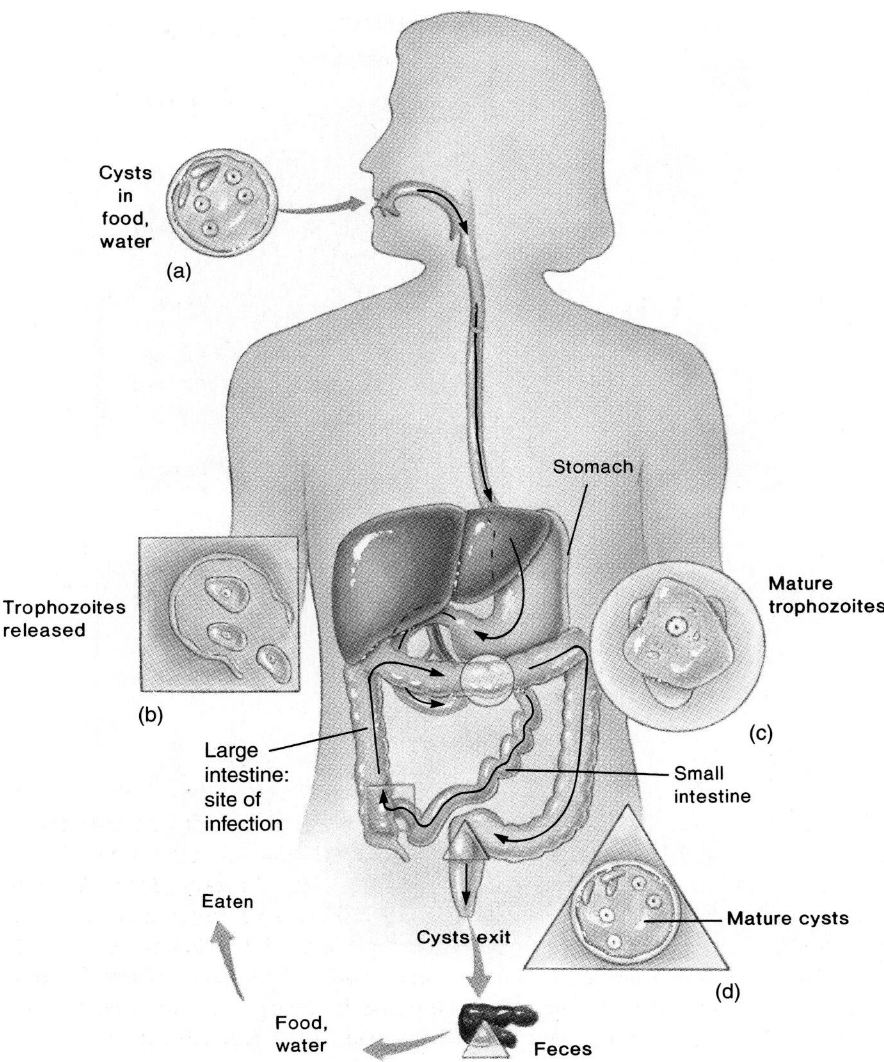

Figure 5.34

Stages in the infection and transmission of amebic dysentery, showing the route of infection and the appearance of *Entamoeba histolytica*. (*a*) Cysts are eaten. (*b*) Trophozoites (amebas) emerge from cysts. (*c*) Trophozoites invade the large intestinal wall. (*d*) Mature cysts are released in the feces.

Transmission occurs from bug to mammal and from mammal to bug, but usually not from mammal to mammal, except across the placenta during pregnancy. The general phases of this cycle are presented in figure 5.33.

The trypanosome trophozoite multiplies in the intestinal tract of the reduviid bug and is harbored in the feces. During the night, the bug seeks a host and bites the mucous membranes, usually of the eye, nose, or lips. As it fills with blood, the bug soils the bite with feces containing the trypanosome. Ironically, the victims themselves inadvertently contribute to the entry of the microbe by scratching the bite wound. The trypanosomes ultimately become established and multiply in muscle and white blood cells. Periodically, these parasitized cells rupture, releasing large numbers of new trophozoites into the blood. Eventually, the trypanosome can spread to many systems, including the lymphoid organs, heart, liver, and brain. Manifestations of the resultant disease range from mild to very severe, and include fever, inflammation, and heart and brain damage. In many cases, the disease has an extended course and can cause death. Unfortunately, an effective treatment has yet to be developed. The insect phase of development is completed when a reduviid bug takes a blood meal from an infected mammal, thus establishing the microbe in the digestive tract of its insect host.

Infective Amebas: Entamoeba Several species of amebas cause disease in humans, but probably the most common disease is amebiasis, or amebic dysentery,* caused by Entamoeba histolytica. This microbe is widely distributed in the world, from northern zones to the tropics, and is nearly always associated with humans. Amebic dysentery is the fourth most common protozoan infection in the world. This microbe has a life cycle quite different from the trypanosomes in that it does not involve multiple hosts and a blood-sucking vector. It lives part of its cycle as trophozoite and part as a cyst. Because the cyst is the more resistant form and can survive in water and soil for several weeks, it is the more important stage for transmission. The primary way that people become infected is by ingesting food or water contaminated with human feces.

Figure 5.34 shows the major features of the amebic dysentery cycle, starting with the ingestion of cysts. The viable, heavy-

*dysentery (dis´-en-ter´´-ee) Any inflammation of the intestine accompanied by bloody stools. It can be caused by a number of factors, both microbial and nonmicrobial.

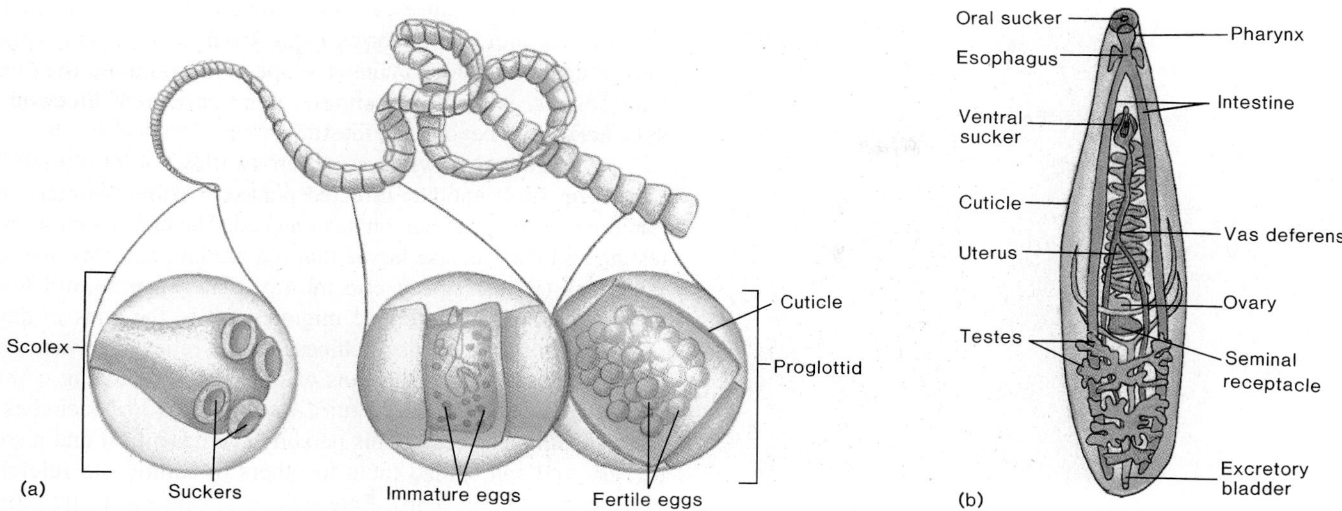

Figure 5.35

Parasitic flatworms. (*a*) A cestode (beef tapeworm), showing the scolex; long, tapelike body; and magnified views of immature and mature proglottids. (*b*) The structure of a trematode (liver fluke). Note the suckers that attach to host tissue and the dominance of reproductive and digestive organs.

walled cyst passes through the stomach unharmed. Once inside the small intestine, the cyst germinates into a large multinucleate ameba that subsequently divides to form small amebas (the trophozoite stage). These trophozoites migrate to the large intestine, colonize the intestinal surface, and begin to feed and grow. From this site, they can penetrate the lining of the intestine and invade the liver, lungs, and skin. Common symptoms include gastrointestinal disturbances such as nausea, vomiting, and diarrhea, leading to weight loss and dehydration. Untreated cases with extensive damage to the organs experience a high death rate. The cycle is completed in the infected human when certain trophozoites in the feces begin to form cysts with several nuclei, which then pass out of the body with fecal matter. Knowledge of the amebic cycle and role of cysts has been a boon in controlling the disease. Important preventive measures include sewage treatment, curtailing the use of human feces as fertilizers, and adequate sanitation of food and water.

THE PARASITIC HELMINTHS

Tapeworms, flukes, and roundworms are collectively called **helminths,** from the Greek word meaning worm. Adult animals are usually large enough to be seen with the naked eye, and they range from the longest tapeworms, measuring up to about 23 m in length, to roundworms less than 1 mm in length. Nevertheless, they are included among microorganisms because the microscope is necessary to see the details of smaller helminths and to identify their eggs and larvae.

On the basis of morphological form, the two major groups of parasitic helminths are the **flatworms** (Phylum Platyhelminthes), with a very thin, often segmented body plan (figure 5.35), and the **roundworms** (Phylum Aschelminthes, also called **nematodes***), with an elongate, cylindrical, unsegmented body plan (figure 5.36). The flatworm group is subdivided into the **cestodes,*** or tapeworms, named for their long, ribbonlike arrangement, and the **trematodes,*** or flukes, characterized by flat, ovoid bodies. Not all flatworms and roundworms are parasites by nature; many live free in soil and water. Both here and in chapter 23, we concern ourselves with the medically important worms (see table 23.4).

GENERAL WORM MORPHOLOGY

All helminths are multicellular animals equipped to some degree with organs and organ systems. In parasites, the most developed organs are those of the reproductive tract, with some degree of reduction in the digestive, excretory, nervous, and muscular systems. In particular groups, such as the cestodes, reproduction is so dominant that the worms are reduced to little more than a series of flattened sacs filled with ovaries, testes, and eggs (microfile 5.3 and figure 5.35*a*). Not all worms have such extreme adaptations as cestodes, but most have a highly developed reproductive potential, thick cuticles for protection, and mouth glands for breaking down the host's tissue.

LIFE CYCLES AND REPRODUCTION

Many worms have complex life cycles that alternate between hosts. Reproduction of individuals is primarily sexual, involving the production of eggs and sperm in the same worm or in separate

*nematode (neem´-ah-tohd) Gr. *nemato,* thread, and *eidos,* form.

*cestode (sess´-tohd) L. *cestus,* a belt, and *ode,* like.

*trematode (treem´-a-tohd) Gr. *trema,* hole. Named for the appearance of having tiny holes.

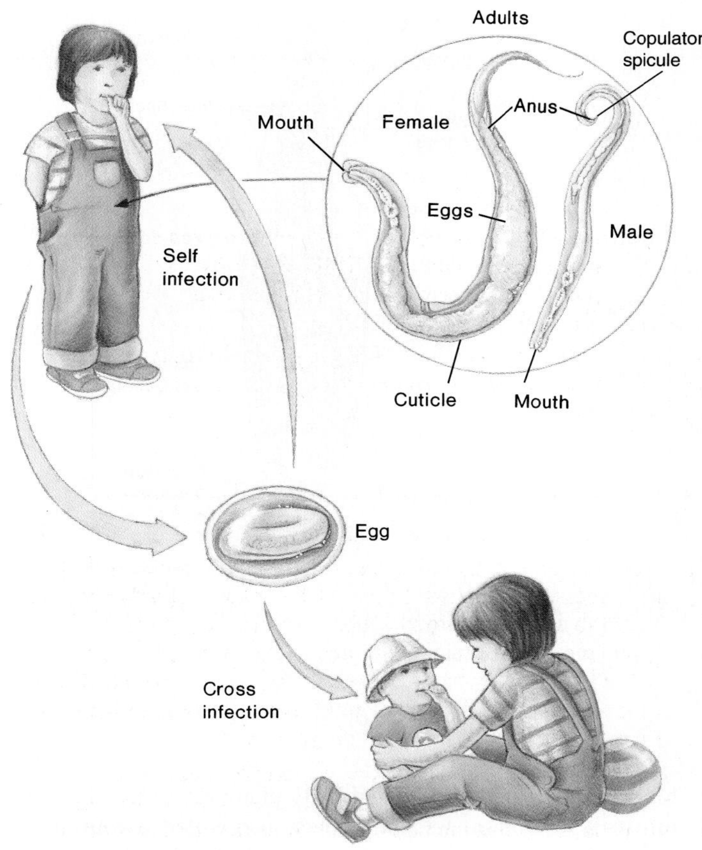

Figure 5.36

The life cycle of the pinworm, a roundworm. Eggs are the infective stage and are transmitted by unclean hands. Children frequently reinfect themselves and also pass the parasite on to others.

male and female worms. Fertilized eggs are usually released to the environment and are provided with a protective shell and extra food to aid their development into larvae. Even so, most eggs and larvae are vulnerable to heat, cold, drying, and predators and are destroyed or unable to reach a new host. To counteract this formidable mortality rate, certain worms have adapted a reproductive capacity that borders on the incredible: A single female *Ascaris*[8] can lay 200,000 eggs a day, and a large female can contain over 25,000,000 eggs at varying stages of development! If only a tiny number of these eggs makes it to another host, the parasite will have been successful in completing its life cycle.

A Helminth Cycle: The Pinworm

To illustrate a helminth cycle in humans, we will use the example of a roundworm, *Enterobius vermicularis,* the pinworm or seatworm. This worm causes a very common infestation of the large intestine (figure 5.36). Worms range from 2 to 12 mm long and have a tapered, curved cylinder shape. The condition they cause, enterobiasis, is usually a simple, uncomplicated infection that does not spread beyond the intestine.

A cycle starts when a person swallows microscopic eggs picked up from another infected person by direct contact or by touching articles that person has touched. The eggs hatch in the intestine and then release larvae that mature into adult worms in the appendix (within about one month). There, male and female worms mate, and the female migrates out to the anus to deposit eggs, which cause intense itchiness that is relieved by scratching. Herein lies a significant means of dispersal: Scratching contaminates the fingers, which, in turn, transfer eggs to bedclothes and other inanimate objects. This person becomes a host and a source of eggs and can spread them to others in addition to reinfesting himself (see figure 5.36). Enterobiasis occurs most often among families and in other close living situations. Its distribution is worldwide among all socioeconomic groups, but it seems to attack younger people more frequently than older ones and females three times more often than males.

Helminth Classification and Identification

The helminths are classified according to their shape; size; the degree of development of various organs; the presence of hooks, suckers, or other special structures; the mode of reproduction; the kinds of hosts; and the appearance of eggs and larvae. They are identified in the laboratory by microscopic detection of the adult worm or its larvae and eggs, which often have distinctive shapes or external and internal structures (see chapter 19). Occasionally, they are cultured in order to verify all of the life stages.

Distribution and Importance of Parasitic Worms

About 50 species of helminths parasitize humans. They are distributed in all areas of the world that support human life. Some worms are restricted to a given geographic region, and many have a higher incidence in tropical areas. This knowledge must be tempered with the realization that jet age travel, along with human migration, are gradually changing the patterns of worm infections, especially of those species that do not require alternate hosts or special climatic conditions for development. The yearly estimate of worldwide cases numbers in the billions, and these are not confined to developing countries. A conservative estimate places 50,000,000 helminth infections in the United States alone. The primary targets are young, malnourished children, whose death rate is increased in cases of gross infestation, multiple worm infections, and impaired host defenses.

8. *Ascaris* is a genus of parasitic intestinal roundworms.

MICROFILE 5.3 BIOLOGICAL ADAPTATION IN THE TAPEWORMS

The adaptation of intestinal helminths serves as a useful model to illustrate the economy, opportunity, and even resourcefulness of parasitic worms. Deep within the dark, airless tunnel of the host's intestine, the worms experience no temperature extremes, no water shortage, and no predators. They literally float in a sea of food that the host has already digested. Because these worms merely have to absorb nutrients and water through their skin, they do not need an extensive digestive tract of their own. Living this easy life is not without problems, however. First, the animals must complete their life cycles by having their eggs safely transmitted to another host. Second, they must avoid being digested or expelled by the host. The first problem is solved by ample reproductive organs, and the second is solved by developing a resistant covering, or cuticle, and by various hooks and suckers for attachment to host tissues.

Chapter Checkpoints

The eucaryotic microorganisms include the Fungi (Myceteae), the Protista (algae and protozoa), and the Helminths (Kingdom Animalia).

The Kingdom Fungi (Myceteae) is composed of nonphotosynthetic haploid species with cell walls of chitin. The fungi are either saprobes or parasites, and may be unicellular, colonial, or multicellular. Forms include yeasts (unicellular budding cells) and molds (filamentous cells called hyphae). Their primary means of reproduction involves asexual and sexual spores.

The protists are mostly unicellular or colonial eucaryotes that lack specialized tissues. There are two major organism types: the Algae and the Protozoa. Algae are photosynthetic organisms that contain chloroplasts with chlorophyll and other pigments. Protozoa are heterotrophs that usually display some form of locomotion. Most are single celled trophozoites and many produce a resistant stage, or cyst.

The Kingdom Animalia has only one group that contains microscopic members. These are the helminths or worms. Parasitic examples include flat worms and round worms.

CHAPTER CAPSULE WITH KEY TERMS

EUCARYOTIC STRUCTURE[9]

Appendages (cilia, flagella), glycocalyx, cell wall, cytoplasmic (or cell) membrane, ribosomes, organelles (nucleus, nucleolus, endoplasmic reticulum, Golgi apparatus, mitochondria, chloroplasts, cytoskeleton, microfilaments, microtubules).

THE KINGDOM FUNGI (MYCETEAE)

Common names of two particular types: macroscopic fungi (mushrooms, bracket fungi, puffballs) and microscopic fungi (yeasts, molds).

Overall Morphology: At the cellular (microscopic) level, typical eucaryotic cell, with thick **cell walls. Yeasts** are single cells that form **buds** and sometimes short chains called **pseudohyphae. Hyphae** are long, tubular filaments that can be septate or nonseptate and grow in a network called a **mycelium:** hyphae are characteristic of the filamentous fungi called **molds;** some fungi (mushrooms) produce multicellular structures such as fleshy fruiting bodies. Fungal cells are largely **nonmotile,** except for some motile gametes.

Nutritional Mode/Distribution: All are **heterotrophic.** The majority are harmless **saprobes** living off organic **substrates** such as dead animal and plant tissues. A few are **parasites,** living on the tissues of other organisms, but none are obligate. They do not contain chlorophyll and are not photosynthetic. Preferred temperature of growth is 20° – 40°C. Distribution is extremely widespread in many habitats.

Reproduction: Primarily through **spores** formed on special **reproductive hyphae.** In **asexual reproduction,** spores are formed through budding, partitioning of a hypha, or in special sporogenous structures; examples are **conidia** and **sporangiospores.** In **sexual reproduction,** spores are formed through fusion of male and female strains and the formation of a sexual hypha of some sort; sexual spores are a basis for classification.

Major Groups: The four important divisions among the terrestrial fungi, given with sexual spore type, are **Zygomycota (zygospores), Ascomycota (ascospores), Basidiomycota (basidiospores),** and **Deuteromycota** (no sexual spores).

Importance: Essential decomposers of plant and animal detritis in the environment with return of valuable nutrients to the ecosystem. Economically beneficial as sources of antibiotics; used in making foods and in genetic studies. Harmful plant pathogens; decompose fruits and vegetables; several fungi cause infections, or **mycoses;** some produce substances that are toxic if eaten.

9. A review comparing the major differences between eucaryotic and procaryotic cells is provided in table 2.6, page 54.

MICROSCOPIC PROTISTS—THE ALGAE

Include plantlike protists, kelps,and seaweeds. Specific groups are the **euglenids, green algae, diatoms, dinoflagellates, brown algae,** and **red seaweeds.**

Overall Morphology: Contain **chloroplasts** with **chlorophyll** and other pigments; cell wall; may or may not have flagella. Microscopic forms are unicellular, colonial, filamentous; macroscopic forms are colonial and multicellular.

Nutritional Mode/Distribution: **Photosynthetic;** most are free-living in the aquatic environment, both fresh water and marine (common component of **plankton**).

Major Groups: Classification according to types of pigments and cell walls. Microscopic algae: Euglenophyta, Chlorophyta, Chrysophyta, Pyrrophyta: Macroscopic algae: Phaeophyta (kelps) and Rhodophyta (sea weeds) are multicellular algae.

Importance: Algae provide the basis of the food web in most aquatic habitats, and they produce a large proportion of atmospheric O_2 through photosynthesis. Some are harvested as a source of cosmetics, food, and medical products. Dinoflagellates cause red tides and give off toxins that can render fish toxic to humans.

THE PROTOZOA

Include animal-like **protists;** unicellular animals such as **amebas, flagellates, ciliates, sporozoa.**

Overall Morphology: Most are unicellular, colonies rare, no multicellular forms; most have locomotor structures, such as **flagella, cilia, pseudopods;** special feeding structures can be present; lack a cell wall; extreme variations in shape. Can exist in **trophozoite,** a motile, feeding stage, or **cyst,** a dormant resistant stage.

Nutritional Mode/Distribution: All are **heterotrophic.** Most are free-living in a moist habitat (water, soil); feed on other microorganisms and organic matter. A number of animal parasites; can be spread from host to host by **insect vectors.**

Reproduction: Asexual by binary fission and **mitosis,** budding; sexual by fusion of free-swimming gametes, conjugation.

Major Groups: Protozoa are subdivided into four groups based upon mode of locomotion and type of reproduction: **Mastigophora,** the flagellates, motile by flagella; **Sarcodina,** the amebas, motile by pseudopods; **Ciliophora,** the ciliates, motile by cilia; **Sporozoa,** all parasites; motility not well developed; produce unique reproductive structures.

Importance: Ecologically important in food webs and decomposing organic matter. Medical significance: hundreds of millions of people are afflicted with one of the many protozoan infections (malaria, trypanosomiasis, amebiasis); protozoan infections can be geographically restricted.

THE HELMINTH PARASITES

Includes parasitic worms, tapeworms, flukes, nematodes.

Overall Morphology: Animal cells; multicellular; individual organs specialized for reproduction, digestion, movement, protection, though some of these are reduced.

Nutritional/Reproductive Mode: Parasitize host fluids, tissues; have mouthparts for attachment to or digestion of host tissues. Most are aerobic, but some can live in anaerobic conditions. Most have well-developed sex organs that produce eggs and sperm. Fertilized eggs go through larval period in or out of host body.

Major Groups: **Flatworms** have highly flattened body; no definite body cavity; digestive tract a blind pouch; male and female sex organs; simple excretory and nervous systems. **Cestodes** (tapeworms) long, flat chains of segments; attach to host's intestine by hooked mouthpart. **Trematodes,** or flukes, are flattened, ovoid, nonsegmented worms with sucking mouthparts. **Roundworms (nematodes)** have round bodies in cross section, a complete digestive tract, a protective surface cuticle, spines and hooks on mouth; excretory and nervous systems poorly developed.

How Transmitted and Acquired: Through ingestion of larvae or eggs in food, soil, water; can be carried by insect vectors.

Importance: Afflict billions of humans. Because these worms can cause illness and death, they have medical and economic impact of staggering proportions.

MULTIPLE-CHOICE QUESTIONS

1. Both flagella and cilia are found primarily in
   a. algae
   b. protozoa
   c. fungi
2. Features of the nuclear envelope include
   a. ribosomes
   b. a double membrane structure
   c. pores that allow communication with the cytoplasm
   d. b and c
   e. all of these
3. In general, if two haploid cells fuse, ____ will result.
   a. a germ cell c. mitosis
   b. a diploid zygote d. meiosis
4. The cell wall is found in which eucaryotes?
   a. fungi c. protozoa
   b. algae d. a and b
5. What is embedded in rough endoplasmic reticulum?
   a. ribosomes c. chromatin
   b. Golgi apparatus d. vesicles
6. Yeasts are ____ fungi, and molds are ____ fungi.
   a. macroscopic, microscopic
   b. unicellular, filamentous
   c. motile, nonmotile
   d. water, terrestrial
7. In general, fungi derive nutrients through
   a. photosynthesis
   b. engulfing bacteria
   c. digesting organic substrates
   d. parasitism
8. A hypha divided into compartments by cross walls is called
   a. nonseptate c. septate
   b. imperfect d. perfect
9. A conidium is a/an ____ spore, and a zygospore is a/an ____ spore.
   a. sexual, asexual

b. free, endo
c. ascomycete, basidiomycete
d. asexual, sexual

10. Algae generally contain some types of
a. spore
b. chlorophyll
c. locomotor organelle
d. toxin

11. All protozoa have a
a. locomotor organelle
b. cyst stage
c. pellicle
d. trophozoite

12. The protozoan trophozoite is the
a. active feeding stage
b. inactive dormant stage
c. infective stage
d. spore-forming stage

13. All mature sporozoa are
a. parasitic
b. nonmotile
c. carried by vectors
d. both a and b

14. Helminth parasites reproduce with
a. spores
b. eggs and sperm
c. mitosis
d. cysts
e. all of these

15. Matching. Select the description that best fits the word in the left column.

____ diatom	a. the cause of malaria
____ *Rhizopus*	b. single-celled alga with silica in its cell wall
____ *Histoplasma*	c. fungal cause of Ohio Valley fever
____ *Cryptococcus*	d. the cause of amebic dysentery
____ euglenid	e. genus of black bread mold
____ dinoflagellate	f. helminth worm involved in pinworm infection
____ *Trichomonas*	g. motile flagellated alga with eyespots
____ *Entamoeba*	h. a yeast that infects the lungs
____ *Plasmodium*	i. flagellated protozoan genus that causes an STD
____ *Enterobius*	j. alga that causes red tides

CONCEPT QUESTIONS

1. Construct a chart indicating the major similarities and differences between procaryotic and eucaryotic cells.
2. a. Which kingdoms of the five-kingdom system contain eucaryotic microorganisms? How do unicellular, colonial, and multicellular organisms differ from each other?
   b. Give examples of each type.
3. a. Describe the anatomy and functions of each of the major eucaryotic organelles.
   b. How are flagella and cilia similar? How are they different?
   c. Compare and contrast the smooth ER, the rough ER, and the Golgi apparatus in structure and function.
4. Trace the synthesis of cell products, their processing, and their packaging through the organelle network.
5. a. Describe the detailed structure of the nucleus.
   b. Why can one usually not see the chromosomes?
   c. When are the chromosomes visible?
   d. What causes them to be visible?
6. a. Define mitosis and explain its function.
   b. What happens to the chromosome number during this process?
   c. How does a diploid organism remain diploid and a haploid organism remain haploid?
7. a. Define meiosis and explain its function.
   b. When does it occur in diploid organisms?
   c. In haploid organisms?
   d. How does it differ from mitosis?
8. Describe some of the ways that organisms use lysosomes.
9. For what reasons would a cell need a "skeleton"?
10. a. Differentiate between the yeast and hypha types of fungal cell.
    b. What is a mold?
    c. What does it mean if a fungus is dimorphic?
11. a. How does a fungus feed?
    b. Where would one expect to find fungi?
12. a. Describe the functional types of hyphae.
    b. Describe the two main types of asexual fungal spores and how they are formed.
    c. What are some types of conidia?
    d. What is the reproductive potential of molds in terms of spore production?
    e. How do mold spores differ from procaryotic spores?
13. a. Explain the importance of sexual spores to fungi.
    b. Describe the three main types, and explain how each is formed by means of a simple diagram.
14. How are fungi classified? Give an example of a member of each fungus division and describe its structure and importance.
15. What is a mycosis? What kind of mycosis is athlete's foot? What kind is coccidioidomycosis?
16. What is a working definition of a "protist"?
17. a. Describe the principal characteristics of algae that separate them from protozoa.
    b. How are algae important?
    c. What causes the many colors in the algae?
    d. Are there any algae of medical importance?
18. a. Explain the general characteristics of the protozoan life cycle.
    b. Describe the protozoan adaptations for feeding.
    c. Describe protozoan reproductive processes.
19. a. Briefly outline the characteristics of the four protozoan groups.
    b. What is an important pathogen in each group?
20. a. Which protozoan group is the most complex in structure and behavior?
    b. In life cycle?
    c. What characteristics set the sporozoa apart from the other protozoan groups?
21. a. Construct a chart that compares the four groups of eucaryotic microorganisms (fungi, algae, protozoa, helminths) in cellular structure.
    b. Indicate whether the group has a cell wall, chloroplasts, motility, or some other distinguishing feature.
    c. Include also the manner of nutrition and body plan (unicellular, colonial, filamentous, or multicellular).

CRITICAL-THINKING QUESTIONS

1. Suggest some ways that one would go about determining if mitochondria and chloroplasts are a modified procaryotic cell.
2. Give the common name of a eucaryotic microbe that is unicellular, walled, non-photosynthetic, nonmotile, and bud-forming.
3. Give the common name of a microbe that is unicellular, nonwalled, motile with flagella, and has chloroplasts.
4. Which group of microbes has long, thin pseudopods and is encased in a hard shell?
5. What general type of multicellular parasite is composed primarily of thin sacs of reproductive organs?
6. a. Name two parasites that are transmitted in the cyst form.
   b. How must a non-cyst-forming pathogenic protozoan be transmitted? Why?
7. You just found an old container of food in the back of your refrigerator. You open it and see a mass of multicolored fuzz. As a budding microbiologist, describe how you would determine what types of organisms are growing on the food.
8. Here is a simple project using mushrooms as art. A unique print can be made by placing a mushroom cap, gill side down, on a piece of paper and leaving it undisturbed for a few days. The spores will fall onto the paper, leaving a perfect pattern of the gills. If carefully lacquered, this can be preserved as a unique and beautiful design.
9. Explain what factors could cause opportunistic mycoses to be a growing medical problem.
10. a. How are bacterial endospores and cysts of protozoa alike?
    b. How do they differ?
11. You have gone camping in the mountains and plan to rely on water present in forest pools and creeks for drinking water. Certain encysted pathogens often live in this type of water, but you do not discover this until you arrive at the campground. How might you treat the water to prevent becoming infected?
12. a. Explain the two levels of parasitism at work in Chagas' disease.
    b. Do you suppose the trypanosome has a parasite, too?
    c. If so, could its parasite have a parasite?
    d. What does this tell you about the prevalence of parasitism and its success as a mode of life?
    e. What is a potential weakness in parasitism?
13. Can you think of a way to determine if a child is suffering from pinworms? Hint: Scotch tape is involved.

INTERNET SEARCH TOPIC

Use the World Wide Web to explore two topics:

1. The endosymbiotic theory of eucaryotic cell evolution. List data from studies that support this idea.
2. The dinoFlagellate toxin epidemic that occurred in 1997. Explain factors that may have accounted for the emergence of this protist.

chapter 6

An Introduction to the Viruses

Virology is the study of an important group of intriguing infectious agents whose major characteristics are so different from those of cellular microorganisms that they must be considered separately. The involvement of viruses in human diseases, ranging from AIDS and the common cold to influenza and cancer, inspires even the elementary microbiology student to develop a basic grasp of their properties. A generous portion of this chapter focuses on the relationship between the virus and its host cell because of the intimate connection between them. Along the way, you will also become acquainted with the unique features of virus structure, physiology, multiplication, cultivation, and identification.

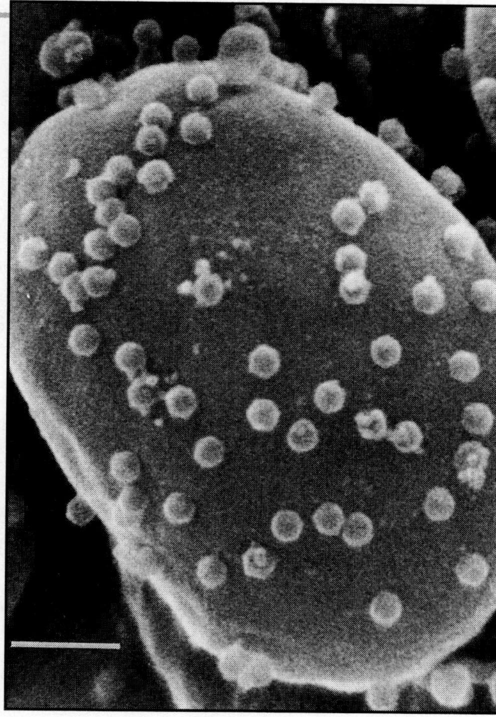

No living thing escapes virus infection. Here, a unicellular alga *(Chlorella)* has been attacked by numerous small particles (viruses) that are injecting their DNA into the cell. In time, the cell will become a factory for new viruses and will be destroyed in the process of lysis. Bar = 0.5 μm

The Search for the Elusive Viruses

The discovery of the light microscope made it possible to see firsthand the agents of many bacterial, fungal, and protozoan diseases. But the techniques for observing and cultivating these relatively large microorganisms were virtually useless for viruses. For many years, the causes of viral infections such as smallpox and polio remained cloaked in mystery, even though it was clear that the diseases were transmitted from person to person. The French bacteriologist Louis Pasteur was certainly on the right track when he postulated that rabies was caused by a "living thing" smaller than bacteria, and in 1884 he was able to develop the first vaccine for rabies. Pasteur also proposed the term **virus** (L. poison) to denote this special group of infectious agents.

The first substantial revelations about the unique characteristics of viruses occurred in the 1890s. First, D. Ivanovski and M. Beijerinck showed that a disease in tobacco was caused by a virus (tobacco mosaic virus). Then, Friedrich Loeffler and Paul Frosch discovered an animal virus that causes foot-and-mouth disease in cattle. These early researchers found that when infectious fluids from host organisms were passed through porcelain filters designed to trap bacteria, the filtrate still remained infectious. This result proved that an infection could be caused by a cell-free fluid containing agents smaller than bacteria and thus first introduced the concept of a *filterable virus.*

Over the succeeding decades, a remarkable picture of the physical, chemical, and biological nature of viruses began to take form. Years of experimentation were required to show that viruses were noncellular particles with a definite size, shape, and chemical composition. Like bacteria, they could be cultured in the laboratory. By the 1950s, virology had grown into a multifaceted discipline that promised to provide much information on disease, genetics, and even life itself (see microfile 6.1).

The Position of Viruses in the Biological Spectrum

Viruses are a unique group of biological entities known to infect every type of cell, including bacteria, algae, fungi, protozoa, plants, and animals. Although the emphasis in this chapter is on animal viruses, much credit for our knowledge must be given to experiments with bacterial and plant viruses. The exceptional and curious nature of viruses prompts numerous questions, including: (1) Are they organisms; that is, are they alive? (2) What are their distinctive biological characteristics? (3) How can particles so small, simple, and seemingly insignificant be capable of causing disease and death? and (4) What is the connection between viruses and cancer? In this chapter, we address these questions and many others.

The unusual structure and behavior of viruses have led to debates about their connection to the rest of the microbial world. One viewpoint holds that viruses are unable to exist independently from the host cell, so they are not living things but are more akin to large, infectious molecules. Another viewpoint proposes that even though viruses do not exhibit most of the life processes of cells (discussed in chapter 2), they can direct them and thus are certainly more than inert and lifeless molecules. Depending upon the circumstances, both views are defensible. In applied virology, this debate has greater philosophical than practical importance because viruses are agents of disease and must be dealt with through control, therapy, and prevention, whether we regard them as living or not. In keeping with their special position in the biological spectrum, it is best to describe viruses as *infectious particles* (rather than organisms) and as either *active* or *inactive* (rather than alive or dead).

Viruses are different from their host cells in size, structure, behavior, and physiology. They are **obligate intracellular parasites,** meaning that they cannot multiply unless they invade a specific host cell and instruct its genetic and metabolic machinery to make and release quantities of new viruses. Because of this characteristic, viruses are capable of causing serious damage and disease. Other unique properties of viruses are summarized in table 6.1.

Table 6.1

Novel Properties of Viruses

- Ultramicroscopic size, ranging from 20 nm up to 450 nm.
- Are not cells; structure is very compact and economical.
- Do not independently fulfill the characteristics of life (see chapter 2).
- Are inactive macromolecules outside of the host cell and active only inside host cells.
- Are geometric; can form crystal-like masses.
- Basic structure consists of protein capsid and nucleic acid.
- Capsid is made of repeating subunits; encloses and protects nucleic acid.
- Nucleic acid can be either DNA or RNA but not both.
- Nucleic acid can be double-stranded DNA, single-stranded DNA, single-stranded RNA, or double-stranded RNA.
- Molecules on virus surface impart high specificity for attachment to host cell.
- Multiply by assembly-line method; do not divide.
  Cycle includes attachment of virus to host cell, penetration by genetic material, production of virus components by the cell, assembly of new viruses, and release from host cell.
- Lack enzymes for most metabolic processes.
- Lack machinery for synthesizing proteins.
- Are obligate intracellular parasites of bacteria, protozoa, fungi, algae, plants, and animals.

Chapter Checkpoints

Viruses are noncellular entities whose properties have been identified through technological advances in microscopy and tissue culture.

Viruses are infectious particles that invade every known type of cell. They are not alive, yet they are able to redirect the metabolism of living cells to reproduce virus particles.

Viral replication inside a cell usually causes death or disease of that cell.

TABLE 6.2

RELATIVE SIZES OF SELECTED CELLS, VIRUSES, AND MOLECULES 

Cells	Largest Diameter (nm)
Red blood cells	7,500
Bacteria	
Streptococcus	750
Rickettsia	250
Viruses	
Vaccinia	210
Herpes simplex	130
Rabies	125
Influenza	85
Adenovirus	75
T2 bacteriophage	65
Poliomyelitis	27
Yellow fever	22
Foot-and-mouth	21
Tobacco mosaic	15 (but 300 in length)
Molecules	
Hemoglobin molecule	15
Egg albumin molecule	10

THE GENERAL STRUCTURE OF VIRUSES

SIZE RANGE

As a group, viruses represent the smallest infectious agents (with some unusual exceptions to be discussed later). Their size relegates them to the realm of the **ultramicroscopic.** This term means that most of them are so minute ($< 0.2\ \mu m$) that an electron microscope is necessary to detect them or to examine their fine structure. They are dwarfed by their host cells: More than 2,000 bacterial viruses could fit into an average bacterial cell, and more than 50 million polioviruses could be accommodated by an average human cell. Animal viruses range in size from the small parvoviruses[1] (around 20 nm [0.02 μm] in diameter) to poxviruses[2] that are as large as small bacteria (up to 450 nm [0.4 μm] in length). Some cylindrical viruses are relatively long (800 nm [0.8 μm] in length) but so narrow in diameter (15 nm [0.015 μm]) that their visibility is still limited without the high magnification and resolution of an electron microscope. Table 6.2 compares the sizes of several viruses with procaryotic and eucaryotic cells and molecules.

1. DNA viruses that cause respiratory infections in humans.
2. A group of large, complex viruses, including smallpox, that cause raised skin swellings called pox.

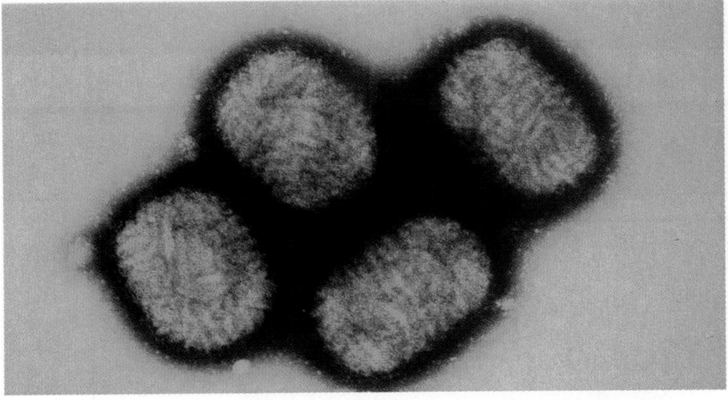
(a)

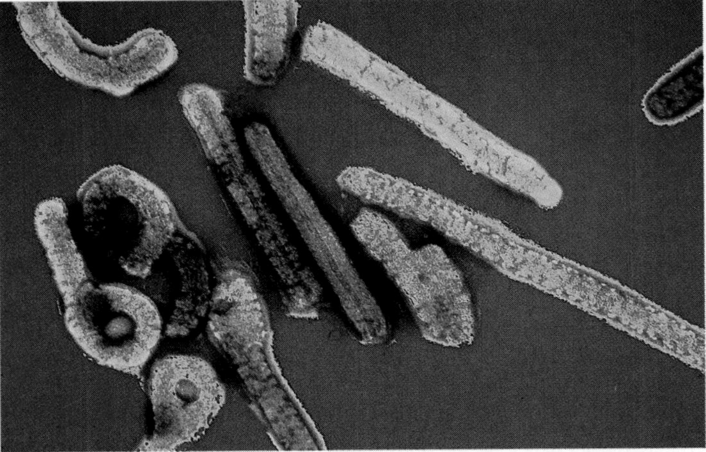
(b)

(c)

Figure 6.1
Methods of viewing viruses. *(a)* Negative staining of an orfvirus (a type of poxvirus), revealing details of its outer coat (122,000×). *(b)* Positive stain of the Ebola virus, a type of filovirus, so named because of its tendency to form long strands. Note the textured capsid. *(c)* Shadowcasting image of a vaccinia virus (17,500×).

Viral architecture is most readily observed through special stains in combination with electron microscopy (figure 6.1). Negative staining uses very thin layers of an opaque salt to outline the shape of the virus against a dark background and to enhance textural features on the viral surface. Internal details are revealed by positive staining of specific parts of the virus such as protein or nucleic acid. The *shadowcasting* technique attaches a virus preparation to a surface and showers it with a dense metallic vapor directed from a certain angle. The thin metal coating over the surface of the virus approximates its contours, and a shadow is cast on the unexposed side.

Unique Viral Constituents: Capsids, Nucleic Acids, and Envelopes

It is important to realize that viruses bear no real resemblance to cells and that they lack any of the protein-synthesizing machinery found in even the simplest cells. Their molecular structure is composed of regular, repeating subunits that give rise to their crystalline appearance. Indeed, many purified viruses can form large aggregates or crystals if subjected to special treatments (figure 6.2). The general plan of virus organization is exceptional in its simplicity and compactness. Viruses contain only those parts needed to invade and control a host cell: an external coating and a core containing one or more nucleic acid strands of either DNA or RNA. This pattern of organization can be represented with a flowchart:

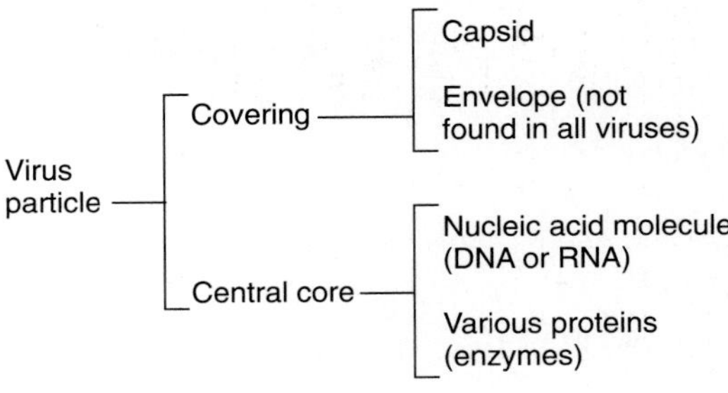

All viruses have a protein **capsid,*** or shell, that surrounds the nucleic acid strand. Together the capsid and the nucleic acid strand are referred to as the **nucleocapsid** (figure 6.3). Members of 13 of the 20 families of animal viruses possess an additional covering external to the capsid called an **envelope,** which is actually a modified piece of the host's cell membrane (figure 6.3*b*). Viruses that lack this envelope are considered **naked nucleocapsids** (figure 6.3*a*). As we shall see later, the enveloped viruses also differ from the naked viruses in the way that they enter and leave a host cell.

*capsid (kap′-sid) L. *capsa,* box.

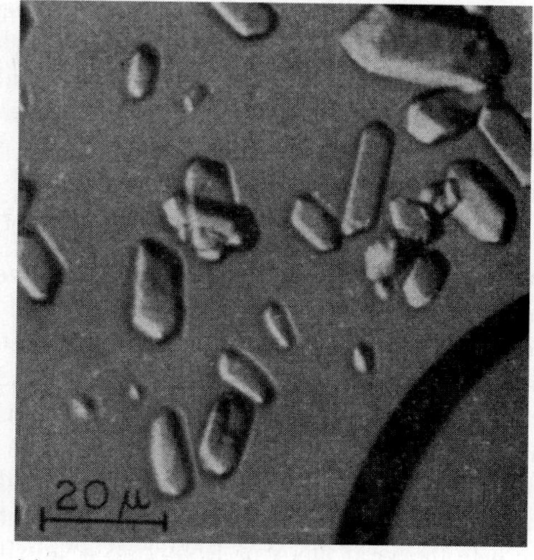

(a)

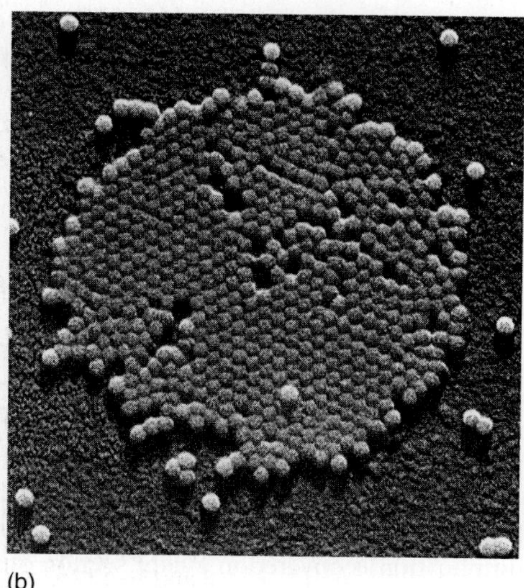

(b)

Figure 6.2
The crystalline nature of viruses. *(a)* Light microscope magnification (1,200×) of purified poliovirus crystals. *(b)* Highly magnified (150,000×) electron micrograph of the capsids of this same virus, demonstrating their highly geometric nature.

(a) **Naked Nucleocapsid Virus**

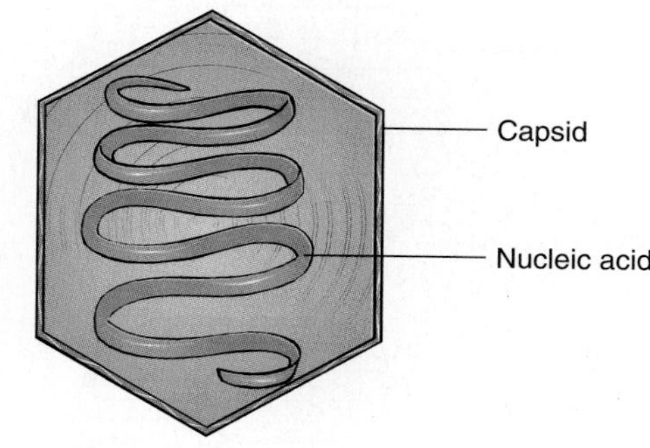

(b) **Enveloped Virus**

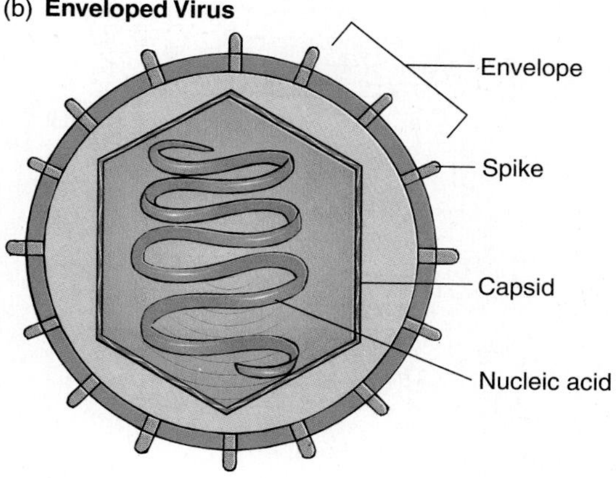

Figure 6.3
Generalized structure of viruses. *(a)* The simplest virus is a naked virus (nucleocapsid) consisting of a geometric capsid assembled around a nucleic acid strand or strands. *(b)* An enveloped virus is composed of a nucleocapsid surrounded by a flexible membrane called an envelope. The envelope usually has special receptor spikes inserted into it.

The Viral Capsid: The Protective Outer Shell

When a virus particle is magnified several hundred thousand times, the capsid appears as the most prominent geometric feature (see figure 6.2*b*). In general, each capsid is constructed from identical building blocks called **capsomers,*** and each capsomer, in turn, is a cluster of smaller protein molecules (called protomers). The capsomers spontaneously self-assemble into the finished capsid. Depending on how the capsomers are shaped and arranged, this binding results in two different types: helical and icosahedral.

The simpler **helical capsids** have rod-shaped capsomers that bond together to form a series of hollow discs resembling a bracelet. During the formation of the nucleocapsid, these discs link with other discs to form a continuous helix into which the nucleic acid strand is coiled (figure 6.4). In electron micrographs, the appearance of a helical capsid varies with the type of virus. The nucleocapsids of naked helical viruses are very rigid and tightly wound into a cylinder-shaped package (figure 6.5 *a,b*). An example is the *tobacco mosaic virus,* which attacks tobacco leaves and gives them a mottled appearance. Enveloped helical nucleocapsids, on the other hand, are more flexible and tend to be arranged as a looser helix within the envelope (figure 6.5*c,d*). This type of morphology is found in several enveloped human viruses, including those of influenza, measles, and rabies.

The capsids of a number of major virus families are arranged in a **polygon*** or **icosahedron***—a three-dimensional, 20-sided figure with 12 evenly spaced corners. Such many sided capsids contain two types of capsomers: triangular hexons that form the flat faces and round pentons that form the corners

*capsomer (kap′-soh-meer) L. *capsa,* box, and *mer*, part.

*polygon; a three-dimensional, many-sided figure.

*icosahedron (eye″-koh-suh-hee′-drun) Gr. *eikosi,* twenty, and *hedra,* side.

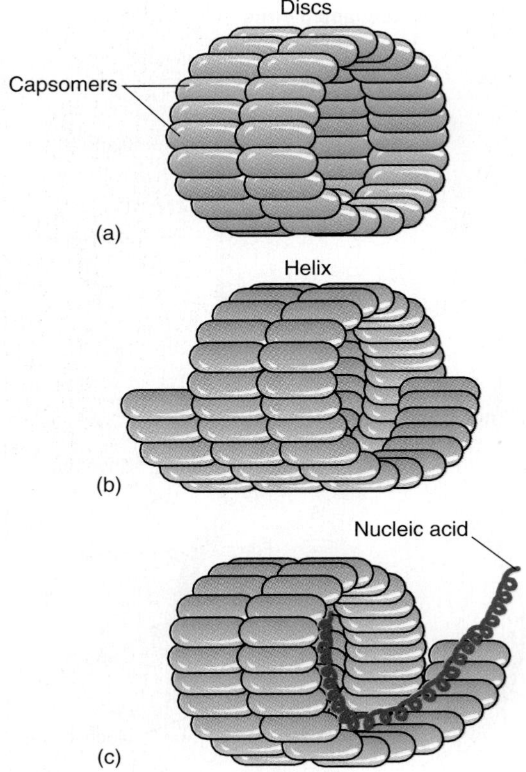

Figure 6.4
Stages in the assembly of helical nucleocapsids. *(a)* Individual capsomers are bound together into discs. *(b)* Several adjacent discs are assembled into a helix. *(c)* A nucleic acid strand is inserted into the core of the helix.

(figure 6.6). During assembly of the virus (discussed later), the nucleic acid is packed into the center of this icosahedron, forming a nucleocapsid. Although the capsids of all polygonal viruses have this sort of symmetry, they can have major variations in the number of capsomers; for example, a poliovirus has 32, and an adenovirus has 240 capsomers. Individual capsomers can look either ring- or dome-shaped, and the capsid itself can appear spherical or cubical (figure 6.7). Another factor that alters the appearance of icosahedral viruses is whether or not they have an outer envelope; contrast a papillomavirus (warts) and its naked nucleocapsid with herpes simplex (cold sores) and its enveloped nucleocapsid (figure 6.7).

The Viral Envelope

When enveloped viruses (mostly animal) are released from the host cell, they take with them a bit of its membrane system in the form of an envelope (see figure 6.19). Some viruses bud off the cell membrane; others leave via the nuclear envelope or the endoplasmic reticulum. Whichever avenue of escape, the viral envelope differs significantly from the host's membranes. In the envelope, some or all of the regular membrane proteins are replaced with special viral proteins. Some proteins form a binding layer between the envelope and capsid of the virus, and others (glycoproteins) remain exposed on the outside of the envelope. These protruding molecules, called **spikes** or **peplomers,*** are essential for the attachment of viruses to the next host cell. Because the envelope is more supple than the capsid, enveloped viruses are pleomorphic and range from spherical to filamentous.

Functions of the Viral Capsid/Envelope

The outermost covering of a virus, whether a capsid or an envelope, is indispensable to viral function because it protects the nucleic acid from the effects of various enzymes and chemicals when the virus is outside the host cell. For example, the capsids of enteric (intestinal) viruses such as polio and hepatitis A are resistant to the acid- and protein-digesting enzymes of the gastrointestinal tract. Capsids and envelopes are also responsible for helping to introduce the viral DNA or RNA into a suitable host cell, first by binding to the cell surface and then by assisting in penetration of the viral nucleic acid (to be discussed in more detail later). In addition, parts of viral capsids and envelopes stimulate the immune system to produce antibodies that can neutralize viruses and protect the host's cells against future infections (see chapters 15, 24, and 25).

Complex Viruses: Atypical Viruses

Two special groups of viruses, termed **complex viruses** (figure 6.8), are more intricate in structure than the helical, icosahedral, naked, or enveloped viruses just described. The **poxviruses** (including the agent of smallpox) are very large viruses that contain a DNA core but lack a regular capsid and have in its place several layers of lipoproteins and coarse surface fibrils. Another group of very complex viruses, the **bacteriophages,*** have a polyhedral head, a helical tail, and fibers for attachment to the host cell. Their mode of multiplication is covered in a later section of this chapter. Figure 6.9 summarizes the morphological types of some common viruses.

Nucleic Acids: At the Core of a Virus

So far, one biological constant is that the genetic information of living cells is carried by nucleic acids (DNA, RNA). Viruses, although neither alive nor cells, are no exception to this rule, but there is a significant difference. Unlike cells, which contain both DNA and RNA, viruses contain either DNA or RNA but not both. Because viruses must pack into a tiny space all of the genes necessary to instruct the host cell to make new viruses, the size of a viral genome is quite small compared with that of a cell. It varies from four genes in hepatitis B virus to hundreds of genes in some herpesviruses. By comparison, the bacterium *Escherichia coli* has approximately 4,000 genes, and a human cell has approximately 100,000 genes.

---

*peplomer (pep′-loh-meer) Gr. *peplos,* envelope, and *mer,* part.

*bacteriophage (bak-teer′-ee-oh-fayj″) From *bacteria,* and Gr. *phagein,* to eat. These viruses parasitize bacteria.

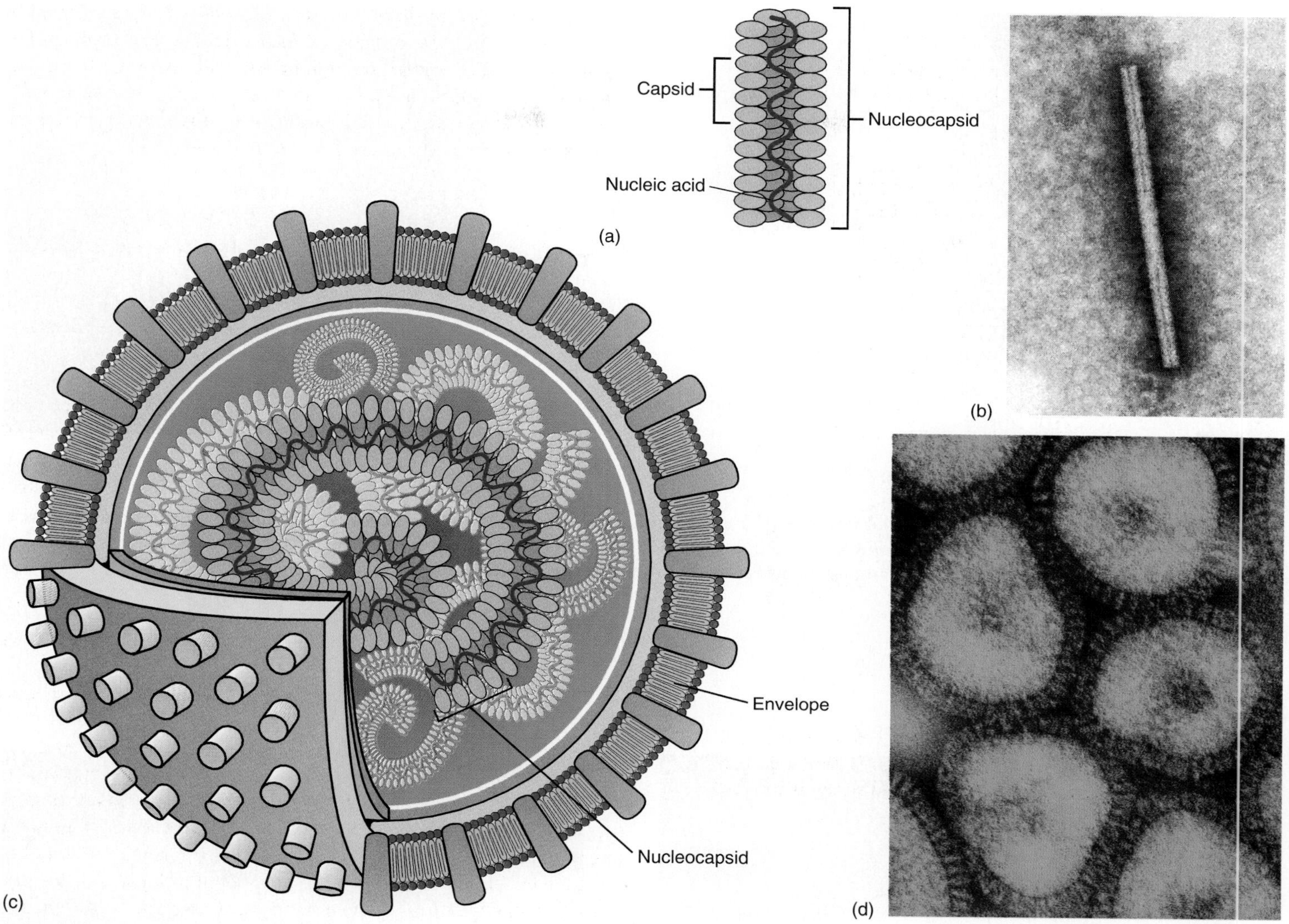

Figure 6.5

Typical variations of viruses with helical nucleocapsids. Naked helical virus (tobacco mosaic virus): *(a)* a schematic view and *(b)* a greatly magnified micrograph. Note the overall cylindrical morphology. Enveloped helical virus (influenza virus): *(c)* a schematic view and *(d)* an electron micrograph of the same virus (350,000×).

In chapter 2 we learned that DNA usually exists as a double-stranded molecule and that RNA is single-stranded. Although most viruses follow this same pattern, a few exhibit distinctive and exceptional forms. Notable examples are the parvoviruses (the cause of erythema contagiosum[3]), which contain single-stranded DNA, and reoviruses (a cause of respiratory and intestinal tract infections), which contain double-stranded RNA. In all cases, these tiny strands of genetic material hold the key to the behavior of a virus. In a very real sense, viruses are **genetic parasites** because they cannot multiply until their nucleic acid has reached the internal habitat of the host cell. Experiments have shown that the nucleic acid alone of some viruses can cause infection when introduced into a host cell.

Other Substances in the Virus Particle

In addition to the protein of the capsid, the proteins and lipids of envelopes, and the nucleic acid of the core, viruses can contain enzymes for specific operations within their host cell. Examples of these enzymes are *polymerases,** for synthesizing DNA and RNA, and enzymes, for digesting host DNA and proteins. Viruses completely lack the genes for synthesis of metabolic enzymes. As we shall see, this deficiency has little consequence, because viruses have adapted to completely take over their hosts' metabolic resources. Some viruses can actually carry away substances from their host cell. For instance, arenaviruses pack along host ribosomes, and retroviruses "borrow" the host's tRNA molecules.

3. A common childhood disease described in chapter 24.

*polymerase (pol-im′-ur-ace) An enzyme that synthesizes a large molecule from smaller subunits.

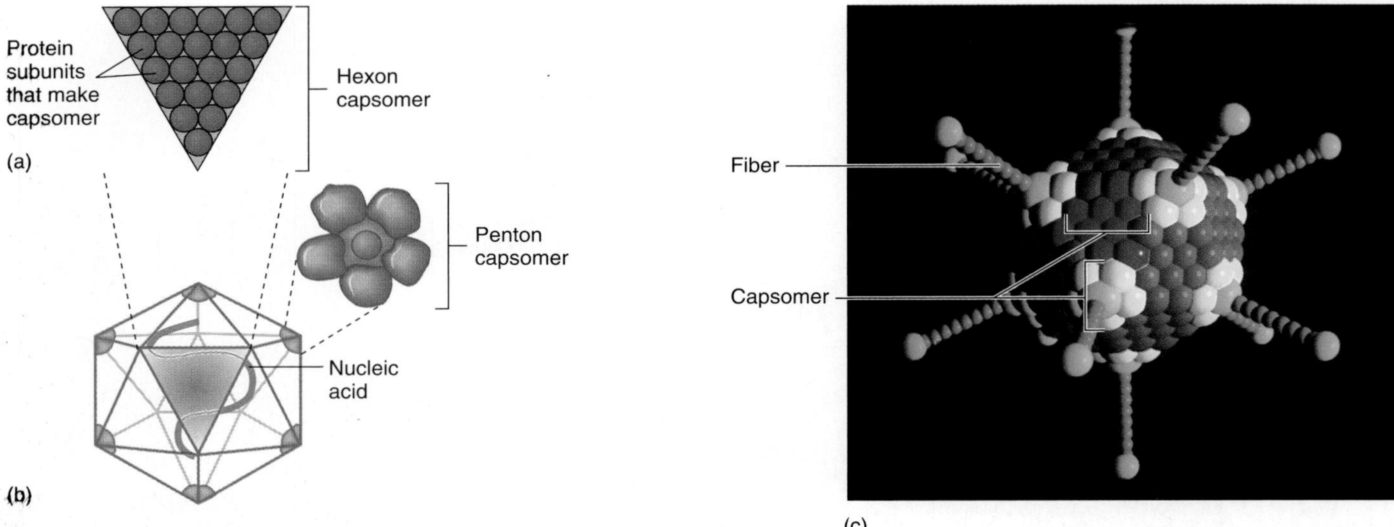

Figure 6.6

Formation and structure of icosahedral viruses. *(a)* The basic building block is a triangular capsomer composed of spherical proteins. In this example, each capsomer contains 21 protein units, although viruses may vary in this feature. *(b)* View of a completed capsid, shown transparent to reveal its 20 triangular panels and 12 corners. *(c)* Three-dimensional model of an adenovirus.

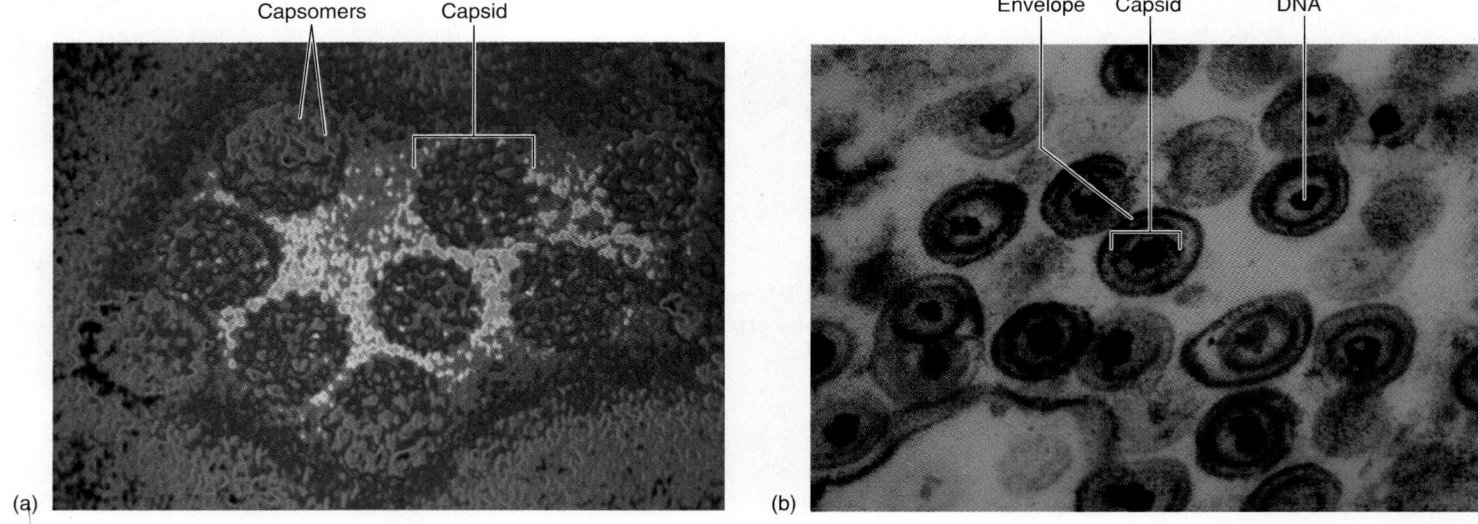

Figure 6.7

Two types of icosahedral viruses, highly magnified. *(a)* False-color micrograph of papillomaviruses with unusual, ring-shaped capsomers. *(b)* Herpes simplex virus, an enveloped icosahedron (300,000×).

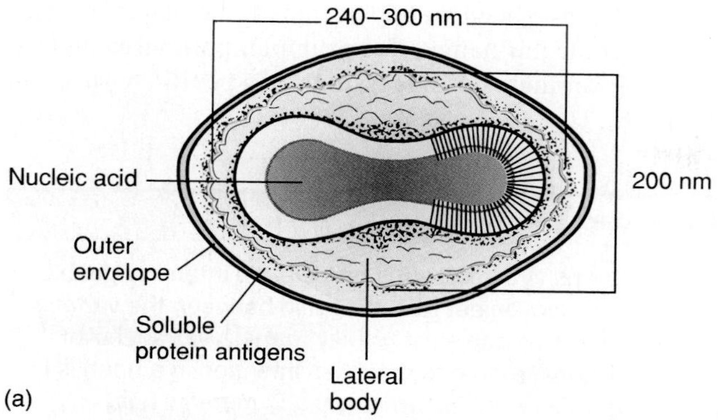

240–300 nm
200 nm
Nucleic acid
Outer envelope
Soluble protein antigens
Lateral body
(a)

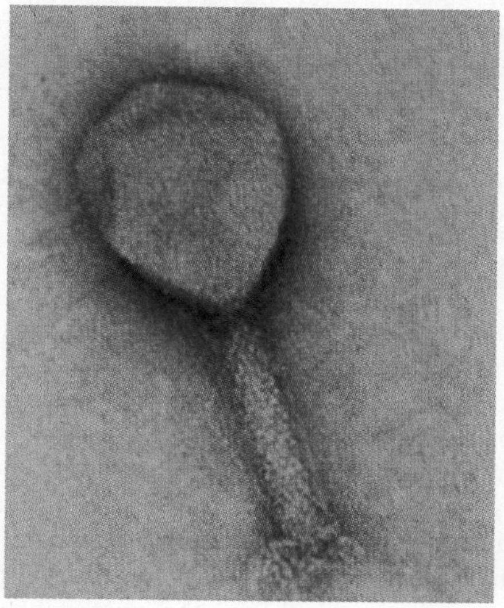

(b)

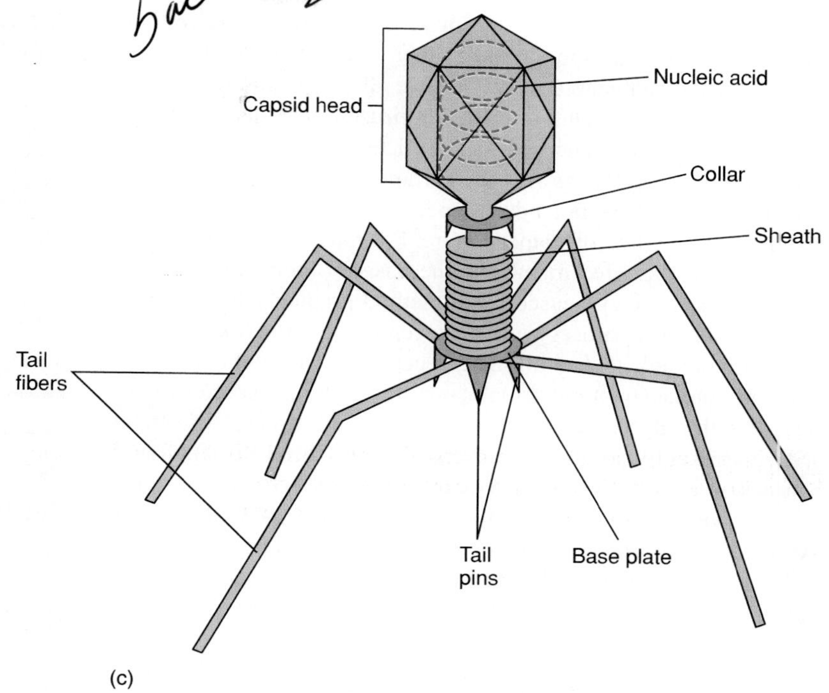

Capsid head
Nucleic acid
Collar
Sheath
Tail fibers
Tail pins
Base plate
(c)

Figure 6.8

Detailed structure of complex viruses. *(a)* The vaccinia virus, a poxvirus. *(b)* Photomicrograph and *(c)* diagram of a T4 bacteriophage. (See text for functions.)

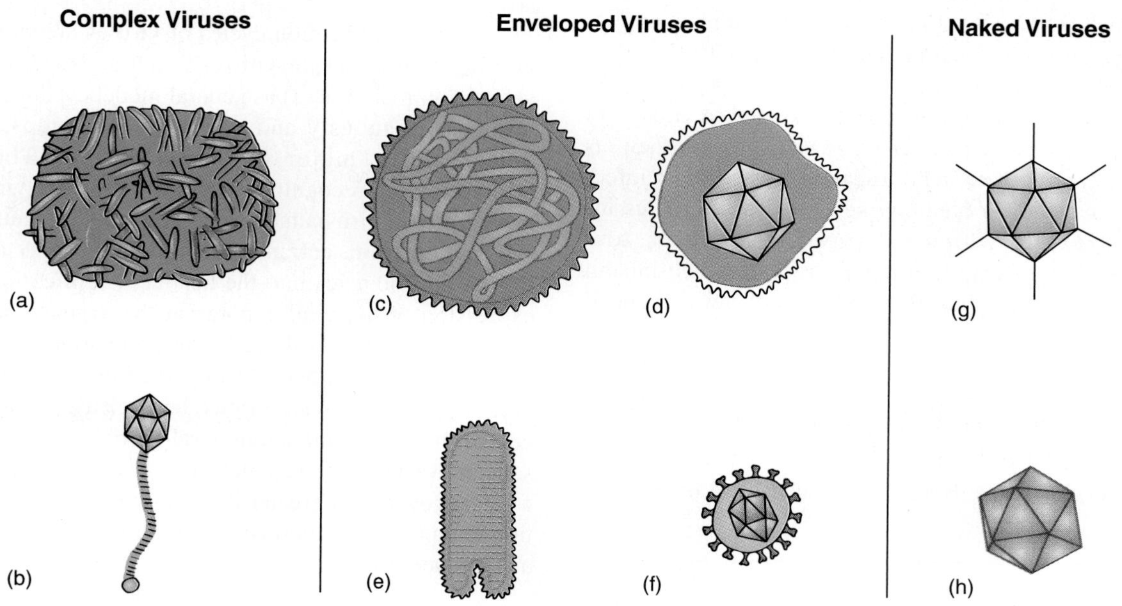

Figure 6.9

Complex viruses: *(a)* poxvirus, a large DNA virus; *(b)* flexible-tailed bacteriophage. Enveloped viruses: *(c)* mumps virus, an enveloped RNA virus with a helical nucleocapsid; *(d)* herpesvirus, an enveloped DNA virus with an icosahedral nucleocapsid; *(e)* rhabdovirus, an RNA virus with a bullet-shaped envelope; *(f)* HIV (AIDS), an RNA retrovirus with an icosahedral capsid. Naked viruses: *(g)* adenovirus, a DNA virus with fibers on the capsid; *(h)* papillomavirus, a DNA virus that causes warts.

HOW VIRUSES ARE CLASSIFIED AND NAMED

Although viruses are not classified as members of the kingdoms discussed in chapter 1, they are diverse enough to require their own classification scheme to aid in their study and identification. In an informal and general way, we have already begun classifying viruses—as animal, plant, or bacterial viruses; enveloped or naked viruses; DNA or RNA viruses; and helical or icosahedral viruses. These introductory categories are certainly useful in organization and description, but the study of specific viruses requires a more standardized method of nomenclature. For many years, the animal viruses were classified mainly on the basis of their hosts and the kind of diseases they caused. This method gradually became less workable as more viruses were discovered and numerous areas of overlap emerged. It was ultimately necessary to adopt a system that took into account the actual nature of the virus particles themselves, with only partial emphasis on host and disease. The main criteria presently used to group viruses are structure, chemical composition, and similarities in genetic makeup.

A widely used scheme for classifying animal viruses first assigns them to one of two superfamilies, either those containing DNA or those containing RNA. DNA viruses can be further subdivided into six families, and RNA viruses into 13 families, for a total of 19 families of animal viruses (table 6.3). Virus families are given a name composed of a Latin root followed by *-viridae.* Characteristics used for placement in a particular family include type of capsid, nucleic acid strand number, presence and type of envelope, overall viral size, and area of the host cell in which the virus multiplies. Some virus families are named for their microscopic appearance (shape and size). Examples include *rhabdoviruses,** which have a bullet-shaped envelope, and *togaviruses,** which have a cloaklike envelope. Anatomical or geographic areas have also been used in naming. For instance, *adenoviruses** were first discovered in adenoides (one type of tonsil), and bunyaviruses (bun-yah) were originally isolated in an area in Africa called Bunyamwera. Viruses can also be named for their effects on the host. *Lentiviruses** tend to cause slow, chronic infections. Acronyms made from blending several characteristics include *picornaviruses,** which are tiny RNA viruses, and reoviruses (or **r**espiratory **e**nteric **o**rphan viruses), which inhabit the respiratory tract and the intestine and are not yet associated with any known disease state.

Each different type of virus is also assigned genus status according to its host, target tissue, and the type of disease it causes (table 6.4). Viral genera are denoted by a special Latinized root followed by the suffix *-virus* (for example, *Enterovirus* and *Herpesvirus*). Because the use of standardized species names has not been widely accepted, the genus or common English vernacular names (for example, poliovirus and rabies virus) predominate in discussions of specific viruses in this text.

A Note on Terminology

Although the terms *virus* and *virus particle* are interchangeable, virologists find it convenient to distinguish between the various states in which a virus can exist. A fully formed, extracellular particle that is virulent (able to establish infection in a host) is called a **virion** (vir′-ee-on). Once its genetic material has entered a host cell, a virus may undergo a multiplication phase called the **lytic** (lih′-tik) **cycle;** in which the host cell is disrupted to release more virions. Some viruses remain in an inactive, or **latent** (lay′-tunt), stage in which the virus does not lyse the host cell.

MODES OF VIRAL MULTIPLICATION

Viruses tend to remain closely associated with their hosts. In addition to providing the viral habitat, the host cell is absolutely necessary for viral multiplication. The process of viral multiplication is an extraordinary biological phenomenon. Viruses have often been aptly described as minute parasites that appropriate the synthetic and genetic machinery of cells. The nature of this cycle dictates viral pathogenicity, transmission, the responses of the immune defenses, and human measures to control viral infections. From these perspectives, we cannot overemphasize the importance of a working knowledge of the relationship between viruses and their host cells.

The multiplication cycles of viruses are similar enough that virologists use certain viruses (such as bacteriophages and enveloped animal viruses) as general models. Although a given cycle occurs continuously and not in discrete steps, it is helpful to demarcate its major sequential events. These events are **adsorption,*** a recognition process between a virus and host cell that results in virus attachment to the external surface of the host cell; **penetration,** entrance of the virion (either a whole virus or just its nucleic acid) into the host cell; **replication,*** copying and expression of the viral genome at the expense of the host's synthetic equipment, resulting in the production of the various virus components; **assembly** and **maturation** of these individual viral parts into whole, intact virions; and **release,** escape from the host cell of the active, infectious viral particles (figure 6.10). We present an overview of the genetic events of virus cycles here, but this topic is covered in greater detail in chapter 9. The following section compares and contrasts the multiplication stages for bacterial and human viruses.

*rhabdovirus (rab″-doh-vy′-rus) Gr. *rhabdo,* little rod.

*togavirus (toh″-guh-vy′-rus) L. *toga,* covering or robe.

*adenovirus (ad″-uh-noh-vy′-rus) G. *aden,* gland.

*lentivirus (len″-tee-vy′-rus) Gr. *lente,* slow. HIV, the AIDS virus, belongs in this group.

*picornavirus (py-kor″-nah-vy′-rus) Sp. *pico,* small, plus RNA.

*adsorption (ad-sorp′-shun) L. *ad.* to, and *sorbere,* to suck. The attachment of one thing onto the surface of another.

*replication (rep-lih-kay′-shun) L. *replicare,* to reply. To make an exact duplicate.

Table 6.3

Animal Virus Families and Their Major Characteristics

Family	Strand Type	Capsid Type	Envelope	Size (diameter in nm)	Common Name of Important Members*
DNA Viruses					
Poxviridae	Double	None	+	130–300	Smallpox virus (Variola); complex virus; brick-shaped
Herpesviridae	Double	Icosahedral	+	150–200	Herpes simplex virus, Varicella zoster virus, Epstein-Barr virus; can become latent**
Adenoviridae	Double	Icosahedral	–	70–90	Human adenoviruses
Papovaviridae	Double	Icosahedral	–	45–55	Human papillomavirus
Hepadnaviridae	Single or double	Icosahedral	+	42	Hepatitis B virus
Parvoviridae	Single	Icosahedral	–	18–26	Parvovirus B19
RNA Viruses					
Picornaviridae	Single	Icosahedral	–	20–30	Hepatitis A virus, poliovirus, coxsackieviruses, rhinoviruses
Calciviridae	Single	Icosahedral	–	35–40	Norwalk virus
Togaviridae	Single	Icosahedral	+	45–70	Rubella virus, western equine encephalitis
Flaviviridae	Single	Icosahedral	+	40–70	Yellow fever virus, Japanese encephalitis virus
Filoviridae	Single	Helical	+	790–970	Ebola and Marburg viruses
Bunyaviridae	Single	Helical	+	90–100	Bunyamwera virus, Hanta virus
Reoviridae	Double	Icosahedral	–	60–80	Human rotavirus, Colorado tick fever virus
Orthomyxoviridae	Single	Helical	+	80–120	Influenza viruses
Paramyxoviridae	Single	Helical	+	125–250	Parainfluenza virus, mumps virus, measles virus
Rhabdoviridae	Single	Helical	+	60–75	Rabies virus
Retroviridae	Single	Icosahedral	+	100	Human immunodeficiency virus (AIDS), oncoviruses***
Arenaviridae	Single	?	+	50–300	Lassa virus; lymphocytic choriomeningitis virus
Coronaviridae	Single	Helical	+	80–130	Human infectious bronchitis and corona viruses
CHINA	—	—	—	—	Chronic Infectious Neuropathic Agents (CHINA)****

*This is only a partial list of the principal viruses that are important to humans. Table 6.4 enlarges on the characteristics of each viral genus.
**A term indicating that the virus is within the host but in a state of inactivity; the virus can become active at a later date.
***From the root *onco,* meaning mass. These are viruses that cause cancer.
****Infectious forms that appear to be viruslike and have only recently been observed microscopically. They all cause progressive, degenerative disease of the nervous system.

The Multiplication Cycle in Bacteriophages

When Frederick Twort and Felix d'Herelle discovered bacterial viruses in 1915, it first appeared that the bacterial host cells were being eaten by some unseen parasite, hence the name bacteriophage was used. Most bacteriophages (often shortened to *phage*) contain double-stranded DNA, though single-stranded DNA and RNA types exist as well. So far as is known, every bacterial species is parasitized by various specific bacteriophages. Probably the most widely studied bacteriophages are those of the intestinal bacterium *Escherichia coli*—especially the T-even (for even-numbered type) phages. Their complex structure has been previously described. They have an icosahedral capsid head containing DNA, a central tube (surrounded by a sheath), collar, base plate, tail pins, and fibers, which in combination make an efficient package for infecting a bacterial cell (see figure 6.8*c*). Momentarily setting aside a strictly scientific and objective tone, it is tempting to think of these extraordinary viruses as minute spacecrafts docking on an alien planet, ready to unload their genetic cargo.

Adsorption: Docking onto the Host Cell Surface

For a successful infection, a phage must meet and stick to a susceptible host cell. Adsorption takes place when certain molecules on the phage tail and fibers bind to specific molecules (receptors) on the cell envelope of the host bacterium (figure 6.10). Among the bacterial structures that serve as phage receptors are surface molecules of the cell wall, pili, and flagella. Attachment is a function of correct fit and chemical attraction; that is, the phage's tail molecules must interact specifically with those of the bacterial receptors. Once the phage is firmly affixed to the cell, it is positioned for penetration.

TABLE 6.4

IMPORTANT HUMAN VIRAL GENERA, COMMON NAMES, AND TYPES OF DISEASES THEY CAUSE

Genus of Virus	Common Name of Genus Members	Name of Disease	Clinical Characteristics in Humans
DNA Viruses			
Orthopoxvirus	Variola major and minor	Smallpox	Pox—pustules on skin
Herpesvirus	Herpes simplex (HSV) I virus	Fever blister, cold sores	Lesions on lips, eyes
	Herpes simplex (HSV) II virus	Genital herpes	Lesions on genitals; damage to newborn infants
	Varicella zoster virus (VZV)	Chickenpox, shingles	Generalized rash (pox)
	Human cytomegalovirus (CMV)	CMV infections	Congenital viral infections, some cases of mononucleosis
	Epstein-Barr virus (EBV)	Infectious mononucleosis	Sore throat, fever, enlargement of lymph glands, Burkitt's lymphoma tumors
Mastadenovirus	Human adenoviruses	Adenovirus infection	Various acute respiratory, enteric, and eye infections
Papillomavirus	Human papillomavirus (HPV)	Several types of warts	Epidermal tumors on skin, mucous membranes
Polyomavirus	JC virus (JCV)	Progressive multifocal leukoencephalopathy (PML)	Generally fatal brain infection
None assigned	Hepatitis B virus (HBV or Dane particle)	Serum hepatiti	Progressive liver infections
RNA Viruses			
Enterovirus	Poliovirus	Poliomyelitis	Infection of nervous system, paralysis
	Coxsackievirus	Several syndromes	Affects several systems: brain, upper respiratory tract, skin
	ECHO* viruses	Several syndromes	Affect several systems: brain, upper respiratory tract, skin
	Hepatitis A virus (HAV)	Infectious hepatitis	Attacks liver cells; jaundice
Rhinovirus	Human rhinovirus	Common cold, bronchitis	Fever, cough, nasal congestion
Calicivirus	Norwalk virus	Viral diarrhea, Norwalk virus syndrome	Acute enteritis
Alphavirus	Eastern equine encephalitis virus	EEE	Fatal encephalitis**
	Western equine encephalitis virus	WEE	Encephalitis
	Ross River virus	Same as virus name	Rash, arthritis
	Yellow fever virus	Yellow fever	Fever, hemorrhage, hepatitis
	St. Louis encephalitis virus	St. Louis encephalitis	Brain infection
Rubivirus	Rubella virus	Rubella (German measles)	Skin, lymph node, joint symptoms; damage to fetus
Flavivirus	Dengue fever virus	Dengue fever	Fever, rash, hemorrhage
Bunyavirus	Bunyamwera viruses	California encephalitis	Fever and viremia***
Filovirus	Ebola, Marburg virus	Ebola fever	Fever and viremia; very severe
Hantavirus	Muerto Canyon virus	Respiratory distress syndrome	Attacks the lungs
Phlebovirus	Rift Valley fever virus	Rift Valley fever	Fever, encephalitis, hemorrhages, blindness
Nairovirus	Crimean–Congo hemorrhagic fever virus (CCHF)	Crimean – Congo hemorrhagic fever	Fever, bleeding in intestine and skin
Orbivirus	Colorado tick fever virus	Colorado tick fever	Fever, encephalitis; attacks bone marrow
Rotavirus	Human rotavirus	Rotavirus gastroenteritis	Vomiting, diarrhea; dehydration in infants
Influenza virus	Influenza virus, type A (Asian, Hong Kong, and Swine influenza viruses)	Influenza or "flu"	Acute infection of the nasopharynx, trachea, and bronchi
	Influenza virus, type B	Influenza or "flu"	
Paramyxovirus	Parainfluenza virus, types 1 – 5	Parainfluenza	Respiratory infections, including croup, common cold
	Mumps virus	Mumps	Infects salivary glands, testes, ovaries, and brain
Morbillivirus	Measles virus	Measles (red)	Fever, rash, cough, nasal discharge
Pneumovirus	Respiratory syncytial virus (RSV)	Common cold syndrome	Pneumonia in infants
Lyssavirus	Rabies virus	Rabies (hydrophobia)	Fatal brain infection
Oncornavirus	Human T-cell leukemia virus (HTLV)	T-cell leukemia	Cancer of T cells
Lentivirus	HIV (human immunodeficiency viruses 1 and 2)	Acquired immunodeficiency syndrome (AIDS)	Loss of immune function
Arenavirus	Lassa virus	Lassa fever	Fever, severe hemorrhage, shock
Coronavirus	Infectious bronchitis virus (IBV)	Bronchitis	Acute infection of upper respiratory tract
	Enteric corona virus	Coronavirus enteritis	Intestinal infections

*An acronymn for **E**nteric **C**ytopathogenic **H**uman **O**rphan, denoting the origin and effect of these viruses on cells.
**An inflammation of the brain caused by an infectious agent or a toxic substance.
***Presence of viruses in the blood.

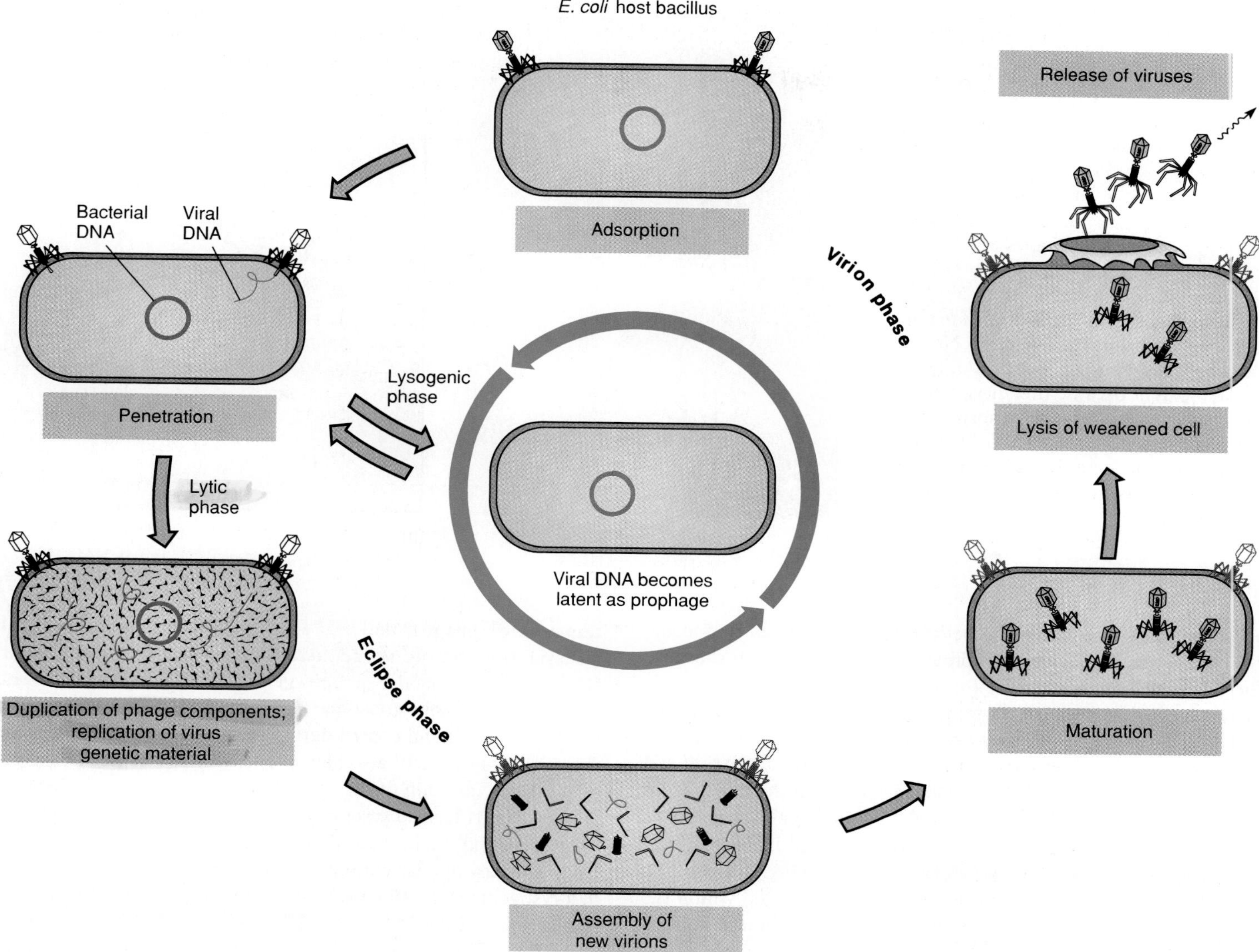

Figure 6-10

Events in the multiplication cycle of T-even bacteriophages. The cycle is divided into the eclipse phase (during which the phage is developing but is not yet infectious) and the virion phase (when the virus is mature and capable of infecting a host). (The text provides further details of this cycle. See figure 6.14 for the steps in the lysogenic phase.)

Bacteriophage Penetration: Entry of the Nucleic Acid

After adsorption, a phage is still on the outside of its host cell and remains inactive unless it can gain entrance to the cell's cytoplasm. The strong, rigid bacterial cell wall is quite an impenetrable barrier, and the entire virus particle is unable to cross it. But the T-even bacteriophages have an exquisite mechanism for injecting nucleic acid across this barrier and into the cell. Like a miniscule syringe, the sheath constricts, pushing the inner tube through the host's cell wall and membrane, forming a passageway for its nucleic acid into the interior of the cell (figure 6.11). Once inside, the viral nucleic acid alone can complete the multiplication cycle. The spent virus shell (or ghost) has fulfilled its principal functions—protection, adsorption, and injection of DNA—and it will remain attached outside to the cell wall fragment even after the cell has lysed.

The Bacteriophage Assembly Line

Entry of the nucleic acid into the cell results in sweeping changes in the bacterial cell's activities. Within a few minutes, the bacterium stops synthesizing its own molecules, and its metabolism shifts to the expression of genes on the viral nucleic acid strand. In T-even bacteriophages, the viral DNA redirects the genetic and metabolic activity of the cell, blocking the utilization of host DNA and ensuring that viral DNA is copied and used to synthesize new viral components (see replication of DNA in chapter 2).

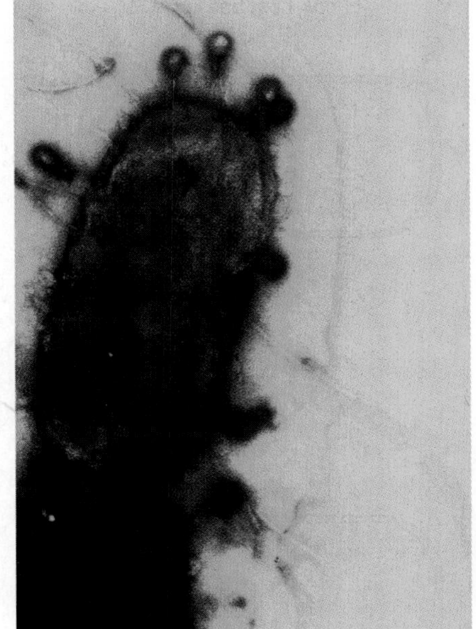

(a)

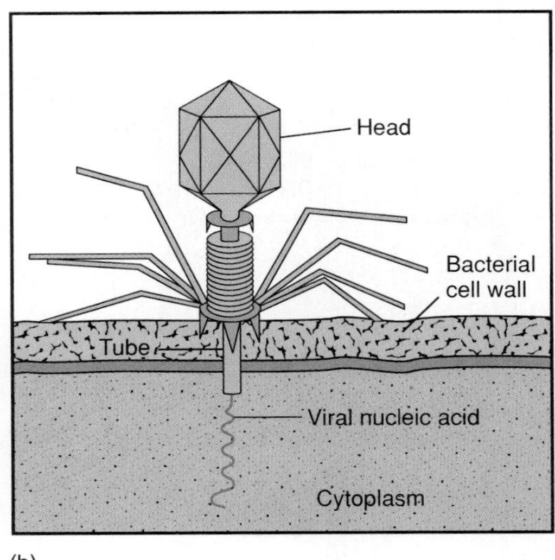

(b)

Figure 6.11
Penetration of a bacterial cell by a bacteriophage. *(a)* A single *Escherichia coli* cell with several phages attached. Note that some heads are opaque and others are transparent because they have injected their nucleic acid. *(b)* After adsorption, the phage plate becomes embedded in the cell wall, and the sheath contracts, pushing the tube through the cell wall and releasing the nucleic acid into the interior of the cell.

Expression of that phage's genetic information gives rise to the following viral molecules: (1) proteins to "seal" the cell (because the host cell is punctured by the virus during injection, and it is advantageous for the virus to repair the hole before the host cell's integrity is destroyed); (2) enzymes for copying the virus genome; (3) proteins that make up the capsid head and parts of the tail; and (4) enzymes that help weaken the cell wall so that the new phages can easily escape. This period of viral synthesis exploits the host's cytoplasmic nutrient resources, its ribosomes, and its energy supplies. The early stages of viral replication are known as *eclipse,* a period during which no mature virions can be detected within the host cell. This corresponds with the period of penetration, replication, and early assembly, and if the cell were disrupted at this time, no infective virus would be released.

As the host cell rapidly produces new phage parts, these parts spontaneously assemble into bacteriophages, similar to a mass production assembly line (figure 6.12). The first parts to assemble are the capsid and tail. Viral DNA is inserted into the capsid before the capsomers are completely joined and the collar and sheath unite with the tail pins. Mature phage particles are formed when the capsid and tail fit together and the fibers are anchored to the tail pins. An average-sized *Escherichia coli* cell can contain up to 200 new phage units at the end of this period. Eventually, the host cell becomes so packed with viruses that it **lyses**—splits open—thereby liberating the mature virions (figure 6.13). This process is hastened by viral enzymes released late in the infection cycle that digest the cell envelope thereby weakening it. Upon release, the virulent phages can spread to other susceptible bacterial cells and begin a new cycle of infection.

Lysogeny: The Silent Virus Infection

The lethal effects of a virulent phage on the host cell present a dramatic view of virus-host interaction. Not all bacteriophages complete the lytic cycle, however. Special DNA phages, called **temperate*** phages, undergo adsorption and penetration in the bacterial host but are not replicated or released. Instead, the viral DNA enters an inactive **prophage*** state, in which it is inserted into the bacterial chromosome. This viral DNA will be retained by the bacterial cell and copied during its normal cell division so that the cell's progeny will also have the temperate phage DNA (figure 6.14). This condition of the host chromosome's carrying bacteriophage DNA is termed **lysogeny.*** Because the viral genome is not expressed, the bacterial cells carrying temperate phages do not lyse, and they appear entirely normal. On occasion, the prophage in a lysogenic cell will be activated and progress directly into viral replication and the lytic cycle. Lysogeny is a less deadly form of parasitism than the full lytic cycle and is thought to be an advancement that allows the virus to spread without killing the host. In a later section and in chapters 9, 17, and 24, we describe a similar relationship that exists between certain animal viruses and human cells.

The Impact of Bacteriophages

The cycle of bacterial viruses illustrates general features of viral multiplication in a very concrete and memorable way. It is fascinating to realize that viruses are capable of lying "dormant" in their host cells, possibly becoming active at some later time. Because of the intimate association between the genetic material of the virus and host, phages occasionally serve as transporters of bacterial genes from one bacterium to another and consequently can play a profound role in bacterial genetics. This phenomenon, called transduction, is responsible for toxin production and drug resistance being transferred between bacteria (see chapters 9 and 12).

*temperate (tem′-pur-ut) A reduction in intensity.

*prophage (pro′fayj) L. *pro,* before, plus phage.

*lysogeny (ly-soj′-uhn-ee) The potential ability to produce phage.

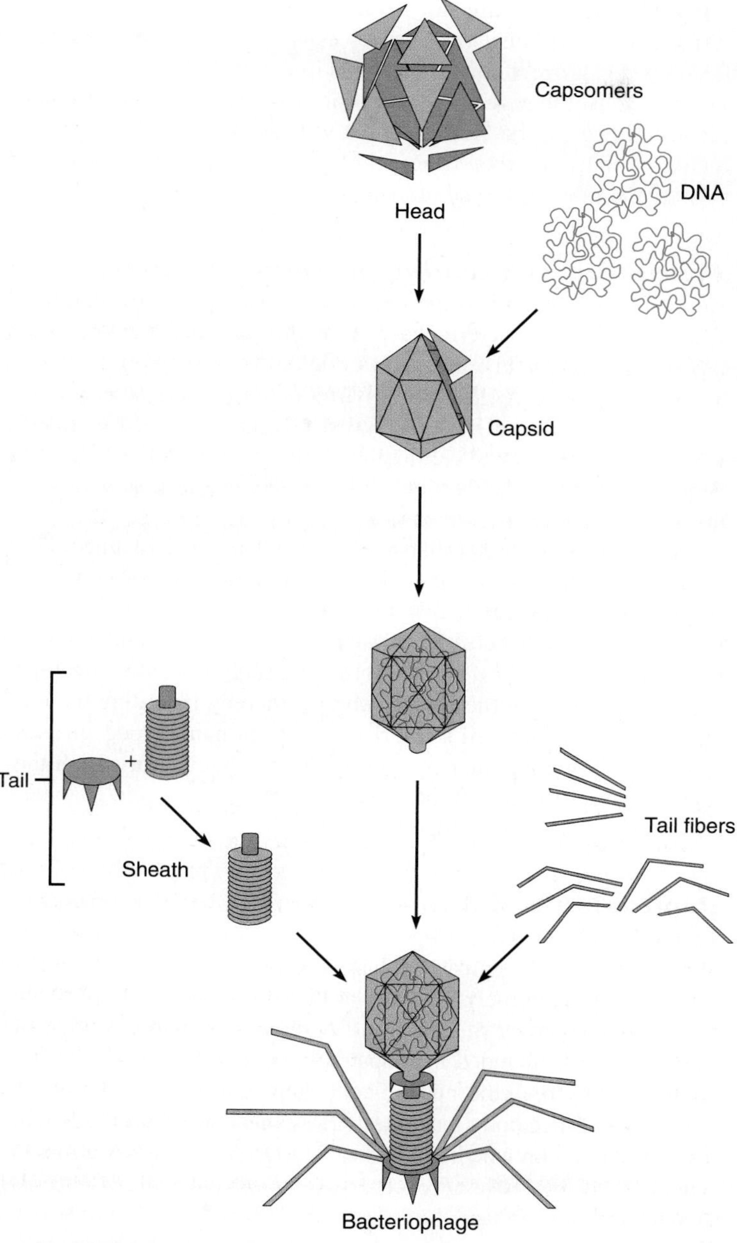

Figure 6.12
Bacteriophage assembly line. First the protein subunits (head, sheath, tail fibers) are synthesized by the host cell. A strand of viral nucleic acid also replicated by the host's machinery is inserted during the later stages of capsid formation. During final assembly, the prefabricated components fit together into whole parts and finally into the finished product.

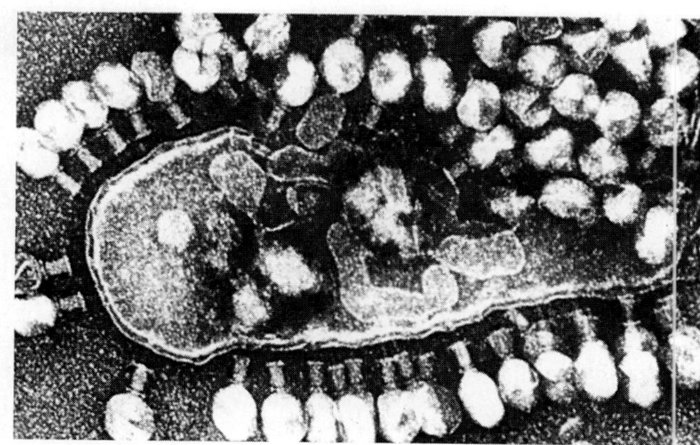

Figure 6.13

A weakened bacterial cell, crowded with viruses, has ruptured and released numerous virions that can then attack nearby susceptible host cells.

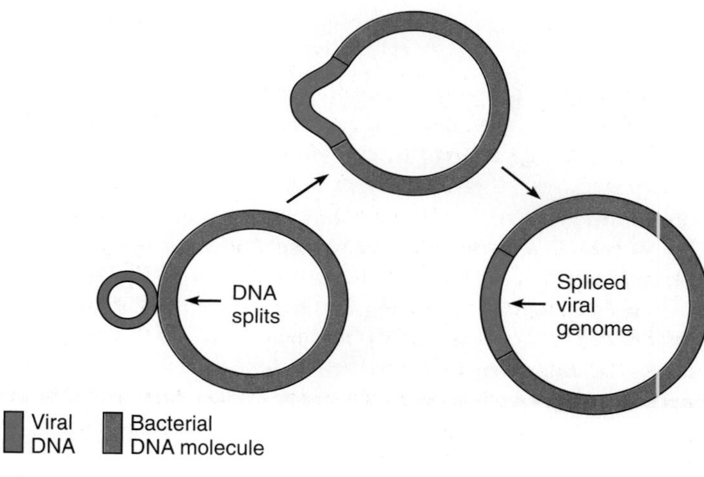

Figure 6.14
The lysogenic state in bacteria. A bacterial DNA molecule can accept and insert viral DNA molecules at specific sites on its genome. This additional viral DNA is duplicated along with the regular genome and can provide adaptive characteristics for the host bacterium.

Multiplication Cycles in Animal Viruses

Now that we have a working concept of viral multiplication in bacteria, let us turn to the animal viruses for yet another model system. Although the basic cycle is very similar to that of bacteriophages, several significant differences exist between the host cells, their viruses, and the multiplication cycles (table 6.5). The general phases in the cycle of animal viruses are adsorption, penetration, uncoating, replication, assembly, and departure from the host cell. The length of the entire multiplication cycle varies from 6 hours in polioviruses to 36 hours in herpesviruses. See figure 6.15 for a comparison of the major phases of two types of animal viruses.

Adsorption and Host Range

Invasion begins when the virus encounters a susceptible host cell and adsorbs specifically to receptor sites on the cell membrane. The membrane receptors that viruses attach to are usually proteins the cell requires for its normal function. For example, the rabies

TABLE 6.5

COMPARISON OF BACTERIOPHAGE AND ANIMAL VIRUS MULTIPLICATION

	Bacteriophage	Animal Virus
Adsorption	Precise attachment of special tail fibers to cell wall	Attachment of spikes, capsid, or envelope to cell surface receptors
Penetration	Injection of nucleic acid through cell wall; no uncoating of nucleic acid	Whole virus enters (is engulfed), or virus surface fuses with cell membrane, nucleic acid is released
Replication and maturation	Occurs in cytoplasm Cessation of host synthesis	Occurs in cytoplasm and nucleus Cessation of host synthesis
	Viral DNA or RNA is replicated and begins to function Viral components synthesized	Viral DNA or RNA is replicated and begins to function Viral components synthesized
Viral persistence	Lysogeny	Latency, chronic infection, cancer
Exit from host cell	Cell lyses when viral enzymes weaken it	Some cells lyse; enveloped viruses bud off host cell membrane
Cell destruction	Immediate	Immediate; delayed in some

virus affixes to the acetylcholine receptor of nerve cells, and the human immunodeficiency virus (HIV or AIDS virus) attaches to the CD4 protein on certain white blood cells. The mode of attachment varies between the two general types of viruses. In enveloped forms such as influenza virus and HIV, glycoprotein spikes bind to the cell membrane receptors. Viruses with naked nucleocapsids (poliovirus, for example), possess surface proteins that adhere to cell membrane receptors (figure 6.16). The actual number of viruses attached to the cell varies with the number of cell receptors and the concentration of viruses, but certain animal cells have as many as 100,000 viral receptors.

Because a virus can invade its host cell only through making an exact fit with a specific host molecule, the scope of hosts it can infect in a natural setting is limited. This limitation, known as the **host range,** may be as restricted as hepatitis B, which infects only liver cells of humans, intermediate like the poliovirus, which infects intestinal and nerve cells of primates (humans, apes, and monkeys), or as broad as the rabies virus, which can infect various cells of all mammals. Cells that lack compatible virus receptors can block adsorption. This explains why, for example, human cells resist infection with the canine hepatitis virus and dog cells are not invaded by the human hepatitis A virus. It also explains why viruses usually have tissue specificities called *tropisms** for certain cells in the body. The hepatitis B virus targets the liver, and the mumps virus targets salivary glands. However, the fact that most viruses can be induced to infect cells that they would not infect naturally makes it possible to cultivate them in the laboratory.

Penetration/Uncoating of Animal Viruses

Animal viruses exhibit some impressive mechanisms for entering a host cell. Unlike bacteriophages, they have no mechanism to inject their nucleic acids. Instead, the flexible cell membrane of the host is penetrated by the whole virus or its nucleic acid (figure 6.17). In penetration by **endocytosis** (figure 6.17*a*), the entire virus is engulfed by the cell and enclosed in a vacuole or vesicle. When enzymes in the vacuole dissolve the envelope and capsid, the virus is said to be **uncoated,** a process that releases the viral nucleic acid into the cytoplasm. The exact manner of uncoating varies, depending on whether the virus is enveloped or complex. Another means of entry involves direct fusion of the viral envelope with the host cell membrane (as in influenza and mumps viruses) (figure 6.17*b*). In this form of penetration, the envelope merges directly with the cell membrane, thereby liberating the nucleocapsid into the cell's interior. A few nonenveloped viruses (such as poliovirus) enter by an obscure mechanism in which the capsid adheres to the cell membrane and the nucleic acid is somehow translocated into the cell.

Replication and Maturation of Animal Viruses: Host Cell As Factory

The synthetic and replicative phases of animal viruses are highly regulated and extremely complex at the molecular level. They are too detailed to cover at this point. A more thorough coverage of viral genetics is included in chapters 9, 24, and 25. As with bacteriophages, the free viral nucleic acid exerts control over the host's synthetic and metabolic machinery. How this control proceeds will vary, depending on whether the virus is a DNA or an RNA virus. In general, the DNA viruses (except poxviruses) enter the host cell's nucleus and are replicated and assembled there (see figure 9.19). With few exceptions (such as retroviruses), RNA viruses are replicated and assembled in the cytoplasm (see figure 9.20).

For an overview of the phases of duplication and assembly, we will use RNA viruses as a model. Almost immediately upon entry, the viral nucleic acid alters the genetic expression of the host and instructs it to synthesize the building blocks for new viruses. First, the RNA of the virus becomes a message for synthesizing viral proteins (translation). Some viruses come equipped with the necessary enzymes for synthesis of viral components; others utilize those of the host. In the next phase, new RNA is synthesized using host nucleotides. Proteins for the capsid, spikes, and viral enzymes are synthesized on the host's ribosomes using its amino acids. Toward the end of the cycle, mature virus particles are constructed from the growing pool of parts. In most in-

*__tropism__ (troh′-pizm) Gr. *trope,* a turn. Having a special affinity for an object or substance.

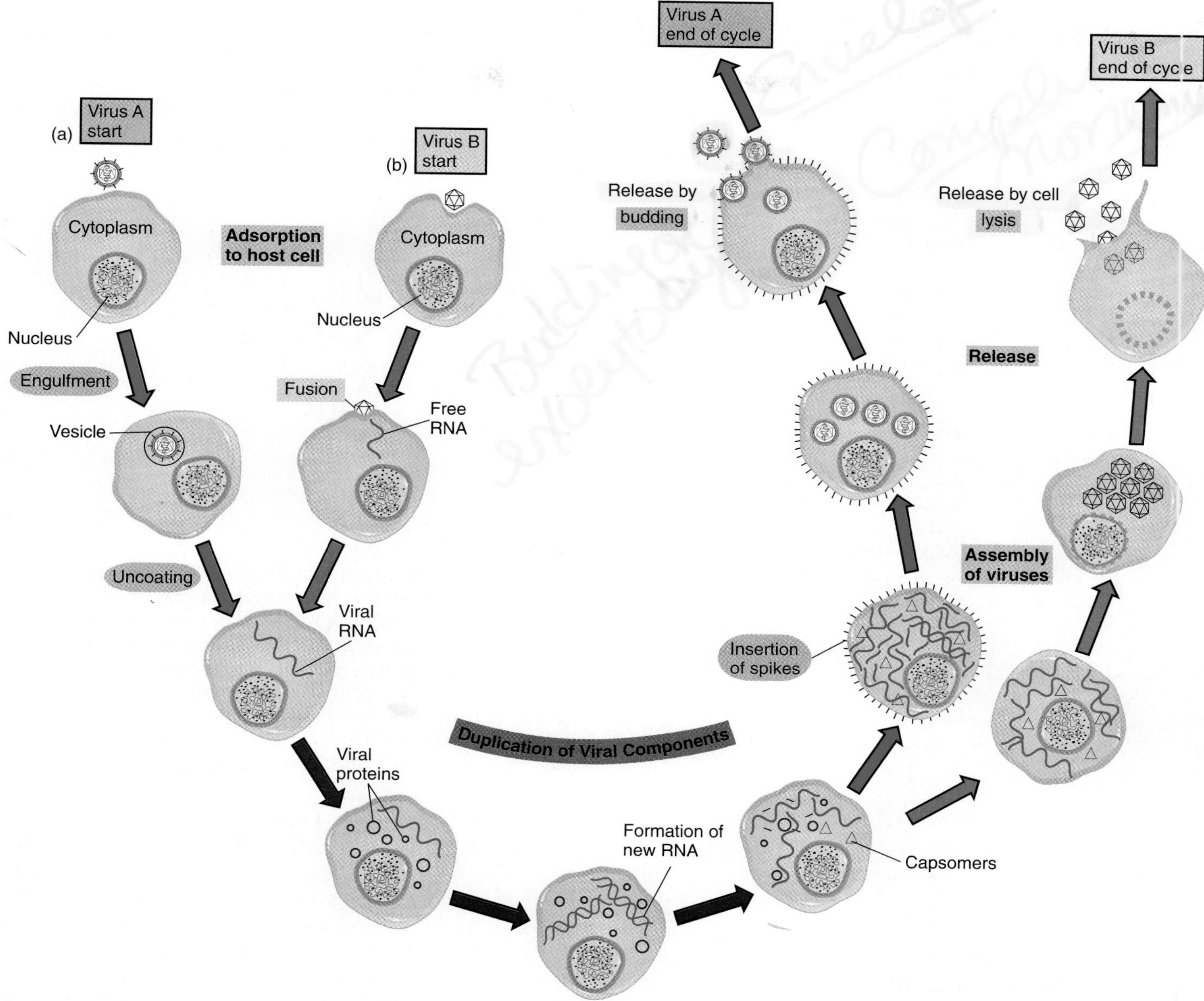

Figure 6.15

Features in the multiplication cycle of animal viruses, comparing two general types of RNA viruses. *(a)* An enveloped virus (*Rubella* virus). *(b)* A naked nucleocapsid (poliovirus). Both viruses adsorb to specific host receptors, but they penetrate the host cell by different mechanisms. Once in the cell, the free RNA molecule instructs the synthesis of new viral components (proteins for capsid and RNA). The two viruses also differ in the mode of release. Virus *(a)* inserts spikes into the host cell membrane and is budded off, taking with it an envelope. Virus *(b)* does not insert spikes and is released by cell disruption.

stances, the capsid is first laid down as an empty shell that will serve as a receptacle for the nucleic acid strand. Electron micrographs taken during this time show cells with masses of viruses, often in crystalline packets (figure 6.18). One important event in enveloped viruses is the insertion of viral spikes into the host's cell membrane so they can be picked up as the virus buds off with its envelope (see figures 6.15 and 6.19).

Release of Mature Viruses

To complete the cycle, assembled viruses leave their host in one of two ways. Nonenveloped and complex viruses that reach maturation in the cell nucleus or cytoplasm are released when the cell dies and lyses spontaneously. Enveloped viruses are liberated by **budding** or **exocytosis** from the membranes of the cytoplasm, nucleus, endoplasmic reticulum, or vesicles. This process is

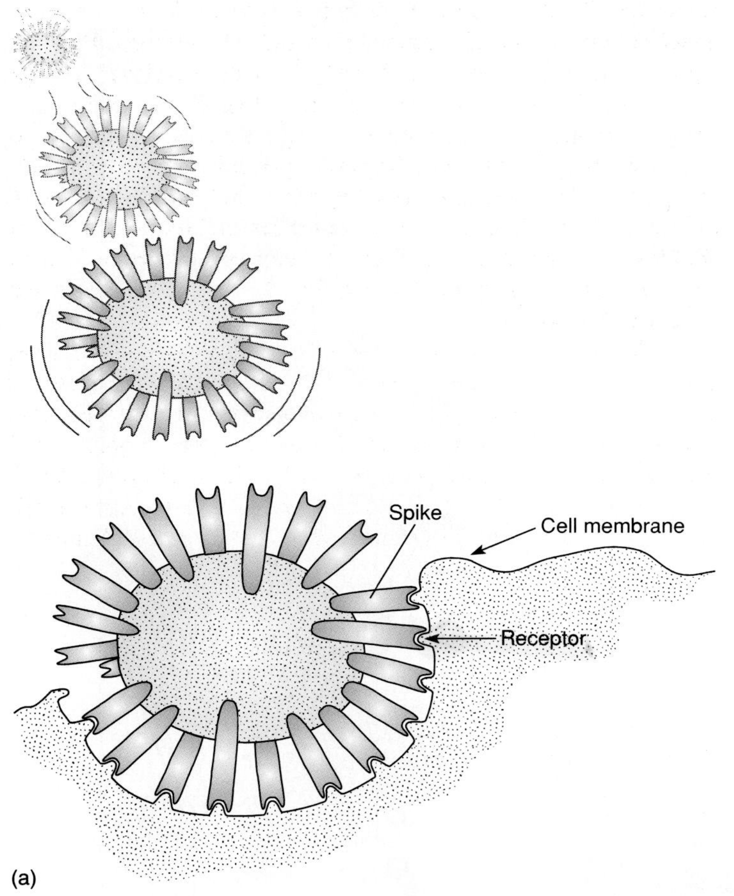

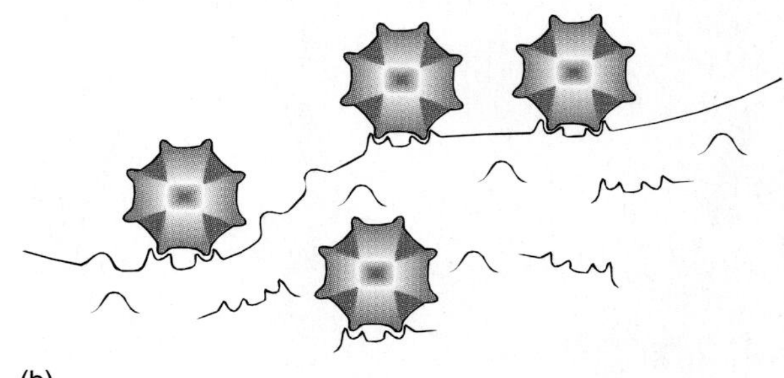

Figure 6.16

The mode by which animal viruses adsorb to the host cell membrane. *(a)* A virus with spikes. The configuration of the spike has a complementary fit for cell receptors. The process in which the virus lands on the cell and plugs into receptors is termed docking. *(b)* A virus with a naked capsid adheres to its host cell by nestling surface molecules on its capsid into the receptors on the host cell's membrane.

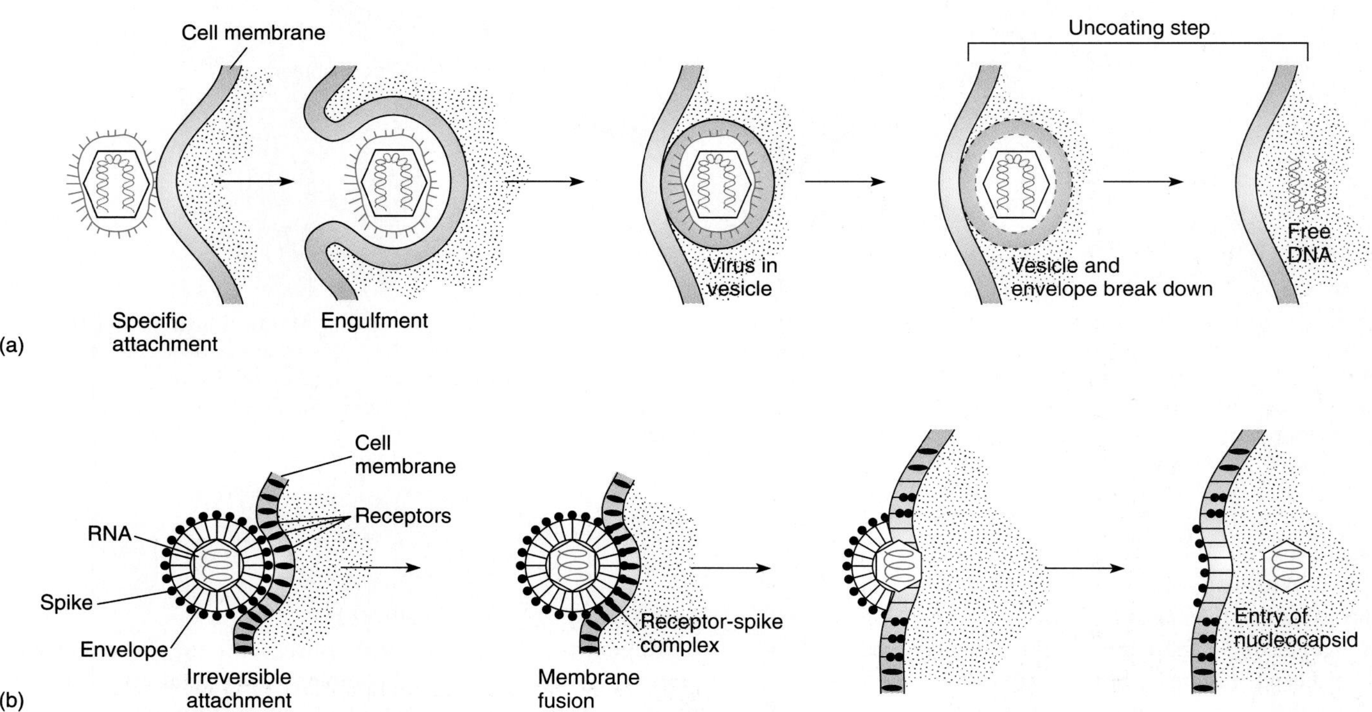

Figure 6.17

Two principal means by which animal viruses penetrate. *(a)* Endocytosis (engulfment) and uncoating of a herpesvirus. *(b)* Fusion of the cell membrane with the viral envelope (mumps virus).

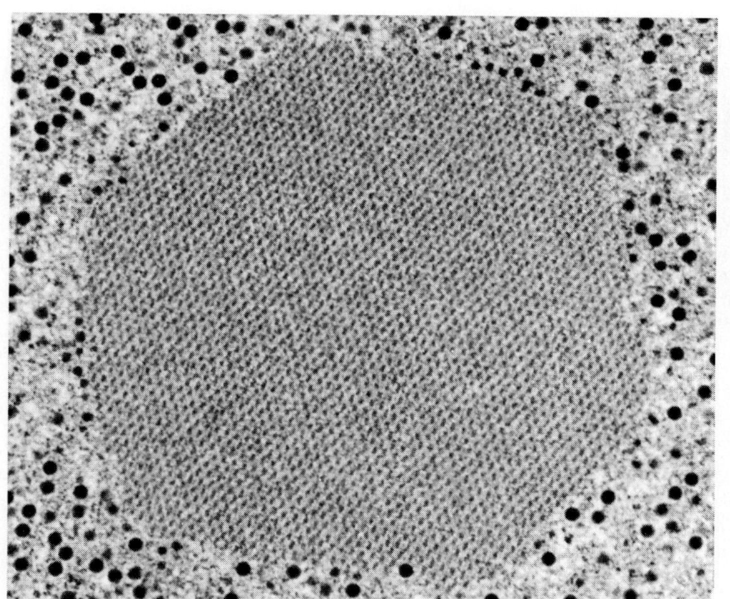

Figure 6.18
Nucleus of a cell, containing a crystalline mass of adenovirus (35,000×).

somewhat like the reverse of adsorption because the nucleocapsid attaches to the inside of the membrane and is budded off. Budding simultaneously completes the formation of the envelope carrying its spikes, if they are present (figure 6.19). Exocytosis of enveloped viruses causes them to be shed gradually, without the sudden destruction of the cell. Regardless of how the virus leaves, most active viral infections are ultimately lethal to the cell because of accumulated damage. Lethal damages include a permanent shutdown of metabolism and genetic expression, destruction of cell membrane and organelles, toxicity of virus components, and release of lysosomes.

The number of viruses released by infected cells is variable, controlled by factors such as the size of the virus and the health of the host cell. As "few" as 3,000 or 4,000 virions are released from a single cell infected with poxviruses, whereas a poliovirus-infected cell can release over 100,000 virions. If even a small number of these virions happens to meet another susceptible cell and infect it, the potential for rapid viral proliferation is immense.

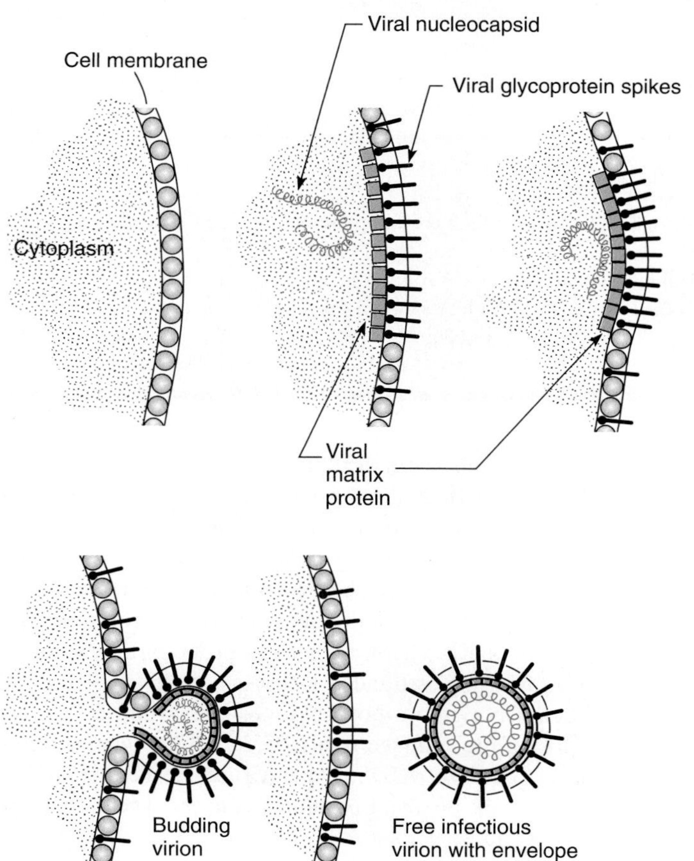

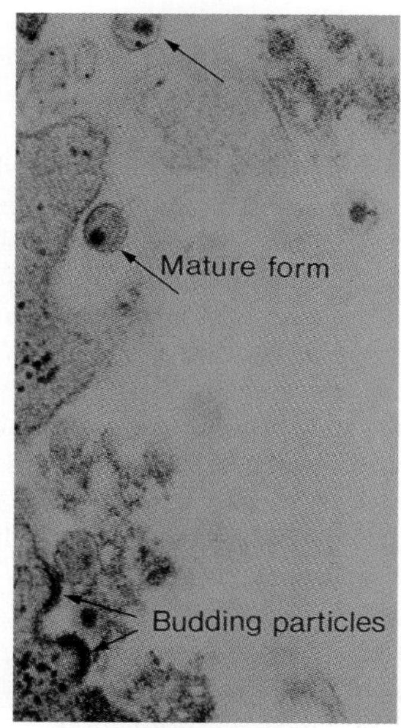

Figure 6.19
(a) Stages in the maturation of an enveloped virus (parainfluenza virus). As the virus is budded off the membrane, it simultaneously picks up an envelope and spikes. *(b)* AIDS virus (HIV) leave their host white blood cell by budding off its surface. (Note early and mature stages.)

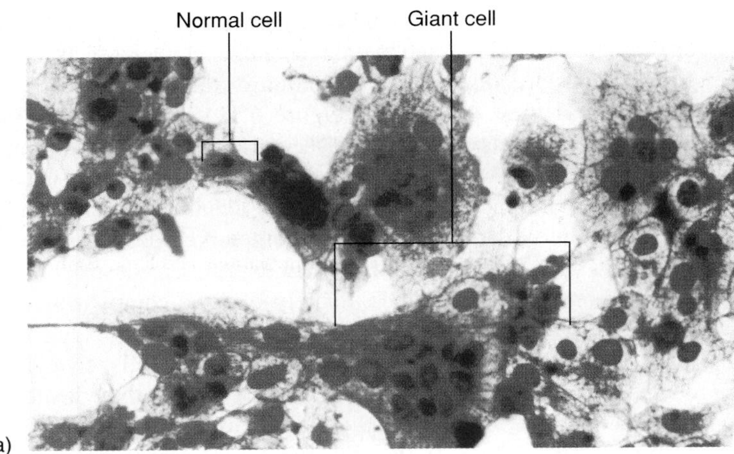

(a)

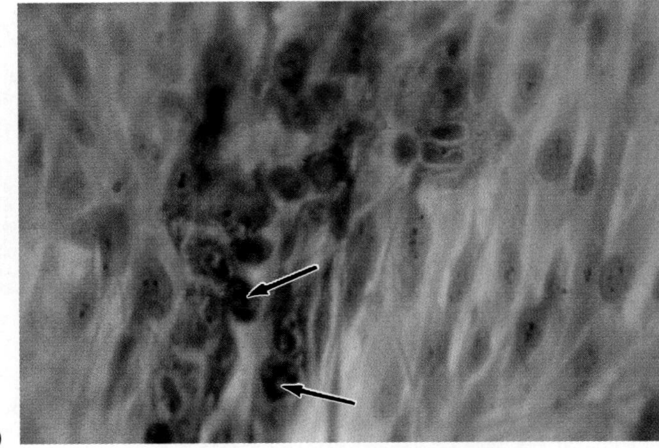
(b)

Figure 6.20

Cytopathic changes in cells and cell cultures infected by viruses. *(a)* Human epithelial cells infected by herpes simplex virus demonstrate the giant size of the cells and their multiple nuclei. *(b)* Human cells infected with cytomegalovirus. Note the inclusion bodies (arrows). Note also that both viruses disrupt the cohesive junctions between cells.

Damage to the Host Cell and Persistent Infections

The short- and long-term effects of viral infections on animal cells are well documented. **Cytopathic effects*** (CPEs) are defined as virus-induced damage to the cell that alters its microscopic appearance. Individual cells can become disoriented, undergo gross changes in shape or size, or develop intracellular changes (figure 6.20*a*). It is common to note *inclusion bodies,* or compacted masses of viruses or damaged cell organelles, in the nucleus and cytoplasm (figure 6.20*b*). Examination of cells and tissues for cytopathic effects is an important part of the diagnosis of viral infections. Table 6.6 summarizes some prominent cytopathic effects associated with specific viruses.

TABLE 6.6

Cytopathic Changes in Selected Virus-Infected Animal Cells

Virus	Response in Animal Cell
Smallpox virus	Cells round up; inclusions appear in cytoplasm
Herpes simplex	Cells become giant with multiple nuclei; nuclear inclusions
Adenovirus	Clumping of cells; nuclear inclusions
Poliovirus	Cell lysis; no inclusions
Togavirus (EEE)	Cell lysis; no inclusions
Reovirus	Cell enlargement; vacuoles and inclusions in cytoplasm
Influenza virus	Cells round up; no inclusions
Rabies virus	No change in cell shape; cytoplasmic inclusions (Negri bodies)
HIV	Giant cells with numerous nuclei (multinucleate)

Although accumulated damage from a virus infection kills most host cells, some cells maintain a carrier relationship, in which the cell harbors the virus and is not destroyed by it. These so-called *persistent infections* can last from a few weeks to much longer (for life, in some cases). The most serious persistent viruses remain in a *chronic latent state,* periodically becoming reactivated. Examples of this are herpes simplex viruses (fever blisters and genital herpes) and herpes zoster virus (chickenpox and shingles), which can go into latency in nerve cells and later emerge under the influence of various stimuli to cause recurrent infections. Specific damage that occurs in viral diseases is covered more completely in chapters 24 and 25.

Some persistent animal viruses enter their host cell and alter its growth and metabolic patterns to such an extent that the cells become cancerous. These viruses are termed *oncogenic,* and their effect on the cell is called *transformation.* A startling feature of these viruses is that their nucleic acid is consolidated into the host DNA like a prophage. Changes acquired by transformed cells include an increased rate of growth; alterations in chromosomes; changes in the cell's surface molecules; and the capacity to divide for an indefinite period, unlike normal animal cells. Mammalian viruses capable of initiating tumors are called *oncoviruses.* Some of these are DNA viruses such as papillomavirus (genital warts are associated with cervical cancer), herpesviruses (Epstein-Barr virus causes Burkitt's lymphoma), and adenoviruses. Retroviruses are an unusual family of RNA viruses that can program the synthesis of DNA using their single-stranded RNA as a template. They can then insert their DNA into a host chromosome, where it can remain for the life of the cell (see figure 25.16*a*). One type of retrovirus is HIV, the cause of AIDS. Other viruses related to HIV—**HTLV** I and II*—are involved in human cancers. These findings

*cytopathic (sy″-toh-path′-ik) Gr. *cyto,* cell, and *pathos,* disease.

*human T-cell lymphotropic viruses; cause types of leukemia.

have spurred a great deal of speculation on the possible involvement of viruses in cancers whose cause is still unknown. Additional information on the connection between viruses and cancer is found in chapters 9, 17, 24, and 25.

Chapter Checkpoints

Virus size range is from 20 nm to 450 nm.

Viruses are composed of an outer protein capsid enclosing either DNA or RNA plus a variety of enzymes. Some viruses also exhibit an envelope around the capsid.

Viral particles are called virions.

Viruses reproduce through five major events that turn a cell into a virus factory that synthesizes viral components, then assembles them into complete virion units. The new virus particles leave the cell host by lysing the cell or by budding.

Bacterial viruses, or bacteriophages, sometimes inadvertently incorporate bacterial DNA into the virion in place of viral DNA, thus providing a means of genetic transfer in bacteria.

Lysogeny is a condition in which viral DNA is inserted into the bacterial chromosome and remains inactive for an extended period. It is replicated right along with the chromosome every time the bacteria divides.

Animal viruses vary significantly from bacteriophage in their methods of adsorption, penetration, site of replication, and method of exit from host cells.

Animal viruses can persist in host tissues as chronic latent infections that can reactivate periodically throughout the host's life. Some persistent animal viruses are oncogenic.

TECHNIQUES IN CULTIVATING AND IDENTIFYING ANIMAL VIRUSES

One problem hampering earlier animal virologists was their inability to propagate specific viruses routinely in pure culture and in sufficient quantities for their studies. Virtually all of the pioneering attempts at cultivation had to be performed in the animal or plant that was the usual host for the virus, even though using the intact host organism often left much to be desired. How could researchers have ever traced the stages of viral multiplication if they had been restricted to the natural host, especially in the case of human viruses? Fortunately, systems of cultivation with broader applications were developed, including *in vivo** inoculation of laboratory-bred animals and embryonic bird tissues and *in vitro** cell (or tissue) culture methods. Such use of substitute host systems permits greater control, uniformity, and wide-scale harvesting of viruses.

*in vivo (in-vee'-voh) L. *vivos,* life. Experiments performed in a living body.

*in vitro (in vee'-troh) L. *vitros,* glass. Experiments performed in test tubes or other artificial environments.

TABLE 6.7

COMMON ANIMAL VIRUSES AND LABORATORY METHODS FOR CULTIVATION

Virus	Cultivated In
Poliovirus	Human and primate tissue culture cells
Rhinovirus (cold virus)	Human embryonic kidney and lung tissue culture
Bunyavirus	Baby mice, mosquitos, human tissue culture
Rubella virus	Monkey cell culture
Influenza, mumps, measles viruses	Human tissue culture, chicken embryos, monkey or calf kidney
Rhabdovirus (rabies)	Mice, human or hamster kidney tissue culture, chicken embryo
HIV	Human lymphocyte cell culture, chimpanzees
Papillomavirus (warts)	Human fetal brain tissue culture
Herpesviruses	Human embryonic fibroblast culture
Poxviruses	Human cell culture, chicken embryo
Hepatitis A	Human cell culture
Hepatitis B	Primates (virus cannot yet be cultivated in cell culture)

The primary purposes of viral cultivation are: (1) to isolate and identify viruses in clinical specimens; (2) to prepare viruses for vaccines; and (3) to do detailed research on viral structure, multiplication cycles, genetics, and effects on host cells. Table 6.7 summarizes some major human viruses and the systems used to culture them.

USING LIVE ANIMAL INOCULATION

Specially bred strains of white mice, rats, hamsters, guinea pigs, and rabbits are the usual choices for animal cultivation of viruses. Invertebrates (insects) or nonhuman primates are occasionally used as well. Because viruses can exhibit some host specificity, certain animals can propagate a given virus more readily than others. Depending on the particular experiment, tests can be performed on adult, juvenile, or newborn animals. The animal is exposed to the virus by injection of a viral preparation or specimen into the brain, blood, muscle, body cavity, skin, or footpads.

USING BIRD EMBRYOS

An **embryo** is an early developmental stage of animals marked by rapid differentiation of cells. Birds undergo their embryonic period within the closed protective case of an egg, which makes an incubating bird egg a nearly perfect system for viral propagation. It is an intact and self-supporting unit, complete with its own sterile environment and nourishment. Furthermore, it furnishes several embryonic tissues that readily support viral multiplication.

Chicken, duck, and turkey eggs are the most common choices for inoculation. The egg must be injected through the

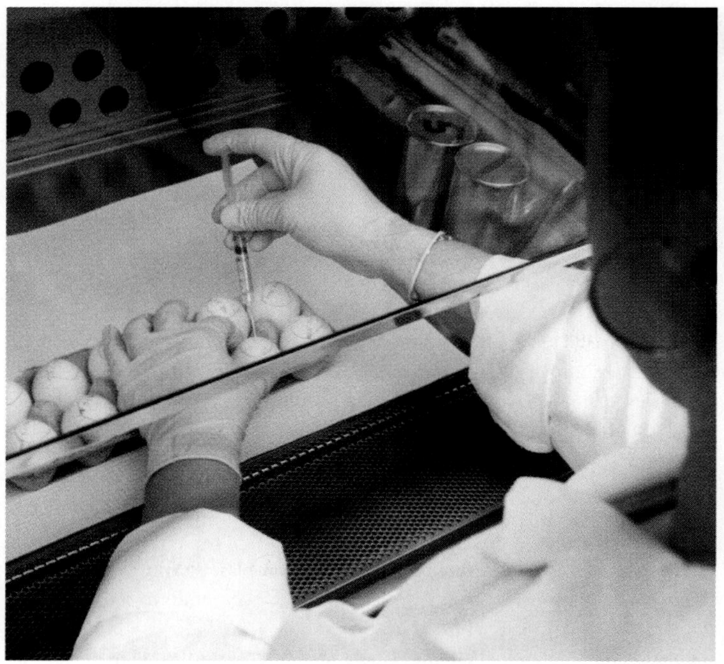

(a)

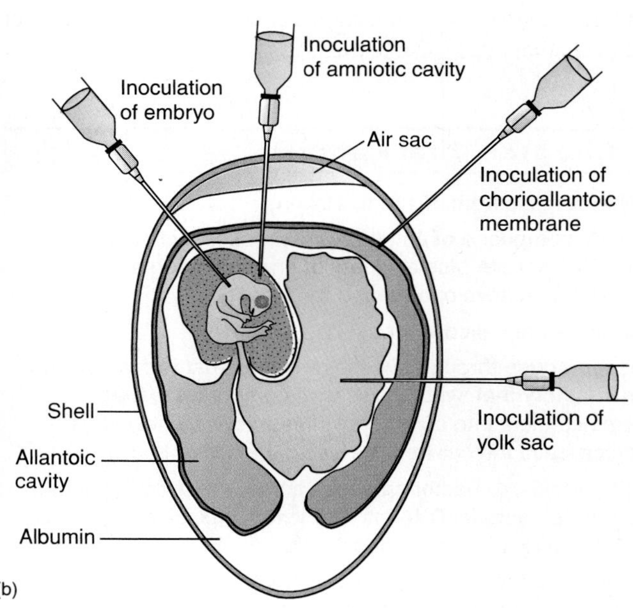

Figure 6.21
(a) A technician inoculates fertilized chicken eggs with viruses in the first stage of preparing vaccines. This process requires the highest levels of sterile and aseptic precautions. *(b)* Cultivating animal viruses in a developing bird embryo. The shell is perforated using sterile techniques, and a virus preparation is injected into a site selected to grow the viruses. Targets include the allantoic cavity, a fluid-filled sac that functions in embryonic waste removal; the amniotic cavity, a sac that cushions and protects the embryo itself; the chorioallantoic membrane, which functions in embryonic gas exchange; the yolk sac, a membrane that mobilizes yolk for the nourishment of the embryo; and the embryo itself.

shell, usually by drilling a hole or making a small window. Because this procedure can violate the asepsis of the egg, rigorous sterile techniques must be used to prevent contamination by bacteria and fungi from the air and the outer surface of the shell. The exact tissue that is inoculated is guided by the type of virus being cultivated and the goals of the experiment (figure 6.21).

Viruses multiplying in embryos may or may not cause effects visible to the naked eye. The signs of viral growth include death of the embryo, defects in embryonic development, and localized areas of damage in the membranes, resulting in discrete, opaque spots called **pocks** (a variant of pox) (figure 6.22). If a virus does not produce overt changes in the developing embryonic tissue, virologists have other methods of detection. Embryonic fluids and tissues can be prepared for direct examination with an electron microscope. Certain viruses can also be detected by their ability to agglutinate[4] red blood cells or by their reaction with an antibody of known specificity that will affix to its corresponding virus, if it is present.

4. To form large masses or clumps.

Using Cell (Tissue) Culture Techniques

The most important early discovery that led to cultivation of viruses was the development of a simple and effective way to grow populations of isolated animal cells in culture. These types of *in vitro* cultivation systems are termed **cell culture** or **tissue culture.** (Although these terms are used interchangeably, cell culture is probably a more accurate description.) So prominent is this method that most viruses are propagated in some sort of cell culture, and much of the virologist's work involves developing and maintaining these cultures. Animal cell cultures are grown in sterile chambers with special media that contain the correct types and proportions of ions and nutrients required by animal cells to survive. The cultured cells grow in the form of a *monolayer,* a single, confluent sheet of cells that supports viral multiplication and permits close inspection of the culture for signs of infection (figure 6.23).

Cultures of animal cells usually exist in the primary or continuous form. *Primary cell cultures* are prepared by placing freshly isolated animal tissue in a growth medium. The cells undergo a series of mitotic divisions to produce a monolayer. Embryonic, fetal,

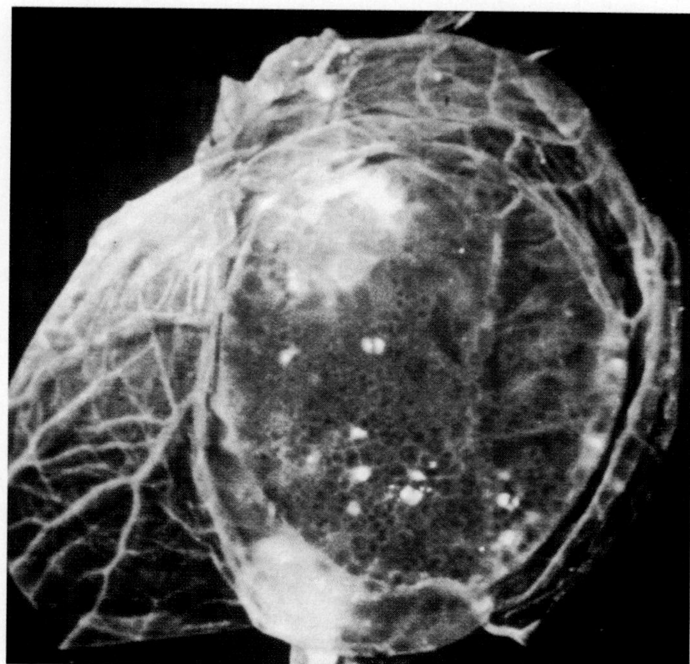

Figure 6.22
Pocks (white patches) developing on a chorioallantoic membrane removed from a chicken embryo. Several viral groups, including poxviruses and herpesviruses (shown here), cause pocks.

adult, and even cancerous tissues from many organs have served as sources of primary cultures. A primary culture retains several characteristics of the original tissue from which it was derived, but this original line generally has a limited existence. Eventually, it will die out or mutate into a line of cells that can grow continuously. These **cell lines** tend to have altered chromosome numbers, grow rapidly, and show changes in morphology, and they can be continuously subcultured, provided they are routinely transferred to fresh nutrient. One very clear advantage of cell culture is that a specific cell line can be available for viruses with a very narrow host range. Strictly human viruses can be propagated in one of several primary or continuous human cell lines, such as embryonic kidney cells, fibroblasts, bone marrow, and heart cells.

One way to detect the growth of a virus in culture is to observe degeneration and lysis of infected cells in the monolayer of cells. The areas where virus-infected cells have been destroyed show up as clear, well-defined patches in the cell sheet called **plaques*** (figure 6.23*c*). This same technique is used to detect and count bacteriophages, because they also produce plaques when grown in soft agar cultures of their host cells (bacteria). A plaque develops when the viruses released by an infected host cell radiate out to nearby host cells. As new cells become infected, they die and release more viruses, and so on. As this process continues, the infection spreads gradually and symmetrically from the original point of infection, causing the macroscopic appearance of round, clear spaces that correspond to areas of dead cells.

*plaque (plak) Fr. *placke,* patch or spot.

Chapter Checkpoints

Animal viruses must be studied in some type of host cell environment such as laboratory animals, bird embryos, or tissue cultures.

Tissue cultures are cultures of host cells grown in special sterile chambers containing correct types and proportions of growth factors using aseptic techniques to exclude unwanted microorganisms.

Virus growth in tissue culture is detected by the appearance of plaques.

MEDICAL IMPORTANCE OF VIRUSES

The number of viral infections that occur on a worldwide basis is nearly impossible to predict accurately. Certainly, viruses are the most common cause of acute infections that do not result in hospitalization, especially when one considers widespread diseases such as colds, hepatitis, chickenpox, influenza, herpes, and warts. If one also takes into account prominent viral infections found only in certain regions of the world, such as Nile Valley fever, Rift Valley fever, and yellow fever, the total could easily exceed several billion cases each year. Although most viral infections do not result in death, some, such as rabies, AIDS, and Ebola infection have very high mortality rates, and others can lead to long-term debility (polio, neonatal rubella). Current research is focused on the possible connection of viruses to several chronic afflictions of unknown cause, such as type I diabetes and multiple sclerosis.

Special Viruslike Infectious Agents

Not all noncellular infectious agents have typical viral morphology. One group of most unusual forms, even smaller and simpler than viruses, is implicated in chronic, persistent diseases in humans and animals. The infection has a long period of latency (usually several years) before the first clinical signs appear. Signs range from mental derangement to loss of muscle control, and the diseases are progressive and universally fatal.

Creutzfeldt-Jakob disease afflicts the central nervous system of humans and causes gradual degeneration and death. Cases in which medical workers developed the disease after handling autopsy specimens seem to indicate that it is transmissible, but by an unknown mechanism. Several animals (sheep, monkeys, mice) are victims of a similar transmissible disease called *scrapie.*

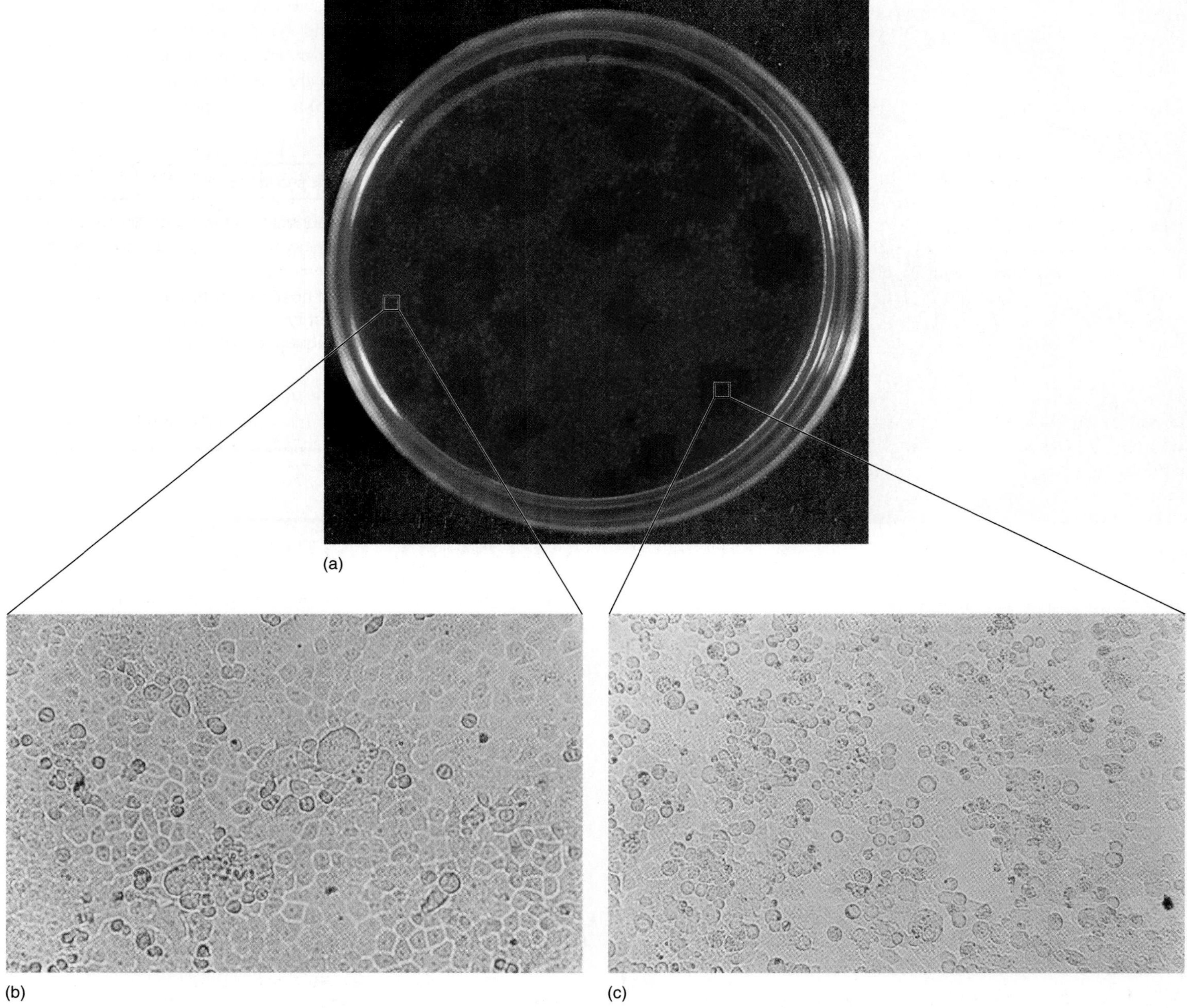

Figure 6.23

Appearance of normal and infected cell cultures. *(a)* Macroscopic view of a Petri dish containing a monolayer (single layer of attached cells) of monkey kidney cells. Clear spaces in culture indicate sites of virus growth (plaques). *(b)* Microscopic views of normal, undisturbed cell layer and *(c)* plaques, which consist of cells disrupted by viral infection.

A prion (see below) disease called bovine spongiform encephalopathy, or "mad cow disease," was recently the subject of fears and a crisis in Europe when researchers found evidence that the disease could be acquired in humans who consumed contaminated beef. This was the first incidence of prion disease transmission from animals.

A common feature of these conditions is the deposition of distinct fibrils in the brain tissue (see figure 25.30). Some researchers have hypothesized that these fibrils are the agents of the disease and have named them *prions* (**pro**teinacious **in**fectious particles). Whether these fibrils actually represent an infectious agent is currently controversial. If continued study supports this hypothesis, the discovery that they are composed primarily of protein (no nucleic acid) will certainly revolutionize our ideas of what can constitute an infectious agent. Other investigators have linked the pathology of scrapie to small pieces of nucleic acid coated with protein, which they have termed *virinos.* (See chapter 25 for further discussion of this phenomenon.)

Some fascinating viruslike agents in human disease are defective forms called satellite viruses that are actually dependent on

MICROFILE 6.1 VIRUSES, FETUSES, AND NEONATES

The common childhood viral infections such as measles, rubella, and chickenpox are usually mild and free of severe complications—provided they occur in healthy, normal children whose immune defenses are active. Sometimes even innocuous viral infections can become life-threatening in highly vulnerable fetuses or newborns. One of the most dangerous of these is the rubella virus. If a mother develops rubella during pregnancy, the virus can cross the placenta and cause abnormal development in the fetus. These intrauterine infections often result in congenital defects such as cardiac disease, cataracts, deafness, abnormal brain development, and even death (see figure 25.9). Herpesviruses such as chickenpox and cytomegalovirus can also permanently damage or kill a fetus.

Herpes simplex II, the cause of genital herpes in adults, does not usually cross the placenta and infect the fetus, but it can be picked up by the neonate through contact with the mother's birth canal. The disease it causes varies from a short-lived skin infection to extreme brain damage. Hepatitis B virus also can be acquired from the mother's birth canal. The potential risk is so great that pregnant mothers must be carefully monitored and screened for the presence of these viruses.

other viruses! Two remarkable examples are the adeno-associated virus (AAV), which can replicate only in cells infected with adenovirus, and the delta agent, a naked strand of RNA that is expressed only in the presence of the hepatitis B virus and can worsen the severity of liver damage.

Plants are also parasitized by viruslike agents called **viroids** that differ from ordinary viruses by being very small (about one-tenth the size of an average virus) and being composed of only naked strands of RNA, lacking a capsid or any other type of coating. Viroids are significant pathogens in several economically important plants, including tomatoes, potatoes, cucumbers, citrus trees, and chrysanthemums.

DETECTION AND CONTROL OF VIRAL INFECTIONS

Life-threatening viral diseases such as AIDS, hantavirus pulmonary syndrome, influenza, and those that pose a serious risk to the fetus (rubella or cytomegalovirus; microfile 6.1) demand rapid detection and correct diagnosis so that appropriate therapy and other control measures can be prescribed. Viruses tend to be more difficult to identify than cellular microbes. Physicians often use the overall clinical picture of the disease (specific signs) to guide diagnosis, but exact virus identification from clinical specimens (blood, respiratory secretions, feces, throat, and vaginal secretions) is more satisfactory (figure 6.24*a,b*). One direct, rapid method is to examine a specimen for the presence of the virus or signs of cytopathic changes in cells or tissues (see herpesviruses, figure 24.9). This may involve immunofluorescence techniques or direct examination with an electron microscope (figure 6.24*c*). Samples can also be screened for the presence of indicator molecules (antigens) from the virus itself. A new technique that is rapidly becoming standard procedure is the polymerase chain reaction (PCR), which can detect and amplify even minute amounts of viral DNA or RNA in a sample (figure 6.24*e*). This method greatly facilitated identification of the first documented cases of American hantavirus infection in New Mexico. In certain infections, definitive diagnosis will require the use of cell culture, embryos, or animals, but this method can be time-consuming and slow to give results (figure 6.24*f*). Frequently, it is helpful as a screening test to detect specific antibodies that indicate signs of virus infection from patient's blood. It is the main test for HIV infection (figure 6.24*d*). Details of viral diagnosis are provided in chapters 10, 16, 24, 25.

TREATMENT OF ANIMAL VIRAL INFECTIONS

The novel nature of viruses has at times been a major impediment to effective therapy. Because viruses are not bacteria, antibiotics aimed at disrupting procaryotic cells do not work on them. On the other hand, many antiviral drugs block virus replication by targeting the function of host cells, and can cause severe side effects. An example is azidothymidine (AZT), a drug used to treat AIDS, which has severe toxic side effects such as immunosuppression and anemia. In addition, most antiviral drugs prevent the virus from replicating or maturing and therefore must be capable of entering human cells to have an effect. Another compound that shows some potential for treating and preventing viral infections is a naturally occurring human cell product called *interferon* (see chapter 14). Vaccines that stimulate immunity are an extremely valuable tool but are available for only a limited number of viral diseases (see chapter 16).

Chapter Checkpoints

Viruses are easily responsible for several billion cases of infections each year. It is conceivable that many chronic diseases of unknown cause will be connected to viral agents.

Other noncellular agents of disease are the so-called slow viruses, which are not viruses at all, but protein fibers called **prions;** virions, extremely small lengths of protein-coated nucleic acid; and satellite viruses, which require larger viruses to cause disease.

Diagnosis of viral disease agents is done directly through culture, electron microscopy, immunofluorescence microscopy, and by the PCR technique. Indirect diagnosis is done by testing patient serum for host antibody to viral agents.

Viral infections are difficult to treat because the drugs that attack the viral replication cycle also cause serious side effects in the host.

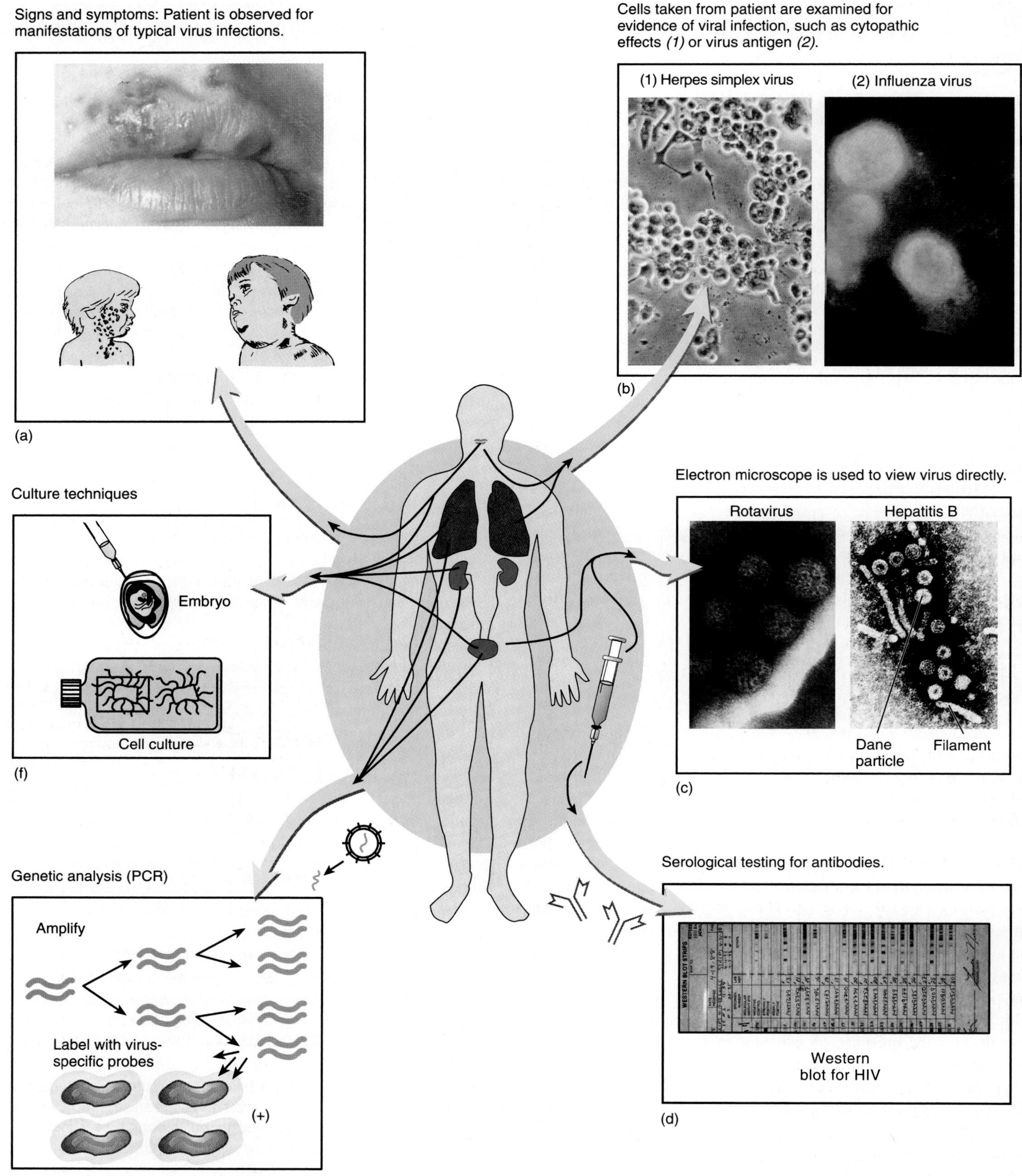
Signs and symptoms: Patient is observed for manifestations of typical virus infections.
(a)
Cells taken from patient are examined for evidence of viral infection, such as cytopathic effects (1) or virus antigen (2).
(1) Herpes simplex virus
(2) Influenza virus
(b)
Electron microscope is used to view virus directly.
Rotavirus
Hepatitis B
Dane particle
Filament
(c)
Serological testing for antibodies.
Western blot for HIV
(d)
Genetic analysis (PCR)
Amplify
Label with virus-specific probes
(+)
(e)
Culture techniques
Embryo
Cell culture
(f)

Figure 6.24
Summary of methods used to diagnose virus infections.
© Electron micrograph by Derek Garbellini.

CHAPTER CAPSULE WITH KEY TERMS

Other names based on host cell or disease: general—**virions, viral particles;** specific—**bacteriophage, poxvirus, herpesvirus,** etc.

OVERALL MORPHOLOGY OF VIRUSES

Viruses are infectious particles and **not cells;** lack protoplasm, organelles, locomotion of any kind; are large, complex molecules; can be **crystalline** in form. A virus particle is composed of a nucleic acid core (**DNA** or **RNA,** not both) surrounded by a geometric protein shell, or **capsid;** combination called a **nucleocapsid;** capsid is **helical,** or **icosahedral** in configuration; many are covered by a membranous **envelope** containing viral protein **spikes; complex viruses** have additional external and internal structures.

Shapes/Sizes: Cuboidal, spherical, cylindrical, brick- and bullet-shaped. Smallest infectious forms range from the largest poxvirus (0.45 μm or 450 nm) to the smallest viruses (0.02 μm or 20 nm).

Nutritional and Other Requirements: Lack enzymes for processing food or generating energy; are tied entirely to the host cell for all needs **(obligate intracellular parasites).**

DISTRIBUTION/HOST RANGE

Viruses are known to parasitize all types of cells, including bacteria, algae, fungi, protozoa, animals, and plants. Each viral type is limited in its **host range** to a single species or group, mostly due to specificity of adsorption of virus to specific host receptors.

CLASSIFICATION

The two major types of viruses are **DNA** and **RNA viruses.** These are further subdivided into families, depending on shape and size of capsid, presence or absence of an envelope, whether double- or single-stranded nucleic acid, and antigenic similarities. **Animal viruses:** Common names of major DNA animal viruses include poxviruses (smallpox), herpesviruses, adenoviruses, papillomavirus (wart), hepatitis B virus, parvoviruses. RNA animal viruses include poliovirus, hepatitis A virus, rhinovirus (common cold), encephalitis viruses, yellow fever virus, rubella virus, influenza virus, mumps virus, measles virus, rabies viruses, and HIV (AIDS virus).

MULTIPLICATION CYCLE

A unique method of multiplication occurs—no division is involved. Viruses make multiple copies of themselves through: (1) attachment, or **adsorption,** to a host cell; they must have special binding molecules for receptors on the host cell; (2) **penetration** of a cell by means of whole virus or only nucleic acid; (3) assuming control of the cell's genetic and metabolic components, thus (4) programming synthesis of new viral parts; (5) directing **assembly** of these parts into new virus particles, and (6) **releasing** complete, mature viruses. This can proceed through **lysis** of cell or **budding** of viruses. Bacteriophages and animal viruses differ in how they enter and leave the cell. Some viruses go into a latent or **lysogenic** phase in which they integrate into the DNA of the host cell and later may become active and produce a lytic infection.

METHOD OF CULTIVATION

The need for an intracellular habitat makes it necessary to grow viruses in living cells, either in the intact host animal, in bird embryos, or in isolated cultures of host cells **(cell culture).**

MEDICAL IMPORTANCE

Viruses attach to specific target hosts or cells. They cause a variety of infectious diseases, ranging from mild respiratory illness (common cold) to destructive and potentially fatal conditions (rabies, AIDS). Some viruses can cause birth defects and cancer in humans and other animals.

Agricultural: Hundreds of cultivated plants and domestic animals are susceptible to viral infections, often with adverse economic and ecologic repercussions.

Research: Because of their simplicity, viruses have become an invaluable tool for studying basic genetic principles.

IDENTIFICATION

By means of **cytopathic effects** in host cells, direct examination of viruses or their components in samples, analyzing the patient's body fluids for antibodies against viruses, and symptoms.

UNIQUE FEATURES

Viruses exist at a level between living things and nonliving molecules. They consist of a capsid and nucleic acid, either DNA or RNA, not both; lack metabolism and respiratory enzymes; multiply inside host cells using the assembly line of the host's synthetic machinery; are genetic parasites; can pass through fine filters; are ultramicroscopic and crystallizable. Some viruses are enveloped, others are naked; some have unique nucleic acid structure (single-stranded DNA and double-stranded RNA); some animal viruses can become **latent** and cause cancer.

MULTIPLE-CHOICE QUESTIONS

1. A virus is a tiny infectious
   a. cell c. particle
   b. living thing d. nucleic acid
2. Viruses are known to infect
   a. plants c. fungi
   b. bacteria d. all organisms
3. The capsid is composed of protein subunits called
   a. spikes c. virions
   b. protomers d. capsomers
4. The envelope of an animal virus is derived from the ____ of its host cell.
   a. cell wall c. glycocalyx
   b. cell membrane d. receptors
5. The nucleic acid of a virus is
   a. DNA only
   b. RNA only
   c. both DNA and RNA
   d. either DNA or RNA
6. The general steps in a viral multiplication cycle are
   a. adsorption, penetration, replication, maturation, and release
   b. endocytosis, uncoating, replication, assembly, budding
   c. adsorption, uncoating, duplication, assembly, and lysis
   d. endocytosis, penetration, replication, maturation, and exocytosis

7. A prophage is an early stage in the development of a/an
   a. bacterial virus c. lytic virus
   b. poxvirus d. enveloped virus

8. The nucleic acid of animal viruses enters the host cell through
   a. translocation c. endocytosis
   b. fusion d. all of these

9. In general, RNA viruses multiply in the cell ____, and DNA viruses multiply in the cell ____.
   a. nucleus, cytoplasm
   b. cytoplasm, nucleus
   c. vesicles, ribosomes
   d. endoplasmic reticulum, nucleolus

10. Enveloped viruses carry surface receptors called
   a. buds c. fibers
   b. spikes d. sheaths

11. Viruses that persist in the cell and cause recurrent disease are considered
   a. oncogenic c. latent
   b. cytopathic d. resistant

12. Examples of cytopathic effects of viruses are
   a. inclusion bodies c. multiple nuclei
   b. giant cells d. all of these

13. Viruses cannot be cultivated in:
   a. tissue culture c. live mammals
   b. bird embryos d. blood agar

14. Clear patches in cell cultures that indicate sites of virus infection are called
   a. plaques c. colonies
   b. pocks d. prions

CONCEPT QUESTIONS

1. a. Describe 10 *unique* characteristics of viruses (can include structure, behavior, multiplication).
   b. After consulting table 6.2, what additional statements can you make about viruses, especially as compared with cells?
2. a. What does it mean to be an obligate intracellular parasite?
   b. What is another way to describe the sort of parasitism exhibited by viruses?
3. a. Characterize the viruses according to size range.
   b. What does it mean to say that they are ultramicroscopic?
   c. That they are filterable?
4. a. Describe the general structure of viruses.
   b. What is the capsid, and what is its function?
   c. How are the two types of capsids constructed?
   d. What is a nucleocapsid?
   e. Give examples of viruses with the two capsid types.
   f. What is an enveloped virus, and how does the envelope arise?
   g. Give an example of a common enveloped human virus.
   h. What are spikes, how are they formed, and what is their function?
5. a. What are bacteriophages, and what is their structure?
   b. What is a tobacco mosaic virus?
   c. How are the poxviruses different from other animal viruses?
6. a. Since viruses lack metabolic enzymes, how can they synthesize necessary components?
   b. What are some enzymes with which the virus is equipped?
7. a. How are viruses classified? What are virus families?
   b. How are generic and common names used?
   c. Looking at table 6.3, how many different viral diseases can you count?
8. a. Compare and contrast the main phases in the lytic multiplication cycle in bacteriophages and animal viruses.
   b. When is a virus a virion?
   c. What is necessary for adsorption?
   d. Why is penetration so different in the two groups?
   e. What is eclipse?
   f. In simple terms, what does the virus nucleic acid do once it gets into the cell?
   g. What is involved in assembly?
9. a. What is a prophage or temperate phage?
   b. What is lysogeny?
10. a. What dictates the host range of animal viruses?
   b. What are two ways that animal viruses penetrate the host cell?
   c. What is uncoating?
   d. Describe the two ways that animal viruses leave their host cell.
11. a. Describe several cytopathic effects of viruses.
   b. What causes the appearance of the host cell?
   c. How might it be used to diagnose viral infection?
12. a. What does it mean for a virus to be persistent or latent, and how are these events important?
   b. Briefly describe the action of an oncogenic virus.
13. a. Describe the three main techniques for cultivating viruses.
   b. What is the advantage of using cell culture?
   c. The disadvantages?
   d. What is a disadvantage of using live intact animals or embryos?
   e. What is a cell line?
   f. A monolayer?
   g. How are plaques formed?
14. a. What is the principal effect of the agent of Creutzfeldt-Jakob disease?
   b. How is the agent different from viruses?
   c. What is a viroid?
15. Why are virus diseases more difficult to treat than bacterial diseases?
16. Circle the viral infections from this list: cholera, rabies, plague, cold sores, whooping cough, tetanus, genital warts, gonorrhea, mumps, Rocky Mountain spotted fever, syphilis, rubella, rat bite fever.

Critical-Thinking Questions

1. a. What characteristics of viruses could be used to characterize them as life forms?
   b. What makes them more similar to lifeless molecules?
2. a. Comment on the possible origin of viruses. Is it not curious that the human cell welcomes a virus in and hospitably removes its coat?
   b. How do spikes play a part in the action of the host cell?
3. a. If viruses that normally form envelopes were prevented from budding, would they still be infectious?
   b. If the RNA of an influenza virus were injected into a cell by itself, could it cause a lytic infection?
4. a. The end result of most viral infections is death of the host cell.
   a. If this is the case, how can we account for such differences in the damage that viruses do (compare the effects of the cold virus with those of the rabies virus)?
   b. What is one adaptation of viruses that does not immediately kill the host cell?
5. a. Given that DNA viruses can actually be carried in the DNA of the host cell's chromosomes, comment on what this phenomenon means in terms of inheritance in the offspring.
   b. Could some cancers be explained on the basis of inheritance of a latent virus?
6. You have been given the problem of constructing a building on top of a mountain. It will be impossible to use huge pieces of building material. Using the virus capsid as a model, can you think of a way to construct the building?
7. The AIDS virus attacks only specific types of human cells, such as certain white blood cells and nerve cells. Can you explain why a virus can enter some types of human cells but not others?
8. a. Consult table 6.4 to determine which viral diseases you have had and which ones you have been vaccinated against.
   b. Which viruses would you investigate as possible oncoviruses?
9. One early problem in cultivating the AIDS virus was the lack of a cell line that would sustain indefinitely *in vitro,* but eventually one was developed. What do you expect were the stages in developing this cell line?
10. a. If you were involved in developing an antiviral drug, what would be some important considerations? (Can a drug "kill" a virus?)
    b. How could multiplication be blocked?
11. a. Is there such a thing as a "good virus"?
    b. Explain why or why not. Consider both bacteriophages and viruses of eucaryotic organisms.
12. Why is an embryonic or fetal viral infection so harmful?
13. How are computer program viruses analogous to real viruses?

Internet Search Topic

Look up the term *outbreak* on an Internet search engine. Make a list of worldwide outbreaks of infectious diseases and determine how many of them are viral in origin. Use microfile 1.1 as a reference in this search.

7 chapter

Elements of Microbial Nutrition, Ecology, and Growth

With its widely fluctuating physical, chemical, and biological conditions, the earth provides an amazing array of habitats. Numerous environmental factors combine in these habitats to dictate just what kinds of microbial residents can adapt and survive there. Factors that exert the greatest influence are (1) nutrients and energy sources, (2) ambient temperature, (3) amount of moisture, (4) presence or absence of gases, (5) osmotic pressure, (6) pH (acidity or alkalinity), (7) presence or absence of light or radiation, and (8) other organisms. Some microbes are unable to survive the prevailing conditions in a habitat; some migrate to other habitats; and others remain and flourish. Microbes are so immensely adaptable and plastic that they have proliferated into every type of habitat and *niche,** from the most temperate to the most extreme (microfile 7.1).

*niche (nitch) Fr. *nichier,* to nest. The role of an organism in its ecological setting.

Biotechnology researchers probe for samples of thermophilic microorganisms in a hot spring in Yellowstone National Park.

MICROFILE 7.1 LIFE IN THE EXTREMES

Any extreme habitat—whether hot, cold, salty, acidic, alkaline, high pressure, arid, oxygen-free, or toxic—is likely to harbor microorganisms that have made special adaptations to their conditions. Although in most instances the inhabitants are archaea and bacteria, certain fungi, protozoans, and algae are also capable of living in harsh habitats. Microbiologists have termed such remarkable organisms **extremophiles.**

HOT AND COLD

Perhaps the most extreme habitats are hot springs, geysers, and ocean vents, all of which support flourishing microbial populations. Temperatures in these regions range from 50°C to well above the boiling point of water, with some ocean vents even approaching 350°C. Many heat-adapted microbes are highly primitive archaea whose genetics and metabolism are extremely modified for this mode of existence. Some of the brilliant colors surrounding the sulfur springs and other hot regions of Yellowstone National Park are caused by the growth of cyanobacteria. A unique ecosystem based on hydrogen sulfide–oxidizing bacteria exists in the hydrothermal vents lying along deep oceanic ridges (see microfile 7.8). Heat-adapted bacteria even plague home water heaters and the heating towers of power and industrial plants.

A large part of the earth exists at cold temperatures. Microbes settle and grow throughout the Arctic and Antarctic, and in the deepest parts of the ocean, in temperatures that hover near the freezing point of water (2°–10°C). Several species of algae and fungi thrive on the surfaces of snow and glaciers (see figure 7.11). More surprising still is a strategy of bacteria and algae adapted to the sea ice of Antarctica. Although the ice appears to be completely solid, it is honeycombed by various-sized pores and tunnels filled with liquid water. These frigid microhabitats harbor a virtual microcosm of planktonic life, including predators (fish and shrimp) that live on these algae and bacteria.

SALT, ACIDITY, ALKALINITY

The growth of most microbial cells is inhibited by high amounts of salt; for this reason, salt is a common food preservative. Yet whole communities of salt-dependent bacteria and algae occupy habitats in oceans, salt lakes, and inland seas, some of which are saturated with salt (30%). Most of these microbes have demonstrable metabolic requirements for high levels of minerals such as sodium, potassium, magnesium, chlorides, or iodides. Because of their salt-loving nature, some species are pesky contaminants in salt–processing plants, pickling brine, and salted fish.

Highly acidic or alkaline habitats are not common, but acidic bogs, lakes, and alkaline soils contain a specialized microbial flora. A few species of algae and bacteria can actually survive at a pH near that of concentrated hydrochloric acid. They not only require such a low pH for growth, but particular bacteria (for example, *Thiobacillus*) actually help maintain the low pH by releasing strong acid.

OTHER FRONTIERS TO CONQUER

It was once thought that the region far beneath the soil and upper crust of the earth's surface was sterile. However, new work with deep core samples (from 330 m down) indicates a vast microbial population in these zones. Myriad bacteria, protozoa, and fungi exist in this moist clay, which is high in minerals and complex organic substrates. An extremely rich community of bacteria was discovered in porous rock several kilometers beneath the continental shelf. This finding is all the more remarkable because the bacteria serve as the main nutrient source for an abundance of animals inhabiting the surrounding ocean.

Numerous species have carved a niche for themselves in the depths of mud, swamps, and oceans, where oxygen gas and sunlight cannot penetrate. The predominant living things in the deepest part of the oceans (10,000 m or below) are pressure- and cold-loving microorganisms. Even parched zones in sand dunes and deserts harbor a hardy brand of microbes, and thriving bacterial populations can be found in petroleum, coal, and mineral deposits containing copper, zinc, gold, and uranium. Although one might think that formaldehyde, alcohol, airplane fuel, hydrogen sulfide gas, and carbolic acid would be harmful, these substances are food for some microbes. Even artificial habitats such as asphalt, wet cement, and old rubber tires soon develop microbial colonists.

As a rule, a microbe that has adapted to an extreme habitat will die if placed in a moderate one. And, except for rare cases, none of the organisms living in these extremes are pathogens, because the human body is a hostile habitat for them.

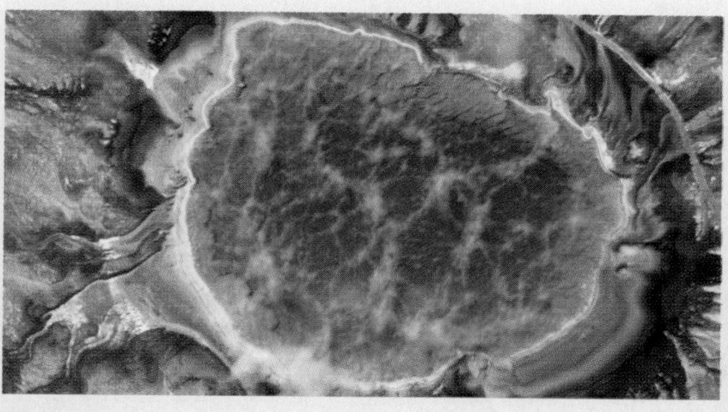

(a)

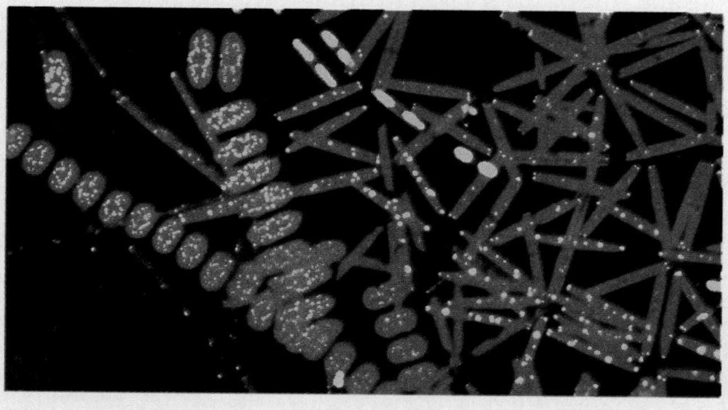

(b)

Examples of extreme habitats. (a) *The grand prismatic spring in Yellowstone National Park. Bright colors encircling this geyser arise from abundant cyanobacterial pigments. Obvious variations in color are due to varying species of bacteria that differ in heat tolerance. Only non-photosynthetic (colorless) forms can live toward the hottest center of the pool.* (b) *Life on ice: diatoms living in Antarctic sea ice tunnels.*

MICROBIAL NUTRITION

Nutrition* is a process by which chemical substances called **nutrients** are acquired from the environment and used in cellular activities such as metabolism and growth. With respect to nutrition, microbes are not really so different from humans. Bacteria living in mud on a diet of inorganic sulfur or protozoa digesting wood in a termite's intestine seem to show radical adaptations, but even these organisms require a constant influx of certain substances from their habitat. In general, all living things require a source of elements such as carbon, hydrogen, oxygen, phosphorus, potassium, nitrogen, sulfur, calcium, iron, sodium, chlorine, magnesium, and certain other elements. (A helpful mnemonic device for the necessary elements is **CHOPKINS CaFe, Sodium Chl**oride **M**ighty **g**ood!) But the ultimate source of a particular element, its chemical form, and how much of it the microbe needs are all points of variation (table 7.1). Any substance, whether in an elemental or molecular, form that must be provided to an organism is called an **essential nutrient.** Once absorbed, nutrients are processed and transformed into the chemicals of the cell (microfile 7.2).

Two categories of essential nutrients are **macronutrients** and **micronutrients.** Macronutrients are required in relatively large quantities and play principal roles in cell structure and metabolism. Examples of macronutrients are proteins, carbohydrates, and other molecules that contain carbon, hydrogen, and oxygen. Micronutrients, or **trace elements,** such as manganese, zinc, and nickel are present in much smaller amounts and are involved in enzyme function and maintenance of protein structure. What constitutes a micronutrient can vary from one microbe to another and often must be determined in the laboratory. This determination is made by deliberately omitting the substance in question from a growth medium to see if the microbe can grow in its absence.

Another way to categorize nutrients is according to their carbon content. An **inorganic nutrient** is an atom or simple molecule that contains a combination of atoms other than carbon and hydrogen. The natural reservoirs of inorganic compounds are mineral deposits in the crust of the earth, bodies of water, and the atmosphere. Examples include metals and their salts (magnesium sulfate, ferric nitrate), gases (oxygen, carbon dioxide), and water (table 7.2). The molecules of **organic nutrients** contain carbon and hydrogen atoms and are usually the products of living things. They range from the simplest organic molecule, methane (CH_4), to large polymers (carbohydrates, lipids, proteins, and nucleic acids). The source of nutrients is extremely varied: Some microbes (sulfur-reducing bacteria) obtain their nutrients strictly from inorganic sources, and others require a combination of organic and inorganic sources (see table 7.1). Parasites capable of invading and living on the human body derive all essential nutrients from host tissues, tissue fluids, secretions, and wastes.

* **nutrition;** L. *nutrire,* nourishment.

TABLE 7.1

NUTRIENT REQUIREMENTS AND SOURCES FOR TWO ECOLOGICALLY DIFFERENT BACTERIAL SPECIES

	Thiobacillus thioxidans	*Mycobacterium tuberculosis*
Habitat	Sulfur springs	Human lung
Nutritional Type	Chemoautotroph (sulfur-oxidizing bacteria)	Chemoheterotroph (parasite; cause of tuberculosis in humans)
Essential Element	**Provided Principally by**	
Carbon	Carbon dioxide (CO_2) in air	Minimum of L-glutamic acid (an amino acid), glucose (a simple sugar), albumin (blood protein), oleic acid (a fatty acid), and citrate
Hydrogen	H_2O, H^+ ions	Nutrients listed above, H_2O
Nitrogen	Ammonium (NH_4)	Ammonium, L-glutamic acid
Oxygen	Phosphate (PO_4), sulfate (SO_4), O_2	Oxygen gas (O_2) in atmosphere
Phosphorus	PO_4	PO_4
Sulfur	Elemental sulfur (S), SO_4	SO_4
Miscellaneous Nutrients		
Vitamins	None required	Pyridoxine, biotin
Inorganic salts	Potassium, calcium, iron, chloride	Sodium, chloride, iron, zinc, calcium, copper, magnesium
Principal energy source	Oxidation of sulfur (inorganic)	Oxidation of glucose (organic)

CHEMICAL ANALYSIS OF MICROBIAL CYTOPLASM

Examining the chemical composition of a bacterial cell can indicate its nutritional requirements. Table 7.3 lists the major contents of the intestinal bacterium *Escherichia coli.* Some of these components are absorbed in a ready-to-use form, and others must be synthesized by the cell from simple nutrients. Several important features of cell composition as it is related to nutrition can be summarized:

- Water content is the highest of all the components (70%).
- Proteins are the next most prevalent chemical.
- About 97% of the dry cell weight is composed of organic compounds.

MICROFILE 7.2 DINING WITH AN AMEBA

An ameba gorging itself on bacteria could be compared to a person eating a bowl of vegetable soup, because its nutrient needs and uses are fundamentally similar to a human's. Most food is a complex substance that contains many different types of nutrients. Some smaller molecules such as sugars can be absorbed directly by the cell; larger food debris and molecules must first be ingested and broken down into a size that can be absorbed. As nutrients are taken in, they add to a dynamic pool of inorganic and organic compounds dissolved in the cytoplasm. This pool will provide raw materials to be assimilated into the organism's own specialized proteins, carbohydrates, lipids, and other macromolecules used in growth and metabolism.

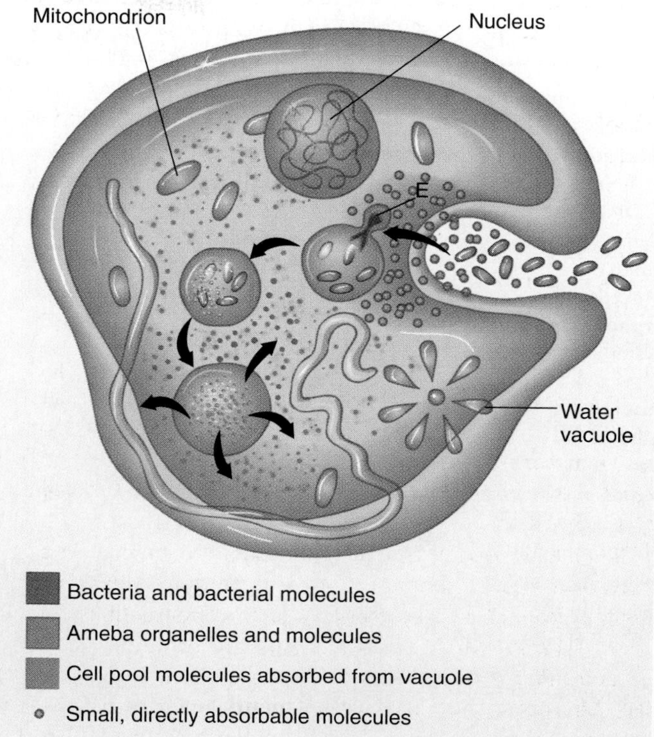

Steps in obtaining and incorporating nutrients. Food particles are surrounded and enclosed by pseudopods into a vacuole that fuses with a lysozyme containing digestive enzymes (E). Smaller subunits of digested macromolecules are transported out of the vacuole into the cell pool and are used in the metabolic and synthetic activities of the cell.

TABLE 7.2

PRINCIPAL INORGANIC RESERVOIRS OF ELEMENTS

Element	Inorganic Environmental Reservoir
Carbon	CO_2 in air; CO_3^{-2} in rocks and sediments
Oxygen	O_2 in air, certain oxides, water
Nitrogen	N_2 in air; NO_3^-, NO_2^-, NH_4^+ in soil and water
Hydrogen	Water, H_2 gas, mineral deposits
Phosphorus	Mineral deposits (PO_4^{-3}, H_3PO_4)
Sulfur	Mineral deposits, volcanic sediments (SO_4^{-2}, H_2S, S)
Potassium	Mineral deposits, the ocean (KCl, K_3PO_4)
Sodium	Mineral deposits, the ocean (NaCl, NaSi)
Calcium	Mineral deposits, the ocean ($CaCO_3$, $CaCl_2$)
Magnesium	Mineral deposits, geologic sediments ($MgSO_4$)
Chloride	The ocean (NaCl, NH_4Cl)
Iron	Mineral deposits, geologic sediments ($FeSO_4$)
Manganese, molybdenum, cobalt, nickel, zinc, copper, other micronutrients	Various geologic sediments

- About 96% of the cell is composed of six elements (represented by CHNOPS).
- Bioelements are needed in the overall scheme of cell growth, but most of them are available to the cell as compounds and not as pure elements (table 7.3).
- A cell as "simple" as *E. coli* contains on the order of 5,000 different compounds, yet it needs to absorb only a few types of nutrients to synthesize this great diversity. These include $(NH_4)_2SO_4$, $FeCl_2$, NaCl, trace elements, glucose, KH_2PO_4, $MgSO_4$, $CaHPO_4$, and water.

SOURCES OF ESSENTIAL NUTRIENTS

The elements that make up nutrients ultimately exist in an environmental inorganic reservoir of some type. These reservoirs serve not only as a permanent, longtime source of these elements but also can be replenished by the activities of organisms. In fact, as we shall see in chapter 26, the framework of microbial nutrition underlies the nutrient cycles of all life on the earth.

For convenience, the following section on nutrients is organized by element. You will no doubt notice that some categories overlap and that many of the compounds furnish more than one element.

TABLE 7.3

ANALYSIS OF THE CHEMICAL COMPOSITION OF AN *ESCHERICHIA COLI* CELL

	% Total Weight	% Dry Weight
Organic Compounds		
Proteins	15	50
Nucleic acids		
RNA	6	20
DNA	1	3
Carbohydrates	3	10
Lipids	2	Not determined
Miscellaneous	2	Not determined
Inorganic Compounds		
Water	70	
All others	1	3
Elements		
Carbon (C)		50
Oxygen (O)		20
Nitrogen (N)		14
Hydrogen (H)		8
Phosphorus (P)		3
Sulfur (S)		1
Potassium (K)		1
Sodium (Na)		1
Calcium (Ca)		0.5
Magnesium (Mg)		0.5
Chloride (Cl)		0.5
Iron (Fe)		0.2
Manganese (Mn), zinc (Zn), molybdenum (Mo), copper (Cu), cobalt (Co), zinc (Zn)		0.3

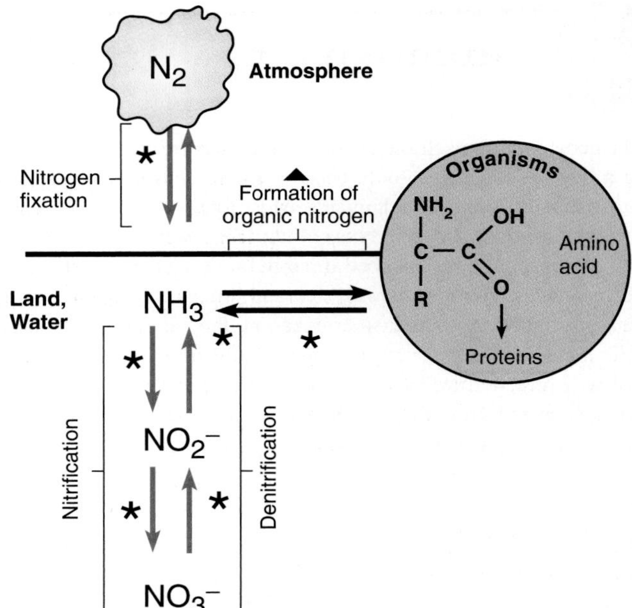

Figure 7.1
A simplified scheme of nitrogen compounds and their uses by the biological community.

Carbon Sources

It seems worthwhile to emphasize a point about the *extracellular source* of carbon as opposed to the *intracellular function* of carbon compounds. Although a distinction is made between the type of carbon compound cells absorb as nutrients (inorganic or organic), the majority of carbon compounds involved in the normal structure and metabolism of all cells are organic.

Microorganisms called **heterotrophs*** obtain carbon principally in the form of organic matter from the bodies of other organisms and are consequently dependent on other life forms. Among the common organic molecules that can satisfy this requirement are proteins, carbohydrates, lipids, and nucleic acids. In most cases, these nutrients provide several other elements as well. Some organic nutrients available to heterotrophs already exist in a form that is simple enough for absorption (for example, monosaccharides and amino acids), but many larger molecules must be digested by the cell before absorption. Moreover, heterotrophs vary in their capacities to use various organic carbon sources. Some are restricted to a few organic substrates, whereas others (certain *Pseudomonas* bacteria, for example) are so versatile that they can metabolize more than 100 different organic substrates.

The sole carbon source of microorganisms called **autotrophs*** is carbon dioxide (CO_2), an inorganic gas that currently makes up about 0.034% of the earth's atmosphere. Because autotrophs have the special capacity to convert CO_2 into organic compounds, they are not as nutritionally dependent on other living things. In a later section, we enlarge on the topic of nutritional types as based on carbon and energy sources.

Nitrogen Sources

The main reservoir of nitrogen is nitrogen gas (N_2), which makes up about 79% of the earth's atmosphere. This element is indispensable to the structure of proteins, DNA, RNA, and ATP. Such nitrogenous compounds are the primary source for heterotrophs, but to be useful, they must first be degraded into their basic building blocks (proteins into amino acids; nucleic acids into nucleotides). Some bacteria and algae utilize inorganic nitrogenous nutrients (NO_3^-, NO_2^-, or NH_3). A small number of bacteria can transform N_2 into compounds usable by living things through the process of nitrogen fixation (figure 7.1; also see chapter 26). Regardless of the initial form in which the inorganic nitrogen enters the cell, it must

* heterotroph (het´uhr-oh-trohf) Gr. *hetero,* other, and *troph,* to feed. See microfile 7.4.

* autotroph (aw´-toh-trohf) Gr. *auto,* self, and *troph,* to feed.

MICROFILE 7.3 METAL IONS AND INFECTIOUS DISEASE

A link exists between the course of an infection and the presence of certain metal ions. The ion best known for this effect is iron (Fe^{+2}). Several fascinating studies have revealed that the balance of iron in a mammalian host's body can actually stimulate or inhibit infections. When the iron concentration is increased, faster growth has been observed in bacteria that cause gonorrhea, meningitis, enteric fevers, and gas gangrene. It has been suggested that some hosts have defense mechanisms for depriving an infectious agent of iron as a means of slowing the microbe's growth.

A low concentration of iron in the host has also been found to tip the balance in favor of the microbe, thus promoting infection. For example, the diphtheria bacillus produces its toxin in response to a low-iron environment, and the agent of bubonic plague (*Yersinia pestis*) is more virulent when iron-starved.

Other metals have also been implicated in infectious processes. Toxic shock syndrome is most common in young women who use highly absorbant tampons. Evidently, when these tampons absorb large amounts of magnesium from the vaginal mucus, resident *Staphylococcus aureus* bacteria respond to the lower amount of this ion by producing a toxin with very destructive effects.

first be converted to NH_3, the only form that can be directly combined with carbon to synthesize amino acids and other compounds.

Oxygen Sources

Because oxygen is a major component of organic compounds such as carbohydrates, lipids, and proteins, it plays an important role in the structural and enzymatic functions of the cell. Oxygen is likewise a common component of inorganic salts such as sulfates, phosphates, nitrates, and water. Free gaseous oxygen (O_2) makes up 20% of the atmosphere. It is absolutely essential to the metabolism of many organisms, as we shall see later in this chapter and in chapter 8.

Hydrogen Sources

Hydrogen is a major element in all organic compounds and several inorganic ones, including water (H_2O), salts ($Ca(OH)_2$), and certain naturally occurring gases (H_2S, CH_4, and H_2). These gases are both used and produced by microbes. Hydrogen performs the following overlapping roles in the biochemistry of cells: (1) maintaining pH, (2) forming hydrogen bonds between molecules, and (3) serving as the source of free energy in oxidation-reduction reactions of respiration (see chapter 8).

Phosphorus (Phosphate) Sources

The main inorganic source of phosphorus is phosphate (PO_4^{-3}), derived from phosphoric acid (H_3PO_4) and found in rocks and oceanic mineral deposits. Phosphate is a key component of nucleic acids and is thereby essential to the genetics of cells and viruses. Because it is found in the nucleotide ATP, it also serves in cellular energy transfers. Other phosphate-containing compounds are phospholipids in cell membranes and coenzymes such as NAD and NADP (see chapter 8). Phosphate can be so scarce in the environment that its rarity severely limits growth, which explains why bacteria such as *Corynebacterium* concentrate it in metachromatic granules.

Sulfur Sources

Sulfur is widely distributed throughout the environment in mineral form. Rocks and sediments (such as gypsum) can contain sulfate (SO_4^{-2}), sulfides (FeS), hydrogen sulfide gas (H_2S), and elemental sulfur (S). The amino acids cysteine (sis´-tee-een) and methionine (meth-eye´-oh-neen) and certain vitamins contain sulfur in their structure. The sulfur-containing amino acids can contribute to the structural stability and shape of proteins by forming unique linkages called disulfide bonds (see figure 2.22).

Other Nutrients Important in Microbial Metabolism

The remainder of important elements are mineral ions. **Potassium** is essential to protein synthesis and membrane function. **Sodium** is important for some types of cell transport. **Calcium** is a stabilizer of the cell wall and endospores of bacteria. It is also combined with carbonate (CO_3^{-2}) in the formation of shells by foraminiferans and radiolarians. **Magnesium** is a component of chlorophyll and a stabilizer of membranes and ribosomes. **Iron** is an important component of the cytochrome pigments of cell respiration. Zinc is an essential regulatory element for eukaryotic genetics. It is a major component of "zinc fingers"—binding factors that help enzymes adhere to specific sites on DNA. Copper, cobalt, nickel, molybdenum, manganese, silicon, iodine, and boron are needed in small amounts by some microbes but not others. A discovery with important medical implications is that metal ions can directly influence certain diseases by their effects on microorganisms (microfile 7.3).

Growth Factors: Essential Organic Nutrients

Few microbes are as versatile as *Escherichia coli* in assembling molecules from scratch. Many fastidious bacteria lack the genetic and metabolic mechanisms to synthesize every organic compound they need for survival. An organic compound such as an amino acid, nitrogen base, or vitamin that cannot be synthesized by an organism and must be provided as a nutrient is a **growth factor.** For example, although all cells require 20 different amino acids for proper assembly of proteins, many cells cannot synthesize them all. Those that must be obtained from food are called **essential amino acids.** A notable example of the need for growth factors occurs in *Haemophilus influenzae,* a bacterium that causes meningitis and respiratory infections in humans. It can grow only when hemin (factor X), NAD (factor V), thiamine and pantothenic acid (vitamins), uracil, and cysteine are provided by another organism or a growth medium.

MICROFILE 7.4 THE LANGUAGE OF MICROBIAL ADAPTATIONS

Although initially you may feel inundated with new words, there is really no mystery to the terminology of microbial adaptations. One can easily decipher the ecological or nutritional description of a microbe by knowing a few prefixes and suffixes. In terms of nutrition, the prefixes *auto-* (self), *hetero-* (other), *photo-* (light), and *chemo-* (chemical) and the suffix *-troph* (to feed) can be put together in various ways to succinctly describe a microbe's carbon and energy sources. For example, an alga is a *photoautotroph* because it "feeds itself using the sun"; a parasitic worm is a *chemoheterotroph* that "feeds on the chemicals of others." One can describe other nutritional patterns by combining *sapro-* (rotten) with *-troph* and *halo-* (salt) with *-phile* (to love).

Other useful roots and words for characterizing adaptations are *thermo-* (heat), *psychro-* (cold), *meso-* (intermediate), *aero-* (air), *anaero-* (no air), or *baro-* (pressure) combined with suffixes *-phile* or *-obe* (living). Thus, a *thermophile* "loves heat" and an *aerobe* "lives in air."

Certain modifying terms help to express how microbes deal with nutritional or ecological changes. Microbes that are *obligate* or *strict* are restricted to a narrow range of growth conditions or life-style. Thus, viruses are *obligate* intracellular parasites, and humans (and most animals) are *strict* aerobes; that is, they cannot survive without air. Other microorganisms are less restricted and can adjust to a wider range of growth conditions. An all-around adjective for describing such a microbe is *facultative.** One must be careful in interpreting this term; it should be used in the sense of "can adapt to." For example, a *facultative halophile* can adapt to higher than normal salt concentrations, but it does not normally require salt or live in a high-salt habitat. Likewise, a *facultative anaerobe* can metabolize and grow without air, but it is more efficient if air is present. The roots *-tolerant* and *-duric* denote survival but not growth under certain conditions. A *thermoduric* bacterium can survive heat for a short period, but it does not ordinarily grow at hot temperatures.

* **facultative** (fak´-uhl-tay-tiv) Gr. *facult,* capability or skill, and *tatos,* most.

HOW MICROBES FEED: NUTRITIONAL TYPES

The earth's limitless habitats and microbial adaptations are rivaled by an elaborate menu of nutritional schemes. Fortunately, most organisms show consistent trends and can be described by a few general categories (table 7.4) and a few selected terms (microfile 7.4). The main determinants of a microbe's nutritional type are its sources of carbon and energy. In a previous section, microbes were defined as autotrophs, whose primary carbon source is inorganic carbon (CO_2), and heterotrophs, which are dependent on organic carbon compounds. In terms of energy source, microbes that photosynthesize are generally classified as **phototrophs,** and those that oxidize chemical compounds are **chemotrophs.** The terms for carbon and energy source are often merged into a single word for convenience (table 7.4). The categories described here are meant to describe only the major nutritional groups and do not include unusual exceptions.

TABLE 7.4

NUTRITIONAL CATEGORIES OF MICROBES BY CARBON AND ENERGY SOURCE

Category	Carbon Source	Energy Source	Example
Autotroph	CO_2	Nonliving environment	
Photoautotroph	CO_2	Sunlight	Photosynthetic organisms, such as algae, plants, cyanobacteria
Chemoautotroph	CO_2	Simple inorganic chemicals	Only certain bacteria, such as methanogens, vent bacteria
Heterotroph	Organic	Other organisms or sunlight	
Photoheterotroph	Organic	Sunlight	Nonsulfur bacteria
Chemoheterotroph	Organic	Metabolic conversion of the nutrients from other organisms	Protozoa, fungi, many bacteria, animals
Saprobe	Organic	Metabolizing the organic matter of dead organisms	Fungi, bacteria (decomposers)
Parasite	Organic	Utilizing the tissues, fluids of a live host	Various parasites and pathogens; can be bacteria, fungi, protozoa, animals

Autotrophs and Their Energy Sources

Autotrophs derive energy from one of two possible nonliving sources: sunlight (photoautotrophs) and chemical reactions involving simple inorganic chemicals (chemoautotrophs). **Photoautotrophs** are photosynthetic; that is, they capture the energy of light rays and transform it into chemical energy that can be used in cell metabolism (microfile 7.5). Because photosynthetic organisms (algae, plants, some bacteria) produce organic molecules that can be used by themselves and heterotrophs, they form the basis for most food webs. Their role as primary producers of organic matter is discussed in chapter 26.

A significant type of bacteria called **chemoautotrophs** have an unusual nutritional adaptation that requires neither sunlight nor organic nutrients. Some microbiologists prefer to call them **lithoautotrophs** (rock feeders) in reference to their total reliance

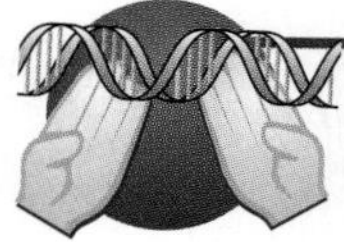

MICROFILE 7.5 SUNLIGHT-DRIVEN ORGANIC SYNTHESIS

Two equations sum up the reactions of photosynthesis in a simple way. The first equation shows a reaction that results in the production of oxygen:

$$CO_2 + H_2O \xrightarrow[\text{by chlorophyll}]{\text{Sunlight absorbed}} (CH_2O)_n{}^* + O_2$$

This oxygenic (oxygen-producing) type of photosynthesis occurs in plants, algae, and cyanobacteria (see figure 4.34). The function of chlorophyll is to capture light energy. Carbohydrates produced by the reaction can be used by the cell to synthesize other cell components, and they also become a significant nutrient for heterotrophs that feed on them. The production of oxygen is vital to maintaining this gas in the atmosphere.

A second equation shows a photosynthetic reaction that does not result in the production of oxygen:

$$CO_2 + H_2S \xrightarrow[\text{by bacteriochlorophyll}]{\text{Sunlight absorbed}} (CH_2O)^n + S + H_2O$$

This anoxygenic (no oxygen produced) type of photosynthesis is found in bacteria such as purple and green sulfur bacteria. Note that the type of chlorophyll (bacteriochlorophyll, a substance unique to these microbes), one of the reactants (hydrogen sulfide gas), and one product (elemental sulfur) are different from those in the first equation. These bacteria live in the anaerobic regions of aquatic habitats.

*$(CH_2O)_n$ is shorthand for a carbohydrate.

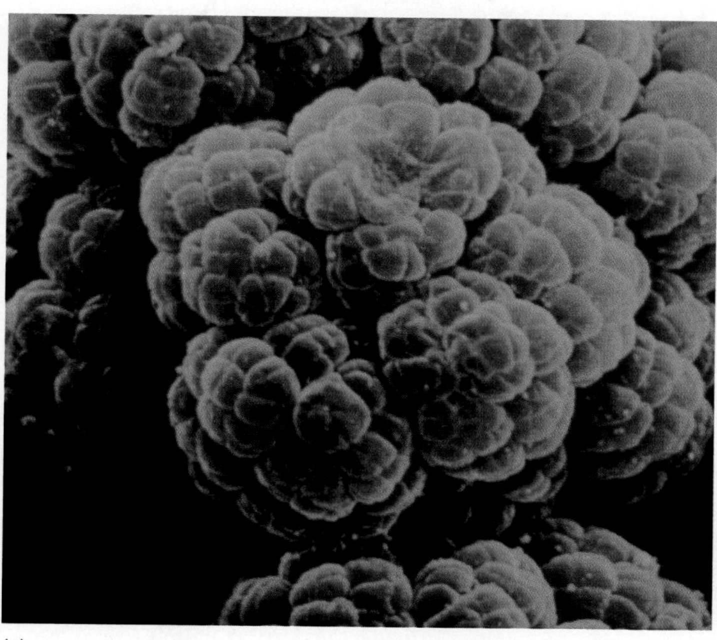

(a)

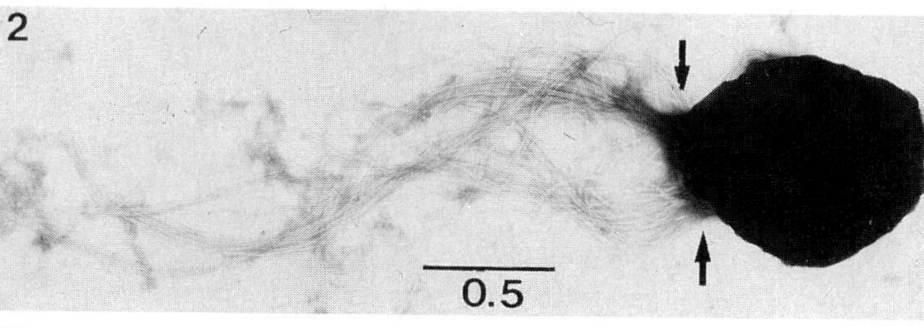

(b)

Figure 7.2
Methane-producing archaea. Members of this group are primitive procaryotes with unusual cell walls and membranes. (*a*) A small colony of *Methanosarcina*. (*b*) *Methanococcus junnuschii,* a motile archaea that inhabits hot vents in the seafloor and uses hydrogen. The genome of this procaryote was recently deciphered and found to contain unique genetic sequences.

on inorganic minerals. These bacteria derive energy in diverse and rather amazing ways. In very simple terms, they remove electrons from inorganic substrates such as hydrogen gas, hydrogen sulfide, sulfur, or iron and combine them with carbon dioxide and hydrogen. This reaction provides simple organic molecules and a modest amount of energy to drive the synthetic processes of the cell. Chemoautotrophic bacteria play an important part in recycling inorganic nutrients. For an example of chemoautotrophy and its importance to deep-sea communities, see microfile 7.8.

An interesting group of chemoautotrophs are *methanogens,** which produce methane (CH_4) from hydrogen gas and carbon dioxide (figure 7.2):

$$4H_2 + CO_2 \rightarrow CH_4 + 2H_2O$$

Methane, sometimes called "swamp gas," is formed in anaerobic, hydrogen-containing microenvironments of soil, swamps, mud, and even the intestines of some animals. Many methanogens are archaea that live in extreme habitats such as ocean vents and hot springs, up to 125°C. Methane can be harvested and used as an inexpensive energy source in certain industries. Biogas generators are devices primed with a mixed population of microbes (including methanogens) and fueled with various waste materials that can supply enough methane to drive a steam generator. Methane also plays a role as one of the greenhouse gases that is currently an environmental concern (see chapter 26).

Heterotrophs and Their Energy Sources

The majority of heterotrophic microorganisms are **chemoheterotrophs** that derive both carbon and energy from organic compounds. Processing these organic molecules by respiration or fermentation releases energy in the form of ATP. An example of chemoheterotrophy is *aerobic respiration,* the principal energy-

* **methanogen** (meth´-an-oh-gen) From *methane,* a colorless, odorless gas, and *gennan,* to produce.

yielding reaction in all animals, most protozoa and fungi, and many bacteria. It can be simply represented by the equation:

$$\text{Glucose or other monosaccharide } [(CH_2O)_n] + O_2 \rightarrow CO_2 + H_2O + \text{Energy stored as ATP}$$

You might notice that this reaction is complementary to photosynthesis. Here, glucose and oxygen are reactants, and carbon dioxide is given off. Indeed, the earth's balance of both energy and metabolic gases is greatly dependent on this relationship. Chemoheterotrophic microorganisms belong to one of two main categories that differ in how they obtain their organic nutrients: **Saprobes**[1] are free-living microorganisms that feed primarily on organic detritus from dead organisms, and **parasites** ordinarily derive nutrients from the cells or tissues of a host. Most microbes of biomedical importance belong to these two categories.

Saprobic Microorganisms Saprobes occupy a niche as decomposers of plant litter, animal matter, and dead microbes. If not for the work of decomposers, the earth would gradually fill up with organic material, and the nutrients they contain would not be recycled. Although some saprobes can phagocytose detritus particles and digest them intracellularly, saprobes with rigid cell walls (fungi and bacteria, which make up the majority) are incapable of engulfing large food particles. To compensate, they release enzymes to the extracellular environment and digest the food particles into smaller molecules that can pass freely into the cell (figure 7.3). Saprobes exist in great array. *Obligate saprobes* exist strictly on organic matter in soil and water and are unable to adapt to the body of a live host. This group includes many free-living protozoa, fungi, and budding and gliding bacteria. Apparently, there are fewer of these strict species than was once thought, and many supposedly nonpathogenic saprobes can infect a susceptible host. When a saprobe infects a host, it is considered a *facultative parasite.* Because infection by a saprobe usually occurs when the host is compromised, we call the microbe an *opportunistic pathogen.* For example, although its natural habitat is soil and water, *Pseudomonas aeruginosa* frequently causes infections in hospital patients. The yeast *Cryptococcus neoformans* causes a severe lung and brain infection in AIDS patients, yet its natural habitat is the soil.

Parasitic Microorganisms Parasites live in or on the body of a host, which they usually harm to some degree. Parasites range from viruses to helminth worms, and they can live on the body (ectoparasites), in the organs and tissues (endoparasites), or even within cells (intracellular parasites, the most extreme type). Although there are several degrees of parasitism, the more successful parasites generally have no fatal effects and eventually evolve to a less harmful relationship with their host.

Parasites inclined to cause damage to tissues (disease) or even death are called **pathogens.** In general, relatively new parasites (such as HIV, the virus that causes AIDS) are more pathogenic and more severe in their effects, whereas older pathogens give rise to milder diseases that mainly cause discomfort (a cold

[1] Synonyms are *saprotroph* and *saprophyte.* We prefer to use the terms *saprobe* and *saprotroph* because they are more consistent with other terminology and because the term *saprophyte* is a holdover from the time when bacteria and fungi were considered plants.

Digestion in Bacteria and Fungi

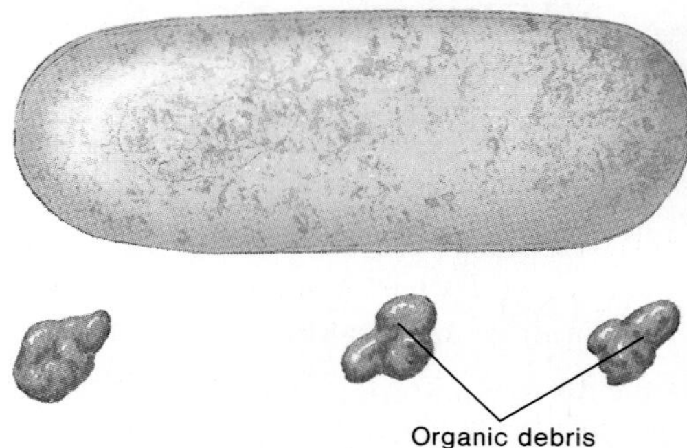
Organic debris
(a) Walled cell is inflexible.

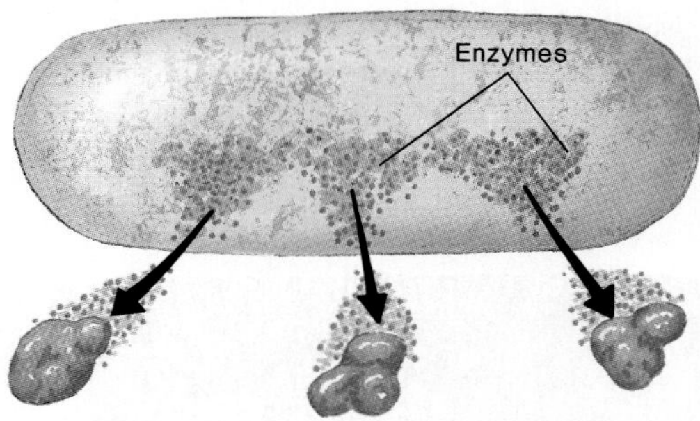
Enzymes
(b) Enzymes are transported across the wall.

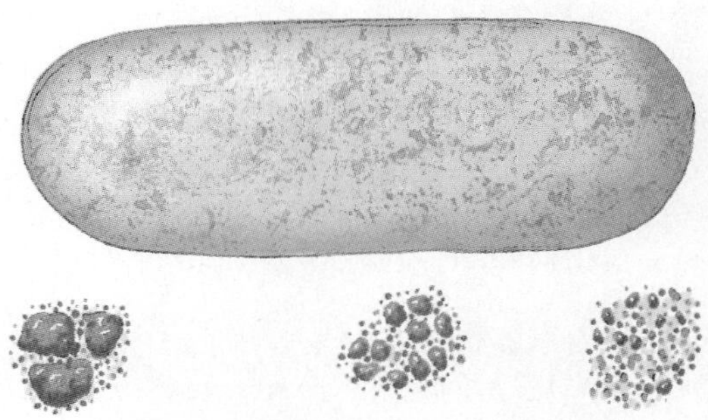
(c) Enzymes hydrolyze the bonds on nutrients.

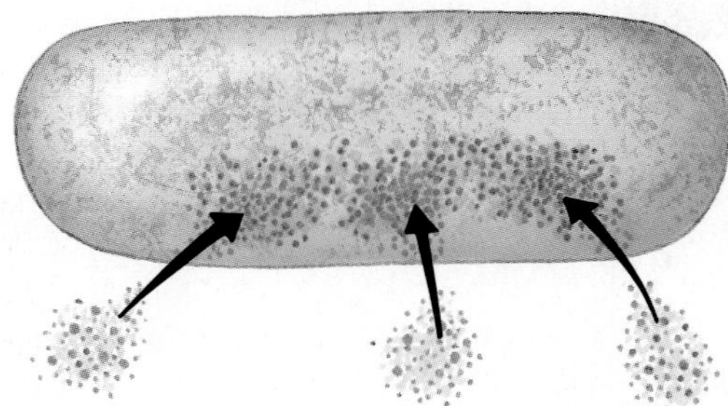
(d) Smaller molecules are transported into the cytoplasm.

Figure 7.3
Extracellular digestion in a saprobe with a cell wall (bacterium or fungus). (*a*) A walled cell is inflexible and cannot engulf large pieces of organic debris. (*b*) In response to a usable substrate, the cell synthesizes enzymes that are transported across the wall into the extracellular environment. (*c*) The enzymes hydrolyze the bonds in the debris molecules. (*d*) Digestion produces molecules small enough to be transported into the cytoplasm.

virus). *Obligate parasites* (for example, the leprosy bacillus and the syphilis spirochete) are unable to grow outside of a living host. Parasites that are less strict can be cultured artificially if provided with the correct nutrients and environmental conditions. Bacteria such as *Streptococcus pyogenes* (the cause of strep throat) and *Staphylococcus aureus* can grow on artificial media as saprobes.

Obligate intracellular parasitism is an extreme but relatively common mode of life. Microorganisms that spend all or part of their life cycle inside a host cell include the viruses, a few bacteria (rickettsias, chlamydias), and certain protozoa (apicomplexa). Contrary to what one might think, the inside of a cell is not completely without hazards, and microbes must overcome some difficult challenges. They must find a way into the cell, keep from being destroyed, not destroy the host cell too soon, multiply, and find a way to infect other cells. Intracellular parasites obtain different substances from the host cell, depending on the group. Viruses are the most extreme, parasitizing the host's genetic and metabolic machinery. Rickettsias are primarily energy parasites; chlamydia are metabolic and nutrient parasites; and the malaria protozoan is a hemoglobin parasite.

Figure 7.4

Simplified analogies comparing the three major types of transport. In passive transport, nutrients exist in a gradient from a high concentration outside the cell to a low concentration inside the cell. The molecules naturally diffuse into the cell without its having to expend energy. In active transport, the cell actively picks up nutrients from a solution in which the nutrients are not in a gradient (energy expenditure is required). In bulk transport, large solids or masses of liquids enter the cell intact by engulfment. The cell actively works during this process.

How Microbes Take In Molecules

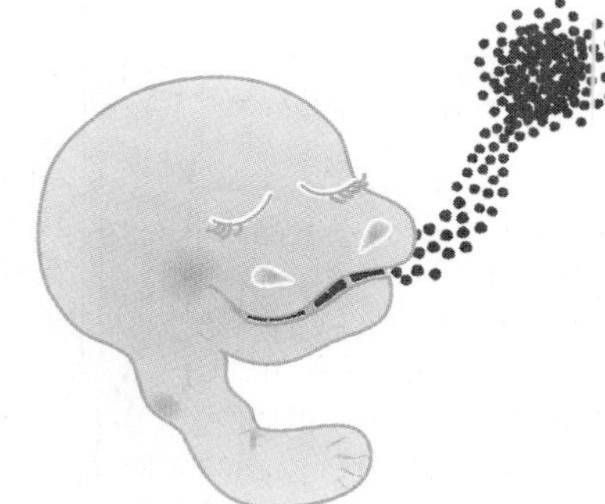

Passive Transport

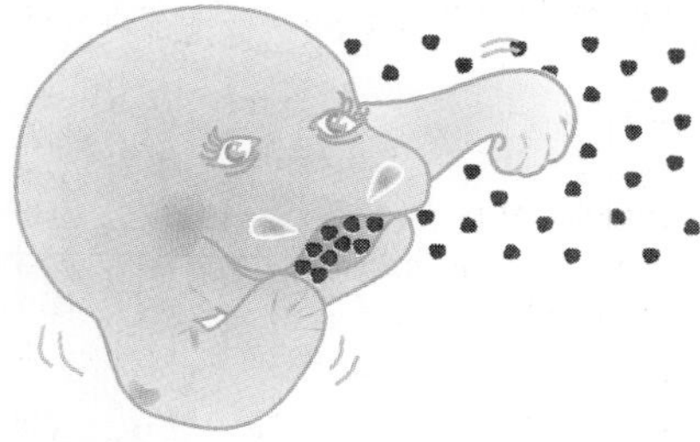

Active Transport

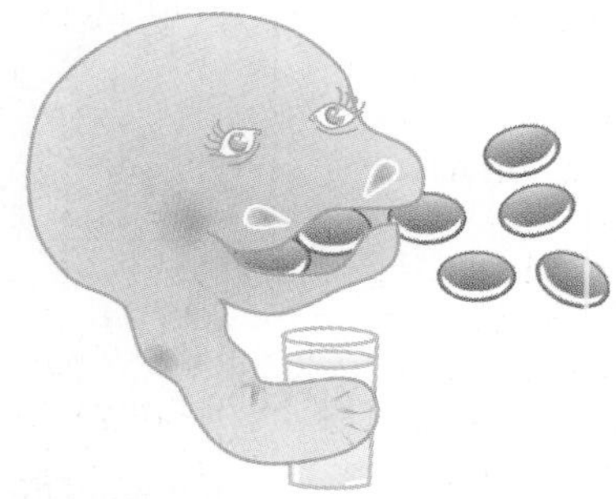

Bulk Transport

TRANSPORT MECHANISMS FOR NUTRIENT ABSORPTION

A microorganism's habitat provides necessary nutrients—some abundant, others scarce—that must still be taken into the cell. Survival also requires that cells transport waste materials into the environment. Whatever the direction, transport occurs across the cell membrane, the structure specialized for this role. This is true even in organisms with cell walls (bacteria, algae, and fungi), because the cell wall is usually too nonselective to screen the entrance or exit of molecules. Three general types of transport are **passive transport,** which follows physical laws that are not unique to living systems and do not generally require direct energy input from the cell; **active transport,** which requires carrier proteins in the membranes of living cells and the expenditure of energy; and **bulk transport,** the movement of large masses of material across membranes (figure 7.4).

Passive Transport: Diffusion, Osmosis

The driving force of passive transport is atomic and molecular movement—the natural tendency of atoms and molecules to be in constant random motion. The existence of this motion is evident in Brownian movement (see chapter 4) of small particles suspended in liquid. It can be demonstrated by a variety of simple observations. A drop of perfume released into one part of a room is soon smelled in another part, or a lump of sugar in a cup of tea spreads through the whole cup without stirring. This phenomenon of molecular movement, in which atoms or molecules move in a gradient from an area of higher density or concentration to an area of lesser density or concentration, is **diffusion*** (figure 7.5). Although passive transport requires molecules to be arranged in a gradient, diffusion remains an important way for cells to obtain freely diffusible materials (oxygen, carbon dioxide, and water) and to release wastes to the environment.

Diffusion of water through a selectively permeable membrane, a process called **osmosis,*** is also a physical phenomenon that is easily demonstrated in the laboratory with nonliving materials. It provides a model of how cells deal with various solute concentrations in aqueous solutions (figure 7.6). In an osmotic system, the membrane is *selectively,* or *differentially, permeable,* having passageways that allow free diffusion of water but can block certain other dissolved molecules. When this membrane is placed between solutions of differing concentrations and the solute is not diffusible (protein, for example), then under the laws of diffusion, water will diffuse at a faster rate from the side that has more water to the side that has less water. As long as the concentrations of the solutions differ, one side will experience a net loss of water and the other a net gain of water, until equilibrium is reached and the rate of diffusion is equalized.

* diffusion (dih-few´-zhun) L. *dis,* apart, and *fundere,* to pour.

* osmosis (oz-moh´-sis) Gr. *osmos,* impulsion, and *osis,* a process.

How Molecules Diffuse In Aqueous Solutions

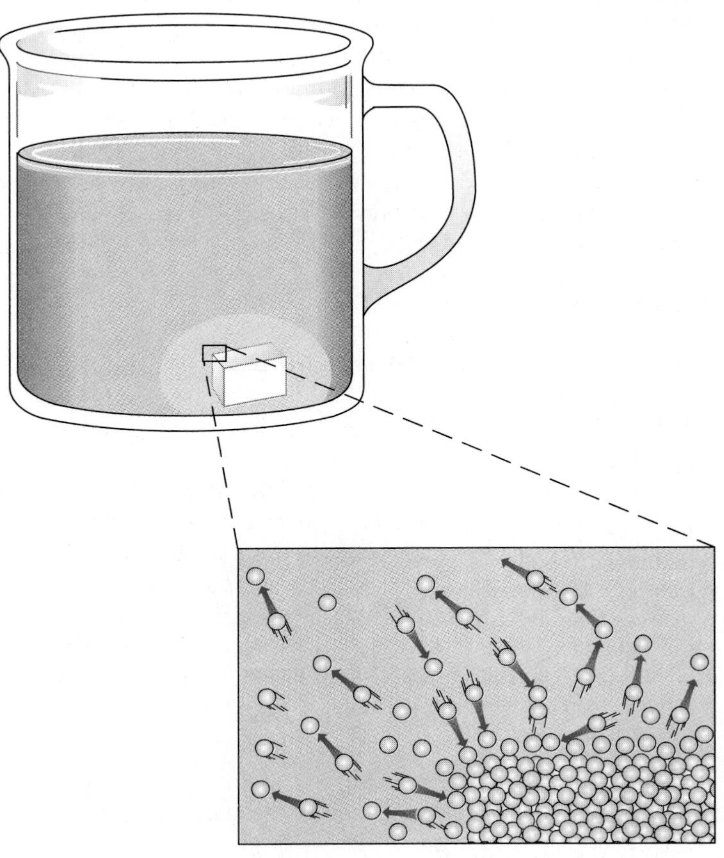

Figure 7.5
Diffusion of molecules in aqueous solutions. A high concentration of sugar exists in the cube at the bottom of the liquid. An imaginary molecular view of this area shows that sugar molecules are in a constant state of motion. Those at the edge of the cube diffuse from the concentrated area into more dilute regions. As diffusion continues, the sugar will spread evenly throughout the aqueous phase and eventually there will be no gradient. At that point, the system is said to be in equilibrium.

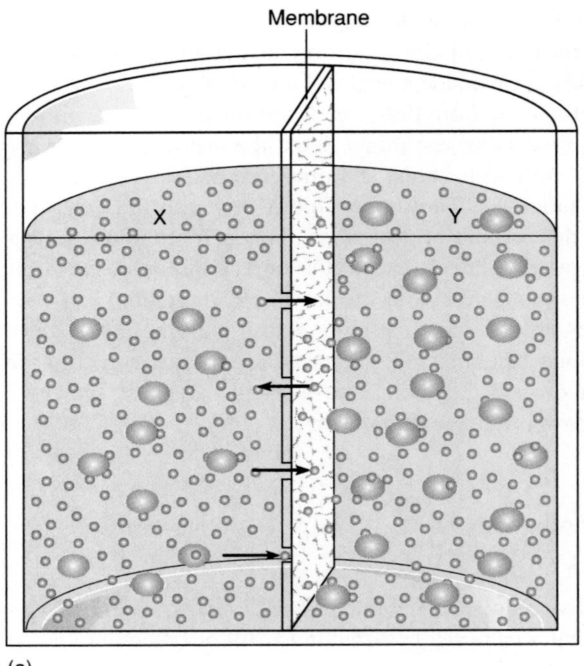

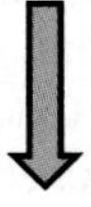

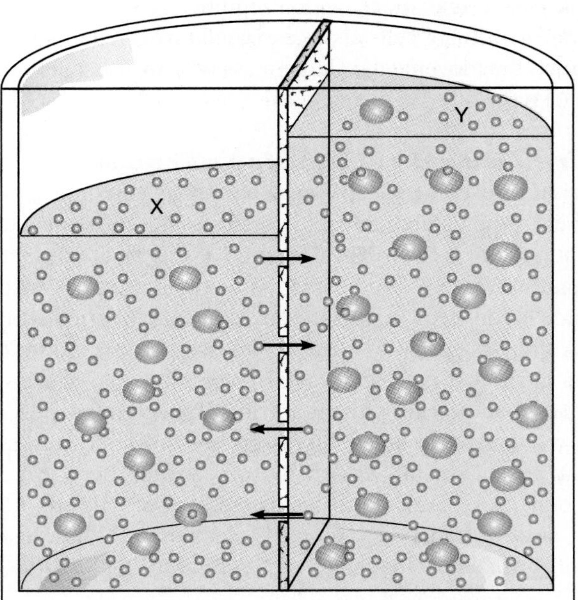

Figure 7.6
Osmosis, the diffusion of water through a selectively permeable membrane. (*a*) A membrane has pores that allow the ready passage of water but not large solute molecules from one side to another. Placement of this membrane between solutions of different solute concentrations (X = less concentrated and Y = more concentrated) results in a diffusion gradient for water. Water molecules undergo diffusion and move across the membrane pores in both directions. Because there is more water in solution X, the opportunity for a water molecule to successfully hit and go through a pore is greater for X than for Y. The result will be a net movement of water from X to Y. (*b*) The level of solution on the Y side rises as water continues to diffuse in. This process will continue until equilibration occurs and the rate of diffusion of water is equal on both sides.

Be aware of an important point about the dynamics of osmosis and the solutions involved. Whereas the primary considerations are the concentration and direction of movement of water (solvent), we also must know the solute concentration, because it will dictate how much water is present to diffuse. For example, a more concentrated 70% solution contains 70 parts of solute and 30 parts of water. By comparison, a less concentrated 5% solution contains 5 parts of solute and 95 parts of water. When these two solutions are placed on opposite sides of a membrane, water will diffuse at a faster rate from the 5% solution into the 70% one.

Osmosis in living systems is similar to the model shown in figure 7.6, largely because living membranes are selectively permeable. They generally block the entrance and exit of larger molecules and permit free diffusion of water. Because most cells are surrounded by some free water, the amount of water entering or leaving has a far-reaching impact on cellular activities and survival. This osmotic relationship between cells and their environment is determined by the relative concentrations of the solutions on either side of the cell membrane (table 7.5). Such systems can be compared using the terms isotonic, hypotonic, and hypertonic.

Under **isotonic*** conditions, the environment is equal in concentration to the cell's internal environment, and, because diffusion of water proceeds at the same rate in both directions, there is no net change in cell volume. Isotonic solutions are the most stable environments for cells, because they are already in osmotic equilibrium with the cell. Parasites living in host tissues are most likely to be living in isotonic habitats.

Under **hypotonic*** conditions, the solute concentration of the external environment is lower than that of the cell's internal environment. Pure water provides the most hypotonic environment for cells because it has no solute. The net direction of osmosis is from the hypotonic solution into the cell, and cells without walls swell and can burst.

Hypertonic* conditions are also out of balance with the tonicity of the cell's cytoplasm, but in this case, the environment has a higher solute concentration than the cytoplasm. Because a hypertonic environment will force water to diffuse out of a cell, it is said to have high *osmotic pressure* or potential. Hypertonic solutions are lethal to many microbes. Salt water and concentrated sugar solutions are examples of such solutions, and food preservation with salt and sugar takes advantage of their hypertonic effect.

Adaptations to Osmotic Variations in the Environment

Let us now see how specific microbes have adapted osmotically to their environments. In general, isotonic conditions pose little stress on cells, so survival depends on counteracting the adverse effects of hypertonic and hypotonic environments. A bacterium and an ameba living in fresh pond water are examples of cells that live in constantly hypotonic conditions. The rate of water diffusing across the cell membrane into the cytoplasm is rapid and constant, and the cells would die without compensatory mechanisms. The majority of bacterial cells compensate by having a cell wall that protects them from bursting even as the cytoplasm membrane becomes *turgid** with water. The ameba's adaptation is an anatomical and physiological one that requires the constant expenditure of energy. It has a water, or contractile, vacuole that siphons excess water back out into the habitat like a tiny pump (see figure 5.30). A microbe living in a high-salt environment (hypertonic) has the opposite problem and must either restrict its loss of water to the environment or increase the salinity of its internal environment. Halobacteria living in the Great Salt Lake and the Dead Sea actually absorb salt to make their cells isotonic with the environment, thus they have a physiological need for a high-salt concentration in their habitats.

A form of passive transport called **facilitated diffusion** (figure 7.7) also requires a concentration gradient of the transported molecule and does not expend energy, but it is more specific than other passive systems. The only molecules to be transported are those that can bind to specialized membrane proteins. Examples of facilitated diffusion include yeast transport of sugars, bacterial transport of glycerol, and the transport of calcium into endospores by spore-forming bacteria.

Active Transport: Bringing in Molecules Against a Gradient

Free-living microbes exist under relatively nutrient-starved conditions and cannot rely completely on slow and rather inefficient passive transport mechanisms. To ensure a constant supply of nutrients and other required substances, microbes must capture those that are in extremely short supply and actively transport them into the cell. Features inherent in **active transport systems** are (1) the transport of nutrients against the natural diffusion gradient or in the same direction as the natural gradient but at a rate faster than by diffusion alone, (2) the presence of specific membrane proteins (permeases and pumps; figure 7.8*a*), and (3) the expenditure of energy. Examples of substances transported actively are monosaccharides, amino acids, organic acids, phosphates, and metal ions. Some freshwater algae have such efficient active transport systems that an essential nutrient can be found in intracellular concentrations 200 times that of the habitat.

Active transport also occurs within intracellular membranes such as the mitochondrion, the chloroplast, and the endoplasmic reticulum. This behavior is particularly important in mitochondrial ATP formation (see the discussion of oxidative phosphorylation in chapter 8) and protein synthesis. One special type of active transport, **group translocation,** couples the transport of a nutrient with its conversion to a substance that is immediately useful inside the cell (figure 7.8*b*). This method is used by certain bacteria to

* **isotonic** (eye-soh-tahn′-ik) Gr. *iso,* same, and *tonos,* tension.
* **hypotonic** (hy-poh-tahn′-ik) Gr. *hypo,* under, and *tonos,* tension.
* **hypertonic** (hy-pur-tahn′-ik) Gr. *hyper,* above, and *tonos,* tension.

* **turgid** (ter′-jid) A condition of being swollen or congested.

TABLE 7.5

CELL RESPONSES TO SOLUTIONS OF DIFFERING OSMOTIC CONTENT

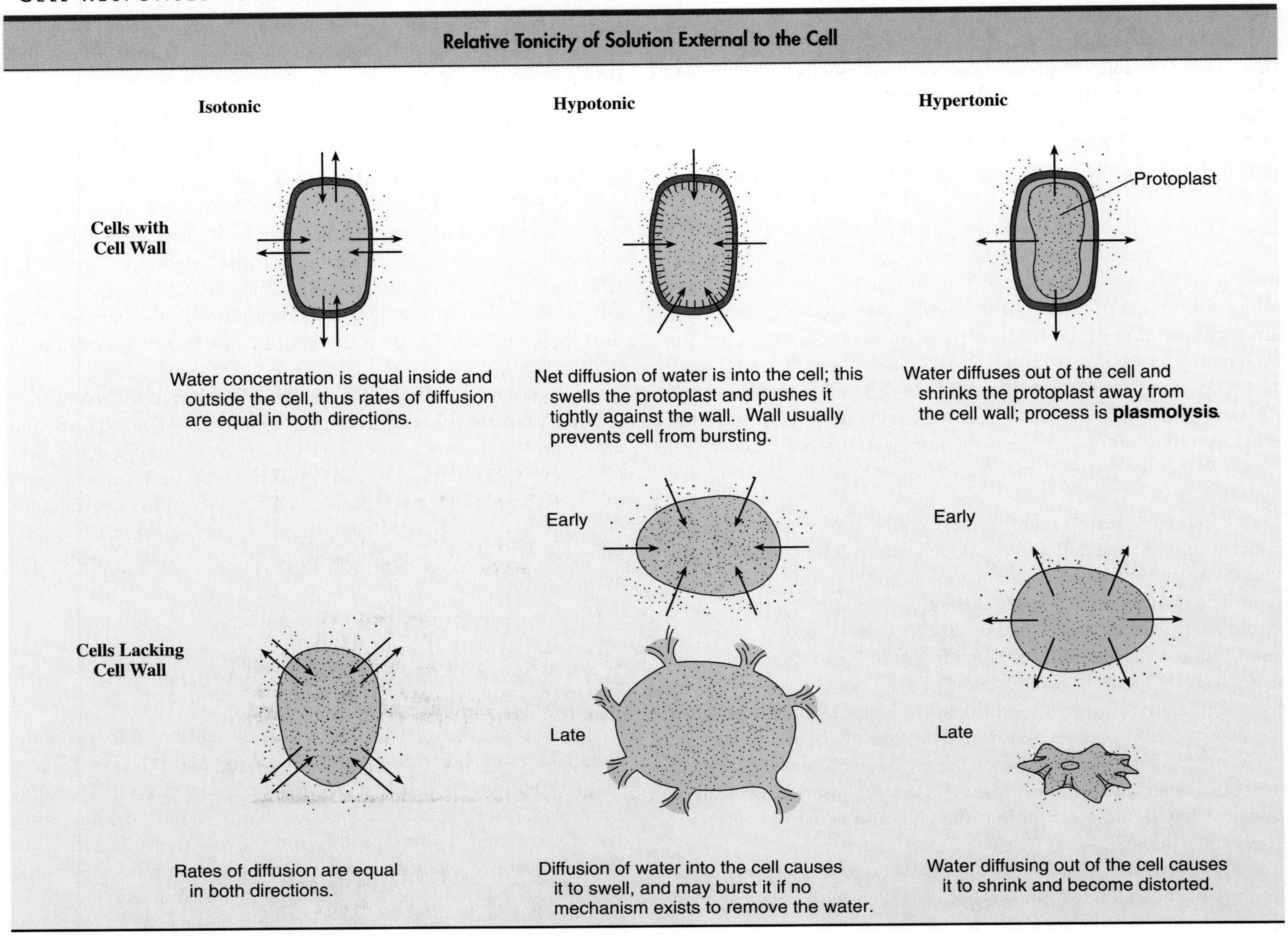

	Relative Tonicity of Solution External to the Cell		
	Isotonic	**Hypotonic**	**Hypertonic**
Cells with Cell Wall	Water concentration is equal inside and outside the cell, thus rates of diffusion are equal in both directions.	Net diffusion of water is into the cell; this swells the protoplast and pushes it tightly against the wall. Wall usually prevents cell from bursting.	Protoplast Water diffuses out of the cell and shrinks the protoplast away from the cell wall; process is **plasmolysis**.
Cells Lacking Cell Wall	Rates of diffusion are equal in both directions.	Early Late Diffusion of water into the cell causes it to swell, and may burst it if no mechanism exists to remove the water.	Early Late Water diffusing out of the cell causes it to shrink and become distorted.

→ Direction of net water movement.

transport sugars (glucose, fructose) while simultaneously adding molecules such as phosphate that prepare them for the next stage in metabolism.

Bulk Transport: Eating and Drinking by Cells

Some cells transport large molecules, particles, liquids, or even other cells across the cell membrane. Because the cell usually expends energy to carry out this transport, it is also a form of active transport. The substances transported do not pass physically through the membrane but are carried into the cell by **endocytosis.*** This involves the engulfment of the substance followed by vacuole or vesicle formation (figure 7.9). Amebas and certain white blood cells ingest large solid matter by **phagocytosis;*** liquids, such as oils or large molecules in solution, enter the cell through **pinocytosis.***

* endocytosis (en´´-doh-cy-toh´-sis) Gr. *endo,* in; *cyte,* cell; and *osis,* a process.
* phagocytosis (fag´´-oh-cy-toh´-sis) Gr. *phagein,* to eat.
* pinocytosis (pin´´-oh-cy-toh´-sis) Gr. *pino,* to drink.

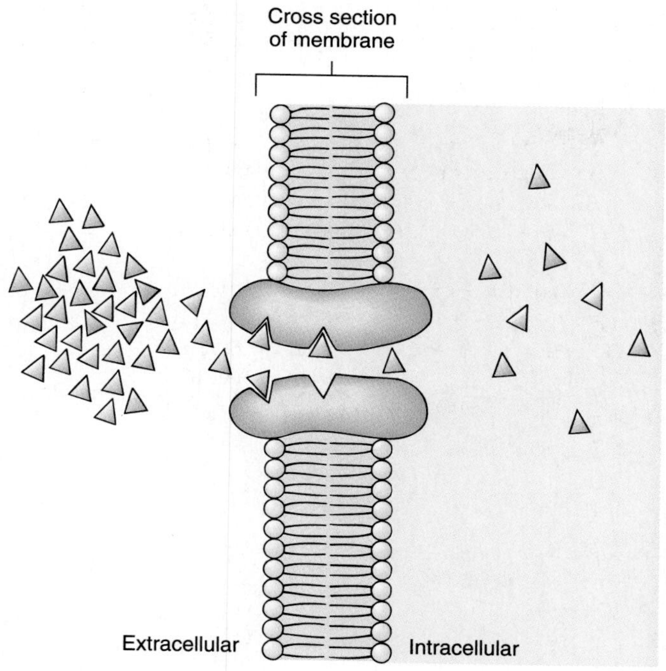

Figure 7.7

Facilitated diffusion involves the attachment of a molecule to a specific protein carrier. Bonding of the molecule causes a conformational change in the protein that facilitates the molecule's passage into the cell. Because molecules are carried to the cell by a diffusion gradient, this is considered passive transport.

Chapter Checkpoints

Nutrition is a process by which all living organisms obtain substances from their environment to convert to metabolic uses.

Although the chemical form of nutrients varies widely, all organisms require six bioelements: carbon, hydrogen, oxygen, nitrogen, phosphorus, and sulfur to survive, grow, and reproduce.

Nutrients are categorized by the amount required (macronutrients or micronutrients), by chemical structure (organic or inorganic), and by their importance to the organism's survival (essential or nonessential).

Microorganisms are classified both by the chemical form of their nutrients and the energy sources they utilize.

Nutrient requirements of microorganisms determine their respective niches in the food webs of major ecosystems.

Nutrients are transported into microorganisms by two kinds of processes: those requiring expenditure of energy and those that occur passively.

The molecular size and concentration of a nutrient determine which method of transport is used.

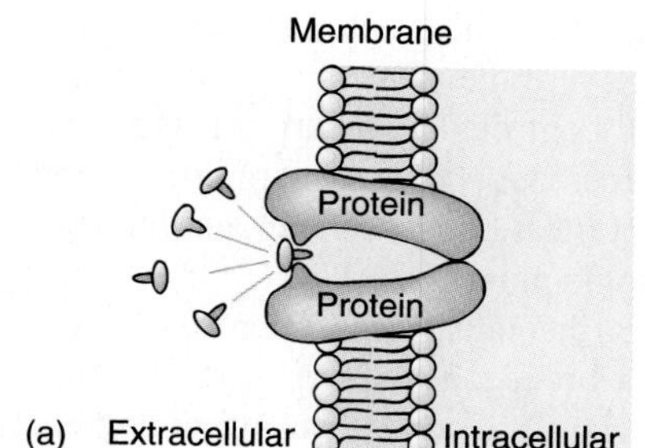

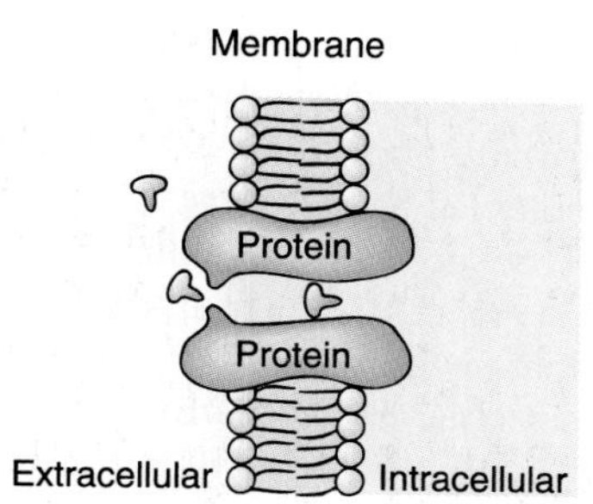

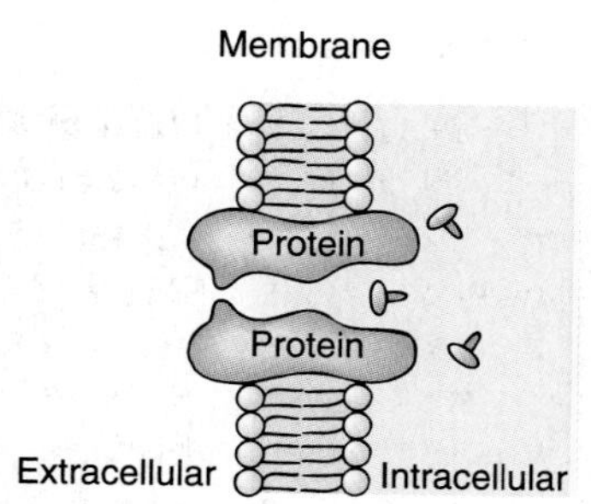

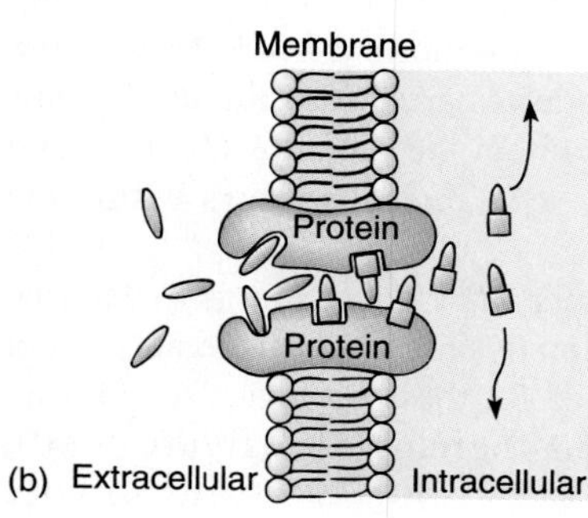

Figure 7.8

In active transport mechanisms, energy is expended to transport the molecule across the cell membrane. (*a*) Molecular channels. The membrane proteins (permeases) have attachment sites for essential nutrient molecules. As these molecules bind to the permease, they are pumped into the cell's interior through special channels or pores in the membrane protein. Microbes have these systems for transporting various ions (sodium, iron) and small organic molecules. (*b*) In group translocation, the molecule is actively captured, but, along the route of transport, it is chemically altered. By coupling transport with synthesis, the cell conserves energy.

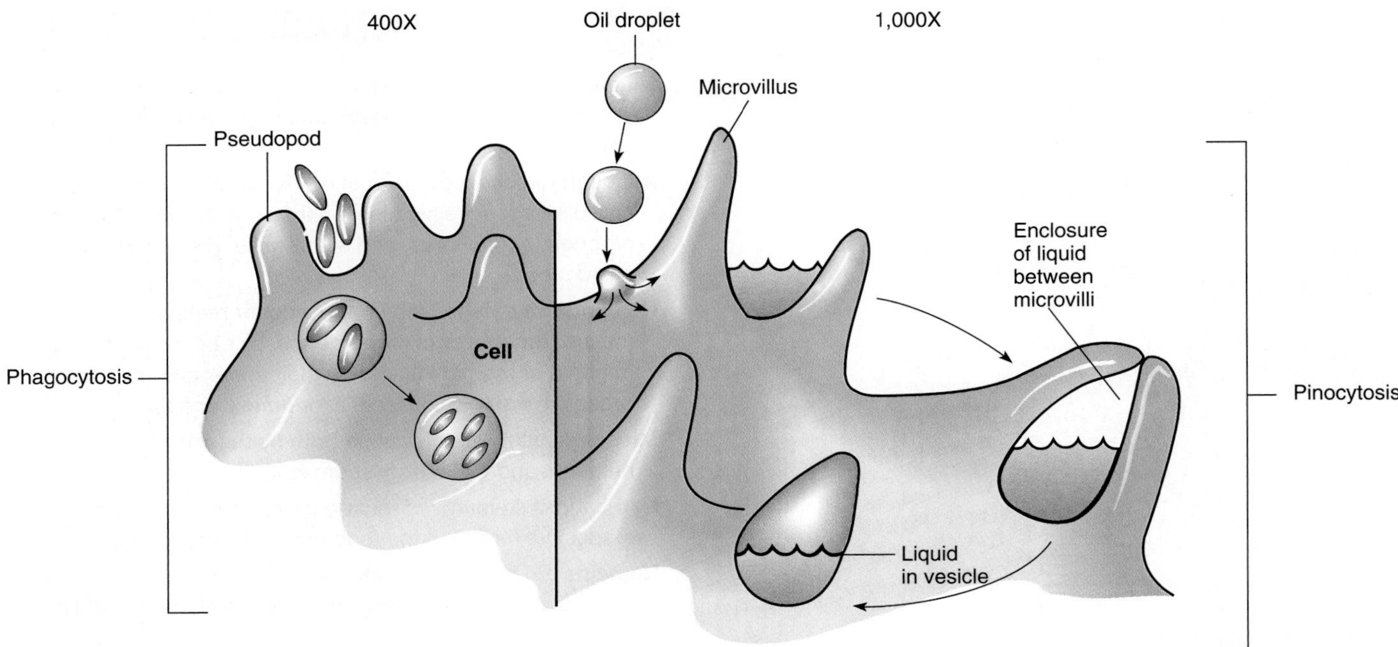

Figure 7.9

Endocytosis (phagocytosis and pinocytosis). Solid particles are phagocytosed by large cell extensions called pseudopods, and they are pinocytosed into vesicles by very fine cell protrusions called microvilli. Oil droplets fuse with the membrane and are released directly into the cell.

ENVIRONMENTAL FACTORS THAT INFLUENCE MICROBES

Microbes are exposed to a wide variety of environmental factors in addition to nutrients. One aim in ecology is to account for the ways that microorganisms deal with or adapt to such factors as heat, cold, gases, acid, radiation, osmotic and hydrostatic pressures, and even other microbes. Adaptation is a complex adjustment in biochemistry or genetics that enables long-term survival and growth. The nature of these adjustments will become more evident in chapters 8 and 9. For now, it can be stated that environmental factors fundamentally affect the function of metabolic enzymes. Thus, survival in a changing environment is largely a matter of whether the enzyme systems of microorganisms can adapt to alterations in their habitat. Incidentally, one must be careful to differentiate between growth in a given condition and tolerance, which implies survival without growth.

TEMPERATURE ADAPTATIONS

Microbial cells are unable to control their temperature and therefore assume the ambient temperature of their natural habitats. Their survival is dependent on adapting to whatever temperature variations are encountered in that habitat. The range of temperatures for microbial growth can be expressed as three *cardinal temperatures.* The **minimum temperature** is the lowest temperature that permits a microbe's continued growth and metabolism; below this temperature, its activities are inhibited. The **maximum temperature** is the highest temperature at which growth and metabolism can proceed. If the temperature rises slightly above maximum, growth will stop, but if it continues to rise beyond that point, the enzymes and nucleic acids will eventually become permanently inactivated and the cell will die. This is why heat works so well as an agent in microbial control. The **optimum temperature** covers a small range, intermediate between the minimum and maximum, which promotes the fastest rate of growth and metabolism (rarely is the optimum a single point).

Depending on their natural habitats, some microbes have a narrow cardinal range, others a broad one. Some strict parasites will not grow if the temperature varies more than a few degrees below or above the host's body temperature. For instance, the typhus rickettsia multiplies only in the range of 32°–38°C, and rhinoviruses (one cause of the common cold) multiply successfully only in tissues that are slightly below normal body temperature (33°–35°C). In contrast, other parasites are not so limited. Strains of *Staphylococcus aureus* grow within the range of 6°–46°C, and the intestinal bacterium *Enterococcus faecalis* grows within the range of 0°–44°C.

Another way to express temperature adaptation is to describe whether an organism grows optimally in a cold, moderate, or hot temperature range. The terms used for these ecological groups are psychrophile, mesophile, and thermophile (figure 7.10), respectively.

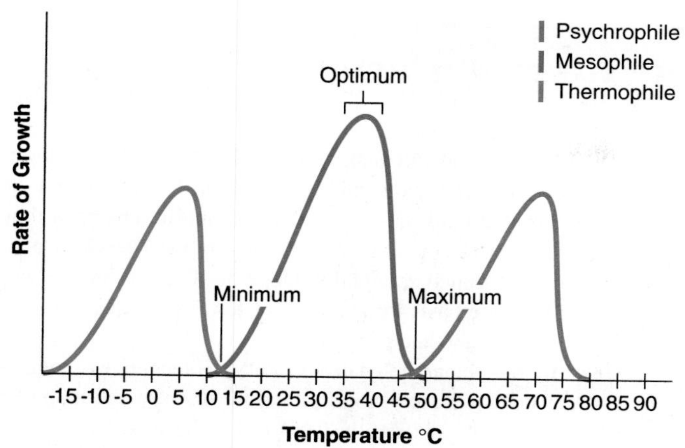

Figure 7.10

Ecological groups by temperature of adaptation. Psychrophiles can grow at or near 0°C and have an optimum below 15°C. As a group, mesophiles can grow between 10°C and 50°C, but their optima usually fall between 20°C and 40°C. Generally speaking, thermophiles require temperatures above 45°C and grow optimally between this temperature and 80°C. The cardinal temperatures are labeled for mesophiles. Note that the extremes of the ranges can overlap to an extent.

(a)

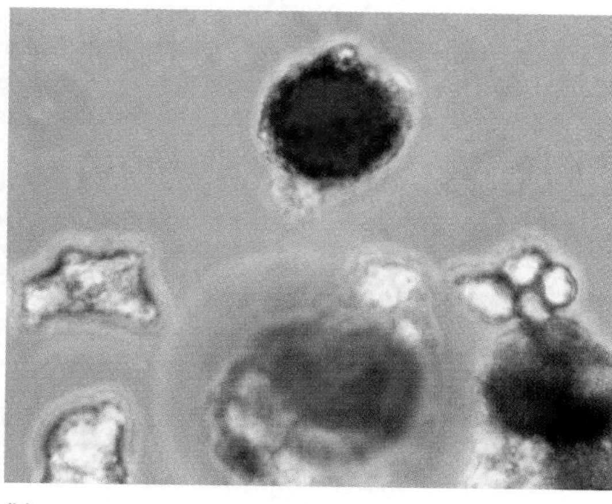

(b)

Figure 7.11

Red snow. (*a*) An early summer snowbank provides a perfect habitat for psychrophilic photosynthetic organisms like *Chlamydomonas nivalis.* (*b*) Microscopic view of this snow alga (actually classified as a "green" alga).

A **psychrophile** (sy´-kroh-fyl) is a microorganism that has an optimum temperature below 15°C and is capable of growth at 0°C. It is obligate with respect to cold and generally cannot grow above 20°C. Laboratory work with true psychrophiles can be a real challenge. Inoculations have to be done in a cold room because room temperature can be lethal to the organisms. Unlike most laboratory cultures, storage in the refrigerator incubates, rather than inhibits, them. As one might predict, the habitats of psychrophilic bacteria, fungi, and algae are snowfields (figure 7.11), polar ice, and the deep ocean. Rarely, if ever, are they pathogenic. True psychrophiles must be distinguished from *psychrotrophs* or *facultative psychrophiles* that grow slowly in cold but have an optimum temperature above 20°C. Bacteria such as *Staphylococcus aureus* and *Listeria monocytogenes* are a concern because they can grow in refrigerated food and cause foodborne illness.

The majority of microorganisms important to the medical microbiologist are **mesophiles** (mez´-oh-fylz), organisms that grow at intermediate temperatures. Although an individual species can grow at the extremes of 10°C or 50°C, the optimum growth temperatures (optima) of most mesophiles fall into the range of 20°–40°C. Organisms in this group inhabit animals and plants as well as soil and water in temperate, subtropical, and tropical regions. Most human pathogens have optima somewhere between 30°C and 40°C (human body temperature is 37°C). *Thermoduric* microbes, which can survive short exposure to high temperatures but are normally mesophiles, are common contaminants of heated or pasteurized foods (see chapter 11). They are often heat-resistant cysts such as *Giardia,* or sporeformers such as *Bacillus* and *Clostridium.*

A **thermophile** (thur´-moh-fyl) is a microbe that grows optimally at temperatures greater than 45°C. Such heat-loving microbes live in soil and water associated with volcanic activity and in habitats directly exposed to the sun. Thermophiles vary in heat requirements, with a general range of growth of 45°–80°C. Most eucaryotic forms cannot survive above 60°C, but a few thermophilic bacteria called hyperthermophiles, grow between 80°C and 100°C (currently thought to be the temperature limit endured by enzymes and cell structures). These thermophiles are so heat-tolerant that researchers use an autoclave to isolate them in culture. Many thermophiles are endospore-forming bacteria, and a small number are pathogens. Currently, there is intense interest in thermal microorganisms by biotechnology companies (microfile 7.6).

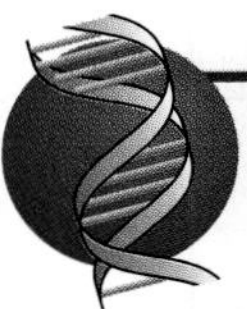

MICROFILE 7.6 CASHING IN ON "HOT" MICROBES

The smoldering thermal springs in Yellowstone National Park are more than just one of the geologic wonders of the world. They are also a hotbed of some of the most unusual microorganisms in the world. The thermophiles thriving at temperatures near the boiling point are the focus of serious interest from the scientific community. For many years, biologists have been intrigued that any living thing could function at such high temperatures. Such questions as these come to mind: Why don't they melt and disintegrate, why don't their proteins coagulate, how can their DNA possibly remain intact?

One of the earliest thermophiles to be isolated was *Thermus aquaticus*. It was discovered by Thomas Brock in Yellowstone's Mushroom Pool in 1965 and was registered with the American Type Culture Collection. Interested parties studied this species and discovered that it has extremely heat-stable proteins and nucleic acids, and its cell membrane does not break down readily at high temperatures. Later, an extremely heat-stable DNA-replicating enzyme was isolated from the species.

What followed is a riveting example of how pure research for the sake of understanding and discovery also offered up a key ingredient in a multimillion dollar process. Developers of the polymerase chain reaction (PCR), a versatile tool for making multiple copies of DNA fragments, found that the technique would work only if they performed the test at temperatures around 65°–72°C. The mesophilic enzymes they tested were destroyed at such high temperatures, but the *Thermus* DNA-copying enzyme functioned splendidly. Once the PCR technique was perfected, it became the basis for a variety of test procedures in forensics and gene detection and analysis.

Spurred by this remarkable success story, biotechnology companies have descended on Yellowstone, a treasure trove of thermophilic habitats, in the hopes of capturing bacteria and archaea with still other unique enzymes that have applications in molecular biology (see chapter opening photo). These industries are looking to thermophiles as a means of developing high-temperature industrial fermentations and techniques for bioremediation. This quest has also brought attention to questions such as: Who owns these microbes, and can their enzymes be patented? Should some of the profits made possible by Yellowstone thermophiles go to the government and National Park Service? Although no final decisions on microbial management have been made, it is obvious that we must begin to regard microbes as much a natural resource as forests and wildlife.

GAS REQUIREMENTS

The atmospheric gases that most influence microbial growth are O_2 and CO_2. Of these, oxygen has the greatest impact on microbial adaptation. Not only is it an important respiratory gas, but it is also a powerful oxidizing agent that exists in many toxic forms. In general, microbes fall into one of three categories: those that use oxygen and can detoxify it; those that can neither use oxygen nor detoxify it; and those that do not use oxygen but can detoxify it.

How Microbes Process Oxygen

As oxygen enters into cellular reactions, it is transformed into several toxic products. Singlet oxygen ($^1O_2$) is an extremely reactive molecule produced by both living and nonliving processes. Notably, it is one of the substances produced by phagocytes to kill invading bacteria (see chapter 14). The build-up of singlet oxygen and the oxidation of membrane lipids and other molecules can damage and destroy a cell. The highly reactive superoxide ion (O_2^-), peroxides (H_2O_2), and hydroxyls (OH^-) are other destructive metabolic by-products of oxygen. To protect themselves against damage, most cells have developed enzymes that go about the business of scavenging and neutralizing these chemicals. The complete conversion of superoxide ion into harmless oxygen requires a two-step process and at least two enzymes:

Step 1. $O_2^- + O_2^- + 2H^+ \xrightarrow[\text{dismutase}]{\text{Superoxide}} H_2O_2 \text{ (hydrogen peroxide)} + O_2$

Step 2. $H_2O_2 + H_2O_2 \xrightarrow{\text{Catalase}} 2H_2O + O_2$

In this series of reactions (essential for aerobic organisms), the superoxide ion is first converted to hydrogen peroxide and normal oxygen by the action of an enzyme called superoxide dismutase. Because hydrogen peroxide is also toxic to cells (it is a disinfectant and antiseptic), it will be degraded by the enzyme catalase into water and oxygen. If a microbe is not capable of dealing with toxic oxygen by these or similar mechanisms, it is forced to live in habitats free of oxygen.

With respect to oxygen requirements, several general categories are recognized. An **aerobe*** (aerobic organism) grows well in the presence of normal atmospheric oxygen and possesses the enzymes needed to process toxic oxygen products. An organism that cannot grow without oxygen is an **obligate aerobe.** Most fungi and protozoas, as well as many bacteria (genera *Micrococcus* and *Pseudomonas*), have strict requirements for oxygen in their metabolism.

A **facultative anaerobe** is an aerobe that does not require oxygen for its metabolism and is capable of growth in the absence of oxygen. This type of organism metabolizes by aerobic respiration when oxygen is present, but, in its absence, it adopts an anaerobic mode of metabolism such as fermentation. Facultative anaerobes usually possess catalase and superoxide dismutase. A large number of bacterial pathogens fall into this group (for example, gram-negative **enteric*** bacteria and staphylococci). A **microaerophile** (myk´´-roh-air´-oh-fyl) does not grow at normal atmospheric tensions of oxygen but requires a small

*aerobe (air´-ohb) Although the prefix means air, it is used in the sense of oxygen.

*enteric (en-terr´-ik) Gr. *enteron,* intestine. A family of bacteria that live in the large intestines of animals.

MICROFILE 7.7 ANAEROBIC INFECTIONS

Even though human cells use oxygen and oxygen is found in the blood and tissues, some body sites present anaerobic pockets or microhabitats where colonization or infection can occur. It is now believed that many areas are maintained in an anaerobic state by the coexistence of aerobic and facultative organisms that use up the oxygen. As long as oxygen users continue to metabolize, so can anaerobes. One region that supports a mixed community and is an important site of anaerobic infections is the oral cavity. Dental caries are partly due to the complex actions of aerobic and anaerobic bacteria, and most gingival infections consist of similar mixtures of oral bacteria that have invaded damaged gum tissues. Another common site for anaerobic infections is the large intestine, a relatively oxygen-free habitat that harbors a rich assortment of strictly anaerobic bacteria such as *Bacteroides.* Anaerobic infections can accompany abdominal surgery, antimicrobic therapy, and traumatic injuries (gas gangrene and tetanus).

amount of it in metabolism. *Actinomyces israelii,* the cause of lumpy jaw in humans, and *Treponema pallidum,* the cause of syphilis, are examples of microaerophiles. Most organisms in this category live in a habitat (soil, water, or the human body) that provides small amounts of oxygen but is not directly exposed to the atmosphere.

An **anaerobe** (anaerobic microorganism) does not grow in normal atmospheric oxygen, and it lacks the metabolic enzyme systems for using oxygen in respiration. Because **strict,** or **obligate, anaerobes** also lack the enzymes for processing toxic oxygen, they cannot tolerate any free oxygen in the immediate environment and will die if exposed to it. Strict anaerobes live in highly reduced habitats, such as deep muds, lakes, oceans, soil, and even the bodies of animals (microfile 7.7). Among the more important anaerobic pathogens are some species of *Clostridium, Bacteroides, Fusobacterium,* and the protozoan *Trichomonas.* Growing anaerobic bacteria usually requires special media, methods of incubation, and handling chambers that exclude oxygen (figure 7.12*a*).

Aerotolerant anaerobes do not utilize oxygen but can survive in its presence. These anaerobes are not killed by oxygen, mainly because they possess alternate mechanisms for breaking down peroxides and superoxide. Certain lactobacilli and streptococci use manganese ions or peroxidases to perform this task. Determining the oxygen requirements of a microbe from a biochemical standpoint can be a very time-consuming process. Often it is illuminating to perform culture tests with reducing media (those that contain an oxygen-absorbing chemical). One such technique demonstrates oxygen requirements by the location of growth in a tube of fluid thioglycollate (figure 7.13).

Although all microbes require some carbon dioxide in their metabolism, *capnophiles** grow best at a higher CO_2 tension than is normally present in the atmosphere. This becomes important in the initial isolation of some pathogens from clinical specimens, notably *Neisseria* (gonorrhea, meningitis), *Brucella* (undulant fever), and *Streptococcus pneumoniae.* Incubation is carried out in a CO_2 incubator, sealed plastic pouch, or candle jar that provides 3% to 10% CO_2 (see figure 7.12*b*).

(a)

(b)

Figure 7.12
Culturing techniques for anaerobes and capnophiles. (*a*) A special anaerobic environmental chamber makes it possible to handle strict anaerobes without exposing them to air. It also has provisions for incubation and inspection in a complete O_2-free system. (*b*) The candle jar. This simple, yet useful, method involves lighting a candle and immediately closing the jar. After a few moments, the flame goes out because of depleted O_2. Meanwhile, the combustion process has added CO_2 to the air as well.

* **capnophile** (kap -noh-fyl) Gr. *kapnos,* smoke.

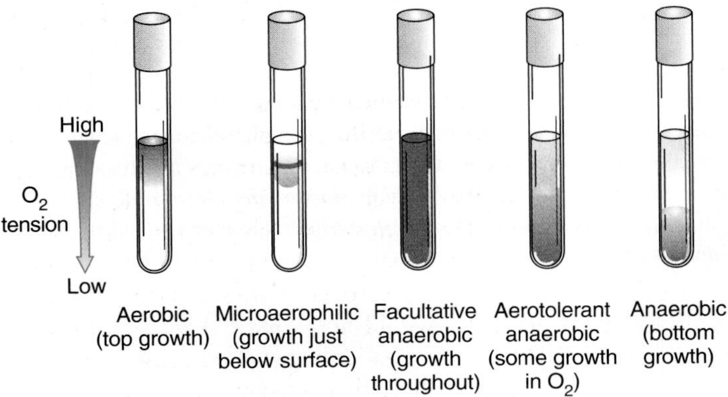

Figure 7.13
Use of thioglycollate broth to demonstrate oxygen requirements. Thioglycollate is a chemical that absorbs O_2 gas and renders it unavailable to bacteria. Oxygen from the air continues to be dissolved in the medium and absorbed; its progress can be shown by the red dye resazurin. When a series of tubes is inoculated with bacteria that differ in O_2 requirements, the relative position of growth provides some indication of their adaptations to oxygen use.

Effects of pH

Microbial growth and survival are also influenced by the pH of the habitat. The pH was defined in chapter 2 as the degree of acidity or alkalinity (basicity) of a solution. It is expressed by the pH scale, a series of numbers ranging from 0 to 14. The pH of pure water (7.0) is neutral, neither acidic nor basic. As the pH value decreases toward 0, the acidity increases, and as the pH increases toward 14, the alkalinity increases. The majority of organisms do not live or grow in high or low pH habitats, because acids and bases can be highly damaging to enzymes and other cellular substances. The optimum pH range for most microorganisms is between 6 and 8, and most human pathogens grow optimally at a pH of 6.5 to 7.5.[2]

A few microorganisms live at pH extremes. Obligate *acidophiles* include *Euglena mutabalis,* an alga that grows in acid pools between 0 and 1.0 pH, and *Thermoplasma,* an archaea that lacks a cell wall, lives in hot coal piles at a pH of 1 to 2, and will lyse if exposed to pH 7. Because many molds and yeasts tolerate moderate acid, they are the most common spoilage agents of pickled foods. Alkalinophiles live in hot pools and soils that contain high levels of basic minerals (up to pH 10.0). Bacteria that decompose urine create alkaline conditions, since ammonium (NH_4^+, an alkaline ion) can be produced when urea (a component of urine) is digested. Metabolism of urea is one way that the ulcer bacterium *Helicobacter pylori* can survive the acidity of the stomach.

Osmotic Pressure

Although most microbes exist under hypotonic or isotonic conditions, a few, called **halophiles** (hay´-loh-fylz) live in habitats with a high solute concentration. *Obligate halophiles* such as *Halobacterium* and *Halococcus* inhabit salt lakes, ponds, and other hypersaline habitats. They grow optimally in solutions of 25% NaCl but require at least 9% NaCl (combined with other salts) for growth. These archaea have significant modifications in their cell walls and membranes and will lyse in hypotonic habitats. *Facultative halophiles* are remarkably resistant to salt, even though they do not normally reside in high-salt environments. For example, *Staphylococcus aureus* can grow on NaCl media ranging from 0.1% up to 20%. Although it is common to use high concentrations of salt and sugar to preserve food (jellies, syrups, and brines), many osmophilic bacteria and fungi actually thrive under these conditions and are common spoilage agents.

2. The pH of human blood is normally 7.4.

Miscellaneous Environmental Factors

Various forms of electromagnetic radiation (ultraviolet, infrared, visible light) stream constantly onto the earth from the sun. Some microbes (phototrophs) can use visible light rays as an energy source, but non-photosynthetic microbes tend to be damaged by the toxic oxygen products produced by contact with light. Some microbial species produce yellow carotenoid pigments to protect against the damaging effects of light by absorbing and dismantling toxic oxygen. Other types of radiation that can damage microbes are ultraviolet and ionizing rays (X rays and cosmic rays). In chapter 11, we will see just how these types of energy are applied in microbial control.

Descent into the ocean depths subjects organisms to increasing hydrostatic pressure. Deep-sea microbes called **barophiles** exist under pressures that range from a few times to over 1,000 times the pressure of the atmosphere. These bacteria are so strictly adapted to high pressures that they will rupture when exposed to normal atmospheric pressure.

Because of the high water content of cytoplasm, all cells require water from their environment to sustain growth and metabolism. Water is the solvent for cell chemicals, and it is needed for enzyme function and digestion of macromolecules. A certain amount of water on the external surface of the cell is required for the diffusion of nutrients and wastes. Even in apparently dry habitats, such as sand or dry soil, the particles retain a thin layer of water usable by microorganisms. Dormant, dehydrated cell stages (for example, spores and cysts) tolerate extreme drying because of the inactivity of their enzymes.

Ecological Associations Among Microorganisms

Up to now, we have considered the importance of nonliving environmental influences on the growth of microorganisms. Another profound influence comes from other organisms that share (or sometimes are) their habitats. In all but the rarest instances, microbes live in shared habitats, which give rise to complex and fascinating ecological associations. Some associations are between similar or dissimilar types of microbes; others involve multicellular organisms such as animals or plants. Interactions can have beneficial, harmful, or no particular effects on the organisms involved; they can be obligatory or nonobligatory to the members; and they often involve nutritional interactions. The following outline provides an overview of the major types of microbial associations:

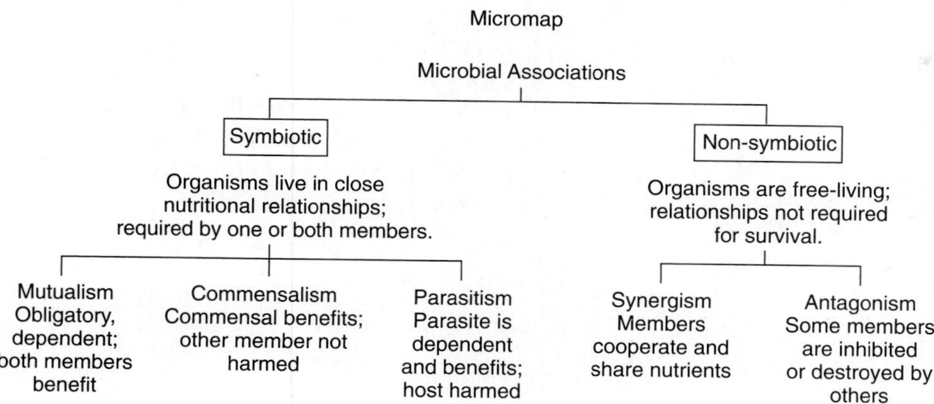
Micromap
Microbial Associations
Symbiotic
Organisms live in close nutritional relationships; required by one or both members.
Non-symbiotic
Organisms are free-living; relationships not required for survival.
Mutualism
Obligatory, dependent; both members benefit
Commensalism
Commensal benefits; other member not harmed
Parasitism
Parasite is dependent and benefits; host harmed
Synergism
Members cooperate and share nutrients
Antagonism
Some members are inhibited or destroyed by others

A general term used to denote a situation in which two organisms live together in a close partnership is **symbiosis,*** and the members are termed *symbionts.* Three main types of symbiosis occur. **Mutualism** exists when organisms live in an obligatory but mutually beneficial relationship. This association is rather common in nature because of the survival value it has for the members involved. Microfile 7.8 gives several examples to illustrate this concept. In the other symbiotic relationships—commensalism and parasitism—the relationship tends to be unequal, meaning it benefits one member and not the other, and it can be obligatory.

In a relationship known as **commensalism,*** the member called the **commensal** receives benefits, while its coinhabitant is neither harmed nor benefited. A classic commensal interaction between microorganisms called *satellitism* arises when one member provides nutritional or protective factors to the other. One example of nutritional satellitism is observed when one microbe provides a growth factor that another one needs (figure 7.14). Some microbes can break down a substance that would be toxic or inhibitory to another microbe. Relationships between humans and resident commensals that derive nutrients from the body are discussed in a later section.

In an earlier section, we introduced the concept of **parasitism** as an interrelationship in which the host organism provides the parasitic microbe with nutrients and a habitat. Multiplication of the parasite usually harms the host to some extent. As this relationship evolves, the host can even develop tolerance for or dependence on a parasite, at which point we call the relationship commensalism or mutualism.

Synergism* is an interrelationship between two or more free-living organisms that benefits them but is not necessary for their survival. Together, the participants cooperate to produce a result that none of them could do alone. An example of synergism is observed in the exchange between soil bacteria and plant roots (see chapter 26).

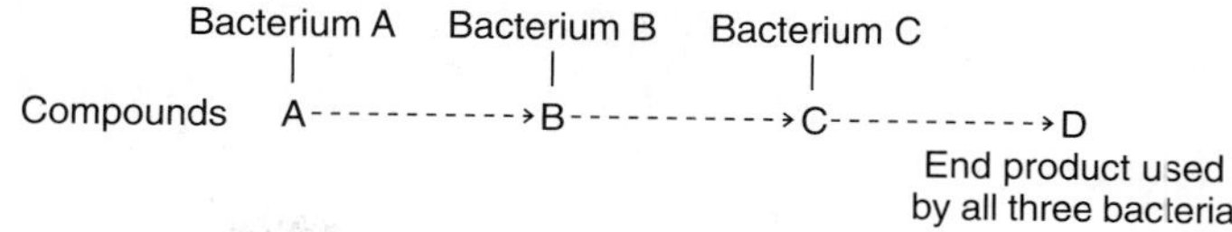
Bacterium A
Bacterium B
Bacterium C
Compounds
A
B
C
D
End product used by all three bacteria

The plant provides various growth factors, and the bacteria help fertilize the plant by supplying it with minerals. Other synergists metabolize a compound sequentially, with a resulting end product that can be used by all members. In synergistic infections, a combination of organisms (sometimes as many as 10 species) can produce tissue damage that a single organism would not cause alone. Gum disease, dental caries, and gas gangrene involve mixed infections by bacteria interacting synergistically.

Antagonism* is an association between free-living species that arises when members of a community compete. In this interaction, one microbe secretes chemical substances into the surrounding environment that inhibit or destroy another microbe in the same habitat. The first microbe may gain a competitive advantage by increasing the space and nutrients available to it. Interactions of this type are common in the soil, where mixed communities often compete for space and food. Antagonistic substances are produced by a wide variety of bacteria and fungi. *Antibiosis**—the production of inhibitory compounds called antibiotics—is actually a form of antagonism. Hundreds of naturally occurring antibiotics have been isolated from microorganisms and used as drugs to control diseases (see chapter 12). *Bacteriocins** are another class of antimicrobial proteins that are toxic to bacteria other than the ones that produced them.

Interrelationships Between Microbes and Humans

The human body is a rich habitat for symbiotic bacteria, fungi, and a few protozoa. Microbes that normally live on the skin, in the alimentary tract, and in other sites are called the *normal microbial flora* (see chapter 13). These residents participate in commensal, parasitic, and synergistic relationships with their human hosts. For example, certain bacteria living symbiotically in the intestine produce some vitamins, and species of symbiotic *Lactobacillus* residing in the vagina help maintain an environment that protects against infection by other microorganisms. Hundreds of commensal species "make a living" on the body without either harming or benefiting it. For example, *Staphylococcus epidermidis* and *Malassezia furfur* reside in the outer dead regions of the skin; oral microbes feed on the constant flow of nutrients in the mouth; and billions of bacteria live on the wastes in the large intestine. Because the normal flora and the body are in a constant state of change, these relationships are not absolute, and a commensal can convert to a parasite by invading body tissues and causing disease.

* **symbiosis** (sim′′-bye-oh′-sis) Gr. *syn,* together, and *bios,* to live.
* **commensalism** (kuh-men′-sul-izm) L. *com,* together, and *mensa,* table.
* **synergism** (sin′-ur-jizm) Gr. *syn,* together; *erg,* work; and *ism,* process.
* **antagonism** (an-tag′-oh-nizm) Gr. *antagonistes,* an opponent.
* **antibiosis** (an′′-tee-by-oh′-sis) Gr. *anti,* against, and *bios,* life.
* **bacteriocin** (bak-teer′-ee-oh-sin) Gr. *bakterion,* little rod, and *ios,* poison.

MICROFILE 7.8 LIFE TOGETHER: MUTUALISM

A tremendous variety of mutualistic partnerships occur in nature. These associations gradually evolve over millions of years as the participating members come to rely on some critical substance or habitat that they share. Recall that microbiologists have strong evidence that eucaryotic cells arose through a form of metabolism between two kinds of procaryotic cells.

Protozoan cells often receive growth factors from symbiotic bacteria and algae that, in turn, are nurtured by the protozoan cell. One peculiar ciliate propels itself by affixing symbiotic bacteria to its cell membrane to act as "oars." These relationships become so obligatory that some amebas and ciliates require mutualistic bacteria for survival. This kind of relationship is especially striking in the complex mutualism of termites, which harbor protozoans specialized to live only inside them. The protozoans, in turn, contain endosymbiotic bacteria. Wood eaten by the termite gets processed by the protozoan and bacterial enzymes, and all three organisms thrive.

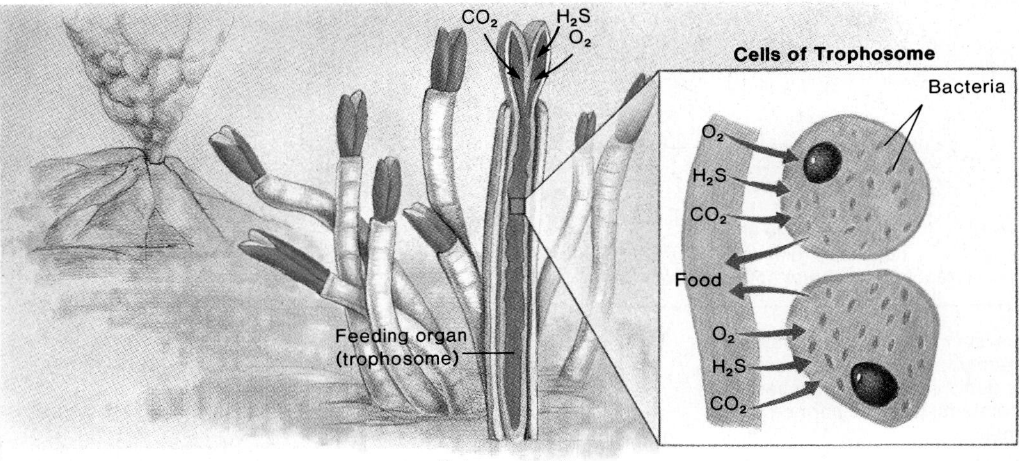

A view of a vent community based on mutualism and chemoautotrophy. The giant tube worm Riftia *houses bacteria in its specialized feeding organ, the trophosome. Raw materials in the form of dissolved inorganic molecules are provided to the bacteria through the worm's circulation. With these, the bacteria produce usable organic food that is absorbed by the worm.*

SYMBIOSIS BETWEEN MICROBES AND ANIMALS

Microorganisms carry on symbiotic relationships with animals as diverse as sponges, worms, and mammals. Bacteria and protozoa are essential in the operation of the rumen (a complex, four-chambered stomach) of cud-chewing mammals. These mammals produce no enzymes of their own to break down the cellulose that is a major part of their diet, but the microbial population harbored in their rumens does. The complex food materials are digested through several stages, during which time the animal regurgitates and chews the partially digested plant matter (the cud) and occasionally burps methane produced by the microbial symbionts.

THERMAL VENT SYMBIONTS

Another fascinating symbiotic relationship has been found in the deep thermal ridges (vents) in the seafloor, where geologic forces spread the crustal plates and release heat and gas. These vents are a focus of tremendous biological and geologic activity. Discoveries first made in the late 1970s demonstrated that the basis of the energy chain in this community is not the sun, because the vents are too deep for light to penetrate (2,600 m). Instead, this ecosystem is based on a massive chemoautotrophic bacterial population that oxidizes the abundant hydrogen sulfide (H_2S) gas given off by the volcanic activity there.

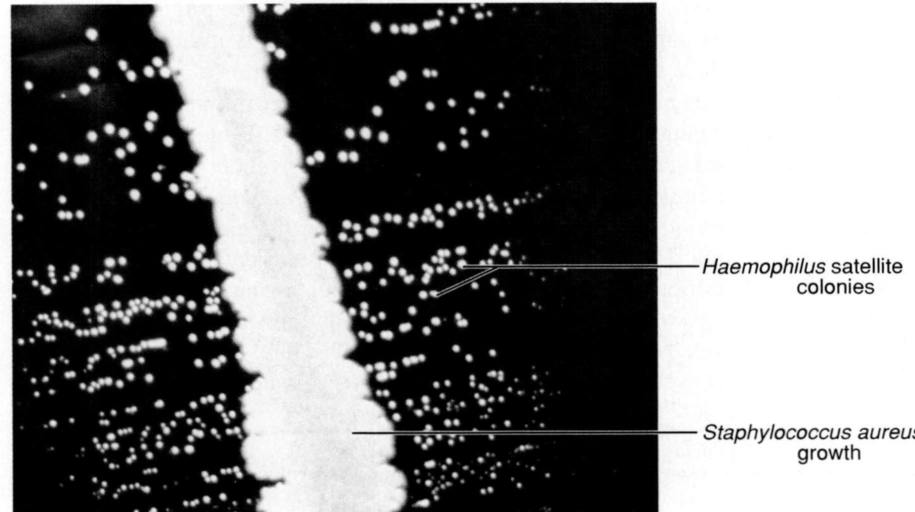

Figure 7.14
Satellitism, a type of commensalism between two microbes. In this example on blood agar, *Staphylococcus aureus* provides growth factors of *Haemophilus influenzae,* which grows as tiny satellite colonies near the streak of *Staphylococcus.* By itself, *Haemophilus* could not grow on blood agar.

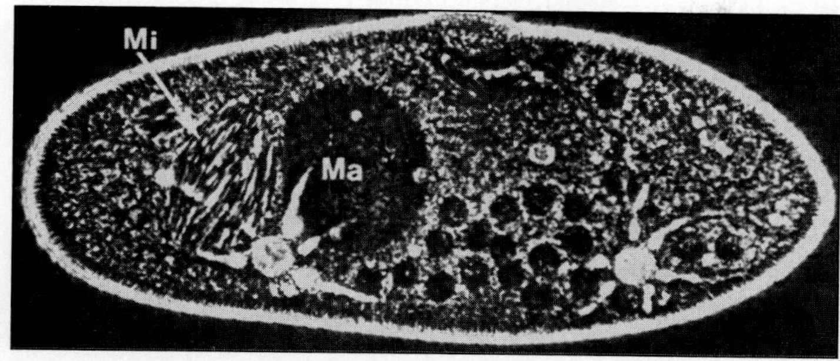

(a)

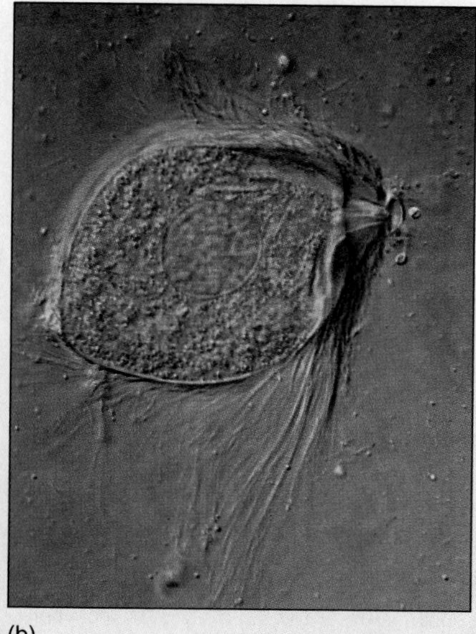

(b)

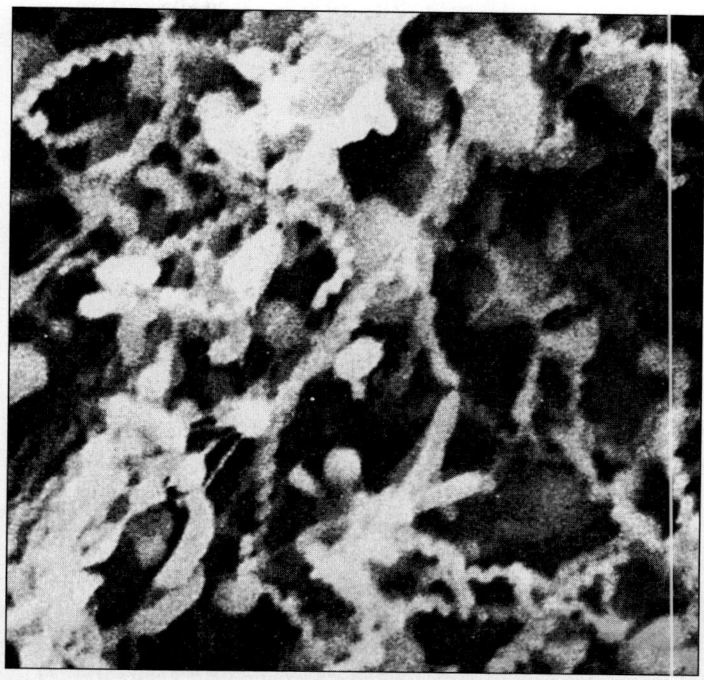

(c)

(a) The endosymbiotic bacteria of Paramecium. *This familiar pond protozoan harbors a bacterium in its micronucleus (Mi) that swells it to 50 times its normal size and makes it as large as the macronucleus (Ma). (b) The real culprits in the final breakdown of wood are species of protozoa in a termite's gut. Here we see a trichonymph filled with tiny particles of wood (135×). Also note the fringe of flagella radiating over its body. (c) Scanning electron micrograph of microbial flora of a sheep's rumen (5,500×). This sample is extremely mixed, with pockets of rod- and spiral-shaped bacteria.*

Chapter Checkpoints

The environmental factors that control microbial growth are temperature, pH, moisture, radiation, gases, and other microorganisms.

Environmental factors control microbial growth by their influence on microbial enzymes.

Three cardinal temperatures for a microorganism describe its temperature range and the temperature at which it grows best. These are the minimum temperature, the maximum temperature, and the optimum temperature.

Microorganisms are classified by their temperature requirements as psychrophiles, mesophiles, or thermophiles.

Most eucaryotic microorganisms are aerobic, but bacteria vary widely in their oxygen requirements from facultative to anaerobic.

Microorganisms live in associations with other species that range from mutually beneficial symbiosis to parasitism and antagonism.

THE STUDY OF MICROBIAL GROWTH

Microbes that are provided with nutrients and the required environmental factors become metabolically active and grow. Growth takes place on two levels. On one level, a cell synthesizes new cell components and increases its size; on the other level, the number of cells in the population increases. This capacity for multiplication, increasing the size of the population by cell division, has tremendous importance in microbial control, infectious disease, and biotechnology. In the following section, we will focus primarily on the characteristics of bacterial growth that are generally representative of single-celled microorganisms.

THE BASIS OF POPULATION GROWTH: BINARY FISSION

The division of a bacterial cell occurs mainly through **binary,** or **transverse, fission;** *binary* means that one cell becomes two, and *transverse* refers to the position of the division plane forming across the width of the cell. During binary fission, the parent cell enlarges, duplicates its chromosome, and forms a central transverse septum that divides the cell into two daughter cells. This process is repeated at intervals by each new daughter cell in turn, and with each successive round of division the population increases. The stages in this continuous process are shown in greater detail in figures 7.15 and 7.16.

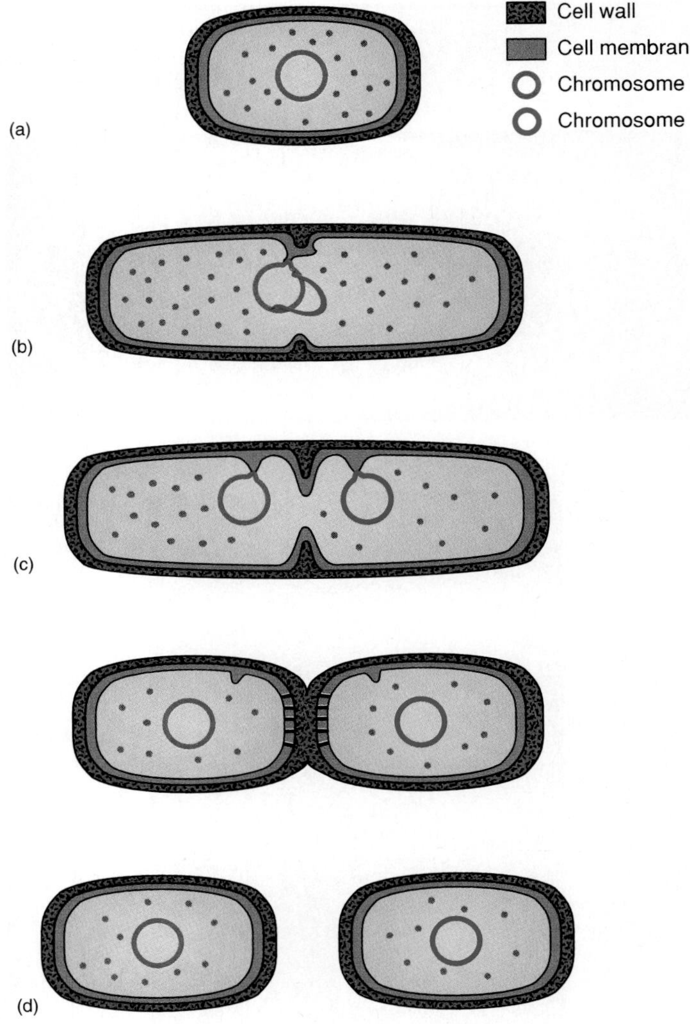

Figure 7.15

Steps in binary fission. (*a*) A parent cell prepares for division by (*b*) enlarging its cell wall, cell membrane, and overall volume. Midway in the cell, the wall develops notches that will eventually form the transverse septum, and the duplicated chromosome becomes affixed to a special membrane site. (*c*) The wall septum grows inward, and the chromosomes are pulled toward opposite cell ends as the membrane enlarges. Other cytoplasmic components (ribosome granules) are distributed (randomly) to the two developing cells. (*d*) The septum is synthesized completely through the cell center, and the cell membrane patches itself so that there are two separate cell chambers. At this point the daughter cells are divided. Some species will separate completely as shown here, and others will remain attached.

THE RATE OF POPULATION GROWTH

The time required for a complete fission cycle—from parent cell to two new daughter cells—is called the **generation,** or **doubling, time.** The term *generation* has a similar meaning as it does in humans. It is the period between an individual's birth and the time of producing offspring. In bacteria, each new fission cycle or generation increases the population by a factor of 2, or doubles it. Thus, the initial parent stage consists of 1 cell, the first generation consists of 2 cells, the second 4, the third 8, then 16, 32, 64, and so on (figure 7.16). As long as the environment remains favorable, this doubling effect can continue at a constant rate. With the passing of each generation, the population will double, over and over again. The length of the generation time is a measure of the growth rate of an organism.

Compared with the growth rates of most other living things, bacteria are notoriously rapid. The average generation time is 30–60 minutes under optimum conditions. The shortest generation times average 5–10 minutes, and the longest generation times require days. For example, *Mycobacterium leprae,* the cause of leprosy, has a generation time of 10–30 days—as long as in some animals. Most pathogens have relatively short doubling times. *Salmonella enteritidis* and *Staphylococcus aureus,* bacteria that cause food-borne illness, double in 20 to 30 minutes, which is why leaving food at room temperature even for a short period has caused many a person to be suddenly stricken with an attack of food-borne disease. In a few hours, a population of these bacteria can easily grow from a small number of cells to several million.

Figure 7.16 shows several quantitative characteristics of growth: (1) The cell population size can be represented by the number 2 with an exponent (2^1, 2^2, 2^3, 2^4); (2) the exponent increases by one in each generation; and (3) the number of the exponent is also the number of the generation. This growth pattern is termed **exponential.** Population growth expressed this way is **geometric,** with a constantly increasing slope. Because these populations often contain very large numbers of cells, it is useful to express them by means of exponents or logarithms (see appendix B). The data from a growing bacterial population are graphed by plotting the number of cells as a function of time. The cell number can be represented logarithmically or arithmetically. Plotting the logarithm number over time provides a straight line in-

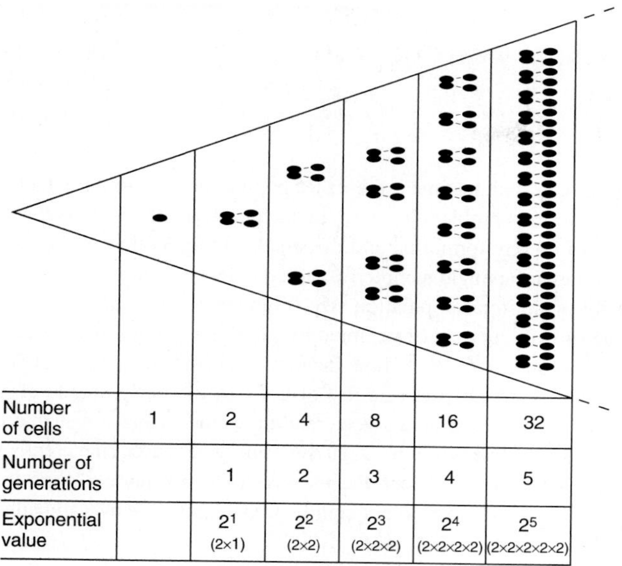

Number of cells
1
2
4
8
16
32
Number of generations
1
2
3
4
5
Exponential value
2^1 (2×1)
2^2 (2×2)
2^3 (2×2×2)
2^4 (2×2×2×2)
2^5 (2×2×2×2×2)
(a)

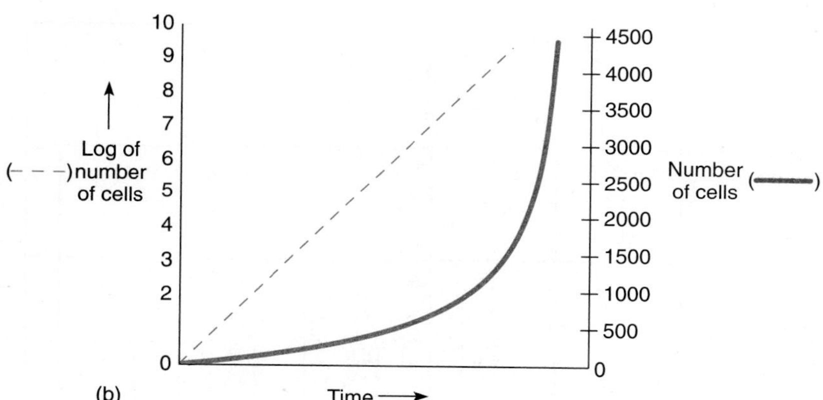

Figure 7.16
The mathematics of population growth. (*a*) Starting with a single cell, if each product of reproduction goes on to divide in a binary fashion, the population doubles with each new division cycle or generation. This process can be represented by logarithms (2 raised to an exponent) or simple numbers. (*b*) Plotting the logarithm of the cells produces a straight line indicative of exponential growth, whereas plotting the cell numbers arithmetically gives a curved slope.

dicative of exponential growth. Plotting the data arithmetically gives a constantly curved slope. In general, logarithmic graphs are preferred because accurate cell number is easier to read, especially during early growth phases.

Predicting the number of cells that will arise during a long growth period (yielding millions of cells) is based on a relatively simple concept. One could use the method of addition 2 + 2 = 4; 4 + 4 = 8; 8 + 8 = 16; 16 + 16 = 32, and so on or a method of multiplication (for example, $2^5 = 2 \times 2 \times 2 \times 2 \times 2$), but it is easy to see that for 20 or 30 generations, this calculation could take reams of paper and be very tedious. An easier way to calculate the size of a population over time is to use an equation such as:

$$N_f = (N_i)2^n$$

In this equation, N_f is the total number of cells in the population at some point in the growth phase, N_i is the starting number, the exponent n denotes the generation number and 2^n represents the number of cells in that generation. If we know any two of the values, the other values can be calculated. Let us use the example of *Staphylococcus aureus* to calculate how many cells (N_f) will be present in an egg salad sandwich after it sits in a warm car for 4 hours. We will assume that N_i is 10 (number of cells deposited in the sandwich while it was being prepared). To derive n, we need to divide 4 hours (240 minutes) by the generation time (we will use 20 minutes). This calculation comes out to 12, so 2^n is equal to 2^{12}. Referring to a table on the powers of numbers or using a calculator, we find that 2^{12} is 4,096.

$$\text{Final number } (N_f) = 10 \times 4{,}096$$
$$= 40{,}960 \text{ cells in the sandwich}$$

This same equation, with modifications, is used to determine the generation time, a more complex calculation that requires knowing the number of cells at the beginning and end of a growth period. Such data are obtained through actual testing by a method discussed in the following section.

THE POPULATION GROWTH CURVE

In reality, a population of bacteria does not maintain its potential growth rate and does not double endlessly, because in most systems numerous factors prevent the cells from continuously dividing at their maximum rate. Quantitative laboratory studies indicate that a population typically displays a predictable pattern, or **growth curve,** over time. The method traditionally used to observe the population growth pattern is a viable count technique, in

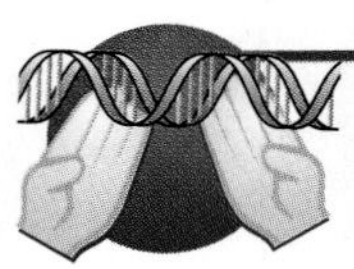

MICROFILE 7.9 STEPS IN A VIABLE PLATE COUNT—BATCH CULTURE METHOD

A growing population is established by inoculating a flask containing a known quantity of sterile liquid medium with a few cells of a pure culture. The flask is incubated at that bacteria's optimum temperature, and timing is begun. The population size at any point in the growth cycle is quantified by removing a tiny measured sample of the culture from the growth chamber and plating it out on a solid medium to develop isolated colonies. This procedure is repeated at evenly spaced intervals (every hour for 24 hours).

Evaluating the samples involves a common and important principle in microbiology: One colony on the plate represents one cell or colony-forming unit (CFU) from the original sample. Because the CFU of some bacteria is actually composed of several cells (consider the clustered arrangement of *Staphylococcus,* for instance), using a colony count can underestimate the exact population size to an extent. This is not a serious problem because, in such bacteria, the CFU is the smallest unit of colony formation and dispersal. Multiplication of the number of colonies in a single sample by the container's volume gives a fair estimate of the total population size (number of cells) at any given point. The growth curve is determined by graphing the number for each sample in sequence for the whole incubation period (see figure 7.17).

Because of the scarcity of cells in the early stages of growth, some samples can give a zero reading even if there are viable cells in the culture. The sampling itself can remove enough viable cells to alter the tabulations, but since the purpose is to compare relative trends in growth, these factors do not significantly change the overall pattern.

Flask inoculated

Samples taken at equally spaced intervals (0.1 ml)

500 ml	60 min	120 min	180 min	240 min	300 min	360 min	420 min	480 min	540 min	600 min
	0.1 ml									
Sample is diluted in liquid agar medium and poured or spread over surface of solidified medium										
Plates are incubated, colonies are counted	None									
Number of colonies (CFU) per 0.1 ml	0*	1	3	7	13	23	45	80	135	230
Total cell population in flask	0*	5,000	15,000	35,000	65,000	115,000	225,000	400,000	675,000	1,150,000

* Zero CFUs only means that too few cells are present to be assayed.

The viable plate count, a technique for determining population size and rate of growth.

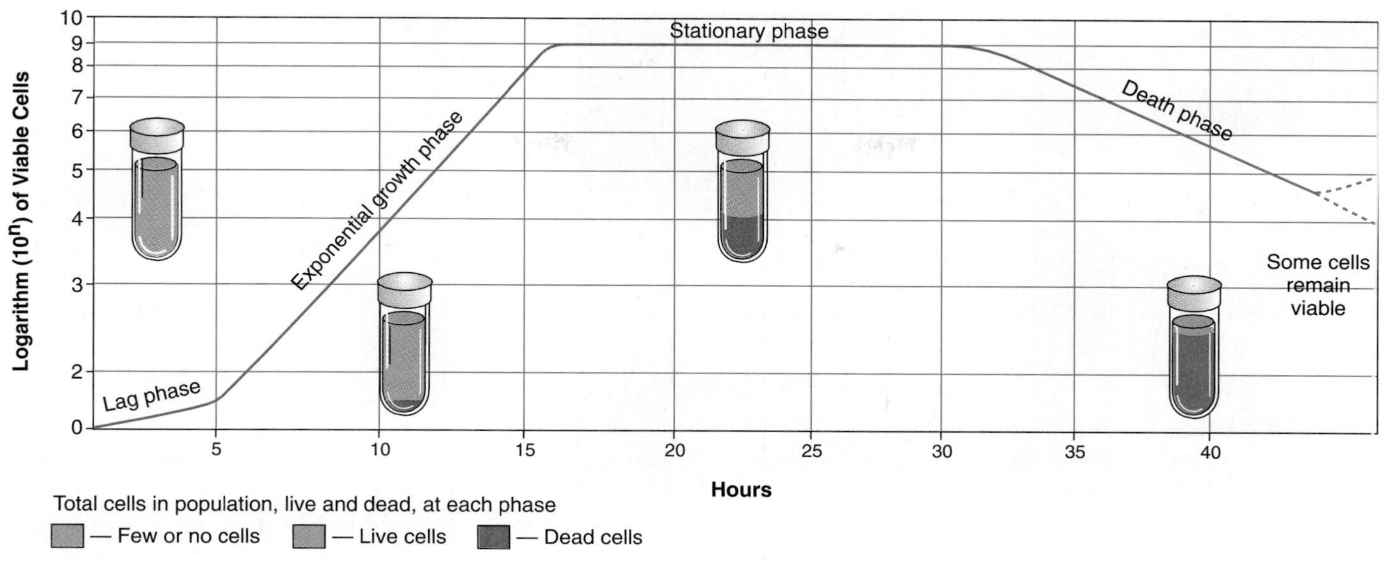

Figure 7.17

The growth curve in a bacterial culture. On this graph, the number of viable cells expressed as a logarithm (log) is plotted against time. See text for discussion of the various phases. Note that with a generation time of 30 minutes, the population has risen from 10 (10^1) cells to 1,000,000,000 (10^9) cells in only 16 hours.

which the total number of live cells is counted over a given time period. In brief, this method entails (1) placing a tiny number of cells into a sterile liquid medium; (2) incubating this culture over a period of several hours; (3) sampling the broth at regular intervals during incubation; (4) plating each sample onto solid media; and (5) counting the number of colonies present after incubation. Microfile 7.9 gives the details of this process.

STAGES IN THE NORMAL GROWTH CURVE

The system of batch culturing described in microfile 7.9 is *closed,* meaning that nutrients and space are finite and there is no mechanism for the removal of waste products. Data from an entire growth period of 3–4 days typically produce a curve with a series of phases termed the lag phase, the exponential growth (log) phase, the stationary phase, and the death phase (figure 7.17).

The **lag phase** is a relatively "flat" period on the graph when the population appears not to be growing or is growing at less than the exponential rate. Growth lags primarily because: (1) the newly inoculated cells require a period of adjustment, enlargement, and synthesis; (2) the cells are not yet multiplying at their maximum rate; and (3) the population of cells is so sparse or dilute that the sampling misses them. The length of the lag period varies somewhat from one population to another.

The cells reach the maximum rate of cell division during the **exponential growth (log) phase,** a period during which the curve increases geometrically. This phase will continue as long as cells have adequate nutrients and the environment is favorable.

At the **stationary growth phase,** the population enters a survival mode in which cells stop growing or grow slowly. The curve levels off because the rate of cell inhibition or death balances out the rate of multiplication. The decline in the growth rate is caused by depleted nutrients and oxygen, excretion of organic acids and other biochemical pollutants into the growth medium, and an increased density of cells.

As the limiting factors intensify, cells begin to die in exponential numbers (literally perishing in their own wastes), and they are unable to multiply. The curve now dips downward as the **death phase** begins. The speed with which death occurs depends on the relative resistance of the species and how toxic the conditions are, but it is usually slower than the exponential growth phase. Viable cells often remain many weeks and months after this phase has begun. In the laboratory, refrigeration is used to slow the progression of the death phase so that cultures will remain viable as long as possible.

Practical Importance of the Growth Curve

The tendency for populations to exhibit phases of rapid growth, slow growth, and death has important implications in microbial control, infection, food microbiology, and cultural technology. Antimicrobial agents such as heat and disinfectants rapidly accelerate the death phase in all populations, but microbes in the exponential growth phase are more vulnerable to these agents than are those that have entered the stationary phase. In general, actively growing cells are more vulnerable to conditions that disrupt cell metabolism and binary fission.

Growth patterns in microorganisms can account for the stages of infection (see chapter 13). Microbial cells produced during the exponential phase are far more numerous and virulent than those released at a later stage of infection. A person shedding bacteria in the early and middle stages of an infection is more likely to spread it to others than is a person in the late stages. The course of an infection is also influenced by the relatively faster rate of multiplication of the microbe, which can overwhelm the slower growth rate of the host's own cellular defenses.

Understanding the stages of cell growth is crucial for work with cultures. Sometimes a culture that has reached the stationary

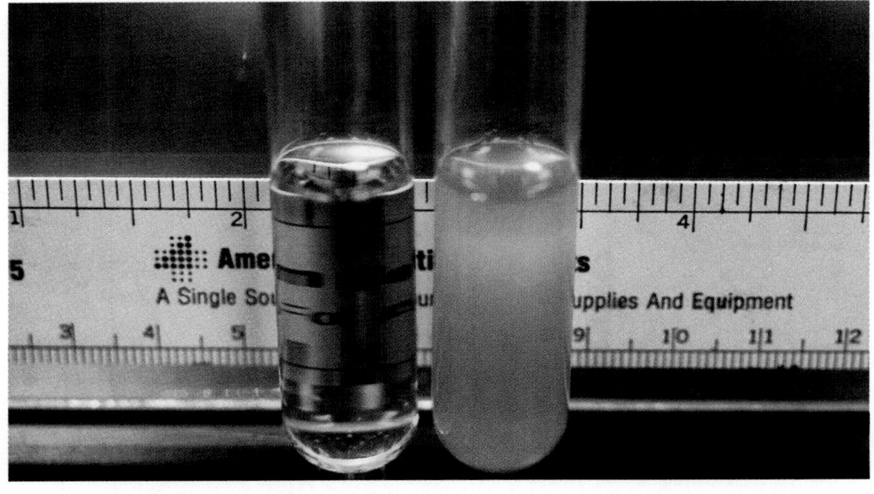

(a)

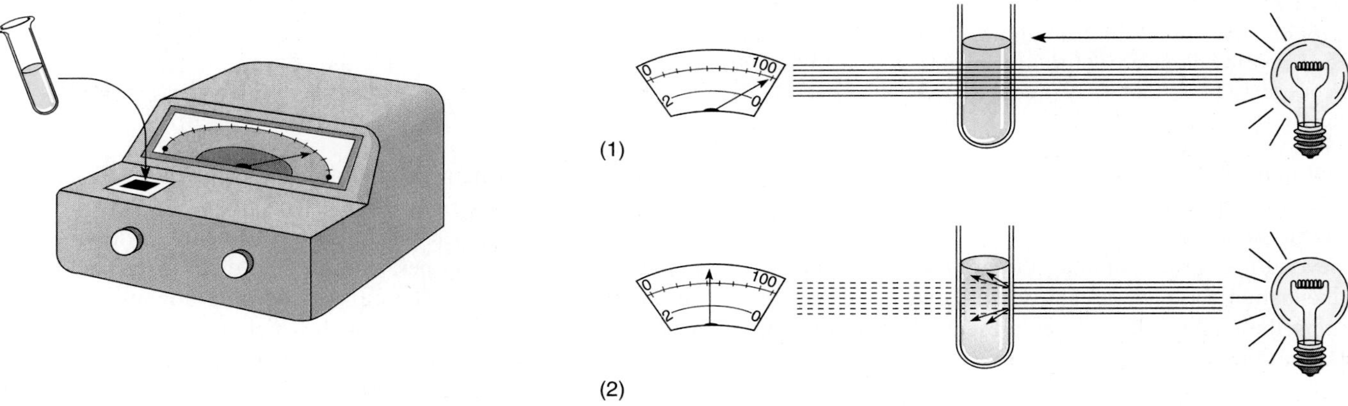

(b)

Figure 7.18

Turbidity measurements as indicators of growth. (*a*) Holding a broth to the light is one method of checking for gross differences in cloudiness (turbidity). The broth on the left is transparent, indicating little or no growth; the broth on the right is cloudy and opaque, indicating heavy growth. (*b*) The eye is not sensitive enough to pick up fine degrees in turbidity; more sensitive measurements can be made with a spectrophotometer. (*1*) A tube with no growth transmits more light and gives a higher reading. (*2*) In a tube with growth, the cells scatter the light so that transmittance is reduced and the reading is lower. This technique provides only a relative measure of growth; it cannot determine actual numbers or differentiate between dead and live cells.

phase is incubated under the mistaken impression that enough nutrients are present for the culture to multiply. In most cases, it is unwise to continue incubating a culture beyond the stationary phase, because doing so will reduce the number of viable cells and the culture could die out completely. It is also preferable to do stains (an exception is the spore stain) and motility tests on young cultures, because the cells will show their natural size and correct reaction, and motile cells will have functioning flagella.

For certain research or industrial applications, closed batch culturing with its four phases is inefficient. The alternative is an automatic growth chamber called the *chemostat,* or continuous culture system. This device can admit a steady stream of new nutrients and siphon off used media and old bacterial cells, thereby stabilizing the growth rate and cell number. The chemostat is very similar to the industrial fermenters used to produce vitamins and antibiotics (chapter 26). It has the advantage of maintaining the culture in a biochemically active state and preventing it from entering the death phase.

Other Methods of Analyzing Population Growth

Microbiologists have developed several alternative ways of analyzing bacterial growth qualitatively and quantitatively. One of the simplest methods for estimating the size of a population is through turbidometry. This technique relies on the simple observation that a tube of clear nutrient solution loses its clarity and becomes cloudy, or **turbid,** as microbes grow in it. In general, the greater the turbidity, the larger the population size, which can be measured by means of sensitive instruments (figure 7.18*b*).

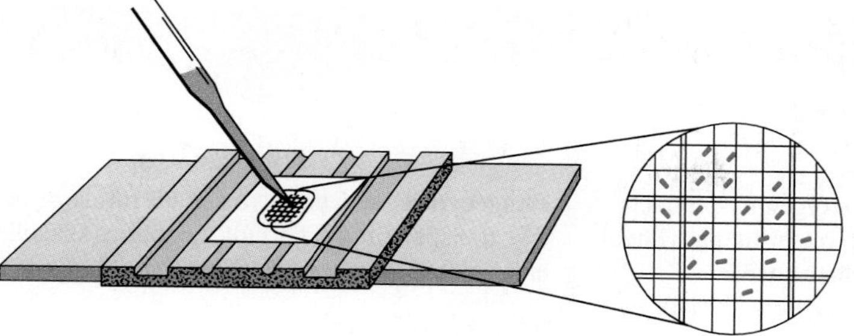

Figure 7.19

Direct microscopic count of bacteria. A small sample is placed on the grid under a cover glass. Individual cells, both living and dead, are counted. This number can be used to calculate the total count of a sample.

Enumeration of Bacteria

Turbidity readings are useful for evaluating relative amounts of growth, but if a more quantitative evaluation is required, the viable colony count described previously or some other enumeration (counting) procedure is necessary. The **direct,** or **total, cell count** involves counting the number of cells in a sample microscopically (figure 7.19). This technique, very similar to that used in blood cell counts, employs a special microscope slide (cytometer) calibrated to accept a tiny sample that is spread over a premeasured grid. The cell count from a cytometer can be used to estimate the total number of cells in a larger sample (for instance, of milk or water). One inherent inaccuracy in this method is that no distinction can be made between dead and live cells, both of which are included in the count.

Counting can be automated by sensitive devices such as the Coulter counter, which electronically scans a culture as it passes through a tiny pipette. As each cell flows by, it is detected and registered on an electronic sensor (figure 7.20*a*). A flow cytometer works on a similar principle, but in addition to counting, it can measure cell size and even differentiate between live and dead cells. When used in conjunction with fluorescent dyes and antibodies, it has been used to differentiate between gram-positive and gram-negative bacteria. It is being adapted for use as a rapid method to identify pathogens in patient specimens and to compare bacterial species on the basis of genetic differences such as guanine and cytosine content (figure 7.20*b*).

Chapter Checkpoints

Microbial growth refers both to increase in cell size and increase in number of cells in a population.

The generation time is a measure of the growth rate of a microbial species. It varies in length according to environmental conditions.

Microbial cultures in a nutrient-limited environment exhibit four distinct stages of growth: the lag phrase, the exponential growth (log) phase, the stationary phase, and the death phase.

Microbial cell populations show distinct phases of growth in response to changing nutrient and waste conditions.

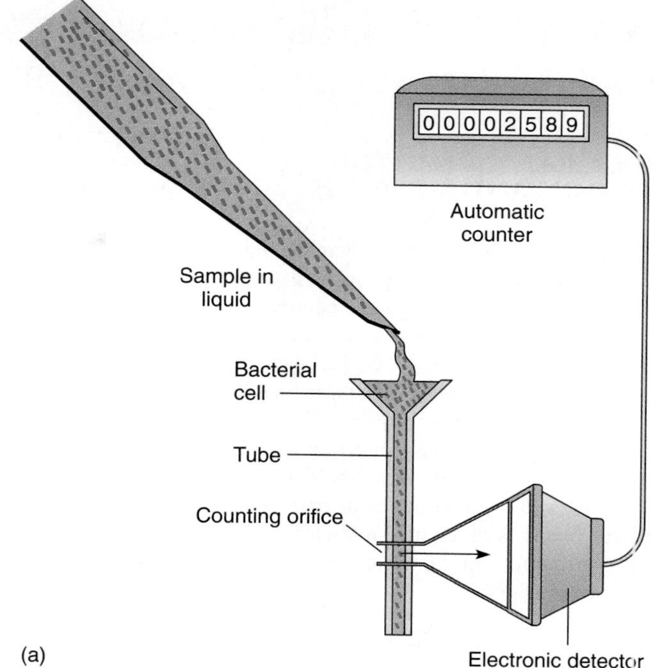

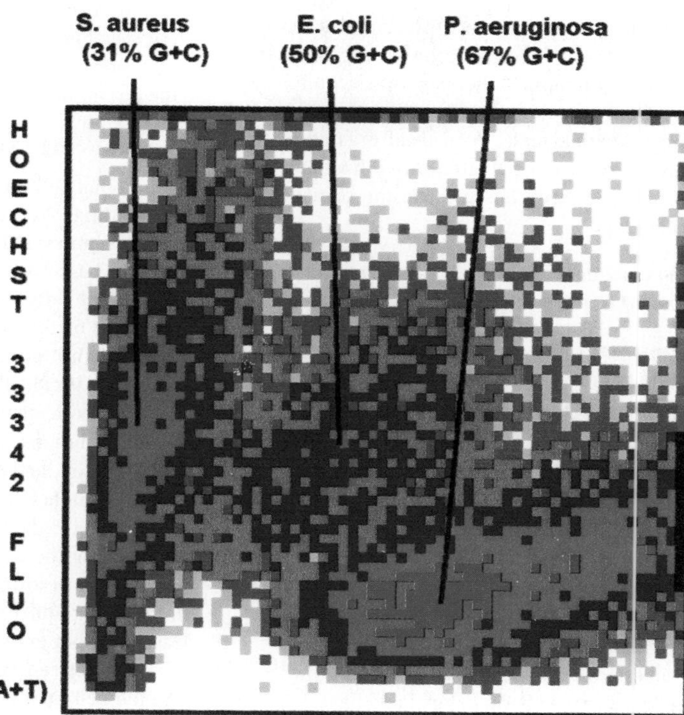

Figure 7.20

(*a*) Coulter counter. As cells pass through this device, they trigger an electronic sensor that tallies their numbers. (*b*) A variation on this machine, called a flow cytometer, can record the number, size, and types of cells by tagging them with fluorescent substances and passing them through a beam of light. Shown here is a fluorescence signature of three species, which are differentiated on the basis of guanine and cytosine percentages.

CHAPTER CAPSULE WITH KEY TERMS

MICROBIAL NUTRITION, ECOLOGY, AND GROWTH

Nutrition consists of taking in chemical substances (nutrients) and assimilating and extracting energy from them. Substances required for survival are **essential nutrients**—usually containing the elements (C, H, N, O, P, S, Na, Cl, K, Ca, Fe, Mg). Essential nutrients are considered as **macronutrients** (required in larger amounts) or **micronutrients** (trace elements required in smaller amounts—Zn, Mn, Cu). Nutrients are classed as either inorganic or organic. A **growth factor** is an organic nutrient (amino acid and vitamin) that cannot be synthesized and must be provided.

NUTRITIONAL CATEGORIES

An **autotroph** depends on carbon dioxide for its carbon needs. If its energy needs are met by light, it is a **photoautotroph,** but if it extracts energy from inorganic substances, it is a **chemoautotroph.** A **heterotroph** acquires carbon from organic molecules. A **saprobe** is a decomposer that feeds upon dead organic matter, and a **parasite** feeds from a live host and usually causes harm. Disease-causing parasites are pathogens.

ENVIRONMENTAL INFLUENCES ON MICROBES

Temperature: An organism exhibits **optimum, minimum,** and **maximum temperatures.** Organisms that cannot grow above 20°C but thrive below 15°C and continue to grow even at 0°C are known as **psychrophiles. Mesophiles** grow from 10°C to 50°C, having temperature optima from 20°C to 40°C. The growth range of **thermophiles** is 45°–80°C.

Oxygen Requirements: The ecological need for free oxygen (O_2) is based on whether a cell can handle toxic by-products such as superoxide and peroxide. **Aerobes** grow in normal atmospheric oxygen and have enzymes to handle toxic oxygen by-products. An aerobic organism capable of living without oxygen if necessary is a **facultative anaerobe.** An aerobe that prefers a small amount of oxygen but does not grow under anaerobic conditions is a **microaerophile. Strict (obligate) anaerobes** do not use free oxygen and cannot produce enzymes to dismantle reactive oxides. They are actually damaged or killed by oxygen. An **aerotolerant anaerobe** cannot use oxygen for respiration, yet is not injured by it.

Effects of pH: Acidity and alkalinity affect the activity and integrity of enzymes and the structural components of a cell. Optimum pH for most microbes ranges approximately from 6 to 8. **Acidophiles** prefer lower pH, and **alkalinophiles** prefer higher pH.

Other Environmental Factors: Electromagnetic radiation and barometric pressure affect microbial growth. A **barophile** is adapted to life under high pressure (bottom dwellers in the ocean, for example).

TRANSPORT MECHANISMS

A microbial cell must take on nutrients from its surroundings by transporting them across the cell membrane. **Diffusion** is a form of passive transport based on a concentration gradient. **Osmosis** is diffusion of water through a selectively permeable membrane. A special form of passive transport that permits movement of selected substances is called **facilitated diffusion.**

Osmotic changes that affect cells are **hypotonic** solutions, which contain a lower solute concentration, and **hypertonic** solutions, which contain a higher solute concentration. **Isotonic** solutions have the same solute concentration as the inside of the cell. A **halophile** thrives in hypertonic surroundings, and an **obligate halophile** requires a salt concentration of at least 15%, but grows optimally in 25%.

In **active transport,** substances are taken into the cell by a process that consumes energy. **Phagocytosis** and **pinocytosis** are forms of active transport in which bulk quantities of solid and fluid material are taken into the cell.

MICROBIAL INTERACTIONS

Microbes coexist in varied relationships in nature. Types of **symbiosis** are **mutualism,** a reciprocal, obligatory, and beneficial relationship between two organisms, and **commensalism,** when an organism benefits from but does not harm the other organism in the relationship (satellitism is a special form of commensalism.). Parasitism occurs between a host and an infectious agent. **Synergism** is a mutually beneficial but not obligatory coexistence. **Antagonism** entails competition, inhibition, and injury directed against the opposing organism. Special cases of antagonism are antibiotic and bacteriocin production.

MICROBIAL GROWTH

The splitting of a parent bacterial cell to form a pair of similar-sized daughter cells is known as **binary,** or **transverse, fission.** The duration of each division is called the **generation,** or **doubling, time.** A population theoretically doubles with each generation, so the growth rate is **exponential,** and each cycle increases in **geometric progression.** A **growth curve** is a graphic representation of a closed population over time. Plotting a curve requires an estimate of live cells, called a **viable count.** The initial flat period of the curve is called the **lag phase,** followed by the **exponential growth phase,** in which viable cells increase in logarithmic progression. Adverse environmental conditions combine to inhibit the growth rate, causing a plateau, or **stationary growth phase.** In the **death phase,** nutrient depletion and waste build-up cause increased cell death.

Cell numbers can be counted directly by a microscope counting chamber, Coulter counter, or flow cytometer. Cell growth can also be determined by turbidometry and a total cell count.

MULTIPLE-CHOICE QUESTIONS

1. An organic nutrient essential to an organism's metabolism that cannot be synthesized itself is termed a/an
   a. trace element c. growth factor
   b. micronutrient d. essential nutrient
2. The source of the necessary elements of life are
   a. an inorganic environmental reservoir
   b. the sun
   c. rocks
   d. the air
3. An organism that can synthesize all its required organic components from CO_2 using energy from the sun is a
   a. photoautotroph
   b. photoheterotroph
   c. chemoautotroph
   d. chemoheterotroph
4. An obligate halophile requires high
   a. pH c. salt
   b. temperature d. pressure
5. Chemoautotrophs can survive on ____ alone.
   a. minerals c. minerals and CO_2
   b. CO_2 d. methane
6. A pathogen would most accurately be described as a
   a. parasite c. saprobe
   b. commensal d. symbiont
7. A substance that would be moved by passive transport is
   a. sugar c. amino acids
   b. water d. food detritus
8. A cell exposed to a hypertonic environment will ____ by osmosis.
   a. gain water
   b. lose water
   c. neither gain nor lose water
   d. burst
9. Active transport of a substance across a membrane requires
   a. a gradient
   b. the expenditure of ATP
   c. water
   d. diffusion
10. Environmental factors such as temperature and pH exert their effect on the ____ of microbial cells.
    a. membranes c. enzymes
    b. DNA d. cell wall
11. Psychrophiles would be expected to grow
    a. in hot springs
    b. on the human body
    c. at refrigeration temperatures
    d. at low pH
12. Superoxide ion is toxic to strict anaerobes because they lack
    a. catalase c. dismutase
    b. peroxidase d. oxidase
13. The time required for a cell to undergo binary fission is called the
    a. exponential growth rate
    b. growth curve
    c. generation time
    d. lag period
14. In a viable plate count, each ____ represents a ____ from the sample population.
    a. cell, colony
    b. colony, cell
    c. hour, generation
    d. cell, generation
15. During the ____ phase, the rate of new cells being added to the population has slowed down.
    a. stationary c. lag
    b. death d. exponential growth

CONCEPT QUESTIONS

1. Differentiate between micronutrients and macronutrients.
2. Briefly describe the general function of the bioelements CHNOPS in the cell. Define growth factors, and give examples of them.
3. Name some functions of metallic ions in cells.
4. Compare autotrophs and heterotrophs with respect to the form of carbon-based nutrients they require.
5. Describe the nutritional strategy of two types of chemoautotrophs (lithotrophs) in the chapter.
6. Briefly fill in the following table

	Source of Carbon	Usual Source of Energy	Example
Photoautotroph			
Photoheterotroph			
Chemoautotroph			
Chemoheterotroph			
Saprobe			
Parasite			

7. Compare the effects of isotonic, hypotonic, and hypertonic solutions on an ameba and on a bacterial cell. If a cell is in a hypotonic environment, what is the condition of its cytoplasm relative to the environment?

8. Look at the following diagrams and predict which direction osmosis will take place. Use arrows to show the net direction of osmosis. Is one of these microbes a halophile? Which one?

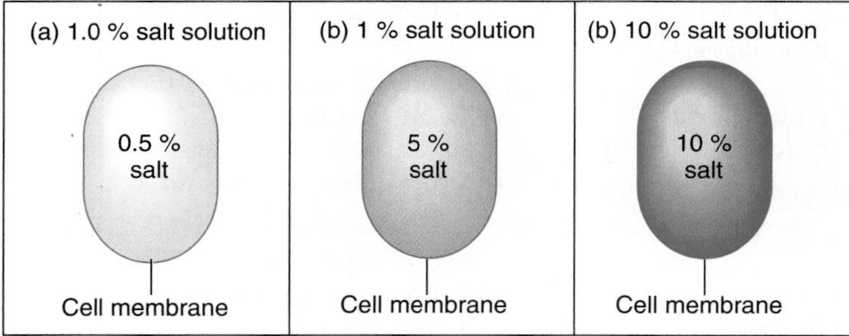

9. Why are most pathogens mesophilic?
10. What are the ecological roles of psychrophiles and thermophiles?
11. What is the natural habitat of a facultative parasite? of strict saprobe?
12. Name three groups of obligate intracellular parasites.
13. Classify a human with respect to oxygen requirements.
14. a. What might be the habitat of an aerotolerant anaerobe?
    b. Where in the body are anaerobic habitats apt to be found?
15. Where do superoxide ions and hydrogen peroxide originate?
16. a. Define symbiosis and differentiate among mutualism, commensalism, synergism, parasitism, and antagonism, using examples.
    b. How are parasitism and antagonism similar and different?
    c. Are any of these relationships obligatory?
17. Explain the relationship between colony counts and colony-forming units. Why can one use the number of colonies as an index of population size?
18. Why is growth called exponential? What makes it a geometric progression?
19. Explain what is happening to the population at points A, B, C, and D.

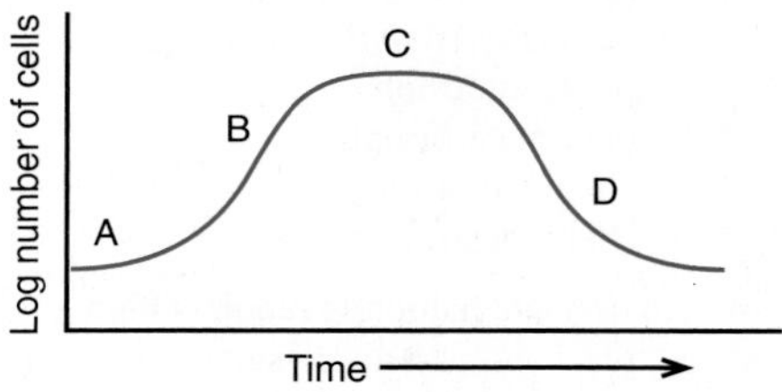

20. Why are we concerned with differentiating live versus dead cells in populations?

Critical-Thinking Questions

1. a. Is there a microbe that could grow on a medium that contains only the following compounds dissolved in water: $CaCO_3$, $MgNO_3$, FeC_{12}, $ZnSO_4$, and glucose? Defend your answer.
   b. Check the last entry in the cell composition summary on page 194. Are all needed elements present here? Where is the carbon?
2. Describe how one might determine the nutrient requirements of a microbe from Mars. If, after exhausting the nutrient schemes, it still does not grow, what other factors might one take into account?
3. a. Explain what ultimately determines whether a microorganism can adapt to a certain habitat.
   b. Give two examples of ways that microbes become modified to survive.
4. a. How can osmotic pressure and pH be used in preserving foods?
   b. How do they affect microbes?
5. a. How might one effectively treat anaerobic infections using gas?
   b. Explain how it would work.
6. How can you explain the observation that unopened milk will spoil even while refrigerated?
7. What would be the effect of a fever on a thermophilic pathogen?
8. Patients with acidotic diabetes are especially susceptible to fungal infections. Can you explain why?
9. Using the concept of synergism, can you describe a way to grow a fastidious microbe?
10. Describe a way to isolate an antibiotic-producing bacterium.
11. a. If an egg salad sandwich sitting in a warm car for 4 hours develops 40,960 bacterial cells, how many more cells would result with just one more hour of incubation?
    b. With 10 hours of incubation?
    c. What would the cell count be after 4 hours if the initial bacterial dose were 100?
    d. What do your answers tell you about using clean techniques in food preparation (other than aesthetic considerations)?
12. Why is an older culture needed for spore staining?
13. Should biotechnology companies be allowed to isolate and own microorganisms taken from the earth's habitats and derive profits from them? Why or why not?

Internet Search Topic

Look up the term **extremophile** on a search engine. Find examples of microbes that have adapted to various extremes. Is there a species that exists in several extremes simultaneously?

chapter 8

MICROBIAL METABOLISM: The Chemical Crossroads of Life

This chapter discusses four interrelated concepts that are all part of metabolism: the nature of enzymes, the function of enzymes, the flow of energy in the cell, and the pathways that govern nutrient processing. Because a common metabolic thread passes among all organisms, a study of metabolism increases our understanding of life processes in general. Our present knowledge of human cell physiology and the success of modern drug therapy were made possible through studies of microbial metabolism. Such diverse areas as biotechnology and clinical diagnosis depend on microbial metabolism. The manufacture of many types of antibiotics, organic acids, alcohols, hormones, and vitamins is dependent on microorganisms. Pathogenic microorganisms are frequently identified by metabolic and enzymatic tests that are specific chemical "fingerprints" for each species.

A vat of wine in the early stages of fermentation. The foam on top is created by CO_2 gas given off by yeast as it metabolizes sugar.

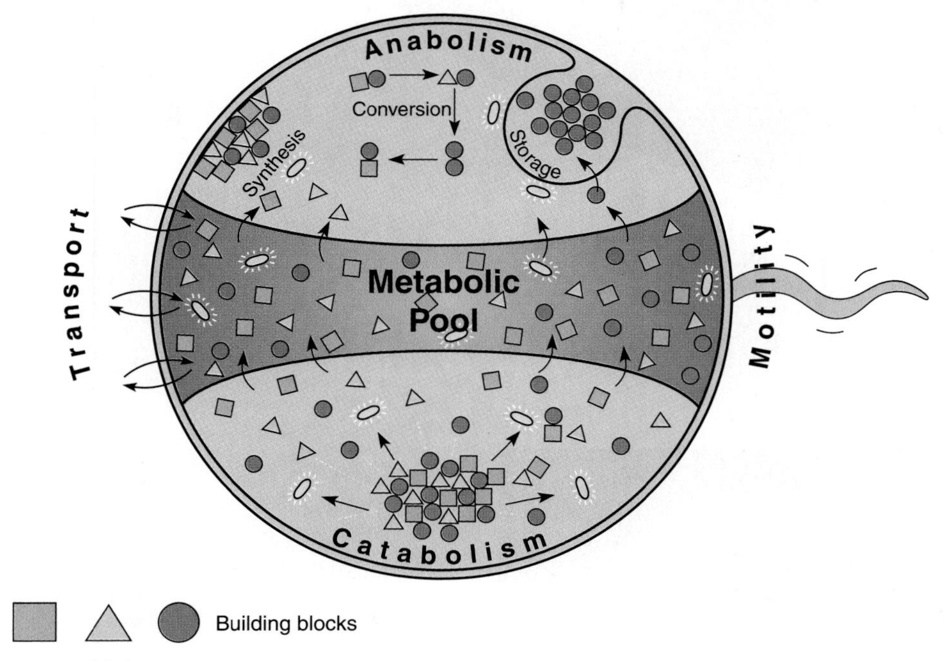

Figure 8.1
Summary of metabolic functions. Energy is released in catabolic reactions and used in activities such as transport, motility, and anabolic pathways.

THE METABOLISM OF MICROBES

Metabolism, from the Greek term *metaballein,* meaning change, pertains to all chemical reactions and physical workings of the cell. Although metabolism entails thousands of different reactions, most of them fall into one of two general categories. **Anabolism,*** sometimes also called *biosynthesis,* is any process that results in synthesis of cell molecules and structures. It is a building and bond-making process that forms larger molecules from smaller ones, and it usually requires the input of energy. **Catabolism*** is the opposite, or complement, of anabolism. Catabolic reactions are degradative; they break bonds, convert larger molecules into smaller components, and often produce energy. The linking of anabolism to catabolism ensures the efficient completion of many thousands of cellular processes.

Metabolism is a cyclical self-regulatory process that maintains the stability of the cell (figure 8.1). Its actions provide a dynamic pool of chemical building blocks and give rise to enzymes and structural components of the cell. Metabolism of nutrients can extract energy in the form of adenosine triphosphate (ATP), or other high-energy compounds, that can be channeled into such processes as biosynthesis, transport, growth, and motility. As we will see, **metabolites**—compounds given off by the complex networks of metabolism—often serve multipurpose roles in the economy of the cell. Metabolism is highly organized and responsive to fine controls. For instance, there are mechanisms for diminishing or ceasing the production of a substance that is not in demand and for diverting excess nutrients into storage. The details of metabolic schemes and pathways are indeed complex. Having such an intricate chemical organization makes cell function and survival possible.

ENZYMES: CATALYZING THE CHEMICAL REACTIONS OF LIFE

A microbial cell could be viewed as a microscopic factory, complete with basic building materials, a source of energy, and a "blueprint" for running its extensive network of metabolic reactions. But the chemical reactions of life, even when highly organized and complex, cannot proceed without a special class of molecules called **enzymes.*** Enzymes are a remarkable example of **catalysts,*** chemicals that increase the rate of a chemical reaction without becoming part of the products or being consumed in the reaction. Do not make the mistake of thinking that an enzyme creates a reaction. Because of the great energy of some molecules, a reaction would occur spontaneously at some point even without an

* anabolism (ah-nab′-oh-lizm) Gr. *anabole,* a throwing up.
* catabolism (kah-tab′-oh-lizm) Gr. *katabole,* a throwing down.
* enzyme (en′-zyme) Gr. *en,* in, and *syme,* leaven. Named for catalytic agents first found in yeasts.
* catalyst (kat′-uh-list) Gr. *katalysis,* dissolution.

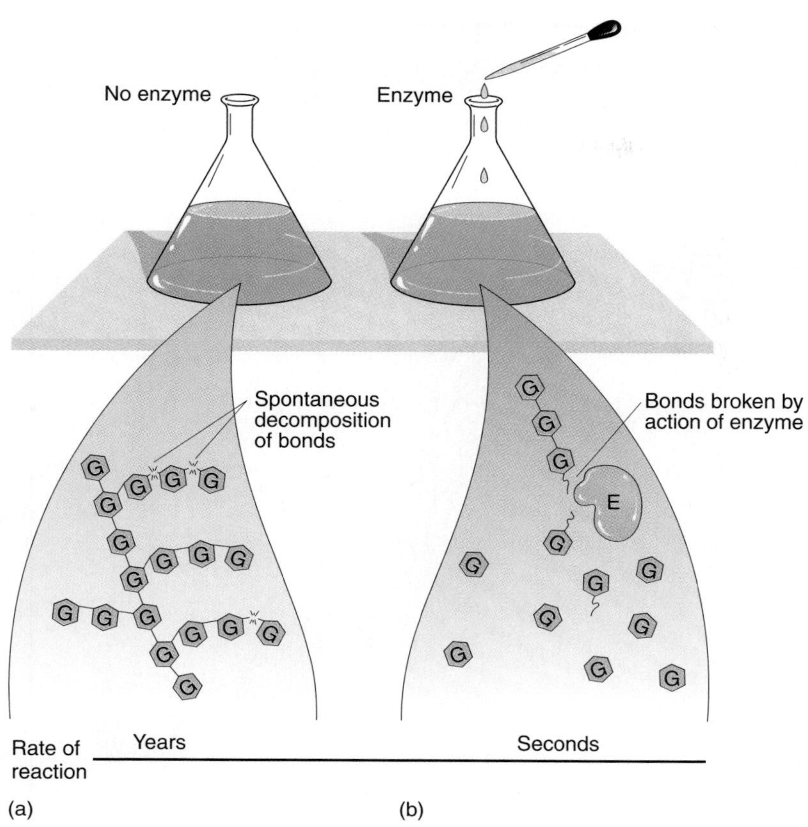

Figure 8.2
Comparison of a noncatalyzed reaction with a catalyzed reaction. *(a)* A starch molecule will spontaneously decompose into glucose after many years in the absence of an enzyme. *(b)* Addition of an enzyme can speed up the reaction so that it happens in a few seconds (E).

enzyme (figure 8.2). But such uncatalyzed metabolic reactions do not occur fast enough to sustain life processes, and thus enzymes, which speed up the rate of reactions, are indispensable to life. Other characteristics of enzymes are summarized in table 8.1.

How Do Enzymes Work?

We have said that an enzyme speeds up the rate of a metabolic reaction, but just how does it do this? During a chemical reaction, reactants are converted to products by bond formation or breakage. A certain amount of energy is required for every such reaction, which limits its rate. This resistance to a reaction, which must be overcome for a reaction to proceed, is measurable and is called the **energy of activation** (microfile 8.1). In the laboratory, overcoming this initial resistance can be achieved by (1) increasing thermal energy (heating), which increases molecular velocity, (2) increasing the concentration of reactants, or (3) adding a catalyst. In most living systems, the first two alternatives are not feasible, because elevating the temperature is potentially harmful and higher concentrations of reactants are not practical. This leaves only the action of catalysts, and enzymes fill this need efficiently and potently.

At the molecular level, an enzyme physically promotes a reaction by serving as a physical site upon which one or more reactant molecules—the **substrate** (or substrates)—can be positioned for various interactions. Although an enzyme becomes physically attached to the substrate and participates directly in bonding, it does not become a part of the products, is not used up by the reaction, and can function over and over again. In fact, studies have shown that a single molecule of an enzyme can catalyze the reactions of several million substrate molecules. To further visualize the roles of enzymes in metabolism, we must next look at their structure.

TABLE 8.1

CHECKLIST OF ENZYME CHARACTERISTICS

- Act as organic catalysts to speed up the rate of cellular reactions
- Are composed of protein and may require cofactors
- Have unique characteristics such as shape, specificity, and function
- Enable metabolic reactions to proceed at a speed compatible with life
- Provide a reactive site for target molecules called substrates
- Associate closely with substrates but do not become integrated into the reaction products
- Are not used up or permanently changed by the reaction
- Lower the activation energy required for a chemical reaction to proceed (microfile 8.1)
- Can be recycled, thus function in extremely low concentrations
- Are limited by particular conditions of temperature and pH
- Can be regulated by feedback and genetic mechanisms

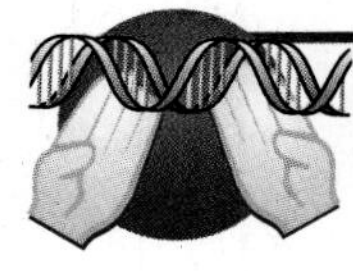

MICROFILE 8.1 ENZYMES AS BIOCHEMICAL LEVERS

An analogy will allow us to envision the relationship of enzymes to the energy of activation. A large boulder sitting precariously on a cliff's edge contains a great deal of potential energy, but it will not fall to the ground and release this energy unless something disturbs it. It might eventually be disturbed spontaneously as the cliff erodes below it, but that could take a very long time. Moving it with a crowbar (to overcome the resistance of the boulder's weight) will cause it to tumble down freely by its own momentum. If we relate this analogy to chemicals, the reactants are the boulder on the cliff, the activation energy is the energy needed to move the boulder, the enzyme is the crowbar, and the product is the boulder at the bottom of the cliff. Like the crowbar, an enzyme permits a reaction to occur rapidly by lowering the energy obstacle. (Although this analogy concretely illustrates the concept of the energy of activation, it is imperfect in that the enzyme, unlike the person with the crowbar, does not actually add any energy to the system.)

This phenomenon can be represented by plotting the energy of the reaction against the direction of the reaction. The curve of reaction is shown in the accompanying graph. It is evident that there is an energy "hill" called the energy of activation (E_{act}). That must be overcome for the reaction to proceed. When a catalyst is present, it lowers the E_{act}. It does this by providing the reactants with an energy "shortcut" by which they can progress to the final state. Note that the products are at the same final energy state with or without an enzyme.

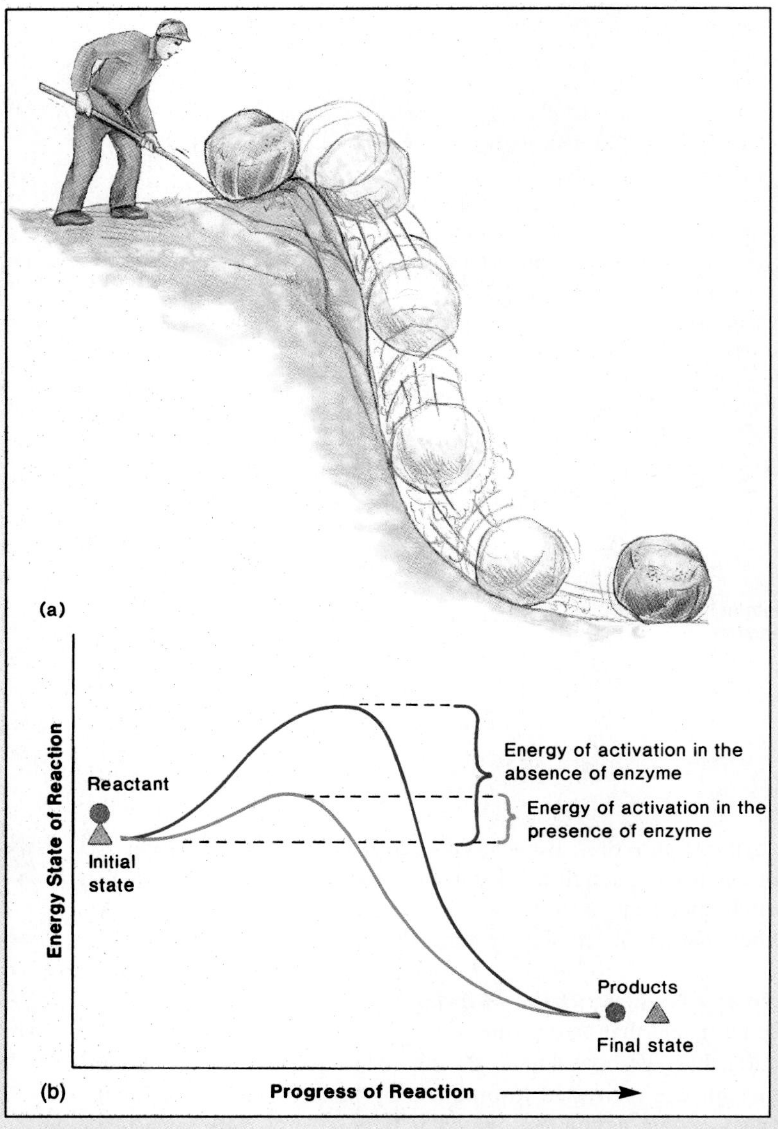

Analogy demonstrating the influence of enzymes on chemical reactions. (a) A boulder can represent potential energy available for a chemical reaction. (b) Graph of chemical reaction, with and without an enzyme. Energy (called energy of activation) is required in both cases to convert a reactant molecule to products. But in an enzyme-catalyzed reaction, the enzyme significantly lowers this energy of activation and allows the reaction to proceed more readily and rapidly.

Enzyme Structure

Enzymes can be classified as simple or conjugated. **Simple** enzymes consist of protein alone, whereas **conjugated** enzymes (figure 8.3) contain protein and nonprotein molecules (with some exceptions; microfile 8.2). A conjugated enzyme, sometimes referred to as a **holoenzyme,*** is a combination of a protein, now called the **apoenzyme,** and one or more **cofactors** (table 8.2). Cofactors are either organic molecules, called **coenzymes,** or inorganic elements (metal ions). In some enzymes, the cofactor is loosely associated with the apoenzyme by noncovalent bonds; in others it is tenaciously linked by covalent bonds.

Apoenzymes: Specificity and the Active Site

Apoenzymes range in size from small polypeptides with about 100 amino acids and a molecular weight of 12,000 to large polypeptide conglomerates with thousands of amino acids and a molecular

* holoenzyme (hol-oh-en′-zyme) Gr. *holos,* whole.

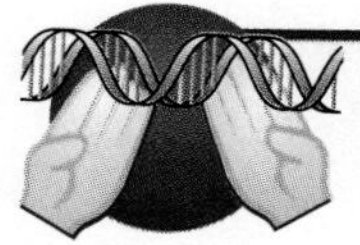

MICROFILE 8.2 UNCONVENTIONAL ENZYMES

The secrets of the cell never cease to surprise biologists. Just when it seemed that the protein nature of biological catalysts and enzymes had been well established, researchers discovered a novel type of RNA in both procaryotes and eucaryotes that also acts as a catalyst. Even more startling is the fact that the target of these molecules, called *ribozymes,* is RNA itself. In their basic mechanisms, ribozymes are really very similar to protein-based enzymes: they have a very specialized active site that binds to a certain place on a substrate and changes that substrate in some way. It appears that the precise function of RNA-based enzymes is somewhat more restricted than that of protein-based enzymes. They help process the genetic code by cleaving RNA molecules, but they do not handle the diversity of substrates that regular protein enzymes do.

Meanwhile, back at the lab, immunologists were studying enzymatic possibilities from yet another perspective. They were already well acquainted with the concept of a lock-and-key specific fit because of the way antibodies lock onto their targets. This knowledge led to the inevitable question: Can antibodies, which are also proteins, act as enzymes, too? The answer is a qualified yes—if the antibody is carefully chosen. These catalytic antibodies, or *abzymes,* have extreme specificity for their substrate and can speed up a reaction, form an abzyme-substrate complex, participate in bond breaking, and release a product. What is exciting about the use of abzymes is that, unlike enzymes, antibodies are very easy to fine-tune to fit almost any possible substrate configuration and can be manufactured in large amounts using monoclonal methods (see chapter 15). The applications for medicine and industry are considerable. For instance, an abzyme could be designed to do double duty—both attaching to a target molecule and enzymatically destroying it. This technology has the potential for treating cancers, infections, abnormal blood clots, and other medical problems very specifically. The possible use of abzymes to purify drugs and inactivate viruses is already being explored.

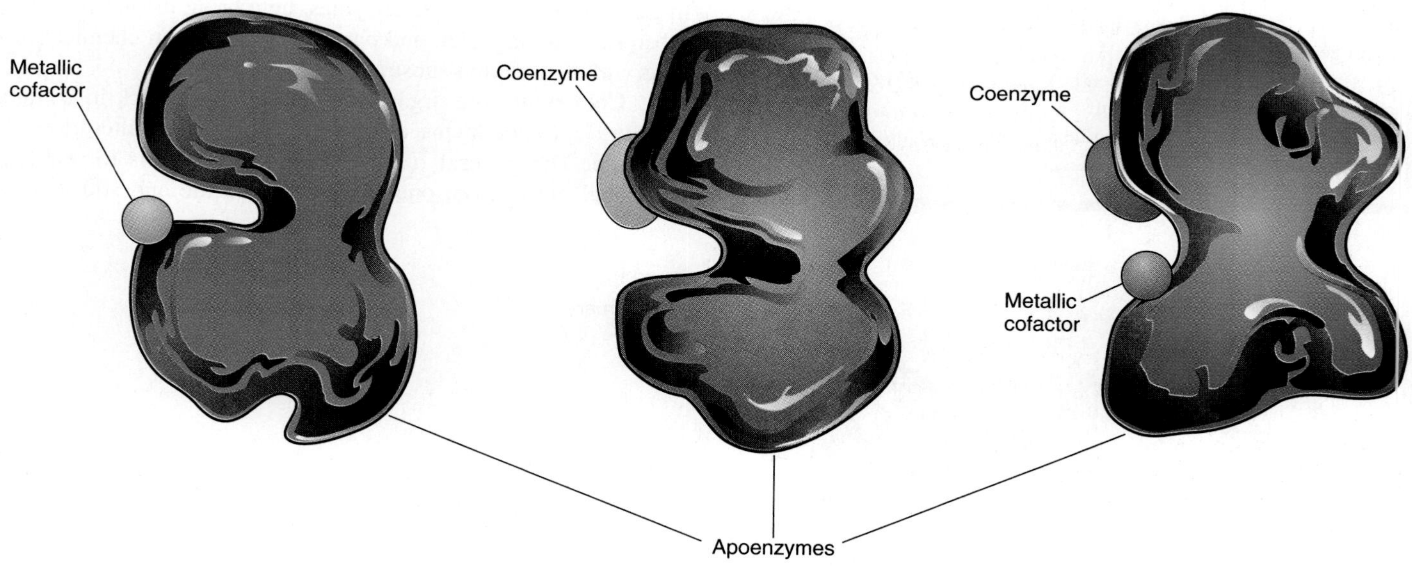

Figure 8.3
Conjugated enzymes are composed of various combinations. All have an apoenzyme (polypeptide or protein) component and one or more cofactors.

weight of over one million. Like all proteins, an apoenzyme exhibits levels of molecular complexity called the primary, secondary, tertiary, and, in larger enzymes, quaternary organization (figure 8.4). As we saw in chapter 2, the first three levels of structure arise from a single polypeptide chain that has undergone an automatic folding process and achieved stability by forming disulfide and other types of bonds. Folding causes the surface of the apoenzyme to acquire three-dimensional surface features. Folding and the surface features that result are related to enzyme specificity in that (1) the apoenzyme of each enzyme differs from others in its primary structure (the type and sequence of amino acids);[1] (2) nuances in polypeptide folding produce a particular three-dimensional configuration (tertiary structure); and (3) surface features of the tertiary structure provide a unique and specific site (usually a crevice or groove) for attachment of substrate molecules. The site that

1. As specified by the genetic blueprint in DNA (see chapter 9).

TABLE 8.2

SELECTED ENZYMES, CATALYTIC ACTIONS, AND COFACTORS

Enzyme	Action	Cofactor
Catalase	Breaks down hydrogen peroxide	Iron (Fe)
Oxidase	Adds electrons to oxygen	Iron, copper (Cu)
Hexokinase	Transfers phosphate to glucose	Magnesium (Mg)
Urease	Splits urea into ammonium	Nickel (Ni)
Nitrate reductase	Reduces nitrate to nitrite	Molybdenum (Mo)
DNA polymerase complex	Synthesis of DNA	Zinc (Zn) and Mg
Enzyme	**Coenzyme and the Vitamin Required for Function**	
Glycine synthetase	Tetrahydrofolate requires folic acid	
Transaminase	Pyridoxal phosphate requires vitamin B_6	
Pyruvate dehydrogenase	Coenzyme A requires pantothenic acid	
Flavin dehydrogenase	Flavin adenine dinucleotide (FAD) requires riboflavin	
Various other dehydrogenases	Nicotinamide adenine dinucleotide (NAD) requires niacin	

accepts a substrate is called the **active,** or **catalytic, site,** and there can be from one to several such sites (figure 8.4).

Enzyme-Substrate Interactions

For a reaction to take place, a temporary enzyme-substrate union must occur at the active site. This interaction has been described as a complementary lock-and-key fit (figure 8.5). The bonds formed between the substrate and enzyme are weak and, of necessity, easily reversible. Once the temporary enzyme-substrate complex has formed, appropriate reactions occur on the substrate, often with the aid of a cofactor, and a product is formed and released. The enzyme can then attach to another substrate molecule and repeat this action. Although enzymes can potentially catalyze reactions in both directions, most examples in this chapter will depict them working in one direction only.

Cofactors: Supporting the Work of Enzymes

In chapter 7 we learned that microorganisms require specific metal ions called trace elements and certain organic growth factors. In many cases, the need for these substances arises from their roles as cofactors. The **metallic cofactors,** including iron, copper, magnesium, manganese, zinc, cobalt, selenium, and many others, participate in precise functions between the enzyme and its substrate. In general, metals activate enzymes, help bring the active site and substrate close together, and participate directly in chemical reactions with the enzyme-substrate complex.

Coenzymes are organic compounds that work in conjunction with an apoenzyme to perform a necessary alteration of a substrate. The general function of a coenzyme is to remove a functional group from one substrate molecule and add it to an-

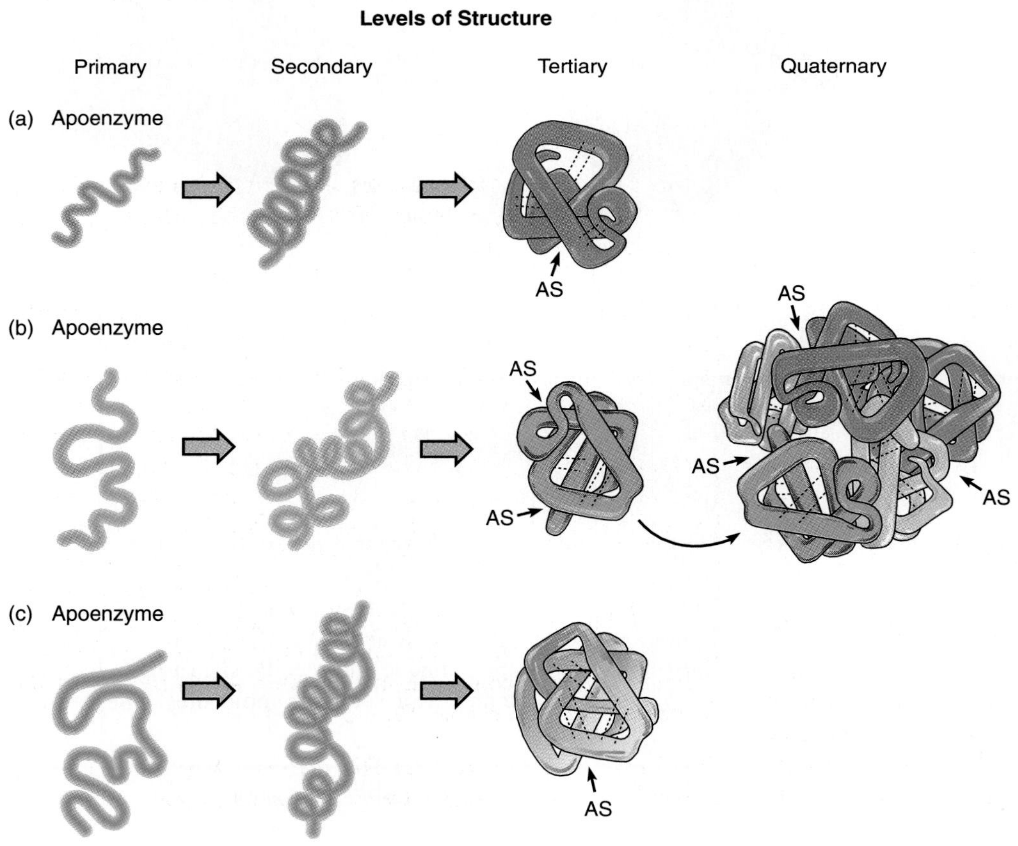

Figure 8.4
How the active site and specificity of the apoenzyme arise. As the polypeptide forms intrachain bonds and folds, it assumes a three-dimensional (tertiary) state with numerous surface features. Because each different polypeptide folds differently, each apoenzyme will have differently shaped active sites (AS). Some enzymes have more than one active site; others have sites to attach cofactors and regulatory compounds; and more complex enzymes have a quaternary structure consisting of several polypeptides bound by weak forces, as in *(b)*.

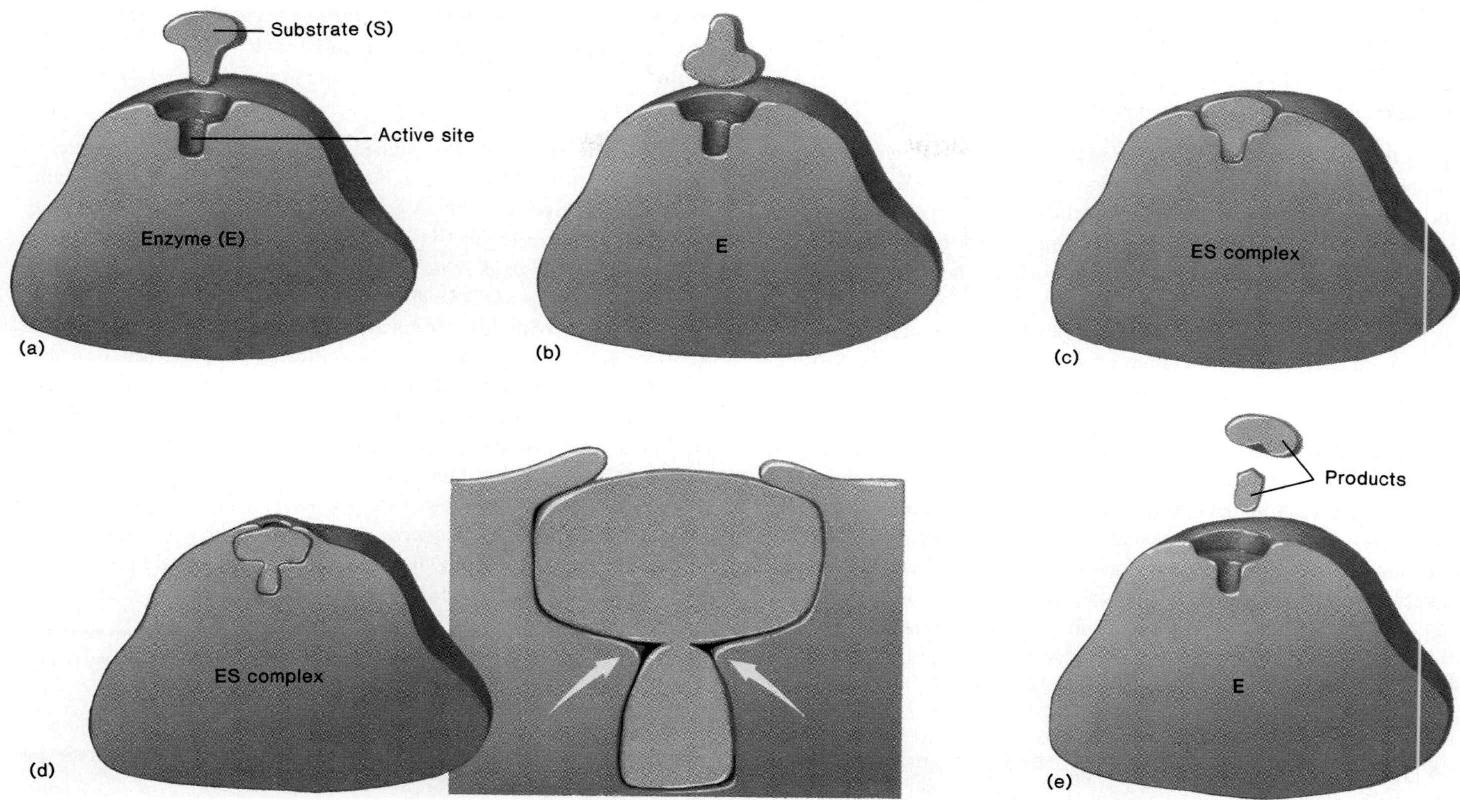

Figure 8.5

Enzyme-substrate reactions: fit, proximity, and orientation. When the enzyme and substrate come together, the substrate (S) must be close and in correct position with respect to the enzyme (E). *(a)* and *(b)* show incorrect proximity and orientation; (c) shows a correct fit, position, and proximity. *(d)* When the ES complex is formed, it enters a transition state. During this temporary but tight interlocking union, the enzyme tightly compresses the substrate as it participates directly in breaking or making bonds. *(e)* Once the reaction is complete, the enzyme releases the products.

other, thereby serving as a transient carrier of this group (figure 8.6). The specific activities of coenzymes are many and varied. In a later section of this chapter, we shall see that coenzymes carry and transfer hydrogen atoms, electrons, carbon dioxide, and amino groups. Because coenzymes are **vitamins** or contain vitamins, it becomes even clearer why vitamins are important to nutrition and are required as growth factors in the majority of living things. Vitamin deficiencies prevent the complete holoenzyme from forming. Consequently, both the chemical reaction and the structure or function dependent upon that reaction are compromised.

Classification of Enzyme Functions

Enzymes are classified and named according to characteristics such as site of action, type of action, and substrate (microfile 8.3).

Location and Regularity of Enzyme Action Enzymes exhibit several patterns of performance. After initial synthesis in the cell, **exoenzymes** are transported extracellularly, where they break down (hydrolyze) large food molecules or harmful chemicals. Examples of exoenzymes are cellulase, amylase, and penicillinase. By contrast, **endoenzymes** are retained intracellularly and function there. Most enzymes of the metabolic pathways are of this variety (figure 8.7).

In terms of their presence in the cell, enzymes are not all produced in equal amounts or at equal rates. Some, called **constitutive enzymes** (figure 8.7*c*), are always present and in relatively constant amounts, regardless of the amount of substrate. The enzymes involved in utilizing glucose, for example, are very important in metabolism and thus are constitutive. An **induced** (or inducible) **enzyme** (figure 8.7*d*) is not constantly present and is produced only when its substrate is present. Induced enzymes are present in amounts ranging from a few molecules per cell to several thousand times that many, depending on the metabolic requirement. This property of selective synthesis of enzymes prevents a cell from wasting energy by making enzymes that will not be used immediately. The induction of enzymes constitutes an important metabolic control discussed later in this section of the chapter and again in chapter 9.

Synthesis and Hydrolysis Reactions A growing cell is in a frenzy of activity, constantly synthesizing proteins, DNA, and

MICROFILE 8.3 THE ENZYME NAME GAME

Most metabolic reactions require a separate and unique enzyme. Up to the present time, researchers have discovered and named over 5,000 of them. Given the complex chemistry of the cell and the extraordinary diversity of living things, many more enzymes probably remain to be discovered. A standardized system of nomenclature and classification was developed to prevent discrepancies.

In general, an enzyme name is composed of two parts: a prefix or stem word derived from a certain characteristic—usually the substrate acted upon or the type of reaction catalyzed, or both—followed by the end *-ase.*

The system classifies the enzyme in one of the following six classes, on the basis of its general biochemical action: (1) *Oxidoreductases* transfer electrons from one substrate to another, and *dehydrogenases* transfer a hydrogen from one compound to another. (2) *Transferases* transfer functional groups from one substrate to another. (3) *Hydrolases* cleave bonds on molecules with the addition of water. (4) *Lyases* add groups to or remove groups from double-bonded substrates. (5) *Isomerases* change a substrate into its **isomeric* form.** (6) *Ligases* catalyze the formation of bonds with the input of ATP and the removal of water.

Each enzyme is also assigned a systematic name that indicates the specific reaction it catalyzes, but some of these names are too cumbersome to be used routinely. A common name based on the substrate or some other notable feature, is easier to use and identify. For example, *hydrogen peroxide oxidoreductase,* is more commonly known as *catalase,* and *mucopeptide N-acetylmuramoylhydrolase* is called (thankfully) *lysozyme.* With this system, an enzyme that digests a carbohydrate substrate is a *carbohydrase;* more specifically, *amylase,* acts on starch (amylose is a major component of starch). The enzyme *maltase* digests the sugar maltose. An enzyme that hydrolyzes peptide bonds of a protein is a *proteinase, protease,* or *peptidase,* depending on the size of the protein substrate. Some fats and other lipids are digested by *lipases.* DNA is hydrolyzed by *deoxyribonuclease,* generally shortened to *DNase.* A *synthetase* or *polymerase* bonds together many small molecules into large molecules. Other examples of enzymes are presented in the table below. (See also table 8.2).

*An isomer is a compound that has the same molecular formula as another compound but differs in arrangement of the atoms.

TABLE 8.A
A SAMPLING OF ENZYMES, THEIR SUBSTRATES, AND THEIR REACTIONS

Common Name	Systematic Name	Enzyme Class	Substrate	Action
Cellulase	1,4 β-glycosidase	Hydrolase	Cellulose	Cleaves 1, 4 β-glycosidic linkages
Lactase	β-D-galactosidase	Hydrolase	Lactose	Breaks lactose down into glucose and galactose
Penicillinase	Beta-lactamase	Hydrolase	Penicillin	Hydrolyzes beta-lactam ring
Lipase	Triacylglycerol acylhydrolase	Hydrolase	Triglycerides	Cleaves bonds between glycerol and fatty acids
DNA polymerase	DNA nucleotidyl-transferase	Transferase	DNA nucleosides	Synthesizes a strand of DNA using the complementary strand as a model
Hexokinase	ATP-glucose phosphotransferase	Transferase	Glucose	Catalyzes transfer of phosphate from ATP to glucose
Aldolase	Fructose diphosphate aldolase	Lyase	Fructose diphosphate	Catalyzes the conversion of the substrate to two 3-carbon fragments
Lactate dehydrogenase	Same as common name	Oxidoreductase	Pyruvic acid	Catalyzes the conversion of pyruvic acid to lactic acid
Oxidase	Cytochrome oxidase	Oxidoreductase	Molecular oxygen	Catalyzes the reduction (addition of electrons and hydrogen) to O_2

RNA; forming storage polymers such as starch and glycogen; and assembling new cell parts. Such anabolic reactions require enzymes (ligases) to form covalent bonds between smaller substrate molecules. Also known as *condensation reactions,* synthesis reactions typically require ATP and always release one water molecule for each bond made (figure 8.8*a*). Catabolic reactions involving energy transactions, remodeling of cell structure, and digestion of macromolecules are also very active during cell growth. These processes require enzymes that reduce substrates to a series of smaller molecules by breaking intramolecular bonds. Because the breaking of each bond requires the input of a water molecule, digestion is often termed a *hydrolysis* reaction* (figure 8.8*b*).

* **hydrolysis** (hy-drol′-uh-sis) Gr. *hydro,* water, and *lysis,* setting free.

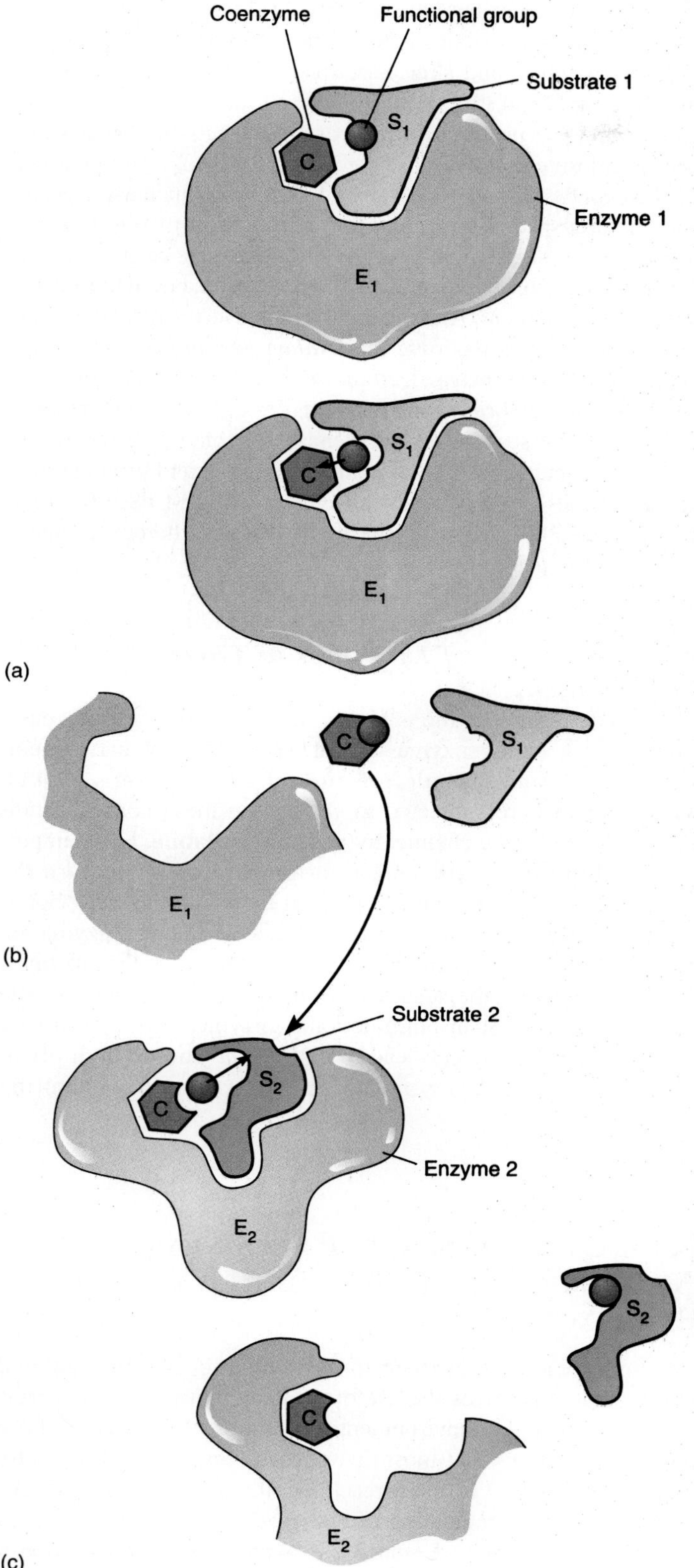

Figure 8.6
The carrier functions of coenzymes. *(a)* A coenzyme carrier in position to remove a functional group from substrate 1 (S_1). *(b)* Coenzyme with the attached functional group and the substrate that donated the group are released. *(c)* The coenzyme carries the functional group to a second enzyme-substrate complex and passes the group to a second substrate (S_2), which is then released.

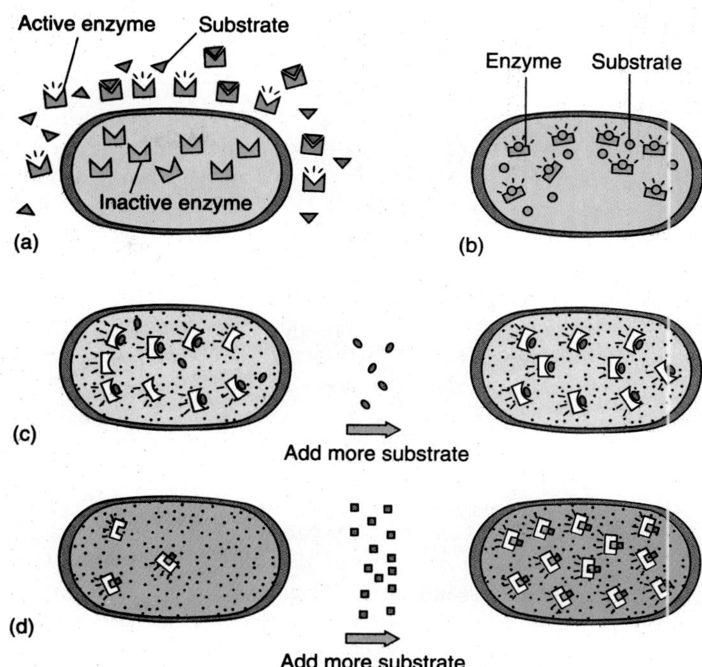

Figure 8.7
Types of enzymes, as described by their location of action and quantity. *(a)* Exoenzymes function extracellularly. *(b)* Endoenzymes function intracellularly. *(c)* Constitutive enzymes are present in constant amounts in a cell. The addition of more substrate does not increase the numbers of these enzymes. *(d)* Induced enzymes are normally present in trace amounts, but their quantity can be increased a thousandfold by the addition of substrate.

Transfer Reactions by Enzymes Other enzyme-driven processes that involve the simple addition or removal of a functional group are important to the overall economy of the cell. Oxidation-reduction and other transfer activities are examples of these types of reactions.

Some atoms and compounds readily give or receive electrons and participate in oxidation (the loss of electrons) or reduction (the gain of electrons). The compound that loses the electrons is **oxidized,** and the compound that receives the electrons is **reduced.** Such oxidation-reduction (redox) reactions are common in the cell and indispensable to the energy transformations discussed later in this chapter. Important components of cellular redox reactions are oxidoreductases, which remove electrons from one substrate and add them to another, and their coenzyme carriers, nicotinamide adenine dinucleotide (NAD; see figure 8.14) and flavin adenine dinucleotide (FAD).

Other enzymes play a role in the molecular conversions necessary for the economical use of nutrients by directing the transfer of functional groups from one molecule to another. For example,

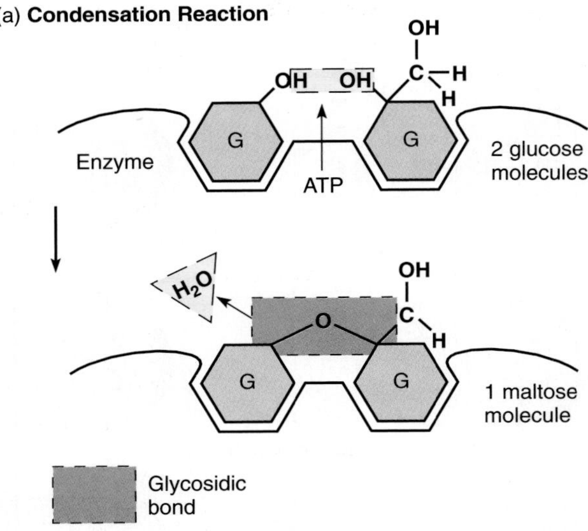

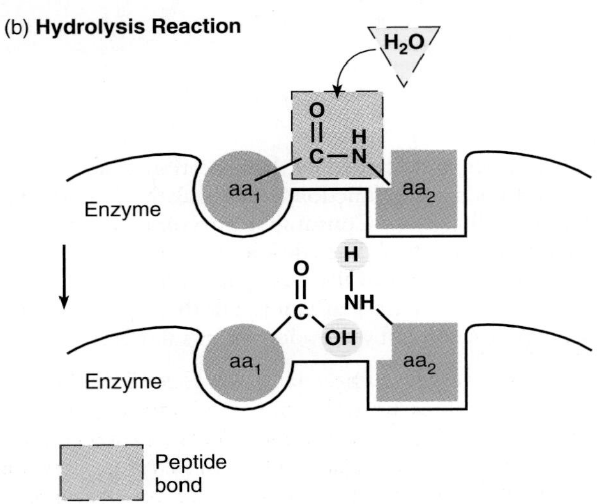

Figure 8.8
Examples of enzyme-catalyzed synthesis and hydrolysis reactions. *(a)* Condensation reaction. Forming a glycosidic bond between two glucose molecules to generate maltose requires the removal of a water molecule and energy from ATP. *(b)* Hydrolysis reaction. Breaking a peptide bond between two amino acids requires a water molecule that adds as an H and OH to the amino acids.

aminotransferases convert one type of amino acid to another by transferring an amino group (see figure 8.28); *phosphotransferases* participate in the transfer of phosphate groups and are involved in energy transfer; *methyltransferases* move a methyl (CH_3) group from substrate to substrate; and *decarboxylases* (also called carboxylases) catalyze the removal of carbon dioxide from organize acids in several metabolic pathways.

The Role of Microbial Enzymes in Disease Many bacterial pathogens secrete unique exoenzymes that help them avoid host defenses or promote their multiplication in tissues. Because these enzymes contribute to pathogenicity, they are referred to as virulence factors, or toxins in some cases. *Streptococcus pyogenes* (a cause of throat and skin infections) produces a streptokinase that digests blood clots and apparently assists in invasion of wounds. It also produces a protease that accounts for a severe form of "flesh eating" disease (see microfile 18.2). The lipases of *Staphylococcus aureus* (the cause of skin boils) increase the virulence of this species by promoting its invasion of oil-producing glands on the skin. *Pseudomonas aeruginosa,* a respiratory and skin pathogen, produces elastase and collagenase, which digest elastin and collagen. These increase the severity of certain lung diseases and burn infections. *Clostridium perfringens,* an agent of gas gangrene, synthesizes lecithinase C, a lipase that profoundly damages cell membranes and accounts for the tissue death associated with this disease. The collagenases of some parasitic worms digest the principal protein of connective tissue and promote invasion of tissues. However, not all enzymes digest tissues; some, such as penicillinase, inactivate penicillin and thereby protect a microbe from its effects.

The Sensitivity of Enzymes to Their Environment

The activity of an enzyme is highly influenced by the cell's environment. In general, enzymes operate only under the natural temperature, pH, and osmotic pressure of an organism's habitat. When enzymes are subjected to changes in these normal conditions, they tend to be chemically unstable, or **labile.** Low temperatures inhibit catalysis, and high temperatures denature the apoenzyme. **Denaturation** is a process by which the weak bonds that collectively maintain the native shape of the apoenzyme are broken. This disruption causes extreme distortion of the enzyme's shape and prevents the substrate from attaching to the active site (see figure 11.4). Such nonfunctional enzymes block metabolic reactions and thereby can lead to cell death. Low or high pH or certain chemicals (heavy metals, alcohol) are also denaturing agents.

Regulation of Enzymatic Activity and Metabolic Pathways

Metabolic reactions proceed in a systematic, highly regulated manner that maximizes the use of available nutrients and energy. The cell responds to environmental conditions by adopting those metabolic reactions that most favor growth and survival. Because enzymes are critical to these reactions, the regulation of metabolism is largely the regulation of enzymes by an elaborate system of checks and balances. Before we examine some of these control systems, let us take a look at some general features of metabolic pathways.

Metabolic Pathways

Metabolic reactions rarely consist of a single action or step. More often, they occur in a multistep series or pathway, with each step catalyzed by an enzyme. An individual reaction is shown in various ways, depending on the purpose at hand (figure 8.9). The product of one reaction is often the reactant (substrate) for the next, forming a linear chain of reactions. Many pathways have branches that provide alternate methods for nutrient processing. Others take a cyclic form, in which the starting molecule is regenerated to initiate another turn of the cycle (for example, the TCA cycle; see figure 8.21). Pathways generally do not stand alone; they are interconnected and merge at many sites.

Every pathway has one or more enzyme pacemakers (usually the slowest enzyme in the series) that sets the rate of a pathway's progression. These enzymes respond to various control signals and, in so doing, determine whether a pathway proceeds. Regulation of pacemaker enzymes proceeds on two fundamental levels. Either the enzyme itself is directly inhibited or activated or the amount of the enzyme in the system is altered (decreased or increased). Factors that affect the enzyme directly provide a means for the system to be finely controlled or tuned, whereas regulation at the genetic level enzyme synthesis) is a coarser control.

Direct Controls on the Behavior of Enzymes

Competitive Inhibition In **competitive inhibition**, other molecules with a structure similar to the normal substrate can occupy the enzyme's active site. Although these molecular mimics have the proper fit for the site, they cannot be further acted upon. The presence of one of these mimics effectively prevents an enzyme from attaching to its usual substrate and blocks its activity, output, and possibly the rest of that pathway. This mechanism is common in natural metabolic pathways. It is also an important mode of action by some drugs for treating infections (see chapter 12).

Feedback Control The term *feedback* is borrowed from the field of engineering, where it describes a process in which the product of a system is *fed back* into the system to keep it in balance. Feedback is positive or negative, depending upon whether the process is stimulated (+) or inhibited (−) by the product. **Negative feedback** is common in electrical appliances such as hot plates or furnaces with automatically controlled temperatures. These devices possess a means of controlling the temperature (a heating coil or flame) and a temperature sensor (thermostat) that is set to a desired point and regulates the turning on or off of the heat source. In the case of a hot plate, when the set temperature has been reached, the thermostat switches the heating elements off. A subsequent fall below the set temperature causes the heating element to be turned on again.

Many enzymatic reactions are controlled by a similar negative mechanism, whereby the end product being fed back into the system *negates* (cancels) an enzyme's activity:

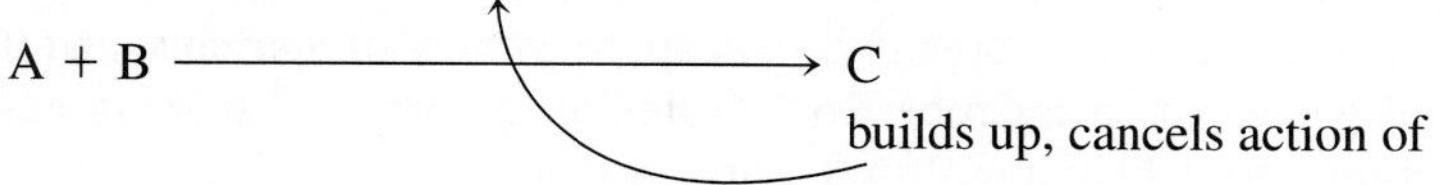

Figure 8.9
In general, metabolic pathways consist of a linked series of individual chemical reactions that produce intermediary metabolites and lead to a final product. These pathways occur in several patterns, including linear, cyclic, and branched. Anabolic pathways involved in biosynthesis result in a more complex molecule, each step adding on a functional group, whereas catabolic pathways involve the dismantling of molecules and can generate energy. Virtually every reaction in a series involves a specific enzyme.

At high levels of substrate and scant levels of end product, the enzyme works freely, but as the end product builds up, it stops the action of that enzyme. It usually targets the first enzyme in a metabolic pathway. The mechanisms of negative feedback—that is, how the product actually stops the action of an enzyme—can best be understood if we look at the allosteric behavior of enzymes.

We have previously shown that globular proteins of enzymes are bulky and possess numerous surface features and that one part of an enzyme's terrain (the active site) accommodates the substrate. **Allosteric* enzymes** have an additional **regulatory site**

* **allosteric** (al-oh-stair′-ik) Gr. *allos*, other, and *steros*, solid. Literally, "another space."

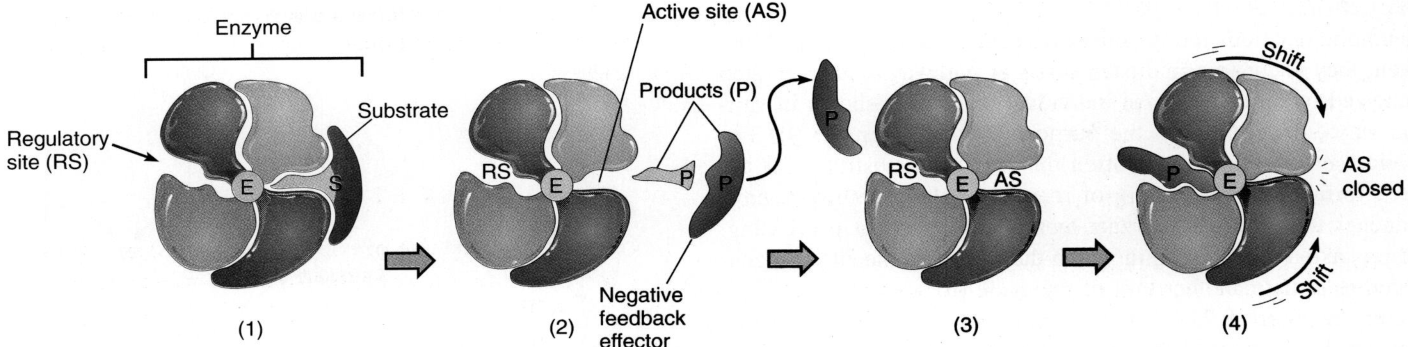

Figure 8.10

Enzyme control by negative feedback in system. Allosteric enzymes have a quaternary structure with more than one site of attachment. *(1,2)* The enzyme complex normally attaches to the substrate at the active site (AS) and releases products (P). *(3)* One product can function as a negative-feedback effector by fitting into a regulatory site (RS) at a different location. *(4)* The entrance of P into RS causes a confirmational shift of the enzyme that closes the active site. The enzyme cannot catalyze further reactions with substrate as long as the product is inserted in the regulatory site.

for the attachment of molecules other than substrate. This regulatory site usually exists on a different polypeptide unit of a quaternary structure. When an end product molecule fits into this regulatory site, the enzyme's active site is so distorted that it can no longer bind to its substrate. Although this distortion does not denature the enzyme and is quite reversible, allosteric inhibitors can temporarily stop the action of that enzyme (figure 8.10). This mechanism is termed **feedback inhibition.** When at some point the end product is used up and more of it is needed, the enzyme will be released from inhibition and can resume catalysis.

Controls on Enzyme Synthesis

Controlling enzymes by controlling their synthesis is effective because enzymes do not last indefinitely. Some wear out, some are deliberately degraded, and others are diluted with each cell division. For catalysis to continue, enzymes eventually must be replaced. This cycle works into the scheme of the cell, where replacement of enzymes can be regulated according to cell demand. The mechanisms of this system are genetic in nature; that is, they require regulation of DNA and the protein synthesis machinery, topics we shall encounter once again in chapter 9.

Enzyme repression is a means to stop further synthesis of an enzyme somewhere along its pathway. As the level of the end product from a given enzymatic reaction has built to excess, the genetic apparatus responsible for replacing these enzymes is automatically suppressed (figure 8.11). The response time takes longer than feedback inhibition, but its effects are more enduring.

A response that resembles the inverse of feedback repression is **enzyme induction.** In this process, enzymes appear (are induced) only when suitable substrates are present; that is, the synthesis of enzyme is induced by its substrate (see figure 9.21). Both mechanisms are important genetic control systems in bacteria.

A classic model of enzyme induction occurs in the response of *Escherichia coli* to certain sugars. For example, if a particular strain of *E. coli* is inoculated into a medium whose principal carbon source is lactose, it will produce lactase to hydrolyze it into glucose and galactose. If the bacterium is subsequently inoculated into a medium containing only sucrose as a carbon source, it will cease synthesizing lactase and begin synthesizing sucrase. This response enables the organism to adapt to variety of nutrients, and it also prevents a microbe from "spinning its wheels," making enzymes for which no substrates are present.

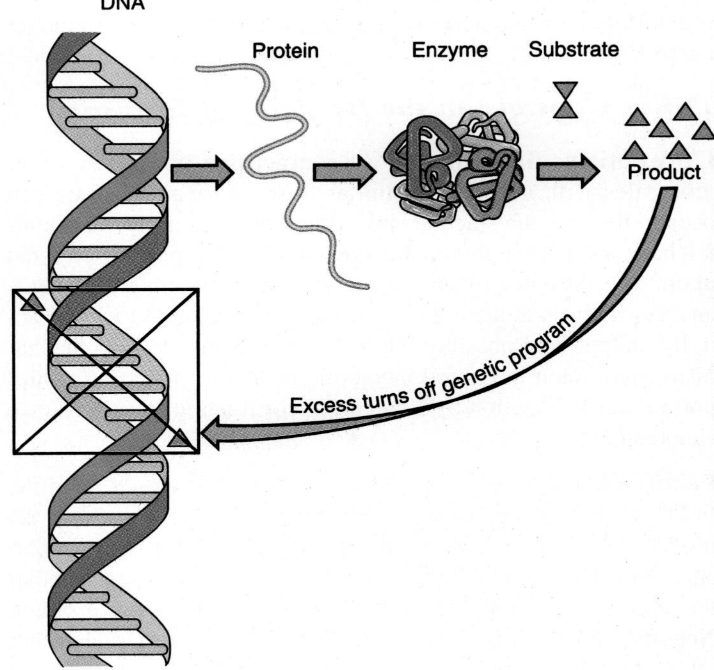

Figure 8.11

One type of genetic control of enzyme synthesis: feedback/enzyme repression. The enzyme is synthesized continuously until enough product has been made, at which time the excess product reacts with a site on DNA that regulates the enzyme's synthesis, thereby inhibiting further enzyme production.

Chapter Checkpoints

Metabolism includes all the biochemical reactions that occur in the cell. It is a cyclical, self-regulating complex of interdependent processes that encompasses many thousands of chemical reactions.

Anabolism is the energy-requiring subset of metabolic reactions, which synthesize large molecules from smaller ones.

Catabolism is the energy-releasing subset of metabolic reactions, which degrade or break down large molecules into smaller ones.

Enzymes are proteins that catalyze all biochemical reactions by forming enzyme-substrate complexes. The physical distortion of the substrate by an enzyme makes possible both bond-forming and bond-breaking reactions, depending on the pathway involved.

Enzymes are classified and named according to the kinds of reactions they catalyze.

To function effectively, enzymes require specific conditions of temperature, pH, and osmotic pressure.

Enzyme activity is regulated by processes of feedback inhibition, induction, and repression, which, in turn, respond to availability of substrate and concentration of end products, as well as to other environmental factors.

THE PURSUIT AND UTILIZATION OF ENERGY

In chapter 7 we surveyed the constant supply of nutrients required for the cellular machine to function. But none of the major cell processes—biosynthesis, movement, transport, or growth—could proceed without the constant input and expenditure of some form of usable **energy.** Except for certain chemoautotrophic bacteria, the ultimate source of energy is the sun, but only photosynthetic autotrophs can tap this source directly. This capacity of photosynthesis to convert solar energy into chemical energy provides both a nutritional and an energy basis for all heterotrophic living things. Although energy exists in several general forms, only chemical energy can routinely operate cell transactions at the level of the cytoplasm (microfile 8.4). The general topic of this section is energy as it is related to the oxidation of fuels, the role of redox carriers, and the generation of ATP.

THE ENERGY IN ELECTRONS

The energy of the cell resides generally in the bonds between atoms and specifically in valence electrons that atoms can exchange. When these bonds remain untapped or unchanged and are not spent doing cellular work, they have what is called **potential energy.** When bond energy is freed for cellular work, it is termed **kinetic energy.** Kinetic energy can be directed toward synthesis, movement, and growth.

Not all cellular reactions are equal with respect to energy. Some release energy, and others require it to proceed. If the cellular reaction:

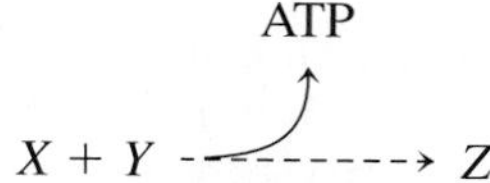

proceeds with the release of energy, it is termed **exergonic.*** Energy released from some exergonic reactions can be stored as bond energy in ATP.

If the reaction:

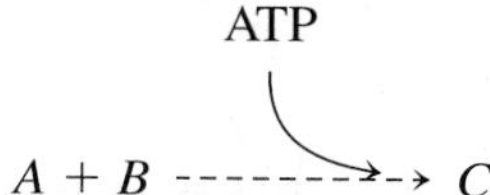

requires the input of energy, it is **endergonic.*** Endergonic reactions can be driven forward by energy given off by ATP.

Summaries of metabolism make it seem that cells "create" energy from nutrients, but they do not. What they actually do is extract chemical energy already present in nutrient fuels and apply that energy toward useful work in the cell. Indeed, a cell can be likened to a gasoline engine that releases energy as it burns the fuel and uses the energy to perform work. The engine does not actually produce energy, but it converts the potential energy of the fuel to kinetic energy of work.

The processes by which cells handle energy are well understood, and most of them are explainable in basic biochemical terms. At the simplest level, cells possess specialized enzyme systems that entrap the energy present in the bonds of nutrients as they are progressively broken (figure 8.12). During exergonic reactions, energy released by electrons is stored in various high-energy phosphate molecules such as ATP. As we shall see, the ability of ATP to temporarily store and release the energy of chemical bonds fuels endergonic cell reactions. Before discussing ATP, let us examine the process behind electron transfer: redox reactions.

A CLOSER LOOK AT BIOLOGICAL OXIDATION AND REDUCTION

We stated earlier that energy extraction in biological systems involves **redox reactions.** Such reactions always occur in pairs, with an electron donor and an electron acceptor, which constitute a *conjugate pair,* or *redox pair.* Written in equation form:

$$\frac{\text{Electron donor}}{\text{(oxidized)}} \rightarrow \bar{e} \quad \frac{\text{Electron acceptor}}{\text{(reduced)}} + \text{Energy}$$

This transfer often involves an intermediate electron carrier.

This process salvages electrons along with their inherent energy, but it also releases some kinetic energy, leaving the reduced

* **exergonic** (ex-er-gon′-ik) Gr. *exo,* without, and *ergon,* work. In general, catabolic reactions are exergonic.

* **endergonic** (en-der-gon′-ik) Gr. *endo,* within, and *ergon,* work. Anabolic reactions tend to be endergonic.

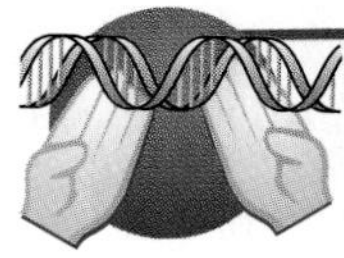

MICROFILE 8.4 ENERGY IN BIOLOGICAL SYSTEMS

Energy is the capacity to do work or to cause change. Energy commonly exists in various forms: (1) thermal, or heat, energy from molecular motion; (2) radiant (wave) energy from visible light or other rays; (3) electrical energy from a flow of electrons; (4) mechanical energy from a physical change in position; (5) atomic energy from reactions in the nucleus of an atom; and (6) chemical energy present in the bonds of molecules. Cells are far too fragile to rely in any constant way on thermal or atomic energy for cell transactions. Cells are largely chemical entities, so it is only logical that chemicals are the basis of cellular energetics.

Fortunately for cells, energy is convertible from one form to another. Light energy is convertible to heat energy; heat is convertible to light; light to chemical; chemical to mechanical; and so on. This phenomenon allows for the biological world to use energy effectively and efficiently. For example, photosynthetic microbes trap the energy in visible light and transform it into the chemical energy of nutrients. This chemical energy can be transformed to the mechanical energy of flagellar movement, or it can even produce light. Conversions do not occur without some energy being lost as heat (in amounts compatible with life). But even heat serves a useful purpose by maintaining a temperature optimal for enzyme function.

(a)

(b)

Photobacteria contain a system of enzymes that expend energy to produce light. (a) *When a compound called luciferin is acted on by luciferase, these bacteria emit a fair amount of visible light. The possible function of light emission in bacteria is very speculative,* (b) *Certain marine fishes carry a type of these bacteria symbiotically in a sac beneath the eye, which apparently serves as a light organ to attract other fish and to facilitate predation. This is one of the more extreme examples of biological energy conversion.*

compound with less energy than the oxidized one. The released energy can **phosphorylate,** or add an inorganic phosphate (P_i), to ADP or to some other compound, a process that captures this energy and stores it in a high-energy molecule (ATP, for example). In many cases, the cell does not handle electrons as discrete entities but rather as parts of an atom such as hydrogen. For simplicity's sake, we will continue to use the term *electron transfer,* but keep in mind that hydrogens are involved in the transfer process. The removal of hydrogens (one hydrogen atom consists of a single proton and a single electron) from a compound during a redox reaction is called **dehydrogenation.** The job of handling these protons and electrons falls to one or more carriers, which function as short-term repositories for the electrons until they can be used in ATP synthesis (figure 8.13). As we shall see, dehydrogenations are an essential supplier of electrons for the respiratory electron transport system.

Electron Carriers: Molecular Shuttles

Electron carriers resemble shuttles that are alternately loaded and unloaded, repeatedly accepting and releasing electrons and hydrogens to facilitate the transfer of redox energy. Most carriers are coenzymes that transfer both electrons and hydrogens, but some transfer electrons only. The most common carrier is NAD (nicotinamide adenine dinucleotide), which carries hydrogens (and a pair of electrons) from dehydrogenation reactions (figure 8.14). Reduced NAD can be represented in various ways. Because 2 hydrogens are removed, the actual carrier state is NADH + H^+, but this is somewhat cumbersome, so we will represent it with the shorter NADH. In catabolic pathways, electrons are extracted and carried through a series of redox reactions until the **final electron acceptor** at the end of a particular pathway is reached. In aerobic metabolism, this acceptor is molecular oxygen; in anaerobic metabolism, it is some other inorganic or organic compound. Other common redox carriers are FAD, NADP (NAD phosphate), coenzyme A, and the compounds of the respiratory chain, which are fixed into membranes (see figure 8.23).

ADENOSINE TRIPHOSPHATE: METABOLIC MONEY

In what ways do cells extract chemical energy from electrons, store it, and then tap the storage sources? To answer these questions, we must look more closely at the powerhouse molecule, adenosine triphosphate. ATP has also been described as metabolic

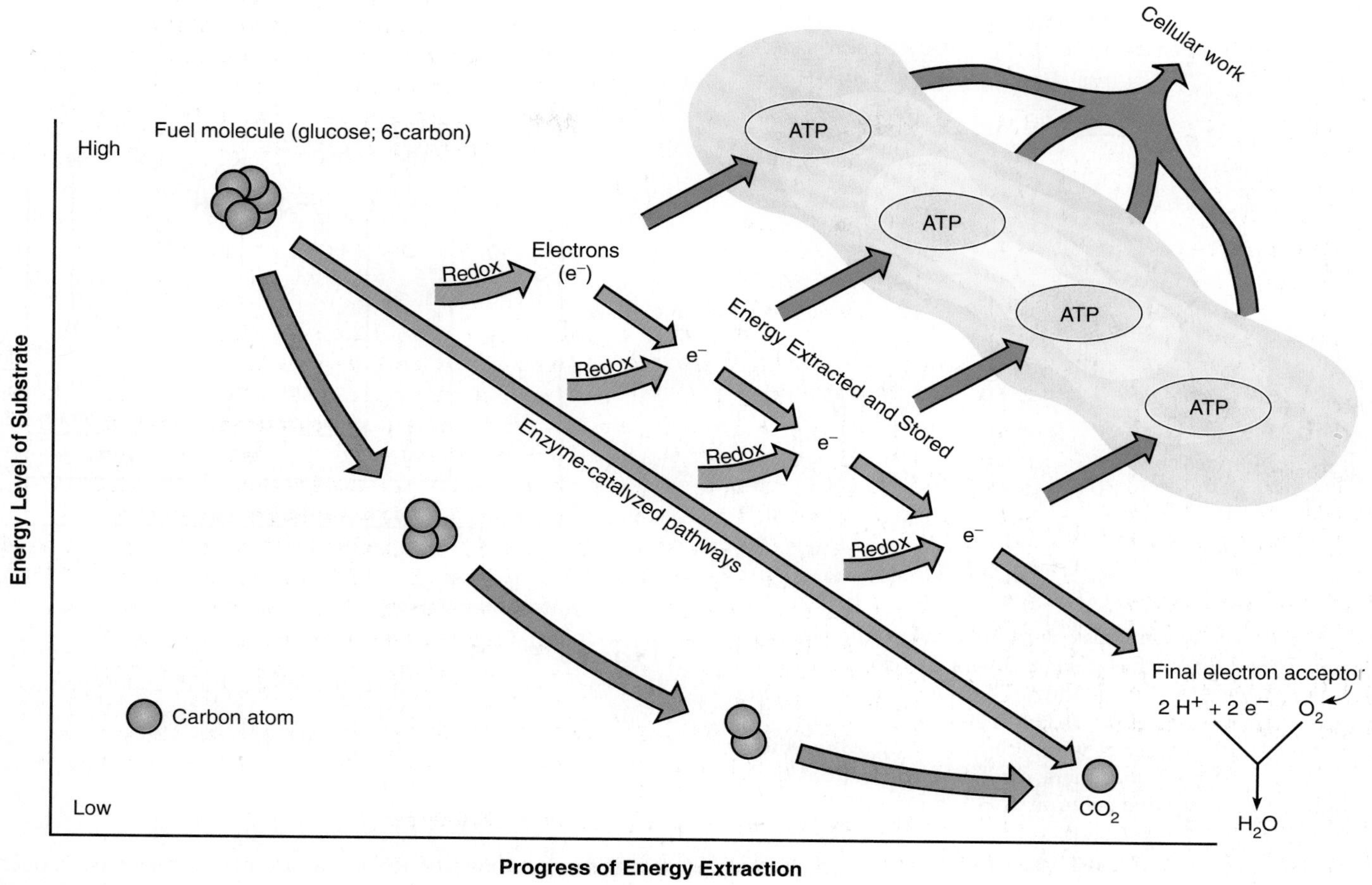

Figure 8.12
A simplified model of the cell's energy machine. The central events of cell energetics include the release of energy during the systemic dismantling of a fuel such as glucose. This is achieved by redox reactions that give off electrons, followed by the shuttling of these electrons to sites in the cell where their energy can be transferred to ATP.

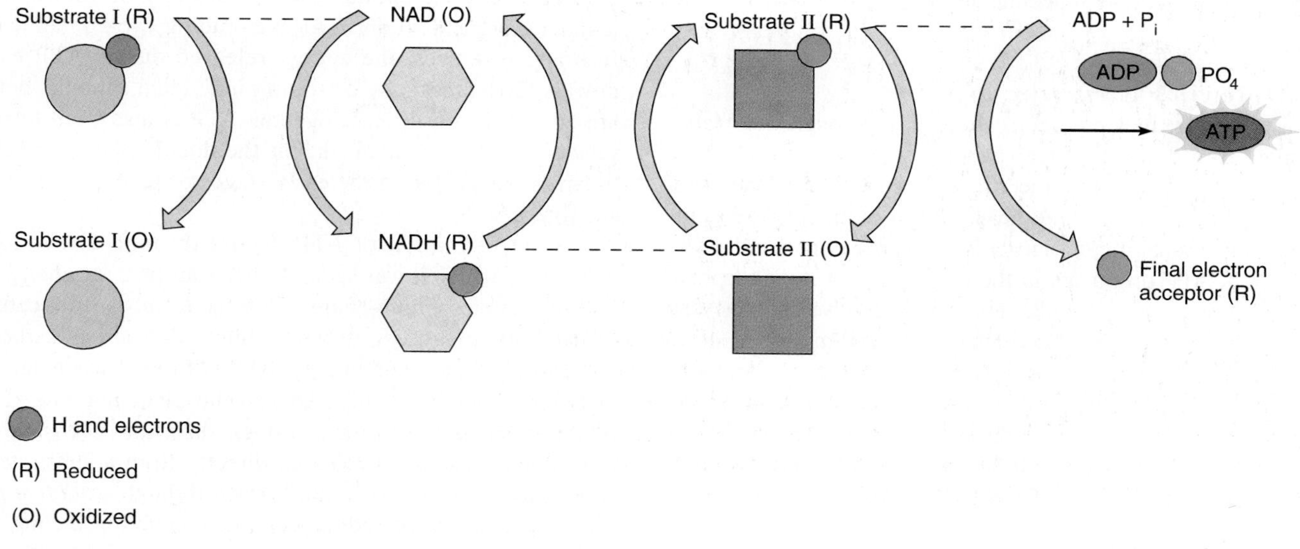

Figure 8.13
The role of electron carriers in redox reactions. Electrons and hydrogens are transferred by paired or coupled reactions in which the donor compound is oxidized (loses electrons) and the acceptor is reduced (gains electrons). A molecule such as NAD can serve as a transient intermediate carrier of the electrons between substrates. During this series of reactions, some energy given off by electron transfer is used to synthesize ATP. To complete the reaction, the electrons are passed to a final electron acceptor.

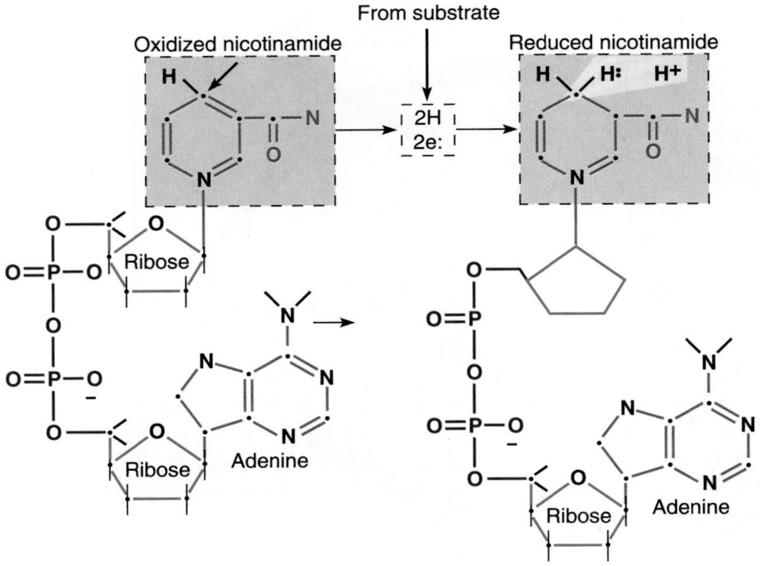

Figure 8.14
Details of NAD reduction. This coenzyme contains the vitamin nicotinamide and the purine adenine attached to double ribose phosphate molecules (a dinucleotide). The principal site of action is on the nicotinamide (boxed area). Hydrogens and electrons donated by a substrate interact with a carbon on the top of the ring (arrow). One hydrogen bonds there, carrying two electrons, and the other hydrogen is carried in solution as H^+ (a proton).

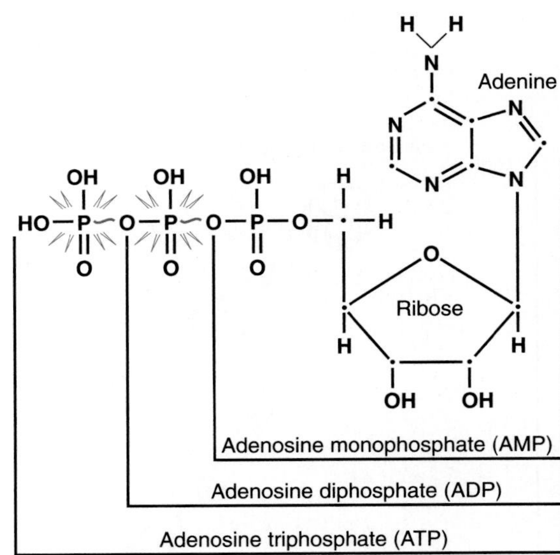

~ Bond that releases energy when broken

Figure 8.15
The structure of adenosine triphosphate (ATP) and its partner compounds, ADP and AMP.

money because it can be earned, banked, saved, spent, and exchanged. As a temporary energy repository, ATP provides a connection between energy-yielding catabolism and all other cellular activities that require energy. Some clues to its energy-storing properties lie in its unique molecular structure (figure 8.15).

The Molecular Structure of ATP

ATP is a three-part molecule consisting of a nitrogen base (adenine) linked to a 5-carbon sugar (ribose), with a chain of three phosphate groups bonded to the ribose (figure 8.15). The type, arrangement, and especially the proximity of atoms in ATP combine to form a compatible but unstable high-energy molecule. The high energy of ATP originates in the orientation of the phosphate groups, which are relatively bulky and carry negative charges. The proximity of these repelling electrostatic charges imposes a strain that is most acute on the bonds between the last two phosphate groups. When ATP is formed, this instability is accommodated by the distributing of the bond stress through the rest of the molecule. The strain on the phosphate bonds accounts for the energetic quality of ATP, because removal of the terminal phosphates releases the bond energy.

Breaking the bonds between two successive phosphates yields adenosine diphosphate (ADP) and adenosine monophosphate (AMP). It is worthwhile noting how economically a cell manages its pool of adenosine nucleotides. AMP derivatives help form the backbone of RNA and are also a major component of certain coenzymes (NAD, FAD, and coenzyme A).

The Metabolic Role of ATP

Cellular reactions involving ATP are necessarily intertwined and balanced. ATP expenditure must inevitably be followed by ATP regeneration, and so on in a never-ending cycle in an active cell. In many instances, the energy released during ATP hydrolysis powers biosynthesis by activating individual subunits before they are enzymatically linked together. ATP is also used to prepare a molecule for catabolism such as the double phosphorylation of a 6-carbon sugar during the early stages of glycolysis (figure 8.16; see figure 8.19).

The formation of ATP occurs through a reversal of the process by which it was spent. In heterotrophs, the energy infusion that regenerates a high-energy phosphate comes from certain steps of catabolic pathways, in which nutrients such as carbohydrates are degraded and yield energy. ATP is formed when substrates or electron carriers add a high-energy phosphate bond to ADP. Some ATP molecules are formed through *substrate-level phosphorylation;* that is, energy is released directly from a substrate to ADP (figure 8.17). Other ATPs are formed through *oxidative phosphorylation,* a series of redox reactions occurring during the final phase of the respiratory pathway (see figure 8.23). Phototrophic organisms have a system of *photophosphorylation,* in which the ATP is formed through a series of sunlight-driven reactions (see chapter 26).

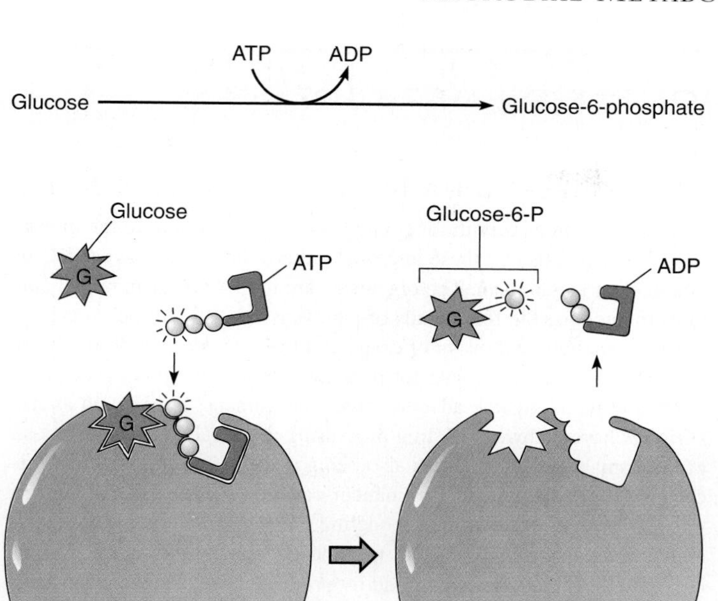

Figure 8.16

An example of phosphorylation of glucose by ATP. The first step in catabolizing glucose is the addition of a phosphate from ATP by an enzyme called glucokinase. This use of high-energy phosphate as an activator is a recurring feature of many metabolic pathways.

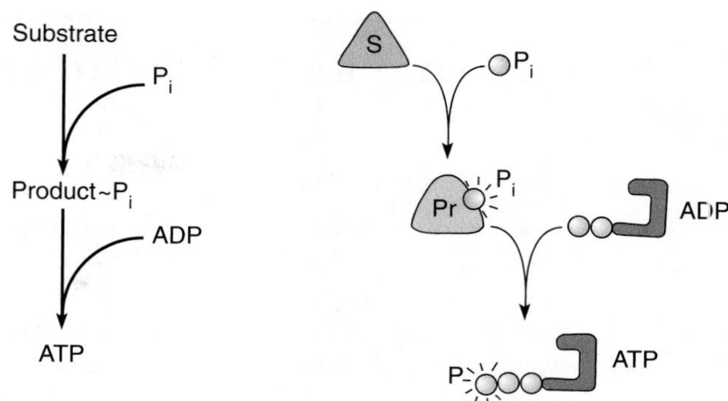

Figure 8.17

ATP formation at the substrate level shown in outline and illustrated form. The inorganic phosphate (P_i) and the substrate form a bond with high potential energy. In a reaction catalyzed by ATP synthase, the phosphate is transferred to ADP, thereby producing ATP.

Chapter Checkpoints

All metabolic processes require the constant input and expenditure of some form of usable energy. Chemical energy is the currency that runs the metabolic processes of the cell, but many forms of energy are involved in cell metabolism.

Chemical energy is obtained from the electrons of nutrient molecules through catabolism. It is used to perform the cellular "work" of biosynthesis, movement, membrane transport, and growth.

Energy is extracted from nutrient molecules by redox reactions. A redox pair of substances passes electrons and hydrogens between them. The donor substance loses electrons, becoming oxidized. The acceptor substance gains electrons, becoming reduced.

ATP is the energy molecule of the cell. It donates free energy to anabolic reactions and is continuously regenerated by three phosphorylation processes: substrate phosphorylation, oxidative phosphorylation, and (in certain organisms) photophosphorylation.

PATHWAYS OF BIOENERGETICS

One of the scientific community's greatest achievements was deciphering the biochemical pathways of cells. Initial work with bacteria and yeasts, followed by studies with animal and plant cells, clearly demonstrated metabolic similarities and strongly supported the concept of the universality of metabolism (microfile 8.5). The study of the production and use of energy by cells is called **bioenergetics.** The two major types of metabolic pathways covered by this science are catabolic routes that gradually degrade nutrients and anabolic or biosynthetic routes that are involved in cell growth. Although these pathways are interconnected and interdependent and most individual reactions are reversible, anabolic pathways are not simply reversals of catabolic ones. At first glance it might seem more economical to use identical pathways, but having different enzymes and divided pathways allows anabolism and catabolism to proceed simultaneously without interference. For simplicity, we shall focus our discussion on the most common catabolic pathways that will illustrate general principles of other pathways as well.

CATABOLISM: AN OVERVIEW OF NUTRIENT BREAKDOWN AND ENERGY RELEASE

The primary catabolism of fuels that results in energy release in many organisms proceeds through a series of three coupled pathways: (1) **glycolysis,*** also called the Embden-Meyerhof-Parnas (EMP) pathway, (2) the **tricarboxylic acid cycle** (TCA), also

* **glycolysis** (gly-kol′-ih-sis) Gr. *glykys,* sweet, and *lysis,* a loosening.

MICROFILE 8.5 MICROBIAL MODELS OF METABOLISM

ESCHERICHIA COLI

Because of its ease of cultivation, metabolic versatility, and relative nonpathogenicity, *Escherichia coli* has been the species of choice from the earliest studies in metabolism and genetics, and it was a source of major discoveries in these areas. Many of the known biochemical pathways and metabolic enzymes were worked out with it. A picture that emerged from these and ensuing studies on other organisms was that the chemical activities of cells, ranging from *E. coli* to human cells, were fundamentally the same. Because so much information was compiled on the physiology and genetics of *E. coli* over time, it automatically became the model for techniques in biotechnology and genetic engineering. Although bacterial species in the genera *Bacillus* and *Salmonella,* the yeast *Saccharomyces,* and cell cultures are more recent contenders for large roles in research and biotechnology, the colon bacillus continues to be a valuable subject.

MICROBES—THE MODERN GUINEA PIGS?

The metabolism of certain fungi and bacteria is so similar to the metabolism of mammals that these microbes are now being used as models of mammalian metabolism. The organisms are used alone or in mixed cultures to decipher the fine details of pathways that are still obscure or to test the metabolic pathways of drugs and toxic substances. In the latter case, researchers are looking for metabolic breakdown products of new drugs and their possible adverse effects on humans. Fungi such as *Aspergillus* have pathways of drug processing remarkably similar to those of the human liver. Progress is also being made in determining how microbes process pollutants and cancer-causing chemicals. Among the greatest benefits of microbial modeling are that the number of experimental animals can be reduced, that the microbes are easy to grow in large numbers, and that metabolic by-products can be amassed in bulk amounts.

known as the citric acid or Krebs cycle,[2] and (3) the **respiratory chain** (electron transport and oxidative phosphorylation). Each segment of the pathway is responsible for a specific set of actions on various products of glucose. The interconnections of these pathways in aerobic respiration are represented in figure 8.18, and their reactions are summarized in table 8.3. In the following sections, we will observe each of these pathways in greater detail.

ENERGY STRATEGIES IN MICROORGANISMS

Nutrient processing is extremely varied, especially in bacteria, yet in most cases it is based on three basic catabolic pathways. In previous discussions, microorganisms were categorized according to their requirement for oxygen gas, and this requirement is related directly to their mechanisms of energy release. As we shall see, **aerobic respiration** is a series of reactions (glycolysis, the TCA cycle, and the respiratory chain) that converts glucose to CO_2 and gives off energy. It relies on free oxygen as the final acceptor for electrons and hydrogens and produces a relatively large amount of ATP. Aerobic respiration is characteristic of many bacteria, fungi, protozoa, and animals, and it is the system we will emphasize here. Facultative and aerotolerant anaerobes may use only the glycolysis scheme to incompletely oxidize or **ferment** glucose. In this case, oxygen is not required, organic compounds are the final electron acceptors, and a relatively small amount of ATP is produced. Some strictly anaerobic microorganisms metabolize by means of **anaerobic respiration.** This system involves the same three pathways as aerobic respiration, but it does not use molecular oxygen as the final electron acceptor. Aspects of fermentation and anaerobic respiration of covered in subsequent sections of this chapter.

Aerobic Respiration

Aerobic respiration is a series of enzyme-catalyzed reactions in which electrons are transferred from fuel molecules such as glucose to oxygen as a final electron acceptor. This pathway is the principal energy-yielding scheme for aerobic heterotrophs, and it provides both ATP and metabolic intermediates for many other pathways in the cell, including those of protein, lipid, and carbohydrate synthesis.

Aerobic respiration in microorganisms can be summarized by an equation:

$$\text{Glucose } (C_6H_{12}O_6) + 6\ O_2 + 38\ \text{ADP} + 38\ P_i \rightarrow$$
$$6\ CO_2 + 6\ H_2O + 38\ \text{ATP}$$

The seeming simplicity of the equation for aerobic respiration conceals its complexity. Fortunately, we do not have to present all of the details to address some important concepts concerning its reactants and products. These ideas are (1) the steps in the oxidation of glucose, (2) the involvement of coenzyme carriers and the final electron acceptor, (3) where and how ATP originates, (4) where carbon dioxide originates, (5) where oxygen is required, and (6) where water originates.

Glucose: The Starting Compound Carbohydrates such as glucose are good fuels because these compounds are highly reduced; that is, they are superior hydrogen and electron donors. The enzymatic withdrawal of hydrogen from them also removes electrons that are unstable and rich in energy. The end products of the conversion of these carbon compounds are energy-rich ATP and energy-poor carbon dioxide and water. Polysaccharides

2. The EMP pathway is named for the biochemists who first outlined its steps. TCA refers to the involvement of several organic acids containing three carboxylic acid groups, citric acid being the first tricarboxylic acid formed; Krebs is in honor of Sir Hans Krebs who, with F. A. Lipmann, delineated this pathway, an achievement for which they won the Nobel Prize in 1953.

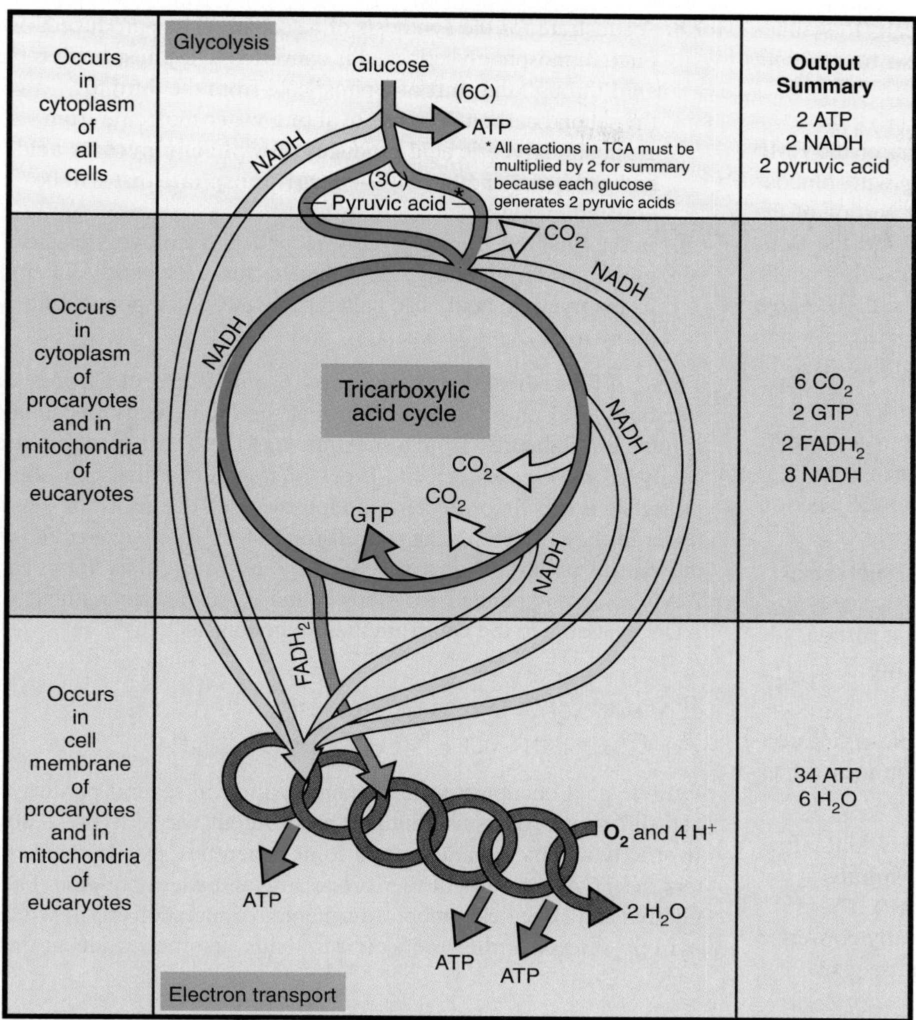

Figure 8.18

Overview of the flow, location, and products of pathways in aerobic respiration. Glucose is degraded through a gradual stepwise process to carbon dioxide and water. Glycolysis proceeds in the absence of oxygen, divides the glucose to two 3-carbon fragments, and produces a small amount of ATP. The tricarboxylic acid cycle receives these 3-carbon pyruvic acid fragments and processes them through redox reactions that extract the electrons and hydrogens. These are shuttled into electron transport to be used in ATP synthesis. CO_2 is an important product of the TCA cycle. Transport of electrons in the third phase generates a large amount of ATP; in the final step, the electrons and hydrogens are received by oxygen, which forms water. Both TCA and electron transport require oxygen to proceed.

TABLE 8.3

METABOLIC STRATEGIES AMONG HETEROTROPHIC MICROORGANISMS

Scheme	Pathways Involved	Final Electron Acceptor	Net Products	Chief Microbe Type
Aerobic respiration	Glycolysis, TCA cycle, electron transport	O_2	38 ATP, CO_2, H_2O	Aerobes; facultative anaerobes
Anaerobic metabolism				
Fermentative	Glycolysis	Organic molecules	2 ATP, CO_2, ethanol, lactic acid*	Facultative, aerotolerant, strict anaerobes
Respiration	Glycolysis, TCA cycle, electron transport	Various inorganic salts (NO_3^-, SO_4^{-2}, CO_3^{-3})	CO_2, ATP, organic acids, H_2S, CH_4, N_2	Anaerobes; some facultatives

*The products of microbial fermentations are extremely varied and include organic acids, alcohols, and gases.

(starch, glycogen) and disaccharides (maltose, lactose) are stored sources of glucose for the respiratory pathways. Although we use glucose as the main starting compound, other hexoses (fructose, galactose) and fatty acid subunits can enter the pathways of aerobic respiration as well (see figure 8.27).

Glycolysis: The Starting Lineup

An anaerobic process called **glycolysis** enzymatically converts glucose through several steps into **pyruvic acid.** Depending on the organism and the conditions, it may be only the first phase of aerobic respiration, or it may serve as the primary metabolic pathway (fermentation). Glycolysis provides a significant means to synthesize a small amount of ATP anaerobically and also to generate pyruvic acid, an essential intermediary metabolite.

Steps in the Glycolytic Pathway Glycolysis, or the EMP pathway, proceeds along nine linear steps, starting with glucose and ending with pyruvic acid (figure 8.19). The first portion of the EMP pathway involves activation of the substrate, and the steps following involve oxidation reactions of the glucose fragments, the synthesis of ATP, and the formation of pyruvic acid. *Although each step of metabolism is catalyzed by a specific enzyme, we will not mention it for most reactions.* The following outline lists the principal steps of glycolysis.

1. **Glucose** is phosphorylated by means of an **ATP** to activate it for the subsequent oxidations. The product is **glucose-6-phosphate.** (Numbers in chemical names refer to the position of the phosphate on the carbon skeleton.).
2. Glucose-6-phosphate is converted to its isomer, **fructose-6-phosphate.**
3. Another **APT** is spent in phosphorylating the first carbon of fructose-6-phosphate, which yields **fructose 1,6-diphosphate.**

Up to this point, no energy has been released, no oxidation-reduction has occurred, and, in fact, 2 ATPs have been used. In addition, the molecules remain in the 6-carbon state.

4. Now doubly activated, fructose 1,6-diphosphate is split into two 3-carbon fragments: **glyceraldehyde-3-phosphate (G-3-P)** and dihydroxyacetone phosphate (DHAP). These molecules are isomers, and DHAP is enzymatically converted to G-3-P, which is the more reactive form for subsequent reactions.

The effect of this step is to double every subsequent reaction, because where there was once a single molecule, there are now two to be fed into the remaining pathways.

5. Each molecule of glyceraldehyde-3-phosphate becomes involved in the single oxidation-reduction reaction of glycolysis, a reaction that sets the scene for ATP synthesis. Two reactions occur simultaneously and are catalyzed by the same enzyme. First, the coenzyme **NAD** picks up or removes hydrogen from G-3-P, forming **NADH.** This step is followed by the addition of an **inorganic phosphate** (P_i) to form a high-energy bond on the third carbon of the G-3-P substrate. The product of these reactions is **diphosphoglyceric acid (DPGA).**

In aerobic organisms, the NADH formed during this step will undergo further reactions in the electron transport system, where the final H acceptor will be oxygen and additional ATPs will be generated (see figure 8.23). In organisms that ferment glucose anaerobically, the NADH will be oxidized back to NAD, and the hydrogen acceptor will be an organic compound (see figure 8.25).

6. One of the **high-energy phosphates** of DPGA is donated to **ADP,** resulting in a molecule of **ATP.** The product of this reaction is **3-phosphoglyceric acid.**

7,8. A substrate for the synthesis of a second ATP is produced. First, 3-phosphoglyceric acid is converted to **2-phosphoglyceric acid** through the shift of a phosphate from the third to the second carbon. The removal of a water molecule from 2-phosphoglyceric acid produces **phosphoenolpyruvic acid** and generates another **high-energy phosphate** that will be donated to ADP.

9. In the final reaction of glycolysis, phosphoenolpyruvic acid gives up its high-energy phosphate to form a **second ATP** and forms **pyruvic acid,** also called pyruvate, a compound with many roles in metabolism.

The 2 ATPs formed during steps 6 and 9 are examples of substrate-level phosphorylation in that the high-energy phosphate is transferred directly from a substrate to ADP. These reactions are catalyzed by kinases, special types of transferase that can phosphorylate a substrate. Because both molecules that arose in step 4 undergo these reactions, an overall total of 4 ATPs is generated in the partial oxidation of a glucose to 2 pyruvic acids. However, 2 ATPs were expended for steps 1 and 3, so the net number of ATPs available to the cell from these reactions is 2.

Pyruvic Acid—A Central Metabolite

Pyruvic acid occupies an important position in several pathways, and different organisms handle it in different ways (figure 8.20). In strictly aerobic organisms and some anaerobes, pyruvic acid enters the TCA cycle for further processing and energy release. Facultative anaerobes can adopt a fermentative metabolism, in which pyruvic acid is further reduced into acids or other products (see page 242).

The Tricarboxylic Acid Cycle—A Carbon and Energy Wheel

As we have seen, the anaerobic oxidation of glucose yields a comparatively small amount of energy and gives off pyruvic acid. Pyruvic acid is still energy-rich, containing a number of extractable hydrogens and electrons to power ATP synthesis, but this can be achieved only through the work of the second and third phases of respiration, in which pyruvic acid is converted to CO_2 and H_2O (see figure 8.18). The tricarboxylic acid cycle converts the 3-carbon pyruvic acid to a 2-carbon fragment that is cycled through a number of organic acids as it is completely degraded. This process takes place in the mitochondrial matrix in eucaryotes and in the cytoplasm of bacteria. Coupled closely to the TCA cycle is the respiratory chain, a system that transports the hydrogens and electrons from other cycles, siphons off their energy into ATP, and donates them to oxygen.

In the first step of the **tricarboxylic acid** cycle, pyruvic acid is converted to a starting compound for that cycle (figure 8.21). This step involves the first oxidation-reduction reaction of this phase of respiration, and it also removes the first carbon dioxide molecule. This highly complex reaction involves a cluster of enzymes and **coenzyme A** that participate in the dehydrogenation

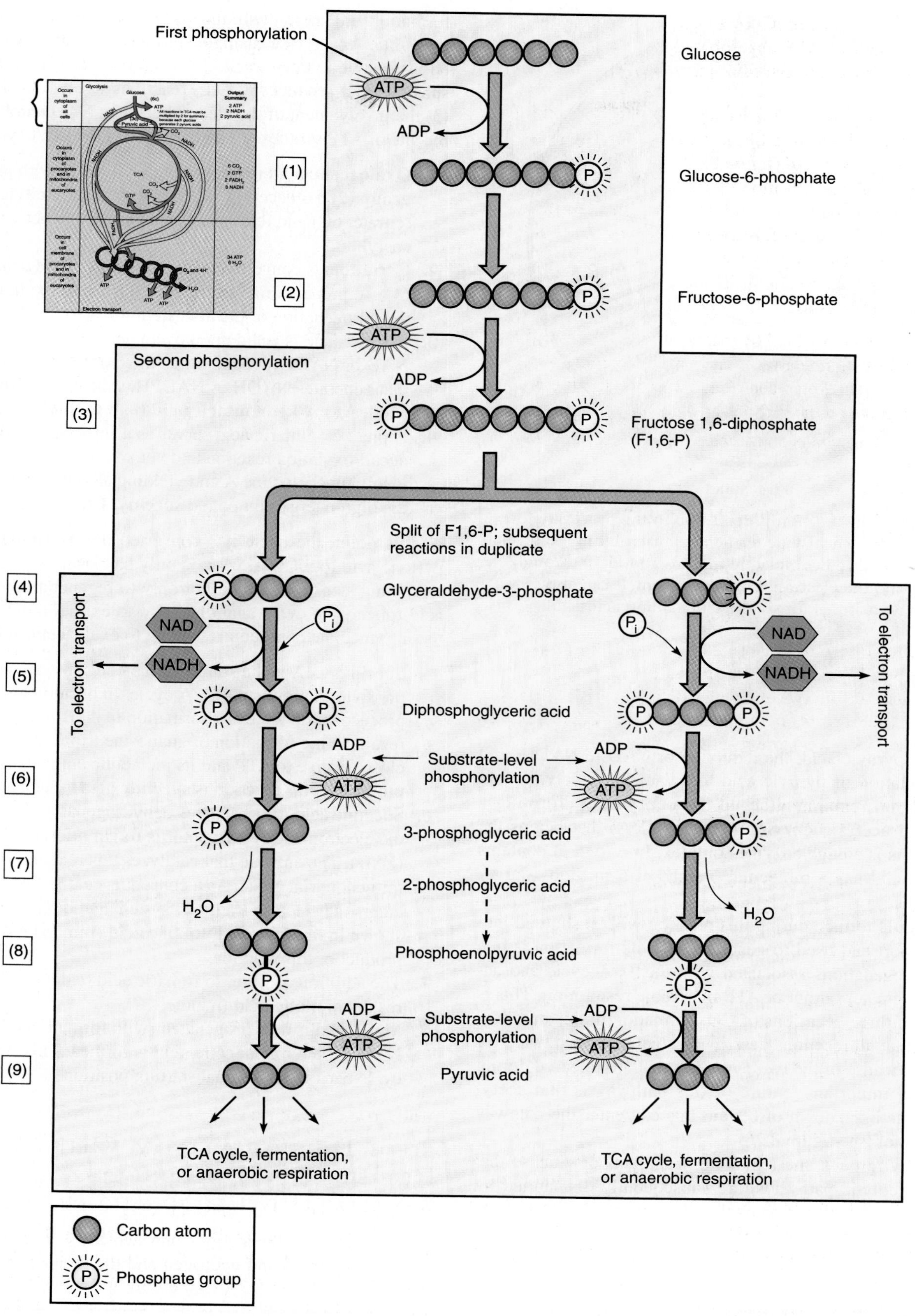

Figure 8.19

The reactions of the glycolysis system (Embden-Meyerhof-Parnas pathway), described in the text.

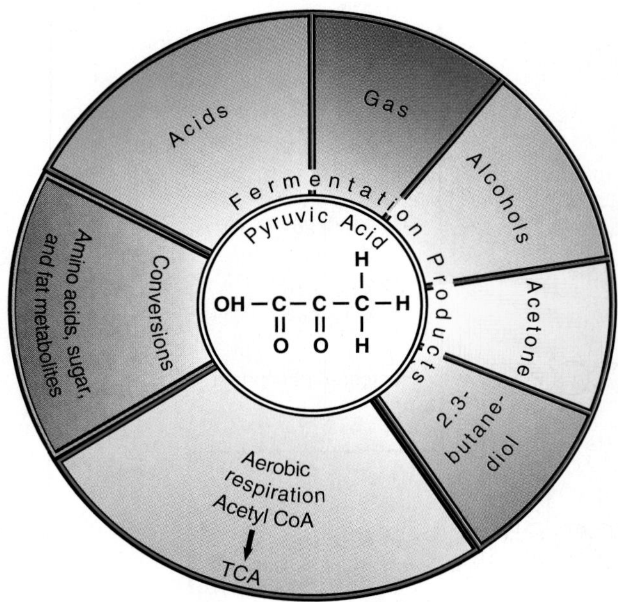

Figure 8.20
Pyruvic acid (pyruvate) is an important hub in the processing of nutrients by microbes. It may be fermented anaerobically to several end products or oxidized completely to CO_2 and H_2O through the TCA cycle and the electron transport system. It can also serve as a source of raw material for synthesizing amino acids and carbohydrates.

(oxidation) of pyruvic acid, the reduction of NAD to NADH, and the decarboxylation of pyruvic acid to a 2-carbon **acetyl** group. The acetyl group remains attached to coenzyme A, forming a complex called **acetyl coenzyme A** (acetyl CoA) that begins the TCA cycle. This is a huge enzyme complex. In *E. coli,* it contains four vitamins and has a molecular weight of 6 million—larger than a ribosome!

The NADH formed during this reaction will be shuttled into electron transport and used to generate ATP; its formation is one of five dehydrogenations associated with the TCA cycle and accounts for the greater output of ATP in aerobic respiration. This is also the first of three occasions of CO_2 formation in this pathway. Be reminded that all reactions described actually happen twice for each glucose because of the two pyruvates that are released during glycolysis. An important feature of this pathway is that acetyl groups from the breakdown of certain fats can enter the pathway at this same point (see figure 8.26).

The acetyl groups, the remaining fragments of glucose that started the glycolytic pathways, are subsequently dismantled by first combining with a 4-carbon oxaloacetate molecule to form citric acid, a 6-carbon molecule. Although this seems a cumbersome way to dismantle such a small molecule, it is necessary in biological systems for extracting larger amounts of energy.

As we have seen, a cyclic pathway is one in which the starting compound is regenerated at the end (see figure 8.9). The tricarboxylic acid cycle has eight steps, beginning with citric acid formation and ending with the organic acid **oxaloacetic acid**[3] (figure 8.21). As we take a single spin around the TCA cycle, it will be instructive to keep track of (1) the numbers of carbons of each substrate and product (#C), (2) reactions where CO_2 is generated, (3) the involvement of the electron carriers NAD and FAD, and (4) the site of ATP synthesis. The reactions in the TCA cycle are:

1. **Oxaloacetic acid** (oxaloacetate; 4C) reacts with the **acetyl group** (2C) on acetyl CoA, thereby forming **citric acid** (citrate; 6C) and releasing coenzyme A to react with another acetyl.
2. Citric acid is converted to its isomer, **isocitric acid** (isocitrate; 6C), to prepare this substrate for the decarboxylation and dehydrogenation of the next step.
3. Isocitric acid is acted upon by an enzyme complex including NAD or NADP (depending on the organism), in a reaction that generates NADH or NADPH, splits off a carbon dioxide, and leaves **α-ketoglutaric acid** (α-ketoglutarate; 5C).
4. Alpha-ketoglutaric acid serves as a substrate for the last decarboxylation reaction and yet another redox reaction involving coenzyme A and yielding NADH. The product is the high-energy compound **succinyl CoA** (4C).

At this point, the cycle has converted the 3 original carbons of pyruvic acid to 3 CO_2s, yet it is only half over. As it turns out, the remaining steps are needed not only to regenerate the oxaloacetic acid to start the cycle again but also to extract more energy from the intermediate compounds leading to oxaloacetic acid.

5. Succinyl CoA is the source of the one substrate level phosphorylation is the TCA cycle. In bacteria and plants, it proceeds with the usual formation of ATP. In mammals, it results in the formation of guanosine triphosphate (GTP), a close relative to ATP and its metabolic equivalent. The product of this reaction is **succinic acid** (succinate; 4C).
6. Succinic acid next becomes dehydrogenated, but in this case, the electron and H^+ acceptor is **flavin adenine dinucleotide, (FAD).** The enzyme that catalyzes this reaction is found in the mitochondrial crista. As a consequence, the $FADH_2$ directly enters the electron transport system and bypasses the first step shown in figure 8.23. **Fumaric acid** (fumarate; 4C) is the product of this reaction.
7. The addition of water to fumaric acid (called hydration) results in **malic acid** (malate; 4C).
8. Malic acid is dehydrogenated (with formation of a final NADH), and **oxaloacetic acid** is formed. This step brings the cycle back to its original starting position.

The Respiratory Chain: Electron Transport and Oxidative Phosphorylation

We now come to the energy chain, which is the final "processing mill" for electrons and hydrogen and the major generator of ATP.

3. In biochemistry, the organic acids are represented interchangeably as the acid (oxaloacetic acid) or as its salt (oxaloacetate).

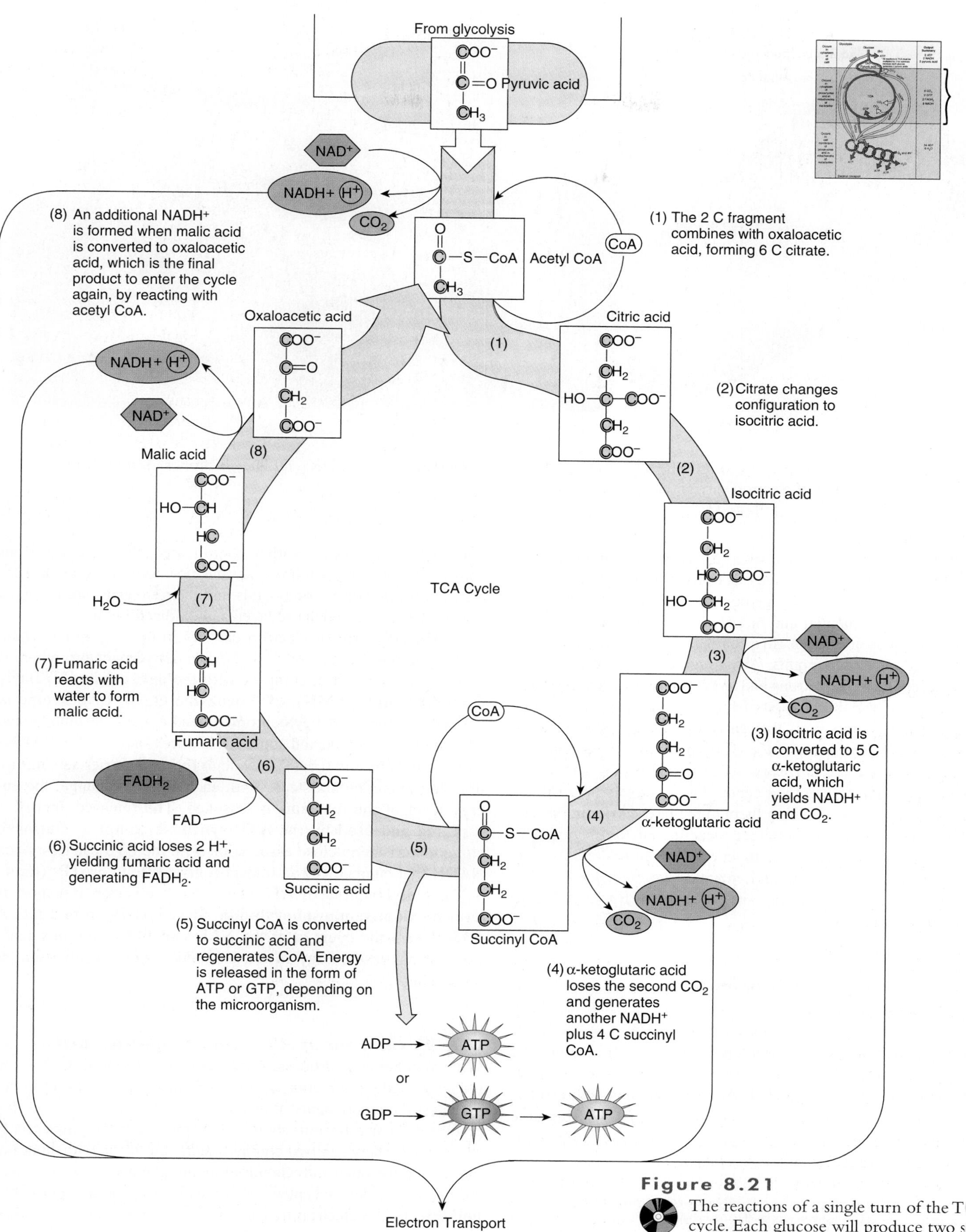
From glycolysis
Pyruvic acid
NAD+
NADH+ H+
CO2
Acetyl CoA
CoA
(1) The 2 C fragment combines with oxaloacetic acid, forming 6 C citrate.
(8) An additional NADH+ is formed when malic acid is converted to oxaloacetic acid, which is the final product to enter the cycle again, by reacting with acetyl CoA.
Oxaloacetic acid
Citric acid
(2) Citrate changes configuration to isocitric acid.
Malic acid
Isocitric acid
TCA Cycle
H2O
(7) Fumaric acid reacts with water to form malic acid.
Fumaric acid
(3) Isocitric acid is converted to 5 C α-ketoglutaric acid, which yields NADH+ and CO2.
α-ketoglutaric acid
FADH2
FAD
(6) Succinic acid loses 2 H+, yielding fumaric acid and generating FADH2.
Succinic acid
Succinyl CoA
(5) Succinyl CoA is converted to succinic acid and regenerates CoA. Energy is released in the form of ATP or GTP, depending on the microorganism.
(4) α-ketoglutaric acid loses the second CO2 and generates another NADH+ plus 4 C succinyl CoA.
ADP
ATP
or
GDP
GTP
Electron Transport

Figure 8.21

The reactions of a single turn of the TCA cycle. Each glucose will produce two spins.
Adapted from Purves and Orions

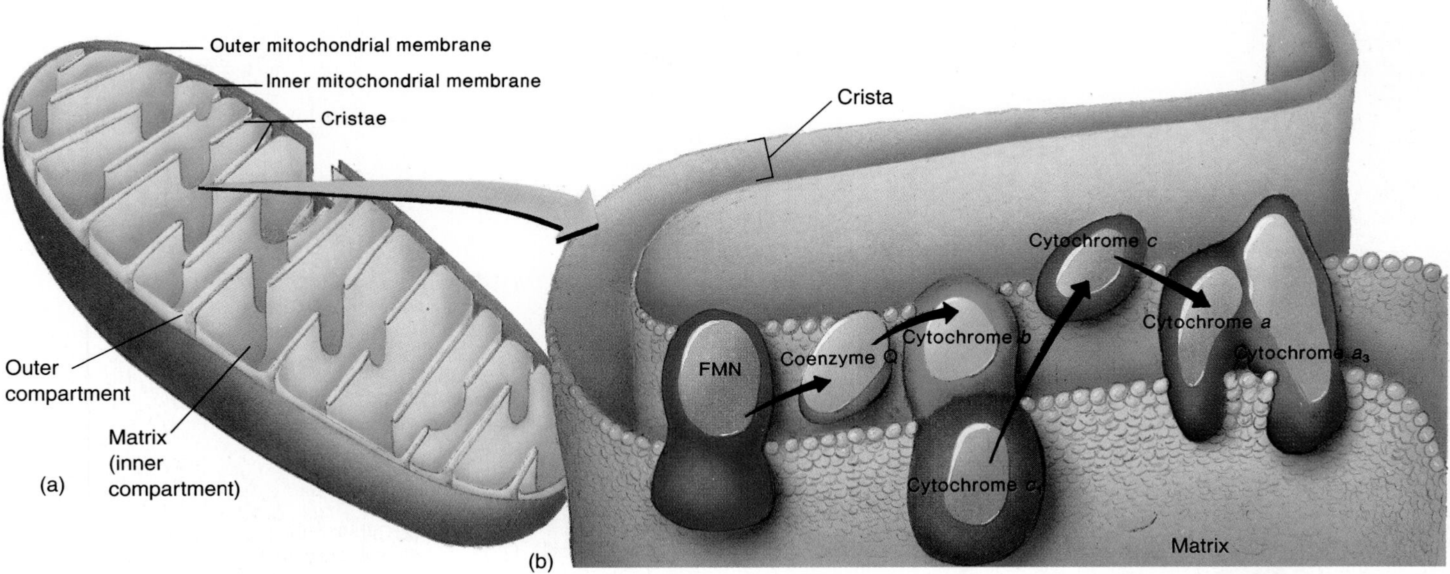

Figure 8.22

(a) Location of the electron transport scheme with respect to cristae membranes and matrix. *(b)* The electron transport chain, showing orientation of electron carriers along the crista membrane.

Overall, the electron transport system (ETS) consists of a chain of special redox carriers that receive electrons from reduced carriers (NADH, $FADH_2$) generated by glycolysis and the TCA cycle and shuttle them in a sequential and orderly fashion (see figure 8.18). The flow of electrons down this chain is highly energetic and gives off ATP at various points. At its end, an enzyme catalyzes the final acceptance of electrons and hydrogen by oxygen, producing water. Some variability exists from one organism to another, but the principal compounds that carry out these complex reactions are NADH dehydrogenase, flavoproteins, coenzyme Q *(ubiquinone),** and *cytochromes.** The cytochromes contain a tightly bound metal atom at their center that is actively involved in accepting electrons and donating them to the next carrier in the series. The highly compartmentalized structure of the respiratory chain is an important factor in its function. Note in figure 8.22 that the electron transport carriers and enzymes are embedded in the inner mitochondrial membranes in eucaryotes. Bacteria carry them in the cell membrane.

Elements of Electron Transport: The Energy Cascade

The principal questions about the electron transport system are: How are the electrons passed from one carrier to another in the series? How is this progression coupled to ATP synthesis? and Where and how is oxygen utilized? Although the biochemical details of this process are rather complicated, the basic reactions consist of a number of redox reactions now familiar to us. In general, the seven carrier compounds and their enzymes are arranged in linear sequence and are reduced and oxidized in turn.

The sequence of electron carriers in the respiratory chain of most aerobic organisms is (1) NADH dehydrogenase, which is closely associated in a complex with the next carrier, (2) flavin mononucleotide (FMN), (3) coenzyme Q, (4) cytochrome *b*, (5) cytochrome c_1, (6) cytochrome *c*, and (7) cytochromes *a* and a_3, which are complexed together. Conveyance of the NADHs from glycolysis and the TCA cycle to the first carrier sets in motion the remaining six steps. With each redox exchange, the energy level of the reactants is lessened. The released energy is captured and used by the **ATP synthase** complex, stationed along the cristae in close association with the ETS carriers. Each NADH that enters electron transport gives rise to 3 ATPs (figure 8.23). This coupling of ATP synthesis to electron transport is termed **oxidative phosphorylation.** Since $FADH_2$ from the TCA cycle enters the cycle after the NAD and FMN complex reactions, it has less energy to release, and 2 ATPs result from its processing.

The Formation of ATP and Chemiosmosis

What biochemical processes are involved in coupling electron transport to the production of ATP? We have observed that in eucaryotes, the components of electron transport are embedded in a precise sequence on mitochondrial membranes. They essentially are stationed between two compartments: the inner mitochondrial matrix and the outer mitochondrial membrane and cytoplasm. According to a widely accepted concept called the **chemiosmotic hypothesis,** as the electron transport carriers shuttle electrons, they

* ubiquinone (yoo-bik′-wih-nohn) L. *ubique,* everywhere. A type of chemical, similar to vitamin K, that is very common in cells.

* cytochrome (sy′-toh-krohm) Gr. *cyto,* cell, and *kroma,* color. Cytochromes are pigmented, iron-containing molecules similar to hemoglobin.

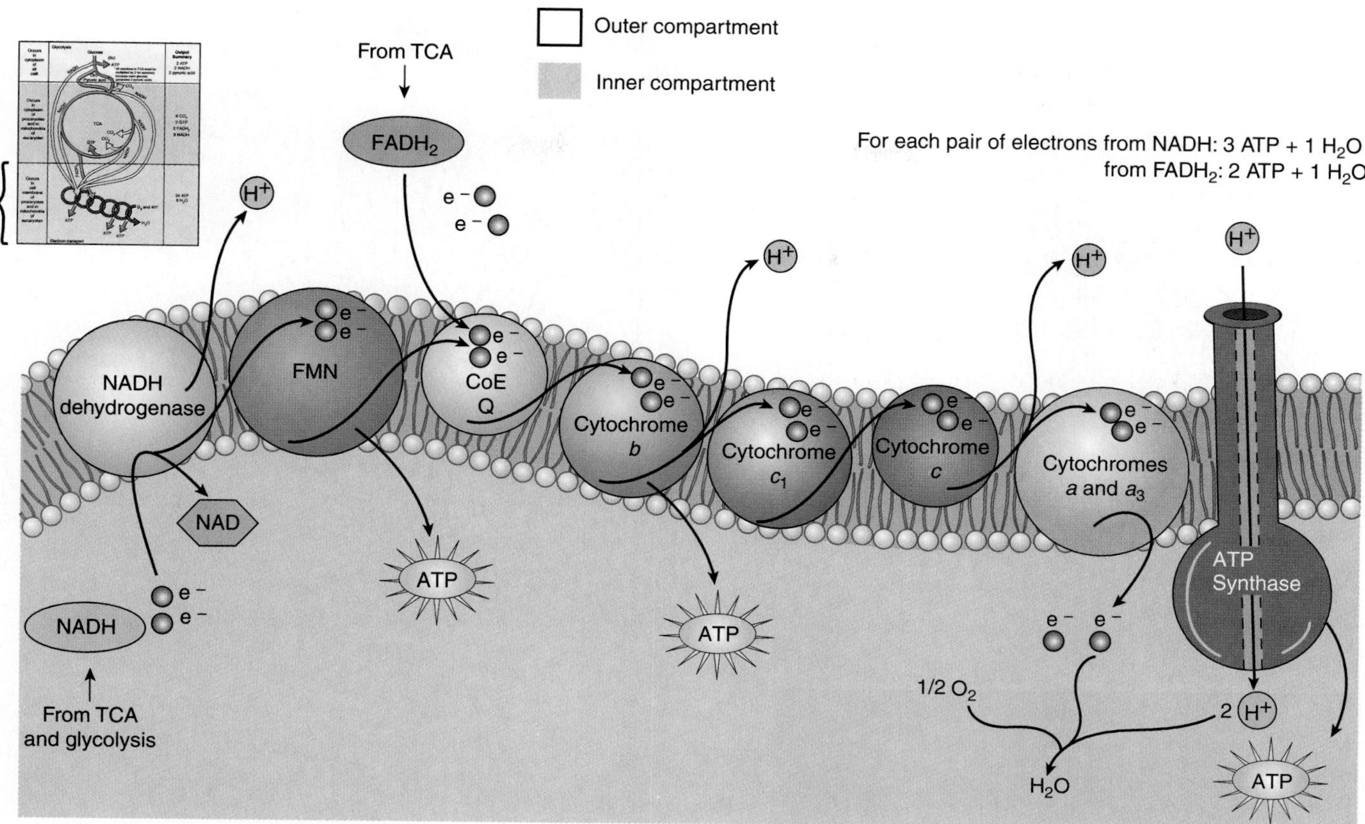

Figure 8.23

The electron transport system and oxidative phosphorylation. Starting at NADH dehydrogenase, electrons brought in from the TCA cycle by NADH are passed along the chain of electron transport carriers. Each adjacent pair of transport molecules undergoes a redox reaction. At 3 sites along this chain, this electron transport is coupled to H^+ movement across the membrane. These actions drive ATP synthesis as shown in figure 8.24.

actively pump hydrogen ions (protons) into the outer compartment of the mitochondrion. This process sets up a concentration gradient of hydrogen ions called the *proton motive force* (PMF). The PMF generates a difference in charge between the outer membrane compartment (+) and the inner membrane compartment (−) (figure 8.24*a*).

Separating the charge has the effect of a battery, which can temporarily store potential energy. This charge will be maintained by the impermeability of the inner cristae membranes to H^+. The only site where H^+ can diffuse into the inner compartment is at the ATP synthase complex, which sets the stage for the final processing of H^+ leading to ATP synthesis (figure 8.24*c*).

ATP synthase is a complex enzyme composed of two large units, F_0 and F_1. Elegant research by Paul Boyer and associates has clarified the intriguing actions of this enzyme. It is embedded in the membrane but can rotate around like a motor and trap chemical energy. As the H^+ ions flow through the F_0 center of the enzyme by diffusion, the F_1 compartments pull in ADP and P_i. Rotation causes a three-dimensional change in the enzyme that bonds these two molecules, thereby releasing ATP into the inner compartment (figure 8.24*c*). The enzyme is then rotated back to the start position and will continue the process. For this work Boyer won the Nobel Prize in 1997.

Bacterial ATP synthesis occurs by means of this same overall process. However, bacteria have the ETS stationed in the cell membrane, and the direction of the proton movement is from the cytoplasm to the periplasmic space. This difference will affect the amount of ATP produced (discussed in the next section). In both cell types, the chemiosmotic theory has been supported by tests showing that oxidative phosphorylation is blocked if the mitochondrial or bacterial cell membranes are disrupted.

Potential Yield of ATPs from Oxidative Phosphorylation

The total of five NADHs (four from the TCA cycle and one from glycolysis) can be used to synthesize:

15 ATPs for ETS (5 × 3 per electron pair)

and

15 × 2 = 30 ATPs per glucose

The single FADH produced during the TCA cycle results in:

2 ATPs per electron pair

and

2 × 2 = 4 ATPs per glucose

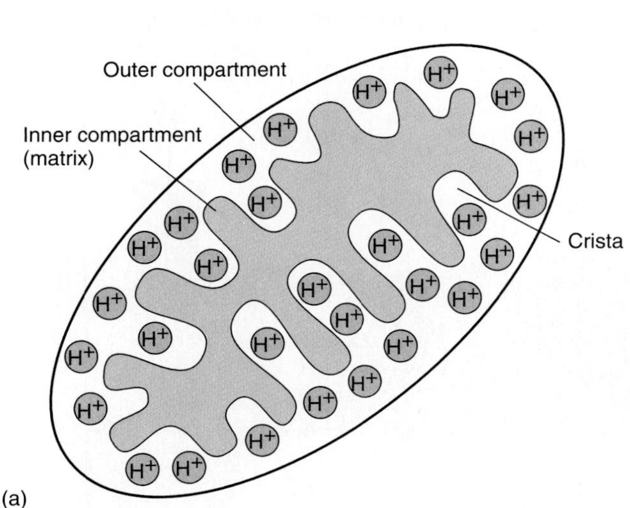

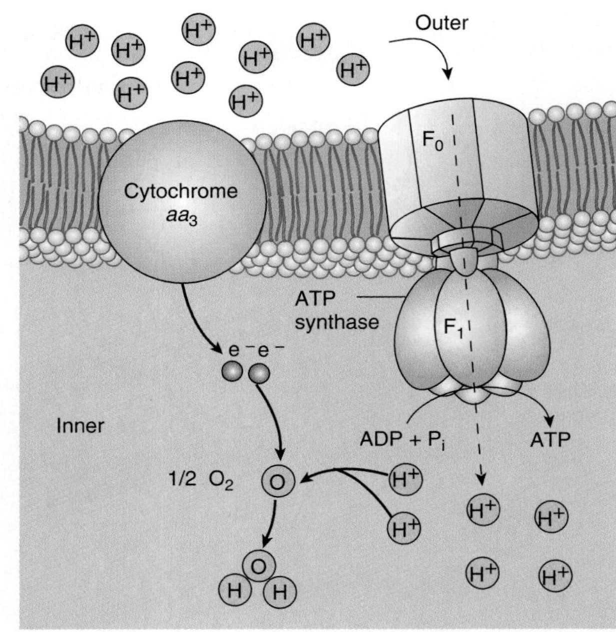

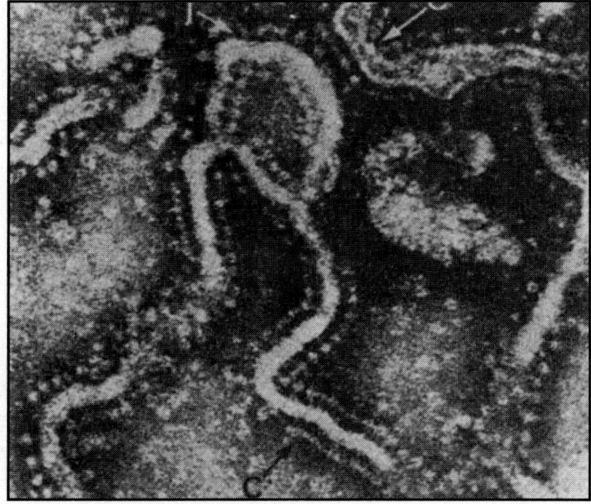

Figure 8.24

Chemiosmosis—the force of ATP synthesis. *(a)* As the carriers in the mitochondrial cristae transport electrons, they also actively pump H^+ ions (protons) to the outer compartment, producing a chemical and charge gradient between the outer and inner mitochondrial compartments. *(b)* Photomicrograph of isolated cristae captures the ATP synthase complex as hundreds of tiny "lollipops" (arrow) projecting from its surface. *(c)* The distribution of electric potential across the membrane drives the synthesis of ATP by ATP synthase. The rotation of this enzyme couples diffusion of H^+ to the inner compartment with the bonding of ADP and P_i. The final event of electron transport is the reaction of the electrons with the accumulated H^+ and O_2 to form metabolic H_2O. This step is catalyzed by cytochrome oxidase (cytochrome *aa*$_3$).

(b) Garret and Grisham, Biochemistry *(Saunders, 1995), p. 647.*

Table 8.4 summarizes the total of ATP and other products for the entire aerobic pathway. These totals are the potential yields possible but may not be fulfilled by many organisms.

Summary of Aerobic Respiration

Originally, we presented a summary equation for respiration. We are now in a position to tabulate the input and output of this equation at various points in the pathways and sum up the final ATP. Close examination of table 8.4 will review several important facets of aerobic respiration:

1. The total yield of ATP is 40: 4 from glycolysis, 2 from the TCA cycle, and 34 from electron transport. However, since 2 ATPs were expended in early glycolysis, this leaves a maximum of **38 ATPS.**

The actual totals may be lower in certain eucaryotic cells because energy is expended in transporting the NADH across the mitochondrial membrane. Certain aerobic bacteria come closest to achieving the full total of 38 because they lack mitochondria and thus do not have to use ATP in transport of NADH across membranes.

2. Six carbon dioxide molecules are generated during the TCA cycle.
3. Six oxygen molecules are consumed during electron transport.
4. Six water molecules are produced in electron transport and 2 in glycolysis, but because 2 are used in the TCA cycle, this leaves a net number of 6.

TABLE 8.4

SUMMARY OF AEROBIC RESPIRATION FOR ONE GLUCOSE MOLECULE

	Glycolysis*	Net Output	TCA Cycle*	Net Output	Respiratory Chain	Net Output	Total Net Output per Glucose
ATP produced	2 × 2 =	4	1 × 2 =	2**	17 × 2 =	34	40 − 2 (used) = 38
ATP used	2		0		0		
NADH produced	1 × 2 =	2	4 × 2 =	8	0		10
FADH produced	0		1 × 2 =	2	0		2
CO_2 produced	0		3 × 2 =	6	0		6
O_2 used	0		0		3 × 2 =	6	
H_2O produced	2		0		3 × 2 =	6	8 − 2 (used) = 6
H_2O used	0		2		0		

*Products are multiplied by 2 because the first figure represents the amount for only one trip through the pathway, and two molecules make this trip for each glucose.

**The product is maybe GTP, but this will be converted to ATP.

The Terminal Step

The terminal step, during which oxygen accepts the electrons, is catalyzed by cytochrome aa_3, also called cytochrome oxidase. This large enzyme complex is specifically adapted to receive electrons from cytochrome *c*, pick up hydrogens from solution, and react with oxygen to form a molecule of water (figure 8.24*c*). This reaction, though in actuality more complex, is summarized as follows:

$$2\ H + 2\ e^- + \frac{1}{2}O_2 \rightarrow H_2O$$

Most eucaryotic aerobes have a fully functioning cytochrome system, but bacteria exhibit wide-ranging variations in this part of the system. Some species lack one or more of the redox steps; others have several alternative electron transport schemes. Because many bacteria lack cytochrome c oxidase, this variation can be used to differentiate among certain genera of bacteria. The oxidase test (see figure 18.29) can detect this enzyme. It is used to help identify members of the genera *Neisseria* and *Pseudomonas* and some species of *Bacillus.* Another variation in the cytochrome system is evident in certain bacteria (*Klebsiella, Enterobacter*) that can grow even in the presence of cyanide because they lack cytochrome oxidase. Cyanide will cause rapid death in humans and other eucaryotes because it blocks cytochrome oxidase, thereby completely terminating aerobic respiration, but it is harmless to these bacteria.

A potential side reaction of the respiratory chain in aerobic organisms is the incomplete reduction of oxygen to superoxide ion ($:O_2-$) and hydrogen peroxide (H_2O_2). Because these toxic oxygen products (previously discussed in chapter 7) can be very damaging to cells, aerobes have various neutralizing enzymes (*superoxide dismutase* and *catalase*). One exception is the genus *Streptococcus,* which can grow well in oxygen yet lacks both cytochromes and catalase. The tolerance of these organisms to oxygen can be explained by the neutralizing effects of a special peroxidase. The lack of cytochromes, catalase, and peroxidases in anaerobes as a rule limits their ability to process free oxygen and contributes to its toxic effects on them.

Alternate Catabolic Pathways

Certain bacteria follow a different pathway in carbohydrate catabolism. The **phosphogluconate pathway** (also called the hexose monophosphate shunt) provides ways to anaerobically oxidize glucose and other hexoses, to release ATP, to produce large amounts of NADPH, and to process pentoses (5-carbon sugars). This pathway, common in heterolactic fermentative bacteria, yields various end products, including lactic acid, ethanol, and carbon dioxide. Furthermore, it is a significant intermediate source of pentoses for nucleic acid synthesis.

ANAEROBIC RESPIRATION

Some bacteria have evolved an anaerobic respiratory system that functions like the aerobic cytochrome system except that it utilizes oxygen-containing salts, rather than free oxygen, as the final electron acceptor. Of these, the nitrate (NO_3) and nitrite (NO_2) reduction systems are best known. The reaction in species such as *Escherichia coli* is represented as:

$$NO_3^- + H_2 \xrightarrow{\text{Nitrite reductase}} NO_2^- + H_2O$$

An enzyme catalyzes the removal of oxygen from nitrate, leaving nitrite and water as products. A test for this reaction is one of the physiological tests used in identifying bacteria.

Some species of *Pseudomonas* and *Bacillus* possess enzymes that can further reduce nitrite to nitric oxide (NO), nitrous oxide (N_2O), and even nitrogen gas (N_2). This process, called *denitrification,* is a very important step in recycling nitrogen in the biosphere (see chapter 26). Other oxygen-containing nutrients reduced anaerobically by various bacteria are carbonates and sulfates. None of the anaerobic pathways produce as much ATP as aerobic respiration.

THE IMPORTANCE OF FERMENTATION

Of all the results of pyruvate metabolism, probably the most varied is fermentation (see figure 8.20). Technically speaking, **fermentation*** is the incomplete oxidation of glucose or other carbohydrates in the absence of oxygen, a process that uses organic compounds as the terminal electron acceptors and yields a small amount of ATP. Over time, the term *fermentation* has acquired several looser connotations. Originally, Pasteur called the microbial action of yeast during wine production *ferments* (see microfile 8.6), and to this day, biochemists use the term in reference to the production of ethyl alcohol by yeasts acting on glucose and other carbohydrates. Fermentation is also what bacteriologists call the formation of acid, gas, and other products by the action of various bacteria on pyruvic acid. The process is a common metabolic strategy among bacteria. Industrial processes that produce chemicals on a massive scale through the actions of microbes are also called fermentations (see chapter 26). Each of these usages is acceptable for one application or another.

It may seem that fermentation would yield only meager amounts of energy (2 ATPs maximum per glucose). What actually happens, however, is that many bacteria can grow as fast as they would in the presence of oxygen. This rapid growth is made possible by an increase in the rate of glycolysis. From another standpoint, fermentation permits independence from molecular oxygen and allows colonization of anaerobic environments. It also enables microorganisms with a versatile metabolism to adapt to variations in the availability of oxygen. For them, fermentation provides a means to grow even when oxygen levels are too low for aerobic respiration.

Bacteria that digest cellulose in the rumens of cattle (as described in chapter 7) are largely fermentative. After initially hydrolyzing cellulose to glucose, they ferment the glucose to organic acids, which are then absorbed as the bovine's principal energy source. Even human muscle cells can undergo a form of fermentation that permits short periods of activity after the oxygen supply in the muscle has been exhausted. Muscle cells convert pyruvic acid into lactic acid, which allows anaerobic production of ATP to proceed for a time. But this cannot go on indefinitely, and, after a few minutes, the accumulated lactic acid causes muscle fatigue.

Products of Fermentation

Alcoholic beverages (wine, beer, whiskey) are perhaps the most prominent among fermentation products; others are solvents (acetone, butanol), organic acids (lactic, acetic), dairy products, and many other foods. Derivatives of proteins, nucleic acids, and other organic compounds are fermented to produce vitamins, antibiotics, and even hormones such as hydrocortisone.

Fermentation products can be grouped into two general categories: alcoholic fermentation products and acidic fermentation products (figure 8.25). **Alcoholic fermentation** occurs in yeast species that have metabolic pathways for converting pyruvic acid

* **fermentation** (fur-men-tay′-shun) L. *fervere,* to boil, or *fermentatum,* leaven or yeast.

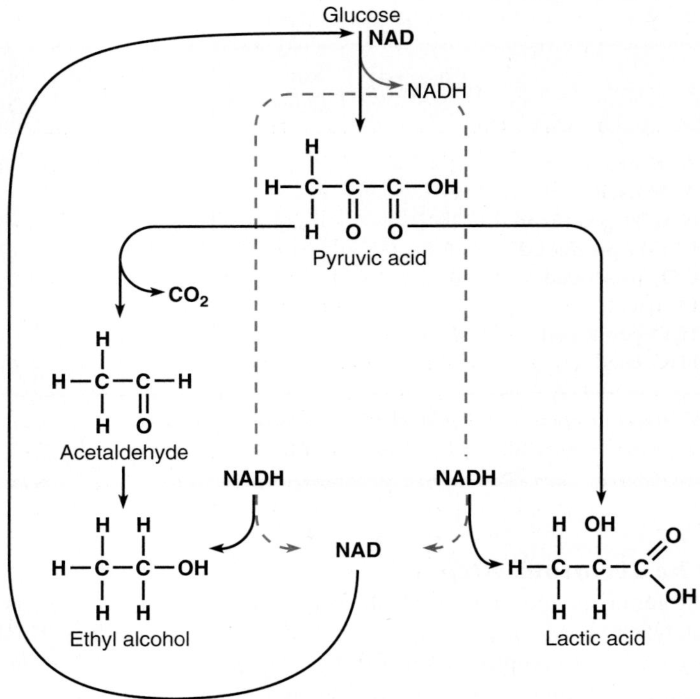

Figure 8.25
The chemistry of fermentation systems that produce acid and alcohol. In both cases, the final electron acceptor is an organic compound. In yeasts, pyruvic acid is decarboxylated to acetaldehyde, and the NADH given off in the glycolytic pathway reduces it to ethyl alcohol. In homolactic fermentative bacteria, pyruvic acid is reduced by NADH to lactic acid. Both systems regenerate NAD to feed back into glycolysis.

to ethanol. This process involves a decarboxylation of pyruvic acid to acetaldehyde, followed by a reduction of the acetaldehyde to ethanol. In oxidizing the NADH formed during glycolysis, this step regenerates NAD, thereby allowing the glycolytic pathway to continue. These processes are crucial in the production of beer and wine, though the actual techniques for arriving at the desired amount of ethanol and the prevention of unwanted side reactions are important tricks of the brewer's trade (microfile 8.6). Note that the products of alcoholic fermentation are not only ethanol but also CO_2, a gas that accounts for the bubbles in champagne and beer.

Alcohols other than ethanol can be produced during bacterial fermentation pathways. Certain clostridia produce butanol and isopropanol through a complex series of reactions. Although this process was once an important source of alcohols for industrial use, it has been largely replaced by a nonmicrobial petroleum process.

The pathways of **acidic fermentation** are extremely varied (see figure 8.25). Lactic acid bacteria ferment pyruvate in the same way that humans do—by reducing it to lactic acid. If the product of this fermentation is mainly lactic acid, as in certain

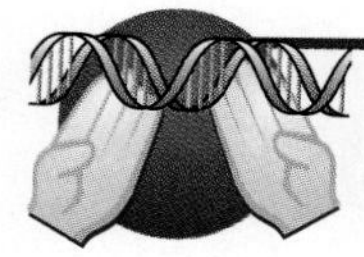

MICROFILE 8.6 PASTEUR AND THE WINE-TO-VINEGAR CONNECTION

The microbiology of alcoholic fermentation was greatly clarified by Louis Pasteur after French wine makers hired him to uncover the causes of periodic spoilage in wines. Especially troublesome was the conversion of wine to vinegar and the resultant sour flavor. Up to that time, wine formation had been considered strictly a chemical process. After extensively studying beer making and wine grapes, Pasteur concluded that wine, both fine and not-so-fine, was the result of microbial action on the juices of the grape and that wine "disease" was caused by contaminating organisms that produced undesirable products such as acid. Although he did not know it at the time, the bacterial contaminants responsible for the acidity of the spoiled wines were likely to be *Acetobacter* or *Gluconobacter* introduced by the grapes, air, or wine-making apparatus. These common gram-negative genera further oxidized ethanol to acetic acid and are presently used in commercial vinegar production. The following formula shows how this is accomplished:

$$\text{H}-\underset{\text{H}}{\overset{\text{H}}{\text{C}}}-\underset{\text{H}}{\overset{\text{H}}{\text{C}}}-\text{OH} \longrightarrow \text{H}-\underset{\text{H}}{\overset{\text{H}}{\text{C}}}-\text{C}\begin{matrix}{\nearrow}\text{O} \\ {\searrow}\text{H}\end{matrix}$$

Ethanol → Acetic acid

Pasteur's far-reaching solution to the problem is still with us today—mild heating, or *pasteurization,* of the grape juice to destroy the contaminants, followed by inoculation of the juice with a pure yeast culture. It also introduced the concept that a single microbe could be responsible for both a desired product and a disease. The topic of wine making is explored further in chapter 26.

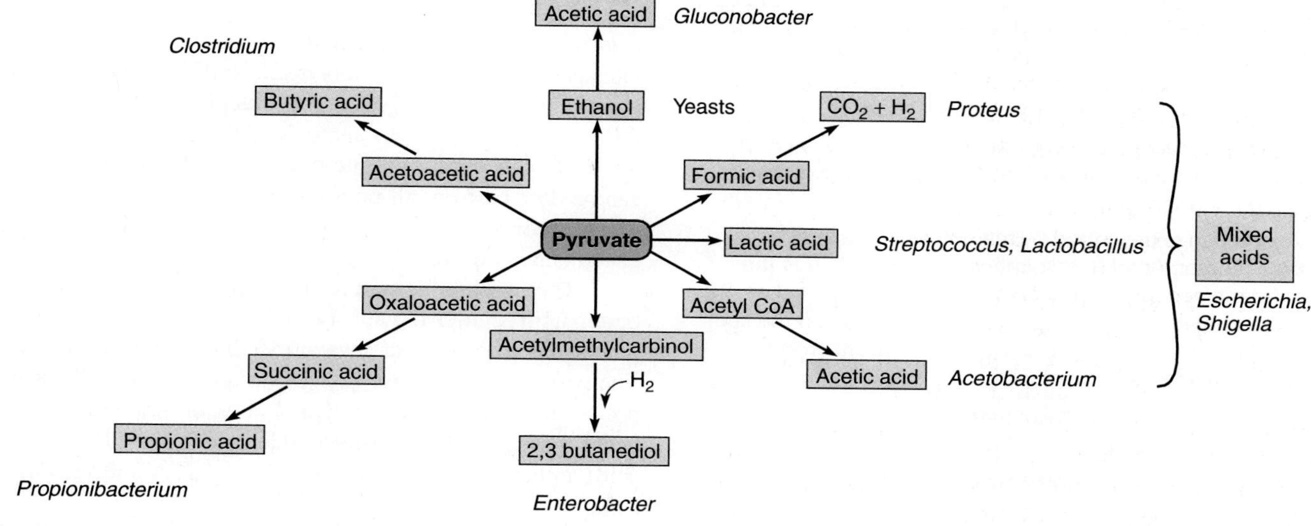

Figure 8.26
Miscellaneous products of pyruvate fermentation and the bacteria involved in their production.

species of *Streptococcus* and *Lactobacillus,* it is termed *homolactic.* The souring of milk is due largely to the production of this acid by bacteria. When glucose is fermented to a mixture of lactic acid, acetic acid, and carbon dioxide, as is the case with *Leuconostoc* and other species of *Lactobaccilus,* the process is termed *heterolactic fermentation.*

Many members of the family Enterobacteriaceae (*Escherichia, Shigella,* and *Salmonella*) possess enzyme systems for converting pyruvic acid to several acids simultaneously (figure 8.26). **Mixed acid fermentation** produces a combination of acetic, lactic, succinic, and formic acids, and it lowers the pH of a medium to about 4.0. *Propionibacterium* produces primarily propionic acid, which gives the characteristic flavor to Swiss cheese while fermentation gas (CO_2) produces the holes. Some members also further decompose formic acid completely to carbon dioxide and hydrogen gases. Because enteric bacteria commonly occupy the intestine, this fermentative activity accounts for the accumulation of some types of gas—primarily CO_2 and H_2—in the intestine. Some bacteria oxidize the organic acids and produce the neutral end product 2,3-butanediol (microfile 8.7).

We have provided only a brief survey of fermentation products, but it is worth noting that microbes can be harnessed to synthesize a variety of other substances by varying the raw materials provided them. In fact, so broad is the meaning of the word

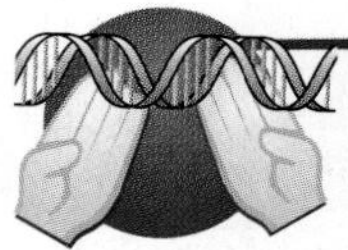

MICROFILE 8.7 FERMENTATION AND BIOCHEMICAL TESTING

The knowledge and understanding of the fermentation products of given species are important not only in industrial production but also in identifying bacteria by biochemical tests. Fermentation patterns in enteric bacteria, for example, are an important identification tool. Specimens are grown in media containing various carbohydrates, and the production of acid or acid and gas is noted (see figure 3.12). For instance, *Escherichia* ferments the milk sugar lactose, whereas *Shigella* and *Proteus* do not. On the basis of its gas production during glucose fermentation, *Escherichia* can be further differentiated from *Shigella,* which ferments glucose but does not generate gas. Other enteric bacteria are separated on the basis of whether glucose is fermented to mixed acids or to 2, 3-butanediol. *Escherichia coli* produces mixed acids, whereas *Enterobacter* and *Serratia* form primarily 2,3-butanediol. These are part of the IMViC testing described in chapter 20. The M refers to the methyl red test for mixed acids, and the V refers to the Voges-Proskauer test for the butanediol pathway.

fermentation that the large-scale industrial syntheses by microorganisms often utilize entirely different mechanisms from those described here, and they even occur aerobically, particularly in antibiotic, hormone, vitamin, and amino acid production (see chapter 26).

Chapter Checkpoints

Bioenergetics describes metabolism in terms of production, utilization, and transfer of energy by cells.

Catabolic pathways release energy through three pathways: glycolysis, the tricarboxylic acid cycle, and the respiratory electron transport system.

Cellular respiration is described by the nature of the final electron acceptor. Aerobic respiration implies that O_2 is the final hydrogen acceptor. Anaerobic respiration implies that some other molecule is the final hydrogen acceptor. If the final hydrogen acceptor is an organic molecule, the anaerobic process is considered fermentation.

Carbohydrates are preferred cell energy sources because they are superior hydrogen (electron) donors.

Glycolysis is the catabolic process by which glucose is oxidized into two molecules of pyruvic acid, with a net gain of 2 ATP. This process is also called substrate phosphorylation.

The tricarboxylic acid cycle converts the 3-carbon pyruvic acid to three CO_2 molecules and transfers its electrons to redox carriers for energy harvesting.

The electron transport chain generates free energy through sequential redox reactions collectively called oxidative phosphorylation. This energy is used to generate up to 36 ATP for each glucose molecule catabolized.

BIOSYNTHESIS AND THE CROSSING PATHWAYS OF METABOLISM

Our discussion now turns from catabolism and energy extraction to anabolic functions and biosynthesis. In this section we will overview aspects of intermediary metabolism, including amphibolic pathways, the synthesis of simple molecules, and the synthesis of macromolecules.

THE FRUGALITY OF THE CELL—WASTE NOT, WANT NOT

It must be obvious by now that cells have mechanisms for careful management of carbon compounds. Rather than being dead ends, most catabolic pathways contain strategic molecular intermediates that can be diverted into anabolic pathways. In this way, a given molecule can serve multiple purposes, and the maximum benefit can be derived from all nutrients and metabolites of the cell pool. The property of a system to function in both catabolism and anabolism is **amphibolism.***

The amphibolic nature of intermediary metabolism can be better appreciated if we look at some examples of the branch points and crossroads of several metabolic pathways (figure 8.27). The pathways of glucose catabolism are an especially rich "metabolic marketplace." The principal sites of amphibolic interaction occur during glycolysis (glyceraldehyde-3-phosphate and pyruvic acid) and the TCA cycle (acetyl coenzyme A and various organic acids).

Amphibolic Sources of Cellular Building Blocks

Glyceraldehyde-3-phosphate can be diverted away from glycolysis and converted into precursors for amino acid, carbohydrate, and triglyceride (fat) synthesis. (A precursor molecule is a compound that is the source of another compound.) Earlier we noted the numerous directions that pyruvic acid catabolism can take. In terms of synthesis, pyruvate also plays a pivotal role in providing intermediates for amino acids. In the event of an inadequate glucose supply, it serves as the starting point in glucose synthesis from various metabolic intermediates, a process called *gluconeogenesis.**

* amphibolism (am-fee-bol′-izm) Gr. *amphi,* two-sided, and *bole,* a throw.

* gluconeogenesis (gloo″-koh-nee″-oh-gen′-uh-sis) Literally, the formation of "new" glucose.

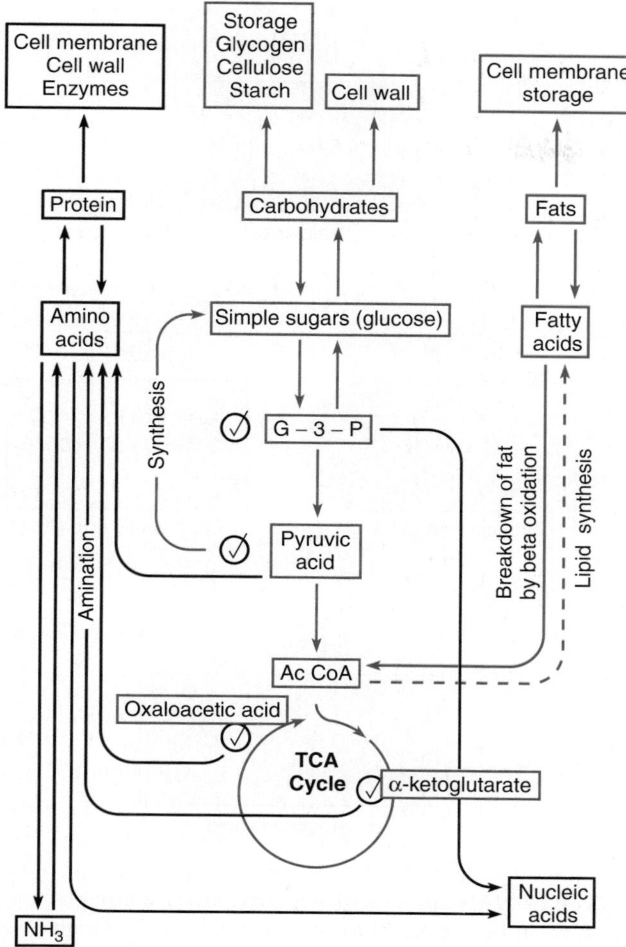

Figure 8.27
A summary of metabolic interactions. Many intermediate compounds serve as amphibolic branch points (✓). With comparatively small modifications, these compounds can be converted into other compounds and enter a different pathway. Note that catabolism of glucose (center) furnishes numerous intermediates for amino acid, fat, nucleic acid, and carbohydrate synthesis. Likewise, molecules arising from fat metabolism can be channeled into the TCA cycle.

The acetyl group that starts the TCA cycle is another extremely versatile metabolite that can be fed into a number of synthetic pathways. This 2-carbon fragment can be converted as a single unit into one of several amino acids, or a number of these fragments can be condensed into hydrocarbon chains that are important bases for fatty acid and lipid synthesis. Note that the reverse is also true—fats can be degraded to acetyl and thereby enter the TCA cycle at acetyl coenzyme A. This aerobic process, called *beta oxidation,* can provide a large amount of energy. Oxidation of a 6-carbon fatty acid yields 50 ATP, compared with 38 for a 6-carbon sugar.

Two metabolites of carbohydrate catabolism that the TCA cycle spins off, oxaloacetic acid and α-ketoglutaric acid, are essential intermediates in the synthesis of certain amino acids. This occurs through **amination,** the addition of an amino group to a carbon skeleton (figure 8.28*a*). A certain core group of amino acids can then be used to synthesize others. Amino acids and carbohydrates can be interchanged through transamination (figure 28*b*).

Pathways that synthesize the nitrogen bases (purines, pyrimidines), which are components of DNA and RNA, originate in amino acids and so can be dependent on intermediates from the TCA cycle as well. Because the coenzymes NAD, NADP, FAD, and others contain purines and pyrimidines similar to the nucleic acids, their synthetic pathways are also dependent on amino acids. During times of carbohydrate deprivation, organisms can likewise convert amino acids to intermediates of the TCA cycle by deamination (removal of an amino group) and thereby derive energy from proteins. Deamination results in the formation of nitrogen waste products such as ammonium or urea (figure 8.28*c*).

Formation of Macromolecules

Monosaccharides, amino acids, fatty acids, nitrogen bases, and vitamins—the building blocks that make up the various macromolecules and organelles of the cell—come from two possible sources. They can enter the cell from the outside as nutrients, or they can be synthesized through various cellular pathways. The degree to which an organism can synthesize its own building blocks is determined by its genetic makeup, a factor that varies tremendously from group to group. In chapter 7 we found that autotrophs require only CO_2 as a carbon source, a few minerals to synthesize all cell substances, and no organic nutrients. Some heterotrophic organisms (*E. coli,* yeasts) are also very efficient in that they can synthesize all cellular substances from minerals and one organic carbon source such as glucose. Compare this with a strict parasite that has few synthetic abilities of its own and drains most precursor molecules from the host.

Whatever their source, once these building blocks are added to the metabolic pool, they are available for synthesis of polymers by the cell. The details of synthesis vary among the types of macromolecules, but all of them involve the formation of bonds by specialized enzymes and the expenditure of ATP (see figure 8.8*a*).

Amino Acids, Protein Synthesis, and Nucleic Acid Synthesis

Proteins account for a large proportion of a cell's constituents. They are essential components of enzymes, the cell membrane, the cell wall, and cell appendages. As a general rule, 20 amino acids are needed to make these proteins (see chapter 2). Although some organisms (*E. coli,* for example) have pathways that will synthesize all 20 amino acids, others (especially animals) lack some or all of the pathways for amino acid synthesis and must acquire the essential ones from their diets. Protein synthesis itself is a complex process that requires a genetic blueprint and the operation of intricate cellular machinery, as we shall see in chapter 9.

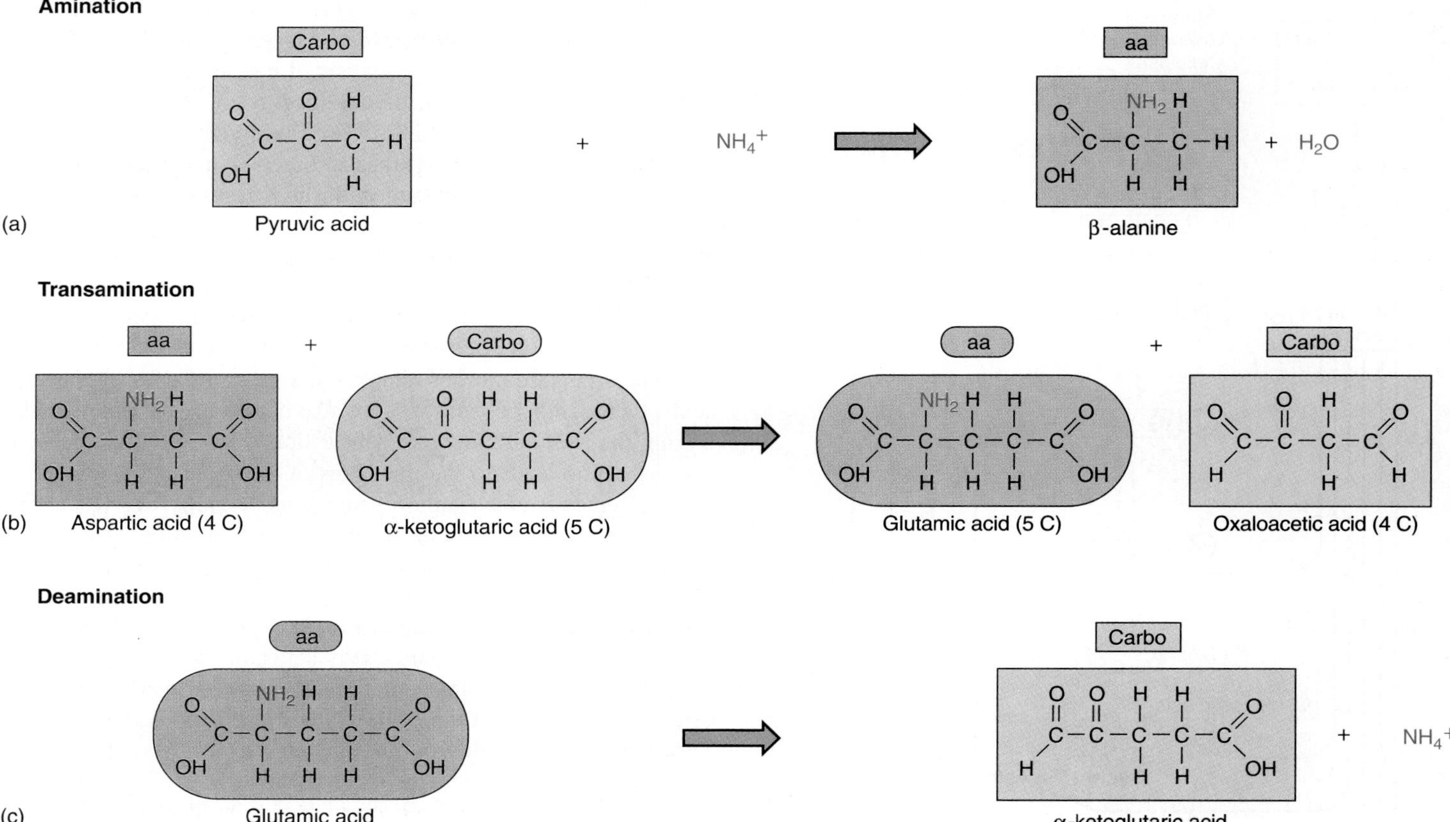

Figure 8.28

Reactions that produce and convert amino acids. All of them require energy as ATP or NAD and specialized enzymes. *(a)* Through amination (the addition of an ammonium molecule [amino group]), a carbohydrate can be converted to an amino acid. *(b)* Through transamination (transfer of an amino group from an amino acid to a carbohydrate fragment), metabolic intermediates can be converted to amino acids that are in low supply. *(c)* Through deamination (removal of an amino group), an amino acid can be converted to a useful intermediate of carbohydrate catabolism. This is how proteins are used to derive energy. Ammonium is one waste product.

DNA and RNA are important for the hereditary continuity of cells and the overall direction of protein synthesis. Because nucleic acid synthesis is a major topic of genetics and is closely allied to protein synthesis, it will likewise be covered in chapter 9.

Carbohydrate Biosynthesis

The role of glucose in bioenergetics is so crucial that its biosynthesis is ensured by several alternative pathways. Certain structures in the cell depend on an adequate supply of glucose as well. It is the major component of the cellulose cell walls of some eucaryotes and of certain storage granules (starch, glycogen). Monosaccharides other than glucose are important in the synthesis of bacterial cell walls. Peptidoglycan contains a linked polymer of muramic acid and glucosamine. Carbohydrates (deoxyribose, ribose) are also an essential building block in nucleic acids. Polysaccharides are the predominant components of cell surface structures such as capsules and the glycocalyx, and they are commonly found in slime layers (dextran).

Chapter Checkpoints

"Intermediary metabolism" refers to the metabolic pathways that connect anabolic and catabolic reactions.

Amphibolic compounds are the "crossroads compounds" of intermediary metabolism. They not only participate in catabolic pathways but also are precursor molecules to biosynthetic pathways.

Biosynthetic pathways utilize building-block molecules from two sources: the environment and the cell's own catabolic pathways. Microorganisms construct macromolecules from these monomers using ATP and specialized enzymes particular to each species.

Proteins are essential macromolecules in all cells because they function as structural constituents, enzymes, and cell appendages.

Carbohydrates are crucial as energy sources, cell wall constituents, and components of nucleotides.

CHAPTER CAPSULE WITH KEY TERMS

MICROBIAL METABOLISM

Metabolism is the sum of cellular chemical changes; it involves scores of reactions that interlink in linear or branched pathways. Metabolism is a complementary process consisting of **anabolism,** the reactions that convert small molecules into macromolecules, and **catabolism,** in which large molecules are degraded. Intermediary molecules called **metabolites** are generated, and their levels are regulated.

ENZYMES

Control is exercised by regulating the activity of organic **catalysts,** or **enzymes.** Enzymes facilitate reactions by lowering the **energy of activation,** but they are not consumed and can be reused. The majority of cellular reactions are catalyzed, and each enzyme acts specifically upon its assigned metabolite, called the **substrate.**

Enzyme Structure: Depending upon its composition, an enzyme is either **conjugated** or **simple.** A conjugated enzyme consists of a protein component called the **apoenzyme** and one or more activators called **cofactors.** Some cofactors are organic molecules called **coenzymes,** and others are inorganic elements, typically metal ions. To function, a conjugated enzyme must be a complete **holoenzyme** with all its parts. Simple enzymes are composed solely of protein.

Enzyme-Substrate Interactions: Metallic cofactors impart greater reactivity to the enzyme-substrate complex. Coenzymes such as **NAD** (nicotinamide adenine dinucleotide) are transfer agents that pass functional groups from one substrate to another. Coenzymes usually contain **vitamins.** Substrate attachment occurs in a special pocket called the **active,** or **catalytic, site.** In order to fit, a substrate must conform to the active site of the enzyme. This configuration is determined by the amino acid content, sequence, and folding of the apoenzyme. Thus, enzymes are **substrate-specific.**

Enzyme Classification: Enzyme names consist of a prefix derived from the type of reaction or the substrate and the ending *-ase.* By convention, enzymes may also be classified according to their location of action. Thus, an **exoenzyme** is secreted, but an **endoenzyme** is not. Moreover, a **constitutive enzyme** is regularly found in a cell, whereas an **induced enzyme** is synthesized only if its substrate is present.

Types of Enzyme Function: Metabolic reactions vary. Sometimes water is released in anabolism to forge new covalent bonds; these are called **condensation reactions. Hydrolysis reactions** occurring in catabolism involve addition of water. **Functional groups** may be added, removed, or traded in many reactions. In energy reactions, valence electrons are often released accompanied by protons (H^+). Compounds yielding electrons are **oxidized,** whereas those gaining electrons are **reduced.**

Enzyme Sensitivity: Enzymes are **labile** (unstable) and function only within narrow operating ranges of temperature and pH, and they are especially vulnerable to **denaturation.** Enzymes are vital metabolic links and thus constitute easy targets for many harmful physical and chemical agents.

REGULATION OF ENZYMATIC ACTIVITY

Regulation involves key enzymes located at or near the beginning of a pathway. A substance that resembles the normal substrate and can occupy the same active site is said to exert **competitive inhibition.** In **feedback** (end product) **control,** the concentration of the output at the end of a pathway blocks the action of a key enzyme. A special mechanism called **feedback inhibition** operates by means of **allosteric enzymes,** which have special **regulatory sites** away from the active site. When secondary substrates attach, the enzyme's active site is distorted. Another mechanism, **feedback repression,** operates through enzyme replacement. Depending upon the status of the end product under control, the replenishment of key enzymes is inhibited at the genetic level. Enzyme control through **induction** conserves cell resources by producing enzymes only when the appropriate substrate is present.

BIOENERGETICS—HOW CELLS PRODUCE AND USE ENERGY

Potential and **kinetic energy** both have the capacity to perform work. Energy is consumed in **endergonic reactions** and is released in **exergonic reactions.** The freed energy is associated with electrons that can be temporarily captured and transferred in high-energy molecules such as ATP. Extracting energy requires a series of **electron carriers** arrayed in a downhill **redox chain** between electron donors and electron acceptors. In **oxidative phosphorylation,** energy is transferred to inorganic phosphate, causing it to form high-energy compounds such as ATP.

Products of Pyruvic Acid Oxidation in TCA and Electron Transport: Acetyl-coenzyme A is a product of pyruvic acid processing. This compound undergoes further oxidation and decarboxylation in the TCA cycle, which generates ATP, CO_2, and H_2O. Important intermediary metabolites are pyruvate, oxaloacetate, citrate, isocitrate, α-ketoglutarate, succinate, fumarate, and malate. The respiratory chain completes energy extraction. Important redox carriers of the electron transport system are NAD, **FAD** (flavin adenine dinucleotide), coenzyme Q, and cytochromes. The **chemiosmotic hypothesis** is a conceptual model that explains the origin and maintenance of electropotential gradients across a membrane that leads to ATP synthesis, by **ATP synthase.**

Glycolysis *and the* **TCA Cycle:** Carbohydrates, such as glucose, are energy-rich because they can yield a large number of electrons per molecule. Glucose is dismantled in two stages. **Anaerobic respiration,** or **glycolysis,** is the degradation of glucose to pyruvic acid in the absence of oxygen. Pyruvic acid is processed in **aerobic respiration** via the **tricarboxylic acid (TCA) cycle** and its associated respiratory chain. Whereas oxygen is the **final electron acceptor** in aerobic respiration, in anaerobic respiration sulfate, nitrate, or nitrite serve this function. Important intermediates in glycolysis are glucose-6-phosphate, fructose-1,6-diphosphate, glyceraldehyde-3-phosphate, diphosphoglyceric acid, phosphoglyceric acids, phosphoenolpyruvate, and pyruvic acid.

Fermentation is anaerobic respiration in which both the electron donor and final electron acceptors are organic compounds. Fermentation enables anaerobic and facultative microbes to survive in environments devoid of oxygen. Production of alcoholic brews, vinegar, and certain industrial solvents relies upon fermentation. Fermentation products also play a part in identifying some bacteria. The **phosphogluconate pathway** is an alternative anaerobic pathway for hexose oxidation that also provides for the synthesis of NADPH and pentoses.

Versatility of Glycolysis and TCA Cycle: Glycolysis and the TCA cycle are bidirectional, or **amphibolic, pathways.** Metabolites of these pathways double as building blocks and sources of energy. Intermediates that are convertible into amino acids through **amination** contribute to protein synthesis. Amino acids can be **deaminated** and used as energy sources. Components for purines and pyrimidines are derived from amino acid pathways. Two-carbon acetate molecules from pyruvate **decarboxylation** are units available for fatty acid synthesis. Glyceraldehyde-3-phosphate is a backbone for fatty acid attachment. The combination yields triglyceride, a typical storage fat, especially in times of carbohydrate abundance. Reversing the process, as in **beta oxidation** of fatty acids, taps fuel from storage sites.

MULTIPLE-CHOICE QUESTIONS

1. ____ is another term for biosynthesis.
   a. catabolism c. metabolism
   b. anabolism d. catalyst
2. Catabolism is a form of metabolism in which ____ molecules are converted into molecules.
   a. large, small c. amino acid, protein
   b. small, large d. food, storage
3. An enzyme ____ the activation energy required for a chemical reaction.
   a. increases c. lowers
   b. converts d. catalyzes
4. An enzyme
   a. becomes part of the final products
   b. is nonspecific for substrate
   c. is consumed by the reaction
   d. is heat and pH labile
5. An apoenzyme is where the ____ is located
   a. cofactor c. redox reaction
   b. coenzyme d. active site
6. Many coenzymes are
   a. metals c. proteins
   b. vitamins d. substrates
7. To digest cellulose in its environment, a fungus produces a/an
   a. endoenzyme c. catalase
   b. exoenzyme d. polymerase
8. In negative feedback control of enzymes, a build-up in the amount of ____ decreases the activity in the enzyme.
   a. substrate c. product
   b. reactant d. ATP
9. Energy in biological systems is primarily
   a. electrical c. radiant
   b. chemical d. mechanical
10. Energy is carried from catabolic to anabolic reactions in the form of ____.
    a. ADP
    b. high-energy ATP bonds
    c. coenzymes
    d. inorganic phosphate
11. Exergonic reactions
    a. release potential energy
    b. consume energy
    c. form bonds
    d. occur only outside the cell
12. A reduced compound is
    a. NAD c. NADH
    b. FAD d. ADP
13. Most oxidation reactions in microbial bioenergetics involve the
    a. removal of electrons and hydrogens
    b. addition of electrons and hydrogens
    c. addition of oxygen
    d. removal of oxygen
14. Products of glycolysis are
    a. ATP c. CO_2
    b. H_2O d. both a and b
15. Fermentation of a glucose molecule gives off a net number of ____ ATPs.
    a. 4 c. 40
    b. 2 d. 0
16. Complete oxidation of glucose in aerobic respiration yields a net output of ____ ATP.
    a. 40 c. 38
    b. 6 d. 2
17. The compound that enters the TCA cycle from glycolysis is
    a. citric acid
    b. oxaloacetic acid
    c. pyruvic acid
    d. acetyl coenzyme A
18. The $FADH_2$ formed during the TCA cycle enters the electron transport system at which site?
    a. NADH dehydrogenase
    b. cytochrome
    c. coenzyme Q
    d. ATP synthase
19. ATP synthase complexes can generate ____ ATPs for each NADH that enters electron transport.
    a. 1 b. 3
    b. 2 d. 4

CONCEPT QUESTIONS

1. Show diagrammatically the interaction of holoenzyme and its substrate and general products that can be formed from a reaction.
2. Give the general name of the enzyme that:
   a. synthesizes ATP; digests RNA
   b. carries out the transformation from DHAP to G-3-P (step 4 in glycolysis)
   c. catalyzes the formation of acetyl from pyruvic acid (just before step 1 in the TCA cycle);
   d. reduces pyruvic acid to lactic acid; and reduces nitrate to nitrate
3. a. Explain what an allosteric enzyme is and how negative feedback works. Two steps in glycolysis are catalyzed by allosteric enzymes. These are: (1) phosphofructokinase, which catabolizes step 3, and (2) pyruvate kinase, which catabolizes step 9.
   b. Suggest what metabolic products might regulate these enzymes.
   c. How might one place these regulators in figure 8.19?
4. Explain how oxidation of a substrate proceeds without oxygen.
5. In the following redox pairs, which compound is reduced and which is oxidized?
   a. NAD and NADH
   b. $FADH_2$ and FAD
   c. lactic acid and pyruvic acid
   d. NO_3 and NO_2
   e. ethanol and acetaldehyde
6. a. Discuss the relationship of anabolism to catabolism
   b. of ATP to ADP
   c. of glycolysis to fermentation
   d. of electron transport to oxidative phosphorylation

7. a. What is meant by the concept of the "final electron acceptor"?
   b. What are the final electron acceptors in aerobic, anaerobic, and fermentative metabolism?
8. Name the major ways that substrate-level phosphorylation is different from oxidative phosphorylation.
9. Compare the location of glycolysis, TCA cycle, and electron transport in procaryotic and eucaryotic cells.
10. a. Outline the basic steps in glycolysis, indicating where ATP is used and given off.
    b. Where does NADH originate, and what is its fate in an aerobe?
    c. What is its fate in a fermentative organism?
11. a. What is the source of ATP in the TCA cycle?
    b. How many ATPs are formed from the original glucose molecule?
12. How many ATPs would be formed as a result of aerobic respiration if cytochrome oxidase were missing from the respiratory chain (as is the case with many bacteria)?
13. a. Summarize the chemiosmotic theory of ATP formation.
    b. What is unique about the actions of ATP synthase?
14. How are aerobic and anaerobic respiration different?
15. Compare the general equation for aerobic metabolism with table 8.4 and verify that all figures balance.
16. Water is one of the end products of aerobic respiration. Where in the metabolic cycles is it formed, and where is it used?
17. Briefly outline the use of certain metabolites of glycolysis and TCA in amphibolic pathways.

CRITICAL-THINKING QUESTIONS

1. Using the concept of fermentation, describe the microbial (biochemical) mechanisms that cause milk to sour.
2. *Trichomonas vaginalis* is a protozoan agent of a sexually transmitted disease in humans. It lives on the vaginal or urinary mucous membranes, feeding off dead cells and glycogen. Astonishingly, this eucaryote completely lacks mitochondria. Predict what sort of metabolism it must have.
3. Explain some of the main functions of vitamins and why they are essential growth factors in human and microbial metabolism.
4. Explain the reasons that bacteria can generate more ATP per glucose than eucaryotic cells can.
5. Beer production requires an early period of rapid aerobic metabolism of glucose by yeast. Given that anaerobic conditions are necessary to produce alcohol, can you explain why this step is necessary?
6. Draw a model of ATP synthase in 3 dimensions, showing how it works. Where in the mitochondrion does the ATP supply collect?
7. Microorganisms are being developed to control man-made pollutants and oil spills that are metabolic poisons to animal cells. A promising approach has been to genetically engineer bacteria to degrade these chemicals. What is actually being manipulated in these microbes?

INTERNET SEARCH TOPIC

Look up fermentation on the WWW, and outline some of the products made by this process.

9 chapter

MICROBIAL GENETICS

The study of modern genetics is really a study of the language of the cell, a special language found in deoxyribonucleic acid—DNA. In this chapter we shall investigate the structure and function of this molecule and explore how it is copied, how its language is interpreted into useful cell products, how it is controlled, how it changes, how it is transferred from one cell to another, and its applications in microbial genetics.

The search for the structure of the genetic material is one of the most compelling stories in biology. See microfile 9.1 for a short history of this saga.

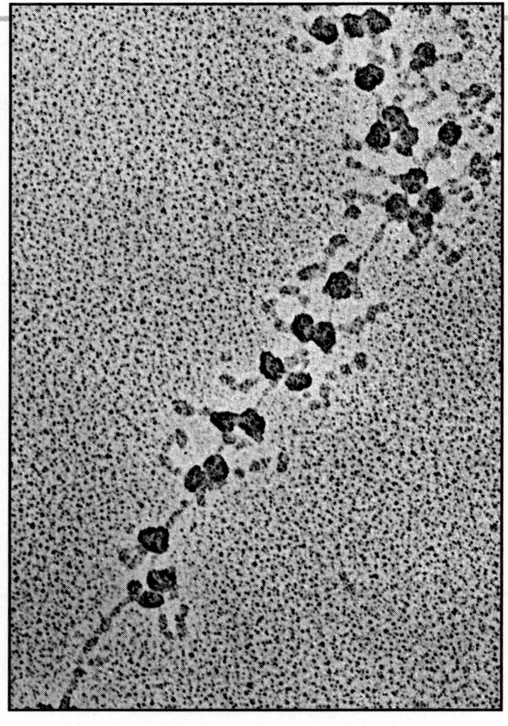

The protein assembly line. Photomicrograph of a chain of ribosomes (larger, spherical objects) linked together by a single messenger RNA (elongate strand). Translation of the mRNA can produce multiple copies of the same protein (smaller granules). See figure 9.17.

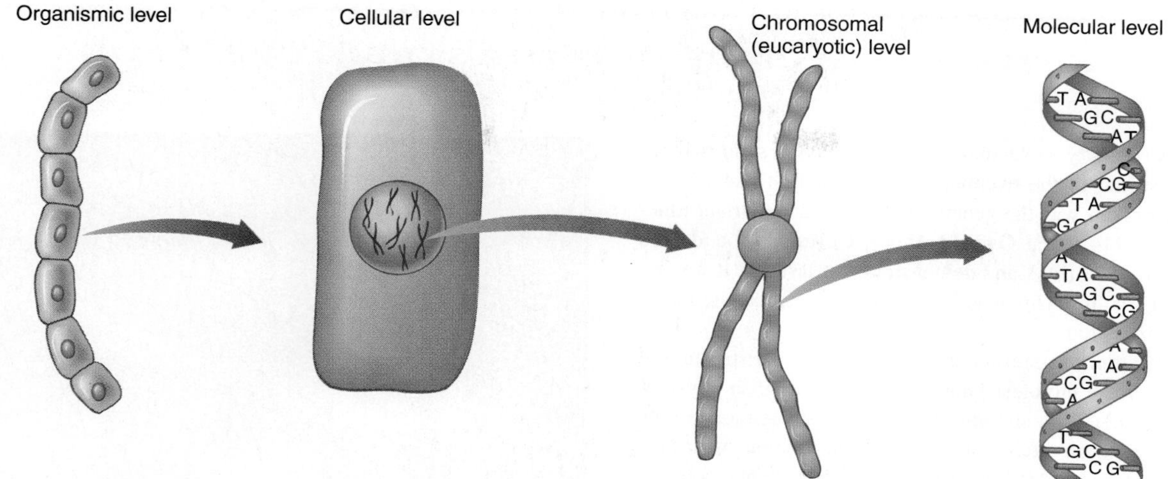

Figure 9.1
Levels of genetic study. The operations of genetics can be observed at the organismic, cellular, chromosome with genes, and molecular (DNA) levels.

INTRODUCTION TO GENETICS AND GENES: UNLOCKING THE SECRETS OF HEREDITY

Genetics* is the study of the inheritance, or **heredity,*** of living things. It is a wide-ranging science that explores the transmission of biological properties (traits) from parent to offspring, the expression and variation of those traits, the structure and function of the genetic material, and how this material changes. The study of genetics exists on several levels (figure 9.1). Organismic genetics observes the heredity of the whole organism or cell; chromosomal genetics examines the characteristics and actions of chromosomes; and molecular genetics deals with the biochemistry of the genes. All of these levels are useful areas of exploration, but in order to understand the expressions of microbial structure, physiology, mutations, and pathogenicity, we need to examine the operation of genes at the cellular and molecular levels. The study of microbial genetics provides a greater understanding of human genetics and an increased appreciation for the astounding advances in genetic engineering we are currently witnessing (chapter 10).

THE NATURE OF THE GENETIC MATERIAL

For a species to survive, it must have the capacity of self-replication. In single-celled microorganisms, reproduction involves the division of the cell by means of binary fission, budding, or mitosis, but these forms of reproduction involve a more significant activity than just simple cleavage of the cell mass. Because the genetic material is responsible for inheritance, it must be accurately duplicated and separated into each daughter cell to ensure its normal function. This genetic material is a long, encoded molecule of DNA that can be studied on several levels. Before we look at how DNA is copied, let us explore the organization of this genetic material, proceeding from the general to the specific.

The Levels of Structure and Function of the Genome

The **genome** is the sum total of genetic material of a cell. Although most of the genome exists in the form of chromosomes, genetic material can appear in nonchromosomal sites as well (figure 9.2). For example, bacteria and some fungi contain tiny extra pieces of DNA (plasmids), and certain organelles of eucaryotes (the mitochondria and chloroplasts) are equipped with their own genetic programs. Genomes of cells are composed exclusively of DNA, but viruses contain either DNA or RNA as the principal genetic material. Although the specific genome of an individual organism is unique, the general pattern of nucleic acid structure and function is similar among all organisms.

In general, a **chromosome** is a discrete cellular structure composed of a neatly packaged elongate DNA molecule. The chromosomes of eucaryotes and bacterial cells differ in several respects. The structure of eucaryotic chromosomes consists of a DNA molecule tightly wound around histone proteins (see microfile 9.3), whereas a bacterial chromosome (chromatin body) is condensed and secured into a packet by means of histoneike proteins. Eucaryotic chromosomes are located in the nucleus; they vary in number from a few to hundreds; they can occur in pairs (diploid) or singles (haploid); and they appear elongate. In contrast, the single chromosome of bacterial cells exists free in the cytoplasm, is haploid, and usually has a closed-circle configuration.

* **genetics** L. *genesis,* birth, generation.

* **heredity** L. *hereditas,* heirship. All the characteristics genetically inherited by an organism.

MICROFILE 9.1 DECIPHERING THE STRUCTURE OF DNA

The search for the primary molecules of heredity was a serious focus throughout the first half of the twentieth century. At first many biologists thought that protein was the genetic material. An important milestone occurred in 1944 when Oswald Avery, Colin MacLeod, and Maclyn McCarty purified DNA and demonstrated at last that it was indeed the blueprint for life. This was followed by an avalanche of research, which continues today.

One area of extreme interest concerned the molecular structure of DNA. In 1951, American biologist James Watson and English physicist Francis Crick collaborated on solving the DNA puzzle. Although they did little of the original research, they attacked the problem with brilliance, perception, and considerable energy. The existing data from experiments with DNA intrigued them. It had been determined by Erwin Chargaff that any model of DNA structure would have to contain deoxyribose, phosphate, purines, and pyrimidines arranged in a way that would provide variation and a simple way of copying itself. Watson and Crick spent long hours constructing models with cardboard cutouts and kept alert for any and every bit of information that might give them an edge. Two English biophysicists, Maurice Wilkins and Rosalind Franklin, had been painstakingly collecting data on X-ray crystallographs of DNA for several years. With this technique, molecules of DNA bombarded by X rays produce a photographic image that can predict the three-dimensional structure of the molecule. After being allowed to view certain of Wilkins and Franklin's data privately, Watson and Crick noticed an unmistakable pattern: the molecule appeared to be a double helix. Gradually, the pieces of the puzzle fell into place, and a final model was assembled—a model that explained all of the qualities of DNA, including how it is copied. Although Watson and Crick were rightly hailed for the clarity of their solution, it must be emphasized that their success was due to the considerable efforts of a number of English and American scientists. This historic discovery showed that the tools of physics and chemistry have useful applications in biological systems, and it also spawned ingenious research in all areas of molecular genetics.

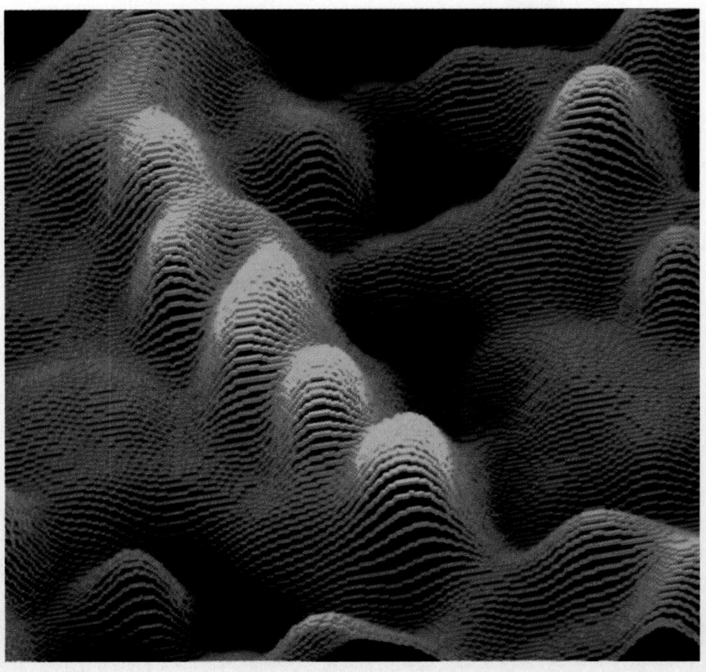

The first direct glimpse at DNA's structure. This false-color scanning tunneling micrograph of calf thymus gland DNA (2,000,000×) brings out the well-defined folds in the helix.

Since the discovery of the double helix in 1953, an extensive body of biochemical, microscopic, and crystallographic analysis has left little doubt that the model first proposed by Watson and Crick is correct. Newer techniques using scanning tunneling microscopy produce three-dimensional images of DNA magnified 2 million times. These images verify the helical shape and twists of DNA represented by models.

The chromosomes of all cells are subdivided into basic informational packets called genes. A **gene** can be defined from more than one perspective. In classical genetics, the term refers to the fundamental unit of heredity responsible for a given trait in an organism. In the molecular and biochemical sense, it is a site on the chromosome that provides information for a certain cell function. More specifically still, it is a certain segment of DNA that contains the necessary code to make a protein or RNA molecule. This last definition of a gene will be emphasized in this chapter. For an analogy that clarifies the relationship of the genome, chromosomes, genes, and DNA, see microfile 9.2.

The Size and Packaging of Genomes

Genomes vary greatly in size. The smallest viruses have four or five genes; the bacterium *Escherichia coli* has a single chromosome containing about 4,000 genes, and a human cell packs about 100,000 genes into 46 chromosomes (the human genome is relatively modest in size compared with the genomes of some plants and other animals). The total length of DNA relative to cell size is notorious: the chromosome of *E. coli* would measure about 1 mm if unwound and stretched out linearly, and yet this fits within a cell that measures only about 2 μm across, making the DNA 500 times as long as the cell (figure 9.3). Still, the bacterial chromosome takes up only about one-third to one-half of the cell's volume. Likewise, if the sum of all DNA contained in the 46 human chromosomes were unraveled and laid end to end, it would measure about 6 feet. This means that the DNA is about 180,000 times longer than a cell 10 μm wide and a million times longer than the width of the nucleus. How can such elongated genomes fit into the miniscule volume of a cell, and, in the case of eucaryotes, into an even smaller compartment, the nucleus? The answer lies in the regular coiling of the DNA chain (microfile 9.3).

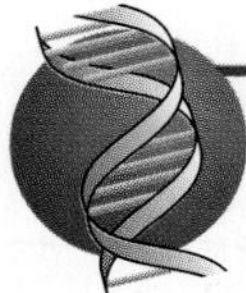

MICROFILE 9.2 THE VIDEO GENOME

The relationship of the genome, chromosomes, and genes to one another and to DNA is analogous to a collection of videotapes of family events. The whole library of tapes (genome) contains several individual cassettes (chromosomes); the tape on the cassette is divided sequentially into several separate events (genes), which may be selected and played to generate a picture on the television screen (product). This analogy works in several other ways:

1. At all levels (library, cassette, event), the basic informational unit is still the tape itself (the DNA molecule).
2. Like a tape, the DNA molecule can be copied, spliced, and edited.
3. Somewhat like the spools of a cassette, the long DNA of a chromosome is carefully wrapped up so that it is compact, easy to read, and will not get tangled (see microfile 9.3).
4. For the tape (DNA code) to be translated into images (cell product), a tape player (special cell machinery) is necessary.
5. The tape, like the DNA molecule, makes sense only if played in a certain direction.
6. The entire collection of tapes, like the genome, is a store of information. Not all of it is being "played" at any one time.

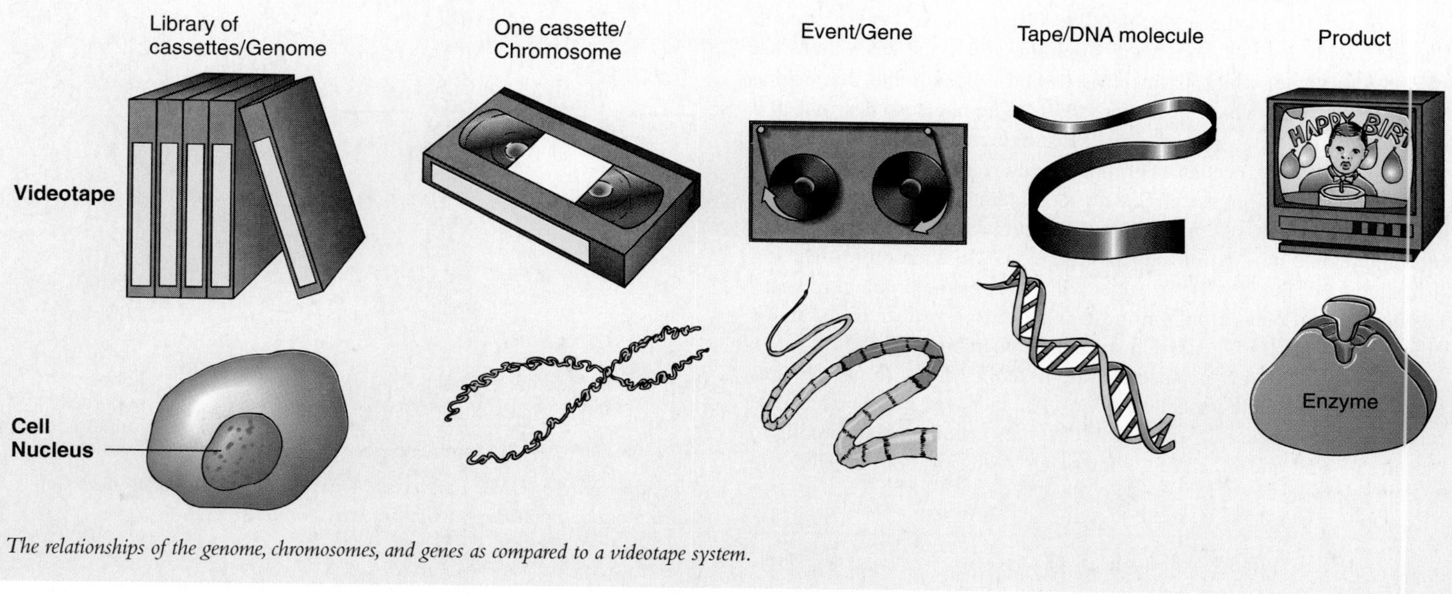

The relationships of the genome, chromosomes, and genes as compared to a videotape system.

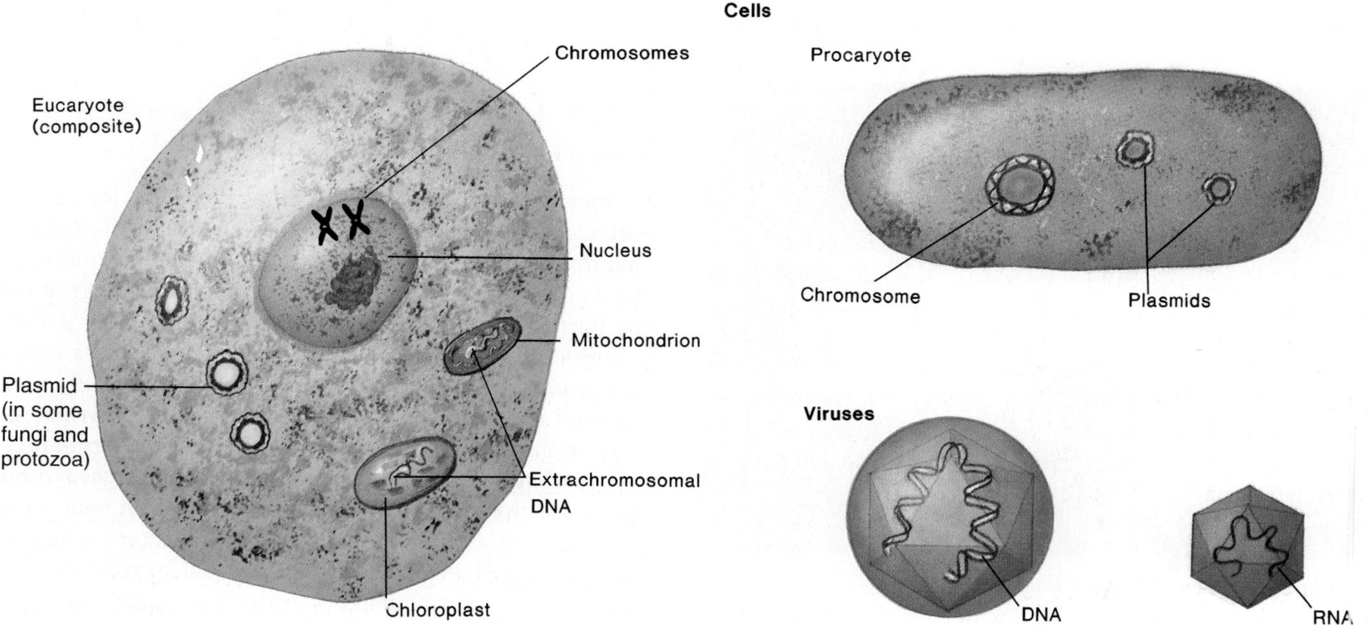

Figure 9.2
The general location and forms of the genome in microbes.

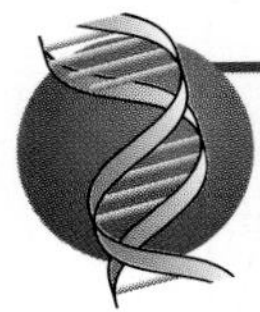

MICROFILE 9.3 THE PACKAGING OF DNA: WINDING, TWISTING, AND COILING

The analogy of DNA to a cassette tape is imperfect because the DNA molecule is not perfectly wound around a spool. Packing the mass of DNA into the cell involves further levels of DNA structure called supercoils or superhelices. In the simpler system of procaryotes, the chromosome, a continuous circle, is twisted around itself by the action of a special enzyme called a *topoisomerase** (specifically DNA *gyrase*).* This enzyme folds, splices, and holds DNA, thereby introducing a reversible series of twists into the molecule. The system in eucaryotes is more complex, with three or more levels of coiling. First, the DNA molecule of a chromosome, which is linear, is wound twice around the histone proteins, creating a chain of *nucleosomes.** The nucleosomes fold in a spiral formation upon one another. Most experts believe that an even greater supercoiling occurs when this spiral arrangement further twists on its radius into a giant spiral with loops radiating from the outside. This extreme degree of compactness is what makes the eucaryotic chromosome visible during mitosis (see figure 5.6). In addition to reducing the volume occupied by DNA, supercoiling solves the problem of keeping the chromosomes from getting tangled during cell division, and it protects the code from massive disruptions due to breakage.

From a different perspective, highly coiled DNA is far too condensed to be available for cell activities. Another type of topoisomerase (also called a helicase or untwisting enzyme) uncoils the supercoils to permit copying and other functions of DNA during cell division and protein synthesis.

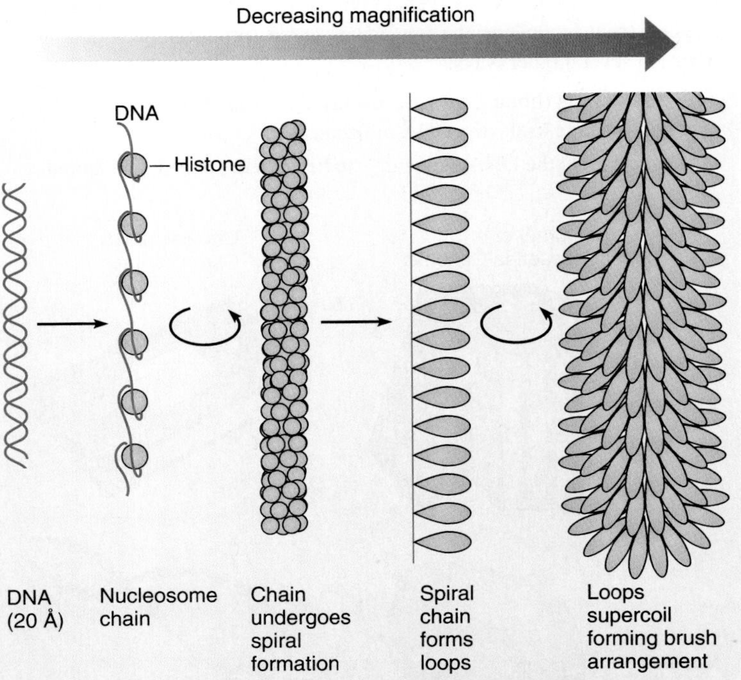

The packaging of DNA. Eucaryotic DNA undergoes several orders of coiling and supercoiling, which greatly condense it.

* topoisomerase (tah″-poh-eye-saw′-mur-ayce) Any enzyme that changes the configuration of DNA.

* gyrase (jy′-rayce) L. *gyros,* ring or circle. A bacterial enzyme that produces supercoils.

* nucleosomes "Nucleus bodies" arranged like beads on a chain.

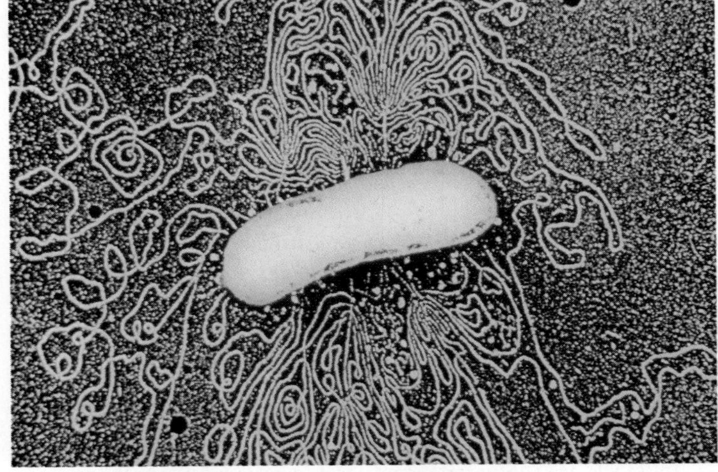

Figure 9.3

An *Escherichia coli* bacillus disrupted to release its DNA molecule. The cell has spewed out its single, uncoiled DNA strand into the surrounding medium.

THE DNA CODE: A SIMPLE YET PROFOUND MESSAGE

Examining the function of DNA at the molecular level requires an even closer look at its structure. To do this we will imagine being able to magnify a small piece of a gene about 5 million times. What such fine scrutiny will disclose is one of the great marvels, and for many years, mysteries, of biology. Our first view of DNA in chapter 2 revealed that it is a gigantic molecule, a type of nucleic acid, with two polynucleotide strands combined into a double helix (see figure 2.25). Except in some viruses that contain single-stranded DNA, the general structure is universal. It consists of a **deoxyribose sugar-phosphate** backbone attached to **nitrogenous base** cross-pieces. The sugar and phosphate molecules alternate, creating a repetitive and regular molecular skeleton. Two phosphates are bonded covalently to one deoxyribose sugar. One of the bonds is to the number 5′ (read "five prime") carbon on deoxyribose, and the other is to the 3′ carbon, which confers a certain order and direction on each strand (figure 9.4).

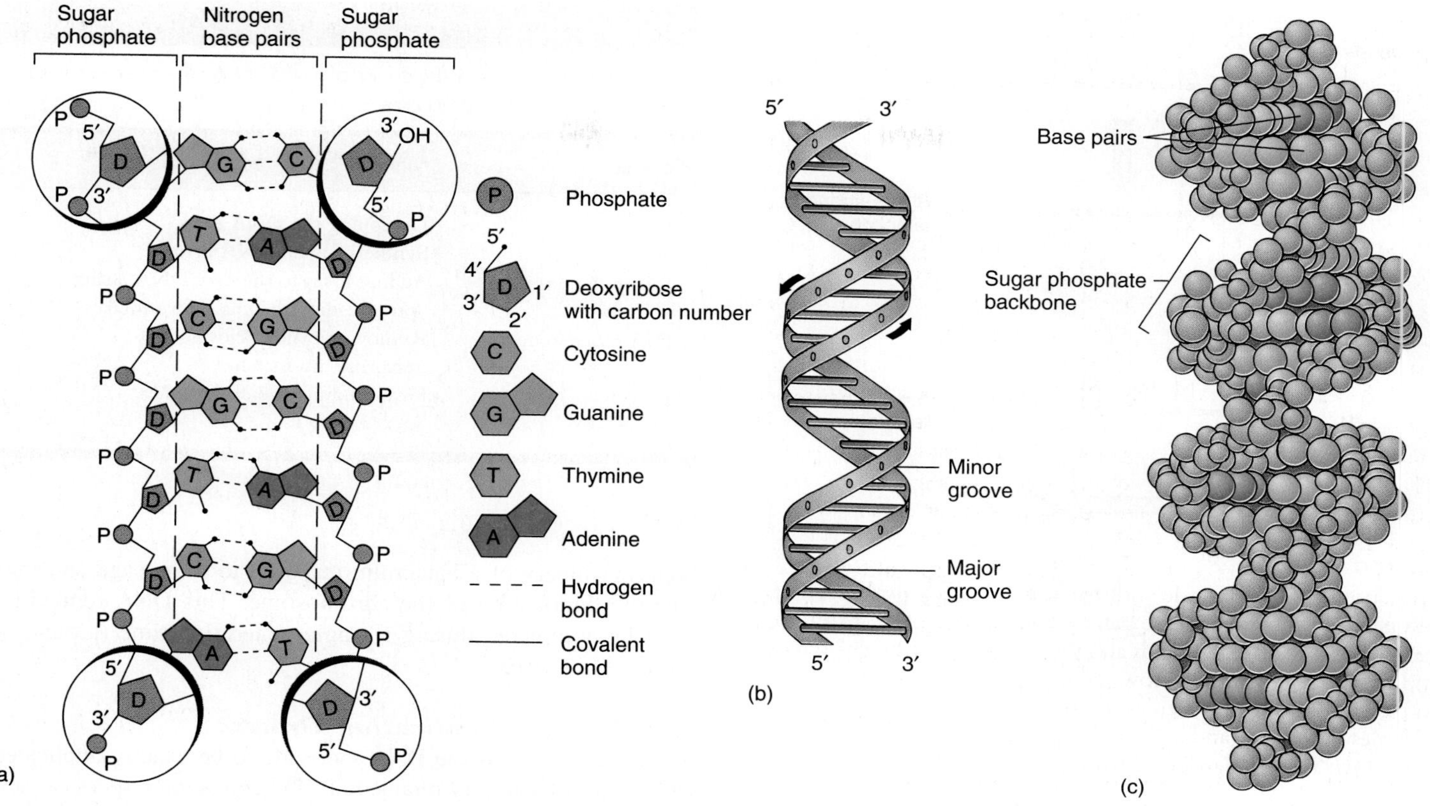

Figure 9.4
Three views of DNA structure. (*a*) A highly schematic nonhelical model, showing sugar-phosphate backbones, details of deoxyribose bonds with phosphate (5´ or 3´), and bonds with sugar (1´). Purine-pyrimidine base pairs are shown with hydrogen bonds; G–C have three of these bonds; A–T have two. (*b*) Simplified model that highlights the antiparallel arrangement and the major and minor grooves. (*c*) Space-filling model that more accurately depicts the structure of DNA.

The nitrogenous bases, **purines** and **pyrimidines,** attach by covalent bonds at the 1´ position of the sugar. They span the center of the molecule and pair with appropriate complementary bases from the other side of the helix. The paired bases are so aligned as to be joined by hydrogen bonds. Such weak bonds are easily broken, allowing the molecule to be "unzipped" into its complementary strands. Later we will see that this feature is of great importance in gaining access to the information encoded in the nitrogenous base sequence. Pairing of purines and pyrimidines is not random; it is dictated by the chemical and physical reality that a given base readily forms hydrogen bonds with certain other bases. Thus, in DNA, the purine **adenine** (A) pairs with the pyrimidine **thymine** (T), and the purine **guanine** (G) pairs with the pyrimidine **cytosine** (C). New research also indicates that the bases are attracted to each other in this pattern because each has a complementary three-dimensional shape that matches its pair. Although the base-pairing partners generally do not vary, the sequence of base pairs along the DNA molecule can assume any order, resulting in an infinite number of possible nucleotide sequences.

Other important considerations of DNA structure concern the nature of the double helix itself. The halves are not parallel or oriented in the same direction. One side of the helix runs in the opposite direction of the other, in an *antiparallel arrangement* (figure 9.4*b*). The order of the bond between the carbon on deoxyribose and the phosphates is used to keep track of the direction of the two sides of the helix. Thus, one helix runs from the 5´ to 3´ direction and the other runs from the 3´ to 5´ direction. This characteristic is a significant factor in DNA synthesis and decoding. As apparently perfect and regular as the DNA molecule may seem, it is not exactly symmetrical. The torsion in the helix and the stepwise stacking of the nitrogen bases produce two different-sized surface features, the major and minor grooves (figure 9.4*b*).

The Significance of DNA Structure

The nitrogen bases influence DNA in two major ways:

1. **Maintenance of the code during reproduction.** The constancy of base-pairing guarantees that the code will be retained during cell growth and division. When the two strands are separated, each one provides a **template** (pattern or model) for the replication (exact copying) of a new molecule (figure 9.5). Because the sequence of one strand automatically gives the sequence of its partner, the code can be duplicated with fidelity.

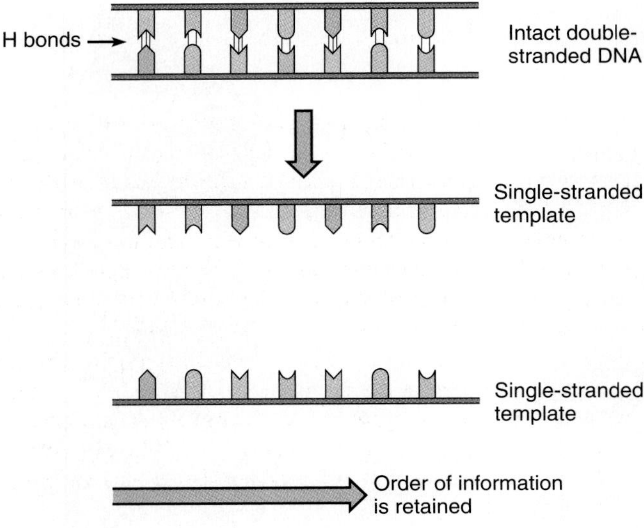

Figure 9.5
Implications of DNA's code and structure. The easy disruption of hydrogen bonds between base pairs is an important factor in the separation of the double strands and the formation of single-stranded templates. Each strand may now serve as the pattern for forming two double-stranded molecules. Yet the covalent bonds between molecules on each strand are not easily broken, and the order of bases, which is the language, remains intact.

2. **Providing variety.** The order of bases along the length of the DNA strand constitutes the genetic program, or the language, of the DNA code. Adding to our earlier definitions, the message present in a gene is a precise arrangement of these bases, and the genome is the collection of all DNA bases that, in an ordered combination, are responsible for the unique qualities of each organism.

It is tempting to ask how such a seemingly simple code can account for the extreme differences among forms as diverse as a virus, *E. coli,* and a human. The English language, based on 26 letters, can create an infinite variety of words, but how can this complex genetic language be based on just four nitrogen base "letters"? Put in mathematical terms, for a segment of DNA only 1,000 bases long, there are $4^{1,000}$ different genetic combinations. Because the complexity of even the simplest bacterial genome is many times greater than this, one must conclude that the potential for variation is virtually infinite.

DNA Replication: Preserving the Code and Passing It On

It has been established that the sequence of bases along the length of a gene constitutes the language of DNA. For this language to be preserved for hundreds of generations, it will be necessary for the genetic program to be duplicated and passed on to each offspring. This process of duplication is called **DNA replication.** In the following example, we will show replication in bacteria, but, with some exceptions, it also applies to the process as it works in eucaryotes and some viruses. Early in the division cycle, the metabolic machinery of a bacterium responds to a message and initiates the duplication of the chromosome. This DNA replication must be completed during a single generation time (around 20 minutes in *E. coli*).

Table 9.1

Some Enzymes Involved in DNA Replication and Their Functions

Enzyme	Function
Helicase	Unzipping the DNA helix
Primase	Synthesizing an RNA primer
DNA polymerase III	Adding bases to the new DNA chain; proofreading the chain for mistakes
DNA polymerase I	Removing primer, closing gaps, repairing mismatches
Ligase	Final binding of nicks in DNA
Gyrase	Supercoiling

The Overall Replication Process

What features allow the DNA molecule to be exactly duplicated, and how is its integrity retained? DNA replication requires a careful orchestration of the actions of 30 different enzymes (partial list in table 9.1), which separate the strands of the existing DNA molecule, copy its template, and produce two complete daughter molecules. A simplified version of replication is shown in figure 9.6 and includes the following: (1) uncoiling the parent DNA molecule, (2) unzipping the hydrogen bonds between the base pairs, thus separating the two strands and exposing the nucleotide sequence of the helix to serve as **templates,** and (3) synthesis of two double strands by attachment of the correct complementary nucleotides to each single-stranded template. It is worth noting that each daughter molecule will be identical to the parent in composition, but neither one is completely new; the strand that serves as a template is an original parental DNA strand. The preservation of the parent molecule in this way, termed *semiconservative replication,* helps explain the reliability and fidelity of replication.

Refinements and Details of Replication

The circular bacterial DNA molecule replicates by means of a special configuration called a **replicon.** Replication begins at a precise initiation site containing a certain repeated sequence called a palindrome, which forms the origin of replication (microfile 9.4). Here, enzymes called helicases ("unzipping enzymes") untwist the helix and break its hydrogen bonds. This process is best seen in animation; the molecule does not unzip like a common clothing zipper but rotates as it is untwisted.

Replication begins when an *RNA primer* is synthesized and enters at the initiation site. The primer provides DNA polymerase III with a short strand made of RNA that signals the point to begin synthesis. This is necessary because the polymerase can only extend from a preexisting strand (the primer). Because the bacterial DNA molecule is circular, opening of the circle forms

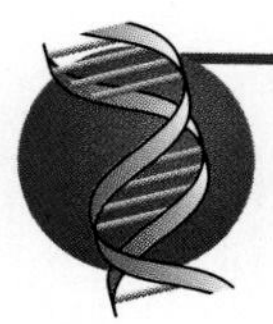

MICROFILE 9.4 PALINDROMES—WORD GAMES WITH THE LANGUAGE OF DNA

Intriguing components of DNA structure are the numerous palindromic sequences present throughout the molecule. In language, a palindrome is a word, phrase, or sentence that reads the same both forward and backward: for example, **radar; madam I'm Adam; too hot to hoot;** and **poor Dan is in a droop.** A DNA palindrome, also called an *inverted repeat,* might read as follows:

GCTAGC
CGATCG

Unlike words, a DNA palindrome occurs not in a single line but in the order of bases in the two complementary strands, with the top strand being read from left to right and its complement being read from right to left.

DNA palindromes vary in size from a few bases to several hundred and appear to have a number of functions. They can, for instance, be regulatory, providing a starting site for DNA replication or serving as a binding site for enzymes and other molecules that govern genetic expression. Inverted repeats can also permit loops to form in supercoiling DNA, thereby relieving tension on the molecule. Of great significance is the discovery of special enzymes of bacterial origin called **restriction endonucleases,** which hydrolyze, or *nick,* DNA internally at sites of palindromic sequences. These enzymes recognize foreign DNA and are capable of snipping it apart at these sites. In the bacterial cell, this ability protects against the incompatible DNA of bacteriophages or plasmids. In the biotechnologist's lab, the enzymes can be used to cleave DNA at desired sites and are a must for the techniques of recombinant DNA technology (see chapter 10).

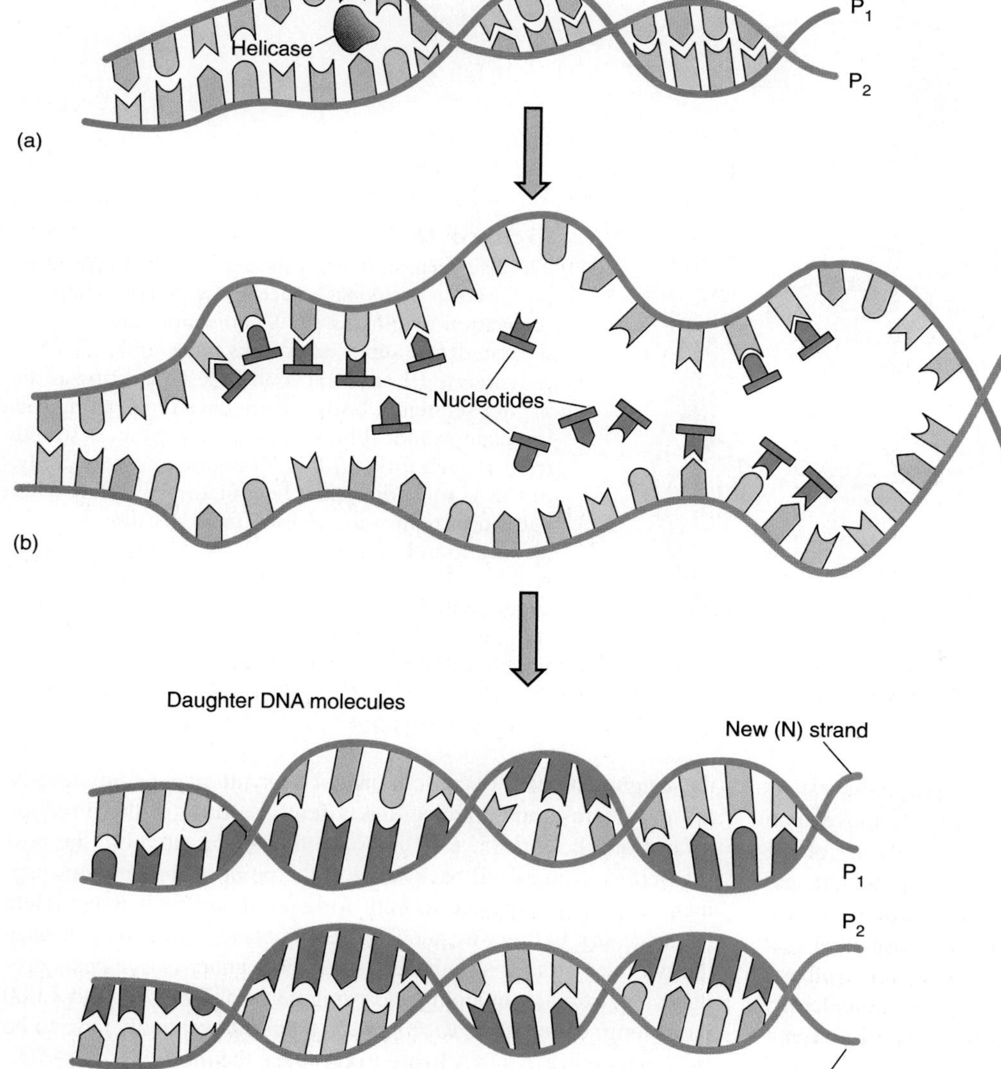

Figure 9.6

Simplified steps in semiconservative replication of DNA. (*a*) A helicase unwinds the double helix into two parent strands (P_1 and P_2). (*b*) The replication of new complementary strands proceeds through the action of an enzyme that attaches nucleotides, using the exposed strands as templates. (*c*) Each completed daughter molecule contains one strand that is newly synthesized and one of the original parent strands.

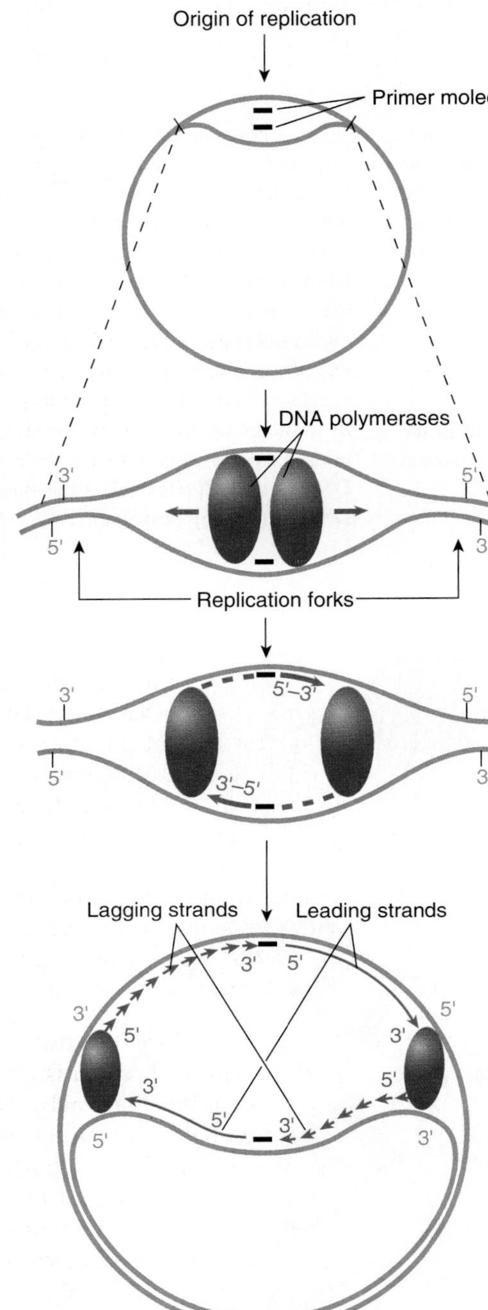

Origin of replication
Primer molecules
(a) Replication origin. Short RNA primers are positioned to start replication.
DNA polymerases
3'
5'
5'
3'
(b) Strands separate, two polymerase molecules attach at origin. Arrows indicate direction of replication.
Replication forks
3'
5'-3'
5'
(c) At primer sequence, each polymerase synthesizes 2 strands at the replication forks.
5'
3'-5'
3'
Lagging strands
Leading strands
3'
5'
3'
5'
5'
3'
3'
5'
5'
3'
5'
3'
(d) Enzyme orientation results in one new strand that grows continuously in 5'→3' direction and one new strand that grows by short segments in the 5'→3' direction.

Figure 9.7
The bacterial replicon: a model for DNA synthesis. (*a*) Circular DNA has a special origin site where replication originates. (*b*) When strands are separated, two replication forks form, and a DNA polymerase III enters at each fork. (*c*) Starting at the primer sequence, both polymerases move along the template strands (blue), synthesizing the new strands (red) at each fork. (*d*) DNA polymerase works only in the 5´ to 3´ direction, necessitating a different pattern of replication at each fork. Because the leading strand orients in the 5´ to 3´ direction, it will be synthesized continuously. The lagging strand, which orients in the opposite direction, can only be synthesized in short sections, 5´ to 3´, which are later linked together.

two **replication forks,** each containing its own polymerases. As replication proceeds, these forks open up and gradually move apart (figure 9.7*b*). Simple models show that further elongation of the molecules proceeds by the simple addition of complementary nucleotides to each parent strand continuously as the fork is opened up. But this is not actually what happens on both strands at that fork. Because DNA polymerase is correctly oriented for synthesis *only* in the 5´ to 3´ direction of the new molecule (red) strand, only this one strand, the **leading strand,** can be synthesized continuously. The strand with the opposite orientation (3´ to 5´) is termed the **lagging strand.** Because it cannot be synthesized continuously, the polymerase adds nucleotides a few at a time in the direction away from the fork (5´ to 3´). As the fork opens up a bit, the next segment is synthesized backwards to the point of the previous segment, a process repeated at both forks until synthesis is complete (figure 9.7*d*). In this way, the DNA polymerase is able to synthesize the two new strands simultaneously. This manner of synthesis produces one strand containing short fragments of DNA (100 to 1,000 bases long) called *Okazaki fragments* that will eventually have to be spliced together by DNA ligase to complete the molecules.

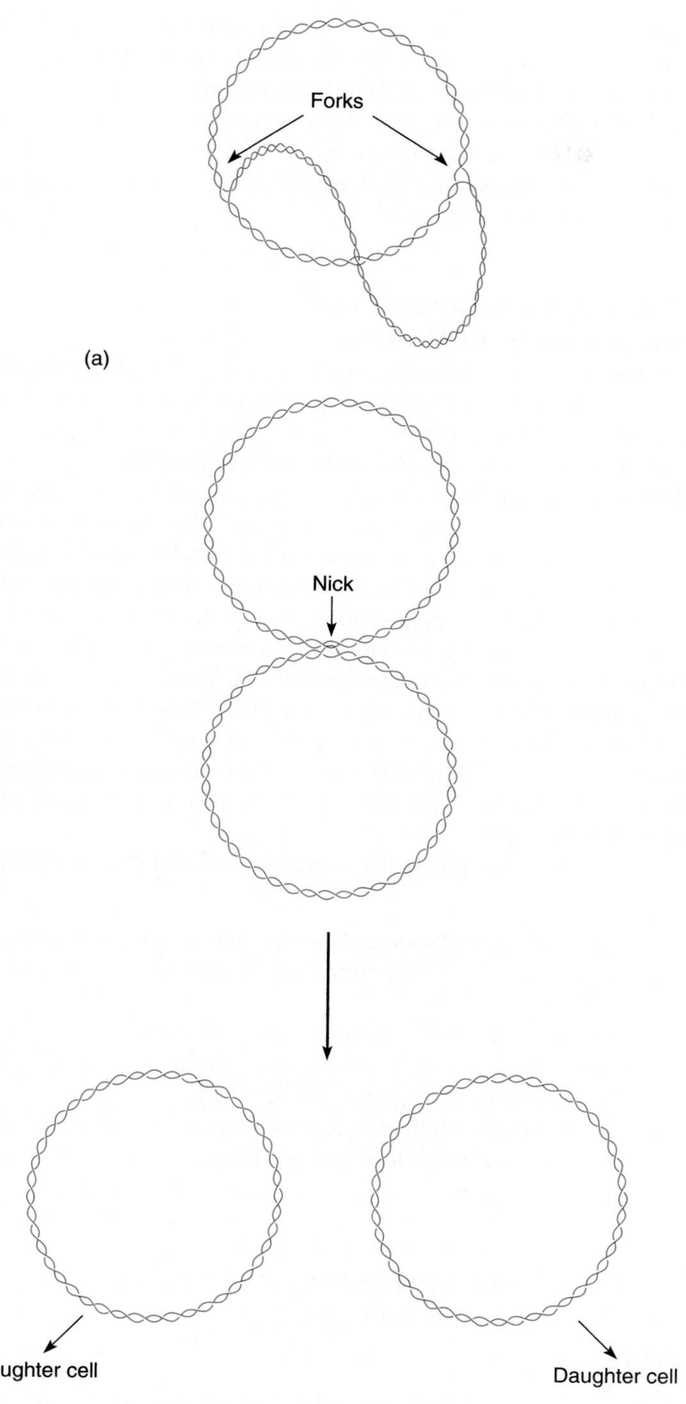

Figure 9.8
Completion of chromosome replication in bacteria. (*a*) The theta, or "eye," stage of replication, in which one strand loops down as it grows in length. (*b*) Nicking, separation, repair, and release of two completed molecules that will be separated into daughter cells during binary fission.

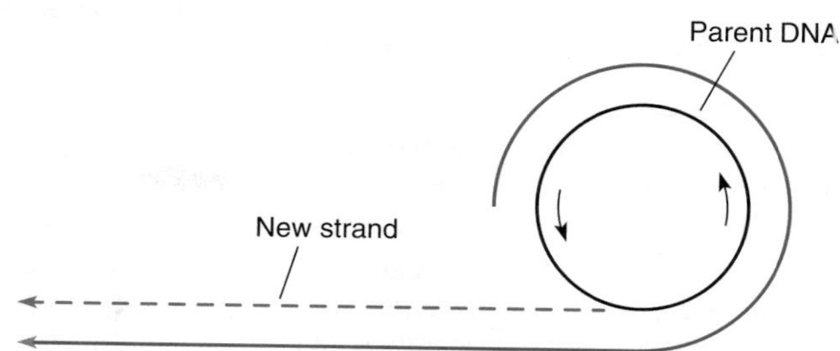

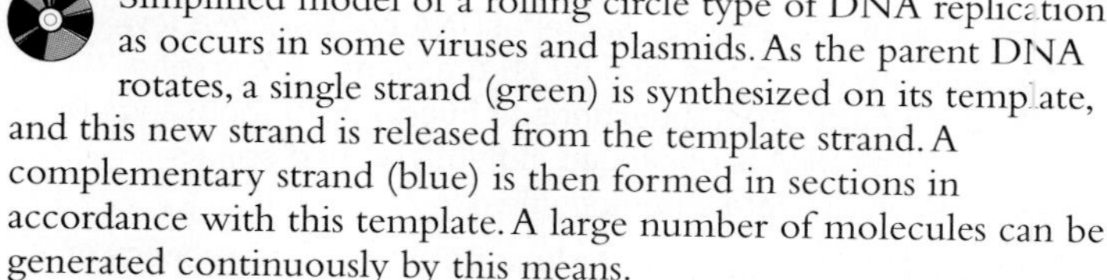

Figure 9.9
Simplified model of a rolling circle type of DNA replication as occurs in some viruses and plasmids. As the parent DNA rotates, a single strand (green) is synthesized on its template, and this new strand is released from the template strand. A complementary strand (blue) is then formed in sections in accordance with this template. A large number of molecules can be generated continuously by this means.

Elongation and Termination of the Daughter Molecules As replication proceeds along both forks, the replicon acquires the look of a half-open eye, or the Greek letter *theta* (θ), due to one duplicating strand that droops down (figure 9.8*a*). The addition of bases proceeds at an astonishing pace, estimated in some bacteria at 750 bases per second at each fork! When the replication forks come full circle and meet, ligases move along the lagging strand to begin the initial linking of the fragments and to complete synthesis and separation of the two circular daughter molecules (figure 9.8*b*). The DNA polymerase III removes the RNA primers used to initiate DNA synthesis.

Other Variations on the Theme The bidirectional replicon as described for bacteria is also the mechanism of many viruses and some bacterial plasmids. The eucaryotic chromosome is replicated simultaneously at numerous sites along the molecule, a process that speeds up duplication of the larger eucaryotic chromosome. Other aspects of the process, however, are very similar in all groups. A novel form of DNA synthesis called **rolling circle** occurs in some bacterial viruses and plasmids (figure 9.9). Although rather complicated in practice, it involves the synthesis of DNA in a single direction, with the new strand of DNA rolling out from the circle of DNA like paper from a roll of tissue.

Like any language, DNA is occasionally "misspelled" when an incorrect base is added to the growing chain. Studies have shown that such mistakes are made once in approximately 100,000 to 1,000,000 bases, but most of these are corrected. If not corrected, they become mutations and can lead to serious cell dysfunction and even death. Because continued cellular integrity is very dependent on accurate replication, cells have evolved their own proofreading function for DNA. DNA polymerase III, the enzyme that elongates the molecule, can also detect incorrect, unmatching bases, excise them, and replace them with the correct base. DNA polymerase I can also proofread the molecule and remove the original RNA primers that positioned replication.

Chapter Checkpoints

Nucleic acids are molecules that contain the blueprints of life in the form of genes. DNA is the blueprint molecule for all cellular organisms. The blueprints of viruses, however, can be either DNA or RNA.

The total amount of DNA in an organism is termed its genome. The genome of each species contains a unique arrangement of genes that define its appearance (phenotype), metabolic activities, and pattern of reproduction.

The genome of procaryotes is quite small compared with the genomes of eucaryotes. Bacterial DNA consists of a few thousand genes in one circular chromosome. Eucaryotic genomes range from thousands to hundreds of thousands of genes. Their DNA is packaged in tightly wound spirals arranged in discrete chromosomes.

DNA copies itself just before cellular division by the process of semiconservative replication. Semiconservative replication means that each "old" DNA strand is the template upon which each "new" strand is synthesized.

The circular bacterial chromosome is replicated at two forks as directed by DNA polymerase III. At each fork, two new strands are synthesized—one continuously and one in short fragments, and mistakes are proofread and removed.

APPLICATIONS OF THE DNA CODE: TRANSCRIPTION AND TRANSLATION

We have explored how the genetic message in the DNA molecule is conserved through replication. Now we must consider the precise role of DNA in the cell. Given that the sequence of bases in DNA is a genetic code, just what is the nature of this code and how is it utilized by the cell? Although the DNA library is full of critical information, the molecule itself does not perform cell processes directly. Its stored information is conveyed to RNA molecules, which carry out instructions. The concept that genetic information flows from DNA to RNA to protein is a central theme of molecular biology (figure 9.10). More precisely, it states that the master code of DNA is first copied onto an RNA molecule called a messenger through **transcription,** and the RNA message is decoded by special cell components into proteins during **translation.** The principal exceptions to this pattern are found in RNA viruses, which convert RNA to other RNA, and in retroviruses, which convert RNA to DNA.

THE GENE-PROTEIN CONNECTION

Genes fall into three basic categories: *structural genes* that code for proteins, genes that code for RNA, and *regulatory genes* that control gene expression. The sum of an organism's structural genes constitutes its distinctive genetic makeup, or **genotype,*** which, when expressed in the individual, creates traits (certain structures or functions) referred to as the **phenotype.*** Just as a person inherits a combination of genes (genotype) that gives a certain eye color or height (phenotype), a bacterium inherits genes that direct the formation of a flagellum, and a virus contains genes for its capsid structure.

The Triplet Code and the Relationship to Proteins

Several questions invariably arise concerning the relationship between genes and cell function. For instance, how does gene structure lead to the expression of traits in the individual, and what features of gene expression cause one organism to be so distinctly different from another? For answers, we must turn to the correlation between gene and protein structure. We know that each structural gene is a linear sequence of nucleotides that codes for a protein. Because each protein is different, each gene must also differ somehow in its composition. In fact, the language of DNA exists in the order of groups of three consecutive bases called **triplets** on one DNA strand (figure 9.11). Thus, one gene differs from another in its composition of triplets. An equally important part of this concept is that each triplet represents a code for a particular amino acid. When the triplet code is transcribed and translated, it dictates the type and order of amino acids in a polypeptide (protein) chain.

The final key points that connect DNA and protein function are:

1. A protein's primary structure—the order and type of amino acids in the chain—determines its characteristic shape and function.
2. Proteins ultimately determine phenotype, the expression of all aspects of cell function and structure. Put more simply, living things are what their proteins make them.
3. DNA is mainly a blueprint that tells the cell which kinds of proteins to make and how to make them.

THE MAJOR PARTICIPANTS IN TRANSCRIPTION AND TRANSLATION

Transcription, the formation of RNA using the DNA code, and translation, the synthesis of proteins, are highly complex. A number of components participate: most prominently, messenger RNA, transfer RNA, ribosomes, several types of enzymes, and a storehouse of raw materials. After first examining each of these components, we shall see how they come together in the assembly line of the cell.

* **genotype** (jee´-noh-typ) Gr. *gennan,* to produce, and *typos,* type.

* **phenotype** (fee´-noh-typ) Gr. *phainein,* to show. The physical manifestation of gene expression.

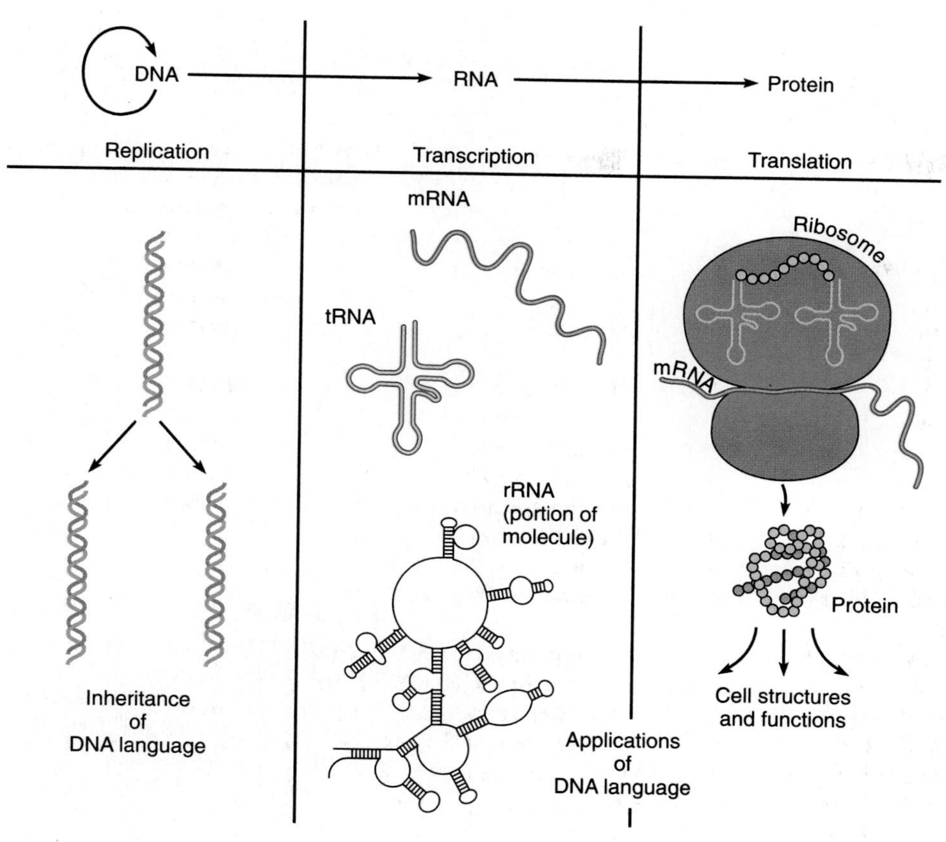

Figure 9.10

Summary of the flow of genetic information in cells. This biological theme, sometimes succinctly summed up as "DNA makes RNA makes protein," indicates the role of DNA as storehouse and distributor of genetic information and shows the involvement of other participants in carrying out its instructions.

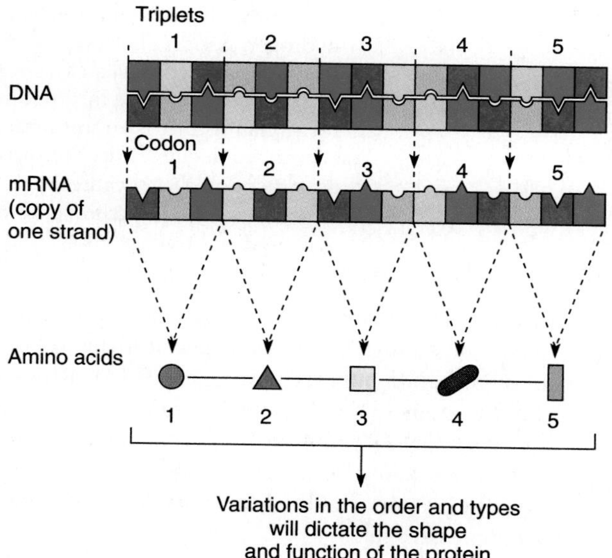

Figure 9.11

The DNA molecule is a continuous chain of base pairs, but the sequence must be interpreted in groups of three base pairs (a triplet). Each triplet as copied into mRNA codons will translate into one amino acid; consequently, the ratio of base pairs to amino acids is 3:1.

RNAs: Tools in the Cell's Assembly Line

Ribonucleic acid is an encoded molecule like DNA, but its general structure is different in several ways: (1) It is a **single-stranded molecule;** that is, it is a single helix. This is not to say that it does not form a secondary structure (hairpin loops) or a tertiary structure, but these structures still arise from a single strand. (2) RNA contains **uracil,** instead of thymine, as the complementary base-pairing mate for adenine. This does not change the inherent DNA code in any way because the uracil still follows the pairing rules. (3) Although RNA, like DNA, contains a backbone that consists of alternating sugar and phosphate molecules, the sugar in RNA is **ribose** rather than deoxyribose. The many functional types of RNA range from small regulatory pieces to large structural ones (table 9.2). All types of RNA are formed through transcription of a DNA gene, but only mRNA is further translated into another type of molecule (protein).

Messenger RNA: Carrying DNA's Message

Messenger RNA **(mRNA)** is a transcript (copy) of a structural gene or genes complementary to DNA. It is synthesized by a process similar to DNA replication, and the complementary base-pairing rules ensure that the code will be faithfully copied in the mRNA transcript. The message of this transcribed strand is displayed in a series of triplets called **codons** (figure 9.12*a*). The details of transcription and the function of mRNA in translation will be covered shortly.

Transfer RNA: The Key to Translation

Transfer RNA **(tRNA)** is also a copy of a DNA code; however, it differs from mRNA. It contains sequences of bases that form hydrogen bonds with complementary sections of the same tRNA strand. At these bonds, the molecule bends back upon itself into several *hairpin loops* (figure 9.12*b*), giving the molecule a

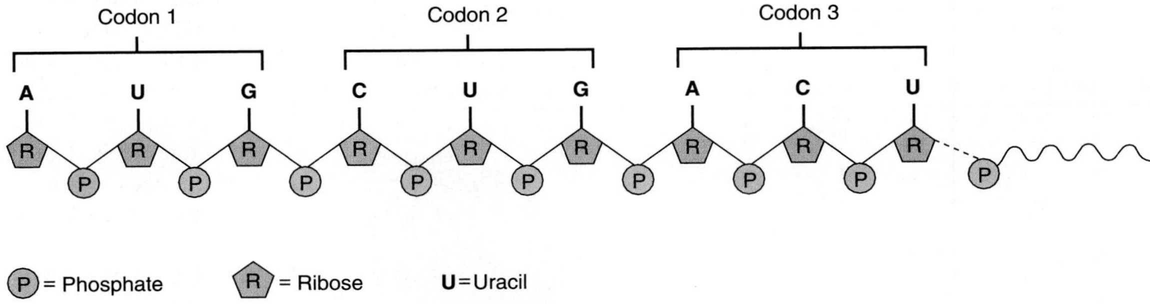

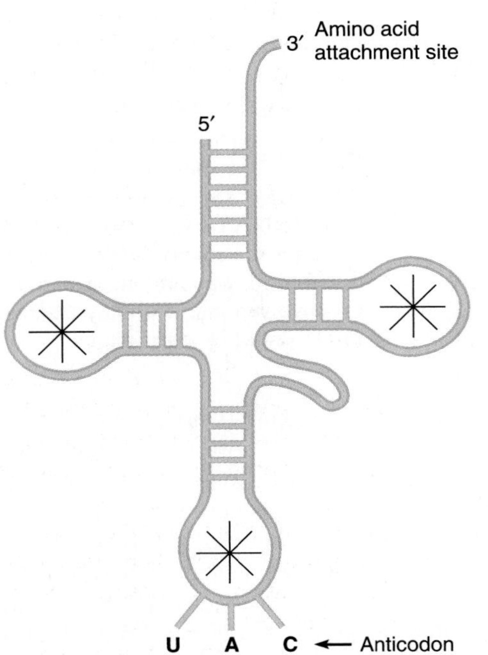

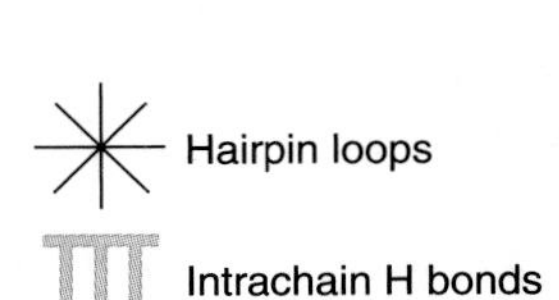

Figure 9.12
Characteristics of messenger and transfer RNA. (*a*) A short piece of messenger RNA (mRNA) illustrates the general structure of RNA: single-strandedness, repeating phosphate-ribose sugar backbone; single nitrogen bases with uracil instead of thymine. (*b*) Transfer RNA (tRNA) can loop back on itself to form intrachain hydrogen bonds. The result of the secondary structure is a cloverleaf structure, shown here in simplified form. The tRNA is an important adaptor molecule. At its bottom is an anticodon that specifies the attachment of a particular amino acid at the 3′ trailing end.

secondary *cloverleaf* structure that folds even further into a complex, three-dimensional helix. This compact molecule is an adaptor that converts RNA language into protein language. The bottom loop of the cloverleaf exposes a triplet, the **anticodon,** that both designates the specificity of the tRNA and complements mRNA's codons. At the opposite end of the molecule is an acceptor position for the amino acid that is specific for that tRNA's anticodon. For each of the 20 amino acids (see table 2.5), there is at least one specialized type of tRNA to carry it. The charging of the tRNA takes place in two enzyme-driven steps: First an ATP activates the amino acid, and then this group binds to the acceptor end of the tRNA. Because tRNA is the molecule that will convert the master code on mRNA into a protein, the accuracy of this step is crucial.

The Ribosome: A Mobile Molecular Factory for Translation

The procaryotic (70S) ribosome is a particle composed of tightly packaged ribosomal RNA and protein. Figure 9.10 shows a small section of rRNA. Electron microscope and biochemical studies have revealed that the ribosomal particle is made up of two subunits that, when joined together, form a special niche to hold the components of protein synthesis (see figure 9.14). Ribosomes contribute both enzymatic and attachment functions for mRNA and tRNA. A metabolically active bacterial cell can accommodate up to 20,000 of these miniscule factories—all actively engaged in reading the genetic program, taking in raw materials, and emitting proteins at an impressive rate.

Transcription: The First Stage of Gene Expression

How does DNA present its code? At a segment corresponding to a gene, the DNA helices are unwound by an enzyme. This unwinding exposes the nitrogen base triplets so that they can serve as a template for synthesis of an mRNA strand. Only one strand of the DNA—the **template strand** (also called the sense strand)—contains meaningful instructions for synthesis of a functioning polypeptide. In most cases, the triplet on DNA that signals the

TABLE 9.2

TYPES OF RIBONUCLEIC ACID

RNA Type	Contains Codes For	Function in Cell	Translated
Messenger (mRNA)	Sequence of amino acids in protein	Carries the DNA master code to the ribosome	Yes
Transfer (tRNA)	A cloverleaf tRNA to carry amino acids	Brings amino acids to ribosome	Helps in translation
Ribosomal (rRNA)	Several large structural rRNA molecules	Forms the major part of a ribosome	No
Primer	An RNA that can begin DNA replication	Primes DNA	No
Ribozymes	RNA enzymes Parts of splicer enzymes	Remove introns from other RNAs	No

start of transcription is TAC, which will transcribe as AUG, the start codon on mRNA. The nontranscribed strand is called the *antisense strand.* The strand of DNA that serves as a template varies from one gene to another (figure 9.13).

Transcription proceeds through several stages, directed by a huge and very complex enzyme system, **RNA polymerase.** During initiation, RNA polymerase recognizes a segment of the DNA called the *promoter region* that lies near the beginning of the gene segment to be transcribed (figure 9.13). This region is usually some version of a palindrome such as TATATA or TATA. It is here that the polymerase begins building the mRNA chain. During elongation, which proceeds in the 3′ to 5′ direction, nucleotide building blocks are assembled in accordance with the DNA template, except that uracil (U) is placed as adenine's complement. As elongation continues (at the rate of 40 nucleotides added per second in bacteria), the part of DNA already transcribed is rewound into its original helical form. At termination, the polymerases recognize another code that signals the separation and release of the mRNA strand, called the **transcript**. How long is the mRNA? The very smallest mRNA might consist of 100 bases; an average-sized mRNA might consist of 1,200 bases; and a large one, of several thousand.

TRANSLATION: THE SECOND STAGE OF GENE EXPRESSION

In translation, all of the elements needed to synthesize a protein, from the mRNA to the amino acids, are brought together on the ribosomes (figure 9.14). The entire process proceeds through the stages of initiation, elongation, termination, and protein folding and processing.

Beginning Stages of Translation: Initiation

The mRNA molecule leaving the DNA transcription site is short-lived and must immediately be translated. In fact, in procaryotes the mRNA is usually translated as it is being transcribed, which greatly speeds up the entire process. Ribosomal subunits nearby are ready to assemble, accept the mRNA, and bind to its 5′ end. One way to envision the attachment of the RNA is to return to our videotape analogy and imagine the end of the mRNA being strung through the ribosome somewhat like a tape through the head of a tape player so that it will be read on the proper frame. The mRNA usually has a short leader that does not encode part of the protein (like the first part of a tape before the movie begins). As the ribosome begins to scan the mRNA, it encounters the START codon, which is almost always **AUG** (and rarely GUG).

With the mRNA message in place on the assembled ribosome, the next step in translation involves entrance of tRNAs with their amino acids (figure 9.15). The pool of cytoplasm around the region contains a complete array of tRNAs, previously charged by having the correct amino acid attached. The step in which the complementary tRNA meets with the mRNA code is guided by the two sites on the large subunit of the ribosome called the **P site** (left) and the **A site** (right).[1] Think of these sites as shallow depressions in the larger subunit of the ribosome, each of which accommodates a tRNA.

The Master Genetic Code: Look to mRNA

By convention, the master genetic code is represented by the mRNA codons and the amino acids they specify (table 9.3). Except in a very few cases, this code is universal, whether for procaryotes, eucaryotes, or viruses. It is worth noting that once the triplet code on mRNA is known, the original DNA sequence, the complementary tRNA code, and the types of amino acids in the protein are automatically known (figure 9.16). One cannot predict with any certainty from protein structure to mRNA and DNA sequence because of a factor called **degeneracy,** meaning that more than one codon can translate into a given amino acid.

In table 9.3, the mRNA codons and their corresponding amino acid specificities are given. Because there are 64 different triplet codes[2] and only 20 different amino acids, it is not surprising that some amino acids are represented by several codons. For example, leucine and serine can each be represented by any of six different triplets, and only tryptophan and methionine are represented by a single codon. In such codons as leucine only the first two nucleotides are required to encode the correct amino acid, and the third nucleotide does not effect a difference. This property, called *wobble* is thought to permit some variation or mutation without harming the message.

The Beginning of Protein Synthesis

With mRNA serving as a model, the stage is finally set for actual protein assembly. The correct **tRNA** (labeled 1 on figure 9.15) enters the P site and binds to the **start codon (AUG)** presented by the mRNA. Rules of pairing dictate that the anticodon of this tRNA must be complementary to the mRNA codon AUG, thus the tRNA with anticodon UAC will first occupy site P. It happens that the amino acid carried by the initiator tRNA is formyl *methionine* ($_f$Met; table 9.3), though in many cases, it may not remain a permanent part of the finished protein.

1. P stands for peptide site; A stands for aminoacyl site (an aminoacyl group is the charged amino acid).
2. $64 = 4^3$ (the 4 different codons in all possible combinations of 3).

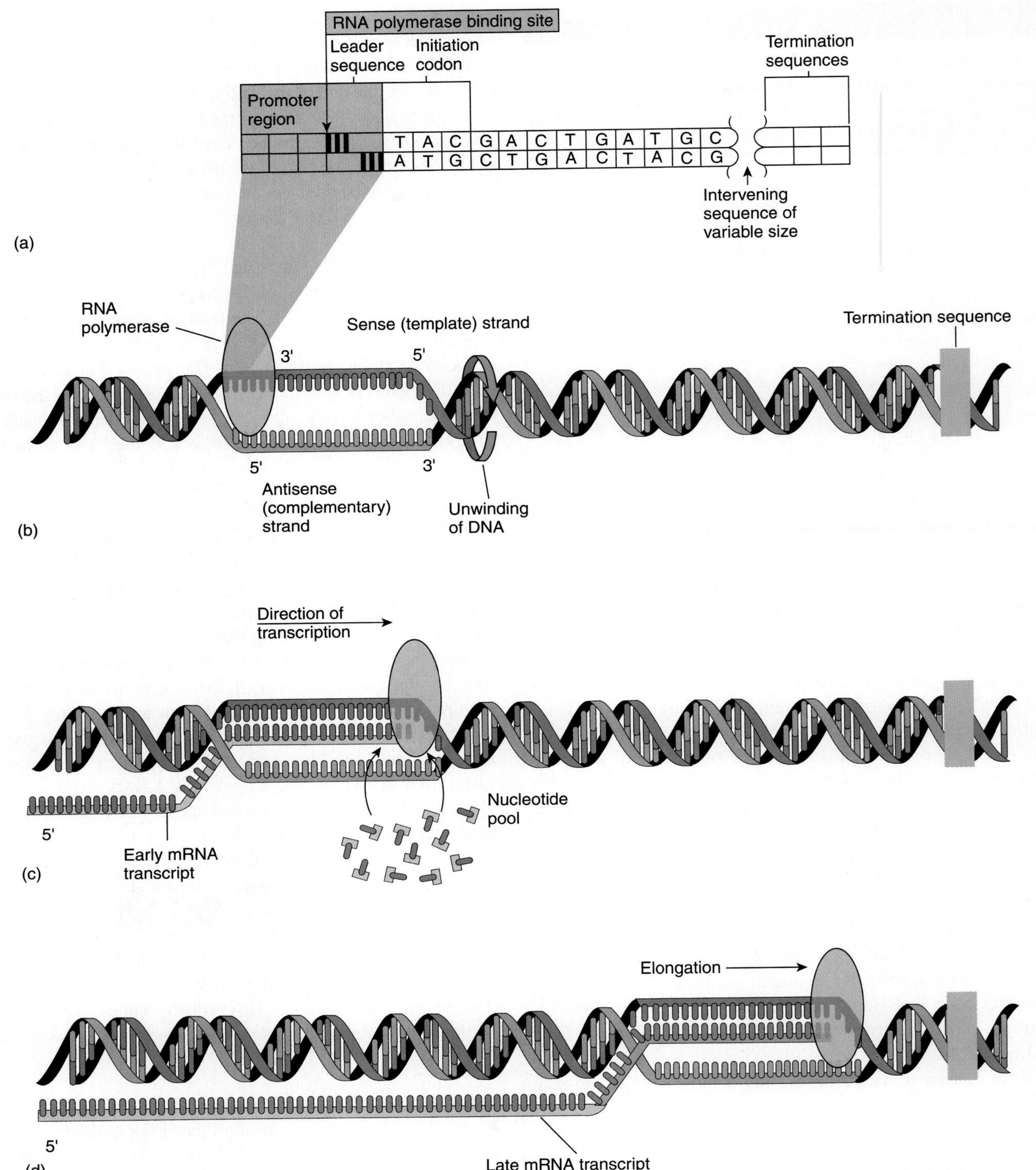

Figure 9.13

The major events in mRNA synthesis or transcription. (*a*) Overall view of a gene. Each gene contains a specific promoter region and a leader sequence for guiding the beginning of transcription. This is followed by the region of the gene that codes for a polypeptide and ends with a series of terminal sequences that stop translation. (*b*) DNA is unwound at the promoter to allow for the entry of RNA polymerase and the reading of the DNA codes. Only one strand, called the sense, or template, strand, is copied by the RNA polymerase. This strand runs in the 3´ to 5´ direction. (*c*) As the RNA polymerase moves along the strand, it adds complementary nucleotides as dictated by the DNA template, forming the single-stranded mRNA that reads in the 5´ to 3´ direction. (*d*) The polymerase continues transcribing until it reaches a termination site and the mRNA transcript is released for translation. Note that the section of the DNA that has been transcribed is rewound into its original configuration.

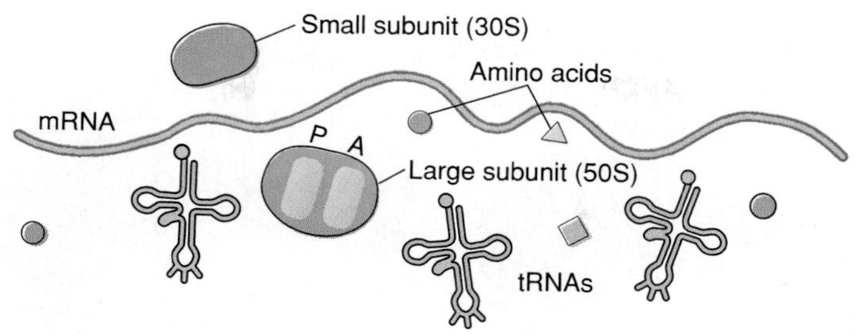

Figure 9.14
Participants in translation. P and A are sites on the ribosome, each of which holds a tRNA. See Figure 9.15.

TABLE 9.3

THE GENETIC CODE: CODONS OF mRNA THAT SPECIFY A GIVEN AMINO ACID

First Position	Second Position	Third Position: U	Third Position: C	Third Position: A	Third Position: G
U	U	UUU	UUC	UUA	UUG
		Phenylalanine		Leucine	
	C	UCU	UCC	UCA	UCG
			Serine		
	A	UAU	UAC	UAA STOP**	UAG STOP**
		Tyrosine			
	G	UGU	UGC	UGA STOP*	UGG Tryptophan
		Cysteine			
C	U	CUU	CUC	CUA	CUG
			Leucine		
	C	CCU	CCC	CCA	CCG
			Proline		
	A	CAU	CAC	CAA	CAG
		Histidine		Glutamine	
	G	CGU	CGC	CGA	CGG
			Arginine		
A	U	AUU	AUC	AUA	START AUG
		Isoleucine			Methionine*
	C	ACU	ACC	ACA	ACG
			Threonine		
	A	AAU	AAC	AAA	AAG
		Asparagine		Lysine	
	G	AGU	AGC	AGA	AGG
		Serine		Arginine	
G	U	GUU	GUC	GUA	GUG
			Valine		
	C	GCU	GCC	GCA	GCG
			Alanine		
	A	GAU	GAC	GAA	GAG
		Aspartic acid		Glutamine	
	G	GGU	GGC	GGA	GGG
			Glycine		

*This codon initiates translation.

**For these codons, which give the orders to stop translation, there are no corresponding tRNAs and no amino acids.

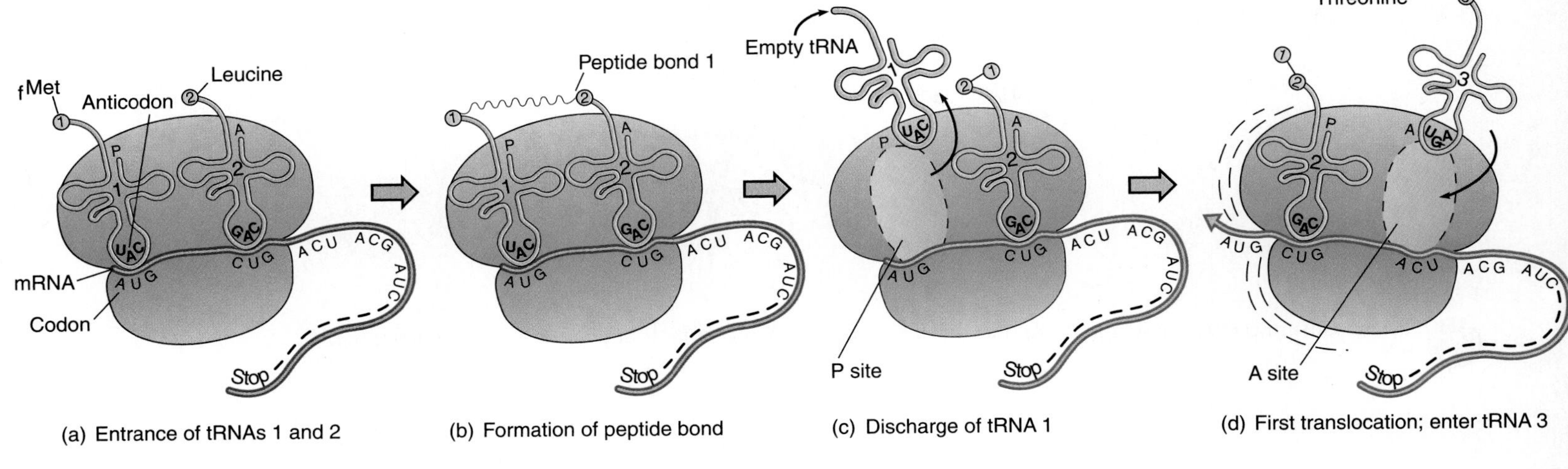

(a) Entrance of tRNAs 1 and 2 (b) Formation of peptide bond (c) Discharge of tRNA 1 (d) First translocation; enter tRNA 3

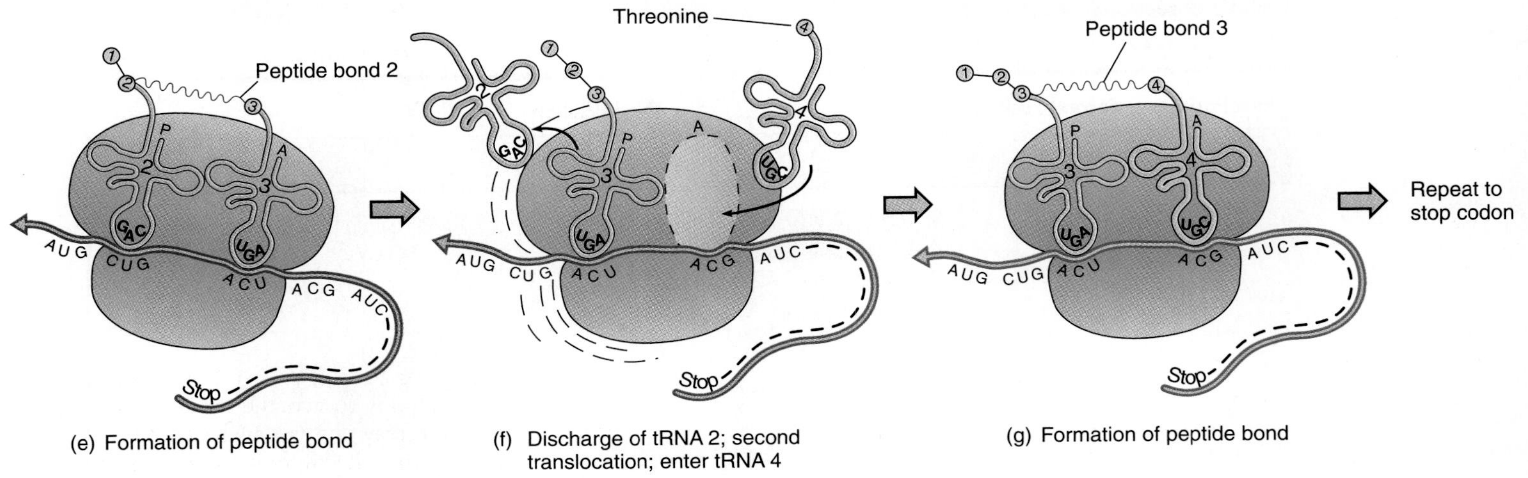

(e) Formation of peptide bond (f) Discharge of tRNA 2; second translocation; enter tRNA 4 (g) Formation of peptide bond

Figure 9.15
The events in protein synthesis.

Continuation and Completion of Protein Synthesis: Elongation and Termination

To keep track of the dynamic process of protein assembly, you will want to remain aware that the ribosome shifts its "reading frame" to the right along the mRNA from one codon to the next. As each new codon is brought into the head, a complementary tRNA is brought to the A position, a peptide bond is formed between the amino acids on the P and A tRNAs, and the polypeptide grows in length (figure 9.15).

The first step in elongation occurs with the filling of the A site by a second **tRNA** (2 on figure 9.15). The identity of this tRNA and its amino acid is dictated by the second mRNA codon.

Once tRNA 2 has been brought to the A site in accordance with the second codon on mRNA, the two adjacent tRNAs are now in favorable proximity for a peptide bond to form between the amino acids (aa) they carry. The $_f$Met is transferred from the first tRNA to aa 2, resulting in two coupled amino acids called a dipeptide (figure 9.15*b*).

For the next step to proceed, some room must be made on the ribosome, and the next codon in sequence must be brought into position for reading. This process is accomplished by *translocation,* the enzyme-directed shifting of the ribosome to the right along the mRNA strand, which causes the blank tRNA (1) to be discharged from the ribosome (figure 9.15*c*) and tRNA 2 with the dipeptide attached to be brought into P position. Site A is tem-

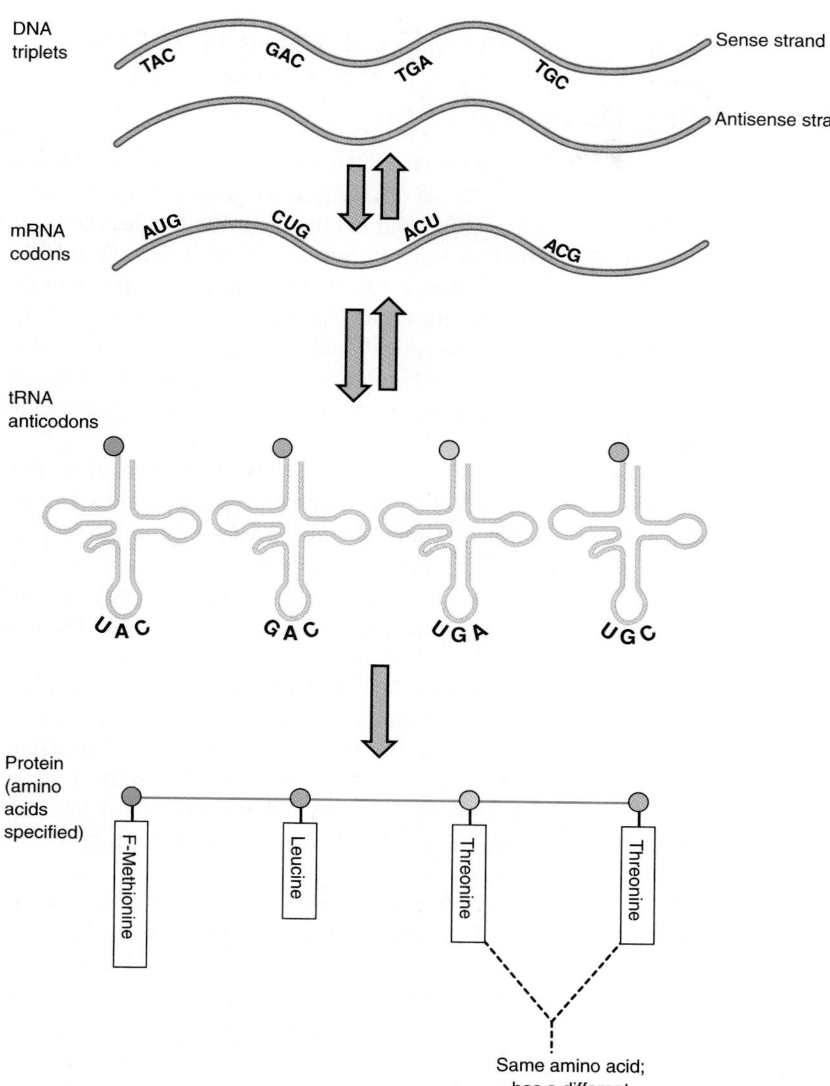

Figure 9.16
If the DNA code is known, the mRNA codon can be surmised. If a codon is known, the anticodon can be determined, but it may not be possible to determine codon sequence from the anticodons because of the degeneracy of the code. It is also not possible to determine the exact codon or anticodon from protein structure.

porarily left empty. The tRNA that has been released is now free to drift off into the cytoplasm and become recharged with an amino acid for later additions to this or another protein.

The stage is now set for the insertion of tRNA 3 at site A as directed by the third mRNA codon (figure 9.15*d*). This insertion is followed once again by peptide bond formation between the dipeptide and aa3 (making a tripeptide), splitting of the peptide from tRNA 2, and translocation: tRNA 2 is released, mRNA is shifted to the next position, tRNA 3 moves to position P, and codon 4 and tRNA 4 enter into place. From this point on, peptide elongation proceeds repetitively by this basic series of actions out to the end of the mRNA.

The termination of protein synthesis is not simply a matter of reaching the end codon on mRNA. It is brought about by the presence of at least one special codon occurring just after the codon for the last amino acid. Terminating codons—UAA, UAG, and UGA—do not translate into either a tRNA or an amino acid. Although they are often called **nonsense codons,** they carry a necessary and useful message: *Stop* here. When this codon is reached, a special enzyme breaks the bond between the final tRNA and the finished polypeptide chain, releasing it from the ribosome.

Before newly made proteins can carry out their structural or enzymatic roles, they often require finishing touches. Even before the peptide chain is released from the ribosome, it begins folding upon itself to achieve its biologically active tertiary conformation. Other alterations called *posttranslational* modifications may be necessary. Some proteins must have the starting amino acid (formyl methionine) clipped off; proteins destined to become complex enzymes have cofactors added; and some join with other completed proteins to form quaternary levels of structure.

The machinelike operation of transcription and translation is overwhelming in its precision. Protein synthesis in bacteria is both efficient and rapid. At 37°C, 12–17 amino acids per second are added to a growing peptide chain. An average protein consisting of about 400 amino acids requires less than half a minute for complete synthesis. Further efficiency is gained when the translation of mRNA starts while transcription is still occurring (figure 9.17). A single mRNA is long enough to be fed through more than one ribosome, thus several polypeptides may be synthesized from the same mRNA transcript arrayed in tandem procession along a chain of ribosomes. This **polyribosomal complex** is indeed an assembly line for mass production of proteins.

Protein synthesis consumes an enormous amount of energy. The equivalent of 3 ATPs is used in forming each peptide bond. An ATP is needed to charge a tRNA; one GTP is needed for the binding of a tRNA; and another GTP is spent on shifting the mRNA. Nearly 1,200 ATPs are required just for an average-sized protein.

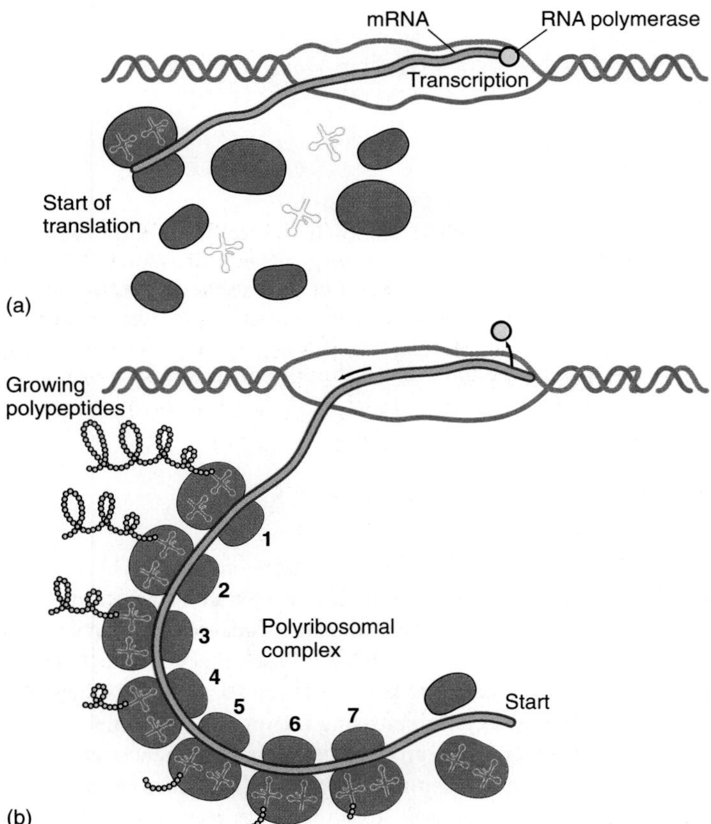

Figure 9.17
Speeding up the protein assembly line in bacteria. (*a*) The mRNA transcript encounters ribosomal parts immediately as it leaves the DNA. (*b*) The ribosomal factories assemble along the mRNA in a chain, each ribosome reading the message and translating it into protein. Many products will thus be well along the synthetic pathway before transcription has even terminated.

Eucaryotic Transcription and Translation: Similar Yet Different

There are two basic differences between procaryotic and eucaryotic gene expression. These differences are primarily in the cellular location of transcription and in how the mRNA is handled. Eucaryotic DNA lies in the nucleus, so that is where mRNA originates. To be translated, it has to pass through the pores in the nuclear envelope to the ribosomes. Simultaneous transcription and translation as exhibited in procaryotes is not possible. Protein synthesis in eucaryotes is very similar in many respects, but it appears to be slower. In developing red blood cells, only two amino acids per second are added at 37°C.

We have given the simplified definition of a gene that works well for procaryotes, but many eucaryotic genes are not colinear[3]—meaning that they do *not* exist as an uninterrupted series of triplets coding for a protein. A eucaryotic gene contains the code for a protein, but located along this code are one to several intervening sequences of bases, called **introns,** that do not code for product. Introns are interspersed between coding regions, called **exons,** that will be translated into product (figure 9.18). We can use words as examples. A short section of colinear procaryotic gene might read TOM SAW OUR DOG DIG OUT; a eucaryotic gene that codes for the same portion would read TOM SAW XZKP FPL OUR DOG QZWVP DIG OUT. The recognizable words are the exons, and the gibberish represents the introns.

This unusual genetic architecture, sometimes called a **split gene,** requires further processing before translation (figure 9.18). Transcription of the entire gene with both exons and introns occurs first. Next a type of RNA called small nuclear ribonucleoprotein, or *snRNP,* recognizes the exon-intron junctions and enzymatically cuts through them. The action of this "splicer enzyme" loops the introns into lariat-shaped pieces, excises them, and joins the exons end to end. By this means, a strand of mRNA with no intron material is produced. This strand can then proceed to the cytoplasm to be translated.

At first glance, this system seems to be a cumbersome way to make a transcript, and the value of this extra genetic baggage is still the subject of much debate. Several different types of introns have been discovered, some of which do code for proteins. One particular intron discovered in yeast gives the code for a reverse transcriptase, leading to speculation that cells have their own mechanisms for producing DNA from RNA. Some experts hypothesize that introns serve as a "sink" for extra bits of genetic material that could be available for splicing into existing genes, thus promoting genetic change and evolution (see microfile 10.3).

3. Colinearity means that the base sequence can be read directly into a series of amino acids.

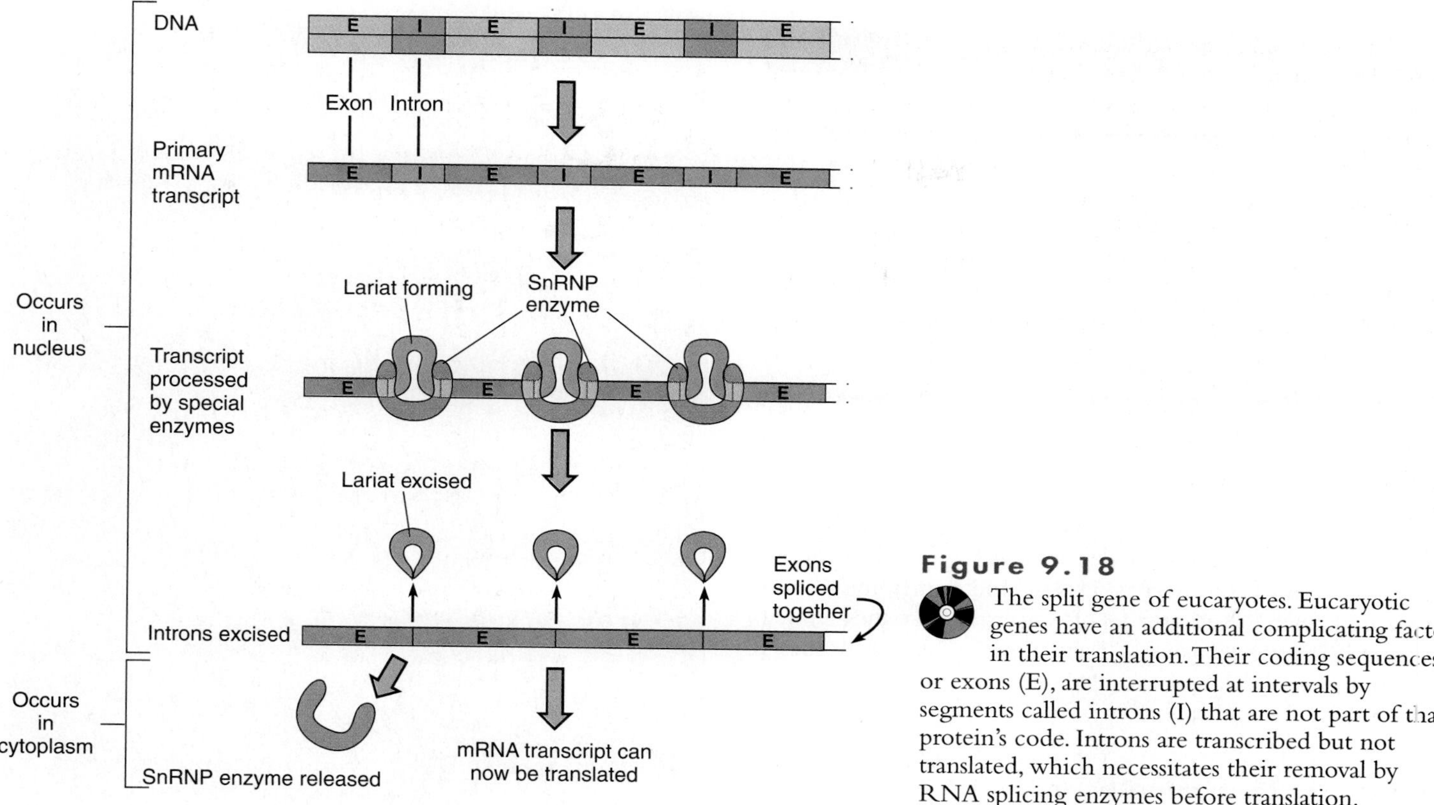

Figure 9.18

The split gene of eucaryotes. Eucaryotic genes have an additional complicating factor in their translation. Their coding sequences, or exons (E), are interrupted at intervals by segments called introns (I) that are not part of that protein's code. Introns are transcribed but not translated, which necessitates their removal by RNA splicing enzymes before translation.

THE GENETICS OF ANIMAL VIRUSES

The genetic nature of viruses was described in chapter 6. Viruses essentially consist of one or more pieces of DNA or RNA enclosed in a protective coating. Above all, they are genetic parasites that require access to their host cell's genetic and metabolic machinery to be replicated, transcribed, and translated, and they also have the potential for genetically changing the cells. Because they contain only those genes needed for the production of new viruses, the genomes of viruses tend to be very compact and economical. In fact, this very simplicity makes them excellent subjects for the study of gene function.

The genetics of viruses is quite diverse (see chapters 24 and 25). In many viruses, the nucleic acid is linear in form; in others it is circular. The genome of most viruses exists in a single molecule, though in a few, it is segmented into several smaller molecules. Most viruses contain normal double-stranded (ds) DNA or single-stranded (ss) RNA, but other patterns are seen: in ssDNA viruses, dsRNA viruses, and retroviruses, which work backward by making dsDNA from ssRNA. In some instances, viral genes overlap one another, and in a few DNA viruses, both strands contain a sense message.

A few generalities can be stated about viral genetics. In all cases, the viral nucleic acid penetrates the cell and is introduced into the host's gene-processing machinery at some point. In successful infection, an invading virus instructs the host's machinery to synthesize large numbers of new virus particles by a mechanism specific to a particular group. Replication of the DNA molecule of DNA animal viruses occurs in the nucleus, where the cell's DNA replication machinery lies (except in the poxviruses); the genome of most RNA viruses is replicated in the cytoplasm (except in the orthomyxoviruses [influenza]). In all viruses, viral mRNA is translated into viral proteins on host cell ribosomes using host tRNA. In the next section, we will briefly observe some major patterns of genetic replication (table 9.4).

Replication, Transcription, and Translation of dsDNA Viruses

Replication of dsDNA viruses is divided into phases (figure 9.19). During the early phase, viral DNA enters the nucleus, where several genes are transcribed into a messenger RNA. This transcript moves into the cytoplasm to be translated into viral proteins (enzymes) needed to replicate the viral DNA; this replication occurs in the nucleus. The host cell's own DNA polymerase is often involved, though some viruses (herpes, for example) have their own. During the late phase, other parts of the viral genome are transcribed and translated into proteins required to form the capsid and other structures. The new viral genomes and capsids are assembled, and the mature viruses are released by budding or cell disintegration.

Double-stranded DNA viruses interact directly with the DNA of their host cell. In some viruses, the viral DNA becomes silently *integrated* into the host's genome by insertion at a particular site on the host genome (figure 9.19). This integration may

TABLE 9.4

PATTERNS OF GENETIC FLOW IN VIRUSES

DNA Viruses
dsDNA → dsDNA (semiconservative)
ssDNA → dsDNA → ssDNA (only one virus group)
RNA Viruses
(+) ssRNA → (–) ssRNA → (+) ssRNA
(–) ssRNA → (+) ssRNA → (–) ssRNA
ssRNA → ssDNA → dsDNA → ssRNA (retroviruses)
dsRNA → ssRNA → dsRNA (conservative)

later lead to the transformation[4] of the host cell into a cancer cell and the production of a tumor. Several DNA viruses, including hepatitis B (HBV), the herpesviruses, and papillomaviruses (warts), are known to be initiators of cancers and are thus termed *oncogenic.** The mechanisms of transformation and oncogenesis appear to involve special genes called oncogenes that can regulate cellular genomes (see chapter 17).

Replication, Transcription, and Translation of RNA Viruses

RNA viruses exhibit several differences from DNA viruses. Their genomes are smaller and less stable; they enter the host cell already in an RNA form; and the virus cycle occurs entirely in the cytoplasm for most viruses. RNA viruses can have one of the following genetic messages: a positive-sense genome (+) that comes ready to be translated into proteins, a negative-sense genome (–) that must be converted to positive before translation, a positive-sense genome (+) that can be converted to DNA, or a dsRNA genome.

Positive-Sense Single-Stranded RNA Viruses Positive-sense RNA viruses must first replicate a negative strand as a master template to produce more positive strands. Shortly after the virus uncoats in the cell, its positive strand is translated into a large protein that is soon cleaved into individual functional units, one of which is a polymerase that initiates the replication of the viral strand (figure 9.20). Replication of a single-stranded positive-sense strand is done in two steps. First, a negative strand is synthesized using the parental positive strand as a template by the usual base-pairing mechanism. The resultant negative strand becomes a master template against which numerous positive daughter strands are made. Further translation of the viral genome produces large numbers of structural proteins for final assembly and maturation of the virus. Examples of these viruses are poliovirus and hepatitis A virus.

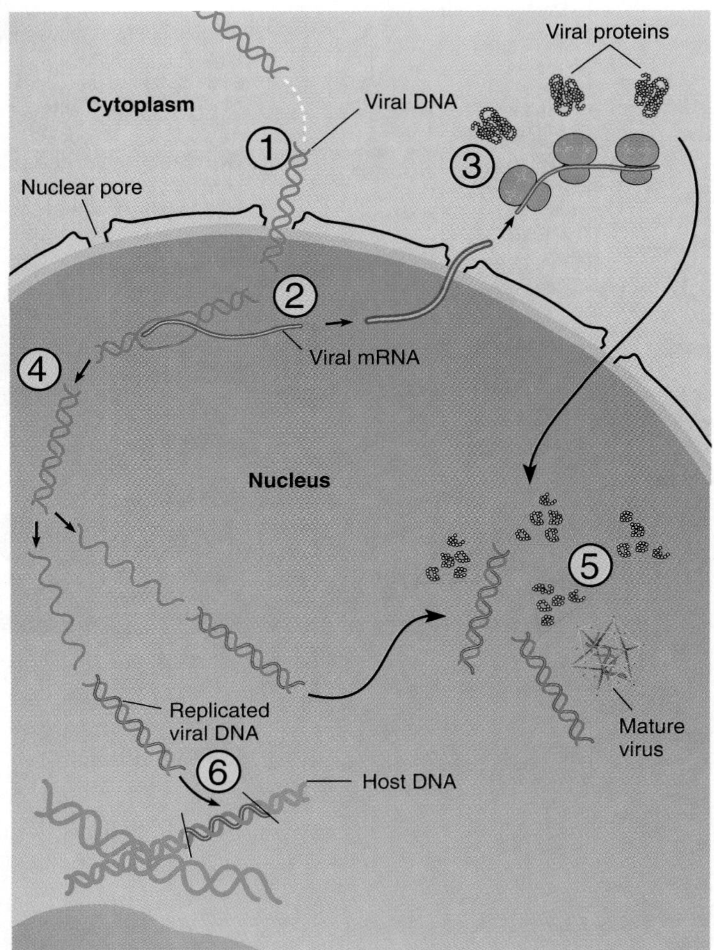

Figure 9.19

Genetic stages in the multiplication of double-stranded DNA viruses. The virus penetrates the host cell and releases DNA, which (*1*) enters the nucleus and (*2*) is transcribed. Other events are: (*3*) Viral mRNA is translated into structural proteins; proteins enter the nucleus. (*4*) Viral DNA is replicated repeatedly in the nucleus. (*5*) Viral DNA and proteins are assembled into a mature virus in the nucleus. (*6*) Because it is double-stranded, the viral DNA can insert itself into host DNA (latency).

RNA Viruses with Reverse Transcriptase: Retroviruses A most unusual class of viruses has a unique capability to reverse the order of the flow of genetic information. Thus far in our discussion, all genetic entities have shown the patterns DNA → DNA, DNA → RNA, or RNA → RNA. Retroviruses, including HIV, the cause of AIDS, and HTLV I, a cause of one type of human leukemia, synthesize DNA using their RNA genome as a template (see figure 25.16). They accomplish this by means of an enzyme, **reverse transcriptase,** that comes packaged with each virus particle. This enzyme synthesizes a single-stranded DNA against the viral RNA template and then directs the formation of a

4. The process of genetic change in a cell, leading to malignancy.

* **oncogenic** (ahn″-koh-jen′-ik) Gr. *onkos,* mass, and *gennan,* to produce. Refers to any cancer-causing process. Viruses that do this are termed oncoviruses.

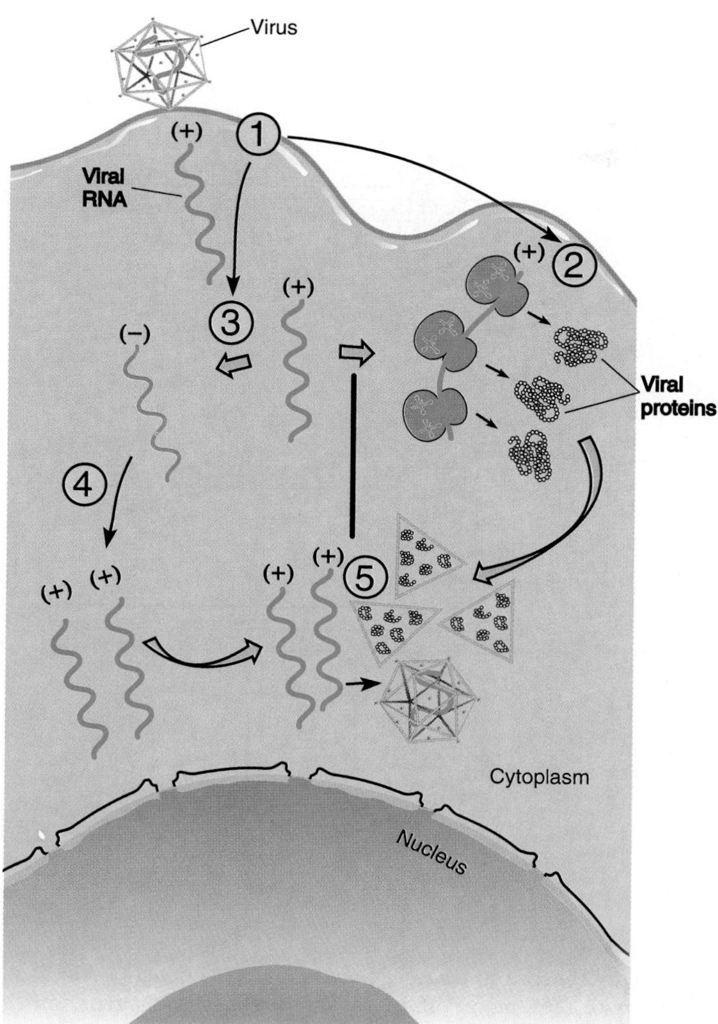

Figure 9.20
Replication of positive-sense single-stranded RNA viruses. In general, these viruses do not enter the nucleus. (*1*) Penetration and uncoating of viral RNA. (*2*) Because it is positive in sense and single-stranded, the RNA can be directly translated on host cell ribosomes into various necessary viral proteins. (*3*) A negative genome is synthesized against the positive template to produce large numbers of positive genomes for final assembly. (*4*) The negative template is then used to synthesize a series of positive replicates. (*5*) RNA strands and proteins assemble into mature viruses.

complementary strand of this ssDNA, resulting in a double strand of viral DNA. The dsDNA strand enters the nucleus, where it can be integrated into the host genome and transcribed by the usual mechanisms into new viral ssRNA. Translation of the viral RNA yields viral proteins for final virus assembly. The capacity of a retrovirus to become inserted into the host's DNA as a provirus has several possible consequences. In some cases, these viruses are oncogenic and are known to transform cells and produce tumors. It allows the AIDS virus to remain latent in an infected cell until a stimulus activates it to continue a productive cycle.

Chapter Checkpoints

Information in DNA genes is converted to actual substances by the process of transcription and translation. The substances produced are structural proteins and enzymes. Enzymes, in turn, control an organism's metabolic activities.

DNA triplets specify the same mRNA codons, regardless of the species in which they occur. The language of DNA and mRNA extends into the noncellular world of viruses as well.

The processes of transcription and translation are similar but not identical for procaryotes and eucaryotes. Eucaryotes transcribe DNA in the nucleus, remove its introns, and translate it in the cytoplasm. Bacteria transcribe and translate simultaneously because there are no nucleus and introns. Bacterial polyribosomes attach themselves directly to the mRNA as it is released by its DNA template.

Viruses replicate by utilizing the transcription and translation processes of cellular organisms. They invade host cells and "force" them to transcribe and translate viral genes into new virus particles. In some cases, viruses connect themselves to the host genome and are passed on to its progeny, some of which will become virus factories many generations later.

The genetic material of viruses can be either DNA or RNA occurring in single, double, positive-, or negative-sense strands. Viral DNA contains just enough information to force a cell to become a virus factory.

GENETIC REGULATION OF PROTEIN SYNTHESIS AND METABOLISM

In chapter 8 we surveyed the metabolic reactions in cells and the enzymes involved in those reactions. At that time, we mentioned types of metabolic regulation that are genetic in origin. These control mechanisms ensure that not all genes are active at all times. In this way, enzymes will be produced to appropriately reflect nutritional status and prevent the waste of energy and materials in dead-end synthesis. Genetic function in procaryotes is regulated by a specific unit of DNA called an **operon.*** Operons contain various control genes (regulators, promoters, and operators) that govern the operation of related structural genes. Such gene regulation responds to external stimuli and usually occurs at the level of transcription. Operons act in one of two ways: (1) The operon can be turned on (*induced*) by the substrate of the enzyme for which the structural genes code, or (2) the operon can be turned off (*repressed*) by the product its enzymes synthesize. Although operons are the primary regulators in bacteria, comparable devices probably govern the much more complicated eucaryotic metabolism as well.

* operon (op´-ur-on) L. *opera,* exertion. A regulatory site for gene expression.

The Lactose Operon: A Model for Inducible Gene Regulation in Bacteria

The best understood cell system for explaining control through genetic induction is the **lactose (*lac*) operon.*** This concept, first postulated in 1961 by François Jacob and Jacques Monod, accounts for the regulation of lactose metabolism in *Escherichia coli.* Many other operons with similar modes of action have since been identified, and together they furnish convincing evidence that the environment of a cell can have great impact on gene expression.

The lactose operon is made up of three segments, or *loci,** of DNA: (1) the **regulator,** composed of the gene that codes for a protein capable of repressing the operon (a **repressor**); (2) the **control locus,** composed of two genes, the **promoter** (recognized by RNA polymerase) and the **operator,** a sequence where transcription of the structural genes is initiated; and (3) the **structural locus,** made up of three genes, each coding for a different enzyme (β-galactosidase, permease, and transacetylase) needed to catabolize lactose (figure 9.21). In bacteria, structural genes required for the metabolism of a nutrient tend to be arranged in procession. This is an efficient strategy that permits genes for a particular metabolic pathway to be induced or repressed in unison. The enzymes of the operon are of the inducible sort mentioned in chapter 8. The promoter, operator, and structural components lie in contiguous order, but the regulator can be at a distant site.

In inductive systems like the *lac* operon, the operon is normally in an *off mode* and does not initiate enzyme synthesis when the appropriate substrate is absent (figure 9.21*a*). How is the operon maintained in this mode? The key is in the repressor protein that is coded by the regulatory locus. This relatively large and pliant molecule has sites for allosteric binding, one for the operator and another for lactose. In the absence of lactose, this repressor binds with the operator locus, thereby blocking the transcription of the structural genes lying downstream. Think of the repressor as a lock on the operator, and if the operator is locked, the structural genes cannot be transcribed.

If lactose is added to the cell's environment, it triggers several events that turn the operon *on.* Because lactose is ultimately responsible for stimulating protein synthesis, it is called an *inducer.* The binding of lactose to the repressor protein causes a conformational change in the repressor that dislodges it from the operator segment (figure 9.21*b*). The control segment that was previously inactive is now unlocked, and RNA polymerase can now bind to the promoter. The structural genes are transcribed in a single unbroken transcript coding for all three enzymes. During translation, however, each gene is used to synthesize a separate protein.

As lactose is depleted, further enzyme synthesis is not necessary, so the order of events reverses. At this point there is no longer sufficient lactose to inhibit the repressor, hence the repressor is again free to attach to the operator. The operator is locked, and transcription of the structural genes and protein synthesis related to lactose both stop.

* loci (loh´-sy) sing. locus (loh´-kus) L. *locus,* a place. The site on a chromosome occupied by a gene.

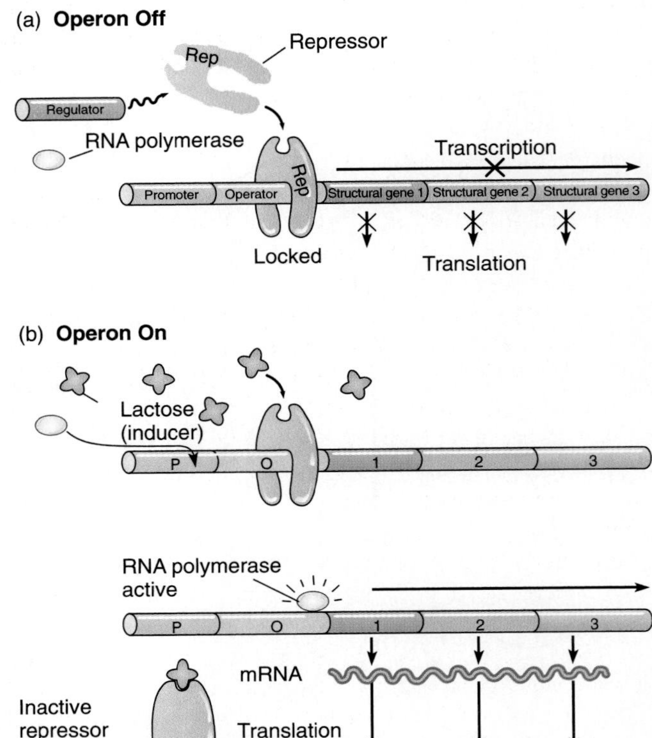

Figure 9.21
The lactose operon in bacteria: how inducible genes are controlled by substrate. (*a*) Operon off. In the absence of lactose, a repressor protein (the product of a regulatory gene) attaches to the operator gene of the operon. This effectively locks the operator and prevents any transcription of structural genes downstream (to its right). Suppression of transcription (and, consequently, of translation) prevents the unnecessary synthesis of enzymes for processing lactose. (*b*) Operon on. Upon entering the cell, the substrate (lactose) becomes a genetic inducer by attaching to the repressor, which loses its grip and falls away. The operator is now free to initiate transcription, and enzymatic products of translated genes perform the necessary reactions on their lactose substrate.

A fine but important point about the *lac* operon is that it functions only in the absence of glucose. Glucose is the nutrient preferentially metabolized by constitutive enzymes, and if it is present in the medium, lactose cannot enter the cell.

A Repressible Operon

Bacterial systems for synthesis of amino acids, purines, and pyrimidines and many other processes work on a slightly different principle—that of repression. Similar factors such as repressor proteins, operators, and a series of structural genes exist for this operon, but with some important differences. Unlike the *lac* operon, this operon is normally in the *on* mode and will be turned *off* only when this nutrient is no longer required. The nutrient plays an additional role as a **corepressor** needed to block the action of the operon.

A growing cell that needs the amino acid arginine (arg) effectively illustrates the operation of a repressible operon. Under these conditions, the *arg* operon is set to *on,* and arginine is being actively synthesized through the action of its enzymatic products (figure 9.22*a*). In an active cell, the arginine will be used up and will not accumulate. In this instance, the repressor will remain inactive because there is too little free arginine to activate it. As the cell's metabolism begins to slow down, however, the synthesized arginine will no longer be used up and will accumulate. The free arginine is then available to act as a corepressor by attaching to the repressor. This reaction produces a functioning repressor that locks the operator and stops further transcription and arginine synthesis (figure 9.22*b*).

Analogous gene control mechanisms in eucaryotic cells are not as well understood, but it is known that genes can be turned on by intrinsic regulatory segments similar to operons. Some molecules, called transcription factors, insert on the grooves of the DNA molecule and enhance transcription of specific genes. Examples include zinc "fingers" and leucine "zippers." Evidence of certain kinds of regulation (or its loss) can be detected in cancer cells, where genes called oncogenes abnormally override built-in genetic controls. An oncogene can exert its control by causing the formation of a chemical that permanently activates a part of the genome controlling cell growth. A cell with no constraints on the number of cell divisions can grow out of all normal bounds into tumors and leukemias (see chapter 17).

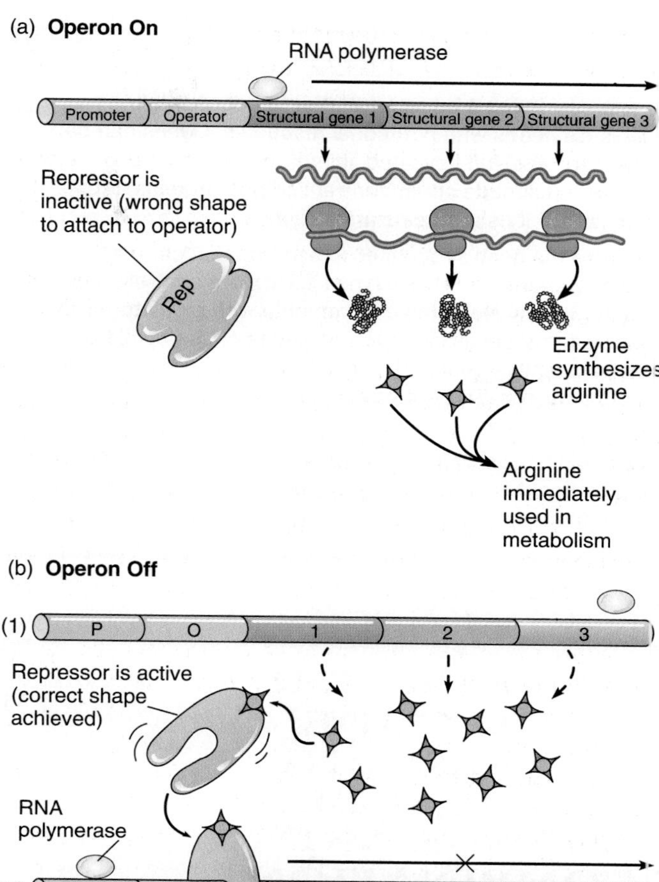

Figure 9.22
Repressible operon: control of a gene through excess nutrient. (*a*) Operon on. A repressible operon remains on when its nutrient products (here, arginine) are in great demand by the cell because the repressor remains inactive at low nutrient levels. (*b*) Operon off. The operon is repressed when (*1*) arginine builds up and, serving as a corepressor, reacts with the repressor, thereby activating it, and (*2*) the repressor complex affixes to the operator and blocks the RNA polymerase and further transcription of genes for arginine synthesis.

ANTIBIOTICS THAT AFFECT TRANSCRIPTION AND TRANSLATION

Naturally occurring cell nutrients are not the only agents capable of modifying gene expression. Some infection therapy is based on the concept that certain drugs react with DNA, RNA, and ribosomes and thereby alter genetic expression (see chapter 12). Treatment with such drugs is based on an important premise: that growth of the infectious agent will be inhibited by blocking its protein-synthesizing machinery selectively, without disrupting the cell synthesis of the patient receiving the therapy.

Drugs that inhibit protein synthesis exert their influence on transcription or translation. For example, the rifamycins used in therapy for tuberculosis bind to RNA polymerase, blocking the initiation step of transcription, and are selectively more active against bacterial RNA polymerase than the corresponding eucaryotic enzyme. Actinomycin D binds to DNA and halts mRNA chain elongation, but its mode of action is not selective for bacteria. For this reason, it is very toxic and never used to treat bacterial infections, though it can be applied in tumor treatment.

The ribosome is a frequent target of antibiotics that inhibit ribosomal function and ultimately protein synthesis. The value and safety of these antibiotics again depend upon the differential susceptibility of procaryotic and eucaryotic ribosomes. One problem with drugs that selectively disrupt procaryotic ribosomes is that the mitochondria of humans contain a procaryotic type of ribosome, and these drugs may inhibit the function of the host's mitochondria. One group of antibiotics (including erythromycin and spectinomycin) prevents translation by interfering with the attachment of mRNA to ribosomes. Chloramphenicol, lincomycin, and tetracycline bind to the ribosome in a way that prevents the elongation of the polypeptide, and aminoglycosides (such as streptomycin) inhibit peptide initiation and elongation. It is interesting to note that these drugs have served as important tools to explore genetic events because they can arrest specific stages in these processes.

Chapter Checkpoints

Gene expression must be orchestrated to coordinate the organism's needs with nutritional resources. Genes can be turned "on" and "off" by specific molecules, which expose or hide their nucleotide codes for transcribing proteins. Most, but not all, of these proteins are enzymes.

Operons are the gene sequences in bacteria that code for specific proteins (usually enzymes). They are controlled by regulator genes. Nutrients can combine with regulator gene products to turn a set of structural genes on (inducible genes) or off (repressible genes). The *Lac* operon is an example of an inducible operon. The *Arg* operon is an example of a repressible operon.

The rifamycins, tetracyclines, and aminoglycosides are classes of antibiotics that are effective because they interfere with transcription and translation processes in microorganisms.

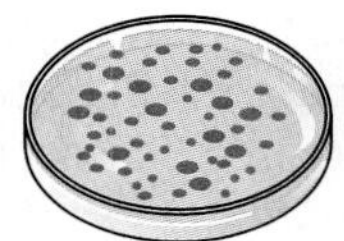

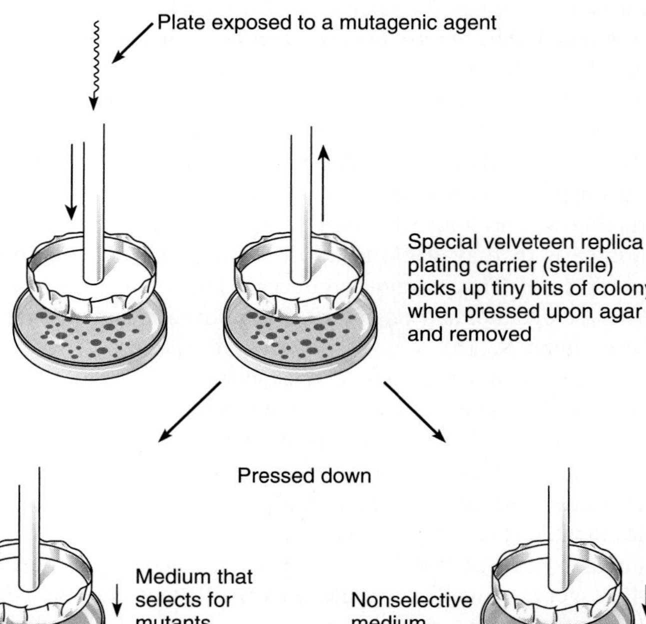

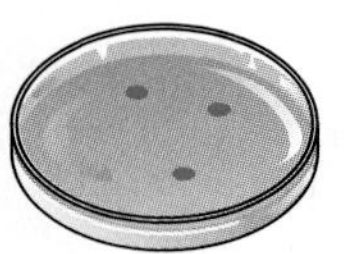

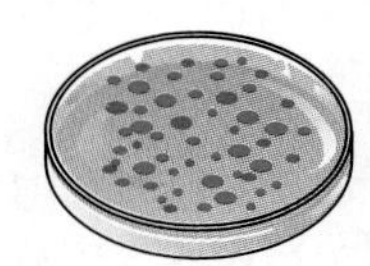

Figure 9.23

The general basis of replica plating, a method developed by Joshua Lederberg for detecting and isolating mutant strains of microorganisms.

CHANGES IN THE GENETIC CODE: MUTATIONS AND INTERMICROBIAL EXCHANGE AND RECOMBINATION

As precise and predictable as the rules of genetic expression seem, permanent changes do occur in the genetic code. Indeed, genetic change is the driving force of evolution. In microorganisms, such changes appear first in phenotype alterations, such as the appearance or disappearance of anatomical or physiological traits. For example, a normally colored bacterium can lose its ability to form pigment, or a strain of the malarial parasite can develop resistance to a drug. Phenotypic changes of this type are ultimately due to changes in the genotype. Any permanent, inheritable change in the genetic information of the cell is a **mutation.** On a strictly molecular level, a mutation is an alteration in the nitrogen base sequence of DNA. It can involve the loss of base pairs, the addition of base pairs, or a rearrangement in the order of base pairs. Do not confuse this with genetic recombination, in which microbes transfer whole segments of genetic information between themselves.

A microorganism that exhibits a natural, nonmutated characteristic is known as a **wild type,** or wild strain. If a microorganism bears a mutation, it is called a **mutant strain.** Mutant strains can show variance in morphology, nutritional characteristics, genetic control mechanisms, resistance to chemicals, temperature preference, and nearly any type of enzymatic function. Mutant strains are very useful for tracking genetic events, unraveling genetic organization, and pinpointing genetic markers. A classic method of detecting and isolating mutant strains is through nutritional manipulation of a culture. For example, in a culture of a wild-type bacterium that is lactose-positive (meaning it has the necessary enzymes for fermenting this sugar), a small number of mutant cells have become lactose-negative, having lost the capacity to ferment this sugar. If the culture is plated on a medium containing indicators for fermentation, each colony can be observed for its fermentation reaction, and the negative strain isolated. Another standard method of detecting and isolating microbial mutants is by replica plating (figure 9.23).

CAUSES OF MUTATIONS

A mutation is described as spontaneous or induced, depending upon its origin. A **spontaneous mutation** is a random change in the DNA arising from mistakes in replication or the detrimental effects of natural background radiation (cosmic rays) on DNA. The frequency of spontaneous mutations has been measured for a

TABLE 9.5

SELECTED MUTAGENIC AGENTS AND THEIR EFFECTS

Agent	Effect
Chemical	
Nitrous acid, bisulfite	Removes an amino group from some bases
Mustard gas	Causes cross-linkage of DNA strands
Acridine dyes	Cause frameshifts due to insertion between base pairs
Nitrogen base analogs	Compete with natural bases for sites on replicating DNA
Radiation	
Ionizing (gamma rays, X rays)	Form free radicals that cause single or double breaks in DNA
Ultraviolet	Causes cross-links between adjacent pyrimidines

number of organisms. Mutation rates vary tremendously from one mutation in 10^5 replications (a high rate) to one mutation in 10^{10} replications (a low rate). The rapid rate of bacterial reproduction allows these mutations to be observed more readily in bacteria than in animals.

Induced mutations result from exposure to known **mutagens,** which are primarily physical or chemical agents that interact with DNA in a disruptive manner (table 9.5). The carefully controlled use of mutagens has proved a useful way to induce mutant strains of microorganisms for study.

Chemical mutagenic agents such as acridine dyes insert completely across the DNA helices between adjacent bases to produce a frameshift mutation and distort the helix (figure 9.24*a*). Analogs[5] of the nitrogen bases (5-bromodeoxyuridine and 2-aminopurine, for example) are chemical mimics of natural bases that are incorporated into DNA during replication. Addition of these abnormal bases leads to mistakes in base-pairing. Many chemical mutagens are also carcinogens, or cancer-causing agents (see the discussion of the Ames test in a later section of this chapter).

Physical agents that alter DNA are primarily types of radiation. High-energy gamma rays and X rays create reactive free radicals in the cell. Because of its role as a genetic resource, DNA is highly vulnerable to the effects of radicals, and it accumulates breaks that may not be repairable. Ultraviolet (UV) radiation induces abnormal bonds between adjacent pyrimidines that prevent normal replication (see figure 11.10). The overall effects of radiation depend upon the length of exposure and its intensity. Exposure to large doses of radiation can be fatal, which is why radiation is so effective in microbial control; it can also be carcinogenic in animals. Ionizing radiation such as X-ray therapy has been linked to thyroid cancer, and exposure to nuclear radiation is a known cause of leukemia.

5. An analog is a chemical structured very similarly to another chemical except for minor differences in functional groups.

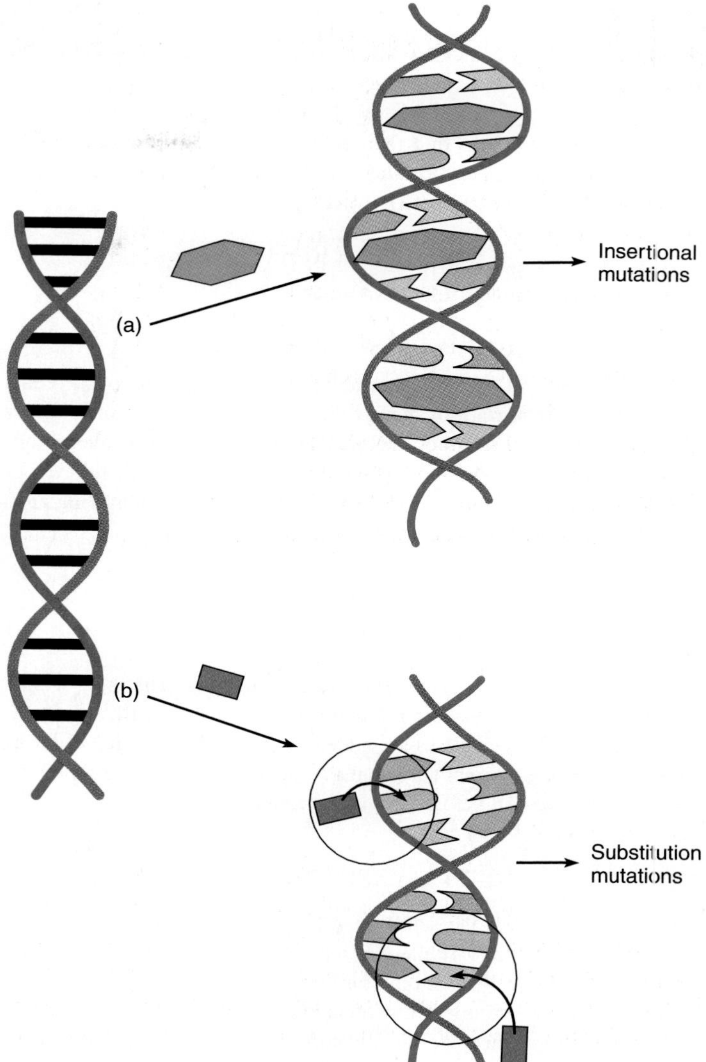

Figure 9.24
Two types of point mutation, showing modes of action of chemical mutagenic agents. (*a*) Some mutagens insert into the helix and interfere with its replication. (*b*) Other chemicals act as imposters. Owing to their structural similarity to a regular nitrogenous base, they can fit in its place, but they cannot function like the regular base.

CATEGORIES OF MUTATIONS

Mutations range from large mutations, in which whole chromosomes are lost or large genetic sequences are inserted (see later section on transposons), to small ones that affect only a single base on a gene. These latter mutations, which involve addition, deletion, or substitution of a few bases, are called **point mutations.** The effects of some point mutations are described in microfile 9.5 and table 9.6.

To understand how a change in DNA influences the cell, remember that the DNA code appears in a particular order of triplets (three bases) that is transcribed into mRNA codons, each of which

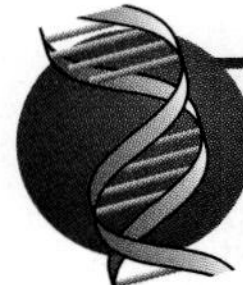

MICROFILE 9.5 MUTATIONS: MISTAKES IN DNA LANGUAGE

The nature of mutations and their effects on the product can best be demonstrated by a sentence analogy, with letters representing the bases. Let us say that the message of DNA reads:

DIDPAMSEETADSUP?

Or, if set in word triplets (like DNA) to make sense:

DID PAM SEE TAD SUP?

Two kinds of mutation can move the reading frame of the code. Such **frameshift mutations** disrupt the natural order of the message by shifting the frame forward or back, thus setting the whole sequence off. In **deletion,** one (or more) base is removed, and the bases all shift left of the deletion point to form new triplets. In our example, removing an E in SEE creates the following frameshift:

In this example, the message still means something. This sort of missense mutation could result in a functional protein in the cell.

If a new base or bases are added to the DNA message, as during **insertion,** the base combinations must shift right of the insertion point. Thus, if X is inserted before SEE in the original sentence, the following sentence results:

DID PAM XSE ETA DSU P?

This change removes the sense from both the words and the sentence and is an example of a frameshift that could lead to a useless protein and possibly alter cell function. The location of a frameshift can be significant in that it determines how many amino acids will be changed. In general, one occurring near the beginning of a gene will have greater effect than one near the end because it increases the chance for a nonsense mutation to occur by shifting more bases, and changing more amino acids.

Other types of mutations do not necessarily shift the frame in the manner of our previous examples because they are limited to one base or codon and do not affect those in surrounding positions. In an **inversion,** for example, the order of bases is switched, and only one or possibly two codons are altered. Consider inversion of the letters in SUP:

DID PAM SEE TAD SPU?

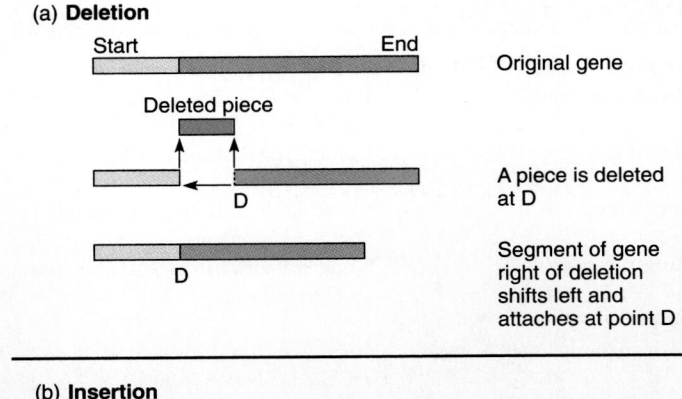

(a) **Deletion**

Start End Original gene

Deleted piece

D A piece is deleted at D

D Segment of gene right of deletion shifts left and attaches at point D

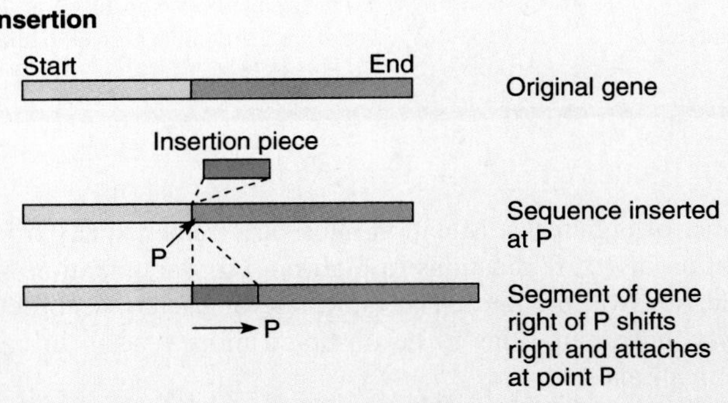

A diagrammatic scheme of two types of mutations that shift the frame of reference. (a) *Deletion.* (b) *Insertion. (Double-stranded DNA is represented here as a bar, for easier visualization.)*

With base **substitution,** one base is removed and replaced with another. This, again, can cause a different codon and amino acid to be added to the protein during translation. If an N were substituted for the final P in the original sentence, the code would read:

DID PAM SEE TAD SUN?

Inversions and substitutions generally have subtler effects than frameshifts. For more work with mutations, see the critical-thinking questions at the end of this chapter.

specifies an amino acid. A permanent alteration in the DNA that is copied faithfully into mRNA and translated can change the structure of the protein. A change in a protein can likewise change the morphology or physiology of a cell. Most mutations have a harmful effect on the cell, leading to cell dysfunction or death; these are called lethal mutations. Neutral mutations produce neither adverse nor helpful changes. A small number of mutations are beneficial in that they provide the cell with a useful change in structure or physiology.

A change in the code that leads to placement of a different amino acid is called a **missense mutation.** A missense mutation can do one of the following: (1) create a faulty, nonfunctional protein, (2) produce a different but functional protein, or (3) cause no significant alteration in protein function.

A **nonsense mutation,** on the other hand, changes a normal codon into a stop codon that does not code for an amino acid and stops the production of the protein wherever it occurs. A nonsense mutation almost always results in a nonfunctional protein. A **silent**

TABLE 9.6

CLASSIFICATION OF MAJOR TYPES OF MUTATIONS

Type of Mutation	Description	Result	Manifestation
Categories Based on Alteration of Base Sequence in DNA			
Frameshift	Addition or loss of one or two bases in a gene	Change of reading frame such that codon triplets no longer read in correct register	Produces entirely new genetic code from point of change; most result in mass changes in amino acids and nonexistent or incomplete protein
	Deletion	Removal of bases; triplet codes are offset by loss	May be caused by chromosome breakage and loss of segments
	Insertion	Addition of bases; triplet codes are offset by gain	May be caused by chemical mutagens, such as ethmidium bromide, that insert between bases
Substitution	A wrong base is put in place of a correct base	Error in base pairing; change in codon	Different amino acid; different protein; may be caused by chemical mutagens, such as nitrous oxide, that convert one base to another
Inversion	Adjacent bases exchange positions	Error in base pairing; change in one or two codons	Different amino acids; cause change in protein sequence and function
Categories Based on Overall Effect of Mutation			
Silent	Change in base that causes no alteration in codon	GGT to GGA still gives proline	No alteration in amino acid or protein
Missense	Change in base that causes alteration in one codon with consequences that range from none to severe, depending upon the nature of the amino acid change	Leaky mutation substitutes different amino acid but does not greatly affect protein function: e.g., TAA mutated to GAA substitutes leucine for isoleucine	
		Harmful mutation changes amino acid and protein structure so that function is not competent: e.g., CTT to CAT substitutes valine for glutamic acid, changing the folding pattern of the protein	
Nonsense	Change in base that causes the development of a STOP codon somewhere along the gene	Will always interrupt protein synthesis and cause a nonfunctional, incomplete protein; outcome can be lethal if this is an essential gene with only a few copies: e.g., *substitution:* AAC to ATC goes from leucine to STOP; *Inversion:* TAT to ATT goes from tyrosine to STOP	

mutation alters a base but does not change the amino acid and thus has no effect. Because of the degeneracy of the code, ACU, ACC, ACG, and ACA all code for threonine, so a mutation that changes only the last base will not alter the sense of the message in any way. A **back-mutation** occurs when a gene that has undergone mutation reverses (mutates back) to its original base composition.

REPAIR OF MUTATIONS

Earlier we indicated that DNA has a proofreading mechanism to repair mistakes in replication that might otherwise become permanent. Because mutations are potentially life-threatening, the cell has additional systems for finding and repairing DNA that has been damaged by various mutagenic agents and processes. Most ordinary DNA damage is resolved by enzymatic systems specialized for finding and fixing such defects.

DNA that has been damaged by ultraviolet radiation can be healed by **photoactivation** or **light repair.** This repair mechanism requires visible light and a light-sensitive enzyme, DNA photolyase, which can detect and attach to the damaged areas (sites of abnormal pyrimidine binding). Ultraviolet repair mechanisms are successful only for a relatively small number of UV mutations. Cells cannot repair severe, widespread damage and will die. In humans, the genetic disease *xeroderma pigmentosa* is due to nonfunctioning genes for the enzyme photolyase. Persons suffering from this rare disorder develop severe skin cancers; this relation provides strong evidence for a link between cancer and mutations.

Mutations can be excised by a series of enzymes that remove the incorrect bases and add the correct ones. This process is known as *excision repair.* First, enzymes break the bonds between the bases and the sugar-phosphate strand at the site of the error. A different enzyme subsequently removes the defective bases one at a time, leaving a gap that will be filled in by DNA polymerase I and ligase (figure 9.25). A repair system can also locate **mismatched bases** that were missed during proofreading: for example, C mistakenly paired with A, or G with T. The base must be replaced soon after the mismatch is made, or it will not be recognized by the repair enzymes.

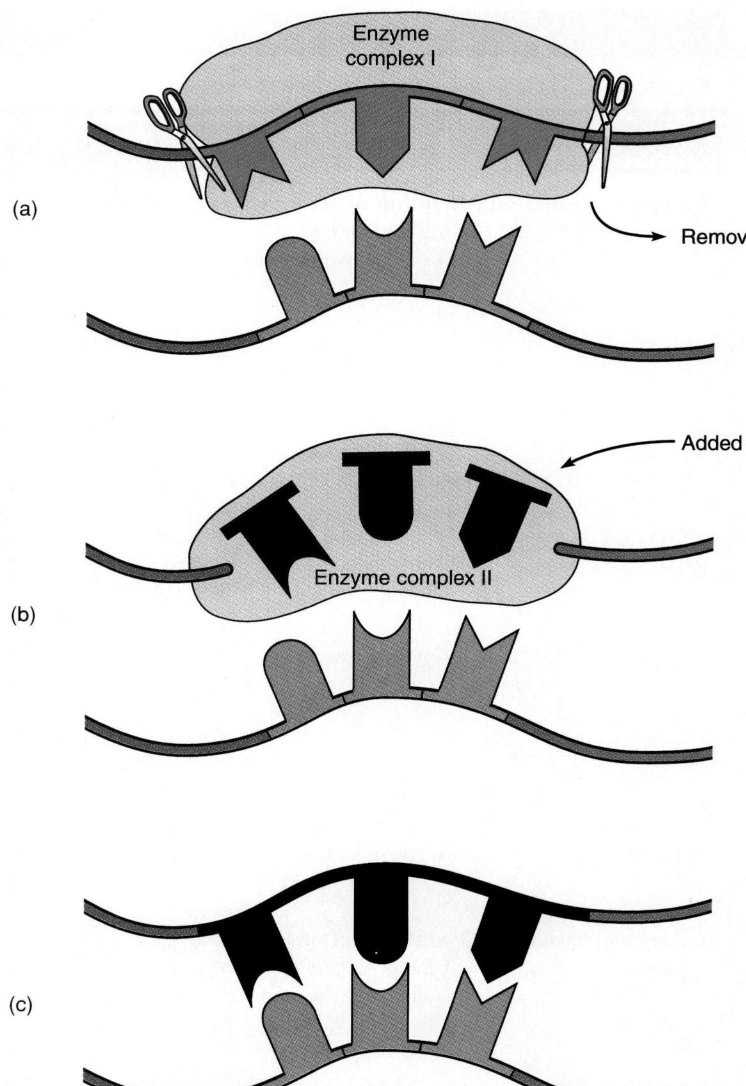

Figure 9.25

Excision repair of mutation by enzymes. (*a*) The first enzyme complex recognizes one or several incorrect bases and removes them. (*b*) The second complex (DNA polymerase I and ligase) places correct bases and seals the gaps. (*c*) Repaired DNA.

The Ames Test

New agricultural, industrial, and medicinal chemicals are constantly being added to the environment, and exposure to them is widespread. The discovery that many such compounds are mutagenic and that up to 83% of these mutagens are linked to cancer is significant. Although animal testing has been a standard method of detecting chemicals with carcinogenic potential, a more rapid screening system called the **Ames test**[6] is also commonly used. In this ingenious test, the experimental subjects are bacteria whose gene expression and mutation rate can be readily observed and monitored. The premise is that any chemical capable of mutating bacterial DNA can similarly mutate mammalian (and thus human) DNA and is therefore potentially hazardous.

One indicator organism in the Ames test is a mutant strain of *Salmonella typhimurium*[7] that has lost the ability to synthesize the amino acid histidine, a defect highly susceptible to back-mutation because the strain also lacks DNA repair mechanisms. Mutations that cause reversion to the wild strain, which is capable of synthesizing histidine, occur spontaneously at a low rate. A test agent is considered a mutagen if it enhances the rate of back-mutation beyond levels that would occur spontaneously. One variation on this testing procedure is outlined in figure 9.26. The Ames test has proved invaluable for screening an assortment of environmental and dietary chemicals for mutagenesis and carcinogenicity without resorting to more expensive and time-consuming animal studies.

6. Named for its creator, Bruce Ames.
7. *S. typhimurium* inhabits the intestine of poultry and causes food poisoning in humans. It is used extensively in genetic studies of bacteria.

Positive and Negative Effects of Mutations

Many mutations are not repaired. How the cell copes with them depends on the nature of the mutation and the strategies available to that organism. Mutations that are permanent and heritable will be passed on to the offspring of organisms and new viruses and will be a long-term part of the gene pool. Some mutations are harmful to organisms; others provide adaptive advantages.

If a mutation leading to a nonfunctional protein occurs in a gene for which there is only a single copy, as in haploid or simple

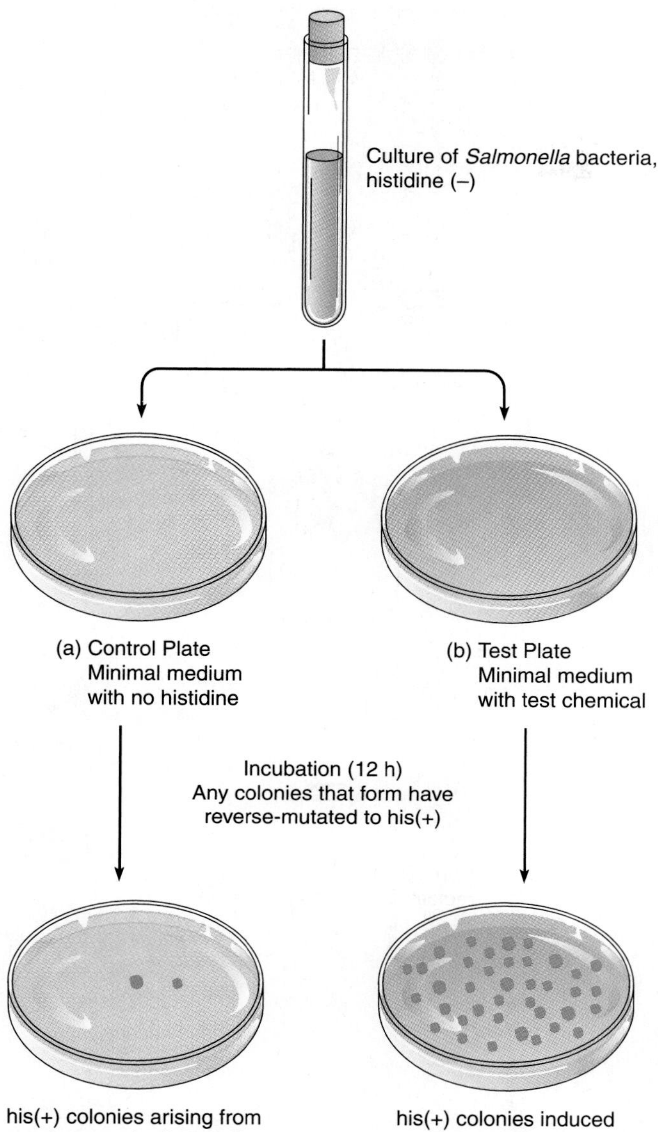

Figure 9.26

The Ames test uses a strain of *Salmonella typhimurium* that cannot synthesize histidine (his−), lacks the enzymes to repair DNA so that mutations show up readily, and has leaky cell walls that permit the ready entrance of chemicals. Many potential carcinogens (benzanthracene and aflatoxin, for example) are mutagenic agents only after being acted on by mammalian liver enzymes, so an extract of these enzymes is added to the test medium. (*a*) In the control setup, bacteria are plated on a histidine-free medium containing liver enzymes but lacking the test agent. (*b*) The experimental plate is prepared the same way except that it contains the test agent. After incubation, plates are observed for colonies. Any colonies developing on the plates are due to a back-mutation in a cell, which has reverted it to a his(+) strain. The degree of mutagenicity of the chemical agent can be calculated by comparing the number of colonies growing on the control plate with the number on the test plate. Chemicals that produce an increased incidence of back-mutation are considered carcinogens.

organisms, the cell will probably die. This happens when certain mutant strains of *E. coli* acquire mutations in the genes needed to repair damage by UV radiation. Mutations of the human genome affecting the action of a single protein (mostly enzymes) are responsible for more than 400 diseases. Microfile 9.6 discusses one such disease, sickle-cell anemia.

Although most spontaneous mutations are not beneficial, a small number contribute to the success of the individual and the population by creating variant strains with alternative ways of expressing a trait. Microbes are not "aware" of this advantage and do not direct these changes; they simply respond to the environment they encounter. Those organisms with beneficial mutations can more readily adapt, survive, and reproduce. In the long-range view, mutations and the variations they produce are the raw materials for change in the population and, thus, for evolution.

Mutations that create variants occur frequently enough that any population contains mutant strains for a number of characteristics, but as long as the environment is stable, the population will remain stable. When the environment changes, however, it can become hostile for the survival of certain individuals, and only those microbes bearing protective mutations can **adapt** to the new environment and survive. In this way, the environment **naturally selects** certain mutant strains that will reproduce, give rise to subsequent generations, and, in time, be the dominant strain in the population. Through these means, any change that confers an advantage during selection pressure will be retained by the population. One of the clearest models for this sort of selection and adaptation is acquired drug resistance in bacteria (see chapter 12). One of the fascinating mechanisms bacteria have developed for increasing their adaptive capacity is genetic exchange (called genetic recombination).

INTERMICROBIAL DNA TRANSFER AND RECOMBINATION

Genetic recombination, or hybridization through sexual reproduction, is an important foundation of genetic variation in eucaryotes. Although bacteria have no exact equivalent to sexual reproduction, they exhibit a primitive means for sharing or recombining parts of their genome. An event in which one bacterium donates DNA to another bacterium is a type of genetic transfer termed **recombination.** The properties of bacteria that influence recombination are the presence of extrachromosomal DNA and the versatility bacteria display in interchanging genes. The end result of recombination is a new strain different from both the donor and the original recipient strain. Genetic exchanges have tremendous effects on the genetic diversity of bacteria, and unlike spontaneous mutations, are generally beneficial to them. They provide additional genes for resistance to drugs and metabolic poisons, new nutritional and metabolic capabilities, and increased virulence and adaptation to the environment in general.

Genetic Recombination in Bacteria

DNA transmitted between bacteria involves **plasmids,** closed circular molecules of DNA separate from chromosomes, or small chromosomal fragments that have escaped from a lysed cell. Plasmids are not necessary to basic bacterial survival, but they can be

MICROFILE 9.6 THE MAGNITUDE OF A MUTATION: SICKLE-CELL ANEMIA

Sickle-cell anemia provides insight into the dramatic repercussions of a single base change in the DNA molecule. This change is magnified across all levels of the biological spectrum: When a codon is changed, an amino acid is substituted. This substitution makes the protein and the cells containing it abnormal. Ultimately, the tissues are harmed, the organs malfunction, the individual is drastically impaired; and even the population itself can be affected.

Persons with this severe and life-threatening illness have inherited two recessive genes (homozygous) for an abnormal type (S) of hemoglobin (Hb), the principal oxygen-carrying protein of the blood. In certain populations at some time in the past, a point mutation occurred on the gene that encodes one polypeptide of the hemoglobin molecule. Because this mutation was germinal (that is, it affected the gametes), it was inherited and appeared in all cells of certain offspring. The red blood-forming cells in which the hemoglobin genes are active are most affected initially. The exact site of this mutation is the sixth codon of the gene, specifically CTT, which would ordinarily cause the insertion of glutamic acid (GAA). The mutation causes substitution of a single base (A) in the center, yielding CAT (and a GUA transcript), which is the code for valine. Insertion of valine in place of glutamic acid during translation causes the polypeptide to behave differently.

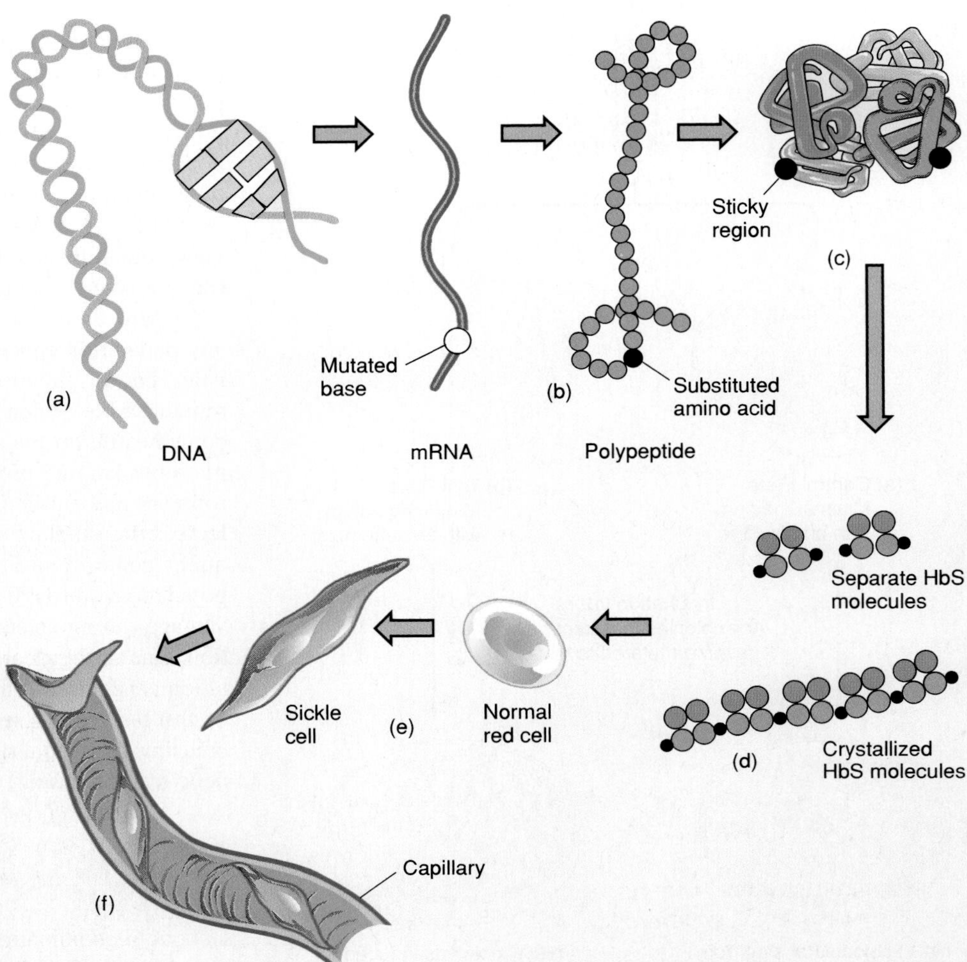

The scope of mutational effects in organisms: sickle-cell anemia in humans. (a) *Substitution of a single base creates a mutation on the hemoglobin gene that causes a different amino acid to be added during hemoglobin synthesis.* (b) *Abnormal HbS polypeptide folds differently, and* (c), *in its quaternary structure, exposes sticky regions.* (d) *HbS molecules form long crystalline strands instead of remaining separate.* (e) *Distortion of RBCs into sickle cells.* (f) *Sickle cells get hung up and cannot circulate in capillaries, leading to anemia and organ damage.*

Such a change seems insignificant—only one base—yet the consequences are profound. Hemoglobin S bears "sticky" spots that cause mutual attraction among the individual hemoglobin molecules. They crystallize into long chains that build up in the red blood cells. Although red blood cells are ordinarily disc-shaped, these chains distort them into elongate, crescent- or sickle-shaped cells. The fine capillary network becomes plugged by these oddly shaped cells, blocking circulation in the tissues and precipitating a crisis that severely damages vital organs. Fragile sickle cells are subject to bursting, which instigates anemia and other side effects. These symptoms first appear in infancy and recur throughout the person's life, which is generally shortened.

The highest incidence of the HbS mutation exists in the populations of Central Africa, though it is also found among Mediterranean, Arabic, and Indian peoples. Sickle-cell anemia occurs in about 1 in 400 births among Afro-Americans. Carriers of a single HbS gene (heterozygous) do not develop a full case of sickle-cell anemia but, in fact, actually have resistance to malaria, giving them a survival advantage over normal noncarriers. Thus, a mechanism that maintains the carriers simultaneously conserves the sickle-cell anemia gene and thus keeps the gene in the population.

TABLE 9.7

TYPES OF INTERMICROBIAL EXCHANGE

Mode	Requirements	Direct or Indirect*	Genes Transferred
Conjugation	Sex pilus on donor Fertility plasmid in donor Both donor and recipient alive Gram-negative cells	Direct	Drug resistance; resistance to metals; toxin production; enzymes; adherence molecules; degradation of toxic substances; uptake of iron
Transformation	Free donor DNA (fragment) Live, competent recipient cell	Indirect	Polysaccharide capsule; unlimited with cloning techniques
Transduction	Donor is lysed bacterial cell Defective bacteriophage is carrier of donor DNA Live, competent recipient cell of same species as donor	Indirect	Toxins; enzymes for sugar fermentation; drug resistance

*Direct means the donor and recipient are in contact during exchange; indirect means they are not.

genetically useful and confer greater versatility or adaptability on the recipient cell. In addition to being transferrable, donated DNA can be integrated into the chromosome of the recipient and replicated and transmitted to progeny during cell division. Plasmids and integrated genes are transcribed and translated along with the regular chromosome. Genetic recombination occurs at a low rate in natural populations, and it can also be induced in the laboratory. In chapter 10 we will see that bacteria, viruses, and some yeasts are capable of receiving and using the DNA of other organisms that is grafted into a plasmid.

Depending upon the mode of transmission, the means of genetic recombination is called conjugation, transformation, or transduction. **Conjugation*** requires the attachment of two related species through a pilus and the presence of a special plasmid. **Transformation*** entails the transfer of naked DNA and requires no special vehicle. **Transduction*** is DNA transfer mediated through the action of a bacterial virus (table 9.7).

Conjugation: Bacterial Sex Conjugation is mode of sexual mating in which a plasmid or other genetic material is transferred by a donor to a recipient cell via a specialized appendage. It occurs primarily in gram-negative bacteria. The donor cell possesses a plasmid **(fertility,** or **F, factor)** that allows it to synthesize a **sex pilus,** or **conjugative pilus.** The recipient cell is a related species or genus that has a recognition site on its surface. A cell's role in conjugation is denoted by F^+ for the cell that has the F plasmid and by F^- for the cell that lacks it. Contact is made when a sex pilus grows out from the F^+ cell, attaches to the surface of the F^- cell, contracts, and draws the two cells together (see figure 4.9). Although the physical details of transmission are still somewhat obscure, the replicated DNA passes across or through the bridge formed by the pilus. Conjugation is a very conservative process, in that the donor bacterium generally retains a copy of the genetic material being transferred.

According to findings with *E. coli,* conjugative transfer exhibits three patterns:

1. The donor (F^+) cell makes a copy of its F factor and transmits this to a recipient (F^-) cell. The F^- cell is thereby changed into an F cell capable of producing a sex pilus and conjugating with other cells (figure 9.27*a*). No additional donor genes are transferred at this time.
2. In high-frequency recombination (Hfr) donors, the fertility factor has been integrated into the F^+ donor chromosome. The term *high-frequency recombination* was adopted to denote that a cell with an integrated F factor transmits its chromosomal genes at a higher frequency than other cells. The F factor can direct a more comprehensive transfer of part of the donor chromosome to a recipient cell. This transfer occurs through duplication of the DNA by means of the rolling circle mechanism. One strand of DNA is retained by the donor and the other strand is transported across to the recipient cell (figure 9.27*b*). The F factor may not be transferred during this process. The transfer of an entire chromosome takes about 100 minutes, but the pilus bridge between cells is ordinarily broken before this time, and rarely is the entire genome of the donor cell transferred.
3. In a specialized form of conjugation called **sexduction,** an F factor originally integrated into the chromosome of a donor bacterium becomes free and carries away a segment of donor chromosomal DNA. This plasmid contains additional genes that can be acquired by a recipient cell during conjugation. In this case, the new genes remain in the plasmid and are not integrated into the recipient's chromosome.

Conjugation has great biomedical importance. Special **resistance (R) plasmids,** or **factors,** that bear genes for resisting antibiotics and other drugs are commonly shared among bacteria through conjugation. Transfer of R factors can confer multiple resistance to antibiotics such as tetracycline, chloramphenicol,

* conjugation (kahn″-jew-gay′-shun) L. *conjugatus,* yoked together.

* transformation (trans-for-may′-shun) L. *trans,* across, and *formatio,* to form. This term is also used to mean the cancerous (malignant) conversion of cells.

* transduction (trans-duk′-shun) L. *transducere,* to lead across.

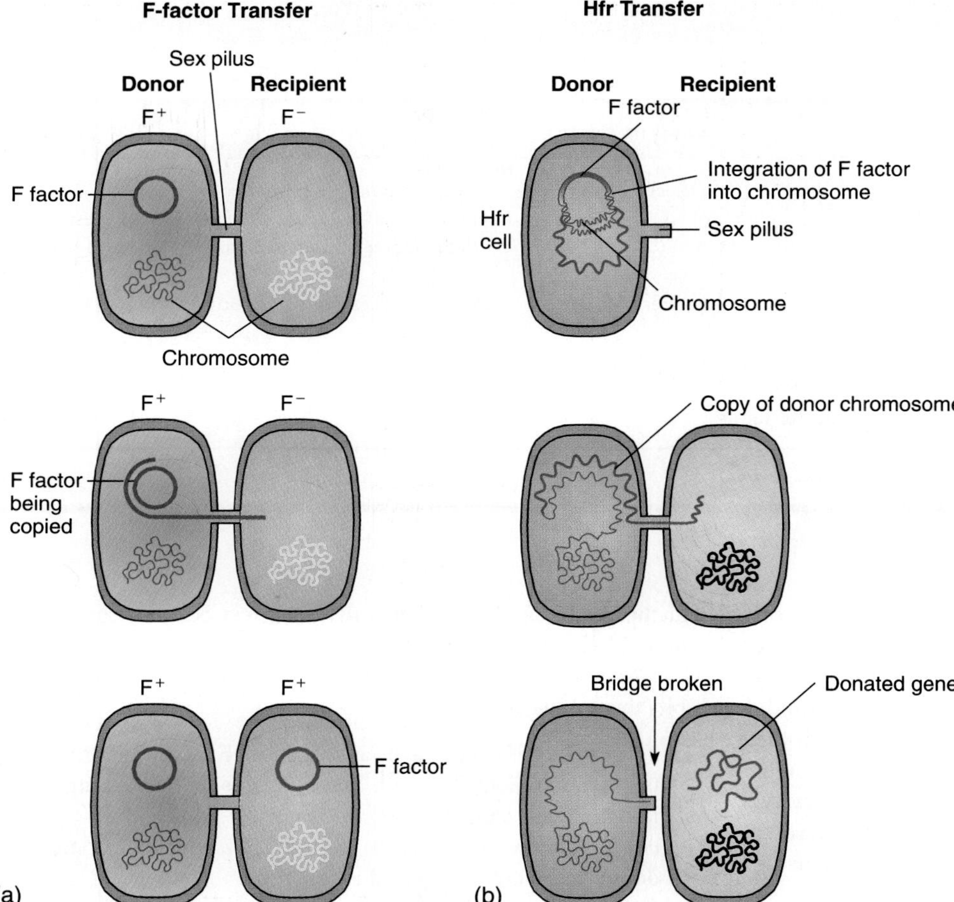

Figure 9.27

Conjugation: genetic transmission through direct contact. The sex pilus forms a connection between the cells (donor on the left, recipient on the right). Whether the donated genes actually pass through the bridge remains controversial. (*a*) Transfer of the F factor, or conjugative plasmid. A cell must have this plasmid to transfer chromosomal genes. (*b*) High-frequency (Hfr) transfer involves transmission of chromosomal genes from a donor cell to a recipient cell. The donor chromosome is duplicated and transmitted in part to a recipient cell.

streptomycin, sulfonamides, and penicillin. This phenomenon is discussed further in chapter 12. Other types of R factors carry genetic codes for resistance to heavy metals (nickel and mercury) or for synthesizing virulence factors (toxins, enzymes, and adhesion molecules) that increase the pathogenicity of the bacterial strain. Conjugation studies have also provided an excellent way to map the bacterial chromosome.

Transformation: Capturing DNA from Solution

One of the cornerstone discoveries in microbial genetics was made in the late 1920s by the English biochemist Frederick Griffith working with *Streptococcus pneumoniae* and laboratory mice. The pneumococcus exists in two major strains based on the presence of the capsule, colonial morphology, and pathogenicity. Encapsulated strains bear a smooth (S) colonial appearance and are virulent; strains lacking a capsule have a rough (R) appearance and are nonvirulent (see figure 4.12). (Recall that the capsule protects a bacterium from the phagocytic host defenses.) To set the groundwork, Griffith showed that when a mouse was injected with a live, virulent (S) strain, it soon died (figure 9.28*a*). When another mouse was injected with a live nonvirulent (R) strain, the mouse remained alive and healthy (figure 9.28*b*). Next he tried a variation on this theme. First, he heat-killed an S strain and injected it into a mouse, which remained healthy (figure 9.28*c*). Then came the ultimate test: Griffith injected both dead S cells and live R cells into a mouse, with the result that the mouse died from a pneumococcal blood infection (figure 9.28*d*). If killed bacterial cells do not come back to life and the nonvirulent strain was harmless, why did the mouse die? Although he did not know it at the time, Griffith had demonstrated that dead S cells, while passing through the body of the mouse, broke open and released some of their DNA (by chance, that part containing the genes for making a capsule). A few of the live R cells subsequently picked up this loose DNA and were **transformed** by it into virulent, capsule-forming strains.

Later studies supported the concept that a chromosome released by a lysed cell breaks into fragments small enough to be accepted by a recipient cell and that DNA, even from a dead cell, retains its code. This nonspecific acceptance by a bacterial cell of small fragments of soluble DNA from the surrounding environment is termed **transformation.** Transformation is apparently fa-

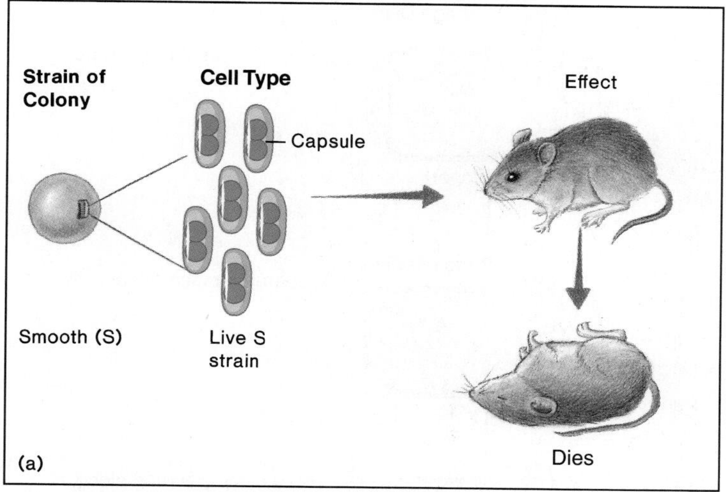

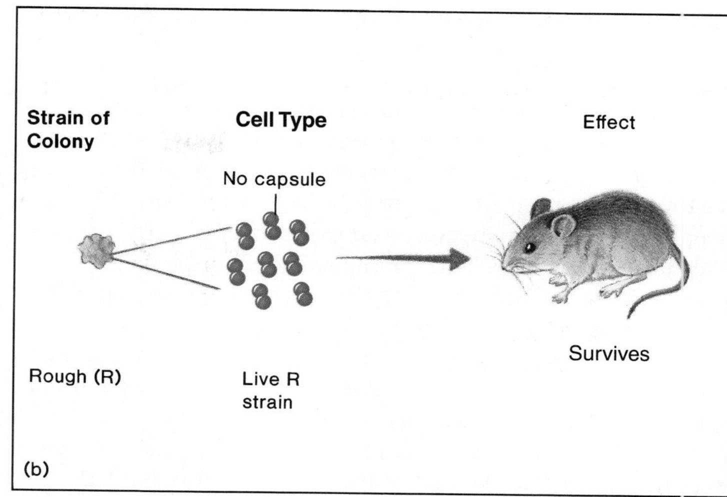

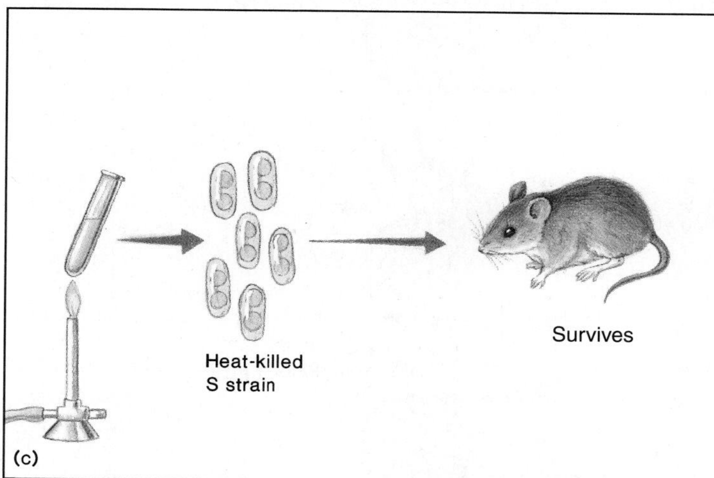

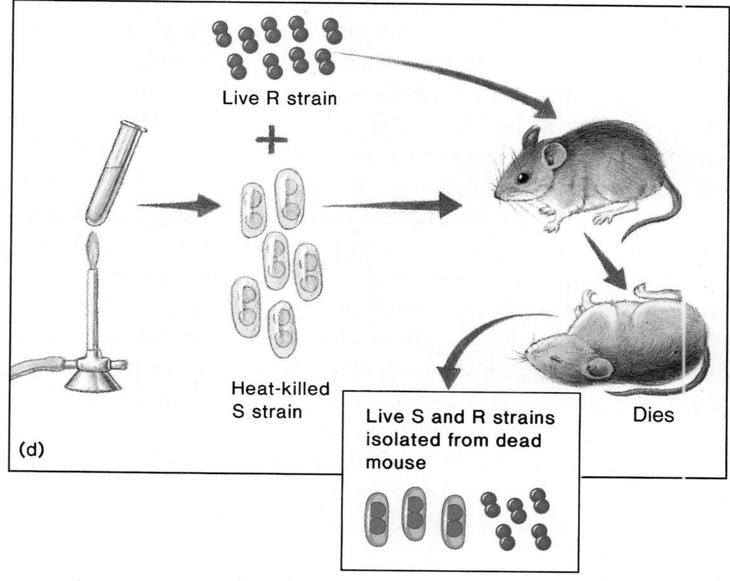

Figure 9.28

Griffith's classic experiment in transformation. In essence, this experiment proved that DNA released from a killed cell can be acquired by a live cell. The cell receiving this new DNA is genetically transformed; in this case, from a nonvirulent strain to a virulent one.

cilitated by special DNA-binding proteins on the cell wall that capture DNA from the surrounding medium. Cells that are capable of accepting genetic material through this means are termed *competent.* The new DNA is processed by the cell membrane and transported into the cytoplasm, where it is inserted into the bacterial chromosome. Transformation is a natural event among both gram-positive streptococci and gram-negative *Haemophilus* and *Neisseria* species. In addition to genes coding for the capsule, bacteria also exchange genes for antibiotic resistance and bacteriocin synthesis in this way.

Because transformation requires no special appendages, and the donor and recipient cells do not have to be in direct contact, the process is useful for certain types of recombinant DNA technology. With this technique, foreign genes from a completely unrelated organism are inserted into a plasmid, which is then introduced into a competent bacterial cell through transformation. These recombinations can be carried out easily in a test tube, and human genes can be experimented upon and even expressed outside the human body by placing them in a microbial cell. This same phenomenon in eucaryotic cells, termed *transfection,* is an essential aspect of genetically engineered yeasts, bird embryos, and mice, and it has been proposed as a future technique for curing genetic diseases in humans. These topics are covered in more detail in chapter 10.

Transduction: The Case of the Piggyback DNA Bacteriophages (bacterial viruses) have been previously described as destructive bacterial parasites. Infection by a virus does not always kill the host cell, however, and viruses can in fact serve as genetic vectors (an entity that can bring foreign DNA into a cell). The process by which a bacteriophage serves as the carrier of DNA from a donor cell to a recipient cell is **transduction.** Although it occurs naturally in a broad spectrum of bacteria, the participating bacteria in a single transduction event must be the same species because of the specificity of viruses for host cells. The events in transduction are shown in figure 9.29.

There are two versions of transduction. In *generalized transduction,* random fragments of disintegrating host DNA are taken up by the phage during assembly. Virtually any gene from the bacterium can be transmitted through this means. In *specialized transduction,* a highly specific part of the host genome is regularly incorporated into the virus. This specificity is explained by the prior existence of a temperate prophage inserted in a fixed site on the bacterial chromosome. When activated, the prophage DNA separates from the bacterial chromosome, carrying a small segment of host genes with it. During a lytic cycle, these specific viral-host gene combinations are incorporated by the viral particles and carried to another bacterial cell.

Several cases of specialized transduction have biomedical importance. The virulent strains of bacteria such as *Corynebacterium diphtheriae, Clostridium* spp., and *Streptococcus pyogenes* all produce toxins with profound physiological effects, whereas nonvirulent strains do not produce toxins. It turns out that virulence is due entirely to lysogenic conversion, in which a bacteriophage introduces genes that code for toxins. Only those bacteria infected with a temperate phage are toxin formers. Other instances of transduction are seen in staphylococcal transfer of drug resistance and in the transmission of gene regulators in gram-negative rods (*Escherichia, Salmonella*).

Transposons: "This Gene Is Jumpin'"

One type of genetic transferral of great interest involves **transposons.** These elements have the distinction of shifting from one part of the genome to another and so are termed "jumping genes." When the idea of their existence in corn plants was first postulated by the geneticist Barbara McClintock, it was greeted with some skepticism, because it had long been believed that the locus of a given gene was set and that genes did not or could not move around. Now it is evident that jumping genes are widespread among procaryotic and eucaryotic cells and viruses.

All transposons share the general characteristic of traveling from one location to another on the genome—from one chromosomal site to another, from a chromosome to a plasmid, or from a plasmid to a chromosome (figure 9.30*a*). Because transposons occur in plasmids, they can also be transmitted from one cell to another in bacteria and a few eucaryotes. Some transposons replicate themselves before jumping to the next location, and others simply move without replicating first.

Transposons can be recognized by the palindromic sequences at each end, some of which are hundreds of bases long (figure 9.30*b*). These regions permit discovery and removal of the

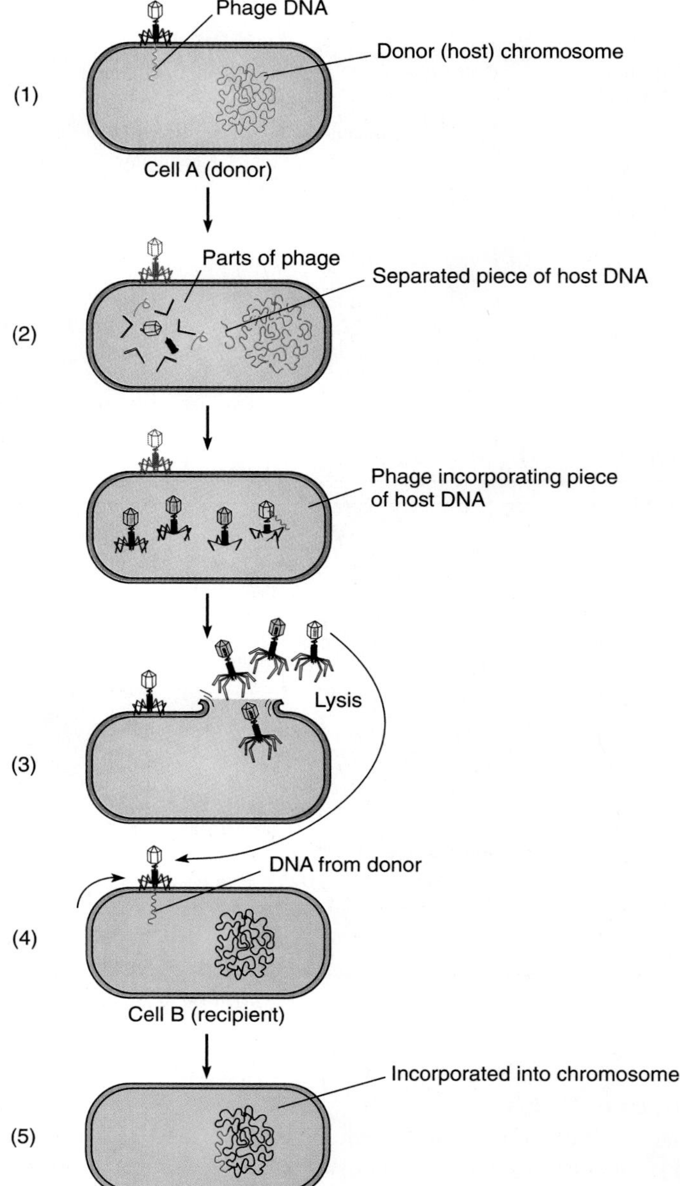

Figure 9.29

Generalized transduction: genetic transfer by means of a virus carrier. (*1*) A phage infects cell A (the donor cell) by normal means. (*2*) During replication and assembly, a phage particle incorporates a segment of bacterial DNA by mistake. (*3*) Cell A then lyses and releases the mature phages, including the genetically altered one. (*4*) The altered phage adsorbs to and penetrates another host cell (cell B), injecting the DNA from cell A rather than viral nucleic acid. (*5*) Cell B receives this donated DNA, which recombines with its own DNA. Because the virus is defective (biologically inactive as a virus), it is unable to complete a lytic cycle. The transduced cell survives and can use this new genetic material.

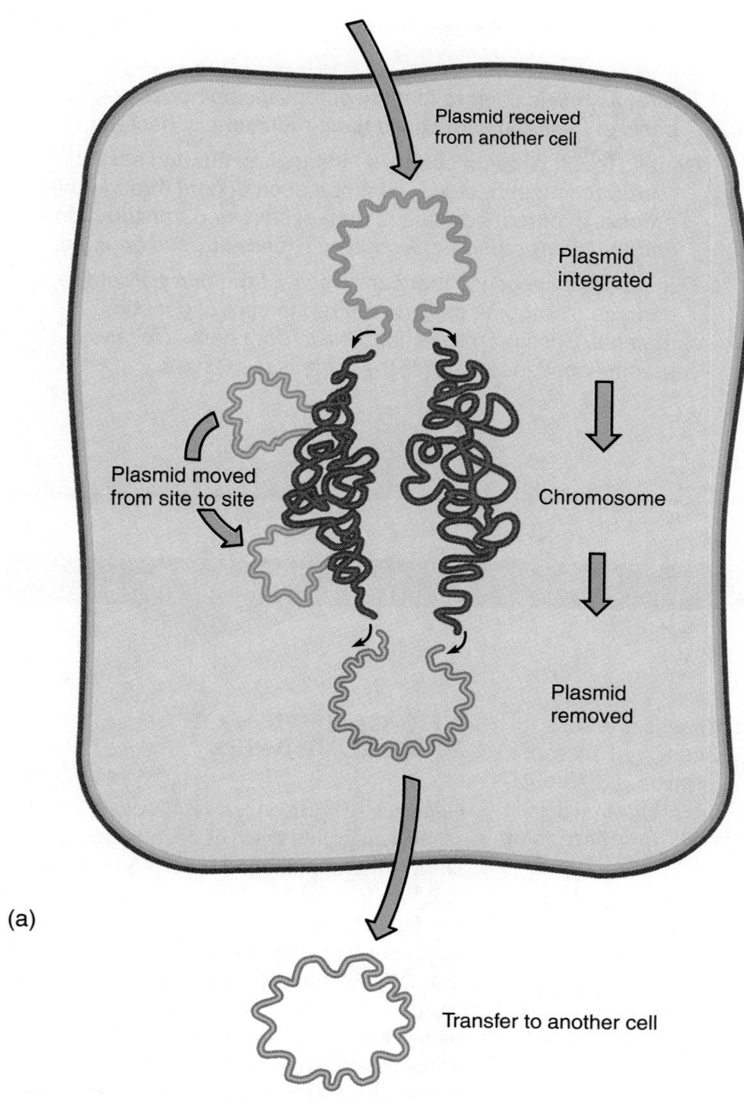

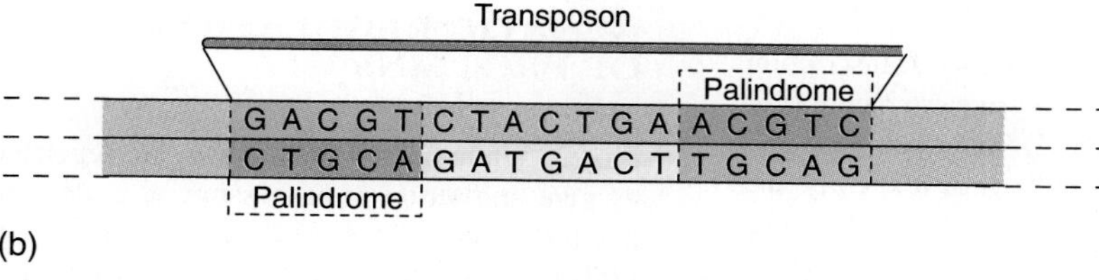

Figure 9.30
Transposons: shifting segments of the genome. (*a*) Potential mechanisms in the movement of transposons in bacterial cells. (*b*) Schematic view of a transposon flanked by palindromic sequences.

transposon sequence, and they dictate the site of insertion into DNA when the transposon relocates.

The overall effect of transposons—to scramble the genetic language—can be beneficial or adverse, depending upon such variables as where insertion occurs in a chromosome, what kinds of genes are relocated, and the type of cell involved. On the beneficial side, transposons are known to be involved in (1) the creation of different genetic combinations, necessary for the high levels of variation in antibodies and receptor molecules in cells (see chapter 15); (2) changes in traits such as colony morphology, pigmentation, pili, and antigenic characteristics; (3) replacement of damaged DNA; and (4) the intermicrobial transfer of drug resistance (in bacteria). On the negative side, rearrangement of DNA that leads to mutations such as deletions, insertions, translocations, inversions, and chromosome breakage can be disruptive and even lethal (see microfile 10.3). Retroviruses such as the AIDS virus behaving as transposons randomly insert in the genome in ways that can lead to severe cell dysfunction (see chapters 17 and 25).

Chapter Checkpoints

Changes in the genetic code can occur by two means: mutation and recombination. Mutation means a change in the nucleotide sequence of the organism's genome. Recombination means the addition of genes from an outside source, such as a virus or another cell.

Mutations can be either spontaneous or induced by exposure to some external mutagenic agent.

All cells have enzymes that repair damaged DNA. When the degree of damage exceeds the ability of the enzymes to make repairs, mutations occur.

Mutation-induced changes in DNA nucleotide sequencing range from a single nucleotide to addition or deletion of large sections of genetic material.

The Ames test is used to identify potential carcinogens on the basis of their ability to cause back-mutations in bacteria.

Genetic recombination occurs in eucaryotes through sexual reproduction. In bacteria, recombination occurs through the processes of transformation, conjugation, and transduction and the incorporation of lysogenic viruses into the genome.

Transposons are genes that can relocate from one part of the genome to another, causing rearrangement of genetic material. Such rearrangements have both beneficial and harmful consequences for the organism involved.

Chapter Capsule with Key Terms

GENES AND THE GENETIC MATERIAL

Genetics is the study of **heredity,** and the **genome** is the sum total of genetic material of a cell. A **chromosome** is composed of DNA in all organisms; **genes** are specific segments of this linear molecule. Genes code for specific peptides such as enzymes, antibodies, or structural proteins.

GENE STRUCTURE AND REPLICATION

A gene consists of DNA, **antiparallel** strands of repeating deoxyribose sugar–phosphate units attached to nitrogenous bases of purine or pyrimidine. Complementary base-pairing (adenine-thymine and cytosine-guanine) ensures genetic fidelity in DNA synthesis called **replication.** Replication is **semiconservative** and requires enzymes such as **helicase, primase, polymerase, ligase,** and **gyrase.** These components in conjunction with the chromosome being duplicated constitute a **replicon.** The unzipped strands of DNA function as **templates.** Synthesis proceeds along two **replication forks,** each with a **leading strand** and a **lagging strand.** In bacteria, the replicon resembles **theta,** and in certain viruses, the replicon is called a **rolling circle.**

GENE FUNCTION

Processing of genetic information proceeds from DNA to RNA and to protein. RNA synthesis is called **transcription,** and protein synthesis is called **translation.** An organism's genetic makeup, or **genotype,** dictates all its traits, called its **phenotype.** The genetic code is organized into **triplets** of nitrogen bases, and each triplet (codon) corresponds to a particular amino acid in a protein.

Types of RNA: Unlike DNA, RNA is single-stranded and contains **uracil** instead of thymine and **ribose** instead of deoxyribose. Chief forms are messenger, transfer, and ribosomal RNA **(mRNA, tRNA,** and **rRNA).** Triplets of bases on mRNA called **codons** convey genetic information for protein structure. The corresponding **anticodon** on tRNA ensures delivery of the appropriate amino acid. Ribosomes are the staging sites where all translation components come together.

Transcription and Translation: Transcription begins when **RNA polymerase** recognizes a **promoter** region on DNA and elongation proceeds in the 5′ to 3′ direction. Only one DNA strand **(template)** is copied. The mRNA strand goes to a ribosome for translation. Translation begins at the **AUG** or **start codon.** Protein assembly proceeds as tRNAs enter the ribosome with anticodons complementary to the codons of the mRNA. Peptide bonds are formed between amino acids on adjacent tRNAs. This process continues until the **nonsense (stop) codon** on mRNA is reached.

Eucaryotic Gene Expression: Eucaryotes have **split genes** interrupted by noncoding regions called **introns** that separate coding segments called **exons.** The synthesis of the final mRNA transcript requires splicing to delete stretches that correspond to introns.

THE GENETICS OF ANIMAL VIRUSES

Genomes of viruses can be linear or circular; segmented or not; made of double-stranded (ds) DNA, single-stranded (ss) DNA, ssRNA, or dsRNA. In general, DNA viruses replicate in the nucleus, RNA viruses in the cytoplasm. Retroviruses synthesize dsDNA from ssRNA. The DNA of some viruses can be silently integrated into the host's genome. Integration by **oncogenic** viruses can lead to **transformation** of the host cell into an immortal cancerous cell. RNA viruses have strand polarity (positive- or negative-sense genome) and double-strandedness.

GENETIC REGULATION OF PROTEINS

Protein synthesis and metabolism are regulated by **gene induction** or **repression,** as controlled by an **operon.** An operon is a DNA unit of **regulatory genes** (made up of **regulators, promoters,** and **operators**) that controls the expression of **structural genes** (which code for enzymes and structural peptides). **Inducible operons** such as the **lactose operon** are normally *off* but are turned *on* by a lactose inducer. **Repressible operons** govern anabolism and are usually *on,* but can be shut *off* when the end product is no longer needed.

Certain antibiotics affect transcription and translation. Adverse modes of action include interference with RNA polymerase, RNA elongation, and ribosomal activity.

GENE MUTATION

Genome changes in microbes come from **mutations** and intermicrobial genetic exchanges. The term **wild type,** or strain, denotes the original form; **mutant strain** refers to the altered version. Mutations are **spontaneous** if they occur randomly and **induced** if they are due to directed chemical or physical agents called **mutagens. Point mutations** entail addition, removal, or substitution of a few bases. A **missense mutation** leads to amino acid substitution, and a **nonsense mutation** arrests peptide synthesis without amino acid insertion. A **silent mutation** causes base substitution without amino acid substitution. A **back-mutation** is a reversion to the original base composition.

Repair of Mutations: Enzymes can identify and repair mutations. Some enzymes locate **mismatched** bases and engage in repairs. DNA damaged by ultraviolet light can be corrected by **photoactivation** or **light repair.** The **Ames test** is a mutation-screening method based on the susceptibility of mutant *Salmonella typhimurium* to a back-mutation. It is used to measure the mutagenicity of chemicals.

Effects of Mutations: Mutations can benefit organisms through adaptive advantage, but some are lethal. Intermicrobial transfer and genetic recombination permit gene sharing between bacteria. In **conjugation,** one bacterium donates a plasmid (**fertility,** or **F, factor**) to a compatible recipient via a **sex,** or **conjugative, pilus.** Conjugation resulting in transfer of **resistance,** or **R, plasmids** is biomedically important. In bacterial **transformation,** naked DNA is transferred without special carriers. It was first demonstrated in smooth and rough pneumococcal strains. **Transduction** is transfer of host DNA by bacteriophages. Drug resistance and virulence genes can be transferred by this route. **Transposons** are large DNA sequences that regularly depart and reinsert into chromosomal and plasmid sites within the same cell or between cells.

MULTIPLE-CHOICE QUESTIONS

1. What is the smallest unit of heredity?
   a. chromosome c. codon
   b. gene d. nucleotide

2. The nitrogen bases in DNA are bonded to the
   a. phosphate c. ribose
   b. deoxyribose d. hydrogen

3. DNA replication is semiconservative because the ____ strand will become half of the ____ molecule.
   a. RNA, DNA c. sense, mRNA
   b. template, finished d. codon, anticodon

4. In DNA, adenine is the complementary base for ____ , and cytosine is the complement for ____ .
   a. guanine, thymine c. thymine, guanine
   b. uracil, guanine d. thymine, uracil

5. The base pairs are held together primarily by
   a. covalent bonds b. hydrogen bonds
   c. ionic bonds d. gyrases

6. Why must the lagging strand of DNA be replicated in short pieces?
   a. because of limited space
   b. otherwise, the helix will become distorted
   c. the DNA polymerase can synthesize in only one direction
   d. to make proofreading of code easier

7. Messenger RNA is formed by ____ of a gene on the DNA template strand.
   a. transcription c. translation
   b. replication d. transformation

8. Transfer RNA is the molecule that
   a. contributes to the structure of ribosomes
   b. adapts the genetic code to protein structure
   c. transfers the DNA code to mRNA
   d. provides the master code for amino acids

9. As a general rule, the template strand on DNA will always begin with
   a. TAC c. ATG
   b. AUG d. UAC

10. The *lac* operon is usually in the ____ position and is activated by a/an ____ molecule.
    a. on, repressor c. on, inducer
    b. off, inducer d. off, repressor

11. The repressible operon is important in regulating ____ .
    a. amino acid synthesis
    b. DNA replication
    c. sugar metabolism
    d. ATP synthesis

12. For mutations to have an effect on populations of microbes, they must be
    a. inheritable c. beneficial
    b. permanent d. a and b
    e. all of the above

13. Which of the following characteristics is *not* true of a plasmid?
    a. it is a circular piece of DNA
    b. it is required for normal cell function
    c. it is found in bacteria
    d. it can be transferred from cell to cell

14. Which genes can be transferred by all three methods of intermicrobial transfer?
    a. capsule production
    b. toxin production
    c. F factor
    d. drug resistance

15. Which of the following would occur through specialized transduction?
    a. acquisition of Hfr plasmid
    b. transfer of genes for toxin production
    c. transfer of genes for capsule formation
    d. transfer of a plasmid with genes for degrading pesticides

16. Multiple matching. Fill in the blanks with all the letters of the words on at the right that apply.

____ genetic transfer that occurs after the donor is dead
____ carries the codon
____ carries the anticodon
____ a process synonymous with mRNA synthesis
____ bacteriophage participate in this transfer
____ a process requiring an F^+ pilus
____ duplication of the DNA molecule
____ process in which transcribed DNA code is deciphered into a polypeptide
____ involves plasmids

a. replication
b. tRNA
c. conjugation
d. ribosome
e. transduction
f. mRNA
g. transcription
h. transformation
i. translation
j. none of these

CONCEPT QUESTIONS

1. Compare the genetic material of eucaryotes, bacteria, and viruses in terms of general structure, size, and mode of replication.
2. Briefly describe how DNA is packaged to fit inside a cell.
3. Give the base sequence of the complementary strand of the following strand of DNA: 5´-ATCGGCTACGTTCAC-3´
4. Describe what is meant by the antiparallel arrangement of DNA.
5. Name several characteristics of DNA structure that enable it to be replicated with such great fidelity generation after generation.
6. Explain the following relationship: DNA formats RNA, which makes protein.
7. What message does a gene provide? How is the language of the gene expressed?
8. If a protein is 3,300 amino acids long, how many nucleotide pairs long is the gene sequence that codes for it?
9. a. What about a palindrome makes it recognizable by a restriction endonuclease?
   b. Draw a short sequence of a palindrome.
10. Compare the structure and functions of DNA and RNA.
11. a. Where does transcription begin?
    b. What are the sense and antisense strands of DNA?
    c. Is the same strand of DNA always transcribed?
12. Compare and contrast the actions of DNA and RNA polymerase.
13. What are the functions of start and nonsense codons?
14. The following sequence represents triplets on DNA:

    TAC CAG ATA CAC TCC CCT GCG ACT

    a. Give the mRNA codons and tRNA anticodons that correspond with this sequence, and then give the sequence of amino acids in the polypeptide.
    b. Provide another mRNA strand that can be used to synthesize this same protein.
15. a. Summarize how bacterial and eucaryotic cells differ in gene structure, transcription, and translation.
    b. Discuss the roles of exons and introns.
16. Compare DNA viruses with RNA viruses in their general methods of nucleic acid synthesis and viral replication.
17. a. Compare and contrast the *lac* operon with a repressible operon system.
    b. How is the *lac* system related to the feedback control of enzymes mentioned in chapter 8?
18. What is the premise of the Ames test?
19. Describe the principal types of mutations. Give an example of a mutation that is beneficial and one that is lethal or harmful.
20. a. Compare conjugation, transformation, and transduction on the basis of general method, nature of donor, and nature of recipient.
    b. List some examples of genes that can be transferred by intermicrobial transfer.
21. By means of a flowchart, show the possible jumps that a transposon can make. Show the involvement of viruses in its movement.

CRITICAL-THINKING QUESTIONS

1. A simple test you can do to demonstrate the coiling of DNA in bacteria is to open a large elastic band, stretch it taut, and twist it. First it will form a loose helix, then a tighter helix, and finally, to relieve stress, it will twist back upon itself. Further twisting will result in a series of knotlike bodies; this is how bacterial DNA is condensed.

2. To envision the replicon of bacteria, take a very thick rubber band and nick it longitudinally at some point. This will be the site of transcription origin. Slice the rubber band for a few millimeters and label the orientation of the two beginning strands (5´ to 3´ and 3´ to 5´). Label one strand leading and one strand lagging. This will also produce the two forks of replication and allow you to play with the production of the two new strands. Snip toward one fork to imagine how the polymerase will enter to add the correct bases. To represent the bases, add one paper clip to the leading strand in sequence toward the fork on the leading strand as each snip is made. Add clips to the lagging strand in the appropriate direction for that strand, but show that it is discontinuous. Do the same on the other fork, and continue around until you have two separate strands. About halfway through, look for the theta configuration.
3. Knowing that retroviruses operate on the principle of reversing the direction of transcription from RNA to DNA, propose a drug that might possibly interfere with their replication.
4. Using the piece of DNA in concept question 14, show a deletion, an insertion, a substitution, and an inversion. Which ones are frameshift mutations? Are any of your mutations nonsense? Missense? (Use the universal code to determine this.)
5. Using tables 9.3 and 9.6, go through the steps in mutation of a codon followed by its transcription and translation that will give the end result in silent, missense, and nonsense mutations.
6. Explain the principle of "wobble" and find four amino acids that are encoded by wobble bases (table 9.3). Suggest some benefits of this phenomenon to microorganisms.
7. Suggest a reason for having only one strand of DNA serve as a source of useful genetic information. What could be some possible functions of the coding or antisense strand?
8. The enzymes required to carry out transcription and translation are themselves produced through these same processes. Speculate which may have come first in evolution—proteins or nucleic acids—and explain your choice.
9. Why can one not reliably predict the sequence of nucleotides on mRNA or DNA by observing the amino acid sequence of proteins?
10. Speculate on the manner in which transposons can be involved in cancer.

INTERNET SEARCH TOPIC

Look up variations on DNA structure by entering the terms Z-DNA and B-DNA. Compare these forms of DNA with the one presented in this chapter.

10 chapter

Genetic Engineering: A Revolution in Molecular Biology

Imagine having the power to cure genetic diseases, to identify an infectious agent or diagnose early cancer in a few hours from a single cell, and to witness the entire human genome readout by a computer. Such sensational prospects at one time may have seemed fantastic, but today they are realities. These feats and many others have been placed within our grasp by **genetic engineering*** and **biotechnology.** Scientists from diverse disciplines now have at their disposal a variety of powerful molecular tools to manipulate, alter, and analyze the genetic material of microbes, plants, animals, and viruses. This relatively young science has ushered in a new era in the history of humankind, and it has spawned an explosion of medical, agricultural, and industrial applications.

The genetic revolution has forced every person to become DNA-literate. The media keep us constantly updated on genetic news events ranging from provocative to profound: mice that glow in the dark, sheep that make moth-proof wool, deciphering the ancient DNA of fossils, using viruses to cure genetic diseases.

We have been challenged to consider the accuracy of DNA fingerprinting evidence in court and the safety of eating genetically altered fruits and animal products. The ability to change genetic blueprints and create totally new types of organisms is so alarming that a group of scientists and religious leaders have strongly urged the government to legislate against genetic alteration of human gametes and embryos. Some environmental activists have even gone so far as to attempt to block the release of engineered microbes and plants into natural habitats. Such controversies will require serious **bioethical*** considerations (microfile 10.1).

*Sometimes called bioengineering—defined as the direct, deliberate modification of an organism's genome. Biotechnology is a more encompassing term that includes the use of DNA, genes, or genetically altered organisms in commercial production.

*bioethics: A field that relates biological issues to human conduct and moral judgment.

A medical milestone. These two young ladies—Ashanti DeSilva (right) and Cynthia Cutshall (left)—are the world's first recipients of therapy for an inborn genetic disease. They received gene therapy for adenosine deaminase (ADA) deficiency, an immune disease that usually causes premature death. Although they are not yet cured, their health and quality of life have been greatly improved by genetic engineering.

BASIC ELEMENTS AND APPLICATIONS OF GENETIC ENGINEERING

Information on genetic engineering and its biotechnological applications is growing at such an expanding rate that some new discovery or product is disclosed almost on a daily basis. To keep this subject somewhat manageable, we will present essential concepts and applications that are both conventional and as up-to-date as possible. As an organizational aid, the topics are divided into the following six major sections, designated with a roman numeral: I. genetic engineering tools and techniques; II. recombinant DNA technology; III, recombinant products; IV. recombinant (transgenic) microbes, plants, and animals; V. genetic medical treatments; and VI. genome analysis. Figure 10.1 summarizes the main categories to be covered in these topics.

Chapter Checkpoints

The genetic revolution has produced a wide variety of industrial technologies that translate and radically alter the blueprints of life. The potential of biotechnology promises not only improved quality of life and enhanced economic opportunity but also serious ethical dilemmas. Increased public understanding is essential for developing appropriate guidelines for responsible use of these revolutionary techniques.

TOOLS AND TECHNIQUES OF GENETIC ENGINEERING

Genetic studies over the last 30 years have provided a number of insights into certain measurable and predictable activities of nucleic acids. Unraveling this information also had a practical benefit, which was the development of highly sophisticated laboratory tools and techniques. Traditional methods of genetic analysis have involved observing phenotypic inheritance in a test subject. But with these newer techniques, discovery is more likely to involve seeing actual DNA patterns on electrophoresis gels, locating the exact gene on a chromosome, and producing computer-generated gene sequences. Each tool or technique described in the next several sections does not stand alone but is usually an important part of an overall methodology.

DNA: THE MARVELOUS MOLECULAR TOY

DNA as an object of study holds many surprises and delights. Its structure is like a child's building blocks, and its language is a special code that imparts information on life processes. It can be wound up and untwisted like a rubber band, opened into two strands like a zipper, diced up into shorter pieces, and spliced back together.

One useful property of DNA is that is readily **anneals,*** or changes, its binding properties in response to heating and cooling. Exposure to temperatures just below boiling (90°–95°C) causes DNA to become temporarily denatured (figure 10.2). When heat breaks the hydrogen bonds holding the double helix together, it separates longitudinally into two strands. Each strand displays its nucleotide code so that the DNA in this form can be subjected to tests or replicated. When heating is followed by slow cooling, two single DNA strands rejoin (renature) by hydrogen bonds at complementary sites. As we shall see, annealing is a necessary feature of the polymerase chain reaction (PCR) and nucleic acid probes described later.

Enzymes for Dicing, Splicing, and Reversing Nucleic Acids

The polynucleotide strands of DNA can also be clipped crosswise at selected positions by means of enzymes called **restriction endonucleases** (see microfile 9.4). So far, hundreds of restriction endonucleases have been discovered in bacteria. Each type has a known sequence of 4 to 10 base pairs as its target, so sites of cutting can be finely controlled. These enzymes have the unique property of recognizing and snipping inverted repeats called palindromes (figure 10.3*a*). Examples of three common endonucleases with their sites of cutting are:

*__anneal__ (ah-neel′) O. E. *anaelan,* to kindle. A process of heating and slowly cooling a substance to change its properties.

MICROFILE 10.1 BIOETHICS CONFRONTS BIOENGINEERING

No one would minimize the real and potential benefits to be derived from genetic engineering. Yet, in our endorsement of this technology, we must also at least consider its possible adverse impact. Some alarmists feel that tampering with the genes of humans and other organisms is tantamount to playing God, and they predict dire sociological, ecological, and medical consequences. Jeremy Rifkin of the Foundation on Economic Trends voiced his group's fear:

> We are embarking on a very potentially troublesome journey, where we begin to reduce all other animals on this planet to genetically engineered products. . . . We will increasingly think of ourselves as just gene codes and blueprints and programs that can be tinkered with.

Other critics argue that creating entirely new organisms and releasing them into the earth's habitats may upset the balance of nature. They point out that the same techniques that are used to engineer plants to resist herbicides and bacteria to make insulin could also create a virus that inserts into the human genome and causes cancer or other diseases.

The possible consequences of this technology are far-reaching. For instance, is the world heading toward a time when normal children will be given genetically engineered growth hormone to increase size? A controlled study was conducted to assess the effects of recombinant growth hormone on the development of children who are healthy but somewhat small for their age. The purposes of these clinical trials are both to assess the safety of the drug and to determine whether it can have substantial effects on the normal body. Data from these trials indicate that hormone therapy does increase height if given early, but the increases are not dramatic. The perception of small size as a "disease" that needs to be treated is very distressing to some medical experts, and the treatment is currently limited to treating children who are deficient in this hormone. Yet supporters feel that the use of engineered hormones is really no different from cosmetic surgery and should be available to people who want it.

Detailed analysis of the genetic blueprint of human beings is expected to be completed around the year 2005. This will not only provide a greater understanding of what causes genetic diseases, but it will also allow screening for any defective genes that could lead to disease. Such a possibility will also give rise to numerous ethical and legal questions. Opponents have already objected vehemently to genetic screening as an invasion of privacy, not unlike requiring genetic "social security numbers." Might such tests be demanded as a means of determining one's fitness for a job or insurance coverage? Most experts view the likelihood of mandatory genetic screening as very remote.

It is far more likely that the ability to profile an individual's genetic makeup will have beneficial effects. Children with serious or life-threatening defects may require prompt therapy to correct or alleviate the effects of the disease. In pinpointing the risk for cancer and other serious diseases, screening will help people make meaningful life-style and medical decisions. In addition, genetic counseling would be greatly improved since parents with a complete genetic profile could determine in advance what potentially harmful genes their children might inherit.

One research group recently separated a human embryo into several identical clone embryos, while another group successfully implanted mouse sperm cells with marker genes that were passed to their offspring. Tests are now available to determine the sex of an eight-cell-stage embryo and whether it carries certain genetic diseases. Will these techniques be used to promote sex selection, or will they influence decisions to terminate pregnancy on the basis of genetic defects? Although it has not yet been attempted in humans, gene line therapy (correcting genetic defects in the zygote or early embryo) will be a distinct possibility in the future. The most vocal critics imagine a day when parents will be able to create superbabies by requesting that certain favorable genes be added to their children's genomes. Others, however, feel that gene line therapy would be the ultimate solution to correcting genetic defects that diminish the quality of life.

An inevitable outcome of bioengineering has been its potential for profit. By 2000, over a hundred bioengineered products are expected to earn as much as $20 billion in revenues. With thousands of biotechnology companies worldwide, each actively developing various drugs,

Endonuclease	EcoRI	HindIII	HaeIII
Cutting pattern	G¦A A T T C C T T A A¦G	A¦A G C T T T T C G A¦A	G G¦C C C C¦G G

Endonucleases are named by combining the first letter of the bacterial genus, the first two letters of the species, and the endonuclease number. Thus, EcoRI is the first endonuclease found in *Escherichia coli,* and HindIII is the third endonuclease discovered in *Haemophilus influenzae* type d.

Endonucleases are necessary for some of the steps in gene removal and insertion as carried out in the recombinant DNA techniques described in a subsequent section. Those such as HaeIII make straight, blunt cuts on DNA. But more often (EcoRI and HindIII, for example), the enzymes make staggered symmetrical cuts that leave short tails called "sticky ends." Such adhesive tails will base-pair with complementary tails on other chromosomes or open plasmids. This effect makes it possible to splice genes into specific sites. Circularization into a plasmid can occur when the complementary tails are on opposite ends of the same linear DNA fragment (figure 10.3*b*).

Fragments of DNA produced by restriction endonucleases are termed *restriction fragment length polymorphisms* (RFLPs) because of the way the endonuclease clips the DNA into pieces of various lengths. They are polymorphic, meaning they vary considerably from one individual to another, and they are inherited in predictable patterns. Hundreds of cleavage sites that produce RFLPs are sprinkled throughout genomes. Because RFLPs serve as a type of *genetic marker,* they can help locate specific sites along a DNA strand and are thus useful in preparation of gene maps and DNA profiles (see figure 10.18 and microfile 10.4).

research tools, or industrial applications, this industry will continue to expand. Research and development on new drugs and treatments often involves billion dollar monetary investments. At the present time, pharmaceutical companies are testing and requesting approval for over 400 new biotech medicines. Most of these are for treating cancer, AIDS, vaccines, and gene therapy.

Most biotechnology companies would like to have rights to exclusive marketing of the drugs, procedures, animals, plants, and microbes they have developed. A limited number of patents have been issued for unique and commercially useful organisms and drugs. Currently, one of the most provocative questions is: Who owns the human genome? Several companies are attempting to patent human databases for genetic diseases such as breast cancer. So far, companies have obtained patent protection on colon cancer and obesity genes because the DNA was isolated and characterized, and it existed in a form different from nature. Such patents have been used to develop test kits to detect defective genes and to assess risk, but later they may be used to develop therapies.

Protecting the public's best interest will require that bioethicists and representatives of many disciplines—science, law, medicine, business, government, and education—continue to discuss the basic moral and economic issues and to develop guidelines for the future. The main problem may well be whether laws can keep up with the rapid advances in the field. There is no doubt that bioengineering will eventually touch all of our lives and even change the ways we regard the world.

Another enzyme called a **ligase** is necessary to seal the sticky ends together by rejoining the phosphate-sugar bonds cut by endonucleases. Its main application is in final splicing of genes into plasmids and chromosomes.

An enzyme called **reverse transciptase** is best known for its role in the replication of the AIDS virus and other retroviruses. It also provides geneticists with a valuable tool for converting RNA into DNA. Copies called **copy DNA,** or **cDNA,** can be made from messenger, transfer, ribosomal, and other forms of RNA. The technique provides a valuable means of synthesizing eucaryotic genes from mRNA transcripts. The advantage is that the synthesized gene will be free of the intervening sequences (introns) that can complicate the management of eucaryotic genes in genetic engineering. Copy DNA can also be used to analyze the nucleotide sequence of RNAs, such as those found in ribosomes and transfer RNAs (see figure 4.29).

Visualizing DNA with the Naked Eye

One way to produce a readable pattern of DNA fragments is through **gel electrophoresis** (figure 10.4). In this technique, samples are placed in compartments (wells) in a soft agar gel and subjected to an electrical current. Flow of electricity from the negative pole to the positive pole causes the DNA pieces to migrate in the gel substrate toward the positive pole. The rate of movement is based primarily on the size of the fragments. The larger fragments move more slowly and remain nearer the top of the gel, whereas the smaller fragments migrate faster and are positioned farther from the wells. The positions of DNA fragments are determined by staining the gel or developing it with probes and comparing the pattern against a known standard (see microfile 10.4). Electrophoresis patterns can be quite distinctive and are very useful in characterizing DNA fragments and comparing the degree of genetic similarities among samples (so-called genetic fingerprints).

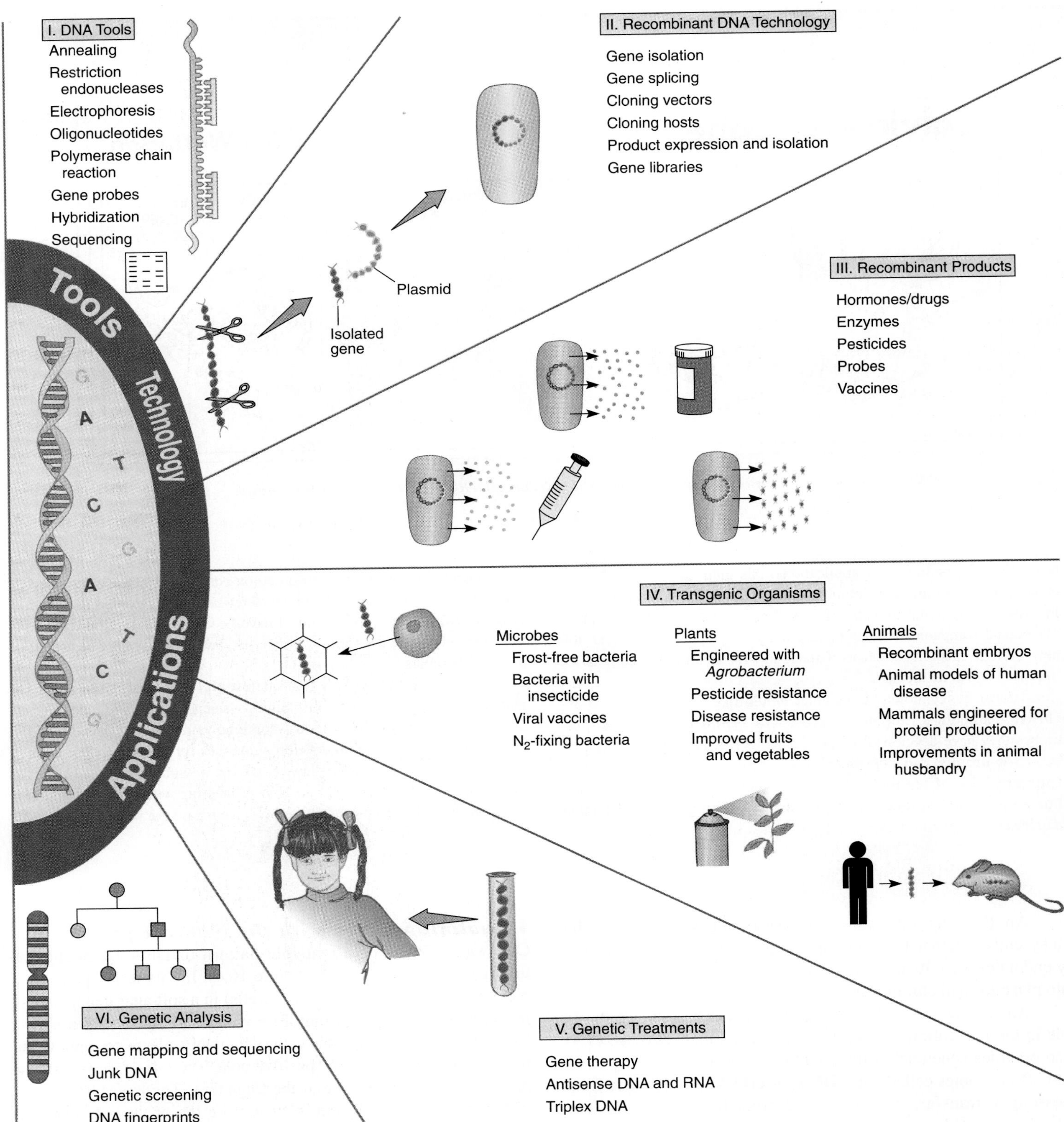

Figure 10.1
Foundations and applications of genetic engineering and DNA technology. Each roman numeral heading corresponds to a major section covered in the text.

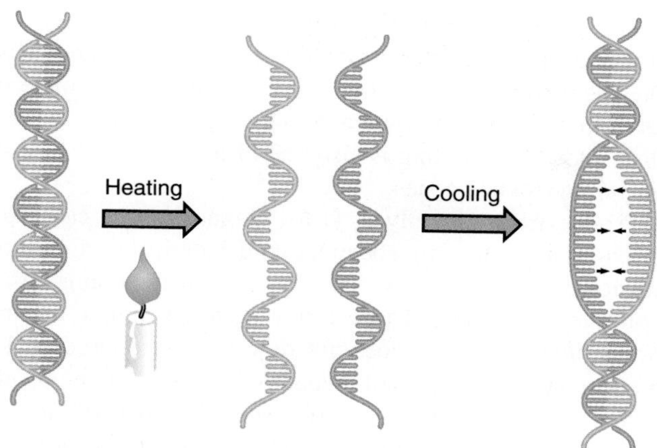

Figure 10.2
DNA annealing process. DNA acts somewhat like molecular "Velcro." Heat will denature and separate the strands, but when cooled the two strands rejoin at complementary sites. The two strands need not be from the same organisms as long as they have matching sites.

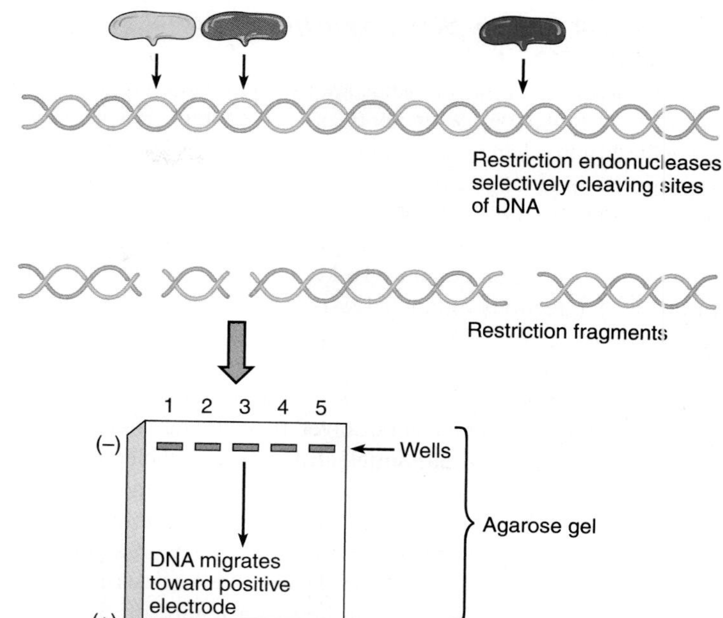

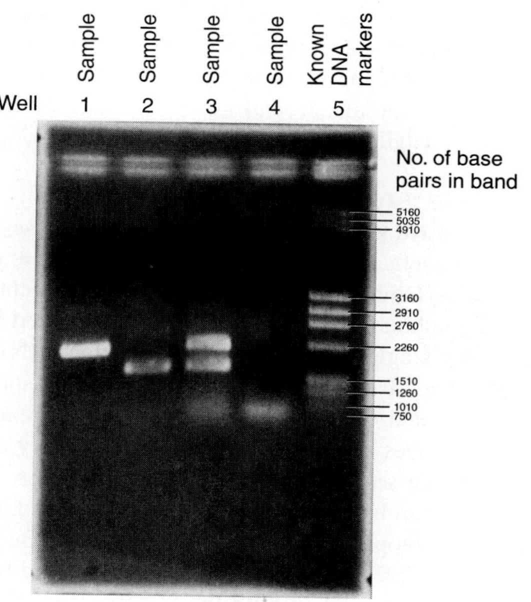

Figure 10.4
Visualizing DNA macroscopically with electrophoresis. After cleavage into fragments, DNA is placed into wells on one end of an agarose gel. When an electrical current is passed through the gel (from the negative pole to the positive pole), the DNA, being negatively charged, migrates toward the positive pole. The larger fragments, measured in numbers of base pairs, migrate more slowly and remain higher in the gel than the smaller (shorter) fragments. Development with radioactivity, fluorescence, or dyes reveals a separation pattern of different fragments of DNA. This pattern, or fingerprint, can be unique for a given DNA. The nature of a given DNA band can be determined by a known set of markers (lane 5) specific for size or DNA code. This result is called a ladder in reference to its appearance.

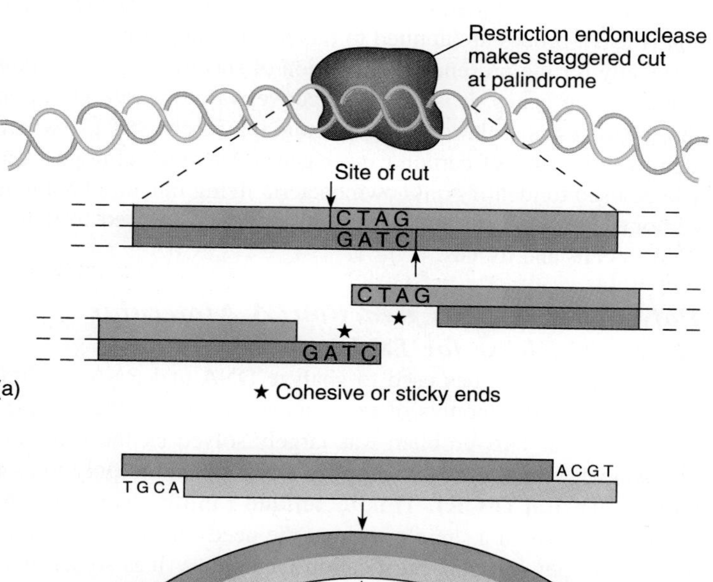

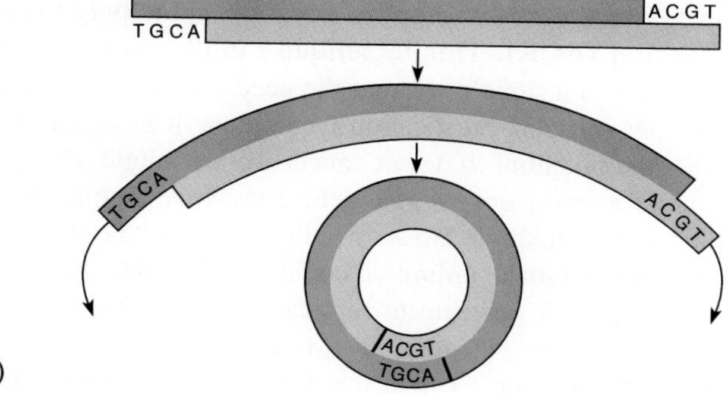

Figure 10.3
Action of restriction endonucleases and circularization of plasmids. (*a*) The restriction endonuclease recognizes and cleaves DNA at the site of the palindromic sequence. Cleavage produces staggered tails called sticky ends that can accept complementary tails for gene splicing. (*b*) Endonucleases can also produce fragments with complementary ends that are attracted to each other, thereby circularizing the molecule into a plasmid.

Methods Used to Size, Synthesize, and Sequence DNA

The relative sizes of nucleic acids are usually denoted by the number of base pairs (bp) or nucleotides they contain. For example, the palindromic sequences recognized by endonucleases are usually 4 to 10 bp in length, and an average gene in *E. coli* is approximately 1,300 bp, or 1.3 kilobases (kb), and its entire genome is approximately 4.7 million base pairs (Mb). The DNA of the human mitochondrion contains 16 kb, and the Epstein-Barr virus (a cause of infectious mononucleosis) has 172 kb. Humans have approximately 3.5 billion base pairs (Bb) arrayed along 46 chromosomes.

Large segments of DNA or RNA are difficult to handle and analyze, but this problem can be solved by using far shorter pieces of DNA or RNA called **oligonucleotides.*** Oligonucleotides vary in length from 2 to 200 bp, although the most common ones are about 20 to 30 bp. Although they can be isolated from cells, for most applications oligonucleotides are tailor-made on a gene machine. DNA synthesizers add the desired nucleotides in a controlled series of steps but are limited to a length of about 200 nucleotides.

An enormously useful type of information about a given gene or genome is the precise sequence of the base pairs it contains. This information was very slow in coming until machines called *sequencers* were developed. A sequencer automatically determines the order of nucleotides in DNA fragments from 500 to 10 million base pairs, and it can read out at a rate of thousands of nucleotides a day. Automated sequencers have been critical in producing the most detailed genetic maps (microfile 10.2).

Nucleic Acid Hybridization and Probes

In biology, a **hybrid** is defined as an offspring that arises from the union of two genetically different parents. Two different nucleic acids can also **hybridize** by uniting at their complementary sites. All different combinations are possible: single-stranded DNA can unite with other single-stranded DNA or RNA, and RNA can hybridize with other RNA. This property has been the inspiration for specially formulated oligonucleotide tracers called **gene probes.** Hybridization probes have practical value because they can detect specific nucleotide sequences in unknown samples. So that areas of hybridization can be visualized, the probes carry radioactive labels, which are isotopes that emit radiation, or luminescent labels, which give off visible light. Reactions can be revealed by placing photographic film in contact with the test reaction. Fluorescent probes contain dyes that can be visualized with ultraviolet radiation, and enzyme-linked probes give off colored dyes.

When probes hybridize with an unknown sample of DNA or RNA, they tag the precise area and degree of hybridization and help determine the nature of nucleic acid present in a sample. In a method called the *Southern blot,*[1] DNA fragments are first separated by electrophoresis and then denatured and immobilized on a special filter. A DNA probe is then incubated with the sample, and wherever this probe encounters the segment for which it is complementary, it will attach and form a hybrid. Development of the hybridization pattern will show up as one or more bands (figure 10.4). This method is a sensitive and specific way to isolate fragments from a complex mixture and to find specific gene sequences on DNA. Southern blotting is also one of the important first steps for preparing isolated genes.

Probes are commonly used for diagnosing the cause of an infection from a patient's specimen and identifying a culture of an unknown bacterium or virus. The method for doing this was first outlined in chapter 4 (see figure 4.28). A simple and rapid method called a *dot blot* does not require electrophoresis. DNA from a test sample is isolated, denatured, placed on an absorbent filter, and combined with a microbe-specific probe (figures 10.5 and 24.21). The blot is then developed and observed for areas of hybridization. Commercially available diagnostic kits are now on the market for identifying intestinal pathogens such as *E. coil, Salmonella, Campylobacter, Shigella, Clostridium difficile,* rotaviruses, and adenoviruses. Other bacterial probes exist for *Mycobacterium, Legionella, Mycoplasma,* and *Chlamydia;* viral probes are available for herpes simplex and zoster, papilloma (genital warts), hepatitis A and B, and AIDS. DNA probes have also been developed for human genetic markers and some types of cancer (figure 10.5*f*).

With another method, called *fluorescent in situ hybridization* (FISH), probes are applied to intact cells and observed microscopically for the presence and location of specific genetic marker sequences on genes. It is a very effective way to locate genes on chromosomes (see the opening photograph in chapter 17, which identifies the sites of certain cancer genes). In situ techniques can also be used to identify unknown bacteria living in natural habitats without having to culture them, and they can be used to detect RNA in cells and tissues.

Polymerase Chain Reaction: A Molecular Xerox Machine for DNA

Some of the techniques used to analyze DNA and RNA are limited by the small amounts of test nucleic acid or by the complex steps involved. This problem was largely solved by the invention of a simple, versatile way to amplify DNA called the **polymerase chain reaction (PCR).** This technique rapidly increases the amount of DNA in a sample without the need for making cultures or carrying out complex purification techniques. It is so sensitive that it holds the potential to detect cancer from a single cell or to diagnose an infection from a single gene copy. It is comparable to being able to pluck a single DNA "needle" out of a "haystack" of other molecules and make unlimited copies of the DNA. The rapid rate of PCR makes it possible to replicate a target DNA from a few copies to billions of copies in a few hours. It can amplify DNA fragments that consist of a few base pairs to whole genes containing several thousand base pairs.

Initiating the reaction requires a few specialized ingredients (figure 10.6). **Primers** are synthetic oligonucleotides of a known sequence of 15–30 bases. They serve as landmarks to indicate where DNA amplification will begin. It is essential to have some information on what is being amplified in order to select appropriate primers. So that DNA templates do not recombine and interfere with DNA synthesis, processing must be carried out at a rela-

1. Named for its developer, E. M. Southern. The "Northern" blot is a similar method used to analyze RNA.

*oligonucleotides (aw″-lig-oh-noo′-klee-oh-tydz) Gr. *oligo,* few or scant.

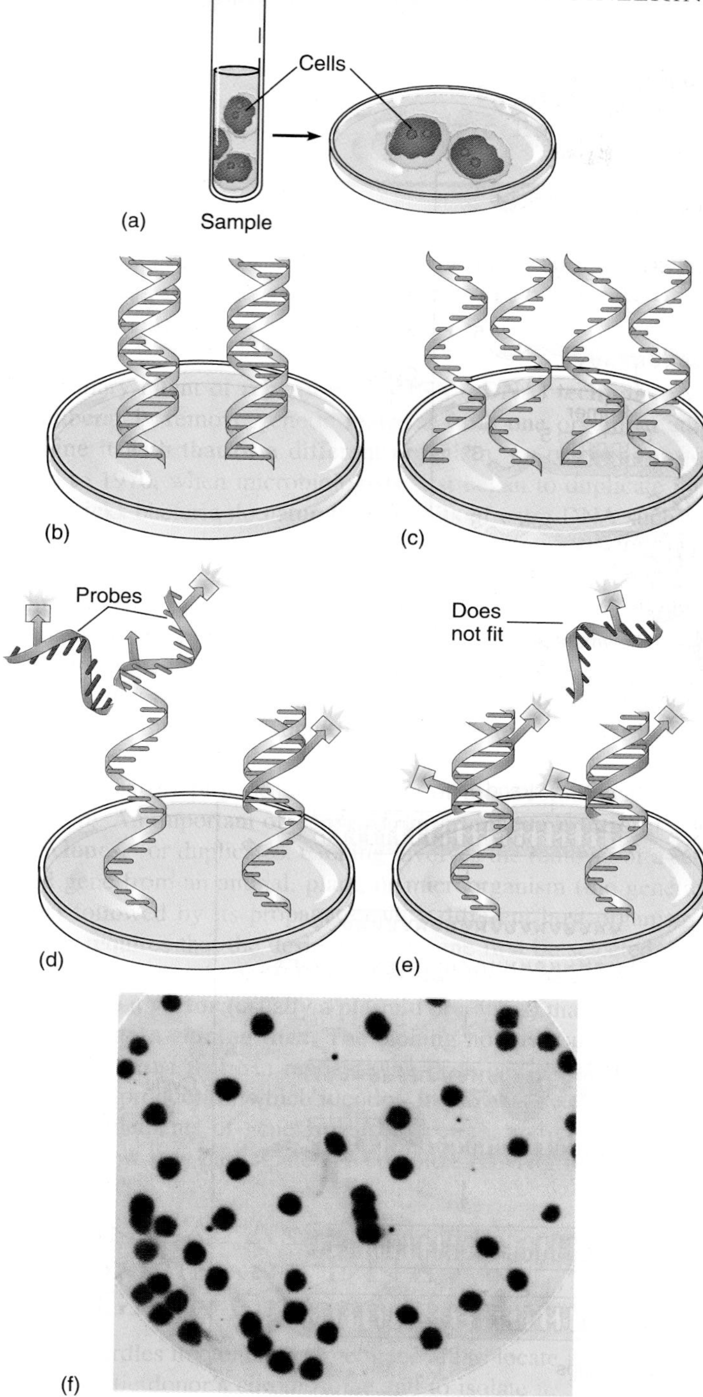

Figure 10.5
Conducting a DNA probe or dot blot technique. (Cells and DNA are greatly disproportionate to show reaction.) *(a)* A sample of the test specimen (cells) is bound on a support matrix in a Petri plate. *(b)* The cells are disrupted to release their DNA. *(c)* The DNA is denatured to separate into its two strands, which are immobilized on the matrix. *(d)* Labeled probes, consisting of a short single strand of radioactive or fluorescent nucleotides, hybridize with any segments on the unknown test strands to which they are complementary. *(e)* Unreacted probes are rinsed away, leaving only those that have attached. *(f)* Developed hybridization test demonstrates colonies with positive reactions (dark spots) to a luminescent probe. The probe is specific to the TGF-β1 gene of humans, that regulates cell growth.

tively high temperature. This necessitates the use of special **DNA polymerases** isolated from thermophilic bacteria. Examples of these unique enzymes are Taq polymerase obtained from *Thermus aquaticus* (see microfile 7.6) and Vent polymerase from *Thermococcus litoralis.* Another useful component of PCR is a machine called a thermal recycler that automatically initiates the cyclic temperature changes.

The PCR technique operates by repetitive cycling of three basic steps: denaturation, priming, and extension. The first step involves heating target DNA to 94°C to separate it into two strands. Next, the system is cooled to 65°C and primers are added in a concentration that favors binding to their complementary strand of test DNA. This reaction prepares the two DNA strands, now called **amplicons,** for synthesis. In the third phase, which proceeds at 72°C, DNA polymerase and raw materials in the form of nucleotides are added. Beginning at the free end of the primers on both strands, the polymerases extend the molecule by adding appropriate nucleotides and produce two complete strands of DNA.

It is through repetition of these same steps that DNA becomes amplified. When the DNAs formed in the first cycle are denatured, they become amplicons to be primed and extended in the second cycle. Each subsequent cycle converts the new DNAs to amplicons and doubles the number of copies. The number of cycles required to produce a million molecules is 20, but it could theoretically be carried out to 40 cycles. One significant advantage of this technique has been its natural adaptability to automation. A PCR machine can perform 20 cycles on nearly 100 samples in 2 or 3 hours.

The amplified DNA comes in a form ready for electrophoresis, probe identification, and sequencing. PCR can be adapted to analyze RNA by initially converting an RNA sample to DNA with reverse transcriptase. This cDNA can then be amplified by PCR in the usual manner. It is by such means that ribosomal RNA and messenger RNA are readied for sequencing. The polymerase chain reaction has found prominence as a powerful workhorse of molecular biology, medicine, and biotechnology. It often plays an essential role in gene mapping, the study of genetic defects and cancer, forensics, taxonomy, and evolutionary studies.

For all of its advantages, PCR has some problems. A serious concern is the introduction and amplification of nontarget DNA from the surrounding environment. Such contamination can be minimized by using equipment and rooms dedicated for DNA analysis maintained with the utmost degree of cleanliness. Problems with contaminants can also be reduced by using gene-specific primers and treating samples with special enzymes that can degrade the contaminating DNA before it is amplified.

Chapter Checkpoints

Genetic engineering utilizes a wide range of methods that physically manipulate DNA for purposes of visualization, sequencing, hybridizing, and identifying specific sequences. The tools of genetic engineering include specialized enzymes, gel electrophoresis, DNA sequencing machines, and gene probes. The polymerase chain reaction (PCR) technique amplifies small amounts of DNA into larger quantities for further analysis.

*clone (klohn) Gr. *klon,* young shoot or twig. A clone can also be defined as an organism that is genetically identical to a parent.

3. DNA can be synthesized artificially on a machine. The size of DNA that can be synthesized is limited.

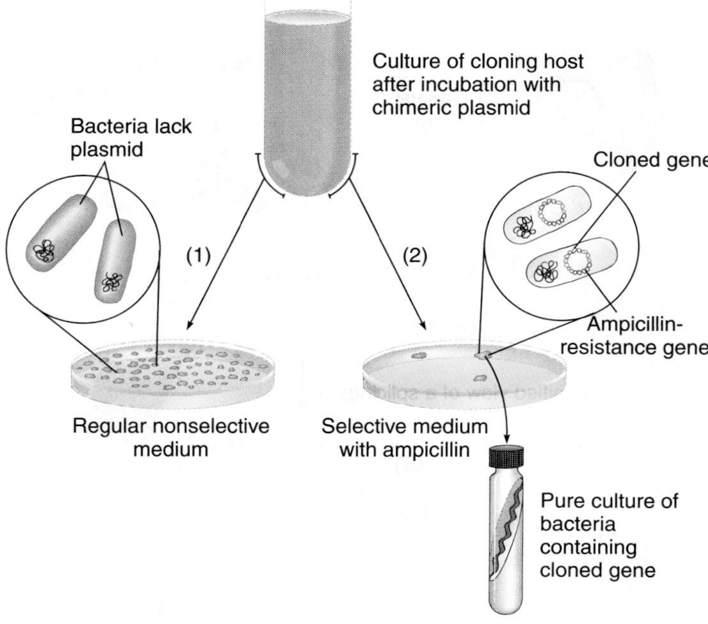

Figure 10.10
One method for screening clones of bacteria that have been transformed with the donor gene. *(1)* Plating the culture on nonselective medium will not separate the transformed cells from normal cells, which lack the plasmid. *(2)* Plating on selective medium containing ampicillin will permit only cells containing the plasmid to multiply. Colonies growing on this medium that carry the cloned gene can be used to make a culture for gene libraries, industrial production, and other processes. (Plasmids are shown disproportionate to relative size.)

Following this procedure, the chimera is introduced by transformation into the cloning host, a special laboratory strain of *E. coli* that lacks any extra plasmids that could complicate the expression of the gene. Because the recombinant plasmid enters only some of the cloning host cells, it is necessary to locate these recombinant clones. Cultures are plated out on medium containing ampicillin, and only those clones that carry the plasmid with ampicillin resistance can form colonies (figure 10.10). These recombinant colonies are selected from the plates and cultured. As the cells multiply, the plasmid is replicated along with the cell's chromosome. In a few hours of growth, there can be billions of cells, each containing the interferon gene.

The bacteria's ability to express the eucaryotic gene is ensured, because the plasmid has been modified with the necessary transcription and translation leader sequences. As the *E. coli* culture grows in a special medium under optimal conditions, it transcribes and translates the interferon gene, synthesizes the peptide, and secretes it into the growth medium. At the end of the process, the cloning cells and other chemical and microbial impurities are removed from the medium. Final processing to excise a terminal amino acid from the peptide yields the interferon product in a relatively pure form (figure 10.9). The scale of this procedure can range from test tube size to gigantic industrial vats that can manufacture thousands of gallons of product (see figure 26.40).

Although the process we have presented here produces interferon, some variation of it can be used to mass produce a variety of hormones, enzymes, and agricultural products such as pesticides.

Chapter Checkpoints

Recombinant DNA techniques combine DNA from different sources to produce microorganism "factories" that produce hormones, enzymes, and vaccines on an industrial scale.

Cloning is the process by which genes are removed from the original host and duplicated for transfer into a cloning host by means of cloning vectors.

Plasmids, bacteriophages, and cosmids are types of cloning vectors used to transfer recombinant DNA into a cloning host.

Cloning hosts are simply organisms that readily accept recombinant DNA, grow easily, and synthesize large quantities of specific gene product.

BIOCHEMICAL PRODUCTS OF RECOMBINANT DNA TECHNOLOGY

Recombinant DNA technology is used by pharmaceutical companies to manufacture medications that cannot be manufactured by any other means. Diseases such as diabetes and dwarfism, caused by the lack of an essential hormone, are treated by replacing the missing hormone. Porcine and bovine insulin were once the only forms available to treat diabetes, even though such animal products can cause allergic reactions in sensitive individuals. In contrast, dwarfism cannot be treated with animal growth hormones, so the only source of human growth hormone (HGH) was originally from the pituitaries of cadavers. At one time, not enough HGH was available to treat the thousands of children in need.

Recombinant technology changed the outcome of these and many other conditions by enabling large-scale manufacture of life-saving hormones and enzymes of human origin. Recombinant human insulin can now be prescribed for diabetics, and recombinant HGH can now be administered to children with dwarfism, a premature aging disease called progeria, or failure to grow. Other protein-based hormones, enzymes, and vaccines produced through rDNA are summarized in table 10.2. In addition to the obvious medical applications, numerous agricultural and industrial products are being developed at a rapid rate. An especially valuable aspect is that the techniques of genetic engineering can be combined with other areas of biotechnology to manufacture commercial quantities of substances such as enzymes (figure 10.11).

Nucleic acid products also have a number of medical applications. A new development in vaccine formulation involves using microbial DNA as a stimulus for the immune system. So far, ani-

TABLE 10.2

CURRENT PROTEIN PRODUCTS FROM RECOMBINANT DNA TECHNOLOGY

Immune Treatments

Interferons—peptides used to treat some types of cancer, multiple sclerosis, and viral infections such as hepatitis and genital warts
Interleukins—types of cytokines that regulate the immune function of white blood cells; used in cancer treatment
Orthoclone—an immune suppressant in transplant patients
Macrophage colony stimulating factor (GM-CSF)—used to stimulate bone marrow activity after bone marrow grafts
Tumor necrosis factor (TNF)—used to treat cancer

Hormones

Erythropoietin (EPO)—a peptide that stimulates bone marrow used to treat some forms of anemia
Tissue plasminogen activating factor (tPA)—can dissolve potentially dangerous blood clots
Hemoglobin A—form of artificial blood to be used in place of real blood for transfusions
Factor VIII—needed as replacement blood-clotting factor in type A hemophilia
Relaxin—an aid to childbirth
Proleukin—a drug to treat kidney cancer

Enzymes

rH DNase (pulmozyme)—a treatment that can break down the thick lung secretions of cystic fibrosis
Antitrypsin—replacement therapy to benefit emphysema patients
PEG-SOD—a form of superoxide dismutase that minimizes damage to brain tissue after severe trauma

Vaccines

Vaccines for hepatitis B and *Haemophilus influenzae* b meningitis
Experimental malaria and AIDS vaccines based on recombinant surface antigens

Miscellaneous

Bovine growth hormone or bovine somatotropin (BST)—given to cows to increase milk production
Apolipoprotein—to deter the development of fatty deposits in the arteries and to prevent strokes and heart attacks
Spider silk—a light, tough fabric for parachutes and bulletproof vests

mal tests using DNA vaccines for AIDS and influenza indicate that this may be a breakthrough in vaccine design. Recombinant DNA could also be used to produce DNA-based drugs for the types of gene and antisense therapy discussed in the genetic treatments section.

Chapter Checkpoints

Bioengineered hormones, enzymes, and vaccines are safer and more effective than similar substances derived from animals. Recombinant DNA drugs and vaccines are useful alternatives to traditional treatments for disease.

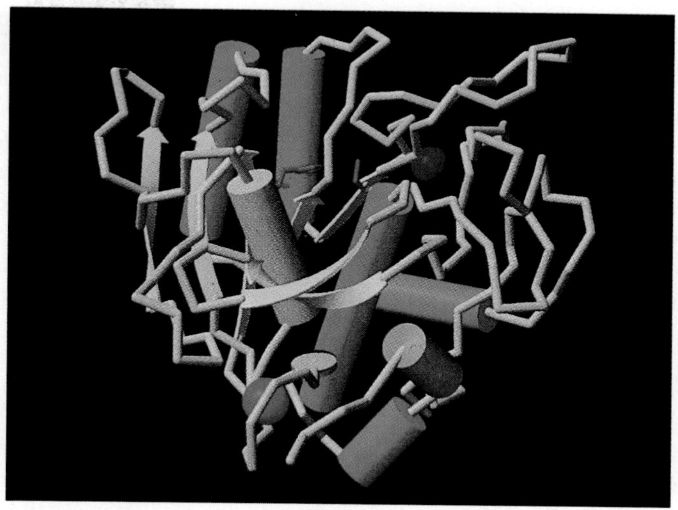

Figure 10.11
Model of a protein, subtilisin, produced by the soil bacterium *Bacillus amyloliquefaciens* that has been altered by genetic engineering. This "designed" molecule has enhanced ability to break down stains and other substances. Procter & Gamble plan to add this enzyme to laundry detergents in the near future.

RECOMBINANT ORGANISMS: HOW TO IMPROVE ON NATURE

The process of artificially introducing foreign genes into organisms is termed *transfection,* and the recombinant organisms produced in this way are called *transgenic.* Foreign genes have been inserted into a variety of microbes, plants, and animals through recombinant DNA techniques developed especially for them. Transgenic "designer" organisms are available for a variety of biotechnological applications. Because they are unique life forms that would never have otherwise occurred, they can be patented.

RECOMBINANT MICROBES: MODIFIED BACTERIA AND VIRUSES

One of the first practical applications of rDNA in agriculture was to create a genetically altered strain of the bacterium *Pseudomonas syringae.* The wild strain ordinarily contains an ice nucleation gene that promotes ice or frost formation on moist plant surfaces. Genetic alteration of the frost gene using recombinant plasmids created a different strain that could prevent ice crystals from forming. A commercial product called Frostban has been successfully applied to stop frost damage in strawberry and potato crops. A strain of *Pseudomonas fluoresces* has been engineered with the gene from a bacterium *(Bacillus thuringiensis)* that codes for an insecticide. These recombinant bacteria are released to colonize plant roots and help destroy invading insects. All releases of recombinant microbes must be approved by the Environmental Protection Agency (EPA) and are closely monitored. So far, extensive studies have shown that such microbes do not proliferate in the environment and probably pose little harm to humans.

Microbiologists have invented an unusual means of assaying drug resistance in the pathogen *Mycrobacterium tuberculosis.* They engineered a bacteriophage with the genes for luciferase, an enzyme from fireflies that causes light emission. Live *M. tuberculosis* infected with these viruses light up; dead or dying cells do not. The test is conducted by incubating cultures with the usual drugs used in treatment and observing the viability of the culture. Dying cells indicate that the drug is working, and a healthy culture indicates drug resistance. The advantage to this type of test is that results can be obtained in a few days rather than the several weeks usually required (figure 10.12).

Viruses can be genetically engineered for transfection purposes. Several viral vectors have been developed for gene therapy (see figure 10.16) and for an experimental AIDS vaccine (see figure 25.21). In an agricultural application, a strain of virus pathogenic to cabbage looper insects was released into a cabbage patch. It was engineered to spontaneously self-destruct after a certain period.

Another very significant bioengineering interest has been to create microbes to bioremediate disturbed environments. Biotechnologists have already developed and tested several types of bacteria that clean up oil spills and degrade pesticides and toxic substances (see chapter 26).

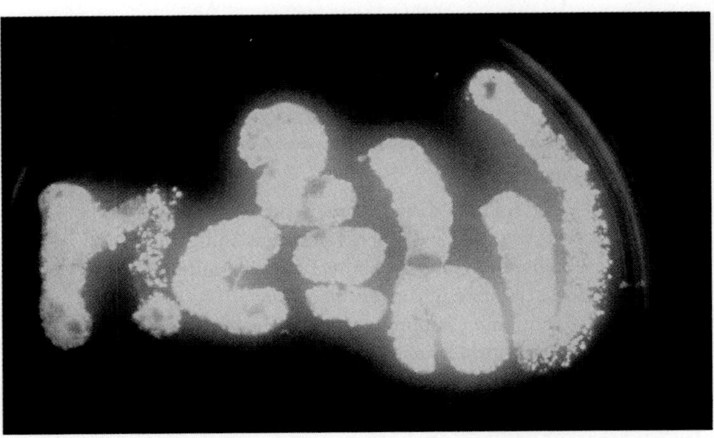

Figure 10.12
A special strain of *Mycobacterium tuberculosis* engineered to glow in the presence of light. This transgenic bacterium is the basis of a new test for determining whether a patient is infected with drug-resistant tuberculosis.

Transgenic Plants: Improving Crops and Foods

Two unusual species of bacteria in the genus *Agrobacterium* are the original genetic engineers of the plant world. These pathogens live in soil and can invade injured plant tissues. Inside the wounded tissue, the bacteria transfer a discrete DNA fragment called T-DNA into the host cell. The DNA is integrated into the plant cell's chromosome, where it directs the synthesis of nutrients to feed the bacteria and hormones to stimulate plant growth. In the case of *Agrobacterium tumefaciens,* increased growth causes development of a tumor called *crown gall disease,* a mass of undifferentiated tissue on the stem. The other species, *A. rhizogenes,* attacks the roots and transforms them into abnormally overgrown "hairy roots."

The capacity of these bacteria, especially *A. tumefaciens,* to transfect host cells can be attributed to a large plasmid termed **Ti** (tumor-inducing). This plasmid inserts into the genomes of the infected plant cells and transforms them (figure 10.13). Even after the bacteria in a tumor are dead, the plasmid genes remain in the cell nucleus and the tumor continues to grow. This plasmid is a perfect vector for inserting foreign genes into plant genomes. The procedure involves removing the Ti plasmid, inserting a previously isolated gene into it, and returning it to *Agrobacterium.* Infection of the plant by the recombinant bacteria automatically transfers the plasmid with the foreign gene into the plant cells.

The insertion of genes with a Ti plasmid works primarily to engineer important dicot plants such as potatoes, tomatoes, cotton, and grapes. Examples of agriculturally important transgenic plants are herbicide-resistant soybean and tobacco plants (figure 10.14*a);* cotton plants with a built-in insecticide; potatoes and tomatoes with resistance to bacterial pathogens; squash and melons that resist viral pathogens; pea seeds that can resist the destructive effects of weevils during storage (figure 10.14*b);* and a starchier potato that absorbs less grease during frying.

For plants that cannot be transformed with *Agrobacterium,* seed embryos have been successfully implanted with additional genes using gene guns—small "shotguns" that shoot the genes into plant embryos with tiny gold bullets. This technique was used to create wheat that resists pesticides, corn that resists a blight fungus, and viral-resistant rice (figure 10.14*c).* Agricultural officials had approved more than 30 plants for field testing through late 1997. This movement toward release of engineered plants into the environment has led to some controversy. Many plant geneticists and ecologists are seriously concerned that transgenic plants will share their genes for herbicide and virus resistance with natural plants, leading to "superweeds" that could flourish and become indestructible. The Department of Agriculture is carefully regulating all releases of transgenic plants.

Another strategy for engineering plants is to insert a strand of antisense RNA that turns off the expression of a target gene (see figure 10.17, which illustrates the antisense concept). Researchers at the University of California at Davis transformed a strain of tomatoes with antisense RNA to block the gene that controls fruit ripening. Plants with this gene produce normal fruits that lack certain enzymes necessary for ripening. As a result, the tomato does not ripen too rapidly, and it can be left on the vine to develop its natural flavor. It can be induced to ripen using ethylene gas. A California company, Calgene, is employing antisense technology to develop a rapeseed (canola) oil plant with more saturated fat, making it suitable for making margarine without artificial hydrogenation. Traditional recombinant techniques have been used to create tobacco plants that synthesize human antibodies. These "plantibodies" can be used in diagnostic tests and treating infectious diseases.

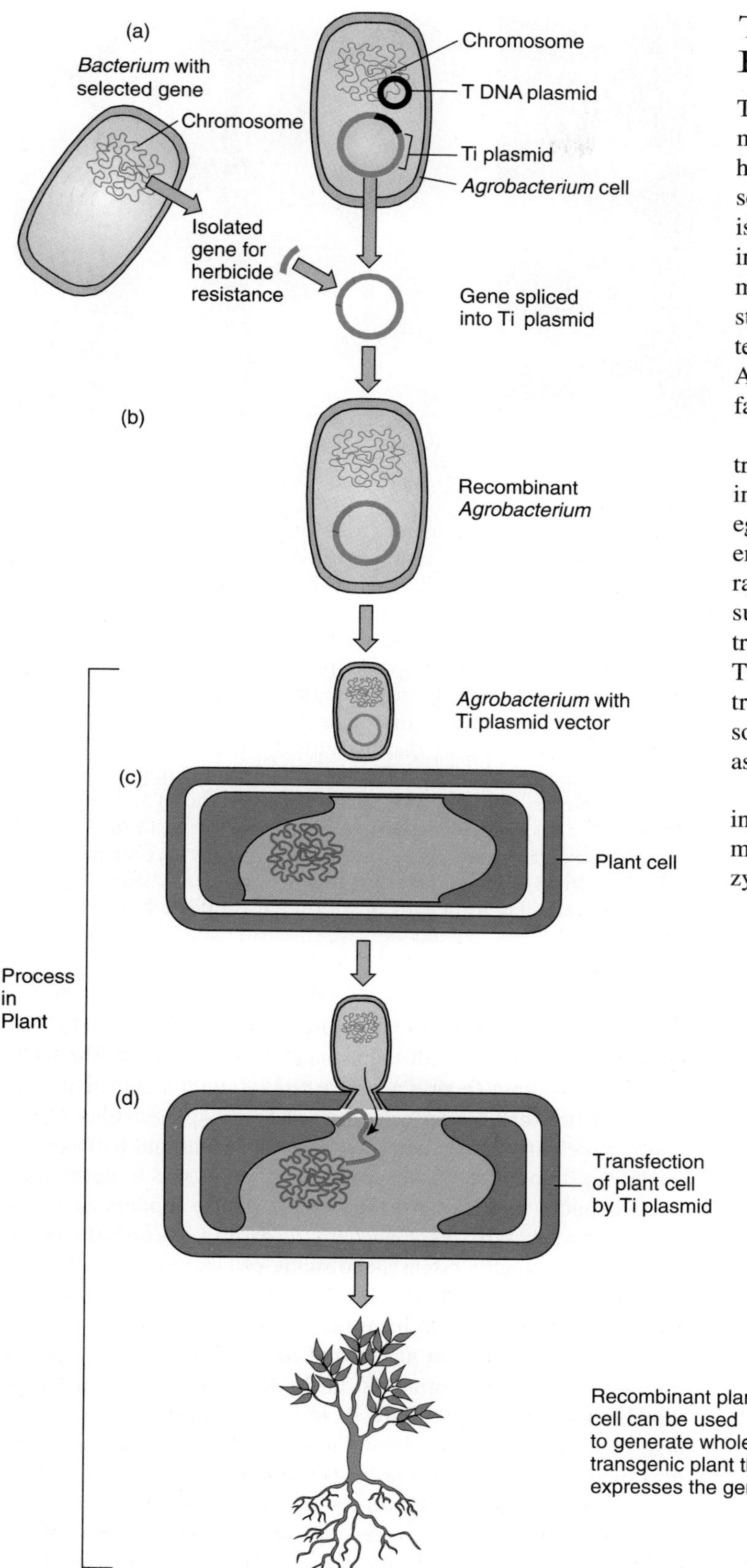

Transgenic Animals: Engineering Embryos

The inclination to be genetically engineered also exists for animals. In fact, animals are so amenable to transfection that several hundred strains of transgenic animals have been introduced by research and industry. One reason for this movement toward animals is that, unlike bacteria and yeasts, they can express human genes in organs and organ systems that are very similar to those of humans. This advantage has led to the design of animal models to study human genetic diseases and then to use these natural systems to test new genetic therapies before they are used in humans. Animals can also be engineered to become "factories" to manufacture human proteins and other products.

The most effective way to insert genes into animals is by transfecting the fertilized egg or early embryo (gene line engineering). Foreign genes can be delivered into an egg by pulsing the egg with high voltage. This electric stimulus enables the gene to enter through small pores in the cell membrane. The technique is rather difficult to control, and only a small number of embryos survive. The success rate does not have to be high, because a transgenic animal will usually pass the genes on to its offspring. The dominant animals in the genetic engineer's laboratory are transgenic forms of mice. Mice are so easy to engineer that in some ways they have become as prominent in genetic engineering as *E. coli.*

In some early experiments, fertilized mouse embryos were injected with human genes for growth hormone, producing supermice twice the size of normal mice (figure 10.15*a*). Other mouse zygotes were recently implanted with the genes for luminescence.

Figure 10.13

Bioengineering of plants using a natural tumor-producing bacterium called *Agrobacterium tumifaciens.* (*a*) The large plasmid (Ti) of this bacterium can be used as a cloning vector for foreign genes that code for herbicide or disease resistance. (*b*) The recombinant plasmids are taken up by the *Agrobacterium* cells, which multiply and copy the foreign gene. (*c*) Genetically engineered *Agrobacterium* is inoculated into a culture of target plant cells and infects the cells. (*d*) Fusion of the bacterium with the plant cell wall permits entrance of the Ti plasmid and incorporation of the herbicide gene into the plant chromosome. Mature plants can be grown from single cells, and these transgenic plants will express the new gene.

(a)

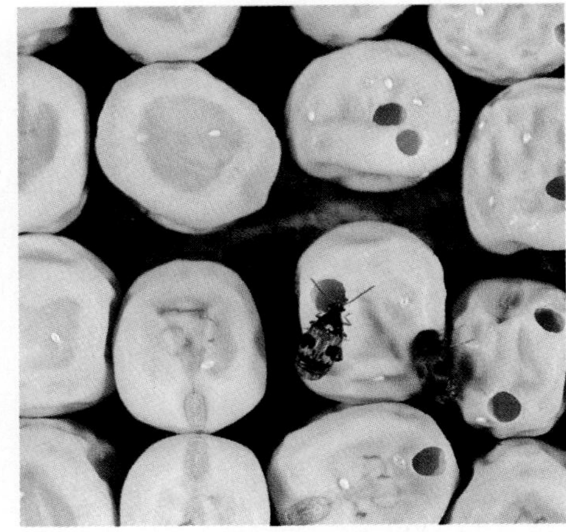

(b)

(c)

Figure 10.14

(*a*) Herbicide protection in transgenic (transformed) plants. Tobacco plants in the upper row have been transformed with a gene that provides protection against buctril, a systemic herbicide. Plants in the lower row are normal and not transformed. Both groups were sprayed with buctril and allowed to sit for 6 days. (The control plants at the beginning of each row were sprayed with a blank mixture lacking buctril.) Note the relative survival in the transgenic and normal plants. (*b*) Pest protection for seeds. Pea plants on the left were engineered with a gene that prevents digestion of the seed starch. This gene keeps tiny insects called weevils from feeding on the seeds. Plants on the right are normal controls with holes created by weevil pests. (*c*) Virus resistance in rice. A new form of rice plant was transfected with genes for resistance to the rice stripe virus (right). The virus is a significant pathogen that severely stunts young plants and reduces crop yields (plant on left is infected control).

The newborn mice that glow in the dark are expected to grow into adults with luminescent mucous membranes. These mice will be used to trace development and gene expression (figure 10.15*b*). Japanese researchers have successfully transferred whole human chromosomes into embryonic mice. They plan to use these mice to study gene control, protein synthesis, and genetic defects involving extra chromosomes.

The so-called knock-out mouse has become a standard way of producing animals with tailor-made genetic defects. The technique involves transfecting a mouse embryo with a defective gene and cross-breeding the progeny through several generations. Eventually, mice will be born with two defective genes and will express the disease. One of the first successful attempts was to develop an animal model of cystic fibrosis. Other mouse models exist for hardening of the arteries, Gaucher's disease (a lysosomal storage disease), Alzheimer's disease, and sickle-cell anemia.

Animal husbandry has been quick to adopt some of these techniques. Pig embryos have been implanted with the bovine growth hormone gene in an attempt to increase the pig's growth rate and decrease fat formation in the meat. The transgenic pigs did indeed have less fat (figure 10.15*c*), but, unfortunately, the foreign gene appeared to contribute to severe health problems in the pigs such as arthritis and heart and kidney disease.

The pharmaceutical industry has already engineered a number of animals to manufacture commercial quantities of human proteins. A strain of transgenic pig can synthesize certain clotting factors and human hemoglobin as a blood substitute. Also, several companies have successfully harnessed the synthetic abilities of mammary glands. Sheep have been designed to give off alpha antitrypsin (an emphysema treatment) and clotting factors in their

(a)

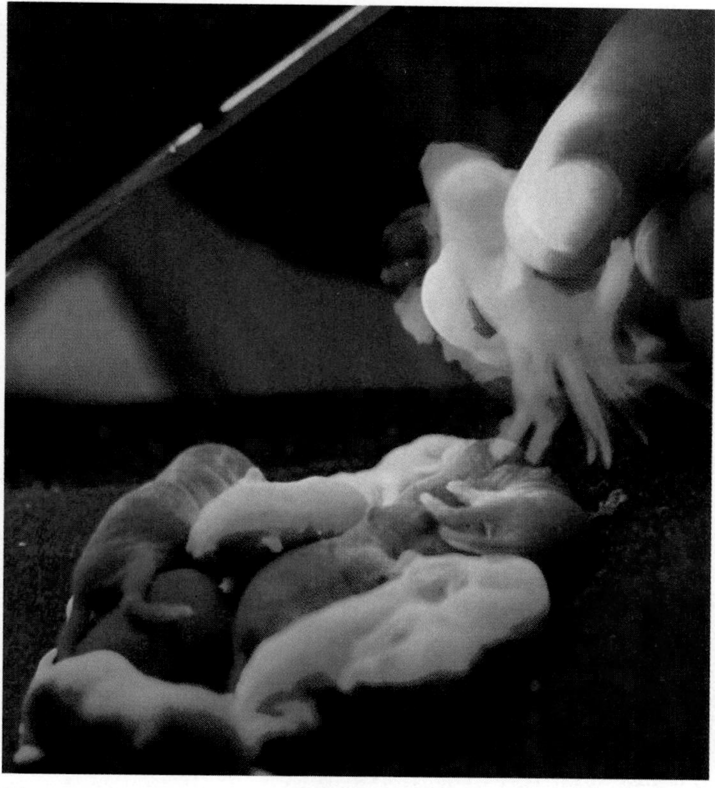
(b)

(c)

Figure 10.15
(*a*) These two mice are genetically the same strain, except that the one on the right has been transfected with the human growth hormone gene (shown in its circular plasma vector). The new gene spurred an increase in this mouse's growth rate until it dwarfed its normal-sized companion. (*b*) Radiant rodents. Japanese researchers have successfully transfected mouse embryos with bioluminescence genes taken from jellyfish. It is hoped that this technology can be used for markers or tracers for studying cell and organ development and diseases such as cancer. The mice glow in the presence of UV radiation and will be monitored to determine if this trait is passed to offspring. (*c*) Comparison of the meat from a pig into which bovine growth hormone genes were introduced (left) and a normal pig (right). A section taken at the same rib area of the two pigs indicates significantly reduced fat content in the transgenic pig. The two animals weighed about the same.

milk, and a strain of transgenic goats produce cystic fibrosis membrane protein in their milk. Transgenic cows have been cloned and genetically customized to synthesize human albumin. Harvard researchers have developed techniques for cultivating the tissues of sheep, rats, and rabbits in special molds to produce organ replacements. The cloned body parts—hearts, kidneys, and bladder—functioned well when transplanted. These techniques will be tested in humans to correct birth defects. Plans for the future include using pigs as living factories to grow tissues and organs for transplantation.

Chapter Checkpoints

Transfection is the process by which foreign genes are introduced into organisms. The transfected organism is termed *transgenic.* Transgenic microorganisms are genetically designed for medical diagnosis, crop improvement, pest reduction, and bioremediation. Transgenic animals are genetically designed to model genetic therapies, improve meat yield, or synthesize specific biological products.

GENETIC TREATMENTS: INTRODUCING DNA INTO THE BODY

GENE THERAPY

One result of the research with transgenic animals is **gene therapy,** a technique for replacing a faulty gene with a normal one in people with fatal or extremely debilitating genetic diseases. The inherent benefit of this therapy is to permanently cure the physiological dysfunction by repairing the genetic defect. There are two strategies for this therapy. In *ex vivo* therapy, the normal gene is cloned in vectors such as retroviruses (mouse leukemia virus) or adenoviruses that are infectious but relatively harmless. Tissues removed from the patient are incubated with these genetically modified viruses to transfect them with the normal gene. The transfected cells are then reintroduced into the patient's body by transfusion (figure 10.16). In contrast, the *in vivo* type of therapy skips the intermediate step of incubating excised patient tissue. Instead, the naked DNA or a virus vector is directly introduced into the patient's tissues.

So far, therapy has been hampered by a number of difficulties. The systems for delivering the gene into target cells are not completely successful, and, the rate of productive transfection by the viruses is very low (10% or less of the target cells). Some experts fear that the virus could insert the DNA into a site that turns on an oncogene or could activate other dormant viruses. Alternative delivery methods are being tested to deliver the product more precisely. These include gene guns similar to those previously described for plants as well as liposomes and dendrimers. Liposomes are tiny spheres with an exterior lipid bilayer and aqueous center that can carry genes across cell membranes. Dendrimers are very larger, branched hydrocarbon molecules that have proved very effective at transferring genes into cells.

Other problems have to do with incorporation, retention, and expression of the DNA by the cell. In some cells, the DNA appears to be picked up and held free in the nucleus. The transfected cells can even translate the gene for a limited time, but they may not duplicate or pass it on during cell division, and these cells often have a limited lifespan. One solution tailored for diseases of blood cells is to genetically engineer the long-lived stem cells in bone marrow. If enough stem cells take up and retain the replacement gene, it will be inherited by new blood cells for the life of the patient.

The first gene therapy experiment in humans was initiated in 1990 by researchers at the National Institutes of Health. The subject was a 4-year-old girl suffering from a severe immunodeficiency disease caused by the lack of the enzyme adenosine deaminase (ADA). She was transfused with her own blood cells into which the functional ADA gene had been inserted by viruses. Later, another child was also given the same type of therapy. Both children showed remarkable improvement and continue to be healthy but are not yet cured (chapter-opening figure). This treatment is not permanent and must be repeated, but it does reduce the severity of the disease. In trials for cystic fibrosis (CF) therapy, adenoviruses carrying a normal CF gene are delivered directly into the nose and lungs. So far, the results of this therapy have been less successful.

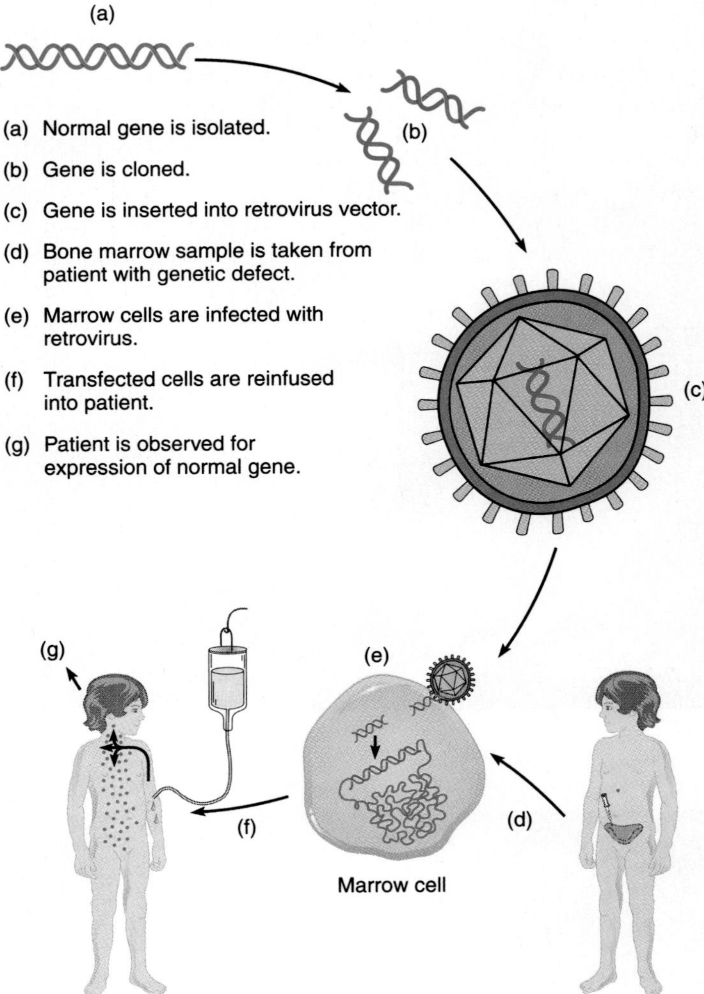

(a)
(a) Normal gene is isolated.
(b) Gene is cloned.
(c) Gene is inserted into retrovirus vector.
(d) Bone marrow sample is taken from patient with genetic defect.
(e) Marrow cells are infected with retrovirus.
(f) Transfected cells are reinfused into patient.
(g) Patient is observed for expression of normal gene.
(b)
(c)
(g)
(e)
(f)
(d)
Marrow cell

Figure 10.16
Protocol for the *ex vivo* type of gene therapy in humans. *(a)* A normal human gene is identified and isolated. *(b)* The gene is cloned. *(c)* The gene is inserted into an appropriate vector such as a retrovirus. *(d)* Harvested bone marrow sample from a patient with a genetic defect is *(e)* infected with retrovirus. *(f)* The transfected cells are then dripped back into the patient. *(g)* The patient is monitored for improved function due to replacement of missing protein (enzyme or other factor).

Other single-gene defects such as hemophilia and sickle-cell anemia may be treated with similar techniques. Clinical trials are under way or are being planned for several hundred patients with genetic diseases. Some of these tests are summarized in table 10.3. Medical researchers have been successful in infusing growth factor genes into patients with vascular disease. This therapy corrected the damage by growing new vasculature and improving circulation. Another new technique introduced by doctors at the National Eye Institute is to use a small skin graft to deliver a gene locally into tissue. If it succeeds, certain eye defects could be corrected without surgery on the eye.

TABLE 10.3

SAMPLE OF HUMAN GENETIC DEFECTS TARGETED FOR GENE THERAPY

Disease	Nature of Defect	Gene Replacement Strategy
Adenosine deaminase (ADA) deficiency	Severe immunodeficiency due to destruction of lymphocytes by toxic metabolic product.	ADA gene inserted into bone marrow white blood cells using virus vector or liposomes.
Hypercholesterolemia	Lack of receptors for low-density lipoprotein causes build-up of cholesterol in blood and liver disease.	Liver cells transfected with correct LDL receptor gene.
Cystic fibrosis	Lack of necessary protein for conducting chloride ions across membranes causes build-up of mucus and secretions in lungs, pancreas, other organs.	CFTR gene placed in adenovirus or liposomes and inoculated into nasal cavity or lungs.
Sickle-cell anemia beta-thalassemia	Lack of correct gene to make the normal B globin protein of hemoglobin.	Hemoglobin A (HbA) gene placed in bone marrow cells by vector.
Hemophilia B	Lack of gene to produce factor VIII, needed for blood clotting.	Normal gene for factor VIII incorporated into marrow cells.
Gaucher's disease	Deficiency in an enzyme that is needed to metabolize glucocerebroside leads to its build-up in lysosomes.	Insertion of normal glucocerebrosidase gene into marrow cells.
Muscular dystrophy	Lack of gene for producing dystrophin necessary for normal muscle development.	Dystrophin gene delivered into muscle tissue by one of several methods.
AIDS	Infection by HIV destroys important immune cells (T cells) and causes severe loss of immune function.	Fibroblasts are infected with gene that stimulates an immune response against HIV; other test uses ribozyme enzyme to block virus activity.
Malignant melanoma	Severest form of skin cancer that does not respond well to traditional therapy.	Insertion of B7 gene into tumor cells to boost the natural white blood cell response against the cancer.
Acute myelogenous leukemia	Blood disease caused by overexpression of a gene.	Genetic modification of cells by replacement of defective gene; also addition of antisense DNA to block the actions of cancer genes.

The ultimate sort of gene therapy is gene line therapy, in which genes are inserted into an egg, sperm, or early embryo. Although tests in animals point to its possible effectiveness, its use on humans is far too controversial and has been banned by the U.S. government for the time being.

ANTISENSE AND TRIPLEX DNA TECHNOLOGY: GENETIC MEDICINES

Up to now we have considered the use of recombinant hormones and other proteins to replace a missing or dysfunctional protein and gene therapy to restore a missing or defective gene. Biomedical researchers have been exploring yet a third type of therapy with genetic drugs called **antisense** and **triplex agents.** With this approach, oligonucleotides are delivered into cells to block transcription or translation of a specific DNA site, thereby bypassing the problem gene and protein altogether.

Antisense DNA: Targeting Messenger RNA

Note that the term *antisense* as used in molecular genetics does *not* mean "no sense." It is used to describe any nucleic acid strand with a base sequence that is complementary to the sense or translatable strand. For example, DNA contains a sense strand that is transcribed and a matching antisense strand that is not usually transcribed. We can also apply this terminology to RNA. Messenger RNA is considered the translatable sense strand, and a strand with the matching sequence of nucleotides would be the antisense strand. To illustrate: If an mRNA sequence read AUGCGAGAC, then an antisense RNA strand for it would read UACGCUCUG. In practice, most antisense agents are made of DNA because the technology for large-scale RNA synthesis is less feasible. The therapy takes advantage of the readiness with which an antisense DNA can hybridize with RNA. It is worth noting that the concept of antisense nucleic acids did not originate with genetic engineers. It appears to be one mechanism by which organisms naturally control gene expression.

When an adequate dose of antisense DNA is delivered across the cell membrane into the cytoplasm and nucleus, it binds to specific sites on any mRNAs that are the targets of therapy. As a result, the reading of that mRNA transcript on ribosomes will be blocked and the gene product will not be synthesized (figure 10.17*a*).

From the beginning of its development, antisense technology was hailed as a revolution in clinical pharmacology. On the positive side, animal experiments show that antisense drugs are relatively

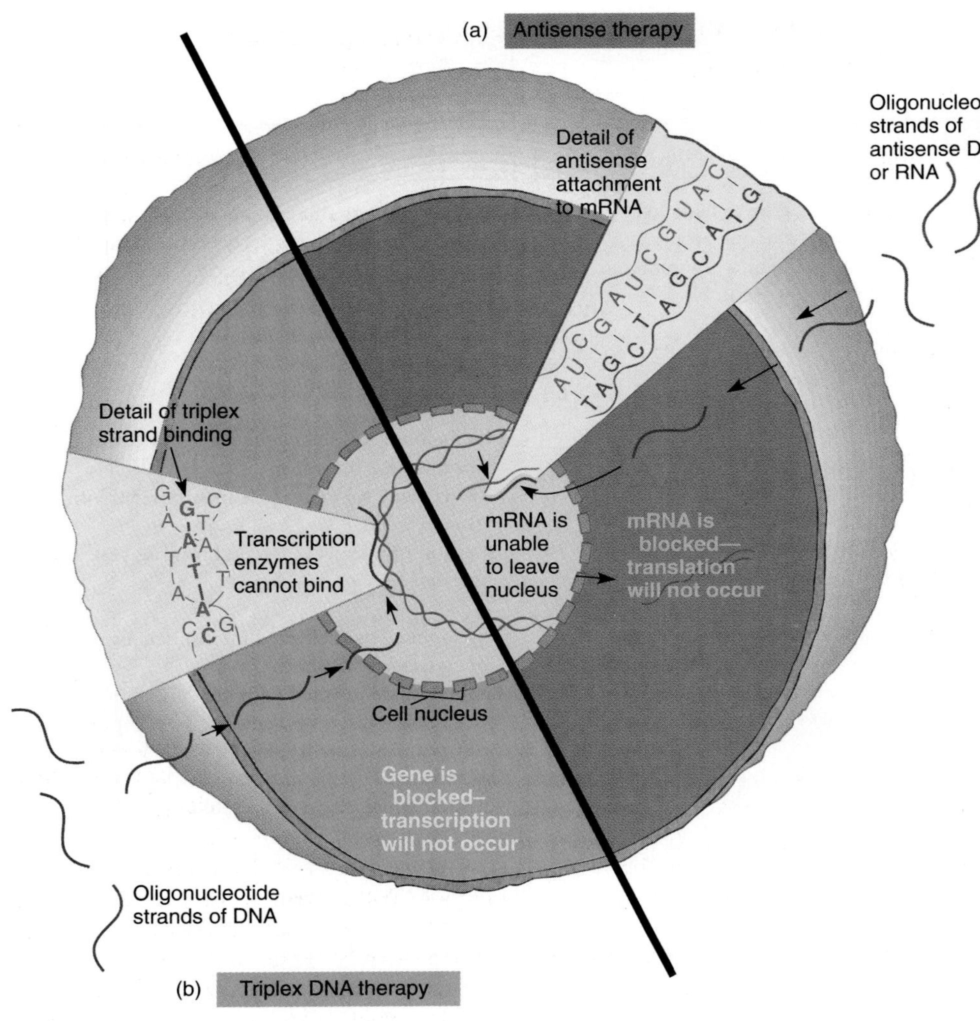

Figure 10.17
Mechanisms of antisense DNA and triplex DNA. These genetic medicines are designed to prevent the expression of an undesirable gene or virus. Both use oligonucleotides of known sequence that can bind with genetic targets that are the cause of disease. The site of action is in the nucleus. (*a*) When antisense strands enter the nucleus, they bind to complementary areas on an existing mRNA molecule. Such bound RNA will be unavailable for translation on the cell's ribosomes and no protein will be made. (*b*) Triplex DNA strands attach by base-pairing to regulatory sites and other sites on the chromosome itself. This action masks the points of attachment for enzymes of transcription, and mRNA cannot be made. Without mRNA, there will be no translation or protein formed.

nontoxic and safe to use, and cell culture tests indicate that they can retard synthesis of an undesirable protein. But, as with most new therapies, there are also some stumbling blocks. One is the drugs' high cost and the need to administer them throughout the course of a person's life (of course, this is one feature that attracted drug companies in the first place). Another drawback has been the inability to get a significant dose of the agent into cells. Whether the DNA could become a permanent part of the genome of cells and lead to cancer and other diseases is another concern.

It is also clear that this therapy will not work for many genetic-based diseases such as cystic fibrosis and sickle-cell anemia that are caused by the lack of a functioning gene. However, it promises to be most effective in preventing diseases caused by undesirable gene expression and by viruses. Diseases that might be amenable to this therapy are cancers, autoimmune diseases such as multiple sclerosis, infections such as hepatitis B and AIDS, and Alzheimer's disease. Human clinical trials are under way for blocking the replication of human papillomavirus (the cause of genital warts and cervical cancer) and the AIDS virus. Trials of acute and chronic myelogenous leukemias, caused by the abnormal expression of a cancer gene, are also being carried out (table 10.3). The alteration of plant genes with antisense agents has already proved itself in agricultural applications.

Triplex DNA: Three's a Crowd

Although the usual structure of DNA is a helix composed of two strands, researchers have known for years that DNA can exist in other naturally and artificially induced formats. One of these unusual variations, **triplex DNA,** is a triple helix formed when a third strand of DNA inserts into the major groove of the molecule. The inserted strand forms hydrogen bonds with the purine bases on one of the adjacent strands (figure 10.17*b*). It seems logical that the template of DNA that has an extra strand wedged into its structure can be relatively inaccessible to normal transcription. Indeed, it is theorized that this is yet another natural means for cells to control their genetic expression.

The potential therapeutic applications of triplex DNA are still in their infancy. So far, oligonucleotides have been synthesized to form triplex DNA that interacts with regulatory sequences in genes. It can interrupt the action of transcription factors and prevent RNA polymerase from forming mRNA transcripts. Studies performed on cell cultures revealed that the genes coding for oncogenes, viruses, and the receptor for interleukin-2 (an immune stimulant) could be blocked with triplex DNA. In the clinical trials to come, this therapy will no doubt be plagued with some of the same problems mentioned for the antisense approach.

Chapter Checkpoints

Gene therapy is the replacement of faulty host genes with functional genes by use of cloning vectors such as viruses, liposomes, and dendrimers. This type of transfection can be used to treat genetic disorders and acquired disease.

Antisense DNA and triplex DNA are used to block expression of undesirable host genes in plants and animals as well as those of intracellular parasites.

GENETIC MAPS, FINGERPRINTS, AND FAMILY TREES

DNA technology has opened the way to substantial discoveries in basic genetics. It has provided a means of deriving a complete genetic knowledge of any organism or virus. Such knowledge will in turn lead to understanding of diverse biological phenomena, including development, virulence, disease, and evolution. Its influence is being felt in numerous other fields such as forensics (medical law), genetic pedigreeing, anthropology, neurobiology, and archeology.

"We finished the genome map, now we can't figure out how to fold it!"

Cartoon by John Chase. Reprinted by permission.

GENOME MAPPING AND SCREENING: AN ATLAS OF THE GENOME

Gene mapping is the process of delineating the relative order and positions of genes on the chromosomes of living things. The least detailed maps are chromosomal ones that show approximate locations of genes. More detailed *physical maps* indicate the order of genes to some degree of specificity. *Sequence maps,* which require tedious analysis of DNA, identify the exact order of base pairs making up a gene or other chromosomal site. The first genomes to be sequenced were viruses and bacteria. By 1998, 190 viruses, 55 bacteria, and several eucaryotes have had their DNA sequenced.

In 1986, the federal government began an ambitious effort to map the human genome at an estimated cost of $3 billion. This gigantic task, called the Human Genome Project, is expected to take 20 years. Within the first 10 years, a large number of genes were already mapped to a given chromosome; approximately 10,000 genes of the 100,000 genes in the human library were correlated with a specific location, and about 3% of the human genome has been completely sequenced (microfile 10.2). When finished, the full map will be like a "human gene dictionary" or "encyclopedia of humans" composed of 23 volumes, one for each chromosome pair. This text would be tedious reading for all but the most powerful computers. Interestingly, only a small fraction of the genome actually codes for real phenotypic genes, and the rest is something quite remarkable (microfile 10.3).

The outcome of the genome project promises to forever change the human condition. With a map of the human genome, the genetic background of the 5,000 genetic diseases could eventually be disclosed, clearing the way for extensive genetic screening. It would not only enable families to know in advance their risks for certain diseases but would also guide decisions on correcting or alleviating them through gene and other forms of therapy. Map sites have been discovered for approximately 450 of these diseases (partial list in table 10.4). As new genes are discovered on a weekly basis, it becomes increasingly possible to be tested for a wide variety of genetic diseases. This testing goes hand in hand with methods of genetic analysis described in the next section.

DNA FINGERPRINTING: A UNIQUE PICTURE OF A GENOME

Although DNA is based on a structure of nucleotides, the exact way these nucleotides are combined is unique for each organism. It is now possible to apply DNA technology in a manner that emphasizes these differences and arrays the entire genome in a pattern for comparison. **DNA fingerprinting** (also called DNA typing or profiling) is best known as a tool of forensic science (microfile 10.4). But it can also identify hereditary relationships, the risk for hereditary diseases, and the patterns of inheritance in conditions such as Huntington's disease, cystic fibrosis, and Alzheimer's disease. One application of this technique is to identify the risk for genetic disease as represented by Alzheimer's disease (figure 10.18). Test DNA from various family members is

MICROFILE 10.2 THE HUMAN HIGHWAY: MAPPING THE GENOME

Life is so short . . . and DNA is so long.

Comment of F. E. Blattner, head of the *E. coli* Genome Project, upon finally finishing the long-awaited sequencing of this bacterium in 1997

Imagine trying to develop a detailed map from New York to Los Angeles one inch at a time, with few known landmarks to use as references. This is something akin to what molecular geneticists are doing in mapping the human genome. When it is finally finished, its sequence of 100,000 genes and 3.5 billion base pairs will fill a million pages of text. How can this feat be accomplished? If it had continued at the pace of the first sequencing attempts, it would have taken 200 years. Now it is being accomplished cooperatively at 10 major U.S. centers and hundreds of smaller public and private sites and includes thousands of individuals in 50 countries.

The key to developing an accurate map is analogous to first setting up markers along the DNA that can be used as points of reference (see figure). Developing these resolution maps involves a process called "walking a chromosome." For this, the investigators clone large segments (about 125 kb) of DNA in BAC or YAC vectors (already discussed) and screen these clones for equally spaced unique sequences that are found only in human DNA. These markers are called sequence tagged sites, or STSs, and expressed sequence tags, or ESTs. Known tagged sequences at opposite ends of a cloned site can be compared with other clones to find areas of overlap. This process will require at least 30,000 of these markers to be laid down in accurate order to serve as landmarks. The result is something called a "contig (for contiguous) map" that gives a fairly precise positioning of the cloned fragments of DNA and, eventually, of exact gene locations.

The contig map can pave the way for the next step, which will be to actually sequence the DNA in the fragments to form a structural map. This will be an even more technology-intensive undertaking. The current sequencing is being carried out in large-scale centers using giant sequencing machines that collectively can determine the order of millions of base pairs in a few days. This rate must be maintained in order to meet the completion date sometime in the year 2005.

Analyzing and storing this massive amount of new data requires specialized computers. Because it is information of both a biological and mathematical content, a whole new discipline has grown up around managing these data: *bioinformatics.* The job of bioinformaticians will be to analyze and classify genes, determine protein sequences, and ultimately determine the function of the genes. It is possible that by 2010, we can have a detailed understanding of normal and pathologic phenotypes. In a sense, the human genome map (genotype) will become a phenotype.

If you are interested in discovering more about this project, there are dozens of Web sites on the Internet, for example, for Human genome centers go to ⟨http://www.ornl.gov/hgmis/CENTERS.HTML⟩, or type Genbank into a search engine such as Alta Vista.

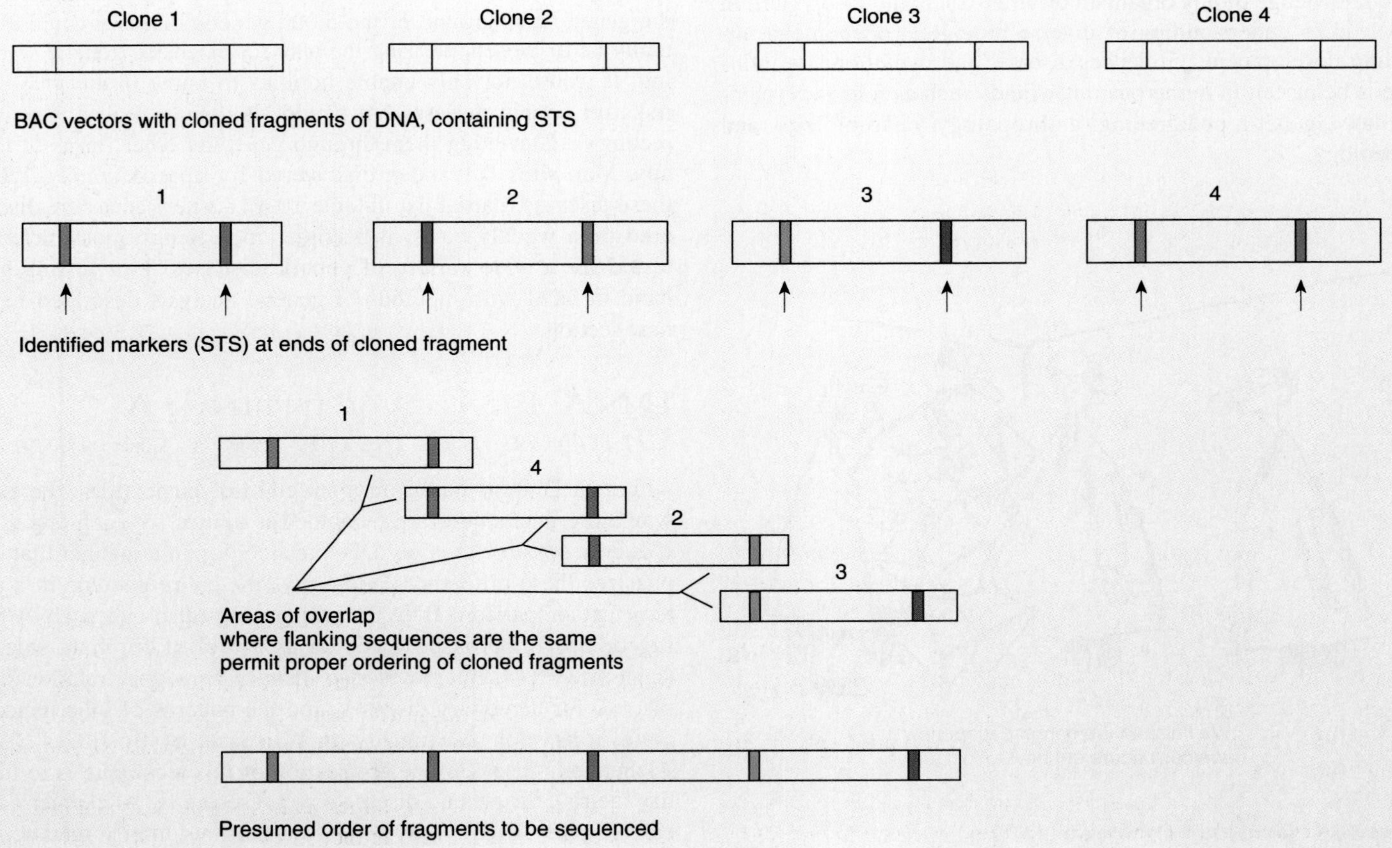

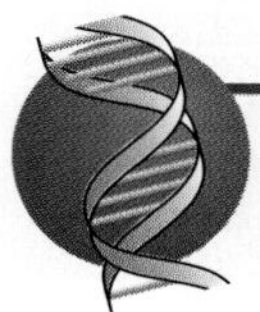

MICROFILE 10.3 GENETIC JUNK SHOP YIELDS TREASURES

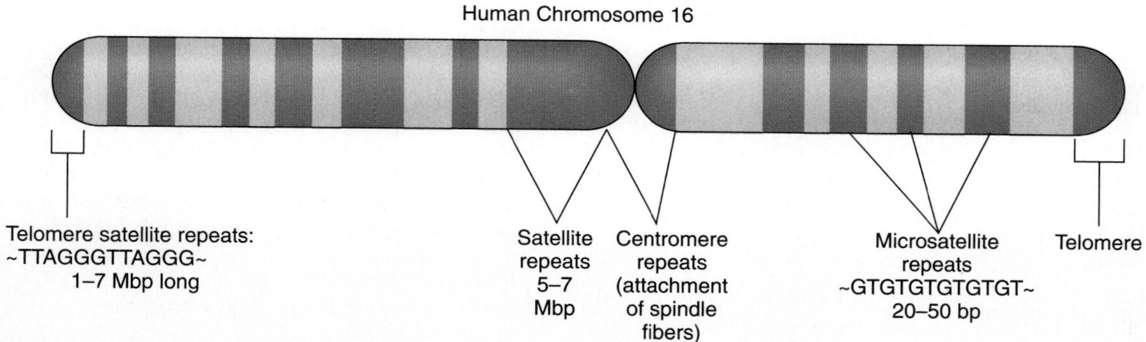

Examples of areas on the chromosome that are devoted to support DNA, most of which consist of highly repetitive nucleotides.

In addition to shedding light on human genetic diseases and evolution, human genome mapping has provided some genetic surprises. Perhaps the most startling discovery was that the majority of chromosomal DNA is not made up of genes coding for cell products such as protein or tRNA. Thus, when the total number of genes is estimated to be 100,000, this figure accounts for only about 3% of the total amount of DNA in chromosomes! The other 97% was initially considered functionless and was given the nickname "junk DNA." Many molecular geneticists were perplexed by this idea and argued that it did not make biological sense to have most of the genome composed of useless DNA. It turns out that they are correct; the so-called junk DNA actually carries important information. Initial analyses show that this category of DNA may function as chromosome stabilizers, gene regulators, attachment sites for mitotic fibers, and in ribosome assembly.

One subgroup of support DNA* is composed of short sections of highly repetitive DNA, meaning that the same nucleotide sequence is repeated 10 to several thousand times. Three identifiable types are: (1) *Satellite DNA* consisting of 1–7 million base pairs located mainly at the ends and centers of chromosomes. It appears to maintain the integrity of chromosomes. (2) *Minisatellite DNA* is a series of 200–3,000 base pair repeats scattered throughout the chromosomes. One type, called variable number tandem repeats (VNTRs), is an important type of restriction fragment used in mapping genes and tracing inheritance. (3) *Microsatellites* are even tinier (20–50 bp) repetitive blocks of DNA. Although the exact function of the last two types has yet to be discovered, it has been shown that mutations in them result in cancers of various types.

Another sort of DNA "filler" familiar from chapter 9 is *introns.* These are short segments interspersed between the exon regions of genes in most eucaryotic and a few procaryotic cells that do not code for protein. They are transcribed along with the exons but are then snipped out of the mRNA transcript before translation. We now know that introns do in fact serve some purposes. For example, small nucleolar RNAs are introns that appear to play a role in ribosome assembly. Other introns probably belong to a newly discovered class of RNAs that regulate the expression of genes.

Some other mysterious types of DNA are SINEs (short interspersed elements) and LINEs (long interspersed elements). SINEs are 300 bp sections that can be repeated up to 500,000 times. LINEs are longer (7,000 bp), highly repetitive segments. Both interspersed elements are considered to be nonfunctional blocks of DNA perhaps left over from infection by retroviruses or other intranuclear viruses. They behave as transposons, jumping around in the genome and causing disruptions and disease if they insert in the middle of a gene. A cancer of nervous tissue called neurofibromatosis is caused by a SINE. One type of hemophilia has been associated with insertion of a LINE that disrupts the factor VIII gene.

**Although the term* junk *DNA is still being used, we are awaiting the invention of a better term.*

clipped by restriction endonucleases, electrophoresed, and reacted with gene-specific probes that will show exactly which restriction sites (genetic markers) the person possesses. The genetic markers are not the actual gene that causes the disease, but they are located nearby and are inherited in exactly the same pattern. The genetic family tree for many diseases can be worked out by comparing fingerprint patterns in a family that is known to carry and express the gene. Futurists predict that a DNA fingerprint will be made for every child at birth and kept on file, much as footprints and regular fingerprints have been used until now.

This same method of analysis can be used for clarifying human paternity and maternity, the pedigree (genetic ancestry) of domestic animals, and the genetic diversity of animals bred in zoos. In conjunction with PCR, it enables analysis of even small fragments of DNA, and from this ability highly novel applications have sprung.

In 1995 the Soviet government tested the remains of what they believed were the family of Tsar Nicolas, executed in 1918. The samples were compared with the DNA of known family members. The analysis confirmed that the bodies were indeed those of the Romanovs. The same technique also proved that a woman claiming to be their missing daughter, Anastasia, could not have been related—putting that legend finally to rest.

The family of Jesse James had the body they claimed to be his exhumed to prove it was indeed the outlaw. DNA results confirmed that it was.

Table 10.4

Selected Gene Loci Involved in Human Disease Mapped to Specific Sites on Chromosomes*

Gene/Defect	Chromosome of Location	Effect of Disease
Huntington's disease	4	Degenerative neurological syndrome with onset of brain atrophy and muscular loss in midlife.
Marfan's syndrome	15	Defect in formation of connective tissue affects development of skeleton, heart, and eyes.
Retinitis pigmentosa	X	Missing or damaged gene causes degeneration of retinal light receptors and blindness.
Tay-Sachs disease	5, 15	Dysfunctional gene for an enzyme causes build-up of toxic metabolic products, leading to mental retardation, seizures, blindness in infancy.
Duchenne's muscular dystrophy	X	Defect causes gradual deterioration in muscle fibers; the major form of childhood dystrophy.
Cystic fibrosis	7	Defect in gene for electrolyte transport causes malfunction in secretion affecting lungs and pancreas; most common known lethal mutation in humans.
Sickle-cell anemia	11	Gene for B globulin has single-base mutation that causes hemoglobin to polymerize and deform red blood cells.
Amyotrophic lateral sclerosis	21, 2	Defective superoxide dismutase gene progressively destroys motor neurons, leading to muscle paralysis.
Fragile X syndrome	X	Abnormal gene in the X chromosome causes mental impairment; most common in males.
Familial hypercholesterolemia	19	Mutated gene for the low-density lipoprotein receptor causes build-up of cholesterol and rapid onset of artery disease.
Hemophilia A and B	X	Type A due to a genetic mutation in clotting factor VIII; type B due to mutation in gene for factor IX; both types suffer from mild to severe clotting deficiencies and bleeding.
Type I diabetes	6, 11	Defect in some part of multigene complex leads to destruction of insulin-secreting cells in pancreas.
Neurofibromatosis 1 and 2	17, 22	Gene causes deforming tumors of the nerve sheaths of the peripheral and central nervous systems.
Familial Alzheimer's disease	14	Most common cause of dementia, associated with degeneration of cerebral cortex and leading to loss of cognitive functions, beginning about age 45 (figure 10.18).
Hemochromatosis	6	Defect in intestine cells compromises iron absorption and causes damage to the liver, heart, and other organs
Various oncogenes	Various	Familial colon cancer, breast cancer (see microfile 17.8, basal cell carcinoma.)

*Because the loci for these diseases have been mapped, it is also possible to do genetic tests to determine if the gene has been inherited.

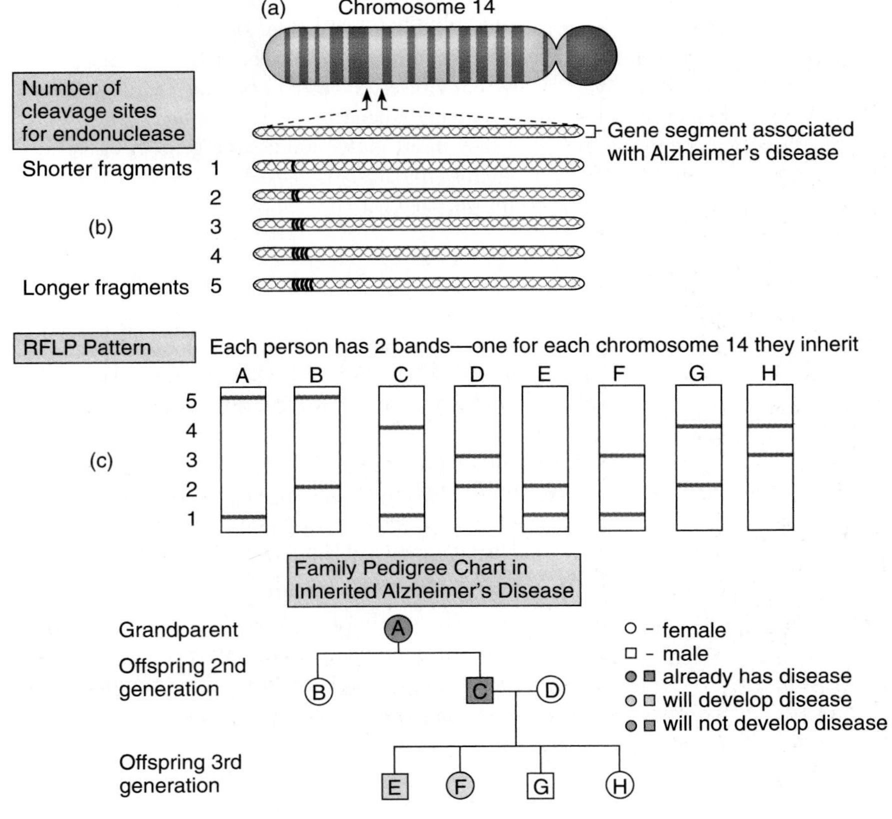

Figure 10.18
Pedigree analysis based on genetic screening for familial type Alzheimer's disease. (*a*) The site for this disease is on chromosome 14. (*b*) Gene mapping has determined that this chromosome segment has five variations, depending on the number and size of cleavage sites. (*c*) A Southern blot using the restriction fragments (RFLPs) is performed on a family with three generations. Each person will have two bands that correspond with each of the chromosomes they have inherited. The grandmother (A) with pattern 1,5 and her son (C) with pattern 1,4 both already have the disease. Patterns for his sister (B), his wife (D), and their four children (E, F, G, and H) are also shown. Comparing the band patterns of A and B with those of other family members makes it possible to predict which children will inherit the gene for Alzheimer's disease. Because the band shared by A and C is 1, we know that this is the variant associated with Alzheimer's disease. Children E (1, 2) and F (1, 3) are at high risk for developing the disease, and children G (2, 4) and H (2, 3) are not. It is also evident that this form of Alzheimer's disease is inherited as a dominant trait.

MICROFILE 10.4 DNA FINGERPRINTS: THE BAR CODES OF LIFE

The elegant technique for producing genetic fingerprints was devised in the mid-1980s by Alex Jeffreys of Great Britain. It involves taking a sample of cells, breaking the cells open and isolating the DNA, and exposing it to endonucleases that cleave the DNA into fragments. The number and size of these fragments vary with the types of endonucleases used and the genetic makeup of the individual DNA. Each restriction endonuclease cleaves the DNA only in certain locations, which can vary among the population. If a collection of different endonucleases is used, the patterns of cleavage can be so varied that each individual's pattern of fragments is unique. Even siblings in the same family can have quite different DNA fingerprint patterns.

The fingerprint pattern is developed by methods previously discussed and summarized here. The resulting pattern has been aptly compared to the bar scanning codes on grocery store items. The more probes that are used, the more individualized the bar code pattern.

One of the first uses of the technique was in forensic medicine. Besides real fingerprints, criminals often leave a bit of other evidence at the site of a crime—a hair, a piece of skin or fingernail, semen, blood, or saliva. Because it can be combined with amplifying techniques such as PCR, DNA fingerprinting provides a way to test even specimens that have only minute amounts of DNA, or even old specimens. Although contamination can be a problem in interpretation, new refinements in technique have made this less of a problem than it once was.

DNA fingerprinting has been an important form of evidence in several thousand trials and has helped both in conviction and exoneration of suspects. If enough markers are used, it is possible to give probabilities in the billions and even trillions that the DNA matches evidence at the crime scene. DNA fingerprinting is thus considered excellent supportive evidence in trials. The FBI is currently establishing a standardized method to set up a national database of DNA fingerprints, and several states allow DNA fingerprinting of felons in prison (so-called DNA dogtags). DNA traces are so pervasive in the environment that nearly any object a person has touched will contain small amounts of his DNA. In a most amazing application of this knowledge, forensic scientists were able to positively identify the culprits in the World Trade Center bombing and the unibomber, Ted Kazynski, by DNA recovered from the back of postage stamps they had licked!

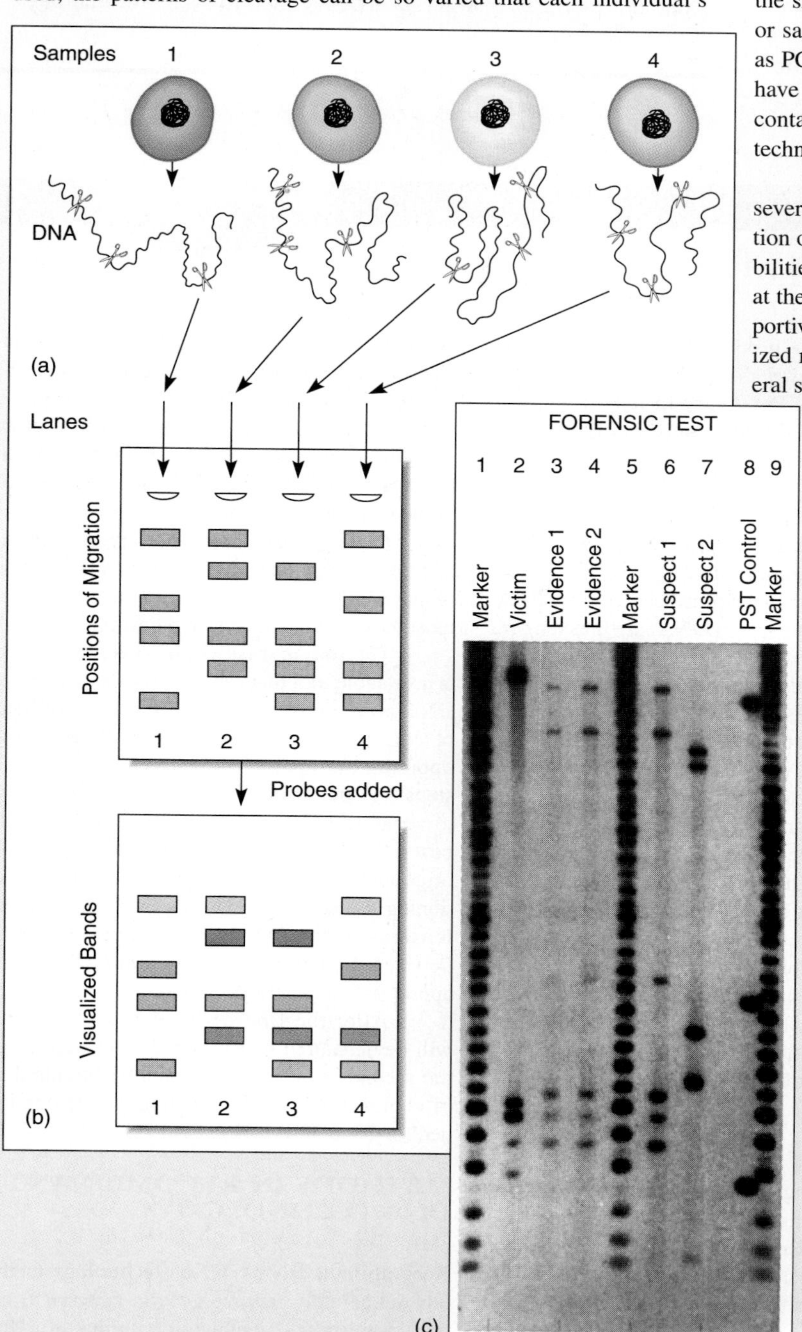

A different kind of fingerprinting using DNA. (a) *Cells from different samples are treated with a substance to release their DNA. All DNA samples are exposed to the same endonuclease, which snips them at specific sites into various-sized fragments.* (b) *Electrophoresing the fragments of DNA sorts them by size (larger fragments at the top, smaller ones at the bottom). The relative positions of these fragments are made visible by labeled DNA probes designed to attach to specific DNA markers. The developed gel (this one uses radioactivity) appears with a series of bands that correspond to the sample's pattern.* (c) *An actual DNA fingerprint used in a rape trial. Control lanes with known markers are in lanes 1, 5, 8, and 9. The second lane contains a sample of DNA from the victim's blood. Evidence samples 1 and 2 (lanes 3 and 4) contain semen samples taken from the victim. Suspects 1 and 2 (lanes 6 and 7) were tested. Can you tell by comparing evidence and suspect lanes which individual committed the rape? (It was suspect 1, who was convicted partly on the basis of this evidence.)*

The U.S. military has begun keeping DNA profiles on their personnel. When the remains of several "unknown soldiers" who had died in the Persian Gulf War were too mutilated and mixed to be properly identified, the DNA content was analyzed. A comparison with known DNA profiles of soldiers missing in action allowed the remains to be identified and separated.

Because even fossilized DNA can remain partially intact, ancient DNA is being studied for comparative and evolutionary studies. Anthropologists have traced the migrations of ancient people by analyzing the mitochondrial DNA found in bone fragments. Mitochondrial DNA (mtDNA) is less subject to degradation than is chromosomal DNA and can be used as an evolutionary time clock. A group of evolutionary biologists have been able to recover Neanderthal mtDNA and use it to show that this group of ancient primates could not have been an ancestor of humans. Biologists studying canine origins have also been able to show that the wolf is the ancestral form for modern canines.

Chapter Checkpoints

DNA technology has advanced understanding of basic genetic principles that have significant applications in a wide range of disciplines, particularly medicine, evolution, forensics, and anthropology.

The Human Genome Project has not only identified sites of specific human genes but has also clarified the functions of introns, transposons, and other types of support DNA.

DNA fingerprinting is a technique by which individuals are identified for purposes of medical diagnosis, genetic ancestry, and forensics.

Mitochrondrial DNA analysis is being used to trace evolutionary origins in animals and plants.

CHAPTER CAPSULE WITH KEY TERMS

Sciences that manipulate, alter, and analyze the genetic material of microbes, plants, animals, and viruses are **genetic engineering** (bioengineering) and **biotechnology.** Genetic engineering is defined as the direct, deliberate modification of an organism's genome, and biotechnology is the use of DNA or genetically altered organisms in commercial production.

In addition to biology and related fields, the influence of applied genetics is felt in numerous other areas such as forensics (medical law), agriculture, industry, anthropology, and archeology. These powerful and useful tools promise to change many areas of law, commerce, and medicine.

TECHNIQUES IN GENETIC ENGINEERING (REFER ALSO TO THE SUMMARY IN FIGURE 10.1)

Several properties of nucleic acids make them amenable to alteration and manipulation.

DNA readily **anneals** (becomes temporarily denatured) when heat breaks the hydrogen bonds holding the double helix together, and it separates into two strands. Slow cooling causes single DNA strands to rejoin (renature) at complementary sites.

DNA can also be clipped crosswise at specific sites by means of bacterial enzymes called **restriction endonucleases.** These enzymes leave short tails called sticky ends that base-pair with complementary tails on linear DNA or plasmids. Endonucleases allow splicing of genes into specific sites and create fragments that circularize into a plasmid.

Fragments produced by restriction endonucleases—*restriction fragment length polymorphisms* (RFLPs)—vary in length and are inherited in predictable patterns that make them useful *markers* of unique genetic characteristics.

Ligases rejoin sticky ends made by endonucleases for final splicing of genes into plasmids and chromosomes.

Reverse transcriptase can convert RNA into DNA. Copies, or **cDNA,** of messenger, transfer, and ribosomal RNA provide a valuable means of synthesizing eucaryotic genes from mRNA transcripts.

Gel electrophoresis can produce a readable pattern of DNA fragments. When samples in gel are subjected to an electrical current, the DNA pieces migrate in the gel substrate toward the positive pole, forming a pattern based on fragment size. The pattern is analyzed by comparing it against known standards to characterize genetic similarities.

Oligonucleotides are short fragments of DNA or RNA usually made by DNA synthesizers. The exact sequence of the base pairs in DNA is automatically analyzed by *sequencers.* Sequencing is critical for detailed gene analysis.

NUCLEIC ACID HYBRIDIZATION AND PROBES

The ability of single-stranded nucleic acids to **hybridize** or join together at complementary sites makes it possible to use **gene** or **hybridization probes** to detect specific nucleotide sequences.

Hybridization probes can identify unknown samples or isolate precise fragments from a complex mixture. In the *Southern blot* method, probes are used to label a gene or nucleotide sequence on an electrophoretic gel; the *dot blot* is a rapid, direct probe of a sample. Probes are used for diagnosing the cause of an infection from a patient's specimen and identifying a culture of an unknown bacterium or virus with a microbe-specific probe.

In situ hybridization uses probes on intact cells to visualize the presence and location of specific nucleic acids such as genes on chromosomes or RNA in cells and tissues.

The **polymerase chain reaction (PCR)** is a valuable tool that can amplify the amount of DNA in a sample from a few copies to billions of copies in a few hours. The PCR technique operates by repetitive cycling of three basic steps: (1) **denaturation**—heating target DNA to separate it into two strands, called **amplicons;** (2) addition of **primers** (synthetic oligonucleotides) to serve as guides for positioning the start of DNA amplification; (3) **extension**—thermostable DNA polymerase synthesizes the complementary strand by adding appropriate nucleotides.

After the first cycle, the DNA strands will be denatured to be primed and extended in the second cycle, thereby doubling the number of copies to four. This doubling can be continued for 20 to 30 cycles.

METHODS IN RECOMBINANT DNA TECHNOLOGY

Recombinant DNA (rDNA) technology methods deliberately remove genetic material from one organism and combine it with that of a different organism. It has numerous applications, including isolation of genes, genetic engineering to mass produce proteins, and creation of genetically transformed organisms.

An important objective of rDNA is the formation of genetic **clones,** or duplicates. Cloning involves the selection and removal of a foreign gene from an animal, plant, or microorganism (the genetic donor) followed by its propagation in a different host organism. Insertion of the foreign gene is accomplished by means of a **vector** (usually a plasmid or virus) that will carry the DNA into a **cloning host,** usually a bacterium or yeast.

A target gene can be isolated or prepared by (1) analysis of the genetic donor's chromosomes through Southern blotting and screening techniques; (2) synthesis by reverse transcription from an mRNA transcript; and (3) synthesis by machines. Cloned genes are maintained in vectors, producing genomic libraries of donor genes.

Characteristics of Cloning Vectors and Cloning Hosts

The best recombinant vectors are plasmids and bacteriophages that carry a significant piece of donor DNA and can be readily transferred into appropriate host sells. Examples include *E. coli* plasmids; a modified phage vector, the *Charon* phage; hybrid vectors formed by merging a plasmid and a phage, called a *cosmid.* Other systems for cloning large pieces of foreign DNA are *yeast artificial chromosomes* (YACs) and *bacterial artificial chromosomes* (BACs).

The best cloning hosts have a rapid growth rate, are nonpathogenic, have a well-delineated, simple genome, can function with plasmids and virus vectors, and will replicate and express foreign genes. Common cloning hosts are special strains of *E. coli* and *Saccharomyces cerevisiae.*

Recombinant Process

In the first cloning step, the foreign gene is excised with an endonuclease and spliced into a plasmid that has been cut with the same endonuclease so that the terminal nucleotides of the two will properly mate; the bonds are sealed by a ligase. The recombinant plasmid, or *chimera* can be introduced by transformation into the proper bacterial cloning host. Recombinant clones are identified by their antibiotic resistance.

Progeny cells containing the chimera gene can express the foreign gene by synthesizing the polypeptide it codes for and secreting it. After final processing a relatively pure product is left. This process may be done on an industrial scale to mass produce a variety of hormones, enzymes, and agricultural products.

BIOCHEMICAL PRODUCTS OF RECOMBINANT DNA TECHNOLOGY

Recombinant DNA techniques give rise to biochemical products: genetically recombined microbes, plants, and animals; medicines for gene therapy; and methods for genomic analysis.

Recombinant DNA technology is used to manufacture medications for diseases, such as diabetes and dwarfism, caused by the lack of an essential hormone. Various recombinant hormones are available to treat medical conditions. A summary list of other products is given in table 10.2

Recombinant Organisms

Artificially introducing foreign genes into organisms is termed *transfection,* and recombinant organisms are called *transgenic.* Foreign genes have been used to engineer unique microbes, plants, and animals for a variety of biotechnological applications.

Recombinant Bacteria and Viruses

Recombinant DNA is used to genetically alter *Pseudomonas syringae* to make a commercial product called Frostban that can prevent ice crystals from forming on plants in the field. A natural insecticide gene was added to *Pseudomonas fluorescens* to colonize plant roots. Biotechnologists have developed bacteria to clean up oil spills and degrade pollutants.

Viral vectors have been developed for testing drug resistance in *Mycobacterium tuberculosis,* gene therapy, an experimental AIDS vaccine, and to introduce resistance genes into crops.

TRANSGENIC PLANTS

Agrobacterium bacteria can invade plant cells and integrate their DNA into the genome and cause a tumor called *crown gall disease.* This ability makes them valuable for inserting foreign genes into plant genomes through a special **Ti plasmid** using rDNA techniques. Infection of the plant with the recombinant bacteria automatically transfers the foreign gene into its cells. Transfection of dicot plants has led to built-in pesticide and pathogen resistance in many species. Other plants are transfected with gene guns that forcefully impel genes into plant embryos.

Other developments in plant genetics include tomatoes that do not ripen too rapidly, plants that are pest- and virus-resistant and tobacco plants that can synthesize human antibodies.

TRANSGENIC ANIMALS

Several hundred strains of transgenic animals have been introduced by research and industry to study genetic diseases, to test new genetic therapies, and to become "factories" to manufacture proteins and other products.

Common transgenic animals are mice transfected by inserting genes into the embryo (gene line engineering). Mice have been engineered with human genes for growth hormone and to create animal models for human diseases. Other transgenic animals are pigs that synthesize human hormones and hemoglobin, and sheep and goats that produce various human hormones in their milk.

GENETIC TREATMENTS

Gene Therapy

Gene therapy can possibly cure genetic diseases by replacing a faulty gene. The *ex vivo* method uses virus vectors containing the normal gene to transfect human tissues in a test tube. The transfected cells are then reintroduced into the patient's body. In *in vivo* therapy, the body is directly infected with virus vectors. Some benefit has occurred in children with a severe immunodeficiency disease called adenosine deaminase (ADA) deficiency and some with cystic fibrosis (CF). Similar trials are under way to treat such diseases as hemophilia, sickle-cell anemia, and even some types of cancer.

Antisense and Triplex DNA Technology

Antisense and **triplex agents** are oligonucleotide drugs delivered into cells to block undesirable expression of genes. Antisense refers to a nucleic acid strand that is complementary to the sense, or translatable, strand. Antisense drugs (usually DNA) are chemically modified agents that bind to a target mRNA and interfere with its reading on ribosomes. Experimental antisense drugs show some success in treating certain cancers, autoimmune diseases, infections, and Alzheimer's disease.

Triplex DNA is a triple helix formed when a third strand of DNA forms hydrogen bonds with the purine bases on one of the helices. This extra strand can make the DNA template inaccessible to normal transcription. Most drugs based on triplex DNA are made to block gene sequences that regulate cancer genes, viruses, and immune reactions.

GENOME MAPPING AND SCREENING

Gene mapping is a way to demarcate the nature of a genome. Location, or physical maps, indicate sites of the genes on chromosomes, and sequence maps provide the order of base pairs in a gene. The Human Genome Project is a long-term attempt to map the 100,000 genes in the human genome. The completed map will present detailed genetic information on inherited diseases and will allow genetic screening, enabling families to know their risks for certain diseases.

Most DNA in humans does not code for proteins. This "junk DNA" is important in stabilizing chromosomes and regulating genes. Much of this support DNA is highly repetitive DNA in which the same nucleotide sequence is repeated 10 to several thousand times. Examples include *satellite DNA* and *introns.*

DNA FINGERPRINTING

The exact way DNA nucleotides are combined is unique for each individual, and DNA technology can be used to array the genome in patterns for comparison. It involves releasing DNA from cells, isolating it, and exposing it to restriction endonucleases that cleave the DNA into a set of relatively unique fragments. After the fragments have been separated by gel electrophoresis, probes are applied to highlight specific restriction landmarks (RFLPs). The pattern of bands is called a **DNA fingerprint.** This technique can identify hereditary relationships and inheritance patterns of genetic diseases. It is also used to keep genetic records in the military and to analyze ancient DNA for comparative and evolutionary studies. Anthropologists and historians have applied the technology in tracing the possible origins of humans.

The technique is an important tool for analyzing evidence in forensics. Although still surrounded by some controversy, DNA fingerprinting is considered by most geneticists to be excellent supportive evidence in trials.

MULTIPLE-CHOICE QUESTIONS

1. Which gene is incorporated into plasmids to detect recombinant cells?
   a. restriction endonuclease
   b. virus receptors
   c. a gene for antibiotic resistance
   d. reverse transcriptase

2. Which of the following is *not* essential to carry out the polymerase chain reaction?
   a. primers
   b. DNA polymerase
   c. gel electrophoresis
   d. high temperature

3. Which of the following must be present to make a nucleic acid hybrid?
   a. Taq polymerase c. a plasmid
   b. a labeled probe d. RNA

4. What do we call the synthetic unit of the polymerase chain reaction?
   a. annealer c. amplicon
   b. ligase d. primer

5. The function of ligase is to
   a. rejoin segments of DNA
   b. make longitudinal cuts in DNA
   c. synthesize cDNA
   d. break down ligaments

6. The pathogen of plant roots that is used as a cloning host is
   a. *Pseudomonas*
   b. *Agrobacterium*
   c. *Escherichia coli*
   d. *Saccharomyces cerevisiae*

7. Which of the following is a site that could be clipped by an endonuclease?
   a. ATCGATCG
TAGCTAGC   c. ATATATA
TATATAT
   b. AAGCTT
TTCGAA   d. ACCAT
TGGTA

8. The antisense DNA strand that complements mRNA AUGCGCGAC is
   a. UACGCUCUG
   b. GTCTCGCAT
   c. TACGCTCTG
   d. DNA cannot complement mRNA

9. Which DNA fragment will be closest to the top (negative pole) of an electrophoretic gel?
   a. 450 bp c. 5 kb
   b. 3,560 bp d. 1,500 bp

10. Which of the following is a primary participant in cloning an isolated gene?
    a. restriction endonuclease
    b. vector
    c. host organism
    d. all of these

11. For which of the following would a nucleic acid probe *not* be needed?
    a. locating a gene on a chromosome
    b. developing a Southern blot
    c. identifying a microorganism
    d. constructing a recombinant plasmid

12. Satellite DNA
    a. contains important metabolic genes
    b. is composed of highly repetitive sequences
    c. is useless junk
    d. causes cancer

13. Match the term with its description:
    ______ nucleic acid probe
    ______ antisense strand
    ______ sense strand
    ______ reverse transcriptase
    ______ Taq polymerase
    ______ triplex DNA
    ______ primer
    ______ restriction endonuclease
    a. enzyme that transcribes RNA into DNA
    b. DNA molecule with an extra strand inserted
    c. the nontranslated strand of DNA or RNA
    d. enzyme that snips DNA at palindromes
    e. oligonucleotide that initiates the PCR
    f. strand of nucleic acid that is transcribed or translated
    g. thermostable enzyme for synthesizing DNA
    h. oligonucleotide used in hybridization

CONCEPT QUESTIONS

1. Define genetic engineering and biotechnology, and summarize the important purposes of these fields. Review the use of the terms *genome, chromosome, gene, DNA,* and *RNA* from chapter 9.

2. a. Describe the processes involved in denaturing and renaturing of DNA.
   b. What is useful about this procedure?
   c. Why is it necessary to denature the DNA in the Southern blot test?
   d. How would the Southern blot be used with PCR?

3. a. Given the shorthand for the following restriction endonucleases: *Arthrobacter luteus* I; *Providencia stuartii* I; *Bacillus amyloliquefaciens* H I
   b. What kind of cut will HaeIII make?
   c. Using nucleotide letters, show how a piece of DNA can be cut to circularize it.
   d. What are restriction length polymorphisms, and how are they used?

4. Explain how electrophoresis works and the general way that DNA is sized. Estimate the size of the DNA fragment in base pairs in the first lane of the gel in figure 10.4. Define oligonucleotides, explain how they are formed, and give three uses for them.
5. a. Briefly describe the functions of DNA synthesizers and sequencers.
   b. How would you make a copy of DNA from an mRNA transcript?
   c. Show how this process would look, using base notation.
   d. What is this DNA called?
   e. Why would it be an advantage to synthesize eucaryotic genes this way?
6. a. Explain the meaning of this shorthand to represent the polymerase chain reaction:

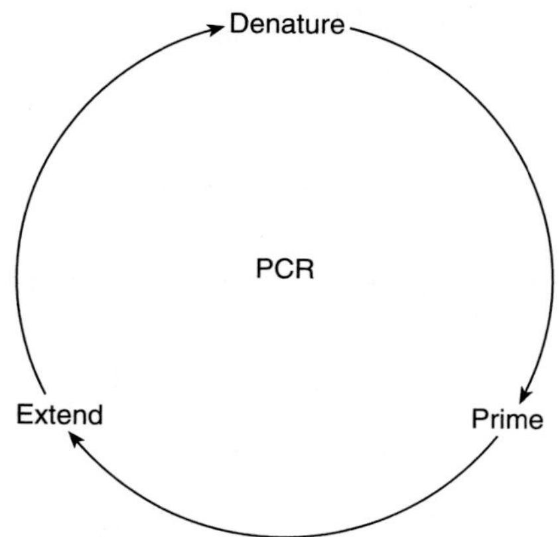

   b. Describe the effects of temperature change in PCR.
   c. Exactly what are the functions of the primer and Taq polymerase?
   d. Explain why the PCR is unlikely to amplify contaminating bacterial DNA in a sample of human DNA.
7. a. What characteristics of plasmids and bacteriophages make them good cloning vectors?
   b. Name several types of vectors, and explain what benefits they have.
   c. List the types of genes that they can contain.
8. a. Describe the principles behind recombinant DNA technology.
   b. Outline the main steps in cloning a gene.
   c. Once cloned, how can this gene be used?
   d. Characterize several ways that recombinant DNA technology can be used.
9. a. What characteristics of bacteria make them good cloning hosts?
   b. What is one way to determine whether a bacterial culture has received a recombinant plasmid?
10. a. What is transfection, and what are transgenic organisms?
    b. Explain how *Agrobacterium* is used to transfect plants.
    c. Describe a method for transfecting animals.
    d. Summarize some of the uses for transgenic plants and animals.
11. a. What is the main difference between *ex vivo* and *in vivo* gene therapy?
    b. Does the virus vector used in gene therapy replicate itself in the host cell?
    c. Why would this not be a good idea?
    d. What are some of the main problems with gene therapy?
12. a. Describe the molecular mechanisms by which a DNA antisense molecule could work as a genetic medicine.
    b. Do the same for triplex DNA.
    c. Are the therapies permanent?
    d. Why or why not?
13. a. What is a gene map?
    b. Show by a diagram how chromosomal, physical, and sequence maps are different?
    c. Which organisms are being mapped, and what uses will these maps perform?
    d. Why is the human genome map going to require 20 years?
    e. What are some possible effects of knowing the genetic map of humans?
14. a. Described what a DNA fingerprint is and why and how restriction fragments can be used to form a unique DNA pattern.
    b. Discuss briefly how DNA fingerprinting is being used routinely by biomedicine, the law, the military, and human biology.

CRITICAL-THINKING QUESTIONS

1. a. Give an example of a benefit of genetic engineering to society and a possible adverse outcome.
   b. Give an example of an ecological benefit and a possible adverse side effect.
2. a In reference to microfile 10.1, what is your opinion of Jeremy Rifkin's analysis of the dangers associated with genetic engineering?
   b. Most of us would agree to growth hormone therapy for a child with dwarfism, but what is an answer to the parents who want to give growth hormones to their 8-year-old son so that he will be tall enough to play basketball?
3. a. If gene probes, fingerprinting, and mapping could make it possible for you to know of future genetic diseases in you or one of your children, would you wish to use this technology to find out?
   b. What if it were used as a screen for employment or insurance?
4. a. Can you think of a reason that bacteria make restriction endonucleases?
   b. What is it about the endonucleases that prevents bacteria from destroying their own DNA?
5. a. Describe how a virus might be genetically engineered to make it highly virulent.
   b. Can you trace the genetic steps in the development of a tomato plant that has become frost-free from the addition of a flounder's antifreeze genes? (Hint: You need to use *Agrobacterium.*)
   c. What is a different approach to preventing frost on tomatoes?
6. a. Give three different methods to treat or cure cystic fibrosis using the techniques of genetic engineering.
   b. Outline a way that gene therapy could be used to treat a patient with genetic defects due to the presence of extra genes (trisomies as described in chapter 17).
7. You have obtained a blood sample in which only red blood cells are left to analyze.
   a. Can you conduct a DNA analysis of this blood?
   b. If no, explain why.
   c. If yes, explain what you would use to analyze it and how to do it.

8. The way that PCR amplifies DNA is similar to the doubling in a population of growing bacteria; a single DNA strand is used to synthesize 2 DNA strands, which become 4, then 8, then 16, etc. If a complete cycle takes 3 minutes,
   a. how many strands of DNA would theoretically be present after 10 minutes?
   b. after 30 minutes?
   c. after 1 hour?
9. a. Design an antisense DNA drug for this gene: TACGGCTATATTCCGGGC
   b. Design an antisense RNA drug for the same gene.
   c. How would a triplex drug for this gene work?
   d. Describe how antisense DNA works to block the replication of viruses.
10. a. How would you regard the release of a bioengineered bacterium in your own backyard?
    b. Describe any moral, ethical, or biological problems associated with eating tomatoes from an engineered plant or pork from a transgenic pig.
    c. What are the moral considerations of using transgenic animals to manufacture various human products?
11. You are on a jury to decide whether a person committed a homicide and you have to weigh DNA fingerprinting evidence. Two different sets of fingerprints were done: one that tested 5 markers and one that tested 10. Both sets match the defendant's profile.
    a. Which one is more reliable and why?
    b. How important is the fingerprint if you are told that the fingerprint pattern occurs in one person out of 10,000 in the general population?
    c. Would knowing that the defendant lived in the same apartment building as the victim have any effect on your decision?
12. a. How do you suppose the fish and game department using DNA evidence, could determine whether certain individuals had poached a deer or if a particular mountain lion had eaten part of a body.
    b. Can you think of some reasons that it would *not* be possible to recreate dinosaurs using the technology we have described in this chapter?
13. a. Look at the pedigree chart for Alzheimer's disease. How many genes for this disease must be inherited for it to be expressed?
    b. What kinds of offspring could be produced if a person with the genotype of 1,3 married a person with genotype 1,5?
    c. If cystic fibrosis has the same general profile of inheritance as Alzheimer's disease, yet it is recessive (requires two genes to be expressed), what would be the future of the offspring in the pedigree chart?
14. Who actually owns the human genome?
    a. Make cogent arguments on various sides of the question.
    b. Explain the steps required in producing a structural map (base sequence) of DNA.

INTERNET SEARCH TOPIC

a. Locate information on mitochondrial DNA and determine how geneticists use it for their study.
b. Look up HUGO (Human Genome Organization) and find out how far along the genome sequencing project has progressed.

chapter 11

Physical and Chemical Control of Microbes

The natural condition of humanity is to share surroundings with a large, diverse population of microorganisms. The complete exclusion of microbes from the environment is not only impossible but of questionable value. However, certain circumstances require constant and concerted efforts to exclude them. Such daily activities as cleaning, refrigeration, and cooking are, by nature, antimicrobial processes. The medical, dental, and commercial methods that prevent the spread of infectious agents, hinder spoilage, and make products safe are routine and very broad in scope. These methods for destroying, removing, and inhibiting microbes are the subjects of this and the following chapter.

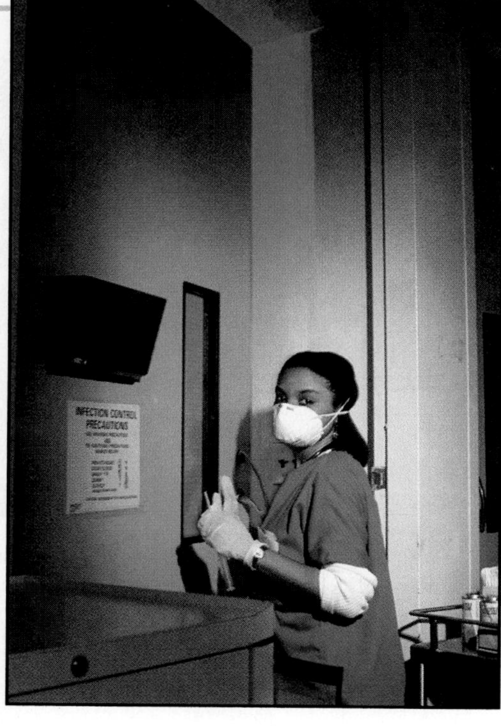

A hospital isolation room is the focus of many forms of microbial control to prevent infection. Health care workers are charged with executing asepsis, sterile precautions, and stringent disinfection procedures. Some rooms are designed with airflow systems that regulate the movement of air into and out of the room.

MICROFILE 11.1 MICROBIAL CONTROL IN ANCIENT TIMES

No one knows for sure when humans first applied methods that could control microorganisms, but perhaps the discovery and use of fire in prehistoric times was the starting point. We do know that records describing simple measures to control decay and disease appear from civilizations that existed several thousand years ago. We know, too, that these ancient people had no concept that germs caused disease, but they did have a mixture of religious beliefs, skills in observing natural phenomena, and, possibly, a bit of luck. This combination led them to carry out simple and sometimes rather hazardous measures that contributed to the control of microorganisms.

Salting, smoking, pickling, and drying foods and exposing food, clothing, and bedding to sunlight were prevalent practices among early civilizations. The Egyptians showed surprising sophistication and understanding of decomposition by embalming the bodies of their dead with strong salts and pungent oils. They introduced filtration of wine and water as well. The Greeks and Romans burned clothing and corpses during epidemics, and they stored water in copper and silver containers. The armies of Alexander the Great reportedly boiled their drinking water and buried their wastes. Burning sulfur to fumigate houses and applying sulfur as a skin ointment also date from this approximate era.

During the great plague pandemic of the Middle Ages, it was commonplace to bury corpses in mass graves, burn the clothing of plague victims, and ignite aromatic woods in the houses of the sick in the belief that fumes would combat the disease. In a desperate search for some sort of protection, survivors wore peculiar garments and anointed their bodies with herbs, strong perfume, and vinegar. These attempts may sound foolish and antiquated, but it now appears that they may have had some benefits. Burning wood releases formaldehyde, which could have acted as a disinfectant; herbs, perfume, and vinegar contain mild antimicrobial substances. Each of these early methods, although somewhat crude, laid the foundations for microbial control methods that are still in use today.

Illustration of protective clothing used by doctors in the 1700s to avoid exposure to plague victims. The beaklike portion of the hood contained volatile perfumes to protect against foul odors and possibly inhaling "bad air."

CONTROLLING MICROORGANISMS

Much of the time in our daily existence, we take for granted tap water that is drinkable, food that is not spoiled, shelves full of products to eradicate "germs," and drugs to treat infections. Controlling our degree of exposure to potentially harmful microbes is a monumental concern in our lives, and it has a long and eventful history (microfile 11.1).

GENERAL CONSIDERATIONS IN MICROBIAL CONTROL

The methods of microbial control belong to the general category of *decontamination* procedures, in that they destroy or remove contaminants. In microbiology, contaminants are microbes present at a given place and time that are undesirable or unwanted. Most decontamination methods employ either **physical agents** such as heat or *radiation* or **chemical agents** such as disinfectants and antiseptics. This separation is convenient, but the categories overlap in some cases; for instance, radiation can cause damaging chemicals to form, or chemicals can generate heat. A flowchart (figure 11.1) summarizes the major applications and aims in microbial control.

RELATIVE RESISTANCE OF MICROBIAL FORMS

The primary targets of microbial control are microorganisms capable of causing infection or spoilage that are constantly present in the external environment and on the human body. This targeted population is rarely simple or uniform; in fact, it often contains mixtures of microbes with extreme differences in resistance and harmfulness. Contaminants that can have far-reaching effects if not adequately controlled include bacterial vegetative cells and en-

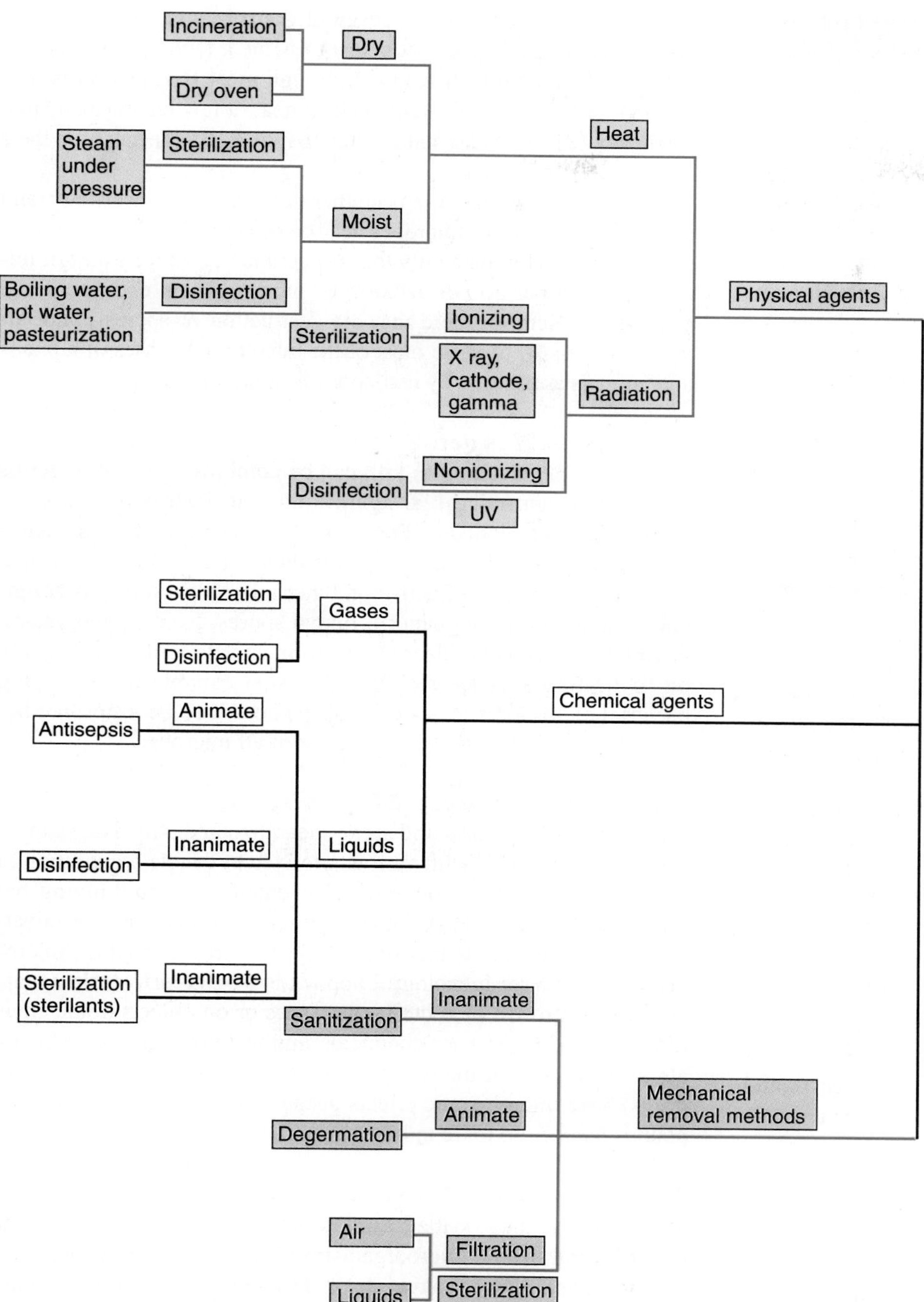

Figure 11.1
Flowchart of microbial control measures.

dospores, fungal hyphae and spores, yeasts, protozoan trophozoites and cysts, worms, insects and their eggs, and viruses. The following scheme compares the general resistance of these forms to physical and chemical methods of control:

Highest resistance
Bacterial endospores

Moderate resistance
Protozoan cysts; some fungal sexual spores (zygospores); some viruses. In general, naked viruses are more resistant than enveloped forms. Among the most resistant viruses are the hepatitis B virus and the poliovirus. Particular vegetative bacteria that have higher resistance are *Mycobacterium tuberculosis, Staphylococcus aureus,* and *Pseudomonas* species.

Least resistance
Most bacterial vegetative cells; ordinary fungal spores and hyphae; enveloped viruses; yeasts; and trophozoites

Actual comparative figures on the requirements for destroying various groups of microorganisms are shown in table 11.1. Because bacterial endospores are the most resistant microbial

TABLE 11.1

RELATIVE MICROBIAL RESISTANCE TO PHYSICAL AND CHEMICAL AGENTS

Method	Spores*	Vegetative Forms*	Relative Resistance**
Heat (moist)	120°C	80°C	1.5×
Radiation (X-ray) dosage	0.4 Mrad	0.1 Mrad	4×
Ultraviolet rays (exposure time)	1.5 h	10 min	9×
Sterilizing gas (ethylene oxide)	1,200 mg/l	700 mg/l	1.7×
Sporicidal liquid (2% glutaraldehyde)	3 h	10 min	18×

*Values are based on methods (concentration, exposure time, intensity) that are required to destroy the most resistant pathogens in each group.

**The greater resistance of spores versus vegetative cells given as an average figure.

entities, their destruction constitutes the goal of processes that sterilize (see definition in following section), because any process that kills them will invariably kill all less resistant forms. Other methods of control (disinfection, antisepsis) act primarily upon microbes that are less hardy than endospores.

TERMINOLOGY AND METHODS OF MICROBIAL CONTROL

Through the years, a growing terminology has emerged for describing and defining measures that control microbes. To complicate matters, the everyday use of some of these terms can at times be vague and inexact. For example, occasionally one may be directed to "sterilize" or "disinfect" a patient's skin, even though this usage does not fit the technical definition of either term. To lay the groundwork for the concepts in microbial control to follow, we present here a series of concepts, definitions, and usages in antimicrobial control.

Sterilization

Sterilization is a process that destroys or removes all viable microorganisms, including viruses. Any material that has been subjected to this process is said to be **sterile.*** These terms should be used only in the strictest sense for methods that have been proved to sterilize. An object cannot be slightly sterile or almost sterile—it is either sterile or not sterile. Control methods that sterilize are generally reserved for inanimate objects, because sterilizing parts of the human body would call for such harsh treatment that it would be highly dangerous and impractical. As we shall see in a subsequent chapter, many internal parts of the body—the brain, muscles, and liver, for example—are naturally free of microbes.

Sterilized products—surgical instruments, syringes, and commercially packaged foods, just to name a few—are frequently essential to human well-being. Although most sterilization is performed with a physical agent such as heat, a few chemicals called *sterilants* can be classified as sterilizing agents because of their ability to destroy spores.

At times, sterilization is neither practicable nor necessary, and only certain groups of microbes need to be controlled. Some antimicrobial agents eliminate only the susceptible vegetative states of microorganisms but do not destroy the more resistant endospore and cyst stages. Keep in mind that the destruction of spores is not always a necessity, because most of the infectious diseases of humans and animals are caused by non-spore-forming microbes.

Microbicidal Agents

The root *-cide,* meaning to kill, can be combined with other terms to define an antimicrobial agent aimed at destroying a certain group of microorganisms. For example, a **bactericide** is a chemical that destroys bacteria except for those in the endospore stage. It may or may not be effective on other microbial groups. A **fungicide** is a chemical that can kill fungal spores, hyphae, and yeasts. A **virucide** is any chemical known to inactivate viruses, especially on living tissue. A **sporicide** is an agent capable of destroying bacterial endospores. A sporicidal agent can also be a sterilant because it can destroy the most resistant of all microbes.

Agents That Cause Microbistasis

The Greek words *stasis* and *static* mean to stand still. They can be used in combination with various prefixes to denote a condition in which microbes are temporarily prevented from multiplying but are not killed outright. Although killing or permanently inactivating microorganisms is the usual goal of microbial control, microbistasis does have meaningful applications. **Bacteriostatic** agents prevent the growth of bacteria on tissues or on objects in the environment, and *fungistatic* chemicals inhibit fungal growth. Materials used to control microorganisms in the body (antiseptics and drugs) have microbistatic effects because many microbicidal compounds can be too toxic to human cells.

Germicides, Disinfection, Antisepsis

A **germicide,*** also called a *microbicide,* is any chemical agent that kills pathogenic microorganisms. A germicide can be used on inanimate (nonliving) materials or on living tissue, but it ordinarily cannot kill resistant microbial cells. Any physical or chemical agent that kills "germs" is said to have **germicidal** properties.

The related term, **disinfection,*** refers to the use of a physical process or a chemical agent (a **disinfectant**) to destroy vegetative pathogens but not bacterial endospores. It is important to note that disinfectants are normally used only on inanimate objects because, in the concentrations required to be effective, they can be toxic to human and other animal tissue. Disinfection processes also remove the harmful products of microorganisms (toxins) from materials. Examples of disinfection include applying a solution of 5% bleach

*sterile (ster´-ill) Gr. *steira,* barren. (This has another, older meaning that connotes the inability to produce offspring.)

*germicide (jer´-mih-syd) L. *germen,* germ, and *caedere,* to kill. Germ is a common term for a pathogenic microbe.

*disinfection (dis″-in-fek´-shun) L. *dis,* apart, and *inficere,* to corrupt.

to an examining table, boiling food utensils used by a sick person, and immersing thermometers in an iodine solution between uses.

In modern usage, **sepsis** is defined as the growth of microorganisms or the presence of microbial toxins in blood and other tissues. The term **asepsis*** refers to any practice that prevents the entry of infectious agents into sterile tissues and thus prevents infection. Aseptic techniques commonly practiced in health care range from sterile methods that exclude all microbes to **antisepsis.*** In antisepsis, chemical agents called **antiseptics** are applied directly to exposed body surfaces (skin and mucous membranes), wounds, and surgical incisions to destroy or inhibit vegetative pathogens. Examples of antisepsis include preparing the skin before surgical incisions with iodine compounds, swabbing an open root canal with hydrogen peroxide, and ordinary handwashing with a germicidal soap.

Methods That Reduce the Numbers of Microorganisms

Several applications in commerce and medicine do not require actual sterilization, disinfection, or antisepsis but are based on reducing the levels of microorganisms so that the possibility of infection or spoilage is greatly decreased. Restaurants, dairies, breweries, and other food industries consistently handle large numbers of soiled utensils that could readily become sources of infection and spoilage. These industries must keep microbial levels to a minimum during preparation and processing. **Sanitization*** is any cleansing technique that mechanically removes microorganisms (along with food debris) to reduce the level of contaminants. A **sanitizer** is a compound such as soap or detergent used to perform this task.

Cooking utensils, dishes, glass bottles, cans, and used clothing that have been washed and dried may not be completely free of microbes, but they are considered safe for normal use (sanitary). Air sanitization with ultraviolet lamps reduces airborne microbes in hospital rooms, veterinary clinics, and laboratory installations. It is important to note that some sanitizing processes (such as dishwashing machines) are rigorous enough to sterilize objects, but this is not true of all sanitization methods.

It is often necessary to reduce the numbers of microbes on the human skin through **degermation.** This process usually involves scrubbing the skin or immersing it in chemicals, or both. It also emulsifies oils that lie on the outer cutaneous layer and mechanically removes potential pathogens on the outer layers of the skin. Examples of degerming procedures are the surgical handscrub, the application of alcohol wipes to the skin, and the cleansing of a wound with germicidal soap and water. The concepts of antisepsis and degermation clearly overlap, since a degerming procedure can simultaneously be antisepic, and vice versa.

WHAT IS MICROBIAL DEATH?

Death is a phenomenon that involves the permanent termination of an organism's vital processes. Signs of life in complex organisms such as animals are self-evident, and death is made clear by loss of nervous function, respiration, or heartbeat. In contrast, death in microscopic organisms that are composed of just one or a few cells is often hard to detect, because they reveal no conspicuous vital signs to begin with. Lethal agents (such as radiation and chemicals) do not necessarily alter the overt appearance of microbial cells. Even the loss of movement in a motile microbe cannot be used to indicate death. This fact has made it necessary to develop special qualifications that define and delineate microbial death.

The destructive effects of chemical or physical agents occur at the level of a single cell. As the cell is continuously exposed to an agent such as intense heat or toxic chemicals, various cell structures become dysfunctional, and the entire cell can sustain irreversible damage. At present, the most practical way to detect this damage is to determine if a microbial cell can still reproduce when exposed to a suitable environment. If the microbe is so disturbed metabolically and structurally that it is unable to multiply, then it is no longer viable. *The permanent loss of reproductive capability, even under optimum growth conditions, has become the accepted microbiological definition of death.*

Factors That Affect Death Rate

The ability to define microbial death has tremendous theoretical and practical importance. With this concept, workers in the field of microbial control have defined the conditions required to destroy microorganisms and have pinpointed the manner by which antimicrobial agents kill cells. Standards of sterilization and disinfection in medicine, dentistry, and industry have been established largely by means of rigorous experimentation. Hundreds of testing procedures have been developed for evaluating physical and chemical agents. Appendix C presents a general breakdown of certain types of antimicrobial tests, along with the kinds of variables that must be controlled.

The cells of a pure culture show marked variations in susceptibility to a given microbicidal agent. Death of the whole population is not instantaneous but occurs in a logarithmic manner and requires a certain time of exposure (figure 11.2*a*). The most susceptible cells (younger, actively growing cells) die immediately, whereas less susceptible cells (older, inactive ones) have greater resistance and require longer exposure times. Eventually, a point is reached at which survival of any cells is highly unlikely; this point is equivalent to sterilization.

The effectiveness of a particular agent is governed by several factors besides time. The following additional factors influence the action of antimicrobial agents:

1. The number of microorganisms (figure 11.2*b*). A higher load of contaminants requires more time to destroy.
2. The nature of the microorganisms in the population (figure 11.2*c*). In most actual circumstances of disinfection and sterilization, the target population is not a single species of microbe but a mixture of bacteria, fungi, spores, and viruses, presenting an even greater spectrum of microbial resistance.
3. The temperature and pH of the environment.
4. The concentration (dosage, intensity) of the agent. For example, UV radiation is most microbicidal at 260 nm;

*asepsis (ay-sep´-sis) Gr. *a,* no or none, and *sepsis,* decay.
*antisepsis *anti,* against.
*sanitization (san″-ih-tih-zay´-shun) L. *sanitas,* health.

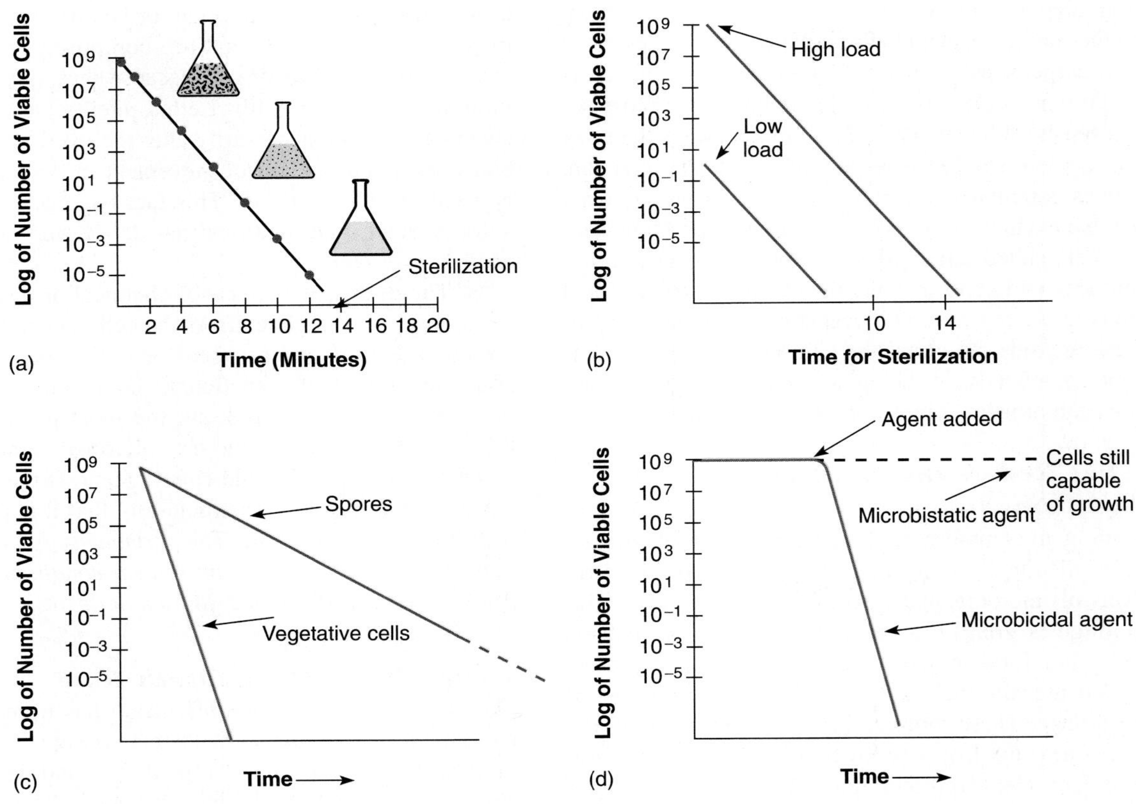

Figure 11.2
Factors that influence the rate at which microbes are killed by antimicrobial agents. (*a*) Length of exposure to the agent. During exposure to a chemical or physical agent, all cells of a microbial population, even a pure culture, do not die simultaneously. Over time, the number of viable organisms remaining in the population decreases logarithmically, giving a straight-line relationship on a graph. The point at which the number of survivors is infinitesimally small is considered sterilization. (*b*) Effect of the microbial load. (*c*) Relative resistance of spores versus vegetative forms. (*d*) Action of the agent, whether destructive or inhibitory.

most disinfectants are more active at higher concentrations.

5. The mode of action of the agent (figure 11.2*d*). How does it kill or inhibit the microorganism?
6. The presence of solvents, interfering organic matter, and inhibitors. Large amounts of saliva, blood, and feces can inhibit the actions of disinfectants and even of heat.

The influence of these factors will be discussed in greater detail in subsequent sections.

How Antimicrobial Agents Work: Their Modes of Action

An antimicrobial agent's adverse effect on cells is known as its *mode* (or *mechanism*) *of action.* Determining an agent's precise mode of action can be very difficult, and we do not yet completely understand how agents such as radiation and heat achieve their effects. Many antimicrobial agents affect more than one cellular target and may inflict both primary and secondary damages that eventually lead to cell death. Agents can be classified according to the following degrees of selectiveness: (1) agents that are less selective in their scope of destructiveness, inflict severe damage on many cell parts, and are generally very biocidal (heat, radiation, some disinfectants); (2) moderately selective agents with intermediate specificity (certain disinfectants and antiseptics); and (3) more selective agents (drugs) whose target is usually limited to a specific cell structure or function and whose effectiveness is restricted only to certain microbes.

The cellular targets of physical and chemical agents fall into four general categories: (1) the cell wall, (2) the cell membrane, (3) cellular synthetic processes (DNA, RNA), and (4) proteins.

The Effects of Agents on the Cell Wall

The cell wall maintains the structural integrity of bacterial and fungal cells. Several types of chemical agents damage the cell wall by blocking its synthesis, digesting it, or breaking down its surface. A cell deprived of a functioning cell wall becomes fragile and is lysed very easily. Examples of this mode of action include some antimicrobial drugs (penicillins) that interfere with the synthesis of the cell wall in bacteria (see figure 12.3). Detergents and alcohol can also disrupt cell walls, especially in gram-negative bacteria.

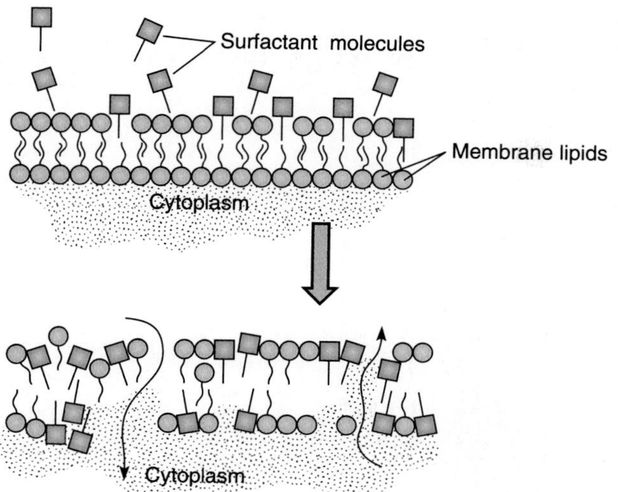

Figure 11.3
Mode of action of surfactants on the cell membrane. Surfactants inserting in the lipoidal layers disrupt it and create abnormal channels that alter permeability and cause leakage both into and out of the cell.

How Agents Affect the Cell Membrane

All microorganisms have a cell membrane composed of lipids and proteins, and even some viruses have an outer membranous envelope. As we learned in previous chapters, a cell's membrane provides a two-way system of transport. If this membrane is disrupted, a cell loses its selective permeability and can neither prevent the loss of vital molecules nor bar the entry of damaging chemicals. Loss of those abilities leads to cell death. Detergents called **surfactants*** work as microbicidal agents because they lower the surface tension of cell membranes. Surfactants are polar molecules with hydrophilic and hydrophobic regions that can physically bind to the lipid layer and penetrate the internal hydrophobic region of membranes. In effect, this process "opens up" the once tight interface, leaving leaky spots that allow injurious chemicals to seep into the cell and important ions to seep out (figure 11.3).

Agents That Affect Protein and Nucleic Acid Synthesis

Microbial life depends upon an orderly and continuous supply of proteins to function as enzymes and structural molecules. As we saw in chapter 9, these proteins are synthesized on the ribosomes through a complex process called translation. Any agent that interferes with accurate translation also prevents synthesis of a complete, functioning protein. For instance, the antibiotic chloramphenicol binds to the ribosomes of bacteria in a way that stops peptide bonds from forming. In its presence, many bacterial cells are inhibited from forming proteins required in growth and metabolism and are thus inhibited from multiplying. Most of the agents that block protein synthesis are drugs used in antimicrobial therapy. These drugs will be discussed in greater detail in chapter 12.

The nucleic acids are likewise necessary for the continued functioning of microbes. DNA must be regularly replicated and

*surfactant (sir-fak´-tunt) A word derived from **surf**ace-**act**ing age**nt**.

Figure 11.4
Modes of action affecting protein function. The functional three-dimensional, or native, state maintained by specific bonds creates active sites to react with substrate. Some agents denature the protein by breaking all or some secondary and tertiary bonds. Results are (*a*) complete unfolding or (*b*) random bonding and incorrect folding. (*c*) Some agents react with protein functional groups and can block the active site or interfere with bonding.

transcribed in growing cells, and any agent that either impedes these processes or changes the genetic code is potentially antimicrobial. Some agents bind irreversibly to DNA, preventing both transcription (formation of RNA) and translation; others are mutagenic agents. Gamma, ultraviolet, or X radiation causes mutations that result in permanent inactivation of DNA. Chemicals such as formaldehyde and ethylene oxide also interfere with DNA and RNA function.

Agents That Alter Protein Function

A microbial cell contains large quantities of proteins that function properly only if they remain in a normal three-dimensional configuration called the *native state.* The antimicrobial properties of some agents arise from their capacity to disrupt, or **denature,** proteins. In general, denaturation occurs when the bonds that maintain the secondary and tertiary structure of the protein are broken. Breaking these bonds will cause the protein to unfold or create random, irregular loops and coils (figure 11.4). One way that proteins can be denatured is through coagulation by moist heat (the same reaction seen in the irreversible solidification of the white of an egg when boiled). Chemicals such as strong organic solvents (alcohols, acids) and phenolics also coagulate proteins. Other antimicrobial agents, such as metallic ions, attach to the active site

MICROFILE 11.2 HOW CAN YOU STERILIZE THE WORLD?

Most human beings manage to remain healthy despite the fact that they live in continual intimate contact with microorganisms. We really do not have to be preoccupied with microbes every minute or feel overly concerned that the things we touch, drink, or eat are sterile, as long as they are somewhat clean and free of pathogens. For most of us, resistance to infection is well maintained by our numerous host defenses; but such is not the case for everyone.

In the last 20 years, largely owing to increasingly sophisticated medical advances, many babies with various immune deficiency diseases have survived. These children have poorly developed immune responses and may be in constant danger from infections by common, even "innocuous" microbes that exist all around us (see chapter 17). Being delivered by cesarian section (to maintain sterility) and then being isolated in a sterile environment enables them to survive.

Many of these children have received successful bone marrow transplants, a procedure that offers them freedom and a normal life—if they can overcome one more problem. For many months after the transplant, as the children are gradually developing immunities, they cannot be exposed to the usual levels of microbial contaminants. At first they are kept in strict isolation, and all personnel and visitors are antiseptically scrubbed and gowned before entering the room. Later, the children are allowed to go home, but with many restrictions and preventive routines.

The parents of severely immunocompromised children are under tremendous pressure to keep their surroundings as germ-free as possible; at times, they must feel compelled to sterilize the world. Everything that passes the child's lips, touches the hands or skin, or could be inhaled in the air requires evaluation. It is frequently necessary for both parents and child to be gowned and masked. Direct physical contact, outside visitors, and playmates must be discouraged. Toys and hands are fastidiously and frequently treated with germicides, and eating utensils, as well as all food and water, are sterilized. The house requires constant and thorough disinfection and vacuuming, and even then, the child cannot crawl on the floors. Every little sniffle can be a cause for panic. As the number of severely compromised patients increases because of AIDS and other diseases, we will continue to see dramatic changes in life support systems for protecting them.

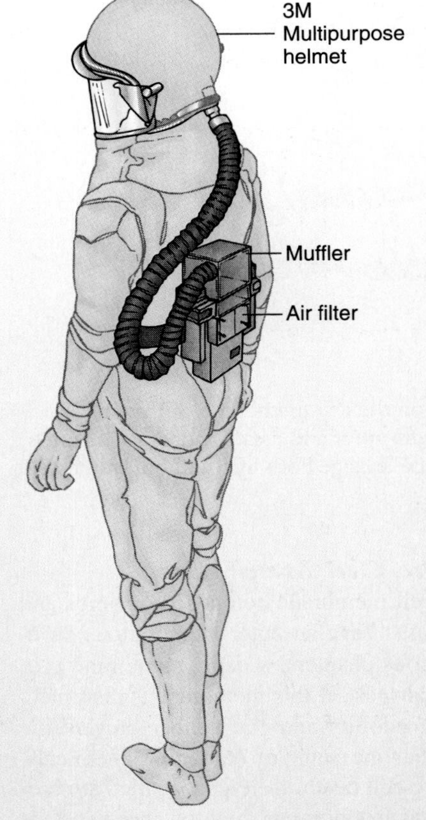

A special space suit designed for severely compromised patients to wear when outside of their sterile rooms delivers sterile air but allows patient mobility. This suit is also a method for protecting microbiologists working with highly infectious agents. *Source: Thomas L. Talbot and Philip A. Pizzo, Department of Biomedical Engineering & Instrumentation Branch, National Institutes of Health, Bethesda, MD.*

of the protein and prevent it from interacting with its correct substrate. Regardless of the exact mechanism, such losses in normal protein function can promptly arrest metabolism.

Practical Concerns in Microbial Control

Numerous considerations govern the selection of a workable method of microbial control. The following are among the most pressing concerns: (1) Does the application require sterilization, or is disinfection adequate? In other words, must spores be destroyed, or is it necessary to destroy only vegetative pathogens? (2) Is the item to be reused or permanently discarded? If it will be discarded, then the quickest and least expensive method should be chosen. (3) If it will be reused, can the item withstand heat, pressure, radiation, or chemicals? (4) Is the control method suitable for a given application? (For example, ultraviolet radiation is a good sporicidal agent, but it will not penetrate solid materials.) Or, in the case of a chemical, will it leave an undesirable residue? (5) Will the agent penetrate to the necessary extent? (6) Is the method cost- and labor-efficient, and is it safe?

A remarkable variety of substances can require sterilization. They run the gamut from durable solids such as rubber to sensitive liquids such as serum, and from air to tissue grafts. Hundreds of situations requiring sterilization confront the network of persons involved in health care, be it technician, nurse, doctor, or manufacturer, and no universal method works well in every case.

Considerations such as cost, effectiveness, and method of disposal are all important. For example, the disposable plastic items such as catheters and syringes that are used in invasive medical procedures have the potential for infecting the tissues. These must be sterilized during manufacture by a nonheating method (gas or radiation), because heat can damage delicate plastics. After these items have been used, it is often necessary to destroy or decontaminate them before they are discarded because of the potential risk to the handler (needlesticks). Steam sterilization, which is quick and sure, is a sensible choice at this point, because it does not matter if the plastic is destroyed. One of the most demanding situations in microbial control arises when extremely compromised patients must be kept in germ-free environments (microfile 11.2). Health care workers are held to very high standards of infection prevention (see universal precautions, appendix D).

Chapter Checkpoints

- Microbial control methods involve the use of physical and chemical agents to eliminate or reduce the numbers of microorganisms from a specific environment.
- Microbial control methods are used to prevent the spread of infectious agents, retard spoilage, and keep commercial products safe.
- The population of microbes that cause spoilage or infection varies widely in species composition, resistance, and harmfulness, so microbial control methods must be adjusted to fit individual situations.
- The type of microbial control is indicated by the terminology used. Sterilization agents destroy all viable organisms, including viruses. Antisepsis, disinfection, and sanitization agents reduce the numbers of viable microbes to a specified level.
- Antimicrobic agents are described according to their ability to destroy or inhibit microbial growth. Microbicidal agents cause microbial death. They are described by what they are *-cidal* for: sporocides, bactericides, fungicides, viricides.
- An antiseptic agent is applied to living tissue to destroy or inhibit microbial growth.
- A disinfectant agent is used on inanimate objects to destroy vegetative pathogens but not bacterial endospores.
- Sanitization reduces microbial numbers on inanimate objects to safe levels by physical or chemical means.
- Degermation refers to the process of mechanically removing microbes from the skin.
- Microbial death is defined as the permanent loss of reproductive capability in microorganisms.
- Antimicrobic agents attack specific cell sites to cause microbial death or damage. Any given antimicrobic agent attacks one of four major cell targets: the cell wall, the cell membrane, biosynthesis pathways for DNA or RNA, or protein (enzyme) function.

METHODS OF PHYSICAL CONTROL

Microorganisms have adapted to the tremendous diversity of habitats the earth provides, even severe conditions of temperature, moisture, pressure, and light. For microbes that normally withstand such extreme physical conditions, our attempts at control would probably have little effect. Fortunately, the vast majority of

TABLE 11.2

COMPARISON OF TIMES AND TEMPERATURES TO ACHIEVE STERILIZATION WITH MOIST AND DRY HEAT

	Temperature	Time to Sterilize
Moist Heat	121°C	15 min
	125°C	10 min
	134°C	3 min
Dry Heat	121°C	600 min
	140°C	180 min
	160°C	120 min
	170°C	60 min

microbes that must be controlled are not adapted to such extremes and are readily controlled by abrupt changes in environment. Most prominent among antimicrobial physical agents is heat. Other less widely used agents include radiation, filtration, ultrasonic waves, and even cold. The following sections will examine some of these methods and explore their practical applications in medicine, commerce, and the home.

HEAT AS AN AGENT OF MICROBIAL CONTROL

A sudden departure from a microbe's temperature of adaptation is likely to have a detrimental effect on it. As a rule, elevated temperatures (exceeding the maximum) are microbicidal, whereas lower temperatures (below the minimum) tend to have inhibitory or microbistatic effects. The two physical states of heat used in microbial control are moist and dry. **Moist heat** occurs in the form of hot water, boiling water, or steam (vaporized water). In practice, the temperature of moist heat usually ranges from 60° to 135°C. As we shall see, the temperature of steam can be regulated by adjusting its pressure in a closed container. The expression **dry heat** denotes air with a low moisture content that has been heated by a flame or electric heating coil. In practice, the temperature of dry heat ranges from 160°C to several thousand degrees Celsius.

Mode of Action and Relative Effectiveness of Heat

In addition to their physical state, moist and dry heat differ in their efficiency. At a given temperature, moist heat works several times faster than dry heat. At the same length of exposure, moist heat kills cells at a lower temperature than dry heat (table 11.2). Both forms of heat disrupt important cell components, but it appears that their specific modes of action are different. Exposure to moist heat generally coagulates and therefore denatures cell proteins. The greater effectiveness of moist heat has been attributed to this particular mode of action, because protein (enzyme) denaturation occurs more rapidly and at a lower temperature if moisture is present. Components such as the membrane, ribosomes, DNA, and RNA are also damaged by moist heat.

TABLE 11.3

THERMAL DEATH TIMES OF VARIOUS ENDOSPORES

Organism	Temperature	Time of Exposure to Kill Spores
Moist Heat		
Bacillus subtilis	121°C	1 min
B. stearothermophilis	121°C	12 min
Clostridium botulinum	120°C	10 min
C. tetani	105°C	10 min
Dry Heat		
Bacillus subtilis	121°C	120 min
B. stearothermophilis	140°C	5 min
B. xerothermodurans	200°C	139 h
Clostridium botulinum	120°C	120 min
C. tetani	100°C	60 min

TABLE 11.4

AVERAGE THERMAL DEATH TIMES OF VEGETATIVE STAGES OF MICROORGANISMS

Microbial Type	Temperature	Time (Min)
Non-spore-forming pathogenic bacteria	58°C	28
Non-spore-forming nonpathogenic bacteria	61°C	18
Vegetative stage of spore-forming bacteria	58°C	19
Fungal spores	76°C	22
Yeasts	59°C	19
Heat inactivation of viruses		
Nonenveloped	57°C	29
Enveloped	54°C	22
Protozoan trophozoites	46°C	16
Protozoan cysts	60°C	6
Worm eggs	54°C	3
Worm larvae	60°C	10

Dry heat also has several effects on microbes. For instance, it can oxidize cells and, at extremely high temperatures, reduce them to ashes. It dehydrates cell components and can denature proteins and DNA, but proteins are more stable in dry heat than in moist heat, and higher temperatures are required to inactivate them.

Heat Resistance and Thermal Death of Spores and Vegetative Cells

Bacterial endospores exhibit the greatest resistance, and vegetative states of bacteria and fungi are the least resistant to both moist and dry heat. Destruction of spores usually requires temperatures above boiling (table 11.3), although resistance varies widely. In boiling water (100°C), the spores of *Bacillus anthracis* (the agent of anthrax) can be destroyed in a few minutes, whereas the spores of some thermophilic and anaerobic species can require several hours.

Vegetative cells also vary in their sensitivity to heat, though not to the same extent as spores (table 11.4). Among bacteria, the death times with moist heat range from 50°C for 3 minutes (*Neisseria gonorrhoeae*) to 60°C for 60 minutes (*Staphylococcus aureus*). It is worth noting that vegetative cells of sporeformers are just as susceptible as vegetative cells of non-sporeformers and that pathogens are neither more nor less susceptible than nonpathogens. Other microbes, including fungi (yeasts, molds, and some of their spores), protozoa, and worms, are rather similar in their sensitivity to heat. Viruses are surprisingly resistant to heat, with a tolerance range extending from 55°C for 2–5 minutes (adenoviruses) to 60°C for 600 minutes (hepatitis virus). For practical purposes, all non-heat-resistant forms of bacteria, yeasts, molds, protozoa, worms, and viruses are destroyed by exposure to 80°C for 20 minutes.

Practical Concerns in the Use of Heat: Thermal Death Measurements

Adequate sterilization requires that both temperature and length of exposure be considered. As a general rule, higher temperatures allow shorter exposure times, and lower temperatures require longer exposure times. A combination of these two variables constitutes the **thermal death time,** or TDT, defined as the shortest length of time required to kill all microbes at a specified temperature. The TDT has been experimentally determined for the microbial species that are common or important contaminants in various heat-treated materials. Another way to compare the susceptibility of microbes to heat is the thermal death point (TDP), defined as the lowest temperature required to kill all microbes in a sample in 10 minutes.

Many perishable substances are processed with moist heat. Some of these products are intended to remain on the shelf at room temperature for several months or even years. The chosen heat treatment must render the product free of agents of spoilage or disease. At the same time, the quality of the product and the speed and cost of processing must be considered. For example, in the commercial preparation of canned green beans, one of the cannery's greatest concerns is to prevent growth of the agent of botulism. From several possible TDTs for *Clostridium botulinum* spores, the cannery must choose one that kills all spores but does not turn the beans to mush. Out of these many considerations emerges an optimal TDT for a given processing method. Commercial canneries heat low-acid foods at 121°C for 30 minutes, a treatment that sterilizes these foods. Because of such strict controls in canneries, cases of botulism due to commercially canned foods are rare.

Common Methods of Moist Heat Control

The four ways that moist heat is employed to sterilize or disinfect are (1) steam under pressure, (2) live, nonpressurized steam, (3) boiling water, and (4) pasteurization.

Steam Under Pressure A temperature of 100°C is the highest that steam can reach under normal atmospheric pressure at sea level. This pressure is measured at 15 pounds per square inch (psi), or 1 atmosphere. In order to raise the temperature of steam above this point, it must be pressurized in a closed chamber. This phenomenon is explained by the physical principle that governs the behavior of gases under pressure. When a gas is compressed, its temperature rises in direct relation to the amount of pressure. So, when the pressure is increased to 5 psi above normal atmospheric pressure, the temperature of steam rises to 109°C. When the pressure is increased to 10 psi above normal, its temperature will be 115°C, and at 15 psi (a total of 2 atmospheres), it will be 121°C. It is not the pressure by itself that is killing microbes, but the increased temperature it produces.

Such pressure-temperature combinations can be achieved only with a special device that can subject pure steam to pressures greater than 1 atmosphere. Health and commercial industries use an **autoclave** for this purpose, and a comparable home appliance is the pressure cooker. Autoclaves have a fundamentally similar plan: a cylindrical metal chamber with an airtight door on one end and racks to hold materials (figure 11.5). Its construction includes a complex network of valves, pressure and temperature gauges, and ducts for regulating and measuring pressure and conducting the steam into the chamber. Sterilization is achieved when the steam condenses against the objects in the chamber and gradually raises their temperature.

Experience has shown that the most efficient pressure-temperature combination for achieving sterilization is 15 psi, which yields 121°C. It is possible to use higher pressure to reach higher temperatures (for instance, increasing the pressure to 30 psi raises the temperature 11°C), but doing so will not significantly reduce the exposure time and can harm the items being sterilized. It is important to avoid overpacking or haphazardly loading the chamber, which prevents steam from circulating freely around the contents and impedes the full contact that is necessary. The duration of the process is adjusted according to the bulkiness of the items in the load (thick bundles of material or large flasks of liquid) and how full the chamber is. The range of holding times varies from 10 minutes for light loads to 40 minutes for heavy or bulky ones; the average time is 20 minutes.

The autoclave is a superior choice to sterilize heat-resistant materials such as glassware, cloth (surgical dressings), rubber (gloves), metallic instruments, liquids, paper, some media, and some heat-resistant plastics. If the items are heat-sensitive (plastic Petri dishes) but will be discarded, the autoclave is still a good choice. However, the autoclave is ineffective for sterilizing substances that repel moisture (oils, waxes, powders).

Intermittent Sterilization Selected substances that cannot withstand the high temperature of the autoclave can be subjected to **intermittent sterilization,** also called *tyndallization.*[1] This technique requires a chamber to hold the materials and a reservoir for boiling water. Items in the chamber are exposed to free-flowing steam for 30–60 minutes. This temperature is not sufficient to reliably kill spores, so a single exposure will not suffice. On the assumption that surviving spores will germinate into less resistant vegetative cells, the items are incubated at appropriate temperatures for 23–24 hours, and then again subjected to steam treatment. This cycle is repeated for 3 days in a row. Because the temperature never gets above 100°C, highly resistant spores that do not germinate may survive even after 3 days of this treatment.

Intermittent sterilization is used most often to process heat-sensitive culture media, such as those containing sera, egg, or carbohydrates (which can break down at higher temperatures) and some canned foods. It is probably not effective in sterilizing items such as instruments and dressings that provide no environment for spore germination, but it certainly can disinfect them (see next section).

Boiling Water: Disinfection A simple boiling water bath or chamber can quickly decontaminate items in the clinic and home. Because a single processing at 100°C will not kill all resistant cells, this method can be relied on only for disinfection and not for sterilization. When materials are exposed to boiling water for 30 minutes, all non-spore-forming pathogens will be killed, including resistant species such as the tubercle bacillus and staphylococci. Probably the greatest disadvantage with this method is that the items can be easily recontaminated when removed from the water. Boiling is also a recommended method of disinfecting unsafe drinking water. In the home, boiling water is a fairly reliable way to sanitize and disinfect materials for babies, food preparation, and utensils, bedding, and clothing from the sickroom.

Pasteurization: Disinfection of Beverages Fresh beverages such as milk, fruit juices, beer, and wine are easily contaminated during collection and processing. Because microbes have the potential for spoiling these foods or causing illness, heat is frequently used to reduce the microbial load and destroy pathogens. **Pasteurization** (see microfile 8.6) is a technique in which heat is applied to liquids to kill potential agents of infection and spoilage, while at the same time retaining the liquid's flavor and food value. Ordinary pasteurization techniques require special vats and heat exchangers that expose the liquid to 71.6°C for 15 seconds (flash method) or to 63°–66°C for 30 minutes (batch method). The first method is preferable because it is less likely to change flavor and nutrient content, and it is more effective against certain resistant pathogens such as *Coxiella* and *Mycobacterium.* Although these treatments inactivate most viruses and destroy the vegetative stages of 97–99% of bacteria and fungi, they do not kill endospores or *thermoduric* species (mostly nonpathogenic lactobacilli, micrococci, and yeasts). Milk is not sterile after regular pasteurization. In fact, it can contain 20,000 microbes per milliliter or more, which explains why even an unopened carton of milk will eventually spoil. Newer techniques can also produce *sterile milk* that has a storage life of 3 months. This milk is processed with ultrahigh temperature (UHT)–134°C for 1–2 seconds (see chapter 26).

1 Named for the British physicist John Tyndall who did early experiments with sterilizing procedures.

(a)

One important aim in pasteurization is to prevent the transmission of milk-borne diseases from infected cows or milk handlers. The primary targets of pasteurization are non-spore-forming pathogens: *Salmonella* species (a common cause of food infection), *Campylobacter jejeuni* (acute intestinal infection), *Listeria monocytogenes* (listeriosis), *Brucella* species (undulant fever), *Coxiella burnetii* (Q fever), *Mycobacterium bovis* and *M. tuberculosis,* and several enteric viruses. A California epidemic of listeriosis afflicted hundreds of people and resulted in 40 deaths, mostly among infants and elderly persons. Epidemiologists searching for the infection source were first led to a particular type of fresh, soft cheese that had been eaten by the patients and was contaminated with *Listeria.* Later evidence linked the outbreak to milk taken from infected cows and inadequately pasteurized prior to cheesemaking.

Pasteurization also has the advantage of extending milk storage time, and it can also be used by some wineries and breweries to stop fermentation and destroy contaminants.

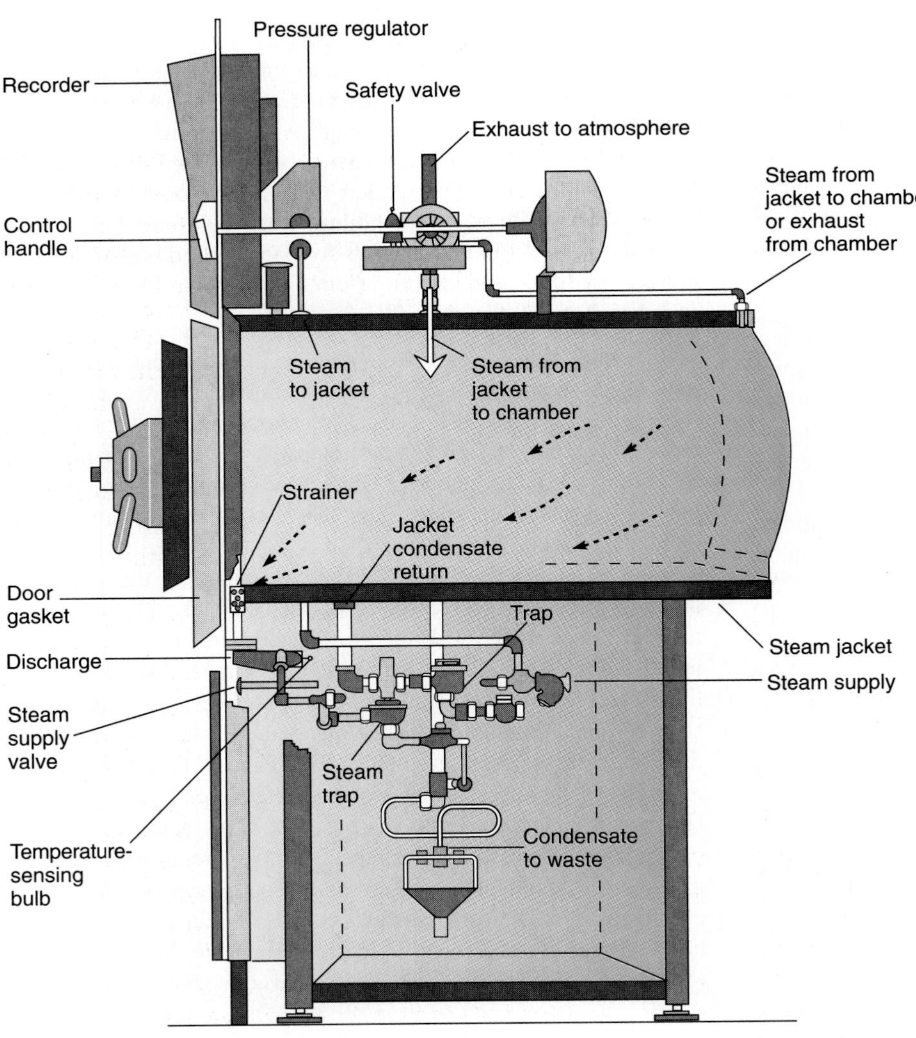

(b)

Figure 11.5

Steam sterilization with the autoclave. (*a*) A large automatic autoclave used in sterilization by drug companies. (*b*) Cutaway section, showing autoclave components.

(*b*) *From John J. Perkins,* Principles and Methods of Sterilization in Health Science, *2nd ed., 1969. Courtesy of Charles C. Thomas, Publisher, Springfield, Illinois.*

Dry Heat: Hot Air and Incineration

Dry heat is not as versatile or as widely used as moist heat, but it has several important sterilization applications. The temperatures and times employed in dry heat vary according to the particular method, but, in general, they are greater than with moist heat. **Incineration** in a flame or electric heating coil is perhaps the most rigorous of all heat treatments. The flame of a Bunsen burner reaches 1,870°C at its hottest point, and furnace/incinerators operate at temperatures of 800°–6,500°C. Direct exposure to such intense heat ignites and reduces the microbes, and sometimes their vehicles, to ashes and gas.

Incineration of microbial samples on inoculating loops and needles using a Bunsen burner is a very common practice in the microbiology laboratory. This method is fast and effective, but it is also limited to metals and heat-resistant glass materials. Incinerators (figure 11.6) are regularly employed in hospitals and research labs for complete destruction and disposal of infectious materials such as syringes, needles, cultural materials, dressings, bandages, bedding, animal carcasses, and pathology samples.

The hot-air oven provides another means of dry-heat sterilization. The so-called **dry oven** is usually electric (occasionally gas) and has coils that radiate heat within an enclosed compartment. Heated, circulated air contacts the items, transferring its heat in the process. Sterilization is accomplished by exposure of items to 150°–180°C for 2 to 4 hours, which ensures thorough heating of the objects and destruction of spores.

The dry oven is used in laboratories and clinics for heat-resistant items that do not sterilize well with moist heat. Substances appropriate for dry ovens are glassware, powders, and oils that steam does not penetrate well and metallic instruments that can be corroded by steam. This method is not suitable for plastics, cotton, and paper, which may burn at the high temperatures, or for solutions, which will dry out. Another limitation is the time required for it to work.

Figure 11.6
Infrared incinerator with shield to prevent spattering of microbial samples during flaming.

THE EFFECTS OF COLD AND DESICCATION

The principal benefit of cold treatment is to slow growth of cultures and microbes in food during processing and storage. *It must be emphasized that cold merely retards the activities of most microbes.* Although it is true that some microbes are killed by cold temperatures, most are not adversely affected by gradual cooling, long-term refrigeration, or deep-freezing. In fact, freezing temperatures, ranging from −70°C to −135°C, provide an environment that can preserve cultures of bacteria, viruses, and fungi for long periods. Some microbes (psychrophiles) grow very slowly even at freezing temperatures and can continue to secrete toxic products. Unawareness of these facts is probably responsible for numerous cases of food poisoning from frozen foods that have been defrosted at room temperature and then inadequately cooked. Pathogens able to survive several months in the refrigerator are *Staphylococcus aureus, Clostridium* species (sporeformers), *Streptococcus* species, and several types of yeasts, molds, and viruses. A recent outbreak of *Salmonella* food infection from homemade ice cream attests to the unreliability of freezing temperatures to kill pathogens.

Vegetative cells directly exposed to normal room air gradually become dehydrated, or *desiccated.** More delicate pathogens such as *Streptococcus pneumoniae,* the spirochete of syphilis, and *Neisseria gonorrhoeae* can die after a few hours of air-drying, but many others are not killed and some are even preserved. Endospores of *Bacillus* and *Clostridium* are viable for thousands of years under extremely arid conditions. Staphylococci and streptococci, if protected by dried secretions (pus, saliva), and the tubercle bacillus, when surrounded by droplets of sputum, can remain viable in air and dust for lengthy periods. Many viruses (especially nonenveloped) and fungal spores can also withstand long periods of desiccation. Desiccation can be a valuable way to preserve foods such as meats, vegetables, and fruits because it greatly reduces the amount of water available to support microbial growth.

It is interesting to note that a combination of freezing and drying—*lyophilization**—is a common method of preserving microorganisms and other cells in a viable state for many years. Pure cultures are frozen instantaneously and exposed to a vacuum that rapidly removes the water (it goes right from the frozen state into the vapor state). This method avoids the formation of ice crystals that would damage the cells. Although not all cells will survive this process, enough of them do to permit future reconstitution of that culture.

As a general rule, chilling, freezing, and desiccation should not be construed as methods of disinfection or sterilization because their antimicrobial effects are erratic and uncertain, and one cannot be sure that pathogens subjected to them have been killed.

*desiccate (des′-ih-kayt) To dry at normal environmental temperatures.

*lyophilization (ly-off″-il-ih-za′-shun) Gr. *lyein,* to dissolve, and *philein,* to love.

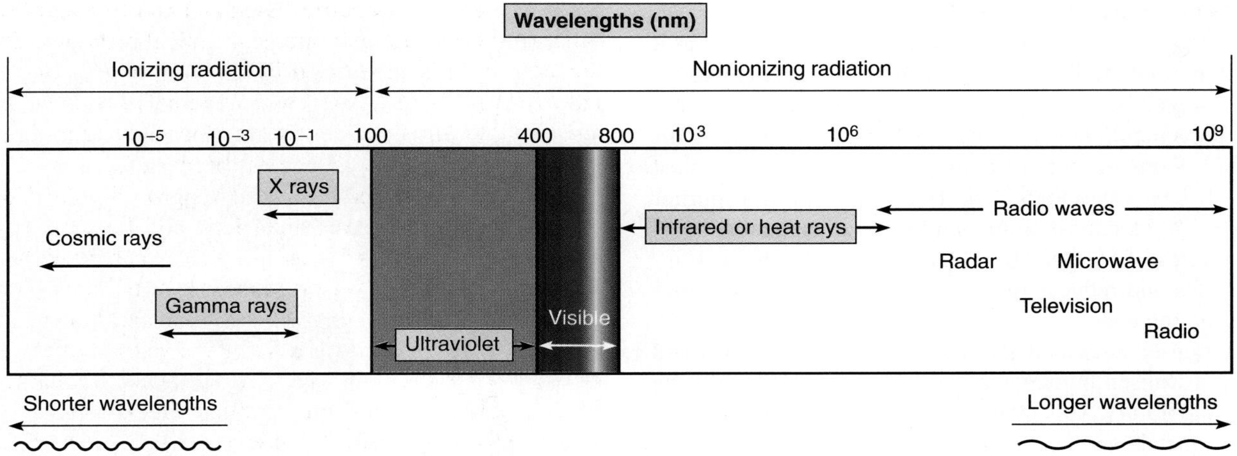

Figure 11.7
Electromagnetic radiation used in chemical control. Short wavelengths of gamma and X rays are the most manageable and practical forms of ionizing radiation. Nonionizing radiation, such as ultraviolet rays in the range of 240 to 280 nm and infrared (heat) rays, are fair microbicidal agents. Other waves are not routinely used in microbial control.

Radiation As a Microbial Control Agent

Another type of energy that exerts antimicrobial effects is radiation. For our purposes, **radiation** is defined as energy emitted from atomic activities and dispensed at high velocity through matter or space. Radiation can behave as waves or as particles, depending upon the conditions. Here we will consider the wavelike character of electromagnetic radiation and the particulate character of particle radiation. The wavelengths of **electromagnetic radiation** range from high-energy, short-wavelength gamma rays at one extreme to low-energy, very long radio waves at the other (figure 11.7). For the most part, only electromagnetic radiations at the gamma ray, X-ray (also called roentgen ray), and ultraviolet ray (UV) levels are suitable for microbial control. **Particle radiation** consists of subatomic particles such as electrons, protons, and neutrons that have been freed from the atom. The only particle radiation feasible for antimicrobial applications is the high-speed electron (also called a β particle or cathode ray).

Visible light does exert some antimicrobial effects, but it is not predictable enough in its effects to control all microbes. Infrared rays can be destructive by virtue of the heat they produce, but they have fewer applications than other heating methods. Microwaves also apparently kill microorganisms through heat production, but the antimicrobial effects of long radio waves are minimal.

Modes of Action of Ionizing Versus Nonionizing Radiation

The actual physical effects of radiation on microbes can be understood by visualizing the process of *irradiation,* or bombardment with radiation, at the cellular level (figure 11.8). When a cell is bombarded by certain waves or particles, its molecules absorb some of the available energy, leading to one of two consequences: (1) If the radiation ejects orbital electrons from an atom, it causes ions to form. This is the effect of **ionizing radiation.** One of the most sensitive targets for ionizing radiation is the DNA molecule, which can sustain mutations on a broad scale. Secondary lethal effects appear to be chemical changes in organelles and the production of toxic substances. Gamma rays, X rays, and high-speed electrons are all ionizing in their effects. (2) **Nonionizing radiation,** best exemplified by UV, excites atoms by raising them to a higher energy state, but it does not ionize them. This atomic excitation, in turn, leads to abnormal linkages within molecules such as DNA and is thus a source of mutations (see chapter 9).

Ionizing Radiation: Gamma Rays, X Rays, and Cathode Rays

Over the past several years, ionizing radiation has become safer and more economical to use, and its applications have mushroomed. It is a highly effective alternative for sterilizing materials that are sensitive to heat or chemicals. Because it sterilizes in the absence of heat, irradiation is a type of *cold sterilization.* Devices that emit ionizing rays include gamma-ray machines containing radioactive cobalt, X-ray machines similar to those used in medical diagnosis, and cathode-ray machines that operate like the vacuum tube in a television set. Items are placed in these machines and irradiated for a short time with a carefully chosen dosage. The dosage is measured in rads (**r**adiation **a**bsorbed **d**ose). Depending on the application, exposure ranges from 0.5 to 5 megarads (Mrad; a megarad is equal to 1,000,000 rads). Although all ionizing radiations can penetrate solids and liquids, gamma rays are most penetrating, X rays are intermediate, and cathode rays least penetrating. The microbial forms with the greatest resistance to radiation are certain gram-positive cocci (*Deinococcus*), because of their extremely efficient DNA repair mechanisms, and bacterial spores,

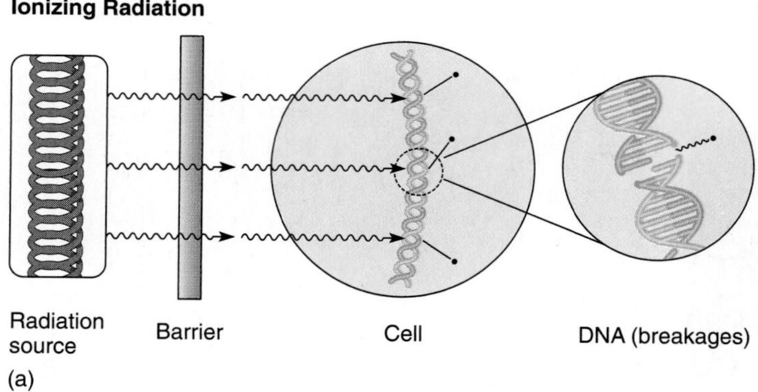

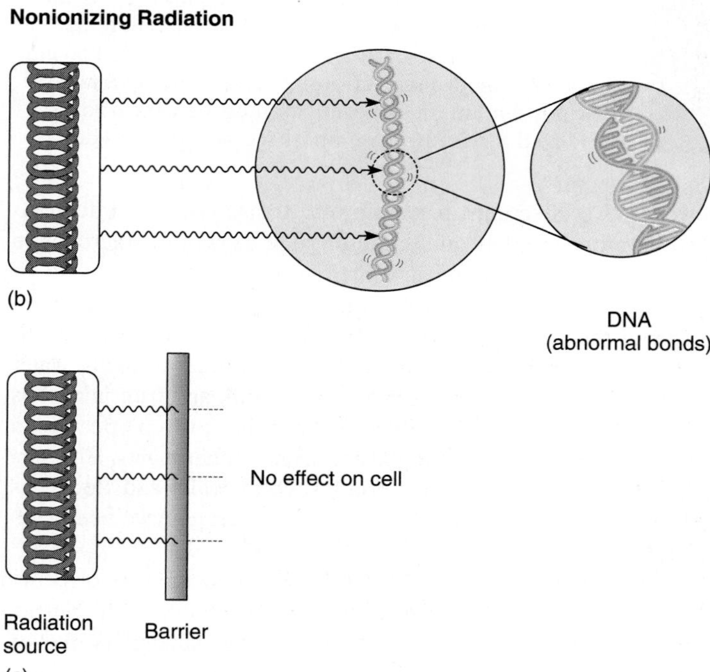

Figure 11.8

Cellular effects of irradiation. (*a*) Ionizing radiation can penetrate a solid barrier, bombard a cell, enter it, and dislodge electrons from molecules. Breakage of DNA creates massive mutations. (*b,c*) Nonionizing radiation. (*b*) Nonionizing radiation enters a cell, strikes molecules, and excites them. The effect on DNA is mutation by formation of abnormal bonds. (*c*) A solid barrier cannot be penetrated by nonionizing radiation.

because of their density. Viruses are second in resistance, followed by yeasts and molds. Vegetative bacterial cells, protozoa, worms, and insects are the most sensitive to radiation.

Applications of Ionizing Radiation

Even though various agencies have been treating food with ionizing radiation for over 40 years, it remains a controversial method. Consumer organizations are concerned that radiation poses a danger both in potential toxicity and in radioactivity that it can then impart to foods. Two known problems that can occur with irradiated food are (1) changes in flavor and nutrition and (2) the possibility of introducing undesirable chemical reactions. These disadvantages have been reduced by carrying out the process in very cold, oxygen-free chambers with extremely low doses of radiation (figure 11.9). Irradiated food cannot be sold to consumers without clear labeling that this method has been used.

Ionizing radiation is currently approved in the United States for cutting down the microbial load and for disinfecting cured meats, some spices, and seasonings. It is also used on fresh pork to destroy *Trichinella* worms (the cause of trichinosis); on chicken carcasses and beef to control *Salmonella* and *E. coli*; on wheat to kill insect eggs; and on fresh fruits and vegetables to cut down on surface microbes that can increase the rate of spoilage. See chapter 26 for further discussion on irradiation of food.

Sterilizing medical products with ionizing radiation is a rapidly expanding field. Drugs, vaccines, medical instruments (especially plastics), syringes, sutures, surgical gloves, and tissues such as bone, skin, and heart valves for grafting all lend themselves to this mode of sterilization. Its main advantages include speed, high penetrating power (it can sterilize materials through outer packages and wrappings), and the absence of heat. Its main disadvantages are potential dangers to machine operators from exposure to radiation and possible damage to some materials.

Nonionizing Radiation: Ultraviolet Rays

Sunlight is the natural source of ultraviolet rays, which accounts for its microbicidal effects. Absorption of most of the sun's UV waves by the atmosphere prevents its full intensity from being felt and thus limits its practical use. Ultraviolet radiation ranges in wavelength from approximately 100 nm to 400 nm. It is most lethal from 240 nm to 280 nm (with a peak at 260 nm). In everyday practice, the source of UV radiation is the germicidal lamp, which generates radiation at 254 nm. Owing to its lower energy state, UV radiation is not as penetrating as ionizing radiation. Because UV radiation passes readily through air, slightly through liquids, and only poorly through solids, the object to be disinfected must be directly exposed to it for full effect.

As UV radiation passes through a cell, it is initially absorbed by DNA. Specific molecular damage occurs on the pyrimidine bases (thymine and cytosine), which form abnormal linkages with each other called *pyrimidine dimers* (figure 11.10). These bonds occur between adjacent bases on the same DNA strand and interfere with normal DNA replication and transcription. The results are inhibition of growth and cellular death. In addition to altering DNA, UV radiation also disrupts cells by generating toxic photochemical products (free radicals). Ultraviolet rays are a powerful tool for destroying fungal cells and spores, bacterial vegetative cells, protozoa, and viruses. Bacterial spores are about 10 times more resistant to radiation than are vegetative cells, but they can be killed by increasing the time of exposure.

Applications of UV Radiation UV radiation is usually directed at disinfection rather than sterilization. Germicidal lamps can cut down on the concentration of airborne microbes as much as 99%. They are used in hospital rooms, operating rooms, schools, food preparation areas, nursing homes, and military housing. Ultraviolet disinfection of air has proved effective in reducing

Radiation room
Chamber with radiation shield
Conveyor system with pallets of sterilized materials
Radioactive source
Automatic Pallet Irradiator.
(a)

(b)

Figure 11.9
(*a*) An irradiation machine that uses radioactive cobalt 60 as a gamma radiation source to sterilize fruits, vegetables, meats, fish, and spices. Although this method has stirred some controversy regarding its safety, it is gaining in acceptance because of its ability to increase shelf-life and reduce food-borne infections. (*b*) Regulations dictate that this symbol for radioactivity must be affixed to all irradiated materials (here, a package of garlic powder).

postoperative infections, preventing the transmission of infections by respiratory droplets, and curtailing the growth of microbes in food-processing plants and slaughterhouses.

Ultraviolet irradiation of liquids requires special equipment to spread the liquid into a thin, flowing film that is exposed directly to a lamp. This method can be used to treat drinking water (figure 11.11) and to purify other liquids (milk and fruit juices) as an alternative to heat. Ultraviolet treatment has proved effective in freeing vaccine antigens and plasma from contaminants. The surfaces of solid, nonporous materials such as walls and floors, as well as meat, nuts, tissues for grafting, and drugs have been successfully disinfected with UV.

One major disadvantage of UV is its poor powers of penetration through solid materials such as glass, metal, cloth, plastic, and even paper. Another drawback to UV is the damaging effect of overexposure on human tissues, including sunburn, retinal damage, cancer, and skin wrinkles.

Sound Waves in Microbial Control

High-frequency sound (sonic) waves beyond the sensitivity of the human ear are known to disrupt cells. These frequencies range from 15,000 to more than 200,000 cycles per second (supersonic to ultrasonic). Such vibrations transmitted through a water-filled chamber (sonicator) induce pressure changes and minute, bubble-like cavities in the liquid, a phenomenon called **cavitation.** The short-lived bubbles swell and collapse, creating intense points of turbulence that can stress and burst cells in the vicinity (figure 11.12). Gram-negative rods are most sensitive to ultrasonic vibrations, and gram-positive cocci, fungal spores, and bacterial spores are most resistant to them. Sonication also forcefully dislodges foreign matter from objects. Heat generated by sonic waves (up to 80°C) also appears to contribute to the antimicrobial action. Ultrasonic devices are used in dental and

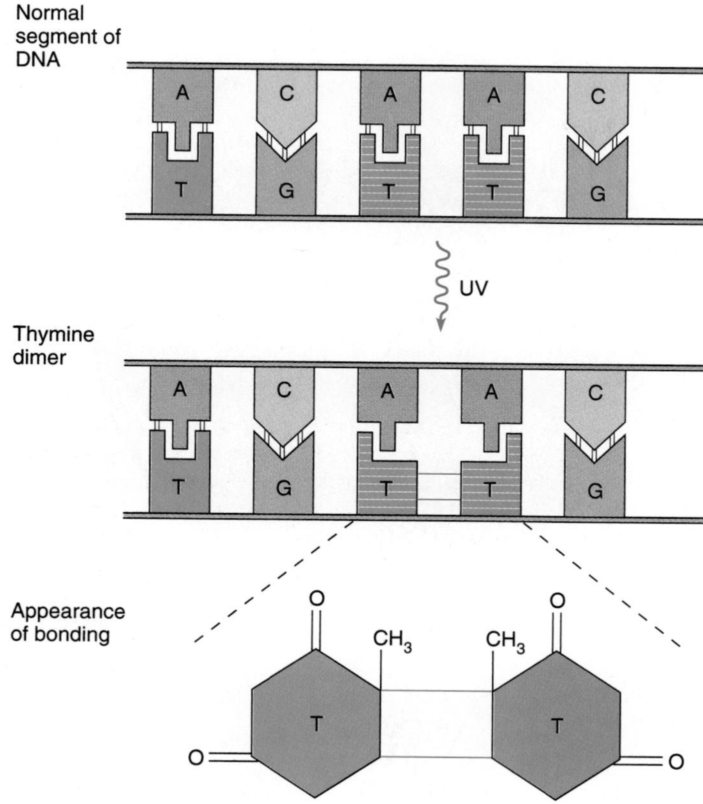

Figure 11.10
Formation of pyrimidine dimers by the action of ultraviolet (UV) radiation. This shows what occurs when two adjacent thymine bases on one strand of DNA are induced by UV rays to bond laterally with each other. The result is a thymine dimer. These dimers can also occur between adjacent cytosines and the thymine and cytosine bases. If they are not repaired, dimers can prevent that segment of DNA from being correctly replicated or transcribed. Massive dimerization is lethal to cells.

some medical offices to clear debris and saliva from instruments before sterilization and to clean dental restorations. However, most sonic machines are not predictable enough to be used in disinfection or sterilization. Other types of ultrasonic devices are available for medical diagnosis and for removing plaque and calculus from teeth.

Sterilization by Filtration: Techniques for Removing Microbes

Filtration is an effective method to remove microbes from air and liquids. In practice, a fluid is strained through a filter with openings large enough for the fluid to pass through but too small for microorganisms to pass through (figure 11.13).

Most modern microbiological filters are thin membranes of cellulose acetate, polycarbonate, and a variety of plastic materials (Teflon, nylon) whose pore size can be carefully controlled and standardized. Ordinary substances such as charcoal, diatomaceous earth, or unglazed porcelain are also used in some applications. Viewed microscopically, most filters are perforated by very precise, uniform pores (figure 11.13*b*). The pore diameters vary from coarse (8 μm) to ultrafine (0.02 μm), permitting selection of the minimum particle size to be trapped. Those with the smallest pore diameters permit true sterilization by removing viruses, and some will even remove large proteins. A sterile liquid filtrate is typically produced by suctioning the liquid through a sterile filter into a presterilized container. These filters are also used to separate mixtures of microorganisms and to enumerate bacteria in water analysis (see chapter 26).

Applications of Filtration Sterilization Filtration sterilization is used to prepare liquids that cannot withstand heat, including serum and other blood products, vaccines, drugs, IV

Figure 11.11
An ultraviolet (UV) treatment system for disinfection of wastewater. Water flows through racks of UV lamps and is exposed to 254 nm UV radiation. This system has a capacity of several million gallons per day and can be used as an alternative to chlorination.

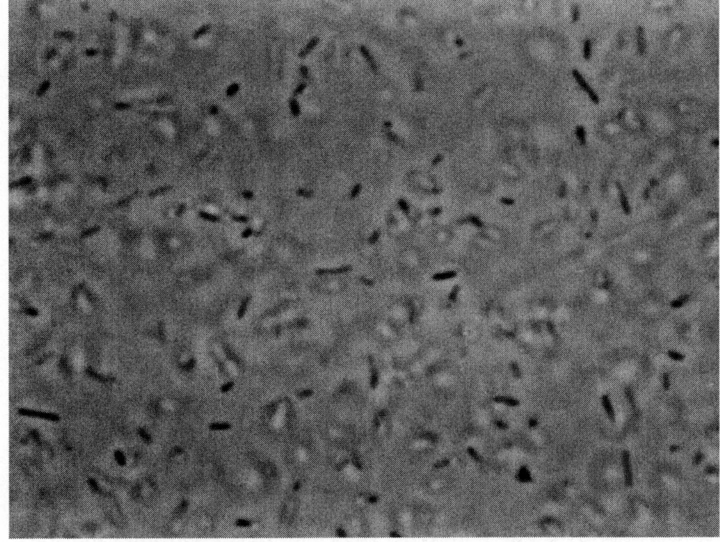

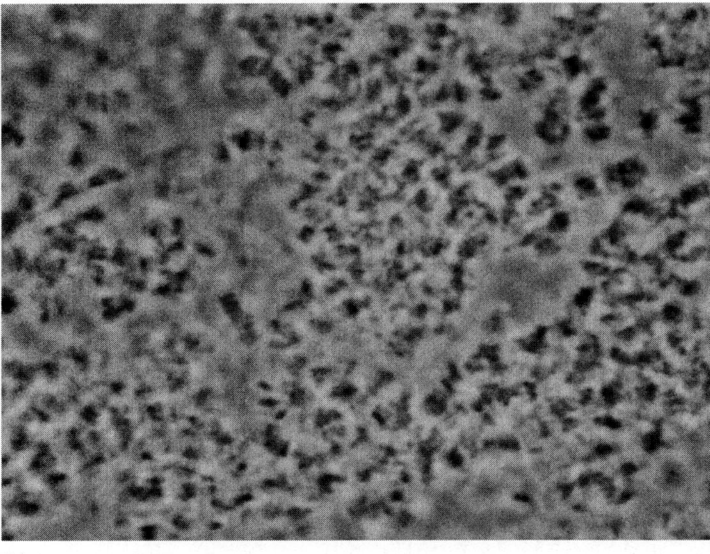

Figure 11.12
(*a*) Phase-contrast microscopic views of *Escherichia coli* cells before being subjected to ultrasonic vibrations (sonication). (*b*) Masses of fragmented bacteria after sonication.

fluids, enzymes, and media. Filtration has been employed as an alternative to sterilize milk and beer without altering their flavor. It is also an important step in water purification. Its usage extends to filtering out particulate impurities (crystals, fibers, and so on) that can cause severe reactions in the body. It has the disadvantage of not removing soluble molecules (toxins) that can cause disease. Filtration is also an efficient means of removing airborne contaminants that are a common source of infection and spoilage. High efficiency particulate air (HEPA) filters are widely used to provide a flow of sterile air to hospital rooms and sterile rooms.

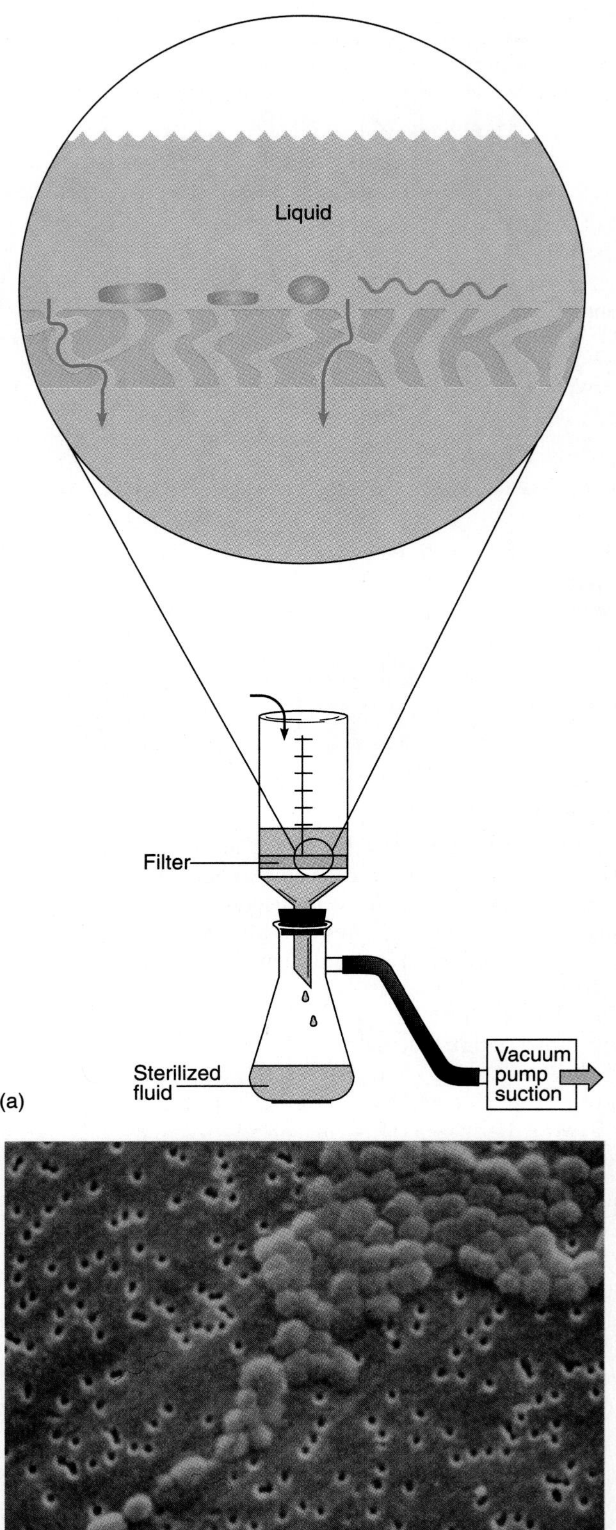

Figure 11.13
Membrane filtration. (*a*) Vacuum assembly for achieving filtration of liquids through suction. Inset shows filter as seen in cross section, with tiny passageways (pores) too small for the microbial cells to enter but large enough for liquid to pass through.
(*b*) Scanning electron micrograph of filter, showing relative size of pores and bacteria trapped on its surface (5,900×).

Chapter Checkpoints

- Physical methods of microbial control include heat, cold, radiation, and drying.
- Heat is the most widely used method of microbial control. It is used in combination with water (moist heat) or as dry heat (oven, flames).
- The thermal death time (TDT) is the shortest length of time required to kill all microbes at a specific temperature. The TDT is longest for spore-forming bacteria and certain viruses.
- The thermal death point (TDP) is the lowest temperature at which all microbes are killed in a specified length of time (10 minutes).
- Autoclaving, or steam sterilization, is the process by which steam is heated under pressure to sterilize a wide range of materials in a comparatively short time (minutes to hours). It is effective for most materials except water-resistant substances such as oils, waxes, and powders.
- Boiling water and pasteurization of beverages disinfect but do not sterilize materials.
- Dry heat is microbicidal under specified times and temperatures. Flame heat, or incineration, is microbicidal. It is used when total destruction of microbes and materials is indicated.
- Chilling, freezing, and desiccation are microbistatic but not microbicidal. They are *not* considered true methods of disinfection because they are not consistent in their effectiveness.
- Ionizing radiation or cold sterilization by gamma rays and X rays is used to sterilize medical products, meats, and spices. It damages DNA and cell organelles by producing disruptive ions.
- Ultraviolet light, or nonionizing radiation, has limited penetrating ability. It is therefore restricted to disinfecting air and certain liquids.
- Ultrasound is microbistatic to most microbes, but it is microbicidal to gram-negative bacteria. It is used primarily to reduce microbial load from inanimate objects.
- Sterilization by filtration removes microbes from heat-sensitive liquids and circulating air. The pore size of the filter determines what kinds of microbes are removed.

CHEMICAL AGENTS IN MICROBIAL CONTROL

Chemical control of microbes probably emerged as a serious science in the early 1800s, when physicians used chloride of lime and iodine solutions to treat wounds and to wash their hands before surgery. At the present time, approximately 10,000 different antimicrobial chemical agents are manufactured; probably 1,000 of them are used routinely in the allied health sciences and the home. There is a genuine need to avoid infection and spoilage, but the abundance of products available to "kill germs, disinfect, antisepticize, clean and sanitize, deodorize, fight plaque, and purify the air" indicates a preoccupation with eliminating microbes from the environment that, at times, seems excessive (microfile 11.3).

Antimicrobial chemicals occur in the liquid, gaseous, or even solid state and vary from disinfectants and antiseptics to sterilants and preservatives (chemicals that inhibit the deterioration of substances). Liquid agents contain water, alcohol, or a mixture of the two, with various dissolved solutes. Solutions containing pure water as the solvent are termed *aqueous,* whereas those with pure alcohol or water-alcohol mixtures are termed *tinctures.*

CHOOSING A MICROBICIDAL CHEMICAL

The choice and appropriate use of antimicrobial chemical agents is of constant concern in medicine and dentistry. Although actual clinical practices of chemical decontamination vary widely, some desirable qualities in a germicide have been identified, including: (1) rapid action even in low concentrations, (2) solubility in water or alcohol and long-term stability, (3) broad-spectrum microbicidal action without being toxic to human and animal tissues, (4) penetration of inanimate surfaces to sustain a cumulative or persistent action, (5) resistance to becoming inactivated by organic matter, (6) noncorrosive or nonstaining properties, (7) sanitizing and deodorizing properties, and (8) inexpensiveness and ready availability. As yet, no chemical can completely fulfill all of those requirements, but glutaraldehyde and hydrogen peroxide approach this ideal. At the same time, we should question the rather overinflated claims made about certain commercial agents such as mouthwashes and disinfectant air sprays.

Germicides are evaluated in terms of their effectiveness in destroying microbes on medical and dental materials. The three levels of chemical decontamination procedures are *high, intermediate,* and *low* (table 11.5). High-level germicides kill endospores, and, if properly used, are sterilants. Materials that necessitate high-level control are medical devices—for example, catheters, heart-lung equipment, and implants—that are not heat-sterilizable and are intended to enter body tissues during medical procedures. Intermediate-level germicides kill fungal (but not bacterial) spores, resistant pathogens such as the tubercle bacillus, and viruses. They are used to disinfect items (respiratory equipment, endoscopes) that come into intimate contact with the mucous membranes but are noninvasive. Low levels of disinfection eliminate only vegetative bacteria, vegetative fungal cells, and some viruses. They are required for materials such as electrodes, straps, and furniture that touch the skin surfaces but not the mucous membranes.

FACTORS THAT AFFECT THE GERMICIDAL ACTIVITY OF CHEMICALS

Factors that control the effect of a germicide include the nature of the microorganisms being treated, the nature of the material being treated, the degree of contamination, the time of exposure, and

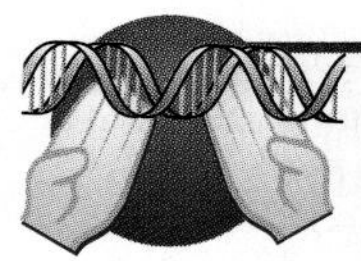

MICROFILE 11.3 PATHOGEN PARANOIA: "THE ONLY GOOD MICROBE IS A DEAD MICROBE"

The sensational publicity over outbreaks of infections such as flesh-eating disease, tuberculosis, and microbial food poisoning has monumentally influenced the public view of microorganisms. An estimated 8,000 such stories have sprinkled the news services over the years 1995–1997. On the positive side, this glut of information has improved people's awareness of the importance of microorganisms. And, certainly, such knowledge can be seen as beneficial when it leads to well-reasoned and sensible choices, such as using greater care in handwashing, food handling, and personal hygiene. But sometimes a little knowledge can be dangerous. The trend also seems to have escalated into an obsessive fear of "germs" lurking around every corner and a fixation on eliminating microbes from the environment and the human body.

As might be expected, commercial industries have found a way to capitalize on those fears. Since 1996, nearly 200 new products that contain antibacterial or germicidal chemicals have been marketed. A widespread array of cleansers and commonplace materials that could possibly control microorganisms have already had antimicrobic chemicals added. First it was hand soaps and dishwashing detergents, and eventually the list grew to include shampoos, laundry aids, hand lotions, foot pads for shoes, deodorants, sponges and scrub pads, kitty litter, acne medication, cutting boards, garbage bags, toys, and toothpaste. It is likely that many more products will be introduced in the near future.

By far, the prevalent chemical agent routinely added to these products is a phenolic called *triclosan* (Irgasan). This substance is fairly mild and nontoxic and does indeed kill most pathogenic bacteria. However, it does not reliably destroy viruses or fungi and has been linked to cases of skin rashes due to hypersensitivity.

Medical experts are concerned that the widespread overuse of these antibacterial chemicals could favor the survival and growth of resistant strains of bacteria. Many of these pathogens could cause serious infection and disease, especially in children and adults with weakened host defenses.

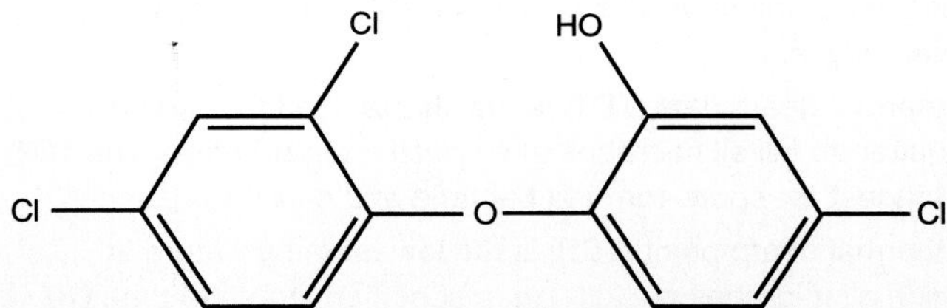

The molecular structure of triclosan, also known as Irgasan and Ster-Zac, a phenol-based chemical that destroys bacteria by disrupting cell walls and membranes.

Another unfortunate result of the negative news on microbes is how it fosters the feeling that all microbes are harmful, overlooking the idea of their contributions to health. Overuse of environmental germicides promises to reduce the natural contact with microbes that is required to maintain the normal resident flora and stimulate immunities. Constant use of these agents could shift the balance in the normal flora of the body by killing off harmless or beneficial microbes.

Infectious disease specialists urge a happy medium approach. Instead of filling the home with questionable germicidal products, they encourage cleaning with traditional soaps and detergents, reserving more potent products to reduce the spread of infection among household members.

the strength and chemical action of the germicide (table 11.6). Standardized procedures for testing the effectiveness of germicides are summarized in appendix C. The modes of action of most germicides involve the cellular targets discussed in an earlier section of this chapter: proteins, nucleic acids, the cell wall, and the cell membrane.

A chemical's strength or concentration is expressed in various ways, depending upon convention and the method of preparation. The content of many chemical agents can be expressed by more than one notation. In dilutions, a small volume of the liquid chemical (solute) is diluted in a larger volume of solvent to achieve a certain ratio. For example, a common laboratory phenolic disinfectant such as Amphyl is usually diluted 1:200; that is, one part of chemical has been added to 200 parts of water by volume. Solutions such as chlorine that are effective in very high dilutions are expressed in parts per million (ppm). In percent solutions, the solute is added to water by weight or volume to achieve a certain percentage in the solution. Alcohol, for instance, is used in percentages ranging from 50% to 95%. In general, solutions of low dilution or high percentage have more of the active chemical (are more concentrated) and tend to be more germicidal, but expense and potential toxicity can necessitate using the minimum strength that is effective.

Another factor that contributes to germicidal effectiveness is the length of exposure. Most compounds require adequate contact time to allow the chemical to penetrate and to act on the microbes present. The composition of the material being treated must also be considered. Smooth, solid objects are more reliably disinfected than are those with pores or pockets that can trap soil. An item contaminated with common biological matter such as serum, blood, saliva, pus, fecal material, or urine presents a problem in disinfection. Large amounts of organic material can hinder

TABLE 11.5

QUALITIES OF CHEMICAL AGENTS USED IN ALLIED HEALTH

Agent	Target Microbes	Level of Activity	Toxicity	Comments
Chlorine	Sporicidal (slowly)	Intermediate	Gas is highly toxic; solution irritates skin	Inactivated by organics; unstable in sunlight
Iodine	Sporicidal (slowly)	Intermediate	Can irritate tissue; toxic if ingested	Iodophors* are milder forms
Phenolics	Some bacteria, viruses, fungi	Intermediate to low	Can be absorbed by skin; can cause CNS damage	Poor solubility; expensive
Alcohols	Most bacteria, viruses, fungi	Intermediate	Toxic if ingested; a mild irritant; dries skin	Inflammable, fast-acting
Hydrogen peroxide,* stabilized	Sporicidal	High	Toxic to eyes; toxic if ingested	Improved stability; works well in organic matter
Quaternary ammonium compounds	Some bactericidal, virucidal, fungicidal activity	Low	Irritating to mucous membranes; poisonous if taken internally	Weak solutions can support microbial growth; easily inactivated
Soaps	Certain very sensitive species	Very low	Very bland; few if any toxic effects	Used for removing soil
Mercurials	Weakly microbistatic	Low	Highly toxic if ingested, inhaled, absorbed	Easily inactivated
Silver nitrate	Bactericidal	Low	Toxic, irritating	Discolors skin
Glutaraldehyde*	Sporicidal	High	Can irritate skin; toxic if absorbed	Not inactivated by organic matter; unstable
Formaldehyde	Sporicidal	High to intermediate	Very irritating; fumes damaging, carcinogenic	Slow rate of action
Ethylene oxide gas*	Sporicidal	High	Very dangerous to eyes, lungs; carcinogenic	Explosive in pure state; good penetration; materials must be aerated
Dyes	Weakly bactericidal, fungicidal	Low	Low toxicity	Stain materials, skin
Chlorhexidine*	Most bacteria, some viruses, fungi	Low to intermediate	Low toxicity	Fast-acting, mild, has residual effects

*These forms approach the ideal by having some of the following characteristics: broad spectrum, low toxicity, fast action, penetrating abilities, residual effects, stability, potency in organic matter, and solubility.

the penetration of a disinfectant and, in some cases, form bonds that reduce its activity. Adequate cleaning of instruments and other reusable materials ensures that the germicide or sterilant will better accomplish the job for which it was chosen.

GERMICIDAL CATEGORIES ACCORDING TO CHEMICAL GROUP

Several general groups of chemical compounds are widely used for antimicrobial purposes in medicine and commerce (see table 11.5). Prominent agents include halogens, heavy metals, alcohols, phenolic compounds, oxidizers, aldehydes, detergents, and gases. These groups will be surveyed in the following section from the standpoint of each agent's specific forms, modes of action, indications for use, and limitations.

The Halogen Antimicrobial Chemicals

The **halogens*** are fluorine, bromine, chlorine, and iodine, a group of nonmetallic elements that commonly occur in minerals, seawater, and salts. Although they can exist in either the ionic (halide) or nonionic state, most halogens exert their antimicrobial effect primarily in the nonionic state, not the halide state (chloride, iodide, for example). Because fluorine and bromine are difficult and dangerous to handle, and are no more effective than chlorine and iodine, only the latter two are used routinely in germicidal preparations. These elements are highly effective components of disinfectants and antiseptics because they are microbicidal and not just microbistatic, and they are sporicidal with longer exposure. For these reasons halogens are the active ingredients in nearly one-third of all antimicrobial chemicals currently marketed.

*halogens (hay´-loh-jenz) Gr. *halos,* salt, and *gennan,* to produce.

TABLE 11.6

REQUIRED CONCENTRATIONS AND TIMES FOR CHEMICAL DESTRUCTION OF SELECTED MICROBES

Organism	Concentration	Time
Agent: Aqueous Iodine		
*Staphylococcus aureus**	2%	2 min
*Escherichia coli***	2%	1.5 min
Enteric viruses	2%	10 min
Agent: Chlorine		
*Mycobacterium tuberculosis**	50 ppm	50 sec
Entamoeba cysts (protozoa)	0.1 ppm	150 min
Hepatitis A virus	3 ppm	30 min
Agent: Phenol		
Staphylococcus aureus	1:85 dil	10 min
Escherichia coli	1:75 dil	10 min
Agent: Ethyl Alcohol		
Staphylococcus aureus	70%	10 min
Escherichia coli	70%	2 min
Poliovirus	70%	10 min
Agent: Hydrogen Peroxide		
Staphylococcus aureus	3%	12.5 sec
Neisseria gonorrhoeae	3%	0.3 sec
Herpes simplex virus	3%	12.8 sec
Agent: Quaternary Ammonium Compound		
Staphylococcus aureus	450 ppm	10 min
*Salmonella typhi***	300 ppm	10 min
Agent: Silver Ions		
Staphylococcus aureus	8 μg/ml	48 h
Escherichia coli	2 mg/ml	48 h
Candida albicans (yeast)	14 mg/ml	48 h
Agent: Glutaraldehyde		
Staphylococcus aureus	2%	<1 min
Mycobacterium tuberculosis	2%	<10 min
Herpes simplex virus	2%	<10 min
Agent: Ethylene Oxide Gas		
Streptococcus faecalis	500 mg/l	2–4 min
Influenza virus	10,000 mg/l	25 h
Agent: Chlorhexidine		
Staphylococcus aureus	1:10 dil	15 sec
Escherichia coli	1:10 dil	30 sec

*Gram-positive vegetative bacterium.

**Gram-negative vegetative bacterium.

Chlorine and Its Compounds Chlorine has been used for disinfection and antisepsis for approximately 200 years. The major forms used in microbial control are liquid and gaseous chlorine (Cl_2), hypochlorites (OCl), and chloramines (NH_2Cl). In solution, these compounds combine with water and release hypochlorous acid (HOCl), which oxidizes the sulfhydryl (S—H) group on the amino acid cysteine and interferes with disulfide (S—S) bridges on numerous enzymes. The resulting denaturation of the enzymes is permanent and suspends metabolic reactions. Chlorine kills not only bacterial cells and endospores but also fungi and viruses. The major limitations of chlorine compounds are: (1) They are ineffective if used at an alkaline pH; (2) excess organic matter can greatly reduce their activity; and (3) they are relatively unstable, especially if exposed to light.

Chlorine Compounds in Disinfection and Antisepsis Gaseous and liquid chlorine are used almost exclusively for large-scale disinfection of drinking water, sewage, and wastewater from such sources as agriculture and industry. Chlorination to a concentration of 0.6 to 1.0 parts of chlorine per million parts of water will ensure that water is safe to drink. This rids the water of most pathogenic vegetative microorganisms without unduly affecting its taste (some persons may debate this).

Hypochlorites are perhaps the most extensively used of all chlorine compounds. The scope of applications is broad, including sanitization and disinfection of food equipment in dairies, restaurants, and canneries and treatment of swimming pools, spas, drinking water, and even fresh foods. Hypochlorites are used in the allied health areas to treat wounds and to disinfect equipment, bedding, and instruments. Common household bleach is a weak solution (5%) of sodium hypochlorite that serves as an all-around disinfectant, deodorizer, and stain remover.

Chloramines (dichloramine, halazone) are being employed more frequently as an alternative to pure chlorine in treating water supplies. Because standard chlorination of water is now believed to produce unsafe levels of cancer-causing substances such as trihalomethanes, some water districts have been directed by federal agencies to adopt chloramine treatment of water supplies. Chloramines also serve as sanitizers and disinfectants and for treating wounds and skin surfaces.

Iodine and Its Compounds Iodine is a pungent black chemical that forms brown-colored solutions when dissolved in water or alcohol. The two primary iodine preparations are *free iodine* in solution (I_2) and *iodophors.* Iodine rapidly penetrates the cells of microorganisms, where it apparently disturbs a variety of metabolic functions by interfering with the hydrogen and disulfide bonding of proteins (similar to chlorine). All classes of microorganisms are killed by iodine if proper concentrations and exposure times are used. Iodine activity is not as adversely affected by organic matter and pH as chlorine is.

Applications of Iodine Solutions **Aqueous iodine** contains 2% iodine and 2.4% sodium iodide; it is used as a topical antiseptic before surgery and occasionally as a treatment for burned and infected skin. A stronger iodine solution (5% iodine and 10% potassium iodide) is used primarily as a disinfectant for plastic items, rubber instruments, cutting blades, and thermometers. **Iodine tincture** is a 2% solution of iodine and sodium iodide in 70% alcohol that can be used in skin antisepsis. Because iodine can be extremely irritating to the skin and toxic when absorbed, strong aqueous solutions and tinctures (5–7%) are no longer considered safe for routine antisepsis. Iodine tablets are available for disinfecting water during emergencies or destroying pathogens in impure water supplies.

Iodophors are complexes of iodine and a neutral polymer such as a polyvinylalcohol. This formulation allows the slow release of free iodine and increases its degree of penetration. These compounds have largely replaced free iodine solutions in medical antisepsis because they are less prone to staining or irritating tissues. Common iodophor products marketed as Betadine, Povidone (PVP), and Isodine contain 2–10% of available iodine. They are used to prepare skin and mucous membranes for surgery and injections, in surgical handscrubs, to treat burns, and to disinfect equipment and surfaces. A recent study showed that Betadine solution is an effective means of preventing eye infections in newborn infants, and it may replace antibiotics and silver nitrate as the method of choice.

Phenol and Its Derivatives

Phenol (carbolic acid) is an acrid, poisonous compound derived from the distillation of coal tar. First adopted by Joseph Lister in 1867 as a surgical germicide, phenol was the major antimicrobial chemical until other phenolics with fewer toxic and irritating effects were developed. Solutions of phenol are now used only in certain limited cases, but it remains one standard against which other phenolic disinfectants are rated. Substances chemically related to phenol are often referred to as phenolics. Hundreds of these chemicals are now available.

Phenolics consist of one or more aromatic carbon rings with added functional groups (figure 11.14). Among the most important are alkylated phenols (cresols), chlorinated phenols, and bisphenols. The mode of action of these compounds has been thoroughly studied. In high concentrations, they are cellular poisons, rapidly disrupting cell walls and membranes and precipitating proteins; in lower concentrations, they inactivate certain critical enzyme systems. The phenolics are strongly microbicidal and will destroy vegetative bacteria (including the tubercle bacillus), fungi, and most viruses (not the hepatitis B virus), but they are not reliably sporicidal. Their continued activity in the presence of organic matter and their detergent actions contribute to their usefulness. Unfortunately, the toxicity of many of the phenolics makes them too dangerous to use as antiseptics.

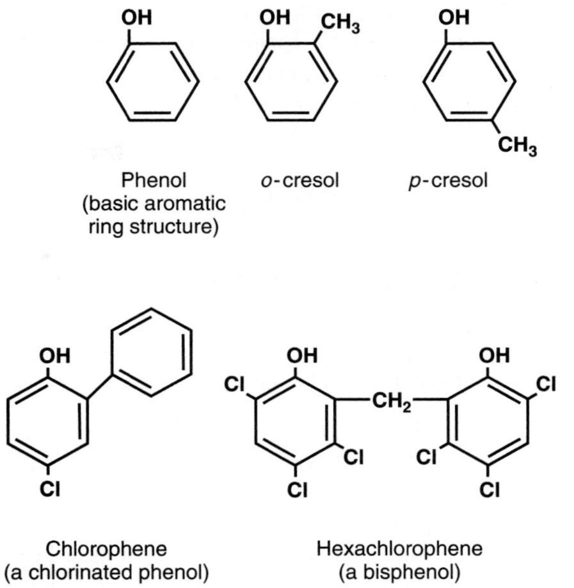

Applications of Phenolics Phenol itself is still used for general disinfection of drains, cesspools, and animal quarters, but it is seldom applied as a medical germicide. The cresols are simple phenolic derivatives that are combined with soap for intermediate or low levels of disinfection in the hospital. Lysol and creolin, in a 1–3% emulsion, are common household versions of this type.

The bisphenols are also widely employed in commerce, clinics, and the home. One type, orthophenyl phenol, is the major ingredient in disinfectant aerosol sprays. This same phenolic is also found in some proprietary compounds (Amphyl, O-syl) often used in hospital and laboratory disinfection. One particular bisphenol, hexachlorophene, was once a common additive of cleansing soaps (pHisoHex) used in the hospital and home. When it was discovered that hexachlorophene is absorbed through the skin and can cause neurological damage, it was no longer available without a prescription. It is occasionally used to control outbreaks of skin infections.

Perhaps the most widely used phenolic is **triclosan,** chemically known as dichlorophenoxyphenol (see microfile 11.3). It is the antibacterial compound added to dozens of products, from soaps to kitty litter. It acts as both a disinfectant and antiseptic and is broad-spectrum in its effects.

Phenol (basic aromatic ring structure) | *o*-cresol | *p*-cresol

Chlorophene (a chlorinated phenol) | Hexachlorophene (a bisphenol)

Figure 11.14

Some phenolics. All contain a basic aromatic ring, but they differ in the types of additional compounds such as Cl and CH_3.

Chlorhexidine

The compound chlorhexidine (Hibiclens, Hibitane) is a complex organic base containing chlorine and two phenolic rings. It is somewhat related to the cationic detergents in its mode of action, being both a surfactant and a protein denaturant. At moderate to high concentrations, it is bactericidal for both gram-positive and gram-negative bacteria but inactive against spores. Its effects on viruses and fungi vary. It possesses distinct advantages over many other antiseptics because of its mildness, low toxicity, and rapid action, and it is not absorbed into deeper tissues to any extent. Alcoholic or aqueous solutions of chlorhexidine are now commonly used for handscrubbing, preparing skin sites for surgical incisions and injections, and whole body washing. Chlorhexidine solution also serves as an obstetric antiseptic, a neonatal wash, a wound degermer, a mucous membrane irrigant, and a preservative for eye solutions.

Alcohols As Antimicrobial Agents

Alcohols are colorless hydrocarbons with one or more —OH functional groups. Of several alcohols available, only ethyl and isopropyl are suitable for microbial control. Methyl alcohol is not very microbicidal, and more complex alcohols are either poorly soluble in water or too expensive for routine use. Alcohols are employed alone in aqueous solutions or as solvents for tinctures (iodine, for example). Alcohol's mechanism of action depends in part upon its concentration. Concentrations of 50% and higher dissolve membrane lipids, disrupt cell surface tension, and compromise membrane integrity. Alcohol that has entered the protoplasm denatures proteins through coagulation, but only in alcohol-water solutions of 50–95%. Absolute alcohol (100%) dehydrates cells and inhibits their growth but is generally not a protein coagulant.

Although useful in intermediate- to low-level germicidal applications, alcohol does not destroy bacterial spores at room temperature. Alcohol can, however, destroy resistant vegetative forms, including tubercle bacilli and fungal spores, provided the time of exposure is adequate. Alcohol is generally more effective in inactivating enveloped viruses than the more resistant nonenveloped viruses such as poliovirus and hepatitis A virus.

Applications of Alcohols Ethyl alcohol, also called ethanol or grain alcohol, is known for being germicidal, nonirritating, nontoxic, and inexpensive. Solutions of 70% to 95% are routinely used as skin degerming agents because the surfactant action removes skin oil, soil, and some microbes sheltered in deeper skin layers. One limitation to its effectiveness is the rate at which it evaporates. Ethyl alcohol is occasionally used to disinfect electrodes, face masks, and thermometers, which are first cleaned and then soaked in alcohol for 15–20 minutes. Isopropyl alcohol, sold as rubbing alcohol, is even more microbicidal and less expensive than ethanol, but these benefits must be weighed against its toxicity. It must be used with caution in disinfection or skin cleansing, because inhalation of its vapors can adversely affect the nervous system.

Hydrogen Peroxide and Related Germicides

Hydrogen peroxide (H_2O_2) is a colorless, caustic liquid that decomposes in the presence of light, metals, or catalase into water and oxygen gas. Peroxide solutions have been used for about 50 years. Early formulations were unstable and inhibited by organic matter, but manufacturing methods now permit synthesis of H_2O_2 so stable that even dilute solutions retain activity through several months of storage.

The germicidal effects of hydrogen peroxide are due to the direct and indirect actions of oxygen. One toxic product of oxygen in cells is hydroxyl free radicals (.OH), which, like the superoxide radical (see chapter 7), are highly reactive. Although most microbial cells produce catalase to inactivate the metabolic hydrogen peroxide, it cannot neutralize the amount of hydrogen peroxide entering the cell during disinfection and antisepsis. Hydrogen peroxide is bactericidal, virucidal, and fungicidal and, in higher concentrations, sporicidal.

Applications of Hydrogen Peroxide As an antiseptic, 3% hydrogen peroxide serves a variety of needs, including skin and wound cleansing, bedsore care, and mouthwashing. It is especially useful in treating infections by anaerobic bacteria because of the lethal effects of oxygen on these forms. Hydrogen peroxide is also a versatile disinfectant for soft contact lenses, surgical implants, plastic equipment, utensils, bedding, and room interiors. Hydrogen peroxide solutions and vapors are the latest method in flash sterilization of food packaging equipment and other industrial processes.

Other compounds with effects similar to those of hydrogen peroxide are ozone (O_3), occasionally used to disinfect air and water, and peracetic acid, a potent oxidizing agent sometimes used to inhibit fungi on fruits and to sterilize pacemakers and other medical devices.

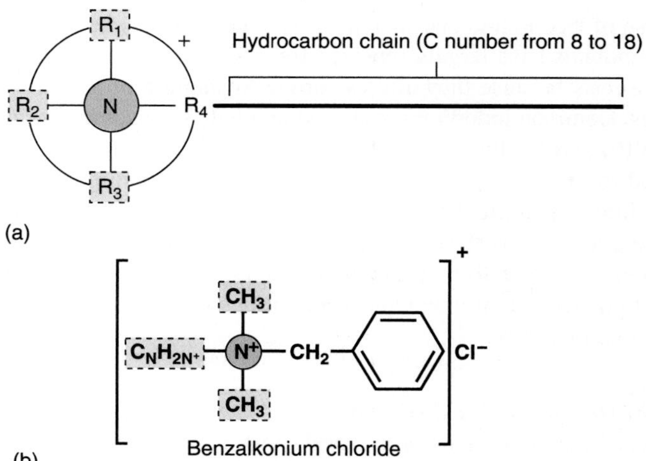

Figure 11.15
(*a*) In general, detergents are polar molecules with a positively charged head and at least one long, uncharged hydrocarbon chain. The head contains a central nitrogen nucleus with various alkyl (R) groups attached. (*b*) The structure of a common quaternary ammonium detergent, benzalkonium chloride.

Chemicals with Surface Action: Detergents

Detergents are complex organic substances that, as a group, are often termed surfactants. Chemical types are nonionic, anionic, and cationic detergents. The nonionic (uncharged) detergents are not really effective germicides. Most anionic (negatively charged) detergents have limited microbicidal power. Soaps belong to this group. Cationic detergents are the most effective of the three, especially the quaternary ammonium compounds (usually shortened to *quats*).

All detergents have a general molecular structure that includes a long-chain hydrocarbon residue and a highly charged, or polar, group (figure 11.15). This configuration makes them soluble in lipid-based structures, and they act as surfactants. These properties can have several effects, but chief among them is the disruption of the cell membrane and the loss of its selective permeability. Quaternary ammonium compounds kill microbes by causing leakage of microbial cytoplasm, precipitating proteins, and inhibiting metabolism. Because of their ability to interact with surfaces, detergents make good wetting agents, cleansing agents, and emulsifiers.

The range of activity of detergents is varied. Quaternary ammonium compounds, if used at medium concentrations, are effective against some gram-positive bacteria, viruses, fungi, and algae. In low concentrations, they have only microbistatic effects. The quats are ineffective against the tubercle bacillus, hepatitis virus, *Pseudomonas,* and spores at any concentration. Their activity can be greatly reduced in the presence of organic matter, and they function best in alkaline solutions. As a result of their limitations, quats are rated only for low-level disinfection in the clinical setting.

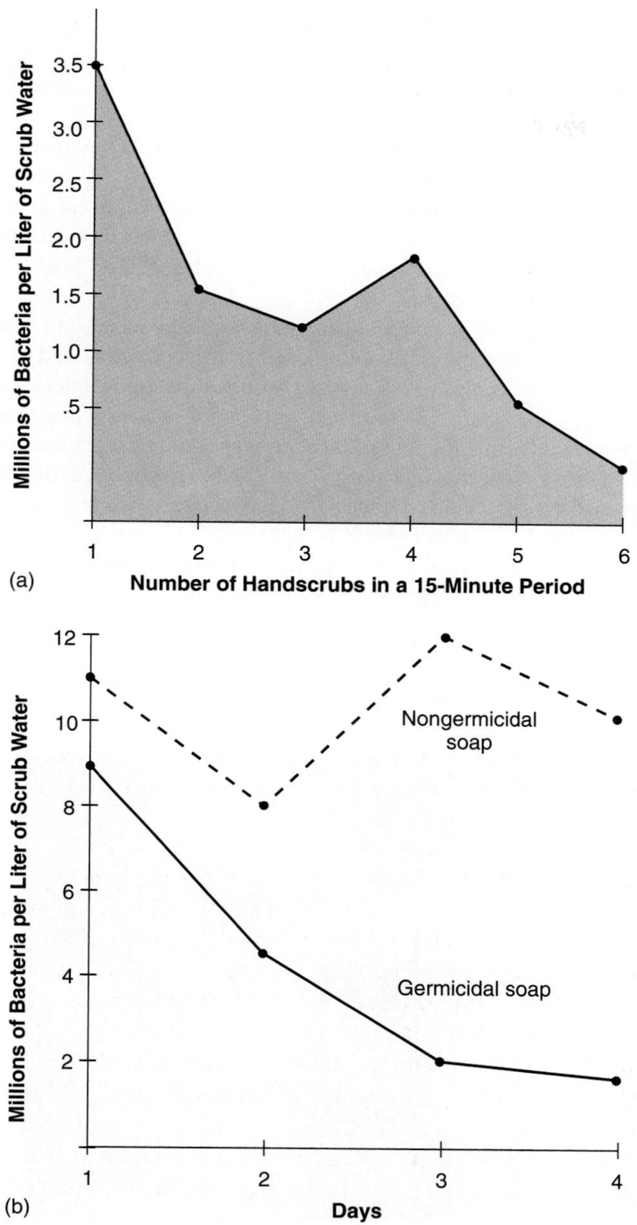

Figure 11.16

Graphs showing effects of handscrubbing. (*a*) Reduction of microbes on the hands during a single (15-minute) episode of handwashing with nongermicidal soap. The increase in the fourth scrub is due to a high level of resident microbes uncovered in that skin layer. (*b*) Comparison of scrubbing over several days with a nongermicidal soap versus a germicidal soap. Germicidal soap has persistent effects on skin over time, keeping the microbial count low. Without germicide, soap does not show this sustained effect.

Applications of Detergents and Soaps Quaternary ammonium compounds include benzalkonium chloride, Zephiran, and cetylpyridinium chloride (Ceepryn). In dilutions ranging from 1:100 to 1:1,000, quats are mixed with cleaning agents to simultaneously disinfect and clean floors, furniture, equipment surfaces, and restrooms. Their detergent properties and low toxicity make quats among the best sanitizers. They are used to clean restaurant eating utensils, food-processing equipment, dairy equipment, and clothing. They are common preservatives for ophthalmic solutions and cosmetics. Their level of disinfection is far too low for disinfecting medical instruments.

Soaps are alkaline compounds made by combining the fatty acids in oils (coconut, castor, cottonseed, linseed) with sodium or potassium salts. In usual practice, soaps are only weak microbicides, and they destroy only highly sensitive forms such as the agents of gonorrhea, meningitis, and syphilis. The common hospital pathogen *Pseudomonas* is so resistant to soap that various species grow abundantly in soap dishes. Soaps function primarily as cleansing agents and sanitizers in industry and the home. The superior sudsing and wetting properties of soaps help to mechanically remove large amounts of surface soil, greases, and other debris that contains microorganisms. Soaps gain greater germicidal value when mixed with agents such as alcohol (called green soap), chlorhexidine, or iodine. They can be used for cleaning instruments before heat sterilization, degerming patients' skin, routine handwashing by medical and dental personnel, and preoperative handscrubbing. Vigorously brushing the hands with germicidal soap over a 15-minute period is an effective way to remove dirt, oil, and surface contaminants as well as some resident microbes, but it will never sterilize the skin (microfile 11.4 and figure 11.16).

Heavy Metal Compounds

Various forms of the metallic elements mercury, silver, gold, copper, arsenic, and zinc have been applied in microbial control over several centuries. These are often referred to as heavy metals because of their relatively high atomic weight. However, from this list, only preparations containing mercury and silver still have any significance as germicides. Although some metals (zinc, iron) are actually needed in small concentrations as cofactors on enzymes, the higher molecular weight metals (mercury, silver, gold) can be very toxic, even in minute quantities (parts per million). This property of having antimicrobial effects in exceedingly small amounts is called an *oligodynamic* action* (figure 11.17). Heavy metal germicides contain either an inorganic or an organic metallic salt, and they come in the form of aqueous solutions, tinctures, ointments, or soaps.

Mercury, silver, and most other metals exert microbicidal effects by binding onto functional groups, including sulfhydryls, hydroxyls, amines, and phosphates. This binding inactivates proteins and rapidly brings metabolism to a standstill (see figure 11.4*b*). This mode of action can destroy many types of microbes, including vegetative bacteria, fungal cells and spores, algae, protozoa, and viruses (but not endospores).

Unfortunately, there are several drawbacks to using metals in microbial control: (1) Metals are very toxic to humans if ingested, inhaled, or absorbed through the skin, even in small quantities, for the same reasons that they are toxic to microbial cells; (2) they commonly cause allergic reactions; (3) large quantities of biological fluids and wastes neutralize or depress their actions; and (4) microbes can develop resistance to metals. Health and environmental considerations have dramatically reduced the use of metallic antimicrobic compounds in medicine, dentistry, commerce, and agriculture.

***oligodynamic** (ol″-ih-goh-dy-nam′-ik) Gr. *oligos,* little, and *dynamis,* power.

MICROFILE 11.4 THE QUEST FOR STERILE SKIN

More than a hundred years ago, before sterile gloves were a routine part of medical procedures, the hands remained bare during surgery. Realizing the danger from microbes, medical practitioners attempted to sterilize the hands of surgeons and their assistants to prevent surgical infections. Several stringent (and probably very painful) techniques involving strong chemical germicides and vigorous scrubbing were practiced. Here are a few examples.

In Schatz's method, the hands and forearms were first cleansed by brisk scrubbing with liquid soap for 3–5 minutes, then soaked in a saturated solution of permanganate at a temperature of 110°F until they turned a deep mahogany brown. Next, the limbs were immersed in saturated oxalic acid until the skin became decolorized. Then, as if this were not enough, the hands and arms were rinsed with sterile limewater and washed in warm bichloride of mercury for 1 minute.

Or, there was Park's method (more like a torture). First, the surfaces of the hands and arms were rubbed completely with a mixture of cornmeal and green soap to remove loose dirt and superficial skin. Next, a paste of water and mustard flour was applied to the skin until it began to sting. This potion was rinsed off in sterile water, and the hands and arms were then soaked in hot bichloride of mercury for a few minutes, during which the solution was rubbed into the skin.

Another method once earnestly suggested for getting rid of microorganisms was to expose the hands to a hot-air cabinet to "sweat the germs" out of skin glands. Pasteur himself advocated a quick flaming of the hands to maintain asepsis.

The old dream of sterilizing the skin was finally reduced to some basic realities: The microbes entrenched in the epidermis and skin glands cannot be completely eradicated even with the most intense efforts, and the skin cannot be sterilized without also seriously damaging it. Because this is true for both medical personnel and their patients, the chance always exists that infectious agents can be introduced during invasive medical procedures. Fortunately, the Goodyear Tire and Rubber Company developed rubber gloves in 1890, making it possible to place a sterile barrier around the hands. Of course, this did not mean that skin cleansing and antiseptic procedures were abandoned or downplayed. A thorough scrubbing of the skin, followed by application of an antiseptic, is still needed to remove the most dangerous source of infections—the superficial contaminants constantly picked up from whatever we touch.

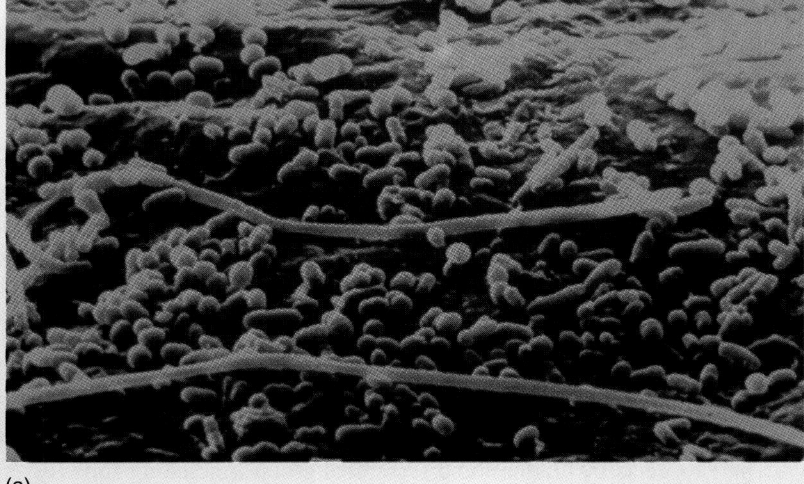

(a)

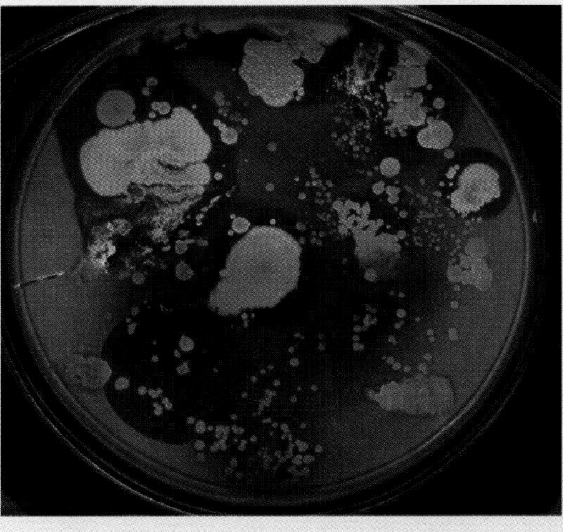

(b)

Microbes on normal unwashed hands, (a) *A scanning electron micrograph of a piece of skin from a fingertip shows clusters of bacteria perched atop a fingerprint ridge (47,000×).* (b) *Heavy growth of microbial colonies on a plate of blood agar. This culture was prepared by passing an open sterile plate around a classroom of 30 students and having each one touch its surface. After incubation, a mixed population of bacteria and fungi appeared. Clear zones in agar can be indicative of pathogens.*

Applications of Heavy Metals Weak (0.001% to 0.2%) organic mercury tinctures such as thimerosal (Merthiolate) and nitromersol (Metaphen) are fairly effective antiseptics and infection preventives, but they should never be used on broken skin because they are harmful and can delay healing. The organic mercurials also serve as preservatives in cosmetics and ophthalmic solutions. Mercurochrome, that old staple of the medicine cabinet, is now considered among the poorest of antiseptics.

A silver compound with several applications is silver nitrate ($AgNO_3$) solution. Crede introduced it in the late nineteenth century for preventing gonococcal infections in the eyes of newborn infants who had been exposed to an infected birth canal. This preparation is not used as often now because many pathogens are resistant to it. It has been replaced by antibiotics in most instances.

Solutions of silver nitrate (1–2%) are used as topical germicides on mouth ulcers and occasionally root canals, although this

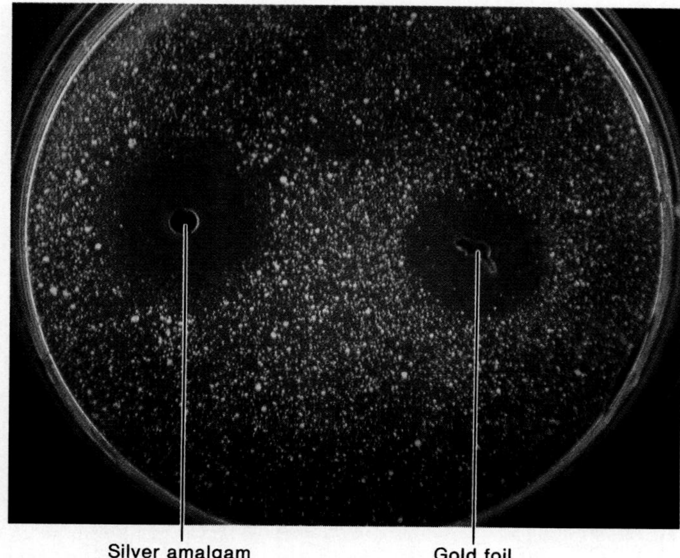

Figure 11.17
Demonstration of the oligodynamic action of heavy metals. A pour plate inoculated with saliva has small fragments of heavy metals pressed lightly into it. During incubation, clear zones indicating growth inhibition developed around both fragments. The slightly larger zone surrounding the amalgam (used in tooth fillings) probably reflects the synergistic effect of the silver and mercury it contains.

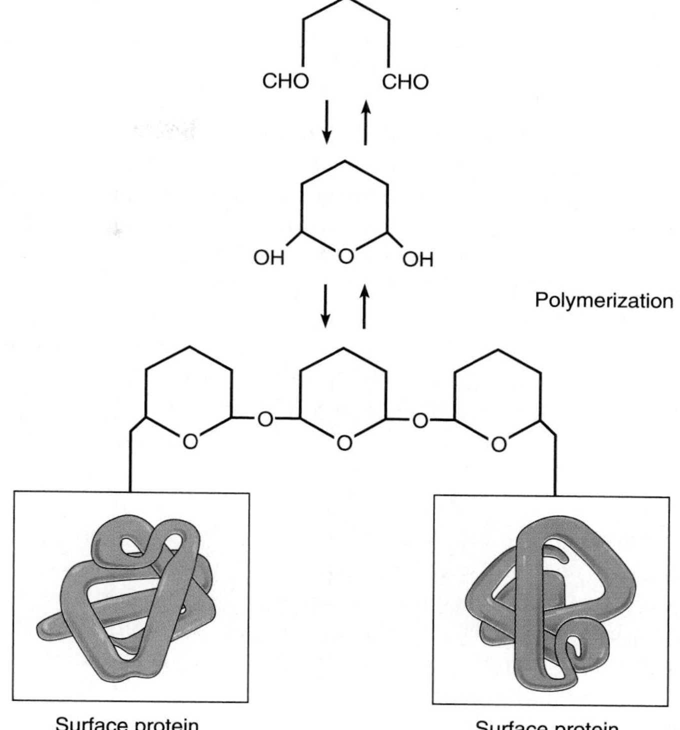

Figure 11.18

Actions of glutaraldehyde. The molecule forms a closed ring that can polymerize. When these alkylating polymers react with amino acids, they cross-link and inactivate proteins.

can discolor teeth. Silver sulfadiazine ointment, when added to dressings, effectively prevents infection in second- and third-degree burn patients. Colloidal silver preparations are much milder and less toxic than inorganic salts but are restricted to use as mild germicidal ointments or rinses for the mouth, nose, eyes, and vagina.

Aldehydes As Germicides

Organic substances bearing a —CHO functional group (a strong reducing group) on the terminal carbon are called aldehydes. Several common substances such as sugars and some fats are technically aldehydes. The two aldehydes used most often in microbial control are **glutaraldehyde** and **formaldehyde.**

Glutaraldehyde is a yellow acidic liquid with a mild odor. The molecule's two aldehyde groups favor the formation of polymers. The mechanism of activity against microbes is not completely understood, but glutaraldehyde appears to cross-link protein molecules on the cell surface through alkylation of amino acids, a process in which a hydrogen atom on an amino acid is replaced by the glutaraldehyde molecule itself (figure 11.18). It can also irreversibly disrupt the activity of enzymes within the cell. Glutaraldehyde is a rapid, broad-spectrum antimicrobial chemical and one of the few chemicals officially accepted as a sterilant and high-level disinfectant. It kills spores in 3 hours and fungi and vegetative bacteria (even *Mycobacterium* and *Pseudomonas*) in a few minutes. Viruses, including the most resistant forms, appear to be inactivated after relatively short exposure times. It retains its potency even in the presence of organic matter, is noncorrosive, does not damage plastics, and is less toxic or irritating than formaldehyde. Its principal disadvantage is that it is somewhat unstable, especially with increased pH and temperature.

Formaldehyde is a sharp, irritating gas that readily dissolves in water to form an aqueous solution called *formalin.* Full saturation of formaldehyde (37%) produces a solution of 100% formalin. The chemical is microbicidal through its attachment to nucleic acids and functional groups of amino acids. Formalin is an intermediate- to high-level disinfectant, although it acts more slowly than glutaraldehyde. Formaldehyde's extreme toxicity (it is classified as a carcinogen) and irritating effects on the skin and mucous membranes greatly limit its clinical usefulness.

Applications of the Aldehydes Glutaraldehyde is a milder substitute for formalin in chemically sterilizing materials that are damaged by heat. Commercial products (Cidex, Glutarol) diluted to 2% are used to sterilize respiratory therapy equipment, hemostats, fiberoptic endoscopes (laparoscopes, arthroscopes), and anesthetic devices. Glutaraldehyde is employed in dental offices for practical disinfection of instruments (when heat sterilization is not possible or necessary) because of its ability to destroy the hepatitis B virus. It can also be used to preserve vaccines, sanitize poultry carcasses, and degerm cow's teats.

Formalin tincture (8%) has limited use as a disinfectant for surgical instruments, and 1% formalin solution is still used to disinfect some reusable kidney dialysis instruments. Any object that comes into intimate contact with the body must be thoroughly rinsed with sterile water to remove the formalin residue. It is, after all, one of the active ingredients in embalming fluid.

Gaseous Sterilants and Disinfectants

Processing inanimate substances with chemical vapors, gases, and aerosols provides a versatile alternative to heat or liquid chemicals. Currently, those vapors and aerosols having the broadest applications are ethylene oxide (ETO), propylene oxide, and betapropiolactone (BPL).

Ethylene oxide is a colorless substance that exists as a gas at room temperature. It is very explosive in air, a feature that can be eliminated by combining it with a high percentage of carbon dioxide or fluorocarbon. Like the aldehydes, ETO is a very strong alkylating agent, and it reacts vigorously with guanine molecules of DNA and functional groups of proteins. Through these mechanisms, it blocks both DNA replication and enzymatic actions. Ethylene oxide is the only gas generally accepted for chemical sterilization because, when employed according to strict procedures, it is a sporicide. A specially designed ETO sterilizer called a *chemiclave,* a variation on the autoclave, is equipped with a chamber, gas ports, and temperature, pressure, and humidity controls (figure 11.19). Ethylene oxide is rather penetrating but relatively slow-acting, requiring from 90 minutes to 3 hours. Some items absorb ETO residues and must be aerated for several hours after exposure to ensure dissipation of as much residual gas as possible. For all of its effectiveness, ETO has some unfortunate features. Its explosiveness makes it dangerous to handle; it can damage the lungs, eyes, and mucous membranes if contacted directly; and it is rated as a carcinogen by the government.

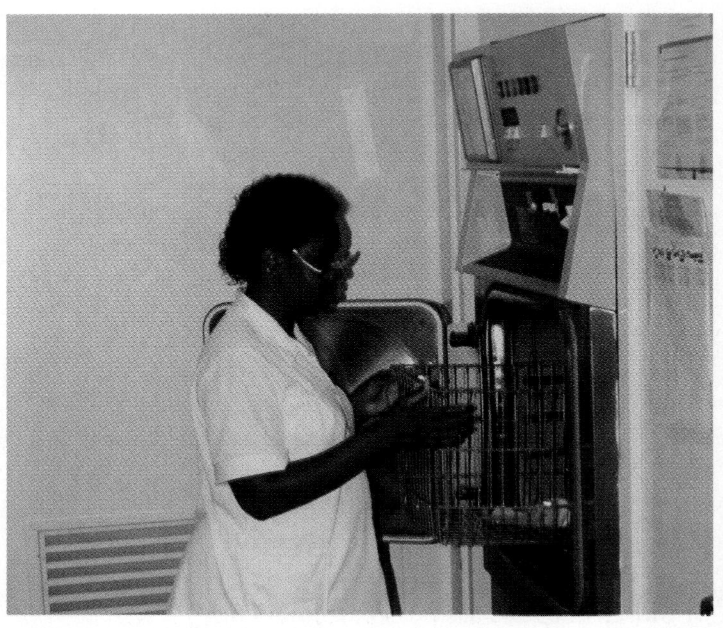

(a)

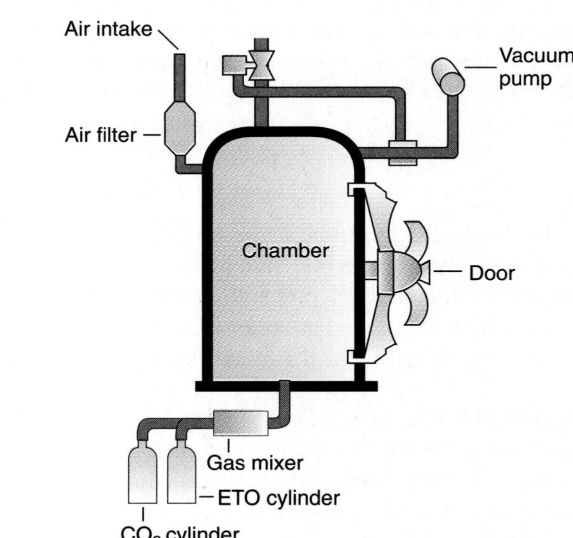

(b)

Figure 11.19
(*a*) A chemiclave, an automatic ethylene oxide sterilizer. (*b*) The machine is equipped with gas canisters containing ethylene oxide (ETO) and carbon dioxide, a chamber to hold items, and mechanisms that evacuate gas and introduce air.

Applications of Gases and Aerosols Ethylene oxide (carboxide, cryoxide) is an effective way to sterilize and disinfect plastic materials and delicate instruments in hospitals and industries. It can safely sterilize prepackaged heart pacemakers, artificial heart valves, surgical supplies, syringes, and disposable Petri dishes. Ethylene oxide has been used extensively to disinfect sugar, spices, dried foods, and drugs. It has also been considered as a possible means to sterilize blood, serum, and culture media.

Propylene oxide is a close relative of ETO, with similar physical properties and mode of action, although it is less toxic. Because it breaks down into a relatively harmless substance, it is safer than ETO for sterilization of foods (nuts, powders, starches, spices).

Betapropiolactone (BPL) is a nonexplosive liquid that is rapidly microbicidal but nonpenetrating. This feature and its toxicity to animals greatly restrict its applications. Betapropiolactone is used for disinfecting rooms and instruments, sterilizing bone and arterial grafts, and inactivating viruses in vaccines.

Dyes As Antimicrobial Agents

Dyes are important in staining techniques and as selective and differential agents in media; they are also a primary source of certain drugs used in chemotherapy. Because aniline dyes such as crystal violet and malachite green are very active against gram-positive species of bacteria and various fungi, they are incorporated into solutions and ointments to treat skin infections (ringworm, for example). The yellow acridine dyes, acriflavine and proflavine, are sometimes utilized for antisepsis and wound treatment in medical and veterinary clinics. For the most part, dyes will continue to have limited applications because they stain and have a narrow spectrum of activity.

Acids and Alkalies

Conditions of very low or high pH can destroy or inhibit microbial cells; however, most inorganic acids and alkalies are far too corrosive, caustic, and hazardous to use in microbial control. Aqueous solutions of ammonium hydroxide remain a common component of detergents, cleansers, and deodorizers. Organic acids are widely used in food preservation because they prevent spore germination and bacterial and fungal growth and because they are generally regarded as safe to eat. Acetic acid (in the form of vinegar) is a pickling agent that inhibits bacterial growth; propionic acid is commonly incorporated into breads and cakes to retard molds; lactic acid is added to sauerkraut and olives to prevent spoilage by anaerobic bacteria (especially the clostridia), and benzoic and sorbic acids are added to beverages, syrups, and margarine to inhibit yeasts.

Chapter Checkpoints

Chemical agents of microbial control are classified by their physical state and chemical nature.

Chemical agents can be either microbicidal or microbistatic. They are also classified as high-, medium-, or low-level germicides.

Factors that determine the effectiveness of a chemical agent include the type and numbers of microbes involved, the material involved, the strength of the agent, and the exposure time.

Halogens are effective chemical agents at both microbicidal and microbistatic levels. Chlorine compounds disinfect water, food, and industrial equipment. Iodine is used as either free iodine or iodophor to disinfect water and equipment. Iodophors are also used as antiseptic agents.

Phenols are strongly microbicidal agents used in general disinfection. Milder phenol compounds, the bisphenols, are also used as antiseptics.

Alcohols dissolve membrane lipids and destroy cell proteins. Their action depends upon their concentration, but they are generally only microbistatic.

Surfactants are of two types: detergents and soaps. They reduce cell membrane surface tension, causing membrane rupture. Cationic detergents, or quats, are low-level germicides limited by the amount of organic matter present and the microbial load.

Aldehydes are potent sterilizing agents and high-level disinfectants that irreversibly disrupt microbial enzymes.

Ethylene oxide is the only accepted gaseous sterilant. It is effective for batch sterilizing medical materials, but it is explosive, toxic to tissues, and carcinogenic.

CHAPTER CAPSULE WITH KEY TERMS

PHYSICAL METHODS FOR CONTROLLING MICROORGANISMS

MOIST HEAT

Hot water or steam may be used in **disinfection** to kill vegetative cells or **sterilization** to kill spores. Mode of action: Denaturation of proteins, destruction of membranes and DNA.

For Sterilization: (1) **Autoclave** uses steam under pressure (15 psi/121°C/10–40 min). Applied on heat-resistant materials that steam can penetrate, including cloth, glassware, metallic items, media, rubber, certain foods. Not recommended for oils, powders, heat-sensitive materials.

(2) **Intermittent sterilization** uses free-flowing, unpressurized steam, 100°C applied for 30–60 minutes on 3 successive days. Used for substances that cannot be autoclaved, especially media with egg, serum, sugars, and certain canned foods.

For Disinfection: (1) **Boiling** water at 100°C used to destroy pathogens (not spores) on dishes, clothing, utensils, instruments, glassware, bedding; does not kill spores; materials may become recontaminated.

(2) **Pasteurization** is application of heat less than 100°C to liquids; Employs the *flash method* to heat liquid to 71°C for 15 seconds. Can destroy non-spore-forming milk-borne pathogens, such as *Salmonella* and *Listeria* and/or reduce spoilage by lowering the overall microbial count; also applied to beer, wine, fruit juices. Only reduces microbial content; many thermodurics survive.

DRY HEAT

Hot air for sterilization or **decontamination.**

Mode of Action: Depending on the temperature, combustion to ashes, oxidation, dehydration, coagulation of proteins.

(1) **Incineration:** *Open flame*—materials are placed for a few seconds in a very hot (800°–1,800°C) flame. Bunsen burner used to flame tips of inoculating implements, test tubes.

(2) *Incineration in furnace:* Incinerator chamber equipped with very high temperature flame (600°–1,200°C) to burn items and microbes to ashes. Used for disposing of hospital and industrial wastes that are disposable; may add pollutants to atmosphere.

(3) **Dry oven:** Materials exposed to 150°–180°C for 2–4 hours. Used for sterilizing

empty glassware, metal instruments, needles, oils, waxes, powders. Too vigorous for liquids containing water, plastics, paper, cloth.

Cold Temperatures

Refrigeration (0°–15°C) or freezing (below 0°C), for microbial inhibition or preservation.

Mode of Action: **Microbistasis,** which slows the growth rate of most microorganisms. Used for maintenance of foods, increasing shelf-life of drugs and chemicals, preserving the viability of microbial cultures. Not a mode of disinfection or sterilization; many pathogens survive chilling and freezing.

Drying/Desiccation

Gradual withdrawal of water from cells by exposure to room air.

Mode of Action: Concentration of cellular solutes, leading to metabolic inhibition. Used in preparing dehydrated foods. Not reliable for pathogens, especially if protected by saliva, pus, blood; not really effective for infection control.

Radiation/Irradiation

Energy in the form of waves **(electromagnetic)** and particles that can be transmitted through space. Used in **cold sterilization.**

Ionizing: High-energy, short waves and particles that can dislodge electrons from atoms. Types are *gamma rays, X rays, cathode rays* (high-speed electrons) used in **irradiation.** *Mode of Action:* Direct damage to DNA chain, causing breaks and mutations. Used to sterilize heat-sensitive medical materials such as plastics, drugs, instruments, vaccines; also to disinfect and increase storage time of fresh products; very penetrating. Somewhat more expensive and dangerous than other methods.

Nonionizing: Moderate energy, medium-length *ultraviolet waves* excite, but do not ionize, atoms. *Mode of Action:* Acts on DNA molecule, forming **dimers** between **adjacent pyrimidines;** production of toxic products. UV lamps used for disinfection of air in medicine and industry; also for treating water, sera, vaccines, drugs; disinfection of solid surfaces in food preparation. Does not penetrate glass, paper, plastic, wood, metal; can cause damage to human tissues (skin, eyes).

Sound Waves

Very high-frequency waves—super and ultrasonic. *Mode of Action:* Creates vibration and turbulence that disrupts cells. Uses are limited; mainly for cleaning and disinfection of instruments.

Filtration

Physical removal of microbes from liquids and air with fine filters with tiny pores. The fluid passes through while the microbes are trapped. Can be used to remove viruses. Filters include cellulose acetate, glass, plastics; can sterilize heat-sensitive liquids such as vaccines, blood products, drugs, media, milk, water; also for removing microbes from air in hospital rooms, isolation units, drug and food preparation areas.

CHEMICAL CONTROL OF MICROORGANISMS

General Uses: **Disinfectants, antiseptics, sterilants,** preservatives, **sanitizers, degermers.**

Physical States: Solutions come in **aqueous** form (chemicals dissolved in water) and **tincture** form (chemicals dissolved in alcohol); solutions expressed as dilutions, percents, and parts per million. In gases, chemical is in vapor or aerosol phase.

Halogens

Chlorine: Forms are Cl_2, hypochlorites, chloramines. *Mode of Action:* Denaturation of proteins by disrupting disulfide bonds. *Spectrum of Action:* All microbial types: sporicidal with adequate time. *Applications:* Elemental chlorine (1 ppm) controls pathogenic vegetative pathogens in water and wastewater; hypochlorites (chlorine bleach) are used extensively for disinfection and sanitization; chloramines are clinical disinfectants and antiseptics and alternative water disinfection agents. *Limitations:* Chemical action may be retarded by high levels of organic matter; may be unstable at basic pHs and in light.

Iodine: Forms are free iodine (I_2) and iodophors (iodine complexed to organic polymers). *Mode of Action:* Interference with protein interchain bonds, causing denaturation. *Spectrum of Action:* Broad—sporicidal. *Applications:* Weak solutions—topical antiseptic for prepping skin; iodine tincture, 2%, for high degree of surgical asepsis; strong solutions (5–7%)—disinfection of plastic, rubber, glass, some metal instruments.

Limitations: Iodine solutions stain, corrode, and have a strong odor. Stronger solutions are too irritating and toxic to use on tissue.

Iodophors—Aqueous solutions with 2–10% iodine (Betadine, Povidone) are milder components of medical and dental degerming (handscrubbing, skin prepping) agents, disinfectants, and ointments.

Phenolics

Forms: Cresols and bisphenols. *Mode of Action:* Disruption of cell membrane, protein precipitators. *Spectrum of Action:* **Bactericidal, fungicidal, virucidal,** but not sporicidal. *Applications:* Disinfectants for biological wastes; phenol is an older toxic disinfectant; 1–3% cresol in soap (Lysol) is a common housekeeping disinfectant and cleaner. Orthophenyl phenol is a milder phenolic used in air sprays and some disinfectants (Amphyl). Triclosan is a common antibacterial additive to soaps, detergents, and numerous household products. *Limitations:* Most are too toxic for use as antiseptics and are relatively insoluble.

Chlorhexidine

Forms: Hibiclens, Hibitane. *Mode of Action:* Surfactant, protein denaturant. *Spectrum of Action:* Bactericidal, some antiviral and antifungal effects, not sporicidal. *Applications:* An aqueous or alcohol solution is a skin degerming agent for preoperative scrubs, skin cleaning, and burns. Product is relatively mild, nontoxic, and fast-acting.

Alcohols

Forms: Ethyl and isopropyl, usually in solutions of 50–95%. *Modes of Action:* **Surfactant** that dissolves membrane lipids and coagulates proteins; pure (100%) alcohol slowly dehydrates cells. *Spectrum of Action:* Narrow; on bacterial vegetative cells and fungi; not sporicidal. *Applications:* Ethyl alcohol (70–95%) is a germicide in the clinic, laboratory, and home for skin degerming and low-level disinfection and sanitization; isopropyl alcohol has similar uses, though it is more toxic and less safe. *Limitations:* Does not inactivate polio and hepatitis B virus.

Hydrogen Peroxide

Forms: Weak (3%) to strong (25%) hydrogen peroxide solutions. *Mode of Action:* Produces highly active hydroxyl-free radical that damages proteins and DNA molecules; decomposes to water and O_2 gas, which can be toxic. *Spectrum of Action:* Broad; strong solutions are sporicidal. *Applications:* 3% hydrogen peroxide is used for skin and wound antisepsis, care of mucous membrane infections; also for disinfection of equipment, utensils; 6–25% can be used to sterilize equipment. *Limitations:* Decomposes in presence of light and catalase.

Detergents

Forms: Primarily cationic **(quaternary ammonium compounds—quats)** and soaps. *Mode of Action:* Surfactant that disrupts membranes and alters permeability. *Spectrum of Action:* Somewhat erratic; have limited bactericidal and fungicidal action; are not sporicidal. *Applications:* Benzalkonium and cetylpyridinium chlorides are environmental disinfectants and cleansers in clinics and the food industry; also preservatives for pharmaceuticals. Soaps are not very microbicidal but function in the mechanical removal of grease and soil on skin, utensils, and environmental surfaces. *Limitations:* Do not destroy the tubercle bacillus or hepatitis B virus; are inactivated by large quantities of organic matter.

HEAVY METAL COMPOUNDS

Mercury and silver solutions and tinctures. *Mode of Action:* **Oligodynamic action,** precipitate proteins. *Spectrum of Action:* Broad, but not sporicidal. *Applications:* Organic mercurial tinctures (thimerosal, nitromersol) are skin antiseptics and infection treatments. 1–2% silver nitrate solutions may be used as a topical antiseptic; silver sulfadiazine ointment can prevent burn infections; colloidal silver preparations are used for mouth and eye rinses. *Limitations:* Metal solutions are highly toxic, cause allergies, and are neutralized by organic substances.

ALDEHYDES

Forms: **Glutaraldehyde** and **formaldehyde** solutions (formalin). *Mode of Action:* Alkylation of amino and nucleic acids. *Spectrum of Action:* Very broad; glutaraldehyde may be used as a **sterilant.** *Applications:* Glutaraldehyde in 2% solutions (Cidex) is used for sterilizing heat-sensitive instruments in the medical and dental office, also for some types of environmental disinfection; formalin in 1–8% has limited use in disinfection of some instruments, rooms, and as a preservative. *Limitations:* Glutaraldehyde is somewhat unstable; formaldehyde is toxic and irritating.

GASES

Forms: **Ethylene oxide** (ETO), propylene oxide, and betapropiolactone. *Mode of Action:* Alkylation of nucleic acids and proteins. *Spectrum of Action:* All are sporicidal, but only ETO is an approved **sterilant.** *Applications:* Ethylene oxide is mixed with carbon dioxide in a special chamber to sterilize plastic medical and industrial supplies and to treat spices and dried foods; propylene oxide can disinfect food (nuts, starch); betapropiolactone is used to disinfect rooms and some biological materials. *Limitations:* ETO is somewhat dangerous because of its explosiveness and toxicity; it is slow-acting. Betapropiolactone is toxic, with poor penetration.

DYES

Forms: Aniline (crystal violet) and acridine (acriflavine). *Mode of Action:* Bind to nucleic acids and proteins. *Spectrum of Action:* Mostly inhibitory for gram-positive bacteria and fungi. *Applications:* Primarily as treatments for skin infections and components of media.

ACIDS AND ALKALIES

Mode of Action: Very low or high pHs disrupt biological molecules. *Applications:* Organic acids (benzoic, propionic, acetic) are food preservatives; sodium and ammonium hydroxide are detergents and cleaning agents. *Limitations:* Corrosive and caustic.

MULTIPLE-CHOICE QUESTIONS

1. A microbicidal agent does what?
   a. kills spores
   b. inhibits microorganisms
   c. is ineffective against viruses
   d. destroys microbes

2. Microbial control methods that kill ____ are able to sterilize.
   a. viruses c. endospores
   b. the tubercle bacillus d. cysts

3. Any process that destroys the non-spore-forming contaminants on inanimate objects is
   a. antisepsis c. sterilization
   b. disinfection d. degermation

4. Sanitization is a process by which
   a. the microbial load on objects is reduced
   b. objects are made sterile with chemicals
   c. utensils are scrubbed
   d. skin is debrided

5. An example of an agent that lowers the surface tension of cells is
   a. phenol c. alcohol
   b. chlorine d. formalin

6. High temperatures ____ and low temperatures ____.
   a. sterilize, disinfect
   b. kill cells, inhibit cell growth
   c. denature proteins, burst cells
   d. speed up metabolism, slow down metabolism

7. The temperature-pressure combination for an autoclave is
   a. 100°C and 4 psi c. 131°C and 9 psi
   b. 121°C and 15 psi d. 115°C and 3 psi

8. Microbes that are the targets of pasteurization:
   a. *Clostridium botulinum*
   b. *Mycobacterium* species
   c. *Salmonella* species
   d. both b and c

9. Ionizing radiation removes ____ from atoms.
   a. protons c. electrons
   b. waves d. ions

10. The primary mode of action of nonionizing radiation is to
    a. produce superoxide ions
    b. make pyrimidine dimers
    c. denature proteins
    d. break disulfide bonds

11. An effective method of sterilizing heat-sensitive liquids is
    a. UV radiation
    b. exposure to ozone
    c. beta propiolactone
    d. filtration

12. ____ is the iodine antiseptic of choice for wound treatment.
    a. Eight percent tincture
    b. Five percent aqueous
    c. Iodophor
    d. Potassium iodide solution

13. A chemical with sporicidal properties is
    a. phenol
    b. alcohol
    c. quaternary ammonium compound
    d. glutaraldehyde

14. Silver nitrate is used
    a. in antisepsis of burns
    b. as a mouthwash
    c. to treat genital gonorrhea
    d. to disinfect water

15. Detergents are
    a. high-level germicides
    b. low-level germicides
    c. excellent antiseptics
    d. used in disinfecting surgical instruments

16. Which of the following is an approved sterilant?
    a. chlorhexidine
    b. betadyne
    c. ethylene oxide
    d. ethyl alcohol

CONCEPT QUESTIONS

1. a. Explain the effect of a tuberculocide.
   b. What would be the effect of a pseudomonicide?
   c. What does a virustatic agent do?
2. Compare sterilization with disinfection and sanitization. Describe the relationship of the concepts of sepsis, asepsis, and antisepsis.
3. a. Briefly explain how the type of microorganisms present will influence the effectiveness of exposure to antimicrobial agents.
   b. Explain how the numbers of contaminants can influence the measures used to control them.
4. a. Precisely what is microbial death?
   b. Why does a population of microbes not die instantaneously when exposed to an antimicrobial agent?
5. Why are antimicrobial processes inhibited in the presence of extraneous organic matter?
6. Describe four modes of action of antimicrobial agents, and give a specific example of how each works.
7. a. Summarize the nature, mode of action, and effectiveness of moist and dry heat.
   b. Compare their effects on vegetative cells and spores.
8. How can the temperature of steam be raised above 100°C? Explain the relationship involved.
9. a. What do you see as a basic flaw in tyndallization?
   b. In boiling water devices?
   c. In incineration?
   d. In ultrasonic devices?
10. What are several microbial targets of pasteurization?
11. Explain why desiccation and cold are not reliable methods of disinfection.
12. a. What are some advantages of ionizing radiation as a method of control?
    b. Some disadvantages?
13. a. What is the precise mode of action of ultraviolet radiation?
    b. What are some disadvantages to its use?
14. What are the superior characteristics of iodophors over free iodine solutions?
15. a. What does it mean to lower the surface tension?
    b. What cell parts will be most affected by surface active agents?
16. a. Name one chemical for which the general rule that a higher concentration is more effective is *not* true.
    b. What is a sterilant?
17. Name the principal sporicidal chemical agents.
18. Why is hydrogen peroxide solution so effective against anaerobes?
19. Give the uses and disadvantages of the heavy metal chemical agents, glutaraldehyde, and the sterilizing gases.
20. What does it mean to say that a chemical has an oligodynamic action?

CRITICAL-THINKING QUESTIONS

1. What is wrong with this statement: "The patient's skin was sterilized with alcohol"? What would be the more correct wording?
2. For each item on the following list, give a reasonable method of sterilization. You cannot use the same method more than three times; the method must sterilize, not just disinfect; and the method must not destroy the item or render it useless. After considering a workable method, think of a method that would not work. Note: Where an object containing something is given, you must sterilize everything (jar and Vaseline). Some examples of methods are autoclave, ethylene oxide gas, dry oven, ionizing radiation.

   room air
   inside of a refrigerator
   blood in a syringe
   wine
   serum
   a jar of Vaseline
   a pot of soil
   a child's toy
   plastic Petri dishes
   fruit in plastic bags
   heat-sensitive drugs
   rubber gloves
   cloth dressings
   disposable syringes
   leather shoes from a thrift shop
   talcum powder
   a cheese sandwich
   milk
   human hair (for wigs)
   orchid seeds
   a flask of nutrient agar
   metal instruments
   the world (can you?)
3. a. Graph the data on tables 11.3 and 11.4, plotting the time on the Y axis and the temperature on the X axis for three different organisms.
   b. What conclusions can you draw regarding the effects of time and temperature on thermal death?
   c. Is there any difference between the graph for a sporeformer and the graph for a non-sporeformer?
4. Can you think of situations in which the same microbe would be considered a serious contaminant in one case and completely harmless in another?
5. What microbial control methods around the house require the use of disinfection rated as high level? As moderate level? As low level?
6. Devise an experiment that will differentiate between bacteriocidal and bacteriostatic effects.
7. a. Define cold sterilization.
   b. Name three totally different methods that qualify for this definition.
8. There is quite a bit of concern that chlorine used as a water purification chemical presents serious dangers. Can you think of some alternative methods to purify vast water supplies and yet keep them safe from contamination?
9. The shelf-life and keeping qualities of fruit and other perishable foods are greatly enhanced through irradiation, saving industry and consumers billions of dollars. How would you personally feel about eating a piece of irradiated fruit? How about spices that had been sterilized with ETO?
10. The microbial levels in saliva are astronomical (<10^6 cells per ml, on average). What effect do you think a mouthwash can have against this high number? Comment on the claims made for such products.
11. Can you think of some innovations the health care community can use to deal with medical waste and its disposal that prevent infection but are ecologically sound?

chapter 12

Drugs, Microbes, Host:
The Elements of Chemotherapy

The subject of this chapter is antimicrobial chemotherapy—the use of chemicals to control or prevent infection. As you will see, the administration of a drug has an impact beyond simply the action of the drug on the microbe. It also includes any effects microbes exert against the drug and, because drugs work inside the body, interactions between the drug and the patient's tissues. The principal areas to be covered here are general concepts of chemotherapy; mechanisms of drug action; drug resistance; antibacterial, antifungal, antihelminth, and antiviral drugs; adverse human reactions to drugs; methods of drug testing; and factors in drug selection.

Crystals of azidothymidine (AZT) magnified 25 times. This drug is one of several important therapies for slowing the progress of AIDS.

MICROFILE 12.1 FROM WITCHCRAFT TO WONDER DRUGS

Early human cultures relied on various types of primitive medications such as potions, salves, poultices, and mudplasters. Many were concocted by medicine men from plant, animal, and mineral products that had been found, usually through trial and error or accident, to have some curative effect upon ailments and complaints. In one ancient Chinese folk remedy, a fermented soybean curd was applied to skin infections. The Greeks used wine and plant resins (myrrh and frankincense), rotting wood, and various mineral salts to treat diseases. Some folk medicines were occasionally effective, but most of them were probably witches' brews that either had no effect or were even harmful. It is interesting that the Greek word *pharmakeutikos* originally meant the practice of witchcraft. These ancient remedies were handed down from generation to generation, but it was not until the Middle Ages that a specific disease was first treated with a specific chemical. Dosing syphilitic patients with inorganic arsenic and mercury compounds may have proved the ancient axiom: Graviora quaedum sunt remedia periculus. ("Some remedies are worse than the disease.")

An enormous breakthrough in the science of drug therapy came with the proof of the germ theory of infection by Robert Koch (see chapter 1). This allowed disease treatment to focus on a particular microbe, which in turn opened the way for Paul Ehrlich to formulate the first theoretical concepts in chemotherapy in the late 1800s. Ehrlich had observed that specific dyes often affixed themselves to specific microorganisms and not to animal tissues. This observation led to the profound idea that if a drug was properly selective in its actions, it would zero in on and destroy a microbial target and leave human cells unaffected. He decided to test this theory in treating syphilis. First, he worked out the structure of an arsenic-based drug that was very toxic to the spirochete of syphilis but, unfortunately, to humans as well. Ehrlich gradually and systematically altered this parent molecule, creating numerous derivatives. Finally, on the 606th try, he arrived at a compound he called *salvarsan.* This drug had some therapeutic merit and was used for a few years, but it eventually had to be discontinued because it was still not selective enough in its toxicity. Ehrlich's work had laid important foundations for many of the developments to come.

Another pathfinder in early drug research was Gerhard Domagk, whose discoveries in the 1930s launched a breakthrough in therapy that marked the true beginning of broad-scale usage of antimicrobic drugs. Experimenting with numerous synthetic dyes, Domagk showed that the red dye prontosil was active against certain bacterial infections in animals, even though it was inactive against that same infectious agent in a test tube. It turned out that prontosil was chemically changed by the body into an entirely different compound, with specific activity against bacteria. This new substance was sulfonamide—the first *sulfa* drug. In a short time, the structure of this drug was determined, and it became possible to synthesize it on a wide scale and to develop scores of other sulfonamide drugs. Although these drugs had immediate applications in therapy (and still do), still another fortunate discovery was needed before the golden age of antibiotics could really blossom.

The discovery of antibiotics dramatically demonstrates how developments in science and medicine often occur through a combination of accident, persistence, collaboration, and vision. In the London laboratory of Alexander Fleming in 1928, a plate of *Staphylococcus aureus* became contaminated with the mold *Penicillium notatum.* Observing these plates, Fleming noted that the colonies of *Staphylococcus* were evidently being destroyed by some activity of the nearby *Penicillium* colonies. Struck by this curious phenomenon, he extracted from the fungus a compound he called penicillin and showed that it was responsible for the inhibitory effects.

Fleming lost interest in penicillin, and nothing was done with it for a decade, until the pressures of impending war spurred Howard Florey and Ernst Chain to develop methods for industrial production of penicillin in England. (At that time, war fatalities due to infectious diseases were higher than for other causes.) Clinical trials of penicillin, conducted in 1941, ultimately proved its effectiveness, and cultures of the mold were brought to the United States for an even larger-scale effort. When penicillin was made available to the world's population, it seemed a godsend at first, but in time, because of extreme overuse and misunderstanding of its capabilities, it also became the model for one of the most serious drug problems—namely, drug resistance. Fortunately, by the 1950s the pharmaceutical industry had entered an era of drug research and development that soon made penicillin only one of a large assortment of antimicrobic drugs.

PRINCIPLES OF ANTIMICROBIAL THERAPY

A hundred years ago in the United States, one out of three children was expected to die of an infectious disease before the age of five. Early death or severe lifelong debilitation from scarlet fever, diphtheria, tuberculosis, meningitis, and many other bacterial diseases was a fearsome yet undeniable fact of life to most of the world's population. The introduction of modern drugs to control infections in the 1930s began as a medical revolution that has added significantly to the lifespan and health of humans. It is no wonder that for many years, antibiotics in particular were regarded as the miracle cure-all for infectious diseases. In later discussions, we will evaluate this misconception in light of the shortcomings of chemotherapy. Although antimicrobic drugs have greatly reduced the incidence of certain infections, they have definitely not eradicated infectious disease and probably never will. In fact, in some parts of the world, mortality rates from infectious diseases are as high as before the arrival of antimicrobic drugs. Nevertheless, humans have been taking medicines to try to control diseases for thousands of years (microfile 12.1).

THE TERMINOLOGY OF CHEMOTHERAPY

Any chemical used in treatment, relief, or **prophylaxis*** of disease is defined as a **chemotherapeutic drug** or agent. When chemotherapeutic drugs are given as a means to control infection, the practice is termed **antimicrobial chemotherapy.*** Antimicro-

*prophylaxis (proh′′-fih-lak′-sis) Gr. *prophylassein,* to keep guard before. A process that prevents infection or disease in a person at risk.

*chemotherapy (kee′′-moh-ther′-uh-pee) Gr. *chemieia,* chemistry, and *therapeia,* service to the sick. Use of drugs to treat disease.

Table 12.1

Characteristics of the Ideal Antimicrobic Drug

- Selectively toxic to the microbe but nontoxic to host cells
- Microbicidal rather than microbistatic
- Relatively soluble and functions even when highly diluted in body fluids
- Remains potent long enough to act and is not broken down or excreted prematurely
- Not subject to the development of antimicrobial resistance
- Complements or assists the activities of the host's defenses
- Remains active despite the presence of large volumes of organic materials
- It is readily delivered to the site of infection
- Its cost is not excessive
- Does not disrupt the host's health by causing allergies or predisposing the host to other infections

bial drugs (also termed anti-infective drugs) are a special class of compounds capable even in high dilutions of destroying or inhibiting microorganisms. The origin of modern antimicrobial drugs is varied. Some, called **antibiotics,** are substances produced by the natural metabolic processes of some microorganisms that can inhibit or destroy other microorganisms. Other antimicrobial substances, termed **synthetic** drugs, are derived in the laboratory from dyes or other organic compounds. Although separation into these two categories has been traditional, they tend to overlap, because many antibiotics are now chemically altered in the laboratory *(semisynthetic).* The current trend is to use the term **antimicrobic** for all antimicrobial drugs, regardless of origin.

Antimicrobial drugs vary in their scope of activity. The so-called **narrow-spectrum** agents are effective against a limited array of different microbial types. Examples are bacitracin, an antibiotic whose inhibitory effects extend mainly to certain gram-positive bacteria, or griseofulvin, which is used chiefly in fungal skin infections. **Broad-spectrum** agents are active against a wider range of different microbes. The targets of antibiotics in the tetracycline group, for example, are a variety of gram-positive and gram-negative bacteria, rickettsias, mycoplasmas, and even protozoa. The concept of spectrum of activity has also become blurred to some extent, because an agent can be induced by artificial means to change its spectrum.

Drug, Microbe, Host—Some Basic Interactions

The broad concept of anti-infective therapy revolves around three interacting factors: the drug, the microorganism, and the infected host. To put it simply, the drug should destroy the infectious agent without harming the host's cells. This process can be visualized more tangibly as a series of stages (figure 12.1): (1) The drug is administered to the host via a designated route—primarily by mouth (oral), by injection into a vein (intravenous), by injection into a muscle (intramuscular), by applying to the skin surface (topical), and occasionally, by injection into the skin or body cavity. (2) The drug is dissolved in body fluids. (3) The drug is delivered to the infected area (extracellular or intracellular). (4) The drug destroys the infectious agent or inhibits its growth. (5) The drug is eventually excreted or broken down by the host's organs, ideally without harming them. Among the factors that affect the outcome of drug therapy are those that aid or block drug action and those that influence the behavior of microbe or host tissues. Favorable features of antimicrobial drugs in relation to the infectious agents and host are presented in table 12.1.

The Origins of Antimicrobic Drugs

Nature is undoubtedly the most prolific producer of antimicrobial drugs. Antibiotics, after all, are common metabolic products of aerobic spore-forming bacteria and fungi. By inhibiting the growth of other microorganisms in the same habitat (antagonism), antibiotic producers presumably enjoy less competition for nutrients and space. The greatest numbers of antibiotics are derived from bacteria in the genera *Streptomyces* and *Bacillus* and molds in the genera *Penicillium* and *Cephalosporium.* Chemists have sought to decipher the structure of the more useful antibiotics in order to produce related semisynthetic compounds with specialized features (microfile 12.2). They have even tried to synthesize antibiotics from simply compounds.

Chapter Checkpoints

Antimicrobial chemotherapy involves the use of drugs to control infection on or in the body.

Antimicrobial drugs are produced either synthetically or from natural sources. They inhibit or destroy microbial growth in the infected host. Antibiotic drugs are the subset of antimicrobics produced by the natural metabolic processes of microorganisms.

Antimicrobial drugs are classified by their range of effectiveness. Broad-spectrum antimicrobics are effective against many types of microbes. Narrow-spectrum antimicrobics are effective against a limited group of microbes.

Antimicrobial therapy involves three interacting factors: the drug, the microbe, and the infected host.

Two genera of bacteria and two genera of fungi are the primary sources of most antibiotics. The molecular structures of these compounds can be chemically altered to form additional semisynthetic antimicrobics.

Characteristic Interactions Between Drug and Microbe

The primary effect of antimicrobic drugs is either to disrupt the cell processes or structures of bacteria, fungi, and protozoa or to inhibit virus replication. Most of the drugs used in chemotherapy

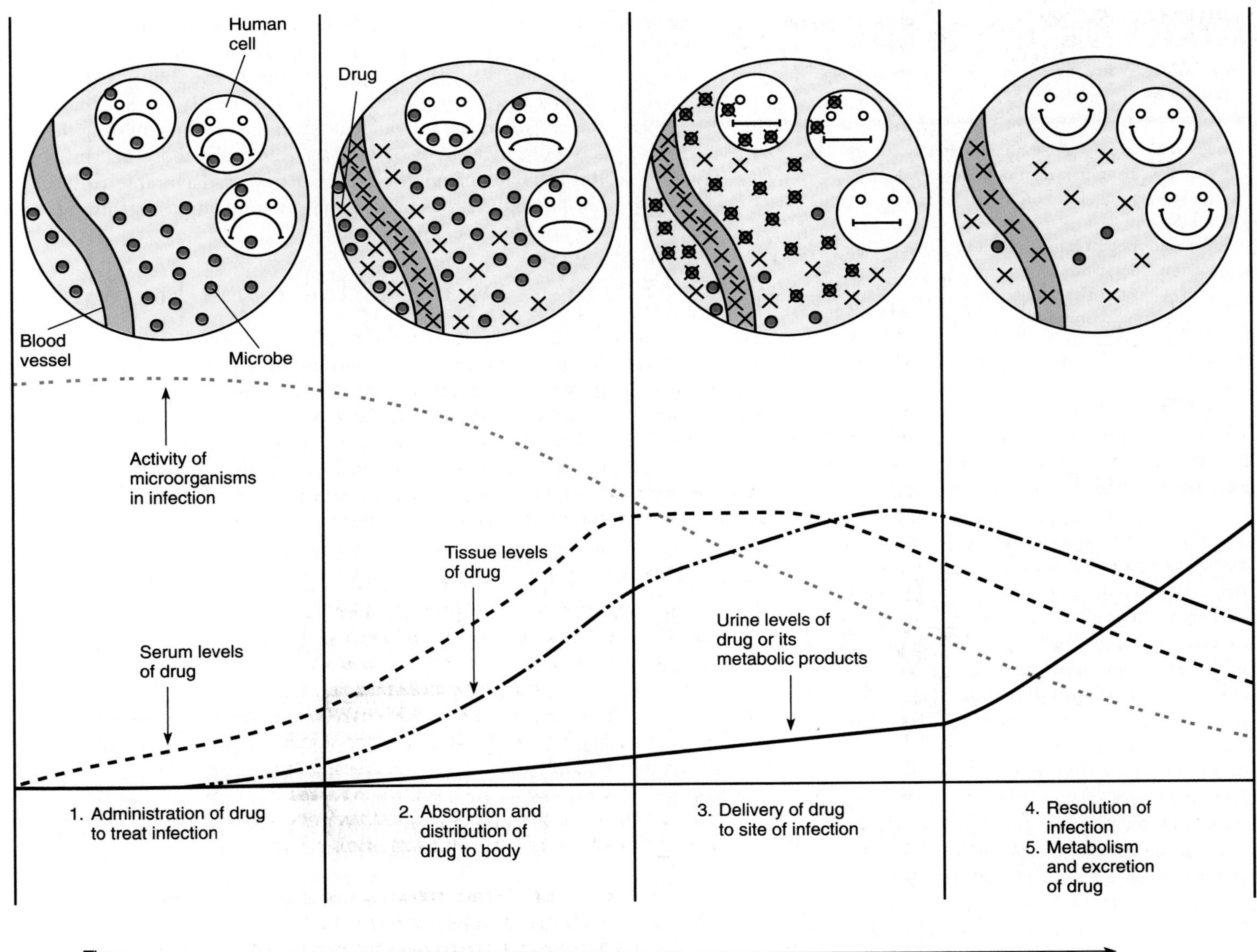

Figure 12.1
The course of events in chemotherapy; interaction between patient, drug, and infectious agent.

interfere with the function of enzymes required to synthesize or assemble macromolecules, or they destroy structures already formed in the cell. Preferably, such drugs should be **selectively toxic,** which means they produce adverse effects on microbial cells without simultaneously damaging host tissues. This concept of selective toxicity is central to chemotherapy, and the best drugs are those that block the actions or synthesis of molecules in microorganisms but not in vertebrate cells. Examples of drugs with selective toxicity are those that block the synthesis of the cell wall in bacteria (penicillins). They have low toxicity and few direct effects on human cells because human cells lack a wall and are thus neutral to this action of the antibiotic. Among the most toxic to human cells are drugs that act upon a common structure such as the cell membrane (amphotericin B, for example). In a later section, we will address the circumstances in which the ideal of selective toxicity is not met.

Mechanisms of Drug Action

Antimicrobial drugs injure microbes through several different mechanisms. Microbicidal drugs lyse and kill microorganisms by inflicting direct damage upon specific cellular targets. Microbistatic drugs interfere with cell division, and thus inhibit reproduction. Drugs that inhibit growth are not considered directly responsible for subsequent microbial death; their primary importance is to keep the microbes at bay and prevent their spread, thus allowing the host defenses an opportunity to destroy and remove the infectious agent.

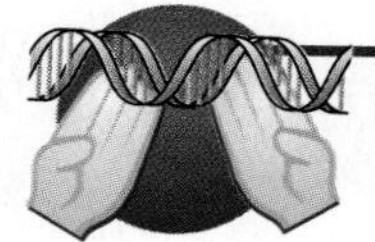

MICROFILE 12.2 A MODERN QUEST FOR DESIGNER DRUGS

Extraordinary time and effort were expended in the discovery, testing, and marketing of the first truly significant antibiotic (penicillin). Once this monumental event had transpired, the world immediately witnessed a scientific scramble to find more antibiotics. This search was advanced on several fronts. Hundreds of investigators began the laborious task of screening samples from soil, dust, muddy lake sediments, rivers, estuaries, oceans, plant surfaces, compost heaps, sewage, skin, and even the hair and skin of animals for antibiotic-producing bacteria and fungi. This intense effort has paid off over the past 50 years, because more than 10,000 antibiotics were eventually discovered (though surprisingly, only a relatively small number have been applicable to chemotherapy). Finding a new antimicrobic substance is only a first step. The complete pathway of drug development from discovery to therapy takes at least 9 years at a cost of several billion dollars.

Antibiotics are products of fermentation pathways that occur in many spore-forming bacteria and fungi. The true role of antibiotics in the lives of these microbes continues to be somewhat mysterious, although the evolutionary preservation of genes for antibiotic production does point to an important function for them. Some experts theorize that antibiotic-releasing microorganisms can inhibit or destroy nearby competitors or predators; others feel that these compounds play a part in spore formation. Whatever benefit the microbes derive, the existence of these compounds has been extremely profitable for humans. Every year, the pharmaceutical industry farms vast quantities of microorganisms and harvests their products to treat diseases caused by other microorganisms. Researchers have facilitated the work of nature by selecting mutant species that yield more abundant or useful products, by varying the growth medium, or by altering the procedures for large-scale industrial production (see chapter 26).

Another approach in the drug quest is that of organic synthesis—the chemical manipulation of molecules by adding or removing functional groups. Drugs produced in this way are designed to have advantages over other, related drugs. With the **semisynthetic** method, a natural product of the microorganism is joined with various preselected functional groups. The antibiotic is reduced to its basic molecular framework (called the nucleus), and to this nucleus specially selected side chains (R groups) are added. A case in point is the metamorphosis of the semisynthetic penicillins. The nucleus is an inactive penicillin derivative called aminopenicillanic acid, which has an opening on the number 6 carbon for addition of R groups. A particular carboxylic acid (R group) added to this nucleus can "fine-tune" the penicillin, giving it special characteristics. For instance, some R groups will make the product resistant to penicillinase (methicillin), some confer a broader activity spectrum (ampicillin), and others make the product acid-resistant (penicillin V). Cephalosporins and tetracyclines have likewise been subjected to semisynthetic conversions. The potential for using bioengineering techniques to design drugs seems almost limitless, and, indeed, several drugs have already been produced by manipulating the genes of antibiotic producers.

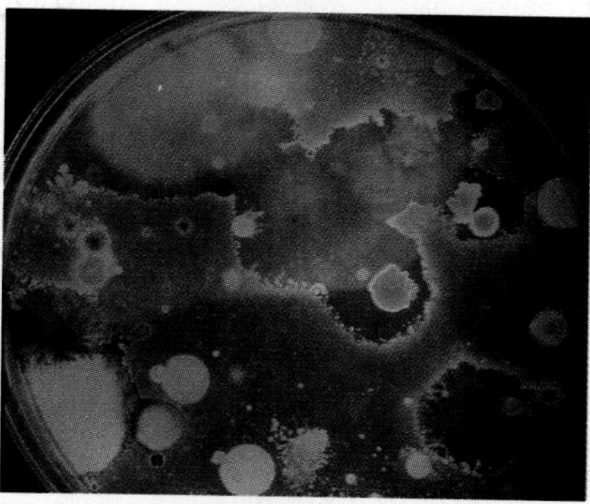

A plate with several discrete colonies of soil bacteria was sprayed with a culture of Escherichia coli *and incubated. Zones of inhibition (clear areas with no growth) surrounding several colonies indicate species that produce antibiotics.*

Synthesizing new types of penicillins. (a) *The original penicillin G molecule is a fermentation product of* Penicillium chrysogenum *that appears somewhat like a split-level house with a removable carport on one end. This house without the carport is the basic nucleus called aminopenicillanic acid. (b–d) Various new fixtures (R groups) can be added in place of the carport according to need. These R groups will produce different penicillins:* (b) *methicillin;* (c) *ampicillin; and* (d) *penicillin V.*

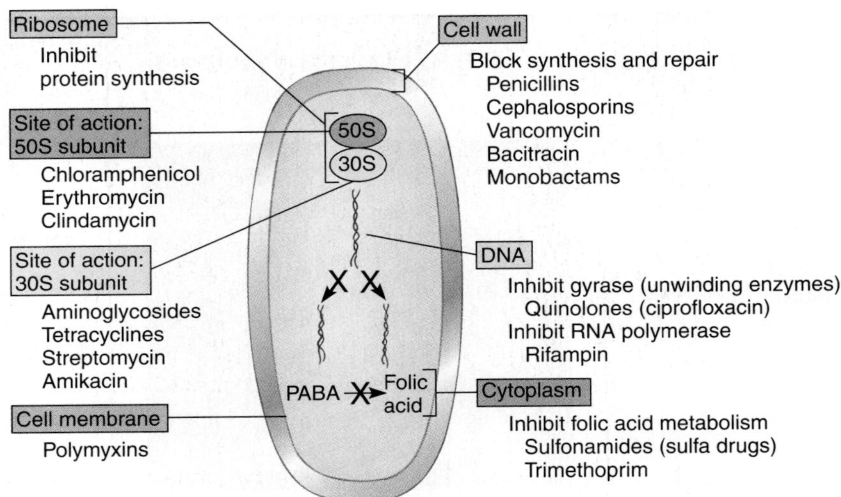

Figure 12.2
Primary sites of action of antimicrobic drugs on bacterial cells.

Antimicrobial drugs function specifically in one of the following ways (figure 12.2): (1) They inhibit cell wall synthesis; (2) they inhibit nucleic acid synthesis; (3) they inhibit protein synthesis; or (4) they interfere with the function of the cell membrane. These categories are not completely discrete, and some effects can overlap. In chapter 11 we presented an overall picture of antimicrobial mechanisms; here, we include selected examples of antimicrobics that will illustrate the general ways that drugs act upon microbial cells. It is worth noting that much knowledge of cellular structure and function has emerged as an added bonus from research on the effects of drugs.

Antimicrobic Drugs That Affect the Bacterial Cell Wall

The cell walls of most bacteria contain a rigid girdle of peptidoglycan. This structure, which is many layers thick in gram-positive species and quite thin in gram-negative ones, protects the cell against rupture from hypotonic environments. Cells actively engaged in enlargement or binary fission must constantly synthesize new peptidoglycan and transport it to its proper place in the cell envelope. Drugs such as penicillins and cephalosporins react with one or more of the enzymes required to complete this process, causing the cell to develop weak points at growth sites and to become osmotically fragile (figure 12.3). Antibiotics that produce this effect are considered bactericidal, because the weakened cell is subject to lysis. It is essential to note that most of these antibiotics are active only in young, actively growing cells, because old, inactive, or dormant cells do not synthesize peptidoglycan. (One exception is a new class of antibiotics called the penems.)

Cycloserine inhibits the formation of the basic peptidoglycan subunits, and vancomycin hinders the elongation of the peptidoglycan. The beta-lactams (penicillins and cephalosporins) bind to peptidases that are essential to cross-link the glycan molecules, thereby interrupting the completion of the cell wall (figure 12.4). Penicillins that do not penetrate the outer membrane are less effective against gram-negative bacteria, but broader-spectrum penicillins and cephalosporins can cross the cell walls of gram-negative species.

Antimicrobic Drugs That Affect Nucleic Acid Synthesis

As you will recall from chapter 9, the metabolic pathway that generates DNA and RNA molecules is a long, enzyme-catalyzed series of reactions. Like any complicated process, it is subject to breakdown at many different points along the way, and inhibition at any given point in the sequence can block subsequent events. Antimicrobial drugs interfere with nucleic acid synthesis by blocking synthesis of nucleotides, inhibiting replication, or stopping transcription. Because functioning DNA and RNA are required for proper translation as well, the effects on protein metabolism can be far-reaching.

Sulfonamides One of the more extensively studied modes of action is that of the **sulfonamides** (sulfa drugs). Because these synthetic drugs interfere with an essential metabolic process in bacteria, they represent a model for **competitive inhibition.** They act as structural or **metabolic analogs*** that mimic the natural substrate of an enzyme and vie for its active site. In practice, sulfa drugs are very similar to the natural metabolic compound PABA (para-aminobenzoic acid) required by bacteria to synthesize folic acid. Folic acid, in turn, is a component of the coenzyme tetrahydrofolic acid, which participates in the synthesis of purines and certain amino acids. A sulfonamide molecule has extreme affinity for the PABA site on the enzyme that synthesizes folic acid; thus it can successfully compete in a "chemical race" with PABA for the same sites (figure 12.5). Sulfonamides ultimately cause an inadequate supply of folic acid for purine production, which invariably halts nucleic acid synthesis and prevents bacterial cells from multiplying.

Sulfonamides are valuable in therapy because they inhibit bacteria and certain protozoa, but not mammalian cells. This one-sided inhibition stems from a basic nutritional difference between humans and microorganisms. Although humans require folic acid for nucleic acid synthesis as much as bacteria do, humans cannot

*analog (an´-uh-log) A compound whose configuration closely resembles another compound required for cellular reactions.

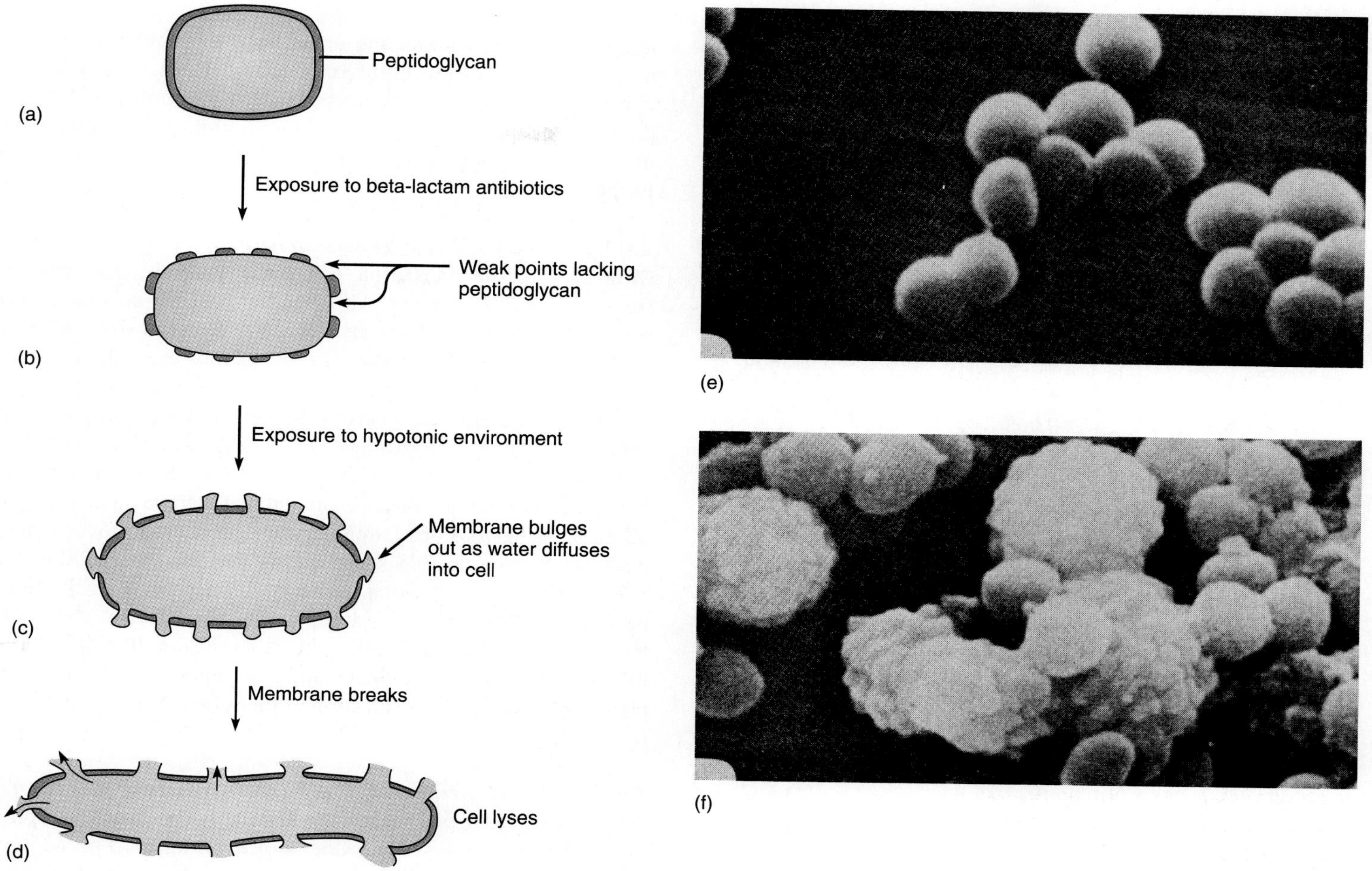

(a) Peptidoglycan

Exposure to beta-lactam antibiotics

(b) Weak points lacking peptidoglycan

Exposure to hypotonic environment

(c) Membrane bulges out as water diffuses into cell

Membrane breaks

(d) Cell lyses

(e)

(f)

Figure 12.3

The consequences of exposing a growing cell to antibiotics that prevent cell wall synthesis. *(a)* Coccus is exposed to beta-lactam agent (cephalosporin). *(b)* Weak points develop where peptidogylcan is incomplete. *(c)* The weakened cell is exposed to a hypotonic environment. *(d)* The cell lyses. *(e)* Scanning electron micrograph of bacterial cells in their normal state (10,000×). *(f)* Scanning electron micrograph of the same cells in the drug-affected state, showing surface bulges (10,000×).

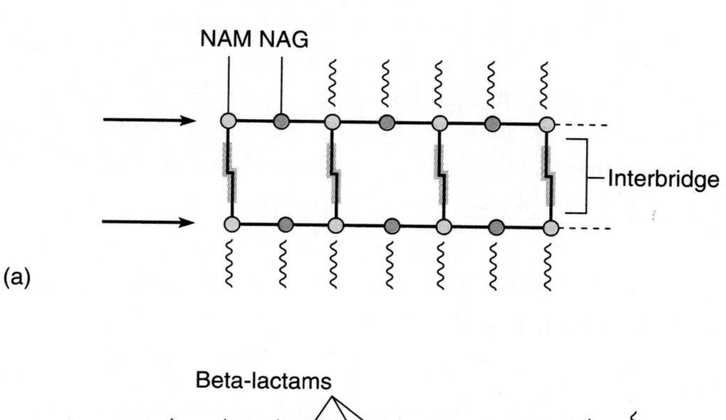

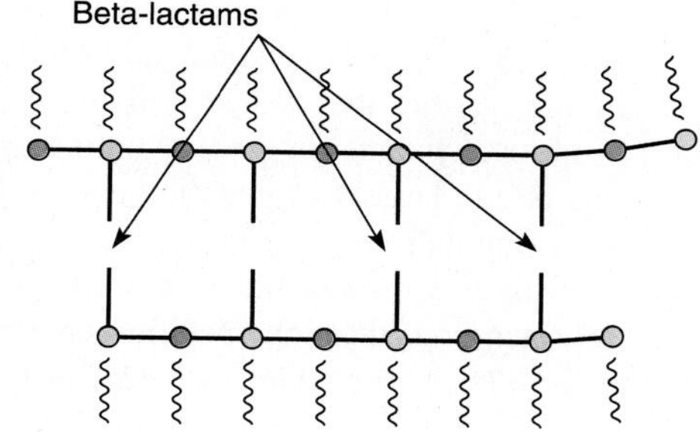

Figure 12.4

The mode of action of beta-lactam antibiotics on the bacterial cell wall. *(a)* Intact peptidogylcan has alternating NAM (N-acetyl muramic acid) and NAG (N-acetyl glucosamine) glycans cross-linked by peptide interbridges. *(b)* The beta-lactam antibiotics block the peptidases required to attach the interbridges, thereby greatly weakening the cell wall meshwork.

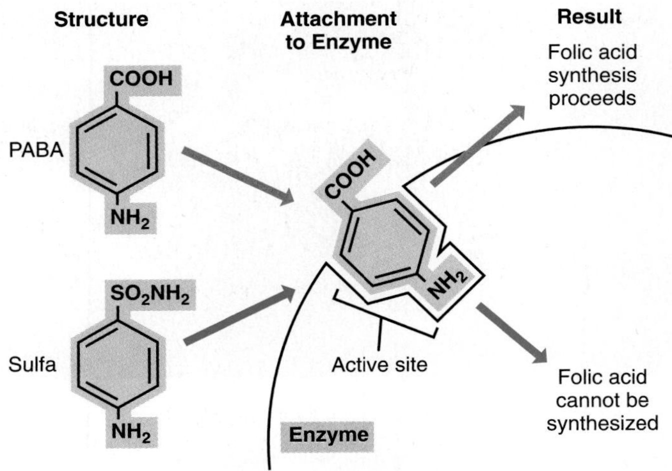

Figure 12.5

Competitive inhibition of a pathway in nucleic acid synthesis. Sulfonamides are structural analogs of PABA, a chemical required to synthesize folic acid. The similar configuration of the two enables sulfonamides to bind to the active site on an enzyme involved in folic acid synthesis. Although it binds, sulfa is not the appropriate substrate for further synthesis.

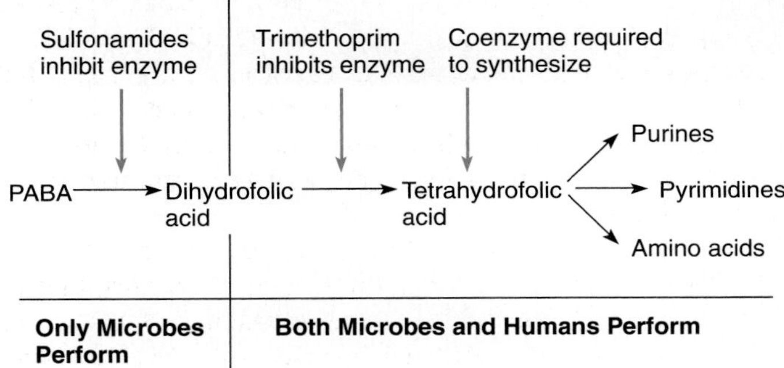

Figure 12.6

A scheme for the synthesis and function of folic acid and points of inhibition by antimetabolites. Sulfa drugs (sulfonamides) inhibit the first step, and trimethoprim blocks the second; together, they have a synergistic action that inhibits the entire pathway leading to nitrogen base and amino acid synthesis.

synthesize folic acid; it is an essential nutrient (vitamin) that must come from the diet. Human cells lack this special enzymatic system for incorporating PABA into folic acid, so their metabolism cannot be inhibited by sulfa drugs.

Other Nucleic Acid Inhibitors Trimethoprim is a metabolic analog that competes for a site on the enzyme needed in folic acid synthesis (figure 12.6). Blocking this enzyme interrupts the synthesis of purines and pyrimidines. Trimethoprim is often given simultaneously with sulfonamides to achieve a **synergistic** effect. In pharmacology, this refers to an additive effect achieved by two drugs working together, thus requiring a lower dose of each.

Other antimicrobics inhibit DNA synthesis. Chloroquine (an antimalarial drug) binds and cross-links the double helix. The newer broad-spectrum quinolones inhibit DNA unwinding enzymes or helicases, thereby stopping DNA transcription. Antiviral drugs that are analogs of purines and pyrimidines (including idoxuridine and acyclovir) insert in the viral nucleic acid and block further replication.

Drugs That Block Translation

Most inhibitors of translation, or protein synthesis, react with the ribosome-mRNA complex. Although human cells also have ribosomes, the ribosomes of eucaryotes are different in size and structure from those of procaryotes, so for the most part, these antimicrobics have a selective action against bacteria. One potential therapeutic consequence of drugs that bind to the procaryotic ribosome is the damage they can do to eucaryotic mitochondria, which contain a procaryotic type of ribosome. Two possible targets of ribosomal inhibition are the 30S subunit and the 50S subunit (figure 12.7). Aminoglycosides (streptomycin, gentamicin, for example) insert on sites on the 30S subunit and cause the misreading of the mRNA, leading to abnormal proteins. Tetracyclines block the attachment of tRNA on the A acceptor site and effectively stop further synthesis. Other antibiotics attach to sites on the 50S subunit in a way that prevents the formation of peptide bonds (chloramphenicol) or inhibits translocation of the subunit during translation (erythromycin).

Drugs That Disrupt Cell Membrane Function

A cell with a damaged membrane invariably dies from disruption in metabolism or lysis and does not even have to be actively dividing to be destroyed. The antibiotic classes that damage cell membranes have specificity for a particular microbial group, based on differences in the types of lipids in their cell membranes.

Polymyxins interact with membrane phospholipids, distort the cell surface, and cause leakage of proteins and nitrogen bases, particularly in gram-negative bacteria (figure 12.8). The polyene antifungal antibiotics (amphotericin B and nystatin) form complexes with the sterols on fungal membranes; these complexes cause abnormal openings and seepage of small ions. Unfortunately, this selectivity is not exact, and the universal presence of membranes in microbial and animal cells alike means that most of these antibiotics can be quite toxic to humans.

THE ACQUISITION OF DRUG RESISTANCE

The wide-scale use of antimicrobics soon led to microbial **drug resistance,** an adaptive response in which microorganisms begin to tolerate an amount of drug that would ordinarily be inhibitory. The development of mechanisms for circumventing or inactivating antimicrobic drugs is due largely to genetic versatility and adaptability of microbial populations. The property of drug resistance can be intrinsic as well as acquired. Intrinsic drug resistance exists naturally and is not acquired through specific genetic changes. In our context, the term *drug resistance* will mean resistance acquired by microbes that had formerly been sensitive to the drug.

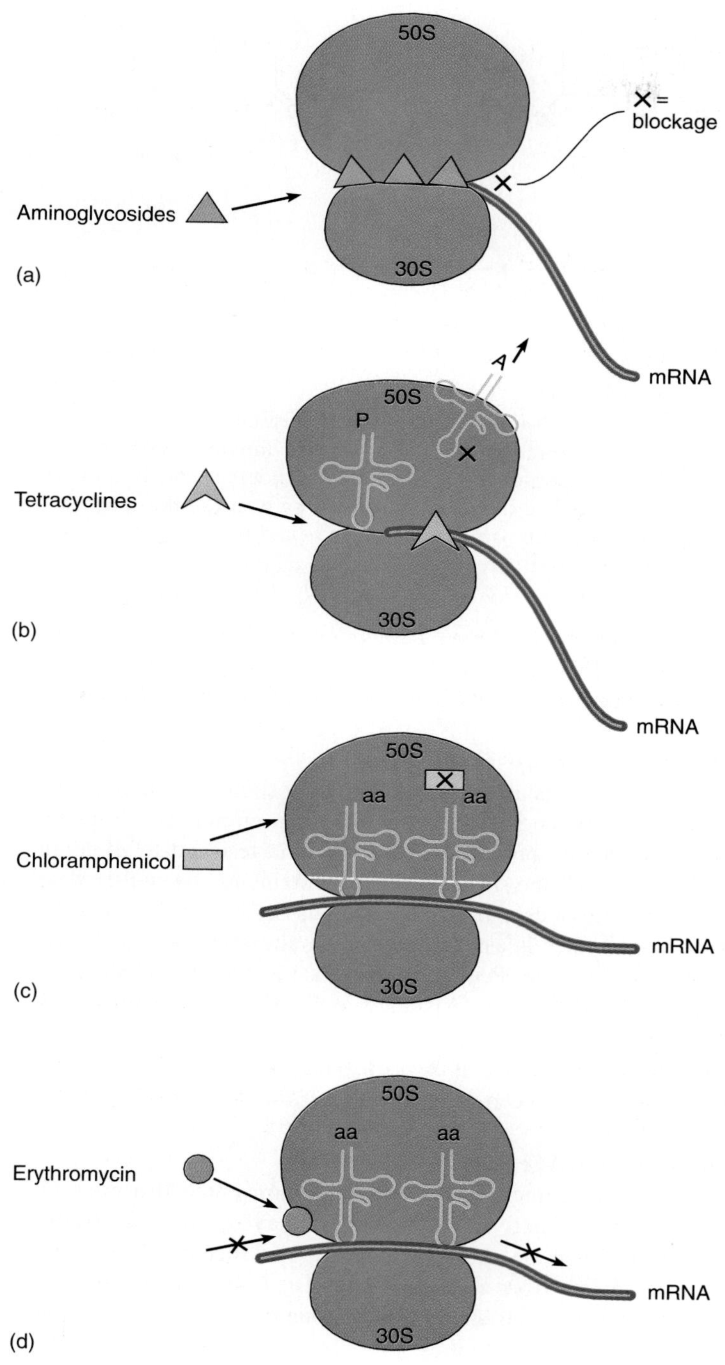

Figure 12.7

(a–d) Sites of inhibition on the procaryotic ribosome and major antibiotics that act on these sites. All have the general effect of blocking protein synthesis. Blockage sites are indicated by X.

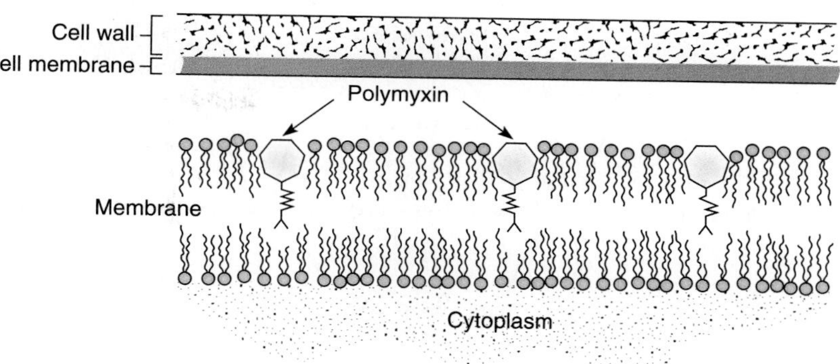

Figure 12.8

The detergent action of polymyxin. After passing through the cell wall of gram-negative bacteria, polymyxin binds to the cell membrane and disrupts its structure.

How Does Drug Resistance Develop?

The genetic events most often responsible for drug resistance are either chromosomal mutations or transfer of extrachromosomal DNA from a resistant species to a sensitive one. Chromosomal drug resistance usually results from spontaneous random mutations in bacterial populations. The chance that such a mutation will be advantageous is minimal, and the chance that it will confer resistance to a specific drug is lower still. Nevertheless, given the huge numbers of microorganisms in any population and the constant rate of mutation, such mutations do occur. The end result varies from slight changes in microbial sensitivity, which can be overcome by larger doses of the drug, to compete loss of sensitivity.

Resistance associated with intermicrobial transfer originates from plasmids called *resistance factors,* or **R factors,** that are transferred through conjugation, transformation, or transduction. Studies have shown that plasmids encoded with drug resistance, like mutations occurring on chromosomes, are naturally present in microorganisms before they have been exposed to the drug. Such traits are "lying in wait" for an opportunity to be expressed and to confer adaptability on the species. Many bacteria also maintain transposable drug resistance sequences (transposons) that are duplicated and inserted from one plasmid to another or from a plasmid to the chromosome. Chromosomal genes and plasmids containing codes for drug resistance are faithfully replicated and inherited by all subsequent progeny. This sharing of resistance genes accounts for the rapid proliferation of drug-resistant species (figure 12.9).

Specific Mechanisms of Drug Resistance

In general, a microorganism loses its sensitivity to a drug by expressing genes that stop the action of the drug. Gene expression takes the form of: (1) synthesis of enzymes that inactivate the drug, (2) decrease in cell permeability and uptake of the drug, (3) change in the number or affinity of the drug receptor sites, or (4) modification of an essential metabolic pathway. Some bacteria can become resistant indirectly by lapsing into dormancy, or, in the case of penicillin, by converting to a cell-wall-deficient form (L form) that penicillin cannot affect.

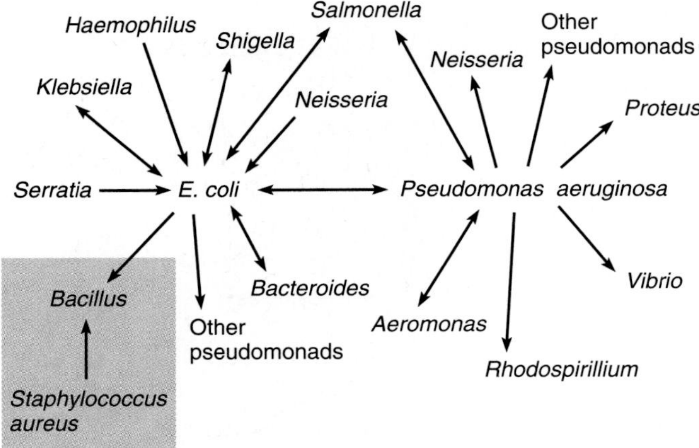

Figure 12.9
The promiscuous exchange of drug resistance is shown by this network of bacterial genera that have received and/or donated plasmids for drug resistance through conjugation and transduction. (All are gram-negative, except *Bacillus* and *Staphylococcus.)* This phenomenon is responsible for the growing numbers of drug-resistant microbes.
Source: Data from Young and Mayer, Review of Infectious Diseases, *1:55, 1979.*

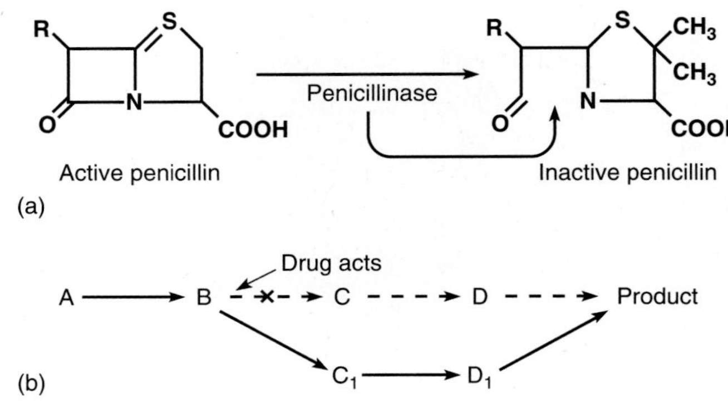

Drug Inactivation Mechanisms Microbes inactivate drugs by producing enzymes that permanently alter drug structure. One example, bacterial exoenzymes called **beta-lactamases,*** hydrolyze the *beta-lactam* ring structure of some penicillins and cephalosporins. Two beta-lactamases—*penicillinase* and *cephalosporinase*—disrupt the structure of certain penicillin or cephalosporin molecules (figure 12.10*a*). So many strains of *Staphylococcus aureus* produce penicillinase that regular penicillin is rarely a possible therapeutic choice. Now that some strains of *Neisseria gonorrhoeae,* called **PPNG,**[1] have also acquired penicillinase, alternative drugs are required to treat gonorrhea. A large number of other gram-negative species are inherently resistant to some of the penicillins and cephalosporins because of naturally occurring beta-lactamases.

Decreased Drug Permeability or Increased Drug Transport The resistance of some bacteria can be due to a mechanism that prevents the drug from entering the cell and acting on its target. For example, the outer membrane of the cell wall of certain gram-negative bacteria is a natural blockade for some of the penicillin drugs. Resistance to the tetracyclines can arise from plasmid-encoded proteins that pump the drug out of the cell. Resistance to the aminoglycoside antibiotics is a special case in which microbial cells have lost the capacity to transport the drug intracellularly.

Many bacteria possess **multidrug resistant** (MDR) **pumps** that actively transport drugs and other chemicals out of cells. These pumps are proteins encoded by plasmids and chromosomes. They are stationed in the cell membrane and expel molecules by a proton-motive force similar to ATP synthesis. They confer drug resistance on many gram-positive pathogens *(Staphylococcus, Streptococcus)* and gram-negative pathogens *(Pseudomonas, E. coli).* Because they lack selectivity, one type of pump can expel a broad array of antimicrobic drugs, detergents, and other toxic substances.

Figure 12.10
Examples of mechanisms of acquired drug resistance. *(a)* Inactivation of a drug like penicillin by penicillinase, an enzyme that cleaves a portion of the molecule and renders it inactive. *(b)* The drug has blocked the usual metabolic pathway, so the microbe circumvents it by using an alternate, unblocked pathway that achieves the required outcome.

Change of Drug Receptors

Because most drugs act on a specific target such as protein, RNA, DNA, or membrane structure, microbes can circumvent drugs by altering the nature of this target. In bacteria resistant to rifampin and streptomycin, the structure of key proteins has been altered so that these antibiotics can no longer bind. Erythromycin and clindamycin resistance is associated with an alteration on the 50S ribosomal binding site. Penicillin resistance in *Streptococcus pneumoniae* and methicillin resistance in *Staphylococcus aureus* is related to an alteration in the binding proteins in the cell wall. Fungi can become resistant by decreasing their synthesis of ergosterol, the principal receptor for certain antifungal drugs.

Changes in Metabolic Patterns

The action of antimetabolites can be circumvented if a microbe develops an alternative metabolic pathway or enzyme (figure 12.10*b*). Sulfonamide and trimethoprim resistance develops when microbes deviate from the usual patterns of folic acid synthesis. Fungi can acquire resistance to flucytosine by completely shutting off certain metabolic activities.

Natural Selection and Drug Resistance

So far, we have been considering drug resistance at the cellular and molecular levels, but its full impact is felt only if this resistance occurs throughout the cell population. Let us examine how this might happen and its long-term therapeutic consequences. Recall that any large population of microbes is likely to contain a few individual cells that are already drug-resistant because of prior mutations or transfer of plasmids (figure 12.11*a*). As long as the drug is not present in the habitat, the numbers of these resistant forms will remain low because they have no particular growth advantage. But if the population is subsequently exposed to this drug (figure 12.11*b*), sensitive individuals are inhibited or destroyed, and resistant forms survive and prolifer-

1. Penicillinase-producing *Neisseria gonorrhoeae.*
*beta-lactam (bay'-tuh-lak'-tam) Molecular structure shown in figure 12.12.

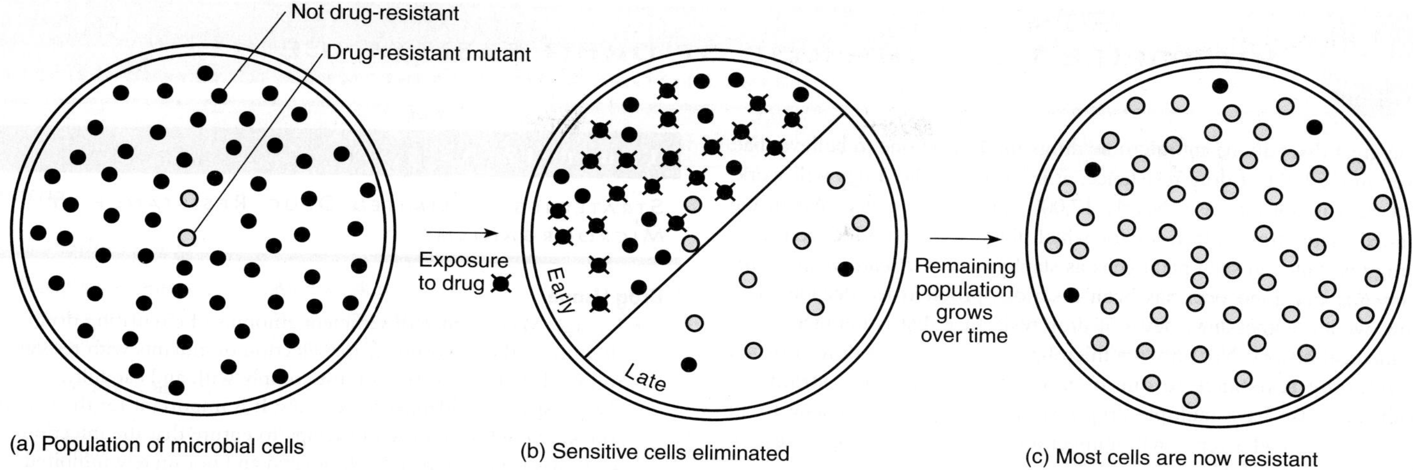

Figure 12.11

Schematic detailing the events in natural selection for drug resistance. *(a)* Populations of microbes can harbor some members with a prior mutation that confers drug resistance. *(b)* Environmental pressure (here, the presence of the drug) selects for survival of these mutants so that *(c)* they eventually become the dominant members of the population.

ate. During subsequent population growth, all offspring of these resistant microbes will inherit this drug resistance. In time, the replacement population will have a preponderance of the drug-resistant forms and can even become completely resistant (figure 12.11*c*). In ecological terms, the environmental factor (in this case, the drug) has put selection pressure on the population, allowing the more "fit" microbe (the drug-resistant one) to survive, and the population has evolved to a condition of drug resistance. Natural selection for drug-resistant forms is apparently a common phenomenon. It takes place most frequently in various natural habitats, laboratories, and medical environments, but it occasionally occurs within the bodies of humans and animals during drug therapy (microfile 12.3).

Chapter Checkpoints

- Microorganisms are termed drug-resistant when they are no longer inhibited by an antimicrobic to which they were previously sensitive.
- Drug resistance is genetic; microbes develop or acquire genes that code for methods of inactivating or escaping the antimicrobic. Resistance is selected for in environments where antimicrobics are present in high concentrations, such as in hospitals.
- Microbial drug resistance develops through random mutation and through acquisition of resistance genes from other microorganisms.
- Varieties of microbial drug resistance include drug inactivation, decreased drug uptake, decreased drug receptor sites, and modification of metabolic pathways formerly attacked by the drug.
- Widespread indiscriminate prescribing of antimicrobics has resulted in an explosion of microorganisms resistant to all common drugs, including the following: *Streptococcus, Staphylococcus,* all the Enterobacteriaceae, *Salmonella, Shigella,* and *Mycobacterium tuberculosis.*

SURVEY OF MAJOR ANTIMICROBIC DRUG GROUPS

Scores of antimicrobic drugs are marketed in the United States, and new ones are being discovered or developed every year. Although the medical and pharmaceutical literature contains a dizzying array of names for antimicrobics, most of them are variants of a small number of drug families. About 260 different antimicrobial drugs currently classified are grouped in 20 drug families. Drug reference books may give the impression that there are 10 times that many because various drug companies assign different trade names to the very same generic drug. Ampicillin, for instance, is sold as Ampen, Amcill, Omnipen, PenA, Principen, and nearly 50 other names. Most antibiotics are useful in controlling bacterial infections, though we shall also consider a number of antifungal, antiviral, and antiprotozoan drugs. Table 12.2 summarizes some major infectious agents, the diseases they cause, and the drugs indicated to treat them (drugs of choice).

Antibacterial Drugs

Penicillin and Its Relatives

The **penicillin** group of antibiotics, named for the parent compound, is a large, diverse group of compounds, most of which end in the suffix *-cillin.* Although penicillins could be completely synthesized in the laboratory from simple raw materials, it is more practical and economical to obtain natural penicillin through microbial fermentation. The natural product can then be used either in unmodified form or to make semisynthetic derivatives. *Penicillium chrysogenum* is the major source of the drug. All penicillins consist of three parts: a thiazolidine ring, a beta-lactam ring, and a variable side chain that dictates its microbicidal activity (figure 12.12).

TABLE 12.2

CONTINUED

Infectious Agent	Typical Infection	Drug of Choice	Alternative Drug*
Fungi			
Superfical Mycoses (Dermatophytoses)			
Candida albicans	Superficial candidiasis, intestinal candidiasis	Ketoconazole	Nystatin (topical)
Epidermophyton	Athlete's foot	Topical miconazole	Other azoles
Microsporon	Ringworm	Topical clotrimazole	Other azoles
Trichophyton	Athlete's foot	Oral griseofulvin	Topical azoles
Systemic Mycoses			
Aspergillus	Aspergillosis	Amphotericin B and flucytosine	
Blastomyces	Blastomycosis	Ketoconazole	Amphotericin B
Candida albicans	Candidiasis	Amphotericin B and Flucytosine	Fluconazole
Coccidioides immitis	Valley fever	Amphotericin B	Azoles
Cryptococcus neoformans	Cryptococcosis	Amphotericin B and fluconazole	Flucytosine
Pneumocystis carinii	Pneumonia (PCP)	SxT, pentamidine	
Sporothrix schenckii	Sporotrichosis	Iodides	Itraconazole
Protozoa			
Balantidium coli	Dysentery	Oxytetracycline	Lodoquinol
Entamoeba histolytica	Amebiasis	Metronidazole/tetracycline	Paromomycin
Giardia lamblia	Giardiasis	Quinacrine	Metronidazole
Plasmodium	Malaria	Chloroquine/primaquine	Quinine
Toxoplasma gondii	Toxoplasmosis	Pyrimethamine/sulfadiazine	
Trichomonas vaginalis	Trichomoniasis	Metronidazole	
Trypanosoma cruzi	Chagas' disease	Nitrifurimox****	
T. brucei	Sleeping sickness	Suramin****	Pentamidine
Helminths			
Ascaris	*Ascariosis*	*Mebendazole/pyrantel*	*Piperazine*
Cestodes	Tapeworm	Niclosamide	Praziquantel
Schistosoma	Schistosomiasis	Praziquantel	Metrifonate
Various fluke infections		Praziquantel	Tetrachlorethylene/bithionol
Various roundworm infections		Mebendazole/thiabendazol	Piperazine
Viruses			
Herpesvirus	Genital herpes, oral herpes, shingles	Acyclovir/vidarabine	Ganciclovir
HIV	AIDS	AZT/protease inhibitors	ddI, ddC, d4T
Orthomyxovirus	Type A influenza	Amantadine	Rimantidine

*Given in the case of patient allergy, microbial resistance, or some other medical cancer.
**( ) Usually given in combination.
***A combination of sulfamethoxazole and trimethoprim.
****Available in the United States only from the Drug Service of the Centers for Disease Control.

tered. The first compounds in this group were isolated in the late 1940s from the mold *Cephalosporium acremonium.* However, it was not until 20 years later that drug researchers discovered a mutant strain of this species that could be used in drug manufacture. Cephalosporins are similar to penicillins in their beta-lactam structure that can be synthetically altered (figure 12.13) and in their mode of action. The generic names of these compounds are often recognized by the presence of the root *cef, ceph,* or *kef* in their names.

Subgroups and Uses of Cephalosporins The cephalosporins are versatile. They are relatively broad-spectrum, resistant to penicillinases, and cause fewer allergic reactions than penicillins. Although some cephalosporins are given orally, many are

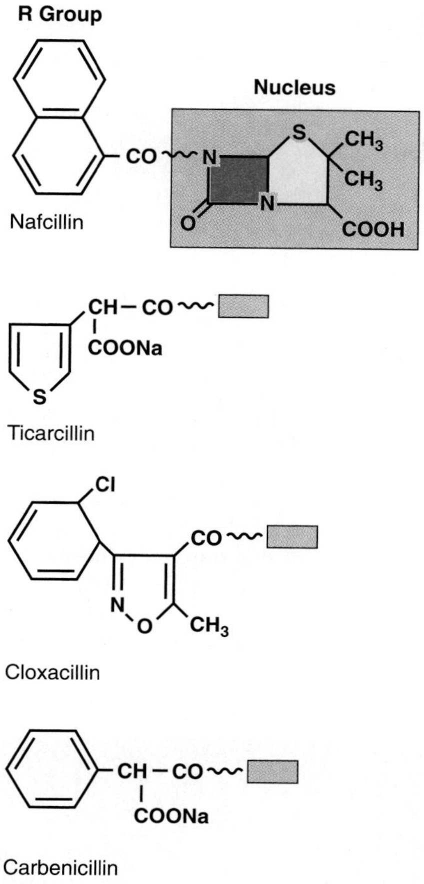

Figure 12.12

Chemical structure of penicillins. All penicillins contains a thiazolidine ring (yellow) and a beta-lactam ring (red), but each differs in the nature of the side chain (R group), which is also responsible for differences in biological activity.

poorly absorbed from the intestine and must be administered **parenterally,*** by injection into a muscle or a vein.

Three generations of cephalosporins exist, based upon their antibacterial activity. First-generation cephalosporins such as cephalothin and cefazolin are most effective against gram-positive cocci and are mildly active against a few gram-negative bacteria. Second-generation forms include cefaclor and cefonacid, which are more effective than the first-generation forms in treating infections by certain gram-negative bacteria such as *Enterobacter, Proteus,* and *Haemophilus.* Third-generation cephalosporins such as cephalexin (Keflex) and cefotaxime, are broad-spectrum with especially well-developed activity against enteric bacteria that produce beta-lactamases. Ceftriaxone (rocephin) is a new semisynthetic broad-spectrum drug for treating a wide variety of respiratory, skin, urinary, and nervous system infections.

Other Beta-Lactam Antibiotics

Related antibiotics include imipenem, a broad-spectrum drug for infections with aerobic and anaerobic pathogens. It is active in very small concentrations and can be taken by mouth with few side effects. Aztreonam, isolated from the bacterium *Chromobacterium violaceum,* is a newer narrow-spectrum drug for treating pneumonia, septicemia, and urinary tract infections by gram-negative aerobic bacilli.

The Aminoglycoside Drugs

Antibiotics composed of two or more amino sugars and an aminocyclitol (6-carbon) ring are referred to as **aminoglycosides** (figure 12.14). These complex compounds are exclusively the

*parenterally (par-ehn´-tur-ah-lee) Gr. *para,* beyond, and *enteron,* intestine. A route of drug administration other than the gastrointestinal tract.

TABLE 12.3

CHARACTERISTICS OF SELECTED PENICILLIN DRUGS

Name	Spectrum of Action	Uses, Advantages	Disadvantages
Penicillin G	Narrow	Best drug of choice when bacteria are sensitive; low cost; low toxicity	Can be hydrolyzed by penicillinase; allergies occur; requires injection
Penicillin V	Narrow	Good absorption from intestine; otherwise, similar to penicillin G	Hydrolysis by penicillinase; allergies
Oxacillin, dicloxacillin	Narrow	Not susceptible to penicillinase; good absorption	Allergies; expensive
Methicillin, nafcillin	Narrow	Not usually susceptible to penicillinase	Poor absorption; allergies; growing resistance
Ampicillin	Broad	Works on gram-negative bacilli	Can be hydrolyzed by penicillinase; allergies; only fair absorption
Amoxicillin	Broad	Gram-negative infections; good absorption	Hydrolysis by penicillinase; allergies
Carbenicillin	Broad	Same as ampicillin	Poor absorption; used only parenterally
Azlocillin, mezlocillin Ticarcillin	Very broad	Effective against *Pseudomonas* species; low toxicity compared with aminoglycosides	Allergies; susceptible to many beta-lactamases

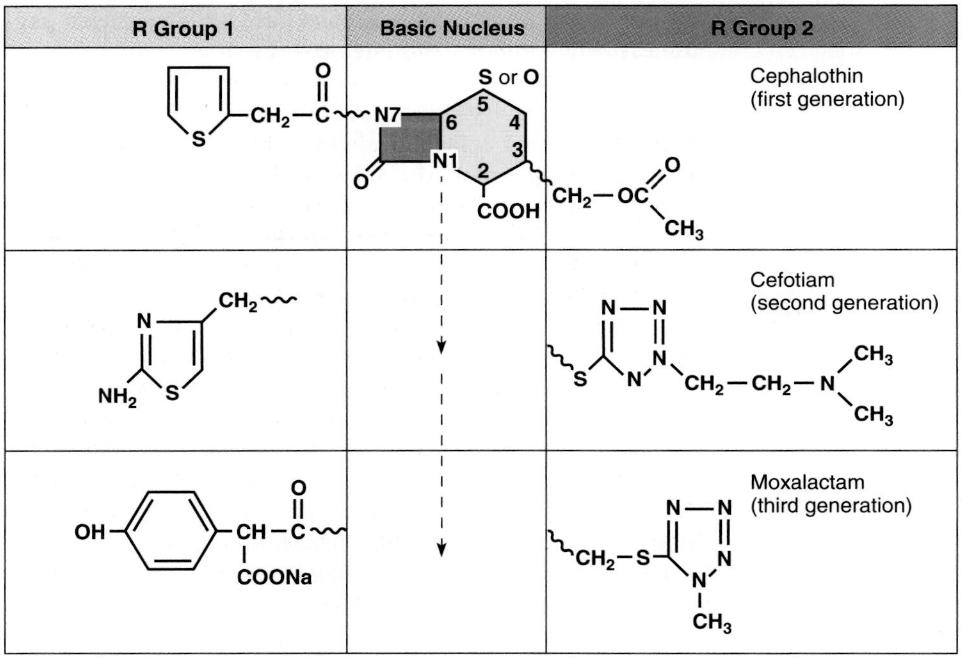

R Group 1	Basic Nucleus	R Group 2
CH_2—C(=O)— (thiophene ring)	S or O; N7, 6, 5, 4, 3, 2, N1; COOH	Cephalothin (first generation); CH_2—OC(=O)CH_3
NH_2, N, S, CH_2 (aminothiazole ring)	↓	Cefotiam (second generation); S, N—N, N, N, CH_2—CH_2—N(CH_3)(CH_3)
OH—(benzene ring)—CH(COONa)—C(=O)—	↓	Moxalactam (third generation); CH_2—S—, N—N, N, N, CH_3

Figure 12.13
The structure of cephalosporins. Like penicillin, they have a beta-lactam ring (red) but they have a different main ring (yellow). However, unlike penicillins, they have two sites for placement of R groups (at positions 3 and 7). This makes possible several generations of molecules with greater versatility in function and complexity in structure.

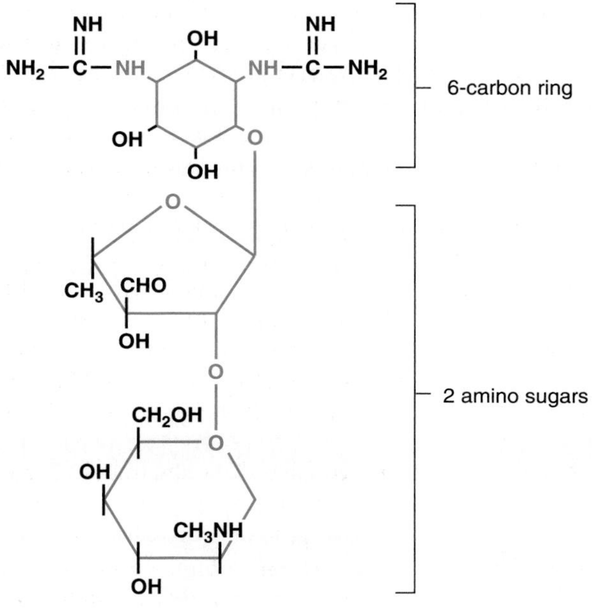

Figure 12.14
The structure of streptomycin, showing the general arrangement of an aminoglycoside (colored portions of molecule).

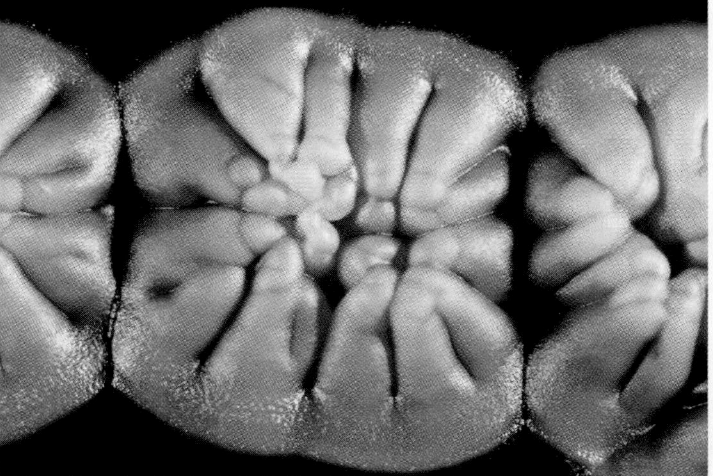

Figure 12.15
A colony of *Streptomyces,* one of nature's most prolific antibiotic producers.

products of various species of soil *actinomycetes** in the genera *Streptomyces* (figure 12.15) and *Micromonospora.*

Subgroups and Uses of Aminoglycosides The aminoglycosides have a relatively broad antimicrobial spectrum because they inhibit protein synthesis. They are especially useful in treating infections caused by aerobic gram-negative rods and certain gram-positive bacteria. Streptomycin is among the oldest of the drugs and has gradually been replaced by newer forms with less mammalian toxicity. It is still the antibiotic of choice for treating bubonic plague and tularemia and is considered a good antituberculosis agent. Gentamicin is less toxic and is widely administered for infections caused by gram-negative rods (*Escherichia, Pseudomonas, Salmonella,* and *Shigella*). Two relatively new aminoglycosides, tobramycin and amikacin, are also used for gram-negative bacillary infections and have largely replaced kanamycin.

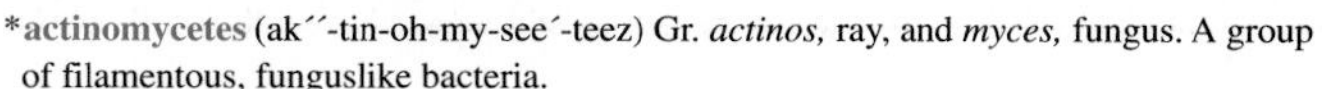

*actinomycetes (ak′′-tin-oh-my-see′-teez) Gr. *actinos,* ray, and *myces,* fungus. A group of filamentous, funguslike bacteria.

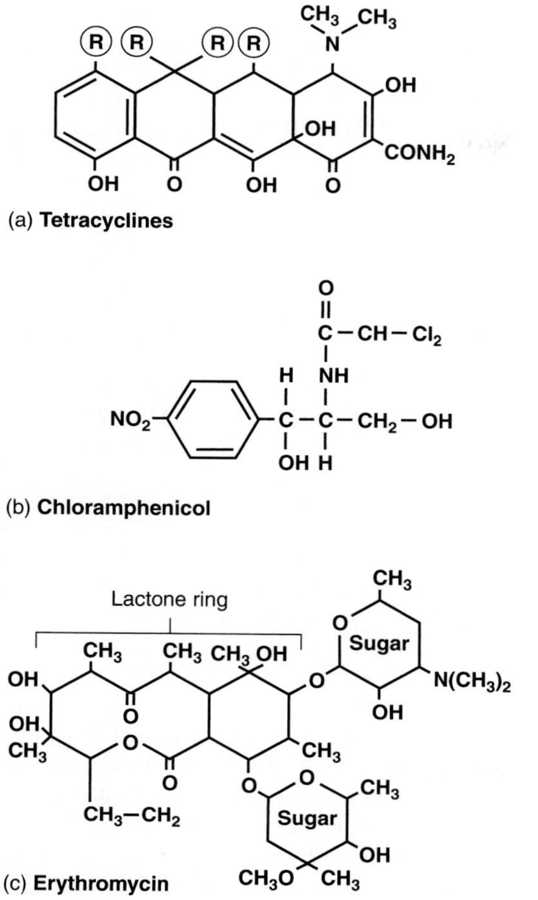

(a) Tetracyclines

(b) Chloramphenicol

(c) Erythromycin

Figure 12.16

Structures of miscellaneous broad-spectrum antibiotics.
(a) Tetracyclines. These are named for their regular group of four rings. The several types vary in structure and activity by substitution at the four R groups. *(b)* Chloramphenicol. *(c)* Erythromycin, an example of a macrolide drug. Its central feature is a large lactone ring to which two hexose sugars are attached.

Tetracycline Antibiotics

In 1948, a yellow colony of *Streptomyces* isolated from a soil sample gave off a substance, aureomycin, with strong antimicrobic properties. This antibiotic was used to synthesize its relatives terramycin and tetracycline. These natural parent compounds and semisynthetic derivatives are called, collectively, the **tetracyclines** (figure 12.16*a*). Their action of binding to ribosomes and blocking protein synthesis accounts for the broad-spectrum effects in the group.

Subgroups and Uses of Tetracyclines The scope of microorganisms inhibited by tetracyclines includes gram-positive and gram-negative rods and cocci, aerobic and anaerobic bacteria, mycoplasmas, rickettsias, and spirochetes. Tetracycline compounds such as doxycycline and minocycline are administered orally to treat several sexually transmitted diseases, Rocky Mountain spotted fever, typhus, *Mycoplasma* pneumonia, cholera, leptospirosis, acne, and even some protozoan infections. Although generic tetracycline is low in cost and easy to administer, its side effects—namely, gastronintestinal disruption and deposition in hard tissues—can limit its use (see table 12.4).

Chloramphenicol

Originally isolated in the late 1940s from *Streptomyces venezuelae,* **chloramphenicol** is a potent broad-spectrum antibiotic with a unique nitrobenzene structure (figure 12.16*b*). Its primary effect on cells is to block peptide bond formation and protein synthesis. It is one type of antibiotic that is no longer derived from the natural source but is entirely synthesized through chemical processes. Although this drug is fully as broad-spectrum as the tetracyclines, it is so toxic to human cells that its uses are restricted. A small number of people undergoing long-term therapy with this drug incur irreversible damage to the bone marrow that usually results in a fatal form of aplastic anemia.[2] Its administration is now limited to typhoid fever, brain abscesses, some forms of meningitis, and certain life-threatening infections for which an alternative therapy is not available. Chloramphenicol should never be given in large doses repeatedly over a long time period, and the patient's blood must be monitored during therapy.

Erythromycin, Clindamycin, Vancomycin, Rifamycin

Erythromycin is a macrolide antibiotic first isolated in 1952 from a strain of *Streptomyces* isolated from a Philippine soil sample. Its structure consists of a large lactone ring with sugars attached (figure 12.16*c*). This drug is relatively broad-spectrum and of fairly low toxicity. Its mode of action is to block protein synthesis by attaching to the ribosome. It is administered orally as the drug of choice for *Mycoplasma* pneumonia, legionellosis, *Chlamydia* infections, pertussis, and diphtheria and as a prophylactic drug prior to intestinal surgery. It also offers a useful substitute for dealing with penicillin-resistant streptococci and gonococci and for treating syphilis and acne. Newer semisynthetic macrolides include *clarithromycin* and *azithromycin.* Both drugs are useful for middle ear, respiratory, and skin infections. Both drugs have also been approved for *Mycobacterium* (MAC) infections in AIDS patients. Clarithromycin has additional applications in controlling infectious stomach ulcers.

Clindamycin is a broad-spectrum antibiotic related to lincomycin. The tendency of clindamycin to cause adverse reactions in the gastronintestinal tract limits its applications to (1) serious infections in the large intestine and abdomen due to anaerobic bacteria (*Bacteroides* and *Clostridium*), that are unresponsive to other antibiotics, (2) infections with penicillin-resistant staphylococci, and (3) acne medications applied to the skin.

Vancomycin is a narrow-spectrum antibiotic most effective in treating staphylococcal infections in cases of penicillin and methicillin resistance or in patients with an allergy to penicillins. It has also been chosen to treat *Clostridium* infections in children and endocarditis (infection of the lining of the heart) caused by *Enterococcus faecalis.* Because it is very toxic and hard to administer, vancomycin is usually restricted to the most serious, life-threatening conditions.

2. A failure of the blood-producing tissue that results in very low levels of red and white blood cells.

Another product of the genus *Streptomyces* is rifamycin, which is altered chemically into **rifampin.** It is somewhat limited in spectrum because the molecule cannot pass through the cell envelope of many gram-negative bacilli. It is mainly used to treat infections by several gram-positive rods and cocci and a few gram-negative bacteria. Rifampin figures most prominently in treating mycobacterial infections, especially tuberculosis and leprosy, but it is usually given in combination with other drugs to prevent development of resistance. Rifampin is also recommended for prophylaxis in *Neisseria meningitidis* carriers and their contacts, and it is occasionally used to treat *Legionella, Brucella,* and *Staphylococcus* infections.

The Bacillus Antibiotics: Bacitracin and Polymyxin

Bacitracin is a narrow-spectrum peptide antibiotic produced by a strain of the bacterium *Bacillus subtilis.* Since it was first isolated, its greatest claim to fame has been as a major ingredient in a common drugstore antibiotic ointment (Neosporin) for combating superficial skin infections by streptococci and staphylococci. For this purpose, it is usually combined with neomycin (an aminoglycoside) and polymyxin (below).

Bacillus polymyxa is the source of the **polymyxins,** narrow-spectrum peptide antibiotics with a unique fatty acid component that contributes to their detergent activity (see figure 12.8). Only two polymyxins—B and E (also known as colistin)—have any routine applications, and even these are limited by their toxicity to the kidney. Either drug can be indicated to treat drug-resistant **Pseudomonas aeruginosa** and severe urinary tract infections caused by other gram-negative rods.

New Classes of Antibiotics

For 20 years new antibiotics have been formulated from the traditional drug classes. Now, two additional drug classes have been introduced. **Fosfomycin** trimethamine is a phosphoric acid agent being used to treat urinary tract infections caused by enteric bacteria. **Synercid** is a combined antibiotic from the streptogramin group of drugs. It is being hailed as the solution to many cases of drug-resistant bacteria. It is effective against *Staphylococcus* species that cause endocarditis and surgical infections and against resistant strains of *Streptococcus.* Many experts are urging physicians to use these medications only when no other drugs are available to reduce the rate of drug resistance.

SYNTHETIC ANTIBACTERIAL DRUGS

The synthetic antimicrobics as a group do not originate from bacterial or fungal fermentations. Some were developed from aniline dyes, and others were originally isolated from plants. Although they have been largely supplanted by antibiotics, several types are still useful.

The Sulfonamides, Trimethoprim, and Sulfones

The very first modern antimicrobic drugs were the **sulofnamides,** or sulfa drugs, named for para-aminobenzenesulfonamide (sulfanilamide; figure 12.17). Although thousands of sulfonamides

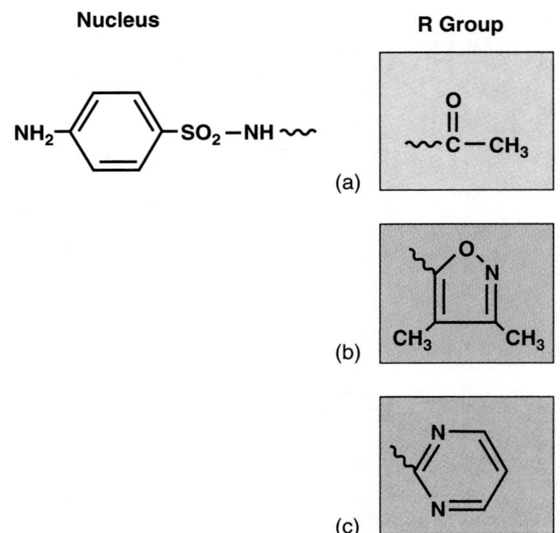

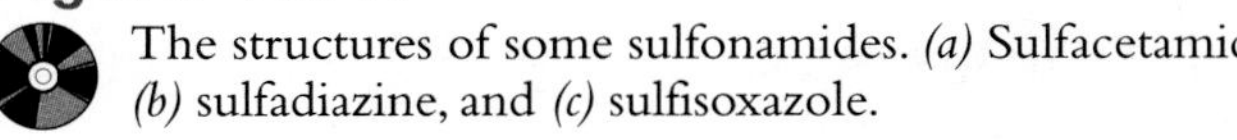

Figure 12.17

The structures of some sulfonamides. *(a)* Sulfacetamide, *(b)* sulfadiazine, and *(c)* sulfisoxazole.

have been formulated, only a few have gained any importance in chemotherapy. Because of its solubility, sulfisoxazole is the best agent for treating shigellosis, acute urinary tract infections, and certain protozoan infections. Silver sulfadiazine ointment and solution are prescribed for treatment of burns and eye infections. In many cases, sulfamethoxazole is given in combination with **trimethoprim** (Septra, Bactrim) to take advantage of the synergistic effect of the two drugs. This combination is one of the primary treatments for *Pneumocystis carinii* pneumonia (PCP) in AIDS patients.

Sulfones are compounds chemically related to the sulfonamides but lack their broad-spectrum effects. This lack does not diminish their importance as key drugs in treating leprosy. The most active form is dapsone, usually given in combination with rifampin and clofazamine (an antibacterial dye) over long periods.

Miscellaneous Antibacterial Agents **Isoniazid** (INH) has been in use since 1952 to treat tuberculosis. It is bactericidal to *Mycobacterium tuberclosis,* but only against growing cells. oral doses are indicated for both active tuberculosis and prophylaxis in cases of a positive TB test. Ethambutol, a closely related compound, is effective in treating the early stages of tuberculosis.

Nitrofurantoin, a derivative of furan sugars, is a fairly broad-spectrum synthetic drug most effective in gram-negative urinary tract infections because it is excreted by the kidneys and is concentrated in the urine.

Much excitement has been generated by the new class of synthetic drugs chemically related to quinine called **fluoroquinolones.** These drugs exhibit several ideal traits, including potency and broad spectrum. Even in minimal concentrations, quinolones inhibit a wide variety of gram-positive and gram-negative bacterial species. In addition, they are readily absorbed from the intestine and less subject to microbial resistance than other drugs. The principal quinolones, norfloxacin and

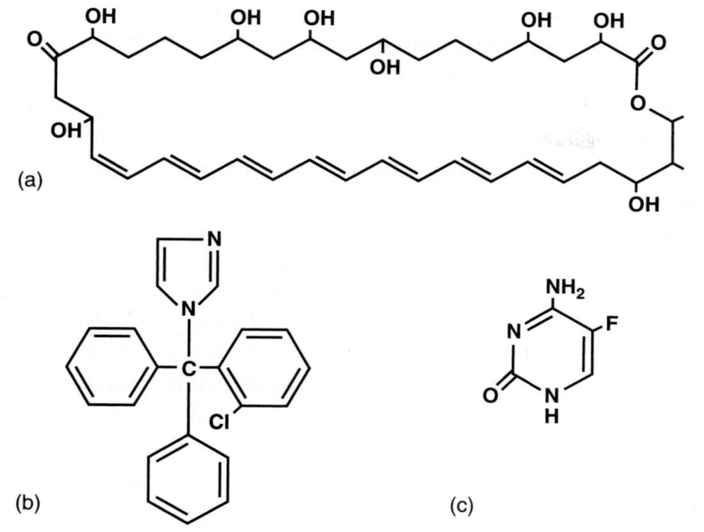

Figure 12.18
Some antifungal drug structures. *(a)* Polyenes. The example shown is amphotericin B (proposed structure), a complex steroidal antibiotic that inserts into fungal cell membranes. *(b)* Clotrimazole, one of the azoles. *(c)* Flucytosine, a structural analog of cytosine that contains fluoride.

ciprofloxacin, have been successful in therapy for urinary tract infections, sexually transmitted diseases, gastrointestinal infections, osteomyelitis, respiratory infections, and soft tissue infections. Newer drugs in this category are sparfloxacin and levofloxacin. These agents are especially recommended for pneumonia, bronchitis, and sinusitis. Side effects that limit the use of quinolones include seizures and other brain disturbances.

Agents to Treat Fungal Infections

Because the cells of fungi are eucaryotic, they present special problems in chemotherapy. For one, the great majority of chemotherapeutic drugs useful in treating bacterial infections are generally ineffective in combating fungal infections, and for another, the similarities between fungal and human cells often mean that drugs toxic to fungal cells are capable of harming human tissues. A few agents with special antifungal properties have been developed for treating systemic and superficial fungal infections. Four main drug groups currently in use are the macrolide polyene antibiotics, griseofulvin, synthetic azoles, and flucytosine (figure 12.18).

Macrolide polyenes, represented by **amphotericin B** (named for its acidic and basic—amphoteric—properties) and **nystatin** (for New York State, where it was discovered), have a structure that mimics the lipids in some cell membranes. Both compounds were originally isolated from species of *Streptomyces.* Amphotericin B (Fungizone) is by far the most versatile and effective of all antifungals. Not only does it work on most fungal infections, including skin and mucous membrane lesions caused by *Candida albicans,* but it is one of the few drugs that can be injected to treat systemic fungal infections such as histoplasmosis and cryptococcus meningitis. Nystatin is used only topically or orally to treat candidiasis of the skin and mucous membranes, but it is not useful for subcutaneous or systemic fungal infections or for ringworm.

Griseofulvin is an antifungal product especially active in certain dermatophyte infections such as athlete's foot. The drug is deposited in the epidermis, nails, and hair, where it inhibits fungal growth. Because complete eradication requires several months and griseofulvin is relatively nephrotoxic, this therapy is given in only the most extreme cases.

The **azoles** are broad-spectrum antifungal agents with a complex ringed structure. The most effective drugs are ketoconazole, intraconazole, fluconazole, clotrimazole, and miconazole. Ketoconazole is used orally and topically for cutaneous mycoses, vaginal and oral candidiasis, and some systemic mycoses. Itraconazole and fluconazole can be used in selected patients for AIDS-related mycoses such as aspergillosis and cryptococcus meningitis. Clotrimazole and miconazole are used mainly as topical ointments for infections in the skin, mouth, and vagina.

Flucytosine is an analog of cytosine that was first developed for tumor therapy in mammals. Although not an effective anticancer drug, it turned out to be useful in combating fungi. Its best features are its rapid absorption after oral therapy and its property of entering the blood and cerebrospinal fluid. Alone, it can be used to treat certain cutaneous mycoses. Now that many fungi are resistant to flucytosine, it must be combined with amphotericin B to effectively treat systemic mycoses.

Antiparasitic Chemotherapy

The enormous diversity among protozoan and helminth parasites and their corresponding therapies reaches far beyond the scope of this textbook; however, a few of the more common drugs will be surveyed here and again in chapter 23 (table 23.5). Presently, a small number of approved and experimental drugs are used to treat malaria, leishmaniasis, trypanosomiasis, amebic dysentery, and helminth infections, but the need for new and better drugs has spurred considerable research in this area.

Antimalarial Drugs: Quinine and Its Relatives

Quinine, extracted from the bark of the cinchona tree, was the principal treatment for malaria for hundreds of years. After World War II, however, it was replaced by the synthesized quinolines, mainly chloroquine and primaquine, which had less toxicity to humans. Because there are several species of *Plasmodium* (the malaria parasite) and many stages in its life cycle, no single drug is universally effective for every species and stage, and each drug is restricted in application. For instance, primaquine eliminates the liver phase of infection, and chloroquine suppresses acute attacks associated with infection of red blood cells. Chloroquine is taken alone for prophylaxis and suppression of acute forms of malaria. Primiquine is administered to patients with relapsing cases of malaria. Although quinine chemotherapy was abandoned for a time because of its toxicity, the development of chloroquine-resistant *Plasmodium* in South America and Southeast Asia restored quinine to a role in treating drug-resistant infections.

Chemotherapy for Other Protozoan Infections

A widely used amebicide, metronidazole (Flagyl), is effective in treating mild and severe intestinal infections and hepatic disease caused by *Entamoeba histolytica.* Given orally, it also has applications for infections by *Giardia lamblia* and *Trichomonas vaginalis.* Other drugs with antiprotozoan activities are quinicrine (a quinine-based drug), sulfonamides, and tetracyclines.

Antihelminthic Drug Therapy

Treating helminth infections has been one of the most difficult and challenging of all chemotherapeutic tasks. Flukes, tapeworms, and roundworms are much larger parasites than other microorganisms and, being animals, have greater similarities to human physiology. Also, the usual strategy of using drugs to block their reproduction is usually not successful in eradicating the adult worms. The most effective drugs immobilize, disintegrate, or inhibit the metabolism of all stages of the life cycle.

Mebendazole and thiabendazole are broad-spectrum antiparasitic drugs used in several roundworm and tapeworm intestinal infestations. These drugs work locally in the intestine to inhibit the function of the microtubules of worms, eggs, and larvae. Inhibiting the function of microtubules interferes with their glucose utilization and disables them. The compounds pyrantel and piperazine paralyze the muscles of intestinal roundworms. Niclosamide destroys the scolex and the adjoining proglottids of tapeworms, thereby loosening the worm's holdfast. In these forms of therapy, the worms are unable to maintain their grip on the intestinal wall and are expelled along with the feces by the normal peristaltic action of the bowel. Two newer antihelminth drugs are praziquantel, a treatment for various tapeworm and fluke infections, and ivermectin, a veterinary drug now used for strongyloidiasis and oncocercosis in humans.

Antiviral Chemotherapeutic Agents

The treatment of viral infections with chemotherapeutic agents is still in its infancy. Traditionally, viral infections have been prevented by vaccination, but useful vaccines have not been developed for many viruses. Epidemics of AIDS, genital warts, and influenza, not to mention the common cold, continue to emphasize the need for effective antiviral drugs. The existing antiviral agents have some limitations such as narrow spectrum and acting only intracellularly. Many of them are toxic as well.

Most compounds have their effects on the completion of the virus cycle. Three major modes of action are: (1) barring complete penetration of the virus into the host cell, (2) blocking the transcription and translation of viral molecules, and (3) preventing the maturation of viral particles. Although antiviral drugs protect uninfected cells by keeping viruses from being synthesized and released, most are unable to destroy extracellular viruses or those in a latent state.

Several antiviral agents mimic the structure of nucleotides and compete for sites on replicating DNA. The incorporation of these synthetic nucleotides inhibits further DNA synthesis. **Acyclovir** (Zovirax) is a synthetic purine compound that blocks DNA synthesis in a small group of viruses, particularly the herpes groups of viruses. In the topical form, it is most effective in controlling the primary attack of facial or genital herpes. Intravenous or oral acyclovir therapy can reduce the severity of primary and recurrent genital herpes episodes. One relative called Famciclovir is used to treat shingles and chickenpox caused by the herpes zoster virus. Another, gancyclovir, is approved to treat cytomegalovirus infections of the eye. An analog of adenine, *vidarabine,* is also effective against the herpesviruses. **Ribavirin** is a guanine analog used in aerosol form to treat life-threatening infections by the respiratory syncytial virus (RSV) in infants and some types of viral hemorrhagic fever.

Azidothymidine (AZT or Zidovudine) is a thymine analog used exclusively to treat AIDS patients. This drug is specific for HIV by preventing the natural action of the viral reverse transcriptase and blocking further DNA synthesis and viral replication. It is indicated for patients with full-blown AIDS and for those in the earlier phases of disease. Although it is not a cure, AZT does slow the course of the disease in most instances. Other approved anti-AIDS drugs that also act as nucleotide analogs are didanosine (ddI), zalcitabine (ddC), and stavudine (d4T). These drugs are used as alternatives to AZT or in combination with it to improve the clinical condition of the patient.

The most recent additions to AIDS treatment regimens are **protease inhibitors** such as saquinavir and ritonivir. These drugs have the benefit of preventing the assembly of functioning viral particles. When used in combination with nucleotide analogs, protease inhibitors can greatly slow the progression of AIDS. See chapter 25 for further coverage of this topic.

Amantadine and its relative, rimantidine, are amines restricted almost exclusively to treating infections by influenza A virus. Because their action appears to inhibit the uncoating of the viral RNA, these drugs must be given rather early in an infection. In addition to these antiviral drugs, dozens of other agents are under investigation. For a novel approach for controlling viruses see microfile 12.4.

A sensible alternative to artificial drugs has been the natural product of infected cells, **interferon** (IFN). Interferon is a carbohydrate-containing protein produced primarily by fibroblasts and leukocytes in response to various immune stimuli. The discovery of interferon's potential application in viral infections and cancer therapy prompted a tremendous push for research on its biological effects. We now know that it is a versatile part of animal host defenses, having a broad spectrum of activities and great import in natural immunities. (Its mechanism is discussed in chapter 14.)

The first investigations of interferon's antiviral activity were limited by the extremely minute quantities that could be extracted from human blood. Several types of interferon are currently produced by the recombinant DNA technology techniques outlined in chapter 10. Extensive clinical trials have tested its effectiveness in viral infections and cancer. Some of the known therapeutic benefits include: (1) reducing the time of healing and some of the complications in certain infections (mainly of herpesviruses), (2) preventing or reducing some symptoms of cold and papillomaviruses (warts), (3) slowing the progress of certain cancers, including bone cancer and cervical cancer, and certain leukemias and lymphomas. A form of interferon is approved for treating a rare cancer called hairy-cell leukemia, hepatitis C (a viral liver infection), and genital warts. It is presently being tested as a nasal spray to control the common cold virus and as therapy for Kaposi's sarcoma in AIDS patients.

MICROFILE 12.4 HOUSEHOLD REMEDIES—FROM APPLES TO ZINC

Who would have thought that drinking a glass of apple juice, eating a clove of garlic, or sneezing into a facial tissue might nip a viral infection in the bud? A series of research findings from the past few years seems to point to a possible role for these and other humble medicinal aids. Apple juice, fruit juices, and even tea contain natural antiviral substances, thought to be tannic acid or other organic acids, that kill the poliovirus and coxsackievirus. Drinking beverages that contain these substances can help prevent the passage of those viruses into the intestine (their usual site of entry). Could this be a reason that "an apple a day keeps the doctor away"?

Specialists in human rhinoviruses suggest that the most important route of transmission of cold viruses is through hand-to-hand contact. Researchers with Kimberly-Clark used this information to develop a special tissue impregnated with iodine and citric acid. If used by patients to catch sneezes, coughs, and secretions, these tissues proved quite effective in retarding the spread of the cold virus and even in reducing the severity of the infection in some cases. Controlled studies now support the benefits of zinc ions to control the common cold. An attractive hypothesis to explain how it works is that the zinc attaches to cell receptors and blocks the attachment of cold viruses. Various tablets and lozenges are now sold over the counter as cold deterrents.

The therapeutic benefits of certain foods are often surprising. Yogurt made with live cultures has been shown to contain a natural antibiotic. This may explain the benefits of eating yogurt to control yeast infections of the gastrointestinal tract and vagina. Research indicates that garlic extract also contains active ingredients that clearly inactivate several types of animal viruses. If all else fails, one should not overlook the recuperative powers of chicken soup, sometimes known as "Jewish penicillin." This, too, has been found in controlled scientific tests to shorten the length and relieve the symptoms of colds, though the active ingredients have not been isolated. It appears that a timely trip to the kitchen cabinet could be as beneficial as one to the medicine cabinet. Might this be what is meant by "feeding a cold and starving a fever"?

Chapter Checkpoints

Antimicrobics are classified into 20 major drug families, based on their chemical composition, source of origin, and their site of action.

The majority of antimicrobics are effective against bacteria, but a limited number are effective against protozoa, helminths, fungi, and viruses.

Penicillins, cephalosporins, bacitracin, vancomycin, and cycloserines block cell wall synthesis, primarily in gram-positive bacteria.

Aminoglycosides and tetracyclines block protein synthesis in procaryotes.

Sulfonamides, trimethoprim, isoniazid, nitrofurantoin, and the fluoroquinolones are synthetic antimicrobics effective against a broad range of microorganisms. They block steps in the synthesis of nucleic acids.

Fungal antimicrobials, macrolide polyenes, griseofulvin, aoles, and fluorocytosine must be monitored carefully because of the potential toxicity to the infected host. They promote lysis of cell membranes.

There are fewer antiparasitic drugs than antibacterial drugs because parasites are eucaryotes like their human hosts and they have several life stages, some of which can be resistant to the drug.

Antihelminth drugs immobilize or disintegrate infesting helminths or inhibit their metabolism in some manner.

Antiviral drugs interfere with viral replication by blocking viral entry into cells, blocking the replication process, or preventing the assembly of viral subunits into complete virions.

Many antiviral agents are analogs of nucleotides. They inactivate the replication process when incorporated into viral nucleic acids.

Although interferon is effective *in vivo* against certain viral infections, commercial interferon is not currently effective as a broad-spectrum antiviral agent.

CHARACTERISTICS OF HOST-DRUG REACTIONS

Although selective antimicrobial toxicity is the ideal constantly being sought, chemotherapy by its very nature involves contact with foreign chemicals that can harm human tissues. In fact, estimates indicate that at least 5% of all persons taking an antimicrobic drug experience some type of serious adverse reaction to it. The major **side effects** of drugs fall into one of three categories: direct damage to tissues through toxicity, allergic reactions, and disruption in the balance of normal microbial flora. The damage incurred by antimicrobial drugs can be short-term and reversible or permanent, and it ranges in severity from cosmetic to lethal. Table 12.4 summarizes drug groups and their major side effects.

TOXICITY TO ORGANS

Certain drugs adversely affect the following organs: the liver (hepatotoxic drugs), kidneys (nephrotoxic drugs), gastrointestinal tract, cardiovascular system and blood-forming tissue, nervous system (neurotoxic drugs), respiratory tract, skin, bones, and teeth.

TABLE 12.4

MAJOR ADVERSE TOXIC REACTIONS TO COMMON DRUG GROUPS

Antimicrobic Drug	Primary Tissue Affected	Primary Damage or Abnormality Produced
Antibacterials		
Penicillin G	Skin	Rash
	Brain	Seizures*
Carbenicillin	Platelets	Abnormal bleeding
Ampicillin	GI tract	Diarrhea and enterocolitis**
Cephalosporins	Platelet function	Inhibition of prothrombin synthesis
	White blood cells	Decreased circulation
	Kidney	Nephritis
Tetracyclines	GI tract	Diarrhea and enterocolitis
	Liver	Damage to hepatocytes*
	Teeth, bones	Gray to brown discoloration of tooth enamel in fetuses and children from birth through age 8
	Skin	Reactions to sunlight (photosensitization)
Chloramphenicol	Bone marrow	Injury to red and white blood cell precursors; blood cell deficiencies
Aminoglycosides (streptomycin, gentamicin, amikacin)	GI tract, hair cells in cochlea, vestibular cells, neuromuscular, kidney tubules	Diarrhea and enterocolitis; malaborption; loss of hearing, dizziness, respiratory failure; loss of filtration ability
Isoniazid	Liver	Hepatitis
	Brain	Seizures
	Skin	Dermatitis
Sulfonamides	Kidney	Formation of crystals; blockage of urine flow
	Red blood cells	Hemolysis
	Platelets	Reduction in number
Polymyxin	Kidney	Damage to membranes of tubule cells
	Neuromuscular	Weakened muscular responses
Quinolones (ciprofloxacin, norfloxacin)	Nervous system, bones, GI tract	Headache, dizziness, tremors, GI distress
Rifampin	Liver	Damage to hepatic cells
	Skin	Dermatitis
Antifungals		
Amphotericin B	Kidney	Disruption of tubular filtration
Flucytosine	White blood cells	Decreased number
Antiprotozoan Drugs		
Metronidazole	GI tract	Nausea, vomiting
Chloroquine	GI tract	Vomiting
	Brain	Headache
	Skin	Itching
Antihelminthics		
Niclosamide	GI tract	Nausea, abdominal pain
Pyrantel	GI tract	Irritation
	Brain	Headache, dizziness
Antivirals		
Acyclovir	Brain	Seizures, confusion
	Skin	Rash
Amantadine	Brain	Nervousness, lightheadedness
	GI tract	Nausea
AZT	Bone marrow	Immunosuppression, anemia

*A very rare reaction.
**An inflammation of the intestinal tract.

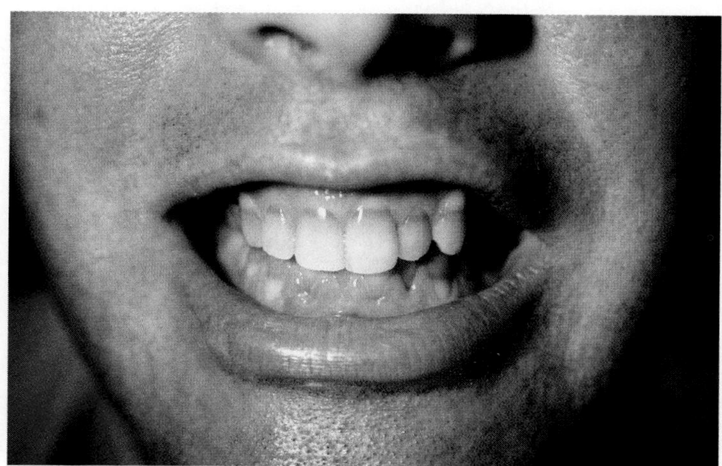

Figure 12.19
An adverse effect of tetracycline given to young children is the permanent discoloration of tooth enamel.

Because the liver is responsible for metabolizing and detoxifying foreign chemicals in the blood, it can be damaged by a drug or its metabolic products. Injury to liver cells can result in enzymatic abnormalities, fatty liver deposits, hepatitis, and liver failure. The kidney, too, excretes drugs and their metabolites. Some drugs irritate the nephron tubules, creating changes that interfere with their filtration abilities. Drugs such as sulfonamides crystallize in the kidney pelvis and form stones that can obstruct the flow of urine.

The most common complaint associated with oral antimicrobial therapy is diarrhea, which can progress to severe intestinal irritation or colitis. Although some drugs directly irritate the intestinal lining, the usual gastrointestinal complaints are caused by disruption of the intestinal microflora (discussed in a subsequent section).

Many drugs given for parasitic infections are toxic to the heart, causing irregular heartbeats and even cardiac arrest in extreme cases. Chloramphenicol can severely depress blood-forming cells in the bone marrow, resulting in either a reversible or a permanent (fatal) anemia. Some drugs hemolyze the red blood cells, others reduce white blood cell counts, and still others damage platelets or interfere with their formation, thereby inhibiting blood clotting.

Certain antimicrobics act directly on the brain and cause seizures. Others, such as aminoglycosides, damage nerves (very commonly, the 8th cranial nerve), leading to dizziness, deafness, or motor and sensory disturbances. When drugs block the transmission of impulses to the diaphragm, respiratory failure can result.

The skin is a frequent target of drug-induced side effects. The skin response can be a symptom of drug allergy or a direct toxic effect. Some drugs interact with sunlight to cause photodermatitis, a skin inflammation. Tetracyclines are contraindicated (not advisable) for children from birth to 8 years of age because they bind to the enamel of the teeth, creating a permanent gray to brown discoloration (figure 12.19). Pregnant women should avoid tetracyclines because they cross the placenta and can be deposited in the developing fetal bones and teeth.

Allergic Responses to Drugs

One of the most frequent drug reactions is heightened sensitivity, or *allergy.* This reaction occurs because the drug acts as an antigen (a foreign material capable of stimulating the immune system) and stimulates an allergic response. This response can be provoked by the intact drug molecule or by substances that develop from the body's metabolic alteration of the drug. In the case of penicillin, for instance, it is not the penicillin molecule itself that causes the allergic response but a major product, *benzlpenicilloyl.* Allergic reactions have been reported for every major type of antimicrobic drug, but the penicillins account for the greatest number of antimicrobic allergies, followed by the sulfonamides.

People who are allergic to a drug become sensitized to it during the first contact, usually without symptoms. Once the immune system is sensitized, a second exposure to the drug can lead to a reaction such as a skin rash (hives), respiratory inflammation, and, rarely, anaphylaxis, an acute, overwhelming allergic response that develops rapidly and can be fatal. (This topic is discussed in greater detail in chapter 17.)

Suppression and Alteration of the Microflora by Antimicrobics

Most normal, healthy body surfaces, such as the skin, large intestine, outer openings of the urogenital tract, and oral cavity, provide numerous habitats for a virtual "garden" of microorganisms. These normal colonists or residents, called the **flora*** or microflora, consist mostly of harmless or beneficial bacteria, but some can be potential pathogens. Although we shall defer a more detailed discussion of this topic to chapter 13, here we focus on the general effects of drugs on this population.

If a broad-spectrum antimicrobic is introduced into the host to treat infection, it will destroy microbes regardless of their roles in the ecological balance, affecting not only the targeted infectious agent but also many others in sites far removed from the original infection (figure 12.20). In some cases, the result of this therapy is the destruction of beneficial resident species and the subsequent survival and overgrowth of opportunistic residents or contaminants. This complication is called a **superinfection.**

Some common examples demonstrate how a disturbance in microbial flora leads to replacement flora and superinfection. A broad-spectrum cephalosporin used to treat a urinary tract infection by *Escherichia coli* will cure the infection, but it will also destroy the lactobacilli in the vagina that normally maintain a protective acidic environment there. The drug has no effect, however, on *Candida albicans,* a yeast that also resides in normal vaginas. Released from the inhibitory pH normally provided by lactobacilli, the yeasts

*__flora__ (flor´-ah) Gr. *flora,* the goddess of flowers. The microscopic life present in a particular location.

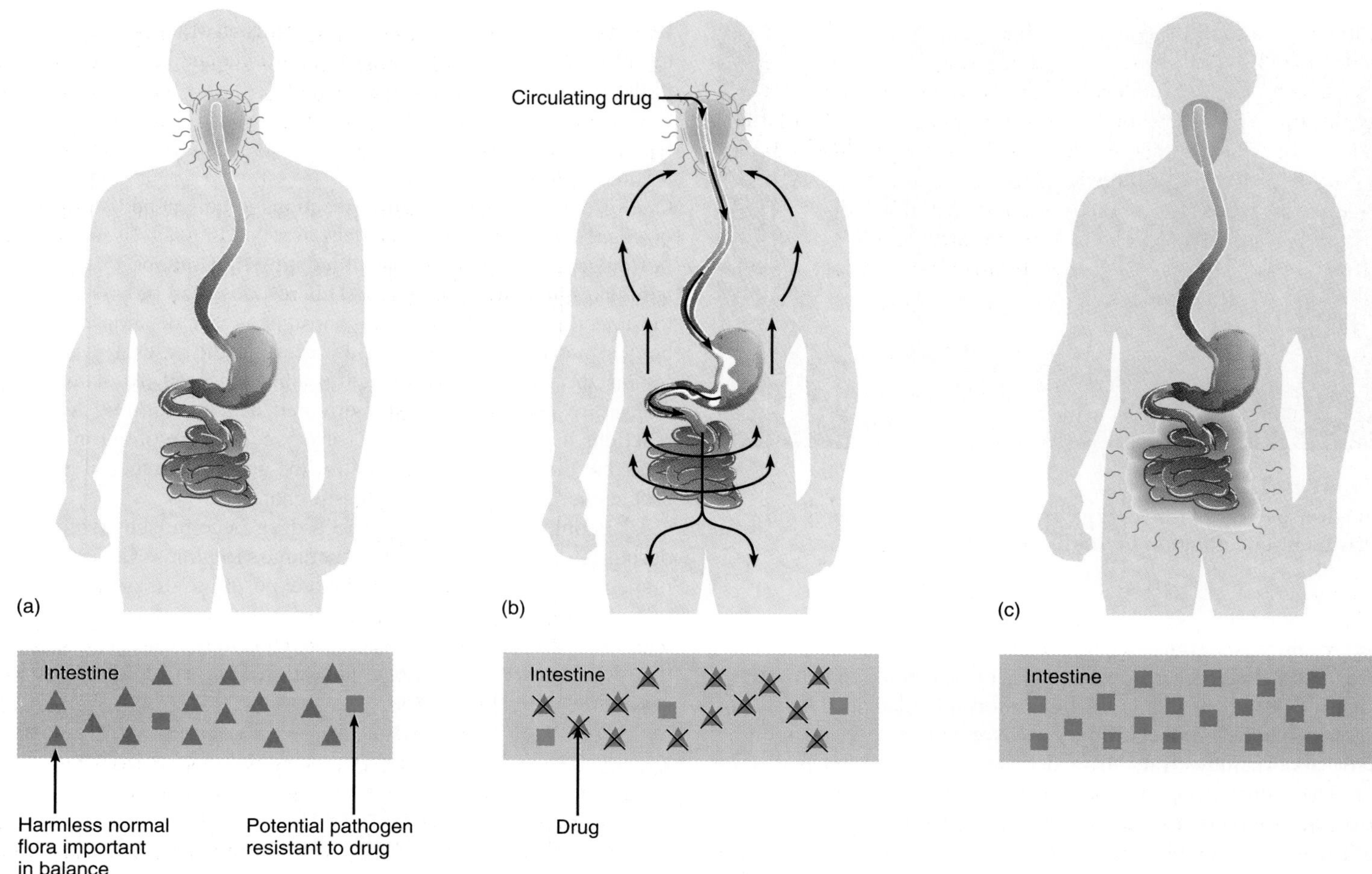

Figure 12.20
The role of antimicrobics in disrupting microbial flora and causing superinfections. *(a,b)* A primary infection in the throat is treated with an oral antibiotic that enters the intestine and is absorbed into the circulation. *(c)* The primary infection is cured, and the intestine is superinfected by drug-resistant pathogens that survived and proliferated in the intestine.

proliferate and cause an infection. *Candida* can cause similar superinfections of the oropharynx (thrush) and the large intestine.

Oral therapy with tetracyclines, clindamycin, and broad-spectrum penicillins and cephalosporins is associated with a serious and potentially fatal condition known as *antibiotic-associated colitis* (pseudomembranous colitis). This condition is due to the overgrowth in the bowel of *Clostridium difficile,* a spore-forming bacterium that is resistant to the antibiotic. It invades the intestinal living and releases toxins that induce diarrhea, fever, and abdominal pain.

Chapter Checkpoints

The three major side effects of antimicrobic are toxicity to organs, allergic reactions, and problems resulting from suppression or alteration of normal flora.

Antimicrobic that destroy most but not all normal flora allow the unaffected normal flora to overgrow, causing a superinfection.

CONSIDERATIONS IN SELECTING AN ANTIMICROBIC DRUG

Before actual antimicrobic therapy can begin, it is important that at least three factors be known: (1) the nature of the microorganism causing the infection; (2) the degree of the microorganism's susceptibility (also called sensitivity) to various drugs; and (3) the overall medical condition of the patient.

IDENTIFYING THE AGENT

Identification of infectious agents from body specimens should be attempted as soon as possible. It is especially important that such specimens be taken before the antimicrobic drug is given, just in case the drug eliminates the infectious agent. Direct examination of body fluids, sputum, or stool is a rapid initial method for detecting and perhaps even identifying bacteria or fungi. A doctor often begins the therapy on the basis of such immediate findings. The choice of drug will be based on experience with drugs that are known to be effective against the microbe; this is called the "informed best guess." For instance, if a sore throat appears to be caused by *Streptococcus pyogenes,* the physician might prescribe penicillin, because this species seems to be universally sensitive to it so far. If the infectious agent is not or cannot be isolated, epidemiologic statistics may be required to predict the most likely agent in a given infection. For example, *Haemophilus influenzae* accounts for the majority of cases of meningitis in children, followed by *Streptococcus pneumoniae* and *Neisseria meningitidis.*

Testing for the Drug Susceptibility of Microorganisms

Testing is essential in those groups of bacteria commonly showing resistance, primarily *Staphylococcus* species, *Neisseria gonorrhoeae, Streptococcus pneumoniae,* and *Enterococcus faecalis,* and the aerobic gram-negative enteric bacilli. However, not all infectious agents require antimicrobial sensitivity testing. Drug testing in fungal or protozoan infections is difficult and is often unnecessary. When certain groups, such as a group A streptococci and all anaerobes (except *Bacteroides),* are known to be uniformly susceptible to penicillin G, testing may not be necessary unless the patient is allergic to penicillin.

Selection of a proper antimicrobial agent begins by demonstrating the *in vitro* activity of several drugs against the infectious agent by means of standardized methods. In general, these tests involve exposing a pure culture of the bacterium to several different drugs and observing the effects of the drugs on growth. The *Kirby-Bauer* technique is an agar diffusion test that provides useful semiquantitative data on antimicrobic susceptibility. In this test, the surface of a plate of special medium is covered completely with the test bacterium, and small discs containing a premeasured amount of antimicrobic are dispensed onto the bacterial lawn. After 18 to 24 hours of incubation at 37°C, the zone of inhibition surrounding the discs is measured and compared with a standard for each drug (figure 12.21 and table 12.5). The profile of antimicrobic sensitivity, or *antibiogram,* provides data for drug selection. An advantage of the Kirby-Bauer procedure is that many drugs can be tested simultaneously in a single plate; however, it is less effective for bacteria that are anaerobic, highly fastidious, or slow-growing *(Mycobacterium).*

More sensitive and quantitative results can be obtained with tube dilution tests. First the antimicrobic is diluted serially in containers of broth, and then each tube is inoculated with a small uniform sample of pure culture. After incubation for 18–24 hours, the tubes are examined for growth (turbidity). The smallest concentration of drug that visibly inhibits growth is called the **minimum inhibitory concentration,** or MIC. The MIC is useful in determining the smallest effective dosage of a drug and in providing a comparative index against other antimicrobics (figure 12.22 and table 12.6). In many clinical laboratories, these antimicrobic testing procedures are automated (figure 12.22*b*).

Applying the Results of Drug Susceptibility Tests The results of antimicrobic sensitivity tests guide the physician's choice of a suitable drug. If therapy has already commenced, it is imperative to determine if the tests bear out the use of that particular drug. Once therapy has begun, it is important to observe the patient's clinical response, because the *in vitro* activity of the drug is not always correlated with its *in vivo* effect. When antimicrobic treatment fails, the failure is due to (1) the inability of the drug to diffuse into that body compartment (the brain, joints, skin); (2) a few resistant cells in the culture that did not appear in the sensitivity test; or (3) an infection caused by more than one pathogen (mixed), some of which are resistant to the drug. If therapy does fail, a different drug, combined therapy, or a different method of administration must be considered.

Medical Considerations in Anti-Infective Chemotherapy

Many factors influence the choice of an antimicrobic drug besides microbial sensitivity to it. The nature and spectrum of the drug, its potential adverse effects, and the condition of the patient can be critically important. When several antimicrobic drugs are available for treating an infection, final drug selection advances to a new series of considerations. In general, it is better to choose the narrowest-spectrum drug of those that are effective. This decreases the potential for superinfections and other adverse reactions.

Because drug toxicity is of concern, it is best to choose the one with high selective toxicity for the infectious agent and low human toxicity. This concept describes the **therapeutic index,** the ratio of a drug's toxic dose to its minimum effective (therapeutic) dose. The closer these two figures are (the smaller the ratio), the greater is the potential for toxic drug reactions. For example, a drug that has a therapeutic index of:

$$\frac{10\ \mu\text{g/ml: toxic dose}}{9\ \mu\text{g/ml (MIC)}} = 1.1$$

is a riskier choice than one with a therapeutic index of:

$$\frac{10\ \mu\text{g/ml}}{1\ \mu\text{g/ml}} = 10$$

Drug companies recommend dosages that will inhibit the microbes but not adversely affect patient cells. When a series of drugs being considered for therapy have similar MICs, the drug with the highest therapeutic index usually has the widest margin of safety.

The physician must also take a careful history of the patient to discover any preexisting medical conditions that will influence the activity of the drug or the response of the patient. A history of allergy to a certain class of drugs should preclude the administration of that drug. Underlying liver or kidney disease will ordinarily necessitate the modification of drug therapy because these organs play such an important part in metabolizing or excreting the drug. Infants, the elderly, and pregnant women require special precautions. For example, age can diminish gastrointestinal absorption and organ function, and most antimicrobic drugs cross the placenta and could affect fetal development.

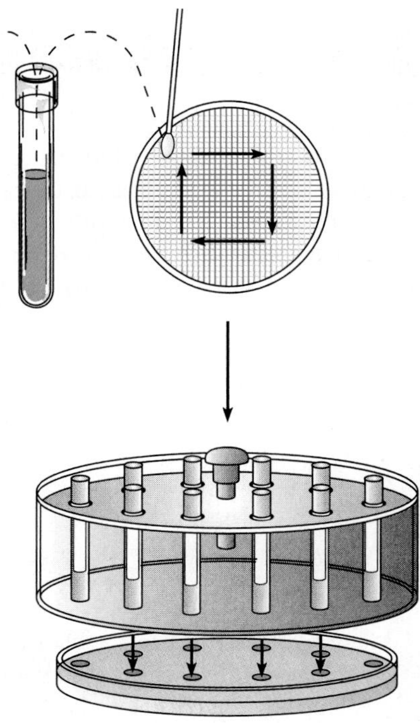
(a)

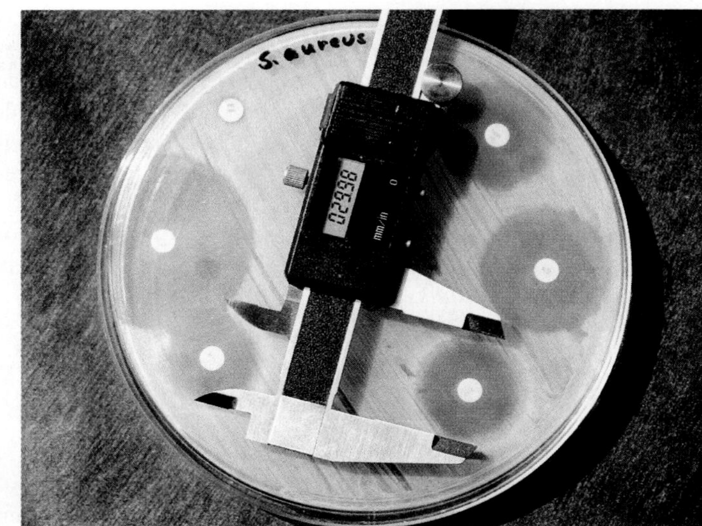

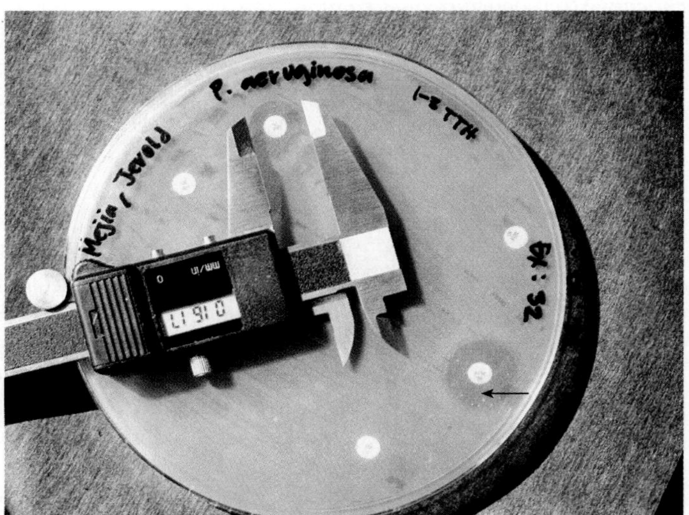

(b)

Figure 12.21

Method and interpretation of disc diffusion. *(a)* Steps in plate preparation and measurement. During incubation, antimicrobics become increasingly diluted as they diffuse out of the disc into the medium. *(b)* Interpretation of results. If the test bacterium is sensitive to a drug, it will not grow, and a zone of inhibition develops around the disc. The larger the size of this zone, the greater is the bacterium's sensitivity to the drug. The diameter of each zone is measured to the nearest whole millimeter and evaluated for susceptibility or resistance by means of a comparative standard (see table 12.5). Drug resistance can be detected by the complete lack of a zone around the disc or by tiny colonies within the zone of inhibition (arrow). This figure compares two species, *Staphylococcus aureus* (top) and *Pseudomonas aeruginosa,* which vary considerably in their sensitivity to drugs.

TABLE 12.5

RESULTS OF KIRBY-BAUER TEST

Drug	Zone Sites (mm) Required For: Susceptibility (S)	Zone Sites (mm) Required For: Resistance (R)	Actual Result (mm) for Lab Strain of *Staphylococcus aureus*	Evaluation
Bacitracin	>13	<8	15	S
Chloramphenicol	>18	<12	20	S
Erythromycin	>18	<13	25	S
Gentamicin	>13	<12	16	S
Kanamycin	>18	<13	20	S
Neomycin	>17	<12	12	R
Penicillin G	>29	<20	10	R
Polymyxin B	>12	<8	10	R
Streptomycin	>15	<11	11	R
Vancomycin	>12	<9	15	S
Tetracycline	>19	<14	25	S

TABLE 12.6

COMPARATIVE MICs (μG/ML) FOR FIVE COMMON DRUGS AND SEVEN PATHOGENS

Bacterium	Penicillin G	Ampicillin	Sulfamethoxazole	Tetracycline	Cefaclor
Staphylococcus aureus	4.0	0.05	3.0	0.3	4.0
Enterococcus faecalis	3.6	1.6	100.0	0.3	60.0
Neisseria gonorrhoeae	0.5	0.5	5.0	0.8	2.0
Escherichia coli	100.0	12.0	3.0	6–50.0	3.0
Pseudomonas aeruginosa	>500.0	>200.0		>100.0	
Salmonella species.	12.0	6.0	10.0	1.0	0.8
Clostridium	0.16			3.0	12.0

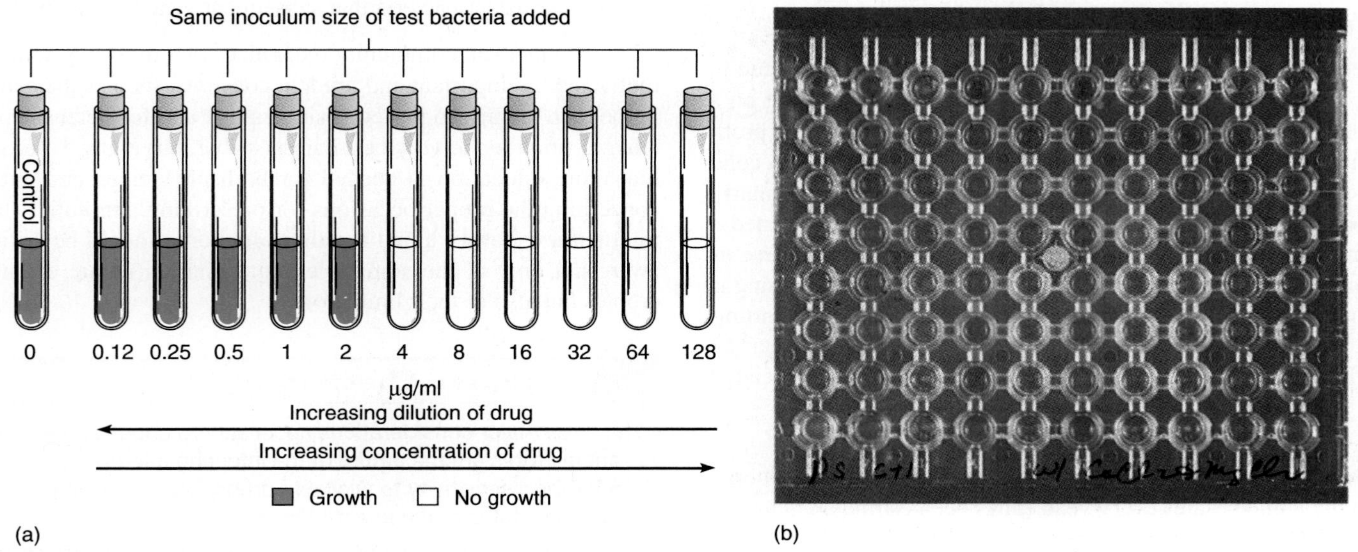

Figure 12.22

Tube dilution test for determining the minimum inhibitory concentration (MIC). *(a)* The antibiotic is diluted serially through tubes of liquid nutrient from right to left. All tubes are inoculated with an identical sample of a test bacterium and then incubated. The first tube on the left is a control containing only the microbe and nutrient. The turbidity of each successive tube is compared with the control. The dilution of the first tube in the series that shows no growth (no turbidity) is the MIC. *(b)* An automated method for determining the MIC of an isolated pathogen. A machine automatically dilutes and dispenses the drug and inoculates the test microbe into a multiple-chambered plate. After incubation, the clear wells are those showing inhibition; turbid wells indicate growth and lack of inhibition by the drug at that dilution.

The current intake of other drugs must be carefully scrutinized, because incompatibilities can result in increased toxicity or failure of one or more of the drugs. For example, the combination of aminoglycosides and cephalosporins increases nephrotoxic effects; antacids reduce the absorption of isoniazid; and the interaction of tetracycline or rifampin with oral contraceptives can abolish the contraceptive's effect. Some drugs (penicillin with certain aminoglycosides, or amphotericin B with flucytosine) act synergistically, so that reduced doses of each can be used in combined therapy. Other concerns in choosing drugs include any genetic or metabolic abnormalities in the patient, the site of infection, the route of administration, and the cost of the drug.

The Art and Science of Choosing an Antimicrobic Drug Even when all the information is in, the final choice of a drug is not always easy or straightforward. Consider the case of an elderly alcoholic patient with pneumonia caused by *Klebsiella* and complicated by diminished liver and kidney function. All drugs must be given parenterally because of prior damage to the gastrointestinal lining and poor absorption. Drug tests show that the infectious agent is sensitive to third-generation cephalosporins, gentamicin, imipenem and azlocillin. The patient's history shows previous allergy to the penicillins, so these would be ruled out. Drug interactions occur between alcohol and the cephalosporins, which are also associated with serious bleeding in elderly patients, so this may not be a good choice. Aminoglycosides such as gentamicin are nephrotoxic and poorly cleared by damaged kidneys. Imipenem causes intestinal discomfort, but it has less toxicity and would be a viable choice.

In the case of a cancer patient with severe systemic *Candida tropicalis* infection, there will be fewer criteria to weigh. Intravenous amphotericin B alone or in combination with flucytosine is about the only possibility, despite the drug's nephrotoxicity and other possible adverse side effects. In a life-threatening situation, in which a dangerous chemotherapy is perhaps the only chance for survival, the choices are reduced and the priorities are different.

An Antimicrobic Drug Dilemma

We began this chapter with a view of the exciting strides made in chemotherapy during the past few years, but we must end it on a note of qualification and caution. There is now a worldwide problem in the management of antimicrobic drugs, which rank second only to some nervous system drugs in overall usage. The remarkable progress in treating many infectious diseases has spawned a view of antimicrobics as a "cure-all" for infections as diverse as the common cold and acne. And, although it is true that nothing is as dramatic as curing an infectious disease with the correct antimicrobic drug, in many instances, drugs have no effect or can be harmful. The depth of this problem can perhaps be appreciated better with a few statistics:

1. Roughly 200 million prescriptions for antibiotics are written in the United States every year. It has been estimated that nearly half of these prescriptions are inappropriate because the infection is viral in origin. Many drugs are also misprescribed as to type, dosage, or length of therapy. Such overuse of antimicrobics is also known to increase the development of antimicrobial resistance. Not only can inappropriate prescriptions harm the patient, but they waste billions of dollars.
2. There is a tendency to use a "shotgun" antimicrobial therapy for minor infections, which involves administering a broad-spectrum drug instead of a more specific narrow-spectrum one. This practice can lead to superinfections as well as toxic reactions. Tetracyclines and chloramphenicol are still prescribed routinely for infections that would be treated more effectively with narrower spectrum, less toxic drugs.
3. Drugs are often prescribed without benefit of culture or susceptibility testing, even when such testing is clearly warranted.
4. More expensive newer drugs are chosen when a less costly older one would be just as effective. Among the most expensive drugs are cephalosporins and the longer-acting tetracyclines, yet these are among the most commonly prescribed antibiotics.
5. Tons of excess antimicrobic drugs produced in this country are exported to other countries, where controls are not as strict. Nearly 200 different antibiotics are sold over the counter in Latin America and Asian countries. It is common to self-medicate without understanding or medical indication. Drugs used in this way are largely ineffectual, but, worse yet, they may inadvertently account for the emergence of some drug-resistant bacteria that subsequently cause epidemics.

The medical community recognizes that most physicians are motivated by important and prudent concerns, such as the need for immediate therapy to protect a sick patient and for defensive medicine to provide the very best care possible, but many experts feel that more education is needed for both physicians and patients concerning the proper occasions for prescribing antibiotics. In the final analysis, every allied health professional should be critically aware not only of the admirable and utilitarian nature of antimicrobics but also of their limitations.

Chapter Checkpoints

- The three major considerations necessary to choose an effective antimicrobic are the nature of the infecting microbe, the microbe's sensitivity to available drugs, and the overall medical status of the infected host.
- The Kirby-Bauer test identifies antimicrobics that are effective against a specific infectious bacterial isolate.
- The MIC (minimum inhibitory concentration) identifies the smallest effective dose of an antimicrobic toxic to the infecting microbe.
- The therapeutic index is a ratio of the amount of drug toxic to the infected host and the MIC. The smaller the ratio, the greater the potential for toxic host-drug reactions.
- The effectiveness of antimicrobic drugs is being compromised by several alarming trends: inappropriate prescription, use of broad-spectrum instead of narrow-spectrum drugs, use of higher-cost drugs, sale of over-the-counter antimicrobics in other countries, and lack of sufficient testing before prescription.

CHAPTER CAPSULE WITH KEY TERMS

ANTIMICROBIAL CHEMOTHERAPY

Purposes of Chemotherapeutic Drugs: Treatment of infections, control of microbes in the body.

Chemical Nature: Aromatic peptides, sugars, amino acids, nucleotides.

Categories of Antimicrobics: **Antibiotics** are chemicals derived from bacteria and molds, natural or **semisynthetic** (part natural, part synthetic); **synthetic drugs** are derived completely from industrial processes. Drugs may be **narrow-spectrum** or **broad-spectrum,** microbicidal or microbistatic.

Scope: Antibacterial (largest number), antifungal, antiprotozoan, antihelminthic, antiviral.

Ideal Qualities of Antimicrobics: **Selective toxicity,** meaning high toxicity to microorganisms, low toxicity to vertebrates; potency unaltered by dilution; stability and solubility in tissue fluids; lack of disruption to host's immune system or microflora; exempt from drug resistance.

Route of Administration: By mouth; **parenteral** (injection into vein, muscle); topical on skin, mucous membranes.

Major Adverse Results of Chemotherapy: Toxicity ranging from slight, short-term damage to permanent debilitation; allergic reactions; disruption of normal microbial flora of body; **superinfections** by resistant species; **drug resistance** (selection of strains of microorganisms genetically resistant through mutation of intermicrobial transfer of resistance factors).

Special Clinical Approaches: **Prophylaxis,** administering antimicrobic drugs to prevent infections in highly susceptible persons; combined therapy, administering two or more antimicrobics simultaneously to circumvent drug resistance or to achieve **synergism,** the additive or magnified effectiveness of certain drugs working together.

Stages in Selection of Proper Drug: Identification of microbe *in vitro;* testing antimicrobial sensitivity or susceptibility (determining the **MIC);** assessing the **therapeutic index.** Final drug selection weighs potential effectiveness, toxicity, *in vivo* effects, spectrum, and the medical condition of the patient.

Perspectives in Antimicrobic Abuse: Drugs are overprescribed, overproduced, and used inappropriately on a worldwide basis, with unfortunate medical and economic consequences.

ANTIBACTERIAL ANTIBIOTICS

PENICILLINS

Types/Source: **Beta-lactam**–based drugs; from ***Penicillium chrysogenum*** mold; natural form is penicillin G; semisynthetic forms (ampicillin, carbenicillin, methicillin, nafcillin) vary in specific therapeutic applications.

Mode of Action: Bactericidal; blocks the completion of the cell wall, causes weak points and cell rupture.

Specific Uses/Spectra: Penicillins G and V are narrow-spectrum—mostly gram-positive and a few gram-negative (for example, gonococcus) bacteria. Semisynthetics are moderate to broad in spectrum, effective in infections by specified gram-positive and gram-negative species.

Problems in Therapy: Not highly toxic, but common cause of allergic reactions; some can cause superinfections; bacterial resistance to some forms of penicillin occurs through **beta-lactamase** (for example, **penicillinase).**

CEPHALOSPORINS

Types/Source: Natural and semisynthetic forms from *Cephalosporium acremonium* mold.

Mode of Action: Similar to penicillins; inhibit peptidogylcan synthesis.

Specific Uses/Spectra: First generation is narrow spectrum, against gram-positive cocci; second generation is narrow-spectrum, for some gram-negative rods; third generation is broad-spectrum, especially for gram-negative enteric rods and gram-positive cocci.

Problems in Therapy: Adverse blood and kidney reactions; superinfections; allergic reactions; bacterial resistance through **cephalosporinases.**

AMINOGLYCOSIDES

Types/Source/Spectrum: Mostly narrow-spectrum, from ***Streptomyces.***

Mode of Action: Interference with the bacterial ribosome, inhibition of protein synthesis.

Specific Uses/Spectra: Streptomycin, narrow-spectrum, not in common usage, except for tuberculosis therapy; gentamicin, standard therapy for gram-negative enteric infections; tobramycin and amikacin are alternate drugs.

Problems in Therapy: Toxic reactions to 8th cranial nerve, kidney damage, intestinal disturbances; drug resistance.

TETRACYCLINES AND CHLORAMPHENICOL

Source/Spectrum: Very broad-spectrum drugs originally from species of *Streptomyces* (now semi- or fully synthetic).

Mode of Action: Interfere with translation (protein synthesis).

Specific Uses: Tetracyclines (e.g., tetracycline and minocycline) used for rickettsial infections, *Mycoplasma* pneumonia, cholera, acne, and some sexually transmitted diseases. Chloramphenicol (Chloromycetin) limited by toxicity; indicated for serious infections where there is no alternative.

Problems in Therapy: Tetracycline may lead to hepatotoxicity, gastric disturbance, discoloration of tooth enamel in children, superinfections; chloramphenicol may damage bone marrow.

ERYTHROMYCIN, CLINDAMYCIN, VANCOMYCIN, RIFAMPIN

Modes of Action: Erythromycin and clindamycin disrupt protein synthesis; vancomycin interferes with early cell wall synthesis; rifampin inhibits RNA synthesis.

Specific Uses/Spectra: Erythromycin is broad-spectrum, for *Legionella, Chlamydia, Mycoplasma,* and some penicillin-resistant cocci; clindamycin used for intestinal infections by anaerobes; vancomycin applied in life-threatening staphylococcal infections; rifampin used for tuberculosis and leprosy.

Problems in Therapy: Vancomycin is neurotoxic; clindamycin and erythromycin can harm the GI tract; rifampin is hepatotoxic; resistant bacteria occur for all.

BACITRACIN AND POLYMYXIN

Spectrum: Narrow-spectrum.

Mode of Action: Bacitracin prevents synthesis of the cell wall of gram-positive bacteria; polymyxin has detergent action that disrupts cell membrane of gram-negative bacteria.

Specific Uses: Bacitracin used in ointments with neomycin for skin infections; polymyxins used to treat *Pseudomonas* infection or in skin ointments.

Problems in Therapy: Polymyxin can cause nephrotoxic and neuromuscular reactions; bacitracin is useful only for topical applications.

SYNTHETIC ANTIBACTERIAL DRUGS

SULFONAMIDES (SULFA DRUGS)

Types/Spectrum: Originally derived from prontosil; all types have similar basic structure; commonest is sulfisoxazole; relatively broad-spectrum.

Mode of Action: Acts as an **antimetabolite,** a **metabolic analog** that causes **competitive inhibition,** resulting in blockage in nucleic and amino acid synthesis.

Specific Uses: Urinary tract infections, nocardiosis, burn and eye infections; often combined with trimethoprim.

Problems in Therapy: Formation of crystals in kidney and allergy.

MISCELLANEOUS

Trimethoprim, medication for urinary, respiratory, and gastrointestinal infections; a competitive inhibitor in nucleic acid synthesis; can cause bone marrow damage. Dapsone, a major antileprosy drug, combined with rifampin to block drug resistance. **Isoniazid** (INH), an antituberculosis drug; blocks synthesis of cell wall of mycobacteria; may damage liver. **Fluoroquinolones** (ciprofloxacin), promising new broad-spectrum drugs.

DRUGS FOR FUNGAL INFECTIONS

Polyenes: Amphotericin B, nystatin, antibiotics that disrupt cell membrane by detergent action. Amphotericin is a key drug in systemic fungal infections; nystatin is used for skin and mucous membrane candidiasis. Both are commonly nephrotoxic.

Azoles: Synthetic drugs that interfere with membrane synthesis; ketoconazole, miconazole, and clotrimazole for cutaneous and membrane infections; can cause liver damage.

Flucytosine: Synthetic inhibitor of DNA synthesis; used alone or in combination with amphotericin for systemic mycoses; may lower WBC count; fungal resistance.

DRUGS FOR PROTOZOAN INFECTIONS

Quinines: Dominant types are chloroquine, primaquine, and quinine for managing malaria; choice depends upon sensitivity and stage in cycle of *Plasmodium;* can cause intestinal symptoms and eye disturbances; resistance and complexity of life cycle are main hurdles.

Others: Metronidazole (Flagyl) for amebiasis, giardiasis, trichomonas infections; suramin, melarsoprol, indicated in treatment of African trypanosomiasis; nitrifurimox, for acute South American trypanosomiasis.

DRUGS FOR HELMINTH INFECTIONS

Mebendazole, thiabendazole, and praziquantel, all-purpose agents in treating intestinal roundworm, tapeworm, and some fluke infestations; pyrantel and piperazine, primarily for intestinal roundworms; niclosamide, for tapeworms. Taken orally, cure occurs only upon incapacitation or death of worms and eggs followed by their expulsion in feces.

DRUGS FOR VIRAL INFECTIONS

Most antivirals function intracellularly to block virus multiplication; main drawbacks are lack of diversity and toxicity to host. **Acyclovir,** idoxuridine, and vidarabine, synthetic nitrogen bases that block synthesis of viral components in herpesviruses; **amantadine,** restricted to treating influenza A infections; **AZT** and **protease inhibitors** anti-AIDS drugs, **interferon,** a naturally occurring protein that can be useful in reducing symptoms of some viral infections and treating a few cancers.

MULTIPLE-CHOICE QUESTIONS

1. A compound synthesized by bacteria or fungi that destroys or inhibits the growth of other microbes is a/an
   a. synthetic drug
   b. antibiotic
   c. antimicrobic drug
   d. competitive inhibitor

2. Which statement is *not* an aim in the use of drugs in antimicrobial chemotherapy? The drug should:
   a. have selective toxicity
   b. be active even in high dilutions
   c. be broken down and excreted rapidly
   d. be microbicidal

3. Drugs that prevent the formation of the bacterial cell wall are
   a. quinolones
   b. beta-lactams
   c. tetracyclines
   d. aminoglycosides

4. Sulfonamide drugs initially disrupt which process?
   a. folic acid synthesis
   b. transcription
   c. PABA synthesis
   d. protein synthesis

5. Microbial resistance to drugs is acquired through
   a. conjugation
   b. transformation
   c. transduction
   d. all of these

6. R factors are______that contain a code for ______.
   a. genes, replication
   b. plasmids, drug resistance
   c. transposons, interferon
   d. plasmids, conjugation

7. When a patient's immune system becomes reactive to a drug, this is an example of
   a. superinfection
   b. drug resistance
   c. allergy
   d. toxicity

8. An antibiotic that disrupts the normal flora can cause
   a. the teeth to turn brown
   b. aplastic anemia
   c. a superinfection
   d. hepatotoxicity

9. Most antihelminth drugs function by
   a. weakening the worms so they can be flushed out by the intestine
   b. inhibiting worm metabolism
   c. blocking the absorption of nutrients
   d. inhibiting egg production

10. An example of an antiviral drug that can prevent a viral nucleic acid from being replicated is
    a. azidothymidine
    b. acyclovir
    c. amantadine
    d. both a and b
11. Which of the following effects do antiviral drugs *not* have?
    a. killing extracellular viruses
    b. stopping virus synthesis
    c. inhibiting virus maturation
    d. blocking virus receptors
12. Which of the following modes of action would be most selectively toxic?
    a. interrupting ribosomal function
    b. dissolving the cell membrane
    c. preventing cell wall synthesis
    d. inhibiting DNA replication
13. The MIC is the ______ of a drug that is required to inhibit growth of a microbe.
    a. largest concentration
    b. standard dose
    c. smallest concentration
    d. lowest dilution
14. An antimicrobic drug with a ______ therapeutic index is a better choice than one with a ______ therapeutic index.
    a. low, high
    b. high, low

CONCEPT QUESTIONS

1. Differentiate between antibiotics and synthetic drugs.
2. a. Differentiate between narrow-spectrum and broad-spectrum antibiotics.
   b. Can you determine why some drugs have narrower spectra than others? (Hint: Look at their mode of action.)
   c. How might one determine whether a particular antimicrobic is broad- or narrow-spectrum?
3. a. What is the major source of antibiotics?
   b. What appears to be the natural function of antibiotics?
4. a. Using the following diagram as a guide, briefly explain how the three factors in drug therapy interact.

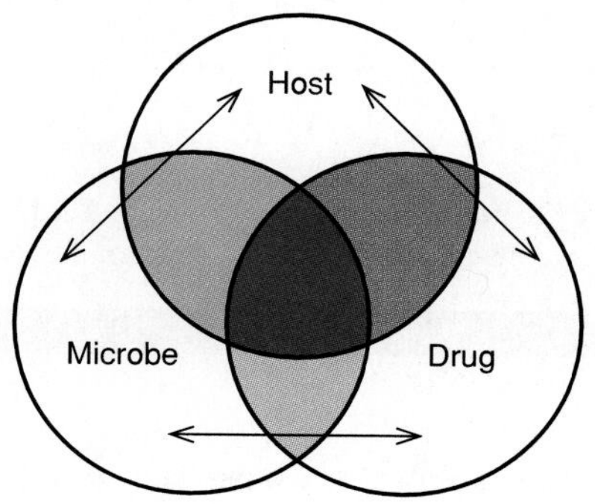

   b. What drug characteristics will make treatment most effective?
   c. What are the major aims of new antimicrobic drugs?
   d. Which of your answers to question c do you think is the most important?
5. a. Explain the major modes of action of antimicrobial drugs, and give an example of each.
   b. What is competitive inhibition?
   c. What is the basic reason that a metabolic analog molecule can inhibit metabolism?
   d. What are the long-term effects of drugs that block transcription?
   e. Why would a drug that blocks translation on the ribosomes of bacteria also affect human cells?
   f. Why do drugs that act on bacterial and fungal membranes generally have high toxicity?
6. a. Explain the phenomenon of drug resistance from the standpoint of microbial genetics (include a description of R factors).
   b. How can one test for drug resistance?
   c. Multiple drug resistance is becoming increasingly common in microorganisms. Explain how one bacterium can acquire resistance to several drugs.
7. a. Explain four general ways that microbes evade the effects of drugs.
   b. What is the effect of beta-lactamase?
8. What causes mutated or plasmid-altered strains of drug-resistant microbes to persist in a population?
9. Construct a chart that summarizes the modes of action and applications of the major groups of antibacterial drugs (antibiotics and synthetics), antifungal drugs, antiparasitic drugs, and antiviral drugs.
10. a. Explain why there are so few antifungal, antiparasitic, and antiviral drugs.
    b. What effect do nitrogen-base analogs have upon viruses?
    c. Summarize the origins and biological actions of interferon.
11. a. Generally overview the adverse effects of antimicrobic drugs on the host.
    b. On what basis can one explain allergy to drugs?
    c. Describe the stages in a superinfection.
12. a. Outline the steps in antimicrobic susceptibility testing.
    b. Compare the interpretation of the Kirby-Bauer technique with the MIC technique.
    c. What is the therapeutic index, and how is it used?

Critical-Thinking Questions

1. Describe the events that occur when an oral drug is taken to treat (1) a skin infection or (2) meningitis (an infection of the meninges of the brain).
2. Occasionally, one will hear the expression that a microbe has become "immune" to a drug.
   a. What is a better way to explain what is happening?
   b. Explain a simple test one could do to determine if drug resistance was developing in a culture.
3. a. Can you think of additional ways that drug resistance can be prevented?
   b. What can health care workers do?
   c. What can one do on a personal level?
4. Drugs are often given to surgical patients, to dental patients with heart disease, or to healthy family members exposed to contagious infections.
   a. What word would you use to describe this use of drugs?
   b. What is the purpose of this form of treatment?
   c. Explain some potential undesired effects of this form of therapy.
5. a. Your pregnant neighbor has been prescribed a daily dose of oral tetracycline for acne. Do you think this therapy is advisable for her? Why or why not?
   b. A woman has been prescribed a broad-spectrum oral cephalosporin for a strep throat. What are some possible consequences in addition to cure of the infected throat?
   c. A man has a severe case of gastroenteritis that is negative for bacterial pathogens. A physician prescribes an oral antibacterial drug in treatment. What are your opinions of this therapy?
6. You have been directed to take a sample from a growth-free portion of the zone of inhibition in the Kirby-Bauer test and inoculate it onto a plate of nonselective medium.
   a. What does it mean if growth occurs on the new plate?
   b. What if there is no growth?
7. In cases in which it is not possible to culture or drug test an infectious agent (such as middle ear infection), how would the appropriate drug be chosen?
8. Using the results in tables 12.5 and 12.6 and reviewing drug characteristics, choose an antimicrobic for each of the following situations (explain your choice):
   a. for an adult patient suffering from *Mycoplasma* pneumonia
   b. for a child with meningitis (drug must enter into cerebrospinal fluid)
   c. for a patient with allergy to erythromycin
   d. for a urinary tract infection by *Enterococcus*
   e. for gonorrhea
9. What factors can play a part in drug synergism?
10. How would you personally feel about being told by a physician that your infection cannot be cured by an antibiotic, and that the best thing to do is go home, drink a lot of fluids, and take aspirin or other symptom-relieving drugs?
11. Refer to the tube dilution test shown in figure 12.22*a,* and give the MIC of the drug being tested.
12. Explain why it could be helpful to use combined therapy in treating HIV infection.

Internet Search Topic

1. Look up multidrug efflux pumps on a search engine and determine how these molecules confer drug resistance on bacteria.
2. Locate information on new classes of antibiotics such as mode of action and indications for use.

chapter 13

Microbe-Human Interactions: Infection and Disease

The human body exists in a state of dynamic equilibrium with microorganisms. In the healthy individual, this balance is maintained as a peaceful coexistence and lack of disease. But on occasion, the balance tips in favor of the microorganism, and an infection or disease results. In this chapter, we explore each component of the host-parasite relationship, beginning with the nature and function of normal flora, moving to the stages of infection and disease, and closing with a study of epidemiology and the spread of disease. These topics will set the scene for the next two chapters, which deal with the ways the host defends itself against assault by microorganisms.

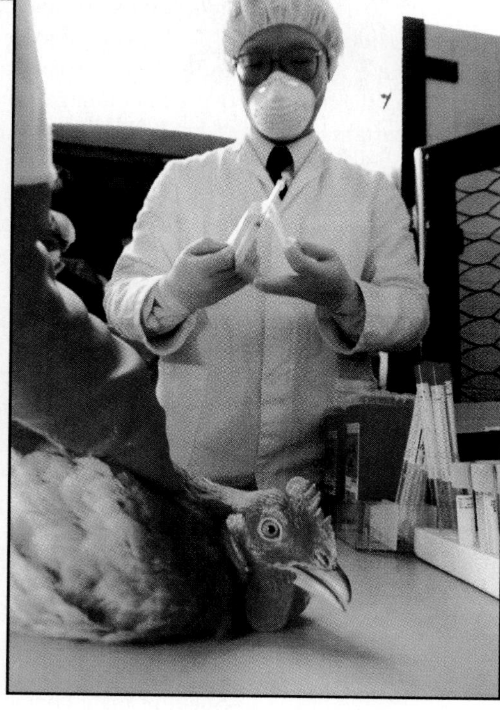

A Chinese virologist prepares to take blood from a chicken to screen for infection with a strain of influenza. This deadly virus was responsible for widespread illness in chickens and some human deaths in 1997. Public health officials mandated the slaughter of several million chickens to prevent a possible epidemic.

MICROFILE 13.1 SOME POTENTIAL RESULTS OF MICROBE-HUMAN INTERACTIONS

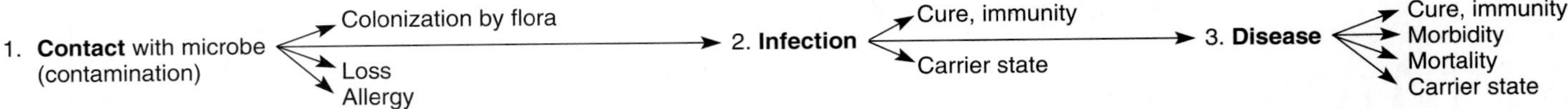

1. Microbes are first acquired on the exposed areas of the body through contact with other living things or the environment. **The instances and consequences of microbe-human contact vary with the type of microbe, the condition of the body, and the actions of the host:**

a. After initial contact, some microbes take up residence as part of the normal flora, a diverse yet relatively stable collection of microbes that are important in maintaining the host equilibrium. (Note that even normal flora may, under certain circumstances, cause infection.)

b. Some microbes with greater infectious potential may evade the host defenses, infiltrate the body, and cause an infection; that is, they act as pathogens. Many factors determine whether an infection takes place, but, in general, the greater the **virulence* of the microbe, the more likely it is that infection will take place.**

c. Some microbes (the transients) stay on the body for only a short time and are destroyed by the host defenses or removed by an action such as cleaning or antisepsis.

d. Some microbes or microbial products result in hypersensitivity reactions, or allergy.

2. In the event of infection, two outcomes are possible:

a. The immune system may arrest the infection before injury to tissues and organs occurs.

b. The infectious agent is not arrested and becomes entrenched in tissues, where its **pathologic*** effects cause some degree of damage.

3. If infection proceeds to disease, several outcomes are possible:

a. The immune system may eventually arrest the microbe and stop the disease process.

b. Damaged tissues and organs may lead to dysfunction or **morbidity.***

c. Damage may be severe enough to cause death, or mortality.

d. The microbe may be harbored inconspicuously for varying lengths of time (carrier state).

*virulence (veer-yoo-lents) L. *virulentia,* virus, poison.

*pathologic (path″-uh-loj′-ik) A disease state caused by structural and functional damage to tissues.

*morbidity (mor-bih′-dih-tee) L. *morbidis,* sick. A condition of being diseased.

THE HUMAN HOST

In previous chapters, several of the basic interrelationships between humans and microorganisms were considered. Most of the microbes inhabiting the human body benefit from the nutrients and protective habitat it provides. From the human point of view, these relationships run the gamut from mutualism to commensalism to parasitism and can have beneficial, neutral, or harmful effects (see page 209). A common characteristic of all microbe-human relationships, regardless of where they lead, is that they begin with contact.

CONTACT, INFECTION, DISEASE—A CONTINUUM

The body surfaces are constantly exposed to microbes. Some microbes become implanted there as colonists (normal flora), some are rapidly lost (transients), and others invade the tissues. Such intimate contact with microbes inevitably leads to **infection,** a condition in which pathogenic microorganisms penetrate the host defenses, enter the tissues, and multiply. When the cumulative effects of the infection damage or disrupt tissues and organs, a **disease** results. A disease is defined as any deviation from health. There are hundreds of different diseases caused by such factors as infections, diet, genetics, and aging. In this chapter, however, we will discuss only **infectious disease**—the disruption of a tissue or organ caused by microbes or their products.

The pattern of the host-parasite relationship can be viewed as a series of stages that begins with contact, progresses to infection, and ends in disease (microfile 13.1). Because of numerous factors relating to host resistance and degree of pathogenicity, not all contacts lead to infection and not all infections lead to disease. In fact, contamination without infection and infection without disease are the rule. Before we consider further details of infection and disease, let us examine the fascinating relationship between humans and their resident flora.

TABLE 13.1

SITES THAT HARBOR A NORMAL FLORA

Skin and its contiguous mucous membranes
Upper respiratory tract
Gastrointestinal tract (various parts)
Outer opening of urethra
External genitalia
Vagina
External ear canal
External eye (lids, conjunctiva)

TABLE 13.2

STERILE (MICROBE-FREE) ANATOMICAL SITES AND FLUIDS

All Internal Tissues and Organs
Heart and circulatory system
Liver
Kidneys and bladder
Lungs
Brain and spinal cord
Muscles
Bones
Ovaries/testes
Glands (pancreas, salivary, thyroid)
Sinuses
Middle and inner ear
Internal eye
Fluids Within an Organ or Tissue
Blood
Urine in kidneys, ureters, bladder
Cerebrospinal fluid
Saliva prior to entering the oral cavity
Semen prior to entering the urethra
Amniotic fluid surrounding the embryo and fetus

RESIDENT FLORA: THE HUMAN AS A HABITAT

With its constant source of nourishment and moisture, relatively stable pH and temperature, and extensive surfaces upon which to settle, the human body provides a favorable habitat for an abundance of microorganisms. In fact, it is so favorable that, cell-for-cell, microbes outnumber human cells ten to one! The large and mixed collection of microbes adapted to the body has been variously called the **normal resident flora,** or *indigenous* flora,* though some microbiologists prefer to use the terms *microflora* and *commensals.* The normal residents include an array of bacteria, fungi, protozoa, and, to a certain extent, viruses and arthropods.

Acquiring Resident Flora

Like a virgin planet being settled by alien beings, humans begin to acquire microflora from their earliest contacts with their immediate environment. Development of the flora proceeds in a relatively orderly and characteristic way. Of the seemingly limitless influx of microbes, only some are capable of persisting in and on the body. The vast majority are destroyed by host defenses or are removed. Those that remain survive by adapting to a particular microhabitat—one that fills their requirements for food, moisture, presence or lack of oxygen, layers of dead cells, and even other microorganisms. In general, most anatomical surfaces directly exposed to the environment can and do harbor microbes. As indicated in table 13.1, these surfaces are mainly the skin and mucous membranes, parts of the inner surface of the gastrointestinal tract, and openings to the cutaneous surface from the urinary, respiratory, and reproductive tracts. Moist areas through which food passes tend to have the largest and most diverse populations. By contrast, organs and fluids inside the body cavity and the central nervous system remain free of normal flora during life because host defense mechanisms maintain their sterility (table 13.2). Although microorganisms can transiently enter these sites, they do not normally become established there.

Interactions between host and indigenous flora have evolved over millions of years into complex and dynamic relationships that directly and indirectly influence many aspects of body function. For the most part, effects of normal flora are protective or nutritional. For instance, the flora can modify the microhabitat by altering pH and oxygen tension; excreting chemicals such as fatty acids, gases, alcohol, and antibiotics; or creating barriers. Such reactions can favor colonization by more beneficial types of microbes. They can also inhibit potential pathogens by restricting their population size and activity through antagonism and competition for nutrients. Certain enteric bacteria release residual vitamins, creating a supplementary dietary source, though the significance of this source to humans remains in question. Other expected functions of the normal flora are discussed in a later section on germ-free animals.

Although relatively stable, the flora can fluctuate to a limited extent with general health, age, variations in diet, hygiene, hormones, and drug therapy. Most of the microorganisms are not serious pathogens and remain harmless as long as they do not penetrate superficial skin and mucosal barriers. However, a few species have the potential to become medically important if the balance of the flora shifts. This imbalance is seen in the overgrowth of *Clostridium difficile* or *Candida albicans* in patients undergoing antibiotic therapy. After being temporarily disturbed, the normal balance of flora is restored by surviving residents harbored in protected areas and by contact with the environment. Members of the flora can also become infectious if the health of the host is compromised (as by cancer or AIDS), allowing opportunists (such as the agents of pneumococcal and *Pneumocystis* pneumonia) to overcome the weakened host defenses (see table 13.5). It is notable that some humans harbor organisms that can be pathogenic to others. As a consequence,

*indigenous (in-dih´-juh-nus) Belonging or native to.

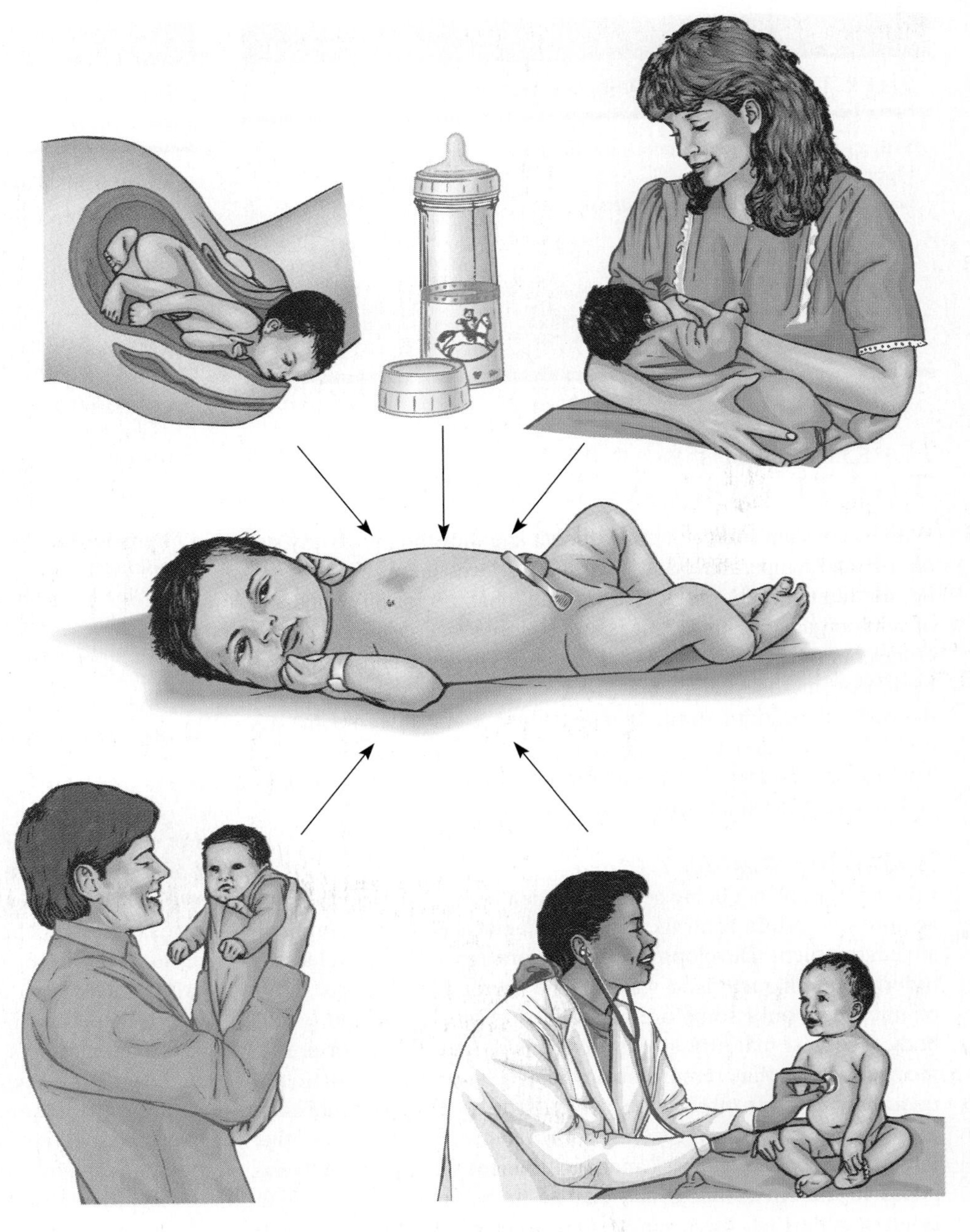

Figure 13.1
The origins of flora in newborns.

the flora is an important reservoir for bacterial pathogens such as *Staphylococcus aureus, Corynebacterium diphtheriae,* and *Neisseria meningitidis.*

Initial Colonization of the Newborn

The uterus and its contents are normally sterile during embryonic and fetal development and remain essentially germ-free until just before birth. The event that first exposes the infant to microbes is the breaking of the fetal membranes, at which time microbes from the mother's vagina can enter the womb. Comprehensive exposure occurs during the birth process itself, when the baby unavoidably comes into intimate contact with the birth canal (figure 13.1). Within 8 to 12 hours after delivery, the newborn typically has been colonized by bacteria such as streptococci, staphylococci, and lactobacilli, acquired primarily from its mother. The nature of the flora initially colonizing the large intestine depends upon whether the baby is bottle- or breast-fed. Bottle-fed infants (receiving milk or a milk-based formula) tend to acquire a mixed population of coliforms, lactobacilli, enteric streptococci, and staphylococci. In contrast, the intestinal flora of breast-fed infants consists primarily of *Bifidobacterium* species whose growth is favored by a growth factor from the milk. This bacterium metabolizes sugars into acids that protect the infant from infection by certain intestinal pathogens. The skin, gastrointestinal tract, and portions of the respiratory and

TABLE 13.3

LIFE ON HUMANS: SITES CONTAINING WELL-ESTABLISHED FLORA AND REPRESENTATIVE EXAMPLES

Anatomic Sites	Common Genera	Remarks
Skin	**Bacteria:** *Staphylococcus, Micrococcus, Corynebacterium, Propionibacterium, Mycobacterium*	Microbes live only in upper dead layers of epidermis, glands, and follicles; dermis and layers below are sterile.
	Fungi: *Malassezia* yeast	Dependent on skin lipids for growth.
	Arthropods: *Demodix* mite	Present in sebaceous glands and hair follicles.
Gastrointestinal Tract		
Oral cavity	**Bacteria:** *Streptococcus, Neisseria, Veillonella, Staphylococcus, Fusobacterium, Lactobacillus, Bacteroides, Corynebacterium, Actinomyces, Eikenella, Treponema, Haemophilus*	Colonize the epidermal layer of cheeks, gingiva, pharynx; surface of teeth; found in saliva in huge numbers.
	Fungi: *Candida* species	Can cause thrush.
	Protozoa: *Trichomonas tenax, Entamoeba gingivalis*	Frequent the gingiva of persons with poor oral hygiene.
Large intestine and rectum	**Bacteria:** *Bacteroides, Fusobacterium, Eubacterium, Bifidobacterium, Clostridium,* fecal streptococci, *Peptococcus, Lactobacillus,* coliforms (*Escherichia, Enterobacter*)	Sites of lower gastrointestinal tract other than large intestine and rectum have sparse or nonexistent flora. Flora consists predominantly of strict anaerobes; other microbes are aerotolerant or facultative.
	Fungi: *Candida*	
	Protozoa: *Entamoeba coli, Trichomonas hominis*	Feed on waste materials in the large intestine.
Upper Respiratory Tract	Microbial population exists in the nasal passages, throat, and pharynx; owing to proximity, flora is similar to that of oral cavity	Trachea and bronchi have a sparse population; smaller breathing tubes and alveoli have no normal flora and are essentially sterile.
Genital Tract	**Bacteria:** *Lactobacillus, Streptococcus, Corynebacterium, Escherichia, Mycobacterium*	In females, flora occupies the external genitalia and vaginal and cervical surfaces; internal reproductive structures normally remain sterile. Flora responds to hormonal changes during life.
	Fungi: *Candida*	Cause of yeast infections.
Urinary Tract	**Bacteria:** *Staphylococcus, Streptococcus,* coliforms	In females, flora exists only in the first portion of the urethral mucosa; the remainder of the tract is sterile. In males, the entire reproductive and urinary tract is sterile except for a short portion of the anterior urethra.

genitourinary tract all continue to be colonized as contact continues with family members, hospital personnel, the environment, and food.

Milestones that contribute to development of the adult pattern of flora are eruption of teeth, weaning, and introduction of the first solid food. Although exposure to microbes is unavoidable and even necessary for the maturation of the infant's flora, contact with pathogens is dangerous, because the neonate is not yet protected by a full complement of flora, and owing to its immature immune defenses, is extremely susceptible to infection.

INDIGENOUS FLORA OF SPECIFIC REGIONS

Although we tend to speak of the flora as a single unit, it is a complex mixture of hundreds of species, differing somewhat in quality and quantity from one individual to another. Studies of the flora have shown that most people harbor certain specially adapted bacteria, fungi, and protozoa (table 13.3).

Flora of the Human Skin

The skin is the largest and most accessible of all organs. Its major layers are the epidermis, an outer layer of dead cells continually being sloughed off and replaced, and the dermis, which lies atop the subcutaneous layer of tissue (figure 13.2*a*). Depending on its location, skin also contains hair follicles and several types of glands, and the outermost surface is covered with a protective, waxy cuticle that can help microbes adhere. The normal flora resides only in or on the dead cell layers, and, except for in follicles and glands, it does not extend into the dermis or subcutaneous levels. The nature of the population varies according to site. Oily, moist skin supports a more prolific flora than dry skin. Humidity, occupational exposure, and clothing also

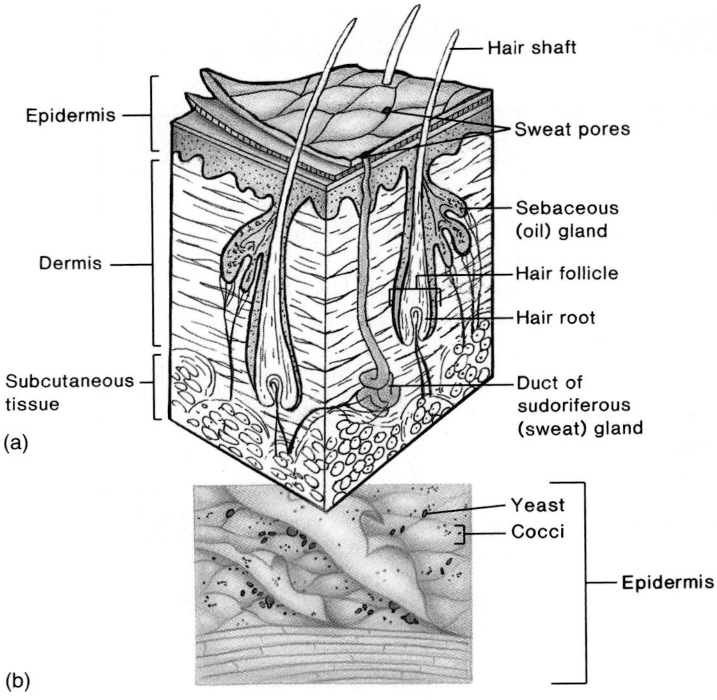

Figure 13.2

The landscape of the skin. (*a*) The epidermis along with associated glands and follicles (colored) provide rich and diverse habitats. Noncolored regions (dermis and subcutaneous layers) are free of microbial flora. (*b*) A highly magnified view of the skin surface reveals beds of rod- and coccus-shaped bacteria and yeasts beneath peeled-back skin flakes.

influence its character. Transition zones where the skin joins with the mucous membranes of the nose, mouth, and external genitalia harbor a particularly rich flora.

Ordinarily, there are two cutaneous populations. The **transient population,** or exposed flora, clings to the skin surface but does not ordinarily grow there. It is acquired by routine contact, and it varies markedly from person to person and over time. The transient flora includes any microbe a person has picked up, often species that do not ordinarily live on the body, and it is greatly influenced by the hygiene of the individual.

The **resident population** lives and multiplies in deeper layers of the epidermis and in glands and follicles (figure 13.2*b*). The composition of the resident flora is more stable, predictable, and less influenced by hygiene than is the transient flora. The normal skin residents consist primarily of bacteria (notably *Staphylococcus, Corynebacterium,* and *Propionibacterium*) and yeasts. Moist skin folds, especially between the toes, tend to harbor fungi, whereas lipophilic mycobacteria and staphylococci are prominent in sebaceous[1] secretions of the axilla, external genitalia, and external ear canal. One species, *Mycobacterium smegmatis,* lives in the cheesy secretion, or *smegma,* on the external genitalia of men and women.

1. From sebum, a lipid material secreted by glands in the hair follicles.

Flora of the Gastrointestinal Tract

The gastrointestinal (GI) tract receives, moves, digests, and absorbs food; it also removes waste. It encompasses the oral cavity, esophagus, stomach, small intestine, large intestine, rectum, and anus. Stating that the GI tract harbors flora may seem to contradict our earlier statement that internal organs are sterile, but it is not an exception to the rule. How can this be true? In reality, the GI tract is a long, hollow tube (with numerous pockets and curves), bounded by the mucous membranes of the oral cavity on one extreme and those of the anus on the other. Because the innermost surface of this tube is exposed to the environment, it is topographically outside the body, so to speak (figure 13.3). This architecture permits ingested materials to be processed before they cross the mucosal epithelium, but, at the same time, it creates numerous niches for microbes.

The shifting conditions of pH, oxygen tension, and differences in microscopic anatomy of the GI tract are reflected by the variations in or distribution of the flora (figure 13.4). Some microbes remain attached to the mucous epithelium or its associated structures, and others dwell in the *lumen.** Although the abundance of nutrients invites microbial growth, the only areas that harbor appreciable permanent flora are the oral cavity, large intestine, and rectum. The esophagus undergoes wavelike contractions (peristalsis), a process that constantly flushes microorganisms; the stomach acid inhibits most microbes; and peristalsis and digestive enzymes help exclude flora from all but the terminal segment of the small intestine.

Flora of the Mouth The oral cavity has a unique flora that is among the most diverse and abundant of the body. Microhabitats, including the cheek epithelium, gingiva, tongue, floor of the mouth, and tooth enamel, provide numerous adaptive niches for hundreds of different species to colonize. The most common residents are aerobic *Streptococcus* species—*S. sanguis, S. salivarius, S. mitis*—that colonize the smooth superficial epithelial surfaces. Two species, *S. mutans* and *S. sanguis,* make a major contribution to dental caries by forming sticky dextran slime layers in the presence of simple sugars. The adherence of dextrans to the tooth surface establishes a medium that attracts other bacteria (see figure 21.30). Over time, acidic products given off by the growing bacteria etch the enamel, causing caries.

Once the teeth have erupted, an anaerobic habitat is established in the gingival crevice that favors the colonization of anaerobic bacteria that can be involved in dental caries and periodontal[2] infections. The instant that saliva is secreted from ducts into the oral cavity, it becomes laden with resident and transient flora. Saliva normally has a high bacterial count (up to 5×10^9 cells per milliliter), a fact that tends to make mouthwashes rather ineffective and a human bite very dangerous (microfile 13.2).

Flora of the Large Intestine The flora of the intestinal tract has complex and profound interactions with the host. The large intestine (cecum and colon) and the rectum harbor a huge

2. Situated or occurring around the tooth.

*lumen (loo´-men) The space within a tubular structure.

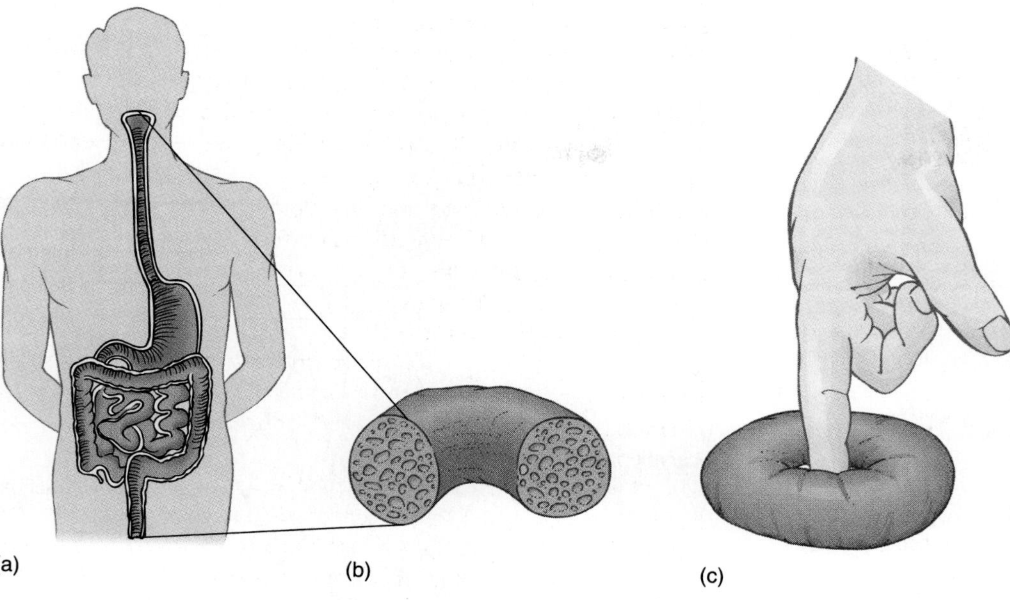

Figure 13.3
(*a*) The body is like an elongate doughnut, perforated from one end to the other by a long tube, the gastrointestinal (GI) tract. (*b*) Things inside the lumen are not inside the body, (*c*) just as something inside the doughnut hole is not inside the donut itself.

population of microbes (10^8–10^{11} per gram of feces) (figure 13.5). So abundant and prolific are these microbes that they constitute 10–30% of the fecal volume. Even an individual on a long-term fast passes feces consisting primarily of bacteria.

Because of the state of the intestinal environment, the predominant fecal flora are strictly anaerobic bacteria (*Bacteroides, Bifidobacterium, Fusobacterium,* and *Clostridium*). **Coliforms*** such as *Escherichia coli, Enterobacter,* and *Citrobacter* are present in smaller numbers. Many species ferment waste materials in the feces, generating vitamins (B_{12}, vitamin K, pyridoxine, riboflavin, and thiamine) and acids (acetic, butyric, and propionic acids) of potential value to the host. Occasionally significant are bacterial digestive enzymes that convert disaccharides to monosaccharides or promote steroid metabolism.

Intestinal bacteria contribute to intestinal odor by producing *skatole,** amines, and gases (CO_2 H_2, CH_4, and H_2S). Intestinal gas is known in polite circles as *flatus,* and the expulsion of it as *flatulence.* Some of the gas arises through the action of bacteria on dietary carbohydrate residues from vegetables such as cabbage, corn, and beans. The bacteria produce an average of 8.5 liters of gas daily, but only a small amount is ejected in flatus. Combustible gases occasionally form an explosive mixture in the presence of oxygen that has reportedly ignited during intestinal surgery and ruptured the colon!

Late in childhood, members of certain ethnic groups lose the ability to secrete the enzyme lactase. When they ingest milk or other lactose-containing dairy products, lactose is acted upon instead by intestinal bacteria, and severe intestinal distress can result. The recommended treatment for this deficiency is avoiding these foods or eating various lactose-free substitutes.

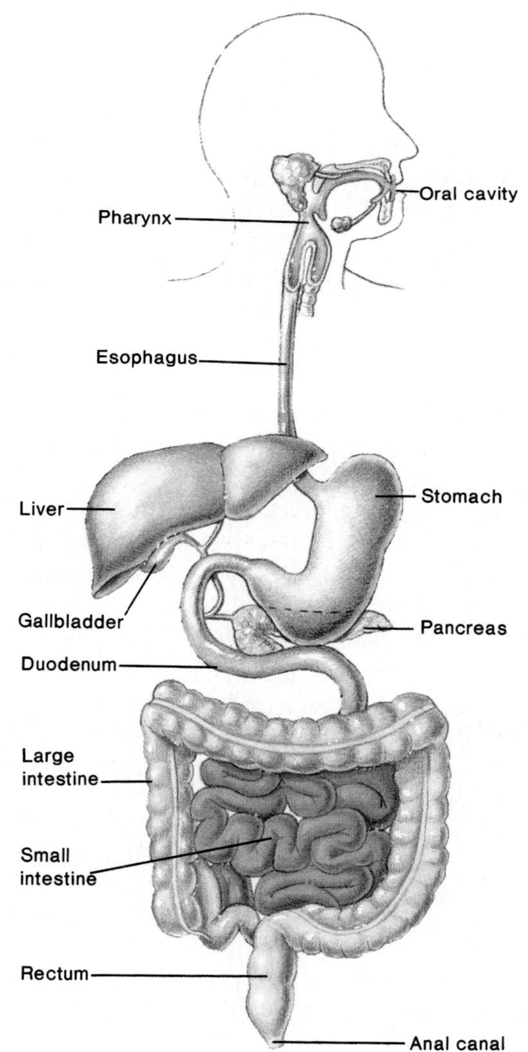

Figure 13.4
Areas of the gastrointestinal tract that shelter resident flora are high lighted in color. Noncolored areas do not regularly harbor residents.

*coliform (koh´-lih-form) L. *colum,* a sieve. Gram-negative, facultatively anaerobic, and lactose-fermenting microbes.

*skatole (skat´-ohl) Gr. *skatos,* dung. One chemical that gives feces its characteristic stench.

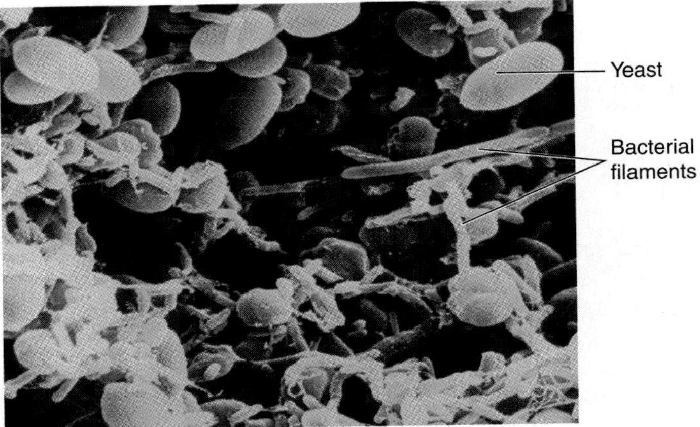

Figure 13.5
Electron micrograph of the mucous membrane of the large intestine. Microcolonies of yeasts and long, filamentous bacteria located in pockets are visible.

Flora of the Respiratory Tract

The first microorganisms to colonize the upper respiratory tract (nasal passages and pharynx) are predominantly oral streptococci. Inhaled air regularly contains microbes that are filtered out, destroyed, or expelled, although some can adapt to specific regions of this habitat. *Staphylococcus aureus* preferentially resides in the nasal entrance, nasal vestibule, and anterior nasopharynx, and *Neisseria* species take up residence in the mucous membranes of the nasopharynx behind the soft palate (figure 13.6). Lower still are assorted streptococci and species of *Haemophilus* that colonize the tonsils and lower pharynx. Conditions lower in the respiratory tree (bronchi and lungs) are unfavorable habitats for permanent residents.

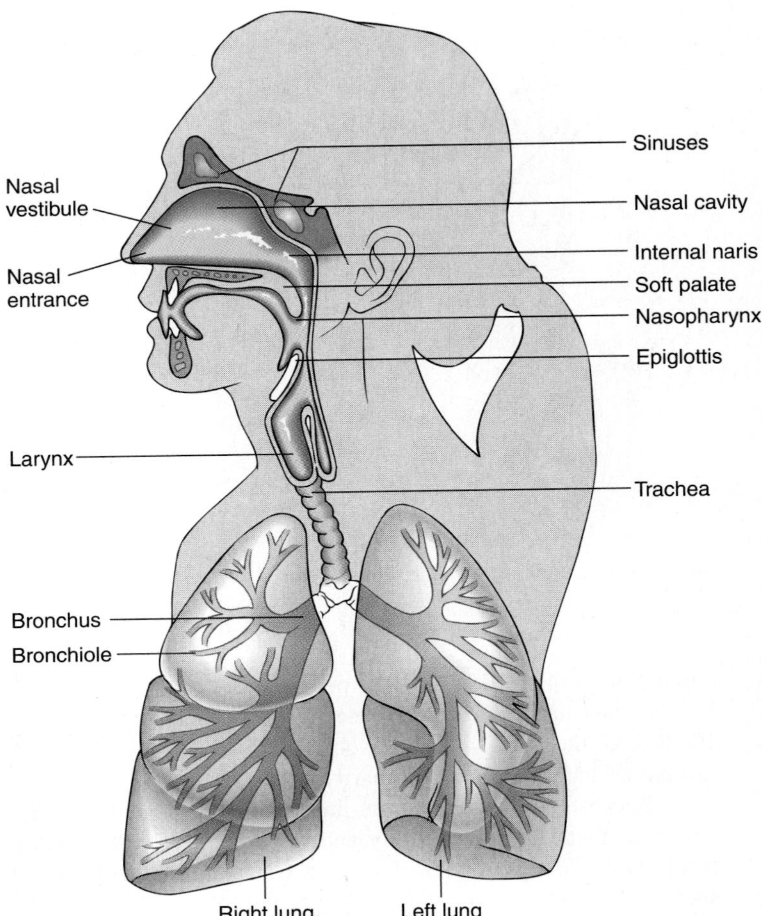

Figure 13.6
Colonized regions of the respiratory tract. The moist mucous blanket of the nasopharynx has a well-entrenched flora (pink). Some colonization occurs in the upper trachea, but lower regions of bronchi, bronchioles, and lungs lack resident microbes.

Flora of the Genitourinary Tract

The regions of the genitourinary tract that harbor microflora are the vagina and outer opening of the urethra in females and the anterior urethra in males (figure 13.7). The internal reproductive organs are kept sterile through physical barriers such as the cervical plug and other host defenses. The kidney, ureter, bladder, and upper urethra are presumably kept sterile by urine flow and regular bladder emptying. Because the urethra in women is so short (about 3.5 cm long), it can form a passage for bacteria to the bladder and lead to urinary tract infections. The principal residents of the urethra are nonhemolytic streptococci, staphylococci, corynebacteria, and, occasionally, coliforms.

The vagina presents a notable example of how changes in physiology can greatly influence the composition of the normal flora. An important factor influencing these changes in women is the hormone estrogen. Estrogen normally stimulates the vaginal mucosa to secrete the carbohydrate glycogen, which certain bacteria (primarily *Lactobacillus* species) ferment, thus lowering the pH to about 4.5. Before puberty, a girl produces little estrogen, little glycogen, and has a vaginal pH of about 7. These conditions favor the establishment of diphtheroids,[3] staphylococci, streptococci, and some coliforms. As hormone levels rise at puberty, the vagina begins to deposit glycogen, and the flora shifts to the acid-producing lactobacilli. It is thought that the acidic pH of the vagina during this time prevents the establishment and invasion of microbes with potential to harm a developing fetus. The estrogen-glycogen effect continues, with minor disruptions, throughout the childbearing years until menopause, when the flora returns to a mixed population similar to that of prepuberty. These transitions are not abrupt, but occur over several months to years.

3. Any nonpathogenic species of *Corynebacterium*.

Studies with Germ-Free Animals

For years, questions lingered about how essential the microbial flora is to normal life and what functions various members of the flora might serve. The need for animal models to further investigate these questions led eventually to development of laboratory strains of **germ-free,** or **axenic,** mammals and birds. The techniques and facilities required for producing and maintaining a germ-free colony are exceptionally rigorous. After the young

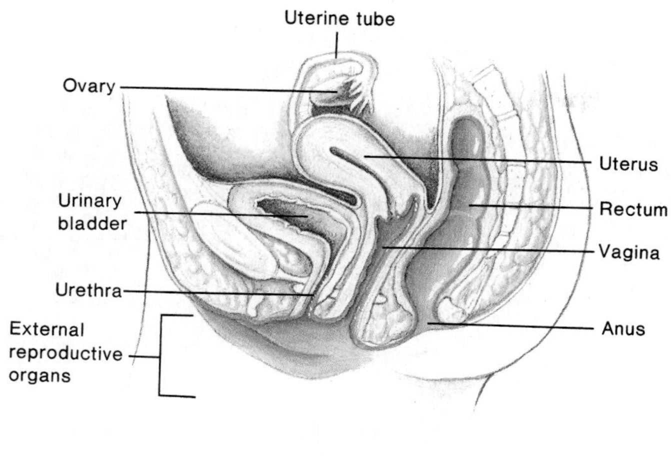

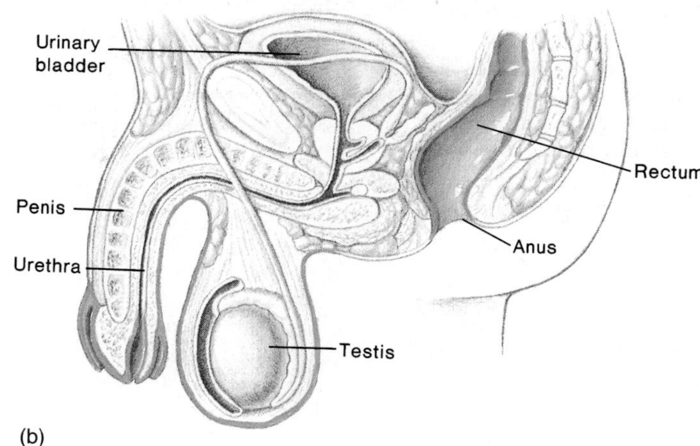

Figure 13.7

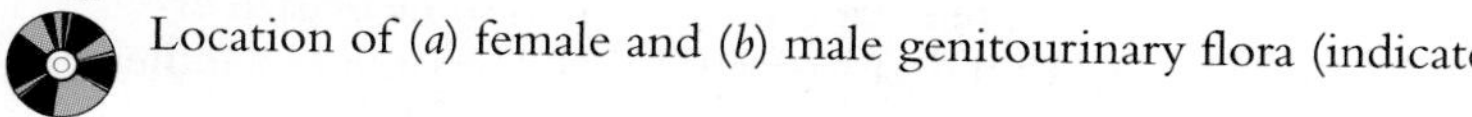
Location of (*a*) female and (*b*) male genitourinary flora (indicated by color).

mammals are taken from the mother aseptically by cesarian section, they are immediately transferred to a sterile isolator or incubator (figure 13.8). The newborns must be fed by hand through gloved ports in the isolator until they can eat on their own, and all materials entering their chamber must be sterile. Rats, mice, rabbits, guinea pigs, monkeys, dogs, hamsters, and cats are some of the mammals raised in the germ-free state.

Experiments with germ-free animals are of two basic varieties: (1) general studies on how the lack of normal microbial flora influences the nutrition, metabolism, and anatomy of the animal and (2) *gnotobiotic** studies, in which the germ-free subject is inoculated either with a single type of microbe to determine its individual effect or with several known microbes to determine interrelationships. Results are validated by comparing the germ-free group with a conventional, normal control group. Table 13.4 summarizes some major conclusions arising from studies with germ-free animals.

A dramatic characteristic of germ-free animals is that they live longer and have fewer diseases than normal controls, as long as they remain in a sterile environment. From this standpoint, it is clear that the flora is not needed for survival and may even be the source of infectious agents. At the same time, it is also clear that axenic life is highly impractical. Additional studies have revealed important facts about the effect of the flora on various organs and systems. For example, the flora contributes significantly to the development of the immune system. When germ-free animals are placed in contact with normal control animals, they gradually develop a flora similar to that of the controls. However, germ-free subjects are less tolerant of microorganisms and can die from infections by relatively harmless species. This susceptibility is due to the immature character of the immune system of germ-free animals. These animals have a reduced number of certain types of white blood cells and slower antibody response.

*gnotobiotic (noh´´toh-by-ah´-tik) Gr. *gnotos*, known, and *biota*, the organisms of a region.

Gnotobiotic experiments have clarified the dynamics of several infectious diseases. Perhaps the most striking discoveries were made in the case of oral diseases. For years, the precise involvement of microbes in dental caries had been ambiguous. Studies with germ-free rats, hamsters, and beagles confirmed that caries development is influenced by heredity, a diet high in sugars, and poor oral hygiene. Even when all these predisposing factors are present, however, germ-free animals still remain free of caries unless they have been inoculated with specific bacteria. Further discussion on dental diseases is found in chapter 21.

The ability of known pathogens to cause infection can also be influenced by normal flora, sometimes in opposing ways. Studies have indicated that germ-free animals are highly susceptible to experimental infection by the enteric pathogens *Shigella* and

Figure 13.8
Sterile enclosure for rearing and handling germ-free laboratory animals.

TABLE 13.4

EFFECTS AND SIGNIFICANCE OF EXPERIMENTS WITH GERM-FREE SUBJECTS

Effect in Germ-Free Animal	Significance
Enlargement of the cecum; other degenerative diseases of the intestinal tract of rats, rabbits, chickens	Microbes are needed for normal intestinal development
Vitamin deficiency in rats	Microbes are a significant nutritional source of vitamins
Underdevelopment of immune system in most animals	Microbes are needed to stimulate development of certain host defenses
Absence of dental caries and periodontal disease in dogs, rats, hamsters	Normal flora are essential in caries formation and gum disease
Heightened sensitivity to enteric pathogens (*Shigella, Salmonella, Vibrio cholerae*) and to fungal infections	Normal flora are antagonistic against pathogens
Lessened susceptibility to amebic dysentery	Normal flora facilitate the completion of the life cycle of the ameba in the gut

Vibrio, whereas normal animals are less susceptible, presumably because of their protective flora. In marked contrast, *Entamoeba histolytica* (the agent of amebic dysentery) is more pathogenic in the normal animal than in the germ-free animal. One explanation for this phenomenon is that *E. histolytica* must feed on intestinal bacteria to complete its life cycle.

Chapter Checkpoints

- Humans are contaminated with microorganisms from the moment of birth onward. An infection is a condition in which contaminating microorganisms overcome host defenses, multiply, and cause disease, damaging tissues and organs.
- The resident or normal flora includes bacteria, fungi, and protozoa.
- There are two types of cutaneous populations of microbes: the transients, which cling to but do not grow on the superficial layers of the skin, and the more permanent residents which reside in the deeper layers of the epidermis and its glands.
- The flora of the alimentary canal is confined primarily to the mouth, large intestine, and rectum.
- The flora of the respiratory tract extends from the nasal cavity to the lower pharynx.
- The flora of the genitourinary tract is restricted to the urethral opening in males and to the urethra and vagina in females.

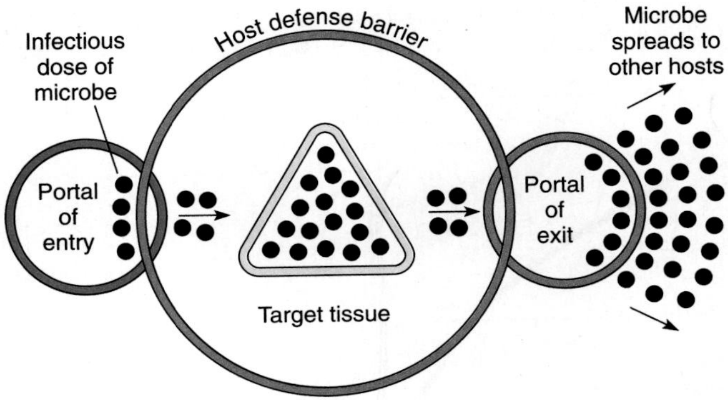

Figure 13.9
An overview of the events in infection. An adequate dose of an infectious agent overcomes a defense barrier and enters the sterile tissues of a host through one of several portals of entry. From here, it moves or is carried to a specific organ or tissue, called the target tissue, where it multiplies and usually causes some degree of damage. The exit of the pathogen through the same or another portal can facilitate its transmission to another host.

THE PROGRESS OF AN INFECTION

When the association between host and parasite tilts in the direction of infection and disease, a particular series of events is set into motion. The microbe enters the body, attaches itself, invades (crosses host barriers), multiplies in a target tissue, and finally is released to the exterior by various pathways (figure 13.9).

The type and severity of an infection depend on numerous factors, most of which are related to the pathogenicity of the microbe and the condition of the host. *Pathogenicity,* we learned earlier, is a broad concept that describes a microorganism's potential to cause an infection or disease. Pathogenic microbes have been traditionally divided into two categories, depending upon the nature of the microbe-host relationship. **True pathogens** (primary pathogens) are capable of causing infection and disease in healthy persons with normal immune defenses. Examples include the influenza virus, plague bacillus, rabies virus, malarial protozoan, and other microorganisms with well-developed qualities of virulence (discussed next). Primary pathogens are generally associated with a distinct, recognizable disease. The degree of pathogenicity of true pathogens ranges from weak to potent; for instance, infection with the cold virus is mild and nonfatal, whereas infection with the rabies virus is nearly 100% fatal.

Opportunistic pathogens cause disease when the host's defenses are compromised or when they become established in a part of the body that is not natural to them. Opportunists are not considered pathogenic to the normal, healthy person and do not usually have well-developed virulence properties. It could be argued that most microbes have some capacity for pathogenicity. For example, a severely compromised host may be challenged by a large inoculum in a vulnerable site. Another side of this idea is that anyone who becomes infected may be temporarily compromised to an

TABLE 13.5

FACTORS THAT WEAKEN HOST DEFENSES AND INCREASE SUSCEPTIBILITY TO INFECTION*

Old age
Extreme youth (infancy, prematurity)
Malnutrition
Genetic defects in immunity
Acquired defects in immunity (AIDS)
Physical and mental stress
Organ transplant
Cancer
Chemotherapy/immunosuppressive drugs
Anatomical defects
Diabetes
Liver disease
Surgery

*These conditions compromise defense barriers or immune responses.

extent. Factors that greatly **predispose** a person to infections both primary and opportunistic are shown in table 13.5.

Recognizing that classifying pathogens as either true or opportunistic may be an oversimplification, the Centers for Disease Control and Prevention has adopted a system of biosafety categories for pathogens based on their degree of pathogenicity and the relative danger in handling them. With this system, presented in appendix D, pathogens are assigned a level or class. Microbes not known to cause disease in humans (such as *Micrococcus luteus*) are given class 1 status; moderate risk agents such as *Staphylococcus aureus* are assigned to class 2; readily transmitted and virulent agents such as *Mycobacterium tuberculosis* are in class 3; and the highest risk microbes (deadly pathogens such as rabies and Ebola fever viruses) are in class 4.

Differences in pathogenicity can be accounted for on the basis of virulence. Virulence takes into account the ability of microbes to invade a host and produce toxins (toxigenicity). These properties are called **virulence factors.** Virulence can be due to single or multiple factors. In some microbes, the causes of virulence are clearly established, but in others they are not. Although virulence is sometimes used interchangeably with pathogenicity, it is actually not a synonym. Pathogenicity is a more general term for comparing degrees of microbial infectiousness. In the following section, we examine the effects of virulence factors while simultaneously outlining the stages in the progress of an infection.

THE PORTAL OF ENTRY: GATEWAY TO INFECTION

To initiate an infection, a microbe enters the tissues of the body by a characteristic route, the **portal of entry,** usually a cutaneous or membranous boundary. The source of the infectious agent can be **exogenous,** originating from a source outside the body (the environment or another person or animal), or **endogenous,** already existing on or in the body (normal flora or latent infection).

For the most part, the portals of entry are the same anatomical regions that also support normal flora: the skin, gastrointestinal tract, respiratory tract, and urogenital tract. The majority of pathogens have adapted to a specific portal of entry, one that provides a habitat for further growth and spread. This adaptation can be so restrictive that if certain pathogens enter the "wrong" portal, they will not be infectious. For instance, inoculation of the nasal mucosa with the influenza virus invariably gives rise to the flu, but if this virus contacts only the skin, no infection will result. Likewise, contact with athlete's foot fungi in small cracks in the toe webs can induce an infection, but inhaling the fungus spores will not infect a healthy individual. Occasionally, an infective agent can enter by more than one portal. For instance, *Mycobacterium tuberculosis* enters through both the respiratory and gastrointestinal tracts, and *Corynebacterium diphtheriae* can infect by way of the throat and the skin. Common pathogens in the genera *Streptococcus* and *Staphylococcus* have adapted to invasion through several portals of entry such as the skin, urogenital tract, and respiratory tract.

Infectious Agents That Enter the Skin

The skin is a very common portal of entry. The actual sites of entry are usually nicks, abrasions, and punctures (many of which are tiny and inapparent) rather than smooth, unbroken skin. *Staphylococcus aureus* (the cause of boils), *Streptococcus pyogenes* (an agent of impetigo), the fungal dermatophytes, and agents of gangrene and tetanus gain access through damaged skin. The viral agent of cold sores (herpes simplex, type 1) enters through the mucous membranes near the lips.

Some infectious agents create their own passageways into the skin using digestive enzymes. For example, certain helminth worms burrow through the skin directly to gain access to the tissues. Other infectious agents enter through bites. The bites of insects, ticks, and other animals offer an avenue to a variety of viruses, rickettsias, and protozoa. Artificial means for breaching the skin barrier are contaminated hypodermic needles by intravenous drug abusers (microfile 13.2) and the insertion of indwelling catheters.

Although the conjunctiva, the outer protective covering of the eye, is ordinarily a relatively good barrier to infection, bacteria such as *Haemophilus aegyptius* (pinkeye), *Chlamydia trachomatis* (trachoma), and *Neisseria gonorrhoeae* have special affinity for this membrane. Foreign bodies in the eye or minor injuries can serve as a portal for herpes simplex virus and *Acanthamoeba* (a protozoan that can contaminate homemade contact lens solutions).

The Gastrointestinal Tract As Portal

The gastrointestinal tract is the portal of entry for pathogens contained in food, drink, and other ingested substances. They are adapted to survive digestive enzymes and abrupt pH changes. Most enteric pathogens possess specialized mechanisms for entering and localizing in the mucosa of the small or large intestine. The best-known enteric agents of disease are gram-negative rods in the genera *Salmonella, Shigella, Vibrio,* and certain strains of *Escherichia coli* (see microfile 20.3). Viruses that enter through the gut are poliovirus, hepatitis A virus, echovirus, and rotavirus. Important enteric protozoans are *Entamoeba histolytica* (amebiasis) and

MICROFILE 13.2 BITES AND NEEDLES—A TRAUMATIC PORTAL OF ENTRY

That venerable tale of "man bites dog" turns the tables in more ways than one, because a human bite is probably far more dangerous than a dog bite. The main reason for its potency is the presence in human saliva of huge numbers of relatively virulent bacteria. A bite also creates a ragged, deep wound favorable to anaerobes. The most common lacerating injuries occur in fistfights, when the knuckles of the hitter are bashed against the teeth of his opponent. The inoculated wound rapidly advances to a deep ulcer that can spread and cause damage to the joints if left untreated. Infections also occur in bite wounds inflicted by children or psychiatric patients, in persons who self-inflict a wound by nervously chewing the insides of their cheeks, and in dental personnel accidentally inoculated with saliva.

A tragic and growing societal problem is the drug abuser who frequently and willingly forms a portal of entry into his own skin and veins with a hypodermic syringe. There is no denying that those who inject heroin, cocaine, or amphetamines are in dire jeopardy from the drug itself, but many drug-abuse practices increase chances for infection and are just as life-threatening. A fiend looking for the perfect system to transmit infectious agents might well use the behavioral patterns of the drug addict as a model. Drugs can be injected into the skin (skin popping) or into the veins (IV). Some IV users "boot" the drug by drawing out a small amount of blood and mixing it with drugs before reinjecting it—a sure way to contaminate the syringe. Most serious of all, "needles" are often shared and even rented with meager or no attempts at disinfection. Considering that a single "shooting gallery" can play host to hundreds of users every day, and that many participants hop from one gallery to another, the potential for spread of infection is tremendous. Programs have been adopted by many cities to distribute kits with sterile syringes and bleach for disinfecting them.

Users who inject drugs are predisposed to a disturbing list of well-known diseases: hepatitis, AIDS, tetanus, tuberculosis, osteomyelitis, and malaria. A resurgence of some of these infections is directly traceable to drug use. Contaminated needles often contain bacteria from the skin or environment that induce heart disease (endocarditis), lung abscesses, and chronic infections of the injection site.

Giardia lamblia (giardiasis). Although the anus is not a typical portal of entry, it becomes one in people who practice anal sex.

The Respiratory Portal of Entry

The oral and nasal cavities are also the gateways to the respiratory tract, the portals of entry to the greatest number of pathogens. The extent to which an agent is carried into the respiratory tree is based primarily on its size. In general, small cells and particles are inhaled more deeply than larger ones. Infectious agents with this portal of entry include the bacteria of streptococcal sore throat, meningitis, diphtheria, and whooping cough and the viruses of influenza, measles, mumps, rubella, chickenpox, and the common cold. Pathogens that are inhaled into the lower regions of the respiratory tract (bronchioles and lungs) can cause **pneumonia,** an inflammatory condition of the lung. Bacteria (*Streptococcus pneumoniae*, *Klebsiella, Mycoplasma*) and fungi (*Cryptococcus* and *Pneumocystis*) are a few of the agents involved in pneumonias. All types of pneumonia are on the increase owing to the greater susceptibility of AIDS patients to them. Other agents causing unique recognizable lung diseases are *Mycobacterium tuberculosis* and fungal pathogens such as *Histoplasma.*

Urogenital Portals of Entry

The urogenital tract is the portal of entry for pathogens that are contracted by sexual means (intercourse or intimate direct contact). The older term, *venereal disease*, has been replaced by the more descriptive term **sexually transmitted disease** (STD; microfile 13.3). The microbes of STDs enter the skin or mucosa of the penis, external genitalia, vagina, cervix, and urethra. Some can penetrate an unbroken surface; others require a cut or abrasion. The once predominant sexual diseases syphilis and gonorrhea have been supplanted by a large and growing list of STDs headed by genital warts, chlamydia, and herpes. Evolving sex practices have increased the incidence of STDs that were once uncommon, and diseases that were not originally considered STDs are now so classified.[4] Other common sexually transmitted agents are HIV (AIDS virus), *Trichomonas* (a protozoan), *Candida albicans* (a yeast), and hepatitis B virus.

Pathogens That Infect During Pregnancy and Birth

The placenta is an exchange organ formed by maternal and fetal tissues that separates the blood of the developing fetus from that of the mother yet permits diffusion of dissolved nutrients and gases to the fetus. The placenta is ordinarily an effective barrier against microorganisms in the maternal circulation. However, a few microbes such as the syphilis spirochete can cross the placenta, enter the umbilical vein, and spread by the fetal circulation into the fetal tissues (figure 13.10).

Other infections such as herpes simplex occur perinatally when the child is contaminated by the birth canal. The common infections of fetus and neonate are grouped together in a unified cluster, known by the acronym **STORCH,** that medical personnel must monitor. STORCH stands for **s**yphilis, **t**oxoplasmosis, **o**ther diseases (hepatitis B, AIDS, and chlamydia), **r**ubella, **c**ytomegalovirus, and **h**erpes simplex virus. The most serious complications of STORCH infections are spontaneous abortion, congenital abnormalities, brain damage, prematurity, and stillbirths.

4. Amebic dysentery, scabies, salmonellosis, and *Strongyloides* worms are examples.

Microfile 13.3 What's Love Got to Do with It?

The word *venereal* is derived from Venus, the Latin name of the goddess of love. For centuries, sexually oriented diseases were considered a curse from God, with the greater share of blame attributed to women. During biblical times, so great was the condemnation of women afflicted with sex-associated diseases that many of them were executed. Attempts to treat or control infection during the early twentieth century were squelched by clergymen protesting that, "If God had wanted to eradicate venereal disease, he would not have given it to women in the first place." Posters from World Wars I and II and even into the 1950s flaunt the role of women in these diseases and depict them in the most debasing ways. Just as it is clear that STDs can be stigmatizing, it is equally clear that men are profoundly involved in their transmission. The term *sexually transmitted disease* recognizes this fact.

A poster dating from the early 1950s unfairly places blame on women for the spread of STDs.

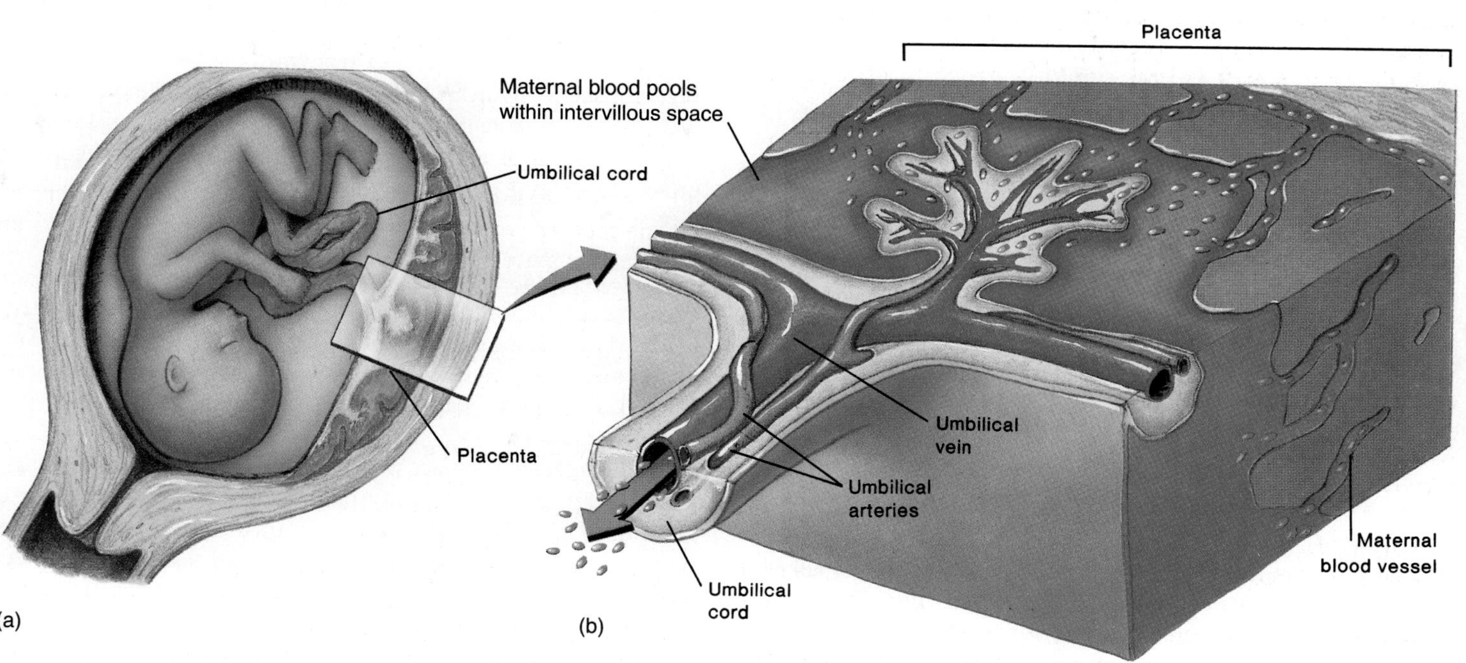

Figure 13.10
(*a*) Transplacental infection of the fetus. (*b*) In a closer view, microbes are shown penetrating the maternal blood vessels and entering the blood pool of the placenta. They then invade the fetal circulation by way of the umbilical vein.

The Size of the Inoculum

Another factor crucial to the course of an infection is the quantity of microbes in the inoculating dose. For most agents, infection will proceed only if a minimum number, called the *infectious dose* (ID), is present. This number has been determined experimentally for many microbes. In general, microorganisms with smaller infectious doses have greater virulence. The ID varies from the astonishing number of 1 rickettsial cell in Q fever to about 10 infectious cells in tuberculosis, giardiasis, and coccidioidomycosis. The ID is 1,000 cells for gonorrhea and 10,000 cells for typhoid fever in contrast to 1,000,000,000 cells in cholera. Lack of an infectious dose will generally not result in an infection. But if the quantity is far in excess of the ID, the onset of disease can be extremely rapid. Even weakly pathogenic species can be rendered more virulent with a large inoculum.

Chapter Checkpoints

Microbial infections result when a microorganism penetrates host defenses, multiplies, and damages host tissue. The *pathogenicity* of a microbe refers to its ability to cause infection or disease. The *virulence* of a pathogen refers to the degree of damage it inflicts on the host tissues.

True pathogens cause infectious disease in healthy hosts, whereas *opportunistic pathogens* become infectious only when the host immune system is compromised in some way.

The site at which a microorganism first contacts host tissue is called the *portal of entry.* Most pathogens have one preferred portal of entry, although some have more than one.

The respiratory system is the portal of entry for the greatest number of pathogens.

The *infectious dose,* or ID, refers to the minimum number of microbial cells required to initiate infection in the host. The ID varies widely among microbial species.

Fimbriae, flagella, hooks, and adhesive capsules are types of adherence factors by which pathogens physically attach to host tissues.

Mechanisms of Invasion and Establishment of the Pathogen

Following entry of the pathogen, the next stage in infection requires that the pathogen (1) bind to the host, (2) penetrate its barriers, and (3) become established in the tissues. How the pathogen achieves these ends greatly depends upon its specific biochemical and structural characteristics.

How Pathogens Attach

Adhesion is a process by which microbes gain a more stable foothold at the portal of entry. Because this often involves a specific interaction between molecules on the microbial surface and receptors on the host cell, adhesion can determine the specificity of a pathogen for its host organism and in some cases the specificity of a pathogen for a particular cell type. Once attached, the pathogen is poised advantageously to invade the sterile body compartments. Bacterial pathogens attach most often by mechanisms such as fimbriae (pili), flagella, and adhesive slimes or capsules; viruses attach by means of specialized receptors (figure 13.11). Protozoa can infiltrate by means of their organelle of locomotion; for example, *Balantidium coli* is said to penetrate by the boring action of its cilia. In addition, parasitic worms are mechanically fastened to the portal of entry by suckers, hooks, and barbs (see chapter 23). Adhesion methods of various microbes and the diseases they lead to are shown in table 13.6.

How Virulence Factors Contribute to Tissue Damage

Virulence factors from a microbe's perspective are simply adaptations it uses to invade and establish itself in the host. These same factors determine the degree of tissue damage that occurs. Effects vary from pathogens that invade the tissues but multiply with relatively minor effects, such as cold viruses, to very harmful pathogens (the tetanus bacillus and AIDS virus, for instance) that severely damage and even kill the host. For convenience, we will divide the virulence factors into exoenzymes, toxins, and antiphagocytic factors (figure 13.12). Although this distinction is useful, there is often a very fine line between enzymes and toxins because many substances called toxins actually function as enzymes.

Extracellular Enzymes Many pathogenic bacteria, fungi, protozoa, and worms secrete **exoenzymes** that break down and inflict damage on tissues. Other enzymes dissolve the host's defense barriers and promote the spread of microbes to deeper tissues.

Examples of enzymes are (1) Mucinase, which digests the protective coating on mucous membranes, is a factor in amebic dysentery. (2) Keratinase, which digests the principal component of skin and hair, is secreted by fungi that cause ringworm. (3) Collagenase digests the principal fiber of connective tissue and is an invasive factor of *Clostridium* species and certain worms. (4) Hyaluronidase digests hyaluronic acid, the ground substance that cements animal cells together. This enzyme is an important virulence factor in staphylococci, clostridia, streptococci, and pneumococci.

Some enzymes react with components of the blood. Coagulase, an enzyme produced by pathogenic staphylococci, causes clotting of blood or plasma. By contrast, the bacterial kinases (strepto-kinase, staphylokinase) do just the opposite, dissolving fibrin clots and expediting the invasion of damaged tissues. In fact, one form of streptokinase (streptase) is marketed as a therapy to dissolve blood clots in patients with problems with thrombi and emboli.[5]

5. These conditions are intravascular blood clots that can cause circulatory obstructions.

TABLE 13.6

ADHESION PROPERTIES OF MICROBES

Microbe	Disease	Adhesion Mechanism
Neisseria gonorrhoeae	Gonorrhea	Fimbriae attach to genital epithelium
Escherichia coli	Diarrhea	Well-developed K antigen capsule
Shigella	Dysentery	Fimbriae can attach to intestinal epithelium
Vibrio	Cholera	Glycocalyx anchors microbe to intestinal epithelium
Treponema	Syphilis	Tapered hook embeds in host cell
Mycoplasma	Pneumonia	Specialized tip at ends of bacteria fuse tightly to lung epithelium
Pseudomonas aeruginosa	Burn, lung infections	Fimbriae and slime layer
Streptococcus pyogenes	Pharyngitis, impetigo	Lipotechoic acid and M-protein anchor cocci to epithelium
Streptococcus mutans, S. sobrinus	Dental caries	Dextran slime layer glues cocci to tooth surface
Influenza virus	Influenza	Viral spikes react with receptor on cell surface
Poliovirus	Polio	Capsid proteins attach to receptors on susceptible cells
HIV	AIDS	Viral spikes adhere to white blood cell receptor
Giardia lamblia (protozoan)	Giardiasis	Small suction disc on underside attaches to intestinal surface

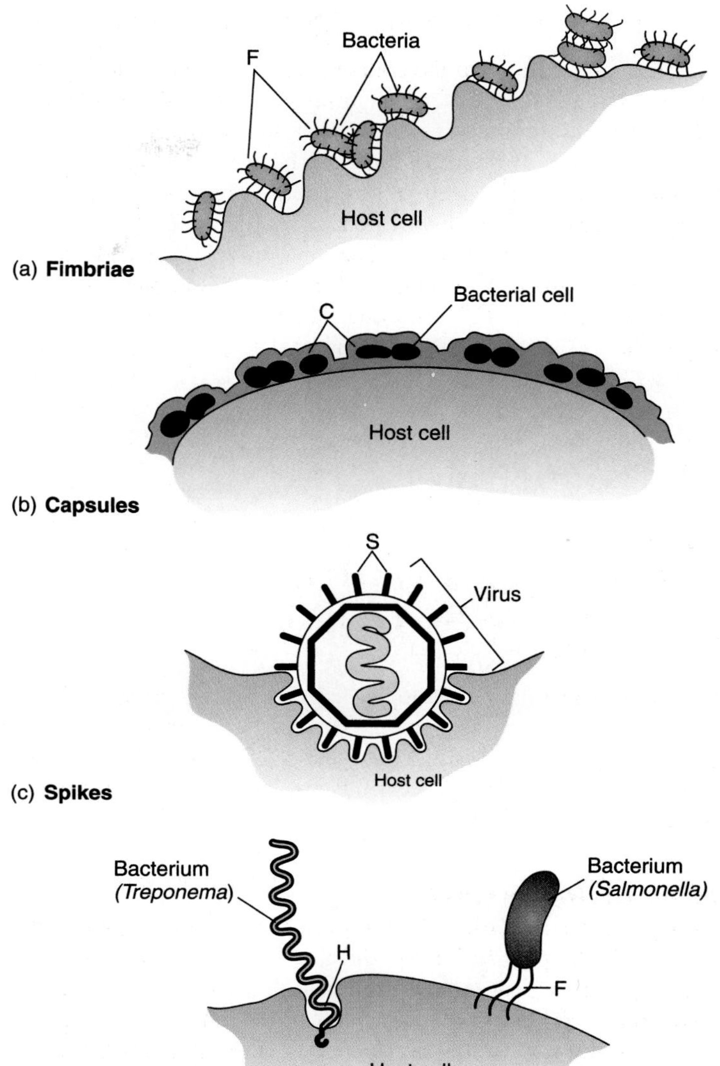

Figure 13.11

Mechanisms of adhesion by pathogens (see table 13.6 for specific examples). (*a*) Fimbriae (F), minute bristle-like appendages. (*b*) Adherent extracellular capsules (C) made of slime or other sticky substances. (*c*) Viral envelope spikes (S). (*d*) Specialized cell hooks (H) or flagella (F).

Bacterial Toxins: A Potent Source of Cellular Damage

A **toxin** is a specific chemical product of microbes, plants, and some animals that is poisonous to other organisms. **Toxigenicity,** the power to produce toxins, is a genetically controlled characteristic of many species and is responsible for the adverse effects of a variety of diseases generally called **toxinoses.** A type of toxinosis in which the toxin is spread by the blood from the site of infection is a **toxemia** (tetanus and diphtheria, for example), whereas one caused by ingestion of toxins is an **intoxication** (botulism). A toxin is named according to its specific target of action: Neurotoxins act on the nervous system; enterotoxins act on the intestine; hemotoxins lyse red blood cells; and nephrotoxins damage the kidneys.

A more traditional scheme classifies toxins according to their origins (figure 13.13). An unbound toxin molecule secreted by a living bacterial cell into the infected tissues is an **exotoxin.** A toxin that is not secreted but is released only after the cell is damaged or lysed is an **endotoxin.** Other important differences between the two groups, summarized in table 13.7, are generally chemical and medical in nature.

Exotoxins are proteins with a strong specificity for a target cell and extremely powerful, sometimes deadly, effects. They generally affect cells by damaging the cell membrane and initiating lysis or by disrupting intracellular function. **Hemolysins*** are a class of bacterial exotoxin that disrupts the cell membrane of red blood cells (and some other cells too). This damage causes the red blood cells to **hemolyze**—to burst and release hemoglobin pigment. Hemolysins that increase pathogenicity include the strep-

*hemolysin (hee-mahl′-uh-sin) Gr. *haima,* blood, and *lysis,* dissolution. A substance that causes hemolysis.

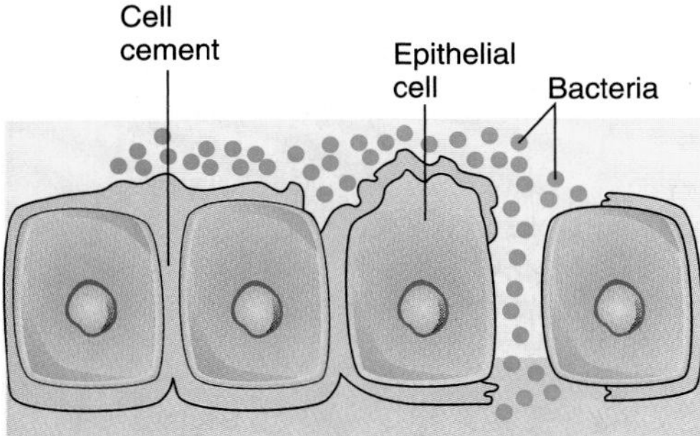

(a) **Exoenzymes**

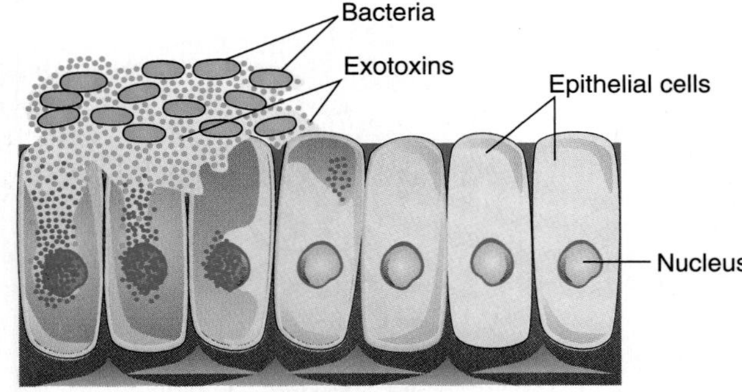

(b) **Toxins**

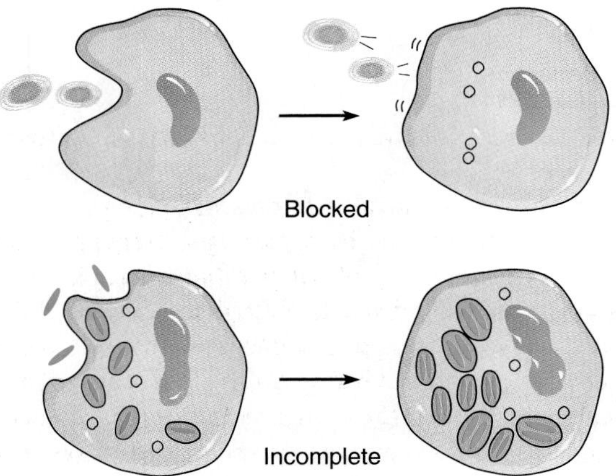

(c) **Phagocytosis**

Figure 13.12

The function of exoenzymes, toxins, and phagocyte blockers in invasiveness. (*a*) Exoenzymes. Bacteria produce extracellular enzymes that dissolve extracellular barriers and penetrate through or between cells to underlying tissues. (*b*) Toxins (primarily exotoxins) secreted by bacteria diffuse to target cells, which are poisoned and disrupted. (*c*) Blocked (top) or incomplete (bottom) phagocytosis. Phagocytosis is ineffective for several reasons. For example, some pathogens have a protective coating that makes them hard to engulf, or the phagocyte ingests them but the microbes can still multiply.

tolysins of *Streptococcus pyogenes* and the alpha (α) and beta (β) toxins of *Staphylococcus aureus.* When colonies of bacteria growing on blood agar produce hemolysin, distinct zones appear around the colony. The pattern of hemolysis is often used to identify bacteria and determine their degree of pathogenicity (see chapter 18).

The toxins of diphtheria, tetanus, and botulism, among others, attach to a particular target cell, become internalized, and interrupt an essential cell pathway. The consequences of cell disruption depend upon the target. One toxin of *Clostridium tetani* blocks the action of certain spinal neurons; the toxin of *Clostridium botulinum* prevents the transmission of nerve-muscle stimuli; pertussis toxin inactivates the respiratory cilia; and cholera toxin provokes profuse salt and water loss from intestinal cells. More details of the pathology of exotoxins are found in the chapters that treat the specific diseases.

Endotoxins belong to a class of chemicals called lipopolysaccharides (LPS), which are part of the outer membrane of gram-negative cell walls (see microfile 20.1). When gram-negative bacteria cause infections, some of them eventually lyse and release these LPS molecules into the infection site or into the circulation. Endotoxins differ from exotoxins in having a variety of systemic effects on tissues and organs. Depending upon the amounts present, endotoxins can cause fever, inflammation, hemorrhage, and diarrhea. Blood infection by gram-negative bacteria such as *Salmonella, Shigella, Neisseria meningitidis,* and *Escherichia coli* are particularly dangerous, in that it can lead to fatal endotoxic shock.

How Microbes Escape Phagocytosis *Antiphagocytic factors* are another type of virulence factor used by some pathogens to avoid certain white blood cells called phagocytes. These cells would ordinarily engulf and destroy pathogens by means of enzymes and other antibacterial chemicals (see chapter 14). The antiphagocytic factors of resistant microorganisms help them to circumvent some part of the phagocytic process (figure 13.12*c*). The most aggressive strategy involves bacteria that kill phagocytes outright. Species of both *Streptococcus* and *Staphylococcus* produce **leukocidins,** substances that are toxic to white blood cells. Some microorganisms secrete an extracellular surface layer (slime or capsule) that makes it physically difficult for the phagocyte to engulf them. *Streptococcus pneumoniae, Salmonella typhi, Neisseria meningitidis,* and *Cryptococcus neoformans* are notable examples. Some bacteria are well adapted to survival inside phagocytes after ingestion. For instance, pathogenic species of *Legionella, Mycobacterium,* and many rickettsias are readily engulfed but are capable of avoiding further destruction. The ability to survive intracellularly in phagocytes has special significance because it provides a place for the microbes to hide, grow, and be spread throughout the body.

Establishment, Spread, and Pathologic Effects

Aided by virulence factors, microbes eventually settle in a particular target organ and continue to cause damage at the site. The type and scope of injuries inflicted during this process account for the typical stages of an infection (microfile 13.4), the

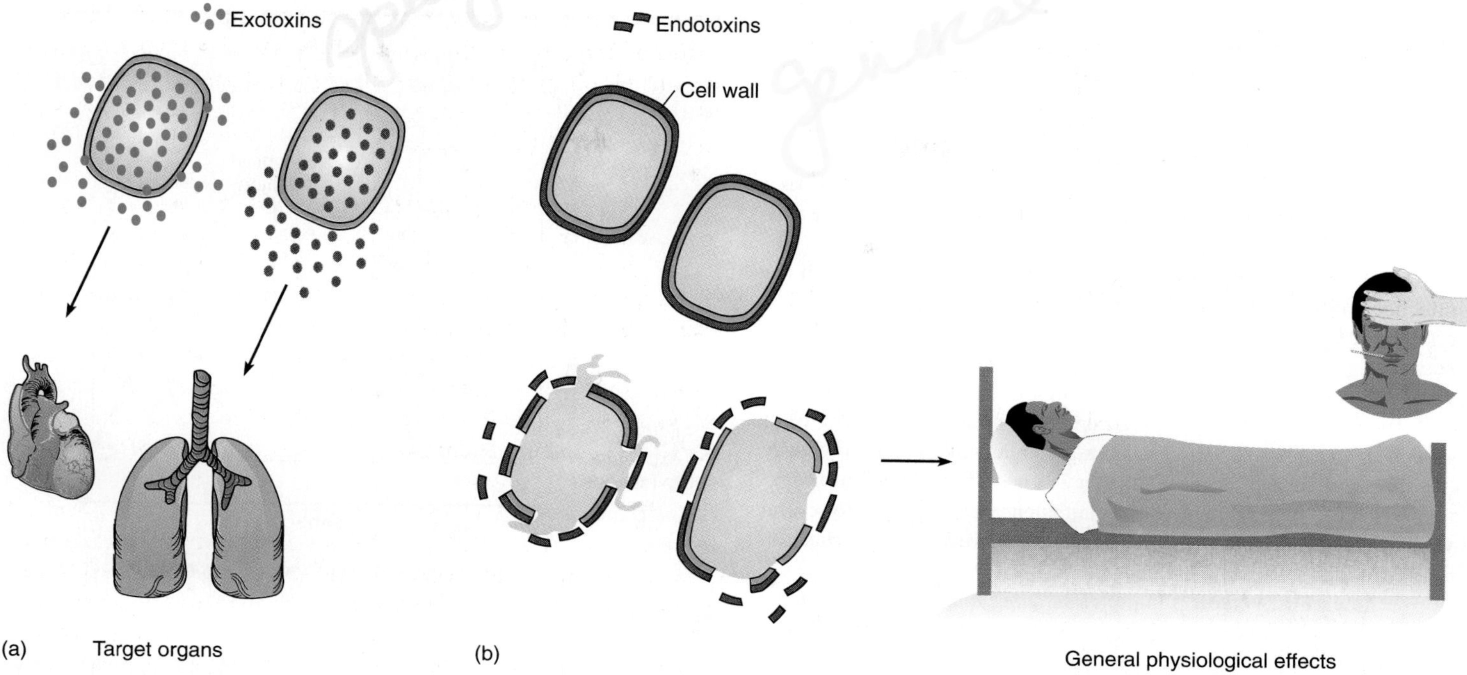

Figure 13.13

The origins and effects of circulating exotoxins and endotoxins. (*a*) Exotoxins, given off by live cells, have highly specific targets and physiological effects. (*b*) Endotoxins, given off when the cell wall of gram-negative bacteria disintegrates, have more generalized physiological effects.

Table 13.7

Differential Characteristics of Bacterial Exotoxins and Endotoxins

Characteristic	Exotoxins	Endotoxins
Toxicity	Toxic in minute amounts	Toxic in high doses
Effects on the Body	Specific to a cell type	Systemic: fever, inflammation
Chemical Composition	Polypeptides	Lipopolysaccharide of cell wall
Heat Denaturation at 60°C	Unstable	Stable
Toxoid Formation	Convert to toxoid*	Do not convert to toxoid
Immune Response	Stimulate antitoxins**	Do not stimulate antitoxins
Fever Stimulation	Usually not	Yes
Manner of Release	Secreted from live cell	Released by cell during lysis
Typical Sources	A few gram-positive and gram-negative	All gram-negative bacteria

*A toxoid is an inactivated toxin used in vaccines.

**An antitoxin is an antibody that reacts specifically with a toxin.

patterns of the infectious disease, and its manifestations in the body.

In addition to the adverse effects of enzymes, toxins, and other factors, multiplication by a pathogen frequently weakens host tissues. Pathogens can obstruct tubular structures such as blood vessels, lymphatic channels, fallopian tubes, and bile ducts. Accumulated damage can lead to cell and tissue death, a condition called **necrosis.** Although viruses do not produce toxins or destructive enzymes, they destroy cells by multiplying in and lysing them. Many of the cytopathic effects of viral infection arise from the impaired metabolism and death of cells (see chapter 6).

Patterns of Infection Patterns of infection are many and varied (figure 13.14). In the simplest situation, a **localized infection,** the microbe enters the body and remains confined to a specific tissue (figure 13.14*a*). Examples of localized infections are boils, fungal skin infections, and warts.

Many infectious agents do not remain localized but spread from the initial site of entry to other tissues. In fact, spreading is necessary for pathogens, such as rabies and hepatitis A virus, whose target tissue is some distance from the site of entry. The rabies virus travels from a bite wound along nerve tracts to its target in the brain, and the hepatitis A virus moves from the intestine to the liver via the circulatory system. When an infection spreads to several sites and tissue fluids, usually in the bloodstream, it is called a **systemic infection** (figure 13.14*b*). Exam-

MICROFILE 13.4 THE CLASSIC STAGES OF CLINICAL INFECTIONS

As the body of the host responds to the invasive and toxigenic activities of a parasite, it passes through four distinct phases of infection and disease: the incubation period, the prodromium, the period of invasion, and the convalescent period.

The **incubation period** is the time from initial contact with the infectious agent (at the portal of entry) to the appearance of the first symptoms. During the incubation period, the agent is multiplying at the portral of entry but has not yet caused enough damage to elicit symptoms. Although this period is relatively well defined and predictable for each microorganism, it does vary according to host resistance, degree of virulence, and distance between the target organ and the portal of entry (the farther apart, the longer the incubation period). Overall, an incubation period can range from several hours in pneumonic plague to several years in leprosy. The majority of infections, however, have incubation periods ranging between 2 and 30 days.

The earliest notable symptoms of infection appear as a vague feeling of discomfort, such as head and muscle aches, fatigue, upset stomach, and general malaise. This short period (1-2 days) is known as the **prodromal stage.** The infectious agent next enters a **period of invasion,** during which it multiplies at high levels, exhibits its greatest toxicity, and becomes well established in its target tissue. This period is often marked by fever and other prominent and more specific signs and symptoms, which can include cough, rashes, diarrhea, loss of muscle control, swelling, jaundice, discharge of exudates, or severe pain, depending on the particular infection. The length of this period is extremely variable.

As the patient begins to respond to the infection, the symptoms decline—sometimes dramatically, other times slowly. During the recovery that follows, called the **convalescent period,** the patient's strength

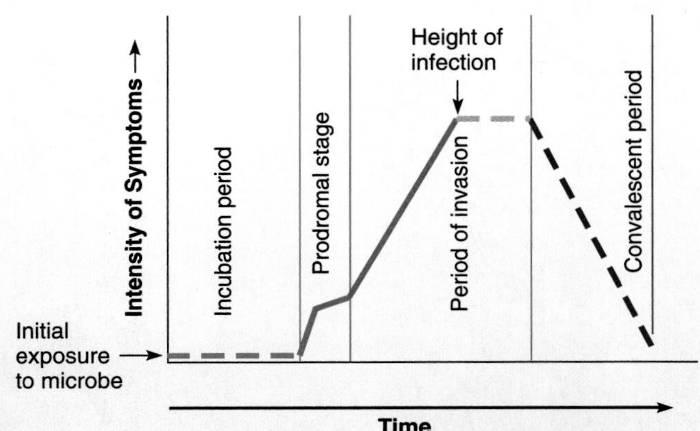

Stages in the course of infection and disease. Dashed lines represent periods with a variable length; the fastigium is the height of the disease.

and health gradually return owing to the healing nature of the immune response. An infection that results in death is called **terminal.** This term is particularly applicable if the infection is the immediate cause of death in a patient already suffering from a degenerative disease or cancer. Thus, an alcoholic might succumb to terminal pneumonia, or an AIDS patient will actually die from a secondary infection.

The transmissibility of the microbe during these four stages must be considered on an individual basis. A few agents are released mostly during incubation (measles, for example); many are released during the invasive period (*Shigella*); and others can be transmitted during all of these periods (hepatitis B).

ples of systemic infections are viral diseases (measles, rubella, chickenpox, and AIDS); bacterial diseases (brucellosis, anthrax, typhoid fever, and syphilis); and fungal diseases (histoplasmosis and cryptococcosis). Infectious agents can also travel to their targets by means of nerves (as in rabies) or cerebrospinal fluid (as in meningitis).

A **focal infection** is said to exist when the infectious agent breaks loose from a local infection and is carried into other tissues (figure 13.14*c*). This pattern is exhibited by tuberculosis or by streptococcal pharyngitis, which gives rise to scarlet fever. In the condition called toxemia,[6] the infection itself remains localized at the portal of entry but the toxins produced by the pathogens are carried by the blood to the actual target tissue. In this way, the target of the bacterial cells can be different from the target of their toxin (see discussions of tetanus and diphtheria in chapter 19).

An infection is not always caused by a single microbe. In a **mixed infection,** several agents establish themselves simultaneously at the infection site (figure 13.14*d*). In some mixed or synergistic infections, the microbes cooperate in breaking down a tissue. In other mixed infections, one microbe creates an environment that enables another microbe to invade. Gas gangrene, wound infections, dental caries, and human bite infections tend to be mixed.

Some diseases are described according to a sequence of infection. When an initial, or **primary, infection** is complicated by another infection caused by a different microbe, the second infection is termed a **secondary infection** (figure 13.14*e*). This pattern often occurs in a child with chickenpox (primary infection) who may scratch his pox and infect them with *Staphylococcus aureus* (secondary infection). The secondary infection need not be in the same site as the primary infection, and it usually indicates altered host defenses.

Infections that come on rapidly, with severe but short-lived effects, are called **acute infections.** Infections that progress and persist over a long period of time are **chronic infections. Subacute infections** do not come on as rapidly as acute infections or persist as long as chronic ones. Microfile 13.5 illustrates other common terminology used to describe infectious diseases.

6. Not to be confused with toxemia of pregnancy, which is a metabolic disturbance and not an infection.

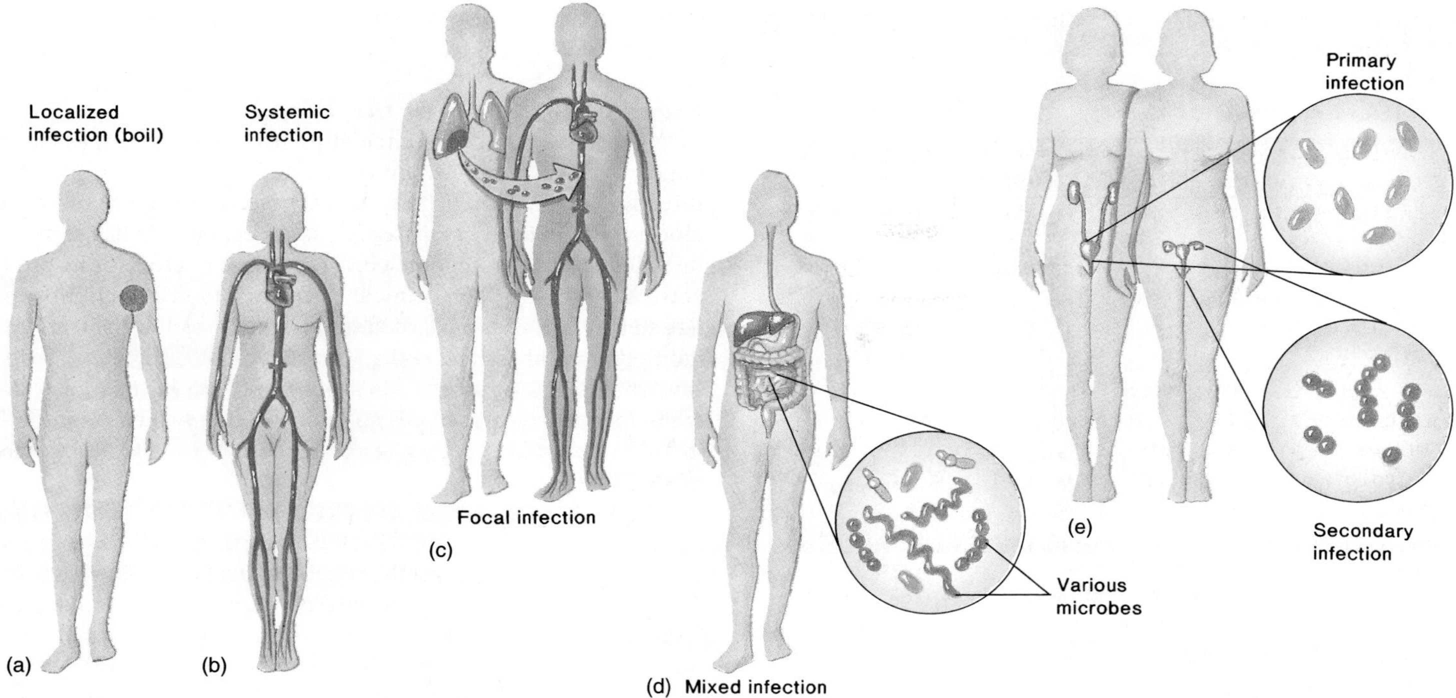

Figure 13.14
The occurrence of infections with regard to location, type of microbe, and length of time. (*a*) A localized infection, in which the pathogen is restricted to one specific site. (*b*) Systemic infection, in which the pathogen spreads through circulation to many sites. (*c*) A focal infection occurs initially as a local infection, but circumstances cause the microbe to be carried to other sites systemically. (*d*) A mixed infection, in which the same site is infected with several microbes at the same time. (*e*) In a primary-secondary infection, an initial infection is complicated by a second one in a different location (usually) and caused by a different microbe.

MICROFILE 13.5 A QUICK GUIDE TO THE TERMINOLOGY OF INFECTION AND DISEASE

Words in medicine have great power and economy. A single technical term can often replace a whole phrase or sentence, thereby saving time and space in patient charting. The beginning student may feel overwhelmed by what seems like a mountain of new words. However, learning a few root words and a fair amount of anatomy can help you learn many of these words and even deduce the meaning of unfamiliar ones. Some examples of medical shorthand follow.

The suffix **-itis** means an inflammation and, when affixed to the end of an anatomical term, indicates an inflammatory condition in that location. Thus, meningitis is an inflammation of the meninges surrounding the brain; encephalitis is an inflammation of the brain itself; hepatitis involves the liver; vaginitis, the vagina; gastroenteritis, the intestine; and otitis media, the middle ear. Although not all inflammatory conditions are caused by infections, many infectious diseases inflame their target organs.

The suffix **-emia** is derived from the Greek word *haeima,* meaning blood. When added to a word, it means "associated with the blood." Thus, septicemia means sepsis (infection) of the blood; bacteremia, bacteria in the blood; viremia, viruses in the blood; and fungemia, fungi in the blood. It is also applicable to specific conditions such as toxemia, gonococcemia, and spirochetemia.

The suffix **-osis** means "a disease or morbid process." It is frequently added to the names of pathogens to indicate the disease they cause: for example, listeriosis, histoplasmosis, toxoplasmosis, shigellosis, salmonellosis, and borreliosis. A variation of this suffix is *-iasis,* as in trichomoniasis and candidiasis.

The suffix **-oma** comes from the Greek word *onkomas* (swelling) and means tumor. Although the root is often used to describe cancers (sarcoma, melanoma), it is also applied in some infectious diseases that cause masses or swellings (tuberculoma, leproma).

TABLE 13.8

COMMON SIGNS AND SYMPTOMS OF INFECTIOUS DISEASES

Signs	Symptoms
Fever	Chills
Septicemia	Pain, ache, soreness, irritation
Microbes in tissue fluids	Nausea
Chest sounds	Malaise, fatigue
Skin eruptions	Chest tightness
Leukocytosis	Itching
Leukopenia	Headache
Swollen lymph nodes	Nausea
Abscesses	Abdominal cramps
Tachycardia (increased heart rate)	Anorexia (lack of appetite)
Antibodies in serum	Sore throat

SIGNS AND SYMPTOMS: WARNING SIGNALS OF DISEASE

When an infection causes **pathologic** changes leading to disease, it is often accompanied by a variety of signs and symptoms. A **sign** is any objective evidence of disease as noted by an observer; a **symptom** is the subjective evidence of disease as sensed by the patient. In general, signs are more precise than symptoms, though both can have the same underlying cause. For example, an infection of the brain might present with the sign of bacteria in the spinal fluid and symptom of headache. Or a streptococcal infection might produce a sore throat (symptom) and inflamed pharynx (sign). Disease indicators that can be sensed and observed can qualify as either a sign or a symptom. When a disease can be identified or defined by a certain complex of signs and symptoms, it is termed a **syndrome.** Signs and symptoms with considerable importance in diagnosing infectious diseases are shown in table 13.8.

Signs and Symptoms of Inflammation

The earliest symptoms of disease result from the activation of the body defense process called **inflammation.*** The inflammatory response includes cells and chemicals that respond nonspecifically to disruptions in the tissue. This subject is discussed in greater detail in chapter 14, but it is worth noting here that signs and symptoms of infection are caused by the mobilization of this system. Some common symptoms of inflammation include fever, pain, soreness, and swelling. Signs of inflammation include **edema,*** the accumulation of fluid in an afflicted tissue; *granulomas* and *abscesses,* walled-off collections of inflammatory cells and microbes in the tissues; and *lymphadenitis,* swollen lymph nodes. The skin is also a common focus of signs and symptoms (microfile 13.6).

Signs of Infection in the Blood

Changes in the number of circulating white blood cells, as determined by special counts, are considered to be signs of possible infection. *Leukocytosis** is an increase in the level of white blood cells, whereas *leukopenia** is a decrease. Other signs of infection revolve around the occurrence of a microbe or its products in the blood. The clinical term for blood infection, **septicemia,** refers to a general state in which microorganisms are multiplying in the blood and are present in large numbers. When small numbers of bacteria or viruses are found in the blood, the correct terminology is **bacteremia** or **viremia,** which means that these microbes are present in the blood but are not necessarily multiplying.

During infection, a normal host will invariably show signs of an immune response in the form of antibodies in the serum or some type of sensitivity to the microbe. This fact is the basis for several serological tests used in diagnosing infectious diseases such as AIDS or syphilis. Such specific immune reactions indicate the body's attempt to develop specific immunities against pathogens. We will concentrate on this role of the host defenses in the next two chapters.

Infections That Go Unnoticed

It is rather common for an infection to produce no noticeable symptoms, even though the microbe is active in the host tissue. In other words, although infected, the host does not manifest the disease. Infections of this nature are known as **asymptomatic,** *subclinical,* or *inapparent* because the patient experiences no symptoms or disease and does not seek medical attention. However, it is important to note that most infections are attended by some sort of sign. In the section on epidemiology, we further address the significance of subclinical infections in the transmission of infectious agents.

THE PORTAL OF EXIT: VACATING THE HOST

Earlier, we introduced the idea that a parasite is considered *unsuccessful* if it kills its host. A parasite is equally unsuccessful if it does not have a provision for leaving its host and moving to other susceptible hosts. With few exceptions, pathogens depart by a specific avenue called the **portal of exit** (figure 13.15). In most cases, the pathogen is shed or released from the body through secretion, excretion, discharge, or sloughed tissue. The usually very high number of infectious agents in these materials increases both virulence and the likelihood that the pathogen will reach other hosts. In many cases, the portal of exit is the same as the portal of entry, but a few pathogens use a different route. As we see in

*inflammatory (in-flam´-uh-tor´´-ee) L. *inflammatio,* to set on fire.

*edema (uh-dee´-muh) Gr. *oidema,* swelling.

*leukocytosis (loo´´-koh-sy-toh´-sis) From *leukocyte,* a white blood cell, and the suffix *-osis.*

*leukopenia (loo´´-koh-pee´-nee-uh) From *leukocyte* and *penia,* a loss or lack of.

MICROFILE 13.6 RASHES, RUSHES, AND FLUSHES

Rashes and other skin eruptions are common in many diseases, and because they tend to mimic each other, it can be difficult to differentiate among diseases on this basis alone. The general term for any damaged or dysfunctional body area is **lesion.*** Skin lesions form at the point of microbial entry, at its exit, or when it induces an inflammatory reaction. Skin lesions can be restricted to the epidermis and its glands and follicles, or they can extend into the dermis and subcutaneous regions. The lesions of some infections undergo characteristic changes in appearance during the course of disease and thus fit more than one category (see smallpox, figure 24.1).

A small, flat-colored skin lesion is a *macule.** The color may be red, tan, or white. Crops of these spots are early signs of measles and rubella. A raised lesion composed of solid tissue is a *papule;** it varies in texture from smooth to rough and can be any number of colors. Often, papules are a later stage in the development of a macule, producing the so-called maculopapular rash. A thicker papule that penetrates into the dermis is a *nodule.** Leprosy and several fungus infections produce nodular lesions.

Many lesions contain some sort of liquid or semiliquid material called an exudate. A *vesicle** is a small blister that contains clear, yellowish fluid, and a *bulla** is a large, fluid-filled blister. Vesicles are characteristic of herpes simplex, impetigo, and chickenpox (a pock is one type of vesicle); bullae are seen in some types of staphylococcal skin infections. If the space in the lesion is filled with pus (a mixture of fluid, white blood cells, tissue debris, and bacteria), it is a *pustule.** Acne lesions are types of pustules. Vesicles, bullae, and pustules have a fragile surface that can burst and release the exudate, which dries and forms a *crust.*

When a lesion results from skin being sloughed, the process is known as *erosion* or *ulceration.* Erosion is a superficial loss of the epidermis, whereas an ulcer is a deeply penetrating open sore. The chancre of syphilis and the eschar of anthrax are erosive types of lesions; large, draining ulcers appear in leishmaniasis and some worm infections.

*lesion (lee´-zhun) L. *laesio,* to hurt.
*macule (mak´-yool) L. *maculatus,* spotted.
*papule (pap´-yool) L. *papula,* pimple.
*nodule (nawd´-yool) L. *nodulus,* little knot.
*vesicle (ves´-ik-ul) L. *vesicula,* small bladder.
*bulla (byoo´-lah) L. *bulla,* bubble.
*pustule (pust´-yool) L. *pustula,* blister or pimple.

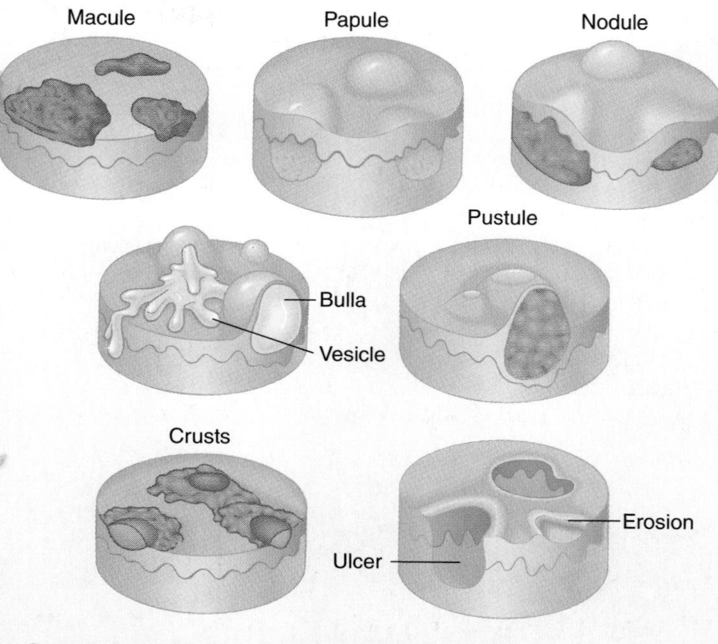

Cutaway views of the skin, comparing various lesions.

Some skin manifestations are due to alterations in circulation. *Erythema** is a confluent reddening of the skin due to increased blood flow. The rash of scarlet fever and the peculiar migrating skin eruption of Lyme disease are types of erythema. Hemorrhaging into the skin produces a brown to purple discoloration called *purpura,** as seen in bubonic plague and meningococcal meningitis.

*erythema (air-ih-thee´-mah) Gr. *erythema,* flush upon the skin.
*purpura (pur´-pur-ah) L. *purpura,* purple.

the next section, the portal of exit concerns epidemiologists because it greatly influences the dissemination of infection in a population.

Respiratory and Salivary Portals

Mucus, sputum, nasal drainage, and other moist secretions are the media of escape for the pathogens that infect the lower or upper respiratory tract. The most effective means of releasing these secretions are coughing and sneezing (see figure 13.22), although they can also be released during talking and laughing. Tiny particles of liquid released into the air form aerosols or droplets that can spread the infectious agent to other people. The agents of tuberculosis, influenza, measles, and chickenpox most often leave the host through airborne droplets. Droplets of saliva are the exit route for several viruses, including those of mumps, rabies, and infectious mononucleosis.

Skin Scales

The outer layer of the skin and scalp are constantly being shed into the environment. A large proportion of household dust is actually composed of skin scales. A single person can shed several billion skin cells a day, and some persons, called shedders, disseminate massive numbers of bacteria into their immediate surroundings. Skin lesions and their exudates can serve as portals of exit in warts, fungal infections, boils, herpes simplex, smallpox, and syphilis.

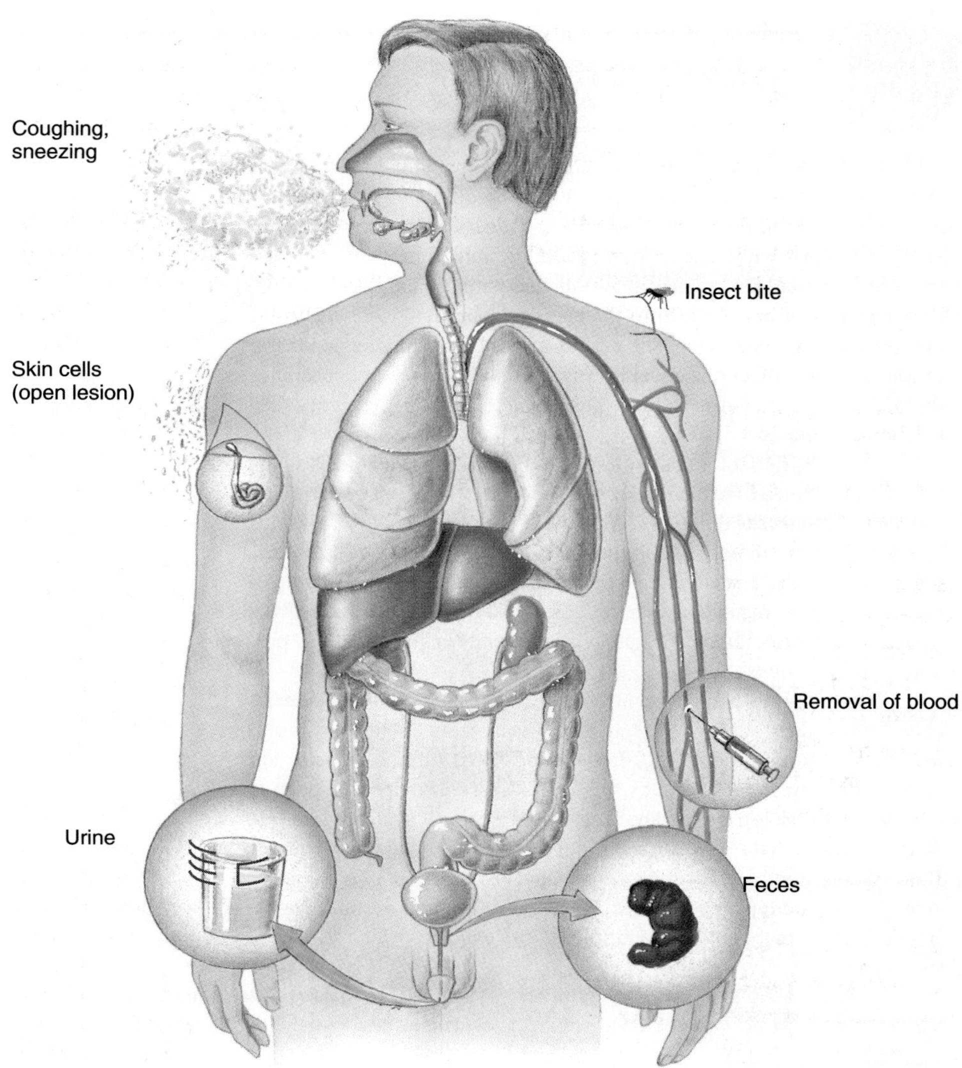

Figure 13.15
Major portals of exit of infectious diseases.

Fecal Exit

Feces is a very common portal of exit. Some intestinal pathogens grow in the intestinal mucosa and create an inflammation that increases the motility of the bowel. This increased motility speeds up peristalsis, resulting in diarrhea, and the more fluid stool provides a rapid exit for the pathogen. A number of helminth worms release cysts and eggs through the feces. Feces containing pathogens are a public health problem when allowed to contaminate drinking water or when used to fertilize crops (see chapter 23).

Urogenital Tract

A number of agents involved in sexually transmitted infections leave the host in vaginal discharge or semen. This is also the source of neonatal infections such as herpes simplex, *Chlamydia,* and *Candida albicans,* which infect the infant as it passes through the birth canal. Less commonly, certain pathogens that infect the kidney are discharged in the urine: for instance, the agents of leptospirosis, typhoid fever, tuberculosis, and schistosomiasis.

Removal of Blood or Bleeding

Although the blood does not have a direct route to the outside, it can serve as a portal of exit when it is removed or released through a vascular puncture made by natural or artificial means. Blood-feeding insects such as mosquitos and fleas are common transmitters of pathogens (see microfile 21.5). The AIDS and hepatitis viruses are transmitted by shared needles or through small gashes in a mucous membrane caused by sexual intercourse. Blood donation is also a means for certain microbes to leave the host, though this means of exit is now unusual because of close monitoring of the donor population and blood used for transfusions.

The Persistence of Microbes and Pathologic Conditions

The apparent recovery of the host does not always mean that the microbe has been completely removed or destroyed by the host defenses. After the initial symptoms in certain chronic infectious diseases, the infectious agent retreats into a dormant state called **latency.** Throughout this latent state, the microbe can periodically become active and produce a recurrent disease. The viral agents of herpes simplex, herpes zoster, hepatitis B, AIDS, and Epstein-Barr can persist in the host for long periods. The agents of syphilis, typhoid fever, tuberculosis, and malaria also enter into latent stages. The person harboring a persistent infectious agent may or may not shed it during the latent stage. If it is shed, such persons are chronic carriers who serve as sources of infection for the rest of the population.

Some diseases leave **sequelae*** in the form of long-term or permanent damage to tissues or organs. For example, meningitis can result in deafness, a strep throat can lead to rheumatic heart disease, Lyme disease can cause arthritis, and polio can produce paralysis.

Chapter Checkpoints

Exoenzymes, toxins, and antiphagocytic factors are the three main types of *virulence factors* pathogens utilize to combat host defenses and damage host tissue.

Exotoxins and endotoxins differ in their chemical composition and tissue specificity.

Antiphagocytic factors produced by microorganisms include leukocidins, capsules, and factors that resist digestion by white blood cells.

Patterns of infection vary with the pathogen or pathogens involved. They range from local and focal to systemic.

A mixed infection is caused by two or more microorganisms simultaneously.

Infections can be characterized by their sequence as primary or secondary and by their duration as either acute or chronic.

An infectious disease is characterized by both objective signs and subjective symptoms.

Infectious diseases that are asymptomatic or subclinical nevertheless produce clinical signs.

The portal of exit by which a pathogen leaves its host is usually but not always the same as the portal of entry.

The portals of exit and entry determine how pathogens spread in a population.

Some pathogens persist in the body in a latent state; others cause long-term diseases called *sequelae*.

EPIDEMIOLOGY: THE STUDY OF DISEASE IN POPULATIONS

So far, our discussion has revolved primarily around the impact of an infectious disease in a single individual. Let us now turn our attention to the effects of diseases on the community (microfile 13.7)—the realm of **epidemiology.*** By definition, this term involves the study of the frequency and distribution of disease and other health-related factors in defined human populations. It involves many disciplines—not only microbiology, but anatomy, physiology, immunology, medicine, psychology, sociology, ecology, and statistics—and it considers many diseases other than infectious ones, including heart disease, cancer, drug addiction, and mental illness. The epidemiologist is a medical sleuth who collects clues on the causative agent, pathology, sources and modes of transmission and tracks the numbers and distribution of cases of disease in the community. In fulfilling these demands, the epidemiologist asks, who, when, where, how, why, and what? about diseases. The outcome of these studies helps public health departments develop prevention and treatment programs and establish a basis for predictions.

WHO, WHEN, AND WHERE? TRACKING DISEASE IN THE POPULATION

Epidemiologists are concerned with all of the factors covered earlier in this chapter: virulence, portals of entry and exit, and the course of disease. But they are also interested in **surveillance**—that is, collecting, analyzing, and reporting data on the rates of occurrence, mortality, morbidity, and transmission of infections. Surveillance involves keeping data for a large number of diseases seen by the medical community and reported to public health authorities. By law, certain **reportable,** or notifiable, diseases must be reported to authorities; others are reported on a voluntary basis.

A well-developed network of individuals and agencies at the local, district, state, national, and international levels keeps track of infectious diseases. Physicians and hospitals report all notifiable diseases that are brought to their attention. Case reporting can focus on a single individual or collectively on group data.

Local public health agencies first receive the case data and determine how they will be handled. In most cases, health officers investigate the history and movements of patients to trace their prior contacts and to control the further spread of the infection as soon as possible through drug therapy, immunization, and education. In sexually transmitted diseases, patients are asked to name their partners so that these persons can be notified, examined, and treated. It is very important to maintain the confidentiality of the persons in these reports. The principal government agency responsible for keeping track of infectious diseases nationwide is the Centers for Disease Control and Prevention (CDC) in Atlanta, Georgia, which is a part of the United States Public Health Service. The CDC publishes a weekly notice of diseases (the *Morbidity and Mortality Report*) that provides weekly and cumulative summaries of the case rates and deaths for about 48 notifiable diseases, highlights important and unusual diseases, and presents data concerning disease occurrence in the major regions of the United States (see endpapers of this book). Ultimately, the CDC shares its statistics on disease with the World Health Organization (WHO) for worldwide tabulation and control.

Epidemiologic Statistics: Frequency of Cases

The **prevalence** of a disease is the total number of existing cases with respect to the entire population. It is a cumulative statistic, usually represented as the percentage of the population having a particular disease at any given time. Disease **incidence** measures

*sequelae (suh-kwee´-lee) L. *sequi,* to follow.

*epidemiology (ep´´-ih-dee-mee-ahl´-uh-gee) Gr. *epidemios,* prevalent.

Pertussis (whooping cough)—by year, United States, 1957–1997

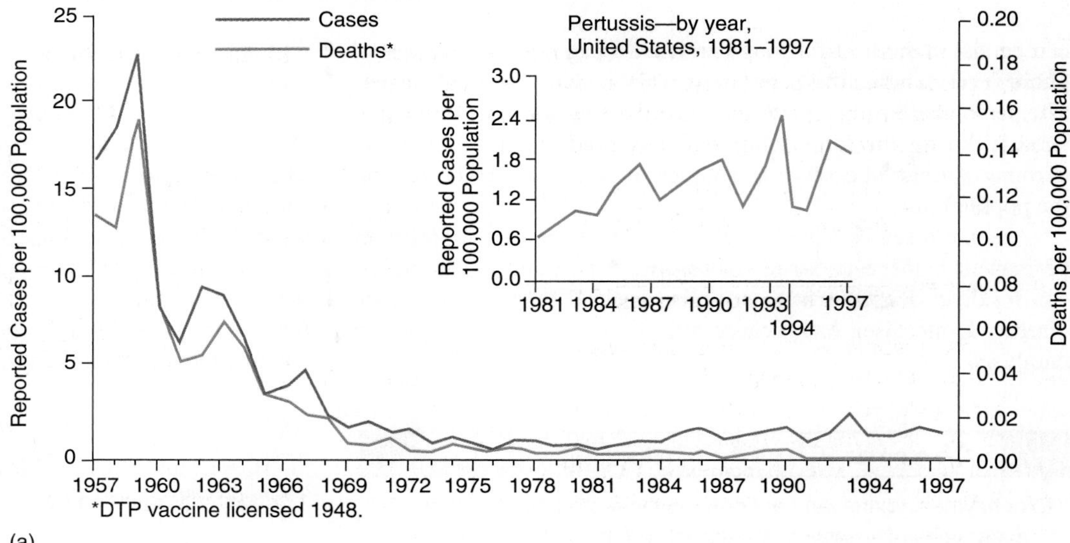

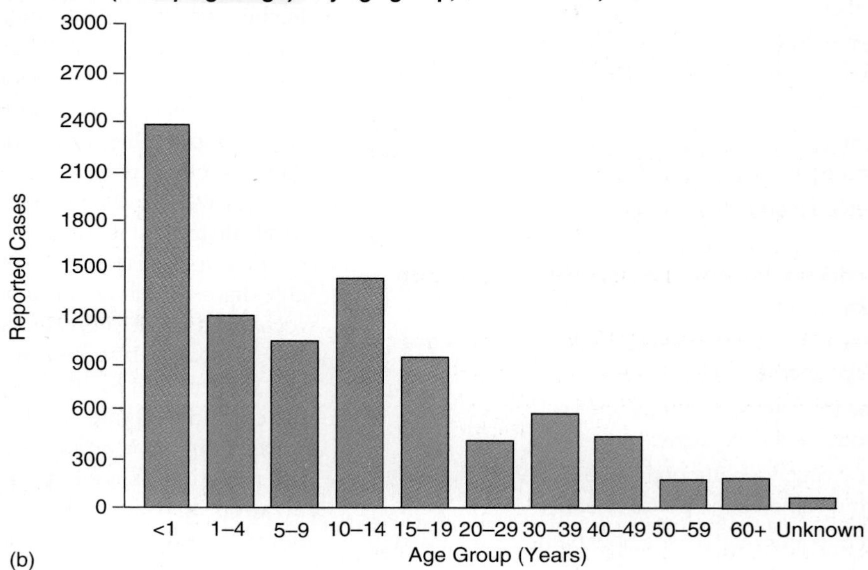

Number of reported tuberculosis cases—United States,* 1975–1997

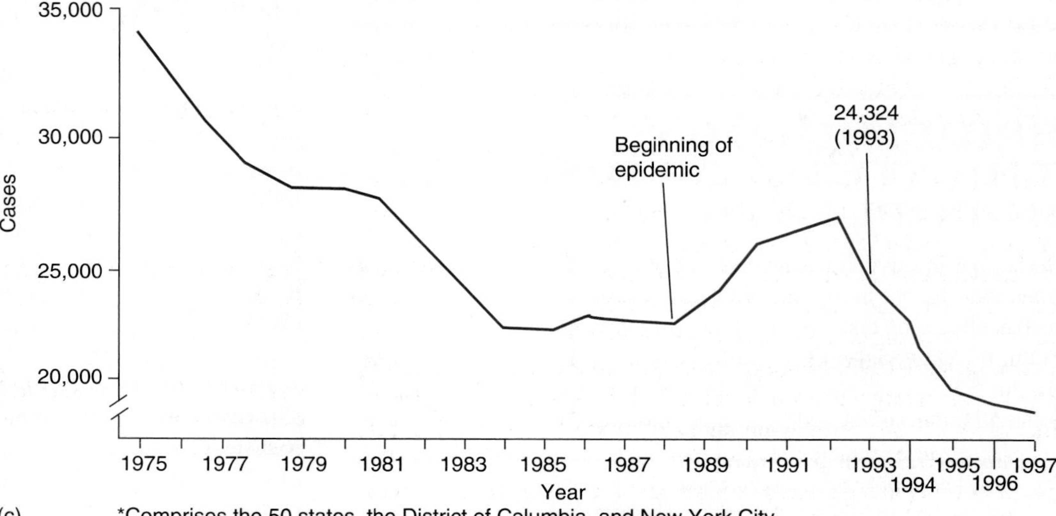

Figure 13.16
Methods for analyzing epidemiologic data. (*a*) Case rates (incidence) of pertussis, 1987–1997, per 100,000 population. The inset magnifies the recent incidence of epidemics (peaks on the graph) that appear to coincide with lapses in population immunity. (*b*) A histogram graphs data of the total case rate of pertussis according to the age group affected. Notice that the majority of cases occur in children 1 to 4 years of age. (*c*) Reported incidence of tuberculosis from 1975 to 1997 traces the emergence of an epidemic in 1988, which abruptly changed a significant pattern of decline for over 10 years.
Source: Data from Morbidity and Mortality Weekly Report, *Vol. 45 No. 53, October 31, 1997. U.S. Department of Health and Human Services, Centers for Disease Control and Prevention, Atlanta, GA.*

the number of new cases over a certain time period, as compared with the general healthy population. This statistic, also called the case, or morbidity, rate, indicates both the rate and the risk of infection. The equations used to figure these rates are:

$$\text{Prevalence} = \frac{\text{Total number of cases in population}}{\text{Total number of persons in population}} \times 100 = \%$$

$$\text{Incidence} = \frac{\text{Number of new cases}}{\text{Number of healthy persons}} = \text{Ratio}$$

As an example, let us use a classroom of 50 students exposed to a new strain of influenza. Before exposure, the prevalence and incidence in this population are both zero (0/50). If in one week, 5 out of the 50 people contract the disease, the prevalence is 5/50 = 10%, and the incidence is 1 in 9 (5 cases compared with 45 healthy persons). If after 2 weeks, 5 more students contract the flu, the prevalence becomes 5 + 5 = 10/50 = 20%, and the incidence becomes 1 in 8 (5/40). When dealing with large populations, the incidence is usually given in numbers of cases per 1,000 or 100,000 population.

The changes in incidence and prevalence are usually followed over a seasonal, yearly, and long-term basis and are helpful in predicting trends (figure 13.16). Statistics of concern to the epidemiologist are the rates of disease with regard to sex, race, or geographic region. Also of importance is the **mortality rate,** which measures the total number of deaths in a population due to a certain disease. Over the past century, the overall death rate from infectious diseases has dropped, although the number of persons afflicted with infectious diseases (the **morbidity rate**) has remained relatively high.

Monitoring statistics also makes it possible to define the frequency of a disease in the population (figure 13.17). An infectious disease that exhibits a relatively steady frequency over a long time period in a particular geographic locale is **endemic** (figure 13.17*a*). For example, Lyme disease is endemic to certain areas of the United States where the tick vector is found. A certain number of new cases are expected in these areas every year. When a disease is **sporadic,** occasional cases are reported at irregular intervals in unpredictable locales (figure 13.17*b*). Tetanus and diphtheria are reported sporadically in the United States (fewer than 50 cases a year). When statistics indicate that the prevalence of an endemic or sporadic disease is increasing beyond what is expected for that population, the pattern is described as an **epidemic** (figure 13.17*c*). (See figure 13.16 for an idea of how epidemics look on a graph.) An epidemic exists when an increasing trend is observed in a particular population. The time period is not defined—it can range from hours in food poisoning to years in syphilis—nor is an exact percentage of increase needed before an outbreak can qualify as an epidemic. Several epidemics occur every year in the United States. A recent example is tuberculosis. From 1975 until 1987, the case rate declined from 34,000 to a historic low of 22,000. Beginning in 1988, a resurgence in TB cases occurred that continued until 1993. Many of these new cases occurred in AIDS patients and many were due to multidrug-resistant strains. The incidence of TB has once again declined to the lowest level in history, in the U.S., but it is increasing on a worldwide scale. The spread of an epidemic across continents is a **pandemic,** as exemplified by AIDS and influenza (figure 13.17*d*).

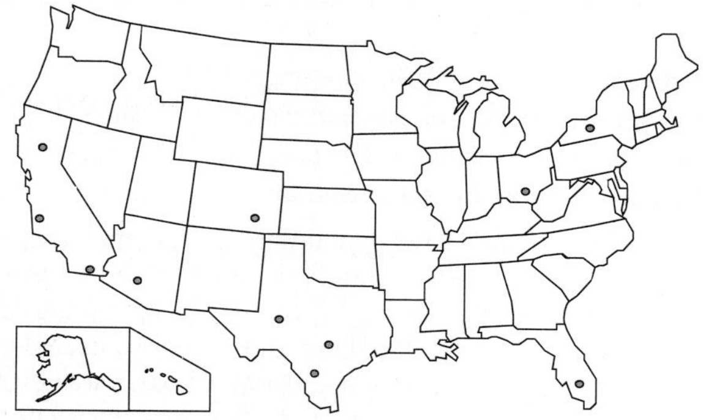

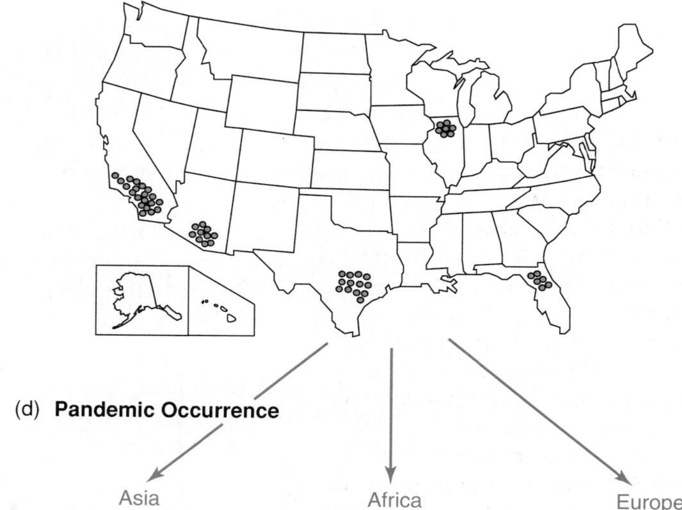

Figure 13.17

Patterns of infectious disease occurrence. (*a*) In endemic occurrence, cases are concentrated in one area at a relatively stable rate. (*b*) In sporadic occurrence, a few cases are dispersed over a wide area. (*c*) An epidemic is an increased number of cases that often appear in geographic clusters. (*d*) Pandemic occurrence means that an epidemic ranges over more than one continent.

MICROFILE 13.7 CHRONOLOGY OF A MYSTERY EPIDEMIC

In the mid-1980s, CDC epidemiologist Dr. Scott Holmberg was caught up in an engrossing and circuitous search for the source of an epidemic of salmonellosis in the Midwest. Although he encountered numerous false clues, dead ends, and the slimmest of leads, several key discoveries eventually helped him solve the mystery. Here is a diary of the investigative pathway Holmberg took:

Incident: Outbreak of food infection, a cluster of 11 cases—**Cluster 1**

Where: Southern Minnesota.

Etiologic agent: *Salmonella newport.*

Unusual factors: All persons were infected with a species not usually found in that part of the Midwest. The bacteria were resistant to amoxicillin and penicillin.

Patient condition: Very ill, requiring hospitalization.

Patient history: All but two patients were unrelated; all had been taking oral amoxicillin or penicillin just prior to the attack.

Investigative questions and course:

1. Was the antibiotic contaminated? Patients' leftover drugs and the pharmacy supply were tested thoroughly for contamination; a single colony of *Salmonella* was found in one capsule. Should medical authorities and the public be warned? Conclusion: Capsule was probably contaminated by fingers of pill taker—**Dead end 1.**
2. What other factors might be involved? Recent histories of patients were taken, but no common travel experience was discovered, and no remarkable food habits were noted—**Dead end 2.**
3. Dr. Holmberg made a personal visit to the homes of stricken persons, interviewed them in depth, and inspected their homes closely. Again, no common source was discovered—**Dead end 3.**
4. Next Holmberg reexamined the interacting facts of which he was certain:
   Salmonella newport infections were associated with antibiotic therapy. This appeared to be an important clue. Did the microbe become resistant in the patient? What was its source? Clearly, there was a third unknown factor.
5. New technology was applied. A DNA analysis of *S. newport* isolates indicated that all strains were exactly the same; the source, whatever it was, had to be a common one.
6. A series of inquiries to state health departments in neighboring states uncovered a cluster of cases of drug-resistant *S. newport* reported from South Dakota, with

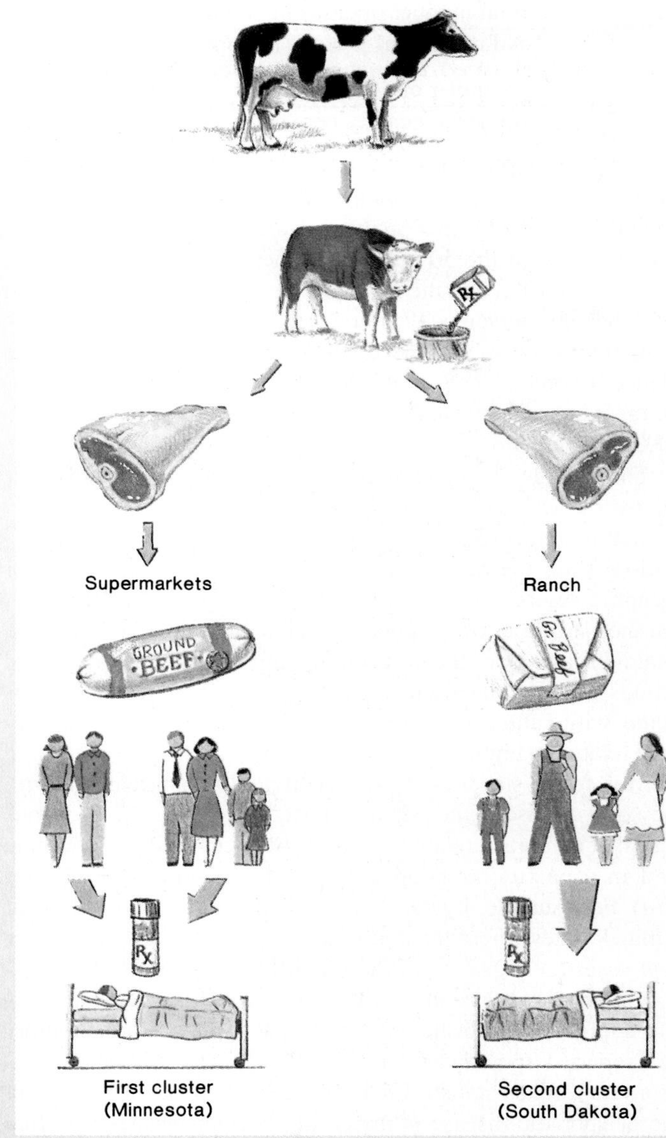

Reconstruction of an epidemic of salmonellosis.

One important epidemiologic truism might be called the "iceberg effect." Regardless of case reporting and public health screening, a large number of cases of infection in the community go undiagnosed and unreported. (For a list of reportable diseases in the United States, see table 13.9.) In the instance of salmonellosis, approximately 40,000 cases are reported each year. Epidemiologists estimate that the actual number is more likely somewhere between 400,000 and 4,000,000 (see figure 20.15). The iceberg effect can be even more lopsided for sexually transmitted diseases or for infections that are not brought to the attention of reporting agencies.

one death—**Cluster 2.** All patients were relatives, but they evidently had not shared a common source. The inquiry revealed one important bit of evidence: One patient was a dairy farmer.

7. Scrutiny of the farm disclosed a history of epidemic calf diarrhea. State officials had isolated the same strain of *S. newport* from one calf. But it was concluded that milk could not have been a source, because other victims in the cluster had not gotten milk from this dairy—**Dead end 4?**
8. Further discussion with the farmer revealed that his uncle on an adjacent farm had a beef herd that supplied meat (ground beef) to the families striken by infection. It was also revealed that dairy and beef herds had mingled through a broken fence and that the beef cattle had eaten feed mixed with antibiotics that would have selected for antibiotic-resistant strains of *Salmonella.*
9. Dr. Holmberg next followed the course of the cattle connection: The South Dakota farmer had sold his beef cattle to a broker who subsequently sold them to a slaughterhouse. The slaughterhouse made ground beef from some of the meat, which was shipped to markets in Minnesota from which the patients in Cluster 1 had bought ground beef. Eventually, 18 cases were linked to this one source.
10. The ultimate source of *S. newport* in the cattle was never discovered. But several loose ends were tied up. It was concluded that:
    - Low doses of antibiotics helped maintain drug-resistant *Salmonella* in the herd.
    - During slaughter, meat contaminated with the bacterium was made into ground beef, which served as a very effective dispersal agent because meat from several cows was mixed.
    - Ground beef was not adequately cooked before consumption. The microbe came to be harbored in the intestines of the patients asymptomatically.
    - Antibiotic therapy for an unrelated infection created a selective environment that led to the overgrowth of *Salmonella* and development of disease.

Besides being a brilliant bit of epidemiologic detective work, this discovery created a furor among agriculture and health officials, because it clearly revealed that feeding antibiotics to animals could select for strains of resistant species that could be transmitted to humans.

TABLE 13.9

NOTIFIABLE DISEASES IN THE UNITED STATES*

AIDS	Lymphogranuloma venereum
Amebiasis	Malaria
Anthrax	Measles (rubeola)
Arboviral infections	Meningococcal infections
Aseptic meningitis	Mumps
Botulism	Pertussis
Brucellosis	Plague
Chancroid	Poliomyelitis
Chickenpox	Psittacosis
Cholera	Rabies
Chlamydiosis	Rheumatic fever
Diphtheria	Rocky Mountain spotted fever
Encephalitis	Rubella
Enterovirus	Salmonellosis
Escherichia coli 0157:H7	Shigellosis
Gonorrhea	Syphilis
Hansen's disease (leprosy)	Tetanus
Haemophilus influenzae	Toxic shock syndrome
Hepatitis A	Trichinosis
Hepatitis B	Tuberculosis
Hepatitis, other	Tularemia
Influenza	Typhoid fever
Legionellosis	Typhus
Leptospirosis	Yellow fever
Lyme disease	

*Depending on the state, some of these diseases are reported only if they occur at epidemic levels; others must be reported on a case-by-case basis. You can request a list of reportable infectious diseases in your state by calling the State Department of Health.

Investigative Strategies of the Epidemiologist

Initial evidence of a new disease or an epidemic in the community is fragmentary. A few sporadic cases are seen by physicians and are eventually reported to authorities. However, it can take several reports before any alarm is registered. Epidemiologists and public health departments must piece together odds and ends of data from a series of apparently unrelated cases and work backward to reconstruct the epidemic pattern. A completely new disease requires even greater preliminary investigation, because the infectious agent must be isolated and linked directly to the disease (see the discussion of Koch's postulates in a later section of this chapter). All factors possibly impinging on the disease are scrutinized. Investigators search for *clusters* of cases indicating spread between persons or a public (common) source of infection; they also look at possible contact with animals, contaminated food, water, and public facilities and at human interrelationships or changes in community structure. Out of this maze of case information, the investigators hope to recognize a pattern that indicates the source of infection so that they can quickly move to control it (microfile 13.7).

RESERVOIRS: WHERE PATHOGENS PERSIST

In order for an infectious agent to continue to exist and be spread, it must have a permanent place to reside. The **reservoir** is the primary habitat in the natural world from which a

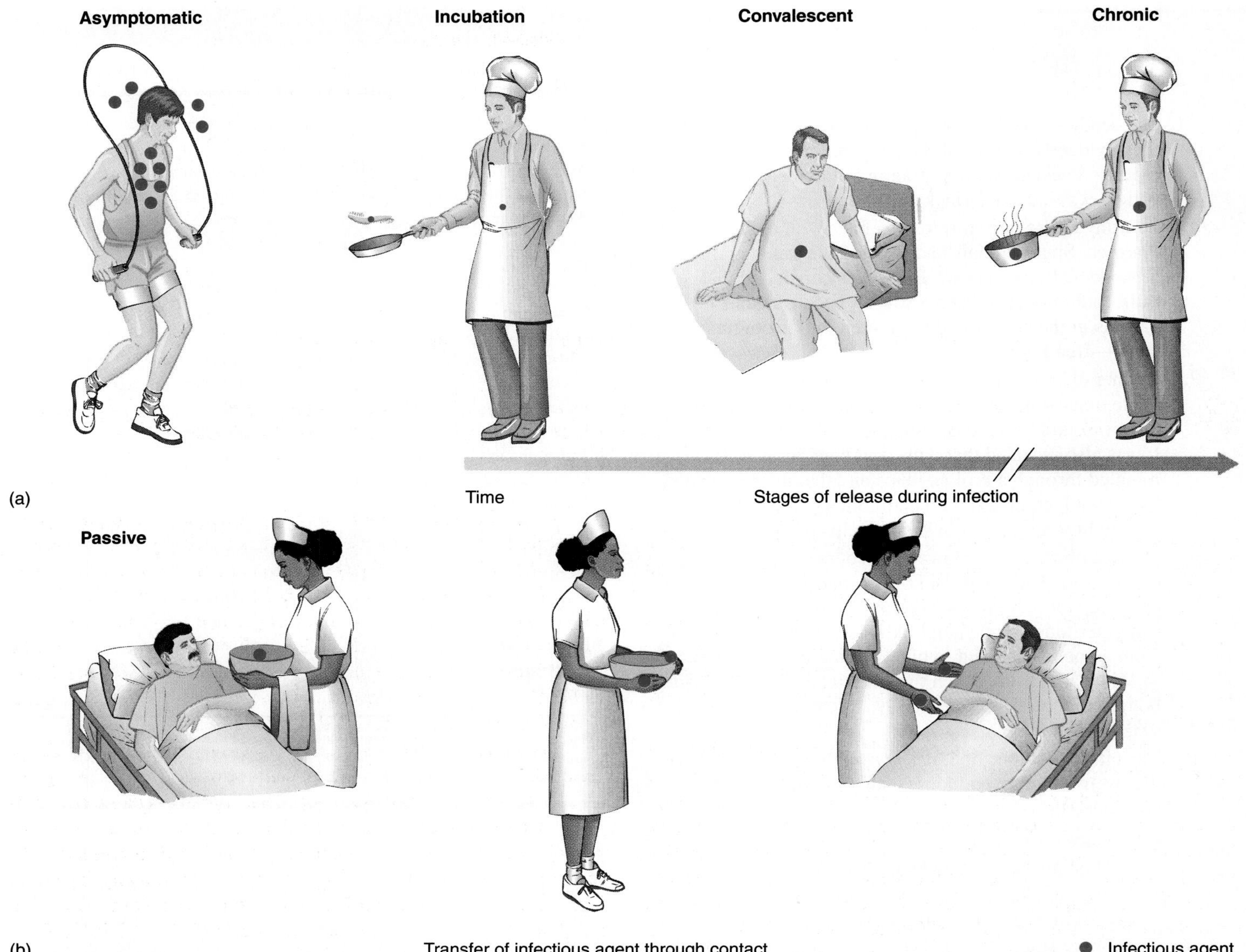

Figure 13.18

Types of carriers. (*a*) An asymptomatic carrier is infected without symptoms. Incubation carriers are in the early stages of infection; convalescent carriers are in late stages of recovery; chronic carriers sequester the microbe for long periods after the infection is over. (*b*) A passive carrier is contaminated but not infected.

pathogen originates. Often it is a human or animal carrier, although soil, water, and plants are also reservoirs. The reservoir can be distinguished from the infection **source,** which is the individual or object from which an infection is actually acquired. In diseases such as syphilis, the reservoir and the source are the same (the human body). In the case of hepatitis A, the reservoir (a human carrier) is usually different from the source of infection (contaminated food).

Living Reservoirs

Many pathogens continue to exist and spread because they are harbored by members of a host population. Persons or animals with frank symptomatic infection are obvious sources of infection, but a **carrier** is, by definition, an individual who *inconspicuously* shelters a pathogen and spreads it to others without any notice. Although human carriers are occasionally detected through routine screening (blood tests, cultures) and other epidemiologic devices, they are unfortunately very difficult to discover and control. As long as a pathogenic reservoir is maintained by the carrier state, the disease will continue to exist in that population, and the potential for epidemics will be a constant threat. The duration of the carrier state can be short- or long-term, and actual infection of the carrier may or may not be involved.

Several situations can produce the carrier state. **Asymptomatic** (apparently healthy) **carriers** are indeed infected, but as previously indicated, they show no symptoms (figure 13.18*a*). A few asymptomatic infections (gonorrhea and genital warts, for instance) can carry out their entire course without

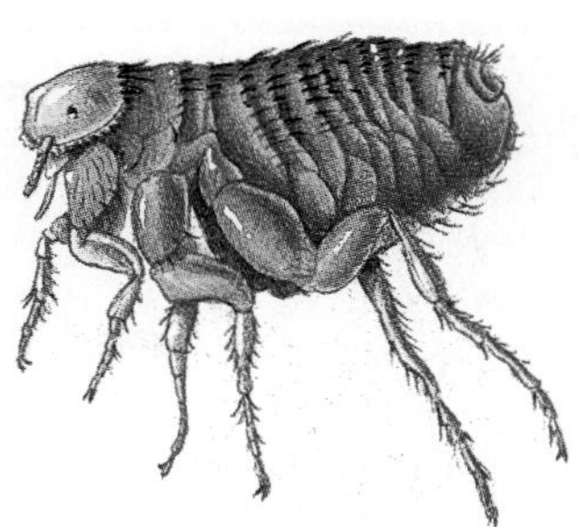

(a) Biological vectors are infected.

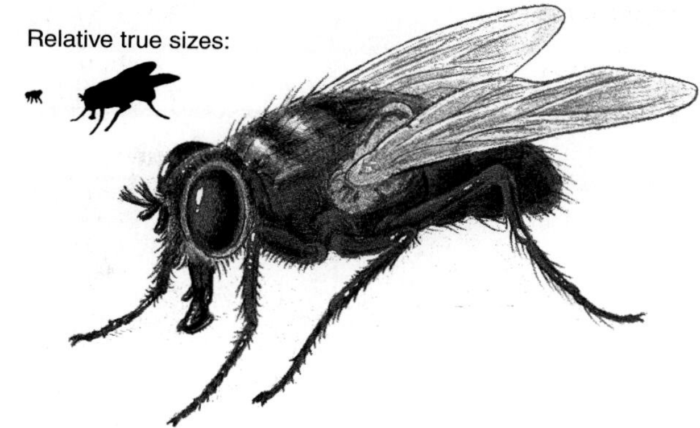

(b) Mechanical vectors are not infected.

Figure 13.19

Two types of vector. (*a*) Biological vectors serve as hosts during pathogen development. They include the flea, a carrier of bubonic plague and murine typhus. (*b*) Mechanical vectors such as the house fly ingest filth and transport pathogens on their feet and mouthparts.

overt manifestations. Other asymptomatic carriers, called *incubation carriers,* spread the infectious agent during the incubation period. For example, AIDS patients can harbor and spread the virus for months and years before their first symptoms appear. Recuperating patients without symptoms are considered *convalescent carriers* when they continue to shed viable microbes and convey the infection to others. Diphtheria patients, for example, spread the microbe for up to 30 days after the disease has subsided.

An individual who shelters the infectious agent for a long period after recovery because of the latency of the infectious agent is a *chronic carrier.* Patients who have recovered from tuberculosis, hepatitis, and herpes infections frequently carry the agent chronically. About one in 20 victims of typhoid fever continues to harbor *Salmonella typhi* in the gallbladder for several years, and sometimes for life. The most infamous of these was Typhoid Mary, a cook who created an epidemic in the early 1900s (see microfile 20.4).

The **passive carrier** state is of great concern during patient care (see a later section on nosocomial infections). Medical and dental personnel who must constantly handle materials that are heavily contaminated with patient secretions and blood are at risk for picking up pathogens mechanically and accidentally transferring them to other patients (figure 13.18*b*). Proper handwashing, handling of contaminated materials, and aseptic techniques greatly reduce this likelihood.

Animals As Reservoirs and Sources Up to now, we have lumped animals with humans in discussing living reservoirs or carriers, but animals deserve special consideration as vectors of infections. The word **vector** is used by epidemiologists to indicate a live animal that transmits an infectious agent from one host to another. (The term is sometimes misused to include any object that spreads disease.) The majority of vectors are arthropods such as fleas, mosquitos, flies, and ticks, although larger animals can also spread infection—for example, mammals (rabies), birds (psittacosis), or lower vertebrates (salmonellosis).

By tradition, vectors are placed into one of two categories, depending upon the animal's relationship with the microbe (figure 13.19). A **biological vector** actively participates in a pathogen's life cycle, serving as a site in which it can multiply or complete its life cycle. A biological vector communicates the infectious agent to the human host by biting, aerosol formation, or touch. In the case of biting vectors, the animal can (1) inject infected saliva into the blood (the mosquito), (2) defecate around the bite wound (the flea) (figure 13.19*a*), or (3) regurgitate blood into the wound (the tsetse fly). More detailed discussions of the roles of biological vectors are found in chapters 20, 21, 23, and 25.

Mechanical vectors are not necessary to the life cycle of an infectious agent and merely transport it without being infected. The external body parts of these animals become contaminated when they come into physical contact with a source of pathogens. The agent is subsequently transferred to humans indirectly by an intermediate such as food or, occasionally, by direct contact (as in certain eye infections). House flies have habits that suit them to the role of mechanical vector (figure 13.19*b*). Their mouthparts are adapted for feeding on decaying garbage and feces, and while they are feeding their feet and mouth parts can easily become contaminated. They also regurgitate juices onto food to soften and digest it. Flies spread more than 20 bacterial, viral, protozoan, and worm infections. Other nonbiting flies transmit tropical ulcers, yaws, and trachoma (see chapter 21). Cockroaches, which have similar unsavory habits, play a role in the mechanical transmission of fecal pathogens as well as contributing to allergy attacks in asthmatic children.

Many vectors and animal reservoirs spread their own infections to humans. An infection indigenous to animals but naturally transmissible to humans is a **zoonosis.*** In these types of infections, the human is essentially a dead-end host and does not contribute to the natural persistence of the microbe. Some zoonotic infections (rabies, for instance) can have multihost involvement, and others can have very complex cycles in the wild (see plague in chapter 20). Zoonotic spread of disease is promoted by close associations of humans with animals, and

*zoonosis (zoh´´-uh-noh´-sis) Gr. *zoion,* animal, and *nosos,* disease.

TABLE 13.10

ANIMAL HOSTS AND THEIR COMMON ZOONOTIC INFECTIONS

Disease	Domestic Animals					Wild Animals			
	Cats	*Dogs*	*Cattle*	*Horses*	*Poultry*	*Arthropods*	*Birds*	*Rodents*	*Primates*
Viruses									
Rabies	+	+	+	–	–	–	–	+	+
Yellow fever	–	–	–	–	–	+	–	+	+
Viral fevers	–	–	–	–	–	+	–	+	–
Hantavirus	–	–	–	–	–	–	–	+	–
Influenza	–	–	–	–	+	–	+	–	–
Bacteria									
Q fever	–	–	+	–	–	–	+	+	–
Rocky Mountain spotted fever	–	+	–	–	–	+	–	–	–
Psittacosis	–	–	–	–	+	–	+	–	–
Leptospirosis	+	+	+	+	–	–	–	+	+
Anthrax	+	+	+	+	–	–	–	+	–
Brucellosis	–	+	+	–	–	–	–	–	–
Listeriosis	–	+	+	+	–	–	+	+	–
Plague	+	+	–	–	–	+	–	+	–
Salmonellosis	+	+	+	+	+	+	+	+	+
Tularemia	+	+	–	+	–	+	+	+	–
Miscellaneous									
Ringworm	+	+	+	+	–	–	–	+	+
Toxoplasmosis	+	–	+	–	–	–	+	+	–
Trypanosomiasis	+	+	+	–	–	+	–	+	+
Larval migrans	+	+	–	–	–	–	–	–	–
Trichinosis	–	–	–	+	–	–	–	–	–
Tapeworm	–	–	+	–	–	–	–	–	–
Scabies (mange)	+	+	+	+	+	–	–	–	–

people in animal-oriented or outdoor professions are at greatest risk. At least 150 zoonoses exist worldwide; the most common ones are listed in table 13.10. It is worth noting that zoonotic infections are impossible to completely eradicate without also eradicating the animal reservoirs. Attempts have been made to eradicate mosquitos and certain rodents, but it is inconceivable that such extreme measures would ever be tried on wild birds or mammals.

Nonliving Reservoirs

It is clear that microorganisms have adapted to nearly every habitat in the biosphere. They thrive in soil and water and often find their way into the air. Although most of these microbes are saprobic and cause little harm and considerable benefit to humans, some are opportunists and a few are regular pathogens. Because human hosts are in regular contact with these environmental sources, acquisition of pathogens from natural habitats is of diagnostic and epidemiologic importance.

Soil harbors the vegetative forms of bacteria, protozoa, helminths, and fungi, as well as their resistant or developmental stages such as spores, cysts, ova, and larvae. Regular bacterial pathogens include the anthrax bacillus and species of *Clostridium* that are responsible for gas gangrene, botulism, and tetanus. Pathogenic fungi in the genera *Coccidioides* and *Blastomyces* are spread by spores in the soil and dust. The invasive stages of the hookworm *Necator* occur in the soil. Natural bodies of water carry fewer nutrients than soil does but still support a number of pathogenic species such as *Legionella, Cryptosporidium,* and *Giardia.*

How and Why? The Acquisition and Transmission of Infectious Agents

Infectious diseases can be categorized on the basis of how they are acquired. A disease is **communicable** when an infected host can transmit the infectious agent to another host and establish infection in that host. (Although this is standard terminology, one must realize that it is not the disease that is communicated, but the microbe. Also be aware that the word *infectious* is sometimes used interchangeably with the word *communicable,* but this is not precise usage.) The transmission of the agent can be direct or indirect, and the ease with which the disease is transmitted varies considerably from one agent to another. If the agent is highly transmissible, especially through direct contact, the disease is

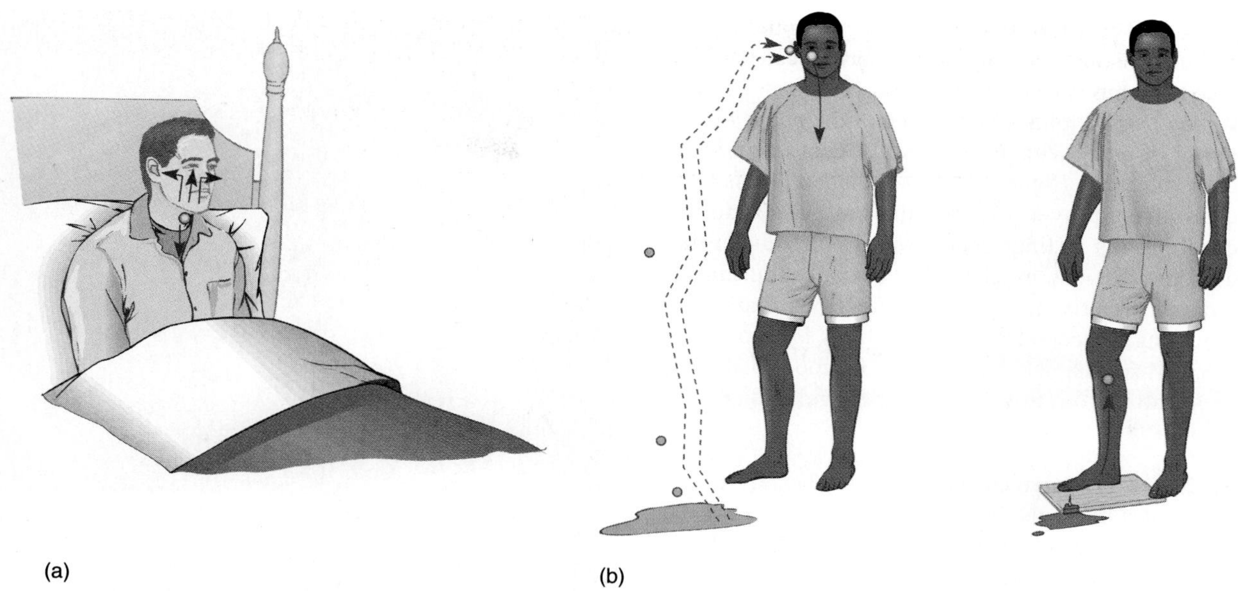

Figure 13.20
Non-communicable infections can be acquired (*a*) from normal flora and (*b*) from contact with microbes that live in the soil and water. Note that these infections are not acquired from another infected person, either directly or indirectly.

contagious. Influenza and measles move readily from host to host and thus are contagious, whereas leprosy is only weakly communicable. Because they can be spread through the population, communicable diseases will be our main focus in the following sections.

In contrast, a **non-communicable** infectious disease does *not* arise through transmission of the infectious agent from host to host. The infection and disease are acquired through some other, special circumstance. Non-communicable infections occur primarily when a compromised person is invaded by his own microflora (as with certain pneumonias, for example) or when an individual has accidental contact with a facultative parasite that exists in a nonliving reservoir such as soil (figure 13.20). Some examples are certain mycoses acquired through inhalation of fungal spores and tetanus, in which *Clostridium tetani* spores from a soiled object enter a cut or wound. Persons thus infected do not become a source of disease to others.

Patterns of Transmission in Communicable Diseases

The routes or patterns of disease transmission are many and varied. The spread of diseases is by direct or indirect contact with animate or inanimate objects and can be horizontal or vertical. The term *horizontal* means the disease is spread through a population from one infected individual to another; *vertical* signifies transmission from parent to offspring via the ovum, sperm, placenta, or milk. The extreme complexity of transmission patterns among microorganisms makes it very difficult to generalize. However, for easier organization, we will divide microorganisms into two major groups, as shown in figure 13.21: transmission by direct contact or transmission by indirect routes (with the latter category divided into vehicle and airborne transmission).

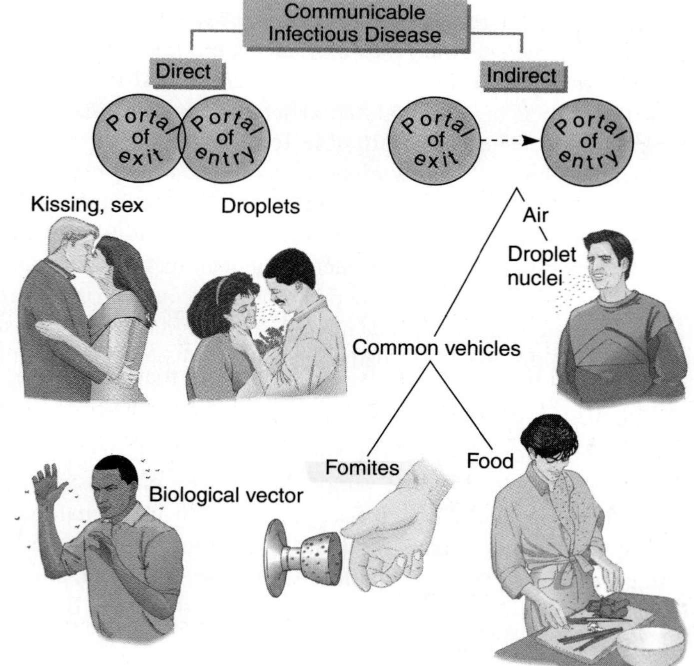

Figure 13.21
Summary of how communicable infectious diseases are acquired. (*a*) Direct mechanisms of transmission involve physical contact between two hosts. (*b*) Indirect modes of transmission require some object or material to transfer the infectious agent between hosts.

Modes of Direct Transmission In order for microbes to be directly transferred, some type of contact must occur between the skin or mucous membranes of the infected person and that of the infectee. It may help to think of this route as the portal of exit meeting the portal of entry without the involvement of an intermediate object or substance (figure 13.21*a*). Included in this category are fine droplets sprayed directly upon a person during sneezing or coughing (as distinguished from droplet nuclei that are transmitted some distance by air). Most sexually transmitted diseases are spread directly. In addition, infections that result from kissing, nursing, placental transfer, or bites by biological vectors are direct. Most obligate parasites are far too sensitive to survive for long outside the host and can be transmitted only through direct contact.

Routes of Indirect Transmission For microbes to be considered indirectly transmitted, the infectious agent must pass from an infected host to an intermediate conveyor and from there to another host. This form of communication is especially pronounced when the infected individual contaminates inanimate objects, food, or air through his activities (figure 13.21*b*). The transmitter of the infectious agent can be either openly infected or a carrier.

Indirect Spread by Vehicles: Contaminated Materials The term **vehicle** specifies any inanimate material commonly used by humans that can transmit infectious agents. A *common vehicle* is a single material that serves as the source of infection for many individuals. Some specific types of vehicles are food, water, various biological products (such as blood, serum, and tissue), and fomites. A **fomite*** is an inanimate object that harbors and transmits pathogens. The list of possible fomites is as long as your imagination allows. Probably highest on the list would be objects commonly in contact with the public such as doorknobs, telephones, push buttons, and faucet handles that are readily contaminated by touching. Shared bed linens, handkerchiefs, toilet seats, toys, eating utensils, clothing, personal articles, and syringes are other examples. Although paper money is impregnated with a disinfectant to inhibit microbes, pathogens are still isolated from bills as well as coins.

Outbreaks of food poisoning often result from the role of food as a common vehicle. The source of the agent can be soil, the handler, or a mechanical vector. In the type of transmission termed the *oral-fecal route,* a fecal carrier with inadequate personal hygiene contaminates food during handling, and an unsuspecting person ingests it. Hepatitis A, amebic dysentery, shigellosis, and typhoid fever are often transmitted this way. Because milk provides a rich growth medium for microbes, it is a significant means of transmitting pathogens from diseased animals, infected milk handlers, and environmental sources of contamination. The agents of brucellosis, tuberculosis, Q fever, salmonellosis, and listeriosis are transmitted by contaminated milk. Water that has been contaminated by feces or urine can carry pathogens such as *Salmonella, Vibrio* (cholera) and viruses (hepatitis A, polio), and pathogenic protozoans (*Giardia, Cryptosporidium*).

*fomite (foh´-myt) L. *fomes,* tinder.

Figure 13.22

The explosiveness of a sneeze. Special photography dramatically captures droplet formation in an unstifled sneeze. Even the merest attempt to cover a sneeze with one's hand will reduce this effect considerably. When such droplets dry and remain suspended in air, they are droplet nuclei.

Indirect Spread by Airborne Route: Droplet Nuclei and Aerosols Unlike soil and water, outdoor air cannot provide nutritional support for microbial growth and seldom transmits airborne pathogens. On the other hand, indoor air (especially in a closed space) can serve as an important medium for the suspension and dispersal of certain respiratory pathogens via droplet nuclei and aerosols. **Droplet nuclei** are dried microscopic residues created when microscopic pellets of mucus and saliva are ejected from the mouth and nose. They are generated forcefully in an unstifled sneeze or cough (figure 13.22) or mildly during other vocalizations. Although the larger beads of moisture settle rapidly, smaller particles evaporate and remain suspended for longer periods. Droplet nuclei are implicated in the spread of hardier pathogens such as the tubercle bacillus and the influenza virus. **Aerosols** are suspensions of fine dust or moisture particles in the air that contain live pathogens. Q fever is spread by dust from animal quarters and psittacosis by aerosols from infected birds. An unusual outbreak of coccidioidomycosis (a lung infection) occurred during the 1994 southern California earthquake. Epidemiologists speculate that disturbed hillsides and soil gave off clouds of dust containing the spores of *Coccidioides.*

Nosocomial Infections: The Hospital As a Source of Disease

Infectious diseases acquired as a result of a hospital stay are known as **nosocomial*** infections. This concept seems incongruous at first thought, because a hospital is regarded as a place to

*nosocomial (nohz´´-oh-koh´-mee-al) Gr. *nosos,* disease, and *komeion,* to take care of. Originating from a hospital or infirmary.

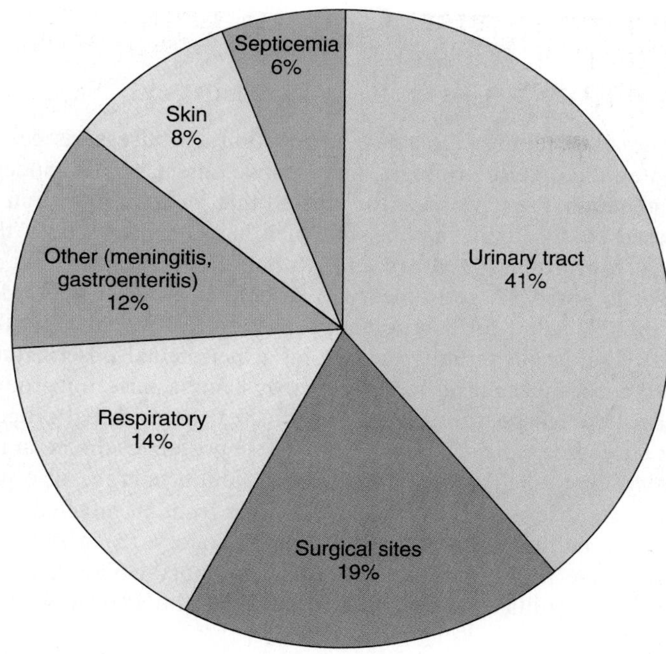

Figure 13.23
Most common nosocomial infections. Relative frequency by body site.

get treatment for a disease, not a place to acquire a disease. Yet it is not uncommon for a surgical patient's incision to become infected or a burn patient to develop a case of pneumonia in the clinical setting. The rate of nosocomial infections can be as low as 0.1% or as high as 20% of all admitted patients at any one time, but the average is about 5%. In light of the number of admissions, this amounts to from 2 to 4 million cases a year, which result in 20,000 to 40,000 deaths. Nosocomial infections cost time and money. By one estimate, they amount to 8 million days of hospitalization a year at an additional cost of $4 to $4.5 billion.

So many factors unique to the hospital environment are tied to nosocomial infections that a certain number of infections are virtually unavoidable. After all, the hospital both attracts and creates compromised patients, and it serves as a collection point for pathogens. Some patients become infected when surgical procedures or lowered defenses permit resident flora to invade their bodies. Other patients acquire infections directly or indirectly from fomites, medical equipment, other patients, medical personnel, visitors, air, and water. The health care process itself increases the likelihood that infectious agents will be transferred from one patient to another. Treatments using reusable instruments such as respirators and thermometers constitute a possible source of infectious agents. Indwelling devices such as catheters, prosthetic heart valves, grafts, drainage tubes, and tracheostomy tubes form a ready portal of entry and habitat for infectious agents. An additional problem is the tendency for drug-resistant strains of microorganisms to develop in hospitals, thereby further complicating treatment.

The most common nosocomial infections involve the urinary tract, the respiratory tract, and surgical incisions (figure 13.23). Gram-negative intestinal flora (*Escherichia coli, Klebsiella, Pseudomonas*) are cultured in more than half of patients with nosocomial infections (see figure 20.7). Gram-positive bacteria (staphylococci and streptococci) and yeasts make up most of the remainder. True pathogens such as *Mycobacterium tuberculosis, Salmonella,* hepatitis B, and influenza virus can be transmitted in the clinical setting as well.

The potential seriousness and impact of nosocomial infections have required hospitals to develop committees that monitor infectious outbreaks and develop guidelines for infection control and aseptic procedures. *Surgical asepsis* involves a high level of disinfection, antisepsis, and sterilization as would be required in surgery. Instruments, dressings, sponges, and all other supplies coming into contact with the patient are sterilized. Personnel must be fully covered in sterile garments, and room surfaces and air must be thoroughly disinfected. *Medical asepsis* includes any practice that lowers the load of infectious microbes in patients, personnel, and the hospital environment. Included are handwashing, decontamination procedures, and *isolation* of patients. In isolation, various barriers (gloves, mask, gown) and other techniques are used in the patient's room to prevent the entry or exit of infectious agents on health care workers, visitors, and into the surroundings. The aim of four categories of isolation is to contain the spread of infectious agents from infected patients or carriers, and one category (reverse isolation) prevents highly compromised patients from coming into contact with pathogens from the outside environment. Table 13.11 summarizes guidelines for the major types of isolation.

An essential member of the infection control team is the *infection control officer.* This person oversees all hospital personnel and procedures to minimize the spread of infection. The infection control officer's responsibilities include keeping track of infections in the wards, determining possible epidemics, relaying this information to the rest of the team, seeking out breaches in asepsis in nursing care or surgery, and training others in aseptic techniques.

Control procedures critical to reducing nosocomial infections are proper handwashing and surgical scrub techniques; proper disposal procedures for contaminated substances and blood; patient isolation; antibiotic prophylaxis; strict sterilization, disinfection, and sanitization procedures; restricting infected personnel; and immunizing personnel.

Another high-risk group for infections in the hospital are patient caretakers. The very nature of their work exposes them to needlesticks (a type of inoculation infection), infectious secretions and blood, and physical contact with the patient. The same practices that interrupt the routes of infection in the patient can also protect the health worker. It is for this reason that most hospitals have adopted universal precautions that recognize that all secretions from all persons in the clinical setting are potentially infectious and that transmission can occur in either direction. See appendix D for a full description of these precautions.

TABLE 13.11

LEVELS OF ISOLATION USED IN CLINICAL SETTINGS

Type of Isolation*	Protective Measures**	To Prevent Spread Of
Enteric precautions	Gowns and gloves must be worn by all persons having direct contact with patient; masks not required; special precautions taken for disposing of feces and urine	Diarrheal diseases; *Shigella, Salmonella,* and *Escherichia coli* gastroenteritis; cholera; hepatitis A; rotavirus; and giardiasis
Respiratory precautions	Private room with closed door is necessary; gowns and gloves not required; masks usually indicated; items contaminated with secretions must be disinfected	Tuberculosis, measles, mumps, meningitis, pertussis, rubella, chickenpox
Drainage and secretion precautions	Gowns and gloves required for all persons; masks not needed; contaminated instruments and dressings require special precautions	Staphylococcal and streptococcal infections; gas gangrene; herpes zoster; burn infections
Strict isolation	Private room with closed door required; gowns, masks, and gloves must be worn by all persons; contaminated items must be wrapped and sent to central supply for decontamination	Mostly highly virulent or contagious microbes; includes diphtheria, anthrax, some types of pneumonia, extensive skin and burn infections, disseminated herpes simplex and zoster
Reverse isolation (also called protective isolation)	Same guidelines as for strict isolation; room may be ventilated by unidirectional or laminar airflow filtered through a high-efficiency particulate air (HEPA) filter that removes most airborne pathogens; infected persons must be barred	Used to protect patients extremely immunocompro- mised by cancer therapy, surgery, genetic defects, burns, prematurity, or AIDS and therefore vulnerable to opportunistic pathogens

*Precautions are based upon the primary portal of entry and communicability of the pathogen.

**In all cases, visitors to the patient's room must report to the nurses' station before entering the room; all visitors and personnel must wash their hands upon entering and leaving the room.

WHAT IS THE CAUSE? USING KOCH'S POSTULATES TO DETERMINE ETIOLOGY

An essential aim in the study of infection and disease is determining the precise **etiologic,** or causative, agent. In our modern technological age, we take for granted that a certain infection is caused by a certain microbe, but such has not always been the case. More than a century ago, Robert Koch realized that in order to prove the germ theory of disease he would have to develop a standard for determining causation that would stand the test of scientific scrutiny. Out of his experimental observations on the transmission of anthrax in cows came a series of proofs, called **Koch's postulates,** that established the principal criteria for etiologic studies (figure 13.24). These postulates direct an investigator to (1) find evidence of a particular microbe in every case of a disease, (2) isolate that microbe from an infected subject and cultivate it artificially in the laboratory, (3) inoculate a susceptible healthy subject with the laboratory isolate and observe the resultant disease, and (4) reisolate the agent from this subject.

Valid application of Koch's postulates requires attention to several critical details. Each isolated culture must be pure, observed microscopically, and identified by means of characteristic tests; the first and second isolate must be identical; and the pathologic effects, signs, and symptoms of the disease in the first and second subject must be the same. Once established, these postulates were rapidly put to the test, and within a short time, they had helped determine the causative agents of tuberculosis, diphtheria, and plague. Today, most infectious diseases have been directly linked to a known infectious agent.

Koch's postulates continue to play an essential role in modern epidemiology. Every decade, new diseases challenge the scientific community and require application of the postulates. Prominent examples are toxic shock syndrome, AIDS, Lyme disease, and Legionnaires' disease (named for the American Legion members who first contracted a mysterious lung infection in Philadelphia).

Koch's postulates are reliable for most infectious diseases, but they cannot be completely fulfilled in certain situations. For example, some infectious agents are not readily isolated or grown in the laboratory. If one cannot elicit a similar infection by inoculating it into an animal, it is very difficult to prove the etiology. In the past, scientists have attempted to circumvent this problem by using human subjects (microfile 13.8).

A small but vocal group of critics has claimed that the postulates have not been adequately carried out for AIDS and thus, that HIV cannot be claimed as the causative agent, despite the fact that there is overwhelming evidence from observing infected humans and primates that it is. Cases of accidental infection through exposure to blood have yielded much proof. In all cases, subjects developed an early virus syndrome, carried high virus levels, and later developed AIDS. One study with baboons revealed that they do indeed develop symptoms of AIDS when infected with a variant of HIV.

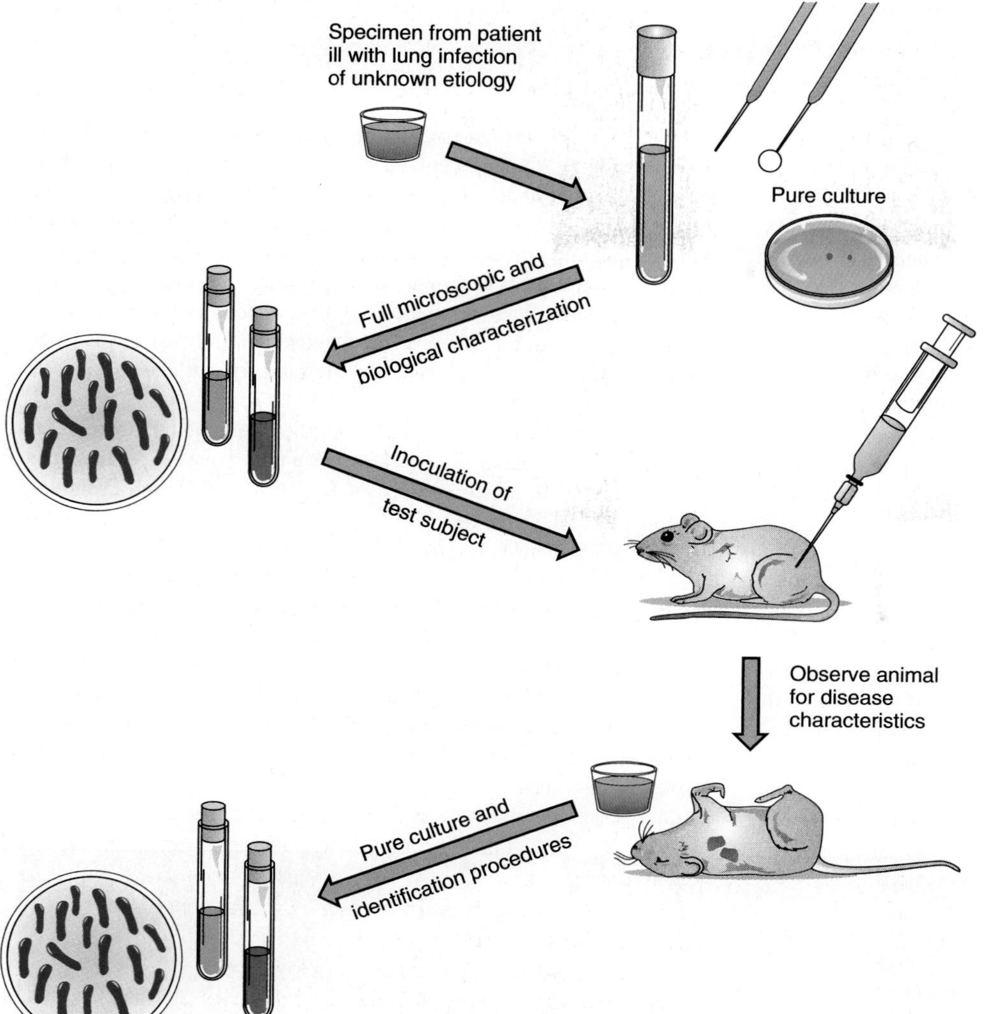

Figure 13.24
Koch's postulates: Is this the etiologic agent? The microbe in the initial and second isolations and the disease in the experimental animal must be identical for the postulates to be satisfied.

Chapter Checkpoints

Epidemiology is the study of the determinants and distribution of all diseases in populations. The study of infectious disease in populations is just one aspect of this field.

Data on specific, reportable diseases is collected by local, national, and worldwide agencies.

The *prevalence* of a disease is the percentage of existing cases in a given population. The disease *incidence,* or *morbidity rate,* is the ratio of newly infected to uninfected members of a population.

The disease frequency is described as sporadic, epidemic, pandemic, or endemic.

The primary habitat of a pathogen is called its reservoir. A human reservoir is also called a carrier.

Animals can be either reservoirs or vectors of pathogens. An infected animal is a biological vector. Uninfected animals, especially insects, that transmit pathogens mechanically are called mechanical vectors.

Soil and water are nonliving reservoirs for pathogenic bacteria, protozoa, fungi, and worms.

A communicable disease can be transmitted from an infected host to others, but not all infectious diseases are communicable.

The spread of infectious disease from person to person is called horizontal transmission. The spread of infectious disease from parent to offspring is called vertical transmission.

Infectious diseases are spread by either direct or indirect routes of transmission. Vehicles of indirect transmission include soil, water, food, droplet nuclei, and fomites (inanimate objects).

Nosocomial infections are acquired in a hospital from surgical procedures, equipment, personnel, and exposure to drug-resistant microorganisms.

Causative agents of infectious disease must be isolated and identified according to Koch's postulates.

MICROFILE 13.8 THE HISTORY OF HUMAN GUINEA PIGS

In this day and age, human beings are not used as subjects for determining the cause of infectious disease, but in earlier times they were. A long tradition of human experimentation dates well back into the eighteenth century, with the subject frequently the experimenter himself. In some studies, mycologists inoculated their own skin and even that of family members with scrapings from fungal lesions to demonstrate that the disease was transmissible. In the early days of parasitology, it was not uncommon for a brave researcher to swallow worm eggs in order to study the course of his infection and the life cycle of the worm.

In a sort of reverse test, a German colleague of Koch's named Max von Petenkofer believed so strongly that cholera was *not* caused by a bacterium that he and his assistant swallowed cultures of the vibrio. Fortunately for them, they did not acquire serious infection. Many self-experimenters have not been so fortunate.

One of the most famous cases is that of Jesse Lazear, a Cuban physician who worked with Walter Reed on the etiology of yellow fever in 1900. Dr. Lazear was convinced that mosquitos were directly involved in the spread of yellow fever, and, by way of proof, he allowed himself and two volunteers to be bitten by mosquitos infected with the blood of yellow fever patients. Although all three became ill as a result of this exposure, Dr. Lazear's sacrifice was the ultimate one—he died of yellow fever. Years later, paid volunteers were used to completely fulfill the postulates.

Dr. Lazear was not the first martyr in this type of cause. Fifteen years previously, a young Peruvian medical student, Daniel Carrion, attempted to prove that a severe blood infection, Oroya fever, had the same etiology as *verucca peruana,* an ancient disfiguring skin disease. After inoculating himself with fluid from a skin lesion, he developed the severe form and died, becoming a national hero. In his honor, the disease now identified as bartonellosis is also sometimes called Carrion's disease. Eventually, the microbe (*Bartonella bacilliformis*) was isolated, a monkey model was developed, and the sand fly was shown to be the vector for this disease.

For many years, syphilis and gonorrhea were thought to be different stages of the same disease because of an unfortunate experiment (see microfile 18.3). In an ironic twist of fate, Fritz Schaudinn eventually proved the causation of syphilis, only to die later from amebic dysentry—which he had acquired from self-experimentation.

The incentive for self-experimentation has continued, but present-day researchers are more likely to test experimental vaccines than dangerous microbes. Jonas Salk injected himself with his polio vaccine before allowing it to be used on others. Recently, a group of physicians, researchers, and AIDS activists have enthusiastically volunteered to be injected by a live, weakened AIDS virus to speed the process of vaccine development.

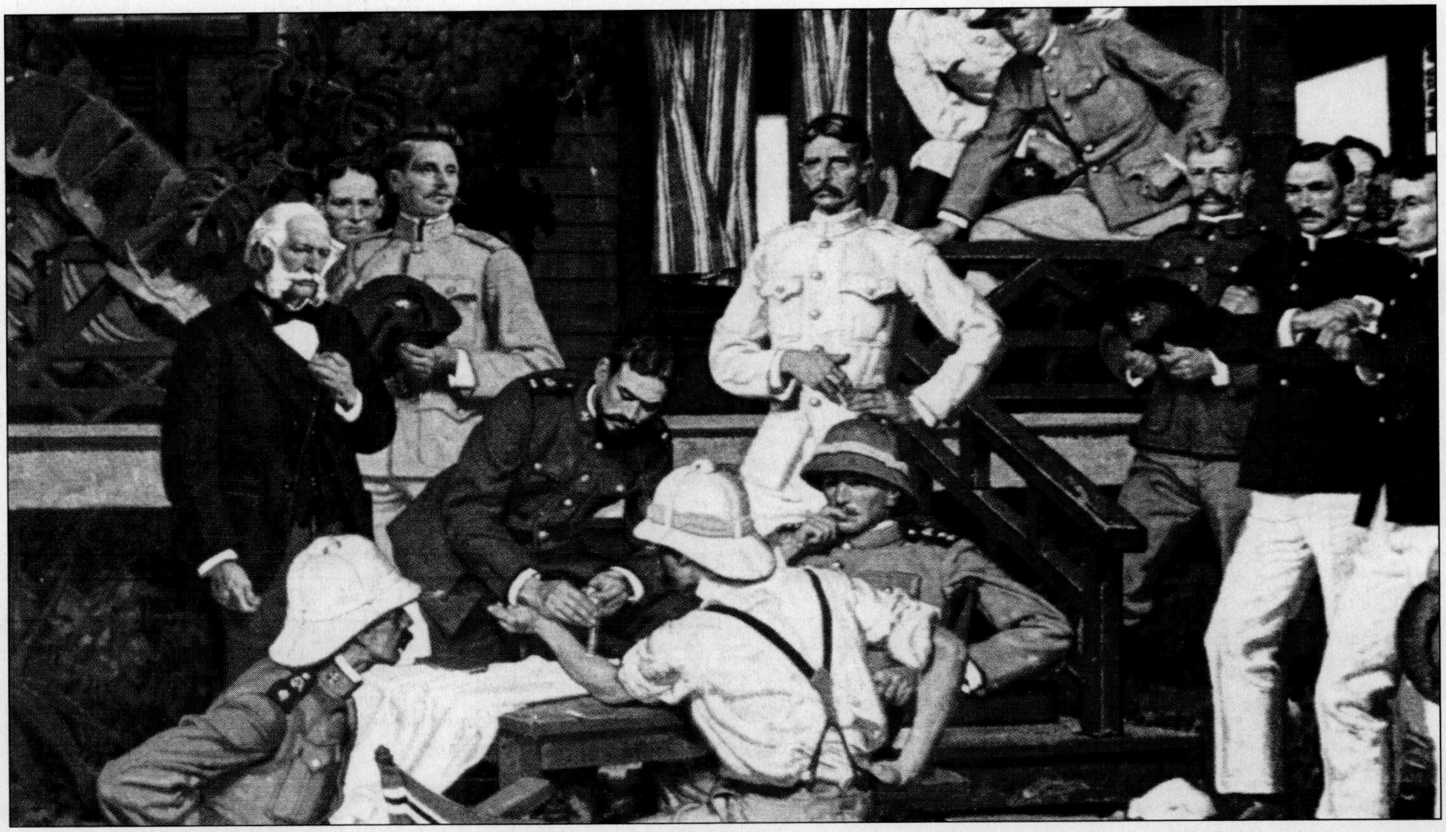

Painting of Dr. Jesse Lazear exposing the arm of James Carroll to a mosquito infected with the yellow fever virus. Human volunteers contributed significantly to our understanding of the transmission of yellow fever and many other diseases.

CHAPTER CAPSULE WITH KEY TERMS

CONTACT-INFECTION-DISEASE: THE HOST-PARASITE RELATIONSHIP

The human body is constantly in contact (contaminated) with microbes. Some are pathogens that may cause an **infection** by circumventing the host defense system, entering normally sterile tissues, and multiplying there. When infections lead to a disruption in tissues, **infectious disease** results. The outcome is highly variable, but most contacts do not result in infection, and most infections do not lead to disease.

THE BODY AS A HABITAT

The **resident flora** or **microflora** is a huge and rich mixed population of microorganisms residing on body surfaces exposed to the environment, including the skin, mucous membranes, parts of the gastrointestinal tract, urinary tract, reproductive tract, and upper respiratory tract. Anatomical sites lying within the body cavity (organs) and fluids (blood, urine) in those sites do not harbor flora.

Colonization: Begins just prior to birth and continues over an individual's life; variations occur in response to individual differences in age, diet, hygiene, and health.

Role of Flora: Bacteria may maintain a balance in the normal conditions; some produce vitamins; studies with **axenic** animals (free of any normal flora) show that flora contribute to the development of the immune and gastrointestinal systems and also to some diseases (dental caries). Normal flora are sometimes agents of infection.

FACTORS AFFECTING THE COURSE OF INFECTION AND DISEASE

Pathogenicity and Virulence: **Pathogenicity** is the property of microorganisms to cause infection and disease. **Virulence** is the precise means by which the microbe invades and damages host tissues; it helps define the degree of pathogenicity. Pathogenicity varies with a microbe's ability to invade or harm host tissues and with the condition of host defenses. A **true pathogen** produces **virulence factors** that allow it to readily evade host defenses and to harm host tissues. True pathogens can infect normal, healthy hosts with intact defenses. An **opportunistic pathogen** is not highly virulent but can invade and cause disease in persons whose host defenses are compromised by *predisposing conditions* such as age, genetic defects, medical procedures, and underlying organic disease.

Mechanisms of Infection and Disease: Entry, adherence, invasion, multiplication, and disruption of target tissues. The **portal of entry** is the route by which microbes enter the tissues, primarily via skin, alimentary tract, respiratory tract **(pneumonia),** urogenital tract **(sexually transmitted diseases),** or placenta. Pathogens that come from outside the body are **exogenous;** those that originate from normal flora are **endogenous.** The size of the *infectious dose* is of great importance. In the process of **adhesion,** a microbe attaches to the host cell by means of fimbriae, flagella, capsule, or receptors; this puts it in advantageous position for invasion.

Virulence Factors: Enzymes, **toxins, antiphagocytic factors. Exoenzymes** digest host epithelial tissues; disrupt tissues; aid invasion. **Toxigenicity** is a microbe's capacity to produce toxins at site of multiplication; may affect local or distant targets. **Toxinoses** are diseases caused by toxins that damage structure or function of host cells; **toxemia** refers to toxins absorbed into the blood; **intoxication** means ingestion of toxins. An **exotoxin** is a protein secreted by living bacteria with powerful effects on a specific organ. Examples are **hemolysins** and tetanus and diphtheria toxins. An **endotoxin** is the lipopolysaccharide portion of a gram-negative cell wall released when a bacterial cell dies; more generalized and weaker in its toxicity; one cause of fever. Antiphagocytic chemicals include leukocidins (white blood cell poisons) and capsules.

EFFECTS ON TARGET ORGAN/SPREAD OF INFECTION

Patterns of Infection: Stages in infection/disease are **incubation period,** the period from contact with infectious agent until appearance of first symptoms; **prodromium,** a short period of initial, vague symptoms; **period of invasion,** a variable period during which microbe multiplies in high numbers and causes severest symptoms; and **convalescent period,** a period of recovery, with decline of symptoms.

Types of Infections/Diseases: **Localized infection,** microbe remains in isolated site; **systemic infection,** microbe is spread through the tissues by circulation; **focal infection,** microbe spreads from local site to entire body (systemic); **mixed infection,** several microbes cause one type of infection simultaneously; **primary infection,** the initial infection in a series; **secondary infection,** a second infection that complicates a primary infection; **septicemia** and **bacteremia** refer to microbes in the blood; **acute infection** appears suddenly, has a short course, is relatively severe; **chronic infection** persists over a long period of time; **subacute infection,** has a pattern between acute and chronic.

Signs and Symptoms: Manifestations of disease, indicators of **pathologic** effects on target organs. A **sign** is objective, measurable evidence noted by an observer. Examples include septicemia, change in number of white blood cells; skin **lesions;** inflammation; **necrosis,** lysis or death of tissue. A **symptom** is a subjective effect of disease as sensed by patient. Examples are pain, fatigue, and nausea. A **syndrome** is a disease that manifests as a predictable complex of symptoms; infections that do not show symptoms are called *asymptomatic, subclinical,* or *inapparent.* Through the **portal of exit,** microbe is released with bodily secretions and discharges so that it may have access to new host; portals include respiratory droplets from sneezing, coughing; saliva; skin; feces; urogenital tract (urine, mucus, semen); blood. A microbe may become dormant **(latent)** and cause recurrent infections. Damaging effects that remain in organs and tissues after infection are **sequelae.**

EPIDEMIOLOGY

Epidemiology is a science that determines the factors influencing causation, frequency, and distribution of disease in a community; epidemiologists are involved in **surveillance** of reportable diseases in populations; consider measures to protect the public health; are concerned with disease statistics such as **prevalence** (the total number of cases), **incidence** (the number of new cases), **morbidity** (general health of the population), and **mortality** (death).

Frequency of Disease: **Endemic,** a disease constantly present in a certain geographic area; **sporadic,** disease occurs occasionally with no predictable pattern; **epidemic,** sudden outbreak of disease in which numbers increase beyond expected trends; **pandemic,** worldwide epidemic.

ORIGIN OF PATHOGENS

The **reservoir** is a place where the pathogen ultimately originates (its habitat); **source** of infection refers to the immediate origin of an infectious agent; **carrier** is an individual that inconspicuously shelters a pathogen and spreads it to others; **asymptomatic carrier** is infected without symptoms. **Incubation carriers** carry early in disease; **convalescent carriers** carry in last phases of recovery; **chronic carriers** carry for long periods after recovery; **passive carriers** are uninfected but convey infectious agents from infected persons to uninfected ones by hand and instrument contact.

VECTORS/ZOONOSES

A **vector** is an animal that transmits pathogens; **biological vector,** an alternate animal host (mosquito, flea) that assists in completion of life cycle of microbe; **mechanical vector,** animal not host in microbial life cycle, transmits by contaminated body parts (house fly); **zoonosis,** an infection for which animals are natural reservoir and host; can be transmitted to humans.

ACQUISITION OF INFECTION

Communicable infectious disease occurs when pathogen is transmitted from host to host directly or indirectly; **contagious diseases** are readily transmissible through direct contact; **non-communicable diseases** are not spread from host to host; acquired from one's own flora (pneumonia) or from a nonliving environmental reservoir (tetanus).

Direct Transmission: Infectious agent is spread through direct contact of portal of exit with portal of entry (STDs, herpes simplex).

Indirect Transmission: A material **(vehicle)** contaminated with pathogens serves as intermediate source of infection; **fomite,** inanimate object contaminated with pathogens (public facilities, personal items); food may be a vehicle; **droplet nuclei,** airborne dried particles containing infectious agents, formed by sneezing, coughing.

Nosocomial Infections: Infectious diseases that originate in the hospital or clinical setting. Common among surgical and chronically ill patients; hospitals monitor various asepsis procedures to help reduce the number of infections; isolation of patients and other universal precautions are necessary controls.

Koch's Postulates: A series of criteria that must be followed to determine the **etiologic** (causative) agent of disease.

MULTIPLE-CHOICE QUESTIONS

1. The best descriptive term for the resident flora is
   a. commensals c. pathogens
   b. parasites d. mutualists
2. Resident flora is commonly found in the
   a. stomach c. salivary glands
   b. kidney d. urethra
3. Resident flora is absent from the
   a. pharynx c. intestine
   b. lungs d. hair follicles
4. Virulence factors include
   a. toxins c. capsules
   b. enzymes d. all of these
5. The specific action of hemolysins is to
   a. damage white blood cells
   b. cause fever
   c. damage red blood cells
   d. cause leukocytosis
6. The ____ is the time that lapses between encounter with a pathogen and the first symptoms.
   a. prodromium
   b. period of invasion
   c. period of convalescence
   d. period of incubation
7. A short period early in a disease that manifests with general malaise and achiness is the
   a. period of incubation
   b. prodromium
   c. sequela
   d. period of invasion
8. The presence of a few bacteria in the blood is termed
   a. septicemia c. bacteremia
   b. toxemia d. a secondary infection
9. A ____ infection is acquired in a hospital.
   a. subclinical c. nosocomial
   b. focal d. zoonosis
10. A/an ____ is a passive animal transporter of pathogens.
    a. zoonosis
    b. biological vector
    c. mechanical vector
    d. asymptomatic carrier
11. An example of a non-communicable infection is:
    a. measles c. pneumonia
    b. leprosy d. tetanus
12. A general term that refers to an increased white blood cell count is
    a. leukopenia c. leukocytosis
    b. inflammation d. leukemia
13. A positive antibody test for HIV would be a ____ of infection
    a. sign c. syndrome
    b. symptom d. sequela
14. Which of the following would *not* be a portal of entry?
    a. the meninges c. skin
    b. the placenta d. small intestine
15. Which of the following is *not* a condition of Koch's postulates?
    a. isolate the causative agent of a disease
    b. cultivate the microbe in a lab
    c. inoculate a test animal to observe the disease
    d. test the effects of pathogen on humans

CONCEPT QUESTIONS

1. Differentiate between contamination, infection, and disease. What are the possible outcomes in each?
2. How are infectious diseases different from other diseases?
3. Name the general body areas that are sterile. Why is the inside of the intestine not sterile like many other organs?
4. What causes variations in the flora of the newborn intestine?
5. What factors influence how the flora of the vagina develops?
6. Why must axenic young be delivered by cesarian section?
7. Explain several ways that true pathogens differ from opportunistic pathogens.
8. a. Distinguish between pathogenicity and virulence.
   b. Define virulence factors, and give examples of them in gram-positive and gram-negative bacteria, viruses, and parasites.
9. Describe the course of infection from contact with the pathogen to its exit from the host.

10. a. Explain why most microbes are limited to a single portal of entry.
    b. For each portal of entry, give a vehicle that carries the pathogen and the course it must travel to invade the tissues.
    c. Explain how the portal of entry could differ from the site of infection.
11. Differentiate between exogenous and endogenous infections.
12. a. What factors possibly affect the size of the infectious dose?
    b. Name five factors involved in microbial adhesion.
13. Which body cells or tissues are affected by hemolysins, leukocidins, hyaluronidase, kinases, tetanus toxin, pertussis toxin, and enterotoxin?
14. Compare and contrast: systemic versus local infections; primary versus secondary infections; infection versus intoxication.
15. What is the difference between signs and symptoms? (First put yourself in the place of a patient with an infection and then in the place of a physician examining you. Describe what you would feel and what the physician would detect upon examining the affected area.)
16. a. What are some important considerations about the portal of exit?
    b. Name some examples of infections and their portals of exit.
17. Complete the table:

	Exotoxins	Endotoxins
Chemical make-up		
General source		
Degree of toxicity		
Effects on cells		
Symptoms in disease		
Examples		

18. a. Outline the science of epidemiology and the work of an epidemiologist.
    b. Using the following statistics, based on number of reported cases, can you determine which show endemic, sporadic, or epidemic patterns? How can you determine each type?

United States Region	Disease Statistics	
Meningitis	**1995**	**1997**
Northeast	131	182
East Coast	824	854
Central	761	897
Southeast	213	294
West	810	822
Total cases	2739	3049
Leprosy (Hansen's disease)		
Northeast	4	0
East Coast	18	10
Central	7	8
Southeast	22	16
West	78	91
Total cases	129	125
Cholera		
Northeast	2	1
East Coast	9	2
Central	6	2
Southeast	4	1
West	16	4
Total cases	37	10

    c. Explain what would have to occur for these diseases to have a pandemic distribution.
19. Distinguish between mechanical and biological vectors, giving one example of each.
20. a. Explain what it means to be a carrier of infectious disease.
    b. Describe four ways that humans can be carriers.
    c. What is epidemiologically and medically important about carriers in the population?
21. a. Explain the precise difference between communicable and non-communicable infectious diseases.
    b. Between direct and indirect modes of transmission.
    c. Between vectors and vehicles as modes of transmission.
22. a. Nosocomial infections can arise from what two general sources?
    b. From this chapter and figure 20.7 outline the major agents involved in nosocomial infections.
    c. Do they tend to be true pathogens or opportunists?
    d. Outline the two types of hospital asepsis and define isolation.
    e. What is the work of an infection control officer?
23. a. List the main features of Koch's postulates.
    b. Why is it so difficult to prove them for some diseases?

CRITICAL-THINKING QUESTIONS

1. a. Discuss the relationship between the vaginal residents and the colonization of the newborn.
   b. Can you think of some serious medical consequences of this relationship?
   c. Why would normal flora cause some infections to be more severe and other infections to be less severe?
2. If the following patient specimens produced positive cultures when inoculated and grown on appropriate media, indicate whether this result indicates a disease state and why or why not:

Urine	Throat
Lung biopsy	Feces
Saliva	Blood
Cerebrospinal fluid	Urine from bladder
Liver biopsy	Semen

   What are the important clinical implications of positive blood or cerebrospinal fluid?
3. Pretend that you have been given the job of developing a colony of germ-free cockroaches.
   a. What will be the main steps in this process?
   b. What possible experiments can you do with these animals?
4. If healthy persons are resistant to infection with opportunists or weak pathogens, what is the expected result if a compromised person is exposed to a true pathogen?

5. Explain how the endotoxin gets into the bloodstream of a patient with endotoxic shock.
6. You are a physician following the course of an infection in a small child who has been exposed to scarlet fever. Describe the main events that occur, what is happening during each stage, and the causes of each sign and symptom.
7. Describe each of the following infections using correct technical terminology. (Descriptions may fit more than one category.)

   Caused by needlestick in dental office

   Pneumocystis pneumonia in AIDS patient

   Bubonic plague from rat flea bite

   Diphtheria

   Acute necrotizing gingivitis

   Syphilis of long duration

   Large numbers of gram-negative rods in the blood

   A boil on the back of the neck

   An inflammation of the meninges

   Scarlet fever
8. a. Using figure 13.16, determine the years of pertussis outbreaks or epidemics from 1957 to 1997.
   b. Do the epidemics occur at regular intervals?
   c. Account for the dramatic decrease in cases of whooping cough beginning in 1960.
   d. Which age group is most affected by this disease?
9. Name 10 fomites that you came into contact with today.
10. Describe what parts of Koch's postulates were unfulfilled by Dr. Lazear's experiment described in microfile 13.8.
11. a. Suggest several reasons why urinary tract, respiratory tract, and surgical infections are the most common nosocomial infections.
    b. Name several measures that health care providers must exercise at all times to prevent or reduce nosocomial infections.

INTERNET SEARCH TOPIC

Use an Internet search engine to look up the following topics:

Pathogens on money

Tuskeegee syphilis experiment

Koch's postulates verified for AIDS and HIV

Write a short summary of what you discover on your searches.

chapter 14

The Nature of Host Defenses

The survival of the host depends upon an elaborate network of defenses that keeps harmful microbes and other foreign materials from penetrating the body. Should they penetrate, additional host defenses are summoned to prevent them from becoming established in tissues. Defenses involve barriers, cells, and chemicals, and they range from nonspecific to specific and from inborn to acquired. This chapter introduces the main lines of defense intrinsic to all humans. Topics included in this survey are the anatomical and physiological systems that detect, recognize, and destroy foreign substances and the general adaptive responses that account for an individual's long-term immunity or resistance to infection and disease.

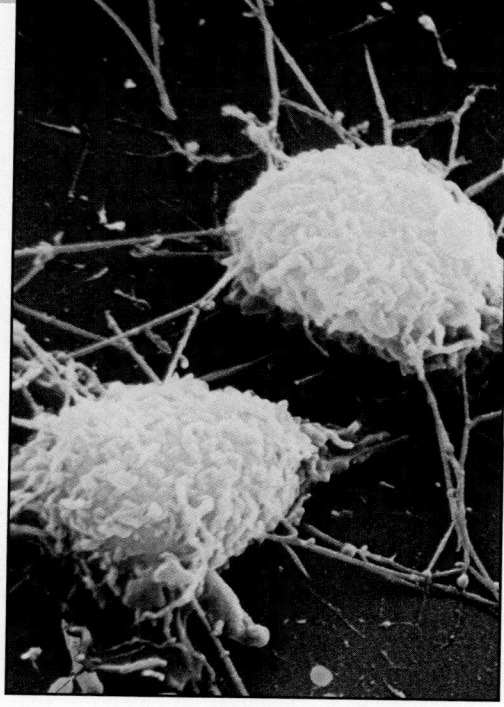

Like an octopus, a macrophage sends out long, tentacled pseudopods to capture its *Escherichia coli* prey. False-color scanning electron micrograph (22,000×).

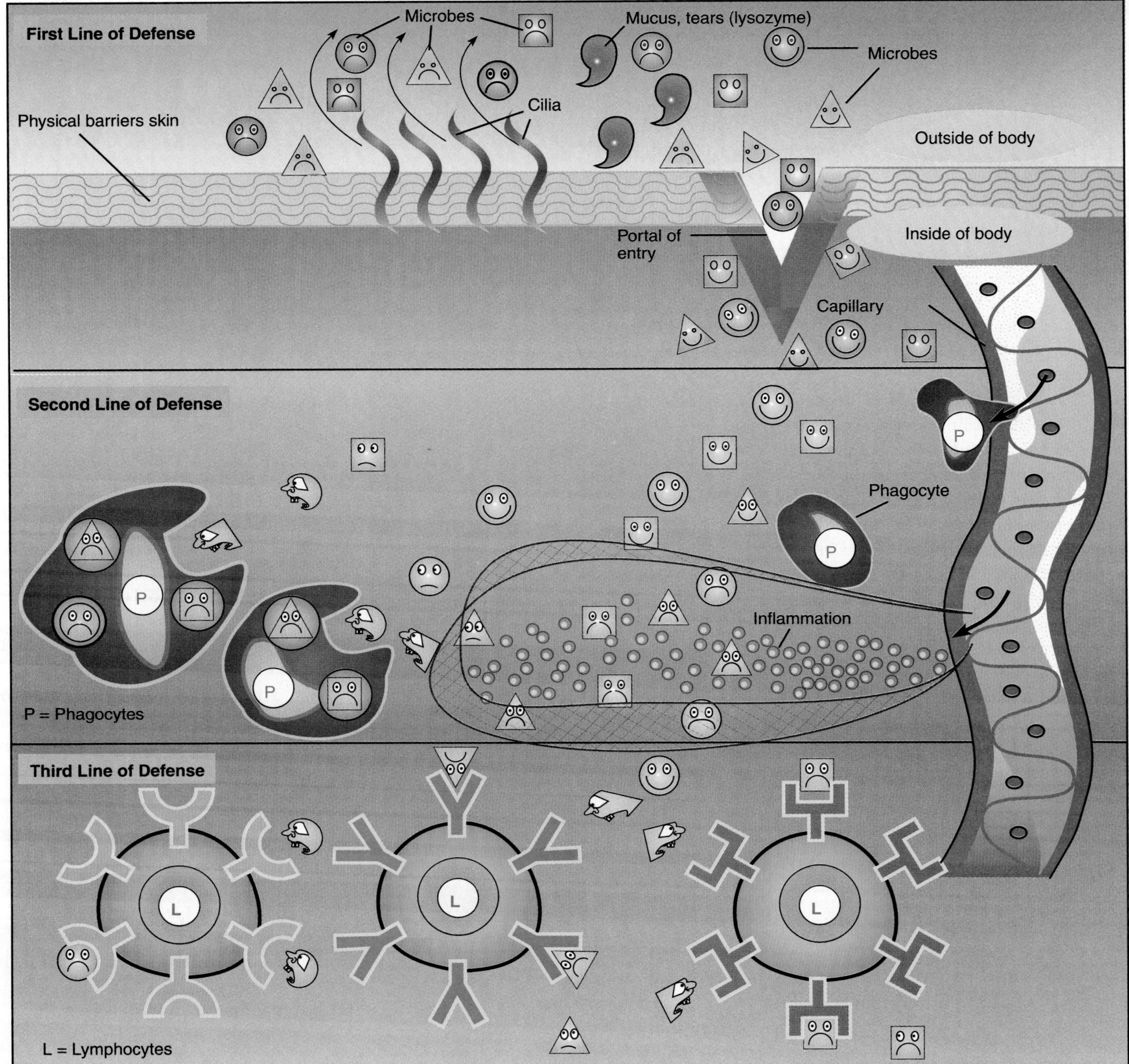

Figure 14.1

The levels of host defense. The first line of defense consists of nonspecific physical and chemical barriers that prevent the entrance of infectious agents into the tissues. The second line of defense consists of white blood cells (phagocytes) and chemical defenses (inflammation) that remove and destroy infectious agents that have entered the tissues. The third line of defense is based on white blood cells that show extreme specificity for their target microbes. This defense must be brought into play for each different microbe that is encountered, and it accounts for the immunities to diseases that develop following infection.

DEFENSE MECHANISMS OF THE HOST IN PERSPECTIVE

In chapter 13 we explored the host-parasite relationship, with emphasis on the role of microorganisms in disease. In this chapter we examine the other side of the relationship—that of the host defending itself against microorganisms. As previously stated, in light of the unrelenting contamination and colonization of humans, it is something of a miracle that we are not constantly infected and diseased. This does *not* happen because of a remarkable, fascinating, and amazingly complex system of defenses. In the war against all sorts of invaders, microbial and otherwise, the body erects a series of barriers, sends in an army of cells, and emits a flood of chemicals to protect tissues from harm.

The host defenses embrace a multilevel network of innate, nonspecific components and specific **immunities*** referred to as the first, second, and third lines of defense (figure 14.1). The interaction and cooperation of these three levels of defense normally provide complete protection against infection. The **first line of defense** includes any barrier that blocks invasion at the portal of entry. This mostly nonspecific line of defense limits access to the internal tissues of the body. However, it is not considered a true immune response because it does not involve recognition of or response to a specific foreign substance. The **second line of defense** is a slightly more internalized system of protective cells and fluids that includes inflammation and phagocytosis. It acts rapidly at both the local and systemic levels once the first line of defense has been circumvented. The highly specific **third line of defense** is acquired on an individual basis as each foreign substance is encountered by white blood cells called lymphocytes. The reaction with each different microbe produces unique protective substances and cells that can come into play if that microbe is encountered again. The third line of defense provides long-term immunity. A final summary chart that outlines the levels of host defenses can be found in figure 15.24.

The human body is armed with various levels of defense that do not operate in a completely separate fashion; most defenses overlap and are even redundant in some of their effects. This "immunological overkill" literally bombards microbial invaders with an entire assault force, making their survival unlikely. Because of the interwoven nature of host defenses, it is difficult to introduce one part without also discussing another, so we will set the scene by presenting general concepts and then tackle the specifics later in this chapter and chapter 15.

Barriers at the Portal of Entry: A First Line of Defense

A number of defenses are a normal part of the body's anatomy and physiology. These natural, inborn, nonspecific defenses can be divided into physical, chemical, and genetic barriers that impede the entry of microbes at the site of first contact (figure 14.2).

*immunity (im-yoo´-nih-tee) Gr. *immunis,* free, exempt. A state of resistance to infection.

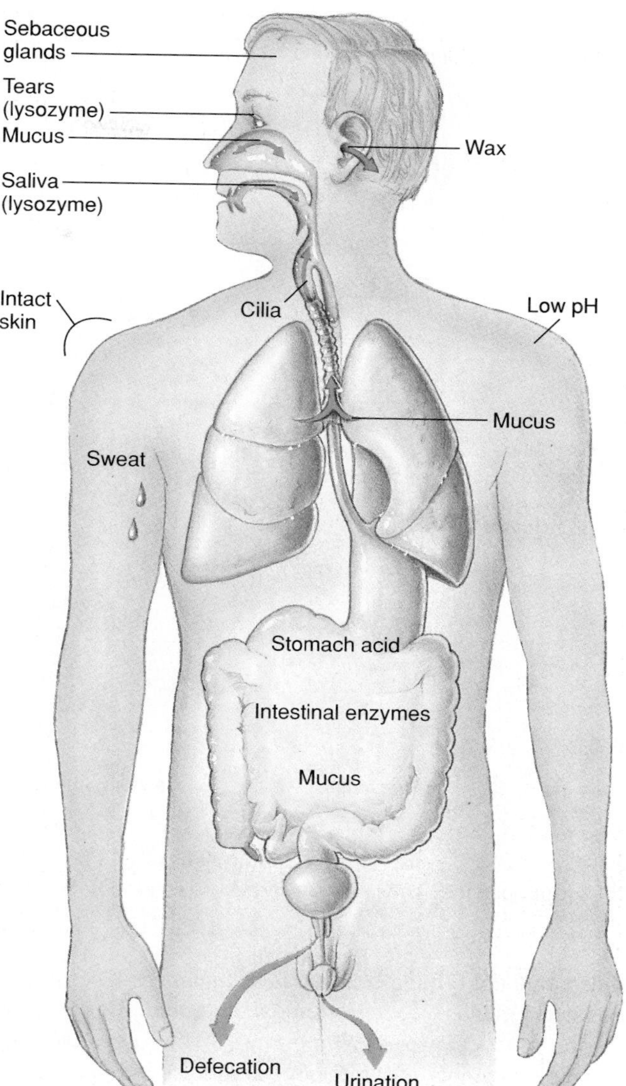

Figure 14.2

The primary physical and chemical defense barriers.

Physical or Anatomical Barriers at the Body's Surface

The skin and mucous membranes of the respiratory and digestive tracts have several built-in defenses. The outermost layer (stratum corneum) of the skin is composed of epithelial cells that have been cornified and keratinized, meaning that they have become compacted, cemented together, and impregnated with an insoluble protein, keratin. The result is a thick, tough layer that is highly impervious and waterproof. Few pathogens can penetrate this unbroken barrier, especially in regions such as the soles of the feet or the palms of the hands, where the stratum corneum is much thicker than on other parts of the body. Other cutaneous barriers include hair follicles and skin glands. The hair shaft is periodically

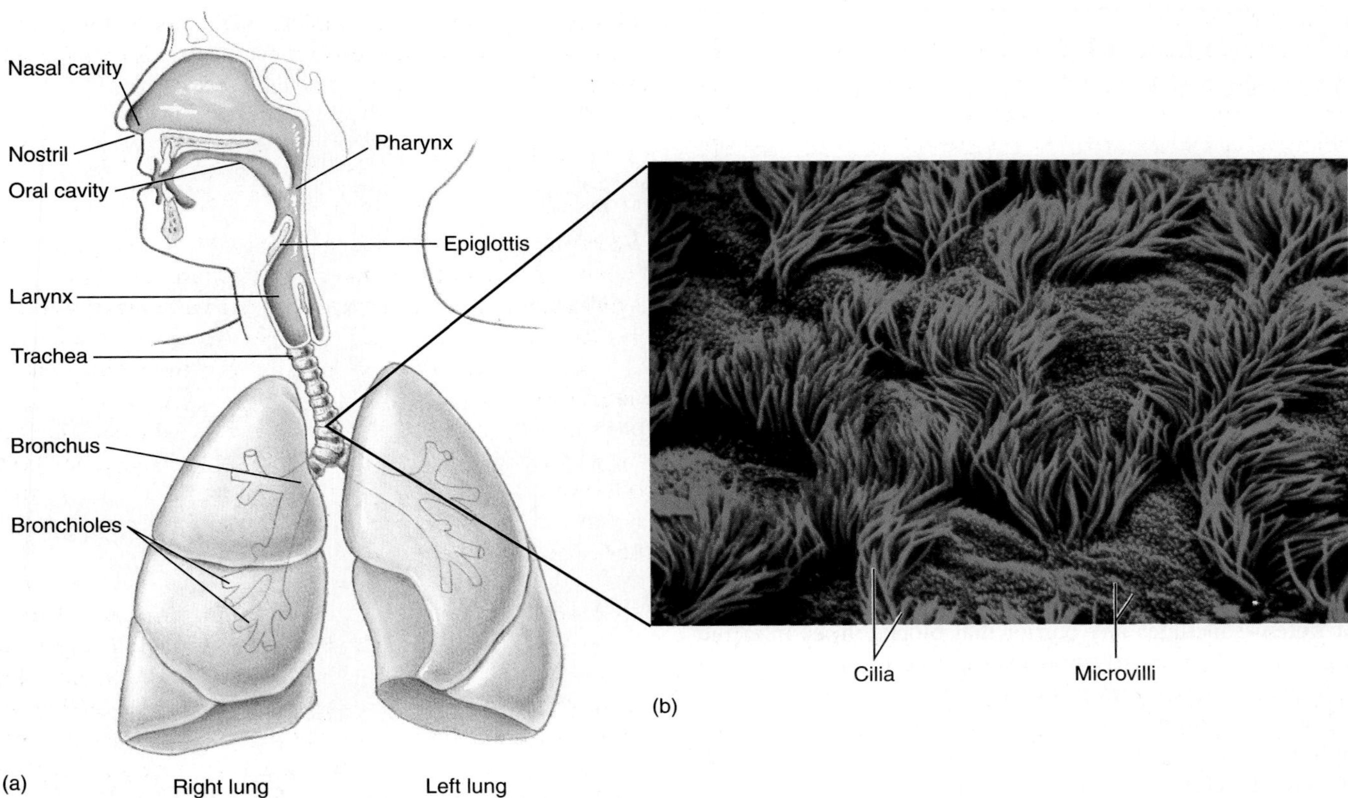

Figure 14.3

The ciliary defense of the respiratory tree. (*a*) The epithelial lining of the airways contains a brush border of cilia to entrap and propel particles upward toward the pharynx. (*b*) Tracheal mucosa (5,000×).

extruded, and the follicle cells are *desquamated.** The flushing effect of sweat glands also helps remove microbes.

The mucocutaneous membranes of the digestive, urinary, and respiratory tracts and of the eye are moist and permeable. Despite the normal wear and tear upon these epithelia, damaged cells are rapidly replaced. The mucous coat on the free surface of some membranes impedes the entry of bacteria. Blinking and *lacrimation** flush the eye's surface with tears and rid it of irritants. The constant flow of saliva helps carry microbes into the harsh conditions of the stomach. Vomiting and defecation also evacuate noxious substances or microorganisms from the body.

The respiratory tract is constantly guarded from infection by elaborate and highly effective adaptations. Nasal hair traps larger particles. Rhinitis, the copious flow of mucus and fluids that occurs in allergy and colds, exerts a flushing action. In the respiratory tree (primarily the trachea and bronchi), a ciliated epithelium conveys foreign particles entrapped in mucus toward the pharynx to be either expelled or swallowed (figure 14.3). This so-called ciliary escalator propels entrapped particles at a rate of 10 to 30 mm/h. Irritation of the nasal passage reflexly initiates a sneeze, which expels a large volume of air at high velocity. Similarly, the acute sensitivity of the bronchi, trachea, and larynx to foreign matter triggers coughing, which ejects irritants.

The genitourinary tract derives partial protection from the continuous trickle of urine through the ureters and from periodic bladder emptying that flushes the urethra.

The composition and protective effect exerted by microflora were discussed in chapter 13. Even though the resident flora does not constitute an anatomical barrier, its presence can block the access by pathogens to epithelial surfaces and can create an unfavorable environment for pathogens.

Nonspecific Chemical Defenses

The skin and mucous membranes offer a variety of chemical defenses. Sebaceous secretions exert an antimicrobial effect, and specialized glands such as the meibomian glands of the eyelids lubricate the conjunctiva with an antimicrobial secretion. An additional defense in tears and saliva is **lysozyme,** an enzyme that hydrolyzes the peptidoglycan in the cell wall of bacteria. The high lactic acid and electrolyte concentrations of sweat and the skin's acidic pH and fatty acid content are also inhibitory to many microbes. Likewise, the hydrochloric acid in the stomach renders protection against many pathogens that are swallowed, and the intestine's digestive juices and bile are potentially destructive to microbes. Even semen contains an antimicrobial chemical (spermine) that inhibits bacteria, and the vagina has a protective acidic pH maintained by normal flora.

*desquamate (des´-kwuh-mayt) L. *desquamo,* to scale off. The casting off of epidermal scales.

*lacrimation (lak´´-rih-may´-shun) L. *lacrimatio,* tears.

Genetic Defenses

Some hosts are genetically immune to the diseases of other hosts. One explanation for this phenomenon is that some pathogens have such great specificity for one host species that they are incapable of infecting other species. One way of putting it is: "Humans can't acquire distemper from cats, and cats can't get mumps from humans." This specificity is particularly true of viruses, which can invade only by attaching to a specific host receptor. But it does not hold true for zoonotic infectious agents that attack a broad spectrum of animals. Genetic differences in susceptibility can also exist within members of one species. Humans carrying a gene or genes for sickle-cell anemia are resistant to malaria (see microfile 9.6). Genetic differences also exist in susceptibility to tuberculosis, leprosy, and certain systemic fungal infections.

The vital contribution of barriers is clearly demonstrated in people who have lost them or never had them. Patients with severe skin damage due to burns are extremely susceptible to infections; those with blockages in the salivary glands, tear ducts, intestine, and urinary tract are also at greater risk for infection. But as important as it is, the first line of defense alone is not sufficient to protect against infection. Because many pathogens find a way to circumvent the barriers by using their virulence factors (discussed in chapter 13), a whole new set of defenses—inflammation, phagocytosis, specific immune responses—are brought into play.

Chapter Checkpoints

The multilevel, interconnecting network of host protection against microbial invasion is organized into three lines of defense. The first line consists of physical and chemical barricades provided by the skin and mucous membranes. The second line encompasses all the nonspecific cells and chemicals found in the tissues and blood. The third line, the specific immune response, is customized to react to specific antigens of a microbial invader. This response immobilizes and destroys the invader every time it appears in the host.

INTRODUCING THE IMMUNE SYSTEM

Immunology is the study of all biological, chemical, and physical events surrounding immune phenomena. This field has mushroomed into an exciting, precedent-setting area of molecular biology that dominates the progress of many areas of biology and medicine. Several recent Nobel prizes in medicine were presented for work in immunology. Breakthroughs in the areas of cancer, AIDS, or therapy emerge from the immunologic research community on a constant basis.

A study of immunities necessitates examining a number of interrelated concepts. For example, because immune reactions actually happen at the molecular level, we must understand the concepts of foreignness, surveillance, recognition, and specificity. In order to discuss phagocytosis, we must know something about blood cells; in turn, in order to discuss inflammation, we must understand the roles of body fluids and chemical mediators (cytokines); and in order to understand specific immunities, we must know something about all of these concepts and several related ones.

THE MOLECULAR BASIS OF IMMUNE RESPONSES

Whenever foreign material such as an infectious agent enters the tissues, the cells of the immune system are rapidly enlisted in a formidable molecular interchange. The main players in these reactions are molecules called **markers**[1] that protrude from the cell surface like minuscule signposts announcing that cell or molecule's identity. The identity of a given marker is based on its **specificity,** a unique configuration that dictates the kinds of immune responses it can elicit. Markers can also be **receptors,** molecules that bind specifically with complementary molecules in ways that signal, communicate, and trigger reactions inside the cell.

Surveillance, Recognition, and Destruction

The body has a fleet of white blood cells strategically located to explore continually through its tissues, compartments, and fluids like billions of tiny fingers, feeling and evaluating what is there and searching for any markers that are new or different. This process of scouting the tissues for foreign receptors and other possibly threatening particles is **surveillance** (figure 14.4). At the same time that white blood cells survey the tissues, they also evaluate and differentiate the molecules they discover by a process of **recognition.** Inherent in recognition is the capacity of the probing cells to sort out the **natural markers of the body** (that is, **self**) from the **foreign markers** (or **nonself**) (see microfile 14.1). This is an essential step, because the immune system is programmed to react as if anything foreign is potentially harmful and should be earmarked for destruction, whereas self is not. In this hunt for things that are not part of self, the immune system has a "sense of touch" so profound and exquisite that it often boggles the mind. When nonself is discovered and recognized, a whole battalion of responses is brought into play to **entrap and destroy** (kill, neutralize) the invading offender (figure 14.4).

Chapter Checkpoints

The immune system operates first as a surveillance system that discriminates between the host's self identity markers and the nonself identity markers of foreign cells. When it recognizes that a marker or antigen is foreign, or nonself, the immune system tailors its response specifically to each different antigen. As far as the immune system is concerned, if an antigen is not self, it is foreign, does not belong, and must be destroyed.

1. The term *marker* is also employed in genetics in a different sense—that is, to denote a detectable characteristic of a particular genetic mutant. A genetic marker may or may not be a surface marker.

MICROFILE 14.1 WHAT IS FOREIGN? WHAT IS NOT?

Understanding how the immune system functions requires a discussion of **foreignness.** In the most general sense, a foreign material is something that can be recognized or distinguished as not being a natural part of an organism's body. The properties of recognition and identification of nonself are traits of primitive origin, present to a limited extent in most organisms, and often essential to survival. When an ameba seeks out and ingests bacteria, it is showing a primitive capacity to distinguish that which is foreign (and edible). A similar recognition occurs in the life cycle of slime molds, whose separate ameboid cells migrate toward one another and aggregate into a large, many-celled stage. Even sponges, the most primitive animals, have a recognition system sufficiently developed to distinguish the cells of other species. If two species of sponges are finely divided and mixed, the cells reaggregate only with cells from their original species. This property is even more refined among the higher invertebrates and vertebrates, which have specialized blood cells to conduct surveillance and defense.

An even more specific term for a foreign, nonself molecule is **antigen.*** An antigen is a substance, often a surface marker, that evokes a specific immune response. As we shall see later in this chapter and in the next, antigens are responsible for eliciting specific reactions that occur during infection, vaccination, and allergies.

*antigen (an´-tih-jen) From *antibody* + *gen,* to produce. Originally, any substance that elicits the production of antibodies, but the term is now used to include other kinds of reactions.

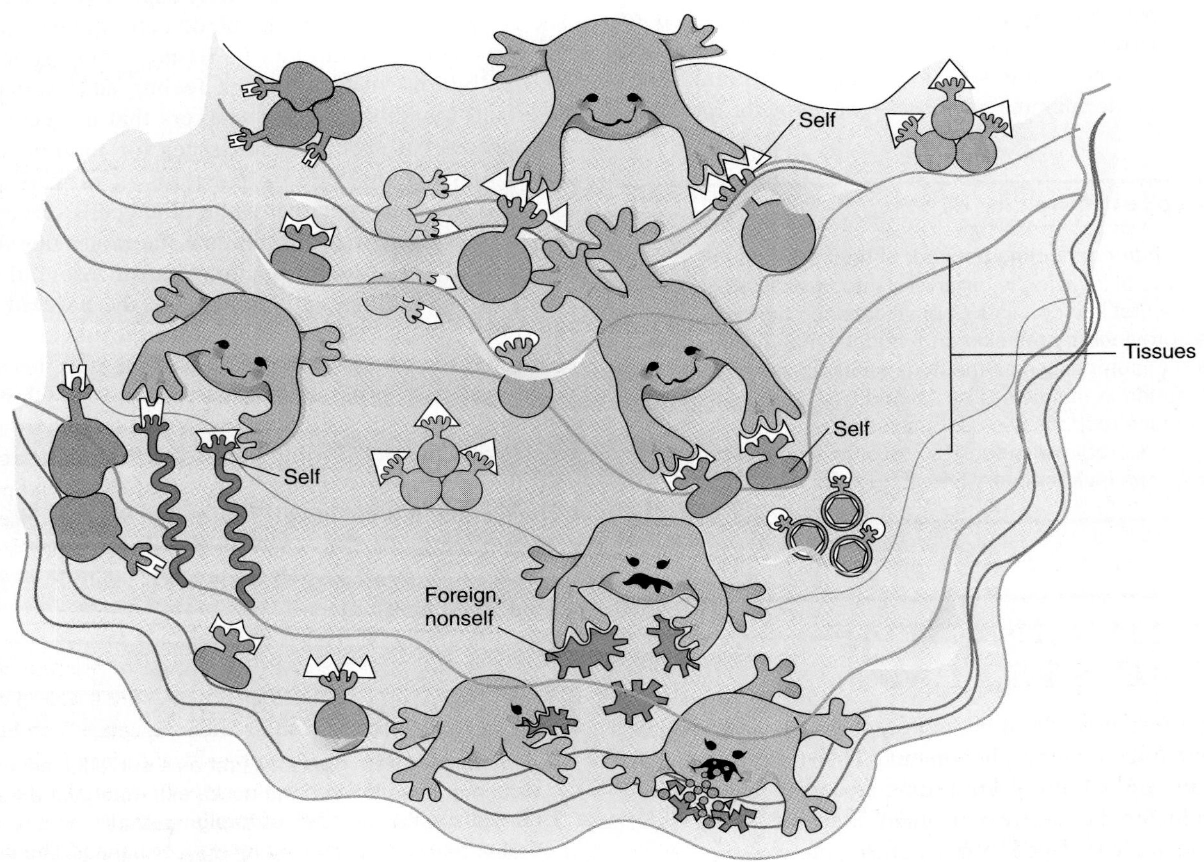

Figure 14.4
Search, recognize, and destroy is the mandate of the immune system. White blood cells are equipped with a very sensitive sense of "touch." As they sort through the tissues, they feel surface markers that help them determine what is self and what is not. When self markers are recognized, no response occurs. However, when nonself is detected, a reaction to destroy it is mounted.

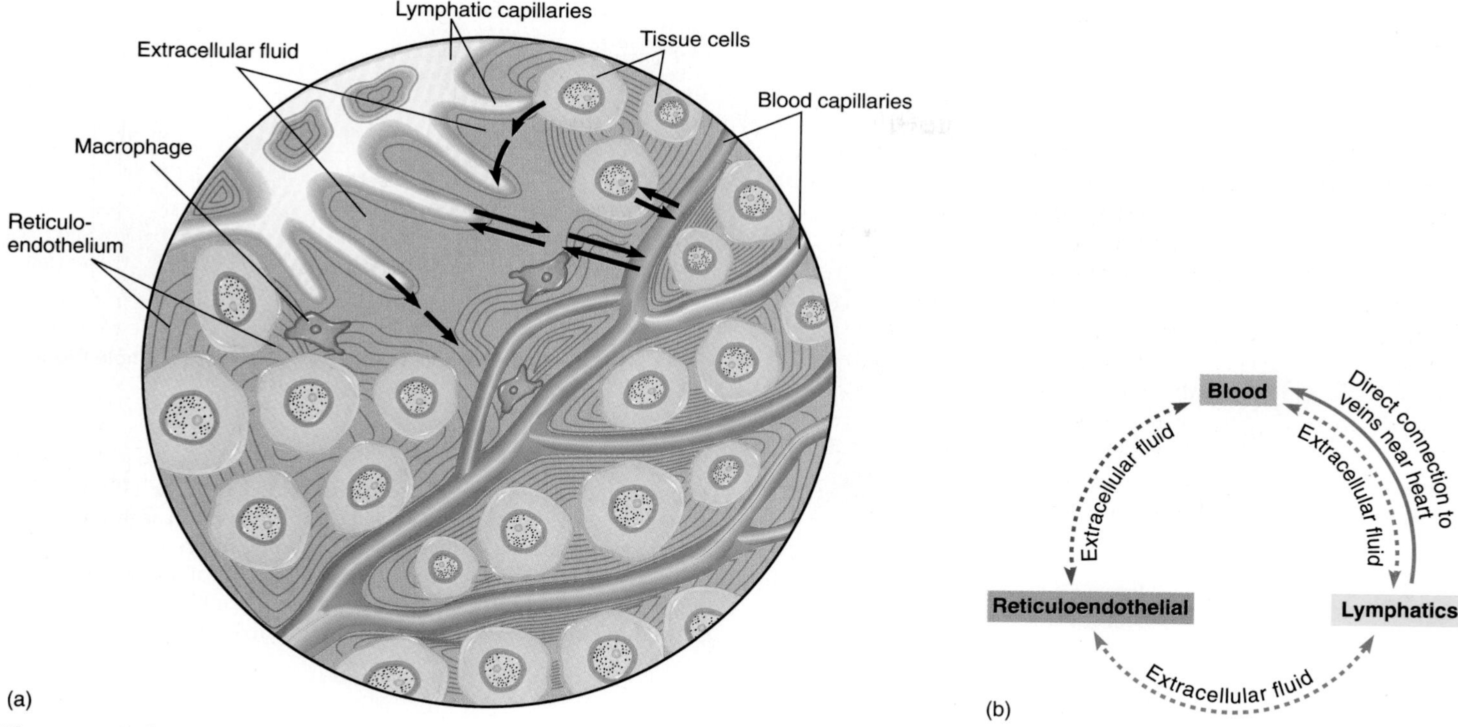

Figure 14.5
The body compartments are separate but connected. (*a*) The meeting of the major fluid compartments at the microscopic level. (*b*) Schematic view of the main fluid compartments and how they form a continuous cyclic system of exchange. Reactions in one section are rapidly broadcast to the others.

SYSTEMS INVOLVED IN IMMUNE DEFENSES

Unlike many systems, the **immune system** does not exist in a single, well-defined site; rather, it encompasses a large, complex, and diffuse network of cells and fluids that permeate every organ and tissue. It is this very arrangement that promotes the surveillance and recognition processes that help screen the body for harmful substances.

The body is partitioned into several fluid-filled spaces called the intracellular, extracellular, lymphatic, cerebrospinal, and circulatory compartments. Although these compartments are physically separated, they have numerous connections. Their structure and position permit extensive interchange and communication (figure 14.5). Four main body compartments that dominate in immune function are (1) the *reticuloendothelial system (RES)*,* (2) the spaces surrounding tissue cells that contain *extracellular fluid (ECF)*, (3) the *bloodstream,* and (4) the *lymphatic system.* In the following section, we consider the anatomy of these main compartments and how they interact in the second and third lines of defense.

*reticuloendothelial (reh-tik′′-yoo-loh-en′′-doh-thee′-lee-al) L. *reticulum,* a small net, and *endothelium,* lining of the blood vessel.

THE COMMUNICATING BODY COMPARTMENTS

For effective immune responsiveness, the activities in one fluid compartment must be conveyed to other compartments. Let us see how this occurs by viewing tissue at the microscopic level (figure 14.5*a*). At this level, clusters of tissue cells are in direct contact with the reticuloendothelial system (RES) and the extracellular fluid (ECF). Other compartments (vessels) that penetrate at this level are blood and lymphatic capillaries. Not only can cells and chemicals that originate in the RES and ECF diffuse or migrate into the blood and lymphatics; any products of a lymphatic reaction can be transmitted directly into the blood through the connection between these two systems. In addition, certain cells and chemicals originating in the blood can move through the vessel walls into the extracellular spaces and migrate into the lymphatic system.

The flow of events among these systems depends on where an infectious agent or foreign substance first intrudes. A typical progression might begin in the extracellular spaces and RES, move to the lymphatic circulation, and ultimately end up in the bloodstream. Regardless of which compartment is first exposed, an immune reaction in any one of them will eventually be communicated to the others at the microscopic level. An obvious benefit of such an integrated system is that no cell of the body is far removed from competent protection, no matter how isolated. Let us take a closer look at each of these compartments.

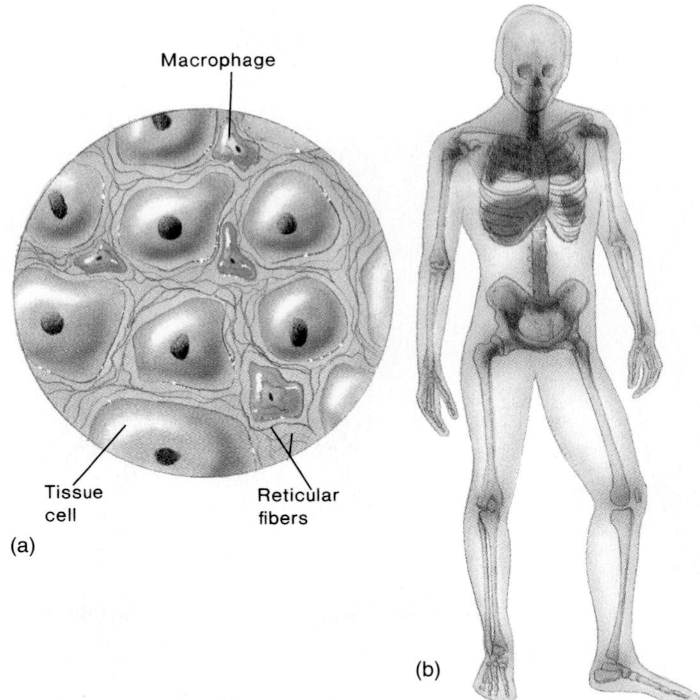

Figure 14.6

The reticuloendothelial system occurs as a pervasive, continuous connective tissue framework throughout the body. (*a*) This system begins at the microscopic level with a fibrous support network (reticular fibers) enmeshing each cell. This web connects one cell to another within a tissue or organ and provides a niche for macrophages, which can crawl within and between tissues using the fibers as guidelines. (*b*) The degrees of shading in the body indicate variations in phagocyte concentration (darker = greater).

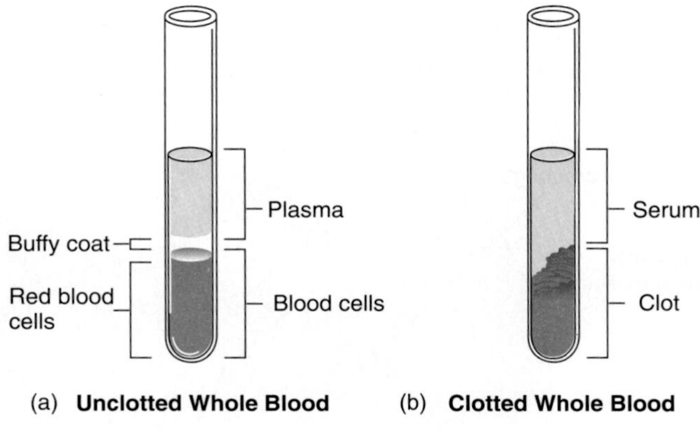

Figure 14.7

The macroscopic composition of whole blood. (*a*) When blood containing anticoagulants is allowed to sit for a period, it stratifies into a clear layer of plasma, a thin layer of off-white material called the buffy coat (which contains the white blood cells), and a layer of red blood cells in the bottom, thicker layer. (*b*) When blood is drawn and allowed to clot, a clear layer of serum is formed above the clot.

Immune Functions of the Reticuloendothelial System

The tissues of the body are permeated by a support network of connective tissue fibers, or a *reticulum,* that originates in the cellular basal lamina, interconnects nearby cells, and meshes with the massive connective tissue network surrounding all organs. The reticular network is intrinsic to the immune function because it provides a route of passage for phagocytes and forms a continuous network through which materials can enter the extracellular fluid surrounding tissues (figure 14.6). The term **reticuloendothelial system** has traditionally been used, although those who prefer to emphasize the system's phagocytic properties have termed it the **mononuclear phagocyte system.** The system is heavily endowed with white blood cells called macrophages waiting to attack passing foreign intruders as they arrive in the skin, lungs, liver, lymph nodes, spleen, and bone marrow.

Origin, Composition, and Functions of the Blood

The circulatory system consists of the circulatory system proper, which includes the heart, arteries, veins, and capillaries that circulate the blood, and the lymphatic system, which includes lymphatic vessels and lymphatic organs (lymph nodes) that circulate lymph (see figure 14.15). As we shall see, these two circulations parallel, interconnect with, and complement one another.

The substance that courses through the arteries, veins, and capillaries is **whole blood,** a liquid connective tissue consisting of **blood cells** (formed elements) suspended in **plasma** (ground substance). One can visualize these two components with the naked eye when a tube of *unclotted* blood is allowed to sit or is spun in a centrifuge. The cells' density causes them to sediment into an opaque layer at the bottom of the tube, leaving the plasma, a clear, yellowish fluid, on top (figure 14.7). The terms *plasma* and *serum* are sometimes mistakenly interchanged, and though these substances have the same basic origin, they differ in one important way. Because **serum** is the fluid extruded from *clotted* blood and the clotting proteins enter into the formation of the clot, serum does not contain the clotting proteins that plasma does. An easy way to remember this is: "Plasma can clot and serum cannot." Their composition is otherwise very similar.

Fundamental Characteristics of Plasma Plasma contains hundreds of different chemicals produced by the liver, white blood cells, endocrine glands, and nervous system and absorbed from the digestive tract. The main component of this fluid is water (92%), and the remainder consists of proteins such as albumin and globulins (including antibodies), other immunochemicals, fibrinogen and other clotting factors, hormones, nutrients (glucose, amino acids, fatty acids), ions (sodium, potassium, calcium, magnesium, chloride, phosphate, bicarbonate), dissolved gases (O_2 and CO_2), and waste products (urea). These substances support the normal physiological functions of nutrition, development, protection, homeostasis, and immunity. We return to the subject of plasma and its function in immune interactions later in this chapter and in chapter 15.

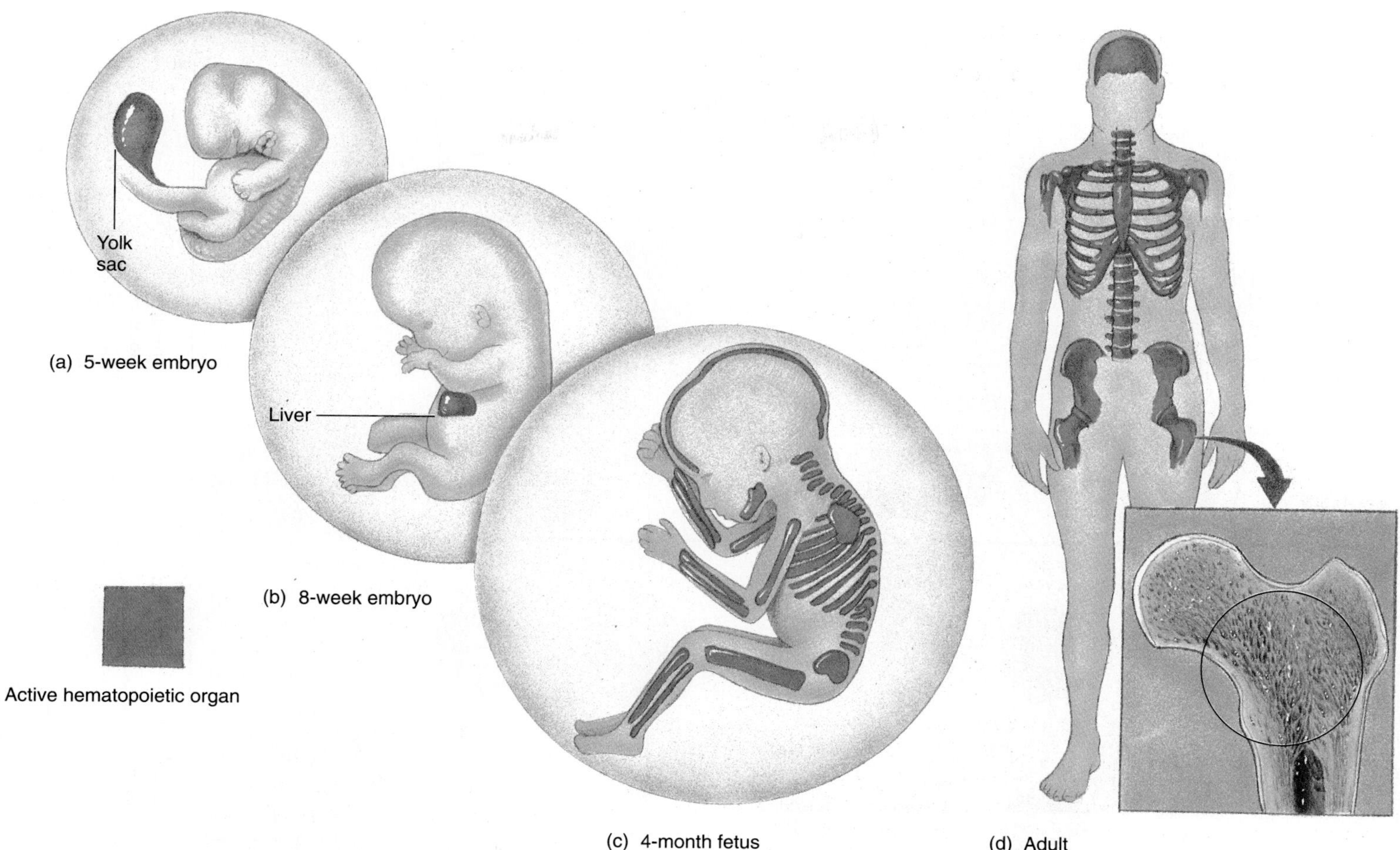

Figure 14.8

Stages in hemopoiesis. The sites of blood cell production change as development progresses from (*a,b*) yolk sac and liver in the embryo, to (*c*) extensive bone marrow sites in the fetus and (*d*) selected bone marrow sites in the child and adult. (*Inset*) Red marrow occupies the spongy bone (circle).

A Survey of Blood Cells The production of blood cells, or **hemopoiesis,*** begins early in embryonic development in the yolk sac (an embryonic membrane). Later it is taken over by the liver and lymphatic organs, and it is finally assumed entirely and permanently by the red bone marrow (figure 14.8). Although much of a newborn's red marrow is devoted to hemopoietic function, the active marrow sites gradually recede, and by the age of 4 years, only the ribs, sternum, pelvic girdle, flat bones of the skull and spinal column, and proximal portions of the humerus and femur are devoted to blood cell production.

The relatively short life of blood cells demands a rapid turnover that is continuous throughout a human lifespan. The primary precursor of new blood cells is a pool of undifferentiated cells called pluripotential **stem cells**[2] maintained in the marrow. During development, these stem cells proliferate and *differentiate*—meaning that immature or unspecialized cells develop the specialized form and function of mature cells. The primary lines of cells that arise from this process produce red blood cells (RBCs, or erythrocytes), white blood cells (WBCs, or leukocytes), and platelets (thrombocytes). The white blood cell lines are programmed to develop into several secondary lines of cells during the final process of differentiation (figure 14.9). These committed lines of WBCs are largely responsible for immune function.

The **white blood cells,** or **leukocytes,*** are traditionally evaluated by their reactions with Wright stain, a mixture of dyes that differentiates cells by color and morphology (figure 14.10). When this stain used on blood smears is evaluated by the light microscope, the leukocytes appear either with or without noticeable colored granules in the cytoplasm and, on that basis, are divided into two groups: **granulocytes** and **agranulocytes.** Greater magnification reveals that even the agranulocytes have tiny granules in their cytoplasm, so some hematologists categorize leukocytes on the basis of the appearance of the nucleus. **Polymorphonuclear***

2. Cells that can develop into several different types of blood cells; unipotential cells have already committed to a specific line of development.

*hemopoiesis (hee″-moh-poy-ee′-sis) Gr. *haima,* blood, and *poiesis,* a making. Also called hematopoiesis.

*leukocyte (loo′-koh-syte) Gr. *leukos,* white, and *kytos,* cell. The whiteness of unstained WBCs is best seen in the white layer, or buffy coat, of sedimented blood.

*polymorphonuclear Gr. *poly,* many; *morph,* shape; and *nuclear,* nucleus.

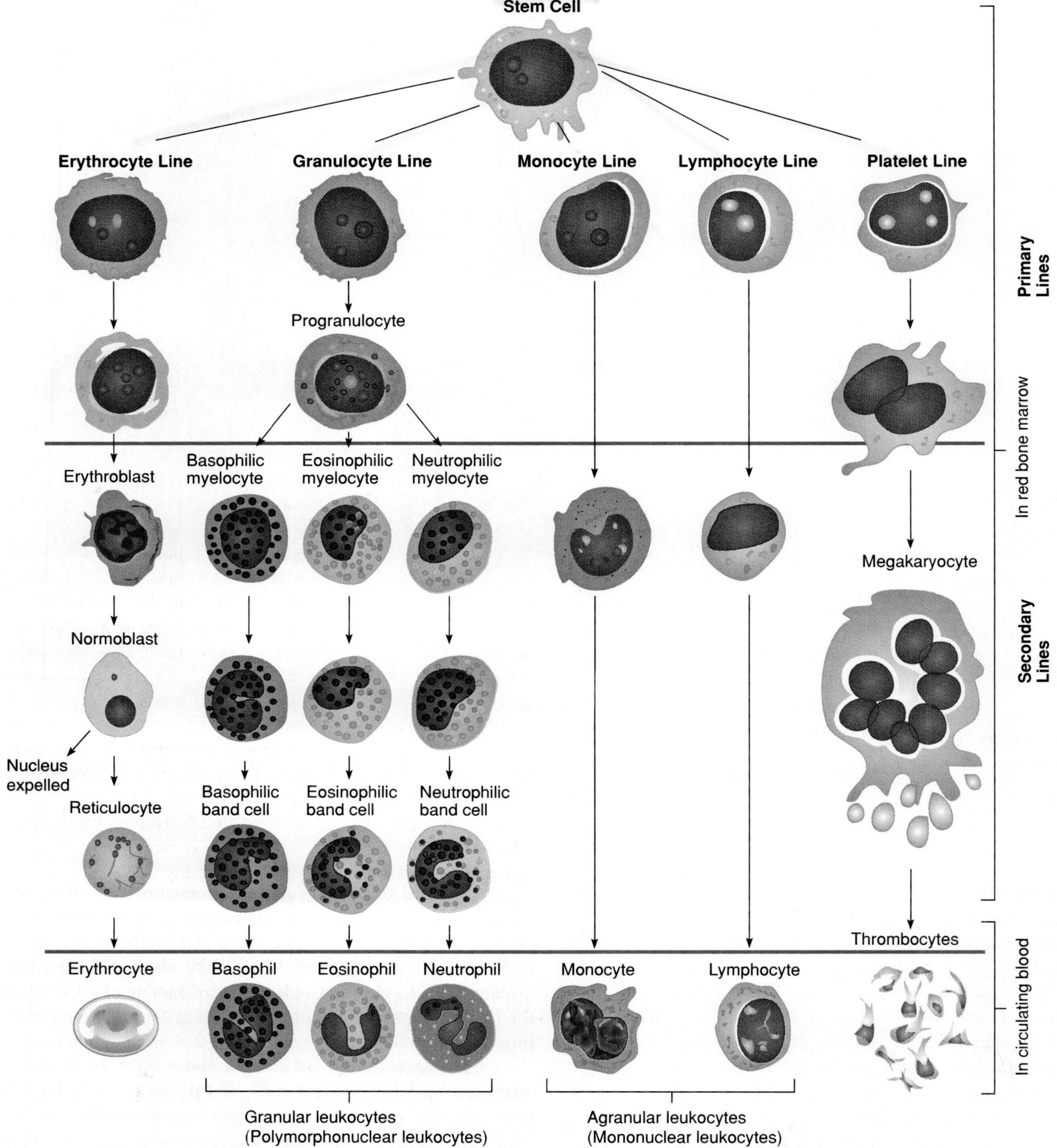

Figure 14.9
The development of blood cells and platelets. Each cell type in circulating blood (bottom row) is ultimately derived from an undifferentiated stem cell in the red marrow. During differentiation, the stem cell gives rise to several cell lines that become more and more specialized. Mature cells are released into the circulatory system.

leukocytes (granulocytes) have a lobed nucleus, and **mononuclear leukocytes** (agranulocytes) have an unlobed, rounded nucleus. Note that the cells fall into the same groups regardless of the system used (figure 14.9).

Granulocytes The types of polymorphonuclear leukocytes present in the bloodstream are neutrophils, eosinophils, and basophils. All three are known for prominent cytoplasmic granules that stain with some combination of acidic dye (eosin) or basic

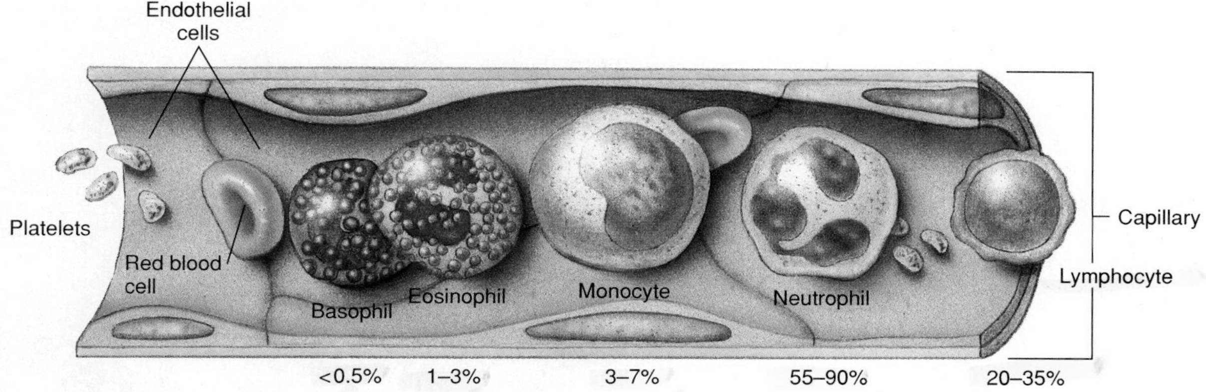

Figure 14.10

The microanatomy and circulating cells of the bloodstream. A cutaway view of a capillary reveals a histological picture of thin, tile-like endothelial cells that form a thin, one-cell-thick conduit. The cell types and relative proportions of circulating white blood cells are also shown.

dye (methylene blue). Although these granules are useful diagnostically, they also function in numerous physiological events.

Neutrophils* are distinguished from other leukocytes by their conspicuous, lobed nuclei and by their fine, pale lavender granules. In cells newly released from the bone marrow, the nuclei are horseshoe-shaped, but as they age, they form multiple lobes (up to five). Polymorphonuclear neutrophils (PMNs) make up 55% to 90% of the circulating leukocytes, about 25 billion cells in the circulation at any given moment. The main work of the neutrophils is in phagocytosis.[3] Their high numbers in both the blood and tissues suggest a constant challenge from resident microflora and environmental sources. Most of the cytoplasmic granules carry digestive enzymes and other chemicals that degrade the phagocytosed materials (see the discussion of phagocytosis later in this chapter). The average neutrophil lives only about 8 days, spending much of this time in the tissues and only about 6–12 hours in circulation.

Eosinophils* are readily distinguished in a Wright stain preparation by their larger, orange to red (eosinophilic) granules and bilobed nucleus. They are much more numerous in the bone marrow and the spleen than in the circulation, contributing only 1% to 3% of the total WBC count. The role of the eosinophil in the immune system is not fully defined, though several functions have been suggested. Their granules contain peroxidase, lysozyme, and other digestive enzymes. Eosinophils appear to be weakly phagocytic for bacteria, foreign particles, and antigen-antibody complexes. High levels of eosinophils are observed in infections by helminth parasites and fungi, suggesting that they are directed against larger eucaryotic parasites (figure 14.11). Much evidence is accumulating on the role of eosinophils as mediators in immediate allergies such as asthma and anaphylaxis (see chapter 17).

Basophils* are characterized by pale-stained, constricted nuclei and very prominent dark blue to black granules. They are the scarcest type of leukocyte, making up less than 0.5% of the total circulating WBCs in a normal individual. Basophils share some morphological and functional similarities with widely distributed tissue cells called **mast cells.*** Although these two cell types were once regarded as identical, mast cells are nonmotile elements bound to connective tissue around blood vessels, nerves, and epithelia, and basophils are motile elements derived from bone marrow. With regard to immune function, both types act primarily in immediate allergy and inflammation. Their granules release chemicals such as histamine, serotonin, heparin, and several enzymes with pronounced physiological effects. As we shall see in chapter 17, many allergy symptoms are directly attributable to the effects of these chemicals on tissues and organs.

Agranulocytes Agranular leukocytes have globular, nonlobed nuclei and lack prominent cytoplasmic granules when viewed with the light microscope. The two general types are monocytes and lymphocytes.

Lymphocytes are the second most common WBC in the blood, comprising 20% to 35% of the total circulating leukocytes. The fact that their overall number throughout the body is among the highest of all cells indicates how important they are to immunity. One estimate suggests that about one-tenth of all adult body cells are lymphocytes, exceeded only by erythrocytes and fibroblasts. The entire collection of lymphocytes for a 150-pound human would weigh about 2.2 pounds. In a stained blood smear, most lymphocytes are small, spherical cells with a

3. The neutrophil is sometimes called a microphage, or "small eater."

*__neutrophil__ (noo´-troh-fil) L. *neuter,* neither, and *philos,* to love. The granules are neutral and do not react markedly with either acidic or basic dyes. In clinical reports they are often called "polys" or PMNs for short.

*__eosinophil__ (ee´´-oh-sin´-oh-fil) Gr. *eos,* dawn, rosy, and *philos,* to love. Eosin is a red, acidic dye attracted to the granules.

*__basophil__ (bay´-soh-fil) Gr. *basis,* foundation. The granules attract to basic dyes.

*__mast__ From Ger. *mast.* food. Early cytologists thought these cells were filled with food vacuoles.

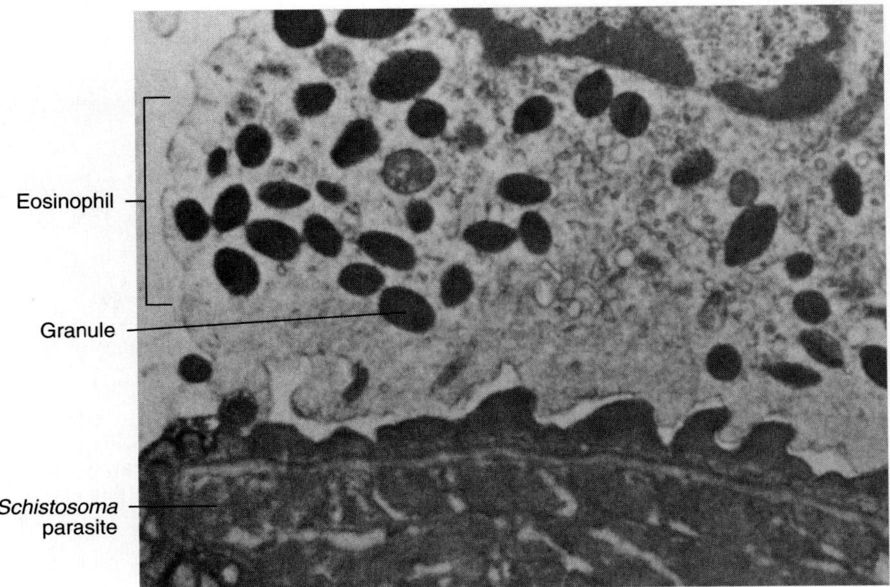

Figure 14.11
An electron micrograph section of an eosinophil attacking a parasitic worm (*Schistosoma*). The eosinophil releases the contents from its prominent granules, which initially disrupts the worm's outer surface and later destroys it completely.

uniformly dark blue, rounded nucleus surrounded by a thin fringe of clear, pale blue cytoplasm, although in tissues, they can become much larger and can even mimic monocytes in appearance (see figure 14.9). Lymphocytes exist as two functional types—the bursal-equivalent, or **B lymphocytes** (**B cells,** for short), and the thymus-derived, or **T lymphocytes** (**T cells,** for short). B cells were first demonstrated in and named for a special lymphatic gland of chickens called the *bursa of Fabricius,* the site for their maturation in birds. In humans B cells mature in special bone marrow sites, but humans do not have a bursa. T cells mature in the thymus gland in all birds and mammals. Both populations of cells are transported by the bloodstream and lymph and move about freely between lymphoid organs and connective tissue.

Lymphocytes are the key cells of the third line of defense and the specific immune response (figure 14.12). When stimulated by foreign substances (antigens), lymphocytes are transformed into activated cells that neutralize and destroy that foreign substance. The contribution of B cells is mainly in **humoral immunity,**[4] defined as protective molecules carried in the fluids of the body. When activated B cells divide, they form specialized *plasma cells,* which produce **antibodies,*** large protein molecules that interlock with an antigen and participate in its destruction. Activated T cells engage in a spectrum of immune functions characterized as **cell-mediated immunity** (CMI) in which T cells help, suppress, and modulate immune functions and kill foreign cells. The action of both classes of lymphocytes accounts for the recognition and memory typical of immunity. So important are lymphocytes to the defense of the body that a large portion of the next chapter is devoted to their reactions.

4. In reference to the humors, the liquids of the body. Humoral immunity includes antibodies, complement, and interferon.

*antibody (an´-tih-bahd´´-ee) Gr. *anti,* against, and O.E., *bodig,* body.

Monocytes* are generally the largest of all white blood cells and the third most common in the circulation (3–7%). As they mature, the nucleus becomes oval- or kidney-shaped—indented on one side, off-center, and often contorted with fine wrinkles. The pale blue cytoplasm holds many fine vacuoles containing digestive enzymes. Monocytes are discharged by the bone marrow into the bloodstream, where they live as phagocytes for a few days. Later they leave the circulation to undergo final differentiation into **macrophages*** (see figure 14.20). Unlike many other WBCs, the monocyte-macrophage series is relatively long-lived and retains an ability to multiply. Macrophages are among the most versatile and important of cells. Most immune reactions involve them either directly or indirectly. In general, they are responsible for (1) many types of specific and nonspecific phagocytic and killing functions (they assume the job of cellular housekeepers, mopping up the messes created by infection and inflammation); (2) processing foreign molecules and presenting them to lymphocytes; and (3) secreting biologically active compounds that assist, mediate, attract, and inhibit immune cells and reactions. We touch upon these functions in several ensuing sections.

Erythrocyte and Platelet Lines The **erythrocyte*** line of cells goes through a process in which the nucleus is finally extruded. The resultant red blood cells are simple, biconcave sacs of hemoglobin that transport oxygen and carbon dioxide to and from the tissues (see figure 14.10). These are the most numerous of circulating blood cells, appearing in Wright stains as small pink circles. Red blood cells do not ordinarily have immune functions, though they can be the target of immune reactions (see chapter 17).

*monocyte (mon´-oh-syte) From *mono,* one, and *cytos,* cell.

*macrophage (mak´-roh-fayj) Gr. *macro,* large, and *phagein,* to eat. They are the "large eaters" of the tissues.

*erythrocyte (eh-rith´-roh-syte) Gr. *erythros,* red. The red color comes from hemoglobin.

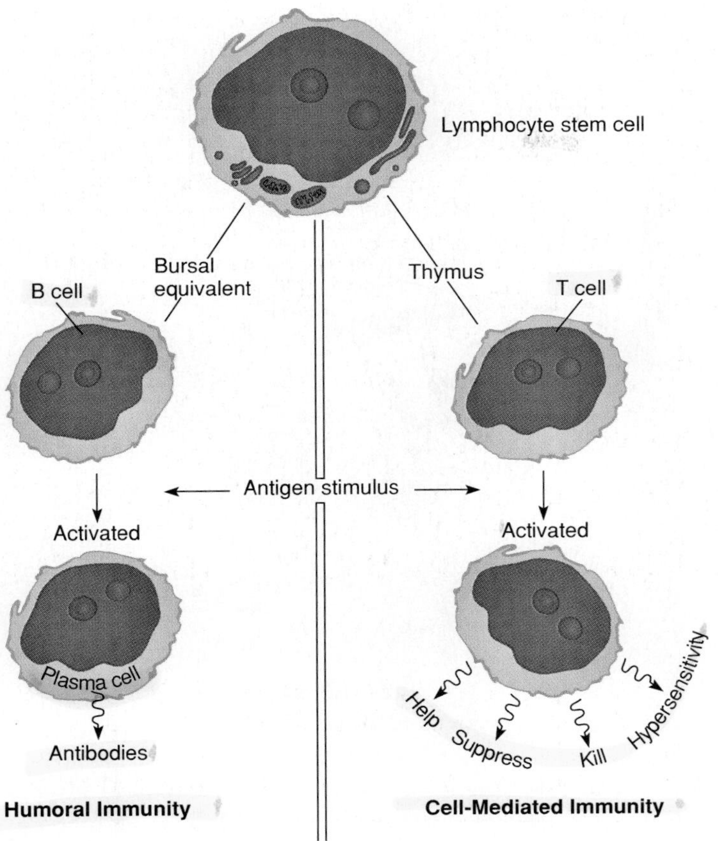

Figure 14.12
Summary of the general development and functions of lymphocytes, which are the cornerstone of specific immune reactions. B cells and T cells arise from the same stem cell but later diverge into two cell lines. Their appearances are similar, and one cannot differentiate them on the basis of a Wright stain. Note the relatively large nucleus: cytoplasm ratio and the lack of granules.

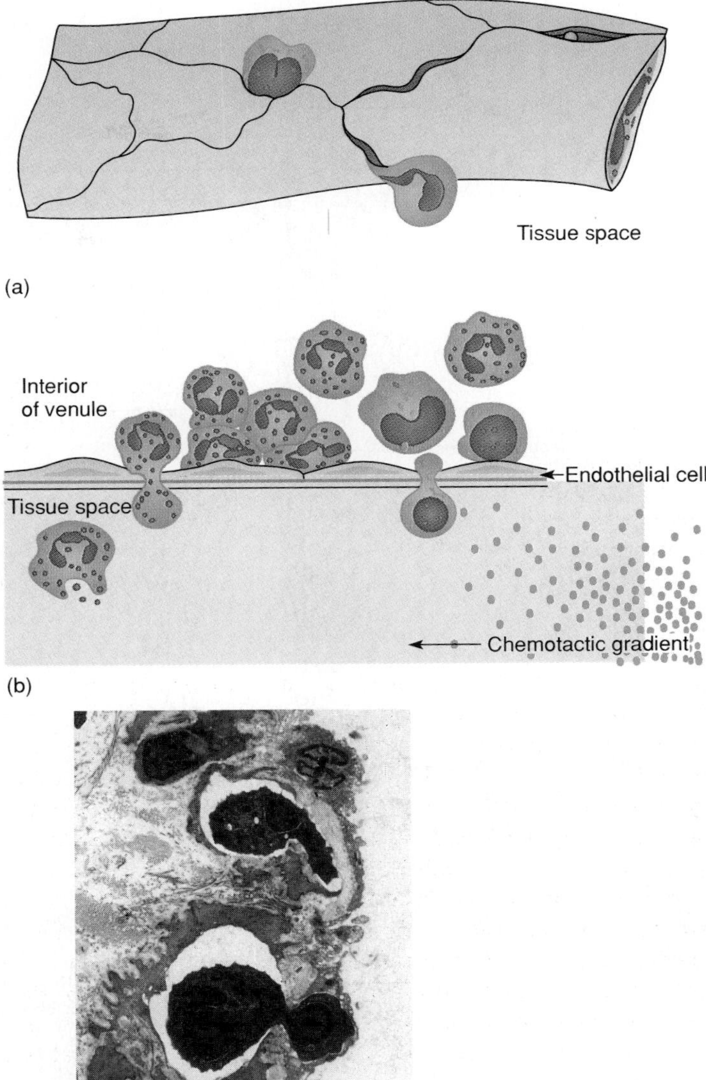

Figure 14.13
Diapedesis and chemotaxis of leukocytes. (*a*) View of a venule depicts white blood cells squeezing themselves between spaces in the blood vessel wall through diapedesis. (*b*) This process, shown in cross section, indicates how the pool of leukocytes adheres to the endothelial wall. From this site, they are poised to migrate out of the vessel into the tissue space. (*c*) This photograph captures neutrophils in the process of diapedesis.

Platelets,* or *thrombocytes,** are formed elements in circulating blood that are *not* whole cells. They originate in the bone marrow when a giant multinucleate cell called a *megakaryocyte* disintegrates into numerous tiny, irregular-shaped pieces, each containing bits of the cytoplasm and nucleus (see figure 14.9). In Wright stains, platelets are blue-gray with fine red granules and are readily distinguished from cells by their small size (1–3 μm in diameter). Platelets function primarily in hemostasis (plugging broken blood vessels to stop bleeding) and in releasing chemicals that act in blood clotting and inflammation.

Unique Dynamic Characteristics of White Blood Cells

Many lymphocytes and phagocytes make regular journeys from the blood and lymphatics to the tissues and back again to the circulation as part of the constant surveillance of the compartments. In order for these WBCs to complete this circuit, they adhere to the inner walls of the smaller blood vessels. From this position, they are poised to migrate out of the blood into the tissue spaces by a process called **diapedesis.*** Upon contact with the small spaces between the patchwork of endothelial cells that line blood vessels, these WBCs extrude themselves between the spaces and slip into the nearby extracellular region (figure 14.13). Their actively motile nature and plastic shape enable them to do this. This process occurs primarily in postcapillary venules (little veins) whose endothelial cells have loose junctions between them, and it happens without injury to the cells or vessel wall.

*platelet (playt´-let) Gr. *platos,* flat, and *let,* small.
*thrombocyte (throm´-boh-syte) Gr. *thrombos,* clot.

*diapedesis (dye´´-ah-puh-dee´-sis) Gr. *dia,* through, and *pedan,* to leap.

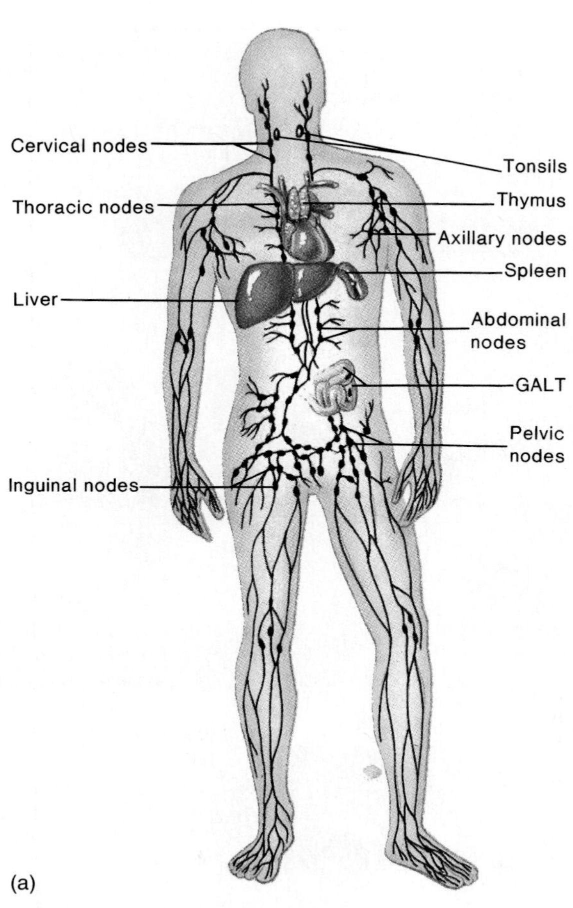

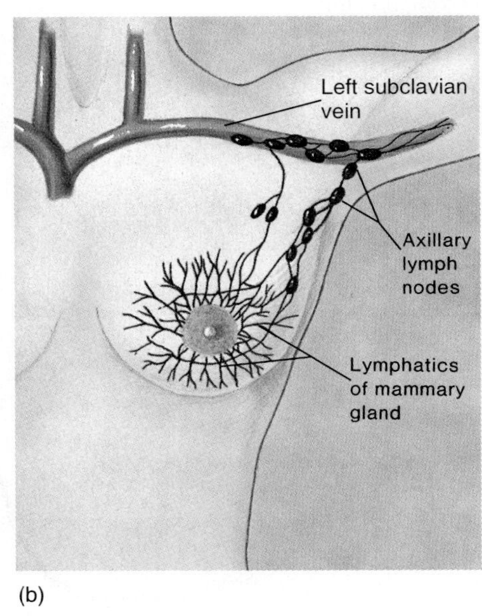

Figure 14.14

General components of the lymphatic system. (*a*) This system consists of an anastomosing, branching network of vessels that permeate most tissues of the body. Note the especially high concentration of this network in the hands, feet, and breasts. (*b*) Organs that form part of this system are lymph nodes, clustered at major drainage points (armpit, groin, intestine). Other lymphatic structures are the spleen, part of the small intestine (gut-associated lymphoid tissue, GALT), the thymus, and the tonsils.

An important factor in the migratory habits of these WBCs is **chemotaxis.*** This is defined as the tendency of cells to migrate in the direction of a specific chemical stimulus given off at a site of injury or infection (see inflammation and phagocytosis later in this chapter). Through this means, cells swarm from many compartments to the site of infection and remain there to perform general and specific immune functions. These basic properties are absolutely essential for the sort of intercommunication and deployment of cells required for most immune reactions.

Components and Functions of the Lymphatic System

The lymphatic system is a compartmentalized network of vessels, cells, and specialized accessory organs (figure 14.14). It begins in the farthest reaches of the tissues as tiny capillaries that transport a special fluid (lymph) through an increasingly larger tributary system of vessels and filters (lymph nodes), and it leads to major vessels that drain back into the regular circulatory system. Some major functions of the lymphatic system are (1) to provide an auxiliary route for the return of extracellular fluid to the circulatory system proper,[5] (2) to act as a "drain-off" system for the inflammatory response, and (3) to render surveillance, recognition, and protection against foreign materials through a system of lymphocytes, phagocytes, and antibodies.

Lymphatic Fluid Lymph is a plasmalike liquid carried by the lymphatic circulation. It is formed when certain blood components move out of the blood vessels into the extracellular spaces and diffuse or migrate into the lymphatic capillaries. Thus, the composition of lymph parallels that of serum in many ways. It is made up of water, dissolved salts, and 2% to 5% protein (especially antibodies and albumin). Like blood, it also transports numerous white blood cells (especially lymphocytes) and miscellaneous materials such as fats, cellular debris, and infectious agents that have gained access to the tissue spaces. Unlike blood, red blood cells are not normally found in lymph.

*chemotaxis (kee-moh-tak´-sis) NL *chemo*, chemical, and *taxis*, arrangement.

5. The importance of this function is most evident in cases of impaired lymphatic drainage as seen for example, in patients with filariasis, a roundworm infection (see figure 23.22).

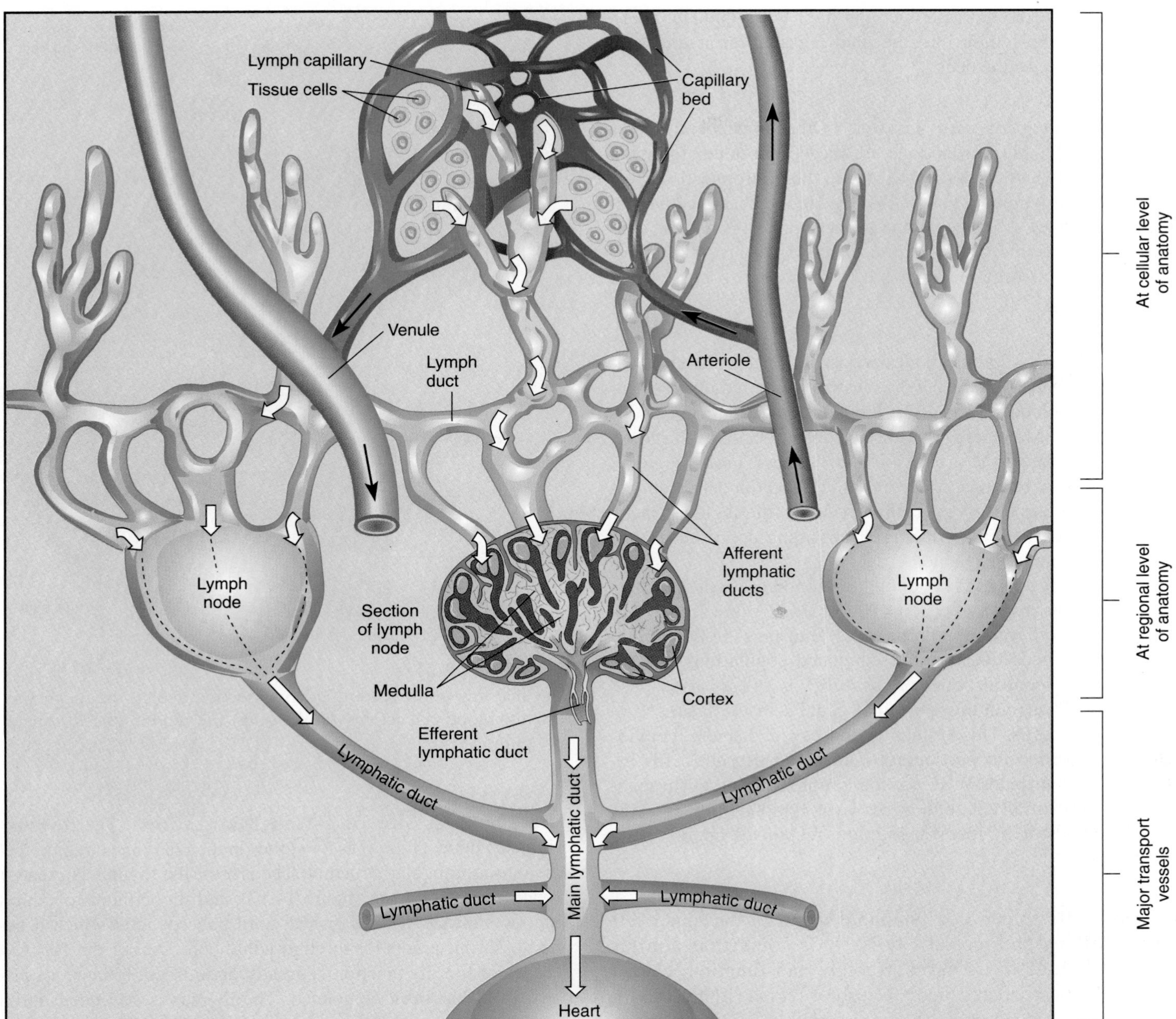

Figure 14.15
The circulatory scheme of the lymphatic vessels. The lymphatic capillaries originate at the finest level of the tissues as tiny, fingerlike projections that absorb excess tissue fluid and transport it into a network of fewer and larger vessels (ducts). These ducts drain their fluid (lymph) into the lymph nodes. The center insert shows the anatomy of a lymph node. Its afferent lymphatic ducts penetrate into the sinuses. As the lymph percolates through these sinuses, it is filtered and exposed to populations of B and T lymphocytes that are segregated into specific sites. After filtration, the lymph leaves the node by a single afferent duct from which it is transported into a larger series of drainage vessels. These vessels empty their load of lymph into large main ducts near the heart. By this means, lymphatic chemicals and cells are returned to the blood.

Lymphatic Vessels The system of vessels that transports lymph is constructed along the lines of regular blood vessels. The tiniest ones, lymphatic capillaries, accompany the blood capillaries and permeate all parts of the body except the central nervous system and certain organs such as bone, placenta, or thymus. The thin capillary walls make them permeable to extracellular fluid. The density of lymphatic vessels is particularly high in the hands, feet, and around the areolae of the breasts. Unlike the bloodstream, the lymphatic system does not circulate cyclically, but in one direction—from the extremities toward the heart (figure 14.15). The lymphatic capillary network feeds into a series of fewer but larger vessels that, in turn, drain finally into two main ducts (the thoracic duct and the right lymphatic duct), which empty the lymph into the main circulation at the large veins at the base of the neck.

Lymphoid Organs and Tissues Other organs and tissues that perform lymphoid functions are the lymph nodes (glands), thymus, spleen, and clusters of tissues in the gastrointestinal tract (gut-associated lymphoid tissue; GALT) and the pharynx (the tonsils, for example). A trait common to these organs is a loose connective tissue framework that houses aggregations of lymphocytes, the important class of white blood cells mentioned previously.

Lymph Nodes Lymph nodes are small, encapsulated, bean-shaped organs stationed, usually in clusters, along lymphatic channels and large blood vessels of the thoracic and abdominal cavities (figure 14.14). Major aggregations of nodes occur in the loose connective tissue of the armpit (axillary nodes), groin (inguinal nodes), and neck (cervical nodes). Both the location and architecture of these nodes clearly specialize them for filtering out materials that have entered the lymph and providing appropriate cells and niches for immune reactions.

A view of a single sectioned lymph node reveals its filtering and cellular response systems (figure 14.15). Incoming lymphatic vessels (afferent lymphatic ducts) transport the lymph into sinuses in the node that contain segregated populations of lymphocytes. The central zone, or *medulla** is the location of T cells, and the surrounding germinal centers in the *cortex** are packed with B cells. This system of sinuses and discrete lymphocyte zones filters out particulate materials (microbes, for instance) and contributes WBCs to the lymph as it passes through. Many of the initial encounters between lymphocytes and microbes that result in specific immune responses occur in the lymph nodes.

Spleen The spleen is a lymphoid organ in the upper left portion of the abdominal cavity. It is somewhat similar to a lymph node in its basic structure and function, except that the spleen circulates blood instead of lymph (figure 14.16). It consists of a connective tissue network of vascular sinuses called the *red pulp,* which is rich in macrophages, neutrophils, and erythrocytes. Nestled within the red pulp are compact regions of lymphocytes called the *white pulp.* B cells and T cells occupy separate predetermined areas of the white pulp. The spleen serves as an important station for phagocytosis of foreign matter and immune reactions against bacteria and removes and breaks down worn erythrocytes. Although adults whose spleens have been surgically removed can live a relatively normal life, asplenic children are extremely immunocompromised.

*medulla (meh-dul´-ah) L. *medius,* middle or marrow.
*cortex (kor´-teks) L. *cortex,* bark, rind, shell.

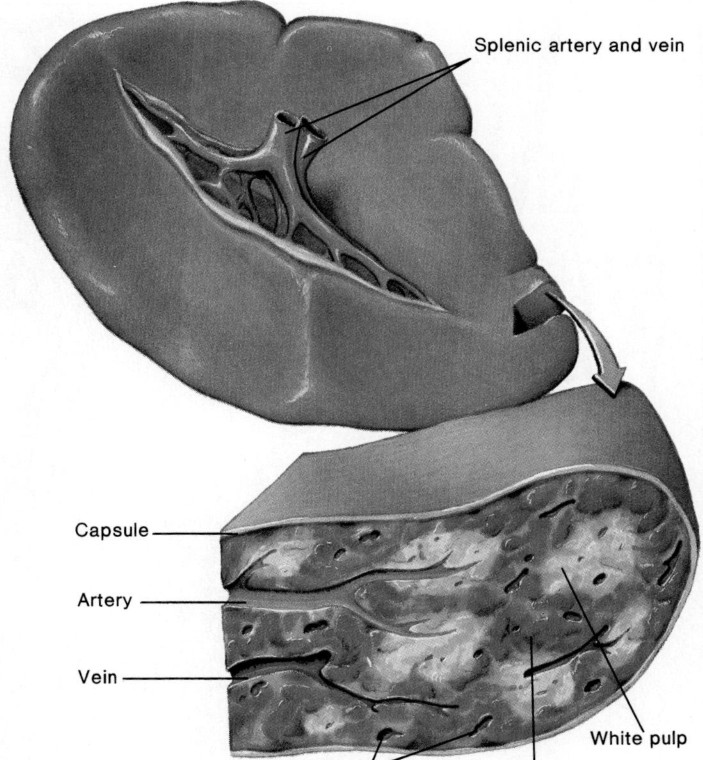

Figure 14.16
The anatomy of the spleen. Enlarged section depicts the major sites of white blood cell activity (white pulp) and red blood cell storage (red pulp).

The Thymus: Site of T-Cell Maturation The **thymus*** originates in the embryo as two lobes in the pharyngeal region that fuse into a triangular structure. The size of the thymus is greatest proportionately at birth (figure 14.17), and it continues to exhibit high rates of activity and growth until puberty, after which it begins to shrink gradually through adulthood. During the last few decades of life, its function is greatly diminished, because its primary work has been completed. The thymus is sectioned into a medulla, composed of special epithelial cells, and a cortex, containing undifferentiated lymphocytes called thymocytes. Under the influence of thymic hormones, thymocytes develop specificity and are released into the circulation as mature T cells. The T cells subsequently migrate to and settle in other lymphoid organs (for example, the lymph nodes and spleen), where they occupy the specific sites described previously.

The thymus gland was once thought to have no important function. Medical science was so mistaken about its significance that children's necks were sometimes irradiated to "cure" a condition called "enlarged thymus." Experiments in the early 1960s fi-

*thymus (thigh´-mus) Gr. *thymos,* soul, mind.

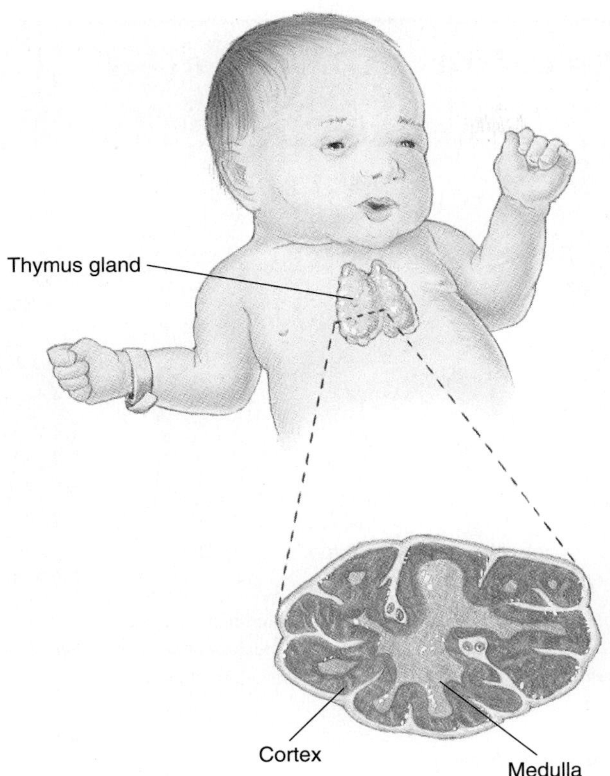

Figure 14.17
The thymus gland. Immediately after birth, the thymus is a large organ that nearly fills the region over the midline of the upper thoracic region. In the adult, however, it is proportionately smaller (to compare, see figure 14.14*a*). Section shows the main anatomical regions of the thymus.

nally clarified its link to lymphocyte development. If the thymus is surgically removed in neonates, they become severely immunodeficient and fail to thrive, and babies born without a functional thymus are vulnerable to a disease called DiGeorge syndrome (see page 542). Adults have developed enough mature T cells that removal of the thymus or reduction in its function has milder effects. Do not confuse the thymus with the thyroid gland, which is located nearby but has an entirely different function.

Miscellaneous Lymphoid Tissue At many sites on or just beneath the mucosa of the gastrointestinal and respiratory tracts lie discrete bundles of lymphocytes. The positioning of this diffuse system provides an effective first-strike potential against the constant influx of microbes and other foreign materials in food and air. In the pharynx, a ring of tissues called the **tonsils** provides an active source of lymphocytes. The breasts of pregnant and lactating women also become temporary sites of antibody-producing lymphoid tissues (see colostrum, microfile 15.5). The intestinal tract houses the best-developed collection of lymphoid tissue, called **gut-associated lymphoid tissue,** or **GALT.** Examples of GALT include the appendix, the lacteals (special lymphatic vessels stationed in each intestinal villus), and *Peyer's patches,* compact aggregations of lymphocytes in the ileum of the small intestine. GALT provides immune functions against intestinal pathogens and is a significant source of some types of antibodies.

Chapter Checkpoints

The immune system is a complex collection of fluids and cells that penetrate every organ, tissue space, fluid compartment, and vascular network of the body. The four major subdivisions of this system are the RES, the ECF, the blood vascular system, and the lymphatic system.

The RES, or reticuloendothelial system, is a network of connective tissue fibers inhabited by macrophages ready to attack and ingest microbes invading the first and second lines of defense.

The ECF, or extracellular fluid, compartment surrounds all tissue cells and is penetrated by both blood and lymph vessels, which bring all components of the second and third line of defense to attack infectious microbes.

The blood contains both specific and nonspecific defenses. Nonspecific cellular defenses include the granulocytes and macrophages. The two components of the specific immune response are the T lymphocytes, which provide specific cell-mediated immunity, and the B lymphocytes, which produce specific antibody or humoral immunity.

The lymphatic system has three functions: (1) It returns tissue fluid to general circulation; (2) it carries away excess fluid in inflamed tissues; (3) it concentrates and processes foreign invaders and initiates the specific immune response. Important sites of lymphoid tissues are lymph nodes, spleen, thymus, tonsils, and GALT.

NONSPECIFIC IMMUNE REACTIONS OF THE BODY'S COMPARTMENTS

Now that we have introduced the principal anatomical and physiological framework of the immune system, let us address some mechanisms that play important roles in host defenses: inflammation, phagocytosis, interferon, and complement. Because of the generalized nature of these defenses, they contain elements that are nonspecific, but they also support and interact with the specific immune responses described in chapter 15.

MICROFILE 14.2 RUBOR, CALOR, TUMOR, DOLOR: GROSS SIGNS OF INFLAMMATION

The classic signs and symptoms of inflammation have been known for centuries and are characterized succinctly by four Latin terms: *rubor, calor, tumor,* and *dolor.* Rubor (redness) is caused by increased circulation and vasodilation in the injured tissues; calor (warmth) is the heat given off by the increased flow of blood; tumor (swelling) is caused by increased fluid escaping into the tissues; and dolor (pain) is caused by the stimulation of nerve endings. Combined, these events often cause the temporary loss of function of the afflicted tissue (*functio laesa*). Many of these reactions are unpleasant and at times can seem more harmful than protective. It is clear, however, that they serve both to warn that injury has taken place and to set in motion responses that save the body from further injury.

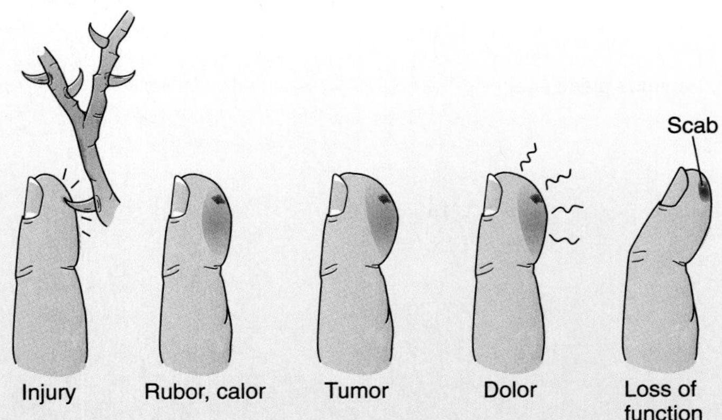

The response to injury. This classic checklist encapsulates the reactions of the tissues to an assault. Each of the events is an indicator of one of the mechanisms of inflammation described in succeeding sections of this chapter.

The Inflammatory Response: A Complex Concert of Reactions to Injury

At its most general level, the inflammatory response is a reaction to any traumatic event in the tissues. It is so very commonplace that most of us manifest inflammation in some way every day. It appears in the nasty flare of a cat scratch, the blistering of a burn, the painful lesion of an infection, and the symptoms of allergy. It is readily identifiable by a classic series of signs and symptoms (microfile 14.2). Factors that can elicit inflammation include trauma from infection (the primary emphasis here), tissue injury or necrosis due to physical or chemical agents, and specific immune reactions. Although the details of inflammation are very complex, its chief functions can be summarized as follows: (1) to mobilize and attract immune components to the site of the injury, (2) to set in motion mechanisms to repair tissue damage and localize and clear away harmful substances, and (3) to destroy microbes and block their further invasion (figure 14.18). The inflammatory response is a powerful defensive reaction, a means for the body to maintain stability and restore itself after an injury. But when it is chronic, it has the potential to actually *cause* tissue injury, destruction, and disease (see microfile 14.4).

The Stages of Inflammation

The process leading to inflammation is a dynamic, predictable sequence of events that can be acute, lasting from a few minutes or hours, to chronic, lasting for days, weeks, or years. Once the initial injury has occurred, a chain reaction takes place at the site of damaged tissue, summoning beneficial cells and fluids into the injured area. As an example, we will look at an injury at the microscopic level and observe the flow of major events (figure 14.18).

Vascular Changes: Early Inflammatory Events

Following an injury, some of the earliest changes occur in the vasculature (arterioles, capillaries, venules) in the vicinity of the damaged tissue. These changes are controlled by nervous stimulation and **chemical mediators** called **cytokines*** released by blood cells, tissue cells, and platelets in the injured area. Some of these mediators are *vasoactive*—that is, they affect the endothelial cells and smooth muscle cells of blood vessels, and others are **chemotactic factors,** also called **chemokines,** that affect white blood cells. Other cytokines cause fever, stimulate lymphocytes, prevent virus spread, and cause allergic symptoms (microfile 14.3 and figure 14.19). Although the constriction of arterioles is stimulated first, it lasts for only a few seconds or minutes and is followed in quick succession by the opposite reaction, vasodilation. The overall effect of vasodilation is to increase the flow of blood into the area, which facilitates the influx of immune components and also causes redness and warmth.

Edema: Leakage of Vascular Fluid into Tissues

Some vasoactive substances cause the endothelial cells surrounding postcapillary venules to contract and form gaps through which blood-borne components exude into the extracellular spaces. The fluid part that escapes is called the *exudate.* Accumulation of this fluid in the tissues gives rise to local swelling and hardness called **edema.** The edematous exudate contains varying amounts of plasma proteins, such as globulins, albumin, the clotting protein

*cytokine (sy´-toh-kyne) Gr. *cytos,* cell, and *kinein,* to move. A chemical stimulant released from one tissue that induces activity in another tissue.

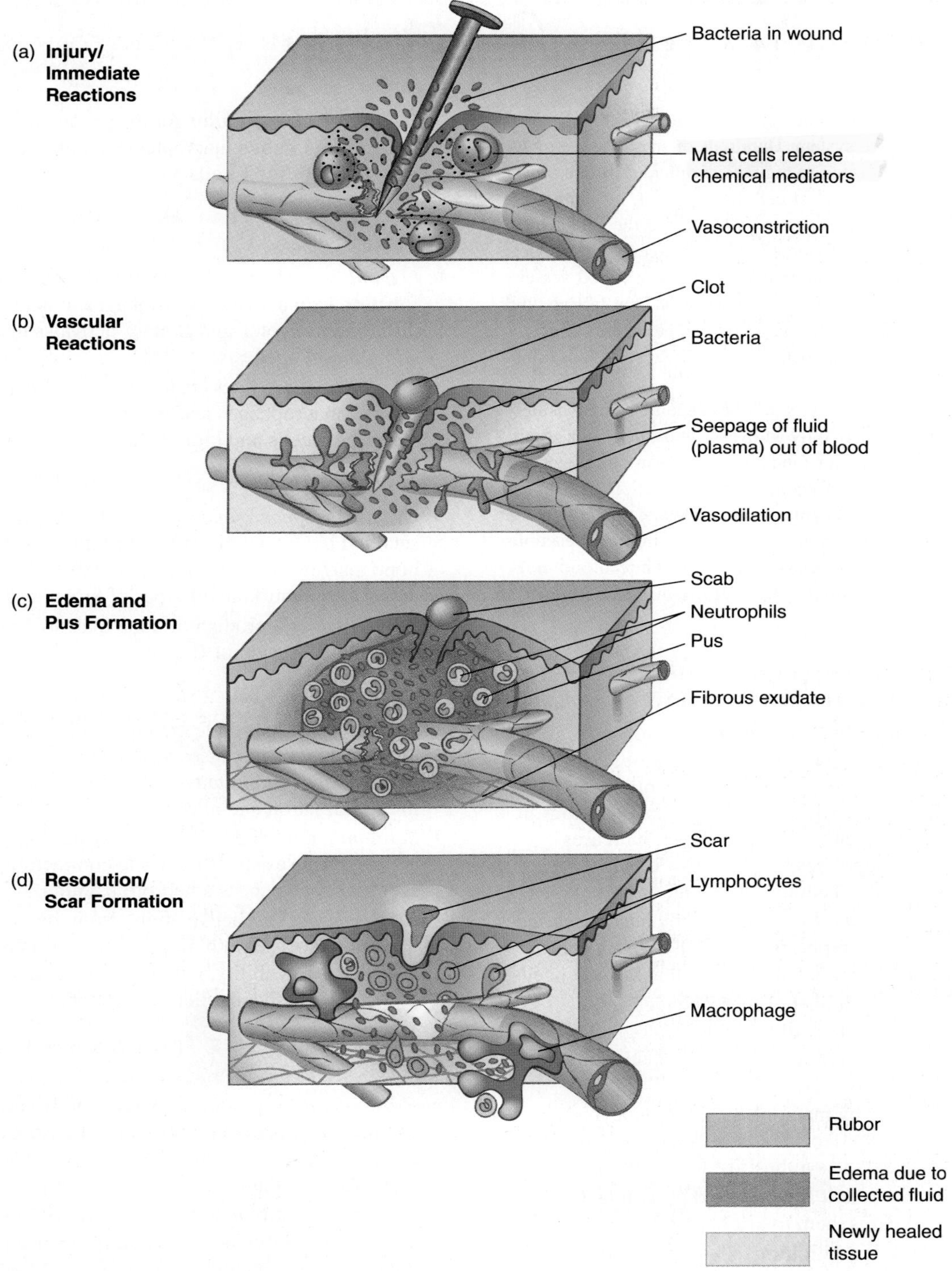

Figure 14.18

The major events in inflammation. (*a*) Injury → Reflex narrowing of the blood vessels (vasoconstriction) lasting for a short time → Release of chemical mediators into area. (*b*) Increased diameter of blood vessels (vasodilation) → Increased blood flow → Increased vascular permeability → Leakage of fluid (plasma) from blood vessels into tissues (exudate formation). (*c*) Edema → Infiltration of site by neutrophils and accumulation of pus. (*d*) Macrophages and lymphocytes → Repair, either by complete resolution and return of tissue to normal state or by formation of scar tissue.

MICROFILE 14.3 THE DYNAMICS OF THE CYTOKINE NETWORK

Just as the nervous system is coordinated by a complex communications network, so too is the immune system. Hundreds of small, active molecules are constantly being secreted to regulate, stimulate, suppress, and otherwise control the many aspects of cell development, inflammation, and immunity. These substances, called *cytokines,* are the products of several types of cells, including monocytes, macrophages, lymphocytes, fibroblasts, mast cells, platelets, and endothelial cells of blood vessels. Their effects may be local or systemic, short-term or long-lasting, and nonspecific or specific. Although the cytokines can be naturally protective, they may also contribute to the pathology of diseases such as infections, cancer, and allergy.

In recent times, the field of cytokines has become so increasingly complex that we can include here only an overview of the major groups of important cytokines.* The major functional types can be categorized into (1) cytokines that mediate nonspecific immune reactions such as inflammation and phagocytosis, (2) cytokines that regulate the growth and activation of lymphocytes, (3) cytokines that activate immune reactions during inflammation, (4) hemopoeisis factors for white blood cells, (5) vasoactive mediators, and (6) miscellaneous inflammatory mediators.

NONSPECIFIC MEDIATORS OF INFLAMMATION AND IMMUNITY

- *Tumor necrosis factor (TNF),* a substance from macrophages that in low concentrations increases chemotaxis and phagocytosis and stimulates other cells to secrete inflammatory cytokines. In higher concentrations, it is an endogenous pyrogen that induces fever, increases blood coagulation, suppresses bone marrow metabolism, and suppresses appetite, leading to wasting, or cachexia.
- *Interferon (IFN), alpha and beta,* produced by leukocytes and fibroblasts, inhibits virus replication and cell division and increases the action of certain lymphocytes that kill other cells.
- *Interleukin (IL) 1,*** a product of macrophages and epithelial cells that has many of the same biological activities as TNF, such as inducing fever and activation of certain white blood cells.
- *Interleukin-6,* secreted by macrophages, lymphocytes, and fibroblasts. Its primary effects are to stimulate the growth of B cells and to increase the synthesis of liver proteins.
- *Interleukin-5,* a very potent activator and growth factor for eisinophils that prepares them to kill hemonths.
- *Various chemokines*. By definition, chemokines are cytokines that stimulate the movement and migration of white blood cells (chemotactic factors). Included among these are complement C5A, interleukin 8, and platelet factor. There are 30 different known chemokines.

CYTOKINES THAT REGULATE LYMPHOCYTE GROWTH IN ACTIVATION

- *Interleukin-2,* the primary growth factor from T cells. Interestingly, it acts on the same cells that secrete it. It stimulates mitosis and secretion of other cytokines. In B cells, it is a growth factor and stimulus for antibody synthesis.
- *Interleukin-4,* a stimulus for the production of allergy antibodies; inhibits macrophage actions; and is also a growth factor for T cells.

CYTOKINES THAT ACTIVATE SPECIFIC IMMUNE REACTIONS

- *Gamma interferon,* a T-cell derived mediator whose primary function is to activate macrophages. It also promotes the differentiation of T and B cells, activates neutrophils, and stimulates diapedesis.
- *Interleukin-5* activates eosinophils and B cells; *interleukin-10* inhibits macrophages and B cells; and *interleukin-12* activates T cells and killer cells.

CYTOKINES INVOLVED IN HEMOPOEISIS

- *Granulocyte-macrophage colony stimulating factor (GM-GSF),* secreted by T cells directly at the site of action such as the bone marrow and sites of inflammation. It stimulates growth and differentiation in all types of blood cells.
- *Interleukin-7,* a product of bone marrow stem cells that aids in the development of B cells.

VASOACTIVE MEDIATORS

- *Histamine,* a vasoactive mediator produced by mast cells and basophils that causes vasodilation, increased vascular permeability, and mucus production. It functions primarily in inflammation and allergy.
- *Serotonin,* a mediator produced by platelets and intestinal cells that causes smooth muscle contraction, inhibits gastric secretion, and acts as a neurotransmitter.
- *Bradykinin,* a vasoactive amine from the blood or tissues that stimulates smooth muscle contraction and increases vascular permeability, mucus production, and pain. It is particularly active in allergic reactions.

MISCELLANEOUS INFLAMMATORY MEDIATORS

- *Prostaglandins,* produced by most body cells; complex chemical mediators that can have opposing effects (for example, dilation or constriction of blood vessels) and are powerful stimulants of inflammation and pain.
- *Leukotrienes* stimulate the contraction of smooth muscle and enhance vascular permeability. They are implicated in the more severe manifestations of immediate allergies (constriction of airways).
- *Platelet-activating factor,* a substance released from basophils, causes the aggregation of platelets and the release of other chemical mediators during immediate allergic reactions.

**For an excellent detailed review of the subject, see* Clinical Microbiology Reviews, *October 1997, pp. 742–780.*

**Interleukin *is a term that refers to a group of small peptides originally isolated from leukocytes. There are currently 17 known interleukins. We now know that other cells besides leukocytes can synthesize them and that they have a variety of biological activities. Functions of some selected examples will be presented in chapter 15.*

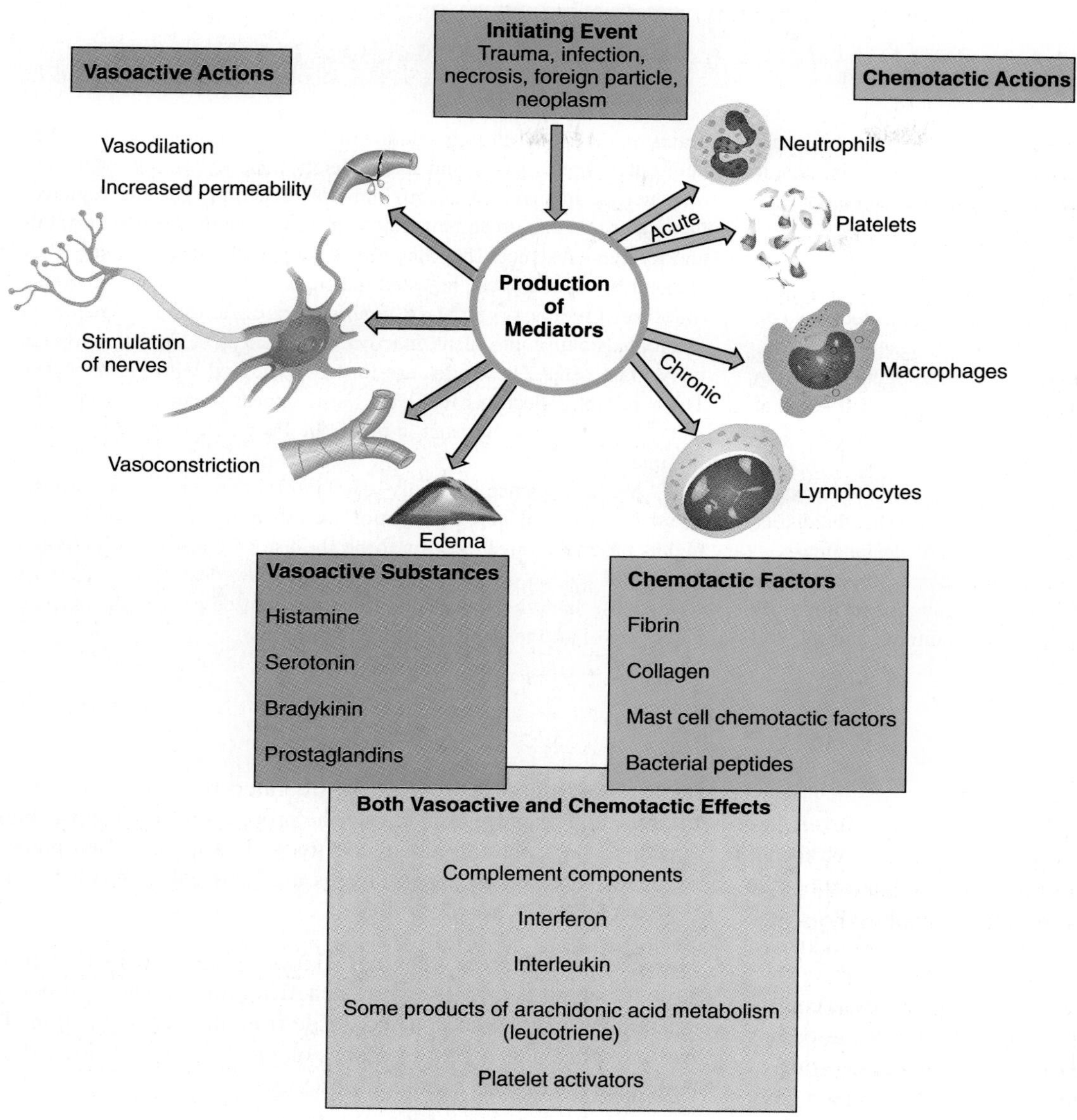

Figure 14.19 Chemical mediators of the inflammatory response and their effects.

fibrinogen, blood cells, and cellular debris. Depending upon its content, the exudate varies from serous (clear) to serosanguinous (containing red blood cells) to purulent (containing pus). In some types of edema, the fibrinogen is converted to fibrin threads that enmesh the injury site. Within an hour, multitudes of neutrophils responding chemotactically to special signaling molecules converge on the injured site (see figure 14.18*c*).

The Benefits of Edema and Chemotaxis Both the formation of edematous exudate and the infiltration of neutrophils are physiologically beneficial activities. The influx of fluid dilutes toxic substances, and the fibrin clot can effectively trap microbes and prevent their further spread. The neutrophils that aggregate in the inflamed site are immediately involved in phagocytosing and destroying bacteria, dead tissues, and particulate matter (by mechanisms discussed in a later section on phagocytosis). In some types of inflammation, accumulated phagocytes contribute to **pus,** a whitish mass of cells, liquefied cellular debris, and bacteria. Certain bacteria (streptococci, staphylococci, gonococci, and meningococci) are especially powerful attractants for neutrophils and are thus termed **pyogenic,** or pus-forming, bacteria.

Late Reactions of Inflammation Sometimes a mild inflammation can be resolved by edema and phagocytosis. Inflammatory reactions that are more long-lived attract a collection of monocytes, lymphocytes, and macrophages to the reaction site. Clearance of pus, cellular debris, dead neutrophils, and damaged tissue is performed by macrophages, the only cells that can engulf and dispose of such large masses. At the same time, B lymphocytes react with foreign molecules and cells by producing specific antimicrobial proteins (antibodies), and T lymphocytes kill intruders directly. Late in the process the tissue is completely repaired, if possible, or replaced by connective tissue in the form of a scar (figure 14.18*d*). If the inflammation cannot be relieved or resolved in this way, it can become chronic and create a long-term pathologic condition (microfile 14.4).

MICROFILE 14.4 WHEN INFLAMMATION GETS OUT OF HAND

Not every aspect of inflammation is protective or results in the proficient resolution of tissue damage. As one looks over a list of diseases, it is rather striking how many of them are due in part or even completely to an overreactive or dysfunctional inflammatory response.

Some "itis" reactions mentioned in chapter 13 are a case in point (see microfile 13.5). Inflammatory exudates that build up in the brain in African trypanosomiasis, cryptococcosis, and other brain infections can be so injurious to the nervous system that impairment is permanent. Frequently, an inflammatory reaction that walls off the pathogen leads to an abscess, a swollen mass of neutrophils and dead, liquefied tissue that can harbor live pathogens in the center. Abscesses are a prominent feature of staphylococcal, amebic, and enteric infections.

Other pathologic manifestations of chronic diseases—for example, the tubercles of tuberculosis, the lesions of late syphilis, the disfiguring nodules of leprosy, and the cutaneous ulcers of leishmaniasis—are due to an aberrant tissue response called *granuloma formation* (see figure 19.17). Granulomas develop not only in response to microbes but also in response to inanimate foreign bodies (sutures and mineral grains that are difficult to break down). This condition is initiated when neutrophils ineffectively and incompletely phagocytose the pathogens or materials involved in an inflammatory reaction. The macrophages then enter to clean up and attempt to phagocytose the dead neutrophils and foreign substances, but they fail to completely manage them. They respond by storing these ingested materials in vacuoles and becoming inactive. Over a given time period, large numbers of adjacent macrophages fuse into giant, inactive multinucleate cells called foreign body giant cells. These sites are further infiltrated with lymphocytes. The resultant collections make the tissue appear granular—hence, the name. A granuloma can exist in the tissue for months, years, or even a lifetime.

Medical science is rapidly searching for new applications for the massive amount of new information on inflammatory mediators. One highly promising area appears to be the use of chemokine inhibitors that could reduce chemotaxis and the massive, destructive influx of leukocytes. Such therapy could ultimately be used for certain cancers, hardening of arteries, and Alzheimer's disease.

Fever: An Adjunct to Inflammation

An important systemic component of inflammation is **fever,** defined as an abnormally elevated body temperature. Although fever is a nearly universal symptom of infection, it is also associated with certain allergies, cancers, and other organic illnesses. Fevers whose causes are unknown are called fevers of unknown origin, or FUO.

The body temperature is normally maintained by a control center in the hypothalamus. This thermostat regulates the body's heat production and heat loss and sets the core temperature at around 37°C (98.6°F), with slight fluctuations (1°F) during a daily cycle. Fever is initiated when a circulating substance called **pyrogen*** resets the hypothalamic thermostat to a higher setting. This change signals the musculature to increase heat production and peripheral arterioles to decrease heat loss through vasoconstriction (microfile 14.5). Fevers range in severity from low-grade (37.7°–38.3°C or 100°–101°F) to moderate (38.8°–39.4°C, or 102°–103°F) to high (40.0°–41.1°C, or 104°–106°F). Fevers of 106° F and above are dangerous, but they are rare and are usually due to organic disease, not infectious agents. Pyrogens are described as *exogenous* (coming from outside the body) or *endogenous* (originating internally). Exogenous pyrogens are products of infectious agents such as viruses, bacteria, protozoa, and fungi. One well-characterized exogenous pyrogen is endotoxin, the lipopolysaccharide found in the cell walls of gram-negative bacteria. Blood, blood products, vaccines, or injectable solutions can also contain exogenous pyrogens. A product labeled "nonpyrogenic" is safe to use because it contains no fever-causing contaminants. Endogenous pyrogens are liberated by monocytes, neutrophils, and macrophages during the process of phagocytosis and appear to be a natural part of the immune response. Two potent pyrogens released by macrophages are interleukin-1 and tumor necrosis factor.

Benefits of Fever The association of fever with infection strongly suggests that it serves a beneficial role, a view still being debated but gaining acceptance. Aside from its practical and medical importance as a sign of a physiological disruption, increased body temperature has additional benefits:

- Fever inhibits multiplication of temperature-sensitive microorganisms such as the poliovirus, cold viruses, herpes zoster virus, systemic and subcutaneous fungal pathogens, *Mycobacterium* species, and the syphilis spirochete.
- Fever impedes the nutrition of bacteria by reducing the availability of iron. It has been demonstrated that during fever, the macrophages stop releasing their iron stores, which could retard several enzymatic reactions needed for bacterial growth.
- Fever increases metabolism and stimulates immune reactions and naturally protective physiological processes. It speeds up hematopoiesis, phagocytosis, and specific immune reactions.

Treatment of Fever With this revised perspective on fever, whether to suppress it or not can be a difficult decision. Some advocates feel that a slight to moderate fever in an otherwise healthy person should be allowed to run its course, in light of its potential benefits and minimal side effects. All medical experts do agree that high and prolonged fevers, or fevers in patients with cardiovascular disease, seizures, and respiratory ailments, are risky and must be treated immediately

*pyrogen (py´-roh-jen) Gr. *pyr,* fire, and *gennan,* produce. As in funeral pyre and pyromaniac.

MICROFILE 14.5 SOME FACTS ABOUT FEVER

Fever is such a prevalent reaction that it is a prominent symptom of hundreds of diseases. For thousands of years, people believed fever was part of an innate protective response. Hippocrates offered the idea that it was the body's attempt to burn off a noxious agent. Sir Thomas Sydenham wrote in the seventeenth century: "Why, fever itself is Nature's instrument!" So widely held was the view that fever could be therapeutic that pyretotherapy (treating disease by inducing an intermittent fever) was once used to treat syphilis, gonorrhea, leishmaniasis (a protozoan infection), and cancer. This attitude fell out of favor when drugs for relieving fever (aspirin) first came into use in the early 1900s, and an adverse view of fever began to dominate.

Changing Views of Fever In recent times, the medical community has returned to the original concept of fever as more healthful than harmful. Experiments with vertebrates indicate that fever is a universal reaction, even in cold-blooded animals such as lizards and fish. A study with *febrile** mice and frogs indicated that fever increases the rate of antibody synthesis. Work with tissue cultures showed that increased temperatures stimulate the activities of T cells and increase the effectiveness of interferon. Artificially infected rabbits and pigs allowed to remain febrile survive at a higher rate than those given suppressant drugs. Fever appears to enhance phagocytosis of staphylococci by neutrophils in guinea pigs and humans.

Hot and Cold. Why Do Chills Accompany Fever? Fever almost never occurs as a single response; it is usually accompanied by chills. What causes this oddity—that a person flushed with fever periodically feels cold and trembles uncontrollably? The explanation lies in the natural physiological interaction between the thermostat in the hypothalamus and the temperature of the blood. As long as the hypothalamus reads the blood's temperature as cooler than the set point of the thermostat, mechanisms for increasing the body temperature will continue. For example, if the thermostat has been set (by pyrogen) at 102°F but the blood temperature is 99°F, the muscles are stimulated to contract involuntarily (shivering) as a means of producing heat. In addition, the vessels in the skin constrict, creating a sensation of cold, and the piloerector muscles in the skin cause "goose bumps" to form.

*febrile (fee´-bril) L. *febris*, fever. Feverish.

with suppressant drugs. The classic therapy for fever is an *antipyretic** drug such as aspirin or acetaminophen (Tylenol) that lowers the setting of the hypothalamic center and restores normal temperature. Any physical technique that increases heat loss (tepid baths, for example) can also help reduce the core temperature.

PHAGOCYTES: THE EVER-PRESENT BUSYBODIES OF INFLAMMATION AND SPECIFIC IMMUNITY

By any standard, a phagocyte represents an impressive piece of living machinery, meandering through the tissues to seek, capture, and destroy a target. The general activities of phagocytes are (1) to survey the tissue compartments and discover microbes, particulate matter (dust, carbon particles, antigen-antibody complexes), and injured or dead cells; (2) to ingest and eliminate these materials; and (3) to extract immunogenic information (antigens) from foreign matter. It is generally accepted that all cells have some capacity to engulf materials, but *professional phagocytes* do it for a living. The three main types of phagocytes are neutrophils, monocytes, and macrophages.

Granulocytic Phagocytes: Neutrophils and Eosinophils

As previously stated, neutrophils are general-purpose phagocytes that react early in the inflammatory response to bacteria and other foreign materials and to damaged tissue (see figure 14.18). A common sign of bacterial infection is a high neutrophil count in the blood (neutrophilia), and neutrophils are also a primary component of pus. Eosinophils are attracted to sites of parasitic infections and antigen-antibody reactions, though they play only a minor phagocytic role.

*antipyretic (an″-tih-py-reh′-tik) L. *anti*, against, and *pyretos*, fever. An agent that relieves fever.

Macrophage: King of the Phagocytes

After emigrating out of the bloodstream into the tissues, monocytes are transformed by various inflammatory mediators into macrophages. This process is marked by an increase in size and by enhanced development of lysosomes and other organelles (figure 14.20). At one time, macrophages were classified as either fixed (adherent to tissue) or wandering, but this terminology can be misleading. All macrophages retain the capacity to move about. Whether they reside in a specific organ or wander depends upon their stage of development and the immune stimuli they receive. Specialized macrophages called *histiocytes* migrate to a certain tissue and remain there during their lifespan. Examples are alveolar (lung) macrophages, the Kupffer cells in the liver, Langerhans cells in the skin (figure 14.21), and macrophages in the spleen, lymph nodes, bone marrow, kidney, bone, and brain. Other macrophages do not reside permanently in a particular tissue and drift nomadically throughout the RES. Not only are macrophages dynamic scavengers, but they also process foreign substances and prepare them for reactions with B and T lymphocytes (see chapter 15).

Mechanisms of Phagocytic Discovery, Engulfment, and Killing

Although the term *phagocytosis* literally means the engulfment of particles by cells, phagocytes actually endocytose both particulate and liquid substances. But phagocytosis is more than just the

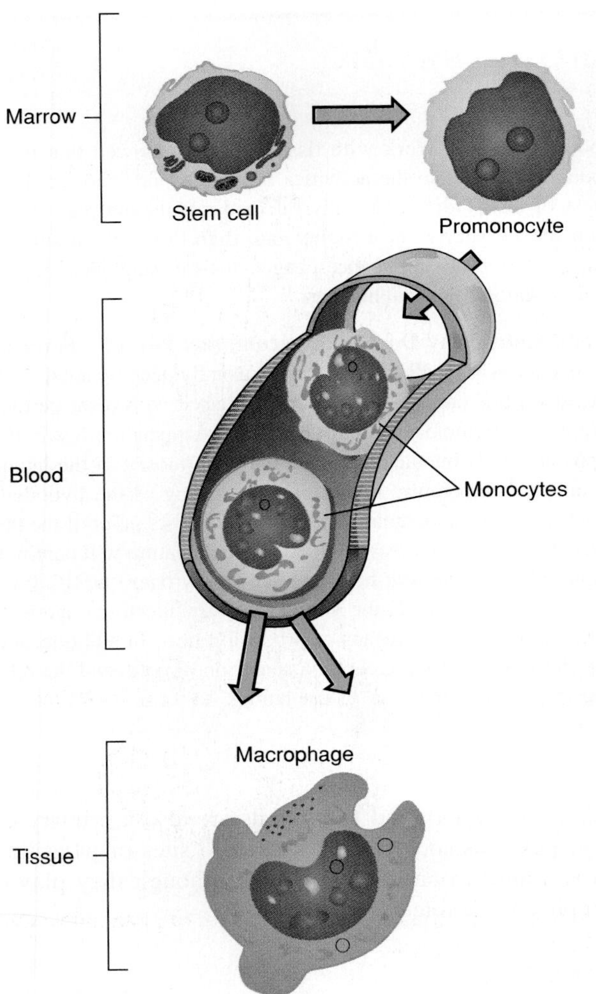

Figure 14.20
The developmental stages of monocytes and macrophages. The cells progress through maturational stages in the bone marrow and peripheral blood. Once in the tissues, a macrophage can remain nomadic or take up residence in a specific organ.

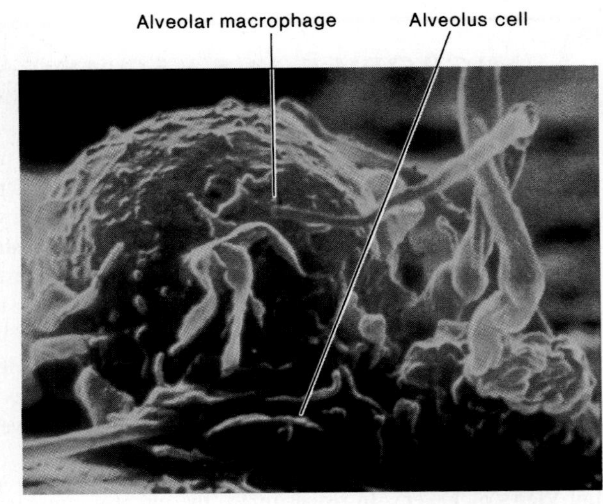

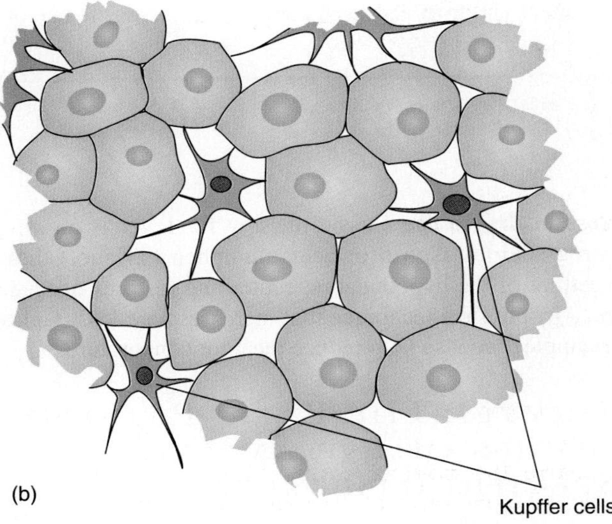

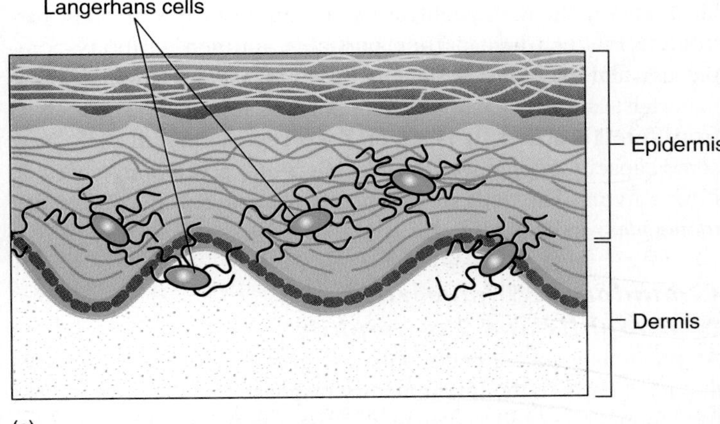

Figure 14.21
Sites containing macrophages. (*a*) Scanning electron micrograph view of a lung with an alveolar macrophage. (*b*) Liver tissue with Kupffer cells. (*c*) Langerhans cells deep in the epidermis.

physical process of engulfment, because phagocytes also actively attack and dismantle foreign cells with a wide array of antimicrobial substances. The events in phagocytosis include chemotaxis, ingestion, phagolysosome formation, destruction, and excretion (figure 14.22).

Chemotaxis and Ingestion Phagocytes move into a region of inflammation with a deliberate sense of direction, attracted by a gradient of stimulant products from the parasite and host tissue at the site of injury. On the scene of an inflammatory reaction, they often trap cells or debris against the fibrous network of connective tissue or the wall of blood and lymphatic vessels. Phagocytosis is often accompanied by *opsonization* (discussed again in chapter 15). This is a process that coats the surface of microorganisms with antibodies or complement, thereby facilitating recognition and engulfment. Once the phagocyte has made contact with its prey, it extends pseudopods that enclose the cells or particles in a pocket and internalize them in a vacuole called a *phagosome* (chapter opening photograph).

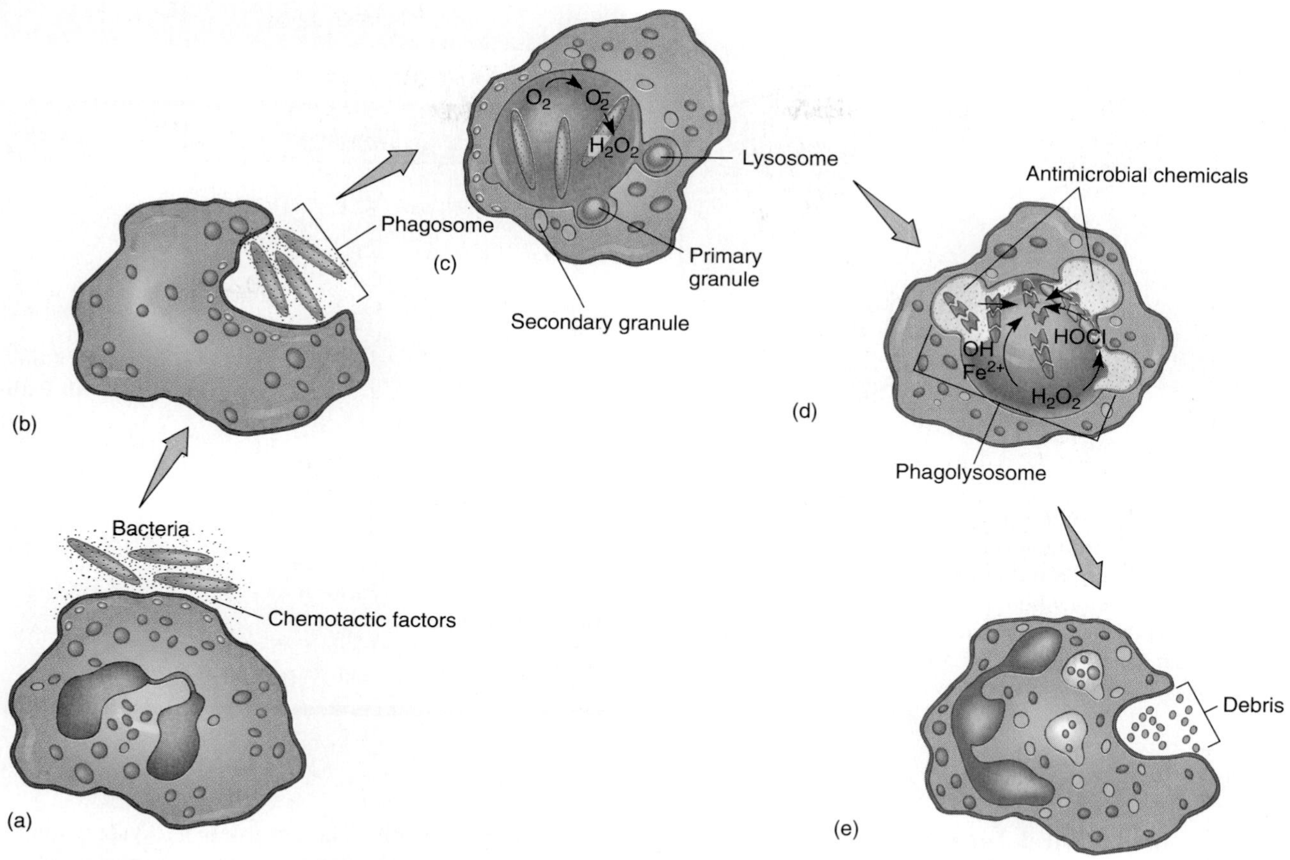

Figure 14.22

The phases in phagocytosis. (*a*) Chemotaxis. (*b*) Contact and ingestion (forming a phagosome). (*c*) Formation of phagolysosome (granules fuse with phagosome). (*d*) Killing, digestion of the microbe. (*e*) Release of debris.

Phagolysosome Formation and Killing In a short time, *lysosomes* migrate to the scene of the phagosome and fuse with it to form a *phagolysosome.* Other granules containing antimicrobial chemicals are released into the phagolysosome, forming a potent brew designed to poison and then dismantle the ingested material. The destructiveness of phagocytosis is evident by the death of bacteria within 30 minutes after contacting this battery of antimicrobial substances.

Destruction and Elimination Systems Two separate systems of destructive chemicals await the microbes in the phagolysosome. The oxygen-dependent system elaborates several substances that were described in chapters 7 and 11. Myeloperoxidase, an enzyme found in granulocytes, forms halogen ions (OCl^-) that are strong oxidizing agents. Other products of oxygen metabolism such as hydrogen peroxide, the superoxide anion (O_2^-), activated or so-called singlet oxygen ($^1O_2$), and the hydroxyl free radical (.OH) separately and together have formidable killing power. Other mechanisms that come into play are the liberation of lactic acid, lysozyme, and *nitric oxide* (NO), a powerful mediator that kills bacteria and inhibits viral replication. Cationic proteins that injure bacterial cell membranes and a number of proteolytic and other hydrolytic enzymes complete the job. The small bits of undigestible debris are exocytosed and released.

Contributors to the Body's Chemical Immunity

Interferon: Antiviral Cytokines and Immune Stimulants

Interferon (IFN) was described in chapter 12 as a small protein produced naturally by certain white blood and tissue cells that is used in therapy against certain viral infections and cancer. Although the interferon system was originally thought to be directed exclusively against viruses, it is now known to be involved also in defenses against other microbes and in immune regulation and intercommunication. Three major types are *alpha interferon,* a product of lymphocytes and macrophages; *beta interferon,* a product of fibroblasts and epithelial cells; and *gamma interferon,* a product of T cells.

All three classes of interferon are produced in response to viruses, RNA, immune products, and various antigens. Their biological activities are extensive. In all cases, they bind to cell surfaces and induce changes in genetic expression, but the exact results vary. In addition to antiviral effects discussed in the next section, all three IFNs can inhibit the expression of cancer genes and have tumor suppressor effects. Alpha and beta IFN stimulate phagocytes, and gamma IFN is an immune regulator of macrophages and T and B cells.

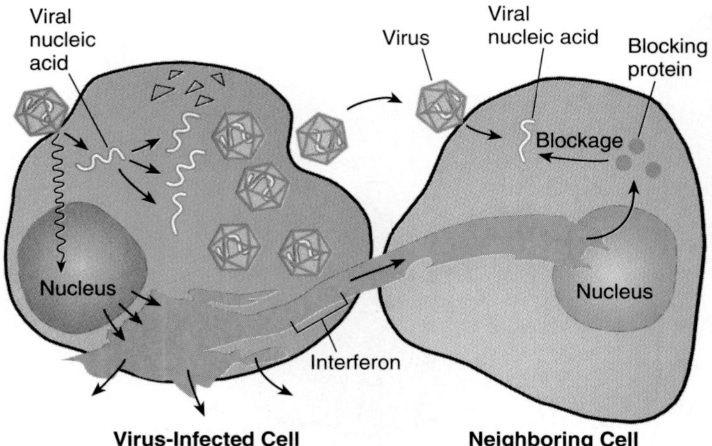

Figure 14.23
The antiviral activity of interferon. When a cell is infected, its nucleus is triggered to transcribe and translate the interferon (IFN) gene. The interferon diffuses out of the infected cell into nearby (uninfected) cells, where it enters the nucleus. Here IFN activates a gene for synthesizing a peptide that blocks viral replication. Note that the original cell is not protected by IFN and that IFN does not prevent viruses from invading the protected cells.

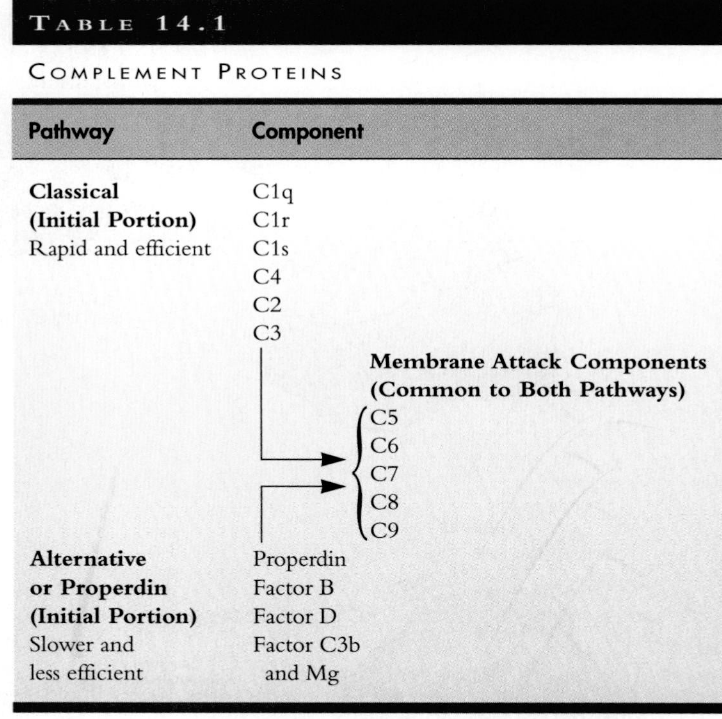

TABLE 14.1

COMPLEMENT PROTEINS

Pathway	Component	
Classical (Initial Portion) Rapid and efficient	C1q C1r C1s C4 C2 C3	**Membrane Attack Components (Common to Both Pathways)** C5 C6 C7 C8 C9
Alternative or Properdin (Initial Portion) Slower and less efficient	Properdin Factor B Factor D Factor C3b and Mg	

Characteristics of Antiviral Interferon The binding of a virus to the receptors of an infected cell sends a signal into the cell nucleus that activates the genes coding for interferon (figure 14.23). As interferon is synthesized, it is rapidly secreted by the cell into the extracellular spaces. The action of antiviral interferon is indirect; it does not kill or inhibit the virus directly. After diffusing to nearby, uninfected cells and entering them, IFN activates a gene complex that codes for another protein. This second protein, not interferon itself, interferes with the multiplication of viruses. Interferon is not virus-specific, so its synthesis in response to one type of virus will also protect against other types. Because this inhibitory protein is the direct inhibitor of viruses, it could be a valuable treatment for AIDS and other virus infections in the future.

Other Roles of Interferon Interferons are also important immune regulatory cytokines that activate or instruct the development of white blood cells. For example, alpha interferon produced by T lymphocytes activates a subset of cells called natural killer (NK) cells. In addition, one type of beta interferon plays a role in the maturation of B and T lymphocytes and in inflammation. Gamma interferon inhibits cancer cells, stimulates B lymphocytes, activates macrophages, and enhances the effectiveness of phagocytosis.

Complement: A Versatile Backup System

Among its many overlapping functions, the immune system has another complex and multiple-duty system called **complement (C factor)** that, like inflammation and phagocytosis, is brought into play at several levels. The complement system, named for its property of "completing" immune reactions, consists of 20 blood proteins that work in concert to destroy bacteria and certain viruses. The sources of complement factors are liver hepatocytes, lymphocytes, and monocytes. Some knowledge of this important system will help in your understanding of topics in chapters 15 and 16.

The concept of a cascade reaction is helpful in understanding how complement functions. A cascade reaction is a sequential physiological response like that of blood clotting, in which the first substance in a chemical series activates the next substance, which activates the next, and so on, until a desired end product is reached. Complement exhibits two schemes, the *classical pathway* and the *alternative pathway,* which differ in how they are activated and in speed and efficiency. The two pathways merge at the final stages into a common pathway with a similar end result (figure 14.24). Since the complement numbers (C1–C9) are based on the order of their discovery, be aware that factors C1–C4 do not appear in numerical order during activation (table 14.1).

Overall Stages in the Complement Cascade In general, the complement cascade includes the three stages of *initiation, amplification and cascade,* and *membrane attack.* At the outset, an initiator (such as microbes, cytokines, and antibodies; table 14.2) reacts with the first complement chemical, which propels the reaction on its cascading course. This process develops a recognition site on the surface of the target cell where the initial C components will bind. Through a stepwise series, each component reacts with another on or near the recognition site. In the C2–C5

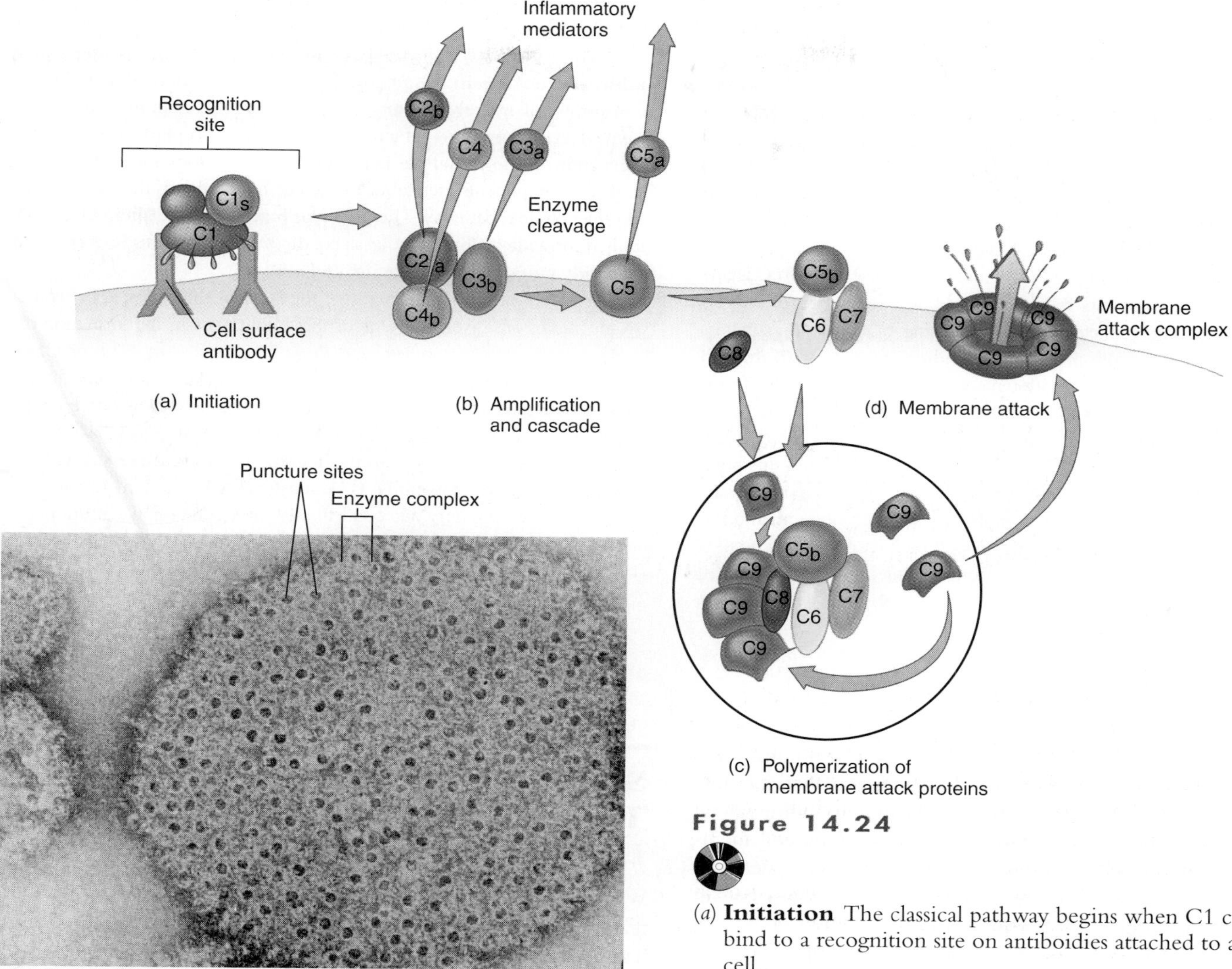

Figure 14.24

(*a*) **Initiation** The classical pathway begins when C1 components bind to a recognition site on antiboidies attached to a foreign cell.

(*b*) **Amplification and Cascade** The C1 complex is an enzyme that activates a second series of components, C4 and C2. When these have been enzymatically cleaved into separate molecules, they become a second enzyme complex that activates C3. At this same site, C3 binds to C5 and cleaves it to form a product that is tightly bound to the membrane.

(*c*) **Polymerization.** C5 is a reactive site for the final assembly of an attack complex. In series, C6, C7, and C8 aggregate with C5 and become integrated into the membrane. They form a substrate upon which the final component, C9, can bind. Up to 15 of these units ring the central core of the complex.

(*d*) **Membrane attack.** The final product of these reactions is a large donut shaped enzyme that punctures small pores through the membrane, leading to cell lysis.

(*e*) An electron micrograph (187,000×) of a cell reveals multiple puncture sites over its surface. The lighter, ringlike structures are the actual enzyme complex.

TABLE 14.2

SUBSTANCES THAT ACTIVATE THE COMPLEMENT PATHWAYS

Activators in the Classical Pathways	Activators in the Alternative Pathway
Complement-fixing antibodies: IgG, IgM	Cell wall components; e.g., yeast and bacteria
	Viruses; e.g., influenza virus
Bacterial lipopolysaccharide	Parasites; e.g., *Schistosoma*
Pneumococcal C-reactive protein	Fungi; e.g., *Cryptococcus*
Retroviruses	Some tumor cells
Polynucleotides	X-ray opaque media, dialysis membranes
Mitochondrial membranes	

MICROFILE 14.6 HOW COMPLEMENT WORKS

The **classical pathway** is a part of the specific immune response covered in chapter 15. It is initiated when antibody, called complement-fixing antibody, attaches to antigen on the surface of a membrane. The first chemical, C1, is a large complex of three molecules, C1q, C1r, and C1s (see table 14.1). When the C1q subunit has recognized and bound two or more antibodies, the C1r subunit cleaves the C1s proenzyme, and an activated enzyme emerges. During amplification, the C1s enzyme has as its primary targets proenzymes C4 and C2. Through the enzyme's action, C4 is converted into C4a and C4b, and C2 is converted into C2a and C2b. C4b and C2a fragments remain attached as an enzyme, C3 convertase, whose substrate is factor C3. The cleaving of C3 yields subunits C3a and C3b. C3b has the property of binding strongly with the cell membrane in close association with the component C5, and it also forms an enzyme complex with C4b-C2a that converts C5 into two fragments, C5a and C5b. C5b will form the nucleus for the membrane attack complex. This is the point at which the two pathways merge. From this point on, C5b reacts with C6 and C7 to form a stable complex inserted in the membrane. Addition of C8 to the complex causes the polymerization of several C9 molecules into a giant ring-shaped membrane attack complex that bores ring-shaped holes in the membrane, which lyse the target cell or destroy the virus (figure 14.24*e*).

The **alternative pathway,** sometimes called the **properdin*** **pathway,** is not specific to a particular microbe. It can be initiated by a wide variety of microbes, tumors, and cell walls. It requires a different group of serum proteins—factors B, D, and P (properdin), C3b, and magnesium in the initiation and amplification phases rather than C1, C2, or C4 components (see table 14.1). The remainder of the steps occur as in the classical pathway. The principal function of the alternative system is to provide a slower but less specific means of lysing foreign cells (especially gram-negative bacteria) and viruses.

You will notice that at many of the steps of the classical pathway, two molecules are given off. One of these continues in the formation of the membrane attack complex, and the other (C2a, C4a, C3a, or C5a) goes on to become a cytokine or stimulant of inflammation and other immune reactions. Complement components also behave as one type of opsonin that promotes phagocytosis. Complement can participate in inflammation and allergy by causing liberation of vasoactive substances from mast cells and basophils. C3a and C5a are so potent in this response that injecting only one quadrillionth of a gram elicits an immediate flare-up at the site.

* **properdin** (proh´-pur-din) L. *pro,* before, and *perdere,* to destroy.

series, enzymatic cleavage produces several inflammatory cytokines. Other details of the pathways differ, but whether classical or alternative, the functioning end product is a large, multienzyme *membrane attack complex* that can digest holes in the cell membranes of bacteria, cells, and enveloped viruses, thereby destroying them (figure 14.24*a–d*). The two pathways are described in more detail in microfile 14.6.

Chapter Checkpoints

Nonspecific immune reactions are generalized responses to invasion, regardless of the type. These include inflammation, phagocytosis, interferon, and complement.

The four symptoms of inflammation are rubor (redness), calor (heat), tumor (edema), and dolor (pain).

Fever is another component of nonspecific immunity. It is caused by both endogenous and exogenous pyrogens. Fever increases the rapidity of the host immune responses and reduces the viability of many microbial invaders.

Macrophages are activated monocytes. Along with neutrophils (PMNs), they are the key phagocytic agents of nonspecific response to disease.

The plasma contains complement, a nonspecific group of chemicals that works with the third line of defense to attack foreign cells.

SPECIFIC IMMUNITIES: THE THIRD AND FINAL LINE OF DEFENSE

When host barriers and nonspecific defenses fail to control an infectious agent, a person with a normal functioning immune system has an extremely substantial mechanism to resist the pathogen—the third, specific line of immunity. This aspect of immunity is the resistance developed after contracting childhood ailments such as chickenpox or measles that provides long-term protection against future attacks. This sort of immunity is not innate, but adaptive; it is acquired only after an immunizing event such as an infection. The absolute need for acquired or adaptive immunity is impressively documented in children who have genetic defects in this system or in AIDS patients who have lost it. Even with heroic measures to isolate the patient, combat infection, or restore lymphoid tissue, the victim is constantly vulnerable to life-threatening infections.

Acquired specific immunity is the product of a dual system that we have previously mentioned—the B and T lymphocytes. During fetal development, these lymphocytes undergo a selective process that specializes them for reacting only to one specific antigen. During this time, **immunocompetence,** the ability of the body to react with myriad foreign substances, develops. An infant is born with the theoretical potential to acquire millions of different immunities.

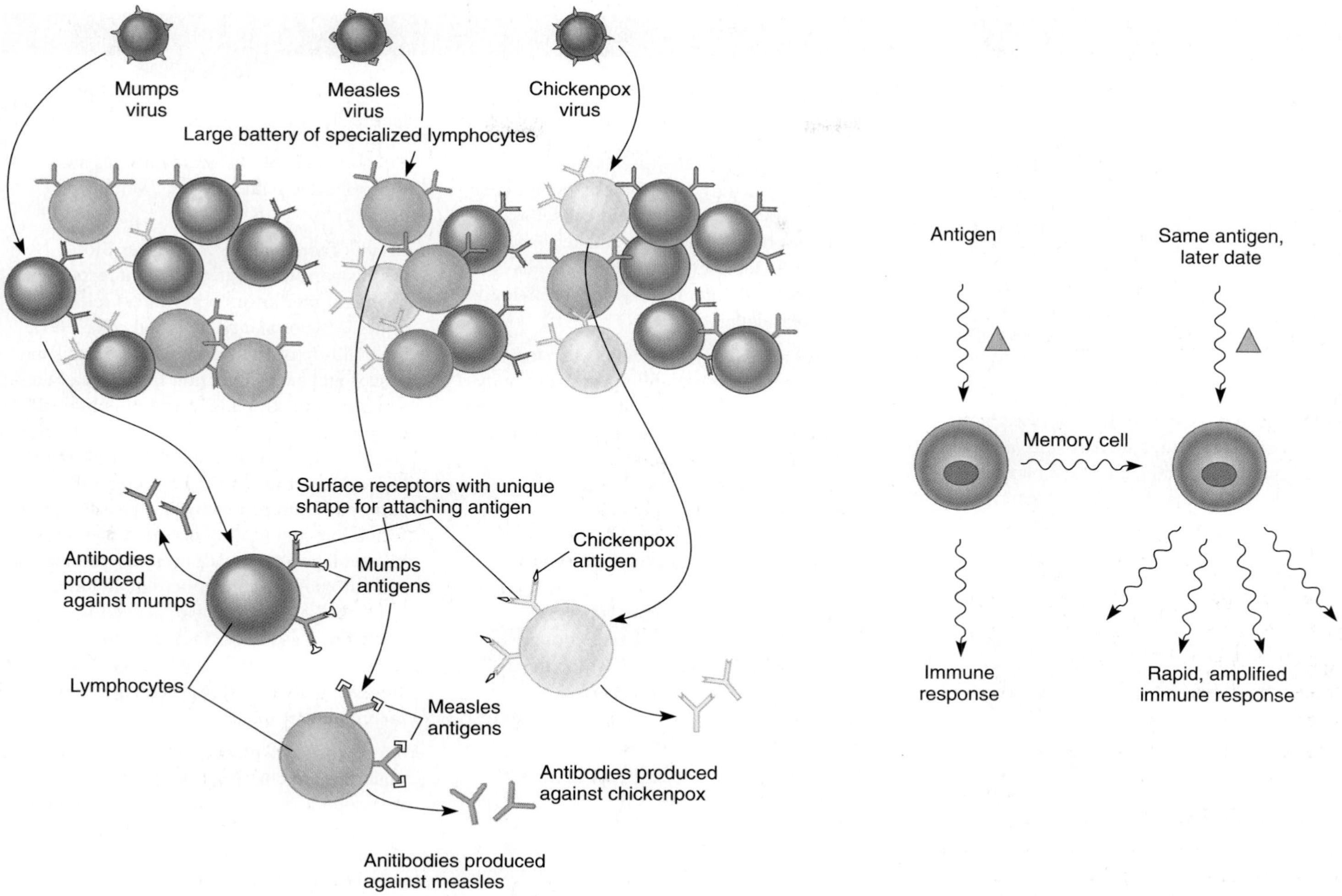

(a) Specificity: Viruses and other infectious agents contain antigen molecules that are specific to a single type of lymphocyte. One result of binding will be the production of virus-specific antibodies.

(b) Memory: First contact with antigen creates a memory and quick recall upon second and other future contacts with that antigen.

Figure 14.25

The characteristics of acquired immunity: (*a*) specificity and (*b*) memory.

Two features that most characterize this third line of defense are **specificity** and **memory.** Unlike mechanisms such as anatomical barriers or phagocytosis, acquired immunity is highly selective. For example, the antibodies produced during an infection against the chickenpox virus will function against that virus and not against the measles virus (figure 14.25*a*). The property of memory pertains to the brisk mobilization of lymphocytes that have been programmed to "recall" their first engagement with the invader and rush to the attack once again (figure 14.25*b*). This complex and fascinating response will be covered more extensively in chapter 15.

Chapter Checkpoints

B and T lymphocytes are the agents of the specific immune response. Unlike the nonspecific responses, it develops after exposure to a specific antigen and is tailor-made to it. In addition, the B and T cells "remember" the antigen and respond much more rapidly on subsequent exposures.

CHAPTER CAPSULE WITH KEY TERMS

THE THREE LEVELS OF HOST DEFENSES

The **first line of defense** is composed of barriers that block the pathogen at the portal of entry; the **second line of defense** includes generalized protective cells and fluids in tissues; and the **third line of defense** includes specific immune reactions with microbes that are required for survival.

First-Line Defenses: Physical barriers are anatomical structures such as skin, mucous membranes, and cilia; chemical barriers include fatty acids, lysozyme in tears, saliva, and gastric acidity; genetic barriers can be a lack of susceptibility to an infectious agent due to the specialization of a microbe for an exact host.

IMMUNITY/IMMUNOLOGY

Immunity is specific resistance acquired to an infectious agent. Functions of the immune system are: **surveillance** by specialized **white blood cells (WBCs)** for **markers** or receptors (molecules on surface of cells) in tissues; **recognition** by WBCs that detect and differentiate normal markers **(self)** from **foreign** markers or **antigens (nonself)**. WBCs are programmed to destroy and remove any foreign material, while self is usually unaffected.

Components of Immunity: The immune system is a diffuse network of cells, fibers, chemicals, fluids, tissues, and organs that permeates the body. The structure at the cellular level includes separate but interconnected compartments: the **reticuloendothelial (mononuclear phagocyte) system,** a continuous network of fibers and phagocytes that surrounds the tissues and organs; the **extracellular fluid (ECF),** a liquid environment in which all cells are bathed; the **lymphatic system,** a series of vessels and organs that carry *lymph* from tissues; and the **bloodstream,** which circulates blood to all organs. The constant communication among the compartments ensures that a reaction in one will be transmitted to another.

CIRCULATORY SYSTEM: BLOOD AND LYMPHATICS

Composition of Whole Blood: **Plasma** is a clear, complex liquid that contains nutrients, ions, gases, hormones, antibodies, albumin, and waste products dissolved in water. **Serum** is plasma minus the clotting factors. **Blood cells** are formed by **hemopoiesis** in particular bone marrow sites. From **stem cells,** three main lines of cells are differentiated—white blood cells **(leukocytes),** red blood cells **(erythrocytes),** and megakaryocytes that give rise to **platelets.**

Functions of Leukocytes: Leukocytes, the primary cells of host defenses and immunity, are either **granulocytes (polymorphonuclear leukocytes)** or **agranulocytes (mononuclear leukocytes).** Granulocytes contain distinct granules in cytoplasm and include **neutrophils,** which function as phagocytes; **eosinophils,** which function in worm and fungal infections; **basophils,** which are involved in allergic responses, along with tissue **mast cells.** Agranulocytes lack noticeable granules and include **monocytes** and **lymphocytes.** Monocytes function as blood phagocytes and give rise to **macrophages** in tissues. Two main types of lymphocytes are **B cells,** which produce **antibodies** as part of **humoral immunities,** and **T cells,** which participate in **cell-mediated immunities** (CMIs). Leukocytes are motile by ameboid motion, can insert between the endothelial cells of small blood vessels **(diapedesis)** and enter surrounding tissues, and can respond to tissue injury or infection by migrating toward chemical signals **(chemotaxis).**

LYMPHATIC SYSTEM

The lymphatic system begins as fine capillaries in tissues that gradually join together into larger vessels that eventually drain into the blood circulation. The vessels transport **lymph,** a fluid that contains serum components and white blood cells. Lymphatic organs include **lymph nodes,** compact filters where lymphocytes aggregate and where immune challenges occur; the **spleen,** a blood filter and repository of immune cells; and the **thymus,** where T cells mature. Other lymphatic tissue includes tonsils, gut-associated lymphoid tissue **(GALT),** and Peyer's patches.

GENERALIZED IMMUNE REACTIONS

Inflammatory Response: This complex system responds to tissue injury (infection, burn, allergy) by mobilizing the immune system against pathogens, repairing damage, and clearing infection. Its common signs and symptoms are redness, heat, swelling, and pain.

Stages of Inflammatory Response: Blood vessels narrow and then dilate in response to **chemical mediators (cytokines)** released by injured tissues and white blood cells. Next, the build-up of fluid from **edema** swells the tissues and keeps infection from spreading, and attract neutrophils to engulf debris and microbes. WBCs, microbes, debris, and fluid collect to form **pus;** macrophages clean up the residue of inflammation; lymphocytes carry out immune reactions such as antibody formation; and healing occurs. Long-term inflammation can result in injury and disease (granuloma). **Fever,** an increase in body temperature above normal, is due to **pyrogens,** substances that alter the temperature setting in the brain. Fever can slow microbial multiplication and stimulate the immune response.

Phagocytosis is a process whereby foreign materials are engulfed and destroyed. Neutrophils engulf small particles, microbes, molecules; macrophages are larger cells that scavenge large packets of cellular debris and extract antigenic information. Macrophages live in a specific tissue or organ (liver, lung, skin) or are free and wandering. After materials are engulfed by the cell into a **phagosome** vacuole, **lysosomes** containing powerful chemicals unite with the phagosome and destroy its contents.

IMPORTANT CYTOKINES AND RELATED DEFENSES

Interferon (IFN) is a family of proteins produced by leukocytes and fibroblasts in response to infection, cancer, or various immune signals. *Alpha* and *beta* interferon are natural infection fighting substances, and *gamma* interferon is an immune activator and regulator. All three also function as antiviral and anticancer cytokines. IFN works by inhibiting the synthesis of viral proteins and by increasing the cellular immune defenses. **Complement** is a complex chemical defense system that destroys certain pathogens and produces chemical mediators. It involves chemicals called complement **(C factor)** that act in cascade fashion: one component activates the next in line, which activates the next, and so on. Two major pathways are: (1) classical, which involves activation of complement by specific antibody; and (2) alternative, which is a nonspecific reaction to infections. The result of complement activation is a huge enzyme, the **membrane attack complex,** that can kill cells and inactivate viruses by digesting holes in their membranes.

CHARACTERISTICS OF ACQUIRED IMMUNITIES

Immunocompetent individuals possess a third line of defense that is acquired only after direct exposure to an infectious agent. It is made possible by a large repertoire of lymphocytes present from birth that mounts an individualized response to each different infectious agent. These responses are extremely **specific** in their effects and give rise to **immunologic memory,** which provides protection in the case of reexposure to the same pathogen.

MULTIPLE-CHOICE QUESTIONS

1. An example of a nonspecific chemical barrier to infection is:
   a. unbroken skin c. cilia in trachea
   b. lysozyme in saliva d. all of these
2. Which nonspecific host defense is associated with the trachea?
   a. lacrimation c. desquamation
   b. ciliary lining d. lactic acid
3. Which of the following blood cells function primarily as phagocytes?
   a. eosinophils c. lymphocytes
   b. basophils d. neutrophils
4. Which of the following is not a lymphoid tissue?
   a. spleen c. lymph nodes
   b. thyroid gland d. GALT
5. What is included in GALT?
   a. thymus c. tonsils
   b. Peyer's patches d. breast lymph nodes
6. Monocytes are ____ leukocytes that develop into ____ .
   a. granular, phagocytes
   b. agranular, mast cells
   c. agranular, macrophages
   d. granular, T cells
7. Which of the following inflammatory signs specifies pain?
   a. tumor c. calor
   b. dolor d. rubor
8. An example of an inflammatory cytokine that stimulates vasodilation is
   a. histamine c. complement C5a
   b. collagen d. interferon
9. ____ is an example of an inflammatory cytokine that stimulates chemotaxis.
   a. Endotoxin c. Fibrin clot
   b. Serotonin d. Interleukin-2
10. An example of an exogenous pyrogen is
    a. interleukin-1 c. interferon
    b. complement d. endotoxin
11. ____ interferon is secreted by ____ and is involved in destroying viruses.
    a. Gamma, fibroblasts
    b. Beta, lymphocytes
    c. Alpha, lymphocytes
    d. Beta, fibroblasts
12. Which of the following substances is *not* produced by phagocytes to destroy engulfed microorganisms?
    a. myeloperoxidase
    b. superoxide anion
    c. hydrogen peroxide
    d. bradykinin
13. Which of the following is the end product of the complement system?
    a. properdin
    b. cascade reaction
    c. membrane attack complex
    d. complement factor C9
14. Immunologic memory refers to the ability of the immune system to
    a. recognize millions of different antigens
    b. react with millions of different antigens
    c. migrate from the blood vessels into the tissues
    d. recall a previous immune response
15. Which subset of WBCs accounts for acquired, specific immunity?
    a. monocytes c. T cells
    b. B cells d. both b and c

CONCEPT QUESTIONS

1. a. Explain the functions of the three lines of defense.
   b. Which is the most essential to survival?
   c. What is the difference between nonspecific host defenses and immune responses?
2. a. Describe the main elements of the process through which the immune system distinguishes self from nonself.
   b. How is surveillance of the tissues carried out?
   c. What is responsible for it?
   d. What does the term *foreign* mean in reference to the immune system?
   e. Define the term *antigen*.
3. a. Trace the complete cycle of a bacterium through the immune compartments, starting with the blood, the RES, the lymphatics, and the ECF. (You will start and end at the same point.)
   b. What is the direct connection between the bloodstream and the lymphatic circulation?
4. a. What are the main components of the reticuloendothelial system?
   b. Why is it also called the mononuclear phagocyte system?
   c. How does it communicate with tissues and the vascular system?
5. a. Prepare a simplified outline of the cell lines of hemopoiesis.
   b. What is a stem cell?
   c. Review the anatomical locations of hemopoiesis at various stages of development.
6. a. Differentiate between granulocytes and agranulocytes.
   b. Describe the main cell types in each group, their functions, and their incidence in the circulation.
7. a. What is the principal function of lymphocytes?
   b. Differentiate between the two lymphocyte types and between humoral and cell-mediated immunity.
8. a. Why are platelets called formed elements?
   b. What are their functions?
   c. How are they associated with inflammation?
9. a. Explain the processes of diapedesis and chemotaxis, and show how they interrelate.

b. Referring to question 3, explain how these processes can account for the movements of leukocytes through the fluid compartments.

10. a. What is lymph, and how is it formed?
 b. Why are white cells but not red cells normally found in it?
 c. What are the functions of the lymphatic system?
 d. Explain the filtering action of a lymph node.
 e. What is GALT, and what are its functions?

11. Differentiate between the terms *pyogenic* and *pyrogenic*. Use examples to explain the impact they have on the immune response.

12. Briefly account for the origins and actions of the major types of inflammatory mediators (cytokines).

13. a. Describe the events that give rise to macrophages.
 b. What types are there, and what are their principal functions?

14. a. Outline the major phases of phagocytosis.
 b. In what ways is a phagocyte a tiny container of disinfectants?

15. a. Briefly describe the three major types of interferon, their sources, and their biological effects.
 b. Describe the mechanism by which interferon acts as an antiviral compound.

16. a. Describe the general complement reaction in terms of a cascade.
 b. What is the end result of complement activation?
 c. What are some other functions of complement components?

17. What are immunocompetence, immunologic specificity, and immunologic memory?

Critical-Thinking Questions

1. Suggest some reasons that there is so much redundancy of action and there are so many interacting aspects of immune responses.

2. a. What is probably missing in children born without a functioning lymphocyte system?
 b. What is the most important component extracted in bone marrow transplants?

3. a. What is the likelihood that plants have some sort of immune protection?
 b. Explain your reasoning.

4. a. What substances contained in plasma are absent from serum? (Which of the two would be used to perform a clotting test?)
 b. If one wishes to give antibodies to treat infections, which is better to use, plasma or serum, or does it matter?

5. A patient's chart shows an increase in eosinophil levels.
 a. What does this cause you to suspect?
 b. What if the basophil levels are very high?

6. How can adults continue to function relatively normally after surgery to remove the thymus, tonsils, spleen, or lymph nodes?

7. a. What actions of the inflammatory and immune defenses account for swollen lymph nodes and leukocytosis?
 b. What is pus, and what does it indicate?
 c. In what ways can edema be beneficial?
 d. In what ways is it harmful?

8. An obsolete treatment for syphilis involved inducing fever by deliberately infecting patients with the agent of relapsing fever. A recent but failed AIDS treatment involved applying heat to the body to induce hyperthermia. Can you provide some possible explanations behind these peculiar forms of treatment?

9. Patients with a history of tuberculosis often show scars in the lungs and experience recurrent infection. Account for these effects on the basis of the inflammatory response.

10. Explain why the alternative pathway in complement action is nonspecific and the classical pathway is more specific in its effects, even though they arrive at the same end product.

11. *Shigella, Mycobacterium,* and numerous other pathogens have developed mechanisms that prevent them from being killed by phagocytes.
 a. Suggest two or three factors that help them avoid destruction by the powerful antiseptics in macrophages.
 b. In addition, suggest the potential implications that these infected macrophages can have in the development of disease.

12. Macrophages perform the final job of removing tissue debris and other products of infection. Indicate some of the possible effects when these scavengers cannot successfully complete the work of phagocytosis.

13. Account for the several symptoms that occur in the injection site when one has been vaccinated against influenza.

14. a. Knowing that fever is potentially both harmful and beneficial, what are some possible guidelines for deciding whether to suppress it or not?
 b. What is the specific target organ of a fever-suppressing drug?

Internet Search Topic

Type in the term *cytokine* on a search engine and look for a cytokine Web page. Find information relating to the number of currently known substances, and describe two medical applications of cytokine therapy.

chapter 15

The Acquisition of Specific Immunity and Its Applications

The specific reactions of the human immune system represent some of the most exciting, useful, and challenging concepts in microbiology. A detailed exploration of the inner workings of this remarkable, finely tuned system will reveal not only how it defends against microorganisms and destroys cancer cells but also how it is involved in diseases as diverse as hay fever, diabetes, and AIDS.

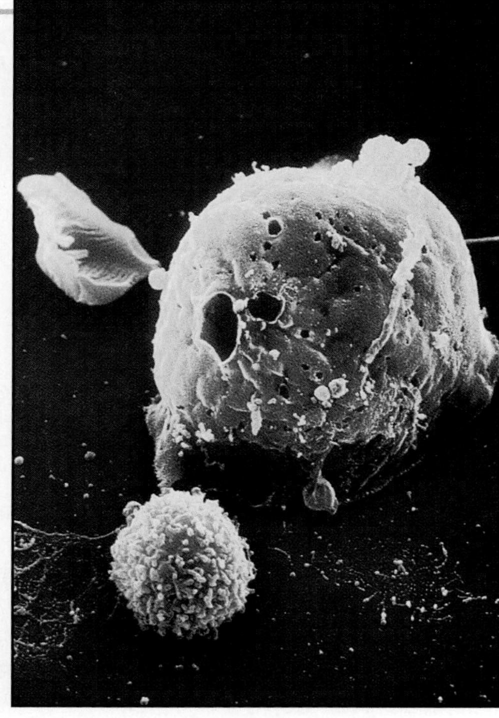

A cytotoxic T cell (lower cell with fine projections) attacks a tumor cell (upper cell). The substances secreted by the T cell perforate the cancer cell with several holes, causing it to eventually burst and die.

FURTHER EXPLORATIONS INTO THE IMMUNE SYSTEM

In chapter 14 we described the capacity of the immune system to survey, recognize, and react to foreign cells and molecules, and we overviewed the characteristics of nonspecific host defenses, blood cells, phagocytosis, inflammation, and complement. In addition, we introduced the concepts of acquired immunity and specificity. In this chapter we take a closer look at those topics.

The elegance and complexity of immune function are largely due to lymphocytes working closely together with macrophages. To simplify and clarify the network of immunologic development and interaction, we present it here as a series of five stages, with each stage covered in a separate section (figure 15.1). The principal stages include the following: I. lymphocyte development and differentiation; II. the processing of antigens; III. the challenge of B and T lymphocytes by antigens; IV. B lymphocytes and the production and activities of antibodies; and V. T-lymphocyte responses. Notice that as each stage is covered, we will simultaneously be following the sequence of an immune response.

THE DUAL NATURE OF SPECIFIC IMMUNE RESPONSES

In the following overview, Roman numerals correspond with the subsequent text sections that cover these topics in more detail.

I. Development of the Dual Lymphocyte System

Lymphocytes are central to immune responsiveness. They undergo a sequential development that begins in the embryonic yolk sac and shifts to the liver and bone marrow. Although all lymphocytes arise from the same basic stem cell type, at some point in development, they diverge into two distinct types. Final maturation of B cells occurs in specialized bone marrow sites and that of T cells occurs in the thymus. This process commits each individual B cell or T cell to one specificity. Both cell types subsequently migrate to precise, separate areas in the lymphoid organs (for instance, nodes and spleen, as described in chapter 14) to serve as a lifelong continuous source of immune responsiveness.

II. Entrance and Processing of Antigens and Clonal Selection

When foreign cells, or antigens, enter a fluid compartment of the body, they immediately encounter a network consisting of the lymphatics, blood, and the reticuloendothelial system (RES). In these sites the antigenic materials are met by a battery of cells that work together to screen, entrap, and eliminate them. Certain specialized macrophages are usually the first cells to recognize and react. They ingest and process antigens and present them to the lymphocytes that are specific for that antigen. This selection process is the trigger that activates the lymphocytes. In most cases, the response of B cells also requires the additional assistance of special classes of T cells.

III. B and III T. Activation of Lymphocytes and Clonal Expansion

When challenged by antigen, both B cells and T cells further differentiate and proliferate. The multiplication of a particular lymphocyte creates a clone, or group of genetically identical cells, some of which are memory cells that will ensure future reactiveness against that antigen. Because the B-cell and T-cell responses depart notably from this point in the sequence, they will be summarized separately.

IV. Products of B Lymphocytes: Antibody Structure and Functions

The active progeny of a dividing B-cell alone are called plasma cells. These cells are programmed to synthesize and secrete antibodies into the tissue fluid. When these antibodies attach to the antigen for which they are specific, the antigen is marked for destruction or neutralization. Because secreted antibody molecules circulate freely in the tissue fluids, lymph, and blood, the immunity they provide is humoral.

V. How T Cells Respond to Antigen: Cell-Mediated Immunity (CMI)

T-cell types and responses are extremely varied. When activated (sensitized) by antigen, a T cell gives rise to one of several types of progeny, each involved in a cell-mediated immune function. The four main classes of T cells are: (1) helper cells that assist in immune reactions, (2) suppressor cells that suppress immune reactions, (3) cytotoxic, or killer, cells that destroy specific target cells, and (4) delayed hypersensitivity cells that function in certain allergic reactions. Although T cells secrete cytokines that help destroy antigen or regulate immune responses, they do not produce antibodies.

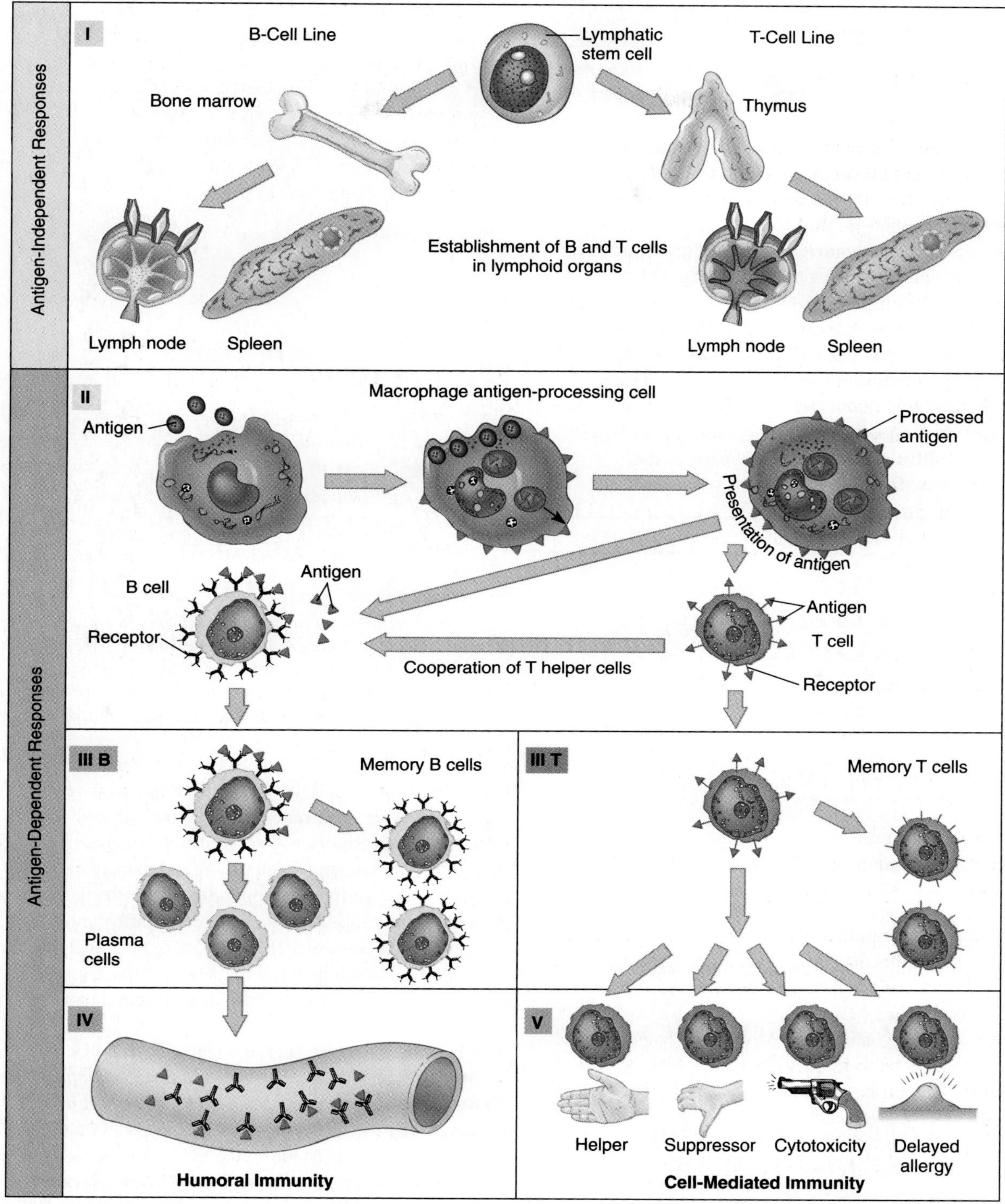

Figure 15.1
Overview of the stages of lymphocyte development and function. 1. Development of B- and T-lymphocyte specificity and migration to lymphoid organs. II. Antigen processing by macrophage and presentation to lymphocytes; assistance to B cells by T cells. III B and III T. Lymphocyte activation, clonal expansion, and formation of memory B and T cells. IV. Humoral immunity. B-cell line produces antibodies to react with the original antigen. V. Cell-mediated immunity. Activated T cells perform various functions, depending on the signal and type of antigen.

Chapter Checkpoints

Acquired specific immunity is an elegant but complex matrix of interrelationships between lymphocytes and macrophages consisting of several stages.

Stage I. Lymphocytes originate in hemopoietic tissue but go on to diverge into two distinct types: B cells, which produce antibody, and T cells, which produce cytokines that initiate and coordinate the entire immune response.

Stage II. Macrophages detect invading foreign antigens and present them to lymphocytes, which recognize the antigen and initiate the specific immune response.

Stage III. Lymphocytes proliferate, producing clones of progeny that include groups of responder cells and memory cells.

Stage IV. Activated B lymphocytes become plasma cells that produce and secrete large quantities of antibodies.

Stage V. Activated T lymphocytes differentiate into one of four subtypes, which regulate and participate directly in the specific immune responses.

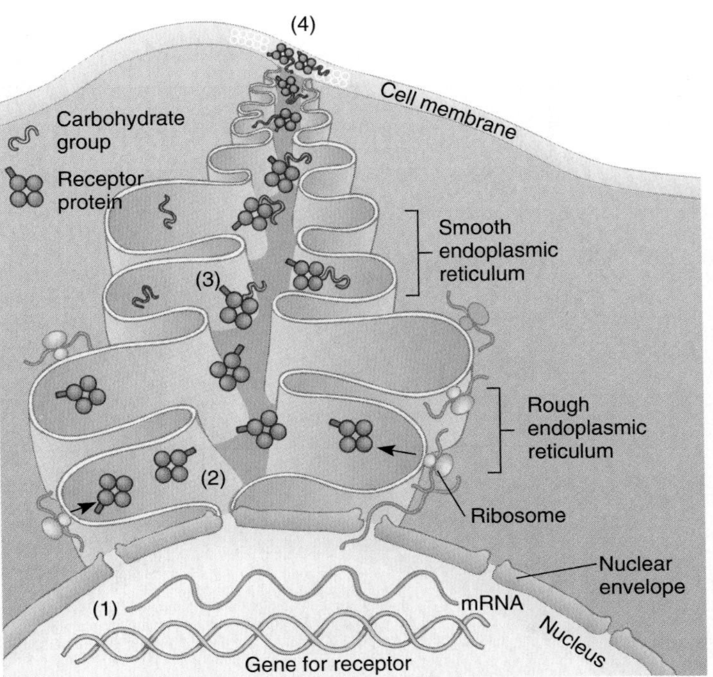

Figure 15.2
Receptor formation in a developing cell. (*1*) Gene coding for the receptor is transcribed. (*2*) In the endoplasmic reticulum, mRNA is translated into the protein portion of the receptor. (*3*) A carbohydrate side chain is added, forming glycoprotein. (*4*) The finished receptor is transported to and inserted into the cell membrane.

ESSENTIAL PRELIMINARY CONCEPTS FOR UNDERSTANDING IMMUNE REACTIONS OF SECTIONS I–V

Before we examine lymphocyte development and function in greater detail, we must initially review concepts such as the unique structure of molecules (especially proteins), the characteristics of cell surfaces (membranes and envelopes), the ways that genes are expressed, and immune recognition and identification of self and nonself. Ultimately, the shape and function of protein receptors and markers protruding from the surfaces of cells are the result of genetic expression, and these molecules are responsible for specific immune recognition and, thus, immune reactions.

Markers on Cell Surfaces Involved in Recognition of Self and Nonself

Chapter 14 touched on the fundamental idea that cell markers or receptors confer specificity and identity. A given cell can express several different receptors, each type playing a distinct and significant role in detection, recognition, and cell communication. Major functions of receptors are: (1) to perceive and attach to nonself or foreign molecules (antigens), (2) to promote the recognition of self molecules, (3) to receive and transmit chemical messages among other cells of the system, and (4) to aid in cellular development. Because of their importance in the immune response, we will concentrate here on the major receptors of lymphocytes and macrophages.

How Are Receptors Formed?

The nature of cell receptors is dictated by which specific elements of the genome are active during development. As a cell matures—be it liver or brain cell, lymphocyte or macrophage—certain genes that code for the cell receptors will be transcribed and translated into protein products with a distinctive shape, specificity, and function. This receptor is modified and packaged by the endoplasmic reticulum and Golgi apparatus. It is ultimately inserted into the cell membrane so as to be accessible to antigens, other cells and chemical mediators (figure 15.2). Note that the receptors of many cells, including lymphocytes and macrophages, are glycoproteins with additional carbohydrate fragments added.

Major Histocompatibility Complex

One set of genes that codes for human cell receptors is the **major histocompatibility complex (MHC).** This gene complex gives rise to a series of glycoproteins (called MHC antigens) found on all cells except red blood cells. Because these marks were first identified in humans on the surface of white blood cells, the MHC is also known as the **human leukocyte antigen (HLA)** system. This receptor complex plays a vital role in recognition of self by the immune system and in rejection of foreign tissue. Genes that regulate and code for the MHC of humans are located on the sixth chromosome, clustered in a multigene complex of three subgroups called class I, class II, and class III (figure 15.3).

The functions of the three MHC groups have been identified. Class I genes code for markers that regulate acceptance or rejection of tissue grafts, which is how the term *histo-* (tissue) *compatibility* (acceptance) originated. The system is rather complicated in its details, but in general, each human being inher-

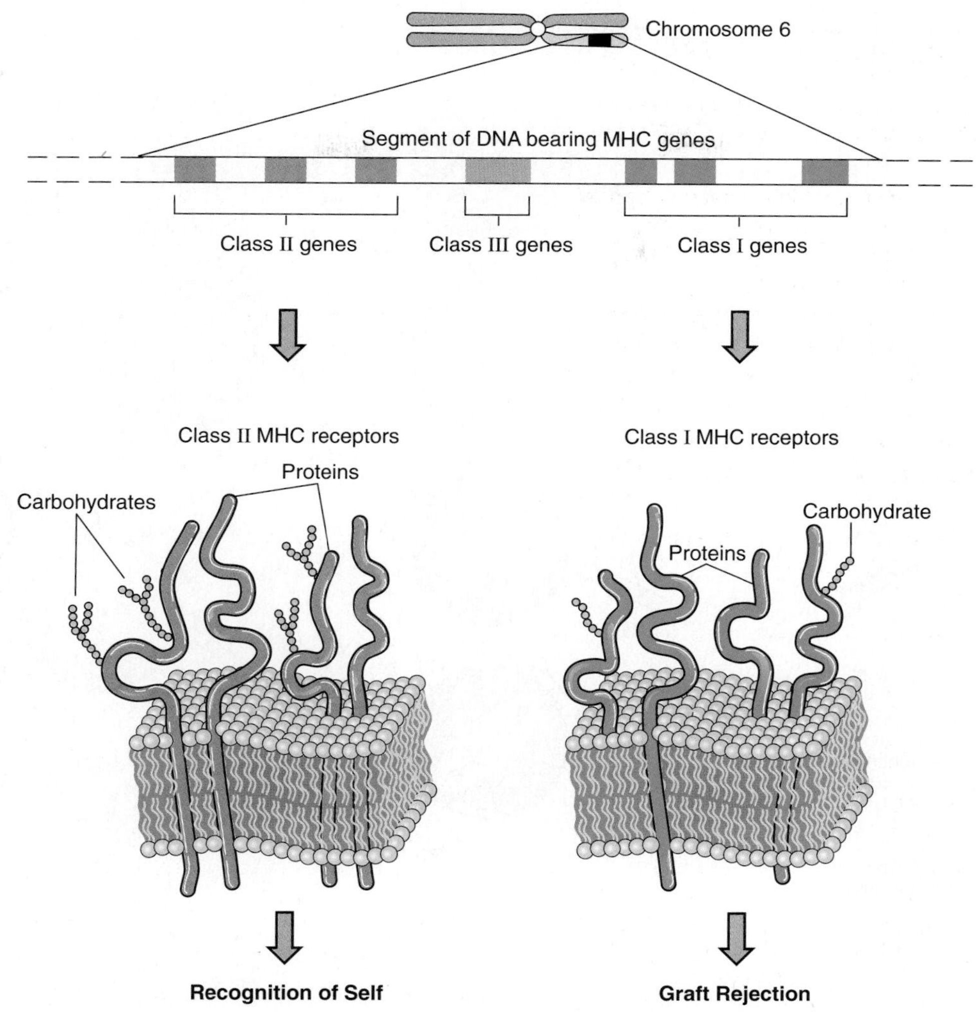

Figure 15.3

Glycoprotein receptors of the human major histocompatibility (human leukocyte antigen) gene complex (MHC).

its a particular combination of class I MHC (HLA) genes in a relatively predictable fashion. Although millions of different combinations and variations of these genes are possible among humans, the closer the relationship, the greater the probability for similarity in MHC profile (see figure 17.19). Individual differences in the exact inheritance of MHC genes, however, make it highly unlikely that even closely related persons will express an identical MHC profile. This fact introduces an important recurring theme: Although humans are genetically the same species, the cells of each individual express molecules that are foreign (antigenic) to other humans. This fact necessitates testing for HLA and other antigens when blood is transfused and organs are transplanted (see graft rejection, chapter 17).

Class II MHC genes regulate immune responses. This system of genes and receptors is located primarily on macrophages and B cells, and it functions in cooperative immune responses to antigens mounted by these cells and also T cells. Unlike the other two classes, which code for molecules that are inserted into cell surfaces, class III MHC genes code for certain secreted complement components such as C2 and C4 (described in chapter 14).

Lymphocyte Receptors and Specificity to Antigen

The part lymphocytes play in immune surveillance and recognition emphasizes the essential role of their receptors. Although they possess MHC antigens for recognizing self, they also carry receptors on their membranes that recognize antigens. Antigen molecules exist in great diversity; there are potentially millions and even billions of unique types. The many sources of antigens include microorganisms as well as an awesome array of chemical compounds in the environment. One of the most fascinating questions in immunology is: How can the lymphocyte receptors be varied to react with such a large number of different antigens? After all, it is generally accepted that there will have to be a different lymphocyte receptor for each unique antigen. Some questions that naturally follow are: How can a cell accommodate enough genetic information to respond to millions or even billions of antigens? When, where, and how does the capacity to distinguish native from foreign tissue arise? To answer these questions, we must first introduce a central theory of immunity.

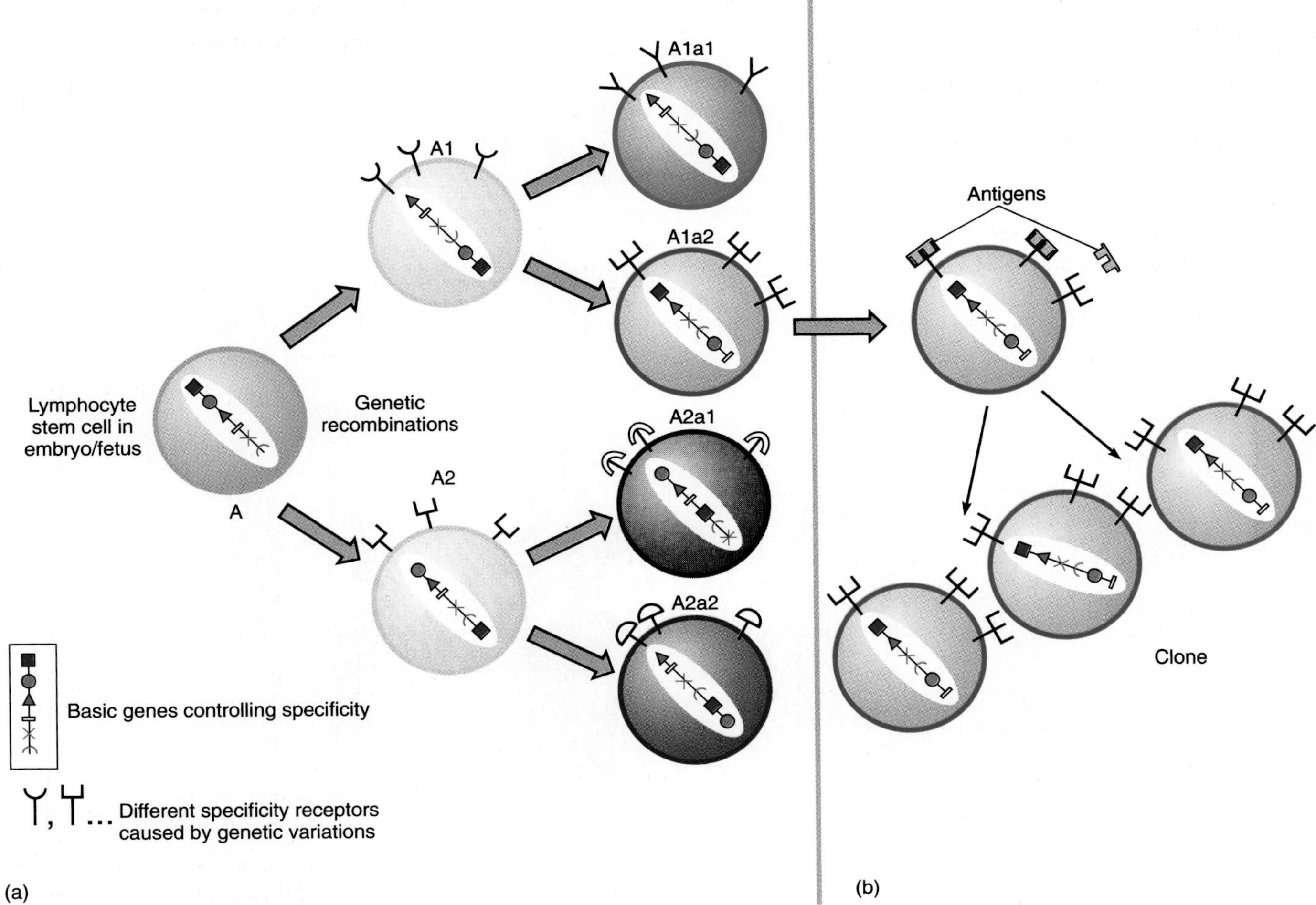

Figure 15.4

Clonal selection theory, the origins of lymphocyte specificity and proliferation. (*a*) During the development of stem cells, the genes that program for specificity to antigen are rearranged so that lymphocyte daughter cells inherit differing genetic programs. These alterations are expressed in the nature of the protein and the shape of the receptor. The potential for genetic and receptor variation is so great that hundreds of millions of genetically different clones can be formed. The receptor is important for providing a specific recognition and reaction site for an antigen, but this part of development does not require the actual presence of the antigen. (*b*) Clonal selection and expansion (antigen is present). The complete repertoire of lymphocytes has developed by the time a child is born, so that any antigen entering the system will automatically encounter and select its prespecified lymphocyte and stimulate it to expand into a genetically identical group or clone, each of which can respond to the same antigen. Note that all receptors on a single lymphocyte are identical.

The Origin of Diversity and Specificity in the Immune Response

The Clonal Selection Theory and Lymphocyte Development

Research findings have shown that lymphocytes use only about 500 genes to produce a tremendous variety of specific receptors. To understand how this occurs, we turn to a widely accepted concept called the **clonal selection theory.** According to this theory, undifferentiated lymphocytes undergo genetic mutations and recombinations while they proliferate in the embryo (figure 15.4). This process leads to extreme variations in the expression of their receptor specificity. The mechanism might be pictured as follows: As an immature, undifferentiated lymphocyte (A) divides, its receptor genes recombine randomly to produce differing genetic codes in the two daughter cells. Because of gene shuffling, daughter cell A1 receives a combination of genes and a specificity that are different from those of daughter cell A2. By the same random shuffling, the progeny of A1 and A2 will likewise receive a different genetic program, and so on, for millions of different lymphocytes. The ultimate impact of these genetic variations is that the amino acid composition of the receptor, and thus its shape and specificity, will be different for each lymphocyte type. Each genetically distinct group of lymphocytes that possesses the same specificity is called a **clone.** Although B cells and T cells have different kinds of receptors on their surfaces, the general principles of genetics and acquisition of specificity are the same for both.

As a result of gene reassortment on such a massive scale, the lymphoid system eventually can develop (in theory) at least a billion different clones of lymphocytes. So, by the time of birth, the ability to react with a tremendous variety of antigens is already in place in the lymphoid tissue. Two important generalities of immune responsiveness one can derive from the clonal selection theory are that: (1) lymphocyte specificity is preprogrammed, existing in the genetic makeup before an antigen has ever entered the system, and (2) each genetically different type of lymphocyte expresses only a single specificity.

According to another part of the clonal selection theory, any lymphocyte that could possibly mount a harmful response against *self* molecules is eliminated or suppressed. Innmunologists call this feature **tolerance to self.** Some immune diseases (called autoimmune) are believed to arise when the immune system loses this tolerance and attacks self (see chapter 17).

The final part of the clonal selection theory proposes that the first introduction of each distinct type of antigen into the immune system selects a genetically distinct lymphocyte and causes it to expand into a clone of cells that can react to that antigen.

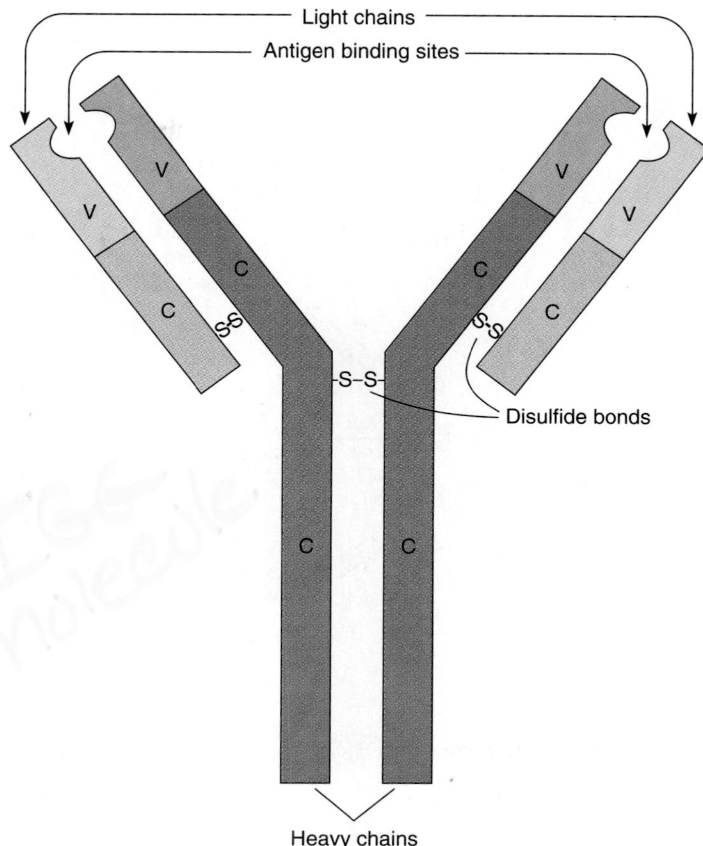

Figure 15.5

Simplified structure of an immunoglobulin molecule. The main components are four polypeptide chains—two identical light chains and two identical heavy chains bound by disulfide bonds as shown. Each chain consists of a variable region (V) and a constant region (C). The variable regions of light and heavy chains form a binding site for antigen.

The Specific B-Cell Receptor: An Immunoglobulin Molecule

In the case of B lymphocytes, the receptor genes that undergo the recombination described are those governing **immunoglobulin*** **(Ig)** synthesis. Immunoglobulins are large glycoprotein molecules that serve as the specific receptors of B cells and as antibodies. The basic immunoglobulin molecule is a composite of four polypeptide chains: a pair of identical heavy (H) chains and a pair of identical light (L) chains (figure 15.5). One light chain is bonded to one heavy chain, and the two heavy chains are bonded to one another with disulfide bonds, creating a symmetrical, Y-shaped arrangement. The ends of the forks formed by the light and heavy chains contain pockets, called the **antigen binding sites.** It is these sites that can be highly variable in shape to fit a wide range of antigens. This extreme versatility is due to **variable regions (V)** where amino acid composition is highly varied from one clone of B lymphocytes to another. The remainder of the light chains and heavy chains consist of **constant regions (C)** whose amino acid content does not vary greatly from one antibody to another. Although we will subsequently discuss immunoglobulins and their function as antibodies, for now, we will concentrate on the genetics that explain the origins of lymphocyte specificity.

Development of the Receptors During Lymphocyte Maturation

The genes that code for immunoglobulins lie on three different chromosomes. An undifferentiated lymphocyte has about 150 different genes that code for the variable region of light chains and a total of about 250 genes for the variable and **diversity regions (D)** of the heavy chains. It has only a few genes for coding for the constant regions and a few for the joining regions (J) that join segments of the molecule together. Owing to genetic recombination during development, only the selected (V and D) receptor genes are active in the mature cell, and all the other V and D genes are inaction (figure 15.6). This is how the singular specificity of lymphocytes arises.

It is helpful to think of the immunoglobulin genes as blocks of information that code for a polypeptide lying in sequence along a chromosome. During development of each lymphocyte, the genetic blocks are independently segregated, randomly rearranged, and assembled as follows:

- For a heavy chain, a variable region gene and diversity region gene are randomly selected from among the hundreds available and spliced to one joining region gene and one constant region gene.
- For a light chain, one variable, one joining, and one constant gene are spliced together.
- After transcription and translation of each gene complex into a polypeptide, a heavy chain combines with a light chain to form half an immunoglobulin; two of these combine to form a completed monomer (figure 15.6).

*immunoglobulin (im´´-yoo-noh-glahb´-yoo-lin) The technical name for an antibody.

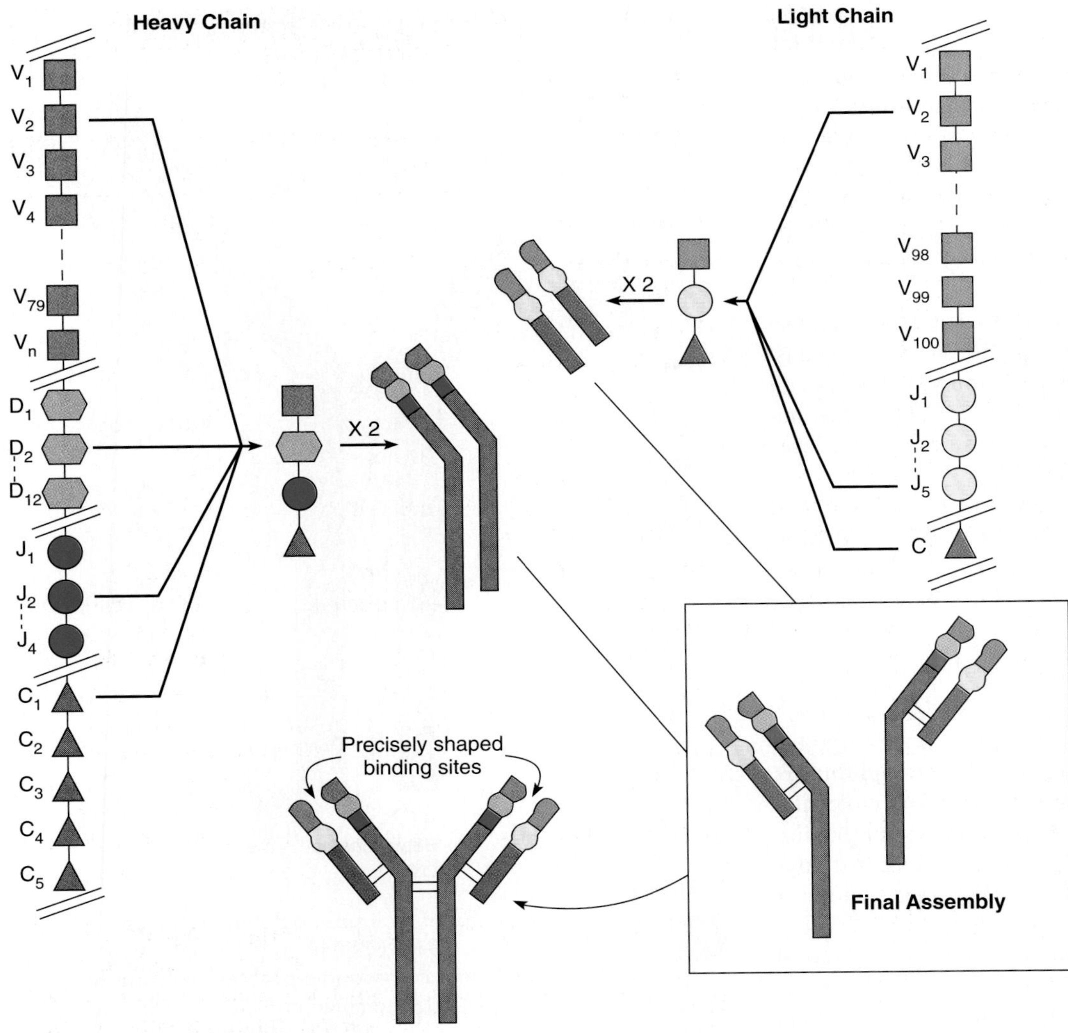

Figure 15.6

A simplified look at immunoglobulin genetics. The phenomenon of B-cell differentiation could be compared to a "cutting and pasting" process, in which the final gene that codes for a heavy or light chain is assembled by splicing blocks of genetic material from several regions. (*Left*) The heavy-chain gene is composed of genes from four separate segments (V, D, J, and C) that are transcribed and translated to form a single polypeptide chain. (*Right*) The light-chain genes are put together like heavy ones, except that the final gene is spliced from three gene groups (V, J, and C). During final assembly, first the heavy and light chains are bound, and then the heavy-light combinations are connected to form the immunoglobulin molecule.

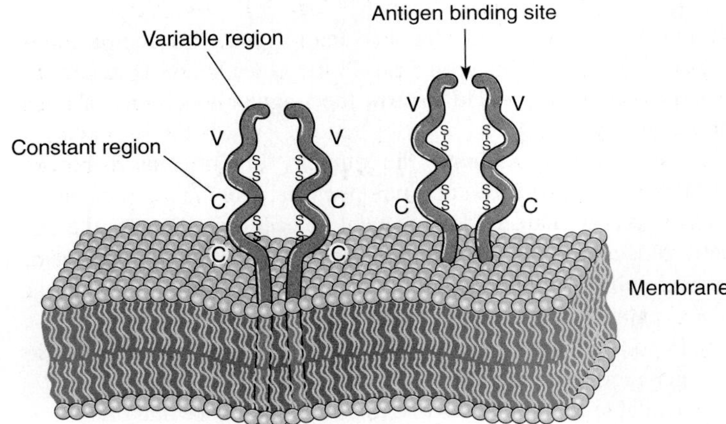

Figure 15.7

Proposed structure of the T-cell receptor for antigen.

Once synthesized, the immunoglobulin product is transported to the cell membrane and inserted there to act as a receptor that expresses the specificity of that cell and to react with an antigen as shown in figure 15.2. The first receptor on most B cells is a small form of IgM, and mature B cells carry IgD receptors (see table 15.2). It is notable that for each lymphocyte, the genes that were selected for the variable region, and thus for its specificity, will be locked in for the rest of the life of that lymphocyte and its progeny. We will discuss the different classes of Ig molecules further in section IV.

T-Cell Receptors

The T-cell receptor for antigen belongs to the same immunoglobulin superfamily as the B-cell receptor. It is similar to B cells in being formed by genetic modification, having variable and constant regions, being inserted into the membrane, and having an antigen-combining site formed from two parallel polypeptide chains (figure 15.7). Unlike the immunoglobulins, the T-cell receptor is relatively small and does not appear to have a humoral function. The several other receptors of T cells are described in a later section.

Chapter Checkpoints

The surfaces of all cell membranes contain identity markers known as protein receptors. These cell markers function in identification, communication, and cell development. They are also self identity markers.

Human cell markers are genetically determined by MHCs. MHC I codes for identity markers on all host cells. MHC II codes for immune receptors on macrophages and B cells. MHC III codes for secretion of complement C2 and C4.

The clonal selection theory explains that during prenatal development, both B and T cells develop millions of genetically different clones through independent segregation, random reassortment, and mutation. Together these clones possess enough genetic variability to respond to many millions of different antigens. Each clone, however, can respond to only one specific antigen.

TABLE 15.1

CONTRASTING PROPERTIES OF B-CELL AND T-CELL LINES

	B Cells	T Cells
Site of maturation	Bone marrow	Thymus
Nature of surface markers	Immunoglobulin	Several CD receptors
Texture of surface	Rough	Smoother
Circulation in blood	Low numbers	High numbers
Rosette formation with plain sheep RBCs	No	Yes
Distribution in lymphatic organs	Cortex (in follicles)	Paracortical (interior to the follicles)
Product of antigenic stimulation	Plasma cells and memory cells	Sensitized T lymphocytes of several types and memory cells
General functions	Production of antibodies	Cells function in helping, suppressing, killing, delayed hypersensitivity; synthesize cytokines

THE LYMPHOCYTE RESPONSE SYSTEM IN DEPTH

Now that you have a working knowledge of some factors in the development of immune specificity, let us look at each stage of an immune response as originally outlined in figure 15.1

I. Development of the Dual Lymphocyte System: The Stages in Origin, Differentiation, and Maturation

Development generally follows a similar general pattern in both types of lymphocytes (figure 15.8). Starting in embryonic and fetal stages, stem cells in the yolk sac, liver, and bone marrow give rise to immature lymphocytes that are released into the circulation. Because these undifferentiated cells cannot yet react with antigens, they must first undergo developmental changes at some specific anatomical location (table 15.1). This maturation occurs along two separate lines that will characterize all future responses. The fully mature B and T lymphocytes are released and ultimately take up residence in various lymphoid organs. Lymphocyte differentiation and immunocompetence are basically complete by the late fetal or early neonatal period.

Specific Happenings in B-Cell Maturation

The site of B-cell maturation was first discovered in birds, which have an organ in the intestine called the bursa. For some time, the human bursal equivalent was not established. Now it is known to be certain bone marrow sites that harbor *stromal cells.* These huge cells nurture the lymphocyte stem cells and provide hormonal signals that initiate B-cell development. As a result of gene modification and selection, hundreds of millions of distinct B cells develop. These naive lymphocytes "home" to specific sites in the lymph nodes, spleen, and gut-associated lymphoid tissue (GALT). Here they will come into contact with antigens throughout life. In addition to having immunoglobulins as surface receptors, a fully differentiated B cell has a distinctively rough appearance under high magnification, with numerous microvillus projections.

Specific Happenings in T-Cell Maturation

The maturation of T cells and the development of their specific receptors are directed by the thymus gland and its hormones. Due to the complexity of T-cell function, there are at least seven classes of T-cell receptors or markers, termed the CD cluster (abbreviated from common determinant—CD1, CD2, etc.). Some receptors (figure 15.7) recognize antigen molecules; others recognize receptors on B cells, other T cells, and macrophages. Like B cells, T cells also migrate to lymphoid organs and occupy specific sites. Additional characteristics typical of mature T cells are smaller and fewer microvilli under high magnification and the property of rosette formation when mixed with normal sheep red blood cells (see figure 16.13).

II. Entrance and Processing of Antigens and Clonal Selection

Having reviewed the characteristics of lymphocytes, let us now examine the properties of antigens, the substances that cause them

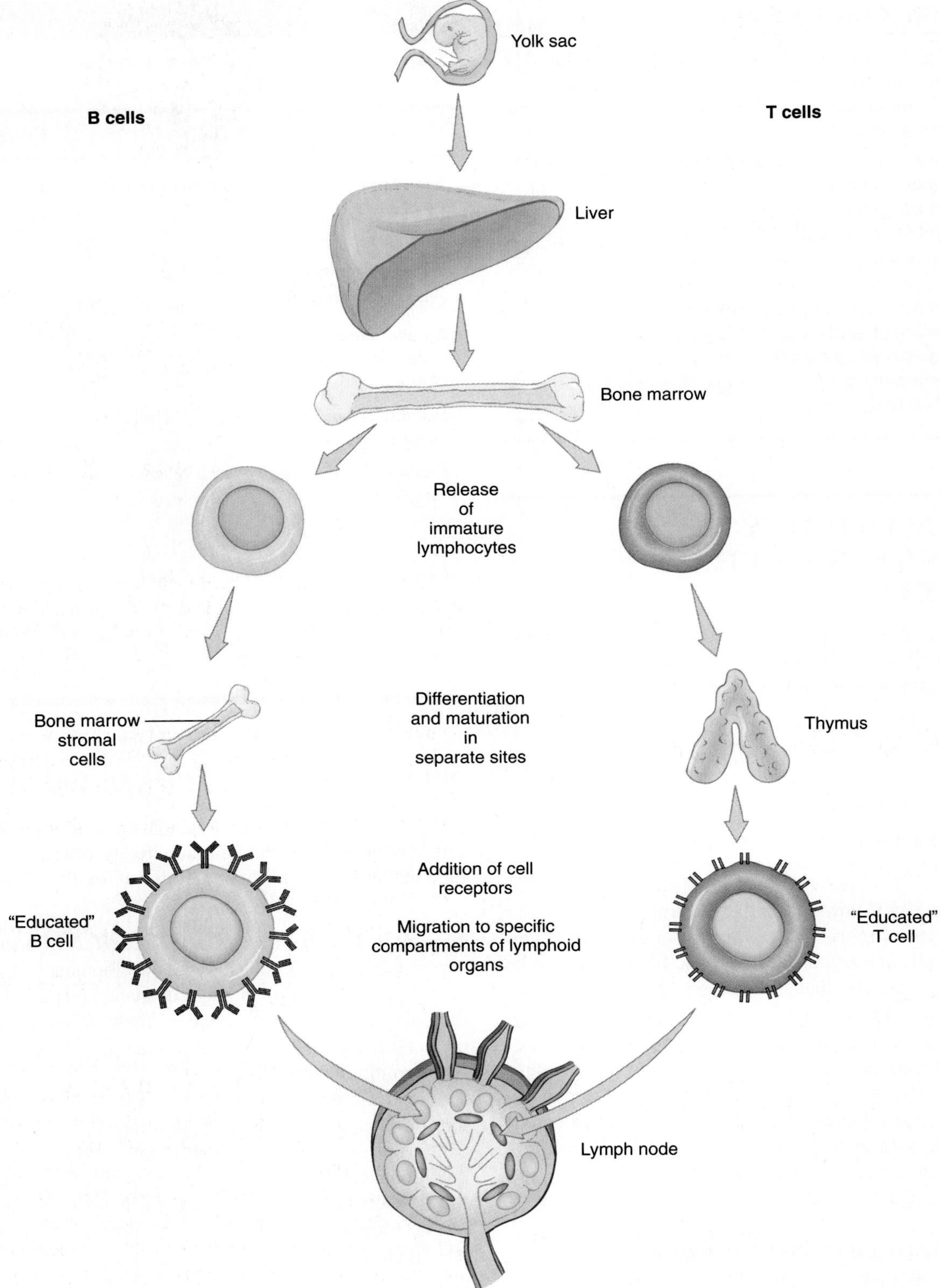

Figure 15.8
Major stages in the development of B and T cells.

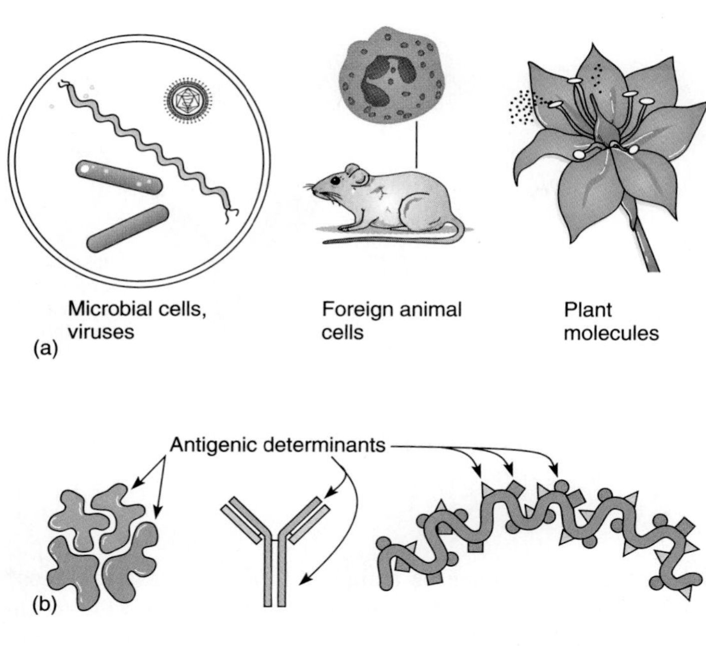

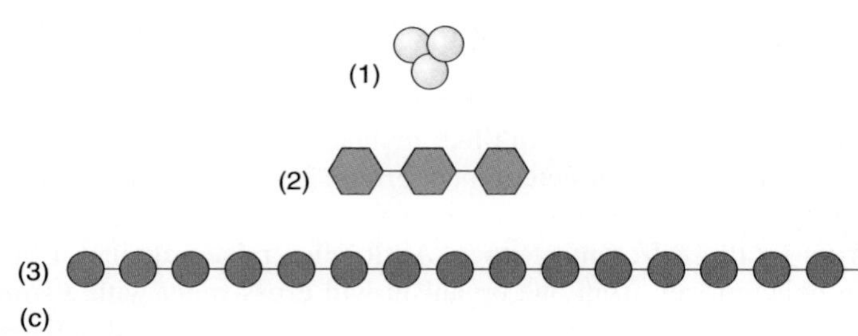

Figure 15.9
Characteristics of antigens. (*a*) Whole cells and viruses make good immunogens. (*b*) Complex molecules with several antigenic determinants make good immunogens. (*c*) Poor immunogens include small molecules not attached to a carrier molecule (*1*), simple molecules, (*2*), and large but repetitive

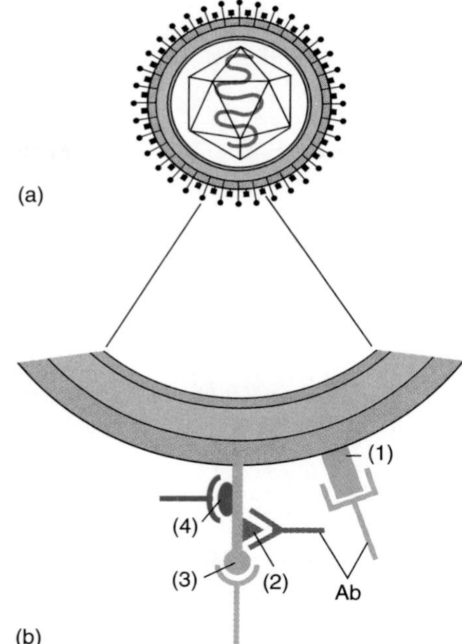

Figure 15.10
Mosaic antigens. (*a*) Microbes such as viruses present various sites that serve as separate antigenic determinants. (*b*) Inset indicates that each determinant (*1, 2, 3, 4*) will stimulate a different lymphocyte and antibody response.

to react. As we reported in chapter 14, an **antigen (Ag)**[1] is a substance that provokes an immune response in specific lymphocytes. The property of behaving as an antigen is called **antigenicity.** The term *immunogen* is another term of reference for a substance that can elicit an immune response. To be perceived as an antigen or immunogen, a substance must meet certain requirements in foreignness, shape, size, and accessibility.

Characteristics of Antigens

One important characteristic of an antigen is that it be perceived as **foreign,** meaning that it is not a normal constituent of the body. Whole microbes or their parts, cells, or substances that arise from other humans, animals, plants, and various molecules all possess this quality of foreignness and thus are potentially antigenic to the immune system of an individual (figure 15.9). Molecules of complex composition such as proteins and protein-containing compounds prove to be more immunogenic than repetitious polymers composed of a single type of unit. Most materials that serve as antigens fall into the following chemical categories:

- Proteins and polypeptides (enzymes, albumin, antibodies, hormones, exotoxins)
- Lipoproteins (cell membranes)
- Glycoproteins (blood cell markers)
- Nucleoproteins (DNA complexed to proteins, but not pure DNA)
- Polysaccharides (certain bacterial capsules) and lipopolysaccharides

Effects of Molecular Shape and Size To initiate an immune response, a substance must also be large enough to "catch the attention" of the surveillance cells. Molecules with a molecular weight (MW) of less than 1,000 are seldom complete antigens, and those between 1,000 MW and 10,000 MW are weakly so. Complex macromolecules approaching 100,000 MW are the most immunogenic, a category also dominated by large proteins. Note that large size alone is not sufficient for antigenicity; glycogen, a polymer of glucose with a highly repetitious structure, has an MW over 100,000 and is not normally antigenic, whereas insulin, a protein with an MW of 6,000, can be antigenic.

A lymphocyte's capacity to discriminate differences in molecular shape is so fine that it recognizes and responds to only a portion of the antigen molecule. This molecular fragment, called the **antigenic determinant,** is the primary signal that the molecule is foreign (figure 15.10). The particular tertiary structure and

1. Originally, **anti**body **gen**erator.

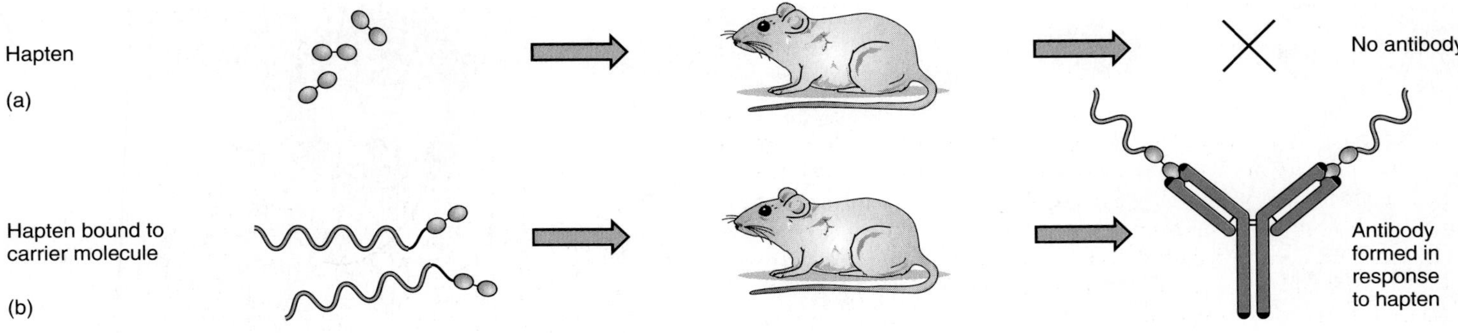

Figure 15.11
The hapten-carrier phenomenon. (*a*) Haptens are too small to be discovered by an animal's immune system; no response. (*b*) A hapten bound to a large molecule will serve as an antigenic determinant and stimulate a response and an antibody that is specific for it.

shape of this determinant must conform like a key to the receptor "lock" of the lymphocyte, which then responds to it. Certain amino acids accessible at the surface of proteins or protruding carbohydrate side chains are typical examples. Many foreign cells and molecules are very complex antigenically, with numerous determinants, each of which will elicit a separate and different lymphocyte response. Examples of these multiple, or *mosaic, antigens* include bacterial cells containing cell wall, membrane, flagellar, capsular, and toxin antigens; and viruses, which express various surface and core antigens (figure 15.10).

Small foreign molecules that consist only of a determinant group and are too small by themselves to elicit an immune response are termed **haptens.** However, if such an incomplete antigen is linked to a larger carrier molecule, the combination develops immunogenicity (figure 15.11). The carrier group contributes to the size of the complex and enhances the proper spatial orientation of the determinative group, while the hapten serves as the antigenic determinant. Haptens includes such molecules as drugs, metals, and ordinarily innocuous household, industrial, and environmental chemicals. Many haptens develop antigenicity in the body by combining with large carrier molecules such as serum proteins (see allergy in chapter 17).

Special Types of Antigens So far, we have emphasized the role of microbial antigens, but tissues and proteins (including enzymes and antibodies) from other humans and animals are also antigenic. On occasion, even a part of the body can take on the character of an antigen. During lymphocyte differentiation, immune tolerances to self tissue occurs, but a few anatomical sites can contain sequestered (hidden) molecules that escape this assessment. Such molecules, called **autoantigens,** can occur in tissues (eye, thyroid gland, for example) that are walled off early in embryonic development before the surveillance system is in complete working order. Because tolerance to these substances has not yet been established, they can subsequently be mistaken as foreign; this mechanism appears to account for some types of autoimmune diseases such as rheumatoid arthritis (chapter 17).

Because each human being is genetically and biochemically unique (except for identical twins), the proteins and other molecules of one person can be antigenic to another. **Alloantigens*** **(isoantigens)** are cell surface markers and molecules that occur in some members of the same species but not in others. Alloantigens are the basis for an individual's blood group (see chapter 17) and major histocompatibility profile, and they are responsible for incompatibilities that can occur in blood transfusion or organ grafting.

On the other hand, different organisms can possess some molecules, called **heterophile,*** or heterogenetic, antigens, with the same or a similar determinant group. These antigens stimulate a response from the same lymphocyte clone even when they are from totally different sources. Antibodies raised against a heterophile antigen from one organism will **cross-react** with a similar or identical antigen from another source. Examples of antigens of different origins with heterophilic determinants are carbohydrate residues on the surfaces of bacteria and red blood cells, the antigens of Group A streptococci and human heart tissue, and cardiolipin, a phospholipid present in a wide assortment of living things. Heterophile antigens may play a part in diseases such as rheumatic fever and in false positive diagnostic tests (as occur in syphilis).

Some bacterial toxins, called *superantigens,* are potent stimuli for T cells. Their presence in an infection activates T cells at a rate 100 times greater than ordinary antigens. The result can be an overwhelming release of cytokines and cell death. Such diseases as toxic shock syndrome (chapter 18) and certain autoimmune diseases (chapter 17) are associated with this class of antigens.

Antigens that evoke allergic reactions, called **allergens,** will be characterized in detail in chapter 17.

Host Response to Antigens: A Cooperative Affair

The basis for most immune responses is the encounter between antigens and white blood cells. Microbes and other foreign substances enter most often through the respiratory or gastrointestinal

---

*alloantigen, isoantigen (al´-oh, eye´-soh) Gr. *allos,* other, and *iso,* same.

*heterophile (het´-ur-oh-fyl) Gr. *hetero,* other, and *phile,* to love.

MICROFILE 15.1 AN IMMUNOLOGIC BARGAIN BASEMENT

Realizing the great variety and quantity of antigens that find their way into the body, one might wonder how this microscopic flotsam and jetsam can ever meet up with the right lymphocyte. After all, the body's fluid compartments are relatively large, and antigens are constantly entering through some portal or another. The events of this awesome sorting process could be likened to a rummage sale in a microscopic department store. The store (lymph node) is crowded with large numbers of all types of customers (lymphocytes), and it contains all sorts of merchandise (antigens). Picture the mounds of antigenic merchandise arrayed on tables and thousands of lymphocyte customers milling around, pawing through the piles of materials with their receptor hands to find just the right specificity (size, fit, color, shape, and brand). The sales personnel (macrophages and T helper cells) display the merchandise, help in the selection, and expedite the sale (contact of the antigen with its matching lymphocyte). Like customers, some lymphocytes become "experienced shoppers." On subsequent shopping sprees, they retain a memory of previous sales and are even more efficient.

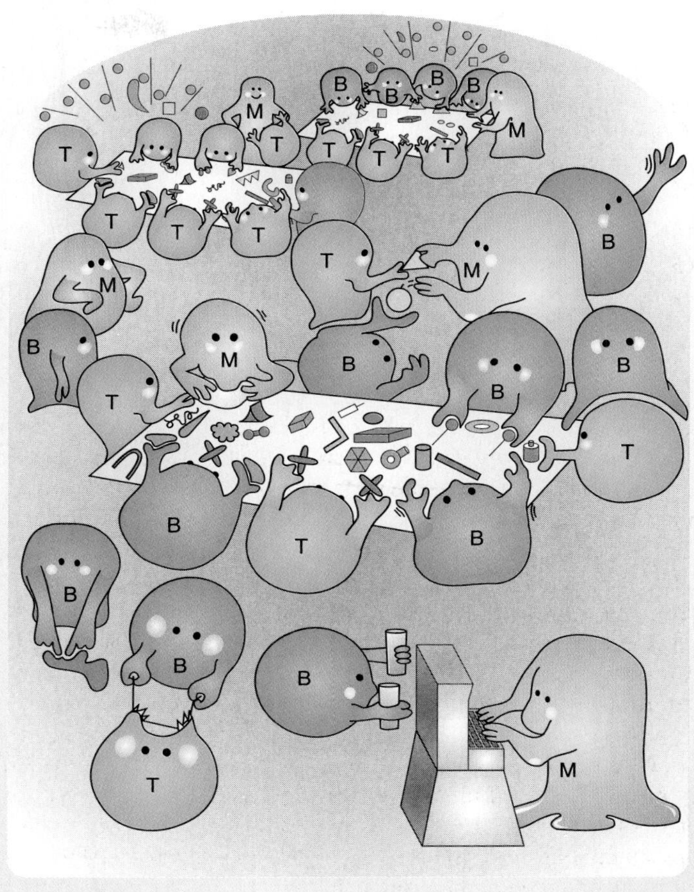

mucosa and less frequently through other mucous membranes, the skin, or across the placenta. Antigens introduced intravenously become localized in the liver, spleen, bone marrow, kidney, and lung. If introduced by some other route, antigens are carried in lymphatic fluid and concentrated by the lymph nodes. The lymph nodes and spleen are important in concentrating the antigens and circulating them thoroughly through all areas populated by lymphocytes so that they come into contact with the proper clone. To help imagine this process, see the analogy in microfile 15.1.

The Role of Macrophages: Antigen Processing and Presentation

In most immune reactions, the antigen must be further acted upon and formally presented to lymphocytes by special macrophages or other cells called **antigen-processing cells** (APCs). These large **dendritic*** cells engulf the antigen and modify is so that it will be more immunogenic and recognizable to lymphocytes. After processing is complete, the antigen is moved to the surface of the APC and bound to the MHC receptor so that it will be readily accessible to the lymphocytes during presentation (figure 15.12).

Presentation of Antigen to the Lymphocytes and Its Early Consequences

For lymphocytes to respond to the APC-bound antigen, certain conditions must be met. **T-cell dependent antigens,** usually protein-based, require recognition steps between the macrophage, antigen, and lymphocytes. The first cells on the scene to assist the macrophage in activating B cells and other T cells are a special class of **helper T cells (T_H).** This class of T cell bears a receptor that binds simultaneously with the class II MHC receptor on the macrophage and with one site on the antigen (figure 15.12). Once identification has occurred, a cytokine, **interleukin* 1 (IL-1),** produced by the macrophage, activates this T helper cell. The T_H cell, in turn, produces a different cytokine, **interleukin-2 (IL-2),** that stimulates a general increase in activity of committed B and T cells (see microfile 15.4). The manner in which B and T cells

*dendritic (den-drih-tik) Gr. *dendron,* tree. In reference to the long, branchlike extensions of the cell membrane.

*interleukin (in″-tur-loo′-kin) A chemical that carries signals between white blood cells.

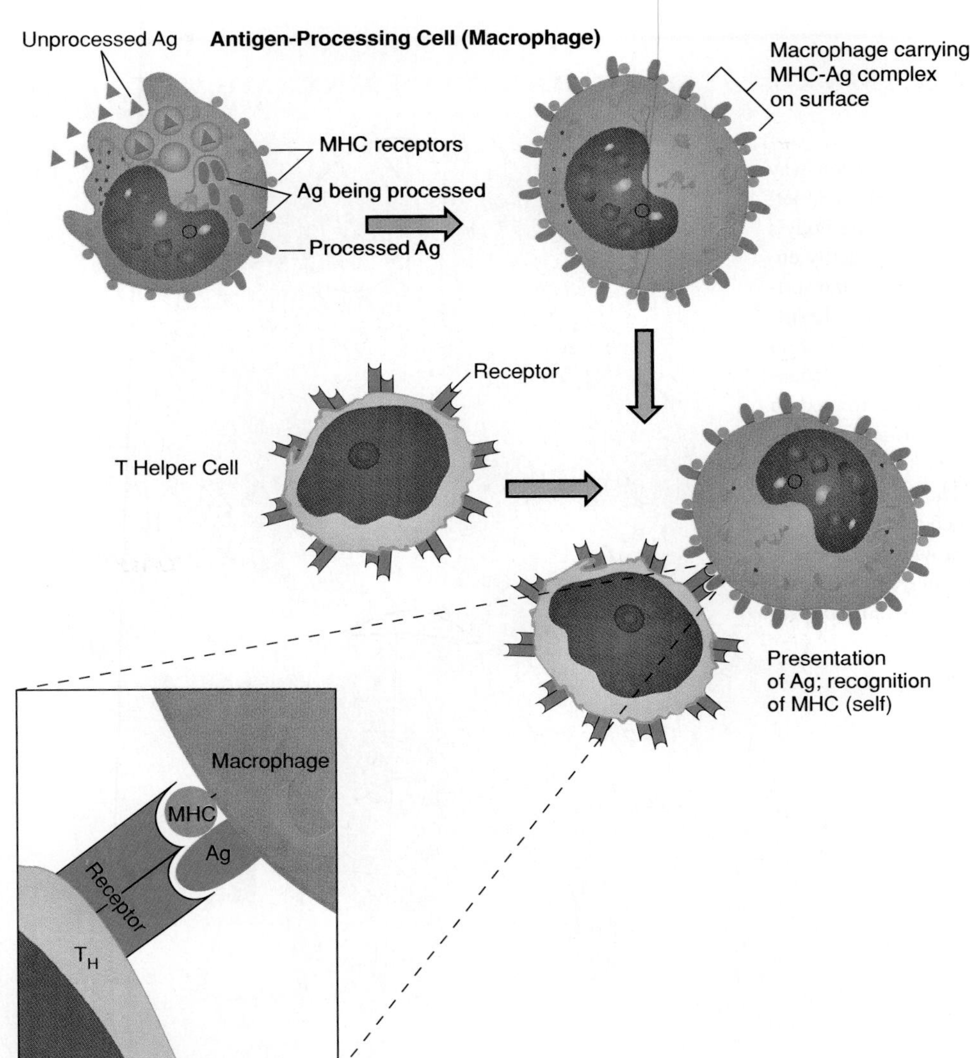

Figure 15.12
Cell cooperation between a macrophage and a T cell during the initial response to antigen. Inset of macrophage-lymphocyte identification shows that the macrophage-receptor complex fits precisely with the receptor of the T_H cell so that there is recognition of both self and nonself.

subsequently become activated by the macrophage–T helper cell complex and their individual responses to antigen will be addressed separately in sections III B and IV, and III T and V.

A few antigens can trigger a response from B lymphocytes without the cooperation of macrophages or T helper cells. These **T-cell-independent** antigens are usually simple molecules such as carbohydrates with many repeating and invariable determinant groups. Examples include lipopolysaccharide from the cell wall of *Escherichia coli* and polysaccharide from the capsule of *Streptococcus pneumoniae.* Because so few antigens are of this type, most B-cell reactions require helper T cells.

III B. Activation of B Lymphocytes: Clonal Expansion and Antibody Production

The immunologic activation of most B cells requires a series of events (figure 15.13):

1. **Binding of antigen and clonal selection.** In this case, a precommitted B cell of a particular clonal specificity picks up the antigen itself or is presented with the antigen by the macrophage complex so that the antigen is bound to the B-cell receptors. Further recognition of self, involving the binding of a B-cell MHC receptor to the T_H cell, also occurs at this point.
2. **Instruction by chemical mediators.** The B cell receives developmental signals from macrophages and T cells (interleukins 6, 2), and various other cytokines (growth factors).
3. The combination of these stimuli on the membrane receptors causes a signal to be transmitted internally to the B-cell nucleus.
4. This event triggers the cell to begin **blast transformation.** A blast is an enlarged, highly active cell whose DNA synthesis and organelle bulk are greatly increased in preparation for mitotic divisions.
5. A stimulated B cell multiplies through successive mitotic divisions and produces a large population of genetically identical daughter cells. Some cells that stop short of becoming fully differentiated are **memory cells,** which remain for long periods to react with that same antigen at a

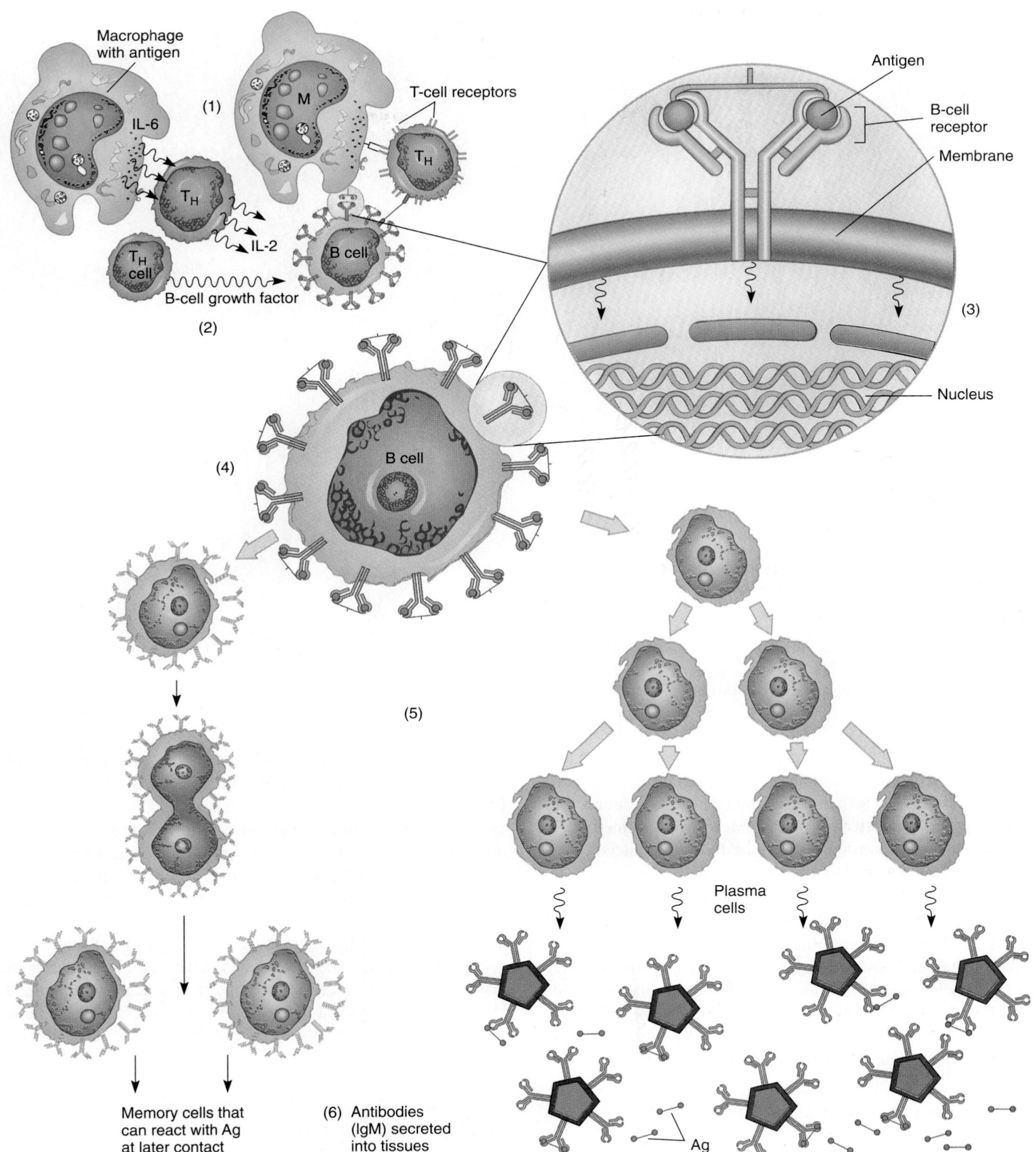

Figure 15.13

Events in B-cell activation. (*1*) Clonal selection. Antigen is presented to the B cell by a macrophage; activated T cells facilitate recognition by binding to both the macrophage and the B cell and (*2*) produce growth factors that stimulate the B cell. (*3, inset*) The binding of antigen on the immunoglobulin receptors of the B cell transmits a signal to the cell nucleus to begin activation of the B cell. (*4*) The B cell is transformed into a blast cell. (*5*) The blast cell undergoes mitotic divisions and clonal expansion into memory and plasma cells. (*6*) Plasma cells secrete antibodies.

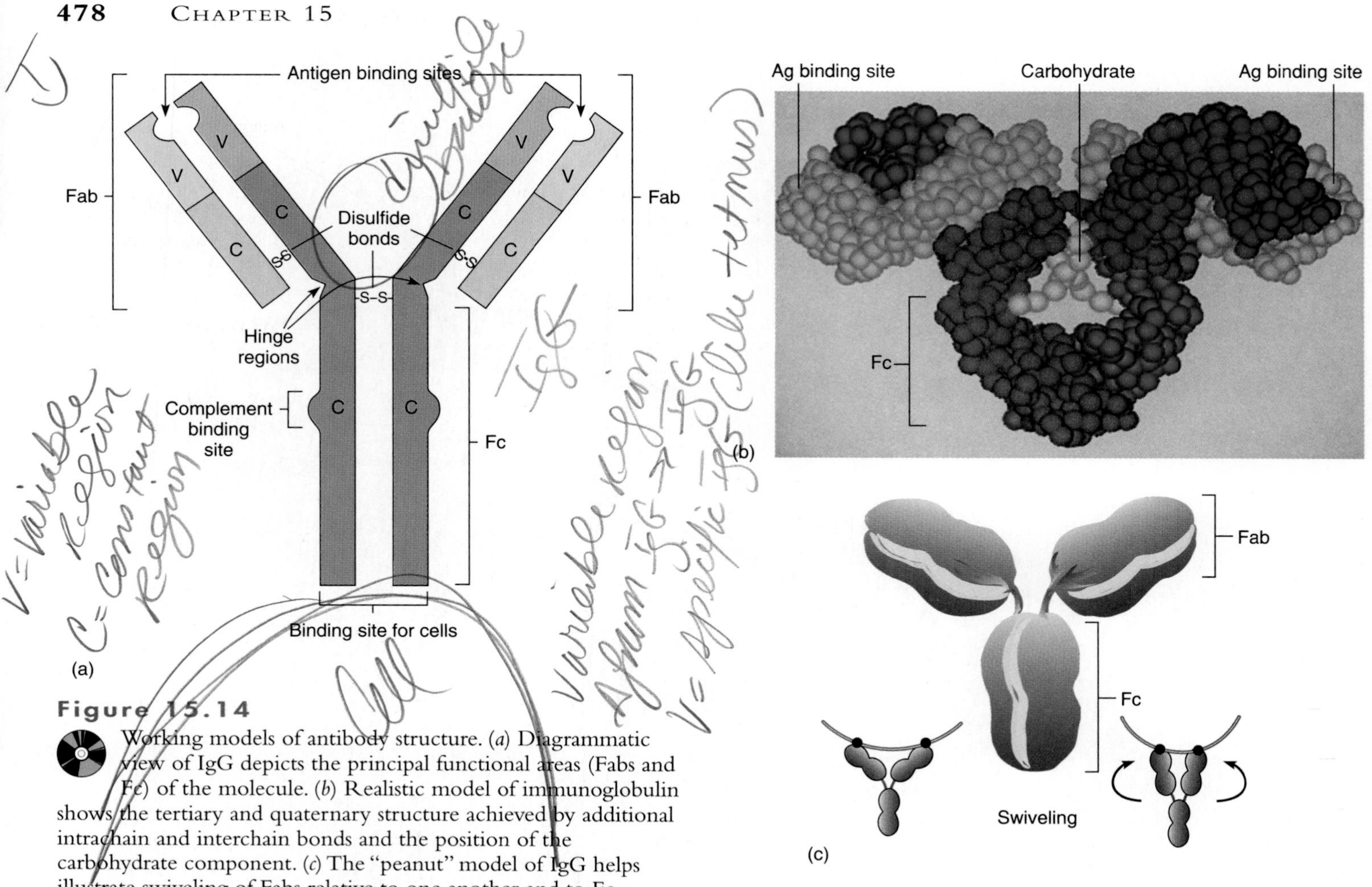

Figure 15.14

Working models of antibody structure. (*a*) Diagrammatic view of IgG depicts the principal functional areas (Fabs and Fc) of the molecule. (*b*) Realistic model of immunoglobulin shows the tertiary and quaternary structure achieved by additional intrachain and interchain bonds and the position of the carbohydrate component. (*c*) The "peanut" model of IgG helps illustrate swiveling of Fabs relative to one another and to Fc.

later time. This reaction also expands the clone size, so that subsequent exposure to that antigen provides more cells with that specificity. This expansion of the clone size accounts for the increased memory response. By far the most numerous progeny are large, specialized, terminally differentiated B cells called **plasma cells.**

6. The primary action of plasma cells is to secrete into the surrounding tissues copious amounts of antibodies with the same specificity as the original receptor (figure 15.13). Although an individual plasma cell can produce around 2,000 antibodies per second, production does not continue indefinitely because of regulation from the T suppressor (T_S) class of cells. The plasma cells do not survive for long and deteriorate after they have synthesized antibodies.

IV. Products of B Lymphocytes: Antibody Structure and Function

The Structure of Immunoglobulins

Earlier we saw that a basic immunoglobulin (Ig) molecule contains four polypeptide chains connected by disulfide bonds. Let us view this structure once again, using an **IgG molecule** as a model. Two functionally distinct segments called *fragments* can be differentiated. The two "arms" that bind antigen are termed **antigen binding fragments (Fabs),** and the rest of the molecule is the **crystallizable fragment (Fc),** so called because it has been crystallized in pure form. The distal end of each Fab fragment (consisting of the variable regions of the heavy and light chains) folds into a groove that will accommodate one antigenic determinant. The presence of a special *hinge* region at the site of attachment between the Fab and Fc fragments allows swiveling of the Fab fragments. In this way they can change their angle to accommodate nearby antigen sites that vary slightly in distance and position. The Fc fragment is involved in binding to various cells and molecules of the immune system itself. Figure 15.14 shows three views of antibody structure, and microfile 15.2 suggests a working model.

Antibody–Antigen Interactions and the Function of the Fab

The site on the antibody where the antigenic determinant inserts is composed of a *hypervariable region* whose amino acid content can be extremely varied. Antibodies differ somewhat in the exactness of this groove for antigen, but a certain complementary fit is necessary for the antigen to be held effectively (figure 15.15). The specificity of antigen binding sites for antigens is very similar to enzymes and substrates (in fact, some antibodies

MICROFILE 15.2 THE HEADLESS ANTIBODY

A vivid (and we hope memorable) analogy compares basic immunoglobulin structure to a human body standing with the legs together and the arms raised. It is a fair model of the structure if you pretend that the head is gone and realize that no body part corresponds to a section of the light chains. The arms (Fab fragments) are connected to the body by hinges (the hinge regions) that can move the arms up and down. The body and legs (Fc fragments) can be planted solidly in one spot. The structure is symmetrical; that is, one side is a mirror image of the other. The model is also functionally dynamic. It reminds us that the basic immunoglobulin molecule has two antigen (Ag) binding sites (hands, which could be likened to the hypervariable region that binds antigen). Changing the positions of the fingers and thumb allows unlimited variations in the three-dimensional shape of a hand so it can grasp an infinite variety of shapes, just as the Ag binding sites can be varied.

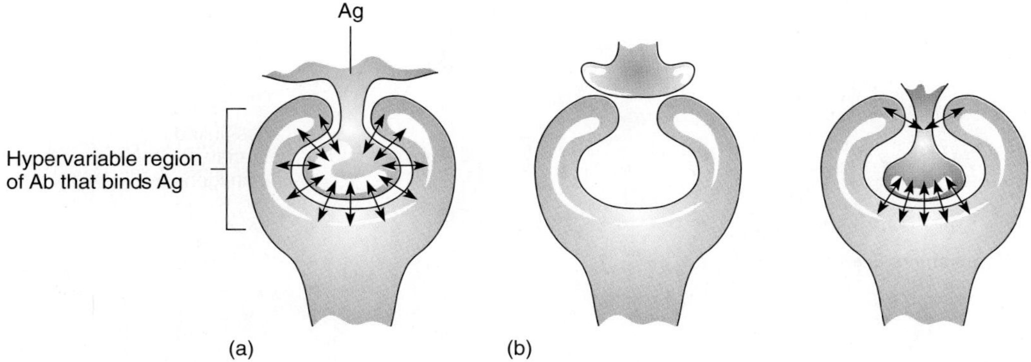

Figure 15.15
Antigen-antibody binding. The union of antibody (Ab) and antigen (Ag) is characterized by a certain degree of fit and is supported by weak linkages such as hydrogen bonds and electrostatic attraction. (*a*) In a snug fit such as that shown here, there is great opportunity for attraction and strong attachment. The strength of this union confers high affinity. (*b*) Examples of potential interactions of other antigens with this same antibody. The first Ag clearly cannot be accommodated. The second (purple) antigen is not a perfect fit, but it can bind to the antibody.

are used as enzymes; see microfile 8.2). So specific are some immunoglobulins for antigen that they can distinguish between a single functional group of a few atoms. Because the specificity of the Fab sites is identical, an Ig molecule can bind antigenic determinants on the same cell or on two separate cells and thereby link them.

The principal activity of an antibody is to unite with, immobilize, call attention to, or neutralize the antigen for which it was formed (figure 15.16). Antibodies called **opsonins** stimulate **opsonization,*** a process in which microorganisms or other particles are coated with specific antibodies so that they will be more readily recognized by phagocytes, which dispose of them. Opsonization has been likened to putting handles on a slippery object to provide phagocytes a better grip. The capacity for antibodies to aggregate, or *agglutinate,* antigens is the consequence of their cross-linking cells or particles into large clumps. This is a principle behind certain immune tests discussed in chapter 16. The interaction of an antibody with complement can result in the specific rupturing of cells and some viruses. In **neutralization** reactions, antibodies fill the surface receptors on a virus or the active site on a molecule to prevent it from functioning normally. **Antitoxins** are a special type of antibody that neutralize bacterial exotoxins. It should be noted that not all antibodies are protective; some neither benefit nor harm, and a few actually cause diseases (see autoimmunity in chapter 17).

*opsonization (ahp′′-son-uh-zay′-shun) Gr. *opsonein,* to prepare food.

Functions of the Crystallizable Fragment: Interactions with Self

Although the Fab fragments bind antigen, the Fc fragment has a different binding function. In most classes of immunoglobulin, the proximal end of Fc contains an effector molecule that can bind to certain receptors on the membrane of cells, such as macrophages, neotrophils, eosinophils, mast cells, basophils, and lymphocytes. The effect on an antibody's Fc fragment binding to a cell receptor depends upon that cell's role. In the case of opsonization, the attachment of antibody to foreign cells and viruses exposes the Fc fragments to phagocytes. Certain antibodies have receptors on the Fc portion for fixing complement, and in some immune reactions, the binding of Fc causes the release of cytokines. For example, the antibody of allergy (IgE) binds to basophils and mast cells, which causes the release of allergic mediators such as histamine (see chapter 17). The size and amino acid composition of Fc also determine an antibody's permeability, its distribution in the body, and its class.

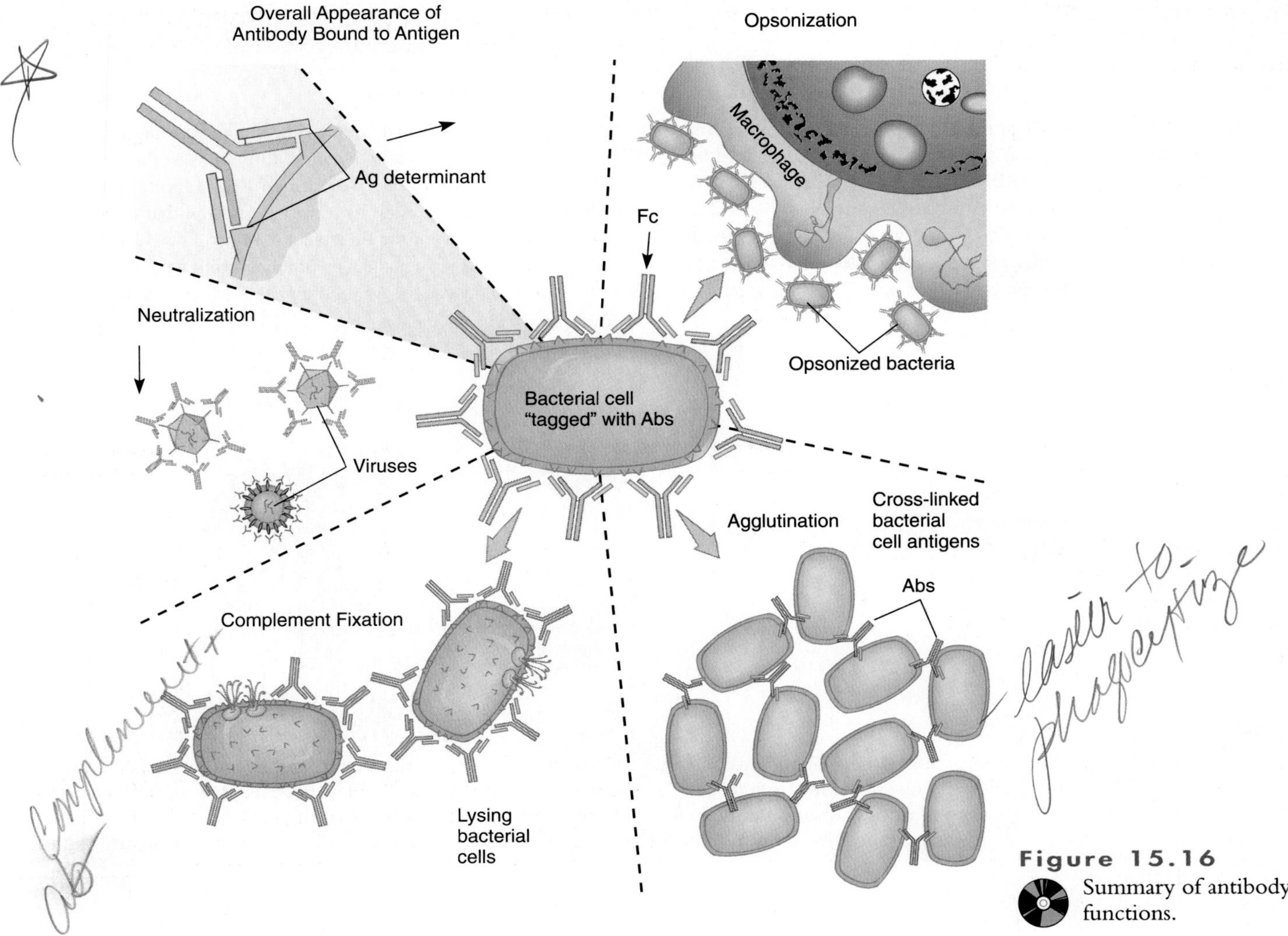

Figure 15.16
Summary of antibody functions.

Accessory Molecules on Immunoglobulins

All antibodies contain molecules in addition to the basic polypeptides. Varying amounts of carbohydrates are affixed to the constant regions in most instances (table 15.2). Two additional accessory molecules are the *J chain* that joins the monomers of IgA and IgM, and the *secretory component,* which helps move Ig across mucous membranes. These proteins occur only in certain immunoglobulin classes.

The Classes of Immunoglobulins

Immunoglobulins exist as structural and functional classes called *isotypes* (compared and contrasted in table 15.2). The differences in these classes are due primarily to variations in the Fc fragment and its accessory molecules. The classes are differentiated with shorthand names (Ig, followed by a letter: IgG, IgA, IgM, IgD, IgE).

The structure of **IgG** has already been presented. It is a monomer produced by memory cells responding the second time to a given antigenic stimulus. It is by far the most prevalent antibody circulating throughout the tissue fluids and blood. It has numerous functions: It neutralizes toxins, opsonizes, and fixes complement, and it is the only antibody capable of crossing the placenta.

The two forms of **IgA** are: (1) a monomer that circulates in small amounts in the blood and (2) a dimer that is a significant component of the mucous and serous secretions of the salivary glands, intestine, nasal membrane, breast, lung, and genitourinary tract. The dimer, called **secretory IgA,** is formed in a plasma cell by two monomers attached by a J piece. To facilitate the transport of IgA across membranes, a secretory piece is later added by the gland cells themselves. IgA coats the surface of these membranes and appears free in saliva, tears, colostrum, and mucus. It confers the most important specific tool immunity to enteric, respiratory, and genitourinary pathogens. Its contribution in protecting newborns who derive it passively from nursing is mentioned in microfile 15.5.

IgM (M for *macro)* is a huge molecule composed of five monomers (making it a pentamer) attached by the Fc receptors to

TABLE 15.2

CHARACTERISTICS OF THE IMMUNOGLOBULIN (Ig) CLASSES

	IgG	IgA (dimer only)	IgM	IgD	IgE
	Monomer	**Dimer, Monomer**	**Pentamer**	**Monomer**	**Monomer**
Number of antigen binding sites	2	4 2	10	2	2
Molecular weight	150,000	170,000–385,000	900,000	180,000	200,000
Percent of total antibody in serum	80%	13%	6%	1%	0.002%
Average life in serum (days)	23	6	5	3	2.5
Crosses placenta?	Yes	No	No	No	No
Fixes complement?	Yes	No	Yes	No	No
Fc binds to	Phagocytes	Phagocytes	B lymphocytes	B lymphocytes	Mass cells and basophils
Biological function	Long-term immunity; memory antibodies	Secretory antibody; on mucous membranes	Produced at first response to antigen; can serve as B-cell receptor	Receptor on B cells	Antibody of allergy; worm infections

C = carbohydrate.

J = J chain.

a central J chain. With its 10 binding sites, this molecule has tremendous avidity for antigen (*avidity* means the capacity to bind antigens). It is the first class synthesized by a plasma cell following its first encounter with antigen. Its complement-fixing and opsonizing qualities make it an important antibody in many immune reactions. It circulates mainly in the blood and is far too large to cross the placental barrier.

IgD is a monomer found in miniscule amounts in the serum, and it does not fix complement, opsonize, or cross the placenta. Its main function is to serve as a receptor for antigen on B cells, usually along with IgM. It seems to be the triggering molecule for B-cell activation, and it can also play a role in immune suppression.

IgE is also an uncommon blood component unless one is allergic or has a parasitic worm infection. Its Fc region interacts with receptors on mast cells and basophils. Its biological significance is to stimulate an inflammatory response through the release of potent physiological substances by the basophils and mast cells. Because inflammation would enlist blood cells such as eosinophils and lymphocytes to the site of infection, it would certainly be one defense against parasites. Unfortunately, IgE has another, more insidious effect—that of mediating anaphylaxis, asthma, and certain other allergies (see chapter 17).

Evidence of Antibodies in Serum

Regardless of the site where antibodies are first secreted, a large quantity eventually ends up in the blood by way of the body's communicating networks. If one submits a sample of **antiserum** (serum containing specific antibodies) to electrophoresis, the major groups of proteins migrate in a pattern consistent with their mobility and size (figure 15.17). The albumins show up in one band, and the globulins in four bands called alpha-1 (α_1), alpha-2 (α_2), beta (β), and gamma (γ) globulins. Most of the globulins represent antibodies, which explains how the term *immunoglobulin* was derived. **Gamma globulin** is composed primarily of IgG, whereas β and α_2 globulins are a mixture of IgG, IgA, and IgM. As we will see in chapter 16, the gamma globulin fraction of serum is important in immune therapies.

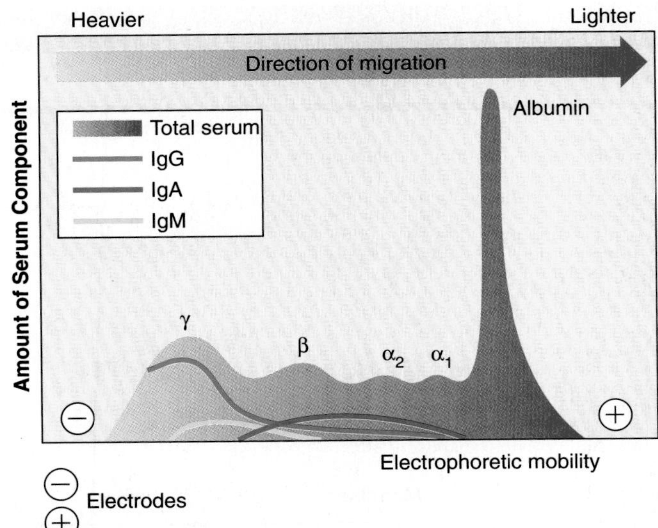

Figure 15.17
Pattern of human serum after electrophoresis. When antiserum is subjected to electrical current, the various proteinaceous components are separated into bands. The relative sizes of the molecules can be discerned because heavier molecules migrate more slowly than light ones.

Monitoring Antibody Production over Time: Primary and Secondary Responses to Antigens

We can learn a great deal about how the immune system reacts to an antigen by studying the levels of antibodies in serum over time (figure 15.18). This level is expressed quantitatively as the **titer,*** or concentration of antibodies. Upon the first exposure to an antigen, the system undergoes a **primary response.** The earliest part of this response, the *latent period,* is marked by a lack of antibodies for that antigen, but much activity is occurring. During this time, the antigen is being concentrated in lymphoid tissue, and is being processed by the correct clones of B lymphocytes. As plasma cells synthesize antibodies, the serum titer increases to a certain plateau and then tapers off to a low level over a few weeks or months. When the class of antibodies produced during this response is tested, an important characteristic of the response is uncovered. It turns out that, early in the primary response, most of the antibodies are the IgM type, which is the first class to be assembled by B cells. Later, the class of the antibodies (but not their specificity) is switched to IgG.

When the immune system is exposed again to the same immunogen within weeks, months, or even years, a **secondary response** occurs. The rate of antibody synthesis, the peak titer, and the length of antibody persistence are greatly increased over the primary response. The rapidity and amplification seen in this response are attributable to the memory B cells that were formed during the primary response. Because of its association with recall, the secondary response is also called the **anamnestic* response.** The advantage of this response is evident: It provides a quick and potent strike against subsequent exposures to infectious agents. This memory effect forms the basis for giving **boosters**—additional doses of vaccine—to increase the serum titer.

Monoclonal Antibodies: Useful Products from Cancer Cells

The value of antibodies as tools for locating or identifying antigens is well established. For many years, antiserum extracted from human or animal blood was the main source of antibodies for tests and therapy, but most antiserum has a basic problem: It is **polyclonal,** containing a mixture of antibodies with multiple specificities derived from several clones. This characteristic is to be expected, because several immune reactions may be occurring simultaneously, and even a single species of microbe can stimulate several different types of antibodies. Certain applications in immunology require a pure preparation of **monoclonal antibodies** (MABs) that originate from a single clone and have a single specificity.

The technology for producing monoclonal antibodies arose from hybridization of cancer cells and plasma cells *in vitro* (figure 15.19). A central part of this technique arose from the discovery that tumors isolated from multiple *myelomas** in mice consist of identical plasma cells. These monoclonal plasma cells were found to secrete a strikingly pure form of antibodies with a single specificity and to continue to divide indefinitely. Immunologists recognized the potential in these plasma cells and devised a **hybridoma** approach to creating MABs. The basic idea behind this approach is to hybridize or fuse a myeloma cell with a normal plasma cell from a mouse spleen to create an immortal cell that secretes a supply of functional antibodies with a single specificity.

The introduction of this technology has the potential for numerous biomedical applications. Monoclonal antibodies have provided immunologists with excellent standardized tools for studying the immune system and for expanding disease diagnosis and treatment. Most of the successful applications thus far use MABs in *in vitro* diagnostic testing and research. Although injecting monoclonal antibodies to treat human disease is an exciting prospect, so far this therapy has been stymied because most MABs are of mouse origin, and many humans will develop hypersensitivity to them. The development of human MABs and other novel approaches using genetic engineering is currently under way (microfile 15.3).

III T. ACTIVATION OF T LYMPHOCYTES AND V. HOW T CELLS RESPOND TO ANTIGEN: CELL-MEDIATED IMMUNITY

During the time that B cells have been actively responding to antigens, the T-cell limb of the system has been similarly en-

*titer (ty´-tur) Fr. *titre,* standard. One method for determining titer is shown in figure 16.5.

*anamnestic (an-am-ness´-tik) Gr. *anamnesis,* a recalling.

*myeloma (my-uh-loh´-muh) Gr. *myelos,* marrow, and *oma,* tumor. A malignancy of the bone marrow.

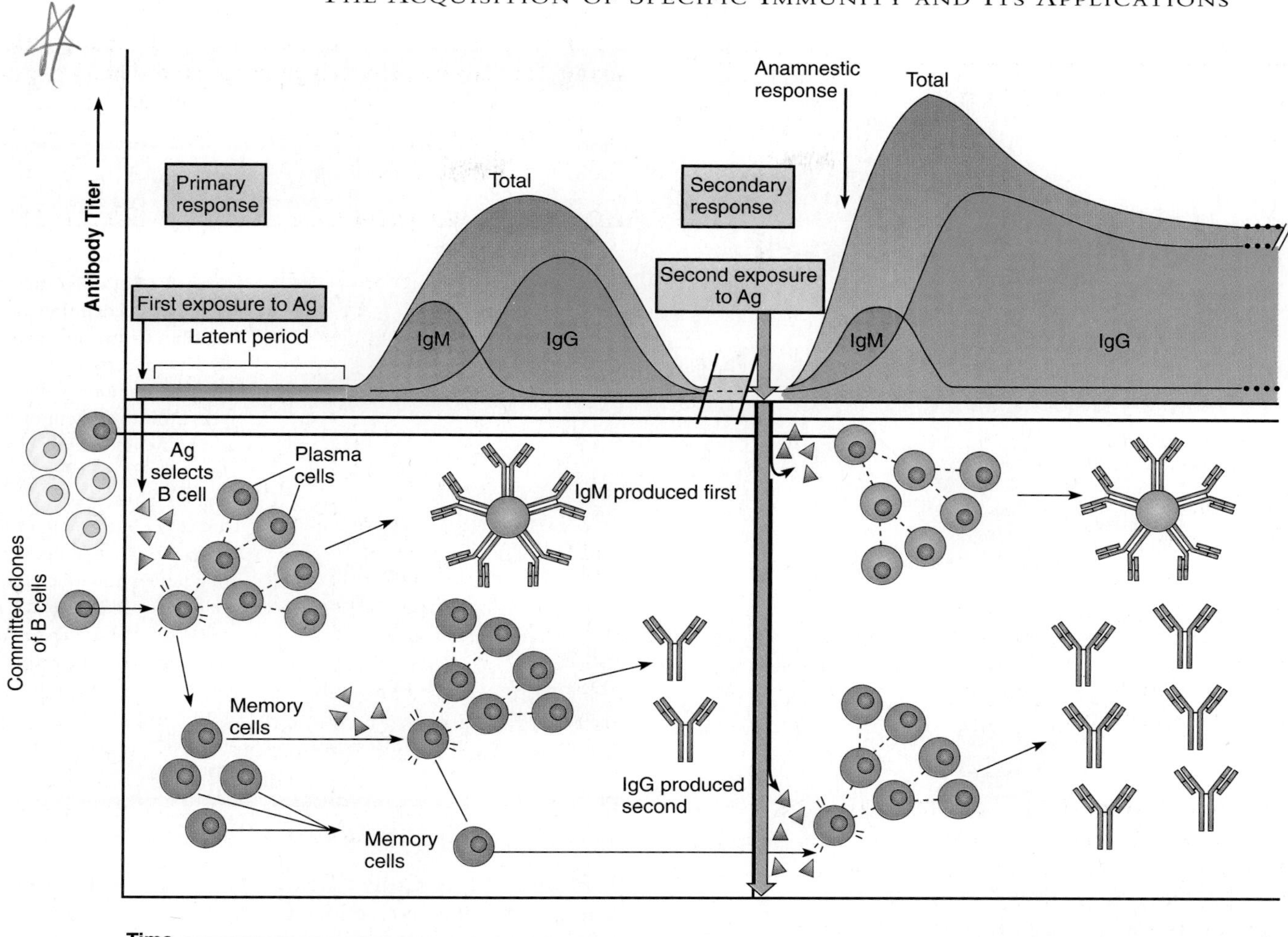

Figure 15.18

Primary and secondary responses to antigens. (*Top*) The pattern of antibody titer and subclasses as monitored during initial and subsequent exposure to the same antigen. (*Bottom*) A view of the B-cell responses that account for the pattern. Depicted as clonal selection, clonal expansion, production of memory cells, and the predominant antibody class occurring at first and second contact with antigen (Ag).

gaged. The responses of T cells, however, are **cell-mediated** immunities, which require the direct involvement of T lymphocytes throughout the course of the reaction. These reactions are among the most complex and diverse in the immune system and involve several subsets of T cells whose particular actions are dictated by CD receptors. All mature T cells have CD2 (the cause of rosetting) and CD2, but **CD4** and **CD8** are found only on certain classes (table 15.3). T cells are restricted; that is, they require some type of MHC (self) recognition before they can be activated, and all produce cytokines with a spectrum of biological effects (microfile 15.4).

T cells have notable differences in function from B cells. Rather than making antibodies to control foreign antigens, the whole T cell acts directly in contact with the antigen. They also stimulate other T cells, B cells, and phagocytes.

The Activation of T Cells and Their Differentiation into Subsets

The mature T cells in lymphoid organs are primed to react with antigens that have been processed and presented to them by macrophages. A T cell is initially **sensitized** when antigen is bound to its receptor. By mechanisms not yet fully characterized, sensitization leads to the final differentiation of the cell into one of four functionally specialized subsets: helper, suppressor, cytotoxic, or delayed hypersensitivity T cells (table 15.3 and figure 15.20). As with B cells, activated T cells transform into lymphoblasts in preparation for mitotic divisions, and they divide into one of the subsets of effector cells and memory cells that can interact with the antigen upon subsequent contact. Memory T cells are some of the longest-lived blood cells known (70 years in one well-documented case).

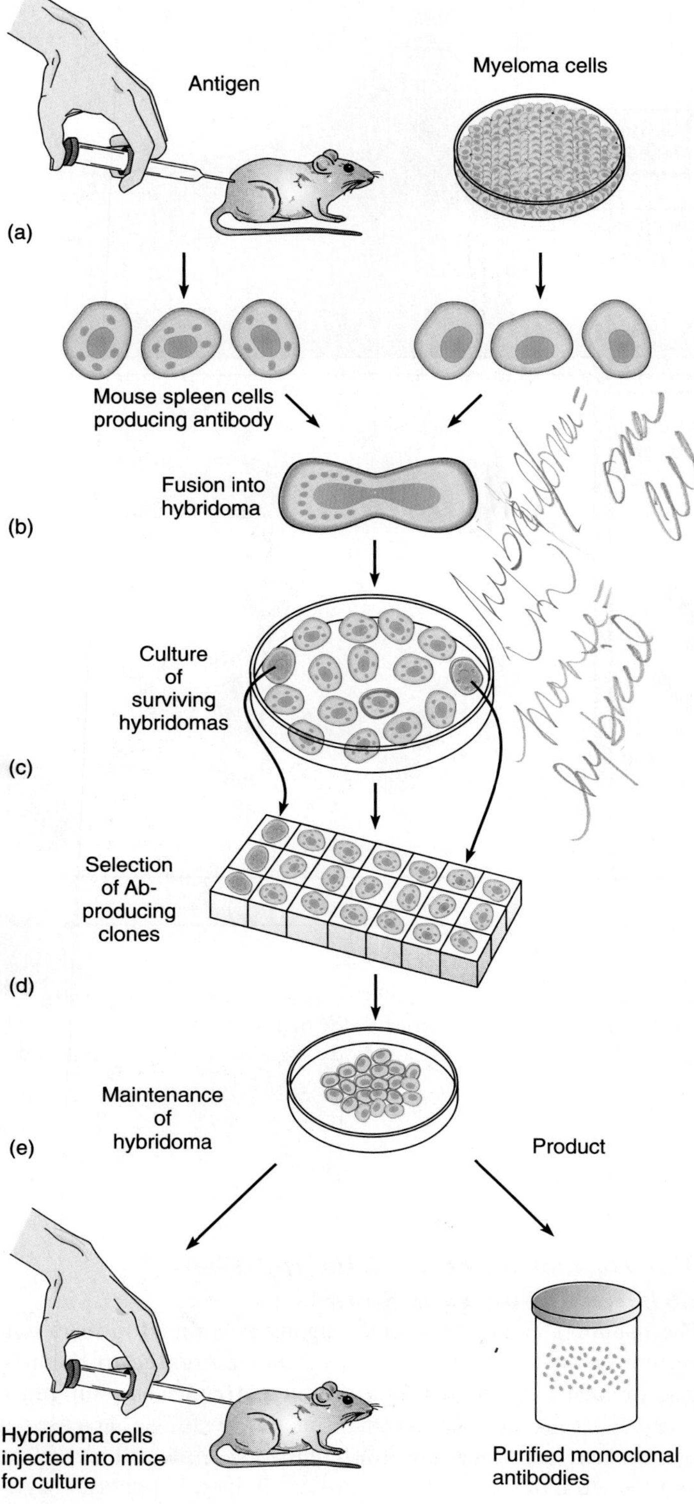

Figure 15.19
Summary of the technique for producing monoclonal antibodies by hybridizing myeloma tumor cells with normal plasma cells. (*a*) A normal mouse is inoculated with an antigen having the desired specificity, and plasma cells are isolated from its spleen. A special strain of mouse provides the myeloma cells. (*b*) The two cell populations are mixed with polyethylene glycol, which causes some cells in the mixture to fuse and form hybridomas. (*c*) Surviving cells are cultured and separated into individual wells. (*d*) Tests are performed on each hybridoma to determine the specificity of the antibody (Ab) it secretes. (*e*) A hybridoma with the desired specificity is grown in tissue culture; antibody product is then isolated and purified. The hybridoma is maintained in a susceptible mouse for future use.

TABLE 15.3

CHARACTERISTICS OF SUBSETS OF T CELLS

Subset	Shorthand Designations	Functions/ Important Features
T helper cells	T_H, T_4	Assist B cells in recognition of antigen; assist other subsets of T cells in recognition and reaction to antigen; identified by CD4 receptors
T suppressor cells	T_S, T_8	Regulate immune reactions; cells limit the extent of antibody production; block some T-cell activity; carry CD5 and CD8 receptors
Cytotoxic (killer) cells	T_C, T_K	Destroy a target foreign cell by lysis; important in destruction of complex microbes, cancer cells, virus-infected cells; graft rejection; allergy; also have CD5 and CD8 receptors
Delayed hypersensitivity cells	T_D, T_{DTH}	Responsible for allergies occurring several hours or days after contact; skin reactions as in tuberculin test

T Helper (T_H) Cells Helper cells play a central role in assisting with immune reactions to antigens, including those of B cells and other T cells. They do this directly by receptor contact and indirectly by releasing cytokines (interleukin-2, B-cell growth factor) that stimulate lymphocyte development and monitor both specific and nonspecific immune reactions. T helper cells are the most prevalent type of T cell in the blood and lymphoid organs, making up about 65% of this population. The severe depression of this class of T cells (with CD4 receptors) by HIV is what largely accounts for the immunopathology of AIDS.

T Suppressor (T_S) Cells An essential part of the immune mechanism is to restrict rampant, uncontrolled immune responses that could be inappropriate or destructive. Although immunosuppression reactions have not yet been completely characterized, they are known to involve T suppressor cells. These cells regulate the production of antibodies by plasma cells and can inhibit the actions of T_H cells. It is thought that certain autoimmune diseases and cancers are influenced by the abnormal function of T_S cells.

MICROFILE 15.3 MONOCLONAL ANTIBODIES: VARIETY WITHOUT LIMIT

Imagine releasing millions of tiny homing pigeons into a molecular forest and having them navigate directly to their proper roost, and you have some sense of what monoclonal antibodies can do. Laboratories use them to identify antigens, receptors, and antibodies; to differentiate cell types (T cells versus B cells) and cell subtypes (different sets of T cells); to diagnose diseases such as cancer and AIDS; and to identify bacteria and viruses.

A number of promising techniques have been directed toward using monoclonals as drugs. When genetic engineering is combined with hybridoma technology, the potential for producing antibodies of almost any desired specificity and makeup is possible. For instance, *chimeric* MABs (monoclonal antibodies) that are part human and part mouse have been produced through splicing antibody genes. It is even possible to design an antibody molecule that has antigen binding sites with different specificities. With this technology, extra molecular groups can be added to increase antibody affinity or to give antibodies destructive powers. Monoclonals can be hybridized with plant or bacterial toxins to form **immunotoxin** complexes that attach to a target cell and poison it. The most exciting prospect of this therapy is that it can destroy a specified cancer cell and not harm normal cells. Monoclonals are currently employed in treatment of cancers such as lymphoma and colon cancers. One such drug, called Rituxan, binds to the cancer cells and triggers their death.

Such antibodies could also be used to suppress allergies, antoimmunities, and graft rejection. Drugs such as OKT3 and Orthoclone are currently used to prevent the rejection of organ transplants by incapacitating cytotoxic T cells. Now that researchers have genetically engineered plants and mice to produce human MABs, therapeutic uses will continue to expand.

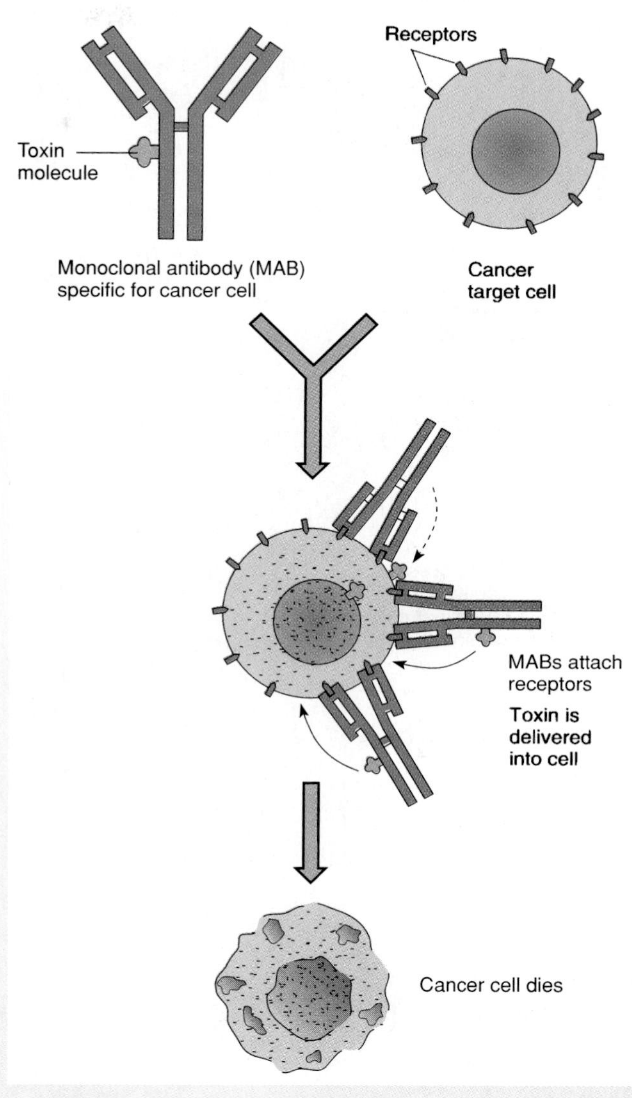

The mechanism of an immunotoxin. A potential therapy for cancer and immune dysfunctions is the use of a monoclonal antibody specific for a certain tumor that carries a potent toxin molecule. The antibody would circulate to cancer cells and deliver the toxin to them. Normal body cells would be unharmed.

MICROFILE 15.4 LYMPHOKINES: CHEMICAL PRODUCTS OF T CELLS

Although the immunities of T cells are usually thought of as cell-mediated, one must not overlook the fact that T cells are also prolific chemical factories. The products of sensitized T cells that communicate with and act upon other cells are a type of cytokine called **lymphokines.** Lymphokines have diverse effects. Some (interferon and interleukin) regulate immune reactions; some mediate inflammation (chemotactic factors); and others (lymphotoxins) kill whole cells. In the previous chapter, we mentioned the role of **gamma interferon** in regulating B cells and T cells. It also activates natural killer cells (NK) and stimulates macrophages. **Interleukin-2** was discussed in conjunction with T helper cells and its role as a growth promoter and general stimulus for lymphocyte activity. **Interleukin-3** is a powerful stimulus for stem cell development in the bone marrow. Several other interleukins (IL-4, 5, 6, 7) promote proliferation, differentiation, and secretion of B and T cells.

The mechanisms by which cytotoxic lymphocytes destroy their whole cell targets is just being understood. It appears that they seek out the foreign cell's membrane and secrete various metabolic products into it. The compounds that damage the target cell are termed **lymphotoxins.** These molecular poisons can disrupt the cell membrane by forming pores (see chapter opening photo). They also precipitate the lysis of the cell nucleus and initiate cell death (see figure 15.21).

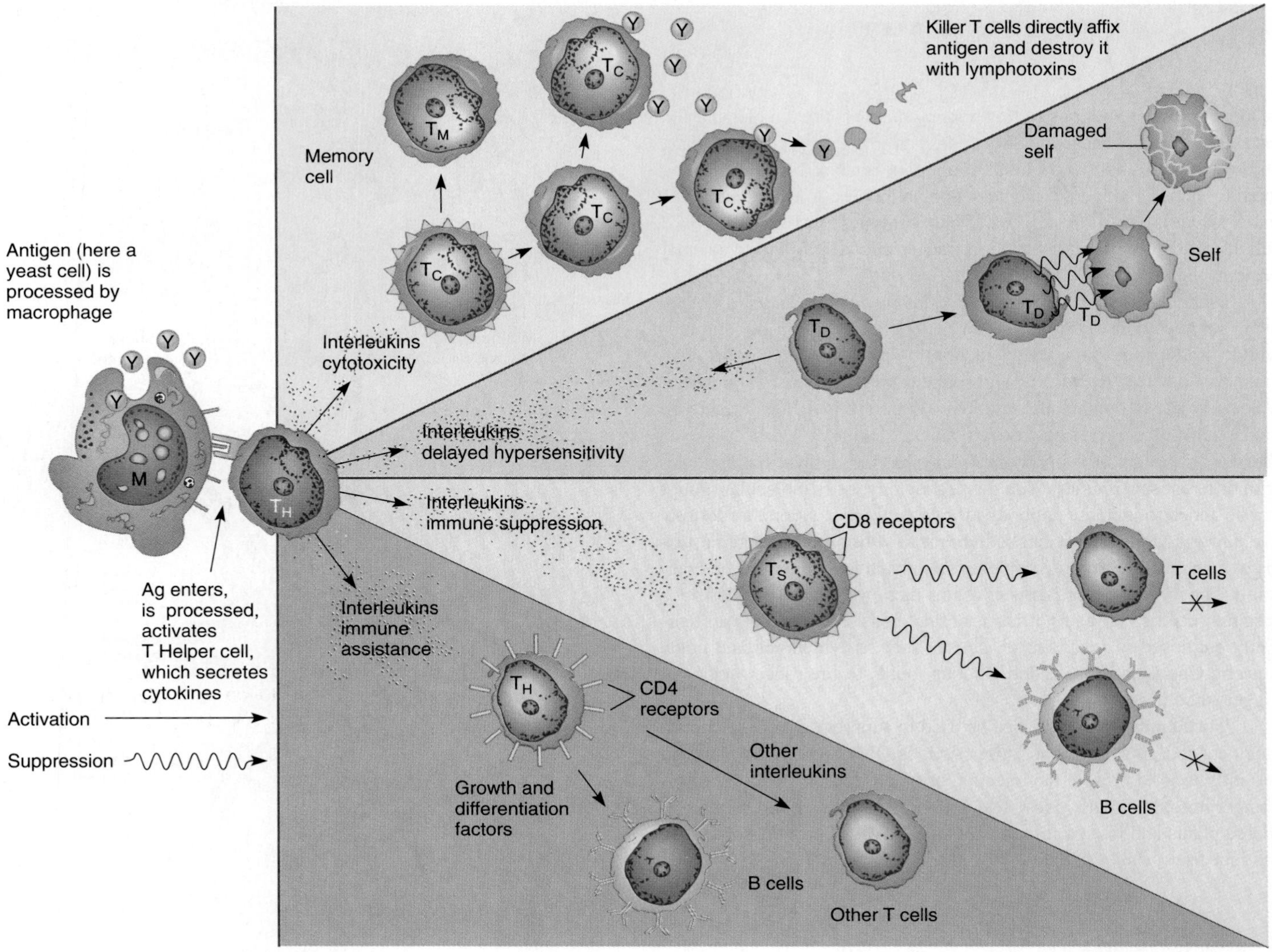

Figure 15.20

Overall scheme of T-cell activation and differentiation into one of our major subsets of T cells: T_C cells destroy certain microbes and foreign cells; T_D cells react with tissues and cause a type of hypersensitivity that damages self; T_S cells inhibit the actions of B and T cells; T_H cells assist in the actions of B and T cells.

Cytotoxic T (T_C) Cells: Cells That Kill Other Cells

Cytotoxicity is the capacity of certain T cells to kill a specific target cell. It is a fascinating and powerful property that accounts for much of our immunity to foreign cells and cancer, and yet, under some circumstances, it can lead to disease. For a *killer T cell* to become activated, it must recognize foreign receptors on a target cell and mount a direct attack upon it. After activation, the T_C cell delivers a dose of several cytokines that severely injures the target cell membrane (figure 15.21). This release of cytokines is followed by target cell death through a process called *apoptosis* (ah-poh-toh´-sis). The apoptosis is genetically programmed and results in destruction of the nucleus and complete cell lysis.

Target cells that T_C cells can destroy:

- Fungi, protozoa, and complex bacteria (mycobacteria).
- Virally infected cells (figure 15.21). Cytotoxic cells recognize these because of telltale virus receptors expressed on their surface. Cytotoxic defenses are an essential protection against viruses.
- Cancer cells. T cells constantly survey the tissues and immediately attack any abnormal cells they encounter (figure 15.22). The importance of this function is clearly demonstrated in the susceptibility of T-cell-deficient people to cancer (chapter 17).
- Cells from other animals and humans. Cytotoxic CMI is the most important factor in **graft rejection.** In this instance, the

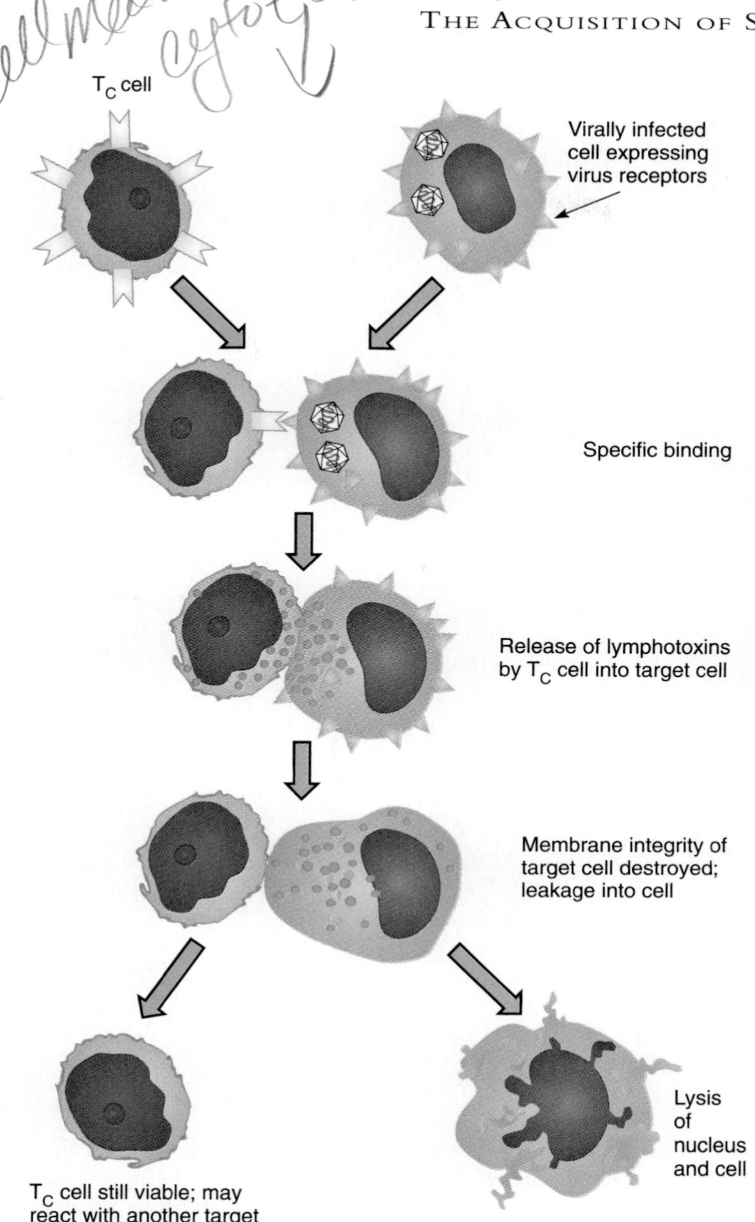

Figure 15.21

Stages of cell-mediated cytotoxicity and the action of lymphotoxins on virus-infected target cells.

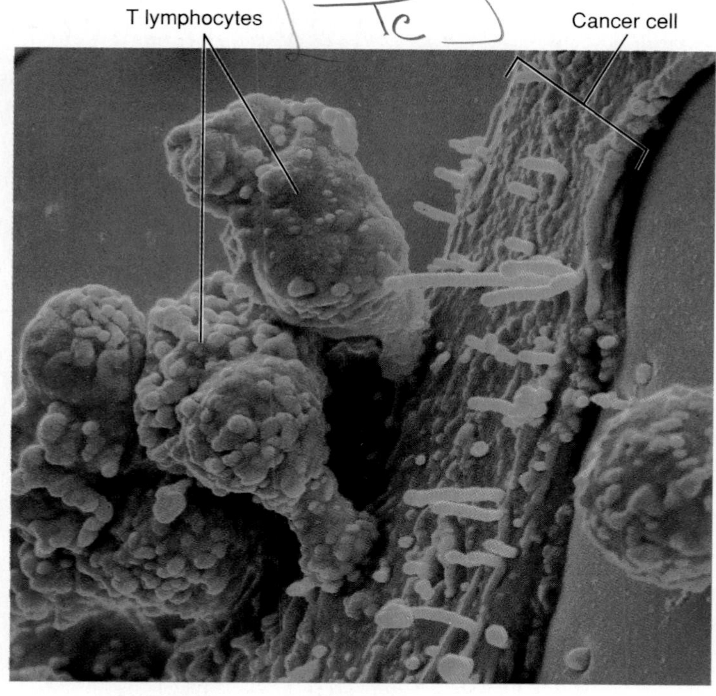

Figure 15.22

A large, crablike cancer cell is attacked by killer T cells, which release lymphotoxins directly into the cancer and eventually leave behind a lifeless shell.

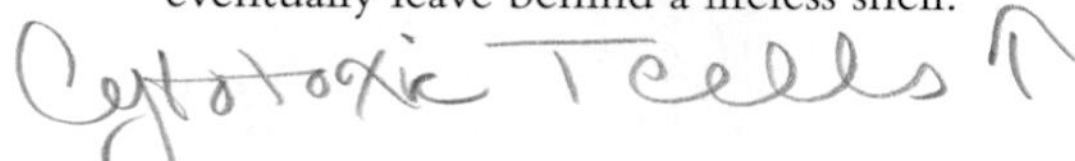

T_C cells attack the foreign tissues that have been implanted into a recipient's body.

Other Types of Killer Cells **Natural killer (NK)** cells are a type of lymphocyte related to T cells that lack specificity for antigens. They circulate through the spleen, blood, and lungs and are probably the first killer cells to attack cancer cells and virus-infected cells. They destroy such cells by similar mechanisms as T cells. Their activities are acutely sensitive to cytokines such as interleukin-12 and type 1 interferon.

Delayed Hypersensitivity T (T_D) Cells Although immediate allergies such as hay fever and anaphylaxis are mediated by antibodies, certain delayed responses to allergens (the tuberculin reactions for example) are initiated by special T cells. These reactions are discussed in chapter 17.

A Practical Scheme for Classifying Specific Immunities

The means by which humans acquire immunities can be conveniently encapsulated within four interrelated categories: active, passive, natural, and artificial.

Active immunity occurs when an individual receives an immune stimulus (antigen) that activates the B and T cells,

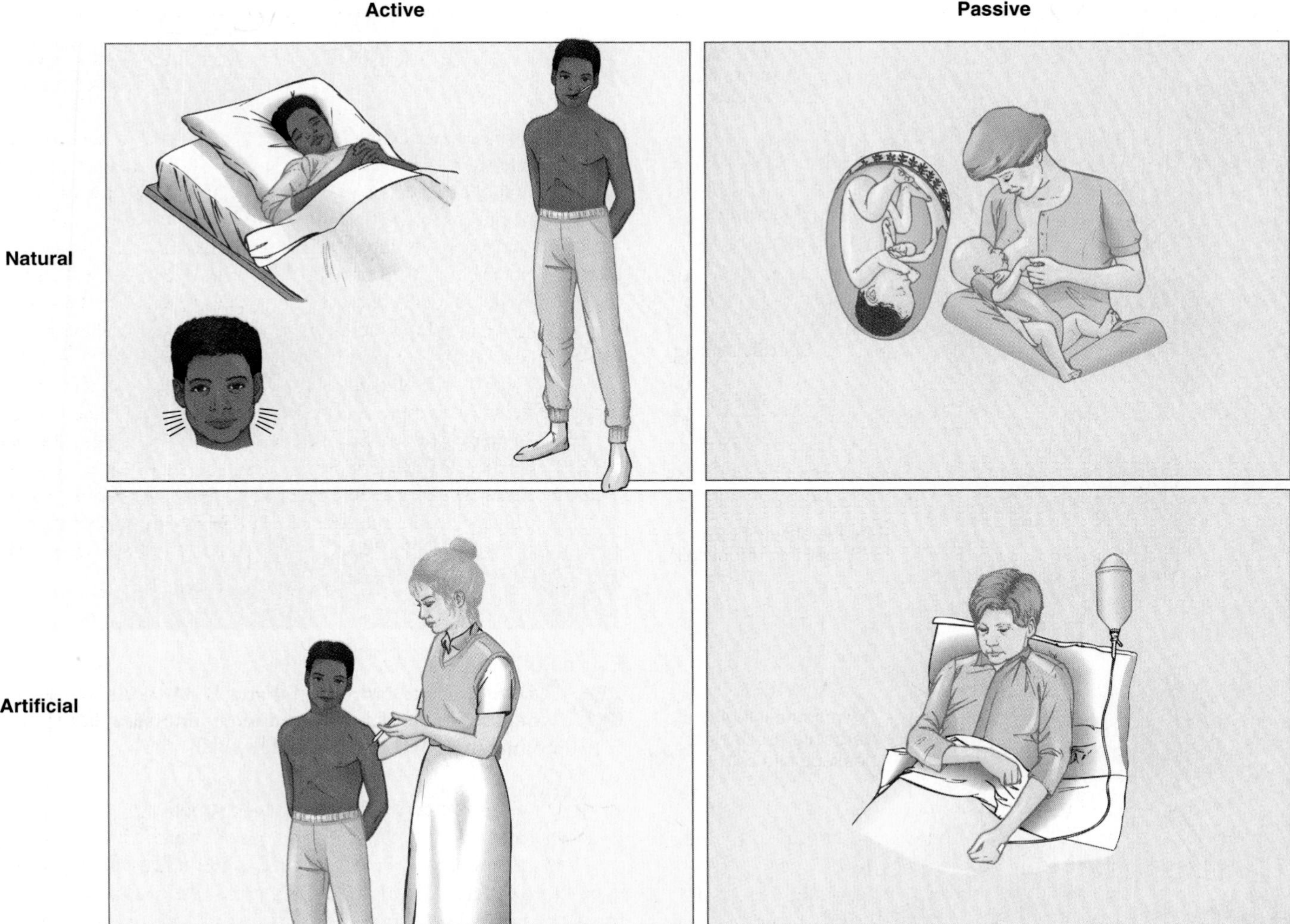

Figure 15.23
Categories of acquired immunities. Natural immunities, which occur during the normal course of life, are either active (acquired from an infection and then recovering) or passive (antibodies donated by the mother to her child). Artificial immunities are acquired through medical practices and can be active (vaccinations with antigen, to stimulate an immune response) or passive (immune therapy with a serum containing antibodies).

causing the body to produce immune substances such as antibodies. Active immunity is marked by several characteristics: (1) It is an essential attribute of an immunocompetent individual; (2) it creates a memory that renders the person ready for quick action upon reexposure to that same antigen; (3) it requires several days to develop; and (4) it lasts for a relatively long time, sometimes for life. Active immunity can be stimulated by natural or artificial means.

Passive immunity occurs when an individual receives immune substances (antibodies) that were produced actively in the body of another human or animal donor. The recipient is protected for a time even though he or she has not had prior exposure to the antigen. It is characterized by: (1) lack of memory of the original antigen, (2) lack of production of new antibodies against that disease, (3) immediate onset of protection, and (4) short-term effectiveness, because antibodies have a limited period of function, and ultimately, the recipient's body disposes of them. Passive immunity can also be natural or artificial in origin.

Natural immunity encompasses any immunity acquired during the normal biological experiences of an individual that does not involve medical intervention.

Artificial immunity is protection from infection obtained through medical procedures. This type of immunity is induced by immunization with vaccines and immune serum.

Figure 15.23 illustrates the various possible combinations of acquired immunities.

MICROFILE 15.5 BREAST FEEDING: THE GIFT OF ANTIBODIES

An advertising slogan from the past claims that cow's milk is "nature's most nearly perfect food." One could go a step further and assert that human milk is nature's *perfect* food for young humans. Clearly, it is loaded with essential nutrients, not to mention being available on demand from a readily portable, hygienic container that does not require refrigeration or warming. But there is another and perhaps even greater benefit. During lactation, the breast becomes a site for the proliferation of lymphocytes that produce IgA, a special class of antibody that protects the mucosal surfaces from local invasion by microbes. The very earliest secretion of the breast, a thin, yellow milk called *colostrum,* is very high in IgA. These antibodies form a protective coating in the gastrointestinal tract of a nursing infant that guards against infection by a number of enteric pathogens *(Escherichia coli, Salmonella,* poliovirus, rotavirus). Protection at this level is especially critical because an infant's own IgA and natural intestinal barriers are not yet developed. As with immunity in utero, the necessary antibodies will be donated only if the mother herself has active immunity to the microbe through a prior infection or vaccination.

The benefits of nursing have been known since time immemorial. In the Middle Ages, women of privilege who did not want to nurse often bore huge families of 20 or more children and dispatched each to the care of a wet-nurse for its first two years of life. In an era when infant mortality was extremely high, the children often had little contact with the birth mother until their survival was assured. In more recent times, the ready availability of artificial formulas and the changing life-styles of women have reduced the incidence of breast feeding. Where adequate hygiene and medical care prevail, bottle-fed infants get through the critical period with few problems, because the foods given them are relatively sterile and they have received protection against some childhood infections in utero. Mothers in developing countries with untreated water supplies or poor medical services are strongly discouraged from using prepared formulas, because they can actually inoculate the baby's intestine with pathogens from the formula. Millions of neonates suffer from severe and life-threatening diarrhea that could have been prevented by the hygienic protection of nursing.

Natural Active Immunities: Getting the Infection

After recovering from infectious disease, a person may be actively resistant to reinfection for a period that varies according to the disease. In the case of childhood viral infections such as measles, mumps, and rubella, this natural active stimulus provides lifelong immunity. Other diseases result in a less extended immunity of a few months to years (such as pneumococcal pneumonia and shigellosis), and reinfection is possible. Even a subclinical infection can stimulate natural active immunity. This probably accounts for the fact that some people are immune to an infectious agent without ever having been noticeably infected with or vaccinated for it.

Natural Passive Immunity: Mother to Child

Natural, passively acquired immunity occurs only as a result of the prenatal and postnatal, mother-child relationship. During fetal life, IgG antibodies circulating in the maternal bloodstream are small enough to pass or be actively transported across the placenta. Antibodies against tetanus, diphtheria, pertussis, and several viruses regularly cross the placenta. This natural mechanism provides an infant with a mixture of many maternal antibodies that can protect it for the first few critical months outside the womb, while its own immune system is gradually developing active immunities. Depending upon the microbe, passive protection lasts anywhere from a few months to a year. But eventually, the infant's body clears the antibody. Most childhood vaccinations are timed so that there is no lapse in protection against common childhood infections.

Another source of natural passive immunity comes to the baby by way of mother's milk (microfile 15.5). Although the human infant acquires 99% of natural passive immunity in utero and only about 1% through nursing, the milk-borne antibodies provide a special type of intestinal protection that is not forthcoming from transplacental antibodies.

Artificial Immunity: Immunization

Immunization is any clinical process that produces immunity in a subject. Because it is often used to give advance protection against infection, it is also called *immunoprophylaxis.* The use of these terms is sometimes imprecise, thus it should be stressed that active immunization, in which a person is administered antigen, is synonymous with vaccination, and that passive immunization, in which a person is given antibodies, is a type of immune therapy.

Vaccination: Artificial Active Immunization

The term **vaccination** originated from the Latin word *vacca* (cow), because the cowpox virus was used in the first preparation for active immunization against smallpox (see microfile 16.1). Vaccination exposes a person to a specially prepared microbial (antigenic) stimulus, which then triggers the immune system to produce antibodies and lymphocytes to protect the person upon future exposure to that microbe. As with natural active immunity, the degree and length of protection vary. Commercial vaccines are currently available for about 26 diseases. Methods of vaccine antigen selection and modes of vaccination are discussed more fully in chapter 16.

Immunotherapy: Artificial Passive Immunization

In **immunotherapy,** a patient at risk for acquiring a particular infection is administered a preparation that contains specific antibodies against that infectious agent. In the past, these therapeutic substances were obtained by vaccinating animals

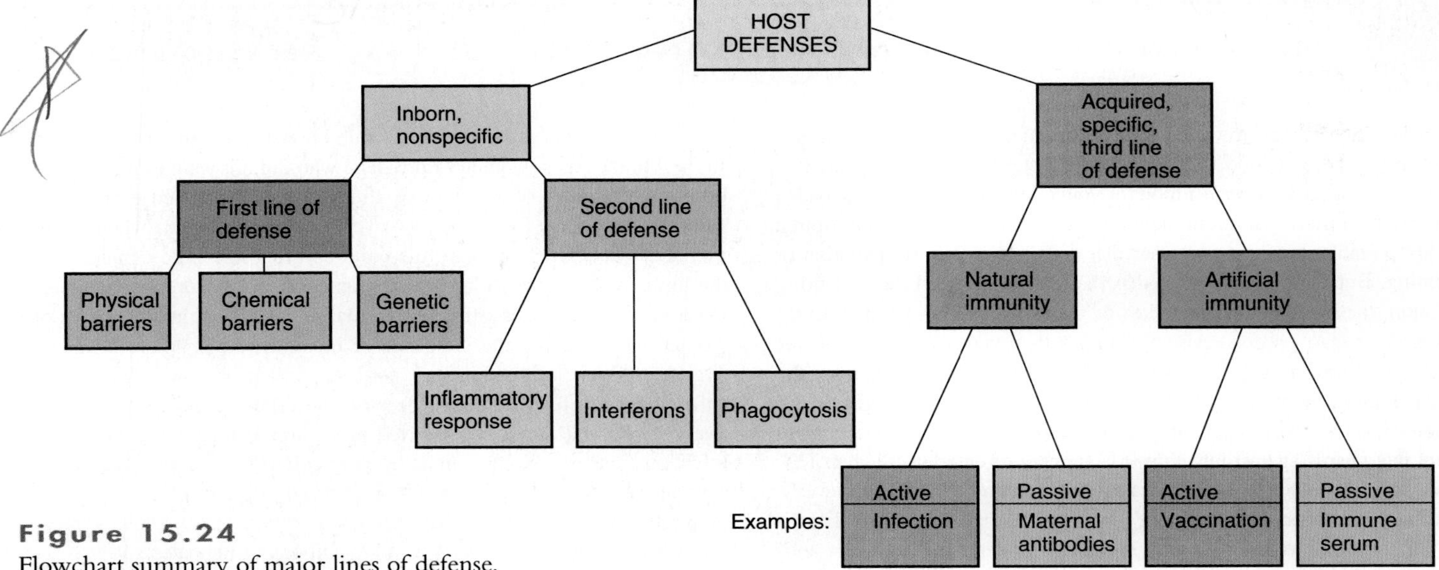

Figure 15.24
Flowchart summary of major lines of defense.

(horses in particular), then taking blood, and extracting the serum. However, horse serum is now used only in limited situations because of the potential for hypersensitivity to it. Pooled human serum from donor blood (gamma globulin) and **immune serum globulins** containing high quantities of antibodies are more frequently used. Immune serum globulins are used to protect people who have been exposed to hepatitis, measles, and rubella. More specific immune serum, obtained from patients recovering from a recent infection, is useful in preventing and treating hepatitis B, rabies, pertussis, and tetanus.

A final outline summarizing the system of host defenses covered in chapter 14 and 15 is presented in figure 15.24. This chart will serve as a resource to review major aspects of immunity as well as a guide in answering certain questions (see concept question 17).

Chapter Checkpoints

Immature lymphocytes released from hemopoietic tissue migrate to one of two sites for further development. B cells mature in the stromal cells of the bone marrow. T cells mature in the thymus.

Antigens or immunogens are proteins or other complex molecules of high molecular weight that trigger the immune response in the host.

Lymphocytes respond to a specific portion of an antigen called the antigenic determinant. A given microorganism has many such determinants, all of which stimulate individual specific immune responses.

Haptens are molecules that are too small to trigger an immune response alone but can be immunogenic when they attach to a larger substance, such as host serum protein.

Autoantigens and allergens are types of antigens that cause damage to host tissue as a consequence of the immune response.

Macrophages or other antigen-processing cells (APCs) bind foreign antigen to their cell surfaces for presentation to lymphocytes. Physical contact between the APC, T cells, and B cells activates these lymphocytes to proceed with their respective immune responses.

B cells produce five classes of antibody: IgM, IgG, IgA, IgD and IgE. IgM and IgG predominate in plasma. IgA predominates in body secretions. IgD binds to B cells as an antigen receptor. IgE binds to tissue cells, promoting inflammation.

Antibodies bind physically to the specific antigen that stimulates their production, thereby immobilizing the antigen and enabling it to be destroyed by other components of the immune system.

The anamnestic response means that the second exposure to antigen calls forth a much faster and more vigorous response than the first.

Monoclonal, or pure, antibodies can be produced commercially by fusing a plasma cell with a myeloma cell to produce an immortal hybridoma.

T cells do not produce antibodies. Instead they produce different cytokines that play diverse roles in the immune response. Each subject of T cell produces a particular cytokine that stimulates lymphocytes or destroys foreign cells.

Active immunity means that your body produces antibodies to a disease agent. If you contract the disease, you can develop natural active immunity. If you are vaccinated, your body will produce artificial active immunity.

In passive immunity, you receive antibodies from another person. Natural passive immunity comes from the mother. Artificial passive immunity is administered medically.

Chapter Capsule with Key Terms

SPECIFIC IMMUNITY AND ITS APPLICATIONS[3]

Development of Lymphocyte Specificity/Receptors

Acquired immunity involves the reactions of B and T lymphocytes to foreign molecules, or **antigens.** Before they can react, each lymphocyte must undergo differentiation into its final functional type by developing protein receptors for antigen, the specificity of which is genetically controlled and unique for each type of lymphocyte.

The **clonal selection theory** explains this process. Genetic recombination and mutation during embryonic and fetal development produce millions of different lymphocyte clones, each bearing a different antigen receptor. This provides a huge lymphocyte repertoire required to react with millions of antigens. **Tolerance to self,** the elimination of any lymphocyte clones that can attack self, occurs during this time.

The receptors on B cells are **immunoglobulin (Ig)** molecules, and receptors on T cells are smaller glycoprotein molecules. Other receptors needed in recognition are governed by the **major histocompatibility (MHC)** gene complex, which is also referred to as the **human leukocyte antigen (HLA)** complex. Expression of these genes gives rise to receptors on most cells that govern cell communication and recognition of self and antigens.

B-Cell Maturation: Immature B stem cells originate in the yolk sac, liver, and bone marrow and differentiate into mature cells under the influence of special stromal cells in the bone marrow. Mature cells acquire Ig receptors and migrate to predetermined sites in lymphoid organs, ready to react with antigen.

T-Cell Maturation: Immature T stem cells originate in the same areas of the embryo and fetus but mature under the influence of the thymus gland. Specificity is acquired through addition of CD receptors, and mature cells migrate to different sites in lymphoid organs.

Introduction of Antigen/Immunogens

An antigen (Ag) is any substance that stimulates an immune response. Requirements for **antigenicity** include foreignness (recognition as nonself), large size, and complexity of cell or molecule. Foreign cells and large complex molecules (over 10,000 MW) are most antigenic; foreign molecules less than 1,000 MW (**haptens**) are not antigenic unless attached to a larger carrier molecule. The **antigenic determinant** is the small molecular group of the foreign substance that is actually involved in recognition and reactions by lymphocytes. Cells, viruses, and large molecules can have numerous antigenic determinants.

Special categories of antigens include **autoantigens,** molecules on self tissues for which tolerance is inadequate; **alloantigens,** cell surface markers of one individual that are antigens to another of that same species; **heterophile antigens,** molecules from unrelated species that bear similar antigenic determinants; *superantigens,* complex bacterial toxins; and **allergen,** the antigen that provokes allergy.

Cooperation in Immune Reactions to Antigen

T-cell-dependent antigens must be processed by special macrophages, the **antigen-processing cells** (APCs). An APC alters the antigen and attaches it to its MHC receptor for presentation to lymphocytes. Antigen presentation involves a direct collaboration among the macrophage, a T helper (T_H) cell, and an antigen-specific B or T cell. A match occurs between the MHC receptors of the APC and T_H cell, the antigen, and the lymphocyte, and cytokines are released. **Interleukin-1** from the APC activates the T_H cells, and **interleukin-2** produced by the T_H cell, activates B and other T cells.

B-Cell Activation and Antibody Production

When B cells receive the antigen and are stimulated by B-cell growth and differentiation factors, they undergo **blast formation** in preparation for mitosis and **clonal expansion.** Divisions give rise to **plasma cells** that secrete antibodies and **memory cells** that can react to that same antigen later.

Nature of Antibodies (Immunoglobulins): A single immunoglobulin molecule (monomer) is a large Y-shaped protein molecule consisting of four polypeptide chains. It contains two identical fragments **(Fab)** with ends that form the active site that binds antigen. Each different antibody has identical Fabs, and each has a unique specificity for a particular antigen. The single fragment of the antibody **(Fc)** binds to self.

Antigen-Antibody (Ag-Ab) Reactions: In **opsonization,** antibodies tag the antigen to make it more readily phagocytosed; in **neutralization,** antibodies block the active site on a toxin **(antitoxin)** or a receptor on a virus; some antibodies can *agglutinate* (aggregate) antigens; some antibodies fix complement and cause destruction of cells; the Fc portion can bind to various body cells and mediate inflammation and allergy.

The five **antibody classes,** which differ in size and function, are **IgG,** the major Ab in circulation that crosses the placenta; **IgA,** secretory Ab, present on mucous membranes and secretions of them; **IgM,** a large pentamer formed during first response to Ag; **IgD,** primarily a receptor on B cells; **IgE,** antibody of certain allergies.

Antibodies in Serum (Antiserum): Serum antibodies can be identified through electrophoresis and quantified by testing the **titer** (levels of antibodies) over time. The first introduction of an Ag to the immune system produces a **primary response,** with a gradual increase in Ab titer that gradually recedes. The second contact with the same Ag occurs rapidly and produces a larger titer called the **secondary,** or **anamnestic, response,** due to memory cells produced during initial response.

Monoclonal antibodies are artificially induced antibodies formed by fusing a mouse B cell with a cancer cell. The antibodies have a single specificity to Ag and are used in diagnosis of disease, identification of microbes, and therapy.

T Cells and Cell-Mediated Immunity (CMI)

T cells function by coming into direct contact with antigens and foreign cells. After presentation of Ag by a macrophage and activation by T_H cells, a sensitized T cell will become one of four functional types. Mitosis leads to a proliferation of activated and long-lasting memory T cells. Depending upon its **CD** receptor type: (1) **T helper cells** (T_H) assist other T cells and B cells, directly and by means of cytokines; (2) **T suppressor cells** (T_S) limit the actions of other T cells and B cells; (3) **cytotoxic,** or **killer, T cells** (T_C) seek out and destroy large, complex foreign or abnormal cells (microbes, cancer, grafted tissues, virus-infected cells); and (4) **delayed hypersensitivity cells** (T_D) cause a form of hypersensitivity. T cells secrete a series of cytokines **(interleukin,**

3. Also review figure 15.1.

interferon, lymphotoxins) that destroy antigen or stimulate reactions.

Classification of Acquired Immunity

Immunities acquired through B and T lymphocytes can be classified by a simple system. **Natural immunity** is acquired as part of normal life experiences, whereas **artificial immunity** is acquired through medical procedures such as **immunization. Active immunity** results when a person is challenged with antigen that stimulates production of protective substances (antibodies). It creates memory, takes time, and is lasting. In **passive immunity,** preformed protective substances (antibodies) are donated to an individual. It does not create memory, acts immediately, and is short term.

Combinations of acquired immunity are **natural active,** acquired upon infection and recovery; **natural passive,** acquired by child through placenta and breast milk; **artificial active (vaccination),** inoculation with a selected antigen prevents future infections; **artificial passive,** administration of **immune serum** or globulin to treat or prevent infection.

Multiple-Choice Questions

1. The primary B-cell receptor is:
   a. IgD
   b. IgA
   c. IgE
   d. IgG
2. In humans, B cells mature in the ______ and T cells mature in the ______.
   a. GALT, liver
   b. bursa, thymus
   c. bone marrow, thymus
   d. lymph nodes, spleen
3. Small, simple molecules are ______ antigens
   a. poor
   b. never
   c. good
   d. heterophilic
4. An example of a mosaic antigen is
   a. albumin
   b. a lipopolysaccharide molecule
   c. a virus
   d. a hapten
5. Which type of cell actually secretes antibodies?
   a. T cells
   b. macrophages
   c. plasma cells
   d. monocytes
6. The cross-linkage of antigens by antibodies is known as
   a. opsonization
   b. a cross-reaction
   c. agglutination
   d. complement fixation
7. The greatest concentration of antibodies is found in the ______ fraction of the serum.
   a. gamma globulin
   b. albumin
   c. beta globulin
   d. alpha globulin
8. ______ is a measurement of the relative amount of immunoglobulin in the serum.
   a. Gamma globulin
   b. Secretory antibody
   c. Antibody titer
   d. Blast formation
9. A cytokine that stimulates the activity of B and T cells is
   a. interleukin-2
   b. opsonin
   c. lymphotoxin
   d. interleukin-1
10. ______ T cells assist in the functions of certain B cells and other T cells.
    a. Sensitized
    b. Cytotoxic
    c. Helper
    d. Natural killer
11. T_C cells are important in controlling
    a. virus infections
    b. allergy
    c. autoimmunity
    d. all of these
12. Vaccination is synonymous with ______ immunity.
    a. natural active
    b. artificial passive
    c. artificial active
    d. natural passive
13. Fusion between a plasma cell and a tumor cell creates a
    a. lymphoblast
    b. hybridoma
    c. natural killer cell
    d. myeloma
14. T cells are the source of which cytokines?
    a. interleukin
    b. interferon
    c. lymphotoxin
    d. a and b
    e. all of the choices
15. Multiple matching. Place all possible matches in the space at the left.

____ IgG	a. Found in mucous secretions
____ IgA	b. A monomer
____ IgD	c. A dimer
____ IgE	d. Has greatest number of Fabs
____ IgM	e. Is primarily a surface receptor for B cells
	f. Major Ig of primary response to Ag
	g. Major Ig of secondary response to Ag
	h. Crosses the placenta
	i. Fixes complement

Concept Questions

1. a. What function do receptors play in specific immune responses?
   b. How can receptors be made to vary so widely?
2. Describe the major histocompatibility complex, and explain how it participates in immune reactions.
3. a. Evaluate the following statement: Each different lymphocyte type must have a unique receptor to react with antigen.
   b. How many different Ags might one be expected to meet up with during life?
4. a. What constitutes a clone of lymphocytes?
   b. Explain the clonal selection theory of antibody specificity and diversity.
   c. During development, when is antigen not needed, and why is it not needed?
   d. When is antigen needed?
   e. Why must the body develop tolerance to self?

5. a. Trace the development of the B-cell receptor from gene to cell surface.
   b. What is the structure of the receptor?
   c. What is the function of the variable regions?

6. a. Trace the origin and development of B lymphocytes; of T lymphocytes.
   b. What is happening during lymphocyte maturation?

7. Describe three ways that B cells and T cells are similar and at least five major ways in which they are different.

8. a. What is an antigen or immunogen?
   b. What is the antigenic determinant?
   c. How do foreignness, size, and complexity contribute to antigenicity?
   d. What is a mosaic antigen?
   e. Why are haptens by themselves not antigenic, even though foreign?
   f. How can they be made to behave as antigens?

9. a. Differentiate among autoantigens, alloantigens, and heterophile antigens.
   b. Explain briefly what importance each has in immune reactions.

10. a. Describe the actions of an antigen-processing cell.
    b. What is the difference between a T-cell-dependent and T-cell-independent response?

11. a. Trace the immune response system, beginning with the entry of a T-cell-dependent antigen, antigen processing, presentation, the cooperative response among the macrophage and lymphocytes, and the reactions of activated B and T cells.
    b. What are the actions of interleukins-1 and -2?

12. a. On what basis is a particular B-cell clone selected?
    b. How are B cells activated, and what events are involved in blast formation?
    c. What happens when B cells are activated?
    d. What are the functions of plasma cells, clonal expansion, and memory cells?

13. a. Describe the structure of immunoglobulin.
    b. What are the functions of the Fab and Fc portions?
    c. Describe four or five ways that antibodies function in immunity.
    d. Describe the attachment of Abs to Ags. (What eventually happens to the Ags?)

14. a. Contrast the primary and secondary response to Ag.
    b. Explain the type, order of appearance, and amount of immunoglobulin in each response and the reasons for them.
    c. What causes the latent period? The anamnestic response?
    d. Explain how monoclonal and polyclonal antibodies are different.
    e. Outline the basic steps in production of monoclonal antibodies.
    f. Describe several possible applications of monoclonals in medicine.

15. a. Why are the immunities involving T cells called cell-mediated?
    b. How do T cells become sensitized?
    c. Summarize the function of each category of T cell and the types of receptors with which they are associated. Define cytokines, and provide some examples of them.
    d. How do cytotoxic cells kill their target?
    e. Why would the immune system naturally require suppression?
    f. What is a natural killer cell, and what are its functions?

16. a. Contrast active and passive immunity in terms of how each is acquired, how long it lasts, whether memory is triggered, how soon it becomes effective, and what immune cells and substances are involved.
    b. Name at least two major ways that natural and artificial immunities are different.

17. Multiple matching (summarizes information from chapters 14 and 15).
    In the blanks on the left place the letters of all of the host defenses and immune responses in the right column that can fit the description.

____ vaccination for tetanus	a. active
____ lysozyme in tears	b. passive
____ immunization with horse serum	c. natural
____ in utero transfer of antibodies	d. artificial
____ booster injection for diphtheria	e. acquired
____ recovery from a case of mumps	f. innate, inborn
____ colostrum	g. chemical barrier
____ interferon	h. mechanical barrier
____ action of neutophils	i. genetic barrier
____ injection of gamma globulin	j. specific
____ recovery from a case of mumps	k. nonspecific
____ edema	l. inflammatory response
____ humans having protection from canine distemper virus	m. second line of defense
	n. none of these
____ stomach acid	
____ cilia in trachea	
____ asymptomatic chickenpox	
____ complement	

CRITICAL-THINKING QUESTIONS

1. What is the advantage of having lymphatic organs screen the body fluids, directly and indirectly?

2. Cells contain built-in suicide genes to self-destruct by apoptosis under certain conditions. Can you explain why development of the immune system might depend in part of this sort of adaptation?

3. Double-stranded DNA is a large, complex molecule, but it is not generally immunogenic unless it is associated with proteins or carbohydrates. Can you think why this might be so? (Hint: How universal is DNA?)

4. How would you go about producing monoclonal antibodies that would participate in the destruction of cancer cells but would not kill normal human cells?

5. Explain how it is possible for people to give a false positive reaction in blood tests for syphilis, AIDS, and infectious mononucleosis.

6. Describe the cellular/microscopic pathology in the immune system of AIDS patients that results in opportunistic infections and cancers.
7. Explain why most immune reactions result in a polyclonal collection of antibodies.
8. a. Why does rising titer of Abs indicate infection?
   b. Why do some vaccinations require three or four boosters?
9. a. What event must occur for passive immunity to exist at all?
   b. Armed with this knowledge, suggest an effective way to immunize a fetus.
   c. Is this method uniformly safe?
   d. What would be an even safer way to ensure that fetuses get necessary antibodies?
10. a. Combine information on the functions of different classes of Ig to explain the exact mechanisms of natural passive immunity (both transplacental and colostrum-induced).
    b. Why are these sorts of immunity short-term?
11. Using the pattern along the lines of microfiles 15.1 and 15.2, develop an analogy that compares the clonal selection theory of antibody diversity and specificity to buying clothes right off the rack versus having them tailor-made.
12. Using words and arrows, complete a flow outline of an immune response, beginning with entrance of antigen; include processing, cell interaction, involvement of cytokines, and the end results for B and T cells.

INTERNET SEARCH TOPICS:

1. Access information on lymphokine-activated killer cells (LAK) and tumor-infiltrating leukocytes (TIL) as clinical therapy approaches for cancer. Describe the therapy and its effectiveness for various cancers.
2. Look up apoptosis; find 3 or 4 papers that feature research on this topic. What are some potential applications of the concept of programmed cell death? What causes the cells to die?

chapter 16

IMMUNIZATION AND IMMUNE ASSAYS

An expanded knowledge of immune function has yielded significant breakthroughs in medical technology that are related to manipulating and monitoring the immune system. Among the practical benefits of immunology have been to develop vaccines and other immune treatments against common infectious diseases such as hepatitis and diphtheria that once caused untold sickness and death. Another valuable application of this technology is testing the blood for signs of infection or disease. It is routine medical practice to diagnose such infections as HIV, syphilis, hepatitis B, and rubella by means of special immunologic analysis. Both separately and in combination, these methods have made sweeping contributions to individual and community health, treating disease, and controlling the spread of disease. Many hopes for the survival of humankind lie in our ability to harness the amazing workings of the immune system.

Detail from "The Cowpock," an 1808 etching that caricatured the worst fears of the English public concerning Edward Jenner's smallpox vaccine.

PRACTICAL APPLICATIONS OF IMMUNOLOGIC FUNCTION

A knowledge of the immune system and its responses to antigens has provided extremely valuable biomedical applications in two major areas: (1) use of antiserum and vaccination to provide artificial protection against disease and (2) diagnosis of disease through immunologic testing.

IMMUNIZATION: METHODS OF MANIPULATING IMMUNITY FOR THERAPEUTIC PURPOSES

The concept of artificially induced immunity was introduced in chapter 15. Methods that actively or passively immunize people are widely used in disease prevention and treatment. When a patient is given preformed antibodies, the immunization is a form of passive therapy. When a patient is vaccinated with a microbe or its antigens, the immunization is considered active and primarily preventive.

Immunotherapy: Artificial Passive Immunity

The first attempts at passive immunization involved the transfusion of horse serum containing antitoxins to prevent tetanus and to treat patients exposed to diphtheria. Since then, antisera from animals have been replaced with products of human origin that function with various degrees of specificity. **Immune serum globulin (ISG),** sometimes called gamma globulin, contains immunoglobulin extracted from the pooled blood of at least 1,000 human donors. Each lot of serum presents a broad cross section of IgG (and some IgM) antibodies, and the content varies from lot to lot. The method of processing ISG concentrates the antibodies to increase potency and eliminates potential pathogens (such as the hepatitis B and HIV viruses). It is a treatment of choice in preventing measles and hepatitis A and in replacing antibodies in immunodeficient patients. Most forms of ISG are injected intramuscularly to minimize adverse reactions, and the protection it provides lasts 2 to 3 months.

A preparation called **specific immune globulin (SIG)** is derived from a more defined group of donors. Companies that prepare SIG obtain serum from patients who are convalescing and in a hyperimmune state after such infections as pertussis, rabies, tetanus, chickenpox, and hepatitis B. These globulins are preferable to ISG because they contain higher titers of specific antibodies obtained from a smaller pool of patients. Although useful for prophylaxis in persons who have been exposed or may be exposed to infectious agents, these sera are often limited in availability.

When a human immune globulin is not available, antisera and antitoxins of animal origin can be used. Sera produced in horses are available for diphtheria, botulism, and spider and snake bites. Unfortunately, the presence of horse antigens can stimulate allergies such as serum sickness or anaphylaxis (see chapter 17). Although donated immunities only last a relatively short time, they act immediately and can protect patients for whom no other useful medication or vaccine exists.

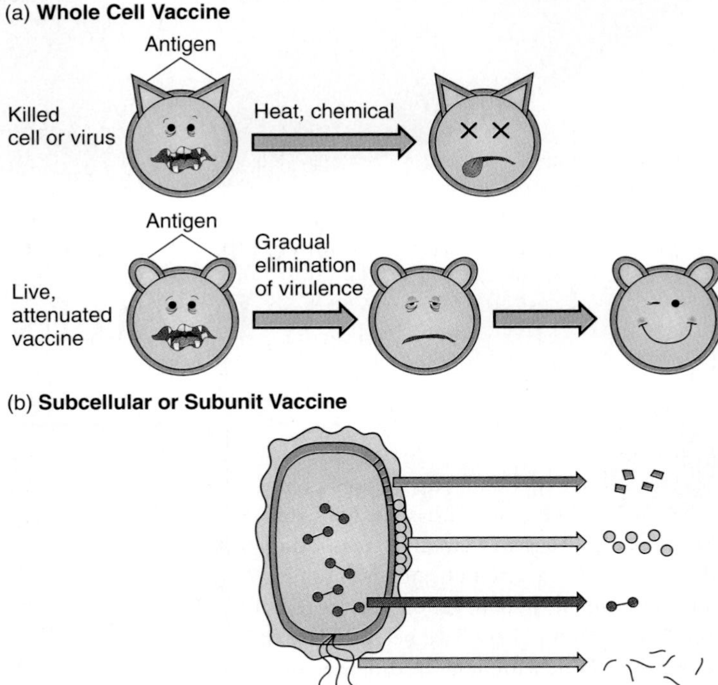

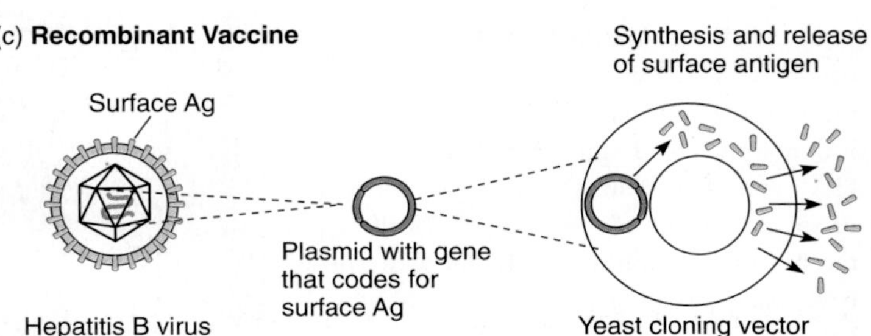

Figure 16.1

Strategies in vaccine design. (*a*) Whole cells, killed or attenuated. (*b*) Subcellular or subunit vaccines are made by disrupting the microbe to release various molecules or cell parts that can be isolated and purified. (*c*) Recombinant vaccines are made by isolating a gene for antigenicity from the pathogen (here a hepatitis virus) and splicing it into a plasmid. Insertion of the recombinant plasmid into a cloning host (yeast) results in the production of large amounts of viral surface antigen to use in vaccine preparation.

Most passive immunization involves the administration of antibodies, but occasionally T cells are also given. For example, after the HLA antigens between the donor and recipient have been closely matched, sensitized T cells called *transfer factor* can be provided for immunodeficient patients chronically infected with *Candida albicans.*

Artificial Active Immunity: Vaccination

Active immunity can be conferred artificially by **vaccination**—exposing a person to material that is antigenic but not pathogenic. The discovery of vaccination was one of the farthest reaching and most important developments in medical science (microfile 16.1).

MICROFILE 16.1 THE LIVELY HISTORY OF ACTIVE IMMUNIZATION

The basic notion of immunization has existed for thousands of years. It probably stemmed from the observation that persons who had recovered from certain communicable diseases rarely if ever got a second case. Undoubtedly, the earliest crude attempts involved bringing a susceptible person into contact with a diseased person or animal. The first recorded attempt at immunization occurred in sixth century China. It consisted of drying and grinding up smallpox scabs and blowing them with a straw into the nostrils of vulnerable family members. By the tenth century, this practice had changed to the deliberate inoculation of dried pus from the smallpox pustules of one patient into the arm of a healthy person, a technique later called **variolation** (variola is the smallpox virus). This method was used in parts of the Far East for centuries before Lady Mary Montagu brought it to England in 1721. Although the principles of the technique had some merit, unfortunately, many recipients and their contacts died of smallpox. This outcome vividly demonstrates a cardinal rule for a workable vaccine: It must contain an antigen that will provide protection but not cause the disease. Variolation was so controversial that any English practitioner caught doing it was charged with a felony.

Eventually, this human experimentation paved the way for the first really effective vaccine, developed by the English physician Edward Jenner in 1796 (see chapter 24). Jenner conducted the first scientifically controlled study, one that had a tremendous impact on the advance of medicine. His work gave rise to the words **vaccine** and **vaccination** (from L., *vacca,* cow), which now apply to any immunity obtained by inoculation with selected antigens. Jenner was inspired by the case of a dairymaid who had been infected by a pustular infection called cowpox. This is a related virus that afflicts cattle but causes a milder condition in humans. She explained that she and other milkmaids had remained free of smallpox. Other residents of the region expressed a similar confidence in the cross-protection of cowpox. To test the effectiveness of this new vaccine, Jenner prepared material from human cowpox lesions and inoculated a young boy. When challenged 2 months later with an injection of crusts from a smallpox patient, the boy proved immune. Jenner's discovery—that a less pathogenic agent could confer protection against a more pathogenic one—is especially remarkable in view of the fact that microscopy was still in its infancy and the nature of viruses was unknown. At first, the use of the vaccine was regarded with some fear and skepticism (see chapter opener). When his method proved successful and word of its significance spread, it was eventually adopted in many other countries. At various times, the method of producing the vaccine was changed, and somewhere along the line, the original virus mutated into a unique strain (*vaccinia* virus). But the essence of Jenner's method—scratching the vaccine into the skin with a sharp tool—remained. Now that smallpox is no longer a threat, smallpox vaccination has been essentially discontinued.

Other historical developments in vaccination included using heat-killed bacteria in vaccines for typhoid fever, cholera, and plague and techniques for using neutralized toxins for diphtheria and tetanus. Throughout the history of vaccination, there have been vocal opponents and minimizers, but numbers do not lie: Whenever a vaccine has been introduced, the prevalence of that disease has declined.

It profoundly reduced the prevalence and impact of many infectious diseases that were once common and often deadly. In this section, we survey the principles of vaccine preparation and important considerations surrounding vaccination in a community. (Vaccines are also given specific consideration in later chapters on bacterial and viral diseases.)

Principles of Vaccine Preparation A vaccine must be considered from the standpoints of antigen selection, effectiveness, ease in administration, safety, and cost. In natural immunity, an infectious agent stimulates appropriate B and T lymphocytes and creates memory clones. In artificial active immunity, the objective is to obtain this same response with a modified version of the microbe or its components. A safe and effective vaccine should mimic the natural protective response, not cause a serious infection or other disease, have long-lasting effects in a few doses, and be easy to administer. Most vaccine preparations contain one of the following antigenic stimulants (figure 16.1): (1) killed whole cells or inactivated viruses, (2) live, attenuated cells or viruses, (3) antigenic components of cells or viruses, or (4) genetically engineered microbes or microbial antigens. A survey of the major licensed vaccines and their indications is presented in table 16.1.

Large, complex antigens such as intact cells or viruses are very effective immunogens. Depending on the vaccine, these are either killed or attenuated. **Killed, whole vaccines** are prepared by cultivating the desired strain or strains of a bacterium or virus and treating them with formalin, radiation, or some other agent that does not destroy antigenicity. One type of vaccine for the bacterial diseases pertussis and typhoid fever are of this type (see chapter 20). Salk polio vaccine and rabies vaccine contain killed whole viruses. Because the microbe does not multiply, killed vaccines often require a larger dose and more boosters to be effective. Although one might think that vaccines containing dead microorganisms would be quite safe, they are not without adverse side effects.

A number of vaccines are prepared from **live, attenuated*** microbes. **Attenuation** is any process that substantially lessens or negates the virulence of viruses or bacteria. It is usually achieved by modifying the growth conditions or manipulating microbial genes in a way that eliminates virulence factors. Attenuation methods include long-term cultivation, selection of mutant strains that grow at colder temperatures (cold mutants), passage of the microbe through unnatural hosts or tissue culture, and removal of virulence genes. The vaccine for tuberculosis (BCG) was obtained after 13 years of subculturing the agent of bovine tuberculosis (see chapter 19). Vaccines for measles, mumps, polio (Sabin), and rubella contain live, nonvirulent viruses. The advantages that favor

*attenuated (ah-ten´-yoo-ayt-ed) L., *attenuare,* to thin. Able to multiply, but nonvirulent.

TABLE 16.1

CURRENTLY APPROVED VACCINES

Disease	Route of Administration	Recommended Usage/Comments
Contain Killed Whole Bacteria		
Cholera	Subcutaneous (SQ) injection	For travelers; effect not long-term
Pertussis	Intramuscular (IM) injection	For newborns and children; newer acellular vaccine can reduce side effects
Typhoid	SQ and IM	For travelers only; efficacy variable
Plague	SQ	For exposed individuals and animal workers; variable protection
Contain Live, Attenuated Bacteria		
Tuberculosis (BCG)	Intradermal (ID) injection	For high-risk occupations only; protection variable
Subcellular Vaccines (Capsular Polysaccharides)		
Meningitis (meningococcal)	SQ	For protection in high-risk infants, military recruits; short duration
Meningitis (*Haemophilus influenzae*)	IM	For infants and children; may be administered with DPT
Pneumococcal pneumonia	IM or SQ	Important for people at high risk: the young, elderly, and immunocompromised; moderate protection
Pertussis	IM	For newborns and children; contains recombinant protein antigens
Toxoids (Formaldehyde-Inactivated Bacterial Exotoxins)		
Diphtheria	IM	A routine childhood vaccination; highly effective in systemic protection
Tetanus	IM	A routine childhood vaccination; highly effective
Botulism	IM	Only for exposed individuals such as laboratory personnel
Contain Killed Whole Viruses		
Poliomyelitis (Salk)	IM	Routine childhood vaccine; effective
Rabies	IM	For victims of animal bites or otherwise exposed; effective
Influenza	IM	For high-risk populations; requires constant updating for new strains; immunity not durable
Hepatitis A	IM	Protection for travelers, institutionalized people
Contain Live, Attenuated Viruses		
Adenovirus infection	Oral	For immunizing military recruits
Measles (rubeola)	SQ	Routine childhood vaccine; very effective
Mumps (parotitis)	SQ	Routine childhood vaccine; very effective
Poliomyelitis	Oral	Routine childhood vaccine; very effective
Rubella	SQ	Routine childhood vaccine; very effective
Chickenpox (varicella)	SQ	Routine childhood vaccine; immunity can diminish over time; approved in 1995
Yellow fever	SQ	Travelers, military personnel in endemic areas
Rotavirus	Oral	Immunization of newborn infants
Subunit Viral Vaccines		
Hepatitis B	IM	For medical, dental, laboratory personnel and others at risk
Influenza	IM	See influenza above
Recombinant Vaccines		
Hepatitis B	IM	Medical, dental, laboratory personnel, newborns, others at risk
Pertussis	IM	See subcellular above

live preparations are: (1) Viable microorganisms can multiply and produce infection (but not disease) like the natural organism; (2) they confer long-lasting protection; and (3) they usually require fewer doses and boosters than other types of vaccines. Disadvantages of using live microbes in vaccines are that they require special storage facilities, can be transmitted to other people, and can mutate back to a virulent strain (see polio, chapter 25).

If the exact antigenic determinants that stimulate immunity are known, it is possible to produce a vaccine based on a selected component of a microorganism. These vaccines for bacteria are called **subcellular** or **acellular vaccines.** For viruses they are called **subunit vaccines.** The antigen used in these vaccines may be taken from cultures of the microbes, produced by rDNA technology, or synthesized chemically.

Examples of component antigens currently in use are the capsules of the pneumococcus and meningococcus, the protein surface antigen of anthrax, and the surface proteins of hepatitis B virus. A special type of vaccine is the **toxoid,*** which consists of a purified bacterial exotoxin that has been chemically denatured. By eliciting the production of antitoxins that can neutralize the natural toxin, toxoid vaccines provide protection against toxinoses such as diphtheria and tetanus.

New Vaccine Strategies

Despite considerable successes, dozens of bacterial, viral, protozoan, and fungal diseases still remain without a functional vaccine. Of all of the challenges facing vaccine specialists, probably the most difficult has been choosing a vaccine antigen that is safe and that properly stimulates immunity. Currently, much attention is being focused on newer strategies for vaccine preparation that employ antigen synthesis, recombinant DNA, and gene cloning technology.

When the exact composition of an antigenic determinant is known, it is possible to synthesize it. This ability permits preservation of antigenicity while greatly increasing antigen purity and concentration. The malaria vaccine currently being used in areas of South America and Africa is composed of three synthetic peptides from the parasite. Several biotechnology companies are exploring the possibility of using plants to synthesize microbial proteins, potentially leading to mass production of edible vaccine antigens.

Some of the genetic engineering concepts introduced in chapter 10 offer novel approaches to vaccine development. These methods are particularly effective in designing vaccines for obligate parasites that are difficult or expensive to culture, such as the syphilis spirochete or the malaria parasite. This technology provides a means of isolating the genes that encode various microbial antigens, inserting them into plasmid vectors, and cloning them in appropriate hosts. The outcome of recombination can be varied as desired. For instance, the cloning host can be stimulated to synthesize and secrete a protein product (antigen), which is then harvested and purified (figure 16.1*c*). This is how certain vaccines for hepatitis B rotavirus and Lyme disease are prepared. Antigens from the agents of syphilis, *Schistosoma,* and influenza have been similarly isolated and cloned and are currently being considered as potential vaccine material.

Another ingenious technique using genetic recombination has been nicknamed the *Trojan horse* vaccine. The term derives from an ancient legend in which the Greeks sneaked soldiers into the fortress of their Trojan enemies by hiding them inside a large, mobile wooden horse. In the microbial equivalent, genetic material from a selected infectious agent is inserted into a live carrier microbe that is nonpathogenic. In theory, the recombinant microbe will multiply and express the foreign genes, and the vaccine recipient will be immunized against the microbial antigens. *Vaccinia,* the virus originally used to vaccinate for smallpox, and adenoviruses have proved practical agents for this technique. Vaccinia is used as the carrier in one of the experimental vaccines for AIDS (see figure 25.21), herpes simplex 2, leprosy, and tuberculosis.

DNA vaccines are being hailed as the most promising of all of the newer approaches to immunization. The technique in these formulations is very similar to gene therapy as described in figure 10.17, except in this case, microbial (not human) DNA is inserted into a plasmid vector and inoculated into a recipient (figure 16.2*a*). The expectation is that the human cells will take up some of the plasmids and express the microbial DNA in the form of proteins. Because these proteins are foreign, they will be recognized during immune surveillance and cause B and T cells to be sensitized and form memory cells.

Experiments with animals have shown that these vaccines are very safe and that only a small amount of the foreign antigen need be expressed to produce effective immunity. Another advantage to this method is that any number of potential microbial proteins can be expressed, making the antigenic stimulus more complex and improving the likelihood that it will stimulate both antibody and cell-mediated immunity. At the present time, over 30 DNA-based vaccines are being tested in animals. Vaccines for Lyme disease, hepatitis C, herpes simplex, influenza, tuberculosis, papillomavirus, and malaria are undergoing animal trials, most with encouraging results. In one dramatic test, chimpanzees vaccinated with nucleic acid from human immunodeficiency virus were protected from infection by a large virus challenge. Rapid development of inexpensive human vaccines is expected to follow.

Vaccine effectiveness relies, in part, on the production of antibodies that closely fit the natural antigen. Realizing that such reactions are very much like a molecular jigsaw puzzle, researchers have proposed an entirely new concept in vaccines. The *anti-idiotype vaccine* is based on the principle that the antigen binding (variable) region, or *idiotype,** of a given antibody (A) can be antigenic to a genetically different recipient and can cause that recipient's immune system to produce antibodies (B; also called anti-idiotypic antibodies) specific for the variable region on antibody A (figure 16.2*b*). The purpose for making anti-idiotypic antibodies is that they will display an identical configuration as the desired antigen and can be used in vaccines. This method avoids giving a microbial antigen, thus reducing the potential for dangerous side effects. Using monoclonal antibodies, this approach has been used to mimic the surface antigen of hepatitis B virus and *Trypanosoma* with some success.

Route of Administration and Side Effects of Vaccines

Most vaccines are injected by subcutaneous, intramuscular, or intradermal routes. Oral vaccines are available for only three diseases (table 16.1), but they have some distinct advantages. An oral dose of a vaccine can stimulate protection (IgA) on the mucous membrane of the portal of entry. Oral vaccines are also easier to give, more readily accepted, and well tolerated. An influenza vaccine given intranasally is another method that shows promise. Some vaccines require the addition of a special binding substance, or *adjuvant.** An adjuvant is any compound that enhances

***toxoid** (tawks´-oyd) Toxinlike.

***idiotype** (id´-ee-oh-type) Gr. *idios,* own, peculiar. Another term for the antigen binding site.

***adjuvant** (ad´-joo-vunt) L. *adjuvare,* to help.

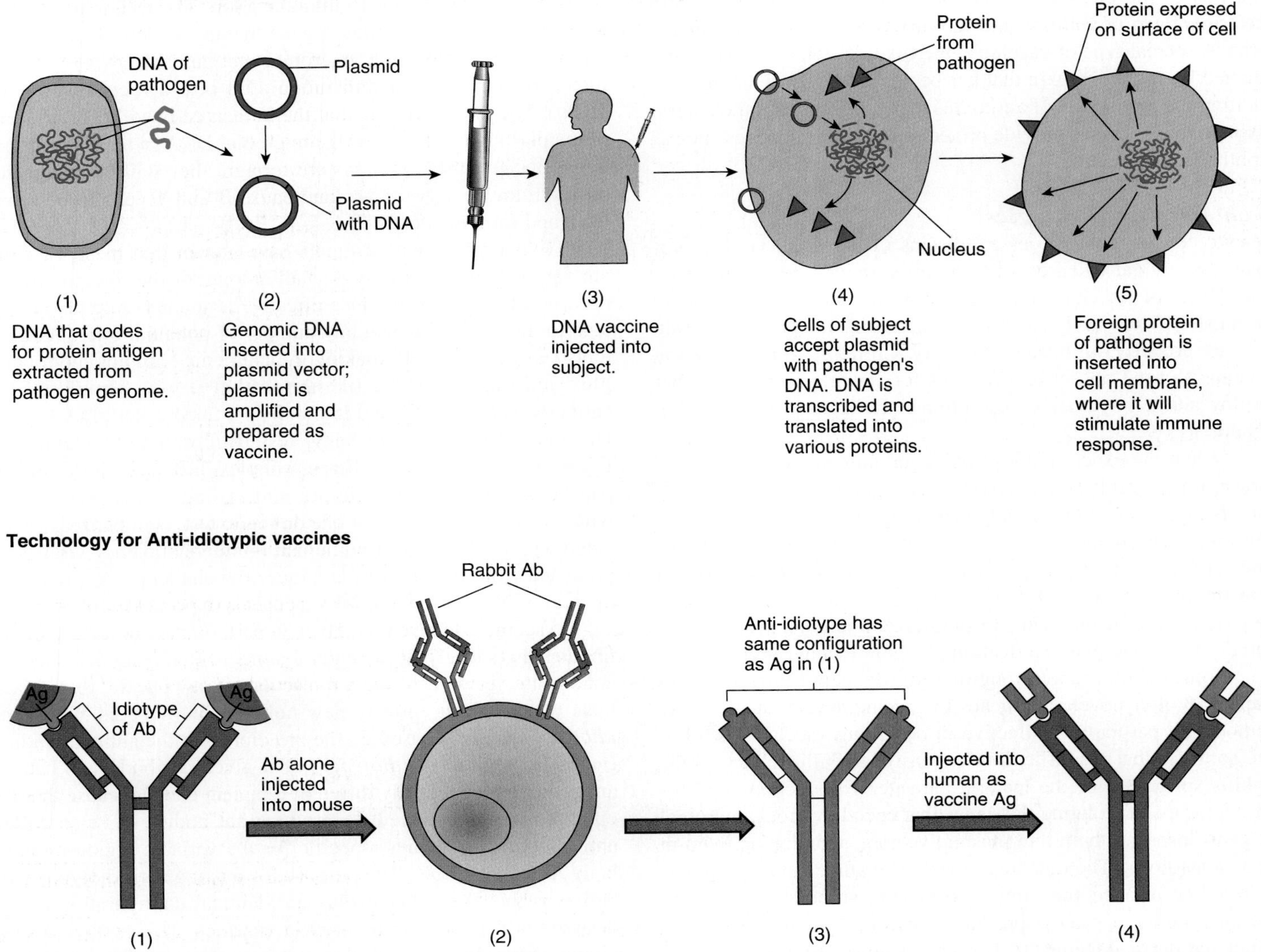

Figure 16.2

Other technologies useful in preparing vaccines. (*a*) DNA vaccines contain all or part of the pathogen's DNA, which is used to "infect" recipient's cells. Processing of the DNA leads to production of an antigen protein that can stimulate a specific response against that pathogen. (*b*) Anti-idiotypic vaccines use antibodies as foreign proteins to mimic the structure of the natural antigen. See text for details.

immunogenicity and prolongs antigen retention at the injection site. The adjuvant precipitates the antigen and holds it in the tissues so that it will be released gradually. Its gradual release, presumably, facilitates contact with macrophages and lymphocytes. Common adjuvants are alum (aluminum hydroxide salts), Freund's adjuvant (emulsion of mineral oil, water, and extracts of mycobacteria), and beeswax.

Vaccines must go through many years of trials in experimental animals and humans before they are licensed for general use. Even after they have been approved, like all therapeutic products, they are not without complications. The most common of these are local reactions at the injection site, fever, allergies, and other adverse reactions. Relatively rare reactions (about 1 case out of 220,000 vaccinations) are panencephalitis (from measles vaccine), back-mutation to a virulent strain (from polio vaccine), disease due to contamination with dangerous viruses or chemicals, and neurological effects of unknown cause (from pertussis and swine flu vaccines). Some patients experience allergic reactions to the medium (eggs or tissue culture) rather than to vaccine antigens. Some recent studies have attempted to link childhood vaccinations to later development of diabetes and asthma. Whether this is an accurate conclusion awaits further study.

TABLE 16.2

RECOMMENDED REGIMEN AND INDICATIONS FOR VACCINATION

Routine Schedules for Children and Adults

Vaccine	Birth	2 Months	4 Months	6 Months	12 Months	15 Months	18 Months	4–6 Years	11–12 Years	Adults	Comments
Mixed vaccines											
Diphtheria, Pertussis, Tetanus (DPT)		DTP	DTP	DTP		DTP one dose		DTP	Td or T		DPT is same as DTP; Td is tetanus/diphtheria; T is tetanus alone—given as booster every 10 years.
Measles,* Mumps, Rubella (MMR)						MMR		MMR or	MMR		First dose varies with disease incidence; booster given either 4–6 or 11–12 years.
Haemophilus influenzae type b (Hib)		Hib	Hib	Hib		Hib					Schedule depends upon source of vaccine; given with DPT as TriHIBit
Single vaccines											
Poliovirus (OPV)		OPV	OPV	OPV		one dose		OPV			Similar schedule to DPT; oral vaccine
Hepatitis B (HB)	HB option 1	HB option 2		HB option 3							Option depends upon condition of infant; 3 doses given
Chickenpox (CPV)						CPV one dose				CPV 2 doses	Cannot be given to children <1 year old

Shaded bars indicate that a dose can be given once during this time frame

*Measles vaccine (Meruvax) can be given alone to children during epidemics or to adults immunized before 1970.

Used in Cases of Specific Risk Due to Occupational or Other Exposure

Vaccine	Group Targeted
Hepatitis B	Health care personnel; people exposed through life-style
Pneumococcal	Elderly patients, children with sickle-cell anemia
Influenza, polio, tuberculosis (BCG)	Hospital, laboratory, health care workers
Rabies, plague, Lyme disease	People whose jobs involve contact with animals (veterinarians, forest rangers); known or suspected exposure to rabid animal
Cholera, hepatitis B, measles, yellow fever, meningococcal meningitis, polio, rabies, typhoid, plague, hepatitis A	Travelers to endemic regions, including military recruits (varies with geographic destination)

*May vary slightly from region to region, depending on the incidence of disease.

Professionals involved in giving vaccinations must understand their inherent risks but also realize that the risks from the infectious disease almost always outweigh the chance of an adverse vaccine reaction. The greatest caution must be exercised in giving live vaccines to immunocompromised and pregnant patients, the latter because of possible risk to the fetus.

To Vaccinate: Why, Whom, and When?

Vaccination confers long-lasting, sometimes lifetime, protection in the individual, but an equally important effect is to protect the public health. Vaccination is an effective method of establishing **herd immunity** in the population. According to this concept, individual immune to a communicable infectious disease will not be carriers, and therefore the occurrence of that microbe will be reduced. With a larger number of immune individuals in a population (herd), it will be less likely that an unimmunized member of the population will encounter the agent. In effect, collective immunity through mass immunization confers indirect protection on the nonimmune (such as children). Herd immunity maintained through immunization is as important force in averting epidemics.

Vaccination is recommended for all typical childhood diseases for which a vaccine is available and for people in certain special circumstances (health workers, travelers, military personnel). Table 16.2 outlines a general schedule for childhood immunization and the indication for special vaccines. Not only are some vaccines mixtures of antigens, but in certain cases several vaccines are also administered simultaneously. Examples are military recruits who receive as many as 15 injections within a few minutes and children who receive boosters for DPT and polio at the same time they receive the MMR vaccine. Experts doubt that immune interference (inhibition of one immune response by another) is a significant problem in these instances, and the mixed vaccines are carefully balanced to prevent this eventuality. The main problem with simultaneous administration is that side effects can be amplified.

Chapter Checkpoints

Knowledge of the specific immune response has two practical applications: (1) commercial production of antisera and vaccines and (2) development of rapid, sensitive methods of disease diagnosis.

Artificial passive immunity usually involves administration of antiserum, and occasionally B and T cells. Antibodies collected from donors (human or otherwise) are injected into people who need protection immediately. Examples include ISG (immune serum globulin) and SIG (specific immune globulin).

Artificial active agents are vaccines that provoke a protective immune response in the recipient but do not cause the actual disease. Vaccination is the process of challenging the immune system with a specially selected antigen. Examples are: (1) killed or inactivated microbes, (2) live, attenuated microbes, (3) subunits of microbes, and (4) genetically engineered microbes or microbial parts.

Vaccination programs seek to protect the individual directly through raising the antibody titer and indirectly through the development of herd immunity.

Serological and Immune Tests: Measuring the Immune Response *in Vitro*

The antibodies formed during an immune reaction are important in combating infection, but they hold additional practical value. Characteristics of antibodies (such as their quantity or specificity) can reveal the history of a patient's contact with microorganisms or other antigens. This is the underlying basis of **serological testing. Serology** is the branch of immunology that traditionally deals with *in vitro* diagnostic testing of serum. Serological testing is based on the familiar concept that antibodies have extreme specificity for antigens, so when a particular antigen is exposed to its specific antibody, it will fit like a hand in a glove. The ability to visualize this interaction by some means provides a powerful tool for detecting, identifying, and quantifying antibodies—or for that matter, antigens. The scheme works both ways, depending on the situation. One can detect or identify an unknown antibody using a known antigen, or, if the antigen is unknown, one can use an antibody of known specificity to help detect or identify it (figure 16.3). Modern serological testing has grown into a field that tests more than just serum. Urine, cerebrospinal fluid, whole tissues, and saliva can also be used to determine the immunologic background of patients. These and other immune tests are helpful in confirming a suspected diagnosis or in screening a certain population for disease.

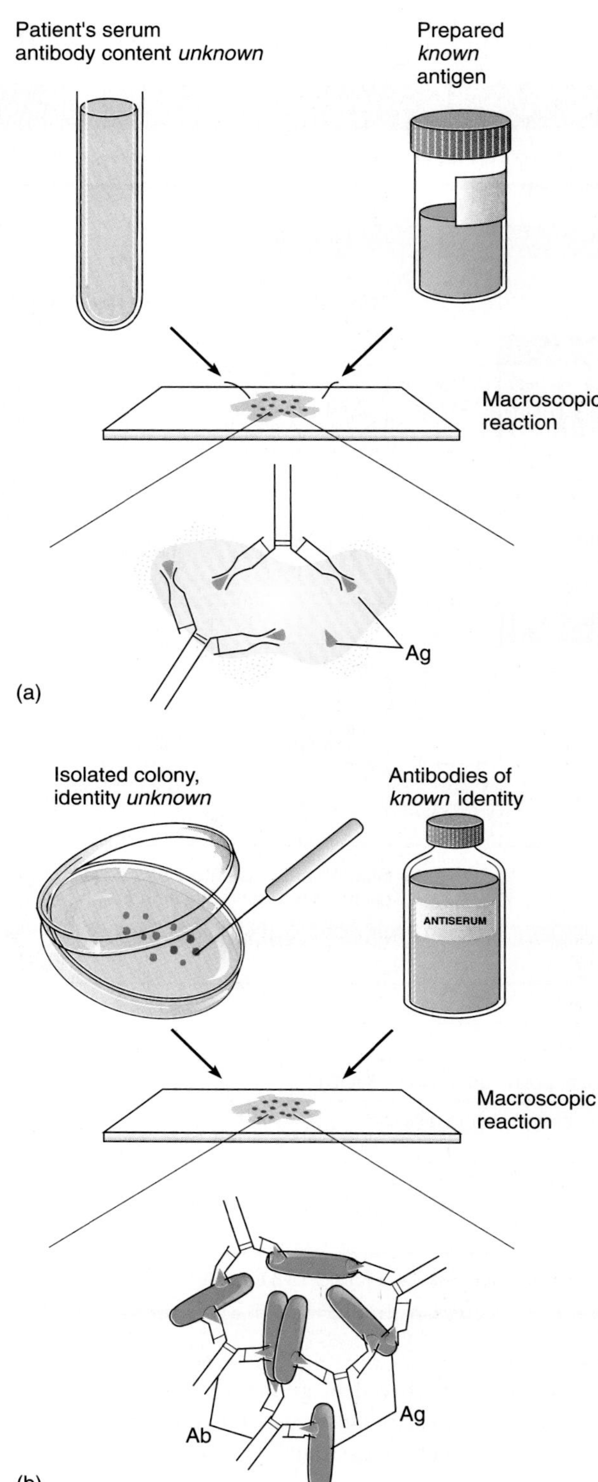

Figure 16.3
Basic principles of testing using antibodies and antigens. (*a*) In serological diagnosis of disease, a blood sample is scanned for the presence of antibody using an antigen of known specificity. A positive reaction is usually evident in some readily visible reaction, such as color change or clumping, that indicates a specific interaction between antibody and antigen. (The reaction at the molecular level is rarely observed.) (*b*) Identification of an unknown microbe using serum containing antibodies of known specificity, a procedure known as serotyping. Microscopically or macroscopically observable reactions indicate a correct match between antibody and antigen.

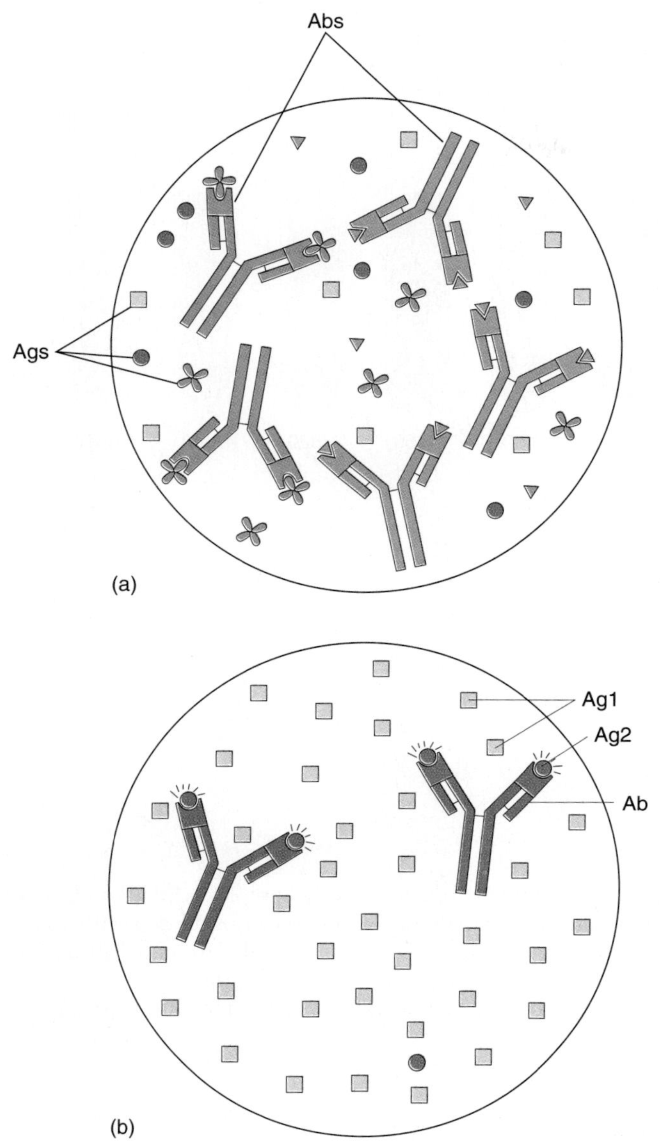

Figure 16.4

Specificity and sensitivity in immune testing. (*a*) This test shows specificity in which an antibody (Ab) attaches with great exactness with only one type of antigen (Ag). (*b*) Sensitivity is demonstrated by the fact that Ab can locate Ag, even when it is greatly diluted.

General Features of Immune Testing

The strategies of immunologic tests are diverse, and they underline some of the brilliant and imaginative ways that antibodies and antigens can be used as tools. We will summarize them under the headings of agglutination, precipitation, immunodiffusion, complement fixation, fluorescent antibody tests, and immunoassay tests. First we will overview the general characteristics of immune testing, and we will then look at each type separately.

The most effective serological tests have a high degree of specificity and sensitivity (figure 16.4). **Specificity** is the property of a test to focus upon only a certain antibody or antigen and not to react with unrelated or distantly related ones. **Sensitivity** means that the test can detect even very small amounts of antibodies or antigens that are the targets of the test. New systems using monoclonal antibodies have greatly improved specificity, and those using radioactivity, enzymes, and electronics have improved sensitivity.

Visualizing Antigen-Antibody Interactions The primary basis of most tests is the binding of an antibody (Ab) to a specific molecular site on an antigen (Ag). Because this reaction cannot be readily seen without an electron microscope, tests involve some type of endpoint reaction visible to the naked eye or with regular magnification that tells whether the result is positive or negative. In the case of large antigens such as cells, Ab binds to Ag and creates large clumps or aggregates that are visible macroscopically or microscopically (figure 16.5*a*). Smaller Ag-Ab complexes that do not result in readily observable changes will require special indicators in order to be visualized. Endpoints are often revealed by dyes or fluorescent reagents that can tag molecules of interest. Similarly, radioactive isotopes incorporated into antigens or antibodies constitute sensitive tracers that are detectable with photographic film.

An antigen-antibody reaction can be used to read a **titer.** Knowing the relative quantity of antibodies permits different samples to be standardized and compared. Titer is determined by serially diluting a sample in tubes or in a multiple-welled microtiter plate and mixing it with antigen (figure 16.5*b*). Titer is expressed as the highest dilution of serum that produces a visible reaction with an antigen. The more a sample can be diluted and yet still react with antigen, the greater is the concentration of antibodies in that sample and the higher is its titer. Interpretation of testing results is discussed in microfile 16.2.

Agglutination and Precipitation Reactions

The essential differences between agglutination and precipitation are in size, solubility, and location of the antigen. In agglutination, the antigens are whole cells such as red blood cells or bacteria with determinant groups on the surface. In precipitation, the antigen is a soluble molecule. In both instances, when Ag and Ab are optimally combined so that neither is in excess, one antigen is interlinked by several antibodies to form an insoluble, three-dimensional aggregate so large that it cannot remain suspended and it settles out (figure 16.5*a*).

Agglutination Testing Agglutination is discernible because the antibodies called *agglutinins* cross-link the antigens or *agglutinogens* to form visible clumps. Agglutination tests are performed routinely by blood banks to determine AGO and Rh (Rhesus) blood types in preparation for transfusions. In this test, antisera containing antibodies against the blood group antigens on red blood cells are mixed with a small sample of blood and read for the presence or absence of clumping (see figure 17.9). The *Widal test* is an example of a tube agglutination test for diagnosing salmonelloses and undulant fever. In addition to detecting specific antibody, it also gives the serum titer (figure 16.5*b*).

Numerous variations of agglutination testing exist. The Rapid Plasma Reagin (RPR) test is one of several tests commonly used to test for antibodies to syphilis. The cold agglutinin

Agglutination

Antigenic determinant

Whole cell

Agglutinogen

Agglutinin

Microscopic appearance of clumps

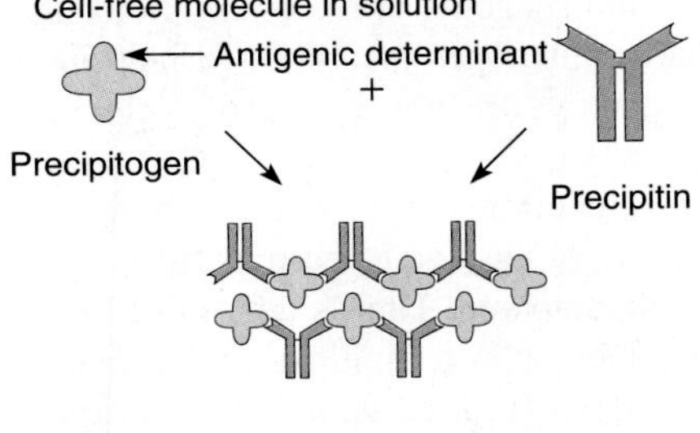

(a) Microscopic appearance of precipitate

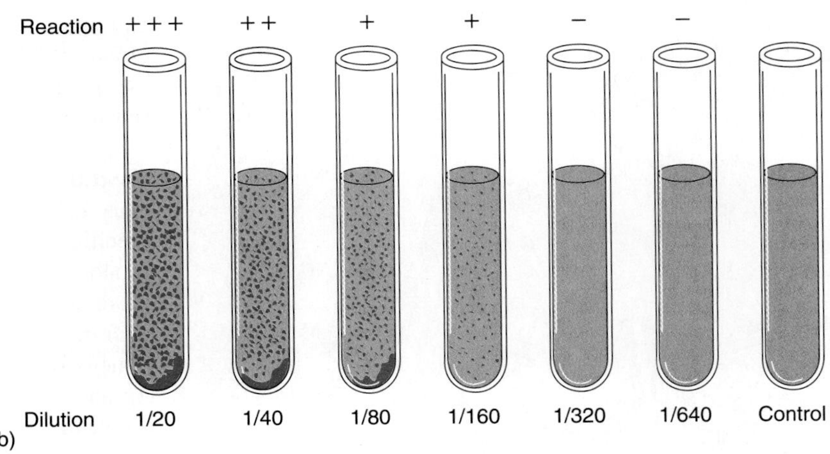

Figure 16.5

(*a*) Cellular/molecular view of agglutination and precipitation reactions that produce visible antigen-antibody complexes. (Although IgG is shown as the Ab, IgM is also involved in these reactions.) (*b*) The tube agglutination test. A sample of patient's serum is diluted with saline through a series of tubes. The dilution is made in a way that halves the number of antibodies in each subsequent tube. An equal amount of the antigen (here, dead bacterial cells) is added to each tube. The control tube has antigen, but no serum. After incubation and centrifugation, each tube is examined for agglutination clumps as compared with the control, which will be cloudy and clump-free. The titer is defined as the dilution of the last tube in the series that shows agglutination.

test, named for antibodies that react only at lower temperatures (4° − 20° C), was developed to diagnose *Mycoplasma* pneumonia. The *Weil-Felix reaction* is an agglutination test sometimes used in diagnosing rickettsial infections.

In some tests, special agglutinogens have been prepared by affixing antigen to the surface of an inert particle. In *latex agglutination* tests, the inert particles are tiny latex beads. Kits using latex beads are available for assaying pregnancy hormone in the urine, identifying *Candida* yeasts and bacteria (staphylococci, streptococci, and gonococci), and diagnosing rheumatoid arthritis.

In *viral hemagglutination* testing, agglutinogen is a red blood cell that reacts naturally with certain viral antigens. The RBCs are mixed with a known virus and a patient's serum of unknown content (figure 16.6). The test is interpreted differently than other agglutination tests because it is based on a competition between the RBCs and the antibodies for the virus antigens. If the patient's serum does *not* contain antibodies specific to the virus, the virus reacts with the RBCs instead and agglutinates them. Thus, agglutination here indicates no antibodies and a seronegative result. On the other hand, if the specific antibodies for the virus are present, they attach to the virus particles, the RBCs remain free, and there is no agglutination. As performed in the wells of microtiter plates, the difference is apparent to the naked eye (figure 16.6). Several viral diseases (measles, rubella, mumps, mononucleosis, and influenza) can be diagnosed with this test.

Precipitation Tests In precipitation reactions, the soluble antigen (*precipitogen*) is precipitated (made insoluble) by an antibody (*precipitin*). This reaction is observable in a test tube in which antiserum has been carefully laid over an antigen solution (figure 16.7*a*). At the point of contact, a cloudy or opaque zone forms.

One example of this technique is the VDRL (Veneral Disease Research Lab) test that also detects antibodies to syphilis.

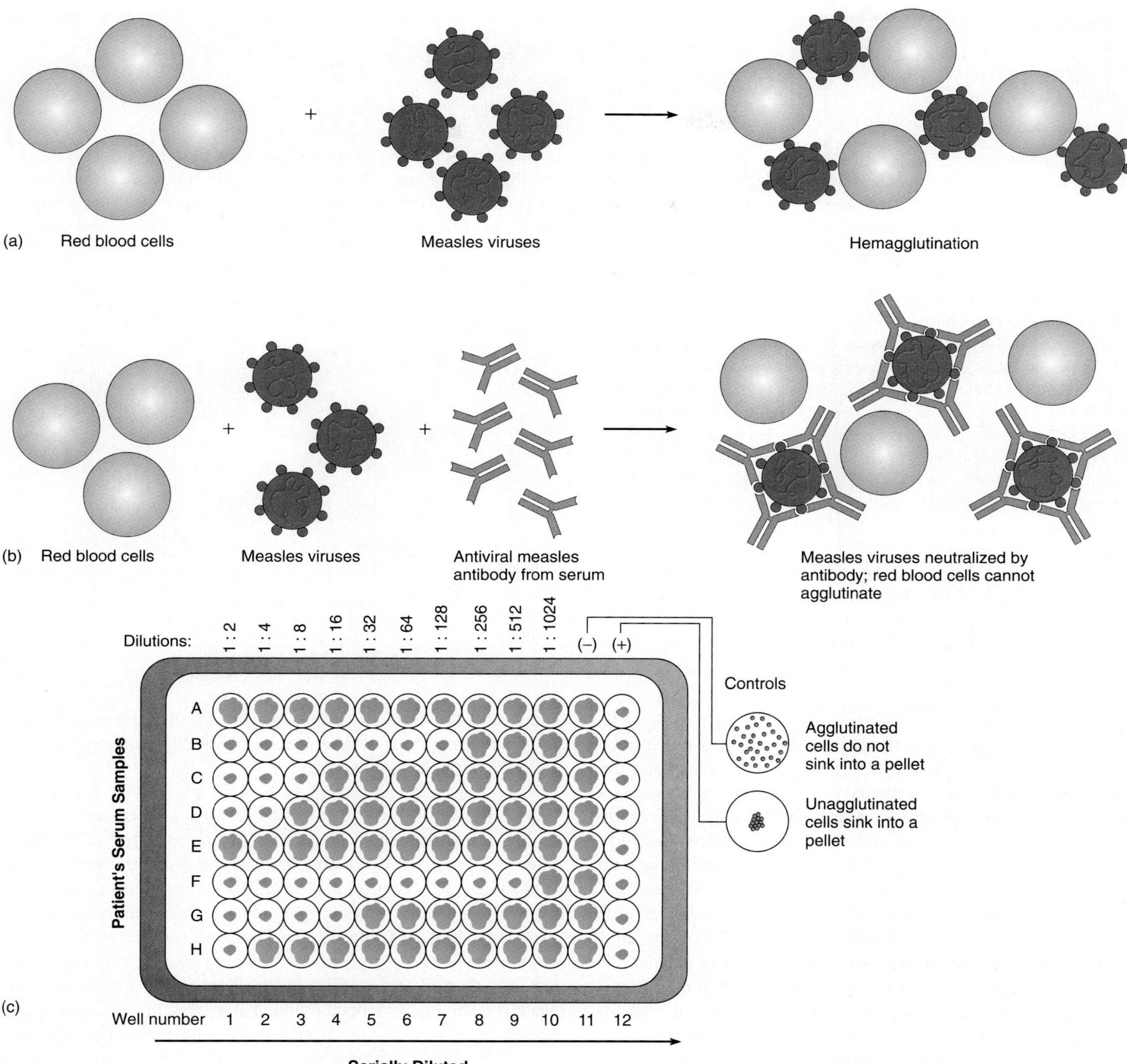

Figure 16.6

Theory and interpretation of viral hemagglutination. (*a*) In an antibody-free system, the natural tendency of measles virus to combine and interlink red blood cells causes an agglutination reaction. (*b*) In a test system using antibodies specific to the measles virus, these Abs will react with the viruses and prevent them from binding to red blood cells, thereby blocking agglutination. Thus, no agglutination means a positive reaction for antibody. (*c*) In an actual test, a multiwell microtiter plate is set up to run dilutions of several patients' sera. In a negative test, there will be agglutination, which appears as complexes covering the bottom of the well; in a positive test, the unagglutinated red blood cells fall into a small pile (pellet) in the well's bottom. (This test allows the reading of titer as well.)

MICROFILE 16.2 WHEN POSITIVE IS NEGATIVE: HOW TO INTERPRET SEROLOGICAL TEST RESULTS

What if a patient's serum gives a positive reaction—is **seropositive**—in a serological test? In most situations, it means that antibodies specific for a particular microbe have been detected in the sample. But one must be cautious in proceeding to the next level of interpretation. The mere presence of antibodies does not necessarily indicate that the patient has a disease but only that he or she has possibly had contact with a microbe or its antigens through infection or vaccination. In screening tests for determining a patient's history (rubella, for instance), knowing that a certain titer of antibodies is present can be significant, because it shows that the person has some protection. However, when the test is being used to diagnose current disease, a series of tests to show a rising titer of antibodies is necessary. The accompanying figure will indicate how such a test can clarify the status of patients who have symptoms of diseases that mimic other diseases. Lyme disease, for instance, can be mistaken for arthritis or viral infections. In the first group, note that the antibody titer against *Borrelia burgdorferi* increased steadily over a 6-week period. A control group that shared similar symptoms did not exhibit a rise in titer for this microbe.

Another important consideration in testing is the occasional appearance of biological **false positives.** These are results in which a patient's serum shows a positive reaction, even though, in reality, he is not or has not been infected by the microbe. False positives such as those in syphilis and AIDS testing arise when antibodies or other substances present in the serum **cross-react** with the test reagents, producing a positive result. Such false results may require retesting by a method that greatly minimizes cross-reactions.

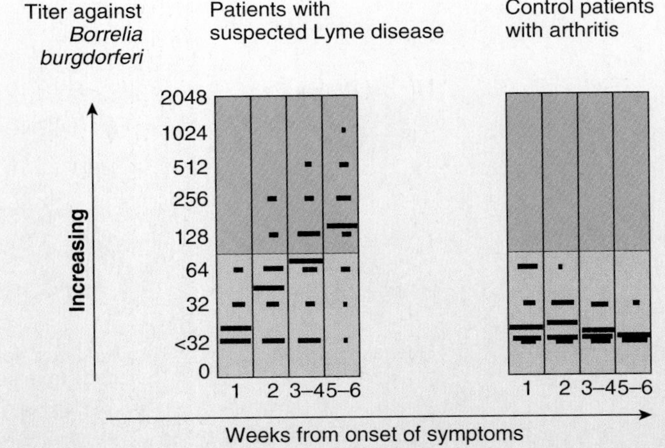

Although it is a good screening test, it contains a heterophilic antigen (cardiolipin) that may give rise to false positive results. Although precipitation is a useful detection tool, the precipitates are so easily disrupted in liquid media that most precipitation reactions are carried out in agar gels. These substrates are sufficiently soft to allow the reactants (Ab and Ag) to freely diffuse, yet firm enough to hold the Ag-Ab precipitate in place. One technique with applications in microbial identification and diagnosis of disease is the **double diffusion** (Ouchterlony) method. It is called double diffusion because it involves diffusion of both antigens and antibodies. The test is performed by punching a pattern of small wells into an agar medium and filling them with test antigens and antibodies. A band forming between two wells indicates that antibodies from one well have met and reacted with antigens from the other well. Variations on this technique provide several results (figure 16.7*b,c*): (1) Antigens placed in the center well can be used to assess the content of sera; (2) antiserum placed in the center well can help identify antigens placed in the outer wells; (3) identification reactions between the contents of the outer wells can be determined.

Immunoelectrophoresis constitutes yet another refinement of diffusion and precipitation in agar. With this method, a serum sample is first electrophoresed to separate the serum proteins as previously shown (see figure 15.17). After antibodies that can react with these serum samples are placed in a trough parallel to the direction of migration, reaction arcs specific for blood proteins develop (figure 16.8). This test is widely used to detect disorders in the production of antibodies. *Counterimmunoelectrophoresis,* which uses an electrical current to speed up the migration of antibody and antigen, is a newer technique for identifying bacterial and viral antigens in blood.

The Western Blot For Detecting Proteins

A variation of immunoelectrophoresis is the basis for the **Western blot** test. This test is a counterpart of the Southern blot test for identifying DNA (see figure 10.4). It is a very specific and sensitive way to identify or verify a particular protein (antibody or antigen) in a sample (figure 16.9). First, the test material is electrophoresed in a gel to separate out particular bands. The gel is then transferred to a special blotter that binds the reactants in place. The blot is developed by incubating it with a solution of antibody or antigen that has been labeled with radioactive, fluorescent, or luminescent labels. Sites of specific binding will appear as a pattern of bands that can be compared with known positive and negative samples. This is currently the verification test for people who are antibody-positive for HIV in the ELISA test (described in a later section), because it tests more types of antibodies and is less subject to misinterpretation than are other antibody tests. The technique has significant applications for detecting microbes and their antigens in specimens.

Complement Fixation

An antibody that requires (fixes) complement to complete the lysis of its antigenic target cell is a **lysin** or **cytolysin.** When lysins act in conjunction with the intrinsic complement system on red blood cells, the cells hemolyze (lyse and release their hemoglo-

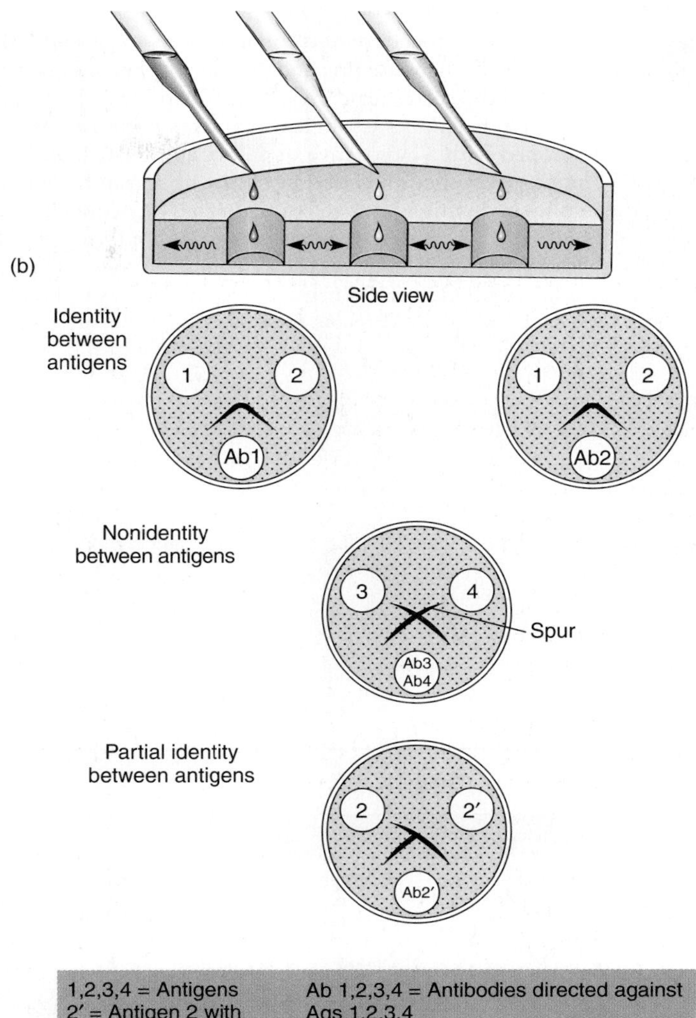

Figure 16.7

Precipitation reactions. (*a*) A tube precipitation test for streptococcal group antigens. Specific antiserum has been placed in the bottom of the tubes and antigen solution carefully overlaid to form a zone of contact. The left-hand tube has developed a heavy band of precipitate indicative of a positive reaction between antibody (Ab) and antigen (Ag). The right-hand tube is negative. (*b*) In one method of setting up a double diffusion test, wells are punctured in soft agar, and Abs and Ags are added in a pattern. As the contents of the wells diffuse toward each other, a number of reactions can result. When a well containing Ab is placed near two Ag wells, lines of precipitate that form between the Ab–Ag wells are indicative of various characteristics of the test antigens. A line indicates a distinct Ag–Ab reaction between the two wells; a spur indicates lack of identify. (*c*) In the top example, the smooth V between wells 1 and 2 and the Ab show that they are the same antigen. When two lines cross each other to form a spur, as in the middle figure, this indicates complete nonidentity of Ags 3 and 4. If a spur occurs on one side only (bottom figure), there is partial serological similarity between the two antigens.

(a) *From Gillies and Dodds,* Bacteriology Illustrated, *5th ed., fig. 14. Reprinted by permission of Longman Group Ltd.*

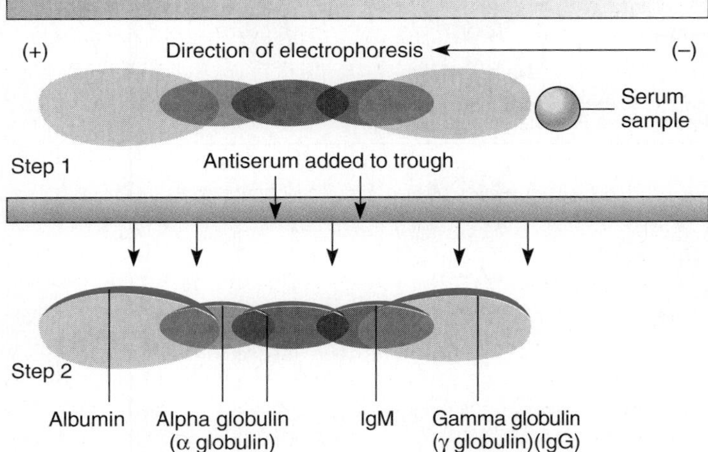

Figure 16.8

Immunoelectrophoresis of normal human serum. *Step 1.* Proteins are separated by electrophoresis on a gel. *Step 2.* To identify the bands and increase visibility, antiserum containing antibodies specific for serum proteins is placed in a trough and allowed to diffuse toward the bands. This diffusion produces a pattern of numerous arcs representing major serum components.

bin). This lysin-mediated hemolysis is the basis of a group of tests called **complement fixation,** or CF (figure 16.10). Complement fixation testing uses four components—antibody, antigen, complement, and sensitized sheep red blood cells—and it is conducted in two stages. In the first stage, the test antigen is allowed to react with the test antibody (at least one must be of known identify) in the absence of complement. If the Ab-Ag are specific for each other, they form complexes. To this mixture, purified complement proteins from guinea pig blood are added. If antibody and antigen have complexed during the previous step, they attach, or **fix,** the complement to them, thus preventing it from participating in further reactions. The extent of this complement fixation is determined in the second stage; sheep RBCs are mixed with sheep-specific lysins. This mixing will produce an indicator complex that can also fix complement. Contents of the stage 1 tube are mixed with the stage 2 tube and observed for hemolysis, which can be observed with the naked eye as a clearing of the solution. If hemolysis *does not* occur, it means that the complement was used up by the first stage Ab-Ag complex and that the unknown antigen or antibody was indeed present. This result is considered positive. If hemolysis *does* occur, it means that unfixed complement from

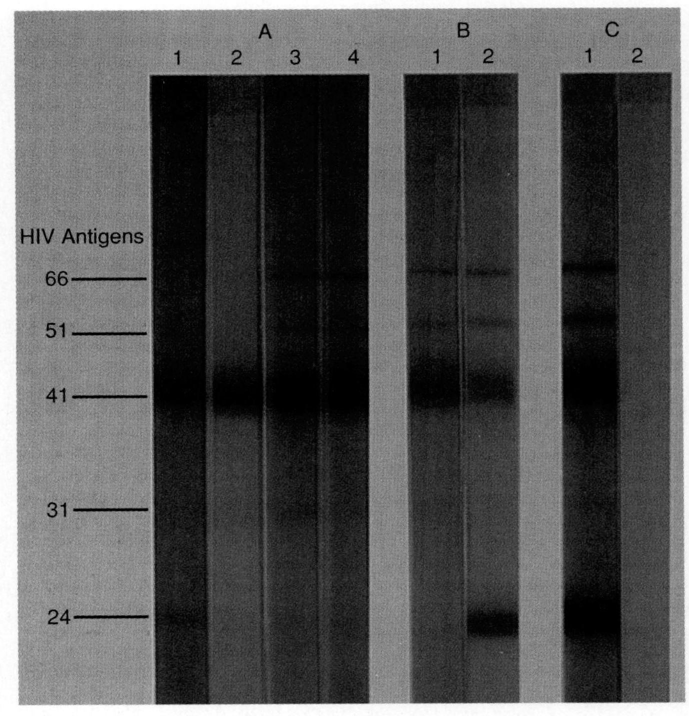

Figure 16.9

The Western blot procedure tests for antibodies to specific HIV antigens. The test strips are prepared by electrophoresing several of the major HIV surface and core antigens and then blotting them onto special filters. The test strips are incubated with patient's serum and developed with a radioactive or colorimetric label. Sites where HIV antigens have bound antibodies show up as bands. Patients' sera are then compared with a positive control strip (C1) containing antibodies for all HIV antigens. In this series, A strips are from patients with fully developed AIDS, B strips are from patients in the early phase of HIV disease, and strip C2 is HIV negative (no antibodies to HIV).
Courtesy of Philip D. Markham, Advanced BioScience Laboratories, Inc.

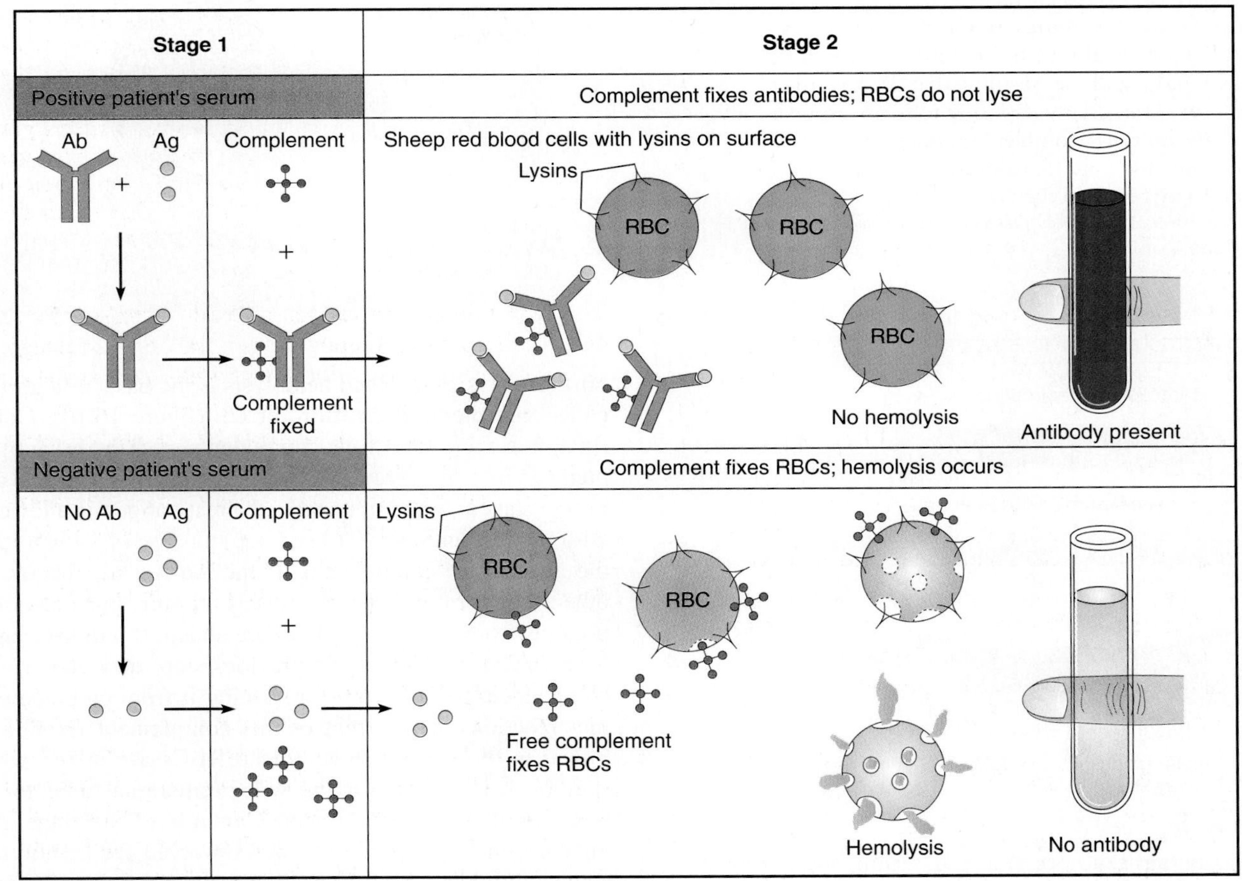

Figure 16.10

Complement fixation test. In this example, a patient's serum is being tested for antibodies to a certain infectious agent. Note that macroscopically, unhemolyzed red blood cells present a turbid solution that one cannot see through. Hemolysis creates a transparent pink solution.

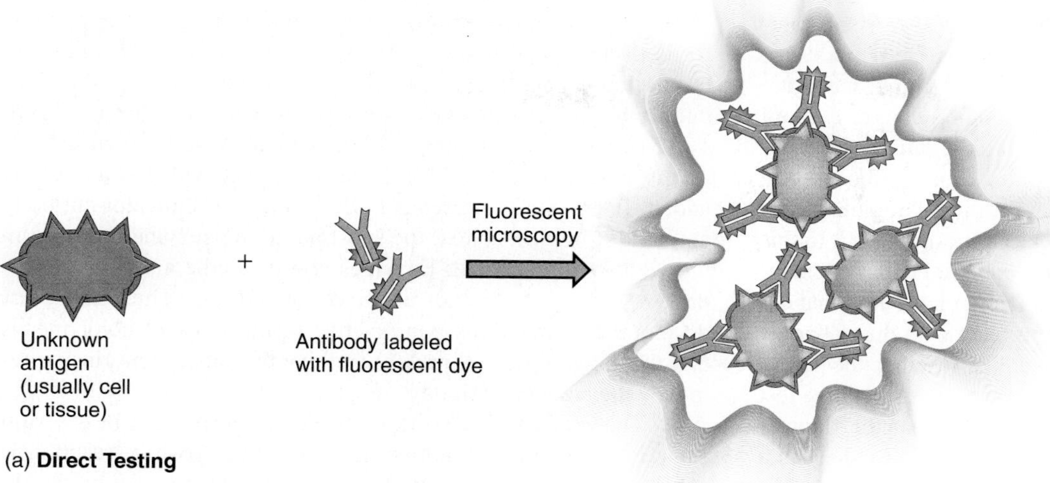

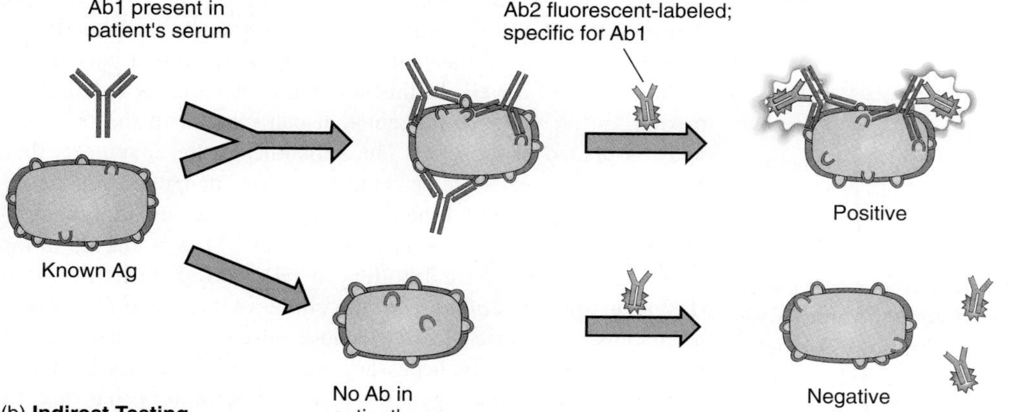

Figure 16.11

Immunofluorescence testing. The success of these techniques is contingent upon the accumulation of sufficient labeled antibody (Ab) in one site that it will show up as fluorescing cells or masses. (*a*) Direct: Unidentified antigen (Ag) is directly tagged with fluorescent Ab. (*b*) Indirect: Ag of known identity is used to assay unknown Ab; a positive reaction occurs when the second Ab (with fluorescent dye) affixes to the first Ab.

tube 1 reacted with the RBC complex instead, thereby causing lysis of the sheep RBCs. This result is negative for the antigen or antibody that was the target of the test. Complement fixation tests are somewhat complicated, yet they are invaluable in diagnosing certain viral, rickettsial, fungal, and parasitic infections.

The antistreptolysis O (ASO) titer test measures the levels of antibody against the streptolysin toxin, an important hemolysin of group A streptococci. It employs a technique related to complement fixation. A serum sample is exposed to known suspensions of streptolysin and then allowed to incubate with RBCs. Lack of hemolysis indicates antistreptolysin antibodies in the patient's serum that have neutralized the streptolysin and prevented hemolysis. This is an important verification procedure for scarlet fever, rheumatic fever, and other related streptococcal syndromes (see chapter 18).

Miscellaneous Serological Tests

A test that relies on changes in cellular activity as seen microscopically is the *Treponema pallidum immobilization* (TPI) test for syphilis. The impairment or loss of motility of the *Treponema* spirochete in the presence of test serum and complement indicates that the serum contains anti-*Treponema pallidum* antibodies (see chapter 21). In *toxin neutralization* tests, a test serum is incubated with the microbe that produces the toxin. If the serum inhibits the growth of the microbe, one can conclude that antitoxins are present.

Serotyping is an antigen-antibody technique for identifying, classifying, and subgrouping certain bacteria into categories called *serotypes,* using antisera for cell antigens such as the capsule, flagellum, and cell wall. It is widely used in typing *Salmonella* species and strains and is the basis for the numerous serotypes of streptococci. The Quellung test, which identifies serotypes of the pneumococcus involves a precipitation reaction in which antibodies react with the capsular polysaccharide. Although the reaction makes the capsule seem to swell, it is actually creating a zone of Ab-Ag complex on the cell's surface (see figure 18.20).

Fluorescent Antibodies and Immunofluorescence Testing

The property of dyes such as fluroescein and rhodamine to emit visible light in response to ultraviolet radiation was discussed in chapter 3. This property of fluorescence has found numerous applications in diagnostic immunology. The fundamental tool in immunofluorescence testing is a fluorescent antibody—a monoclonal antibody labeled by a fluorescent dye (fluorochrome).

The two ways that fluorescent antibodies can be used are shown in figure 16.11. In *direct testing,* an unknown test specimen or antigen is fixed to a slide and exposed to a fluorescent antibody solution of known composition. If the antibodies are complementary to antigens in the material, they will bind to it. After the slide is rinsed to remove unattached antibodies, it is observed with the fluorescent microscope. Fluorescing cells or specks indicate the presence of Ab-Ag complexes and a positive result. These tests are valuable for identifying and locating antigens on the surfaces of cells or in tissues and in identifying the disease agents of syphilis (see figure D, page 559), gonorrhea, chlamydiosis, whooping cough, Legionnaires' disease, plague, trichomoniasis, meningitis, and listeriosis.

In *indirect testing* methods, the fluorescent antibodies are *anti-isotypic* antibodies made to react with the Fc region of another antibody (remember that antibodies can be antigenic). In this scheme, an antigen of known character (a bacterial cell, for example) is combined with a test serum of unknown antibody content. The fluorescent antibody solution that can react with the unknown antibody is applied and rinsed off to visualize whether the serum contains antibodies that have affixed to the antigen. A positive test shows fluorescing aggregates or cells, indicating that the fluorescent antibodies have combined with the unlabeled antibodies. In a negative test, no fluorescent complexes will appear. This technique is frequently used to diagnose syphilis (FTA-ABS) and various viral infections discussed in chapters 24 and 25.

Immunoassays: Tests of Great Sensitivity

The elegant tools of the microbiologist and immunologist are being used increasingly in athletics, criminology, government, and business to test for trace amounts of substances such as hormones, metabolites, and drugs. But traditional techniques in serology are not refined enough to detect a few molecules of these chemicals. Extremely sensitive alternative methods that permit rapid and accurate measurement of trace antigen or antibody are called **immunoassays.** Examples of the technology for detecting an antigen or antibody in minute quantities include radioactive isotope labels, enzyme labels, and sensitive electronic sensors. Many of these tests are based on specifically formulated monoclonal antibodies.

Radioimmunoassy (RIA) Antibodies or antigens labeled with a radioactive isotope can be used to pinpoint minute amounts of a corresponding antigen or antibody. Although very complex in practice, these assays compare the amount of radioactivity present in a sample before and after incubation with a known, labeled antigen or antibody. The labeled substance competes with its natural, nonlabeled partner for a reaction site. Large amounts of a bound radioactive component indicate that the unknown test substance was not present. The amount of radioactivity is measured with an isotope counter or a photographic emulsion (autoradiograph). Radioimmunoassay has been employed to measure the levels of insulin and other hormones and to diagnose allergies, chiefly by the radioimmunosorbent test (RIST) for measurement of IgE in allergic patients and the radioallergosorbent test (RAST) to standardize allergenic extracts (chapter 17).

Enzyme-Linked Immunosorbent Assay ELISA The **ELISA test,** also known as enzyme immunoassay (EIA), contains an enzyme-antibody complex that can be used as a color tracer for antigen-antibody reactions. The enzymes used most often are horseradish peroxidase and alkaline phosphatase, both of which release a dye (chromogen) when exposed to their substrate. This technique also relies on a solid support such as a plastic microtiter plate that can *adsorb* (attract on its surface) the reactants (figure 16.12).

As with immunofluorescence, this technique is applicable for both direct and indirect assays. In *direct ELISA,* or sandwich, tests, a known antibody is absorbed to the bottom of a well and incubated with a solution containing unknown antigen (figure 16.12*a*). After excess unbound components have been rinsed off, an enzyme-antibody indicator that can react with the antigen is added. If antigen is present, it will attract the indicator-antibody and hold it in place. Next, the substrate to the enzyme is placed in the wells and incubated. Enzymes affixed to the antigen will hydrolyze the substrate and release a colored dye. Thus, any color developing in the wells is a positive result. Lack of color means that the antigen was not present and that the subsequent rinsing removed the enzyme-antibody complex.

The *indirect ELISA* test can detect antibodies in a serum sample. As with other indirect tests, the final positive reaction is achieved by means of an antibody-antibody reaction. Like the direct ELISA, the indicator antibody is complexed to an enzyme that produces a color change with positive serum samples (figure 16.12*b*). The starting reactant is a known antigen that is adsorbed to the surface of a well. To this, an unknown serum is added. After rinsing, an enzyme-Ab reagent that can react with the test antibody is placed in the well. The substrate to the enzyme is then added, and the wells are scanned for color changes. Color development indicates that all the components reacted and that the antibody was present in the patient's serum. This is the common screening test for the antibodies to HIV (AIDS virus), various rickettsial species, *Salmonella,* the cholera vibrio, and *Helicobacter,* a cause of gastric ulcers. Because false positives can occur, a verification test may be necessary (such as Western blot for HIV).

A newer technology uses electronic monitors that directly read out antibody-antigen reactions. Without belaboring the technical aspects, these systems contain computer chips that sense the minute changes in electrical current given off when an antibody binds to antigen. The potential for sensitivity is extreme; it is thought that amounts as small as 12 molecules of a substance can be detected in a sample. In another procedure, antibody-substrate molecules are incubated with sample and then exposed to the enzyme alkaline phosphatase. If the antibody is bound, the enzyme reacts with the substrate and causes visible light to be emitted. The light can be detected by machines or photographic films.

Tests That Differentiate T Cells and B Cells

So far we have concentrated on tests that identify antigens and antibodies in samples, but techniques also exist that differentiate between B cells and T cells and can quantify subsets of each. Information on the types and numbers of lymphocytes in blood and other samples is a common way to evaluate immune dysfunctions such as those in AIDS, immunodeficiencies, and cancer. A simple method for identifying T cells is to mix them with untreated sheep red blood cells. Receptors on the T cell bind the RBCs into a flowerlike cluster called a *rosette formation* (figure 16.13*a*). Rosetting can also occur in B cells if one uses Ig-coated bovine RBCs or mouse erythrocytes.

Indirect fluorescent techniques have been developed that cannot only differentiate between T cells and B cells but also can subgroup them (figure 16.13*b*). These subgroup tests utilize

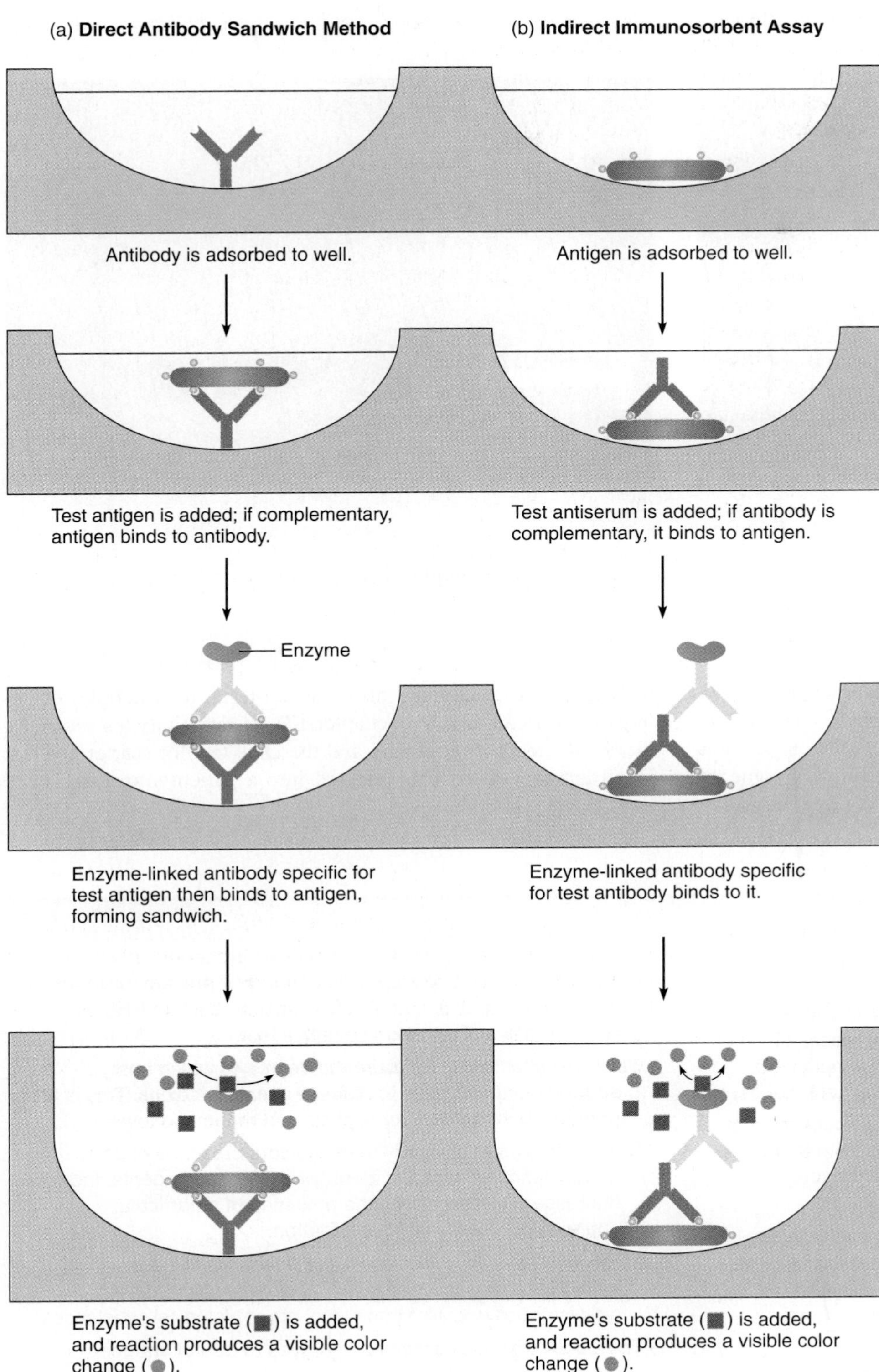

(a) Direct Antibody Sandwich Method
(b) Indirect Immunosorbent Assay
Antibody is adsorbed to well.
Antigen is adsorbed to well.
Test antigen is added; if complementary, antigen binds to antibody.
Test antiserum is added; if antibody is complementary, it binds to antigen.
Enzyme
Enzyme-linked antibody specific for test antigen then binds to antigen, forming sandwich.
Enzyme-linked antibody specific for test antibody binds to it.
Enzyme's substrate (■) is added, and reaction produces a visible color change (●).
Enzyme's substrate (■) is added, and reaction produces a visible color change (●).

Figure 16.12 Methods of ELISA testing. The success of these tests depends on adequate rinsing between steps to remove unreacted or nonspecific components. (*a*) Direct: Antibody sandwich method. (*b*) Indirect: Immunosorbent assay. Current screening for HIV infection is performed as in (*b*).

monoclonal antibodies produced in response to specific cell markers. B-cell tests categorize different stages in B-cell development and are very useful in characterizing B-cell cancers. Tests that can help differentiate the CD4, CD8, and other T-cells subsets are important in monitoring AIDS and other immunodeficiency diseases.

In Vivo Testing

Probably the first immunologic tests were performed not in a test tube but on the body itself. A classic example of one such technique is the *tuberculin test,* which uses a small amount of purified protein (PPD) from *Mycobacterium tuberculosis* injected into the skin. The appearance of a red, raised lesion can indicate previous

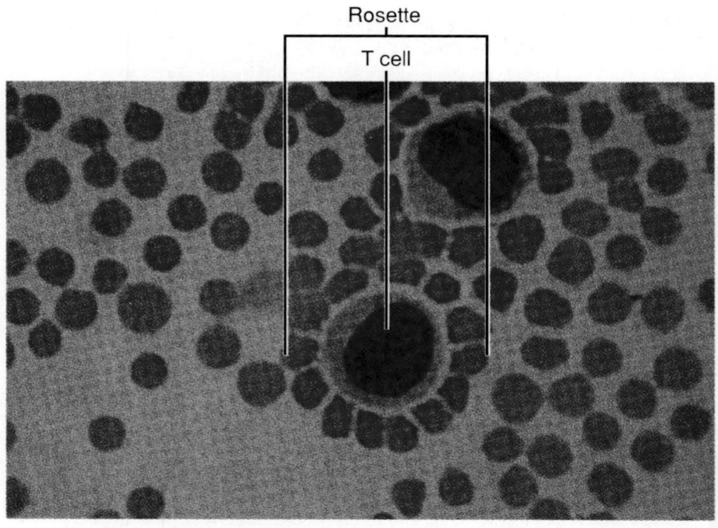

(a)

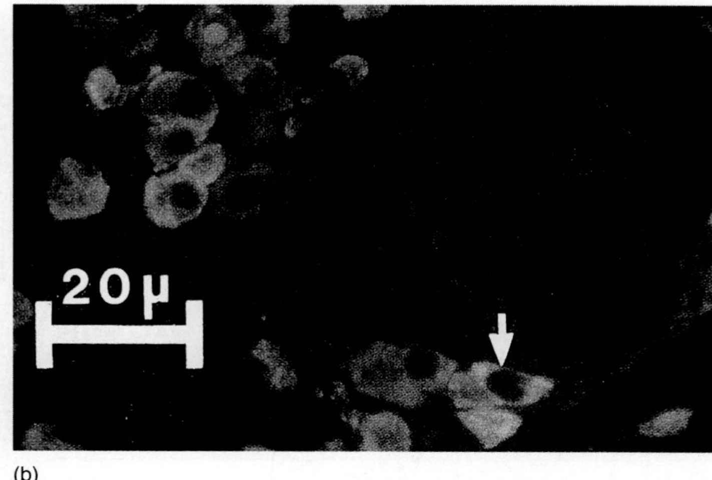

(b)

Figure 16.13
Tests for characterizing T cells and B cells. (*a*) Photomicrograph of rosette formation that identifies T cells. (*b*) Plasma cells highlighted by fluorescent antibodies.

exposure to tuberculosis (see figure 17.18). In practice, *in vivo* tests employ principles similar to serological tests, except in this case an antigen or an antibody is introduced into a patient to elicit some sort of visible reaction. Like the tuberculin test, some of these diagnostic skin tests are useful for evaluating infections due to fungi (coccidioidin and histoplasmin tests, for example) or allergens (see skin testing in chapter 17). In sensitivity tests such as the Schick test for diphtheria and the Dick test for scarlet fever, a small amount of toxin is injected into a patient's arm to detect neutralizing antibodies.

Chapter Checkpoints

Serological tests can test for either antigens or antibodies. Most are *in vitro* assessments of antigen-antibody reactivity from a variety of body fluids. The basis of these tests is an antigen-antibody reaction made visible through the processes of agglutination, precipitation, immunodiffusion, complement fixation, fluorescent antibody, and immunoassay techniques.

One measurement is the *titer,* described as the concentration of antibody in serum. It is the highest dilution of serum that gives a visible antigen-antibody reaction. The higher the titer, the greater the level of antibody present.

Agglutination reactions occur between antibody and antigens bound to cells. This results in visible clumps caused by large antibody-antigen complexes. In viral hemagglutination testing, the antibody reacts with the antigen and inhibits it from agglutinating red blood cells.

In precipitation reactions, soluble antigen and antibody react to form insoluble, visible precipitates. Precipitation reactions can also be visualized by adding radioactive or enzyme markers to the antigen-antibody complex.

In immunoelectrophoresis techniques such as the Western blot, proteins that have been separated by electrical current are identified by labeled antibodies. HIV infections are verified with this method.

Complement fixation involves a two-part procedure in which complement fixes to a specific antibody if present, or to red blood cell antigens, if antibody is absent. Lack of RBC hemolysis is indicative of a positive test.

Serological tests can measure the degree to which host antibody binds directly to disease agents or toxins. This is the principle behind tests for syphilis and rheumatic fever.

Direct fluorescent antibody tests indicate presence of an antigen and are useful in identifying infectious agents. Indirect fluorescent tests indicate the presence of a particular antibody and can diagnose infection.

Immunoassays can detect very small quantities of antigen, antibody, or other substances. Radioimmunoassay uses radioisotopes to detect trace amounts of biological substances.

The ELISA test uses enzymes and dyes to detect antigen-antibody complexes. It is widely used to detect viruses, bacteria, and antibodies in HIV infection.

Technicians use precise assays to differentiate between B and T cells and to identify subgroups of these cells for disease diagnosis.

In vivo serological testing, such as the tuberculin test, involves subcutaneous injection of antigen to elicit a visible antigen-antibody response in the host.

CHAPTER CAPSULE WITH KEY TERMS

BIOMEDICAL APPLICATIONS OF SPECIFIC IMMUNITY

Immunization: Producing immunity by medical intervention. Passive immunotherapy includes administering immune serum globulin and specific immune globulins pooled from donated serum to prevent infection and disease in those at risk; antisera and antitoxins from animals are occasionally used.

Active immunization is synonymous with **vaccination;** provides an antigenic stimulus that does not cause disease but can produce long-lasting, protective immunity. **Vaccines** are made with: (1) **killed** whole cells or inactivated viruses that do not reproduce but are antigenic; (2) **live, attenuated** cells or viruses that are able to reproduce but have lost virulence; (3) **subcellular** or **subunit** components of microbes such as surface antigen or neutralized toxins **(toxoids);** and (4) **genetic engineering** techniques, including cloning of antigens, recombinant attenuated microbes, and DNA-based vaccines. Boosters (additional doses) are often required. Vaccination increases **herd immunity,** protection provided by mass immunity in a population.

Serological/Immune Testing: **Serology** is a science that attempts to detect signs of infection in a patient's serum such as antibodies specific for a microbe. The basis of serological tests is that Abs specifically bind to Ag *in vitro.* If an Ag of known identity is added to an unknown serum sample, it will affix to the antibody. The reverse is also true; known antibodies can be used to detect and type antigens. These Ag-Ab reactions are visible in the form of obvious clumps and precipitates, color changes, or the release of radioactivity. Test results are read as positive or negative; desirable properties of tests are **specificity** and **sensitivity.**

Types of Tests: In **agglutination** tests, antibody cross-links whole cell antigens, forming complexes that settle out and form visible clumps in the test chamber; examples are tests for blood type, some bacterial diseases, and viral diseases. **Double diffusion precipitation** tests involve the diffusion of Ags and Abs in a soft agar gel, forming zones of **precipitation** where they meet.

In **immunoelectrophoresis,** migration of serum proteins in gel is combined with precipitation by antibodies. One example, the **Western blot** test, separates antigen into bands. After the gel is affixed to a blotter, it is reacted with a test specimen and developed by radioactivity or with dyes.

Complement fixation tests detect **lysins**—antibodies that fix complement and can lyse target cells. It involves first mixing test Ag and Ab with complement and then with sensitized sheep RBCs. If the complement is fixed by the Ag-Ab, the RBCs remain intact and the test is positive. If RBCs are hemolyzed, specific antibodies are lacking.

In **direct assays,** known marked Ab is used to detect unknown Ag (microbe). In **indirect testing,** known Ag reacts with unknown Ab and the reaction is made visible by a second Ab that can affix to and identify the unknown Ab. **Immunofluorescence testing** uses **fluorescent antibodies** (FABs tagged with fluorescent dye) either directly or indirectly to visualize cells or cell aggregates that have reacted with the FABs.

Immunoassays are highly sensitive tests for Ag and Ab. In **radioimmunoassay,** Ags or Abs are labeled with radioactive isotopes and traced. The **enzyme-linked immunosorbent assay (ELISA)** can detect unknown Ag or Ab by direct or indirect means. A positive result is visualized when a colored product is released by an enzyme-substrate reaction. Tests are also available to differentiate B cells from T cells and their subtypes. With *in vivo* testing, Ags are introduced into the body directly to determine the patient's immunologic history.

MULTIPLE-CHOICE QUESTIONS

1. A living microbe with reduced virulence that is used for vaccination is considered
   a. a toxoid c. denatured
   b. attenuated d. an adjuvant
2. A vaccine that contains parts of viruses is called
   a. subcellular c. subunit
   b. recombinant d. attenuated
3. Widespread immunity that protects the population from the spread of disease is called
   a. seropositivity
   b. cross-reactivity
   c. epidemic prophylaxis
   d. herd immunity
4. DNA vaccines contain ____ DNA that stimulates cells to make ____ antigens.
   a. human, RNA
   b. microbial, protein
   c. human, protein
   d. microbial, polysaccharide
5. Administration of immune serum globulin is a form of ____ immunization that ____.
   a. active, prevents infection
   b. passive, provides long-term immunity
   c. therapeutic, prevents disease
   d. prophylactic, stimulates the immune system.
6. In agglutination reactions, the antigen is a ____; in precipitation reactions, it is a ____.
   a. soluble molecule, whole cell
   b. whole cell, soluble molecule
   c. bacterium, virus
   d. protein, carbohydrate
7. Which reaction requires complement?
   a. hemagglutination
   b. precipitation
   c. hemolysis
   d. toxin neutralization
8. A patient with a ____ titer of antibodies to an infectious agent has greater protection than a patient with a ____ titer.
   a. high, low b. low, high

9. Direct immunofluorescence tests use a labeled antibody to identify ____.
   a. an unknown microbe
   b. an unknown antibody
   c. fixed complement
   d. precipitins
10. The Western blot test can identify
    a. unknown antibodies
    b. unknown antigens
    c. specific DNA
    d. both a and b
11. An example of an *in vivo* serological test is
    a. indirect immunofluorescence
    b. radioimmunoassay
    c. tuberculin test
    d. complement fixation
12. Multiple matching of vaccines. First look up the disease for which the vaccine is intended (if not apparent), then list the letters of all choices that fit (if there is more than one possible answer).
    ____ BCG
    ____ Measles
    ____ Pertussis
    ____ Tetanus
    ____ Mumps
    ____ Diphtheria
    ____ Hepatitis B
    ____ Hib
    ____ Polio
    ____ Rubella
    ____ Pneumococcal

    a. Uses live, attenuated microbes
    b. Based on a toxoid
    c. Subunit vaccine
    d. Used in combined vaccine
    e. Does not require boosters
    f. Uses killed, whole cells
    g. Uses killed, whole viruses
    h. Routinely given in childhood
    i. Mainly for people at high risk
    j. Made by genetic engineering

CONCEPT QUESTIONS

1. a. Name three products used in passive artificial immunization.
   b. What are the primary reasons for using these substances?
   c. What are some disadvantages of this form of immunization?
   d. What is the difference between immunization used for prophylaxis and that used for treatment?
2. a. Outline the strategies for developing vaccines, and give specific examples for each method.
   b. By what means are microorganisms attenuated?
   c. What is the purpose of an adjuvant?
   d. Describe the way that a Trojan horse vaccine works.
3. a. What are the advantages and disadvantages of a killed vaccine; a live, attenuated vaccine; a subunit vaccine; a recombinant vaccine; and a DNA vaccine?
   b. Use an outline to explain how an inoculation with tetanus toxoid will protect a person the next time he or she steps on a dirty piece of glass.
4. a. Describe the concept of herd immunity.
   b. How does vaccination contribute to its development in a community?
   c. Give some possible explanations for the recent epidemics of diphtheria and whooping cough.
5. a. What is the basis of serology and serological testing?
   b. Differentiate between specificity and sensitivity.
   c. Describe several general ways that Ag-Ab reactions are detected.
6. a. What does seropositivity mean?
   b. In what ways is the antibody titer of serum important?
   c. What causes false positive tests?
7. a. Explain how agglutination and precipitation reactions are alike.
   b. In what ways are they different?
   c. Make a drawing of the manner in which antibodies cross-link the antigens in agglutination and precipitation reactions.
   d. Give examples of several tests that employ the two reactions.
8. a. What is meant by complement fixation? What are cytolysins?
   b. What is the purpose of using sheep red blood cells in this test?
9. a. Explain the differences between direct and indirect procedures in serological or immunoassay tests
   b. How is fluorescence detected?
   c. How is the reaction in a radioimmunoassay detected?
   d. How does a positive reaction in an ELISA test appear?
10. a. Briefly describe the principles and give an example of the use of a specific test using immunoelectrophoresis, Western blot, complement fixation, fluorescence testing (direct and indirect), and immunoassays (direct and indirect ELISA).
    b. Explain a rapid microscopic method for differentiating T cells from B cells.

CRITICAL-THINKING QUESTIONS

1. Describe the relationship between an antitoxin, a toxin, and a toxoid.
2. It is often said that a vaccine does not prevent infection; rather, it primes the immune system to undergo an immediate response to prevent an infection from spreading. Explain what is meant by this statement, and outline what is happening at the cellular molecular level from the time of vaccination until infection with the infectious agent actually occurs.
3. At least three boosters are given for DPT vaccines.
   a. Explain what each subsequent booster does, and why more than one is needed.
   b. Which features of the immune system allow it to react efficiently with 10–15 different vaccine antigens simultaneously?
4. Explain how to design a vaccine that could:
   a. induce protective IgA in the intestine
   b. give immunity to dental caries
   c. protect against the liver phase of the malaria parasite
   d. be derived from a single microbe and immunize against two different infectious diseases

5. a. Suggest several reasons that it could be risky to administer a vaccine containing a live, attenuated DNA virus.
   b. Explain what is involved in making a DNA vaccine.
   c. Explain why measles vaccine does not reliably protect infants if it is given before 12 months of age.
6. a. Determine the vaccines you have been given and those for which you will require periodic boosters.
   b. Suggest vaccines you may need in the future.
7. When traders and missionaries first went to the Hawaiian Islands, the natives there experienced severe disease and high mortality rates from smallpox, measles, and certain STDs.
   a. Explain what factors are involved in the sudden prevalence of disease in previously unexposed populations.
   b. Explain the ways in which vaccination has been responsible for the eradication of diseases such as smallpox and polio.
8. Why do some tests for antibody in serum (such as for HIV and syphilis) require backup verification with additional tests at a later data?
9. a. Look at figure 16.5*b*. What is the titer as shown?
   b. If the titer had been 1/40, what interpretation would be made as to the immune status of the patient?
   c. What would it mean if a test 2 weeks later revealed a titer of 1 : 1280?
   d. What would it mean if no agglutination had occurred in any tube?
   e. From figure 16.6*c*, can you tell which patients have measles antibody and which do not?
   f. What are the titers of the Ab-positive patients?
10. Why do we interpret positive hemolysis in the complement fixation test to mean negative for the test substance?
11. Explain how an immunoassay method could use monoclonal antibodies to differentiate between B and T cells and between different subsets of T cells.

INTERNET SEARCH TOPIC

1. Type the search words *DNA vaccine* into an Internet browser. Find at least two recent vaccine trials that are listed and describe their purposes and outcome.
2. Find descriptions of current tests for HIV infection such as ELISA, Western Blot, and antigen tests; determine the indications for use and their relative sensitivity and specificity.

17 chapter

Disorders in Immunity

Humans possess a powerful and intricate system of defense, which by its very nature also carries the potential for causing injury and disease. In most instances, a defect in immune function is expressed in commonplace, but miserable, symptoms such as those of hay fever and dermatitis. But abnormal or undesirable immune functions are also actively involved in debilitating or life-threatening diseases such as asthma, anaphylaxis, rheumatoid arthritis, graft rejection, and cancer.

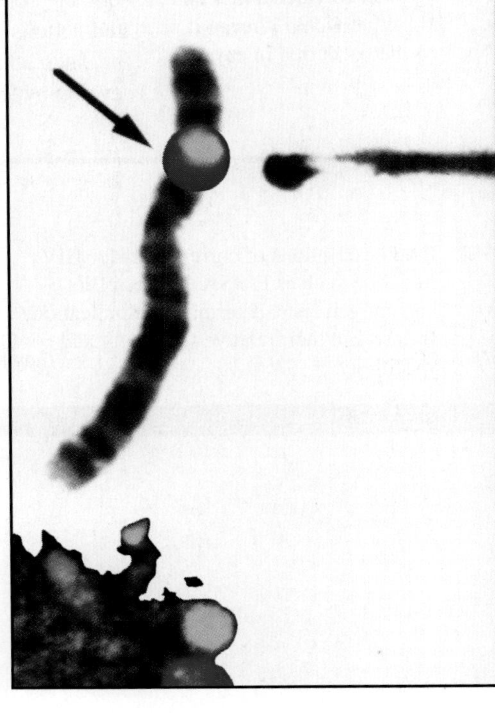

Human chromosome 2 stained by FISH (fluorescent in-situ hybridization) to pinpoint a genetic marker for hereditary nonpolyposis colon cancer. This is one of several common human cancers for which special genetic screening tests are being developed.

THE IMMUNE RESPONSE: A TWO-SIDED COIN

With few exceptions, our previous discussions of the immune response have centered around its numerous beneficial effects. The precisely coordinated system that seeks out, recognizes, and destroys an unending array of foreign materials is clearly protective, but it also presents another side—a side that promotes rather than prevents disease. In this chapter, we will survey **immunopathology,** the study of disease states associated with overreactivity or underreactivity of the immune response (figure 17.1). In the cases of **allergies** and **autoimmunity,** the tissues are innocent bystanders attacked by excessive immunologic functions. In **grafts** and **transfusions,** a recipient reacts to the foreign tissues and cells of another individual. In **immunodeficiency disease,** immune function is incompletely developed, suppressed, or destroyed. **Cancer** falls into a special category, because it is both a cause and an effect of immune dysfunction. As we shall see, one fascinating by-product of studies of immune disorders has been our increased understanding of the basic workings of the immune system.

OVERREACTIONS TO ANTIGENS: ALLERGY/ HYPERSENSITIVITY

The term **allergy*** means a condition of altered reactivity or exaggerated immune response that is manifested by inflammation. Although it is sometimes used interchangeably with **hypersensitivity,** some experts refer to immediate reactions such as hay fever as allergies and to delayed reactions as hypersensitivities. Allergic individuals are acutely sensitive to repeated contact with antigens, called **allergens,** that do not noticeably affect nonallergic individuals. Although the general effects of hyperactivity are detrimental, we must be aware that it involves the very same types of immune reactions as those at work in protective immunities. These include humoral and cell-mediated actions, the inflammatory response, phagocytosis, and complement. Such an association means that all humans have the potential to develop hypersensitivity under particular circumstances.

*allergy (al´ -er-jee) Gr. *allos,* other, and *ergon,* work.

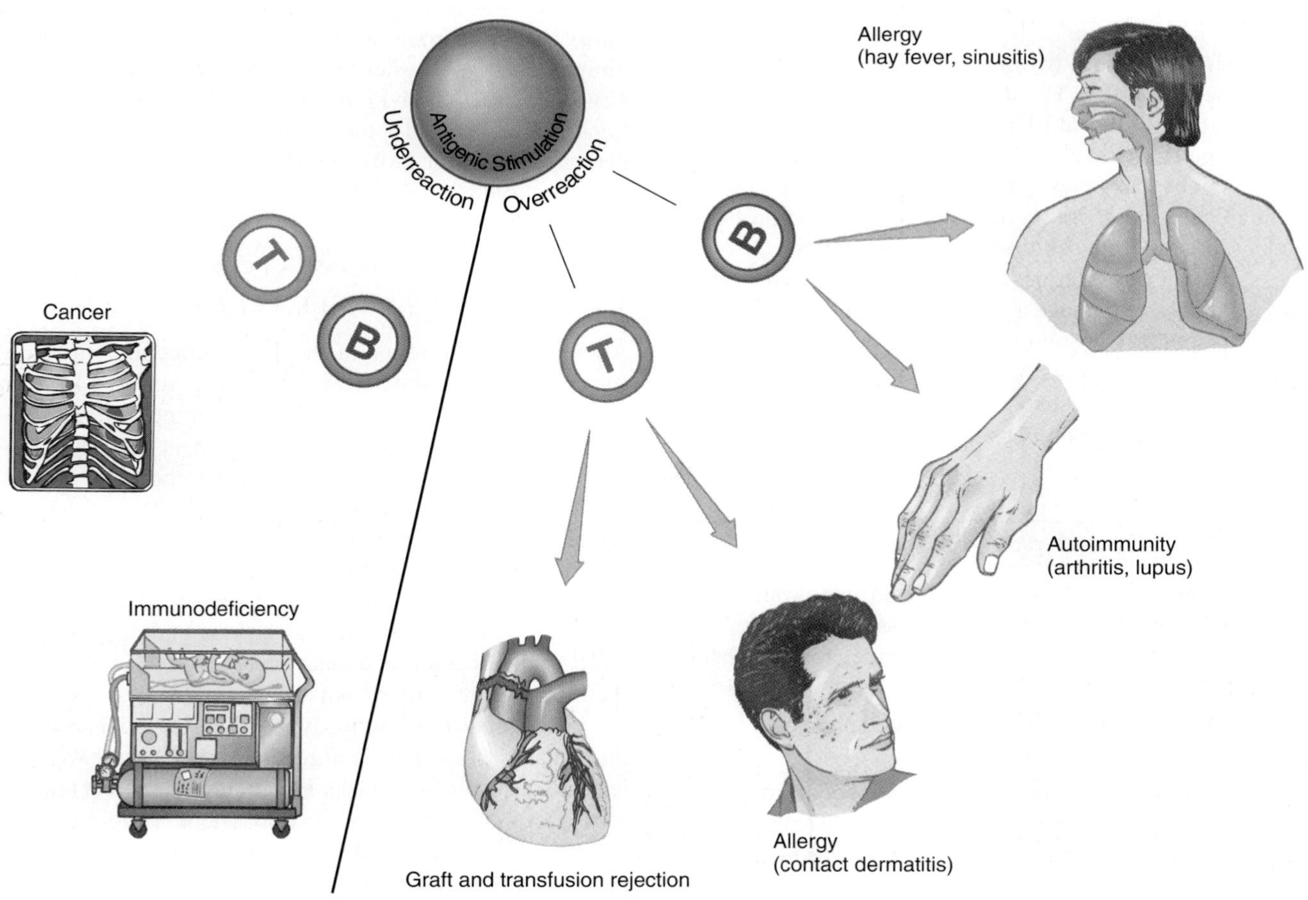

Figure 17.1
The immune system in health and disease. Just as the system of T cells and B cells provides necessary protection against infection and disease, the same system can cause serious and debilitating conditions by overreacting or underreacting to immune stimuli.

TABLE 17.1

HYPERSENSITIVITY STATES

Type	Antigen Source	Systems Involved	Examples
I	Exogenous	Immediate allergy: IgE-mediated; involves mast cells, basophils	Anaphylaxis, atopic allergies such as hay fever, asthma
II	Exogenous, endogenous	IgG, IgM antibodies act upon cells with complement and cause cell lysis; includes some autoimmune diseases	Blood group incompatibility; pernicious anemia; myasthenia gravis
III	Exogenous, endogenous	Immune complex; antibody-mediated inflammation; circulating IgG complexes deposited in basement membranes of target organs; includes some autoimmune diseases	Systemic lupus erythematosus; rheumatoid arthritis; serum sickness; rheumatic fever
IV	Exogenous, endogenous	Delayed hypersensitivity, mediated by T cells, NK cells, and macrophages	Infection reactions; contact dermatitis; graft rejection; granulomas and skin reactions

Originally, allergies were defined as either immediate or delayed, depending upon the time lapse between contact with the allergen and onset of symptoms. Subsequently, they were differentiated as humoral versus cell-mediated. But as information on the nature of the allergic immune response accumulated, it became evident that, although useful, these schemes oversimplified what is really a very complex spectrum of reactions. The most widely accepted classification, first introduced by immunologists P. Gell and R. Coombs, includes four major categories: type I (atopic and anaphylaxis allergies), type II (cytotoxic autoimmunities), type III (immune complex), and type IV (delayed hypersensitivity) (table 17.1). In general, types I, II, and III involve a B cell–immunoglobulin response, and type IV involves a T-cell response. The antigens that elicit these reactions can be exogenous, originating from outside the body (microbes, pollen grains, and foreign cells and proteins), or endogenous, arising from self tissue (autoimmunities).

One of the reasons allergies are easily mistaken for infections is that both involve damage to the tissues and thus trigger the inflammatory response (see figure 14.18). Many symptoms and signs of inflammation (redness, heat, skin eruptions, edema, and granuloma) are prominent features. But unlike inflammation from infection, allergic inflammation is uncontrolled and usually accelerates tissue damage.

Chapter Checkpoints

- Immunopathology is the study of diseases associated with excesses and deficiencies of the immune response. Such diseases include allergies, autoimmunity, grafts, transfusions, immunodeficiency disease, and cancer.
- An allergy or hypersensitivity is an exaggerated immune response that injures or inflames tissues.
- There are four categories of hypersensitivity reactions: type I (atopy and anaphylaxis), type II (cytotoxic), type III (immune complex reactions), and type IV (delayed hypersensitivity reactions).
- Antigens that trigger hypersensitivity reactions are allergens. They can be either exogenous (originate outside the host) or endogenous (involve the host's own tissue).

TYPE I ALLERGIC REACTIONS: ATOPY AND ANAPHYLAXIS

All type I allergies share a similar physiological mechanism, are immediate in onset, and are associated with exposure to specific antigens. However, it is convenient to recognize two subtypes: **atopy*** is any chronic local allergy such as hay fever or asthma; **anaphylaxis*** is a systemic, often explosive reaction that involves airway obstruction and circulatory collapse. In the following sections, we will consider the epidemiology of type I allergies, allergens and routes of inoculation, mechanisms of disease, and specific syndromes.

EPIDEMIOLOGY AND MODES OF CONTACT WITH ALLERGENS

Allergies exert profound medical and economic impact. Allergists (physicians who specialize in treating allergies) estimate that about 10% to 30% of the population is prone to atopic allergy. It is generally acknowledged that self-treatment with over-the-counter medicines accounts for significant underreporting of cases. The 35 million people afflicted by hay fever (15–20% of the population) spend about half a billion dollars annually for medical treatment. The monetary loss due to employee debilitation and absenteeism is immeasurable. The majority of type I allergies are relatively mild, but certain forms such as asthma and anaphylaxis may require hospitalization and cause death. About 2 million people in the United States suffer from asthma.

The predisposition for type I allergies is inherited. Be aware that what is hereditary is a generalized *susceptibility,* not the allergy to a specific substance. For example, a parent who is allergic to ragweed pollen can have a child who is allergic to cat hair. The prospect of a child's developing atopic allergy is at least 25% if one parent is atopic, increasing up to 50% if grandparents or siblings are also afflicted. The actual basis for atopy appears to be a genetic

*atopy (at′-oh-pee) Gr. *atop,* out of place.

*anaphylaxis (an″-uh-fih-lax′-us) Gr. *ana,* excessive, and *phylaxis,* protection.

Table 17.2

Common Allergens, Classified by Portal of Entry

Inhalants	Ingestants	Injectants	Contactants
Pollen Dust Mold spores Dander Animal hair Insect parts Formalin Drugs Enzymes	Food (chocolate, wheat, eggs, milk, nuts, strawberries, fish) Food additives Drugs (aspirin, penicillin)	Hymenopteran venom (bee, wasp) Drugs Vaccines Serum Enzymes Hormones	Drugs Cosmetics Heavy metals Detergents Formalin Rubber Glue Solvents Dyes

program that favors allergic antibody (IgE) production, increased reactivity of mast cells, and increased susceptibility of target tissue to allergic mediators. Allergic persons often exhibit a combination of syndromes, such as hay fever, eczema, and asthma.

Other factors that affect the presence of allergy are age, infection, and geographic locale. New allergies tend to crop up throughout an allergic person's life, especially as new exposures occur after moving or changing life-style. In some persons, atopic allergies last for a lifetime; others "outgrow" them, and still others suddenly develop them later in life. Some features of allergy are not yet completely explained.

The Nature of Allergens and Their Portals of Entry

As with other antigens, allergens have certain immunogenic characteristics. Not unexpectedly, proteins are more allergenic than carbohydrates, fats, or nucleic acids. Some allergens are haptens, non-proteinaceous substances with a molecular weight of less than 1,000 that can form complexes with carrier molecules in the body (see figure 15.11). Organic and inorganic chemicals found in industrial and household products, cosmetics, food, and drugs are commonly of this type. Table 17.2 lists a number of common allergenic substances.

Allergens typically enter through epithelial portals in the respiratory tract, gastrointestinal tract, and skin. The mucosal surfaces of the gut and respiratory system present a thin, moist surface that is normally quite penetrable. The dry, tough keratin coating of skin is less permeable, but access still occurs through tiny breaks, glands, and hair follicles. It is worth noting that the organ of allergic expression may or may not be the same as the portal of entry.

Pollen in Los Angeles
Outlook for January
Trees: Alder and walnut are blooming extremely early this year. Black walnut is the worst allergy offender of the walnut family.
Weeds: Weed season has finally ended.
Grasses: Grasses will be mostly dormant until late February, early March.
Molds: Molds will continue to be sporadic and with major increases following periods of precipitation.
Weekly Levels for December
Major
Moderate
Minor
Week 1
Week 2
Week 3
Week 4
Trees
Weeds
Grasses
Mold spores

Airborne environmental allergens such as pollen, house dust, dander (shed skin scales), or fungal spores are termed *inhalants.* Each geographic region harbors a particular combination of airborne substances that varies with the season and humidity (figure 17.2*a*). Pollen, the most common offender, is given off seasonally by the reproductive structures of pines and flowering plants (weeds, trees, and grasses). Unlike pollen, mold spores are released throughout the year and are especially profuse in moist

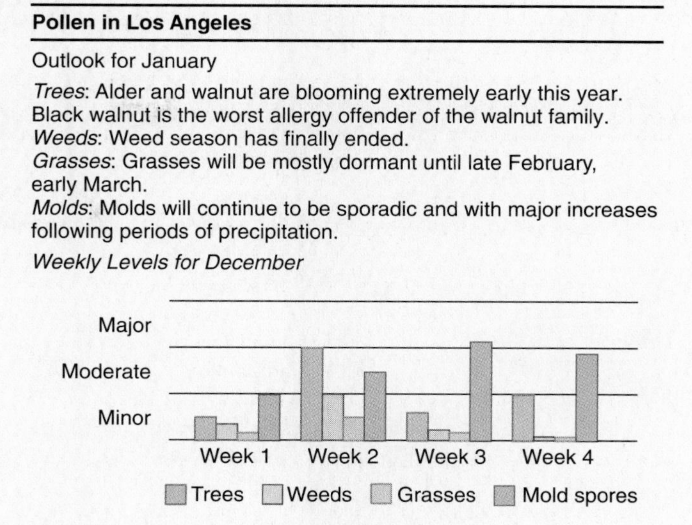

(a)

(b)

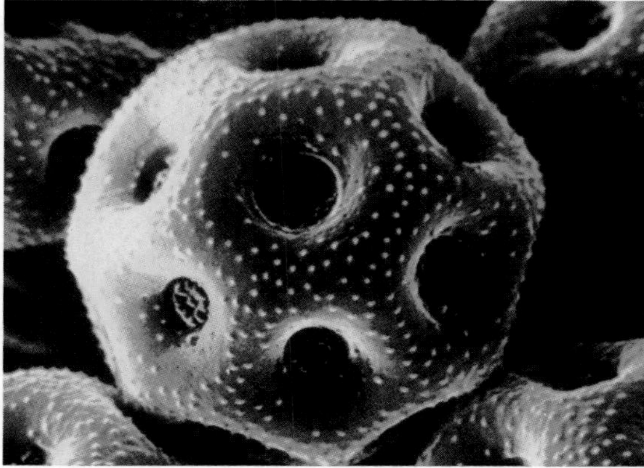

(c)

Figure 17.2

Monitoring airborne allergens. (*a*) The air in heavily vegetated places with a mild climate is especially laden with allergens. Although the seasons change, pollen and mold spores are present year-round. For example, in just one month in southern California, weed pollen subsided to near zero and mold spore levels doubled. (*b*) Because the dust mite *Dermatophagoides* feeds primarily on human skin cells in house dust, these mites are found in abundance in bedding and carpets. Airborne mite feces and particles from their bodies are an important source of allergies. (*c*) Scanning electron micrograph of a single pollen grain from a rose (6,000×). Millions of these are released from a single flower.

(a) *Source: Data from the* Asthma and Allergy Foundation of America, *Los Angeles, CA.*

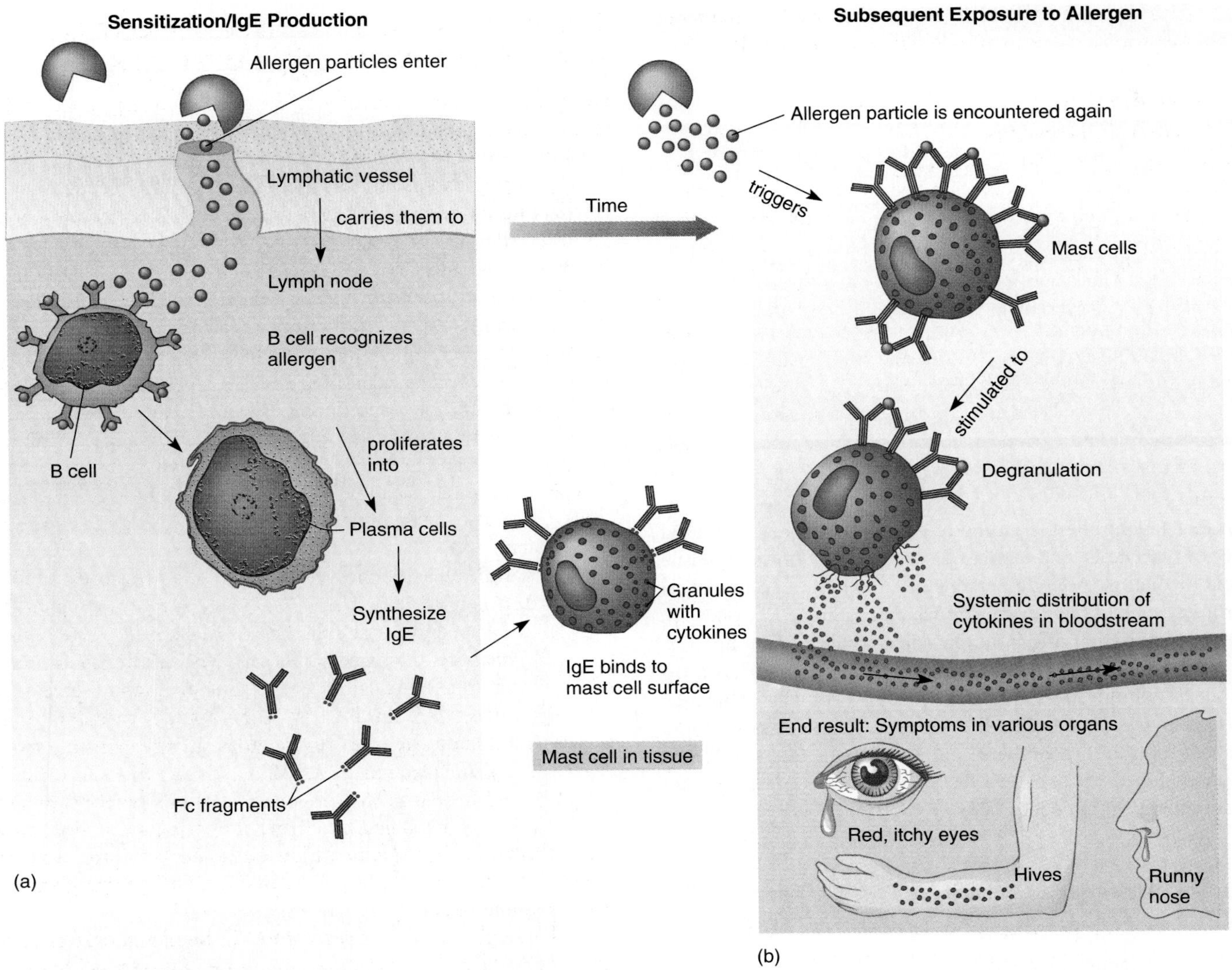

Figure 17.3

A schematic view of cellular reactions during the type I allergic response. (*a*) Sensitization (initial contact with sensitizing dose). (*b*) Provocation (later contacts with provocative dose).

areas of the home and garden. Airborne animal hair and dander, feathers, and the saliva of dogs and cats are common sources of allergens. The component of house dust that appears to account for most dust allergies is not soil, lint, or other debris, but the decomposed bodies of tiny mites that commonly live in this dust (figure 17.2*b*). Some people are allergic to their work, in the sense that they are exposed to allergens on the job. Florists, bakers, beauty operators, woodworkers, food processors, farmers, bookbinders, drug processors, leather tanners, welders, and plastics manufacturers work in environments that aggravate inhalant allergies.

Allergens that enter by mouth, called *ingestants,* often cause food allergies. *Injectant* allergies are an important adverse side effect of drugs or other substances used in diagnosing, treating, or preventing disease. A natural source of injectants is venom from stings by hymenopterans, a family of insects that includes honey bees and wasps. *Contactants* are allergens that enter through the skin. Many contact allergies are of the type IV, delayed variety discussed later in this chapter.

Mechanisms of Type I Allergy: Sensitization and Provocation

What causes some people to sneeze and wheeze every time they step out into the spring air, while others suffer no ill effects? In order to answer this question, we must examine what occurs in the tissues of the allergic individual that does not occur in the normal person. In general, type I allergies develop in stages (figure 17.3). The initial encounter with an allergen provides a **sensitizing dose** that primes the immune system for a subsequent encounter with that allergen but generally elicits no signs or symptoms. The memory cells and immunoglobulin are then ready to react with a subsequent **provocative dose** of the same allergen. It is this dose that precipitates the signs and symptoms of allergy. Despite numerous anecdotal reports of people becoming allergic upon first contact with an allergen, it is generally believed that these individuals unknowingly had contact at some previous time. Fetal exposure to allergens from the mother's bloodstream is one possibility, and foods can be a prime source of "hidden" allergens such as penicillin.

The Physiology of IgE-Mediated Allergies

During primary contact and sensitization, the allergen penetrates the portal of entry (figure 17.3*a*). When large particles such as pollen grains, hair, and spores encounter a moist membrane, they release molecules of allergen that pass into the tissue fluids and lymphatics. The lymphatics then carry the allergen to the lymph nodes, where specific clones of B cells recognize it, are activated, and proliferate into plasma cells. These plasma cells produce **immunoglobulin E** (IgE), the antibody of allergy also known as *reagin.** IgE is different from other immunoglobulins in having an Fc receptor region with great affinity for mast cells and basophils. The binding of IgE to these cells in the tissues sets the scene for the reactions that occur upon repeated exposure to the same allergen (see figures 17.3*b* and 17.6).

The Role of Mast Cells and Basophils

The most important characteristics of mast cells and basophils relating to their roles in allergy are:

1. Their ubiquitous location in tissues. Mast cells are located in the connective tissue of virtually all organs, but particularly high concentrations exist in the lungs, skin, gastrointestinal tract, and genitourinary tract. Basophils circulate in the blood but migrate readily into tissues.
2. Their capacity to bind IgE during sensitization (figure 17.3). Each cell carries 30,000 to 100,000 cell receptors that attract 10,000 to 40,000 IgE antibodies.
3. Their cytoplasmic granules (secretory vesicles), which contain physiologically active cytokines (histamine, serotonin—introduced in chapter 14).
4. Their tendency to **degranulate** (figures 17.3*b* and 17.4), or release the contents of the granules into the tissues when properly stimulated by allergen.

Let us now see what occurs when sensitized cells are challenged with allergen a second time.

The Second Contact with Allergen

After sensitization, the IgE-primed mast cells can remain in the tissues for years. Even after long periods without contact, a person can retain the capacity to react immediately upon reexposure. The next time allergen molecules contact these sensitized cells, they bind across adjacent receptors and stimulate degranulation. As chemical mediators are released, they diffuse into the tissues and bloodstream. Cytokines give rise to numerous local and systemic reactions, many of which appear quite rapidly (figure 17.3*b*). The symptoms of allergy are not caused by the direct action of allergen on tissues but by the physiological effects of mast cell mediators on target organs.

***reagin** (ree´-ah-jin) Derived from *react* or *reaction.*

Cytokines, Target Organs, and Allergic Symptoms

Numerous substances involved in mediating allergy (and inflammation) have been identified. The principal mast cell and basophil cytokines are histamine, serotonin, leukotriene, platelet-activating factor, prostaglandins, and bradykinin (figure 17.4). These chemicals, acting alone or in combination, account for the tremendous scope of allergic symptoms. For some theories pertaining to this function of the allergic response, see microfile 17.1. Targets of cytokines include the skin, upper respiratory tract, gastrointestinal tract, and conjunctiva. The general responses of these organs include rashes, itching, redness, rhinitis, sneezing, diarrhea, and shedding of tears. Systemic targets include smooth muscle, mucous glands, and nervous tissue. Because smooth muscle is responsible for regulating the size of blood vessels and respiratory passageways, changes in its activity can profoundly alter blood flow, blood pressure, and respiration. Pain, anxiety, agitation, and lethargy are also attributable to the effects of cytokines on the nervous system.

Histamine* is the most profuse and the fastest-acting of the cytokines. It is a potent stimulator of smooth muscle, glands, and eosinophils. Histamine's actions on smooth muscle vary with location. It *constricts* the smooth muscle layers of the small bronchi and intestine, thereby causing labored breathing and increased intestinal motility. In contrast, histamine *relaxes* vascular smooth muscle and dilates arterioles and venules. It is responsible for the *wheal* and flare* reaction in the skin (see figure 17.6*a*), pruritis (itching), and headache. More severe reactions (such as anaphylaxis) can be accompanied by edema and vascular dilation, which lead to hypotension, tachycardia, circulatory failure, and, frequently, shock. Salivary, lacrimal, mucous, and gastric glands are also histamine targets.

Although the role of **serotonin*** in human allergy is uncertain, its effects appear to complement those of histamine. In experimental animals, serotonin increases vascular permeability, capillary dilation, smooth muscle contraction, intestinal peristalsis, and respiratory rate, but it diminishes central nervous system activity.

Before the specific types were identified, **leukotriene*** was known as the "slow-reacting substance of anaphylaxis," or SRSA, for its property of inducing gradual contraction of smooth muscle. This type of leukotriene is responsible for the prolonged bronchospasm, vascular permeability, and mucus secretion of the asthmatic individual. Other leukotrienes stimulate the activities of polymorphonuclear leukocytes.

Platelet-activating factor is a lipid released by basophils, neutrophils, monocytes, and macrophages that causes platelet aggregation and lysis. The physiological response to stimulation by this factor is similar to that of histamine, including increased vascular permeability, pulmonary smooth muscle contraction, pulmonary edema, hypotension, and a wheal and flare response in the skin.

***histamine** (his´-tah-meen) Gr. *histio,* tissue, and amine.

***wheal** (weel) A smooth, slightly elevated, temporary welt that is surrounded by a flushed patch of skin (flare).

***serotonin** (ser´´-oh-toh´-nin) L. *serum,* whey, and *tonin,* tone.

***leukotriene** (loo´´-koh-try´-een) Gr. *leukos,* white blood cell, and *triene,* a chemical suffix.

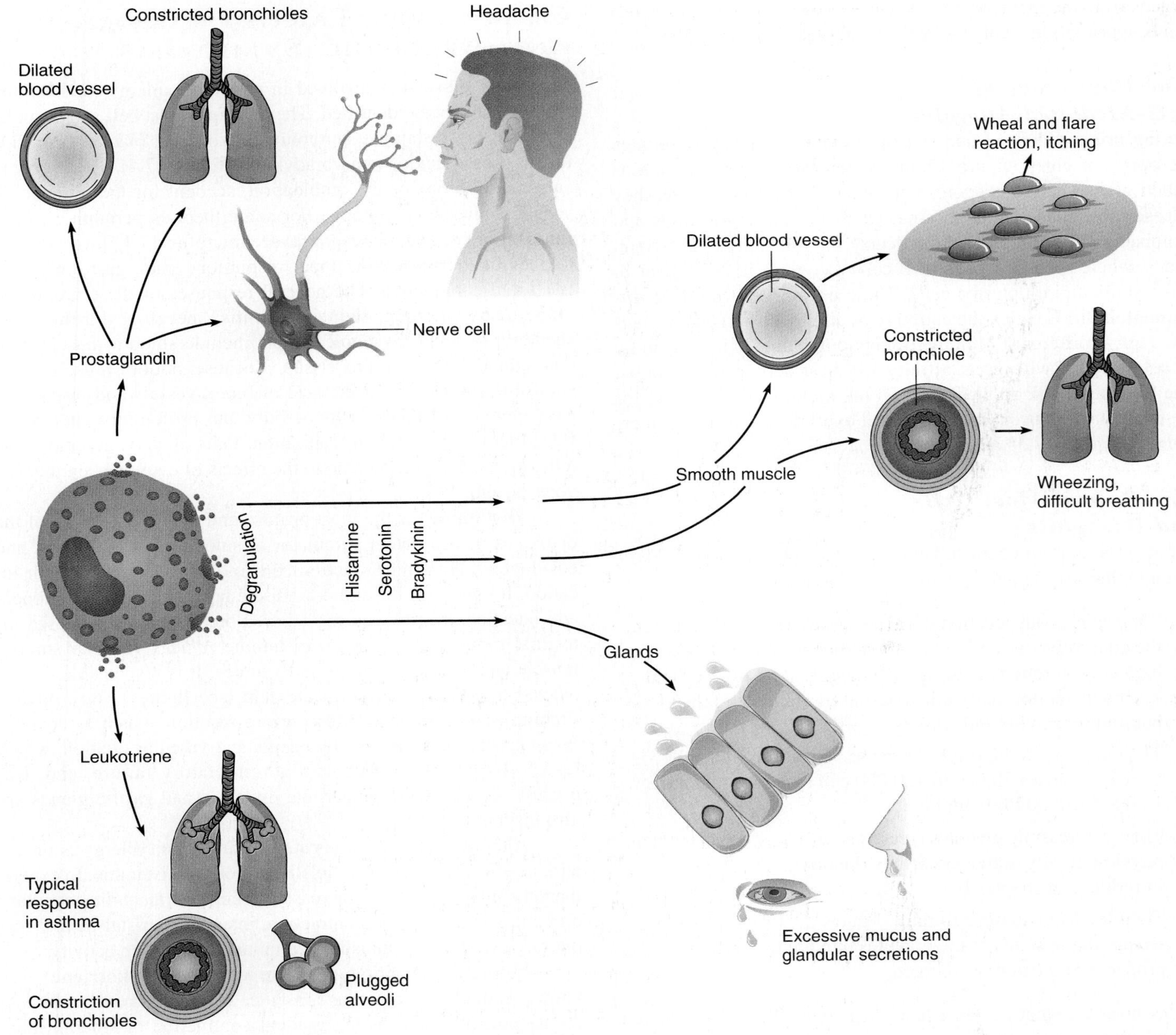

Figure 17.4

The spectrum of reactions to inflammatory cytokines and the common symptoms they elicit in target tissues and organs. Note the extensive overlapping effects.

Prostaglandins* are a group of powerful inflammatory agents. Normally, these substances regulate smooth muscle contraction (for example, they stimulate uterine contractions during delivery). In allergic reactions, they are responsible for vasodilation, increased vascular permeability, increased sensitivity to pain, and bronchoconstriction. Certain anti-inflammatory drugs work by preventing the actions of prostaglandins.

Bradykinin* is related to a group of plasma and tissue peptides known as kinins that participate in blood clotting and chemotaxis. In allergy, it causes prolonged smooth muscle contraction of the bronchioles, dilatation of peripheral arterioles, increased capillary permeability, and increased mucous secretion.

*__prostaglandin__ (pross´´-tah-glan´-din) From prostate gland. The substance was originally isolated from semen.

*__bradykinin__ (brad´´-ee-kye´-nin) Gr. *bradys,* slow, and *kinein,* to move.

MICROFILE 17.1 OF WHAT VALUE IS ALLERGY?

Why would humans and other mammals evolve an allergic response that is capable of doing so much harm and even causing death? It is unlikely that this limb of immunity exists merely to make people miserable; it must have a role in protection and survival. What are the underlying biological functions of IgE, mast cells, and the array of potent cytokines? Analysis has revealed that, although allergic persons have high levels of IgE, trace quantities are present even in the sera of nonallergic individuals, just as mast cells and inflammatory cytokines are also part of normal human physiology. It is generally believed that one important function of this system is to defend against helminth worms that are ubiquitous human parasites. In chapter 14, we learned that cytokines serve valuable inflammatory functions, such as increasing blood flow and vascular permeability to summon essential immune components to an injured site. They are also responsible for increased mucous secretion, gastric motility, sneezing, and coughing, which help expel noxious agents. The difference is that, in allergic persons, the quantity and quality of these reactions are excessive and uncontrolled.

Specific Diseases Associated with IgE- and Mast Cell–Mediated Allergy

The mechanisms just described are basic to hay fever, allergic asthma, food allergy, drug allergy, eczema, and anaphylaxis. In this section, we cover the main characteristics of these conditions, followed by methods of detection and treatment.

Atopic Diseases

Hay fever is a generic term for **allergic rhinitis,*** a seasonal reaction to inhaled plant pollen or molds, or a chronic, year-round reaction to a wide spectrum of airborne allergens or inhalants (see table 17.2). The targets are typically respiratory membranes, and the symptoms include nasal congestion; sneezing; coughing; profuse mucous secretion; itchy, red, and teary eyes; and mild bronchoconstriction.

Asthma* is a respiratory disease characterized by episodes of impaired breathing due to severe bronchoconstriction. The airways of asthmatic people are exquisitely responsive to minute amounts of inhalant allergens, food, or other stimuli, such as infectious agents. The symptoms of asthma range from occasional, annoying bouts of difficult breathing to fatal suffocation. Labored breathing, shortness of breath, wheezing, cough, and ventilatory *rales** are present to one degree or another. The respiratory tract of an asthmatic person is chronically inflamed and severely overreactive to allergic cytokines, especially leukotrienes and serotonin from pulmonary mast cells. Other pathologic components are thick mucous plugs in the air sacs and lung damage that can result in long-term respiratory compromise (see figure 17.4). An imbalance in the nervous control of the respiratory smooth muscles is apparently involved in asthma, and the episodes are influenced by the psychological state of the person, which strongly supports a neurological connection.

The number of asthma sufferers in the United States is estimated at 10 million, with nearly one-third of them children. For reasons that are not completely understood, asthma is on the increase, and deaths from it have doubled since 1982, even though effective agents to control it are more available now than they have ever been before. A recent study of inner-city children has correlated high levels of asthma to contact with cockroach antigens in their living quarters. Nearly 40% of children age 10 or younger showed extreme sensitivity to the droppings and remains of these insects.

Atopic dermatitis is an intensely itchy inflammatory condition of the skin, sometimes also called **eczema.*** Sensitization occurs through ingestion, inhalation, and, occasionally, skin contact with allergens. It usually begins in infancy with reddened, vesicular, weeping, encrusted skin lesions. It then progresses in childhood and adulthood to a dry, scaly, thickened skin condition (figure 17.5). Lesions can occur on the face, scalp, neck, and inner surfaces of the limbs and trunk. The itchy, painful lesions cause considerable discomfort, and they are often predisposed to secondary bacterial infections. An anonymous writer once aptly described eczema as "the itch that rashes" or "one scratch is too many but one thousand is not enough."

Food Allergy

The ordinary diet contains a vast variety of compounds that are potentially allergenic. It is generally believed that food allergies are due to a digestive product of the food or to an additive (preservative or flavoring). Although the mode of entry is intestinal, food allergies can also affect the skin and respiratory tract. Gastrointestinal symptoms include vomiting, diarrhea, and abdominal pain. In severe cases, nutrients are poorly absorbed, leading to growth retardation and failure to thrive in young children. Other manifestations of food allergies include eczema, hives, rhinitis, asthma, and occasionally, anaphylaxis. Classic food hypersensitivity involves IgE and degranulation of mast cells, but not all reactions involve this mechanism (see microfile 17.2). The most common food allergens come from peanuts, fish, cow's milk, eggs, shellfish, and soybeans.

Drug Allergy

Modern chemotherapy has been responsible for many medical advances. Unfortunately, it has also been hampered by the fact that drugs are foreign compounds capable of stimulating allergic

*rhinitis (rye-nye´-tis) Gr. *rhis,* nose, and *itis,* inflammation.

*asthma (az´-muh) The Greek word for gasping.

*rales (rails) Abnormal breathing sounds.

*eczema (eks´-uh-mah; also ek-zeem´-uh) Gr. *ekzeo,* to boil over.

MICROFILE 17.2 "PLEASE HOLD THE ADDITIVES"

A number of adverse reactions to food additives can provoke the symptoms of allergy. In some cases, a true allergy to the substance exists; in others, the reaction is an unusual intolerance of unknown origin.

Sodium metabisulfite (sulfite) is sometimes added to wine to prevent spoilage or to vegetables to prevent browning. Sensitive persons exposed to this agent undergo severe, asthmalike attacks and occasionally even anaphylaxis. "Hot dog headache" is experienced by some persons after ingesting nitrate and nitrite preservatives commonly used in processed meats (hot dogs, bacon, sausages). Tartrazine yellow dye 5, used to color foods, pills, tablets, and capsules can cause symptoms of asthma, rhinitis, or hives in sensitive people. Monosodium glutamate flavor enhancer is responsible for the so-called Chinese restaurant syndrome, a sensation of burning, tightness, or numbness in the chest, neck, and face that begins shortly after the first few bites of food and lasts 2 to 3 hours.

Food sensitivity can also arise from microbial contaminants. Poisoning from negligently handled and stored seafood has been traced to bacterial contaminants that produce histamine-like compounds. Certain plants naturally contain small amounts of active compounds that exert a pharmacologic effect. For example, coffee has caffeine, tea has theophylline, and cocoa has theobromine. In larger quantities, these compounds can produce insomnia, headache, nervousness, tachycardia, nausea, abdominal pain, and diarrhea. Some fresh fruits and vegetables and processed foods (chocolate, cheese, and dried meats) contain vasoactive amines (histamine, epinephrine, norepinephrine) that can cause headache and other symptoms in sensitive individuals.

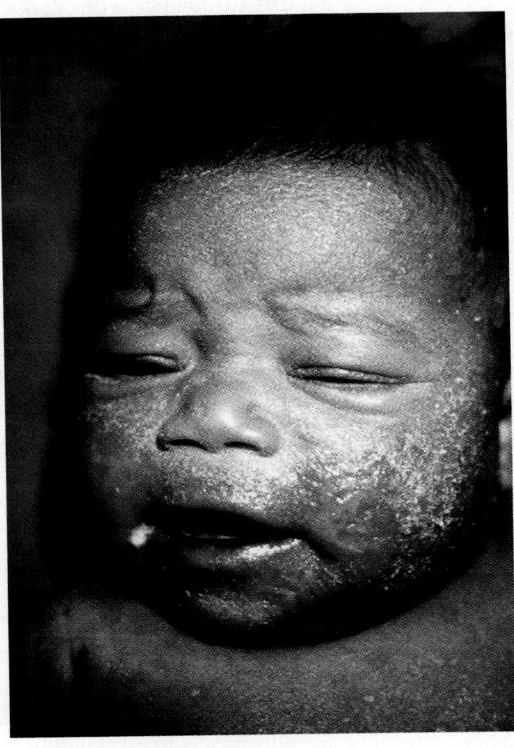

Figure 17.5
Atopic dermatitis, or eczema. Vesicular, encrusted lesions are typical in afflicted infants. This condition is prevalent enough to account for 1% of pediatric care.

reactions. In fact, allergy to drugs is one of the most common side effects of treatment (present in 5–10% of hospitalized patients). Depending upon the allergen, route of entry, and individual sensitivities, virtually any tissue of the body can be affected, and reactions range from mild atopy to fatal anaphylaxis. Compounds implicated most often are antibiotics (penicillin is number one in prevalence), synthetic antimicrobics (sulfa drugs), aspirin, opiates, and anaesthetics. The actual allergen is not the intact drug itself but a hapten given off when the liver processes the drug. Penicillin allergy has also been traced to contamination of meat, milk, and other foods and to exposure to *Penicillium* mold in the environment.

Anaphylaxis: An Overpowering Systemic Reaction

The term **anaphylaxis** or **anaphylactic shock** was first used to denote a reaction of animals injected with a foreign protein. Although the animals showed no response during the first contact, upon reinoculation with the same protein at a later time, they exhibited acute symptoms—itching, sneezing, difficult breathing, prostration, and convulsions—and many died in a few minutes. Two clinical types of anaphylaxis are distinguished in humans. *Cutaneous anaphylaxis* is the wheal and flare inflammatory reaction to the local injection of allergen. *Systemic anaphylaxis,* on the other hand, is characterized by sudden respiratory and circulatory disruption that can be fatal in a few minutes. In humans, the allergen and route of entry are variable, though bee stings and injections of antibiotics or serum are implicated most often. Bee venom is a complex material containing several allergens and enzymes that can create a sensitivity that can last for decades after exposure.

The underlying physiological events in systemic anaphylaxis parallel those of atopy, but the concentration of cytokines and the strength of the response are greatly amplified. The immune system of a sensitized person exposed to a provocative dose of allergen responds with a sudden, massive outpouring of allergic cytokines into the tissues and blood, which act rapidly on the target organs. Anaphylactic persons have been known to die in 15 minutes from complete airway blockage.

Diagnosis of Allergy

Because allergy mimics infection and other conditions, it is important to determine if a person is actually allergic. If possible or necessary, it is also helpful to identify the specific allergen or allergens. Allergy diagnosis involves several levels of tests, including nonspecific, specific, *in vitro,* and *in vivo* methods.

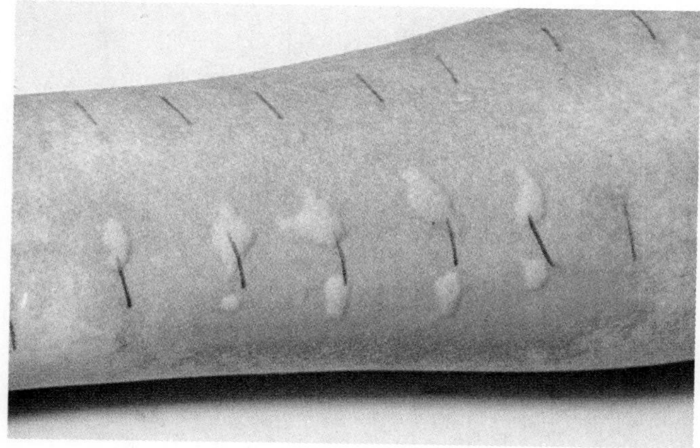

(a)

Figure 17.6
A method for conducting an allergy skin test. The forearm (or back) is mapped and then injected with a selection of allergen extracts. The allergist must be very aware of potential anaphylaxis attacks triggered by these injections. (*a*) Close-up of skin wheals showing a number of positive reactions (dark lines are measurer's marks). (*b*) Standardized skin tests for some common environmental allergens with a legend for assessing them.

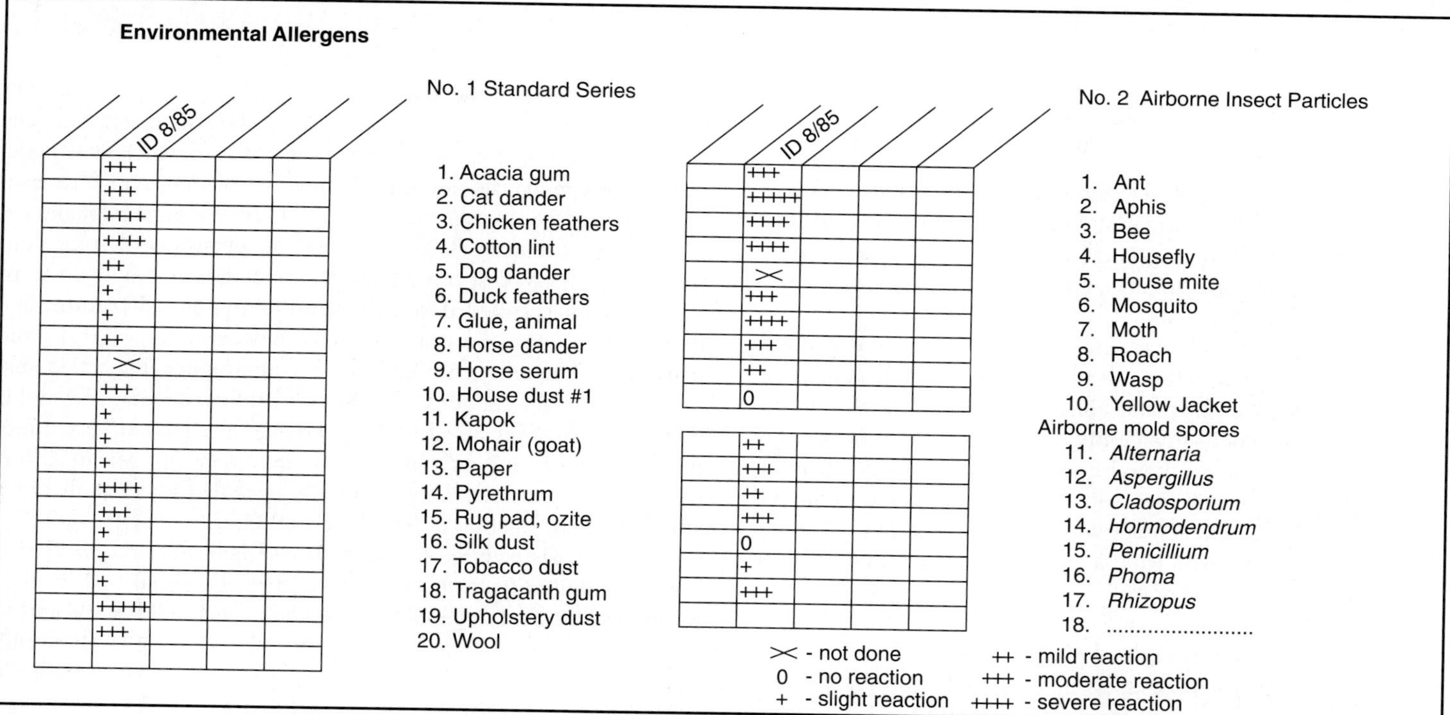

(b)

A new test that can distinguish whether a patient has experienced an allergic attack measures elevated blood levels of tryptase, an enzyme released by mast cells that increases during an allergic response. Several types of specific *in vitro* tests can determine the allergic potential of a patient's blood sample. The leukocyte histamine-release test measures the amount of histamine released from the patient's basophils when exposed to a specific allergen. Serological tests that use radioimmune assays (see chapter 16) to reveal the quantity and quality of IgE are also clinically helpful.

Skin Testing

A useful *in vivo* method to detect precise atopic or anaphylactic sensitivities is skin testing. With this technique, a patient's skin is injected, scratched, or pricked with a small amount of a pure allergen extract. Hundreds of these allergen extracts are obtainable from pharmaceutical companies, including common airborne allergens (plant and mold pollen) and more unusual allergens (mule dander, theater dust, parakeet feathers). Unfortunately, skin tests for food allergies using food extracts are unreliable in most cases. In patients with numerous allergies, the allergist maps the skin on the inner aspect of the forearms or back and injects the allergens intradermally according to this predetermined pattern (figure 17.6*a*). Approximately 20 minutes after antigenic challenge, each site is appraised for a wheal response indicative of histamine release. The diameter of the wheal is measured and rated on a scale of 0 (no reaction) to 4+ (greater than 15 mm). Figure 17.6*b* shows skin test results for a person with extreme inhalant allergies.

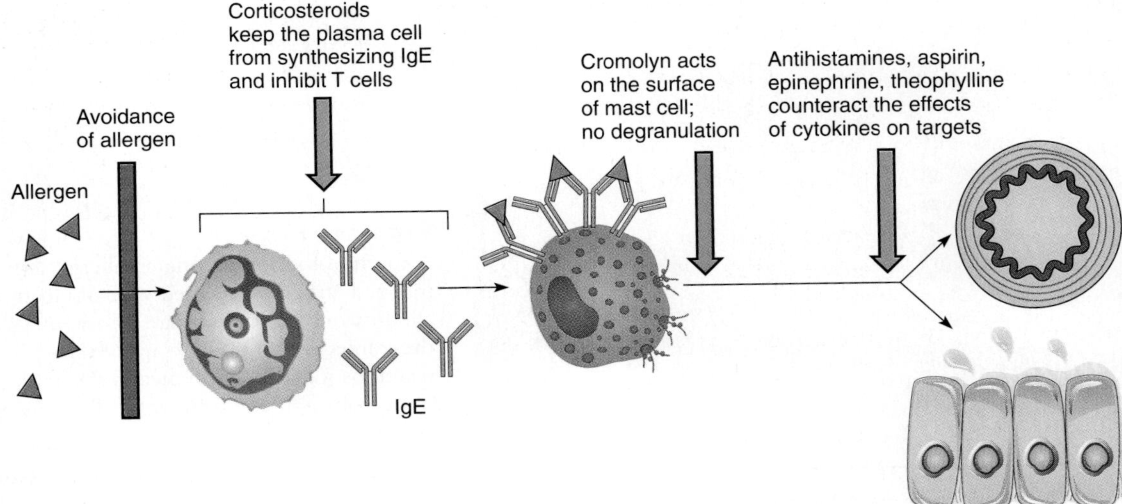

Figure 17.7
Strategies for circumventing allergic attacks.

Treatment and Prevention of Allergy

In general, the methods of treating and preventing type I allergy involve (1) avoiding the allergen, though this may be very difficult in many instances; (2) taking drugs that block the action of lymphocytes, mast cells, or cytokines; and (3) undergoing desensitization therapy.

It is not possible to completely prevent initial sensitization, since there is no way to tell in advance if a person will develop an allergy to a particular substance. The practice of delaying the introduction of solid foods apparently has some merit in preventing food allergies in children, though even breast milk can contain allergens ingested by the mother. Although rigorous cleaning and air conditioning can reduce contact with airborne allergens, it is not feasible to isolate a person from all allergens, which is the reason drugs are so important in control.

Therapy to Counteract Allergies

The aim of antiallergy medication is to block the progress of the allergic response somewhere along the route between IgE production and the appearance of symptoms (figure 17.7). Oral anti-inflammatory drugs such as corticosteroids inhibit the activity of lymphocytes and thereby reduce the production of IgE, but they also have dangerous side effects and should not be taken for prolonged periods. In addition, some drugs block the degranulation of mast cells and reduce the levels of inflammatory cytokines. The most effective of these are diethylcarbamazine and cromolyn. Asthma sufferers can find relief from two new drugs that block synthesis of leukotriene and prevent its attachment to target cell membranes.

Widely used medications for preventing symptoms of atopic allergy are *antihistamines,* the active ingredients in most over-the-counter allergy-control drugs. Antihistamines interfere with histamine activity by binding to histamine receptors on target organs. Most of them have major side effects, however, such as drowsiness. Newer antihistamines lack this side effect because they do not cross the blood-brain barrier. Other drugs that relieve inflammatory symptoms are aspirin and acetaminophen, which reduce pain by interfering with prostaglandin, and theophylline, a bronchodilator that reverses spasms in the respiratory smooth muscles. Persons who suffer from anaphylactic attacks are urged to carry at all times injectable epinephrine (adrenaline) and an identification tag indicating their sensitivity. An aerosol inhaler containing epinephrine can also provide rapid relief. Epinephrine reverses constriction of the airways and slows the release of allergic cytokines.

Approximately 70% of allergic patients benefit from controlled injections of specific allergens as determined by skin tests. This technique, called **desensitization** or **hyposensitization,** is a therapeutic way to prevent reactions between allergen, IgE, and mast cells. The allergen preparations contain pure, preserved suspensions of plant antigens, venoms, dust mites, dander, and molds (but so far, hyposensitization for foods has not proved very effective). The immunologic basis of this treatment is open to differences in interpretation. One theory suggests that injected allergens stimulate the formation of high levels of allergen-specific IgG (figure 17.8). It has been proposed that these IgG *blocking antibodies* remove allergen from the system before it can bind to IgE, thus preventing the degranulation of mast cells. Other evidence points to the possibility that allergen delivered in this fashion combines with IgE and prevents it from reacting with the mast cells. It has been further postulated that the therapy induces specific clones of suppressor T cells that block the production of IgE by B cells.

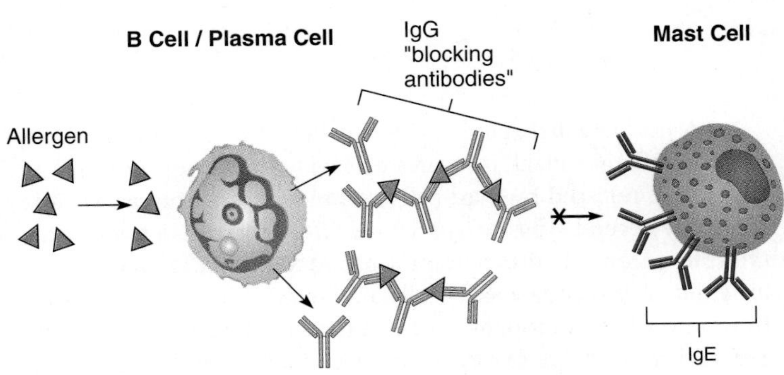

Figure 17.8
The blocking antibody theory for allergic desensitization. An injection of allergen causes IgG antibodies to be formed instead of IgE; these blocking antibodies cross-link and effectively remove the allergen before it can react with the IgE in the mast cell.

Chapter Checkpoints

Type I hypersensitivity reactions result from excessive IgE production in response to an exogenous antigen.

The two kinds of type I hypersensitivities are atopy, a chronic, local allergy, and anaphylaxis, a systemic, potentially fatal allergic response.

The predisposition to type I hypersensitivities is inherited, but age, geographic locale, and infection also influence allergic response.

Type I allergens include inhalants, ingestants, injectants, and contactants.

The portals of entry for type I antigens are the skin, respiratory tract, gastrointestinal tract, and genitourinary tract.

Type I hypersensitivities are set up by a **sensitizing dose** of allergen and expressed when a second **provocative dose** triggers the allergic response. The time interval between the two can be many years.

The primary participants in type I hypersensitivities are IgE, basophils, mast cells, and agents of the inflammatory response.

Allergies are diagnosed by a variety of *in vitro* and *in vivo* tests that assay specific cells, IgE, and local reactions.

Allergies are treated by medications that interrupt the allergic response at certain points. Allergic reactions can often be prevented by desensitization therapy.

TYPE II HYPERSENSITIVITIES: REACTIONS THAT LYSE FOREIGN CELLS

The diseases termed type II hypersensitivities are a complex group of syndromes that involve complement-assisted destruction (lysis) of cells by antibodies (IgG and IgM) directed against those cells' surface antigens. This category includes transfusion reactions and some types of autoimmunities (discussed in a later section). The cells targeted for destruction are often red blood cells, but other cells can be involved.

HUMAN BLOOD TYPES

Chapters 14 and 15 described the functions of unique surface receptors or markers on cell membranes. Ordinarily, these receptors play essential roles in transport, recognition, and development, but they become medically important when the tissues of one person are placed into the body of another person. Blood transfusions and organ donations introduce isoantigens (molecules that differ in the same species) on donor cells that are recognized by the lymphocytes of the recipient. These reactions are not really immune dysfunctions as allergy and autoimmunity are. The immune system is in fact working normally, but it is not equipped to distinguish between the desirable foreign cells of a transplanted tissue and the undesirable ones of a microbe.

TABLE 17.3

CHARACTERISTICS OF ABO BLOOD GROUPS

Genotype	Phenotype A or B* RBC Antigen	Prevalence in Population**	Serum Content of Antibodies
OO	Neither	Most common	Both anti-a and anti-b
AA, AO	A	Second most common	Anti-b
BB, BO	B	Third most common	Anti-a
AB	AB	Least common	Neither antibody

*Capital letters generally denote antigen; lowercase denotes antibody.
**True of most large populations of mixed racial and ethnic groups.

THE BASIS OF HUMAN ABO ISOANTIGENS AND BLOOD TYPES

The existence of human blood types was first demonstrated by an Austrian pathologist, Karl Landsteiner, in 1904. While studying incompatibilities in blood transfusions, he found that the serum of one person could clump the red blood cells of another. Landsteiner identified four distinct types, subsequently called the ABO blood groups.

Like the MHC antigens on white blood cells, the ABO isoantigen markers on red blood cells are genetically determined and composed of glycoproteins. These ABO antigens are inherited as two (one from each parent) of three alternative *alleles:** A, B, or O. A and B alleles are dominant over O and codominant with one another. As table 17.3 indicates, this mode of inheritance gives rise to four blood types (phenotypes), depending on the particular combination of genes. Thus, a person with an *AA* or *AO* genotype has **type A** blood; genotype *BB* or *BO* gives **type B;** genotype *AB* produces **type AB;** and genotype *OO* produces **type O.** Some important points about the blood types are: (1) They are named for the dominant antigen(s); (2) the RBCs of type O persons have antigens, but not A and B antigens; and (3) tissues other than RBCs carry A and B antigens. Aspects of the molecular biology of the RBC markers are discussed in microfile 17.3.

ANTIBODIES AGAINST A AND B ANTIGENS

Although an individual does not normally produce antibodies in response to his or her own RBC antigens, the serum can contain antibodies that react with blood of another antigenic type, even though contact with this other blood type has *never* occurred. These preformed antibodies account for the immediate and intense quality of transfusion reactions. As a rule, type A blood contains antibodies (anti-b) that react against the B antigens on type B and

*allele (ah-leel´) Gr. *allelon*, of one another. An alternate form of a gene for a given trait.

MICROFILE 17.3 THE ORIGIN OF ABO ANTIGENS

The A and B genes each code for an enzyme that adds a terminal carbohydrate to RBC receptors during maturation. RBCs of type A contain an enzyme that adds N-acetylgalactosamine to the receptor; RBCs of type B have an enzyme that adds D-galactose; RBCs of type AB contain both enzymes that add both carbohydrates; and RBCs of type O lack the genes and enzymes to add a terminal molecule.

The genetics of ABO antigens were once used to rule out paternity. For example, if a man is type A, the mother type O, and the child type B, we know this man could not have fathered this child. However, this same logic cannot prove paternity. If the child is type A instead, it is possible for the man to be the father, but so could some other man with blood type A. Highly sensitive methods based on specific and variable MHC antigens and DNA fingerprinting have been developed to gather more precise evidence of paternity or maternity (in cases of kidnapping or adoption, for instance).

The genetic/molecular basis for the A and B antigens (receptors) on red blood cells. In general, persons with blood types A, B, and AB inherit a gene for the enzyme that adds a certain terminal sugar to the basic RBC receptor. Type O persons do not have such an enzyme and lack the terminal sugar.

AB red blood cells. Type B blood contains antibodies (anti-a) that react with A antigen on type A and AB red blood cells. Type O blood contains antibodies against both A and B antigens. Type AB blood does not contain antibodies against either A or B antigens[1] (table 17.3). What is the source of these anti-a and anti-b antibodies? It appears that they develop in early infancy because of exposure to certain heterophile antigens that are widely distributed in nature. These antigens are surface molecules on bacteria and plant cells that mimic the structure of A and B isoantigens. Exposure to these sources stimulates the production of corresponding antibodies.[2]

Clinical Concerns in Transfusions

The presence of ABO antigens and a, b antibodies underlie several clinical concerns in giving blood transfusions. First, the individual blood types of donor and recipient must be determined. By use of a standard technique, drops of blood are mixed with antisera that contain antibodies against the A and B antigens and are then observed for the evidence of agglutination (figure 17.9).

Knowing the blood types involved makes it possible to determine which transfusions are safe to do. The general rule of compatibility is that the RBC antigens of the donor must not be agglutinated by antibodies in the recipient's blood (figure 17.10). The ideal practice is to transfuse blood that is a perfect match (A to A, B to B). But even in this event, blood samples must be cross-matched before the transfusion because other blood group incompatibilities can exist. This test involves mixing the blood of the donor with the serum of the recipient to check for agglutination.

Under certain circumstances (emergencies, the battlefield), the concept of universal transfusions can be used. To appreciate how this works, we must apply the rule stated in the previous paragraph. Type O blood lacks A and B antigens and will not be agglutinated by other blood types, so it could theoretically be used in any transfusion. Hence, a person with this blood type is called a *universal donor.* Because type AB blood lacks agglutinating antibodies, an individual with this blood could conceivably receive any type of blood. Type AB persons are consequently called *universal recipients.* Although both types of transfusions involve antigen-antibody incompatibilities, these are of less concern because of the dilution of the donor's blood in the body of the recipient. Additional RBC markers that can be significant in transfusions are the Rh, MN, and Kell antigens (see next sections).

1. Why would this be true? The answer lies in the first sentence of the paragraph.
2. Evidence comes from germ-free chickens which do not have antibodies against the isoantigens of blood types, whereas normal chickens possess these antibodies.

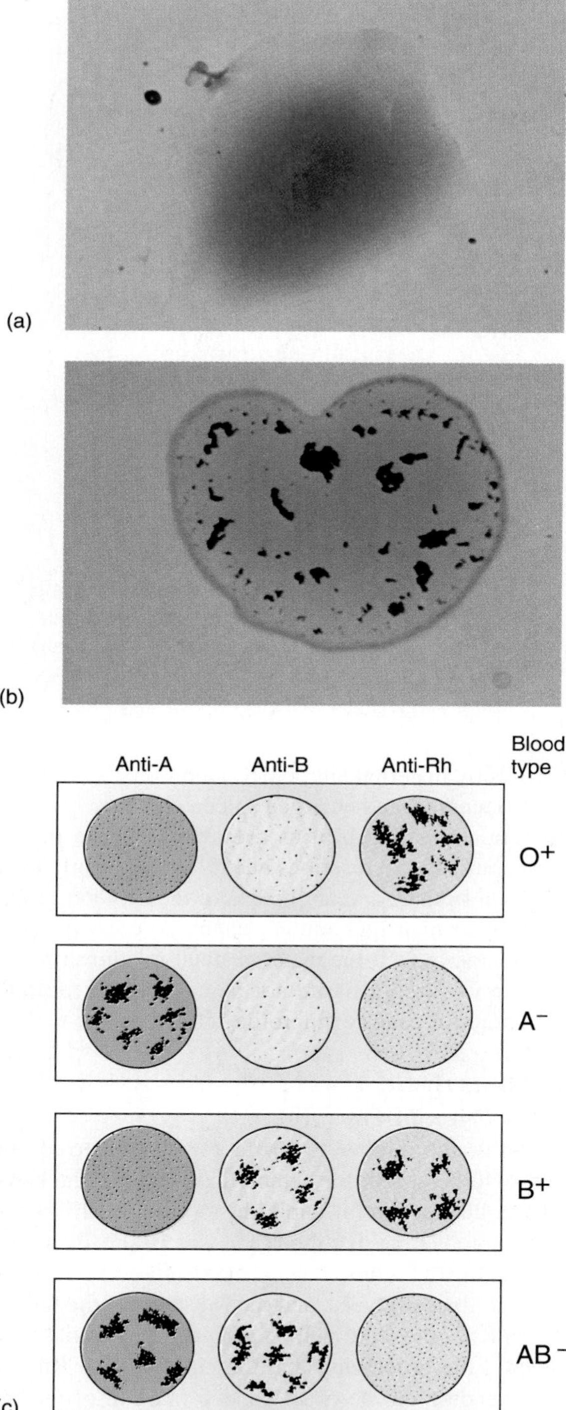

Figure 17.9

Interpretation of blood typing. In this test, a drop of blood is mixed with a specially prepared antiserum known to contain antibodies against the A, B, or Rh antigens. (*a*) If that particular antigen is not present, the red blood cells in that droplet do not agglutinate and form an even suspension. (*b*) If that antigen is present, agglutination occurs and the RBCs form visible clumps. (*c*) Several patterns and their interpretations. Anti-A, anti-B, and anti-Rh are shorthand for the antiserum applied to the drops. (In general, O^+ is the most common blood type, and AB^- is the rarest.)

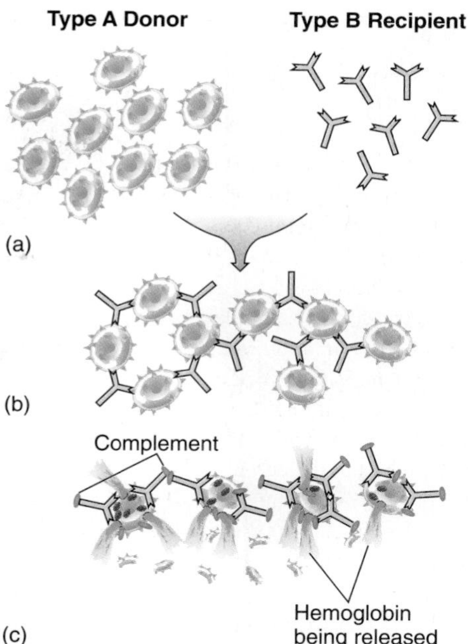

Figure 17.10

Microscopic view of a transfusion reaction. (*a*) Incompatible blood. The red blood cells of the type A donor contain antigen A, while the serum of the type B recipient contains anti-a antibodies that can agglutinate donor cells. (*b*) Agglutination particles can block the circulation in vital organs. (*c*) Activation of the complement by antibody on the RBCs can cause hemolysis and anemia. This sort of incorrect transfusion is very rare because of the great care taken by blood banks to ensure a correct match.

Transfusion of the wrong blood type causes various degrees of adverse reaction. The severest reaction is massive hemolysis when the donated red blood cells react with recipient antibody and trigger the complement cascade (figure 17.10). The resultant destruction of red cells leads to systemic shock and kidney failure brought on by the blockage of glomeruli (blood filtering apparatus) by cell debris. Death is a common outcome. Other reactions caused by RBC destruction are fever, anemia, and jaundice. A transfusion reaction is managed by immediately halting the transfusion, administering drugs to remove hemoglobin from the blood, and beginning another transfusion with red blood cells of the correct type.

The Rh Factor and Its Clinical Importance

Another RBC isoantigen of major clinical concern is the **Rh factor** (or D antigen). This factor was first discovered in experiments exploring the genetic relationships among animals. Rabbits inoculated with the RBCs of rhesus monkeys produced an antibody that also reacted with human RBCs. Further tests showed that this monkey antigen (termed Rh for rhesus) was present in about 85% of humans and absent in the other 15%. The details of Rh inheritance are more complicated than those of ABO, but in simplest terms, a person's Rh type results from a combination of two possible alleles—a dominant one that produces the factor and a recessive one that does not. A person inheriting at least one Rh gene

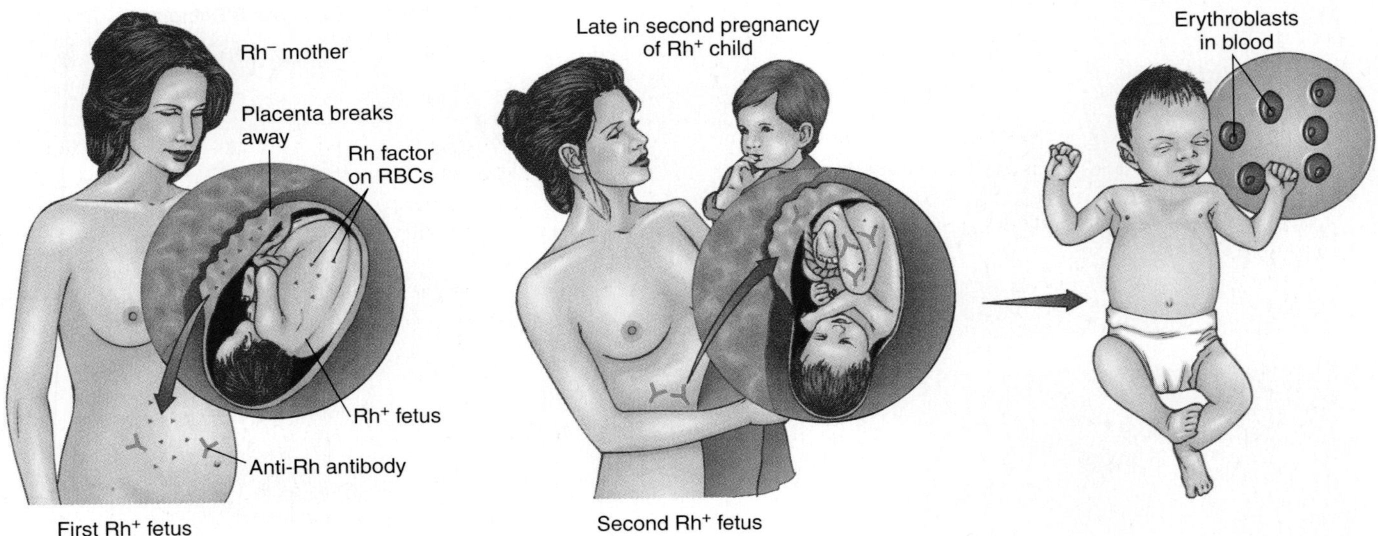

Figure 17.11
The development and aftermath of Rh sensitization. Initial sensitization of the maternal immune system to fetal Rh factor occurs during delivery, when the placenta tears away. The child will escape hemolytic disease in most instances, but the mother, now sensitized, will be capable of an immediate reaction to a second Rh^+ fetus and its Rh factor antigen. At that time, the mother's anti-Rh antibodies pass into the fetal circulation and elicit severe hemolysis in the fetus and neonate.

will be Rh^+; only those persons inheriting two recessive genes are Rh^-. This factor is denoted by a symbol above the blood type, as in O^+ or B^- (see figure 17.9*c*). However, unlike the ABO antigens, exposure to normal flora does not sensitize Rh^- persons to the Rh factor. The only ways one can develop antibodies against this factor are through placental sensitization or transfusion.

Hemolytic Disease of the Newborn and Rh Incompatibility

The potential for placental sensitization occurs when a mother is Rh^- and her unborn child is Rh^+. The obvious intimacy between mother and fetus makes it possible for fetal RBCs to leak into the mother's circulation during childbirth, when the detachment of the placenta creates avenues for fetal blood to enter the maternal circulation. The mother's immune system detects the foreign Rh factors on the fetal RBCs and is sensitized to them by producing antibodies and memory B cells. The first Rh^+ child is usually not affected because the process begins so late in pregnancy that the child is born before maternal sensitization is completed. However, the mother's immune system has been strongly primed for a second contact with this factor in a subsequent pregnancy (figure 17.11).

In the next pregnancy with an Rh^+ fetus, fetal blood cells escape into the maternal circulation late in pregnancy and elicit a memory response. The fetus is at risk when the maternal anti-Rh antibodies cross the placenta into the fetal circulation, where they affix to fetal RBCs and cause complement-mediated lysis. The outcome is a potentially fatal **hemolytic disease of the newborn** (HDN) called *erythroblastosis fetalis* (eh-rith″-roh-blas-toh′-sis fee-tal′-is). This term is derived from the presence of immature nucleated RBCs called erythroblasts in the blood. They are released into the infant's circulation to compensate for the massive destruction of RBCs by maternal antibodies. Additional symptoms are severe anemia, jaundice, and enlarged spleen and liver.

Maternal-fetal incompatibilities are also possible in the ABO blood group, but adverse reactions occur less frequently than with Rh sensitization because the antibodies to these blood group antigens are IgM rather than IgG and are unable to cross the placenta in large numbers. In fact, the maternal-fetal relationship is a fascinating instance of foreign tissue not being rejected, despite the extensive potential for contact (microfile 17.4).

Preventing Hemolytic Disease of the Newborn

Once sensitization of the mother to Rh factor has occurred, all other Rh^+ fetuses will be at risk for hemolytic disease of the newborn. Prevention requires a careful family history of an Rh^- pregnant woman. It can predict the likelihood that she is already sensitized or is carrying an Rh^+ fetus. It must take into account other children she has had, their Rh types, and the Rh status of the father. If the father is also Rh^-, the child will be Rh^- and free of risk, but if the father is Rh^+, the probability that the child will be Rh^+ is 50% or 100%, depending on the exact genetic makeup of the father. If there is any possibility that the fetus is Rh^+, the mother must be passively immunized with antiserum containing antibodies against the Rh factor (*Rh_o (D) immune globulin*, or *RhoGAM*).[3] This antiserum, injected at 28–32 weeks and again immediately after delivery, reacts with any fetal RBCs that have escaped into the maternal circulation, thereby preventing the sensitization of the mother's immune system to Rh factor (figure 17.12). RhoGAM must be given with each pregnancy that involves an Rh^+ fetus. It is

3. Immunoglobulin fraction of human anti-Rh serum, prepared from pooled human sera.

MICROFILE 17.4 WHY DOESN'T A MOTHER REJECT HER FETUS?

Think of it: Even though mother and child are genetically related, the father's genetic contribution guarantees that the fetus will contain molecules that are antigenic to the mother. In fact, with the recent practice of implanting one woman with the fertilized egg of another woman, the surrogate mother is carrying a fetus that has no genetic relationship to her. Yet, even with this essentially foreign body inside the mother, dangerous immunologic reactions such as Rh incompatibility are rather rare. In what ways do fetuses avoid the surveillance of the mother's immune system? The answer appears to lie in the placenta and embryonic tissues. The fetal components that contribute to these tissues are not strongly antigenic, and they form a barrier that keeps the fetus isolated in its own antigen-free environment. The placenta is coated with a thick layer that prevents the passage of maternal cells, and it can also function as a sponge to absorb, remove, and inactivate circulating antigens.

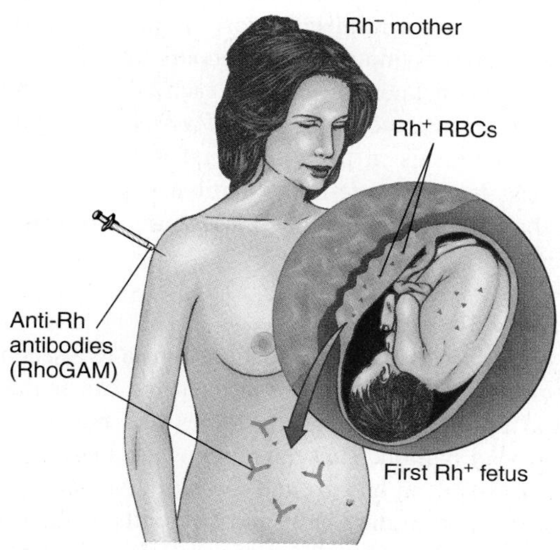

Figure 17.12
Prevention of erythroblastosis fetalis with anti-Rh immune globulin (RhoGAM). Injecting a mother who is at risk with RhoGAM during her first Rh$^+$ pregnancy helps to inactivate and remove the fetal Rh factor before her immune system can react with it and develop sensitivity.

ineffective if the mother has already been sensitized by a prior Rh$^+$ fetus or an incorrect blood transfusion, which can be determined by a serological test. As in ABO blood types, the Rh factor should be matched for a transfusion, although it is acceptable to transfuse Rh$^-$ blood if the Rh type is not known.

OTHER RBC ANTIGENS

Although the ABO and Rh systems are of greatest medical significance, about 20 other red blood cell isoantigen groups have been discovered. Examples are the *MN, Ss, Kell,* and *P* blood groups. Because of incompatibilities that these blood groups present, transfused blood is screened to prevent possible cross-reactions. The study of these blood antigens (as well as ABO and Rh) has given rise to other useful applications. For example, they can be useful in forensic medicine (crime detection), studying ethnic ancestry, and tracing prehistoric migrations in anthropology. Many blood cell antigens are remarkably hardy and can be detected in dried blood stains, semen, and saliva. Even the 2,000-year-old mummy of King Tutankhamen has been typed A_2MN!

Chapter Checkpoints

- Type II hypersensitivity reactions occur when preformed antibodies react with foreign cell–bound antigens. The most common type II reactions occur when transfused blood is mismatched to the recipient's ABO type. The donor cells are destroyed by complement fixation of host IgG and IgM to ABO antigen sites.
- Type II hypersensitivities are stimulated by antibodies formed against red blood cell (RBC) antigens or against other cell-bound antigens following prior exposure.
- Complement, IgG, and IgM antibodies are the primary cytokines of type II hypersensitivities.
- The concepts of universal donor (type O) and universal recipient (type AB) apply only under emergency circumstances. Cross-matching donor and recipient blood is necessary to determine which transfusions are safe to perform.
- Type II hypersensitivities can also occur when Rh$^-$ mothers are sensitized to Rh$^+$ RBCs of their unborn babies and the mother's anti-Rh antibodies cross the placenta, causing hemolysis of the newborn's RBCs. This is called hemolytic disease of the newborn, or erythroblastosis fetalis.

TYPE III HYPERSENSITIVITIES: IMMUNE COMPLEX REACTIONS

Type III hypersensitivity involves the reaction of soluble antigen with antibody and the deposition of the resulting complexes in basement membranes of epithelial tissue. It is similar to type II, because it involves the production of IgG and IgM antibodies after repeated exposure to antigens and the activation of complement. Type III differs from type II because its antigens are not attached to the surface of a cell. The interaction of these antigens with antibodies produces free-floating complexes that can be deposited in the tissues, causing an **immune complex reaction** or disease. This category includes therapy-related disorders (serum sickness and the Arthus reaction) and a number of autoimmune diseases (such as glomerulonephritis and lupus erythematosus).

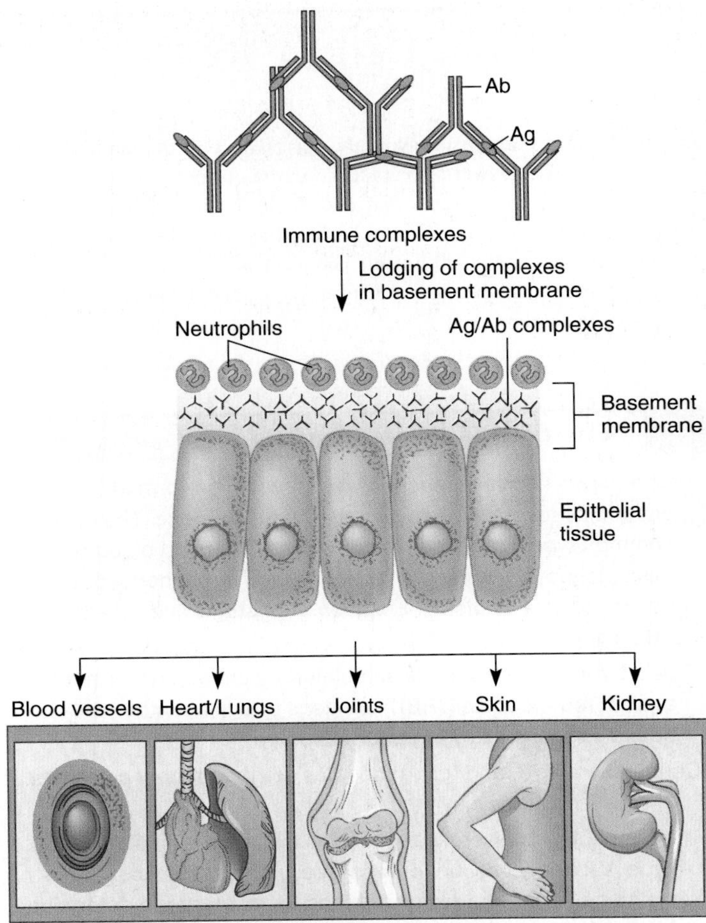

Figure 17.13

The background of immune complex disease. In general, circulating immune complexes become lodged in the basement membrane of the epithelia and cause vascular damage and organ malfunction.

Mechanisms of Immune Complex Disease

After initial exposure to a profuse amount of antigen, the immune system produces large quantities of antibodies that circulate in the fluid compartments. When this antigen enters the system a second time, it reacts with the antibodies to form antigen-antibody complexes (figure 17.13). These complexes summon various inflammatory components such as complement and neutrophils, which would ordinarily eliminate Ag-Ab complexes as part of the normal immune response. In an immune complex disease, however, these complexes are so abundant that they deposit in the basement membranes[4] of epithelial tissues and become inaccessible. In response to these events, neutrophils release lysosomal granules that digest tissues and cause a destructive inflammatory condition. The symptoms of type III hypersensitivities are due in great measure to this pathologic state.

Types of Immune Complex Disease

During the early tests of immunotherapy using animals, hypersensitivity reactions to serum and vaccines were common. In addition to anaphylaxis, two syndromes, the **Arthus reaction**[5] and **serum sickness,** were identified. These syndromes are associated with certain types of passive immunization (especially with animal serum).

Serum sickness and the Arthus reaction are like anaphylaxis in requiring sensitization and preformed antibodies. Characteristics that set them apart are: (1) They depend upon IgG, IgM, or IgA (precipitating antibodies) rather than IgE; (2) they require large doses of antigen (not a miniscule dose as in anaphylaxis); and (3) they have delayed symptoms (a few hours to days). The Arthus reaction and serum sickness differ from each other in some important ways. The Arthus reaction is a *localized* dermal injury due to inflamed blood vessels in the vicinity of any injected antigen. Serum sickness is a *systemic* injury initiated by antigen-antibody complexes that circulate in the blood and settle into membranes at various sites.

The Arthus Reaction

The Arthus reaction is usually an acute response to a second injection of vaccines (boosters) or drugs at the same site as the first injection. In a few hours, the area becomes red, hot to the touch, swollen, and very painful. These symptoms are mainly due to the destruction of tissues in and around the blood vessels and the release of histamine from mast cells and basophils. Although the reaction is usually self-limiting and rapidly cleared, intravascular blood clotting can occasionally cause necrosis and loss of tissue.

Serum Sickness

Serum sickness was named for a condition that appeared in soldiers after repeated injections of horse serum to treat tetanus. It can also be caused by injections of animal hormones and drugs. The immune complexes enter the circulation, are carried throughout the body, and are eventually deposited in blood vessels of the kidney, heart, skin, and joints (figure 17.13). The condition can become chronic, causing symptoms such as enlarged lymph nodes, rashes, painful joints, swelling, fever, and renal dysfunction.

An Inappropriate Response Against Self, or Autoimmunity

The immune diseases we have covered so far are all caused by foreign antigens. In the case of **autoimmunity,** an individual actually develops hypersensitivity to himself. This pathologic process accounts for **autoimmune diseases,** in which **autoantibodies** and, in certain cases, T cells mount an abnormal attack against self antigens. The scope of autoimmune diseases is extremely varied. In general, they can be differentiated as *systemic,* involving sev-

4. Basement membranes are basal partitions of epithelia that normally filter out circulating antigen-antibody complexes.

5. Named after Maurice Arthus, the physiologist who first identified this localized inflammatory response.

TABLE 17.4

SELECTED AUTOIMMUNE DISEASES

Disease	Target	Characteristics
Systemic lupus erythematosus (SLE)	Systemic	Inflammation of many organs; antibodies against red and white blood cells, platelets, clotting factors, nucleus
Rheumatoid arthritis and ankylosing spondylitis	Systemic	Vasculitis; frequent target is joint lining; antibodies against other antibodies (rheumatoid factor)
Scleroderma	Systemic	Excess collagen deposition in organs; antibodies formed against many intracellular organelles
Hashimoto's thyroiditis	Thyroid	Destruction of the thyroid follicles
Graves' disease	Thyroid	Antibodies against thyroid-stimulating hormone receptors
Pernicious anemia	Stomach lining	Antibodies against receptors prevent transport of vitamin B_{12}
Myasthenia gravis	Muscle	Antibodies against the acetylcholine receptors on the nerve-muscle junction alter function
Type I diabetes	Pancreas	Antibodies stimulate destruction of insulin-secreting cells
Type II diabetes	Insulin receptor	Antibodies block attachment of insulin
Multiple sclerosis	Myelin	T cells and antibodies sensitized to myelin sheath destroy neurons
Goodpasture's syndrome (glomerulonephritis)	Kidney	Antibodies to basement membrane of the glomerulus damage kidneys
Rheumatic fever	Heart	Antibodies to group A *Streptococcus* cross-react with heart tissue

eral major organs, or *organ-specific,* involving only one organ or tissue. They usually fall into the categories of type II or type III hypersensitivity, depending upon how the autoantibodies bring about injury. Some major autoimmune diseases, their targets, and basic pathology are presented in table 17.4.

Genetic and Gender Correlation in Autoimmune Disease

In most cases, the precipitating cause of autoimmune disease remains obscure, but we do know that susceptibility is determined by genetics and influenced by gender. Cases cluster in families, and even unaffected members tend to develop the autoantibodies for that disease. More direct evidence comes from studies of the major histocompatibility gene complex. Particular genes in the class I and II major histocompatibility complex (see figure 15.3) coincide with certain autoimmune diseases. For example, autoimmune joint diseases such as rheumatoid arthritis and ankylosing spondylitis are more common in persons with the B-27 HLA type; systemic lupus erythematosus, Graves' disease, and myasthenia gravis are associated with the B-8 HLA antigen. Why autoimmune diseases (except ankylosing spondylitis) afflict more females than males also remains a mystery. Females are more susceptible during childbearing years than before puberty or after menopause, suggesting a possible hormonal relationship.

The Origins of Autoimmune Disease

Very low titers of autoantibodies in otherwise healthy individuals suggest some normal function for them. A moderate, regulated amount of autoimmunity is probably required to dispose of old cells and cellular debris. Disease apparently arises when this regulatory or recognition apparatus goes awry. Attempts to explain the origin of autoimmunity include the following theories.

The *sequestered antigen theory* explains that during embryonic growth, some tissues are immunologically privileged; that is, they are sequestered behind anatomical barriers and cannot be scanned by the immune system (figure 17.14*a*). Examples of these sites are regions of the central nervous system, which are shielded by the meninges and blood-brain barrier; the lens of the eye, which is enclosed by a capsule; and antigens in the thyroid and testes, which are sequestered behind an epithelial barrier. Eventually the antigen becomes exposed by means of infection, trauma, or deterioration, and is perceived by the immune system as a foreign substance.

According to the *clonal selection theory,* the immune system of a fetus develops tolerance by eradicating all self-reacting lymphocyte clones, called *forbidden clones,* while retaining only those clones that react to foreign antigens. Some of these clones may survive, and since they have not been subjected to this tolerance process, they will attack tissues with self antigens.

The *theory of immune deficiency* proposes that mutations in the receptor genes of some lymphocytes render them reactive to self or that a general breakdown in the normal T-suppressor function sets the scene for inappropriate immune responses.

Some autoimmune diseases appear to be caused by *molecular mimicry,* in which microbial antigens bear molecular determinants similar to normal human cells. An infection could cause formation of antibodies that can cross-react with tissues. This is one purported explanation for the pathology of rheumatic fever.

Autoimmune disorders such as type I diabetes and multiple sclerosis are likely triggered by *viral infection.* Viruses can noticeably alter cell receptors, thereby causing immune cells to attack the tissues bearing viral receptors (figure 17.14*b*).

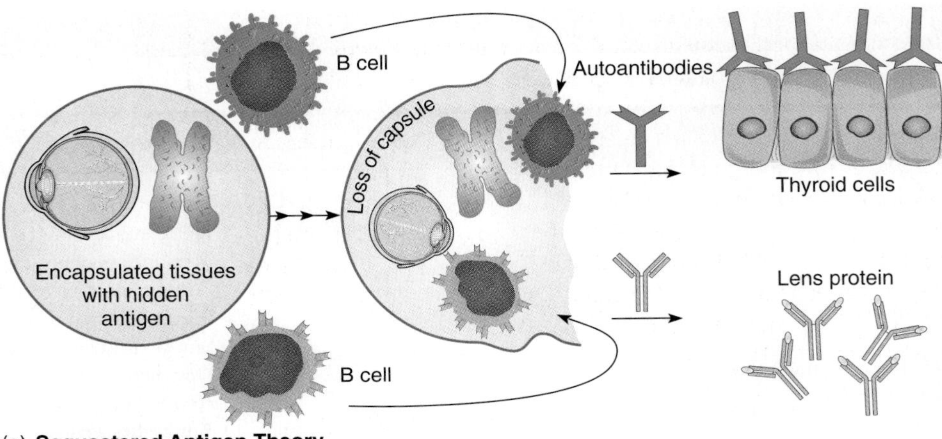

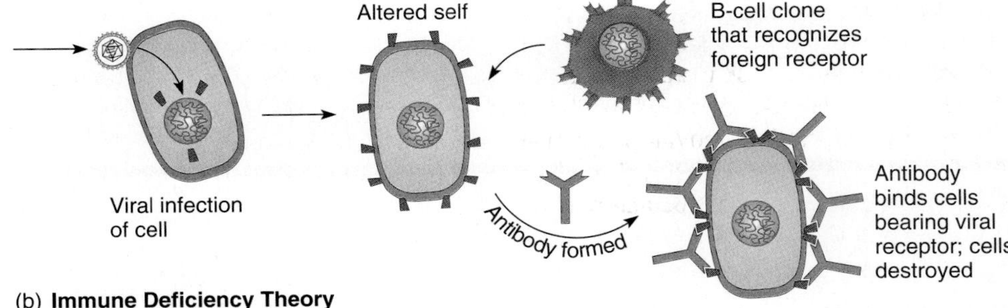

Figure 17.14
Possible explanations for autoimmunity. (*a*) Self is sequestered and is later incorrectly identified as an antigen by B lymphocytes. (*b*) Self is altered by viral infection, which causes an immune response against the perceived foreign cell.

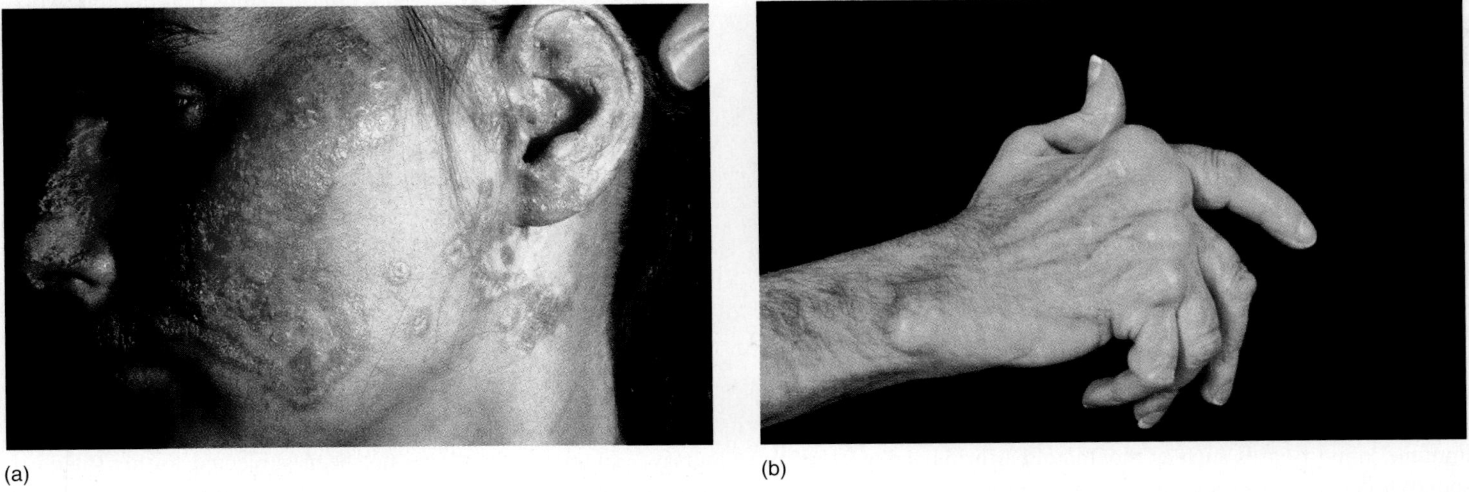

Figure 17.15
Common autoimmune diseases. (*a*) Systemic lupus erythematosus. One symptom is a prominent rash across the bridge of the nose and on the cheeks. These papules and blotches can also occur on the chest and limbs. (*b*) Rheumatoid arthritis commonly targets the synovial membrane of joints. Over time, chronic inflammation causes thickening of this membrane, erosion of the articular cartilage, and fusion of the joint. These effects severely limit motion and can eventually swell and distort the joints.

Examples of Autoimmune Disease

Systemic Autoimmunities One of the severest chronic autoimmune diseases is **systemic lupus erythematosus*** (SLE, or lupus). This name originated from the characteristic rash that spreads across the nose and cheeks in a pattern suggesting the appearance of a wolf (figure 17.15*a*). Although the manifestations of the disease vary considerably, all patients produce autoantibodies against a great variety of organs and tissues. The organs most involved are the kidneys, bone marrow, skin, nervous system, joints, muscles, heart, and GI tract. Antibodies to intracellular materials such as the nucleoprotein of the nucleus and mitochondria are also common.

*systemic lupus erythematosus (sis-tem´-ik loo´-pis air´´-uh-theem-uh-toh´-sis) L. *lupus,* wolf.

In SLE, autoantibody-autoantigen complexes appear to be deposited in the basement membranes of various organs. Kidney failure, blood abnormalities, lung inflammation, myocarditis, and skin lesions are the predominant symptoms. One form of chronic lupus (called discoid) is influenced by exposure to the sun and primarily afflicts the skin. The etiology of lupus is still a puzzle. It is not known how such a generalized loss of self-tolerance arises, though viral infection or loss of T-cell suppressor function are suspected. The fact that women of childbearing years account for 90% of cases indicates that hormones may be involved. The diagnosis of SLE can usually be made with blood tests. Antibodies against the nucleus (ANA) and various tissues (detected by indirect fluorescent antibody or radioimmune assay techniques) are common, and a positive test for the lupus factor (an anticoagulant factor) is also very indicative of the disease. Therapy involves anti-inflammatory drugs such as aspirin and cortisone.

Rheumatoid arthritis,* another systemic autoimmune disease, incurs progressive, debilitating damage to the joints. In some patients, the lung, eye, skin, and nervous system are also involved. In the joint form of the disease, autoantibodies form immune complexes that bind to the synovial membrane of the joints and activate phagocytes and stimulate release of cytokines. Chronic inflammation leads to scar tissue and joint destruction. The joints in the hands and feet are affected first, followed by the knee and hip joints (figure 17.15*b*). The precipitating cause in rheumatoid arthritis is not known, though infectious agents such as Epstein-Barr virus have been suspected. The most common feature of the disease is the presence of an IgM antibody, called rheumatoid factor (RF), directed against other antibodies. This does not cause the disease but is used mainly in diagnosis. Some relief can be achieved with anti-inflammatory agents, immunosuppressive drugs, and gold salt injections in some individuals.

Autoimmunities of the Endocrine Glands On occasion, the thyroid gland is the target of autoimmunity. The underlying cause of **Graves' disease** is the attachment of autoantibodies to receptors on the follicle cells that secrete the hormone thyroxin. The abnormal stimulation of these cells causes the overproduction of this hormone and the symptoms of hyperthyroidism. In **Hashimoto's thyroiditis,** both autoantibodies and T cells are reactive to the thyroid gland, but in this instance, they reduce the levels of thyroxin by destroying follicle cells and by inactivating the hormone. As a result of these reactions, the patient suffers from hypothyroidism.

The pancreas and its hormone, insulin, are other autoimmune targets. Insulin, secreted by the beta cells in the pancreas, regulates and is essential to the utilization of glucose by cells. **Diabetes mellitus** is caused by a dysfunction in insulin production or utilization (figure 17.16). Type I diabetes (also termed insulin-dependent diabetes) is associated with autoantibodies and sensitized T cells that damage the beta cells. A complex inflammatory reaction leading to lysis of these cells greatly reduces the amount of insulin secreted. In a form of type II diabetes (non-insulin-dependent diabetes), sufficient insulin is usually produced, but cellular receptors for insulin are reduced in number or availability. This reduction may be due to autoantibodies that compete with insulin for receptor binding sites.

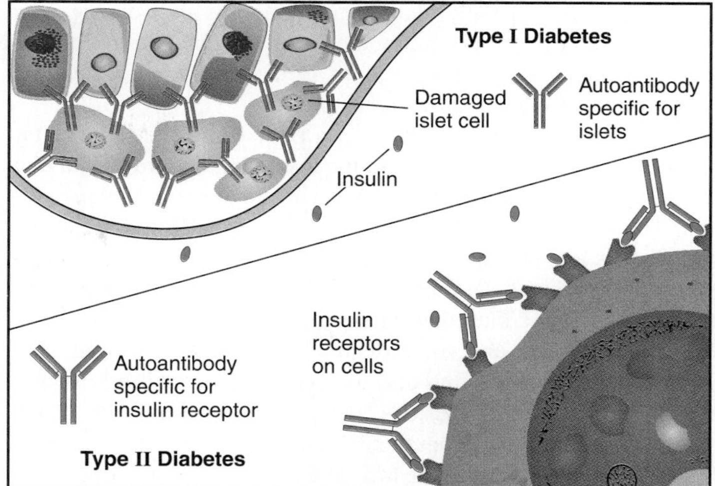

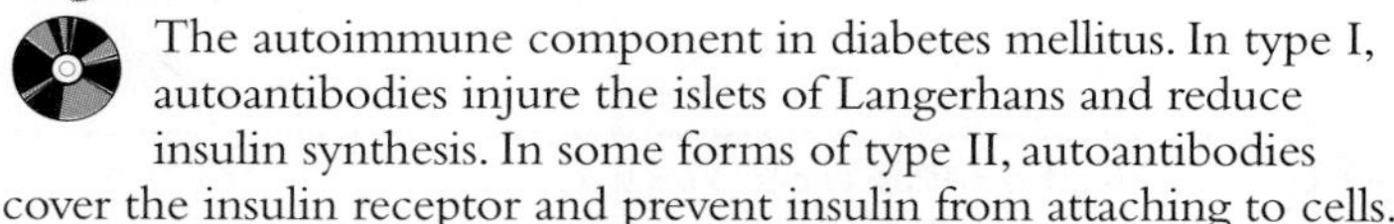

Figure 17.16

The autoimmune component in diabetes mellitus. In type I, autoantibodies injure the islets of Langerhans and reduce insulin synthesis. In some forms of type II, autoantibodies cover the insulin receptor and prevent insulin from attaching to cells.

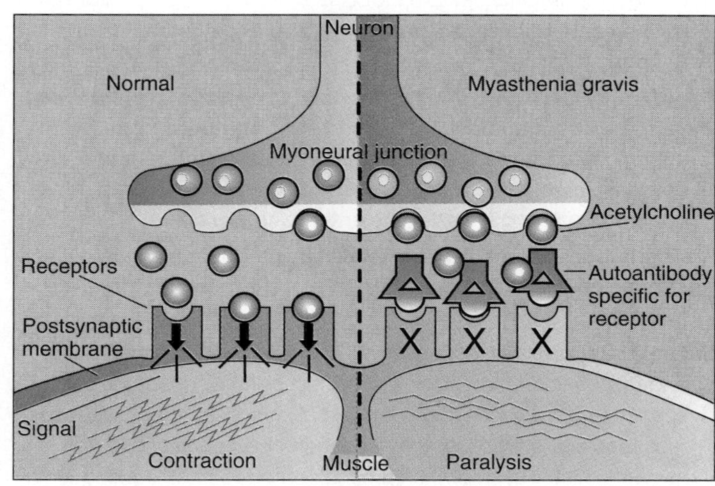

Figure 17.17

Proposed mechanisms for involvement of autoantibodies in myasthenia gravis. Antibodies developed against receptors on the post-synaptic membrane block them so that acetylcholine cannot bind and muscle contraction is inhibited.

Neuromuscular Autoimmunities **Myasthenia gravis*** is named for the pronounced muscle weakness that is its principal symptom. Although the disease afflicts all skeletal muscle, the first effects are usually felt in the muscles of the eyes and throat. Eventually, it can progress to complete loss of muscle function and death. The classic syndrome is caused by autoantibodies binding to the receptors for acetylcholine, a chemical required to transmit a nerve impulse across the synaptic junction to a muscle (figure 17.17). The immune attack so severely damages the muscle cell membrane that transmission is blocked and paralysis ensues.

---

*rheumatoid arthritis (roo´-muh-toyd ar-thry´tis) Gr. *rheuma,* a moist discharge, and *arthron,* joint.

*myasthenia gravis (my´´-us-thee´-nee-uh grah´-vis) Gr. *myo,* muscle, *astheneia,* weakness, and *gravida,* heavy.

Current treatment usually includes immunosuppressive drugs and therapy to remove the autoantibodies from the circulation. Experimental therapy using immunotoxins to destroy lymphocytes that produce autoantibodies shows some promise.

Multiple sclerosis* (MS) is a paralyzing neuromuscular disease associated with lesions in the insulating myelin sheath that surrounds neurons in the white matter of the central nervous system. The underlying pathology involves damage to the sheath by both T cells and autoantibodies that severely compromises the capacity of neurons to send impulses. The principal motor and sensory symptoms are muscular weakness and tremors, difficulties in speech and vision, and some degree of paralysis. Most MS patients first experience symptoms as young adults, and they tend to experience remissions (periods of relief) alternating with recurrences of disease throughout their lives. Convincing evidence from studies of the brain tissue of MS patients points to a strong connection between the disease and infection with human herpesvirus 6 (chapter 24). The disease can be treated passively with monoclonal antibodies that target T cells, and a vaccine containing the myelin protein has shown beneficial effects. Immunosuppressants such as cortisone and interferon B may also alleviate symptoms.

Chapter Checkpoints

- Type III hypersensitivities are induced when a profuse amount of antigen enters the system and results in large quantities of antibody formation.
- Type III hypersensitivity reactions occur when large quantities of antigen react with host antibody to form large, soluble immune complexes that settle in tissue cell membranes, causing chronic destructive inflammation. The reactions appear hours or days after the antigen challenge.
- The mediators of type III hypersensitivity reactions include soluble IgA, IgG, or IgM, and agents of the inflammatory response.
- Two kinds of type III hypersensitivities are **local** (Arthus) reactions and **systemic** (serum sickness). Arthus, or local, reactions occur at the site of injected drugs or booster immunizations. Systemic reactions occur when repeated antigen challenges cause systemic distribution of the immune complexes and subsequent inflammation of joints, lymph nodes, and kidney tubules.
- Autoimmune hypersensitivity reactions occur when autoantibodies or host T cells mount an abnormal attack against self antigens. Autoimmune antibody responses can be either local or systemic type II or type III hypersensitivity reactions. Autoimmune T-cell responses are type IV hypersensitivity reactions.
- Susceptibility to autoimmune disease appears to be influenced by gender and by genes in the MHC complex.
- Autoimmune disease may be an excessive response of a normal immune function, the appearance of sequestered antigens, "forbidden" clones of lymphocytes that react to self antigens, or the result of alterations in the immune response caused by infectious agents, particularly viruses.
- Examples of autoimmune diseases include systemic lupus erythematosus, rheumatoid arthritis, diabetes mellitus, myasthenia gravis, and multiple sclerosis.

TYPE IV HYPERSENSITIVITIES: CELL-MEDIATED (DELAYED) REACTIONS

The adverse immune responses we have covered so far are explained primarily by B-cell involvement and antibodies. But type IV hypersensitivity involves primarily the T-cell branch of the immune system. Type IV immune dysfunction has traditionally been known as delayed hypersensitivity because the symptoms arise one to several days following the second contact with an antigen. In general, type IV diseases result when T cells respond to self tissues or transplanted foreign cells. Examples of type IV hypersensitivity include delayed allergic reactions to infectious agents, contact dermatitis, and graft rejection.

Delayed-Type Hypersensitivity

Infectious Allergy

A classic example of a delayed-type hypersensitivity occurs when a person sensitized by tuberculosis infection is injected with an extract (tuberculin) of the bacterium *Mycobacterium tuberculosis.* The so-called **tuberculin reaction** is an acute skin inflammation at the injection site appearing within 24 to 48 hours. So useful and diagnostic is this technique for detecting present or prior tuberculosis that it is the chosen screening device (figure 17.18*a*). Other infections that use similar skin testing are leprosy, syphilis, histoplasmosis, toxoplasmosis, and candidiasis. This form of hypersensitivity arises from time-consuming cellular events involving the $\mathbf{T_D}$ class of cells. After these cells receive processed microbial antigens from macrophages, they release broad-spectrum cytokines that attract inflammatory cells to the site—particularly mononuclear cells, fibroblasts, and other lymphocytes. In a chronic infection (tertiary syphilis, for example), extensive damage to organs can occur through granuloma formation.

Contact Dermatitis

The most common delayed allergic reaction, **contact dermatitis,** is caused by exposure to resins in poison ivy or poison oak (microfile 17.5), to simple haptens in household and personal articles (jewelry, cosmetics, elasticized undergarments), and to certain drugs. Like immediate atopic dermatitis, the reaction to these allergens requires a sensitizing and a provocative dose. The allergen first penetrates the outer skin layers, is processed by Langerhans cells (skin macrophages), and is presented to T cells. When subsequent exposures attract lymphocytes and macrophages to this area, these cells give off enzymes and inflammatory cytokines that severely damage the epidermis in the immediate vicinity. This response accounts for the intensely itchy papules and blisters that are the early symptoms (figure 17.18*b*). As

*sclerosis (skleh-roh´-sis) Gr. *sklerosis,* hardness.

MICROFILE 17.5 PRETTY, PRICKLY, POISONOUS PLANTS

As a cause of allergic contact dermatitis (affecting about 10 million people a year), nothing can compare with a single family of beautiful but pesky plants. At least one of these plants—either poison ivy, poison oak, or poison sumac—flourishes in the forests, woodlands, or along the trails of most regions of America. The allergen in these plants, an oil called urushiol, has such extreme potency that a pinhead-sized amount could spur symptoms in 500 people, and it is so long-lasting that botanists must be careful when handling 100-year-old plant specimens. Although degrees of sensitivity vary among individuals, it is estimated that half of all Americans are potentially hypersensitive to this compound. Some people are so acutely sensitive that even the most miniscule contact, such as handling pets or clothes that have touched the plant or breathing vaporized urushiol, can trigger an attack.

Humans first become sensitized by contact during childhood. Individuals at great risk (firefighters, hikers) are advised to determine their degree of sensitivity using a skin test, so that they can be adequately cautious and prepared. Some odd remedies include skin potions containing bleach, buttermilk, ammonia, hair spray, and meat tenderizer. Drinking milk from goats that have grazed on poison ivy is believed by some to be an effective desensitizing method. Allergy researchers are capitalizing on this idea by testing oral vaccines containing a form of urushiol, which seem to work in experimental animals. An effective method using poison ivy desensitization injection is currently available to people with extreme sensitivity.

Poison ivy

Poison oak

Poison sumac

(a)

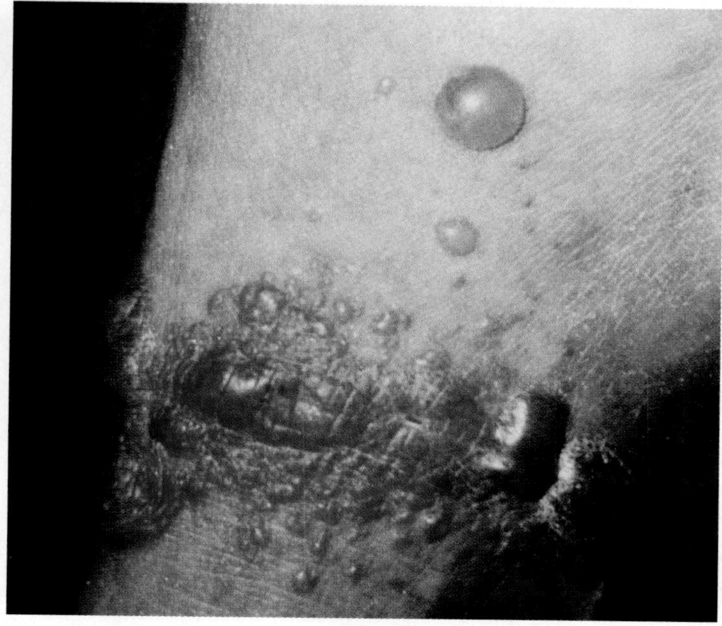

(b)

Figure 17.18

(*a*) Positive tuberculin test. Intradermal injection of tuberculin extract in a person sensitized to tuberculosis yields a slightly raised red bump greater than 10 mm in diameter. (*b*) Contact dermatitis from poison oak, showing various stages of involvement: blisters, scales, and thickened patches.

healing progresses, the epidermis is replaced by a thick, horny layer. Depending upon the dose and the sensitivity of the individual, the time from initial contact to healing can be a week to 10 days.

T Cells and Their Role in Organ Transplantation

Transplantation or grafting of organs and tissues is a common medical procedure. Although it is life-giving, this technique is plagued by the natural tendency of lymphocytes to seek out foreign tissues and mount a campaign to reject them. The bulk of the damage that occurs in graft rejections can be attributed to expression of cytotoxic T cells and other killer cells. This section will cover the mechanisms involved in graft rejection, tests for transplant compatibility, reactions against grafts, prevention of graft rejection, and types of grafts.

The Genetic and Biochemical Basis for Graft Rejection

In chapter 15, we discussed the role of major histocompatibility (MHC or HLA) genes and receptors in immune function. In general, the genes and receptors in MHC classes I and II are extremely important in recognizing self and in regulating the immune response. These receptors also set the events of graft rejection in motion. The MHC genes of humans are inherited from among a large pool of genes, so the cells of each person can exhibit variability in the pattern of cell surface molecules (figure 17.19). The pattern is identical in different cells of the same person and can be similar in related siblings and parents, but the more distant the relationship, the less likely that the MHC genes and receptors will be similar. When donor tissue (a graft) displays surface receptors of a different MHC class, the T cells of the recipient (called the host) will recognize its foreignness and react against it.

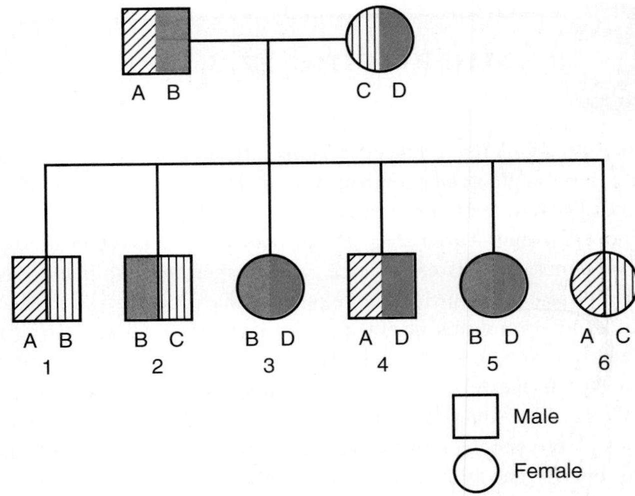

Figure 17.19
The pattern of inheritance of MHC (HLA) genes. A simplified version of the human leukocyte antigen (HLA) complex in a family. In this example, there are two genes in the complex, and each parent has a different set of genes (A/B and C/D). A child can inherit one of four different combinations. Out of six children, two sets (1 and 6, 3 and 5) have identical HLA genes and are good candidates for exchange grafts. Children sharing one gene (for example, 1, 4, and 6 share antigen A) are close matches, but two pairs of children (2 and 4, 3 and 6) do not match at all.

T Cell–Mediated Recognition of Foreign MHC Receptors

Host Rejection of Graft When the T cells of a host recognize foreign class II MHC receptors on the surface of grafted cells, they release interleukin-2 as part of a general immune mobilization. Receipt of this stimulus amplifies helper and cytotoxic T cells specific to the foreign antigens on the donated cells. The cytotoxic cells bind to the grafted tissue and secrete lymphokines that begin the rejection process within 2 weeks of transplantation (figure 17.20*a*). Late in this process, antibodies also formed against the graft tissue speed its loss. A final blow is the destruction of the vascular supply, promoting death of the grafted tissue.

Graft Rejection of Host In certain severe immunodeficiencies, the host cannot or does not reject a graft. But this failure may not protect the host from serious damage, because graft incompatibility is a two-way phenomenon. Some grafted tissues (especially bone marrow) contain an indigenous population called passenger lymphocytes. This makes it quite possible for the graft to reject the host, causing **graft versus host disease (GVHD)** (figure 17.20*b*). Since any host tissue bearing MHC receptors foreign to the graft can be attacked, the effects of GVHD are widely systemic and toxic. A papular, peeling skin rash is the most common symptom. Other organs affected are the liver, intestine, muscles, and mucous membranes. GVHD occurs in approximately 30% of bone marrow transplants within 100–300 days of the graft. A relatively high percentage of recipients die from its effects.

Classes of Grafts

Grafts are generally classified according to the genetic relationship between the donor and the recipient. Tissue transplanted from one site on an individual's body to another site on his body is known as an **autograft.** Typical examples are skin replacement in burn repair and the use of a vein to fashion a coronary artery bypass. In an **isograft (syngeneic graft),** tissue from an identical twin is used. Because isografts do not contain foreign antigens, they are not rejected, but this type of grafting has obvious limitations. **Allografts (homografts),** the most common type of grafts, are exchanges between genetically different individuals belonging to the same species (two humans). A close genetic correlation is sought for most allograph transplants (see next section). A **xenograft** is a tissue exchange between individuals of different species. Until rejection can be better controlled, most xenografts are experimental or for temporary therapy only.

Avoiding and Controlling Graft Incompatibility

Graft rejection can be averted or lessened by directly comparing the tissue of the recipient with that of potential donors. Several **tissue matching** procedures are used. In the *mixed lymphocyte reaction*

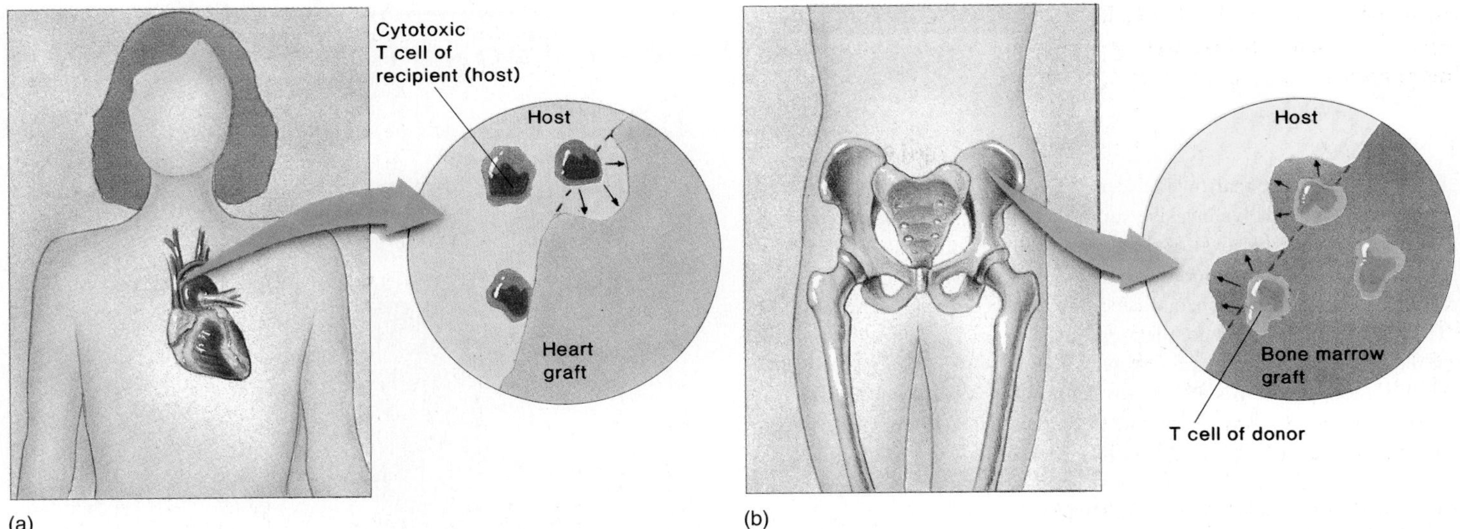

Figure 17.20
Potential reactions in transplantation. (*a*) The host's immune system (primarily cytotoxic T cells) encounters the cells of the donated organ (heart) and rejects the organ by secreting cytokines. (*b*) Grafted tissue (bone marrow) contains endogenous T cells that recognize the host's tissues as foreign and mount a cytokine attack. The recipient will develop symptoms of graft versus host disease.

(*MLR*), lymphocytes of the two individuals are mixed and incubated. If an incompatibility exists, some of the cells will become activated and proliferate. *Tissue typing* is similar to blood typing, except that specific antisera are used to disclose the HLA antigens on the surface of lymphocytes. In most grafts (one exception is bone marrow transplants), the ABO blood type must also be matched. Although a small amount of incompatibility is tolerable in certain grafts (liver, heart, kidney), a closer match is more likely to be successful, so the closest match possible is sought.

Drugs That Suppress Allograft Rejection Despite an international computerized hotline for matching recipients with donors and a greater availability of viable organs than in the past, an ideal match between donor and recipient is still the exception rather than the rule, and some sort of immunosuppressive therapy to overcome rejection is usually required. Rejection can be controlled with agents such as cyclosporin A[6] methotrexate, prednisone, and a monoclonal antibody OKT3. Except for cyclosporin A, intervention with drugs can be complicated by general suppression of the immune system (especially T cells) and frequent opportunistic infections.

Cyclosporin A is a polypeptide isolated from a fungus. It has dramatically improved the survival rate of allograft patients (kidney, heart, liver, and bone marrow) and has reduced the incidence of fatal infections. Although its action is not entirely understood, cyclosporin appears to block the activation of T helper cells and interfere with the release of interleukin-2. What makes this drug so valuable is that it does not inhibit important lymphoid cells and phagocytes, and the body is better able to ward off infections. Its adverse effects of kidney toxicity and increased blood pressure can be reduced by adjusting the dose and monitoring blood levels of the drug. Because of its ability to inhibit undesirable T-cell activity, cyclosporin is also being used to treat autoimmune diseases such as type I diabetes and rheumatoid arthritis. Newer drugs aim to block the binding of IL-2 on T cells.

Types of Transplants

Today, transplantation is a recognized medical procedure whose benefit is reflected in several thousand transplants each year. It has been performed on every major organ, including parts of the brain. The most frequent transplant operations involve skin, heart, kidney, coronary artery, cornea, and bone marrow. The sources of organs and tissues are live donors (kidney, skin, bone marrow, liver), cadavers (heart, kidney, cornea), and fetal tissues. In the past decade, we have witnessed some unusual types of grafts. For instance, the fetal pancreas has been implanted as a potential treatment for diabetes, and fetal brain tissues for Parkinson's disease. Part of a liver has been transplanted from a live parent to a child, and parents have donated a lobe from their lungs to help restore function in their children with severe cystic fibrosis.

Bone marrow transplantation is a rapidly growing medical procedure. Thousands of transplants have been performed for patients with immune deficiencies, aplastic anemia, leukemia and other cancers, and radiation damage. This procedure is extremely expensive, costing up to $200,000 per patient. Before bone marrow from a closely matched donor can be infused (microfile 17.6), the patient is pretreated with chemotherapy and whole-body irradiation, a procedure designed to destroy his own blood stem cells and thus prevent rejection of the new marrow cells. Within 2 weeks to a month after infusion, the grafted cells are established in the host. Because donor lymphoid cells can still cause GVHD, anti-rejection drugs may be necessary. An amazing consequence of bone marrow transplantation is that a recipient's blood type may change to the blood type of the donor.

6. Marketed as Sandimmune.

MICROFILE 17.6 THE MECHANICS OF BONE MARROW TRANSPLANTATION

In some ways, bone marrow is the most exceptional form of transplantation. It does not involve invasive surgery in either the donor or recipient, and it permits the removal of tissue from a living donor that is fully replaceable. While the donor is sedated, a bone marrow/blood sample is aspirated by inserting a special needle into an accessible marrow cavity. The most favorable sites are the crest and spine of the ilium (major bone of the pelvis). During this procedure, which lasts 1 to 2 hours, 3% to 5% of the donor's marrow is withdrawn in 20 to 30 separate extractions. Between 500 and 800 ml of marrow is removed. The donor may experience some pain and soreness, but there are rarely any serious complications. In a few weeks, the depleted marrow will naturally replace itself. Implanting the harvested bone marrow is rather convenient, because it is not necessary to place it directly into the marrow cavities of the recipient. Instead, it is dripped intravenously into the circulation, and the new marrow cells automatically settle in the appropriate bone marrow regions. The survival and permanent establishment of the marrow cells are increased by administering various growth factors and stem cell stimulants to the patient.

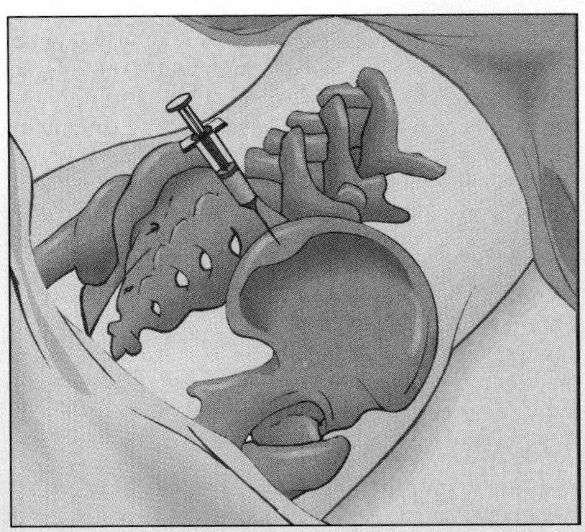

Removal of a bone marrow sample for transplantation. Samples are removed by inserting a needle into the spine or crest of the ilium. (The ilium is a prolific source of bone marrow.)

Chapter Checkpoints

Type IV hypersensitivity reactions occur when cytotoxic T cells attack either self tissue or transplanted foreign cells. Type IV reactions are also termed delayed hypersensitivity reactions because they occur hours to days after the antigenic challenge.

Type IV hypersensitivity reactions are mediated by T lymphocytes and are carried out against foreign cells that show both a foreign MHC and a nonself receptor site.

Examples of type IV reactions include the tuberculin reaction, contact dermatitis, and mismatched organ transplants (host rejection and GVHD reactions).

The four classes of transplants or grafts are determined by the degree of MHC similarity between graft and host. From most to least similar these are: autografts, isografts, allografts, and xenografts.

Graft rejection can be minimized by tissue matching procedures, immunosuppressive drugs, and use of tissues that do not provoke a type IV response.

IMMUNODEFICIENCY DISEASES: HYPOSENSITIVITY OF THE IMMUNE SYSTEM

It is a marvel that development and function of the immune system proceed as normally as they do. On occasion, however, an error occurs and a person is born with or develops weakened immune responses. In many cases, these very "experiments" of nature have provided penetrating insights into the exact functions of certain cells, tissues, and organs because of the specific signs and symptoms shown by the immunodeficient individuals. The predominant consequences of immunodeficiencies are recurrent, overwhelming infections, often with opportunistic microbes. Immunodeficiencies fall into two general categories: *primary diseases,* present at birth (congenital) and usually stemming from genetic errors, and *secondary diseases,* acquired after birth and caused by natural or artificial agents (table 17.5).

PRIMARY IMMUNODEFICIENCY DISEASES

Deficiencies affect both specific immunities such as antibody production and less-specific ones such as phagocytosis. Consult figure 17.21 to survey the places in the normal sequential development of lymphocytes where defects can occur and the possible consequences. In many cases, the deficiency is due to an inherited abnormality, though the exact nature of the abnormality is not known for a number of diseases. Because the development of B cells and T cells departs at some point, an individual can lack one or both cell lines. It must be emphasized, however, that some deficiencies affect other cell functions. For example a T-cell deficiency can affect B-cell function because of the role of T helper cells. In some deficiencies, the lymphocyte in question is completely absent or is present at very low levels, whereas in others, lymphocytes are present but do not function normally.

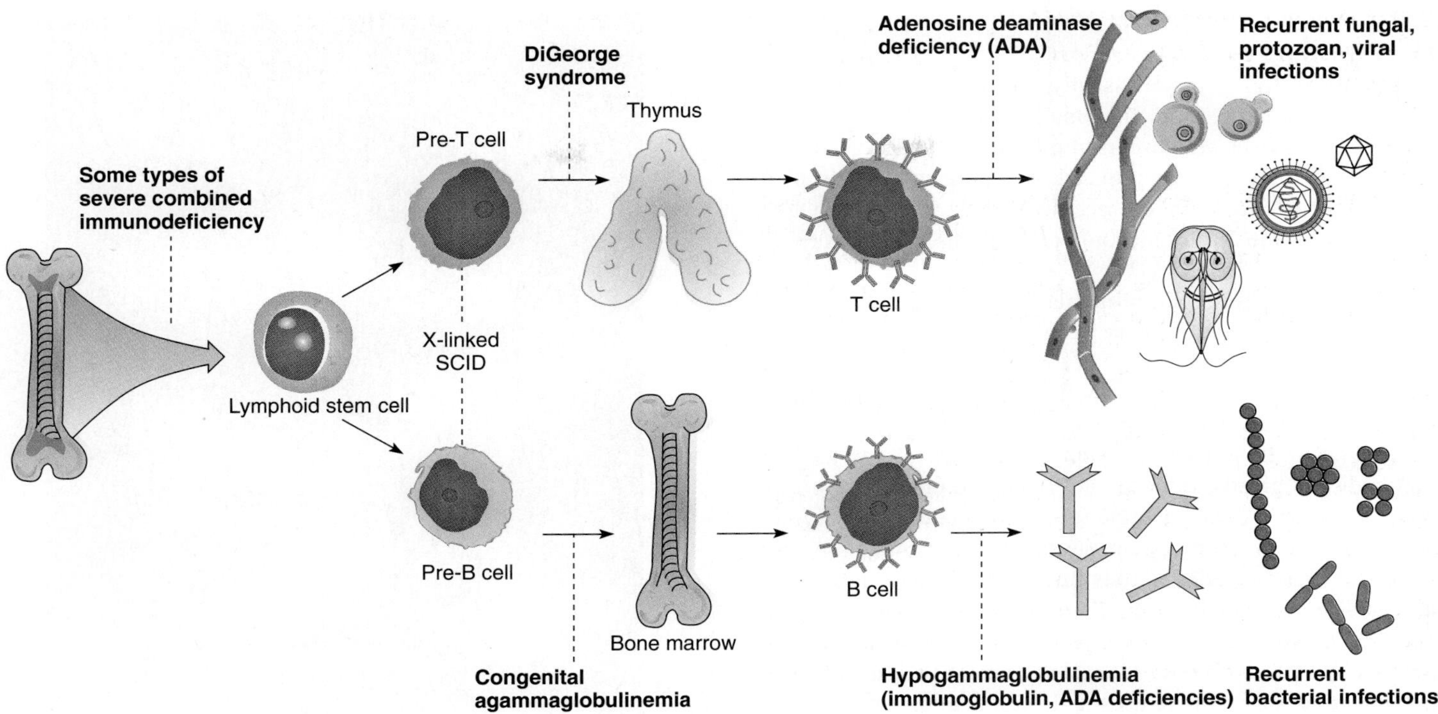

Figure 17.21

The stages of development and the functions of B cells and T cells, whose failure causes immunodeficiencies. Dotted lines represent the phases in development where breakdown can occur.

TABLE 17.5

GENERAL CATEGORIES OF IMMUNODEFICIENCY DISEASES

Primary Immune Deficiencies (Genetic)	Secondary Immune Deficiencies (Acquired)
B-Cell Defects (Low Levels of Antibodies)	**From Natural Causes**
Agammaglobulinemia (X-linked, non-sex-linked)	Infection: AIDS, leprosy, tuberculosis, measles
Hypogammaglobulinemia	Other disease: cancer, diabetes
Selective immunoglobulin deficiencies	Nutrition deficiencies
T-Cell Defects (Lack of All Classes of T Cells)	Stress
Thymic aplasia (DiGeorge syndrome)	Pregnancy
Chronic mucocutaneous candidiasis	Aging
Combined B-Cell and T-Cell Defects (Usually Caused by Lack or Abnormality of Lymphoid Stem Cell)	**From Immunosuppressive Agents**
Severe combined immunodeficiency disease (SCID)	Irradiation
Adenosine deaminase (ADA) deficiency	Severe burns
Wiskott-Aldrich syndrome	Steroids (cortisones)
Ataxia-telangiectasia	Drugs to treat graft rejection and cancer
Phagocyte Defects	Removal of spleen
Chédiak-Higashi syndrome	
Chronic granulomatous disease of children	
Lack of surface adhesion molecules	
Complement Defects	
Lacking one of C components	
Hereditary angioedema	
Associated with rheumatoid diseases	

Clinical Deficiencies in B-Cell Development or Expression

Genetic deficiencies in B cells usually appear as an abnormality in immunoglobulin expression. In some instances, only certain immunoglobulin classes are absent; in others, the levels of all types of immunoglobulins (Ig) are reduced. A significant number of B-cell deficiencies are X-linked recessive traits, meaning that the gene occurs on the X chromosome and the disease appears primarily in male children.

The term **agammaglobulinemia** literally means the absence of gamma globulin, the primary fraction of serum that contains immunoglobulins. Because it is very rare for Ig to be completely absent, some physicians prefer the term **hypogammaglobulinemia.**[7] Both sex-linked and autosomal[8] recessive types of this rare syndrome occur. In both cases, mature B cells are absent, the level of antibodies is greatly reduced, and lymphoid organs are incompletely developed. T-cell function in these patients is usually normal. The symptoms of recurrent, serious bacterial infections usually appear about 6 months after birth. The bacteria most often implicated are pyogenic cocci, *Pseudomonas,* and *Haemophilus influenzae,* and the most common infection sites are the lungs, sinuses, meninges, and blood. Many Ig-deficient patients can have recurrent infections with viruses and protozoa, as well. Patients often manifest a wasting syndrome and have a reduced lifespan, but modern therapy has improved their prognosis. The current treatment for this condition is passive immunotherapy with immune serum globulin and continuous antibiotic therapy.

The lack of a particular class of immunoglobulin is a relatively common condition. Although genetically controlled, its underlying mechanisms are not yet clear. **IgA deficiency** is the most prevalent, occurring in about one person in 600. Such persons have normal quantities of B cells and other immunoglobulins, but they are unable to synthesize IgA. Consequently, they lack protection against local microbial invasion of the mucous membranes and suffer recurrent respiratory and gastrointestinal infections. The usual treatment using Ig replacement does not work, because conventional preparations are high in IgG, not IgA.

Clinical Deficiencies in T-Cell Development or Expression

Due to their critical role in immune defenses, a genetic defect in T cells results in a broad spectrum of disease, including severe opportunistic infections, wasting, and cancer. In fact, a dysfunctional T-cell line is usually more devastating than a defective B-cell line because T helper cells are required to assist in most specific immune reactions. The deficiency can occur anywhere along the developmental spectrum, from thymus to mature, circulating T cells.

Abnormal Development of the Thymus The most severe of the T-cell deficiencies involve the congenital absence or immaturity of the thymus gland. Thymic aplasia, or **DiGeorge syndrome,** results when the embryonic third and fourth pharyngeal pouches fail to develop. Some cases are associated with a deletion in chromosome 22. Children with this syndrome not only lack thymus activity but also manifest congenital heart defects and hypoparathyroidism (figure 17.22). The accompanying lack of cell-mediated immunity makes them highly susceptible to persistent infections by fungi, protozoa, and viruses. Common, usually benign childhood infections such as chickenpox, measles, or mumps can be overwhelming and fatal in these children. Even vaccinations using attenuated microbes pose a danger. Other symptoms of thymic failure are growth retardation, wasting of the body, unusual facial characteristics, and an increased incidence of lymphatic cancer. These children can have reduced antibody levels, and they are unable to reject transplants. The major therapy for them is a transplant of thymus tissue.

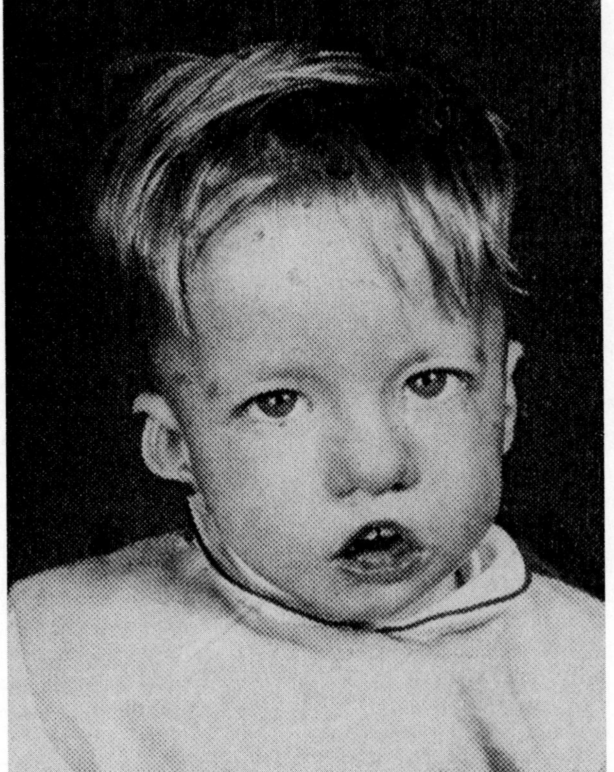

Figure 17.22
Facial characteristics of a child with DiGeorge syndrome. Typical defects include low-set, deformed earlobes; wide-set, slanted eyes; a small, bowlike mouth; and the absence of a philtrum (the vertical furrow between the nose and upper lip).

Severe Combined Immunodeficiencies: Dysfunction in B and T Cells

Severe combined immunodeficiencies (SCIDs) are the most dire and potentially lethal of the immunodeficiency diseases because they involve dysfunction in both lymphocyte systems. Some SCIDs are due to the complete absence of the lymphocyte stem cell in the marrow; others are attributable to the dysfunction of B cells and T cells later in development. Infants with SCID usually manifest the T-cell deficiencies within days after birth by developing candidiasis, sepsis, pneumonia, or systemic viral infections. This debilitating condition appears to have several forms. In the

7. Hypo-denoting a lowered level of Ig.
8. Meaning that the gene is located on a chromosome other than a sex chromosome.

MICROFILE 17.7 AN ANSWER TO THE BUBBLE BOY MYSTERY

David Vetter, the most famous SCID child, lived all but the last 2 weeks of his life in a sterile environment to isolate him from the microorganisms that could have quickly ended his life. When medical tests performed before birth had indicated that David might inherit this disease, he was delivered by cesarian section and immediately placed in a sterile isolette. From that time, he lived in various plastic chambers—ranging from room-size to a special suit that allowed him to walk outside. Remarkably, he developed into a well-adjusted child, even though his only physical contact with others was through special rubber gloves. When he was 12, his doctors decided to attempt a bone marrow transplant that might allow him to live free of his bubble prison. David was transplanted with his sister's bone marrow, but the marrow harbored a common herpesvirus called Epstein-Barr virus. Because he lacked any form of protective immunities against this oncogenic virus, a cancer spread rapidly through his body. Despite the finest medical care available, David died a short time later from the metastatic cancer.

In 1993, after several years of study, researchers discovered the basis for David's immunodeficiency. He had inherited an X-linked form (called XSCID) that arises from a defective genetic code in the receptors for interleukin-2, interleukin-4, and interleukin-7. The defect prevents the receptors on T cells and B cells from receiving the appropriate interleukin signals for growth, development, and reactivity. The end result is that both cytotoxic immunities and antibody-producing systems are shut down, leaving the body defenseless against infections and cancer.

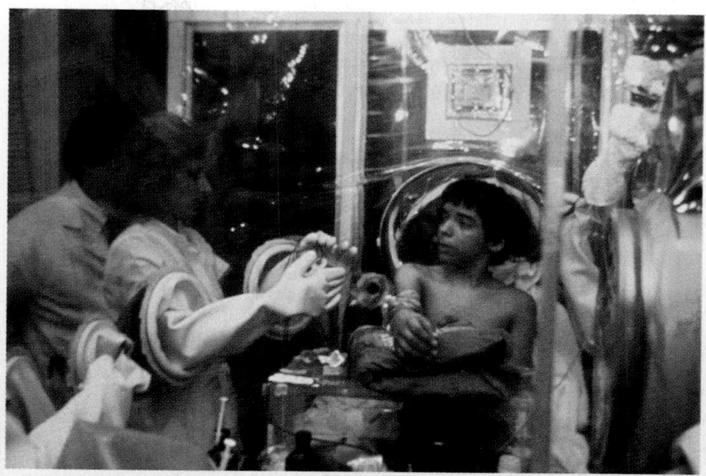

David Vetter, the boy in the plastic bubble.

two most common forms, Swiss-type agammaglobulinemia and thymic alymphoplasia, the numbers of all types of lymphocytes are extremely low, the blood antibody content is greatly diminished, and the thymus and cell-mediated immunity are poorly developed. Both diseases are due to a genetic defect in the development of the lymphoid cell line.

A rarer form of SCID is **adenosine deaminase (ADA) deficiency,** which is caused by an autosomal recessive defect in the metabolism of adenosine. In this case, lymphocytes develop, but a metabolic product builds up abnormally and selectively destroys them. Infants with ADA deficiency are subject to the recurrent infections and severe wasting that occur in other SCID syndromes. A small number of SCID cases are due to a developmental defect in receptors for B and T cells. An X-linked deficiency in interleukin receptors was responsible for the disease of David, the child in the "plastic bubble" (microfile 17.7). Another newly identified condition, *bare lymphocyte syndrome,* is caused by the lack of genes that code for class II MHC receptors.

Because of their profound lack of specific adaptive immunities, SCID children require the most rigorous kinds of aseptic techniques to protect them from opportunistic infections. Aside from life in a sterile plastic bubble, the only serious option for their longtime survival is total replacement or correction of dysfunctional lymphoid cells. Some infants can benefit from fetal liver or bone marrow grafts. Although transplanting compatible bone marrow has been about 50% successful in curing the disease, it is complicated by graft versus host disease. The condition of some ADA deficient patients has been partly corrected by periodic transfusions of red blood cells containing large amounts of ADA. A more lasting treatment for ADA has been found in gene therapy—insertion of a gene to replace the defective gene (see figure 10.16).

SECONDARY IMMUNODEFICIENCY DISEASES

Secondary acquired deficiencies in B cells and T cells are caused by one of four general agents: (1) infection, (2) organic disease, (3) chemotherapy, or (4) radiation.

The most recognized infection-induced immunodeficiency is **AIDS.** This syndrome is caused when several types of immune cells, including T helper cells, monocytes, macrophages, and antigen presenting cells, are infected by the human immunodeficiency virus (HIV). It is generally thought that the depletion of T helper cells and functional impairment of immune responses ultimately account for the cancers and opportunistic protozoan, fungal, and viral infections associated with this disease. See chapter 25 for an extensive discussion of AIDS. Other infections that can deplete immunities are measles, leprosy, and malaria.

Cancers that target the bone marrow or lymphoid organs can be responsible for extreme malfunction of both humoral and cellular immunity (see next section). In leukemia, the massive number of cancer cells compete for space and literally displace the normal cells of the bone marrow and blood. Plasma cell tumors produce large amounts of nonfunctional antibodies, and thymus gland tumors cause severe T-cell deficiencies.

An ironic outcome of life-saving medical procedures is the possible suppression of a patient's immune system. For instance, some immunosuppressive drugs that prevent graft rejection by T cells can likewise suppress beneficial immunities. Although

Chapter Checkpoints

Immunodeficiency diseases occur when the immune response is reduced or absent.

Primary immune diseases are genetically induced deficiencies of B cells, T cells, the thymus gland, or combinations of these. Secondary immune diseases are caused by infection, organic disease, chemotherapy, or radiation.

The best-known infection-induced immunodeficiency is AIDS.

radiation and anticancer drugs are the first line of therapy for many types of cancer, both agents are extremely damaging to the bone marrow and other body cells.

CANCER: CELLS OUT OF CONTROL

The term **cancer** comes from the Latin word for crab and presumably refers to the appendage-like projections that a spreading tumor develops. A cancer is defined as the new growth of abnormal cells. The disease is also known by the synonym *neoplasm,** or by more specific terms that usually end in the suffix *-oma.* **Oncology,** the field of medicine that specializes in cancer, is growing with such great rapidity that we can only survey major concepts as they relate to immunology, including the basic characteristics, classification, and origins of cancer.

Characteristics and Classification of Tumors and Cancers

An abnormal growth or tumor is generally characterized as benign or malignant. A **benign tumor** is a self-contained mass within an organ that does not spread into adjacent tissues. The mass is usually slow-growing, rounded, and not greatly different from its tissue of origin. Benign tumors do not ordinarily cause death unless they grow into a critical space such as the heart valves or brain ventricles. The feature that most distinguishes a **malignant tumor** (cancer) is uncontrolled growth of abnormal cells within normal tissue. As a general rule, cancer originates in cells such as skin and bone marrow that have retained the capacity to divide, while mature cells that have lost this power (neurons, for instance) do not commonly become cancerous.

Most cancerous growths show other characteristics, including (1) disorganized behavior and independence from surrounding normal tissues, (2) permanent loss of cell differentiation, and (3) expression of special markers on their surface. The cells of a malignant tumor also do not remain encapsulated but tend to spread both locally and distantly to other organs. As the initial (primary) tumor grows, it invades nearby tissues, enters lymphatic and blood vessels, and establishes secondary tumors in remote sites. This property of spreading is called **metastasis,** and the cancer is said to *metastasize.* Malignant tumors range in effects from those seen in pancreatic cancer, which spread rapidly and are almost always fatal, to others as occur in basal cell carcinoma of the skin, which are much less aggressive, more treatable, and seldom fatal.

Another tumor-classifying scheme relies on the tissue or cell of primary origin. Although cancers are commonly referred to simply as bladder cancer, breast cancer, or liver cancer, their technical names more clearly define the tumor origin. In general, cancers originating from epithelial tissues are called **carcinomas,*** and those originating from mesenchyme (embryonic connective tissue) are **sarcomas.*** Combining these terms with the tissue of origin supplies the full descriptive name. For example, adenocarcinoma develops in glandular tissue such as the pancreas or thyroid; squamous cell carcinoma arises in the skin epidermis; retinoblastoma occurs in the retina; lymphosarcoma in the lymph nodes; hepatoma (hepatic sarcoma) in the liver; and melanoma in the skin melanocytes. A special name, leukemia, denotes cancer of the blood-forming tissues. Considering the large number of cell types in a given organ, the great diversity of tumors (more than 100 clinical types) is to be expected.

Epidemiology of Cancer

One-third of all U.S. citizens will develop cancer some time during their lifetime. Cancer is the second leading cause of death in humans (following heart and circulatory disease). Annually, approximately 850,000 new cases are diagnosed, and 500,000 people die from various cancers. Cancer rates also vary according to gender, race, ethnic group, age, geographic region, occupational exposure to carcinogens, diet, and life-style. Figure 17.23 compiles the most recent statistics on cancer incidence in the United States for mixed populations. Cancers currently on the increase are melanoma (presumably due to increased levels of UV radiation entering the atmosphere), Kaposi's sarcoma (in AIDS patients), thyroid cancer, esophageal cancer, and prostate cancer (because of an improved method of detection). Lung cancer accounts for the largest number of deaths in both sexes. The second most common cause of cancer deaths is prostate cancer in men and breast cancer in women.

It is increasingly evident that susceptibility to certain cancers is inherited. At least 30 neoplastic syndromes that run in families have been described. A particularly stunning discovery is that some cancers in animals are actually due to endogenous viral genomes passed from parent to offspring in the gametes. Many cancers reflect an interplay of both hereditary and environmental factors. An example is lung cancer, which is associated mainly with tobacco smoking. But not all smokers develop lung cancer, and not all lung cancer victims smoke tobacco.

Proposed Mechanisms of Cancer

Cancer is clearly a complex disease associated with numerous physical, chemical, and biological agents (table 17.6). For years, there was no clear unifying explanation for how so many different

*neoplasm (nee´-oh-plazm) Gr. *neo,* new, and *plasm,* formation.

*carcinoma (kar´´-sih-noh´-mah) Gr. *karkinos,* crab.

*sarcoma (sar-koh´-mah) Gr. *sarcos,* flesh, and *oma,* tumor. Sometimes referred to as soft-tissue tumors.

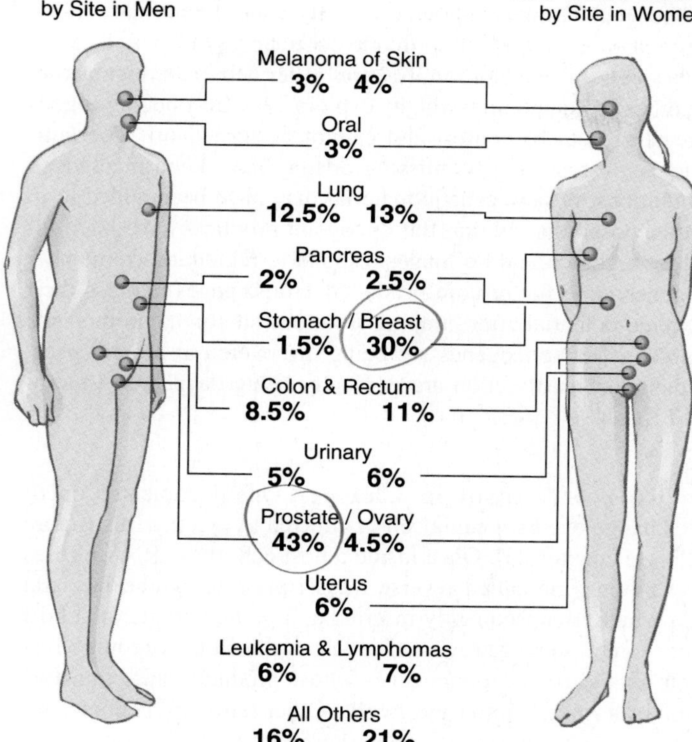

Figure 17.23
Incidence of cancer in men and women as of 1997. Note that breast cancer is the most common type in women, and prostate cancer predominates in men. Data (unofficial) are from the American Cancer Society.

factors could cause normal cells to become cancerous. But recent research findings link all cancers to genetic damage or inherited genetic predispositions that alter the function of certain genes found normally in all cells. The evidence for the interrelationship between genes and cancer is derived from the following observations: (1) Cancer cells often have damaged chromosomes; (2) a specific alteration in a gene can lead to cancer; (3) the predisposition for some cancers is inherited; (4) rates of cancer are highest in individuals who cannot repair damaged DNA; (5) mutagenic agents cause cancer; (6) cells contain genes that can be transformed to cancer-causing oncogenes; and (7) tumor-suppressor genes (anti-oncogenes) exist in the normal genome. These discoveries and other advances in cancer research have provided the seeds for a general unifying theory for cancer.

Oncogene: A Normal Gene Gone Haywire? Cancer is another experiment of nature that continues to inform us about normal cell function. In the section on gene regulation in chapter 9, we learned that normal cell operations are strictly controlled by regulatory genes. Extracellular chemical signals are received by receptors, which transmit them to a regulatory complex in the nucleus that switches the appropriate genes on or off. During the rapid growth of a young individual, cells are extremely active, and the switch that initiates cell division is operating much of the time. By contrast, most

TABLE 17.6

SELECTED CANCER-TRIGGERING AGENTS AND THEIR DISEASES

Agent	Type of Exposure	Cancer Type
Physical Agents		
Ionizing radiation	Occupational exposure, therapy	Leukemia, many others
UV radiation	Sunlight, suntanning lamps	Skin
Asbestos	Industrial, building insulation	Lung
Chemical Agents		
Tobacco	Smoking, chewing	Oral, lung
Alcohol	Ingestion	Oral, esophageal, liver
Arsenic	Manufacturing, mining	Lung, skin, liver
Benzene	Various industrial processes	Leukemia
Polycyclic hydrocarbons	Medicines, industrial use (coal tar derivatives)	Lung, skin
Vinyl chloride	Manufacture of plastic	Liver
Aflatoxin	Fungus-contaminated foods	Liver
Aromatic amines	Manufacturing processes	Bladder
Heavy metal dust	Industry	Lung, nasal
Anabolic steroids	Medication	Liver
Estrogens	Medication	Vaginal, liver
Radon gas	Mines, homes	Lung
Biological Agents (Acquired Through Infection)		
Retrovirus	T-cell leukemia	
Papillomaviruses	Cervical cancer	
Polyoma viruses	Various tumors in mice and hamsters	
Epstein-Barr virus	Burkitt's lymphoma; nasopharyngeal carcinoma	
Hepatitis B virus	Liver cancer	

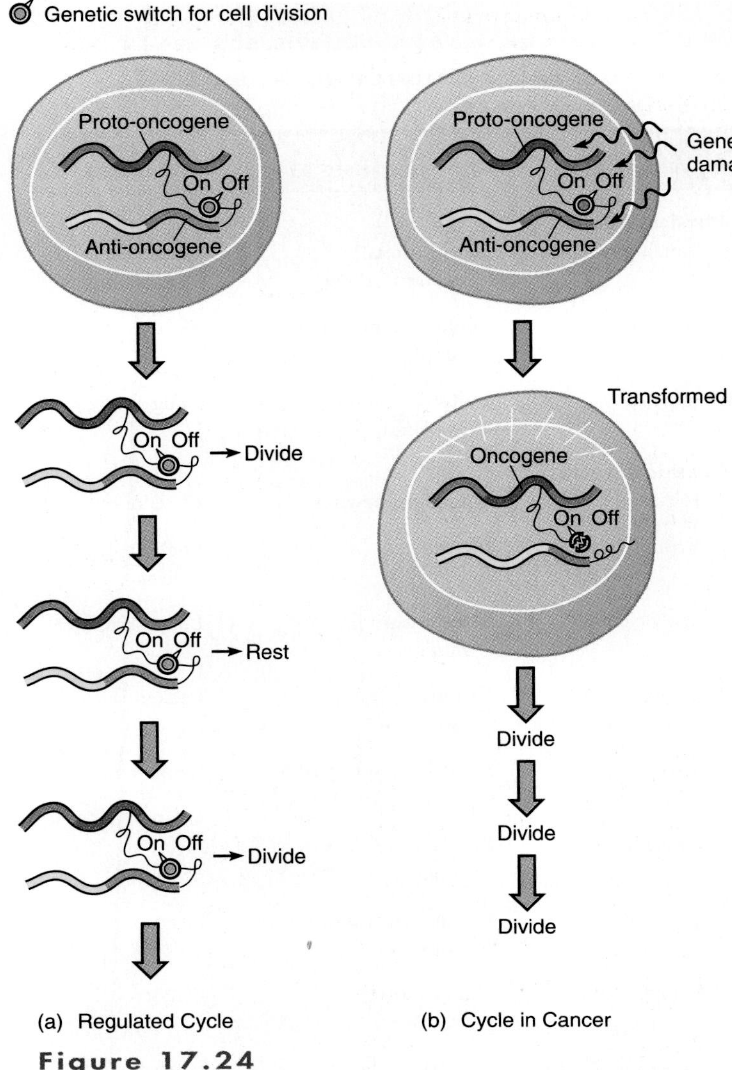

Figure 17.24

Common pathway for neoplasias. (*a*) Normal cell cycle and genetic controls. (*b*) Event that stimulates transformation into a cancer cell. Any agent that can alter the genetic control mechanisms can cause a cell to divide at incorrect times and in incorrect places.
Source: Data from the American Cancer Society.

adult cells exist in a nondividing state, except for divisions involved in repair and maintenance. This state of balance ensures that cells divide at a rate compatible with normal development and function.

Cancer results when the system that organizes cell division is short-circuited. Current theories explain that the gene complex controlling cell division contains a gene termed a **proto-oncogene** that regulates the onset of mitosis. The proto-oncogene is, in turn, regulated by another gene called a **tumor-suppressor gene,** or **anti-oncogene,** that prevents the proto-oncogene from acting continuously (figure 17.24*a*) and thereby keeps the cell division cycle operating normally. It is now clear that genetic disruptions (mutation, chromosomal alterations) can affect the normal actions of the proto-oncogenes and tumor-suppressor genes. If a genetic error keeps the proto-oncogene switched on, it converts to an **oncogene.** The oncogene is then programmed to override the normal mitotic controls and cause the cell to divide continuously (figure 17.24*b*). This process of cell immortalization is termed **transformation.** A number of happenings might explain how an oncogene goes permanently out of control and cannot be turned off. The anti-oncogene system may be missing or may have been inactivated by mutations, or new genetic material may have been added to it. Another possibility is that the oncogene products have been altered by mutation and no longer respond to regulation (remember that genes code for protein products). Some genes express their oncogenic potential after jumping from one site on a chromosome to another. New oncogenes are being discovered at a rapid pace, and their modes of action are gradually being decphered (microfile 17.8).

The Role of Viruses in Cancer One documented oncogenic change occurs when a cell is infected by a retrovirus (figure 17.25; see chapter 25). Once inside a host cell, these RNA viruses utilize an enzyme called reverse transcriptase to synthesize viral DNA, which is subsequently inserted at a particular site on a host chromosome. Some retroviruses, such as the Rous sarcoma virus of chickens, carry viral oncogenes whose products cause transformation of host cells into cancer cells. Other retroviral genomes insert on host chromosomes at regulatory sites; their insertion at those sites can lead to activation of the cell's own proto-oncogenes. One type of cancer in humans, T-cell leukemia, is attributable to insertional mutagenesis by a retrovirus. DNA viruses are also involved in cancer. Human papillomavirus (HPV), which causes genital warts and is strongly implicated in cervical carcinoma, inserts its genome directly into a chromosome in a way that overrides the usual cellular growth controls. Epstein-Barr virus (EBV), a cause of infectious mononucleosis and Burkitt's lymphoma induces a chromosomal alteration that can interfere with cell death.

Chromosomal Damage and Cancer A number of cancers have been associated with gross chromosomal changes or other aberrations that occur during mitotic divisions (figure 17.26). In the form of chromosomal disruption called *translocation* (figure 17.26*a*), a segment of one chromosome is transferred to another chromosome that is not its mate. The abnormal placement of the translocated genes is believed to act as a strong oncogenic stimulus. Burkitt's lymphoma (see figure 24.13), a malignant tumor of lymphocytes, is associated with the translocation of part of chromosome 8 to chromosome 14. Most cases of myelogenous leukemia are due to the translocation of a proto-oncogene on chromosome 9 to a different site on chromosome 22.

Certain cancers appear to result from a mistake in mitosis that produces cells bearing three rather than the usual pair of a particular chromosome. Such *trisomies* (figure 17.26*b*) occurring with chromosome 8 and 12 can lead to leukemia, and an extra copy of chromosome 7 is associated with melanoma. Several types of leukemia can be traced to the *loss* of a whole chromo-

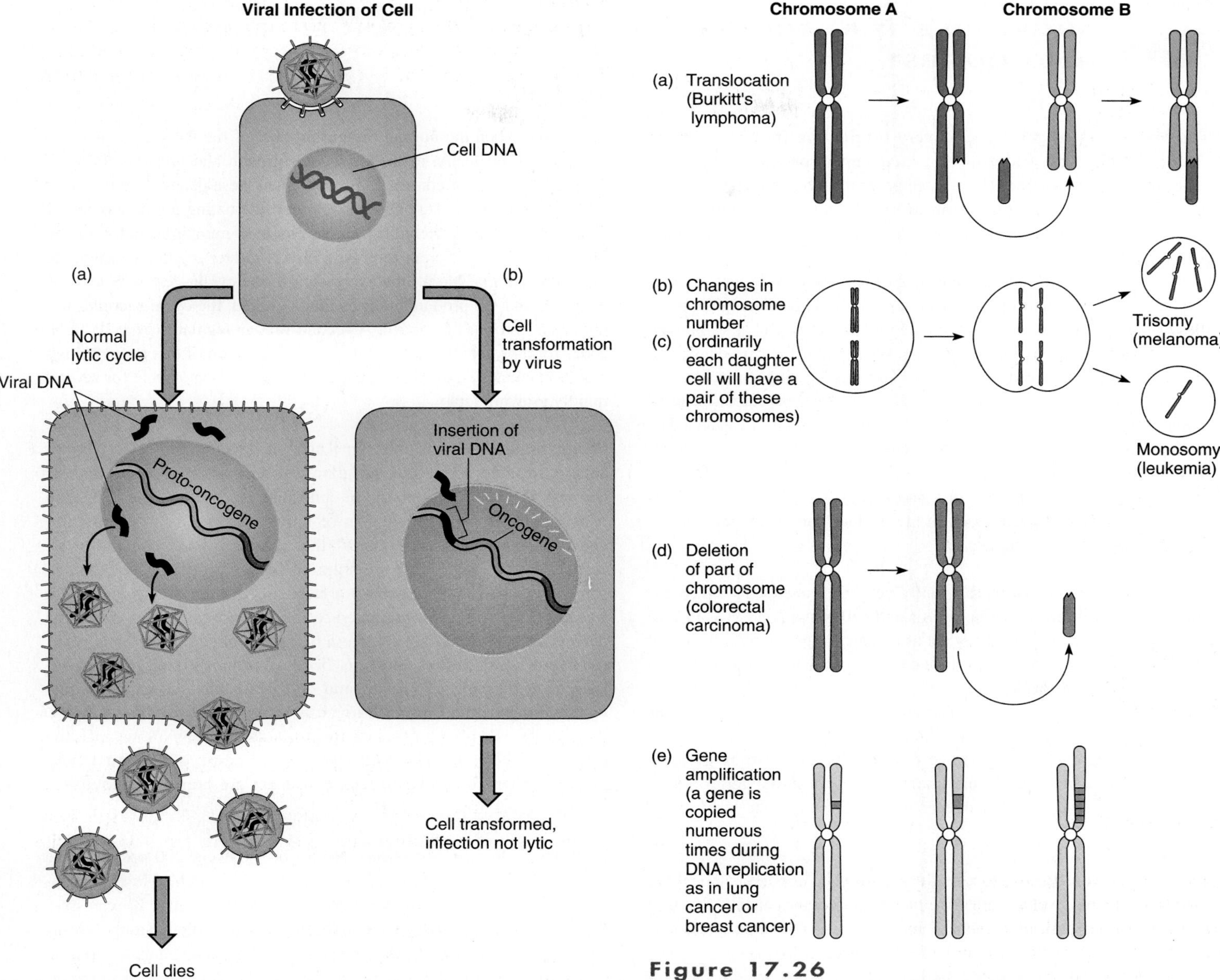

Figure 17.25
Outcomes of viral infection. (*a*) A virus can go through the normal lytic cycle and destroy its host cell. (*b*) A possible mechanism for viral induction of cancer. DNA viruses and retroviruses can insert DNA into the host's genome. If it inserts into a sensitive site (near a proto-oncogene), the cell can be converted into a tumor cell.

Figure 17.26
(*a–e*) Alterations in chromosomes that can be implicated in some cancers.

some (figure 17.26*c*). The loss of part of a chromosome during cell division (chromosomal deletion) is also an important factor in some cancers. A large number of colorectal carcinomas manifest a deletion of a small part of chromosome 17 (figure 17.26*d*), and retinoblastoma occurs through deletions on chromosome 13 (see microfile 17.8).

Sufficient evidence indicates that some cells become cancerous because of *gene amplification,* (figure 17.26*e*), a process whereby numerous extra copies of a gene are produced during DNA replication. When the copied gene is a proto-oncogene, the likelihood of oncogenic transformation is greatly increased. Certain tumors arising in squamous cell tissue or tissue of the lung, brain, and breast are linked to gene amplification.

The Role of Carcinogens What is the true etiologic role of chemical and physical carcinogenic agents (see table 17.6)? It

MICROFILE 17.8 HOT ON THE TRAIL OF CANCER CLUES AND CURES*

The outpouring of genetic discovery originating from the Human Genome Project has had a monumental impact on identification of genes involved in cancers. Remarkable progress is being made in identifying cancer genes, mapping them to specific areas of chromosomes, and developing special markers for these genes (see chapter opening photo).

What has followed naturally from this new knowledge on exact locations of genes has been a tremendous activity in disclosing the ways these genes cause cancer. It is becoming clear that cancer development involves many complex interactions between genes, their products, and external signals received by the cell. There is no single mechanism. Consider the fact that at least eight different genetic defects lead to colon cancer and nearly that many are implicated in breast cancers. The one unifying pathway for all of these cancers, however, is a genetic alteration that disrupts the normal cell division cycle and transforms a normal cell into a cancer cell. In a large number of cases, cancer cells appear when the cell cycle clock in the nucleus can no longer respond appropriately to growth signals.

The Role of Oncogenes in the Cell Cycle. Cell chemicals called cyclins that control the phases in mitosis and cell growth can explain how some types of cancer arise. They act as chemical signals to promote cell division in a timely and orderly fashion. One of them, cyclin D1, is important for the early growth phase of cells. This substance turns on a set of enzymes called kinases that stimulate the replication of DNA. It turns out that the gene that codes for cyclin D1 (also called bc11) is a type of oncogene. When a defect in the gene causes the cyclin to be overproduced or produced at the wrong time, the cell goes into a nonstop division mode. Several other oncogenes apparently act at the level of the cyclins.

The Role of Tumor-Suppressor Genes. Another important model for cancer formation involves defective tumor-suppressor genes (TSGs), of which there are presently about a dozen known. One example is the human retinoblastoma (Rb) gene on chromosome 13, which prevents retinal cancer and other cancers. Individuals who have inherited defective genes or whose normal genes have been mutated by carcinogens can develop retinal tumors, osteosarcoma (a form of bone cancer), breast cancer, and a form of lung cancer. Molecular biologists determined that when the normal protein product of the Rb gene binds to a specific site on DNA, it blocks cellular growth and prevents the cells from growing out of control. However, when these genes are mutated or damaged, there is no suppression. Another fascinating tumor-suppressor gene, p53, becomes altered by simple missense mutations and has been implicated in 51 different tumors. p53 is a master regulator switch for the cell cycle that blocks the expression of the cyclin genes. When it is defective and the production of cyclin is altered, the cycle becomes unregulated and out of control. A second master regulator gene, P-TEN, codes for a protein to slow cell division. Disruption of this gene through mutation creates an aggressive cancer cell that is responsible for several rapidly growing tumors.

Programmed Cell Suicide. One normal developmental activity of cells is to undergo natural, programmed death, or *apoptosis*. This is one way that the body gets rid of abnormal, diseased, or old cells, and it accounts for various aspects of aging. Apoptosis is genetically coded by special "suicide genes" that are part of the genome of all cells. Now it appears that a disruption in this process may figure in some types of cancer. Because a cancer cell is an abnormal cell, it would ordinarily be programmed to die by apoptosis. But if genetic defects keep the suicide gene from being switched on, the cells never receive a death signal and thus become immortalized. Recent research on the role of viruses and cell death has shown that some viruses, such as the Epstein-Barr virus, contain a gene called C-MYC that interferes with the onset of apoptosis and thereby allows infected cells to survive. The EB virus is a known cause of Burkitt's lymphoma. Other viruses that block cell death are human adenoviruses and papillomaviruses (warts).

Inability to Repair Mutations. Nearly one in every 200 people in the Western Hemisphere carries a defective gene that can lead to colon cancer, making this one of the most common genetic defects. If expressed, the defective gene leads to a condition called hereditary nonpolyposis colon cancer (HNPCC). The cause of this cancer is yet another variation on the theme of genetic defects. In this case, the defect occurs in the re-

**For an excellent survey on cancer, see Scientific American, September, 1996.*

appears that carcinogens genetically alter or damage DNA in a way that activates or deregulates a proto-oncogene and transforms it into an oncogene. Although physical agents such as X rays and UV radiation have well-developed powers to mutate or damage DNA (see figure 11.8), most chemical carcinogens operate in a stepwise fashion that requires at least two different stimuli. The first chemical, the *initiating stimulus,* is metabolized by the liver into a more reactive chemical that permanently alters the DNA of a particular target cell. This is the carcinogenic chemical targeted by the Ames test (see chapter 9). Before the altered cell becomes neoplastic, it must be acted upon by a second chemical stimulus, called a *promoter* or *co-carcinogen,* that completes the transformation to cancer.

Progression of Cancer

Transformation is marked by the appearance of aberrant histology and physiology in the malignant cell and its progeny. Cancer cells are frequently pleomorphic. They may appear as giant cells with bizarre shapes; show an abnormal, vacuolated cytoplasm; or possess enlarged, multiple nuclei. They express surface markers that

TABLE 17.A

SELECTED CANCER GENES

Cancer Gene*	Associated with Cancer of	Chromosomal Location	Genetic Mechanism
p53	Colon, breast	17	Defective tumor suppressor/ apoptosis blocker
C-MYC	Burkitt's lymphoma	8	Defective apoptosis gene
Rb	Retinoblastoma	13	Defective tumor suppressor
NF1	Neurofibromatosis	17	Defective tumor suppressor
BRCA1 and 2	Breast	13	Defective tumor suppressor
HNPCC	Colon	2	Defective DNA mismatch repair
bcl-2	Lymphoma	18	Oncogene/apoptosis blocker
neu/HER2	Breast, prostate	17	Oncogene
Cyclin D1/bc11	B-cell lymphoma, benign parathyroid, breast cancers, and esophageal	11	Oncogene
Ras	Lung, pancreas, liver	Various	Oncogene

*Some genes are named for the cancer they cause, and others are named for the gene, enzyme, or origin of the cancer.

pair mechanisms for DNA. The genes that ordinarily regulate the repair of mismatched nucleotides are defective and can no longer detect and repair mutations. This condition leads to accumulated mutations and increases the likelihood of cancer.

Many discoveries in the basics of cancer genetics are already finding applications. In the future, it may be possible to use gene therapy to cure some types of cancer. Biotechnology and drug companies are already in the process of developing specialized therapies and cancer drugs based on molecular genetics (see chapter 10). Now that the exact locations and DNA sequences of oncogenes and tumor-suppressor genes are known, genetic probes and other markers are being developed at a rapid rate. The time is rapidly approaching when routine screening for susceptibility to many cancers will be as commonplace as routine medical tests. Other important developments are rapidly occurring in immunotherapy using monoclonal antibodies that inactivate receptors on cancer cells. This type of therapy has been used successfully on breast cancer. A vaccine that stimulates the immune system to attack melanoma cancer cells also shows promise

may or may not be normal. Some markers are surface receptors for receiving stimuli for *growth factors,* small polypeptides involved in cellular development that can favor cancerous growth.[9] Some tumors secrete chemical factors that stimulate growth of additional circulatory vessels, an effect not unlike "self-feeding." An important feature of malignant cells is the loss of cohesiveness that keeps normal cells aggregated, so that they readily become dislodged from a tumor, metastasize into the circulation, and are carried to distant sites. Such invasive tumor cells also have the property of binding to and disrupting connective tissue barriers surrounding tissues and organs.

9. It is felt that certain oncogenes are the source of these growth factors.

THE FUNCTION OF THE IMMUNE SYSTEM IN CANCER

What role does the immune system play in controlling cancer? A concept that readily explains the detection and elimination of cancer cells is **immune surveillance.** Many experts postulate that cells with cancer-causing potential arise constantly in the body but that the immune system ordinarily discovers and destroys these cells, thus keeping cancer in check. Experiments with mammals have amply demonstrated that components of cell-mediated immunity interact with tumors and their antigens. The primary types of cells that operate in surveillance and destruction of tumor cells are cytotoxic T cells, natural killer (NK) cells, lymphokine-activated killer cells (LAK), and macrophages. It appears that these cells recognize abnormal or foreign surface markers on the tumor cells and destroy them by mechanisms shown in figure 15.21. Antibodies help destroy tumors by interacting with macrophages and natural killer cells. This involvement of the immune system in destroying cancerous cells has inspired several new therapies (see microfile 17.8).

How do we account for the commonness of cancer in light of this powerful range of defenses against tumors? To an extent, the answer is simply that, as with infection, the immune system can and does fail. In some cases, the cancer may not be immunogenic enough; it may retain self markers and not attract the attention of the surveillance system. In other cases, the tumor antigens may have mutated to escape detection. As we saw earlier, patients with immunodeficiencies such as AIDS and SCID are more susceptible to various cancers because they lack essential T-cell or cytotoxic functions. It may turn out that most cancers are associated with some sort of immunodeficiency, even a slight or transient one.

Chapter Checkpoints

Cancer is caused by genetic transformation of normal host cells into malignant cells. These transformed cells perform no useful function but, instead, grow unchecked and interfere with normal tissue function.

Benign tumors are self-contained and slow-growing and do not differ greatly from their tissues of origin.

Malignant tumors are invasive, fast-growing, and very different from their tissues of origin. They metastasize into organs remote from the site of origin.

Possible causes of cancerous transformation include irregularities in mitosis, genetic damage, activation of oncogenes, and infection by retroviruses.

Cytotoxic T cells, NK cells, LAK cells, and macrophages identify and destroy transformed cancerous cells by recognizing and attaching to foreign surface markers on the transformed cells. Cancerous cells survive when these mechanisms fail.

Immunotherapy offers the most promise in treating cancers, compared with surgery, radiation, or conventional chemotherapy.

Chapter Capsule with Key Terms

IMMUNOPATHOLOGY

The study of disease states involving the malfunction of the immune system is called **immunopathology. Allergy,** or **hypersensitivity,** is an exaggerated, adverse expression of certain immune responses. Abnormal responses to foreign antigens are characteristic of **immune complex disease;** undesirable reactions to foreign tissues are graft reactions. **Autoimmunity** involves abnormal responses to self antigens. A deficiency or loss in immune function is called **immunodeficiency.**

ALLERGY/HYPERSENSITIVITY

An allergic reaction is a heightened immune response to antigens involving misdirected but natural mechanisms of humoral immunity, cellular immunity, phagocytosis, and inflammation. The four types of immune reactions are classified according to the type of lymphocyte involved, type of antigen, and nature of damage.

TYPE I HYPERSENSITIVITIES

Immediate-onset allergies involve contact with **allergens,** antigens that affect certain people; susceptibility is inherited; allergens enter through four portals: Inhalants are breathed in (pollen, dust); ingestants are swallowed (food, drugs); injectants are inoculated (drugs, bee stings); contactants react on skin surface (cosmetics, glue).

Mechanism: On first contact with allergen, specific B cells react with allergen and form a special antibody class called **IgE,** which affixes by its Fc receptor to **mast cells, basophils.** This **sensitizing dose** primes the allergic response system. Upon subsequent exposure (the **provocative dose**), the same allergen binds to the IgE-mast cell complex, causing **degranulation,** release of intracellular granules containing cytokines (**histamine,** serotonin, leukotriene, prostaglandin), with physiological effects such as vasodilation and bronchoconstriction; symptoms are rash, itching, redness, increased mucous discharge, pain, swelling, and difficulty in breathing.

Specific Diseases: An **atopic allergy** is a local reaction to an allergen. **Allergic rhinitis (hay fever)** is a seasonal respiratory allergy; **asthma** is a chronic respiratory condition; **atopic dermatitis (eczema)** is characterized by an itchy skin rash; **food allergy** involves respiratory, cutaneous, and skin reactions to common foodstuffs.

Systemic anaphylaxis is an acute, extreme reaction to allergens that results in severe respiratory and circulatory symptoms; death may occur through compromised respiration and circulatory collapse.

Diagnosis of Allergy: Histamine release test on basophils; serological assays for IgE; skin testing, which injects allergen into the skin, and mirrors the degree of reaction.

Control of Allergy: Drugs are used to control the action of histamine, inflammation, and release of cytokines from mast cells; desensitization therapy involves the administration of purified allergens.

TYPE II HYPERSENSITIVITIES

Type II reactions involve the interaction of antibodies, foreign cells, and complement, leading to lysis of the foreign cells. In transfusion reactions, humans may become sensitized to special isoantigens on the surface of the red blood cells of other humans. The **ABO blood groups** are genetically controlled: Type A blood has A antigens on the RBCs; type B has B antigens; type AB has both A and B antigens; and type O has neither antigen. People produce antibodies against A or B antigens they are lacking; these antibodies can react with these antigens if the wrong blood type is transfused. **Rh factor** is another RBC antigen that becomes a problem if a mother lacking the factor is sensitized by a fetus having the factor; a second fetus can receive antibodies she has made against the factor and develop **hemolytic disease of the newborn.** Prevention involves therapy with **Rh immune globulin.**

TYPE III IMMUNE COMPLEX REACTIONS

Exposure to a large quantity of soluble foreign molecules (serum, drugs) stimulates antibodies that react with antigens to produce large, insoluble Ag-Ab complexes; trapping these **immune complexes** in various organs and tissues incites a damaging inflammatory response. **Arthus reaction** is a local reaction to a series of injected antigens in the same body site; may lead to tissue destruction. **Serum sickness** is a systemic disease resulting from repeated injections of foreign proteins; high levels of circulating immune complexes are deposited in various organs and cause tissue damage.

Autoimmunity: In certain type II and III hypersensitivities, the immune system has lost tolerance to self molecules (autoantigens) and forms autoantibodies and sensitized T cells against them. Disruption of function can be systemic or organ specific; diseases are genetically determined and more common in females.

Systemic lupus erythematosus (SLE) is a chronic, systemic disease in which antibodies are deposited in the kidney, skin, lungs, and heart. **Rheumatoid arthritis** is a chronic systemic autoimmunity in the joints; appears to be associated with immune complexes that cause chronic inflammation and scar tissue. **Endocrine autoimmunities** include Graves' disease, Hashimoto's thyroiditis, and types I and II **diabetes mellitus. Myasthenia gravis** is an immune attack upon the myoneural junction, with muscle paralysis. In **multiple sclerosis,** T cells and antibodies damage the myelin sheath of nerve cells; accompanied by motor and sensory loss.

TYPE IV CELL-MEDIATED HYPERSENSITIVITY

A delayed response to antigen involving the activation of and damage by T cells.

Delayed Allergic Response: Skin response to allergens, including infectious agents. Example is **tuberculin reaction;** contact dermatitis is caused by exposure to poison plants (ivy, oak) and simple environmental molecules (metals, cosmetics); cytotoxic T cells acting on allergen elicit a skin reaction.

Graft Rejection: Reaction of cytotoxic T cells directed against foreign cells of a grafted tissue; involves recognition of foreign HLA by T cells and rejection of tissue. Host may reject graft; graft may reject host. Types of grafts include: **autograft,** from one part of body to another; **isograft,** grafting between identical twins; **allograft,** between two members of same species; **xenograft,** between two different species. All major organs (heart, kidney, bone marrow, skin) may be successfully transplanted. Allografts require **tissue match** (HLA antigens must correspond); rejection is controlled with drugs.

Immunodeficiency Diseases: Components of the immune response system are absent; involve B and T cells, phagocytes, complement. **Primary immunodeficiency** is genetically based, congenital; defect in inheritance leads to lack of B-cell activity, T-cell activity, or both. **B-cell defect** is called **agammaglobulinemia;** patient lacks antibodies; serious recurrent bacterial infections result. In Ig deficiency, one of the classes of antibodies is missing or deficient. In **T-cell defects,** the thymus is missing or abnormal. In **DiGeorge syndrome,** the thymus fails to develop; afflicted children experience recurrent infections with eucaryotic pathogens and viruses; immune response is generally underdeveloped. In **severe combined immunodeficiency (SCID),** both limbs of the lymphocyte system are missing or defective; no adaptive immune response exists; fatal without replacement of bone marrow or special facilities. **Secondary (acquired) immunodeficiency** is due to damage after birth (infections, drugs, radiation). AIDS is the most common of these; T helper cells are main target; deficiency manifests in numerous opportunistic infections and cancers.

CANCER AND THE IMMUNE SYSTEM

Cancer is characterized by overgrowth of abnormal tissue, also known as a *neoplasm.* It appears to arise from malfunction of immune surveillance. **Tumors** can be **benign** (a nonspreading local mass of tissue) or **malignant** (a cancer) that spreads (metastasizes) from the tissue of origin to other particular sites by means of the circulation; malignant tumors may be **carcinomas,** originating from epithelial tissue, or **sarcomas,** originating from embryonic connective tissue. Cancers occur in nearly every cell type (except mature, nondividing cells).

Cancer cells appear to share a common basic mechanism involving some type of gene alteration that turns a normal gene **(proto-oncogene)** into an **oncogene;** this **transforms** the cell and triggers uncontrolled growth and abnormal structure and function. Gene alterations may be due to chromosome disruptions during cell division, viral infection, or the actions of chemical and physical **carcinogens.** Malignant cancer cells display special markers, respond to growth factors, and lose their cohesiveness. Treatment is with surgery, chemicals, radiation, and immune therapy.

MULTIPLE-CHOICE QUESTIONS

1. Pollen is which type of allergen?
   a. contactant c. injectant
   b. ingestant d. inhalant
2. B cells are responsible for which allergies?
   a. asthma c. tuberculin reactions
   b. anaphylaxis d. both a and b
3. Which allergies are T cell–mediated?
   a. type I c. type III
   b. type II d. type IV
4. The contact with allergen that results in symptoms is called the
   a. sensitizing dose
   b. degranulation dose
   c. provocative dose
   d. desensitizing dose
5. Production of IgE and degranulation of mast cells are involved in
   a. contact dermatitis c. Arthus reaction
   b. anaphylaxis d. both a and b
6. The direct, immediate cause of allergic symptoms is the action of
   a. the allergen directly on smooth muscle
   b. the allergen on B lymphocytes
   c. allergic mediators released from mast cells and basophils
   d. IgE on smooth muscle
7. Theoretically, type ____ blood can be donated to all persons because it lacks ____
   a. AB, antibodies c. AB, antigens
   b. O, antigens d. O, antibodies
8. An example of a type III immune complex disease is
   a. serum sickness c. graft rejection
   b. contact dermatitis d. atopy
9. Type II hypersensitivities are due to
   a. IgE reacting with mast cells
   b. activation of cytotoxic T cells
   c. IgG-allergen complexes that clog epithelial tissues
   d. complement-induced lysis of cells in the presence of antibodies
10. Production of autoantibodies may be due to
    a. emergence of forbidden clones of B cells
    b. production of antibodies against sequestered tissues
    c. infection-induced change in receptors
    d. all of these are possible
11. Rheumatoid arthritis is an ____ that affects the ____ .
    a. immunodeficiency disease, muscles
    b. autoimmune disease, nerves
    c. allergy, cartilage
    d. autoimmune disease, joints
12. A positive tuberculin skin test is an example of
    a. a delayed-type allergy
    b. acute contact dermatitis
    c. autoimmunity
    d. eczema
13. Which disease would be most similar to AIDS in its pathology?
    a. X-linked agammaglobulinemia
    b. SCID
    c. ADA deficiency
    d. DiGeorge syndrome
14. A general feature common to all cancers is a ____ that leads to ____ .
    a. carcinogenic agent, carcinoma
    b. mutation, gene amplification
    c. genetic defect, uncontrolled cell growth
    d. proto-oncogene, anti-oncogene
15. The property whereby cancer cells display abnormal anatomy and physiology is called
    a. metastasis c. carcinogenic
    b. oncogene d. malignancy
16. A cancer associated with viral infections includes
    a. cervical carcinoma
    b. Burkitt's lymphoma
    c. T-cell leukemia
    d. all of these

CONCEPT QUESTIONS

1. a. Define allergy and hypersensitivity.
   b. What accounts for the reactions that occur in these conditions?
   c. What does it mean when a reaction is immediate or delayed?
   d. Give examples of each type.
2. Describe several factors that influence types and severity of allergic responses.
3. a. How are atopic allergies similar to anaphylaxis?
   b. How are they different?
4. a. How do allergens gain access to the body?
   b. What are some examples of allergens that enter by these portals?
5. a. Trace the course of a pollen grain through sensitization and provocation in type I allergies.
   b. Include in the discussion the role of mast cells, basophils, IgE, and allergic mediators.
   c. Outline the target organs and symptoms of the principal atopic diseases and their diagnosis and treatment.
6. a. Describe the allergic response that leads to anaphylaxis. Include its usual causes, how it is diagnosed and treated, and two effective physiological targets for treatment.
   b. Explain how hyposensitization is achieved and suggest two mechanisms by which it might work.
7. a. What is the mechanism of type II hypersensitivity?
   b. Why are the tissues of some people antigenic to others?
   c. Would we be concerned about this problem if it were not for transfusions?
   d. What is the actual basis of the four ABO and Rh blood groups?
   e. Where do we derive our natural hypersensitivities to the A or B antigens that we do not possess?
   f. How does a person become sensitized to Rh factor? List consequences.
8. Explain the rules of transfusion. Illustrate what will happen if type A blood is accidently transfused into a type B person.
9. a. Contrast type II and type III hypersensitivities with respect to type of antigen, antibody, and manifestations of disease.
   b. What is immune complex disease?
   c. Differentiate between the Arthus reaction and serum sickness.
10. a. Explain the pathologic process in autoimmunity.
    b. Draw diagrams that explain five mechanisms to describe how autoimmunity develops.
    c. Describe four major types of autoimmunity, comparing target organs and symptoms.

11. Compare and contrast type I (atopic) and type IV (delayed) hypersensitivity as to mechanism, symptoms, eliciting factors, and allergens.

12. a. What is the molecular/cellular basis for a host rejecting the graft tissue?
    b. Account for a graft rejecting the host.
    c. Compare the four types of grafts.
    d. What does it mean to say that two tissues constitute a close match?
    e. Describe the procedure involved in a bone marrow transplant.

13. a. In general, what causes primary immunodeficiencies?
    b. Acquired immunodeficiencies?
    c. Why can T-cell deficiencies have greater impact than B-cell deficiencies?
    d. What kinds of symptoms accompany a B-cell defect?
    e. A T-cell defect?
    f. Combined defects?
    g. Give examples of specific diseases that involve each type of defect.

14. a. Name several medical conditions that require immunosuppressive drugs.
    b. Name some immunosuppressive drugs, and explain what they do.

15. a. Define cancer, and give some synonyms.
    b. Differentiate between a benign tumor and a malignant tumor, and give examples.
    c. How are cancers technically differentiated?

16. a. Describe a possible mechanism to account for transformation of a normal cell to a cancer cell.
    b. Describe the possible genetic and environmental factors that increase oncogenesis.
    c. How do infectious agents such as viruses cause cancer?

17. Relate how the immune system is involved in cancer.

18. Matching: Choose the description that best fits the cytokines.

___ histamine
___ prostaglandin
___ bradykinin
___ leukotriene
___ platelet factor

a. a peptide involved in blood clotting and chemotaxis
b. a lipid that aggregates and lyses thrombocytes
c. causes prolonged bronchospasm and mucous secretion in the lungs
d. increases inflammation and sensitivity to pain
e. a potent stimulus for smooth muscle and glandular secretion

CRITICAL-THINKING QUESTIONS

1. a. Discuss the reasons that the immune system is sometimes called a double-edged sword.
   b. Suggest a possible function of allergy.

2. A 3-week-old neonate develops severe eczema after being given penicillin therapy for the first time. Can you explain what has happened?

3. Can you explain why a person is allergic to strawberries when he eats them, but shows a negative skin test to them?

4. a. Where in the course of type I allergies do antihistamine drugs work?
   b. Cortisone?
   c. Desensitization?

5. a. Although we call persons with type O blood universal donors and those with type AB blood universal recipients, what problems might accompany transfunctions involving O donors or AB recipients?
   b. What is the truly universal donor and recipient blood type? (Include Rh type.)
   c. Can you explain how a person could have a genotype for type A or B blood but a phenotype for type O?

6. Why would it be necessary for an Rh^- woman who has had an abortion, miscarriage, or an ectopic pregnancy to be immunized against the Rh factor?

7. a. Describe three circumstances that might cause antibodies to develop against self tissues.
   b. Can you explain how people with autoimmunity could develop antibodies against intracellular components (nucleus, mitochondria, and DNA)?

8. Would a person show allergy to poison oak upon first contact? Explain.

9. a. Looking at figure 17.19, predict which family members would be good allograft pairs and which would be incompatible.
   b. Why might a graft be rejected even between closely matched siblings?
   c. Transplanting a baboon heart into a baby is what kind of graft?

10. Why are primary immunodeficiencies considered experiments of nature?

11. Why are defective genes found on the X chromosome expressed most often in males?

12. a. Explain why babies with agammaglobulinemia do not develop opportunistic infections until about 6 months after birth.
    b. Explain why people with B-cell deficiencies can benefit from artificial passive therapy. Explain whether vaccination would work for them.

13. a. Why are SCID children unable to reject grafts?
    b. What would be the major problem in most bone marrow transplantations for these children?
    c. Draw a general outline of the events that cause the "bubble boy syndrome."
    d. Why did David acquire cancer?

14. In what ways can cancer be both a cause and a symptom of immunodeficiency?

15. What features of cancer cells account for metastasis?

16. In what ways could cancer be inherited?

17. If a chemical is found to be a mutagen by the Ames test, what other factors will be required for it to induce cancer?

18. Most cancer therapy removes or destroys the metastatic cells. Can you think of a genetic prevention for cancer? (Hint: You may have to refer to chapter 10.)

INTERNET SEARCH TOPICS

Go to a search engine and locate web sites involving these areas of interest:

1 Angiogenesis inhibitor drugs and tumor suppressor gene therapy and their mechanism of action. What is the current status of their use in treating cancer?
2. Poison ivy or oak hypersensitivity. How can you identify and treat it?
3. Bone marrow transplants. Locate information on registering to donate marrow and the process surrounding the transplantation.

Introduction to Identification Techniques in Medical Microbiology

Introduction to Identification Techniques in Medical Microbiology

PREFACE TO THE SURVEY OF MICROBIAL DISEASES

The next eight chapters cover the most clinically significant bacterial, fungal, parasitic, and viral diseases in the world, with an emphasis on those most prevalent in North America. This survey will encompass the morphology, physiology, and virulence of the microbes and the epidemiology, pathology, treatment, and prevention of the diseases they cause. The bacteria are divided into the gram-positive and gram-negative cocci (chapter 18); the gram-positive bacilli (chapter 19); the gram-negative bacilli (chapter 20); miscellaneous bacteria and dental infections (chapter 21); fungi (chapter 22); protozoan and helminth parasites (chapter 23); DNA viruses (chapter 24); and RNA viruses (chapter 25).

Several explanatory concepts can be found in previous chapters. Information on general methods of specimen processing is in figures 3.1 and 6.24 and in chapters 4 and 10. Other references include media (chapter 3), chemotherapy (chapter 12), and host-parasite relations and epidemiology (chapter 13). Particular attention is called to microfile 13.5 on the terminology of disease, microfile 13.6 on skin lesions, and the discussion of vaccines and serological tests (chapter 16). An extensive reference that lists organisms and diseases by site of infection is located in appendix F.

The investigation and diagnosis of these diseases depend to a considerable degree on a series of basic laboratory techniques. This section summarizes the methods used by clinical microbiologists and other medical personnel to collect, isolate, and identify infectious agents.

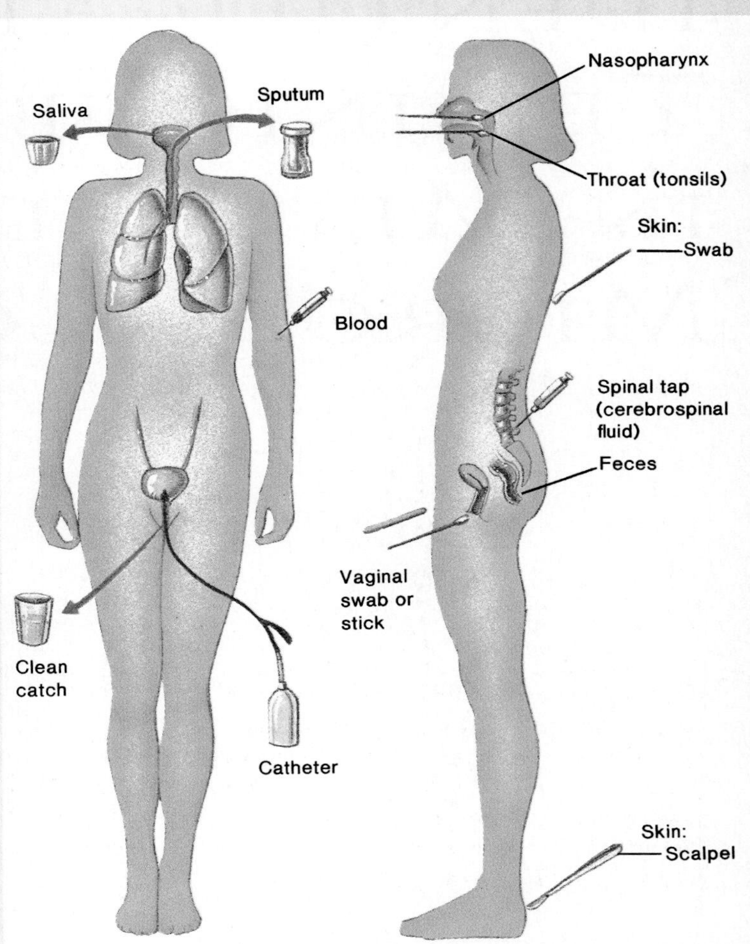

Figure A
Sampling sites and methods of collection for clinical laboratories.

ON THE TRACK OF THE INFECTIOUS AGENT: SPECIMEN COLLECTION

Regardless of the method of diagnosis, specimen collection is the common point that guides the health care decisions of every member of a clinical team. Indeed, the success of identification and treatment depends on how specimens are collected, handled, and stored. Specimens can be taken by a medical technologist, nurse, physician, or even by the patient himself. However, it is imperative that general aseptic procedures be used, including sterile sample containers and other tools to prevent contamination from the environment or the patient. Figure A delineates the most common sampling sites and procedures.

In sites that normally contain resident microflora, care should be taken to sample only the infected site and not surrounding areas. For example, throat and nasopharyngeal swabs should not touch the tongue, cheeks, or saliva. Saliva is an especially undesirable contaminant because it contains millions of bacteria per milliliter, most of which are normal flora. Saliva samples are occasionally taken for dental diagnosis by having the patient expectorate into a container. Depending on the nature of the lesion, skin can be swabbed or scraped with a scalpel to expose deeper layers. The mucous lining of the vagina, cervix, or urethra can be sampled with a swab or applicator stick.

Urine is taken aseptically from the bladder with a thin tube called a catheter. Another method, called a "clean catch," is taken by washing the external urethra and collecting the urine midstream. The latter method inevitably incorporates a few normal flora into the sample, but these can usually be differentiated from pathogens in an actual infection. Sputum, the mucous secretion that coats the lower respiratory surfaces, especially the lungs, is discharged by coughing or taken by catheterization to avoid contamination with saliva. Sterile materials such as blood, cerebrospinal fluid, and tissue fluids must be taken by sterile needle aspiration. Antisepsis of the puncture site is extremely important in these cases. Additional sources of specimens are the vagina, eye, ear canal, nasal cavity (all by swab), and diseased tissue that has been surgically removed (biopsied).

After proper collection, the specimen is promptly transported to a lab and stored appropriately (usually refrigerated) if it must be held for a time. Nonsterile samples in particular, such as urine, feces, and sputum, are especially prone to deterioration at room temperature. Special swab and transport systems are designed to collect the specimen and maintain it in stable condition for several hours. These devices contain nonnutritive maintenance media (so the microbes do not grow), a buffering system, and an anaerobic environment to prevent possible destruction of oxygen-sensitive bacteria.

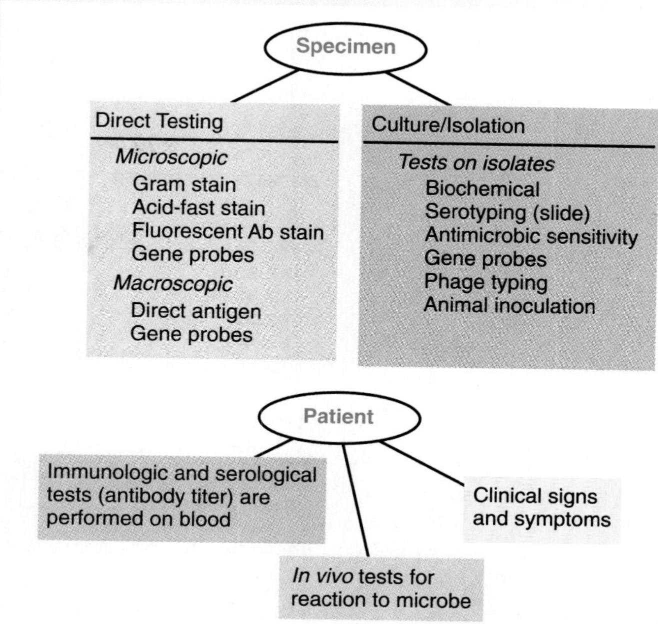

Figure B
A scheme of specimen isolation and identification.

Overview of Laboratory Techniques

The routes taken in specimen analysis are the following: (1) direct tests using microscopic, immunologic, or other specific methods that provide immediate clues as to the identity of the microbe or microbes in the sample and (2) cultivation, isolation, and identification of pathogens using a wide variety of general and specific tests (figure B). Most test results fall into two categories: **presumptive data,** which place the isolated microbe (isolate) in a preliminary category such as a genus, and more specific, **confirmatory data,** which provide more definitive evidence of a species. Some tests are more important for some groups of bacteria than for others. The total time required for analysis ranges from a few minutes in a streptococcal sore throat to several weeks in tuberculosis.

Results of specimen analysis are entered in a summary patient chart (figure C) that can be used in assessment and treatment regimens. The type of antimicrobic drugs chosen for testing varies with the type of bacterium isolated.

Some diseases are diagnosed without the need to identify microbes from specimens. Serological tests on a patient's serum can detect signs of an antibody response. One method that clarifies whether a positive test indicates current or prior infection is to take two samples several days apart to see if the antibody titer is rising. Skin testing can pinpoint a delayed allergic reaction to a microorganism (see chapters 16 and 17). These tests are also important in screening the general population for exposure to an infectious agent such as rubella or tuberculosis.

Because diagnosis is both a science and an art, the ability of the practitioner to interpret signs and symptoms of disease can be very important. AIDS, for example, is usually diagnosed by serological tests and a complex of signs and symptoms without ever isolating the virus. Some diseases (athlete's foot, for example) are diagnosed purely by the typical presenting symptoms and may require no lab tests at all.

Immediate Direct Examination of Specimen

Direct microscopic observation of a fresh or stained specimen is one of the most rapid methods of determining presumptive and sometimes confirmatory characteristics (see figures 18.26 and 22.25). Stains most often employed for bacteria are the Gram stain (see microfile 4.1) and the acid-fast stain (see figure 19.15). For many species these ordinary stains are useful, but they do not work with certain organisms. Direct fluorescence antibody (DFA) tests can highlight the presence of the microbe in patient specimens by means of labeled antibodies (figure D). DFA tests are particularly useful for bacteria, such as the syphilis spirochete, that are not readily cultivated in the laboratory or if rapid diagnosis is essential for the survival of the patient.

Another way that specimens can be analyzed is through *direct antigen testing,* a technique similar to direct fluorescence in that known antibodies are used to identify antigens on the surface of bacterial isolates. But in direct antigen testing, the reactions can be seen with the naked eye. Quick test kits that greatly speed clinical diagnosis are available for *Staphylococcus aureus, Streptococcus pyogenes* (see figure 18.14), *Neisseria gonorrhoeae, Haemophilus influenzae,* and *Neisseria meningitidis.* However, when the microbe is very sparse in the specimen, direct testing is like looking for a needle in a haystack, and more sensitive methods are necessary.

BACT NUMBER ______________

Summit Health Ltd **ST. LUKE MEDICAL CENTER**
2632 E. WASHINGTON BLVD. - BIN 7021
PASADENA, CA 91109-7021
J.R. CRAIG, M.D., Ph. D.
ROY H. MOFFATT, M.D.

MICROBIOLOGY UNIT

DATE, TIME & PERSON COLLECTING	SPECIMEN NUMBER	ANTIBIOTIC THERAPY	TENTATIVE DIAGNOSIS

SOURCE OF SPECIMEN

☐ THROAT
☐ SPUTUM
☐ STOOL
☐ CERVIX
☐ AEROSOL INDUCED SPUTUM
☐ WOUND - SPECIFY SITE ______
☐ OTHER - SPECIFY ______
☐ BLOOD
☐ URINE - CLEAN CATCH
☐ URINE - CATH
☐ BRONCHIAL WASHING

TEST REQUEST

☐ GRAM STAIN
☐ ROUTINE CULTURE
☐ SENSITIVITY
☐ MIC
☐ ANAEROBIC CULTURE
☐ R/O GROUP A STREP
☐ WRIGHT STAIN (WBC)
☐ ACID FAST SMEAR
☐ ACID FAST CULTURE
☐ FUNGUS WET MOUNT
☐ FUNGUS CULTURE
☐ PARASITE STUDIES
☐ OCCULT BLOOD
☐ PCR ANALYSIS

DO NOT WRITE BELOW THIS LINE -# FOR LAB USE ONLY

GRAM STAIN (4+ NUMEROUS; 3+ MANY; 2+ MODERATE; 1+FEW; 0 NONE SEEN)
COCCI: GRAM POS.______ GRAM NEG.______ W B C______
BACILLI: GRAM POS.______ GRAM NEG.______ EPITHELIAL CELLS______
INTRACELLULAR & EXTRACELLULAR GRAM-NEGATIVE DIPLOCOCCI______
YEAST______ ☐ No organisms seen.

FUNGUS: WET MOUNT ☐ No mycotic elements or budding structures seen.
☐ ______
CULTURE ☐ ______

AFB: SMEAR ☐ No acid fast bacilli seen.
☐ ______
CULTURE ☐ ______

PARASITE STUDIES: DIRECT:______
CONCENTRATE:______
PERMANENT:______

OCCULT BLOOD:
APPEARANCE OF STOOL:______
OCCULT BLOOD:______

COLONY COUNT: Urine organisms/ml. ☐ ______ ☐ > 100,000

MISCELLANEOUS RESULTS:
☐ NO GROWTH IN: ☐ 2 DAYS ☐ 3 DAYS ☐ 5 DAYS ☐ 7 DAYS
☐ NORMAL FLORA ISOLATED
☐ NO ENTEROPATHOGENS ISOLATED
☐ SPUTUM UNACCEPTABLE FOR CULTURE — REPRESENTS SALIVA — NEW SPECIMEN REQUESTED
☐ URINE > 2 COLONY TYPES PRESENT REPRESENT CONTAMINATION — NEW SPECIMEN REQUESTED

CULTURE RESULTS

1+ FEW 3+ MANY
2+ MODERATE 4+ NUMEROUS

ANAEROBES: ☐ BACTEROIDES ☐ CLOSTRIDIUM ☐ PEPTOSTREPTOCOCCUS ☐ ☐
ENTERICS: ☐ ESCHERICHIA COLI ☐ ENTEROBACTER ☐ KLEBSIELLA ☐ PROTEUS ☐
STAPHYLOCOCCUS: ☐ AUREUS ☐ EPIDERMIDIS ☐ SAPROPHYTICUS
STREPTOCOCCUS: ☐ GROUP A ☐ GROUP B ☐ GROUP D ENTEROCOCCI ☐ GROUP D NON ENTEROCOCCI ☐ PNEUMONIAE ☐ VIRIDANS ☐
YEAST: ☐ CANDIDA ☐
OTHER ISOLATES: ☐ PSEUDOMONAS ☐ HAEMOPHILUS ☐ GARDNERELLA VAGINALIS ☐ NEISSERIA ☐ CL. DIFFICILE ☐

SENSITIVITY TESTS

NOTE: Bacteria with intermediate susceptibility may not respond satisfactorily to therapy.

	AMIKACIN	AMPICILLIN	BETA LACTAMASE PRODUCTION	CARBENICILLIN	CEFAZOLIN	CEFOTAXIME	CEFOXITIN	CEFUROXIME	CHLORAMPHENICOL	CLINDAMYCIN	ERYTHROMYCIN	GENTAMICIN	METHICILLIN	METRONIDAZOLE	NITROFURANTOIN	PENICILLIN	IMMUNOLOGY	TETRACYCLINE	TOBRAMYCIN	TRIMETHO PRIM SULFAME-THOXAZOLE	VANCOMYCIN	CIPROFLOX
A																						
B																						
C																						

☐ COMMENTS: ______

DATE ______ TECHNOLOGIST ______

706-30A

Figure C
Example of a clinical form used to report data on patients' specimens.

Cultivation of Specimen

Isolation Media Such a wide variety of media exist for microbial isolation that a certain amount of preselection must occur, based on the nature of the specimen. In cases where the suspected pathogen is present in small numbers or is easily overgrown, the specimen can be initially enriched with specialized media. In specimens such as urine and feces that have high bacterial counts and a diversity of species, selective media are used (see figure 20.9). In most cases, specimens are also inoculated into differential media that define such characteristics as reactions in blood (blood agar) and fermentation patterns (mannitol salt and MacConkey agar). A patient's blood is usually cultured

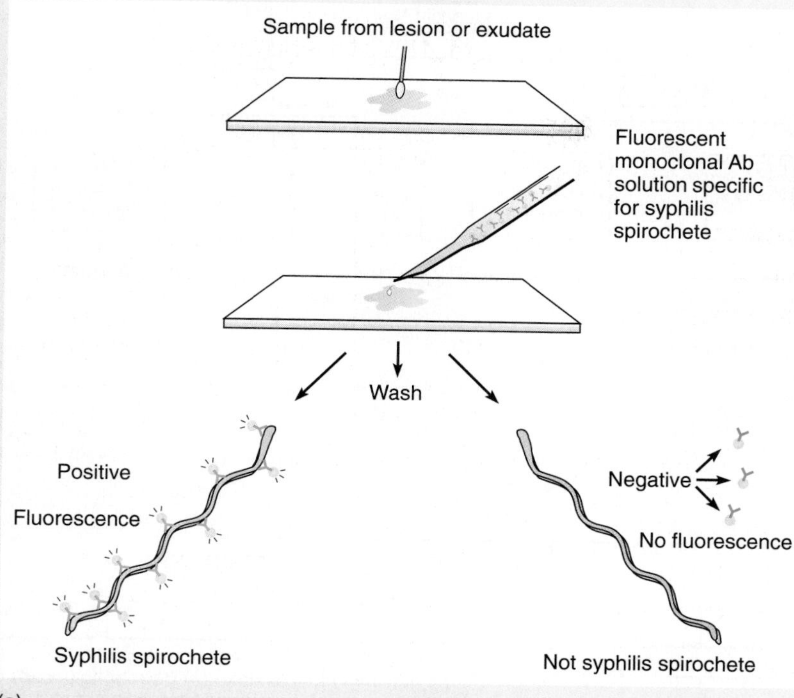

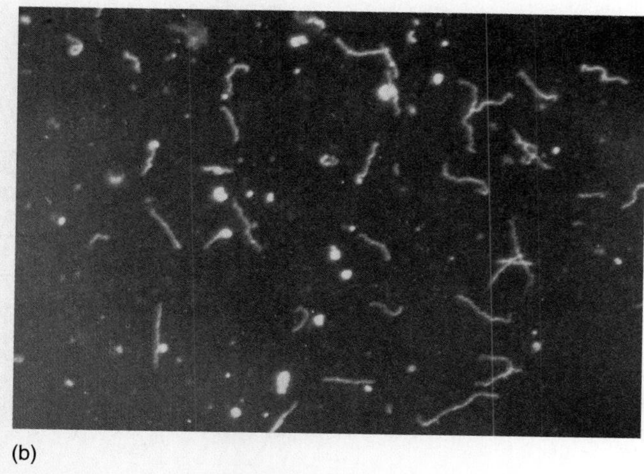

Figure D
(*a*) Direct fluorescent antigen test results for *Treponema pallidum,* the syphilis spirochete, and an unrelated spirochete. (*b*) Photomicrograph of this technique used on a blood sample from a syphilitic patient.

in a special bottle of broth that can be periodically sampled for growth. Numerous other examples of isolation, differential, and biochemical media were presented in chapter 3 and appear in various sections of chapters 18, 19, 20, and 21. So that subsequent steps in identification will be as accurate as possible, all work must be done from isolated colonies or pure cultures, because working with a mixed or contaminated culture gives misleading and inaccurate results. From such isolates, clinical microbiologists obtain information about a pathogen's microscopic morphology and staining reactions, cultural appearance, motility, oxygen requirements, and biochemical characteristics.

Biochemical Testing The physiological reactions of bacteria to nutrients and other substrates provide excellent indirect evidence of the types of enzyme systems present in a particular species. Many of these tests are based on the following scheme:

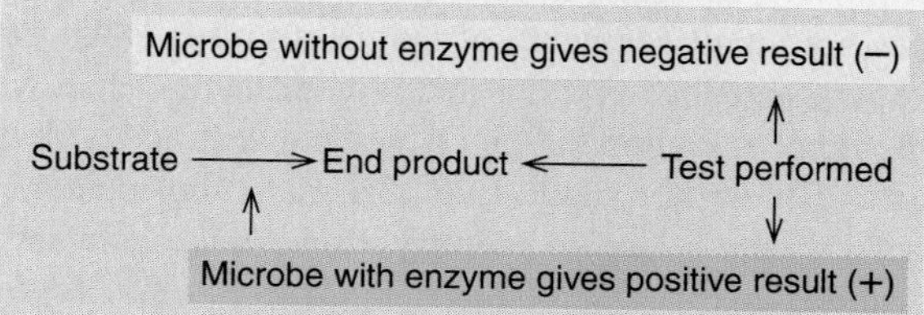

The microbe is cultured in a medium with a special substrate and then tested for a particular end product. The presence of the end product indicates that the enzyme is expressed in that species; its absence means it lacks the enzyme for utilizing the substrate in that particular way. These types of reactions are particularly meaningful in bacteria, which are haploid and generally express their genes for utilizing a given nutrient.

Among the prominent biochemical tests are carbohydrate fermentation (acid and/or gas); hydrolysis of gelatin, starch, and other polymers; enzyme actions such as catalase, oxidase, and coagulase; and various by-products of metabolism. Examples of these tests are presented in chapters 18, 19, and 20. Many are presently performed with rapid, miniaturized systems that can simultaneously determine up to 23 characteristics in small individual cups or spaces (see figures 18.7 and 20.10). An important plus, given the complexity of biochemical profiles, is that such systems are readily adapted to computerized analysis.

Common schemes for identifying bacteria are somewhat artificial but convenient. They are based on easily recognizable characteristics such as motility, oxygen requirements, Gram stain reactions, shape, spore formation, and various biochemical reactions. Schemes can be set up as flowcharts (figure E) or keys that trace a route of identification by offering pairs of opposing characteristics (positive versus negative, for example) from which to select. Eventually, an endpoint is reached, and the name of a genus or species that fits that particular combination of characteristics appears. Diagnostic tables that provide more complete information are preferred by many laboratories because variations from the general characteristics used on the flowchart can be misleading. Both systems are used in this text.

Miscellaneous Tests When morphological and biochemical tests are insufficient to complete identification, other tests come into play. Commercial **serotyping** kits are important for identifying isolates in the genera *Salmonella, Shigella, Streptococcus* and *Staphylococcus.* In general, a bit of culture is mixed with specific, animal-derived antisera or monoclonal antibodies that agglutinate the cells if they are complementary (technique shown in figure 16.3).

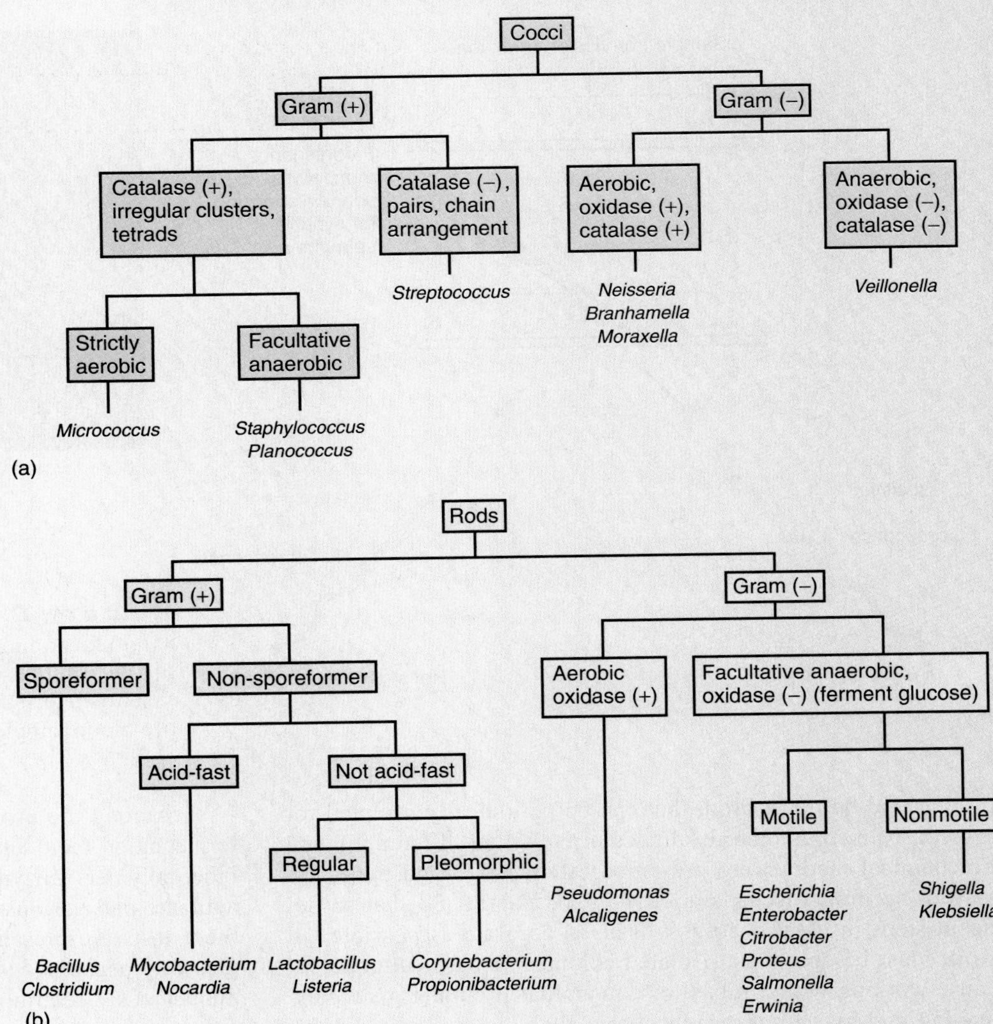

Figure E
Flowchart to separate primary genera of gram-positive and gram-negative (*a*) cocci and (*b*) rods involved in human diseases.

Bacteria host viruses called bacteriophages that are very species- and strain-specific. Such selection by a virus for its host is useful in typing some bacteria, primarily *Staphylococcus* and *Salmonella.* The technique of **phage typing** involves inoculating a lawn of cells onto a Petri dish, mapping it off into blocks, and applying a different phage to each block. Cleared areas corresponding to lysed cells indicate sensitivity to that phage. Phage typing is chiefly used for tracing strains of bacteria in epidemics.

Animals must be inoculated to cultivate bacteria such as *Mycobacterium leprae* and *Treponema pallidum,* whereas avian embryos and cell cultures are used to grow rickettsias, chlamydias, and viruses. Animal inoculation is also occasionally used to test bacterial or fungal virulence.

Modern molecular biology has provided some extremely specific and sensitive systems for identifying microbes. The value of such molecular testing is that it measures similarities or differences in the actual molecular and genetic structure of microbes, not just in morphological appearance or physiological reactions.

The uses of these methods are expanding rapidly. Some of them have been discussed and illustrated in prior chapters and others will be covered with individual infectious agents. **Gene probes** and hybridization techniques based on unique arrangments of nucleotides in DNA are gaining in importance (see figures 4.28 and F). Probes are small segments of single-stranded DNA or RNA with a known sequence that are mixed with nucleic acid from an unknown microbe. Hybridization of the known strand with the unknown strand is highly suggestive of a matching identity. Gene probes are currently used to identify *Escherichia coli* and other gram-negative bacteria, *Mycobacterium, Mycoplasma, Legionella, Chlamydia, Treponema* and numerous viruses.

Many nucleic acid assays use the polymerase chain reaction (PCR). This method can amplify DNA present in samples even in tiny amounts, which greatly improves the sensitivity of the test (see figure 10.6). PCR tests are being used or developed for a wide variety of bacteria, viruses, protozoa, and fungi.

Antimicrobic sensitivity tests are not only important in determining the drugs to be used in treatment (see figure 12.21), but

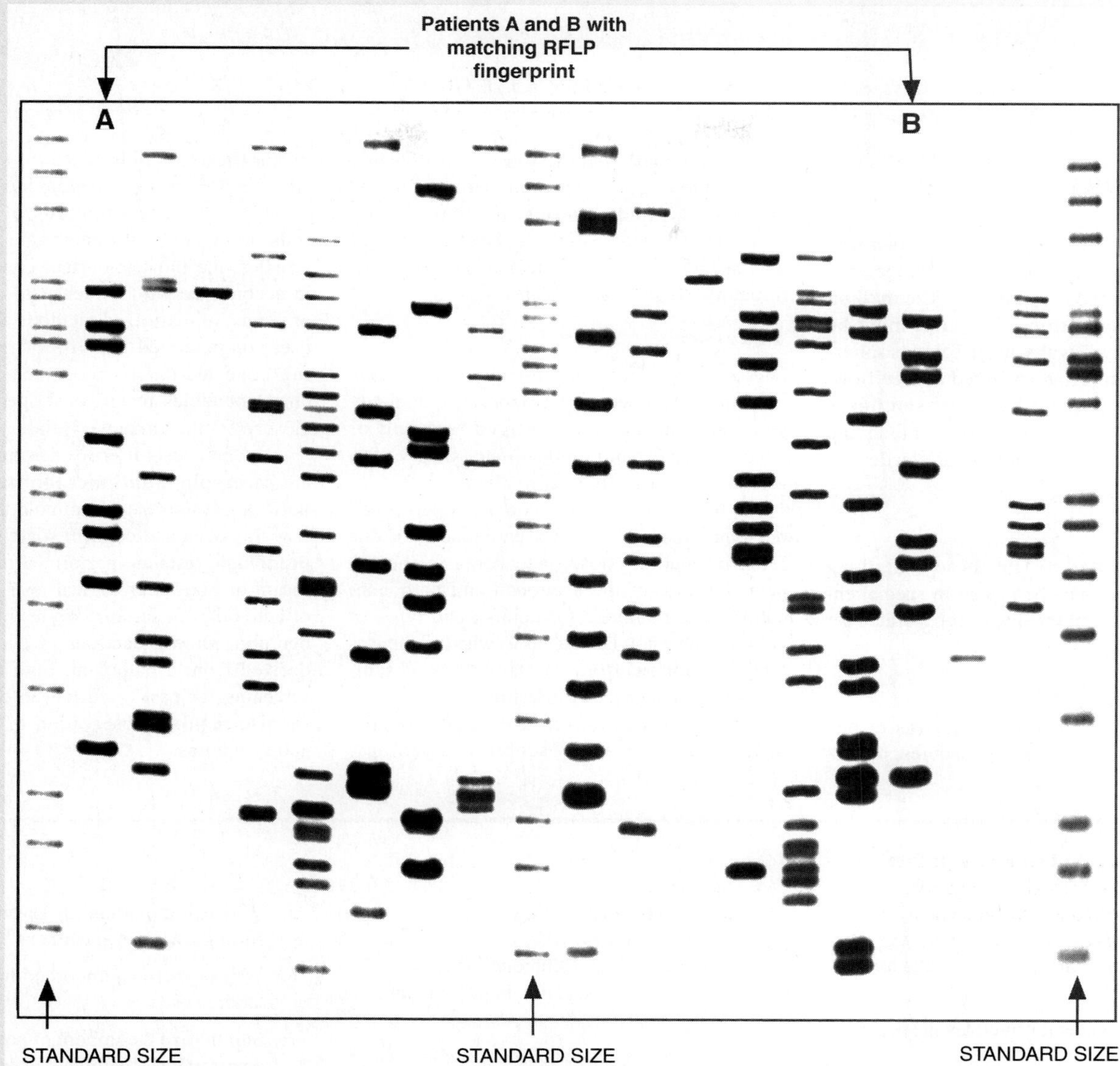

Figure F
A DNA fingerprint called a restriction fragment length polymorphism (RFLP) is shown for *Mycobacterium tuberculosis.* The pattern shows results for various strains isolated from 17 patients. Bands were developd by DNA hybridization using probes specific to genes from several *M. tuberculosis* strains. Lanes 1, 10, and 20 provide size markers for reference. Patients A and B are infected with the same common strain of the pathogen.

the patterns of sensitivity can also be used in presumptive identification of some species of *Streptococcus* (see figure 18.14*a*), *Pseudomonas,* and *Clostridium.* Antimicrobics are also used as selective agents in many media.

Determining Clinical Significance of Cultures Questions that can be difficult but necessary to answer in this era of debilitated patients and opportunists are: Is an isolate clinically important, and how do you decide whether it is a contaminant or just part of the normal flora? The number of microbes in a sample is one useful criterion. For example, a few colonies of *Escherichia coli* in a urine sample can simply indicate normal flora, whereas several hundred can mean active infection. In contrast, the presence of a single colony of a true pathogen such a *Mycobacterium tuberculosis* in sputum or an opportunist in sterile sites such as cerebrospinal fluid or blood is highly suggestive of its role in disease. Furthermore, the repeated isolation of a relatively pure culture of any microorganism can mean it is an agent of disease, though care must be taken in this diagnosis. Another problem facing the laboratory technician is that of differentiating a pathogen from species in the normal flora that are similar in morphology from their more virulent relatives.

REVIEW OF CLINICAL LABORATORY TECHNOLOGY WITH KEY TERMS

CONCERNS OF THE CLINICAL LABORATORY

The lab provides technical support in determining the causative agents of disease and the antimicrobics to be used for treatment. The **method of specimen collection** is crucial; must be done aseptically; swabs used for nasopharynx, vagina, skin; urine collected as free flow; sterile urine requires a catheter; sputum is coughed up or aspirated from lung; blood and spinal fluid taken by puncture of sterile site with needle.

HANDLING

Samples transported to lab; those containing fragile pathogens must be placed in special environment, processed rapidly, or refrigerated for a short time.

LABORATORY PROTOCOLS

Identification of microbe and diagnosis of disease require a wide array of procedures, including (1) direct specimen testing; (2) cultivation, and isolation, (3) examination of biochemistry, antigenicity, and genetics of microbe; and (4) examination of patient. Collected data fall into categories of presumptive evidence or confirmatory evidence of a species.

TESTING

Direct tests include microscopic examination of stained specimen, direct fluorescent antibody tests, and macroscopic antigen tests, all of which provide rapid clinical data. **Cultivation** is usually required for confirmation.

Initial isolation of pathogen is performed on some types of selective, enrichment, or differential media, chosen according to the expected nature of the specimen and pathogen; cultures are examined for number and types of colonies; decision is made as to whether the isolates are normal flora, contaminants of sampling, or actual cause of infection.

Isolated colonies are examined for microscopic and macroscopic morphology. Further tests include motility, special stains, and the presence of enzyme systems for processing nutrients; hundreds of biochemical tests are available to determine these characteristics; the most important ones are selected in accordance with evidence of a given group or genus; miniaturized methods are in common use; computerized analysis of results is helpful; collected data is processed through flowcharts or tables to arrive at species that most closely fits the unknown isolate.

Other tests used in determining species are **serotyping, antimicrobic sensitivity, genetic analysis;** use of animals may be necessary to isolate and confirm some microbes. **Immunologic tests** are performed on the patient's serum or body: Serological tests give evidence of antibodies in serum; skin testing (such as tuberculin) shows reactions of body to proteins derived from a pathogen; both are helpful in screening for past or current infection; diagnosis also requires observation of clinical signs and symptoms.

REVIEW QUESTIONS

1. Why do specimens need to be taken aseptically even when nonsterile sites are being sampled and selective media are to be used?
2. Explain the general principles in specimen collection.
3. a. What is involved in direct specimen testing?
   b. In presumptive and confirmatory tests?
   c. In cultivating and isolating the pathogen?
   d. In biochemical testing?
   e. In gene probes?
   f. Explain which techniques are more sensitive and specific, in general, and why this would be the case.
4. Differentiate between the serological tests used to identify isolated cultures of pathogens and those used to diagnose disease from patients' serum.
5. Why is it important to prevent microbes from growing in specimens?
6. Why is speed so important in the clinical laboratory?
7. Summarize the important points in determining if a clinical isolate is involved in infection.

chapter 18

The Cocci of Medical Importance

Gram-positive and gram-negative cocci are among the most significant infectious agents of humans. Because these bacteria tend to stimulate pus formation, they are often referred to collectively as the **pyogenic cocci.** The most common infectious species in this group belong to four genera: *Staphylococcus, Streptococcus, Enterococcus,* and *Neisseria.*

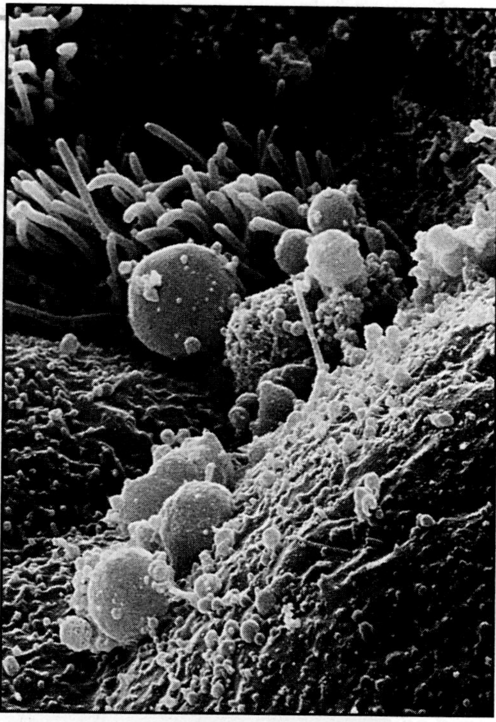

Several tiny spherical cells of the pneumococcus are firmly attached to epithelial cells of the respiratory tract. Early toxic damage to the cells and loss of cilia are evident. This pathogen is a frequent resident of the human pharynx, but it is also the major cause of bacterial pneumonia (3,750×).

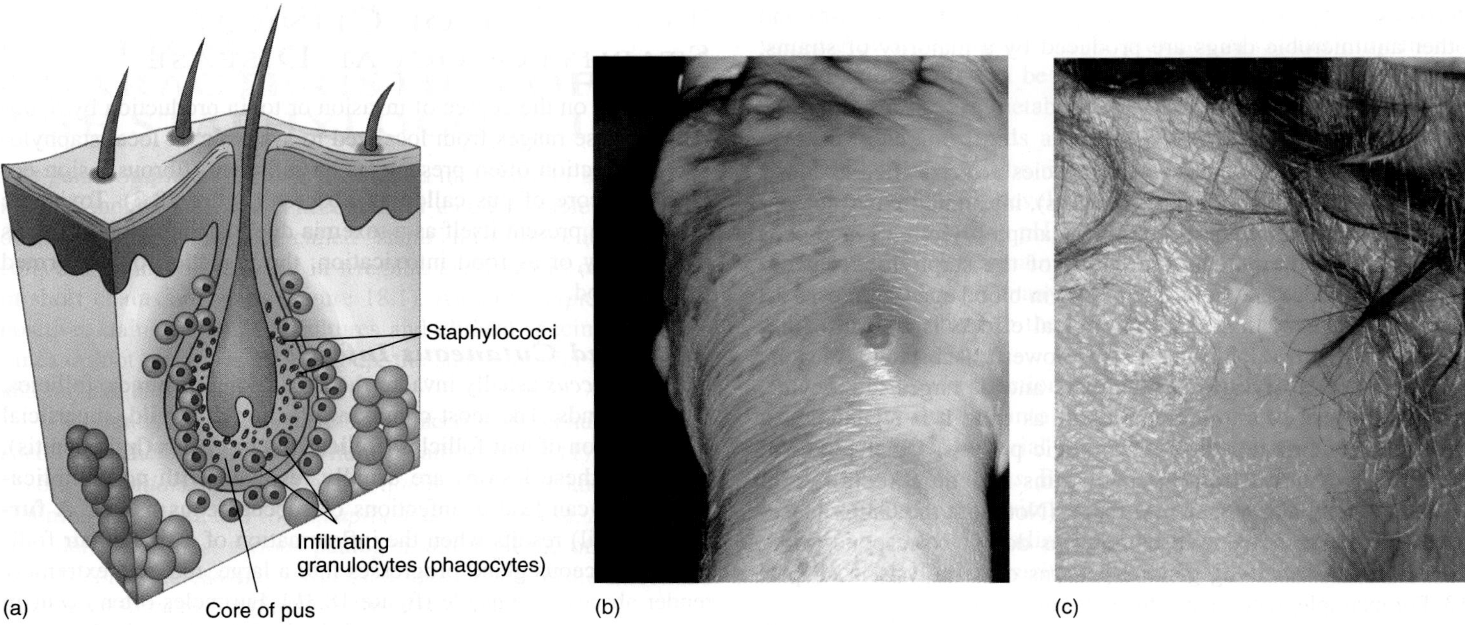

Figure 18.3

Cutaneous lesions of *Staphylococcus aureus.* Fundamentally, all are skin abscesses that vary in size, depth, and degree of tissue involvement. (*a*) Sectional view of a boil or furuncle, a single pustule that develops in a hair follicle or gland and is the classic lesion of the species. The inflamed infection site becomes abscessed when masses of phagocytes, bacteria, and fluid are walled off by fibrin. (*b*) A furuncle on the back of the hand. (*c*) A carbuncle on the back of the neck. Carbuncles are massive deep lesions that result from multiple, interconnecting furuncles. Swelling and rupture into the surrounding tissues can be marked.

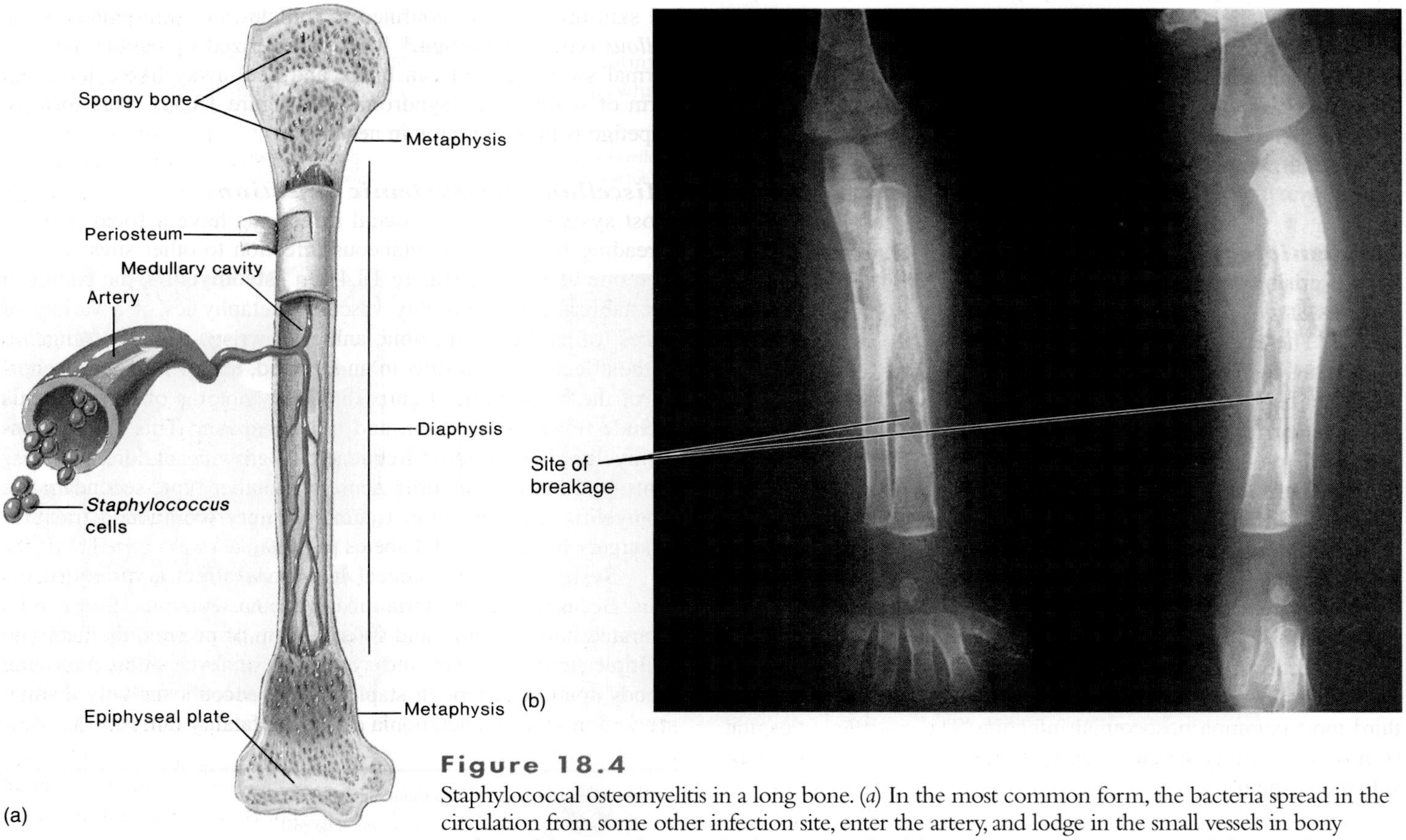

Spongy bone
Metaphysis
Periosteum
Medullary cavity
Artery
Diaphysis
Staphylococcus cells
Epiphyseal plate
Metaphysis
(a)
Site of breakage
(b)

Figure 18.4

Staphylococcal osteomyelitis in a long bone. (*a*) In the most common form, the bacteria spread in the circulation from some other infection site, enter the artery, and lodge in the small vessels in bony pockets of the metaphysis or diaphysis. Growth of the cells cause inflammation and damage that are manifest as swelling and necrosis. (*b*) Two X-ray views of a ruptured ulna caused by osteomyelitis.

MICROFILE 18.1 TAMPONS AND TSS

Toxic shock syndrome (TSS) was first identified as a discrete clinical entity in 1978 in young women using vaginal tampons, even though the disease has probably existed for much longer. At first, the precise link between TSS and the use of tampons was a mystery. Then Harvard Medical School researchers discovered that ultra-absorbent brands of tampons bind magnesium ions, thereby creating a habitat for increased colonization and growth of vaginal *S. aureus* and increased TSS toxin production. The toxin enters the bloodstream and causes a series of reactions, including fever, vomiting, rash, and renal, liver, blood, and muscle involvement, which are sometimes fatal. A small percentage of infections occurs in women using diaphragm or sponge-type contraceptives. Cases of TSS have also been reported in children, men, and nonmenstruating women and in patients recovering from nasal surgery. The evidence implicating ultra-absorbent tampons as a major factor in the disease was so compelling that they were taken off the market in 1981. From the highest yearly count of 880 TSS cases in 1981, the number gradually decreased to about 130 cases in 1997.

An explosive form of toxic shock syndrome is also caused by group A streptococci. This TSS is apparently more common in persons with a history of viral infections, trauma, and surgery.

cases occur in infants and children suffering from cystic fibrosis and measles; this form of pneumonia is also one of the most serious complications of influenza in elderly patients.

Staphylococcal bacteremia causes a high mortality rate among hospitalized patients with chronic disease. Its primary origin is bacteria that have been released from an infection site or from colonized medical devices (catheters, shunts). Circulating bacteria transported to the kidneys, liver, and spleen often form abscesses and give off toxins into the circulation. One consequence of staphylococcal bacteremia is a fatal form of endocarditis associated with the colonization of the heart's lining, cardiac abnormalities, and rapid destruction of the valves. Infection of the joints can produce a deforming arthritis (pyoarthritis). A severe form of meningitis (accounting for about 15% of cases) occurs when *S. aureus* invades the cranial vault.

Toxigenic Staphylococcal Disease

Disorders due strictly to the toxin production of *S. aureus* are food intoxication, scalded skin syndrome, and toxic shock syndrome (microfile 18.1). Enterotoxins produced by certain strains are responsible for the most common type of food poisoning in the United States (see chapter 26). This illness is associated with eating foods such as custards, sauces, cream pastries, processed meats, chicken salad, or ham that have been contaminated by handling and then left unrefrigerated for a few hours. Because of the high salt tolerance of *S. aureus,* even foods containing salt as a preservative are not exempt. The toxins produced by the multiplying bacteria do not noticeably alter the food's taste or smell. Enterotoxins are heat-stable (inactivation requires 100°C for at least 30 minutes), so heating the food after toxin production may not prevent disease. The ingested toxin acts upon the gastrointestinal epithelium and stimulates nerves, with acute symptoms of cramping, nausea, vomiting, and diarrhea that appear in 2 to 6 hours. Recovery is rapid, usually within 24 hours.

Children with infection of the umbilical stump or eyes are susceptible to a toxemia called **staphylococcal scalded skin syndrome** (SSSS). Upon reaching the skin, this toxin induces a painful, bright red flush over the entire body that first blisters and then causes desquamation of the epidermis (figure 18.5). The vast majority of SSSS cases have been described in infants and children under the age of 4. This same exfoliative toxin causes the local reaction in bullous impetigo, which can afflict people of all ages.

HOST DEFENSES AGAINST *S. AUREUS*

Despite regular close contact throughout life, humans have a well-developed resistance to staphylococcal infections. Studies performed on human volunteers established that an injection of several hundred thousand staphylococcal cells into unbroken skin was not sufficient to cause abscess formation. Yet, when sutures containing only a few hundred cells were sewn into the skin, a classic lesion rapidly formed, indicating the aggravating effect of a foreign body in infection. Specific antibodies are produced against most of the staphylococcal antigens and intracellular substances, though none of these antibodies appears to effectively immunize a person against reinfection for long periods (except for the antitoxin to SSSS toxin). The most powerful defense lies in the phagocytic response by neutrophils and macrophages. When aided by the opsonic action of complement, phagocytosis effectively disposes of staphylococci that have gained access to tissues. Inflammation and cell-mediated immunity that stimulate abscess formation also help contain staphylococci and prevent their further spread.

THE OTHER STAPHYLOCOCCI

Because they lack the enzyme coagulase, other species in the genus *Staphylococcus* are called the **coagulase-negative staphylococci** (CNS). Several species are included in this group, some of human origin and others of nonhumans, mammalian origin. Although this group was once considered clinically nonsignificant, its importance has greatly increased over the past 20 years. Coagulase-negative staphylococci currently account for a large proportion of nosocomial and opportunistic infections in immunocompromised patients.

The normal habitat of ***Staphylococcus epidermidis*** is the skin and mucous membranes. *Staphylococcus hominis* lives in skin areas high in apocrine glands, and *S. capitis* populates the scalp, face, and external ear. From these sites, the bacteria are positioned to enter breaks in the protective skin barrier. Infections usually occur after surgical procedures, such as insertion of shunts, catheters, and various prosthetic devices that require incisions through the skin. It appears that any foreign object supports colonization and proliferation of these bacteria, which develop thick adherent capsules. It is particularly alarming that the rate of these infections has increased owing

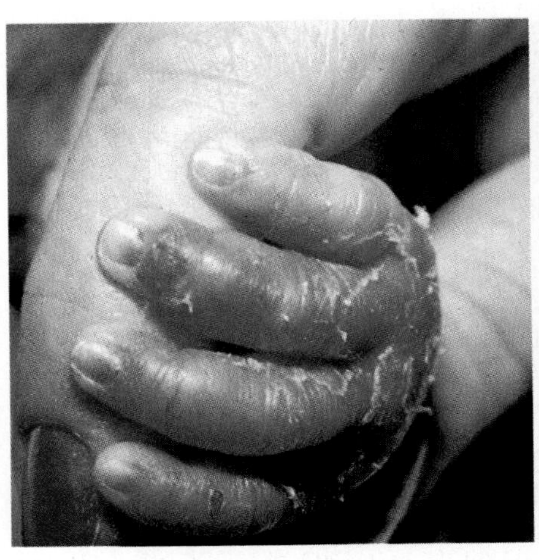

(a)

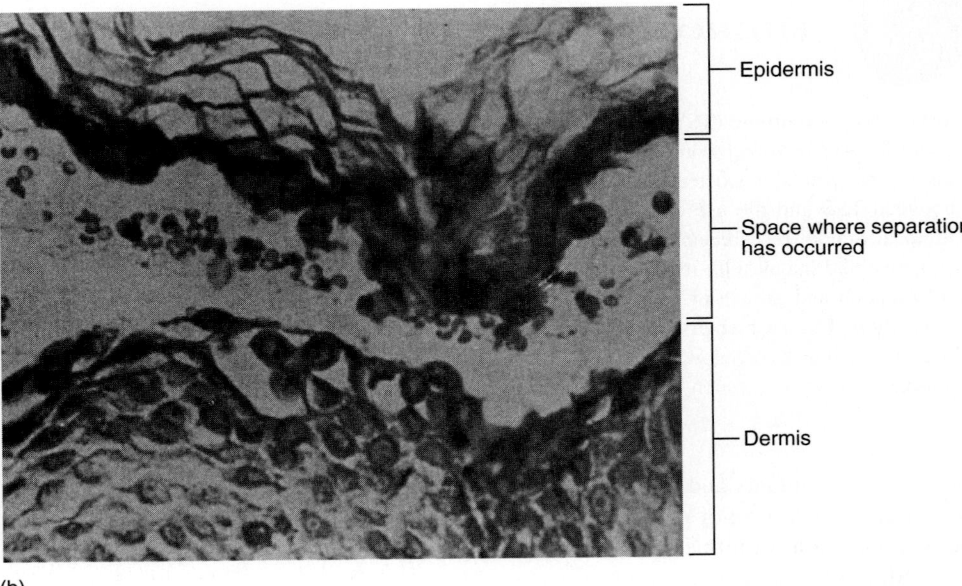

(b)

Figure 18.5

Staphylococcal scalded skin syndrome (SSSS) in a newborn child. (*a*) Exfoliative toxin produced in local infections causes blistering and peeling away of the skin to expose a red underlayer. (*b*) Photomicrograph of a segment of skin affected with SSSS. The point of epidermal shedding, or desquamation, is at the dermis. The lesions will heal well because the level of separation is so superficial.

to technological advances in maintaining patients. Although *S. epidermidis* is not as invasive or toxic as *S. aureus,* it can also cause endocarditis, bacteremia, and urinary tract infections.

Less is known about the distribution and epidemiology of ***S. saprophyticus.*** It is an infrequent resident of the skin, lower intestinal tract, and vagina, but its primary importance is in urinary tract infections. For unknown reasons, urinary *S. saprophyticus* infection is found almost exclusively in sexually active adolescent women and is the second most common cause of urinary infections in this group.

Identification of *Staphylococcus* in Clinical Samples

Staphylococci are frequently isolated from pus, tissue exudates, sputum, urine, and blood. To prevent inadvertently culturing staphylococci from the environment or from normal residents, great care in collection must be exercised. Primary isolation is achieved by inoculation on sheep or rabbit blood agar or in heavily contaminated specimens, on selective media such as mannitol salt agar. In addition to culturing, the specimen can be Gram stained and observed for irregular clusters of gram-positive cocci. Because differentiating among gram-positive cocci is not possible using colonial and morphological characteristics alone, other presumptive tests are required (see figure E in "Introduction to Identification Techniques in Medical Microbiology"). The production of catalase, an enzyme that breaks down hydrogen peroxide accumulated during oxidative metabolism, can be used to differentiate the staphylococci, which produce it, from the streptococci, which do not (figure 18.6*a*). The property of *Staphylococcus* to grow anaerobically and to ferment sugars separates it from *Micrococcus,* a nonpathogenic genus that is a common specimen contaminant.

One key technique for separating *S. aureus* from other species of *Staphylococcus* is the coagulase test (figure 18.6*b*). By definition, any isolate that coagulates plasma is *S. aureus,* and all others are coagulase-negative. Rapid multitest systems are used routinely to collect other physiological information (figure 18.7). Table 18.1 lists some of the characteristics that further separate the five species most often involved in human infections. An important confirming identification of *S. aureus* can be made with a latex bead agglutination test. It is based on the surface protein (A) that can bind to IgG antibodies.

Differentiating the common coagulase-negative staphylococci from one another relies primarily upon the novobiocin resistance of *S. saprophyticus,* mannose fermentation in *S. epidermidis,* the lack of urease production in *S. capitis,* and lack of anaerobic growth in *S. hominis.*

Clinical Concerns in Staphylococcal Infections

The staphylococci are notorious in their acquisition of resistance to new drugs, and they continue to defy attempts at medical control. Because resistance to common drugs is likely, antimicrobic susceptibility tests are essential in selecting a correct therapeutic agent (see chapter 12).

Ninety-five percent of strains of *S. aureus* have acquired genes for penicillinase, which makes them resistant to the traditional drugs—penicillin and ampicillin. Even more problematic are strains of MRSA (**m**ethicillin-**r**esistant *Staphylococcus* ***au**reus)* that carry multiple resistance to a wide range of antimicrobics, including methicillin, gentamicin, cephalosporins, tetracycline, erythromycin, and even quinolones. A few strains have

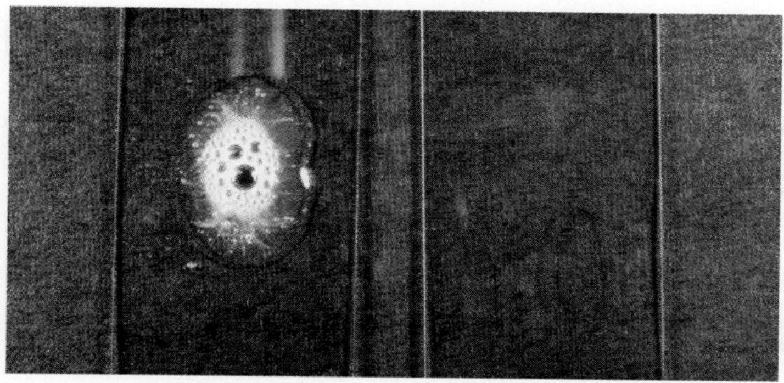

(a)

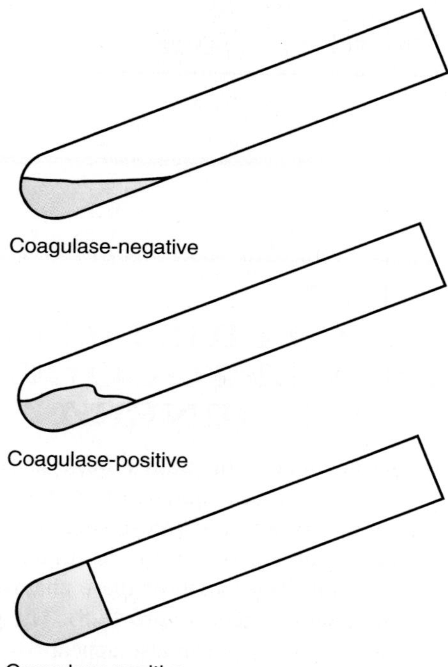

(b)

Figure 18.6

(*a*) Tests for differentiating the genus *Staphylococcus* from *Streptococcus* and for (*b*) identifying *Staphylococcus aureus.* (*a*) In the *slide catalase test,* adding 3% solution of hydrogen peroxide to a small sample of isolate will cause the sample to bubble vigorously if catalase is present and remain undisturbed if catalase is lacking. (*b*) Staphylococcal coagulase is an enzyme that reacts with factors in plasma to initiate clot formation. In the coagulase test, a tube of plasma is inoculated with an isolate. If it remains liquid and flows, the test is negative. If the plasma develops a lump or becomes completely clotted, the test is positive.

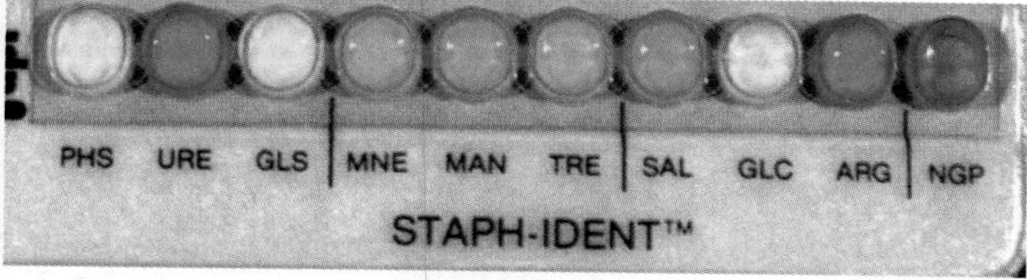

All tests: positive

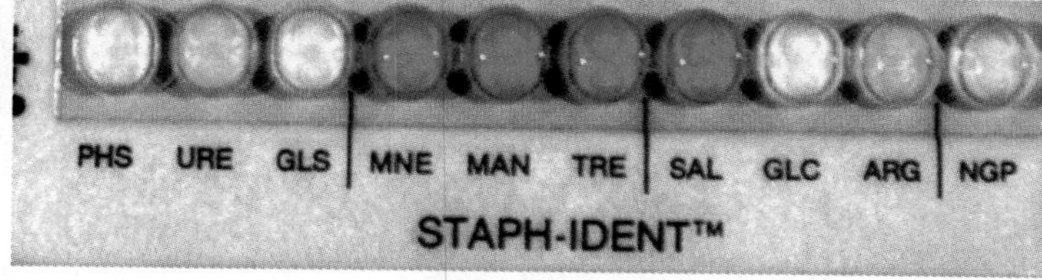

Inoculated: all negative

Figure 18.7

Miniaturized test system used in further identification of *Staphylococcus* isolates. Microbiologists can separate 19 species with this system. The cupules contain substrates that detect phosphatase production (PHS), urea hydrolysis (URE), glucosidase production (GLS), mannose fermentation (MNE), mannitol fermentation (MAN), trehalose fermentation (TRE), salicin fermentation (SAL), glucuronidase production (GLC), arginine hydrolysis (ARG), and galactosidase production (NGP).

acquired resistance to all major drug groups except vancomycin. Most of these isolates are found in hospitals, where they account for 40% of staph infections. Now an even greater cause for concern are reports of vancomycin resistance. Infectious disease specialists are monitoring this new trend and exploring potential drug replacements such as synercid (see chapter 12).

Treatment of Staph Infections

Clinical experience has shown that abscesses will not respond to therapy unless they are surgically perforated and cleared of pus and foreign bodies. Severe systemic conditions such as endocarditis, septicemia, osteomyelitis, pneumonia, and toxemia respond slowly and require intensive, lengthy therapy by oral or injected drugs or some combination of the two. No useful vaccine has yet been marketed, but a trial is currently underway to test the effectiveness of a vaccine to control infections with the two most common strains of *S. aureus* in high risk surgical patients.

Prevention of Staph Infections

The principal reservoir of the pathogenic staphylococci can never be eliminated. As long as there are humans, there will be carriers and infections. It is difficult to block the colonization of the human body, but one can minimize the opportunity for nosocomial infection by careful hygiene and adequate cleansing of surgical incisions and burns. Controlling the increasing incidence of nosocomial outbreaks, especially among people at greatest risk (surgical patients and neonates) is a real challenge for hospital staff. Patients or hospital personnel who are asymptomatic carriers or have frank infections constitute the most common source. The burden of effective deterrence again falls to meticulous handwashing, proper disposal of infectious dressings and discharges, isolation of people with open lesions, and attention to indwelling catheters and needles.

When possible, other preventive measures must be instituted. For example, hospital workers who carry *S. aureus* nasally may be barred from nurseries, operating rooms, and delivery rooms. Occasionally, carriers are treated for several months with a combination of antimicrobic drugs (rifampin and dicloxacillin, for example). A startling experimental technique is *interference therapy,* the deliberate insertion of a small inoculum of a less virulent strain of *S. aureus* into the nasal mucosa or umbilicus of a newborn. If successful, this procedure promotes colonization by this strain and inhibits establishment of more virulent staphylococci, but it must be used with caution.

TABLE 18.1

SEPARATION OF CLINICALLY IMPORTANT SPECIES OF *Staphylococcus*

Test	*S. aureus*	*S. epidermidis*	*S. saprophyticus*	*S. capitis*	*S. hominis*
Coagulase	+*	–	–	–	–
Anaerobic growth	+	+	+	+	–
Urease production	+	+	+	–	+
Mannitol fermentation (aerobic)	+*	–	W	+	–
Mannose fermentation	+	+	–	+	–
Trehalose fermentation	+	–	+	–	+/–
β-hemolysis by α-toxin	+	–	–	–	–
Produces DNase and RNase	+	–	–	–	–
Sensitive to lysostaphin	+	–	–	–	–
Susceptible to novobiocin	+	+	–	+	+
Pathogenicity	Primary	Opportunistic	Opportunistic	Opportunistic	Opportunistic

*A few strains test negative for this.

W = May occur, but weakly.

Chapter Checkpoints

The genus *Staphylococcus* contains 31 species. Most of these are human commensals, but *S. aureus, S. epidermidis,* and *S. saprophyticus* and others can be pathogenic.

All species in the genus *Staphylococcus* are gram-positive facultative anaerobes that tolerate extremes of temperature, osmotic pressure, and drying. All are catalase-positive.

Coagulase-positive staphylococci are by definition *S. aureus,* the most resistant of all non-spore-forming pathogenic bacteria.

S. aureus produces an impressive array of virulence factors that enable it to resist phagocytosis, destroy host tissue, and invade the blood. These include coagulase, hemolysins, leukocidins, enzymes that digest host tissue, and toxins that attack the epidermis, vascular smooth muscle, and the intestinal epithelium. These virulence factors enable *S. aureus* to cause both local and systemic infections.

Abscesses are localized staphylococcal infections. They occur predominantly in hair follicles and glands.

S. aureus causes disease mainly when the host is weakened by poor health, injury, disease, surgery, or immunodeficiency.

S. epidermidis and *S. saprophyticus* are coagulase-negative staphylococci that cause opportunistic infections. *S. epidermidis* causes nosocomial skin incision infections, and *S. saprophyticus* causes urinary tract infections.

Overuse and inappropriate prescription of antimicrobials have selected for multiple drug-resistant strains of *S. aureus* in hospitals, causing serious problems in treatment. All *S. aureus* isolates should therefore be checked for antimicrobic sensitivity so that an effective therapeutic can be prescribed.

Consistent practice of universal precautions by *all* hospital staff is necessary to prevent nosocomial staphylococcal infections.

GENERAL CHARACTERISTICS OF THE STREPTOCOCCI AND RELATED GENERA

The genus ***Streptococcus**** includes a large and varied group of bacteria. Some are normal residents or agents of disease in humans and animals; others are free-living in the environment. Members of this group are known for the arrangement of cocci in long, beadlike chains. The length of these chains varies, and it is common to find them in pairs (figure 18.8). The general shape of the cells is spherical, but they can also appear ovoid or rodlike, especially in actively dividing young cultures.

Streptococci are non-spore-forming and nonmotile (except for an occasional flagellated strain), and they can form capsules and slime layers. They are facultative anaerobes that ferment a variety of sugars, usually with the production of lactic acid (homofermentative). Streptococci do not form catalase, but they do have a peroxidase system for inactivating hydrogen peroxide, which allows their survival even in the presence of oxygen. Most parasitic forms are fastidious in nutrition and require enriched media for cultivations. Colonies are usually small, nonpigmented, and glistening. Most members of the genus are quite sensitive to drying, heat, and disinfectants and seldom develop drug resistance, though pneumococci and enterococci are notable exceptions.

Species of *Streptococcus* have traditionally been classified according to a system developed by Rebecca Lancefield in the 1930s. She discovered that the cell wall carbohydrates (antigens) of various cultures stimulated formation of antibodies with differ-

**Streptococcus* Gr. *streptos,* winding, twisted. The chain arrangement is the result of division in only one plane.

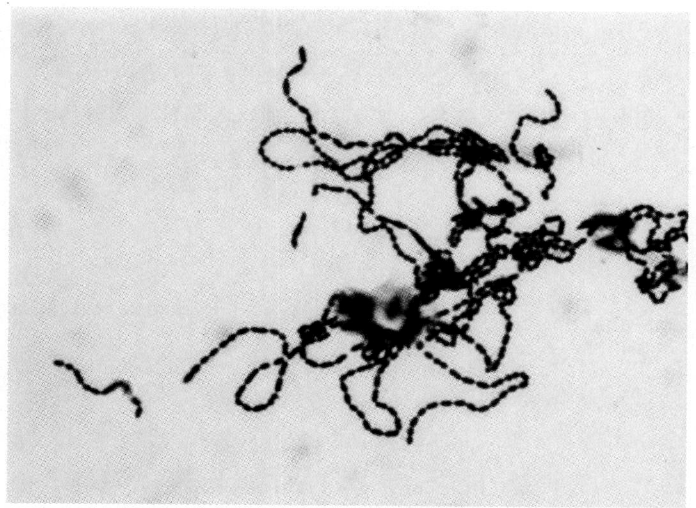

Figure 18.8

One of the truly great sights in microbiology—a freshly isolated *Streptococcus* with its long, intertwining chains. These occur only in liquid media and are especially well-developed if nutrients are in limited supply (1,000×).

ing specificities. She characterized 14 different groups using an alphabetic system (A, B, C). An alternative division method of grouping is based on their reaction in blood agar (figure 18.9). Those producing a zone of β-hemolysis on sheep blood agar are members of Lancefield groups A, B, C, G, and certain strains of D, and those producing α-hemolysis are *S. pneumoniae* and the *viridans** streptococci. Several species of nonhemolytic cocci exist, but most of them are of less clinical significance. Table 18.2 summarizes the streptococcal groups, their habitats, and their pathogenicity.

Despite the large number of streptococcal species, human disease is most often associated with ***S. pyogenes, S. agalactiae,*** and **viridans streptococci, *S. pneumoniae,*** and ***Enterococcus faecalis.*** The latter species was formerly called *Streptococcus faecalis* until DNA hybridization tests indicated that it is a distinct genus. Several members of groups C and G are commensals of domestic animals that also colonize and infect humans. In fact, streptococci are notorious for being shared between humans and their pets.

β-HEMOLYTIC STREPTOCOCCI: *STREPTOCOCCUS PROGENES*

By far the most serious streptococcal pathogen of humans is *Streptococcus pyogenes,* the main representative of group A. It is a relatively strict parasite, inhabiting the throat, nasopharynx, and occasionally the skin of humans. The involvement of this species in severe disease is partly due to the substantial array of surface antigens, toxins, and enzymes it generates, although even highly virulent strains do not produce all of the toxins and enzymes.

*viridans (vih´-rih-denz) L. *viridis,* green. This term comes from the green color that develops around colonies that produce α-hemolysis.

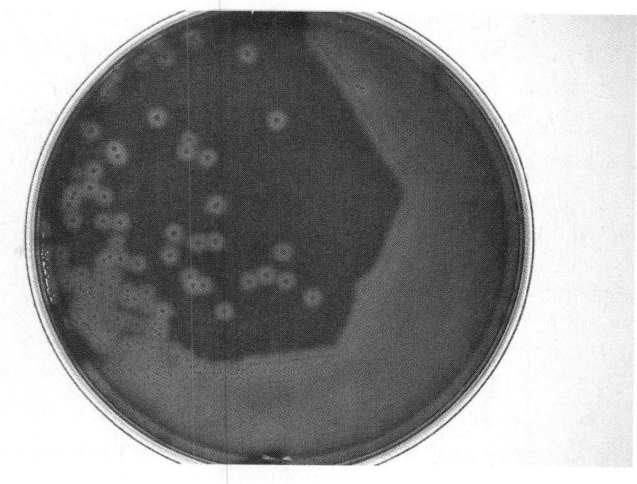

(a)

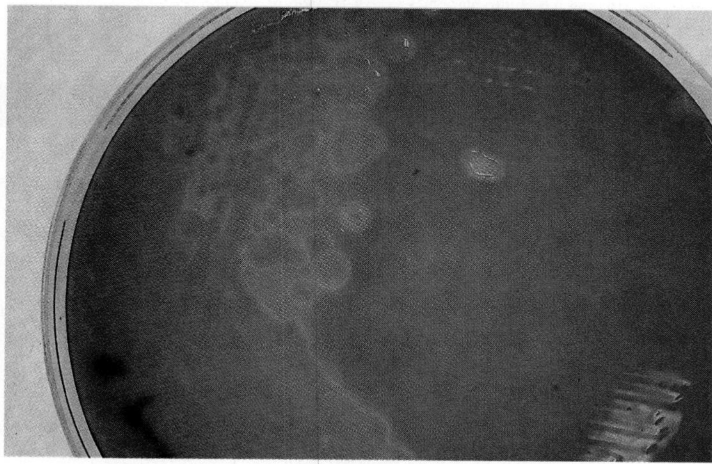

(b)

Figure 18.9

Hemolysis patterns on blood agar may be used to separate streptococci into major subgroups. (*a*) Colonies of *Streptococcus pyogenes* showing β-hemolysis, which forms clear zones with completely lysed red blood cells. (*b*) Colonies of *S. pneumoniae* demonstrate the less-discrete greenish zones typical of α-hemolysis.

Cell Surface Antigens and Virulence Factors

Streptococci display numerous surface antigens (figure 18.10). C-carbohydrates, specialized polysaccharides or techoic acids found on the surface of the cell wall, are the basis for Lancefield groups. Their apparent contribution to pathogenesis is to protect the bacterium from being dissolved by the lysozyme defense of the host. Lipotechoic acid accounts for the adherence of *S. pyogenes* to epithelial cells in the skin or pharynx. Another type-specific molecule is the *M-protein,* of which about 80 different subtypes exist. This substance is the main component of fimbriae, the spiky surface projections that contribute to virulence by resisting phagocytosis and improving adherence. A capsule is formed by most *S. pyogenes* strains, but it remains attached to the cell surface only during the rapid growth phase of a culture and is probably not as important to virulence as are other factors.

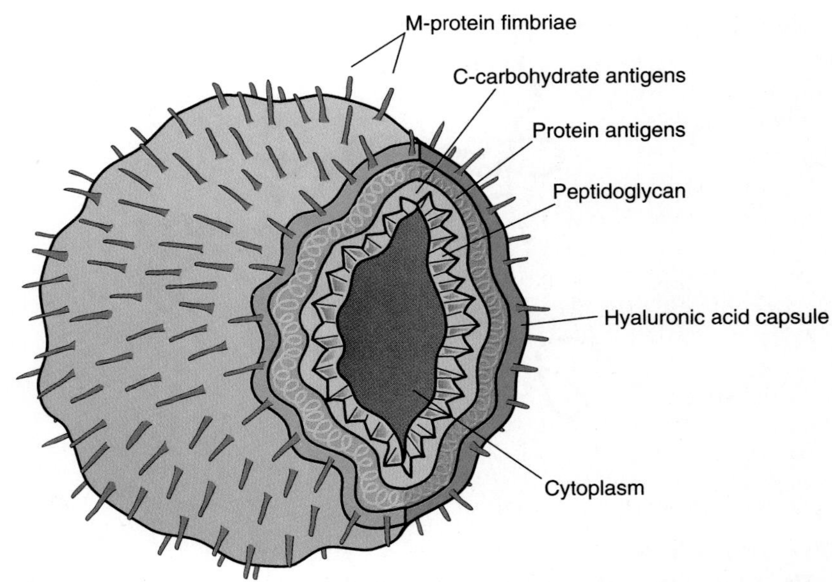

Figure 18.10

Cutaway view of group *A Streptococcus.* The outermost fringe consists of fimbriae composed in part of M-protein, group-specific substance. Other layers making up the cell envelope are the capsule, protein antigens, and C-carbohydrate antigens.

TABLE 18.2

MAJOR SPECIES OF *Streptococcus* AND RELATED GENERA

Species	Lancefield Group	Hemolysis Type	Habitat	Pathogenicity to Humans
S. pyogenes	**A**	Beta (β)	**Human throat**	**Skin, throat infections, scarlet fever**
S. agalactiae	**B**	β	**Human vagina, cow udder**	**Neonatal, wound infections**
S. equisimilis	C	β	Swine, cows, horses	Pharyngitis, endocarditis
S. equi, S. zooepidemicus	C	β	Various mammals	Rare, in abscesses
S. dysgalactiae	C	β	Cattle	Rare
Enterococcus faecalis	**D**	α, β, N	**Human, animal intestine**	**Endocarditis, UTI***
E. faecium, E. durans	D	Alpha (α)	Human, animal intestine	Similar to *E. faecalis*
S. bovis	D	N	Cattle	Subacute endocarditis, bacteremia
S. anginosus	F, G, L	β	Humans, dogs	Endocarditis, URT** infections
S. sanguis	**H**	α	**Human oral cavity**	**Endocarditis, dental caries**
S. salivarius	K	N	Human saliva	Endocarditis
Lactococcus lactis	N	V	Dairy products	Very rare
S. mutans	**NI*****	N	**Human oral cavity**	**Dental caries**
S. uberis, S. acidominimus	NI	V	Domestic mammals	Rare
S. mitior	O, M	α	Human oral cavity	Tooth abscess, endocarditis
S. milleri	F	N	URT	Endocarditis, organ abscess
S. pneumoniae	**NI**	α	**Human RT**	**Bacterial pneumonia**

Note: Species in bold type are the most significant sources of human infection and disease.

N = none V = varies

*Urinary tract infection.

**Upper respiratory tract.

***No group C-carbohydrate identified.

Major Extracellular Toxins Group A streptococci owe several pathologic properties to the effects of hemolysins called **streptolysins.** The two types are streptolysin O (SLO) and streptolysin S (SLS).[2] Although both types cause β-hemolysis of sheep blood agar, SLS produces the major form of surface hemolysis, whereas SLO produces hemolysis only in deep (anaerobic) colonies. Both hemolysins rapidly injure many cells and tissues, including leukocytes, liver, and heart muscle.

A key toxin in the development of scarlet fever (discussed in a later section of this chapter) is **(erythrogenic*) pyrogenic**

2. In SLO, O stands for oxygen because the substance is inactivated by oxygen. SLO is produced by most strains of *S. pyogenes.* In SLS, S stands for serum because the substance has an affinity for serum proteins. SLS is oxygen-stable.

*erythrogenic (en-rith′′-roh-jen′-ik) Gr. *erythros,* red, and *gennan,* to produce.

MICROFILE 18.2 "FLESH-EATER" FUELS UNFOUNDED FEARS

Strep infections are "occupational diseases of childhood" that usually follow a routine and uncomplicated course. The greatest cause for concern are those few occasions when such infections erupt into far more serious ailments. One dramatic example is **necrotizing fasciitis*** a rare disease that has been known for hundreds of years. In 1994, several small outbreaks of the disease in England and the United States received heavy publicity by the press, and the so called "flesh-eating disease" caused by "killer bacteria" suddenly became a household name. Needless to say, the threat was largely overstated by the media. Part of the reason for the horror associated with this disease is that it can begin with an innocuous cut in the skin and spread rapidly into nearby tissue. In addition, it has a real potential for disfigurement and death.

There is really no mystery to the pathogenesis of necrotizing fasciitis. It begins very much like impetigo and other skin infections: Streptococci on the skin are readily introduced into small abrasions or cuts, where they begin to grow rapidly. The name "flesh-eating" for this effect is really a misnomer. These strains of group A streptococci have great toxigenicity and invasiveness by releasing special enzymes and toxins. Their enzymes digest the connective tissue in skin and their toxins poison the epidermal and dermal tissue. As the flesh is killed, it separates and sloughs off, forming a pathway for the bacteria to spread into deeper tissues such as muscle. More dangerous infections involve a mixed infection with anaerobic bacteria and systemic spread of the toxin to other organs. It is true that some patients have lost parts of their limbs and faces, and others have suffered amputation, but early diagnosis and treatment can prevent these complications. Fortunately, even virulent strains of *Streptococcus pyogenes* are not highly drug-resistant.

More mysterious, however, is the background epidemiology involved in these virulent new outbreaks. The incidence of aggressive infections had been gradually declining throughout the antibiotic era, but then in the latter 1980s clusters began cropping up again. Bacteria isolated from cases in the Rocky Mountains and the East appeared to be new, more virulent strains. Infectious disease experts theorize that these new strains of streptococci are probably mutants that have acquired toxin genes from infecting viruses. An explosive combination that led to several childhood deaths in California was that of chickenpox lesions secondarily infected by streptococci. Elsewhere, reports of strep throat and even scarlet fever and rheumatic fever are on the rise.

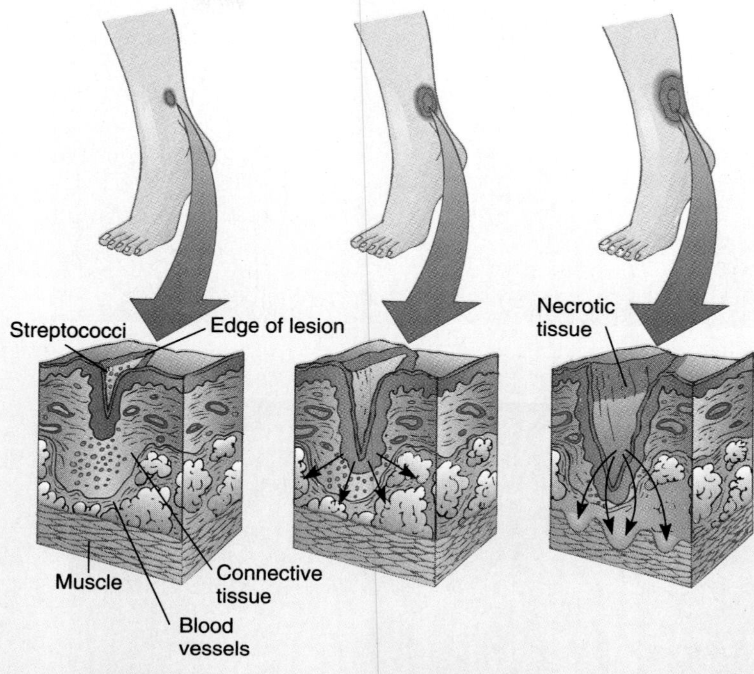

The phases of Streptococcus pyogenes*–induced necrotizing fasciitis.*

*(nee´-kroh-ty´´-zing fass´´-ee-eye´-tis) Gr. *nekrosis,* deadness, and L. *fascia,* the connective tissue sheath around muscles and other organs.

toxin. This toxin is responsible for the bright red rash typical of this disease, and it also induces fever by acting upon the temperature regulatory center. Only lysogenic strains of *S. pyogenes* that contain genes from a temperate bacteriophage can synthesize this toxin.

Some of the streptococcal toxins (pyrogenic and streptolysin O) contribute to increased tissue injury by acting as *superantigens.* These toxins are strong stimulants of monocytes and T lymphocytes. When activated, these cells proliferate and produce *tumor necrosis factor,* which in high quantities leads to vascular injury. This is the likely mechanism for the severe pathology of shock syndrome and necrotizing faciitis.

Major Extracellular Enzymes Several enzymes for digesting macromolecules are given off by the group A streptococci, though their role in pathogenesis is somewhat obscure. Streptokinase, similar to staphylokinase, activates a pathway leading to the digestion of fibrin clots and may play a role in invasion. Hyaluronidase breaks down the binding substance in connective tissue and promotes spreading of the pathogen into the tissues, while streptodornase (DNase) liquefies purulent discharges by hydrolyzing DNA.

Epidemiology and Pathogenesis of S. pyogenes

Streptococcus pyogenes has traditionally been linked with a diverse spectrum of infection and disease. Before the era of antibiotics, it accounted for a major portion of serious human infections and deaths from such diseases as rheumatic fever and puerperal sepsis. Although its importance has diminished in the past 50 years, recent epidemics have reminded us of its potential for sudden and serious illness (microfile 18.2).

Humans are the only significant reservoir for *S. pyogenes.* Healthy or subclinical carriers of virulent strains constitute about 5–15% of the population. Infection is generally transmitted through

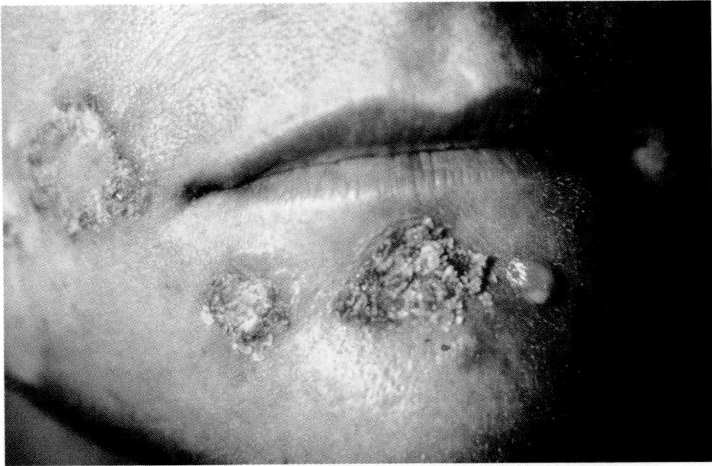

(a)

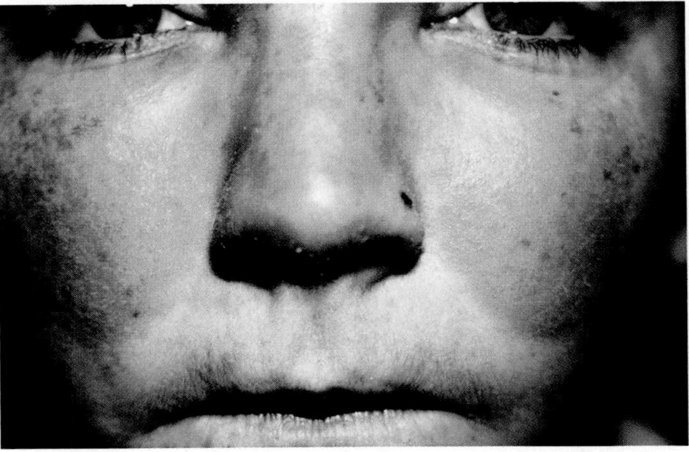

(b)

Figure 18.11

Streptococcal skin infections. *(a)* Impetigo lesions on the face. *(b)* Erysipelas of the face. Although it is a superficial infection, the inflammatory reaction spreads horizontally from the initial point of entry. Tender, red, puffy lesions have a sharp border that can burst and release fluid.

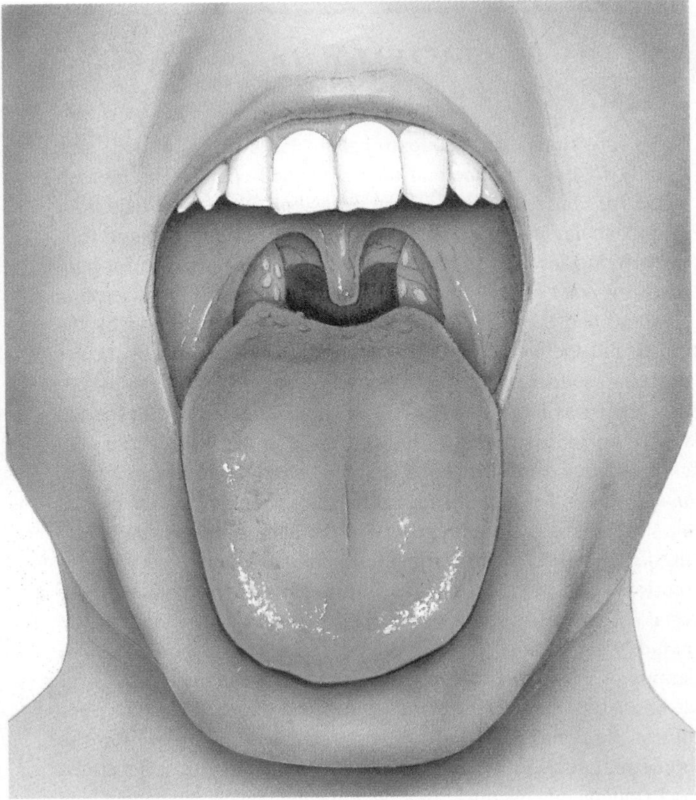

Figure 18.12

The appearance of the throat in pharyngitis and tonsillitis. The tongue, pharynx, and tonsils (if present) become bright red and suppurative. Whitish pus nodules may also appear on the tonsils.

direct contact, droplets, and, occasionally, food or formites. The bacteria invade at periods of lowered host resistance, primarily through the skin and pharynx. The incidence and types of infections are altered by climate, season, and living conditions. Skin infections occur more frequently during the warm temperatures of summer and fall, and pharyngeal infections increase in the winter months. Children 5–15 years of age are the predominant group affected by both types of infections. In addition to local cutaneous and throat infections, *S. pyogenes* can give rise to a variety of systemic infections and progressive sequelae if not properly treated.

Skin Infections When virulent streptococci invade a nick in the skin or the mucous membranes of the throat, an inflammatory primary lesion is produced. Pyogenic infections appearing after local invasion of the skin are pyoderma or erysipelas; those developing in the throat are pharyngitis or tonsillitis.

Pyoderma, or streptococcal impetigo, is marked by burning, itching papules that break and form a highly contagious yellow crust (figure 18.11*a*). Impetigo often occurs in epidemics among school children, and it is also associated with insect bites, poor hygiene, and crowded living conditions. A slightly more invasive form of skin infection is **erysipelas.*** The pathogen usually enters through a small wound or incision on the face or extremities and eventually spreads to the dermis and subcutaneous tissues. Early symptoms are edema and redness of the skin near the portal of entry, fever, and chills. The lesion begins to spread outward, producing a slightly elevated edge that is noticeably red, hot, and often vesicular (figure 18.11*b*). Depending on the depth of the lesion and how the infection progresses, cutaneous lesions can remain superficial or produce long-term systemic complications. Severe cases involving large areas of skin are occasionally fatal.

Most people associate streptococci with the condition called strep throat or, more technically, **streptococcal pharyngitis** (tonsillitis). Estimates indicate that most humans will acquire this particular infection at some time during their lifetime. The organism multiplies in the tonsils or pharyngeal mucous membranes, causing redness, edema, enlargement, and extreme tenderness that make swallowing difficult and painful (figure 18.12). These symptoms can be accompanied by fever, headache, nausea, and abdom-

*erysipelas (er′′-ih-sip′-eh-las) Gr. *erythros,* red, and *pella,* skin.

inal pain. Other signs include a *purulent** exudate over the tonsils, swollen lymph nodes, and occasionally white, pus-filled nodules on the tonsils.

Throat infection can lead to **scarlet fever** (scarlatina) when it involves a strain of *S. pyogenes* carrying a prophage that codes for pyrogenic toxin. Systemic spread of this toxin results in high fever and a bright red, diffuse rash over the face, trunk, inner arms and legs, and even the tongue. Within 10 days, the rash and fever usually disappear, often accompanied by desquamation (sloughing) of the epidermis. Many cases of pharyngitis and even scarlet fever are mild and uncomplicated, but on occasion they elicit an inflammatory reaction that leads to severe sequelae.

Systemic Infections The dissemination of streptococci into the lymphatics and the blood can give rise to septicemia, an infrequent complication limited to people who are already weakened by underlying disease. Another relatively rare infection is *S. pyogenes* pneumonia, which accounts for less than 5% of bacterial pneumonias and is usually secondary to influenza or some other pulmonary disorder. A recent invasive infection, called *streptococcal toxic shock syndrome,* has appeared in the United States and Europe. It begins as a profound bacteremia and deep tissue infection and rapidly progresses to multiple organ failure. Even with treatment, around 30% of patients die. At one time, a major cause of maternal death was childbed, or *puerperal,** fever caused by *S. pyogenes* introduced into the vagina by contaminated hands. From this site, it spreads rapidly into uterine tissue disrupted by childbirth and causes massive abdominal sepsis. Modern obstetric techniques and antimicrobic therapy have greatly reduced this cause of postpartum disease. Presently, most uterine infections occurring after childbirth are due to normal vaginal flora and not exogenous pathogens.

Long-Term Complications of Group A Infections

The principal sequelae that appear within a few weeks after group A streptococcal infections are (1) **rheumatic* fever (RF),** a delayed inflammatory condition of the joints, heart, and subcutaneous tissues, and (2) **acute glomerulonephritis (AGN),** a disease of the kidney glomerulus and tubular epithelia. The actual mechanisms for these streptococcal sequelae have yet to be completely explained, though several theories have been proposed. It is possible that tissue injury in both diseases is tied directly to streptococcal invasion of tissues or to the action of toxins such as SLO and SLS. According to the autoimmunity theory, antibodies formed against the streptococcal cell wall and membranes cross-react with molecules in the host that are similar to these bacterial antigens. In the heart, these antibody-antigen reactions trigger inflammation and injury of cardiac tissues, and in the kidney, immune complexes deposited in the epithelia inflame and damage the filtering apparatus (see chapter 17).

Rheumatic fever usually follows an overt or subclinical case of streptococcal pharyngitis or tonsillitis in children. Its major

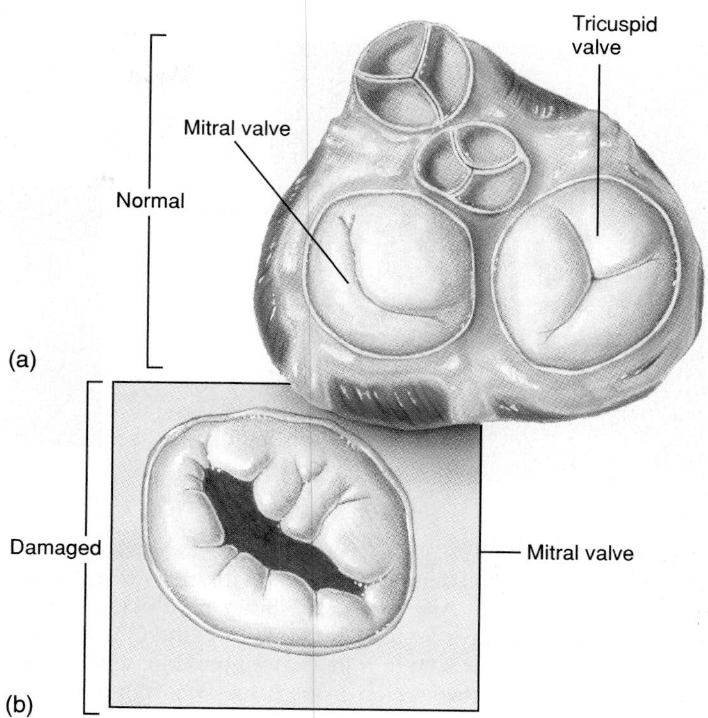

Figure 18.13

The cardiac complications of rheumatic fever. Pathologic processes of group A streptococcal infection can extend to the heart. In this example, it is believed that cross-reactions between streptococcal-induced antibodies and heart proteins have a gradual destructive effect on the atrioventricular valves (especially the mitral valve) or semilunar valves. Scarring and deformation change the capacity of the valves to close and shunt the blood properly. (*a*) Normal valves, viewed from above. (*b*) Inset reveals scar tissue on a damaged mitral valve.

clinical features are carditis, and abnormal electrocardiogram, painful arthritis, chorea, nodules under the skin, and fever.[3] The course of the syndrome extends from 3 to 6 months, usually without lasting damage. In patients with severe carditis, however, extensive damage to the heart valves and muscle can occur (figure 18.13). Although the degree of permanent damage does not usually reveal itself until middle age, it is often extensive enough to require the replacement of damaged valves. Heart disease due to RF is on the wane because of medical advances in diagnosing and treating streptococcal infections, but it is still responsible for thousands of deaths each year in the United States and many times that number in the rest of the world.

In AGN, the kidney cells are so damaged that they cannot adequately filter blood. The first symptoms are nephritis (appearing as swelling in the hands and feet and low urine output), increased blood pressure, and occasionally heart failure. Urine samples are extremely abnormal, with high levels of red and white blood cells and protein. AGN can clear up spontaneously, can become chronic and lead to kidney failure later, or can be immediately fatal.

*purulent (puh´-roo-lent) L. *purulentus,* inflammation. In reference to pus.

*puerperal (poo-er´-per-al) L. *puer,* child, and *parere,* to bear. From puerperium, the period of confinement after labor and birth.

*rheumatic (roo-mat´-ik) Gr. *rheuma,* flux. Involving inflammation of joints, muscles, and connective tissues.

3. Carditis is inflammation of heart tissues; chorea is a nervous disorder characterized by involuntary, jerky movements.

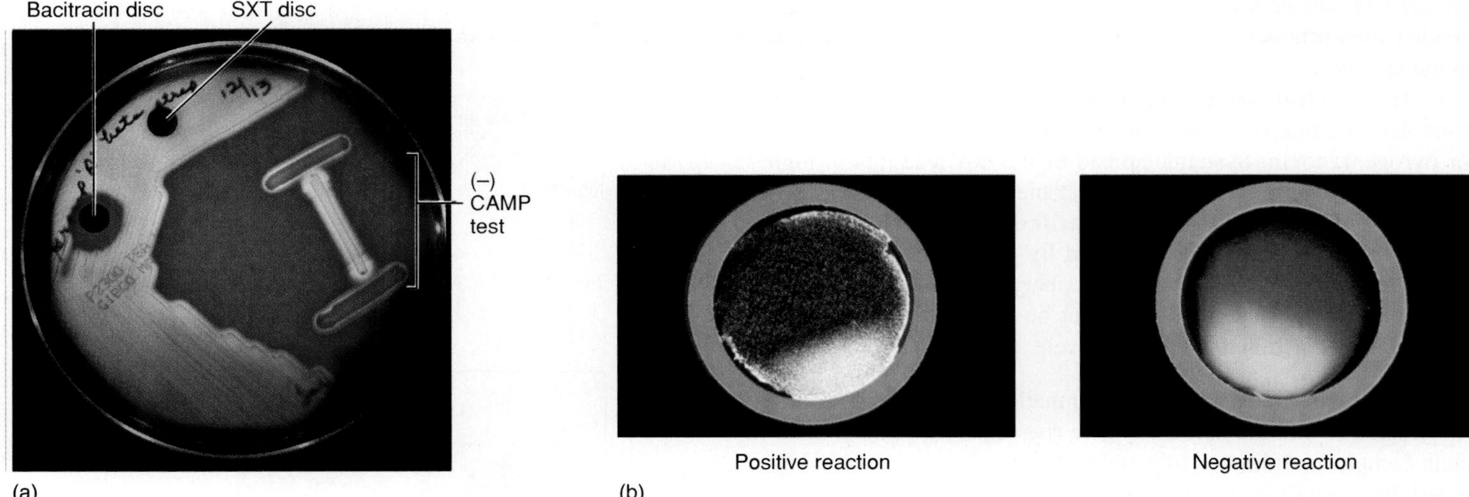

Figure 18.14

Streptococcal tests. (*a*) Bacitracin disc test. With very few exceptions, only *Streptococcus pyogenes* is sensitive to a minute concentration (0.02 μg) of bacitracin. Any zone of inhibition around the B disc is interpreted as a presumptive indication of this species. (Note: Group A streptococci are negative for SXT sensitivity and the CAMP test.) (*b*) A rapid, direct test kit for diagnosis of group A infections. With this method, a patient's throat swab is introduced into a system composed of latex beads and monoclonal antibodies. (*Left*) In a positive reactions, the C-carbohydrate on group A streptococci produces visible clumps. (*Right*) A smooth, milky reaction is negative.

Two antibodies can give long-term protection against group A infections. One is the type-specific antibody produced in response to the M-protein, though it will not prevent infections with a different strain (which explains why a person can have recurring cases of strep throat). The other is a neutralizing antitoxin against the erythrogenic toxin that prevents the fever and rash of scarlet fever. Serological tests for anti-streptococcal antibodies are used primarily to detect continuing or recent streptococcal infection and to judge a patient's susceptibility to rheumatic fever and glomerulonephritis.

Group B: *Streptococcus agalactiae*

Several other species of β-hemolytic *Streptococcus* in groups B, C, and D live among the normal flora of humans and other mammals and can be isolated in clinical specimens from diseased human tissue. The group B streptococci (GBS), represented by the species *S. agalactiae,* demonstrate clearly how the distribution of a parasite can change in a relatively short time. It regularly resides in the human vagina, pharynx, and large intestine. A strain found in cattle is a frequent cause of bovine mastitis.[4] The major result of human colonization has been a sudden increase in serious infections in newborns and compromised people.

Streptococcus agalactiae has been chiefly implicated in neonatal, puerperal, wound, and skin infections and in endocarditis. People suffering from diabetes and vascular disease are particularly susceptible to wound infections. Because of its location in the vagina, GBS can be transferred to the infant during delivery, sometimes with dire consequences. An early-onset infection develops a few days after birth and is accompanied by sepsis, pneumonia, and high mortality (50%). A later complication comes on in 2 to 6 weeks, with symptoms of meningitis—fever, vomiting, and seizures. This bacterial pathogen is the most prevalent cause of neonatal pneumonia, sepsis, and meningitis in the United States and Europe. Because most cases occur in the hospital, personnel must be aware of the risk of passively transmitting this pathogen, especially in the neonatal and surgical units. Pregnant women should be screened for colonization in the third trimester and immunized with immune globulin. A group B strep vaccine is currently undergoing clinical trials.

4. Inflammation of the mammary gland.

Group D: Enterococci and Groups C and G Streptococci

Enterococcus faecalis, E. faecium, and *E. durans* are called the enterococci because they are normal colonists of the human large intestine. Two other members of Group D, *Streptococcus bovis* and *S. equinus,* are nonenterococci that colonize other animals and occasionally humans. Infections caused by *E. faecalis* arise most often in elderly patients undergoing surgery and affect the urinary tract, wounds, blood, the endocardium, the appendix, and other intestinal structures. Enterococci are emerging as serious nosocomial opportunists, primarily because of the rising incidence of multi-drug-resistant strains and the ease with which they are transferred from person to person.

Groups C and G are common flora of domestic animals but are frequently isolated from the human upper respiratory tract. Occasionally, they imitate group A strep in causing pharyngitis and glomerulonephritis. More often, they cause bacteremia and disseminated deep-seated infections in severely compromised patients.

TABLE 18.3

SCHEME FOR DIFFERENTIATING β-HEMOLYTIC STREPTOCOCCI

Characteristic	Group A (*S. pyogenes*)	Group B (*S. agalactiae*)	Groups C/G (*S. equisimilis*)	Group D (*Enterococcus faecalis*)
Bacitracin sensitivity	+	−	−	−
CAMP factor*	−	+	−	−
Esculin** hydrolysis in 40% bile	−	−	−	+
SxT sensitivity***	−	−	+	−
Growth at 45°C	−	−	−	+
Growth in 6.5% salt	−	+	−	+
Hippurate hydrolysis	−	+	−	−
PYR test****	+	−	−	+

*Name is derived from the first letters of its discoverers. CAMP is a diffusable substance of group B, which lyses sheep red blood cells in the presence of staphylococcal hemolysin.
**A sugar that can be split into glucose and esculetin by the group D streptococci.
***Sulfa and trimethoprim. The test is performed (like bacitracin) with discs containing this combination drug.
****PYR for L pyrrolidonyl-β-naphthylamide. This tests for an enzyme that is found mainly in group A and enterococci.

LABORATORY IDENTIFICATION TECHNIQUES

Because streptococcal disease can come on rapidly, it is beneficial to routinely culture throat and skin infections in children. The failure to recognize group A streptococcal infections (even very mild ones) can have devastating effects. Rapid cultivation and diagnostic techniques that will ensure proper treatment and prevention measures are essential. Several companies have developed rapid diagnostic test kits to be used in clinics or offices to detect group A streptococci from pharyngeal swab samples. These tests, based on monoclonal antibodies that react with the C-carbohydrates of group A, are highly specific and sensitive (figure 18.14*b*). They are primarily useful as a method for screening patients who test positive.

Complete identification usually necessitates cultivating a specimen on sheep blood agar plates and occasionally enrichment media. Group A streptococci are by far the most common β-hemolytic isolates in human lesions, but lately an increased number of infections by group B streptococci and the existence of β-hemolytic enterococci have made it important to use differentiation tests. (Groups C and G are also β-hemolytic, but they are most often associated with infections in other mammals and are only infrequently found in humans.) A positive bacitracin disc test (figure 18.14*a*) provides important evidence of group A. Other important characteristics that separate the various β-hemolytic streptococci are presented in table 18.3.

Group B streptococci are differentiated from groups A and D by the CAMP reaction and hippurate hydrolysis, both of which are positive for group B and negative for groups A and D (figure 18.15). *Enterococcus faecalis* (the predominant enterococcus) differs from the streptococci in its resistance to heat, 6.5% salt, and low concentrations of penicillin G. For these reasons, certain strains can be confused with *Staphylococcus aureus,* but a Gram stain and catalase test help differentiate these two. Because *E. faecalis* is sometimes found in the pharynx, strains can also be mistaken for *Streptococcus pyogenes,* but the ability of *E. faecalis,* to hydrolyze esculin in 40% bile salt and its resistance to bacitracin help distinguish it. All groups can also be differentiated with specific agglutimation tests, available in commercial kits.

TREATMENT AND PREVENTION OF STREPTOCOCCAL INFECTIONS

Antimicrobic therapy is aimed at curing infection and preventing complications. All strains of *S. pyogenes* continue to be very sensitive to penicillin or one of its derivatives. In cases of pharyngitis, small children receive an intramuscular injection of 600,000 units of benzathine penicillin to achieve the necessary circulating levels for a period of 10 days; larger children receive 900,000 units, and adults 1.2 million units. An alternative therapy often used for impetigo is oral penicillin V taken for 10 days. If a patient is allergic to the penicillins, erythromycin (if the strain is sensitive) or a cephalosporin is prescribed.

The only certain way to arrest rheumatic fever or acute glomerulonephritis is to treat the preceding infection, because once these two pathologic states have developed, there are no specific treatments. Developing a safe vaccine for group A streptococci is a desirable but elusive goal. The antibodies produced against a particular M-protein can protect against further infection and disease by that strain, but not the other 79 strains. The M-proteins used in vaccines must be separated from other antigens that cause undesirable toxic side effects.

Some physicians recommend that people with a history of rheumatic fever or recurring strep throat receive continuous, long-term penicillin prophylaxis. Mass prophylaxis has also been indicated for young people exposed to epidemics in boarding schools, military camps, and other institutions and for known carriers. Tonsillectomy (removal of the tonsils) has been used extensively in the past, but with questionable effectiveness, because these bacteria can infect other parts of the throat. In hospitals, carriers of

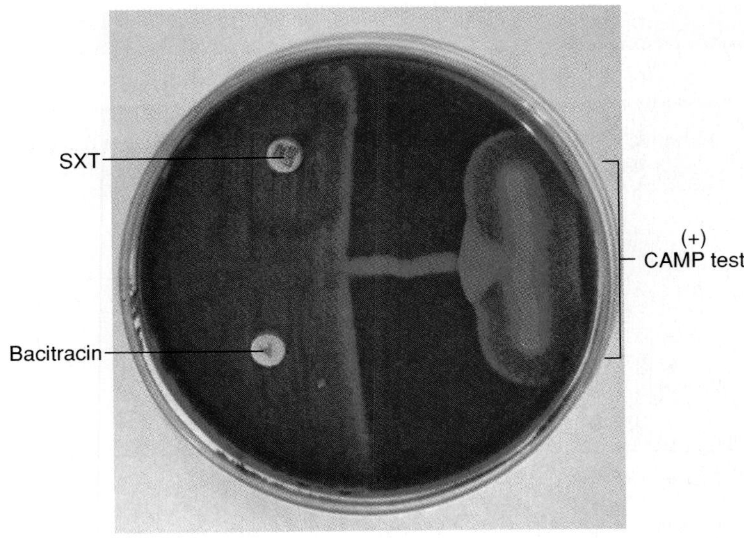

(a)

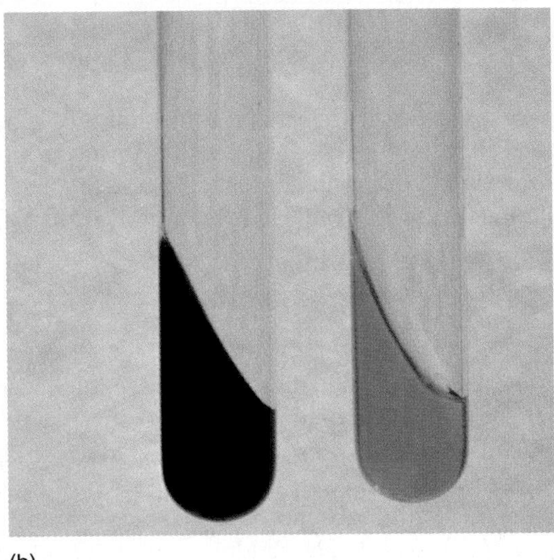

(b)

Figure 18.15
Tests for characterizing other β-hemolytic streptococci. (*a*) In the CAMP test, a special strain of β-hemolytic *Streptococcus aureus* that reacts synergistically with group B *Streptococcus* (GBS) is streaked between two smears of the sample. If the isolate is GBS, increased areas of hemolysis appear where the smears meet. (Note that group B streptococci are negative for bacitracin and SXT tests.) (*b*) Group D streptococci are identified in part by growing in bile and hydrolyzing esculin. One of the by-products of this reaction causes the precipitation of an iron salt that is brown to black. Here, the tube on the left is positive, and the one on the right is negative.

S. pyogenes should not be allowed to work with surgical, obstetric, and immunocompromised patients. Patients with group A infections must be isolated, and high-level precautions must be practiced in handling infectious secretions.

The antibiotic treatment of choice for group B streptococcal infection is penicillin G; alternatives are vancomycin and cephalosporins. Some physicians advocate routine penicillin prophylaxis in colonized mothers and infants, but others fear that this practice can increase the rate of allergies and infections by resistant strains. Artificial passive immunization with human immunoglobulins is currently being considered for treatment and prevention in mothers and infants at high risk. Treating enterococcal infection usually requires combined therapy with ampicillin and aminoglycoside (gentamicin) to take advantage of the synergism of these drugs and to overcome antimicrobic resistance.

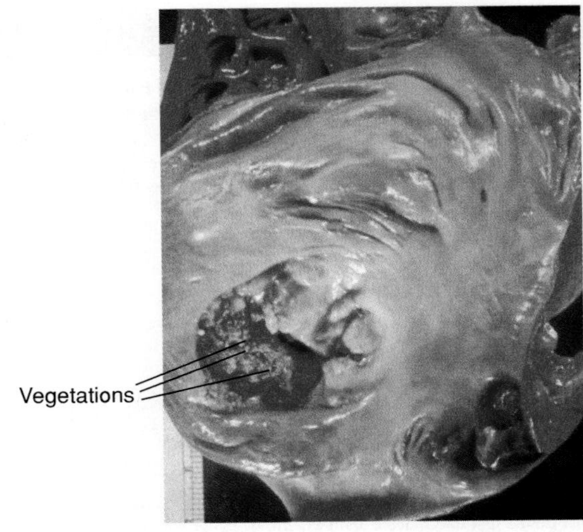

Figure 18.16
A view of the heart in subacute bacterial endocarditis. The tricuspid valve has nodular vegetations and distortions on the surface of the leaflets, which constantly release bacteria into the circulation.

α-Hemolytic Streptococci: The Viridans Group

The viridans category encompasses a large and complex group of human streptococci not entirely groupable by Lancefield serology. They are the most numerous and widespread residents of the oral cavity (gingival, cheeks, tongue, saliva) and are also found in the nasopharynx, genital tract, and skin. These species can cause serious systemic infections, although most of them are opportunists and lack the full complement of toxins and enzymes that occur in group A. The one characteristic shared by all species *(Streptococcus mitis, S. mutans, S. milleri, S. salivarius, S. sanguis)* is α-hemolysis. Other characteristics used to distinguish them are too numerous to include here.

Because viridans streptococci are not highly invasive, their entrance into tissues usually occurs through dental or surgical instrumentation and manipulation. These organisms are constant inhabitants of the gums and teeth, and even chewing hard candy or brushing the teeth can provide a portal of entry for them. Dental procedures can lead to bacteremia, meningitis, abdominal infection, and tooth abscesses. But the most important complication of all viridans streptococcal infections is **subacute endocarditis.** In this condition, blood-borne bacteria settle on areas of the heart lining or valves that have previously been injured by rheumatic fever, valve surgery, or the like. Colonization of these surfaces leads to thick layers of bacteria and fibrin called *vegetations* (figure 18.16). As the disease progresses, vegetations increase in size, constantly releasing masses of bacteria into the circulation. These

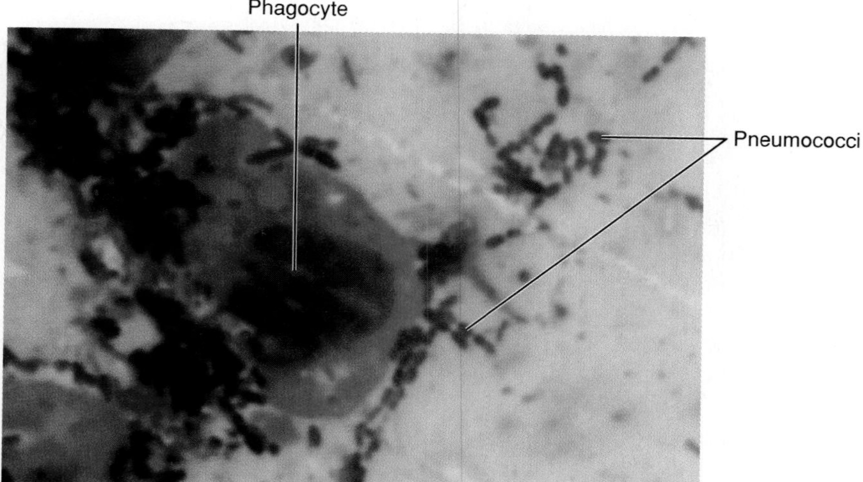

Figure 18.17

This Gram stain of sputum from a pneumonia patient indicates the morphology of *Streptococcus pneumoniae* as small, pointed diplococci. Large cells interspersed among the pneumococci are phagocytes (600×).

masses, or emboli, can travel to the lungs and brain, creating a crisis in circulation and damage to those organs. Because of its insidious and somewhat concealed course, endocarditis is considered subacute rather than acute or chronic. Symptoms and signs range from fever, heart murmur, and emboli to weight loss and anemia.

Endocarditis is diagnosed almost exclusively by blood culture, and repeated blood samples positive for bacteremia are highly suggestive of it. The goal in treatment is to completely destroy the microbes in the vegetations, which usually can be accomplished by long-term therapy with penicillin G. Because persons with preexisting heart conditions are at high risk for this disease, they usually receive prophylactic antibiotics prior to dental and surgical procedures.

Another very common dental disease involving viridans streptococci is dental caries. In the presence of sugar, *A. mutans* and *S. sanguis* produce slime layers made of glucose polymers that adhere tightly to tooth surfaces. These sticky polysaccharides are the basis for plaque, the adhesive white material on teeth that becomes coinfected with other bacteria and fosters dental disease (see chapter 21).

STREPTOCOCCUS PNEUMONIAE: THE PNEUMOCOCCUS

High on the list of significant human pathogens is ***Streptococcus pneumoniae,*** a unique species that was formerly called *Diplococcus* until its genetic similarity to the streptococci was demonstrated. Because it causes 60–70% of all bacterial pneumonias, *S. pneumoniae* is also referred to as the **pneumococcus.** Gram stains of sputum specimens from pneumonia patients reveal small lancet-shaped cells arranged in pairs and short chains (figure 18.17). Cultures require complex media such as blood or chocolate agar, and they produce smooth or mucoid colonies and α-hemolysis (see figures 4.12 and 18.9*b*). Growth is improved by the presence of 5–10% CO_2, and cultures die in the presence of oxygen because they lack catalase and peroxidases.

All pathogenic (smooth) strains form rather large capsules, this being their major virulence factor. Rough strains lack a capsule and are nonvirulent. Capsules help the streptococci escape phagocytosis, which is the major host defense in pyogenic infections. They contain a polysaccharide antigen called the specific soluble substance (SSS) that varies chemically among the pneumococcal types and stimulates antibodies of varying specificity. So far, 84 different capsular types (specified by numerals 1, 2, 3, . . .) have been identified, using a technique called the *Quellung* test or capsular swelling reaction (see figure 18.20 and under laboratory diagnosis).

Epidemiology and Pathology of the Pneumococcus

From 5% to 50% of all people carry *S. pneumoniae* as part of the normal flora in the nasopharynx. Although infection is often acquired endogenously from one's own flora, it occasionally occurs after direct contact with respiratory secretions or droplets from carriers. *Streptococcus pneumoniae* is very delicate and does not survive long out of its habitat. Factors that favor development of pneumonia are old age, the season (rate of infection is highest in the winter), other diseases (people with underlying lung disease or viral infections have weakened defenses), and living in institutions, which increases the chance of contact with infected people.

The Pathology of Pneumonia Healthy people commonly inhale microorganisms into the respiratory tract without serious consequences because of the host defenses present there. Pneumonia is likely to occur when mucus containing a load of bacterial cells is aspirated from the pharynx into lungs. The lungs of susceptible individuals have temporarily or permanently lost their defenses. Passing into the bronchioles and alveoli, the pneumococci multiply and induce an overwhelming inflammatory response. This is marked by the release of a torrent of edematous fluids into the lungs. In a form of pneumococcal pneumonia termed **lobar pneumonia,** this fluid accumulates in the alveoli along with red and white blood cells. As the infection and inflammation spread rapidly through the lung, the patient can actually "drown" in his own secretions. If this mixture of exudate, cells, and bacteria

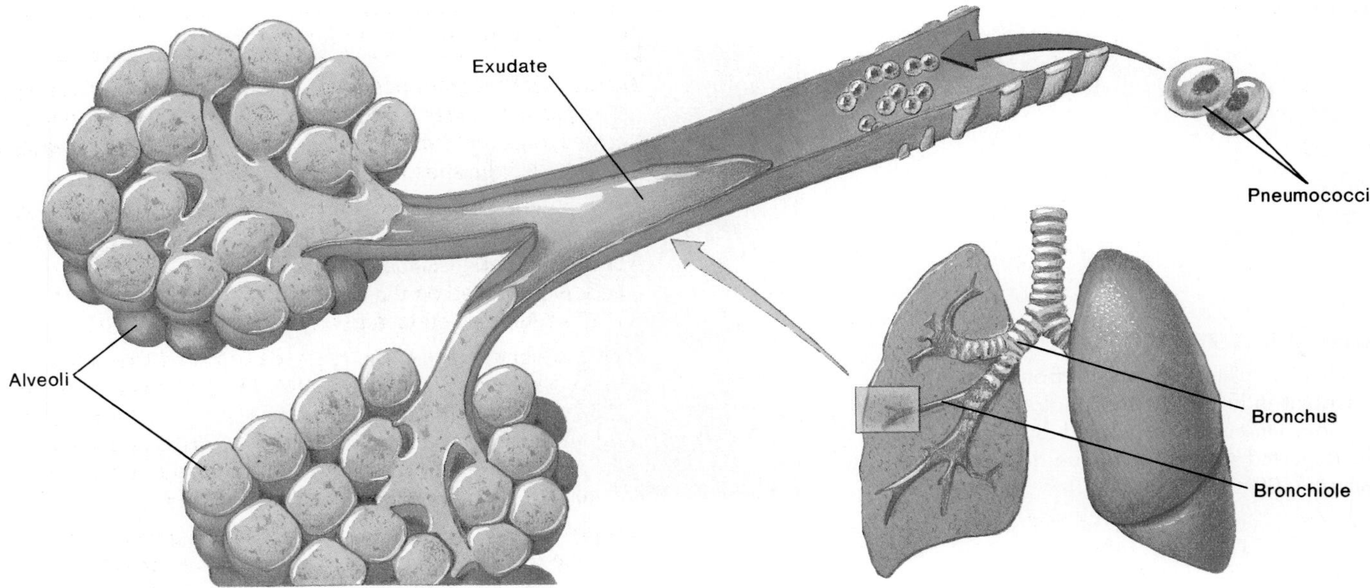

Figure 18.18
The course of bacterial pneumonia. As the pneumococcus traces a pathway down the respiratory tree, it provokes intense inflammation and exudate formation. The blocking of the bronchioles and alveoli by consolidation of inflammatory cells and products is evident.

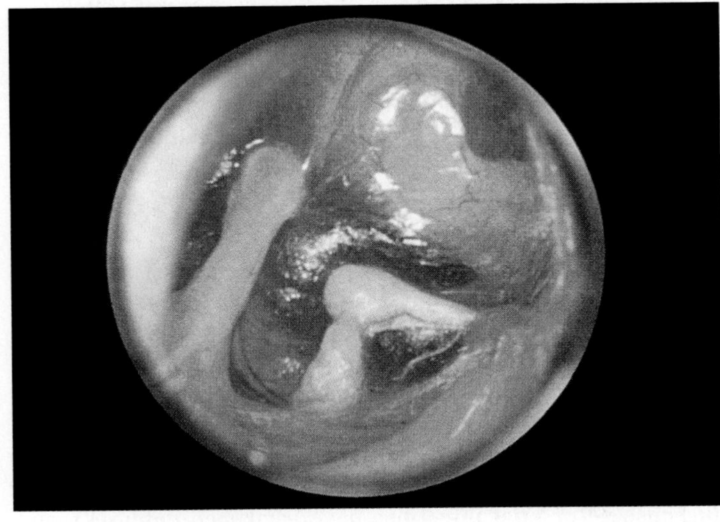

Figure 18.19
Pneumococcal otitis media. The acute inflammation and internal pressure of a middle ear infection are evident in the redness and bulging of the tympanum as seen during otoscope examination. Occasionally, the eardrum breaks as shown here. (White objects are the auditory bones.)

solidifies in the air spaces, a condition known as *consolidation** occurs (figure 18.18). In infants and the elderly, the areas of infection are usually spottier and centered more in the bronchi than in the alveoli (bronchial pneumonia).

Symptoms of pneumococcal pneumonia are chills, shaking, rapid breathing, and fever. The patient can experience severe pain in the chest wall, cyanosis (due to compromised oxygen intake), a cough that produces rusty-colored (bloody) sputum, and abnormal breathing sounds. Systemic complications of pneumonia are pleuritis and endocarditis, but pneumococcal bacteremia and meningitis are the greatest danger to the patient.

In young children, *S. pneumoniae* is a common agent of upper respiratory tract infections that can spread to the meninges and cause meningitis. It is even more common for this agent to gain access to the chamber of the middle ear by way of the eustachian tube and cause a middle ear infection called **otitis media.** This occurs readily in children under 2 years because of their relatively short eustachian tubes. Otitis media is the third most common childhood disease in the United States, and the pneumococcus accounts for the majority of cases. The severe inflammation in the small space of the middle ear induces acutely painful earaches and sometimes even temporary deafness (figure 18.19).

Healthy people have high natural resistance to the pneumococcus. The natural mucous and ciliary responses of the respiratory tract help flush out transient organisms. Respiratory phagocytes are also essential in the eradication process, but this system works only if the capsule is coated by opsonins in the presence of complement. After recovery, the individual will be immune to future infections by that particular pneumococcal type. This immunity is the basis for a successful vaccine (see subsequent section on prevention).

Laboratory Cultivation and Diagnosis A specimen is usually necessary before diagnosing pneumococcal infections because a number of other microbes can cause pneumonia, sepsis, and meningitis. Blood cultures, sputum, pleural fluid, and spinal fluid are common specimens. A Gram-stained specimen is very instructive (figure 18.17), and presumptive identification can often be made by this method alone. Also highly definitive is the Quel-

*consolidation L. *consolidatio,* to make firm.

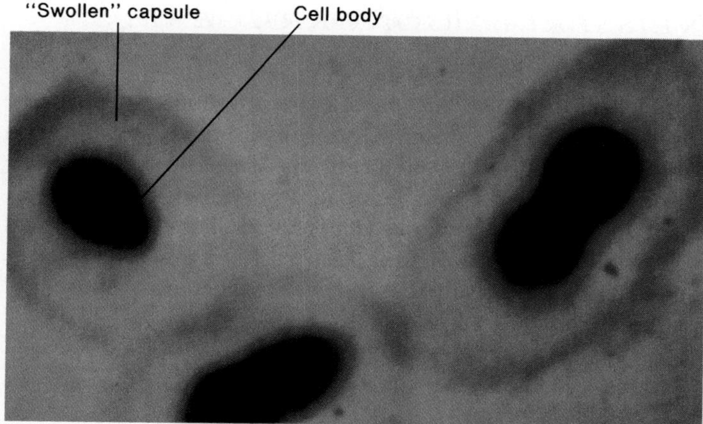

Figure 18.20
A positive Quellung, or capsular swelling, test with anticapsular precipitins, is confirmatory for *Streptococcus pneumoniae.* It can also be used to identify the precise capsular serotype. The reaction of antibodies with the capsular polysaccharide intensifies the capsule.

lung reaction, a serological test in which sputum is mixed with anti-capsular antisera and observed microscopically. Precipitation of antibodies on the surface of the capsule gives the appearance of swelling (figure 18.20). Alpha-hemolysis helps differentiate the pneumococcus from non-viridans streptococci. Diagnosis can then be confirmed by testing for the agent's sensitivity to the drug optochin, a positive bile solubility, and positive insulin fermentation tests.

Treatment and Prevention of Pneumococcal Infections The treatment of choice for pneumococcal infections has traditionally been a large daily dose of penicillin G or penicillin V. In recent times, however, clinics have reported increased cases of drug-resistant strains. Drug sensitivity testing is therefore essential. Alternative drugs are cephalosporins, vancomycin, quinolones, and erythromycin. Daily penicillin prophylaxis has recently been suggested to protect children with sickle-cell anemia against recurrent pneumococcal infections. Untreated cases in these children have a mortality rate up to 30%.

Although antibiotics arrest the course of pneumonia, active immunity is important in preventing recurrences. This is the one streptococcal disease for which effective vaccination is available. Polyvalent vaccines (Pneumovax and Pnu-immune) that contain capsular antigens of 23 of the most frequently encountered serotypes are indicated for patients at particularly high risk, including those with sickle-cell anemia, lack of a spleen, congestive heart failure, lung disease, diabetes, kidney disease, and advanced age. These vaccines are effective for about 5 years in 60–70% of those vaccinated. Immunization is not indicated in normal, healthy infants or in seriously immunosuppressed individuals.

Children suffering from chronic otitis media may require a procedure in which drainage tubes are inserted into the eardrum to remove trapped fluid that promotes infection.

Chapter Checkpoints

Streptococci are gram-positive cocci arranged in chains and pairs. The genus includes both harmless commensals and formidable pathogens. All *Streptococcus* species are catalase-negative, which distinguishes them from the staphylococci.

Streptococci are classified immunologically by their cell wall antigens (Lancefield groups) and also by their hemolysis of red blood cells. Clinical identification of *Streptococcus* species is based on the sensitivity to certain drugs, presence of the CAMP reaction, and other biochemical tests.

Streptococci that use hemolysins to completely hemolyze red blood cells are terms beta-hemolytic. Lancefield groups A, B, C, G, and some members of group D are all beta-hemolytic. Alpha-hemolytic streptococci are found in several Lancefield groups. Alpha-hemolysis is caused by a hemolysin that partially clears red blood cells in blood agar cultures.

S. pyogenes is a group A beta-hemolytic pathogen. It is extremely pathogenic because of its many virulence factors such as M-protein, tissue-digesting enzymes, and streptolysin S and O, which attack leukocytes, kidney, and heart muscle. Long-term sequelae of *S. pyogenes* infections are rheumatic fever and acute glomerulonephritis (AGN).

Local infections caused by *S. pyogenes* include the skin infections impetigo and erysipelas, as well as pharyngitis ("strep throat").

The group B pathogen *S. agalactiae* is a common opportunist agent of wound, skin, and neonatal infections.

The group D species *Enterococcus faecalis* causes opportunistic infections of wounds, blood, endocardium, and the urinary and gastrointestinal tracts.

Streptococci from groups C and G, usually animal flora, are increasingly isolated in pharyngitis, AGN, and bacteremias in immunocompromised patients.

Alpha-hemolytic streptococci of the viridans group cause dental caries as well as subacute bacterial endocarditis following dental surgery or injury to the oral mucosa.

Alpha-hemolytic *S. pneumoniae* causes 60–70% of all bacterial pneumonias. Young children, the elderly, and immunocompromised individuals are particularly susceptible. However, immunization against *S. pneumoniae* gives effective protection to these groups.

Most streptococcal infections can be cured with penicillin G, except for the enterococci and pneumococci. They require combination drug therapy because of developing resistance to antimicrobic drugs.

THE FAMILY NEISSERIACEAE: GRAM-NEGATIVE COCCI

Members of the Family Neisseriaceae are residents of the mucous membranes of warm-blooded animals. Most species are relatively innocuous commensals, but two are primary human pathogens

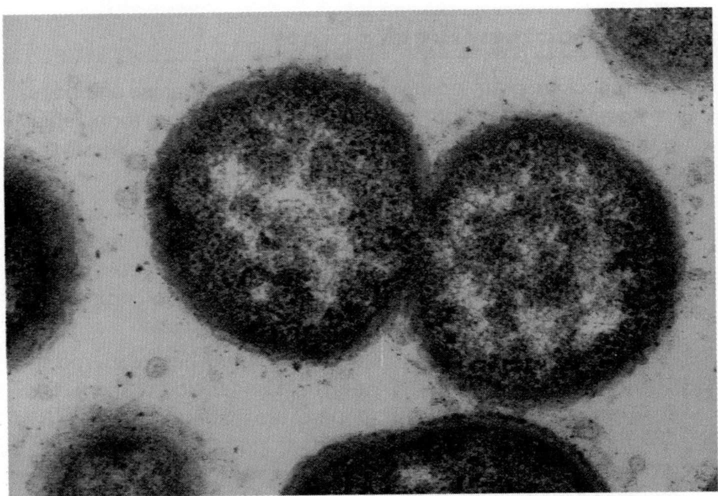

Figure 18.21
This transmission electron micrograph of *Neisseria* (52,000×) clearly indicates how the diplococci form.

with far-reaching medical impact. The genera contained in this group are *Neisseria, Moraxella,* and *Acinetobacter.* Of these, *Neisseria* has the greatest clinical significance.

A distinguishing feature of the ***Neisseria*** is their cellular morphology. Rather than being perfectly spherical, the cells are bean-shaped and paired, with their flat sides touching (figure 18.21). None develop flagella or spores, but capsules can be found on the pathogens. The cells are typically gram-negative, possessing an outer membrane in the cell wall and, in many cases, pili.

Most *Neisseria* are strict parasites that do not survive long outside the host, particularly where hostile conditions of drying, cold, acidity, or light prevail. *Neisseria* species are aerobic or microaerophilic and have an oxidative form of metabolism. They produce catalase, enzymes for fermenting various carbohydrates, and the enzyme cytochrome oxidase that can be used at a general level in identification. The pathogenic species, *N. gonorrhoeae* and *N. meningitidis,* require complex enriched media and grow best in an atmosphere containing additional CO_2. We shall concentrate on other features of the pathogenic *Neisseria* in the following sections.

*NEISSERIA** *GONORRHOEAE*: THE GONOCOCCUS

Gonorrhea* has been known as a sexually transmitted disease since ancient times. Its name originated with the Greek physician Claudius Galen, who thought that it was caused by an excess flow of semen. For a fairly long period in history, gonorrhea was confused with syphilis (microfile 18.3). Later, microbiologists went on to cultivate *N. gonorrhoeae,* also known as the **gonococcus,** and to prove conclusively that it alone was the etiologic agent of gonorrhea.

Factors Contributing to Gonococcal Pathogenicity

The virulence of the gonococcus is due chiefly to the presence of pili and other surface molecules that promote mutual attachment of cocci to each other and invasion and infection of epithelial tissue. In addition to their role in adherence, pili also seem to slow phagocytosis by macrophages and neutophils. Another contributing factor in pathogenicity is a protease that cleaves the secretory antibody (IgA) on mucosal surfaces and keeps it from working.

Epidemiology and Pathology of Gonorrhea

Gonorrhea is a strictly human infection that occurs worldwide and ranks among the top five sexually transmitted diseases in prevalence. Although about 300,000 cases are reported in the United States each year, it is estimated that the actual incidence is much higher—in the millions if one counts asymptomatic infections. Most cases occur in young people (18–24 years old) with multiple sex partners. Figures on the prevalence of gonorrhea and syphilis over the past 60 years show a fluctuating pattern, apparently corresponding to periods of social and political upheaval, when promiscuity tends to increase (figure 18.22). One interesting effect occurred during the sexual revolution of the 1960s, when reliance on oral contraceptives rather than condoms to prevent pregnancy increased the transmission of the gonococcus. Although gonorrhea has long occurred in epidemic proportions, present trends indicate a decline in the number of cases to the levels of the early 1960s.

**Neisseria* (ny-serr´-ee-uh) After the German physician Albert Neisser who first observed the agent of gonorrhea in 1879.

*gonorrhea Gr. *gonos,* seed, and *rhein,* to flow.

MICROFILE 18.3 AN UNFORTUNATE SELF-EXPERIMENT

Throughout the Middle Ages, gonorrhea and syphilis were thought to be different manifestations of the same disease. In 1767, a dedicated and daring English physician named John Hunter attempted to determine if this were so with a shocking and unfortunate experiment. He played the parts of both guinea pig and scientist by inoculating himself with pus from a gonorrhea patient. Dr. Hunter did not know that his patient happened to be simultaneously infected with gonorrhea and syphilis. Not only did the poor doctor acquire both infections, but his findings continued to foster the old and totally incorrect belief that the two diseases were one and the same.

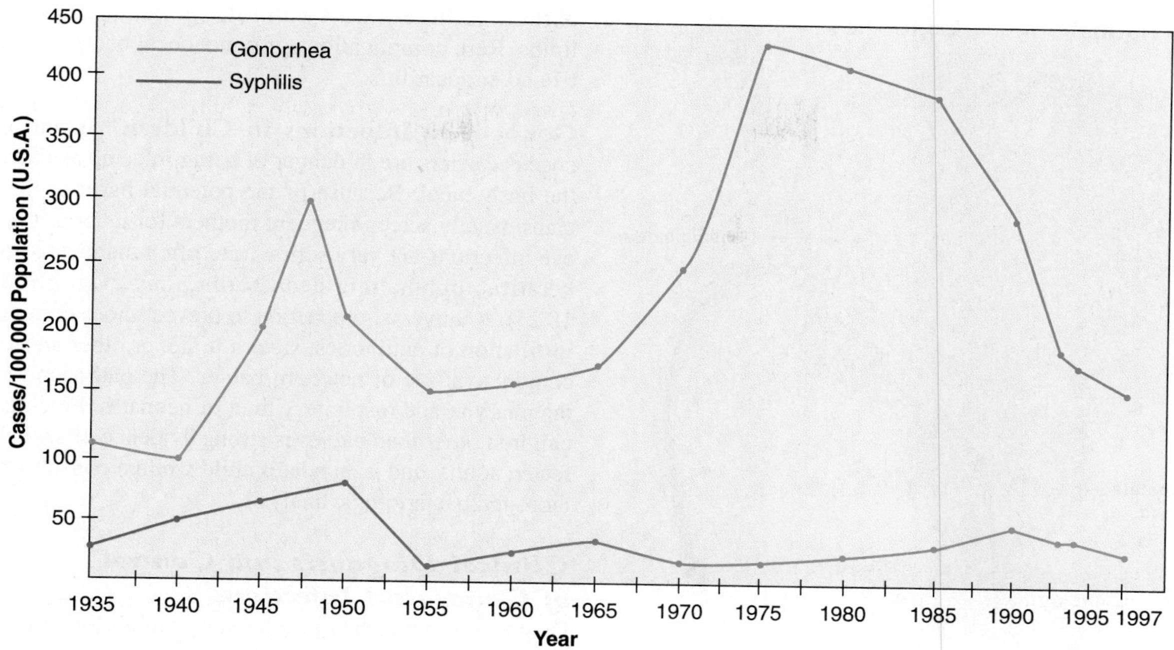

Figure 18.22
Comparative graph of two reportable infectious STDs. Gonorrhea was once the most common reportable STD in the United States. Notice the rise in cases (epidemics) corresponding with two periods: 1940–1946 (World War II) and 1965–1975 (the Vietnam War). This pattern is correlated with times of social upheaval and changing sexual mores. Epidemiologists also theorize that increased use of oral contraceptives contributed to the second epidemic period. The current trend for both diseases is a decrease in prevalence.

For various reasons, not every person coming in contact with a virulent strain of the gonococcus becomes infected. Studies done with male volunteers revealed that an infectious dose can range from 100 to 1,000 colony-forming units. *Neisseria gonorrhoeae* does not survive for more than one or two hours out of the body and is most infectious when transferred directly to a suitable mucous membrane. Survival is extremely unlikely on fomites. Except for neonatal infections, the gonococcus spreads through some form of sexual contact. The pathogen comes in contact with an appropriate portal of entry that is genital or extragenital (rectum, eye, or throat). After attaching to the epithelial surface by pili, the bacteria invade the underlying connective tissue. In 2 to 6 days, this process results in an inflammatory reaction that may or may not produce noticeable symptoms. The infection is asymtomatic in approximately 10% of males and 50% of females. It is this reservoir that is most important in the persistence and spread of the pathogen. The following sections survey the several categories of gonorrhea.

Genital Gonorrhea in the Male Infection of the urethra elicits urethritis, painful urination, and a yellowish discharge, though a relatively large number of cases are asymptomatic. In most cases, infection is limited to the distal urogenital tract, but it can occasionally spread from the urethra to the prostate gland and epididymis (figure 18.23). Scar tissue formed in the spermatic ducts during healing of an invasive infection can render the individual infertile. This outcome is becoming increasingly rarer with improved diagnosis and treatment regimens.

Genitourinary Gonorrhea in the Female The proximity of the genital and urinary tract openings increases the likelihood that both systems can be infected during sexual intercourse. A mucopurulent or bloody vaginal discharge occurs in about half the cases, along with painful urination if the urethra is affected. Major complications occur when the infection ascends from the vagina and cervix to higher reproductive structures such as the uterus and fallopian tubes (figure 18.24). One disease resulting from this progression is **salpingitis,*** also known as **PID** (pelvic inflammatory disease), a condition characterized by fever, abdominal pain, and tenderness. It is not unusual for the microbe to become involved in mixed infections with anaerobic bacteria. The build-up of scar tissue from these infections can block the fallopian tubes, causing sterility and ectopic pregnancies.

Extragenital Gonococcal Infections in Adults Extragenital sexual transmission and carriage of the gonococcus are not uncommon. Anal intercourse can lead to proctitis, and oral copulation can result in pharyngitis and gingivitis. These complications appear most often in homosexual males. Careless personal hygiene can account for self-inoculation of the eyes and a serious form of conjunctivitis. In 1% of gonorrhea cases, the gonococcus enters the bloodstream and is disseminated to the joints and skin. Involvement of the wrist and ankle can lead to

*salpingitis (sal′′-pin-jy′-tis) Gr. *salpinx,* tube, and *itis,* inflammation. An inflammation of the fallopian tubes.

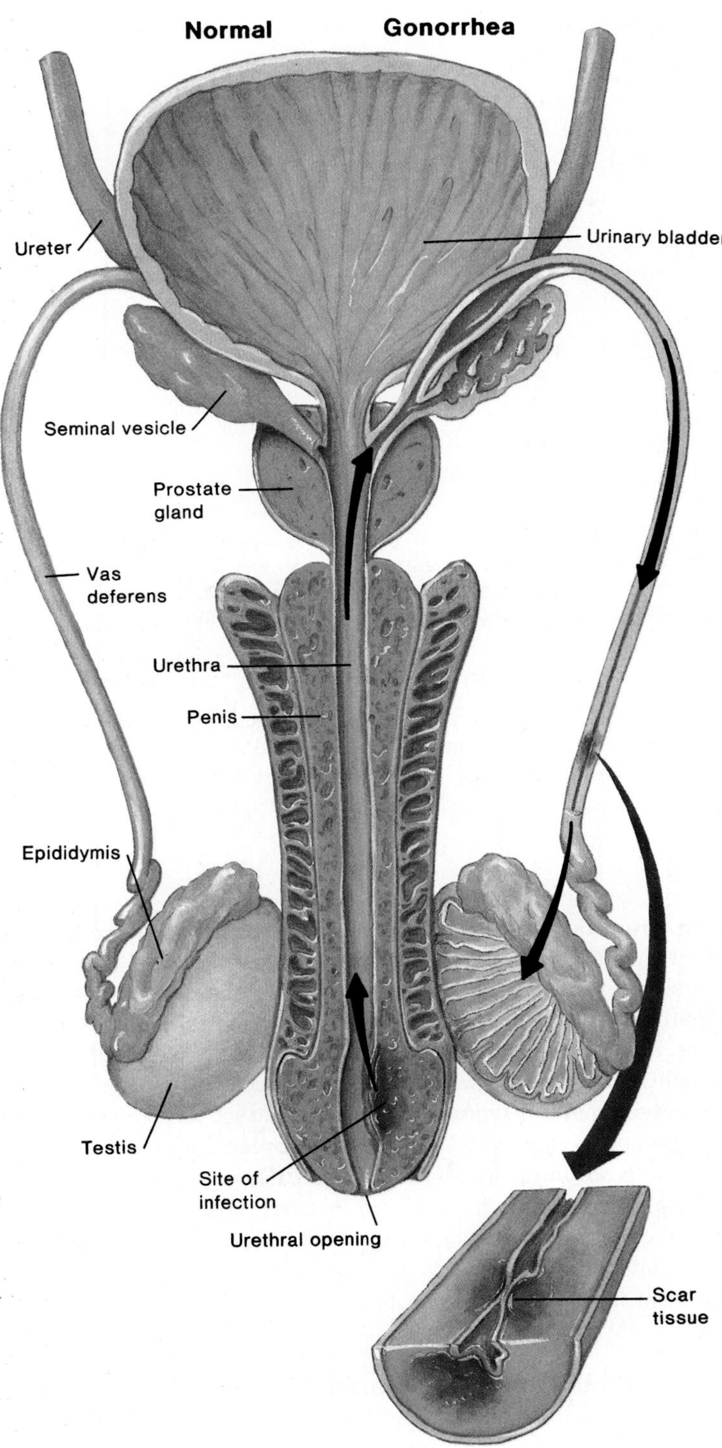

Figure 18.23
(*Left*) Frontal view of the male reproductive tract. (*Left side*) The normal, uninfected state. The sperm-carrying ducts are continuous from the testis to the urethral opening. (*Right side*) The route of ascending gonorrhea complications. Infection begins at the tip of the urethra, ascends the urethra through the penis, and passes into the vas deferens. Occasionally, it can even enter the epididymides and testes. (*Inset*) Damage to the ducts carrying sperm can create scar tissue and blockage, which reduce sperm passage and can lead to sterility.

chronic arthritis and a painful, sporadic, papular rash on the limbs. Rare complications of gonococcal bacteremia are meningitis and endocarditis.

Gonococcal Infections in Children Infants born to gonococcus carriers are in danger of being infected as they pass through the birth canal. Because of the potential harm to the fetus, physicians usually screen pregnant mothers for its presence. Gonococcal eye infections are very serious and often manifest sequelae such as keratitis, ophthalmia neonatorum, and even blindness (figure 18.25). A universal precaution to prevent those complications is the instillation of antibiotics, silver nitrate, or other antiseptics into the conjunctival sac of newborn babies. The pathogen may also infect the pharynx and respiratory tract of neonates. Finding gonorrhea in children other than babies is strong evidence of sexual abuse by infected adults, and it mandates child welfare consultation along with thorough bacteriologic analysis.

Clinical Diagnosis and Control of Gonococcal Infections

Diagnosing gonorrhea is relatively straightforward. The presence of gram-negative diplococci in neutrophils from urethral, vaginal, cervical, or eye exudates is especially diagnostic because gonococci tend to be engulfed and remain viable within phagocytes (figure 18.26). This simple procedure can provide at least presumptive evidence of gonorrhea. It is most successful in males and least successful in asymptomatic infection. Other tests used to identify *Neisseria gonorrhoeae* and differentiate it from related species are discussed in a later section.

Several hundred thousand new cases of gonorrhea occur every year in the United States. Of these cases, 10–20% are caused by drug-resistant strains called penicillinase-producing *N. gonorrhoeae* (PPNG) or tetracycline-resistant (TRNG). A fairly large proportion of cases are complicated by a concurrent STD such as chlamydiosis. An effective solution is combined therapy with a broad-spectrum cephalosporin (cephtriaxone) plus tetracycline (doxycycline). An alternative therapy combines a quinoline (ciprofloxacin) with tetracycline. Other drugs with some effectiveness are spectinomycin and erythromycin.

Although gonococcal infections stimulate local production of antibodies and activate the complement system, these responses do not produce lasting immunity, and some people experience recurrent infections.

Gonorrhea is a reportable infectious disease, which means that any physician diagnosing it must forward the information to a public health department. The follow-up to this finding involves tracing sexual partners to offer prophylactic antibiotic therapy. There is a pressing need to seek out and treat asymptomatic carriers and their sexual contacts, but complete control of this group is nearly impossible. Other control measures include education programs that emphasize the effects of all STDs and promote safer sexual practices such as the use of condoms. Two developments would help revolutionize the management of this disease. The first would be a serological screening test to detect antibodies to the gonococcus in a patient's serum (much like that used for syphilis), thus making it possible to detect asymptomatic carriers. The second would be the perfection of a pilus vaccine to immunize the high-risk population. The latter possibility is the subject of intensive research.

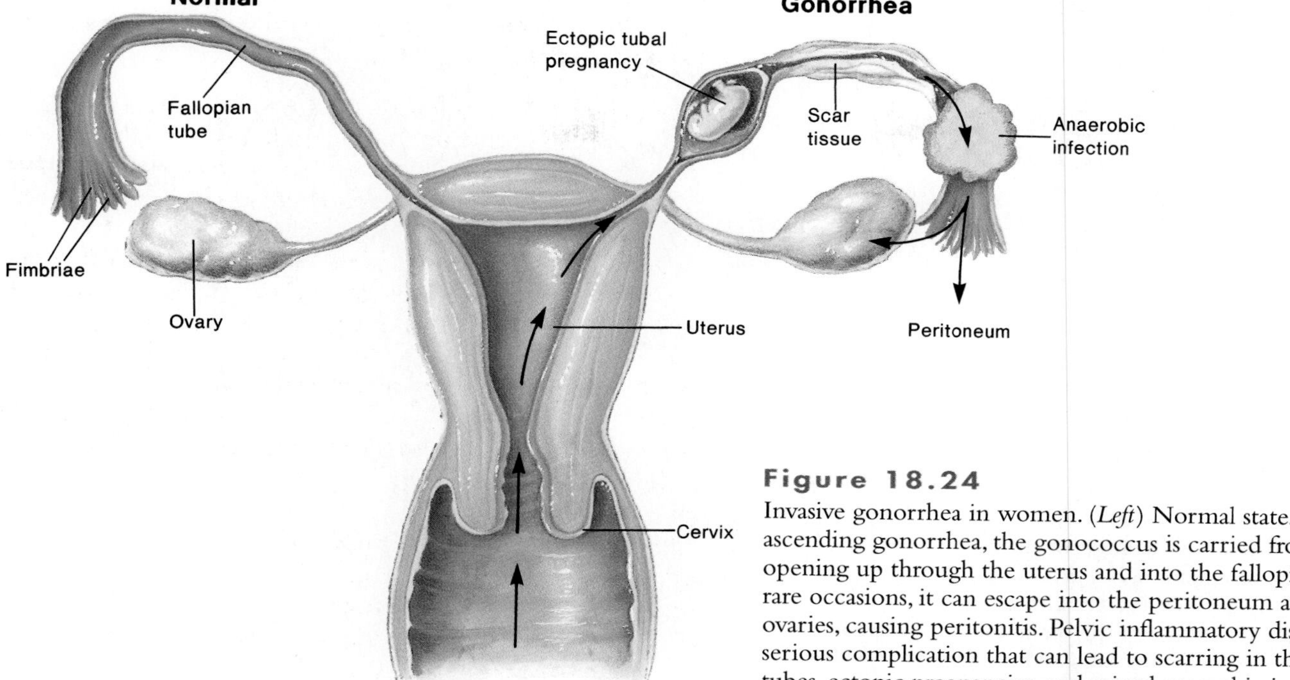

Figure 18.24
Invasive gonorrhea in women. (*Left*) Normal state. (*Right*) In ascending gonorrhea, the gonococcus is carried from the cervical opening up through the uterus and into the fallopian tubes. On rare occasions, it can escape into the peritoneum and invade the ovaries, causing peritonitis. Pelvic inflammatory disease (PID) is a serious complication that can lead to scarring in the fallopian tubes, ectopic pregnancies, and mixed anaerobic infections.

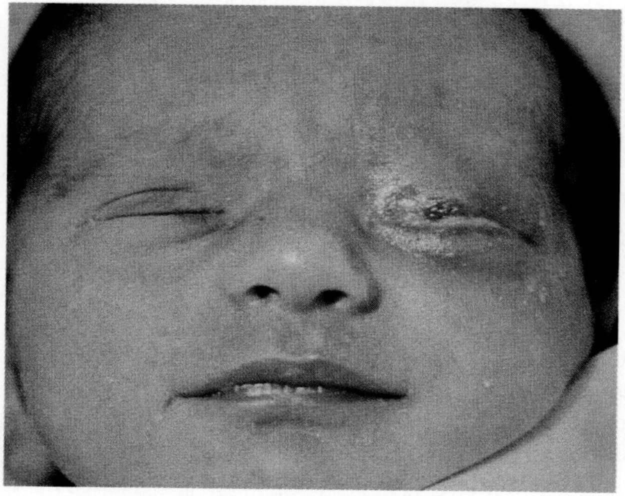

Figure 18.25
Gonococcal ophthalmia neonatorum in a week-old infant. The infection is marked by intense inflammation and edema; if allowed to progress, it causes damage that can lead to blindness. Fortunately, this infection is completely preventable and treatable.

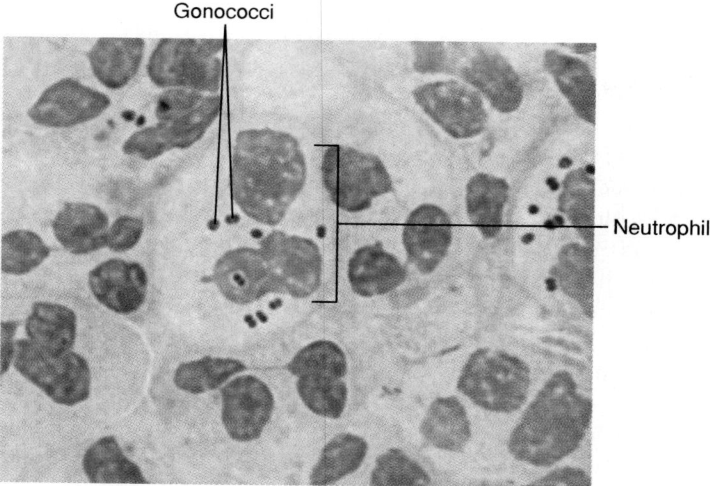

Figure 18.26
Gram stain of urethral pus from a patient with gonorrhea (1,000×). Note the intracellular (phagocytosed) gram-negative diplococci in polymorphonuclear leukocytes (neutophils).

NEISSERIA MENINGITIDIS: THE MENINGOCOCCUS

Another serious human pathogen is ***Neisseria meningitidis,*** a bacterium known commonly as the **meningococcus** and usually associated with epidemic cerebrospinal meningitis. Important factors in meningococcal invasiveness are a polysaccharide capsule, pili, and IgA protease. Although 12 different strains of capsular antigens exist, serotypes A, B, and C are responsible for most cases of infection. Another virulence factor with potent pathologic effects is the lipopolysaccharide (endotoxin) released from the cell wall when the microbe lyses.

Epidemiology and Pathogenesis of Meningococcal Disease

The diseases of *N. meningitidis* have a sporadic or epidemic incidence in late winter or early spring. The continuing reservoir of infection is humans who harbor the pathogen in the nasopharynx. The carriage state, which can last from a few days to several months, exists in 3% to 30% of the adult population and can

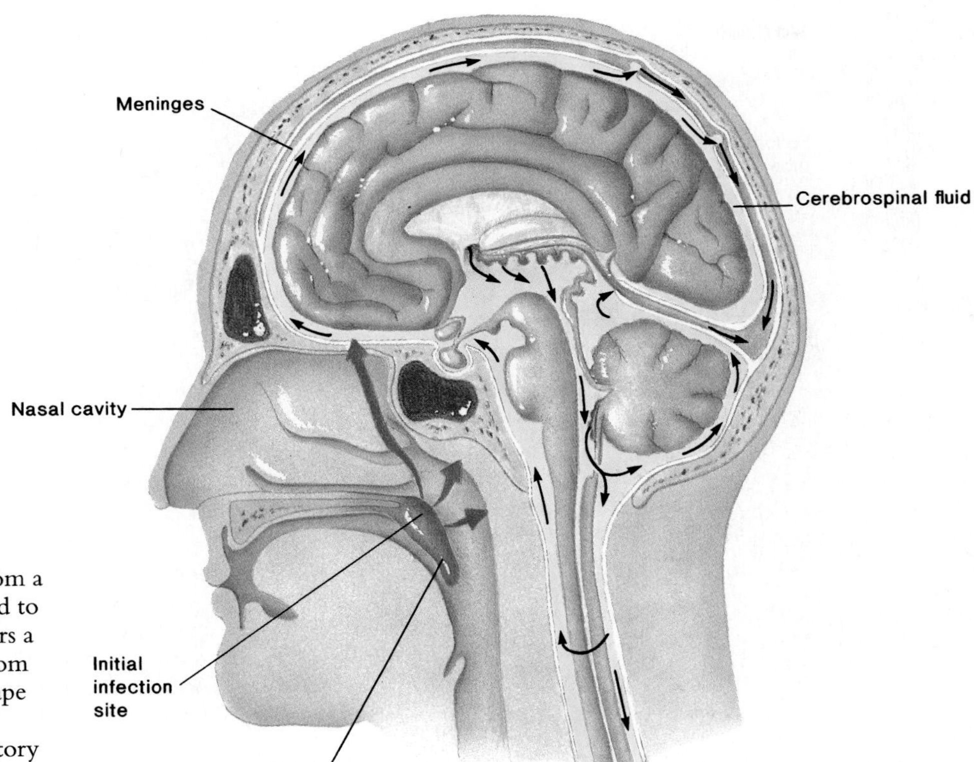

Figure 18.27
Dissemination of the meningococcus from a nasopharyngeal infection. Bacteria spread to the roof of the nasal cavity, which borders a highly vascular area at the base of the brain. From this location, they can enter the blood and escape into the cerebronspinal fluid. Infection of the meninges leads to meningitis and an inflammatory purulent exudate over the brain surface.

exceed 50% in institutional settings. The scene is set for transmission when carriers live in close quarters with nonimmune individuals, as might be expected in families, day care facilities, and military barracks. The highest risk groups are young children (6–36 months old) and older children and young adults (10–20 years old). *Neisseria meningitidis* is the most frequent cause of **meningitis** in all age groups except older patients, who are more frequently infected by *Haemophilus influenzae* (see figure 20.22). At the present time, meningococcal meningitis is on the rise. Several dozen clusters of cases have been reported from 1991 to 1997.

Because this bacterium does not survive long in the environment, meningococci are usually acquired through close contact with secretions or droplets. Upon reaching its portal of entry in the nasopharynx, the meningococcus attaches there with pili. In many people, this can result in simple asymptomatic colonization. In the more vulnerable individual, however, the meningococci are engulfed by epithelial cells of the mucosa and penetrate into the nearby blood vessels, thereby damaging the epithelium and causing pharyngitis.

Bacteria entering the blood vessels rapidly permeate the meninges and produce symptoms of meningitis, the most common complication in children. It is marked by fever, sore throat, headache, stiff neck, convulsions, and vomiting. The most serious complications of meningococcal pharyngitis are due to meningococcemia (figure 18.27). The pathogen sheds endotoxin into the generalized circulation, which is a potent stimulus for certain white blood cells. Damage to the blood vessels caused by cytokines leads to vascular collapse, hemorrhage, and crops of lesions called *petechiae** on the trunk and appendages. In a small number of cases, meningococcemia becomes a fulminant disease with a high mortality rate. It has a violent onset, with fever (higher than 40°C), chills, delirium, severe widespread *ecchymoses,* shock, and coma (figure 18.28). Generalized intravascular clotting, cardiac failure, damage to the adrenal glands, and death can occur within a few hours.

*petechiae (pee-tee´-kee-ee) Small, nonraised, round purple spots caused by hemorrhage into the skin. Larger spots are called **ecchymoses** (ek´´-ih-moh´-seez).

Clinical Diagnosis of Meningococcal Disease

Suspicion of bacterial meningitis constitutes a medical emergency, and differential diagnosis must be done with great haste and accuracy, since complications can come on so rapidly and with such lethal consequences. Cerebrospinal fluid, blood, or nasopharyngeal samples are stained and observed directly for the typical gram-negative diplococci. Cultivation may be necessary to differentiate the bacterium from other species (see subsequent section on laboratory diagnosis). Specific rapid tests are also available for detecting the capsular polysaccharide or the cells directly from specimens without culturing.

Immunity, Treatment, and Prevention of Meningococcal Infection

The infection rate in most populations is about 1%, so well-developed natural immunity to the meningococcus appears to be the rule. This is due to a sort of natural immunization that occurs during the early years of life as one is exposed to the meningococcus and its close relatives. Resistance is due to opsonizing antibodies that develop against the capsular polysaccharides in groups A and C and against membrane antigens in group B. Because even treated meningococcal disease has a mortality rate of up to 15%, it is vital that chemotherapy begin as soon as possi-

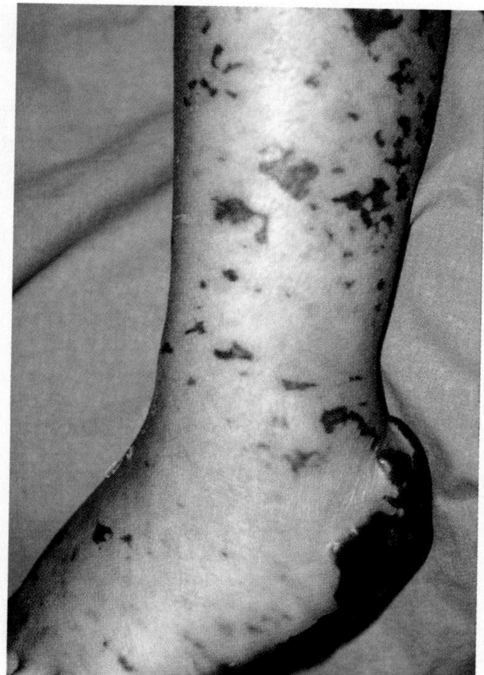

Figure 18.28
The appearance of meningococcemia. The blotches on the leg and foot are due to subcutaneous hemorrhages. This condition can occur anywhere on the body, including the mucous membranes and conjunctiva. Endotoxins released during blood infection are thought to be largely responsible for this pathologic state.

ble with one or more drugs. It can even be given while tests for the causative agent are under way. Penicillin G is the most potent of the drugs available for meningococcal infections; it is generally given in high doses intravenously. If the patient cannot tolerate penicillin, intravenous chloramphenicol is the second choice. Patients may also require treatment for shock and intravascular clotting.

When family members, medical personnel, or children in day care have come in close contact with infected people, preventive therapy with rifampin or tetracycline may be warranted. Meningococcal vaccines that contain specific purified capsular antigens are available to protect high-risk groups, especially during epidemics. Group A vaccine protects all ages, but group C vaccine is useful only for individuals over 2 years of age, and a group B vaccine is not yet available.

DIFFERENTIATING PATHOGENIC FROM NONPATHOGENIC *NEISSERIA*

It is usually necessary to differentiate true pathogens from normal *Neisseria* that also live in the human body and can be present in infectious fluids. Immediately after collection, specimens are streaked on Modified Thayer-Martin medium (MTM) or chocolate agar and incubated in a high CO_2 atmosphere (figure 18.29*a*). Presumptive identification of the genus is obtained by a Gram stain and oxidase testing on isolated colonies (figure 18.29*b*). Further testing may be necessary to differentiate the two pathogenic

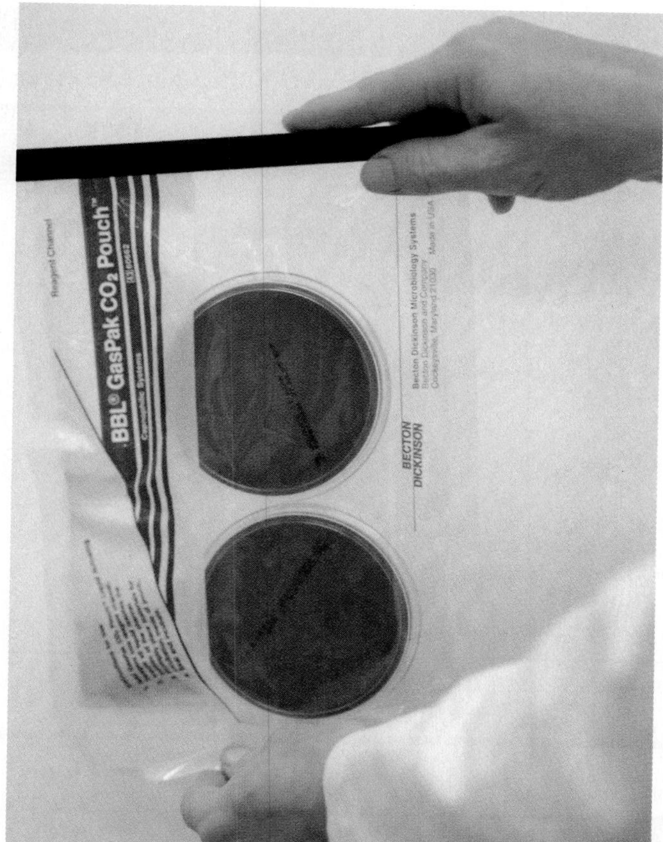

(a)

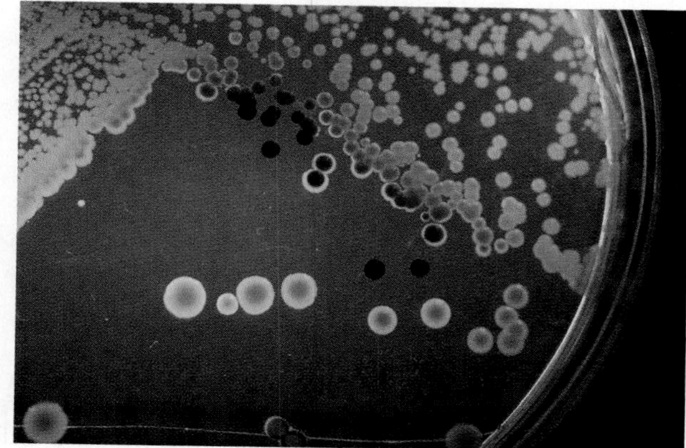

(b)

Figure 18.29
Selected laboratory techniques for *Neisseria* species. (*a*) A pouch system provides an individualized compartment for incubating specimens. Plates of inoculated chocolate agar have been sealed in the pouch after insertion of a CO_2-generating solution. This method favors increased survival and isolation of sensitive pathogens. (*b*) The oxidase test. A drop of oxidase reagent is placed on a suspected *Neisseria* or *Branhamella* colony. If the colony reacts with the chemical to produce a purple to black color, it is oxidase-positive; those that remain white to tan are oxidase-negative. Because several species of gram-negative rods are also oxidase-negative. Because several species of gram-negative rods are also oxidase-positive, this test is presumptive for these two genera only if a Gram stain has verified the presence of gram-negative cocci.

TABLE 18.4

SCHEME FOR DIFFERENTIATING GRAM-NEGATIVE COCCI AND COCCOBACILLI

Characteristic or Test	*Neisseria. gonorrhoeae*	*N. meningitidis*	*N. lactamica**	*N. sicca*	*Branhamella catarrhalis*	*Moraxella* spp.	*Acinetobacter* spp.
Oxidase	+	+	+	+	+	+	−
Growth on MTM	+	+	+	−	+/−	−	−
Growth on nutrient agar	−	−	−	+	+	+	+
Growth on blood agar at 22°C	−	−	+/−	+/−	+	+	+
Yellow pigment	−	−	+	+/−	−	−	−
Require increased CO_2 during isolation	+	−	−	−	−	−	−
Sugar fermentation							
Maltose	−	+	+	+	−	−	+/−
Sucrose	−	−	−	+	−	−	+/−
Lactose	−	−	+	−	−	−	+/−
Nitrate reduction	−	−	−	−	+	+/−	−
Nitrite reduction	−	+/−	++	+	+	−	−
Capsule	+	+	−	+/−	+	+/−	+/−
Human pathogenicity	+	+	+	UDC**	Opportunistic	UDC	Opportunistic

*A weak pathogen, found in the nasopharynx of children and easily mistaken for *N. meningitidis.*

**Stands for Usually Dismissed as Contaminants, because pathogenicity is very rare.

Chapter Checkpoints

Neisseria are fastidious, nonmotile, gram-negative, coffee bean–shaped cocci that are facultative anaerobes and live as commensals in the mucous membranes of mammals. The two significant pathogens in this group are *Neisseria gonorrhoeae* and *Neisseria meningitidis.*

N. gonorrhoeae is the causative agent of gonorrhea, a sexually transmitted disease that causes infections of the male and female reproductive tracts. Genital gonorrhea in males causes urethritis and painful urination. Genital gonorrhea in females can lead to PID and ectopic pregnancies. It also causes sterility, proctitis, and pharyngitis in the sexually active, as well as gonococcal eye infections of the newborn.

Gonorrhea is a reportable STD because it is one of the top five STDs worldwide. It is spread by direct contact with infected carriers, who are often asymptomatic. Humans are the only reservoirs. It is especially prevalent in 18–24 year olds. Gonorrhea is treated by combined drug therapy because of widespread resistance to penicillin and frequent concurrent infection by other STD agents, particularly chlamydia.

Virulence factors of *N. gonorrhoeae* include pili, a capsule, and a protease that inactivates IgA.

N. meningitidis is the causative agent of epidemic cerebrospinal meningitis. It is transmitted by respiratory secretions or droplets from infected carriers.

Meningococcus virulence factors include lipopolysaccharide (endotoxin), IgA protease, pili, and a capsule. Toxins cause cardiac failure, vascular collapse, and clotting disorders. Immunization is available to protect at-risk groups.

species from one another, from other oxidase-positive infectious species, and from normal flora of the oropharynx and genitourinary tract that can be confused with the pathogens. Sugar fermentation, growth patterns, nitrate reduction, and pigment production are useful differentiation tests at this point (table 18.4). Several rapid method identification kits have been developed for this purpose.

OTHER GRAM-NEGATIVE COCCI AND COCCOBACILLI

The genera *Branhamella, Moraxella,* and *Acinetobacter* are included in the same family as *Neisseria* because of morphological and biochemical similarities. Most species are either relatively harmless commensals of humans and other mammals or are saprobes living in soil and water. In the past few years, however, one species in particular has emerged as a significant opportunist in hosts with disturbed immune functions. This species *Branhamella (Moraxella) catarrhalis,** is found in the normal human nasopharynx and can cause purulent disease. It is associated with several clinical syndromes such as meningitis, endocarditis, otitis media, bronchopulmonary infections, and neonatal conjunctivitis. Adult patients with leukemia, alcoholism, malignancy, diabetes, or rheumatoid disease are the most susceptible to it.

Because *Branhamella* is morphologically similar to the gonococcus and the meningococcus, biochemical methods are needed to discriminate it. Key characteristics are the complete lack of carbohydrate fermentation and positive nitrate reduction in *B. catarrhalis.* Treatment of infection by this organism is with erythromycin or cephalosporins, because so many strains of this species produce penicillinase.

*Moraxella** species are short, plump rods rather than cocci, and some exhibit a twitching motility. These bacteria are widely distributed on the mucous membranes of domestic mammals and humans and are generally regarded as weakly pathogenic or nonpathogenic. Some species are rarely implicated in ear infections and conjunctivitis in humans.

*Branhamella catarrhalis (bran´´-hah-mel´-ah cah-tahr-al´-is) After Sarah Branham, a bacteriologist working on the genus *Neisseria,* and L. *catarrhus,* to flow. It is also known as *Moraxella.*

*Moraxella (moh-rak-sel´-uh) From Victor Morax, a Swiss ophthalmologist who worked with the genus.

The genus *Acinetobacter** is similar morphologically to *Moraxella.* It is a small, paired, gram-negative cell that varies in shape from a true rod to a coccus. However, it is quite different from the members of this family in several ways. For example, its habitat is soil, water, and sewage; it is nonfastidious; and it is oxidase-negative. *Acinetobacter* is a common contaminant in clinical specimens and is ordinarily not considered clinically significant. In rare cases, however, it is clearly an agent of disease. Most infections are nosocomial in origin, affecting traumatized or debilitated people with indwelling catheters or other instrumentation. Septicemia, meningitis, endocarditis, pneumonia, and urinary tract infections have been reported.

Chapter Checkpoints

Other Neisseriaceae implicated in infectious disease are *Branhamella* and *Acinetobacter.* Most members, such as *Moraxella* species are harmless commensals or saprobes, but *Branhamella catarrhalis* causes opportunistic infections in immunocompromised individuals. *Acinetobacter* is a common agent in soil and water and is increasingly isolated in nosocomial infections.

*Acinetobacter (ass´´-ih-nee-toh-bak´-tur) Gr. *a,* none, *kinetos,* movement, and *bacterion,* rod.

Chapter Capsule with Key Terms

THE MEDICALLY IMPORTANT COCCI

Genus *Staphylococcus*

Nonmotile, non-spore-forming cocci arranged in irregular clusters; facultative anaerobes; fermentative; salt-tolerant, catalase-positive ***Staphylococcus aureus*** produces a number of virulence factors. Enzymes include **coagulase** (a confirmatory chacteristic), hyaluronidase, staphylokinase, nuclease, and penicillinase. Toxins are β-hemolysins, leukocidin, enterotoxin, exfoliative toxin, toxic shock syndrome toxin. Microbe carried in nasal vestibule, nasopharynx, skin.

Infections: Target is skin; local **abscess** occurs at site of invasion of hair follicle, gland; manifestations are **folliculitis, furuncle, carbuncle,** and **bullous impetigo.** Other common infections are **osteomyelitis,** a focal infection of bone, bacteremia, leading to endocarditis, and pneumonia.

Toxic Disease: Food intoxication due to enterotoxin; **staphylococcal scaled skin syndrome (SSSS),** a skin condition that causes desquamation; **toxic shock syndrome,** toxemia in women due to infection of vagina, associated with wearing tampons.

Principal Coagulase-Negative Staphylococci: ***S. epidermidis,*** a normal resident of skin and follicles; an opportunist and one of the most common causes of nosocomial infections, chiefly in surgical patients with indwelling medical devices or implants; ***S. saprophyticus*** is a urinary pathogen.

Treatment: S. aureus has multiple resistance to antibiotics, especially penicillin, ampicillin, and methicillin; drug selection requires sensitivity testing; cephalosporins often used; abscesses require debridement and removal of pus; extreme resistance of staphylococci to harsh environmental conditions makes control difficult, requiring high level of disinfection and antisepsis; no vaccines available.

Streptococci

A large, varied group of bacteria (about 25 species), containing the genera ***Streptococcus*** and ***Enterococcus.*** Cocci are in chains of various lengths; nonmotile, non-spore-forming; often encapsulated; fermentative; catalase-negative; most pathogens fastidious and sensitive to environmental exposure. Classified into **Lancefield groups** (A–R) according to the type of serological reactions of the cell wall carbohydrate; also characterized by type of hemolysis. The most important sources of human disease are β-hemolytic ***S. pyogenes*** (group A), ***S. agalactiae*** (group B), ***Enterococcus faecalis*** (group D) and α-hemolytic ***S. pneumoniae*** and the **viridans streptococci.**

β-Hemolytic Streptococci: ***S. pyogenes*** is the most serious pathogen of family; produces several virulence factors, including C-carbohydrates, M-protein (fimbriae), streptokinase, hyaluronidase, DNase, **hemolysins** (SLO, SLS), **pyogenic toxin.**

Microbe resides in nasopharynx of carriers; transmitted through close contact; invades skin and mucous membranes. **Skin infections** include **pyoderma** (strep **impetigo**), **erysipelas** (deeper, spreading skin infection). Systemic conditions include **strep throat** or **pharyngitis,** severe inflammation of throat membranes; may lead to toxemia, called **scarlet fever**—generalized flushing of skin and high fever due to erythrogenic toxin; also causes pneumonia, and puerperal fever.

Sequelae caused by immune response to streptococcal toxins include **rheumatic fever,** a delayed allergy that damages heart valves, joints and **glomerulonephritis,** an inflammation that leads to malfunction or destruction of kidney tubules.

Streptococcus agalactiae (group B), a cow pathogen that is increasingly found in the human vagina; causes neonatal, puerperal, wound, skin infections, particularly in debilitated persons. ***Enterococcus faecalis*** and other enteric group D species are normal flora of intestine; cause opportunistic urinary, wound, and surgical infections. Group A and group B are treated primarily with some type of penicillin (G is very effective); sensitivity testing may be necessary for enterococci; no vaccines available.

α-Hemolytic Streptococci: The **viridans streps** *S. Mitis, S. salivarius, S. mutans,* and *S. sanguis* constitute oral flora in saliva; principal infections are **subacute endocarditis,** mass colonization of heart valves following dental procedures; *mutans* and *sanguis* species are the main contributors to plaque and dental disease.

S. pneumoniae, the **pneumococcus,** has heavily encapsulated, lancet-shaped diplococci; capsule is an important virulence factor—84 types; reservoir is nasopharynx of normal healthy carriers.

The pneumococcus is the most common cause of bacterial pneumonia; attacks patients with weakened respiratory defenses; entrance of bacteria into lungs initiates acute, massive inflammatory response that fills lungs (and bronchioles) with fluid; **consolidation** of fluid leads to **lobar pneumonia;** respiration severely compromised, oxygen exchange inefficient; other symptoms include fever/chills, cyanosis, cough; **otitis media,** inflammation of middle ear, is common in children; vaccination available for patients at risk.

GRAM-NEGATIVE COCCI

Primary pathogens are in the genus ***Neisseria,*** common residents of mucous membranes. *Neisseria* species are bean-shaped diplococci that may be encapsulated and are pilated and oxidase-positive, non-spore-forming, nonmotile; pathogens are fastidious and do not survive long in the environment.

Neisseria gonorrhoeae: The **gonococcus,** cause of **gonorrhea;** microbe invades mucous membranes by attaching with **pili;** tends to be located intracellularly in pus cells; among the top five most common STDs; may be transmitted from mother to newborn; asymptomatic carriage is common in both sexes.

Symptoms of **gonorrhea in males** are urethritis, discharge; may infect deeper reproductive structures; causes scarring and infertility. Symptoms of **gonorrhea in females** include vaginitis, urethritis. Ascending infection may lead to **salpingitis (PID),** mixed anaerobic infection of abdomen; common cause of sterility and ectopic tubal pregnancies due to scarred fallopian tubes. **Extragenital infections** may be anal, pharyngeal, conjunctivitis, septicemia, arthritis.

Infections in newborns cause eye inflammation, occasionally infection of deeper tissues and blindness; can be prevented by prophylaxis immediately after birth.

Preferred treatment is a combination of cephalosporin and tetracycline due to increasing **PPNG** (penicillin-resistant) strains; no vaccine exists; safe sex practices, lack of promiscuity are important controls.

Neisseria meningitidis: The **meningococcus,** most common cause of **meningitis** in United States. Agent is a common resident of the nasopharynx; invades when resistance is lowered; is spread by close contact; bacterium adheres by capsule and pili. Disease begins when bacteria enter bloodstream, pass into the cranial circulation, multiply in meninges; very rapid onset; initial symptoms are neurologic; **endotoxin** released by pathogen causes hemorrhage and shock; can be fatal; treated with penicillin and/or chloramphenicol; vaccines exist for groups A and C.

Other Gram-Negative Cocci and Coccobacilli: ***Branhamella catarrhalis,*** common member of throat flora; is an opportunist in cancer, diabetes, alcoholism. ***Moraxella,*** short rods that colonize mammalian mucous membranes. ***Acinetobacter,*** gram-negative coccobacilli that occasionally cause nosocomial infections.

MULTIPLE-CHOICE QUESTIONS

1. Which of the following are pyogenic cocci:
   a. *Streptococcus*
   b. *Staphylococcus*
   c. *Neisseria*
   d. all of these

2. The coagulase test is used to differentiate *Staphylococcus aureus* from
   a. other staphylococci
   b. streptococci
   c. micrococci
   d. enterococci

3. The symptoms in scarlet fever are due to
   a. streptolysin
   b. coagulase
   c. pyrogenic toxin
   d. alpha toxin

4. Penicillin-resistant *Staphylococcus* can be treated with______without sensitivity testing.
   a. ampicillin
   b. erythromycin
   c. methicillin
   d. no antibiotic

5. The most severe streptococcal diseases are caused by
   a. group B streptococci
   b. group A streptococci
   c. pneumococci
   d. enterococci

6. Rheumatic fever damages the ______, and glomerulonephritis damages the______
   a. skin, heart
   b. joints, bone marrow
   c. heart valves, kidney
   d. brain, kidney

7. ______hemolysis is the partial lysis of red blood cells due to bacterial hemolysins.
   a. Gamma
   b. Alpha
   c. Beta
   d. Delta

8. An effective vaccine exists to prevent infections from
   a. *Staph, aureus*
   b. *Strep. pyogenes*
   c. *N. gonorrhoeae*
   d. *Strep. pneumoniae*

9. Viridans streptococci commonly cause
   a. pneumonia
   b. meningitis
   c. subacute endocarditis
   d. otitis media

10. Which genus of bacteria has pathogens that can cause blindness and deafness?
    a. *Streptococcus*
    b. *Staphylococcus*
    c. *Neisseria*
    d. *Branhamella*

11. An important test for identifying *Neisseria* is
    a. production of oxidase
    b. production of catalase
    c. sugar fermentation
    d. beta-hemolysis

12. A complication of genital gonorrhea in both men and women is
    a. infertility
    b. pelvic inflammatory disease
    c. arthritis
    d. blindness

13. The skin blotches in meningitis are due to
    a. skin invasion by *N. meningitidis*
    b. blood clots
    c. erysipelas
    d. endotoxins in the blood

14. Which infectious agent of those covered in the chapter would most likely be acquired from a contaminated doorknob?
    a. *Staphylococcus aureus*
    b. *Streptococcus pyogenes*
    c. *Neisseria meningitidis*
    d. *Streptococcus pneumoniae*

CONCEPT QUESTIONS

1. Differentiate between the pathologies in staphylococcal infections and toxinoses.

2. Explain how the actions of each of the following make all of them virulence factors:
   a. hemolysins
   b. leukocidin
   c. kinases
   d. hyaluronidase

3. a. Describe the normal habitat of *Staphylococcus aureus.*
   b. Distinguish between the four main "staph" skin infections.
   c. Define the term *abscess,* and describe at least two kinds that are caused by staph species.

4. What does it mean to say osteomyelitis is a focal infection?

5. What conditions favor staph food poisoning?

6. What conditions favor toxic shock syndrome?

7. a. Describe the principal role of the coagulase-negative staphylococci in disease.
   b. Describe the usual habitats of these staphylococci.

8. Compare the symptoms of streptococcal and staphylococcal impetigo.

9. a. Describe the major group A streptococcal infections.
   b. Why is a "strep throat" a cause for concern?

10. Discuss the apparent pathology at work in rheumatic fever and acute glomerulonephritis.

11. a. How are group B streptococci important?
    b. How is the genus *Enterococcus* different from the genus *Streptococcus?*
    c. What is the medical significance of enterococci and group C and G streptococci?

12. a. How does lobar pneumonia arise?
    b. How is it different from bronchial pneumonia?
    c. What causes the patient to get a blue tinge to his mucous membranes?
    d. Explain how children acquire otitis media.

13. a. Compare and contrast the characteristics of male and female genital gonorrhea.
    b. Describe at least two extragenital complications of gonorrhea.
    c. Explain how neonates become infected with *N. gonorrhoeae* and describe the disease symptoms.

14. a. Describe the pathway that *N. meningitidis* takes from infection in the nasopharynx to the brain.
    b. Describe the specific virulence factors that cause symptoms of meningitis.

15. Single matching. Only one description in the right-hand column fits a word in the left-hand column.

___ furuncle	a. complete red blood cell lysis
___ osteomyelitis	b. substance involved in heart valve damage
___ coagulase	c. dissolves blood clots
___ pyrogenic toxin	d. enzyme of pathogenic *S. aureus*
___ rheumatic fever	e. cutaneous infection of group A streps
___ β-hemolysis	f. solidification of pockets in lung
___ consolidation	g. unique pathologic feature of *N. gonorrhoeae*
___ viridans streptococci	h. a boil
___ erysipelas	i. cause of tooth abscesses
___ endocarditis	j. focal infection of long bones
___ streptolysin	k. heart colonization by viridans streps
___ streptokinase	l. cause of scarlet fever
	m. long-term sequelae of strep throat

16. Multiple matching. Match the bacterium in the left-hand column with its characteristic. More than one characteristic may fit each bacterium.

___ *Staphylococcus aureus*	a. bacitracin sensitivity
___ *Streptococcus pyogenes*	b. diplococcus
___ *Streptococcus pneumoniae*	c. causes pneumonia
___ *Neisseria gonorrhoeae*	d. complication is glomerulonephritis
___ *Neisseria meningitidis*	e. infects female reproductive tract
___ *Enterococcus faecalis*	f. resident of nasopharynx
	g. is in clusters
	h. is in chains
	i. has a capsule
	j. catalase production

CRITICAL-THINKING QUESTIONS

1. a. What is the probable significance of isolating large numbers of *Streptococcus mitis* from a throat swab sample?
   b. What is the significance of isolating five colonies of *Streptococcus pyogenes* in a throat culture?
   c. What is the significance of isolating three *Staphylococcus epidermidis* colonies from a swab culture of an open wound?
   d. Of isolating 100 colonies of *S. epidermidis* from a swab from an in-dwelling catheter?
   e. How important is the isolation of 10 colonies of *N. meningitidis* from the nasopharynx?
   f. What if it is isolated from the spinal fluid? Give one possible explanation for finding *Enterococcus faecalis* in the blood.
2. Why is it so important to differentiate *S. aureus* from coagulase-negative staphylococci?
3. You have been handed the problem of diagnosing gonorrhea from a single test. Which one will you choose and why?
4. How do the gram-positive and gram-negative diplococci differ in exact morphology?
5. Why do you suppose there are no useful vaccines for most of the pyogenic cocci?
6. You have been called upon to prevent outbreaks of SSSS in the nursery of a hospital where this strain of *S. aureus* has been isolated.
   a. What will be your main concerns?
   b. What procedures will you establish to address these concerns?
   c. Suppose that a pediatrician on the staff is found to be a nasal carrier of MRSA. How will you deal with this discovery?
7. Explain why pneumonia occurs most often in the elderly and the immunosuppressed.
8. How would an adult get gonococcal conjunctivitis?
9. You have been given the assignment of obtaining some type of culture that could help identify the cause of otitis media in a small child.
   a. What area might you sample?
   b. Why is this disease so common in children?
10. a. Name the three species of pyogenic cocci most commonly implicated in neonatal disease?
    b. Explain how infants might acquire such diseases.
11. Comment on the practice of kissing pets on the face or allowing them to lick or kiss one's face. What are some possible consequences of this?
12. a. For which pyogenic cocci is penicillin *not* a good choice for treatment?
    b. Itemize alternative drugs for pyogenic cocci.
13. Explain this statement: The prevalence of gonorrhea is higher in women but the incidence of gonorrhea is higher in men.
14. a. Explain why it is so important to take a detailed cardiovascular and infection history of dental patients.
    b. What is the policy on preventing complications of dental work in patients with heart disease?
15. Case study 1. An elderly man with influenza acquires a case of pneumonia. Gram-positive cocci isolated from his sputum give β-hemolysis on blood agar; the infection is very difficult to treat; it is unreactive to the Quellung test. Later it is shown that the man shared a room with a patient with bone infection. Isolates from both infections were the same.
    a. What is the probable species?
    b. Justify your answer.
    c. What treatment will be necessary?
16. Case study 2. A person with an inflamed cut on his head went to a physician, who then discovered that microbes had invaded the clot and spread into underlying tissues. Patchy areas developed around the lesion, leading to redness and edema. After a brief hospitalization, the patient died.
    a. What was the probable condition, and which species and virulence factors were involved?
    b. What should the treatment have been?
17. Case study 3. A child is brought to the emergency room in a semiconscious state with a high fever. Earlier he had complained of a stiff neck and headache. A tap of spinal fluid is performed and tested. A Gram stain reveals numerous gram-negative cocci or coccobacilli and WBCs, and a diagnosis of meningitis is made. Owing to the patient's serious condition, immediate drug therapy is necessary. Two species could cause this condition (see table A-8).
    a. Which drug would work on both infections?
    b. What are some rapid tests used to differentiate the two bacteria?

INTERNET SEARCH TOPIC

1. Look up RAP (RNA-activating protein) therapy and determine its actions on *Staphylococcus aurens* and its potential therapeutic benefits.
2. Locate information on MRSA, VISA, VRE, and PPNG. What do the acronyms stand for and what is the current medical status of these drug-resistant cocci?

chapter 19

The Gram-Positive Bacilli of Medical Importance

In terms of sheer numbers, rod-shaped bacteria dominate the field of pathogenic bacteriology. Included among the diseases caused by bacilli are ancient, deadly, and fascinating ones such as bubonic plague, typhoid fever, tetanus, and leprosy, as well as newly emerging ones such as listeriosis and Legionnaires' disease. In chapters 19 and 20 we will characterize and differentiate genera and medically important species using a simple modification of the ninth edition of *Bergey's Manual of Systematic Bacteriology.* This system divides the bacilli by Gram reaction and subgroups them by morphological characteristics and oxygen utilization (see figure E, page 548). Each section covers unique features of the pathogens and the epidemiology, pathology, and control measures of the principal diseases.

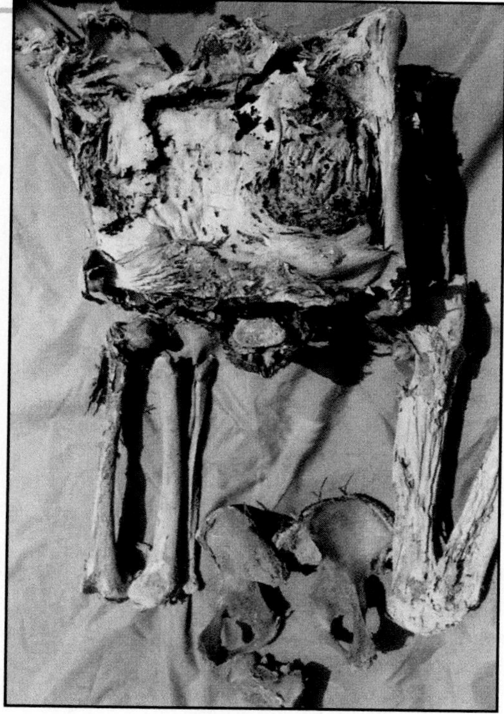

Ancient tuberculosis. A 900-year-old Peruvian mummy was tested for DNA from the tubercle bacillus by means of a highly sensitive PCR technique. Discovery of TB DNA laid to rest the theory that Columbus brought the disease to the Americas from Europe.

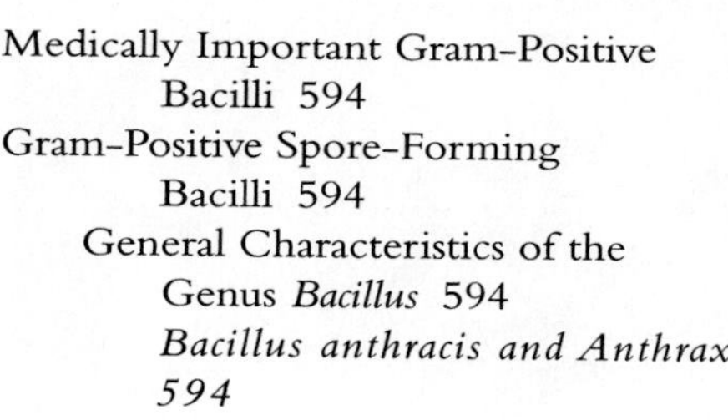

MEDICALLY IMPORTANT GRAM-POSITIVE BACILLI

The gram-positive bacilli can be subdivided into three general groups based on the presence or absence of endospores and the characteristic of acid-fastness. Further levels of separation correspond to oxygen requirements and cell morphology. This scheme can be organized as follows:

Endospore-forming bacilli
- Aerobic: *Bacillus*
- Anaerobic: *Clostridium*

Non-endospore-forming bacilli
- Regular in morphology
  - *Listeria*
  - *Erysipelothrix*
- Irregular in morphology
  - Aerobic: *Corynebacterium*
  - Anaerobic: *Propionibacterium*

Acid-fast bacilli
- *Mycobacterium*
- *Nocardia*

Non-acid-fast branching filamentous bacilli
- Actinomyces

GRAM-POSITIVE SPORE-FORMING BACILLI

Most endospore-forming bacteria are gram-positive, motile, rod-shaped forms in the genera *Bacillus, Clostridium,* and *Sporolactobacillus.* An endospore is a dense survival unit that develops in a vegetative cell in response to nutrient deprivation (figure 19.1; see figure 4.21). The extreme resistance to heat, drying, radiation, and chemicals accounts for the survival, longevity, and ecological niche of sporeformers, and it is also relevant to their pathogenicity.

GENERAL CHARACTERISTICS OF THE GENUS *BACILLUS*

The genus *Bacillus* includes a large assembly of mostly saprobic bacteria widely distributed in the earth's habitats. *Bacillus* species are aerobic and catalase-positive, and, though they have varied nutritional requirements, none is fastidious. The group is noted for its versatility in degrading complex macromolecules, and it is also a common source of antibiotics. Because the primary habitat of many species is the soil, spores are continuously dispersed by means of dust into water and onto the bodies of plants and animals. Despite their ubiquity, the two species with primary medical importance are *B. anthracis,* the cause of anthrax, and *B. cereus,* the cause of one type of food poisoning.

Bacillus anthracis *and Anthrax*

Bacillus anthracis is among the largest of all bacterial pathogens, composed of block-shaped, angular nonmotile rods 3–5 μm long and 1–1.2 μm wide and central spores that develop under all

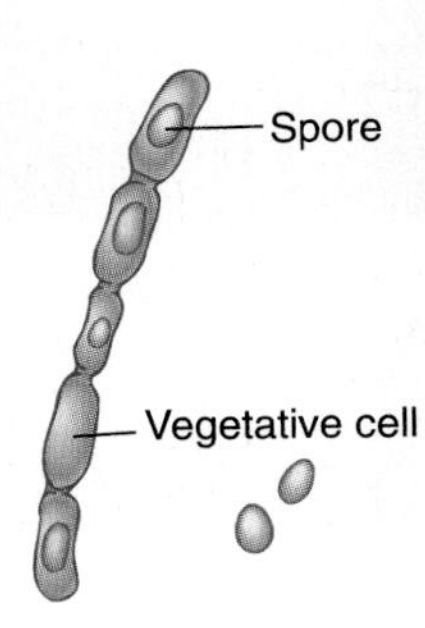

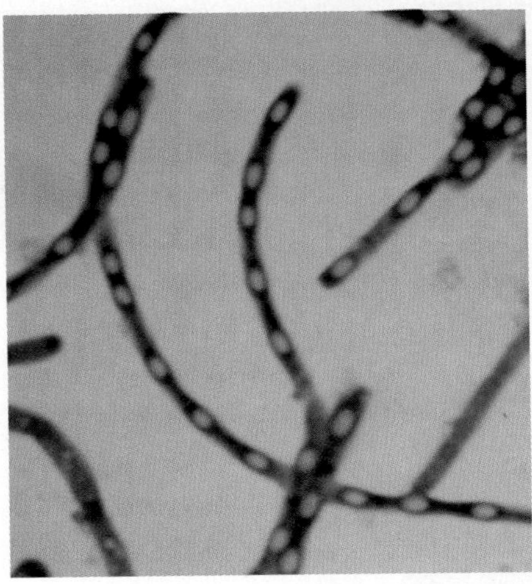

(a)

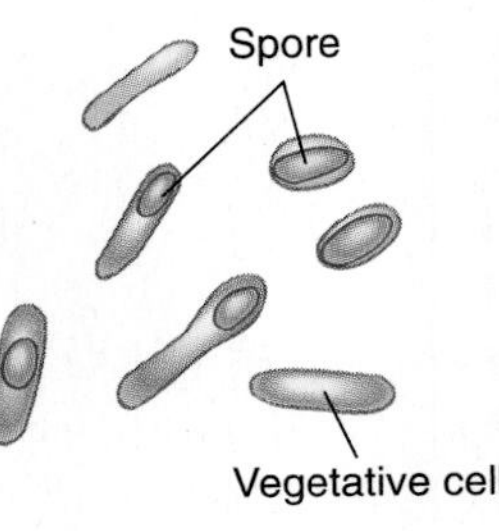

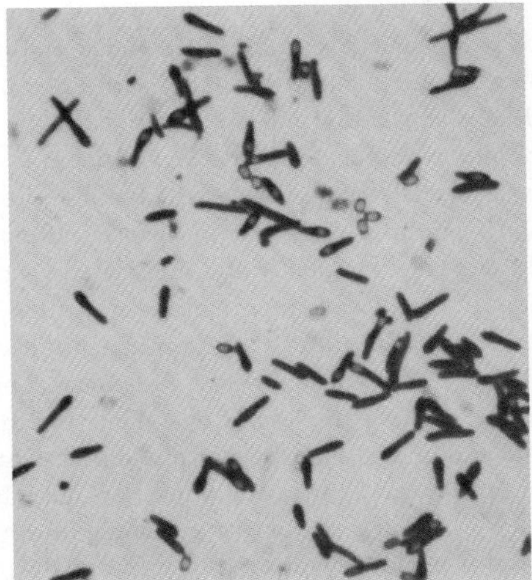

(b)

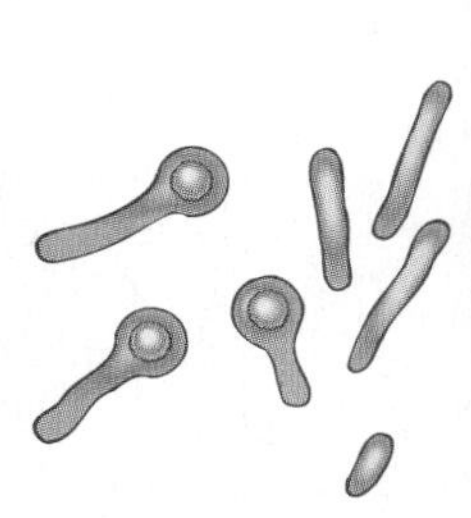
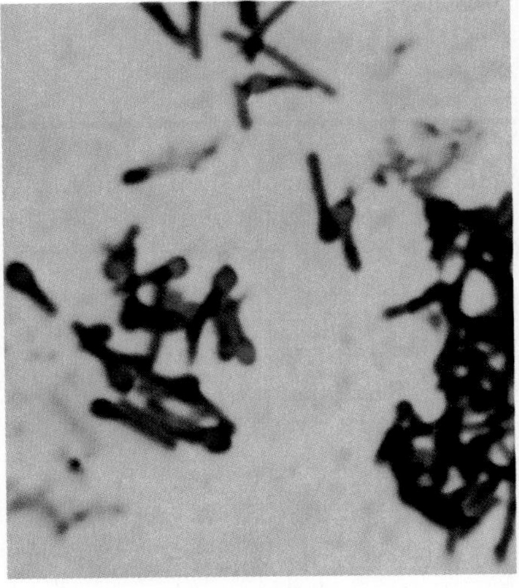

(c)

Figure 19.1

Examples of endospore-forming pathogens. (*a*) The morphological appearance of *Bacillus anthracis,* showing centrally placed endospores and a streptobacillus arrangement (600×). (*b*) A smear of *Clostridium perfringens* with central to terminal spores (410×). (*c*) *Cl. tetani* (710×). Its typical tennis racquet morphology is created by terminal spores that swell the sporangium.

MICROFILE 19.1 A BACILLUS THAT COULD KILL US

Mechanisms of biological warfare are frequently based on the pathogenic potential of microorganisms. Because spore-forming bacteria such as *Bacillus* are so hardy, they have held a particular attraction to scientists as research subjects and, in one case, as the basis of a deadly bomb.

Late in World War II, the mounting outrage over German bombings of London and the pressure for retaliation forced the British and American governments to collaborate on a powerful means of striking back. Given the knowledge that pneumonic anthrax was 100% fatal in a short time period, a highly secret council designed a bomb containing the spores of *Bacillus anthracis.* In the original plan, the detonation of these deadly projectiles over six German cities would have caused a shower of spores to be inhaled by humans and animals. Some estimates predicted that this could have wiped out the population over hundreds of square miles. Late in the war, a plant in Indiana actually began to manufacture the bombs, but by this time, the war had started to wind down. How close the Allies came to dropping these bombs was not disclosed.

During the Cold War, the U.S. Army's Special Operations Division conducted a chilling experiment using *Bacillus subtilis.* The plan was to stimulate a possible method of germ warfare by releasing spore aerosols into the air of a crowded Washington, D.C. airport and bus terminal. Agents carried suitcases to spray the air while hundreds of unsuspecting people inhaled "large and acceptably uniform" doses, presumably with no ill effects.

This study and several others like it were based on the disturbing assumption that *B. subtilis* is a harmless air contaminant. It should be clear even to beginning students that many species of bacteria are opportunists and are capable of causing disease, especially if present in large numbers. In fact, *B. subtilis* can cause lung and blood infections in people with weakened immunity. Although flawed, the test did fulfill its original purpose. It was determined that if smallpox virus had been used instead of *B. subtilis,* 300 people in 93 cities would have become infected and that similar exposures at terminals in several other large cities would have spread the disease rapidly to thousands of others. It was also shown that this contamination procedure could be accomplished without detection.

The potential threat from biological warfare is not only a concern of the past. Even in the late 1990s, the government of Iraq was stockpiling a huge arsenal of germ agents, including anthrax spores and botulinum toxin (another product of a gram-positive sporeformer). Iraq and other countries have amassed several tons of anthrax and other toxic agents that could be incorporated into bombs or missiles to be spread over long distances. International agencies are attempting to monitor and force elimination of the deadly agents.

growth conditions except in the living body of the host (figure 19.1*a*). Its virulence factors include a polypeptide capsule and exotoxins that in varying combinations produce edema and cell death. For centuries, **anthrax*** has been known as zoonotic disease of herbivorous livestock (sheep, cattle, goats). It has an important place in the history of medical microbiology because it was Robert Koch's model for developing his postulates in 1877, and, later, Louis Pasteur used the disease to prove usefulness of vaccination.

The anthrax bacillus is a facultative parasite that undergoes its cycle of vegetative growth and sporulation in the soil. Animals become infected while grazing on grass contaminated with spores. When the pathogen is returned to the soil in animal excrement or carcasses, it can sporulate and become a long-term reservoir of infection for the animal population. The majority of anthrax cases are reported in livestock from Africa, Asia, and the Middle East. Most recent cases in the United States have occurred in textile workers handling imported animal hair or hide or products made from them. People who work with infected animals (veterinarians, livestock handlers) are also at slight risk. Because of effective control procedures, however, the number of cases in the United States is extremely low (fewer than 10 a year).

The circumstances of human infection depend upon the portal of entry. The most common and least dangerous of all forms is **cutaneous anthrax,** caused by spores entering the skin through small cuts and abrasions. Germination and growth of the pathogen in the skin are marked by the production of a papule that becomes increasingly necrotic and later ruptures to form a painless, black **eschar*** (figure 19.2). Handlers of raw wool or hides are at risk of acquiring **pulmonary anthrax** (woolsorter's disease) by inhaling airborne spores. Bacilli grow in the lungs and release exotoxins that produce toxemia with wide-ranging pathologic effects, including capillary thrombosis and cardiovascular shock. The development of septicemia can cause death in a few hours. So fatal is this form of infection that anthrax spores have been a serious choice for biological warfare weapons (microfile 19.1). Gastrointestinal anthrax acquired from contaminated meat is another rare but dangerous form of the disease.

Methods of Anthrax Control Active cases of anthrax are treated with penicillin or tetracycline, but therapy does not lessen the effects of toxemia, so people can still die. A vaccine containing live spores and a toxoid prepared from a special strain of *B. anthracis* are used to protect livestock in areas of the world where anthrax is endemic. Humans can be vaccinated with the purified toxoid if they have occupational contact with livestock or products such as hides and bone or are members of the military. Effective vaccination requires six inoculations given over $1\frac{1}{2}$ years, with yearly boosters. Animals that have died from anthrax must be burned or chemically decontaminated before burial to prevent establishing the microbe in the soil, and imported items containing animal hides, hair, and bone should be gas-sterilized.

*anthrax (an´-thraks) Gr. *anthrax,* carbuncle.

*eschar (ess´-kar) Gr. *eschara,* scab.

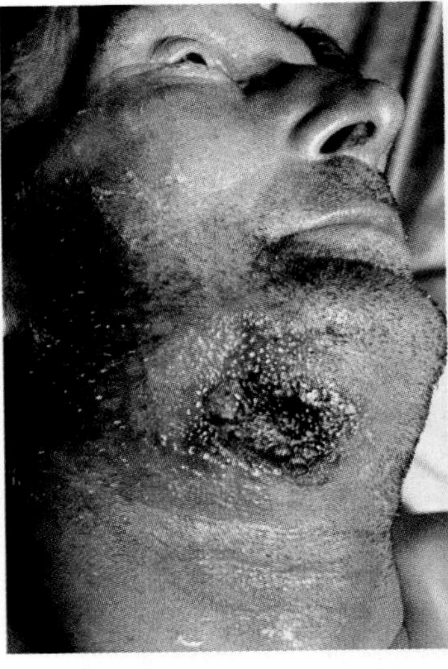

(a)

(b)

Figure 19.2

Cutaneous anthrax. (*a*) In early stages, the tissue around the site of invasion is inflamed and edematous. (*b*) A later stage reveals a thick, necrotic lesion called the eschar. It usually separates and heals spontaneously.

TABLE 19.1

IMPORTANT SPECIES OF *CLOSTRIDIUM*

Species	Chief Importance to Humans	Description of Role
Cl. perfringens	Pathogen	Principal cause of gas gangrene and myonecrosis; common agent in enterotoxigenic food poisoning
Cl. novyi	Pathogen	Second most frequent cause of gas gangrene
Cl. septicum	Pathogen	Third most frequent cause of gas gangrene
Cl. tetani	Pathogen	Cause of tetanus
Cl. botulinum	Pathogen	Cause of botulism
Cl. difficile	Opportunist	Involved in antibiotic-associated colitis
Cl. iodophilum	Industrial uses	Produces organic acids and alcohols for commercial use
Cl. acetobutylicum	Industrial uses	Produces acids, alcohols, and benzene
Cl. butyricum	Industrial uses	Produces butyric acid in butter and cheese
Cl. cellobiofavum	Industrial uses	Digests cellulose

Other Bacillus *Species Involved in Human Disease*

Bacillus cereus is a common airborne and dust-borne contaminant that multiplies very readily in cooked foods such as rice, potato, and meat dishes. The spores survive short periods of cooking and reheating; when the food is stored at room temperature, the spores germinate and release enterotoxins. Ingestion of toxin-containing food causes nausea, vomiting, abdominal cramps, and diarrhea. There is no specific treatment, and the symptoms usually disappear within 24 hours.

For many years, most common airborne *Bacillus* species were dismissed as harmless contaminants with weak to nonexistent pathogenicity. However, infections by these species are increasingly reported in immunosuppressed and intubated patients and in drug addicts who do not use sterile needles and syringes. An important contributing factor is that spores are abundant in the environment and the usual methods of disinfection and antisepsis are powerless to control them.

THE GENUS *CLOSTRIDIUM*

Another genus of gram-positive, spore-forming rods that is widely distributed in nature is ***Clostridium.****** It is differentiated from *Bacillus* on the basis of being anaerobic and catalase-negative. The large genus (over 120 species) is extremely varied in its habitats. Saprobic members reside in soil, sewage, vegetation, and organic debris, and commensals inhabit the bodies of humans and other animals. Infections caused by pathogenic species are not normally communicable but occur when spores are introduced into injured skin.

Clostridia cells produce oval or spherical spores that often swell the vegetative cell (see figure 19.1*b*,*c*). Spores are produced only under anaerobic conditions. Their nutrient requirements are complex, and they can decompose a variety of substrates. They can also synthesize organic acids, alcohols, and other solvents through fermentation. This capacity makes some clostridial species essential tools of the biotechnology industry (see chapter 26). Other extracellular products, primarily exotoxins, play an important role in various clostridial diseases. A number of *Clostridium* species are implicated in serious human disease. Some of these species and their medical and economic importance are listed in table 19.1.

The Role of Clostridia in Infection and Disease

Clostridial disease can be divided into (1) wound and tissue infections, including myonecrosis, antibiotic-associated colitis, and tetanus, and (2) food intoxication of the perfringens and botulism varieties. Most of these diseases are caused when soluble exotoxins, some of which are highly potent, act on specific cellular targets (microfile 19.2).

Gas Gangrene

The majority of clostridial soft tissue and wound infections are caused by ***Clostridium perfringens,*** *Cl. novyi,* and *Cl. septicum.* The spores of these species can be found in soil, on human skin,

**Clostridium* (klaw-strid′-ee-um) Gr. *closter,* spindle.

MICROFILE 19.2 TOXICITY IN THE EXTREME

Imagine a substance so poisonous that a single microgram of it is a mass lethal dose for 200,000 mice, and a cup of it in concentrated form could quite possibly kill all the humans on the planet. This toxin is 100,000 times more powerful than rattlesnake venom and a million times more potent than strychnine. Although this formidable chemical might sound like the science fiction creation of a mad scientist, it actually exists as a product of the common soil bacterium *Clostridium botulinum.* Curiously, the genus *Clostridium* is the source of two of the most deadly toxins known, botulin and tetanospasmin (from *Cl. tetani*). These two toxins share several other characteristics. For instance, both are polypeptides coded by plasmids and acquired through transduction by a virus. Both are neurotoxins that act on nerve cell processes involved in muscular activity, and both cause loss of voluntary muscle control (paralysis).

But here the similarities end. The diseases these two toxins cause are acquired through generally different models. In tetanus, the toxin is formed in the tissues, and in the prominent form of botulism, the toxin is formed in food and then ingested. The mechanism by which the two toxins cause paralysis is also different. Tetanospasmin blocks the activity of inhibitory neurons in the central nervous system, and botulin acts on mononeurons in the peripheral nervous system. Tetanospasmin causes spastic (rigid) paralysis due to hyperactive muscles and sustained contraction (see figure 19.7). In contrast, botulin causes flaccid paralysis by interfering with the transmission of impulses to muscles and blocking contraction (see figure 19.9).

Despite the extreme toxicity of botulin, a form of it called botox has been approved as a drug for treating a variety of muscle ailments. It can be used in extreme dilutions to relax muscles that are overactive. It has been used to successfully treat cross- or wall-eye, blepharospasm, wry neck, spastic vocal cords, and various types of tremors. It minimizes the spasms and pain and improves normal muscle function when administered every 3–4 months. Botox is even being used by cosmetic surgeons to weaken facial muscles that cause frown creases and wrinkles on the forehead.

Muscle fibers
Clostridium
Gas-filled spaces
(a)
(b)

Figure 19.3

(*a*) A microscopic analysis of clostridial myonecrosis, showing a histological section of gangrenous skeletal muscle in cross section. (*b*) A schematic drawing of the same section. The growth of *Clostridium perfringens* (plump rods) has caused gas formation and separation of the fibers.

(a) *From N.A. Boyd et al.,* Journal of Medical Microbiology, *5:459, 1972. Reprinted by permission of Longman Group, Ltd.*

and in the human intestine and vagina. The disease they cause has the common name **gas gangrene*** in reference to the gas produced by the bacteria growing in the tissue. It is technically termed anaerobic cellulitis or *myonecrosis.** The conditions that predispose a person to gangrene are surgical incisions, compound fractures, diabetic ulcers, septic abortions, puncture and gunshot wounds, and crushing injuries contaminated by spores from the body or the environment.

Because clostridia are not highly invasive, infection requires damaged or dead tissue that supplies growth factors and an anaerobic environment. The low oxygen tension results from an interrupted blood supply and the presence of aerobic bacteria that deplete oxygen. Such conditions stimulate spore germination, rapid vegetative growth in the dead tissue, and release of exotoxins. *Clostridium perfringens* produces several physiologically active toxins; the most potent one, *alpha toxin* (lecithinase c), causes red blood cell rupture, edema, and tissue destruction (figure 19.3). Additional virulence factors that enhance tissue destruction are collagenase, hyaluronidase and DNase. The gas formed in tissues, due to fermentation of muscle carbohydrates, can also destroy muscle structure.

Extent and Symptoms of Infection Two forms of gas gangrene have been identified. In **anaerobic cellulitis,** the bacteria spread within damaged necrotic muscle tissue, producing toxin and gas, but the infection remains localized and does not spread into healthy tissue. The pathology of true **myonecrosis** is more destructive. Toxins produced in large muscles, such as the thigh, shoulder, and buttocks, diffuse into nearby healthy tissue and cause local necrosis there. This damaged tissue then serves as a focus for continued clostridial growth, toxin formation, and gas production. The disease can progress through an entire limb or body area, destroying tissues as it goes (figure 19.4). Initial symptoms of pain, edema, and a bloody exudate in the lesion are followed by fever, tachycardia, and blackened necrotic tissue filled

*gas gangrene (gang´-green) Gr. *gangraina,* an eating sore. A necrotic condition associated with the release of gases.

*myonecrosis (my´´-oh-neh-kro´sis) Gr. *myo,* muscle, and *necros,* dead.

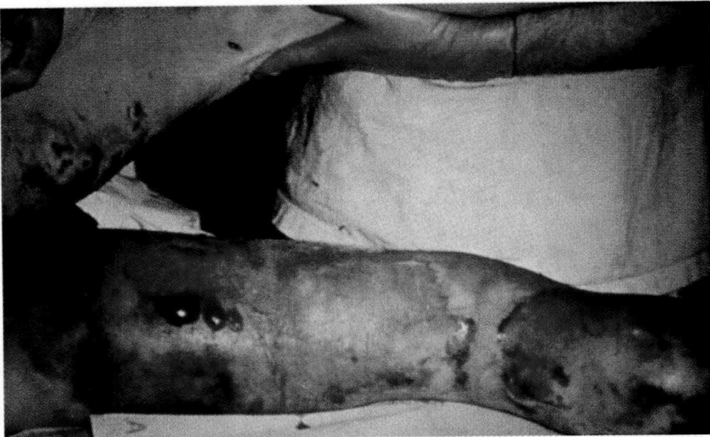

Figure 19.4
The clinical appearance of myonecrosis in a compound fracture of the leg. Necrosis has traveled from the main site of the break to other areas of the leg. Note the explosive nature of the infection, with blackening, general tissue destruction and bubbles on the skin caused by gas formation in underlying tissue.

with bubbles of gas. Gangrenous infections of the uterus due to septic abortions and clostridial septicemia are particularly serious complications. If treatment is not indicated early, the disease is invariably fatal.

Treatment and Prevention of Gangrene *Debridement** of diseased tissue eliminates the conditions that promote the spread of gangrenous infection. This is most difficult in the intestine or body cavity, where only limited amounts of tissue can be removed. Surgery is supplemented by large doses of a broad-spectrum cephalosporin (cefoxitin) or penicillin to control infection. Hyperbaric oxygen therapy, in which the affected part is exposed to an increased oxygen gas tension in a pressurized chamber, can also lessen the severity of infection (figure 19.5). The increased oxygen content of the tissues presumably blocks further bacterial multiplication and toxin production. Extensive myonecrosis of a limb may call for surgical removal, or amputation.

One of the most effective ways to prevent clostridial wound infections is immediate and rigorous cleansing and surgical repair of deep wounds, decubitus ulcers (bedsores), compound fractures, and infected incisions, along with prophylactic antibiotic therapy. A newer promising technique is a form of gene therapy that introduces a gene for stimulating increased growth of blood vessels. The resultant flow of blood into the infection site helps reduce the spread of the pathogen and improves healing of the gangrenous tissue. Because there are so many different antigen subtypes in this group, active immunization is not possible.

Antibiotic-Associated Colitis

A nosocomial infection called **antibiotic-associated** (or pseudomembranous) **colitis** is the second most common intestinal infection after salmonellosis in industrialized countries. The disease is caused by ***Clostridium difficile,*** a minor but normal resident of the intestine that was once considered relatively harmless. In most instances, this infection is traced to therapy with broad-spectrum antibiotics such as ampicillin, clindamycin, and cephalosporins, and it is a major cause of diarrhea in hospitals.

Although the mechanisms of its pathogenesis are not completely understood, drug-resistant *Cl. difficile* seems to superinfect the intestine when the normal flora has been disrupted by antibiotics. Although this bacterium is relatively noninvasive, it produces enterotoxins that cause necrosis of the intestinal epithelium. The predominant symptom is diarrhea commencing late in therapy or even after therapy has stopped. More severe cases exhibit abdominal cramps, fever, and leukocytosis. The colon is inflamed and gradually sloughs off loose, membranelike patches called pseudomembranes consisting of fibrin and cells (figure 19.6). If the condition is not arrested, cecal perforation and death can result.

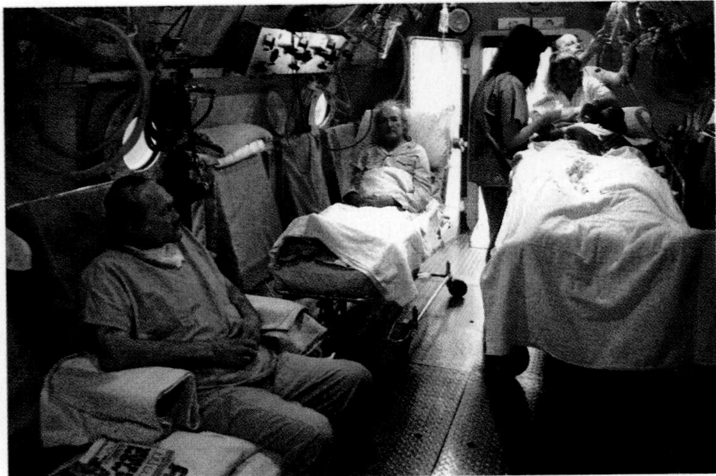

Figure 19.5
A large decompression chamber is being used to treat patients with wound infections and diabetic ulcers. This hyperbaric therapy involves breathing higher levels of oxygen to inhibit anaerobic infection and to promote healing.
Los Angeles Times, December 6, 1993, Metro Section, p. 1. Photographer Rolando Otero.

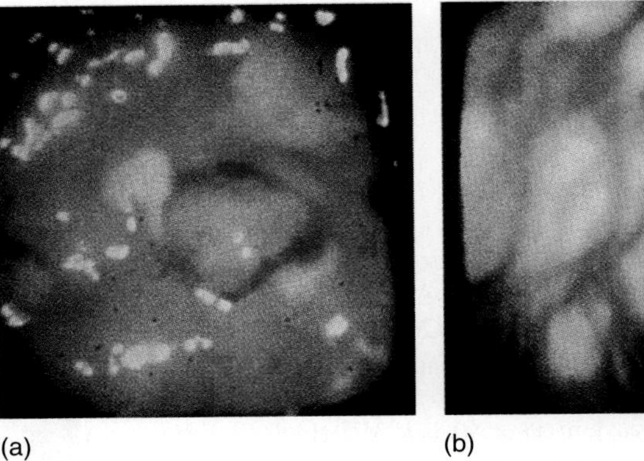

(a) (b)

Figure 19.6
Antibiotic-associated colitis, as revealed by a sigmoidoscope, an instrument capable of photographing the interior of the colon. (*a*) A mild form with diffuse, inflammatory patches. (*b*) Heavy yellow plaques, or pseudomembranes, typical of more severe cases.

***debridement** (dih-breed´-ment) Surgical removal of dead or damaged tissue.

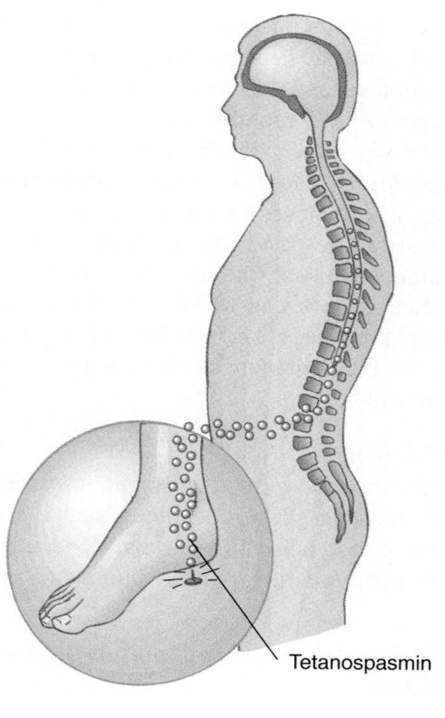

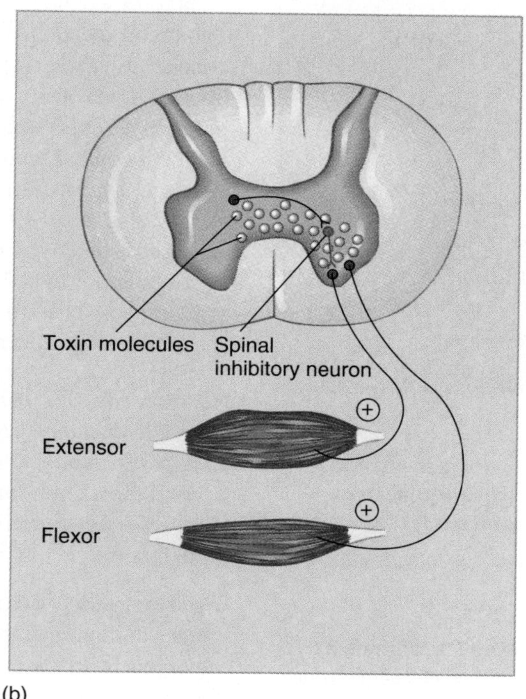

Figure 19.7

The events in tetanus. (*a*) After traumatic injury, bacilli infecting the local tissues secrete tetanospasmin, which is absorbed by the peripheral axons and is carried to the target neurons in the spinal column. (*b*) In the spinal cord, the toxin attaches to the junctions of regulatory neurons that inhibit inappropriate contraction. Released from inhibition, the muscles, even opposing members of a muscle group, receive constant stimuli and contract uncontrollably. (*c*) Muscles contract spasmodically, without regard to regulatory mechanisms or conscious control. Note the clenched jaw typical of risus sardonicus.

Mild, uncomplicated cases respond to withdrawal of antibiotics and replacement therapy for lost fluids and electrolytes. More severe infections are treated with oral vancomycin of metronidazole for several weeks until the intestinal flora returns to normal. Because infected persons often shed large numbers of spores in their stools, increased precautions are necessary to prevent spread of the agent to other patients who may be on antimicrobic therapy. Some new techniques on the horizon are vaccination with *Cl. difficile* toxoid and restoration of normal flora by means of a mixed culture of lactobacilli and yeasts.

Tetanus, or Lockjaw

Tetanus* is a neuromuscular disease whose alternate name, **lockjaw,** refers to an early effect of the disease on the jaw muscle. The etiologic agent, ***Clostridium tetani,*** is a common resident of cultivated soil and the gastrointestinal tracts of animals. Spores usually enter the body through accidental puncture wounds, burns, umbilical stumps, frostbite, and crushed body parts.

The incidence of tetanus is low in North America. Most cases occur among geriatric patients and intravenous drug abusers. The incidence of neonatal tetanus—predominantly the result of an infected umbilical stump or circumcision—is higher in cultures that apply dung, ashes, and mud to these sites to arrest bleeding or as a customary ritual. The disease accounts for several hundred thousand infant deaths a year.

The Course of Infection and Disease The risk for tetanus occurs when spores of *Clostridium tetani* are forced into injured tissue. But the mere presence of spores in a wound is not sufficient to initiate infection because the bacterium is unable to invade damaged tissues readily. It is also a strict anaerobe, and the spores cannot become established unless tissues at the site of the wound are necrotic and poorly supplied with blood, conditions that favor germination.

As the vegetative cells grow, various metabolic products are released into the infection site. Of these, the most serious is **tetanospasmin,** a potent neurotoxin that accounts for the major symptoms of tetanus. The toxin spreads to nearby motor nerve endings in the injured tissue, binds to them, and travels by axons to the ventral horns of the spinal cord (figure 19.7). In the spinal column, the toxin binds to specific target sites on the spinal neurons that are responsible for inhibiting skeletal muscle contraction. The toxin inhibits the release of neurotransmitter, and only a small amount is required to initiate the symptoms. The incubation period varies from 4 to 10 days, and shorter incubation periods signify a more serious condition.

*tetanus (tet´-ah-nus) Gr. *tetanos,* to stretch.

Figure 19.8

Baby with neonatal tetanus, showing spastic paralysis of the paravertebral muscles, which locks the back into a rigid, arched position. Also note the abnormal flexion of the arms and legs.

Tetanospasmin alters the usual regulation mechanisms for muscle contraction. As a result, the muscles are released from normal inhibition and begin to contract uncontrollably. Powerful muscle groups are most affected, and the first symptoms are clenching of the jaw, followed in succession by extreme arching of the back, flexion of the arms, and extension of the legs (figure 19.8). Lockjaw confers the bizarre appearance of *risus sardonicus* (sarcastic grin) that looks eerily as though the person is smiling (see figure 19.7*c*). These contractions are intermittent and extremely painful and may be forceful enough to break bones, especially the vertebrae. Death is most often due to paralysis of the respiratory muscles and respiratory collapse. The fatality rate, ranging from 10% to 70%, is highest in cases involving delayed medical attention, a short incubation time, or head wounds. Full recovery requires a few weeks, and other than transient stiffness, no permanent damage to the muscles usually remains.

Treatment and Prevention of Tetanus Tetanus treatment is aimed at deterring the degree of toxemia and infection and maintaining patient homeostasis. A patient with clinical appearance suggestive of tetanus should immediately receive antitoxin therapy with human tetanus immune globulin (TIGH). Tetanus antitoxin (TAT) from horses may be used but it is less acceptable because of possible allergic reactions. Although the antitoxin inactivates circulating toxin, it will not counteract the effects of toxin already bound to neurons. Other treatment methods include thoroughly cleansing and removing the afflicted tissue, controlling infection with penicillin or tetracycline, and administering muscle relaxants. The patient may require the assistance of a respirator, and a tracheostomy[1] is sometimes performed to prevent respiratory complications such as aspiration pneumonia or lung collapse.

Tetanus is one of the world's most preventable diseases, chiefly because of an effective vaccine containing tetanus toxoid. During World War II, only 12 cases of tetanus occurred among 2,750,000 wounded soldiers who had been previously vaccinated. The recommended vaccination series for 1- to 3-month-old babies consists of three injections given 2 months apart, followed by booster doses about 1 and 4 years later. Children thus immunized probably have protection for 10 years. Additional protection against neonatal tetanus may be achieved by vaccinating pregnant women, whose antibodies will be passed to the fetus. Toxoid should also be given to injured persons who have never been immunized, have not completed the series, or whose last booster was received more than 10 years previously. The vaccine can be given simultaneously with passive TIGH immunization to achieve immediate and long-term protection.

1. The surgical formation of an air passage by perforation of the trachea.

Clostridial Food Poisoning

Two *Clostridium* species are involved in food poisoning. ***Clostridium perfringens,*** type A, accounts for a mild illness that is the second most common form of food poisoning worldwide, whereas ***Clostridium botulinum*** produces a rarer but more serious intoxication.

Perfringens Food Poisoning *Clostridium perfringens* spores contaminate many kinds of food, but those most frequently involved in disease are animal flesh (meat, fish) and vegetables (beans) that have not been cooked thoroughly enough to destroy the spores. When these foods are cooled, spores germinate, and the germinated cells multiply, especially if the food is left unrefrigerated. If the food is eaten without adequate reheating, live *Cl. perfringens* cells enter the small intestine and release enterotoxin. The toxin, acting upon epithelial cells, initiates acute abdominal pain, diarrhea, and nausea in 8 to 16 hours. Recovery is rapid, and deaths are extremely rare. *Clostridium perfringens* also causes an enterocolitis infection similar to that caused by *Cl. difficile.* This infectious type of diarrhea is acquired from contaminated food, or it may be transmissible by inanimate objects.

Botulinum Food Poisoning **Botulism*** is an intoxication associated with eating poorly preserved foods, though it can also occur as an infection. Until recent times, it was relatively prevalent and commonly fatal, but modern techniques of food preservation and medical treatment have reduced both its incidence and its fatality rate. However, botulism is a common cause of death in livestock that have grazed on contaminated food and in aquatic birds that have eaten decayed vegetation.

Clostridium botulinum is a spore-forming anaerobe that commonly inhabits soil and water and, occasionally, the intestinal tract of animals. It is distributed worldwide, but occurs most often in the Northern Hemisphere. The species has eight distinctly different types (designated A, B, C_α, and C_β, D, E, F, and G), which vary in distribution among animals, regions of the world, and type of exotoxin. Human disease is usually associated with types A, B, E, and F, and animal disease with types A, B, C, D, and E.

The Route of Pathogenesis There is a high correlation between cultural dietary preferences and food-borne botulism. In the

*botulism (boch´-oo-lizm) L. *botulis,* sausage. The disease was originally linked to spoiled sausage.

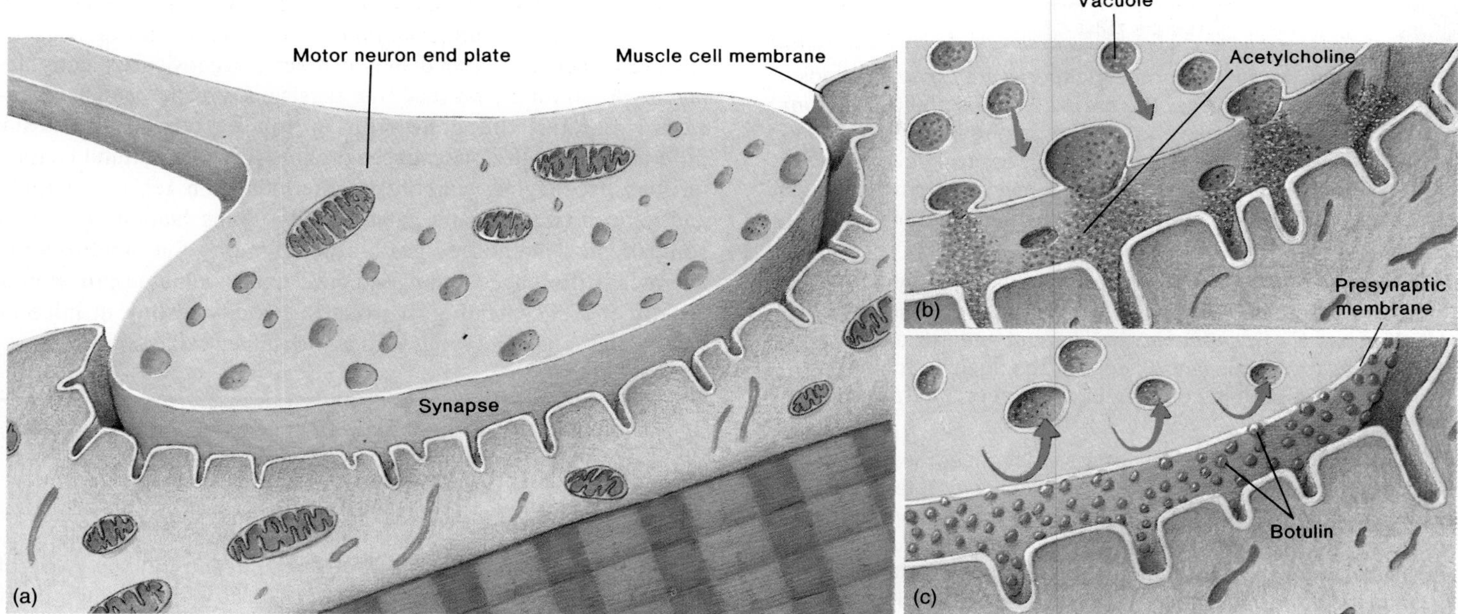

Figure 19.9

The physiological effects of botulism toxin (botulin). (*a*) The relationship between the motor neuron and the muscle at the neuromuscular junction. (*b*) In the normal state, acetylcholine released at the synapse crosses to the muscle and creates an impulse that stimulates muscle contraction. (*c*) In botulism, the toxin enters the motor end plate and attaches to the presynaptic membrane, where it blocks release of the transmitter, prevents impulse transmission, and keeps the muscle from contracting.

United States, the disease is often associated with low-acid vegetables (green beans, corn), fruits, and occasionally meats, fish, and dairy products. Most botulism outbreaks occur in foods that have been home-processed, including canned vegetables, smoked meats, and cheese spreads. The demand for prepackaged convenience foods such as vacuum-packed cooked vegetables and meats has created a new source of risk, but most commercially canned foods are held to very high standards of preservation and are only rarely a source of botulism.

The factors in food processing that lead to botulism are dependent upon several circumstances. Spores are present on the vegetables or meat at the time of gathering and are difficult to remove completely. When contaminated food is bottled and steamed in a pressure cooker that does not reach reliable pressure and temperature, some spores survive (botulinum spores are highly heat-resistant). At the same time, the pressure is sufficient to evacuate the air and create anaerobic conditions. Storage of the bottles at room temperature favors spore germination and vegetative growth. One of the products of metabolism is **botulin,** the most potent microbial toxin known (see microfile 19.2).

Bacterial growth may not be evident in the appearance of the bottle or can or in the food's taste or texture, and only minute amounts of toxin may be present. Swallowed toxin enters the small intestine and is absorbed into the lymphatics and circulation. From there, it travels to its principal site of action, the neuromuscular junctions of skeletal muscles (figure 19.9). The effect of botulin is to prevent the release of the neurotransmitter substance, acetylcholine, that initiates the signal for muscle contraction. The usual time before onset of symptoms is 12–72 hours, depending on the size of the dose. Neuromuscular symptoms first affect the muscles of the head and include double vision, difficulty in swallowing, and dizziness, but there is no sensory or mental lapse. Although nausea and vomiting can occur at an early stage, they are not common. Later symptoms are descending muscular paralysis and respiratory compromise. In the past, death resulted from stoppage of respiration, but mechanical respirators have reduced the fatality rate to about 10%.

Infant and Wound Botulism In rare instances, *Cl. botulinum* causes infection and toxemia. In these special circumstances, called infant and wound botulism, the spores germinate in the body and produce an infection.

Infant botulism was first described in the late 1970s in children between the ages of 2 weeks and 6 months who had ingested spores. It is currently the most common type of botulism in the United States, with approximately 80–100 cases reported annually. The exact food source is not always known, although raw honey has been implicated in some cases, and the spores are common in dust and soil. Apparently, the immature state of the neonatal intestine and microbial flora allows the spores to gain a foothold, germinate, and give off neurotoxin. As in adults, babies exhibit flaccid paralysis, usually with a weak sucking response, generalized loss of tone (the "floppy baby syndrome"), and respiratory complications. Although adults can also ingest botulinum spores in contaminated vegetables and other foods, the adult intestinal tract normally inhibits this sort of infection.

Chapter Checkpoints

Gram-positive bacilli that form endospores are common in soil and water environments. The genera *Bacillus* and *Clostridium* include highly toxigenic and virulent pathogens. *Cl. tetani* and *Cl. botulinum* produce the most deadly toxins known. *B. anthracis* toxin is also deadly in systemic infections. Microbial control of these pathogens must include methods that destroy their highly resistant spores.

Species in the genus *Bacillus* are all aerobic, although a few are facultative anaerobes. Most are also motile. Significant pathogens are *B. anthracis,* the agent of anthrax, a zoonosis of domestic animals, and *B. cereus,* the agent of food poisoning. Important commercial species of *Bacillus* produce antibiotics.

Species in the genus *Clostridium* are strict anaerobes. Significant pathogens are *Cl. tetani,* the agent of tetanus, *Cl. botulinum,* the agent of botulism, *Cl. perfringens* and other species, cause of gas gangrene and myonecrosis, and *Cl. difficile,* an opportunist that infects the intestine after antibiotic therapy. Commercially important clostridia are industrial producers of organic acids and alcohols.

In **wound botulism,** the spores enter a wound or puncture much as in tetanus, but the symptoms are similar to those of food-borne botulism. Increased cases of this form of botulism are being reported in injecting drug users. The rate of infection is highest in people who inject black tar heroin into the skin.

Treatment and Prevention of Botulism Differentiating botulism from other neuromuscular conditions requires testing of the food samples, sampling the patient's blood and pumping the gastrointestinal tract. The Centers for Disease Control and Prevention provides a source of type A, B, and E trivalent horse antitoxins that must be administered early for greatest effectiveness. Patients are also managed with respiratory and cardiac support systems. Infectious botulism is treated with penicillin to control the microbe's growth and toxin production.

Botulism will always be a potential threat for people who consume home-preserved foods. Preventing it depends on educating the public about the proper methods of preserving and handling canned foods. Pressure cookers should be tested for accuracy in sterilizing, and home canners should be aware of the types of foods and conditions likely to cause botulism. Although acidic foods (tomatoes, fruits) are traditionally thought to inhibit the microbe, recent findings indicate that acid content alone may not be sufficient to prevent bacterial growth and toxin production. Other effective preventives include addition of preservatives such as sodium nitrite, salt, or acid. Bulging cans or bottles that look or smell spoiled should be discarded, and all home-bottled foods should be boiled for 10 minutes before eating, because the toxin is heat-sensitive and is rapidly inactivated at 100°C. Military researchers are in the process of testing a genetically engineered toxoid that could protect against all strains of the pathogen. It is likely that this will be included in the routine immunizations given to armed service personnel.

Differential Diagnosis of Clostridial Species

Although clostridia are common isolates, their clinical significance is not always immediately evident. Diagnosis frequently depends on the microbial load, the persistence of the isolate on resampling, and the condition of the patient. Laboratory differentiation relies on testing morphological and cultural characteristics, exoenzymes, carbohydrate fermentation, reaction in milk, and toxin production and pathogenicity. Some laboratories use a sophisticated method of gas chromatography that analyzes the chemical differences among species. Other valuable procedures are direct ELISA testing of isolates, toxicity testing in mice or guinea pigs, and serotyping with antitoxin neutralization tests.

GRAM-POSITIVE REGULAR NON-SPORE-FORMING BACILLI

The non-spore-forming gram-positive bacilli are a mixed group of genera subdivided on the basis of morphology and staining characteristics. One loose aggregate of seven genera is characterized as **regular** because they stain uniformly and do not assume pleomorphic shapes. Regular genera include *Lactobacillus, Listeria, Erysipelothrix, Kurthia, Caryophanon, Bronchothrix,* and *Renibacterium. Lactobacillus* is widely distributed in the environment and is a common resident of the intestinal tract and vagina of humans. Some species are also important in processing dairy products. Only rarely are they pathogens. The most significant pathogens in this group are *Listeria monocytogenes* and *Erysipelothrix rhusiopathiae.*

AN EMERGING FOOD-BORNE PATHOGEN: *LISTERIA MONOCYTOGENES*

*Listeria monocytogenes** ranges in morphology from coccobacilli to long filaments in palisades formation (table 19.2). Cells show tumbling with one to four flagella and do not produce capsules or spores. *Listeria* is not fastidious and is resistant to cold, heat, salt, pH extremes, and bile. It is also β-hemolytic.

Epidemiology and Pathology of Listeriosis

The distribution of *L. monocytogenes* is so broad that its reservoir has been difficult to determine. It has been isolated all over the world from water, soil, plant materials, and the intestines of healthy mammals (including humans), birds, fish, and invertebrates. Apparently, the primary reservoir is soil and water, while animals, plants, and food are secondary sources of infection. Most cases of **listeriosis** are associated with ingesting contaminated milk, cheeses, ice cream, poultry, and meat. Recent epidemics have spurred an in-depth investigation into the prevalence of *L. monocytogenes* in dairy products. Both domestic and imported cheeses were subjected to detailed bacteriological analysis. The

**Listeria monocytogenes* (lis-ter´-ee-ah) For Joseph Lister, the English surgeon who pioneered antiseptic surgery; (mah´´-noh-sy-toj´-uh-neez) For its effect on monocytes.

pathogen was isolated in 1% to 3.5% of the milk, cheeses, and ice creams tested. Cheeses made from raw milk and aged for several months are of special concern because *Listeria* bacilli readily survives such processing and can grow during storage.

Except in cases of pregnancy, human-to-human transmission is probably not a significant factor. A predisposing factor in listeriosis seems to be the weakened condition of host defenses in the intestinal mucosa, since studies have shown that immunocompetent individuals are rather resistant to infection.

Listeriosis in normal adults is often a mild or subclinical infection with nonspecific symptoms of fever, diarrhea, and sore throat. However, listeriosis in immunocompromised patients, fetuses, and neonates usually affects the brain and meninges and results in septicemia. For reasons that are not well understood, pregnant women are highly susceptible to a mild form of the disease, which is transmitted to the infant prenatally when the microbe crosses the placenta or postnatally through the birth canal (figure 19.10). Intrauterine infections are widely systemic and usually result in premature abortion and fetal death. Neonatal infections that localize in the meninges cause extensive damage to the nervous system if not treated at an early stage.

Diagnosis and Control of Listeriosis

Diagnosing listeriosis is hampered by the difficulty in isolating it. However, the chances of isolation can be improved by using a procedure called cold enrichment, in which the specimen is held at 4°C and periodically plated onto media, but this procedure can take 4 weeks. *Listeria monocytogenes* can be differentiated from the nonpathogenic *Listeria* and from other bacteria to which it bears a superficial resemblance by characteristics listed in table 19.2. Rapid diagnostic kits using ELISA, immunofluorescence, and gene probe technology are now available for direct testing of dairy products and cultures. Antibiotic therapy should be started as soon as listeriosis is suspected. Ampicillin and trimethoprim-sulfamethoxazole are the first choices, followed by erythromycin. Prevention can be improved by adequate pasteurization temperatures and by cooking foods that are suspected of being contaminated with animal manure or sewage. Cold storage is not an effective control measure because the microbe can grow at most refrigeration temperatures.

ERYSIPELOTHRIX RHUSIOPATHIAE: A ZOONOTIC PATHOGEN

Epidemiology, Pathogenesis, and Control

Erysipelothrix rhusiopathiae* is a gram-positive rod widely distributed in animals and the environment. Its primary reservoir appears to be the tonsils of healthy pigs. It is also a normal flora of other vertebrates and is commonly isolated from sheep, chickens, and fish. It can persist for long periods in sewage, seawater, soils, and foods. The pathogen causes epidemics of swine erysipelas and sporadic infections in other domestic and wild animals. Humans at greatest risk for infection are those who handle animals, carcasses, and meats, such as slaughterhouse workers, butchers, veterinarians, farmers, and fishermen.

The common portal of entry in human infections is a scratch or abrasion on the hand or arm. The microbe multiplies at the invasion site to produce a disease known as ***erysipeloid,*** characterized by swollen, inflamed, dark red lesions that burn and itch (figure 19.11). Although the lesions usually heal without complications, rare cases of septicemia and endocarditis do arise. Inflamed red sores on the hands of people in high-risk occupations suggest erysipeloid, but the lesions must be cultured for confirmatory diagnosis. The condition is treated with penicillin or erythromycin. Swine erysipelas can be prevented by vaccinating pigs, but the vaccine does not protect humans. Animal handlers can lower their risk by wearing protective gloves.

Chapter Checkpoints

Regular, non-spore-forming gram-positive bacilli are so named because of their consistent shape and staining properties. The seven genera of this group are otherwise diverse in habitat, biochemical properties, and size. Significant pathogens in this group are *Listeria monocytogenes,* the agent of listeriosis, and *Erysipelothrix rhusiopathiae,* agent of erysipelas in both humans and animals. *Lactobacillus* species are important commercial producers of many dairy products.

GRAM-POSITIVE IRREGULAR NON-SPORE-FORMING BACILLI

The **irregular,** non-spore-forming bacilli tend to be pleomorphic and to stain unevenly. Of the 20 genera in this category, *Corynebacterium, Mycobacterium,* and *Nocardia* have the greatest clinical significance. These three genera are grouped together because of similar morphological, genetic, and biochemical traits. They produce catalase and possess mycolic acids and a unique type of peptidoglycan in the cell wall. The following sections discuss the primary diseases associated with *Corynebacterium,* a similar genus called *Propionibacterium, Mycobacterium, Actinomyces,* and *Nocardia.*

CORYNEBACTERIUM DIPHTHERIAE

Although several species of *Corynebacterium** are important, most human disease is associated with ***C. diphtheriae.*** In general morphology, this bacterium is a straight or somewhat curved rod that tapers at the ends, but thin spots in the cell wall cause it to develop pleomorphic club, filamentous, and swollen shapes. Older cells are filled with metachromatic (polyphosphate) granules and can occur in side-by-side palisades arrangement (figure 19.12).

*_Erysipelothrix rhusiopathiae_ (er′′-ih-sip′-eh-loh-thriks) Gr. *erythros,* red, *pella,* skin, and *thrix,* filament. (ruz′′-ee-oh-path′-ee-eye) Gr. rhusios, red, and *pathos,* disease.

*_Corynebacterium_ (kor-eye′′-nee-bak-ter′-ee-um) Gr. *koryne,* club, and *bakterion,* little rod.

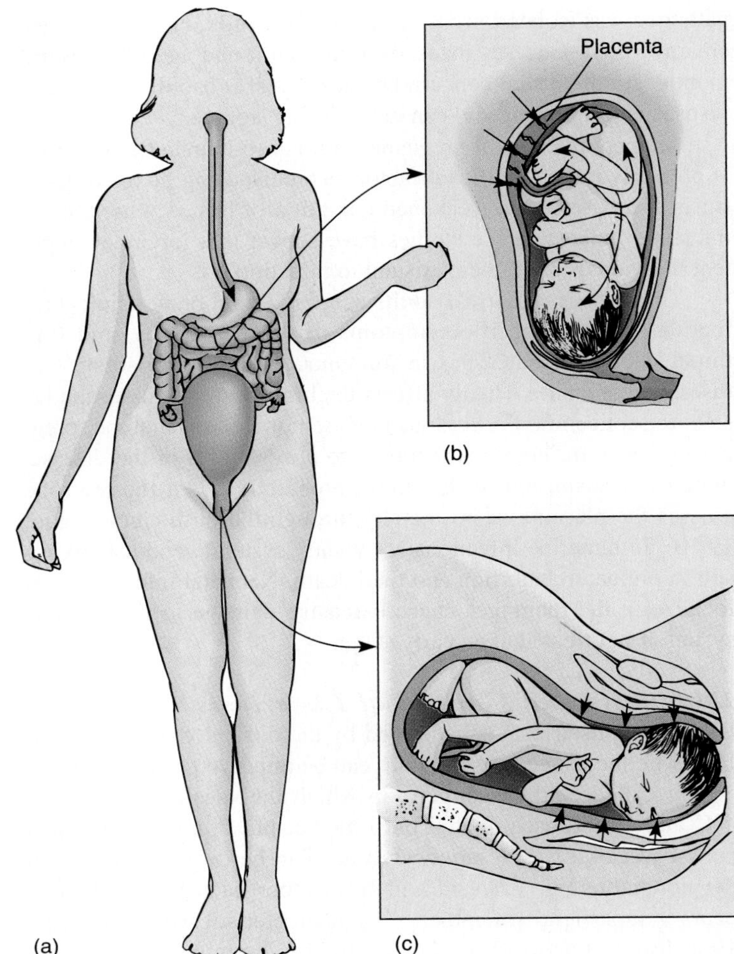

Figure 19.10
Principal routes of infection in listeriosis. (*a*) *Listeria monocytogenes* enters by the gastrointestinal route of adults and infects the intestine. From this site, it can gain access to the circulation. (*b*) In pregnant women, one complication occurs when the pathogen crosses the placenta and initiates a serious fetal infection or abortion. (*c*) An asymptomatic vaginal infection can also transmit infection to the neonate at the time of birth. Infection is manifested as symptoms of meningitis within a few hours or days.

TABLE 19.2

CHARACTERISTICS FOR DIFFERENTIATING *L. MONOCYTOGENES* FROM SIMILAR GRAM-POSITIVE BACTERIA

	Shape	Arrangement	Motility (20°–25°C)	Catalase	Morphology
Listeria monocytogenes	Rods, coccobacilli	Single, in chains	+	+	
Listeria spp.	Rods, coccobacilli	Single, in chains	+	+	
Streptococcus (esp. group D)	Cocci (spheres, ovals)	Pairs, in chains	–	–	
Corynebacterium	Pleomorphic rods	Palisades, single	–	+	
Lactobacillus	Straight rods	Single, in chains	–	–	
Erysipelothrix	Long, slender rods	Filaments	–	–	

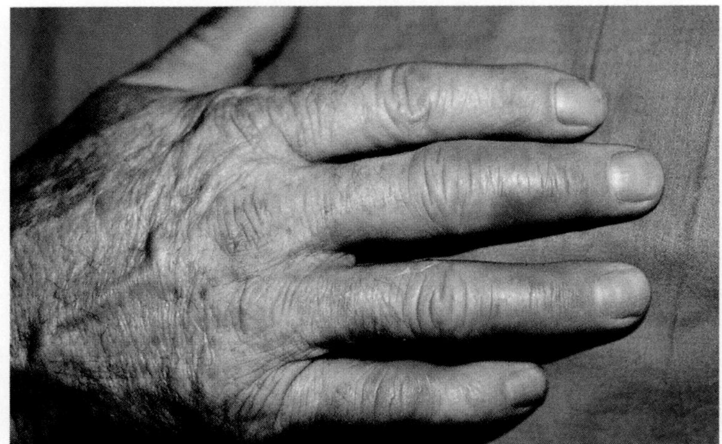

Figure 19.11
Erysipeloid on the hand of an animal handler.

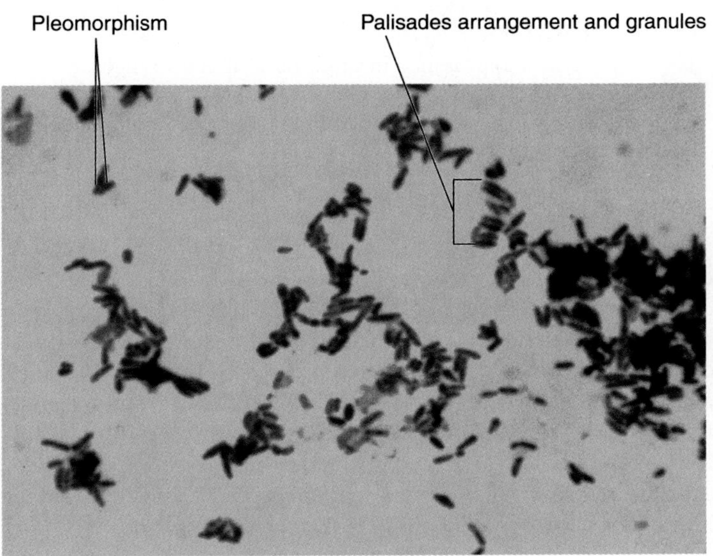

Figure 19.12
A photomicrograph of *Corynebacterium diphtheriae,* showing pleomorphism (especially club forms), metachromatic granules, and palisades arrangement (600×).

Epidemiology of Diphtheria

For hundreds of years, **diphtheria*** was a significant cause of morbidity and mortality, but in the last 50 years, both the number of cases and the fatality rate have steadily declined throughout the world. The current rate for the entire United States is 0.01 cases per million population (figure 19.13). Recent outbreaks of the cutaneous form of the disease have occurred among Native American populations and the homeless. Because many populations harbor a reservoir of healthy carriers, the potential for diphtheria is constantly present. In the mid-1990s, an epidemic occurred in areas of the former Soviet Union, where reduced vaccination programs have put the community at risk. Most cases occur in nonimmunized children from 1 to 10 years of age living in crowded, unsanitary situations.

*diphtheria (dif-thee´-ree-ah) Gr. *diphthera,* membrane.

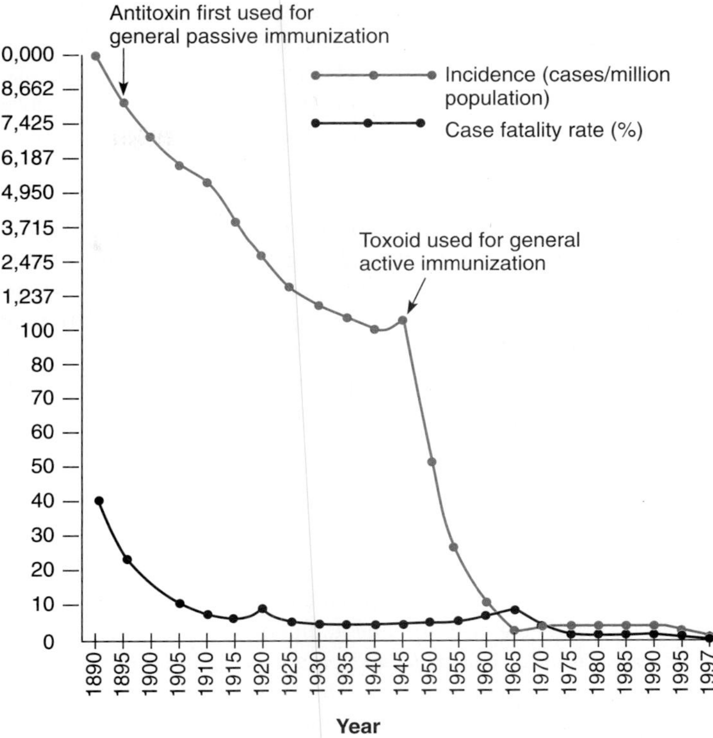

Figure 19.13
The incidence and case fatality rates for diphtheria in the United States during the last 100 years. This graph clearly documents the steady decline in cases, starting with the routine use of antitoxin in 1895 and the widespread use of the toxoid vaccine that began in 1945. The percentage of fatalities also dropped dramatically, long before antimicrobic drugs were available. The current residual level of cases is due to a small number of unimmunized carriers in the population.

Pathology of Diphtheria

Exposure to the diphtheria bacillus usually results from close contact with the droplets of human carriers or active infections and occasionally with fomites or contaminated milk. The clinical disease proceeds in two stages: (1) local infection by *Corynebacterium diphtheriae* and (2) toxin production and toxemia. The most common location of primary infection is in the upper respiratory tract (tonsils, pharynx, larynx, and trachea). *Cutaneous diphtheria* is usually a secondary infection manifesting as deep, erosive ulcers that are slow to heal. The bacterium becomes established by means of virulence factors that assist in its attachment and growth. The cells are not ordinarily invasive and usually remain localized at the portal of entry. This form of the disease is on the rise in the United States.

Diphtherotoxin and Toxemia Although infection is necessary for disease, the cardinal determinant of pathogenicity is the production of **diphtherotoxin.** According to studies, this exotoxin is produced only by toxigenic strains of *C. diphtheriae* that carry the structural gene for toxin production acquired from bacteriophages during transduction (see chapter 9). This cytotoxin consists of two polypeptide fragments. Fragment B binds to and is endocytosed by mammalian target cells in the heart and nervous system. Fragment A interacts metabolically with factors in the cytoplasm and arrests protein synthesis.

The toxin affects the body on two levels. Locally, it produces an inflammatory reaction, low-grade fever, sore throat, nausea, vomiting, enlarged cervical lymph nodes, and severe swelling in the neck. One life-threatening complication is the **pseudomembrane,** a greenish-gray film consisting of solidified fibrous exudate cells, and fluid that develops in the pharynx (figure 19.14). The pseudomembrane is so leathery and tenacious that attempts to pull it away result in bleeding, and if it forms in the airways, it can cause asphyxiation.

The most dangerous systemic complication is **toxemia,** which occurs when the toxin is absorbed from the throat and carried by the blood to certain target organs, primarily the heart and nerves. The action of the toxin on the heart causes myocarditis and abnormal EKG patterns. Cranial and peripheral nerve involvement can cause muscle weakness and paralysis. Although toxic effects are usually reversible, patients with inadequate treatment often die from asphyxiation, respiratory complications, or heart damage.

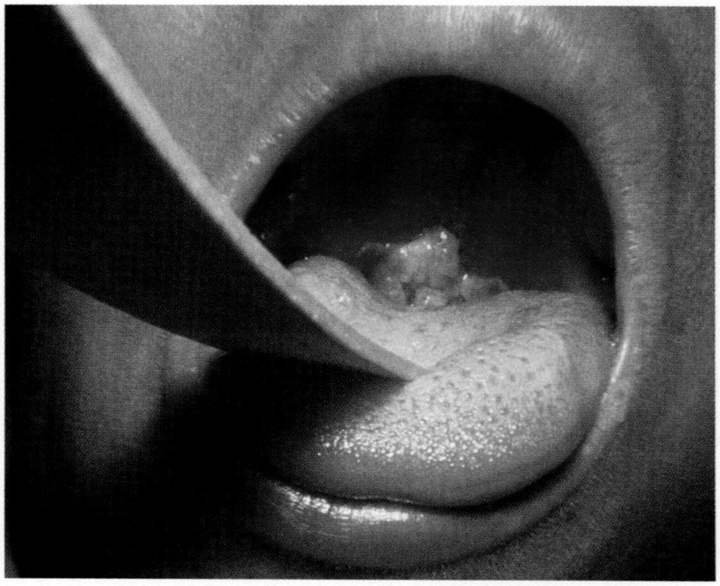

Figure 19.14

The clinical appearance in diphtheria infection includes gross inflammation of the pharynx and tonsils marked by grayish patches (a pseudomembrane) and swelling over the entire area.

Diagnostic Methods for the Corynebacteria

Diphtheria has such great potential for harm that often the physician must make a presumptive diagnosis and begin treatment before the bacteriological analysis is complete. A gray membrane and swelling in the throat are somewhat indicative of diphtheria, although several diseases present a similar appearance. A Gram stain of a membrane or throat specimen can help rule out diphtheria. Epidemiologic factors such as living conditions, travel history, and immunologic history (a positive Schick test) can also aid in initial diagnosis.

A simple stain of *C. diphtheriae* isolates with alkaline methylene blue reveals cells with marked pleomorphism and granulation. Other methods include tests for toxicity and a PCR analysis of the culture. It is important to differentiate *C. diphtheriae* from "diphtheroids"—similar species often present in clinical materials that are not primary pathogens. *Corynebacterium xerosis** normally lives in the eye, skin, and mucous membranes and is an occasional opportunist in eye and postoperative infections. *Corynebacterium pseudodiphtheriticum,** a normal inhabitant of the human nasopharynx, can colonize natural and artificial heart valves.

Treatment and Prevention of Diphtheria

The adverse effects of toxemia are treated with diphtheria antitoxin (DAT) derived from horses. Prior to injection, the patient must be tested for allergy to horse serum and be desensitized if necessary. The infection is treated with antibiotics from the penicillin or erythromycin family. Bed rest, heart medication, and tracheostomy or bronchoscopy to remove the pseudomembrane may be indicated. Diphtheria can be easily prevented by a series of vaccinations with toxoid, usually given as part of a mixed vaccine against tetanus and pertussis called the DPT. Currently recommended are three vaccinations, starting at 6–8 weeks of age, followed by a booster at 15 months and again at school age. Older children and adults who are not immune to the toxin can be immunized with two doses of diphtheria-tetanus (Td) vaccine.

The Genus *Propionibacterium*

Propionibacterium* resembles *Corynebacterium* in morphology and arrangement, but it differs by being aerotolerant or anaerobic and nontoxigenic. The most prominent species is ***P. acnes,**** a common resident of the *pilosebaceous** glands of human skin and occasionally the upper respiratory tract. The primary importance of this bacterium is its relationship with the familiar **acne vulgaris** lesions of adolescence. Acne is a complex syndrome influenced by genetic and hormonal factors as well as by the structure of the epidermis, but it is also an infection (microfile 19.3). *Propionibacterium* is occasionally involved in infections of the eye and artificial joints.

**xerosis* (zee-roh´-sis) Gr. *xerosis,* parched skin.

**pseudodiphtheriticum* (soo´´-doh-dif-ther-it´-ih-kum) Gr. *pseudes,* false, and *diphtheriticus,* of diphtheria.

**Propionibacterium* (pro´´-pee-on´´-ee-bak-tee´-ree-um) Gr. *pro,* before, *prion,* fat, and *backterion,* little rod. Named for its ability to produce propionic acid.

**acnes* (ak´-neez) Gr. *akme,* a point. (The original translators incorrectly changed an *m* to an *n*.)

**pilosebaceous* (py´´-loh-see-bay´-shus) L. *pilus,* hair, and *sebaceous,* tallow.

MICROFILE 19.3 ACNE: THE MICROBIAL CONNECTION

Normally, each pilosebaceous gland is a self-contained system for protecting, softening, and lubricating the skin *(1)*. It contains one or more sebaceous glands that continuously release an oily secretion called sebum into the hair follicle. As hair and skin grow, the dead epidermal cells and sebum work their way upward and are discharged from the pore onto the skin surface.

The skin structure of people prone to pimples and acne traps the mass of sebum and dead cells and clogs the pores. If the skin at the surface swells over the pore's entrance, a closed comedo, or whitehead, results *(2)*; if the pore remains open to the surface but is blocked with a plug of sebum, it is called an open comedo, or blackhead *(3)*. The dark tinge of the blackhead is caused by the accumulation of the pigment melanin not to uncleanliness. An added factor is overproduction of sebum when the sebaceous gland is stimulated by hormones (especially male). *Propionibacterium* bacteria growing in the follicle release lipases to digest this surplus of oil. The combination of digestive products (fatty acids) and bacterial antigens stimulates an intense local inflammation that bursts the follicle *(4)*. In time, the lesion erupts on the surface as a papule or pustule *(5)*. In some cases, secondary bacterial infections cause deeply scarred lesions. Because acne depends in part on an infection, it can be suppressed with topical and oral antibiotics such as clindamycin, erythromycin, or tetracycline. Now after 30 years of widespread use of antimicrobic drugs for treating acne, the bacteria have developed resistance that complicates therapy. Other types of therapy involve chemicals that enhance skin removal (benzoyl peroxide) and slow the production of sebum (Retin A and Accutane).

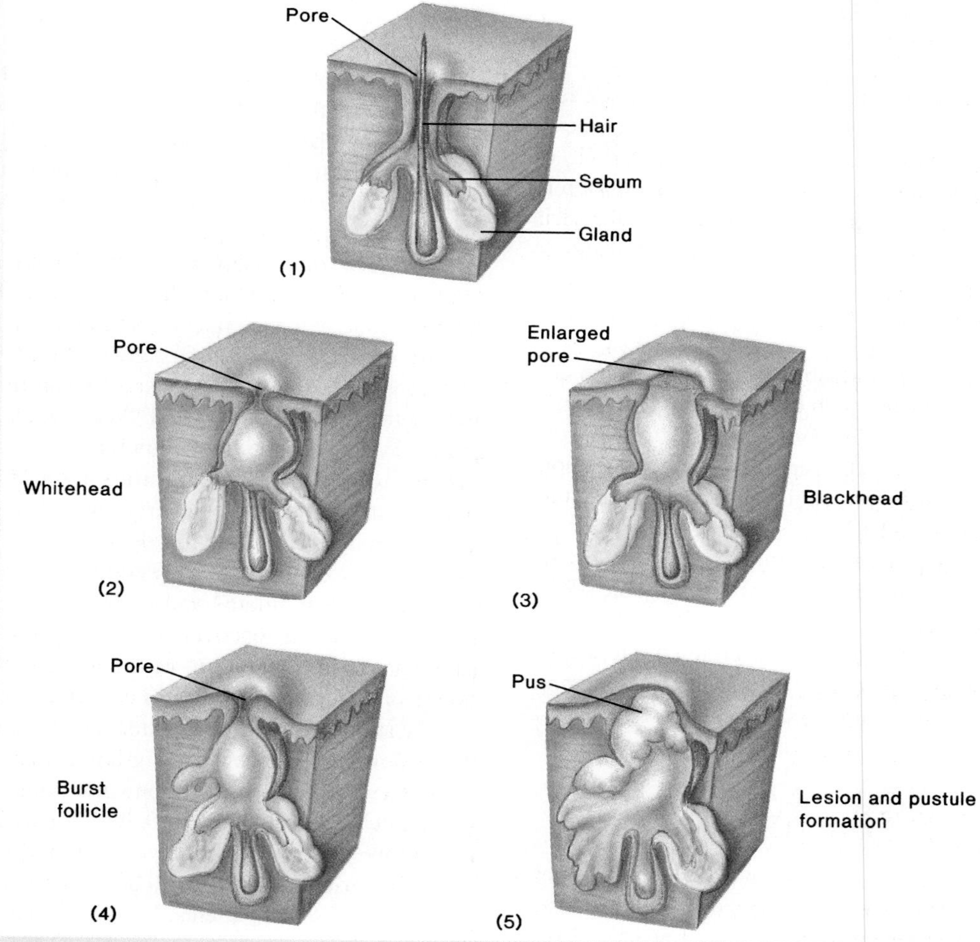

The stages in the formation of an acne lesion.

Chapter Checkpoints

Irregular, non-spore-forming gram-positive bacilli are so named because of irregularities in both morphology and staining properties. The inhabit soil, water, and animal bodies. Three genera contain important pathogens: *Corynebacterium, Propionibacterium,* and *Mycobacterium.*

C. diphtheriae is the agent of diphtheria, an upper respiratory infection. The primary virulence factor is diphtherotoxin, which is toxic to upper respiratory tissue, the peripheral nervous system, and the heart. Immunization provides effective protection.

Propionibacterium acnes is the agent of acne vulgaris, an infection of pilosebaceous glands exacerbated by biochemical changes of adolescence.

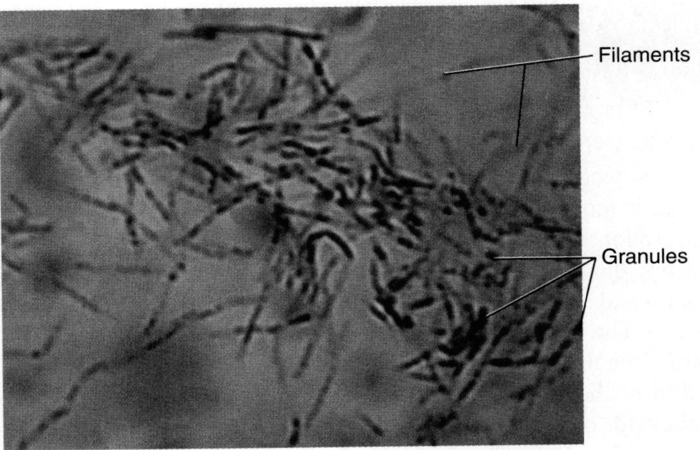

Figure 19.15
The microscopic morphology of mycobacteria (*Mycobacterium tuberculosis* shown here), as illustrated by an acid-fast stain of sputum from a tubercular patient. Note the irregular morphology, granules, and the filamentous forms (700×).

MYCOBACTERIA: ACID-FAST BACILLI

The genus ***Mycobacterium**** is distinguished by its complex layered structure composed of high-molecular-weight mycolic acids and waxes. This high lipid content imparts the characteristic of **acid-fastness** and is responsible for the resistance of the group to drying, acids, and various germicides. The cells of mycobacteria are long, slender, straight, or curved rods with a slight tendency to be filamentous or branching (figure 19.15). Although they usually contain granules and vacuoles, they do not form capsules, flagella, or spores.

Most mycobacteria are strict aerobes that grow well on simple nutrients and media. Compared with other bacteria, the growth rate is generally slow, with generation times ranging from 2 hours to several days. Some members of the genus exhibit colonies containing yellow, orange, or pink carotenoid pigments that require light for development; others are nonpigmented. Many of the 50 mycobacterial species are saprobes living free in soil and water, and several are highly significant human pathogens (table 19.3). Worldwide, millions of people are afflicted with tuberculosis and leprosy. Certain opportunistic species loosely grouped into a category called NTM (**n**on-**t**uberculous **m**ycobacteria) have become an increasing problem in immunosuppressed patients.

MYCOBACTERIUM TUBERCULOSIS: THE TUBERCLE BACILLUS

The *tubercle** bacillus is a long, thin rod that grows in sinuous masses or strands called cords. Unlike many bacteria, it produces no exotoxins or enzymes that contribute to infectiousness. Most strains contain complex waxes and a cord factor (figure 19.16) that contribute to virulence in some yet unexplained way. It is currently thought that protection conferred by the antigens in the cell wall prevents the mycobacteria from being destroyed by the lysosomes or macrophages and thereby contributes to their invasion and persistence as intracellular parasites.

Epidemiology and Transmission of Tuberculosis

Mummies from the Stone Age, ancient Egypt, and Peru provide unmistakable evidence that tuberculosis (TB) is an ancient human disease (see chapter opening figure). Just 100 years ago, tuberculosis was such a prevalent cause of death that it was called "Captain of the Men of Death" and "White Plague." Its epidemiologic patterns vary with the living conditions in a community or an area of the world. Factors that significantly affect a person's susceptibility to tuberculosis are poverty, inadequate nutrition, unsanitary living conditions, underlying debilitation of the immune system, lung damage, and genetics. People in developing countries are often infected as infants and harbor the microbe for many years until the disease is reactivated in young adulthood. Estimates indicate that possibly one-third of the world's population and 15 million people in the United States carry the TB bacillus.

Cases in the United States show a strong correlation with the age, sex, and recent immigration history of the patient. The highest case rates occur in nonwhite males over 30 years of age and nonwhite females over 60. The highest carrier rates occur in new immigrants from certain areas of Indochina, Central and South America, and Africa. These factors, along with the large population of AIDS patients, have contributed to a shift in the epidemiology of tuberculosis (microfile 19.4).

The agent of tuberculosis is transmitted almost exclusively by fine droplets of respiratory mucus suspended in the air. The tubercle bacillus is very resistant and can survive for 8 months in fine aerosol particles. Although larger particles become trapped in the mucus and expelled, tinier ones can be inhaled into the bronchioles and alveoli. This effect is especially pronounced among

*Mycobacterium (my´´-koh-bak-tee´-ree-um) Gr. *myces,* fungus, and *bakterion,* a small rod.

*tubercle (too´-ber-kul)L. *tuberculum,* a swelling or knob.

Table 19.3

Differentiation of Important Mycobacterium Species

Species	Primary Habitat	Disease in Humans	Treatment	Rate of Growth*	Pigmentation**
M. tuberculosis	Humans	Tuberculosis (TB)	Combined drugs	S	NP
M. bovis	Cattle	Tuberculosis	Same as TB	S	NP
M. ulcerans	Humans	Skin ulcers	Surgery, grafts	S	NP
M. kansasii	Not clear	Opportunistic lung infection	Difficult, similar to TB	S	PP
M. marinum	Water, fish	Swimming pool granuloma	Tetracycline, rifampin	S	PP
M. scrofulaceum	Soil, water	Scrofula	Removal of lymph nodes	S	PS
M. avium–M. intracellulare **complex**	Birds	Opportunistic AIDS infection; lung infection like TB	Combined drugs	S	NP
M. fortuitum–M. chelonae **complex**	Soil, water, animals	Wound abscess; postsurgical infection	4–6-drug regimen; surgery	R	NP
M. phlei	Sputum, soil	Not pathogenic	None	R	PS
M. smegmatis	Smegma, soil	Not pathogenic	None	R	Usually NP
M. leprae	Strict parasite of humans	Leprosy	See text	S	Cannot be grown in artificial media

The mycobacteria are grouped into major categories by their growth rate and their pigment production.

*Growth rate is rapid (R), occurring in less than 7 days, or slow (S), occurring in more than 7 days.

**Photochromogens (PP) develop yellow to dark orange pigment in the presence of light; scotochromogens (PS) synthesize pigment in darkness; and nonpigmented forms (NP) have no color.

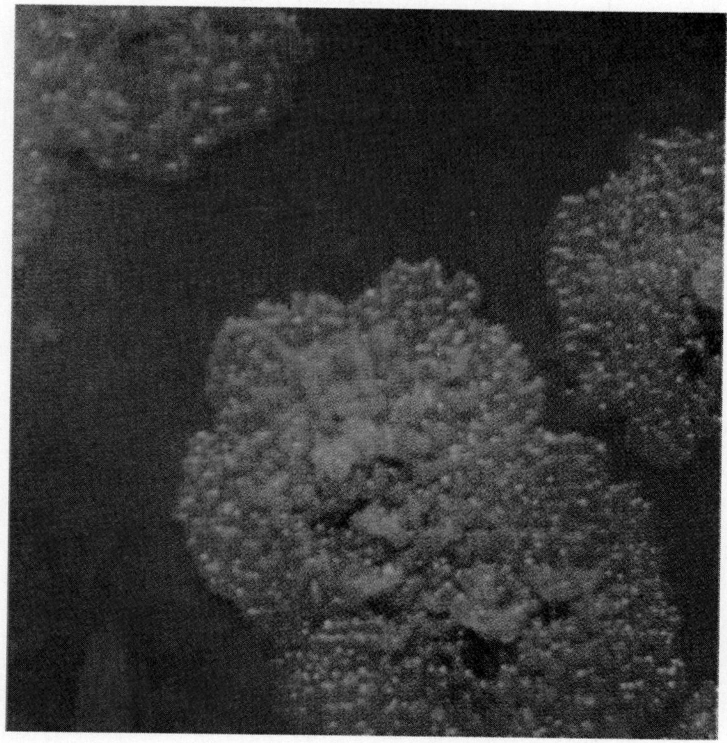

(a)

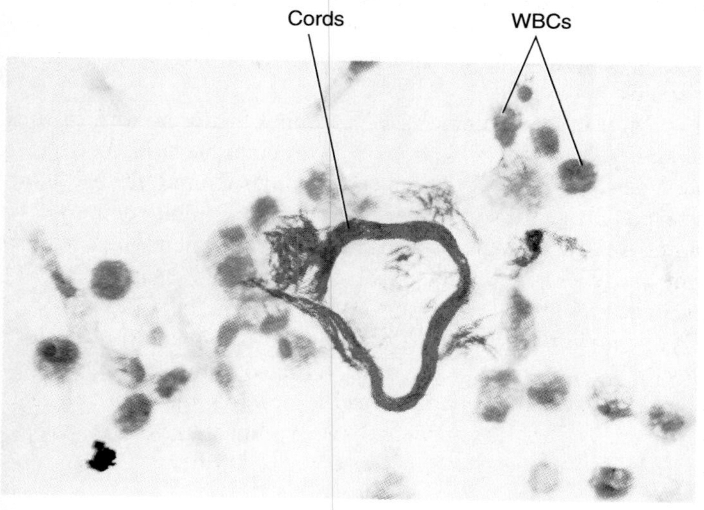

(b)

Figure 19.16

(*a*) The cultural appearance of *Mycobacterium tuberculosis.* Colonies with a typical granular, waxy pattern of growth. (*b*) Cord formation in infected tissue. Red-stained strands are elongate, massed filaments of TB bacilli.

(a) From Gillies and Dodds, Bacteriology Illustrated, *5th ed., Fig. 45, p. 58. Reprinted by permission of Churchill Livingstone.*

MICROFILE 19.4 TUBERCULOSIS: WAKING A SLEEPING GIANT

After 80 years of decline, a tiny wax-coated bacillus that hides in human lungs has reemerged as a serious threat to the world's health. The hope for eradicating tuberculosis is becoming an unlikely possibility. It still remains as the most common infectious cause of death in the world. Approximately 8 million new cases of tuberculosis and 3 million deaths occur every year. Even the United States has experienced recent outbreaks, especially in large urban centers. Ironically, TB had traditionally been curable with drugs, but there was a distinctly unsettling aspect to these new cases. Nearly 14% of isolates were resistant to one or more drugs, and about 2% were multiply resistant **(MRTB),** meaning that they are not sensitive to two or more of the major TB drugs. Patients were actually dying from TB even while undergoing multidrug therapy! Between 1992 and 1997, these drug-resistant strains had spread to 42 states.

What set the population on a collision course for a TB epidemic and increased drug resistance? Major factors that account for the changing epidemiologic picture are homelessness, drug addiction, the HIV epidemic, reduced government support of TB programs, and more cases occurring in new immigrants. Homelessness and drug addiction have the similar effect of increasing host susceptibility and creating conditions that favor the spread of the TB bacillus. These populations are less likely to have access to good medical care and to comply with the necessarily long course of therapy. Fully one-fifth of cases occur in AIDS patients. The mortality rate in this group is very high: up to 80% die within 16 weeks of diagnosis.

The lack of well-funded TB control programs has contributed to the epidemic by discontinuing low-cost treatments, by limiting surveillance, and by not following up on treatment regimens. Experts at the CDC believe that multidrug-resistant strains of TB emerged in those patients with active TB who did not follow appropriate drug therapy. The costs of cutting back and "saving money" on TB prevention for the past 15 years has been staggering both in dollars and human lives. Treating a patient in a hospital with active TB costs $25,000 per episode, compared with a few hundred dollars for preventive medicine. The long-term cost for the excess cases are estimated at $1–$2 billion. The risk for increased MRTB cases will continue well into the end of the century, and only a concerted effort from government, medicine, and research can hope to hold it at bay. Worldwide, only about $11 million is spent on TB programs, when at least $100 million is needed to have an impact.

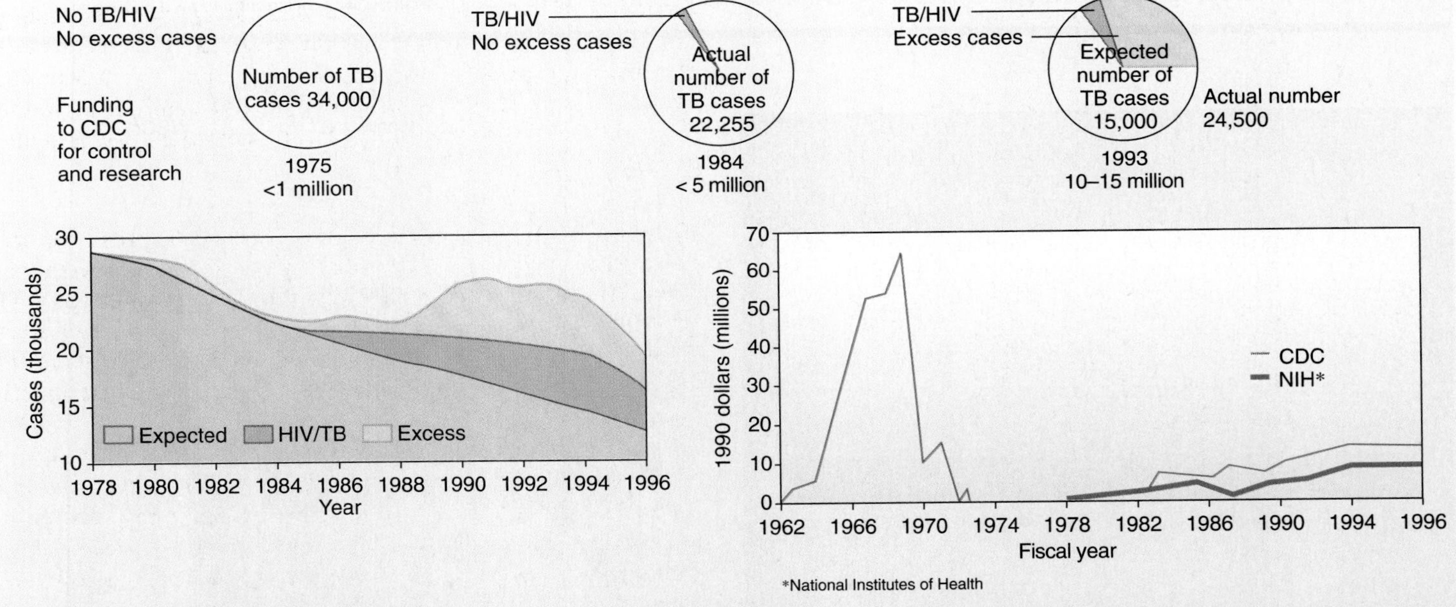

As federal support for TB programs lagged, the number of infected people exceeded estimates. These excess cases began to increase in 1984 and continue to the present day.
Source: Data from Morbidity and Mortality Weekly Report, *U.S. Department of Health and Human Services, Centers for Disease Control and Prevention, Atlanta, GA.*

people sharing closed, small rooms with limited access to sunlight and fresh air.

The Course of Infection and Disease

A clear-cut distinction can be made between infection with the tubercle bacillus and the disease it causes. In general, humans are rather easily infected with the bacillus but are resistant to the disease. Estimates project that only about 5% of infected people actually develop a clinical case of tuberculosis. Untreated tuberculosis progresses slowly and is capable of lasting a lifetime, with periods of health alternating with episodes of morbidity. The majority (85%) of TB cases are contained in the lungs, even though disseminated tubercle bacilli can give rise to tuberculosis in any organ of the body. Clinical tuberculosis is divided into primary tuberculosis, secondary (reactivation or reinfection) tuberculosis, and disseminated tuberculosis.

Primary Tuberculosis The minimum infectious dose for lung infection is around 10 cells. The bacilli are phagocytosed by alveolar macrophages and multiply intracellularly. This period of hidden

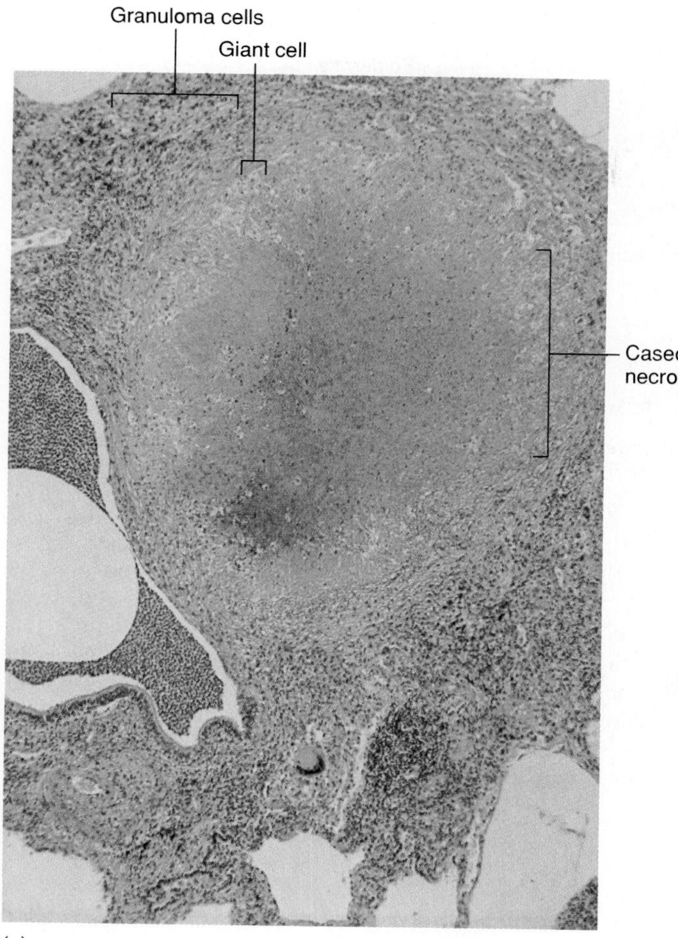

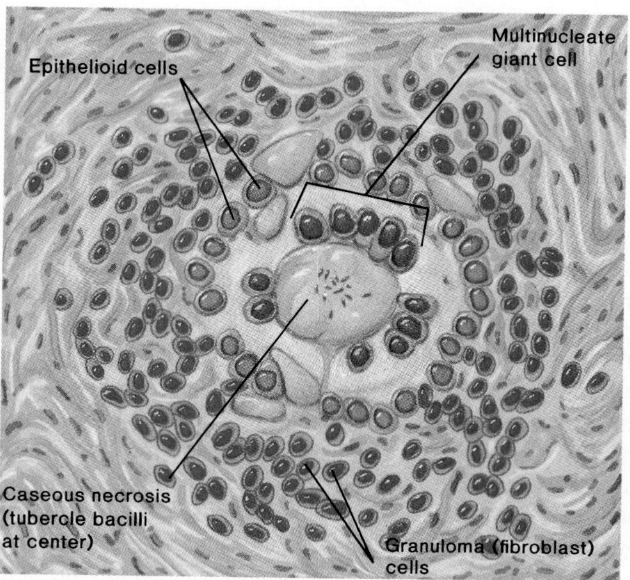

Figure 19.17
Tubercle formation. (*a*) Photomicrograph of a tubercle (16×). (*b*) In this schematic drawing, the massive granuloma infiltrate has obliterated the alveoli and set up a dense collar of fibroblasts, lymphocytes (granuloma cells), and epithelioid cells. The core of this tubercle is a caseous (cheesy) material containing the bacilli.

infection is asymptomatic or accompanied by mild fever, but some cells escape from the lungs into the blood and lymphatics. After 3 to 4 weeks, the immune system mounts a complex, cell-mediated assault against the bacilli. The large influx of mononuclear cells into the lungs plays a part in the formation of specific infection sites called **tubercles.** Tubercles are granulomas that consist of a central core containing TB bacilli and enlarged macrophages and an outer wall made of fibroblasts, lymphocytes, and neutrophils (figure 19.17). Although this response further checks spread of infection and helps prevent the disease, it also carries a potential for damage. Frequently, the centers of tubercles break down into necrotic, *caseous** lesions that gradually heal by calcification (normal lung tissue is replaced by calcium deposits). The response of T cells to *M. tuberculosis* proteins also causes a cell-mediated immune response evident in the **tuberculin reaction,** a valuable diagnostic and epidemiologic tool (see figure 17.18*a*).

Secondary Reactivation Tuberculosis Although the majority of TB patients recover more or less completely from the primary episode of infection, live bacilli can remain dormant and become reactivated weeks, months, or years later, especially in people with weakened immunity. In chronic tuberculosis, tubercles filled with masses of bacilli expand and drain into the bronchial tubes and upper respiratory tract. Gradually, the patient experiences more severe symptoms, including violent coughing, greenish or bloody sputum, low-grade fever, anorexia, weight loss, extreme fatigue, night sweats, and chest pain. It is the gradual wasting of the body that accounts for an older name for tuberculosis—*consumption.* Untreated secondary disease has nearly a 60% mortality rate.

Extrapulmonary Tuberculosis During the course of secondary TB, the bacilli disseminate rapidly to sites other than the lungs. Organs most commonly involved in **extrapulmonary TB** are the regional lymph nodes, kidneys, long bones, genital tract, brain, and meninges. Because of the debilitation of the patient and the high load of tubercle bacilli, these complications are usually grave.

Renal tuberculosis results in necrosis and scarring of the renal medulla and the pelvis, ureters, and bladder. This damage is accompanied by painful urination, fever, and the presence of blood and the TB bacillus in urine. Genital tuberculosis in males damages the prostate gland, epididymis, seminal vesicles, and testes; and in females, the fallopian tubes, ovaries, and uterus. It often affects reproductive function in both sexes.

Tuberculosis of the bone and joints is a common complication. The spine is a frequent site of infection, though the hip, knee, wrist, and elbow can also be involved. Advanced infiltration of the vertebral column produces degenerative changes that collapse the vertebrae (Pott's disease; figure 19.18), resulting in abnormal curvature of the thoracic region (humpback or kyphosis) or of the lumbar region (swayback or lordosis). Neurological damage stemming from compression on nerves can cause extensive paralysis and sensory loss.

*caseous (kay´-see-us) L. *caseus,* cheese. The material formed resembles cheese or curd.

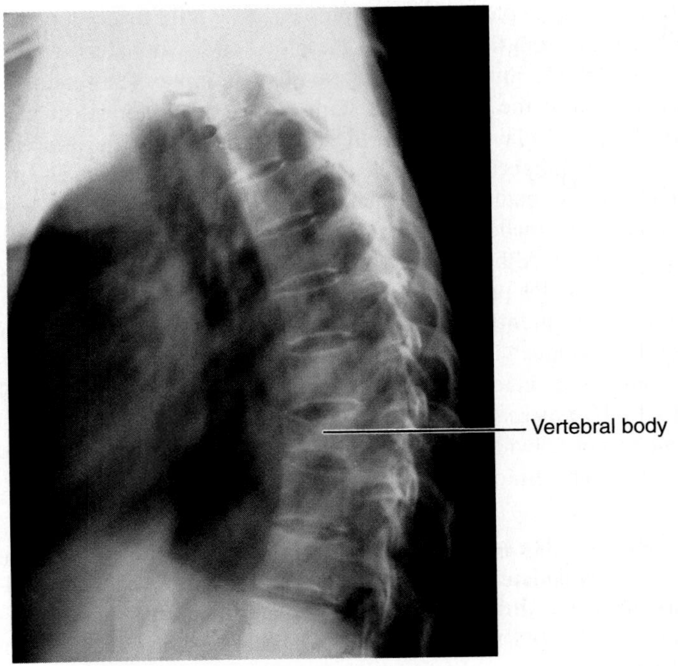

Figure 19.18
Tuberculosis of the spinal column (Pott's disease) as seen by an X ray.

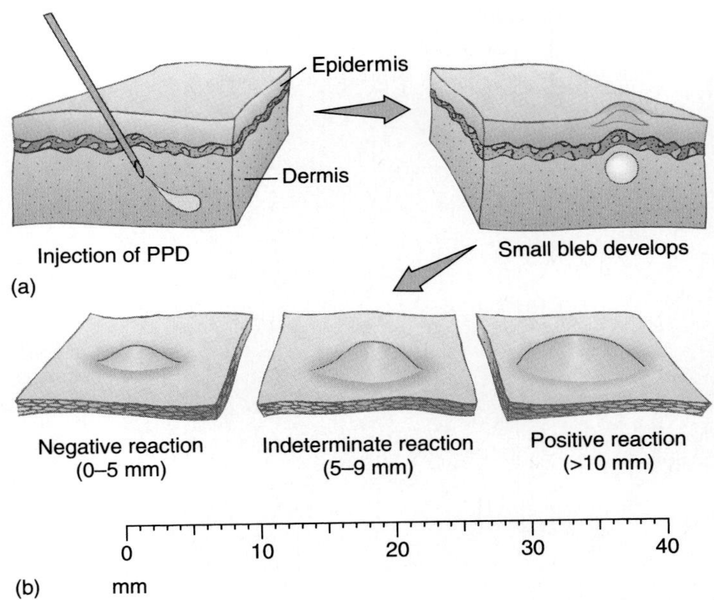

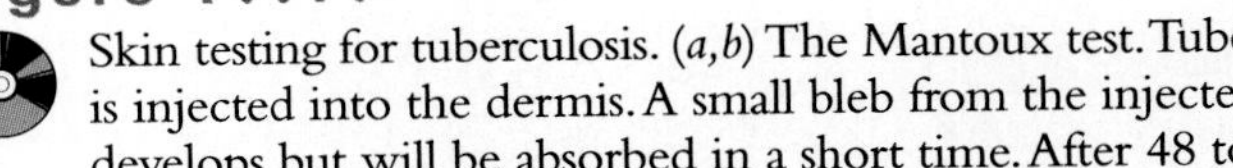

Figure 19.19
Skin testing for tuberculosis. (*a,b*) The Mantoux test. Tuberculin is injected into the dermis. A small bleb from the injected fluid develops but will be absorbed in a short time. After 48 to 72 hours, the skin reaction is rated by the degree (or size) of the reaction. (Views are magnified 3×.) See also figure 17.18*a*.

Tubercular meningitis is the result of an active brain lesion seeding bacilli into the meninges. Over a period of several weeks, the infection of the cranial compartments can create mental deterioration, permanent retardation, blindness, and deafness. Untreated tubercular meningitis is invariably fatal, and even treated cases can have a 30% to 50% mortality rate.

Clinical Methods of Detecting Tuberculosis

Clinical diagnosis of tuberculosis traditionally relies upon four techniques: (1) *in vivo* or tuberculin testing, (2) roentgenography (X rays), (3) direct identification of acid-fast bacilli (AFB) in sputum or some other specimen, and (4) cultural isolation and biochemical testing.

Tuberculin Sensitivity and Testing Because hypersensitivity to tuberculoproteins can persist throughout life, testing for it is an effective way to screen school children, public employees, and health care workers. A positive reaction to these proteins in an unimmunized individual is fairly reliable evidence of recent or past infection or disease. Because vaccination for TB also stimulates delayed hypersensitivity, tuberculin screening is most useful in countries such as the United States where widespread vaccination is not carried out. Physicians should determine a patient's history before giving the tuberculin test.

The test itself involves exposure to a small amount of purified protein derivative (PPD), a standardized solution obtained from culture filtrates of *M. tuberculosis* (figure 19.19*a*). In the **Mantoux test,** 0.1 ml of PPD (equivalent to 5 tuberculin units) is injected intradermally into the forearm to produce an immediate small bleb. The skin is observed and measured for induration after 48 hours and 72 hours. The three reaction levels include: (1) negative reactions, with 5 mm or less induration; (2) intermediate reactions, with 5 mm to 9 mm induration, which indicates doubtful sensitivity and requires retesting; and (3) positive reactions, with 10 mm or more induration (figure 19.19*b*). Because tuberculin can cause a systemic reaction in sensitive individuals, it should not be injected into known tuberculin reactors, as these persons often experience severe ulceration and necrosis at the test site.

A positive reaction can be interpreted in one of the following ways: It may be due to recent contact and a new infection, or it may be reactivation of a prior infection. In these cases, the degree of reaction tends to be more pronounced. False positive reactions are caused by a recent BCG vaccination or an infection with a NTM that cross-reacts with the TB bacillus.

A negative skin test also has more than one possible interpretation. (1) In most cases, it means no contact or infection has occurred (2) It may be too early in the infection for sensitization to appear, indicating a need for retesting after 3 months. (3) Up to 20% of infected subjects are *anergic* and unable to mount a tuberculin reaction. This condition is associated with severe immunocompromise as seen with HIV infection, age, alcoholism, and chronic disease. Skin testing is not a reliable diagnostic indicator in these people.

Roentgenography and Tuberculosis Chest X rays, or roentgenographs, can help verify TB when other tests have given indeterminate results. X-ray films reveal abnormal radiopaque

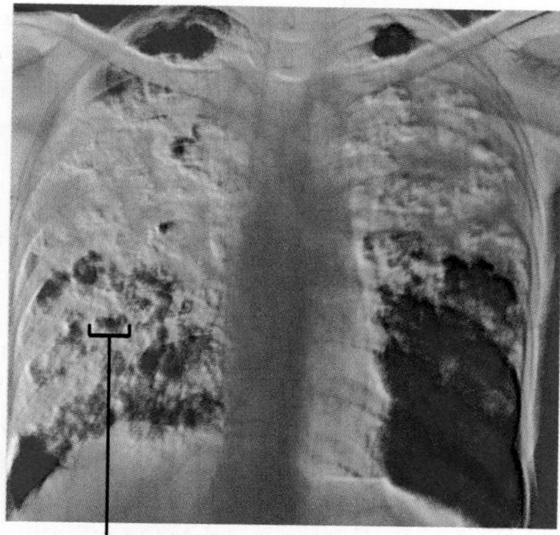

Figure 19.20
Colorized X ray showing a secondary tubercular infection.

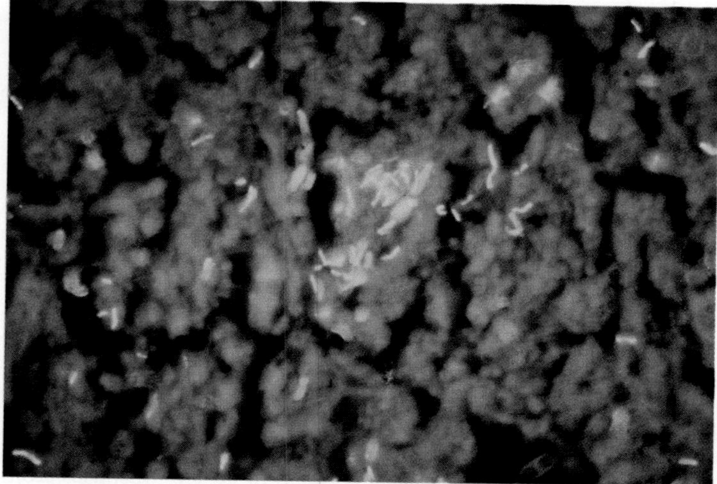

Figure 19.21
A fluorescent acid-fast stain of *Mycobacterium tuberculosis* from sputum. Smears are evaluated in terms of the number of AFB seen per field. This quantity is then applied to a scale ranging from 0 to 4+, 0 being no AFB observed and 4+ being more than 9 AFB per field.

patches whose appearance and location can be very indicative. Primary tubercular infection presents the appearance of fine areas of infiltration and enlarged lymph nodes in the lower and central areas of the lungs. Secondary tuberculosis films show more extensive infiltration in the upper lungs and bronchi and marked tubercles (figure 19.20). Scars from older infections often show up on X rays and can furnish a basis for comparison with which to identify newly active disease.

Acid-Fast Staining The diagnosis of tuberculosis in people with positive skin tests or X rays can be backed up by acid-fast staining of sputum or other specimens. Several variations on the acid-fast stain are currently in use. The Ziehl-Neelsen stain produces bright red acid-fast bacilli (AFB) against a blue background. Fluorescence staining shows luminescent yellow-green bacilli against a dark brown background (figure 19.21). The fluorescent acid-fast stain is becoming the method of choice because it is easier to read and provides a more striking contrast.

Laboratory Cultivation and Diagnosis *Mycobacterium tuberculosis* infection is most accurately diagnosed by isolating and identifying the causative agent in pure culture. Because of the specialized expertise and technology required, this is not done by most clinical laboratories as a general rule. In the United States, the handling of specimens and cultures suspected of containing pathogen is strictly regulated by federal laws.

Diagnosis that differentiates between *M. tuberculosis* and other mycobacteria must be accomplished as rapidly as possible so that appropriate treatment and isolation precautions can be instituted. Because specimens are often contaminated with rapid-growing bacteria that will interfere with the isolation of *M. tuberculosis,* they are pretreated with chemicals to remove contaminants and are plated onto selective egg-potato base media (such as Middlebrook 7H11 or Lowenstein-Jensen media). Cultures are incubated under varying temperature and lighting conditions to clarify thermal and pigmentation characteristics and are then observed for signs of growth. Several newer cultivation schemes have shortened the time to several days instead of the 6–8 weeks once necessary. Several identification techniques use probes to detect specific genes and can confirm positive specimens early in the infection (see figure F on page 561). Rapid diagnosis is particularly important for public health and treatment considerations.

Management of Tuberculosis

Treatment of TB involves administering drugs for a sufficient period of time to kill the bacilli in the lungs, organs, and macrophages, usually 6–24 months. Drug resistance is avoided by a therapeutic regimen that involves combined therapy with at least two drugs selected from a list of eleven, including isoniazid (INH), rifampin, ethambutol, streptomycin, pyrazinamide, thioacetazone, or paraaminosalicylic acid (PAS). The choice of drugs depends upon such considerations as effectiveness, adverse side effects, cost, and special medical problems of the patient. A one-pill regimen called *Rifater* (INH, rifampin, and pyrazinamide) is considered the best combination to effect cure and prevent resistance. If one combination is not working well because of toxicity, drug resistance, or hypersensitivity, a reasonably effective replacement is usually available. The presence of a negative culture or a gradual decrease in the number of AFB on a smear indicates success. However, because permanent cure will not occur if the patient does not comply with drug protocols, there are many relapses. Several programs in the United States have begun using directly observed therapy (DOT) to improve the success of treatment, especially among homeless or isolated patients.

Prevention and Control of Tuberculosis

Although it is essential to identify and treat people with active TB, it is equally important to seek out and treat those in the early stages of infection or at high risk of becoming infected. Treatment groups are divided into tuberculin-positive "converters," who appear to have a reactivated infection, and tuberculin-negative people in high-risk groups such as laboratory workers and the contacts of tubercular patients. The standard prophylactic treatment is a daily dose of isoniazid for a year. In the hospital, the use of UV lamps in air-conditioning systems and negative pressure rooms to isolate TB patients can effectively control the spread of infection.

A vaccine based on the attenuated "bacille Calmet-Guerin" (BCG) strain of *M. bovis* is often given to children in countries that have high rates of tuberculosis. Studies have shown that the success rate of vaccinations is around 80% in children and less than that (20–50%) in adults. The length of protection varies from 5 to 15 years. Because the United States does not have as high an incidence as other countries, BCG vaccination is not generally recommended except among certain health professionals, and military personnel who may be exposed to TB carriers.

Mycobacterium Leprae: The Leprosy Bacillus

Mycobacterium leprae, the cause of leprosy, was first detected in 1873 by a Norwegian physician named Gerhard Hansen, and it is sometimes called Hansen's bacillus in his honor. The general morphology and staining characteristics of the leprosy bacillus are similar to those of other mycobacteria, but it is exceptional in two ways: (1) It is a strict parasite that has not been grown in artificial media or human tissue cultures, and (2) it is the slowest growing of all the species. *Mycobacterium leprae* multiplies within host cells in large packets called globi at an optimum temperature of 30°C.

Leprosy* is a chronic, progressive disease of the skin and nerves known for its extensive medical and cultural ramifications. In ancient times, it was accompanied by stigmatization and ignorance because of the severe disfigurement of the disease and to the belief that it was a divine curse. As if the torture of the disease were not enough, leprosy patients once suffered terrible brutalities, including imprisonment under the most gruesome conditions. The modern view of leprosy is more enlightened. We know that it is not readily communicated and that it should not be accompanied by social banishment. Because of the unfortunate connotations associated with the term *leper* (a person who is shunned or ostracized), more acceptable terms such as leprosy patient, Hansen's patient, or leprotic, are preferred.

Epidemiology and Transmission of Leprosy

Reliable statistics on the worldwide incidence of leprosy are difficult to obtain, but the WHO estimates in excess of 10 million cases, most of them concentrated in areas of Asia, Africa, Central and South America, and the Pacific islands. Leprosy is not restricted to warm climates, however, since it is also reported in Siberia, Korea, and northern China. The disease is endemic to a few limited locales in the United States, including parts of Hawaii, Texas, Louisiana, Florida, and California. The total number of new cases reported nationwide is from 300 to 500 per year, many of which are associated with recent immigrants.

The mechanism of transmission among humans is yet to be fully verified. Theories propose that the bacillus is directly inoculated into the skin through contact with a leprotic, that mechanical vectors are involved, or that inhalation of droplet nuclei is a factor.

Although the human body was long considered the sole host and reservoir of the leprosy bacillus, it is now clear that armadillos harbor a mycobacterial species genetically identical to *M. leprae* and may develop a granulomatous disease similar to leprosy. Whether humans can acquire leprosy from armadillos is not yet known, but it is tempting to speculate that the infection is actually a zoonosis.

Because the leprosy bacillus is not highly virulent, most people who come into contact with it do not develop clinical disease. Indeed, numerous people have lived their entire lives among leprotics without acquiring leprosy. As with tuberculosis, it appears that health and living conditions influence susceptibility and the course of the disease. An apparent predisposing factor is some defect in the regulation of T cells. Mounting evidence also indicates that some forms of leprosy are associated with a specific genetic marker. Long-term household contact with leprotics, poor nutrition, and crowded conditions increase the risks of infection. Many people become infected as children and harbor the microbe through adulthood.

The Course of Infection and Disease

Once *M. leprae* has entered a portal of entry, macrophages successfully destroy the bacilli, and there are no manifestations of disease. But in a small percentage of cases, a weakened or slow macrophage and T-cell response leads to intracellular survival of the pathogen. The usual incubation period varies from 2 to 5 years, with extremes of 3 months to 40 years. The earliest signs of leprosy appear on the skin of the trunk and extremities as small, spotty lesions colored differently from the surrounding skin (figure 19.22). In untreated cases, the bacilli grow slowly in the skin macrophages and Schwann cells of peripheral nerves, and the disease progresses to one of several outcomes.

A useful rating system for leprosy was developed by D.S. Ridley and W. H. Jopling in 1966. At the two extremes are tuberculoid leprosy (TT) and lepromatous leprosy (LL) (table 19.4), and in between are borderline tuberculoid (BT), borderline (BB), and borderline lepromatous (BL). Patients can have more than one form of leprosy simultaneously, and one type can progress to another.

Tuberculoid leprosy, the most superficial form, is characterized by asymmetrical, shallow skin lesions containing very few bacilli (figure 19.22). Microscopically, the lesions appear as thin granulomas and enlarged dermal nerves. Damage to these nerves usually results in local loss of pain reception and feeling. This form has fewer complications and is more easily treated than other types of leprosy.

Lepromatous leprosy is responsible for the disfigurations commonly associated with the disease. It is marked by chronicity

---

*leprosy (lep´-roh-see) Gr. *lepros,* scaly or rough. Translated from a Hebrew word that refers to uncleanliness.

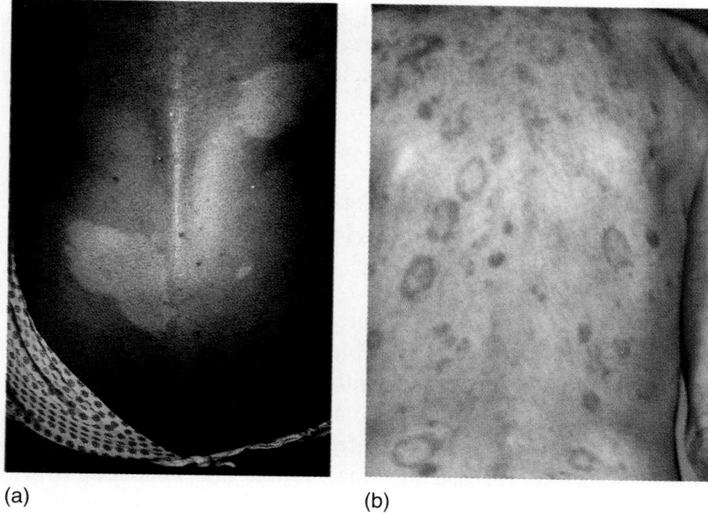

Figure 19.22

Leprosy lesions. Both views are of the tuberculoid form, with shallow, painless lesions. (*a*) Infection in dark-skinned persons manifests as hypopigmented patches or macules. (*b*) In light-skinned individuals, it appears as reddish patches or papules.

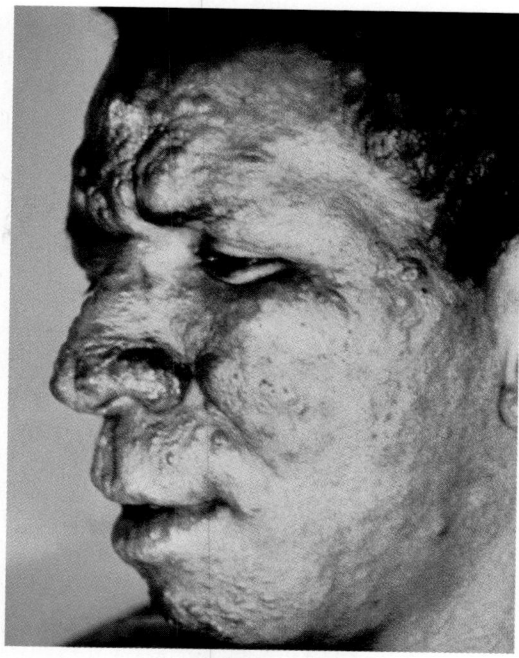

Figure 19.23

A clinical picture of lepromatous leprosy (LL). Infection of the nose, lips, chin, and brows produces moderate facial deformation, typical of lepromas.

TABLE 19.4

THE TWO MAJOR CLINICAL FORMS OF LEPROSY

Tuberculoid Leprosy	Lepromatous Leprosy
Few bacilli in lesions	Many bacilli in lesions
Few shallow skin lesions in many areas	Numerous deeper lesions concentrated in cooler areas of body
Loss of pain sensation in lesions	Sensory loss more generalized; occurs late in disease
No skin nodules	Gross skin nodules
Occasional mutilation of extremities	Mutilation of extremities common
Reactive to lepromin	Not reactive to lepromin
Lymph nodes not infiltrated by bacilli	Lymph nodes massively infiltrated by bacilli
Well-developed cell-mediated (T-cell) response	Poorly developed T-cell response

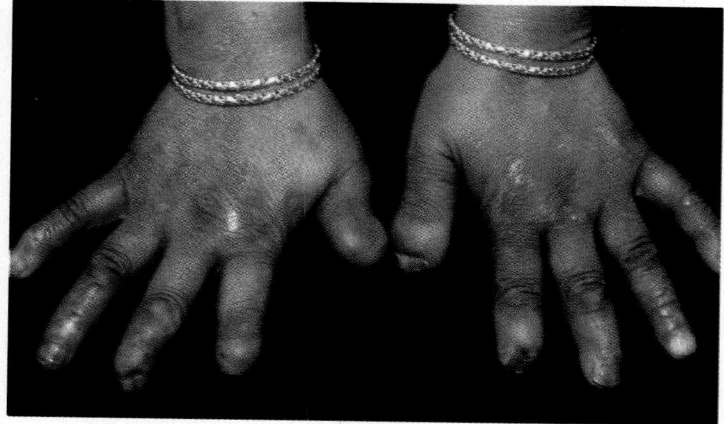

Figure 19.24

Deformation of the hands caused by borderline leprosy. The clawing and wasting are chiefly due to nerve damage that interferes with musculoskeletal activity. Individuals in a later phase of the disease can lose their fingers.

and severe complications due to widespread dissemination of the bacteria. Leprosy bacilli grow primarily in macrophages in cooler regions of the body, including the nose, ears, eyebrows, chin, and testes. As growth proceeds, the face of the afflicted person develops folds and granulomas thickenings, called **lepromas,** which are caused by massive intracellular overgrowth of *M. leprae* (figure 19.23). Advanced LL causes a loss of sensitivity that predisposes the patient to trauma and mutilation, secondary infections, blindness, and kidney or respiratory failure.

The condition of **borderline leprosy** patients can progress in either direction along the scale, depending upon their treatment and immunologic competence. The most severe effect of intermediate forms of leprosy is early damage to nerves that control the muscles of the hands and feet. The subsequent wasting of the muscles and loss of control produces drop food and claw hands (figure 19.24). Sensory nerve damage can lead to trauma and loss of fingers and toes.

Diagnosing Leprosy

Leprosy is diagnosed by a combination of symptomology, microscopic examination of lesions, and patient history. A simple yet

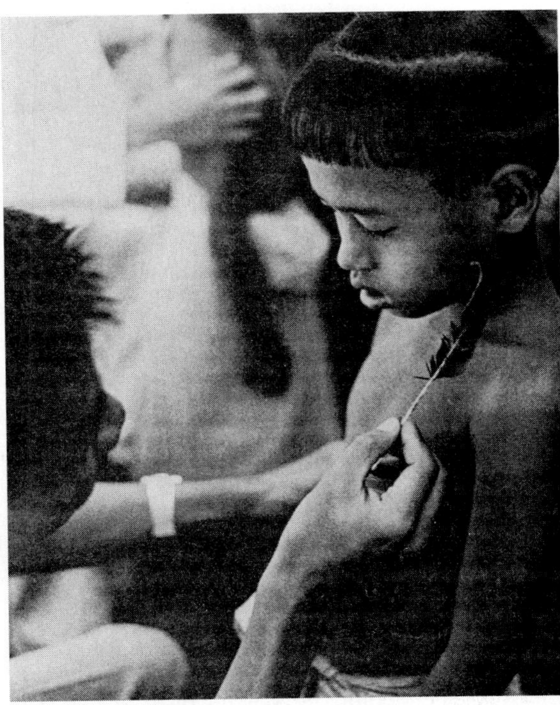

Figure 19.25
The feather test for leprosy. The Burmese child is being tickled over her face and trunk to determine loss of fine sensitivity to touch. This simple technique can provide evidence of early infection.

effective field test in endemic populations is the feather test (figure 19.25). A skin area that has lost sensation and does not itch may be an early symptom. Numbness in the hands and feet, loss of heat and cold sensitivity, muscle weakness, thickened earlobes, and chronic stuffy nose are additional evidence. Laboratory diagnosis relies upon the detection of acid-fast bacilli in smears of skin lesions, nasal discharges, and tissue samples. Knowledge of a patient's prior contacts with leprotics also supports diagnosis. Laboratory isolation of the leprosy bacillus is difficult and not ordinarily attempted.

Treatment and Prevention of Leprosy

Leprosy infection can be controlled by drugs, but therapy is most effective when started before permanent damage to nerves and other tissues has occurred. Because of an increase in resistant strains, multidrug therapy is necessary. Tuberculoid leprosy can be managed with rifampin and dapsone for 6 months. Lepromatous leprosy requires a combination of rifampin, dapsone, and clofazimine until the number of AFB in skin lesions has been substantially reduced (requiring up to 2 years). Then dapsone can be taken alone for an indeterminate period (up to 10 years or more). In the United States, many newly diagnosed leprosy patients are treated at the National Hansen's Disease Center operated by the U.S. Public Health Service in Carville, Louisiana.

Preventing leprosy requires constant surveillance of high-risk populations to discover early cases, chemoprophylaxis of healthy persons in close contact with leprotics, and isolation of leprosy patients. The WHO is currently sponsoring a trial of a vaccine containing killed leprosy bacilli that shows promise in clinical trials.

Infections by Non-Tuberculous Mycobacteria (NTM)

For many years, most mycobacteria were thought to have low pathogenicity for humans (see table 19.3). Saprobic and commensal species are isolated so frequently in soil, drinking water, swimming pools, dust, air, raw milk, and even the human body that both contact and asymtomatic infection appear to be widespread. However, the recent rise in opportunistic and nosocomial mycobacterial infections has demonstrated that many species are far from harmless.

Disseminated Mycobacterial Infection in AIDS

Bacilli of the *Mycobacterium avium* complex (MAC) frequently cause secondary infections in AIDS patients with low T-cell counts. In fact, MAC is the third most common cause of death in these patients, after pneumocystis pneumonia and cytomegalovirus infection. These common soil bacteria usually enter through the respiratory tract, multiply, and rapidly disseminate. In the absence of an effective immune counterattack, the bacilli flood the body systems, especially the blood, bone marrow, bronchi, intestine, kidney, and lifer. Treatment requires two or more combined drugs such as rifabutin, azithromycin, ethambutol, and rifampin. This therapy usually has to be given for several months or even years.

Non-tuberculous Lung Disease Pulmonary infections caused by commensal mycobacteria have symptoms like a milder form of tuberculosis, but they are not communicable. *Mycobacterium kansaii* infection is endemic to urban areas in the midwestern and southwestern United States and parts of England. It occurs most often in adult white males who already have emphysema or bronchitis. *Mycobacterium fortuitum* complex causes postsurgical skin and soft-tissue infection and pulmonary complications in immunosuppressed patients.

Skin, Lymph Node, and Wound Infections An infection by *M. marinum* has been labeled swimming pool granuloma, because it is a hazard of scraping against the rough concrete surfaces lining swimming pools. The disease starts as localized nodule, usually on the elbows, knees, toes, or fingers, which then enlarges, ulcerates, and drains (figure 19.26). The granuloma can clear up spontaneously, but it can also persist and require long-term treatment. *Mycobacterium scrofulaceum* causes an infection of the cervical lymph nodes in children living in the Great Lakes region, Canada, and Japan. The bacterium apparently infects the oral cavity and invades the lymph nodes when ingested with food or milk. In most cases, the infection is without complications, but certain children develop *scrofula,* in which the affected lymph nodes ulcerate and drain.

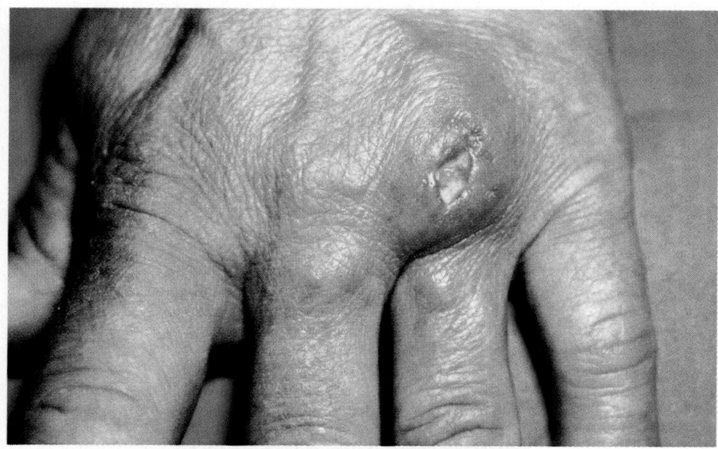

Figure 19.26
A chronic swimming pool granuloma of the hand.

Chapter Checkpoints

The genus *Mycobacterium* is distinguished by its acid-fast cell wall composed of mycolic acids and its slow but persistent rate of growth. Significant pathogens are *M. tuberculosis* and *M. leprae.*

M. tuberculosis is an ancient pathogen that annually kills millions of individuals worldwide. It infects the lungs (pulmonary tuberculosis) and other organs of the body (extrapulmonary tuberculosis). Factors in its pathogenicity include high infectivity, cord factor, ability to escape phagocytosis, ability to stimulate host-delayed hypersensitivity response, and the recent development of multiple drug resistance. Reduction of tuberculosis control programs has also contributed to its status as a major pathogen worldwide.

Tuberculosis is diagnosed by the tuberculin skin test, chest X rays, DNA probes, and direct microscopic identification in clinical specimens. Immunization with BCG vaccine offers 20% to 80% protection in areas with high incidence. Multiple drug therapy is effective in most cases.

M. leprae is the causative agent of leprosy (Hansen's disease). Its infectivity appears to be either direct contact or droplet transmission. Genetic susceptibility and living condition of the host may also be factors. There are two forms of leprosy, both of which destroy epithelial tissue and peripheral nerves. Lepromatous leprosy is the more invasive disfiguring form. Tuberculoid leprosy is the milder, superficial form. Intermediate infections also occur and combine characteristics of both forms. Treatment with combination drug therapy is effective if begun early. Research for an effective vaccine is under way but is not yet available.

Mycobacterium avium complex (MAC), an opportunist responsible for 30% of AIDS-related deaths, is just one of many mycobacterial species increasingly identified as opportunistic and nosocomial agents of infection.

ACTINOMYCETES: FILAMENTOUS BACILLI

A group of bacilli closely related to *Mycobacterium* are the pathogenic **actinomycetes.*** These are nonmotile filamentous rods that may be acid-fast. Certain members produce a mycelium-like growth and spores reminiscent of fungi (figure 19.27*b*). They also produce chronic granulomatous diseases. The main genera of actinomycetes involved in human disease are *Actinomyces* and *Nocardia* (table 19.5).

ACTINOMYCOSIS

Actinomycosis is an endogenous infection of the cervicofacial, thoracic, or abdominal regions by species of *Actinomyces* living normally in the human oral cavity, tonsils, and intestine. The cervicofacial form of disease can be a common complication of tooth extraction, poor oral hygiene, and rampant dental caries. The lungs, abdomen, and uterus are also sites of infection.

In cervicofacial involvement, the bacteria enter a damaged area of the oral mucous membrane and begin to multiply there. Diagnostic signs are swollen, tender nodules in the neck or jaw that give off a discharge containing macroscopic (1–2 mm) sulfur granules (figure 19.27). In most cases, infection remains localized, but bone invasion and systemic spread can occur in people with poor health. Thoracic actinomycosis is a necrotizing lung disorder that an project through the chest wall and ribs. Abdominal actinomycosis is a complication of burst appendices, gunshot wounds, ulcers, or intestinal damage. Uterine actinomycosis has been increasingly reported in women using intrauterine contraceptive devices. Infections are treated with surgical drainage and antibacterial antibiotics (penicillin, erythromycin, tetracycline, and sulfonamides).

Compelling evidence indicates that oral actinomyces play a strategic role in the development of plaque and dental caries. Studies of the oral environment show that *A. viscosus* and certain oral streptococci are the first microbial colonists of the tooth surface. Both groups have specific tooth-binding powers and can adhere to each other and to other species of bacteria (see chapter 21).

NOCARDIOSIS

*Nocardia** is a genus of bacilli widely distributed in the soil. Most species are not infectious, but *N. brasiliensis* is a primary pulmonary pathogen, and *N. asteroides* and *N. caviae* are opportunists. Nocardioses fall into the categories of pulmonary, cutaneous, or subcutaneous infection. Most cases in the United States are reported in patients with deficient immunity, but a few occur in normal individuals.

Pulmonary nocardiosis is a form of bacterial pneumonia with pathology and symptoms similar to tuberculosis. The lung

*actinomycetes (ak´´-tih-noh-my´-seets) Gr. *actinos,* a ray, and *myces,* fungi. Members of the Order Actinomycetales.

**Nocardia* (noh-kar´-dee-ah) After Edmund Nocard, the French veterinarian who first described the genus.

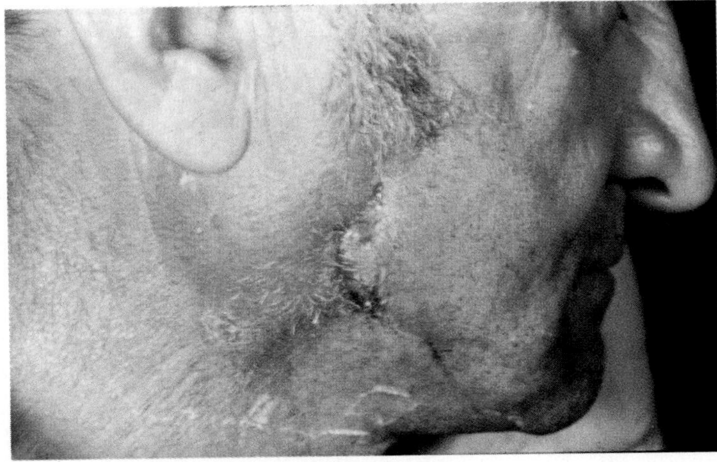

(a)

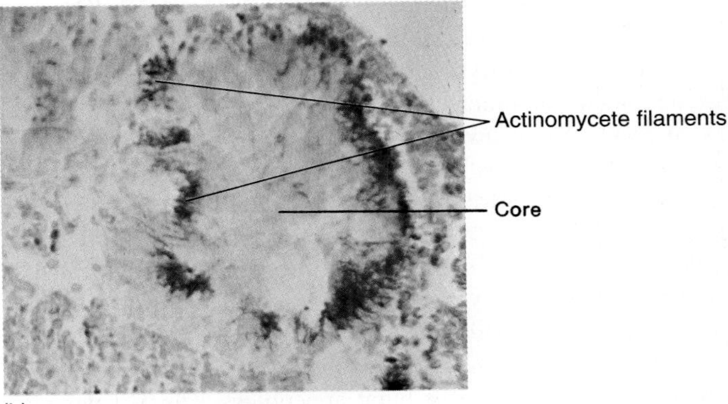

(b)

Figure 19.27
Clinical symptoms and microscopic signs of actinomycosis. (*a*) Early periodontal lesions work their way to the surface in cervicofacial disease. The lesions eventually break out and drain. (*b*) A section of crushed sulfur granule shows filamentous cells surrounding a sulfur-yellow core.

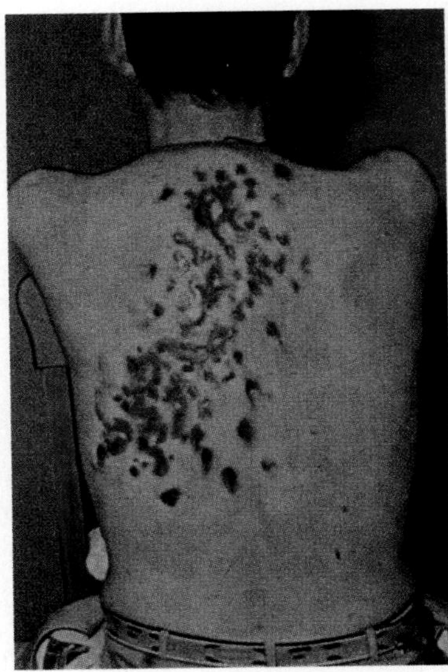

Figure 19.28
Nocardiosis, a case of pulmonary disease that has extended through the chest wall and ribs to the cutaneous surface.

TABLE 19.5

SELECTED ACTINOMYCETES, DISEASES, AND IDENTIFYING CHARACTERISTICS

	Actinomyces	*Nocardia*
Name	*A. israelii*	*N. asteroides* *N. brasiliensis*
Diseases	Actinomycosis	Nocardiosis Mycetoma
Principal characteristics	Branching filaments that fragment to rodlike cell	Filaments fragment to coccoid or elongate cells; can be acid-fast, with aerial spores
Respiration	Anaerobic, microaerophilic	Aerobic
Epidemiology	Worldwide; microbe is part of oral flora, invades through trauma	Worldwide; microbe lives in soil, water, grasses; spores inhaled
Diagnosis	Sulfur granules; short, branching gram-positive filaments	No sulfur granules; acid-fast, coccoid, and bacillary arthrospores

develops abscesses and nodules and can consolidate. Often the lesions extend to the pleura and chest wall and disseminate to the brain, kidneys, and skin (figure 19.28). *Nocardia brasiliensis* is also one of the chief etiologic agents in mycetoma, discussed in chapter 22.

Chapter Checkpoints

Actinomycetes are filamentous, gram-positive bacilli that form a branching mycelium. *Actinomyces* and *Nocardia* are the two pathogenic genera in this group.

Actinomycetes israelli is the causative agent of actinomycosis. This opportunist bacterium infects the mouth, tonsils, and intestine, resulting in nodules that discharge material containing macroscopic sulfur granules. Other actinomycetes are also implicated in dental caries.

Nocardia is a genus of soil bacteria implicated in pulmonary, cutaneous, and subcutaneous infections of immunosuppressed hosts.

Chapter Capsule with Key Terms

THE BACILLI OF MEDICAL IMPORTANCE

Gram-Positive Rods

Endosporeformers: Highly resistant bodies are produced. The aerobic genus ***Bacillus anthracis*** causes **anthrax,** a zoonosis of herbivorous animals; the bacterium inhabits the soil, where it is picked up by grazing animals that become infected and return it to the soil; humans are infected by contact with animals or their other products; the cutaneous form is manifest by **eschar,** a black ulcer at the site of entrance; the pulmonic form, acquired by inhaling spores, is highly fatal. Anthrax is controlled by antibiotics and animal vaccination. A type of food intoxication is caused by *B. cereus.*

The anaerobic genus ***Clostridium*** is common in soil and spread throughout most habitats, even the human body; wound and tissue diseases are due to introduction of the bacterium into damaged tissues; food intoxications arise when pathogenic clostridia grow in food; the basis for disease is powerful **exotoxins** that act upon specific target tissues. ***Cl. perfringens*** causes **gas gangrene,** a soft-tissue and muscle infection **(myonecrosis)** in areas of damage caused by broken limbs, surgery, diabetes, or other injuries; spores germinate in anaerobic tissues and produce gas bubbles; may invade nearby healthy tissue; toxin destroys muscle, necrotizes tissue; treatment is by careful tissue **debridement** and cleansing, hyperbaric oxygen, antibiotics. ***Cl. difficile*** causes **antibiotic-associated colitis,** an acute diarrheal disease associated with disruptions in the GI tract flora due to antimicrobic therapy. ***Cl. tetani*** is the cause of **tetanus** or **lockjaw,** a disease associated with dirty puncture wounds, burns, umbilical stumps, or needles; spores in anaerobic areas of the wound grow and produce a toxin called **tetanospasmin,** which travel to target cells in the CNS; affects muscular coordination by causing severe uncontrollable spasms of large muscles; loss of control may interfere with breathing, cause death; treatment is by antitoxin; control is through vaccination with tetanus toxoid.

Clostridial food poisoning occurs when *Cl. perfringens* is eaten with inadequately cooked meats and fish, producing toxin in the intestine; marked by severe diarrhea, vomiting, not usually fatal. **Botulism,** caused by ***Cl. botulinum,*** is associated with improperly home-canned foods; spores withstand food processing and grow in stored food; **botulin** toxin is released. Botulism is a true food intoxication; ingested toxin enters circulation and acts on myoneural junctions; blocks muscle contraction, leads to flaccid paralysis; treated by antitoxin, maintenance on respirator; control by proper canning, heating of foods prior to eating. **Infant** and **wound botulism** are unusual types of infectious botulism contracted like tetanus.

Non-spore-forming Rods: Straight, regular rods and irregular rods.

Straight, nonpleomorphic rods stain evenly. Genera that are regular in morphology include *Listeria monocytogenes* and *Erysipelothrix rhusiopathiae.* ***L. monocytogenes*** are widely distributed resistant bacteria that cause **listeriosis;** primary habitat appears to be water, soil, and the intestines; most cases are food infections associated with contaminated dairy products and meats; disease is mild, self-limited in young adults but may be severe and complicated in **fetuses,** neonates, the elderly, or the immunocompromised; treated with penicillin; prevention requires proper pasteurization of milk and dairy products. ***E. rhusiopathiae*** is carried by animals (swine) and widely dispersed into the environment; cause of **erysipeloid,** a disease common among occupations that handle animals or animal products; symptom is inflamed red sores on fingers at portal of entry.

Irregular, non-spore-forming rods include ***Corynebacterium*** and ***Propionibacterium. C. diphtheriae*** is a highly pleomorphic aerobic rod with metachromatic granules and palisades arrangement; agent of *diphtheria,* a disease whose reservoir is healthy human carriers; spread by droplets; occurs first as a localized infection of the throat with tenacious **pseudomembrane** that may cause suffocation; later symptoms are due to a classic **toxemia;** toxin spreads in blood to targets; may damage heart and nervous system; treated by antitoxin, penicillin (erythromycin); controlled by vaccination (DPT). ***Propionibacterium acnes*** is an anaerobic rod found in the skin of humans; is regularly associated with **acne vulgaris** lesions; acts on skin oils to create an inflammatory condition that erupts into pimples.

Acid-Fast Bacilli (AFB): Contain large amounts of mycolic acids and waxes; are strict aerobes; slender filamentous rods; widely distributed; tend to be resistant to environmental conditions. ***Mycobacterium tuberculosis*** causes **tuberculosis (TB),** a common lung infection; tubercle bacillus is spread by droplets in close quarters; most susceptible are persons with weakened immunities, poor nutrition, unhealthful living conditions. The disease is divided into (1) **primary pulmonary** infection, often marked by formation of granulomas in the lungs called **tubercles;** (2) **secondary TB,** a severe lung complication that occurs in people with reactivated primary disease; and (3) **disseminated extrapulmonary** infections due to blood-borne TB bacilli carried to bone, kidneys, lymph nodes, or brain; may be fatal; TB diagnosed by tuberculin testing, chest X rays, acid-fast stain for bacilli in sputum, and laboratory cultivation and identification; managed by combined drug therapy, BCG vaccine in some countries.

M. leprae causes **leprosy,** a chronic disease that begins in skin and mucous membranes and progresses into nerves; prevalent in endemic regions throughout the world; spread through direct inoculation from **leprotics;** two forms are (1) **tuberculoid,** a superficial infection without skin disfigurement; it damages nerves and causes loss of pain perception, and (2) **lepromatous,** a deeply nodular, disfiguring form that occurs primarily on the cooler regions of the body; in both forms, mutilation of parts may occur due to loss of pain receptors; treatment is by long-term combined therapy.

NTM (non-tuberculosis mycobacteria) are common environmental species that may be involved in disease. For example, *M. avium* complex (MAC) causes a disseminated disease of AIDS patients and *M. marinum* causes swimming pool granuloma.

Diseases Caused by Actinomycetes

The genera *Actinomyces* and *Nocardia* are filamentous bacteria related to mycobacteria that cause chronic infections of the skin and soft tissues.

Actinomycosis: A. israelii occurs in the human oral cavity, tonsils, intestine; oral disease or surgery may result in cervicofacial infection also abdominal, thoracic, uterine complications.

Nocardiosis: N. brasiliensis causes pulmonary disease similar to TB, with abscesses and nodules that may spread to chest wall; one agent of mycetoma.

MULTIPLE-CHOICE QUESTIONS

1. What is the usual habitat of endospore-forming bacteria that are agents of disease?
   a. the intestine of animals c. water
   b. dust and soil d. foods
2. Most *Bacillus* species are
   a. true pathogens
   b. opportunistic pathogens
   c. nonpathogens
   d. commensals
3. Many clostridial diseases require a/an ____ environment for their development.
   a. living tissue c. aerobic
   b. anaerobic d. low-pH
4. *Clostridium perfringens* causes
   a. myonecrosis
   b. food poisoning
   c. antibiotic-induced colitis
   d. both a and b
5. The action of tetanus exotoxin is on the
   a. neuromuscular junction
   b. sensory neurons
   c. spinal interneurons
   d. cerebral cortex
6. The probable habitat of *Listeria* is
   a. the human intestine c. soil and water
   b. animals d. plants
7. An infection peculiar to swine causes ____ when transmitted to humans.
   a. anthrax c. tuberculosis
   b. diphtheria d. erysipeloid
8. TB is spread by
   a. contaminated fomites
   b. food
   c. respiratory droplets
   d. vectors
9. Diagnosis of tuberculosis is by
   a. X ray c. acid-fast stain
   b. Mantoux test d. all of these
10. The form of leprosy associated with severe disfigurement of the face is
    a. tuberculoid c. borderline
    b. lepromatous d. papular
11. Soil mycobacteria can be the cause of
    a. tuberculosis
    b. leprosy
    c. swimming pool granuloma
12. Caseous lesions containing inflammatory white blood cells are
    a. lepromas c. eschars
    b. pseudomembranes d. tubercles
13. Which infectious agent is an obligate parasite?
    a. *Mycobacterium tuberculosis*
    b. *Corynebacterium diphtheriae*
    c. *Mycobacterium leprae*
    d. *Clostridium difficile*
14. Which infectious agents are facultative pathogens?
    a. *Bacillus anthracis*
    b. *Clostridium tetani*
    c. *Clostridium perfringens*
    d. all of these
15. Which infection(s) would be categorized as a zoonosis?
    a. anthrax c. diphtheria
    b. gas gangrene d. both a and b
16. Actinomycetes are ____ that produce chronic granulomatous diseases.
    a. filamentous bacteria c. molds
    b. yeasts d. cocci

CONCEPT QUESTIONS

1. a. What is the role of spores in infections?
   b. Describe the general distribution of spore-forming bacteria.
2. a. Briefly outline the epidemiology of anthrax; describe the stages in the development of cutaneous anthrax.
   b. Why is pneumonic anthrax so deadly?
   c. What other *Bacillus* species can be involved in infections or diseases?
3. a. What characteristics of *Clostridium* contribute to its pathogenicity?
   b. Compare the toxigenicity of tetanospasmin and botulin.
   c. What predisposes a patient to clostridial infection?
4. a. Outline the epidemiology of the major wound infections and food intoxications of *Clostridium.*
   b. What is the origin of the gas in gas gangrene?
   c. How does hyperbaric oxygen treatment work?
   d. Why is amputation necessary in some cases?
   e. What is debridement, and how does it prevent some clostridial infections?
5. What is the mechanism of antibiotic-associated colitis?
6. a. What causes the jaw to "lock" in lockjaw?
   b. Why do patients with no noticeable infection sometimes present with tetanus?
7. a. Compare the symptomology of botulism and tetanus.
   b. How are the two conditions alike and different?
   c. What is the difference between food, infant, and wound botulism?
8. a. Describe the epidemiology and route of infection in listeriosis.
   b. Why is listeriosis a serious problem even with refrigerated foods?
   c. Which groups are most at risk for serious complications?
9. Why is erysipeloid an occupation-associated infection?
10. a. What are the distinctive morphological traits of *Corynebacterium?*
    b. Differentiate between diphtheria infection and toxemia.
    c. How can the pseudomembrane be life-threatening?
    d. What is the ultimate origin of diphtherotoxin?
11. a. Give the unique characteristics of *Mycobacterium.*
    b. What is the epidemiology of TB?
    c. Differentiate between TB infection and TB disease.
    d. What are tubercles?
    e. What is the course of disseminated disease?
    f. Outline the principles of tuberculin testing, chest X rays, and acid-fast staining.
    g. Why does tuberculosis require combined therapy?

12. a. What characteristics make *M. leprae* different from other mycobacteria?
    b. Differentiate between tuberculoid and lepromatous leprosy.
    c. What causes the deformations?
    d. What causes the mutilations of extremities?
13. a. What is the importance of NTM?
    b. Describe the effects of *Mycobacterium-avium* complex in AIDS patients.
14. a. Describe the bacteria in the actinomycete group, and explain why they are similar to fungi.
    b. Briefly describe two of the common diseases caused by this group.
15. Multiple matching: Choose all descriptions that fit.

____ *Bacillus*	1. is acid fast
____ *Clostridium*	2. cells irregular, pleomorphic
____ *Mycobacterium*	3. can form metachroma tic granules
____ *Corynebacterium*	4. regular-shaped rods
____ *Listeria*	5. forms endospores
	6. is psychrophilic
	7. is primarily anaerobic
	8. is aerobic
	9. is associated with dairy products
	10. cells can be elongate filaments
	11. shows palisades arrangement

16. Match the disease with the principal portal of entry.

____ 1. anthrax	a. skin
____ 2. botulism	b. gastrointestinal tract
____ 3. gas gangrene	c. traumatized tissue
____ 4. antibiotic colitis	d. respiratory tract
____ 5. tetanus	e. urogenital tract
____ 6. diphtheria	f. placenta
____ 7. listeriosis	g. none of these
____ 8. tuberculosis	
____ 9. leprosy	

17. Using the same list of diseases as in question 16, match with symptoms and signs from the following list.
    a. loss of cutaneous sensation ____
    b. double vision ____
    c. necrosis of muscle tissue ____
    d. fever ____
    e. black, cutaneous ulcer ____
    f. severe diarrhea ____
    g. difficulty in breathing ____
    h. heart failure ____
    i. flaccid paralysis of muscles ____
    j. spastic paralysis of muscles ____
    k. severe coughing ____
    l. pseudomembrane ____
    m. skin nodules ____

CRITICAL-THINKING QUESTIONS

1. a. What is the main clinical strategy in preventing gas gangrene?
   b. Why does it work?
2. a. Why is it unlikely that diseases such as tetanus and botulism will ever be completely eradicated?
   b. Name some bacterial diseases in this chapter that could be completely eradicated and explain how.
3. Why is the cause of death similar in tetanus and botulism?
4. a. Why does botulin not affect the senses?
   b. Why does botulism not commonly cause intestinal symptoms?
5. Account for the fact that boiling does not destroy botulism spores but does inactivate botulin.
6. Adequate cooking is the usual way to prevent food poisoning. Why doesn't it work for *perfringens* and *Bacillus* food poisoning?
7. a. Why do patients who survive tetanus and botulism often have no sequelae?
   b. How has modern medicine improved the survival rate for these two diseases?
8. What would be the likely consequence of diphtheria infection alone without toxemia?
9. How can one tell that acne involves an infection?
10. a. Do you think the spittoons of the last century were effective in controlling tuberculosis?
    b. Why or why not?
11. a. What could it mean to say that TB and leprosy are "family diseases"?
    b. What, if anything, can be done about multiple-drug-resistant tuberculosis?
12. Which diseases in this chapter have no real portal of exit?
13. Case history 1: A farm worker sustained a crushing injury to his hand and was taken to a hospital, where he received treatment and tetanus toxoid. During successive weeks, he had several operations to repair the damaged hand and was given antibiotics. After a time, he lost muscle tone and had difficulty talking. Finally, his hand required amputation, and he was given antitoxin.
    a. What do you think the disease was?
    b. Why did the earlier treatments not work?
    c. To what other disease is this similar?
14. Case history 2: 86 people at a St. Patrick's day celebration developed diarrhea, cramps, and vomiting after eating a traditional dinner of corned beef, cabbage, potatoes, and ice cream. Symptoms appeared within 10 hours of eating, on average. Of those afflicted, 85 had eaten corned beef, one had not. The corned beef had been cooked in an oven the day before the event, stored in the refrigerator, and sliced and placed under heat lamps 1½ hours before serving.
    a. Use this information to diagnose the food poisoning agent.
    b. What sorts of lab tests could one do to verify it as this bacterium?
    c. What would you have done to prevent it?
15. Case history 3: An outbreak of gastrointestinal illness was reported in a daycare center. Of 67 people, 14 came down with symptoms of nausea, cramps, and diarrhea within 2–3 hours of eating.

Food items included fried rice, peas, and apple rings. Symptoms occurred in 14 of the 48 who had eaten the fried rice and in none of those who had not eaten it. The rice had been cooked the night before and refrigerated. The next morning it had been fried with leftover chicken and held without refrigeration until lunchtime.

a. On the basis of symptoms and food types, what do you expect the etiologic agent to be?
b. What would you expect to culture from the food?
c. To what other disease is the one similar?
d. How can it be differentiated from the agent and disease in case history 2?

INTERNET SEARCH TOPIC

1. Use a search engine to locate Web sites that cover biological germ warfare. Look for information on how weapons based on microbes are developed and regulated.
2. Search under tuberculin testing to find information on medical guidelines and interpretations for various test outcomes.

chapter 20

The Gram-Negative Bacilli of Medical Importance

The gram-negative bacilli are a large group of non-spore-forming bacteria adapted to a wide range of habitats and niches. The genera in this category are highly diverse in metabolism and pathogenicity. Because the group is so large and complex, and many of its members are not medically important, we have included only those that are human pathogens.

A large number of representative genera are inhabitants of the large intestine (enteric); some are zoonotic; some are adapted to the human respiratory tract; and still others live in soil and water. It is useful to distinguish between the frank (true) pathogens (*Salmonella, Yersinia pestis, Bordetella,* and *Brucella,* for example), which are infectious to the general population, and the opportunists (*Pseudomonas* and coliforms), which are resident flora that cause infection in people with weakened host defenses.

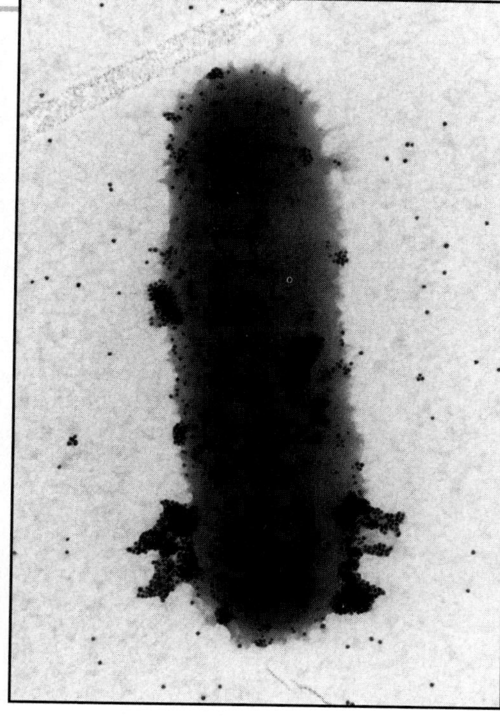

A gram-negative bacterium undergoing lysis following exposure to an antibiotic releases spherical fragments of lipopolysaccharide. This component of the cell wall acts as an endotoxin that seriously disrupts several systems of the body.

TABLE 20.1

SURVEY OF GRAM-NEGATIVE PATHOGENS

Oxygen Requirements	Genus	Species	Disease/Commentary
Aerobes			
	Pseudomonas	*aeruginosa, Fluorescens*	Opportunistic pathogens; invade surgical wounds, burns, lungs
	Brucella	*abortus, melitensis*	Brucellosis (undulant fever)
	Francisella	*tularensis*	Tularemia (rabbit fever)
	Bordetella	*pertussis*	Pertussis (whooping cough)
	Legionella	*pneumophila*	Legionellosis; spread by environmental aerosols
Facultative Anaerobes			
The Family Enterobacteriaceae (Oxidase-Negative)			
	Escherichia	*coli*	Numerous strains; pathogens cause diarrhea; opportunists cause urinary tract infections
	Edwardsiella	*tarda*	Gastroenteritis; opportunistic infections of the lungs, surgical incisions, urinary tract
	Citrobacter		Pathogenic to the immunocompromised, causing opportunistic or secondary infection
	Klebsiella		
	Enterobacter		
	Hafnia		
	Serratia		
	Proteus		
	Providencia		
	Morganella		
	Salmonella		Salmonellosis
		typhi	Typhoid fever
		cholerae-suis	Septicemia, enteric fever
		enteritidis	Gastroenteritis
	Shigella	*dysenteriae*	Bacillary dysentery (shigellosis)
		flexneri	Bacillary dysentery
		boydii	Bacillary dysentery
		sonnei	Bacillary dysentery
	Yersinia	*pestis*	Plague
		pseudotuberculosis	Gastroenteritis; opportunistic
		enterocolitica	Gastroenteritis; opportunistic
The Family Pasteurellaceae (Oxidase-Positive)			
	Pasteurella	*multocida*	Skin abscess; septicemia
	Haemophilus	*influenzae*	Meningitis, epiglottitis
Genera Not Associated with a Family			
	Gardnerella	*vaginalis*	Nonspecific vaginitis
	Eikenella	*corrodens*	Gingival infections
	Streptobacillus	*moniliformis*	Rat bite fever
	Calymmatobacterium	*granulomatis*	Donovanosis or granuloma inguinale
Obligate Anaerobes			
	Bacteroides	*fragilis*	Dominant species in intestine; cause of abdominal infections
		melaninogenicus	Oral soft tissue infections

The gram-negative rods can be organized in several ways. To simplify coverage here, they are grouped into three major categories based on oxygen requirements (table 20.1)

One notable component of all gram-negative rods is a complex cell wall with an outer membrane containing a lipopolysaccharide with properties of an **endotoxin** (see chapter 13). Because endotoxin in the blood can have severe and far-reaching pathophysiologic effects, gram-negative septicemia, a common *iatrogenic** or nosocomial infection, is a cause for great concern (microfile 20.1).

***iatrogenic** (eye-at″-troh-jen′-ik) Gr. *iatros,* doctor. An infection occurring as a result of medical treatment.

AEROBIC GRAM-NEGATIVE NONENTERIC BACILLI

The aerobic gram-negative bacteria of medical importance are a loose assortment of genera including *Pseudomonas,* an opportunistic pathogen; the zoonotic pathogens *Brucella* and *Francisella;* and *Bordetella* and *Legionella,* which are mainly human pathogens.

MICROFILE 20.1 GRAM-NEGATIVE SEPSIS AND ENDOTOXIC SHOCK

One of the most serious consequences of infection with gram-negative bacteria, even if the species involved is not a frank or virulent pathogen, is blood infection. Nationwide, over 100,000 patients a year die from a condition called **septic shock.** Shock is a pathologic state of low blood pressure accompanied by a reduced amount of blood circulating to vital organs, particularly the brain, heart, lungs, and kidneys. Its major symptoms and signs are nausea, tachycardia, cold, clammy skin, and weak pulse. Damage to the organs can elicit respiratory failure, coma, heart failure, and death in a few hours. Although the endotoxins of all gram-negative bacteria can cause shock, the most common clinical cases are due to gram-negative enteric rods.

The adverse factor in gram-negative sepsis is the presence of an endotoxin called lipopolysaccharide (LPS) in the outer membrane of the gram-negative cell wall. Lipopolysaccharide is a potent immune stimulator. The component that accounts for most of the adverse effects is *lipid A* embedded in the external layer of the membrane. This lipid is active only after it is liberated by bacteria that are growing or lysed by host defenses and other factors. In general, lipid A triggers the secretion of interleukins, tumor necrosis factor, and other cytokines by macrophages. In a local infection with only small amounts of endotoxin, the effects of the cytokines are protective. But in massive sepsis, legions of macrophages release profuse amounts into the bloodstream. Some of the cytokines are pyrogenic and induce fever, some activate the complement cascade, and others promote intravascular blood coagulation. The end results are circulation failure, tissue damage, hypotension, and shock.

Ironically, traditional treatments with antibiotics can compound the problem of septic effects because these drugs can actually lyse bacteria (see chapter opening figure). Drug companies are actively developing new drugs that will block the effects of endotoxin or the cytokines. Several experimental drugs contain monoclonal antibodies to inactivate lipid A; one is based on antibodies against the tumor necrosis factor, and another inactivates the interleukin-1 receptor on macrophages. Although none of these drugs is available for therapy yet, these monoclonal products continue to be the subject of intense research and clinical trials.

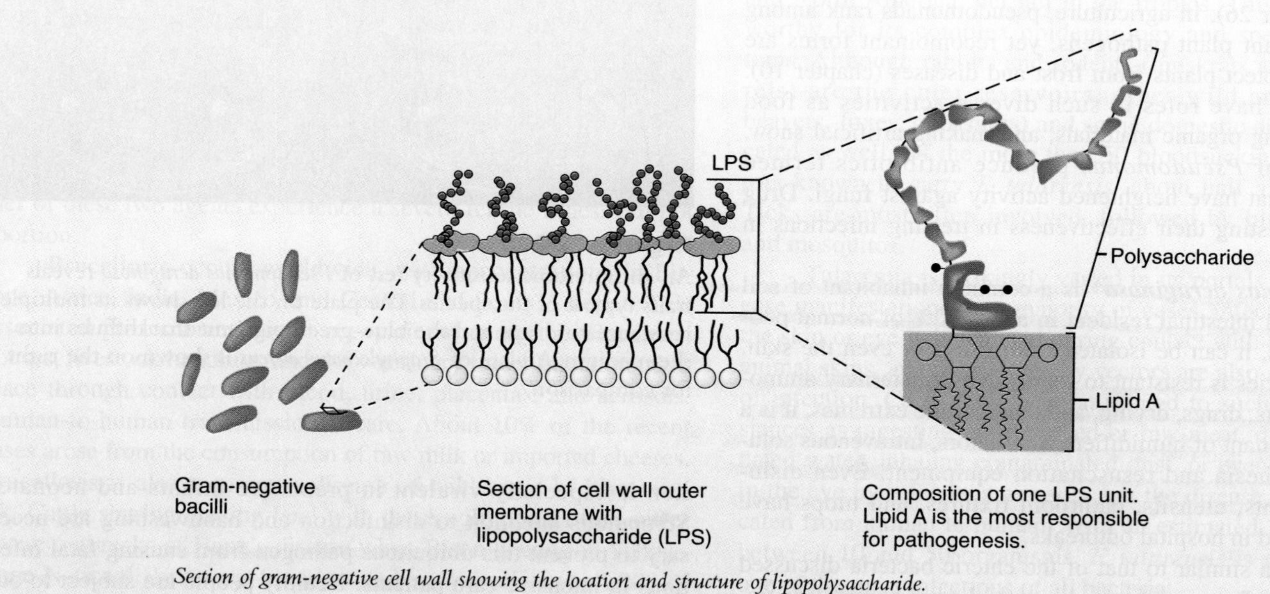

Section of gram-negative cell wall showing the location and structure of lipopolysaccharide.

PSEUDOMONAS: *THE PSEUDOMONADS*

The pseudomonads are a large group of free-living bacteria that live primarily in soil, seawater, and fresh water. They also colonize plants and animals and are frequent contaminants in homes and clinical settings. These small, gram-negative rods have a single polar flagellum (figure 20.1), produce oxidase and catalase, and do not ferment carbohydrates. Although they ordinarily obtain energy aerobically through oxidative metabolism, some species can grow anaerobically if provided with a salt such as nitrate. Many species produce green, brown, red, or yellow pigments that diffuse into the medium and change its color.

*Pseudomonas** species are highly versatile. They can adapt to a wide range of habitats and manage to extract needed energy from minuscule amounts of dissolved nutrients. Most members can grow in a medium containing minerals and one simple organic compound. This adaptability accounts for their constant presence in the environment and their capacity to thrive on hosts. These bacteria are also metabolically versatile. They degrade numerous extracellular substances by means of protease, amylase, pectinase, cellulase, and numerous other enzymes.

**Pseudomonas* (soo″-doh-moh′-nas) Gr. *pseudes,* false, and *monas,* a unit.

Bordetella pertussis and Whooping Cough

Bordetella pertussis* is a minute, encapsulated coccobacillus responsible for **pertussis,** or **whooping cough,** a communicable childhood affliction that causes an acute respiratory syndrome. Contrary to the common impression that the disease is mild and self-limiting, it often causes severe, life-threatening complications in babies. *Bordetella parapertussis* is a closely related species that causes a milder form of the infection.

The primary source of infection is usually other children afflicted with clinical disease. The reservoir appears to be in apparently healthy carriers. Transmission is by direct contact with droplets or inhalation of infectious aerosols. About half of pertussis cases occur in children between birth and 4 years of age. A recent study found evidence that whooping cough is relatively common among older children and adults. It is generally a milder disease, presenting as a chronic cough, and it is often misdiagnosed as a cold or the flu.

Although the introduction of an effective vaccine has decreased the prevalence of pertussis in many countries, it is far from an obsolete disease. In developing countries that cannot routinely vaccinate and in industrialized countries with lax vaccination, pertussis is still a major cause of childhood sickness and death. Even in the United States, the case rate has gradually increased between 1981 and 1997 to a total as high as 1968 (see figure 13.16). Epidemiologists attribute this upsurge to fewer vaccinations amidst concern over its publicized side effects.

The primary virulence factors of *B. pertussis* are (1) receptors that specifically recognize and bind to ciliated respiratory epithelial cells and (2) toxins that destroy and dislodge ciliated cells (a primary host defense). The loss of the ciliary mechanism leads to build-up of mucus and blockage of the airways. The initial phase of pertussis is the *catarrhal stage,* marked by nasal drainage and congestion, sneezing, and occasional coughing. As symptoms deteriorate into the *paroxysmal stage,* the child experiences recurrent, persistent coughing. After fits of 10 to 20 abrupt, hacking coughs, the need for oxygen stimulates a deep inspiration that draws air swiftly through the narrowed larynx and gives off a "whoop." Many of the complications of pertussis are due to compromised respiration.

The standard control measure is early vaccination with pertussis vaccine, usually combined with the diphtheria and tetanus vaccines (DPT). The CDC has recently revised its recommendations for the type of vaccine to be used. An acellular vaccine (aP) is available for all routine immunizations, starting at 6 weeks of age. This vaccine contains toxoid and is well tolerated when given with DT vaccine. The vaccine does not give long-term immunity, which is why adults and older children can have recurrences. Other measures to protect susceptible people during epidemics include erythromycin therapy and isolation of cases.

**Bordetella pertussis* (bor-duh-tel'-uh pur-tus'-is) After Jules Bordet, a discoverer of the agent, and L. *per,* severe, and *tussis,* cough.

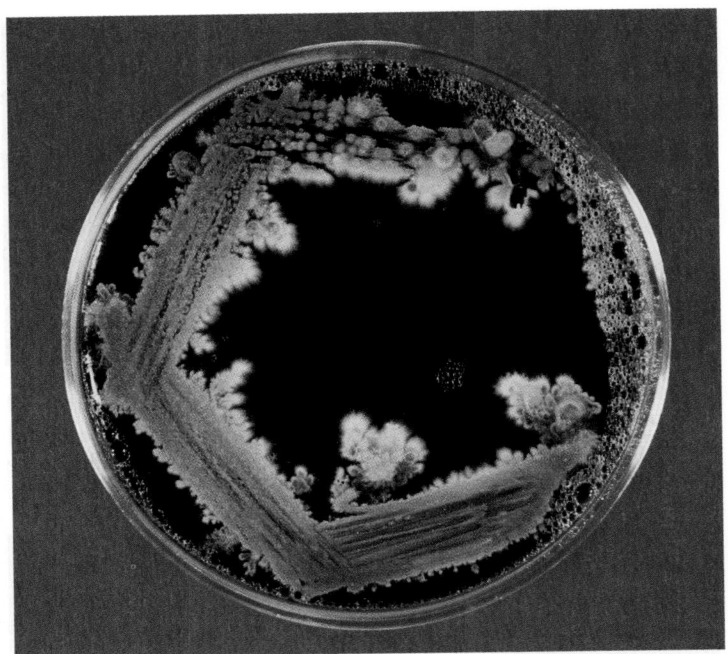

Figure 20.5
Legionella pneumophila colonies on selective charcoal yeast extract medium.

Legionella and Legionellosis

Legionella is a novel bacterium unrelated to other strictly aerobic gram-negative genera. Although the organisms were originally described in the late 1940s, they were not clearly associated with human disease until 1976. The incident that brought them to the attention of medical microbiologists was an explosive and mysterious epidemic of pneumonia that afflicted 200 American Legion members attending a convention in Philadelphia and killed 29 of them. After 6 months of painstaking analysis, epidemiologists isolated the pathogen and traced its source to contaminated air-conditioning vents in the Legionnaires' hotel. The news media latched on to the name **Legionnaires' disease,** and the experts named their new isolate *Legionella* (lee″-jun-ell′-uh).

Legionellas are weakly gram-negative motile rods that range in morphology from cocci to filaments. They have fastidious nutrient requirements and can be cultivated only in special media (figure 20.5) and cell cultures. Several species or subtypes have been characterized, but *L. pneumophila* (lung-loving) is the one most frequently isolated in infections.

Legionella's ability to survive and persist in natural habitats has been something of a mystery, yet it appears to be widely distributed in aqueous habitats as diverse as tap water, cooling towers, spas, ponds, and other fresh waters. Recently, researchers determined that the bacteria live in close associations with free-living amebas (figure 20.6). It is released during aerosol formation and can be carried for long distances. Cases have been traced to supermarket sprayers and the fallout from the Mount St. Helens volcano.

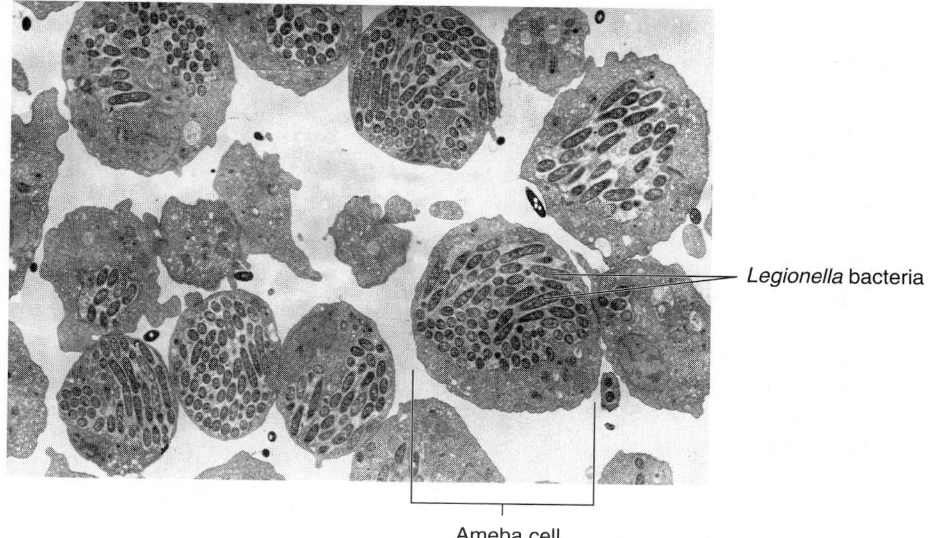

Figure 20.6
Legionella living intracellularly in the ameba *Hartmanella*. Amebas inhabiting natural waters appear to be the reservoir for this pathogen and a means for it to survive in rather hostile environments. The pathogenesis of *Legionella* in humans is likewise dependent on its uptake by and survival in phagocytes.

Studies since 1976 have determined that Legionnaires' disease is not a new disease and that earlier outbreaks occurred but went undiagnosed at the time. Most community cases occur sporadically in conjunction with accidental exposure to a common environmental source of the agent. The disease is now known to be worldwide and is prevalent in males over 50 years of age. Nosocomial infections occur most often in elderly patients hospitalized with diabetes, malignant disease, transplants, alcoholism, and lung disease. However, the disease is not communicable from person to person.

Two major clinical forms of the disease are *Legionnaires' pneumonia* and *Pontiac fever.* The symptoms of both are a rising fever (up to 41°C), cough, diarrhea, and abdominal pain. Legionnaires' pneumonia is the more severe disease, progressing to lung consolidation and impaired respiration and organ function. It has a

Chapter Checkpoints

Medically significant gram-negative bacteria are divided into three groups based on their oxygen requirements. Pathogenic species are aerobic or facultative anaerobes. Opportunistic pathogens occur in all three groups. One common virulence factor in all gram-negative pathogens is endotoxin, the lipopolysaccharide (LPS) component of the cell wall. Endotoxin stimulates excessive release of host defense cytokines, which, in turn, result in high fever, tissue damage, hypotension, and shock.

Medically important **gram-negative aerobic bacilli** include the opportunistic pathogen *Pseudomonas,* the zoonotic pathogens *Brucella* and *Francisella,* as well as the human pathogens *Bordetella* and *Legionella.*

Pseudomonads normally inhabit soil and water. They are metabolically quite versatile and are able to degrade many substances and survive under minimal conditions.

Pseudomonas aeruginosa is a common nosocomial opportunist, infecting human hosts that are either debilitated or have had invasive medical procedures. Infections occur primarily in patients with severe burns, neoplastic disease, or cystic fibrosis. Healthy individuals acquire skin infections from swimming pools, hot tubs, or contaminated sponges or contact lens solutions.

Brucella infections occur primarily through exposure to infected cattle or pigs but also through drinking unpasteurized milk. Brucellosis is a systemic infection characterized by alternating periods of fever, sweating, and chills. The infection is carried by neutrophils to many body organs. Combination drug therapy is effective. Prevention includes animal vaccination and quarantine.

Francisella tularensis is the causative agent of tularemia, an arthropod-borne zoonosis infecting rabbits, rodents, and other mammals. Infection occurs through skin, eye, gastrointestinal tract, and respiratory tract. Tularemia is highly transmissible through infected animals, contaminated water, droplets, dust, or through the bites of infected arthropods. Gentamicin is effective in most cases. Immunization provides protection for occupational risk groups.

Bordetella pertussis is the causative agent of whooping cough, or pertussis, a potentially fatal respiratory infection associated with uncontrollable coughing and transmitted through droplets. Childhood immunization is by far the best protection.

Legionella is the causative agent of legionellosis (Legionnaires' disease) and the milder Pontiac fever. Both are noncommunicable lung infections transmitted by aerosols from air conditioners, cooling towers, and other aquatic habitats. *Legionella* exists in aquatic habitats, possibly in symbiosis with certain amebas. It is a potentially fatal disease if untreated, but it responds to antibiotic therapy.

MICROFILE 20.2 DIARRHEAL DISEASE

Diarrhea is an acute syndrome of the intestinal tract in which the volume, fluid content, and frequency of bowel movements increase. It is usually a symptom of **gastroenteritis** (inflammation of the lining of the stomach and intestine) and can be accompanied by severe abdominal pain. Diarrhea is generated by several pathologic states—most commonly, infection, intestinal disorders, and food poisoning. Although the human large intestine ordinarily harbors a huge microbial population, most bacterial, protozoan, and viral agents of diarrhea are not members of this normal gut flora but are acquired through contaminated food or water.

Infectious diarrhea has two basic mechanisms. In the toxigenic type of disease, bacteria release **enterotoxins** that bind surface receptors of the small intestine. These toxins disrupt the physiology of epithelial cells and cause increased secretion of electrolytes and water loss, a condition called **secretory diarrhea.** The organism itself does not invade the tissues. Secretory diarrhea is characterized by its large volume, and there is little blood in the stool. This is the mechanism of cholera (see chapter 21) and some types of *Escherichia coli* and *Shigella* infectious damage.

In a more invasive diarrheal disease, the microbe invades the wall of the small or large intestine and disrupts its architecture, leading to gross injury. This form is attended by smaller fecal volume, pain in the rectum, blood in the stool, and ulceration of the inner mucosal lining. *Salmonella,* other strains of *Shigella* and *E. coli, Campylobacter,* and *Entamoeba histolytica* are responsible for this type of intestinal disease. Regardless of the cause, the loss of fluid that accompanies diarrhea results in severe dehydration and sometimes death. It accounts for 43% of the world's infectious diseases and 18% of deaths, worldwide. Infants are especially vulnerable to diarrheal illness because of their smaller fluid reserves and inadequate immunities.

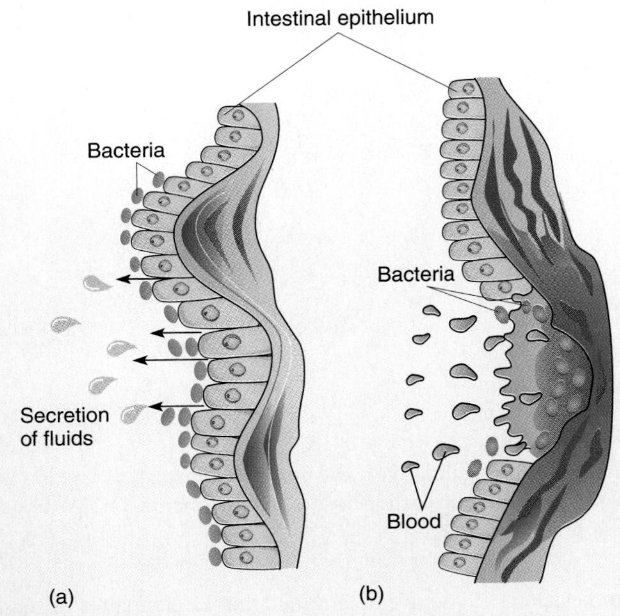

Mechanisms of infectious diarrhea. (a) In a toxigenic infection, the microbe remains on the surface of epithelial cells and secretes toxin into the cells. (b) In an invasive infection, the microbe breaks down epithelial cells and forms ulcerations, loss of the intestinal lining, and bleeding.

See microfile 21.3 to review a successful support treatment that offsets some of the complications of diarrhea.

fatality rate of 3–30%. In contrast, Pontiac fever does not lead to pneumonia and rarely causes death. Legionellosis is diagnosed by symptomology and the patient's history. Laboratory analyses include fluorescent antibody staining of specimens, cultivation on charcoal yeast extract (CYE) agar, and DNA probes. The disease is treated with erythromycin alone or in combination with rifampin. Alternative drugs include azithromycin or quinolones. *Legionella*'s wide dispersal in aquatic habitats makes control difficult, though chlorination and regular cleaning of artificial habitats are of some benefit.

IDENTIFICATION AND DIFFERENTIAL CHARACTERISTICS OF THE ENTEROBACTERIACEAE

One large family of gram-negative bacteria that exhibits a considerable degree of relatedness is the Enterobacteriaceae. Although many members of this group inhabit soil, water, and decaying matter, they are also common occupants of the large bowel of humans and animals. All members are small (the average is 1 μm by 2–3 μm) non-spore-forming rods. The bacteria in this group grow best in the presence of air, but they are facultative and can ferment carbohydrates by an anaerobic pathway. This group is probably the most common one isolated in clinical specimens—both as normal flora and as agents of disease

Enteric pathogens are the most frequent cause of **diarrheal illnesses** (microfile 20.2), which account for an annual mortality rate of 3 million people and an estimated 4 billion infections worldwide. Counting the total number of cases that receive medical attention, along with estimates of unreported, subclinical cases, enteric illness is probably responsible for more morbidity than any other disease. Prominent pathogenic enterics include *Salmonella, Shigella,* and strains of *Escherichia,* many of which are harbored by human carriers.

Enterics (along with *Pseudomonas* species) also account for more than 50% of all isolates in nosocomial infections (figure 20.7). Their widespread involvement in hospital-acquired infections can be attributed to their constant presence in the hospital environment and their survival capabilities. Enterics are commonly isolated from soap dishes, sinks, and invasive devices such as fiberoptic probes, tracheal cannulas, and indwelling catheters. The most important enteric opportunists are *E. coli, Klebsiella, Proteus, Enterobacter, Serratia,* and *Citrobacter.* Interestingly, the

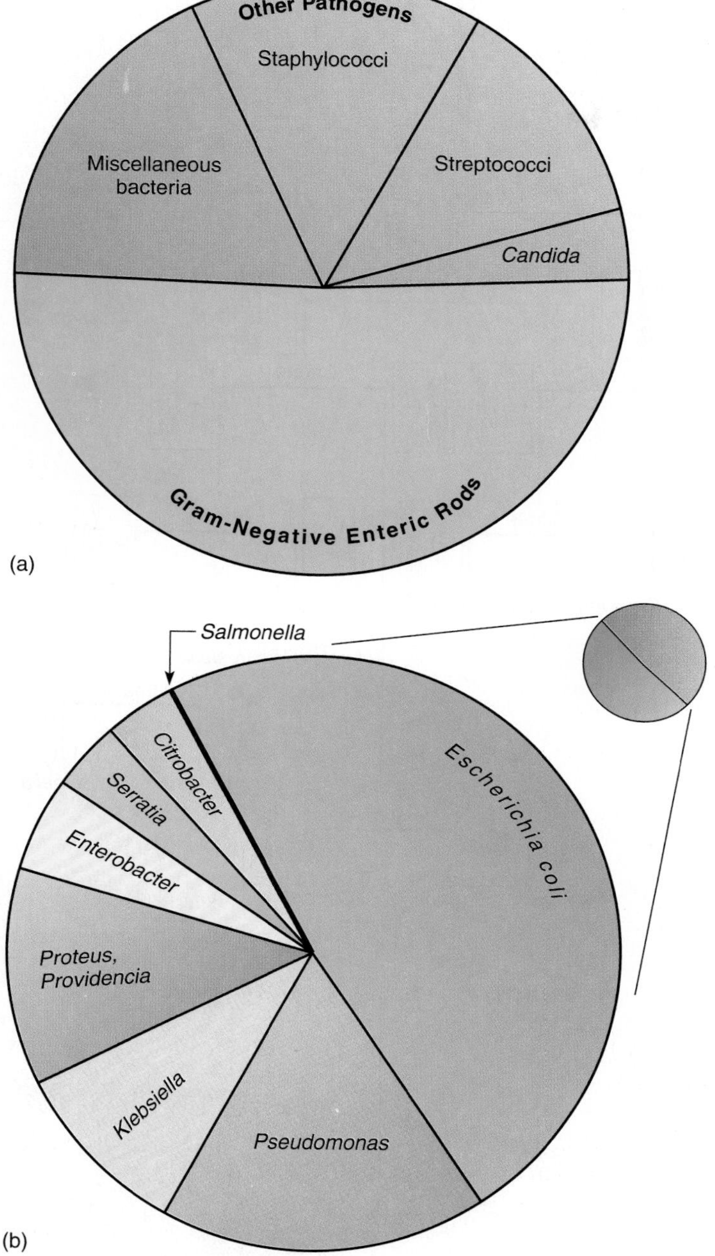

Figure 20.7

Bacteria that account for the majority of nosocomial infections. (*a*) The overall contribution of gram-negative rods versus other agents in hospital infections. (*b*) Comparison of the relative proportion of each genus or species isolated from infections by the gram-negative enteric group.

diseases caused by these agents usually involve systems other than the gastrointestinal tract, such as the lungs and the urinary tract.

The genera in this family share several key properties: They ferment glucose, reduce nitrates to nitrites, and are oxidase-negative. The family is traditionally divided into two subcategories. The **coliforms** include *Escherichia coli* and other gram-negative normal enteric flora that ferment lactose rapidly (within 48 hours). **Noncoliforms** are generally non-lactose-fermenting or slow lactose-fermenting bacteria that are either normal flora or regular pathogens. Although minor exceptions occur, this distinction has considerable practical merit in laboratory, clinical, and epidemiologic terms. The following outline provides a framework for organizing the discussion of the Enterobacteriaceae:

Coliforms: Rapid lactose-fermenting enteric bacteria that are normal flora and opportunistic (some strains of *E. coli* are true pathogens). Examples include
- *Escherichia coli*
- *Klebsiella*
- *Enterobacter*
- *Hafnia*
- *Serratia*
- *Citrobacter*

Noncoliforms: Lactose-negative; may or may not be normal flora
- Examples of opportunistic, normal gut flora
  - *Proteus*
  - *Morganella*
  - *Providencia*
  - *Edwardsiella*
- Examples of pathogenic enterics
  - *Salmonella typhi, S. cholerae-suis, S. enteritidis*
  - *Shigella dysenteriae, Sh. flexneri, Sh. boydii, Sh. sonnei*
  - *Yersinia enterocolitica, Y. pseudotuberculosis*
- Examples of pathogenic nonenterics
  - *Yersinia pestis*

Morphology and staining characteristics are insufficient in themselves to identify the enteric bacilli. Characteristics they share include being non-spore-forming, straight rods that are often motile (exceptions are *Shigella* and *Klebsiella*), catalase-positive, and oxidase-negative. A battery of biochemical tests is commonly used to identify the principal genera and species. Although the details of this process are beyond the scope of this text, a summary will overview the major steps (figure 20.8). Because fecal specimens contain such a high number of normal flora, they can be inoculated first into enrichment media (selenite or GN—gram-negative—broth) that inhibit the normal flora and favor the growth of pathogens. Enrichment and nonfecal specimens are streaked onto several plates of selective, differential media such as MacConkey, EMB, or Hektoen enteric agar and incubated for 24–48 hours. Colonies developing on these media will provide the initial separation into lactose-fermenters and non-lactose-fermenters (figure 20.9; see figure 3.11).

After the reactions of isolated colonies are noted, single colonies are subcultured on triple sugar iron (TSI) agar slants for further analysis. As is evident in figure 20.8, several other tests are required to identify to the level of genus (table 20.2). One series of reactions, the **IMViC** (indole, methyl red, Voges-Proskauer, *i* for pronunciation, and citrate), is a traditional panel that can be used to differentiate among several genera (table 20.3). Commercial identification systems such as the Enterotube (figure 20.10) provide a rapid and compact means of acquiring these and additional data required for species identification. In general, whether an isolate is clinically significant depends upon the specimen's origin.

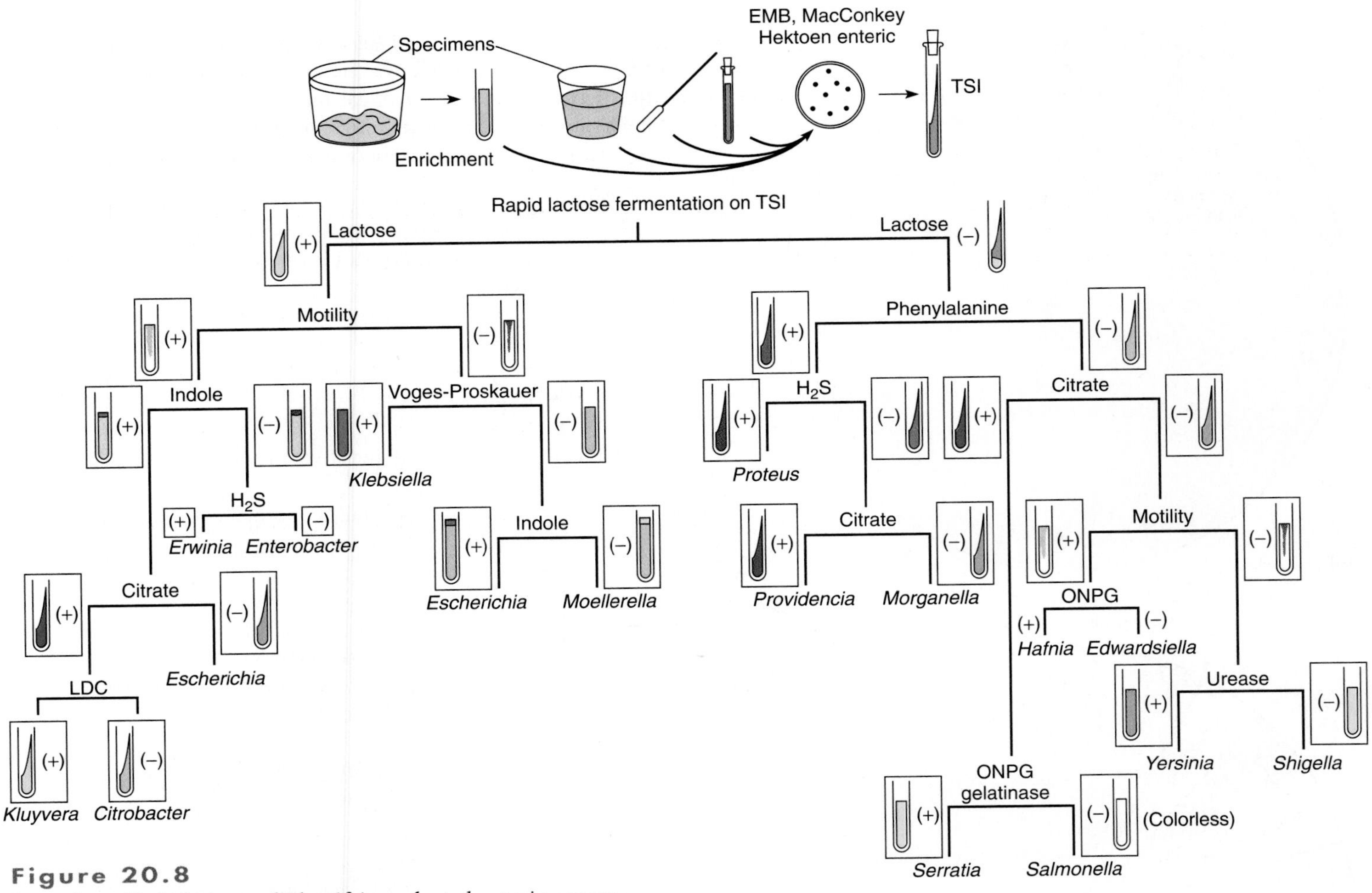

Specimens
Enrichment
EMB, MacConkey
Hektoen enteric
TSI
Rapid lactose fermentation on TSI
(+) Lactose
Lactose (–)
Motility
(+)
(–)
Indole
(+)
(–)
Voges-Proskauer
(+)
(–)
Klebsiella
H2S
(+) (–)
Erwinia Enterobacter
Indole
(+)
(–)
Escherichia
Moellerella
Citrate
(+)
(–)
Escherichia
LDC
(+)
(–)
Kluyvera Citrobacter
Phenylalanine
(+)
(–)
H2S
(+)
(–)
Proteus
Citrate
(+)
(–)
Providencia
Morganella
Citrate
(+)
(–)
Motility
(+)
(–)
ONPG
(+) (–)
Hafnia Edwardsiella
Urease
(+)
(–)
Yersinia
Shigella
ONPG
gelatinase
(+)
(–)
(Colorless)
Serratia
Salmonella

Figure 20.8
Procedures for isolating and identifying selected enteric genera.

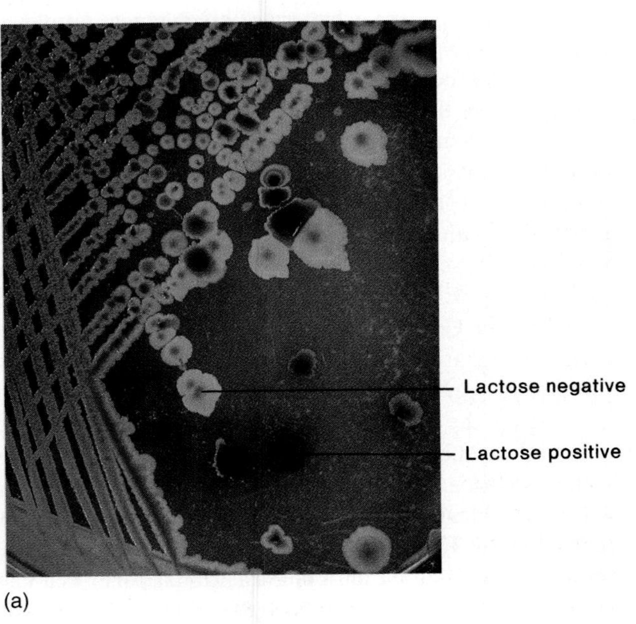

Lactose negative
Lactose positive

(a)

Lactose (+)
Lactose (–)
H2S (+) Lactose (–)

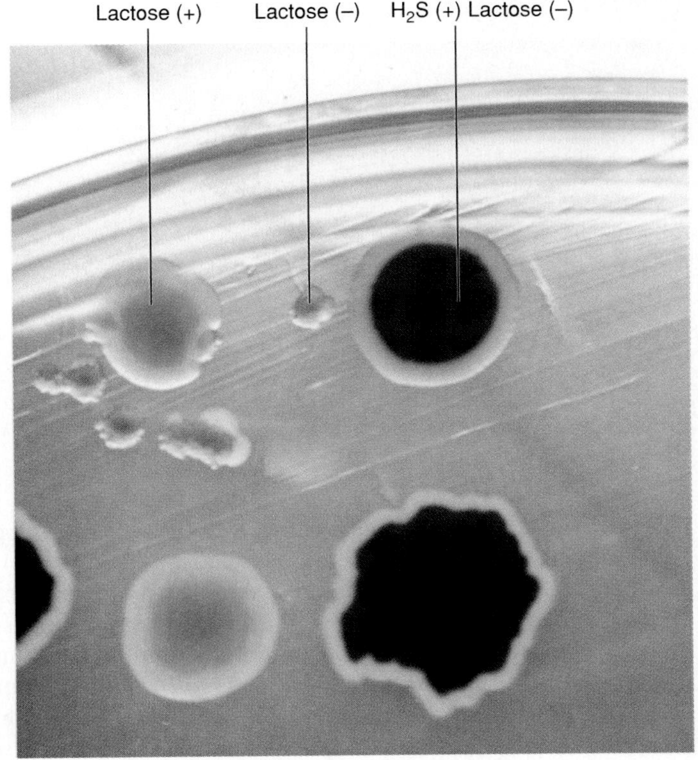

(b)

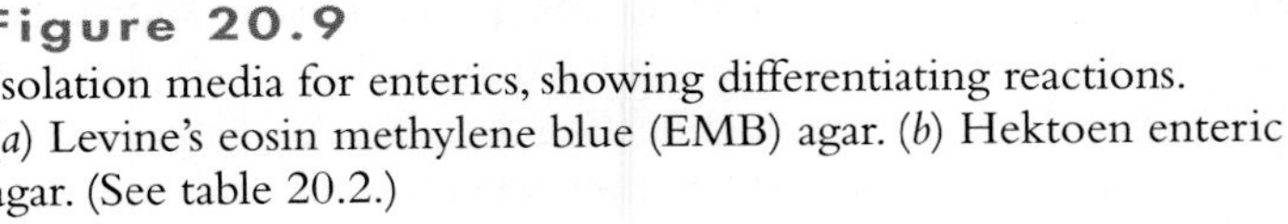

Figure 20.9
Isolation media for enterics, showing differentiating reactions. (*a*) Levine's eosin methylene blue (EMB) agar. (*b*) Hektoen enteric agar. (See table 20.2.)

TABLE 20.2

PROTOCOL IN ISOLATING AND IDENTIFYING SELECTED ENTERIC GENERA*

- **MacConkey agar** contains bile salts and crystal violet to inhibit gram-positive bacteria. It contains lactose and neutral red dye to indicate pH changes and differentiate the rapid lactose-fermenting bacteria from lactose nonfermenters or slow fermenters. Lactose (+) colonies are red to pink from the accumulation of acid acting on the neutral red indicator. Lactose (–) colonies are not colored in this way (see figure 3.10).
- **Levine's EMB agar** also contains bile salts, plus eosin and methylene blue dyes that cause lactose fermenters to develop a dark nucleus and sometimes a metallic sheen over the surface. Nonfermenters are pale lavender and non-nucleated (see figure 20.9*a*).
- **Hektoen enteric agar** also contains bile salts and is especially good for isolating pathogens. The bromothymol blue and acid fuchsin indicators differentiate the lactose fermenters (salmon pink to orange) from the nonfermenters (green, blue-green) and also detect the production of H_2S gas, which turns the center of a colony black (see figure 20.9*b*).
- **Utilization of lactose,** a disaccharide, requires two enzymes—a permease to bring lactose into the cell and β-galactosidase to split it into the monosaccharides glucose and galactose. Rapid fermenters on the initial isolation plates and TSI have both enzymes. Slow fermenters have only galactosidase. **ONPG** is chemical shorthand for a test that detects these slow lactose fermenters, which turn a special ONPG medium yellow. Nonfermenters do not react in either test.
- **TSI** is a nonselective medium that indicates some combination of fermentation reactions (primarily lactose and glucose, and with results from isolation medium, possibly sucrose) by means of a phenol red indicator dye. It also reveals gas and hydrogen sulfide (H_2S) production (see figure 3.21). Hydrogen sulfide is a metabolic product of the reduction of an inorganic or organic sulfur source. H_2S is indicated by a reaction with iron salts to form a black precipitate of ferric sulfide.
- **The indole test** indicates the capacity of an isolate to cleave a compound called indole off the amino acid tryptophan. If cleaving occurs, Kovac's reagent reacts with indole to form a bright red ring at the surface of the tube; if it is negative, the Kovac's remains yellow.
- **The methyl red (MR) test** indicates a form of glucose fermentation in which large amounts of mixed acids accumulate in the medium. The pH is lowered to around 4.2, so that methyl red dye remains red when added to the tube. MR-negative bacteria do not lower the pH to this degree, and the tube is yellow to orange in the presence of the dye.
- **The Voges-Proskauer (VP) test** determines whether the product of glucose fermentation is a neutral metabolite called acetyl-methylcarbinol (acetoin). This substance reacts with Barritt's reagent to form a pink to rosy-red tinge in the medium. With negative results, the tube remains brown to yellow.
- **Citrate media** contain citrate as the only usable carbon source and a pH indicator, bromothymol blue. If an isolate can utilize this carbon source, it grows and produces alkaline by-products, resulting in a conversion of the medium's color from green (neutral) to blue (alkaline).
- **Lysine** and **ornithine decarboxylase (LDC, ODC)** detect the presence of enzymes that remove carbon dioxide from amino acids. The end products are alkaline amines that raise the pH and convert the bromocresol purple to a bright lavender color. Negative results are yellow.
- **Urea media** contain 2% urea, a nitrogenous waste product of mammals, and phenol red pH indicator. Some bacteria produce the enzyme urease, which hydrolyses urea into two molecules of ammonium. Ammonium raises the pH of the medium, and the indicator turns bright pink.
- **Phenylalanine (PA) deaminase,** an enzyme produced primarily by the *Proteus* group, removes an amino group from the amino acid phenylalanine to produce phenylpyruvic acid. This reaction is visualized by adding ferric chloride to the tube, which turns olive green when it reacts with the phenylpyruvic acid.
- **Motility** is tested by a hanging drop slide or motility test medium (see figure 3.21).

*The metabolic basis for some tests is discussed in chapter 8.

TABLE 20.3

THE IMViC TESTS FOR DIFFERENTIATING COMMON OPPORTUNISTIC ENTERICS

Genus	Indole	Methyl Red	Voges-Proskauer	Citrate
Escherichia	+	+	–	–
Citrobacter	+	+	–	–
Klebsiella/ Enterobacter	–	–	V*	+
Serratia	–	V*	+	+
Proteus	+	–	–	+
Providencia	+	+	–	+
Pseudomonas	–	–	–	V*

*V = Species variable.

For example, normal flora isolated from stool specimens do not usually signify infection in that site, but the same organisms isolated from extraintestinal sites in sputum, blood, urine, and spinal fluid are considered likely agents of opportunistic infections.

ANTIGENIC STRUCTURES AND VIRULENCE FACTORS

The gram-negative enterics have complex surface antigens that are important in pathogenicity and are also the basis of immune responses (figure 20.11). By convention, they are designated **H,** the **flagellar antigen; K,** the **capsule** and/or **fimbrial antigen;** and **O,** the **somatic** or **cell wall antigen.**[2] Not all species carry the H and K antigens, but all have O, the lipopolysaccharide implicated

2. H comes from the German word *hauch* (breath) in reference to the spreading pattern of motile bacteria grown on moist solid media. K is derived from the German term *kapsel.* O is for *ohne-hauch,* or nonmotile.

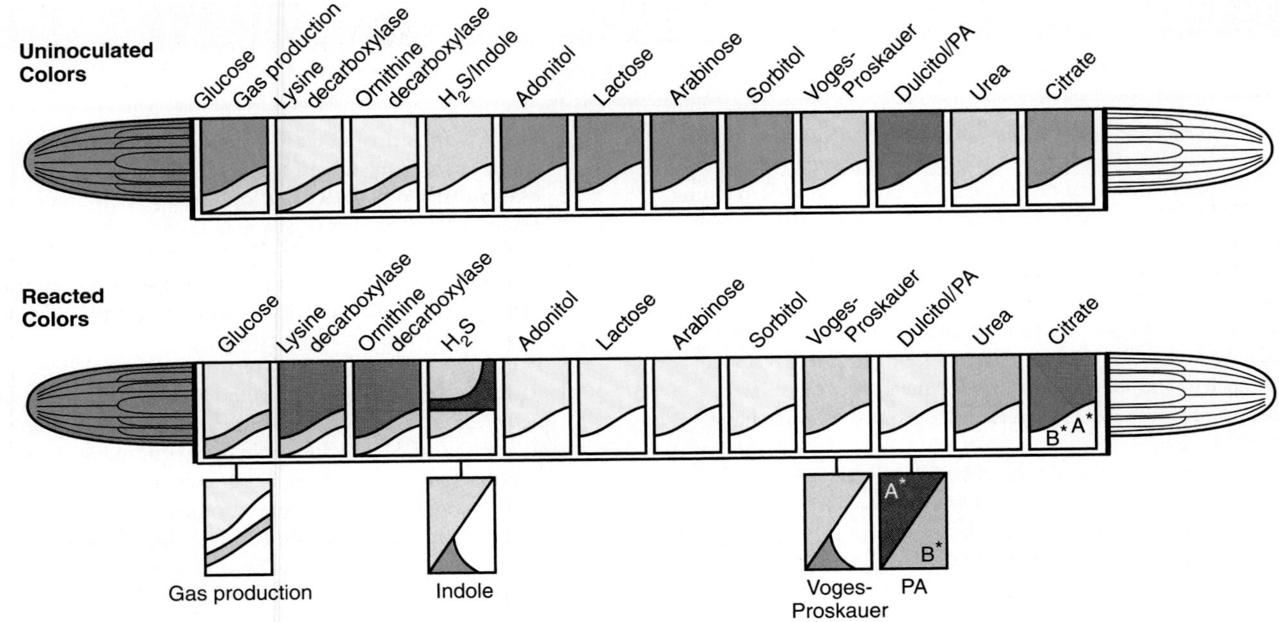

Figure 20.10
BBL Enterotube™ II (Roche Diagnostics), a miniaturized, multichambered tube used for rapid biochemical testing of enterics. An inoculating rod pulled through the length of the tube carries an inoculum to all chambers. Here, an uninoculated tube is compared with one that gives all positive results. For those chambers that give more than one result, the first test is shown on the top (e.g., H_2S) and the second on the bottom (indole).

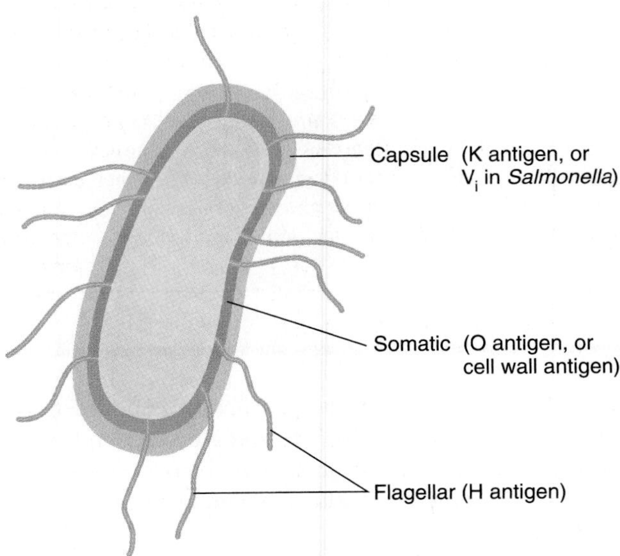

Figure 20.11

Antigenic structures in gram-negative enteric rods. Variations in the composition of these antigens provide the basis for the serological types found in most genera.

in endotoxic shock. Most species of gram-negative enterics exhibit a variety of subspecies or serotypes caused by slight variations in the chemical structure of the HKO antigens. These are usually identified by serotyping, a technique in which specific antibodies are mixed with a culture to measure the degree of agglutination or precipitation (see chapter 16). Not only is serotyping helpful in classifying some species, but it also helps pinpoint the source of outbreaks by a particular serotype, making it an invaluable epidemiologic tool.

The pathogenesis of enterics can be due to endotoxins (see microfile 20.1), exotoxins, and to the pathogen's capacity to overcome host defenses and multiply in the tissues and blood. Although most enterics are commensals, most also have a latent potential to rapidly alter their pathogenicity. As we saw in chapters 9 and 12, gram-negative bacteria freely transfer chromosomal or plasmid-borne genes that can mediate drug resistance, toxigenicity, and other adaptive traits. When this transference occurs in mixed populations of fecal residents (for example, in the intestines of carriers), it is possible for nonpathogenic or less pathogenic residents to acquire greater virulence (see *E. coli* in the next section). The most common virulence genes transferred code for enterotoxins, capsules, hemolysins, and fimbriae that promote colonization of the intestinal or urinary epithelia.

COLIFORM ORGANISMS AND DISEASES

ESCHERICHIA COLI: *THE UNPREDICTABLE ENTERIC BACILLUS*

*Escherichia** *coli* is the best known coliform, largely because of its use as a subject for laboratory studies. Although it is called the colon bacillus and sometimes regarded as the predominant species

**Escherichia* (ess-shur-eek′-ee-uh) After the German physician Theodor Escherich.

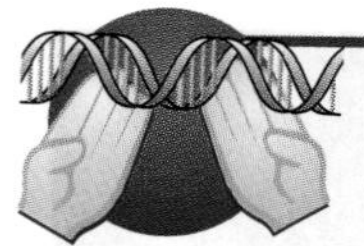

MICROFILE 20.3 *ESCHERICHIA COLI* 0157:H7: A PATHOGEN KNOWN BY ITS NUMBERS

The newest pathogenic strain of *Escherichia coli* has been variously associated with fast-food restaurants, supermarket hamburger, and undercooked beef, but, unlike most new pathogens, it has no catchy nickname. It is known instead by its strain antigen numbers—0 (somatic) type 157 and H (flagellar) type 7. This bacterium is the agent of a spectrum of conditions, ranging from mild gastroenteritis with fever to bloody diarrhea. About 10% of patients develop hemolytic uremic syndrome (HUS), a severe hemolytic anemia that can cause kidney damage and failure. The highest risk is to young children and the elderly and to immunocompromised people. Antibiotic treatments do not seem to help, and there is really no prevention other than avoiding ingestion of the pathogen.

The first human cases of disease were reported in 1982 and since then have been increasingly reported, probably due to improved diagnosis. Approximately 4,000 cases a year are reported in the United States, although the actual number is probably much higher. One of the most prominent recent epidemics occurred in Japan in 1997. Over 9,000 cases were reported, with around 20 fatalities. The source was never entirely determined, but some tests indicated that radish sprouts may have become contaminated during growth or processing.

This new strain appears to have acquired a virulence plasmid for production of Shiga (*Shigella*) toxin that accounts for the more severe disease symptoms. The pathogen originates from the intestine of dairy cattle, and it is transmitted mainly in contaminated beef, water and fresh vegetables. Hamburger is a common vehicle because butcher shops tend to grind meats from several sources together, thereby contaminating hundreds of pounds of hamburger with meat from one animal carrier. In 1997, 25 million pounds of hamburger had to be recalled from supermarkets because of a cluster of cases traced to one large meat plant in Nebraska.

This emerging infection has been an impetus for several changes from the slaughterhouse to the lab to the kitchen. Critics have proposed that the older visual inspection of meat alone is inadequate to detect bacteria and must be supplemented. The primary way to identify 0157:H7 is through culture on selective media and other special laboratory tests (see chapter 3 opener). The extra expense and time involved in such tests would greatly complicate the inspection process and require prolonged storage times. A panel of federal officials has recommended that a rapid chemical test that can detect the microbial load be used on all meat and poultry. A group of Montana researchers has developed a 4-hour test that verifies *E. coli* in samples by means of fluorescent antibodies. It is very sensitive and could be supplied in test kits for slaughterhouses, supermarkets, produce suppliers, or even water companies (see figure).

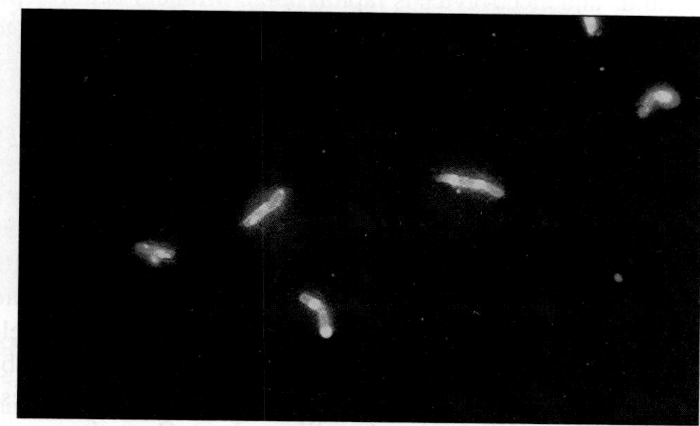

Escherichia coli *cells glow in a sample of raw ground beef mixed with fluorescent antibodies specific for the pathogen.*

Another outgrowth of this newly recognized food-borne pathogen are those small warning labels that now appear on all fresh meat and poultry products to inform the consumer of safe handling instructions (see microfile 20.5). One guideline is to bring the meat up to a temperature of at least 155°F (68°C) for a few minutes. This same temperature would also kill other food-borne pathogens such as *Salmonella.* Certainly, ground beef must be cooked completely through and not consumed when it is still rare or pink. Because of its threat, 0157:H7 was listed as an emerging infectious agent and added to the list of reportable diseases.

in the intestine of humans, *E. coli* is actually outnumbered 9 to 1 by the strictly anaerobic bacteria of the gut (*Bacteroides* and *Bifidobacterium*). Its prevalence in clinical specimens and infections is due to its being the most common aerobic and non-fastidious bacterium in the gut. Another mistaken view is that *E. coli* is a harmless commensal. Although many of the 150 strains are not infectious, some have developed greater virulence through plasmid transfer, and others are opportunists.

Examples of Pathogenic Strains of E. coli

Enterotoxigenic *E. coli* causes a severe diarrheal illness brought on by two exotoxins, termed heat-labile toxin (LT) and heat-stable toxin (ST), that stimulate heightened secretion and fluid loss. In this way it mimics the pathogenesis of cholera (see figure 21.14). This strain of bacteria also has fimbriae that provide adhesion to the small intestine. **Enteroinvasive** *E. coli* causes an inflammatory disease similar to *Shigella* **dysentery*** that involves invasion and ulceration of the mucosa of the large intestine. **Enteropathogenic** strains of *E. coli* are linked to a wasting form of infantile diarrhea whose pathogenesis is not well understood. The newest strain to emerge, *E. coli* 0157:H7, causes a hemorrhagic syndrome that can cause permanent damage to the kidney (microfile 20.3).

Clinical Diseases of E. coli

Most clinical diseases of *E. coli* are transmitted exclusively among humans. Pathogenic strains of *E. coli* are frequent agents of **infantile diarrhea,** the greatest single cause of mortality among babies. In some areas of the world, about 15% to 25% of children 5 years or younger die of diarrhea as either the primary disease or a complication of some other illness. The rate of infection is higher in crowded tropical regions where sanitary facilities are poor and water supplies are contaminated and where adults carry pathogenic strains to which they have developed immunity. The

*dysentery (dis′-en-ter″-ee) Gr. *dys,* bad, and *enteria,* bowels.

MICROFILE 20.4 THE INFAMOUS TYPHOID MARY

Early in this century, a middle-aged Irishwoman named Mary Mallon immigrated to New York State to work as a cook in the houses of the wealthy. What followed is a classic historical example of a chronic carrier state. Seven of the eight families for which Mallon worked over a 7-year period were stricken by typhoid fever, and several of them died. A physician who specialized in diagnosing the disease spent months attempting to trace the source of the epidemic. Water, milk, food, and the environment were all tested for the presence of the typhoid bacillus, but to no avail.

Eventually, the doctor began to consider the possibility that the cook had been the culprit, and he set out to find her. But when finally tracked down, the wary and uncooperative Many ran away. Because of the serious threat she posed to public health, the police were enlisted to capture her and take her to a hospital, where she was imprisoned. When tests for the pathogen proved positive, she was offered two ways to gain her freedom—have her gallbladder removed or stop working as a cook. Mary rejected both choices. She was released after 3 years on the condition that she not ply her trade, but she went right back to kitchen work, shedding bacilli and causing more disease and death. After she was at last apprehended, she spent the remainder of her life as a ward of the hospital and went to her grave stubbornly denying that she had ever been "Typhoid Mary."

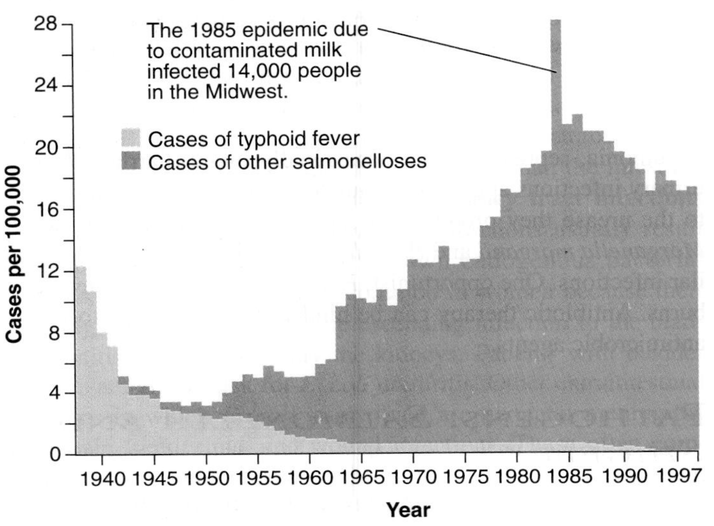

Figure 20.15

Data on the prevalence of typhoid fever and other salmonelloses from 1940 to 1997. The 1985 epidemic due to contaminated milk infected 14,000 people in the Midwest. (Nontyphoidal salmonelloses did occur before 1940, but the statistics are not available.)

Source: Data from Morbidity and Mortality Weekly Report, *January 9, 1998, Vol. 46. Centers for Disease Control and Prevention, Atlanta, GA.*

They are not fastidious, grow readily on most laboratory media, and can survive outside the host in inhospitable environments such as fresh water and freezing temperatures. These pathogens are resistant to chemicals such as bile and dyes (the basis for isolation on selective media) and do not lose virulence after long-term artificial cultivation.

Typhoid fever is so named because it bears a superficial resemblance to typhus, a rickettsial disease, even though the two diseases are otherwise very different. In the United States, the incidence of typhoid fever has remained at a steady rate for the last 30 years, appearing sporadically (figure 20.15). Of the 300–400 cases reported annually, roughly half of them were imported from endemic regions. In other parts of the world, typhoid fever is still a serious health problem responsible for 25,000 deaths each year and probably millions of cases.

The typhoid bacillus usually enters the alimentary canal along with water or food contaminated by feces, although it is occasionally spread by close personal contact. Because humans are the exclusive hosts for *S. typhi,* asymptomatic carriers are important in perpetuating and spreading typhoid. Even 6 weeks after convalescence, the bacillus is still shed by about half of recovered patients. A small number of people chronically carry the bacilli for longer periods in the gallbladder; from this site the bacilli are constantly released into the intestine and feces. This carrier state, combined with unclean personal habits, provides a recipe for disaster (microfile 20.4).

The size of the *S. typhi* dose that must be swallowed to initiate infection is between 1,000 and 10,000 bacilli. At the mucosa of the small intestine, the bacilli adhere and initiate a progressive, invasive infection that leads eventually to septicemia. The typhoid bacillus infiltrates the mesenteric lymph nodes and the phagocytes of the liver and spleen. The periodic escape of bacilli from infected lymphoid tissue gives rise to bacteremia and establishment in organs. Symptoms are fever, diarrhea, and abdominal pain. As the enteric phase progresses, the lymphoid tissue of the small intestine develops areas of ulceration that are vulnerable to hemorrhage and, in a few patients, deep erosion, perforation, and peritonitis occur (figure 20.16). The urinary tract and the liver can develop typhoidal abscesses. Untreated cases last a month, with about a 10% to 15% mortality rate. Prompt treatment greatly reduces the possibility of death.

A preliminary diagnosis of typhoid fever can be based upon the patient's history and presenting symptoms, supported by a rising antibody titer (see chapter 16). However, definitive diagnosis requires isolation of the typhoid bacillus. Chloramphenicol or sulfa-trimethoprim is the drug of choice, though some resistant strains occur. Antimicrobials are usually effective in treating the chronic carrier, but surgical removal of the gallbladder may be necessary in individuals with chronic gallbladder inflammation. Drug companies have produced two new vaccines; one is a live attenuated oral vaccine, and the other is based on the capsular polysaccharide. Both provide temporary protection for travelers and military personnel.

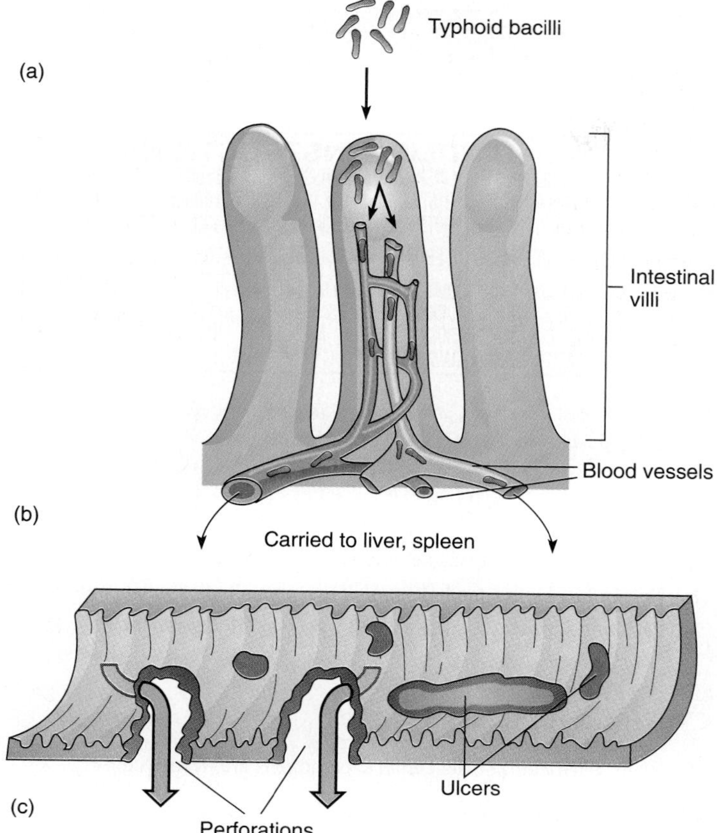

Figure 20.16

The phases of typhoid fever. (*a*) Ingested cells invade the small intestinal lining. (*b*) From this site, they enter the bloodstream, causing septicemia and endotoxemia. (*c*) Infection of the lymphatic tissue of the small intestine can produce varying degrees of ulceration and perforation of the intestinal wall.

Animal Salmonelloses

Salmonelloses other than typhoid fever are termed enteric fevers, *Salmonella* food poisoning, and gastroenteritis. These diseases are usually less severe than typhoid fever and are ascribed to one of the many serotypes of *Salmonella enteritidis*—most frequently, *S. paratyphi A, S. schottmulleri (paratyphi B), S. hirschfeldii (paratyphi C),* and *S. typhimurium.* A close relative, *Arizona*[3] *hinshawii,* is a pathogen found in the intestines of reptiles that parallels the salmonellae in its characteristics and diseases.

Nontyphoidal salmonelloses are more prevalent than typhoid fever and are currently holding steady at around 40,000 to 50,000 cases a year (see figure 20.15). Unlike tyhpoid, all strains are zoonotic in origin, though humans may become carriers under certain circumstances. *Salmonella* are normal intestinal flora in cattle, poultry, rodents, and reptiles. Animal products such as meat and milk can be readily contaminated during slaughter, collection, and processing. There are inherent risks in eating poorly cooked beef or unpasteurized fresh or dried milk, ice cream, and cheese. A particular concern is the contamination of foods by rodent feces. Several outbreaks of infection have been traced to unclean food storage or to food-processing plants infested with rats and mice.

It is estimated that one out of every three chickens is contaminated with *Salmonella,* and other poultry such as ducks and turkeys are also affected. Eggs are a particular problem because the bacteria may actually enter the egg while the shell is being formed in the chicken. The poultry industry is testing a product—Preempt—that is an avirulent strain of *Salmonella* sprayed on newly-hatched chicks to colonize them. This inoculation appears to curb the acquisition of pathogenic strains in a high percentage of treated animals. In general, one should always assume that poultry and poultry products harbor *Salmonella* and handle them accordingly, with clean techniques and adequate cooking. Drug resistance of the salmonellas is on the rise, some of which can be traced to the practice of adding antibiotics to animal feeds.

Most cases are traceable to a common food source such as milk or eggs. Outbreaks have occurred in association with milk, homemade ice cream made with raw eggs, and Caesar's salad. An epidemic at a casino in Las Vegas, Nevada, shows how human carriers may be involved. In this instance, a single food handler transmitted the pathogen to several hundred persons. Some cases may be due to poor sanitation. In 1996, about 60 people became infected after visiting the Komodo dragon (a large lizard) exhibit at the Denver zoo. They apparently neglected to wash their hands after handling the rails and fence of the dragon's cage.

The symptoms, virulence, and prevalence of *Salmonella* infections can be described by means of a pyramid, with typhoid fever at the peak, enteric fevers and gastroenteritis at intermediate levels, and asymptomatic infection at the base. In typhoid and enteric fevers, elevated body temperature and septicemia are much more prominent than gastrointestinal disturbances. Gastroenteritis is manifested by symptoms of vomiting, diarrhea, fluid loss, and mucosal lesions. Depending upon the organ or tissue involved, the lungs, nervous system, and bones can be sites for local infections. In otherwise healthy adults, symptoms spontaneously subside after 2 to 5 days; death is infrequent except in debilitated persons.

Serotyping is expensive, complicated, and not necessary before administering treatment. It is attempted primarily when the source of infection must be traced, as in the case of outbreaks and epidemics. A new probe technology that uses PCR can assay and identify *Salmonella* in less than 24 hours. Treatment for complicated cases of gastroenteritis is similar to that for typhoid fever; uncomplicated cases are handled by fluid and electrolyte replacement. Because the pathogen has an animal reservoir and lacks a practical vaccine for humans total eradication of gastroenteritis caused by nontyphoidal salmonellae seems unlikely, but certain measures can greatly reduce its incidence (microfile 20.5).

Shigella *and Bacillary Dysentery*

*Shigella** causes a common but often incapacitating dysentery called shigellosis, which is marked by crippling abdominal cramps and frequent defecation of watery stool filled with mucus and blood. The etiologic agents (*Shigella dysenteriae, Sh. sonnei, Sh. flexneri,* and *Sh. boydii*) are primarily human parasites, though they can infect apes. All produce a similar disease that can vary in intensity. They are nonmotile, nonencapsulated, and not fastidious

3. Named for the state.

**Shigella* (shih-gel′-uh) After K. Shiga, a Japanese physician.

MICROFILE 20.5 AVOIDING GASTROINTESTINAL INFECTIONS

Most food infections or other enteric illnesses caused by *Salmonella, Shigella, Vibrio, Escherichia coli,* and other enteric pathogens are transmitted by the **4 F's—food, fingers, feces,** and **flies**—as well as by water. This is sometimes rephrased in the saying "Food fussed over with freshly fecaled fingers." Individuals can protect themselves from infection by taking precautions when preparing food and traveling, and communities can protect their members by following certain public health measures.

SAFE HANDLING INSTRUCTIONS

THIS PRODUCT WAS PREPARED FROM INSPECTED AND PASSED MEAT AND/OR POULTRY. SOME FOOD PRODUCTS MAY CONTAIN BACTERIA THAT COULD CAUSE ILLNESS IF THE PRODUCT IS MISHANDLED OR COOKED IMPROPERLY. FOR YOUR PROTECTION, FOLLOW THESE SAFE HANDLING INSTRUCTIONS.

KEEP REFRIGERATED OR FROZEN. THAW IN REFRIGERATOR OR MICROWAVE.

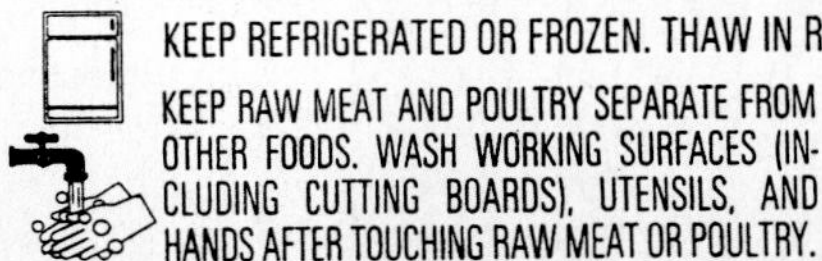

KEEP RAW MEAT AND POULTRY SEPARATE FROM OTHER FOODS. WASH WORKING SURFACES (INCLUDING CUTTING BOARDS), UTENSILS, AND HANDS AFTER TOUCHING RAW MEAT OR POULTRY.

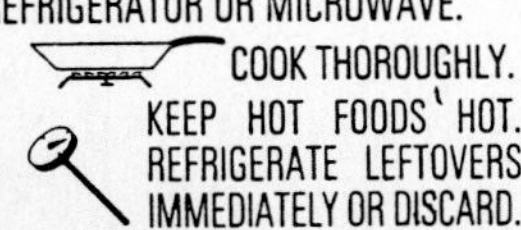

COOK THOROUGHLY.

KEEP HOT FOODS HOT. REFRIGERATE LEFTOVERS IMMEDIATELY OR DISCARD.

Supermarket label emphasizes precautions in handling, storing, and cooking meat and poultry products, as a way to avoid food-borne disease.

IN THE KITCHEN

A little knowledge and good common sense go a long way. Food-related illnesses can be prevented by:

1. Use of good sanitation methods while preparing food. This includes liberal handwashing, disinfecting surfaces, sanitizing utensils, and preventing contamination by mechanical vectors such as flies and cockroaches.
2. Use of temperature. Sufficient cooking of meats, eggs, and seafood is essential. Most microbes are not heat-resistant and will be killed in 10 minutes at 100°C. Refrigeration and freezing prevent multiplication of bacteria in fresh food that is not to be cooked (but remember, it does not kill them).
3. Do not taste uncooked batter made with raw eggs or ground meats before cooking.

WHILE TRAVELING

1. Choose food such as sealed or packaged items that are less likely to be contaminated with enteric pathogens.
2. Drink bottled water whenever possible.
3. Do not brush your teeth or wash fruit or vegetables in tap water.
4. Pack antidiarrheal medicines and water disinfection tablets.

COMMUNITY MEASURES

1. Water purification. Most enteric pathogens are destroyed by ordinary chlorination.
2. Pasteurization of milk.
3. Detection and treatment of carriers.
4. Restriction of carriers from food handling.
5. Constant vigilance during floods and other disasters that result in sewage spillage and contamination of food and drink.

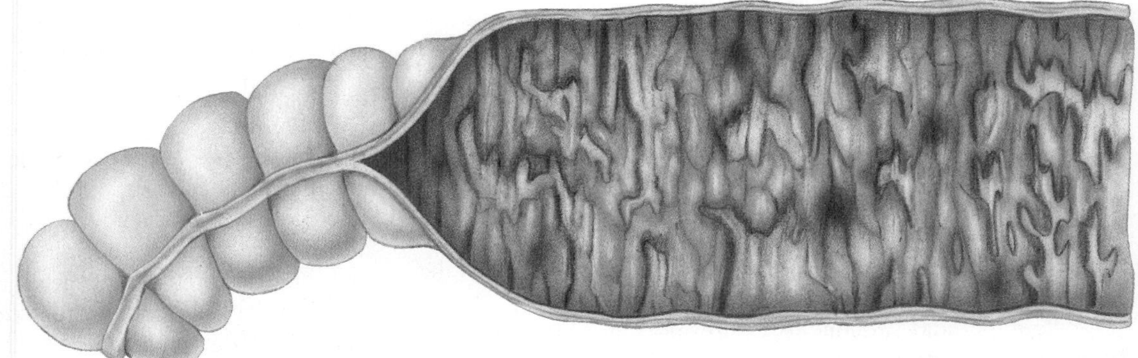

Figure 20.17

The appearance of large intestinal mucosa in *Shigella* (bacillary) dysentery. Note the patches of blood and mucus, the erosion of the lining, and the absence of perforation.

and do not produce H_2S or urease. The shigellae resemble some types of pathogenic *E. coli* so closely that they are placed in the same subgroup.

Although *Sh. dysenteriae* causes the severest form of dysentery, it is uncommon in the United States and occurs primarily in the Eastern Hemisphere. In the past decade, the prevalent agents in the United States have been *Sh. sonnei* and *Sh. flexneri,* which are presently undergoing a dramatic increase. There are currently approximately 25,000 cases reported each year, half of them in children between 1 and 10 years of age. In addition to the usual oral route, shigellosis is also acquired through direct person-to-person contact, largely because of the small infectious dose required (200 cells). The disease is mostly associated with lax sanitation, malnutrition, and crowding and is spread epidemically in daycare centers, prisons, mental institutions, nursing homes, and military camps. As in other enteric infections, a chronic carrier period of weeks to months occurs in some people.

Shigellosis is different from salmonellosis in that *Shigella* invades the villus cells of the large intestine, rather than the small intestine. In addition, it is not as invasive as *Salmonella* and does

Chapter Checkpoints

Enterobacteriaceae are a large family of gram-negative rods. Some are free-living, and others live as enterics, the normal flora of vertebrate intestinal tracts. Enteric species are grouped as coliforms or noncoliforms on the basis of their ability to ferment lactose. Enteric pathogens are identified by growth characteristics on differential media and their reactions to the IMViC test. Virulence factors of pathogenic species include enterotoxins, exotoxins, capsule, fimbriae, hemolysins, and endotoxin.

Pathogenic *Escherichia coli* infections are usually limited to the intestine or urinary tract. An exception is *E. coli* 0157:H7, which causes hemolytic anemia and kidney damage. Virulence factors are exotoxins, which cause excessive fluid loss, and fimbriae that enable it to adhere to the intestinal mucosa. The presence of *E. coli* in food, water, or dairy products indicates fecal contamination and, therefore, the possible presence of enteric pathogens.

Enteric species *Klebsiella, Enterobacter, Serratia,* and *Citrobacter* are not usually harmful to the healthy host, but they can cause serious, even fatal, nosocomial infections when introduced into nonintestinal sites. *Klebsiella* is the most important secondary invader in this group.

The noncoliform enterics *Proteus, Morganella,* and *Providencia* are also opportunistic pathogens that are often involved in septicemia, wound infection, urinary tract infections, and pneumonia.

Salmonella and *Shigella* are noncoliform pathogens that cause intestinal infections with a wide range of severity, depending on the species. *Salmonella* is a hardy genus that is transmitted both directly and also indirectly through contaminated food and water. *Salmonella typhi* causes the most serious disease (typhoid fever). Other species cause salmonella food poisoning, gastroenteritis, and enteric fevers. Humans are the only host for *S. typhi,* but other *Salmonella* pathogens are transmitted through poultry and dairy products, human feces, and other fomites.

Shigellosis, or bacillary dysentery, is a disease of worldwide distribution. It is caused by several species of *Shigella.* Infections are easily initiated because of the small infective dose. Humans are the only reservoir. Virulence factors are endotoxin and enterotoxins that attack the colon.

not perforate the intestine or invade the blood. It enters the intestinal mucosa by means of lymphoid cells in Peyer's patches. Once in the mucosa, it instigates an inflammatory response that destabilizes the epithelium and causes extensive tissue destruction. As it multiplies, it gives off toxins. Endotoxin causes fever, and enterotoxin brings on inflammation of the underlying gut wall layer, degeneration of the villi, and local erosion that causes bleeding and heavy mucous secretion (figure 20.17). Abdominal cramps and pain are caused by the disruption of the muscular function of the intestine. *Shigella dysenteriae* produces a heat-labile exotoxin (shiga toxin) that has a number of effects, including injury to nerve cells and nerves and damage to the intestine.

Diagnosis is complicated by the coexistence of several alternative candidates for bloody diarrhea such as *E. coli* and the protozoans *Entamoeba histolytica* and *Giardia lamblia.* Isolation and identification follow the usual protocols for enterics. Infection is treated by fluid replacement and oral drugs such as ciprofloxacin and sulfa-trimethoprim (SxT) unless drug resistance is detected, in which case ampicillin or cephalosporins are prescribed. Prevention follows the same steps as for salmonellosis (microfile 20.5), and there is no vaccine yet available.

The Enteric *Yersinia* Pathogens

Although formerly classified in a separate category, the genus *Yersinia** has been placed in the Family Enterobacteriaceae on the basis of cultural, biochemical, and serological characteristics. All three species cause zoonotic infections called *yersinioses. Yersinia enterocolitica* and *Y. pseudotuberculosis* are intestinal inhabitants of wild and domestic animals that cause enteric infections in humans, and *Y. pestis* is the nonenteric agent of bubonic plague.

Yersinia enterolitica has been isolated from healthy and sick farm animals, pets, wild animals, and fish throughout the world. Its presence on fruits and vegetables and in drinking water suggests that humans are infected by contaminated food or drink. During the incubation period of about 4 to 10 days, the bacteria invade the small intestinal mucosa, and some cells enter the lymphatics and are harbored intracellularly in phagocytes. Inflammation of the ileum and mesenteric lymph nodes gives rise to severe abdominal pain that mimics appendicitis. *Yersinia pseudotuberculosis* shares many characteristics with *Y. entercolitica,* though infections by the former are more benign and center upon lymph node inflammation rather than mucosal involvement.

NONENTERIC YERSINIA PESTIS AND PLAGUE

The word **plague**[4] conjures up visions of death and morbidity unlike any other infectious disease. Although pandemics of plague have probably occurred since antiquity, the first one that was reliably chronicled killed an estimated 100 million people in the sixth century. The last great pandemic occurred in the late 1800s and was transmitted around the world, primarily by rat-infested ships. The disease was brought to the United States through the port of San Francisco around 1906. Eventually, infected rats mingled with native populations of rodents and gradually spread the populations throughout the West and Midwest. The cause of this dread disease is a rather harmless-looking gram-negative rod called ***Yersinia pestis,*** formerly *Pasteurella pestis,* with unusual bipolar staining and capsules (figure 20.18).

**Yersinia* (yur-sin′-ee-uh) After Alexandre Yersin, a French bacteriologist.

4. From the Latin *plaga,* meaning to strike, infest, or afflict with disease, calamity, or some other evil.

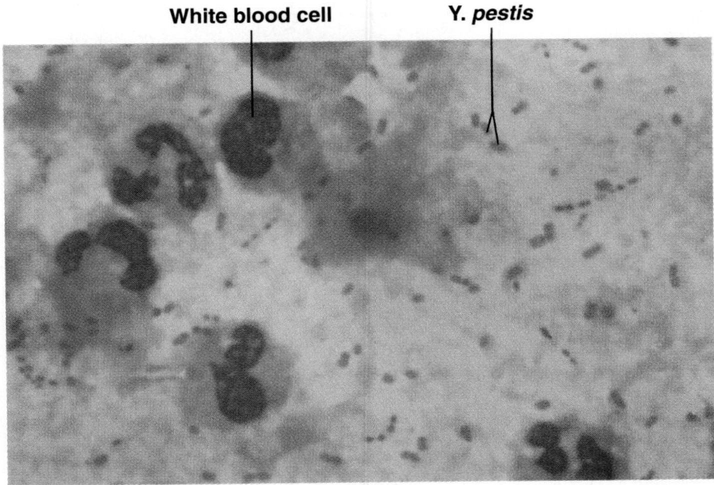

Figure 20.18
A Gram-stained preparation of *Yersinia pestis* in the blood of an infected mouse. The cells exhibit a distinctive bipolar morphology reminiscent of a safety pin.

Virulence Factors

Strains of the plague bacillus are just as virulent today as in the Middle Ages. Some of the virulence factors that coincide with its high morbidity and mortality are capsular and envelope proteins, which protect against phagocytosis and foster intracellular growth. The bacillus also produces coagulase, which clots blood and is involved in clogging the esophagus in fleas and obstructing blood vessels in humans. Other factors that contribute to pathogenicity are endotoxin and a highly potent murine toxin.

The Complex Epidemiology and Life Cycle of Plague

The plague bacillus exists naturally in many animal hosts, and its distribution is extensive, though the incidence of disease has been reduced in most areas. Plague still exists endemically in large areas of Africa, South America, the Mideast, Asia, and the former USSR, where it sometimes erupts into epidemics. Most recently, India experienced its first epidemic in three decades. Several hundred people were infected with about 50 deaths. This new surge in cases was attributed to increased populations of rats following the monsoon floods. In the United States, sporadic cases (usually less than 10) occur as a result of contact with wild and domestic animals. No cases of human-to-human transmission have been recorded since 1924. Persons most at risk for developing plague are veterinarians and people living and working near woodlands and forests.

The epidemiology of plague is among the most complex of all diseases. It involves several different types of vertebrate hosts and flea vectors, and its exact cycle varies from one region to another. A general scheme of the cycle is presented in figure 20.19. Humans can develop plague through contact with wild animals **(sylvatic plague),** domestic or semidomestic animals **(urban plague),** or infected humans.

The Animal Reservoirs The plague bacillus occurs in 200 different species of mammals. The primary long-term *endemic reservoirs* are various rodents such as mice and voles that harbor the organism but do not develop the disease. These hosts spread the disease to other mammals called *amplifying hosts* that become infected with the bacillus and experience massive die-offs during epidemics. These hosts, including the brown rat, ground squirrel, wood rat, black rat, chipmunk, and rabbit, are usually the sources of human plague. The particular mammal that is most important in this process depends on the area of the world. Other mammals (camels, sheep, coyotes, deer, dogs, and cats) can also be involved in the transmission cycle.

Flea Vectors The principal agents in the transmission of the plague bacillus from reservoir hosts to amplifying hosts to humans are fleas. These tiny, blood-sucking insects (see figure 13.19 and microfile 21.5) have a special relationship with the bacillus. After an uninfected flea ingests a blood meal from a plague-ridden animal, the bacilli multiply in its gut. In fleas that effectively transmit the bacillus, the esophagus becomes blocked. Being unable to feed properly, the ravenous flea jumps from animal to animal in a futile attempt to get nourishment. During this process, regurgitated infectious material is inoculated into the bite wound.

When the flea's natural host is available, the flea transmits the bacillus within that population. But many fleas are not host-specific and will attempt to feed on other species, even humans. Depending on the weather conditions, fleas containing viable bacilli can survive up to 3 years in animal habitats. Fleas of rodents such as rats and squirrels are most often vectors in human plague, though occasionally, the human flea is involved. Humans can also be infected by handling infected animals, animal skins, or meat and by inhaling droplets.

Pathology of Plague

The number of bacilli required to initiate a plague infection is small—perhaps 3 to 50 cells. The manifestations of infection lead to bubonic, septicemic, or pneumonic plague. In **bubonic plague,** the plague bacillus multiplies in the flea bite, enters the lymph, and is filtered by the local lymph nodes. This process causes necrosis and swelling of the node called a **bubo,*** typically in the groin and less often in the axilla (figure 20.20). The incubation period lasts 2 to 8 days, ending abruptly with the onset of fever, chills, headache, nausea, weakness, and tenderness of the bubo.

Cases of bubonic plague often progress to massive bacterial growth in the blood termed **septicemic plague.** The release of virulence factors causes disseminated intravascular coagulation, subcutaneous hemorrhage, and purpura that may degenerate into necrosis and gangrene. Because of the visible darkening of the skin, the plague has often been called the "black death." In **pneumonic plague,** infection is localized to the lungs and is highly contagious through sputum and aerosols. Without proper treatment, it is invariably fatal.

*bubo (byoo′-boh) G. *boubon,* the groin.

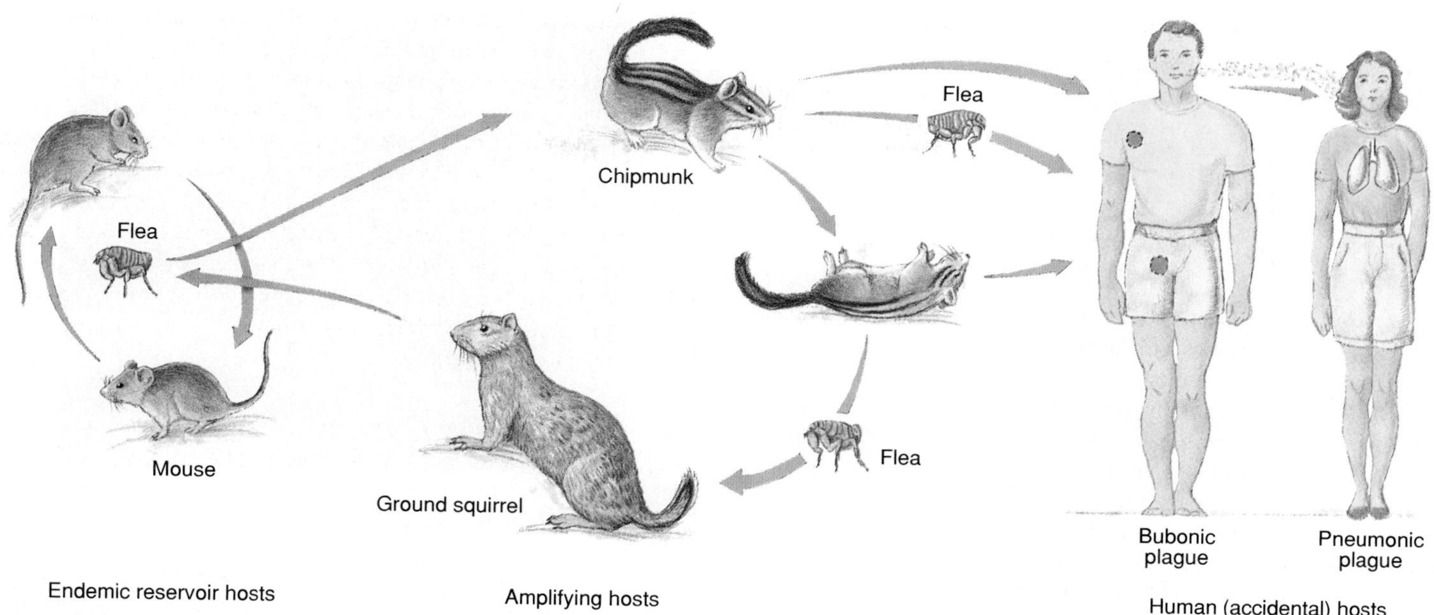

Figure 20.19
The infection cycle of *Yersinia pestis* simplified for clarity.

Diagnosis, Treatment, and Prevention Because death can ensue as quickly as 2 to 4 days after the appearance of symptoms, prompt diagnosis and treatment of plague are imperative. The patient's history, including recent travel to endemic regions, symptoms, and laboratory findings from bubo aspirates, helps establish a diagnosis. Streptomycin, tetracycline, and chloramphenicol are satisfactory treatments. Treated plague has a 90–95% survival rate.

The menace of plague is proclaimed by its status as one of the internationally quarantinable diseases (the others are cholera and yellow fever). In addition to quarantine during epidemics, plague is controlled by trapping and poisoning rodents near urban and suburban communities and by dusting rodent burrows with insecticide to kill fleas. These methods, however, cannot begin to control the reservoir hosts, so the potential for plague will always be present in endemic areas, especially as humans encroach into rodent habitats. A killed or attenuated vaccine that protects against the disease for a few months is given to military personnel, veterinarians, and laboratory workers.

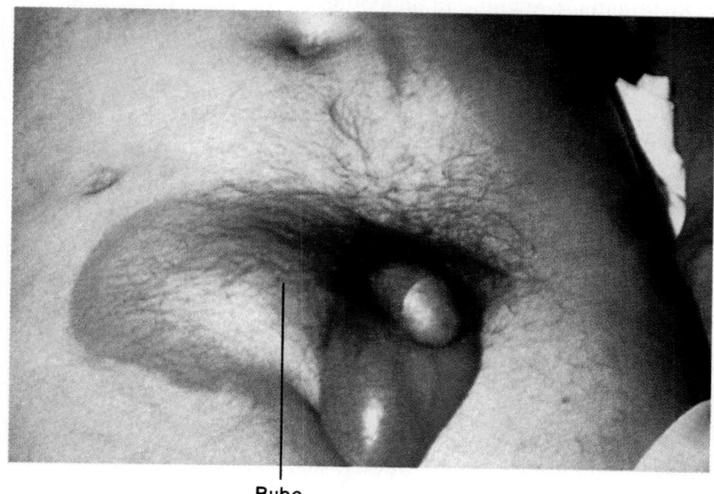

Figure 20.20
A classic inguinal bubo of bubonic plague in a male victim. This hard nodule is very painful and can rupture onto the surface. (The brown discoloration is from iodine.)

OXIDASE–POSITIVE NONENTERIC PATHOGENS

Pasteurella multocida

Pasteurella is a zoonotic genus that occurs as normal flora in animals and is mainly of concern to the veterinarian. Of the six recognized species, ***Pasturella multocida**** is responsible for the broadest spectrum of opportunistic infections. For example, poultry and wild fowl are susceptible to cholera-like outbreaks, and cattle are especially prone to epidemic outbreaks of hemorrhagic septicemia or pneumonia known as "shipping fever." The species has also adapted to the nasopharynx of the household cat and is normal flora in the tonsils of dogs. Because many hosts are domesticated animals with relatively close human contact, zoonotic infections are an inevitable and serious complication.

Animal bites or scratches, usually from cats and dogs, cause a local abscess that can spread to the joints, bones, and lymph nodes. Patients with weakened immune function due to liver cirrhosis or rheumatoid arthritis are at great risk for septicemic complications involving the central nervous system and heart. Patients

**Pasteurella multocida* (pas″-teh-rel′-uh mul-toh-see′-duh) From Louis Pasteur, plus L. *multi,* many, and *cidere,* to kill.

with chronic bronchitis, emphysema, pneumonia, or other respiratory diseases are vulnerable to pulmonary failure. Contrary to the antibiotic resistance of many gram-negative rods, *P. multocida* and other related species are susceptible to penicillin, and tetracycline is an effective alternative.

Haemophilus: The Blood-Loving Bacilli

Haemophilus[5] cells are tiny (0.5 × 0.8 mm) gram-negative pleomorphic rods sometimes confused with the genus *Neisseria* in clinical samples. The members of this group tend to be fastidious and sensitive to drying, temperature extremes, and disinfectants. Even though their name means blood-loving, none of these organisms can grow on blood agar alone without special techniques. Some *Haemophilus* species are normal colonists of the upper respiratory tract or vagina, and others (primarily *H. aegyptius, H. parainfluenzae,* and *H. ducreyi*) are virulent species responsible for conjunctivitis, childhood meningitis, and chancroid.

The characteristic that makes the hemophili fastidious is their requirement for certain factors from blood to complete their metabolic syntheses (figure 20.21). Factor X, hemin, is a necessary component of cytochromes, catalase, and peroxidase. Factor V, nicotinamide adenine dinucleotide (NAD or NADP), is an important coenzyme. These factors are made available to the hemophili in media such as chocolate agar (a form of cooked blood agar; see chapter 3) and Fildes medium.

Haemophilus influenzae was originally named after it was isolated from patients with "flu" about 100 years ago. For over 40 years, it was erroneously proclaimed the causative agent until the real agent, the influenza virus, was discovered. The potential of this species to act as a pathogen remained in question until it was clearly shown to be the agent of **acute bacterial meningitis** in humans. This severe form of meningitis, caused primarily by the b serotype, was once most common in children between 3 months and 5 years of age. The case rates have declined over the past 10 years in this age group likely because of an increased emphasis on vaccination programs. Now the trend is for more cases of *H. influenzae* meningitis in the elderly, and meningococcal infection predominates in young infants and children (figure 20.22). In contrast to *Neisseria* meningitis, *Haemophilus* meningitis is not associated with epidemics in the general population but tends to occur as sporadic cases or clusters in daycare and family settings. It is transmitted by close contact and nose and throat discharges. Healthy adult carriers are the usual reservoirs of the bacillus.

Haemophilus meningitis is very similar to meningococcal meningitis (see chapter 18), with symptoms of fever, vomiting, stiff neck, and neurological impairment. Untreated cases have a fatality rate of nearly 90%, but even with prompt diagnosis and aggressive treatment, 33% of children sustain residual disability, and about 5% must be placed in institutional care. Other important diseases caused by *H. influenzae* are an intense form of epiglottitis common in older children and young adults that may require immediate intubation or tracheostomy to relieve airway obstruction.

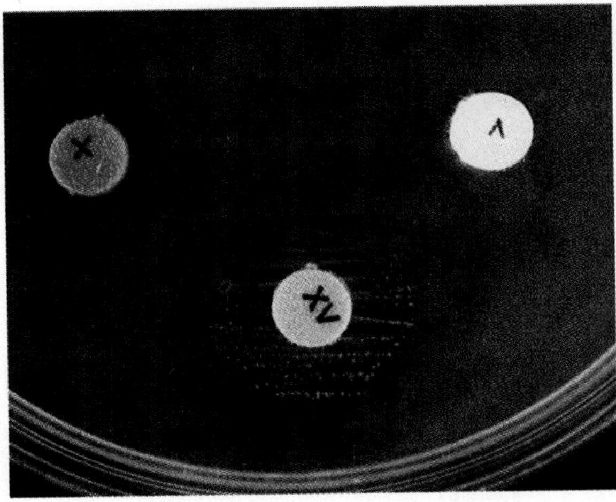

Figure 20.21
Discs containing separate or combined factor X and factor V are used to identify or type *Haemophilus* species. *Haemophilus influenzae* will grow only around the disc that has both factors. Other species (not shown here) require factor X or factor V, but not both.
From Gillies and Dodds, Bacteriology Illustrated, *5th ed. Fig. 78 (right), p. 104. Reprinted by permission of Churchill Livingstone.*

This species is also an agent of otitis media, sinusitis, pneumonia, and bronchitis.

Haemophilus infections are usually treated with a combination of chloramphenicol and ampicillin. Outbreaks of disease in families and daycare centers may necessitate rifampin prophylaxis for all contacts. Routine vaccination with a subunit vaccine (Hib) based on type b polysaccharide is recommended for all children, beginning at age 2 months, with three follow-up boosters. It is available in combination with DPT as TriHiBit.™

Haemophilus aegyptius **(Koch-Weeks bacillus)** is an agent of acute communicable **conjunctivitis,** sometimes called **pinkeye.** The subconjunctival hemorrhage that accompanies infection imparts a bright pink tinge to the sclera (figure 20.23). The disease occurs primarily in children, is distributed worldwide, and is spread through contaminated fingers and shared personal items as well as mechanically by gnats and flies. It is treated with antibiotic eyedrops.

Haemophilus ducreyi is the agent of **chancroid** (soft chancre), a sexually transmitted disease prevalent in the tropics and subtropics that afflicts mostly males. It is transmitted by direct contact with infected lesions and is favored by sexual promiscuity and unclean personal habits. After an incubation period lasting 2 to 14 days, lesions develop in the genital or perianal area. First to appear is an inflammatory macule that evolves into a painful, necrotic ulcer similar to those found in lymphogranuloma venereum and syphilis (see chapter 21). Often the regional lymph nodes develop into bubolike swellings that burst open. Although cotrimoxazole and other antimicrobics are effective, infection often recurs.

Haemophilus parainfluenzae and ***H. aphrophilus,*** members of normal oral and nasopharyngeal flora, are involved in infective endocarditis in adults who have underlying congenital or rheumatic heart disease. Such infections typically stem from routine dental procedures, periodontal disease, or some other oral injury.

5. Also spelled *Hemophilus*.

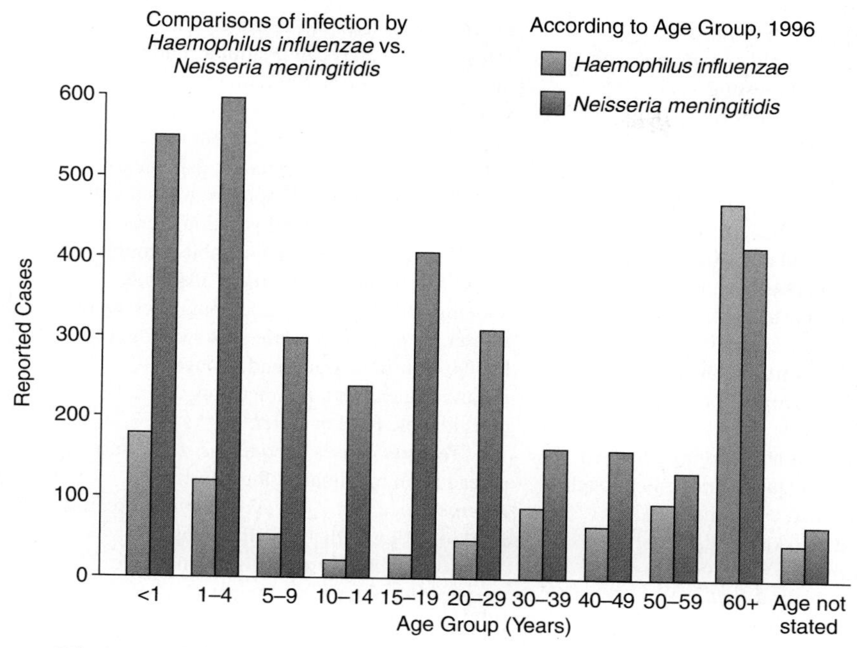

Figure 20.22

Comparisons of meningitis incidence in *Haemophilus influenzae* versus *Neisseria meningitidis.* (Latest statistics available at publication time.)

Source: Data from Morbidity and Mortality Weekly Report; Summary of Notifiable Diseases, U.S., 1996, *Massachusetts Medical Society for the Centers of Disease Control and Prevention, October 31, 1997.*

Figure 20.23

Acute conjunctivitis, or pinkeye, caused by *Haemophilus aegyptius.*

Chapter Checkpoints

Yersinia species are agents of zoonotic infections. *Y. enterocolitica* and *Y. pseudotuberculosis* cause enteric infections, primarily in children. They are transmitted through contaminated food and water. *Y. pestis* causes plague in three forms: bubonic, pneumonic, and septicemic. It is usually transmitted to humans through flea bites. Mice and rodents are the primary reservoirs of this bacillus, but infected fleas spread it to other mammals (amplifying hosts) and from there to humans. Infections respond to antibiotic therapy if treated early, but untreated infections have high mortality rates.

Pasteurella is another genus that causes a wide range of opportunistic zoonoses. *P. multocida* occasionally causes infection in humans through animal bites or scratches.

The genus *Haemophilus* contains both commensals and pathogenic species. *H. influenzae* causes a variety of upper and lower respiratory infections.

Haemophilus meningitis, or acute bacterial meningitis, is the most serious infection. *Haemophilus* infections respond to combination drug therapy. Immunization is available for young children. Other pathogens include *Haemophilus aegyptius,* the causative agent of conjunctivitis, *H. ducreyi,* the agent of soft chancre, an STD; and *H. parainfluenzae* and *H. aphrophilus,* agents of bacterial endocarditis.

CHAPTER CAPSULE WITH KEY TERMS

GRAM-NEGATIVE RODS

A large group of loosely affiliated families and genera, all of which are non-spore-forming; cell walls of most species contain lipopolysaccharide with **endotoxin** affects which causes fever, cardiovascular disruptions, and shock and is one of the gravest complications of gram-negative septicemia.

Aerobic Rods: ***Pseudomonas*** species are among the most widely distributed bacteria; thrive even in hostile conditions or where nutrients are scarce; very versatile and beneficial; medically, ***Ps. aeruginosa*** is a common opportunist of medical treatments when normal defenses are compromised; may attack lungs, skin, burns, urinary tract, eyes, ears; may also infect healthy persons; drug resistance limits treatment choices.

Brucella is a zoonotic genus that causes abortion in cattle, pigs, and goats, and

brucellosis, or **undulant fever,** in humans; bacterium is transmitted through direct contact with infected animals or contaminated animal products and ingestion of raw milk; infection occurs in several systems; marked by organ abscess and fluctuating fever; treated with streptomycin and rifampin; controlled through animal vaccination, pasteurization.

Francisella tularensis causes **tularemia,** or "rabbit fever," a zoonosis of rabbits, rodents, and other wild mammals that spreads to humans through direct contact with animals, bites by vectors (ticks), ingestion of contaminated food or water, or inhalation; very contagious symptoms depend upon portal of entry and organ involved but include skin, lymph nodes, lungs, intestine; treated with gentamicin; vaccine available for risk groups.

Bordetella pertussis causes a strictly human disease called **pertussis,** or **whooping cough;** contagious and prevalent in children under 6 months; may be fatal; infectious by droplets; attachment of pathogen and toxin production destroy cilial defense and produce a cough that occurs in several bursts followed by a sudden inspiration; death may be caused by secondary infections; vaccine (acellular pertussis) is a very effective preventive.

Legionella pneumophila causes **legionellosis,** commonly called **Legionnaires' disease;** unusual agent is a wide-ranging inhabitant of natural water that survives for months in human aquatic environments (cooling towers, air conditioners, taps) and can cause serious lung disease when accidentally inhaled; community infections occur from common environmental sources; hospital epidemics occur due to contaminated air and water supplies.

Facultative Anaerobic Rods: The **Family Enterobacteriaceae** is the largest group of gram-negative **enteric** bacteria; small, often motile, fermentative rods occurring in many habitats but often found in animal intestines; informal division of family is made between **coliforms,** normal flora that are rapid lactose fermenters, and **noncoliforms,** genera that do not or only weakly ferment lactose; enterics are predominant bacteria in clinical specimens; some species cause **diarrheal disease** due to **enterotoxins** acting on the intestinal mucosa or to invasion and disruption of the mucosa; enterics are agents of nonintestinal opportunistic infections or true pathogens; some are opportunists only; identification of group includes series of biochemical and serological tests; primary **antigens** are flagellar (H), cell wall (O), and capsular (K or V_i); drug resistance is well entrenched through plasmids; treatment with drugs requires sensitivity testing.

The most common coliform agent is ***Escherichia coli;*** exists in several forms; **pathogenic strains** have acquired virulence factors for invasiveness and toxigenicity; cause of **infantile diarrhea,** a complication of malnourished babies fed unsanitary food or water; **traveler's diarrhea** occurs in people who pick up a toxigenic strain from water and food in other countries. *E. coli* is the usual cause of **urinary tract infections** (from normal flora) and nosocomial pneumonia and septicemia.

E. coli **0:157 H7** is an emerging pathogen of cattle and humans; acquired primarily through beef, produce and water; causes hemolytic uremic syndrome due to kidney damage.

Other coliforms are ubiquitous in the hospital environment. *Klebsiella, Enterobacter, Serratia,* and *Citrobacter* account for nosocomial infections associated with tracheostomies, respiratory care equipment, endoscopes, and cathethers; other common infections are pneumonia, burn, and incision infections.

Noncoliform infections are caused by opportunists, pathogenic enterics, and pathogenic nonenterics. **Opportunists** include *Proteus, Morganella,* and *Providencia,* which cause nosocomial infections such as urinary tract infections, wound infections, pneumonia, sepsis, and diarrhea.

True enteric pathogens include ***Salmonella*** and ***Shigella.*** *Salmonella* causes **salmonelloses;** most severe disease is **typhoid fever,** caused by ***S. typhi;*** agent is spread only by humans via unclean food or water; often chronically carried bacillus crosses wall of small intestine into circulation, is carried to organs, where it may form abscesses; symptoms include fever, diarrhea, septicemia, and ulceration of small intestine; perforation is a complication; treatment with chloramphenicol; vaccine available. Other species belong to a serotype of *S. enteritidis,* such as *S. typhimurium* or *S. paratyphi A;* bacilli are common flora of cattle, poultry, rats, mice; contaminate meat, milk, and eggs; disease somewhat milder but more prevalent than typhoid fever; main infections are enteric fever or gastroenteritis, depending on severity of symptoms; treatment with antibiotics, oral rehydration.

Shigella causes **shigellosis,** a **bacillary dysentery** characterized by acute painful diarrhea with bloody, mucus-filled stools; *Sh. dysenteriae* causes most severe form, but *Sh. sonnei* and *flexneri* cause similar milder disease; all are primarily human parasites; spread by fingers, feces, food, and flies; bacteria invade large intestine but remain local and do not produce septicemia; damage to intestinal villi causes symptoms; treated with oral antimicrobics and rehydration therapy.

Enteric disease is preventable through cleanliness in processing food, adequate cooking and refrigeration, keeping flies away, awareness of animal carriers, proper toilet habits, control of water and sewage, monitoring carriers, not ingesting questionable food or water.

Yersinia causes *yersinioses,* zoonoses spread from mammals to humans; *Y. enterocolitica* and *Y. pseudotuberculosis* cause food infection with appendicitis-like symptoms. ***Yersinia pestis*** is a nonenteric agent of the **plague,** an ancient virulent disease; with complex epidemiology; agent is maintained by relationship between **endemic hosts** (mice) and **amplifying hosts** (rats, squirrels) and **flea vectors** that carry the bacillus between them; humans enter this cycle by accident, usually through flea bite or contact with infected animal; infected humans may pass the agent to other humans. Forms are (1) **bubonic plague,** in which multiplication at site of bite creates regional lymphatic swelling, or **bubo;** (2) **septicemic plague,** a deadly complication marked by hemorrhage; and (3) **pneumonic plague,** a lung infection spread by aerosols. Treatment includes streptomycin, tetracycline; vector and reservoir control and a vaccine can be preventive.

Nonenteric pathogens include:
(1) ***Pasteurella multocida,*** a zoonosis of cattle, poultry, cats, dogs; spread by bites, scratches, other contact; usual disease is abscess and lymph node swelling;
(2) ***Haemophilus influenzae,*** is the most frequent cause of **acute bacterial meningitis** in children between 3 months and 5 years; sporadic infection occurs primarily in daycare and similar settings; passed through respiratory discharges; disease is typical meningitis, marked by acute neurological complications, high morbidity, and sequelae; treated with chloramphenics ampicillin; vaccination with Hib recommended for children at risk. (3) ***H. aegyptius*** causes a form of conjunctivitis called pinkeye. (4) ***H. ducreyi*** causes the STD known as **chancroid.**

MULTIPLE-CHOICE QUESTIONS

1. A unique characteristic of many isolates of *Pseudomonas* useful in identification is
   a. fecal odor
   b. fluorescent green pigment
   c. drug resistance
   d. motility

2. Human brucellosis is also known as
   a. Bang's disease c. rabbit fever
   b. undulant fever d. Malta fever

3. *Francisella tularensis* has which portal of entry?
   a. tick bite c. respiratory
   b. intestinal d. all of these

4. A classic symptom of pertussis is
   a. labored breathing c. convulsions
   b. paroxysmal coughing d. headache

5. The severe symptoms of pertussis are due to what effect?
   a. irritation of the glottis by the microbe
   b. pneumonia
   c. the destruction of the respiratory epithelium
   d. blocked airways

6. *Escherichia coli* displays which antigens?
   a. capsular c. flagellar
   b. somatic d. all of these

7. Which of the following is *not* an opportunistic enteric bacterium?
   a. *E. coli* c. *Proteus*
   b. *Klebsiella* d. *Shigella*

8. Which of the following represents a major difference between *Salmonella* and *Shigella* infections?
   a. mode of transmission
   b. likelihood of septicemia
   c. the portal of entry
   d. presence/absence of fever and diarrhea

9. Complications of typhoid fever are:
   a. neurological damage
   b. intestinal perforation
   c. liver abscesses
   d. b and c

10. *Shigella* is transmitted by
    a. food c. feces
    b. flies d. all of these

11. The bubo of bubonic plague is a/an
    a. ulcer where the flea bite occurred
    b. granuloma in the skin
    c. enlarged lymph node
    d. infected sebaceous gland

12. *Haemophilus influenzae* requires ____ for growth.
    a. hemin c. blood
    b. NAD d. a and b

13. Circle all diseases for which human vaccines are available:
    a. typhoid fever
    b. shigellosis
    c. *Haemophilus* meningitis
    d. whooping cough
    e. bubonic plague
    f. Legionnaires' disease
    g. *E. coli* infantile diarrhea
    h. brucellosis

14. Which of the following are primarily zoonoses?
    a. tularemia d. brucellosis
    b. salmonellosis e. pasteurellosis
    c. shigellosis f. bubonic plague

15. Match the infectious agent with its disease:
    ____ *Francisella tularensis*
    ____ *Yersinia pestis*
    ____ *Escherichia coli* 0157:H7
    ____ *Shigella* species
    ____ *Salmonella enteritidis*
    ____ *Salmonella typhi*
    ____ *Pseudomonas aeruginosa*
    ____ *Bordetella pertussis*
    ____ *Legionella pneumophila*
    ____ *Haemophilus aegyptius*
    ____ *Haemophilus influenzae*
    ____ *Haemophilus ducreyi*
    ____ *Pasteurella multocida*
    a. dysentery
    b. local abscess
    c. chancroid
    d. enteric fever
    e. whooping cough
    f. meningitis
    g. typhoid fever
    h. hemolytic uremic syndrome
    i. bubonic plague
    j. Pontiac fever
    k. folliculitis
    l. rabbit fever
    m. pinkeye

CONCEPT QUESTIONS

1. Why are bacteria such as *Pseudomonas* and coliforms so often involved in nosocomial infections?

2. a. Briefly describe the human infections caused by *Pseudomonas, Brucella,* and *Francisella.*
   b. How are they similar?
   c. How are they different?

3. a. What is the pathologic effect of whooping cough?
   b. What factors cause it to predominate in newborn infants?

4. What is unusual about *Legionella?* What is the epidemiologic pattern of the disease?

5. a. Describe the chain of events that result in endotoxic shock.
   b. Describe the key symptoms of endotoxemia.
   c. Differentiate between toxigenic diarrhea and infectious diarrhea.

6. a. Describe what each of the following bacteria is: an enteric bacterium, a coliform, and a noncoliform.
   b. Which bacteria in the Family Enterobacteriaceae are true enteric pathogens?
   c. What are opportunists?

7. a. Briefly describe the methods used to isolate and identify enterics.
   b. What is the basis of serological tests, and what is their main use for enterics?

8. a. Explain how *E. coli* can develop increased pathogenicity.
   b. Describe the kinds of infections for which *E. coli* is primarily responsible.
   c. Describe the roles of other coliforms in infections.

9. a. What is salmonellosis?
   b. What is the pattern of typhoid fever?
   c. How does the carrier state occur?
   d. What is the main source of the other salmonelloses?
   e. What kinds of infections do salmonellas cause?

10. What causes the blood and mucus in dysentery?

11. Explain several practices an individual can use to avoid enteric infection and disease at home and when traveling.

12. a. Trace the epidemiologic cycle of plague.
    b. Compare the portal of entry of bubonic plague with that of pneumonic plague.
    c. Why is plague called the black death?
13. Describe the epidemiology and pathology of *Haemophilus influenzae* meningitis.
14. Compare the types of food-related illness discussed in this chapter according to
    a. the Gram reaction of the agent
    b. whether it is a food infection or intoxication
    c. the kinds of food involved
15. a. List the bacteria from this chapter for which general, routine vaccines are given.
    b. For which special groups of bacteria are there vaccines?
    c. For which bacteria are there none?
    d. Why are there no vaccines for these?
16. Briefly outline the zoonotic infections in this chapter, and describe how they are spread to humans.
17. a. Give the portal of entry and target tissues for plague, pertussis, legionellosis, and shigellosis.
    b. Which are primary pulmonary pathogens?
18. Compare and contrast the pathology, diagnosis, and treatment of meningococcal meningitis and *Haemophilus influenzae* meningitis.

CRITICAL-THINKING QUESTIONS

1. What is the logic behind testing for *E. coli* to detect fecal contamination of water?
2. Identify the genera with the following characteristics from figure 20.8:
   a. Lactose−, phenylalanine and urease−, citrate+, ONPG−
   b. Lactose+, motility−, VP−, Indole+
   c. Lactose+, motility, indole−, H_2S+
   d. Lactose−, phenylalanine and urease+, H_2S−, citrate+
3. Given that so many infections are caused by gram-negative opportunists, what would you predict in the future as the number of compromised patients increases, and why do you make these predictions?
4. An infectious dose of a million cells in enteric infections seems like a lot.
   a. In terms of the size and abundance of microbes, what would its appearance be? (Could you see a cluster containing that many cells with the naked eye?)
   b. Refer to chapter 7 on microbial growth cycles. How long would it take an average bacterial species to reach the infectious dose of a million cells starting from a single cell?
   c. Why do enteric diseases require a relatively higher infectious dose than nonenteric diseases?
5. Students in our classes sometimes ask how it is possible for a single enteric carrier to infect 1,000 people at a buffet or for a box turtle to expose someone to food infection. We always suggest that they use their imagination. Provide a detailed course of events that could result from these types of outbreaks.
6. Case study 1. A woman living near a wooded area in a western state discovered her cat carrying a sick mouse. She discarded the mouse, but a neighbor's dog found it and carried it home. In a few days, one child in the neighbor's family got a case of febrile illness that responded to antibiotics. The cat died from an open sore on its neck. Later, the veterinarian who treated the cat developed a fatal pneumonia.
   a. What disease is possible here?
   b. What two possible modes of transmission were demonstrated here?
7. a. Name five bacteria from chapters 19 and 20 that could be used in biological warfare.
   b. What are some possible ways they could be used in warfare?
   c. What is your personal opinion of using microorganisms to gain advantage during war?
8. Case study 2. Several persons working in an exercise gym acquired an acute disease characterized by fever, cough, pneumonia, and headache. Treatment with erythromycin cleared it up. The source was never found, but an environmental focus was suspected.
   a. What do you think might have caused the disease?
   b. People in a different gym got skin lesions after sitting in a redwood hot tub. Which pathogen could have caused that?
9. Case study 3. A 3-year-old severely ill child was admitted to a hospital with symptoms of diarrhea, fever, and malaise. Laboratory testing showed abnormal renal and liver values and anemia. She had no history of previous illness, and her food history was a recent meal of teriyaki beef consumed at a local restaurant. She responded to antibiotics.
   a. What was the probable pathogen?
   b. What was the likely source?
   c. What is the pathologic effect of the pathogen?
10. a. Give your opinion of the U.S. Department of Agriculture's "visual inspection" program to detect pathogenic bacteria in meat.
    b. What is the weakness in this approach?
    c. What are some alternatives to this method of detecting contaminated meat?
11. Looking at the endpapers, determine which of the diseases in this chapter are reportable, which ones are currently increasing in incidence, and which ones are decreasing.

INTERNET SEARCH TOPIC

1. Use the Internet to survey the current recommendations for pertussis vaccination.
2. Explore Internet sites to find information on new methods of inspecting meat and poultry and of improving food safety (such as pasteurizing eggs, preempt).

chapter 21

MISCELLANEOUS BACTERIAL AGENTS OF DISEASE

A number of agents of bacterial infections do not fit the usual categories of gram-positive or gram-negative rods or cocci. This group includes spirochetes and curviform bacteria, obligate intracellular parasites such as rickettsias and chlamydias, and mycoplasmas. This chapter covers not only those agents but also the vectors that many of them are borne by—ticks, lice, and other arthropods. The last section of the chapter surveys oral ecology, dental diseases, and the mixed bacterial infections that are responsible for them.

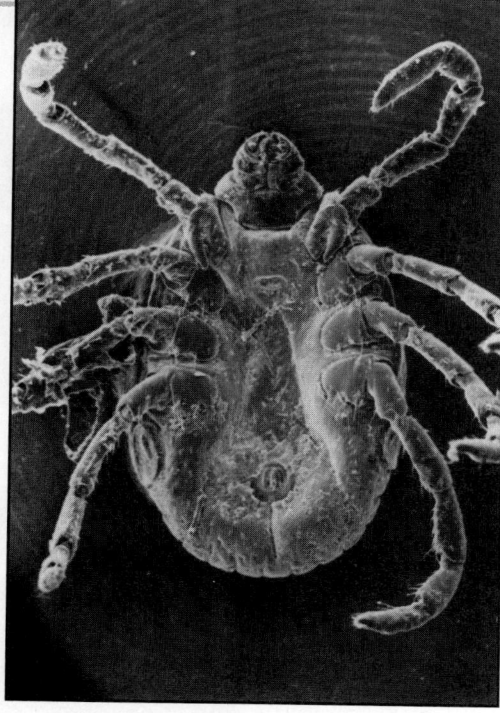

Close-up of the brown dog tick, *Rhiphicephalus sanguineous,* a hard tick that carries the rickettsial agent of Mediterranean spotted fever. All rickettsias are intimately associated with the life cycles of blood-sucking arthropods.

THE SPIROCHETES

Bacteria called spirochetes have a helical form and a mode of locomotion that appear especially striking in live, unstained preparations using the dark-field or phase-contrast microscope (see figure 21.7). Other traits include a typical gram-negative cell wall and a well-developed periplasmic space encloses the flagella (called endoflagella or periplasmic flagella) (figure 21.1*a*). Although internal flagella are constrained somewhat like limbs in a sleeping bag, their flexing propels the cell by rotation and even crawling motions. The spirochetes are classified in the Order Spirochaetales, which contains two families and five genera. The majority of spirochetes are free-living saprobes or commensals of animals and are not primary pathogens. But three genera contain major human pathogens: *Treponema, Leptospira,* and *Borrelia* (figure 21.1*b, c, d,* respectively).

TREPONEMES: MEMBERS OF THE GENUS *TREPONEMA*

Treponemes are thin, somewhat regular, coiled cells that live in the oral cavity, intestinal tract, and perigenital regions of humans and animals. The pathogens are strict parasites with complex growth requirements that necessitate cultivating them in live cells. Their oxygen requirements are minimal; some are strictly anaerobic and others microaerophilic. Diseases caused by *Treponema* are called **treponematoses.** The subspecies *Treponema pallidum pallidum* is responsible for venereal and congenital syphilis; the subspecies *T. p. endemicum* causes nonveneral endemic syphilis, or bejel; and *T. p. pertenue* causes yaws. *Treponema carateum* is the cause of pinta. Infection begins in the skin, progresses to other tissues in gradual stages, and is often marked by periods of healing interspersed with relapses. The major portion of this discussion will center on syphilis, and any mention of *T. pallidum* refers to the subspecies *T. p. pallidum.* Other treponemes of importance are involved in infections of the gingiva (see oral diseases at the end of the chapter).

*Treponema pallidum:** The Spirochete of Syphilis

The origin of syphilis is an obscure yet intriguing topic of speculation. The disease was first recognized at the close of the fifteenth century in Europe, a period coinciding with the return of Columbus from the West Indies, which led some medical scholars to conclude that syphilis was introduced to Europe from the New World. However, a more probable explanation contends that the spirochete evolved from a related subspecies, perhaps an endemic treponeme already present in the Mediterranean basin. The combination of the immunologically naive population of Europe, the European wars, and sexual promiscuity set the stage for worldwide transmission of syphilis that continues to this day.

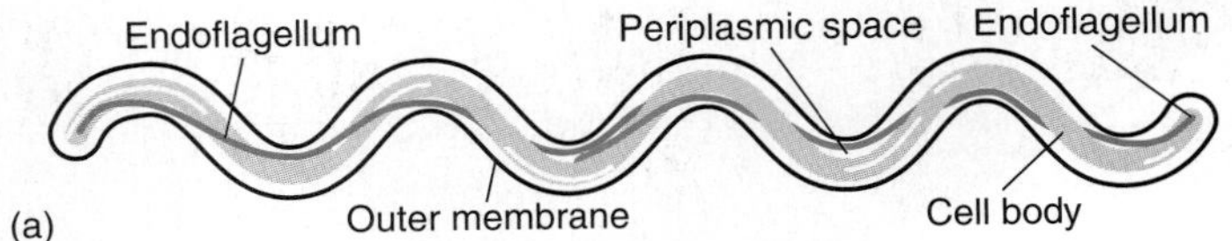

(a)

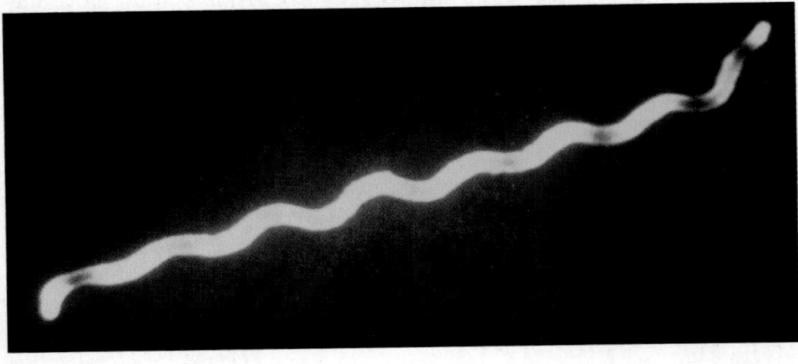
(b)

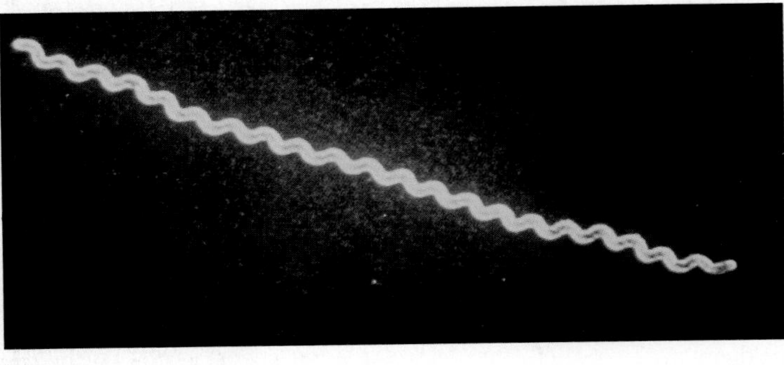
(c)

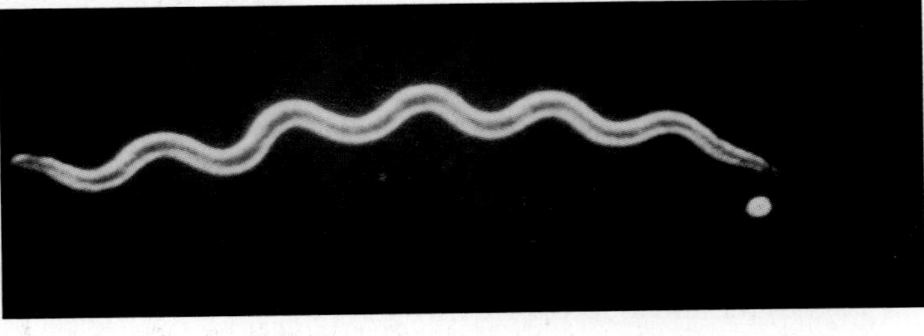
(d)

Figure 21.1
(*a*) Representation of general spirochete morphology with a pair of endoflagella inserted in the opposite poles, lying beneath the outer membrane and within the periplasmic space. In some spirochetes, the free ends overlap, as shown here. (*b–d*) Variations of the basic helical form. (*b*) *Treponema* has 8–20 evenly spaced coils. (*c*) *Leptospira* has numerous fine, regular coils and one or both ends curved. (*d*) *Borrelia* has 3–10 loose, irregular coils.

The term *syphilis* first appeared in a poem entitled "Syphilis sive Morbus Gallicus" by Fracastorius (1530) about a mythical shepherd whose name eventually became synonymous with the disease from which he suffered. Another early Latin name for it was *lues venerea*—literally, the plague of love. Attempting to distance themselves from the disease, various peoples at various times have called syphilis the "Italian disease" or the "French

**Treponema pallidum* (trep"-oh-nee'-mah pal'-ih-dum) Gr. *trepo,* turn, and *nema,* thread; L. *pallidum,* pale. The spirochete does not stain with the usual bacteriological methods.

disease." Another fitting name is the "Great Imitator," which points out that the complex stages of syphilis can easily be mistaken for numerous infectious and noninfectious diseases.

Epidemiology and Virulence Factors of Syphilis

Although infection can be provoked in laboratory animals, the human is evidently the sole natural host and source of *T. pallidum.* It is an extremely fastidious and sensitive bacterium that cannot survive for long outside the host, being rapidly destroyed by heat, drying, disinfectants, soap, high oxygen tension, and pH changes. It survives a few minutes to hours when protected by body secretions and about 36 hours in stored blood. Research with human subjects has demonstrated that the risk of infection from an infected sexual partner is 12% to 30%. Less common modes of transmission are passage to the fetus in utero and laboratory or medical accidents. Syphilitic infection through blood transfusion or exposure to fomites is rare.

Syphilis, like other STDs, has experienced periodic increases during times of social disruption (see figure 18.22). Currently, the number of reported cases is decreasing to the levels of the early 1960s. Because many cases go unreported, the actual incidence is likely to be several times higher than these reports show. Most cases tend to be concentrated in larger metropolitan areas among prostitutes, their contacts, and intravenous drug abusers. Syphilis continues to be a serious problem worldwide, especially in Africa and Asia. Persons with syphilis often suffer concurrent infection with other STDs. Coinfection with the AIDS virus can be an especially deadly combination with a rapidly fatal course.

Pathogenesis and Host Response

Brought into direct contact with mucous membranes or abraded skin, *T. pallidum* binds avidly by its hooked tip to the epithelium (figure 21.2). The number of cells required for infection using human volunteers was established at 57 organisms. At the binding site, the spirochete multiplies and penetrates the capillaries by dissolving the hyaluronic acid between endothelial cells. Within a short time, it moves into the circulation, and the body is literally transformed into a large receptacle for incubating the pathogen—virtually any tissue is a potential target. The slow generation time of this bacterium (about 30 hours) accounts for the slow but progressive nature of the disease.

Specific factors that account for the virulence of the syphilis spirochete appear to be outer membrane proteins. It produces no toxins and does not appear to kill cells directly. Studies have shown that, although phagocytes seem to act against it and several types of antitreponemal antibodies are formed, cell-mediated immune responses are unable to contain it. The primary lesion occurs when the spirochetes invade the spaces around arteries and stimulate an inflammatory response. Organs are damaged when granulomas form at these sites and block circulation.

Clinical Manifestations Untreated syphilis is marked by distinct clinical stages designated as primary, secondary, and tertiary syphilis (table 21.1). It also has latent periods of varying duration during which the disease is quiescent. The spirochete appears in the lesions and blood during the primary and secondary stages, and thus is communicable at these times. It is largely noncommunicable during the tertiary stage, though syphilis can be transmitted during early latency.

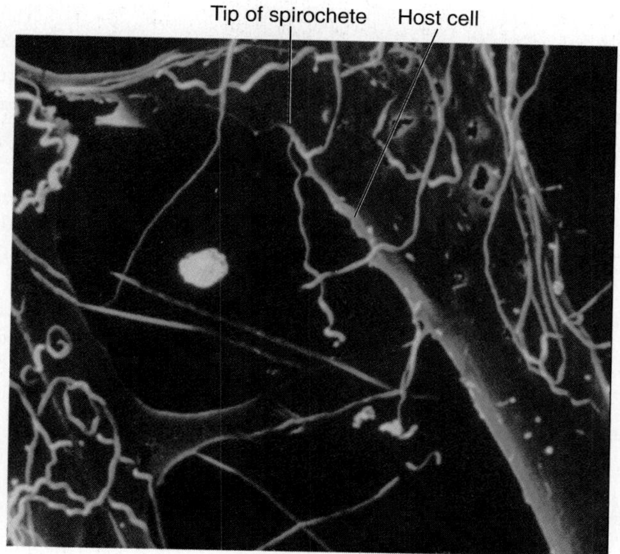

Figure 21.2
Electron micrograph of the syphilis spirochete attached to cells. Notice the hooklike nature of the specialized tip of the spirochete.

Primary Syphilis The earliest indication of syphilis infection is the appearance of a hard **chancre*** at the site of inoculation, after an incubation period that varies from 9 days to 3 months (figure 21.3). The chancre begins as a small, red, hard bump that enlarges and breaks down, leaving a shallow crater with firm margins. The base of the chancre beneath the encrusted surface swarms with spirochetes. Most chancres appear on the internal and external genitalia, but about 20% occur on the lips, oral cavity, nipples, fingers, or around the rectum. Because genital lesions tend to be painless, they may escape notice in some cases. Lymph nodes draining the affected region become enlarged and firm, but systemic symptoms such as fever or headache are virtually absent. The chancre heals spontaneously without scarring in 3 to 6 weeks, but this healing is deceptive, because the spirochete has escaped into the circulation and is entering a period of tremendous activity.

Secondary Syphilis About 3 weeks to 6 months (average is 6 weeks) after the chancre heals, the secondary stage appears. By then, many systems of the body have been invaded, and the signs and symptoms are more profuse and intense. Initially there is fever, headache, and sore throat, followed by lymphadenopathy and a peculiar red or brown rash that breaks out on all skin surfaces, including the palms and the soles (figure 21.4). Like the chancre, the lesions contain viable spirochetes and disappear spontaneously. The major complications develop in the bones, hair follicles, joints, liver, eyes, brain, and kidneys. In most individuals, the symptoms of secondary syphilis disappear in a few weeks, though certain symptoms can linger for months and years.

---

*chancre (shang'-ker) Fr. for canker; from L. *cancer,* crab. An injurious sore.

TABLE 21.1

SYPHILIS: STAGES, SYMPTOMS, DIAGNOSIS, AND CONTROL

Stage	Average Duration	Clinical Setting	Diagnosis	Treatment
Incubation	3 weeks	No lesion; treponemes adhere and penetrate the epithelium; after multiplying, they disseminate	Asymptomatic phase	Not applicable
Primary	2–6 weeks	Initial appearance of chancre at inoculation site; intense treponemal activity in body; chancre later disappears	Dark-field microscopy; VDRL, FTA-ABS, MHA-TP testing	Benzathine penicillin G, 2×10^6 units; aqueous benzyl or procaine penicillin G, 4.8×10^6 units
Primary latency	2–8 weeks	Healed chancre; little scarring; treponemes in blood; few if any symptoms	Serological tests (+)	As above
Secondary	2–6 weeks after chancre leaves	Skin, mucous membrane lesions; hair loss; patient highly infectious; fever, lymphadenopathy; symptoms can persist for months	Dark-field testing of lesions; serological tests	Double doses of penicillins listed above
Latency	6 months–8 or more years	Treponemes quiescent unless relapse occurs; lesions can reappear; seropositivity	Microscopy useless	As above
Tertiary	Variable, up to 20 years	Neural, cardiovascular symptoms; gummas develop in organs; seropositivity	Treponeme may be demonstrated by DNA analysis of tissue	As above

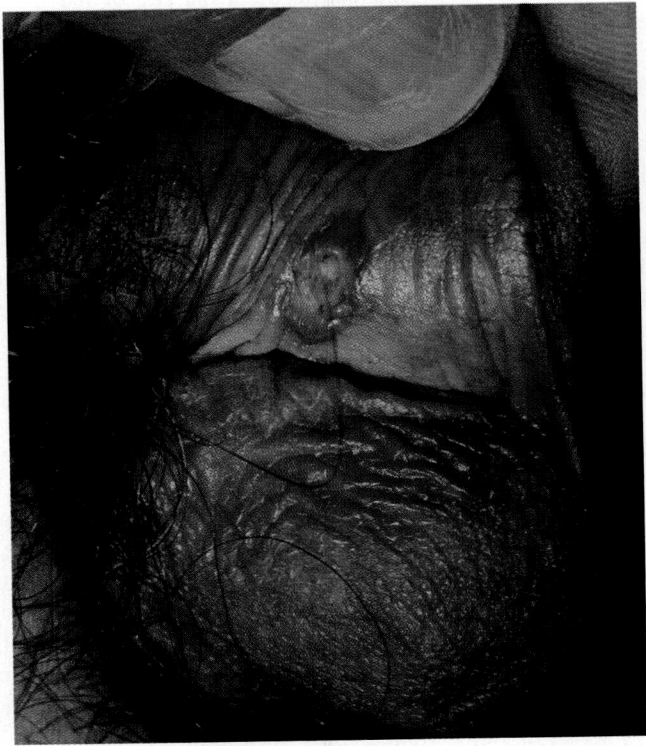

Figure 21.3
A chancre (the lesion of primary syphilis) on the underside of a scrotum. Chancres can appear singley and solitary or in multiple clusters.

Latency and Tertiary Syphilis After resolution of secondary syphilis, about 30% of people infected enter a highly varied latent period that can last for 20 years or longer. Latency is divisible into early and late phases, and though antitreponeme antibodies are readily detected, the parasite itself is not. The final stage of disease, late, or **tertiary, syphilis,** is quite rare today because of widespread use of antibiotics to treat other infections. By the time a patient reaches this phase, the combined action of the latent infection and the body's response to it produces severe pathologic complications. Cardiovascular syphilis results from damage to the small arteries in the aortic wall. As the fibers in the wall weaken, the aorta is subject to distension and fatal rupture. The same pathologic process can damage the aortic valves, resulting in insufficiency and heart failure.

In one form of tertiary syphilis, painful swollen syphilitic tumors called **gummas*** develop in tissues such as the liver, skin, bone, and cartilage (figure 21.5). Gummas are usually benign and only occasionally lead to death, but they can impair function. **Neurosyphilis** can involve any part of the nervous system, but it shows particular affinity for the blood vessels in the brain, cranial nerves, and dorsal roots of the spinal cord. The diverse reactions include severe headaches, convulsions, mental derangement, the Argyll Robertson pupil,[1] atrophy of the optic nerve and blindness. Destruction of parts of the spinal cord can lead to muscle wasting and loss of activity and coordination.

1. Perhaps the most common sign still seen today, this condition is caused by adhesions along the inner edge of the iris that fix the pupil's position into a small, irregular circle.

*gumma (goo'-mah) L. *gummi*, gum. A soft tumorous mass containing granuloma tissue.

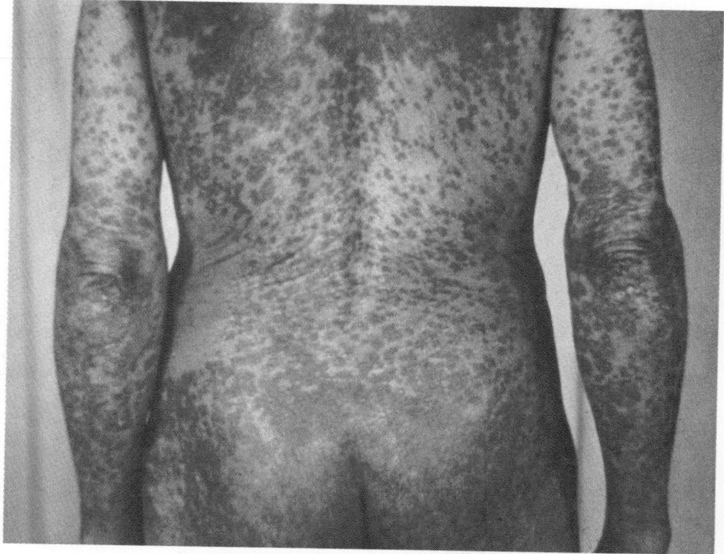

Figure 21.4

The skin rash in secondary syphilis can form on the trunk, arms, and even palms and soles (this latter location is particularly diagnostic). The rash does not hurt or itch and can persist for months.

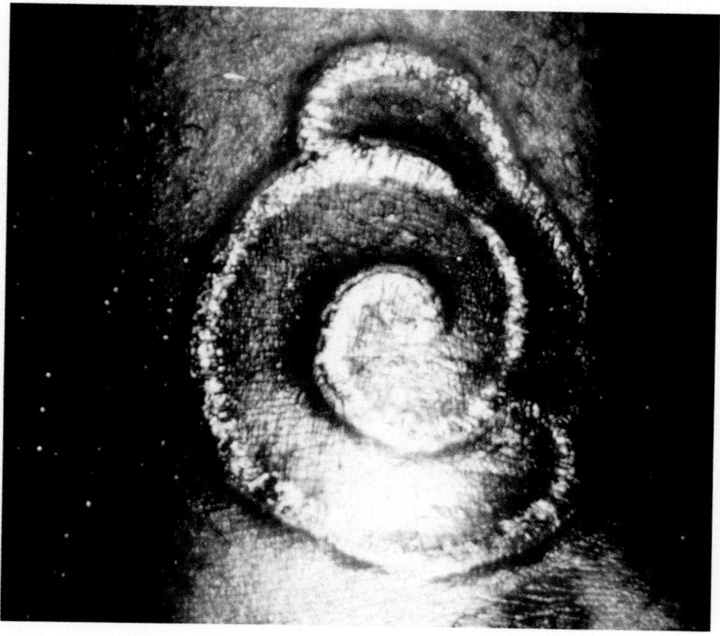

Figure 21.5

The pathology of late, or tertiary, syphilis. A ring-shaped erosive gumma appears on the arm of this patient. Other gummas can be internal.

Congenital Syphilis *Treponema pallidum* can pass from a pregnant woman's circulation into the placenta and can be carried throughout the fetal tissues. An infection leading to **congenital syphilis** can occur in any of the three trimesters, though it is most common in the second and third. The pathogen inhibits fetal growth and disrupts critical periods of development with varied consequences, ranging from mild to the extremes of spontaneous miscarriage or stillbirth. Early congenital syphilis encompasses the period from birth to 2 years of age and is usually first detected 3 to 8 weeks after birth. Infants often demonstrate such signs as nasal discharge (figure 21.6*a),* skin eruptions and loss, bone deformation, and nervous system abnormalities. The late form gives rise to an unusual assortment of stigmata in the bones, eyes, inner ear, and joints and causes the formation of Hutchinson's teeth (figure 21.6*b).* The number of congenital syphilis cases is closely tied to the incidence in adults, and, because it is sometimes not diagnosed, some children will die or sustain lifelong disfiguring disease.

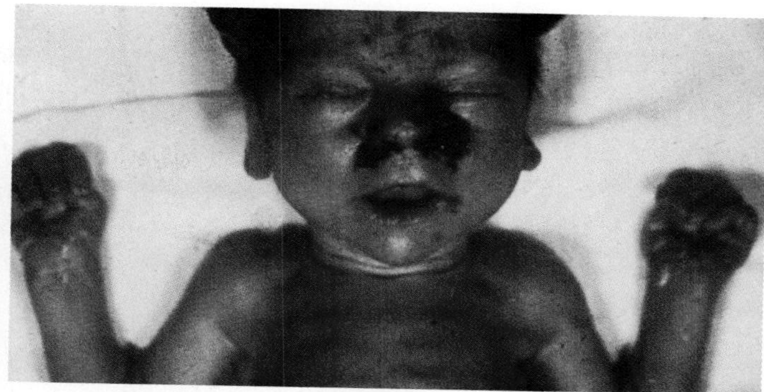

(a)

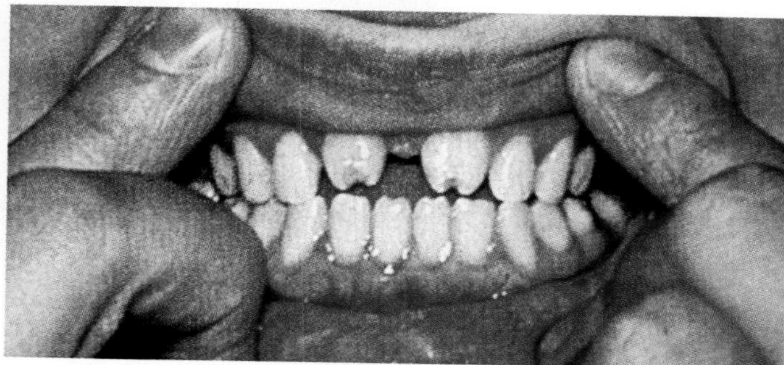

(b)

Figure 21.6

Congenital syphilis. (*a*) An early sign is snuffles, a profuse nasal discharge that obstructs breathing. (*b*) A common characteristic of late congenital syphilis is notched, barrel-shaped incisors (Hutchinson's teeth).

Clinical and Laboratory Diagnosis The pattern of syphilis imposes many complications on diagnosis. Not only do the stages mimic other diseases, but their appearance can be so separated in time as to seem unrelated. The chancre and secondary lesions must be differentiated from bacterial, fungal, and parasitic infections, tumors, and even allergic reactions. Overlapping symptoms of concurrent, sexually transmitted infections such as gonorrhea or chlamydiosis can further complicate diagnosis. The clinician must weigh presenting symptoms, patient history, and microscopic and serological tests in rendering a definitive diagnosis.

One rapid, direct method to diagnose primary, early congenital, and, to a lesser extent, secondary syphilis, is dark-field microscopy of a suspected lesion (figure 21.7). The lesions are

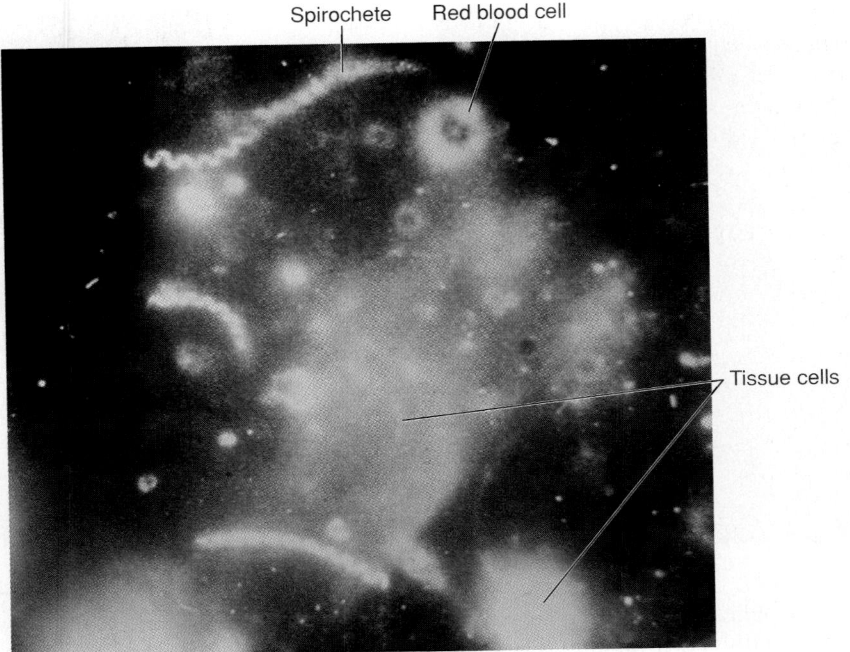

Figure 21.7
Treponema pallidum from a syphilitic chancre, viewed with dark-field illumination. Spirochetes contrast sharply with the red blood cells and tissue cells.

gently squeezed or scraped to extract clear serous fluid. A wet mount prepared from the exudate is then observed for the characteristic size, shape, and motility of *T. pallidum.* A single negative test is insufficient to exclude syphilis, because the patient may have removed the organism by washing, so two follow-up tests without prior washing are recommended. Another microscopic test for discerning the spirochete directly in samples is direct immunofluorescence staining with monoclonal antibodies (see figure D, page 559). Patient samples can also be tested with a DNA probe specific to various spirochete gene sequences (figure 4.28*d*).

Testing Blood for Syphilis If dark-field or direct antigen tests are negative, serological tests provide valuable though indirect diagnostic support. These tests are based upon detection of antibody formed in response to *T. pallidum* infection (see table 21.1). Several of the tests (Rapid Plasma Reagin [RPR], VDRL, Kolmer) are variations on the original test developed by Wasserman using cardiolipin, a natural constituent of many cells, as the antigen. Although anticardiolipin antibodies are not specific for syphilis, the test is an effective way to screen the population for people who may be infected.

Premarital blood tests have traditionally been used as a screening test for syphilis. Blood tests are also suggested for high-risk groups such as homosexuals, male and female prostitutes, people with other STDs, and pregnant women. In the case of a positive result, it is important that a series of serological tests be carried out to detect an elevated antibody titer indicative of active infection, because a single positive test may be due to a prior cured infection. Because the most common screening tests (RPR and VDRL) are based on reactions to a substance found normally in human tissue, biological false positives can occur, especially in patients with autoimmune diseases or impaired immunity. A more specific test is needed for those who are suspected of having a false-positive result.

Typical of these specific tests is the *T. pallidum* microhemagglutination assay (MHA-TP), which employs red blood cells that have been coated with treponemal antigen. Agglutination of the cells by serum indicates antitreponemal antibodies and infection. Another standard test is an indirect immunofluorescent method called the FTA-ABS (Fluorescent Treponemal Antibody Absorbance) test. The test serum is first absorbed with treponemal cells and reacted with antihuman globulin antibody labeled with fluorescent dyes. If antibodies to the treponeme are present, the fluorescence on the outside of these cells is highly visible with a fluorescent microscope (see figure 16.11*b*). The most sensitive and specific of all is the *T. pallidum* immobilization (TPI) test, in which live syphilis spirochetes are mixed with the test serum and observed microscopically for loss of motility. Other tests gaining in use are Western blot and DNA probe–based analyses.

Treatment and Prevention Penicillin G retains its status as a wonder drug in the treatment of all stages and forms of syphilis. It is given parenterally in large doses with benzathine or procaine to maintain a blood level lethal to the treponeme for at least 7 days (see table 21.1). Alternative drugs (tetracycline and erythromycin) are less effective and are indicated only if penicillin allergy has been documented. It is important that all patients be monitored for compliance or possible treatment failure.

The core of an effective prevention program depends upon detection and treatment of the sexual contacts of syphilitic patients. Public health departments and physicians are charged with the task of questioning the patients and tracing their contacts.

All individuals identified as being at risk, even if they show no signs of infection, are given immediate prophylactic penicillin in a single, long-acting dose. The barrier effect of a condom provides superior protection, but washing after sexual intercourse alone is not an adequate protective measure. Protective immunity apparently does arise in humans and in experimentally infected rabbits, which raises the prospect of an effective immunization program in the future. The recent cloning of treponemal surface antigens using recombinant DNA technology promises to improve development of vaccines and new diagnostic testing methods.

Nonsyphilitic Treponematoses

The other treponematoses are ancient diseases that closely resemble syphilis in their effects, though they are rarely transmitted sexually or congenitally. These infections, known as bejel, yaws, and pinta, are endemic to certain tropical and subtropical regions of the world, especially rural areas with unsanitary living conditions. The treponemes that cause these infections are nearly indistinguishable from those of syphilis in morphology and behavior. The diseases are slow and progressive and involve primary, secondary, and tertiary stages. They begin with local invasion by the treponeme into the skin or mucous membranes and its subsequent spread to subcutaneous tissues, bones, and joints. Drug therapy with penicillin, erythromycin, or tetracycline remains the treatment of choice for these treponematoses.

Bejel Bejel is also known as endemic syphilis and nonvenereal childhood syphilis. The pathogen, the subspecies *T. pallidum endemicum,* is harbored by a small reservoir of nomadic and seminomadic people in arid areas of the Middle East and North Africa. It is a chronic, inflammatory childhood disease transmitted by direct contact or shared household utensils and other fomites and is facilitated by minor abrasions or cracks in the skin or a mucous membrane. Often the infection begins as small, moist patches in the oral cavity (figure 21.8*a)* and spreads to the skin folds of the body and to the palms.

Yaws Yaws is a West Indian name for a chronic disease known by the regional names bouba, frambesia tropica, and patek. It is endemic to warm, humid, tropical regions of Africa, Asia, and South America. The microbe, subspecies *T. pallidum pertenue,* is readily spread by direct contact with skin lesions or fomites. Crowded living conditions and poor community or personal hygiene are contributing factors. The earliest sign is a large, abscessed papule called the "mother yaw," usually on the legs or lower trunk. After the initial lesion has healed, a secondary crop of moist nodular tumors appears. These tumors erode the skin, periosteum, and bones but do not penetrate to the viscera (figure 21.8*b.)* Late-stage yaws can result in mutilating ulcerations of the face and extremities. Yaws can be prevented by correcting predisposing conditions, shielding minor skin injuries from mechanical insect vectors, mass treatment, and surveillance for new cases.

Pinta The names *mal del pinto* and *carate* are regional synonyms for pinta, a chronic skin infection caused by *T. carateum.**

(a)

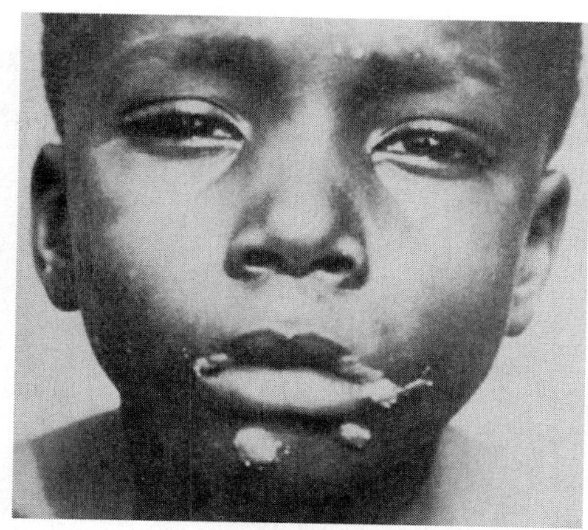

(b)

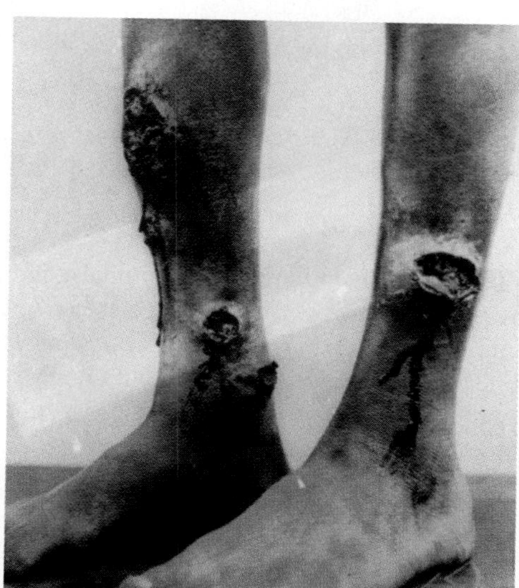

Figure 21.8

Endemic treponematoses. (*a*) Skin and membrane nodules in a young boy with endemic syphilis (bejel). (*b*) The clinical appearance of yaws. The large, draining sores are "mother yaws" that give rise to new lesions.

Transmission evidently requires several years of close personal contact accompanied by poor hygiene and inadequate health facilities. Even though pinta is not currently widespread, the disease is still found in isolated populations inhabiting the tropical forest and valley regions of Mexico and Central and South America. Infection begins in the skin with a dry, scaly papule reminiscent of psoriasis or leprosy. In time, pigmented secondary macules and blanched tertiary lesions appear. Pinta is not life-threatening, but it often creates scars on the afflicted area.

LEPTOSPIRA AND LEPTOSPIROSIS

Leptospires are typical spirochetes marked by tight, regular, individual coils with a bend or hook at one or both ends (figure 21.9). There are only two species in the genus: *Leptospira interrogans,**

**carateum* (kar-uh'-tee-um) From *carate,* the South American name for pinta.

**Leptospira* (lep"-toh-spy'-rah) Gr. *leptos,* slender or delicate, and *speira,* a coil.

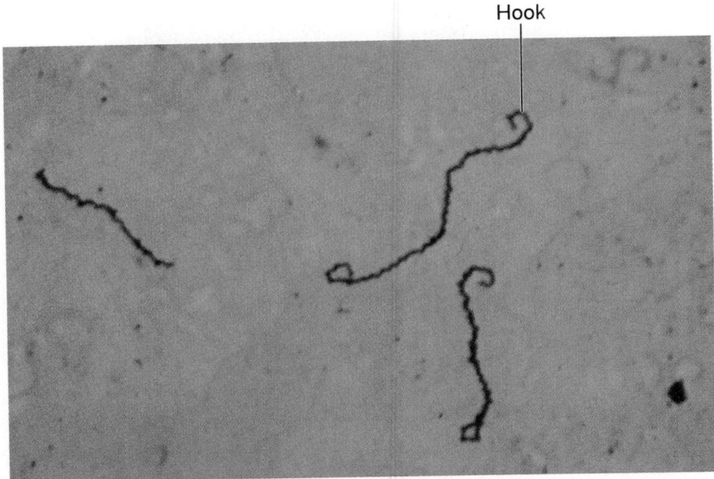

Figure 21.9

Leptospira interrogans, the agent of leptospirosis. Note the curved hook at one end of the spirochete. Its relative, *L. biflexa,* has flexible hooks at both ends.

which causes **leptospirosis** in humans and animals, and *L. biflexa,* a harmless, free-living saprobe. The two species are serologically, genetically, and physiologically distinct. *Leptospira interrogans* demonstrates nearly 200 serotypes distributed among various animal groups, which accounts for the extreme variations in leptospirosis among humans.

Epidemiology and Transmission of Leptospirosis

Leptospirosis is a zoonosis associated with wild animals such as rodents, skunks, raccoons, foxes, and some domesticated animals, particularly horses, dogs, cattle, and pigs. Although these reservoirs are distributed throughout the world, the disease is concentrated mainly in the tropics. Leptospires shed in the urine of an infected animal can survive for several months in neutral or alkaline soil or water. Infection occurs almost entirely through contact of skin abrasions or mucous membranes with animal urine or some environmental source containing urine. Transmission does not seem to occur through ingestion, animal bites, inhalation, or from human to human. In the United States, most of the relatively few cases (50–60) each year are reported among older male children and young adults exposed to polluted water. Soldiers involved in jungle training are also at high risk for infection.

Pathology of Leptospirosis and Host Response

Leptospirosis proceeds in two phases, and its principal targets are the kidneys, liver, brain, and eyes. During the early, or leptospiremic, phase, the pathogen appears in the blood and cerebrospinal fluid. Symptoms are sudden high fever, chills, headache, muscle aches, conjunctivitis, and vomiting. During the second, or immune, phase, the blood infection disappears because of the action of phagocytes, complement, and IgM antibodies. This period is marked by milder fever, headache due to leptospiral meningitis, and *Weil's syndrome,* a cluster of symptoms characterized by kidney invasion, hepatic disease, jaundice, anemia, and neurological disturbances. Long-term disability and even death can result from injury to the kidneys and liver, but they occur primarily with virulent strains and in elderly patients.

Diagnosis, Treatment, and Prevention

A history of environmental exposure, along with presenting symptoms, can support initial diagnosis of leptospirosis, but definitive diagnosis relies on dark-field microscopy of specimens, *Leptospira* culture, and serological tests. Isolation is accomplished by inoculating a specimen into special media or laboratory animals. Because leptospiral infection stimulates a strong humoral response, it is possible to test the patient's serum for its antibody titer. A fast, specific, and effective test called the macroscopic slide agglutination test is most often employed for routine screening. Live or formalinized *L. interrogans* is mixed with the patient's serum and observed for agglutination or lysis with a dark-field microscope.

Treatment with penicillin or tetracycline by the fourth day of illness rapidly reduces symptoms and shortens the course of disease, but after this time, therapy is less effective. Strain-specific vaccines made from killed cells are available for humans, dogs, and cattle, but with so many possible serotypes, these can confer protection only in areas where exposure to a specific endemic strain is anticipated. Vaccination is aimed at those with greatest risk such as combat troops training in jungle regions and animal care and livestock workers. The best controls are to wear protective footwear and clothing and to avoid swimming or wading in livestock watering ponds.

BORRELIA: ARTHROPOD-BORNE SPIROCHETES

Members of the genus ***Borrelia**** are morphologically distinct from other pathogenic spirochetes. They are comparatively larger, ranging from 0.2 to 0.5 μm in width and from 10 to 20 μm in length, and they contain 3 to 10 irregularly spaced and loose coils (see figure 21.1*d*) with an abundance (30 to 40) of periplasmic flagella. The nutritional requirements of *Borrelia* are so complex that the bacterium can be grown in artificial media only with difficulty.

Human infections with *Borrelia,* termed **borrelioses,** are all transmitted by some type of arthropod vector, usually ticks or lice. The two most important human diseases are relapsing fever and Lyme disease.

Epidemiology of Relapsing Fever

Borrelia hermsii, the cause of tick-borne relapsing fever, is carried by soft ticks (see microfile 21.5) of the genus *Ornithodoros.* The mammalian reservoirs of this zoonosis are squirrels, chipmunks, and other wild rodents, and the human is generally an accidental host. The spirochetes mature and persist in the salivary glands and intestines of the tick, and both the bite itself and the subsequent scratching initiate infection. Tick-borne relapsing fever occurs sporadically in the United States, usually in campers, backpackers,

---

**interrogans* (in-terr'-oh-ganz) Its appearance suggests a question mark at one end.

**Borrelia* (boh-ree'-lee-ah) Named after Amédé Borrel, a French bacteriologist.

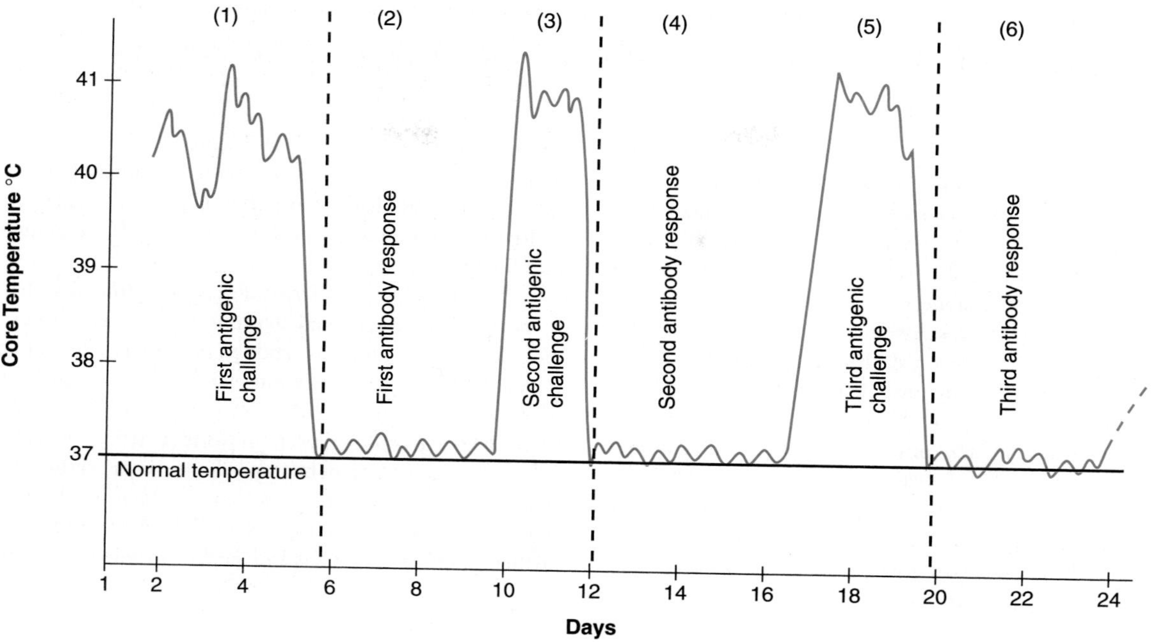

Figure 21.10

The pattern in relapsing fever: (1) Primary infection and fever; (2) initial antibody response with concurrent reduction in symptoms; (3) reinfection with a new antigenic type, causing renewed symptoms; and (4) a second antibody response, producing a second remission. (5,6) This pattern can continue for up to four relapses.

and forestry personnel who frequent the higher elevations of western states. The incidence of infection is higher in endemic areas of the tropics, especially where rodents have easy access to dwellings.

Whenever famine, war, or natural disasters are coupled with poor hygiene, crowding, and inadequate medical attention, epidemics of louse-borne relapsing fever occur. Such conditions favor the survival and spread of the louse vector *Pediculus humanus* (see microfile 21.5), which harbors the spirochete *B. recurrentis* in its body cavity. A host is infected when lice are smashed and accidentally scratched into a wound or the skin. Louse-borne fever is most common in parts of China, Afghanistan, and Africa.

Pathogenesis and the Nature of Relapses The pathologic manifestations are similar in tick- and louse-borne relapsing fever. Even though the blood level can reach an imposing 500,000 borrelias per milliliter, there is little sign of disease during the 2- to 15-day incubation period. The incubation period ends abruptly with the onset of high fever, shaking chills, headache, and fatigue. Later features of the disease include nausea, vomiting, muscle aches, and abdominal pain. Extensive damage to the liver, spleen, heart, kidneys, and cranial nerves occurs in many cases. Half of the patients hemorrhage profusely into organs, and some develop a rash on the shoulders, trunk, and legs. Untreated cases are often lengthy and debilitating and are attended by 5% to 40% mortality.

As the name *relapsing fever* indicates, the fever follows a fluctuating course that is explained by changes in the parasite and the attempts of the immune system to control it (figure 21.10). *Borrelia* have adopted a remarkable strategy for evading the immune system and avoiding destruction. They change surface antigens during growth, so that, in time, the initial antibodies become useless. These antigenically altered cells survive, multiply, and cause a second wave of symptoms. Eventually, the immune system forms new antibodies, but it is soon faced with yet another antigenic form. A single strain has been known to generate 24 distinct serological types, Eventually, cumulative immunity against the variety of antigens develops, and complete recovery can occur.

Diagnosis, Treatment, and Prevention A patient's history of exposure, clinical symptoms, and the presence of *Borrelia* in blood smears are very definitive evidence of borreliosis. Except for pregnant women and children under 7 years old, tetracycline is the treatment of choice. Chloramphenicol, erthyromycin, and doxycycline are also effective antimicrobial agents. Because vaccines are not available, prevention of relapsing fever is dependent upon controlling rodents and avoiding tick bites. Louse-borne relapsing fever is effectively arrested by improving hygiene.

Borrelia burgdorferi *and Lyme Disease*

Lyme disease is the most prominent borreliosis in the United States (microfile 21.1). Sixteen thousand five hundred cases were reported in 1996. Its spirochetal agent, ***Borrelia burgdorferi,**** is transmitted primarily by hard ticks of the genus *Ixodes*. In the northeastern part of the United States, *Ixodes scapularis* (the black-legged deer tick) passes through a complex 2-year cycle that involves two principal hosts (figure 21.11). As a larva or nymph, it feeds on the white-footed mouse, where it picks up the infectious agent. The nymph is relatively nonspecific and will try to feed on nearly any type of vertebrate, thus it is the form most likely to bite

**burgdorferi* (berg-dor'-fer-eye) Named for its discoverer, Dr. Willy Burgdorfer.

MICROFILE 21.1 THE DISEASE NAMED FOR A TOWN

In the 1970s, an enigmatic cluster of arthritis cases appeared in the town and surrounding suburbs of Old Lyme, Connecticut. This phenomenon caught the attention of nonprofessionals and professionals alike, whose persistence and detective work ultimately disclosed the unusual nature and epidemiology of Lyme disease. The story of its unraveling began in the home of Polly Murray, who, along with her family, was beset for years by recurrent bouts of stiff neck, swollen joints, malaise, and fatigue that seemed vaguely to follow a rash from tick bites. When Mrs. Murray's son was diagnosed as having juvenile rheumatoid arthritis, she became skeptical. Conducting her own literature research, she began to discover inconsistencies. Rheumatoid arthritis was described as a rare, noninfectious disease, yet, over an 8-year period, she found that 30 of her neighbors had experienced similar illnesses. Eventually this cluster of cases and several others were reported to state health authorities.

The reports caught the attention of Dr. Allen Steere, a rheumatologist with a CDC background. He was able to forge the vital link between the case histories, the disease symptoms, and the presence of unique spirochetes in ticks preserved by some of the patients. These same spirochetes had been previously characterized in 1981 by Dr. Willy Burgdorfer, though he did not realize their importance at the time (see figure 4.23*c*).

In the years since Lyme disease was formally characterized, retrospective studies showed that this is not really a new disease. The unique bull's-eye rash was reported in Europe at the turn of the century. Recent PCR analysis of tick museum specimens from 50 years ago documents the presence of *Borrelia burgdorferi.* It is now thought that Lyme disease has been present in North America for centuries.

The modern epidemic appears to be the result of an increase in the reservoirs such as deer and mice, which would naturally amplify tick and spirochete numbers. This effect has been compounded by greater mingling of humans and animals in woodland habitats.

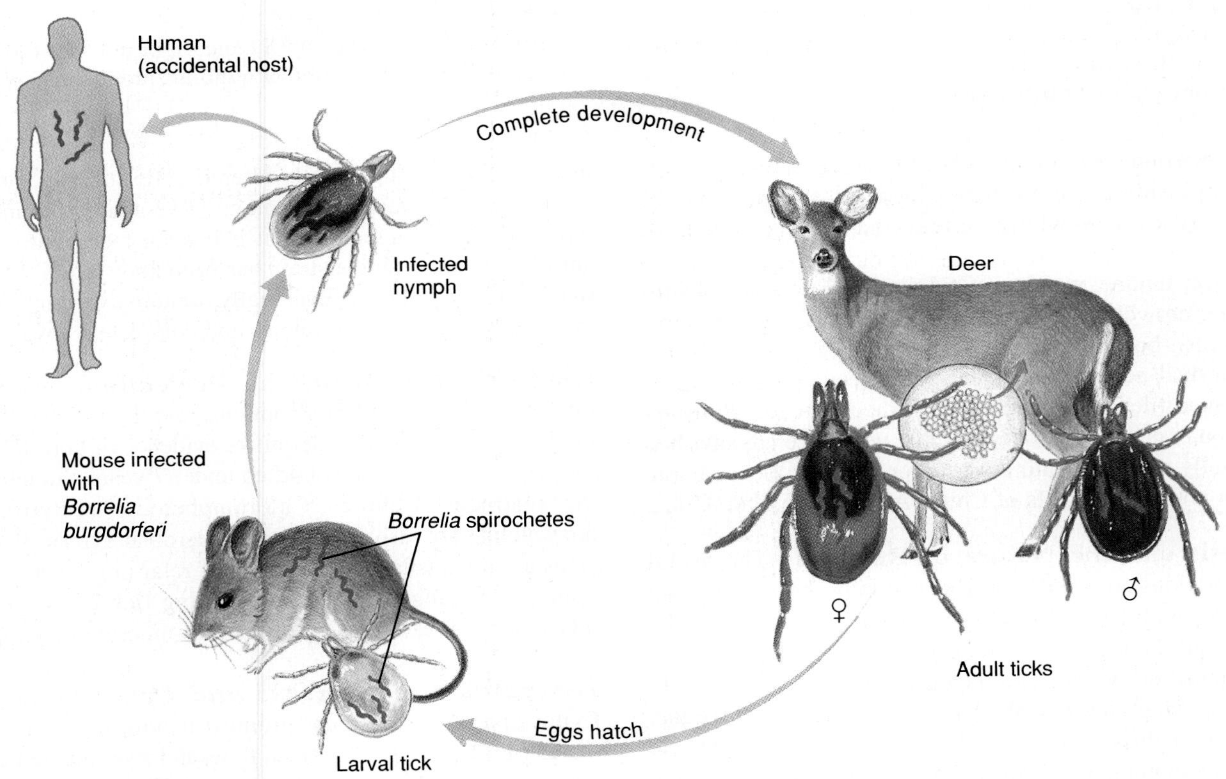

Figure 21.11

The cycle of Lyme Disease in the northeastern United States. The exact reservoir hosts vary from region to region in the United States and worldwide.

humans. The adult tick reproductive phase of the cycle is completed on deer. In California, the transmission cycle involves *Ixodes pacificus* and the dusky-footed woodrat reservoir.

The incidence of Lyme disease is showing a gradual upward trend from 3 cases per 100,000 population in 1990 to 6 in 1996. This may be partly due to improved diagnosis, but it also reflects changes in the numbers of hosts and vectors. The greatest concentrations of Lyme disease are in areas having high mouse and deer populations. Most of the cases recorded since 1978 have occurred in New York, Pennsylvania, Connecticut, New Jersey, Rhode

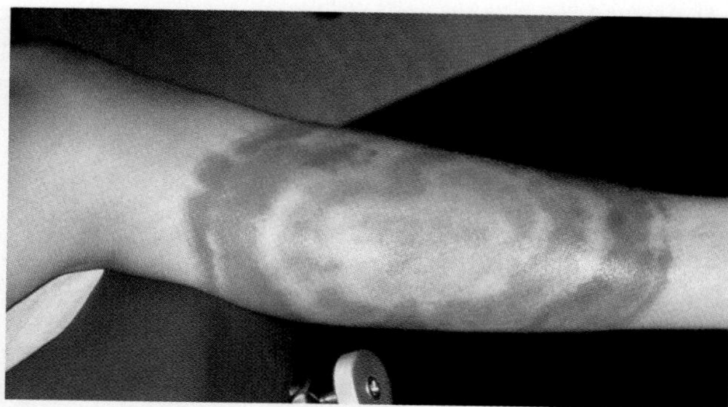

Figure 21.12

Lesions of Lyme disease on the lower leg. Note the flat, reddened rings in the form of a bull's-eye. Primary lesions often give rise to large numbers of secondary lesions in other locations.

Island, and Maryland, though the number in the Midwest and West is growing. Highest risk groups include hikers, backpackers, and people living in newly developed communities near woodlands and forests. Peak seasons are the summer and early fall.

Lyme disease is nonfatal but often evolves into a slowly progressive syndrome that mimics neuromuscular and rheumatoid conditions. An early symptom in 70% of cases is a rash at the site of a larval tick bite. The lesion, called *erythema migrans,* looks something like a bull's-eye, with a raised erythematous ring that gradually spreads outward and a pale central region (figure 21.12). Other early symptoms are fever, headache, stiff neck, and dizziness. If not treated or if treated too late, the disease can advance to the second stage, during which cardiac dysrhythmias and neurological symptoms such as facial palsy develop. After several weeks or months, a crippling polyarthritis can attack joints, especially in the European strain of the agent. Some people acquire chronic neurological complications that are severely disabling.

Diagnosis of Lyme disease can be difficult because of the range of symptoms it presents. Most suggestive are the ring-shaped lesions, isolation of spirochetes from the patient, and serological testing with an ELISA method that tracks a rising antibody titer (see microfile 16.2). Tests for spirochetal DNA in specimens is especially helpful for late-stage diagnosis. Early treatment with tetracycline and amoxicillin is effective, and other antibiotics such as ceftriaxone and azithromycin are used in late Lyme disease therapy. Because dogs can also acquire the disease, a vaccine has been marketed to protect them, and a human vaccine for high-risk populations is currently being tested and will probably be marketed in 1998. So widely distributed are the ticks, its larvae, and reservoir animals that anyone involved in outdoor activities should wear protective clothing, boots, leggings, and insect repellent containing DEET.* Individuals exposed to heavy infestation should routinely inspect their bodies for ticks and remove ticks gently without crushing, preferably with forceps or fingers protected with gloves, because it is possible to become infected by tick feces or body fluids.

*N,N-Diethy-M-toluamide. The active ingredient in OFF! and Cutter repellents.

Chapter Checkpoints

The Order Spirochaetales is a group of spirochetes that are mostly harmless saprobes except for pathogens in the genera *Treponema, Leptospira,* and *Borrelia.*

Many treponemes are part of the normal flora. Significant pathogens in this genus are *T. pallidum pallidum,* the agent of epidemic syphilis, *T. p. pertenue,* the agent of yaws, and *T. p. endemicum,* the agent of endemic syphilis. Other species are agents of oral diseases.

T. pallidum pallidum, a fastidious, obligate parasite, is the agent of venereal syphilis. It is limited to humans. Primary syphilis appears as an infective but painless chancre that usually heals spontaneously. Untreated infection progresses to a systemic condition called secondary syphilis, which is manifest in skin rashes. The spirochetes remain in body organs and can initiate tertiary syphilis after a latency of many years. During this stage, the spirochete can invade the heart, blood vessels, and brain and can form gummas.

Syphilis is difficult to diagnose because of its nonspecific symptoms and the concurrent presence of other STDs. Methods of detection include direct microscopic identification from clinical specimens and serological tests for host antibody (such as the TPI test). Treatment with penicillin G is still effective. Control depends on identification of all sexual contacts and use of barrier contraceptives.

Subspecies of *T. pallidum* are agents of bejel, yaws, and pinta, diseases endemic to specific tropical regions. Their symptoms mimic the stages of syphilis, but all are transmitted nonsexually.

Leptospira interrogans is a hooked spirochete that causes leptospirosis, a tropical zoonosis transmitted through direct contact with the urine of infected animals. It can be diagnosed by serological testing and treated with drugs early in infection.

Spirochetes in the genus *Borrelia* are larger and more loosely coiled than other genera. Pathogens in this group are agents of relapsing fever and Lyme disease. Both are transmitted by arthropods.

The two agents of relapsing fever are *B. hermsii* and *B. recurrentis. B. hermsii* is the agent of tick-borne relapsing fever. It is transmitted by soft ticks to humans from rodents. *B. recurrentis* is transmitted among humans by body lice. Relapsing fever is known by the recurrence of symptoms and the tendency of *Borrelia* to change their surface antigens. The disease can be prevented by control of rodent populations, prevention of tick bites, and good personal hygiene.

Borrelia burgdorferi is the agent of Lyme disease, a slowly progressive systemic infection that can be difficult to diagnose. It is transmitted to humans by hard ticks. The natural hosts are deer and rodents. The most prominent symptom is a bull's-eye rash radiating outward from the bite site. Tetracycline and amoxicillin are effective if administered early.

OTHER CURVIFORM GRAM-NEGATIVE BACTERIA OF MEDICAL IMPORTANCE

Two genera of curved or short spiral rods prominent in medical bacteriology are *Vibrio** and *Campylobacter.** Vibrios are comma-shaped rods with a single polar flagellum. Campylobacters are short spirals that have one or more flagella and a corkscrew type of motility. *Vibrio* is a member of the Family Vibrionaceae, and *Campylobacter* is in the Family Spirillaceae.

THE BIOLOGY OF *VIBRIO CHOLERAE*

A freshly isolated specimen of ***Vibrio cholerae**** reveals quick, darting cells slightly resembling a wiener or a comma (figure 21.13*a*). *Vibrio* shares many cultural and physiological characteristics with members of the Enterobacteriaceae, a closely related family. They are fermentative and grow on ordinary or selective media containing bile at 37°C, but, unlike the enterics, they are oxidase-positive. They possess unique O (somatic) antigens, H (flagella) antigens, and membrane receptor antigens that provide some basis for classifying members of the family.

Epidemiology of Cholera

Epidemic cholera, or Asiatic cholera, has been a devastating disease for centuries. Although the human intestinal tract was once thought to be the primary reservoir, it is now known that the parasite is free-living in certain endemic regions (microfile 21.2). In nonendemic areas such as the United States, the microbe is spread by water and food contaminated by asymptomatic carriers, but it is relatively uncommon. Sporadic outbreaks occasionally occur along the Gulf of Mexico, and the cholera vibrio is sometimes isolated from shellfish in this region. Worldwide, it is still a serious cause of morbidity and mortality, affecting several million people in endemic regions of Asia and Africa.

Pathogenesis of Cholera

After being ingested with food or water, *V. cholerae* encounters the potentially destructive acidity of the stomach. This hostile environment influences the size of the infectious dose (10^8 cells), though certain types of food also shelter the pathogen more readily than others. At the mucosa of the duodenum and jejunum, the vibrios penetrate the mucous barrier using their flagella, adhere to the microvilli of the epithelial cells and multiply there (see figure 21.23*b*). The cells are strictly epipathogens that do not enter the cells or invade the mucosa. The virulence of *V. cholerae* is due entirely to an enterotoxin called *cholera toxin* (CT) that disrupts the normal physiology of intestinal cells. When this toxin binds to specific intestinal receptors, a secondary signaling system is activated. Under the influence of this system, the cells shed large amounts of electrolytes into the intestine, an event that is accompanied by profuse water loss (figure 21.14). Most cases of cholera are mild or self-limited, but in children and weakened individuals, the disease can strike rapidly and violently.

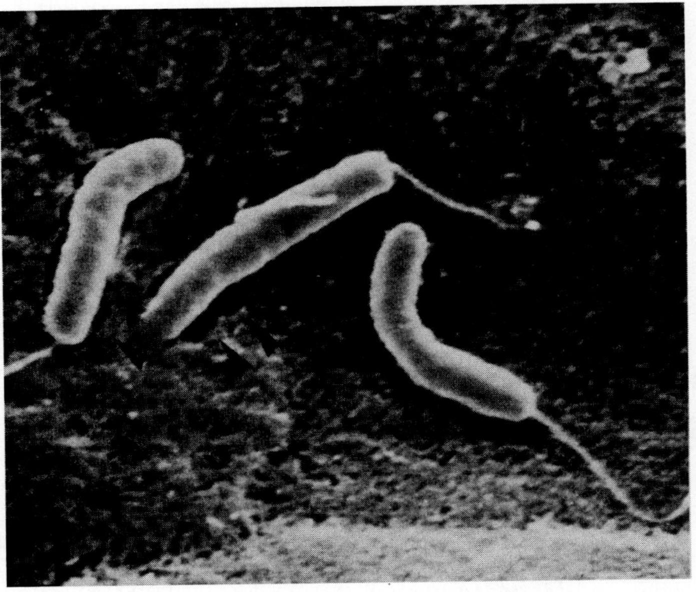

(a)

Vibrios Villus surface

(b)

Figure 21.13 (*a*) *Vibrio cholerae,* showing its characteristic curved shape and single polar flagellum. (*b*) Infectious vibrios burrowing into the surface of intestinal villi.

After an incubation period of a few hours to a few days, symptoms begin abruptly with vomiting, followed by copious watery feces called **secretory diarrhea.** This voided fluid is odorless and contains flecks of mucus, hence the description "rice-water stool." Fluid losses of nearly one liter per hour have been reported in severe cases, and an untreated patient can lose up to 50% of body weight during the course of the disease. The diarrhea causes

Vibrio (vib'-ree-oh) L. *vibrare,* to shake.

Campylobacter (kam"-pih-loh-bak'-ter). Gr. *campylo,* curved and *bacter,* rod.

cholerae (kol'-ur-ee) Gr. *chole,* bile. The bacterium was once named *V. comma* for its comma-shaped morphology.

MICROFILE 21.2 THE MYSTERIOUS EPIDEMIOLOGY OF CHOLERA

Cholera is a strictly human disease, and there are no demonstrable animal reservoirs. The microbe's natural reservoir appears to be the soil and water of two large Indian Rivers, the Ganges and the Brahmaputra. In these endemic areas, transmission of cholera is associated with pilgrimages to the river and religious practices involving mass bathing. Periodically during the past two centuries, this vibrio has migrated by means of waterways and human carriers to areas of Asia, Africa, Europe, the Middle East, and South America, and in the process it has caused at least seven pandemics.

A chain of epidemics struck several South and Central American countries from 1991 to the present. The origin of this latest outbreak is now thought to be a large oil tanker that emptied water into the ocean off the coast of Peru. Within a few months, nearly 700,000 people were stricken and thousands died. Many cases were traced to contaminated shellfish. By 1994, the epidemic had spread to Mexico and Argentina.

The pattern of cholera transmission and the onset of epidemics are greatly influenced by the season of the year and the climate. Cold, acidic, dry environments inhibit the migration and survival of *Vibrio,* whereas warm, monsoon, alkaline, and saline conditions favor them. The most persistent pandemic, which began in 1961 and continues until today, is due to a strain called the *El Tor* biotype. This strain survives longer in the environment, infects a higher number of people, and is more likely to be chronically carried than any other strain.

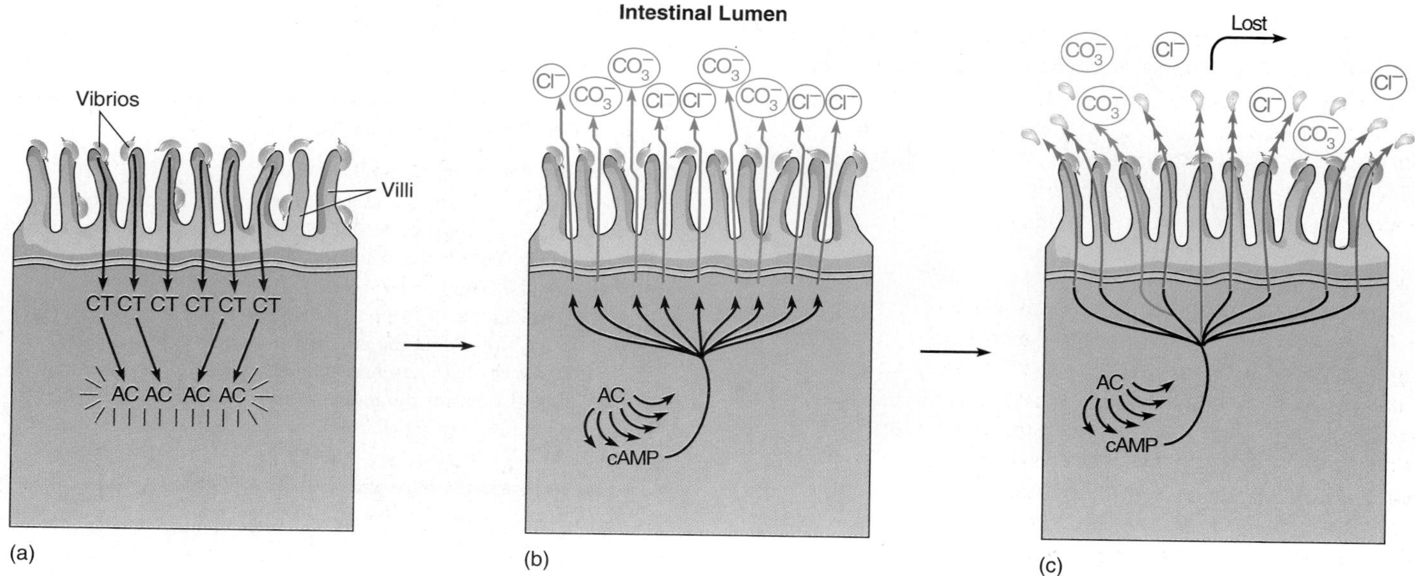

Figure 21.14 (*a*) The specific action of cholera toxin (CT) upon the intestinal epithelial cells heightens the activity of an enzyme called adenyl cyclase (AC). (*b*) This enzyme stimulates abnormally high levels of cAMP (cyclic adenosine monophosphate), a chemical messenger that normally mediates the action of hormones on cells but in higher concentrations promotes removal of anions by the cell membrane. (*c*) Under the constant action of cAMP, the cells begin to secrete large quantities of chloride (Cl^-) and bicarbonate (CO_3^-) ions into the intestinal lumen. Electrolyte loss is followed by water loss from epithelial cells, and the body experiences massive fluid depletion.

loss of blood volume, acidosis from bicarbonate loss, and potassium depletion that predispose the patient to muscle cramps, severe thirst, flaccid skin, and sunken eyes, and, in young children, coma and convulsions. Secondary circulatory consequences can include hypotension, tachycardia, cyanosis, and collapse from shock within 18 to 24 hours. If cholera is left untreated, death can occur in less than 48 hours, and the mortality rate approaches 55%.

Diagnosis and Remedial Measures

During epidemics, clinical evidence is usually sufficient to diagnose cholera. But confirmation of the disease is often required for epidemiologic studies and detection of sporadic cases. *Vibrio cholerae* can be readily isolated and identified in the laboratory from stool samples. Direct dark-field microscopic observation reveals characteristic curved cells with brisk, darting motility as confirmatory evidence. Immobilization or fluorescent staining of feces with group-specific antisera is supportive as well. Difficult or elusive cases can be traced by detecting a rising antitoxin titer in the serum.

The key to cholera therapy is prompt replacement of water and electrolytes, since their loss accounts for the severe morbidity and mortality. This can be accomplished by various rehydration techniques that replace the lost fluid and electrolytes (oral rehydration therapy [ORT]; microfile 21.3).

Cases in which the patient is unconscious or has complications from severe dehydration require intravenous replenishment as well. Only after the patient has been treated for physiological disruption will it be beneficial. Oral antibiotics such as tetracycline and drugs such as trimethoprim-sulfa can terminate the diarrhea in 48 hours, and they also promote recovery and diminish the period of vibrio excretion.

MICROFILE 21.3 ORAL REHYDRATION THERAPY

A sugar-salt solution taken by mouth appears to be a miracle cure for cholera. This relatively simple formulation, developed by the World Health Organization, consists of a mixture of the electrolytes sodium chloride, sodium bicarbonate, potassium chloride, and glucose or sucrose dissolved in water (table 21A). When administered early in amounts ranging from 100 to 400 ml/hour, the solution can restore patients in 4 hours, often bringing them literally back from the brink of death. It works even if the individual has diarrhea, because the particular combination of ingredients is well tolerated and rapidly absorbed. Infants and small children who once would have died now survive so often that the mortality rate for treated cases of cholera is near zero. This therapy has several advantages, especially for countries with few resources. It does not require medical facilities, high-technology equipment, or complex medication protocols. It is also inexpensive, noninvasive, fast-acting, and useful for a number of diarrheal diseases besides cholera.

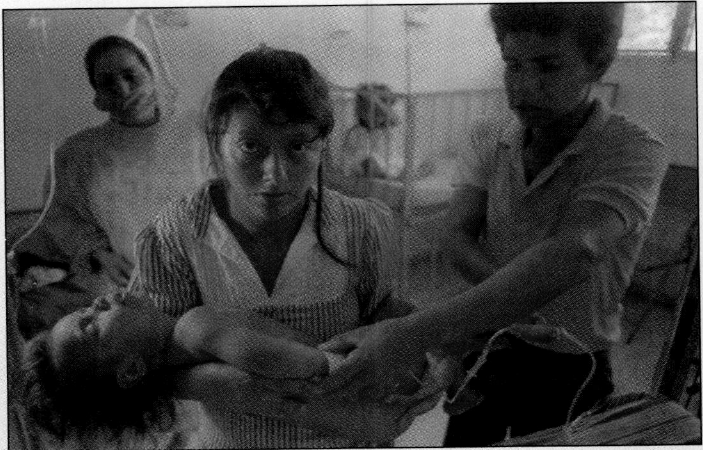

Note the sunken eyes in this child suffering from severe dehydration.

TABLE 21A

RECOMMENDATIONS FOR ORAL REHYDRATION THERAPY (ORT)

Standard Composition of Oral Rehydration Solution (ORS)	Grams per liter (g/l)	Concentration in Millimoles
NaCl	3.5 g/l	90 (Na) Cl (80) total
$NaHCO_3$	2.5 g/l	30 (HCO_3)
KCl	1.5 g/l	20 (K)
Glucose*	20 g/l	111

*Sucrose or rice powder may be substituted if glucose is not available.

Guidelines for Preparation and Treatment

- Components are dissolved in 1 liter of water that has been boiled to disinfect it. It is thoroughly mixed and tasted for saltiness. It should be only slightly salty. This product is similar to a commercial product called Pedialyte or Oralyte.
- Moderately dehydrated children and adults should receive frequent small amounts of ORS at the rate of 4–8 ounces per hour over 6 hours. Plain water can be administered in between ORS administrations.
- To check the patient for the return of normal hydration, test the skin turgor (pinch it to see if a fold remains), and observe whether the eyes are still sunken.

Effective prevention is contingent upon proper sewage disposal and water purification. Detecting and treating carriers with mild or asymptomatic cholera is a serious goal but one that is frequently difficult to attain because of inadequate medical provisions in those countries where cholera is endemic. Vaccines are available for travelers and people living in endemic regions. One contains killed cholera vibrios, but it protects for only 6 months or less. An oral vaccine containing live, attenuated vibrios is an effective alternative.

Vibrio parahaemolyticus and *Vibrio vulnificus*: Pathogens Carried by Seafood

Two additional relatives of *V. cholerae* share its morphology, physiology, and ecological adaptation. *Vibrio parahaemolyticus* and *V. vulnificus* are both salt-tolerant inhabitants of coastal waters and associate with marine invertebrates. In temperate zones, vibrios survive over the winter by settling into the ocean sediment, and, when resuspended by upwelling during the warmer seasons, they become incorporated into the food web, eventually growing on fish, shellfish, and other edible seafood.

Features of V. parahaemolyticus *Gastroenteritis*

Vibrio parahaemolyticus food infection, an acute form of gastroenteritis, was first described in Japan more than 30 years ago. The great majority of cases appear in individuals who have eaten raw, partially cooked, or poorly stored seafood. The vehicles most often implicated are squid, mackerel, sardines, crabs, tuna, shrimp, oysters, and clams. Outbreaks tend to be concentrated along coastal regions during the summer and early fall. The incubation period of nearly 24 hours is followed by explosive, watery diarrhea accompanied by nausea, vomiting, abdominal cramps, and sometimes fever. *Vibrio* toxins cause symptoms that last about 72 hours but can persist for 10 days.

Disease caused by *V. vulnificus* is very similar in symptoms to that caused by *V. parahemolyticus.* It is most often associated with ingesting raw oysters and can have a much more severe outcome in critically ill patients with diabetes or liver

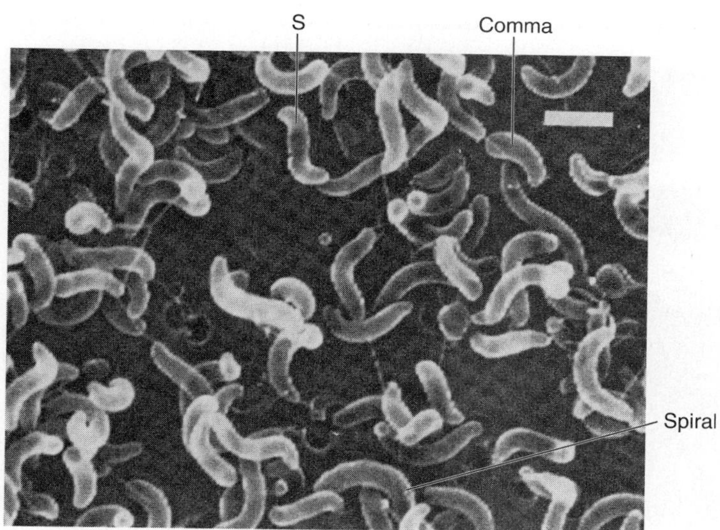

Figure 21.15

Scanning micrograph of *Campylobacter jejuni,* showing comma, S, and spiral forms.

disease. It is the leading cause of death from food-borne illness in some areas.

Treatment of severe gastroenteritis can require fluid and electrolyte replacement, and occasionally antibiotics, and hospitalization may be indicated. Control measures aim to keep the bacterial count in all seafood below the infective dose by continuous refrigeration during transport and storage, sufficient cooking temperatures, and prompt serving. Consumers of raw oysters (or other shellfish) should understand the possible risk of eating raw seafood. In some regions, the USDA mandates that health warnings be posted about this danger in markets and restaurants.

DISEASES OF THE *CAMPYLOBACTER* VIBRIOS

Campylobacters are slender, curved or spiral bacilli propelled by polar flagella at one or both poles, often appearing in S-shaped or gull-winded pairs (figure 21.15). These bacteria tend to be microaerophilic inhabitants of the intestinal tract, genitourinary tract, and oral cavity of humans and animals. Species of *Campylobacter* most significant in medical and veterinary practice are *C. jejuni* and *C. fetus.* A close relative, *Helicobacter* (formerly *Campylobacter) pylori,* is an unusual vibrio implicated in certain types of stomach ulcers (microfile 21.4).

Campylobacter jejuni *Enteritis*

*Campylobacter jejuni** has recently emerged as a pathogen of such imposing proportions that it is considered one of the most important causes of bacterial gastroenteritis worldwide. Early results of epidemiologic and pathologic studies have shown that this species is a primary pathogen transmitted through contaminated beverages and food, especially water, milk, meat, and chicken. The source of a Vermont outbreak of enteritis was traced to a water storage reservoir that supplied most of the community and eventually caused disease in 20% of the population. Cases are extremely common in developing countries because of frequent contact with animal reservoirs and unreliable sanitation.

When ingested, *C. jejuni* cells reach the mucosa at the last segment of the small intestine (ileum) near its junction with the colon; they adhere, burrow through the mucus, and multiply. Symptoms of headache, fever, abdominal pain, and bloody or watery diarrhea commence after an incubation period of 1 to 7 days. The mechanisms of pathology appear to involve a heat-labile enterotoxin called CJT that stimulates a secretory diarrhea like that of cholera.

Diagnosis of *C. jejuni* enteritis requires isolation from fecal samples and occasionally blood samples. More rapid presumptive diagnosis can be obtained from direct examination of feces with a dark-field microscope, which accentuates the characteristic curved rods and darting motility. Resolution of infection occurs in most instances with simple, nonspecific rehydration and electrolyte balance therapy. In more severely affected patients, it may be necessary to administer erythromycin, tetracycline, aminoglycosides, or quinolones. Because vaccines are yet to be developed, prevention depends upon rigid sanitary control of water and milk supplies and care in food preparation.

Traditionally of interest to the veterinarian, *C. fetus* (subspecies *venerealis)* causes a sexually transmitted disease of sheep, cattle, and goats. Its role as an agent of abortion in these animals has considerable economic impact on the livestock industry. About 40 years ago, the significance of *C. fetus* as a human pathogen was first uncovered, though its exact mode of transmission in humans is yet to be clarified. This bacterium appears to be an opportunistic pathogen that attacks debilitated persons or women late in pregnancy. Diseases to which *C. fetus* has been linked are meningitis, pneumonia, arthritis, fatal septicemic infection in the newborn, and occasionally, sexually transmitted proctitis in adults.

jejuni (jee-joo'-nye) L. *jejunum.* The small section of intestine between the duodenum and the ileum.

Chapter Checkpoints

The genera *Vibrio* and *Campylobacter* are curved or short spiral-shaped bacteria that are actively motile. *Vibrio cholerae* is the agent of epidemic, or Asiatic, cholera. It is spread through contaminated water. Its virulence factor is an enterotoxin that causes profuse diarrhea and dehydration. Diagnosis is by microscopic identification of the bacteria in stool specimens or serological tests. Therapy includes prompt replacement of water and electrolytes followed by antibiotics. Preventive measures focus on effective water purification.

V. parahaemolyticus and *V. vulnificus* are agents of seafood gastroenteritis. They are most often acquired by eating raw shellfish that have been exposed to contaminated wastewater.

Campylobacter jejunum is the agent of campylobacter enteritis. This vibrio is transmitted to humans through contaminated water and animal products. Its enterotoxin, CJT, stimulates a diarrhea similar to cholera. Treatment in severe cases is similar to that for cholera.

MICROFILE 21.4 *HELICOBACTER PYLORI:* IT'S ENOUGH TO GIVE YOU ULCERS

Although the human stomach is usually regarded as a hostile habitat, an unusual vibrio, *Helicobacter pylori,* has found its own special niche there. Not only does it thrive in the acidic environment, but evidence has clearly linked it to a variety of gastrointestinal ailments. It is known to cause an inflammatory condition of the stomach lining called gastritis and is implicated in 90% of stomach and duodenal ulcers. It is also an apparent cofactor in the development of a common type of stomach cancer called adenocarcinoma.

The curved cells were first detected by J. Robin Warren in 1979 in stomach biopsies from ulcer patients. He and an assistant, Barry J. Marshall, isolated the microbe in culture and even served as guinea pigs by swallowing a good-sized inoculum to test its effects. Both developed transient gastritis.

Studies have revealed that the pathogen is present in a large proportion of people. It occurs in the stomachs of 25% of healthy middle-aged adults and in more than 60% of adults over 60 years of age. *Helicobacter pylori* is probably transmitted from person to person by the oral-oral or oral-fecal route. It is most common in countries where hygiene and sanitation are not rigorous. It seems to be acquired early in life and carried asymptomatically until its activities begin to damage the digestive mucosa. Because other animals are also susceptible to *H. pylori* and even develop chronic gastritis, it has been proposed that the disease is a zoonosis transmitted from an animal reservoir. Reports on its high frequency in cats suggest that they may be a significant reservoir source. It can also be spread by house flies acting as mechanical vectors.

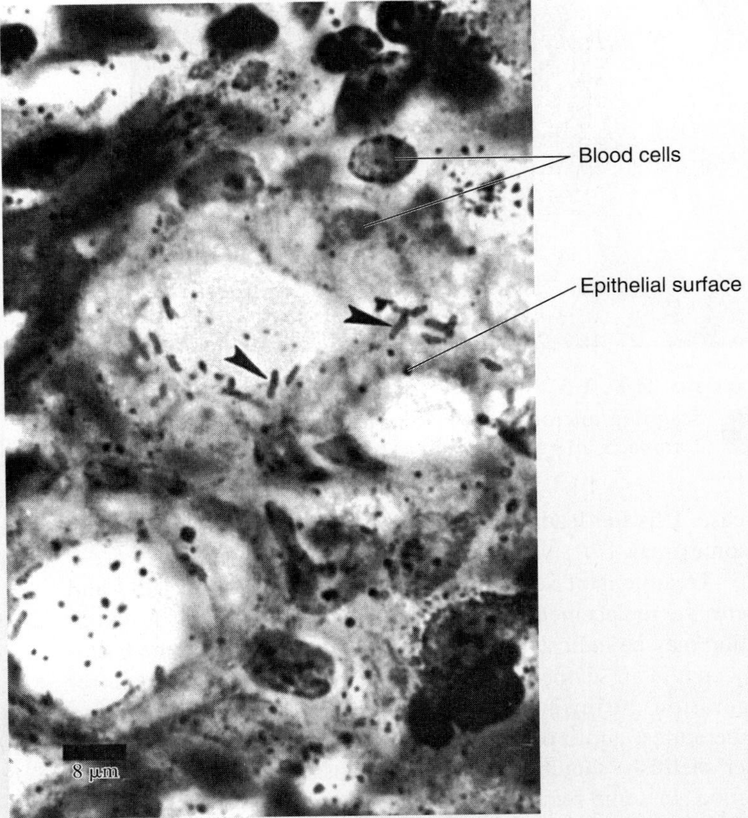

Histological section of stomach showing Helicobacter pylori attached to the epithelial surface (arrows). Note infiltration by blood cells indicative of inflammation.

Other studies have helped to explain how the vibrio takes up residence in the gastrointestinal tract. First, it bores through the outermost mucus that lines the epithelial tissue. Then it attaches to specific binding sites on the cells and entrenches itself. It turns out that one receptor specific for *Helicobacter* is the same receptor as type O blood, which accounts for the higher rate of ulcers in people with this blood type (1.5 to 2 times). Another protective adaptation is the formation of urease, an enzyme that converts urea into ammonium and bicarbonate, both alkaline compounds that can neutralize stomach acid. As the immune system recognizes and attacks the pathogen, infiltrating white blood cells begin to damage the epithelium to some degree, leading to chronic active gastritis. In some people, these lesions lead to deeper erosions and ulcers and, eventually, can lay the groundwork for a cancer to develop.

Helicobacter pylori is isolated primarily from biopsy specimens. Because it gives off large amounts of urease, it can be initially identified by inoculating a rapid urease test (see table 20.2). A less invasive technique is the urea breath test. This test involves swallowing labeled urea that gives off radioactive CO_2 in exhaled air if *H. pylori* is present. A PCR test for microbial DNA in biopsies and feces is also being marketed. The knowledge of this infection has useful applications in treatment. Gastritis and ulcers have traditionally been treated with drugs (Tagamet, Zantac) that suppress symptoms by slowing the secretion of acid in the stomach and must be taken continuously for indefinite periods, and relapses are common. The newest recommended therapy is 2–4 weeks of antimicrobic therapy with clarithromycin and Zantac, which inhibits the formation of stomach acid. This regimen can actually cure the infection permanently.

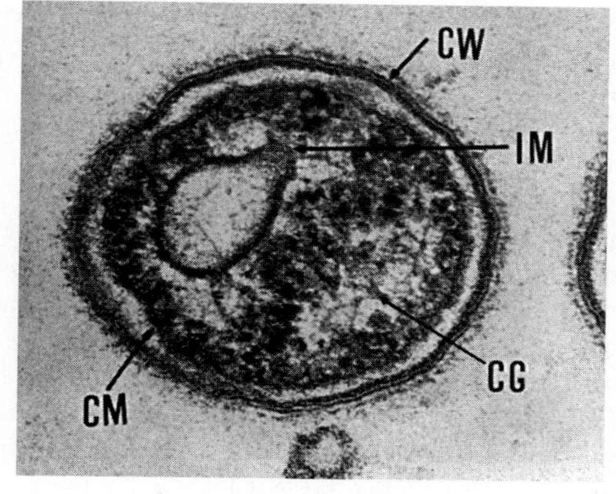

(a)

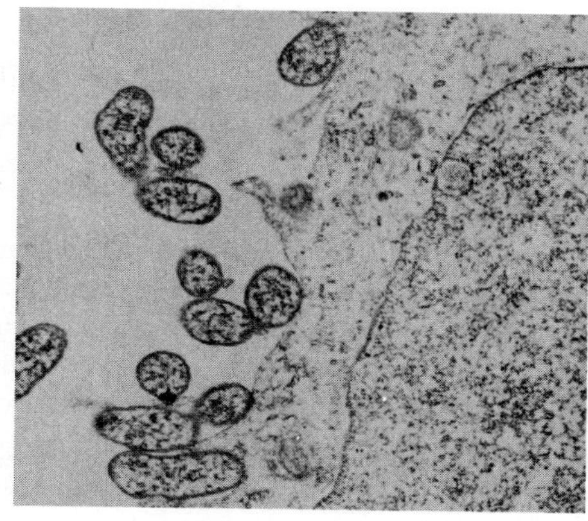
(b)

Figure 21.16

(*a*) The morphology of *Rickettsia.* Several features, including the cell wall (CW), cell membrane (CM), chromatin granules (CG), and mesosome (IM), identify these as tiny, pleomorphic, gram-negative bacteria (185,000×). (*b*) View of rickettsias adhering to the surface of a mouse tissue culture cell.

MEDICALLY IMPORTANT BACTERIA OF UNIQUE MORPHOLOGY AND BIOLOGY

Bacterial groups that exhibit atypical morphology, physiology, and behavior are the following: (1) rickettsias and chlamydias, obligately parasitic gram-negative coccobacilli, and (2) mycoplasmas, highly pleomorphic cell-wall-deficient bacteria.

FAMILY RICKETTSIACEAE

The Family Rickettsiaceae contains about 23 species of pathogens, mostly in the genus *Rickettsia.** Other members include *Ehrlichia** and the recently renamed *Orientia* (formerly *Rickettsia).* These organisms are known commonly as **rickettsiae,** or **rickettsias,** and the diseases they cause are called **rickettsioses.**

Other genera originally included in this Family were *Coxiella** and *Bartonella,** on the basis of similarities in diseases and transmission. Genetic tests using rRNA further clarified their position. *Coxiella* is now grouped in a Family with *Legionella,* and *Bartonella* is allied with *Brucella.*

The rickettsias are all obligate to their host cells and require live cells for cultivation; they also spend part of their life cycle in the bodies of arthropods, which serve as vectors. Rickettsioses are among the most important emerging diseases. Six of the 14 recognized diseases have been identified in the last 12 years.

**Rickettsia* (rik'-ett'-see-ah) After Howard Ricketts, an American bacteriologist who worked extensively with this group.

**Ehrlichia* (ur-lik'-ee-ah) After Paul Ehrlich, a German immunologist.

**Coxiella* (kox"-ee-el'-ah) For H. R. Cox, an American bacteriologist who first isolated this bacterium in the United States.

**Bartonella* (barr"-tun-el'-ah) After A. L. Barton, a Peruvian physician who first described the genus.

Morphological and Physiological Distinctions of Rickettsias

Rickettsias possess a gram-negative cell wall, binary fission, metabolic pathways for synthesis and growth, and both DNA and RNA. They are among the smallest cells, ranging from 0.3 to 0.6 μm wide and from 0.8 to 2.0 μm long. They are nonmotile pleomorphic rods or coccobacilli (figure 21.16).

The precise nutritional requirements of the rickettsias have been difficult to demonstrate because of a close association with host cell metabolism. Their obligate parasitism originates from an inability to metabolize AMP, an important precursor to ADP and ATP, which they must obtain from the host. Rickettsias are generally sensitive to environmental exposure, although *R. typhi* can survive several years in dried flea droppings.

Distribution and Ecology of Rickettsial Diseases

The Role of Arthropod Vectors The rickettsial life cycle depends upon a complex exchange between blood-sucking arthropod[2] hosts and vertebrate hosts (microfile 21.5). Eight tick genera, two fleas, one mite, and one louse are involved in the spread of rickettsias to humans. Humans accidentally enter the zoonotic life cycles through occupational contact with the animals except in the cases of louse-borne typhus and trench fever. Because humans are often incidental or "dead-end" hosts, they are not a regular source of infection. Most vectors apparently harbor rickettsias with no ill effect, but others, like the human body louse, die from typhus infection and do not continuously harbor the pathogen. In certain vectors such as the tick of Rocky Mountain spotted fever, rickettsias are transferred transovarially, from the infected female to her eggs. This event has profound impact; the continuous

2. Ticks and mites are in the Class arachnida, and lice and fleas are in the Class insecta; both in the phylum Arthropoda.

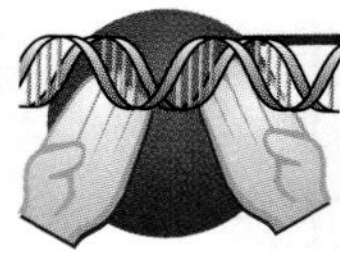

MICROFILE 21.5 OF MICE AND MITES AND LICE AND BITES: THE ARTHROPOD VECTORS OF INFECTIOUS DISEASE

Many bacterial pathogens have evolved with and made complex adaptations to the bodies of arthropods such as ticks and fleas. These same animals happen also to be natural ectoparasites of humans, feeding on blood or tissue fluids. In this role as biological vectors, they are an important source of zoonotic infections in humans.

Ticks Distributed the world over are 810 species of ticks, all of them ectoparasites and about 100 of them vectors of infectious disease. Two families of ticks are differentiated—soft *(argasid)* and hard *(ixodid)*—on the basis of the absence or presence of a dorsal shield called the scutum. Argasids seek specific, sheltered habitats in caves, barns, burrows, or even bird cages, and they parasitize wildlife during nesting seasons and domesticated animals throughout the year. Endemic relapsing fever is transmitted by this type of tick.

In contrast, the ixodids prefer confined habitats, and many have adapted to a wide-ranging life-style, hitchhiking along as their hosts wander through forest, savanna, or desert regions. Depending upon the species, ticks feed during larval, nymph, and adult metamorphic stages. The longevity of ticks is formidable; metamorphosis can extend for 2 years, and adults can survive for 4 years away from a host without feeding. Many tick-borne agents cause mild infection in their primary wildlife hosts but serious infection in livestock and humans. Ixodid ticks are implicated in several rickettsial infections, most particularly, Rocky Mountain spotted and Q fevers and human ehrlichosis. Ticks also play host to *Borrelia* (Lyme disease) and arboviruses.

Mites So abundant are mites that the 30,000 classified species are estimated to represent less than one-tenth of those in existence. Most species are free-living, and only a handful of the parasitic ones are of direct medical importance. The house mouse mite, *Liponyssoides sanguineus,* is a vector for the etiologic agent of rickettsialpox. Although domestic mice and rats are the natural hosts for this mite, it is not specific in its food getting and will readily nibble on humans if the opportunity arises. Larval mites called chiggers develop in the undergrowth of grasslands, forest, fields, gardens, and riverbanks that are frequented by humans. The larvae may escape notice at first, since they are minute and the bite is initially painless. But the bite can become intensely itchy in sensitive individuals. Once the larval stage is passed, the nymph and adult feed on nonvertebrate hosts. Chiggers are the primary reservoir and carriers of *Orientia tsutsugamushi* (scrub typhus).

Fleas Fleas are laterally flattened, wingless insects with well-developed jumping legs and prominent probosci for piercing the skin of warm-blooded animals. They are known for their extreme longevity and resistance, and many are notorious in their nonspecificity, passing with ease from wild or domesticated mammals to humans. In response to mechanical stimulation and warmth, fleas jump onto their targets and crawl about, feeding as they go. A well-known vector is the oriental rat flea, *Xenopsylla cheopis* that transmits *Rickettsia typhi,* the cause of murine typhus. The flea harbors the pathogen in its gut and periodically contaminates the environment with virulent rickettsias by defecating. This same flea is involved in the transmission of plague (see figure 20.19).

Lice Lice are small, flat insects equipped with biting or sucking mouthparts. The lice of humans usually occupy head and body hair *(Pediculus humanus)* or pubic, chest, and axillary hair *(Phthirus pubis).* They feed by gently piercing the skin and sucking blood and tissue fluid. Infection develops when the louse or its feces is inadvertently crushed and rubbed into wounds, skin, eyes, or mucous membranes. Lice are involved in transmitting *R. prowazekii* (epidemic typhus) and *Borrelia recurrentis* (relapsing fever).

inheritance of the microbe through multiple generations of ticks creates a long-standing reservoir.

These arthropods feed on the blood or tissue fluids of their mammalian hosts, but not all of them transmit the rickettsial pathogen through direct inoculation with saliva. Ticks and mites directly inoculate the skin lesion as they feed, but fleas and lice harbor the infectious agents in their intestinal tracts. During their stay on the host's body, these latter insects defecate or are smashed, thereby releasing the rickettsias onto the skin or into a wound. Ironically, scratching the bite helps the pathogen invade deeper tissues.

General Factors in Rickettsial Pathology and Isolation

A common target in rickettsial infections is the endothelial lining of the small blood vessels. The bacteria recognize, enter, and multiply within endothelial cells, causing necrosis of the vascular lining. Among the immediate pathologic consequences are vasculitis, perivascular infiltration by inflammatory cells, vascular leakage, and thrombosis. These pathologic effects are manifested by skin rash, edema, hypotension, and gangrene. Intravascular clotting in the brain accounts for the stuporous mental changes and other neurological symptoms that sometimes occur.

Isolation of most rickettsias from clinical specimens requires a suitable live medium and specialized laboratory facilities, including controlled access and safety cabinets. The usual choices for routine growth and maintenance are the yolk sacs of embryonated chicken eggs, chick embryo cell cultures, and, to a lesser extent, mice and guinea pigs.

SPECIFIC RICKETTSIOSES

Rickettsioses can be differentiated on the basis of their clinical features and epidemiology as: (1) the typhus group; (2) the spotted fever group; and (3) ehrlichosis (table 21.2).

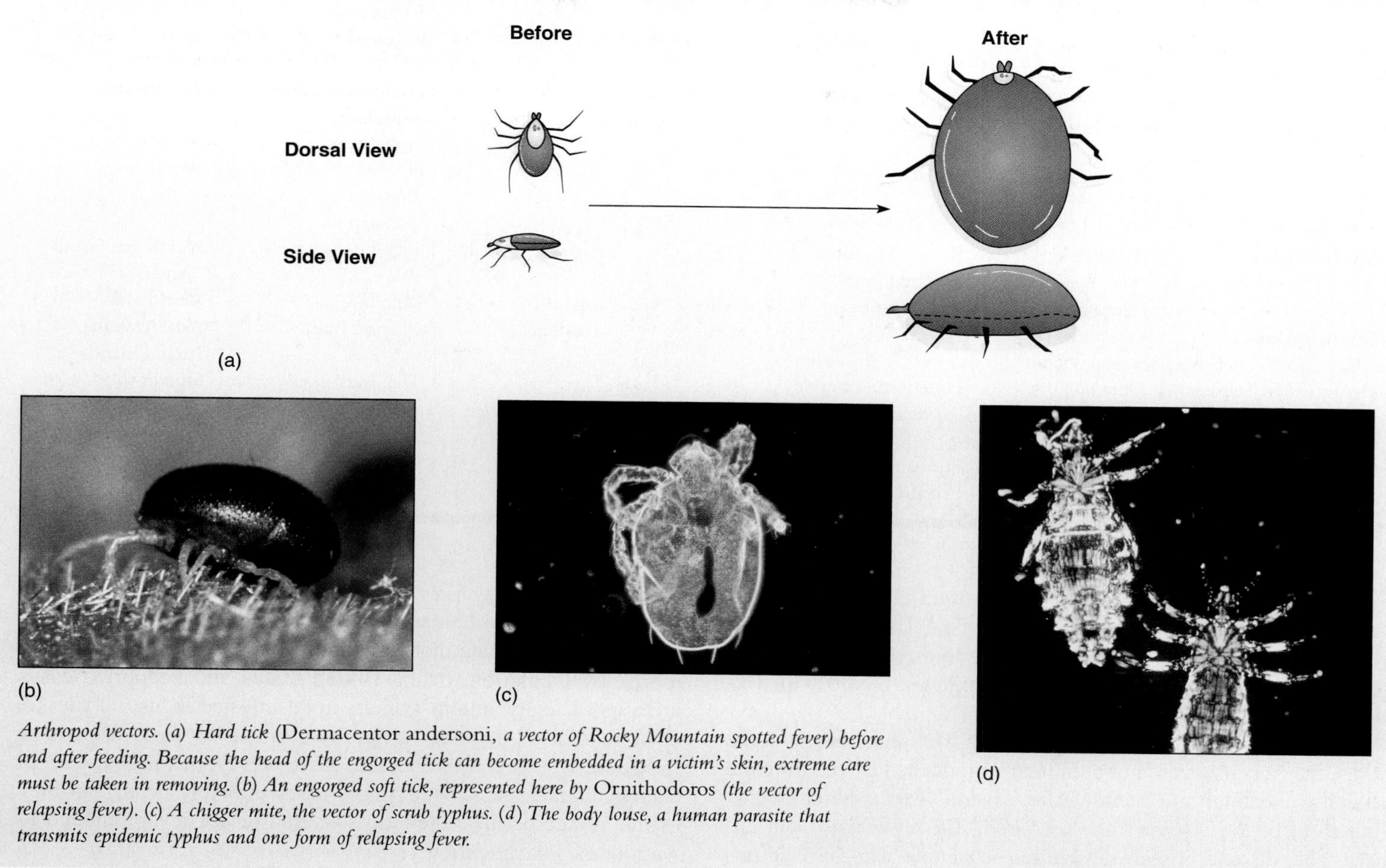

Arthropod vectors. (a) Hard tick (Dermacentor andersoni, *a vector of Rocky Mountain spotted fever) before and after feeding. Because the head of the engorged tick can become embedded in a victim's skin, extreme care must be taken in removing. (b) An engorged soft tick, represented here by* Ornithodoros *(the vector of relapsing fever). (c) A chigger mite, the vector of scrub typhus. (d) The body louse, a human parasite that transmits epidemic typhus and one form of relapsing fever.*

Epidemic Typhus and Rickettsia prowazekii

Epidemic, or louse-borne, typhus has been a constant accompaniment to war, poverty, and famine. The extensive investigations of Dr. Howard Ricketts and Stanislas von Prowazek in the early 1900s led to the discovery of the vector and the rickettsial agent, but not without mortal peril; both men died of the very disease they investigated. *Rickettsia prowazekii* was named in honor of their pioneering efforts.

Epidemiology of Epidemic Typhus* Humans are the sole hosts of human body lice and the only reservoirs of *R. prowazekii.* The louse spreads infection by defecating into its bite wound or other breaks in the skin. Infection of the eye or respiratory tract can take place by direct contact or inhalation of dust containing dried louse feces, but this is a rarer mode of transmission.

Infected lice die from their infection in 1 to 3 weeks, without transmitting the rickettsias to their eggs. The entire metamorphic cycle (egg to adult) can take place upon an infested person's garments or other inanimate objects. Crowding, infrequently changing clothing, and sharing clothing greatly favor transmission. The overall incidence of epidemic typhus in the United States is very low, with no epidemics since 1922. Recent sporadic cases involving transmission from the flying squirrel indicate a possible animal reservoir. Although no longer common in regions of the world with improved standards of living, epidemic typhus presently persists in regions of Africa, Central America, and South America.

Disease Manifestations and Immune Response in Typhus After entering the circulation, rickettsias pass through an intracellular incubation period of 10 to 14 days. The first clinical manifestations are sustained high fever, chills, frontal headache, and muscular pain. Within 7 days, a generalized rash appears, initially on the trunk, and then spreads to the extremities.

*typhus (ty'-fus) Gr. *typhos,* smoky or hazy, underlining the mental deterioration seen in this disease. Typhus is commonly confused with typhoid fever, an unrelated enteric illness caused by *Salmonella typhi.*

TABLE 21.2

CHARACTERISTICS OF MAJOR RICKETTSIAS INVOLVED IN HUMAN DISEASE

Disease Group	Species	Disease	Vector	Primary Reservoir	Mode of Transmission to Humans	Where Found
Typhus	*Rickettsia prowazekii*	Epidemic typhus	Body louse	Humans	Louse feces rubbed into bite; inhalation	Worldwide
	R. typhi (mooseri)	Murine typhus	Flea	Rodents	Flea feces rubbed into skin; inhalation	Worldwide
Spotted Fever	*R. rickettsii*	Rocky Mountain spotted fever	Tick	Small mammals	Tick bite; aerosols	North and South America
	R. akari	Rickettsialpox	Mite	Mice	Mite bite	Worldwide
Scrub Typhus	*Orientia tsutsugamushi*	—	Chigger	Rodents	Chigger bite	Asia, Australia, Pacific Islands
Human Ehrlichosis	*Ehrlichia chaffeensis*	Human monocytic ehrlichosis	Tick		Tick bite	Similar to Rocky Mountain spotted fever
	Ehrlichia species	Human granulocytic ehrlichosis	Tick	Deer, rodents	Tick bite	Unknown

Personality changes, oliguria (low urine output), hypotension, and gangrene complicate the more severe cases. Untreated disease continues for 3 weeks, ending in either rapid recovery or death. Mortality is lowest in children, and as high as 40–60% in patients over 50 years of age.

Recovery usually confers resistance to typhus, but in some cases, the rickettsias are not completely eradicated by the immune response and enter into latency. After several years, a milder recurring form of the disease known as *Brill-Zinsser disease* can appear. This disease is seen most often in people who have immigrated from endemic areas and is of concern mainly because they provide a continuous reservoir of the etiologic agent.

Treatment and Prevention of Typhus The standard chemotherapy for typhus is tetracycline or chloramphenicol. Despite antibiotic therapy, however, the prognosis can be poor in patients with advanced circulatory or renal complications. Eradication of epidemic typhus is theoretically possible by exterminating the vector. Widespread dusting of human living quarters with insecticides has provided some environmental control, and individual treatment with an antilouse shampoo or ointment is also effective. Vaccination against the rickettsial agent provides yet another effective method of control.

Epidemiology and Clinical Features of Endemic Typhus

The agent for **endemic typhus** is *Rickettsia typhi (R. mooseri),* which shares many characteristics with *R. prowazekii* except for pronounced virulence. Synonyms for this rickettsiosis are endemic typhus, murine (mouse) typhus, and flea-borne typhus. The disease occurs in certain parts of Central and South America and in the Southeast, Gulf Coast, and Southwest regions of the United States, where *R. typhi* is regularly harbored by mice and rats. Transmission to the human population is chiefly through infected rat fleas that inoculate the skin, though disease is occasionally acquired by inhalation. In the United States, most reported cases arise sporadically among workers in rat-infested industrial sites. A close relative, *R. felis,* is carried by cats and cat fleas in parts of the Southwest. It causes sporadic cases of a typhus-like disease in California and Texas. Many more cases probably occur, but they escape notice because they are scattered or not severe enough to warrant medical attention.

The clinical manifestations of endemic typhus include fever, headache, muscle aches, and malaise. After 5 days, a skin rash, transient in milder cases, begins on the trunk and radiates toward the extremities. Symptoms dissipate in about 2 weeks. Tetracycline and chloramphenicol are effective therapeutic agents, and various pesticides are available for vector and rodent control.

Scrub Typhus or Tsutsugamushi Disease

Scrub typhus, also called *tsutsugamushi* disease,* was first described nearly two centuries ago in Japan. The illness, caused by *Orientia tsutsugamushi,* is endemic to Southeast Asia, Australia, and India. Current incidence of this disease is difficult to trace, but it appears to occur sporadically in endemic areas.

Scrub typhus rickettsias live naturally in the bodies of chigger mites. Infected mammalian and avian hosts help perpetuate the reservoir. Humans are attacked by chiggers while passing through forests and parkland. At the site of a bite, a distinctive black scab develops in 1 to 3 weeks. Early symptoms of fever, headache, and

**tsutsugamushi* (soot"-soo-gah-moo'-shee) Jap. *tsutsuga,* small and menacing, and *mushi,* a creature.

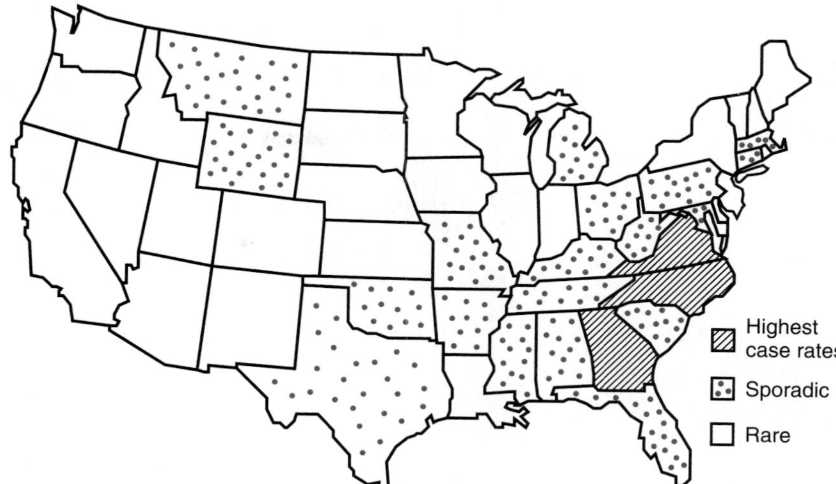

Figure 21.17

The prevalence of Rocky Mountain spotted fever during the last 5 years. Most cases are reported in states on the eastern seaboard and southeastern United States.

muscle aches resemble those of endemic typhus, and, in about half the cases, a generalized rash originates on the trunk and radiates peripherally. Severe cases are accompanied by mental confusion, delirium, pneumonia, and circulatory collapse. Mortality in untreated disease reaches 50%, but with prompt diagnosis and adequate therapy, fatalities can be greatly reduced. Chemotherapy and prevention are nearly identical to that of endemic typhus.

Rocky Mountain Spotted Fever: Epidemiology and Pathology

The rickettsial disease with greatest impact on people living in North America is **Rocky Mountain spotted fever** (RMSF), named for the place it was first seen—the Rocky Mountains of Montana and Idaho. Ricketts identified the etiologic agent *Rickettsia rickettsii* in smears from infected animals and patients and later discovered that it was transmitted by ticks. Despite its geographic name, this disease occurs infrequently in the western United States. The majority of cases are concentrated in the Southeast and eastern seaboard regions (figure 21.17). It also occurs in Canada and Central and South America. Infections occur most frequently among children living in rural and mountain areas, generally in the spring and summer, when the tick vector is most active. The yearly rate of RMSF is 20–40 cases per 10,000 population, with fluctuations coinciding with weather and tick infestations.

The principal reservoirs and vectors of *R. rickettsii* are hard ticks such as the wood tick *(Dermacentor andersoni),* the American dog tick *(D. variabilis,* among others), and the Lone Star tick *(Ambylomma americanum).* The dog tick is probably most responsible for transmission to humans because it is the major vector in the southeastern United States (figure 21.18). The tick passes the pathogen to its offspring, to various domestic and wild mammals, and to humans. Unengorged ticks, picked up as a person brushes against low shrubs or other habitats, can go unnoticed. After wandering on the body, the tick eventually embeds its mouthparts into the skin, feeds, and sheds rickettsias into the bite.

Pathogenesis and Clinical Manifestations of Spotted Fever After 2 to 4 days incubation, the first symptoms are sustained fever, chills, headache, and muscular pain. The distinctive spotted rash usually comes on within 2 to 4 days after the prodromium (figure 21.19). It starts on the wrist and ankles, moves to the arms and legs, converges toward the chest, and eventually covers the entire body. Early lesions are slightly mottled like measles, but later ones are macular, maculopapular, and even petechial. In the severest untreated cases, the enlarged lesions merge and can become necrotic, predisposing to gangrene of the toes or fingertips.

Other grave manifestations of disease are cardiovascular disruption, including hypotension, thrombosis, and hemorrhage. Conditions of restlessness, delirium, convulsions, tremor, and coma are signs of the often overwhelming effects on the central nervous system (see figure 21.18). Fatalities occur in an average of 20% of untreated cases and 5–10% of treated cases. Although death usually occurs late in the disease, there have been instances of mortality within 3 days of onset of symptoms.

Diagnosis, Treatment, and Prevention of Spotted Fever
Any case of Rocky Mountain spotted fever is a cause for great concern and requires immediate treatment, even before laboratory confirmation. Careful clinical observation and patient history are thus essential. Indications sufficiently suggestive to start antimicrobic therapy are the following: (1) a cluster of symptoms, including sudden fever, headache, and rash; (2) recent contact with ticks or dogs; and (3) possible occupational or recreational exposure in the spring or summer. A recent boom to early diagnosis is a method for staining rickettsias directly in a tissue biopsy using fluorescent antibodies. Isolating rickettsias from the patient's blood or tissues is desirable, but it is expensive and requires specially qualified personnel and laboratory facilities. Specimens taken from the rash lesions are suitable for PCR assay, which is very specific and sensitive and can circumvent the need for culture.

Because antibodies appear relatively soon after infection, a change in serum titer detected through an ELISA test can confirm a presumptive diagnosis: The drug of choice for suspected and known cases is tetracycline (doxycycline) administered every day for one week. Chloramphenicol is an alternate choice when diagnosis is unclear. A new vaccine, made with rickettsias grown in chick embryo cell cultures, shows some promise in controlling the disease in humans who are at high risk. Other preventive measures

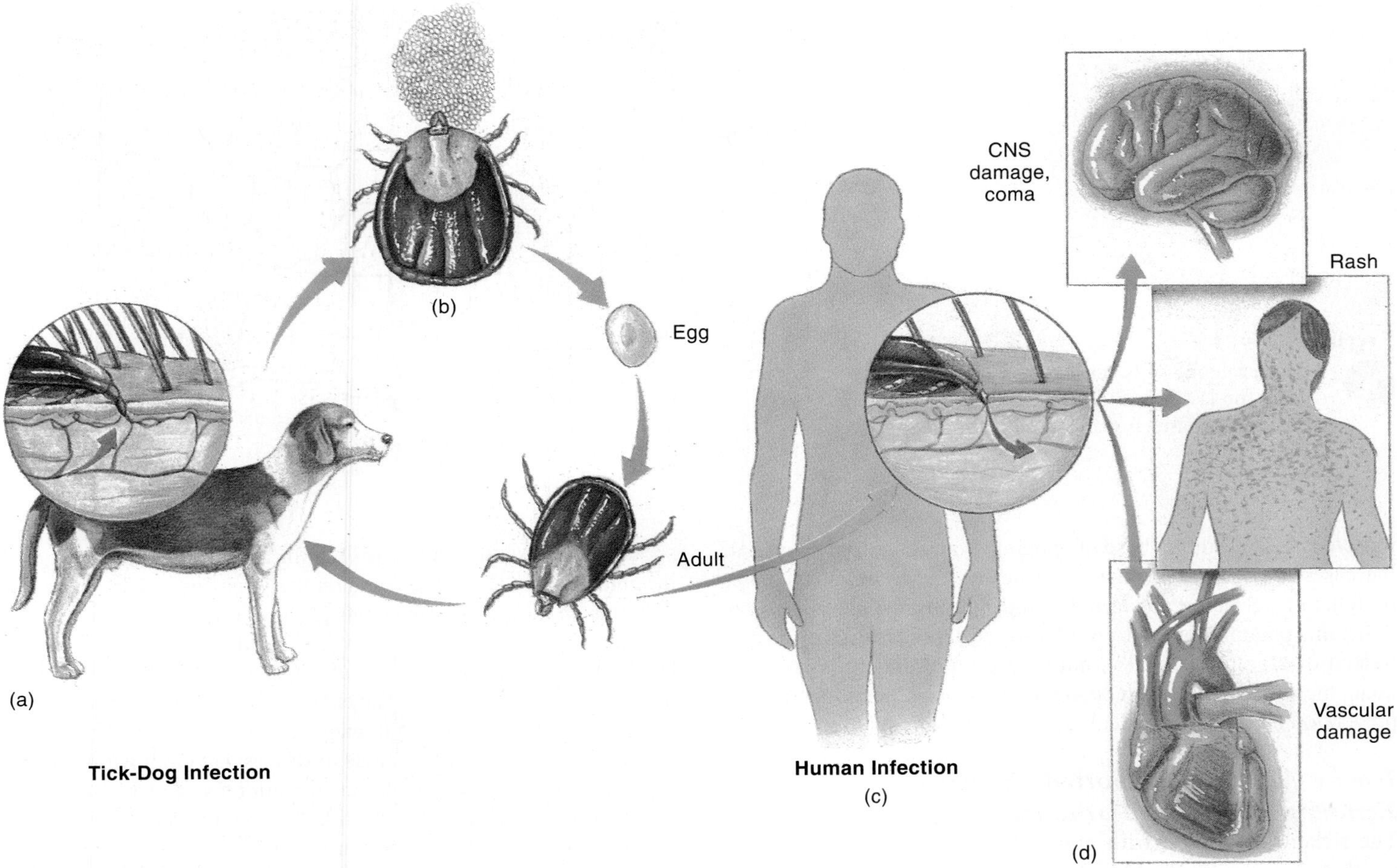

Figure 21.18

The transmission cycle in Rocky Mountain spotted fever. *(a)* Dog and wood ticks are the principal vectors. Ticks are infected from a mammalian reservoir during a blood meal. *(b)* Transovarial passage of *Rickettsia rickettsii* to tick eggs serves as a continual source of infection within the tick population. Infected eggs produce infected adults. *(c)* A tick attaches to a human, embeds its head in the skin, feeds, and sheds rickettsias into the bite. *(d)* Systemic involvement includes severe headache, fever, rash, coma, and vascular damage such as blood clots and hemorrhage.

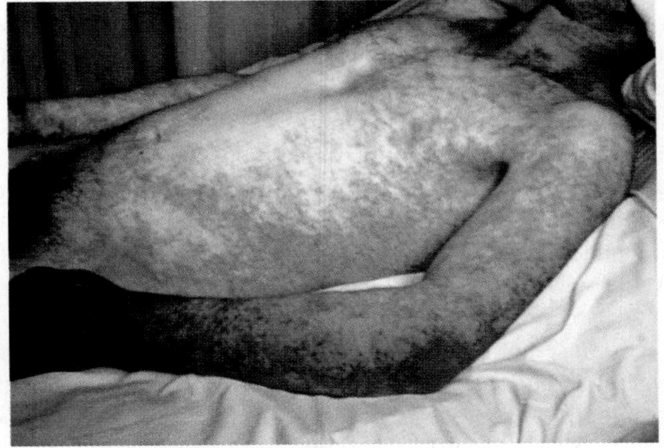

Figure 21.19

Late generalized rash of Rocky Mountain spotted fever. In some cases, lesions become hemorrhagic and predispose to gangrene of the extremities.

follow the pattern for Lyme disease and other tick-borne disease: wearing protective clothing, using insect sprays, and fastidiously removing ticks.

Emerging Rickettsioses

Members of *Ehrlichia* share a number of characteristics with *Rickettsia,* including the strict parasitic existence in host cells and the association with ticks. Although *Ehrlichia* have been identified for years as parasites of horses, dogs, and other mammals, human disease has been known only since 1986. Two emergent pathogens are *E. chaffeensis,* the cause of human monocytic erhlichosis (HME) and an unnamed species that causes human granulocytic erhlichosis (HGE). Both pathogens have a predisposition for growing in white blood cells, hence the names.

Since 1986, approximately 500 cases of HME have been diagnosed and traced to contact with the Lone Star tick *(Ambylomma).* The incidence of HGE since 1993 has been about 50 cases a year. This disease is carried by *Ixodes scapularis* ticks.

The signs and symptoms of the two diseases are similar. They present with an acute febrile state manifesting headache, muscle pain, and rigors. Most patients recover rapidly with no lasting effects, but around 5% of older chronically ill patients die from disseminated infection. Rapid diagnosis is enabled by PCR tests and indirect fluorescent antibody tests. It can be critical to differentiate or detect coinfection with the Lyme disease *Borrelia,* which is carried by the same tick. Doxycycline will clear up most infections within 7 to 10 days.

Chapter Checkpoints

The rickettsias and chlamydias are gram-negative obligate intracellular parasites that lack key metabolic enzymes. The mycoplasmas and L forms are parasites that lack a cell wall.

The rickettsias are extremely small, pleomorphic rods or coccobacilli that cannot synthesize their own AMP. Most have a complex life-style that cycles between arthropod vectors and vertebrate hosts. Rickettsias are usually transmitted by tick bite or tick feces.

Rickettsial diseases are classified into three groups based on the vector and specific clinical characteristics: the typhus group, the spotted fever group, and ehrlichosis. *R. prowazekii* is the agent of epidemic typhus, a systemic infection transmitted by the feces of body lice. Invasion of the vascular endothelium can cause necrosis and hypotension. This disease responds well to antibiotic therapy, and recovery usually provides immunity, but a mild recurrence, Brill-Zinsser disease, sometimes occurs years later. Epidemic typhus is associated with overcrowded living conditions.

R. typhi is the agent of endemic, or murine, typhus, which is less virulent than louse-borne typhus. It is transmitted by the feces of infected rat fleas and enters the host when bites are scratched.

Orientia tsutsugamushi is the agent of scrub typhus, a disease endemic to Asia. It is transmitted by chigger bites. Scrub typhus responds to drug therapy but can be fatal if untreated.

R. rickettsii is the agent of Rocky Mountain spotted fever (RMSF), a potentially fatal disease endemic to North America. It is transmitted by the bites of wood ticks, dog ticks, and the Lone Star tick. RMSF produces an infection that if untreated can cause tissue necrosis and cardiovascular and clotting disorders. This disease usually responds to tetracycline if given promptly.

Two new rickettsias in the genus *Ehrlichia* cause human monocytic and granulocytic ehrlichosis, diseases spread by ticks. Both involve infection of white blood cells, fever, and muscle pains.

DISEASES RELATED TO THE RICKETTSIOSES

Q fever[3] was first described in Queensland, Australia. Its origin was mysterious for a time, until Harold Cox working in Montana and Frank Burnet in Australia discovered the agent later named ***Coxiella burnetti.*** This bacterium is similar to rickettsias in being an intracellular parasite, but it is much more resistant because it produces an unusual type of spore (figure 21.20). It is apparently harbored by a wide assortment of vertebrates and arthropods, especially ticks, which play an essential role in transmission between wild and domestic animals. Humans acquire infection largely by means of environmental contamination and airborne spread. Sources of infectious material include urine, feces, milk, and airborne particles from infected animals. The primary portals of entry are the lungs, skin, conjunctiva, and gastrointestinal tract.

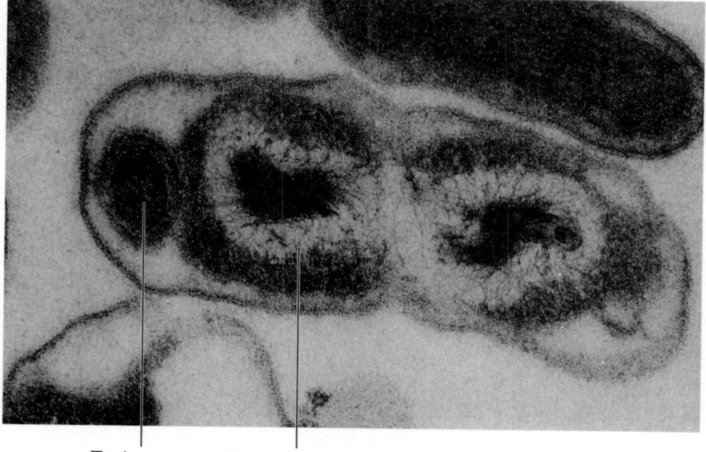

Figure 21.20
The vegetative cells of *Coxiella burnetii* produce unique endospores that are released when the cell disintegrates. Free spores survive outside the host and are important in transmission.

Coxiella burnetti has been isolated from most regions of the world. California and Texas have the highest case rates in the United States, though most cases probably go undetected. People at highest risk are farm workers, meat cutters, veterinarians, laboratory technicians, and consumers of raw milk. The clinical manifestations typical of *Coxiella* infection are abrupt onset of fever, chills, head and muscle ache, but only rarely are skin rashes seen. Disease is sometimes complicated by pneumonitis, hepatitis, and delayed endocarditis, which are occasionally fatal. Mild or subclinical cases resolve spontaneously, and more severe cases respond to tetracycline therapy. A vaccine that gives 5 years of protection is available in many parts of the world and is used on military personnel in the United States. People in contact with livestock and their environment must use precautions in handling animal excrement and secretions.

The Family Bartonellaceae contains the genus ***Bartonella,*** which now includes species formerly known as *Rochalimaea.* These small gram-negative rods are fastidious but not obligate intracellular parasites, and they can be cultured on blood agar. *Bartonella* species are currently considered a group of emerging pathogens. Historically, **trench fever** has been the most prevalent *Bartonella* disease. It first appeared during World War I, when it afflicted at least a million persons. Since 1945, cases have occurred infrequently in endemic regions of Europe, Africa, and

3. For "query," meaning to question, or of unknown origin.

Asia. The causative agent is *Bartonella quintana,* a species that, like epidemic typhus, cycles between humans and lice, but unlike the typhus rickettsia, does not multiply intracellularly and does not kill the louse vector. Highly variable symptoms can include a 5- to 6-day fever (hence 5-day, or quintana, fever); leg pains, especially in the tibial region (shinbone fever); headache; chills; and muscle aches. A macular rash can also occur. The microbe can persist in the blood long after convalescence and is responsible for later relapses.

Bartonella henselae, is the most common agent of **cat-scratch disease** (CSD), an infection connected with a cat scratch or bite. The pathogen can be isolated in over 40% of cats, especially kittens. There are approximately 25,000 cases per year in the United States, 80% of them in children 2–14 years old. The symptoms start after 1–2 weeks, with a cluster of small papules at the site of inoculation. In a few weeks, the lymph nodes along the lymphatic drainage swell and can become pus-filled (figure 21.21). Most infections remain localized and resolve in a few weeks, but drugs such as tetracycline, erythromycin, and rifampin can be effective therapies. The disease can be prevented by thorough degerming of a cat bite or scratch.

Bartonella is currently an important new pathogen in AIDS patients. It is the cause of **bacillary angiomatosus,** a severe cutaneous and systemic infection. The cutaneous lesions arise as reddish nodules or crusts that can be mistaken for Kaposi's sarcoma. Systems most affected are the liver and spleen, and symptoms are fever, weight loss, and night sweats. Treatment is similar to that for CSD.

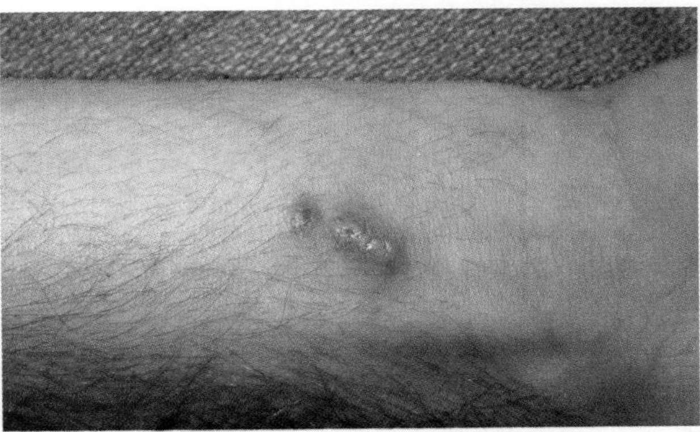

Figure 21.21
Cat-scratch disease. A primary nodule appears at the site of the scratch in about 21 days. In time, large quantities of pus collect, and the regional lymph nodes swell.

Other Obligate Parasitic Bacteria: The Chlamydiaceae

Like rickettsias, the chlamydias are obligate parasites that depend on certain metabolic constituents of host cells for growth and maintenance. They show further resemblance to the rickettsias with their small size, gram-negative cell wall, and pleomorphic morphology, but they are markedly different in several aspects of their life cycle. The species of greatest medical significance are *Chlamydia trachomatis,** a very common pathogen involved in sexually transmitted, neonatal, and ocular disease (trachoma); *C. pneumoniae,* the cause of one type of atypical pneumonia; and *C. psittaci,** a zoonosis of birds and mammals that causes ornithosis in humans.

The Biology of Chlamydia

Chlamydias alternate between two distinct stages: (1) a small, metabolically inactive, infectious form called the **elementary body** that is released by the infected host cell and (2) a larger, noninfectious, actively dividing form called the **reticulate body** that grows within the host cell vacuoles (figure 21.22). Elementary bodies are tiny, dense spheres shielded by a rigid, impervious envelope that ensures survival outside the eucaryotic host cell. Reticulate bodies are finely granulated and have thin cell walls. Studies of the reticulate bodies indicate that they are energy parasites, entirely lacking enzyme systems for catabolizing glucose and other substrates and for synthesizing ATP, though they do possess ribosomes and mechanisms for synthesizing proteins, DNA, and RNA. Reticulate bodies ultimately differentiate into elementary bodies.

Diseases of Chlamydia trachomatis

The reservoir of pathogenic strains of *Chlamydia trachomatis* is the human body. The microbe shows astoundingly broad distribution within the population, often being carried with no symptoms. Elementary bodies are transmitted in infectious secretions, and, although infection can occur in all age groups, disease is most severe in infants and children. The two human strains are the **trachoma** strain, which attacks the squamous or columnar cells of mucous membranes in the eyes, genitourinary tract, and lungs, and the **lymphogranuloma venereum** (LGV) strain, which invades the lymphatic tissues of the genitalia.

Chlamydial Diseases of the Eye The two forms of chlamydial eye disease, ocular trachoma and inclusion conjunctivitis, differ in their patterns of transmission and ecology. **Ocular trachoma,** an infection of the epithelial cells of the eye, is an ancient disease and a major cause of blindness in certain parts of the world. Although a few cases occur yearly in the United States, several million cases occur endemically in parts of Africa and Asia. Transmission is favored by contaminated fingers, fomites, flies, and a hot, dry climate.

The first signs of infection are a mild conjunctival exudate and slight inflammation of the conjunctiva. These are followed by marked infiltration of lymphocytes and macrophages into the infected area. As these cells build up, they impart a pebbled (rough) appearance to the inner aspect of the upper eyelid (figure 21.23*a*). In time, a vascular pseudomembrane of exudate and inflammatory leukocytes forms over the cornea, a condition called *pannus* that lasts a few weeks and usually heals. Corneal damage and impaired vision can be attributed to chronic and secondary infections and deformation of tear ducts and eyelashes. Early treatment of this disease with tetracycline or sulfa drugs is highly effective and prevents all of the complications. It is a tragedy that in this day of

**Chlamydia trachomatis* (klah-mid'-ee-ah trah-koh'-mah-tis) Gr. *chlamys,* a cloak, and *trachoma,* roughness.

**psittaci* (sih-tah'-see) Gr. *psittacus,* a parrot.

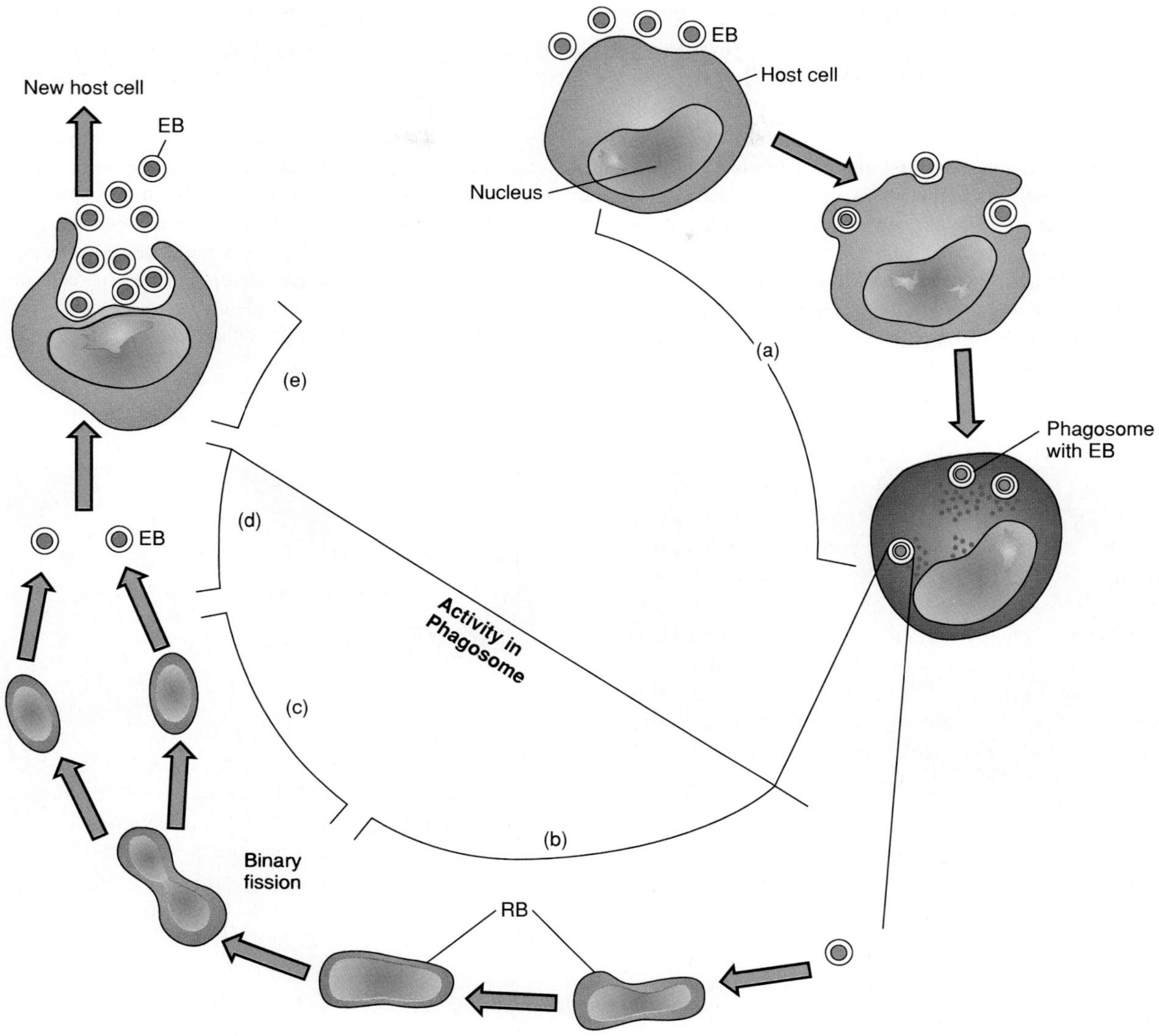

Figure 21.22

The life cycle of *Chlamydia.* (*a*) The infectious stage, or elementary body (EB), is taken into phagocytic vesicles by the host cell. (*b*) In the phagosome, each elementary body develops into a reticulate body (RB). (*c*) Reticulate bodies multiply by regular binary fission. (*d*) Mature RBs become reorganized into EBs. (*e*) Completed EBs are released from the host cell.

preventive medicine, millions of children will develop blindness for want of a few dollars' worth of antibiotics.

Inclusion conjunctivitis is usually acquired through contact with secretions of an infected genitourinary tract. Infantile conjunctivitis develops 5 to 12 days after a baby has passed through the birth canal of its infected mother and is the most prevalent form of conjunctivitis in the United States (100,000 cases per year). The initial signs are conjunctival irritation, a profuse adherent exudate, redness, and swelling (see figure 21.23*b*). Although the disease is generally self-limited and heals spontaneously, trachoma-like scarring occurs often enough to warrant routine prophylaxis of all newborns (as for gonococcal infection). Because the traditional silver nitrate solution is ineffective against *C. trachomatis,* antibiotics such as erthyromycin and tetracycline must be instilled into the eyes.

Most inclusion conjunctivitis in adults has occurred in people with concurrent genital infection. These cases appear to result from self-inoculation or contamination with secretions. Other cases have been traced to contact with swimming pool water and shared towels. The infection is similar to trachoma and can cause corneal scarring if untreated.

Sexually Transmitted Chlamydial Diseases It has been estimated that *C. trachomatis* is carried in the reproductive tract of up to 10% of all people, with even higher rates among the promiscuous. About 70% of infected women harbor it asymptomatically on the cervix, while 10% of infected males show no signs or symptoms. The potential for this disease to cause long-term reproductive damage has initiated its listing as a reportable disease since 1995 (microfile 21.6).

MICROFILE 21.6 REPRODUCTIVE THREATS FROM *CHLAMYDIA*

Chlamydiosis is one of the most prevalent of all sexually transmitted diseases. The official number of cases reported to the CDC is around 500,000 cases per year, but it probably exceeds that level by 10 times. The prevalence of this infection in young, sexually active teenagers is as high as 20% in some regions. Medically and socioeconomically, its clinical significance now eclipses gonorrhea, herpes simplex 2, and syphilis. Far from being simple and non-life-threatening, chlamydial infections, even inapparent ones, have the potential for great harm. When infection ascends to the uterus, fallopian tubes, and peritoneum, the pelvic inflammatory disease can damage the interior of the reproductive organs and cause scar tissue to form. The result is that thousands of young women become sterilized every year, and often this process occurs inconspicuously. Equally serious complications are ectopic pregnancies and cervical infections that lead to prematurity and morbidity in the fetus. Other infections include chlamydial pneumonitis in neonates (30,000 cases per year) and eye disease.

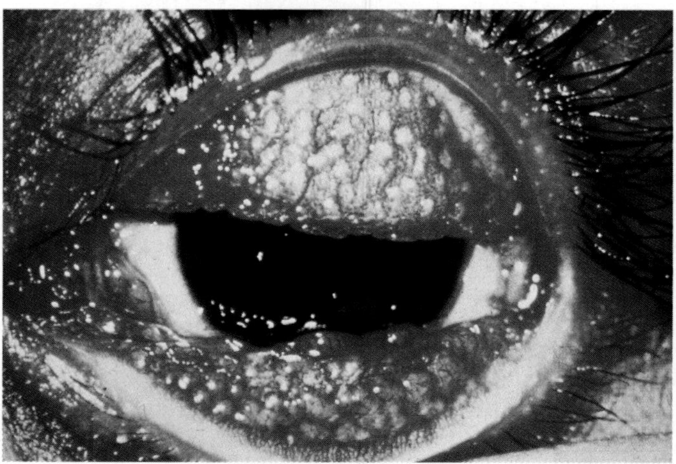

(a)

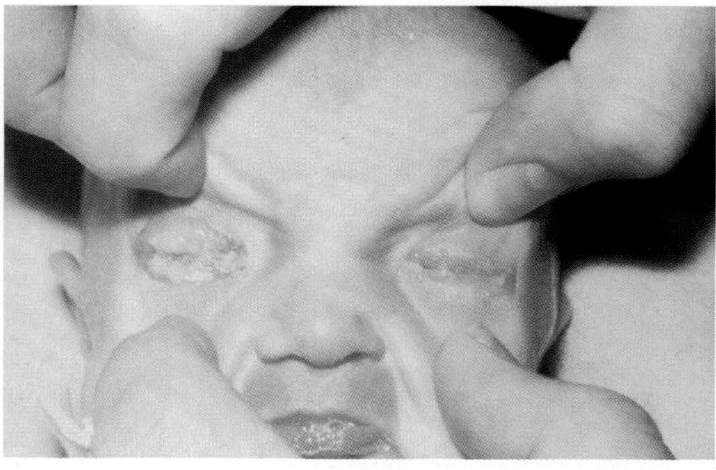

(b)

Figure 21.23

The pathology of primary ocular chlamydia infection. *(a)* Ocular trachoma, an early, pebble-like inflammation of the conjunctiva and inner lid in a child. (Note: The eyelid is being pulled away by the examiner to reveal the lesion.) *(b)* Inclusion conjunctivitis in a newborn. Within 5 to 6 days, an abundant, watery exudite collects around the conjunctival sac. This is currently the most common cause of ophthalmia neonatorum.

A syndrome appearing among males with chlamydial infections is an inflammation of the urethra called **nongonococcal urethritis** (NGU). This diagnosis is derived from the symptoms that mimic gonorrhea yet do not involve gonococci. *Chlamydia* is the most frequent cause of this syndrome. Women with symptomatic chlamydial infection have cervicitis accompanied by a white drainage, endometritis, and salpingitis (pelvic inflammatory disease). As is often the rule with sexually transmitted diseases, chlamydia frequently appears in mixed infections with the gonococcus and other genitourinary pathogens, thereby greatly complicating treatment.

When a particularly virulent strain of *Chlamydia* chronically infects the genitourinary tract, the result is a severe, often disfiguring disease called **lymphogranuloma venereum.**[4] The disease is endemic to regions of South America, Africa, and Asia but occasionally occurs in other parts of the world. Its incidence in the United States is about 500 cases per year. Chlamydias enter through tiny nicks or breaks in the perigenital skin or mucous membranes and form a small, painless vesicular lesion that often escapes notice. Other acute symptoms are headache, fever, and muscle aches. As the lymph nodes near the lesion begin to fill with granuloma cells, they enlarge and become firm and tender (figure 21.24). These nodes, or buboes, can burst and then heal with scarring that obstructs lymphatic channels. Long-term blockage of lymphatic drainage leads to chronic, deforming edema of the genitalia and anus.

Identification, Treatment, and Prevention of Chlamydiosis Because chlamydias reside intracellularly, specimen sampling requires enough force to dislodge some of the cells from the mucosal surface. Genital samples are taken with a swab inserted a few centimeters into the urethra or cervix, rotated, and removed. Although the most reliable diagnosis comes from culture

4. Also called tropical bubo or lymphogranuloma inguinale.

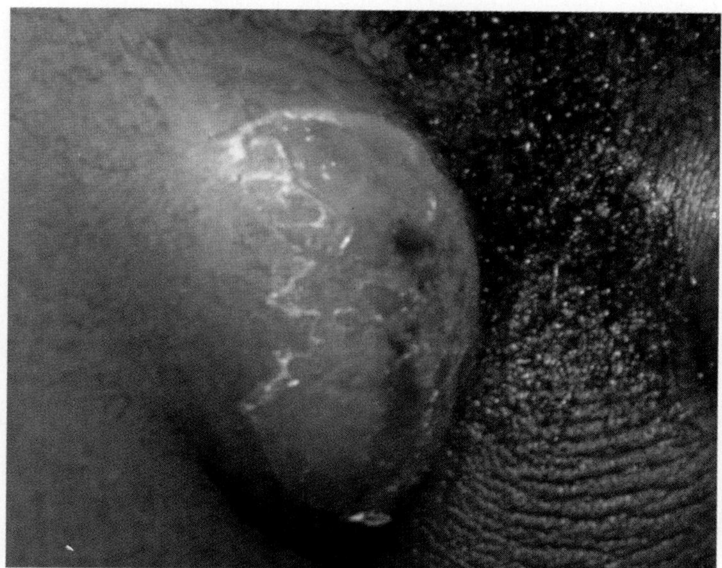

Figure 21.24

The clinical appearance of advanced lymphogranuloma venereum in a man. A chronic local inflammation blocks the lymph channels, causing swelling and distortion of the external genitalia.

in chicken embryos, mice, or tissue culture, this procedure is too costly and time-consuming to be used routinely in STD clinics; however, it is an essential part of diagnosing neonatal infections. The most sensitive and specific tests currently available are a direct assay of specimens using immunofluorescence (figure 21.25*a)* and a PCR-based probe. Methods useful in diagnosing inclusion conjunctivitis are Giemsa or iodine stains (figure 21.25*b)*, but they are not recommended for urogenital specimens because of low sensitivity and the possibility of obtaining false-negative results in asymptomatic patients.

Urogenital chlamydial infections are most effectively treated with drugs that act intracellularly, such as tetracyclines and azithromycin. Penicillin and aminoglycosides are not effective and must not be used. Because of the high carrier rate and the difficulty in detection, prevention of chlamydial infections is a public health priority. As a general rule, sexual partners of infected people should be treated with drug therapy to prevent infection, and sexually active people can achieve some protection with a condom. In addition, screening pregnant women and treating those who test chlamydia-positive helps prevent perinatal disease. Because detection of carriers can fail, routine ocular prophylaxis of newborns is still necessary.

Chlamydia pneumoniae

A strict human pathogen, *Chlamydia pneumoniae,* is distinctly different and not closely related to the other chlamydias. It has been linked to a type of respiratory illness that includes pharyngitis, bronchitis, and pneumonitis. It is usually a mild illness in young adults, though it can cause a severe reaction in asthmatic patients that is responsible for increased rates of death in this group. Compelling evidence is mounting that chronic infection of the arteries by this species is an important contributing factor in the development of hardened arteries and heart disease.

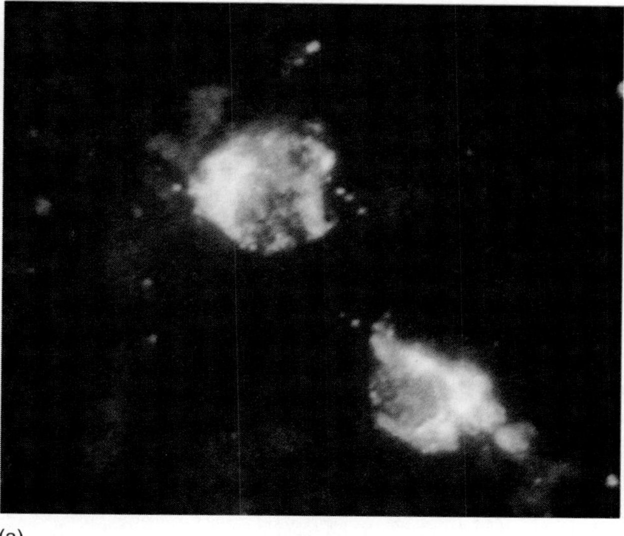

(a)

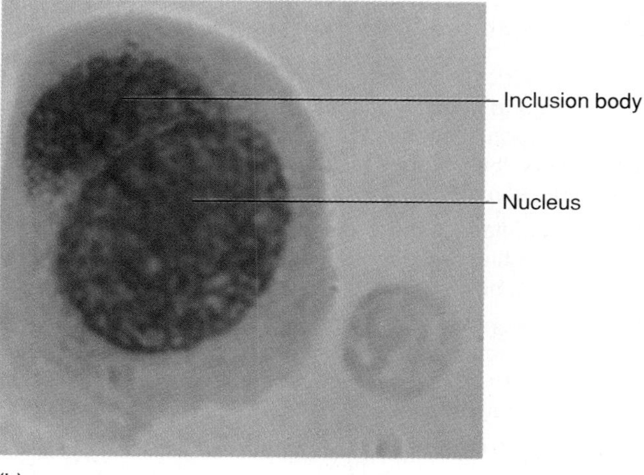

(b)

Figure 21.25

Direct diagnosis of chlamydial infection. *(a)* A specimen stained with monoclonal antibodies bearing a fluorescent dye. Infected cells glow a bright apple green. *(b)* Direct Giemsa stain of the eyelid scraping of a patient suffering from inclusion conjunctivitis. Large inclusion body represents a phagosome packed with chlamydia in various stages of development.

Chlamydia psittaci *and Ornithosis*

The term *psittacosis* was adopted to describe a pneumonia-like illness contracted by people working with imported parrots and other psittacine birds in the last century. As outbreaks of this disease appeared in areas of the world having no parrots, it became evident that other birds could carry and transmit the microbe to humans and other animals. In light of this evidence, the generic term **ornithosis*** has been suggested as a replacement.

*ornithosis (or"-nih-thoh'-sis) Gr. *ornis,* bird. More than 90 species of birds harbor *C. psittaci.*

Ornithosis is a worldwide zoonosis that is carried in a latent state in wild and domesticated birds but becomes active under stressful conditions such as overcrowding. In the United States, poultry have been subject to extensive epidemics that killed as many as 30% of flocks. Infection is communicated to other birds, mammals, and humans by contaminated feces and other discharges that become airborne and are inhaled. Most of the 50–100 yearly human cases in the United States occur among poultry and pigeon handlers.

The symptoms of ornithosis mimic those of influenza and pneumococcal pneumonia. Early manifestations are fever, chills, frontal headache, and muscle aches, and later ones are coughing and lung consolidation. Unchecked, infection can lead to systemic complications involving the meninges, brain, heart, or liver. Although most patients respond well to tetracycline or erythromycin therapy, recovery is often slow and fraught with relapses. Control of the disease is usually attempted by quarantining and giving antimicrobic therapy to imported birds and by taking precautions in handling birds, feathers, and droppings.

Chapter Checkpoints

Coxiella burnetii, the cause of Q fever, is a rickettsia-like agent transmitted to humans primarily by milk, meat, and airborne contamination. It infects a broad range of vertebrates and arthropods. It also forms resistant spores. Prevention includes pasteurization of milk.

Bartonella quintana is the agent of trench fever, a systemic infection lasting 5–6 days. It is transmitted to humans by body lice and can be recurrent.

Bartonella henselae is the agent of cat-scratch disease, a systemic infection that travels from the initial site along the lymph vessels. This disease responds to antibiotic therapy if not resolved spontaneously.

The chlamydias are small, gram-negative, pleomorphic, intracellular parasites that have no catabolic pathways. They exist in two forms: the elementary body, which is the form transmitted between human hosts through direct contact and body secretions, and the reticulate body, which multiplies intracellularly.

Chlamydia trachomatis is the agent of several STDs: NGU (nongonococcal urethritis), pelvic inflammatory disease, and lymphogranuloma venereum. It also causes ocular trachoma, a serious eye infection.

Chlamydia pneumoniae is the agent of respiratory infections in young adults and asthmatics.

Chlamydia psittaci is the agent of ornithosis, an influenza-like disease carried by birds that has serious systemic complications if untreated. Antibiotic therapy is usually successful.

(a)

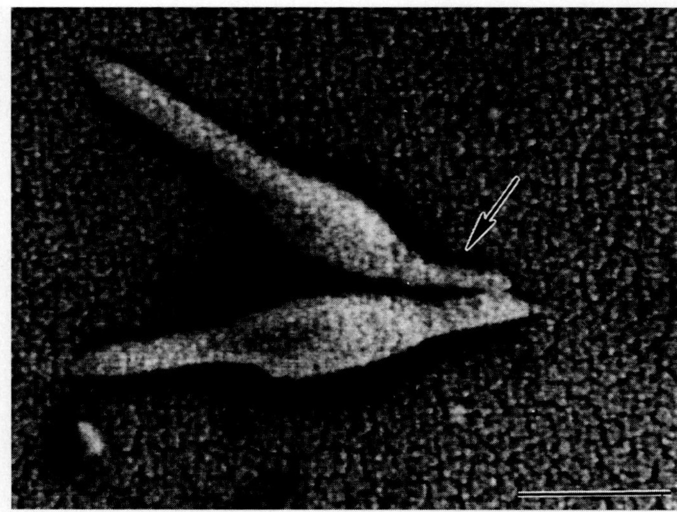

(b)

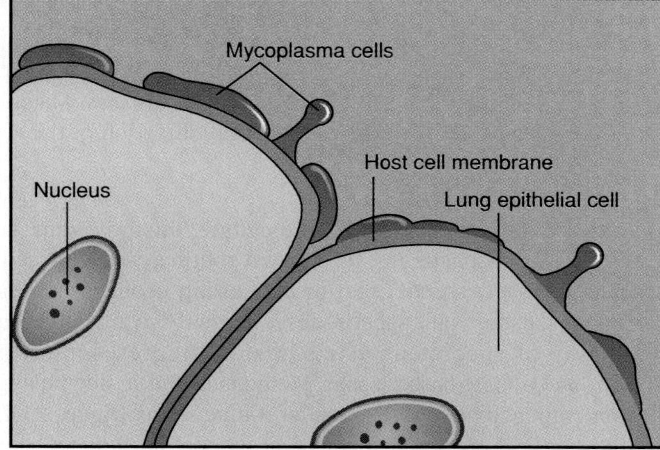

Figure 21.26
The morphology of mycoplasmas. *(a)* A scanning electron micrograph of *Mycoplasma pneumoniae* (bar = 0.5 μm). Note pleomorphic shape and elongate attachment tip (arrow). The cells use this to anchor themselves to host cells. *(b)* Figure depicts how *M. pneumoniae* becomes a membrane parasite that adheres tightly and fuses with the host cell surface. This fusing makes destruction and removal of the pathogen very difficult.

MOLLICUTES AND OTHER CELL-WALL-DEFICIENT BACTERIA

Bacteria in the Class Mollicutes, also called the **mycoplasmas,** are the smallest self-replicating microorganisms. All of them naturally lack a cell wall (figure 21.26*a*), and, except for one genus, all species are parasites of animals and plants. The two most clinically important genera are *Mycoplasma* and *Ureaplasma.* Disease of the respiratory tract has been primarily associated with *Mycoplasma pneumoniae; M. hominis* and *Ureaplasma urealyticum* are implicated in urogenital tract infections.

Biological Characteristics of the Mycoplasmas

Without a rigid cell wall to delimit their shape, mycoplasmas are exceedingly pleomorphic. The small (0.3–0.8 μm), flexible cells assume a spectrum of shapes, ranging from cocci and filaments to doughnuts, clubs, and helices (see figure 4.33). Mycoplasmas are not strict parasites, and they can grow in cell-free media, generate metabolic energy, and synthesize proteins with their own enzymes. However, most are fastidious and require complex media containing sterols, fatty acids, and preformed purines and pyrimidines. Animal species of mycoplasmas are sometimes referred to as membrane parasites because they form intimate associations with membranes of the respiratory and urogenital tracts (figure 21.26*b*). Because mycoplasmas bind to specific receptor sites on cells and adhere so tenaciously that they are not easily removed by usual defense mechanisms, infections are chronic and difficult to eliminate. Mycoplasmas are inhibited by antibiotics that repress protein synthesis but are resistant to penicillin and cephalosporins that block cell wall synthesis.

Mycoplasma pneumoniae *and Atypical Pneumonia*

Mycoplasma pneumoniae is a human parasite that is the most common agent of **primary atypical pneumonia (PAP).** This syndrome is atypical in that its symptoms do not resemble those of pneumococcal pneumonia. Primary atypical pneumonia can also be caused by rickettsias, chlamydias, respiratory syncytial viruses, and adenoviruses. Mycoplasmal pneumonia is transmitted by aerosol droplets among people confined in close living quarters, especially families, students, and the military. Community resistance to this pneumonia is high; only 3–10% of those exposed become infected, and fatalities are rare.

Mycoplasma pneumoniae selectively binds to specific receptors of the respiratory epithelium and inhibits ciliary action. Gradual spread of the bacteria over the next 2 to 3 weeks disrupts the cilia and the epithelium. Even though most mycoplasmas are neither invasive nor particularly toxigenic, the sloughing off of the epithelium causes inflammation and other tissue damage. The first symptoms—fever, malaise, sore throat, and headache—are not suggestive of pneumonia. A cough is not a prominent early symptom, and, when it does appear, it is mostly unproductive. As the disease progresses, nasal symptoms, chest pain, and earache can develop. The lack of acute illness in most patients has given rise to the nickname "walking pneumonia."

Diagnosis Because a culture can take 2 or 3 weeks, early diagnosis of mycoplasma pneumonia is difficult and relies chiefly on close clinical observation. Stains of sputum appear devoid of bacterial cells, leukocyte counts are within normal limits, and X-ray findings are nonspecific. Some physicians find the appearance of a small blister on the tympanic membrane a useful sign. Others use the patient's history to rule out other bacterial and viral agents. Serological tests based on complement fixation, immunofluorescence, and indirect hemagglutination are useful later in the disease.

Tetracycline and erythromycin inhibit mycoplasmal growth and help to rapidly diminish symptoms, but they do not stop the shedding of viable mycoplasmas. Patients frequently experience relapses if treatment is not continued for 14 to 21 days. Preventive measures include controlling contamination of fomites, avoiding contact with droplet nuclei, and reducing aerosol dispersion.

Other Mycoplasmas

Mycoplasma hominis and *Ureaplasma urealyticum* are regarded as weak, sexually transmitted pathogens. They are frequently encountered in samples from the urethra, vagina, and cervix of newborns and adults. These species initially colonize an infant at birth and subsequently diminish through early and late childhood. A second period of colonization and persistence is initiated by the onset of sexual intercourse. Evidence linking genital mycoplasmas to human disease is substantial and growing every year. *Ureaplasma urealyticum* is implicated in some types of nonspecific or nongonococcal urethritis and prostatitis. There is increasing evidence that this mycoplasma plays a role in opportunistic infections of the fetus and fetal membranes. It appears to cause some cases of miscarriage, stillbirth, premature birth, and respiratory infections of newborns. *Mycoplasma hominis* is more often associated with vaginitis, pelvic inflammatory disease, and kidney inflammation.

A newly discovered *Mycoplasma* species has stunned the scientific community and promises to change our view of their pathogenicity. This intracellular pathogen, *M. Fermentens,* was isolated from the spleen, liver, brain, and blood of AIDS patients. At first it appeared to be a potent cofactor accounting for rapid deterioration and cell death in some AIDS patients; however, researchers have now found that it is capable of causing a fatal infection on its own. This mycoplasma has been associated with aggressive flulike illness in people who are negative for HIV infection, and it appears to suppress the immune system on its own. Other researchers are focussing on the possible role of mycoplasmas in a number of autoimmune diseases.

BACTERIA THAT HAVE LOST THEIR CELL WALLS

Exposure of typical walled bacteria to certain drugs (penicillin) or enzymes (lysozyme) can result in wall-deficient bacteria called L forms or L-phase variants (see figure 4.17). L forms are induced or occur spontaneously in numerous species and can even become stable and reproduce themselves, but they are not naturally related to mycoplasmas.

L Forms and Disease

The role of certain L forms in human and animal disease is a distinct possibility, but proving etiology has been complicated because infection is difficult to verify with Koch's postulates. One theory proposes that antimicrobic therapy with cell-wall-active agents induces certain infectious agents to become L forms. In this wall-free state, they resist further treatment with these drugs and remain latent until the therapy ends, at which time they reacquire walls and resume their pathogenic behavior.

Infections with L-phase variants of group A streptococci, *Proteus,* and *Corynebacterium* have been reported, though they are uncommon. In a number of chronic pyelonephritis and endocarditis cases, cell-wall-deficient bacteria have been the only isolates. Research on people with a chronic intestinal syndrome called Crohn's disease has uncovered a strong association with *Mycobacterium paratuberculosis.* After prolonged culture, protoplast wall-deficient bacteria from biopsy specimens converted into walled acid-fast cells. When a PCR technique was used to analyze the DNA of specimens, it was found that 65% of Crohn's patients tested positive for *M. paratuberculosis.*

Chapter Checkpoints

Mycoplasmas are tiny pleomorphic bacteria that lack a cell wall. Although most species are parasitic, mycoplasmas can be cultured on complex artificial media. They are considered membrane parasites because they bind tightly to epithelial linings of the respiratory and urogenital tracts.

Mycoplasma pneumoniae is the agent of primary atypical pneumonia (walking pneumonia).

M. hominis and *Ureoplasma urealyticum* are agents of sexually transmitted infections of the reproductive tract and kidneys, and more recently, of fetal infections.

M. incognitus is a recently identified pathogen that causes systemic infection by suppressing the immune system.

L forms are wall-deficient variants of walled bacteria such as group A streptococci, *Proteus, Mycobacterium,* and *Corynebacterium* that are occasionally involved in diseases.

BACTERIA IN DENTAL DISEASE

The relationship between humans and their oral microflora is a complex, dynamic microecosystem. The mouth contains a diversity of surfaces for colonization, including the tongue, teeth, gingiva, palate, and cheeks, and it provides numerous aerobic, anaerobic, and microaerophilic microhabitats for the estimated 1,000 different oral species with which humans coexist. The habitat of the oral cavity is warm, moist, and greatly enriched by the periodic infusion of food. In most humans, this association remains in balance with little adverse effect, but in people with poor or nonexistent oral hygiene, it teeters constantly on the brink of disease.

The Structure of Teeth and Associated Tissues

Most dental diseases develop on the **tooth** and its surrounding supportive structures, or **periodontium,*** as shown in figure 21.27. A tooth is composed of a **crown** that protrudes above the gum and a **root** that is inserted into a bony socket. The outer surface of the crown is protected by a dense coating of **enamel,** an extremely hard, noncellular material composed of tightly packed rods of calcium phosphate ($CaPO_4$) that cannot be replaced once the tooth has matured. The root is surrounded by a layer of cementum, which anchors the tooth to the periodontal membrane that lines the socket. The major portion of the tooth inside the crown and root is composed of a highly regular calcified material called dentin, and the very core contains a pulp cavity that supplies the living tissues with blood vessels and nerves. The root canal is the portion of the pulp that extends into the roots. Places where the bone protrudes toward the crown are covered by connective tissue and a mucous membrane called the **gingiva,** or gum. Important sites for initial dental infections are the enamel, especially the cusps, and the crevice, or sulcus, formed where the gingiva meets the tooth.

Dental pathology generally affects both hard and soft tissues (figure 21.28). Although both categories of disease are initiated when microbes adhere to the tooth surface and produce dental plaque, their outcomes vary. In the case of dental caries, the gradual breakdown of the enamel leads to invasive disease of the tooth itself, whereas in soft-tissue disease, calcified plaque damages the soft gingival tissues and predisposes them to bacterial invasion. Both diseases are responsible for the loss of teeth, though dental caries are usually implicated in children, and periodontal infections in adults.

Hard-Tissue Disease: Dental Caries

Dental caries* is the most common human disease. It is a complex mixed infection of the dentition that gradually destroys the enamel and often lays the groundwork for the destruction of deeper tissues. It occurs most often on tooth surfaces that are less accessible and harder to clean and on those that provide pockets or crevices where bacteria can cling. Caries commonly develop on enamel pits and fissures, especially those of the occlusal surfaces, though they can also occur on the smoother crown surfaces and subgingivally on the roots.

Over the years, several views have been put forth to explain how dental caries originate. At various times it has been believed that sugar, microbes, or acid cause teeth to rot; however, studies with germ-free animals have now shown that no single factor can account for caries (see figure 1.10 and chapter 13). Caries development occurs in many phases and requires multiple interactions involving the anatomy, physiology, diet, and bacterial flora of the host. The principal stages in the formation of dental caries are pellicle formation, plaque formation, acid production and localization, and enamel etching (figure 21.29*a).*

Plaque Formation

A freshly cleaned tooth is a perfect landscape for colonization by microbes. Within a few moments, it develops a thin, mucous coating called the **acquired pellicle,** which is made up of adhesive salivary proteins. This structure presents a potential substrate upon which certain bacteria first gain a foothold. The most prominent pioneering colonists are the cariogenic genera *Streptococcus* and *Actinomyces.* These gram-positive bacteria have adhesive receptors such as fimbriae and slime layers that allow them to cling to the tooth surfaces and to each other, forming a foundation for the dense, whitish mass called **plaque.*** Fed by a diet high in sucrose, glucose, and certain complex carbohydrates, *Streptococcus mutans* and *S. gordonii* produce a gummy polymer of glucose called dextran that helps them attach to smooth enamel surfaces, holds

*periodontium (per"-ee-oh-don'-shee-um) Gr. *peri,* around, and *odous,* tooth. Gums, bones, and cementum.

*caries (kar'-eez) L. rottenness.

*plaque (plak) Fr. a patch.

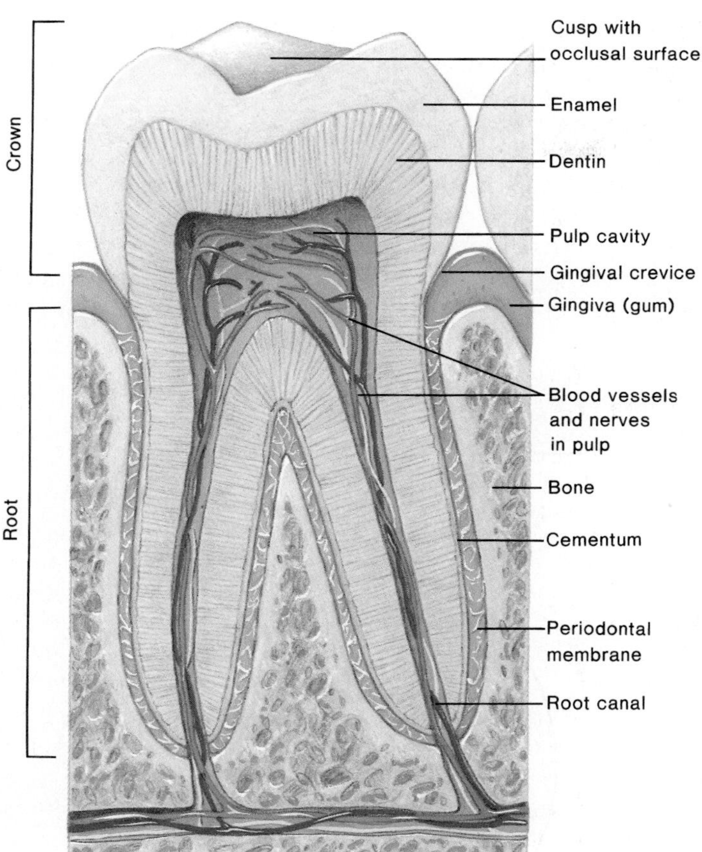

Figure 21.27
The anatomy of a tooth.

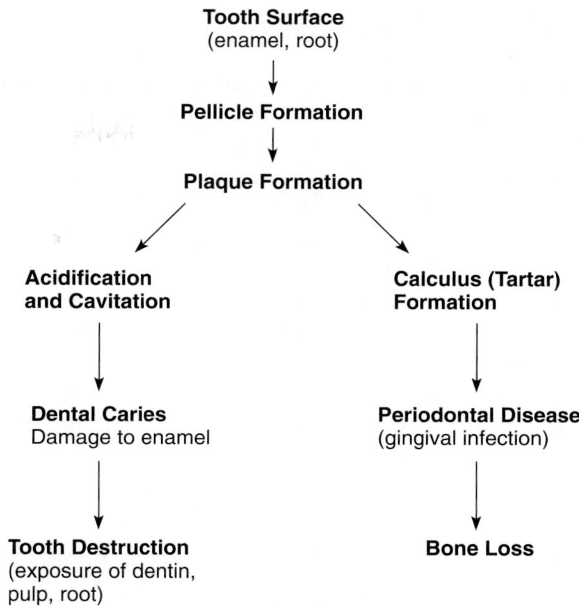

Figure 21.28
Summary of the events leading to dental caries, periodontal disease, and bone loss.

the plaque together, and adds to its bulk. As these primary invaders continue to grow and plaque begins to build up, the scene is set for the invasion and aggregation of additional species of bacteria. Among secondary invaders are species of *Lactobacillus, Bacteroides, Fusobacterium, Neisseria,* and *Treponema.* Plaque can be highlighted by staining the teeth with disclosing tablets (figure 21.30*a*), and a microscopic view reveals a rich and varied network of bacteria and their products along with epithelial cells and fluids (figure 21.30*b*).

Acid Formation and Localization and Etching of the Enamel

If mature plaque is not removed from sites that readily trap food, it usually evolves into a caries lesion. The role of plaque in caries development is related directly to streptococci and lactobacilli, which produce acid as they ferment dietary carbohydrates. If this acid is immediately flushed from the plaque and diluted in the mouth, it has little effect. However, in the denser regions of plaque, the acid can accumulate in direct contact with the enamel surface. Eventually the pH of the environment adjacent to the enamel is lowered below 5, which is acidic enough to begin to dissolve (decalcify) the calcium phosphate of the enamel in that spot. This initial lesion can remain localized in the enamel (first-degree caries) and can be repaired with various inert materials (fillings). Once the deterioration has reached the level of the dentin (second-degree caries), tooth destruction speeds up, and the tooth can be rapidly destroyed. Exposure of the pulp (third-degree caries) is attended by severe tenderness and toothache, and the chance of saving the tooth is diminished (see figure 21.29*b*).

SOFT-TISSUE (PERIODONTAL) DISEASE

Periodontal disease is so common that 97–100% of the population has some degree of it by age 45. Most kinds are due to bacterial colonization and varying degrees of inflammation that occur in response to gingival damage. The most common predisposing condition occurs when the plaque becomes mineralized (calcified) with calcium and phosphate crystals. This process produces a hard, porous substance called **calculus** on both the sub- and supragingival tooth surfaces (figure 21.31). Supragingival calculus is usually not harmful and can even protect the tooth from caries, but its presence in the relatively sheltered gingival sulcus initiates a pathologic lesion that leads to periodontal disease.

Calculus and plaque accumulating in the gingival sulcus irritate the delicate gingival membrane, and the chronic trauma to this area causes a pronounced inflammatory reaction. This event, in turn, creates an avenue of invasion for a variety of normal flora. These bacteria, which are primarily anaerobic and gram-negative, include *Actinobacillus*, *Porphyromonas* and *Bacteriodes* (rods), *Fusobacterium* (spindle-shaped rods), and numerous spirochetes. In response to the mixed infection, the damaged area becomes infiltrated by neutrophils and macrophages, and later by lymphocytes, which cause further inflammation and tissue damage (figure 21.32). The initial signs of **gingivitis** are swelling, loss of normal contour, patches of redness, and increased bleeding of the gingiva. Spaces or pockets of varying depth also develop between the tooth

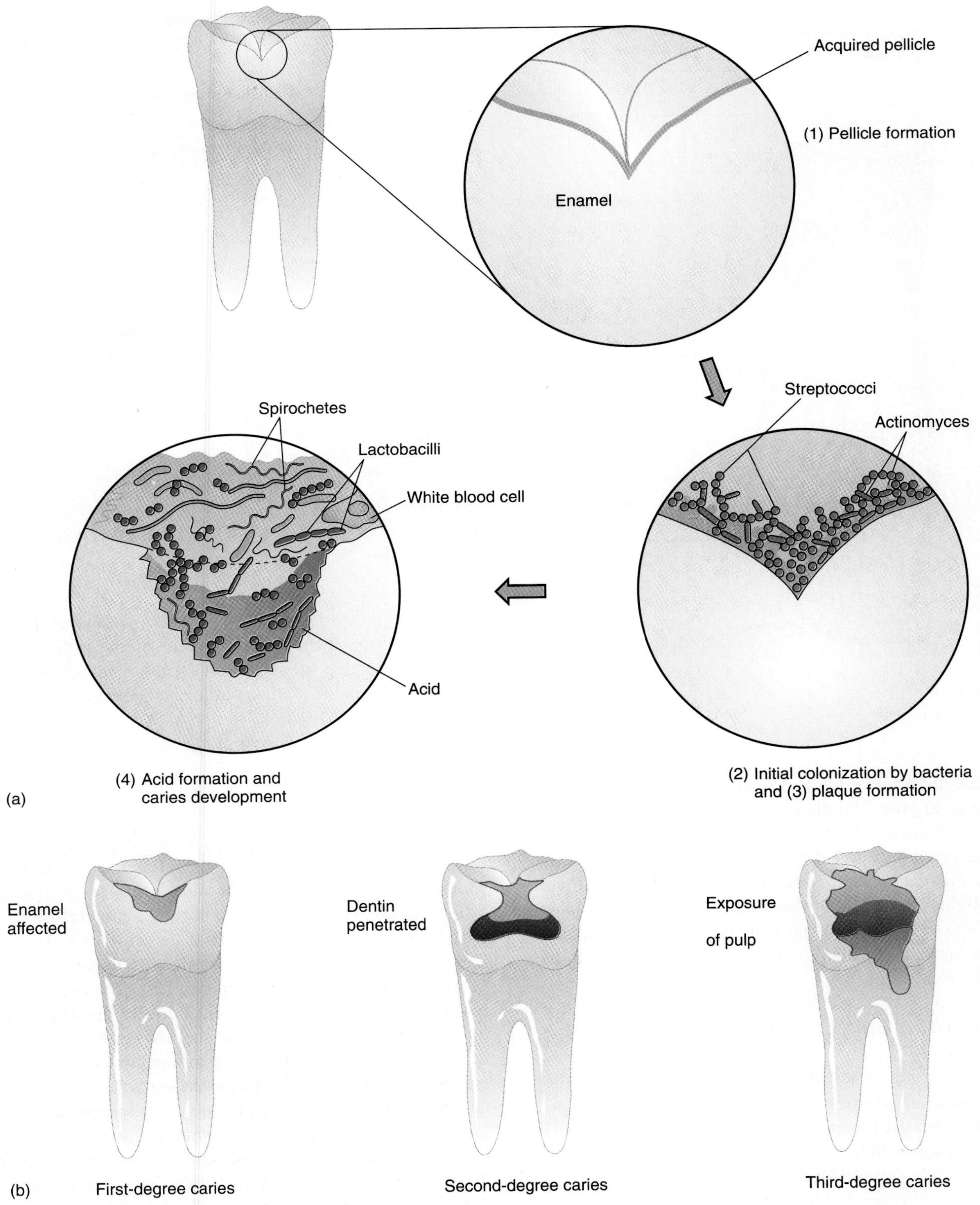

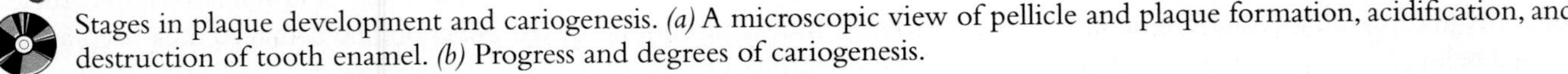

Figure 21.29

Stages in plaque development and cariogenesis. *(a)* A microscopic view of pellicle and plaque formation, acidification, and destruction of tooth enamel. *(b)* Progress and degrees of cariogenesis.

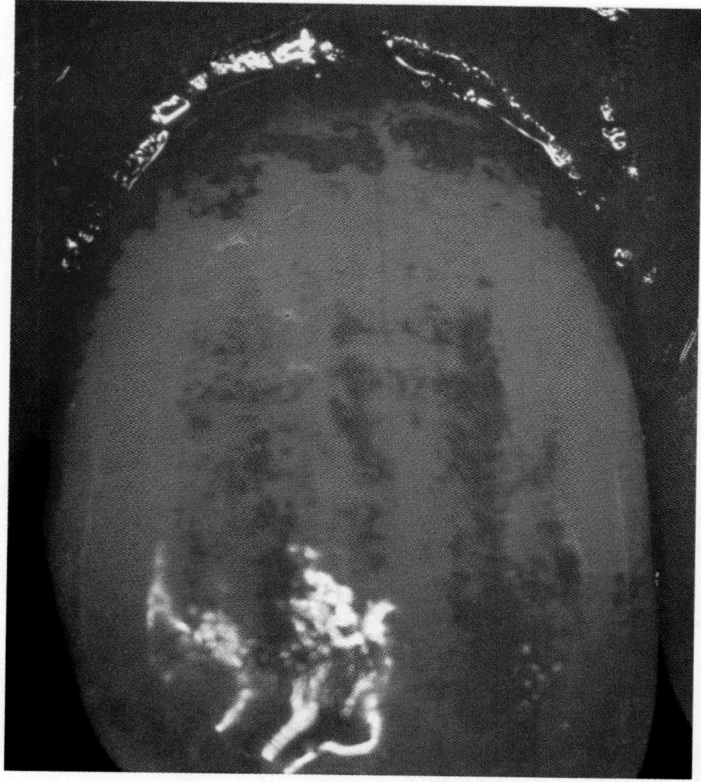

(a)

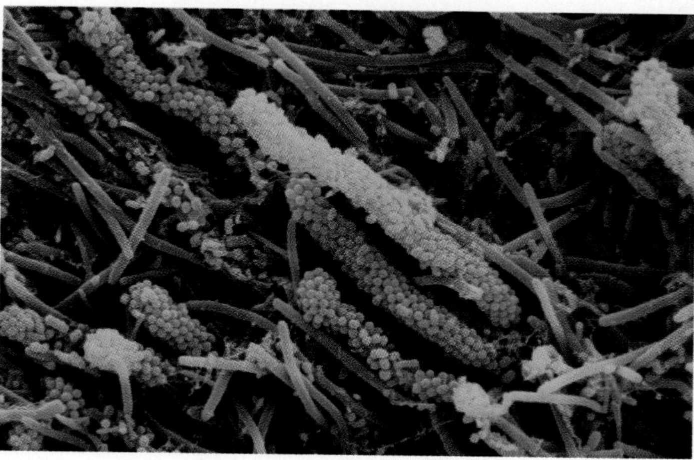

(b)

Figure 21.30
The macroscopic and microscopic appearance of plaque.
(a) Disclosing tablets containing vegetable dye stain heavy plaque accumulations at the junction of the tooth and gingiva.
(b) Scanning electron micrograph of plaque with long filamentous forms and "corn cobs" that are mixed bacterial aggregates.

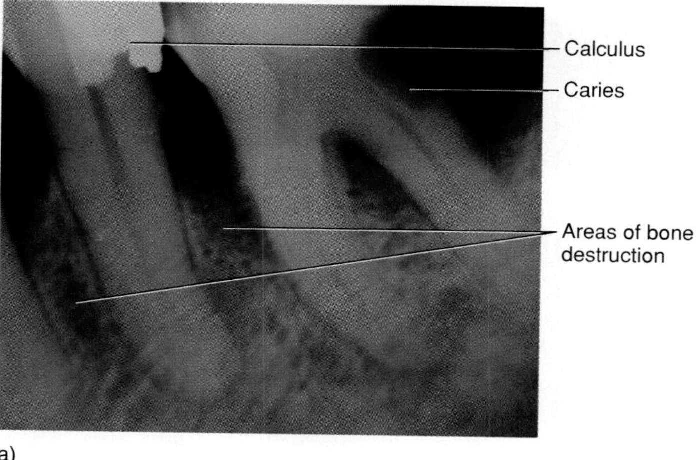

(a)

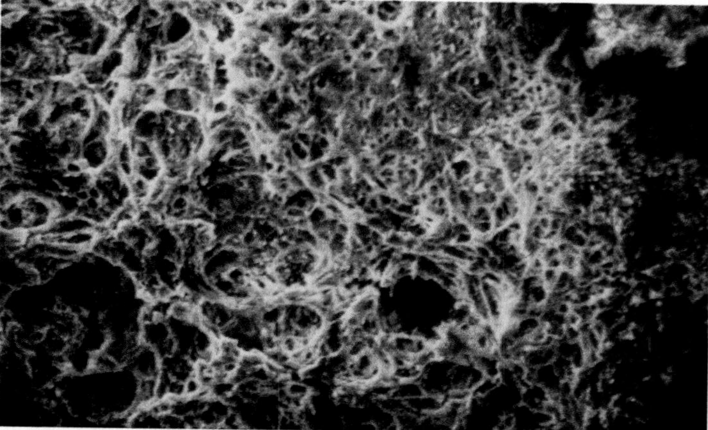

(b)

Figure 21.31
The nature of calculus. *(a)* Radiograph of mandibular premolar and molar, showing calculus on the top and a caries lesion on the right. Bony defects caused by periodontitis affect both teeth. *(b)* Scanning electron micrograph revealing the rough, dense, and porous nature of calculus (500×).

and the gingiva. If this condition persists, a more serious disease called **periodontitis** results. This is the natural extension of the disease into the periodontal membrane and cementum. The deeper involvement increases the size of the pockets and can cause bone resorption severe enough to loosen the tooth in its socket. If the condition is allowed to progress, the tooth can be lost.

Another serious periodontal disease is **acute necrotizing ulcerative gingivitis** (ANUG), formerly called trench mouth or Vincent's disease. This disease is a synergistic infection involving *Treponema denticola* and other spirochetes and fusobacteria that are sufficiently aggressive to invade the gingival tissues and severely damage them (figure 21.32*d*). It develops rapidly and is associated with severe pain, bleeding, pseudomembrane formation, and necrosis. This condition usually results from poor oral hygiene, altered host defenses, or prior gum disease and is not communicable. However, it responds well to broad-spectrum antibiotics.

Factors in Dental Disease

Nutrition and eating patterns have a profound effect on oral diseases. People whose diet is high in refined sugar (sucrose, glucose, and fructose) tend to have more caries, especially if these foods are eaten constantly throughout the day without brushing

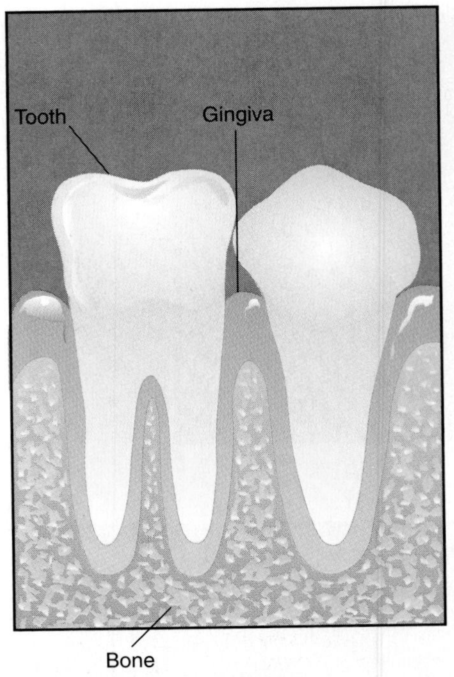

(a) Normal, nondiseased state

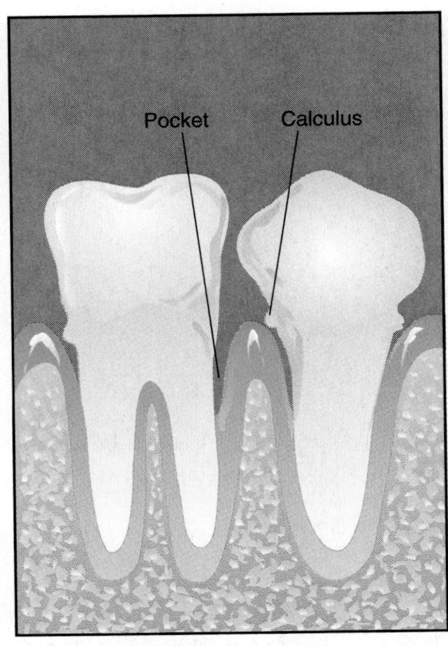

(b) Early gingivitis

(c) Late-stage gingival disease

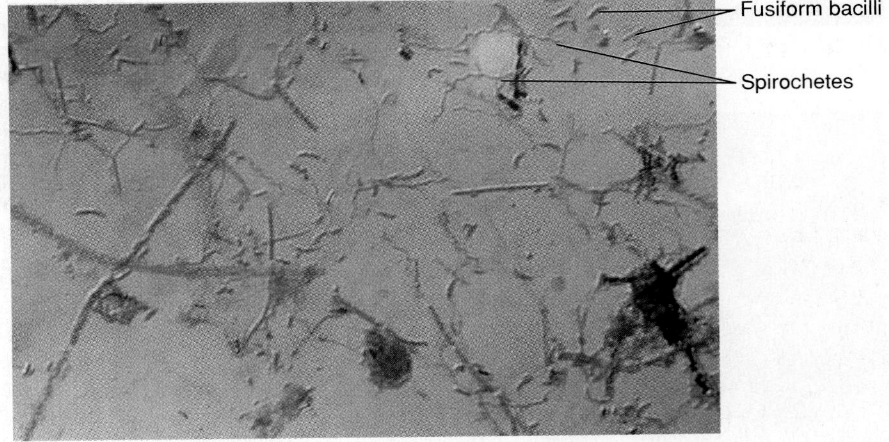

(d)

Figure 21.32
Stages in soft-tissue infection and gingivitis. *(a)* The relationship of tooth, gingiva, and bone in the nondiseased state. *(b)* Calculus build-up and early gingivitis. *(c)* Late-stage disease, tissue destruction, deep pocket formation, and bone loss. *(d)* A sample of exudate from a gingival pocket stained with crystal violet (560×). Note the numerous spirochetes and fusiform bacilli.

the teeth. The practice of putting a baby down to nap with a bottle of fruit juice or formula can lead to rampant dental caries ("nursing bottle caries"). In addition to diet, numerous anatomical, physiological, and hereditary factors influence oral diseases. The structure of the tooth enamel can be influenced by genetics and by environmental factors such as fluoride, which strengthens the enamel bonds. The amount of calcium available to developing fetuses and neonates can also affect tooth development. Saliva that is more fluid in consistency inhibits oral colonization, and its content of inhibitory factors such as antibodies and lysozyme can also help prevent dental disease.

The greatest control of dental diseases is preventive dentistry, including regular brushing and flossing to remove plaque, because stopping plaque build-up automatically reduces caries and calculus production. Mouthwashes are relatively ineffective in controlling plaque formation because of the high bacterial content of saliva and the relatively short acting time of the mouthwash. The federal government has recently blocked advertisements for certain mouthwashes that claim to prevent plaque formation.

Once calculus has formed on teeth, it cannot be removed by brushing but can be dislodged only by special mechanical procedures (scaling) in the dental office. One of the most exciting prospects is the possibility of a vaccine to protect against the primary colonization of the teeth. Some success in inhibiting plaque formation has been achieved in experimental animals with vaccines raised against the whole cells of *Streptococcus mutans* and the fimbriae of *Actinomyces viscosus.*

Chapter Checkpoint

The oral cavity is en ecosystem that provides a broad spectrum of niches for a wide variety of microorganisms. A high sugar diet and poor oral hygiene enhance the growth of bacteria, which cause oral diseases. *Streptococcus, Lactobacillus,* and *Actinomyces* are the main agents of dental caries. Invasion of the gums or gingiva by a mixture of spirochetes and gram-negative bacteria causes many gum diseases. ANUG, or trench mouth, is a severe form of gingivitis that can lead to destruction of connective tissue and bone.

CHAPTER CAPSULE WITH KEY TERMS

MISCELLANEOUS BACTERIAL INFECTIONS

SPIROCHETES

Spirochetes are helical, flexible bacteria that move by periplasmic flagella. Several ***Treponema*** species are obligate parasitic spirochetes with 8–12 regular spirals; best observed under dark-field microscope; cause **treponematoses;** direct observation of treponeme in tissues and blood tests important in diagnosis; cannot be cultivated in artificial media; treatment by large doses of penicillin or tetracycline.

MAJOR TREPONEMATOSES

Syphilis: ***Treponema pallidum*** is a very sensitive pathogen that causes complex progressive disease in adults and children. **Sexually transmitted syphilis** is acquired through close contact with an infected cutaneous surface; untreated sexual disease occurs in stages over long periods; many symptoms mimic other diseases. At site of entrance, multiplying treponemes produce a **primary lesion,** a hard ulcer or **chancre,** which disappears as the microbe becomes systemic. **Secondary syphilis** occurs when spirochete infects many organs; marked by skin rash, fever, damage to mucous membranes. Both primary and secondary syphilis are communicable; after latent period of months to years, the pathogen is established in tissues. Final, non-communicable **tertiary stage** is marked by tumors called **gummas** and life-threatening cardiovascular and neurological effects.

Congenital syphilis is acquired transplacentally; disrupts embryonic and fetal development; survivors may have respiratory, skin, bone, teeth, eye, and joint abnormalities if not treated.

Nonsyphilitic Treponematosis: Slow progressive cutaneous and bone diseases endemic to specific regions of tropics and subtropics; usually transmitted under unhygenic conditions. **Bejel** is a deforming childhood infection of the mouth, nasal cavity, body, and hands; **yaws** occurs from invasion of skin cut, causing a primary ulcer that seeds a second crop of lesions; **pinta** is superficial skin lesion that depigments and scars the skin.

LEPTOSPIRA INTERROGANS

L. interrogans has very regular coils and a prominent hook; causes **leptospirosis,** a worldwide zoonosis acquired through contact with urine of wild and domestic animal reservoirs. Spirochete enters cut, multiplies in blood and spinal fluid, causes muscle aches, headache; second phase is marked by Weil's syndrome, which involves kidneys and liver; requires early treatment with penicillin or tetracycline.

BORRELIA

Borrelia is a loose, irregular spirochete that causes **borreliosis;** infections are vector-borne (mostly by ticks). *B. hermsii* is zoonotic (in wild rodents) and carried by soft ticks; *B. recurrentis* has a strictly human reservoir and is carried by lice; both species cause relapsing fever. Borrelias are introduced into blood and multiply; fever and other symptoms recur because the spirochete repeatedly changes antigenically and forces the immune system to keep adapting; this may occur several times.

Borrelia burgdorferi: A zoonosis carried by mice and spread by a hard tick *(Ixodes)* that lives on deer and mice; causes **Lyme disease,** a syndrome that occurs endemically in several regions of the United States. Tick bite leads to fever and a prominent ring-shaped rash; if allowed to progress, may cause cardiac, neurological and arthritic symptoms; can be controlled by antibiotics and by preventing tick contact.

CURVED BACTERIA (VIBRIOS)

Vibrios are short spirals or sausage-shaped cells with polar flagella.

Vibrio cholerae: Causes **epidemic cholera,** a human disease that originated in Asia but is now distributed worldwide. Organism is free-living in soil and water and is ingested in contaminated food and water; microbe infects the surface of epithelial cells in small intestine, is not invasive; severity of cholera due to potent **cholera toxin** (CT) that causes the cells to release electrolytes and lose water, resulting in **secretory diarrhea;** electrolyte loss and dehydration lead to muscle, circulatory, and neurological symptoms and death; treatment with oral rehydration (electrolyte and fluid replacement) completely restores stability; vaccine available.

Vibrio parahaemolyticus: Causes food infection associated with seafood; organism lives in seawater and is prevalent during warm months; symptoms similar to mild cholera; for prevention, food must be well cooked and refrigerated during storage.

Campylobacter Species: Vibrios with two polar flagella. ***C. jejuni*** is a common cause of severe **gastroenteritis** worldwide; acquired through food or water contaminated by animal feces; enteritis is due to enterotoxin with symptoms like cholera; ***C. fetus*** causes diseases in pregnant women and fatal septicemia in neonates; ***Helicobacter pylori*** may be etiologically involved in diseases of the stomach lining such as gastritis and ulcers.

OBLIGATE INTRACELLULAR PARASITIC BACTERIA

Rickettsias and **chlamydias** are tiny, gram-negative rods or cocci that require certain metabolic elements of the host cell to multiply; diseases treatable with tetracycline and chloramphenicol. ***Rickettsia*** causes **rickettsioses;** most are zoonoses spread by arthropod vectors.

Characteristics of Ectoparasitic, Blood-Sucking Arthropod Vectors: Transmitters of parasites between various vertebrate reservoirs; human usually not a major participant in life cycle; vectors may pass pathogen to offspring transovarially or to hosts through bite, feces, or mechanical injury and scratching.

Ticks are arachnids; hard ticks have a hard shield and soft ticks lack it; ticks devoid of blood are small, can live in environment for many years; cling to host, feed, and inoculate the host with saliva containing the pathogen; carry Rocky Mountain spotted fever, Q fever, borrelioses, viral fevers.

Mites and chiggers are tiny arachnids that live on wild vegetation and nibble on various hosts, including humans; bite is painless, causes itchiness; carry rickettsialpox and scrub typhus.

Fleas are flattened insects with jumping legs and a blood-probing mouth; extremely resistant and nonspecific to host; infect by defecating on injured skin; carry murine typhus, bubonic plague.

Lice are insects that cling to body hair and gently pierce skin; infection occurs when louse is crushed by scratching and rubbed into skin; carry epidemic typhus, trench fever, and relapsing fever.

Epidemic Typhus: Caused by ***Rickettsia prowazekii;*** carried by lice; associated with overcrowding; probably not a zoonosis; disease starts with high fever, chills, headache; rash occurs; Brill-Zinsser disease is a chronic, recurrent form.

Endemic (Murine) Typhus: Zoonosis caused by ***R. typhi,*** harbored by mice and rats; occurs in areas of high flea infestation; not common; like epidemic form, but symptoms milder.

Scrub Typhus: Caused by ***Orientia tsutsugamushi,*** a zoonosis of Japan and Asia; transmitted by chiggers; causes fever, rash; often fatal.

Rocky Mountain Spotted Fever: Etiologic agent is ***R. rickettsii;*** zoonosis carried by dog and wood ticks *(Dermacentor);* most cases on eastern seaboard; infection causes distinct spotted, migratory rash; acute reactions include heart damage, CNS damage, gangrene; can be fatal; disease prevented by blocking ticks' access to body and inspecting for and carefully removing them.

The genus ***Ehrlichia*** contains two species of rickettsias that have recently been discovered in humans. These tick-borne bacteria cause human monocytic and granulocytic ehrlichosis.

Diseases Related to the Rickettsioses

Q Fever: Caused by ***Coxiella burnetii;*** agent has unusual life cycle, with a resistant spore form that can survive out of host; a zoonosis of domestic animals; transmitted by air, dust, unpasteurized milk, ticks; usually inhaled, causing pneumonitis, fever, hepatitis, but no rash.

Bartonella: A closely-related genus, is the cause of trench fever, spread by lice, and cat-scratch disease, a lymphatic infection associated with a clawing injury by cats.

Chlamydia

Organisms in the genus *Chlamydia* pass through a transmission phase involving a hardy **elementary body** and an intracellular **reticulate body** that has pathologic effects.

Chlamydia trachomatis: A strict human pathogen that causes eye diseases and STDs. **Ocular trachoma** is a severe infection that deforms the eyelid and cornea and may cause blindness. **Conjunctivitis** occurs in babies following contact with birth canal; prevented by ocular prophylaxis after birth. *C. trachomatis* causes a very common bacterial STD: nongonococcal urethritis in males, and cervicitis, salpingitis (PID), infertility, and scarring in females; also **lymphogranuloma venereum,** a disfiguring disease of the external genitalia and pelvic lymphatics.

C. pneumoniae causes an atypical pneumonia that is a serious complication in asthma patients.

Chlamydia psittaci: Causes **ornithosis,** a zoonosis transmitted to humans through respiratory discharges and feces of bird vectors; highly communicable among all birds; pneumonia or flulike infection with fever, lung congestion.

Mollicutes/Mycoplasmas

Mycoplasmas are naturally cell-wall-deficient bacteria; highly pleomorphic cells; not obligate parasites but require special lipids; fuse tightly to host membranes during infection, are difficult to dislodge; diseases treated with tetracycline, erythromycin.

Mycoplasma pneumoniae: Causes **primary atypical pneumonia;** pathogen slowly spreads over interior respiratory surfaces; symptoms are not pronounced, may be fever and chest pain, sore throat, no cough; *M. hominis* and Ureaplasma *urealyticum* are weak STDs; normal colonists of most persons; may cause urethritis, PID, and other reproductive tract diseases.

L Forms: Bacteria that normally have cell walls but have transiently lost them through drug therapy. They may be involved in certain chronic diseases.

THE ROLE OF MIXED INFECTIONS IN DENTAL DISEASE

The oral cavity contains hundreds of microbial species that participate in interactions between themselves, the human host, and the nutritional role of the mouth; it is continuously vulnerable to infection and disease.

Hard-Tissue Disease

Dental caries is a slow, progressive infection of irregular areas of enamel surface; begins with colonization of tooth by dextran-forming species of *Streptococcus* and cross adherence with *Actinomyces:* process forms layer of thick, adherent material called **plaque;** this complex layer harbors dense masses of bacteria and extracellular substances; acid formed by agents in plaque is held close against enamel and dissolves the inorganic salts; if unabated, this produces a deep caries lesion, which may invade dentin and root canal and destroy tooth.

Soft-Tissue Disease

Periodontal disease involves **periodontium (gingiva** and surrounding tissues); also begins when plaque forms on tooth (root surface below gingiva) and is mineralized to a hard concretion called **calculus;** this irritates tender gingiva; inflammatory reaction and swelling create **gingivitis** and pockets between tooth and gingiva invaded by bacteria (spirochetes and gram-negative bacilli); immune reaction further damages site; tooth socket may be involved **(periodontitis),** and the tooth may be lost.

MULTIPLE-CHOICE QUESTIONS

1. *Treponema pallidum* is cultured in/on
   a. blood agar c. serum broth
   b. animal tissues d. eggs
2. A gumma is
   a. the primary lesion of syphilis
   b. a syphilitic tumor
   c. the result of congenital syphilis
   d. a damaged aorta
3. The treatment of choice for syphilis is:
   a. tetracycline c. penicillin
   b. antiserum d. sulfa drugs
4. Which of the treponematoses are *not* STDs?
   a. yaws c. syphilis
   b. pinta d. both a and b
5. Lyme disease is caused by_______and spread by_______.
   a. *Borrelia recurrentis,* lice
   b. *Borrelia hermsii,* ticks
   c. *Borrelia burgdorferi,* chiggers
   d. *Borrelia burgdorferi,* ticks
6. Relapsing fever is spread by
   a. lice c. animal urine
   b. ticks d. a and b
7. The primary habitat of *Vibrio cholerae* is
   a. intestine of humans
   b. intestine of animals
   c. natural waters
   d. exoskeletons of crustaceans
8. The best therapy for cholera is
   a. oral tetracycline
   b. oral rehydration therapy
   c. antiserum injection
   d. oral vaccine
9. Rickettsias and chlamydias are similar in being
   a. free of a cell wall
   b. the cause of eye infections
   c. carried by arthropod vectors
   d. obligate intracellular bacteria
10. Which of the following is *not* an arthropod vector of rickettisoses?
    a. mosquito
    b. louse
    c. tick
    d. flea
11. Chlamydiosis caused by *C. trachomatis* attacks which structure(s)?
    a. eye
    b. urethra
    c. fallopian tubes
    d. all of these
12. Ornithosis is a_______infection associated with_______.
    a. rickettsial, parrots
    b. chlamydial, mice
    c. chlamydial, birds
    d. rickettsial, flies
13. Mycoplasmas attack the_______of host cells.
    a. nucleus
    b. cell walls
    c. ribosomes
    d. cell membranes
14. The earliest process that is at the basis of most dental disease is
    a. acquired pellicle
    b. acid release
    c. enamel destruction
    d. plaque accumulation
15. Dental caries are directly due to
    a. microbial acid etching away tooth structures
    b. build-up of calculus
    c. death of tooth by root infection
    d. the acquired pellicle
16. Acute necrotizing ulcerative gingivitis is a_______infection
    a. contagious
    b. mixed
    c. spirochete
    d. systemic

CONCEPT QUESTIONS

1. a. Describe the characteristics of *Treponema pallidum* that are related to its transmission.
   b. Name some factors responsible for the current epidemic of syphilis.
2. a. Describe the stages of untreated syphilis infection. Where does the chancre occur, and what is in it?
   b. What is happening as the chancre disappears?
   c. Which stages are symptomatic, and which are communicable?
   d. What is syphilis latency?
   e. For which tissues does the spirochete have an affinity?
3. Describe the conditions leading to congenital syphilis and the long-term effects of the disease.
4. a. Describe the nonspecific and specific tests for syphilis. What do they test for?
   b. Why are they so important?
5. Outline the general characteristics of bejel, yaws, and pinta.
6. Describe the epidemiology and pathology of leptospirosis.
7. a. Describe the three major borrelioses.
   b. How are arthropods involved?
   c. Why are there relapses in relapsing fever?
   d. What does it mean for a patient to be borrelemic or rickettsemic?
   e. How are these conditions related to the role of vectors in the spread of disease?
8. a. Trace the route of the infectious agent from a tick bite to infection.
   b. Do the same for lice.
9. a. Overview the natural history of Lyme disease and its symptoms.
   b. Explain why it may be mistaken for arthritis, allergy, and neurological diseases.
10. a. Briefly, what is the natural history of cholera?
    b. What is its principal pathologic feature?
11. a. What is secretory diarrhea?
    b. How does oral rehydration therapy work?
12. a. Briefly describe the nature of food infection in species of *Vibrio* and the diseases of *Campylobacter.*
    b. What diseases are *Helicobacter pylori* involved in?
    c. Describe its method of invasion and pathogenesis.

13. a. What do rickettsias and chlamydias derive from the host?
    b. How do antibiotics work to control them?
14. a. Describe the life cycles of *Rickettsia prowazekii, Rickettsia rickettsii,* and *Coxiella burnetii.*
    b. What are the predisposing factors for the types of disease caused by each species?
    c. What makes *Coxiella* unique?
    d. Why are dogs so important in transmission of RMSF?
    e. What are the general symptoms of rickettsial infections?
15. a. Compare the elementary body and the reticulate body of chlamydias.
    b. How are chlamydias transmitted?
    c. Describe the major complications of eye infections and STDs.
16. a. What are the pathologic effects of *Mycoplasma penumoniae?*
    b. Why are the symptoms of mycoplasma pneumonia so mild?
    c. Why doesn't penicillin work on mycoplasma infection?
17. a. In what ways are dental diseases mixed infections?
    b. Discuss the major factors in the development of dental caries and periodontal infections.
18. a. Which diseases in this chapter are zoonoses?
    b. Name them and the major vector involved.
19. a. Find all of the different agents of STDs in this chapter.
    b. Find all of the different agents of pneumonia.
    c. Find all of the agents of gastroenteritis.
20. Matching. Match each disease in the left column with its vector (or vectors) in the right column.

____ leptospirosis	a. wild animals
____ Lyme disease	b. flea
____ murine typhus	c. tick
____ ornithosis	d. birds
____ relapsing fever	e. louse
____ lymphogranuloma venereum	f. mite
____ cat-scratch disease	g. domestic animals
____ epidemic typhus	h. none of these
____ Rocky Mountain spotted fever	
____ Q fever	
____ scrub typhus	
____ cholera	

21. Matching. Match each disease in the left column with its portal of entry in the right column.

____ Q fever	a. skin
____ ornithosis	b. mucous membrane
____ dental caries	c. respiratory tract
____ ANUG	d. urogenital tract
____ mycoplasma	e. eye
____ syphilis	f. oral cavity
____ leptospirosis	g. gastrointestinal tract
____ lymphogranuloma venereum	
____ cholera	
____ Lyme disease	
____ trachoma	
____ *Campylobacter* infection	
____ gastric ulcers	

CRITICAL-THINKING QUESTIONS

1. Why is it so difficult to trace the historical origin of disease, as in syphilis?
2. a. Why does syphilis have such profound effects on the human body?
   b. Why is long-term immunity so difficult to achieve?
3. How can congenital syphilis be prevented?
4. How are the nonsyphilitic treponematoses similar to syphilis?
5. a. In view of the fact that cholera causes the secretion of electrolytes into the intestine, explain what causes the loss of water.
   b. What are the principles of osmosis behind this phenomenon?
6. What would be the best type of vaccine for cholera?
7. a. Explain the general relationships of the vector, the reservoir, and the agent of infection.
   b. Can you think of an explanation for Lyme disease having such a low incidence in the southern United States? (Hint: In this region, the larval stages of the tick feed on lizards, not on mice.)
8. Humans are accidental hosts in many vector-borne diseases. What does this indicate about the relationship between the vector and the microbial agent?
9. a. Why can a louse cause infection only once?
   b. Why must it be crushed in order to cause infection?
   c. What kind of vector can infect many individuals?
   d. How do these vectors infect their hosts?
10. a. Why is arthropod vector control so difficult?
    b. Summarize the methods of preventing arthropod-borne disease.
11. a. Which bacteria presented in this chapter can be cultivated on artificial media?
    b. Which require embryos or cell culture?
12. Name four bacterial diseases for which the dark-field microscope is an effective diagnostic tool.
13. Explain how L forms could be involved in disease.
14. a. In what way is the oral cavity an ecological system?
    b. What causes an imbalance?
    c. What are some logical ways to prevent dental disease besides removing plaque?
15. Case study 1: A journalist returning from a trip experienced severe fever, vomiting, chills, and muscle aches, followed by symptoms of meningitis and kidney failure. Early tests were negative for

septicemia; throat cultures were negative; and penicillin was an effective treatment. Doctors believed the patient's work in the jungles of South America was a possible clue to his disease. What do you think might have been the cause?

16. Case study 2: A man went for a hike in the mountains of New York State and later developed fever and a rash. What two totally different diseases might he have contracted, and what could have been the circumstances of infection?

17. Case study 3: A woman experienced a bout of fever, diarrhea, cramping, and general malaise that lasted for 3 days. She was treated with rehydration therapy but not antibiotics. Cultures and serological tests were negative for any microorganisms tested. The only unusual event that could possibly be linked to her disease was that she had been camping in Michigan and had consumed water from a mountain lake. Name the likely disease.

INTERNET SEARCH TOPIC

1. Go to the World Wide Web to find information on human monocytic and granulocytic ehrlichosis and bartonellosis. Briefly summarize their epidemiology, including the number of new cases, and suggest what factors may be involved in their emergence.
2. Locate information on the possible involvement of chlamydias in arteriosclerosis and of mycoplasmas in autoimmune diseases.

22 chapter

The Fungi of Medical Importance

The eucaryotic microbes collectively called fungi were introduced in chapter 5. The profound importance of fungi stems primarily from their role in the earth's ecological balance and their impact on agriculture. To a lesser—but not minor—extent, the fungi are also medically significant, as agents in human disease, allergies, and mycotoxicoses (intoxications due to ingesting fungal toxins). Diseases resulting from fungal infections, primarily by yeasts and molds, are termed mycoses. In this chapter, we will survey the most prevalent mycotic infections, including systemic, cutaneous, and subcutaneous forms. Other topics to be covered include common respiratory allergies and diseases associated with fungal toxins.

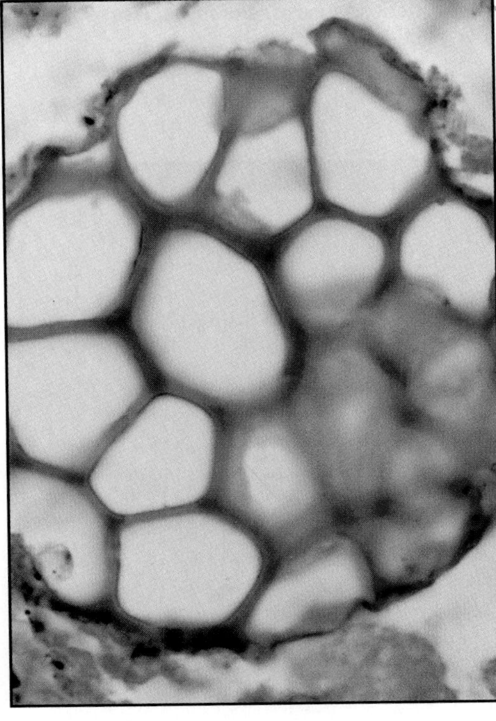

A petrified sporangium of *Coccidioides immitis,* the cause of valley fever, was unearthed in the lungs of a 600–1,000-year-old American Indian skeleton. This is definitive proof of the ancient origins of this disease.

TABLE 22.1

REPRESENTATIVE FUNGAL PATHOGENS, DEGREE OF PATHOGENICITY, AND HABITAT

Microbe	Disease/Infection*	Primary Habitat and Distribution
I. Primary Pathogens		
Histoplasma capsulatum	Histoplasmosis	Soils high in bird guano; Ohio and Mississippi valleys of U.S.; Central and South America; Africa
Blastomyces dermatitidis	Blastomycosis	Presumably soils, but isolation has been difficult; southern Canada; Midwest, Southeast, Appalachia in U.S.; along drainage of major rivers
Coccidioides immitis	Coccidioidomycosis	Highly restricted to alkaline desert soils in southwestern U.S. (California, Arizona, Texas, and New Mexico)
Paracoccidioides brasiliensis	Paracoccidioidomycosis	Soils of rain forests in South America (Brazil, Colombia, Venezuela)
II. Pathogens with Intermediate Virulence		
Sporothrix schenckii	Sporotrichosis	In soil and decaying plant matter; widely distributed
Genera of dermatophytes *Microsporum, Trichophyton, Epidermophyton*	Dermatophytosis (various ringworms or tineas)	Human skin, animal hair, soil throughout the world
III. Secondary Pathogens		
Cryptococcus neoformans	Cryptococcosis	Pigeon roosts and other nesting sites (buildings, barns, trees); worldwide distribution
Candida albicans	Candidiasis	Normal flora of human mouth, throat, intestine, vagina; also normal in other mammals, birds; ubiquitous
Aspergillus spp.	Aspergillosis	Soil, decaying vegetation, grains; common airborne contaminants; extremely pervasive in environment
Pneumocystis carinii	Pneumocystis pneumonia (PCP)	Upper respiratory tract of humans, animals
Genera in Mucorales *Rhizopus, Absidia, Mucor*	Mucormycosis	Soil, dust; very widespread in human habitation

*Specific mycotic infections are usually named by adding *-mycosis, -iasis,* or *-osis* to the generic name of the pathogen.

FUNGI AS INFECTIOUS AGENTS

Molds and yeasts are so widely distributed in air, dust, fomites, and even among the normal flora that humans are incessantly exposed to them. The fact that the planet's surface is totally dusted with spores led W. B. Cooke to christen it "our moldy earth." Fortunately, because of the relative resistance of humans and the comparatively nonpathogenic nature of fungi, most exposures do not lead to overt infection. Of the estimated 100,000 fungal species, only about 300 have been linked to disease in animals, though among plants, fungi are the most common and destructive of all pathogens. Human mycotic disease, or **mycosis,*** is associated with true fungal pathogens that exhibit some degree of virulence or with opportunistic pathogens that take advantage of defective resistance (tables 22.1 and 22.2).

TRUE VERSUS OPPORTUNISTIC PATHOGENS

A **true fungal pathogen** is a species that can invade and grow in a healthy, noncompromised animal host. This behavior is contrary to the metabolism and adaptation of fungi, most of which are

*mycosis (my-koh´-sis) pl. mycoses; Gr. *mykos,* fungi, and *osis,* a disease process.

TABLE 22.2

COMPARISON OF TRUE AND OPPORTUNISTIC FUNGAL INFECTIONS

	True Pathogenic Infections	Opportunistic Infections
Degree of Virulence	Well developed	Limited
Condition of Host	Resistance high or low	Resistance low
Primary Portal of Entry	Respiratory	Respiratory, mucocutaneous
Nature of Infection	Usually primary, pulmonary, and systemic; usually asymptomatic	Varies from superficial skin to pulmonary and systemic; usually symptomatic
Nature of Immunity	Well-developed, specific	Weak, short-lived
Infecting Form	Primarily conidial	Conidial or mycelial
Shows Thermal Dimorphism	Strongly	Not usually
Habitat of Fungus	Soil	Varies from soil to flora of humans and animals
Geographic Location	Restricted to endemic regions	Distributed worldwide

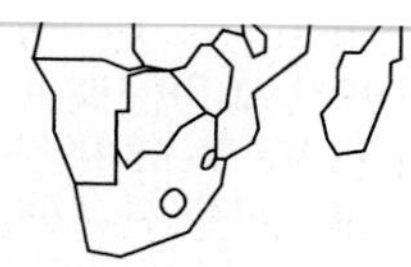

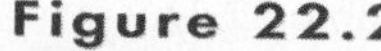

Figure 22.2

Distribution of the four fungal pathogens. The exposure of humans to a given pathogen is greatly influenced by where they live or travel. *Histoplasma* and *Blastomyces* occupy somewhat the same areas of North America, and their incidence is correlated with the course of the Mississippi River and its tributaries. A unique form of *Histoplasma* occurs in Africa and is the only true fungal pathogen on a continent other than the Americas.

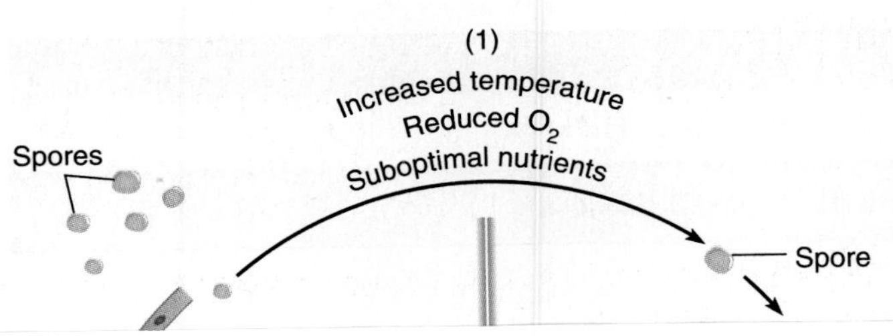

stages in their life cycles, thermal dimorphism is an exception. Opportunistic pathogens vary in their manifestations from superficial and benign colonizations to deep, chronic systemic disease that is rapidly fatal. Mycoses due to opportunists are an increasingly serious problem (microfile 22.1).

Some fungal pathogens exist in a category between true pathogens and opportunists. These species are not inherently

mucous, and cutaneous routes. In general, the agents of primary mycoses have a respiratory portal (spores inhaled from the air); subcutaneous agents enter through inoculated skin (trauma); and cutaneous and superficial agents enter through contamination of the skin surface. Spores, hyphal elements, and yeasts can all be infectious, but spores are most often involved because of their durability and abundance. Spore size is a factor in respiratory infections, because smaller spores are likely to be inhaled more deeply into the respiratory tract, and a large infecting dose increases the severity of disease.

Thermal dimorphism greatly increases virulence by enabling fungi to tolerate the relatively high temperatures and low O_2 tensions of the body. Fungi in the yeast form can be more invasive than hyphal forms because they grow more rapidly and spread through tissues and blood, while those producing hyphae tend to localize along the course of blood vessels and lymphatics. Specific factors contributing to fungal virulence are the subject of much research. Toxinlike substances have been isolated from several species, but their method of damaging the tissues remains unclear. Fungi also produce various adhesion factors and capsules, hydrolytic enzymes, inflammatory stimulants, and allergens, all of which generate strong host responses.

The human body is extremely resistant to establishment of fungi. Among its numerous antifungal defenses are the normal integrity of the skin, mucous membranes, and respiratory cilia, but the most important defenses are cell-mediated immunity, phagocytosis, and the inflammatory reaction. Long-term protective immunity can develop for some of the true pathogens, but for the rest, reinfection is a distinct possibility.

Diagnosis of Mycotic Infections

Satisfactory diagnosis of fungal infections depends mainly upon isolating and identifying the pathogen in the laboratory. Accurate and speedy diagnosis is especially critical to the immunocompromised patient, who must have prompt antifungal chemotherapy. For example, a patient with systemic *Candida* infection can die if the infection is not detected and treated within 5 to 7 days. Because therapy can vary among the pathogenic fungi, identification to the species level is often necessary.

A suitable specimen can be obtained from sputum, skin scrapings, skin biopsies, cerebrospinal fluid, blood, tissue exudates, urine, or vaginal samples, as determined by the patient's symptoms. Routine laboratory procedures include isolation, microscopic and macroscopic examination, histological stains, serology, and animal inoculation (figure 22.3).

Because culture of the sample can require several days, the immediate direct examination of fresh samples is recommended. Wet mounts can be prepared by mixing a small portion of the sample on a slide with saline, water, or potassium hydroxide to clear the specimen of background debris. The relatively large size and unique appearance of many fungi help them stand out. Large round or oval budding cells are evidence of yeasts, whereas thick branching strands suggest hyphae. Nonspecific fluorescent stains or whiteners are valuable for highlighting fungi in tissue samples (figure 22.4).

Fungal elements in samples are most readily detected with methenamine silver, periodic acid-Schiff, hematoxylin and eosin, Giemsa, or Gram stains (see figure 22.24). Isolating the fungal pathogen requires planting the specimen onto three or four types of solid media such as Sabouraud's dextrose agar, mycosel agar, inhibitory mold medium, and brain-heart-infusion agar. These media can be modified by adding one of the following agents: blood to increase the growth of fastidious species, chloramphenicol and gentamicin to inhibit bacteria, or cyclohexamide to slow the growth of fungal contaminants. Cultures are incubated at room temperature, at 30°C, and at 37°C and are observed daily for growth, which can require several weeks in some species. Gross colonial morphology, such as the color and texture of the colony's surface and underside, can be very distinctive. Initial identification of the pathogen can be followed by confirmatory physiological, antigen, and DNA-based testing. Examples of certain tests will be given during discussions of specific diseases later in this chapter.

In vitro tests to detect antifungal antibodies in serum (see figure 22.10) can be a useful diagnostic tool in some infections. *In vivo* skin testing for delayed hypersensitivity to fungal antigens is mostly used to trace epidemiologic patterns. It does not verify ongoing infection, and considerable cross-reactivity exists among *Histoplasma, Blastomyces,* and *Coccidioides.* A negative test in a healthy person, however, can rule out infection by these fungi.

Control of Mycotic Infections

Antimicrobic therapy for fungal infections is covered in chapter 12 and on a disease-by-disease basis in this chapter. Although immunization for most fungal diseases is not considered feasible at this time, T-cell infusions can be used successfully to treat children with candidiasis. Vaccines for coccidioidomycosis and histoplasmosis are being prepared for clinical trials. However, few specific prevention measures exist for fungal infections.

Organization of Fungal Diseases

Fungal infections can be presented in several schemes, none of which is completely satisfactory. Traditional methods are based on taxonomic group, location of infection, and type of pathogen. Chapter 5 presents a taxonomic breakdown of the fungi, and this chapter treats them in the following categories according to the type and level of infection they cause and their degree of pathogenicity: (1) **systemic, subcutaneous, cutaneous,** and **superficial** cutaneous mycoses (the four levels of infection depicted in figure 22.5), and (2) opportunistic mycoses (see table 22.1).

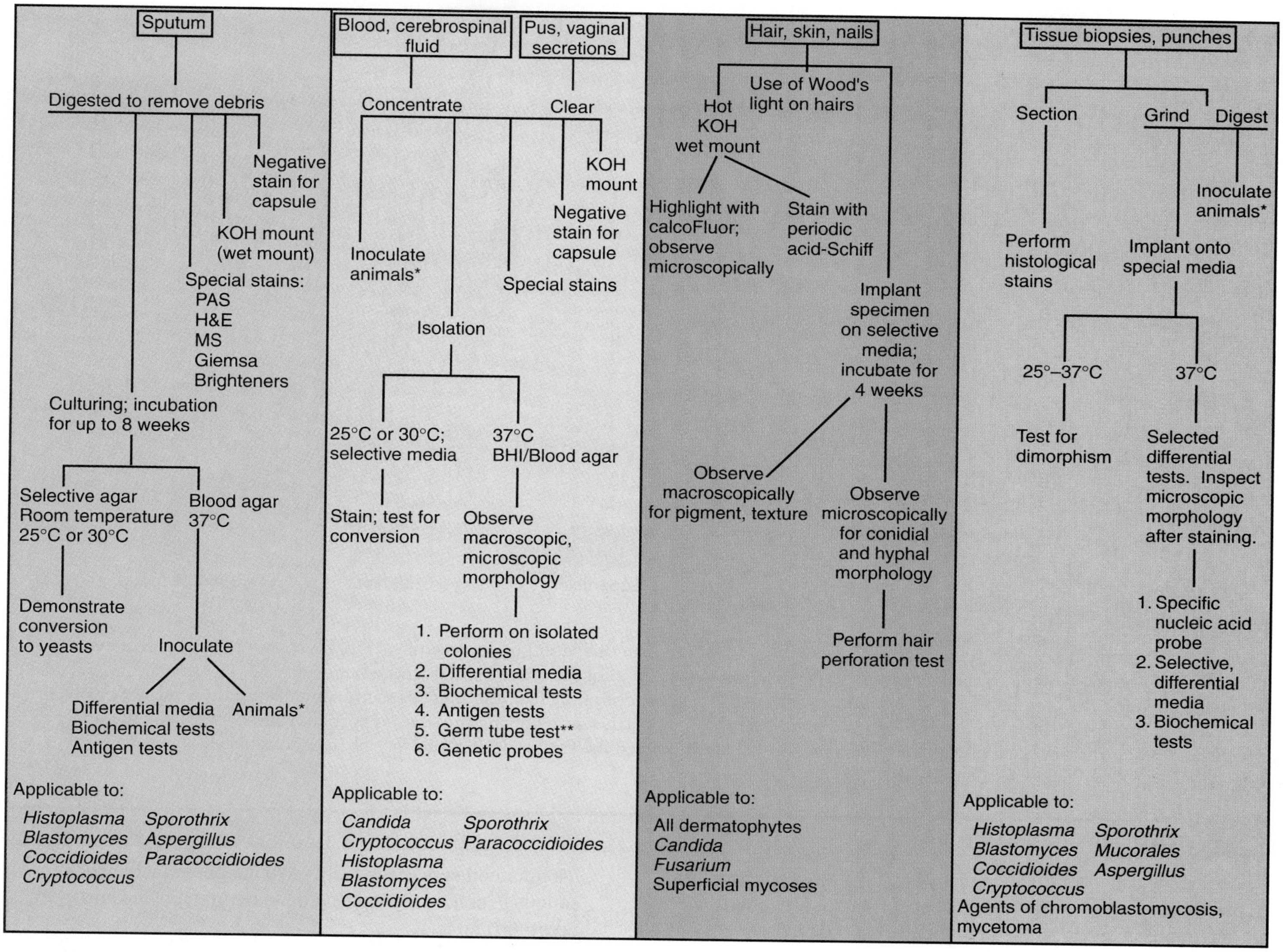

Sputum

Digested to remove debris
- Negative stain for capsule
- KOH mount (wet mount)
- Special stains: PAS, H&E, MS, Giemsa, Brighteners
- Culturing; incubation for up to 8 weeks
  - Selective agar, Room temperature 25°C or 30°C → Demonstrate conversion to yeasts
  - Blood agar 37°C → Inoculate → Differential media, Biochemical tests, Antigen tests; Animals*

Applicable to: *Histoplasma*, *Blastomyces*, *Coccidioides*, *Cryptococcus*, *Sporothrix*, *Aspergillus*, *Paracoccidioides*

Blood, cerebrospinal fluid → Concentrate; **Pus, vaginal secretions** → Clear
- KOH mount
- Negative stain for capsule
- Special stains
- Inoculate animals*
- Isolation
  - 25°C or 30°C; selective media → Stain; test for conversion
  - 37°C BHI/Blood agar → Observe macroscopic, microscopic morphology →
    1. Perform on isolated colonies
    2. Differential media
    3. Biochemical tests
    4. Antigen tests
    5. Germ tube test**
    6. Genetic probes

Applicable to: *Candida*, *Cryptococcus*, *Histoplasma*, *Blastomyces*, *Coccidioides*, *Sporothrix*, *Paracoccidioides*

Hair, skin, nails
- Hot KOH wet mount
  - Highlight with calcoFluor; observe microscopically
  - Stain with periodic acid-Schiff
- Use of Wood's light on hairs
- Implant specimen on selective media; incubate for 4 weeks
  - Observe macroscopically for pigment, texture
  - Observe microscopically for conidial and hyphal morphology → Perform hair perforation test

Applicable to: All dermatophytes, *Candida*, *Fusarium*, Superficial mycoses

Tissue biopsies, punches
- Section → Perform histological stains
- Grind → Implant onto special media
  - 25°–37°C → Test for dimorphism
  - 37°C → Selected differential tests. Inspect microscopic morphology after staining. →
    1. Specific nucleic acid probe
    2. Selective, differential media
    3. Biochemical tests
- Digest → Inoculate animals*

Applicable to: *Histoplasma*, *Blastomyces*, *Coccidioides*, *Cryptococcus*, *Sporothrix*, *Mucorales*, *Aspergillus*, Agents of chromoblastomycosis, mycetoma

Figure 22.3

Methods of processing specimens in fungal disease. *Animal inoculation is performed only to help diagnose systemic mycoses when other methods are unavailable or indeterminant. **Some yeasts, when incubated in serum for 2 to 4 hours, sprout tiny hyphal tubes called germ tubes. *Candida albicans* is identified by this characteristic.

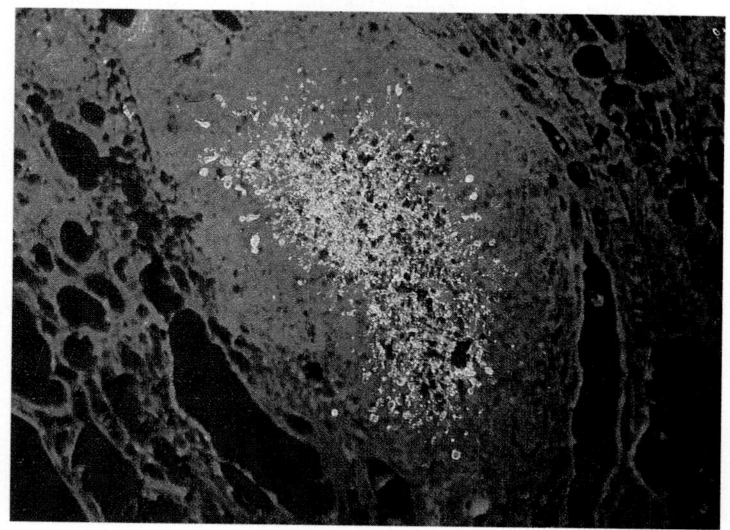

Figure 22.4

Special tissue stains called whiteners or brighteners can amplify the presence of fungal elements in specimens. They bind tightly to the carbohydrates of the fungal surface; when viewed by a fluorescent microscope, they fluoresce with a blue-green light in contrast with the red background. This kidney section of *Candida* is stained with tinopal.

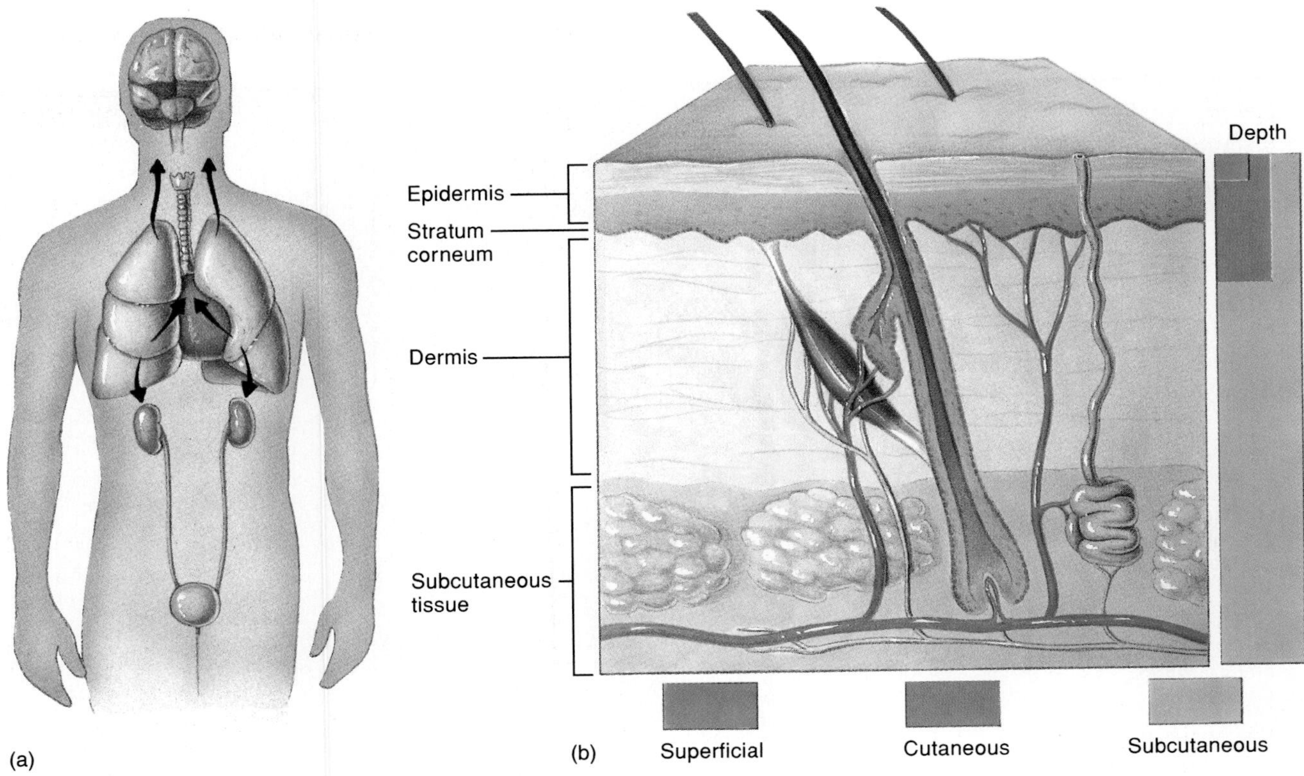

Figure 22.5
Levels of invasion by fungal pathogens. (Some species can invade more than one level.) (*a*) In systemic (deep) mycoses, the fungus disseminates from the lungs or other sites into the circulation. Fungemia leads to infection of the brain, kidneys, and other organs. (*b*) The skin and its attendant structures provide many potential sites for invasion, including the scalp, smooth skin, hair, and mucous membranes. Differing depths of involvement are: superficial, consisting of extremely shallow epidermal colonizations; cutaneous, involving the stratum corneum and occasionally the upper dermis; and subcutaneous, occurring after a puncture wound has introduced the fungus deeper into the subcutaneous tissues.

Chapter Checkpoints

Fungi are widely distributed in many habitats. Most are harmless saprobes, but approximately 300 species cause mycotic infections in humans. The ubiquitous distribution of fungi ensures that most humans will experience one or more mycoses during their lifetime.

Mycotic infectious agents can be either true pathogens or opportunists. Fungi can cause allergies, toxicoses, and specific infections that are seasonally related. Virulence factors include resistant spores, thermal dimorphism, toxin production, and invasive factors. Infective spores are widely distributed through dust and air.

Fungal infections are categorized by their pathogenicity and the level or type of infection they cause. They can be superficial, cutaneous, subcutaneous, and systemic. True pathogens are species that are capable of initiating infection in a healthy host. Such pathogens exhibit thermal dimorphism, exhibiting the yeast form at body temperatures and the mold form at lower temperatures. True pathogens are *Histoplasma, Blastomyces, Coccidioides,* and *Paracoccidioides.*

Opportunists have few if any virulence factors, may or may not have thermal dimorphism, and usually initiate infection in immunodeficient hosts. Examples are *Cryptococcus, Candida,* and *Aspergillus.* Some fungi are not inherently pathogenic but, if introduced subcutaneously, have the potential to cause serious infection. Examples are the dermatophytes and *Sporothrix* species.

Mycoses are not usually communicable, with certain exceptions. The four pathogenic species are endemic to specific ecological regions and may cause epidemics during mass exposure to a common source.

All true pathogens are dimorphic, and all initiate infection through inhalation of spores. The yeast phase develops in the tissues. Host defenses to fungal infection depend on the integrity of all epithelial barriers, a functional inflammatory response, and cell-mediated immunity. Effective treatment of fungal infections requires rapid and accurate diagnosis because drug therapy can vary with the species.

Accurate diagnosis of mycotic agents requires direct microscopic examination of fresh specimens followed by confirmatory isolation on solid media, serological tests for host antibody, and genetic analysis.

Immunization is not usually effective against fungal infections. Preventives are limited to masks and protective clothing to reduce contact with spores. Treatment includes antifungal therapy and, in some cases, surgery.

SYSTEMIC INFECTIONS BY TRUE PATHOGENS

The infections of primary fungal pathogens all follow a similar pattern. They are restricted to certain endemic regions of the world. Infection occurs when soil or other matter containing the fungal conidia is disturbed and the spores are inhaled into the lower respiratory tract. In the lungs, the spores germinate into yeasts or yeastlike cells and produce an asymptomatic or mild **primary pulmonary infection (PPI)** that parallels tuberculosis. In a small number of hosts, this infection becomes systemic and creates severe, chronic lesions. In a few cases, spores are inoculated into the skin, where they form localized granulomatous lesions. All diseases result in immunity that can be long-term and that manifests clinically as an allergic reaction to fungal antigens.

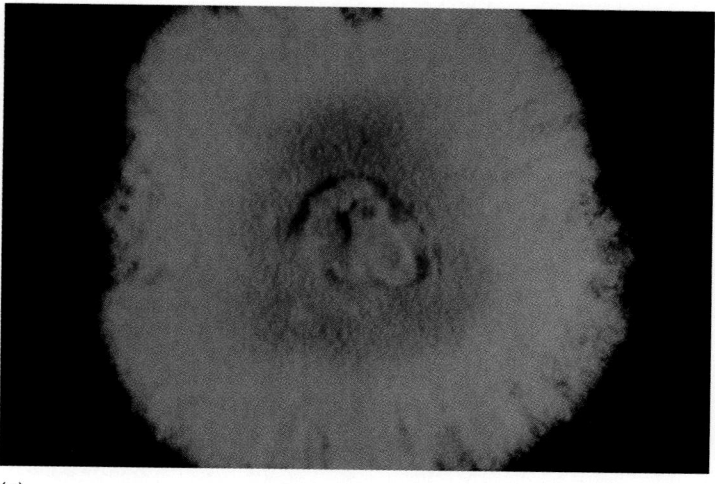

(a)

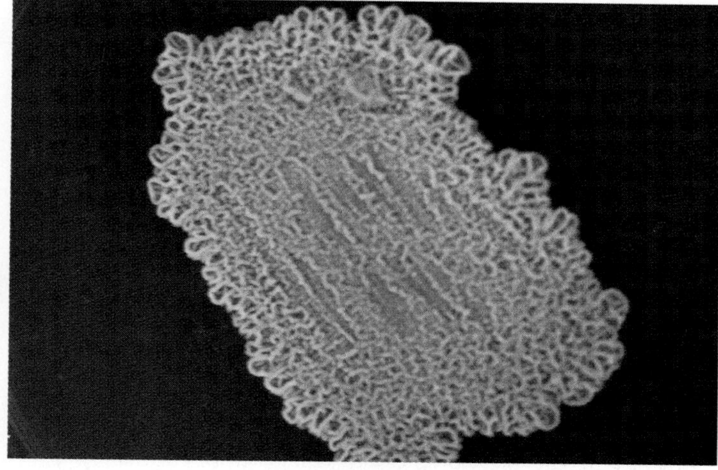

(b)

Figure 22.6

Cellular and cultural characteristics of *Histoplasma capsulatum*. (*a*) A colony at 25°C produces an abundant mycelium. (The spores are shown in figure 22.7.) (*b*) A yeast colony (37°C) is dense and can have a lacy texture.

HISTOPLASMOSIS: OHIO VALLEY FEVER

The most common true pathogen is ***Histoplasma capsulatum,**** the cause of histoplasmosis. This disease has probably afflicted humans since antiquity, though it was not described until 1905 by Dr. Samuel Darling. Through the years, it has been known by various synonyms—Darling's disease, Ohio Valley fever, and reticuloendotheliosis. The impact of histoplasmosis throughout history is not really known. Certain aspects of its current distribution and epidemiology suggest that it has been an important disease as long as humans have practiced agriculture.

Biology and Epidemiology of Histoplasma capsulatum

Histoplasma capsulatum is typically dimorphic. Growth on media below 35°C is characterized by a white or brown, hairlike mycelium, and growth at 37°C on blood agar produces a white, smooth colony (figure 22.6).

Histoplasma capsulatum is discontinuously distributed on all continents except Australia, though its areas of greatest endemicity are the eastern and central regions of the United States (the Ohio Valley). This fungus appears to grow most abundantly in moist soils high in nitrogen content, especially those supplemented by bird and bat *guano.**

A useful tool for determining the distribution of *H. capsulatum* is to inject a fungal extract called **histoplasmin** into the skin and monitor for allergic reactions. Application of this test has verified the extremely widespread distribution of the fungus. In high-prevalence areas such as southern Ohio, Illinois, Missouri, Kentucky, Tennessee, Michigan, Georgia, and Arkansas, 80% to 90% of the population show signs of prior infection. Histoplasmosis prevalence in the United States is estimated at about 500,000 cases per year, with several thousand of them requiring hospitalization and a small number resulting in death.

Histoplasma capsulatum (his''-toh-plaz'-mah kap''-soo-lay'-tum) Gr. *hist.* tissue. and *plasm,* shape; L. *capsula,* small box.

*guano (gwan'-oh) Sp. *huanu,* dung. An accumulation of animal manure.

The spores of the fungus are probably dispersed by the wind and, to a lesser extent, animals. The most striking outbreaks of histoplasmosis occur when concentrations of spores have been dislodged by humans working in parks, bird roosting areas, and old buildings. People of both sexes and all ages incur infection, but adult males experience the majority of cases. The oldest and youngest members of a population are most likely to develop serious disease.

Infection and Pathogenesis of Histoplasma

Histoplasmosis presents the most formidable array of manifestations of any mycosis. It can be benign or severe, acute or chronic, and it can show pulmonary, systemic, or cutaneous lesions. Inhaling a small dose of microconidia into the deep recesses of the

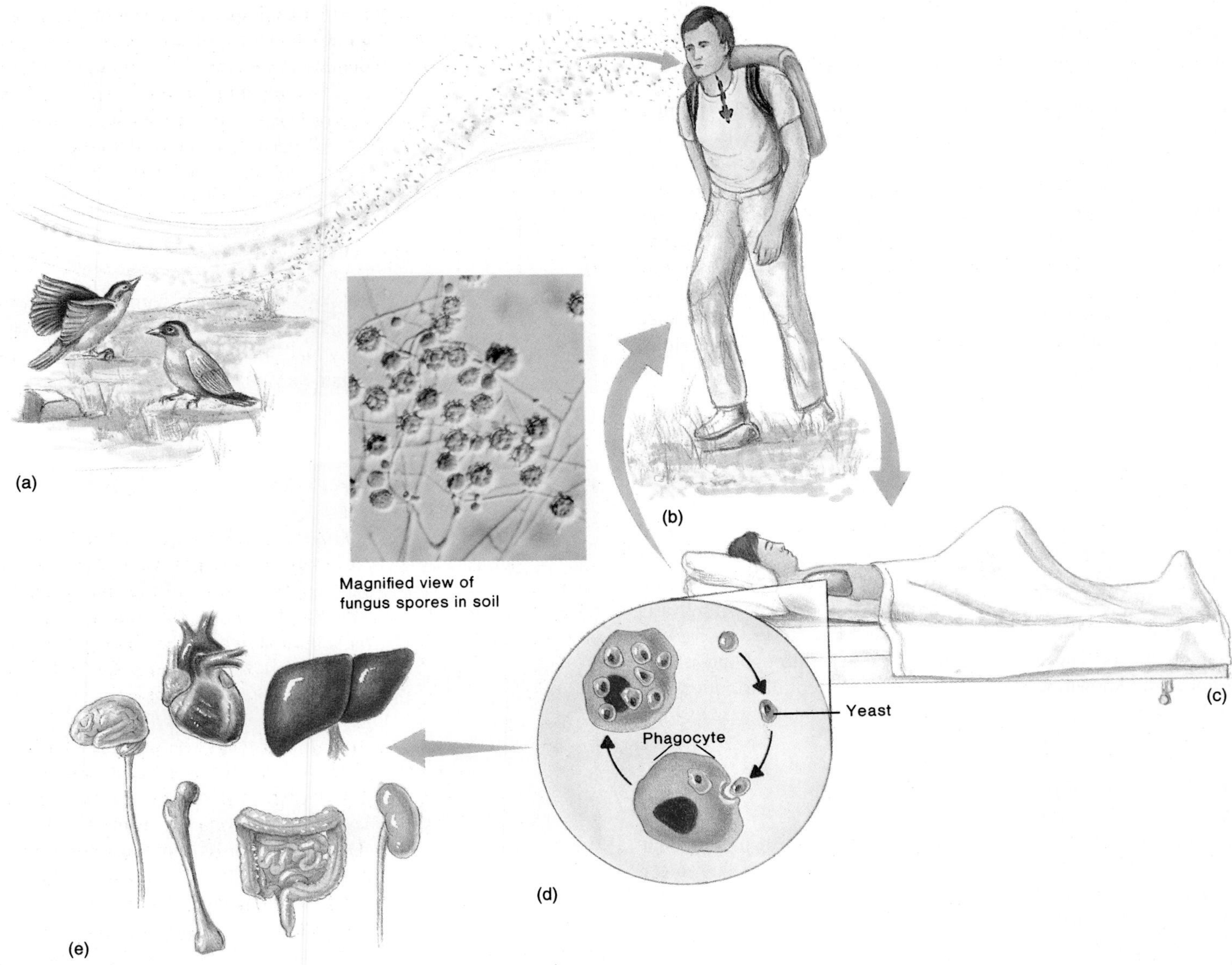

Figure 22.7
Events in histoplasmosis. (*a*) Soil containing bird droppings is whipped up by the wind. (*b*) Microconidia are inhaled. (*c*) The patient develops mild pneumonitis, which might recur. (*d*) In the tissue phase of infection, the yeast phase develops, is phagocytosed, and multiplies by budding intracellularly. Most patients recover without complications, but (*e*) in some cases, phagocytes enter the blood and cause disseminated disease in such sites as the bones, kidney, and liver.

lung establishes a primary pulmonary infection that is usually asymptomatic. Its primary location of growth is in the cytoplasm of phagocytes such as macrophages. Within these cells, it flourishes and is carried to other sites (figure 22.7). Some people experience mild symptoms such as aches, pains, and coughing, but a few develop more severe symptoms, including fever, night sweats, and weight loss. Primary cutaneous histoplasmosis, in which the agent enters via the skin, is very rare.

The most serious systemic forms of histoplasmosis occur in patients with defective cell-mediated immunity such as AIDS patients. The infection may be new or the recurrence of a dormant infection. In children, this can lead to liver and spleen enlargement, anemia, circulatory collapse, and death. Adults with systemic disease can acquire lesions in the brain, intestine, adrenal gland, heart, liver, spleen, lymph nodes, bone marrow, and skin. Persistent colonization of patients with emphysema and bronchitis causes *chronic pulmonary histoplasmosis,* a complication that has signs and symptoms similar to those of tuberculosis.

Diagnosis and Control of Histoplasmosis

Discovering *Histoplasma* in clinical specimens is a substantial diagnostic indicator. Usually it appears as spherical, "fish-eye" yeasts intracellularly in macrophages and occasionally as free yeasts in samples of sputum and cerebrospinal fluid. An essential confirmatory step is to isolate the agent and demonstrate dimorphism, but this step can require up to 12 weeks. Complement fixation and immunodiffusion serological tests can support a diagnosis by showing a rising antibody titer. The histoplasmin test does not indicate concurrent infection and is not useful in diagnosis.

Undetected or mild cases of histoplasmosis resolve without medical management, but chronic or disseminated disease calls for systemic chemotherapy. The principal drug, amphotericin B, is administered in daily intravenous doses for a few days to a few weeks. Under some circumstances, ketoconazole or other azoles are the drugs of choice. Surgery to remove affected masses in the lungs or other organs can also be effective.

Coccidioidomycosis: Valley Fever, San Joaquin Fever, California Disease

Coccidioides immitis** is the etiologic agent of **coccidioidomycosis.** Although the fungus has probably lived in soil for millions of years, human encounters with it are relatively recent and coincide with the encroachment of humans into its habitat (see chapter opening photograph). This unique and fascinating fungus can be the most virulent of all mycotic pathogens.

Biology and Epidemiology of Coccidioides

The morphology of *C. immitis* is very distinctive. At 25°C, it forms a moist, white to brown colony with abundant, branching, septate hyphae. These hyphae fragment into thick-walled, blocklike **arthroconidia** at maturity (inset photo, figure 22.8). On special media incubated at 37°–40°C, an arthrospore germinates into the parasitic phase, a small, spherical cell **(spherule).** This structure swells into a giant sporangium that cleaves internally to form numerous endospores that look like bacterial endospores but have none of their resistance traits (figure 22.8*c*).

Coccidioides immitis occurs endemically in various natural reservoirs and casually in areas where it has been carried by wind and animals. Conditions favoring its settlement in a given habitat include high carbon and salt content and a semiarid, relatively hot climate. The fungus has been isolated in such regions from soils, plants, and a large number of vertebrates. The natural history of *C. immitis* follows a cyclic pattern—a period of dormancy in winter and spring, followed by growth in summer and fall. Growth and spread are greatly increased by cycles of drought and heavy rains, followed by windstorms.

Skin testing has disclosed that the highest incidence of coccidioidomycosis (approximately 100,000 cases per year) is in the southwestern United States, though it also occurs in Mexico and parts of Central and South America. Especially concentrated reservoirs exist in the San Joaquin Valley of California and in southern Arizona. Occasional epidemics are reported in conjunction with soil agitation by humans (excavating and farming). Archeologists digging in ancient dwellings of Native Americans have been common victims of the disease. All persons inhaling the arthrospores probably develop some degree of infection, but certain groups have a genetic susceptibility that gives rise to more serious disease.

A highly unusual outbreak of coccidioidomycosis was traced to the 1994 earthquake in Northridge, California. The CDC reported over 200 cases and one death over a 2-month period following the quake and its aftershocks. Clouds of dust bearing loosened spores were given off by landslides in the mountains north of the epicenter. Local winds then carried the dust into outlying residential areas.

Infection and Pathogenesis of Coccidioidomycosis

The arthrospores of *C. immitis* are lightweight and readily inhaled. In the lung they convert to spherules, which swell, sporulate, burst, and release spores that continue the cycle. In 60% of patients, this primary pulmonary infection is inapparent; in the other 40%, it is accompanied by coldlike symptoms such as fever, chest pain, cough, headaches, and malaise. In uncomplicated cases, complete recovery and lifelong immunity are the rule.

In about five out of a thousand cases, the primary infection does not resolve and progresses with varied consequences. Chronic progressive pulmonary disease is manifested by nodular growths called *fungomas** and cavity formation in the lungs that compromise respiration. Dissemination of the endospores into major organs occasionally takes place in people with impaired cell-mediated immunity. Such disseminated disease impressively demonstrates that when virulence and immunodeficiency coexist in the same infection, the disease can erupt with explosive and sometimes fatal results (figure 22.9).

Diagnosis and Control of Coccidioidomycosis

Diagnosis of coccidioidomycosis is straightforward when the highly distinctive spherules are found in sputum, spinal fluid, and biopsies. This finding is further supported if the typical mycelium and spores are isolated on Sabouraud's agar and if spherules are induced to form. Newer specific antigen tests have been a great boon in identifying and differentiating *Coccidioides* from other fungi. So great is the potential for contamination and infection that cultures are grown in closed tubes or bottles, and are opened in a biological containment hood or killed before inspection. Immunodiffusion (figure 22.10) and latex agglutination tests on serum samples are excellent screens for detecting early infection. Skin tests using an extract of the fungus

**Coccidioides immitis* (kok-sid″-ee-oy′-deez ih′-mih-tis) From *coccidia,* a sporozoan, and L. *immitis,* fierce. The original discoverers thought the microbe looked like a protozoan.

*fungoma (fun-joh′-mah) A fungus tumor or growth.

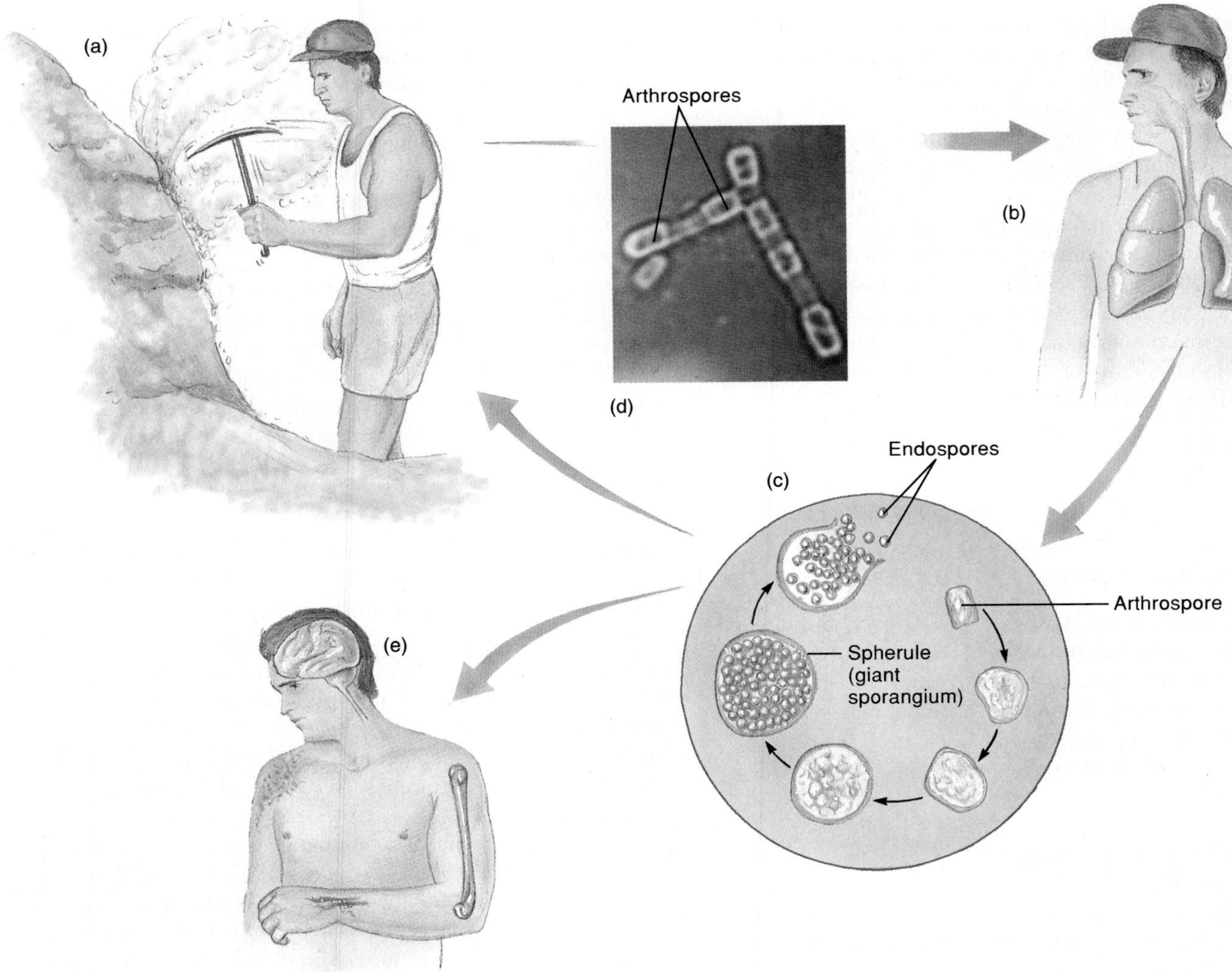

Figure 22.8
Events in coccidioidomycosis. (*a*) A person digging in soil produces aerosol. Inhaled arthrospores (inset photo) establish (*b*) a lung infection. (*c*) Arthrospores develop into spherules that produce endospores; endospores are released in the lungs. (*d*) Immunocompetent persons effectively fight infection and return to health. (*e*) Compromised people can develop meningitis, osteomyelitis, and skin granulomas.

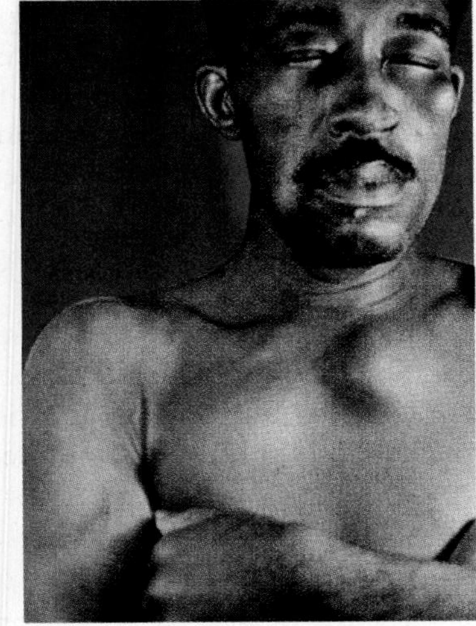

Figure 22.9
Disseminated coccidioidomycosis manifested by subcutaneous abscesses in the chest.

(coccidioidin or spherulin) are of primary importance in epidemiologic studies.

The majority of coccidioidomycosis patients do not require treatment. However, in people with disseminated disease, amphotericin B is administered intravenously. The azole drugs can have some benefit, but they may require a longer term of therapy. Minimizing contact with the fungus in its natural habitat has been of some value. For example, oiling roads and planting vegetation help reduce spore aerosols, and using dust masks while excavating soil prevents workers from inhaling spores.

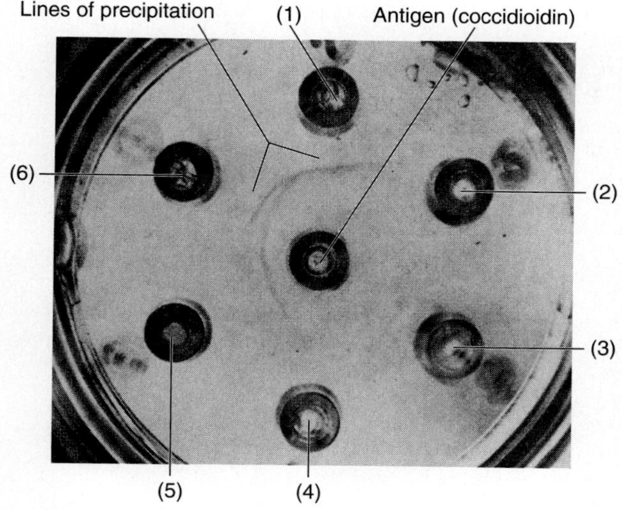

Figure 22.10
Immunodiffusion testing for coccidioidomycosis. Small wells are punched into agar, and patients' sera are placed in numbered wells; the center well contains a known fungal antigen such as coccidioidin. Lines of precipitation forming between the outer well and the inner well indicate a reaction between antibodies in the serum and antigen diffusing from the center well. In this test, reactions occurring for sera 1, 5, and 6 are positive evidence of infection. Subtle differences in these lines can show qualitative differences in the severity of the disease. Wells 2, 3, and 4 indicate the absence of antibodies and infection.

Biology of *Blastomyces dermatitidis*: North American Blastomycosis

Blastomyces dermatitidis,* the cause of **blastomycosis,** is another fungal pathogen endemic to the United States. Alternative names for this disease are Gilchrist's disease, Chicago disease, and North American blastomycosis. The dimorphic morphology of *Blastomyces* follows the pattern of other pathogens. Colonies of the saprobic phase are uniformly white to tan, with a thin, septate mycelium and simple ovoid conidia (figure 22.11*a*). Temperature-induced conversion results in a wrinkled, creamy-white colony that yields large, heavy-walled yeasts with buds nearly as large as the mother cell.

Studies indicate that *B. dermatitidis* inhabits areas high in organic matter such as forest soil, decaying wood, animal manure, and abandoned buildings, but it has been difficult to isolate with regularity. Its life cycle features dormancy during the warmer, dryer times of the year and growth and sporulation during the colder, wetter seasons. In general, blastomycosis occurs from southern Canada to southern Louisiana and from Minnesota to Georgia. Cases have also been reported in Central America, South America, Africa, and the Middle East. Humans, dogs, cats, and horses are the chief targets of infection, which is usually acquired by inhaling conidia-laden dust from living quarters, farm buildings, or forest litter. According to the few statistics available, the frequency of blastomycosis is highest among the middle-aged, males, blacks, and pregnant women.

**Blastomyces dermatitidis* (blas′′-toh-my′-seez der′′-mah-tit′-ih-dis) Gr. *blastos*, germ, *myces*, fungus, *dermato*, skin, and *itis*, inflammation.

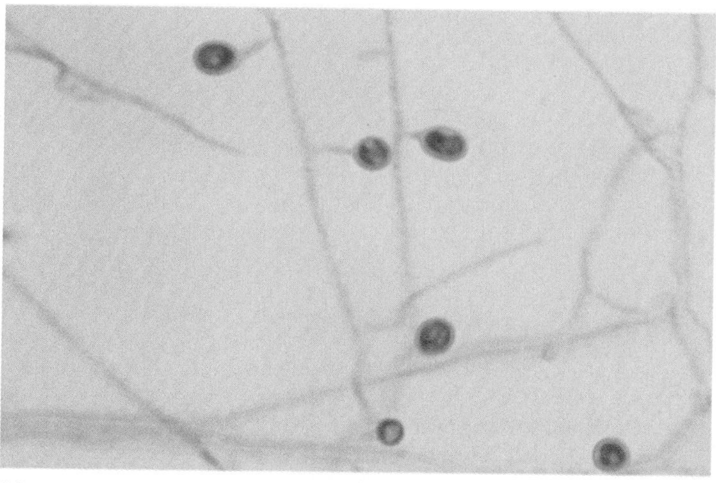

(a)

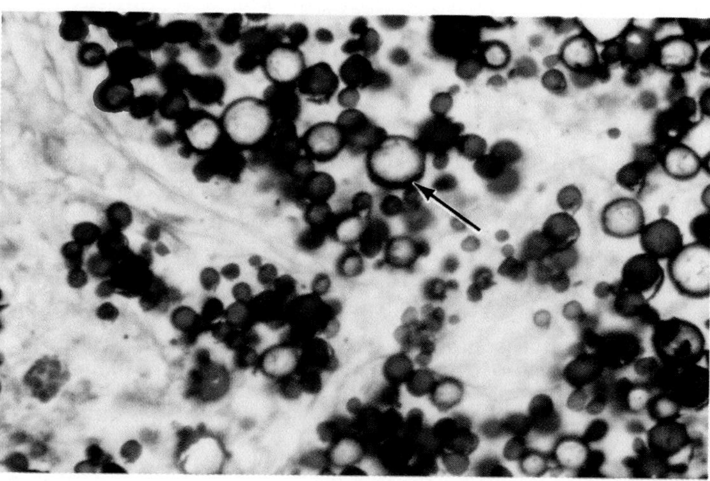

(b)

Figure 22.11

The dimorphic nature of *Blastomyces dermatitidis.* (*a*) Hyphal filaments bear conidia that resemble tiny lollipops. (*b*) The tissue phase as seen in a sputum sample. The arrow points out the very thick cell wall typical of this species.

Infection and Pathology in Blastomycosis

The primary portal of entry of *B. dermatitidis* is the respiratory tract, though it also may enter through accidental inoculation. Inhaling only 10 to 100 conidia is enough to initiate infection. As conidia convert to yeasts and multiply, they encounter macrophages in the lung. A large proportion of primary pulmonary infections are probably symptomatic to some degree.

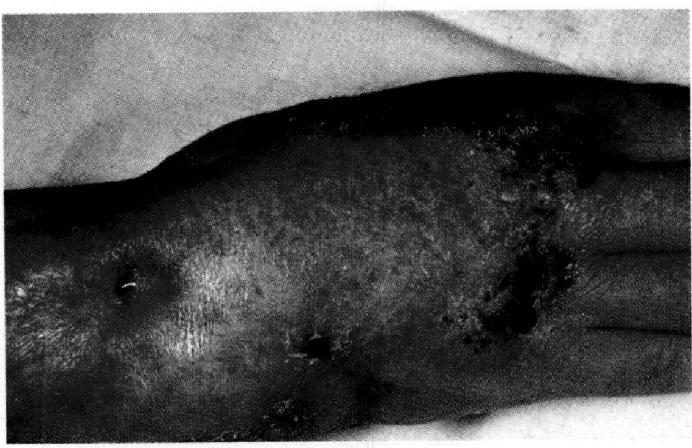

Figure 22.12

Cutaneous blastomycosis in the hand and wrist as a complication of disseminated infection. Note the darkly colored, tumorlike vegetations and scar tissue on the hand.

Mild disease is accompanied by cough, chest pain, hoarseness, and fever. More severe, chronic blastomycosis can progress to the lungs, skin, and numerous other organs. Abscesses and tumorlike nodules developing in the lungs are often mistaken for cancer. Compared with other fungal diseases, the chronic cutaneous form of blastomycosis is rather common. It frequently begins on the face, hand, wrist, or leg as a subcutaneous nodule that erupts to the skin surface (figure 22.12). Yeasts disseminating into the bone induce symptoms of arthritis and osteomyelitis. Involvement of the central nervous system can bring on headache, convulsions, coma, and mental confusion. *Blastomyces* can also spread to the reproductive tract, spleen, liver, and kidneys. Chronic systemic blastomycosis lasting for weeks to years frequently overwhelms the host defenses and kills the patient.

Laboratory Diagnosis and Therapy

Microscopic smears of specimens showing large, ovoid yeasts with broad-based buds can provide the most reliable diagnostic evidence of blastomycosis (see figure 22.11*a*). Cultures prepared on selective and enriched media and incubated at room temperature can require several weeks to develop. Further confirmatory tests are done to verify dimorphism and the presence of antibodies through complement fixation and ELISA tests. Most skin tests for *Blastomyces* are not useful for general testing because of the frequency of false-positive and false-negative results. Although disseminated infections were once nearly 100% fatal, modern drugs have greatly improved the prognosis. The drug of choice for severe disseminated and cutaneous disease is amphotericin B, and milder cases typically respond to several months of dihydroxystilbamidine therapy.

PARACOCCIDIOIDOMYCOSIS

The remaining dimorphic fungal pathogen would be a front-runner in a competition for the most tongue-twisting scientific term. ***Paracoccidioides brasiliensis****** causes **paracoccidioidomycosis,** also known as paracoccidioidal granuloma and South American blastomycosis. Because of its relatively restricted distribution, this disease is the least common of the primary mycoses. *Paracoccidioides* forms a small, nondescript colony with scanty, undistinctive spores at room temperature, but it develops an unusual yeast form at 37°C. The large mother cells sprout small, narrow-necked buds that can radiate around the periphery like spokes on a wheel (figure 22.13).

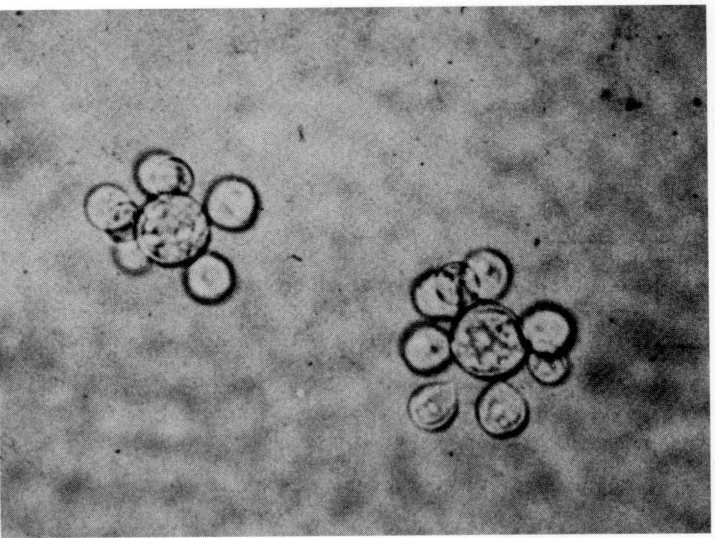

Figure 22.13
The morphology of *Paracoccidioides.* A Gridley stain from a skin lesion (1,200×) reveals the central round mother cell with a series of narrow-necked buds that look like the spokes of a wheel.

The natural history of *Paracoccidioides* has not yet been completely clarified. It has been isolated from the cool, humid soils of tropical and semitropical regions of South and Central America, particularly Brazil, Colombia, Venezuela, Argentina, and Paraguay. People most often afflicted with paracoccidioidomycosis are rural agricultural workers and plant harvesters. A possible animal role in transmission has been postulated but not demonstrated. Other factors believed to influence disease are altered physiology, poor nutrition, and impaired host resistance. Infection is initiated when fungal spores enter the lungs or, occasionally, are inoculated into the skin. Most infections are probably benign, self-limited events that go completely unnoticed. In the minority of patients who do incur progressive systemic disease, the lungs, skin and mucous membranes (especially of the head), and lymphatic organs are frequently involved.

Paracoccidioidomycosis is diagnosed by standard procedures previously described for the other mycoses in this chapter. Closely scrutinizing fresh or stained clinical specimens for the distinctive yeast phase is important, and a cultural follow-up with conversion media is essential. Serological testing can be instrumental in diagnosing and monitoring the course of infection. Principal treatment drugs for disseminated disease, in order of choice, are ketoconazole, amphotericin B, and sulfa drugs.

**Paracoccidioides brasiliensis* (pair″-ah-kok-sid″-ee-oy′-deez brah-sil″-ee-en′-sis) Named for its superficial resemblance to *Coccidioides* and its prevalence in Brazil.

Chapter Checkpoints

Primary fungal pathogens have several common characteristics: (1) Each is endemic to a specific region of the world. (2) All cause primary pulmonary infections by inhalation of spores. (3) In most cases, the infection is not life-threatening. (4) Recovery from infection confers lifelong immunity. True fungal pathogens cause systemic infection in certain susceptible groups of people. The spores of some species can also infect the skin.

Histoplasma capsulatum is the causative agent of histoplasmosis, or Ohio Valley fever. It is endemic to the Ohio Valley in the United States and discontinuously in much of the world. Airborne spores cause pulmonary, systemic, or cutaneous lesions. The severity of the infection depends on the number of spores inhaled and the immunocompetence of the infected host. *H. capsulatum* is identified by "fish-eye" yeast cells in host macrophages.

Coccidioides immitis is the causative agent of coccidioidomycosis, or Valley fever. It is endemic to salty soils in arid regions of the western United States. Airborne spores cause a primary pulmonary infection that is self-limiting in most cases and results in a lifelong immunity. Rare cases progress to chronic pulmonary disease or systemic infections. *C. immitis* is identified by highly distinctive spherules in fresh tissue specimens.

Blastomyces dermatitidis is the causative agent of blastomycosis, or Gilchrist's disease. It is endemic to North America, Africa, and the Middle East. Like other true pathogens, airborne spores cause a primary pulmonary infection, but the skin is also a site of infection. *B. dermatitidis* is identified by its microscopic appearance in fresh specimens and by serological tests for antibody.

Paracoccidioides brasiliensis causes paracoccidioidomycosis, or South American blastomycosis, endemic to regions of Central and South America. In most cases, infections of the lungs and skin are self-limiting, but persistent infections can be debilitating. *P. brasiliensis* is identified by its unusual budding pattern and by serological tests for antibody.

SUBCUTANEOUS MYCOSES

When certain fungi are transferred from soil or plants directly into traumatized skin, they can invade the damaged site. Such infections are termed subcutaneous because they involve tissues within and just below the skin. Most species in this group are greatly inhibited by the higher temperatures of the blood and viscera, and only rarely do they disseminate. Nevertheless, these diseases are progressive and can destroy the skin and associated structures. Mycoses in this category are sporotrichosis, chromoblastomycosis, phaeohyphomycosis, and mycetoma.

The Natural History of Sporotrichosis: Rose-Gardener's Disease

The cause of **sporotrichosis,** ***Sporothrix schenckii,**** is a very common saprobic fungus that decomposes vegetative matter in soil and humus and exhibits both mycelial and yeast phases (figure 22.14; see figure 22.16). *Sporothrix* resides in warm, temperate, and moist areas of the tropics and semitropics, though the incidence of sporotrichosis is highest in Africa, Australia, Mexico, and Latin America. A prick by a rose thorn is the classic origin of the cutaneous form of infection and has given rise to its common name. Most episodes of human infection follow contact with thorns, wood, sphagnum moss, bare roots, bark, and other

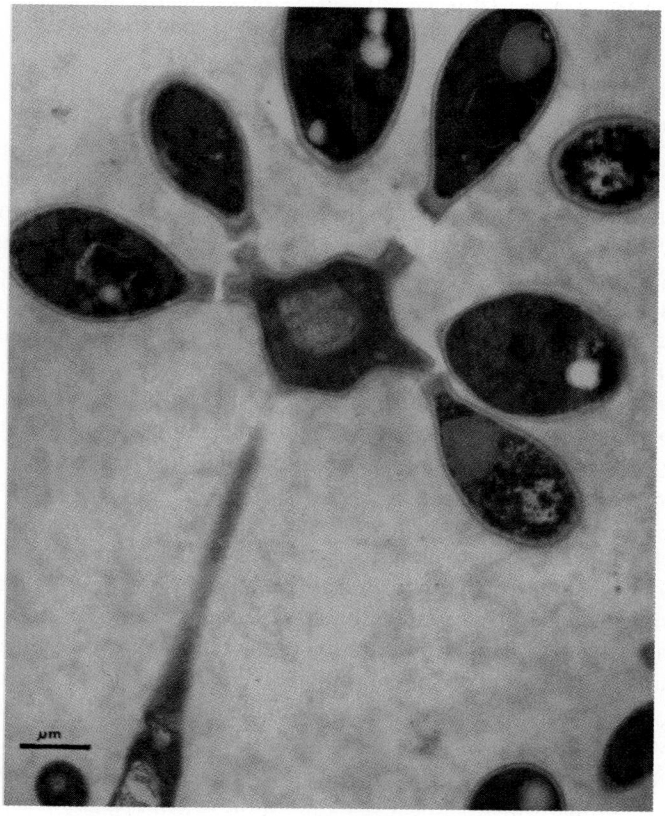

Figure 22.14

The microscopic morphology of *Sporothrix schenckii*. Spores develop as floral clusters borne on conidiophores or as single spores direct from the hyphae.

**Sporothrix schenckii* (spoh´-roh-thriks shenk´-ee-ee) Gr. *sporos,* seed, and *thrix,* hair. Named for B. R. Schenck, who first isolated it.

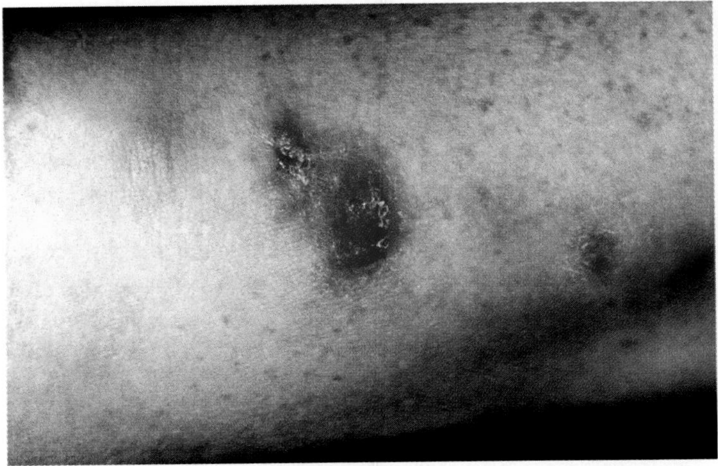

Figure 22.15

The clinical appearance of lymphocutaneous sporotrichosis. A primary sore is accompanied by a series of nodules running along the lymphatic channels of the arm.

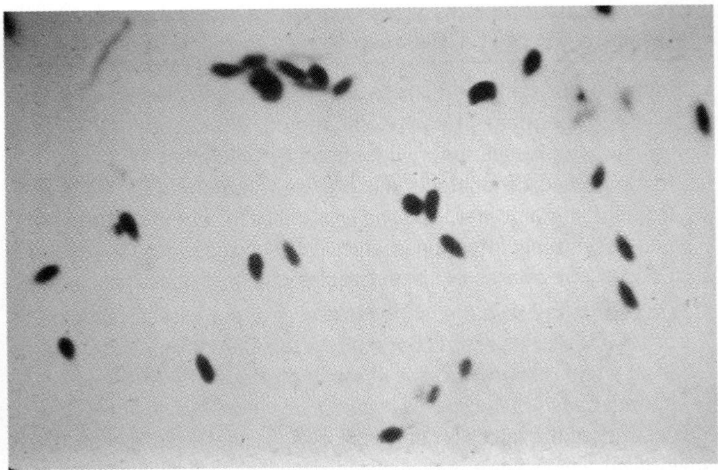

Figure 22.16

A scattering of tiny, plump *Sporothrix schenckii* yeasts with tapered ends.

vegetation. Horticulturists, gardeners, farmers, and basket weavers incur the majority of cases. Overt disease is uncommon unless a person has inadequate immune defenses or is exposed to a large inoculum or a virulent strain. Other mammals, notably horses, dogs, cats, and mules, are also susceptible to sporotrichosis.

Pathology, Diagnosis, and Control of Sporotrichosis

In **lymphocutaneous sporotrichosis,** the fungus grows at the site of penetration and develops a small, hard, nontender nodule within a few days to months. This subsequently enlarges, becomes necrotic, breaks through to the skin surface, and drains (figure 22.15). Infection often progresses along the regional lymphatic channels, leaving a chain of lesions at various stages. If left untreated, this condition persists for several years, though it is contained by the regional lymph nodes and does not spread. When *Sporothrix* conidia are drawn into the lungs, primary pulmonary sporotrichosis can result. Although this infection was once thought to be rare, it is increasingly reported among hospitalized chronic alcoholics.

Discovery of the agent in tissue exudates, pus, and sputum is difficult because there are usually so few cells in the infection. Enzymes that clear the specimen and special stains that highlight the tiny, cigar-shaped yeasts can be applied to enhance their appearance (figure 22.16). Demonstration of the typical macroscopic and microscopic morphology on artificial media confirms the diagnosis. Reliable serological tests are available but are usually not necessary. However, the sporotrichin skin test can be used to determine prior infection. An older but effective drug for sporotrichosis is potassium iodide, given orally in milk or applied topically to open lesions. Amphotericin B and flucytosine can assist in unresponsive cases. The fungus cannot withstand heat, thus local applications of heat packs to lesions have helped resolve some infections. People with occupational exposure should cover their bare limbs as a preventive measure.

Chromoblastomycosis and Phaeohyphomycosis: Diseases of Pigmented Fungi

Chromoblastomycosis* is a progressive subcutaneous mycosis characterized by highly visible *verrucous** lesions. The principal etiologic agents are a collection of widespread soil saprobes containing large amounts of dark pigments. *Fonsecaea pedrosoi, Phialophora verrucosa,* and *Cladosporium carrionii* are among the most common causative species. **Phaeohyphomycosis,*** a closely related infection, differs primarily in the causative species and the appearance of the infectious agent in tissue. Chromoblastomycotic agents produce very large, thick, yeastlike bodies called sclerotic cells, whereas the agents of phaeohyphomycosis remain typically hyphal.

The fungi associated with these infections are of low inherent virulence, and none exhibits thermal dimorphism. Infection occurs when body surfaces, especially the legs and feet, are penetrated by soiled vegetation or inanimate objects. Chromoblastomycosis occurs throughout the world, with peak incidence in the American subtropics and tropics. Most vulnerable to infection are men with rural occupations who go barefoot. After a very long (2- to 3-year) incubation period, a small, colored, warty plaque, ulcer, or papule appears. Because this lesion is generally not painful, many patients do not seek treatment and can go for years watching it progress to a more advanced state (see figure 5.17*d*). Unfortunately, the patient can aggravate the spread of infection on the body and provoke secondary bacterial infections by scratching the nodules.

Chromoblastomycosis is frequently confused with cancer, syphilis, yaws, and blastomycosis. Diagnosis is based on the clinical appearance of the lesions, microscopic examination of biopsied lesions, and the results of culture. Therapy for the condition

---

*chromoblastomycosis (kroh′′-moh-blas′′-toh-my-koh′-sis) Gr. *chroma* color, and *blasto,* germ. The fungal body is highly colored *in vitro.*

*verrucous (ver-oo′-kus) Tough, warty.

*phaeohyphomycosis (fy′′-oh-hy′′-foh-my-coh′-sis) Gr. *phaeo,* brown, and *hypho,* thread.

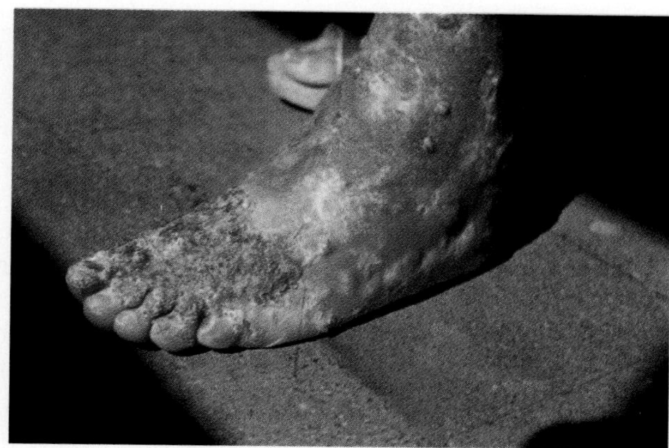

Figure 22.17

Early stages of mycetoma of the foot, caused by *Madurella.* Ulceration, swelling, and scarring are visible.

includes heat, drugs, surgical removal of early nodules, and amputation in advanced cases. Combined topical amphotericin B and thiabendazole or systemic flucytosine have shown some success in arresting the disease.

The infectious diseases termed phaeohyphomycoses are among the strangest oddities of medical science. The etiologic agents are soil fungi or plant pathogens with brown-pigmented mycelia. Victims of the disease are usually highly compromised patients who have become inoculated with these fungi, which are so widespread in household and medical environments that contact is inevitable. The disease process can be extremely deforming, and the microbe is sometimes hard to identify. Some genera of fungi commonly involved in these mycoses are *Alternaria, Aureobasidium, Curvularia, Dreschlera, Exophiala, Phialophora,* and *Wangiella.*The fungi grow slowly from the portal of entry through the dermis and create enlarged, subcutaneous cysts. In patients with underlying conditions such as endocarditis, diabetes, and leukemia, the fungi can spread into the body proper (bones, brain, lungs).

Mycetoma: A Complex Disfiguring Syndrome

Another disease elicited when soil microbes are accidently implanted into the skin is **mycetoma,** a mycosis usually of the foot or hand that looks superficially like a tumor. It is also called *madura foot* for the Indian region where it was first described. About half of all mycetomas are caused by fungi in the genera *Pseudallescheria* or *Madurella,* though filamentous bacteria called actinomycetes (*Nocardia,* discussed in chapter 19) are often etiologically implicated. Mycetomas are endemic to equatorial Africa, Mexico, Latin America, and the Mediterranean, and cases occur sporadically in the United States. Infection begins when bare skin is pierced by a thorn, sliver, leaf, or other type of sharp plant debris. First to appear is a localized abscess in the subcutaneous tissues, which gradually swells and drains (figure 22.17). Untreated cases that spread to the muscles and bones can cause pain and loss of function in the affected body part. Mycetoma has an insidious, lengthy course and is exceedingly difficult to treat.

Chapter Checkpoints

Fungal infections that invade traumatized skin are called subcutaneous mycoses. These localized infections rarely become systemic, but they can be very destructive to the skin and its associated components. Sporotrichosis, chromoblastosis, phaeohyphomycosis, and mycetoma are examples.

Sporothrix schenckii is the causative agent of sporotrichosis, or rose-gardener's disease. When introduced subcutaneously, it produces local lesions with the potential to invade lymphatics.

Chromoblastomycosis and phaeohyphomycosis are both caused by certain pigmented soil saprobes that produce characteristic slow-growing skin lesions of low virulence.

Mycetoma, or madura foot, is caused by filamentous fungi that invade traumatized skin. Lesions appear as abscesses and nodules. The systemic form spreads to bones and muscles and is very difficult to treat.

CUTANEOUS MYCOSES

Fungal infections strictly confined to the nonliving epidermal tissues (stratum corneum) and its derivatives (hair and nails) are termed **dermatophytoses.** Common terms used in reference to these diseases are **ringworm,** because they tend to develop in circular, scaly patches (see figure 22.19), and **tinea*** (early observers thought they were caused by worms). About 39 species in the genera *Trichophyton, Microsporum,* and *Epidermophyton* are involved in various dermatophytoses. The causative agent of a given type of ringworm differs from person to person and from place to place and is not restricted to a particular genus or species (table 22.3).

Characteristics of Dermatophytes

The dermatophytes are so closely related and morphologically similar that they can be difficult to differentiate. Various species exhibit unique macroconidia, microconidia, and unusual types of hyphae. In general, *Trichophyton* produces thin-walled, smooth macroconidia and numerous microconidia; *Microsporum* produces thick-walled, rough macroconidia and sparser microconidia; and *Epidermiphyton* has ovoid, smooth, clustered macroconidia and no microconidia (figure 22.18).

*tinea (tin´-ee-ah) L. a larva or worm.

TABLE 22.3

THE DERMATOPHYTE GENERA AND DISEASES

Genus	Name of Disease	Principal Targets	How Transmitted
Trichophyton	Ringworm of the scalp, body, beard, and nail Athlete's foot	Hair, skin, nails	Human to human, animal to human
Microsporum	Ringworm of scalp Ringworm of skin	Scalp hair Skin; not nails	Animal to human, soil to human, human to human
Epidermophyton	Ringworm of the groin and nail	Skin, nails; not hair	Strictly human to human

Epidemiology and Pathology of Dermatophytoses

The natural reservoirs of dermatophytes are other humans, animals, and the soil (microfile 22.2). Important factors that promote infection are the hardiness of the dermatophyte spores (they can last for years on fomites); hot, sweaty, chafed body parts; and intimate contact. Most infections exhibit a long incubation period (months), followed by localized inflammation and allergic reactions to the fungal proteins. As a general rule, infections acquired from animals and soil cause more severe reactions than do infections acquired from other humans, and infections eliciting stronger immune reactions are resolved faster.

Dermatophytic fungi exist throughout the world. Although some species are endemic, they tend to spread rapidly to nonendemic regions through travel. Dermatophytoses endure today as a serious health concern, not because they are life-threatening, but because of the extreme discomfort, stress, pain, and unsightliness they cause. In many cultures, ringworm and other skin mycoses are regarded as a social disgrace or a sign of uncleanliness.

An Atlas of Dermatophytoses

In the following section, the dermatophytoses are organized according to the area of body they affect, their mode of acquisition, and their pathologic appearance. Both the common English name and body site and its equivalent Latin name (tinea) are given.

Ringworm of the Scalp (Tinea Capitis) This mycosis results from the fungal invasion of the scalp and the hair of the head, eyebrows, and eyelashes (figure 22.19*a,b*). Very common in children, tinea capitis is acquired from other children and adults or from domestic animals. Manifestations range from small, scaly patches (gray patch), to a severe inflammatory reaction (kerion), to destruction of the hair follicle and permanent hair loss. Unless infected hairs are controlled, reinfection is common.

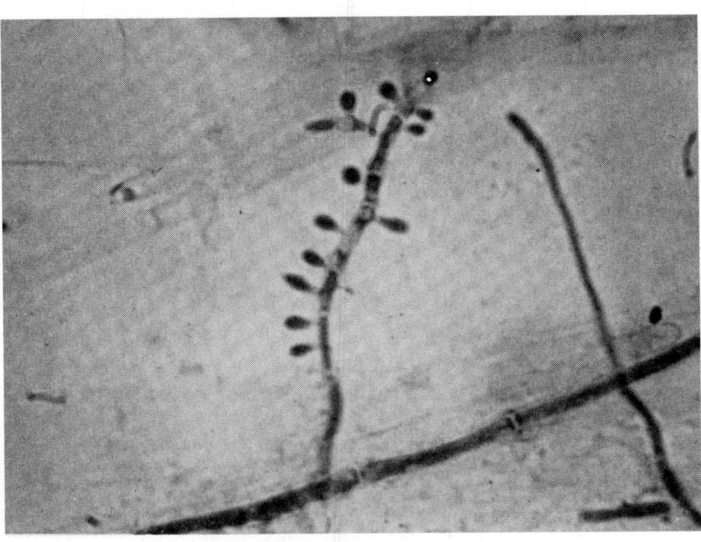

(a)

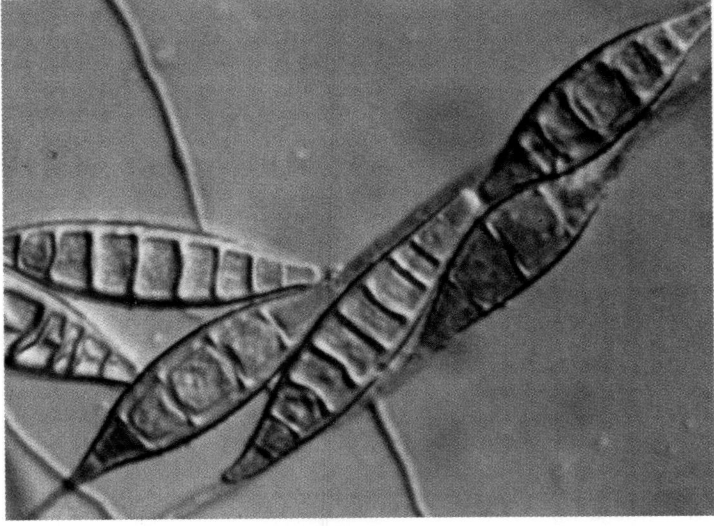

(b)

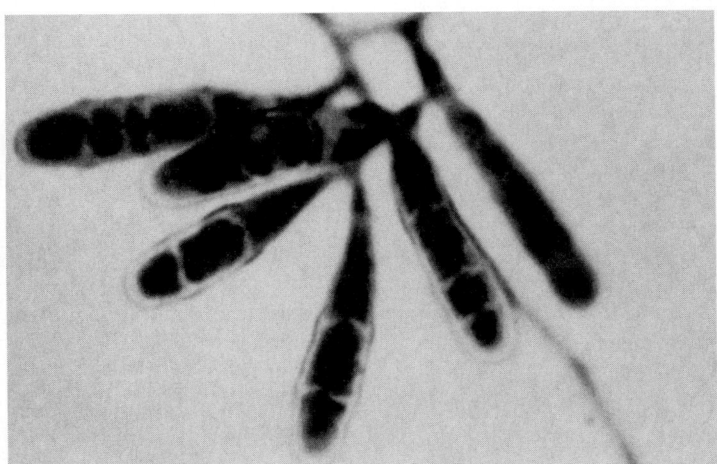

(c)

Figure 22.18

Examples of dermatophyte spores. (*a*) Regular, numerous microconidia of *Trichophyton*. (*b*) Macroconidia of *Microsporum canis,* a cause of ringworm in cats, dogs, and humans. (*c*) Smooth-surfaced macroconidia in clusters characteristic of *Epidermophyton*.

MICROFILE 22.2 THE KERATIN LOVERS

The dermatophytic fungi are especially well adapted to breaking down keratin, the primary protein of the epidermal tissues of vertebrates (skin, nails, hair, feathers, and horns). Their affinity for this compound gives them the name *keratophiles.* Examination of infected hairs indicates that these fungi attach to the hair surface and penetrate into its cortex. In time, it grows along the hair's length to the follicle, where it initiates a skin infection. A study of dermatophyte ecology reveals a gradual evolutionary trend from saprobic soil forms that digest keratin but do not parasitize animals, to soil forms that occasionally parasitize animals, to species that are dependent on live animals. Some species can infect a broad spectrum of animals, and others are specific to one particular animal species or region of the body. One adaptive challenge faced by relatively new fungal parasites is that they are likely to cause severe reactions in the host's skin and to be attacked by the host defenses and eliminated. Thus, the more successful fungi equilibrate with the host by reducing their activity (growth rate, sporulation) to reduce the inflammatory response. Eventually, these dermatophytes become such "good parasites" that they colonize the host for life. A striking example is *Trichophyton rubrum,* a fungus that causes a form of athlete's foot. It has such a tenacious hold and is so hard to cure that its carriers have been called the "*T. rubrum* people."

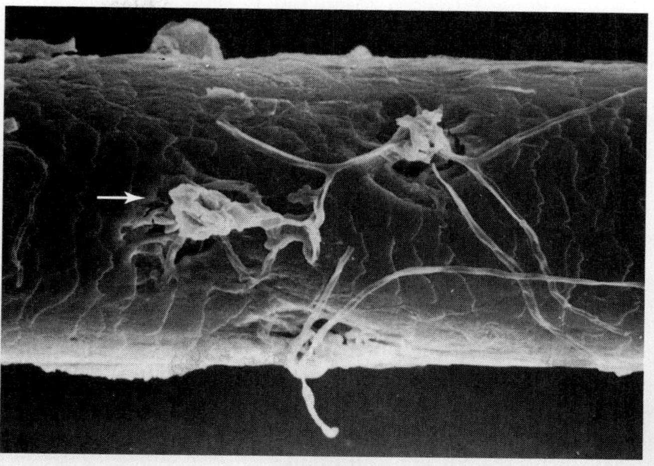

This microscopic view of a human hair shows dermatophyte hyphal growing along the hair and penetrating into its center (arrow).

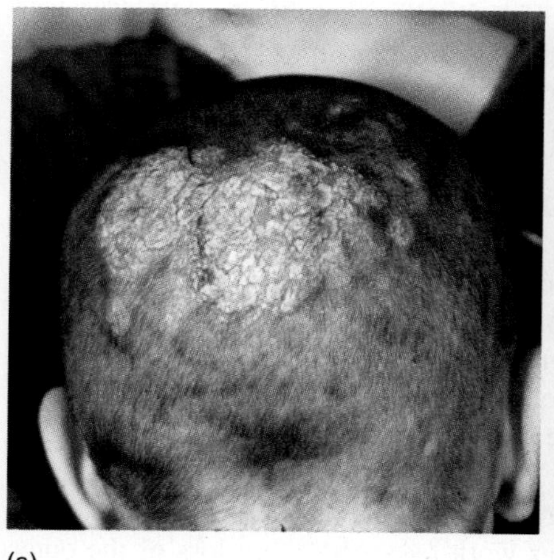

(a)

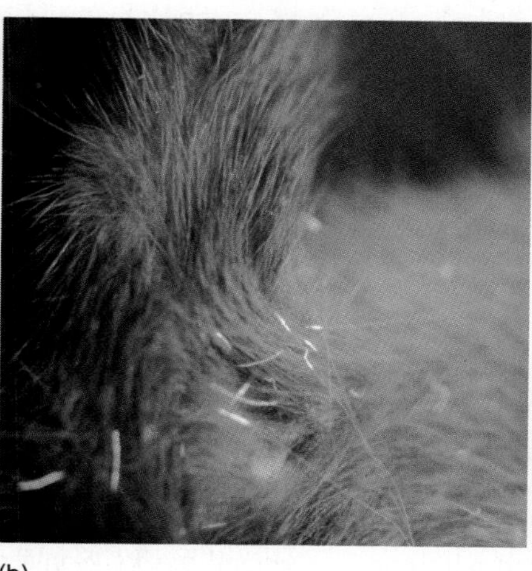

(b)

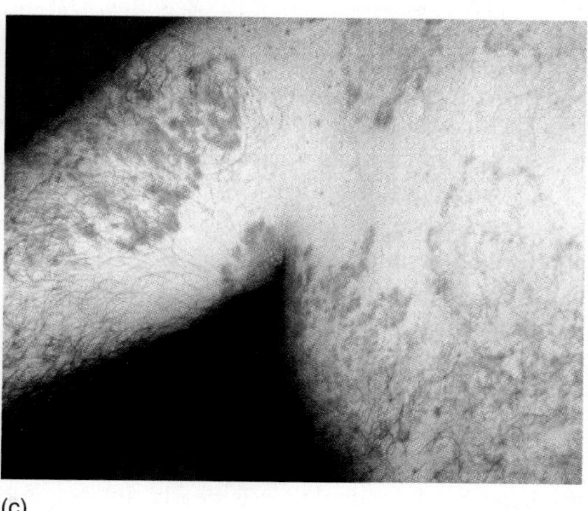

(c)

Figure 22.19

Ringworm lesions on the scalp and body vary in appearance. (*a*) Kerion, with deep exudative involvement and complete hair loss in the affected region. (*b*) A close-up of an infected hair fluorescing under a Wood's light. (*c*) Widespread lesions over the arm and shoulder have a dramatic ringed appearance that results from the gradual spread of inflammation from the center to the newest area of invasion in a circumferential pattern.

Ringworm of the Beard (Tinea Barbae) This tinea, also called "barber's itch," afflicts the chin and beard of adult males. Although once a common aftereffect of unhygienic barbering, it is now contracted mainly from animals.

Ringworm of the Body (Tinea Corporis) This extremely prevalent infection of humans can appear nearly anywhere on the body's glabrous (smooth and bare) skin. The principal sources are other humans, animals, and soil, and it is transmitted primarily by direct contact and fomites (clothing, bedding). The infection usually appears as one or more scaly reddish rings on the trunk, hip, arm, neck, or face (figure 22.19*c*). The ringed pattern is formed when the infection radiates from the original site of invasion into the surrounding skin. Depending on the causal species and the health and hygiene of the patient, lesions vary from mild and diffuse to florid and pustular.

Ringworm of the Groin (Tinea Cruris) Sometimes known as "jock itch," crural ringworm occurs mainly in males on the groin, perianal skin, scrotum, and, occasionally, the penis. The

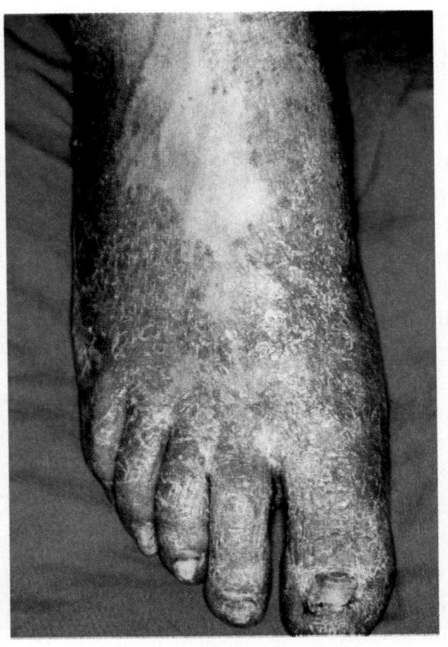

(a)

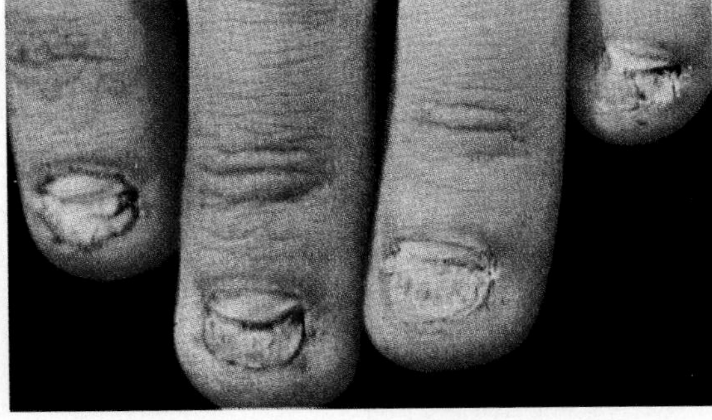

(b)

Figure 22.20
Ringworm of the extremities. (*a*) *Trichophyton* infection spreading over the foot in a "moccasin" pattern. The chronicity of tinea pedis is attributed to the lack of fatty-acid-forming glands in the feet. (*b*) Ringworm of the nails. Invasion of the nail bed causes some degree of thickening, accumulation of cheesy debris, cracking, and discoloration; nails can be separated from underlying structures as shown here. Severe, untreated cases can result in permanent loss of the nails.

fungus thrives under conditions of moisture and humidity created by profuse sweating or tropical climates. It is transmitted primarily from human to human and is pervasive among athletes and persons living in close situations (ships, military quarters).

Ringworm of the Foot (Tinea Pedis) Tinea pedis is known by a variety of synonyms, including athlete's foot and jungle rot. The disease is clearly connected to wearing shoes, because it is uncommon in cultures where the people customarily go barefoot (but as you have seen, bare feet have other fungal risks!). Conditions that encase the feet in a closed, warm, moist environment increase the possibility of infection. Because tinea pedis is a known hazard in shared facilities such as shower stalls, public floors, and locker rooms, many public facilities have rules against bare feet. Infections begin with small blisters between the toes that burst, crust over, and can spread to the rest of the foot and nails (figure 22.20*a*).

Ringworm of the Hand (Tinea Manuum) Infection of the hand by dermatophytes is nearly always associated with concurrent infection of the foot. Lesions usually occur on the fingers and palms of one hand and they vary from white and patchy to deep and fissured.

Ringworm of the Nail (Tinea Unguium) Fingernails and toenails, being masses of keratin, are often sites for persistent fungus colonization. The first symptoms are usually superficial white patches in the nail bed. A more invasive form causes thickening, distortion, and darkening of the nail (figure 22.20*b*). Nail problems caused by dermatophytes are on the rise as more women wear artificial fingernails, which can provide a portal of entry into the nail bed.

Diagnosis of Ringworm

Dermatologists are most often called upon to diagnose dermatophytoses. Occasionally, the presenting symptoms are so dramatic and suggestive that no further testing is necessary, but in most cases, direct microscopic examination and culturing are needed. Diagnosis of tinea of the scalp caused by some species of *Microsporum* is aided by a Wood's light. Illumination by this long-wave ultraviolet lamp causes infected hairs to fluoresce with a greenish glow (see figure 22.19*b*). Samples of hair, skin scrapings, and nail debris treated with heated KOH show a thin, branching fungal mycelium if infection is present. Culturing specimens on selective media and identifying the species can be important diagnostic aids.

Treatment of the Dermatophytoses

Ringworm therapy is based on the knowledge that the dermatophyte is feeding on dead epidermal tissues. These regions undergo constant replacement from living cells deep in the epidermis, so if multiplication of the fungus can be blocked, the fungus will eventually be sloughed off along with the skin or nail. Unfortunately, this takes time. By far the most satisfactory choice for therapy is a topical antifungal agent. Over-the-counter ointments containing tolnaftate, miconazole, or thiabendazine are applied regularly for several weeks. Some drugs work by speeding up loss of the outer skin layer. Intractable infections can be treated with griseofulvin, but placing a patient on this hepatotoxic and nephrotoxic drug for up to 2 years is probably too risky in most cases. Gentle debridement of skin and ultraviolet light treatments can have some benefit.

Chapter Checkpoints

Dermatophytoses are fungal infections of the nonliving epidermis and its derivatives. The causative agents are members of the genera *Trichophyton, Microsporum,* and *Epidermophyton.*

Dermatophytoses are classified according to the body region infected. They are also called ringworm or tinea infections. All three dermatophyte genera are communicable. *Microsporum* is also transmitted from soil to humans.

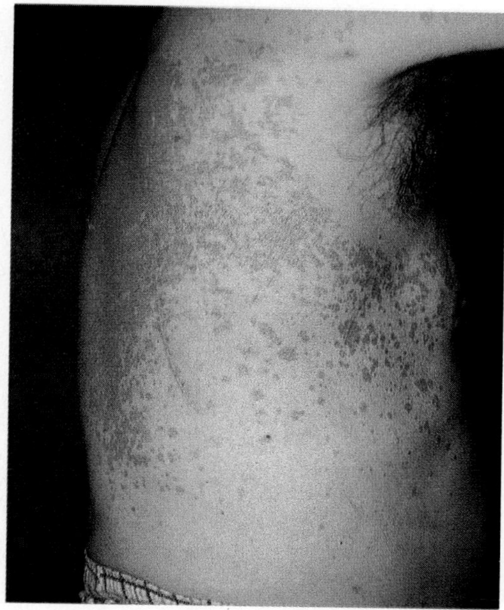

Figure 22.21
Tinea versicolor: mottled, discolored skin pigmentation characteristic of superficial skin infection by *Malassezia furfur.*

SUPERFICIAL MYCOSES

Agents of **superficial mycoses** involve the outer epidermal surface and are ordinarily innocuous infections with cosmetic rather than inflammatory effects. **Tinea versicolor**[1] is caused by the yeast *Malassezia furfur,* a normal inhabitant of human skin that feeds on the high oil content of the skin glands. Even though this yeast is very common (carried by nearly 100% of humans tested), in some people its growth elicits mild, chronic scaling and interferes with production of pigment by melanocytes. The trunk, face, and limbs take on a mottled appearance (figure 22.21). The disease is most pronounced in young people living in the tropics who are frequently exposed to the sun. Other skin conditions in which *M. furfur* is implicated are folliculitis, psoriasis, and seborrheic dermatitis. It is also occasionally associated with systemic infections and catheter-associated sepsis in compromised patients.

*Piedras** are marked by a tenacious, colored concretion forming on the outside surface of hair shafts (figure 22.22). In **white piedra,** caused by *Trichosporon beigelii,* a white to yellow adherent mass develops on the shaft of scalp, pubic, or axillary hair. At times, the mass is invaded secondarily by brightly colored bacterial contaminants, with startling results. **Black piedra,** caused by *Piedraia hortae,* is characterized by dark-brown to black gritty nodules, mainly on scalp hairs. Neither piedra is common in the United States.

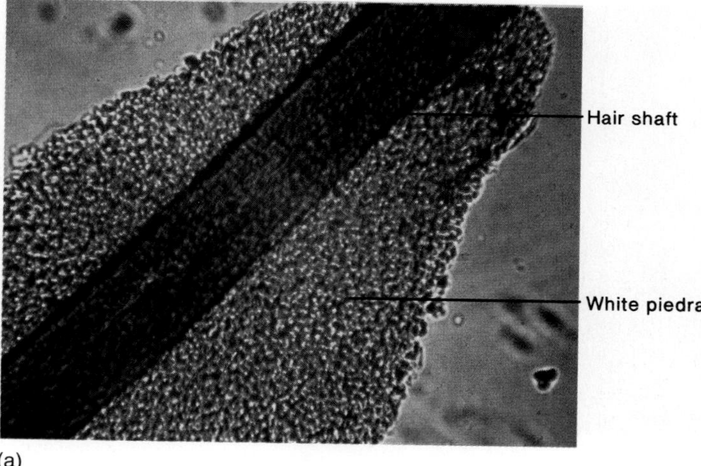

(a)

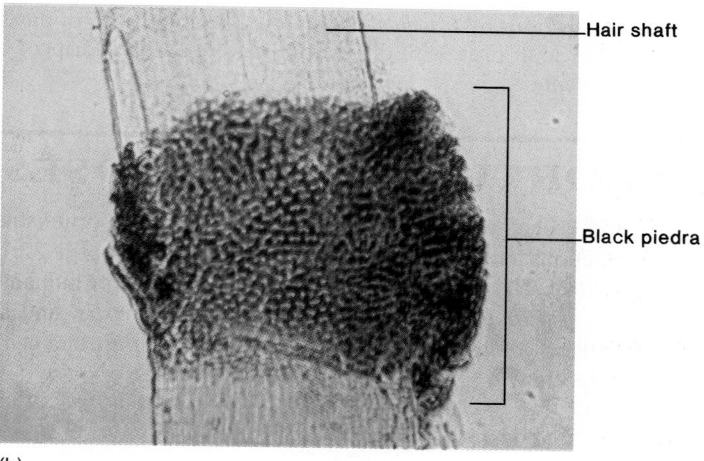

(b)

Figure 22.22
Examples of superficial mycoses. (*a*) Light-colored mass on a hair shaft with white piedra. (*b*) Dark, hard concretion enveloping the hair shaft in black piedra (200×).

Chapter Checkpoints

Superficial mycoses are noninvasive, noninflammatory fungal infections restricted to the hair and outer layer of the epidermis. Superficial mycoses of the hair, or piedra, appear as white or black masses on individual hair shafts. Epidermal infections include tinea versicolor, folliculitis, psoriasis, and seborrheic dermatitis.

1. Versicolor refers to the color variations this mycosis produces in the skin. Also called pityriasis.

*piedra (pee-ay´-drah) Sp. stone. Hard nodules of fungus formed on the hair.

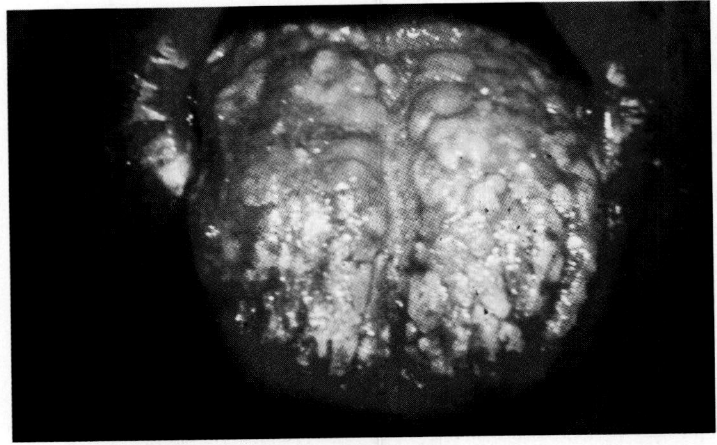
(a)

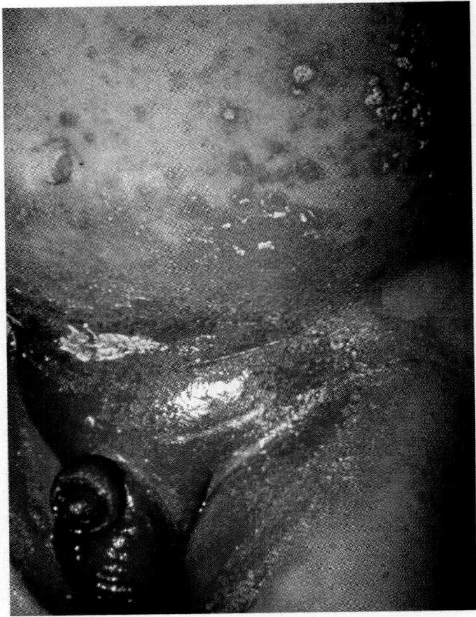
(b)

Figure 22.23
Common infections by *Candida albicans.* (*a*) Chronic thrush of the tongue on an adult male with diabetes. (*b*) Severe candidal diaper rash in an infant.

OPPORTUNISTIC MYCOSES

Earlier in this chapter, we introduced the concept of opportunistic fungal infections and their predisposing factors (see microfile 22.1 and table 22.2). The prevailing opportunistic pathogens of humans are the yeasts *Candida* and *Cryptococcus, Pneumocystis,* and a small number of filamentous fungi, primarily *Aspergillus* and certain zygomycetes.

INFECTIONS BY *CANDIDA*: CANDIDIASIS

Candida albicans,* an extremely widespread yeast, is the major cause of candidiasis (also called candidosis or moniliasis). Manifestations of infection run the gamut from short-lived, superficial skin irritations to overwhelming, fatal systemic diseases. Microscopically, *C. albicans* has budding cells of varying size that may form both elongate pseudohyphae and true hyphae (see figure 22.24*a*); macroscopically, it forms a white- to cream-colored, pasty colony with a yeasty odor.

Epidemiology of Candidiasis

Candida albicans occurs as normal flora in the oral cavity, genitalia, large intestine, and skin of 20% of humans. A healthy person with an unimpaired immune system can hold it in check, but the risk of invasion increases with extreme youth, pregnancy, use of certain antimicrobic drugs, nutritional and organic disease, immunodeficiency, the presence of invasive devices, and trauma. Any situation that maintains the yeast in contact with moist skin provides an avenue for infection. Although candidiasis is usually endogenous and not contagious, it can be spread in nurseries or through surgery, childbirth, and sexual contact. *Candida albicans* and its close relatives account for nearly 80% of nosocomial fungal infections and 30% of deaths from nosocomial infections in general.

Diseases of Candida albicans

Candida albicans causes local infections of the mouth, pharynx, vagina, skin, alimentary canal, and lungs, and it can also disseminate to internal organs. The mucous membranes most frequently involved are the oral cavity and vagina. **Thrush** is a white, adherent, patchy infection affecting the membranes of the mouth, gums, cheeks, or throat, usually in newborn infants and elderly, debilitated patients (figure 22.23). **Vulvovaginal candidiasis (VC),** known more commonly as **yeast infection,** has widespread occurrence in adult women, especially if taking oral antibiotics, diabetic, or pregnant (microfile 22.3). These conditions are thought to increase the glucose content of vaginal secretions, which favors candidal overgrowth and disrupts the balance of normal flora. Binding, impervious undergarments (pantyhose, nylon panties) are probably contributing factors. The chief symptoms of VC are a yellow to white milky discharge, inflammation, painful ulcerations, and itching. The most severe cases spread from the vagina and vulva to the perineum and thighs. Infection of the penis occasionally occurs in males.

Of all the areas of the gastrointestinal tract, *Candida* most often infects the esophagus and the anus. Esophageal candidiasis, which afflicts 70% of AIDS patients, causes painful, bleeding ulcerations; perforations; nausea, and vomiting.

Candidal attack of keratinized structures such as skin and nails, called **onychomycosis,*** is often brought on by predisposing occupational and anatomical factors. People whose occupations require their hands or feet to be constantly immersed in water are at risk for finger and nail invasion. *Intertriginous** infection occurs in moist areas of the body where skin rubs against skin, as beneath the breasts, in the armpit, and between folds of the groin.

Candida albicans (kan´-dih-dah al´-bih-kanz) L. *candidus,* glowing white, and *albus,* white.

*onychomycosis (ahn´´-ih-koh-my-koh´-sis) Gr. *onyx,* nail. Infection of the nails by *Candida* or *Aspergillus.*

*intertriginous (in´´-ter-trij´-ih-nus) L. *inter,* between, and *trigo,* to rub.

MICROFILE 22.3 PREGNANCY, CANDIDIASIS, AND THE NEONATE

Hormonal changes contribute to a high rate of candidiasis in pregnant women (up to 90% of women in the third trimester). Aside from the extreme discomfort of the symptoms, candidal vaginitis poses a threat to the newborn. Most cases of neonatal thrush are traced to contact with the mother's vagina during birth, and cutaneous infection is another common complication. Given the prevalence of vaginal infection during pregnancy, screening for *Candida* is yet another essential prenatal test. Fortunately, treating the mother with an intravaginal drug cream containing an azole or polyene drug is effective and does not jeopardize the fetus.

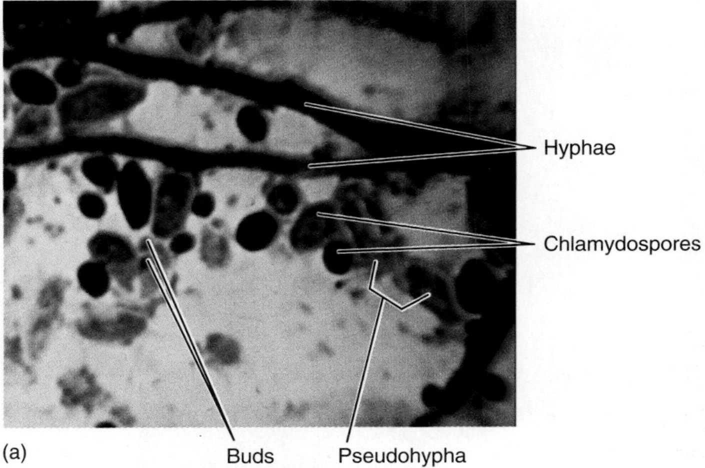

(a)

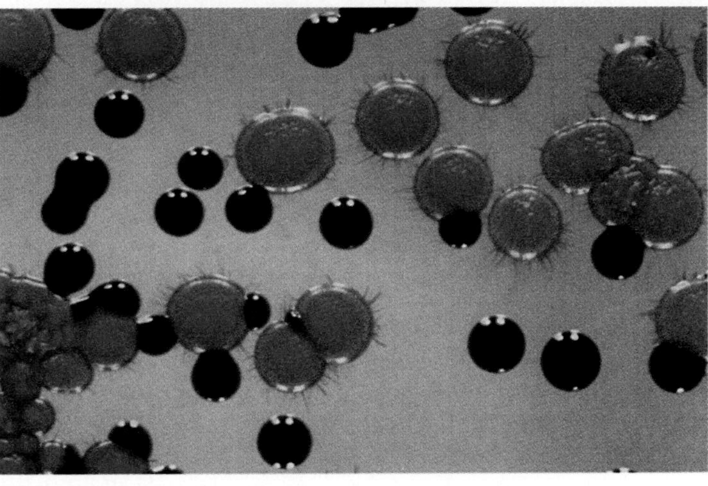

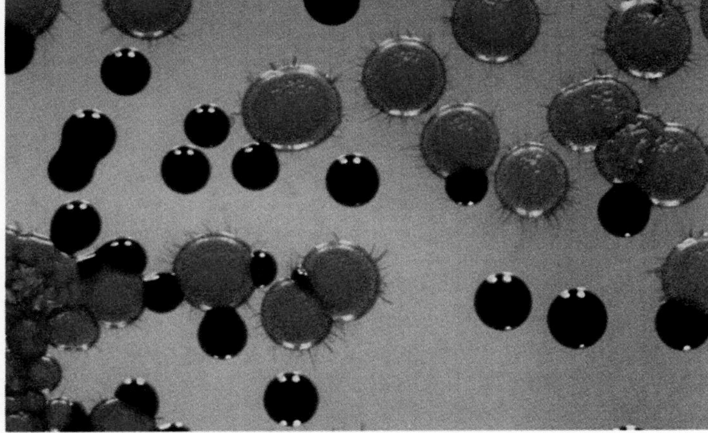

(b)

(c)

Figure 22.24
(*a*) Gram stain of *Candida albicans* in a vaginal smear reveals gram-positive chlamydospores, pseudohypha, and hyphae. In many cases of vaginal candidiasis, infection is detected during a routine Pap smear. (*b*) Colonies of *Candida albicans* appear pale blue on trypan medium, whereas *Cryptococcus* are dark blue. (*c*) Rapid yeast identification system using biochemical reactions to 12 test substances.

Cutaneous candidiasis can also complicate burns and produce a scaldlike rash on the skin of neonates (figure 22.23*b*).

Candidal blood infection in patients chronically weakened by surgery, bone marrow transplants, advanced cancer, or intravenous drug addiction is usually followed by systemic invasion. Although the mechanisms of virulence are not well understood, the presence of *C. albicans* in the blood is such a serious assault that it causes more human mortalities than any other fungal pathogen. Principal targets of systemic infections are the urinary tract, endocardium, and brain. Patients with valvular disease of the heart or indwelling prosthetic devices are vulnerable to candidal endocarditis, usually caused by other species (*C. tropicalis* and *C. parapsilosis*). *Candida krusei* has reportedly caused terminal infections in bone marrow transplant patients and in recipients of anticancer therapy.

Laboratory Techniques in the Diagnosis of Candidiasis

A presumptive diagnosis of *Candida* infection is made if budding yeast cells and pseudohyphae are found in specimens from localized infections (figure 22.24*a*). Specimens are cultured on standard fungal media incubated at 30°C. Identification is complicated by the numerous species of *Candida* and other look-alike yeasts. Growth on a selective, differential medium containing trypan blue can easily differentiate *Candida* species from the yeast *Cryptococcus* (figure 22.24*b*). Confirmatory evidence of *C. albicans* can also be obtained by the germ tube test, the presence of chlamydospores, and multiple panel systems that test for biochemical characteristics (figure 22.24*c*). A sensitive DNA amplification technique has been developed for identifying this species directly from clinical specimens.

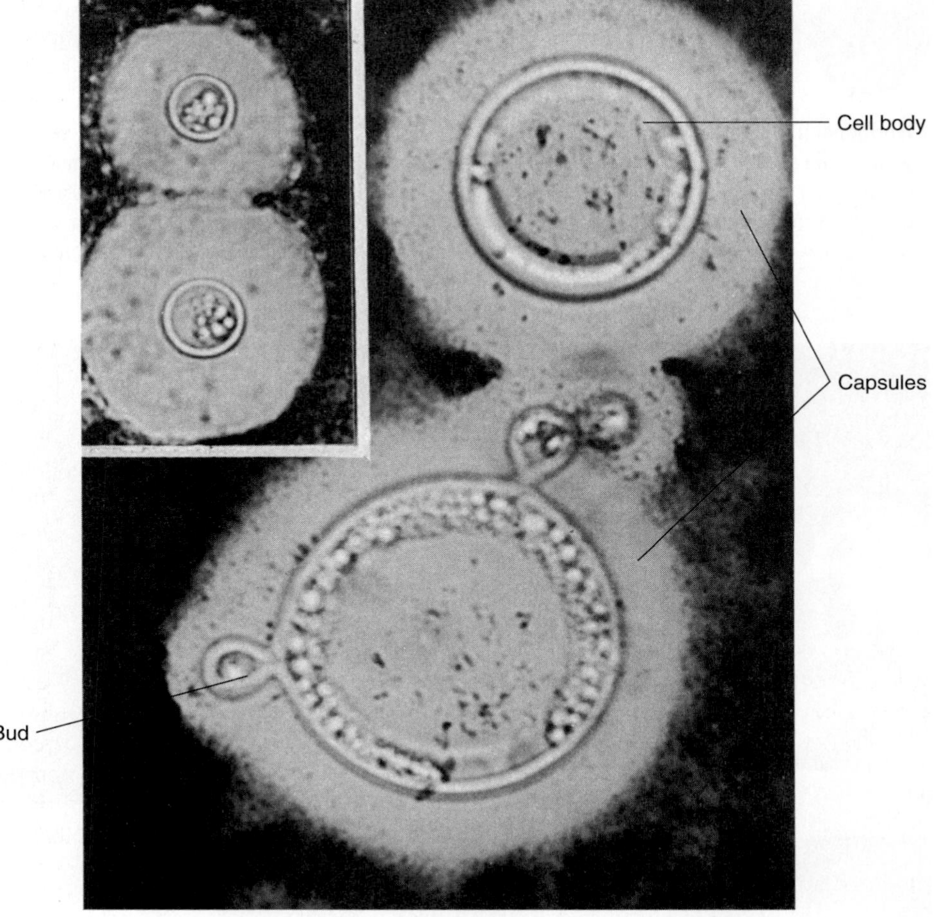

Figure 22.25
Cryptococcus neoformans from infected spinal fluid stained negatively with India ink. Halos around the large spherical yeast cells are thick capsules. Also note the buds forming on one cell. Encapsulation is a useful diagnostic sign for cryptococcosis, although the capsule is fragile and may not show up in some preparations (×150).

Treatment of Candidiasis

Because candidiasis is almost always opportunistic, it will return in many cases if the underlying disease is not also treated. Therapy for superficial mucocutaneous infection consists of topical antifungal agents (azoles and polyenes). A new class of antifungal drugs, terbinafine, has been approved for treating onychomycosis. Amphotericin B with or without flucytosine and fluconazole are usually effective in systemic infections. Recurrent bouts of vulvovaginitis are managed by topical azole drug ointments, now available as over-the-counter drugs. Vaginal candidiasis is sexually transmissible, thus it is very important to treat both sex partners to avoid reinfection.

CRYPTOCOCCOSIS* AND *CRYPTOCOCCUS NEOFORMANS*

Another widespread resident of human habitats is the fungus ***Cryptococcus neoformans.**** This yeast has a spherical to ovoid shape, with small, constricted buds and a large capsule that is important in its pathogenesis (figure 22.25). Its role as an opportunist is supported by evidence that healthy humans have strong resistance to it and that frank infection occurs primarily in debilitated patients. Most cryptococcal infections (*cryptococcoses*) center around the respiratory, central nervous, and mucocutaneous systems.

Epidemiology of C. neoformans

The primary ecological niche of *C. neoformans* is associated with birds. It is prevalent in urban areas where pigeons congregate, and it proliferates in the high nitrogen content of droppings that accumulate on pigeon roosts. Masses of dried yeast cells are readily scattered into the air and dust. The species is sporadically isolated from dairy products, fruits, and the healthy human body (tonsils and skin), where it is considered a contaminant. The highest rates of cryptococcosis occur among patients with AIDS. It causes a severe form of meningitis, which affects at least 10% of them and is frequently fatal. Other conditions that predispose to infection are steroid treatment, diabetes, and cancer. The disease exists throughout the world, with highest prevalence in the United States. Outbreaks have been reported in construction workers exposed to pigeon roosts, but it is not considered communicable among humans.

Pathogenesis of Cryptococcosis

The primary portal of entry for *C. neoformans* is respiratory, and most lung infections are subclinical and rapidly resolved. A few patients with pulmonary cryptococcosis develop fever, cough, and nodules in

*cryptococcosis (krip′′-toh-koh-oh′-sis) Older names are torulosis and European blastomycosis.

**Cryptococcus neoformans* (krip′′-toh-koh′-us nee′′-oh-for′manz) Gr. *kryptos,* hidden, *kokkos,* berry, *neo,* new, and *forma,* shape.

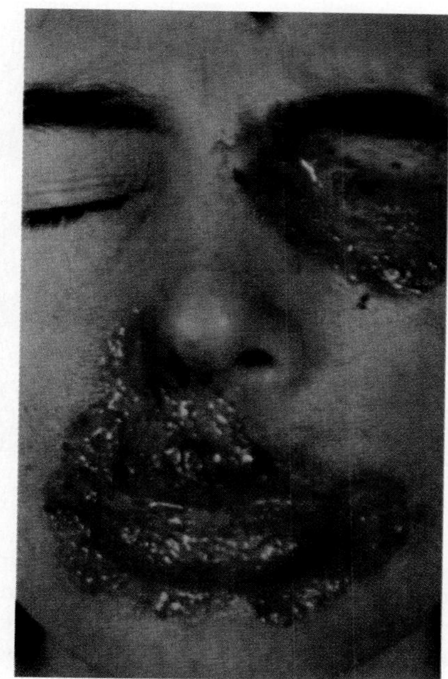

Figure 22.26
A late disseminated case of cutaneous cryptococcosis in which fungal growth produces a gelatinous exudate. The texture is due to the capsules surrounding the yeast cells.

their lungs. The escape of the yeasts into the blood is intensified by weakened host defenses and is attended by severe complications. Sites for which *Cryptococcus* shows an extreme affinity are the brain and meninges. The tumorlike masses formed in these locations can cause headache, mental changes, coma, paralysis, eye disturbances, and seizures. Because of the underlying illness of the host, many cases result in progressive deterioration and death. Some disseminated infections involve the skin, bones, and viscera (figure 22.26).

Diagnosis and Treatment of Cryptococcosis

The first step in diagnosis of cryptococcosis is negative staining of specimens to detect encapsulated budding yeast cells that do not occur as pseudohyphae. Colony development on blood-enriched or selective media requires 2 to 3 days. Two quick screening tests that presumptively differentiate *C. neoformans* from the seven other cryptococcal species are the urease and phenoloxidase tests. Confirmatory results include a negative nitrate assimilation, pigmentation on birdseed agar, and fluorescent antibody tests. Cryptococcal antigen can be detected in a specimen by means of serological tests, and DNA probes can make a positive genetic identification. Systemic cryptococcosis requires immediate treatment with amphotericin B, often combined with flucytosine or fluconazole, over a period of weeks or months.

Pneumocystis carinii and Pneumocystis Pneumonia

Although *Pneumocystis carinii* was discovered in 1909, it remained relatively obscure until it was suddenly propelled into clinical prominence as the agent of ***Pneumocystis carinii* pneumonia (PCP).** PCP is the most frequent opportunistic infection in AIDS patients, most of whom will develop one or more episodes during their lifetimes.

Because this unicellular microbe has characteristics of both protozoa and fungi, its taxonomic status has been somewhat uncertain. Until recently it was considered a type of protozoan because of its cystlike and trophozoite-like phases. But new tests on the sequence of its ribosomal RNA indicate that it is very similar genetically to the yeast *Saccharomyces.* It differs from most other fungi because it lacks ergosterol, has a weak cell wall, and is an obligate parasite.

Unlike most of the human fungal pathogens, little is known about the life cycle or epidemiology of *Pneumocystis.* Evidently the parasite is a common, relatively harmless resident of the upper respiratory tract. Contact with the agent is so widespread that in some populations a majority of people show serological evidence of infection. Until the AIDS epidemic, symptomatic infections by this organism were very rare, occurring only among the elderly or premature patients that were severely debilitated or malnourished. *Pneumocystis* is probably spread in droplet form between humans. In people with intact immune defenses, it is usually held in check by lung phagocytes and lymphocytes, but in those with deficient immune systems, it multiplies intracellularly and extracellularly. The massive numbers of parasites adhere tenaciously to the lung pneumocytes and cause an inflammatory condition. The linings of the lung alveoli slough off and a foamy exudate builds up. Symptoms are nonspecific and include cough, fever, shallow respiration, and cyanosis. The prognosis for untreated disease ranges from 50% to 100% mortality, but even with treatment, many AIDS patients succumb to this infection.

Because human strains of the species have not yet been successfully cultured, PCP must be diagnosed by symptoms and direct examination of lung secretions and tissue. These samples are analyzed by means of stains such as Giemsa, methamine-silver and fluorescent antibodies (figure 22.27). A DNA amplification probe can provide rapid confirmation of infection. Traditional serological tests for antigens and antibodies are not useful diagnostic tools for *Pneumocystis.*

Traditional antifungal drugs are ineffective against *Pneumocystis* pneumonia. The primary treatments are pentamidine and cotrimoxazole (sulfamethoxazole and trimethoprim) given over a 10-day period. This therapy should be applied even if disease appears mild or is only suspected. Pentamidine in an aerosol form is also given as a prophylactic measure in patients with a low T-cell count. The airways of patients in the active stage of infection often must be suctioned to reduce the symptoms, and additional oxygen is sometimes given. Unfortunately, most patients experience numerous recurrences because there are several variant strains of the fungus.

Aspergillosis: Diseases of the Genus *Aspergillus*

Molds of the genus ***Aspergillus****** are possibly the most pervasive of all fungi. The estimated 600 species are widely dispersed in

**Aspergillus* (as´´-per-jil´-us) L. *aspergere,* to scatter.

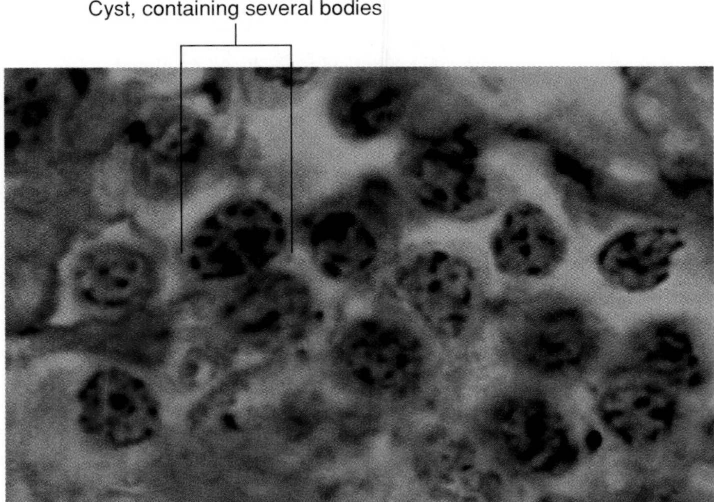

Figure 22.27
Pneumocystis carinii cysts in the lung tissue (stained with methamine silver) of an AIDS patient (500×).

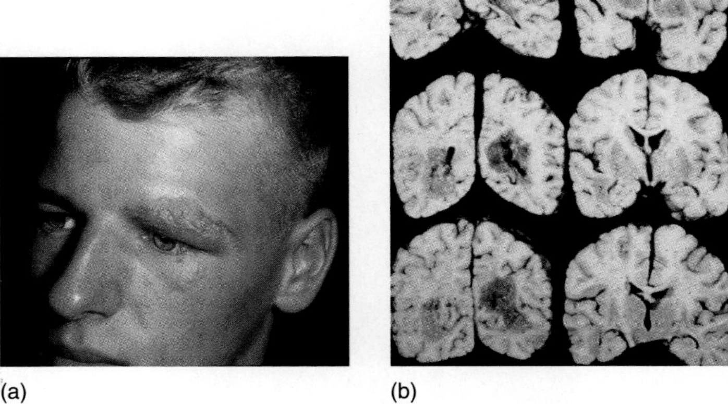

Figure 22.28
Clinical aspects of aspergillosis. (*a*) Generalized conjunctival infection accompanied by intense and painful inflammation. (Long-term contact lens wearers must guard against this infection.) (*b*) Brain abscesses (darkened areas) consistent with disseminated disease.

dust and air and can be isolated from vegetation, food, and compost heaps. Of the eight species involved in human diseases, the thermophilic fungus *A. fumigatus* accounts for the most cases. Aspergillosis is almost always an opportunistic infection, lately posing a serious threat to AIDS, leukemia, and organ transplant patients. It is also involved in allergies and toxicoses (microfile 22.4).

Aspergillus infections usually occur in the lungs. People breathing clouds of conidia from the air of granaries, barns, and silos are at greatest risk. In heavily exposed healthy people, the spores simply germinate in the lungs and form *fungus balls.* Similar benign, noninvasive infections occur in colonization of the sinuses, ear canals, eyelids, and conjunctiva (figure 22.28*a*). In the more invasive form of aspergillosis, the fungus produces a necrotic pneumonia and disseminates to the brain (figure 22.28*b*), heart, skin, and a wide range of other organs. Systemic aspergillosis usually occurs in very ill hospitalized patients with a poor prognosis.

An obvious branching, septate mycelium with characteristic conidial heads in a specimen is presumptive evidence of aspergillosis (figure 22.29). A culture of the specimen that produces abundant colonies and growth at 37°C will rule out its being only an air contaminant. A radioimmunoassay test for rapidly detecting antigen has been effective in diagnosing some cases. Noninvasive disease can be treated by surgical removal of the fungus tumor or by local drug therapy of the lung using amphotericin or nystatin. Combined therapy of amphotericin B and other antifungal agents is the only effective treatment for systemic disease.

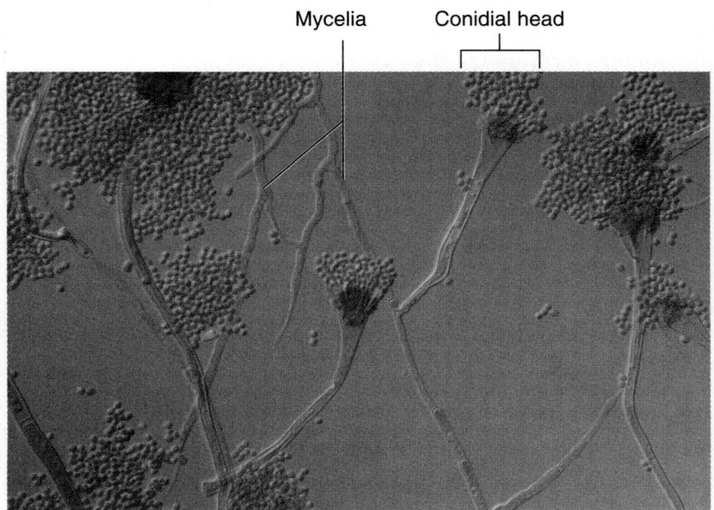

Figure 22.29

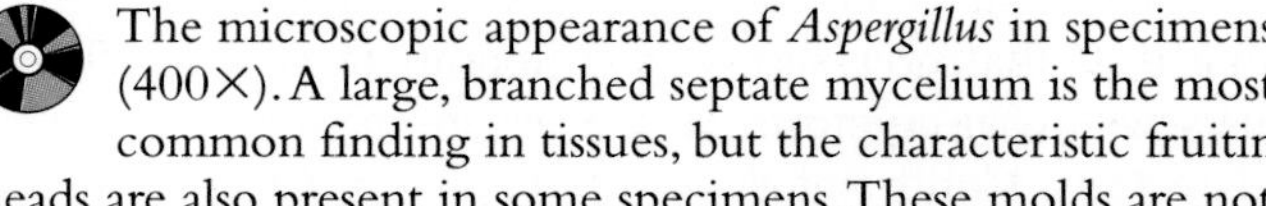

The microscopic appearance of *Aspergillus* in specimens (400×). A large, branched septate mycelium is the most common finding in tissues, but the characteristic fruiting heads are also present in some specimens. These molds are not dimorphic and exist only in the hyphal stage.

Mucormycosis: Zygomycete Infection

The zygomycetes are extremely abundant saprobic fungi found in soil, water, organic debris, and food. Their large, prolific sporangia release multitudes of lightweight spores that literally powder the living quarters of humans and usually do little harm besides rotting fruits and vegetables. But an increasing number of critically ill patients are contracting a disease called *mucormycosis.** The genera most often involved in mucormycoses are *Rhizopus, Absidia,* and *Mucor.*

***mucormycosis** (mew´´-kor-my-koh´-sis) Any infection caused by members of the Order Mucorales. Also called phycomycosis and zygomycosis.

MICROFILE 22.4 FUNGI, FOOD, AND TOXINS

Ingesting fungi (mycophagy) is a common and often pleasant experience. In some cases, one eats them by choice, as with gourmet mushrooms, and quite frequently, they are unintentionally consumed in dairy products such as yogurt and cottage cheese, fruits, vegetables, and many other foods.

Although most of the fungi we eat accidently are harmless, poisonous ones cause toxic illness. A number of mushrooms and molds contain mycotoxins that act on the central nervous system, liver, gastrointestinal tract, bone marrow, or kidney. The primary cause of **mycotoxicosis** is the consumption of wild mushrooms, a centuries-old practice that has taken the life of many an avid mushroom hunter. The deadliest of all mushrooms is *Amanita phalloides,* the so-called death-angel or destroying angel, whose cap contains enough toxin to kill an adult, though this happens rarely. Clearly, collecting and eating wild mushrooms is recommended only for the most expert mycophagist, and the rest of us should be content to gather mushrooms in the grocery store.

Certain poisonous fungi have been significant in history, and others are tied to ancient religious practices. Rye plants infested by *Claviceps purpurea* develop black masses, known as ergot, on the grains. When such rye grains are inadvertently incorporated into food and ingested, absorption of the ergot alkaloids causes *ergotism.* People thus intoxicated exhibit convulsions, delusions, a burning sensation in the hands (St. Anthony's fire), and symptoms similar to LSD (lysergic acid diethylamide) poisoning. This latter symptom occurs because the alkaloids in *Claviceps* and other mind-altering fungi contains the active ingredient lysergic acid. In the Middle Ages, millions of people died from ergotism, but now the disease is very rare. In fact, the active agent in ergot is extracted in minute doses and marketed to treat migraine headaches and to induce uterine contractions.

Some mushrooms have played a part in religious rituals through the ages, and others have been taken for their hallucinogenic effects. European priests used the "fly agaric" mushroom to induce trances, visions, and other derangements in the senses. New World tribes ate *Psilosybe*—"magic mushrooms"—to bring on hallucinations and mild trances. Even today, a mushroom-induced higher consciousness is sought by some people.

Mycotoxicosis is also caused by food contaminated by fungal toxins. One of the most potent of these toxins is *aflatoxin,** a product of *Aspergillus flavus* growing on grains, corn, and peanuts. Although the mold itself is relatively innocuous, the toxin is carcinogenic and hepatotoxic. It is especially lethal to turkeys, ducklings, pigs, and calves that have eaten contaminated feed. In humans, it appears to be a cofactor in liver cancer. The potential hazard aflatoxin poses for domesticated animals and humans necessitates constant monitoring of peanuts, beer, grains, nuts, vegetable oils, animal feed, and even milk. A quick exposure of seeds to ultraviolet radiation is a useful screening test because they glow when contaminated by the toxin.

*aflatoxin (af″-lah-toks′-in) Acronym for *Aspergillus flavus* toxin.

Underlying debilities that seem to increase the risk for mucormycosis are leukemia, burns, and malnutrition. Uncontrolled diabetes is a special risk because it can lead to reduced pH in the blood and tissue fluids that enhances invasion of the fungus. The specific pathology depends on the portal of entry of the fungus, which can be nasal, pulmonary, cutaneous, or gastrointestinal. In rhinocerebral mucormycosis, infection begins in the nasal passages or pharynx and then grows in the blood vessels and up through the palate into the eyes and brain (figure 22.30). Invasion of the lungs occurs most often in leukemics, and gastrointestinal mucormycosis is a complication in children suffering from protein malnutrition (kwashiorkor). Unfortunately, the outlook for most of these patients is not hopeful.

Diagnosis of mucormycosis rests on clinical grounds and biopsies or smears. Very large, thick-walled, nonseptate hyphae scattered through the tissues of a gravely ill diabetic or leukemic patient are very suggestive of mucormycosis. Attempts to isolate the molds from specimens are frequently met with failure, and confusion of the isolates with air contaminants is also a problem. Treatment, which is effective only if commenced early in the infection, consists of surgical removal of infected areas accompanied by high doses of amphotericin B.

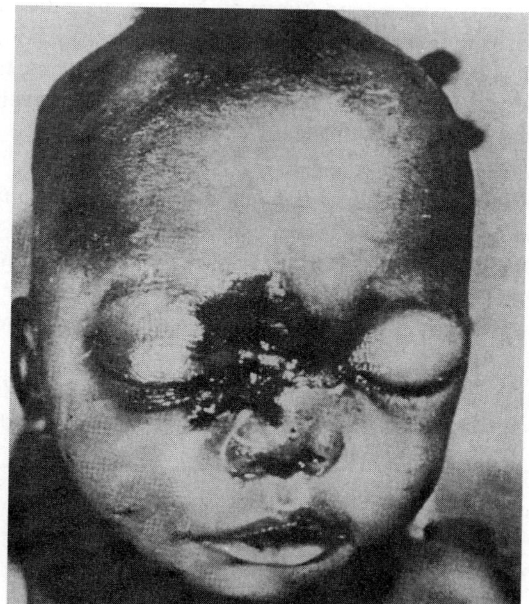

Figure 22.30

Mucormycosis in a malnourished child. This infection occurs in individuals so weakened that they usually have little chance for survival.

MISCELLANEOUS OPPORTUNISTS

It is increasingly clear that nearly any fungus can be implicated in infections when the host's immune defenses are severely compromised. Although it is still relatively unusual to become

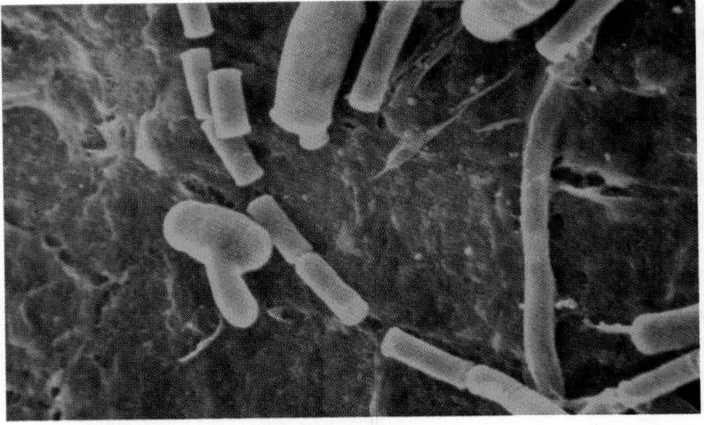

(a)

(b)

Figure 22.31
Common fungi that can cause uncommon infections. (*a*) *Geotrichum candidum* with arthrospores (2,000×). (*b*) *Fusarium* with crescent-shaped conidia (750×).

infected by common airborne fungi, these infections are currently on the rise.

Geotrichosis is a rare mycosis caused by *Geotrichum candidum* (figure 22.31*a*), a mold commonly found in soil, in dairy products, and on the human body. The mold is not highly aggressive and is primarily involved in secondary infection in the lungs of tubercular or highly immunosuppressed patients. Species of *Fusarium* (figure 22.31*b*), a common soil inhabitant and plant pathogen, occasionally infects the eyes, toenails, and burned skin. Mechanical introduction of the fungus into the eye (for example, by contaminated contact lenses) can induce a severe mycotic ulcer. It can also cause a patchy infection of the nail bed somewhat like tinea unguium. *Fusarium* colonization of patients with widespread burns has occasionally led to mortalities. Unusual opportunistic infections have been reported for other commonplace fungi such as *Penicillium, Malassezia,* and the red yeast *Rhodotorula.* Even a rare case of systemic infection by the baker's yeast *Saccharomyces* was recently observed.

Chapter Checkpoints

Candida albicans is the causative agent of a wide variety of opportunistic infections. These range from minor skin mycoses to fatal systemic diseases. *Candida* infections are most likely to occur in hosts with lowered resistance, trauma, or invasive devices or those receiving prolonged antibiotic therapy for bacterial infections.

Cryptococcus neoformans is the causative agent of cryptococcosis, a non-communicable mycosis that infects the respiratory, mucocutaneous, and nervous systems. Cryptococcus is commonly found in bird droppings and is spread through air and dust. Those at greatest risk for serious disease are AIDS patients and other immunocompromised hosts.

Aspergillosis is a fungal infection caused by the spores of *Aspergillus* species. Particularly high concentrations of spores occur in dust from granaries, barns, and silos. Inhaled spores germinate in the lung to form localized "fungus balls." Susceptible hosts can develop systemic infections that have a very poor prognosis.

Pneumocystis carinii is an opportunistic fungus of the upper respiratory tract that causes pneumonia in AIDS patients and other immunosuppressed hosts. Originally classified as a protozoan, it is now known to be related to the yeast *Saccharomyces.*

Mucormycosis is caused by saprobic fungi that colonize debilitated hosts with acidosis. Mucormycosis invades through several portals of entry and progresses rapidly to a systemic infection if not treated early.

The genus *Geotrichum* is involved in rare infections of the lungs, eyes, nail bed, and burned skin.

FUNGAL ALLERGIES AND INTOXICATIONS

It should be amply evident by now that fungi are constantly present in the air we breathe. In some people, inhaled fungal spores are totally without effect; in others, they instigate infection; and in still others, they stimulate hypersensitivity. So significant is the impact of fungal allergens that airborne spore counts are performed by many public health departments as a measure of potential risk (see figure 17.2). Among the significant fungal allergies are the following: (1) asthma, often occurring in seasonal episodes; (2) farmer's lung, a chronic and sometimes fatal allergy of agricultural workers exposed to moldy grasses; (3) teapicker's lung; (4) bagassosis, a condition caused by inhaling moldy dust from processed sugarcane debris; and (5) bark stripper's disease, caused by inhaling spores from logs. The role of fungi as toxic agents is discussed in microfile 22.4.

Chapter Checkpoints

In addition to agents of disease, fungi act as allergens and can produce toxins. Many of the allergies are related to agricultural occupations, but some are also common among the general population. *Claviceps* and *Aspergillus* species produce toxins that are hallucinogenic, hepatotoxic, or carcinogenic.

CHAPTER CAPSULE WITH KEY TERMS

FUNGAL INFECTIONS

GENERAL PRINCIPLES

Fungal diseases are called **mycoses** with names that end in **-osis,** or **-iasis.** Infectious fungi occur in groups based upon the virulence of the pathogen and the level of involvement, whether **systemic, subcutaneous, cutaneous,** or **superficial.**

True or **primary pathogens** have virulence factors that allow them to invade and grow in a healthy host. They are also **thermally dimorphic,** occurring as hyphae in their natural habitat and converting to yeasts while growing as parasites at body temperature (37°C). When spores are inhaled, they initiate **primary pulmonary infection (PPI).** Systemic and cutaneous complications are common, especially in weakened patients. Disease is accidental, not transmissible, and causes a long-term allergic reaction to fungal proteins.

Opportunistic fungal pathogens may be normal flora or common inhabitants of the environment. They lack well-developed virulence and thermal dimorphism and usually attack patients whose immunity has been compromised by surgery, cancer, AIDS, or organic disease. Infections may be local and cutaneous or systemic. Other fungal pathogens such as the **dermatophytes** (skin fungi) are intermediate in virulence. They can infect a healthy person, but are not highly invasive.

Epidemiology: True pathogens are endemic to certain geographic areas; other fungal pathogens are widely distributed; dermatophytoses are most prevalent; infections by the true pathogens are common in endemic regions; opportunistic infections are increasing due to improved medical technology that maintains immunocompromised patients.

Immunity: Primarily nonspecific barriers, inflammation, and cell-mediated; antibodies may be used to detect disease in some cases.

Diagnosis/Identification: Requires microscopic examination of fresh, stained specimens, culturing of pathogen in selective and enriched media, and specific biochemical and *in vitro* serological tests. Skin testing for true and opportunistic pathogens can determine prior disease but is not useful in diagnosis.

Control Measures: Drugs include intravenous amphotericin B and/or flucytosine for local and systemic disease and azoles and nystatin for skin and membrane infections. No useful vaccines exist.

SYSTEMIC MYCOSES CAUSED BY TRUE PATHOGENS

Histoplasmosis (Ohio Valley Fever): The agent is ***Histoplasma capsulatum,*** a dimorphic fungus with macro- and microconidia and small budding yeasts; distributed worldwide, but most prevalent in eastern and central regions of the U.S.; infection common where land or animal excreta have been disturbed and blown by wind. Inhaled spores produce PPI, with flu-like symptoms; may progress in patients with weak defenses to systemic involvement of a variety of organs and chronic lung disease.

Coccidioidomycosis (Valley Fever): ***Coccidioides immitis*** has block-like **arthroconidia** in the free-living stage and swollen **spherules** containing endospores in the lungs; agent lives in alkaline soils and semiarid, hot climates, and is endemic to the southwestern U.S.; infection occurs when soils are disturbed by digging or wind and arthrospores are inhaled. Arthrospores convert to spherules that burst, reinfect lung, and form nodules; dissemination into major organs may cause fatal disease.

Blastomycosis (Chicago Disease): The agent is ***Blastomyces dermatitidis.*** Free-living form has simple conidia; parasitic yeast form is ovoid cell with large bud; distributed in soil of a large section of the midwestern and southeastern U.S.; inhaled conidia convert to yeasts and multiply in lungs; inflammatory response produces cough, fever; may progress to chronic cutaneous nodules and abscesses and bone and nervous system infection.

Paracoccidioidomycosis (South American Blastomycosis): ***Paracoccidioides brasiliensis*** has an unusual, flowerlike yeast form; it is distributed in various regions of Central and South America; infection occurs through inhalation or inoculation of spores; systemic disease not common.

SUBCUTANEOUS MYCOSES

Diseases of tissues in or below the skin.

Sporotrichosis (Rose-Gardener's Disease): Caused by ***Sporothrix schenckii,*** a common saprobic fungus found worldwide that infects the appendages and lungs. The **lymphocutaneous** variety occurs when contaminated plant matter penetrates the skin and the pathogen forms a local nodule, then spreads to nearby lymph nodes.

Chromoblastomycosis: A deforming, deep infection of the tissues of the legs and feet with pigmented fungi following trauma.

Mycetoma: A progressive, tumorlike disease of the hand or foot due to chronic fungal infection; may lead to loss of the body part.

CUTANEOUS MYCOSES

Infections confined to the skin.

Dermatophytoses: Commonly called tinea, ringworm, athlete's foot. Caused by dermatophytes, primarily species in genera *Trichophyton, Microsporum,* and *Epidermophyton.* Keratinized epidermis (skin, hair, nails) is the target. Forms are communicable from other humans, animals, and soil; infection facilitated by broken, moist, chafed skin and by hardiness of spores; reactions to fungi are inflammatory, causing itching and pain.

Ringworm of scalp (tinea capitis) affects scalp and hair-bearing regions of head; hair may be lost. Ringworm of beard (tinea barbae) afflicts the chin and beard of adult males. **Ringworm of body** (tinea corporis) occurs as inflamed red ring lesions anywhere on smooth skin. **Ringworm of groin** (tinea cruris) affects groin and scrotal regions exposed to constant moisture; readily transmissible in close living conditions. **Ringworm of foot** and **hand** (tinea pedis and tinea manuum) are caused by enclosing feet in hot sweaty shoes (athlete's foot); hands become involved when self-inoculated; spread by exposure of feet to public floors; occurs between toes, soft tissues and on nails. **Ringworm of nails** (tinea unguium) is a persistent colonization of the nails of the hands and feet that distorts the nail bed.

SUPERFICIAL MYCOSES

Infections of outermost superficial epidermal structures. **Tinea versicolor** causes mild scaling, mottling of skin; **white piedra** is whitish or colored masses on the long hairs of the body; **black piedra** is dark, hard concretions on scalp hairs.

OPPORTUNISTIC MYCOSES

Candidiasis: Diseases of ***Candida albicans*** and other *Candida* species can grow as yeasts and pseudohyphae; normally resides in mouth, vagina, intestine, skin; infection occurs in cases of lowered resistance (babies, pregnancy, drug therapy, AIDS, trauma); disease acquired endogenously, though may be spread from mother to child and sexually. Diseases include **thrush,** a thick, white, adherent growth on the mucous membranes of mouth and throat, often in neonates; **vulvovaginal yeast infection,** a painful inflammatory condition of the genital area that causes ulceration and whitish discharge; **onychomycosis,** infection of keratinized structures due to long-term occupational exposure; **cutaneous candidiasis,** occurring in moist areas of skin that rub and in neonates and burn patients.

Cryptococcosis: Caused by ***Cryptococcus neoformans,*** a yeast with a well-developed capsule; widespread resident of urban soils and pigeon roosts; dust-borne and inhaled; common infection of AIDS, cancer, or diabetes patients; pulmonary disease causes cough, fever, lung nodules; dissemination to meninges and brain can cause severe neurological disturbance and death; skin may be involved.

Aspergillosis: Aspergillus is a very common airborne soil fungus; inhalation of clouds of spores causes fungus balls in lungs; invasive form occurs in highly debilitated patients and is often fatal.

Mucormycosis: Caused by large molds. *Rhizopus* and *Mucor* invade persons with acidosis (diabetics, malnutrition), which allows these ordinarily harmless air contaminants to grow in the mucous membranes of the nose, eyes, and brain, with deadly consequences.

Mycoses are increasingly caused by common unaggressive fungi such as *Geotrichum, Fusarium, Rhodotorula,* and *Penicillium.*

FUNGAL ALLERGIES AND MYCOTOXICOSES

Airborne fungal spores are common sources of atopic allergies. Ingestion of fungal toxins leads to **mycotoxicoses;** most commonly caused by eating poisonous or hallucinogenic mushrooms or food contaminated with fungal toxins such as ergot or **aflatoxin.**

MULTIPLE-CHOICE QUESTIONS

1. The ability of a fungus to alternate between hyphal and yeast phases in response to temperature is called
   a. sporulation c. dimorphism
   b. conversion d. binary fission
2. Which phase of a fungal life cycle is best adapted to growing in a host's body?
   a. yeast c. conidial
   b. hyphal d. filamentous
3. Primary pathogenic fungi differ from opportunistic fungi in being
   a. more virulent c. more invasive
   b. less virulent d. more transmissible
4. True pathogenic fungi
   a. are transmissible from person to person
   b. cause primary pulmonary infections
   c. are endemic to specific geographic areas
   d. both b and c
5. Example(s) of fungal infections that is /are communicable:
   a. dermatomycosis c. sporotrichosis
   b. candidiasis d. both a and b
6. Histoplasmosis has the greatest endemic occurrence in which region?
   a. Midwest U.S.
   b. Southwest U.S.
   c. Central and South America
   d. North Africa
7. Coccidioidomycosis is endemic to which geographic region in the U.S.?
   a. Southwestern c. Pacific Northwest
   b. Northeastern d. Southeastern
8. Skin testing with antigen is a useful diagnostic procedure for
   a. histoplasmosis c. coccidioidomycosis
   b. candidiasis d. blastomycosis
9. A mycetoma is a
   a. fungal tumor of the lung
   b. collection of pus in the brain caused by *Cryptococcus*
   c. deeply invasive fungal infection of the foot or hand
   d. type of ringworm of the scalp
10. Dermatophytic fungi attack the ____ in the ____
    a. epithelial cells, lungs
    b. melanin, stratum corneum
    c. keratin; skin, nails, hair
    d. bone, foot
11. *Candida albicans* is the cause of ____ , an infection of the ____
    a. onychomycosis, root canal
    b. moniliasis, lungs
    c. thrush, mouth
    d. salpingitis, uterus
12. Cryptococcosis is associated with
    a. blowing winds
    b. plant debris
    c. pigeon droppings
    d. excavation of ancient dwellings
13. Which fungus does *not* commonly cause systemic infection?
    a. *Blastomyces* c. *Cryptococcus*
    b. *Histoplasma* d. *Malassezia*
14. Match the disease with its common name:
    ____ coccidioidomycosis
    ____ histoplasmosis
    ____ blastomycosis
    ____ sporotrichosis
    ____ paracoccidioidomycosis
    ____ dermatophytosis
    ____ candidiasis
    a. tinea
    b. San Joaquin Valley fever
    c. South American blastomycosis
    d. Chicago disease
    e. Rose-gardener's disease
    f. yeast infection
    g. Ohio Valley fever

CONCEPT QUESTIONS

1. a. Explain how true pathogens differ from opportunists in physiology, virulence, types of infection, and distribution.
   b. Give examples of each.
   c. Explain the adaptation and importance of thermal dimorphism.
2. a. Why are most fungi considered facultative parasites and fungal infections considered non-communicable?
   b. Give examples of two communicable mycoses.
3. a. What factors are involved in the pathogenesis of fungi?
   b. What parts of fungi are infectious?
   c. What are the primary defenses of humans against fungi?
4. a. What is the role of skin testing in tracing fungal infections?
   b. What causes cross-reactions?
   c. For which groups is it most useful?
5. a. Give an overview of the major steps in fungal identification.
   b. What is one advantage over bacterial identification?
   c. How are *in vitro* and *in vivo* tests different?
   d. Describe an immunodiffusion test; the histoplasmin test.
   e. What biochemical and genetic tests can be conducted in identification?
6. a. What are the general principles involved in treating fungal diseases with drugs?
   b. Why is amphotericin B so toxic?
   c. Name some topical medications and their uses.
7. a. Differentiate between systemic, subcutaneous, cutaneous, and superficial infections.
   b. What is the initial infection site in true pathogens?
   c. What does it mean if an infection disseminates?
8. a. Briefly outline the life cycles of *Histoplasma capsulatum, Coccidioides immitis,* and *Blastomyces dermatitidis.*
   b. In general, what is the source of each pathogen?
   c. Is disease usually overt or subclinical?
   d. To which organs do these pathogens disseminate?
   e. What causes cutaneous disease? How are cutaneous diseases treated?
9. a. What are the most common subcutaneous infections?
   b. Why are these organisms not more invasive?
   c. Describe the life cycle of *Sporothrix* and the disease it causes.
   d. How could one avoid it? Differentiate between chromoblastomycosis and phaeohyphomycosis.
10. a. What are the dermatophytoses?
    b. What is meant by the term *keratophile?*
    c. What do the fungi actually feed upon?
    d. To what do the terms *ringworm* and *tinea* refer?
    e. What are the three main genera of dermatophytes, and what are their reservoirs?
11. a. What are the five major types of ringworm, their common names, symptoms, and epidemiology?
    b. What conditions predispose an individual to these infections?
12. a. What is the difficulty in ridding the epidermal tissues of certain dermatophytes?
    b. What is necessary in treatment?
13. Briefly describe three superficial mycoses.
14. a. What is the relationship of *Candida albicans* to humans?
    b. What medical or other conditions can predispose a person to candidiasis?
    c. Describe the principal infections of women, adults of both sexes, and neonates.
    d. Why should women be screened for candidiasis?
    e. What are the treatments for localized and systemic candidiasis?
    f. If a woman tests positive for infection, besides treating her, what else should be done?
15. a. Describe the pathology of cryptococcosis.
    b. What specimens could be used to diagnose it?
    c. Which group of people is currently most vulnerable to this disease?
    d. What makes the fungus somewhat unique?
16. a. Briefly list the major diseases caused by *Aspergillus* and zygomycetes.
    b. Why are fungus balls not an infection?
17. Briefly outline the epidemiology and pathology of *Pneumocystis carinii.*

CRITICAL-THINKING QUESTIONS

1. Give reasons most humans are not constantly battling fungal infections, considering how prevalent fungi are in the environment.
2. a. Name several medical conditions that compromise the immunities of patients.
   b. Can you think of some solutions to the problems of opportunistic infections in extremely compromised patients?
   c. Is there any such thing as a harmless contaminant with these patients?
   d. Can medical personnel do anything to lessen their patients' exposure to fungi?
3. Describe the consequences if a virulent fungal pathogen infects a severely compromised patient.
4. Using the general guidelines presented in the "Introduction to Identification Techniques in Medical Microbiology," page 561, how can you tell for sure if a fungal isolate is the actual causative agent or merely a contaminant?
5. a. Observe the reactions in figure 22.10 and describe what is happening at the molecular level to make the lines of precipitation in the immunodiffusion tests.
   b. Make a drawing to show this.
   c. What is happening in the wells without the lines?
6. a. Explain why some dermatophytes are considered good parasites.
   b. Why would infections from soil and animal species be more virulent than ones from fungi adapted to humans?
7. What conditions predispose diabetics to fungal infections?
8. a. What is the function of mushroom and other fungal toxins?
   b. How are these toxins harmful, and how are they used to our advantage?

9. a. Explain how *Pneumocystis carinii,* once classified with the protozoan parasites, came to be considered a fungus.
   b. Describe the characteristics that make it protozoan-like and those that are fungus-like.
   c. What was the definitive reason for changing its taxonomic position?
10. Case study 1. Mr. Jones is admitted to a hospital with severe chest pain, cough, and fever. The lab tests show tiny cocci-like cells in sputum, and he is treated for bacterial pneumonia and released. However, Mr. Jones does not respond to penicillin or cephalosporin, and the lung condition worsens. After a few days, skin bumps begin to appear on the face and chest. A lung biopsy indicates large sporangia filled with tiny spores, and the patient's history reveals a recent trip to Arizona.
    a. On this basis, how would you diagnose the infection?
    b. What other fungal disease might it suggest?
    c. Which drug should have been given?
11. Case study 2. A male researcher wanting to determine the pathogenesis of a certain fungus that often attacks women did the following experiment on himself. He taped a piece of gauze heavily inoculated with the fungus to the inside of his thigh and left it there for several days, periodically moistening it with sterile saline. After a time, he experienced a severe skin reaction that resembled a burn and the skin peeled off.
    a. Suggest a species of fungus that can cause this effect.
    b. What does this experiment tell you about the factors in pathogenesis related to environment and gender?
12. Case study 3. A worker involved in demolishing an older downtown building is the victim of a severe lung infection.
    a. Without any lab data, what two fungal infections do you suspect?
    b. What is your reasoning?

INTERNET SEARCH TOPIC

1. Look up newer antifungal drugs such as fluconazole and terbinafine, and determine their modes of action and therapeutic uses.
2. Find web sites that give information on common fungal infections of AIDS patients and the effects they have on health and survival.

chapter 23

The Parasites of Medical Importance

Parasitology is traditionally the study of eucaryotic parasites. It encompasses protozoa (unicellular animals) and helminths (worms), which are sometimes collectively called the macroparasites, as opposed to the microparasites (bacteria and viruses). This inherently fascinating and practical subject will be presented as a compact survey of the major clinically important protozoa and helminths. This chapter covers elements of their morphology, life cycles, epidemiology, pathogenic mechanisms, and methods of control. Arthropods, which are often included in the field of parasitology, have been previously discussed as vectors for certain viral and bacterial infections in chapters 13, 20, and 21. Other chapters containing information on parasites are chapter 5 (eucaryotes), chapter 7 (nutrition), and chapter 12 (chemotherapy).

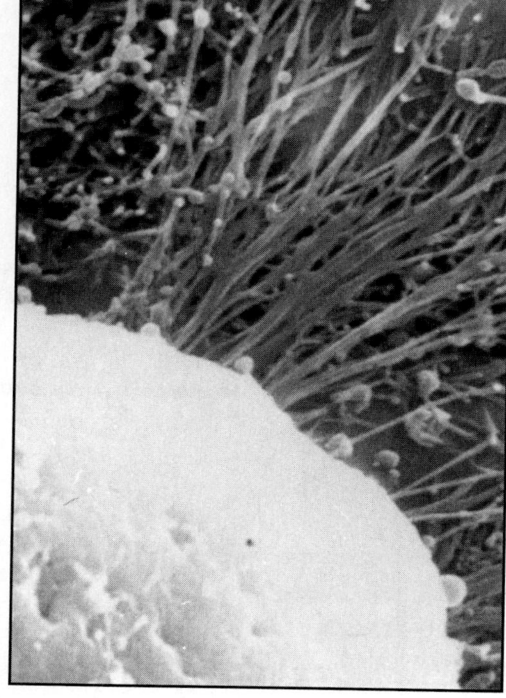

Trophozoite of *Entamoeba histolytica,* the cause of amebic dysentery, displays a fringe of very fine pseudopods it uses to invade and feed on tissue.

THE PARASITES OF HUMANS

Because of their larger size and greater visibility, parasites and their diseases have been known since ancient times. Even so, the science of parasitology is relatively young and still in the early stages of definition at the molecular level. One cannot underestimate the impact of parasites on the everyday lives of humans. A recent World Health Organization listing of the most prevalent tropical diseases indicated that a majority were caused by parasites. In industrialized countries, we are less aware of parasites; nevertheless, they affect most humans in both hidden and overt forms at some time in their lives. The distribution of parasitic diseases is influenced by many factors, including rapid travel, immigration, and the increased number of immunocompromised patients. Because parasites can literally travel around the world, it is not uncommon for exotic parasitic diseases to be discovered even in the United States.

Humans play host to dozens of different parasites. A few are indigenous to humans alone and are spread to other humans through direct contact or contaminated food and water. A considerable number are transmitted among human hosts by arthropod vectors. The remainder are zoonoses that humans acquire from contact with pets or other animals.

The current increase in patients with severe immune suppression is changing the epidemiologic pattern of parasitic diseases. AIDS patients in particular play host to a variety of opportunists that dominate their clinical picture. More and more, protozoa that were once rare and obscure pathogens are emerging as the agents of life-threatening diseases.

Chapter Checkpoints

Parasitic diseases have existed since ancient times but are still a major public health problem today. International travel of people and materials has facilitated the spread of parasites and their vectors worldwide. The increase of immunosuppressed people, particularly those with AIDS, has caused an increase in formerly obscure protozoan infections.

TYPICAL PROTOZOAN PATHOGENS

When we first met the protozoa in chapter 5, they were described as single-celled, animal-like microbes usually having some form of motility. The four commonly recognized groups are sarcodinians (amebas), flagellates, ciliates, and apicomplexans. Although the life cycles of pathogenic protozoa can vary a great deal, most of them propagate by simple asexual cell division of the active feeding cell, called a **trophozoite.** Many species undergo formation of a **cyst** that can survive for periods outside the host and can also function as an infective body (see figure 5.27). Others have a more complex life cycle that includes asexual and sexual phases (see later discussion of apicomplexans). The protozoan parasites and diseases to be surveyed here are listed in table 23.1.

TABLE 23.1

MAJOR PATHOGENIC PROTOZOA, INFECTIONS, AND PRIMARY SOURCES

Protozoan/Disease	Reservoir/Source
Ameboid Protozoa	
Amebiasis: *Entamoeba histolytica*	Human/water and food
Brain infection: *Naegleria, Acanthamoeba*	Free-living in water
Ciliated Protozoa	
Balantidiosis: *Balantidium coli*	Zoonotic in pigs
Flagellated Protozoa	
Giardiasis: *Giardia lamblia*	Zoonotic/water and food
Trichomoniasis: *Trichomonas tenax, T. hominis, T. vaginalis*	Human
Hemoflagellates	
Trypanosomiasis: *Trypanosoma brucei, T. cruzi*	Zoonotic/vector-borne
Leishmaniasis: *Leishmania donovani, L. tropica, L. brasiliensis*	Zoonotic/vector-borne
Apicomplexan Protozoa	
Malaria: *Plasmodium vivax, P. falciparum, P. malariae*	Human/vector-borne
Toxoplasmosis: *Toxoplasma gondii*	Zoonotic/vector-borne
Cryptosporidiosis: *Cryptosporidium*	Free-living/water, food
Isosporosis: *Isospora belli*	Dogs, other mammals
Cyclosporiasis: *Cyclospora cayetanensis*	Water/fresh produce

INFECTIVE AMEBAS

Entamoeba histolytica and Amebiasis

Amebas are widely distributed in aqueous habitats and are frequent parasites of animals, but only a small number of them have the necessary virulence to invade tissues and cause serious pathology. The most significant pathogenic ameba is ***Entamoeba histolytica.**** The relatively simple life cycle of this parasite alternates between a large trophozoite that is motile by means of pseudopods and a smaller, compact, nonmotile cyst (figure 23.1). The trophozoite lacks most of the organelles of other eucaryotes, and it has a large single nucleus that contains a prominent nucleolus called a *karyosome.* Amebas from fresh specimens are often packed with food vacuoles containing digested or whole erythrocytes and bacteria. The mature cyst is encased in a thin, yet tough wall and contains four nuclei as well as distinctive bodies called *chromatoidals,* which are actually dense clusters of ribosomes.

**Entamoeba histolytica* (en″-tah-mee′-bah his″-toh-lit′-ih-kuh) Gr. *ento,* within, *histos,* tissue, and *lysis,* a dissolution.

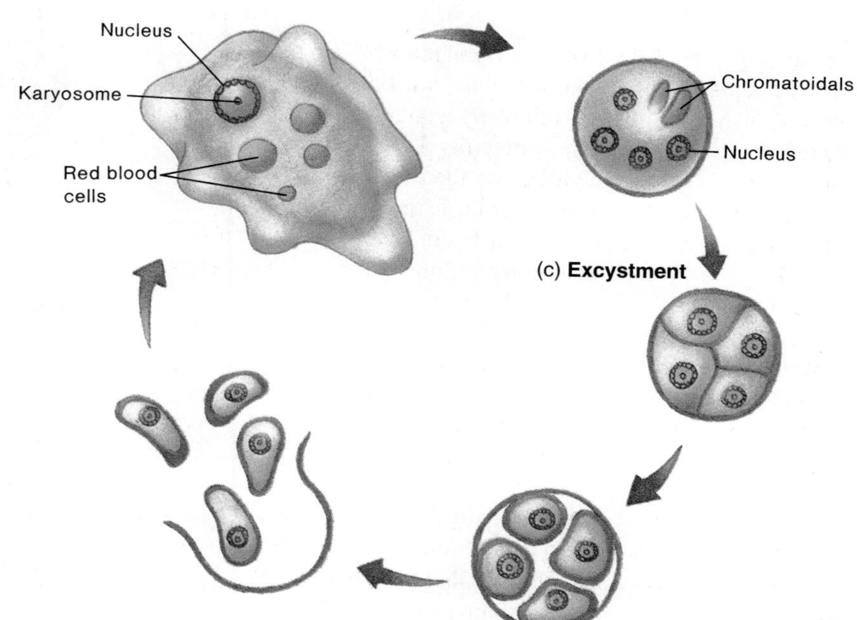

Figure 23.1

Cellular forms of *Entamoeba histolytica*. (*a*) A trophozoite containing a single nucleus, a karyosome, and red blood cells. (*b*) A mature cyst with four nuclei and two blocky chromatoidals. (*c*) Stages in excystment. Divisions in the cyst create four separate cells, or metacysts, that differentiate into trophozoites and are released.

Epidemiology of Amebiasis Humans are the primary hosts of *E. histolytica.* Infection is usually acquired by ingesting food or drink contaminated with cysts released by an asymptomatic carrier. Among the parasitic protozoan diseases, amebic dysentery is surpassed only by malaria in its capacity to cause morbidity and mortality. The ameba is thought to be carried in the intestine of one-tenth of the world's population, and it kills up to 100,000 people a year. Its geographic distribution is partly due to local sewage disposal practices and the degree of soil contamination. Occurrence is highest in tropical regions (Africa, Asia, and Latin America), where night soil[1] or untreated sewage is used to fertilize crops and sanitation of water and food can be substandard. Although the prevalence of the disease is lower in the United States, as many as 10 million people could harbor the agent.

Epidemics of amebiasis are infrequent but have been documented in prisons, hospitals, juvenile care institutions, and communities where water supplies are polluted. Amebic infections can also be transmitted by oral-anal sexual contact among homosexual men.

Life Cycle, Pathogenesis, and Control of *E. histolytica* Amebiasis typically begins when viable cysts survive the oral and gastric secretions and arrive in the small intestine (see figure 5.34). The alkaline pH and digestive juices of this environment stimulate excystment, and from the cyst emerges four trophozoites (figure 23.1*c*). The trophozoites are released in the small intestine and are swept into the *cecum** and large intestine, attaching firmly to the mucosa with fine pseudopods (see chapter opening photo). There, the trophozoites mature, multiply, and actively move about and feed. In about 90% of patients, infection is asymptomatic or very mild, and the trophozoites do not invade beyond the most superficial layer. The severity of the infection appears to vary with the strain of the parasite. Some are more virulent and aggressive than others. Inoculation size, *Entamoeba* strain, diet, and host resistance also play some part in the degree of involvement.

As hinted by its species name, tissue damage is one of the formidable characteristics of untreated *E. histolytica* infection. Clinical amebiasis exists in intestinal and extraintestinal forms. The initial targets of intestinal amebiasis are the cecum, appendix, colon, and rectum. The ameba secretes enzymes that dissolve tissues and actively penetrates deeper layers of the mucosa, leaving erosive ulcerations (figure 23.2). Symptoms associated with this phase of disease are dysentery (bloody, mucus-filled stools), abdominal pain, fever, diarrhea, fatigue, and weight loss. The most life-threatening manifestations of intestinal infection are hemorrhage, perforation, appendicitis, and tumorlike growths called amebomas.

Extraintestinal infection occurs when amebas invade the viscera of the peritoneal cavity. The most common site of invasion is the liver. Here, abscesses containing necrotic tissue and trophozoites develop and cause amebic hepatitis. Another regular complication is pulmonary amebiasis. Less frequent targets of infection are the spleen, adrenals, kidney, skin, and brain. Severe forms of the disease result in about a 10% fatality rate.

Entamoeba is harbored by chronic healthy carriers whose intestines favor the encystment stage of the life cycle. Cyst formation cannot occur in active dysentery because the feces are so

1. Human excrement used as fertilizer.

*cecum (see′-kum) L. *caecus,* blind. The pouchlike anterior portion of the large intestine near the appendix.

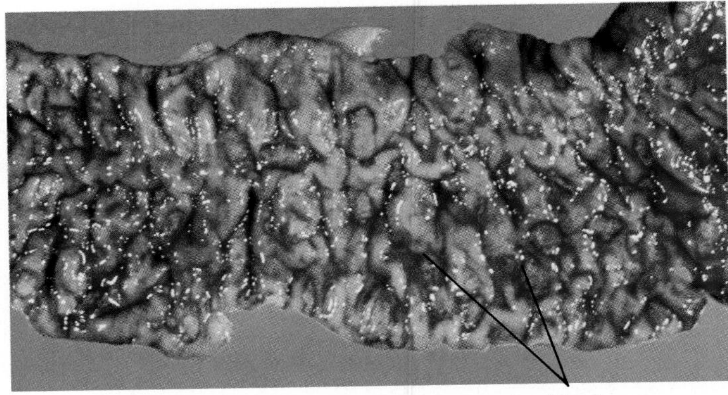

Figure 23.2
Intestinal amebiasis and dysentery of the cecum. Amebic ulcers dot the mucosa in this autopsy specimen.

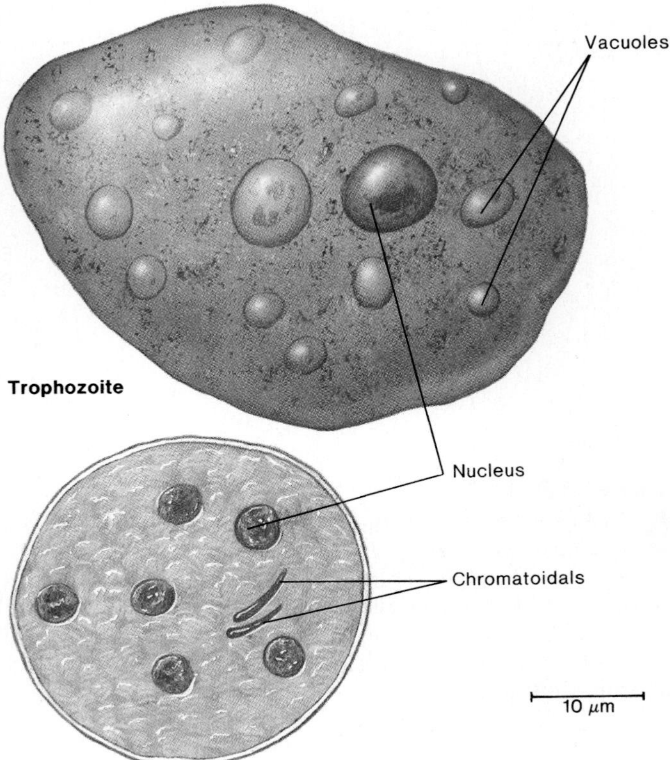

Figure 23.3
Entamoeba coli, a nonpathogenic intestinal ameba, may be mistaken for *E. histolytica,* but it is larger and less motile, with short, blunt pseudopodia. Its cyst has spiny chromatoidals and from one to eight nuclei. Another close relative, *E. dispar,* is also commonly confused with the pathogen.

rapidly flushed from the body, but after recuperation, cysts are continuously shed in feces.

Amebic dysentery has symptoms and signs common to other forms of dysentery, especially bacillary dysentery. Initial diagnosis involves examining a saline suspension of a fecal smear for trophozoites or cysts under high-power magnification and differentiating them from nonpathogenic intestinal amebas that are part of the normal flora (figure 23.3). More definitive identification can be obtained with a stained smear or with a rapid ELISA test using specific monoclonal antibodies. It may be necessary to supplement laboratory findings with clinical data, including symptoms, X rays, and the patient's history, which might indicate possible geographic, occupational, or sexual exposure.

Effective treatment for amebic dysentery usually involves the use of drugs that can act on the parasite in both the feces and the tissues. Drugs such as iodoquinol act in the feces, whereas metronidazole, dehydroemetine, and chloroquine work systematically. In combined intestinal and extraintestinal infections, both iodoquinol and metronidazole are given, and liver abscess alone is treated with chloroquine. Dehydroemetine is used to control symptoms but will not cure the disease. Other drugs are given to relieve diarrhea and cramps, while lost fluid and electrolytes are replaced by oral or intravenous therapy. Infection with *E. histolytica* provokes antibody formation against several antigens, but permanent immunity is unlikely and reinfection can occur. Prevention depends upon concerted efforts similar to those for other enteric diseases (see microfile 20.5). Because regular chlorination of water supplies is insufficient to kill cysts, more rigorous methods such as boiling or iodine are required if contamination has occurred. Several candidates for an amebiasis vaccine are in development. One is based on a combination of three antigens made by recombinant technology and delivered by the oral route.

Amebic Infections of the Brain

Two common free-living protozoans that cause a rare and usually fatal infection of the brain are ***Naegleria fowleri**** and ***Acanthamoeba.**** Both genera are accidental parasites that invade the body only under unusual circumstances. The trophozoite of *Naegleria* is a small, flask-shaped ameba that moves by means of a single, broad pseudopod and has prominent feeding structures, or amebostomes, that lend it a facelike appearance in scanning electron micrographs (figure 23.4). It forms a rounded, thick-walled, uninucleate cyst that is resistant to temperature extremes and mild chlorination. *Acanthamoeba* has a large, ameboid trophozoite with spiny pseudopods and a double-walled cyst.

Both *N. fowleri* and *Acanthamoeba* ordinarily inhabit standing fresh or brackish water, lakes, puddles, ponds, hot springs, moist soil, and even swimming pools and hot tubs. They are especially abundant in warm water with a high bacterial count.

Most cases of *Naegleria* infection reported worldwide occur in healthy young people who have been swimming in warm, natural bodies of fresh water. One epidemic in Australia killed six people after a public water supply had been inadequately disinfected, and in Belgium, an outbreak was reported among people who had bathed in polluted canal water. Infection can begin when amebas are forced into human nasal passages as a result of swimming, diving, or other aquatic activities. Being suddenly inoculated into the abnormal but favorable habitat of the nasal mucosa,

**Naegleria fowleri* (nay-glee′-ree-uh fow′-ler-eye) After F. Nagler and M. Fowler.

**Acanthamoeba* (ah-kah″-thah-mee′-bah) Gr. *acanthos,* thorn.

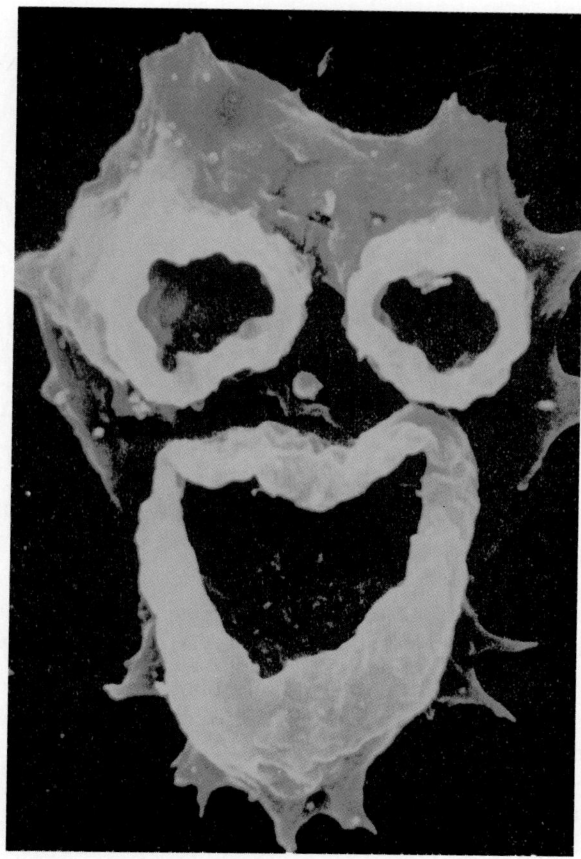

Figure 23.4
Scanning electron micrograph of *Naegleria fowleri* with what looks like a face. The "eyes" and "mouth" are its attachments and feeding structures.

the ameba burrows in, multiplies, and subsequently migrates into the brain and surrounding structures. The result is **primary acute meningoencephalitis,** a rapid, massive destruction of brain and spinal tissue that causes hemorrhage and coma and invariably ends in death within a week or so.

Unfortunately, *Naegleria* meningoencephalitis advances so rapidly that treatment usually proves futile. Studies have indicated that early therapy with amphotericin B, sulfadiazine, tetracycline, or ampicillin alone or in some combination can be of some benefit. Because of the wide distribution of the ameba and its hardiness, no general means of control exists. Public swimming pools and baths must be adequately chlorinated and checked periodically for the ameba.

Acanthamoeba differs from *Naegleria* in its portal of entry, invading broken skin, the conjunctiva, and occasionally the lungs and urogenital epithelia. Although it causes a meningoencephalitis somewhat similar to that of *Naegleria,* the course of infection is lengthier. At special risk for infection are people with traumatic eye injuries, contact lens wearers, and AIDS patients exposed to contaminated water. Ocular infections can destroy the eye or lead to brain infection. However, they can be avoided by carefully tending to injured eyes and using sterile solutions to store and clean contact lenses. Cutaneous and central nervous system infections are complications in AIDS.

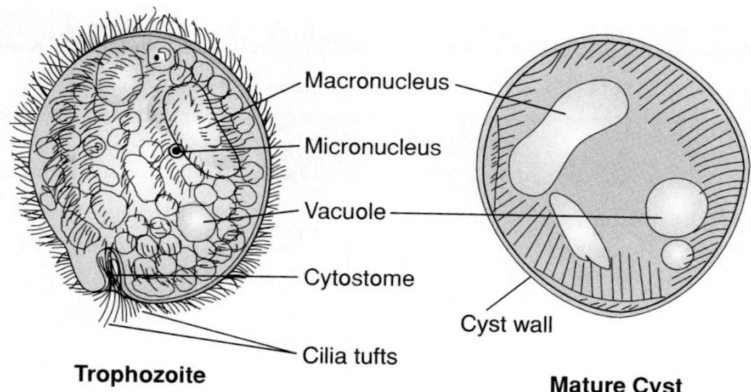

Figure 23.5

The morphological anatomy of a trophozoite and a mature cyst of *Balantidium coli.*

AN INTESTINAL CILIATE: *BALANTIDIUM COLI*

Most ciliates are free-living in various aquatic habitats, where they assume a role in the food web as scavengers or as predators upon other microbes. Several species are pathogenic to animals, but only one, *Balantidium* coli,* occasionally establishes infection in humans. This giant ciliate (about the size of a period on this page) exists in both trophozoite and cyst states (figure 23.5). Its surface is covered by orderly, oblique rows of cilia, and a shallow depression, the cytostome (cell mouth), is at one end. These parasites will grow only under anaerobic conditions in the presence of the host's bacterial flora.

The large intestine of pigs and, to a lesser extent, sheep, cattle, horses, and primates constitutes the natural habitat of *B. coli,* which are shed as cysts in the feces of these animals. Although pigs are the most common source of *balantidiosis,* it can also be spread by contaminated water and from person to person in institutional settings. Healthy humans and animals are resistant to infection, but compromised hosts are vulnerable to attack.

Like cysts of *E. histolytica,* those of *B. coli* resist digestion in the stomach and small intestine and liberate trophozoites that immediately burrow into the epithelium with their cilia. The resultant erosion of the intestinal mucosa produces varying degrees of irritation and injury, leading to nausea, vomiting, diarrhea, dysentery, and abdominal colic. The protozoan rarely penetrates the intestine or enters the blood. The treatment of choice is oral tetracycline, but if this therapy fails, iodoquinol, nitrimidazine, or metronidazole can be given. Preventive measures are similar to those for controlling amebiasis, with additional precautions to prevent food and drinking water from being contaminated with pig manure.

THE FLAGELLATES (MASTIGOPHORANS)

The natural history of flagellated protozoans spans the ecological spectrum from free-living to commensalistic to parasitic. Pathogenic flagellates play an extensive role in human diseases, ranging

**Balantidium* (bal″-an-tid′-ee-um) Gr. A small sac.

TABLE 23.2

FLAGELLATE DISEASES AND TARGET ORGANS

Disease	Major Lesion Site	Flagellate
Trichomoniasis		*Trichomonas* species
Vaginitis, urethritis	Vagina, urethra	*T. vaginalis*
Gingivitis	Opportunist in gums	*T. tenax*
Giardiasis		*Giardia*
Intestinal disease	Duodenum, jejunum	*G. lamblia*
Trypanosomiasis		*Trypanosoma* species
African sleeping sickness	Skin, viscera, brain	*T. brucei*
Chagas' disease	Skin, lymphatics, heart	*T. cruzi*
Leishmaniasis		*Leishmania* species
Oriental sore	Skin	*L. tropica*
Kala-azar	Viscera	*L. donovani*
Espundia	Skin, mucous membranes	*L. brasiliensis, L. mexicana*

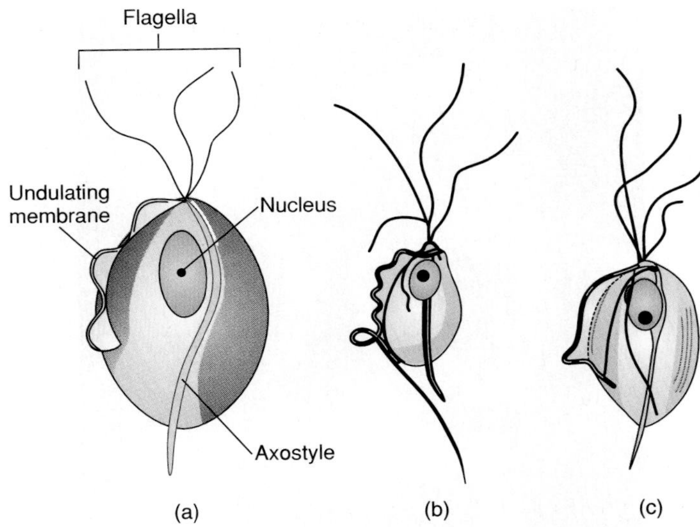

Figure 23.6

The trichomonads of humans. (*a*) *Trichomonas vaginalis,* a urogenital pathogen. (*b*) *T. tenax,* a gingival form (infection is rare). (*c*) *T. hominis,* an intestinal species (infection is rare).

from troublesome but usually self-limited infections (trichomoniasis and giardiasis) to serious and debilitating vector-borne diseases (trypanosomiasis and leishmaniasis). The major flagellate genera and species and their diseases are given in table 23.2.[2]

Trichomonads: Trichomonas* *Species*

Trichomonads are small, pear-shaped protozoa with four anterior flagella and an undulating membrane, which together produce a characteristic twitching motility in live preparations. They exist only in the trophozoite form and do not produce cysts. Three trichomonads intimately adapted to the human body are *Trichomonas vaginalis, T. tenax,* and *T. hominis* (figure 23.6).

The most important species, ***Trichomonas vaginalis,*** is a pathogen of the reproductive tract that causes a sexually transmitted disease called **trichomoniasis.** The reservoir for this protozoan is the human urogenital tract, and about 50% of infected people are asymptomatic carriers. Because *T. vaginalis* has no protective cysts, it is a relatively strict parasite that does not survive for long out of the host. Transmission is primarily through contact between genital membranes, although it can on rare occasions be transmitted through communal bathing, public facilities, and from mother to child. The incidence of infection is highest among promiscuous young women already infected with other sexually transmitted diseases such as gonorrhea or chlamydia. Approximately 3 million new cases occur every year in the United States.

Symptoms and signs of trichomoniasis in the female include a frothy, foul-smelling, green-to-yellow vaginal discharge; vulvitis; cervicitis; and urinary frequency and pain. Severe inflammation of the infection site causes tenderness, edema, chafing, and itching. Males with overt clinical infections often experience irritating persistent or recurring urethritis, with a thin, milky discharge and occasionally prostate infection.

Infection is diagnosed by demonstrating the active swimming trichomonads in a wet film of exudate. Alternative methods to detect asymptomatic infection are a Pap smear or culture of the parasite. The infection has been successfully treated with oral and vaginal metronidazole (Flagyl) for one week, and it is essential to treat both sex partners to prevent "Ping-Pong" reinfection.

Trichomonas tenax is a small trichomonad that ordinarily resides in the oral cavity of 5–10% of humans. Harborage is most common in people with poor oral hygiene or dental disease. *Trichomonas tenax* is the only flagellate in the oral cavity, where it frequents the gingival crevices and areas of calculus, feeding on food debris and sloughed cells. Most dental experts agree that it is not a true pathogen but more likely an opportunist in lesions of gingivitis and periodontal pockets. *Trichomonas hominis* (also *Centatrichomonas)* is a resident of the cecum of a small percentage of humans and great apes, but it is believed to be a harmless commensal and is not associated with disease.

Giardia lamblia *and Giardiasis*

Giardia lamblia* is a pathogenic flagellate first observed by Antonie van Leeuwenhoek in his own feces. For 200 years, it was considered a harmless or weak intestinal pathogen, and only since the 1950s has its prominence as a cause of diarrhea been recognized. In fact, it is the most common flagellate isolated in clinical specimens. Observed straight on, the trophozoite has a unique, symmetrical heart shape with organelles positioned in such a way that they resemble a face (figure 23.7*a*). Four pairs of flagella emerge from the ventral surface, which is concave and acts like a suction cup for attachment to a substrate. *Giardia* cysts are small, compact, and multinucleate (figure 23.7*a).*

2. Many flagellates are named in honor of one of their discoverers—for example, *T. cruzi,* O. Cruz; *Leishmania,* W. Leishman; *donovani,* C. Donovan; *Giardia,* A. Giard and *lamblia,* V. Lambl; and *brucei,* D. Bruce. Geographic location *(gambiense, rhodesiense, tropica, brasiliensis, mexicana)* is another basis for naming. Names are also derived from a certain characteristic of the organism—for example, *tenax,* tenacious; *hominis* (L. *homo,* human); and *vaginalis,* vaginal habitat.

**Trichomonas* (trik″-oh-moh′-nus) Gr. *thrix,* hair, and *monas,* unit.

**Giardia lamblia* (jee′-ard-ee-uh lam′-blee-uh) Also called *G. intestinalis.*

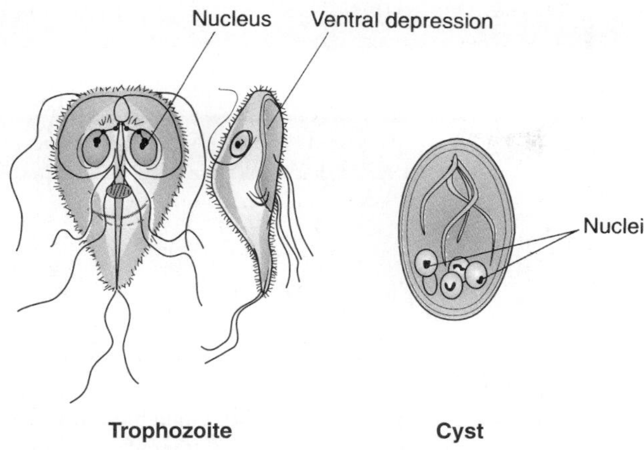

(a)

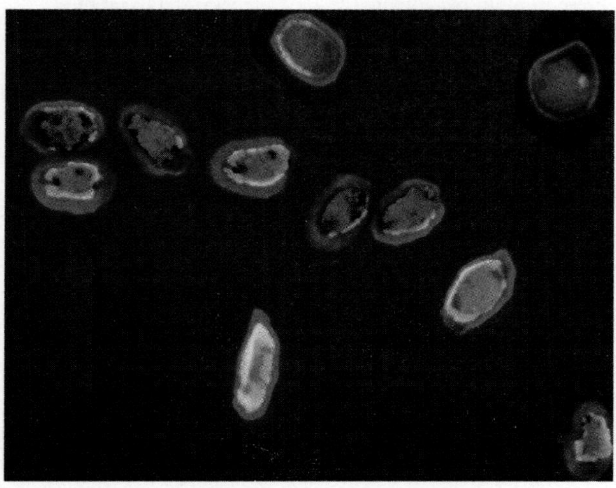

(b)

Figure 23.7

Trophozoites and cysts of *Giardia lamblia*. (*a*) The face of a trophozoite with eyes (paired nuclei). Side view shows the ventral depression and origin of the flagella. Cysts have two or four nuclei. (*b*) A special DNA probe test uses FISH (fluorescent-in situ-hybridization) technology to differentiate *Giardia* species using colored dyes. *G. lamblia* (lower right) glows blue, whereas *G. muris* glows green. In both forms the cell wall antigens are shown in red.

Giardiasis has a complex epidemiologic pattern. The protozoan has been isolated from the intestines of beavers, cattle, coyotes, cats, and human carriers, but the precise reservoir is unclear at this time. Although both trophozoites and cysts escape in the stool, the cysts play a greater role in transmission. Unlike other pathogenic flagellates, *Giardia* cysts can survive for 2 months in the environment. Cysts are usually ingested with water and food or swallowed after close contact with infected people or contaminated objects. In a study with prison volunteers, the cysts were shown to be highly infectious; only 10–100 were required. In tropical climates, giardiasis is most common in children; in industrialized countries, adults are infected as frequently as children.

Outbreaks of giardiasis point to a spectrum of possible modes of transmission. Community water supplies in areas throughout the United States have been implicated as common vehicles of infection. *Giardia* epidemics have broken out in resorts with pristine mountain streams such as Aspen, Colorado, and they have even been traced to chlorinated municipal water supplies in Massachusetts, California, and Washington. In addition, numerous infections have been reported in hikers and campers who used what they thought was clean water from ponds, lakes, and streams in remote mountain areas. Because wild mammals such as muskrats and beavers are intestinal carriers, they could account for cases associated with drinking water from these sources. Obviously, checking water for purity by its appearance is unreliable, because the cysts are too small to be detected.

Giardia is also a well-known cause of travelers' diarrhea, which sometimes does not appear until the person has returned home. Cases of food-borne illness have been traced to carriers who contaminate food through unclean personal habits. Daycare centers have reported outbreaks associated with diaper changing and contact with other types of fomites.

Ingested *Giardia* cysts excyst in the duodenum and travel to the jejunum to feed and multiply. Some trophozoites remain on the surface, while others invade the glandular crypts to varying degrees. The outcome of infection ranges from asymptomatic to severe, chronic giardiasis. Superficial invasion by trophozoites causes damage to the epithelial cells, edema, and infiltration by white blood cells, but these effects are reversible. Typical symptoms include diarrhea, abdominal pain, and flatulence. Diagnosis of giardiasis can be difficult because the organism is shed in feces intermittently. A promising DNA test is now available for detection of the cysts in samples (figure 23.7*b*). The infection can be eradicated with quinacrine or metronidazole.

Now that this parasite is apparently on the increase, many water agencies have had to rethink their policies on water maintenance and testing. The agent is killed by boiling, ozone, and iodine, but, unfortunately, the amount of chlorine used in municipal water supplies does not destroy the cysts. People who must use water from remote sources should assume that it is contaminated and boil or filter it.

Hemoflagellates: Vector-Borne Blood Parasites

Other parasitic flagellates are called hemoflagellates because of their propensity to live in the blood and tissues of the human host. Members of this group, which includes species of ***Trypanosoma****** and ***Leishmania,*** have several distinctive characteristics in common. They are obligate parasites that cause life-threatening and debilitating zoonoses; they are spread by blood-sucking vectors that serve as intermediate hosts; and they are acquired in specific tropical regions. Hemoflagellates have complicated life cycles and undergo morphological changes as they pass between vector and human host. To keep track of these developmental stages, parasitologists have adopted the following set of terms:

amastigote (ay-mas′-tih-goht) Gr. *a-*, without, and *mastix,* whip. The form lacking a free flagellum.

**Trypanosoma* (try′-pan-oh-soh-mah) Gr. *tryanon,* borer, and *some,* body.

TABLE 23.3

CELLULAR AND INFECTIVE STAGES OF THE HEMOFLAGELLATES

Genus/Species	Amastigote	Promastigote	Epimastigote	Trypomastigote
Leishmania	Intracellular in human macrophages	Found in sand fly gut; **infective to humans**	Does not occur	Does not occur
Trypanosoma brucei	Does not occur	Does not occur	Present in salivary gland of tsetse fly	In biting mouthparts of tsetse fly; **infective to humans**
Trypanosoma cruzi	Intracellular in human macrophages, liver, heart, spleen	Occurs	Present in gut of reduviid (kissing) bug	In feces of reduviid bug; **transferred to humans**

promastigote (proh-mas′-tih-goht) Gr. *pro-,* before. The stage bearing a single, free, anterior flagellum.

epimastigote (ep″-ih-mas′-tih-goht) Gr. *epi-,* upon. The flagellate stage, which has a free anterior flagellum and an undulating membrane.

trypomastigote (trih″-poh-mas′-tih-goht) Gr. *trypanon,* auger. The large, fully formed stage characteristic of *Trypanosoma.*

These stages suggest a metamorphic or evolutionary transition from the simple, cystlike amastigote form to the highly complex trypomastigote form. All four stages are present in *T. cruzi,* but *T. brucei* lacks the amastigote and promastigote forms, and *Leishmania* has only the amastigote and promastigote stages (table 23.3). In general, the vector host serves as the site of differentiation for one or more of the stages, and the parasite is infectious to humans only in its fully developed stage.

Trypanosoma *Species and Trypanosomiasis*

Trypanosoma species are distinguished by their infective stage, the trypomastigote, an elongate, spindle-shaped cell with tapered ends that is capable of eel-like, sinuous motility. Two types of **trypanosomiasis** are distinguishable on a geographical basis. *Trypanosoma brucei* is the agent of **African sleeping sickness,** and *T. cruzi* is the cause of **Chagas' disease,** which is endemic to Central and South America. Both exhibit a biphasic life-style alternating between a vertebrate and an insect host. Because the life cycle of *T. cruzi* was discussed in chapter 5, only that of sleeping sickness will be illustrated here.

***Trypanosoma brucei* and Sleeping Sickness** Trypanosomiasis has greatly affected the living conditions of Africans since ancient times. Today, at least 50 million people are at risk, and 20,000 new cases occur each year. It imposes an additional hardship when it attacks domestic and wild mammals. The two variants of sleeping sickness are the Gambian (West African) strain, caused by the subspecies *T. b. gambiense,* and the Rhodesian (East African) strain, caused by *T. b. rhodesiense* (figure 23.8*a*). These geographically isolated types are associated with different ecological niches of the principal tsetse fly vectors. In the West African form, the fly inhabits the dense vegetation along rivers and forests typical of that region, whereas the East African form is adapted to savanna woodlands and lakefront thickets.

The cycle begins when a tsetse fly becomes infected after feeding on an infected reservoir host, such as a wild animal (antelope, pig, lion, hyena), domestic animal (cow, goat), or human (figure 23.8*c*). The trypanosome is taken into the gut with a blood meal, where it multiplies, migrates to the salivary glands, and develops into the infectious stage. A single fly lives up to 3 months and is capable of delivering 50,000 trypanosomes with a single bite, even though only about 500 cells are required for infection. At the site of release, the trypanosome undergoes a series of divisions and produces a sore called the primary chancre. From there, the pathogen moves into the lymphatics and the blood (figure 23.8*b*). Although an immune response occurs, it is counteracted by an unusual adaptation of the trypanosome (microfile 23.1). The onset, severity, and length of the disease vary in the two geographic forms. The Rhodesian form is acute and advances to brain involvement in 3 to 4 weeks and to death in a few months. The Gambian form, on the other hand, has a longer incubation period, is usually chronic, and may not affect the brain for several years.

The principal pathologic effects of trypanosomiasis occur in lymphatics and areas surrounding blood vessels. Symptoms usually include intermittent fever, enlarged spleen, swollen lymph nodes, and joint pain. In both forms, the central nervous system is affected, the initial signs being personality and behavioral changes that are at first subtle but progress to lassitude and sleep disturbances. The disease is commonly called sleeping sickness, but in fact, uncontrollable sleepiness occurs primarily in the day and is followed by sleeplessness at night. Signs of advancing neurological deterioration are muscular tremors, shuffling gait, slurred speech, epileptic-like seizures, and local paralysis. Death results from coma, malnutrition, or secondary infections, and cardiac arrest can occur in the Rhodesian form.

MICROFILE 23.1 THE TRYPANOSOME'S BAG OF TRICKS

The immune system first responds to trypanosome surface antigens by producing immunoglobulins of the IgM type. These antibodies carry out their usual function of tagging the parasite to make it easier to destroy. If this were the end of the matter, sleeping sickness would be a mild infection with few fatalities, but such is not the case. Instead, the trypanosome strikes back with a fascinating genetic strategy in which surviving cells change the structure of their surface glycoprotein antigens. This change in specificity renders the existing IgM ineffective so that the parasite eludes control and multiplies in the blood. When the host fights back by producing IgM against this new antigen, the parasite alters its surface again, the body produces corresponding different IgM, the surface changes again, and so on, in a repetitive cycle that can go on for months or even years. Eventually, the host becomes exhausted and overwhelmed by repeated efforts to catch up with this trypanosome masquerade. This cycle has tremendous impact on the pathology and control of the disease. The presence of the trypanosome in the blood and the severity of symptoms follow a wavelike pattern reminiscent of borreliosis. Development of a vaccine is hindered by the requirement to immunize against at least 100 different antigenic variations.

Sleeping sickness may be suspected if the patient has been bitten by a distinctive fly (usually quite painful) while living or traveling in an endemic area. Trypanosomes are readily demonstrated in the blood, spinal fluid, or lymph nodes. Chemotherapy is most successful if administered prior to nervous system involvement. Brain infection must be treated with a highly toxic arsenic-based drug called melarsoprol, or a less toxic one, difluoromethylornithine (DFMO). Both drugs are expensive and may be beyond the medical resources of many countries.

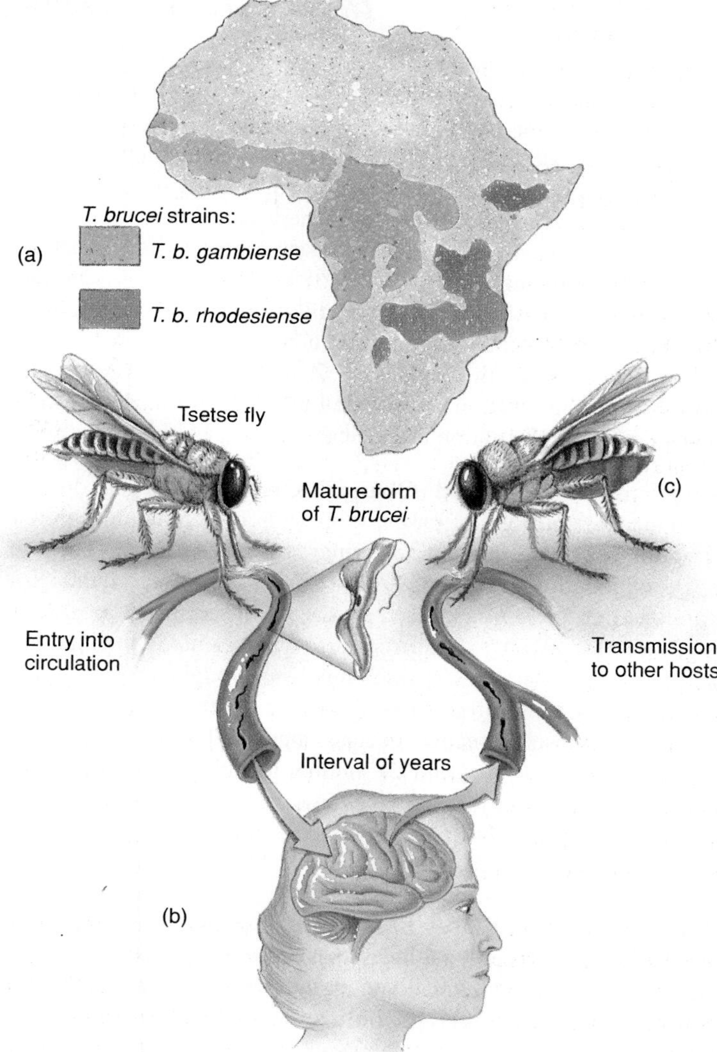

Figure 23.8

(*a*) The distribution of African trypanosomiasis. (*b*) The generalized cycle between humans and the tsetse fly vector. The saliva of a fly infected with *T. brucei* inoculates the human bloodstream. The parasite matures and invades various organs. In time, its cumulative effects cause central nervous system (CNS) damage. (*c*) The trypanosome is spread to other hosts through another fly in whose alimentary tract the parasite completes a series of developmental stages.

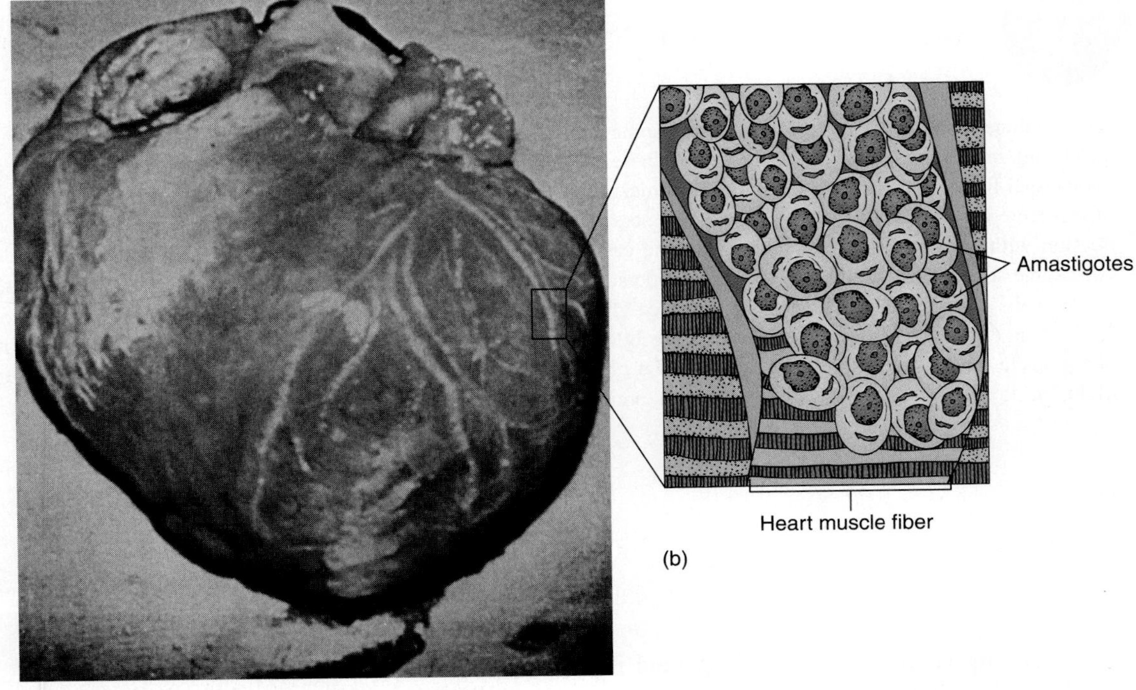

Figure 23.9
Heart pathology in Chagas' disease. (*a*) A greatly enlarged ventricular muscle. (*b*) Section of a heart muscle with a cluster of amastigotes nestled within the fibers.

Control of trypanosomiasis in western Africa, where humans are the main reservoir hosts, involves eliminating tsetse flies by applying insecticides, trapping flies, or destroying shelter and breeding sites. In eastern regions, where cattle herds and large wildlife populations are reservoir hosts, control is less practical because large mammals are the hosts, and flies are less concentrated in specific sites. In some parts of equatorial Africa, *T. b. gambiense* has recently experienced a resurgence. Epidemics are reported in areas of the Sudan and Zaire, where 20–40% of the population is infected. These latest outbreaks are attributed to a civil war and its disruption in medical services.

Trypanosoma cruzi **and Chagas' Disease** Chagas' disease accounts for millions of cases, thousands of deaths, and untold economic loss and human suffering in Latin America. The life cycle of its etiologic agent, *Trypanosoma cruzi* (see chapter 5), parallels that of *T. brucei,* except that the insect hosts are "kissing," or reduviid, bugs that harbor the trypanosome in the hind gut and discharge it in feces. These bugs live throughout Central and South America and often share the living quarters of human and mammalian hosts. Infection occurs when the bug defecates near the site of its own bite, and the human inoculates this wound by accidentally rubbing the bug feces into it. Infection is also possible when bug feces fall onto the mucous membranes or conjunctiva. The disease can even be spread by sexual intercourse, transplacental means, and blood transfusion in some instances. The course of the illness is somewhat similar to African trypanosomiasis, marked by a local lesion, fever, and swelling of the lymph nodes, spleen, and liver. Particular targets are the heart muscle and large intestine, both of which harbor masses of amastigotes. The infection of these organs causes enlargement and severe disruption in function (figure 23.9). Chronically infected people with cardiac complications usually die within 2 years.

In addition to microscopic examination of blood and serological testing, diagnosis of Chagas' disease is sometimes made using a clever technique called xenodiagnosis, in which a germ-free reduviid bug is allowed to feed on the patient and is then observed for signs of infection in a few weeks. Although treatment is sometimes attempted with two drugs, nifurtimox and benzonidazole, they are not effective if the disease has progressed too far, and they have damaging side effects. It has been shown that the prevalence of Chagas' disease can be reduced by vector control. Insecticides are of some benefit, but the bugs are very tough and capable of resisting doses that readily kill flies or mosquitoes. Mass collections of reduviid bugs and housing improvements to keep them out of human dwellings have also been effective. Vaccination has been contemplated, but the development of a successful vaccine lies in the indefinite future.

Leishmania *Species and Leishmaniasis*

Leishmaniasis is a zoonosis transmitted among various mammalian hosts by female **phlebotomine*** flies called sand flies that require a blood meal to produce mature eggs. ***Leishmania*** species have adapted to the insect host by timing their own development with periods of blood-taking (figure 23.10). The protozoa enter the gut in the blood and then multiply and differentiate into promastigotes that migrate to the esophagus and proboscis of the fly.

Leishmaniasis is endemic to regions on or near the equator that provide favorable geographic and biological conditions for sand flies to complete their life cycle. The disease can be transmitted by over 50 species of sand fly, and numerous wild and domesticated animals, especially dogs, compose the rest of the reservoir.

*phlebotomine (flee-bot'-oh-meen). Gr. *phlebo,* vein, and *tomos,* cutting. So called because they feed on blood.

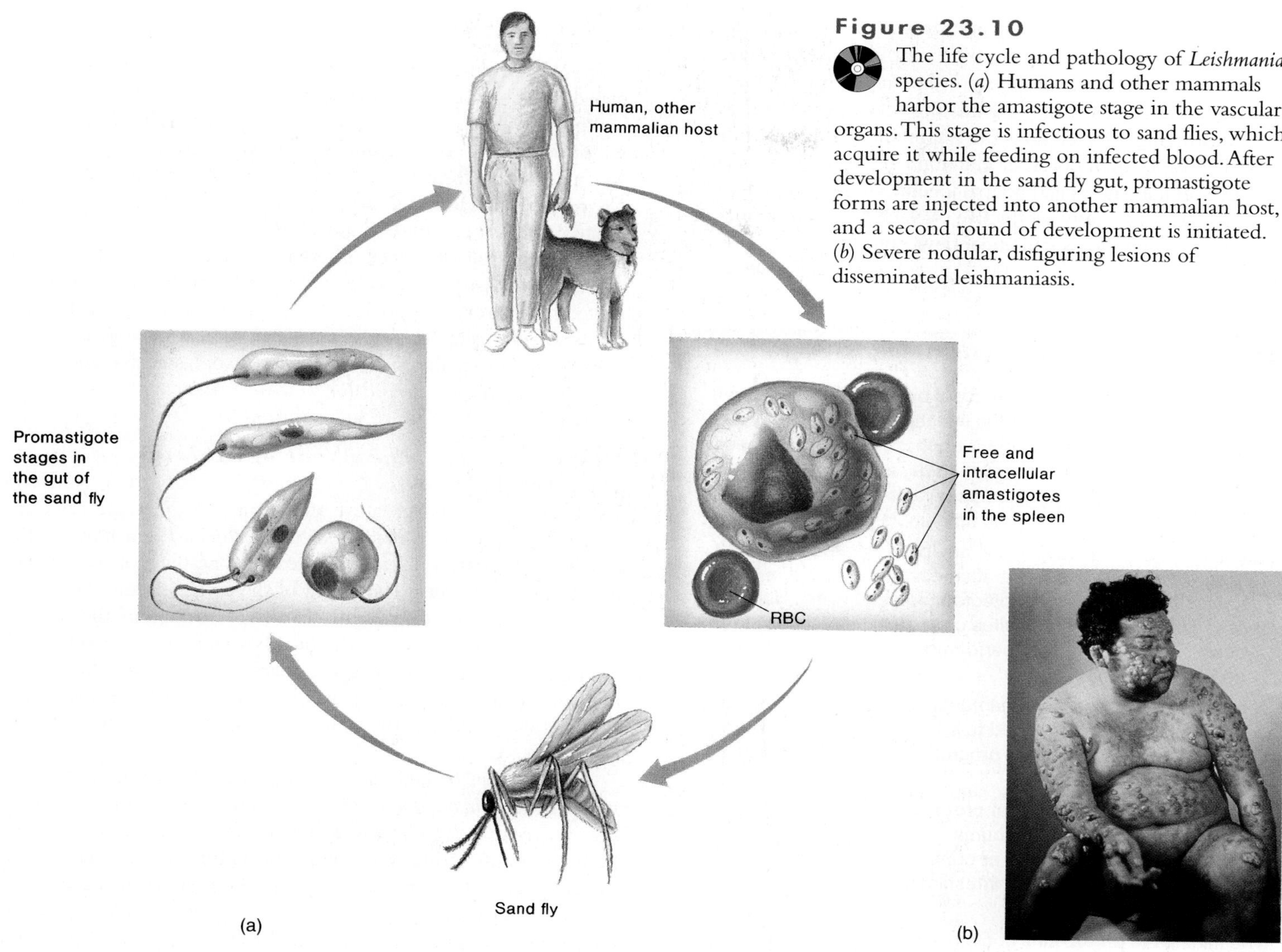

Figure 23.10

The life cycle and pathology of *Leishmania* species. (*a*) Humans and other mammals harbor the amastigote stage in the vascular organs. This stage is infectious to sand flies, which acquire it while feeding on infected blood. After development in the sand fly gut, promastigote forms are injected into another mammalian host, and a second round of development is initiated. (*b*) Severe nodular, disfiguring lesions of disseminated leishmaniasis.

Although humans are usually accidental hosts in the epidemiologic cycle, the fly is said to find human blood particularly appealing. At particular risk are travelers or immigrants who have never had contact with the parasite and lack specific immunities.

The species of *Leishmania* pathogenic to humans are indistinguishable in appearance, even though they are geographically isolated and produce distinct types of diseases. In all cases, infection begins when an infected fly injects promastigotes while feeding. Although nearby macrophages capture the parasite, it is able to resist destruction, convert to an amastigote, and multiply intracellularly in the macrophage. The differences in the nature of the infection depend on the pathway of the macrophage. If infected macrophages remain fixed, the infection is localized in the skin or mucous membranes, but if the infected macrophages migrate, systemic disease occurs.

Cutaneous leishmaniasis (oriental sore, Baghdad boil) is a localized infection of the capillaries of the skin caused by *L. tropica* in certain Mediterranean, African, and Indian regions and by *L. mexicana* in Latin America. A single small, red papule occurs at the site of the bite and spreads laterally into a large "wet" or "dry" ulcer. A form of mucocutaneous leishmaniasis called espundia, caused by *L. brasiliensis* and endemic to parts of Central and South America, affects both the skin and mucous membranes of the head. The primary skin ulcer develops within a few days and after it heals, the protozoa may settle into a chronic infection of the nose, lips, palate, gingiva, and pharynx. Secondary infections of these sites can be very disfiguring (figure 23.10*b*). Systemic or visceral leishmaniasis that occurs in parts of Africa, Latin America, and China is caused by *L. donovani.* The disease appears gradually after 2 to 18 weeks, with a high, spiking, intermittent fever; weight loss; and enlargement of internal organs, especially the spleen, liver, and lymph nodes. The most deadly form, *kala-azar,* is 75% to 95% fatal in untreated cases. Death is due to complications of infection, including the destruction of blood-forming tissues, anemia, secondary infections, and hemorrhage.

Diagnosis is confounded by the similarity of leishmaniasis to several granulomatous bacterial and fungal infections. The organism can usually be detected in the cutaneous lesion or in the bone marrow during visceral disease. Additional tests include cultivating the parasite, serological tests, and leishmanin skin testing.

Chemotherapy with injected pentavalent antimony is usually effective, though pentamidine or amphotericin B may be necessary in resistant cases. Some disease prevention has been achieved by applying insecticides to and eliminating sand fly habitats, as well as by destroying infected dogs (an important reservoir of infection in some areas). Attempts at vaccination have included inoculation with live promastigotes and nonpathogenic strains, but these methods have been generally ineffective. A vaccine based on recombinant DNA technology is still under development.

Chapter Checkpoints

Protozoan infections are categorized by the type of motility, if any, and the complexity of the infectious agent's life cycle. All protozoans propagate via their trophozoite, or active feeding stage. The resting form, or cyst stage, enables them to exist outside their host. Major human parasites are found in the ameba, ciliate, flagellate, and apicomplexan groups.

The most significant pathogenic ameba is *Entamoeba histolytica,* causative agent of amebic dysentery. It is transmitted through cyst-infected food and water. Humans are its only host. The trophozoites can cause ulcerations in the bowel, which can lead to perforation and infection of other peritoneal organs.

Naegleria fowleri and *Acanthamoeba* are free-living amebas that infect humans exposed to contaminated water. Both cause CNS infections that progress rapidly and are usually fatal.

Balantidium coli is the largest protozoan pathogen and the only ciliate pathogenic to humans. Cysts enter the host through contact with pigs or contaminated water. Trophozoites colonize the intestinal mucosa and cause moderate diarrhea.

The flagellated protozoans contain several major pathogens: *Trichomonas, Giardia, Trypanosoma,* and *Leishmania.* The most important trichomonad causes an STD called trichomoniasis. *Giardia* flagellates cause a noninvasive infection of the intestinal mucosa, which can proceed to a severe, chronic form in some cases.

The hemoflagellates are systemic invaders that cause debilitating, life-threatening infections of lymphatics, blood vessels, and specific organs. All pathogens in this group are transmitted by arthropod vectors. All have several morphologically distinctive stages to their life cycles.

African sleeping sickness is caused by *Trypanosoma brucei,* which invades the blood, the lymphatics, and the brain. Tsetse flies are the vector and secondary host. *T. cruzi* is the causative agent of Chagas' disease. The reduviid bug infects the host through its feces deposited at the bite region. Chagas' disease damages the heart and intestine in particular.

Leishmania is the causative agent of leishmaniasis, a zoonosis transmitted by female sand flies that are also the secondary host. Macrophages transmit the infection throughout the human host. The migration of the infected macrophage determines whether the infection is cutaneous or systemic.

APICOMPLEXAN PARASITES

One group of unique protozoa are the apicomplexans, sometimes called the sporozoans. These tiny parasites live on a variety of animal hosts. They are distinct from other protozoans in lacking locomotor organelles in the mature state, though some forms do have a flexing, gliding form of locomotion, and others have flagellated sex cells. Characteristics of the apicomplexan life cycle that add greatly to their complexity are alternations between sexual and asexual phases and between different animal hosts. Most members of the group also form specialized infective bodies that are transmitted by arthropod vectors, food, water, or other means, depending on the parasite. The most important human pathogens belong to *Plasmodium* (malaria), *Toxoplasma* (toxoplasmosis), and *Cryptosporidium* (cryptosporidiosis).

Plasmodium: *The Agent of Malaria*

Throughout human history, including prehistoric times, malaria has been one of the greatest afflictions, in the same ranks as bubonic plague, influenza, and tuberculosis. Even now, as the dominant protozoan disease, it threatens one-third of the world's population every year. The origin of the name is from the Italian words *mal,* bad, and *aria,* air. The superstitions of the Middle Ages explained that evil spirits or mists and vapors arising from swamps caused malaria, because many victims came down with the disease after this sort of exposure. Of course, we now know that a swamp was mainly involved as a habitat for vectors (mosquitoes).

The agent of malaria is an obligate intracellular sporozoan in the genus *Plasmodium,* which contains four species: *P. malariae, P. vivax, P. falciparum,* and *P. ovale.* The human and some primates are the primary vertebrate hosts for these species, which are geographically separate and show variations in the pattern and severity of disease. All forms are spread primarily by the female *Anopheles* mosquito and occasionally by shared needles, blood transfusions, and from mother to fetus. The detailed study of malaria is a large and engrossing field that is far too complicated for this text. We cover only the general aspects of its epidemiology, life cycle, and pathology.

Epidemiology, Life Cycle, and Stages of *Plasmodium* Infection Although malaria was once distributed throughout most of the world, the control of mosquitoes in temperate areas has successfully restricted it mostly to a belt extending around the equator. Despite this achievement, approximately 300 million to 500 million new cases are still reported each year, about two-thirds of them in Africa. The most frequent victims are children and young adults, of whom at least 2 million die annually. The total case rate in the United States is about 1,000 to 2,000 new cases a year, most of which occur in immigrants.

Because the developmental stages of the malarial parasite are exquisitely timed to coincide with the mosquito's life cycle, and the behavior of the mosquito in turn coincides with human infection, we will present the stages in parasites, mosquitoes, and humans as an interactive system (figure 23.11). Development of the malarial parasite is divided into two distinct phases: the asexual phase, carried out in the human, and the sexual phase, carried out in the mosquito.

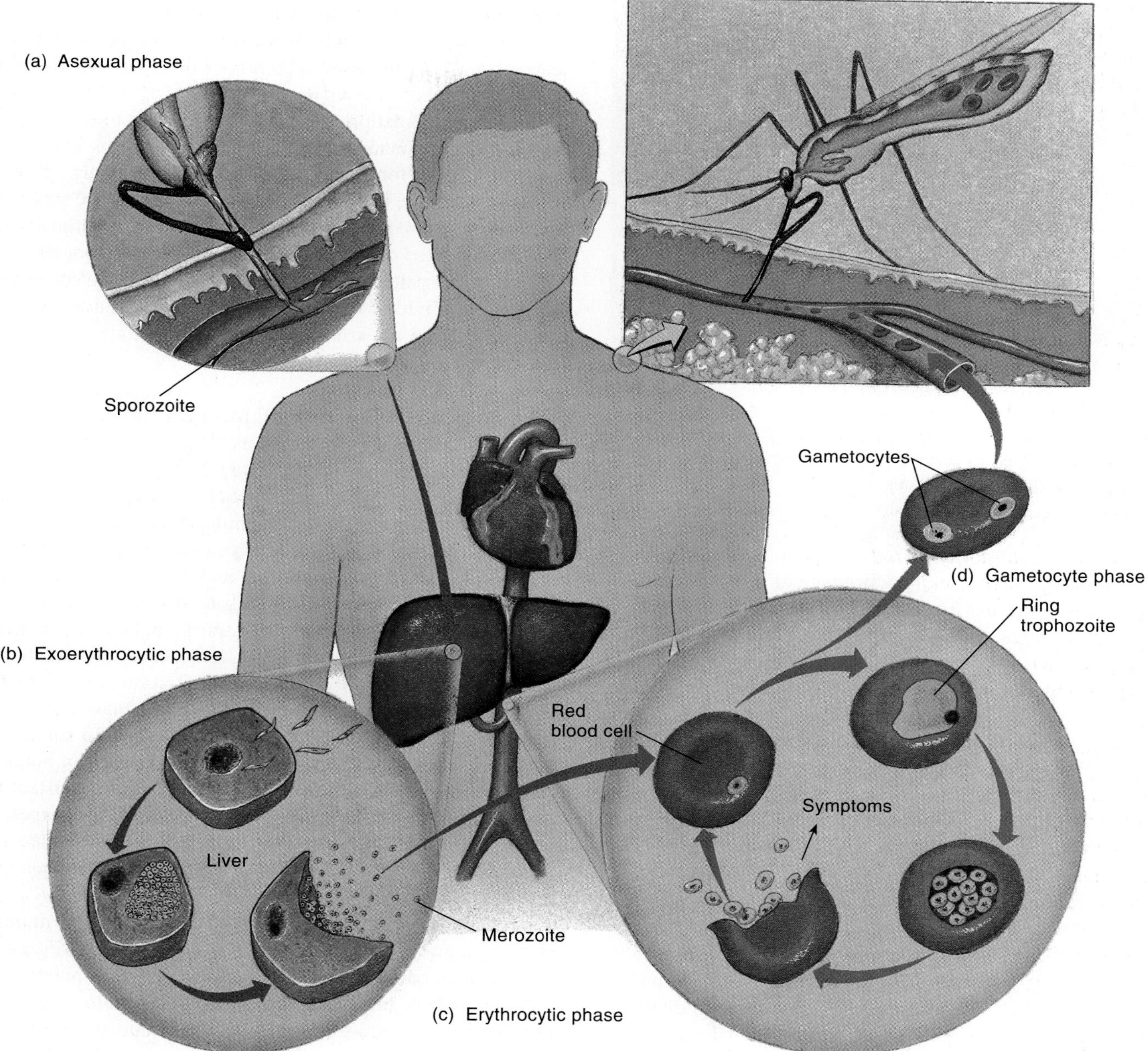

Figure 23.11

The life and transmission cycle of *Plasmodium,* the cause of malaria. (*a*) In the asexual phase in humans, sporozoites enter a capillary through the saliva of a feeding mosquito. (*b*) Exoerythrocytic (liver) phase. Sporozoites invade the liver cells and develop into large numbers of merozoites. (*c*) Erythrocytic phase. Merozoites released into the circulation enter red blood cells. Initial infection is marked by a ring trophozoite; schizogony of the ringed form produces additional merozoites that burst out and infect other red blood cells. (*d*) Gametocytes that develop in certain infected red blood cells are ingested by another mosquito. (*e*) The sexual phase of fertilization and sporozoite formation occurs in the mosquito.

The **asexual phase** and infection begins when an infected female *Anopheles* mosquito extracts the blood necessary to develop her eggs. Immediately before feasting, she injects saliva containing anticoagulant into the capillary she has punctured, an event that also inoculates the blood with motile, spindle-shaped, asexual cells called **sporozoites** (Gr. *sporo,* seed, and *zoon,* animal). The sporozoites circulate through the body, but within an hour, they have taken up residence in the liver. Within liver cells, the process of asexual division called *schizogony* (Gr. *schizo,* to divide, and *gone,* seed) generates numerous daughter parasites, or *merozoites.* This phase of *exoerythrocytic development* lasts from 5 to 16 days, depending upon the species of parasite. When each packed liver cell finally erupts, it releases from 2,000 to 40,000 merozoites into the circulation.

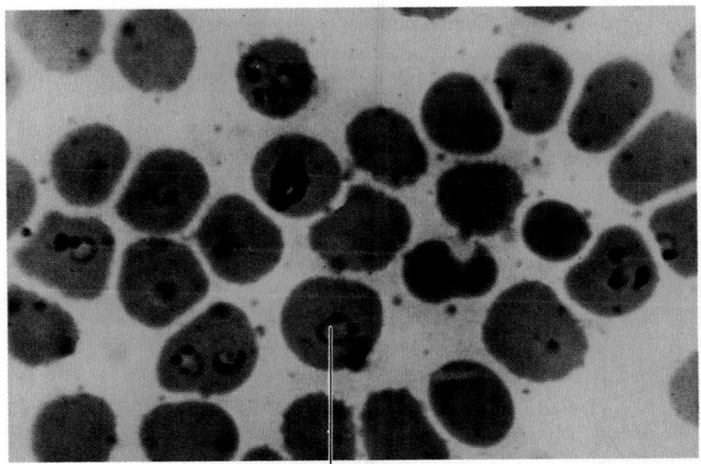

Figure 23.12

The ring trophozoite stage in a *Plasmodium falciparum* infection. A smear of peripheral blood shows several ring forms in some red blood cells.

During the *erythrocytic phase,* merozoites are programmed to infect red blood cells by taking advantage of specific receptors. Inside RBCs, they transform into a circular (ring) trophozoite (figure 23.12). This stage feeds upon hemoglobin, grows, and undergoes multiple divisions to produce a cell called a *schizont,* which is filled with more merozoites. When massively infected red cells burst, the liberated merozoites are freed to infect more red cells. At some point, certain merozoites differentiate into two types of specialized gametes called *macrogametocytes* (female) and *microgametocytes* (male). Because the human does not provide a suitable environment for completion of fertilization and the next phase of development, this is the end of the cycle in humans.

The **sexual phase** (sporogony) results when a mosquito draws infected RBCs into her stomach. In the stomach, the microgametocyte releases several spermlike gametes that fertilize the larger macrogametes. The resultant diploid cell (oocyst) implants into the stomach wall and undergoes multiple meiotic divisions, releasing haploid sporozoites that migrate to the salivary glands and lodge there. This event completes the sexual cycle and makes the sporozoites available for injection when the mosquito feeds on her next victim.

The classic clinical signs and symptoms of malaria result after an incubation period of 10 to 16 days. Prodromal symptoms are malaise, fatigue, vague aches, and nausea followed by bouts of chills, fever, and sweating. These occur first in an irregular pattern, and later at 48- or 72-hour intervals, as a result of the synchronous rupturing of RBCs. The interval, length, and regularity of these symptoms reflect the type of malaria. Patients with falciparum malaria, the most malignant type, often manifest persistent fever, rapid pulse, cough, and weakness for weeks without relief. Complications of malaria are hemolytic anemia from lysed blood cells and organ enlargement and rupture due to accumulated excess cellular debris in the spleen, liver, and kidneys. Additional complications are pulmonary failure, cerebral malaria, and gastrointestinal disturbances. Patients with milder, chronic forms of malaria can achieve spontaneous cure after 3 to 5 years, but falciparum malaria has a high death rate in the acute phase, especially in children. Certain kinds of malaria are subject to relapses because some infected liver cells harbor dormant infective cells for up to 5 years.

Some populations are naturally resistant to *Plasmodium* infection. One dramatic example occurs in Africans who carry one gene that codes for sickle-cell hemoglobin. Although carriers are relatively healthy, they produce some abnormal RBCs that prevent the parasite from deriving sufficient nutrition to multiply. The continued existence of sickle-cell anemia in these geographic areas can be directly attributed to the increased survival of carriers of the gene. Individuals with sickle-cell anemia have even greater protection from malaria; however, the debilitation of sickle-cell disease outweighs its protective effect.

Diagnosis and Control of Malaria Malaria can be diagnosed definitively by the discovery of a typical stage of *Plasmodium* in stained blood smears (figure 23.12). Other indications are knowledge of the patient's residence or travel in endemic areas and such symptoms as recurring chills, fever, and sweating.

As recently as a decade ago, eradicating malaria seemed possible, but since then, morbidity and mortality have increased in several regions. The two main reasons involve the sheer adaptive and survival capacity of both the parasite and the vector. Standard chemotherapy with antimalarial drugs has selected for drug-resistant strains of the malarial parasite, and mosquitoes have developed resistance to some common insecticides used to control them. The current treatments for malaria include some form of quinine. Chloroquine, the least toxic type, is used in nonresistant forms of the disease. In areas of the world where resistant strains of *P. falciparum* and *P. vivax* predominate, a course of mefloquine or quinine is indicated. Eliminating the parasite from the liver and preventing relapses can be managed with long-term therapy with primaquine or proguenil.

Malaria prevention is attempted through long-term mosquito abatement and human prophylaxis. Any standing water that could serve as a breeding site should be eliminated. Combinations of insecticides can be broadcast into the environment to reduce populations of adult mosquitoes, especially near human dwellings. Humans can reduce their risk of infection considerably by using netting, screens, and repellents; remaining indoors at night; and taking weekly doses of prophylactic drugs. People with recent cases of malaria must be excluded from blood donation. Even with massive efforts undertaken by the WHO, the prevalence of malaria in endemic areas is still high.

Malaria and Its Elusive Vaccine Several animal and human models have proved that effective vaccination for malaria is possible, but developing a practical vaccine for broad-scale use has proved more difficult. In fact, malaria is a perfect model to demonstrate the problems of vaccine development. A successful malaria vaccine must be capable of striking a diverse and rapidly changing target. Not only are there four different species, each having different sporozoite, merozoite, and gametocyte stages, but each species can also have different antigenic types of sporozoites and merozites. A sporozoite vaccine would prevent liver infection, whereas a merozoite vaccine would prevent infection of red blood cells and diminish the pathology that is most responsible for morbidity and mortality. The best vaccines would provide long-term protection against all the phases and variants so that if one immunity were to fail, another could come into play.

One type of vaccine already in existence is an irradiated sporozoite vaccine. Although it can protect people from infection for several months, it is impractical because it is effective only when given in the natural way by a mosquito bite. The cloning of major plasmodial surface and core antigens and mapping of the parasite's genome have provided new molecular tools for vaccines. It is believed that challenging the immune system with a spectrum of sporozoite and merozoite antigens will increase the chances of attacking most of the life stages of the parasite. Two of the most promising vaccines currently under development are a recombinant DNA version of the sporozoite surface antigens and one that contains the parasite's DNA.

Coccidian Parasites

The Order Coccidiorida of the Apicomplexa contains a number of members that are parasites of domestic mammals and birds. Several of them are also zoonotic and can cause severe disease in humans. These unusual protozoans exist in several stages, including dormant oocysts and pseudocysts.

Toxoplasma gondii* and Toxoplasmosis** ***Toxoplasma gondii* is an obligate apicomplexan parasite with such extensive cosmopolitan distribution that some experts estimate it affects the majority of the world's population at some time in their lives. In most of these cases, toxoplasmosis goes unnoticed, but disease in the fetus and in immunodeficient people, especially those with AIDS, is severe and often terminal. *Toxoplasma gondii* is a very successful parasite with so little host specificity that it can attack at least 200 species of birds and mammals. However, its primary reservoir and hosts are members of the feline family, both domestic and wild.

To follow the transmission of toxoplasmosis, we must first look at the general stages of *Toxoplasma*'s life cycle in the cat (figure 23.13*a*). The parasite undergoes a sexual phase in the intestine and is then released in feces, where it becomes an infective *oocyst* that survives in moist soil for several months. The parasite can complete its entire life cycle in the cat population. Ingested oocysts release an invasive asexual tissue phase called a tachyzoite that infects many different tissues and often causes disease in the cat. Eventually, these forms enter an asexual cyst state in tissues, called a pseudocyst. Most of the time, the parasite does not cycle in cats alone and is spread to intermediate hosts, usually rodents and birds. The cycle returns to cats when they eat these infected prey animals.

Other vertebrates become a part of this transmission cycle primarily through ingestion of cysts (figure 23.13*b*). Herbivorous animals such as cattle and sheep ingest oocysts that persist in the soil of grazing areas and then develop pseudocysts in their

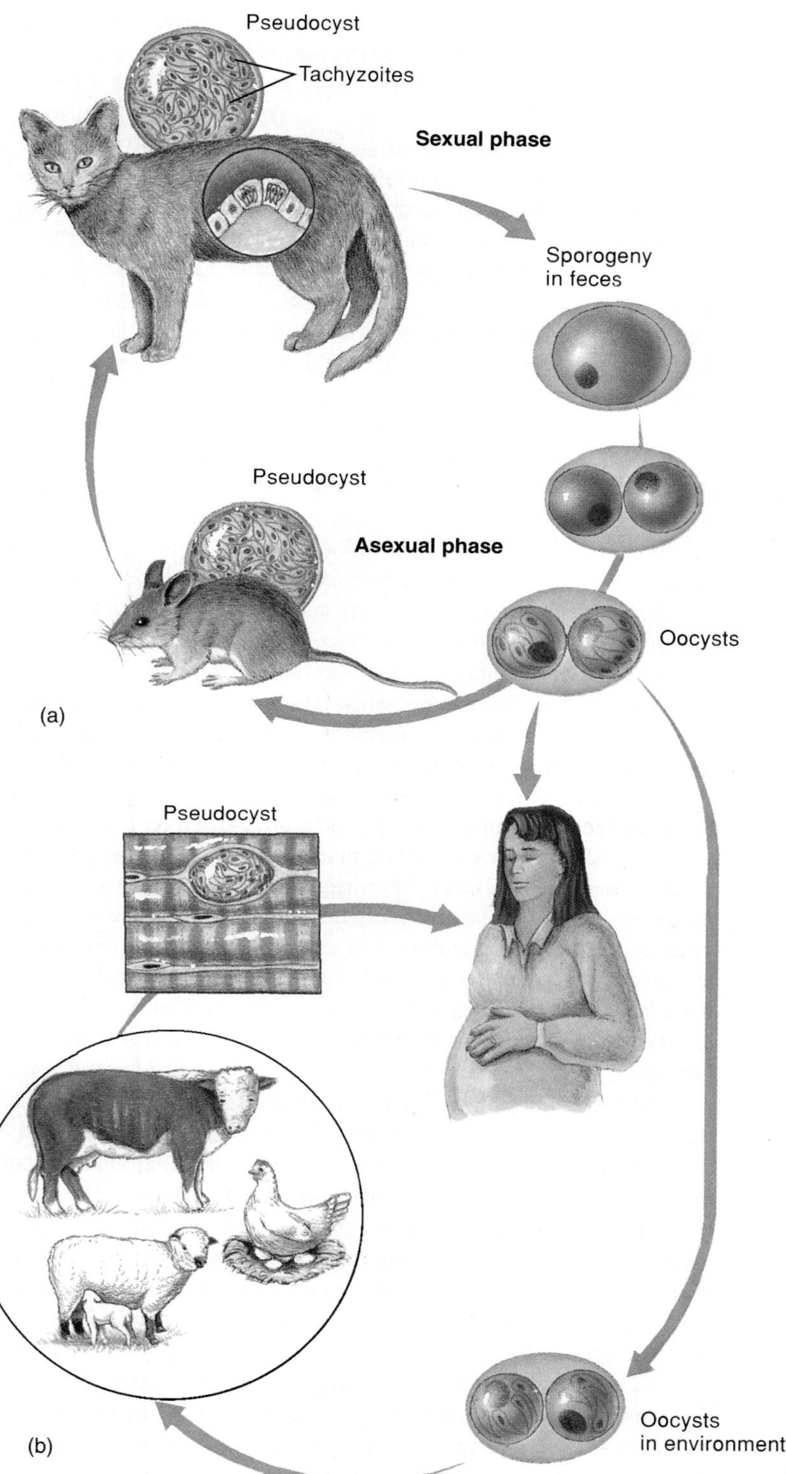

Figure 23.13
The life cycle and morphological forms of *Toxoplasma gondii*. (*a*) The cycle in cats and their prey. (*b*) The cycle in other animal hosts. The zoonosis has a large animal reservoir (domestic and wild) that becomes infected through contact with oocysts in the soil. Humans can be infected through contact with cats or ingestion of pseudocysts in animal flesh. Infection in pregnant women is a serious complication because of the potential damage to the fetus.

**Toxoplasma gondii* (toks″-oh-plaz′-mah gawn′-dee-eye). L. *toxicum*, poison, and Gr. *plasma*, molded; *gundi*, a small rodent.

muscles and other organs. Carnivores such as canines are infected by eating cystic tissues from other carrier animals.

Humans appear to be constantly exposed to the pathogen. The rate of prior infection, as detected through serological tests, can be as high as 90% in some populations. Many cases are caused by ingesting pseudocysts in contaminated meats. A common source is raw or undercooked meat, which is customary fare for certain ethnic groups, such as raw, spiced beef (steak tartare), raw pork (German hackepeter), and a Middle Eastern delicacy of ground lamb called kibbe. The grooming habits of cats spread fecal oocysts on their body surfaces, and unhygienic handling of them presents an opportunity to ingest oocysts. Infection can also occur when occysts are inhaled in air or dust contaminated with cat droppings and when tachyzoites cross the placenta to the fetus.

Most cases of toxoplasmosis are asymptomatic or marked by mild symptoms such as sore throat, lymph node enlargement, and low-grade fever. In patients whose immunity is suppressed by infection, cancer, or drugs, the outlook is grim. Chronic, persistent *Toxoplasma* infection can produce extensive brain lesions and can create fatal disruptions of the heart and lungs. A pregnant woman with toxoplasmosis has a 33% chance of transmitting the infection to her fetus. Congenital infection occurring in the first or second trimester is associated with stillbirth and severe abnormalities such as liver and spleen enlargement, liver failure, hydrocephalus, convulsions, and damage to the retina that can result in blindness.

Toxoplasmosis in AIDS and other immunocompromised patients can result from a primary acute or reactivated latent infection. The acute form is a disseminated, rapidly fatal disease with massive brain involvement. The reactivated form is the most common cause of encephalitis in AIDS patients. Trophozoites and cysts clustered in intracerebral lesions cause seizures, altered mental state, and coma (figure 23.14).

Most cases of toxoplasmosis elude diagnosis because the signs and symptoms mimic infectious mononucleosis or cytomegalovirus infections. Patient history, lymphadenopathy fever, and encephalopathy are helpful in evaluating adult infection. Several serological tests that detect anti-toxoplasma antibodies can be of some benefit, especially those for IgM, which appears early in infection. Disease can also be diagnosed by culture and histological analysis.

The most effective drugs are pyrimethamine and sulfadiazine together or in combination. Because they do not destroy the cyst stage, they must be given for long periods to prevent recurrent infection.

In view of the fact that the oocysts are so widespread and resistant, hygiene is of paramount importance in controlling toxoplasmosis. There is no such thing as a safe form of raw meat, even salted or spiced. Adequate cooking or freezing below −20°C destroys both oocysts and tissue cysts. Oocysts can also be avoided by washing the hands after handling cats or soil that is possibly contaminated with cat feces, especially sandboxes and litter boxes. Pregnant women should be especially attentive to these rules and should never clean out the cat's litter box.

***Sarcocystis* and Sarcocystosis** Although *Sarcocystis* are parasites of cattle, swine, and sheep, some species have an interesting transmission cycle that involves passage through humans as the final host. Domestic animals are intermediate hosts that pick up infective cysts while grazing on grass contaminated by human excrement. Humans are subsequently infected when they eat beef, pork, or lamb containing tissue cysts, but human-to-human transmission does not occur. Sarcocystosis is an uncommon infection mostly caused by ingesting uncooked meats. Initial symptoms of intestinal sarcocystosis are diarrhea, nausea, and abdominal pain that peak in a few hours after ingestion but can persist for about 2 weeks. There is no specific treatment at the present time.

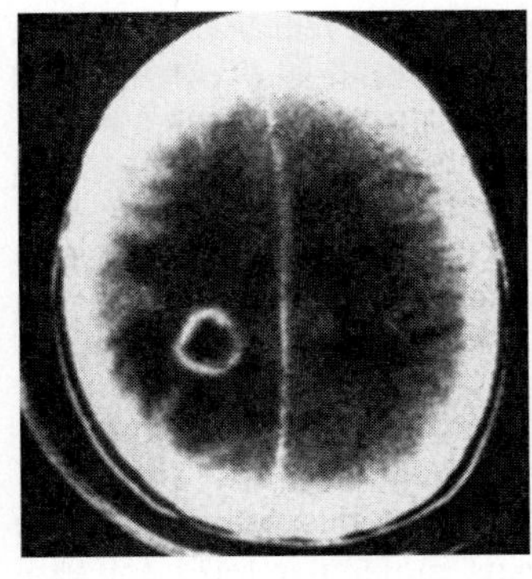
(a)

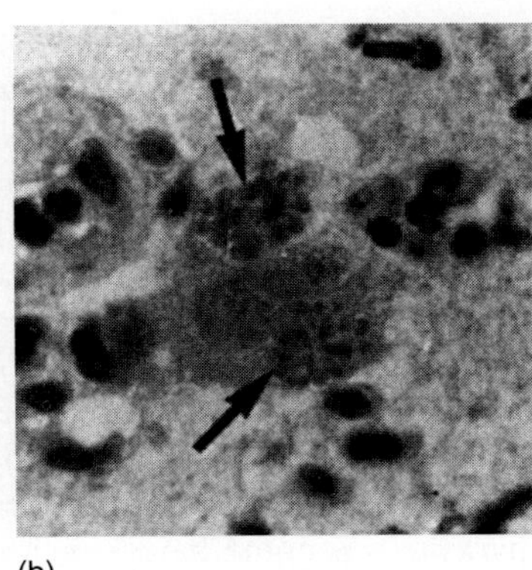
(b)

Figure 23.14
Toxoplasmosis in the brain of an AIDS patient. (*a*) A CT scan of the brain shows a contrast-enhanced image of a cerebral lesion. (*b*) Biopsy tissue stained for *Toxoplasma gondii* indicates clusters of intracellular tachyzoites (arrows).
Courtesy of Dr. Henry W. Muray. From Harrison's Principles of Internal Medicine, *12th ed., McGraw-Hill, Inc. Reproduced with permission of the publisher.*

Cryptosporidium*: A Newly Recognized Intestinal Pathogen** ***Cryptosporidium* is an intestinal apicomplexan that infects a variety of mammals, birds, and reptiles. For many years, **cryptosporidiosis** was considered an intestinal ailment exclusive to calves (neonatal scours), pigs, chickens, and other poultry, but it is clearly a zoonosis as well. The organism exhibits a life cycle similar to that of *Toxoplasma,* with a hardy intestinal oocyst and a tissue phase. In transmission pattern it is very similar to *Giardia.*

Cryptosporidiosis has cosmopolitan distribution. Its highest prevalence is in areas with unreliable water and food sanitation. The carrier state occurs in 3% to 30% of the population in developing countries. The susceptibility of the general public to this pathogen has been amply demonstrated by several large-scale epidemics. In 1993, 370,000 people developed *Cryptosporidium* gastroenteritis from the municipal water supply in Milwaukee, Wisconsin! Other mass outbreaks of this sort have been traced to contamination of the local water reservoir by livestock wastes. Other studies revealed that at least one-third of all fresh surface waters harbor this parasite. Chlorination is not entirely successful in eradicating the cysts, so most treatment plants use filtration to remove them, but even this method can fail.

Infection is initiated by ingestion of oocysts. These give rise to sporozoites that penetrate the intestinal cells and live intracellularly in them. Most cases of infection in healthy people manifest few if any symptoms. The prominent symptoms in patients with

**Cryptosporidium* (krip″-toh-spo-rid′-ee-um). Gr. *cryptos,* hidden, and *sporos,* seed.

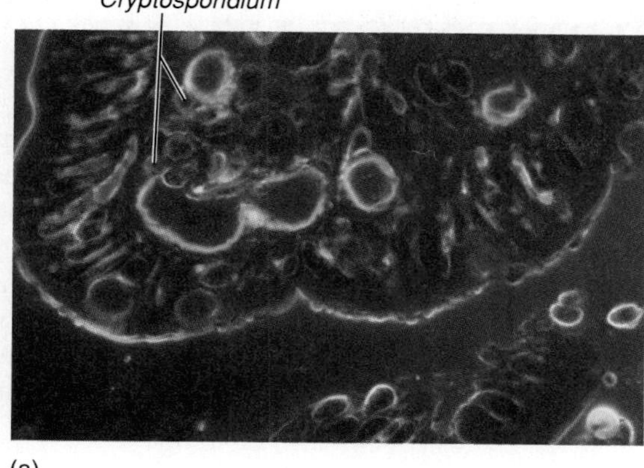

(a)

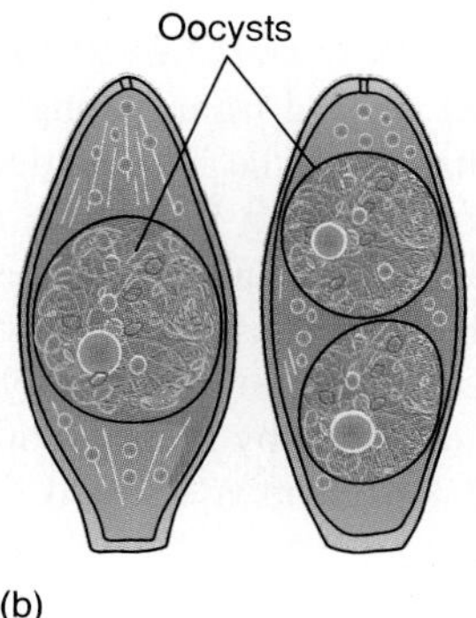

(b)

Figure 23.15

Other apicomplexan parasites. (*a*) An electron micrograph of a *Cryptosporidium* merozoite penetrating an intestinal cell. (*b*) *Isopora belli,* showing oocysts in two stages of maturation.

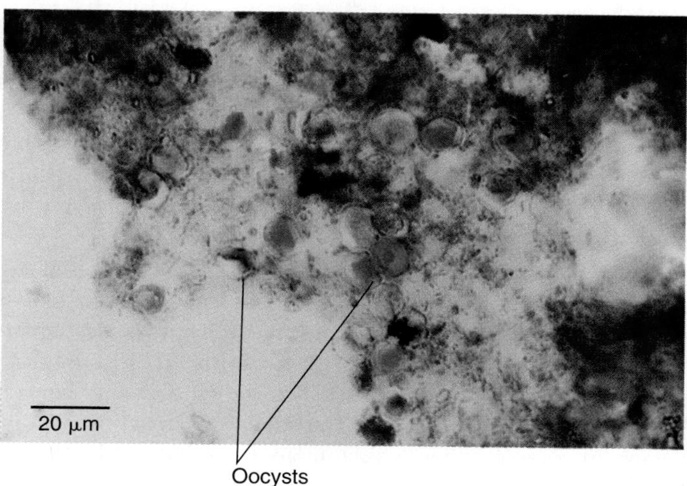

Figure 23.16

An acid-fast stain of *Cyclospora* in a human fecal sample. The large (8–10 μm) cysts stain pink to red and have a wrinkled outer wall. Blue cells in the background are normal intestinal bacteria.

clinical disease mimic other types of gastroenteritis, with headache, sweating, vomiting, severe abdominal cramps, and diarrhea. AIDS patients may experience massive infection of the intestine and uncontrollable diarrhea. Chronic persistent cryptosporidial diarrhea, one criterion for diagnosing AIDS, is a frequent cause of death in these patients. The agent can be detected in fecal samples with indirect immunofluorescence and by acid-fast staining of biopsy tissues (figure 23.15*a*). Treatment is presently hampered by the lack of effective drugs and the tendency of immunodeficient patients to suffer relapses.

Isospora belli* and Coccidiosis** *Isospora* is an intracellular intestinal parasite with a distinctive oocyst stage (figure 23.15*b*). Of several species that colonize mammals, apparently only ***Isospora belli* infects humans. The parasite is transmitted in fecally contaminated food or drink and through laboratory contact. Not much is known about the incidence of infection because it is usually asymptomatic or self-limited. The more prominent symptoms are malaise, nausea and vomiting, diarrhea, fatty stools, abdominal colic, and weight loss. In cases requiring treatment, combined therapy with sulfadiazine and pyrimethamine is usually effective.

Cyclospora cayetanensis* and Cyclosporiasis** ***Cyclospora cayetanensis is an emerging protozoan pathogen related to *Isospora.* The first human cases of infection with this protozoan were seen in 1979, when it was described as a cyanobacterium- or coccidia-like body in specimens. Since that time, hundreds of outbreaks have been reported in the United States and Canada. Its mode of transmission is oral-fecal, and most cases have been associated with consumption of fresh produce and water, presumably contaminated with feces. Cyclosporiasis occurs worldwide, and, although primarily of human origin, it is not spread directly from person to person. One common source of infection in several outbreaks has been raspberries imported from Guatemala. It is not known whether these fruits became contaminated in the field or during handling, but normal washing of the berries did not prevent transmission. Other cases were traced to a salad made with fresh greens and drinking water. The parasite has also been identified as a significant cause of diarrhea in travelers.

The disease is acquired when oocysts enter the small intestine and invade the mucosa. After an incubation period of about one week, symptoms of watery diarrhea, stomach cramps, bloating, fever, and muscle aches appear. Patients with prolonged diarrheal illness experience anorexia and weight loss.

Diagnosis can be complicated by the lack of recognizable oocysts in the feces. Techniques that improve identification of the parasite are examination of fresh preparations under a fluorescent microscope and an acid-fast stain of a processed stool specimen. Large wrinkled oocysts are typical of the species (figure 23.16). A PCR-based DNA test can also be used to identify *Cyclospora* and differentiate it from other coccidian parasites. This form of analysis is more sensitive and can detect protozoan genetic material even in the absence of cysts. Most cases of infection have been effectively controlled with a combination of trimethoprim and sulfamethoxazole (Bactrim, Septra) for a week. Traditional antiprotozoan drugs are not effective. Some cases of disease may be prevented by cooking or freezing food to kill the oocysts.

***Babesia* Species and Babesiosis** Babesiosis, or piroplasmosis, is historically important for two firsts in microbiology.

**Isospora belli* (eye-saus′-poh-rah bel′-eye) Gr. *iso,* equal, and *sporos,* seed. L. *bellum,* war.

Babesia* was the first protozoan found to be responsible for a disease—specifically, redwater fever.[3] This is an acute febrile hemolytic disease of cattle characterized by the destruction of erythrocytes and the presence of hemoglobin in the urine, hence its name. The disease was also the first one associated with an arthropod vector—a tick. Although the morphology and life cycle of *Babesia* are not commonly understood, they are probably like those of the plasmodia.

Human babesiosis is a relatively rare zoonosis caused by a species of *Babesia* from wild rodents that has been accidentally introduced by a hard tick's bite. This infection has been essentially eradicated in the United States, but it remains a problem in various other regions. Most cases occurred in the Northeast, where the *Ixodes* tick vector is located. The infection closely resembles malaria in pathology and symptomology, in that red blood cells burst after they are invaded, and it is diagnosed and treated in a similar fashion.

Chapter Checkpoints

- The apicomplexans are obligate intracellular parasites that lack motility in the mature state. *Plasmodium* and *Babesia* cause serious systemic infections transmitted by mosquitos and ticks. *Toxoplasma* causes mild, systemic infections in healthy hosts. *Cryptosporidium* causes intestinal infections.
- *Plasmodium* species are the causative agents of malaria, the major protozoan disease worldwide. The *Plasmodium* life cycle is timed to coincide with the reproductive cycle of female *Anopheles* mosquitos, which require human blood for the development of their eggs. They transmit *Plasmodium* sporozoites, which complete their development in the liver, invade red blood cells as merozoites, and then differentiate into reproductive gametes. Sexual reproduction is completed in the mosquito. Symptoms are caused by the damage to red blood cells. Prevention consists of mosquito abatement procedures and human prophylaxis. Quinine derivatives are the treatment of choice, but resistant strains are increasing.
- *Toxoplasma gondii* causes a mild, systemic infection of healthy hosts, but fatal infections can occur transplacentally and in the immunosuppressed. Humans are accidental hosts; cats are the primary reservoir. Infection is associated with ingestion of cysts.
- *Cryptosporidium* species cause an intestinal infection in both domestic animals and humans. Infection occurs through contact with infected animal feces, primarily through water supplies contaminated with cysts. Healthy hosts recover spontaneously, but the infection is potentially fatal to immunosuppressed hosts.

A SURVEY OF HELMINTH PARASITES

So far our survey has been limited to relatively small infectious agents, but *helminths,** particularly the adults, are comparatively large, multicellular animals with specialized tissues and organs similar to the hosts they parasitize. The behavior of these parasites and their adaptations to humans are at once fascinating and grotesque. Imagine having long, squirming white worms suddenly expelled from your nose, feeling and seeing a worm creeping beneath your skin, or having a worm slither across the front of your eye, and you have some idea of the conspicuous nature of these parasites.

Worms that parasitize humans are amazingly diverse, ranging from barely visible roundworms (0.3 mm) to huge tapeworms 25 meters long and varying from oblong- and crescent-shaped to whip- or ribbon-shaped. In the introduction to these organisms in chapter 5, we grouped them into three categories: nematodes (roundworms), trematodes (flukes), and cestodes (tapeworms) and discussed basic characteristics of each group. In this section, we present major facets of helminth epidemiology, life cycles, pathology, and control, followed by an accounting of the major worm diseases. Table 23.4 presents a selected list of medically significant helminths.

GENERAL LIFE AND TRANSMISSION CYCLES

The complete life cycle of helminths includes the fertilized egg (embryo), larval, and adult stages. In the majority of helminths, adults derive nutrients and reproduce sexually in a host's body. In nematodes, the sexes are separate and usually different in appearance; in trematodes, the sexes are separate or **hermaphroditic** meaning that male and female sex organs are in the same worm; cestodes are generally hermaphroditic. For a parasite's continued survival as a species, it must complete the life cycle by transmitting an infective form, usually an egg or larva, to the body of another host, either of the same or a different species. By convention, the host in which larval development occurs is the **intermediate (secondary) host,** and adulthood and mating occur in the **definitive (final) host.** A **transport host** is an intermediate host that experiences no parasitic development but is an essential link in the completion of the cycle.

Although individual variations occur in life and transmission cycles, most can be described by one of five basic patterns (figure 23.17). In general, sources for human infection are contaminated food, soil, and water or infected animals, and routes of infection are by oral intake or penetration of unbroken skin. Humans are the definitive hosts for many of the parasites listed in table 23.4, and in about half the diseases, they are also the sole biological reservoir. In other cases, animals or insect vectors serve as reservoirs or are required to complete worm development. In the majority of helminth infections, the worms do not complete the

3. Also known as bovine babesiosis. Texas fever, or tick fever.

**Babesia* (bab-ee′-see″-uh) Named for its discoverer, Victor Babes.

*helminth Gr. *helmins,* worm.

TABLE 23.4

MAJOR HELMINTHS OF HUMANS AND THEIR MODES OF TRANSMISSION

Classification	Common Name of Disease or Worm	Life Cycle Requirement	Spread to Humans By
Nematodes (Roundworms)			
Intestinal Nematodes			
Infective in egg (embryo) stage			Ingestion
Trichuris trichiura	Whipworm	Humans	Fecal pollution of soil with eggs
Ascaris lumbricoides	Ascariasis	Humans	Fecal pollution of soil with eggs
Enterobius vermicularis	Pinworm	Humans	Close contact
Infective in larval stage			
Necator americanus	New World hookworm	Humans	Fecal pollution of soil with eggs
Ancylostoma duodenale	Old World hookworm	Humans	Fecal pollution of soil with eggs
Strongyloides stercoralis	Threadworm	Humans; may live free	Fecal pollution of soil with eggs
Trichinella spiralis	Trichina worm	Pigs, wild mammals	Consumption of meat containing larvae
Tissue Nematodes			Burrowing of larva into tissue
Wuchereria bancrofti	Filarial worm	Humans, mosquitos	Mosquito bite
Onchocerca volvulus	River blindness	Humans, black flies	Fly bite
Loa loa	Eye worm	Humans, mangrove flies or deer flies	Fly bite
Dracunculus medinensis	Guinea worm	Humans and *Cyclops* (an aquatic invertebrate)	Ingestion of water containing *Cyclops*
Trematodes			
Schistosoma japonicum	Blood fluke	Humans and snails	Ingestion of fresh water containing larval stage
S. mansoni	Blood fluke	Humans and snails	Ingestion of fresh water containing larval stage
S. haematobium	Blood fluke	Humans and snails	Ingestion of fresh water containing larval stage
Opisthorchis sinensis	Chinese liver fluke	Humans, snails, fish	Consumption of fish
Fasciola hepatica	Sheep liver fluke	Herbivores (sheep, cattle)	Consumption of water and water plants
Paragonimus westermani	Oriental lung fluke	Humans, mammals, snails, freshwater crabs	Consumption of crabs
Cestodes			
Taenia saginata	Beef tapeworm	Humans, cattle	Consumption of undercooked or raw beef
T. solium	Pork tapeworm	Humans, swine	Consumption of undercooked or raw pork
Diphyllobothrium latum	Fish tapeworm	Humans, fish	Consumption of undercooked or raw fish
Hymenolepis nana	Dwarf tapeworm	Humans	Oral-fecal; close contact

entire life cycle without leaving the host. Unlike microparasites, adult helminths do not continue to multiply in the host, so a more accurate term for their presence in the host is **infestation.**

GENERAL EPIDEMIOLOGY OF HELMINTH DISEASES

The helminths have a far-reaching impact on human health, economics, and culture. It has been said that there are probably more worm infections in humans than there are humans (due to multiple infections in one individual). Although individuals from all societies and regions play host to worms at some time in their lives, the highest case rates occur among children in rural areas of the tropics and subtropics. The most prevalent diseases and those creating the highest mortality and morbidity are schistosomiasis, filariasis, hookworm disease, ascariasis, and onchocerciasis. In most developed countries in temperature climates, the major parasitic infections are pinworm and trichinosis, though exotic parasitic worms are often encountered in immigrants and travelers. The continuing cycle between parasites and humans is often the result of practices such as using human excrement for fertilizer, exposing bare feet to soil or standing water, eating undercooked or raw meat and fish, and defecating in open soil. In some areas of the world parasites have been an accepted part of life for thousands of years.

PATHOLOGY OF HELMINTH INFESTATION

Worms exhibit numerous adaptations to the parasitic habit. They have specialized mouthparts for attaching to tissues and for feeding, enzymes with which they liquefy and penetrate tissues, and a cuticle or other covering to protect them from the host defenses. As we learned in chapter 5, their organ systems are usually reduced to the essentials; getting food and processing it, moving, and reproducing. The signs and symptoms of infestation depend

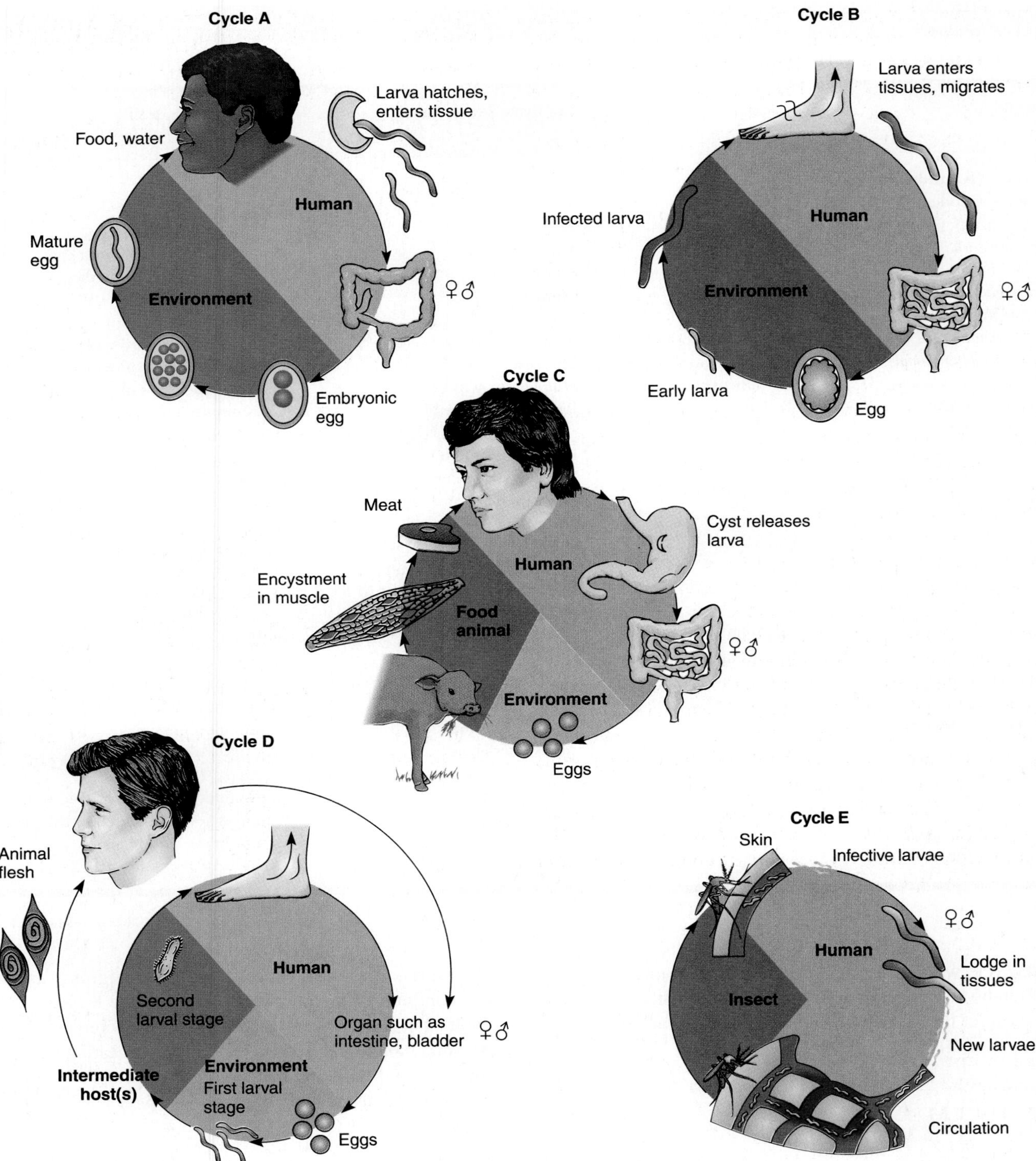

Figure 23.17

Five basic helminth life and transmission cycles. In cycle A, the worm develops in intestine; egg is released with feces into environment; eggs are ingested by new host and hatch in intestine (examples: *Ascaris, Trichuris).* In cycle B, the worm matures in intestine; eggs released with feces; larvae hatch and develop in environment; infection occurs through skin penetration by larvae (example: hookworms). In cycle C, the adult matures in human intestine; eggs are released into environment; eggs are eaten by grazing animals; larval forms encyst in tissue; humans eating animal flesh are infected (example: *Taenia).* In cycle D, eggs are released from human; humans are infected through ingestion or direct penetration by larval phase (examples: *Opisthorchis* and *Schistosoma).* In cycle E, the human is definitive host and carries larval form in blood; insect vector is intermediate host that picks up and transmits larvae through bite (examples: *Wuchereria* and *Onchocerca).*

partly upon the portal of entry. Helminths entering by mouth undergo a period of development in the intestine that is often followed by spread to other organs. Helminths that burrow directly into the skin or are inserted by an insect bite quickly penetrate the circulation and are carried systemically. In both cases, most worms do not remain in a single location, and they migrate by various mechanisms through major systems, inflicting injury as they go.

Adult helminths eventually take up final residence in sites such as the intestinal mucosa, blood vessels, lymphatics, subcutaneous tissue, skin, liver, lungs, muscles, brain, or even the eyes. Traumatic damage occurring in the course of migration and organ invasion is primarily due to tissue feeding, blocked ducts and organs, toxic secretions, and pressure within organs due to high worm load. Pathological effects can include enlargement and swelling of organs, hemorrhage, weight loss, and anemia caused by the parasites' feeding on blood. Despite the popular misconception that intestinal worms cause weight loss by competing for food (the person who can eat everything but never gain weight must have a tapeworm!), weight loss is more likely caused by general malnutrition, malabsorption of food and vitamins due to intestinal damage, lack of appetite, vomiting, and diarrhea.

The mere presence of a parasite alone does not signify disease. In fact, some people can coexist comfortably with parasites and experience few symptoms. More severe manifestations are the rule if the infectious load is high and the health of the patient is poor. One common antihelminthic defense is an increase in granular leukocytes called eosinophils that have a specialized capacity to destroy worms (see chapter 14). Although antibodies and sensitized T cells are also formed, their effects are rarely long-term, and reinfection is always a distinct possibility. The inability of the immune system to completely eliminate worm infections appears to be due to their relatively large size, migratory habits, and inaccessibility. The incompleteness of immunity has hampered the development of vaccines for parasitic diseases and has emphasized the use of chemotherapeutic agents instead.

Table 23.5

Antihelminthic Therapeutic Agents and Indications for Their Use

Drug	Effect	Used For
Piperazine	Paralyzes worm so it can be expelled in feces	Ascariasis, pinworm, hookworm
Pyrantel	Paralyzes worm so it can be expelled in feces	Ascariasis, pinworm, hookworm, trichinosis
Mebendazole	Blocks key step in worm metabolism	Trichuriasis (whipworm), ascariasis, hookworm, trichinosis
Thiabendazole	Blocks key step in worm metabolism	Strongyloidiasis, guinea worm, hookworm, trichinosis
Albendazole	Blocks key step in worm metabolism	Echinococciasis
Diethylcarbamazine	Unknown, but kills microfilariae; does not kill adult worm	Filariasis, loiasis
Niridazole	Alters worm metabolism	Schistosomiasis
Praziquantel	Interferes with worm metabolism	Schistosomiasis, other flukes, tapeworm
Metrifonate	Paralyzes worm	Schistosomiasis
Niclosamide	Inhibits ATP formation in worm; destroys proglottids but not eggs	Tapeworm
Ivermectin	Blocks nerve transmission	Onchocerciasis

Elements of Diagnosis and Control

Diagnosis of helminthic infestation may require several steps. A differential blood count showing eosinophilia and serological tests indicating sensitivity to helminths both provide indirect evidence of worm infestation. A history of travel to the tropics or immigration from those regions is also helpful, even if it occurred years ago, because some flukes and nematodes persist for decades. The most definitive evidence, however, is the discovery of eggs, larvae, or adult worms in stools, sputum, urine, gastric washings, blood, or tissue biopsies. The worms are sufficiently distinct in morphology that positive identification can be based on any stage, including eggs.

Although several useful antihelminthic medications exist, the cellular physiology of the eucaryotic parasites resembles that of humans, and drugs toxic to them can also be toxic to humans. Some antihelminthic drugs act to suppress a metabolic process that is more important to the worm than to the human. Others inhibit the worm's movement and prevent it from maintaining its position in a certain organ. Therapy is also based on a drug's greater toxicity to the relatively smaller and more vulnerable helminths or on the local effects of oral drugs in the intestine. Antihelminthic drugs of choice and their effects are given in table 23.5. In some cases, surgery may be necessary to remove worms or larvae, though this can be difficult if the parasite load is high or not confined.

Preventive measures are aimed at minimizing human contact with the parasite or interrupting its life cycle. In areas where the worm is transmitted by fecally contaminated soil, disease rates are significantly reduced through proper sewage disposal, using sanitary latrines, avoiding human feces as fertilizer, and wearing shoes. If the agent is transmitted by drinking water, disinfection of the water supply by heat or chemicals can be effective. If the larvae are aquatic and invade through the skin, people should avoid direct contact with infested water. Food-borne disease can be avoided by thoroughly washing and cooking vegetables and meats. Also, because adult worms, larvae, and eggs are sensitive to

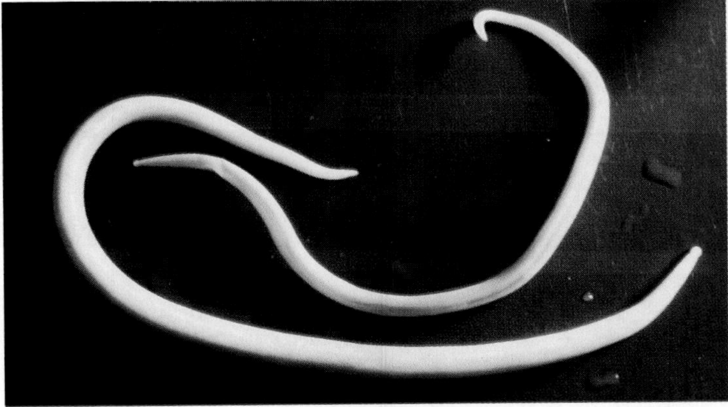

(a)

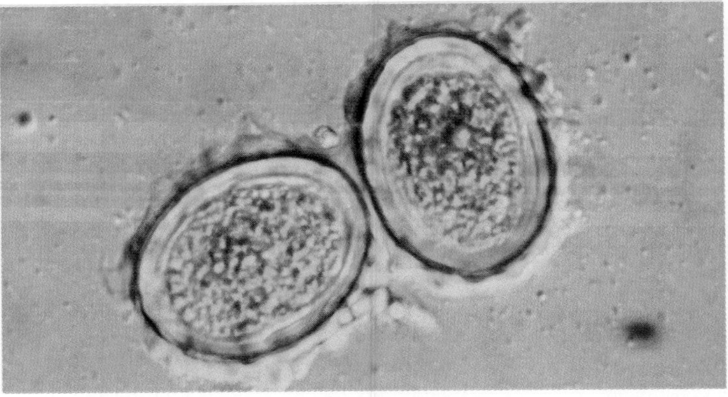

(b)

(c)

Figure 23.18

Ascaris lumbricoides. (*a*) Adult worms. The female is larger; the male has a hooked end. (*b*) A microscopic view of *A. lumbricoides* ova from a fecal sample. (*c*) A mass of 800 worms retrieved from the ileum of a child during an autopsy.

(c) From Rubin and Farbo, Pathology. *Reprinted by permission of J. B. Lippincott Company.*

cold, freezing foods is a highly satisfactory preventive measure. These methods work best if humans are the sole host of the parasite; if they are not, control of reservoirs or vector populations may be necessary.

Nematode (Roundworm) Infestations

Nematodes* are elongate, cylindrical worms that biologists consider one of the most abundant animal groups (a cup of soil or mud can easily contain millions of them). The vast majority are free-living soil and freshwater worms, but a few species are parasitic, of which 50 affect humans. Nematodes bear a smooth, protective outer covering, or cuticle, that is periodically shed as the worms grows. They are essentially headless, tapering to a fine point anteriorly. The sexes are distinct, and the male is often smaller than the female, having a hooked posterior containing spicules for coupling with the female (figure 23.18*a*).

The human parasites are divided into intestinal nematodes, which develop to some degree in the intestine (for example, *Ascaris, Strongyloides,* and hookworms), and tissue nematodes, which spend their larval and adult phases in various soft tissues other than the intestine (for example, various filarial worms).

Intestinal Nematodes (Cycle A)

Ascaris lumbricoides **and Ascariasis** *Ascaris lumbricoides** is a giant (up to 300 mm long) intestinal roundworm that probably causes more infestations than any other worm (estimated at one billion cases worldwide). Although most cases in the United States probably go unreported, the highest number is centered in the southeastern states. *Ascaris* spends its larval and adult stages in humans and releases embryonic eggs in feces, which are then spread to other humans through food, drink, or soiled objects placed in the mouth (figure 23.18). The eggs thrive in warm, moist soils; resist cold and chemical disinfectants; and are killed by sunlight, high temperatures, and drying. After ingested eggs hatch in the human intestine, the larvae embark upon an odyssey in the tissues. First they penetrate the intestinal wall and enter its lymphatic and circulatory drainage, feeding and growing as they go. Swept along by the bloodstream into the

*nematode (nem′-ah-tohd) Gr. *nema,* thread.

**Ascaris lumbricoides* (as′-kah-ris lum″-brih-koy′-dees) Gr. *askaris,* an intestinal worm; L. *lumbricus,* earthworm, and Gr. *eidos,* resemblance.

right ventricle of the heart, they eventually arrive at the capillaries of the lungs. From this point, the larvae migrate up the respiratory tree to the glottis. Worms entering the throat are swallowed and returned to the small intestine, where they reach adulthood and reproduce, producing up to 200,000 fertilized eggs a day.

Even as adults, male and female worms are not attached to the intestine and retain some of their exploratory ways. They are known to invade the biliary channels of the liver and gallbladder, and on occasion the worms emerge from the mouth and nares. Severe inflammatory reactions mark the migratory route, and allergic reactions such as bronchospasm, asthma, or skin rash can occur. Heavy worm loads can retard the physical and mental development of young, malnourished children. One possibility with intestinal worm infestations is self-reinoculation due to poor personal hygiene.

Trichuris trichiura* and Whipworm Infection** The common name for ***Trichuris trichiura*—whipworm—refers to its likeness to a miniature buggy whip. Although the worm is morphologically different from *Ascaris,* it has a similar transmission cycle, with humans as the sole host. Trichuriasis has a high incidence in the tropics and subtropics, where sanitation of feces may be lax and children sometimes ingest dirt. Embryonic eggs deposited in the soil are not immediately infective and must first develop for 3 to 6 weeks in moist, warm soil. Ingested eggs then hatch in the small intestine, where the larvae attach, penetrate the outer wall, and go through several molts. The mature adults move to the large intestine and gain a hold with their long, thin heads, while the thicker tail dangles free in the lumen. Following sexual maturation and fertilization, the females eventually lay 3,000 to 5,000 eggs daily into the bowel. The entire cycle requires about 90 days, and untreated infestation can last up to 2 years.

Worms burrowing into the intestinal mucosa can pierce capillaries, cause localized hemorrhage, and provide a portal of entry for secondary bacterial infection. Light infestation can be asymptomatic, but heavier infestations can cause dysentery, loss of muscle tone, and rectal prolapse, which can prove fatal in children.

Enterobius vermicularis* and Pinworm Infection** The pinworm or seatworm, ***Enterobius vermicularis,* was first discussed in chapter 5 (see figure 5.36). Enterobiasis is the most common worm disease of children in temperate zones. The transmission of the pinworm follows a course similar to that of the other enteric parasites. Freshly deposited eggs have a sticky coating that causes them to lodge beneath the fingernails during scratching and to adhere to fomites. Upon drying, the eggs are readily swept by air currents and settle in house dust. Worms are ingested from contaminated food and drink, and self-inoculation from one's own fingers is common. Eggs hatching in the small intestine release larvae, which migrate to the large intestine, mature, and then mate. Itching is brought on when the mature female emerges from the anus and lays eggs. Although infestation is not fatal and most cases are asymptomatic, the afflicted child can be irritable, nervous, and pallid from lost or disrupted sleep and can sometimes suffer nausea, abdominal discomfort, and diarrhea. A simple, rapid test can be done by pressing a piece of transparent adhesive tape against the anal skin and then applying it to a slide for microscopic evaluation.

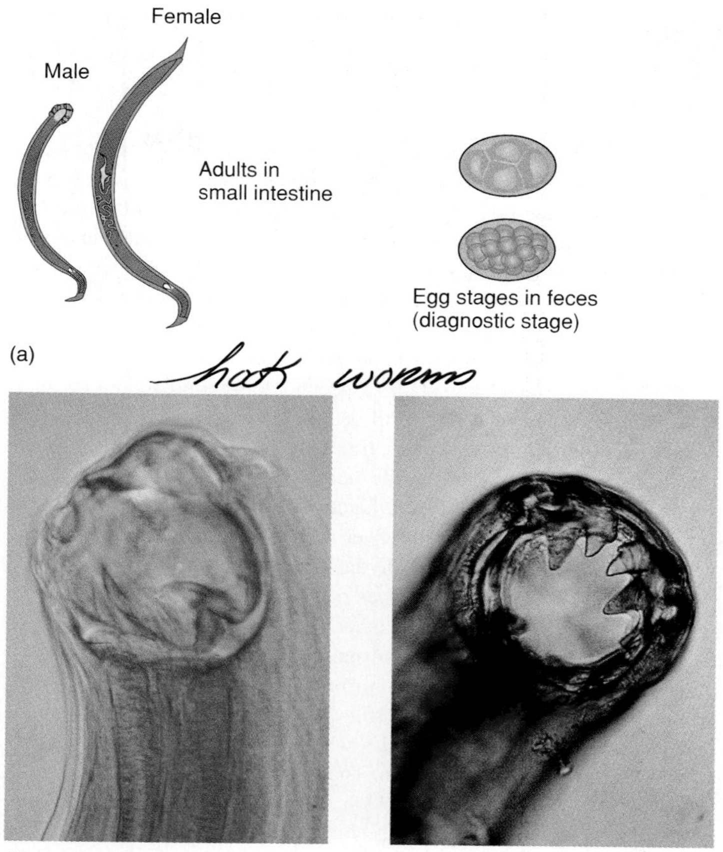

Figure 23.19
(a) Adult and egg stages of *Necator americanus. (b)* Cutting teeth on the mouths of (left) *N. americanus* and (right) *Ancylostoma duodenale.*

Intestinal Helminths (Cycle B)

The Hookworms Two major agents of human hookworm infestation are ***Necator americanus,**** endemic to the New World, and ***Ancylostoma duodenale,**** endemic to the Old World, though the two species overlap in parts of Latin America. Otherwise, with respect to transmission, life cycle, and pathology, they are usually lumped together. Humans are sometimes invaded by cat or dog hookworm larvae, sustaining a dermatitis called creeping eruption (microfile 23.2). The *hook* refers to the adult's oral cutting plates by which it anchors to the intestinal villi and its curved anterior end (figure 23.19). Unlike infection by other intestinal worms, hookworm infection is acquired when larvae hatched outside the body penetrate the skin.

*Trichursi (trik-ur'-is) Gr. *thrix,* hair, and *oura,* tail.

*Enterobius vermicularis (en"-ter-oh'-bee-us ver-mik"-yoo-lah'-ris) Gr. *enteron,* intestine, and *bios,* life; L. *vermiculus,* a small worm.

*Necator americanus (nee-kay'-tor ah-mer"-ih-cah'-nus) L. *necator,* a murderer, and *americanus,* of New World origin.

*Ancylostoma duodenale (an"-kih-los'-toh-mah doo-oh-den-ah'-lee) Gr. *amkylos,* curved or hooked, *stoma,* mouth, and *duodenale,* the intestine.

MICROFILE 23.2 WHEN PARASITES ATTACK THE "WRONG" HOST

Not only do we humans host a number of our own parasites, but occasionally, worms indigenous to other animals attempt to take up residence in us. When this happens, the parasite cannot complete the life cycle in its usual fashion, but in making the attempt, it can cause discomfort and even death in some cases.

One such parasite is *Echinococcus granulosus,* a tiny intestinal tapeworm that causes **hydatid cyst disease.** The parasite's usual definitive host is the dog, though it can be transmitted to other carnivores and herbivores in egg and larval stages. Sheepherders are prone to it because the parasite cycles back and forth between sheepdogs and sheep. Humans become involved through frequent, intimate contact with fur, paws, and tongue, especially while "kissing" their dog. If a person accidentally ingests eggs, the larvae hatch and migrate through the body, eventually coming to rest in the liver and lungs. There they form compact, water-filled packets called **hydatid cysts** (L. *hydatos,* water drop). The bursting of these capsules gives rise to an allergic reaction and also proliferates the worm by seeding other sites.

Another disease that arises from invasion by animal worms is **cutaneous larval migrans,** also described as a creeping eruption caused by both dog and cat hookworm *(Ancylostoma).* It is acquired when humans walk barefoot through areas where dogs and cats defecate, enabling the larvae to enter the skin. The larvae cannot invade completely and so remain just beneath the skin, where they meander, producing dark red tunnels. These lesions become so itchy and inflamed that the person cannot eat or sleep and can contract painful secondary infections. Another migratory infection called **visceral larval migrans** is caused by the dog ascarid *Toxocara* that is ingested in egg form with contaminated soil. The larvae hatch in the intestine and migrate extensively through the liver, lungs, and circulation.

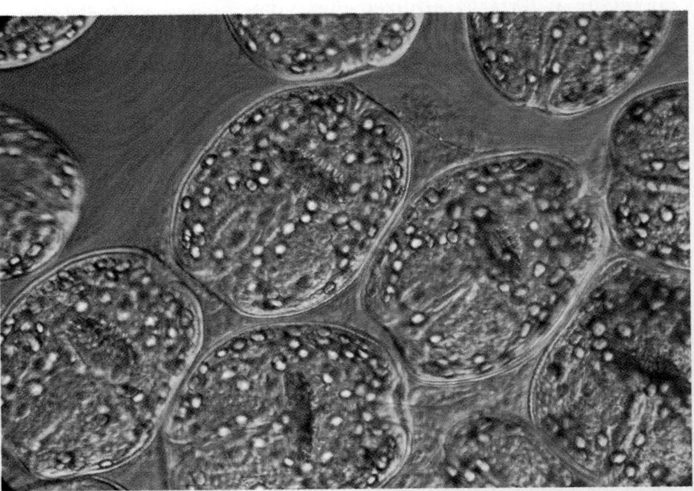

Hydatid cysts from the lung. These fluid-filled spheres are also known as bladder worms.

Likewise, "swimmer's itch" results when larval schistosomes of birds and small mammals penetrate a short distance into the skin and then die. The disintegrating worm sensitizes the humans and causes intense itching and discomfort.

Hookworm infestation remains endemic to the tropics and subtropics and is absent in arid, cold regions. A generation ago, it outranked all other parasitic diseases for morbidity (WHO estimated about 25% of the world's population was affected). Since that time, it has decreased in incidence, especially in the United States. At one time, hookworm plagued the rural South, unfortunately sapping its victims' strength and stigmatizing them as unkempt, irresponsible, and ignorant. The decline of the disease in these regions was due to such simple changes as improved sanitation (indoor plumbing) and generally better standards of living (such as being able to afford shoes).

After being released into warm, moist soil with feces, hookworm eggs hatch into fine *filariform** larvae that instinctively climb onto grass or other vegetation to improve the probability of contact with a host. Ordinarily, the parasite enters sites on bare feet such as hair follicles, abrasions, or the soft skin between the toes, but cases have occurred in which mud splattered on the ankles of people wearing shoes. Infection has even been reported in people handling soiled laundry.

On contact, the hookworm larvae actively burrow into the skin. After several hours they reach the lymphatic or blood circulation and are immediately carried into the heart and lungs. The larvae infiltrate the air spaces and proceed up the bronchi and trachea to the throat. Although some larvae are ejected by spitting, most are swallowed with sputum and arrive in the small intestine, where they anchor, feed on tissues, and mature. Eggs first appear in the stool about 6 weeks from the time of entry, and the untreated infestation can last about 5 years.

The symptoms of infestation correspond to the migratory and feeding activities of the hookworm. Initial penetration with larvae can evoke a localized dermatitis called "ground itch." The transit of the larvae to the lungs is ordinarily brief, but it can cause symptoms of pneumonia and eosinophilia. The potential for injury is greatest during the intestinal phase, when heavy worm burdens can cause nausea, vomiting, cramps, pain, and bloody diarrhea. Because blood loss is significant, iron-deficient anemia develops, and infants are especially susceptible to hemorrhagic shock. Chronic fatigue, listlessness, apathy, and anemia worsen with chronic and repeated infections.

Strongyloides stercoralis* and Strongyloidiasis** The agent of strongyloidiasis, or threadworm infection, is ***Strongyloides stercoralis.* This is an exceptional human nematode because of its

*filariform (fih-lar′-ih-form) L. *filum,* a thread.

**Strongyloides stercoralis* (stron-″-jih-loy′-deez ster″-kor-ah′-lis) Gr. *strongylos,* round, and *stercoral,* pertaining to feces.

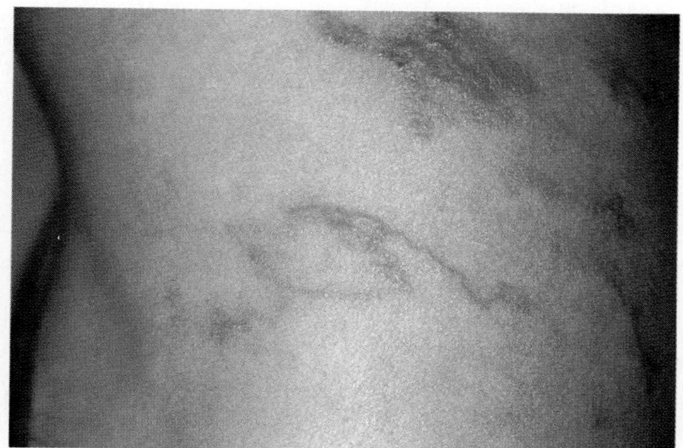

Figure 23.20
A patient with disseminated *Strongyloides* infection. Trails under the skin indicate the migration tracks of the worms.

minute size and its capacity to complete its life cycle either within the human body or outside in moist soil. This nematode shares a similar distribution and life cycle as hookworms and afflicts an estimated 100–200 million people worldwide. Infection occurs when soil larvae that have developed from fecally deposited or free-living eggs penetrate the skin. The worm then enters the circulation, is carried to the respiratory tract and swallowed, and then enters the small intestine to complete development. Although adult *S. stercoralis* lay eggs in the gut just as hookworms do, the eggs can hatch into larvae in the large bowel, penetrate it, and complete their circuitous route through the lungs and back to the small intestine, where they mature into adults. Eggs can likewise exit with feces and go through an environmental cycle. These numerous alternative life cycles greatly increase the chance of transmission, and the unique characteristic of larval development makes chronic infection common.

The first hint of threadworm infection is usually a red, intensely itchy skin rash at the site of entry. Mild migratory activity in an otherwise normal person can escape notice, but heavy worm loads can cause symptoms of pneumonitis and eosinophilia. The physical trauma and irritation that result from nematode activities in the intestine produce bloody diarrhea, liver enlargement, bowel obstruction, and malabsorption. In immunocompromised patients there is a risk of severe, disseminated infestation involving numerous organs (figure 23.20). Hardest hit are AIDS patients, transplant patients on immunosuppressant drugs, and cancer patients receiving irradiation therapy, who can die if not treated promptly.

Trichinella spiralis* and Trichinosis** The life cycle of ***Trichinella spiralis* is spent entirely within the body of a mammalian host, usually a carnivore or omnivore such as a pig, bear, cat, dog, or rat. In nature, the parasite is maintained in an encapsulated (encysted) larval form in the muscles of these animal reservoirs and is transmitted when other animals prey upon them. Humans acquire **trichinosis** (also called trichinellosis) by eating the meat of an infected animal (usually wild or domestic swine or bears), but humans are essentially dead-end hosts (provided corpses are buried and cannibalism is not practiced!) (figure 23.21).

Because all wild and domesticated mammals appear to be susceptible to *T. spiralis,* one might expect human trichinellosis to be common worldwide. But, in actual fact, its incidence in tropical regions may be lower than in certain temperate regions with a higher standard of living. Relatively high rates of infestation occur in the United States (estimated at 200,000 new cases a year) and in Europe. This distribution appears to be related to regional or ethnic customs of eating raw or rare pork dishes or wild animal meats. Bear meat is the source of up to one-third of the cases in the United States. Home or small-scale butchering enterprises that do not carefully inspect pork can spread the parasite, although commercial pork can also be a source. Practices such as tasting raw homemade pork sausage for proper seasoning or serving rare pork or pork-beef mixtures have been responsible for sporadic outbreaks.

Digestion of the cyst envelope in the juices of the stomach or small intestine liberates the larvae. After burrowing into the intestinal mucosa, the larvae reach adulthood and mate. An unusual aspect of this worm's life cycle is that males die and are expelled in the feces, while females burrow into the intestinal lining. The females incubate eggs and shed live larvae that subsequently penetrate the intestine and enter the lympathic channels and blood. All tissues are at risk for destructive invasion, but final development occurs when the coiled larvae are encysted in the skeletal muscle (figure 23.21). At maturity, the cyst is about 1 mm long and can be observed by careful inspection of meat. Although larvae can deteriorate over time, they have also been known to survive for years.

In many cases, there is little sign of *T. spiralis* infestations. In clinical illness, however, the first symptoms mimic influenza or viral fevers, with diarrhea, nausea, abdominal pains, fever, and sweating. The second phase, brought on by the mass migration of larvae and their entrance into muscle, produces puffiness around the eyes, intense muscle and joint pain, shortness of breath, and pronounced eosinophilia. The most serious life-threatening manifestations are heart and brain involvement. In one remarkable case, a woman ate a large amount of raw pork tartare and was dead from massive brain infestation in 48 hours. Although the symptoms eventually subside, a cure is not available once the larvae have encysted in muscles.

The most effective preventive measures for trichinosis are to adequately cook, freeze, or smoke pork and wild meats.

Tissue Nematodes

As opposed to the intestinal worms previously described, tissue nematodes parasitize human blood, lymphatics, or skin. Although these filarial parasites are responsible for grotesque deformities or loss of sight, they are also wonders of evolution and adaptation. The primary examples of tissue roundworms are the filarial worms, *Loa loa* (the eye worm), and *Dracunculus* (the guinea worm).

* *Trichinella spiralis* (trik″-ih-nel′-ah spy-ral′-is) A little, spiral-shaped hair worm.

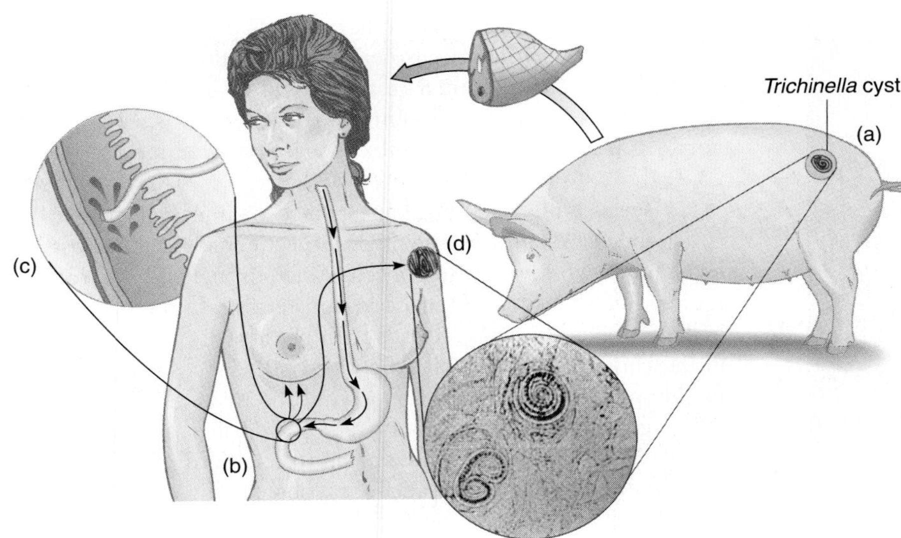

Figure 23.21
The cycle of trichinosis. (*a*) The infective stage is larvae encysted in animal muscle. (*b*) When ingested, the cyst hatches and matures in the intestinal lining. (*c*) Offspring of adult worms burrow through the intestine and penetrate the circulation. (*d*) They eventually form cysts in skeletal muscle that can remain for years. Inset shows a biopsy of human skeletal muscle infected with the coiled larvae of *Trichinella spiralis*.
(d inset) Source: Koneman et al., Diagnostic Microbiology, *4th ed., 1992. Reprinted by permission from J. B. Lippincott Co.*

Filarial Nematodes and Filariasis (Cycle E)

The filarial parasites are long, threadlike worms that can live in host tissues for years. They are characterized by the production of tiny **microfilarial** larvae that circulate in the blood and lymphatics. Their anatomy is relatively simple, but they have a complex biphasic life cycle that alternates between humans and blood-sucking mosquito or fly vector. The two species responsible for most **filariases** are *Wuchereria bancrofti,* the cause of elephantiasis, and *Onchocerca volvulus,* the agent of river blindness. Although there are numerous zoonotic species of filarial worms, these two species exist primarily in humans, and there are no other natural vertebrate reservoirs.

Wuchereria bancrofti* and Bancroftian Filariasis** Filariasis caused by ***Wuchereria bancrofti* afflicts about 90 million people in the hot, humid regions of the globe native to its intermediate host vectors: female *Culex, Anopheles,* and *Aedes* mosquitos. So attuned are the microfilariae to the mosquito feeding cycle that in night-feeding mosquitos, the blood larvae swarm to the cutaneous circulation about 2 hours before and after midnight. In a strain of the worm that is carried by day-feeding mosquitos the parasite is more plentiful in the skin capillaries during the daytime hours. The microfilariae complete development in the flight muscles of the mosquito, and the resultant larvae gather in its mouthparts.

Infection begins when a mosquito bite deposits the infective larvae into the skin. These larvae penetrate the lymphatic vessels, where they develop into long-lived, reproducing adults. Mature female worms expel huge numbers of microfilariae directly into the circulation (figure 23.22*a*).

The first symptoms of larval infestation in humans are inflammation and dermatitis in the skin near the mosquito bite, lymphadenitis, and testicular pain (in males). The most conspicuous and bizarre signs of bancroftian filariasis is the chronic swelling of the scrotum or legs called **elephantiasis.*** This edema is caused by inflammation and blockage of the main lymphatic channels, which prevent return of lymph to the circulation and cause large amounts of fluid to accumulate in the extremities (figure 23.21*b*). The condition can predispose to ulceration and infection. In extreme cases, the patient can become physically incapacitated and suffer from the stigma of having to carry a massive growth. Although travelers may regard with horror the thought of developing this condition, it occurs in individuals in endemic regions only after one to five decades of chronic disease. Acute infection in a young, healthy person is usually subclinical.

Onchocerca volvulus* and River Blindness** ***Onchocerca volvulus* is a filarial worm transmitted by small biting vectors called **black flies.** These voracious flies often attack in large numbers, and it is not uncommon to be bitten several hundred times a day. Onchocerciasis is found in rural settlements along rivers bordered by overhanging vegetation, which are typical breeding sites of the black fly. The life and transmission cycle is similar to that of *Wuchereria,* although the larvae are deposited into the bite wound and develop into adults in the immediate subcutaneous tissues, where disfiguring nodules form within one to two years after initial contact. Microfilariae given off by the adult female migrate into the blood and the eyes. It is the blood phase that infects other feeding black flies.

Some cases of onchocerciasis result in a severe itchy rash that can last for years. As worms degenerate, escaping antigens provoke inflammatory and granulomatous lesions. The other, more serious complication is **river blindness,** arising from infestation of the eye. The worms eventually invade the entire eye, producing much inflammation and permanent damage to the retina and optic nerve. In regions of high prevalence, it is not unusual for an ophthalmologist to see microfilariae wiggling in the anterior chamber

**Wuchereria bancrofti* (voo″-kur-ee′-ree-ah bang-krof′-tee) After Otto Wucherer and J. Bancroft.

*elephantiasis (el″-eh-fan-ty′-ah-sis) Gr. *elephas,* elephant.

**Onchocerca volvulus* (ong″-koh-ser′-kah volv′-yoo-lus) Gr. *onkos,* hook, and *kerkos,* tail; L. *volvere,* to twist.

MICROFILE 23.3 WORMS THAT DESTROY VISION

The parasitic diseases provide numerous examples of human suffering, not the least of which is the tragic legacy of river blindness. In some West African villages, half of the adult population is blind and must be cared for and led around by younger family members, who can also fall victim to the disease. River blindness has been so long a part of the lives of these villagers that it is almost fatalistically accepted as inevitable. The disease has caused social problems as well, because many natives thought the disease was spread by the river water and abandoned their villages. In the past 15 years, the WHO has been successful in controlling the black fly by spraying pesticides along the rivers in some regions. An American drug company (Merck & Co.) is supplying **ivermectin** (Mectizan), a potent antifilarial drug, to treat people at risk for the disease or already infected. This drug kills the blood larvae and prevents the spread of the disease and has already been used successfully to combat parasites in animals. Unfortunately, it does not reverse the blindness or kill the adult worm, so patients must take the drug for several years. A number of African and other world agencies have developed a master plan for complete elimination of onchocerciasis from the endemic regions by 2008. By 1998, 9–10 million people a year had received treatment. With a new antigen test that detects the infection in its early stages, chances for prompter treatment and avoidance of blindness and other complications have improved.

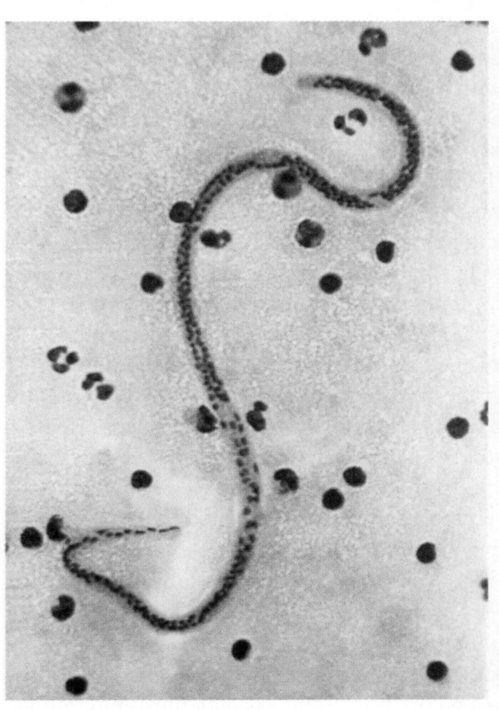

(a)

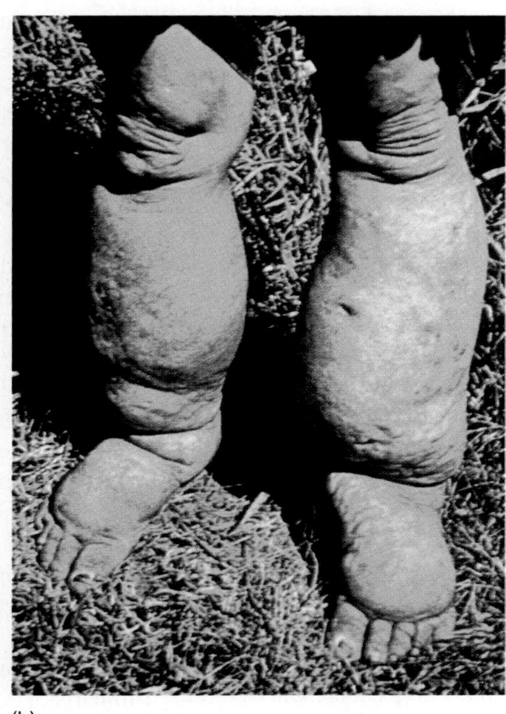

(b)

Figure 23.22

Bancroftian filariasis. (*a*) The appearance of a larva in a blood smear. (*b*) Pronounced elephantiasis in a Costa Rican woman.

during a routine eye checkup. Microfilariae die in several months, but adults can exist for up to 15 years in nodules (microfile 23.3).

***Loa loa:* The African Eye Worm** Another nematode that makes disturbing migratory visits across the eye and under the skin is *Loa loa* (loh′-ah loh′-ah). This worm is larger and more apparent than *O. volvulus.* The agent occurs in parts of western and central Africa and is spread by the bites of dipteran flies. Its life cycle is similar to *Onchocerca* in that the larvae and adults remain more or less confined to the subcutaneous tissues. The first signs of disease are itchy, marble-sized lesions called *calabar swellings* at the site of penetration. These nodules are transient, and the worm lives for only a year or so, but the infestation is often recurrent. Because the nematode is temperature-sensitive, it is drawn to the body surface in response to warmer temperature and is frequently felt slithering beneath the skin or observed beneath the conjunctiva (figure 23.23). One method of treatment is to carefully pull the worm from a small hole in the conjunctiva following application of local anesthesia. Some infections can be suppressed with high doses of diethylcarbamazine.

***Dracunculus medinensis** * **and Dracontiasis** Dracontiasis, caused by the guinea worm, has caused untold misery in residents of India, the Middle East, and central Africa for thousands of years. The parasite is carried by *Cyclops,* an arthropod commonly found in still water. When people ingest water from a contaminated community pond or open well, the larvae are freed in the intestine and

**Dracunculus medinensis* (drah-kung′-kyoo-lus meh-dih-nen′-sis) The little dragon of Medina. Also called guinea worm.

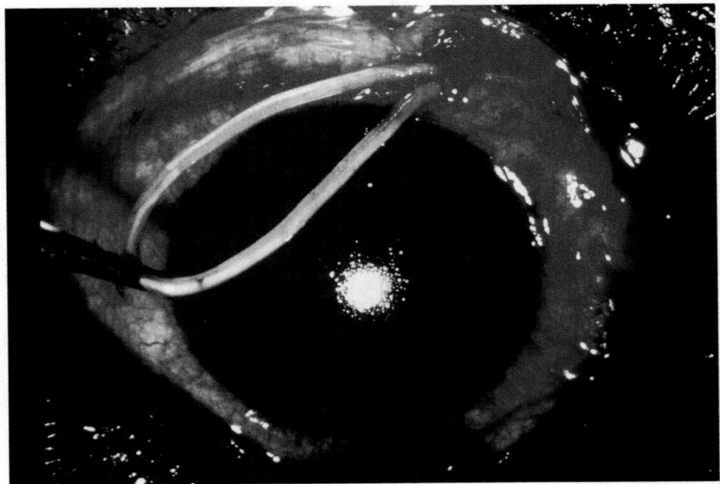

Figure 23.23
Loa loa being surgically removed from an eye.

penetrate into subcutaneous tissues. Although infection often proceeds without notice, the first symptoms coincide with the maturation of the female worm in about a year. Just beneath the skin surface, the worm secretes irritants that provoke an itching and burning blister. When the victim tries to relieve the pain by immersing his feet in well water, the blister erupts and permits the female to expel motile larvae into the water. The cycle is complete when larvae are ingested by *Cyclops,* which are a continuing source of infection.

An international campaign through the WHO has aimed to eradicate the disease in 18 countries. A combined strategy of converting ponds to wells, filtering water through a fine nylon sieve, and treating the water with an insecticide has been very successful. As of 1997, dracontiasis has been eliminated from all parts of the world except for small pockets in Africa.

The Trematodes, or Flukes

The fluke (ME *floke,* flat) is named for the characteristic ovoid or flattened body characteristic of most adult worms, especially liver flukes such as *Fasciola hepatica.* Blood flukes, also called schistosomes, have a more cylindrical body plan. The name **trematode** (Gr. *trema,* a hole) comes from the muscular sucker containing a mouth (hole) at the anterior end of the fluke (see figure 5.35*b).* Flukes have digestive, excretory, neuromuscular, and reproductive systems but lack circulatory and respiratory systems. Except in the blood flukes, both male and female reproductive systems occur in the same individual and occupy a major portion of the body. The trematode life cycles are closely adapted to agricultural practices such as flood irrigating and using raw sewage as fertilizer. In human fluke cycles, animals such as snails or fish are usually the intermediate hosts, and humans are the definitive hosts.

Blood Flukes: The Schistosomes (Cycle D)

Schistosomiasis has afflicted humans for thousands of years. Ancient Egyptian writings that described males "menstruating" were probably referring to blood in the urine caused by renal schistosomiasis. The disease is caused by *Schistosoma* mansoni, S. japonicum,* or *S. haematobium,*[4] species that are morphologically and geographically distinct but share similar life cycle, transmission methods, and general disease manifestations. The disease occurs in 73 countries located in Africa, South America, the Middle East, and the Far East and blood flukes are not native to the United States. Schistosomiasis is a major public health problem, probably affecting 20 million people at any one time.

The schistosome parasite exhibits the stages of egg, miracidium, cercariae, and adult fluke and demonstrates intimate adaptations to both humans and certain species of freshwater snails required to complete its life cycle. The cycle begins when infected humans release eggs into irrigated fields or ponds either by deliberate fertilization with excreta or by defecating or urinating directly into the water. The egg hatches in the water and gives off an actively swimming ciliated larva called a **miracidium*** (figure 23.24*a),* which instinctively swims to a snail and burrows into a vulnerable site, shedding its ciliated covering in the process. In the body of the snail, the miracidium multiplies and transforms into a larger, fork-tailed swimming larva called a **cercaria*** (figure 23.24*b).* Cercariae are given off by the thousands into the water by infected snails.

Upon contact with a human wading or bathing in water, cercariae attach themselves to the skin by ventral suckers and penetrate hair follicles. They pass into small blood and lymphatic vessels and, in about a week, arrive in the liver. Schistosomes achieve sexual maturity in the liver, and the male and female worms remain permanently entwined to facilitate mating (figure 23.24*c).* In time, the pair migrates to and lodges in small blood vessels at specific sites. *Schistosoma mansoni* and *S. japonicum* end up in the mesenteric venules of the small intestine; *S. haematobium* enters the venous plexus of the bladder. While attached to these intravascular sites, the worms feed upon blood, and the female lays spiked eggs that eventually penetrate the intestine or bladder and are voided in feces or urine.

The first symptoms of infestation are itchiness in the vicinity of cercarial entry, followed by fever, chills, diarrhea, and cough. The most severe consequences, associated with chronic infections, are hepatomegaly and liver disease, splenomegaly, bladder obstruction, and blood in the urine. Occasionally, eggs carried into the central nervous system and heart create a severe granulomatous response. Adult flukes can live for many years, and by eluding the immune defenses, cause a chronic affliction.

In parts of Africa where this disease is endemic, epidemiologists are attempting a unique form of biological control. They have introduced into local lakes a species of small fish that feeds on snails. Keeping the snail population in check will curtail development of cercaria, and schistosomiasis can be controlled.

Liver and Lung Flukes (Cycle D)

Several trematode species that infest humans may be of zoonotic origin. The Chinese liver fluke, ***Opisthorchis (Clonorchis) sinensis,**** completes its sexual development in mammals such as cats,

4. The species are named after P. Manson; Japan, and Gr. *haem,* blood, and *obe,* to like.

**Schistosoma* (skis″-toh-soh′-mah) Gr. *schisto,* split, and *soma,* body.

*miracidium (my″-rah-sid′-ee-um) Gr. *meirakidion,* little boy.

*cercaria (sir-kair′-ee-uh) Gr. *kerkos,* tail.

**Opisthorchis sinensis* (oh″-piss-thor′-kis sy-nen′-sis) Gr. *opisthein,* behind, and *orchis,* testis, and *sinos,* oriental.

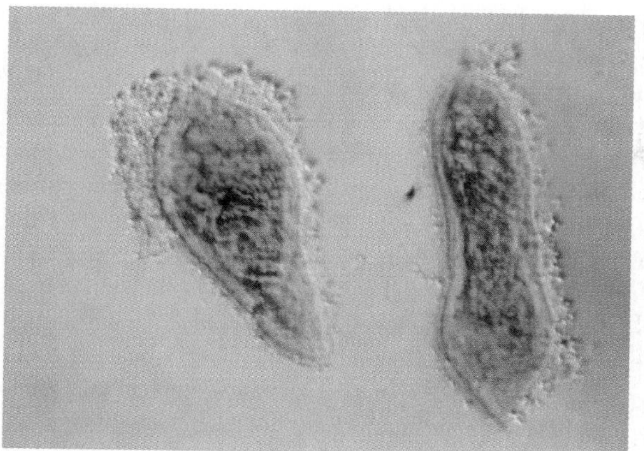
(a)

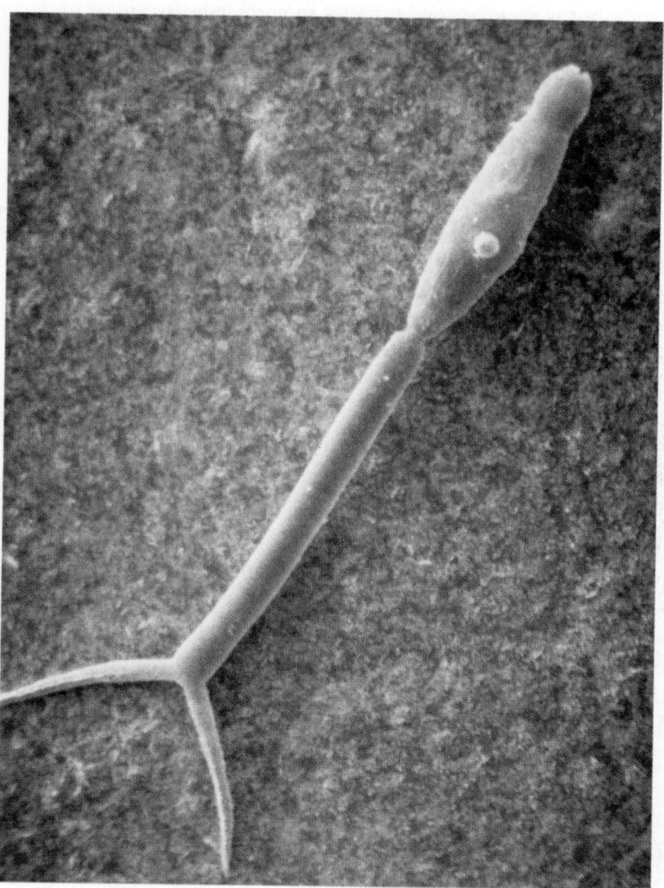
(b)

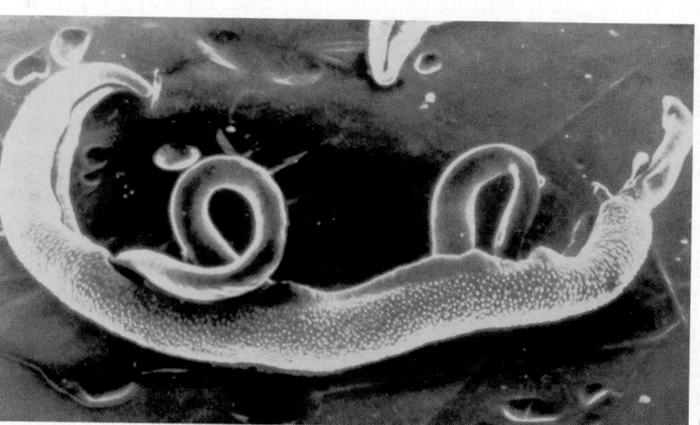
(c)

Figure 23.24
Stages in the life cycle of *Schistosoma*. (*a*) The miracidium phase, which infects the snail. (*b*) The cercaria phase, which is released by snails and burrows into the human host. (*c*) An electron micrograph of normal mating position of adult worms. The male worm holds the female in a groove on his ventral surface.

dogs, and swine. Its intermediate development occurs in snail and fish hosts. Humans ingest cercariae in inadequately cooked or raw freshwater fish and crustaceans. Larvae hatch and crawl into the bile duct, where they mature and shed eggs into the intestinal tract. Feces containing eggs are passed into standing water that harbors the intermediate snail host. The cycle is complete when infected snails release cercariae that invade fish or invertebrates living in the same water. These animals serve as the source of infectious larvae for humans.

The liver fluke ***Fasciola hepatica**** is a common parasite in sheep, cattle, goats, swine, rabbits, and other mammals and is occasionally transmitted to humans (figure 23.25). Periodic outbreaks in temperate regions of Europe and South America are associated with eating wild watercress. The life cycle is very complex, involving the mammal as the definitive host, the release of eggs in the feces, the hatching of the egg in water into miracidia, invasion of freshwater snails, development and release of cercariae, encystment of cercariae on a water plant, and ingestion of the cyst by a mammalian host eating the plant. The cysts release into the intestine young flukes that wander to the liver, lodge in the gallbladder, and develop into the large, leaflike, hermaphroditic adult. Humans develop symptoms of vomiting, diarrhea, hepatomegaly, and bile obstruction only if they are chronically infected by a large number of flukes.

The adult lung fluke ***Paragonimus westermani,**** endemic to the Orient, India, and South America, occupies the pulmonary tissues of various reservoir mammals, usually carnivores such as cats, dogs, foxes, wolves, and weasels. These animals release eggs into water, where they invade the first intermediate host, a snail. Cercariae shed by the snail attack a second intermediate host, a crab or crayfish. Humans accidentally contract infection by eating infected crustaceans that are raw or undercooked. After being released in the intestine, the worms migrate to the lungs, where they can cause cough, pleural pain, and abscess.

***Fasciola hepatica* (fah-see′-oh-lah heh-pat′-ih-kah) Gr. *fasciola,* a band, and *hepatos,* liver.

***Paragonimus westermani* (par″-ah-gon′-ih-mus wes-tur-man′-eye) Gr. *para,* to bear, and *gonimus,* productive; after Westerman, a researcher in flukes.

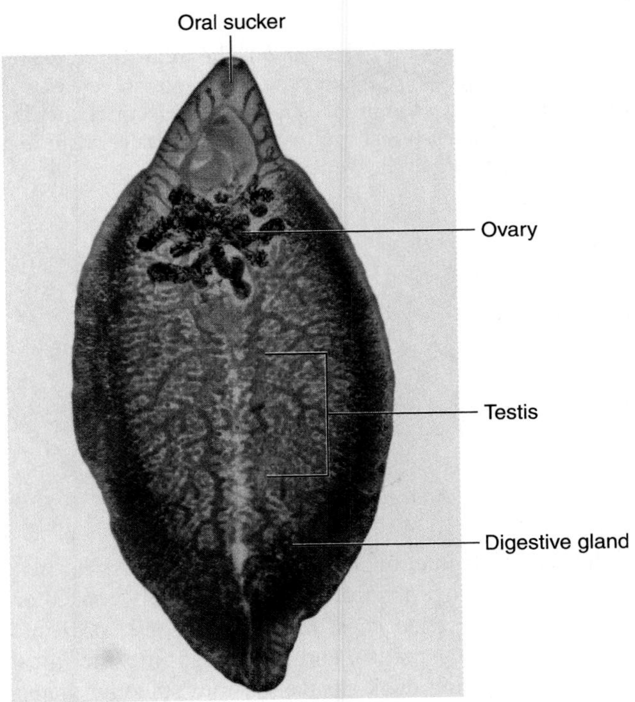

Figure 23.25
Fasciola hepatica, the sheep liver fluke. (2×).

CESTODE (TAPEWORM) INFESTATIONS (CYCLE C)

The **cestodes,*** or tapeworms, are among the most extreme parasitic worms, consisting of little more than a tiny holdfast connected to a chain of flattened sacs that contain the reproductive organs (see microfile 5.3). Several anatomical features of the adult emphasize their adaptations to an intestinal existence (figure 23.26*a,b*). The small **scolex,*** or head of the worm, has suckers or hooklets for clinging to the intestinal wall, but it has no mouth because nutrients are absorbed directly through the body of the worm. The importance of the scolex in anchoring the worm makes it a potential target for antihelminthic drugs. The head is attached by a **neck** to the **strobila,*** a long ribbon composed of individual reproductive segments called **proglottids.*** The contents of proglottids are dominated by male and female reproductive organs, though each one also contains a modest neuromuscular system that permits feeble movements. The neck generates the proglottids, tapering from the smaller, newly formed, and least mature segments nearest the neck to the larger, older ones containing ripe eggs at the distal end of the worm. In some worms, the proglottids are shed intact in feces after detaching from the strobila, and in others, the proglottids break open in the intestine and expel free eggs into the feces.

All tapeworms for which humans are the definitive hosts require an intermediate animal host. The principal human species are *Taenia saginata,* the beef tapeworm, *T. solium,* the swine tapeworm, and *Diphyllobothrium latum,* the fish tapeworm (microfile 23.4). The first two tapeworms cause an important infestation called **taeniasis.** The beef tapeworm follows a cycle C pattern, in which the larval worm is ingested in raw beef, and the pork tapeworm follows both a cycle C and a modified cycle A, in which the eggs are ingested and hatch in the body.

Taenia saginata* is one of the largest tapeworms, composed of up to 2,000 proglottids and anchored by a scolex with suckers. Taeniasis caused by the beef tapeworm is distributed worldwide but is mainly concentrated in areas where raw or undercooked beef is eaten and where cattle ingest proglottids or eggs from contaminated fields and water. After being ingested, the eggs hatch in the small intestine, and the released larvae migrate throughout the organs of the cow (figure 23.26*c*). Ultimately, they encyst in the muscles, becoming **cysticerci,** young tapeworms that are the infective stage for humans. When humans ingest a live cysticercus in beef, it is uncoated and flushed into the intestine, where it firmly attaches by the scolex and develops into an adult tapeworm. So efficient is this egg-laying machine that a single tapeworm is the usual human load. Humans are not known to acquire infection by ingesting the eggs. For such a large organism, it is remarkable how inapparent a tapeworm's presence in the intestine can be. Usually, inflammation is minimal, and infestations often escape detection until the patient discovers proglottids in his stool. At most, some patients complain of vague abdominal pain and nausea that can be relieved by a meal.

Taenia solium differs from *T. saginata* by being somewhat smaller, having a scolex with hooklets and suckers to attach to the intestine, and being infective to humans in both the cysticercus and egg stages. The disease is endemic to regions where pigs are allowed to eat fecally contaminated food and humans consume raw or partially cooked pork (Southeast Asia, South America, Mexico, and eastern Europe). The cycle involving ingestion of cysticerci is nearly identical to that shown in figure 23.26*c*.

When humans ingest pork tapeworm eggs that have gotten into their food or water, a different form of the disease, called **cysticercosis,** results. Although humans are not the usual intermediate hosts, the eggs can still hatch in the intestine, releasing tapeworm larvae that migrate to all tissues. They finally settle down to form peculiar cysticerci, or bladder worms, each of which has a small capsule resembling a little bladder. A connective tissue wrapping contributed by the host completes the formation of the bladder worm. They do the most harm when they lodge in the heart muscle, eye, and brain, especially if a large number of bladder worms occupy tissue space (figure 23.27). It is not uncommon for patients to exhibit seizures, psychiatric disturbances, and other signs of neurological impairment.

---

*cestode (ses′-tohd) L. *caestus,* to strike.

*scolex (skoh′-leks) Gr. *scolos,* worm.

*strobila (stroh-bih′-lah) Gr. *strobilos,* twisted.

*proglottid (proh-glot′-id) Gr. *pro,* before, and *glotta,* the tongue.

**Taenia saginata* (tee′-nee-ah saj-ih-nah′-tah) L. *taenia,* a flat band, *sagi,* a pouch, and *natare,* to swim.

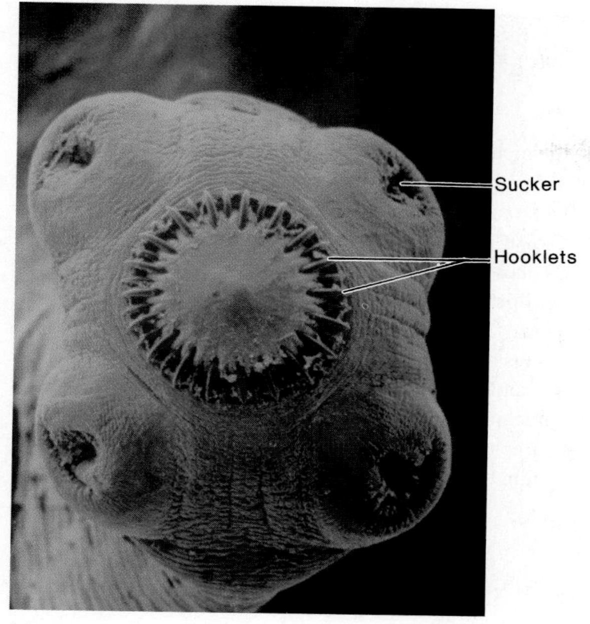

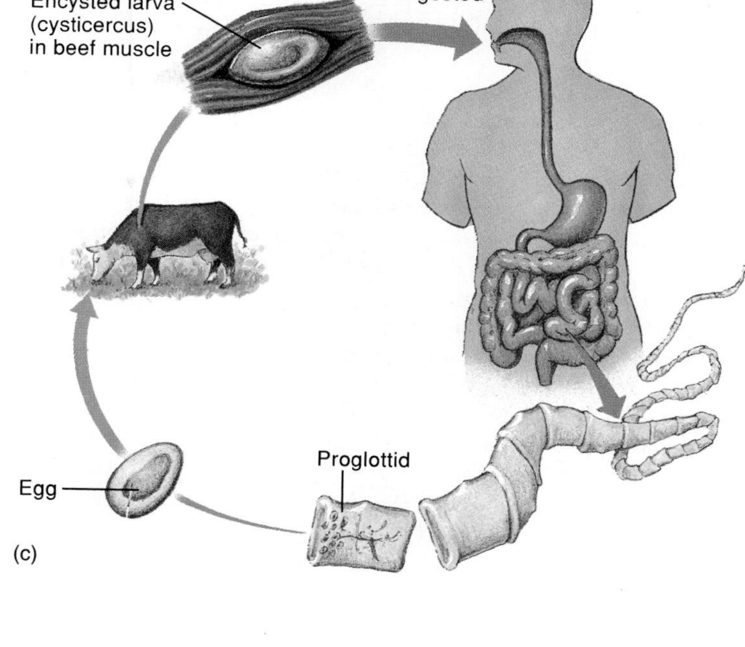

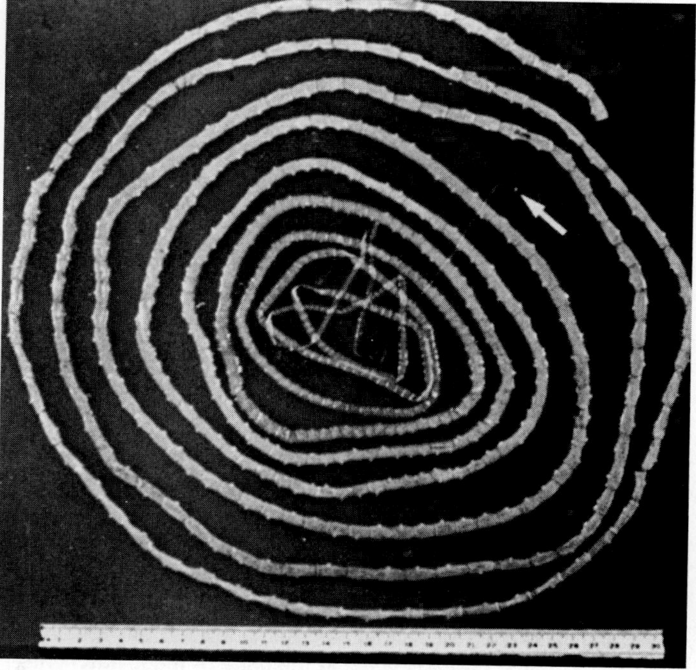

Figure 23.26

Tapeworm infestation in humans. (*a*) Tapeworm scolex showing sucker and hooklets. (*b*) Adult *Taenia saginata*. The arrow points to the scolex; the remainder of the tape, called the strobila, has a total length of 5 meters. (*c*) A generalized diagram of the life cycle of the beef tapeworm *T. saginata*.

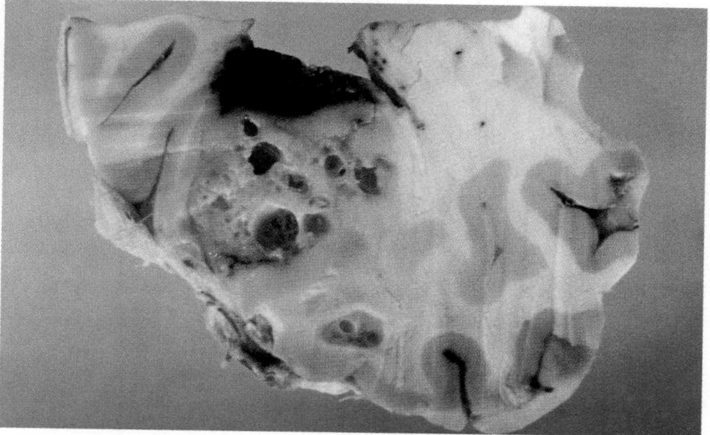

Figure 23.27

Cysticercosis caused by *Taenia solium* in the brain of a 13-year-old girl. A section through the cerebrum contains a cluster of encysted larval worms, which gives a spongy or honeycomb appearance.

MICROFILE 23.4 WAITER, THERE'S A WORM IN MY FISH

Various forms of raw fish or shellfish are part of the cuisine of many cultures. Popular examples are Japanese sashimi and sushi. Although these delicacies are definitely an acquired taste, it is possible for humans to acquire much more than flavor from them, because fish flesh often harbors various types of helminths. Usually humans are not natural hosts for these parasites, so the infestation does not develop fully, but some of the symptoms the worms cause during their sojourn in the body can be quite distressing.

One of the common fish parasites transmitted this way is ***Anisakis,*** a nematode parasite of marine vertebrates, fish, and mammals. The symptoms of anisakiasis, usually appearing within a few hours after a person ingests the half-inch larvae, are due to the attempts of the larvae to burrow into the tissues. The first symptom is often a tingling sensation in the throat, followed by acute gastrointestinal pain, cramping, and vomiting that mimic appendicitis. At times, the cause of these mysterious symptoms is explained when worms are regurgitated or discovered during surgery, but in many cases, they are probably expelled without symptoms. Another worm that can hide in raw or undercooked freshwater fish is the tapeworm *Diphyllobothrium latum,* common in the Great Lakes, Alaska, and Canada. Because humans serve as its definitive host, this worm can develop in the intestine and cause long-term symptoms.

Students always ask about the safety of sushi and seem uncertain about how to eat it and still avoid parasites. One good safeguard is to patronize a reputable sushi bar, because a competent sushi chef carefully examines the fish for the readily visible larvae. Another is to avoid raw salmon and halibut, which are more often involved than tuna and octopus. Fish can also be freed from live parasites by freezing for 5 days at −20°C.

Chapter Checkpoints

Helminths are macroscopic, multicellular worms that have microscopic infective forms: eggs and encysted or free-living larvae. Two major divisions of helminths are the nematodes (roundworms) and platyhelminthes (flatworms). The flatworms are further divided into trematodes (flukes) and cestodes (tapeworms).

Most parasitic helminths have at least two hosts: the definitive host, where sexual reproduction occurs, and the intermediate host, which supports immature or larval forms. The basic patterns of helminth life cycles are characterized by the infectious stage, route of transmission, portal of entry, site of reproduction, portal of exit, and exiting stage. Chemotherapy usually involves drugs that selectively inhibit worm metabolism or weaken their attachment. Preventive measures strive to interrupt the cycle of infection through improved public health conditions.

Nematodes, or roundworms, cause the most severe helminth infestations. They are grouped according to the host site of reproduction as either tissue or intestinal nematodes.

Nematodes of the type A life cycle enter the host as eggs, complete their life cycle in the host, and are transmitted to the environment in new eggs. Examples include *Ascaris, Trichuris,* and *Enterobius.*

Cycle B nematodes enter the host through the skin as larvae. They migrate from the blood to the lungs to the intestine, where they mature and reproduce. Excreted eggs hatch in the environment, producing more larvae. Examples include hookworms *Necator* and *Ancylostoma* and the threadworm *Strongyloides.*

Ascaris lumbricoides, the largest and most common nematode, is the causative agent of ascariasis, and intestinal infestation in which both larvae and adults can invade other body tissues.

Trichuris trichiura, a tropical nematode, is the causative agent of whipworm infection, an intestinal infestation characterized by ulceration and hemorrhage of the colon.

Enterobius vermicularis, a common nematode in temperate zones, causes pinworm infestation of the intestine and is characterized by the appearance of egg-laying females around the anus.

Trichinella spiralis causes trichinosis, a zoonotic disease associated with eating rare pork. Humans are dead-end hosts that suffer from encystment of larvae in muscle, heart, and brain.

Tissue nematodes include the filarial parasites *Wuchereria, Onchocerca,* and *Dracunculus,* all of which cause invasive, debilitating, life-threatening infestations among tropical and subtropical populations.

Trematode pathogens include *Shistosoma, Opisthorchis, Fasciola hepatica,* and *Paragonimus westermani.*

Trematodes exhibit the type D cycle of infection whereby cercarial larvae invade the human host and reproduce in specific organs. Eggs released in the urine or feces hatch into miricidial larvae that invade one or more secondary hosts.

Schistosoma mansonii is a blood fluke that causes a systemic infestation of the bladder and intestine called schistosomiasis.

Cestodes, or tapeworms, are the most extreme parasites, consisting of a feeding attachment to the host and a segmented chain of proglottids containing both male and female reproductive structures. They exhibit the type C reproductive cycle, in which encysted larvae are ingested in the meat of a secondary animal host. The adult develops and reproduces in the intestine and sheds it eggs in the feces. Cestode pathogens include *Taenia saginata, T. solium, Diphyllobothrium latum,* and *Anisakis.*

Chapter Capsule with Key Terms

PARASITOLOGY

Parasitology studies **protozoa** and **helminths** (worms) that live on the body of a host. Parasites are spread from human to human directly or indirectly and from animals and vectors to humans.

PROTOZOAN PATHOGENS

Protozoans propagate only as **trophozoites** (active feeding stage found in host) or alternate between a trophozoite and a **cyst** (dormant, resistant survival body outside of host). Some have complex life cycles with sexual and asexual phases carried out in more than one host.

Infectious Amebas: ***Entamoeba histolytica*** alternates between a large ameboid trophozoite and a multinucleate cyst that is the infectious stage; human is principal host; **amebiasis** occurs worldwide; affects approximately 500 million people in the tropics; cysts released in feces of carriers; spread through unsanitary water and food, agricultural practices, or close contact. Ingested cysts release trophozoites that invade large intestine and cause ulceration; may penetrate into deeper layers and disseminate to extraintestinal sites (liver, spleen, brain). Dysentery is a prominent feature; severe cases may cause fatal intestinal perforation and liver damage; effective drugs are iodoquinol, metronidazole, and chloroquine.

Amebic brain infections are caused by ***Naegleria fowleri*** and ***Acanthamoeba,*** free-living inhabitants of natural waters; **primary acute meningoencephalitis** is acquired through contact with water wherein amebas are flushed into nasal cavity and through traumatic eye damage; infiltration of brain is usually fatal.

Ciliates: The only important ciliate pathogen is ***Balantidium coli,*** a common occupant of the intestines of domestic animals such as pigs and cattle; acquired by humans when cyst-containing food and water are ingested; cyst liberates trophozoite, which erodes intestine and creates intestinal symptoms.

Flagellates: Genus ***Trichomonas*** contains pear-shaped flagellated protozoa that are parasites or commensals of animals. Principal human pathogen is ***T. vaginalis;*** has no cysts; causes an STD, trichomoniasis; disease spread by direct contact (occasionally through fomites); most common in promiscuous people; infects vagina, cervix, urethra, producing frothy yellow discharge; both sex partners treated with metronidazole.

Giardia lamblia is an intestinal flagellate with a unique heart shape and four flagella; has cyst stage; causes **giardiasis;** transmission similar to *Entamoeba;* natural reservoir is apparently intestines of animals; sources of infection are cyst-contaminated fresh water and food; cysts release trophozoites that invade intestinal glands; cause severe diarrhea; may be chronic in immunodeficient patients; contaminated water must be boiled or filtered to eliminate the resistant cyst; quinacrine or metronidazole is usual treatment.

Hemoflagellates: Parasites that occur in blood during infection; cause zoonoses spread by insect vectors; most are exotic, tropical; have complex life cycles, with various **mastigote** phases that develop in the insect and human hosts.

Trypanosoma have tapering, crescent-shaped cells with flagellum and undulating membrane; species that cause **trypanosomiasis** are geographically isolated and have different blood-feeding vectors; diseases endemic to regions where vectors live.

T. brucei causes **African sleeping sickness;** the two strains are Gambian (a chronic disease) and Rhodesian (an acute disease); parasites carried by reservoir hosts such as wild and domestic animals; both are spread by **tsetse fly,** which carries agent in salivary gland and inoculates bite; trypanosome multiplies in blood and alters its surface antigen so that immune system cannot control it; enlargement of spleen, lymph nodes, joints involved; brain damage leads to sleepiness, tremors, paralysis, coma; treatment with suramin not totally effective.

T. cruzi causes **Chagas' disease;** endemic to Latin America; cycle similar to *T. brucei,* except that the kissing, or **reduviid, bug** is vector; infection occurs when bug feces inoculate wound or mucous membrane; invasion is marked by inflammation, organ involvement (especially heart and brain); nifurtimox may give some therapeutic benefit.

Leishmania contains several geographically separate species that cause **leishmaniasis,** a zoonosis transmitted by phlebotomine (sand) flies; reservoirs are wild animals; humans enter cycle when fly inoculates them; macrophages engulf infective cells. Disease may be cutaneous, mucocutaneous, or systemic, with fever, enlarged organs, anemia. Most severe and fatal form is **kala-azar;** therapy is antimony, pentamidine.

Apicomplexans: Tiny obligate intracellular parasites that lack motility in mature stage and have complex sexual and asexual phases; infection occurs through sporozoites, fecal cysts called oocysts, or tissue cysts; most are zoonotic and vector-borne.

Plasmodium consists of four species that cause **malaria;** human is primary host for asexual phase of parasite; female ***Anopheles*** **mosquito** is vector/host for sexual phase; distributed primarily in malaria belt around the equator; 300–500 million cases are estimated worldwide. Infective forms for humans **(sporozoites)** enter blood with mosquito saliva; from blood, they penetrate liver cells, multiply, and form hundreds of **merozoites;** merozoites burst out of liver and enter circulation; multiply in and lyse red blood cells; this cycle continues until sex cells are produced in some RBCs and are picked up by another feeding mosquito; in mosquito gut these gametes produce sporozoites that are harbored in the salivary gland. Symptoms include episodes of chill-fever-sweating, anemia, and organ enlargement. Therapy is chloroquine, quinine, primaquine; a vaccine is being tested.

Toxoplasma gondii is an apicomplexan with wide distribution among birds and mammals; infectious in both tissue cyst and free oocyst phases; primary reservoirs are felines harboring oocysts in the GI tract; feces may spread it among other cats, vertebrates (cattle, pigs), and humans. **Toxoplasmosis** is acquired by ingesting raw or rare meats containing tissue cysts or by accidentally ingesting oocysts from substances contaminated by cat feces. Infection is usually mild and flu-like in older children and adults; more severe in immunodepressed (AIDS) patients or fetuses. **Fetal toxoplasmosis** damages brain, heart, lungs; pregnant women must avoid raw meats and unhygienic handling of cats.

Sarcocytis is a zoonotic parasite transmitted to humans from meat containing live tissue cysts; humans develop diarrhea, can transmit cysts to grazing animals.

Cryptosporidium is a common vertebrate pathogen that exists in both tissue and oocyst phases; most cases of **cryptosporidiosis** are caused by handling animals with infections or by drinking water contaminated with animal feces; infection causes diarrhea, abdominal symptoms similar to giardiasis; is a common chronic, debilitating, wasting infection in AIDS patients; can be fatal.

Isopora is an uncommon human intestinal parasite transmitted through fecal contamination; causes coccidiosis.

Cyclospora causes a diarrheal illness when its oocysts are ingested in fecally contaminated produce and water.

Babesia causes babesiosis, a zoonosis spread by hard ticks; similar to malaria in symptoms and treatment.

HELMINTH PARASITES

Helminths, or worms, are relatively large, multicellular animals with specialized tissues and organs; adaptations to habitat include specialized mouthparts, reduction of organs, protective cuticles, complex life cycles that increase chances of encountering a host. Worms are often long-lived; concentrated in rural semitropics and tropics.

Life cycles involve **adult worms** that mate, produce fertile **eggs** that hatch into **larvae;** larvae mature in several stages to adults; adults live in **definitive host;** eggs and larvae may develop in same host, external environment, or another, **intermediate, host** (an animal). In some worms, the sexes are separate; others are hermaphroditic, with both sexes in same worm. Because worms do not usually increase in number by multiplying in a host, their presence is an **infestation.**

Basic transmission cycles: (1) Egg-laden feces are deposited into soil, water; (2) eggs may be directly infectious when ingested, or (3) go through larval phase in environment that is infectious by direct penetration of skin; (4) animal flesh or plants containing encysted larval worms are ingested; (5) helminth is inoculated into body by insect vector that is intermediate host.

Pathology is due to action of worms on tissues; feeding, breaking down, migrating through tissues, and accumulation of worms and worm products; worms may cause allergies, survive for long periods in the body; **anti-helminthic drugs** such as piperazine, pyrantel, and metrifonate control worms by paralyzing their muscles and causing them to be shed; drugs such as mebendazole, diethylcarbamazine, and praziquantel interfere with metabolism and kill them; other control measures are improved sanitation (not using feces as fertilizer), protective clothing, cooking food adequately, and controlling vectors.

NEMATODES

Roundworms are filamentous, tapered worms; round in cross section; have protective cuticles, circular muscles, digestive tract; separate sexes with well-developed reproductive organs.

Intestinal Nematodes: These worms spend part of their life cycle in the intestine; adults shed eggs from intestine. ***Ascaris lumbricoides*** is indigenous to humans; very common worldwide; ingested eggs hatch in intestine; larvae burrow through intestine into blood, then enter heart, pulmonary circulation, lungs, migrate into pharynx, are swallowed; adult form develops in intestine, produces eggs; symptoms and signs depend upon size of worm burden; may obstruct organs, cause inflammatory reactions.

Trichuris trichiura, the **whipworm,** is a small intestinal pathogen transmitted like *Ascaris* but restricted to intestine throughout development. ***Enterobius vermicularis,*** the **pinworm,** is a very common childhood worm infection confined to intestine; transmitted by close contact, unclean personal hygiene; eggs swallowed, hatch in intestine, develop into adults, release more eggs; main symptom is anal itching due to egg-laying activities of females; self-inoculation is common.

Hookworms: Worms have curved ends and hooked mouths; ***Necator americanus*** (Western Hemisphere) and ***Ancylostoma duodenale*** (Eastern Hemisphere) have similar life cycle, transmission, pathogenesis; very prevalent; human is principal host; eggs shed in feces; hatch into larvae in environment and burrow into skin of lower legs; move into blood, lymph, ending up in lungs; larvae are coughed up, swallowed, and reach adulthood in intestine; eggs released with feces; symptoms are itching at site of penetration, inflammatory reactions, severe intestinal distress, anemia; can sap strength.

Strongyloides stercoralis, the **threadworm,** is a tiny nematode that can complete its life cycle totally in humans or in the external environment (moist soil); larvae emerging from soil-borne eggs penetrate the skin and migrate to lungs, are swallowed, and develop in intestine; unlike most helminths, can reinfect same host without leaving body; symptoms are pneumonic, intestinal; disease most severe in immunosuppressed or AIDS patients.

Trichinella spiralis causes **trichinosis,** a common zoonosis of hogs and bears, in which humans are a dead-end host; acquired from eating undercooked or raw meat containing **encysted larvae;** larvae are released in intestine and develop into adults. Female worms produce larvae that migrate in blood vessels to muscle, heart, and brain, where they encapsulate; disease is flu-like, with fever, aches; may be fatal with high worm loads.

Tissue Nematodes: Principal target organs of maturation are nonintestinal (blood, lympatics, skin). **Filarial worms** are long, threadlike worms with tiny larvae (microfilariae) that circulate in blood and reside in various organs; are spread by various biting insects.

Wuchereria bancrofti causes **bancroftian filariasis,** a tropical infestation spread by various species of mosquitoes; insect deposits larvae into bite; larvae move into lymphatics; after development, adult females shed microfilariae into blood; first symptoms of disease are localized inflammation; chronic blockage of lymphatic circulation causes build-up of fluid in lower extremities, or **elephantiasis,** which may manifest as massive swelling.

Onchocerca volvulus causes **river blindness,** African disease spread by black flies and small, river-associated insects that feed on blood; inoculated larvae develop into adults in skin, forming nodules; in long-term infections, microfilariae migrate into eye and produce inflammation that can blind; new treatment is ivermectin.

Loa loa, the African eye worm, is spread by bite of small flies; worm remains just under skin, where it migrates about; often frequents the conjunctiva.

Dracunculus medinensis, or guinea worm, causes dracontiasis; acquired by persons drinking water containing a small crustacean, *Cyclops,* that hosts the larvae; ingested larvae penetrate intestine and migrate through body; female lays eggs through blister on leg that opens up when victim stands in water.

TREMATODES, OR FLUKES

Flatworms possessing thin, leaflike bodies with suckers and reduced organs; most are hermaphroditic.

Blood flukes and ***Schistosoma*** species; **schistosomiasis** is very prevalent in subtropics and tropics, affecting 20 million people; adult flukes live in humans and release eggs into aquatic environment with feces or urine; in water, they hatch into an early larva, the **miracidium,** that infects freshwater snail; snail releases second larva, called a **cercaria,** that is infective to humans wading in water; larvae penetrate skin, move into liver to mature; adults migrate to intestine or bladder and shed eggs. Acute symptoms are fever, diarrhea; chronic organ enlargement that can be very destructive.

Zoonotic flukes include the Chinese liver fluke ***Opisthorchis sinensis,*** whose definitive hosts are cats, dogs, swine; intermediate hosts are snails and fish; humans are infected when eating fish containing larvae; fluke lives in liver. In another liver fluke, ***Fasciola***

hepatica, herbivores (sheep, cattle) are definitive hosts; snails and aquatic plants are intermediate hosts; humans infected by eating raw aquatic plants; fluke lodges in liver.

CESTODES OR TAPEWORMS

These flatworms have long, very thin, ribbon-like bodies that grip host intestine by **scolex;** budding off slender neck is a chain **(strobila)** of sacs **(proglottids)** that make up the worm body; each proglottid is an independent hermaphroditic unit adapted to absorbing food and making eggs; mature proglottids or eggs shed with feces.

Taenia saginata is the beef tapeworm; humans are definitive hosts; infected by eating raw beef in which the larval form **(cysticercus)** has encysted; larva attaches to small intestine and becomes adult tapeworm; beef animals are infected by grazing on land contaminated with human feces. ***T. solium*** is the pork tapeworm; humans may be infected with cysticerci as in beef tapeworm or by ingesting eggs in food or drink; disease caused by ingesting eggs is **cysticercosis;** larvae hatch and migrate to encyst in many organs and do not reach tapeworm stage but damage organs. ***Diphyllobothrium latum,*** the fish tapeworm, matures in animals and humans; intermediate freshwater hosts are *Cyclops* and various fish; infection acquired from ingesting raw or rare fish.

MULTIPLE-CHOICE QUESTIONS

1. All protozoan pathogens have a ____ phase.
   a. cyst
   b. sexual
   c. trophozoite
   d. blood
2. *Entamoeba histolytica* primarily invades the
   a. liver
   b. large intestine
   c. small intestine
   d. lungs
3. *Giardia* is a/an ____ that invades the ____
   a. flagellate, large intestine
   b. ameba, small intestine
   c. ciliate, large intestine
   d. flagellate, small intestine
4. Blood flagellates are transmitted by
   a. mosquito bites
   b. insect vectors
   c. bug feces
   d. sand fleas
5. *Plasmodium* reproduces sexually in the and asexually in the____.
   a. liver, red blood cells
   b. mosquito, human
   c. human, mosquito
   d. red blood cell, liver
6. In the exoerythrocytic phase of infection, *Plasmodium* invades the
   a. blood cells
   b. heart muscle
   c. salivary glands
   d. liver
7. An oocyst is found in ____ and a pseudocyst is found in____.
   a. humans, cats
   b. cats, humans
   c. feces, tissue
   d. tissue, feces
8. A person can acquire toxoplasmosis from
   a. pseudocysts in raw meat
   b. oocysts in air
   c. cleaning out the cat litter box
   d. all of these
9. All adult helminths produce
   a. cysts and trophozoites
   b. scolex and proglottids
   c. fertilized eggs and larvae
   d. hooks and cuticles
10. The ____ host is where the larva develops, and the____host is where the adults produce fertile eggs.
    a. intermediate, definitive
    b. definitive, intermediate
    c. secondary, transport
    d. primary, secondary
11. Antihelminthic medications work by
    a. paralyzing the worm
    b. disrupting the worm's metabolism
    c. causing vomiting
    d. both a and b
12. A host defense that is most active in worm infestations is
    a. phagocytes
    b. antibodies
    c. killer T cells
    d. eosinophils
13. Currently, the most common nematode infestation worldwide is
    a. hookworm
    b. ascariasis
    c. pinworm
    d. trichinosis
14. Hookworm diseases are spread by
    a. the feces of humans
    b. mosquito bites
    c. contaminated food
    d. microscopic invertebrates in drinking water
15. Trichinosis can only be spread from human to human by
    a. cannibalism
    b. flies
    c. raw pork
    d. contaminated water
16. The swelling of limbs typical of elephantiasis is due to
    a. allergic reaction to the filarial worm
    b. granuloma development due to inflammation by parasites
    c. lymphatic circulation being blocked by filarial worm
    d. heart and liver failure due to infection
17. Match the disease or condition with the causative agent, and indicate for each whether the agent is a protozoan or a helminth.

____ amebic dysentery
____ Chagas' disease
____ tapeworm
____ hookworm
____ African sleeping sickness
____ pinworm
____ filariasis
____ amebic meningoencephalitis
____ malaria
____ toxoplasmois
____ trichinosis
____ whipworm
____ river blindness

a. *Plasmodium vivax*
b. *Onchocerca volvulans*
c. *Enterobius vermicularis*
d. *Toxoplasma gondii*
e. *Trichuris trichiura*
f. *Entamoeba histolytica*
g. *Necator americanus*
h. *Trypanosoma brucei*
i. *Wuchereria bancrofti*
j. *Trypanosoma cruzi*
k. *Taenia saginata*
l. *Naegleria fowleri*
m. *Trichinella spiralis*

18. Match the disease with its primary mode of transmission or acquisition in humans.

____ amebic dysentery	a.	eating poorly cooked beef or pork
____ Chagas' disease	b.	water or food contaminated with animal wastes containing cysts
____ cyclosporiasis	c.	water or food contaminated with human feces containing cysts
____ toxoplasmosis	d.	bite from a reduviid bug
____ giardiasis	e.	bite from a black fly
____ tapeworm	f.	water or food contaminated with human feces containing eggs
____ filariasis	g.	contact with cats or ingesting rare or raw meat
____ schistosomiasis	h.	bite from a mosquito
____ ascariasis	i.	ingesting fecally contaminated water or produce
____ river blindness	j.	freshwater snail vector releases infectious stage

CONCEPT QUESTIONS

1. As a review, compare the four major groups of protozoa according to overall cell structure, locomotion, infective state, and mode of transmission.
2. a. What are the primary functions of the trophozoite and the cyst in the life cycle?
   b. In which diseases is a cyst stage an important part of the infection cycle?
3. a. Why does *Entamoeba* require healthy carriers to complete its life cycle?
   b. What is the role of night soil in transmission of amebiasis?
   c. What is the basic pathology of amebiasis? How and where does it invade?
   d. Why are so many cases asymptomatic?
4. Briefly describe how primary amebic meningoencephalitis is acquired and its outcome.
5. Compare trichomoniasis and amebiasis with respect to transmission, life cycle, and relative hardiness.
6. a. Name ways in which giardiasis and amebic dysentery are similar.
   b. How are they different?
7. a. Compare the infective stages and means of vector transfer in the two types of trypanosomiasis and leishmaniasis.
   b. Are there any stages in the life cycle of the blood flagellates that occur in all three species?
   c. Which phase or phases is/are not ever infective to humans? (Hint: Look at table 23.3.)
8. a. What factors cause the symptoms in trypanosomiasis?
   b. In leishmaniasis?
9. a. What is the name of the infective stage of the malaria parasite in humans?
   b. In mosquitos?
   c. How many times must a female mosquito feed before the parasite can complete the whole cycle?
   d. Where in the human does development take place, and what are the results?
   e. What stage of development is the ring form?
   f. Which events cause the symptoms of malaria?
10. a. Describe the life cycle and host range of *Toxoplasma gondii.*
    b. How are humans infected?
    c. What are the most serious outcomes of infection?
11. Briefly describe the transmission cycle of *Cryptosporidium, Sarcocystis,* and *Babesia.*
12. a. In what ways are helminths different from other parasites, and why are there so many parasitic worm infections worldwide?
    b. What is the nature of human defenses against them?
13. a. Outline the five general cycles, and give examples of helminths that exhibit each cycle.
    b. What are the main portals of entry in worm infestations?
    c. How is damage done by parasites?
14. What are some ways that worms adapt to parasitism, and how are these adaptations beneficial to them?
15. How are nematodes, trematodes, and cestodes different from one another?
16. Where in the body do the helminth adults ultimately reside and produce fertile eggs?
17. a. How do adult *Ascaris* get into the intestine?
    b. How do adult hookworms get into the intestine?
    c. How do microfilariae get into the blood?
18. In what ways is trichinosis different from other worm infections?
19. a. Which worms can be found in the eye?
    b. Which worm can cause blindness?
20. a. What are the stages in *Schistosoma* development?
    b. Which organs are affected by schistosomiasis?
    c. What other organs can be invaded by flukes?
21. a. Describe the structure of a tapeworm.
    b. Name several ways in which tapeworms are spread to humans.
22. a. Which helminths are zoonotic?
    b. Which are strictly human parasites?
    c. Which are borne by vectors?
    d. Which can be STDs?
    e. Which cause intestinal symptoms?

Critical-Thinking Questions

1. Describe some adaptations needed by parasites that enter through the oral cavity.
2. Explain why a person with overt symptoms of intestinal *Entamoeba histolytica* infection is unlikely to transmit infection to others.
3. a. Explain why *Trichomonas vaginalis* is unlikely to be transmitted by casual contact.
   b. What is meant by "Ping-Pong" infection, and why must both sex partners be treated for trichomoniasis?
4. Explain why only female mosquitos are involved in malaria and elephantiasis.
5. Which parasitic diseases could conceivably be spread by contaminated blood and needles?
6. For which diseases can one not rely upon chlorination of water as a method of control? Explain why this is true.
7. a. If a person returns from traveling afflicted with trypanosomiasis or leishmaniasis, is he or she generally infective to others?
   b. Why, or why not?
8. Explain why there is no malaria above 6,000 feet in altitude.
9. Describe some strategies to arrive at an effective malaria vaccine.
10. Suggest a possible way to medically circumvent the antigenic switching of trypanosomes.
11. a. Which diseases end up in the intestine from swallowing larval worms?
    b. From swallowing eggs?
    c. From penetration of larvae into skin?
12. a. What one simple act could in time eradicate dracunculosis?
    b. Hookworm?
    c. Beef tapeworm?
13. a. To achieve a cure for tapeworm, why must the antihelminthic drug either kill the scolex or slacken its grip?
    b. Which tapeworm is pictured in figure 23.26*a?*
    c. How can you tell?
14. Give some reasons why AIDS patients are so susceptible to certain protozoan diseases.
15. Explain why there is no such thing as a safe form of rare or raw meat.
16. Students sometimes react with horror and distress when they discover that cats and dogs carry parasites to humans. Give an example of a disease for each of these animals that can be spread to humans, and explain how to avoid these diseases and still enjoy your pet.
17. a. Why is it necessary for most parasites to leave their host to complete the life cycle?
    b. What are some ways to prevent completion of the life cycle?
    c. What is the benefit to parasites of having numerous hosts?
    d. What are the disadvantages in having more than one host?
18. East Indian natives habitually chew on betel nuts as an alkaloidal stimulant and narcotic. A side effect is phlegm collection, which is eliminated by frequent spitting. The incidence of *Strongyloides* infections is relatively low in this population. Can you account for this? (No, the betel nuts are not effective antihelminthics.)
19. Give some possible reasons why the incidence of cystic hydatid disease is very high among women and children in hot, arid Turkana, Kenya, where dogs form an integral part of tribal life.
20. In New York City, four Orthodox Jewish patients brought into an emergency clinic with seizures and other neurological symptoms tested positive for antibodies to the pork tapeworm, *Taenia solium.* This was embarrassing to them because they do not eat pork for religious reasons. It also became something of a medical mystery, because there were no indications of how they could have become infected. Later it was shown that the patients had recently employed housekeepers and cooks from Latin America who were free of symptoms yet also seropositive for the worm infection. Use this case study to give a plausible explanation for the transmission of the infection.

Internet Search Topic

Conduct an on-line search for emerging protozoan pathogens. Identify five of the major species and make note of the sources of infection and major risk groups.

24 chapter

Introduction to the Viruses of Medical Importance: The DNA Viruses

Viruses are probably the most common infectious agents, though no exact figures exist to support this contention. As a group, they are all obligate parasites and infect not only animals and plants but also other microorganisms. Every time a new virus or viral disease is discovered or a connection is made between a virus and a once-unexplained disease, our understanding of these remarkable infectious particles is increased, and we are reminded of their power. This chapter covers viral diseases of the viral groups most significant to humans and their epidemiology, pathology, and methods of control, with special emphasis on DNA viruses. Chapter 25 covers RNA viruses.

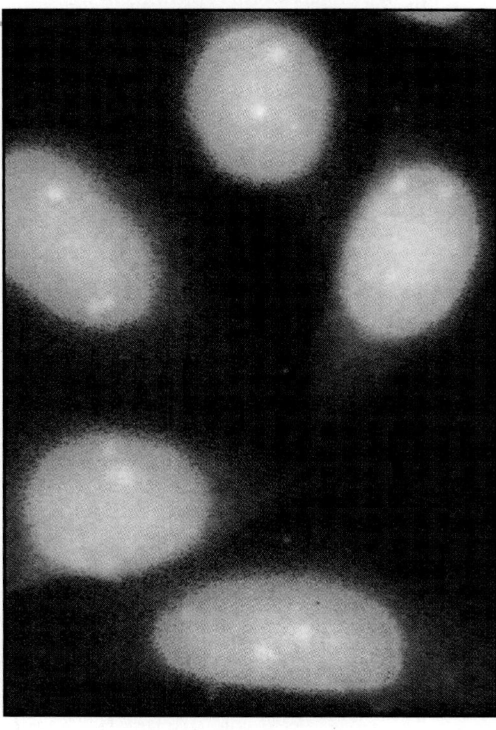

Molecular probes with fluorescent dyes highlight the presence of human papillomavirus (HPV) in the nucleus of infected cells. These fluorescent tags cause integrated viral DNA to appear as small glowing spots.

VIRUSES IN INFECTION AND DISEASE

Viruses are the smallest parasites with the simplest biological structure—essentially small packaged particles of DNA or RNA that depend on the host cell for their qualities of life. Viruses have special adaptations for entering a cell and for reprogramming its genetic and molecular machinery to produce and release new viruses. Animal viruses are divided into families based on the nature of the nucleic acid (DNA or RNA), the type of capsid, and whether or not an envelope is present. All DNA viruses are double-stranded except for the parvoviruses, which have single-stranded DNA. All RNA viruses are single-stranded except for the double-stranded reoviruses. The genome of RNA viruses can be further described as segmented (consisting of more than one molecule) or nonsegmented (consisting of a single molecule). The envelope is derived in part from host cell membranes as the virus buds off the nuclear envelope or cell membrane, and it often contains spikes that interact with the host cell (use tables 24.1 and 25.1 for references to virus structure).

IMPORTANT MEDICAL CONSIDERATIONS IN VIRAL DISEASES

Target Cells

Because receptors on the virus surface react specifically with a molecule on the cell surface, infectiousness is limited to a particular host or cell type in most cases. Various viruses target nearly all types of tissues, including those of the nervous system (polio and rabies), liver (hepatitis), respiratory tract (influenza, colds), intestine (polio), skin and mucous membranes (herpes, pox), and cells of the immune system (AIDS). Most DNA viruses are assembled and budded off the nucleus, whereas most RNA viruses multiply in and are released from the cytoplasm. The presence of assembling and completed viruses in the cell gives rise to various telltale disruptions called cytopathic effects. Most cells productively infected with viruses are destroyed, which accounts for the sometimes severe pathology and loss of function.

Scope Of Infections

Viral disease range from very mild or asymptomatic infections such as colds to severe, deadly syndromes such as rabies and AIDS. Many are so-called childhood diseases that occur primarily in the young and are readily transmitted through droplets. Although we tend to consider these infections as self-limited and inevitable, even measles, mumps, chickenpox, and rubella can manifest severe complications. Many diseases are strictly human in origin, but several such as rabies, yellow fever, and viral encephalitis are zoonoses transmitted by vectors.

The course of viral diseases starts with invasion at the portal of entry by a few virions and a primary infection. In some cases (influenza, colds), the viruses replicate locally and disrupt the tissue, and in others (mumps, polio), the virus enters the blood and creates complications in tissues far from the initial infection. Common manifestations of virus infections are rashes, fever, muscle aches, respiratory involvement, and swollen glands—symptoms that can be too nonspecific to be of use in diagnosis. Many diseases of unknown etiology are thought to have a viral basis. Examples include type I diabetes, multiple sclerosis, and chronic fatigue syndrome (see microfile 24.3).

Protection in viral infections arises from the combined actions of interferon, neutralizing antibodies, and cytotoxic T cells. Because of the nature of the immune response to viruses, infections frequently result in lifelong immunities, and many viruses are adaptable to vaccines.

Latency and Oncogenicity

Most DNA viruses and a few RNA viruses can become permanent residents of the host cell. In some cases, the latent virus alternates between periods of relative inactivity and recurrent infections. Viruses can persist in a more permanent state by splicing their DNA into a site in the host's DNA. This process sometimes transforms the cell into a cancer cell. This potential for oncogenesis is more likely with DNA viruses, though retroviruses, which can convert their RNA to DNA, are also important in cancers. This property of combining with host DNA has been a drawback in developing live, attenuated DNA virus vaccines.

Teratogenicity, Congenital Defects, and Viral Diagnosis

Several viruses can cross the placenta from an infected mother to the embryo or fetus. Their infection of the fetus often causes developmental disturbances and permanent defects in the child that are present at birth (congenital). Among viruses with known **teratogenic*** effects are rubella, cytomegalovirus, and adenovirus. Another risk for infants involves infection at the time of birth, as occurs with hepatitis B and herpes simplex.

Viral diseases are diagnosed by a variety of clinical and laboratory methods, depending on the virus. Some can be diagnosed by symptoms alone; others require isolation in cell or animal culture. Serological testing for antibodies can be of importance in many viral infections. The use of antigen detection assays is on the increase, especially those based on monoclonal antibodies. Rapid nucleic acid probes directed against DNA or RNA are another option for several viruses. These have the detect advantage of being so sensitive that they could conceivably detect the viral nucleic acid in a single infected cell. Review figure 6.24 for a summary of identification techniques.

*teratogenic (tur″-ah-toh-jen′-ik) Gr. *teratos,* monster. Any process that causes physical defects in the embryo or fetus.

TABLE 24.1

DNA VIRUS FAMILIES

Enveloped

***Poxviridae*: Small pox, Molluscum contagiosum**

Complex structure, lack capsid

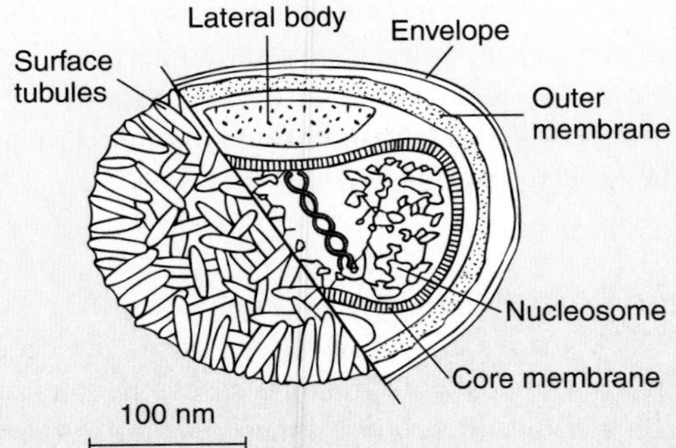

***Herpesviridae*: Cold sores, Mononucleosis**

Bud off nucleus, tend to become latent

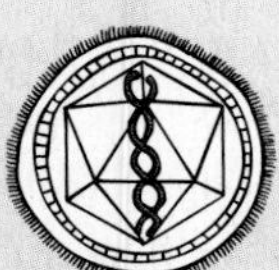

***Hepadnaviridae*: Hepatitis B**

Unusual genome containing both double- and single-stranded DNA

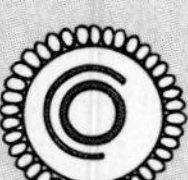

Nonenveloped

***Adenoviridae*: Common cold, Keratoconjunctivitis**

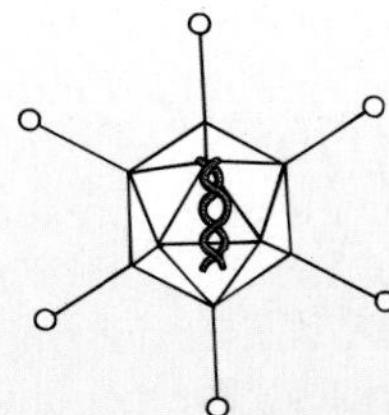

***Papovaviridae*: Common and genital warts, Leucoencephalopathy**

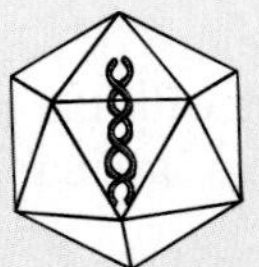

***Parvoviridae:* Erythema contagiosum**

Unusual single-stranded DNA genome

Legend

Genome strandedness

Double (DS) DNA

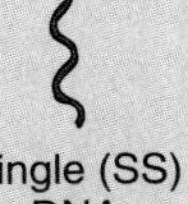
Single (SS) DNA

Source: Poxviridae *from Buller et al., National Institute of Allergy & Infectious Diseases, Department of Health & Human Services.*

Chapter Checkpoints

Viruses are particles of parasitic DNA or RNA. Most DNA viruses are double-stranded, except for the single-stranded parvoviruses. RNA viruses are single-stranded except for the double-stranded reoviruses. Viral infectivity requires a specific viral receptor site on the host cell, but most host tissues possess receptor sites for one or more viruses.

Viral diseases vary in severity, depending on virulence of the virus and age, health, and habitat of the human host. Lifelong immunity develops to some but not all viral agents. Virus infection can be diagnosed by overt symptoms, cultures, antigen detection, and nucleic acid probes.

Most DNA and some RNA viruses can cause chronic infections and combine with the host genome. They also have the potential to activate host oncogenes.

SURVEY OF DNA VIRUS GROUPS

The DNA viruses that cause human diseases are placed into six groups based upon the existence of an envelope, the nature of the genome (double-stranded or single-stranded), size, and target cells (see table 24.1). There are six major groups of DNA viruses. The enveloped DNA viruses include the poxviruses, herpesviruses, and hepadnaviruses. The nonenveloped group includes the adenoviruses, papovaviruses, and parvoviruses.

Enveloped DNA Viruses

Poxviruses: Classification and Structure

Poxviruses produce eruptive skin pustules called **pocks** or **pox** (figure 24.1), which leave small, depressed scars (pockmarks) upon healing. The poxviruses are distinctive because they are the largest and most complex of the animal viruses (see table 24.1), they have the largest genome of all viruses, and they multiply in the cytoplasm in well-defined sites called factory areas, which appear as inclusion bodies in infected cells.

Among the best known poxviruses are **variola,*** the agent of **smallpox,** and vaccinia, a closely related virus used in vaccinations. Other members are of interest to the veterinarian for their role in diseases of domestic or wild animals. A common feature of all poxvirus infections is a specificity for the cytoplasm of epidermal cells and subcutaneous connective tissues. In these sites, they produce the typical lesions and also tend to stimulate cell growth that can lead to tumor formation.

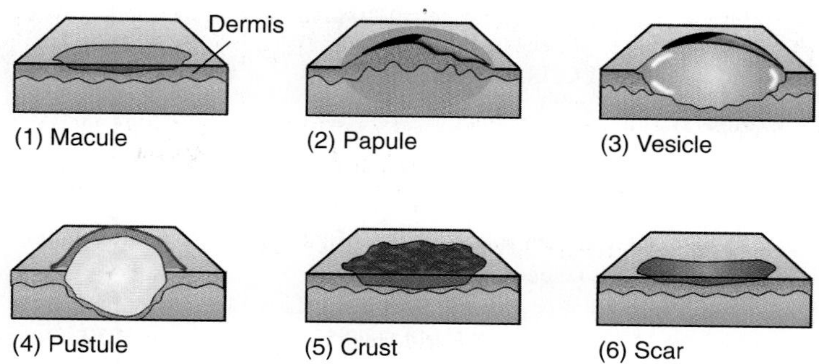

Figure 24.1
Stages in pock development. Because the infection occurs at the skin-forming level (dermis), these pocks will leave a scar.

Smallpox: A Perspective

Largely through the World Health Organization's comprehensive global efforts, smallpox is now a disease of the past. It is one of the few infectious agents that now exists only in government laboratories. (See microfile 24.1 for a discussion of this stunning achievement.) At one time, smallpox was ranked as one of the deadliest infectious diseases, leaving no civilization untouched. When introduced by explorers into naive populations (those never before exposed to the disease) such as Native Americans and Hawaiians, smallpox caused terrible losses and was instrumental in destroying those civilizations. Even as recently as 1967, approximately 10 to 15 million cases occurred worldwide, with thousands of fatalities.

Disease Manifestations Exposure to smallpox usually occurred through inhalation of droplets or skin crusts. Infection was associated with fever, malaise, prostration, and, later, a rash that began in the pharynx, spread to the face, and progressed to the extremities. Initially, the rash was macular, evolving in turn to papular, vesicular, and pustular before eventually crusting over, leaving nonpigmented sites pitted with scar tissue (figure 24.1). The two principal forms of smallpox were variola minor and variola major. Variola major was a highly virulent form that caused toxemia, shock, and intravascular coagulation. People who survived any form of smallpox nearly always developed lifelong immunity.

Laboratory Diagnoses Smallpox was usually diagnosed by clinical signs and symptoms. It must be differentiated from chickenpox, disseminated herpes, vaccinia, monkeypox, and certain nonviral lesions. Scanning stained smears of vesicular fluid for cytoplasmic inclusion bodies or election microscopy provide additional evidence. The virus can be isolated by inoculating the chorioallantoic membrane of chicken eggs with specimen and looking for pocks (see figure 6.22).

Smallpox Vaccination The smallpox vaccine uses a single drop of vaccinia virus punctured into the skin with a double-pronged needle or jetgun. In a successful take, a large, scablike pustule with a reddened periphery develops at the infection site, where it leaves a permanent scar. The vaccine protects people for up to 10 years against variola as well as other poxvirus infections. Smallpox vaccinations have been discontinued for most populations not only because smallpox no longer poses a threat, but also because of the slight risk of vaccinia complications, especially in immunodeficient and allergic people. Vaccination is being considered as a way to control an epidemic of monkeypox in Africa. The vaccinia virus remains an important vehicle for developing genetically engineered vaccines for other diseases (see chapter 16).

Other Poxvirus Diseases

An unclassified poxvirus causes a skin disease called **molluscum contagiosum*** (figure 24.2*a*). This disease is distributed throughout the world, with highest incidence occurring in certain Pacific islands. In endemic regions, it is primarily an infection of children and is transmitted by direct contact and fomites. The moist tropical climate appears to favor skin lesions, which take the form of smooth, waxy nodules on the face, trunk, and limbs. The infection can also be spread by sexual intercourse and is most common in sexually active young people. Most lesions occur on the pubic, genital, and thigh regions (figure 24.2*b*). They are firm, smooth, and translucent; from 2 to 5 mm in diameter; and they may be spread to other body sites through scratching and self-inoculation. AIDS patients suffer from an atypical version of the disease, which attacks the skin of the face and forms giant tumorlike growths.

Treatment requires destruction of the virus by freezing (cryotherapy), electric cautery, and chemical agents applied directly to the lesions.

*variola (ver-ee-oh′-lah) L. *varius,* varied, mottled.

*molluscum contagiosum (mah-lusk′-uhm kahn″-taj-ee-oh-sum) L. *molluscus,* soft. In reference to the soft, rounded cutaneous papules.

MICROFILE 24.1 THE ULTIMATE END TO AN ANCIENT SCOURGE

The thirty-third World Health Assembly "declares solemnly that the world and all its peoples have won freedom from smallpox... an unprecedented achievement in the history of public health..." (Resolution 33-3, May 8, 1980, Geneva, Switzerland)

A concerted vaccination effort against smallpox was first instigated by the United Nations' World Health Organization in 1959. Even seven years later, large numbers of cases were still being reported in Africa, Southeast Asia, Indonesia, and Brazil. In 1966 the WHO launched an even more intensive effort. A medical team of 100,000 people began a meticulous village-by-village campaign to identify and report population reservoirs and to isolate and vaccinate specific high-risk groups or individuals. Over the next 10 years, this strategy constantly diminished the numbers of cases, and, in 1977, the last natural case was reported in Somalia. The main reasons for this program's success are that only humans with active cases of smallpox are infectious and variola is not a latent virus nor is it harbored by healthy carriers. These features made it easier to trace the disease and vaccinate in communities where it was prevalent.

Presently, only the United States and Russia have stocks of viable variola virus. After 10 years of debate concerning the safety of these stocks and the possibility of a laboratory accident, scientists from both countries agreed to steam-sterilize the last batches of virus. When researchers from around the world protested at the deliberate destruction of a unique virus, the decision was put off until several research labs have completed a genome map for further study. A group of representatives from 190 countries has decreed that the remaining stocks of virus will be permanently destroyed on midnight, June 30, 1999.

Ultimately, the victory over smallpox has demonstrated that cooperative efforts and a program of vaccination can succeed, even on a planet of nearly 6 billion people. Hoping for similar success with other agents, the WHO has expanded its drive for immunization against six major childhood killers—measles, polio, diphtheria, whooping cough, tetanus, and tuberculosis. So far, the disease showing greatest promise for total eradication is polio, which has now been declared officially eradicated from the Americas as of August, 1991, and is almost completely controlled in India, Africa, and the Western Pacific.

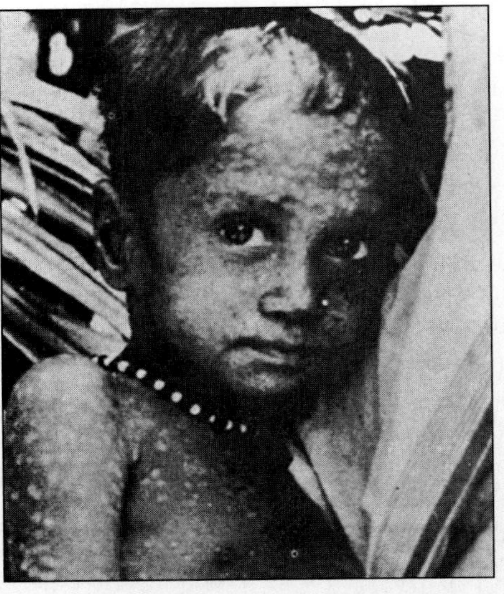

A child in Bangladesh with the last recorded case of variola major (the more serious form of smallpox) in 1975.

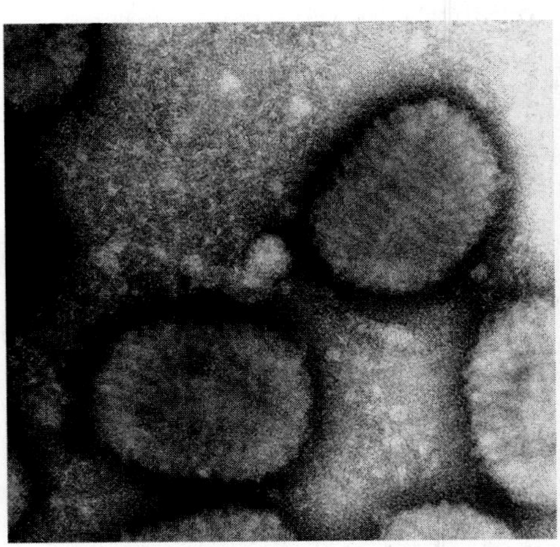

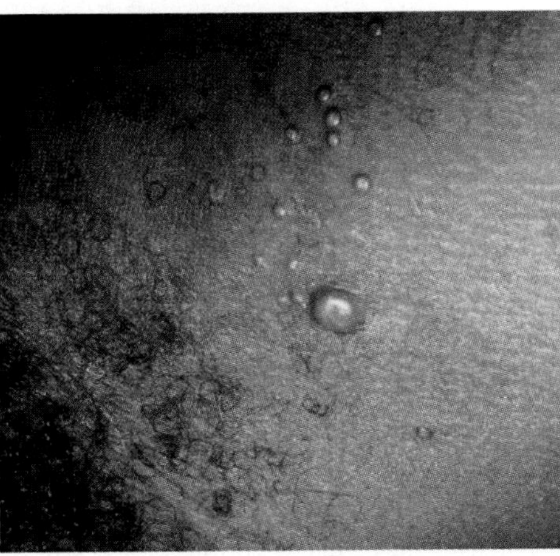

Figure 24.2
Molluscum contagiosum. (*a*) Virus particles have a typical poxvirus structure with prominent surface tubules. These viruses fill the cytoplasm of infected epithelial cells, causing a prominent mass called a molluscum body. (*b*) A sexually acquired infection. The skin eruption takes the form of small, waxy papules in the genital region.

Many mammalian groups host some sort of poxvirus infection. Cowpox, rabbitpox, monkeypox, mousepox, camelpox, buffalopox, and elephantpox are among the types that occur. (But chickenpox is a herpesvirus infection, not a poxvirus, and it does not occur in chickens!). However, only monkeypox and cowpox appear to be capable of infecting humans symptomatically.

The fact that humans are also susceptible to monkeypox virus has produced a new concern in parts of central and western Africa. For the past two years, outbreaks have affected hundreds of people, predominantly children. Historically, this disease occurred sporadically in humans having contact with monkeys, squirrels, or rats—the traditional vectors. But nearly three out of four of these

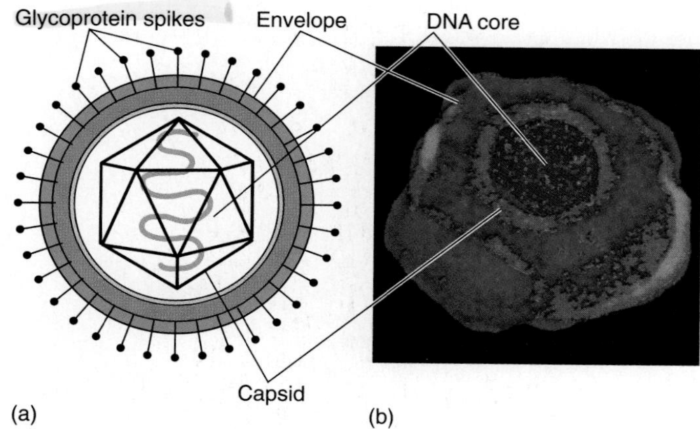

Figure 24.3

(*a*) A schematic of the general structure of herpesvirus alongside (*b*) a color-enhanced view of cytomegalovirus.

latest cases were acquired within communities and families. This changing pattern is thought to be partly due to the discontinuation of smallpox vaccination that had formerly protected against monkeypox as well. Noting that the attack and mortality rates have surpassed any other recorded outbreaks also suggests that the virus is becoming more virulent for humans. The disease manifestations are very similar to smallpox, with skin pocks, fever, and swollen lymph nodes. Because of the potential for this emerging disease to spread rapidly through the susceptible populations, health authorities are reinstating a vaccine program in regions with greatest risk and are keeping close track of new cases.

Although cowpox is an eruptive cutaneous disease that can develop on cows' udders and teats, cows are not the actual reservoirs of this virus. Other mammals such as rodents, cats, and even zoo animals can carry it. Human infection is rare and usually confined to the hands, although the face and other cutaneous sites can be involved.

The Herpesviruses: Common, Persistent Human Viruses

Herpesvirus* was named for the tendency of some herpes infections to produce a creeping rash. It is the common name for a large family whose members include herpes simplex 1 and 2 (HSV), the cause of fever blisters and genital infections; herpes zoster (VZV), the cause of chickenpox and shingles; cytomegalovirus (CMV), which affects the salivary glands and other viscera; Epstein-Barr virus (EBV), associated with infection of the lymphoid tissue; and some recently identified viruses (herpesvirus-6, -7, and -8). Prominent features of the family are its tendency toward viral latency and recurrent infections. This behavior often involves incorporation of viral nucleic acid into the host genome and carries with it the potential for oncogenesis. Two of the viruses in this group have a well-documented association with cancers, and others have suspected involvement.

*herpes (her'-peez) Gr. *herpein,* to creep.

Table 24.2

Comparative Epidemiology and Pathology of Herpes Simplex, Types 1 and 2

	HSV-1	HSV-2
Usual Etiologic Agent of	Herpes labialis Ocular herpes Gingivostomatitis Pharyngitis	Herpes genitalis*
Transmission	Close contact, usually of face	Sexual or close contact
Latency	Occurs in trigeminal ganglion	Occurs primarily in sacral ganglia
Skin Lesions	On face, mouth	On internal, external genitalia, thighs, buttocks
Complications		
Whitlows	Among personnel working on oral cavity	Among obstetric, gynecological personnel
Neonatal Encephalitis	Causes up to 30% of cases**	Causes most cases

*The other herpes simplex type can be involved in this infection, though not as commonly.
**Due to mothers infected genitally by HSV-1 or contamination of the neonate by oral lesions.

Virtually everyone becomes infected with a herpesvirus at some time, usually without adverse effect. The clinical complications of latency and recurrent infections become more severe with advancing age, organ transplants, cancer chemotherapy, or other conditions that compromise the immune defenses. The herpesviruses are among the most common and serious opportunists among AIDS patients. Nearly 95% of this group will experience recurrent bouts of skin, mucous membrane, intestinal, and eye disease from HSV, VZV, CMV, and EBV.

The herpesviruses are among the larger viruses, with a diameter ranging from about 150 nm to 200 nm (figure 24.3). They are enclosed within a loosely fitting envelope that contains glycoprotein spikes. Like other enveloped viruses, herpesviruses are prone to deactivation by organic solvents or detergents and are relatively unstable outside the host's body. The icosahedral capsid houses a core of double-stranded DNA that winds around a proteinaceous spindle in some viruses. Replication occurs primarily within the nucleus, and viral release is usually accompanied by cell lysis.

General Properties of Herpes Simplex Viruses

Humans are susceptible to two varieties of **herpes simplex viruses (HSVs):** HSV-1, characterized by lesions on the oropharynx, and HSV-2, with lesions on the genitalia, differentiated by several characteristics (table 24.2). Even though laboratory mice, guinea pigs, rabbits, hamsters, and other animals can be experimentally infected, humans appear to be the only natural reservoir for herpes simplex virus.

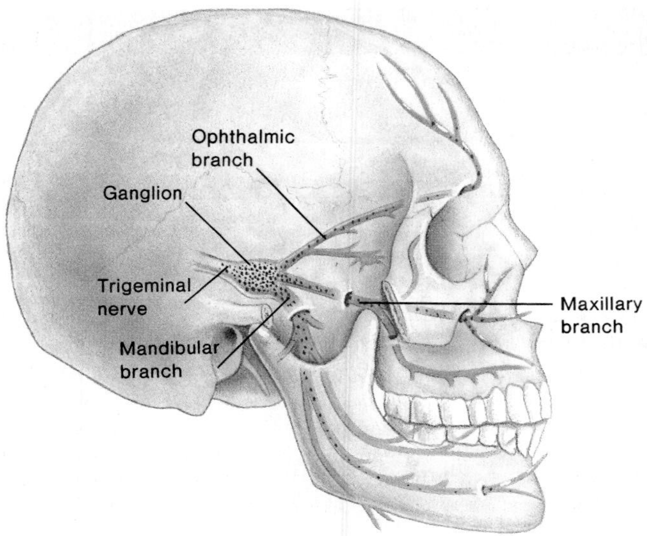

Figure 24.4
The site of latency and routes of recurrence in herpes simplex, type 1. Following primary infection, the virus invades the trigeminal, or fifth cranial, nerve and is harbored in its ganglion. Various provocative stimuli cause it to migrate back to the skin surface by the mandibular, maxillary, or ophthalmic branch.

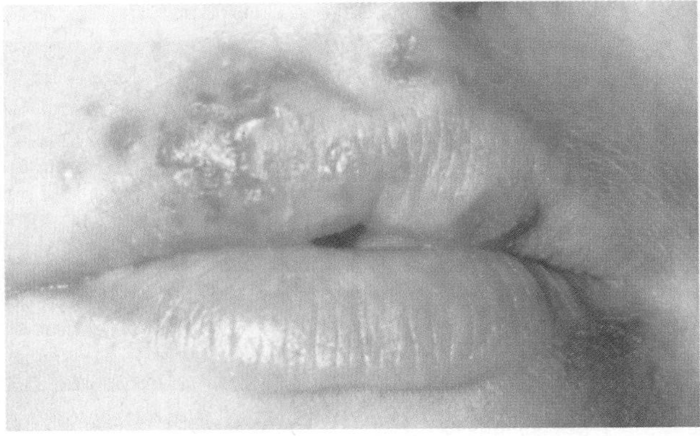

(a)

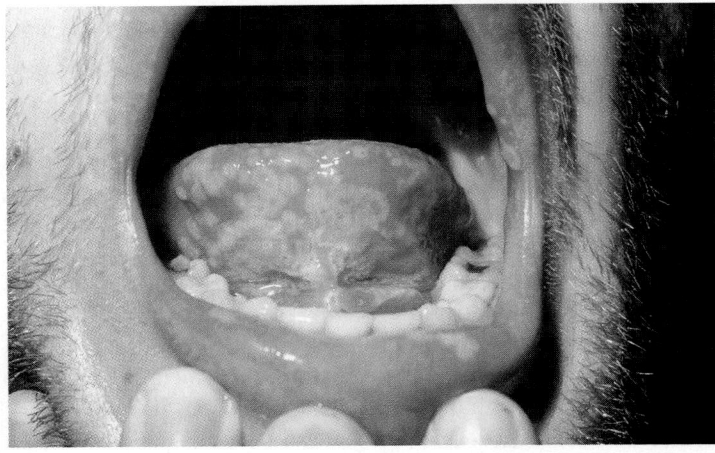

(b)

Figure 24.5
(*a*) Recurrent lesions of herpes labialis (coldsore, fever blister). Tender, itchy papules erupt on the perioral region and progress to vesicles that burst, drain, and scab over. These sores and fluid are highly infectious and should not be touched. (*b*) Primary herpetic gingivostomatitis involving the entire oral mucosa, tongue, cheeks, and lips.

Epidemiology of Herpes Simplex

The herpes simplex viruses have cosmopolitan distribution. Infection occurs in all seasons and among all age groups, and transmission is promoted by direct exposure to secretions containing the virus. The viruses are relatively sensitive to the environment but can remain infectious in moist secretions on inanimate objects for a few hours. People with active lesions are the most significant source of infection, though studies indicate that genital herpes can be transmitted even when no lesions are present.

Primary infections tend to be age-specific: HSV-1 frequently occurs in infancy and early childhood, and, by adulthood, most people exhibit some serological evidence of infection. On the other hand, primary infection with HSV-2 occurs most frequently between the ages of 14 and 29, a pattern that reflects the sexual route of transmission. In fact, genital herpes is one of the most common STDs in the United States. Instances of HSV-1 genital infections and HSV-2 infections of the oral cavity have recently risen. Such cases are probably caused by autoinoculation with contaminated hands or by oral sexual contact.

The Nature of Latency and Recurrent Attacks In approximately 20% to 50% of primary infections, herpes simplex viruses enter the distal regions of sensory neurons and travel to the dorsal root ganglia. Type 1 HSV enters primarily the trigeminal, or fifth cranial, nerve, which has extensive innervations in the oral region (figure 24.4). Type 2 HSV usually becomes latent in the ganglion of the lumbosacral spinal nerve trunk. The exact details of this unusual mode of latency are not yet fully understood, but the virus remains inside the neuron in a nonproliferative stage for a variable time. Recurrent infection is triggered by various stimuli such as fever, UV radiation, stress, or mechanical injury, all of which reactivate the virus. The virus migrates to the body surface and produces a local skin or membrane lesion, often in the same site as a previous infection. The number of such attacks can range from one to dozens each year.

The Spectrum of Herpes Infection and Disease

Herpes simplex infections, often called simply herpes, usually target the mucous membranes. The virus enters cracks or cuts in the membrane surface and then multiplies in basal and epithelial cells in the immediate vicinity. Inflammation, edema, cell lysis, degeneration, and a ballooning effect that produces a characteristic thin-walled vesicle follow. The main diseases of herpes simplex viruses are facial herpes (oral, optic, and pharyngeal), genital herpes, neonatal herpes, and disseminated disease.

Type 1 Herpes Simplex in Children and Adults **Herpes labialis,** otherwise known as **fever blisters** or **coldsores,** is the most common recurrent HSV-1 infection. Vesicles usually crop up on the mucocutaneous junction of the lips or on adjacent skin (figure 24.5*a*). A few hours of tingling or itching precede the

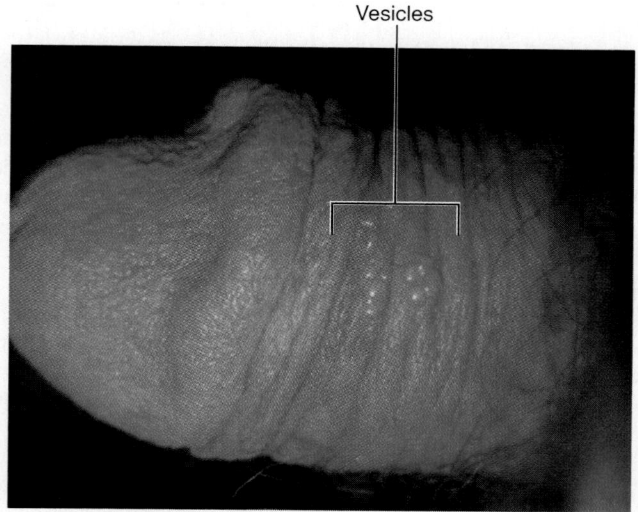

Figure 24.6

Genital herpes. Vesicles start out separate, but become confluent and ulcerate into painful red erosions so tender that inspection is difficult.

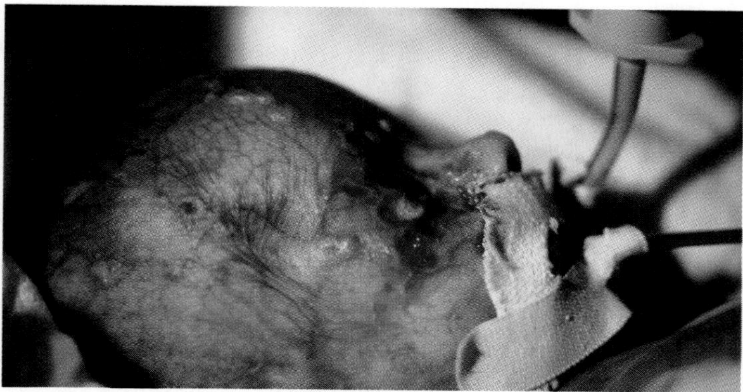

Figure 24.7

Neonatal herpes simplex. This premature infant was born with the classic "cigarette burn" pattern of HSV infection. Babies can be born with the lesions or develop them one to two weeks after birth.

formation of one or more vesicles at the site. Pain is most acute in the earlier stage as vesicles ulcerate; lesions crust over in 2 or 3 days and heal completely in a week.

Although most herpetic infections are mild or asymptomatic, an occasional manifestation known as herpetic *gingivostomatitis* can occur (figure 24.5*b*). This infection of the oropharynx, usually due to HSV-1, strikes young children most frequently. Inflammation of the oral mucosa can involve the gums, tongue, soft palate, and lips, sometimes with ulceration and bleeding. A common complication in adolescents is pharyngitis, a syndrome marked by sore throat, fever, chills, swollen lymph nodes, and difficulty in swallowing.

Herpetic keratitis (also called ocular herpes) is an infective inflammation of the eye in which a latent virus travels into the ophthalmic rather than the mandibular branch of the trigeminal nerve. Preliminary symptoms are a gritty feeling in the eye, conjunctivitis, sharp pain, and sensitivity to light. Some patients develop characteristic branched or opaque corneal lesions as well. In 25% to 50% of cases, keratitis is recurrent and chronic and can lead to visual loss requiring surgery to replace the cornea.

Type 2 Herpes Infections Type 2 herpes infection usually coincides with sexual maturation and an increase in sexual encounters. **Genital herpes** (herpes genitalis) starts out with malaise, anorexia, fever, and bilateral swelling and tenderness in the groin. Then, intensely sensitive vesicles break out in clusters on the genitalia, perineum, and buttocks (figure 24.6). The chief symptoms are urethritis, painful urination, cervicitis, itching, and central nervous system involvement. After a period of days to weeks, the vesicles ulcerate and develop a light-gray exudate before healing. Bouts of recurrent genital herpes in either sex are usually less severe than the original infection and are triggered by menstruation, stress, and concurrent bacterial infection.

Herpes of the Newborn Although HSV infections in healthy adults are annoying, unpleasant, and painful, only rarely are they life-threatening. However, in the neonate and fetus (figure 24.7), HSV infections are very destructive and, in some cases, fatal. Most cases of infection in this group occur when infants are contaminated by the mother's reproductive tract immediately before or during birth (microfile 24.2). Neonatal infections have also been traced to hand transmission from the mother's lesions to the baby. Because type 2 is more often associated with genital infection, it is more frequently involved; however, type 1 infection has similar complications. In infants whose disease is confined to the mouth, skin, or eyes, the mortality rate is 30%, but disease affecting the central nervous system has a 50–80% mortality rate.

Miscellaneous Herpes Infections When either form of herpesvirus enters a break in the skin, a local infection can develop. This route of infection is most often associated with occupational exposure or severely damaged skin. A hazard for health care workers who handle patients or their secretions without hand protection is a disease called herpetic **whitlow.*** Workers in the fields of obstetrics, gynecology, dentistry, and respiratory therapy are probably at greatest risk for contracting it. Whitlows are deep-set, usually occur on one finger, and are extremely painful and itchy (figure 24.8). Afflicted personnel should not work with patients until the whitlow has healed, usually in 2 to 3 weeks.

Life-Threatening Complications Although herpes simplex encephalitis is a rare complication of type 1 infection, it is probably the most common sporadic form of viral encephalitis in the United States. The infection disseminates along nerve pathways to the brain or spinal cord. The effects on the CNS begin with headache and stiff neck and can progress to mental disturbances and coma. The fatality rate in untreated cases is 70%. Patients with underlying immunodeficiency are more prone to

*whitlow (hwit′-loh) MI: *white,* flaw. An abscess on the distal portion of a finger.

MICROFILE 24.2 DEALING WITH PREGNANCY AND HERPES

Because of the danger of herpes to fetuses and newborns and also the increase in the number of cases of genital herpes, it is now standard procedure to screen pregnant women for the herpesvirus early in their prenatal care. Women with a history of recurrent infections must be constantly monitored for any signs of viral shedding, especially in the last 4 weeks of pregnancy. If no evidence of recurrence is seen, vaginal birth is indicated, but any evidence of an outbreak at the time of delivery necessitates a cesarian section. A pregnant woman with active infection and ruptured membranes presents a difficult clinical situation, because these conditions can cause the infant to become infected before birth and negate the protective effect of a cesarian section.

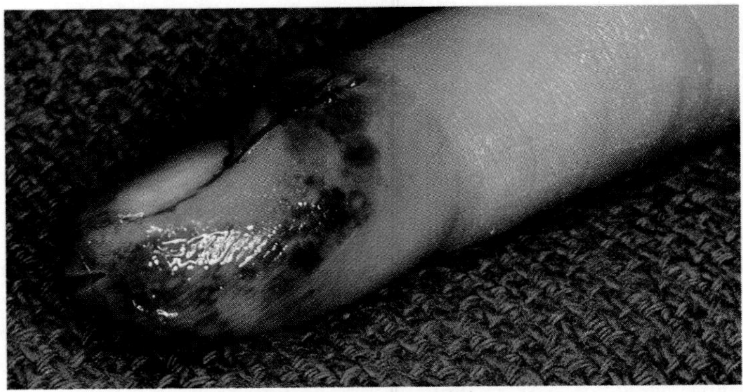

Figure 24.8
Herpetic whitlows. These painful, deepset vesicles can become inflamed and necrotic and are difficult to treat. They occur most frequently in dental and medical personnel or people who carelessly touch lesions on themselves or others.

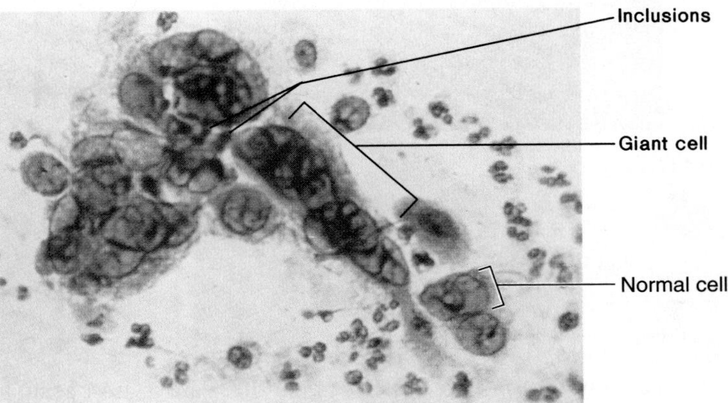

Figure 24.9
Direct cytologic diagnosis of herpesvirus infection. A direct Papanicolaou (Pap) smear of a cervical scraping shows typical enlarged (multinucleate giant) cells and intranuclear inclusions. This technique is not specific for HSV, but most other herpesviruses do not infect the reproductive mucosa.

severe, disseminated herpes infection than are immunocompetent patients. Of greatest concern are patients receiving organ grafts, cancer patients on immunosuppressive therapy, those with congenital immunodeficiencies, and AIDS patients.

Diagnosis, Treatment, and Control of Herpes Simplex

Small, painful, vesiculating lesions on the mucous membranes of the mouth or genitalia, lymphadenopathy, and exudate are typical diagnostic symptoms of herpes simplex. Further diagnostic support is available by examining scrapings from the base of such lesions stained with Giemsa, Wright, or Papanicolaou (Pap) methods (figure 24.9). The presence of multinuclear cells, giant cells, and intranuclear eosinophilic inclusion bodies can help establish herpes infection. This method, however, will not distinguish among HSV-1, HSV-2, or other herpesviruses, which require more specific subtyping.

Laboratory culture and specific tests are essential for diagnosing immunosuppressed and neonatal patients with severe, disseminated herpes infection. A specimen of tissue or fluid is introduced into a primary cell line such as monkey kidney or human embryonic kidney tissue cultures and is then observed for cytopathic effects within 24 to 48 hours. Direct tests on specimens or cell cultures using fluorescent antibodies or DNA probes can differentiate among HSV-1, HSV-2, and closely related herpesviruses. Serological analysis is useful for primary infection but is inconclusive for recurrent illness, because only 10% of patients show increased antibody titers.

Several agents are available for treatment. Acyclovir (Zovirax) is the most effective therapy developed to date. This drug, in use since the early 1980s, is relatively nontoxic yet shows high specificity for the herpes simplex virus. Topical medications applied to primary genital and oral lesions cut the length of infection and reduce viral shedding. In addition, systemic therapy is available for more serious complications. The more toxic drugs idoxuridine and trifluridine have been used for topical treatment of herpes keratitis, and vidarabine has been used for complications of disseminated herpes. Over-the-counter coldsore medications containing menthol, camphor, and local anesthetics lessen pain and may protect against secondary bacterial infections, but they probably do not affect the progress of the viral infection. Some protection in suppressing coldsores can be obtained from the amino acid lysine, taken orally in the earliest phases of recurrence. Although its mechanism is not understood, it can apparently abort the development of the lesions.

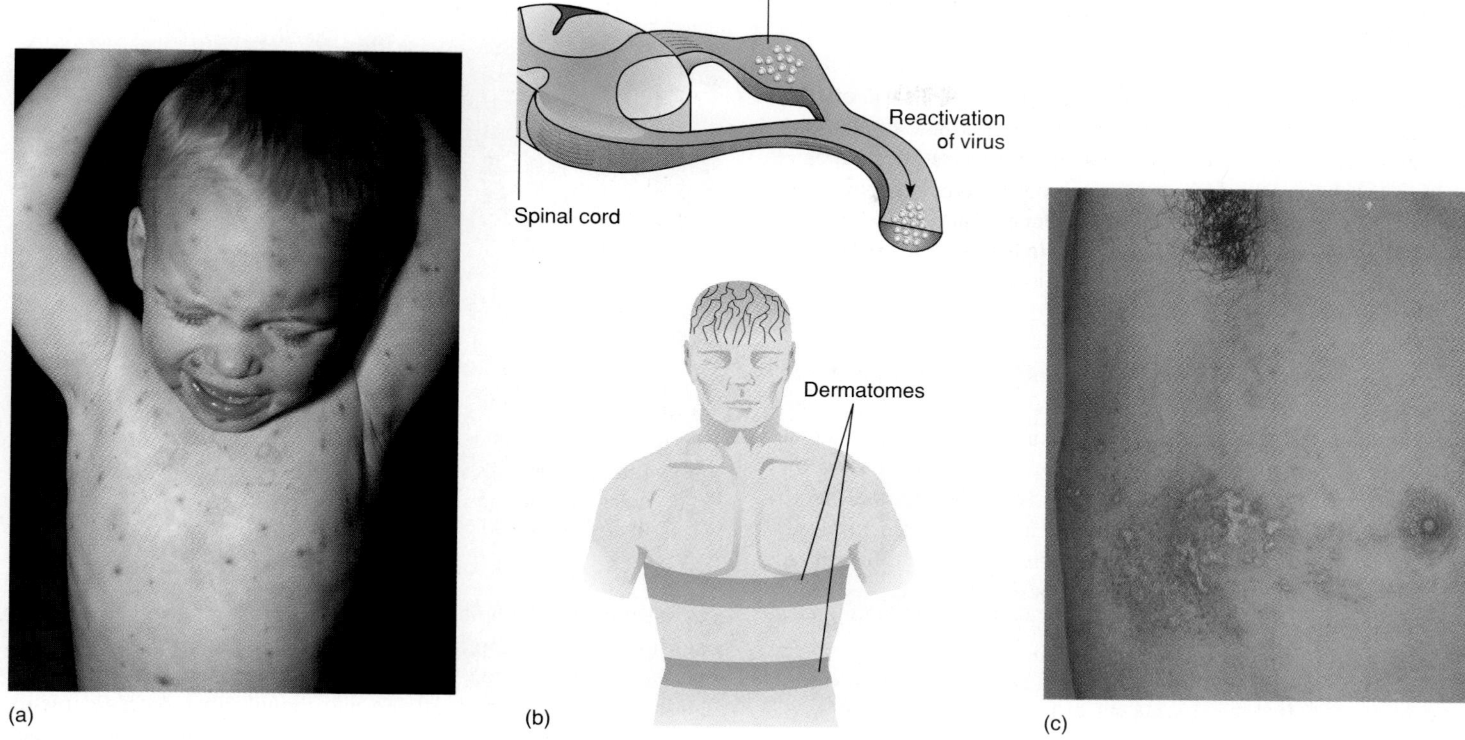

Figure 24.10

The relationship between varicella (chickenpox) and zoster (shingles) and the clinical appearance of each. (*a*) First contact with the virus (usually in childhood) results in a macular, papular, vesicular rash distributed primarily on the face and trunk. (*b*) The virus becomes latent in the dorsal ganglia of nerves that supply dermatomes of mid-thoracic nerves and the cranial nerves that supply facial regions. (*c*) The clinical appearance of shingles. Early symptoms are acute pain in the nerve root and redness of the dermatome, followed by a vesicular-papular rash on the chest and back that is usually asymmetrical and does not cross the midline of the body.

New evidence indicates that a daily dose of oral acyclovir taken for a period of 6 months to one year can be effective in preventing recurrent genital herpes. Although the barrier protection afforded by condoms can diminish the prospects of spreading genital herpes sexually, people with active infection should avoid sexual contact. Mothers with coldsores should observe extreme care in handling their newborns; infants should never be kissed on the mouth; and hospital attendants with active oral herpes infection must be barred from the newborn nursery. Medical and dental workers who deal directly with patients can reduce their risk of exposure by wearing gloves.

The Biology of Varicella-Zoster Virus

Another herpesvirus causes both **varicella*** (commonly known as **chickenpox**) and a recurrent infection called **herpes zoster,*** or **shingles.*** Because the same virus causes two forms of disease, it is known by a composite name, **varicella-zoster virus (VZV).** At one time, the differences in clinical appearance and time separation between varicella and herpes zoster were sufficient to foster the notion that they were caused by different infectious agents. But about 100 years ago, the observation that young children acquired chickenpox from family members afflicted with shingles gave the first hint that the two diseases were related. More recently, doctors followed the diseases in a single patient using Koch's postulates and demonstrated that the virus is the same in both diseases and that zoster is a reactivation of a latent varicella virus (figure 24.10).

Epidemiologic Patterns of VZV Infection Humans are evidently the only natural hosts for varicella-zoster virus. The virus is harbored in the respiratory tract, but it is communicable from both respiratory droplets and the fluid of active skin lesions. Infected persons are most infectious a day or two prior to the development of the rash. The dried scabs are not infectious because the virus is unstable and loses infectivity when exposed to the environment. Patients with shingles are a source of infection for nonimmune children, but the attack rate is significantly lower than with exposure to chickenpox. Only in rare instances will a child develop chickenpox more than once, and even a subclinical case can result in long-term immunity. People who are immune will not acquire chickenpox or shingles from a person who has an active case of either disease.

*varicella (var″-ih-sel′-ah) The Latin diminutive for smallpox or variola.

*zoster (zahs′-tur) Gr. *zoster,* girdle.

*shingles (shing′-gulz) L. *cingulus,* cinch. (Both *zoster* and *shingles* refer to the beltlike encircling nature of the rash.)

Varicella (Chickenpox) The respiratory epithelium serves as the chief portal of entry and the initial site of viral replication for the chickenpox virus, which initially produces no symptoms. Regional lymph nodes support secondary multiplication before the virus is detectable in the blood. After an incubation period of 10 to 20 days, the first symptoms to appear are fever and an abundant rash that begins on the scalp, face, and trunk and radiates in sparse crops to the extremities. Skin lesions progress quickly from macules and papules to itchy vesicles that encrust and drop off, usually healing completely but sometimes leaving a tiny pit or scar (figure 24.10*a*). Lesions number from a dozen to hundreds and are more abundant in adolescents and adults than in young children.

Herpes Zoster (Shingles) In an unknown percentage of individuals, recuperation from varicella is associated with the entry of the varicella-zoster virus into the sensory endings that innervate dermatomes, regions of the skin supplied by the cutaneous branches of nerves, especially the thoracic and trigeminal nerves (figure 24.10*b*). Like HSV, VZV migrates toward the central nervous system and becomes latent in the sensory ganglia. In contrast to herpes simplex, however, zoster generally recurs only once or possibly twice, with a characteristic asymmetrical distribution on the skin of the trunk or head (figure 24.10*c*).

Shingles develops abruptly after reactivation by such stimuli as X-ray treatments, immunosuppressive and other drug therapy, surgery, or developing malignancy. In fact, up to 50% of patients suppressed by radiation therapy, leukemia, or similar malignancies manifest recurrence. The virus is believed to migrate down the ganglion to the skin, where multiplication resumes and produces crops of tender skin vesicles typical of varicella, except that they can persist for weeks. Inflammation of the ganglia and the pathways of intercostal nerves can cause pain and tenderness (radiculitis) that can last for several months. Involvement of cranial nerves can lead to eye inflammation and ocular and facial paralysis.

Diagnosis, Treatment, and Control The cutaneous manifestations of varicella and shingles are sufficiently characteristic for ready clinical recognition. Supportive evidence for a diagnosis of varicella usually comes from a recent history of close contact with a source or with an active case of VZV infection. A diagnosis of shingles is supported by the distribution pattern of skin lesions and a history of varicella. Further laboratory confirmation is achieved by identifying multinucleate giant cells in stained smears prepared from vesicle scrapings. However, unequivocal identification or differentiation of herpes simplex from herpes zoster is best done with fluorescent antibody detection of viral antigen in skin lesions, DNA probe analysis, and culture.

Uncomplicated varicella is self-limited and requires no therapy aside from alleviation of discomfort. Secondary bacterial infection that can cause dangerous complications (especially streptococcal and staphylococcal) is prevented by application of anaesthetic and antimicrobic ointments. The use of aspirin, though, is contraindicated because of the apparent risk of Reye's syndrome in children and young adults (see microfile 25.3). Drugs that can be therapeutically beneficial for systemic disease are intravenous acyclovir or famciclovir and high-dose interferon. Zoster immune globulin (ZIG) and varicella-zoster immune globulin (VZIG) are passive immunotherapies that do not entirely prevent disease but can diminish its complications. Japanese researchers have developed a live, attenuated vaccine that is recommended for routine childhood vaccination. It is safe and thought to protect recipients against shingles later on. A decision is being weighed as to whether it should be combined with the measles, mumps, rubella (MMR) vaccine. Because some parents regard the disease as mild and inevitable, they purposely expose their children to infected individuals as a means of "immunizing" them. This practice is not recommended because of the potential for viral latency.

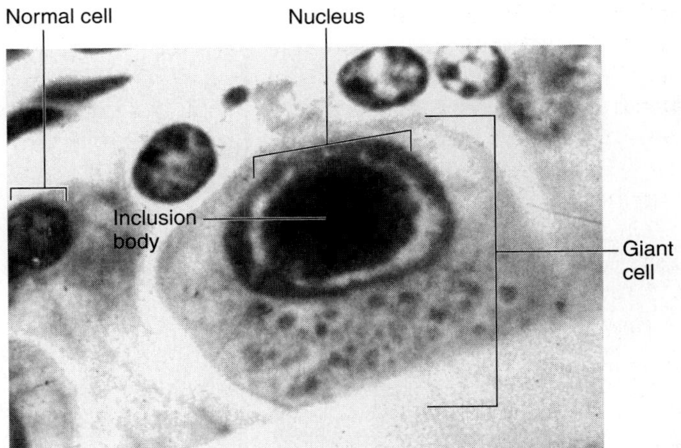

Figure 24.11

Infection of cells with cytomegalovirus shows the cellular enlargement and distortion of the nucleus by a large inclusion (1,500 ×).

The Cytomegalovirus Group

Another group of herpesviruses, the **cytomegaloviruses*** **(CMVs),** are named for their tendency to produce giant cells with nuclear and cytoplasmic inclusions (figure 24.11). These viruses, also termed salivary gland virus and cytomegalic inclusion virus, are among the most ubiquitous pathogens of humans.

Epidemiology of CMV Disease Wherever human populations have been tested, a relatively high percentage (40–100%) have shown the presence of antibodies to CMV developed during a prior infection. Studies in the United States and the United Kingdom revealed infection in 7.5% of newborns, making CMV the most prevalent viral infection in the fetus. Additional surveys that include other countries showed that 20% to 100% of women of childbearing age had CMV antibodies. Other groups with a high carrier rate are IV drug abusers and male homosexuals. Although cytomegalovirus is not noted for being highly communicable, it appears to be transmitted in saliva, respiratory mucus, milk, urine, semen, cervical secretions, and feces. Transmission usually involves intimate exposure such as sexual contact, vaginal birth, transplacental infection, blood transfusion, and organ transplantation. As with other herpesviruses, CMV is commonly carried in a latent state in various tissues.

*cytomegalovirus (sy″-toh-may″-gah-loh-vy′-rus) Gr. *cyto,* cell, and *megale,* big, great.

Infection and Disease Most healthy adults and children with primary CMV infection are asymptomatic. However, three groups that develop a more virulent form of disease are fetuses, newborns, and immunodeficient adults. The incidence of **congenital CMV** infection is high, occurring in about 20% of pregnancies in which the mother has concurrent infection. Although most infected newborns are born without signs, a certain number exhibit enlarged liver and spleen, jaundice, capillary bleeding, microcephaly, and ocular inflammation. In some cases, damage can be so severe and extensive that death follows within a few days or weeks. Babies who survive develop long-term neurological sequelae, including hearing and visual disturbances and even mental retardation. In the United States, about 5,000 babies a year are affected. **Perinatal CMV infection,** which occurs after exposure to the mother's vagina, is chiefly asymptomatic, although pneumonitis and a mononucleosis-like syndrome can develop during the first 3 months after birth.

Cytomegalovirus mononucleosis* is a syndrome characterized by fever and lymphocytosis, somewhat similar to the disease caused by Epstein-Barr virus. Although it can occur in children, it is chiefly an illness of adults. **Disseminated cytomegalovirus** infection can precipitate a medical crisis. It is a common opportunist of AIDS patients, in whom it produces overwhelming systemic disease, with fever, severe diarrhea, hepatitis, pneumonia, and high mortality. One of its more unfortunate targets in these patients is the retina, where spreading lesions can lead to blindness. Most patients undergoing kidney transplantation and half of those receiving bone marrow develop CMV pneumonitis, hepatitis, myocarditis, meningoencephalitis, hemolytic anemia, and thrombocytopenia.

Diagnosis, Treatment, and Prevention In neonates, CMV infection must be differentiated from toxoplasmosis, rubella, and herpes simplex; in adults, it must be distinguished from Epstein-Barr infection and hepatitis. The cytomegalovirus can be isolated from pathologic specimens taken from virtually all organs as well as from epithelial tissue. Cell enlargement and prominent inclusions in the cytoplasm and nucleus are suggestive of CMV. The virus can be cultured and tested with monoclonal antibody against the early nuclear antigen. Direct ELISA tests and DNA probe analysis are also useful to diagnosis. Testing serum for antibodies may fail to diagnose infection in neonates and the immunocompromised, so is less reliable.

Drug therapy is reserved for immunosuppressed patients. The two main drugs are ganciclovir and foscarnet, both of which have toxic side effects and cannot be administered for long periods. Interferon appears, at best, to delay excretion of the virus during therapy, but it shows little or no indication of preventing infection or promoting recovery. None of the other anti-herpes drugs is therapeutically active in CMV infections. The development of a vaccine is hampered by the lack of an animal that can be infected with human cytomegalovirus. One crucial concern is whether vaccine-stimulated antibodies would be protective, since patients already seropositive can become reinfected; another concern is that a vaccine could be oncogenic.

*mononucleosis (mah″-noh-noo″-klee-oh′-sis) Gr. *mono,* one, *nucleo,* nucleus, and *sis,* state.

Epstein–Barr Virus

Although the lymphatic disease known as infectious mononucleosis was first described more than a century ago, its cause was finally discovered through a series of accidental events starting in 1958, when Michael Burkitt discovered in African children an unusual malignant tumor (Burkitt's lymphoma) that appeared to be infectious. Later, Michael Epstein and Yvonne Barr cultured a virus from tumors that showed typical herpesvirus morphology. Proof that the two diseases had a common cause was completed when a laboratory technician accidently acquired mononucleosis while working with the Burkitt's lymphoma virus. The **Epstein-Barr (EBV) virus** shares morphological and antigenic features with other herpesviruses, and, in addition, it contains a circular form of DNA that is readily spliced into the host cell DNA, thus transforming cultured lymphocytes.

Epidemiology of Epstein-Barr Virus This virus is ubiquitous, with a firmly established niche in human lymphoid tissue and salivary glands. Salivary excretion occurs in about 20% of otherwise normal individuals. Direct oral contact and contamination with saliva are the principal modes of transmission, though fomites and arthropod vectors may play a role. Transmission by means of blood transfusions and organ transplants is also possible.

The nature of infection depends on the patient's age at first exposure, socioeconomic level, geographic region, and genetic predisposition. In less-developed regions of the world, infection rates are generally higher and infection occurs at an earlier age. Children living in eastern and west-central Africa are predisposed to Burkitt's lymphoma, and residents of parts of China and North Africa show high levels of nasopharyngeal carcinoma, another tumor linked to EBV. In industrialized countries, exposure to EBV is usually postponed until adolescence and young adulthood. This delay is generally ascribed to improved sanitation and less frequent and intimate contact with other children. Because nearly 70% of college-age Americans have never had EBV infection, this population is vulnerable to infectious mononucleosis, often dubbed the "kissing disease" or "mono." However, by mid-life, 90% to 95% of all people, regardless of geographic region, show serological evidence of prior or current infection.

Diseases of EBV The epithelium of the oropharynx is the portal of entry for EBV during the primary infection. From this site, the virus moves to the parotid gland, which is the major area of viral replication and the source of viremia. After the initial infection, the virus remains latent in the throat and blood for an extended time, periodically being reactivated by some ill-defined mechanism. In some people, the entire course of infection and latency is asymptomatic.

The symptoms of **infectious mononucleosis** are sore throat, high fever, and cervical lymphadenopathy, which develop after a long incubation period (30–50 days). Many patients also have a gray-white exudate in the pharynx (figure 24.12), a skin rash, and enlarged spleen and liver. A notable sign of mononucleosis is sudden leukocytosis, consisting initially of infected B cells and later to T suppressor and cytotoxic T cells. The strong, cell-mediated immune response is decisive in controlling the infection and preventing complications. Some people can become chronically

MICROFILE 24.3 VIRAL INFECTION: THE CAUSE OF CHRONIC FATIGUE?

In 1985, two internists in the small resort area of Lake Tahoe, California, witnessed a large cluster of patients with a common group of non-specific symptoms—disabling fatigue, recurrent colds, memory loss, and enlarged lymph nodes—that lasted for long periods and did not respond to the usual treatments. A series of tests revealed that only one specific sign was shared by all: a high level of antibodies to EBV (Epstein-Barr virus) in the blood. At first, physicians had difficulty reconciling the signs and symptoms with the usual picture of EBV infection, which usually does not occur in epidemic pattern; causes a relatively mild, short-term illness; and is not highly transmissible by casual contact. Investigators from the CDC were called in to help solve the mystery, and they concluded that EBV infection alone could not account for all aspects of the disease.

This mystery disease has since been termed **chronic fatigue syndrome (CFS)** and it is now a recognized disease with hundreds of thousands of diagnosed victims. What has remained mysterious is its exact etiology.

One of the earliest large clinical studies at the University of California, San Francisco, made some break-through findings that supported the probable connection of CFS to a viral infection. The exact causative agent is still a matter of speculation, however. Other possible candidates besides EBV have been herpesvirus-6, a retrovirus (HTLV II), and an enterovirus. It now appears that CFS patients have chronically overactive immune responses, which could readily account for the flu-like nature of the disease and the apparent involvement of viruses. Some hope for control of this devastating syndrome lies in Ampligen, a drug that helps to alleviate symptoms in some patients.

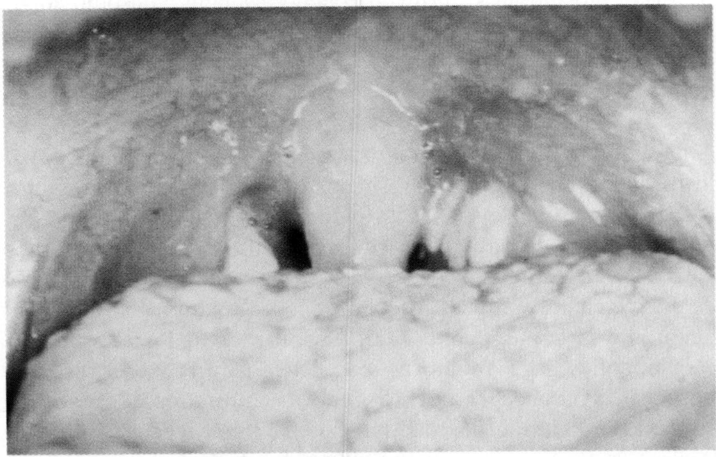

Figure 24.12
Infectious mononucleosis. This view of the pharynx shows the swollen tonsils and throat tissues, a gray-white exudate, and rash on the palate that can be of some use in diagnosis.

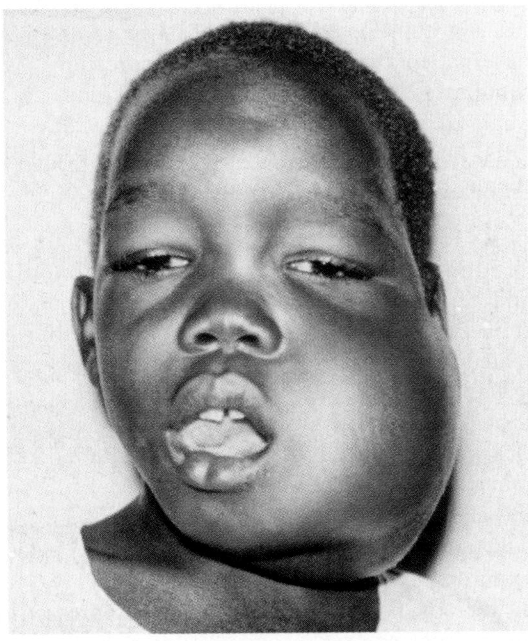

Figure 24.13
Burkitt's lymphoma, a malignancy of B cells commonly associated with chronic Epstein-Barr infection.

infected with EBV, but whether it causes a debilitating illness is controversial (microfile 24.3).

Tumors and Other Complications Associated with EBV **Burkitt's lymphoma** is a B-cell malignancy that usually develops in the jaw and grossly swells the cheek (figure 24.13), though it can also occur in the abdomen. Most cases are reported in central African children 4 to 8 years old, and a small number are seen in North America. The prevalence in Africa may be associated with chronic coinfections with other diseases such as malaria that weaken the immune system and predispose the body to tumors. One theory to explain the neoplastic effect is that persistent EBV infection in young children eventually immortalizes certain B lymphocytes. When this effect is coupled with concurrent malaria infection, the T-cell effector response is overwhelmed and unable to control the overgrowth of these malignant B-cell clones.

Unlike Burkitt's lymphoma, **nasopharyngeal carcinoma** is a malignancy of epithelial cells that occurs in older Chinese and African men. Although little about this disease is understood, high

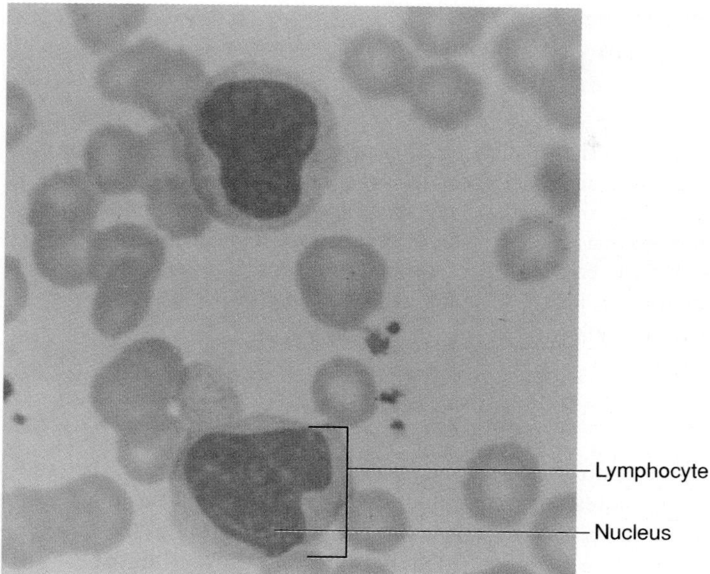

Figure 24.14
Epstein-Barr virus in the blood smear of a patient with infectious mononucleosis. Note the abnormally large lymphocytes containing indented nuclei with light discolorations.

antibody titer to EBV is a consistent finding. It is not known how the virus invades epithelial cells (not its usual target), why the incidence of malignancy is so high in these ethnic groups, or whether dietary carcinogens such as nitrosamines in cured fish are involved.

In general, any person with an immune deficiency (especially of T cells) is highly susceptible to Epstein-Barr virus. Organ transplant patients carry a high risk because their immune systems are weakened by measures to control rejection. The oncogenic potential of the virus is another concern, because it is currently believed to cause some types of lymphatic tumors in kidney transplant patients and is commonly isolated from AIDS-related lymphomas. AIDS patients also develop *hairy leukoplakia,* white adherent plaques on the tongue due to invasion of the epithelium by EBV. Lesions can be reduced by oral acyclovir therapy.

Diagnosis, Treatment, and Prevention Because clinical symptoms in EBV infection are common to many other illnesses, laboratory diagnosis is necessary. A differential blood count that shows lymphocytosis, neutropenia, and large, atypical lymphocytes with lobulated nuclei and vacuolated, basophilic cytoplasm is suggestive of EBV infection (figure 24.14). Serological assays to detect antibody against the capsid and DNA are helpful, as are direct viral antigen tests using probes and labeled antibodies.

The usual treatments for infectious mononucleosis are directed at symptomatic relief of fever and sore throat. Specific antiviral measures for disseminated disease that have shown some promise are intravenous gamma-globulin, interferon, acyclovir, and monoclonal antibodies. Burkitt's lymphoma has usually metastasized by the time of diagnosis, so systemic chemotherapy with cyclophosphamide, vincristine, methotrexate, or prednisolone can induce remission. These measures, along with surgical removal of the tumor, have considerably reduced the mortality rate.

Other Herpesviruses and the Cancer Connection

The latest addition to the herpesvirus family is **human herpesvirus-6 (HHV-6),** also known as human T-lymphotropic virus. It was originally isolated from infected lymphocytes and has characteristics similar to EBV, but it is genetically distinct from the other herpesviruses. The virus can enter and replicate in T lymphocytes, macrophages, and salivary gland tissues. It is probably transmitted by close contact with saliva and other secretions. It is among the most common herpesviruses, with up to 95% prevalence in many human populations. The primary infection usually occurs in babies between 2 and 12 months of age. It is the cause of **roseola** (also known as roseola infantum), an acute febrile disease. It begins with fever that can spike as high as 105°F (40°C) and is followed a few days later by a faint maculopapular rash over the neck, trunk, and buttocks. The disease is usually self-limited, and the children recover spontaneously with support care.

Infected adults present with mononucleosis-like symptoms and may develop lymphadenopathy and hepatitis. Patients with renal or bone marrow transplants often acquire HHV-6 infections that suppress the function of the grafted tissues and are possibly a factor in rejection. Because of the association of HHV-6 with cases of brain infection and encephalitis, it has been considered as a potential cause of chronic neurological disease. Researchers have recently established a link between this virus and multiple sclerosis (MS). Over 70% of MS patients have positive serological signs of infection, and the brain lesions of many of them contain the virus. Although this is not conclusive proof of etiology, it is the first crucial step in connecting this disease to a virus.

The fact that herpesviruses are persistent, can develop latency in the cell nucleus, and have demonstrated oncogenesis has spurred an active search for the roles of these viruses in cancers of unknown cause. One serious candidate for an oncogenic virus is HHV-6. A number of clinical studies have disclosed a significant relationship between it and Hodgkin's lymphoma, oral carcinoma, and certain T-cell leukemias. It must be emphasized that isolating a virus or its antigens from a tumor is not sufficient proof of causation and further evidence is required, but it is highly suggestive of some role in cancer.

In the past 5 years, DNA of a new herpesvirus has been isolated from Kaposi's sarcoma (a common tumor of AIDS patients). This virus was not found in other normal tissues or in patients without sarcoma. It has now been isolated and characterized and is termed *Kaposi's sarcoma-associated herpesvirus* (KSHV), or *herpesvirus-8.* The genetic mechanisms and pathology behind this cancer are an ongoing focus of study. Some researchers have discovered an additional link between KSHV and multiple myeloma, a relatively common cancer of the blood. Initial results from affected patients indicate that the virus lives in dendritic cells (a type of macrophage) that surround hemopoietic cells in the bone marrow. Its role in the myeloma is apparently to stimulate the growth of the malignant cells.

Other mammals are also susceptible to herpesvirus infections. In cases of simian herpesviruses, monkeys can pass the infection to humans, often with fatal results (microfile 24.4).

MICROFILE 24.4 THE DANGERS OF MONKEY HERPESVIRUS

Although herpesviruses tend to cause mild infection in their natural mammalian hosts, interspecies infection can be quite severe and even lethal. A herpesvirus called *Herpesvirus simiae,* or B virus, common among macaque monkeys, produces a localized herpes simplex–like infection in these primates. Macaques are popular subjects for biomedical research and are maintained in laboratories worldwide, so a potential exists for transmission of the virus to humans. Most of the verified human infections have been traced to animal bites or scratches in laboratory workers, although pet monkeys have accounted for a number of cases. Human infection can give rise to a severe neurological disease that is fatal in about 75% of cases. Persons who routinely handle these monkeys must take extreme precautions to avoid contact with them or with their secretions. Because monkeys carry these viruses orally and are natural "biters," they are considered very inappropriate pets, especially for children.

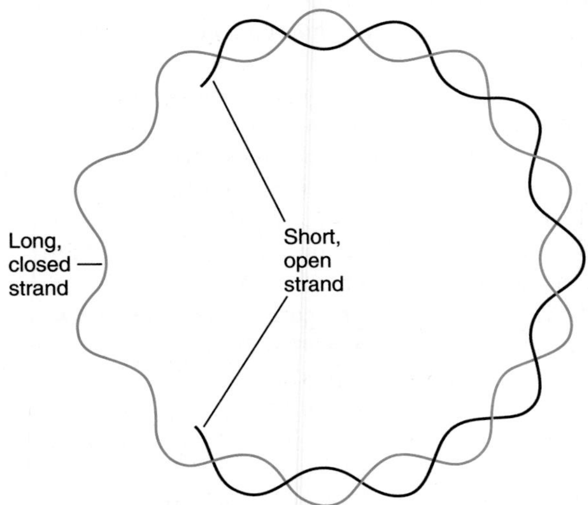

Figure 24.15
The nature of the DNA strand in hepatitis B virus. Like the genomes of most DNA viruses, the genome of HBV is circular, but unlike them, it is not a completely double-stranded molecule. A section of the molecule is composed of single-stranded DNA.

HEPADNAVIRUSES: UNUSUAL ENVELOPED DNA VIRUSES

One group of enveloped DNA viruses, called *hepadnaviruses,** is quite unlike any other so far discovered. These viruses have never been grown in tissue culture and have an unusual genome containing both double- and single-stranded DNA (figure 24.15). Hepadnaviruses show a decided tropism for the liver, where they persist and can be a factor in liver cell carcinoma. One member of the group, **hepatitis B virus (HBV),** causes a common form of hepatitis, and other members cause hepatitis in wood chucks, ground squirrels, and Peking ducks.

General Considerations: What Is Viral Hepatitis?

When certain viruses infect the liver, they cause **hepatitis,** an inflammatory disease marked by necrosis of hepatocytes and a mononuclear response that swells and disrupts the liver architecture. This pathologic change interferes with the liver's excretion of bile pigments such as bilirubin into the intestine. When bilirubin, a greenish-yellow pigment, accumulates in the blood and tissues, it causes **jaundice,*** a yellow tinge in the skin and eyes.

Characteristics of the three principal viruses responsible for human hepatitis are summarized in table 24.3. *Hepatitis A virus (HAV)* is a nonenveloped, single-stranded RNA enterovirus that is the etiologic agent of infectious hepatitis (see chapter 25). In general, HAV disease is far milder, shorter-term, and less virulent than the other forms. Another important agent is *hepatitis C virus* (HCV), an RNA virus in the flavivirus family that causes most cases of infusion hepatitis. HCV is involved in a chronic liver infection that can go undiagnosed, later leading to severe liver damage. It is spread by blood and sexual contact, and is considered one of the most common unreported agents of hepatitis (an estimated 3.5 million cases nationally). Another agent is hepatitis E (HEV), a newly identified RNA virus that causes a disease similar to HAV. It is associated with poor hygiene and sanitation and is spread oro-fecally.

These hepatitis viruses have no relationship to hepatitis B virus (HBV), the only DNA virus that causes hepatitis. The incidence of the major forms is presented in figure 24.16. An unusual virus, hepatitis, D, or delta agent, is a defective RNA virus that cannot produce infection unless a cell is also infected with HBV. Hepatitis D virus has an unusual nucleocapsid that is similar to viroids, and it invades host cells by "borrowing" the outer receptors of HBV. When HBV infection is accompanied by the delta agent, the disease becomes more severe and is more likely to progress to permanent liver damage.

Hepatitis B Virus and Disease

Because a cell culture system for propagating hepatitis B virus is not yet available, by far the most significant source of information about its morphology and genetics has been obtained from viral fragments, intact virions, and antibodies, which are often abundant

*hepadnavirus (hep″-ah-dee″-en-ay-vy′-rus) Gr. *hepatos,* liver, plus DNA and virus.

*jaundice (jon′-dis) Fr. *jaunisse,* yellow. Also called icterus.

TABLE 24.3

PRINCIPAL MORPHOLOGICAL AND PATHOLOGIC FEATURES OF THE MAJOR HEPATITIS VIRUSES

Property/Diseases	HAV	HBV	HCV
Biology			
Nucleic acid	RNA	DNA	RNA
Size	27 nm	42 nm	Various
Protein coat	–	+	+
Cell culture	+	–	–
Envelope	–	+	+
Synonyms	Infectious hepatitis, yellow jaundice	Serum hepatitis	Post-transfusional hepatitis
Epidemiology	Endemic and epidemic	Endemic	Endemic
Case incidence	Decreasing	Decreasing	Decreasing
Reservoir	Active infections	Chronic carrier	Chronic carrier
Transmission	Oral-fecal; water- or food-borne	Overt inoculation from blood, serum; close contact	Inoculation from blood, serum; intimate contact
Incubation Period	2–7 weeks	1–6 months	2–8 weeks
Symptoms	Fever, GI tract disorder	Fever, rash, arthritis	Similar to HBV
Jaundice	1 in 10	Common	Common
Onset/Duration	Acute, short	Gradual, chronic	Acute to chronic
Complications	Uncommon	Chronic active hepatitis, hepatic cancer	Chronic inflammation, cirrhosis
Availability of Vaccine	+	+	–
Diagnostic Tests to Differentiate	+	+	+

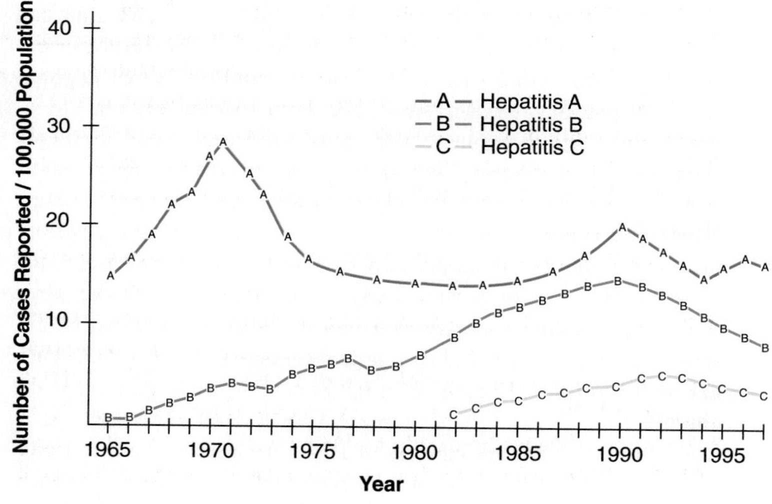

Figure 24.16
The comparative incidence of viral hepatitis in the United States from 1965 to 1997.
Source: Data from the Centers for Disease Control and Prevention, Atlanta, GA.

in patient's blood. One of these blood-borne pieces, the Dane particle, has been suspected as the infectious virus particle, and modern serological and electron microscope studies support that view. Blood-borne viral components are shown in figure 24.17.

Epidemiology of Hepatitis B An important factor in the transmission pattern of hepatitis B virus is that it multiples exclusively in the liver, which continuously seeds the blood with viruses. Electron microscopic studies have revealed up to 10^7 virions per milliliter of infected blood. Even a minute amount of blood (10^{-6} to 10^{-7} ml) can transmit infection. The abundance of circulating virions is so high and the minimal dose so low that such simple practices as sharing a toothbrush or a razor can transmit infection. Over the past 10 years, HBV has also been detected in semen and vaginal secretions, and it is currently believed to be transmitted by these fluids in certain populations. Spread of the

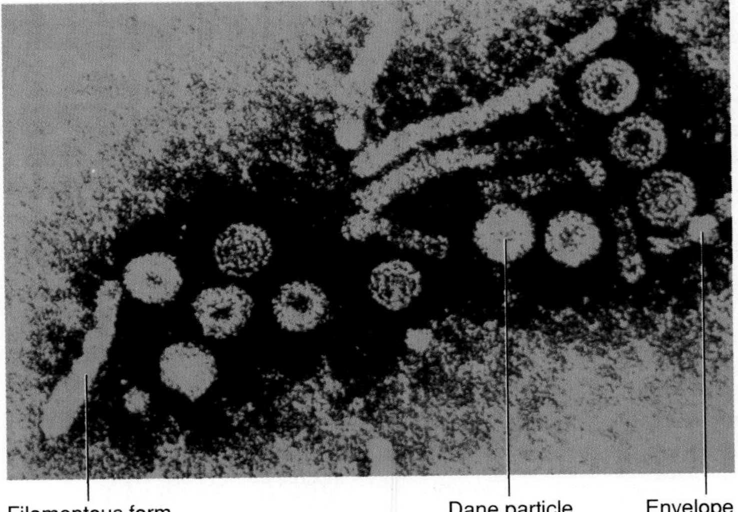

Figure 24.17
Blood from hepatitis B virus–infected patients presents an array of particles, including Dane particles (the intact virus), the envelope (or surface shell), and filaments that are polymers of the envelope antigen.

virus by means of close contact in families or institutions is also well documented.

Hepatitis B is an ancient disease that has been found in all populations, though the incidence and risk are highest among people living under crowded conditions, drug addicts, the sexually promiscuous, and certain occupations, including people who conduct medical procedures involving blood or blood products such as serum (microfile 24.5). The incidence of HBV infection and carriage among drug addicts who routinely share needles is very high. Homosexual males constitute another high-risk group because such practices as anal intercourse can traumatize membranes and permit transfer of virus into injured tissues. Heterosexual intercourse can also spread infection, but is less likely to do so. Infection of the newborn by chronic carrier mothers occurs readily and predisposes to development of the carrier state and increased risk of liver cancer in the child. The role of mosquitos, which can harbor the virus for several days after biting infected persons, has been well-documented in the tropics and in some parts of the United States.

The tropism of HBV for the liver becomes a significant epidemiologic complication in some people (5–10% of HBV-positive people). Persistent or chronic carriers harbor the virus for up to 6 months, and some continue to shed the virus for many years. Worldwide, 300 million people are estimated to carry the virus, but it is most prevalent in Africa, Asia, and the western Pacific and is least prevalent in North America and Europe.

Pathogenesis of Hepatitis B Virus The hepatitis B virus enters the body through a break in the skin or mucous membrane or by injection into the bloodstream. Eventually, it reaches the liver cells (hepatocytes), where it multiplies and releases viruses into the blood (viremia). After an incubation period of 4 to 24 weeks (7 weeks average), some people experience malaise, fever, chills, anorexia, abdominal discomfort, and diarrhea. Surprisingly, the majority of those infected exhibit few overt symptoms and eventually develop an immunity to HBV. In those who do experience disease, the severity of symptoms and the aftermath of hepatic damage vary widely. Fever, jaundice, rashes, and arthritis are common reactions (figure 24.18), and a smaller number of patients develop glomerulonephritis and arterial inflammation. In 90% of patients complete liver regeneration and restored function occur. But for reasons that are not entirely clear, a small number of predisposed older persons develop chronic liver disease in the form of necrosis or cirrhosis (permanent liver scarring and loss of tissue).

The association of HBV with **hepatocellular carcinoma**[1] is based on these observations: (1) Certain hepatitis B antigens are found in malignant cells and are often detected as integrated components of the host genome; (2) persistent carriers of the virus are more likely to develop this cancer; and (3) people from areas of the world with a high incidence of hepatitis B (Africa and the Far East) are more frequently affected. This connection is probably a result of infection early in life and the long-term carrier state. In general, people with chronic hepatitis are 200 times more likely to develop liver cancer, though the exact role of the virus is still the object of molecular analysis.

Diagnosis and Management of Hepatitis B The initial stages of most forms of hepatitis are so similar that differential diagnosis on the basis of symptoms is unlikely. A clinician can usually distinguish hepatitis B from hepatitis A by carefully examining the possible risk factors. Some combination of occupational hazard, drug abuse, age (it occurs primarily in adults), relatively long incubation period, and insidious onset strongly suggests HBV infection. Serological tests are increasingly important. Testing procedures fall into two categories: those that detect virus antigen and those that assay for antibodies. Recent developments in radioimmunoassay and ELISA testing permit detection of the surface antigen very early in the infection. These same tests are essential for screening blood destined for transfusions, semen in sperm banks, and organs intended for transplant. Antibody tests are most valuable in patients who are negative for the antigen.

Mothers who are carriers of HBV are highly likely to transmit infection to their newborns during delivery. It is not currently known whether transplacental HBV infection occurs. Unfortunately, infection with this virus early in life is a known risk factor for chronic liver disease. In light of the increasing incidence of HBV-positive pregnant women, routine testing of all pregnant women is now strongly supported by the CDC and a majority of obstetricians. If adopted, mandatory hepatitis testing would add to the growing number of procedures used to screen prospective mothers for infection with agents that have especially profound

1. HOC. Also called hepatoma, a primary malignant growth of hepatocytes.

MICROFILE 24.5 THE CLINICAL CONNECTION IN HEPATITIS

Hepatitis B is a risk to patients and health care workers alike. Obvious gross contact, in the form of blood transfusions or injections of human plasma, clotting factors, serum, and other products, is a classic route of infection. The virus remains infective for days in dried blood, for months when stored in serum at room temperature, and for decades if frozen. Although it is not inactivated after 4 hours of exposure to 60°C, boiling for the same period can destroy it. Disinfectants containing chlorine, iodine, and glutaraldehyde show potent anti-hepatitis B activity. It should be noted, however, that HBV infection from transfusion is quite rare now because of blood tests for HBV antigens and processing techniques that inactivate any HBV present.

Such diverse procedures as kidney dialysis, reused needles, and acupuncture needles are also known to transmit the HBV virus. Outbreaks of hepatitis B have developed in people vaccinated with a special needle-less gun. These cases were traced to inadequate disinfection of the casing of the gun nozzle, which must be reused continuously between patients. Even cosmetic manipulations such as tattooing and ear or nose piercing can expose a person to infection. The only reliable method for destroying HBV on reusable instruments is through autoclaving.

In regard to occupational exposure, nonimmunized people working with blood, plasma, sera, or materials contaminated with even small amounts of those substances are very vulnerable to infection. Practically any health occupation can be involved, but hematologists, phlebotomists, dialysis personnel, dental workers, surgical staff, and emergency room attendants must take particular precautions. Accidents involving broken blood vials, needlesticks, improper handling of instruments, and splashed blood and serum are common causes of HBV infection in clinical workers.

Other hepatitis viruses, especially HCV, are also closely tied to blood transmission. A blood test can screen blood and organ donations to prevent this source of infection. Currently, because of detection and control measures, all forms of viral hepatitis are declining in the United States.

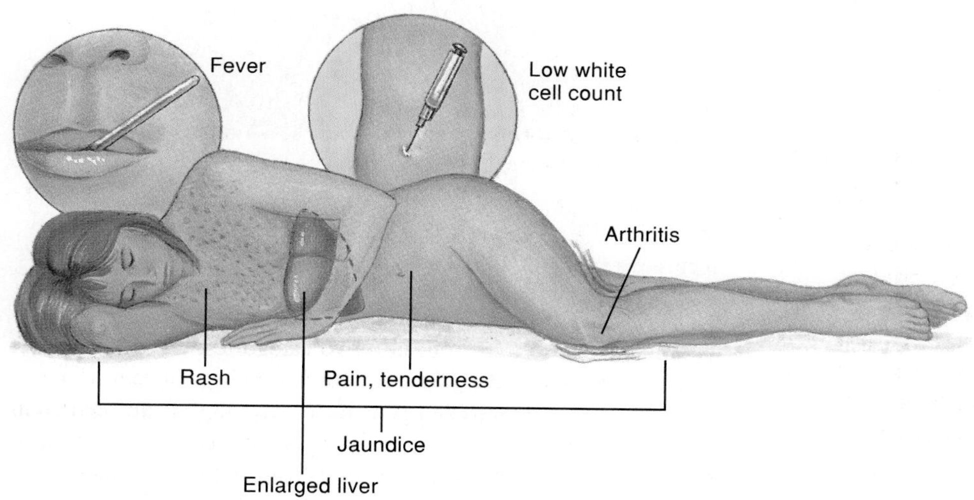

Figure 24.18
The clinical features of hepatitis B. Among the most prominent symptoms are fever, jaundice, and intestinal discomfort. The liver is also enlarged and tender.

risks for fetuses or neonates. These diseases, known collectively as STORCH, are syphilis, toxoplasmosis, other infections, rubella, CMV, and herpes simplex (see chapter 13).

Because most cases of hepatitis B are self-limited, patient care relies on symptomatic treatment and supportive care. An exciting new treatment for chronic infection is the injection of recombinant interferon, which stops virus multiplication and prevents liver damage in up to 60% of patients. Passive immunization with hepatitis B immune globulin (HBIG) gives significant protection to people who have been exposed to the virus through needle puncture, broken blood containers, or skin and mucosal contact with blood. Other groups for whom prophylaxis is highly recommended are the sexual contacts of actively infected people or carriers and neonates born to infected mothers.

The first boon to curbing hepatitis B was a vaccine developed and tested in the late 1970s and committed to general use in 1981. Because HBV cannot be cultivated in tissue cultures or embryos, the source of antigen for this vaccine is the blood of carriers with high levels of the desired surface antigen. The blood is treated to purify and concentrate the antigen and heated to inactivate any intact viruses. Alternative vaccines (Recombivax, Energix) containing the pure surface antigen cloned in yeasts have largely replaced the original vaccine, except in cases of allergy to yeast. Vaccination is a must for medical and dental workers and students, patients receiving multiple transfusions, immunodeficient persons, and cancer patients. The vaccine is also now strongly encouraged for all newborns as part of a routine immunization schedule.

Chapter Checkpoints

Poxviruses are the largest and most complex of all viruses. They infect the skin, producing pustular lesions. Smallpox, which is now eradicated, molluscum contagiosum, monkeypox, and cowpox are the only poxviruses that infect humans.

Herpesviruses are large enveloped viruses that cause recurrent infections and are potentially carcinogenic. Herpes simplex 1 and 2 infect the skin and mucous membranes but can cause encephalitis in neonates. The varicella-zoster virus is the causative agent of chickenpox and shingles. Cytomegalovirus (CMV) causes systemic disease in fetuses, neonates, and immunosuppressed individuals and a type of mononucleosis. Epstein-Barr virus is the causative agent of Burkitt's lymphoma and infectious mononucleosis. Herpesvirus-6 causes roseola and is a factor in multiple sclerosis. Other herpes viruses are implicated in lymphomas, sarcomas, and leukemias.

Hepadnaviruses are the causative agents of liver disease in many animals, including humans. Hepatitis B virus causes the most serious form of human hepatitis (hepatitis B or serum hepatitis). It is spread by direct contact, fomites, and mosquitos. It is potentially lethal in a small percentage of cases, either by direct liver damage or by causing liver cancer. Other forms of viral hepatitis are caused by RNA viruses.

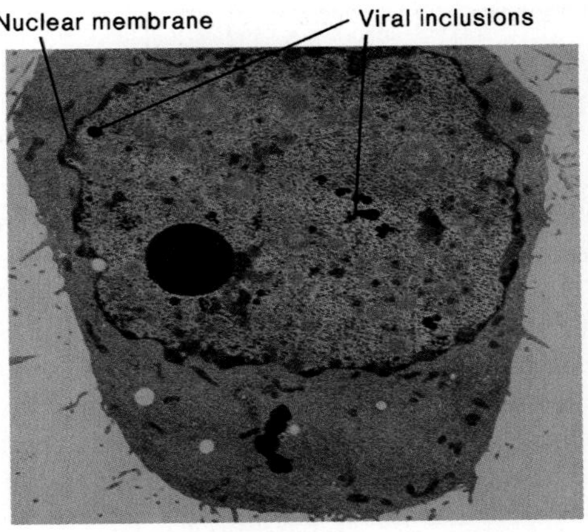

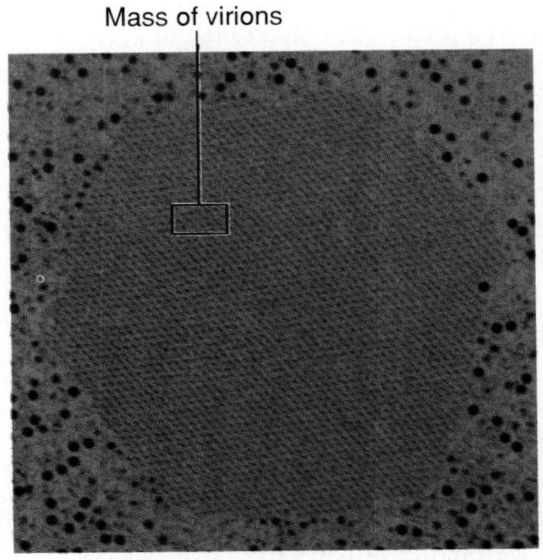

Figure 24.19
(a) Cellular and nuclear alterations in adenovirus infection. Granules of early virus parts accumulating in the nucleus cause inclusions to form (19,000×). *(b)* Magnification of the nucleus shows a crystalline inclusion containing mature virions (35,000×).

NONENVELOPED DNA VIRUSES

THE ADENOVIRUSES

During a study seeking the cause of the common cold, a new virus was isolated from *adenoids** that had been surgically excised from young children. It turned out that this virus, termed **adenovirus,** was not the sole agent of colds but one of several others (described in chapter 25). It was also the first of about 80 strains of nonenveloped, double-stranded DNA viruses to be discovered and eventually classified as adenoviruses. Of this array, about 30 types are associated with human infection, and the rest are animal pathogens. Besides infecting lymphoid tissue, adenoviruses have a preference for the respiratory and intestinal epithelia and the conjunctiva. Adenoviruses produce aggregations of incomplete and assembled viruses that usually lyse the cell (figure 24.19). In cells that do not lyse, the viral DNA may be harbored latently in the nucleus. Experiments show that adenoviruses are oncogenic in animals, but whether they are oncogenic in humans is not known.

Epidemiology of Adenoviruses

Adenoviruses are spread from person to person by means of respiratory and ocular secretions. Most cases of conjunctival infection involve preexisting eye damage and contact with contaminated sources such as swimming pools, dusty working places (shipyards and factories), and unsterilized optical instruments (ocular tonometers used in the glaucoma test). Infection by an adenovirus usually occurs by age 15 in most people, and certain individuals become chronic respiratory carriers. Although the reason is obscure, military recruits are at greater risk for infection than the general population, and respiratory epidemics are common in armed services installations. Civilians, even of the same age group and living in close quarters, seldom experience such outbreaks.

Respiratory, Ocular, and Miscellaneous Diseases of Adenoviruses

The patient infected with an adenovirus is typically feverish, with acute rhinitis, cough, and inflammation of the pharynx, enlarged cervical lymph nodes, and a macular rash (somewhat reminiscent of rubella). Certain adenoviral strains produce an acute follicular lesion of the conjunctiva. Usually one eye is affected by a watery exudate, redness, and partial closure. A deeper and more serious complication is *keratoconjunctivitis,* an inflammation of the conjunctiva and cornea.

* **adenoid** (ad′-eh-noyd) Gr. *adeno,* gland, and *eidos,* form. A lymphoid tissue (tonsil) in the nasopharynx that sometimes becomes enlarged.

Acute hemorrhagic cystitis in children has been traced to an adenovirus. This self-limiting illness of 4 to 5 days duration is marked by hematuria (blood in the urine), frequent and painful urination with occasional episodes of fever, bedwetting, and suprapubic pain. Because the virus apparently infects the intestinal epithelium and can be isolated from some diarrheic stools, adenovirus is also suspected as an agent of infantile gastroenteritis.

Severe cases of adenovirus infection can be treated with interferon in the early stages. An inactivated polyvalent vaccine prepared from viral antigens is an effective preventive measure, especially for military recruits.

Papovaviruses: Papilloma, Polyoma, Vacuolating Viruses

Letters from the terms **pa**pilloma,* **po**lyoma,* and simian **va**cuolating* viruses gave rise to the term **papovavirus.** This family contains two subtypes: the **papillomaviruses,** which are responsible for human and animal papillomas, and the **polyomaviruses,** which include several human (BK and JC viruses) and animal pathogens.

Epidemiology and Pathology of the Human Papillomaviruses

A papilloma is a benign, squamous epithelial growth commonly referred to as a **wart,** or **verruca,** and caused by one of 40 different strains of **human papillomavirus (HPV).** Some types are specific for the mucous membranes; others invade the skin. The appearance and seriousness of the infection vary somewhat from one anatomical region to another. Painless, elevated, rough growths on the fingers and occasionally on other body parts are called **common,** or **seed, warts** (figure 24.20*a*). These commonly occur in children and young adults. **Plantar warts** are deep, painful papillomas on the soles of the feet; flat warts are smooth, skin-colored lesions that develop on the face, trunk, elbows, and knees. A special form of verruca known as **genital warts** is a prevalent STD and is linked to some types of cancer. Warts are transmissible through direct contact with a wart or contaminated fomites, and they can also spread on the same person by autoinoculation. The incubation period varies from 2 weeks to more than a year. Papillomas are common in all sexes, races, and geographic regions.

Genital Warts: An Insidious Papilloma Genital warts are the latest concern among young, sexually active people. Although this disease is not new, case reports have increased to such an extent that it is now regarded as the most common STD in the United States. Studies show that up to 15% of the population are carriers of one of the five types of HPV associated with genital warts (types 6, 11, 16, 18, and 31). The virus invades the external and internal genital membranes, especially in areas prone to friction in the vagina and the head of the penis. Wart morphology ranges from tiny, flat, inconspicuous bumps to giant, branching, cauliflower-like masses called **condylomata acuminata*** (figure 24.20*b*). Infection is also common in the cervix, urethra, and anal skin.

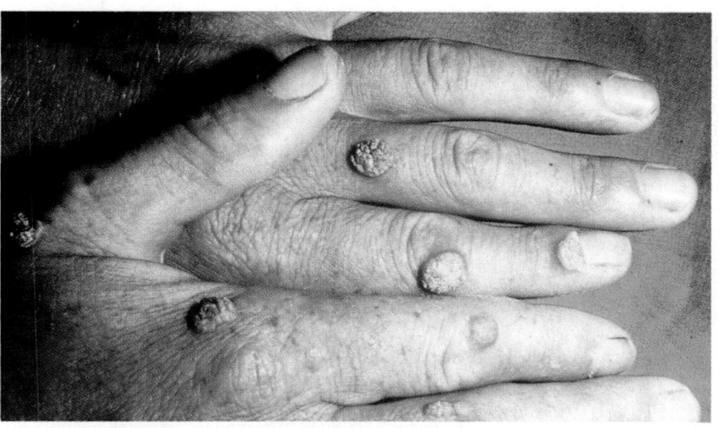

(a)

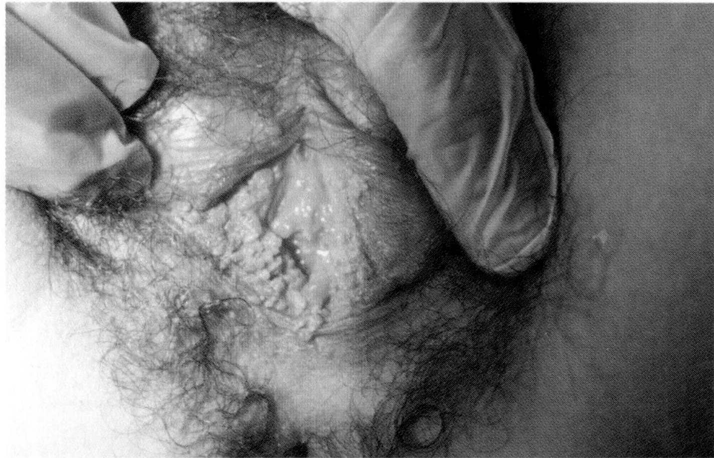

(b)

Figure 24.20

Human papillomas. (*a*) Common warts. Virus infection induces a neoplastic change in the skin and benign growths on the fingers, face, or trunk. One must avoid picking, scratching, or otherwise piercing the lesions to avoid spread and secondary infections. (*b*) Chronic genital warts (condylomata acuminata) that have extended into the labial, perineal, and perianal regions.

The most disturbing aspect of the sudden increase in HPV infection is its strong association with cancer of the reproductive tract, especially the cervix and penis. Two viral types—16 and 18—can be isolated from a majority of female genital cancers and are known to convert some forms of genital warts into malignant tumors. Susceptibility to this infection and its potential for cancer can be greatly reduced by early detection and treatment of both the patient and the sexual partner. Such early diagnosis depends on thorough inspection of the genitals and, in

*papilloma (pap″-il-oh′-mah) L. *papilla,* pimple, and Gr. *oma,* tumor. A benign, nodular, epithelial tumor.

*polyoma (paw″-lee-oh′-mah) L. *poly,* many. A glandular tumor caused by a virus.

*vacuolating (vak-yoo-oh-lay″-ting) L. *vacuus,* empty. The process of forming small spaces or vacuoles in the cytoplasm of the cell.

*condylomata acuminata (kahn″-dee-loh′-mah-tah ah-kyoo″-min-ah′-tah) Gr. *kondyloma,* knob, and L. *acuminatus,* sharp, pointed. Not to be confused with condyloma lata, a broad, flat lesion of syphilis.

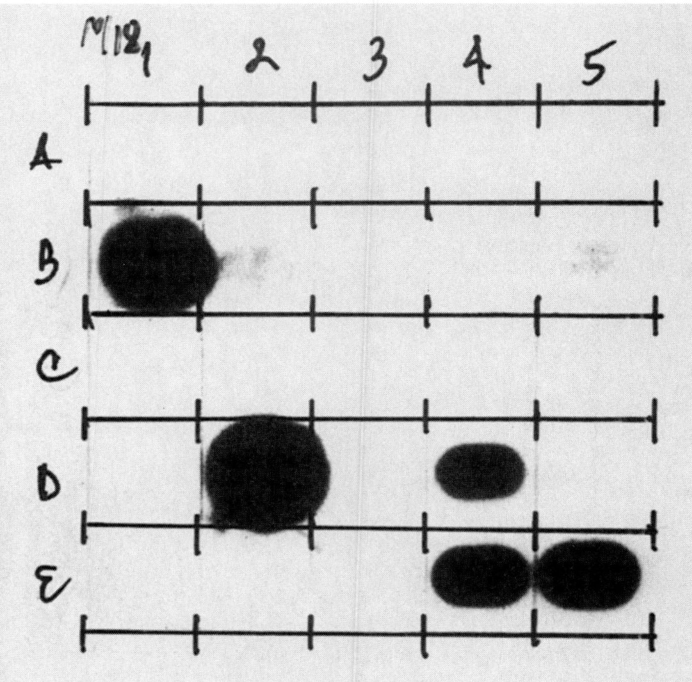

Figure 24.21
Dot blot method for detecting human papillomavirus. DNA is extracted from cervical specimens and blotted onto a special carrier. Labeled DNA probes specific to types 16 and 18 are reacted with the samples and exposed to X-ray film. Results show that patients B1, D2, D4, E4, and E5 carry HPV DNA for these two types and have greater risk for developing cervical cancer.
From Koneman et al., Diagnositc Microbiology, *4th ed. 1992, fig. 20.20, p. 1037* *© J. B. Lippinocott Company.*

women, a Papanicolaou smear to screen for abnormal cervical cells.

Diagnosis, Treatment, and Prevention The warts caused by papillomaviruses are usually distinctive enough to permit reliable clinical diagnosis without much difficulty. However, a biopsy and histological examination can help clarify ambiguous cases. Sensitive DNA probes can detect HPV by means of a dot blot (figure 24.21) or directly in infected cells (chapter opener). These techniques can determine the type of HPV and assess the patient's risk for developing cervical cancer. Although most common warts regress over time, they cause sufficient discomfort and cosmetic concern to require treatment. Treatment strategies for all types of warts include direct chemical application of podophyllin and physical removal of the affected skin or membrane by cauterization, freezing, or laser surgery. Immunotherapy with interferon is effective in many cases. Because treatments may not completely destroy the virus, warts can recur.

The Polyomaviruses

In the mid-1950s the search for a leukemia agent in mice led instead to the discovery of a new virus capable of inducing an assortment of tumors—hence its name, polyomavirus. Among the most important human polyomaviruses are the **JC virus** and **BK virus.**[2] Although many polyomaviruses are fully capable of transforming host cells and inducing tumors in experimental animals, none causes cancer in its natural host.

Epidemiology and Pathology On the basis of serological surveys, it appears that infections by JC virus and BK virus are commonplace throughout the world. Because the majority of infections are asymptomatic or mild, not a great deal is known about their mode of transmission, portal of entry, or target cells, though urine and respiratory secretions are suspected as possible sources of the viruses. The main complications of infection occur in patients who have cancer or have been immunosuppressed. **Progressive multifocal leukoencephalopathy*** **(PML)** is an uncommon but generally fatal infection by JC virus that attacks accessory brain cells and gradually demyelinizes certain parts of the cerebrum (figure 24.22). Infection with BK viruses, usually associated with renal transplants, causes complications in urinary function. The prospects of finding effective therapy for these diseases are bleak. By the time PML is diagnosed, the patient is irreversibly immunocompromised, and extensive brain damage has already occurred. However, BK infection can be prevented by treating renal transplant patients with human leukocyte interferon.

Chapter Checkpoints

Adenoviruses produce inflammatory infections in epithelial cells of the conjunctiva, intestine, respiratory tract, and lymph nodes. Papillomaviruses infect the skin and mucous membranes, producing warts. Genital warts are the leading STD in the United States. The polyomaviruses transform infected host cells into tumors, which can be fatal in immunosuppressed patients.

NONENVELOPED SINGLE-STRANDED DNA VIRUSES: THE PARVOVIRUSES

The **parvoviruses*** **(PV)** are unique among the viruses in having single-stranded DNA molecules. They are also notable for their extremely small diameter (18–26 nm) and genome size. Parvoviruses are indigenous to and can cause disease in several mammalian groups—for example, distemper in cats, an enteric disease in adult dogs, and a potentially fatal cardiac infection in puppies.

2. The initials signify the cancer patients from whom the viruses were first isolated.

*leukoencephalopathy (loo″-koh-en-sef″-uh-lop′-uh-thee) Gr. *leuko,* white, *cephalos,* brain, and *pathos,* disease.

*parvovirus (par″-voh-vy′-rus) L. *parvus,* small.

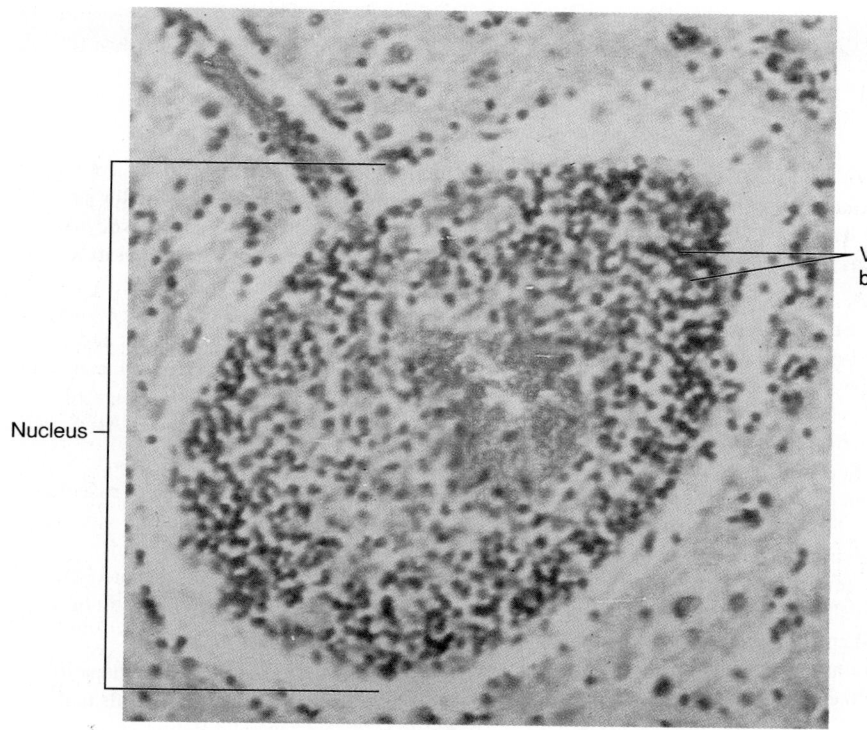

Figure 24.22
Brain pathology in progressive multifocal leukoencephalopathy. Target cells are oligodendrocytes, seen here as an enlarged cell with a giant swollen nucleus and masses of assembling virus particles.

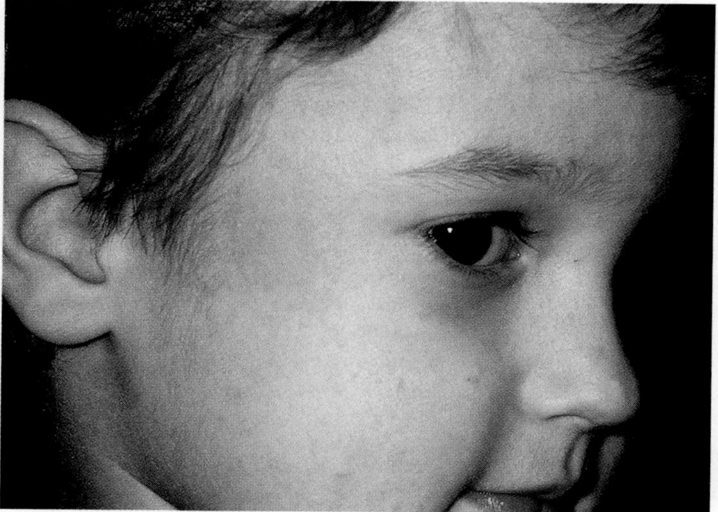

Figure 24.23
Confluent red rash on the check gives this child a "slapped face" appearance. This is one symptom of erythema contagiosum, a common childhood ailment caused by human parvovirus (B19).

The most important human parvovirus is B19, the cause of erythema contagiosum (fifth disease), a common infection in children (figure 24.23). Often the infection goes unnoticed, though the child may have a low-grade fever and a bright red rash on the cheeks. This same virus can be more dangerous in children with immunodeficiency or sickle-cell anemia, because it destroys red blood stem cells. If a pregnant woman contracts infection and transmits it to the fetus, a severe and often fatal anemia can result. One type of parvovirus, the adeno-associated virus (A-AV), is another example of a defective virus that cannot replicate in the host cell without the function of a helper virus (in this case, an adenovirus). The possible impact of this coinfection on humans is still obscure.

Chapter Checkpoints

Parvoviruses are extremely small, resistant, single-stranded DNA viruses that cause diseases in cats, dogs, and humans. The most important human parvovirus virus, B19, is the cause of erythema contagiosum.

Chapter Capsule with Key Terms

GENERAL CHARACTERISTICS OF ANIMAL VIRUSES AND HUMAN VIRAL DISEASES

Viruses are minute parasitic particles consisting of DNA or RNA genomes packaged within a protein capsid; they invade host cells and appropriate the cell's machinery for mass production of new virions, both in cytoplasm and nucleus; viruses exit the host cell by lysis or budding; budded viruses leave with an envelope; actively infected host cells are usually destroyed.

Viruses attack a variety of host and cell types; individual viruses are relatively host/cell specific, due to the need for receptor recognition; target cells include nervous system, blood, liver, skin, and intestine.

Diseases range from mild and self-limited to fatal in effects; symptoms are local, systemic, and depend on the exact tissue target; identification and diagnosis are by culture, microscopy, genetic probes, and serology; immunity is both humoral and cell-mediated; limited number of drugs available for treatment.

DNA and some RNA viruses persist in inactive state (latent) in host cells; some latent viruses cause recurrent infections, and others integrate into host genome and cause cancerous transformation; some viruses are **teratogenic** and cause congenital defects; zoonotic viruses transmitted by vectors cause severe human disease.

DNA VIRUSES

DNA viruses are enveloped or nonenveloped nucleocapsids; most have double-stranded DNA; parvoviruses have single-stranded DNA.

Enveloped Viruses

Poxviruses: Large complex viruses; produce skin lesions called **pox** (pocks). **Variola** is the agent of **smallpox,** the first disease to be eliminated from the world through vaccination; last smallpox case reported in 1977; infection leads to viremia and large deep pustules that may scar; vaccination is by scarification with vaccinia virus but is no longer done routinely, except to protect certain populations against animal poxviruses such as monkeypox.

Molluscum contagiosum, an unclassified poxvirus, causes waxy nodules on skin; transmitted by direct contact; may be an STD in adults. Humans can also be infected by *monkeypox virus,* which causes a disease similar to smallpox.

Herpesviruses: Persistent latent viruses that cause recurrent infections and may be involved in neoplastic transformations; attack skin, mucous membranes, and glands.

Herpes simplex virus (HSV): Types 1 and 2 both create lesions of skin and mucous membranes; migrate into nerve ganglia; are reactivated by anatomical and physiological disruptions; cause serious disseminated disease in newborn infants and immunodeficient patients; cause **whitlows** on fingers of medical and dental personnel. Herpes simplex 1 is transmitted by close contact and droplets and mainly infects lips (coldsores, fever blisters), eyes, and oropharynx. Herpes simplex 2 is sexually transmitted and usually affects the genitalia. Complications occurring in newborns infected at birth with either virus include encephalitis with severe CNS damage. Infections are treated with some form of cyclovir; controlled by means of barriers and care in handling secretions.

Varicella-Zoster Virus (VZV): This herpesvirus causes both **chickenpox (varicella)** and **zoster (shingles);** chickenpox is the primary infection, common in children, zoster is later recurrence of infection by latent virus. Chickenpox is transmitted through droplets; symptoms are fever, poxlike papular rash. In zoster, the virus has migrated into spinal nerves and is reactivated by surgery, cancer, or other stimuli; travels out to skin, causing painful lesions in a pattern on the trunk or head; severe disease treated with acyclovir or famciclovir; immune globulin used to relieve symptoms; vaccine now available.

Cytomegalovirus (CMV): One of the most common herpesvirus; high infection rate in children, pregnant women, drug abusers, and male homosexuals; spread through close contact with body fluids; chronically carried; congenital infection affects the liver, spleen, brain, and eyes; also causes mononucleosis-like syndrome with fever and leukocytosis; disseminated disease affecting many organs is common in AIDS and transplant patients; antiviral drugs may be of some benefit.

Epstein-Barr Virus (EBV): A herpesvirus of the lymphoid and glandular tissue; transmitted by saliva, oral contact; 95% of all persons develop some form of infection, though often asymptomatic; cause of **mononucleosis,** a "kissing disease" common in college students marked by sore throat, fever, swollen lymphoid tissue, leukocytosis; virus is oncogenic; also causes **Burkitt's lymphoma,** a malignancy of the B lymphocytes that swells cheek or abdomen and is prevalent in African children; and causes **nasopharyngeal carcinoma,** cancer in older men in China, Africa. Disseminated EBV infection is a complication of AIDS and transplant patients.

Herpesvirus-6 is the cause of **roseola,** an acute febrile disease in children; virus may be involved in multiple sclerosis and lymphoma. **Herpesvirus-8** is the oncogenic viral agent of Kaposi's sarcoma, a tumor of AIDS patients.

Hepadnaviruses: One cause of viral hepatitis, an inflammatory disease of liver cells that may result from several different viruses. **Hepatitis B virus (HBV)** causes **serum hepatitis,** is most severe; virus has strong affinity for liver cells, is carried and shed into blood; only a tiny infectious dose is required; invades skin or mucous membranes; virus resistant to heat and disinfectants; common clinical risk due to contact with blood; drug addicts, homosexuals are at high risk; may be transmitted from mother to child or by sharing personal articles; infection marked by fever, intestinal disturbance, **jaundice;** chronic carriage may lead to liver disease and liver cancer; managed by serum globulin, interferon; vaccines effective, recommended for health care workers, transfusion patients, other high-risk groups. *Delta agent* is an associated virus that accompanies HBV and causes a more severe disease.

Nonenveloped Viruses

Adenoviruses: Common infectious agents of lymphoid organs, respiratory tract, eyes; spread through close contact with secretions, contaminated vehicles, fomites; diseases include common cold with fever and rash; **keratoconjunctivitis,** a severe eye infection; **cystitis,** an acute urinary infection.

Papovaviruses: Papilloma-and polyomaviruses.

Human papillomaviruses (HPVs) cause skin tumors called **papillomas, verrucas,** or **warts;** spread by close contact with infected skin or fomites. **Common warts** are rough, painless lesions on hands; **plantar warts** are painful, flat, benign tumors on feet, trunk. **Genital warts** are verrucas that start as tiny bumps on membranes or skin of genitals; a very common STD; may progress to large masses called **condylomata acuminata;** chronic infection associated with cervical and penile cancer; treatment by surgery, freezing, cauterization, interferon.

Polyomaviruses include animal tumor viruses and human viruses. **JC virus** causes **progressive multifocal leukoencephalopathy,** a slow destruction of the brain in cancer and immunodeficient patients; **BK virus** causes complications in urinary transplants.

Single-Stranded DNA Viruses: Parvoviruses are the tiniest viruses; cause severe disease in several mammals; **human parvovirus (B19)** causes *erythema contagiosum,* a mild respiratory infection of children that can be dangerous to fetus.

MULTIPLE-CHOICE QUESTIONS

1. Which of the following is *not* caused by a poxvirus?
   a. molluscum contagiosum
   b. smallpox
   c. cowpox
   d. chickenpox

2. In general, zoonotic viral diseases are ____ in humans.
   a. mild, asymptomatic
   b. self-limited
   c. severe, systemic
   d. not contagious

3. Which virus was used in smallpox vaccination?
   a. variola
   b. vaccinia
   c. cowpox
   d. varicella

4. Herpes simplex 1 causes ____, and herpes simplex 2 causes ____.
   a. coldsores, genital herpes
   b. fever blisters, coldsores
   c. canker sores, fever blisters
   d. shingles, stomatitis

5. Herpetic whitlows are ____ infections of the ____.
   a. pox, oral cavity
   b. CMV, lymph nodes
   c. EBV, skin
   d. herpes, fingers

6. Neonatal herpes simplex is usually acquired through
   a. contaminated mother's hands
   b. other babies in the newborn nursery
   c. crossing the placenta
   d. an infected birth canal

7. ____ is an effective treatment for herpes simplex lesions.
   a. Amantadine
   b. Interferon
   c. Acyclovir
   d. Cortisone

8. Which herpes virus is commonly associated with a dangerous fetal infection?
   a. herpes simplex
   b. herpes zoster
   c. EBV
   d. CMV

9. Varicella and zoster are caused by
   a. the same virus
   b. two different strains of VZV
   c. herpes simplex and herpes zoster
   d. CMV and VZV

10. A common sign of hepatitis is
    a. liver cancer
    b. jaundice
    c. anemia
    d. bloodshot eyes

11. A virus associated with chronic liver infection and cancer is
    a. hepatitis C
    b. hepatitis B
    c. hepatitis A
    d. delta agent

12. Benign epithelial growths on the skin of fingers are called
    a. polyomas
    b. verrucas
    c. whitlows
    d. pox

13. Parvoviruses are unique because they contain ____
    a. a double-stranded RNA genome
    b. reverse transcriptase
    c. a single-stranded DNA genome
    d. an envelope without spikes

14. Adenoviruses are the agents of
    a. hemorrhagic cystitis
    b. keratoconjunctivitis
    c. common cold
    d. all of these

15. Matching. Match the virus with the disease or condition. (Some viruses may match more than one condition.)

____ herpes simplex 1
____ variola
____ polyoma
____ varicella
____ herpes simplex 2
____ vaccinia
____ papilloma
____ Epstein-Barr
____ cytomegalovirus
____ parvovirus
____ herpesvirus-6

a. gingivostomatitis
b. condylomata acuminata
c. chickenpox
d. erythema contagiosum
e. fever blisters
f. mononucleosis
g. smallpox
h. genital herpes
i. leukoencephalopathy
j. roseola
k. cancer

CONCEPT QUESTIONS

1. Outline 10 medically important characteristics of viruses.
2. a. What accounts for the affinity of viruses for certain hosts or tissues?
   b. What accounts for the symptoms of viral diseases?
   c. Why are some pathologic states caused by viruses so much more damaging and life-threatening than others?
3. Outline the target organs and general symptoms of DNA viral infections.
4. a. What is the general rule governing the cell locations where DNA and RNA viruses multiply?
   b. What factors determine whether latency occurs or not?
   c. In what special way can people carry DNA viruses?
5. a. Briefly describe the epidemiology and symptoms of smallpox.
   b. Why was it such a great killer?
   c. What is a pock, and how is it formed?
   d. What is the basis for the vaccine given for smallpox?
   e. How is it protective?

6. a. What is the disease molluscum contagiosum?
   b. For what other infection in this chapter could it be mistaken?
   c. Explain why humans would suddenly become more susceptible to monkeypox.
7. What are the common characteristics of herpesviruses?
8. a. Compare the two types of human herpes simplex virus according to the types of diseases they cause, body areas affected, and complications.
   b. Why is neonatal infection accompanied by such severe effects?
   c. Is HSV-1 as severe as HSV-2 in this condition?
   d. What causes recurrent attacks?
9. a. Under what circumstances does one get chickenpox and shingles?
   b. Where do the lesions of shingles occur, and what causes them to appear there?
   c. How are ZIG and VZIG used?
10. a. What are the main target organs of CMV?
    b. How is it transmitted?
    c. What group of people is at highest risk for serious disease?
11. a. What are the main diseases associated with Epstein-Barr virus?
    b. What appears to be the pathogenesis involved in Burkitt's lymphoma?
    c. Why is mononucleosis a disease primarily of college-age Americans?
12. a. Define viral hepatitis, and briefly describe the three main types with regard to causative agent, common name, severity, and mode of acquisition.
    b. What is the effect of the delta agent?
13. a. What is unusual about the genome of hepatitis B virus?
    b. What is the usual source of the virus for study?
    c. What is important about the virus with regard to infectivity and transmission to others?
    d. What groups are most at risk for developing hepatitis B?
14. a. What is the course of hepatitis B infection from portal of entry to portal of exit?
    b. Describe some serious complications.
15. a. What is the nature of the vaccines for hepatitis B?
    b. Who should receive them?
16. a. What are the principal diseases of adenoviruses?
    b. How are they spread?
17. a. Compare the locations and other characteristics of the three types of warts.
    b. What type is most serious and why?
    c. How are warts cured?
18. a. Name the most important human parvovirus and the disease it causes.
19. a. Which DNA viruses are zoonotic?
    b. Which are intestinal viruses?
    c. Which are spread by blood and blood products?
    d. Which are spread by respiratory droplets, kissing, and other nonsexual intimate contact?
20. a. For which DNA viruses are there effective vaccines?
    b. Who receives them?

CRITICAL-THINKING QUESTIONS

1. a. Which DNA viruses have been linked to cancer?
   b. Why are DNA viruses more likely to cause cancer than RNA viruses?
   c. What are some mechanisms that could instigate cancer formation by viruses?
   d. Why is there a danger in using live attenuated DNA viruses for vaccines?
2. a. Why is isolating a virus from a tumor not indicative of its role in disease?
   b. What important indicators could verify that various cancers are caused by a herpes virus?
3. a. What features of variola virus made it possible to eradicate smallpox?
   b. Are any other viruses in this chapter possible candidates for this sort of achievement? Why or why not?
4. Even after the variola virus is extinct, its complete genome will be kept in gene libraries. From what you understand about viruses and their nature, can the virus ever really be extinct? Explain your answer.
5. a. Can you think of some reasons that herpes simplex and zoster viruses are carried primarily in nerve trunks?
   b. What mechanism might account for reactivation of these viruses by various traumatic events?
6. Describe several measures health care workers can take to avoid whitlows.
7. Explain why a baby whose mother has genital herpes is not entirely safe even with a cesarian birth.
8. a. What specific host defenses do immunodeficient, cancer, and AIDS patients lack that make them so susceptible to the viral diseases in this chapter?
   b. Given the ubiquity of most of these viruses, what can health care workers do to prevent transmission of infection to other patients and themselves?
9. What are the medical consequences of transfusing blood labeled Dane (+)?
10. a. Name two different defective viruses that require another virus for function.
    b. Explain how they might interrelate with the host virus.
11. Explain this statement: One acquires chickenpox from others, but one acquires shingles from oneself.
12. A man wants to divorce his wife because he believes her genital herpes could only have been acquired sexually. Her doctor says she has type 1. Can you help them resolve this problem?
13. Weigh the following observation: High titers of antibodies for EBV are found in a leukemia patient; a chronically tired, ill businessman; a healthy military recruit; and an AIDS patient. Comment on the probable significance of antibodies to EBV in human serum.
14. Give some pros and cons of exposing children to chickenpox to infect them and provide subsequent immunity.
15. a. How would you decontaminate a vaccination gun or acupuncture tools used in a series of patients?
    b. What type of virus is most likely to be transmitted by vaccination guns and acupuncture tools?

Internet Search Topic

1. Access a search engine to locate information on the prevalence of herpes simplex 2, genital warts, and molluscum contagiosum as STDs in the United States.
2. Look in online medical sources for up to date information on the involvement of DNA viruses in cancers.

25 chapter

The RNA Viruses of Medical Importance

The RNA viruses are a remarkably diverse group of microbes with a variety of morphological and genetic adaptations and extreme and novel biological characteristics. Viruses are assigned to one of 13 families on the basis of their envelope, capsid, and the nature of their RNA genome (table 25.1). RNA viruses are etiologic agents in a number of serious and prevalent human diseases, including influenza, AIDS, hepatitis, and viral encephalitis. Many experts believe that these viruses and dozens of others have the potential of becoming the most important disease threats of the future (microfile 25.1).

A virologist checks a culture while working in a Level 4 biohazard facility at the CDC. He is completely encased in a "space suit" equipped with its own air supply that protects against accidental exposure to the deadliest pathogens.

TABLE 25.1

RNA VIRUS FAMILIES AND REPRESENTATIVE VIRUSES/DISEASES

Enveloped

Segmented, Single-Stranded, Negative-Sense Genome*

Orthomyxoviridae: Influenza
Bunyaviridae: California encephalitis virus
Hantavirus hemorrhagic fever
Arenaviridae: Hemorrhagic fevers
Lassa fever virus
Argentine hemorrhagic fever virus

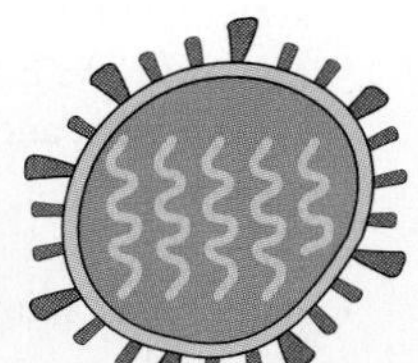

Nonsegmented, Single-Stranded, Negative Sense

Paramyxoviridae: Mumps virus
Measles virus
Respiratory syncytial virus

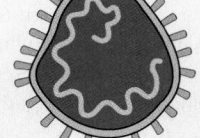

Rhabdoviridae: Rabies virus
Vesicular stomatitis virus

Filoviridae: Ebola fever virus
Marburg virus

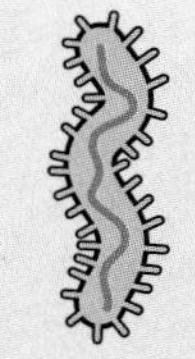

Nonsegmented, Single-Stranded, Positive Sense

Togaviridae: Rubella virus
Western and Eastern equine encephalitis
Flaviviridae: Dengue fever virus
Yellow fever virus

Coronaviridae: Common cold virus

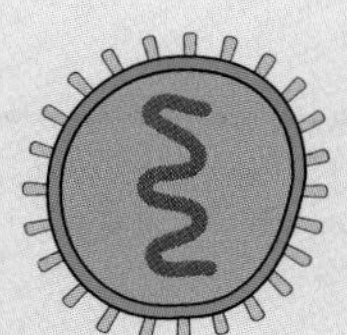

Single-Stranded, Positive Sense, Reverse Transcriptase

Retroviridae: AIDS (HIV)
T-cell leukemia virus
Hairy cell leukemia virus

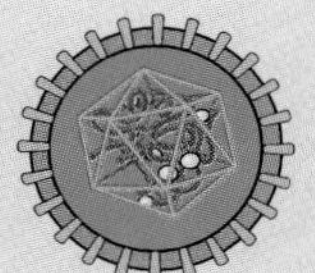

Nonenveloped

Nonsegmented, Single-Stranded, Positive-Sense Genome

Picornaviridae: Polio virus
Hepatitis A virus
Rhinoviruses

Caliciviridae: Norwalk agent

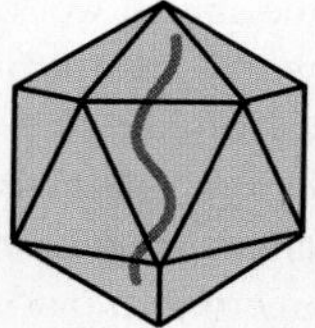

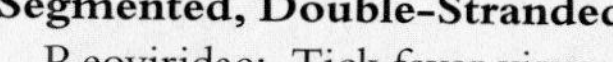

Segmented, Double-Stranded, Positive Sense, Double Capsid

Reoviridae: Tick fever virus
Rotavirus gastroenteritis

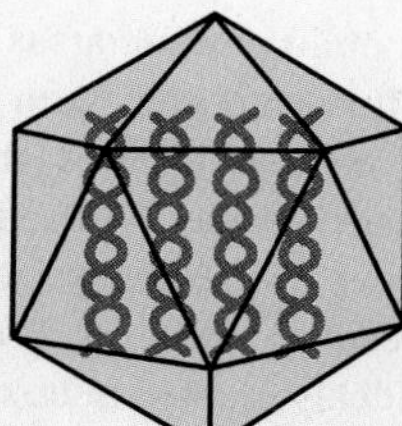

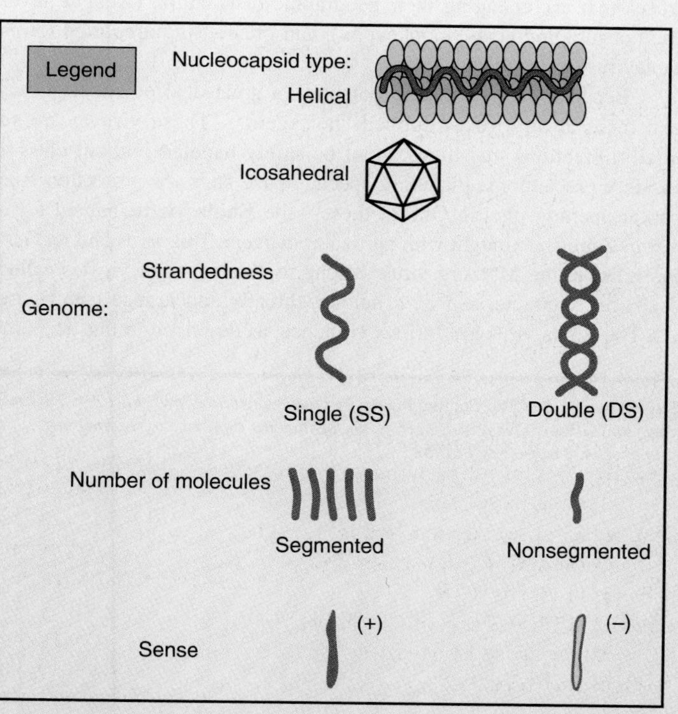

*If the RNA of the virus is in a form ready to be translated by the host's machinery, it is considered a positive-sense genome, and if it is not directly translatable by the host, it is a negative-sense genome.

MICROFILE 25.1 NEW VIRUSES, OLD TRICKS; OLD VIRUSES, NEW TRICKS

Beginning with the emergence of AIDS in the early 1980s, infectious disease specialists expressed their profoundest fears about future epidemics of similar deadly viruses:

> Viruses were mutating at rapid rates: seals were dying in great plagues . . . ; more than 90% of the rabbits of Australia died in a single year following the introduction of a new virus to the land; great influenza pandemics were sweeping through the animal world; megacities were arising in the developing world, creating niches from which "virtually anything might arise"; rain forests were being destroyed, forcing disease-carrying animals and insects into areas of human habitation and raising the very real possibility that lethal, mysterious microbes would, for the first time, infect humanity on a large scale and imperil the survival of the human race.*

Although it is true that we have seemingly vanquished smallpox and polio in some areas of the world, these successes are overshadowed by this emerging viral threat. Over the past 30 years, the world has played host to at least 33 outbreaks of serious virus diseases. The decade of the 1990s is continuing to prove the unpredictability of viruses and reemphasizes this warning. At least once a year for the past 5 years, a new virus has emerged out of nowhere or a familiar virus has suddenly erupted in unusual outbreaks.

A large number of these diseases are zoonotic in origin and become more pathogenic in humans than in their natural hosts. Outbreaks have been traced to viruses of traditional concern (influenza and Ebola); viruses that are changing their geographic distribution (Dengue fever virus) or jumping hosts (monkeypox); and previously unreported forms (hantavirus).

Because of their deadly potential, a great deal of attention has been focused on viruses dubbed "hot agents." These viruses are so lethally infectious that they cannot be safely handled without class 4 biosafety precautions, including special space suits for protection (see chapter opening photo). One of these—the **Ebola virus,** named for a river in Zaire—is fraught with particular dangers. This virus and its German relative, the Marburg virus, belong to a category of viruses called *filoviruses,* characterized by a unique, thready appearance (see figure 6.1). The Ebola virus has furious virulence, as described in this account:

> In this they are like HIV, which also destroys the immune system, but unlike the onset of HIV, the attack by Ebola is explosive. As Ebola sweeps through you, your immune system fails, and you seem to lose your ability to respond to viral attack. . . . Your mouth bleeds, and you bleed around your teeth . . . literally every opening in the body bleeds, no matter how small. Ebola does in ten days what it takes AIDS ten years to accomplish.**

This mysterious disease has erupted several times since 1976. It is thought to originate from monkeys and is spread through contact with body fluids. With no effective treatments, the mortality rate is 80–90%.

In the United States, rabies is rapidly increasing among animals that were previously a minor concern in transmission. In Texas, coyotes have become the reservoir for a rare strain of canine rabies that can be rapidly spread from them to stray dogs and humans. Deaths in southern Texas in 1995 necessitated protective treatment for 1,600 inhabitants of this region.

Other microbiological mysteries involve an unusual virus-like agent found in cattle in England that causes bovine spongiform encephalopathy, or "mad cow disease." This disease, which is transmissible among cows and sheep causes overwhelming deterioration of the brain, frenzy, and death. Several cases of a new variant of human Creutzfeldt-Jakob disease that occurred in England during 1996–1997 were traced to consumption of food products contaminated with the bovine agent. This discovery caused a crisis in the beef industry that resulted in the destruction of most herds of cattle and a ban on exported British beef.

A newly isolated pathogen called Borna disease virus (BDV) is commonly associated with neurological abnormalities in horses, rats, and monkeys. This zoonosis apparently has a wide host range and, when it infects humans, may lead to neuropsychiatric abnormalities such as schizophrenia. The blood of some psychiatric patients contains antibodies against this virus 600 times higher than normal controls.

The fears fostered by these emerging viruses are the possibility of amplification in the population, leading to rapid global transmission, and the unknown, uncontrollable nature of their origins.

**Excerpt from p. 6 of* The Coming Plague *by Laurie Garrett, copyright, 1994; Farrar, Straus, and Giroux. This passage echoes the sentiments expressed by a prominent group of scientists meeting in 1989.*

***Excerpt from pp. 46, 73 of* The Hot Zone *by Richard Preston, copyright 1994, Random House.*

ENVELOPED SEGMENTED SINGLE-STRANDED RNA VIRUSES

THE BIOLOGY OF ORTHOMYXOVIRUSES: INFLUENZA

Orthomyxoviruses are spherical particles with an average diameter of 80–120 nm. They are covered with a lipoprotein envelope that is studded with glycoprotein spikes acquired during viral maturation (see figures 25.1*a* and 6.19). The two glycoproteins that make up the spikes of the envelope and contribute to virulence are hemagglutinin (HA) and neuraminidase (NA). Because **hemagglutinin** spikes combine with a specific carbohydrate molecule found in all eucaryotic cell membranes, the virus has the capacity to bind and clump a variety of animal cells. The name *hemagglutinin* is derived from its particular agglutinating action on red blood cells (see figure 16.6). The property of erythrocyte agglutination is the basis for viral assays that have been used to identify several antigenic types. Hemagglutinin contributes to infectivity by binding to host cell receptors of the respiratory mucosa, a process that facilitates viral penetration. **Neuraminidase** has the principal effect of hydrolyzing the protective mucous coating of the respiratory tract, assisting in viral budding and release, keeping viruses from sticking together, and participating in host cell fusion.

The genome of the influenza virus is known for its extreme variability. It is subject to constant genetic changes that alter its antigenic expression and create new strains of the virus with a different epidemiologic pattern. Depending upon the degree and timing of the change, these patterns are called **antigenic shift** or **antigenic drift.** Antigenic shift is a major alteration in antigenicity occurring when genome segments from different viral strains recombine. A wholly new strain can emerge in a single recombinant event, such as when a human is simultaneously infected by both a human and an animal strain of the virus. By contrast, antigenic drift is a minor change in antigenicity that is caused by small mutations in a single virus strain. As the antigenic variations accumulate over time, the gradual trend is toward a new viral strain (microfile 25.2).

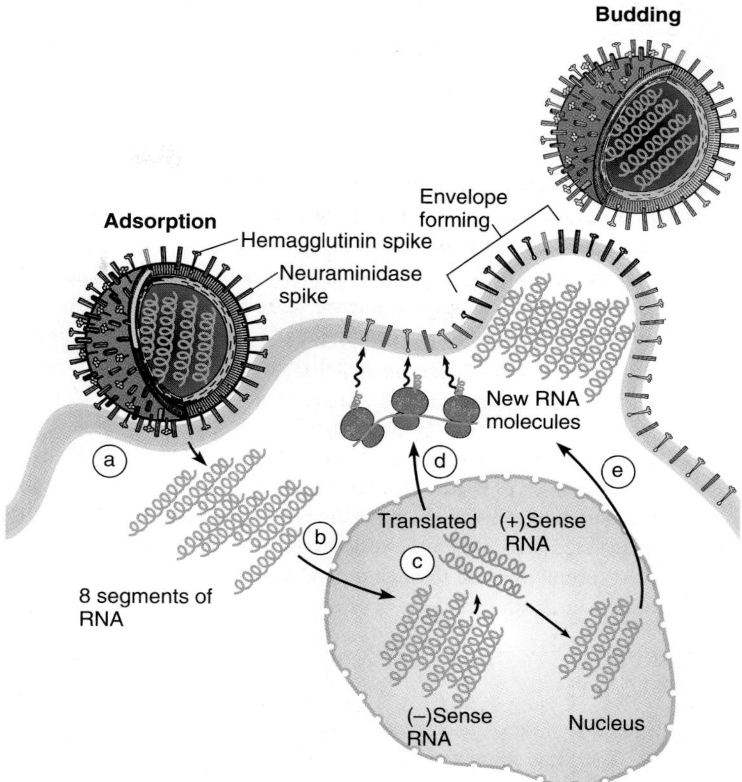

Figure 25.1
Stages in cell invasion and disruption by the influenza virus. (*a*) Virus (structure shown in a cutaway view) adsorbs to a respiratory epithelial cell by hemagglutinin spikes and fuses with the membrane. (*b*) Free RNA moves into the nucleus and (*c*) is transcribed into an mRNA that can be translated. (*d*) mRNA is used to synthesize glycoprotein spikes inserted into the host membrane. (*e*) The (+) sense RNA molecules are used to synthesize (−) sense RNA molecules for mature viruses. Assembly occurs when viral parts gather at the cell membrane and are budded off with an envelope containing spikes.

Epidemiology of Influenza A

Influenza is an acute, highly contagious respiratory illness afflicting people of all ages and marked by seasonal regularity and pandemics at predictable intervals. So convinced were the ancients that this disease was influenced by a celestial alignment of the stars that one of its first names was an Italian derivative of *un influenza di freddo,* alluding to a vague and mysterious influence. This name has remained and is now contracted to the vernacular "flu." Periodic outbreaks of respiratory illness that were probably influenza have been noted throughout recorded history. Of the six pandemics occurring in 1900, 1918, 1946, 1957, 1968, and 1977 the outbreak of 1918 took the greatest toll—at least 20 million deaths. Even today, deaths from influenza or influenzal pneumonia rank among the top 10 causes of death in the United States. Of the three types of influenza viruses, the most virulent, type A, is most commonly associated with human disease.

To characterize each virus strain, virologists have developed a special nomenclature that chronicles the influenza virus type, the presumed animal of origin, the location, and the year of origin. For example, A/duck/Ukraine/63 identifies the strain isolated from domestic or migratory fowl; likewise. A/seal/Mass/1/80 is the causative agent of seal influenza. From this convention come such expressions as "swine flu" and "Hong Kong flu."

Mode of Influenza Transmission

Inhalation of virus-laden aerosols and droplets constitutes the major route of influenza infection, although fomites can play a secondary role. Young children, notorious for their nonchalant hygiene, are the major perpetrators and victims of influenza spread. Transmission is greatly facilitated by crowding and poor ventilation in classrooms, barracks, nursing homes, and military installations in the late fall and winter. In the tropics, infection occurs throughout the year, without a seasonal distribution. Occupational contact with ducks, other poultry, and swine constitutes a special

MICROFILE 25.2 COMMUNITY IMMUNITY, GENETIC CHANGE, AND INFLUENZA

Every fall, the word goes out to expect outbreaks of certain types of flu with exotic names such as Hong King, Australian, or Russian. How can medical experts predict which types of flu to expect in a given year? In order to understand these predictions, we must first examine how changes in the influenza virus influence its epidemiologic pattern. The most important factors in the evolution and fate of influenza virus are the degree of community resistance, antigenic drift, and antigenic shift.

COMMUNITY IMMUNITY

The concept of community, or herd, immunity was introduced in chapter 16. It is defined as the collective resistance of individuals in a population, acquired by active infection, vaccination, and transplacental antibody transfer. Herd immunity works most effectively when the virus strain has recently passed through a population, and it is least effective when a new virus strain appears or is imported from another population.

ANTIGENIC DRIFT AND NEW STRAINS OF VIRUS

The influenza viruses undergo constant genetic variations and antigenic change. The viral components that are most subject to change are the hemagglutinin and neuraminidase spikes, which determine infectivity and induce protective antibodies. Accumulated changes can create viral variants so antigenically different that antibodies from a prior infection are no longer protective. The results are lapses in individual and herd immunity and epidemics.

ANTIGENIC SHIFT AND COINFECTIONS WITH INFLUENZA VIRUS

An event that increases the possibility of antigenic shift is the simultaneous infection of one individual by two different influenza strains. Although the influenza viruses are common in humans, they have also been found in swine and birds. It is possible for humans to become infected with animal strains and vice versa. Occasionally, coinfection of the same animal by two antigenic variants gives rise to completely new viral strains quite different from either original. If such an altered strain is introduced into either animal or human populations, it will spread rapidly because neither group has protective antibodies. Experts have traced the pandemics of 1918, 1957, 1968, and 1977 to strains of a virus that came from pigs.

Scientists at the Armed Forces Institute of pathology have analyzed 79-year-old tissue specimens from soldiers who died in the 1918 pandemic. They were able to isolate enough virus genetic material to determine that genes for spike proteins most closely resemble those of swine. This important finding finally laid to rest the question of where this virulent virus originated, verifying that it was the United States.

More recently, a small but potentially serious outbreak of influenza occurred in Hong Kong. About 20 people developed an infection from the chicken influenza virus, and several of them died. This was the first reported instance of avian influenza in humans. To avert an epidemic, officials ordered the slaughter of 1.4 million chickens.

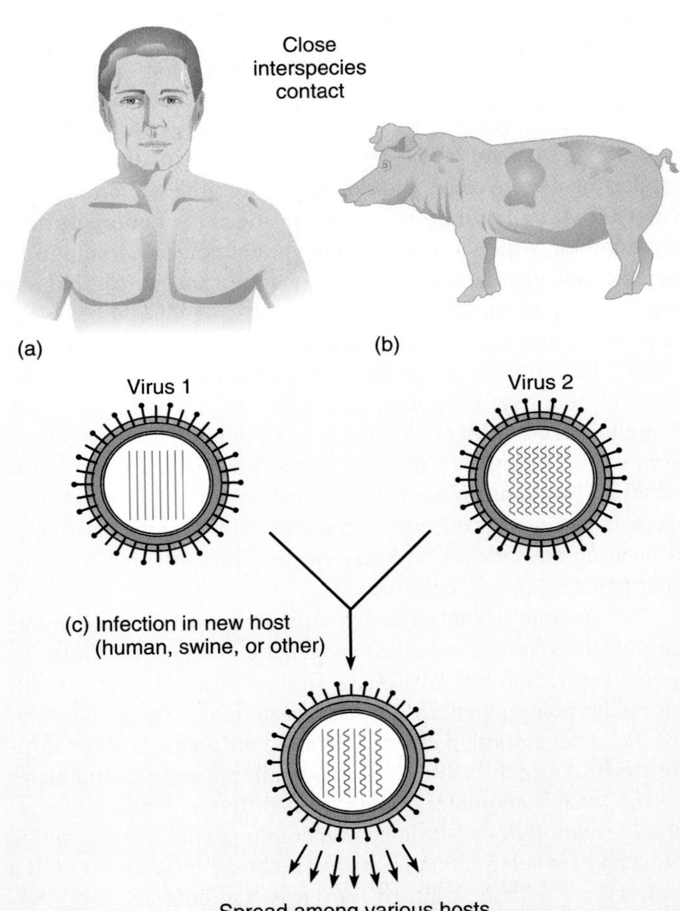

Coinfection of human and pig strains in a single animal. During viral multiplication, the eight-segmented genome of virus 1 can reassort with the eight-segmented genome of virus 2. Theoretically, recombination could produce 256 (2^8) different genome combinations. Only one is shown here.

RECURRENCE OF PANDEMICS

Influenza pandemics are accelerated by the contagiousness of the virus and widespread migrations or travel. They also occur in cycles, with episodes spaced about a decade apart. By compiling extensive serological profiles of populations and the influenza strains to which they are immune, the Centers for Disease Control and Prevention (CDC) has a scientific basis for predicting expected epidemic years and the probable strains involved. One advantage of forecasting the expected viral subtypes is that the most vulnerable people can be vaccinated ahead of time with the appropriate virus strain(s).

MICROFILE 25.3 INFLUENZA AND LONG-TERM MEDICAL COMPLICATIONS

Two major sequelae are associated with influenza virus or its vaccines. One, **Guillain-Barré syndrome,** is a neurological complication that can arise in approximately one in 100,000 vaccine recipients. This syndrome appears to be an autoimmunity induced by viral proteins and marked by varying degrees of demyelination of the peripheral nervous system, leading to weakness and sensory loss. Most patients recover function, but the disease can also be debilitating and fatal. Another deadly neurological complication was associated with the 1970s swine flu vaccinations that killed many recipients. It is now known that such severe reactions are correlated almost exclusively with the H_{swine} antigen, and other vaccines are far safer.

For reasons that are still unknown, a small number of children and young adults infected with influenza and chickenpox virus develop **Reye's syndrome,** a disease that strikes the brain, liver, and kidney and is characterized by the fatty degeneration of those organs. The symptoms are acute and include vomiting, headache, neurological abnormalities, and coma. Although the pathologic mechanisms underlying the disease are not understood, aspirin therapy for influenza and chickenpox significantly increases the risk for Reye's syndrome in children.

high-risk category for disease. The overall mortality rate for influenza A is about 0.1% of cases. Deaths are most common among elderly persons and small children.

Infection and Disease

The influenza virus binds primarily to ciliated columnar cells of the respiratory mucosa (figure 25.1). Infection causes the rapid shedding of these cells along with a load of viruses. About one-half of infections remain asymptomatic; the remainder produce symptoms ranging from inconsequential to lethal. Stripping the respiratory epithelium to the basal layer eliminates protective ciliary clearance and leads to severe inflammation and irritation. The illness is further aggravated by fever, headache, myalgia, pharyngeal pain, shortness of breath, and repetitive bouts of coughing. The viruses tend to remain in the respiratory tract, and viremia is rare. As the normal ciliated columnar epithelium is restored in a week or two, the symptoms usually abate.

Patients with emphysema or cardiopulmonary disease, and those who are very young, elderly, or pregnant, are more susceptible to serious complications. Most often, weakened host defenses and immune function predispose patients to secondary bacterial infections, especially pneumonia caused by *Streptococcus pneumoniae* and *Staphylococcus aureus.* Although less common than bacterial pneumonia, primary influenzal virus pneumonia has the ominous reputation for causing rapid death of people even in their prime. Other complications are bronchitis, meningitis, and neuritis.

Diagnosis, Treatment, and Prevention of Influenza

Except during epidemics, influenza must be differentiated from other acute respiratory diseases. Rapid immunofluorescence tests can detect influenza antigens in a pharyngeal specimen. Serological testing to screen for antibody titer and virus isolation in chick embryos or kidney cell culture are other specific diagnostic methods. Genetic analysis is valuable for determining the origin of the virus.

In most cases, treatment requires little more than measures to alleviate symptoms. Fluids, bed rest, and nonaspirin drugs (microfile 25.3) help to control symptoms. Amantadine, which specifically blocks the uncoating of the influenza virus in the cell, is sometimes prescribed to abate the course of influenza, especially in more severe cases. It can reduce the length of the disease and its symptoms as well as the spread of the virus. Amantadine is given orally, though recent studies have shown that aerosol sprays may be beneficial as well. Because the drug shows some toxicity for the CNS, it must be used with caution.

The standard vaccine contains dead viruses grown in embryonated eggs. They have an overall effectiveness of 70–90%. An attenuated virus vaccine administered nasally is currently undergoing FDA approval. It appears to be more effective than the existing vaccine, and it requires smaller doses that could even be administered at home.

Influenza vaccination is recommended primarily for certain high-risk groups, such as the chronically ill and the elderly, and for people with a high degree of exposure to the public. For complete protection, it is necessary to immunize against all the different strains of influenza viruses occurring in a given year.

BUNYAVIRUSES AND ARENAVIRUSES

Between them, the **bunyavirus*** and **arenavirus*** groups contain hundreds of viruses whose normal hosts are arthropods or rodents. Although none of the viruses are natural parasites of humans, they can be transmitted zoonotically and periodically cause epidemics in human populations. Bunyaviruses are discussed in a later section along with other arboviruses because they are transmitted primarily by insects and ticks. Some of the more important bunyavirus diseases are California encephalitis, spread by mosquitos; Rift Valley fever, an African disease transmitted by sand flies; and Korean hemorrhagic fever associated with hantavirus and

*bunyavirus (bun´-yah-vy´´-rus) From Bunyamwera, an area of Africa where this type of virus was first isolated.

*arenavirus (ah-ree´-nah-vy´´-rus) L. *arena,* sand. In reference to the appearance of virus particles in micrographs.

MICROFILE 25.4 HUNTING FOR HANTAVIRUSES

A bunyavirus with increasing importance is the **hantavirus.** This virus was first associated with a severe kidney disease called Korean hemorrhagic fever reported during the Korean War. Twenty years later, it was isolated from a field mouse taken near the Hantaan River in Korea. Outbreaks of a similar disease called nephropathica epidemica caused by a different virus stain occurred in Russia and Scandinavia in the 1960s and 1970s. Since them, 10 recognized strains of this virus have been reported worldwide, all carried by rodents such as field mice and voles. The case rate is currently estimated at 200,000 people worldwide, and the incidence of hantavirus disease is increasing throughout Europe and Asia.

For many years, American virologists had concerns that similar hantaviruses were to be found in North America. However, strains isolated from rodents in the 1980s could not be linked to any disease outbreaks. The first American cases came to light suddenly in 1993 in the Four Corners area of New Mexico, when a cluster of unusual cases was reported to the CDC. All victims of the infections were healthy young people who were suddenly struck down with high fever, lung edema, and pulmonary failure. Most of them died within a few days. Similar cases were reported sporadically in Arizona, Colorado, and other western states.

Using PCR technology, virologists at the CDC were able to identify the virus in patients' specimens in a short time. The virus was named *Sin Nombre* ("no name") virus, and the disease was named hantavirus pulmonary syndrome (HPS). By late 1997, 174 cases had occurred in 20 states with a 53% fatality rate. Comprehensive studies of the local rodent population revealed that the virus is carried by deer and harvest mice. It is probably spread by dried animal droppings and urine that have become airborne when the rodent nests are uncovered or removed. Although the epidemiology of this virus holds some mysteries, experts predict that the rodent reservoir is very widely distributed throughout the United States, and it is only a matter of time before more cases will emerge.

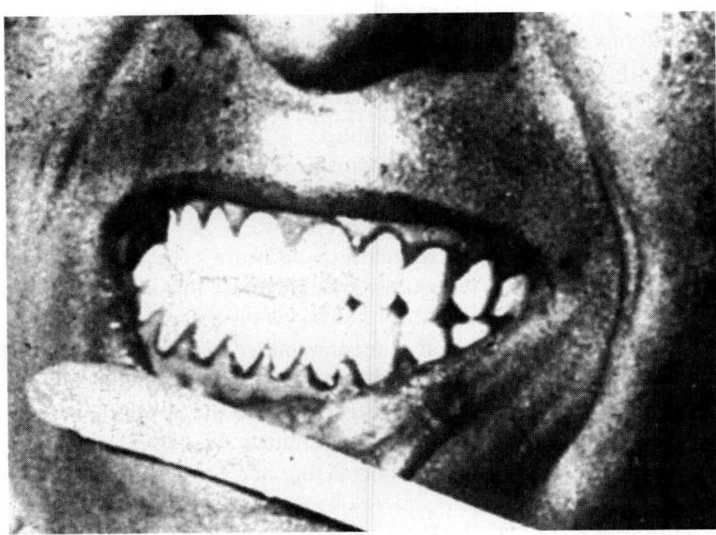

Figure 25.2
Hemorrhage and purpura in a patient with Argentine hemorrhagic fever.

transmitted by rodents. The first cases of hantavirus infection in the United States occurred in 1993 (microfile 25.4).

Four major arenavirus diseases are Lassa fever, found in parts of Africa; Argentine hemorrhagic fever, typified by severe weakening of the vascular bed, hemorrhage, and shock; Bolivian hemorrhagic fever; and lymphocytic choriomeningitis, a widely distributed infection of the brain and meninges. The viruses are closely associated with a rodent host and are continuously shed by the rodent throughout its lifetime. Transmission of the virus to humans is through aerosols and direct contact with the animal or its excreta. These diseases vary in symptomology and severity from mild fever and malaise to complications such as hemorrhage, renal failure, cardiac damage, and shock syndrome (figure 25.2). The lymphocytic choriomeningitis virus can cross the placenta and infect the fetus. Pathologic effects include hydrocephaly, blindness, deafness, and mental retardation. So dangerous are these viruses to laboratory workers that the highest level of containment procedures and sterile techniques must be used in handling them.

Chapter Checkpoints

The enveloped, segmented, single-stranded RNA virus group includes the Orthomyxoviridae, the Bunyaviridae, and the Arenaviridae. Segmented viruses such as influenza (flu) virus undergo small genetic mutations that continuously change their antigenic structure, or **antigenic drift.** This can eventually prevent the development of immunity. When two genetic variants of the flu virus infect a host, they can combine to form a third genetically different strain. This event is termed **antigenic shift.**

Like all enveloped RNA viruses, orthomyxoviruses have two unique glycoprotein virulence factors (receptors) in the envelope, **hemagglutinin** and **neuraminidase.** These assist in viral penetration and release from the host cell. Most orthomyxoviruses are spread by the aerosol route. Infection damages the respiratory epithelium and may lead to pneumonia. Control is with amantidine and yearly vaccination.

Bunyaviruses, grouped among the arboviruses because many of them are spread by insects, cause potentially fatal encephalitis and hemorrhagic fevers. Arenaviruses are typically spread by rodents. They also cause fatal encephalitis and/or hemorrhagic fevers. Humans are considered accidental hosts, infected by direct contact or through aerosol transmission. Specific diseases caused by bunyaviruses include hantavirus pulmonary syndrome and hemorrhagic fever; arenaviruses cause Lassa fever and California encephalitis.

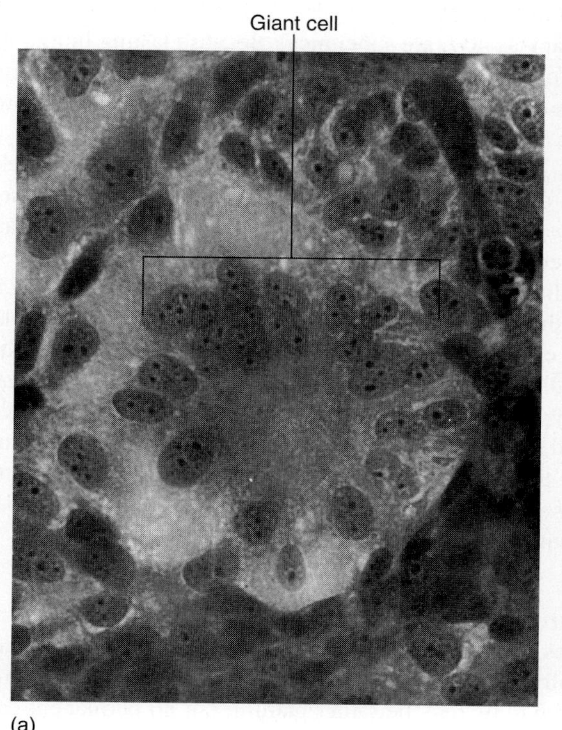

(a)

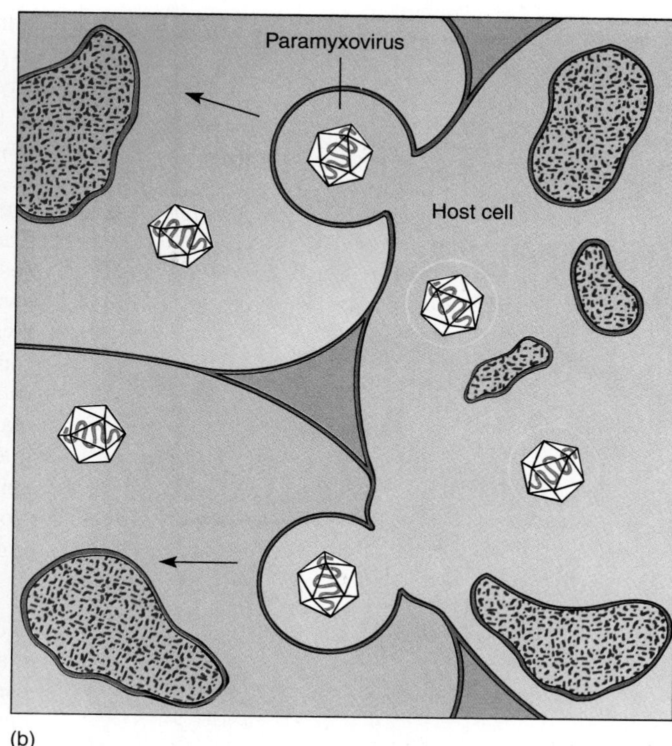

(b)

Figure 25.3
The effects of paramyxoviruses. (*a*) When they infect a host cell, paramyxoviruses induce the cell membranes of adjacent cells to fuse into large multinucleate giant cells, or syncytia. (*b*) This fusion allows direct passage of viruses from an infected cell to uninfected cells by communicating membranes. Through this means, the virus evades antibodies.

ENVELOPED NONSEGMENTED SINGLE-STRANDED RNA VIRUSES

Paramyxoviruses

The important human paramyxoviruses are ***Paramyxovirus*** (parainfluenza and mumps viruses), ***Morbillivirus*** (measles virus), and ***Pneumovirus*** (respiratory syncytial virus), all of which are readily transmitted through respiratory droplets. The envelope of a paramyxovirus possesses HN spikes and F glycoprotein spikes that allow it to infect neighboring cells by a novel mechanism. First, the cell membrane of an infected cell is modified by insertion of the spikes. The HN spikes immediately bind an uninfected neighboring cell, and in the presence of F spikes, the two cells permanently fuse. A chain reaction of multiple cell fusions then produce a *syncytium,** or **multinucleate giant cell,** with cytoplasmic inclusion bodies. This is a diagnostically useful cytopathic effect (figure 25.3).

Epidemiology and Pathology of Parainfluenza

One type of infection by *Paramyxovirus,* called **parainfluenza,** is as widespread as influenza but usually more benign. Primary infection with parainfluenza virus predominates in the late fall and winter. Droplets and respiratory secretions effectively disseminate the virus into the air and onto fomites. Inhalation or inoculation of the mucous membranes by contaminated hands are the usual routes of transmission. Parainfluenzal respiratory disease is seen most frequently in children, most of whom have been infected by the age of 6. Because protective transplacental immunity is frequently missing, babies in their first year are particularly susceptible, and they develop more severe symptoms. Characteristic symptoms of parainfluenza are minor upper respiratory disease (a cold), bronchitis, bronchopneumonia, and laryngotracheobronchitis (croup). *Croup** causes labored and noisy breathing accompanied by a hoarse cough and is most common in infants and young children.

Diagnosis, Treatment, and Prevention of Parainfluenza Presenting symptoms typical of a cold are often sufficient to presume respiratory infection of viral origin. Determining the actual viral agent, however, is difficult and usually unnecessary. Whereas in older children and adults infection is usually self-limited and benign, primary infection in infants can be severe enough to be life-threatening. So far, no specific chemotherapy is available, but supportive treatment with immune serum globulin or interferon can be of benefit.

*syncytium (sin-sish´-yum) Gr. *syn,* together, and *kytos,* cell.

*croup (kroop) Scot. *kropan,* to cry aloud.

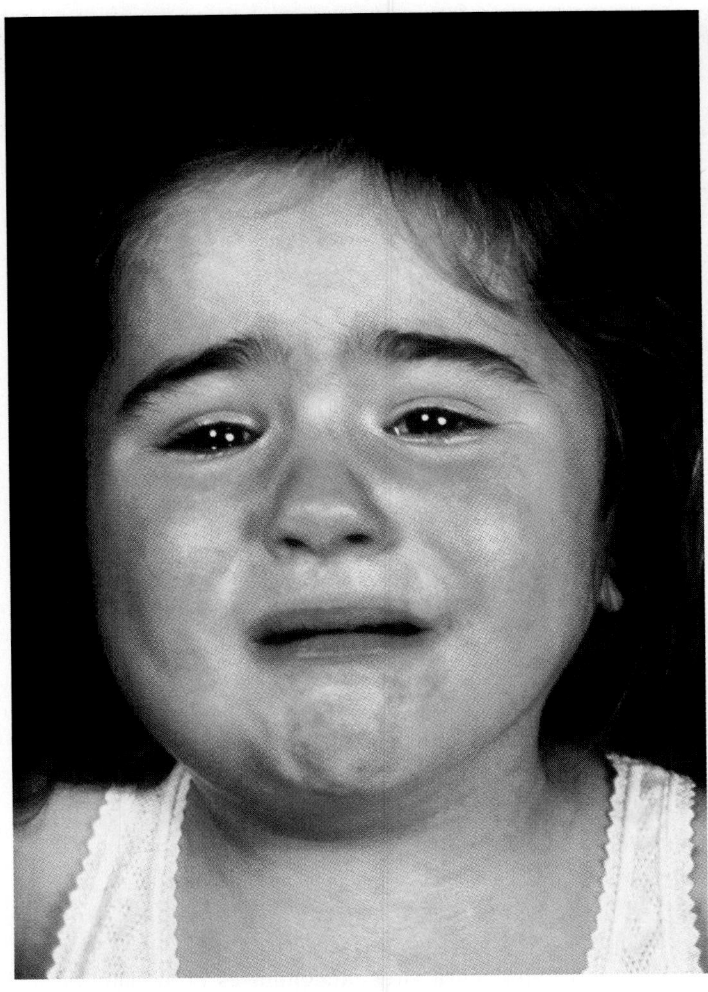

Figure 25.4

The external appearance of swollen parotid glands in mumps (parotitis). Usually both sides are affected, though parotitis affecting one side (as shown here) occasionally develops.

Mumps: Epidemic Parotitis

Another infection caused by *Paramyxovirus* is **mumps** (Old English for lump or bump). So distinctive are the pathologic features of this disease that Hippocrates clearly characterized it several hundred years B.C. as a self-limited, mildly epidemic illness associated with painful swelling at the angle of the jaw (figure 25.4). Also called **epidemic parotitis,** this infection typically targets the parotid salivary glands, but it is not limited to this region. The mumps virus bears morphological and antigenic characteristics similar to the parainfluenza virus and has only a single serological type.

Epidemiology and Pathology of Mumps Humans are the exclusive natural hosts for the mumps virus. It is communicated primarily through salivary and respiratory secretions. Infection occurs worldwide, with epidemic increases in the late winter and early spring in temperate climates. High rates of infection arise among crowded populations or communities with poor herd immunity. Most cases occur in children under the age of 15, and as many as 40% are subclinical. Because lasting immunity follows any form of mumps infection, no long-term carrier reservoir exists in the population. The incidence of mumps is at the lowest point since reporting first began in 1968.

After an average incubation period of 2 to 3 weeks, symptoms of fever, nasal discharge, muscle pain, and malaise develop. These may be followed by inflammation of the salivary glands (especially the parotids), producing the classic gopherlike swelling of the cheeks on one or both sides (figure 25.4). This swelling of the gland and its duct can induce considerable discomfort, especially when the gland is stimulated. Viral multiplication in salivary glands is followed by invasion of other organs, especially glands. Such diverse structures as the testes, ovaries, thyroid gland, pancreas, meninges, heart, and kidney can be involved. Despite the invasion of multiple organs, the prognosis for most infections is complete, uncomplicated recovery, with permanent immunity.

Complication in Mumps In 20% to 30% of young adult males, mumps infection localizes in the epididymis and testis, usually on one side only. The resultant syndrome of orchitis and epididymitis may be rather painful, but no permanent damage occurs. The popular belief that mumps readily causes sterilization of adult males is still held, despite medical evidence to the contrary. Perhaps this notion has been reinforced by the excruciating pain experienced during testicular infection, by the tenderness that continues afterward, and by the partial atrophy of one testis that occurs in about half the cases. Permanent sterility due to mumps is very rare.

In mumps pancreatitis, the virus replicates in beta cells and pancreatic epithelial cells. Viral meningitis, characterized by fever, headache, nausea, vomiting, and stiff neck, is rather common in mumps. It appears 2 to 10 days after the onset of parotitis, lasts for 3 to 5 days, and then dissipates, leaving few or no adverse side effects. On rare occasions, loss of hearing develops abruptly in the course of parotitis, usually affecting only one side. Unfortunately, because the virus replicates in the organ of Corti and causes inflammation and irreversible injury, deafness is usually permanent.

Diagnosis, Treatment, and Prevention of Mumps Mumps can be tentatively diagnosed in a child with swollen parotid glands and a known exposure 2 or 3 weeks previously. Since parotitis is not always present and the incubation period can range from 7 to 23 days, a practical diagnostic alternative is to perform a direct fluorescent test for viral antigen or an ELISA test on the serum.

The general pathology of mumps is mild enough that symptomatic treatment to relieve fever, dehydration, and pain is usually adequate. A live, attenuated mumps vaccine given routinely as part of the MMR vaccine at 15 months of age is a powerful and effective control agent. A separate single vaccine is available for adults who require protection. Although the antibody titer is lower than that produced by wild mumps virus, protection often lasts a decade.

TABLE 25.2

THE TWO FORMS OF MEASLES

	Synonyms	Etiology	Primary Patient	Complications	Skin Rash	Koplik's Spots
Measles	Rubeola, red measles	Paramyxovirus: *Morbillivirus*	Child	SSPE,* pneumonia	Present	Present
German measles	Rubella, 3-day measles, rubeola	Togavirus: *Rubivirus*	Child/fetus	Congenital defects	Present	Absent

*Subacute sclerosing panencephalitis. See microfile 25.5.

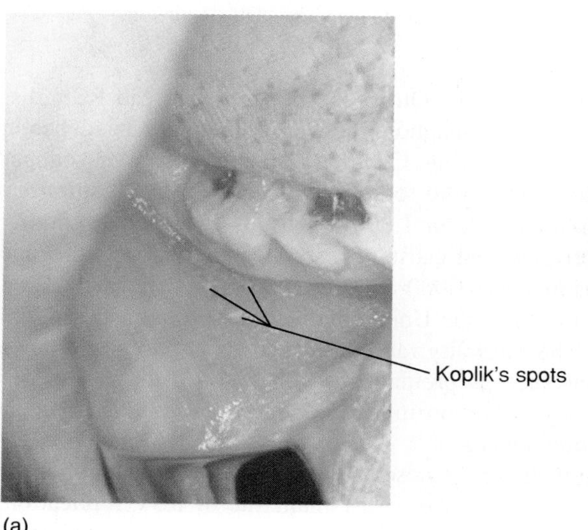

(a)

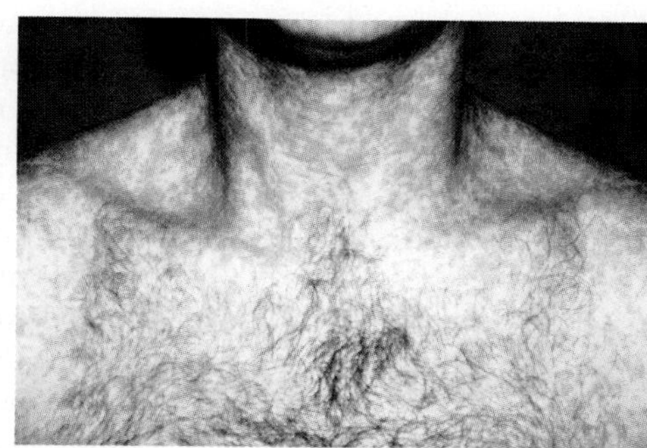

(b)

Figure 25.5 Signs and symptoms of measles. (*a*) Koplik's spots, named for the physician who described them in the last century, are tiny white lesions with a red border that form on the cheek membrane adjacent to the molars just prior to eruption of the skin rash. (*b*) A fully developed measles rash.

Measles: Morbillivirus Infection

Measles, an acute disease caused by Morbillivirus, is also known as **red measles*** and **rubeola.*** Unfortunately, German measles (rubella), an entirely unrelated viral infection, is also called rubeola in some countries. Some differentiating criteria for these two forms of measles are summarized in table 25.2. Despite a tendency to think of measles as a mild childhood illness, several recent outbreaks have reminded us that measles can kill babies and young children, and that it is the seventh most frequent cause of death worldwide.

Epidemiology of Measles Measles is one of the most contagious infectious diseases, transmitted principally by direct contact with respiratory aerosols. Epidemic spread is favored by crowding, low levels of herd immunity, malnutrition, and inadequate medical care. Occasional outbreaks of measles have been linked to the lack of immunization in children or the failure of a single dose of vaccine in many children. There is no reservoir other than humans, and a person is infectious during the periods of incubation, prodromium, and the skin rash. But the virus is not carried for any length of time during convalescence. Only relatively large, dense populations of susceptible individuals can sustain the continuous chain necessary for transmission. The most recent statistics show that measles has declined to an all-time low.

Infection, Disease, and Complications of Measles Infection with the measles virus begins with the invasion of the mucosal lining of the respiratory tract, followed by viremia. The incubation period of nearly 2 weeks ends with symptoms of sore throat, dry cough, headache, conjunctivitis, lymphadenitis, and fever. In a short time, unusual oral lesions called *Koplik's spots* appear as a prelude to the characteristic red maculopapular *exanthem** that erupts on the head and then progresses to the trunk and extremities, until most of the body is covered (figure 25.5). The rash gradually coalesces into red patches that fade to brown.

*measles (mee´-zlz) Dutch *maselen,* spotted.

*rubeola (roo-bee´-oh-lah) L. *ruber,* red.

*exanthem (eg-zan´-thum) Gr. *exanthema,* to bloom or flower. An eruption or rash of the skin.

MICROFILE 25.5 A RARE AND DEADLY OUTGROWTH OF MEASLES

Subacute sclerosing panencephalitis (SSPE)* is a progressive neurological degeneration of the cerebral cortex, white matter, and brain stem. Its incidence is approximately one case in a million measles infections, and it afflicts primarily male children and adolescents. Its relationship to measles remained unsuspected for three decades because of the prolonged incubation period of SSPE. Eventually, researchers discovered intracellular inclusions in infected brain cells that proved identical to cytological features of measles, and finally they were able to isolate the SSPE agent.

The pathogenesis of SSPE appears to involve a defective virus, one that has lost its ability to form a capsid and be released from an infected cell. It can, however, continue to spread unchecked through the brain by cell fusion. Its spread sets into motion inflammatory processes that destroy neurons and accessory cells and break down myelin. The disease is known for the profound intellectual and neurological impairment it brings, beginning with such symptoms as decline in memory, judgment, and motor activity. The course of the disease invariably leads to coma and death in a matter of months or years.

subacute sclerosing panencephalitis (sub-uh-kewt´sklair-oh´-sing pan´´-en-cef´´-uh-ly´-tis) Gr. *skleros*, hard, *pan*, all, and *enkephalos*, brain.

Although in most instances, measles is a self-limited infection, complications can be sufficiently severe to cause death in about 1 in 500 children. Among the complications are laryngitis, bronchopneumonia, and bacterial secondary infections such as otitis media and sinusitis. Children afflicted with leukemia or thymic deficiency are especially predisposed to pneumonia because of their lack of the natural T-cell defense. Undernourished children frequently have severe diarrhea and abdominal discomfort that adds to their debilitation. The gravest complications include the central nervous system (microfile 25.5).

Diagnosis, Treatment, and Prevention of Measles The patient's age, a history of recent exposure to measles, and the season of the year are all useful epidemiologic clues for diagnosis. Clinical characteristics such as dry cough, sore throat, conjunctivitis, lymphadenopathy, fever, and especially the appearance of Koplik's spots and a rash are presumptive of measles.

Treatment relies on reducing fever, suppressing cough, and replacing lost fluid. Complications require additional remedies to relieve neurological and respiratory symptoms and to sustain nutrient, electrolyte, and fluid levels. Antibiotics are sometimes given for bacterial complications, and doses of immune globulin can be therapeutic.

Vaccination is the most practical, economical, and enduring strategy for combating measles. The attenuated viral vaccine is administered by subcutaneous injection and achieves immunity that persists for about 20 years. It must be stressed that because the virus is live, it can cause an atypical infection sometimes accompanied by a rash and fever. Measles immunization is recommended for all healthy children at the age of 15 months (MMR vaccine, with mumps and rubella), and a booster is given before a child enters school. A single antigen vaccine (Meruvax) is also available for older patients who require protection against measles alone. As a general rule, anyone who has had the measles is considered protected, and anyone who received the vaccine before 1980 or whose immunization history is in doubt should be revaccinated.

Respiratory Syncytial Virus: RSV Infections

As its name indicates, **respiratory syncytial virus (RSV),** also called *Pneumovirus,* infects the respiratory tract and produces giant multinucleate cells. Outbreaks of droplet-spread RSV disease occur regularly throughout the world, with peak incidence in the winter and early spring. Children 6 months of age or younger are especially susceptible to serious disease of the respiratory tract. Approximately 5 in 1,000 newborns is affected, making RSV the most prevalent cause of respiratory infection in this age group. An estimated 91,000 children are hospitalized with RSV infection each year in the United States, and approximately 4,500 die from it. The mortality rate is highest for children who have complications such as prematurity, congenital disease, and immunodeficiency. Infection in older children and adults is manifested as a common cold.

The epithelia of the nose and eye are the principal portals of entry, and the nasopharynx is the main site of RSV replication. The first symptoms of primary infection are fever that lasts for 3 days, rhinitis, pharyngitis, and otitis. Infections of the bronchial tree and lung parenchyma give rise to symptoms of croup that include acute bouts of coughing, wheezing, *dyspnea,** and abnormal breathing sounds (rales). In severe cases, compromised breathing can lead to fatalities. Adults and older children with RSV infection can experience cough and nasal congestion but are frequently asymptomatic.

Diagnosis, Treatment, and Prevention of RSV Diagnosis of RSV infection is more critical in babies than in older children or adults. The afflicted child is conspicuously ill, with signs typical of pneumonia and bronchitis. A positive identification of RSV must be made as rapidly as possible so that correct treatment can be started. The best diagnostic procedures are those that demonstrate the viral antigen directly from specimens. The sample must be fresh (the virus is very fragile), and it must contain sufficient infected cells. Direct and indirect fluorescent staining, ELISA testing, and DNA probes are useful methods for evaluating a specimen's viral content.

A highly beneficial new treatment is the administration of RSV immunoglobulin. This serum is obtained from people with high RSV antibody titers, and it greatly reduces complications and the need for hospitalization. An antiviral drug, ribavirin (virazole),

*dyspnea (dysp´-nee-ah) Gr. *dyspnoia,* difficulty in breathing.

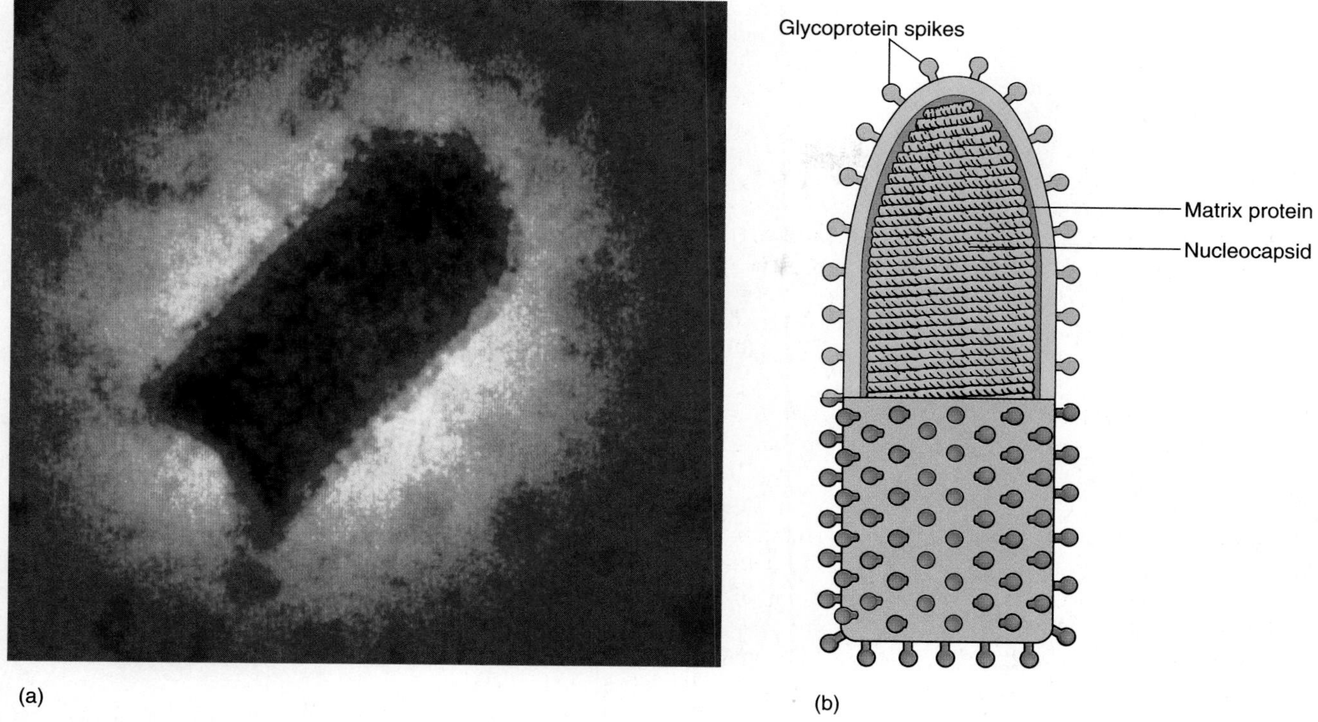

Figure 25.6
The structure of the rabies virus. (*a*) Color-enhanced virion shows internal serrations, which represent the tightly coiled nucleocapsid. (*b*) A schematic model of the virus, showing its major features.

is an alternative therapy administered in aerosol inhaled form. Other supportive measures include drugs to reduce fever, assisting pulmonary ventilation, and treating secondary bacterial infection if present. Technologists are working on an attenuated vaccine to be administered intranasally in order to promote formation of antibodies on the mucosal surface, but so far results have been disappointing.

RHABDOVIRUSES

The most conspicuous *rhabdovirus** is the **rabies*** virus, genus ***Lyssavirus.**** The particles of this virus have a distinctive bullet-like appearance, round on one end and flat on the other. Additional features are a helical nucleocapsid and spikes that protrude through the envelope (figure 25.6). The family contains approximately 60 different viruses, but only the rabies virus and, rarely, some other mammalian lyssaviruses affect humans.

Epidemiology of Rabies

Rabies is a slow, progressive zoonotic disease characterized by a fatal meningoencephalitis. It is distributed nearly worldwide, although 34 countries have remained rabies-free by practicing rigorous animal control. The primary wild reservoirs of the virus are mammals such as wild canines, skunks, raccoons, badgers, wild cats, and bats. These animals can spread the infection to domestic dogs and cats, and both wild and domestic mammals can spread the disease to humans through bites, scratches, and inhalation of droplets. The annual worldwide total for human rabies is estimated at about 30,000 cases, but fewer than 5–10 of these cases occur in the United States. Rabies incidence in wild reservoirs has been on the increase, while domestic dog rabies has declined.

The epidemiology of animal rabies in the United States has varied both chronologically and geographically. The most common wild animal reservoir host has changed from foxes in 1966 to skunks in 1977 and to raccoons in 1993. Regional differences in the dominant reservoir also occur. Bats, skunks, and bobcats are the most common carriers of rabies in California, whereas raccoons are the predominant carriers in the East, and coyotes dominate in Texas.

Infection and Disease

Infection with rabies virus typically begins when an infected animal's saliva enters a puncture site. Occasionally, the virus is inhaled or inoculated orally. The rabies virus remains up to a week at the trauma site, where it multiplies. The virus then gradually enters sensory nerve endings or the neuromuscular junction and advances toward the sensory ganglia, spinal cord, and brain (figure 25.7). Viral multiplication throughout the brain is eventually followed by

*rhabdovirus (rab´-doh-vy´´-rus) Gr. *rhabdos,* rod. In reference to its bullet, or bacillary, form.

*rabies (ray´-beez) L. *rabidus,* rage or fury.

**Lyssavirus* (lye´-suh-vy´´-rus) Gr. *lyssa,* madness.

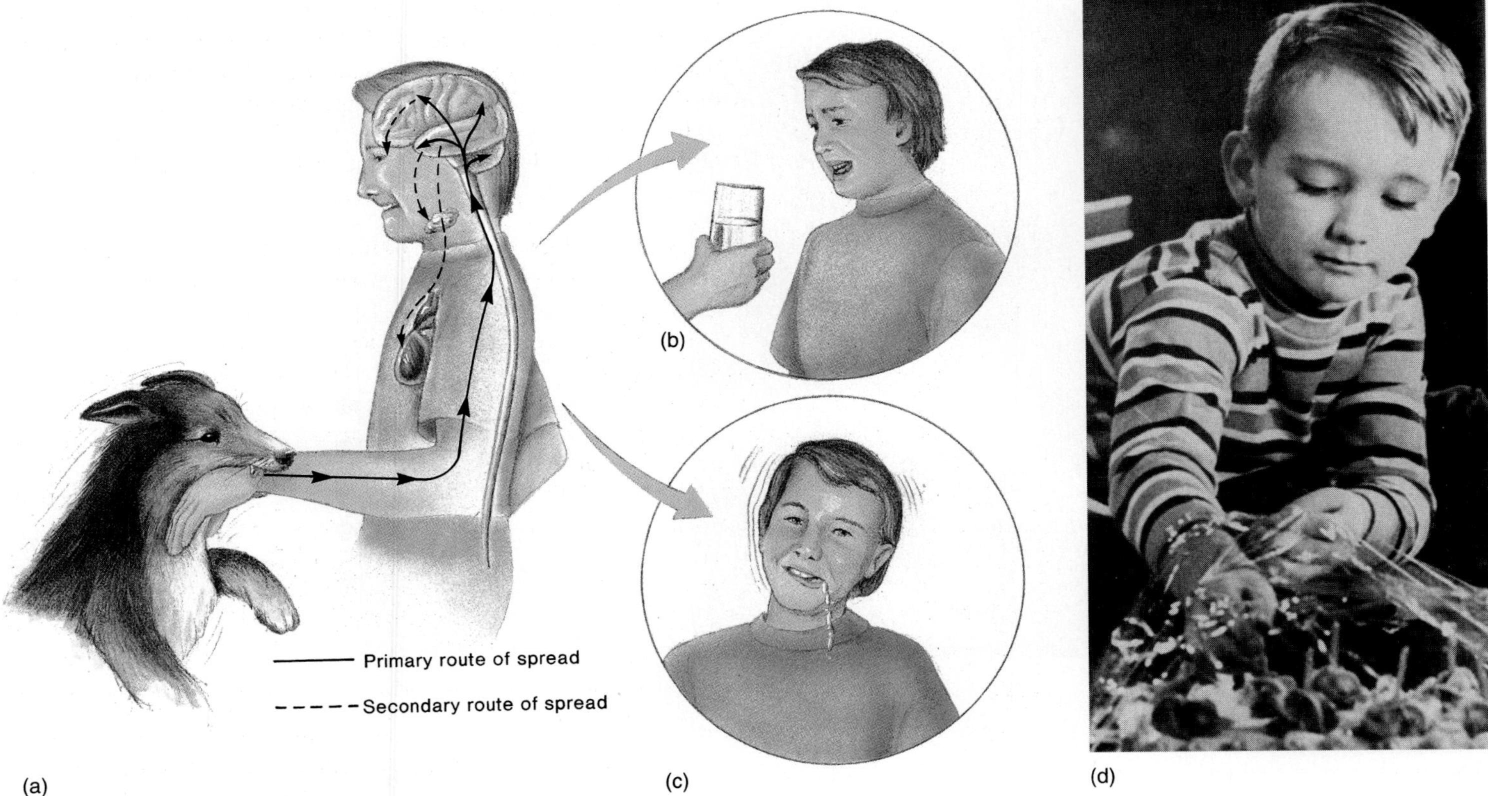

Figure 25.7
A pathologic picture of rabies. (*a*) After an animal bite, the virus spreads to the nervous system and multiplies. (*b*) The classic symptom of hydrophobia is elicited by pain in swallowing. (*c*) Severe neurological symptoms develop, including agitation and twitching. (*d*) A very rare case—a child who survived an attack of rabies.

migration to such diverse sites as the eye, heart, skin, and oral cavity. The infection cycle is completed when the virus replicates in the salivary glands and is shed into the saliva. Clinical rabies proceeds through the identifiable stages of incubation, prodromium, acute neurological phase (encephalitis), coma, and death.

Clinical Phases of Rabies

The average incubation period of rabies is 1 to 2 months with extremes of a week to more than a year. Its length depends upon the wound site, its severity, and the inoculation dose. The incubation period is shorter in facial, scalp, or neck wounds because of greater proximity to the brain. The prodromal phase begins with fever, anorexia, nausea, vomiting, headache, fatigue, and other nonspecific symptoms. Some patients continue to experience pain, burning, prickling, or tingling sensations at the wound site.

In the form of rabies termed *furious,* the first acute signs of neurological involvement are periods of agitation, disorientation, seizures, and twitching. Spasms in the neck and pharyngeal muscles lead to severe pain upon swallowing. As a result, attempts to swallow or even the sight of liquids bring on *hydrophobia* (fear of water). Throughout this phase, the patient is fully coherent and alert. If afflicted with the *dumb* form of rabies, a patient is not hyperactive but paralyzed, disoriented, and stuporous. Ultimately, both forms progress to the coma phase, resulting in death from cardiac or respiratory arrest. Until recently, humans were never known to survive rabies. But three patients have now recovered with minimal aftereffects after receiving intensive, long-term treatment (figure 25.7*d*).

Diagnosis and Management of Rabies

When symptoms appear after a rabid animal attack, the cause-and-effect relationship is vivid, and the disease is readily diagnosed. But the diagnosis can be obscured when contact with an infected animal is not clearly defined or when symptoms are absent or delayed. Anxiety, agitation, and depression can pose as a psychoneurosis; the mental state and pharyngeal spasms resemble tetanus; and encephalitis with convulsions and paralysis mimics a number of other viral infections. Often the disease is diagnosed at autopsy. The laboratory criteria most diagnostic of rabies are intracellular inclusions (Negri bodies) in infected nervous tissue (figure 25.8), isolation and identification of rabies virus in saliva or brain tissue, and demonstration of rabies virus antigens in specimens of the brain, serum, cerebrospinal fluid, or cornea using immunofluorescent methods.

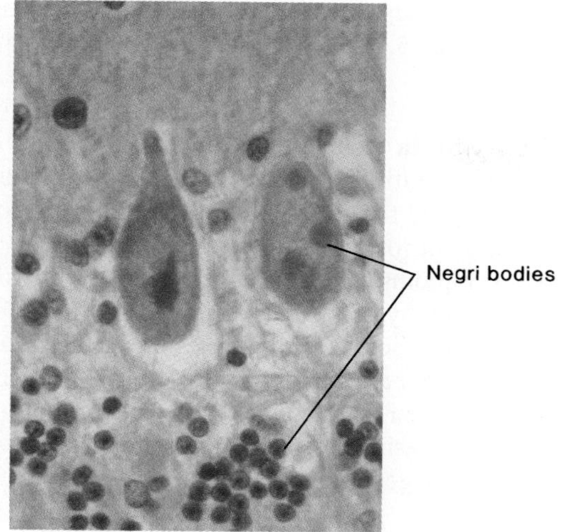

Figure 25.8
Neuronal cells of the cerebellum exhibit distinct intracytoplasmic inclusions called Negri bodies. Although typical of rabies, they are usually not found until the brain is examined in an autopsy.

Postexposure Prophylaxis A bite from a wild or stray animal demands prompt action, including assessment of the animal, meticulous care of the wound, and a specific treatment regimen. A wild mammal, especially a skunk, raccoon, fox, or coyote that bites without provocation is presumed to be rapid, and therapy is immediately commenced. If the animal is captured, brain samples and other tissue are examined for verification of rabies. Apparently healthy domestic animals are observed closely for signs of disease and sometimes quarantined. Preventive therapy is initiated if any signs of rabies appear.

After an animal bite, the wound should be scrupulously washed with soap or detergent and water, followed by debridement and application of an antiseptic such as alcohol or peroxide. Rabies is one of the few infectious diseases for which a combination of passive and active postexposure immunization is therapeutically indicated and successful. Initially the wound is infused with human rabies immune globulin (HRIG) to impede the spread of the virus, and globulin is also injected intramuscularly to provide immediate systemic protection. A full course of human diploid cell vaccine (HDCV) is started simultaneously (microfile 25.6).

Rabies Prevention and Control Measures that effectively prevent and limit rabies are preexposure vaccination and animal control. Three intramuscular injections of HDCV are recommended for high-risk groups such as veterinarians, animal handlers, laboratory personnel, and travelers to endemic areas. Control measures such as vaccination of domestic animals, elimination of strays, and strict quarantine practices have helped reduce the virus reservoir. Several countries have tested a new genetically engineered oral vaccine made with a vaccinia virus that carries the gene for the rabies virus surface antigen. The vaccine is incorporated into bait placed in the habitats of wild reservoir species such as skunks and raccoons. General use of this vaccine in the United States has been started in limited regions.

Chapter Checkpoints

The enveloped, nonsegmented, single-stranded group of RNA viruses includes the Paramyxoviridae and the Rhabdoviridae. Paramyxoviruses cause respiratory infections characterized by the formation of multinucleate giant cells. *Paramyxovirus, Morbillivirus,* and *Pneumovirus* are the principal disease agents in this group. They cause mumps, red measles, and respiratory syncytial virus infections through aerosol transmission.

The most serious of the Rhabdoviridae is the *Lyssavirus,* which causes rabies, a slow progressive, usually fatal, zoonotic disease of the CNS acquired by bites and other close contact with infected mammals.

MICROFILE 25.6 THE DEVELOPMENT OF RABIES VACCINES

The first attempt to prevent rabies by postexposure vaccination came in 1878 when a boy mauled by a rapid dog was brought to Louis Pasteur for treatment with a vaccine never before tried in humans. Pasteur and others had experimented by infecting rabbit brains with canine rabies virus, harvesting the virus from the rabbits' saliva, and inoculating yet other rabbits. This serial passage repeated a hundred times eventually yielded an attenuated, or "fixed," virus. The method of vaccine administration in humans, called the Pasteur treatment, consisted of injecting the patient daily with virus of increasing virulence over a period of 2 to 3 weeks. This technique caused many undesirable side effects, such as severe pain at the injection site, allergies, and even neurological damage.

Later replacements for the Pasteur treatment were phenol-inactivated (Semple) vaccine and the duck embryo vaccine (DEV), prepared from embryonated eggs and given on 21 consecutive days with two boosters. The current vaccine of choice is the **human diploid cell vaccine (HDCV).** This potent inactivated vaccine contains no brain or animal tissue because the virus is cultured in human embryonic fibroblasts. The routine in postexposure vaccination entails intramuscular or intradermal injection on the 1st, 3rd, 7th, 14th, 28th, and 60th days, with two boosters.

OTHER ENVELOPED RNA VIRUSES

Coronaviruses

Coronaviruses[1] are relatively large RNA viruses with distinctive, widely spaced spikes on their envelopes. These viruses are common in domesticated animals and are responsible for epidemic respiratory, enteric, and neurological diseases in pigs, dogs, cats, and poultry. Thus far, two types of human coronaviruses have been characterized. One of these (HCV) is an etiologic agent of the common cold (see microfile 25.11). The same virus is also thought to cause some forms of viral pneumonia and myocarditis. The other human virus is very closely associated with the intestine and may have a role in some human enteric infections. Coronaviruses have also been isolated from the brain, but their role in human neurological disease is uncertain at this time.

Rubivirus, the Agent of Rubella

Togaviruses are nonsegmented, single-stranded RNA viruses with a loose envelope.[2] There are several important members, including *Rubivirus,* the agent of rubella, and certain arboviruses. **Rubella,** or German measles, was first recognized as a distinct clinical entity in the mid-eighteenth century and was considered a benign childhood disease with mild prodromal symptoms and a brief rash. Then, in 1941, the Australian physician Norman Gregg called attention to the *teratogenic* aspect of rubella virus infection. He discovered that cataracts often developed in neonates born to mothers who had contracted rubella in the first trimester. Epidemiologic investigations over the next generation confirmed the link between rubella and many other congenital defects as well.

Epidemiology of Rubella

Rubella is an endemic disease with worldwide distribution. Infection is initiated through contact with respiratory secretions and occasionally urine from infected individuals. The virus is shed during the prodromal phase and up to a week after the rash appears. Because the virus is only moderately communicable, close living conditions are required for its spread. Although epidemics and pandemics of rubella once regularly occurred in 6- to 9-year cycles, the introduction of vaccination has essentially stopped this pattern in the United States. Most cases are reported among adolescents and young adults in military training camps, colleges, and summer camps. The greatest concern is that nonimmune women of childbearing age might be caught up in this cycle, raising the prospect of congenital rubella.

Infection and Disease

Two clinical forms of rubella can be distinguished. Postnatal infection develops in children or adults, and congenital (prenatal) infection of the fetus is expressed in the newborn as various types of birth defects (figure 25.9).

Postnatal Rubella During an incubation period of 2 to 3 weeks, the virus multiplies in the respiratory epithelium, infiltrates local lymphoid tissue, and enters the bloodstream. Despite widespread invasion of organs, half of infections are asymptomatic. Early symptoms include malaise, mild fever, sore throat, and lymphadenopathy. The rash of pink macules and papules first appears on the face and progresses down the trunk and toward the extremities, advancing and resolving in about 3 days. Adult rubella is often accompanied by joint inflammation and pain rather than a rash. Except for an occasional complication, postnatal rubella is generally mild and produces lasting immunity.

Congenital Rubella Transmission of the rubella virus to a fetus in utero can result in a serious complication called **congenital rubella.** The mother is able to transmit the virus even if she is asymptomatic. Fetal injury varies according to the time of infection. It is generally accepted that infection in the first trimester is most likely to induce miscarriage or multiple permanent defects in the newborn such as cardiac abnormalities, ocular lesions, deafness (figure 25.9*b*), and mental and physical retardation. Less drastic sequelae that usually resolve in time are anemia, hepatitis, pneumonia, carditis, and bone infection. Because an infant with congenital rubella sheds virus in its nasal secretions for months, congenitally infected asymptomatic neonates are a concern in the hospital nursery.

Diagnosis of Rubella

Rubella mimics other diseases and is often asymptomatic, so it should never be diagnosed on clinical grounds alone. The confirmatory methods of choice are serological testing and virus isolation systems. Diagnosis of fetal infection by examining cells and amniotic fluid obtained by amniocentesis is a promising technique that awaits further evaluation. IgM tests can determine recent infection, and a rising titer is a clear indicator of continuing rubella infection. Several serological tests for IgM are used, including complement fixation, ELISA, fluorescent assay, and hemagglutination. The latex-agglutination card, a simple, miniaturized test, is a more rapid method for IgM assay.

Prevention of Rubella

Postnatal rubella is generally benign and requires only symptomatic treatment. Because no specific therapy for rubella is available, most control efforts are directed at maintaining herd immunity with attenuated rubella virus vaccine. This vaccine is usually given to children in the combined form (the MMR vaccination) at 15 months and a booster at 4 or 6 years of age. The use of rubella vaccine alone varies according to the age and sex of the individual and whether pregnancy is anticipated or already in progress (microfile 25.7).

1. Named for the resemblance of the viral spikes to a crown of thorns.
2. Reminiscent of a toga.

MICROFILE 25.7 PREVENTION OF CONGENITAL RUBELLA

Preventing congenital rubella is best accomplished by vaccinating all children at an early age so that girls are already immune by the time of sexual maturity. Unfortunately, the vaccine does not always provide long-term protection, and many children are not vaccinated at all. A second important safety net is routine testing for rubella antibodies in women prior to marriage. A titer of 1:8 or greater is interpreted as being protective.

The finding of a negative titer in a postpubertal or pregnant woman requires careful evaluation and management. The current recommendation for nonpregnant, antibody-negative women is immediate immunization. Because the vaccine contains live virus and a teratogenic effect is theoretically possible, the vaccine is administered on the condition that the patient uses contraception for 3 months afterward. Under no circumstances is the vaccine to be given to a pregnant woman. The management of verified rubella in a pregnant woman depends upon whether maternal infection occurred early or late. Therapeutic abortion may be indicated if maternal infection occurred in the first trimester and monitoring of the fetus discloses severe damage.

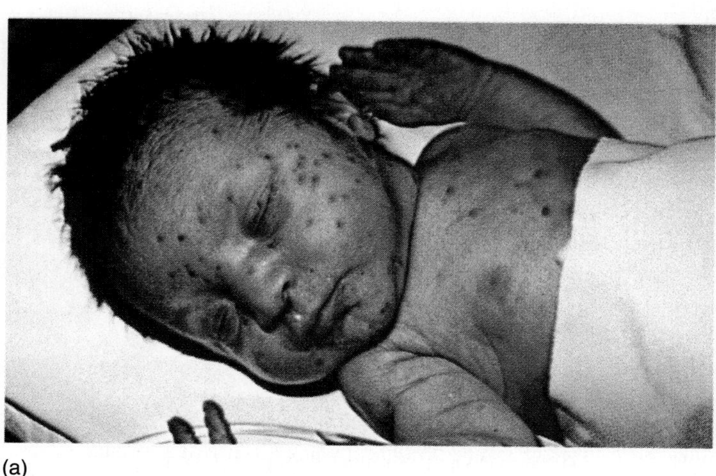

(a)

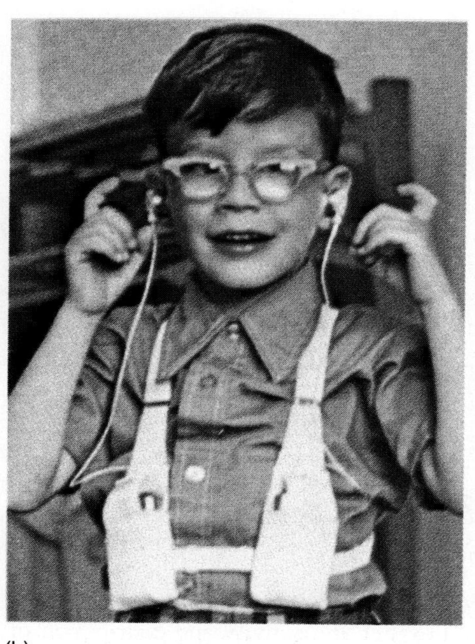

(b)

Figure 25.9
(*a*) An infant born with congenital rubella can manifest a papular and purpurotic rash. (*b*) The sequelae of congenital rubella. Among the permanent defects requiring corrective medical devices are cataracts and deafness. Some children sustain mental retardation and heart defects.

ARBOVIRUSES: VIRUSES SPREAD BY ARTHROPOD VECTORS

Vertebrates are hosts to more than 400 viruses transmitted primarily by arthropods. For simplicity's sake, these viruses are lumped together in a loose grouping called the **arboviruses** (**ar**thropod-**bo**rne **viruses**). The major arboviruses pathogenic to humans are togaviruses *(Alphavirus),* Flaviviruses (Flavivirus), some bunyaviruses (*Bunyavirus* and *Phlebovirus*), and reoviruses *(Orbivirus).* The chief vectors are blood-sucking arthropods such as mosquitos, ticks, flies, and gnats. Most types of illness caused by these viruses are mild, undifferentiated fevers, though some cause severe encephalitides (plural of encephalitis) and life-threatening hemorrhagic fevers. Although these viruses have been assigned taxonomic names, they are more often known by their common names, which are based on geographic location and primary clinical profile (see table 6.4).

EPIDEMIOLOGY OF ARBOVIRUS DISEASE

Wherever there are arthropods there are also arboviruses, so collectively, their distribution is worldwide (figure 25.10). The vectors and viruses tend to be clustered in the tropics and subtropics, but many temperate zones report periodic epidemics. A given arbovirus type may have very restricted distribution, even to a single isolated region, but some range over several continents, and others can spread along with their vectors.

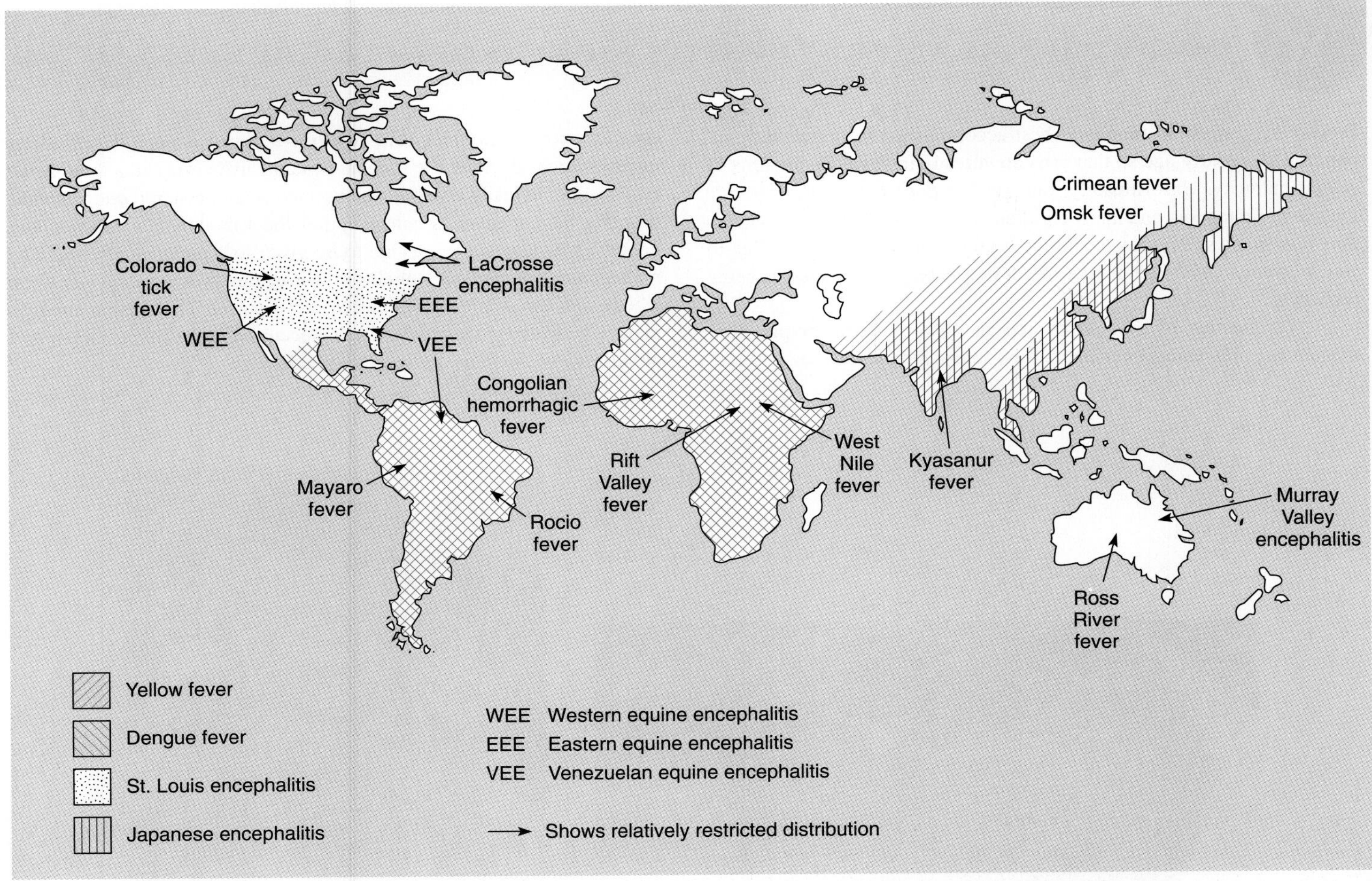

Figure 25.10
Worldwide distribution of major arboviral diseases.

The Influence of the Vector

As one would expect, the activity and distribution of arboviruses are closely tied to the ecology of the vectors. Factors that weigh most heavily are the longevity of the arthropod, the availability of food and breeding sites, and climatic influences such as temperature and humidity. When vectors take a blood meal to finish their reproductive cycle, they acquire the virus and harbor it for varying time periods. Infections show a peak incidence when the arthropod is actively feeding and reproducing, usually from late spring through early fall. Warm-blooded vertebrates also maintain the virus during the cold and dry seasons. Humans can serve as dead-end, accidental hosts, as in equine encephalitis and Colorado tick fever, or they can be a maintenance reservoir, as in dengue fever and yellow fever. The general patterns of arbovirus transmission involving vectors are illustrated in figure 25.11.

Arboviral diseases have a great impact on humans. Although exact statistics are unavailable, it is accepted knowledge that millions of people acquire infections each year and thousands of them die. The uncertain nature of host and viral cycles frequently results in sudden, unexpected epidemics, sometimes with previously unreported viruses. Travelers and military personnel entering endemic areas are at special risk because, unlike the natives of that region, they have no immunity to the viruses.

General Characteristics of Arbovirus Infections

Although the arboviruses have many inherently interesting characteristics, our coverage here will be limited to (1) the chief manifestations of disease in humans, (2) aspects of the most important North American viruses, and (3) methods of diagnosis, treatment and control.

Febrile Illness and Encephalitis

One type of disease elicited by arboviruses is an acute, undifferentiated fever, often accompanied by rash. These infections, typified by dengue fever and Colorado tick fever, are usually mild and self-limited and leave no sequelae. Prominent symptoms are fever up to 40°C, prostration, headache, myalgia, orbital pain, muscle aches, and joint stiffness. About midway through the illness, a maculopapular or petechial rash can erupt over the trunk and limbs.

When the brain, meninges, and spinal cord are invaded, symptoms of viral encephalitis occur. The most common American types are western equine, eastern equine, St. Louis, and California encephalitis. The viruses cycle between wild animals (primarily birds) and mosquitos or ticks, and humans are not usually reservoir hosts. The disease begins with an arthropod bite, the release of the

Figure 25.11
Variations on the arbovirus-vector-human cycle. (*a*) The arthropod vector lives in a habitat separate from humans and spreads the virus among various wild, warm-blooded vertebrates and even to their offspring through transovarial passage. Humans intrude into this cycle by happening on the vector in the wild. (*b*) The infected person then returns to human habitation and serves as a source of viruses for related insects living in the community that may act as biological vectors, or (*c*) wild animal reservoirs and vectors bring the virus into human populations, especially through migration or the encroachment of human habitation into wild areas.

virus into the tissues, and its replication in nearby lymphatics. Prolonged viremia establishes the virus in the brain, where inflammation can cause swelling and damage to various nuclei and nerve tracts as well as to the meninges. Symptoms are extremely variable and can include coma, convulsions, paralysis, tremor, rigidity, loss of coordination, palsies, memory deficits, changes in speech and personality, and heart disorders. In some cases survivors experience some degree of permanent brain damage. Young children and the elderly are most sensitive to injury by arboviruses.

Colorado tick fever (CTF) is the most common tick-borne viral fever in the United States. Restricted in its distribution to the Rocky Mountain states, it occurs sporadically during the spring and summer, corresponding to the time of greatest tick activity and human forays into the wilderness. Two hundred to three hundred endemic cases are reported each year, and deaths are rare.

Western equine encephalitis (WEE) occurs sporadically in the western United States and Canada, first in horses and later in humans. The mosquito that carries the virus emerges in the early summer when irrigation begins in rural areas and breeding sites are abundant. The disease is extremely dangerous to infants and small children, with a case fatality rate of 3% to 7%. But even during epidemic years, usually no more than 100 cases are reported.

Eastern equine encephalitis (EEE) is endemic to an area along the eastern coast of North America and Canada. The usual pattern is sporadic cases, but occasional epidemics can occur in humans and horses. High periods of rainfall in the late summer increase the chance of an outbreak, and disease usually appears first in horses and caged birds. The case fatality rate can be very high (70%).

A disease known commonly as **California encephalitis** is caused by two different viral strains. The California strain occurs occasionally in the western states and has little impact on humans. The LaCrosse strain is widely distributed in the eastern United States and Canada and is a prevalent cause of viral encephalitis in North America. Children living in rural areas are the primary target group, and most of them exhibit mild, transient symptoms. Fatalities are rare.

St. Louis encephalitis (SLE) is the most common of all American viral encephalitides. Cases appear throughout North and South America, but epidemics occur most often in the midwestern and southern states. Inapparent infection is very common, and the total number of cases is probably thousands of times greater than the 200 or so reported each year. People over the age of 55 are most susceptible to life-threatening complications. The seasons of peak activity are spring and summer, depending on the region and species of mosquito. In the east, mosquitos breed in stagnant or polluted water in urban and suburban areas during the summer. In the west, mosquitos frequent spring floodwaters in rural areas.

Hemorrhagic Fevers

Certain arboviruses, principally the yellow fever and dengue fever viruses, can disrupt the vascular bed to such an extent that sudden localized bleeding erupts in the tissues, leading to shock and even death. The exact mechanisms of pathology are obscure, but somehow the virus causes capillary fragility and disrupts the blood clotting system. These hemorrhagic syndromes, which are usually attended by fever and pain, are caused by a variety of viruses, are carried by a variety of vectors, and are distributed globally. Reservoir animals are usually small mammals, although yellow fever and dengue fever can be harbored in the human population.

As a result of intensive studies begun in the 1920s, more is known about **yellow fever** than any other arboviral disease. Although this disease was at one time cosmopolitan in its distribution, mosquito control measures have eliminated it in many countries, including the United States. Two patterns of transmission occur in nature (see figure 25.11). One is an urban cycle between humans and the mosquito *Aedes aegypti,* which reproduces in standing water in cities. The other is a sylvan cycle, maintained between forest monkeys and jungle species of mosquitos. Most cases in the Western Hemisphere occur in the jungles of Brazil, Peru, and Colombia during the rainy season. Infection begins acutely with fever, headache, and muscle pain. The disease is benign in 80–90% of cases, but some patients experience oral hemorrhage, nosebleed, vomiting, jaundice, and liver and kidney damage. The explosive form of the disease leads to death in 50% of cases.

Dengue* fever is caused by a flavivirus and is carried by *Aedes* mosquitos. Although mild infection is the usual pattern, a form called dengue hemorrhagic shock syndrome can be lethal. This disease is also called "breakbone fever" because of the severe pain it induces in the muscles and joints. The illness is endemic to Southeast Asia and India, and several epidemics have occurred in South and Central America, the Caribbean, and Mexico. The Pan American Health Organization has reported an ongoing epidemic of dengue fever in the Americas that has increased to 200,000 cases with 5,500 deaths.

Concern over the possible spread of the disease to the United States had led to CDC to survey mosquito populations in states along the southern border. They discovered a species, the Asian tiger mosquito *(Aedes albopictus)* that maintains breeding colonies in several states. Because the mosquito is a potential vector of dengue fever virus, it is feared that contact of these mosquitos with infected humans could establish the virus in these mosquito populations. So far, no cases of hemorrhagic dengue fever have been reported in the United States, and it is hoped that continued mosquito abatement will block its spread. One group of researchers is currently testing a method to genetically engineer mosquitos so that they are resistant to infection and no longer capable of spreading the virus.

*dengue (den´-gha) A corruption of Sp. "dandy" fever. Other synonyms are dengue shock syndrome and stiffneck fever.

Diagnosis, Treatment, and Control of Arbovirus Infection

Except during epidemics, detecting arboviral infections can be difficult. The patient's history of travel or contact with vectors, along with serum analysis, are highly supportive of a diagnosis. Treatment of the various arbovirus diseases relies completely on support measures to control fever, convulsions, dehydration, shock, and edema.

The most reliable arbovirus vaccine is a live, attenuated yellow fever vaccine that provides relatively long-lasting immunity. Vaccination is a requirement for travelers in the tropics and can be administered to entire populations during epidemics. Live, attenuated vaccines for WEE and EEE are administered to laboratory workers, veterinarians, ranchers, and horses.

Most of the control safeguards for arbovirus disease are aimed at the arthropod vectors. Mosquito abatement by eliminating breeding sites and broadcasting insecticides have been highly effective in restricted urban settings, but if the vectors are widely dispersed through a woodland habitat, these techniques are of little value. In the case of sylvan vectors, the only real preventive is to restrict human contact with the vector through insect repellents and protective clothing. Mosquitos can also be avoided by remaining indoors at night, during the period when they are most active.

Chapter Checkpoints

Enveloped RNA viruses include the Coronaviridae, agents of the common cold; Togaviridae, agents of rubella and Arboviruses, a collection of viruses which are spread by arthropod vectors. They include the togaviruses that cause St. Louis, equine bunyaviruses (agents of hemorrhagic fever), and Flaviviridae, which cause yellow and dengue fevers.

ENVELOPED SINGLE-STRANDED RNA VIRUSES WITH REVERSE TRANSCRIPTASE: RETROVIRUSES

Retroviruses are among the most unusual and disturbing viruses for the following reasons: They have known oncogenic abilities; they cause dire, often fatal diseases; and they are capable of reprogramming a host's DNA in profound ways. The most familiar of these viruses, called *lentiviruses,* is the **human immunodeficiency virus (HIV),** which causes **acquired immune deficiency syndrome (AIDS).** HIV is only one of a whole array of retroviruses associated with cancer and other diseases. Judging from what is currently known about these viruses, they will continue to influence medicine, research, politics, and economics for decades to come.

THE BIOLOGY OF THE RETROVIRUSES

An outstanding characteristic of retroviruses is an unusual enzyme called **reverse transcriptase,** which can convert a single-stranded RNA genome into double-stranded viral DNA. Not only can this newly formed DNA chronically infect host cells, but it also can be incorporated into the host genome as a provirus that can be passed on to progeny cells. Some of the retroviruses are transforming agents; that is, by regulating certain host genes, they convert normal cells into cancer cells (see chapter 17).

The Reversing Viruses

Imagine the surprised group of researchers who discovered the revolutionary properties of retroviruses. These viruses do not follow a central thesis of biology, which maintains that the direction of flow of information and function is from DNA to RNA. Previously, all biological entities, including other viruses, fulfilled this expectation, but this virus has a radical strategy. It comes already equipped with an enzyme that works in the opposite direction by catalyzing the replication of double-stranded DNA from single-stranded RNA. Because these viruses reverse the order of replication, they are termed *retro*viruses (Latin for reversal). This property is apparently found only in them and certain of the DNA viruses.

The association of retroviruses with their hosts can be so intimate that viral genes are permanently integrated into the host genome. In fact, as the technology of DNA probes for detecting retroviral genes is employed, it becomes increasingly evident that retroviruses may be integral parts of chromosomes. So striking is their resemblance to transposons (so-called jumping genes) that some biologists think retroviruses may even have originated from host cells.

NOMENCLATURE AND CLASSIFICATION

Most retroviruses isolated to date cause slow infections that lead to leukemia, tumors, anemia, immune dysfunction, and even neurological diseases in birds and mammals. The first human retrovirus (discovered in 1978) was the T-cell leukemia virus I (HTLV I). This was followed by the isolation of HTLV II and HTLV III. Types I and II are associated with human leukemia or lymphoma; type III is the agent of AIDS. A unified name HIV (human immunodeficiency virus) has been adopted for this virus. There are two types, HIV 1, which is the dominant form in most of the world, and HIV 2, which exists primarily in western Africa.

STRUCTURE AND BEHAVIOR OF RETROVIRUSES

HIV and other retroviruses display structural features typical of enveloped RNA viruses (figure 25.12*a*). The outermost component is a lipid envelope with transmembrane glycoprotein spikes and knobs that mediate viral adsorption to the host cell. Although a few retroviruses are nonspecific, most can infect only host cells that present the required receptors. In the case of the AIDS virus, the target cells possess CD4 receptors and other coreceptors that permit the virus to attach and penetrate several types of leukocytes and tissue cells (figure 25.12*b*).

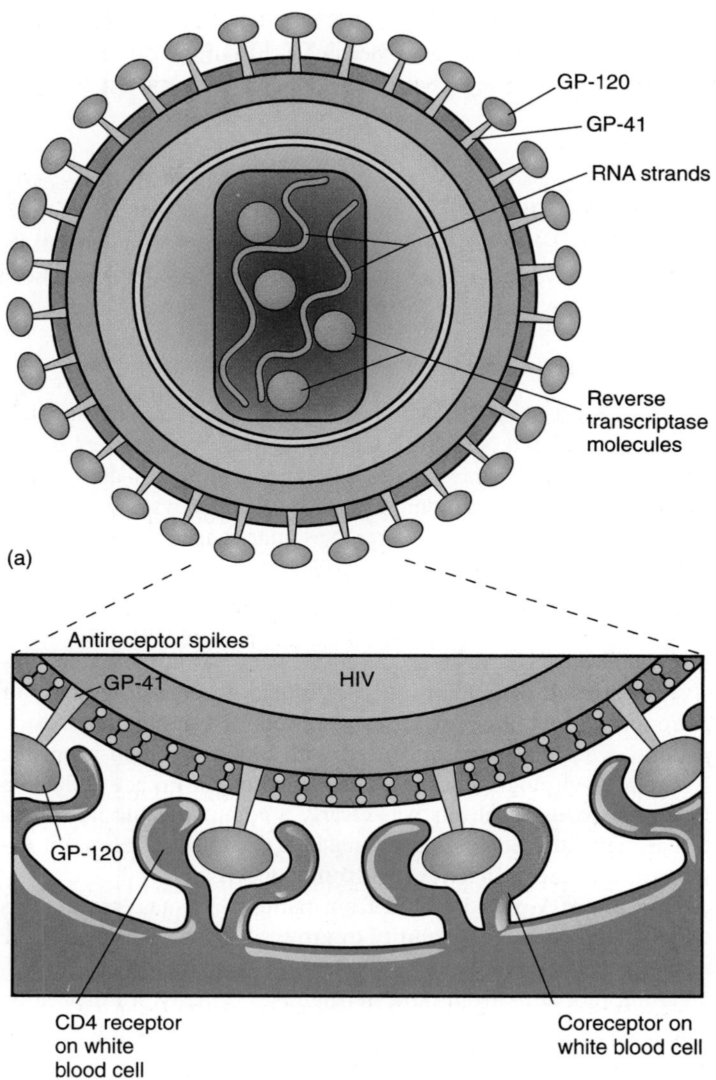

Figure 25.12

(*a*) A cutaway model of HIV. The envelope contains two types of glycoprotein (GP) spikes, two identical RNA strands, and several molecules of reverse transcriptase encased in a protein coating. (*b*) The snug attachment of HIV glycoprotein antireceptors (GP-41 and 120) to their specific receptors on a human cell membrane. These receptors are CD4 and a coreceptor called CCR-5 (fusin) that permit docking with the host cell and fusion with the cell membrane.

Chapter Checkpoints

Retroviruses are unique in their ability to create a double-stranded DNA copy from an RNA template using the enzyme reverse transcriptase. They are also called transforming viruses because the infected cells convert to cancer cells. Retroviruses infect many birds and mammals. Those infecting humans are named HTLV I, II, and HIV and all attack various white blood cells, leading to cancer or immune dysfunction.

ACQUIRED IMMUNE DEFICIENCY SYNDROME (AIDS)

The sudden emergence of AIDS on the world scene in the early 1980s has altered society's collective view of infectious diseases. The AIDS epidemic is not the greatest epidemic ever known, but its fatal effects have focused an enormous amount of public attention, research studies, and financial resources on the virus and its disease.

HISTORICAL BACKGROUND

The first cases of AIDS were seen by physicians in Los Angeles, San Francisco, and New York City. They observed clusters of patients who were all young males with one or more of a complex of symptoms: severe pneumonia caused by *Pneumocystis carinii* (ordinarily a harmless fungus), a rare vascular cancer called *Kaposi's sarcoma;** sudden weight loss; swollen lymph nodes; and general loss of immune function. Another common feature was that all of these young men were homosexuals. Early hypotheses attempted to explain the disease as a consequence of the homosexual lifestyle or a result of immune suppression by chronic drug abuse or infections. Soon, however, cases were reported in nonhomosexual patients who had been transfused with blood or blood products. Eventually, virologists at the Pasteur Institute in France isolated a novel retrovirus. That this was clearly a communicable infectious disease left little doubt, and the medical community termed it **acquired immune deficiency syndrome,** or **AIDS.**

Once HIV was isolated from patients' tissues, researchers were hampered by the inability to grow sufficient amounts of the virus for further study and by the lack of accurate tests for differentiating this virus from other retroviruses. Finally, a functioning line of T cells that remained alive after infection with the new virus became available. This breakthrough was soon followed by the creation of sensitive tests for detecting the virus and the antibody against it. These tests, in turn, made the diagnosis of AIDS and the screening of donor blood possible.

The Origins of AIDS

There are no conclusive answers to the question of where and how HIV originated: significant insight into this question has come from studies comparing the genetics of HIV with various African monkey viruses. The following outline is a proposed family tree for the primate lentiviruses:

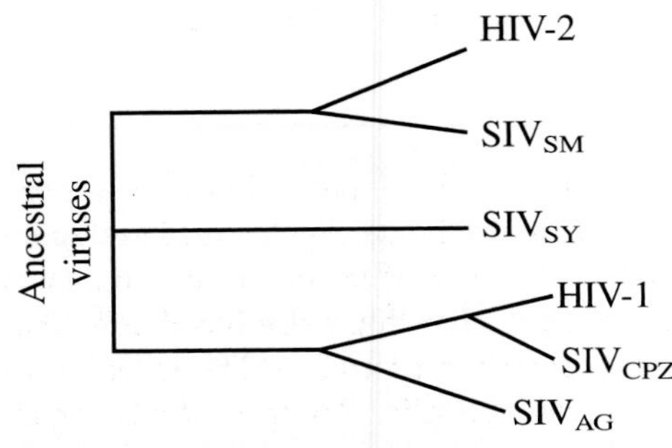

*Kaposi's sarcoma (kah´-poh-seez sahr-koh´-mah) Named for a Viennese dermatologist who first described the condition in the late 1800s.

It now appears that HIV-1 and -2 are not as closely related to each other as they are to simian immunodeficiency viruses (SIV). HIV-1 shares a heritage with SIV_{CPZ}, a virus of chimpanzees. Researchers have discovered a new, atypical strain of HIV that is genetically most similar to a strain of SIV from chimpanzees. HIV-2 is allied with SIV_{SM}, a virus from sooty Mangabey monkeys. The two HIVs originated from separate evolutionary events. Distant relatives are SIV_{SY} (Sykes monkey) virus and SIV_{AG} (African Green monkey) virus. All of these monkey viruses are apparently of ancient origin and do not cause severe disease in their natural hosts.

The first well-documented case of AIDS occurred in an African man in 1959. Samples of his blood yielded genetic material from an early version of HIV. This finding has helped to clarify the time frame for the emergence of HIV. It is theorized that it must have first appeared in the early 1950s, after having jumped from original simian hosts.

It probably remained in small isolated villages, causing sporadic cases and mutating into more virulent strains that were readily transmitted from human to human. When this pattern was coupled with changing social and sexual mores and increased immigration and travel, a pathway was opened up for rapid spread of the virus.

EPIDEMIOLOGY OF AIDS

Changing Statistical Patterns

AIDS has been reported in every country, and parts of Africa and Asia are especially devastated by it (microfile 25.8). Estimates of the number of individuals currently infected with the virus range from 30 to 40 million worldwide, with approximately 1 million (the range is 800,000 to 2 million) in the United States. It is thought that a large number of these people have not yet begun to show symptoms because they are in the latent phase of the disease.

AIDS first became a notifiable disease at the national level in 1984 (figure 25.13). It took 7 years for the first 100,000 cases to appear, but only 2 years for the next 100,000 cases. The disease has continued this epidemic pattern, although it has shown a decrease in incidence since 1994. New therapies that have lengthened the lives of many patients have also slowed the death rate dramatically. AIDS is still the second leading cause of death in men 25–44 years of age and the seventh most frequent overall cause of death nationally.

Although all people appear to be susceptible to the AIDS virus, epidemiologic statistics show that in the United States it is more common among males than females and that whites account for the most cases, followed by blacks, Hispanics, and other groups. The average adult patient is about 35 years of age. If one observes specific sociological groups, however, the sexual, racial, and age statistics can be quite different. (See table 2 in the end papers for a breakdown according to race/ethnicity and sex.) A majority of AIDS patients are concentrated in large metropolitan centers of New York, California, Florida, New Jersey, and Texas.

Modes of AIDS Transmission

It is now clear that the mode of HIV transmission is exclusively through two forms of contact: sexual intercourse and transfer of blood or blood products (figure 25.14). Each can be divided into

MICROFILE 25.8 THE GLOBAL OUTLOOK ON AIDS

AIDS in the Third World shows the same spectrum of illness as is seen in the United States, but it differs in epidemiology. In Africa, the case rate in men and women is about equal, and in some regions, up to 30% of the population may be carrying the AIDS virus. Furthermore, the mode of transmission there is primarily by sexual contact between men and women, and to a lesser extent via blood transfusions and pregnancy. The disease is spreading so rapidly that 16,000 cases occur each day. Especially hard hit are children. In 1997, 450,000 children died of AIDS and 400,000 became infected. Deaths from AIDS are also having a staggering effect on families and communities. By the year 2000, Africa will have 10 million orphaned children, and many villages will be missing an entire generation from 20 to 40 years of age.

Statistics from the rest of the world point to a growing pandemic that involves large parts of Europe, Latin America, India, and Southeast Asia. In some cities, the rate of carriage by prostitutes is over 50%, and 2–3% of healthy military recruits in Thailand show signs of infection. The newly successful drug therapy for controlling the disease is far beyond the reach of most of the world's AIDS patients. It averages $12,000 to $15,000 per patient per year—a figure far beyond the medical resources of these countries.

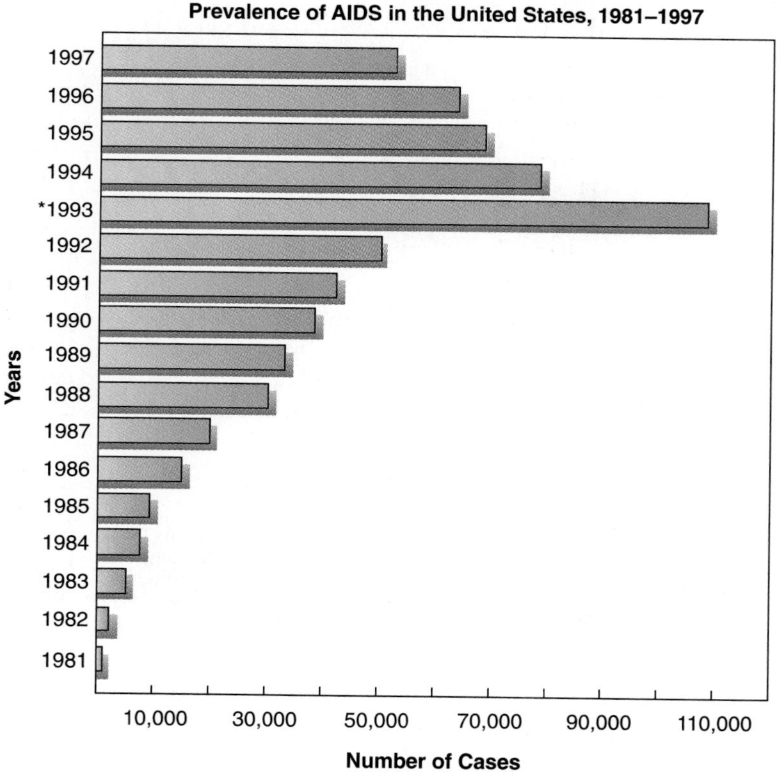

Figure 25.13

Total reported cases of AIDS from 1981 to 1997, based on data from the CDC.* This increase was due to several changes in the case definition of AIDS introduced in January 1993.

Data in part from Summary of Notifiable Diseases, *United States 1996, and* Morbidity and Mortality Weekly Report *No. 52 December 1997. The Centers for Disease Control and Prevention, Atlanta, GA.*

specific high-risk groups that include persons who become infected through a particular life-style or behavior or through medical accidents. The mode of transmission is rather similar to that of hepatitis B virus, except that the AIDS virus does not survive for as long outside the host, and it is far more sensitive to heat and disinfectants. In general, AIDS is spread only by direct and rather specific routes involving intimate contact with an infectious dose. An infection can take place only if the virus crosses the body's epithelial barriers into the fluid compartments. Because the blood of HIV-infected people often harbors high levels of free virus and infected leukocytes, any form of intimate contact involving transfer of blood (trauma, injection) can be a potential source of infection. Semen and vaginal secretions also harbor the virus within white blood cells, thus they are significant factors in sexual transmission. The virus can be isolated from urine, tears, sweat, and saliva, but these fluids contain too few viruses (less than 1 virus per cubic centimeter) to be a source of infection. Because milk contains significant numbers of leukocytes, neonates can become infected through nursing.

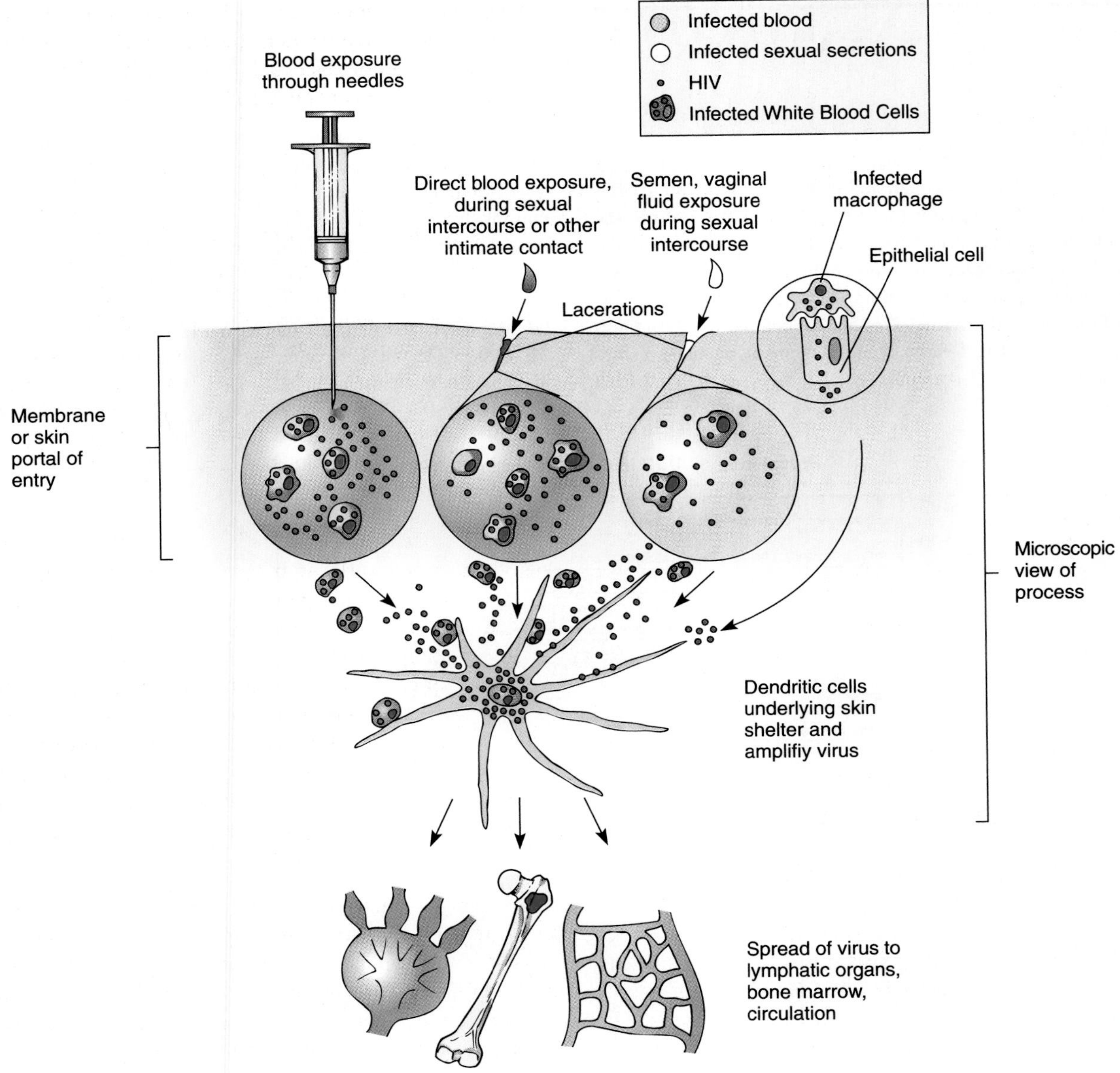

Figure 25.14
Primary sources and suggested routes of infection in AIDS.

Clearing Up Misconceptions: How AIDS Is NOT Transmitted So horrifying is the prospect of acquiring a lethal disease such as AIDS that inaccurate notions and misconceptions rather than established facts have sometimes abounded. Rumors about the role of blood-sucking insects have been common; however, insects such as mosquitos do not serve as biological vectors of the AIDS virus, and the chance of the insect's harboring and delivering an infectious dose mechanically is infinitesimally small. At first it was mistakenly believed that inhaling airborne droplets, shaking hands, handling fomites, and sharing public facilities, food, and swimming pools were possible ways to become infected. However, in-depth studies have indicated that, even within families, infection of healthy persons has occurred only after known exposure to blood. Epidemiologists at the CDC are convinced that if any of these modes of transmission existed, they would have become evident by now.

An Overview of AIDS Cases

The following section will cover definitions of AIDS, the importance of the serological state of patients, the groups most commonly involved, and behavior or events that increase a person's risk. In 1993, the CDC redefined the case definition of AIDS on the basis of changing epidemiologic patterns and medical information. This redesigned system is presented in table 25.3. This scheme places an HIV infected person into a particular clinical category based on the condition of his immune system and the degree and type of symptoms.

TABLE 25.3

REVISED CLASSIFICATION SYSTEM FOR HIV INFECTION AND EXPANDED AIDS SURVEILLANCE CASE DEFINITION FOR ADOLESCENTS AND ADULTS,* 1993

	Clinical Category		
CD4+ T-cell Category	(A) Asymptomatic, acute (primary) HIV or PGL†	(B) Symptomatic, not (A) or (C) conditions	(C) AIDS-indicator conditions**
(1) ≥500/μl	A1	B1	C1
(2) 200–499/μl	A2	B2	C2
(3) <200/μl AIDS-indicator T-cell count	A3	B3	C3

*Persons with AIDs-indicator conditions (category C) as well as those with $CD4^+$ T-lymphocyte counts < 200/μl (category A3 or B3) became reportable as AIDS cases in the United States and Territories, effective January 1, 1993.
†PGL = persistent generalized lymphadenopathy.

Source: Data from *Morbidity and Mortality Weekly Report,* vol. 41. December 18, 1992. The Centers for Disease Control and Prevention, Atlanta, GA.
**See figure 25.18

The Clinical Definition of AIDS

AIDS is defined as a severe immunodeficiency disease arising from infection with HIV and accompanied by some of the following symptoms; life-threatening opportunistic infections, persistent fever, unusual cancers, chronically swollen lymph nodes, weight loss, diarrhea, and neurological disorders. This definition also includes many patients suffering from AIDS-related complex (ARC), once considered a stage in development of the disease. At some point during the infection, the patient tests positive for antibodies to HIV. Seropositivity means only that an infection with HIV has taken place; it does not indicate that the disease exists. However, it is presumed that the antibody-positive person is infectious. For further ramifications of this test, see the later section on diagnosis.

Case Descriptions According to Group, in Order of Prevalence

The following categories summarize the primary exposure and risk categories for HIV infection and AIDS in the United States.

I. Homosexual or Bisexual Males Male homosexuals still account for the majority of AIDS cases (50%) in the United States, though the rate of infection appears to be declining. This demographic group has been so hard hit apparently because of certain types of homosexual practices. Multiple anonymous sexual encounters were once relatively common among gay men frequenting bars and bathhouses. Anal sex, which is known to lacerate the rectal mucosa, could certain provide entry of viruses from semen into the blood. Studies have shown that the passive anal partner in these encounters is the more likely of the two to become infected. In addition, bisexual men provide an avenue for the virus to females and the general heterosexual population.

II. Intravenous Drug Users In large metropolitan areas such as New York City, as many as 60% of intravenous drug addicts can be HIV carriers. Needle sharing is most common in the "shooting galleries" of large eastern cities and is especially risky for cocaine users, whose rate of drug injection can be more frequent than that of users of other injected drugs. Infection from contaminated needles is growing more rapidly than for any other mode of transmission (25% of cases) and is another significant factor in the spread of AIDS to the heterosexual population. A fairly large segment of IV users are also homosexual men. So concerned are public health officials about this trend that campaigns to hand out sterile needles or disinfection kits to drug users have begun in some cities.

III. Heterosexual Sex Partners of HIV Carriers About 8% of AIDS cases arise in people who have had sexual intercourse with an AIDS-infected partner. Individuals in this group include the partners of drug addicts and bisexual men, prostitutes, and spouses of people who have received transfusions or blood factors. The virus spreads readily in either direction, and multiple sexual encounters increase the chance of infection, although even a single encounter with an AIDS patient has been known to transmit the virus (microfile 25.9). The overall rate of infection in the general heterosexual population has increased dramatically in the past few years in adolescent and young adult women. Currently, one-third of new cases occur in women.

IV. Blood Transfusion and Organ Transplant Patients Since routine screening of donated blood for antibodies to the AIDS virus began, transfusions are no longer considered a serious risk. Before then, approximately 2% of AIDS cases arose in patients receiving multiple transfusions, and most new transfusion-related cases can be traced to transfusions given before 1985. Because there can be a lag period of a few weeks to nearly 3 years before antibodies appear in an infected person (figure 25.15*b*), it is remotely possible to be infected through donated blood (estimated at 1 in 500,000 transfusions). As added precautions, potential donors are thoroughly screened, and people anticipating surgery are encouraged to store their own blood in advance. Rarely, organ transplants can carry HIV, so transplant procurement groups are now screening organs for the virus.

V. Patients with Coagulation Disorders Receiving Replacement Factors In this category, which accounts for 1% of all AIDS cases, hemophiliac men predominate. For several years in this country, replacement coagulation products (factor VIII and others) were derived from human plasma and filtered to remove cellular microbial contaminants, but not viruses. Before this route of HIV transmission was discovered, several hundred hemophiliacs became infected. Now replacement infusions are heat-treated to kill any contained viruses, a process that has essentially eliminated risk for this group of people.

MICROFILE 25.9 THE ODDS OF ACQUIRING AIDS THROUGH HETEROSEXUAL CONTACT

A high-risk sexual partner is one who has recently engaged in homosexual activity or intravenous drug use, has lived in a high-prevalence area such as Africa, has had multiple blood transfusions, or has engaged in regular sex with a member of any of these groups. The following estimates on the likelihood of acquiring AIDS come from the Center for AIDS Prevention studies:

	Chance of Acquiring AIDS
One sexual encounter with an HIV-negative person who does not belong to a risk group (condom used)	1 in 5 billion
One sexual encounter with an HIV-negative person who does not belong to a risk group (no condom used)	1 in 500 million
One sexual encounter with a person who has not been tested but who is not at high risk (condom used)	1 in 50 million
One sexual encounter with a person who has not been tested but who is not at high risk (no condom used)	1 in 5 million
One sexual encounter with an HIV-positive person (no condom)	1 in 500
500 sexual encounters with an HIV-positive person (no condom)	3 in 5

From these data, it is apparent that, although the number of contacts and the use of protection are important factors, neither is as important as the choice of sex partners. With AIDS, a person's sexual history takes on crucial importance. Epidemiologists cannot overemphasize the need to screen prospective sex partners and to follow a monogamous sexual life-style. This is one point that has the unanimous support of both scientists and moralists.

VI. Inapparent or Unknown Risk Factors About 8% of all AIDS cases occur in people who apparently belong to none of the risk groups just described. This is the most troublesome category because it opens up questions about some other, unknown route of spread. When the CDC has carefully followed up such cases, a majority of the patients have admitted to prior contact with a prostitute or a history of other STDs. Factors such as patient denial, unavailability, death, or uncooperativeness make it impossible to explain every case.

VII. Congenital and Neonatal AIDS The population of mothers that accounts for the neonatal epidemic are young IV drug addicts or sex partners of drug addicts. The chance of an HIV-positive mother infecting her fetus is 1 in 3, and infected babies develop the disease more rapidly than adults. Recent evidence indicates that treatment of infected mothers with a combination of anti-AIDS drugs can significantly prevent the infection of fetuses. As a result of improved education and treatment, the number of children with AIDS is decreasing.

VIII. Risks Involving Medical and Dental Personnel Health care personnel are not considered a high-risk group, though approximately 100 medical and dental workers are known to have acquired AIDS or become antibody-positive as a result of clinical accidents. A health care worker involved in an accident in which gross inoculation with blood occurs (a needlestick) has about one chance in 500 of becoming infected.

The well-publicized case of a Florida dentist who infected several of his patients has prompted a controversial examination of health workers as a *source* of HIV. Statistics from the CDC indicate that several thousand medical and dental personnel are infected with HIV. With this in mind, lawmakers are considering a federal statute that would force mandatory testing for all health workers. It would also require that HIV-positive physicians and dentists obtain permission from patients and a governing board before performing invasive surgeries.

It must be emphasized that transmission of HIV in either direction will not occur through casual contact or routine nursing procedures and that basic guidelines and universal precautions for infection control (see appendix D) were designed to give full protection for both worker and patient.

The Evolving Picture of AIDS Pathology

Understanding the pathology of AIDS has been aided to an extent by animal models. Studies of macaque monkeys with a simian form of AIDS that is very similar to the human disease have helped outline some of the mechanisms by which retroviruses suppress the immune system. Although chimpanzees can be infected with HIV, they fail to develop the disease. Much promise lies in studying the colonies of transgenic mice that have been implanted with human immune systems. Preliminary findings show that when these special mice are infected with HIV, they develop some symptoms of AIDS.

Initial Infection of Target Cells

The proposed events in transmission and infection are summarized in figure 25.14. It appears that after HIV enters a mucous membrane or the skin through injection, trauma, or direct passage, it travels to dendritic cells, a type of macrophage living beneath the skin. In the dendritic cells, it grows and is shed from the cells without killing them. It is amplified by multiplying in

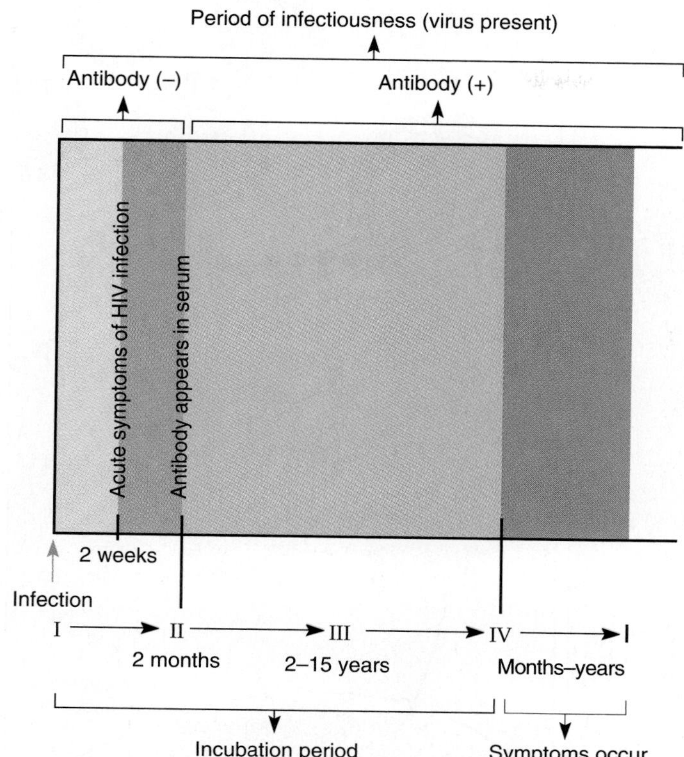

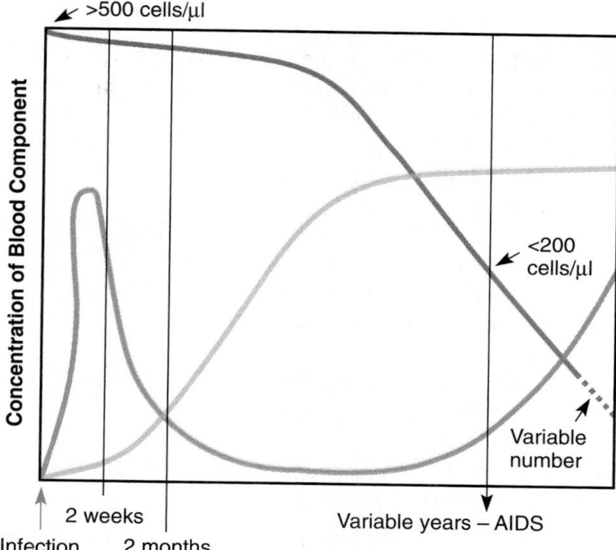

Figure 25.15

(*a*) Stages in HIV infection and disease. (I) Infection with virus. (II) Appearance of antibodies in standard HIV tests. (III) Asymptomatic HIV disease, which can encompass an extensive time period. (IV) Overt symptoms of AIDS include some combination of opportunistic infections, cancers, and general loss of immune function such as a very low T4 lymphocyte count. (*b*) A comparison of blood levels of viruses, antibodies, and T cells covering the same time frame depicted in part (a). Virus levels are high during the initial acute infection and decrease until the later phases of HIV disease and AIDS. Antibody levels gradually rise and remain relatively high throughout phases III and IV. T-cell numbers remain relatively normal until the later phases of HIV disease and full-blown AIDS. Levels below 200 cells μl generally cause the onset of symptoms.

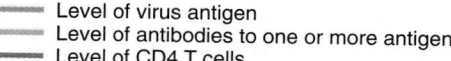

macrophages in the skin, lymph organs, bone marrow, and blood. One of the great ironies of HIV is that it infects and destroys many of the very cells needed to destroy it, including the helper (T4 or CD4) class of lymphocytes, monocytes, macrophages, and even B lymphocytes. The virus is adapted to docking onto its host cell's surface receptors (see figure 25.12). Its main docking point is CD4, but other coreceptors are involved. It binds to CCR-5 on macrophages, which is ordinarily a receiver for cytokine signals. Using the CXCR/4 (fusin) receptor on T cells, it induces viral fusion with the cell membrane and creates syncytia.

Once inside the cell, its transcriptase makes the RNA into DNA. Although initially it can produce a lytic infection, in many cells it enters a latent period in the nucleus of the host cell (figure 25.16), which accounts for the lengthy course of the disease that most often follows. It is during this time that the viral DNA becomes integrated into the nucleus (figure 25.16).

Stages of HIV Infection and Disease

The scope of HIV infection shows a continuous progression from initial contact to full-blown AIDS (see figure 25.15*a*). The initial infection is often attended by vague, mononucleosis-like symptoms that soon disappear. This is followed by a period of asymptomatic infection (HIV disease) with an incubation period that varies in length from 2 to 15 years (the average is about 10). The first antibodies to HIV are usually detectable in serum by the second month following infection, but they are ineffective in neutralizing the virus once it enters the latent period.

Disease can initially present with fever, swollen lymph glands (persistent lymphadenopathy), fatigue, diarrhea, weight loss, and neurological syndromes, as well as opportunistic infections and neoplasms.

Contrary to what is commonly thought, not everyone who becomes infected or is antibody-positive develops AIDS. About 5% of people who are antibody-positive remain free of disease, indicating that functioning immunity to the virus can develop. Any person who remains healthy despite HIV infection is termed a *non-progressor.* When subjected to study, some non-progressors have been found to lack the cytokine receptors that HIV requires. Others are infected by a weakened virus mutant. Individuals who progress to full-blown AIDS have a much brighter prognosis than they had even a few years ago because of the newer therapies discussed in a later section.

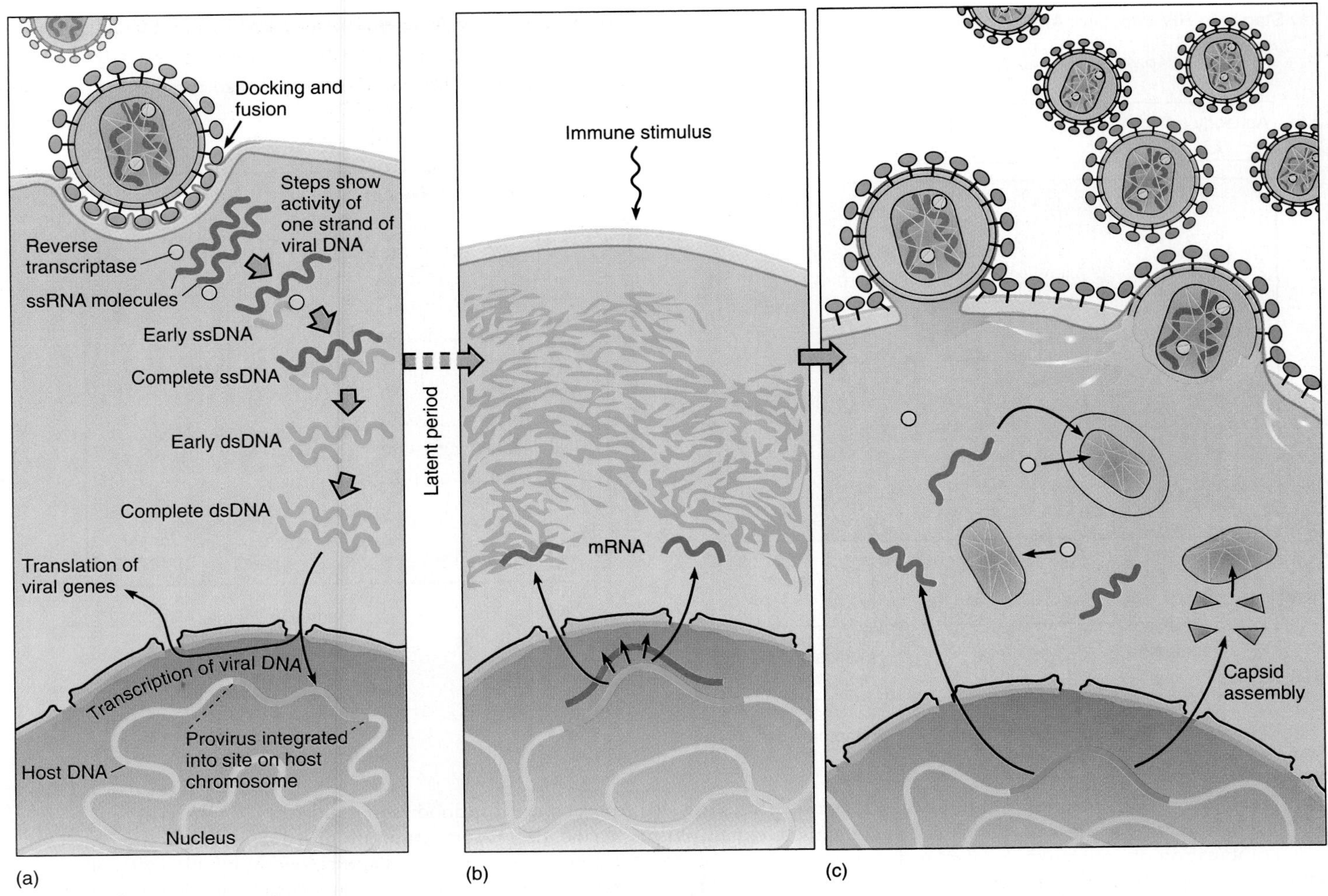

Figure 25.16

The general life cycle of HIV. (*a*) The virus is adsorbed and endocytosed, and the twin RNAs are uncoated. Reverse transcriptase catalyzes the synthesis of a single complementary strand of DNA (ssDNA). This single strand serves as a template for synthesis of a double strand (ds) of DNA. In latency, dsDNA is inserted into the host chromosome as a provirus. (*b*) After an indeterminant time period, various immune activators stimulate the infected cell, causing reactivation of the provirus genes and production of viral mRNA (*c*) Assembly of viruses from translated viral genes and insertion of reverse transcriptase, and envelope receptors. Budding off of mature viruses lyses the infected cell.

The Primary Effects: Harm to T cells and the Brain

In order to understand the outcome of AIDS infection, we must think in terms of an impaired immune system. Over the long period of chronic infection, the infected cells play host to the virus in either a latent or an active phase. Cells with active viruses release large quantities into the circulation and are killed in the process. The death of T cells and other white blood cells results in extreme leukopenia and loss of essential T4-memory clones and stem cells (figure 25.17). The viruses also cause the formation of giant T-cell and other cell syncytia, which allow the spread of viruses directly from cell to cell, followed by mass destruction of the syncytia. As the dose of viruses entering the system increases, a vicious cycle occurs, involving a greater infection rate and death of cells and greater viral burden. Because the T4 lymphocytes have multiple cytotoxic, helper, and inducer functions, their destruction or disability paves the way for invasion by opportunistic agents and cancer. When the T4 cell levels fall below 200 cells/mm^3 (200/μl) of blood, symptoms of AIDS appear (see table 25.3).

Targets of the AIDS virus are not limited to the immune system. The central nervous system is involved in a large proportion of cases. It is believed that infected macrophages cross the blood-brain barrier and release viruses, which then invade nervous tissues. Studies have indicated that some of the viral envelope proteins can have a direct toxic effect on the brain's glial cells and other cells. Other research has shown that some peripheral nerves become demyelinated and the brain becomes inflamed. How the AIDS virus is involved in these pathologic progresses is a topic of intense study and debate.

Secondary Effects That Define AIDS

Opportunistic Infections Among the earliest symptoms of AIDS are debilitating, potentially fatal, and often multiple opportunistic infections (Figure 25.18). In general, the infections involve fungi (chapter 22), protozoa (chapter 23), viruses, or bacteria (chapters 18, 19, and 21) that establish a foothold in the absence of adequate cytotoxic immunity. The most frequent cites of attack are the respiratory, digestive, and nervous systems.

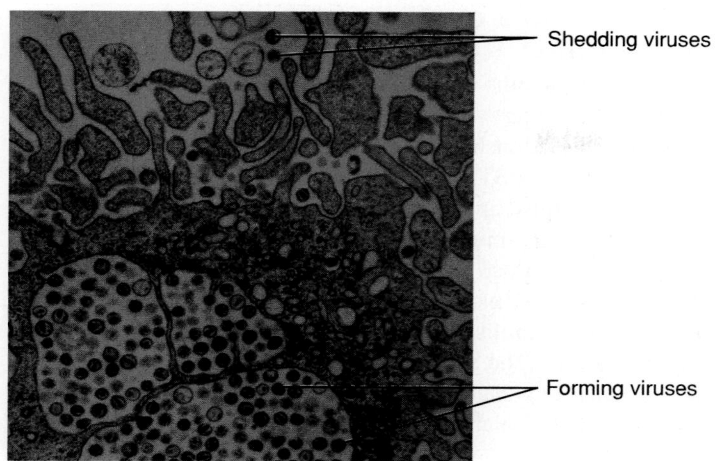

Figure 25.17
Transmission of electron micrograph of a cell section during lytic infection by the AIDS virus. The process of viral shedding, which has virtually disintegrated the cell, accounts for the severe decimation of T4 lymphocytes.

***Pneumocystis carinii* pneumonia (PCP),** caused by a widespread fungus, is the leading cause of morbidity and mortality in AIDS patients. Its symptoms include fever, cough, shortness of breath, thickening of the lung epithelium, and fluid infiltration that compromises respiration. The course of injection can be acute and rapid or slow and progressive. Fungi most frequently involved in AIDS complications are *Cryptococcus neoformans,* a cause of meningitis; *Candida albicans,* which attacks the oral, pharyngeal, and esophageal mucosa; and *Aspergillus* species, which infect the lungs. Protozoan infections common in AIDS patients are toxoplasmosis of the brain (caused by *Toxoplasma gondii*) and *Cryptosporidium* diarrhea.

Common herpesviruses also afflict the AIDS patient. Cytomegalovirus produces a disseminated pneumonia and can infect the brain, eyes, and liver as well. Herpes simplex and herpes zoster can reactivate to produce severe ulceration of the skin, lips, genitalia, and anus. Infections with bacteria such as

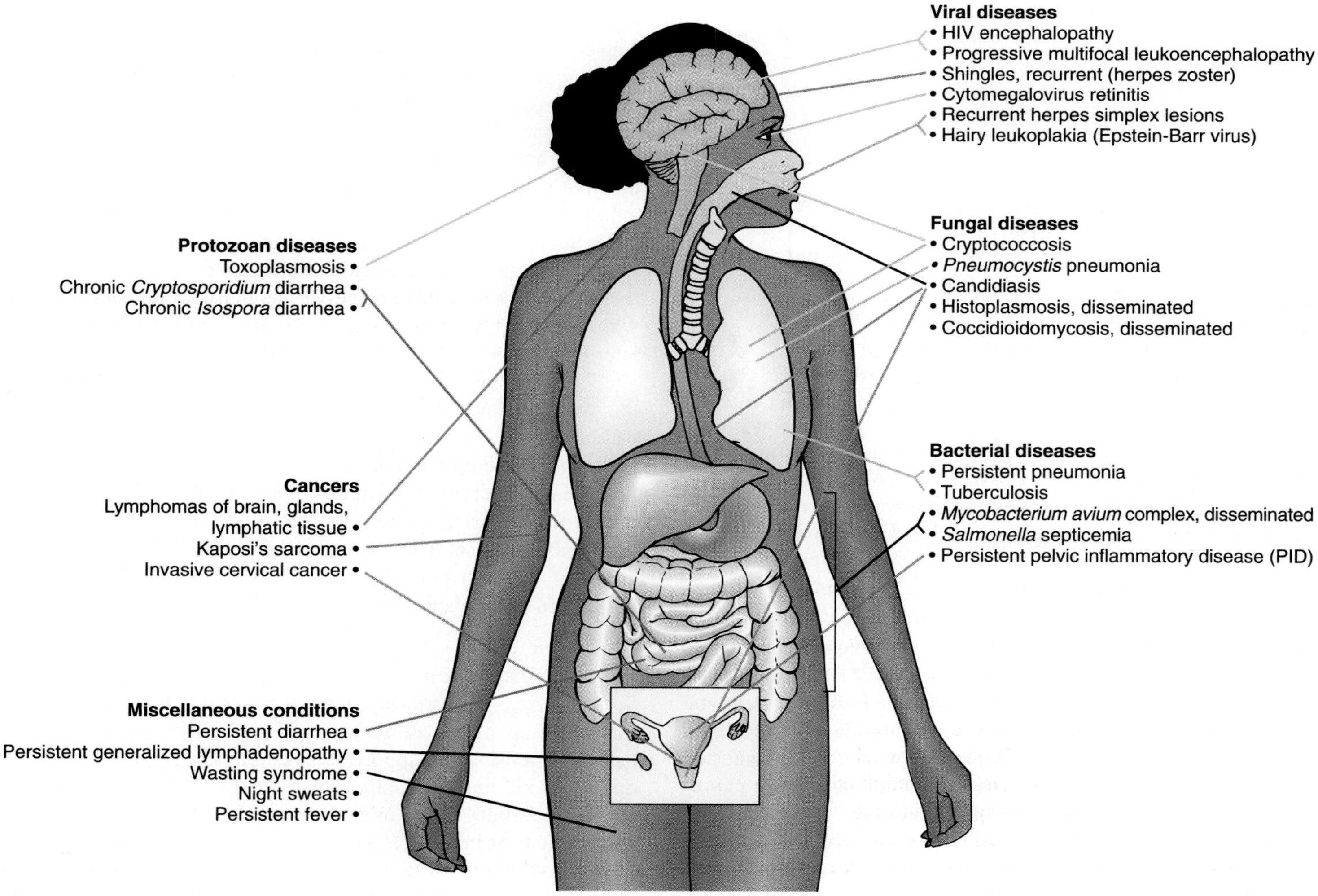

Viral diseases
• HIV encephalopathy
• Progressive multifocal leukoencephalopathy
• Shingles, recurrent (herpes zoster)
• Cytomegalovirus retinitis
• Recurrent herpes simplex lesions
• Hairy leukoplakia (Epstein-Barr virus)

Fungal diseases
• Cryptococcosis
• *Pneumocystis* pneumonia
• Candidiasis
• Histoplasmosis, disseminated
• Coccidioidomycosis, disseminated

Bacterial diseases
• Persistent pneumonia
• Tuberculosis
• *Mycobacterium avium* complex, disseminated
• *Salmonella* septicemia
• Persistent pelvic inflammatory disease (PID)

Protozoan diseases
Toxoplasmosis •
Chronic *Cryptosporidium* diarrhea •
Chronic *Isospora* diarrhea •

Cancers
Lymphomas of brain, glands, lymphatic tissue •
Kaposi's sarcoma •
Invasive cervical cancer •

Miscellaneous conditions
Persistent diarrhea •
Persistent generalized lymphadenopathy •
Wasting syndrome •
Night sweats •
Persistent fever •

Figure 25.18
Opportunistic infections and other diseases used in the expanded case definition of AIDS. One or more of these conditions, in combination with serological tests and CD4 T-cell counts, establishes a system to classify the clinical stage of the disease.

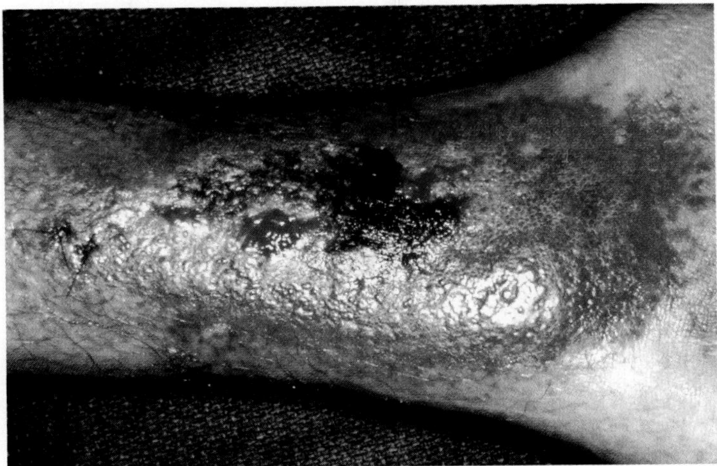

Figure 25.19
Kaposi's sarcoma lesion on the arm. The flat, purple tumors occur in almost any tissue and are frequently multiple.

Mycobacterium are particularly problematic and severe. Tuberculosis in AIDS patients is now at epidemic levels. For reasons that are obscure, several non-tuberculous mycobacteria (*Mycobacterium avium* complex) show a strong affinity for AIDS patients, growing in massive numbers in the bone marrow, liver, spleen, and lymph nodes.

AIDS-Associated Cancers The predisposition of AIDS patients for tumors (Figure 25.18) is due to the loss of natural cancer-killing T cells along with the intrusion of microbes that can cause cancer. Kaposi's sarcoma (KS) accounts for most cancers seen clinically. This sarcoma, which is caused by a newly-identified herpesvirus (see chapter 24), is a nodular purple lesion that develops from endothelial cells in blood vessels of the skin, intestine, and mucous membranes (figure 25.19). Other symptoms that accompany KS are fever, lymphadenopathy, weight loss, and diarrhea. Complications are hemorrhage, perforation, and intestinal obstruction. Additional cancers common to AIDS patients are epithelial carcinomas of the skin, mouth, and rectum, and lymphomas originating from B lymphocytes.

Miscellaneous Conditions AIDS patients experience many nonspecific, disease-related symptoms. Pronounced wasting of body mass that accompanies weight loss, diarrhea, and poor nutrient absorption is a common complaint. Protracted fever, fatigue, sore throat, and night sweats are significant and debilitating. Some of the most virulent complications are neurological. Lesions occur in the brain, meninges, spinal column, and peripheral nerves. Patients with nervous involvement show some degree of withdrawal, persistent memory loss, spasticity, sensory loss, and progressive dementia. Both a rash and generalized lymphadenopathy in several chains of lymph nodes are presenting symptoms in many AIDS patients.

The Search for Cofactors: The Prevalence of Coinfections

From the first months of its discovery, medical science sought to identify contributing cofactors in AIDS besides life-style or transfusions. One relationship that seems certain is the effect of coinfections, especially STDs. Genital ulcers such as those present in chlamydia, syphilis, and warts increase susceptibility to AIDS because breaks in the mucous membrane promote the shedding and entrance of the virus. AIDS patients frequently have concurrent infections with Epstein-Barr virus (mononucleosis), cytomegalovirus, syphilis, tuberculosis, hepatitis B, and mycoplasma (see chapter 21). The severe immune suppression accompanying AIDS appears to enhance the pathogenicity and virulence of both participants and escalate damage.

Diagnosis of HIV Infection and AIDS

Secondary infections and cancers suggestive of AIDS present strong presumptive evidence. When these signs are coupled with a patient's history of predisposing activities and a positive serological test for AIDS antibodies, the diagnosis is confirmed. The ELISA test for AIDS antibodies is highly sensitive, but variations in the quality of testing and the patient's condition can lead to false-positive results in a small number of tests. Newer antibody tests are a rapid latex agglutination test that is 99% accurate and an oral swab test that detects antibodies in the mouth. Positive antibody readings call for the more specific follow-up test known as the Western blot, which detects several different anti-HIV antibodies and can usually rule out a false-positive result (see figure 16.9). False-negative results occur in as many as one in four patients who are indeed infected with HIV. Such silent infections contribute considerably to the unpredictable nature of latency and how to test for it. To rule out the possibility of a false-negative, people who test negative but who practice high-risk behavior should seek a second test at a later date.

Because an infected person has viruses in his blood from the very earliest stages, a more reliable diagnostic approach would be direct discovery of HIV, its genes, or antigens in blood and other specimens. This form of testing can eliminate the false tests inherent in antibody analysis and are a particular boon in monitoring transfusion blood. Several rapid, specific, very sensitive tests using monoclonal antibodies are now available for research purposes, for diagnosing infection in newborns, and as a means of tracing viral load.

AIDS Treatments

So far, there is no cure for AIDS. None of the therapies do more than prolong life or diminish symptoms. Treatment for AIDS patients focuses on supportive care and drugs to control both opportunistic and HIV infections. Drugs used to control PCP are pentamidine, taken in oral or inhaled form; sulfamethoxazole-trimethoprim (SxT); and leucovorin. Drugs that have some control over CMV infections are ganciclovir and foscarnet. Disseminated fungal infections are treated with fluconazole, and Kaposi's sarcoma is relieved by alpha interferon and human chorionic gonadotropin.

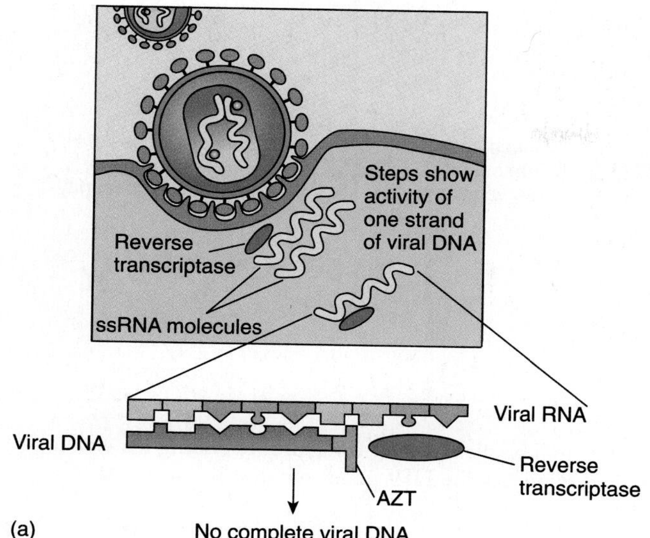

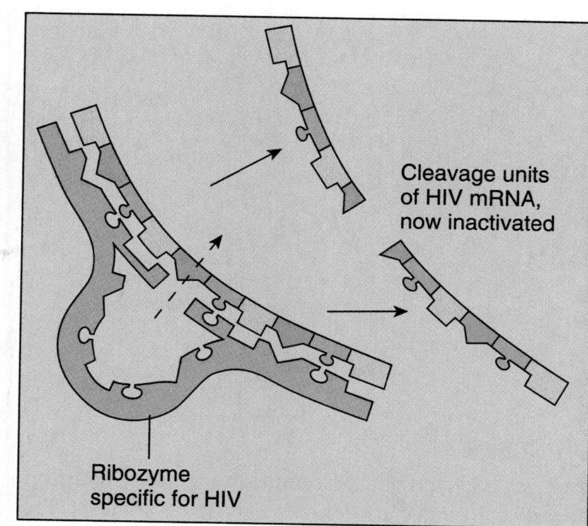

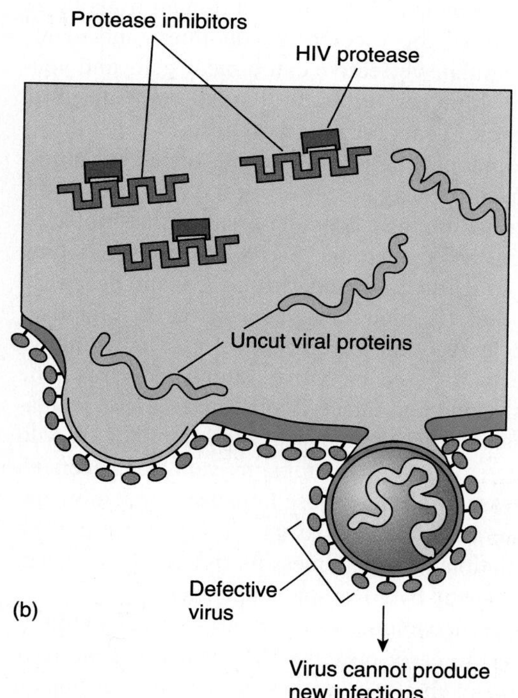

Figure 25.20

Strategies for drugs to treat AIDS. (*a*) A prominent group of drugs (AZT, ddI, ddC) are molecular mimics called nucleoside analogs or reverse transcriptase inhibitors. They are inserted in place of the natural nucleotide by reverse transcriptase but block further action of the enzyme and synthesis of viral DNA (*b*) Protease inhibitors plug into the active sites on HIV protease. This enzyme is necessary to cut elongate HIV protein strands and produce functioning smaller protein units. Because the enzyme is blocked, the proteins remain uncut, and normal viruses are not formed. (*c*) An experimental drug, ribozyme, acts on the mRNA transcript, thereby preventing it from being translated. Ribozymes are special RNA-based enzymes that can snip other RNAs at selected sites. They effectively cleave the viral RNA in half.

AIDS DRUGS: CAUTIOUS HOPE FOR A CURE

Drugs to inhibit infection and replication by AIDS virus have been an intense focus of research. More than 100 drugs are in development or clinical trials, and approximately 50 drugs have been formally approved by the Food and Drug Administration for HIV therapy. The earliest of these drugs are the synthetic nucleoside analogs (reverse transcriptase inhibitors) azidothymidine (AZT), Didanosine (ddI), Zalcitibine (ddC), and Stavudine (d4T). They interrupt the HIV multiplication cycle by being incorporated into the DNA molecule by reverse transcriptase. Because these drugs lack all of the correct binding sites for further DNA synthesis, viral replication and the viral cycle are terminated (figure 25.20*a*). One or two of these drugs is a recommended part of therapy for HIV positive and AIDS patients. Protease inhibitors (figure 25.20b) are an important new class of drugs that block the action of an HIV enzyme involved in the final assembly and maturation of the virus. By combining two reverse transcriptase inhibitors and one protease inhibitor in a three-drug cocktail, the virus is blocked in two different phases of its cycle. The regimen has been successful in reducing viral load to undetectable levels and facilitating the improvement of immune function. It has also reduced the rate of virus drug resistance and the incidence of AIDS deaths. The primary drawbacks to the therapy are high cost, toxic side effects, drug failure due to patient non-compliance, and its inability to completely eradicate the virus.

Additional drugs under consideration block the integration and replication of viral DNA into host cell (antisense and triplex

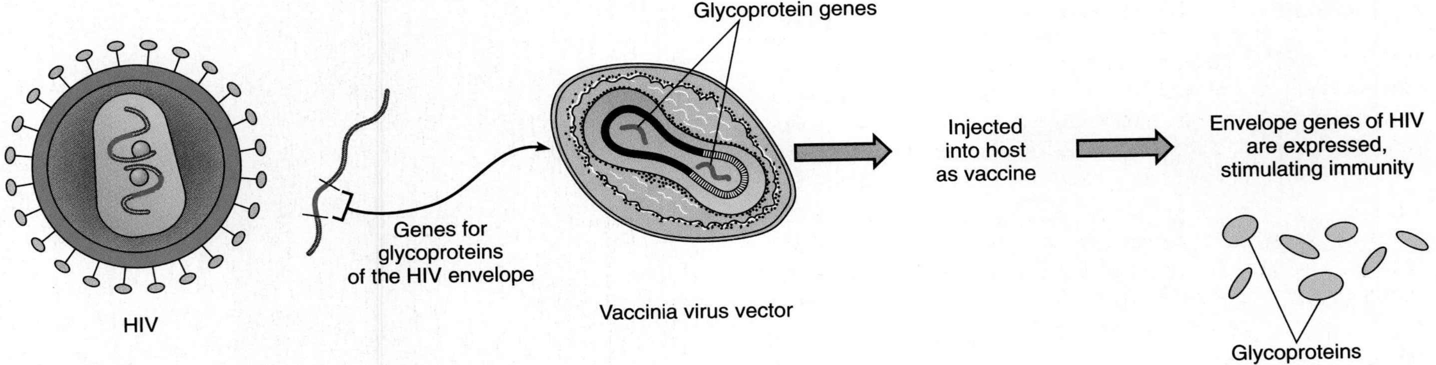

Figure 25.21
"Trojan horse," or viral vector, technique, a novel technique for making an AIDS vaccine. The part of the HIV genome coding for envelope glycoproteins is inserted into a carrier virus (vaccinia). This hybrid virus replicates and expresses the HIV genes when it is injected into a host. Figure is not shown to scale.

agents), inhibit translation of the viral RNA (ribozymes; figure 25.20*c*), and stop virus budding (interferon). Some patients have acquired untested but potentially useful drugs from countries with less stringent drug policies than the United States. To ease this demand for alternative treatments, the FDA has relaxed its usual regulations to allow for compassionate use of experimental AIDS drugs among patients.

AIDS Protection

Measures to prevent direct contact with HIV are an important consideration. There is universal agreement that "safe sex" practices can be very effective. These include the use of condoms with an antimicrobic spermicide, monogamous sex with a long-term sex partner, and avoidance of anal sex. Although avoiding intravenous drugs is an obvious deterrent, many drug addicts do not choose this option. In such cases, risk can be decreased by not sharing syringes and needles or by disinfecting the syringe and needle with hypochlorite solution before use. Anonymous AIDS testing to protect the identity of possible HIV-positive people is currently under way to screen certain populations, such as newborns and hospital, STD, and drug treatment patients, and some states have adopted AIDS testing as part of the marriage license procedure. Testing is mandatory for military recruits. There is some indication that immediate doses of anti-AIDS drugs can deter infection in health care workers accidently exposed to blood during accidents.

Will There Be an AIDS Vaccine?

From the very first years of the AIDS epidemic, the potential for a vaccine has been regarded warily, because the virus presents many seeming insurmountable problems. For example, it becomes latent in cells; its cell surface antigens mutate rapidly; and, although it does elicit immune responses, it is apparently not completely controlled by them. In view of the great need for a vaccine, none of those facts has stopped the medical community from moving ahead.

Most of the 25 vaccines in various stages of development and testing are based on recombinant viruses and viral envelope or core antigens. The principal difficulties in human trials lie in determining the safety of the vaccine in volunteers and determining whether it stimulates effective cytotoxic T cells and neutralizing antibodies. Volunteers are paid a small compensation, and must commit several years for follow-up tests.

Recent developments include a vaccine tested on chimpanzees that immunized them against one strain of the virus and a vaccine used in mice with human immune systems that protected the human cells against HIV infection. A U.S. company is hoping to carry out human trials on a subunit vaccine containing gp120 virus receptors, but there is some disagreement on the effectiveness of this approach. Another vaccine developed in monkeys using an SIV model was based on a live, attenuated virus. The monkeys were healthy and had intact T-cell counts after vaccination and could withstand a challenge with high numbers of wild virus. Using this strategy in humans remains very controversial. An attenuated HIV vaccine could expose humans to mutation and cancer, and the virus could mutate to its virulent form and be spread into the population at large. Unless its risk can be reduced, it is unlikely to be endorsed by government agencies.

In the viral vector technique, vaccinia virus or adenovirus is genetically engineered to carry only the HIV envelope gene (figure 25.21). One form of this vaccine has been tested on human volunteers and has shown some ability to raise a protective response. Studies in monkeys using a DNA plasmid vaccine indicate the potential effectiveness of this approach. Given the challenges of HIV, even the most optimistic experts do not expect a vaccine to be generally available before early next century.

OTHER RETROVIRAL DISEASES IN HUMANS

Other human retroviruses are implicated in various transmissible, fatal lymphocyte malignancies: HTLV I causes adult T-cell leukemia (ATL), a type of lymphosarcoma; HTLV II causes hairy-cell leukemia. The possible mechanisms whereby retroviruses stimulate cancer were covered in chapter 17. One hypothesis says

Chapter Checkpoints

AIDS is caused by the human immunodeficiency virus (HIV). It is the most significant viral pandemic of this century, with 30–40 million estimated cases worldwide and approximately 1 million known cases in the United States. It is acquired through direct contact with blood or sexual fluids in sufficient quantity.

Those at greatest risk for infection include homosexual or bisexual males, IV drug users, heterosexual partners of AIDS carriers, recipients of blood products or tissue transplants, and neonates from infected mothers.

HIV first infects macrophages, where it multiplies and is shed, infecting macrophages, lymphocytes, and other cells that possess the CD4 receptor site. The most significant of these are the T4 (helper) cells that coordinate the immune responses to most infections. Also, the virus mutates rapidly, so that one individual is infected with hundreds of strains. Following infection, the virus often enters a dormant stage lasting 2–15 years.

AIDS is fatal because of overwhelming opportunistic infections resulting from the destruction of T4 lymphocytes and stem cells. Infected macrophages also carry the virus to the CNS, causing neurological degeneration. Opportunistic infections involve fungi, protozoa, bacteria, and other viruses as well as Kaposi's sarcoma, an AIDS-associated cancer.

HIV infection is diagnosed by several tests. The most accurate is the Western blot test, which detects several different antibodies to HIV. There is no cure for AIDS, but improved supportive care has increased the life expectancy for many patients. It is difficult to develop drugs that attack the virus itself because of its high mutation rate.

Protection against HIV includes following safe sex practices and taking increased precautions when handling needles and body fluids. Research continues for an effective AIDS vaccine, but many years of research and testing are necessary. There are also ethical considerations regarding trials of such vaccines in human populations.

that the virus carries an oncogene that is spliced into a host's chromosome during infection and that, when triggered by various carcinogens, immortalizes the cell and frees it from limitations in the number of cell divisions. No oncogenes have been discovered in HTLV I and II, but they do have genes that code for regulatory proteins. One of HTLV's genetic targets seems to be the gene and receptor for interleukin-2, a potent stimulator of T cells.

Epidemiology and Pathology of the Human Leukemia Viruses

Adult T-cell leukemia (ATL) was first described by physicians working with a cluster of patients in southern Japan. Later, a similar clinical disease was described in Caribbean immigrants. In time, it was shown that these two diseases and a cancer called Sézary-cell leukemia were the same disease. ATL was linked to HTLV I when the virus was isolated from patients' T cells. Although more common in Japan, Europe, and the Caribbean, a small number of cases occur in the United States. The disease is not highly transmissible; studies among families show that repeated close or intimate contact is required. Because the virus is thought to be transferred in infected blood cells, blood transfusions and blood products are potential agents of transmission. Intravenous drug users could spread it through needle sharing. A sexual mode of transmission has also been proposed.

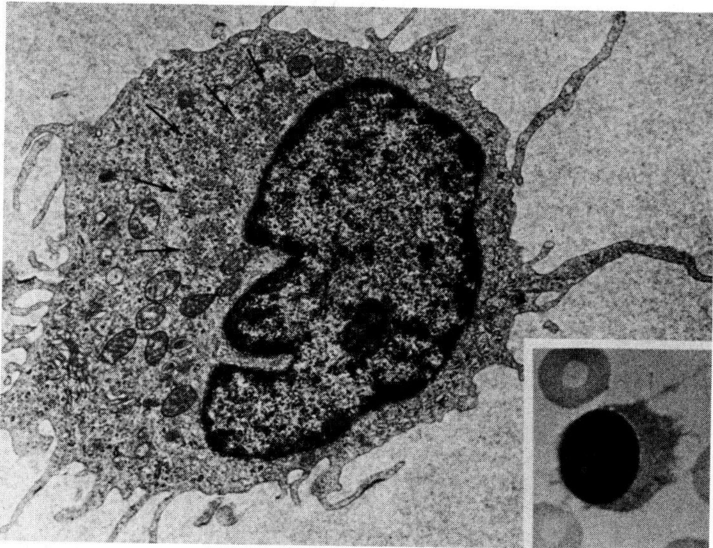

Figure 25.22
Appearance of hairy-cell leukemia, a neoplasm of B lymphocytes caused by HTLV II. Long hairlike processes are evident in an electron microscope section. Arrows indicate cytopathic effects suggestive of viral infection.
From I. Katayarma, C.Y. Li, and L. T.Yam, "Ultrastructure Characteristics of the Hairy Cells of Leukemic Reticuloendotheliosis," American Journal of Pathology, *67:361, 1972. Reprinted by permission of J. B. Lippincott Company.*

ATL is chronic and invariably fatal in its course. Symptoms are extreme, persistent lymphocytosis, with large, atypical lymphocytes. One variant of ATL, mycosis fungoides or cutaneous T-cell lymphoma, is accompanied by dermatitis, with thickened, scaly, ulcerative, or tumorous skin lesions. Other complications are lymphadenopathy and dissemination of the tumors to the lung, spleen, and liver. Many types of treatments with drugs, radiation, or some combination have been tried, but the long-term prognosis remains poor. Since the discovery of this virus, health officials have become concerned that it may be present and transmitted in transfused blood, and they are currently screening blood for this virus along with HIV.

Hairy-cell leukemia (HCL) is a rare form of cancer that has been traced to HTLV II infection. HCL derives its name from the appearance of the afflicted lymphocytes, which have fine cytoplasmic projections or pseudopod-like extensions (figure 25.22). The disease is more common in males than in females and is probably spread through blood and shared syringes and needles. Unlike ATL, this form of leukemia presents with overall leukopenia but with an increased number of neoplastic B lymphocytes. The most serious complication is the infiltration of the spleen, liver, and bone marrow, which interferes with hematopoiesis. Splenectomy, bone marrow transplants, and antineoplastic drugs have been pursued as treatments, but alpha interferon produces the highest rate of remission.

Chapter Checkpoints

Other retroviruses include the human T-cell leukemia viruses (HTLV I and II). They have been identified as the causative agents of ATL (adult T-cell leukemia) and HCL (hairy-cell leukemia).

NONENVELOPED NONSEGMENTED SINGLE-STRANDED RNA VIRUSES: PICORNAVIRUSES AND CALICIVIRUSES

As suggested by the prefix, *picorna* viruses are named for their small (pico) size and their RNA core (figure 25.23). Important representatives include ***Enterovirus*** and ***Rhinovirus,*** which are responsible for a broad spectrum of human neurological, enteric, and other illnesses (table 25.4), and *Cardiovirus,* which infects the brain and heart in humans and other mammals. The following discussion on human picornaviruses centers upon the poliovirus and other related enteroviruses, the hepatitis A virus, and the human rhinoviruses (HRVs).

Poliovirus and Poliomyelitis

Poliomyelitis* (polio) is an acute enteroviral infection of the spinal cord that can cause neuromuscular paralysis. Because it often affects small children, it is also called infantile paralysis. No civilization or culture has escaped the devastation of polio. Drawings made by ancient civilizations hint at its long history (figure 25.24), and such famous persons as Sir Walter Scott and Franklin Delano Roosevelt suffered its complications. Those of us who experienced the epidemics at the middle of this century will never forget the lasting images of polio: crippling paralysis, braces, the iron lung, and the March of Dimes (microfile 25.10).

Epidemiology of Poliomyelitis

The poliovirus has a naked capsid (see figure 25.23*a*) that confers chemical stability and resistance to acid, bile, and detergents. By this means the virus survives the gastric environment and other assaults. This ability to withstand an array of host defenses contributes to its transmission.

Sporadic cases of polio can break out at any time of the year, but its incidence is more pronounced during the summer and fall. The virus is passed within the population through food, water, hands, and objects contaminated with feces. Transmission via mechanical vectors such as flies sometimes occurs, but respiratory droplets are only rarely involved. The efforts of a WHO campaign called National Vaccination Days started in 1988 have significantly reduced the global incidence of polio. By the century's end, all of the last wild polio viruses will have been eradicated. The last

*poliomyelitis (poh´´-lee-oh-my´´-eh-ly´-tis) Gr. *polios,* gray, *myelos,* medulla, and *itis,* inflammation.

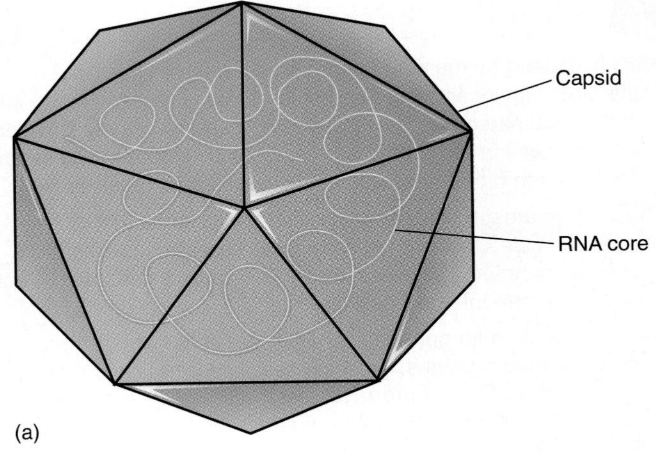

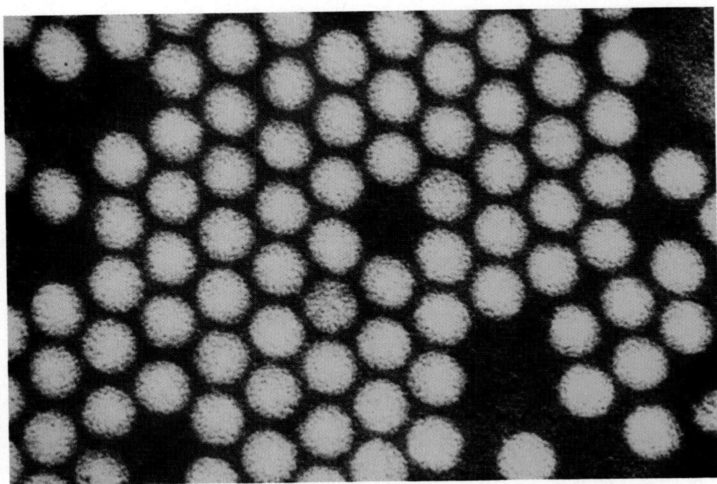

Figure 25.23
(*a*) A poliovirus, a type of picornavirus that is one of the simplest and smallest viruses (30 nm). It consists of an icosahedral capsid shell around a tightly packed molecule of RNA. (*b*) A crystalline mass of stacked poliovirus particles in an infected host cell.

TABLE 25.4

SELECTED CHARACTERISTICS OF THE HUMAN PICORNAVIRUSES

Genus	Representative	Primary Diseases
Enterovirus	Poliovirus	Poliomyelitis
	Coxsackievirus A	Focal necrosis, myositis
	Coxsackievirus B	Myocarditis of newborn
	Echovirus	Aseptic meningitis, enteritis, others
	Enterovirus 72	Hepatitis A
Rhinovirus	Rhinovirus	Common cold
Cardiovirus	Cardiovirus	Encephalomyocarditis
Aphthovirus	Aphthovirus	Foot-and-mouth disease (in cloven-foot animals)

MICROFILE 25.10 THE TRAGIC AFTERMATH OF POLIOMYELITIS

Survivors of the more severe forms of spinal and bulbar polio were left with multiple medical burdens. A patient with respiratory paralysis had to be maintained by a large cylindrical breathing apparatus called the iron lung. Encased and immobilized in these respiratory prisons, many patients could not talk, and their only opportunity to see the world was through a mirror positioned above their prison. Total loss of muscle function meant that even the simplest activities had to be done for them. Maneuvers such as bathing and medical examinations were frightening, because the patient had to be disconnected from the machine. In time, these young people were weaned onto air hoses that entered through a tracheostomy tube or were adapted to portable respirators.

Patients with severe paralysis of the skeletal muscles suffered other plights. The unused muscles began to atrophy, growth was slowed, and severe deformities of the trunk and limbs developed. Common sites of deformities were the spine, shoulder, hips, knees, and feet. Because motor function, but not sensation, was compromised, the crippled limbs were often very painful. Children's spines and limbs were often fused surgically into a rigid state in efforts to prevent this pathology from progressing and even threatening life.

In a desperate attempt to cure polio, health care workers treated patients with gamma globulin, antibiotics, and vitamins, but the epidemic escalated in the early 1950s. As the tragedy and anguish continued to mount, communities were galvanized by a cooperative effort called the March of Dimes. Through the modest donations of millions of schoolchildren and families, enough dimes were collected to underwrite increased research that eventually led to effective vaccines.

What were the prospects for the survivors—those kids in the iron lungs? Because of improved medical care, many became long-term survivors who learned to live with their disabilities. But, as if those impairments were not enough, long-term survivors were also subject to a condition called post-polio syndrome (PPS), an insidious motor neuron disease. PPS manifests as a progressive muscle deterioration that develops in about 25–50% of patients several decades after their original polio attack.

Figure 25.24
Polio in an ancient civilization. This stone tablet from the Eighteenth Dynasty in Egypt depicts Siptah bearing the signs of paralytic polio.

remaining cases of infection are confined to a few pockets in Africa, India, and parts of the Middle East. There have been no cases in the Americas since 1991.

Infection and Disease

After being ingested, polioviruses adsorb to receptors of mucosal cells in the oropharynx and intestine (figure 25.25). Here, they multiply in the mucosal epithelia and lymphoid tissue. Multiplication results in large numbers of viruses being shed into the throat and feces, and some of them leak into the blood.

Most infections are contained as a short-term viremia that is usually asymptomatic. A small number of persons develop mild nonspecific symptoms of fever, headache, nausea, sore throat, and myalgia. If the viremia persists, viruses can be carried to the central nervous system through its blood supply. The virus then spreads along specific pathways in the spinal cord and brain. Being *neurotropic,** the virus infiltrates the motor neurons of the anterior horn of the spinal cord, though it can also attack spinal ganglia, cranial nerves, and motor nuclei (figure 25.26). The so-called major illness, or nonparalytic disease, involves the invasion but not the destruction of nervous tissue. It gives rise to muscle pain and spasm, meningeal inflammation, and vague hypersensitivity.

Paralytic Disease

Invasion of motor neurons causes various degrees of flaccid paralysis over a period of a few hours to several days. Depending on the level of damage to motor neurons, paralysis of the muscles of the legs, abdomen, back, intercostals, diaphragm, pectoral girdle, and bladder can result. In rare cases **bulbar poliomyelitis,** the

*neurotropic (nu´´-roh-troh´-pik) Having an affinity for the nervous system.

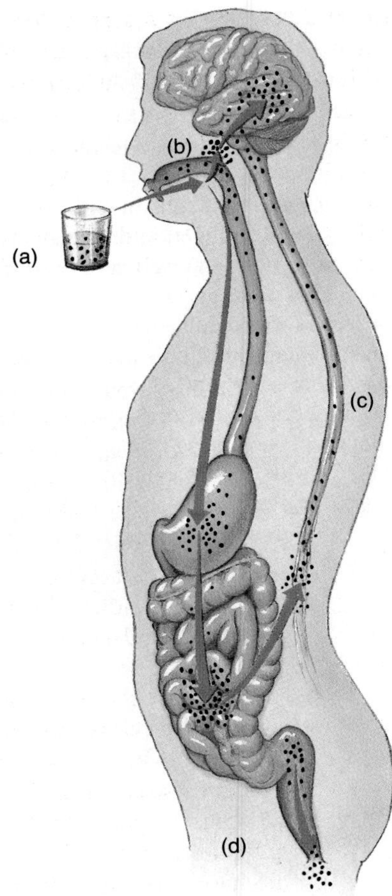

Figure 25.25
The stages of infection and pathogenesis of poliomyelitis. *(a)* First, the virus is ingested and carried to the throat and intestinal mucosa. *(b)* The virus then multiplies in the tonsils. Small numbers of viruses escape to the regional lymph nodes and blood. *(c)* The viruses are further amplified and cross into certain nerve cells of the spinal column and central nervous system. *(d)* Last, the intestine actively sheds viruses.

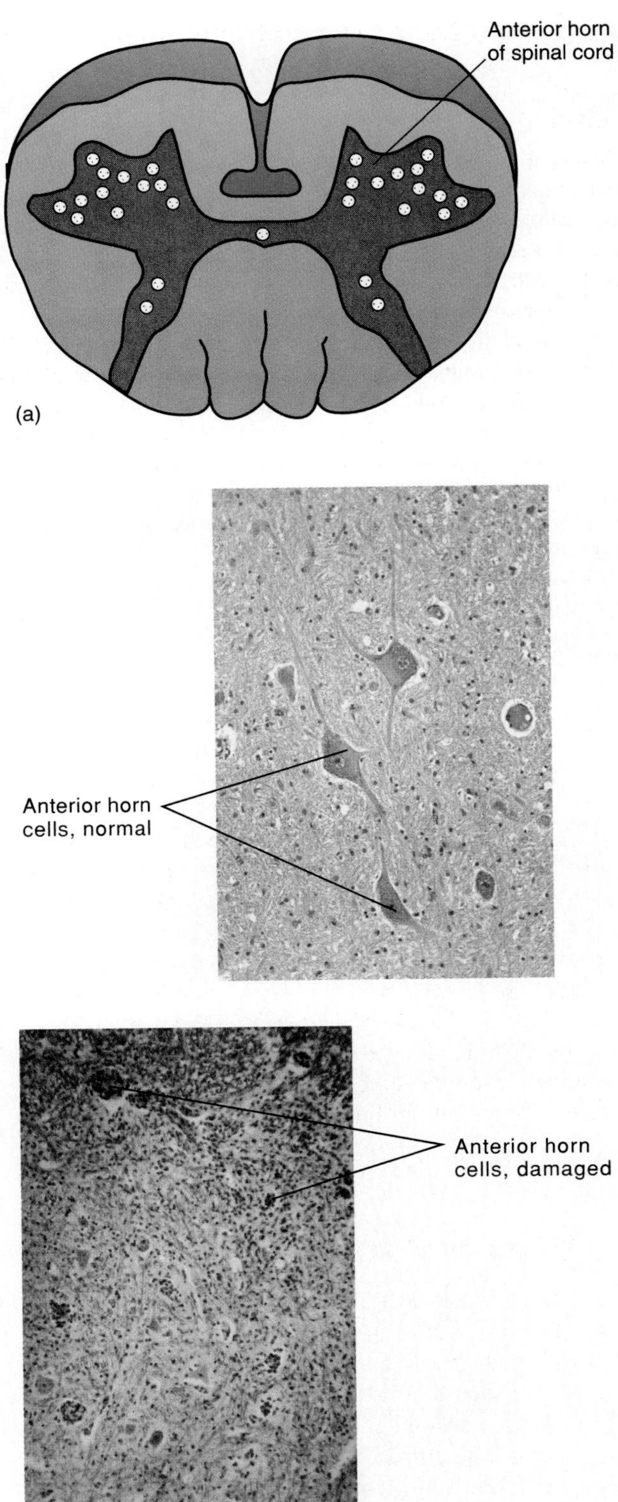

Figure 25.26
Targets of poliovirus. *(a)* A cross section of the spinal column indicates the areas most damaged in spinal poliomyelitis. *(b)* The anterior horn cells of a monkey before (top) and after (bottom) poliovirus infection.

brain stem, medulla, or even cranial nerves are affected. This situation leads to loss of control of cardiorespiratory regulatory centers, palate, pharynx, and vocal cords. The long-term effects of paralytic polio were discussed in microfile 25.10. Some protection is afforded by antibodies to poliovirus in the intestine and tonsils (secretory antibodies) and in the serum.

Diagnosis of Polio

Polio is mainly suspected when epidemics of neuromuscular disease occur in the summer in temperate climates. Poliovirus can usually be isolated by inoculating cell cultures with stool or throat washings in the early part of the disease. The stage of the patient's infection can also be demonstrated by testing serum samples for the type and amount of antibody.

Measures for Treatment and Prevention of Polio

Treatment of polio rests largely on alleviating pain and suffering. During the acute phase, muscle spasm, headache, and associated discomfort can be alleviated by pain-relieving drugs. Respiratory failure may require artificial ventilation maintenance. Because reflex swallowing and coughing can be impaired, drainage or mechanical suction of mucus and secretions is necessary to prevent choking. Prompt physical therapy to diminish crippling deformities and to retrain muscles is recommended after the acute febrile phase subsides.

The mainstay of prevention is vaccination, a measure that is now taken for granted. Polio immunization must be instituted as early in life as possible, usually in four doses starting at about 2 months of age. Adult candidates for immunization are travelers and members of the armed forces. The two forms of vaccine currently in use are inactivated poliovirus vaccine (IPV), known as the Salk[3] vaccine, and oral poliovirus vaccine (OPV), known as the Sabin[4] vaccine. Both are prepared from animal cell cultures and are trivalent (combinations of the three serotypes of the poliovirus). Both vaccines are effective, but one may be favored over the other under certain circumstances.

The Sabin vaccine has been favored in the United States because it is easily administered by mouth, but it is not free of medical complications. It contains an attenuated virus that can multiply in vaccinated people and be spread to others. In very rare instances, the attenuated virus reverts to a neurovirulent strain that causes disease rather than protects against it. Numerous instances of paralytic polio have occurred among children with hypogammaglobulinemia who have been mistakenly vaccinated. There is a tiny risk (about one case in 4 million) that an unvaccinated family member will acquire infection and disease from a vaccinated child. For this reason, some public health officials are calling for the combination of IPV given for the initial two doses, followed by OPV for later boosters. Now that the WHO goal of polio eradication by the year 2000 is nearly achieved, it may be possible to discontinue polio vaccination, as has been done for smallpox.

NONPOLIO ENTEROVIRUSES

Several viruses related to poliovirus commonly cause transient, nonfatal infections. Such viruses as **coxsackieviruses*** A and B, *echoviruses,** and nonpolio enteroviruses, are similar to the poliovirus in many of their epidemiologic and infectious characteristics. The incidence of infection is highest from late spring to early summer in temperate climates and is most frequent in infants and persons living under unhygienic circumstances.

Specific Types of Enterovirus Infection

About 50% to 80% of enteroviral infections are subclinical, and the remainder fall into the category of "undifferentiated febrile illness," characterized by fever, myalgia, and malaise. Symptoms are usually mild and self-limited and last only a few days. The initial phase of infection is intestinal, after which viruses enter the lymph and blood and disseminate to other organs. The outcome of this infection largely depends on the organ affected. An overview of the more severe complications follows.

Important Complications

Children are more prone than adults to lower respiratory tract illness, namely bronchitis, bronchiolitis, croup, and pneumonia. All ages, however are susceptible to the "common cold syndrome" of enteroviruses (microfile 25.11). *Pleurodynia** is an acute disease characterized by recurrent sharp, sudden intercostal or abdominal pain accompanied by fever and sore throat. Although nonpolio enteroviruses are less virulent than the polioviruses, rare cases of coxsackievirus and echovirus paralysis, aseptic meningitis, and encephalitis occur. Even in severe childhood cases involving seizures, ataxia, coma, and other central nervous system symptoms, recovery is usually complete.

Eruptive skin rashes (exanthems) that resemble the rubella rash are other manifestations of enterovirus infection. Coxsackievirus can cause a peculiar pattern of lesions on the hands, feet, and oral mucosa (hand-foot-mouth disease). Rashes are accompanied by fever, headache, and muscle pain. Acute hemorrhagic conjunctivitis is an abrupt inflammation associated with subconjunctival bleeding, serous discharge, painful swelling, and sensitivity to light (figure 25.27). Spread of the virus to the heart in infants can cause extensive damage to the myocardium, leading to heart failure and death in nearly half of the cases. Heart involvement in older children and adults is generally less serious, with symptoms of chest pain, altered heart rhythms, and pericardial inflammation.

Hepatitis A Virus and Infectious Hepatitis

One enterovirus that reacts primarily with the intestinal tract is the **hepatitis A virus (HAV;** enterovirus 72), the cause of infectious, or short-term hepatitis. Although this virus is not related to the hepatitis B virus discussed in chapter 24, it shares its tropism for liver cells. Otherwise, the two viruses are different in almost every respect (see table 24.3). The hepatitis A virus is a cubical picornavirus relatively resistant to heat but sensitive to formalin, chlorine, and ultraviolet radiation. There appears to be only one major serotype of this virus.

Epidemiology of Hepatitis A Hepatitis A virus is spread through the oral-fecal route, but the details of transmission vary from one area to another. In general, the disease is associated with deficient personal hygiene and lack of public health measures. In countries with inadequate sewage control, most outbreaks are associated with fecally contaminated water and food. The United States has a

3. Named for Dr. Jonas Salk, who developed the vaccine in 1954.
4. Named for Dr. Albert Sabin, who had the idea to develop an oral attenuated vaccine in the 1960s.

*coxsackievirus (kok-sak´-ee-vy´´-rus) Named for Coxsackie, New York, where the viruses were first isolated.

*echovirus (ek´-oh-vy´´-rus) An acronym for **e**nteric **c**ytopathic **h**uman **o**rphan **virus.**

*pleurodynia (plur´´-oh-din´-ee-ah) Gr. *pleura,* rib, side, and *odyne,* pain.

MICROFILE 25.11 THE COMMON COLD: A UNIVERSAL HUMAN CONDITION

The common cold touches the lives of humans more than any other viral infection. Everyone instantly recognizes the sensations of a cold—the nasal stuffiness, scratchy throat, headache, sneezing, and coughing. As an infection, the cold is king, afflicting at least half the population every year and accounting for millions of hours of absenteeism from work and school. The reason for its widespread distribution is not that it is more virulent or transmissible than other infections but that symptoms of colds are linked to hundreds of different viruses and viral strains. Among the known causative viruses, in order of importance, are rhinoviruses (which cause about half of all colds), paramyxoviruses, enteroviruses, coronaviruses, reoviruses, and adenoviruses. A given cold can be caused by a single virus type or it can result from a mixed infection.

The name implies a relationship with cold weather or drafts. But studies in which human volunteers with wet heads or feet chilled or exposed to moist, frigid air have failed to support such a link. Most colds occur in the late autumn, winter, and early spring—all periods of colder weather—but this seasonal connection may have more to do with being confined in closed spaces with carriers than with temperature. Studies have shown that certain states such as menstruation, extreme fatigue, and allergic rhinitis can facilitate the onset of a cold.

The most significant single factor in the spread of colds is contamination of hands with mucous secretions. The portal of entry is the mucous membranes of the nose and eyes. Multiplication of the virus produces an exudative inflammation with detachment of the epithelium and mild respiratory symptoms. The most common symptom is a nasal discharge, and the least common is fever, except in infants and children.

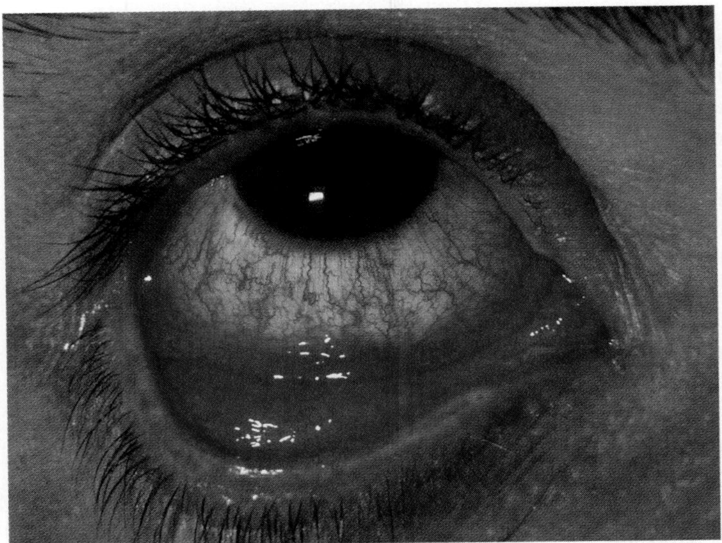

Figure 25.27
Acute hemorrhagic conjunctivitis caused by an enterovirus. In the early phase of disease, the eye is severely inflamed, with the sclera bright red owing to subconjunctival hemorrhage. Later, edema of the lids causes complete closure of the eye.

yearly reported incidence of 25,000–30,000 cases. Most of these result from close institutional contact, unhygienic food handling, eating shellfish, sexual transmission, or travel to other countries. Occasionally, hepatitis A can be spread by blood or blood products. In developing countries, children are the most common targets, because exposure to the virus tends to occur early in life, whereas in North America and Europe, more cases appear in adults. Because the virus is not carried chronically, the principal reservoirs are asymptomatic, short-term carriers or people with clinical disease.

The Course of HAV Infection Swallowed hepatitis A virus multiplies in the small intestine during an incubation period of 2 to 6 weeks. The virus is then shed in the feces for about 2 weeks, and, at some point, it enters the blood and is carried to the liver. Most infections are either subclinical or accompanied by vague, flu-like symptoms. In more overt cases, the presenting symptoms are loss of appetite, nausea, diarrhea, fever, pain and discomfort in the region of the liver, and darkened urine. Jaundice is present in only about one in 10 cases. Occasionally, hepatitis A occurs as a fulminating disease and causes liver damage, but this manifestation is quite rare. Because the virus is not oncogenic and does not predispose to liver cancer, complete uncomplicated recovery results.

Diagnosis and Control of Hepatitis A A patient's history, liver and blood tests, serological tests, and viral identification all play a role in diagnosing hepatitis A and differentiating it from the other forms of hepatitis. Certain liver enzymes are elevated, and leukopenia is common. Diagnosis is aided by detection of anti-HAV IgM antibodies produced early in the infection and tests to identify HA antigen or virus directly in stool samples.

There is no specific treatment for hepatitis A once the symptoms begin. Patients who receive immune serum globulin early in the disease usually experience milder symptoms than patients who do not receive it. Prevention of hepatitis A is based primarily on prophylactic immunization with pooled immune serum globulin. Potential globulin recipients are travelers and armed forces personnel planning to enter endemic areas, contacts of known cases, and occupants of daycare centers and other institutions during epidemics. An inactivated viral vaccine is currently approved and an oral vaccine based on an attenuated strain of the virus is in development. Control of this disease can be improved by sewage treatment, hygienic food handling and preparation, and adequate cooking of shellfish.

MICROFILE 25.12 CONTEMPLATING A CURE FOR THE COMMON COLD

Finding a cure for the common cold has been such a long-standing hope that it sounds almost like a cliché. Why this mania? It is not motivated by the clinical nature of a cold, which is really a rather benign infection. A more likely reason for the search for a "magic cold bullet" is the lost productivity in the workplace and schools, as well as the potential profits from a truly effective cold drug.

One nonspecific approach has been to destroy the virus outright and halt its spread. Special facial tissues impregnated with mild acid have been marketed for use during the cold season. One company claims that its aerosol device, which forces hot air and medication to the nose and throat, can nip colds in the bud. Virus multiplication is inhibited by heat, but whether a virus can be killed by heating cells is another question. After years of controversy, a recent study has finally shown that taking megadoses of vitamin C at the onset of a cold can be beneficial. Zinc lozenges may also help retard the onset of cold symptoms.

The important role of natural interferon in controlling many cold viruses has led to the testing and marketing of a nasal spray containing recombinant interferon. Another company is currently developing a novel therapy based on monoclonal antibodies. These antibodies are raised to the site on the human cell (receptor) to which the rhinovirus attaches. In theory, these antibodies should occupy the cell receptor, competitively inhibit viral attachment, and prevent infection. Experiments in chimpanzees and humans showed that, when administered intranasally, this antibody preparation delayed the onset of symptoms and reduced their severity. It remains to be seen whether any of these products will ever be widely available.

Human Rhinovirus (HRV)

The **rhinoviruses*** are a large group of picornaviruses (more than 110 serotypes) associated with the **common cold** (see microfile 25.11). Although the majority of characteristics are shared with other picornaviruses, two distinctions set rhinoviruses apart. First, they are sensitive to acidic environments such as that of the stomach, and, second, their optimum temperature of multiplication is not normal body temperature, but 33°C, the average temperature in the human nose.

Portrait of Rhinovirus Virologists have subjected rhinovirus (type 14) to a detailed structural analysis. The results of this study provided a striking three-dimensional view of its molecular surface and also explained why immunity to the rhinovirus has been so elusive. The capsid subunits are of two types: protuberances (knobs), which are antigenically diverse among the rhinoviruses, and indentations (pockets), which exist only in two forms (figure 25.28). Because the antigens on the surface are the only ones accessible to the immune system, a successful vaccine would have to contain hundreds of different antigens, so it is not practical. Unfortunately the recessed antigens are too inaccessible for either immune surveillance or antibody fit. These factors make the development of a vaccine unlikely. New knowledge could provide a basis for developing drugs to block the host receptor (microfile 25.12).

Epidemiology and Infection of Rhinoviruses Rhinovirus infections occur in all areas and all age groups at all times of the year. Epidemics caused by a single type arise on occasion, but usually many strains circulate in the population at one time. As mutations increase and herd immunity to a given type is established, newer types predominate. Children are the most successful disseminators of colds and often transmit the virus to the rest of a family. Infected respiratory membranes shed the virus for several days to weeks. People acquire infection from contaminated hands and fomites, and to a lesser extent from droplet nuclei. Although

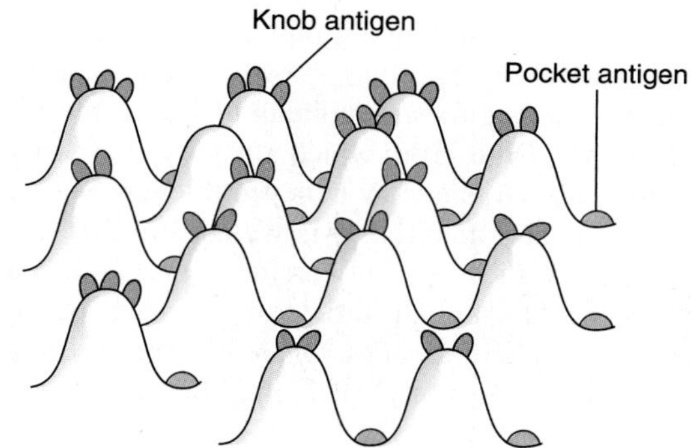

Figure 25.28
(*a*) The surface of a rhinovirus is composed of knobs and pockets. The antigens on the knobs are extremely variable in shape among the scores of viral strains, and those in the pockets do not vary among the strains. (*b*) The knobs are readily accessible to the immune system, and antibodies formed against them will inactivate the virus. But to be fully protected against colds, one would have to produce 100 different kinds of antibodies. Although antibodies against pocket antigens would be universal, the antigens are too inaccessible to react with antibodies.

*rhinovirus (ry´-noh-vy´´-rus) Gr. *rhinos,* nose.

other animals have rhinoviruses, interspecies infections have never been reported. After an incubation period of 1 to 3 days, the patient can experience some combination of headache, chills, fatigue, sore throat, cough, a mild nasal drainage, and atypical pneumonia. Natural host defenses and nasal antibodies have a beneficial local effect on the infection, but immunity is short-lived.

Control of Rhinoviruses The classic therapy is to force fluids and relieve symptoms with various cold remedies and cough syrups that contain nasal decongestants, alcohol, antihistamines, and analgesics. The actual effectiveness of most of these remedies (of which there are hundreds) is rather questionable. See microfile 25.12 for other potential treatments. Owing to the extreme diversity of rhinoviruses, prevention of colds through vaccination will probably never be a medical reality. Some protection is afforded by such simple measures as handwashing and care in handling nasal secretions.

CALICIVIRUSES

*Caliciviruses** are an ill-defined group of enteric viruses found in humans and mammals. The best known human pathogen is the **Norwalk agent,** named for an outbreak of gastroenteritis that occurred in Norwalk, Ohio, from which a new type of virus was isolated. The Norwalk virus is now believed to cause one-third of all cases of viral gastroenteritis. It is transmitted oral-fecally in schools, camps, cruise ships, and nursing homes and through contaminated water and shellfish. Infection can occur at any time of the year and in people of all ages. Onset is acute, accompanied by nausea, vomiting, cramps, diarrhea, and chills; recovery is rapid and complete.

Chapter Checkpoints

Picornaviruses cause a wide variety of viral infections including polio, aseptic meningitis, the common cold, hepatitis A, encephalomyocarditis, and hand-foot-mouth disease. Major groups include *Enterovirus, Rhinovirus, Cardiovirus,* and *Aphthovirus.*

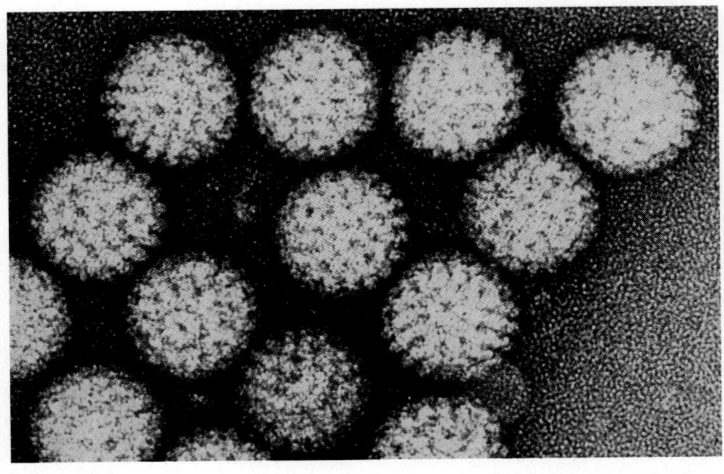

(a)

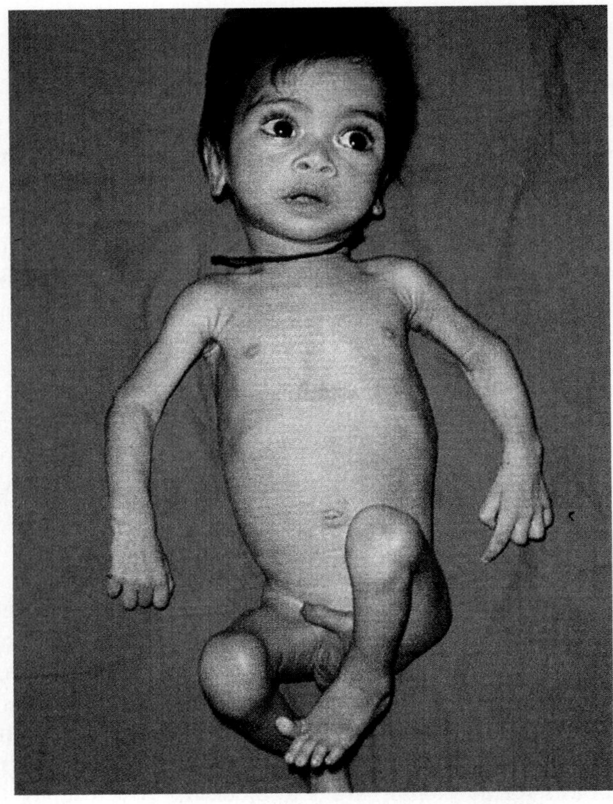

(b)

Figure 25.29
(*a*) A sample of feces from a child with gastroenteritis as viewed by electron microscopy. Note the unique "spoked-wheel" morphology that can be used to identify *Rotavirus.* (*b*) An infant suffering from chronic rotavirus gastroenteritis shows evidence of malnutrition, failure to thrive, and stunted growth.

NONENVELOPED SEGMENTED DOUBLE-STRANDED RNA VIRUSES: REOVIRUSES

Reoviruses have an unusual double-stranded RNA genome and both an inner and an outer capsid (see table 25.1). Two of the best-studied viruses of the group are *Rotavirus* and *Reovirus.* Named for its wheel-shaped capsomer, *Rotavirus** (figure 25.29*a*) is a significant cause of diarrhea in newborn humans, calves, and piglets. Because the viruses are transmitted via fecally contaminated food, water, and fomites, disease is prevalent in areas of the world with poor sanitation. Globally, *Rotavirus* is the primary viral cause of mortality and morbidity resulting from diarrhea, accounting for nearly 50% of all the cases. The effects of infection vary with the nutritional state, general health, and living conditions of the infant. Babies from 6 to 24 months of age lacking maternal antibodies have the greatest risk for fatal disease. These

*calicivirus (kal´-ih-sih-vy´´-rus) L. *calix,* the cup of a flower. These viruses have cup-shaped surface depressions.

*Rotavirus (roh´-tah-vy´´-rus) L. *rota,* wheel.

children present with symptoms of watery diarrhea, fever, vomiting, dehydration, and shock. The intestinal mucosa can be damaged in a way that chronically compromises nutrition, and long-term or repeated infections can retard growth (figure 25.29*b*). In the United States, rotavirus infection is relatively common, but its course is generally mild. Children are treated as in cholera with oral replacement fluid and electrolytes. A live, attenuated oral vaccine has been approved and is currently recommended for immunizing neonates in areas of high prevalence.

*Reovirus** is not considered a significant human pathogen. Adults who were voluntarily inoculated with the virus developed symptoms of the common cold. The virus has been isolated from the feces of children with enteritis and from people suffering from an upper respiratory infection and rash, but most infections are asymptomatic. Although a link to biliary, meningeal, hepatic, and renal syndromes has been suggested, a causal relationship has never been proved.

Chapter Checkpoints

Reoviruses are unusual in possessing both double-stranded RNA and two capsids. The two major groups are rotaviruses and reoviruses. *Rotavirus* is a significant intestinal pathogen for young children worldwide. *Reovirus* causes an upper respiratory infection similar to the common cold.

SLOW INFECTIONS BY UNCONVENTIONAL VIRUSLIKE AGENTS

Diseases such as subacute sclerosing panencephalitis and progressive multifocal leukoencephalopathy, arising from persistent viral infections of the CNS, have been described in previous sections. They are characterized by a long incubation period, and they can progress over months or years to a state of severe neurological impairment. Whereas these slow infections are caused by known viruses with conventional morphology, there is another group of CNS infections for which no typical virus has been isolated.

The **spongiform encephalopathies*** are transmissible, uniformly fatal, chronic infections of the nervous system caused by unusual biological entities that some virologists call unconventional viruses, or virinos. Other virologists have concluded that these diseases are caused by agents called **prions**[5] with extremely atypical chemical and physical properties (table 25.5). The agents cause a degeneration in the brain tissue marked by the formation of vacuoles in nerve cells and a spongy appearance of the cerebrum.

Two diseases of humans are known to be caused by unconventional infectious agents. **Kuru** is an endemic disease that was once prevalent among the natives of the New Guinea highlands but has nearly disappeared since cannibalism was discontinued there. Members of these tribes performed mourning rituals that entailed eating the brains of dead relatives. Because women were the principal participants in these ceremonies, they were the most common victims (figure 25.30*a*). It is believed that the kuru agent enters skin cuts and mucosal surfaces and is then carried to the brain. After an incubation period of 20–30 years, the parts of the brain controlling gait, posture, speech, and eye movement are adversely affected, leading to symptoms of unsteadiness along with shivering tremor (the tribal word for this is *kuru*). As the victims' health gradually deteriorates, they experience profound loss of locomotion and mental deterioration.

TABLE 25.5

PROPERTIES OF THE AGENTS OF SPONGIFORM ENCEPHALOPATHIES

Very resistant to chemicals, radiation, and heat
Do not present virus morphology in electron microscopy of infected brain tissue
Not associated with foreign nucleic acid isolated from infected host cells
Proteinaceous, filterable
Multiply (slowly) in cell culture; no cytopathic effects
Do not elicit inflammatory reaction in host
Do not elicit antibody formation in host
Responsible for plaques and abnormal fibers forming in brain of host
Transmitted only by intimate contact with infected tissues and secretions

Creutzfeldt-Jakob disease (CJD) is another encephalopathy with sporadic, worldwide distribution. It shows a familial pattern of inheritance and appears to be transmitted by intimate contact with tissues of an infected patient. Symptoms include altered behavior, memory loss, impaired senses, delirium, and premature senility. Uncontrollable muscle contractions continue until death, which usually occurs within one year of diagnosis. An autopsy of the brain shows built-up tangled protein fibers (neurofibrillary tangles), enlarged astroglial cells, and vacuoles (figure 25.30*b*). Because the infectious agent is not readily destroyed by usual methods, great care must be taken in handling tissues from CJD patients in a medical setting. Some cases have been transmitted through grafts, accidental inoculation, and handling of brain tissue.

It is becoming clear that a range of neurological diseases have a pathologic picture similar to CJD. Examples are scrapie in sheep and bovine spongiform encephalopathy. It has been suggested that these syndromes be lumped together in the category of transmissible virus dementia (TVD). A common characteristic of these diseases may be that infection or some other factor prevents the transport of a necessary protein neurofilament, which builds up along the neuron and disturbs brain function.

Chapter Checkpoints

Long-term degenerative and incurable CNS infections that cannot be linked to any known viral agent are thought to be caused by *prions,* infectious agents with atypical biological properties that are unusually resistant to destruction. Kuru and Creutzfeldt-Jakob disease are two of several TVDs (transmissible virus dementias) identified to date.

5. A contraction of **pro**teinaceous **in**fectious particle.

***Reovirus** (ree´oh-vy´´-rus) An acronym for **r**espiratory **e**nteric **o**rphan **virus**.

***spongiform encephalopathies** (spunj´-ih-form en-sef´´-uh-lop´-uh-theez)

(a)

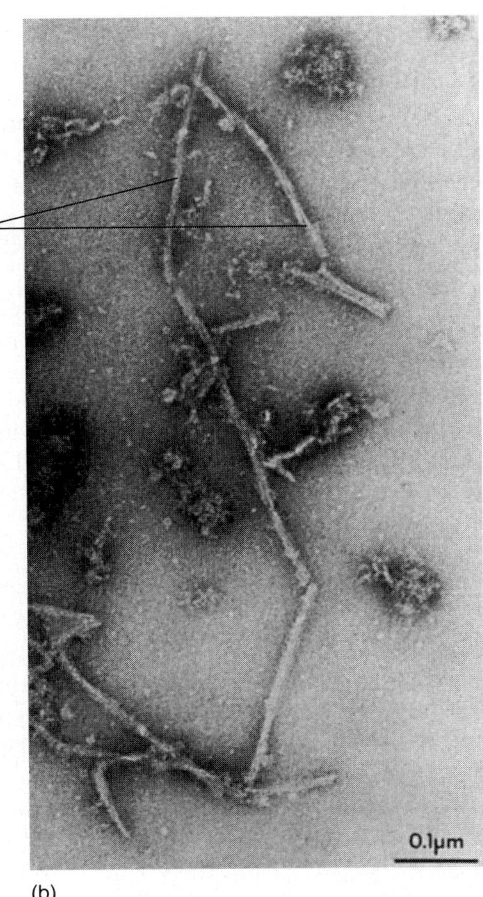

(b)

Figure 25.30
Spongiform encephalopathies. (*a*) New Guinea tribeswomen in the early stages of kuru. Already they require help in maintaining postural stability and exhibit some degree of visual involvement. (*b*) Amyloid fibrils making up the tangles in the brain of a Creutzfeldt-Jakob patient.

CHAPTER CAPSULE WITH KEY TERMS

THE RNA VIRUSES

ENVELOPED SINGLE-STRANDED VIRUSES

Orthomyxoviruses: Influenza viruses have a segmented genome and glycoprotein spikes for binding to the host (hemagglutinin, neuraminidase); are subject to *antigenic shift* and *drift* that creates new viral subtypes. Influenza A is spread in epidemics and pandemics from human to human and from animal (pigs, poultry) to human; virus shed in aerosols in close company; children and elderly most susceptible; clinical disease marked by respiratory symptoms, fever, aches, sore throat; secondary pneumonia is frequent cause of death; long-term complications are Guillain-Barré syndrome and Reye's syndrome; controlled with amantadine and vaccination.

Bunyaviruses are zoonotic viruses widely associated with insects and rodents. One type is the *Hantavirus;* cause of Korean hemorrhagic fever worldwide and *hantavirus pulmonary syndrome,* a newly described disease in the United States; spread by dried animal excreta.

Paramyxoviruses: Nonsegmented genome; glycoprotein spikes facilitate development of **multinucleate giant cells;** are spread through droplets.

Paramyxovirus causes **parainfluenza,** a common, mild form of influenza or coldlike illness, mostly in children; **mumps** or

parotitis, inflammatory infection of salivary glands, sometimes testes, pancreas, and other organs; occurs often in children; can cause deafness; prevented by live, attenuated vaccine (MMR).

Morbillivirus, the viral agent of **measles** or **rubeola,** is an acute, highly contagious respiratory infection that produces a maculopapular rash over the body and in the oral cavity **(Koplik's spots);** usually mild, but can threaten life in a small number of children; other complication is progressive brain disease called **SSPE;** attenuated vaccine administered in combination (MMR) or alone; booster recommended.

Pneumovirus, or **respiratory syncytial virus (RSV)** causes **croup,** an acute respiratory syndrome in newborns that is a problem in hospital nurseries; treated with aerosol ribavirin.

Rhabdoviruses: The main human pathogen is bullet-shaped ***Lyssavirus,*** the cause of **rabies,** a nearly 100% fatal infection of the brain and meninges; a zoonosis carried by wild carnivores; usually contracted through an animal bite or scratch but can be inhaled; virus enters nerves and is carried to spinal cord and brain and eventually to salivary glands; symptoms include both hyperactivity and paralysis, also hydrophobia; death results from cardiac or respiratory arrest; treatment after animal contact requires scrupulous cleaning of wound, passive immunization with rabies immune globulin, and active immunization with **human diploid cell vaccine;** vaccine is also given to prevent infection in domestic animals and humans at risk.

OTHER ENVELOPED VIRUSES

Coronaviruses: Have distinctive crown of spikes; human coronavirus (HCV) is one cause of the common cold.

Togaviruses: Rubella (German measles) **virus** is a **teratogenic** virus that crosses the placenta and causes developmental disruptions; virus spread through close contact, infected mucous droplets; period of respiratory symptoms is followed by rash; usually mild in children; infection during pregnancy can lead to congenital rubella with severe heart and nervous system damage; control by live, attenuated vaccine (MMR), preferably given in childhood; women should be vaccinated prior to pregnancy.

ZOONOTIC VIRUSES

Arenaviruses: Spread by rodent secretions, droppings; example is Lassa fever, a severe and often fatal African hemorrhagic fever.

Arboviruses: Loose grouping of 400 viruses spread by arthropod vectors and harbored by numerous birds and mammals; include togaviruses, bunyaviruses, and reoviruses; diseases occur in conjunction with greatest vector activity; humans usually dead-end host; cause febrile infections (tick-borne fever), mosquito-borne encephalitides in horses and humans **(WEE, EEE),** and in humans and birds (St. Louis encephalitis); hemorrhagic fevers include **yellow fever** and **dengue fever,** sometimes deadly mosquito-borne diseases.

RETROVIRUSES

Retroviruses are enveloped, single-stranded RNA viruses with **reverse transcriptase** that converts their single-stranded RNA into double-stranded DNA genome; insertable into host cell, so viruses are latent and can be oncogenic.

Human Immunodeficiency Virus (HIV, types 1 and 2): Causes **acquired immunodeficiency syndrome (AIDS),** an HIV antibody-positive state with some combination of opportunistic infections, fever, weight loss, chronic lymphadenitis, neoplasms, diarrhea, brain dysfunction. HIV appears to have originated in Africa; an estimated 30–40 million persons are infected worldwide. AIDS has high mortality rate.

Virus occurs in blood, semen, and vaginal secretions; is transmitted through homosexual and heterosexual contact and shared needles; risk groups or behaviors include (1) male homosexuals or bisexuals practicing unsafe anal and or oral sex, (2) intravenous drug abusers, (3) heterosexual partners of drug users and bisexuals not using protection, (4) blood transfusion and organ transplant patients, (5) hemophiliacs (less common now), (6) persons with unknown risk factors, (7) fetuses or newborns of infected mothers.

HIV attacks cells with CD4 receptors, starting in macrophages and moving to T cells and other lymphocytes; initial infection creates a flu-like illness; virus becomes latent, gradually bringing the levels of T-helper cells down and invading the brain; after incubation period of 2–15 years, opportunistic infections occur, including severe intractable ***Pneumocystis carinii* pneumonia (PCP)** and serious diseases by protozoa, fungi, mycobacteria, and viruses; **Kaposi's sarcoma** and lymphoma are common cancers; coinfections with HIV and other agents can hasten the destruction and weakening of immunities; the death rate is over 55%.

Treatment for AIDS includes drug cocktails that combine **azidothymidine** (AZT), ddI, ddC, and d4T and protease inhibitors; drugs must be given for secondary infections; prevention measures are safe sexual practices; numerous vaccines are in clinical trials.

Other Retroviruses and Diseases: HTLV I causes **adult T-cell leukemia** or mycosis fungoides, a fatal cancer; HTLV II causes another cancer with high mortality rate, **hairy-cell leukemia.**

NONENVELOPED RNA VIRUSES

Picornaviruses are the smallest human viruses.

Enteroviruses: Spread primarily by the oral-fecal route. **Poliovirus,** cause of **polio** or infantile paralysis; common in children; virus is resistant, can spread through close contact, flies, contaminated food, water; occasionally leads to paralysis caused by infection and destruction of spinal neurons; bulbar type affects most muscles, requires life support; disease controlled through vaccination with oral vaccine (OPV) or inactivated vaccine (IPV).

Coxsackievirus causes respiratory infection, hand-foot-mouth disease, conjunctivitis. **Hepatitis A virus** causes short-term, milder hepatitis; spread through close institutional contact, contaminated food or shellfish; disease is initially flu-like, followed by liver discomfort and diarrhea; prevented by passive immunization with gamma globulin; vaccines now available.

Rhinovirus: Most prominent cause of common cold; spread by droplets; grows in cooler regions of body (nasal membrane); virus exists in over 110 forms, is difficult to control; symptoms are nasal drainage, cough, sneezing, sore throat; treatment is symptomatic.

Calicivirus: Norwalk agent, a common cause of viral gastroenteritis.

DOUBLE-STRANDED RNA VIRUSES

Reoviruses: Most important one is *Rotavirus*, the cause of severe infantile diarrhea; most common cause of viral enteric disease worldwide; children can die from effects of diarrhea, dehydration, shock.

UNCONVENTIONAL VIRUSLIKE AGENTS

Virinos or **prions** have unusual properties: very resistant, proteinaceous, cause progressive neurological diseases; diseases are the **spongiform encephalopathies,** named for brain's appearance; examples are **kuru,** a disease of cannibals in New Guinea, and **Creutzfeldt-Jakob disease,** a slow deterioration of the brain.

MULTIPLE-CHOICE QUESTIONS

1. Which receptors of the influenza virus are responsible for binding to the host cell?
   a. hemagglutinin c. Type A
   b. neuraminidase d. capsid proteins
2. The primary site of attack in influenza is the
   a. small intestine
   b. respiratory epithelium
   c. skin
   d. meninges
3. Infections with ____ virus causes the development of multinucleate giant cells.
   a. rabies c. pneumo
   b. influenza d. corona
4. Which virus is responsible for Korean hemorrhagic fever?
   a. Ebola virus c. arenavirus
   b. hantavirus d. flavivirus
5. For which disease is a exanthem (skin rash) *not* a symptom?
   a. measles
   b. rubella
   c. Coxsackievirus infection
   d. parainfluenza
6. A common, highly diagnostic sign of measles is
   a. viremia c. sore throat
   b. red rash d. Koplik's spots
7. ____ is a life-threatening disease in infants caused by ____.
   a. Mumps, morbillivirus
   b. Croup, respiratory syncytial virus
   c. Encephalitis, calicivirus
   d. Hairy-cell leukemia, HTLV II
8. Rabies virus has an average incubation period of
   a. 2 to 3 weeks c. 4 to 5 days
   b. 1 to 2 years d. 1 to 2 months
9. For which disease are active and passive immunization given simultaneously?
   a. influenza c. measles
   b. yellow fever d. rabies
10. What property of the retroviruses enables them to integrate into the host genome?
    a. the RNA they carry
    b. presence of glycoprotein receptors
    c. presence of reverse transcriptase
    d. a positive-sense genome
11. Which of the following cells is a target of HIV infection?
    a. dendritic cells
    b. monocytes
    c. helper T cells
    d. all of these can support infection
12. Which of the following conditions is *not* associated with AIDS?
    a. *Pneumocystis carinii* pneumonia
    b. Kaposi's sarcoma
    c. dementia
    d. adult T-cell leukemia
13. Polio and hepatitis A viruses are ____ viruses.
    a. arbo c. cold
    b. enteric d. syncytial
14. Rhinoviruses are the most common cause of
    a. conjunctivitis
    b. gastroenteritis
    c. hand-foot-mouth disease
    d. the common cold
15. Which of the following is *not* a characteristic of the agents of spongiform encephalopathies?
    a. highly resistant
    b. associated with tangled protein fibers in the brain
    c. are naked fragments of RNA
    d. are chronic
16. Multiple matching. Select all descriptions that fit each type of virus.

____ influenza virus	a. enveloped
____ rhabdovirus	b. double-stranded RNA
____ paramyxovirus	c. single-stranded RNA
____ reovirus	d. genome in segments
____ mumps virus	e. icosahedral capsid
____ measles virus	f. helical nucleocapsid
____ poliovirus	g. positive-sense genome
____ retrovirus	h. negative-sense genome
____ hantavirus	

17. Matching. Match the virus with its primary target organ or site of attack. Some viruses may attack more than one organ.

____ HIV	1. brain
____ mumps virus	2. parotid gland
____ measles virus	3. upper respiratory tract
____ hepatitis A virus	4. white blood cells
____ hantavirus	5. kidney
____ western equine encephalitis virus	6. spinal cord
	7. liver
____ enterovirus	8. intestine
____ poliovirus	9. heart
____ rabies virus	10. eye
____ rubella virus	

18. Matching. Use the same list of viruses as in question 17 and match each one with its primary mode of transmission from the list below. Some viruses may have more than one mode.
    a. respiratory droplets
    b. sexual transmission
    c. ingestion (oral-fecal)
    d. arthropod bite
    e. contact with mammals
    f. blood transfusion
    g. fomites

CONCEPT QUESTIONS

1. a. Describe the structure and functions of the hemagglutinin and neuraminidase spikes in influenza virus.
   b. What is unusual about the genome of influenza virus?
   c. Use drawings and words to explain the concepts of antigenic drift and antigenic shift, and explain their impact on the epidemiology of this disease.
   d. How does the CDC anticipate which flu strains will predominate each year?
   e. How do the names for the types of flu originate?
2. a. Explain the course of infection and disease in influenza.
   b. What are the complications?
   c. How is the vaccine prepared, and for which groups is it indicated?
3. a. Give examples of Bunyaviruses and arenaviruses,
   b. Briefly describe the nature of the diseases they cause.
   c. Describe their modes of transmission.
4. a. Describe the steps in the production of multinucleate giant cells during a viral infection.
   b. Which viruses have this effect, and what impact does cell fusion have on the spread of infection?
5. a. Describe the progress of measles symptoms.
   b. What is the cause of death in measles, and what are the most severe complications of measles infections?
   c. Describe methods of treatment and prevention for measles.
6. a. Describe the epidemiologic cycle in rabies.
   b. Which animals in the United States are most frequently involved as carriers?
   c. Describe the route of infection and the viruses' pathologic effects.
   d. Why is rabies so uniformly fatal?
   e. Describe the indications for pre- and postexposure rabies treatment.
   f. What is given in postexposure treatment that is not given in preexposure treatment of rabies?
   g. What makes the latest vaccine less traumatic than earlier ones?
7. a. What is a teratogenic virus?
   b. Which RNA viruses have this potential?
   c. In what ways is rubella different from red measles?
   d. What is the protocol to prevent congenital rubella?
8. a. What are the principal carrier arthropods for arboviruses?
   b. How is the cycle of the virus maintained in the wild?
   c. Describe the symptoms of the encephalitis type of infection (WEE, EEE) and the hemorrhagic fevers (yellow fever).
9. a. What are retroviruses, and how are they different from other viruses?
   b. Give examples of the three principal human retroviruses and the diseases they cause.
   c. Give a comprehensive definition of AIDS, and define categories of HIV disease as characterized by the CDC.
10. a. Briefly explain the activities most likely to spread AIDS and what factors increase the relative risk among certain populations.
    b. Explain why bone marrow transplants are a transmission concern.
    c. Discuss some ways that HIV has *not* been documented to spread.
11. a. Discuss whether AIDS is an all-or-none disease, or a series of gradual pathologic changes.
    b. Differentiate between HIV infection, HIV disease, and AIDS.
    c. What are the primary target tissues and changes?
    d. Explain why only certain T cells are targeted by HIV and what it does to them.
    e. How is the brain affected by infection?
    f. What major factors cause the long latency period in AIDS?
    g. What does it mean to be a non-progressor?
12. a. Describe the changes over time in virus antigen levels, antibody levels, and CD4 T cells in the blood of an HIV-infected individual.
    b. Relate these changes to the progress of the disease, the infectiousness of the person, and the effectiveness of the immune response to the virus.
13. a. List the major opportunistic bacterial, fungal, protozoan, and viral diseases used to define AIDS.
    b. Describe four other secondary diseases or conditions that accompany AIDS.
    c. Explain which effects HIV has that create vulnerability to these particular pathogens and the other conditions.
14. a. If a person is seropositive for HIV, does this mean that the patient has AIDS?
    b. What is the most sensible interpretation of a seronegative result for a monogamous person who has not had a blood transfusion or taken IV drugs?
    c. What is a logical interpretation of a seronegativity in a person who has participated in high-risk behavior?
    d. What is meant by the "window" regarding antibody presence in the blood?
15. a. How do AZT and other nucleotide analogs control the AIDS virus? How do protease inhibitors work?
    b. Are there any actual cures for AIDS?
    c. What is the rationale behind the new "drug cocktail" treatment?
    d. Explain several strategies for vaccine development; include in your answer the primary drawbacks to development of an effective vaccine.
16. a. Describe the epidemiology and progress of polio infection and disease.
    b. What causes the paralysis and deformity?
    c. Compare and contrast the two types of vaccines.
    d. What characteristics of enteric viruses cause them to be readily transmissible?
17. a. Describe some common ways that hepatitis A is spread.
    b. What are its primary effects on the body?
    c. How is it different from hepatitis B?
18. a. What is the common cold, and which groups of viruses can be involved in its transmission?
    b. Why is it so difficult to control?
    c. Discuss the similarities in the symptoms of influenza, parainfluenza, mumps, measles, rubella, RSV, and rhinoviruses.
19. a. For which of the RNA viruses are vaccines available?
    b. Which viruses will probably never have a vaccine developed for them and why?
    c. For which of the RNA viruses are there specific drug treatments?
20. Provide the formal genus name for each of the English vernacular names of viruses listed in multiple-choice matching questions 16 and 17.

CRITICAL-THINKING QUESTIONS

1. a. Explain the relationship between herd immunity and the development of influenzal pandemics.
   b. Why will herd immunity be lacking if antigenic shift or drift occurs?
2. a. Explain how the antibody content of a patient's sera can be used to predict which strain of influenza predominated during a previous epidemic.
   b. Explain the precise way that cross-species influenza infections give rise to new and different strains of virus.
3. Describe the possible connection between viral infection and Guillain-Barré and Reye's syndromes.
4. a. Why do infections such as mumps, measles, polio, rubella, and RSV regularly infect children and not adults?
   b. Measles is considered to be a highly contagious infection. Explain how it is possible to acquire measles from a person who has only been exposed to this infection.
5. AIDS is thought to have originated from simian viruses, and later spread from human populations to the rest of the world.
   a. Give some possible ways that monkeys may have transmitted viruses to humans.
   b. Can you think of a plausible series of events that could explain how it came to be so widespread?
   c. Explain why there can be such a discrepancy between the total number of reported cases of AIDS and the projected number of cases estimated by health authorities.
   d. Using the statistics in Figure 25.13, estimate the total number of cases occurring between 1981 and 1997 and show the percent decrease occurring since the peak incidence.
6. Trace the possible transmission pathways of HIV (source in first person, mode of infection in second person, portals of exit and entry) for each of the following:
   a. From one infected homosexual man to another homosexual man.
   b. From one drug user to another drug user.
   c. From a bisexual man to his wife.
   d. From a prostitute to a contact.
   e. From a monogamous man to his monogamous wife of 50 years.
   f. From one teenager to another.
   g. From a mother to a child.
7. a. It is often stated that full-blown AIDS is always fatal. Is this true? Support your answer.
   b. Provide a possible explanation for the people with AIDS who are long-term survivors or those with HIV infection who are non-progressors.
   c. Is it possible to become immune to HIV?
8. a. Explain the characteristics of the HIV virus and infection that make it such a difficult target for the immune system and for drugs.
   b. Why are coinfections with other microbes so devastating to AIDS patients?
9. a. Explain why a direct antigen test for HIV infection would be preferable to a serological test for antibodies.
   b. What factors have prevented the development of an effective AIDS vaccine?
   c. Give a plausible explanation for the fact that live, attenuated (or even dead) HIV vaccines pose a risk.
10. a. What precautions can a person take to prevent himself or herself from contracting HIV infection?
    b. How can a health care worker prevent possible infection?
11. a. If wild type polio had disappeared from the Western Hemisphere by 1991, how do you explain the 7 cases of polio reported in the United States in 1994?
    b. Provide an explanation for giving the Salk (IPV) vaccine for the first two doses, followed later by Sabin (OPV) vaccine.
12. Various over-the-counter cold remedies control or inhibit inflammation and depress the symptoms of colds. Is this beneficial or not? Support your answer.
13. a. Briefly compare the neurological syndromes caused by measles and AIDS with the slow progressive neurological diseases caused by viruslike agents.
    b. How are they different?
14. Case study 1. Three months after having corneal surgery, a patient reported to his physician with fever and numbness of the face. His throat was tight, and he had difficulty swallowing; he was afraid of showering because of the painful spasms it caused. Within hours, he became paralyzed, lapsed into a coma, and died. Physicians were stumped as to the cause. Can you offer some ideas? What therapy should have been given?
15. Case study 2. Late in the spring, a young man from rural Idaho developed fever, loss of memory, difficulty in speech, convulsions, and tremor and lapsed into a coma. He tested negative for bacterial meningitis and had no contact with dogs or cats. He survived but had long-term mental retardation. What are the possible diseases he might have had, and how would he have contracted them?
16. Case study 3. Biopsies from the liver and intestine of an otherwise asymptomatic 35-year-old male show masses of acid-fast bacilli throughout the tissue. Explain the pathology going on here and provide a preliminary diagnosis.

INTERNET SEARCH TOPICS

1. Look up polio on a search engine and determine the current progress in eradicating this disease.
2. Research the topic of the three-drug cocktail used in treating AIDS and the advantages and disadvantages.

chapter 26

Environmental and Applied Microbiology

The past dozen chapters have focused on the somewhat narrow field of pathogenic microbiology, necessarily overlooking many other interesting and useful aspects of microbiology and microorganisms. This last chapter emphasizes microbial activities that not only are beneficial to humans and other organisms but also help maintain and control the life support systems on the earth. This subject is explored from the standpoints of (1) the natural roles of microorganisms in the environment and their contributions to the ecological balance, including soil, water, and mineral cycles, and (2) the artificial applications of microbes in the food, medical, biochemical, drug, and agricultural industries.

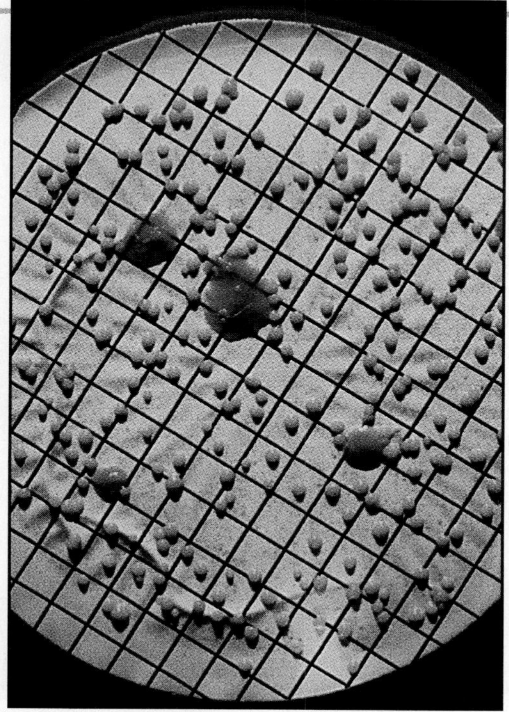

A membrane filter sample of water taken from a hospital tap cultured on media selective for *Mycobacterium*. The major isolate, *M. avium,* is a serious pathogen among AIDS patients. Unusual pathogens like this bacterium as well as protozoa and viruses are appearing in water supplies and indicate weaknesses in traditional treatment methods.

ECOLOGY: THE INTERCONNECTING WEB OF LIFE

The study of microbes in their natural habitats is known as **environmental,** or **ecological, microbiology;** the study of the practical uses of microbes in food processing, industrial production, and biotechnology is known as **applied microbiology.** Separating this chapter into two disciplines is an organizational convenience, since the two areas actually overlap to a considerable degree—largely because most natural habitats have been altered by human activities. Human intervention in natural settings has changed the earth's warming and cooling cycles, increased wastes in soil, polluted water, and altered some of the basic relationships between microbial, plant, and animal life. Now that humans are also beginning to release totally new, genetically recombined microbes into the environment and to alter the genes of plants, animals, and even themselves, what does the future hold? Although this question may be imponderable, we know one thing for certain: Microbes—the most vast and powerful resource of all—will be silently working in nature.

In chapter 7 we first touched upon the widespread distribution of microorganisms and their adaptations to most habitats of the world, from extreme to temperate. Regardless of their exact location or type of adaptation, microorganisms necessarily are exposed to and interact with their environment in complex and extraordinary ways. The science that studies interactions between microbes and their environment and the effects of those interactions on the earth is **microbial ecology.** Unlike studies that deal with the pathologic effects of a single organism or its individual characteristics in the laboratory, ecological studies are aimed at the interactions taking place between organisms and their environment at many levels at any given moment. Therefore, ecology is a broad-based science that merges many subsciences of biology as well as geology, chemistry, and engineering.

Ecological studies take into account both the biotic and the abiotic components of an organism's environment. **Biotic*** factors include any other living or once-living organisms[1] such as symbionts sharing an organism's habitat, parasites, or food substrates. **Abiotic*** factors include any nonliving surroundings such as the atmosphere, soil, water, temperature, and light. A collection of organisms together with its surrounding physical and chemical factors is defined as an **ecosystem** (figure 26.1).

THE ORGANIZATION OF ECOSYSTEMS

The earth initially may seem like a random, chaotic place, but it is actually an incredibly organized, well-tuned machine. Ecological relationships exist at several levels, ranging from the entire

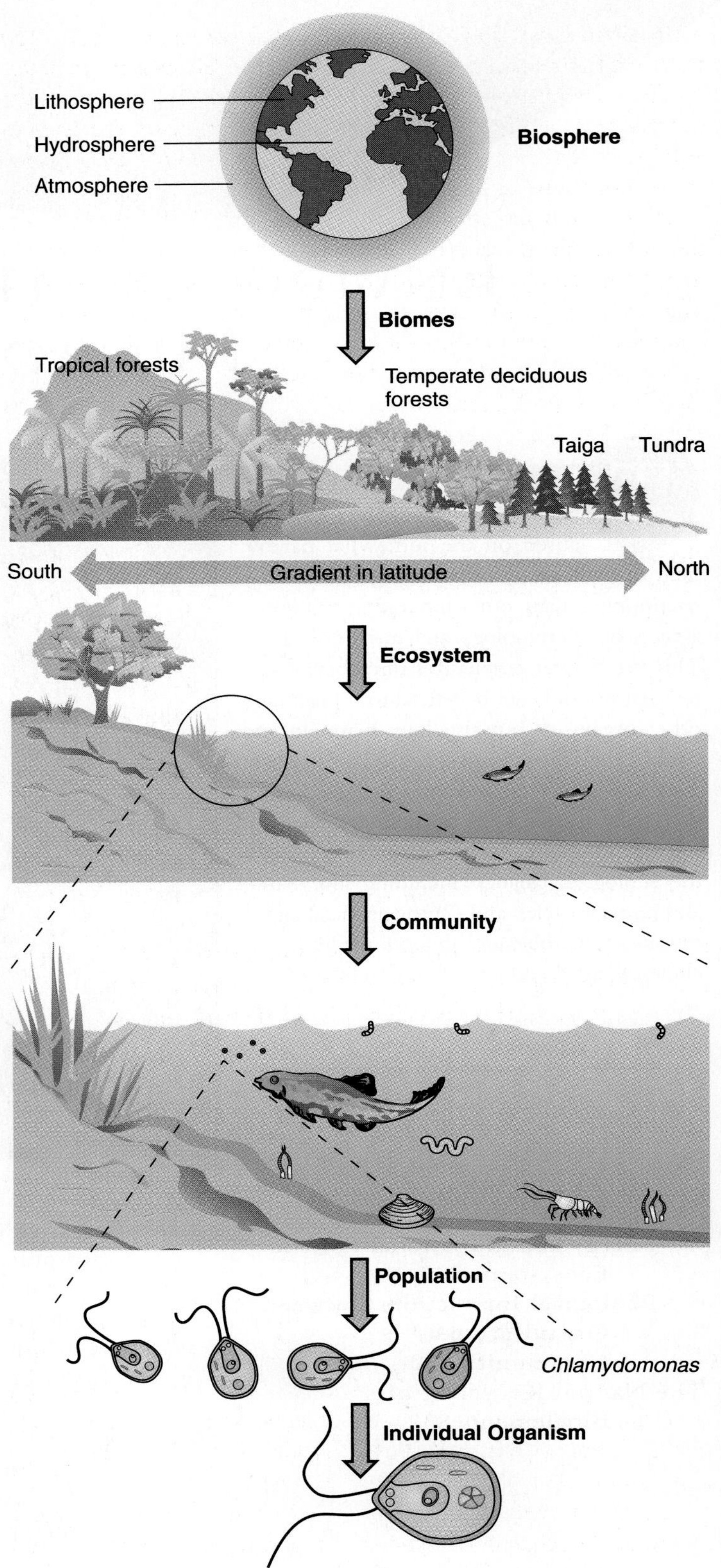

Figure 26.1

Levels of organization in an ecosystem, ranging from the biosphere to the individual organism.

1. Biologists make a distinction between nonliving and dead. A nonliving things has never been alive, whereas a dead thing was once alive but no longer is.

*biotic (by-ah′-tik)

*abiotic (ay″-by-ah′-tik)

earth all the way down to a single organism (figure 26.1). The most all-encompassing of these levels, the **biosphere,** includes the thin envelope of life (about 14 miles deep) that surrounds the earth's surface. This global ecosystem comprises the **hydrosphere** (water), the **lithosphere** (a few miles into the soil), and the **atmosphere** (a few miles into the air). The biosphere maintains or creates the conditions of temperature, light, gases, moisture, and minerals required for life processes. The biosphere can then be naturally subdivided into terrestrial and aquatic realms. The terrestrial realm is usually distributed into particular climatic regions called **biomes** (by′-ohmz), each of which is characterized by a dominant plant form, altitude, and latitude. Particular biomes include grassland, desert, mountain, and tropical rain forest. The aquatic biosphere is generally divisible into freshwater and marine realms.

Biomes and aquatic ecosystems are generally composed of mixed assemblages of organisms that live together at the same place and time and that usually exhibit well-defined nutritional or behavioral interrelationships. These clustered associations are called **communities.** Although most communities are identified by their easily visualized dominant plants and animals, they also contain a complex assortment of bacteria, fungi, algae, protozoa, and even viruses. The basic units of community structure are **populations,** groups of organisms of the same species. The organizational unit of a population is the individual organism, and each organism, in turn, has its own levels of organization (organs, tissues, cells).

Ecosystems are generally balanced, with each organism existing in its particular habitat and niche. The **habitat** is the physical location in the environment to which an organism has adapted. In the case of microorganisms, the habitat is frequently a *microenvironment,* where particular qualities of oxygen, light, or nutrient content are somewhat stable. The **niche** is the overall role that a species (or population) serves in a community. This includes such activities as nutritional intake (what it eats), position in the community structure (what eats it), and rate of population growth. A niche can be broad (such as scavengers that feed on nearly any organic food source) or narrow (microbes that decompose cellulose in forest litter or that fix nitrogen).

Energy and Nutritional Flow in Ecosystems

All living things must obtain nutrients and a usable form of energy from the abiotic and biotic environments. The energy and nutritional relationships in ecosystems can be described in a number of convenient ways. A **food chain** or **energy pyramid** provides a simple summary of the general trophic (feeding) levels, designated as producers, consumers, and decomposers, and traces the flow and quantity of available energy from one level to another (figure 26.2). It is worth noting that microorganisms are the only living things that exist at all three major trophic levels. The nutritional roles of microorganisms in ecosystems are summarized in table 26.1.

Life would not be possible without **producers,** because they provide the fundamental energy source that drives the

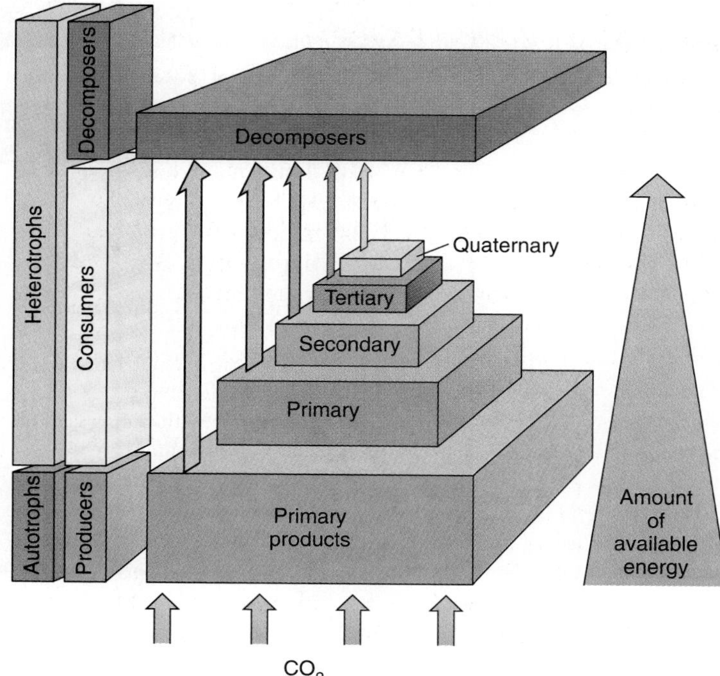

Figure 26.2
A trophic and energy pyramid. The relative size of the blocks indicates the number of individuals that exist at a given trophic level. The arrow on the right indicates the amount of usable energy from producers to top consumers. Both the number of organisms and the amount of usable energy decrease with each trophic level. Decomposers are an exception to this pattern, but only because they can feed from all trophic levels.

trophic pyramid. Producers are the only organisms in an ecosystem that can produce organic carbon compounds such as glucose by assimilating (fixing) inorganic carbon (CO_2) from the atmosphere. Such organisms are also termed autotrophs. Most producers are photosynthetic organisms, such as plants and cyanobacteria, that convert the sun's energy into the energy of chemical bonds. Photosynthesis was introduced in chapter 7, and the chemical steps in the process are covered with the carbon cycle later in this chapter. A small but important amount of CO_2 assimilation is brought about by unusual bacteria called lithotrophs. The metabolism of these organisms derives energy from oxidation-reduction reactions of simple inorganic compounds such as sulfides and hydrogen. In certain ecosystems (see thermal vents, microfile 7.8), lithotrophs are the sole supporters of the energy pyramid.

Consumers feed on other living organisms and obtain energy from bonds present in the organic substrates they contain. The category includes animals, protozoa, and a few bacteria and fungi. A pyramid usually has several levels of consumers, ranging from *primary consumers* (grazers or herbivores), which consume producers; to *secondary consumers* (carnivores), which feed on primary consumers; to *tertiary consumers,* which feed on secondary consumers; and up to *quaternary consumers* (usually the last level), which feed on tertiary consumers. Figure 26.3 shows specific organisms at these levels.

TABLE 26.1

THE MAJOR ROLES OF MICROORGANISMS IN ECOSYSTEMS

Role	Description of Activity	Examples of Microorganisms Involved
Primary producers	Photosynthesis	Algae, cyanobacteria, sulfur bacteria
	Chemosynthesis	Chemolithotrophic bacteria in thermal vents
Consumers	Predation	Free-living protozoa that feed on algae and bacteria; some fungi that prey upon nematodes
Decomposers	Degradation of plant and animal matter and wastes	Soil saprobes (primarily bacteria and fungi) that degrade cellulose, lignin, and other complex macromolecules
	Mineralization of organic nutrients	Soil bacteria that reduce organic compounds to inorganic compounds such as CO_2 and minerals
Cycling agents for biogeochemical cycles	Recycling compounds containing carbon, nitrogen, phosphorus, sulfur	Specialized bacteria that transform elements into different chemical compounds to keep them cycling from the biotic to the abiotic and back to the biotic phases of the biosphere
Parasites	Living and feeding on hosts	Viruses, bacteria, protozoa, fungi, and worms that play a role in population control

Decomposers, primarily microbes inhabiting soil and water, break down and absorb the organic matter of dead organisms, including plants, animals, and other microorganisms. Because of their biological function, decomposers are active at all levels of the food pyramid. Without this important nutritional class of saprobes, the biosphere would stagnate and die. The work of decomposers is to reduce organic matter into inorganic minerals and gases that can be cycled back into the ecosystem, especially for the use of primary producers. This process, also termed *mineralization,* is so efficient that almost all biological products can be reduced by some type of decomposer. Numerous bacterial species decompose cellulose and lignin, which are complex polysaccharides from plant cell walls that account for the vast bulk of detritus in soil and water. Both fungi and bacteria decompose such complex animal compounds as keratin and chitin.

The pyramid illustrates several limitations of ecosystems with regard to energy. Unlike nutrients, which can be passed among trophic levels, recycled, and reused, energy does not cycle. Maintenance of complex interdependent trophic relationships such as those shown in figure 26.3 requires a constant input of energy at the producer level. As energy is transferred to the next level, a large proportion (as high as 90%) of the energy will be lost in a form (primarily heat) that cannot be fed back into the system. Thus, the amount of energy available decreases at each successive trophic level. This energy loss also decreases the actual number of individuals that can be supported at each successive level.

A concrete image of a feeding pathway can be provided by a food chain. Although it is a somewhat simplistic way to describe a community feeding pattern, a food chain helps identify the specific types of organisms that are present at a given trophic level in a natural setting (figure 26.3*a*). Feeding relationships in communities are more accurately represented by a multichannel food chain, or a **food web** (figure 26.3*b*). A food web reflects the actual nutritional structure of a community. It can help to identify feeding patterns typical of herbivores (plant eaters), carnivores (flesh eaters), and omnivores (feed on both plants and flesh).

ECOLOGICAL INTERACTIONS BETWEEN ORGANISMS IN A COMMUNITY

Whenever complex mixtures of organisms associate, they develop various dynamic interrelationships based on nutrition and shared habitat. These relationships, some of which were described in chapter 7, include mutualism, commensalism, parasitism, competition, synergism (cross-feeding), predation, and scavenging (figure 26.4).

Mutually beneficial associations *(mutualism)* such as protozoans living in the termite intestine are so well evolved that the two members require each other for survival. In contrast, *commensalism* is one-sided and independent. Although the action of one microbe favorably affects another, the first microbe receives no benefit. Many commensal unions involve *cometabolism,* meaning that the waste products of the first microbe are useful nutrients for the second one. In *synergism,* two organisms that are usually independent cooperate to break down a nutrient neither one could have metabolized alone. *Parasitism* is an intimate relationship whereby a parasite derives its nutrients and habitat from a host that is usually harmed in the process. In *competition,* one microbe gives off antagonistic substances that inhibit or kill susceptible species sharing its habitat. A *predator* is a form of consumer that actively seeks out and ingests live prey (protozoa that prey on algae and bacteria). *Scavengers* are nutritional jacks-of-all-trades; they feed on a variety of food sources, ranging from live cells to dead cells and wastes.

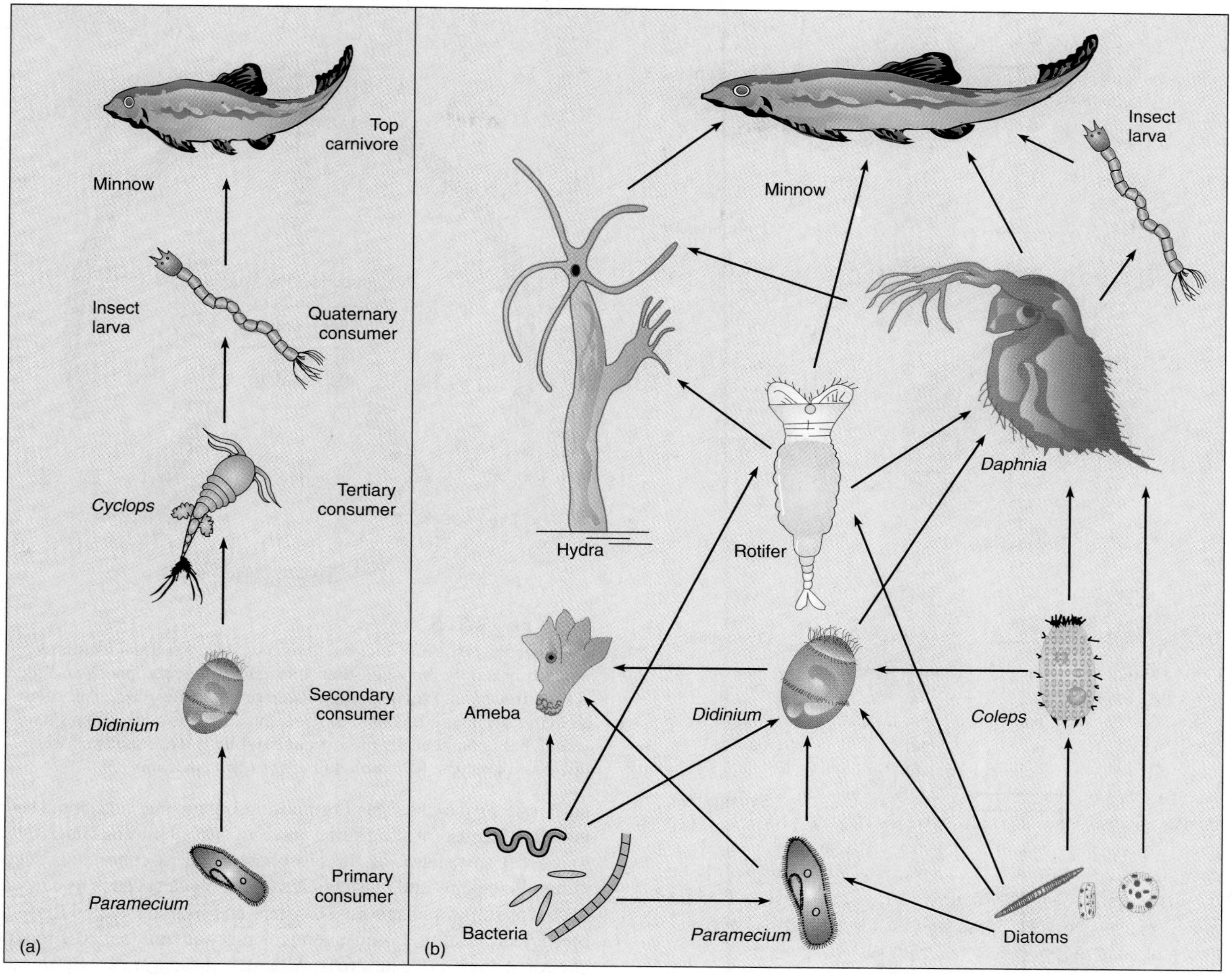

Figure 26.3
Comparison of a food chain and a food web in an aquatic ecosystem. (*a*) A food chain is the simplest way to present specific feeding relationships among organisms, but it may not reflect the total nutritional interactions in a community. (*b*) More complex trophic patterns are accurately depicted by a food web, which traces the multiple feeding options that exist for most organisms. Note: Arrows point toward the consumers.

Chapter Checkpoints

The study of ecology includes both living (biotic) and nonliving components (abiotic) of the earth. Microbial ecology includes ecological microbiology, the study of microbes in their natural habitats, and applied microbiology, their utilization for commercial purposes.

Ecosystems are organizations of living populations in specific habitats. Each ecosystem requires a continuous outside source of energy for survival and a nonliving habitat consisting of soil, water, and air.

A living community is composed of populations that show a pattern of energy and nutritional relationships called a food web. Microorganisms are essential producers and decomposers in any ecosystem.

The relationships between populations in a community are described according to the degree of benefit or harm they pose to one another. These relationships include mutualism, commensalism, predation, parasitism, synergism, scavenging, and competition.

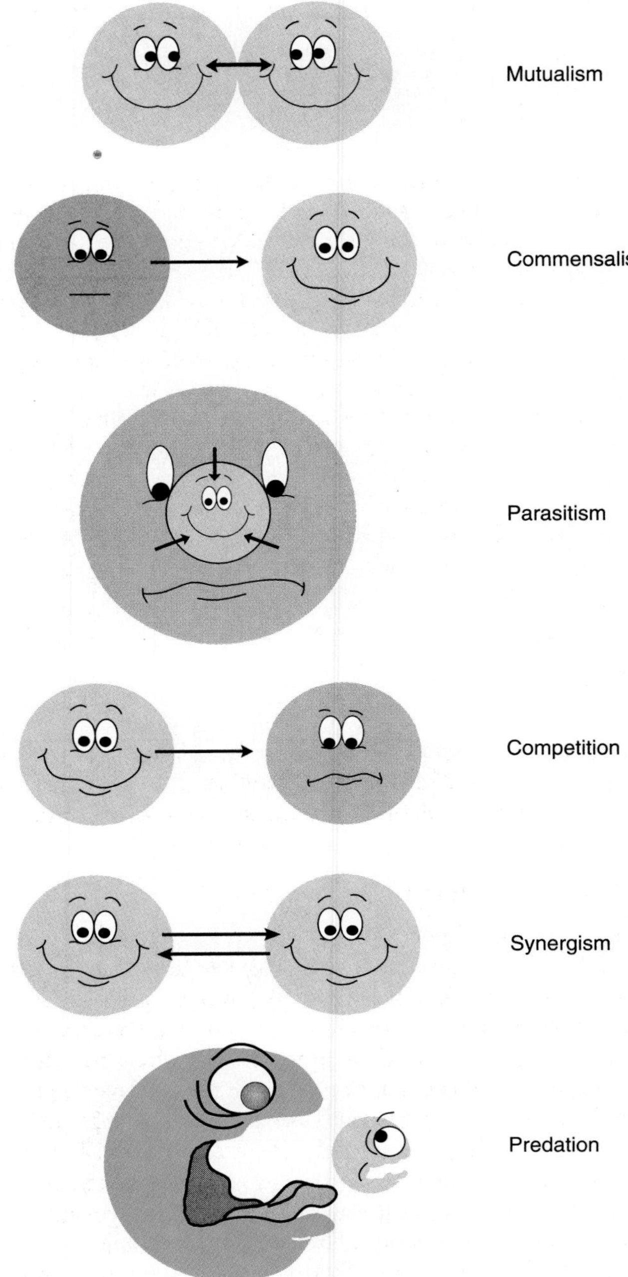

Figure 26.4
Community interactions that have beneficial, harmful, or neutral effects.

THE NATURAL RECYCLING OF BIOELEMENTS

Ecosystems are open with regard to energy because the sun is constantly infusing them with a renewable source. In contrast, the bioelements and nutrients that are essential components of protoplasm are supplied exclusively by sources somewhere in the biosphere and are not being continually replenished from outside the earth. In fact, the lack of a required nutrient in the immediate habitat is one of the chief factors limiting organismic and population growth. Because of the finite source of life's building blocks, the long-term sustenance of the biosphere requires continuous **recycling** of elements and nutrients. Essential elements such as carbon, nitrogen, sulfur, phosphorus, oxygen, and iron are cycled through biological, geologic, and chemical mechanisms called **biogeochemical cycles** (figure 26.5). Although these cycles vary in certain specific characteristics, they share several general qualities, as summarized in the following list:

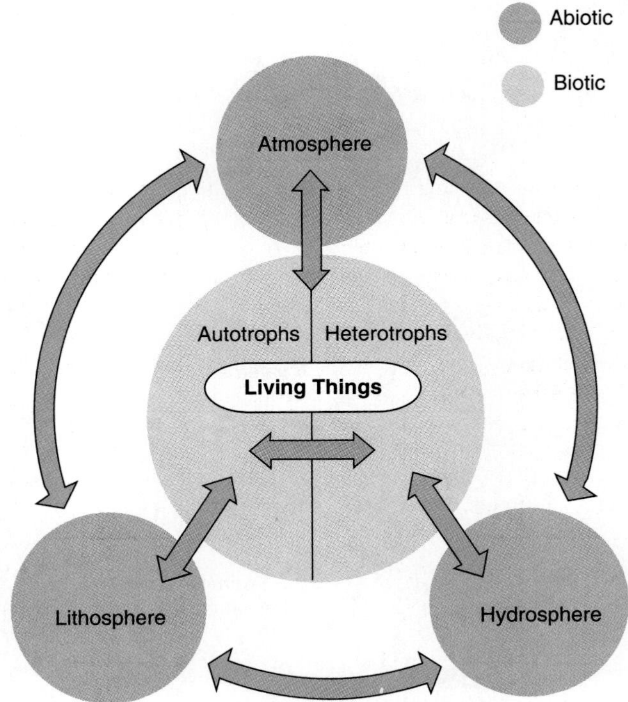

Figure 26.5
The general pattern of biogeochemical cycles. Essential elements constantly recycle between their biotic form in protoplasm and their abiotic (inorganic) form in soil, water, or the atmosphere. Although all living things are involved in these cycles, plants and animals have a somewhat general effect on recycling, while microorganisms act upon and alter the forms of elements in the environment.

- All elements ultimately originate from a nonliving, long-term reservoir in the atmosphere, sedimentary rocks, or water. They cycle in pure form (N_2) or as compounds (PO_4).
- Elements make the rounds between a free, inorganic state in the abiotic environment and a combined, organic state in the biotic environment.
- Recycling maintains a necessary balance of nutrients in the biosphere so that they do not build up or become unavailable.
- Cycles are complex systems that rely upon the interplay of producers, consumers, and decomposers. Often the waste products of one organism become a source of energy or building material for another.
- All organisms participate directly in recycling, but only certain categories of microorganisms have the metabolic pathways for converting inorganic compounds from one nutritional form to another.

MICROFILE 26.1 GAIA: IS THE BIOSPHERE A GIANT, SELF-SUSTAINING ORGANISM?

Reflecting on the remarkable qualities of the earth that support life, British scientist James Lovelock more than 20 years ago became convinced that those qualities could not have arisen out of random accidents. The conventional view had always been that the abiotic realms of the earth existed and evolved independently of living things and that life adapted to the abiotic realm rather than vice versa. But Lovelock proposed that the earth contains a diversity of habitats and niches favorable to life because living things have made it that way. In this view, the biosphere is like a superorganism that behaves as a single unit, is self-regulating, and maintains an equilibrium.

Lovelock called his somewhat controversial idea the **Gaia** (guy′-uh) **Hypothesis** after the mythical Greek goddess of earth. Some scientists have been put off by this unorthodox suggestion of an intent or "wisdom" on the part of the earth, as though it were an organism capable of thought. But Lovelock merely used a fanciful name to express a concept that is not really so incredible: Just as the earth shapes the character of living things, so do living things shape the character of the earth. After all, we know that the chemical compositions of the aquatic environment, the atmosphere, and even the soil would not exist as they do without the actions of living things. Organisms are also very active in the recycling of elements, evaporation and precipitation cycles, formation of mineral deposits, and rock weathering. The Gaia Hypothesis has gradually gained in acceptance as scientists have tested some of its predictions and discovered that many of its claims can be verified.

A wave of new research has uncovered profound evidence of the dominant role played by microbes in the formation and maintenance of the earth's crust, the development of rocks and minerals, and formation of fossil fuels. This revolution in understanding the biological mediation of geologic processes has given rise to a new field called geomicrobiology.

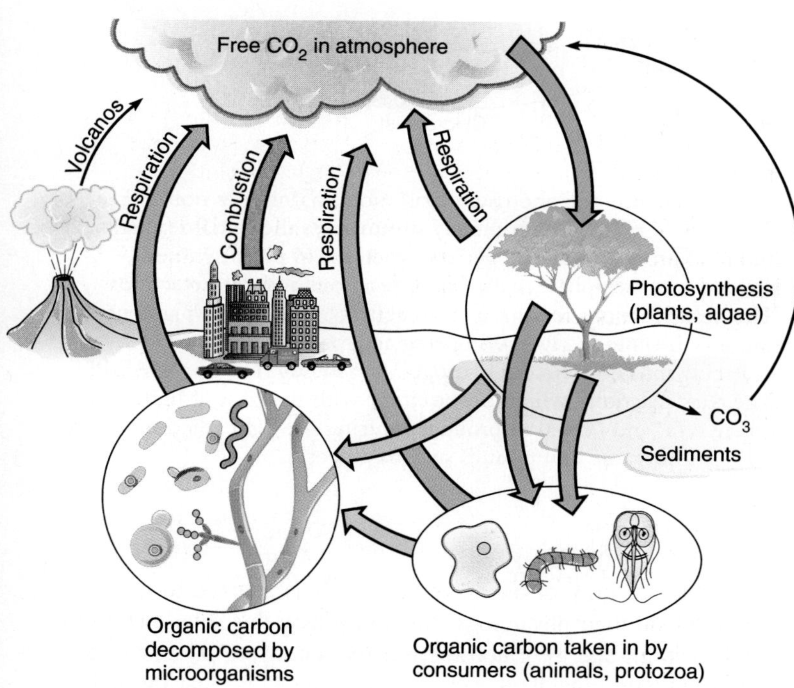

Figure 26.6

The carbon cycle. This cycle traces carbon from the CO_2 pool in the atmosphere to the primary producers (green), where it is fixed into protoplasm. Organic carbon compounds are taken in by consumers (blue) and decomposers (yellow) that produce CO_2 through respiration and return it to the atmosphere (pink). Combustion of fossil fuels and volcanic eruptions also add to the CO_2 pool. Some of the CO_2 is carried into inorganic sediments by organisms that synthesize carbonate (CO_3) skeletons. In time, natural processes acting on exposed carbonate skeletons can liberate CO_2.

In the next several sections we examine how, jointly and over a period of time, the varied microbial activities affect and are themselves affected by the abiotic environment (see also microfile 26.1).

ATMOSPHERIC CYCLES

The Carbon Cycle

Because carbon is the fundamental atom in all biomolecules and accounts for at least one-half of the dry weight of protoplasm, the **carbon cycle** is more intimately associated with the energy transfers and trophic patterns in the biosphere than are other elements. Besides the enormous organic reservoir in the bodies of organisms, carbon also exists in the gaseous state as carbon dioxide (CO_2) and methane (CH_4) and in the mineral state as carbonate (CO_3). In general, carbon is recycled through ecosystems via photosynthesis (carbon fixation), respiration, and fermentation of organic molecules, limestone decomposition, and methane production. A convenient starting point from which to trace the movement of carbon is with carbon dioxide, which occupies a central position in the cycle and represents a large common pool that diffuses into all parts of the ecosystem (figure 26.6). As a general rule, the cycles of oxygen and hydrogen are closely allied to the carbon cycle.

The principal users of the atmospheric carbon dioxide pool are photosynthetic autotrophs (phototrophs) such as plants, algae, and cyanobacteria. An estimated 165 billion tons of organic material per year are produced by terrestrial and aquatic photosynthesis. A smaller amount of CO_2 is used by chemosynthetic autotrophs such as methane bacteria. A review of the general equation for photosynthesis in figure 26.7*a* will reveal that phototrophs use energy from the sun to fix CO_2 into organic compounds such as glucose that can be used in synthesis and respiration. Photosynthesis is also the primary means by which the atmospheric supply of O_2 is regenerated.

Just as photosynthesis removes CO_2 from the atmosphere, other modes of generating energy, such as respiration and fermentation, return it. Recall in the general equation for aerobic respiration (chapter 8) that, in the presence of O_2, organic compounds such as glucose are degraded completely to CO_2 and H_2O, with the release of energy. Carbon dioxide is also released by anaerobic respiration and certain types of fermentation reactions. Other sources of CO_2 are heterotrophic organisms, including consumers, decomposers, and phototrophs, which must respire in the absence of light.

A small but important phase of the carbon cycle involves certain limestone deposits composed primarily of calcium carbonate ($CaCO_3$). Limestone is produced when marine organisms such as mollusks, corals, protozoans, and algae form hardened shells by combining carbon dioxide and calcium ions from the surrounding water. When these organisms die, the durable skeletal components accumulate in marine deposits. As these immense deposits are gradually exposed by geologic upheavals or receding ocean levels, various decomposing agents liberate CO_2 and return it to the CO_2 pool of the water and atmosphere.

The complementary actions of photosynthesis and respiration along with other natural CO_2-releasing processes such as limestone erosion and volcanic activity have maintained a relatively stable atmospheric pool of carbon dioxide. Recent figures show that this balance is being disturbed as humans burn *fossil fuels* and other organic carbon sources. Fossil fuels, including coal, oil, and natural gas, were formed over millions of years through the combined actions of microbes and geologic forces and so are actually a part of the carbon cycle. Humans are so dependent upon this energy source that within the past 25 years, the proportion of CO_2 in the atmosphere has steadily increased from 32 ppm to 36 ppm. Although this increase may seem slight and insignificant, many experts now feel it has the potential to profoundly disrupt the delicate temperature balance of the biosphere (microfile 26.2).

Compared with carbon dioxide, **methane** (CH_4) plays a secondary part in the carbon cycle, though it can be a significant product in anaerobic ecosystems dominated by **methanogens** (methane producers). In general, when methanogens reduce CO_2 by means of various oxidizable substrates, they give off CH_4. Lithotrophic methanogens such as *Methanobacterium* and *Methanococcus* use H_2 as an oxidizing agent, and heterotrophic ones use organic acids such as formate or acetate. The practical applications of methanogens are covered in a subsequent section on sewage treatment, and their contribution to the greenhouse effect is discussed in microfile 26.2.

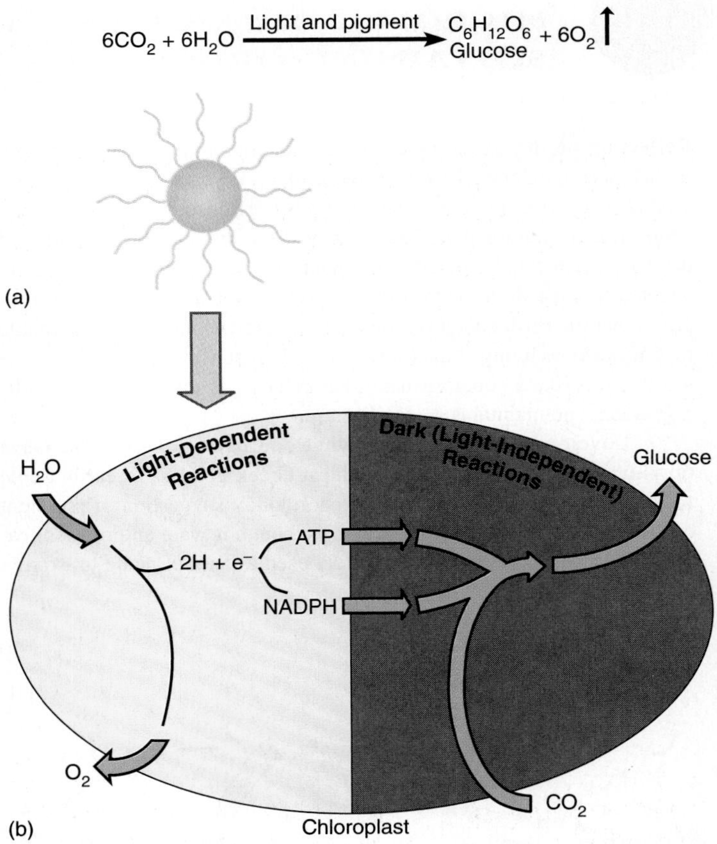

Figure 26.7
(*a*) The equation for photosynthesis. Note that it does not indicate the order of reactions but merely summarizes the initial reactants and final products. (*b*) The general reactions of photosynthesis, divided into two phases called light reactions and dark reactions. The light reactions require light to activate chlorophyll pigment and use the energy given off during activation to split an H_2O molecule into oxygen and hydrogen, producing ATP and NADPH. The dark reactions, which occur either with or without light, utilize ATP and NADPH produced during the light reactions to fix CO_2 into organic compounds such as glucose.

Photosynthesis: The Earth's Lifeline

With few exceptions, the energy that drives all life processes comes from the sun, but this source is directly available only to the cells of photosynthesizers. In the terrestrial biosphere, green plants are the primary photosynthesizers, and in aquatic ecosystems, where 80–90% of all photosynthesis occurs, green, golden-brown, brown, and red algae, as well as cyanobacteria, fill this role. Other photosynthetic procaryotes are green sulfur, purple sulfur, and purple nonsulfur bacteria.

The anatomy of photosynthetic cells is adapted to trapping sunlight, and their physiology effectively uses this solar energy to produce high-energy glucose from low-energy CO_2 and water. Photosynthetic organisms achieve this remarkable feat through a series of reactions, given in the following summary.

- trapping visible light rays by means of special pigments attached to intracellular membranes;
- exciting atoms in the pigment molecule and release of electrons;
- splitting water to release oxygen and hydrogen ions;
- producing ATP with energy extracted from electrons and hydrogen ions; and
- fixing CO_2 into carbohydrates using ATP.

MICROFILE 26.2 GREENHOUSE GASES, FOSSIL FUELS, COWS, TERMITES, AND GLOBAL WARMING

The sun's radiant energy does more than drive photosynthesis; it also helps maintains the stability of the earth's temperature and climatic conditions. As radiation impinges on the earth's surface, much of it is absorbed, but a large amount of the infrared (heat) radiation bounces back into the upper levels of the atmosphere. For billions of years, the atmosphere has been insulated by a layer of gases (primarily CO_2, CH_4, water vapor, and nitrous oxide, N_2O) formed by natural processes such as respiration, decomposition, and biogeochemical cycles. This layer traps a certain amount of the reflected heat yet also allows some of it to escape into space. As long as the amounts of heat entering and leaving are balanced, the mean temperature of the earth will not rise or fall in an erratic or life-threatening way. Although this phenomenon, called the **greenhouse effect,** is popularly viewed in a negative light, it must be emphasized that its function for eons has been primarily to foster life.

The greenhouse effect has recently been a matter of concern because *greenhouse gases* appear to be increasing at a rate that could disrupt the temperature balance. In effect, a denser insulation layer will trap more heat energy and gradually heat the earth. The amount of CO_2 released collectively by respiration, anaerobic microbial activity, fuel combustion, and volcanic activity has increased more than 30% since the beginning of the industrial era. By far the greatest increase in CO_2 production results from human activities such as combustion of fossil fuels, burning forests to clear agricultural land, and manufacturing. Deforestation has the added impact of removing large areas of photosynthesizing plants that would otherwise consume some of the CO_2.

Originally, experts on the greenhouse effect were concerned primarily about increasing CO_2 levels, but it now appears that the other greenhouse gases combined may have a slightly greater contribution than CO_2, and they, too, are increasing. One of these gases, methane (CH_4), released from the gastrointestinal tract of ruminant animals such as cattle, goats, and sheep, has doubled over the past century. The gut of termites also harbors wood-digesting and methane-producing bacteria. Even the human intestinal tract can support methanogens. Other greenhouse gases such as nitrous oxide and sulfur dioxide (SO_2) are also increasing through automobile and industrial pollution.

There is not yet complete agreement as to the extent and effects of global warming. It has been documented that the mean temperature of the earth has increased by 1.6°C since 1860. If the rate of increase continues, by 2020 a rise in the average temperature of 4° to 5°C will begin to melt the polar ice caps and raise the levels of the ocean 2 to 3 feet, but some experts predict more serious effects, including massive flooding of coastal regions, changes in rainfall patterns, expansion of deserts, and long-term climatic disruptions. Early warning signs of global warming are appearing in the Antarctic, where the land mass is breaking up at an increased rate, and in the mass melting of glaciers in many other parts of the world.

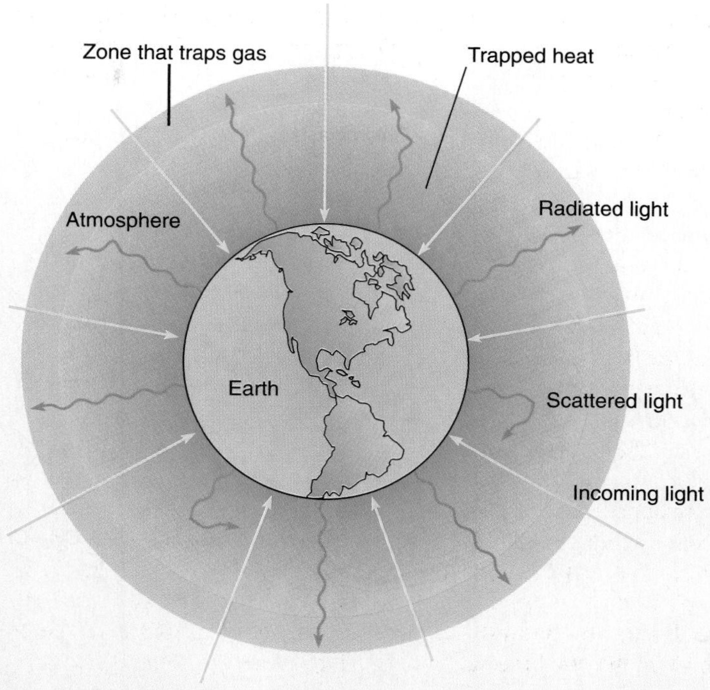

The greenhouse effect. Some of the light radiated or reflected from the earth's surface is trapped in a dense layer of atmospheric gases. Heat radiation trapped in this zone helps maintain the temperature of the atmosphere. Any increase in density of this gas layer could lead to increased entrapment of heat and a gradual rise in the ambient temperature of the earth.

Photosynthesis proceeds in two phases: the **light-dependent reactions,** which proceed only in the presence of sunlight, and the **dark** (light-independent) **reactions,** which proceed regardless of the lighting conditions (light or dark). The summary equation and overall reactions are depicted in figure 26.7.

The Functions of Green Machines Solar energy is delivered in discrete energy packets called **photons** (also called quanta) that travel as waves. The wavelengths of light operating in photosynthesis occur in the visible spectrum between 400 nm (violet) and 700 nm (red). As this light strikes photosynthetic pigments, some wavelengths are absorbed, some pass through, and some are reflected. The activity that has greatest impact on photosynthesis is the absorbance of light by photosynthetic pigments. These include the **chlorophylls,** which are green; **carotenoids,** which are yellow, orange, or red; and **phycobilins,** which are red or blue-green.[2] By far the most important of these pigments are the chlorophylls, which contain a *photocenter* that consists of a magnesium atom held in the center of a complex ringed structure

2. The color of the pigment corresponds to the wavelength of light it reflects.

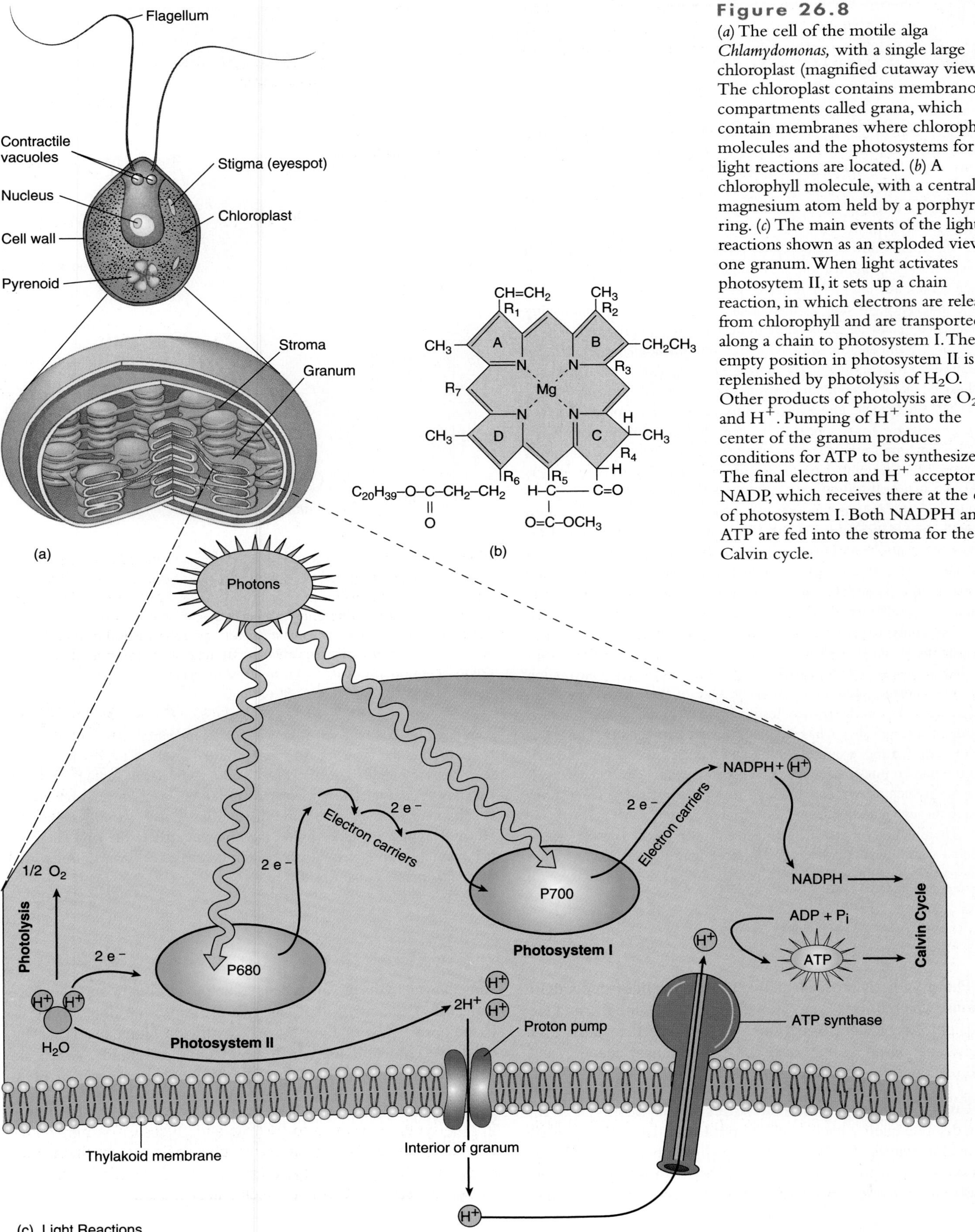

Figure 26.8
(*a*) The cell of the motile alga *Chlamydomonas,* with a single large chloroplast (magnified cutaway view). The chloroplast contains membranous compartments called grana, which contain membranes where chlorophyll molecules and the photosystems for the light reactions are located. (*b*) A chlorophyll molecule, with a central magnesium atom held by a porphyrin ring. (*c*) The main events of the light reactions shown as an exploded view in one granum. When light activates photosytem II, it sets up a chain reaction, in which electrons are released from chlorophyll and are transported along a chain to photosystem I. The empty position in photosystem II is replenished by photolysis of H_2O. Other products of photolysis are O_2 and H^+. Pumping of H^+ into the center of the granum produces conditions for ATP to be synthesized. The final electron and H^+ acceptor is NADP, which receives there at the end of photosystem I. Both NADPH and ATP are fed into the stroma for the Calvin cycle.

called a *porphyrin*. As we will see, the chlorophyll molecule harvests the energy of photons and converts it to electron (chemical) energy. The accessory photosynthetic pigments such as carotenes trap light energy and shuttle it to chlorophyll.

The detailed biochemistry of photosynthesis is beyond the scope of this text, but we will provide an overview to account for the primary process as it occurs in green plants, algae, and cyanobacteria (figure 26.8). The major reactants in light reactions are chlorophyll, photons, and H_2O and the products are ATP, NADPH, and O_2. The reactants in dark reactions are CO_2, ATP, and NADPH, and the products are glucose and other carbohydrates. Many of the basic activities (electron transport and phosphorylation) are biochemically similar to certain pathways of respiration previously covered in chapter 8.

Light Reactions The same systems that carry the photosynthetic pigments are also the sites for the light reactions. They occur in the *thylakoid* membranes of the granum in chloroplasts (figure 26.8*a*) and in specialized lamellae of the cell membranes in procaryotes (see chapter 4). These systems exist as two separate complexes called *photosystem I* (P700) and *photosystem II* (P680)[3] (figure 26.8*c*). Both systems contain chlorophyll and they are simultaneously activated by light, but the reactions in photosystem II help drive photosystem I. Together the systems form a powerful scheme to receive light, transport electrons, pump hydrogen ions, and form ATP and NADPH.

When photons enter the photocenter of the P680 system (PS II), the magnesium atom in chlorophyll becomes excited and releases 2 electrons. The loss of electrons from the photocenter has two major effects: (1) It creates a vacancy in the chlorophyll molecule forceful enough to split an H_2O molecule into hydrogens (H^+) (electrons and hydrogen ions) and oxygen (O_2). This splitting of water, termed **photolysis,** is the ultimate source of the O_2 gas that is an important product of photosynthesis. The electrons released from the lysed water regenerate the photosystem for its next reaction with light. (2) Electrons generated by the first photoevent are immediately boosted through a series of carriers (cytochromes) to the P700 system. At this same time, hydrogen ions accumulate in the internal space of the thylakoid complex, thereby producing an electrochemical gradient.

The P700 (PS I) system has been activated by light so that it is ready to accept electrons generated by the PS II. The electrons it receives are passed along a second transport chain to a complex that uses electrons and hydrogen ions to reduce NADP to NADPH. (Recall that reduction in this sense entails the addition of electrons and hydrogens to a substrate). As noted in chapter 8, a reduced coenzyme such as NADPH is another source of chemical energy and can be used in synthesis.

A second energy reaction involves synthesis of ATP by a chemiosmotic mechanism similar to that shown in figure 8.23. Channels in the thylakoids of the granum actively pump H^+ into the inner chamber, producing a charge gradient. ATP synthase located in this same thylakoid uses the energy from H^+ transport to phosphorylate ADP to ATP. Because it occurs in light, this process is termed **photophosphorylation.** Both NADPH and ATP are released into the stroma of the chloroplast, where they participate in the Calvin cycle.

3. The numbers refer to the wavelength of light to which each system is most sensitive.

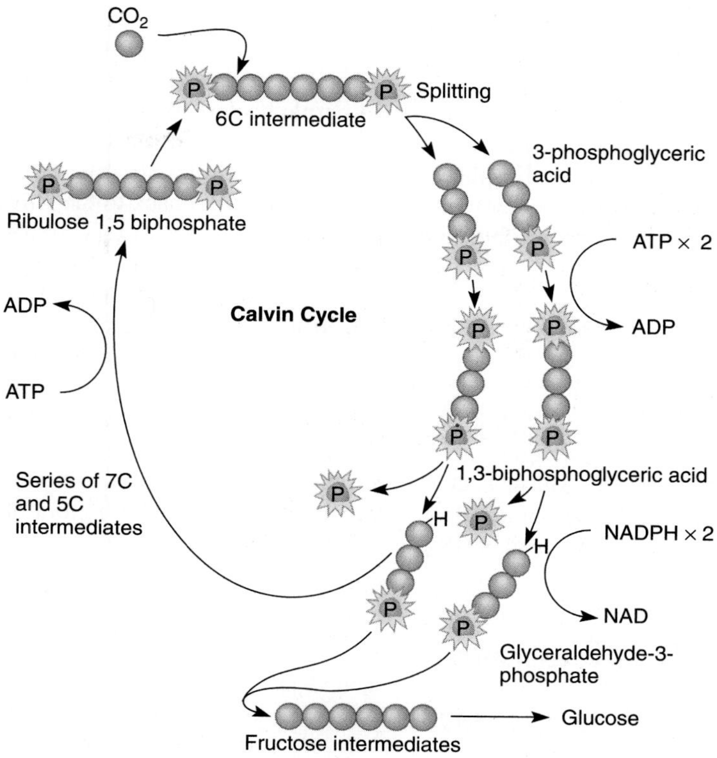

Figure 26.9
The main events of the reactions that do not require light, also called the Calvin cycle. It is during this cycle that carbon is fixed into organic form using the energy (ATP and NADPH) released by the light reactions. The end product, glucose, can be stored as complex carbohydrates, or it can be used in various amphibolic pathways to produce other carbohydrate intermediates or amino acids.

Dark Reactions The subsequent photosynthetic reactions that do not require light occur in the chloroplast stroma or the cytoplasm of cyanobacteria. These dark reactions use energy produced by the light phase to synthesize glucose by means of a cyclic scheme called the *Calvin cycle* (figure 26.9). The cycle begins at the point where CO_2 is combined with a doubly phosphorylated 5-carbon acceptor molecule called ribulose 1,5 biphosphate (RuBP). This process, called **carbon fixation,** generates a 6-C intermediate compound that immediately splits into 2,3 carbon molecules of 3-phosphoglyceric acid (PGA). The subsequent steps use the ATP and NADPH generated by the photosystems to form high-energy intermediates. First, ATP adds a second phosphate to 3-PGA and produces 1,3-biphosphoglyceric acid (BPG). Then, during the same step, NADPH contributes its hydrogen to BPG and one high-energy phosphate is removed. These events give rise to glyceraldehyde-3-phosphate (PGAL). This molecule and its isomer dihydroxyacetone phosphate (DHAP) are key molecules in hexose synthesis leading to fructose and glucose. You may notice that this pathway is very similar to glycolysis, except that it runs in reverse. Bringing the cycle back to regenerate RuBP requires PGAL and several steps not depicted in figure 26.9.

Other Mechanisms of Photosynthesis The **oxygenic,** or oxygen-releasing, photosynthesis that occurs in plants, algae, and cyanobacteria is the dominant type on the earth. Other photosynthesizers such as green and purple bacteria possess bacteriochlorophyll, which is more versatile in capturing light. They have only a cyclic photosystem I, which routes the electrons from the photocenter to the electron carriers and back to the photosystem again. This pathway generates a relatively small amount of ATP, and it may not produce NADPH. As photolithotrophs, these bacteria use H_2, H_2S, or elemental sulfur rather than H_2O as a source of electrons and reducing power. As a consequence, they are **anoxygenic** (non-oxygen-producing), and many are strict anaerobes. Bacteria that use sulfur compounds such as H_2S deposit sulfur residues either intracellularly (*Chromatium,* figure 4.35) or extracellularly, depending on the species. The sulfur bacteria are typically isolated from oxygen-deficient environments, such as mud and water at the bottom of shallow ponds and lakes, where they play a central role in the biogeochemical cycle of sulfur.

The Nitrogen Cycle

Nitrogen (N_2) gas is the most abundant component of the atmosphere, accounting for nearly 79% of air volume. As we will see, this extensive reservoir in the air is largely unavailable to most organisms. Only about 0.03% of the earth's nitrogen is combined (or fixed) in some other form such as nitrates (NO_3), nitrites (NO_2), ammonium (NH_4), and organic nitrogen compounds (proteins, nucleic acids).

The **nitrogen cycle** is relatively more intricate than other cycles because it involves such a diversity of specialized microbes to maintain the flow of the cycle. In many ways, it is actually more of a nitrogen "web" because of the array of adaptations that occur. Higher plants can utilize NO_3^- and NH_4^+; animals must receive nitrogen in organic form from plants or other animals; and microorganisms vary in their source, using NO_2^-, NO_3^-, NH_4^+, N_2, and organic nitrogen. The cycle includes four basic types of reactions: nitrogen fixation, ammonification, nitrification, and denitrification (figure 26.10).

Nitrogen Fixation The biosphere is most dependent upon the only process that can remove N_2 from the air and convert it to a form usable by living things. This process, called **nitrogen fixation,** is the beginning step in the synthesis of virtually all nitrogenous compounds. Nitrogen fixation is brought about primarily by nitrogen-fixing bacteria in soil and water, though a small amount is formed through nonliving factors such as lightning. The nitrogen fixers have developed a unique enzyme system capable of breaking the triple bonds of the N_2 molecule and reducing the N atoms, an anaerobic process that requires the expenditure of considerable ATP. The primary product of nitrogen fixation is the ammonium ion, NH_4^+. Nitrogen-fixing bacteria live free or in a symbiotic relationship with plants. Among the common free-living nitrogen fixers are the aerobic *Azotobacter* and *Azospirillum* and certain members of the anaerobic genus *Clostridium.* Other free-living nitrogen fixers are the cyanobacteria *Anabaena* and *Nostoc.*

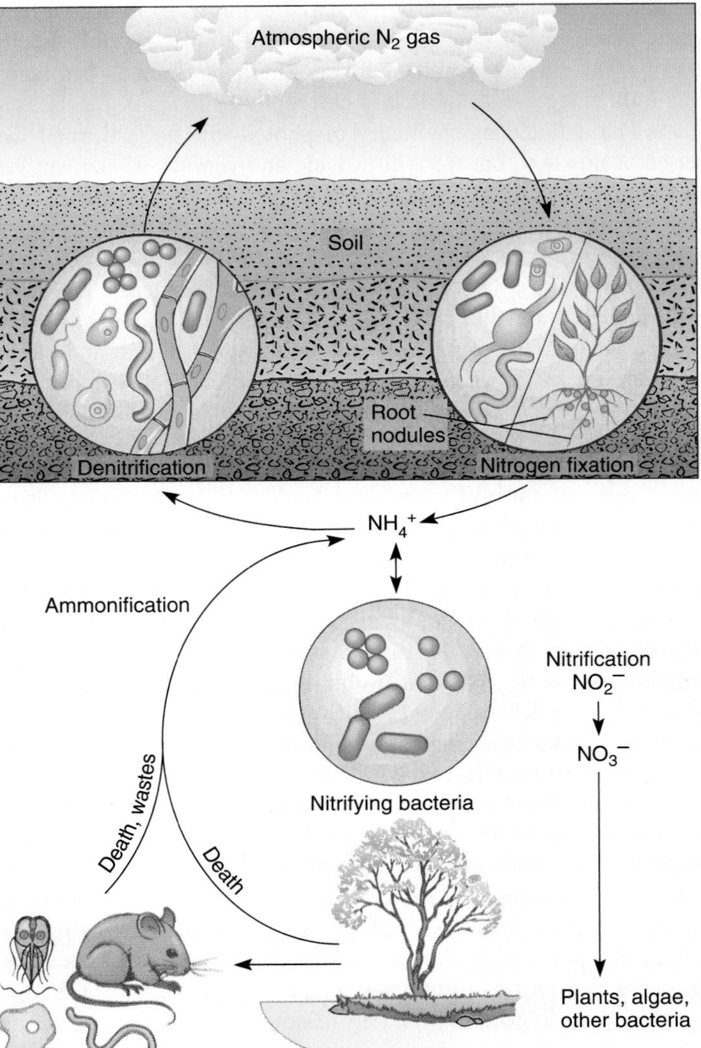

Figure 26.10

The simplified events in the nitrogen cycle. In nitrogen fixation, gaseous nitrogen (N_2) is acted on by nitrogen-fixing bacteria, which give off ammonia (NH_4). Ammonia is converted to nitrite (NO_2^-) and nitrate (NO_3^-) by nitrifying bacteria in nitrification. Plants, algae, and bacteria use nitrates to synthesize nitrogenous organic compounds (proteins, amino acids, nucleic acids). Organic nitrogen compounds are used by animals and other consumers. In ammonification, nitrogenous macromolecules from wastes and dead organisms are converted to NH_4 by ammonifying bacteria. NH_4 can be either directly recycled into nitrates or returned to the atmospheric N_2 form by denitrifying bacteria (denitrification).

Root Nodules: Natural Fertilizer Factories A significant symbiotic association occurs between *rhizobia** (bacteria in the genera *Rhizobium, Bradyrhizobium,* and *Azorhizobium)* and *legumes** (plants such as soybeans, peas, alfalfa, and clover that characteristically produce seeds in pods). The infection of

*rhizobia (ry-zoh′-bee-uh) Gr. *rhiza,* root, and *bios,* to live.

*legume (leg′-yoom) L. *legere,* to gather.

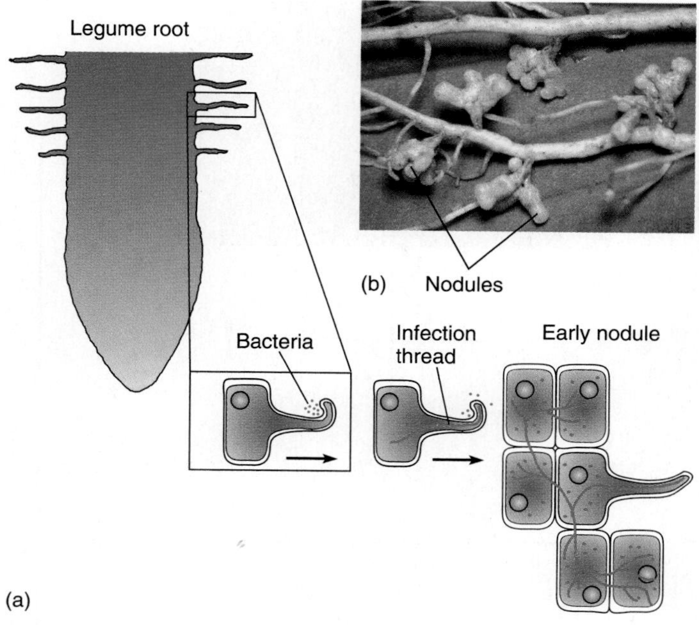

Figure 26.11

Nitrogen fixation through symbiosis. (*a*) Events leading to formation of root nodules. Cells of the bacterium *Rhizobium* attach to a legume root hair and cause it to curl. Invasion of the legume root proper by *Rhizobium* initiates the formation of an infection thread that spreads into numerous adjacent cells. The presence of bacteria in cells causes nodule formation. (*b*) Mature nodules that have developed in a sweet clover plant.

legume roots by these gram-negative, motile, rod-shaped bacteria causes the formation of special nitrogen-fixing organs, called **root nodules** (figure 26.11). Nodulation begins when rhizobia colonize specific sites on root hairs. From there, the bacteria invade deeper root cells and induce the cells to form tumorlike masses. The bacterium's enzyme system supplies a constant source of reduced nitrogen to the plant, and the plant furnishes nutrients and energy for the activities of the bacterium. The legume uses the NH_4 to aminate (add an amino group to) various carbohydrate intermediates and thereby synthesize amino acids and other nitrogenous compounds that are used in plant and animal synthesis (see figure 8.28).

Plant-bacteria associations have great practical importance in agriculture, because an available source of nitrogen is often a limiting factor in the growth of crops. The self-fertilizing nature of legumes makes them valuable food plants in areas with poor soils and in countries with limited resources. It has been shown that crop health and yields can be improved by inoculating legume seeds with pure cultures of rhizobia, because the soil is often deficient in the proper strain for forming nodules (figure 26.12). In the past several years, the idea for converting regular, nonleguminous plants into nitrogen fixers has also been explored. Some researchers attempted to genetically engineer various plants by inserting genes from *Rhizobium* into them, but the plants could not be induced to form nodules or to fix nitrogen. A newer, more promising approach has been to artificially fuse *Rhizobium* to the roots of such plants as rice, rapeseed, and wheat. The nodules they form appear to produce modest amounts of NH_4.

Figure 26.12

Inoculating legume seeds with *Rhizobium* bacteria increases the plant's capacity to fix nitrogen. The legumes in (*a*) were inoculated and are healthy. The poor growth and yellowish color of the uninoculated legumes in (*b*) indicate a lack of nitrogen.

Ammonification, Nitrification, and Denitrification

In the next phase of the nitrogen cycle, nitrogen-containing organic matter is decomposed by various bacteria *(Clostridium, Proteus* for example) that live in the soil and water. Organic detritus consists of large amounts of protein and nucleic acids from dead organisms and nitrogenous animal wastes such as urea and uric acid. The decomposition of these substances splits off amino groups and produces NH_4^+. This process is thus known as **ammonification.** The ammonium released can be reused by certain plants or converted to other nitrogen compounds.

The oxidation of ammonium to NO_2^- and NO_3^- is a process called **nitrification.** This reaction occurs in two phases and involves lithotrophic bacteria in soil and water. In the first phase, certain gram-negative genera such as *Nitrosomonas* and *Nitrosococcus* oxidize NH_4^+ to NO_2^- as a means of generating energy. Nitrite is not useful to most microbes and is rapidly acted upon by a second group of nitrifiers, including *Nitrobacter* and *Nitrococcus,* which perform the final oxidation of NO_2^- to NO_3^-. Nitrates can be assimilated into protoplasm by a variety of organisms (plants, fungi, and bacteria); a large number of bacteria also use it as a source of oxygen.

The nitrogen cycle is complete when nitrogen compounds are returned to the reservoir in the air by a reaction series that converts NO_3^- through intermediate steps to atmospheric nitrogen. The first step, which involves the reduction of nitrate to nitrite, is so common that hundreds of different bacterial species can do it. Other bacteria such as *Bacillus, Pseudomonas, Spirillum,* and *Thiobacillus* can carry out this **denitrification process** to completion as follows:

$$NO_3^- \rightarrow NO_2^- \rightarrow NO \rightarrow N_2O \rightarrow N_2$$

SEDIMENTARY CYCLES

The Sulfur Cycle

The sulfur cycle resembles the nitrogen cycle except that sulfur originates from natural sedimentary deposits in rocks, oceans, lakes, and swamps rather than the atmosphere. Sulfur exists in the elemental form (S) and as hydrogen sulfide gas (H_2S), sulfate (SO_4), and thiosulfate (S_2O_3). Most of the oxidations and reductions that convert one form of inorganic sulfur compound to another are accomplished by bacteria. Plants and many microorganisms can assimilate only SO_4, and animals must have an organic source. Organic sulfur occurs in the amino acids cystine, cysteine, and methionine, which contain sulfhydryl (—SH) groups and form disulfide (S—S) bonds that contribute to the stability and configuration of proteins.

One of the most remarkable contributors to the cycling of sulfur in the biosphere is the genus *Thiobacillus.** These gram-negative, motile rods flourish in mud, sewage, bogs, mining drainage, and brackish springs that can be inhospitable to organisms that require complex organic nutrients. But the metabolism of these specialized lithotrophic bacteria is adapted to extracting energy by oxidizing elemental sulfur, sulfides, and thiosulfate. One species, *T. thiooxidans,* is so efficient at this process that it secretes large amounts of sulfuric acid into its environment, as shown by the following equation:

$$Na_2S_2O_3 + H_2O + O_2 \rightarrow Na_2SO_4 + H_2SO_4 \text{ (sulfuric acid)} + 4S$$

The marvel of this bacterium is its ability to create and survive in the most acidic habitats on the earth. It plays an essential part in the phosphorus cycle, and its relative, *T. ferrooxidans,* participates in the cycling of iron. Other bacteria that can oxidize sulfur to sulfates are the photosynthetic sulfur bacteria mentioned in the section on photosynthesis.

The sulfates formed from oxidation of sulfurous compounds are assimilated into protoplasm by a wide variety of organisms. The sulfur cycle reaches completion when inorganic and organic sulfur compounds are reduced. Bacteria in the genera *Desulfovibrio* and *Desulfuromonas* anaerobically reduce sulfates to hydrogen sulfide and water as the final step in electron transport. Sites in mud and moist soil where they live usually emanate a strong, rotten-egg stench from H_2S and have a blackened film of iron on the surface. Organic sulfur in amino acids can also be reduced to H_2S by a number of saprobic bacteria.

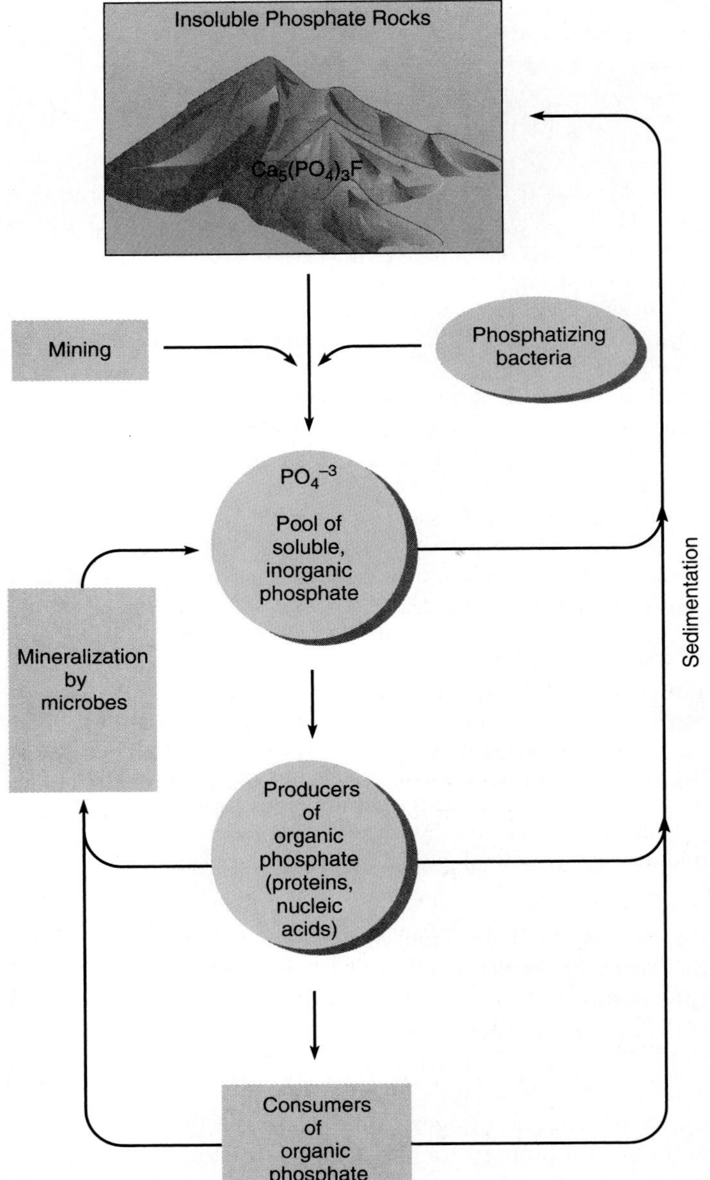

Figure 26.13
The phosphorus cycle. The pool of phosphate existing in sedimentary rocks is released into the ecosystem either naturally by erosion and microbial action or artificially by mining and the use of phosphate fertilizers. Soluble phosphate (PO_4^{-3}) is cycled through producers, consumers, and decomposers back into the soluble pool of phosphate, or it is returned to sediment in the aquatic biosphere.

The Phosphorus Cycle

Phosphorus is an integral component of DNA, RNA, and ATP, and all life depends upon a constant supply of it. It cycles between the abiotic and biotic environments almost exclusively as inorganic phosphate (PO_4) rather than its elemental form (figure 26.13). The chief inorganic reservoir is phosphate rock, which contains the insoluble compound $Ca_5(PO_4)_3F$. Before it can enter biological systems, this mineral must be *phosphatized*—converted into more soluble PO_4^{-3} by the action of acid. Phosphate is released naturally when the sulfuric acid produced by *Thiobacillus* dissolves phosphate rock. Phosphate fertilizers are made commercially by treating mined phosphate rock with sulfuric acid. Soluble phosphate in the soil and water is the principal source for autotrophs, which fix it onto organic molecules and pass it on to heterotrophs in this form. Organic phosphate is returned to the pool of soluble phosphate by decomposers, and it is finally cycled back to the mineral reservoir by slow geologic processes such as sedimentation. Because the low phosphate content of many soils can limit

**Thiobacillus* (thigh″-oh-bah-sil′-us) Gr. *theion,* sulfur. A type of non-photosynthetic sulfur bacteria.

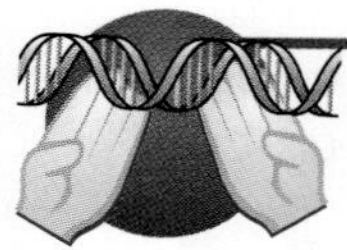

MICROFILE 26.3 BIOGEOCHEMICAL CYCLES AND BIOAMPLIFICATION OF POLLUTANTS

Understanding the action of biogeochemical cycles takes on increased significance when we examine the increasing pollution of the air, soil, and water by humans. The scale of human activity has empowered us to increase the natural cycling rate of such toxic elements as arsenic, chromium, lead, and mercury. Many of the hundreds of thousands of synthetic chemicals introduced into the environment over the past hundred years can also get caught up in cycles. Some of these chemicals will be converted into less harmful substances, but others, such as DDT and heavy metals, persist and flow along with nutrients into all levels of the biosphere. If such a pollutant accumulates in living tissue and is not excreted, it can be **bioamplified,** or concentrated and accumulated by living things, through the natural trophic flow of the ecosystem. Microscopic producers such as bacteria and algae begin the accumulation process. With each new level of the food chain, the consumers gather an increasing amount of the chemical, until the top consumers can contain toxic levels.

The long-term effects of *bioaccumulation* are best demonstrated using the heavy metal mercury as a model. Mercury compounds are used in household antiseptics and disinfectants, agriculture, and industry. Elemental mercury precipitates proteins by attaching to functional groups, and it interferes with the bonding of uracil and thymine. But the elemental form is not as toxic as the organic mercurials such as ethyl or methyl mercury.

The severity of this toxicity was first demonstrated by an incident in Minamata Bay, Japan, during the 1950s. A local plastics manufacturer had drained large quantities of mercury into the bay. In the aquatic sediments, certain mercury-resistant bacteria *(Desulfvibrio)* metabolized the mercury compounds to the highly toxic organic forms. Motile plankton that fed upon these bacteria accumulated the mercury and passed it on to the next consumer level until eventually, fish and shellfish had accumulated high levels of methyl mercury. These foods were a regular part of the diet of local Japanese fishermen and their families. Many adults and older children in the community suffered nerve and brain damage, and fetuses developed severe birth defects. Of the 130 people who were poisoned, 47 died. Since then, no other large-scale cases of toxicity have been traced to eating fish contaminated with methyl mercury. However, recent studies have disclosed increased mercury content in fish taken from oceans and freshwater lakes in North America, and even in canned tuna.

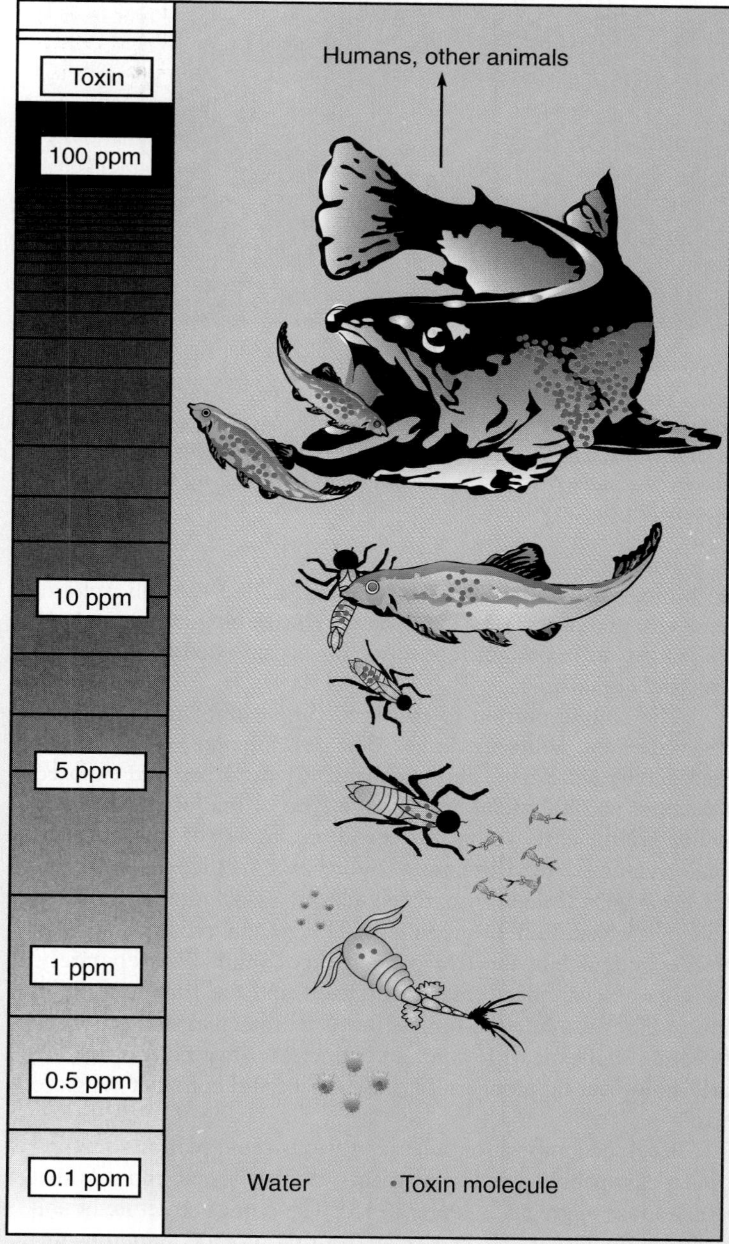

The mechanism for bioamplification of toxic compounds in the food chain.

productivity, phosphate is added to soil to increase agricultural yields. The excess run-off of fertilizer into the hydrosphere is often responsible for overgrowth of aquatic pests (see eutrophication in a subsequent section on aquatic habitats).

The involvement of microbes in cycling elements and compounds has both negative and positive effects. On one hand, they can recycle or create toxic compounds (microfile 26.3); on the other, they can help clean up pesticides (see microfile 26.4) and mine minerals.

SOIL MICROBIOLOGY: THE COMPOSITION OF THE LITHOSPHERE

Descriptions such as "soiled" or "dirty" may suggest that soil is an undesirable, possibly harmful substance. To some people, soil can also appear to be a somewhat homogenous, inert substance. At the microscopic level, however, soil is a dynamic ecosystem that supports complex interactions between numerous geologic, chemical,

Figure 26.14
Rocks covered with patches of crustose and fruticose lichens. Crustose lichens are flat and spreading, whereas fruticose lichens are branching and shrubby. Such microbial associations begin the process of soil formation by invading and breaking up the rocks into smaller particles.

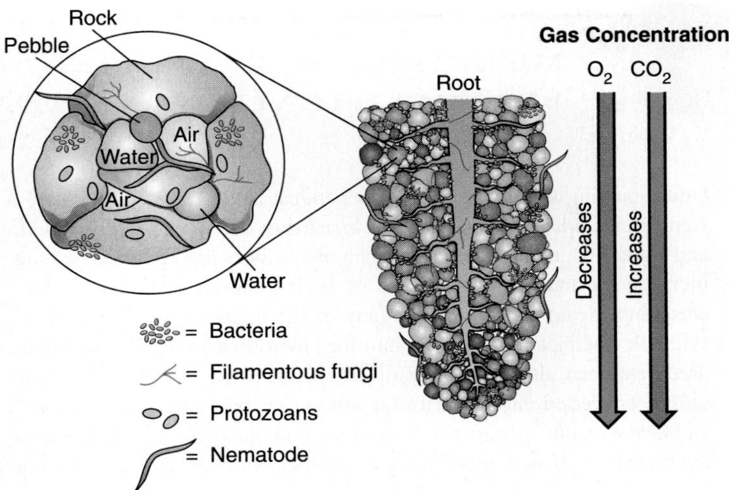

Figure 26.15
The structure of the rhizosphere and the microhabitats that develop in response to soil particles, moisture, air, and gas content.

and biological factors. This rich region, called the lithosphere, teems with microbes, serves a dynamic role in biogeochemical cycles, and is an important repository for organic detritus and dead terrestrial organisms.

The abiotic portion of soil is a composite of mineral particles, water, and atmospheric gas. The development of soil begins when geologic sediments are mechanically disturbed and exposed to weather and microbial action. One type of microbe commonly involved in the early colonization and breakdown of rocks are unusual creatures called **lichens*** (figure 26.14). Lichens are complex organisms formed from the symbiotic association of a fungus and a cyanobacterium or green alga. This mutual relationship is so interdependent that the two participants cannot live separately. The alga or cyanobacterium feeds itself and the fungus through photosynthesis, while the fungus forms a protective shelter for the alga and supplies it with water and minerals. Most lichens that are early colonists or pioneers of rock are of the compact crustose form.

Rock decomposition releases various-sized particles ranging from rocks, pebbles, and sand grains to microscopic morsels that lie in a loose aggregate (figure 26.15). The porous structure of soil creates various-sized pockets or spaces that provide numerous microhabitats. Some spaces trap moisture and form a liquid phase in which mineral ions and other nutrients are dissolved. Other spaces trap air that will provide gases to soil microbes, plants, and animals. Because both water and air compete for these pockets, the water content of soil is directly related to its oxygen content. Water-saturated soils contain less oxygen, and dry soils have more. Gas tensions in soil can also vary vertically. In general, the concentration of O_2 decreases and that of CO_2 increases with the depth of soil. Aerobic and facultative organisms tend to occupy looser, drier soils, whereas anaerobes are adapted to waterlogged, poorly aerated soils.

Within the superstructure of the soil are varying amounts of **humus,*** the slowly decaying organic litter from plant and animal tissues. This soft, crumbly mixture holds water like a sponge. It is also an important habitat for microbes that decompose the complex litter and gradually recycle nutrients. The humus content varies with climate, temperature, moisture and mineral content, and microbial action. Warm, tropical soils have a high rate of humus production and microbial decomposition. Because nutrients in these soils are swiftly released and used up, they do not accumulate. Fertilized agricultural soils in temperate climates build up humus at a high rate and are rich in nutrients. The very low content of humus and moisture in desert soils greatly reduces its microbial flora, rate of decomposition, and nutrient content. Bogs are likewise nutrient-poor, but their lack of nutrients is due to a slow rate of decomposition of the humus caused by high acid content and lack of oxygen. Humans can artificially increase the amount of humus by mixing plant refuse and animal wastes with soil and allowing natural decomposition to occur, a process called *composting.*

Living Activities in Soil

The rich culture medium of the soil supports a fantastic array of microorganisms (bacteria, fungi, algae, protozoa, and viruses). A gram of moist loam soil with high humus content can have a count as high as 10 billion, each competing for its own niche. Some of the most distinctive biological interactions occur in the **rhizosphere,** the zone of soil surrounding the roots of plants that contains associated bacteria, fungi, and protozoa (figure 26.15). Plants interact with soil microbes in a truly synergistic fashion. Studies have shown that a rich microbial community grows in a *biofilm*

*lichens (ly′-kenz) Gr. *leichen,* a tree moss.

*humus (hyoo′-mus) L. earth.

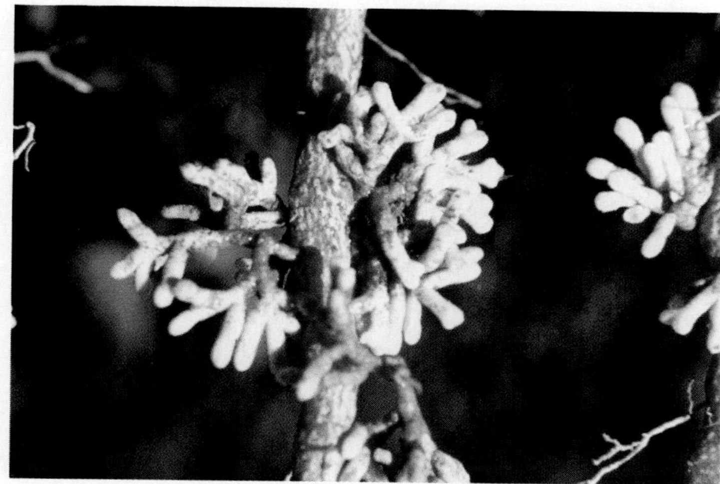

Figure 26.16
Mycorrhizae, symbiotic associations between fungi and plant roots, favor the absorption of water and minerals from the soil.

around the root hairs and other exposed surfaces. Their presence stimulates the plant to exude growth factors such as carbon dioxide, sugars, amino acids, and vitamins. These nutrients are released into fluid spaces, where they can be readily captured by microbes. Bacteria and fungi likewise contribute to plant survival by releasing hormone-like growth factors and protective substances. They are also important in converting minerals into forms usable by plants. We saw numerous examples in the nitrogen, sulfur, and phosphorus cycles.

We previously observed that plants can form close symbiotic associations with microbes to fix nitrogen. Another mutualistic partnership between plant roots and microbes are **mycorrhizae.*** These associations occur when various species of basidiomycetes, ascomycetes, or zygomycetes attach themselves to the roots of vascular plants (figure 26.16). The plant feeds the fungus through photosynthesis, and the fungus sustains the relationship in several ways. By extending its mycelium into the rhizosphere, it helps anchor the plant and increases the surface area for capturing water from dry soils and minerals from poor soils. Plants with mycorrhizae can inhabit severe habitats more successfully than plants without them.

The topsoil, which extends a few inches to a few feet from the surface, supports a host of burrowing animals such as nematodes, termites, and earthworms. Many of these animals are decomposer-reducer organisms that break down organic nutrients through digestion and also mechanically reduce or fragment the size of particles so that they are more readily mineralized by microbes. Aerobic bacteria initiate the digestion of organic matter into carbon dioxide and water and generate minerals such as sulfate, phosphate, and nitrate, which can be further degraded by anaerobic bacteria. Fungal enzymes increase the efficiency of soil decomposition by hydrolyzing complex natural substances such as cellulose, keratin, lignin, chitin, and paraffin.

*mycorrhizae (my″-koh-ry′-zee) Gr. *mykos,* fungus, and *rhiza,* root.

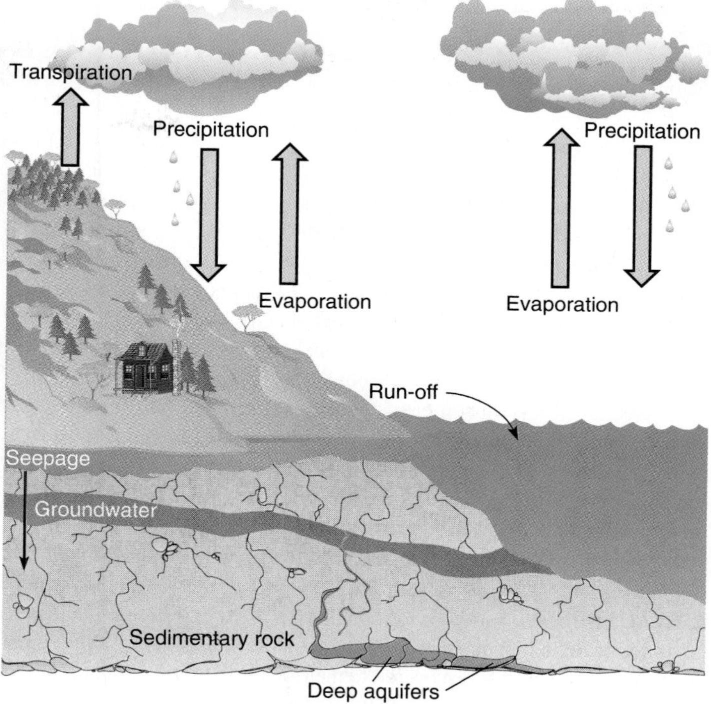

Figure 26.17
The hydrologic cycle. The largest proportion of water cycles through evaporation, transpiration, and precipitation between the hydrosphere and the atmosphere. Other reservoirs of water exist in the groundwater or deep storage aquifers in sedimentary rocks. Plants add to this cycle by releasing water through transpiration, and heterotrophs release it through respiration.

The soil is also a repository for agricultural, industrial, and domestic wastes such as insecticides, herbicides, fungicides, manufacturing wastes, and household chemicals. Microfile 26.4 explores the problem of soil contamination and the feasibility of harnessing indigenous soil microbes to break down undesirable hydrocarbons and pesticides.

Aquatic Microbiology

Water is the dominant compound on the earth. It occupies nearly three-fourths of the earth's surface. In the same manner as minerals, the earth's supply of water is continuously cycled between the hydrosphere, atmosphere, and lithosphere (figure 26.17). The **hydrologic cycle** begins when surface water (lakes, oceans, rivers) exposed to the sun and wind evaporates and enters the vapor phase of the atmosphere. Living things contribute to this reservoir by various activities. Plants lose moisture through transpiration (evaporation through leaves), and all aerobic organisms, from animals to plants to decomposers, give off water during respiration. Airborne moisture accumulates in the atmosphere, most conspicuously as clouds.

Water is returned to the earth through condensation or precipitation (rain, snow). The largest proportion of precipitation falls back into surface waters, where it circulates rapidly between running water and standing water. Only about 2% of water seeps into

MICROFILE 26.4 BIOREMEDIATION: THE POLLUTION SOLUTION?

The soil and water of the earth have long been considered convenient repositories for solid and liquid wastes. Humans have been burying solid wastes for thousands of years, but the process has escalated in the past 50 years. Every year, about 270 metric tons of pollutants, industrial wastes, and garbage are deposited into the natural environment. Often this dumping is done with the mistaken idea that naturally occurring microbes will eventually biodegrade (break down) waste material.

Landfills currently serve as a final resting place for hundreds of castoffs from an affluent society, including yard wastes, paper, glass, plastics, wood, textiles, rubber, metal, paints, and solvents. This conglomeration is dumped into holes and is covered with soil. Although it is true that many substances are readily biodegradable, materials such as plastics and glass are not. Successful biodegradation also requires a compost containing specific types of microorganisms, adequate moisture, and oxygen. The environment surrounding buried trash provides none of these conditions. Large, dry, anaerobic masses of plant materials, paper, and other organic materials will not be successfully attacked by the aerobic microorganisms that dominate in biodegradation. A group at the University of Arizona excavating older landfills discovered to their surprise 25-year-old guacamole and a 16-year-old T-bone steak that were still relatively intact. As we continue to fill up hillsides with waste, the future of these landfills is a prime concern. One of the most serious of these concerns is that they will be a source of toxic compounds that seep into the ground and water.

Pollution of groundwater, the primary source of drinking water for 100 million people in the United States, is an increasing problem. Because of the extensive cycling of water through the hydrosphere and lithosphere, groundwater is often the final collection point for hazardous chemicals released into lakes, streams, oceans, and even garbage dumps. Many of these chemicals are pesticide residues from agriculture (dioxin, selenium, 2,4-D), industrial hydrocarbon wastes (PCBs), and hydrocarbon solvents (benzene, toluene). They are often hard to detect, and, if detected, are hard to remove.

For many years, polluted soil and water were simply sealed off or dredged and dumped in a different site, with no attempt to get rid of the pollutant. But now, with greater awareness of toxic wastes, many Americans are adopting an attitude known as NIMBY (not in my backyard!), and environmentalists are troubled by the long-term effects of contaminating the earth.

In a search for solutions, waste management has turned to **bioremediation**—using microbes to break down or remove toxic wastes in water and soil. Some of these waste-eating microbes are natural soil and water residents with a surprising capacity to decompose even artificial substances. Because the natural, unaided process occurs too slowly, most cleanups are accomplished by commercial bioremediation services that treat the contaminated soil with oxygen, nutrients, and water to increase the rate of microbial action. Through these actions, levels of pesticides such as 2,4-D can be reduced to 96% of their original levels, and solvents can be reduced from one million parts per billion (ppb) to 10 ppb or less (see figure 1.1*a*). Bacteria are also being used to help break up and digest oil spills.

Among the most important bioremedial microbes are species of *Pseudomonas* and *Bacillus* and various toxin-eating fungi. The current quest in bioremediation is for "super bugs" genetically engineered to convert toxic chemicals into CO_2 or other less harmful residues.

(a)

(b)

Can microbes rescue us from our polluted world? (a) A Santa Monica, California shoreline receives a heavy run-off of raw sewage. (b) Solid wastes collect on a beach in the northeastern United States.

One recent isolate from a toxic waste dump is a strain of *Burkholderia cepacia* that contains genes for degrading the chemicals 2,4,S,T and TCDD, major pollutants of industry. Cloning these genes and recombining them in other species can result in genetically engineered bacteria with a high affinity for these compounds.

the earth or is bound in ice, but these are very important reservoirs. Surface water collects in extensive subterranean pockets produced by the underlying layers of rock, gravel, and sand. This process forms a deep **groundwater** source called an **aquifer.** The water in aquifers circulates very slowly and is an important replenishing source for surface water. It can resurface through springs, geysers, and hot vents, and it is also tapped as the primary supply for one-fourth of all water used by humans.

Although the total amount of water in the hydrologic cycle has not changed over millions of years, its distribution and quality have been greatly altered by human activities. Two serious problems have arisen with aquifers. First, as a result of increased well-drilling, land development, and persistent local droughts, the aquifers in many areas have not been replenished as rapidly as they have depleted. As these reserves are used up, humans will have to rely on other delivery systems such as pipelines, dams, and reservoirs, which can further disrupt the cycling of water. Second, because water picks up materials when falling through air or percolating through the ground, aquifers are also important collection points for pollutants. As we will see, the proper management of water resources is one of the greatest challenges of the next century.

The Structure of Aquatic Ecosystems

Surface waters such as ponds, lakes, oceans, and rivers differ to a considerable extent in size, geographic location, and physical and chemical character. Although an aquatic ecosystem is composed primarily of liquid, it is predictably structured, and it contains significant gradients or local differences in composition. Factors that contribute to the development of zones in aquatic systems are sunlight, temperature, aeration, and dissolved nutrient content. These variations create numerous macro- and microenvironments for communities of organisms. An example of zonation can be seen in the schematic section of a freshwater lake in figure 26.18.

A lake is stratified vertically into three zones, or strata. The uppermost region, called the **photic zone,** extends from the surface to the lowest limit of sunlight penetration. Its lower boundary (the compensation depth) is the greatest depth at which photosynthesis can occur. Directly beneath the photic zone lies the **profundal* zone,** which extends from the edge of the photic zone to the lake sediment. The sediment itself, or **benthic* zone,** is composed of organic debris and mud, and it lies directly on the bedrock that forms the lake basin. The horizontal zonation includes the shoreline, or **littoral* zone,** an area of relatively shallow water. The open, deeper water beyond the littoral zone is the **limnetic* zone.**

Marine Environments The marine profile resembles that of a lake, although special characteristics set it apart. The ocean exhibits extreme variations in salinity, depth, temperature, hydrostatic pressure, and mixing. It contains a unique zone where the river meets the sea called an *estuary.* This region fluctuates in salinity, is very high in nutrients, and supports a specialized microbial community. It is often dominated by salt-tolerant species of *Pseudomonas* and *Vibrio.* (At least two species of *Vibrio, parahaemolyticus* and *vulnificus,* are pathogens associated with contaminated seafood.) Another important factor is the tidal and wave action that subjects the coastal habitat to alternate periods of submersion and exposure. The deep ocean, or **abyssal zone,** is poor in nutrients, chiefly because of the absence of sunlight for photosynthesis, and its tremendous depth (up to 10,000 m) makes it oxygen-poor and cold (average temperature 4°C). This zone supports communities with extreme adaptations, including halophilic, psychrophilic, barophilic, and anaerobic.

Aquatic Communities The freshwater environment is a site of tremendous microbiological activity. Microbial distribution is associated with sunlight, temperature, oxygen levels, and nutrient availability. The photic zone, including littoral and limnetic, is the most productive self-sustaining region because it contains large numbers of **plankton,*** a floating microbial community that drifts

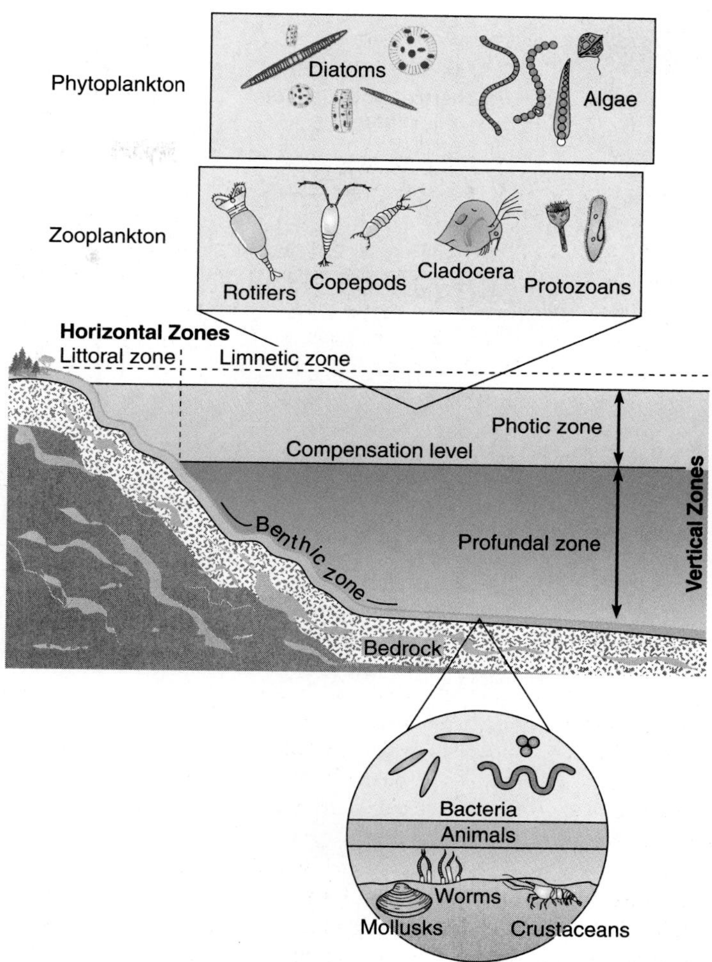

Figure 26.18

Stratification in a freshwater lake. Depth changes in this ecosystem create significant zones that differ in light penetration, temperature, and community structure.

*profundal (proh-fun′-dul) L. *pro,* before, and *fundus,* bottom.

*benthic (ben′-thik) Gr. *benthos,* depth of the sea.

*littoral (lit′-or-ul) L. *litus,* seashore.

*limnetic (lim-neh′-tik) Gr. *limne,* marsh.

*plankton (plang′-tun) Gr. *planktos,* wandering.

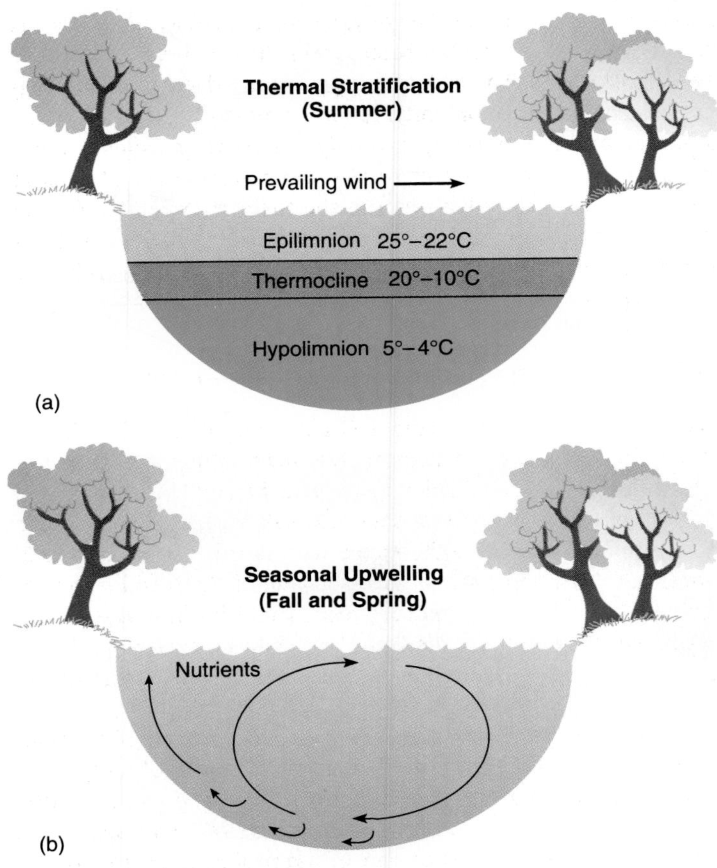

Figure 26.19
Profiles of a lake (*a*) during summer, when it becomes stabilized into three major temperature strata, and (*b*) during fall and spring, when cooling or heating of the water disrupts the temperature strata and causes upwelling of nutrients from the bottom sediments.

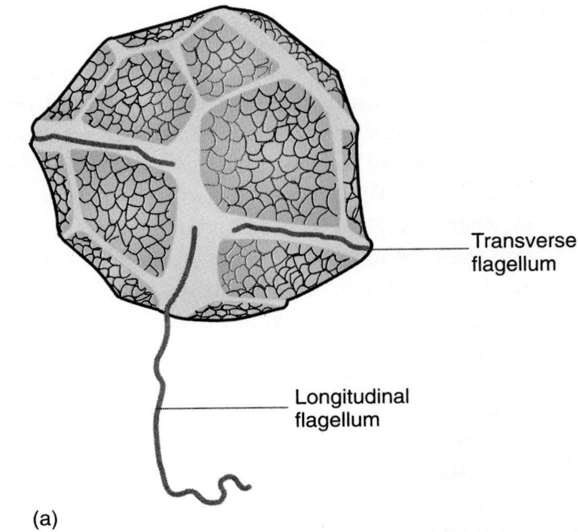

Figure 26.20
Red tides. (*a*) Single-celled red algae called dinoflagellates (*Gymnodinium* shown here) bloom in high-nutrient, warm water and impart a noticeable red color to it, as shown in (*b*). These algae produce a potent muscle toxin that can be concentrated by shellfish through filtration-feeding. When humans eat clams, mussels, or oysters that contain the toxin, they develop paralytic shellfish poisoning. People living in coastal areas are cautioned not to eat shellfish during those months of the year associated with red tides (sometimes characterized as months that lack the letter *r*).

with wave action and currents. A major member of this assemblage is the **phytoplankton,** containing a variety of photosynthetic algae and cyanobacteria (figure 26.18). The phytoplankton provide nutrition for **zooplankton,** microscopic consumers such as protozoa and invertebrates that filter feed, prey, or scavenge. The plankton supports numerous other trophic levels such as larger invertebrates and fish. With its high nutrient content, the benthic zone also supports an extensive variety and concentration of organisms, including aquatic plants, aerobic bacteria and anaerobic bacteria actively involved in recycling organic detritus.

Larger bodies of standing water develop gradients in temperature or thermal stratification, especially during the summer (figure 26.19). The upper region, called the *epilimnion,* is warmest, and the deeper *hypolimnion* is cooler. Between these is a buffer zone, the *thermocline,* that ordinarily prevents the mixing of the two. Twice a year, during the warming cycle of spring and the cooling cycle of fall, temperature changes in the water column break down the thermocline and cause the water from the two strata to mix. Mixing disrupts the stratification and creates currents that bring nutrients up from the sediments. This process, called *upwelling,* is associated with increased activity by certain groups of microbes and is one explanation for the periodic emergence of *red tides* in oceans (figure 26.20). The microorganisms responsible for red tides, dinoflagellates, are highly toxic. A recent outbreak of fish and human disease on the eastern seaboard has been attributed to the overgrowth of certain species of these algae in polluted water (chapter 5).

Because oxygen is not very soluble in water and is rapidly used up by the plankton, its concentration forms a gradient from highest in the epilimnion to lowest in the benthic zone. In general,

Figure 26.21
Heavy surface growth of algae and cyanobacteria in a eutrophic pond.

the amount of O_2 that can be dissolved is dependent on temperature. Warmer strata on the surface tend to carry lower levels of this gas. Of all the characteristics of water, the greatest range occurs in nutrient levels. Pure water recently carried from melting snow into cold mountain ponds and lakes is lowest in nutrients. Such nutrient-deficient aquatic ecosystems, called **oligotrophic,*** support very few microorganisms and often are virtually sterile. Species that can make a living on such starvation rations are *Hyphomicrobium* and *Caulobacter* (see figure 4.37). These bacteria have special stalks that capture even minuscule amounts of hydrocarbons present in oligotrophic habitats. At one time it was thought that viruses were present only in very low levels in aquatic habitats, but researchers have reported finding bacterial virus levels of 125 million particles per milliliter in a water sample from a pristine West German lake. Most of these viruses pose no danger to humans, but as parasites of bacteria, they appear to be a natural control mechanism for these populations.

At the other extreme are waters overburdened with organic matter and dissolved nutrients. Some nutrients are added naturally through seasonal upwelling and disasters (floods or typhoons), but the most significant alteration of natural waters comes from effluents from sewage, agriculture, and industry that contain heavy loads of organic debris or nitrate and phosphate fertilizers. The addition of excess quantities of nutrients to aquatic ecosystems, termed *eutrophication,** often wreaks havoc on the communities involved. The sudden influx of abundant nutrients along with warm temperatures encourages a heavy growth of algae called a bloom (figure 26.21). This heavy algal mat effectively shuts off the O_2 supply to the lake. The oxygen content below the surface is further depleted by aerobic heterotrophs that actively decompose the organic matter. The lack of oxygen greatly disturbs the ecological balance of the community. It causes massive die-offs of strict aerobes (fish, invertebrates), and only anaerobic or facultative microbes will survive. Another serious complication are blooms of certain cyanobacteria that produce the toxin *microcystin.* Several epidemics of fatal toxicosis have decimated populations of birds and fish, and several outbreaks in humans have been linked to toxic blooms.

Water Management to Prevent Disease

Microbiology of Drinking Water Supplies We do not have to look far for overwhelming reminders of the importance of safe water. Recent epidemics of cholera in South America have killed thousands of people, and an outbreak of *Cryptosporidium* in Wisconsin affecting 370,000 people was traced to a contaminated municipal water supply. In a large segment of the world's population, the lack of sanitary water is responsible for billions of cases of diarrheal illness that kills 2 million children each year. In the United States nearly 1 million people develop water-borne illness every year.

Good health is dependent upon a clean, **potable** (drinkable) **water supply.** This means the water must be free of pathogens, dissolved toxins, and disagreeable turbidity, odor, color, and taste. As we shall see, water of high quality does not come easily, and we must look to microbes as part of the problem and part of the solution.

Through ordinary exposure to air, soil, and effluents, surface waters usually acquire harmless, saprobic microorganisms. But along its course, water can also pick up pathogenic contaminants. Among the most prominent water-borne pathogens of recent times are the protozoans *Giardia* and *Cryptosporidium;* the bacteria *Campylobacter, Salmonella, Shigella, Vibrio,* and *Mycobacterium;* and hepatitis A and Norwalk viruses. Some of these agents (especially encysted protozoans) can survive in natural waters for long periods without a human host, whereas others are present only transiently and are rapidly lost. The microbial content of drinking water must be continuously monitored to ensure that the water is free of infectious agents.

Attempting to survey water for specific pathogens can be very difficult and time consuming, so most assays of water purity are more focused on detecting fecal contamination. High fecal levels can mean the water contains pathogens and is consequently unsafe to drink. Thus, wells, reservoirs, and other water sources can be analyzed for the presence of various **indicator bacteria.** These generally nonpathogenic species are intestinal residents of birds and mammals, and they are readily identified using routine lab procedures.

Enteric bacteria most useful in the routine monitoring of microbial pollution are *coliforms* and enteric *streptococci,* which survive in natural waters but do not multiply there. Finding them in high numbers thus implicates recent or high levels of fecal contamination. Environmental Protection Agency standards for water sanitation are based primarily on the levels of coliforms, which are described as gram-negative, lactose-fermenting, gas-producing

*oligotrophic (ahl''-ih-goh-trof'-ik) Gr. *oligo,* small, and *troph,* to feed.

*eutrophication (yoo''-troh-fih-kay'-shun) Gr. *eu,* good.

bacteria such as *Escherichia coli, Enterobacter,* and *Citrobacter.* Fecal contamination of marine waters that poses a risk for gastrointestinal disease is more readily correlated with gram-positive cocci, primarily in the genus *Enterococcus.* Occasionally, coliform bacteriophages and reoviruses (the Norwalk virus) are good indicators of fecal pollution, but their detection is more difficult and more technically demanding.

Water Quality Assays A rapid method for testing the total bacterial levels in water is the **standard plate count.** In this technique, a small sample of water is spread over the surface of a solid medium. The numbers of colonies that develop provide an estimate of the total viable population without differentiating coliforms from other species. This information is particularly helpful in evaluating the effectiveness of various water purification stages. Another general indicator of water quality is the level of dissolved oxygen it contains. It is established that water containing high levels of organic matter and bacteria will have a lower oxygen content because of consumption by aerobic respiration.

Coliform Enumeration Water quality departments employ two standard assays for routine detection and quantification of coliforms. With the **most probable number (MPN)** procedure, coliforms are detected by a series of *presumptive, confirmatory,* and *completed* tests (figure 26.22). The presumptive test involves a series of three subsets of fermentation tubes, each containing different amounts of lactose or lauryl tryptose broth. Each subset contains five tubes that hold an inverted Durham tube to collect gas produced by fermentation. The three subsets are inoculated with water samples of 10, 1.0, and 0.1 ml, respectively. After 24 hours of incubation, the tubes are evaluated for gas production. A positive test for gas formation is presumptive evidence of coliforms, and negative for gas means no coliforms. The number of positive tubes in each subset is tallied, and this set of numbers is applied to a statistical table to estimate the most likely or probable concentration of coliforms (see appendix C table A-1). A confirmatory test for coliforms is made by inoculating another broth from one of the positive tubes. The test is completed by final isolation of the coliform species on selective and differential media, Gram staining the isolate, and reconfirming gas production.

The MPN technique has some notable drawbacks. For one thing, it requires several days to complete. Another problem is that it does not differentiate between coliforms that are commonly found in soil and water *(Enterobacter)* and true *fecal coliforms* that live primarily in the intestines of animals *(Escherichia).* Assays for fecal coliforms require a more specific test such as membrane filtration.

The **membrane filter method** for water analysis is faster, requires fewer steps and media, is less expensive, is more portable, and can process larger quantities of water. This method is more suitable for dilute fluids, such as drinking water, that are relatively free of particulate matter, and it is less suitable for water containing heavy microbial growth or metal inhibitors that bind to the filter. This technique is related to the method described in chapter 11 for sterilizing fluids by filtering out microbial contaminants, except that in this system, the filter containing the trapped microbes is the desired end product. The steps in membrane filtration are diagrammed in figure 26.23*a,b.* After filtration, the membrane filter is removed and placed in a small Petri dish containing selective broth. After incubation, fecal coliform colonies can be counted and often presumptively identified by their distinctive characteristics on these media. More specific rapid techniques for water quality analysis take advantage of substrates that release colored substances into the medium in the presence of lactose-processing enzymes commonly found in coliforms (figure 26.23*c*).

When a test is negative for coliforms, the water is considered generally fit for human consumption. But even slight coliform levels are allowable under some circumstances. For example, municipal waters can have a maximum of 4 coliforms per 100 ml; private wells can have an even higher count. There is no acceptable level for fecal coliforms, enterococci, viruses, or pathogenic protozoans in drinking water. Waters that will not be consumed but are used for fishing or swimming are permitted to have counts of 70 to 200 coliforms per 100 ml. If the coliform level of recreational water reaches 1,000 coliforms per 100 ml, health departments usually bar its usage.

Water and Sewage Treatment Most drinking water comes from rivers, aquifers, and springs. Only in remote, undeveloped, or high mountain areas is this water used in its natural form. In most cities, it must be treated before it is supplied to consumers. Water supplies such as deep wells that are relatively clean and free of contaminants require less treatment than those from surface sources laden with wastes. The stepwise process in water purification as carried out by most cities is shown in figure 26.24. Treatment begins with the impoundment of water in a large reservoir such as a dam or catch basin that serves the dual purpose of storage and sedimentation. The access to reservoirs is controlled to avoid contamination by animal carcasses, wastes, and run-off water. In addition, overgrowth of cyanobacteria and algae that add undesirable qualities to the water is prevented by pretreatment with copper sulfate (0.3 ppm). Sedimentation to remove large particulate matter is also encouraged during this storage period.

Next, the water is pumped to holding ponds or tanks, where it undergoes further settling, aeration, and filtration. The water is filtered first through sand beds or pulverized diatomaceous earth to remove residual bacteria, viruses, and protozoans, and then through activated charcoal to remove undesirable organic contaminants. Pipes coming from the filtration beds collect the water in storage tanks. The final step in treatment is chemical disinfection by bubbling chlorine gas through the tank until it reaches a concentration of 1–2 ppm, but some municipal plants use chloramines (chapter 11) for this purpose. A few pilot plants in the United States are using ozone or peroxide for final disinfection, but these methods are expensive and cannot sustain an antimicrobic effect over long storage times. The final quality varies, but most tap water has a slight odor or taste from disinfection.

Increasingly in the developed world, the same water that serves as a source of drinking water is also used as a dump for

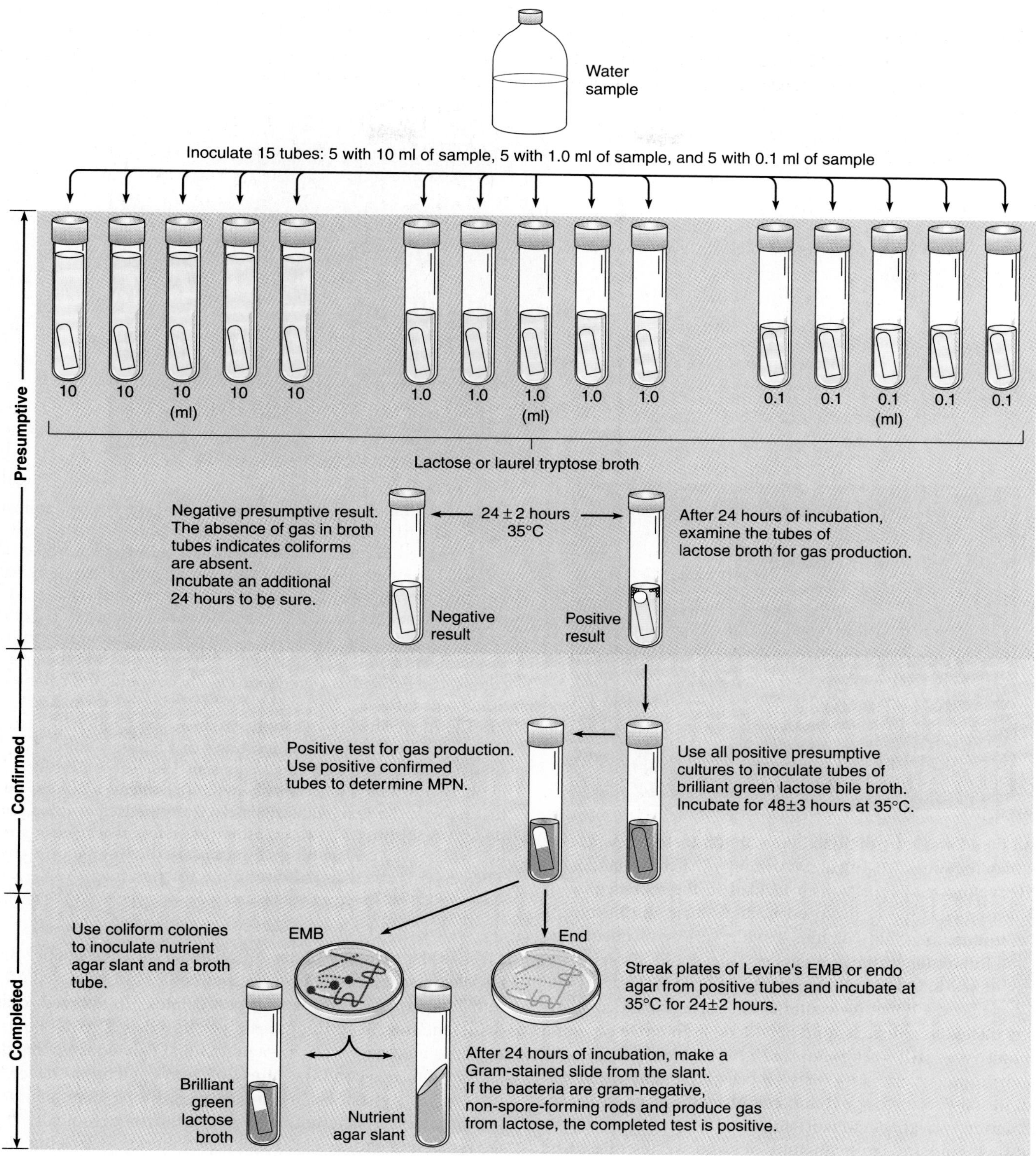

Figure 26.22
The most probable number (MPN) procedure for determining the coliform content of a water sample. In the presumptive test, each set of five tubes of broth is inoculated with a water sample reduced by a factor of 10. After incubation, the sets of tubules are examined and rated for gas production (for instance, 0 means no tubes with gas, 1 means one tube with gas, 2 means two tubes with gas). Applying this result to the MPN table in appendix C will indicate the probable number of cells present in 100 ml of the water sample. Confirmation of coliforms can be achieved by confirmatory tests on additional media, and complete identification can be made through selective and differential media and Gram staining.

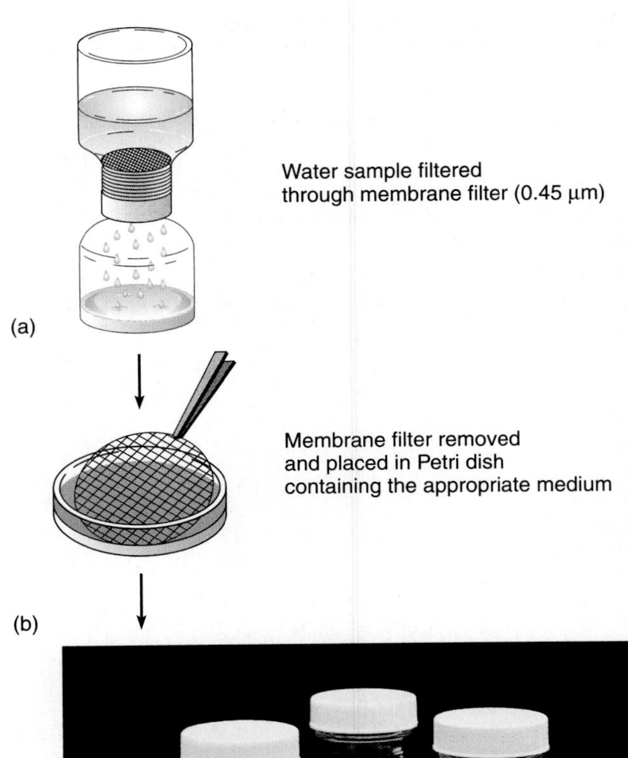
Water sample filtered
through membrane filter (0.45 µm)
(a)
Membrane filter removed
and placed in Petri dish
containing the appropriate medium
(b)

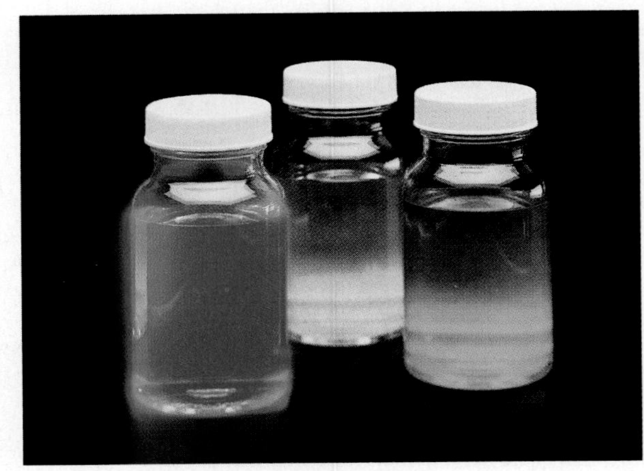
(c)

Figure 26.23

Rapid methods of water analysis for coliform contamination. (*a,b*) Membrane filter technique. (*a*) The water sample is filtered through a sterile membrane filter assembly and collected in a flask. (*b*) The filter is removed and placed in a small Petri dish containing a differential, selective medium such as M-FD endo broth and incubated. On this medium, colonies of *Escherichia coli* often yield a noticeable metallic sheen. The medium permits easy differentiation of various genera of coliforms, and the grid pattern can be used as a guide for rapidly counting the colonies. (*c*) Some tests for water-borne *E. coli* are based on its formation of specialized enzymes to metabolize lactose. The Colilert™ tests shown here utilize synthetic substrates that release a colored substance when the enzymes are present. One substrate, called methyl umbelliferyl gluconuride (MUG), contains a fluorescent dye that is released when glucuronidase acts on the lactose (blue bottle on left). A second test uses a synthetic galactose that releases a yellow dye when galactosidase breaks bonds on lactose (bottle on right). The center bottle is an indicator of no reaction (negative). *(c) Source: Courtesy of IDEXX Company. Colilert is a trademark name of the IDEXX Company.*

solid and liquid wastes. Continued pressure on the finite water resources may require salvaging and recycling of such contaminated water. Recycling has already been tackled in the reclamation of **sewage** water. Sewage is the used wastewater draining out of homes and industries that contains a wide variety of chemicals, debris, and microorganisms. The dangers of typhoid, cholera, and dysentery linked to the unsanitary mixing of household water and sewage have been a threat for centuries. In current practice, some sewage is treated to reduce its microbial load before release, but a large quantity is still being emptied raw (untreated) into the aquatic environment primarily because heavily contaminated waters require far more stringent and costly methods of treatment than are currently available to most cities.

Sewage contains large amounts of solid wastes, dissolved organic matter, and toxic chemicals that pose a health risk. To remove all potential health hazards, treatment typically requires three phases: The primary stage separates out large matter; the secondary stage reduces remaining matter and can remove some toxic substances; and the tertiary stage completes the purification of the water (figure 26.25). Microbial activity is an integral part of the overall process. The systems for sewage treatment are massive engineering marvels (figure 26.26).

In the **primary phase** of treatment, floating bulkier materials such as paper, plastic waste, and bottles are skimmed off. The remaining smaller, suspended particulates are allowed to settle. Sedimentation in settling tanks usually takes 2 to 10 hours and leaves a mixture rich in organic matter. This aqueous residue is carried into a **secondary phase** of active microbial decomposition, or biodegradation. In this phase, a diverse community comprising chiefly bacteria, algae, and protozoa aerobically digests the remaining particles of wood, paper, fabrics, petroleum, and organic molecules. It forms a suspension of material called *sludge* that tends to settle out and slow the process. To hasten aerobic decomposition of the sludge, most processing plants have systems to *activate* it by injecting air, mechanically stirring it, and recirculating it. A large amount of organic matter is mineralized into sulfates, nitrates, phosphates, carbon dioxide, and water. Certain volatile gases such as hydrogen sulfide, ammonia, nitrogen, and methane may also be released. Water from this process is si-

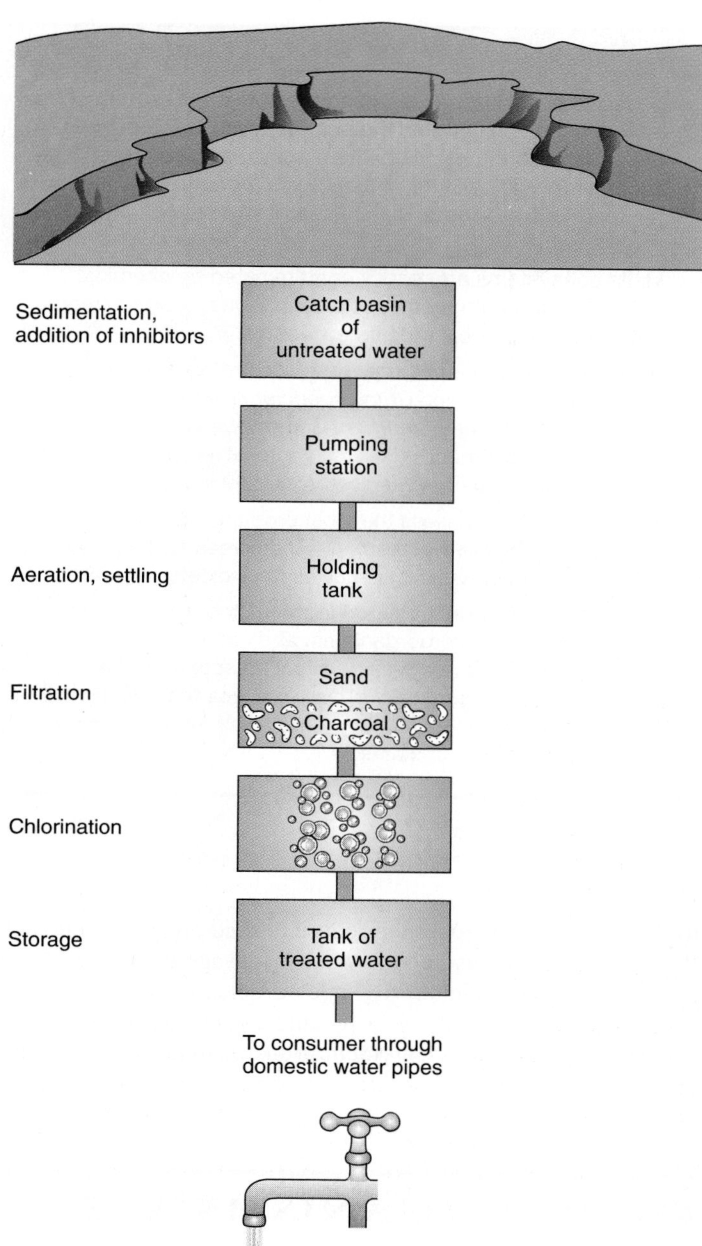

Figure 26.24

The major steps in water purification as carried out by a modern municipal treatment plant.

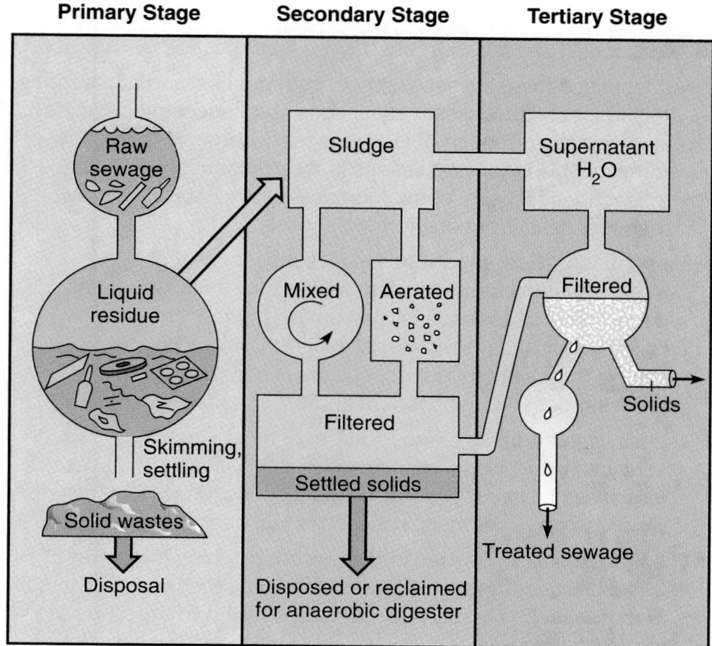

Figure 26.25

The primary, secondary, and tertiary stages in sewage treatment.

Figure 26.26

Aerial photograph of a sewage treatment plant that occupies hundreds of acres and can process several hundred million gallons of water a day.

phoned off and carried to the **tertiary phase,** which involves further filtering and chlorinating prior to discharge. Such reclaimed sewage water is usually used to water golf courses and parks, rather than for drinking.

In some cases, the solid waste that remains after aerobic decomposition is harvested and reused. Its rich content of nitrogen, potassium, and phosphorus makes it a useful fertilizer. But if the waste contains large amounts of nondegradable or toxic substances, it must be disposed of properly. In many parts of the world, the sludge, which still contains significant amounts of simple, but useful organic matter, is used as a secondary source of energy. Further digestion is carried out by microbes in sealed chambers called bioreactors, or **anaerobic digesters.** The digesters convert components of the sludge to swamp gas, primarily methane with small amounts of hydrogen, carbon dioxide, and other volatile compounds. Swamp gas can be burned to provide energy to run the sewage processing facility itself or to power small industrial plants. Considering the mounting waste disposal and energy shortage problems, this technology should gain momentum. (For another microbiological answer to disposal of liquid and solid waste, see microfile 26.4.)

Chapter Checkpoints

Nutrients and minerals necessary to communities and ecosystems must be continuously recycled. These biogeochemical cycles involve transformation of elements from inorganic to organic forms usable by all populations in the community and back again. Specific types of microorganisms are needed to convert many nutrients from one form to another.

Elements of critical importance to all ecosystems that cycle through various forms are carbon, nitrogen, sulfur, phosphorus, and water. Carbon and nitrogen are part of the atmospheric cycle. Sulfur and phosphorus are part of the sedimentary cycling of nutrients. Water circulates in a hydrologic cycle involving both the atmosphere and the lithosphere.

The sun is the primary energy source for most surface ecosystems. Photosynthesis captures this energy and utilizes it for carbon fixation by producer populations. Producers include plants, algae, cyanobacteria, and certain bacterial species.

The lithosphere, or soil, is an ecosystem in which mineral-rich rocks are decomposed to organic humus, the base for the soil community. Soil ecosystems vary according to the kinds of rocks and amount of water, air, and nutrients present. The rhizosphere is the most ecologically active zone of the soil.

Aquatic ecosystems are classified as the photic and profundal zones according to the amount of light penetration. The shoreline is the littoral zone, and the soil at the bottom of the water column constitutes the benthic zone. The food web of the aquatic community is built on phytoplankton and zooplankton. The nature of the aquatic community varies with the temperature, depth, minerals, and amount of light present in each zone.

Aquatic ecosystems are readily contaminated by chemical pollutants and pathogens because of industry, agriculture, and improper disposal of human wastes.

Significant water-borne pathogens include protozoans, bacteria, and viruses. *Giardia* and *Cryptosporidium* are the most significant protozoan pathogens. *Campylobacter, Salmonella,* and *Vibrio* are the most significant bacterial pathogens. Hepatitis A and Norwalk virus are the most significant viral pathogens.

Water quality assays assess the most probable number of microorganisms in a water sample and screen for the presence of enteric pathogens using *E. coli* as the indicator organism.

Wastewater or sewage is treated in three stages to remove organic material, microorganisms, and chemical pollutants. The primary phase removes physical objects from the wastewater. The secondary phase removes the organic matter by biodegradation. The tertiary phase disinfects the water and removes chemical pollutants.

APPLIED MICROBIOLOGY AND BIOTECHNOLOGY

Never underestimate the power of the microbe.
—Jackson W. Foster

The profound and sweeping involvement of microbes in the natural world is inescapable. Although our daily encounters with them usually go unnoticed, human and microbial life are clearly intertwined on many levels. It is no wonder that even long ago humans realized the power of microbes and harnessed them for specific metabolic tasks. The practical applications of microorganisms in manufacturing products or carrying out a particular decomposition process belong to the large and diverse area of **biotechnology.** Biotechnology has an ancient history, dating back nearly 6,000 years to those first observant humans who discovered that grape juice left sitting produced wine or that bread dough properly infused with a starter would rise. Today, biotechnology has become a fertile ground for hundreds of applications in industry, medicine, agriculture, food sciences, and environmental protection and has even come to include the genetic alterations of microbes and other organisms.

Most biotechnological systems involve the actions of bacteria, yeasts, molds, and algae that have been selected or altered to synthesize a certain food, drug, organic acid, alcohol, or vitamin. Many such food and industrial end products are obtained through **fermentation,** a general term used here to refer to the mass, controlled culture of microbes to produce desired organic compounds. It also includes the use of microbes in sewage control, pollution control, and metal mining. A single section cannot cover this diverse area of microbiology in its entirety, but we will touch on some of its more important applications in food technology and industrial processes.

MICROORGANISMS AND FOOD

All human food—from vegetables to caviar to cheese—comes from some other organism, and rarely is it obtained in a sterile, uncontaminated state. Food is but a brief stopover in the overall scheme of biogeochemical cycling. This means that microbes and humans are in direct competition for the nutrients in food, and we must be constantly aware that microbes' fast growth rates give them the winning edge. Somewhere along the route of procurement, processing, or preparation, food becomes contaminated with microbes from the soil, the bodies of plants and animals, water, air, food handlers, or utensils. The final effects depend upon the types and numbers of microbes and whether the food is cooked, preserved, or otherwise processed. In some cases, specific microbes can even be added to food to obtain a desired effect. The effects of microorganisms on food can be classified as detrimental, beneficial, or neutral to humans, as summarized by the following outline:

TABLE 26.2

MAJOR FORMS OF BACTERIAL FOOD POISONING

Disease	Microbe	Foods Involved	Comments
Food Intoxications: Caused by Ingestion of Foods Containing Preformed Toxins			
Staphylococcal enteritis	*Staphylococcus aureus*	Custards, cream-filled pastries, ham, dressings	Very common; symptoms come on rapidly; usually nonfatal
Botulism	*Clostridium botulinum*	Home-canned or poorly preserved low-acid foods	Recent cases involved vacuum-packed foods; can be fatal
Perfringens enterotoxemia	*Clostridium perfringens*	Inadequately cooked meats	Vegetative cells produce toxin within the intestine
Bacillus cereus enteritis	*Bacillus cereus*	Reheated rice, potatoes, puddings, custards	Mimics staphylococcal enteritis; usually self-limited
Food Infections: Caused by Ingestion of Live Microbes That Invade the Intestine			
Salmonellosis	*Salmonella typhimurium* and *S. enteriditis*	Poultry, eggs, dairy products, meats	Very common; can be severe and life-threatening
Shigellosis	Various *Shigella* species	Unsanitary cooked food; fish, shrimp, potatoes, salads	Carriers and flies contaminate food; the cause of bacillary dysentery
Vibrio enteritis	*Vibrio parahaemolyticus*	Raw or poorly cooked seafoods	Microbe lives naturally on marine animals
Listeriosis	*Listeria monocytogenes*	Poorly pasteurized milk, cheeses	Most severe in fetuses, newborns, and the immunodeficient
Campylobacteriosis	*Campylobacter jejuni*	Raw milk; raw chicken, shellfish, and meats	Very common; animals are carriers of other species
Escherichia enteritis	*Escherichia coli*	Contaminated raw vegetables, cheese	Various strains can produce infantile and traveler's diarrhea
	E. coli 0157:H7	Raw or rare beef	The cause of hemolytic uremic syndrome (see chapter 20)

Detrimental Effects
- Food poisoning or food-borne illness
  - Chemical in origin: Pesticides, food additives
  - Biological in origin: Living things or their products
    - Infection: Bacterial, protozoan, worm
    - Intoxication: Bacterial, fungal
- Food spoilage
  - Growth of microbes makes food unfit for consumption; adds undesirable flavors, appearance, and smell; destroys food value

Neutral Effects
- The presence or growth of microbes that do not harm or change the nature of the food

Beneficial Effects
- Food is fermented or otherwise chemically changed by the addition of microbes or microbial products to alter or improve flavor, taste, or texture.
- Microbes can serve as food.

As long as food contains no harmful substances or organisms, its suitability for consumption is largely a matter of taste. But what tastes like bouquet to some may seem like decay to others. The test of whether certain foods are edible is guided by experience and preference. The flavors, colors, textures, and aromas of many cultural delicacies are supplied by bacteria and fungi. Poi, pickled cabbage, Norwegian fermented fish, and limburger cheese are notable examples. If you examine the foods of most cultures, you will find some food that derives its delicious flavor from microbes.

MICROBIAL INVOLVEMENT IN FOOD-BORNE DISEASES

Diseases caused by ingesting food are usually referred to as **food poisoning,** and, although this term is often used synonymously with microbial food-borne illness, not all food poisoning is caused by microbes or their products. Several illnesses are caused by poisonous plant and animal tissues or by ingesting food contaminated by pesticides or other poisonous substances. Table 26.2 summarizes the major types of bacterial food-borne disease.

Food poisoning of microbial origin can be divided into two general categories[4] (figure 26.27). **Food intoxication** results from the ingestion of exotoxin secreted by bacterial cells growing in food. The absorbed toxin disrupts a particular target such as the intestine (if an enterotoxin) or the nervous system (if a neurotoxin). The symptoms of intoxication vary from bouts of vomiting and diarrhea (staphylococcal intoxication) to severely disrupted muscle function (botulism). In contrast, **food infection** is associated with the ingestion of whole, intact microbial cells that proceed to attack the intestine. In some cases, they infect the surface of the intestine, and, in others, they invade the intestine and other body structures. Most food infections manifest some degree of diarrhea and abdominal distress (see microfile 20.2). It is important

4. Although these categories are useful for clarifying the general forms of food poisoning, some diseases are transitional. For example, perfringens intoxication is caused by a toxin, but the toxin is released in the intestinal lumen rather than in the food.

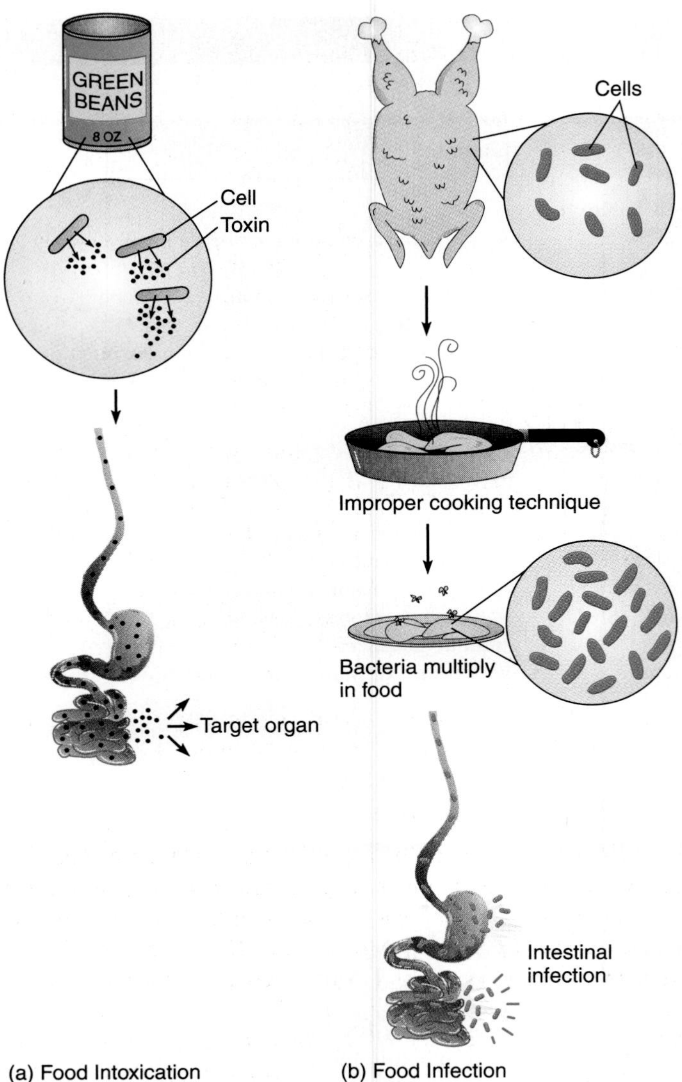

Figure 26.27
Food-borne illnesses of microbial origin. (*a*) Food intoxication. The toxin is produced by microbes growing in the food. After the toxin is ingested, it acts upon its target tissue and causes symptoms. (*b*) Food infection. The infectious agent comes from food or is introduced into it through poor food processing and storage. After the cells are ingested, they invade the intestine and cause symptoms of gastroenteritis.

to realize that disease symptoms in food infection can be initiated by toxins but that the toxins are released by microbes growing in the infected tissue rather than in the food.

Reports of food poisoning are escalating in the United States and worldwide. Outbreaks attributed to common pathogens (*Salmonella, E. coli, Vibrio, Hepatitis A,* and protozoa) have doubled in the past 20 years. The CDC estimates that nearly 10,000 people die each year from some form of food infection. A major factor in this changing pattern is the mass production and distribution of processed food such as raw vegetables, fruits, and meats. Improper handling can lead to gross contamination of these products with soil or animal wastes.

Many reported food poisoning outbreaks occur where contaminated food has been served to large groups of people,[5] but most cases probably occur in the home and are not reported. The most common food-borne illness in the United States is staphylococcal food intoxication. Other relatively common agents of food-associated disease include *Campylobacter, Salmonella, Clostridium perfringens,* and *Shigella.* Food poisoning by *Clostridium botulinum, Bacillus cereus,* and *Vibrio* is less common. The detailed pathology of the diseases listed in table 26.2 was covered in chapters 18, 19, 20, and 21.

Prevention Measures for Food Poisoning and Spoilage

It will never be possible to avoid all types of food-borne illness because of the ubiquity of microbes in air, water, food, and the human body. But most types of food poisoning require the growth of microbes in the food. In the case of food infections, an infectious dose (sufficient cells to initiate infection) must be present, and in food intoxication, enough cells to produce the toxin must be present. Thus, food poisoning or spoilage can be prevented by proper food handling, preparation, and storage (see microfile 20.5). The methods shown in figure 26.28 are aimed at preventing the incorporation of microbes into food, removing or destroying microbes in food, and keeping microbes from multiplying.

Preventing the Incorporation of Microbes into Food Most agricultural products such as fruits, vegetables, grains, meats, eggs, and milk are naturally exposed to microbes. Vigorous washing reduces the levels of contaminants in fruits and vegetables, whereas meat, eggs, and milk must be taken from their animal source as aseptically as possible. Aseptic techniques are also essential in the kitchen. Contamination of foods by fingers can be easily remedied by handwashing and proper hygiene, and contamination by flies or other insects can be stopped by covering foods or eliminating pests from the kitchen. Care and common sense also apply in managing utensils. It is important to avoid cross-contaminating food by using the same cutting board for meat and vegetables, without disinfecting it between uses. The subject of cutting board safety is discussed in microfile 26.5.

Preventing the Survival or Multiplication of Microbes in Food Even with the most hygienic techniques, it is not possible to eliminate all microbes from certain types of food by clean techniques alone, so a more efficient approach preserves the food by some form of physical or chemical method. Hygienically preserving foods is especially important for large commercial companies that process and sell bulk foods and must ensure that products are free from harmful contaminants. Regulations and standards for food processing are administered by two federal agencies: the Food and Drug Administration (FDA) and the United States Department of Agriculture (USDA).

Temperature and Food Preservation Heat is a common way to destroy microbial contaminants or to reduce the load of

5. One-third of all reported cases result from eating restaurant food.

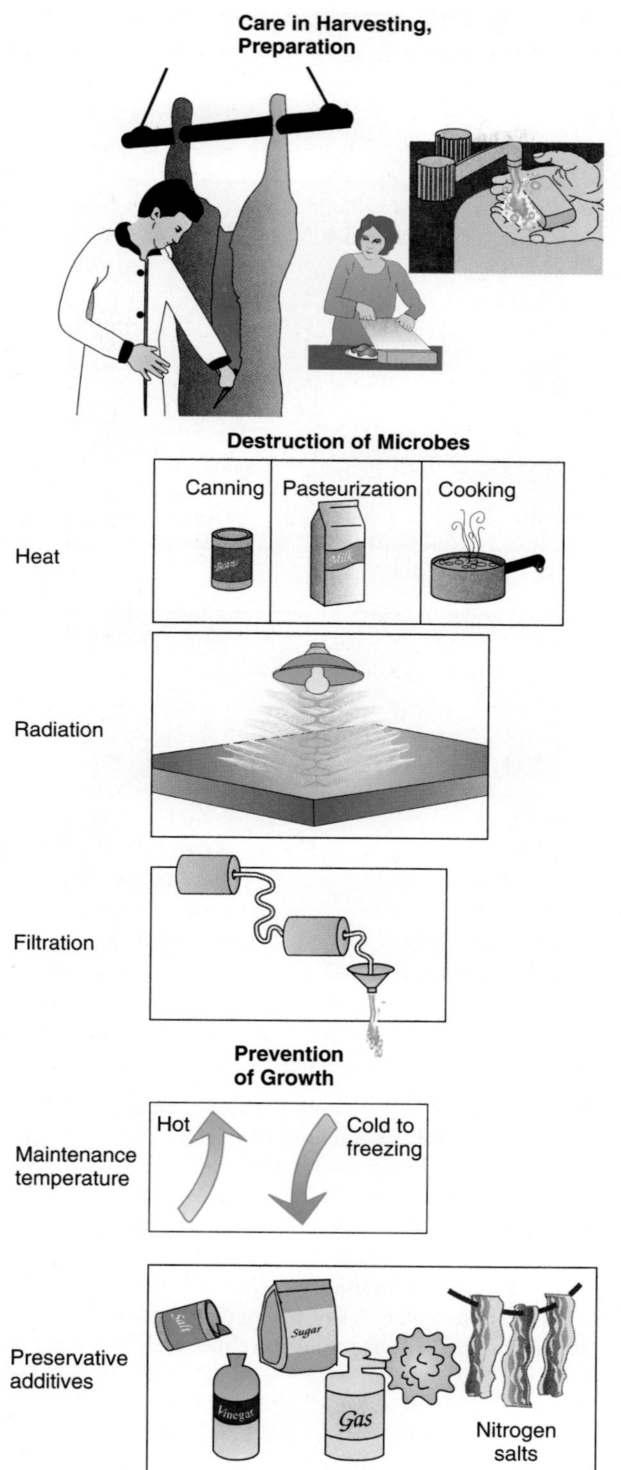

Figure 26.28
The primary methods to prevent food poisoning and food spoilage.

Figure 26.29
A modern flash pasteurizer, a system used in dairies for high-temperature short-time (HTST) pasteurization.
Photo taken at Alta Dena Dairy, City of Industry, California.

microorganisms. Commercial canneries preserve food in hermetically sealed containers that have been exposed to high temperatures over a specified time period. The temperature used depends upon the type of food, and it can range from 60°C to 121°C, with exposure times ranging from 20 minutes to 115 minutes. The food is usually processed at a thermal death time (TDT; see chapter 11) that will destroy the main spoilage organisms and pathogens but will not alter the nutrient value or flavor of the food. For example, tomato juice must be heated to between 121°C and 132°C for 20 minutes to ensure destruction of the spoilage agent *Bacillus coagulans.* Likewise, green beans must be heated to 121°C for 20 minutes to destroy pathogenic *Cl. botulinum.* Most canning methods are rigorous enough to sterilize the food completely, but some only render the food "commercially sterile," which means it contains live bacteria that are unable to grow under normal conditions of storage.

Another use of heat is **pasteurization,** usually defined as the application of heat below 100°C to destroy nonresistant bacteria and yeasts in liquids such as milk, wine, and fruit juices. The heat is applied in the form of steam, hot water, or even electrical current. The most prevalent technology is the *high-temperature short-time (HTST),* or flash method, using extensive networks of tubes that expose the liquid to 72°C for 15 seconds (figure 26.29). An alternative method, ultrahigh temperature (UHT) pasteurization, steams the product until it reaches a temperature of 134°C for at least one second. Although milk processed this way is not actually sterile, it is often marketed as sterile, with a shelf life of up to 3 months. Older methods involve large bulk tanks that hold the fluid at a lower temperature for a long time, usually 62.3°C for 30 minutes.

Cooking temperatures used to boil, roast, or fry foods can render them free or relatively free of living microbes if carried out for sufficient time to destroy any potential pathogens. A quick warming of chicken or an egg is inadequate to kill microbes such as *Salmonella.* In fact, any meat is a potential source of infectious

MICROFILE 26.5 WOOD OR PLASTIC: ON THE CUTTING EDGE OF CUTTING BOARDS

Inquiring cooks have long been curious to have the final word on which type of cutting board is the better choice for food safety. When the USDA first recommended plastic cutting boards in the 1980s, it seemed the logical, reasonable choice. After all, plastic is nonabsorbent and easy to clean, presumably making it less likely to harbor bacteria and other microorganisms on its surface than wood is. But this recommendation was never based on evidence from scientific tests. Recently, two separate research groups turned their attention to this important kitchen question. What emerged from these studies came as rather a surprise—the two groups came up with exactly opposite conclusions.

First came the study by a team of microbiologists from the University of Wisconsin. They experimented with hardwood chopping blocks and acrylic plastic boards inoculated with pathogens such as *Salmonella, Escherichia coli,* and *Listeria monocytogenes.* One of the most unexpected results was that the wooden boards actually killed 99.9% of the bacteria within a few minutes. The team concluded from the lack of viable cells that wood must contain some antibacterial substances, although they were unable to isolate them. The plastic boards did not similarly reduce the numbers of pathogens and they failed to live up to expectations in other ways. For instance, they continued to harbor bacteria if left unwashed for a given time period. If they were scored from extensive use, even after scrubbing with soap and water, they still held live bacteria. In contrast, even heavily used wooden boards did not grow microorganisms and had a far lower bacterial count. The Wisconsin researchers concluded that the grounds for advocating plastic are questionable and that wood is as safe as plastic, if not superior to it.

In the other study, researchers from the Food and Drug Administration performed an electron microscope study of wood. They found that pathogens such as *E. coli* 0157:H7 and *Campylobacter* became trapped in the porous spaces of wooden boards and were able to survive for 2 hours to several days, depending on the moisture content of the wood. They continue to recommend the use of plastic because bacteria trapped in wood would be difficult to remove and could be released during use.

What is a chef to do? Although these contradictory studies seem not to provide a definitive answer, they can serve to emphasize an important point. The solution still exists in simple commonsense guidelines that are the crux of good kitchen practices. It is apparent that both boards can be safe if properly handled and their limitations are taken into account. All boards should be scrubbed with soap and hot water and disinfected between uses, especially if meats, poultry, or fish have been cut on them. Plastic boards should be replaced if their surface has become too roughened with use, and wooden boards must not be left moist for any period of time. A new line of plastic boards containing antibacterial triclosan is expected to improve the safety of cutting boards.

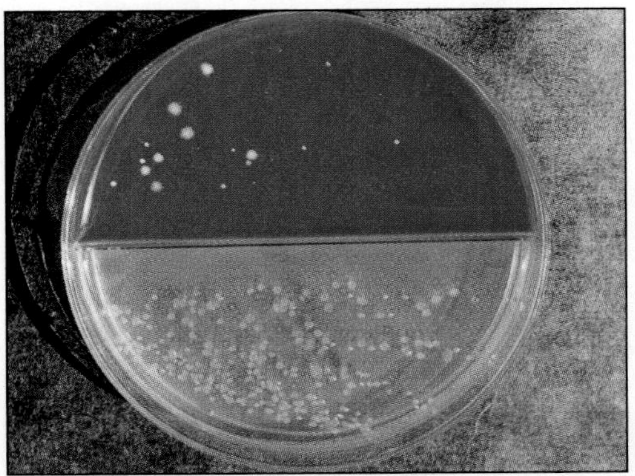

(a)

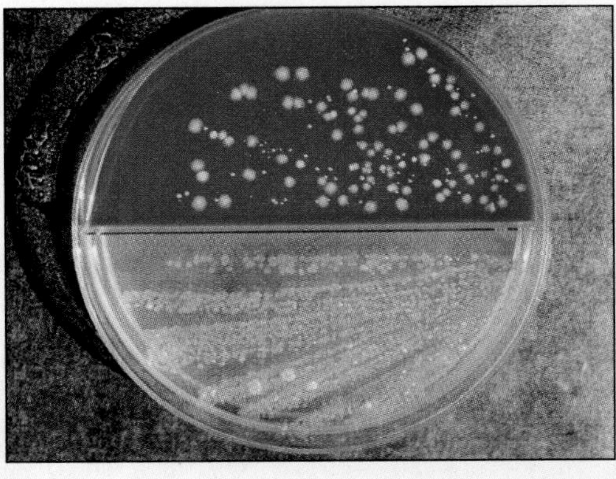

(b)

Double-sided plates of blood agar (top) and MacConkey agar (bottom) after swabbing with samples from cutting boards. The boards were equally contaminated with a fresh chicken carcass and the samples were taken 10 minutes later. Results appear in (a) for the wooden board and in (b) for the plastic board. Note that, in this case, the wooden board yielded significantly fewer colonies on both types of media.

agents and should be adequately cooked. Because most meat-associated food poisoning is caused by nonsporulating bacteria, heating the center of meat to at least 80°C and holding it there for 30 minutes is usually sufficient to kill pathogens. Roasting or frying food at temperatures of at least 200°C or boiling it will achieve a satisfactory degree of disinfection (see guidelines in microfile 20.5).

Any perishable raw or cooked food that could serve as a growth medium must be stored to prevent the multiplication of bacteria that have survived during processing or handling. Because most food-borne bacteria and molds that are agents of spoilage or infection can multiply at room temperature, manipulation of the holding temperature is a useful preservation method (figure 26.30). A good general directive is to store foods at temperatures below 4°C or above 60°C.

Regular refrigeration reduces the growth rate of most mesophilic bacteria by ten times, although some psychrotrophic microbes can continue to grow at a rate that causes spoilage. This factor limits the shelf life of milk, because even at 7°C, a population could go from a few cells to a billion in 10 days. Pathogens

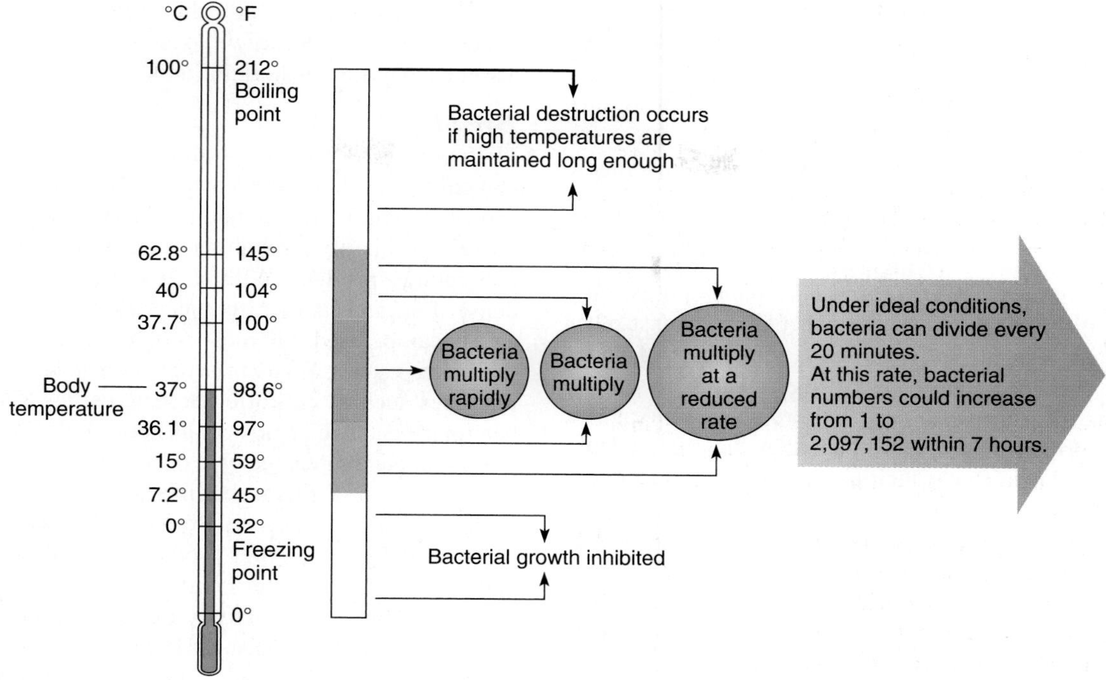

Figure 26.30
Temperatures favoring and inhibiting the growth of microbes in food. Most microbial agents of disease or spoilage grow in the temperature range of 15°–40°C. Preventing unwanted growth in foods in long-term storage is best achieved by refrigeration or freezing (4°C or lower). Preventing microbial growth in foods intended to be consumed warm in a few minutes or hours requires maintaining the foods above 60°C.
From Ronald Atlas, Microbiology: Fundamentals and Applications, *2nd ed., © 1988, p. 475. Reprinted by permission of Prentice-Hall, Upper Saddle River, New Jersey.*

such as *Listeria monocytogenes* and *Salmonella* can also continue to grow in refrigerated foods. Freezing is a longer-term method for cold preservation. Foods can be either slow-frozen for 3 to 72 hours at −15°C to −23°C or rapidly frozen for 30 minutes at −17°C to −34°C. Because freezing cannot be counted upon to kill microbes, rancid, spoiled, or infectious foods will still be unfit to eat after freezing and defrosting. *Salmonella* is known to survive several months in frozen chicken and ice cream, and *Vibrio parahaemolyticus* can survive in frozen shellfish. For this reason, frozen foods should be defrosted rapidly and immediately cooked or reheated. However, even this practice will not prevent staphylococcal intoxication if the toxin is already present in the food before it is cooked.

Foods such as soups, stews, gravies, meats, and vegetables that are generally eaten hot should not be maintained at warm or room temperatures, especially in settings such as cafeterias, banquets, and picnics. The use of a hot plate, chafing dish, or hot water bath will maintain foods above 60°C, well above the incubation temperature of food-poisoning agents.

As a final note about methods to prevent food poisoning, remember the simple axiom: "When in doubt, throw it out."

Radiation Food industries have led the field of radiation sterilization and disinfectant (see chapter 11). Ultraviolet (nonionizing) lamps are commonly used to destroy microbes on the surfaces of foods or utensils, but they do not penetrate far enough to sterilize bulky foods or food in packages. Food preparation areas are often equipped with UV radiation devices that are used to destroy spores on the surfaces of cheese, breads, and cakes and to disinfect packaging machines and storage areas.

Food itself is usually sterilized by gamma or cathode radiation because these ionizing rays can penetrate denser materials. Gamma rays are more dangerous because they originate from a radioactive source such as cobalt-60, but the technology has sophisticated chambers that protect workers from the cobalt. It must also be emphasized that this method does not cause the targets of irradiation to become radioactive.

Concerns have been raised about the possible secondary effects of radiation that could alter the safety and edibility of foods. Experiments over the past 30 years have demonstrated some side reactions that affect flavor, odor, and vitamin content, but it is currently thought that irradiated foods are relatively free of toxic byproducts. The government has currently approved the use of radiation in sterilizing beef, pork, poultry, fish, spices, grain, and some fruits and vegetables (see figure 11.9). Less than 10% of these products are sterilized this way, but outbreaks of food-borne illness have increased its desirability for companies and consumers. It also increases the shelf life of perishable foods, thus lowering their cost.

Other Forms of Preservation The addition of chemical preservatives to many foods can prevent the growth of microorganisms that could cause spoilage or disease. Preservatives include natural chemicals such as salt (NaCl) or table sugar and artificial substances such as ethylene oxide. The main classes of preservatives are organic acids, nitrogen salts, sulfur compounds, oxides, salt, and sugar.

Organic acids, including lactic, benzoic, and propionic acids, are among the most widely used preservatives. They are added to baked goods, cheeses, pickles, carbonated beverages, jams, jellies, and dried fruits to reduce spoilage from molds and some bacteria. Nitrites and nitrates are used primarily to maintain the red color of cured meats (hams, bacon, and sausage). By inhibiting the germination of *Clostridium botulinum* spores, they also prevent botulism intoxication, but their effects against other microorganisms are limited. Sulfite prevents the growth of undesirable molds in dried fruits, juices, and wines and retards discoloration in various foodstuffs. Ethylene and propylene oxide gases disinfect various dried foodstuffs. Because they are carcinogenic and can leave a residue in foods, their use is restricted to fruit, cereals, spices, nuts, and cocoa.

The high osmotic pressure contributed by hypertonic levels of salt plasmolyzes bacteria and fungi and removes moisture from food, thereby inhibiting microbial growth. Salt is commonly added to brines, pickled foods, meats, and fish. However, it does not retard the growth of pathogenic halophiles such as *Staphylococcus aureus,* which grows readily even in 7.5% salt solutions. The high sugar concentrations of candies, jellies, and canned fruits also exert an osmotic preservative effect. Other chemical additives that function in preservation are alcohols and antibiotics. Alcohol is added to flavoring extracts, and antibiotics are approved for treating the carcasses of chickens, fish, and shrimp.

Food can also be preserved by **desiccation,** a process that removes moisture needed by microbes for growth by exposing the food to dry, warm air. Solar drying was traditionally used for fruits and vegetables, but modern commercial dehydration is carried out in rapid-evaporation mechanical devices. Drying is not a reliable microbicidal method, however. Numerous resistant microbes such as micrococci, coliforms, staphylococci, salmonellae, and fungi survive in dried milk and eggs, which can subsequently serve as agents of spoilage and infections.

Microbial Fermentations in Food Products from Plants

In contrast to measures aimed at keeping unwanted microbes out of food are the numerous cases in which they are deliberately added to food. Common substances such as bread, cheese, beer, wine, yogurt, and pickles are the result of **food fermentations.** These reactions actively encourage microbial growth and biochemical activities that impart a particular taste, smell, or appearance to food. The microbe or microbes can occur naturally on the food substrate as in sauerkraut, or they can be added as pure or mixed samples of known bacteria, molds, or yeasts called **starter cultures.** Many food fermentations are synergistic, with a series of microbes acting in concert to convert a starting substrate to the desired end product. Because large-scale production of fermented milk, cheese, bread, alcoholic brews, and vinegar depends upon inoculation with starter cultures, considerable effort is spent selecting, maintaining, and preparing these cultures and excluding contaminants that can spoil the fermentation. Most starting raw materials are of plant origin (grains, vegetables, beans) and, to a lesser extent of animal origin (milk, meat).

Bread

Microorganisms accomplish three functions in bread making: (1) **leavening** the flour-based dough, (2) imparting flavor and odor, and (3) conditioning the dough to make it workable. Leavening is achieved primarily through the release of gas to produce a porous and spongy product. Without leavening, bread dough remains dense, flat, and hard. Although various microbes and leavening agents can be used, the most common ones are various strains of the baker's yeast *Saccharomyces cerevisiae.* Other gas-forming microbes such as coliform bacteria, certain *Clostridium* species, heterofermentative lactic acid bacteria, and wild yeasts can be employed, depending on the type of bread desired.

Yeast metabolism requires a source of fermentable sugar such as maltose or glucose. Because the yeast respires aerobically in bread dough, the chief products of maltose fermentation are carbon dioxide and water rather than alcohol (the main product in beer and wine). Yeast activity can be modified by adjusting the size of the inoculum, the sugar level, temperature, and salt concentration. Other contributions to bread texture come from kneading, which incorporates air into the dough, and from microbial enzymes, which break down flour proteins (gluten) and give the dough elasticity.

Besides carbon dioxide production, bread fermentation generates other volatile organic acids and alcohols that impart delicate flavors and aromas. These are especially well developed in home-baked bread, which is leavened more slowly than commercial bread. Yeasts and bacteria can also impart unique flavors, depending upon the culture mixture and baking techniques used. The tangy, sour flavor of rye bread, for example, comes in part from starter cultures of lactic acid bacteria such as *Lactobacillus plantarum, L. brevis, L. bulgaricus, Leuconostoc mesenteroides,* and *Streptococcus thermophilus.* Sourdough bread gets its unique tang from *Lactobacillus sanfrancisco* (of course!).

Beer and Other Alcoholic Beverages

The production of alcoholic beverages takes advantage of another useful property of yeasts. By fermenting carbohydrates in fruits or grains anaerobically, they produce ethyl alcohol, as shown by this equation (see figure 8.25):

$$C_6H_{12}O_6 \rightarrow 2C_2H_5OH + 2CO_2$$

(yeast + sugar = ethanol + carbon dioxide)

Depending upon the starting materials and the processing method, alcoholic beverages vary in alcohol content and flavor. The principal types of fermented beverages are malt liquors, wines, and spirit liquors.

The earliest evidence of beer brewing appears in ancient tablets by the Sumerians and Babylonians around 6,000 B.C. The starting ingredients for both ancient and present-day versions of beer, ale, stout, porter, and other variations are water, malt (barley grain), hops, and special strains of yeasts. The steps in brewing include malting, mashing, adding hops, fermenting, aging, and finishing (figure 26.31).

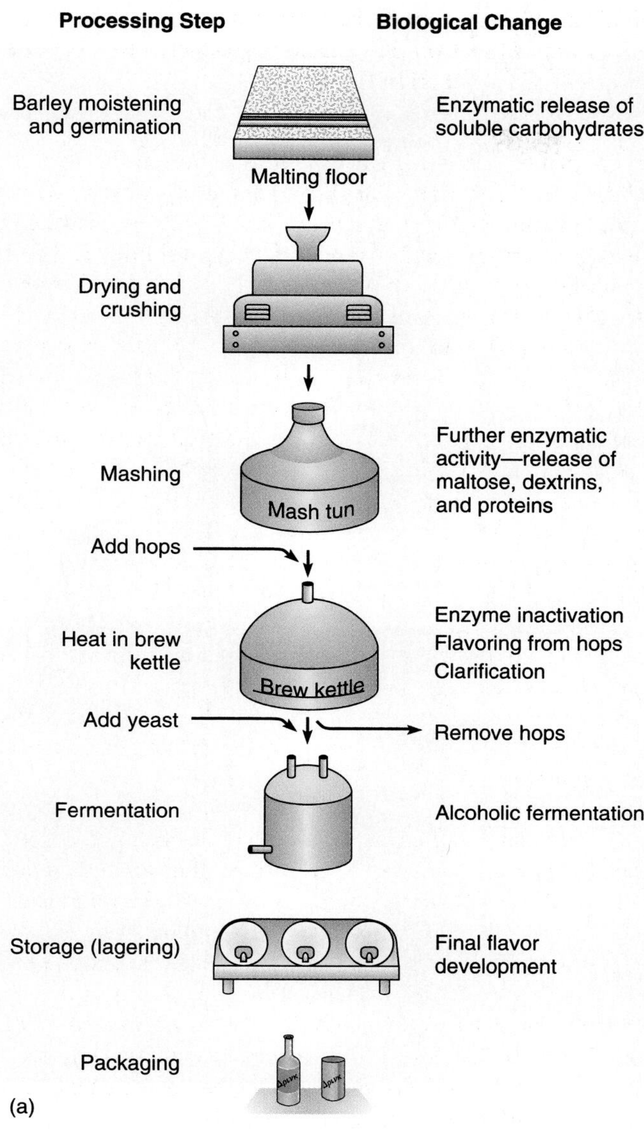

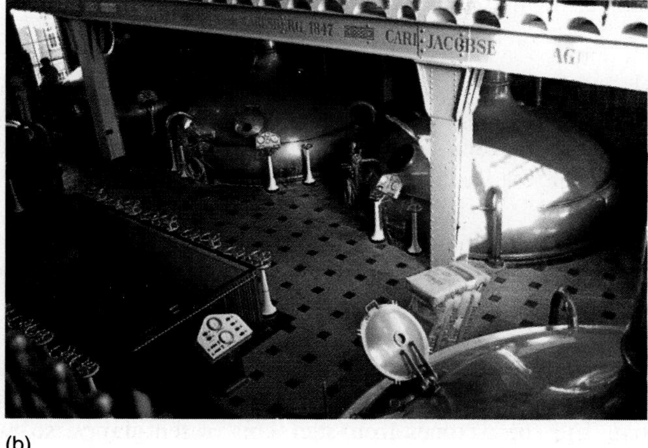

Figure 26.31
(*a*) The general stages in brewing beer, ale, and other malt liquors. (*b*) Fermentation tanks at a commercial brewery.

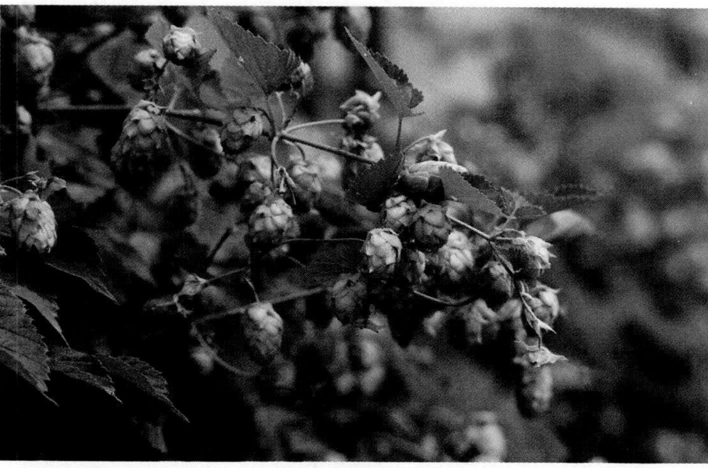

Figure 26.32
Female flowers of hops, the herb that gives beer some of its flavor and aroma.

For brewer's yeast to convert the carbohydrates in grain into ethyl alcohol, the barley must first be sprouted and softened to make its complex nutrients available to microorganisms. This process, called **malting,*** releases amylases that convert starch to maltose and dextrins and proteases that digest proteins. Other sugar and starch supplements added in some forms of beer are corn, rice, wheat, soybeans, potatoes, and sorghum. After the sprouts have been separated, the remaining malt grain is dried and stored in preparation for mashing.

The malt grain is soaked in warm water and ground up to prepare a **mash.** Sugar and starch supplements are then introduced to the mash mixture, which is heated to a temperature of about 65° to 70°C. During this step, the starch is hydrolyzed by amylase, and simple sugars are released. Heating this mixture to 75°C stops the activity of the enzymes. Solid particles are next removed by settling and filtering. **Wort,*** the clear fluid that comes off, is rich in dissolved carbohydrates. It is boiled for about 2.5 hours with **hops,** the dried scales of the female flower of *Humulus lupulus,* a climbing herb (figure 26.32). This boiling phase serves several purposes: (1) It extracts humulus, the bitter acids and resins that give aroma and flavor to the finished product;[6] (2) it caramelizes the sugar and imparts a golden or brown color; (3) it effectively destroys any bacterial contaminants that can destroy flavor; and (4) it concentrates the mixture. The filtered and cooled supernatant is then ready for the addition of yeasts and fermentation.

Fermentation begins when wort is inoculated with a species of *Saccharomyces* that has been specially developed for beer making. Top yeasts such as *Saccharomyces cerevisiae* function at the surface and are used to produce the higher alcohol content of *ales.* Bottom yeasts such as *S. uvarum (carlsbergensis)* function deep in

6. Humulus is an ingredient in beer that provides some rather interesting side effects besides flavor. It is a moderate sedative, a diuretic, and a mild antiseptic.

*malting (mawlt′-ing) Gr. *malz,* to soften.

*wort (wurt) O.E. *wyrt,* a spice.

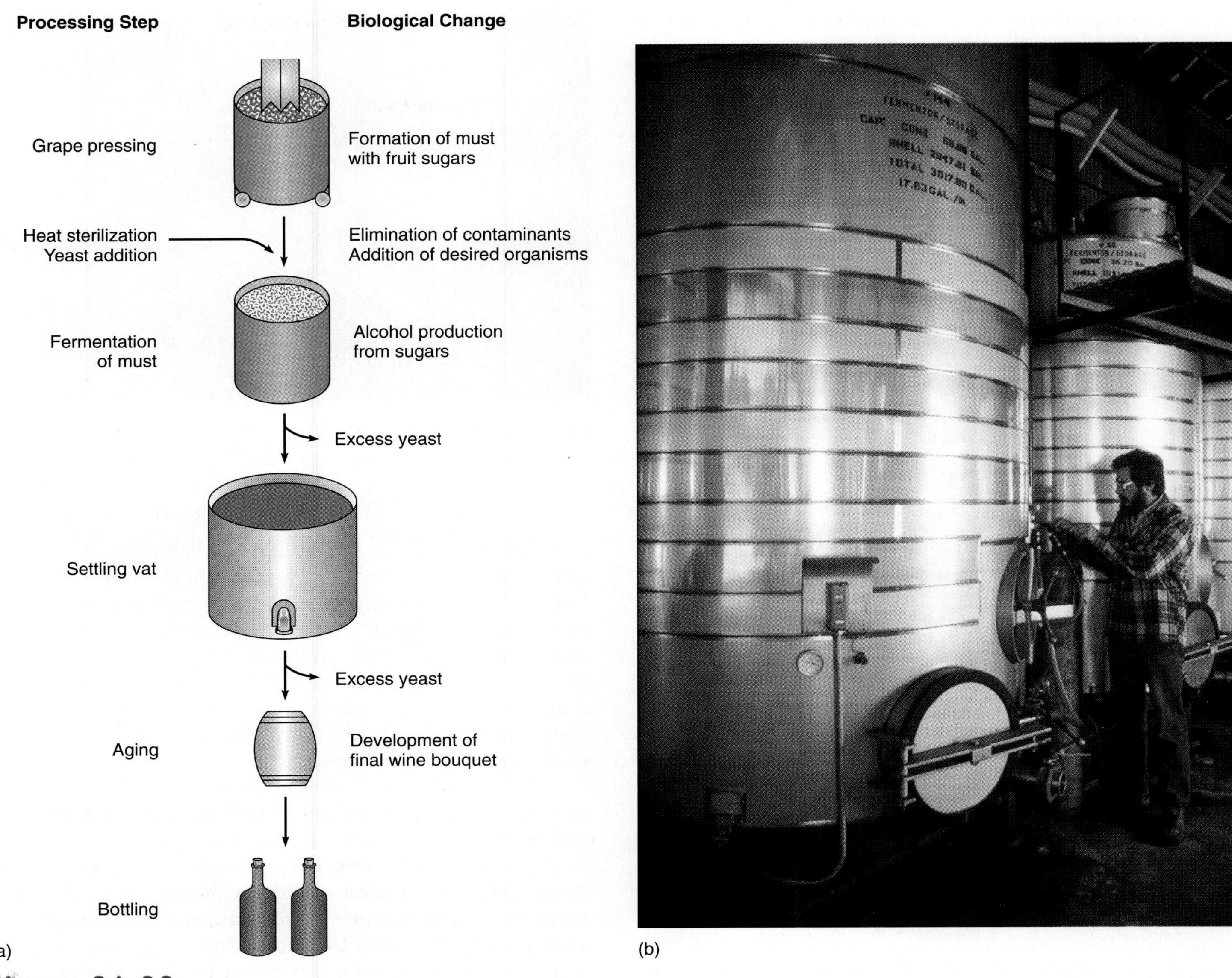

Figure 26.33
(*a*) General steps in wine making. (*b*) Wine fermentation vats in a larger commercial winery.

the fermentation vat and are used to make other beers. In both cases, the initial inoculum of yeast starter is aerated briefly to promote rapid growth and increase the load of yeast cells. Shortly thereafter, an insulating blanket of foam and carbon dioxide develops on the surface of the vat and promotes anaerobic conditions. During fermentation, which lasts 8 to 14 days, the wort sugar is converted chiefly to ethanol and carbon dioxide. The diversity of flavors in the finished product is partly due to the release of small amounts of glycerol, acetic acid, and esters. Fermentation is self-limited, and it essentially ceases when a concentration of 3% to 6% ethyl alcohol is reached.

Freshly fermented, or "green," beer is **lagered,*** meaning it is held for several weeks to months in vats at 0°C. During this maturation period, yeast, proteins, resin, and other materials settle, leaving behind a clear, mellow fluid. Lager beer is subjected to a final cold filtration or pasteurization step to remove any residual yeasts that could spoil it. Finally, it is carbonated with carbon dioxide collected during fermentation. The end product is packaged in kegs, bottles, or cans.

Wine

Wine is traditionally considered any alcoholic beverage arising from the fermentation of grape juice, but practically any fruit can be rendered into wine. The essential starting point is the preparation of **must,** the juice given off by crushed fruit that is used as a substrate for fermentation. In general, grape wines are either *white* or *red.* The color comes from the skins of the grapes, so white wine is prepared either from white-skinned grapes or from red-skinned grapes that have had the skin removed. Red wine comes from the red- or purple-skinned varieties. Steps in making wine include must preparation (crushing), fermentation, storage, and aging (figure 26.33).

*lagered (law'-gurd) Gr. *lager,* to store or age.

For proper fermentation, must should contain 12% to 25% glucose or fructose, so the art of wine making beings in the vineyard. Grapes are harvested when their sugar content reaches 15% to 25%, depending on the type of wine to be made. Grapes from the field carry a mixed microflora on their surface called the *bloom* that can serve as a source of wild yeasts. Some wine makers allow these natural yeasts to dominate, but many wineries inoculate the must with a special strain of *Saccharomyces cerevisiae,* variety *ellipsoideus.* To discourage yeast and bacterial spoilage agents, wine makers sometimes treat grapes with sulfur dioxide or potassium metabisulfite. The inoculated must is thoroughly aerated and mixed to promote rapid aerobic growth of yeasts, but when the desired level of yeast growth is achieved, anaerobic alcoholic fermentation is begun.

The temperature of the vat during fermentation must be carefully controlled to facilitate alcohol production. The length of fermentation varies from 3 to 5 days in red wines and from 7 to 14 days in white wines. The initial fermentation yields ethanol concentrations reaching 7% to 15% by volume, depending upon the type of yeast, the source of the juice, and ambient conditions. From this point, the process varies with the wine. *Dry wines* can undergo a second period of fermentation to reduce the sugar content, whereas *sweet wines* can be immediately placed in storage chambers. The fermented juice (raw wine) is decanted and transferred to large vats to settle and clarify. Before the final aging process, it can then be flash-pasteurized to precipitate proteins and filtered to remove any remaining sediments. Wine is aged in wooden casks for varying time periods (months to years), after which it is bottled and stored for further aging. During aging, nonmicrobial changes produce aromas and flavors (the bouquet) characteristic of a particular wine.

The fermentation processes discussed thus far can only achieve a maximum alcoholic content of 17%, because concentrations above this level inhibit the metabolism of the yeast. The fermentation product must be **distilled** to obtain higher concentrations such as those found in liquors. During distillation, heating the liquor separates the more volatile alcohol from the less volatile aqueous phase. The alcohol is then condensed and collected. The alcohol content of distilled liquors is rated by *proof,* a measurement that is usually two times the alcohol content. Thus, 80 proof vodka contains 40% ethyl alcohol.

Distilled liquors originate through a process similar to wine making, though the starting substrates can be extremely diverse. In addition to distillation, liquors can be subjected to special treatments such as aging to provide unique flavor or color. Vodka, a colorless liquor, is usually prepared from fermented potatoes, and rum is distilled from fermented sugarcane. Assorted whiskeys are derived from fermented grain mashes; rye whiskey is produced from rye mash, and bourbon from corn mash. Brandy is distilled grape, peach, or apricot wine.

Other Fermented Plant Products

Fermentation provides an effective way of preserving vegetables, as well as enhancing flavor with lactic acid and salt. During pickling fermentations, vegetables are immersed in an anaerobic salty solution (brine) to extract sugar and nutrient-laden juices. The salt also disperses bacterial clumps, and its high osmotic pressure inhibits proteolytic bacteria and sporeformers that can spoil the product.

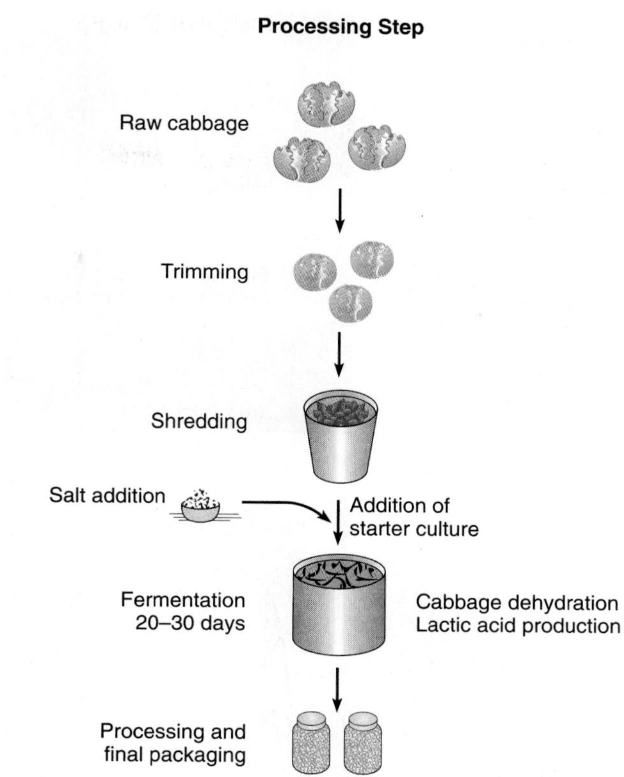

Figure 26.34
Steps in the preparation of sauerkraut.

Sauerkraut is the fermentation product of cabbage (figure 26.34). Cabbage is washed, wilted, shredded, salted, and packed tightly into a fermentation vat. Weights cover the cabbage mass and squeeze out its juices. The fermentation is achieved by natural cabbage microflora or by an added culture. The initial agent of fermentation is *Leuconostoc mesenteroides,* which grows rapidly in the brine and produces lactic acid. It is followed by *Lactobacillus plantarum,* which continues to raise the acid content to as high as 2% (pH 3.5) by the end of fermentation. The high acid content restricts the growth of spoilage microbes.

Fermented cucumber pickles come chiefly in salt (sour, sweet-sour, mixed pickles, and relished) and dill varieties. Salt pickles are prepared by washing immature cucumbers, placing them in barrels of brine, and allowing them to ferment for 6 to 9 weeks. The brine can be inoculated with *Pediococcus cerevisiae* and *Lactobacillus plantarum* to avoid unfavorable qualities caused by natural microflora and to achieve a more consistent product. Fermented dill pickles are prepared in a somewhat more elaborate fashion, with the addition of dill herb, spices, garlic, onion, and vinegar.

Natural vinegar is produced when the alcohol in fermented plant juice is oxidized to acetic acid, which is responsible for the pungent odor and sour taste. Although a reasonable facsimile of vinegar could be made by mixing about 4% acetic acid and a dash of sugar in water, this preparation would lack the traces of various

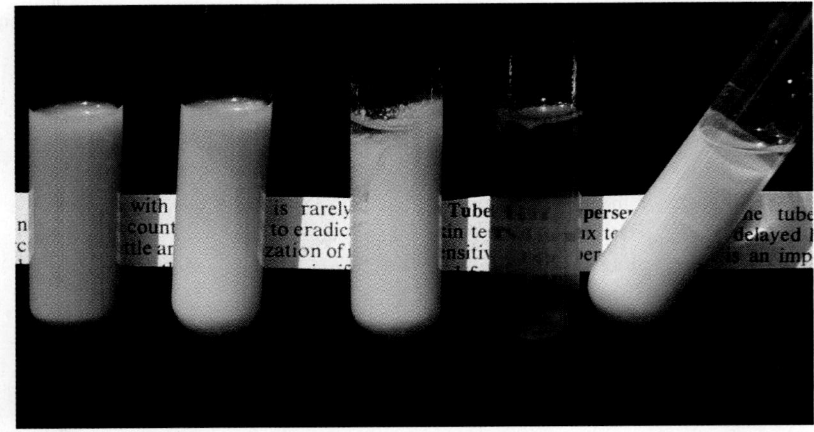

(a)

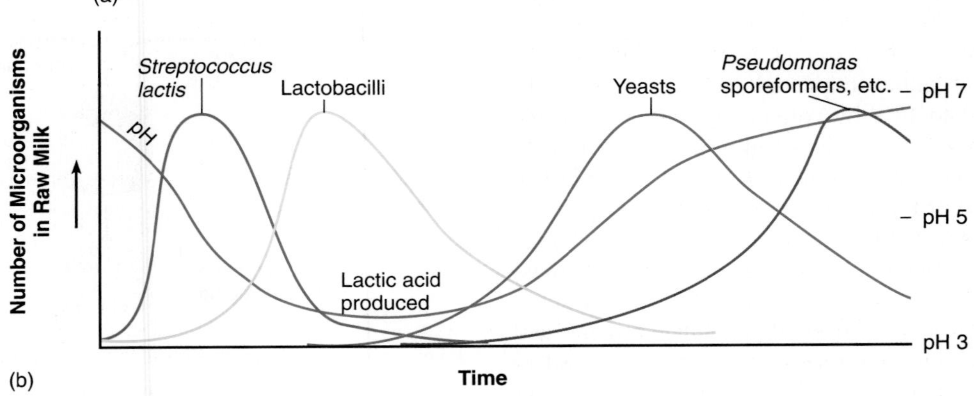

(b)

Figure 26.35

(*a*) Litmus milk is a medium used to indicate pH and consistency changes in milk resulting from microbial action. The first tube is an uninoculated, unchanged control. The second tube has a white, decolorized zone indicative of litmus reduction. The third tube has become acidified (pink), and its proteins have formed a loose curd. In the fourth tube, digestion of milk proteins has caused complete clarification or peptonization of the milk. The fifth tube shows a well-developed solid curd overlaid by a clear fluid, the whey. (*b*) Chart depicting spontaneous changes in the number and type of microorganism and the pH of raw milk as it incubates.

esters, alcohol, glycerin, and volatile oils that give natural vinegar its pleasant character. Vinegar is actually produced in two stages. The first stage is similar to wine or beer making, in which a plant juice is fermented to alcohol by *Saccharomyces.* The second stage involves an aerobic fermentation carried out by acetic acid bacteria in the genera *Acetobacter* and *Gluconbacter.* These bacteria oxidize the ethanol in a two-step process, as shown here:

$$\underset{\text{Ethanol}}{2C_2H_5OH} + \frac{1}{2}O_2 \rightarrow \underset{\text{Acetaldehyde}}{CH_3CHO} + H_2O$$

$$\underset{\text{Acetaldehyde}}{CH_3CHO} + \frac{1}{2}O_2 \rightarrow \underset{\text{Acetic acid}}{CH_3COOH}$$

The abundance of oxygen necessary in commercial vinegar making is furnished by exposing inoculated raw material to air by arranging it in thin layers in open trays, allowing it to trickle over loosely packed beechwood twigs and shavings, or aerating it in a large vat. Different types of vinegar are derived from substrates such as apple cider (cider vinegar), malted grains (malt vinegar), and grape juice (wine vinegar).

Microbes in Milk and Dairy Products

Milk has a highly nutritious composition. It contains an abundance of water and is rich in minerals, protein (chiefly casein), butterfat, sugar (especially lactose), and vitamins. It starts its journey in the udder of a mammal as a sterile substance, but as it passes out of the teat, it is inoculated by the animal's normal flora. Other microbes can be added during milking from the milker, the milking machinery, and collection vessels. Because milk is a nearly perfect culture medium, it is highly susceptible to microbial growth. When raw milk is left at room temperature, a series of bacteria ferment the lactose, produce acid, and alter the milk's content and texture (figure 26.35*a*). This progression can occur naturally, or it can be induced, as in the production of cheese and yogurt.

In the initial stages of milk fermentation, lactose is rapidly attacked by *Streptococcus lactis* and *Lactobacillus* species (figure 26.35*b*). The resultant lactic acid accumulation and lowered pH cause the milk proteins to coagulate into a solid mass called the **curd.** Curdling also causes the separation of a watery liquid called **whey** on the surface. Curd can be produced by microbial action or by an enzyme, **rennin** (casein coagulase), which is isolated from the stomach of unweaned calves.

Cheese

Since 5,000 B.C., various forms of cheese have been produced by spontaneous fermentation of cow, goat, or sheep milk. Present-day, large-scale cheese production is carefully controlled and uses only freeze-dried samples of pure cultures (figure 26.36). These are first inoculated into a small quantity of pasteurized milk to form an active starter culture. This amplified culture is subsequently inoculated into a large vat of milk, where rapid curd development takes place. Such rapid growth is desired because it promotes the overgrowth of the desired inoculum and prevents the activities of undesirable contaminants. Rennin is usually added to increase the rate of curd formation.

After its separation from whey, the curd is rendered to produce one of the 20 major types of soft, semisoft, or hard cheese. The composition of cheese is varied by adjusting water, fat, acid, and salt content. Cottage and cream cheese are examples of the soft, more perishable variety. After light salting and the optional addition of cream, they are ready for consumption without further processing. Other cheeses acquire their character from "ripening," a complex curing process involving bacterial, mold, and enzyme reactions that develop the final flavor, aroma, and other features characteristic of particular cheeses.

The distinctive traits of soft cheeses such as Limburger, Camembert, and Liederkranz are acquired by ripening with a reddish-brown mucoid coating of yeasts, micrococci, and molds. The microbial enzymes permeate the curd and ferment lipids, proteins, carbohydrates, and other substrates. This process leaves assorted acids, salts, aldehydes, ketones, and other by-products that give the finished cheese powerful aromas and delicate flavors. Semisoft varieties of cheese such as Roquefort, blue, or Gorgonzola are infused and aged with a strain of *Penicillium roqueforti* mold. Hard cheeses such as Swiss, cheddar, and Parmesan develop a sharper flavor by aging with selected bacteria. The pockets in Swiss cheese come from entrapped carbon dioxide formed by *Propionibacterium,* which is also responsible for its bittersweet taste.

Other Fermented Milk Products

Yogurt is formed by the fermentation of milk by *Lactobacillus bulgaricus* and *Streptococcus thermophilus.* These organisms produce organic acids and other flavor components and can grow in such numbers that a gram of yogurt regularly contains 100 million bacteria. Live cultures of *Lactobacillus acidophilus* are an important additive to acidophilus milk, which is said to benefit digestion and to help maintain the normal flora of the intestine. Fermented milks such as kefir, koumiss, and buttermilk are a basic food source in many cultures.

MICROORGANISMS AS FOOD

At first, the thought of eating bacteria, molds, algae, and yeasts may seen odd or even unappetizing. We do eat their macroscopic relatives such as mushrooms, truffles, and seaweed, but we are used to thinking of the microscopic forms as agents of decay and disease, or at most, as food flavorings. The consumption of

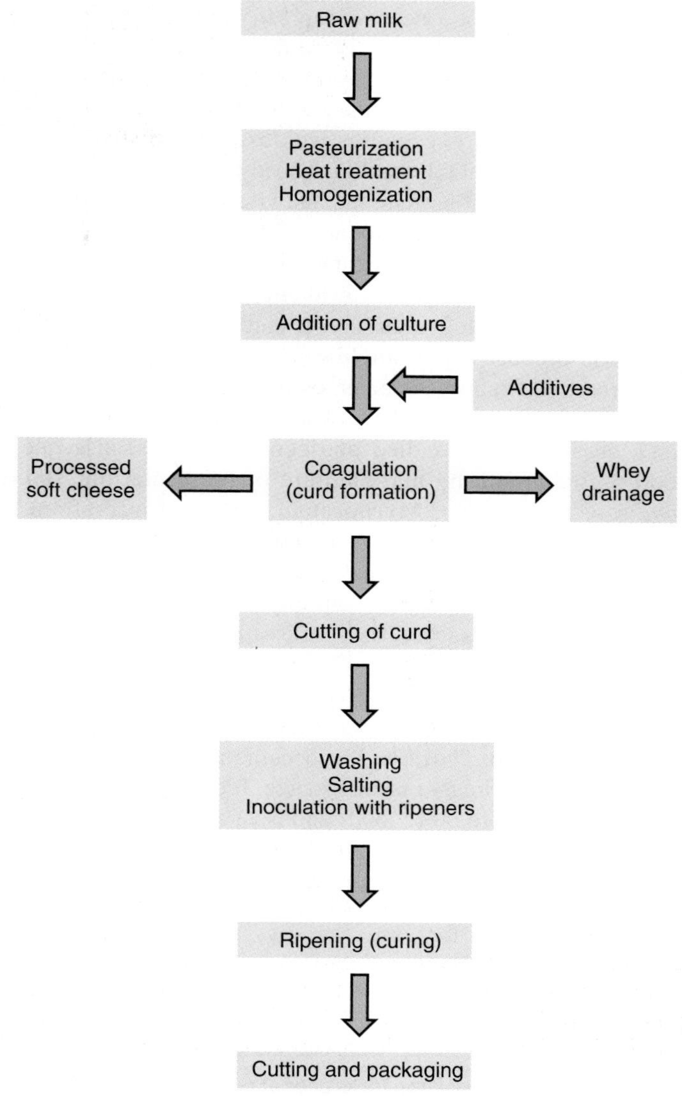

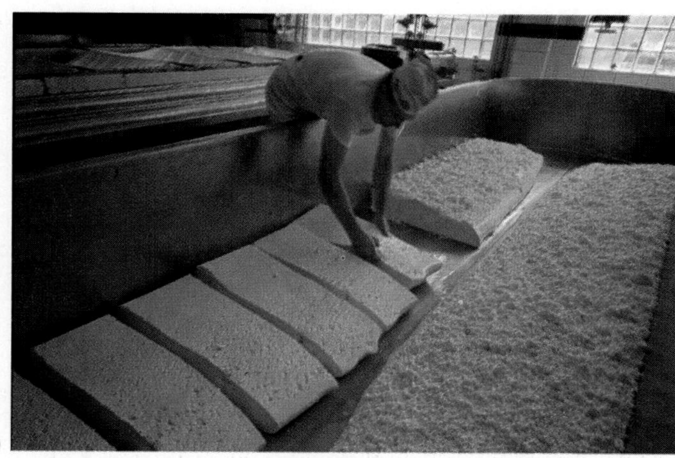

Figure 26.36
(*a*) A flow diagram that summarizes the main events in cheese making. (*b*) The curd-cutting stage in the making of cheddar cheese.

microorganisms is not a new concept. In Germany during World War II, it became necessary to supplement the diets of undernourished citizens by adding yeasts and molds to foods. At present, most countries are able to produce enough food for their inhabitants, but in the future, countries with exploding human populations and dwindling arable land may need to consider microbes as a significant source of protein, fat, and vitamins. Several countries already commercially mass produce food yeasts, bacteria, and, in a few cases, algae. Although eating microbes has yet to win total public acceptance, their use as feed supplements for livestock is increasing. A technology that shows some promise in increasing world food productivity is **single-cell protein (SCP).** This material is produced from waste materials such as molasses from sugar refining, petroleum by-products, and agricultural wastes. In England, an animal feed called **pruteen** is produced at a rate of 3,000 tons a month by mass culture of the bacterium *Methylophilus methylotrophus.* **Mycoprotein,** a product made from the fungus *Fusarium graminearum,* is also sold there. The filamentous texture of this product makes it a likely candidate for producing meat substitutes for human consumption.

Health food stores carry bottles of dark green pellets or powder that are a pure culture of a spiral-shaped cyanobacterium called *Spirulina.* This microbe is harvested from the surface of lakes and ponds, where it grows in great mats. In some parts of Africa and Mexico, *Spirulina* has become a viable alternative to green plants as a primary nutrient source. It can be eaten in its natural form or added to other foods and beverages. The microbe has high productivity and requires very little area or resources to grow. *Spirulina* is nontoxic, high in protein, and has a mild, vegetable-like flavor.

Chapter Checkpoints

The use of microorganisms for practical purposes to benefit humans is called biotechnology.

Microorganisms can compete with humans for the nutrients in food. Their presence in food can be detrimental, beneficial, or of neutral consequence to human consumers. Food poisoning can be an intoxication caused by microbial toxins produced as by-products of microbial decomposition of food. Or it can be a food infection when pathogenic microorganisms in the food attack the human host after being consumed.

Heat, radiation, chemicals, and drying are methods used to limit numbers of microorganisms in food. The type of method used depends on the nature of the food and the type of pathogens or spoilage agents it contains.

Food fermentation processes utilize bacteria or yeast to produce desired components such as alcohols and organic acids in foods and beverages. Beer, wine, yogurt, and cheeses are examples of such processes.

Some microorganisms are used as a source of protein. Examples are single-cell protein, mycoprotein, and *Spirulina.* Microbial protein could replace meat as a major protein source.

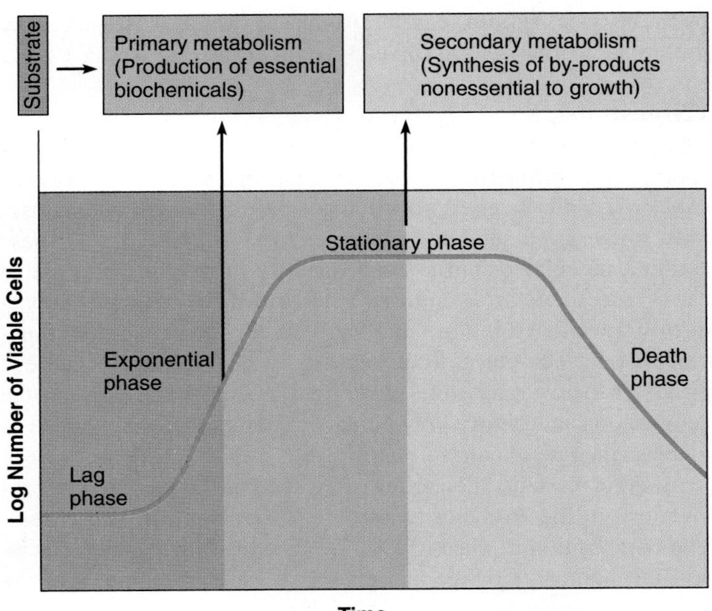

Figure 26.37
The origins of primary and secondary microbial metabolites harvested by industrial processes.

GENERAL CONCEPTS IN INDUSTRIAL MICROBIOLOGY

Virtually any large-scale commercial enterprise that enlists microorganisms to manufacture consumable materials is part of the realm of **industrial microbiology.** Here the term pertains primarily to bulk production of organic compounds such as antibiotics, hormones, vitamins, acids, solvents (table 26.3), and enzymes (table 26.4). Many of the processing steps involve fermentations similar to those described in food technology, but industrial processes usually occur on a much larger scale, produce a specific compound, and involve numerous complex stages. The aim of industrial microbiology is to produce chemicals that can be purified and packaged for sale or for use in other commercial processes. Thousands of tons of organic chemicals worth several billion dollars are produced by this industry every year. To create just one of these products, an industry must determine which microbes, starting compounds, and growth conditions work best. The research and development involved requires an investment of 10 to 15 years and billions of dollars.

The microbes used by fermentation industries have traditionally been naturally occurring (wild) strains of bacteria or molds that carry out a particular metabolic action on a substrate. At an increasing rate, however, these microbes are mutant strains of fungi or bacteria that selectively synthesize large amounts of various metabolic intermediates, or **metabolites.** Two basic kinds of metabolic products are harvested by industrial processes: (1) *Primary metabolites* are produced during the major metabolic pathways and are essential to the microbe's function. (2) *Secondary metabolites* are by-products of metabolism that may not be

TABLE 26.3

INDUSTRIAL PRODUCTS OF MICROORGANISMS

Chemical	Microbial Source	Substrate	Applications
Pharmaceuticals			
Bacitracin	*Bacillus subtilis*	Glucose	Antibiotic effective against gram-positive bacteria
Cephalosporins	*Cephalosporium*	Glucose	Antibacterial antibiotic, broad spectrum
Pencillins	*Penicillium chrysogenum*	Lactose	Antibacterial antibiotics, broad and narrow spectrum
Erythromycin	*Streptomyces*	Glucose	Antibacterial antibiotic, broad spectrum
Tetracycline	*Streptomyces*	Glucose	Antibacterial antibiotic, broad spectrum
Amphotericin B	*Streptomyces*	Glucose	Antifungal antibiotic
Vitamin B_{12}	*Pseudomonas*	Molasses	Dietary supplement
Riboflavin	*Asbya*	Glucose, corn oil	Animal feed supplement
Steroids (hydrocortisone)	*Rhizopus, Cunninghamella*	Deoxycholic acid Stigmasterol	Treatment of inflammation, allergy; hormone replacement therapy
Food Additives and Amino Acids			
Citric acid	*Aspergillus, Candida*	Molasses	Acidifier in soft drinks; used to set jam; candy additive; fish preservative; retards discoloration of crab meat; delays browning of sliced peaches
Lactic acid	*Lactobacillus, Bacillus*	Whey, corn cobs, cottonseed; from maltose, glucose, sucrose	Acidifier of jams, jellies, candies, soft drinks, pickling brine, baking powders
Xanthan	*Xanthomonas*	Glucose medium	Food stabilizer; not digested by humans
Acetic acid	*Acetobacter*	Any ethylene source, ethanol	Food acidifier; used in industrial processes
Glutamic acid	*Corynebacterium, Arthrobacter, Brevibacterium*	Molasses, starch source	Flavor enhancer monosodium glutamate (MSG)
Lysine	*Corynebacterium*	Casein	Dietary supplement (absent in cereals)
Miscellaneous Chemicals			
Ethanol	*Saccharomyces*	Beet, cane, grains, wood, wastes	Additive to gasoline (gasohol)
Acetone	*Clostridium*	Molasses, starch	Solvent for lacquers, resins, rubber, fat, oil
Butanol	*Clostridium*	Molasses, starch	Added to lacquer, rayon, detergent, brake fluid
Gluconic acid	*Aspergillus, Gluconobacter*	Corn steep, any glucose source	Baking powder, glass-bottle washing agent, rust remover, cement mix, pharmaceuticals
Glycerol	Yeast	By-product of alcohol fermentation	Explosive (nitroglycerine)
Dextran	*Klebsiella, Acetobacter, Leuconostoc*	Glucose, molasses, sucrose	Polymer of glucose used as adsorbents, blood expanders, and in burn treatment; a plasma extender; used to stabilize ice cream, sugary syrup, candies
Thuricide	*Bacillus thuringiensis*	Molasses, starch	Used in biocontrol of caterpillars, moths, loopers, and hornworm plant pests

critical to the microbe's function (figure 26.37). In general, primary products are compounds such as amino acids, and organic acids synthesized during the logarithmic phase of microbial growth, and secondary products are compounds such as vitamins, antibiotics, and steroids synthesized during the stationary phase (see chapter 7). Most strains of industrial microorganisms have been chosen for their high production of a particular primary or secondary metabolite. Certain strains of yeasts and bacteria can produce 20,000 times more metabolite than a wild strain of that same microbe.

Industrial microbiologists have several tricks to increase the amount of the chosen end product. First, they can manipulate the growth environment to increase the synthesis of a metabolite. For instance, adding lactose instead of glucose as the fermentation substrate increases the production of penicillin by *Penicillium.* Another strategy is to select microbial strains that genetically lack a feedback system to regulate the formation of end products, thus encouraging mass accumulation of this product. Many syntheses occur in sequential fashion, wherein the waste products of one organism become the building blocks of the next. During these

TABLE 26.4

INDUSTRIAL ENZYMES AND THEIR USES

Enzyme	Source	Application
Amylase	*Aspergillus, Bacillus, Rhizopus*	Flour supplement, desizing textiles, mash preparation, syrup manufacture, digestive aid, precooked foods, spot remover in dry cleaning
Catalase	*Micrococcus, Aspergillus*	To prevent oxidation of foods; used in cheese production, cake baking, irradiated foods
Cellulase	*Aspergillus, Trichoderma*	Liquid-coffee concentrates, digestive aid, increase digestibility of animal feed, degradation of wood or wood by-products
Glucose oxidase	*Aspergillus*	Removal of glucose or oxygen that can decolorize or alter flavor in food preparations as in dried egg products; glucose determination in clinical diagnosis
Hyaluronidase	Various bacteria	Medical use in wound cleansing, preventing surgical adhesions
Keratinase	*Streptomyces*	Hair removal from hides in leather preparation
Lipase	*Rhizopus*	Digestive aid, to develop flavors in cheese and milk products
Pectinase	*Aspergillus, Sclerotina*	Clarifies wine, vinegar, syrups, and fruit juices by degrading pectin, a gelatinous substance; used in concentrating coffee
Penicillinase	*Bacillus*	Removal of penicillin in research
Proteases	*Aspergillus, Bacillus, Streptomyces*	To clear and flavor rice wines, process animal feed, remove gelatin from photographic film, recover silver, tenderize meat, unravel silkworm cocoon, remove spots
Rennet	*Mucor*	To curdle milk in cheese making
Streptokinase	*Streptococcus*	Medical use in clot digestion, as a blood thinner
Streptodornase	*Streptococcus*	Promotes healing by removing debris from wounds and burns

biotransformations, the substrate undergoes a series of slight modifications, each of which gives off a different by-product (figure 26.38). The production of an antibiotic such as tetracycline requires several microorganisms and 72 separate metabolic steps.

FROM MICROBIAL FACTORIES TO INDUSTRIAL FACTORIES

Industrial fermentations begin with microbial cells acting as living factories. When exposed to optimum conditions, they multiply in the trillions, and synthesize large volumes of a desired product. Such mass microbial fermentations are the driving force of industrial microbiology. Producing appropriate levels of growth and fermentation requires cultivation of the microbes in a carefully controlled environment. This process is basically similar to culturing bacteria in a test tube of nutrient broth. It requires a sterile medium containing appropriate nutrients, protection from contamination, provisions for introduction of sterile air or total exclusion of air, and a suitable temperature and pH (figure 26.39).

Many commercial fermentation processes have been worked out on a small scale in a lab and then *scaled up* to a large commercial venture. An essential component for scaling up is a **fermentor,** a device in which mass cultures are grown, reactions take place, and product develops. Some fermentors are large tubes, flasks, or vats, but most industrial types are metal cylinders with built-in mechanisms for stirring, cooling, monitoring, and harvesting product (figure 26.40). Fermentors are made of materials that can withstand pressure and are rust-proof, nontoxic, and leakproof. They range in holding capacity from small, 5-gallon systems used in research labs to larger, 5,000- to 100,000-gallon vessels (figure 26.39), and in some industries, to tanks of 250 million to 500 million gallons.

For optimum yield, a fermentor must duplicate the actions occurring in a tiny volume (a test tube) on a massive scale. Most microbes performing fermentations have an aerobic metabolism, and the large volumes make it difficult to provide adequate oxygen. Fermentors have a built-in device called a *sparger* that aerates the medium to promote aerobic growth. Paddles *(impellers)* located in the central part of the fermentor increase the contact between the microbe and the nutrients by vigorously stirring the fermentation mixture. Their action also maintains its uniformity. The temperature of the chamber is maintained by a cooling jacket.

SUBSTANCE PRODUCTION

The general steps in mass production of organic substances in a fermentor are illustrated in figure 26.41. These can be summarized as: (1) introduction of microbes and sterile media into the reaction chamber, (2) fermentation, (3) *downstream processing* (recovery, purification, and packaging of product), and (4) removal of waste. All phases of production must be carried out aseptically and monitored (usually by computer) for rate of flow and quality of product. The starting raw substrates include crude plant residues, molasses, sugars, fish and meat meals, and whey. Additional chemicals can be added to control pH or to increase the yield. In *batch fermentation,* the substrate is added to the system all at once and taken through a limited run until product is harvested. In *continuous feed systems,* nutrients are continuously fed into the reactor and the product is siphoned off throughout the run.

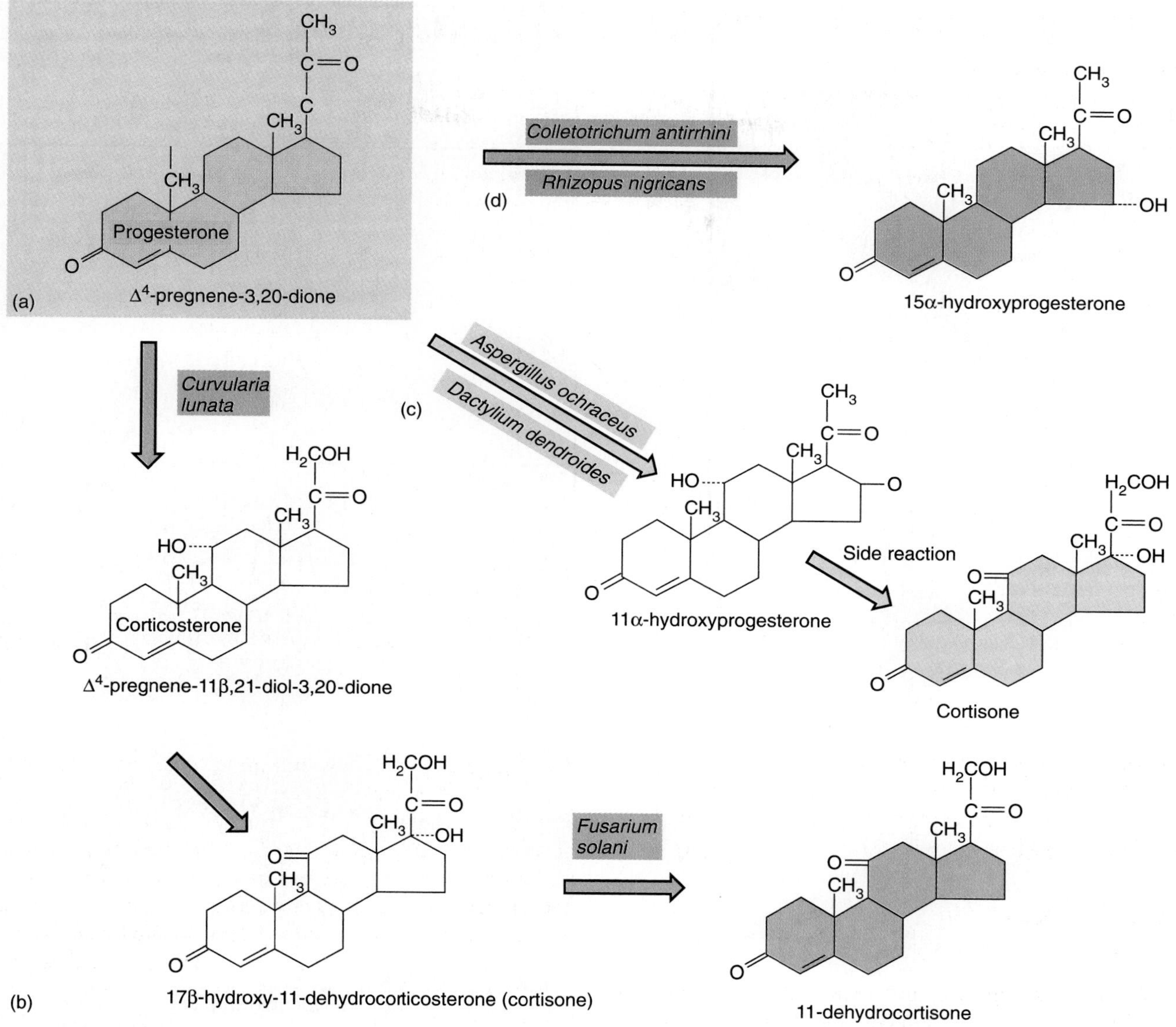

Figure 26.38
An example of biotransformation by microorganisms in the industrial production of steroid hormones. Desired hormones such as cortisone and progesterone can require several steps and microbes, and rarely can a single microbe carry out all the required synthetic steps. (*a*) Beginning with starting compound Δ^4-pregnene-[3,20-dione, (*b*) two species of mold can produce 11-dehydrocortisone in three steps. (*c*) Two other species of mold can convert this starting compound to cortisone in two steps, and (*d*) two species of mold can produce 15α-hydroxyprogesterone in one step.

Ports in the fermentor allow the raw product and waste materials to be recovered from the reactor chamber when fermentation is complete. The raw product is recovered by settling, precipitation, centrifugation, filtration, or cell lysis. Some products come from this process ready to package, whereas others require further purification, extraction, concentration, or drying. The end product is usually in a powder, cake, granular, or liquid form that is placed in sterilized containers. The waste products can be siphoned off to be used in other processes or discarded, and the residual microbes and nutrients from the fermentation chamber can be recycled back into the system or removed for the next run.

Fermentation technology for large-scale cultivation or microbes and production of microbial products is versatile. Table 26.3 itemizes some of the major pharmaceutical substances, food additives, and solvents produced by microorganisms.

Pharmaceutical Products

Health care products derived from microbial biosynthesis are antibiotics, hormones, vitamins, and vaccines. The first mass-produced antimicrobic was penicillin, which came from *Penicillium chrysogenum,* a mold first isolated from a cantaloupe in Wisconsin. The current strain of this species has gone through 40 years of selective mutation and screening to increase its yield. (The original wild *P. chrysogenum* synthesized 60 mg/ml of medium, and the latest isolate yields 85,000 mg/ml). The semisynthetic penicillin derivatives (see microfile 12.2) are produced by

Figure 26.39
A cell culture vessel used to mass produce pharmaceuticals. Such elaborate systems require the highest levels of sterility and clean techniques.

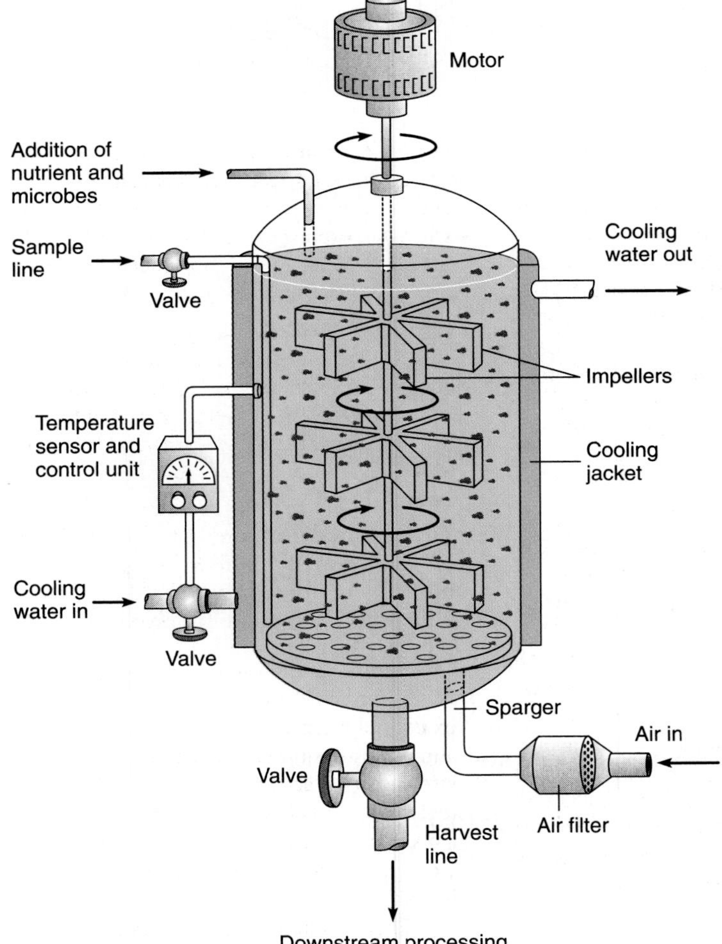

Figure 26.40
A schematic diagram of an industrial fermentor for mass culture of microorganisms. Such instruments are equipped to add nutrients and cultures, to remove product under sterile or aseptic conditions, and to aerate, stir, and cool the mixture automatically.

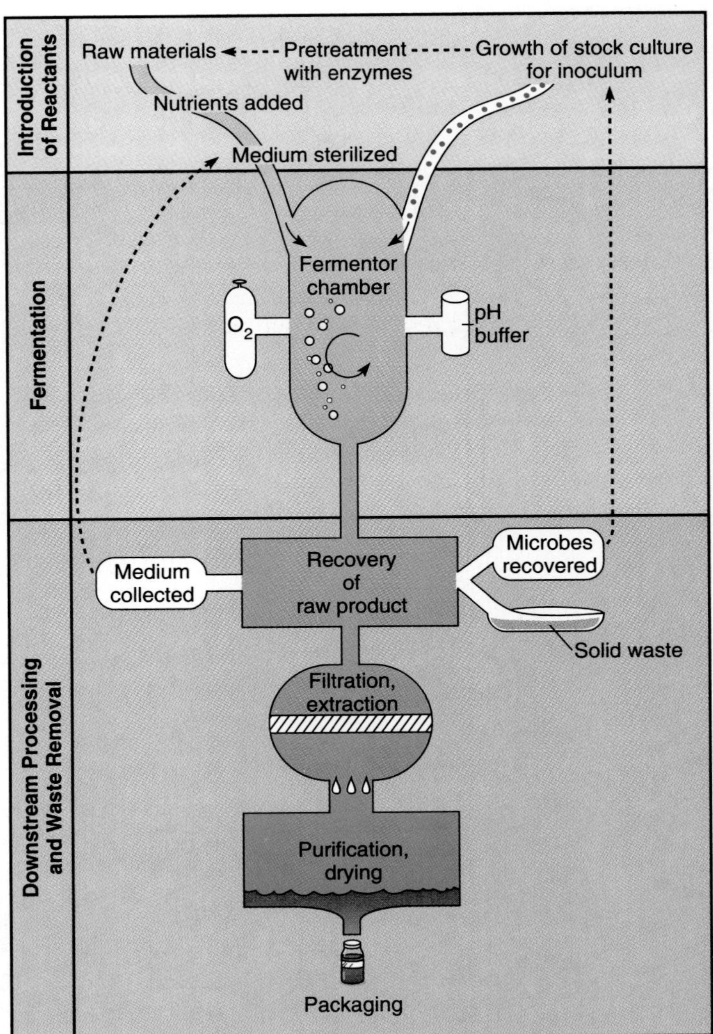

Figure 26.41
The general layout of a fermentation plant for industrial production of drugs, enzymes, fuels, vitamins, and amino acids.

introducing the assorted side-chain precursors to the fermentation vessel during the most appropriate phase of growth. These experiences with penicillin have provided an important model for the manufacture of other antibiotics (see table 26.3).

Several steroid hormones that are clinically therapeutic are produced industrially (see figure 26.39). Corticosteroids of the adrenal cortex, cortisone and cortisol (hydrocortisone), are invaluable for treating inflammatory and allergic disorders, and female hormones such as progesterone or estrogens are the active ingredients in birth control pills. For years, the production of these hormones was tedious and expensive because it involved purifying them from slaughterhouse animal glands or chemical syntheses. In time, it was shown that, through biotransformation, various molds could convert a precursor compound called diogenin into cortisone. By the same means, stigmasterol from soybean oil could be transformed into progesterone.

Some vaccines are also adaptable to mass production fermentation methods. Vaccines for *Bordetella pertussis, Salmonella typhi, Vibrio cholerae,* and *Mycobacterium tuberculosis* are produced in large batch cultures. *Corynebacterium diphtheriae* and *Clostridium tetani* are propagated for the synthesis of their toxins, from which toxoids for the DT vaccines are prepared.

Miscellaneous Products

An exciting innovation has been the development and industrial production of natural *biopesticides* using *Bacillus thuringiensis.* During sporulation, these bacteria produce intracellular crystals that can be toxic to certain insects. When the insect ingests this endotoxin, its digestive tract breaks down and it dies, but the material is relatively nontoxic to other organisms. Commercial dusts are now on the market to suppress caterpillars, moths, and worms on various agricultural crops and trees. A strain of this bacterium is also being considered to control the mosquito vector of malaria and the black fly vector of onchocerciasis.

Enzymes are critical to chemical manufacturing, the agriculture and food industries, textile and paper processing, and even laundry and dry cleaning. The advantage of enzymes is that they are very specific in their activity and are readily produced and released by microbes. Mass quantities of proteases, amylases, lipases, oxidases, and cellulases are produced by fermentation technology (see table 26.4). The wave of the future appears to be custom-designing enzymes to perform a specific task by altering their amino acid content. Other compounds of interest that can be mass-produced by microorganisms are amino acids, organic acids, solvents, and natural flavor and aromatic compounds to be used in air fresheners and foods.

Chapter Checkpoints

Industrial microbiology refers to the bulk production of any organic compound derived from microorganisms. Currently these include antibiotics, hormones, vitamins, acids, solvents, and enzymes.

CHAPTER CAPSULE WITH KEY TERMS

MICROBIAL ECOLOGY

Microbial ecology deals with the interaction between the environment and microorganisms. The environment is composed of **biotic** (living or once-living) and **abiotic** (nonliving) components. The combination of organisms and the environment make up an **ecosystem.**

ECOSYSTEM ORGANIZATION

Living things inhabit only that area of the earth called the **biosphere,** which is made up of the **hydrosphere** (water), the **lithosphere** (soil), and the **atmosphere** (air). The biosphere consists of terrestrial ecosystems of **biomes** (characterized by a dominant plant form, elevation, and latitude) and aquatic ecosystems. Biomes contain **communities,** mixed assemblages of organisms coexisting in the same time and space. Communities consist of **populations,** groups of like organisms usually of the same species. The space within which an organism lives is its **habitat;** its role in community dynamics is its **niche.**

ENERGY AND NUTRIENT FLOW

Organisms consume nutrients and derive energy. Their collective trophic status relative to one another is summarized in a **food** or **energy pyramid.** At the beginning of the chain or pyramid are **producers**—photosynthetic or lithotrophic organisms that can tap a basic nonliving energy source to synthesize large, complex organic compounds from small, simple inorganic molecules. The levels above producer are occupied by **consumers,** organisms that feed upon other organisms. **Decomposers** are consumers that obtain nutrition from the remains of dead organisms from all levels; they help convert organic materials to their inorganic state, a process called **mineralization.** Decomposers are also important nutrient recyclers. Organisms are usually not related in a linear **food chain** fashion; a cross-linked **food web** better reflects their ability to use alternative nutritional sources.

MICROBIAL INTERRELATIONSHIPS

Mutualism involves a beneficial, dependent, two-way partnership; **commensalism** is a one-sided relationship that benefits one partner while neither benefiting nor harming the other; in **cometabolism,** the by-product of one microbe becomes a nutrient for another; in **synergism,** two microbes degrade a substrate cooperatively; in **parasitism,** a host provides habitat and niche and is harmed by the activities of its passenger. **Competition** involves the release of an antagonistic substance by one organism that inhibits another organism. A **predator** is an organism that seeks and ingests live prey.

BIOGEOCHEMICAL CYCLES

The processes by which bioelements and essential building blocks of protoplasm are **recycled** between the biotic and abiotic environments are called **biogeochemical cycles.** They require microorganisms that can remove elements and compounds from a nonliving inorganic reservoir, return them to the food web, and reconvert them to the inorganic state. The **Gaia Hypothesis** is a philosophical explanation for the self-sustaining and self-regulating nature of organisms and the environment.

Atmospheric Cycles: Key compounds in the **carbon cycle** include carbon dioxide, methane, and carbonates. Carbon is **fixed** when autotrophs (photosynthesizers) add carbon dioxide to organic carbon compounds. Most fixed carbon is later returned to the carbon dioxide state by respiration and burning fossil fuels. Volcanic activity is one of several natural ways in which carbonates revert to carbon dioxide. Anaerobic metabolism of **methane** by **methanogens** also contributes significantly to carbon recycling.

The accumulation of a heat-trapping layer of carbon dioxide, methane, nitrous oxide, and water vapor in the upper atmosphere contributes to the **greenhouse effect.** Human activities such as burning fossil fuels and forests have significantly increased the amounts of these gases, which can cause global warming and have a major climatic impact.

Photosynthesis takes place in two stages—**light reactions** (light-dependent) and **dark reactions** (light-independent). During the light reactions, **photons** of solar energy are absorbed by **chlorophyll, carotenoid,** and **phycobilin** pigments in **thylakoid** membranes in chloroplasts or in specialized membrane lamellae. Captured light energy fuels a series of two photosystems that split water by **photolysis** and release oxygen gas and electrons to drive **photophosphorylation.** In this process, released light energy is used to synthesize ATP and NADPH. Dark reactions occur during the **Calvin cycle,** which uses ATP to fix carbon dioxide to a carrier molecule, ribulose 1,5 bisphosphate. The resultant 6-carbon molecule is converted to glucose in a multistep process. The type of photosynthesis commonly found in plants, algae, and cyanobacteria is called **oxygenic** because it liberates oxygen. **Anoxygenic** photosynthesis occurs in photolithotrophs, which use an alternative electron source and do not produce oxygen.

The **nitrogen cycle** requires four processes and several types of microbes. In **nitrogen fixation,** atmospheric N_2 gas (the primary reservoir) is combined with oxygen or hydrogen to form NO_2^-, NO_3^-, or NH_4^+ salts. Microbial fixation is achieved both by free-living soil and symbiotic bacteria *(Rhizobium)* that help form **root nodules** in association with legumes. **Ammonification** is a stage in the degradation of nitrogenous organic compounds (proteins, nucleic acids) by bacteria to ammonium. Some bacteria **nitrify** NH_4^+ by converting it to NO_2 − and then to NO_3^-, which can be incorporated into protoplasm by still other microbes. **Denitrification** is a multistep microbial conversion of various nitrogen salts back to atmospheric N_2.

Cycles in the Lithosphere: In the **sulfur cycle,** environmental sulfurous compounds are converted into useful substrates and returned to the inorganic reservoir through the action of microbes. The major inorganic forms of sulfur are S, H_2S, SO_4, and S_2O_3, and the chief organic forms are in certain amino acids.

The chief compound in the **phosphorus cycle** is phosphate (PO_4) found in certain mineral rocks. Microbial action on this reservoir makes it available to be incorporated into organic phosphate forms such as DNA, RNA, and ATP. Phosphate is often a limiting factor in ecosystems.

Microorganisms often cycle and help **bioamplify** (accumulate) heavy metals and other toxic pollutants that have been added to habitats by human activities, thereby creating potential hazards in the food web.

Soil is a dynamic, complex ecosystem that accommodates a vast array of microbes, animals, and plants coexisting among rich organic debris, water and air spaces, and minerals. The breakdown of rocks and the release of minerals are due in part to the activities of acid-producing bacteria and pioneering **lichens,** symbiotic associations between fungi and cyanobacteria or algae. **Humus** is the rich, moist layer of soil containing plant and animal debris in the process of being decomposed by microorganisms. An important habitat is the **rhizosphere,** a zone of soil around plant roots, which nurtures the growth of microorganisms. **Mycorrhizae** are specialized symbiotic organs formed between specialized fungi and certain plant roots.

Cycles in the Hydrosphere: The three environmental compartments (surface water, atmospheric moisture, and groundwater) are linked through a **hydrologic cycle** that involves evaporation and precipitation. Living things contribute to the cycle through respiration and transpiration. Water that percolates into the earth accumulates in deep underground **aquifers.**

The diversity and distribution of surface water communities are related to sunlight, temperature, aeration, and dissolved nutrients. The three strata occurring vertically are the **photic zone,** the **profundal zone,** and the **benthic zone;** horizontal sections are the **littoral zone** and the **limnetic zone.** Features of the marine environment include tidal and wave action and estuary, intertidal, and abyssal zones that harbor unique microorganisms. **Phytoplankton** and **zooplankton** drifting in the photic zone constitute a microbial community that supports larger invertebrates and marine animals. Temperature gradients in still waters cause stratified layers to form; the **epilimnion** and **hypolimnion** meet at an interface called a **thermocline.** Currents brought on by temperature changes cause **upwelling** of nutrient-rich benthic sediments and outbreaks of abundant microbial growth (red tides). **Oligotrophic** waters contain few nutrients and support a sparser population. Waters artificially spiked with high levels of nutrients lead to **eutrophication** and explosive aquatic overgrowth of algae and other plants.

Water Analysis: Providing **potable water** is central to prevention of water-borne disease. Water is constantly surveyed for certain **indicator bacteria** (coliforms and enterococci) that signal fecal contamination and are easier to detect and longer-lived than pathogens. Assays for possible water contamination include the **standard plate count,** the **most probable number (MPN)** procedure, and **membrane filter tests** to enumerate coliforms.

Water and Sewage Treatment: Drinking water is rendered safe by a purification process that involves storage, sedimentation, settling, aeration, filtration, and disinfection. Sewage or used wastewater can be processed to remove solid matter, dangerous chemicals, and microorganisms before it is released. Treatment occurs in three phases. Microbes biodegrade the waste material or sludge. Solid wastes are further processed in **anaerobic digesters** that can provide combustible methane and fertilizer. Future solutions to the environmental waste problem may involve **bioremediation,** the degradation of pollutants by microbial activities.

APPLIED MICROBIOLOGY AND BIOTECHNOLOGY

Practical application of microbiology in the manufacture of food, industrial chemicals, drugs, and other products is the realm of **biotechnology.** Many of these processes use mass, controlled microbial **fermentations** and bioengineered microorganisms.

FOOD MICROBIOLOGY

Microbes and humans compete for the rich nutrients in food. Although some microbes are present on food as harmless contaminants, others can cause either unfavorable or favorable results.

Food-Borne Disease: Some microbes cause spoilage and **food poisoning.** In **food intoxication,** damage is done when the toxin is produced in food, is ingested, and then acts on a specific body target; examples are staphylococcal and clostridial poisoning. In **food infection,** the intact microbe does damage when it is ingested and invades the intestine; examples are salmonellosis and shigellosis.

Precautions for Food: Microbial growth that leads to spoilage and food poisoning can be avoided by cold or hot maintenance temperatures, sterilizing or disinfecting foods, and preserving foods. High temperature and pressure treatment (canning) is commonly used for foods that are not adversely altered by heat; **pasteurization** is an alternative for disinfecting milk and other heat-sensitive beverages. For short-term needs, refrigeration and freezing can inhibit microbial growth; irradiation can sterilize or disinfect foods for longer-term storage. Alternative preservation methods include additives (salt, nitrites), treatment with ethylene oxide gas, and drying.

Fermentations in Foods: Microbes can impart desirable aroma, flavor, or texture to foods. Bread, alcoholic beverages, some vegetables, and some dairy products are infused with **starter cultures** of pure microbial strains to yield the necessary fermentation products.

Baker's yeast, *Saccharomyces cerevisiae,* is used to **leaven** bread dough by giving off CO_2; bacteria contribute additional flavors.

Beer making involves the following steps: Barley is sprouted **(malted)** to generate digestive enzymes and then dried; malt is ground up and spiked with sugar and starch to yield **mash;** heating of the mash mixture with hops releases nutrients and flavor and produces a **wort;** cooled wort is inoculated with a special strain of yeast that ferments it to a concentration of 3% to 6% alcohol; fermented beer is matured **(lagered)** in large tanks before it is carbonated and packaged.

Wine is usually prepared from grapes, although **must** can be made from any fruit juice. Fermentation of must by special strains of yeast produces wine that is from 7% to 15% alcohol. After storage, wine is bottled and aged. Whiskey, vodka, brandy, and other alcoholic beverages are **distilled** to increase their alcohol content. Vegetable products, including sauerkraut, pickles, and soybean derivatives, can be pickled in the presence of salt or sugar to form an acidic product. Vinegar is produced by fermenting plant juices first to alcohol and then to acetaldehyde and acetic acid.

Most dairy products are produced by microbes acting on the rich assortment of nutrients in milk. In cheese production, milk proteins are coagulated with **rennin** to form solid **curd** that separates from the watery **whey.** The different kinds of cheeses are obtained by varying such components as water, fat, acid, and salt and by aging or ripening with bacteria and yeasts; yogurt and buttermilk are also processed with live cultures.

Mass-cultured microbes such as yeasts, molds, and algae can serve as food. **Single-cell protein,** pruteen, and mycoprotein are currently added to animal feeds; in some countries, large colonies of the cyanobacterium *Spirulina* are harvested to be used as food supplements.

INDUSTRIAL MICROBIOLOGY

Industrial microbiology involves the large-scale commercial production of organic compounds such as antibiotics, vitamins, amino acids, enzymes, and hormones using specific microbes in carefully controlled fermentation settings. Microbes are chosen for their production of a desired **metabolite;** several different species can be used to **biotransform** raw materials in a stepwise series of metabolic reactions; fermentations are conducted in massive culture devices called **fermentors** that have special mechanisms for adding nutrients, stirring, oxygenating, altering pH, cooling, monitoring, and harvesting product; plant organization includes downstream processing (purification and packaging).

MULTIPLE-CHOICE QUESTIONS

1. Which of the following is *not* a major subdivision of the biosphere?
   a. hydrosphere
   b. lithosphere
   c. stratosphere
   d. atmosphere
2. A/an____is defined as a collection of populations sharing a given habitat.
   a. biosphere
   b. community
   c. biome
   d. ecosystem
3. The quantity of available nutrients____ from the lower levels of the energy pyramid to the higher ones.
   a. increases
   b. decreases
   c. remains stable
   d. cycles
4. Photosynthetic organisms convert the energy of____into chemical energy.
   a. electrons
   b. protons
   c. photons
   d. hydrogen atoms
5. Which of the following is considered a greenhouse gas?
   a. CO_2
   b. CH_4
   c. N_2O
   d. all of these
6. The Calvin cycle operates during which part of photosynthesis?
   a. only the light reaction
   b. only the dark reaction
   c. both light and dark reactions
   d. only photosystem I
7. Root nodules contain____, which can____.
   a. *Azotobacter,* fix N_2
   b. *Nitrosomonas,* nitrify NH_3
   c. rhizobia, fix N_2
   d. *Bacillus,* denitrify NO_3
8. Which elements have an inorganic reservoir that exists primarily in sedimentary deposits?
   a. nitrogen
   b. phosphorus
   c. sulfur
   d. b and c
9. The floating assemblage of microbes, plants, and animals that drifts on or near the surface of large bodies of water is the ____ community.
   a. abyssal
   b. benthic
   c. littoral
   d. plankton
10. An oligotrophic ecosystem would be most likely to exist in a/an
    a. ocean
    b. high mountain lake
    c. tropical pond
    d. polluted river
11. Which of the following does not vary predictably with the depth of the aquatic environment?
    a. dissolved oxygen
    b. temperature
    c. penetration by sunlight
    d. salinity
12. Which of the following would be *least* accurate in detecting coliform bacteria in a water sample?
    a. the presumptive MPN test
    b. the standard plate count
    c. the membrane filter method
    d. the confirmatory MPN test
13. Milk is usually pasteurized by
    a. the high-temperature short-time method
    b. ultrapasteurization
    c. batch method
    d. electrical currents
14. The dried, presprouted grain that is soaked to activate enzymes for beer is
    a. hops
    b. malt
    c. wort
    d. mash
15. Substances given off by yeasts during fermentation are
    a. alcohol
    b. carbon dioxide
    c. organic acids
    d. all of these

16. Which of the following is added to facilitate milk curdling during cheese making?
    a. lactic acid
    b. salt
    c. *Lactobacillus*
    d. rennin

17. A food maintenance temperature that generally will prevent food poisoning is
    a. below 4°C
    b. above 60°C
    c. room temperature
    d. both a and b

18. Secondary metabolites of microbes are formed during the____phase of growth.
    a. exponential
    b. stationary
    c. death
    d. lag

CONCEPT QUESTIONS

1. a. Present in outline form the levels of organization in the biosphere. Define the term *biome.*
   b. Compare autotrophs and heterotrophs; producers and consumers.
   c. Where in the energy and trophic schemes do decomposers enter?
   d. Compare the concepts of habitat and niche using *Chlamydomonas* (figure 26.1) as an example.
2. a. Using figure 26.3, point out specific examples of producers; primary, secondary, and tertiary consumers, herbivores; primary, secondary, and tertiary carnivores; and omnivores.
   b. What is mineralization, and which organisms are responsible for it?
3. Using specific examples, differentiate between commensalism, mutualism, parasitism, competition, and scavenging.
4. a. Outline the general characteristics of a biogeochemical cycle.
   b. What are the major sources of carbon, nitrogen, phosphorus, and sulfur?
5. a. In what major forms is carbon found? Name three ways carbon is returned to the atmosphere.
   b. Name a way it is fixed into organic compounds.
   c. What form is the least available for the majority of living things?
   d. Which form is produced during anaerobic reduction of CO_2?
   e. Describe the relationship between photosynthesis and respiration with regard to CO_2 and O_2.
6. a. Tell whether each of the following is produced during the light or dark reactions of photosynthesis: O_2, ATP, NADPH, and glucose.
   b. When are water and CO_2 consumed?
   c. What is the function of chlorophyll and the photosystems?
   d. What is the fate of the ATP and NADH?
   e. Compare oxygenic with nonoxygenic photosynthesis.
7. a. Outline the steps in the nitrogen cycle from N_2 to organic nitrogen and back to N_2.
   b. Describe nitrogen fixation, ammonification, nitrification, and denitrification.
   c. What form of nitrogen is required by plants?
   d. By animals?
8. a. Outline the phosphorus and sulfur cycles.
   b. What are the most important inorganic and organic phosphorus compounds? How is phosphorus made available to plants?
   c. What are the most important inorganic and organic sulfur compounds?
   d. What are the roles of microorganisms in these cycles?
9. a. Describe the structure of the soil and the rhizosphere.
   b. What is humus?
   c. Compare and contrast root nodules with mycorrhizae.
10. a. Outline the modes of cycling water through the lithosphere, hydrosphere, and atmosphere.
    b. What are the roles of precipitation, condensation, respiration, transpiration, surface water, and aquifers?
11. a. Map a section through the structure of a lake.
    b. What types of organisms are found in plankton?
    c. Why is phytoplankton not found in the benthic zone?
    d. What would you expect to find there?
12. a. What causes the formation of the epilimnion, hypolimnion, and thermocline?
    b. What is upwelling?
    c. In what ways are red tides and eutrophic algal blooms similar and different?
13. a. Why must water be subjected to microbiological analysis?
    b. What are the characteristics of good indicator organisms, and why are they monitored rather than pathogens?
    c. Give specific examples of indicator organisms and water-borne pathogens.
    d. Describe two methods of water analysis.
    e. What are the principles behind the most probable number test?
    f. Describe the three phases of sewage treatment.
    g. What is activated sludge?
14. a. Explain the meaning of fermentation from the standpoint of industrial microbiology.
    b. Describe five types of fermentations.
15. a. What is the main criterion regarding the safety of foods?
    b. Differentiate between food infection and food intoxication.
    c. When are microbes on food harmless?
    d. What are the two most common food intoxications?
    e. What are the three most common food infections?
    f. Summarize the major methods of preventing food poisoning and spoilage.
16. a. Which microbes are used as starter cultures in bread, beer, wine, cheeses, and sauerkraut?
    b. Outline the steps in beer making.
    c. List the steps in wine making.
    d. What are curds and whey, and what causes them?
17. a. Describe the aims of industrial microbiology.
    b. Differentiate between primary and secondary metabolites.
    c. Describe a fermentor.
    d. How is it scaled up for industrial use?
    e. What are specific examples of products produced by these processes?

CRITICAL-THINKING QUESTIONS

1. a. What factors cause energy to decrease with each trophic level?
   b. How is it possible for energy to be lost and the ecosystem to still run efficiently?
   c. Are the nutrients on the earth a renewable resource?
   d. Why, or why not?
2. Describe the course of events that could result in a parasitic relationship evolving into a mutualistic one.
3. Give specific examples from biogeochemical cycles that support the Gaia Hypothesis.
4. "Biosphere II" is an experimental, self-contained, self-sustaining ecosystem completely sealed off from the external environment in the Arizona desert. Predict what sorts of organisms must be placed in this habitat for it to function in a balanced manner. (For instance, are humans necessary?)
5. a. Is the greenhouse effect harmful under ordinary circumstances?
   b. What occurrence has made it dangerous to the global ecosystem?
   c. What could each person do on a daily basis to cut down on the potential for disrupting the delicate balance of the earth?
6. Outline a possible genetic engineering method to produce root nodules in nonleguminous plants. (Hint: See chapter 10.)
7. a. If we are to rely on microorganisms to biodegrade wastes in landfills, aquatic habitats, and soil, list some ways that this process could be made more efficient.
   b. Since elemental poisons (heavy metals) cannot be further degraded even by microbes, what is a possible fate of these metals?
   c. What, if anything, can be done about this form of pollution?
8. Why are organisms in the abyssal zone of the ocean necessarily halophilic, psychrophilic, barophilic, and anaerobic?
9. a. What happens to the nutrients that run off into the ocean with sewage and other effluents?
   b. Why can high mountain communities usually dispense with water treatment?
10. Every year supposedly safe municipal water supplies cause outbreaks of enteric illness.
    a. How in the course of water analysis and treatment might these pathogens be missed?
    b. What kinds of microbes are they most likely to be?
11. Describe three important potential applications of bacteria that can make methane.
12. a. Describe four food-preparation and food-maintenance practices in your own kitchen that could expose people to food poisoning and explain how to prevent them.
    b. What is a good general rule to follow concerning suspicious food?
13. a. Describe a controlled experiment that might solve the contradiction between which cutting board is safer to use in the kitchen.
    b. What mistakes might the researchers have made in the experiments described in microfile 26.5?
14. a. What is the purpose of boiling the wort in beer preparation?
    b. What are hops used for?
    c. If fermentation of sugars to produce alcohol in wine is anaerobic, why do wine makers make sure that the early phase of yeast growth is aerobic?
15. Predict the differences in the outcome of raw milk that has been incubated for 48 hours versus pasteurized milk that has been incubated for the same length of time.
16. How are cometabolism and biotransformations of microorganisms harnessed in industrial microbiology?
17. a. Outline the steps in the production of an antibiotic, starting with a newly discovered mold.
    b. Review chapter 10 and describe several ways that recombinant DNA technology can be used in biotechnology processes.

INTERNET SEARCH TOPICS

1. Look up information on bioremediation. Name three new technologies in this field involving microorganisms.
2. Find information on the uses of microorganisms to mine metals such as gold and copper. How can bacteria extract metals from ore?
3. Locate ⟨http://www.fightbac.org⟩ on the Internet. What are the most important factors in reducing food-borne illnesses?

APPENDIX A

IMPORTANT CHARTS

Greek Letters					
α	alpha	ι	iota	ρ	rho
β	beta	κ	kappa	σ	sigma
γ	gamma	λ	lambda	τ	tau
δ	delta	μ	mu	υ	upsilon
ϵ, ε	epsilon	ν	nu	ϕ	phi
ζ	zeta	ξ	xi	χ	chi
η	eta	ο	omicron	ψ	psi
θ	theta	π	pi	ω	omega

UNITS OF MEASUREMENT

Unit	Abbreviation	Size
Length		
meter	m	approximately 39 inches
centimeter	cm	10^{-2} m
millimeter	mm	10^{-3} m
micrometer	μm	10^{-6} m
nanometer	nm	10^{-9} m
angstrom	Å	10^{-10} m
Volume		
liter	L	approximately 1.06 quart
milliliter	ml	10^{-3} L (1 ml = 1 cm^3 = 1 cc)
microliter	μl	10^{-6} L

COMMON NUMERICAL PREFIXES

Prefix	Value	Meaning	Example
giga (G)	1,000,000,000	by a factor of a billion	gigabyte
mega (M)	1,000,000	by a factor of a million	megaton
kilo (K)	1,000	by a factor of a thousand	kilogram
centi (c)	0.01	a hundredth	centimeter
milli (m)	0.001	a thousandth	milligram
micro (μ)	0.000001	a millionth	microliter
nano (n)	0.000000001	a billionth	nanometer
pico (p)	0.000000000001	a trillionth	picogram

TEMPERATURE CONVERSIONS*

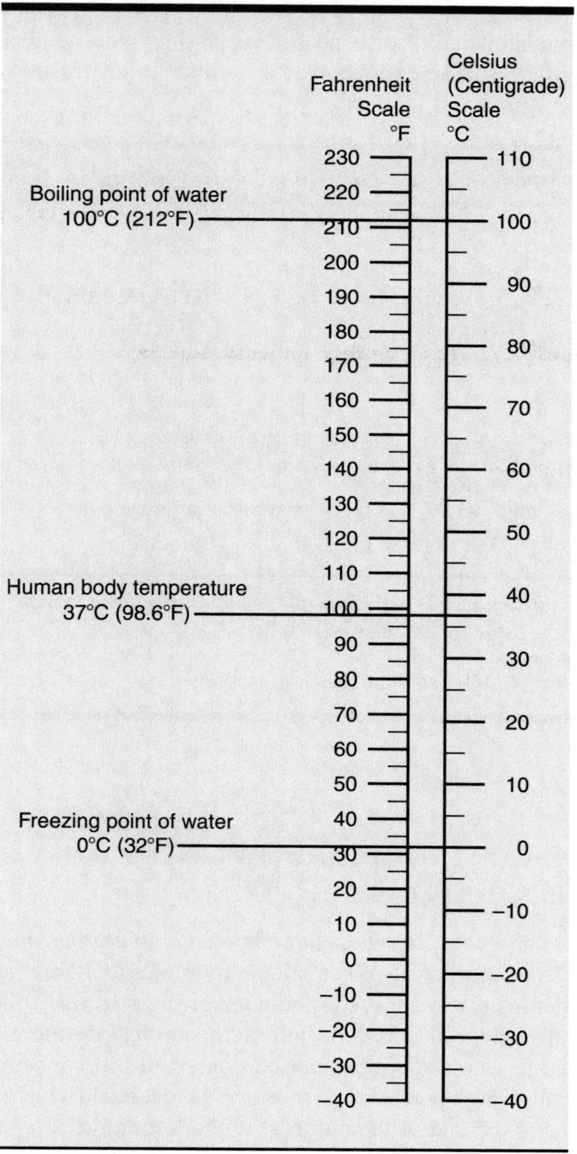

*General rule of conversion: Degrees Fahrenheit (°F) are converted to degrees Celsius (°C) by subtracting 32 from °F and multiplying the result by $\frac{5}{9}$. Degrees Celsius are converted to °F by multiplying °C by $\frac{9}{5}$ and adding 32 to the result.

Appendix B

Exponents

Dealing with concepts such as microbial growth often requires working with numbers in the billions, trillions, and even greater. A mathematical shorthand for expressing such numbers is with exponents. The exponent of a number indicates how many times (designated by a superscript) that number is multiplied by itself. These exponents are also called common *logarithms,* or logs. The following chart, based on multiples of 10, summarizes this system.

Number	Quantity	Exponential Notation*	Number Arrived at By:	One Followed By:
1	One	10^0	Numbers raised to zero power are equal to one	No zeros
10	Ten	10^1**	10×1	One zero
100	Hundred	10^2	10×10	Two zeros
1,000	Thousand	10^3	$10 \times 10 \times 10$	Three zeros
10,000	Ten thousand	10^4	$10 \times 10 \times 10 \times 10$	Four zeros
100,000	Hundred thousand	10^5	$10 \times 10 \times 10 \times 10 \times 10$	Five zeros
1,000,000	Million	10^6	10 times itself 6 times	Six zeros
1,000,000,000	Billion	10^9	10 times itself 9 times	Nine zeros
1,000,000,000,000	Trillion	10^{12}	10 times itself 12 times	Twelve zeros
1,000,000,000,000,000	Quadrillion	10^{15}	10 times itself 15 times	Fifteen zeros
1,000,000,000,000,000,000	Quintillion	10^{18}	10 times itself 18 times	Eighteen zeros

Other large numbers are sextillion (10^{21}), septillion (10^{24}), and octillion (10^{27}).

*The proper way to say the numbers in this column is 10 raised to the *nth* power, where *n* is the exponent. The numbers in this column can also be represented as 1×10^n, but for brevity, the $1 \times$ can be omitted.

**The exponent 1 is usually omitted in numbers at this level.

Converting Numbers to Exponent Form

As the chart shows, using exponents to express numbers can be very economical. When simple multiples of 10 are used, the exponent is always equal to the number of zeros that follow the 1, but this rule will not work with numbers that are more varied. Other large whole numbers can be converted to exponent form by the following operation: First, move the decimal (which we assume to be at the end of the number) to the left until it sits just behind the first number in the series (example: 3568. = 3.568). Then count the number of spaces (digits) the decimal has moved; that number will be the exponent. (The decimal has moved from 8. to 3., or 3 spaces.) In final notation, the converted number is multiplied by 10 with its appropriate exponent: 3568 is now 3.568×10^3.

Rounding Off Numbers

The notation in the previous example has not actually been shortened, but it can be reduced further by rounding off the decimal fraction to the nearest thousandth (three digits), hundredth (two digits), or tenth (one digit). To round off a number, drop its last digit and either increase the one next to it or leave it as it is. If the number dropped is 5, 6, 7, 8, or 9, the subsequent digit is increased by one (rounded up); if it is 0, 1, 2, 3, or 4, the subsequent digit remains as is. Using the example of 3.528, removing the 8 rounds off the 2 to a 3 and produces 3.53 (two digits). If further rounding is desired, the same rule of thumb applies, and the number becomes 3.5 (one digit). Other examples of exponential conversions are as follows:

Number	*Is the Same As*	*Rounded Off, Placed in Exponent Form*
16,825.	$1.6825 \times 10 \times 10 \times 10 \times 10$	1.7×10^4
957,654.	$9.57654 \times 10 \times 10 \times 10 \times 10 \times 10$	9.58×10^5
2,855,000.	$2.855000 \times 10 \times 10 \times 10 \times 10 \times 10 \times 10$	2.86×10^6

NEGATIVE EXPONENTS

The numbers we have been using so far are greater than 1 and are represented by positive exponents. But the correct notation for numbers less than 1 involves negative exponents (10 raised to a negative power, or 10^{-n}). A negative exponent says that the number has been divided by a certain power of 10 (10, 100, 1,000). This usage is handy when working with concepts such as pH that are based on very small numbers otherwise needing to be represented by large decimal fractions—for example, 0.003528. Converting this and other such numbers to exponential notation is basically similar to converting positive numbers, except that you work from left to right and the exponent is negative. Using the example of 0.003528, first convert the number to a whole integer followed by a decimal fraction and keep track of the number of spaces the decimal point moves (example: 0.003528 = 3.528). The decimal has moved three spaces from its original position, so the finished product is 3.528×10^{-3}. Other examples are:

Number	*Is the Same As*	*Rounded Off, Express with Exponents*
0.0005923	$\dfrac{5.923}{10 \times 10 \times 10 \times 10}$	5.92×10^{-4}
0.00007295	$\dfrac{7.295}{10 \times 10 \times 10 \times 10 \times 10}$	7.3×10^{-5}

Appendix C

Methods for Testing Sterilization and Germicidal Processes

Most procedures used for microbial control in the clinical setting are either physical methods (heat, radiation) or chemical methods (disinfectants, antiseptics). These procedures are so crucial to the well-being of patients and staff that their effectiveness must be monitored in a consistent and standardized manner. Particular concerns include the time required for the process, the concentration or intensity of the antimicrobic agent being used, and the nature of the materials being treated. The effectiveness of an agent is frequently established through controlled microbiological analysis. In these tests, a biological indicator (a highly resistant microbe) is exposed to the agent, and its viability is checked. If this known test microbe is destroyed by the treatment, it is assumed that the less resistant microbes have also been destroyed. Growth of the test organism indicates that the sterilization protocol has failed. The general categories of testing include:

Heat The usual form of heat used in microbial control is steam under pressure in an autoclave. Autoclaving materials at a high temperature (121°C) for a sufficient time (15–40 minutes) destroys most bacterial endospores. To monitor the quality of any given industrial or clinical autoclave run, technicians insert a special ampule of *Bacillus stearothermophilus,* a spore-forming bacterium with extreme heat resistance (figure A.1). After autoclaving, the ampule is incubated at 56°C (the optimal temperature for this species) and checked for growth.

Radiation The effectiveness of ionizing radiation in sterilization is determined by placing special strips of dried spores of *Bacillus sphaericus* (a common soil bacterium with extreme resistance to radiation) into a packet of irradiated material.

Filtration The performance of membrane filters used to sterilize liquids may be monitored by adding *Pseudomonas diminuta* to the liquid. Because this species is a tiny bacterium that can escape through the filter if the pore size is not small enough, it is a good indicator of filtrate sterility.

Gas sterilization Ethylene oxide gas, one of the few chemical sterilizing agents, is used on a variety of heat-sensitive medical and laboratory supplies. The most reliable indicator of a successful sterilizing cycle is the sporeformer *Bacillus subtilis,* variety *niger.*

Germicidal assays United States hospitals and clinics commonly employ more than 250 products to control microbes in the environment, on inanimate objects, and on patients. Controlled testing of these products' effectiveness is not usually performed in the clinic itself, but by the chemical or pharmaceutical manufacturer. Twenty-five standardized *in vitro* tests are currently available to assess the effects of germicides. Most of them use a non-spore-forming pathogen as the biological indicator.

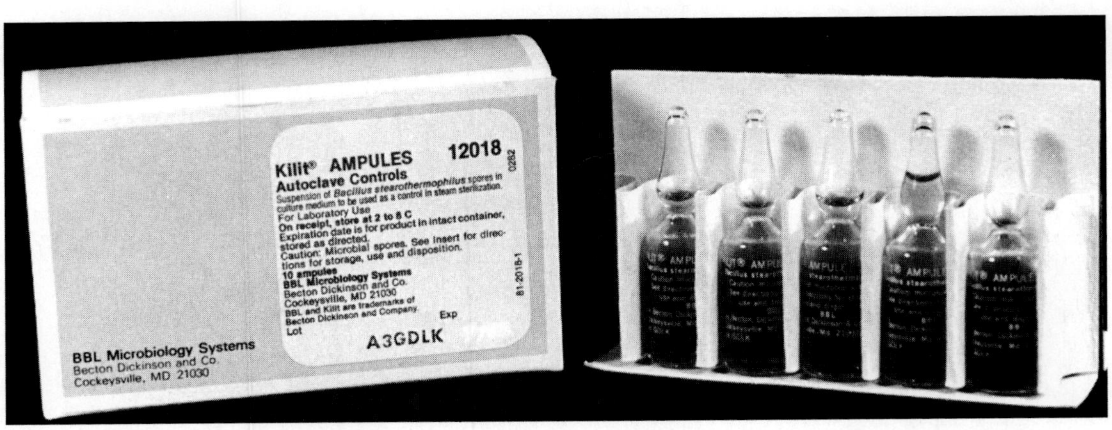

Figure A.1
The sterilizing conditions of autoclave runs may be tested by indicator bacteria in small broth ampules. One ampule is placed in the center of the autoclave chamber along with the regular load. After the process, the ampule is incubated to test for survival.

The Phenol Coefficient (PC)

The older disinfectant phenol has been the traditional standard by which other disinfectants are measured. In the PC test, a water-soluble phenol-based (phenolic) disinfectant (amphyl, lysol) is tested for its bactericidal effectiveness as compared with phenol. These disinfectants are serially diluted in a series of *Salmonella choleraesuis, Staphylococcus aureus,* or *Pseudomonas aeruginosa* cultures. The tubes are left for 5, 10, or 15 minutes and then are subcultured to test for viability. The PC is a ratio derived by comparing the following data:

$$\text{PC} = \frac{\text{Greatest dilution of the phenolic that kills test bacteria in 10 min but not in 5 min}}{\text{Greatest dilution of phenol giving same result}}$$

The general interpretation of this ratio is that chemicals with lower phenol coefficients have greater effectiveness. The primary disadvantage of the PC test is that, because it is restricted to phenolics, it is inappropriate for the vast majority of clinical disinfectants.

Use Dilution Test

An alternative test with broader applications is the use dilution test. This test is performed by drying the test culture (one of those species used in the PC test) onto the surface of small stainless steel cylindrical carriers. These carriers are then exposed to varying concentrations of disinfectant for 10 minutes, removed, rinsed, and placed in tubes of broth. After incubation, the tubes are inspected for growth. The smallest concentration of disinfectant that kills the test organism on 10 carrier pieces is the correct dilution for use.

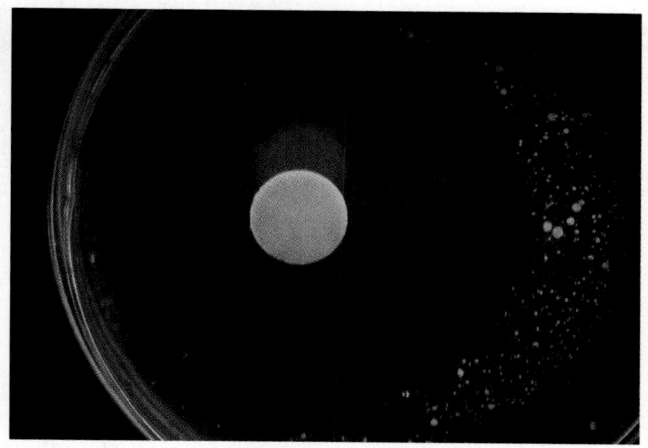

Figure A.2
A filter paper disc containing hydrogen peroxide produces a broad clear zone of inhibition in a culture of *Staphylococcus aureus.*

Filter Paper Disc Method

A quick measure of the inhibitory effects of various disinfectants and antiseptics can be achieved by the filter paper disc method. First, a small (1/2-inch) sterile piece of filter paper is dipped into a disinfectant of known concentration. This is placed on an agar medium seeded with a test organism (*S. aureus* or *P. aeruginosa*), and the plate is incubated. A zone devoid of growth (zone of inhibition) around the disc indicates the capacity of the agent to inhibit growth (figure A.2). Like the antimicrobic sensitivity test, this test measures the minimum inhibitory concentration of the chemical. In general, chemicals with wide zones of inhibition are effective even in high dilutions.

Table A.1

Most Probable Number (Chapter 26) Chart for Evaluating Coliform Content of Water

The Number of Tubes in Series, with Positive Growth at Each Dilution							
10 ml	1 ml	0.1 ml	MPN* per 100 ml of Water	10 ml	1 ml	0.1 ml	MPN* per 100 ml of Water
0	1	0	0.18	5	0	1	3.1
1	0	0	0.20	5	1	0	3.3
1	0	0	0.40	5	1	1	4.6
2	0	0	0.45	5	2	0	4.9
2	0	1	0.68	5	2	1	7.0
2	2	0	0.93	5	2	2	9.5
3	0	0	0.78	5	3	0	7.9
3	0	1	1.1	5	3	1	11.0
3	1	0	1.1	5	3	2	14.0
3	2	0	1.4	5	4	0	13.0
4	0	0	1.3	5	4	1	17.0
4	0	1	1.7	5	4	2	22.0
4	1	0	1.7	5	4	3	28.0
4	1	1	2.1	5	5	0	24.0
4	2	0	2.2	5	5	1	35.0
4	2	1	2.6	5	5	2	54.0
4	3	0	2.7	5	5	3	92.0
5	0	0	2.3	5	5	4	160.0

*Most probable number of cells per sample of 100 ml.

Appendix D

Universal Blood and Body Fluid Precautions

Medical and dental settings require stringent measures to prevent the spread of nosocomial (clinically acquired) infections from patient to patient, from patient to worker, and from worker to patient. But even with precautions, the rate of such infections is rather high. Recent evidence indicates that more than one-third of nosocomial infections could be prevented by consistent and rigorous infection control methods.

Previously, control guidelines were disease-specific, and clearly identified infections were managed with particular restrictions and techniques. With this arrangement, personnel tended to handle materials labeled *infectious* with much greater care than those that were not so labeled. The AIDS epidemic spurred a reexamination of that policy. Because of the potential for increased numbers of undiagnosed HIV-infected patients, the Centers for Disease Control and Prevention laid down more-stringent guidelines for handling patients and body substances. These guidelines have been termed **universal precautions (UPs),** because they are based on the assumption that all patient specimens could harbor infectious agents and so must be treated with the same degree of care. They also include body substance isolation (BSI) techniques to be used in known cases of infection.

It is worth mentioning that these precautions are designed to protect all individuals in the clinical setting—patients, workers, and the public alike. In general, they include techniques designed to prevent contact with pathogens and contamination and, if prevention is not possible, to take purposeful measures to decontaminate potentially infectious materials.

The universal precautions recommended for all health care settings are:

1. Barrier precautions, including masks and gloves, should be taken to prevent contact of skin and mucous membranes with patients' blood or other body fluids. Because gloves can develop small invisible tears, double gloving decreases the risk further. For protection during surgery, venipuncture, or emergency procedures, gowns, aprons, and other body coverings should be worn. Dental workers should wear eyewear and face shields to protect against splattered blood and saliva.
2. More than 10% of health care personnel are pierced each year by sharp (and usually contaminated) instruments. These accidents carry risks not only for AIDS but also for hepatitis B, hepatitis C, and other diseases. Preventing inoculation infection requires vigilant observance of proper techniques. All disposable needles, scalpels, or sharp devices from invasive procedures must immediately be placed in puncture-proof containers for sterilization and final discard. Under no circumstances should a worker attempt to recap a syringe, remove a needle from a syringe, or leave unprotected used syringes where they pose a risk to others. Reusable needles or other sharp devices must be heat-sterilized in a puncture-proof holder before they are handled. If a needlestick or other injury occurs, immediate attention to the wound, such as thorough degermation and application of strong antiseptics, can prevent infection.
3. Dental handpieces should be sterilized between patients, but if this is not possible, they should be thoroughly disinfected with a high-level disinfectant (peroxide, hypochlorite). Blood and saliva should be removed completely from all contaminated dental instruments and intraoral devices prior to sterilization.
4. Hands and other skin surfaces that have been accidently contaminated with blood or other fluids should be scrubbed immediately with a germicidal soap. Hands should likewise be washed after removing rubber gloves, masks, or other barrier devices.
5. Because saliva can be a source of some types of infections, barriers should be used in all mouth-to-mouth resuscitations.
6. Health care workers with active, draining skin or mucous membrane lesions must refrain from handling patients or equipment that will come into contact with other patients. Pregnant health care workers risk infecting their fetuses and must pay special attention to these guidelines. Personnel should be protected by vaccination whenever possible.

Isolation procedures for known or suspected infections should still be instituted on a case-by-case basis.

Table A.2

Primary Biosafety Levels and Agents of Disease

Biosafety Level	Facilities and Practices	Risk of Infection and Class of Pathogens
1	Standard, open bench, no special facilities needed; typical of most microbiology teaching labs; access may be restricted	Low infection hazard; microbes not generally considered pathogens and will not colonize the bodies of healthy persons: *Micrococcus luteus, Bacillus megaterium, Lactobacillus, Saccharomyces*
2	At least Level 1 facilities and practices, plus personnel must be trained in handling pathogens; lab coats and gloves required; safety cabinets may be needed; biohazard signs posted; access restricted	Agents with moderate potential to infect; class 2 pathogens can cause disease in healthy people but can be contained with proper facilities; most pathogens belong to class 2; includes *Staphylococcus aureus, Escherichia coli, Salmonella* spp., *Corynebacterium diphtheriae;* pathogenic helminths; hepatitis A, B, and rabies viruses; *Cryptococcus* and *Blastomyces*
3	Minimum of Level 2 facilities and practices; all manipulation performed in safety cabinets; lab designed with special containment features; only personnel with special clothing can enter; no unsterilized materials can leave the lab; personnel warned, monitored, and vaccinated against infection dangers	Agents can cause severe or lethal disease especially when inhaled; class 3 microbes include *Mycobacterium tuberculosis, Francisella tularensis, Yersinia pestis, Brucella* spp., *Coxiella burnetii, Coccidioides immitis,* and yellow fever, WEE, and AIDS viruses
4	Minimum of Level 3 facilities and practices; facilities must be isolated with very controlled access; clothing changes and showers required for all people entering and leaving; materials must be autoclaved or fumigated prior to entering and leaving lab	Agents being handled are highly virulent microbes that pose extreme risk for morbidity and mortality when inhaled in droplet or aerosol form; most are exotic flaviviruses; arenaviruses, including Lassa fever virus; or filoviruses, including Ebola and Marburg viruses

Laboratory Biosafety Levels and Classes of Pathogens

Personnel handling infectious agents in the laboratory must be protected from possible infection through special risk management or containment procedures. These involve the following: (1) carefully observing standard laboratory aseptic and sterile procedures while handling cultures and infectious samples; (2) using large-scale sterilization and disinfection procedures; (3) restricting eating, drinking, and smoking; and (4) wearing personal protective items such as gloves, masks, safety glasses, laboratory coats, boots, and headgear. Some circumstances also require additional protective equipment such as biological safety cabinets for inoculations and specially engineered facilities to control materials entering and leaving the laboratory in the air and on personnel. Table A.2 summarizes the primary biosafety levels and agents of disease as characterized by the Centers for Disease Control and Prevention.

Appendix E

Answers to Multiple-Choice Questions

Chapter 1
1. d
2. c
3. d
4. a
5. c
6. b
7. d
8. a
9. c
10. c
11. b
12. 1st column
 3, 7, 4, 2
 2nd column
 8, 5, 6, 1

Chapter 2
1. c
2. c
3. b
4. c
5. a
6. e
7. c
8. d
9. a
10. c
11. c
12. a
13. b
14. c
15. c
16. b
17. b
18. d
19. b
20. c
21. c
22. c
23. a
24. d

Chapter 3
1. b
2. c
3. d
4. b
5. d
6. b
7. b
8. c
9. c
10. a
11. b

Chapter 4
1. d
2. a
3. c
4. a
5. c
6. b
7. c
8. d
9. b
10. d
11. c
12. d
13. c
14. b

Chapter 5
1. b
2. d
3. b
4. d
5. a
6. b
7. c
8. c
9. d
10. b
11. d
12. a
13. d
14. b

Chapter 6
1. c
2. d
3. d
4. b
5. d
6. a
7. a
8. d
9. b
10. b
11. c
12. d
13. d
14. a

Chapter 7
1. c
2. a
3. a
4. c
5. c
6. a
7. b
8. b
9. b
10. c
11. c
12. c
13. c
14. b
15. a

Chapter 8
1. b
2. a
3. c
4. d
5. d
6. b
7. b
8. c
9. b
10. b
11. a
12. c
13. a
14. d
15. b
16. c
17. c
18. c
19. c

Chapter 9
1. b
2. b
3. b
4. c
5. b
6. c
7. a
8. b
9. a
10. b
11. a
12. d
13. b
14. d
15. b

Chapter 10
1. c
2. c
3. b
4. c
5. a
6. b
7. b
8. c
9. c
10. d
11. d
12. b

Chapter 11
1. d
2. c
3. b
4. a
5. c
6. b
7. b
8. d
9. c
10. b
11. d
12. c
13. d
14. a
15. b

Chapter 12
1. b
2. c
3. b
4. a
5. d
6. b
7. c
8. c
9. a
10. d
11. a
12. c
13. c
14. b

Chapter 13
1. a
2. d
3. b
4. d
5. c
6. d
7. b
8. c
9. c
10. c
11. d
12. c
13. a
14. a
15. d

Chapter 14
1. b
2. b
3. d
4. b
5. b
6. c
7. b
8. a
9. c
10. d
11. c
12. d
13. c
14. d
15. d

Chapter 15
1. a
2. c
3. a
4. c
5. c
6. c
7. a
8. c
9. a
10. c
11. a
12. c
13. b
14. e

Chapter 16
1. b
2. c
3. d
4. b
5. c
6. b
7. c
8. a
9. a
10. d
11. c

Chapter 17
1. d
2. d
3. d
4. c
5. d
6. c
7. b
8. a
9. d
10. d
11. d
12. a
13. d
14. c
15. b
16. d

Chapter 18
1. d
2. a
3. c
4. d
5. b
6. c
7. b
8. d
9. c
10. c
11. a
12. a
13. d
14. a

Chapter 19
1. b
2. c
3. b
4. d
5. c
6. c
7. d
8. c
9. d
10. b
11. c
12. d
13. c
14. d
15. a
16. a

Chapter 20
1. b
2. b
3. d
4. b
5. c
6. d
7. d
8. b
9. d
10. d
11. c
12. d
13. a, c, d, e
14. a, b, d, e, f

Chapter 21
1. b
2. b
3. c
4. d
5. d
6. d
7. c
8. b
9. d
10. a
11. d
12. c
13. d
14. d
15. a
16. b

Chapter 22
1. c
2. a
3. a
4. d
5. d
6. a
7. a
8. c
9. c
10. c
11. c
12. c
13. d

Chapter 23
1. c
2. b
3. d
4. b
5. b
6. d
7. c
8. d
9. c
10. a
11. d
12. d
13. b
14. a
15. a
16. c

Chapter 24
1. d
2. c
3. b
4. a
5. d
6. d
7. c
8. d
9. a
10. b
11. b
12. b
13. c
14. d

Chapter 25
1. a
2. b
3. c
4. b
5. d
6. d
7. b
8. d
9. d
10. c
11. d
12. d
13. b
14. d
15. c

Chapter 26
1. c
2. b
3. b
4. c
5. d
6. c
7. c
8. d
9. d
10. b
11. d
12. b
13. a
14. b
15. d
16. d
17. d
18. b

APPENDIX F

CLASSIFICATION OF MAJOR MICROBIAL DISEASE AGENTS BY SYSTEM AFFECTED, SITE OF INFECTION, AND ROUTES OF TRANSMISSION

Tables A.3 through A.9 provide a cross reference to accompany the organismic approach of presenting pathogens used in this text. It lists the primary organs and systems that serve as targets for infectious agents and the sources and routes of infection. Many infectious agents affect both the system they first enter and others. (For example polio infects the gastrointestinal tract and also inflicts major damage on the central nervous system.) This reference highlights agents that have major effects on more than one system with an asterisk (*). The chapter that contains detailed information on the listings is shown in parentheses in the far right column.

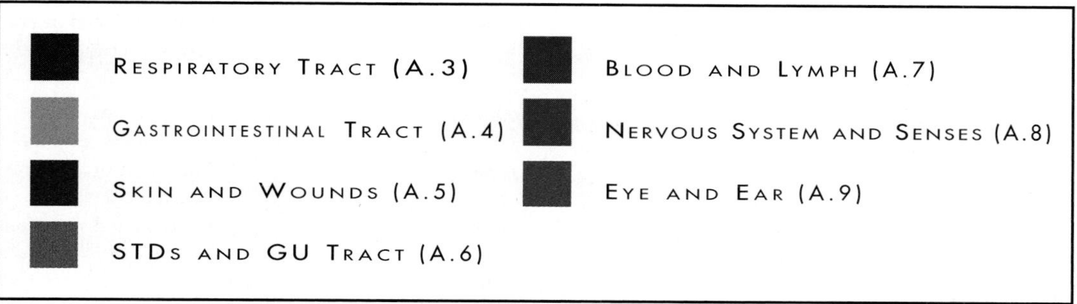

TABLE A.3

INFECTIONS OF THE RESPIRATORY TRACT (RT)

Name of Microbe	Name of Disease	Specific Targets of Infection	Source/Mode of Aquisition	(Ch)
Bacterial Pathogens				
*Streptococcus pyogenes**	Streptococcal pharyngitis	Upper RT membranes	Droplets	(18)
Viridans streptococci	Sinusitis	Upper RT and saliva	Normal flora	(18)
*Bacillus anthracis**	Pulmonary anthrax	Upper and lower RT	Spores inhaled from animals/products	(19)
*Corynebacterium diphtheriae**	Diphtheria	Upper RT membranes; also skin	Droplets	(19)
*Haemophilus influenzae**	Pharyngitis	Upper RT membranes	Droplets	(20)
*Streptococcus pneumoniae**	Pneumococcal pneumonia; otitis media, sinusitis meningitis	Lower RT, lungs; meninges	Droplets, mucus	(18)
*Mycobacterium tuberculosis**	Tuberculosis	Lower RT, lungs	Droplets; droplet nuclei	(19)
Legionella pneumophila	Legionnaire's disease Pontiac fever	Lower RT, lungs	Aerosols from aquatic habitats	(20)
Brucella spp.*	Brucellosis (undulant fever)	Respiratory tract	Aerosols from infected animals	(20)
Bordetella pertussis	Whooping cough	Upper, lower RT	Droplets	(20)
Klebsiella pneumoniae *Pseudomonas aeruginosa*	Opportunistic pneumonias	Lower RT, lungs	Droplets, normal flora, aerosols, fomites	(20)
*Yersinia pestis**	Pneumonic plague	Lower RT, lungs	Droplets, animal aerosols	(20)
Mycoplasma pneumoniae	Primary atypical pneumonia	Lower RT, lungs	Droplets, mucus	(21)

continued

INFECTIONS OF THE RESPIRATORY TRACT (RT)—*CONTINUED*

Name of Microbe	Name of Disease	Specific Targets of Infection	Source/Mode of Aquisition	(Ch)
Chlamydia pneumoniae	Chlamydial pneumonitis	Lower RT, lungs	Droplets, mucus	(21)
Chlamydia psittaci	Ornithosis (parrot fever)	Lower RT, lungs	Aerosols, droplets from birds	(21)
Coxiella burnetii	Q (for query) fever	Lower RT, lungs	Ticks, aerosols from animals, dust	(21)
Nocardia brasiliensis	Nocardiosis (a form of pneumonia)	Lower RT, lungs	Spores inhaled from environment	(19)
Fungal Pathogens (Chapter 22)				
*Histoplasma capsulatum**	Histoplasmosis	Lower RT, lungs	Spores inhaled in dust	
*Blastomyces dermatitidis**	Blastomycosis	Lower RT, lungs	Spores inhaled	
*Coccidioides immitis**	Coccidioidomycosis (Valley fever)	Lower RT, lungs	Spores inhaled	
*Cryptococcus neoformans**	Cryptococcosis	Lower RT, lungs	Yeasts inhaled, can be normal flora	
*Sporothrix schenckii**	Sporothix pneumonia	Lower RT, lungs	Spores on vegetation	
Pneumocystis carinii	Pneumocystis pneumonia (PCP)	Lower RT, lungs	Spores widespread, normal flora	
*Aspergillus fumigatus**	Aspergillosis	Lower RT, lungs	Spores widespread in dust, air	
Parasitic (Protozoan, Helminth) Pathogens (Chapter 23)				
Paragonimus westermani	Adult lung fluke	Lung tissue; intestine	Larvae ingested in raw or undercooked crustacea	
Viral Pathogens				
Influenza viruses	Influenza A, B, C	Mucosal cells of trachea, bronchi	Inhalation of droplets, aerosols from active cases	(25)
Hantavirus	Hantavirus pulmonary syndrome	Lung epithelia	Inhalation of airborne rodent excrement	(25)
Respiratory Syncytial Pneumovirus	RSV pneumonitis, bronchiolitis	Nasopharynx; upper and lower RT	Droplets and secretions from active cases	(25)
Parainfluenza virus Paramyxovirus	Parainfluenza	Upper, lower RT	Direct hand contact with secretions; inhalation of droplets	(25)
Adenovirus	Common cold	Nasopharynx; sometimes lower RT	Close contact with secretions from person with active infection; hand to nose	(24)
Rhinovirus	Common cold	Same as above		(25)
Coronavirus	Common cold	Same as above		(25)
Coxsackie virus	Common cold	Same as above		(25)
Mumps virus*	See Table A.4			
Rubella virus*	See Table A.5			
Measles virus*	See Table A.5			
Chickenpox virus*	See Table A.5			

Table A.4

Diseases of the Gastrointestinal Tract (Food Poisoning and Intestinal, Gastric, Oral Infections)

Name of Microbe	Name of Disease	Specific Targets of Disease	Source/Mode of Aquisition	(Ch)
Bacterial Pathogens				
Staphylococcus aureus	Staph food intoxication	Small intestine	Foods contaminated by human carriers	(18)
Clostridium perfringens	Perfringens food poisoning	Small intestine	Rare meats; toxin produced in in body	(19)
Clostridium difficile	Antibiotic-associated colitis	Large intestine	Spores ingested; can be normal flora; caused by antibiotic therapy	(19)
Listeria monocytogenes	Listeriosis	Small intestine	Water, soil, plants, animals, milk, food infection	(19)
Escherichia coli	Traveler's diarrhea 0157: H7 enteritis	Small intestine	Food, water, rare or raw beef	(20)
*Salmonella typhi**	Typhoid fever	Small intestine	Human carriers; food, drink	(20)
Salmonella species	Food infection, enteric fevers	Small intestine	Cattle, poultry, rodents; fecally contaminated foods	(20)
Shigella spp.	Bacillary dysentery	Large intestine	Human carriers; food, flies	(20)
Yersinia enterocolitica *Y. paratuberculosis*	Yersiniosis (gastroenteritis)	Small intestine, lymphatics	A zoonosis; food and drinking water	
Vibrio cholerae	Epidemic cholera	Small intestine	Natural waters; food contaminated by human carriers	(21)
Helicobacter pylori	Gastritis, gastric intestinal ulcers	Esophagus, stomach, duodenum	Humans, cats (?)	(21)
Campylobacter jejuni	Enteritis	Jejunum segment of small intestine	Water, milk, meat, food infection	(21)
Streptococcus mutans	Dental caries	Tooth surface	Normal flora; saliva	(21)
*Actinomyces israeli**	Actinomycosis	Oral membranes, cervicofacial area	Normal flora; enter damaged tissue	(19)
Fusobacterium, Treponema, Bacteroides	Various forms of gingivitis, periodontitis	Gingival pockets	Normal flora, close contact	(21)
Fungal Pathogens (Chapter 22)				
Aspergillus flavus	Aflatoxin poisoning	Liver	Toxin ingested in food	
*Candida albicans**	Esophagitis	Esophagus	Normal flora	
Parasitic (Protozoan, Helminth) Pathogens (Chapter 23)				
*Entamoeba histolytica**	Amebic dysentery	Caecum, appendix, colon, rectum	Food, water-contaminated fecal cysts	
Giardia lamblia (intestinalis)	Giardiasis	Duodenum, jejunum	Ingested cysts from animal, human carriers	
Balantidium coli	Balantidiosis	Large intestine	Swine, contaminated food	
Cryptosporidium spp.	Cryptosporoidosis	Small and large intestine	Animals, natural waters, contaminated food	
*Toxoplasma gondii**	Toxoplasmosis	Lymph nodes	Cats, rodents, domestic animals; raw meats	
Enterobius vermicularis	Pinworm	Rectum, anus	Eggs ingested from fingers, food	
Ascaris lumbricoides	Ascariasis	Worms burrow into intestinal mucosa	Fecally contaminated food and water containing eggs	
Trichuris trichiura	Trichiuriasis, whipworm	Worms invade small intestine	Food containing eggs	
*Trichinella spiralis**	Trichinosis	Larvae penetrate intestine	Pork and bear meat containing encysted larvae	
*Necator americanus** *Ancylostoma duodenale**	Hookworms	Large and small intestine	Larvae deposited by feces into soil; burrow into feet	
Fasciola hepatica	Sheep liver fluke	Liver and intestine	Watercress	
Opisthorchis sinensis	Chinese liver fluke	Liver and intestine	Raw or undercooked fish or shellfish	
Taenia saginata *T. solium*	Beef tapeworm Pork tapeworm	Large intestine Pork larva can migrate to brain	Rare or raw beef or pork containing larval forms	

continued

Diseases of the Gastrointestinal Tract (Food Poisoning and Intestinal, Gastric, Oral Infections) *Continued*

Name of Microbe	Name of Disease	Specific Targets of Disease	Source/Mode of Aquisition	(Ch)
Viral Pathogens				
Herpes simplex virus, type 1	Cold sores, fever blisters, gingivostomatitis	Skin, mucous membranes of oral cavity	Close contact with active lesions of human	(24)
Mumps virus* *(Paramyxovirus)*	Parotitis	Parotid salivary gland	Close contact with salivary droplets	(25)
Hepatitis A virus	Infectious hepatitis	Liver	Food, water contaminated by human feces; shellfish	(25)
Rotavirus	Rotavirus diarrhea	Small intestinal mucosa	Fecally contaminated food, water, fomites	(25)
Norwalk agent (enterovirus)	Gastroenteritis	Small intestine	Close contact; oral-fecal, water, shellfish	(25)
Poliovirus*	Poliomyelitis	Intestinal mucosa, systemic	Food, water, mechanical vectors	(25)

Table A.5

Infections of Skin and Skin Wounds

Name of Microbe	Name of Disease	Specific Targets of Infection	Source/Mode of Transmission	(Ch)
Bacterial Pathogens				
*Staphylococcus aureus**	Folliculitis Furuncles (boils) Carbuncles Bullous impetigo Scalded skin syndrome	Glands, follicles, glabrous skin	Close contact, nasal and oral droplets; associated with poor hygiene, tissue injury	(18)
*Streptococcus pyogenes**	Pyoderma (impetigo) Erysipelas Necrotizing faciitis	Skin and underlying tissues	Contact with human carrier; invasion of tiny wounds in skin	(18)
*Bacillus anthracis**	Cutaneous anthrax	Necrotic lesion in skin (eschar)	Spores from animal secretions, products, soil	(19)
*Clostridium perfringens**	Gas gangrene	Skin, muscle, connective tissue	Introduction of spores into damaged tissue	(19)
Propionibacterium acnes	Acne	Sebaceous glands	Normal flora; bacterium is lipophilic	(19)
Erysipelothrix rhusiopathiae	Erysipeloid	Epidermal, dermal regions of skin	Normal flora of pigs, other animals; enters in injured tissue	(19)
*Mycobacterium leprae**	Leprosy (Hansen's disease)	Skin, underlying nerves	Long term contact with infected person; skin is inoculated	(19)
Mycobacterium marinum	Swimming pool granuloma	Superficial skin	Skin scraped against contaminated surface	(19)
Pasteurella multocida	Pasteurellosis	Localized skin abcess	Animal bite or scratch	(20)
Bartonella henselae	Cat-scratch disease	Subcutaneous tissues, lymphatic drainage	Cat's claws carry bacteria, inoculate skin	(21)
Fungal Pathogens (Chapter 22)				
Blastomyces dermatitidis	Blastomycosis	Subcutaneous	Accidental inoculation of skin with conidia	
Trichophyton *Microsporum* *Epidermophyton*	Dermatomycoses: tineas, ringworm, athlete's foot	Cutaneous tissues (epidermis, nails, scalp, hair)	Contact with infected humans, animals, spores in environment	
*Sporothrix schenckii**	Sporotrichosis	Epidermal, dermal, subcutaneous	Puncture of skin with sharp plant materials	

continued

INFECTIONS OF SKIN AND SKIN WOUNDS—*CONTINUED*

Name of Microbe	Name of Disease	Specific Targets of Infection	Source/Mode of Transmission	(Ch)
Fungal Pathogens (Chapter 22)—*continued*				
Madurella	Madura foot, mycetoma	Subcutaneous tissues, bones, muscles	Traumatic injury with spore-infested objects	
Cladosporium *Fonsecaea*	Chromoblastomycosis	Deep subcutaneous tissues of legs, feet	Penetration of skin with contaminated objects	
*Candida albicans**	Onychomycosis	Skin, nails	Chronic contact with moisture, warmth	
Malassezia furfur	Tinea versicolor	Superficial epidermis	Normal flora of skin	
Parasitic (Protozoan, Helminth) Pathogens (Chapter 23)				
Leishmania tropica* *L. mexicana*	Cutaneous leishmaniasis	Local infection of skin capillaries	Sand fly salivary gland harbors infectious stage; transmitted through bite	(23)
Dracunculus medinensis	Dracontiasis guinea worm	Starts in intestine, migrates to skin	Microscopic aquatic arthropod harbors larval form; is ingested	
Viral Pathogens				
Varicella zoster virus*	Chickenpox Shingles	Skin of face and trunk; becomes systemic and latent in nerves	Spread by respiratory droplets; recurrences from latent infection	(24)
Herpes virus 6	Roseola infantum	Generalized skin rash	Close contact with droplets	
Parvovirus B 19	Erythema Contagiosum	Skin of cheeks, trunk	Close contact with droplets	
Measles virus* (Morbillivirus)	Measles, rubeola	Skin on head, trunk, extremities	Contact with respiratory aerosols from active case	(25)
Rubella virus* (*Rubivirus*)	Rubella	Skin on face, trunk, limbs	Respiratory droplets, urine of active cases	(25)
Human papillomavirus (See also Table A.6)	Warts, verruca	Skin of hands, feet, body	Contact with warts, fomites	(24)
Molluscum contagiosum	See Table A.6			(24)
Herpes simplex, type 2	See Table A.6			(24)

TABLE A.6

SEXUALLY TRANSMITTED DISEASES (STDs) AND INFECTIONS OF THE GENITOURINARY (GU) TRACT

Name of Microbe	Name of Disease	Specific Targets of Infection	Source/Mode of Acquisition	(Ch)
Bacterial Pathogens				
*Neisseria gonorrhoeae**	Gonorrhea	Vagina, urethra	Mucous secretions; STD	(18)
Haemophilus ducreyi	Chancroid	Genitalia, lymph nodes	Skin lesions; STD	(20)
*Treponema pallidum**	Syphilis	Genitalia; mucous membranes	Chancres, skin membranes lesions; STD	(21)
*Chlamydia trachomatis**	Chlamydiosis (non-gonococcal urethritis) lymphogranuloma venereum	Vagina, urethra, lymph nodes	Mucous discharges; STD	(21)
*Escherichia coli**	Urinary tract infection (UTI)	Urethra, bladder	Normal flora; intestinal tract	(20)
*Leptospira interrogans**	Leptospirosis	Kidney, liver	Animal urine, contact with contaminated water, soil	(21)
Gardnerella vaginalis	Bacterial vaginosis	Vagina, occasionally penis	Mixed infection; normal flora; intimate contact	(20)

continued

Sexually Transmitted Diseases (STDs) and Infections of the Genitourinary (GU) Tract—*Continued*

Name of Microbe	Name of Disease	Specific Targets of Infection	Source/Mode of Acquisition	(Ch)
Fungal Pathogens (Chapter 22)				
Candida albicans	Yeast infections, candidiasis	Vagina, vulva, penis, urethra	Discharge from infected membranes; STD	
Parasitic (Protozoan, Helminth) Pathogens (Chapter 23)				
Trichomonas vaginalis	Trichomoniasis	Vagina, vulva, urethra	Genital discharges, direct contact; STD	
Viral Pathogens				
Molluscum contagiosum poxvirus	Molluscum contagiosum	Skin, mucous membranes	Pox lesions; STD	(24)
Herpes simplex, type 2	Genital herpes	Genitalia, perineum	Contact with vesicles, shed skin cells; STD	(24)
Human papillomavirus	Genital warts, condylomata	Membranes of the vagina, penis	Direct contact with warts; STD; some fomites	(24)
Hepatitis B virus*	Hepatitis B	Liver	Semen, vaginal fluids, blood; STD	(24)
Human immunodeficiency virus (HIV)*	AIDS	White blood cells, brain cells	Blood, semen, vaginal fluids; STD	(25)

Table A.7

Infections Affecting Multiple Systems (Circulatory and Lymphatic)

Name of Microbe	Name of Disease	Specific Targets of Disease	Source/Mode of Acquisition	(Ch)
Bacterial Pathogens				
*Staphylococcus aureus**	Toxic shock syndrome	Vagina, uterus, kidney, liver, blood	Normal flora; grow in environment created by tampons	(18)
	Osteomyelitis	Interior of long bones	Focal from skin infection	
*Streptococcus pyogenes**	Scarlet fever	Skin, brain	Toxemia; complication of strep pharyngitis	(18)
	Rheumatic fever	Heart valves	Autoimmune reaction to strep toxins	
	Glomerulonephritis	Kidney	Immune reaction blocks filtration apparatus	
Streptococcus spp. (viridans)	Subacute endocarditis	Lining of heart and valves	Normal flora invades during dental work; colonizes heart	(18)
*Corynebacterium diphtheriae**	Diphtheria	Heart muscle, nerves, occasionally skin	Droplets from carriers; local infection leads to toxemia	(19)
*Mycobacterium tuberculosis**	Disseminated tuberculosis	Lymph nodes, kidneys, bones, brain, sex organs	Reactivation of latent lung infection	(19)
*Yersinia pestis**	Plague	Lymph nodes, blood	Bite of infected flea	(20)
Francisella tularensis	Tularemia Rabbit fever	Skin, lymph nodes, eyes, lungs	Blood-sucking arthropods (ticks, mosquitos); direct contact with animals and meat; aerosols	(20)
Borrelia recurrentis	Borreliosis, relapsing fever	Liver, heart, spleen, kidneys, nerves	Human lice; influenced by poor hygiene, crowding	(21)
B. hermsii	Same as above	Same as above	Soft tick bite; reservoir is wild rodents	(21)
B. burgdorferi	Lyme disease	Skin, joints, nerves, heart	Larval *Ixodes* ticks; reservoir is deer and rodents	(21)

continued

INFECTIONS AFFECTING MULTIPLE SYSTEMS (CIRCULATORY AND LYMPHATIC)—*CONTINUED*

Name of Microbe	Name of Disease	Specific Targets of Disease	Source/Mode of Acquisition	(Ch)
Bacterial Pathogens—*continued*				
Rickettsia prowazekii	Epidemic typhus	Blood vessels in many organs	Human lice, bites, feces, crowding, lack of sanitation	(21)
R. rickettsii	Rocky Mountain spotted fever	Cardiovascular, central nervous, skin	Wood and dog ticks; picked up in mountain or rural habitats	(21)
R. typhi	Endemic typhus	Similar to epidemic typhus	Rat flea vector; rodent reservoir; occupational contact	(21)
*Coxiella burnetii**	See Table A.3			
Fungal Pathogens (Chapter 22)				
Many pathogenic fungi (*Histoplasma, Coccidioides, Blastomyces,* and *Cryptococcus*) have a systemic component to their infection, especially in immunocompromised patients. Because their primary infection involves the lungs, information on those fungi is listed in Table A.3.				
Parasitic (Protozoan, Helminth) Pathogens (Chapter 23)				
Leishmania spp.	Systemic leishmaniasis (kala azar)	Spleen, liver, lymph nodes	Bite by sand (phlebotamine) fly; reservoir is dogs, rodents, wild carnivores	
Plasmodium spp.	Malaria	Liver, blood, kidney	Female *Anopheles* mosquito; human reservoir, mainly tropical	
*Toxoplasma gondii**	Toxoplasmosis	Pharynx, lymph nodes, nervous system	Cysts from cats and other animals are ingested; many hosts	
Strongyloides stercoralis	Strongiloidiasis Threadworm	Intestine, trachea, lungs, pharynx, subcutaneous, migrate through systems	Larvae deposited into soil or water by feces; burrow into skin	
Schistosoma spp.	Schistosomiasis, blood fluke	Liver, lungs, blood vessels in intestine, bladder	Larval forms develop in aquatic snails; penetrate skin of bathers	
Wuchereria bancrofti	Filariasis Elephantiasis	Lymphatic circulation, blood vessels	Various mosquito species carry larval stage; human reservoir	
Viral Pathogens				
Cytomegalovirus (CMV)	CMV mononucleosis, congenital CMV	Lymph glands, salivary glands	Human saliva, milk, urine, semen, vaginal fluids; intimate contact	(24)
Epstein-Barr virus	Infectious mononucleosis	Lymphoid tissue, salivary glands	Oral contact, exposure to saliva, fomites	(24)
Hepatitis B virus*	See Table A.6			
Flavivirus	Yellow fever	Liver, blood vessels, kidney, mucous membranes	Various mosquito species; human and monkey reservoirs	(24) (25)
	Dengue fever (breakbone fever)	Skin, muscles, and joints	Several mosquito vectors	(25)
Filovirus	Ebola fever	Every organ and tissue except muscle and bone	Close contact with blood and secretions of victim; monkey reservoir?	(25)
Arenavirus	Lassa fever	Blood vessels, kidney, heart	Inhalation of airborne virus from rodent excreta	(25)
HIV*	See Table A.6			
HTLV I	Adult T cell leukemia Sézary-cell leukemia	T cells, lymphoid tissue	Blood, blood products, sexual intercourse, IV drugs	(25)
HTLV II	Hairy-cell leukemia	B cells	Blood, shared needles	(25)

Table A.8

Diseases of the Nervous System

Name of Microbe	Name of Disease	Specific Targets of Disease	Source/Mode of Acquisition	(Ch)
Bacterial Pathogens				
*Neisseria meningitidis**	Meningococcal meningitis	Membranes covering brain, spinal cord	Contact with respiratory secretions or droplets of human carrier	(18)
*Haemophilus influenzae**	Acute meningitis	Meninges, brain	Contact with carrier; normal flora of throat	(20)
Clostridium tetani	Tetanus	Inhibitory neurons in spinal column	Spores from soil introduced into wound; toxin formed	(19)
*Clostridium botulinum**	Botulism	Neuromuscular junction	Canned vegetables, meats; an intoxication	(19)
*Mycobacterium leprae**	Leprosy (Hansen's disease)	Peripheral nerves of skin	Close contact with infected person; enters skin	(19)
Fungal Pathogens (Chapter 22)				
*Cryptococcus neoformans**	Cryptococcal meningitis	Brain, meninges	Pigeon droppings; inhalation of yeasts in air and dust	
Parasitic (Protozoan, Helminth) Pathogens (Chapter 23)				
Naegleria fowleri	Primary acute meningoencephalitis	Brain, spinal cord	Acquired while swimming in fresh, brackish water; ameba invades nasal mucosa	
Acanthamoeba	Meningoencephalitis	Eye, brain	Ameba lives in fresh water; can invade cuts and abrasions	
Trypanosoma brucei	African trypanosomiasis, sleeping sickness	Brain, heart, many organs	Bite of tsetsefly, vector	
Trypanosoma cruzii	South American trypanosomiasis; Chagas' disease	Nerve ganglia, muscle	Feces of reduviid bug rubbed into bite	
Viral Pathogens (Chapters 24 and 25)				
Poliovirus*	Poliomyelitis	Anterior horn cells; motor neurons	Contaminated food, water objects	
Rabies virus	Rabies	Brain, spinal nerves, ganglia	Saliva of infected animal enters bite wound	
Encephalitis viruses	St. Louis encephalitis Western equine encephalitis Eastern equine encephalitis	Brain, meninges, spinal cord	Bite of infected mosquito	
Bunyavirus	California encephalitis	Brain, meninges	Mosquito bite; reservoir is wild birds	
HIV* (See Table A.6)	AIDS, dementia	Destroys brain cells, peripheral nerves	Blood, semen, vaginal fluids; STD	
JC polyomavirus	Progressive multifocal leukoencephalopathy	Oligodendrocytes in cerebrum	Not well defined	
Prions	Spongiform encephalopathies Creutzfeldt-Jacob disease	Brain, neurons	Intimate contact with infected tissues	

TABLE A.9

INFECTIONS OF THE EYE AND EAR

Name of Microbe	Name of Disease	Specific Targets of Infection	Source/Mode of Transmission	(Ch)
Bacterial Pathogens				
*Neisseria gonorrhoeae**	Ophthalmia neonatorum	Conjunctiva, cornea, eyelid	Vaginal mucus enters eyes during birth	(18)
*Streptococcus pneumoniae**	Otitis media	Middle ear	Nasopharyngeal secretions forced into eustachian tube	(18)
Haemophilus ducreyi	Pinkeye	Conjunctiva, sclera	Contaminated fingers, fomites	(20)
*Chlamydia trachomatis**	Inclusion conjunctivitis, trachoma	Conjunctiva, inner eyelid, cornea	Infected birth canal, contaminated fingers	(21)
Fungal Pathogens (Chapter 22)				
Aspergillus fumigatus	Aspergillosis	Eyelids, conjunctiva	Contact lens contaminated with spores	
Fusarium solani	Corneal ulcer, keratitis	Cornea	Eye is accidentally scratched, contact lens is contaminated	
Parasitic (Protozoan, Helminth) Pathogens (Chapter 23)				
Oncocerca volvulus	Onchocerciasis, River blindness	Eye, skin, blood	Black flies inject larvae into bite	
Loa loa	African eye worm	Subcutaneous tissues, conjunctiva, cornea	Bites by dipteran flies	
Viral Pathogens				
Coxsackie virus	Hemorrhagic conjunctivitis	Conjunctive, sclera, eyelid	Contact with fluids, secretions of carrier	(25)

GLOSSARY

A

abyssal zone The deepest region of the ocean, a sunless, high-pressure, cold, anaerobic habitat.

acid-fast A term referring to the property of mycobacteria to retain carbol fuchsin even in the presence of acid alcohol. The staining procedure is used to diagnose tuberculosis.

active immunity Immunity acquired through direct stimulation of the immune system by antigen.

active site The specific region on an apoenzyme that binds substrate. The site for reaction catalysis.

acute Characterized by rapid onset and short duration.

acyclovir A synthetic purine analog that blocks DNA synthesis in certain viruses, particularly the herpes simplex viruses.

adjuvant In immunology, a chemical vehicle that enhances antigenicity, presumably by prolonging antigen retention at the injection site.

aerobe A microorganism that lives and grows in the presence of free gaseous oxygen (O_2).

aflatoxin From *A*spergillus ***fla***vus toxin, a mycotoxin that typically poisons moldy animal feed and can cause liver cancer in humans and other animals.

agammaglobulinemia Also called hypogammaglobulinemia. The absence of or severely reduced levels of antibodies in serum.

agglutination The aggregation by antibodies of suspended cells or similar-sized particles (agglutinogens) into clumps that settle.

agglutinin A specific antibody that cross-links agglutinogen, causing it to aggregate.

agglutinogen An antigenic substance on cell surfaces that evokes agglutinin formation against it.

AIDS **A**cquired **i**mmune **d**eficiency **s**yndrome. The complex of signs and symptoms characteristic of the late phase of human immunodeficiency virus (HIV) infection.

allele A gene that occupies the same location as other alternative (allelic) genes on paired chromosomes.

allergen A substance that provokes an allergic response.

allergy The altered, usually exaggerated, immune response to an allergen. Also called hypersensitivity.

alloantigen An antigen that is present in some but not all members of the same species. Also called isoantigen.

allograft Relatively compatible tissue exchange between nonidentical members of the same species. Also called homograft.

allosteric Pertaining to the altered activity of an enzyme due to the binding of a molecule to a region other than the enzyme's active site.

amastigote The rounded or ovoid nonflagellated form of the *Leishmania* parasite.

Ames test A method for detecting mutagenic and potentially carcinogenic agents based upon the genetic alteration of nutritionally defective bacteria.

amination The addition of an amine ($—NH_2$) group to a molecule.

aminoglycoside A complex group of drugs derived from soil actinomycetes that impairs ribosome function and has antibiotic potential. Example: streptomycin.

amphibolism Pertaining to the metabolic pathways that serve multiple functions in the breakdown, synthesis, and conversion of metabolites.

amphipathic Relating to a compound that has contrasting characteristics, such as hydrophilic-hydrophobic or acid-base.

amphitrichous Having a single flagellum or a tuft of flagella at opposite poles of a microbial cell.

anabolism The energy-consuming process of incorporating nutrients into protoplasm through biosynthesis.

anaerobe A microorganism that grows best, or exclusively, in the absence of oxygen.

analog In chemistry, a compound that closely resembles another in structure.

anamnestic In immunology, an augmented response or memory related to a prior stimulation of the immune system by antigen. It boosts the levels of immune substances.

anaphylaxis The unusual or exaggerated allergic reaction to antigen that leads to severe respiratory and cardiac complications.

anion A negatively charged ion.

antibiotic A chemical substance from one microorganism that can inhibit or kill another microbe even in minute amounts.

antibody A large protein molecule evoked in response to an antigen that interacts specifically with that antigen.

anticodon The trinucleotide sequence of transfer RNA that is complementary to the trinucleotide sequence of messenger RNA (the codon).

antigen Any cell, particle, or chemical that induces a specific immune response by B cells or T cells and can stimulate resistance to an infection or a toxin. *See* immunogen.

antigenic determinant The precise molecular group of an antigen that defines its specificity and triggers the immune response.

antigenic drift Minor antigenic changes in the influenza A virus due to mutations in the spikes' genes.

antigenic shift Major changes in the influenza A virus due to recombination of viral strains from two different host species.

antihistamine A drug that counters the action of histamine and is useful in allergy treatment.

anti-idiotype An anti-antibody that reacts specifically with the idiotype (variable region or antigen-binding site) of another antibody.

antimetabolite A substance such as a drug that competes with, substitutes for, or interferes with a normal metabolite.

antiseptic A growth-inhibiting agent used on tissues to prevent infection.

antiserum Antibody-rich serum derived from the blood of animals (deliberately immunized against infectious or toxic antigen) or from people who have recovered from specific infections.

antitoxin Globulin fraction of serum that neutralizes a specific toxin. Also refers to the specific antitoxin antibody itself.

apoenzyme The protein part of an enzyme, as opposed to the nonprotein or inorganic cofactors.

apoptosis The genetically programmed death of cells that is both a natural process of development and the body's means of destroying abnormal or infected cells.

aquifer A subterranean water-bearing stratum of permeable rock, sand, or gravel.

arthrospore A fungal spore formed by the septation and fragmentation of hyphae.

Arthus reaction An immune complex phenomenon that develops after repeat injection. This localized inflammation results from aggregates of antigen and antibody that bind, complement, and attract neutrophils.

artificial chromosome A large packet of DNA from yeasts (YAC), bacteria (BAC), or humans (HAC) that can be used to carry, transfer, or analyze isolated foreign DNA.

ascospore A spore formed within a saclike cell (ascus) of Ascomycota following nuclear fusion and meiosis.

asepsis A condition free of viable pathogenic microorganisms.

atopy Allergic reaction classified as type I, with a strong familial relationship; caused by allergens such as pollen, insect venom, food, and dander; involves IgE antibody; includes symptoms of hay fever, asthma, and skin rash.

ATP synthase A unique enzyme located in the mitochondrial cristae and chloroplast grana that harnesses the flux of hydrogen ions to the synthesis of ATP.
attenuate To reduce the virulence of a pathogenic bacterium or virus by passing it through a non-native host or by long-term subculture.
autoantibody An "anti-self" antibody having an affinity for tissue antigens of the subject in which it is formed.
autoantigen Molecules that are inherently part of self but are perceived by the immune system as foreign.
autograft Tissue or organ surgically transplanted to another site on the same subject.
autoimmune disease The pathologic condition arising from the production of antibodies against autoantigens. Example: rheumatoid arthritis. Also called autoimmunity.
autosome A chromosome of somatic cells as opposed to a sex chromosome of gametes.
autotroph A microorganism that requires only inorganic nutrients and whose sole source of carbon is carbon dioxide.
axenic A sterile state such as a pure culture. An axenic animal is born and raised in a germ-free environment. *See* gnotobiotic.
axial filament A type of flagellum (called an endoflagellum) that lies in the periplasmic space of spirochetes and is responsible for locomotion. Also called periplasmic flagellum.
azole Five-membered heterocyclic compounds typical of histidine, which are used in antifungal therapy.

B

bacteremia The presence of viable bacteria in circulating blood.
bactericide An agent that kills bacteria.
bacteriocin Proteins produced by certain bacteria that are lethal against closely related bacteria and are narrow spectrum compared with antibiotics; these proteins are coded and transferred in plasmids.
bacteriophage A virus that specifically infects bacteria.
bacteriostatic Any process or agent that inhibits bacterial growth.
barophile A microorganism that thrives under high (usually hydrostatic) pressure.
basement membrane A thin layer (1–6 μm) of protein and polysaccharide found at the base of epithelial tissues.
basidiospore A sexual spore that arises from a basidium. Found in basidiomycota fungi.
basidium A reproductive cell created when the swollen terminal cell of a hypha develops filaments (sterigmata) that form spores.
basophil A motile polymorphonuclear leukocyte that binds IgE. The basophilic cytoplasmic granules contain mediators of anaphylaxis and atopy.
bdellovibrio A bacterium that preys on certain other bacteria. It bores a hole into a specific host and inserts itself between the protoplast and the cell wall. There it elongates before subdividing into several cells and devours the host cell.
benthic zone The sedimentary bottom region of a pond, lake, or ocean.
beta-lactamase An enzyme secreted by certain bacteria that cleaves the beta-lactam ring of penicillin and cephalosporin and thus provides for resistance against the antibiotic. *See* penicillinase.
beta-oxidation The degradation of long-chain fatty acids. Two-carbon fragments are formed as a result of enzymatic attack directed against the second or beta carbon of the hydrocarbon chain. Aided by coenzyme A, the fragments enter the tricarboxylic acid (TCA) cycle and are processed for ATP synthesis.
binary fission The formation of two new cells of approximately equal size as the result of parent cell division.
bioremediation The use of microbes to reduce or degrade pollutants, industrial wastes, and household garbage.
biosphere Habitable regions comprising the aquatic (hydrospheric), soil-rock (lithospheric), and air (atmospheric) environments.
biotechnology The use of microbes or their products in the commercial or industrial realm.
blast cell An immature precursor cell of B and T lymphocytes. Also called a lymphoblast.
blocking antibody The IgG class of immunoglobins that competes with IgE antibody for allergens, thus blocking the degranulation of basophils and mast cells.
B lymphocyte (B cell) A white blood cell that gives rise to plasma cells and antibodies.
booster The additional doses of vaccine antigen administered to increase an immune response and extend protection.
botulin *Clostridium botulinum* toxin. Ingestion of this potent exotoxin leads to flaccid paralysis.
bradykinin An active polypeptide that is a potent vasodilator released from IgE-coated mast cells during anaphylaxis.
broad spectrum A word to denote drugs that affect many different types of bacteria, both gram-positive and gram-negative.
Brownian movement The passive, eratic, nondirectional motion exhibited by microscopic particles. The jostling comes from being randomly bumped by submicroscopic particles, usually water molecules, in which the visible particles are suspended.
bubo The swelling of one or more lymph nodes due to inflammation.
bulla A large, bobble-like vesicle in a region of separation between the epidermis and the subepidermal layer. The space is usually filled with serum and sometimes with blood.

C

Calvin cycle The recurrent photosynthetic pathway characterized by CO_2 fixation and glucose synthesis. Also called the dark reactions.
cancer Any malignant neoplasm that invades surrounding tissue and can metastasize to other locations. A carcinoma is derived from epithelial tissue, and a sarcoma arises from proliferating mesodermal cells of connective tissue.
capsid The protein covering of a virus's nucleic acid core. Capsids exhibit symmetry due to the regular arrangement of subunits called capsomers. *See* icosahedron.
capsomer A subunit of the virus capsid shaped as a triangle or disc.
capsule In bacteria, the loose, gel-like covering or slime made chiefly of simple polysaccharides. This layer is protective and can be associated with virulence.
carbuncle A deep staphylococcal abscess joining several neighboring hair follicles.
carrier A person who harbors infections and inconspicuously spreads them to others. Also, a chemical agent that can accept an atom, chemical radical, or subatomic particle from one compound and pass it on to another.
caseous lesion Necrotic area of lung tubercle superficially resembling cheese. Typical of tuberculosis.
catabolism The chemical breakdown of complex compounds into simpler units to be used in cell metabolism.
catalyst A substance that alters the rate of a reaction without being consumed or permanently changed by it. In cells, enzymes are catalysts.
cation A positively charged ion.
cavitation The formation of a hollow space such as in the lung in tuberculosis.
cecum The intestinal pocket that forms the first segment of the large intestine. Also called the appendix.
cell-mediated The type of immune responses brought about by T cells, such as cytotoxic, suppressor, and helper effects.
cephalosporins A group of broad-spectrum antibiotics isolated from the fungus *Cephalosporium*.
cercaria The free-swimming larva of the schistosome trematode that emerges from the snail host and can penetrate human skin, causing schistosomiasis.
cestode The common name for tapeworms that parasitize humans and domestic animals.
chancre The primary sore of syphilis that forms at the site of penetration by *Treponema pallidum*. It begins as a hard, dull red, painless papule that erodes from the center.
chancroid A lesion that resembles a chancre but is soft and is caused by *Haemophilus ducreyi*.
chemoautotroph An organism that relies upon inorganic chemicals for its energy and carbon dioxide for its carbon. Also called a chemolithotraph.
chemokine Chemical mediators (cytokines) that stimulate the movement and migration of white blood cells.
chemostat A growth chamber with an outflow that is equal to the continuous inflow of nutrient media. This steady-state growth device is used to study such events as cell division, mutation rates, and enzyme regulation.
chemotaxis The tendency of organisms to move in response to a chemical gradient (toward an attractant or to avoid adverse stimuli).
chemotherapy The use of chemical substances or drugs to treat or prevent disease.
chitin A polysaccharide similar to cellulose in chemical structure. This polymer makes up the horny substance of the exoskeletons of arthropods and certain fungi.
chloroplast An organelle containing chlorophyll that is found in photosynthetic eucaryotes.
chromatic aberration Deviant focus of magnification due to refraction of the colored wavelengths that make up white light.
chromatin The genetic material of the nucleus. Chromatin is made up of nucleic acid and stains readily with certain dyes.
chromatin body The bacterial chromosome or nucleoid.

chromophore The chemical radical of a dye that is responsible for its color and reactivity.
chromosome The tightly coiled bodies in cells that are the primary sites of genes.
chronic Any process or disease that persists over a long duration.
clonal selection theory A conceptual explanation for the development of lymphocyte specificity and variety during immune maturation.
clone A colony of cells (or group of organisms) derived from a single cell (or single organism) by asexual reproduction. All units share identical characteristics. Also used as a verb to refer to the process of producing a genetically identical population of cells or genes.
cloning host An organism such as a bacterium or a yeast that receives and replicates a foreign piece of DNA inserted during a genetic engineering experiment.
coagulase A plasma-clotting enzyme secreted by *Staphylococcus aureus*. It contributes to virulence and is involved in forming a fibrin wall that surrounds staphylococcal lesions.
coccobacillus An elongated coccus; a short, thick, oval-shaped bacterial rod.
coccus A spherical-shaped bacterial cell.
codon A specific sequence of three nucleotides in mRNA (or the sense strand of DNA) that constitutes the genetic code for a particular amino acid.
coenzyme A complex organic molecule, several of which are derived from vitamins (e.g., nicotinamide, riboflavin). A coenzyme operates in conjunction with an enzyme. Coenzymes serve as transient carriers of specific atoms or functional groups during metabolic reactions.
cofactor An enzyme accessory. It can be organic, such as coenzymes, or inorganic, such as Fe^{+2}, Mn^{+2}, or Zn^{+2} ions.
cold sterilization The use of nonheating methods such as radiation or filtration to sterilize materials.
coliform A collective term that includes normal enteric bacteria that are gram-negative and lactose-fermenting.
colinear Having corresponding parts of a molecule arranged in the same linear order as another molecule, as in DNA and mRNA.
colony A macroscopic cluster of cells appearing on a solid medium, each arising from the multiplication of a single cell.
colostrum The clear yellow early product of breast milk that is very high in secretory antibodies. Provides passive intestinal protection.
commensalism An unequal relationship in which one species derives benefit without harming the other.
communicable infection Capable of being transmitted from one individual to another.
community The interacting mixture of populations in a given habitat.
complement In immunology, serum protein components that act in a definite sequence when set in motion either by an antigen-antibody complex or by factors of the alternative (properdin) pathway.
conidia Asexual fungal spores shed as free units from the tips of fertile hyphae.
conjugation In bacteria, the contact between donor and recipient cells associated with the transfer of genetic material such as plasmids. Can involve special (sex) pili. Also a form of sexual recombination in ciliated protozoans.
constitutive enzyme An enzyme present in bacterial cells in constant amounts, regardless of the presence of substrate. Enzymes of the central catabolic pathways are typical examples.
contagious Communicable; transmissible by direct contact with infected people and their fresh secretions or excretions.
contaminant An impurity; any undesirable material or organism.
convalescence Recovery; the period between the end of a disease and the complete restoration of health in a patient.
corepressor A molecule that combines with inactive repressor to form active repressor, which attaches to the operator gene site and inhibits the activity of structural genes subordinate to the operator.
crista The infolded inner membrane of a mitochondrion that is the site of the respiratory chain and oxidative phosphorylation.
culture The visible accumulation of microorganisms in or on a nutrient medium. Also, the propagation of microorganisms with various media.
curd The coagulated milk protein used in cheese making.
cyst The resistant, dormant, but infectious form of protozoans. Can be important in spread of infectious agents such as *Entamoeba histolytica* and *Giardia lamblia*.
cysticercus The larval form of certain *Taenia* species, which typically infest muscles of mammalian intermediate hosts. Also called bladderworm.
cystine An amino acid, $HOOC—CH(NH_2)—CH_2—S—S—CH_2—CH(NH_2)COOH$. An oxidation product of two cysteine molecules in which the —SH (sulfhydryl) groups form a disulfide union. Also called dicysteine.
cytochrome A group of heme protein compounds whose chief role is in electron and/or hydrogen transport occurring in the last phase of aerobic respiration.
cytokine A chemical substance produced by white blood cells and tissue cells that regulates development, inflammation, and immunity.
cytopathic effect The degenerative changes in cells associated with virus infection. Examples: the formation of multinucleate giant cells (Negri bodies), the prominent cytoplasmic inclusions of nerve cells infected by rabies virus.
cytotoxic Having the capacity to destroy specific cells. One class of T cells attacks cancer cells, virus-infected cells, and eucaryotic pathogens. *See* killer T cells.

D

debridement Trimming away devitalized tissue and foreign matter from a wound.
decontamination The removal or neutralization of an infectious, poisonous, or injurious agent from a site.
definitive host The organism in which a parasite develops into its adult or sexually mature stage. Also called the final host.
degerm To physically remove surface oils, debris, and soil from skin to reduce the microbial load.
degranulation The release of cytoplasmic granules, as when cytokines are secreted from mast cell granules.
denaturation The loss of normal characteristics resulting from some molecular alteration. Usually in reference to the action of heat or chemicals on proteins whose function depends upon an unaltered tertiary structure.
deoxyribose A 5-carbon sugar that is an important component of DNA.
desquamate To shed the cuticle in scales; to peel off the outer layer of a surface.
diapedesis The migration of intact blood cells between endothelial cells of a blood vessel such as a venule.
differential medium A single substrate that discriminates between groups of microorganisms on the basis of differences in their appearance due to different chemical reactions.
differential stain A technique that utilizes two dyes to distinguish between different microbial groups or cell parts by color reaction.
diffusion The dispersal of molecules, ions, or microscopic particles propelled down a concentration gradient by spontaneous random motion to achieve a uniform distribution.
dimorphic In mycology, the tendency of some pathogens to alter their growth form from mold to yeast in response to rising temperature.
diplococcus Spherical or oval-shaped bacteria, typically found in pairs.
diploid Somatic cells having twice the basic chromosome number. One set in the pair is derived from the father, and the other from the mother.
disaccharide A sugar containing two monosaccharides. Examples: sucrose (fructose + glucose).
disinfection The destruction of pathogenic nonsporulating microbes or their toxins, usually on inanimate surfaces.
droplet nuclei The dried residue of fine droplets produced by mucus and saliva sprayed while sneezing and coughing. Droplet nuclei are less than 5 μm in diameter (large enough to bear a single bacterium and small enough to remain airborne for a long time) and can be carried by air currents. Droplet nuclei are drawn deep into the air passages.

E

ectoplasm The outer, more viscous region of the cytoplasm of a phagocytic cell such as an ameba. It contains microtubules, but not granules or organelles.
eczema An acute or chronic allergy of the skin associated with itching and burning sensations. Typically, red, edematous, vesicular lesions erupt, leaving the skin scaly and sometimes hyperpigmented.
edema The accumulation of excess fluid in cells, tissues, or serous cavities. Also called swelling.
electrolyte Any compound that ionizes in solution and conducts current in an electrical field.
electromagnetic radiation A form of energy that is emitted as waves and is propagated through space and matter. The spectrum extends from short gamma rays to long radio waves.
electron A negatively charged subatomic particle that is distributed around the nucleus in an atom.
electrophoresis The separation of molecules by size and charge through exposure to an electrical current.
electrostatic Relating to the attraction of opposite charges and the repulsion of like charges. Electrical charge remains stationary as opposed to electrical flow or current.

element A substance comprising only one kind of atom that cannot be degraded into two or more substances without losing its chemical characteristics.

ELISA Abbreviation for **e**nzyme-**l**inked **i**mmuno**s**orbent **a**ssay, a very sensitive serological test used to detect antibodies in diseases such as AIDS.

encystment The process of becoming encapsulated by a membranous sac.

endemic disease A native disease that prevails continuously in a geographic region.

endergonic reaction A chemical reaction that occurs with the absorption and storage of surrounding energy. Antonym: exergonic.

endocarditis An inflammation of the lining and valves of the heart. Often caused by infection with pyogenic cocci.

endocytosis The process whereby solid and liquid materials are taken into the cell through membrane invagination and engulfment into a vesicle.

endoenzyme An intracellular enzyme, as opposed to enzymes that are secreted.

endogenous Originating or produced within an organism or one of its parts.

endoplasmic reticulum An intracellular network of flattened sacs or tubules with or without ribosomes on their surfaces.

endospore A small, dormant, resistant derivative of a bacterial cell that germinates under favorable growth conditions into a vegetative cell. The bacterial genera *Bacillus* and *Clostridium* are typical sporeformers.

endotoxin A bacterial intracellular toxin that is not ordinarily released (as is exotoxin). Endotoxin is composed of a phospholipid-polysaccharide complex that is an integral part of gram-negative bacterial cell walls. Endotoxins can cause severe shock and fever.

energy of activation The minimum energy input necessary for reactants to form products in a chemical reaction.

enriched medium A nutrient medium supplemented with blood, serum, or some growth factor to promote the multiplication of fastidious microorganisms.

enteric Pertaining to the intestine.

enteropathogenic Pathogenic to the alimentary canal.

enterotoxin A bacterial toxin that specifically targets intestinal mucous membrane cells. Enterotoxigenic strains of *Escherichia coli* and *Staphylococcus aureus* are typical sources.

enveloped virus A virus whose nucleocapsid is enclosed by a membrane derived in part from the host cell. It usually contains exposed glycoprotein spikes specific for the virus.

enzyme A protein biocatalyst that facilitates metabolic reactions.

enzyme repression The inhibition of enzyme synthesis by the end product of a catabolic pathway.

eosinophil A leukocyte whose cytoplasmic granules readily stain with red eosin dye.

epidemic A sudden and simultaneous outbreak or increase in the number of cases of disease in a community.

epidemiology The study of the factors affecting the prevalence and spread of disease within a community.

epimastigote The trypanosomal form found in the tsetse fly or reduviid bug vector. Its flagellum originates near the nucleus, extends along an undulating membrane, and emerges from the anterior end.

erysipelas An acute, sharply defined inflammatory disease specifically caused by hemolytic *Streptococcus*. The eruption is limited to the skin but can be complicated by serious systemic symptoms.

erysipeloid An inflammation resembling erysipelas but caused by *Erysipelothrix*, a gram-positive rod. The self-limited cellulitis that appears at the site of an infected wound, usually the hand, comes from handling contaminated fish or meat.

erythema An inflammatory redness of the skin.

erythroblastosis fetalis Hemolytic anemia of the newborn. The anemia comes from hemolysis of Rh-positive fetal erythrocytes by anti-Rh maternal antibodies. Erythroblasts are immature red blood cells prematurely released from the bone marrow.

erythrogenic toxin An exotoxin produced by lysogenized group A strains of β-hemolytic streptococci that is responsible for the severe fever and rash of scarlet fever in the nonimmune individual. Also called a pyrogenic toxin.

eschar A dark, sloughing scab that is the lesion of anthrax and certain rickettsioses.

essential nutrient Any ingredient such as a certain amino acid, fatty acid, vitamin, or mineral that cannot be formed by an organism and must be supplied in the diet. A growth factor.

ester bond A covalent bond formed by reacting carboxylic acid with an OH group:

$$(\mathrm{R{-}\overset{\overset{\large O}{\|}}{C}{-}O{-}R'})$$

Olive and corn oils, lard, and butter fat are examples of triacylglycerols—esters formed between glycerol and three fatty acids.

estuary The intertidal zone where a river empties into the sea.

ethylene oxide A potent, highly water-soluble gas invaluable for gaseous sterilization of heat-sensitive objects such as plastics, surgical and diagnostic appliances, and spices. Potential hazards are related to its carcinogenic, metagenic, residual, and explosive nature. Ethylene oxide is rendered nonexplosive by mixing with 90% CO_2 or fluorocarbon.

etiological agent The microbial cause of disease; the pathogen.

eucaryotic cell A cell that differs from a procaryotic cell chiefly by having a nuclear membrane (a well-defined nucleus), membrane-bound subcellular organelles, and mitotic cell division.

eutrophication The process whereby dissolved nutrients resulting from natural seasonal enrichment or industrial pollution of water cause overgrowth of algae and cyanobacteria to the detriment of fish and other large aquatic inhabitants.

exergonic A chemical reaction associated with the release of energy to the surroundings. Antonym: endergonic.

exfoliative toxin A poisonous substance that causes superficial cells of an epithelium to detach and be shed. Example: staphylococcal exfoliatin. Also called an epidermolytic toxin.

exoenzyme An extracellular enzyme chiefly for hydrolysis of nutrient macromolecules that are otherwise impervious to the cell membrane. It functions in saprobic decomposition of organic debris and can be a factor in invasiveness of pathogens.

exon A stretch of eucaryotic DNA coding for a corresponding portion of mRNA that is translated into peptides. Intervening stretches of DNA that are not expressed are called introns. During transcription, exons are separated from introns and are spliced together into a continuous mRNA transcript.

exotoxin A toxin (usually protein) that is secreted and acts upon a specific cellular target. Examples: botulin, tetanospasmin, diphtheria toxin, and erythrogenic toxin.

exponential Pertaining to the use of exponents, numbers that are typically written as a superscript to indicate how many times a factor is to be multiplied. Exponents are used in scientific notation to render large, cumbersome numbers into small workable quantities.

F

facultative Pertaining to the capacity of microbes to adapt or adjust to variations; not obligate. Example: The presence of oxygen is not obligatory for a facultative anaerobe to grow. *See* obligate.

fastidious Requiring special nutritional or environmental conditions for growth. Said of bacteria.

fermentation The extraction of energy through anaerobic degradation of substrates into simpler, reduced metabolites. In large industrial processes, fermentation can mean any use of microbial metabolism to manufacture organic chemicals or other products.

fixation In microscopic slide preparation of tissue sections or bacterial smears, fixation pertains to rapid killing, hardening, and adhesion to the slide, while retaining as many natural characteristics as possible. Also refers to the assimilation of inorganic molecules into organic ones, as in carbon or nitrogen fixation.

fluid mosaic model A conceptualization of the molecular architecture of cellular membranes as a bilipid layer containing proteins. Membrane proteins are embedded to some degree in this bilayer, where they float freely about.

fluorescence The property possessed by certain minerals and dyes to emit visible light when excited by ultraviolet radiation. A fluorescent dye combined with specific antibody provides a sensitive test for the presence of antigen.

folliculitis An inflammatory reaction involving the formation of papules or pustules in clusters of hair follicles.

fomite Virtually any inanimate object an infected individual has contact with that can serve as a vehicle for the spread of disease.

food pyramid A triangular summation of the consumers, producers, and decomposers in a community and how they are related by trophic, number, and energy parameters.

formalin A 37% aqueous solution of formaldehyde gas; a potent chemical fixative and microbicide.

functional group In chemistry, a particular molecular combination that reacts in predictable ways and confers particular properties on a compound. Examples: —COOH, —OH, —CHO.

furuncle A boil; a localized pyogenic infection arising from a hair follicle.

G

Gaia Hypothesis The concept that biotic and abiotic factors sustain suitable conditions for one another simply by their interactions. Named after the mythical Greek goddess of earth.

GALT Abbreviation for **g**ut-**a**ssociated **l**ymphoid **t**issue. Includes Peyer's patches.

gamma globulin The fraction of plasma proteins high in immunoglobulins (antibodies). Preparations from pooled human plasma containing normal antibodies make useful passive immunizing agents against pertussis, polio, measles, and several other diseases.

gene probe Short strands of single-stranded nucleic acid that hybridize specifically with complementary stretches of nucleotides on test samples and thereby serve as a tagging and identification device.

gene therapy The introduction of normal functional genes into people with genetic diseases such as sickle-cell anemia and cystic fibrosis. This is usually accomplished by a virus vector.

genetic engineering A field involving deliberate alterations (recombinations) of the genomes of microbes, plants, and animals through special technological processes.

genome The complete set of chromosomes and genes in an organism.

genotype The genetic makeup of an organism. The genotype is ultimately responsible for an organism's phenotype, or expressed characteristics.

germicide An agent lethal to non-endospore-forming pathogens.

gingivitis Inflammation of the gum tissue in contact with the roots of the teeth.

gluconeogenesis The formation of glucose (or glycogen) from noncarbohydrate sources such as protein or fat. Also called glyconeogenesis.

glycan A polysaccharide.

glycocalyx A filamentous network of carbohydrate-rich molecules that coats cells.

glycolysis The energy-yielding breakdown (fermentation) of glucose to pyruvic or lactic acid. It is often called anaerobic glycolysis because no molecular oxygen is consumed in the degradation.

glycosidic bond A bond that joins monosaccharides to form disaccharides and polymers.

gnotobiotic Referring to experiments performed on germ-free animals.

Golgi apparatus An organelle of eucaryotes that participates in packaging and secretion of molecules.

gonococcus Common name for *Neisseria gonorrhoeae,* the agent of gonorrhea.

Gram stain A differential stain for bacteria useful in identification and taxonomy. Gram-positive organisms appear purple from crystal violet-mordant retention; whereas gram-negative organisms appear red after loss of crystal violet and absorbance of the safranin counterstain.

grana Discrete stacks of chlorophyll-containing thylakoids within chloroplasts.

granulocyte A mature leukocyte that contains noticeable granules in a Wright stain. Examples: neutrophils, eosinophils, and basophils.

granuloma A solid mass or nodule of inflammatory tissue containing modified macrophages and lymphocytes. Usually a chronic pathologic process of diseases such as tuberculosis or syphilis.

greenhouse effect The capacity to retain solar energy by a blanket of atmospheric gases that redirects heat waves back toward the earth.

growth factor An organic compound such as a vitamin or amino acid that must be provided in the diet to facilitate growth. An essential nutrient.

gumma A nodular, infectious granuloma characteristic of tertiary syphilis.

H

habitat The environment to which an organism is adapted.

halophile A microbe whose growth is either stimulated by salt or requires a high concentration of salt for growth.

H antigen The flagellar antigen of motile bacteria. *H* comes from the German word *hauch* that denotes the appearance of speading growth on solid medium.

haploid Having a single set of unpaired chromosomes, such as occurs in gametes and certain microbes.

hapten An incomplete or partial antigen. Although it constitutes the determinative group and can bind antigen, hapten cannot stimulate a full immune response without being carried by a larger protein molecule.

hay fever A form of atopic allergy marked by seasonal acute inflammation of the conjunctiva and mucous membranes of the respiratory passages. Symptoms are irritative itching and rhinitis.

helical Having a spiral or coiled shape. Said of certain virus capsids and bacteria.

helminth A term that designates all parasitic worms.

helper T cell A class of thymus-stimulated lymphocytes that facilitate various immune activities such as assisting B cells and macrophages. Also called a T helper cell.

hemagglutinin A molecule that causes red blood cells to clump or agglutinate. Often found on the surfaces of viruses.

hemolysin Any biological agent that is capable of destroying red blood cells and causing the release of hemoglobin. Many bacterial pathogens produce exotoxins that act as hemolysins.

hemopoieis The process by which the various types of blood cells are formed, such as in the bone marrow.

hepatocyte A liver cell.

herd immunity The status of collective acquired immunity in a population that reduces the likelihood that nonimmune individuals will contract and spread infection. One aim of vaccination is to induce herd immunity.

herpes zoster A recurrent infection caused by latent chickenpox virus. Its manifestation on the skin tends to correspond to dermatomes and to occur in patches that "girdle" the trunk. Also called shingles.

herpetic keratitis Corneal or conjunctival inflammation due to herpes virus type 1.

heterophile antigen An antigen present in a variety of phylogenetically unrelated species. Example: red blood cell antigens and the glycocalyx of bacteria.

heterotroph An organism that relies upon organic compounds for its carbon and energy needs.

hexose A 6-carbon sugar such as glucose and fructose.

histamine A cytokine released when mast cells and basophils release their granules. An important mediator of allergy, its effects include smooth muscle contraction, increased vascular permeability, and increased mucus secretion.

histiocyte Another term for macrophage.

histone Proteins associated with eucaryotic DNA. These simple proteins serve as winding spools to compact and condense the chromosomes.

histoplasmin The antigenic extract of *Histoplasma capsulatum,* the causative agent of histoplasmosis. The preparation is used in skin tests for diagnosis and for conducting surveys to determine the geographic distribution of the fungus.

HLA An abbreviation for **h**uman **l**eukocyte **a**ntigens. This closely linked cluster of genes programs for cell surface glycoproteins that control immune interactions between cells and is involved in rejection of allografts. Also called the major histocompatibility complex (MHC).

holoenzyme An enzyme complete with its apoenzyme and cofactors.

hops The ripe, dried fruits of the hop vine *(Humulus lupulus)* that is added to beer wort for flavoring.

hybridization A process that matches complementary strands of nucleic acid (DNA-DNA, RNA-DNA, RNA-RNA). Used for locating specific sites or types of nucleic acids.

hybridoma An artificial cell line that produces monoclonal antibodies. It is formed by fusing (hybridizing) a normal antibody-producing cell with a cancer cell, and it can produce pure antibody indefinitely.

hydatid cyst A sac, usually in liver tissue, containing fluid and larval stages of the echinococcus tapeworm.

hydration The addition of water as in the coating of ions with water molecules as ions enter into aqueous solution.

hydrolase An enzyme that catalyzes the cleavage of a bond with the additions of —H and —OH (parts of a water molecule) at the separation site.

hydrolysis A process in which water is used to break bonds in molecules. Usually occurs in conjunction with an enzyme.

hypertonic Having a greater osmotic pressure than a reference solution.

hyphae The tubular threads that make up filamentous fungi (molds). This web of branched and intertwining fibers is called a mycelium.

hypogammaglobulinemia An inborn disease in which the gamma globulin (antibody) fraction of serum is greatly reduced. The condition is associated with a high susceptibility to pyogenic infections.

hypotonic Having a lower osmotic pressure than a reference solution.

I

icosahedron A regular geometric figure having 20 surfaces that meet to form 12 corners. Some virions have capsids that resemble icosahedral crystals.

immunity An acquired resistance to an infectious agent due to prior contact with that agent.

immunogen Any substance that induces a state of sensitivity or resistance after processing by the immune system of the body.

immunoglobulin The chemical class of proteins to which antibodies belong.

IMViC Abbreviation for four identification tests: **i**ndole production, **m**ethyl red test, **V**oges-Proskauer test (**i** inserted to concoct a wordlike sound), and **c**itrate as a sole source of carbon. This test was originally developed to distinguish between *Enterobacter aerogenes* (associated with soil) and *Escherichia coli* (a fecal coliform).

incidence In epidemiology, the number of new cases of a disease occurring during a period.

inclusion A relatively inert body in the cytoplasm such as storage granules, glycogen, fat, or some other aggregated metabolic product.

inducible enzyme An enzyme that increases in amount in direct proportion to the amount of substrate present.

infection The entry, establishment, and multiplication of pathogenic organisms within a host.

infectious disease The state of damage or toxicity in the body caused by an infectious agent.

inflammation A natural, nonspecific response to tissue injury that protects the host from further damage. It stimulates immune reactivity and blocks the spread of an infectious agent.

inoculation The implantation of microorganisms into or upon culture media.

interferon Naturally occurring polypeptides produced by fibroblasts and lymphocytes that can block viral replication and regulate a variety of immune reactions.

interleukin A macrophage agent (interleukin-1, or IL-1) that stimulates lymphocyte function. Stimulated T cells release yet another interleukin (IL-2), which amplifies T-cell response by stimulating additional T cells. T helper cells stimulated by IL-2 stimulate B-cell proliferation and promote antibody production.

intoxication Poisoning that results from the introduction of a toxin into body tissues through ingestion or injection.

intron The segments on split genes of eucaryotes that do not code for polypeptide. They can have regulatory functions. *See* exon.

in utero Literally means "in the uterus"; pertains to events or developments occurring before birth.

in vitro Literally means "in glass," signifying a process or reaction occurring in an artificial environment, as in a test tube or culture medium.

in vivo Literally means "in a living being," signifying a process or reaction occurring in a living thing.

iodophor A combination of iodine and an organic carrier that is a moderate-level disinfectant and antiseptic.

ionization The aqueous dissociation of an electrolyte into ions.

ionizing radiation Radiant energy consisting of short-wave electromagnetic rays (X ray) or high-speed electrons that cause dislodgment of electrons on target molecules and create ions.

irradiation The application of radiant energy for diagnosis, therapy, disinfection, or sterilization.

isograft Transplanted tissue from one monozygotic twin to the other; transplants between highly inbred animals that are genetically identical.

isolation The separation of microbial cells by serial dilution or mechanical dispersion on solid media to achieve a clone or pure culture.

isotonic Two solutions having the same osmotic pressure such that, when separated by a semipermeable membrane, there is no net movement of solvent in either direction.

isotope A version of an element that is virtually identical in all chemical properties to another version except that their atoms have slightly different atomic masses.

J

jaundice The yellowish pigmentation of skin, mucous membranes, sclera, deeper tissues, and excretions due to abnormal deposition of bile pigments. Jaundice is associated with liver infection, as with hepatitis B virus and leptospirosis.

J chain A small molecule with high sulfhydryl content that secures the heavy chains of IgM to form a pentamer and the heavy chains of IgA to form a dimer.

K

Kaposi's sarcoma A malignant or benign neoplasm that appears as multiple hemorrhagic sites on the skin, lymph nodes, and viscera and apparently involves the metastasis of abnormal blood vessel cells. It is a clinical feature of AIDS.

keratitis Inflammation of the cornea.

keratoconjunctivitis Inflammation of the conjunctiva and cornea.

killer T cells A T lymphocyte programmed to directly affix cells and kill them. *See* cytotoxic.

Koch's postulates A procedure to establish the specific cause of disease. In all cases of infection: (1) the agent must be found; (2) inoculations of a pure culture must reproduce the same disease in animals; (3) the agent must again be present in the experimental animal; and (4) a pure culture must again be obtained.

Koplik's spots Tiny red blisters with central white specks on the mucosal lining of the cheeks. Symptomatic of measles.

L

labile In chemistry, molecules or compounds that are chemically unstable in the presence of environmental changes.

lacrimation The secretion of tears, especially in profusion.

lager The maturation process of beer, which is allowed to take place in large vats at a reduced temperature.

lagging strand The newly forming 5′ DNA strand that is discontinuously replicated in segments (Okazaki fragments).

lag phase The early phase of population growth during which no signs of growth occur.

latency The state of being inactive. Example: a latent virus or latent infection.

leading strand The newly forming 3′ DNA strand that is replicated in a continuous fashion without segments.

leaven To lighten food material by entrapping gas generated within it. Example: the rising of bread from the CO_2 produced by yeast or baking powder.

lesion A wound, injury, or some other pathologic change in tissues.

leukocidin A heat-labile substance formed by some pyogenic cocci that impairs and sometimes lyses leukocytes.

leukocytosis An abnormally large number of leukocytes in the blood, which can be indicative of acute infection.

leukopenia A lower than normal leukocyte count in the blood that can be indicative of blood infection or disease.

leukotriene An unsaturated fatty acid derivative of arachidonic acid. Leukotriene functions in chemotactic activity, smooth muscle contractility, mucous secretion, and capillary permeability.

L form L-phase variants; wall-less forms of some bacteria that are induced by drugs or chemicals. These forms can be involved in infections.

limnetic zone The deep-water region beyond the shoreline.

lipase A fat-splitting enzyme. Example: Triacylglycerol lipase separates the fatty acid chains from the glycerol backbone of triglycerides.

lipopolysaccharide A molecular complex of lipid and carbohydrate found in the bacterial cell wall. The lipopolysaccharide (LPS) of gram-negative bacteria is an endotoxin with generalized pathologic effects such as fever.

lithotroph An autotrophic microbe that derives energy from reduced inorganic compounds such as N_2S.

littoral zone The shallow region along a shoreline.

lophotrichous Describing bacteria having a tuft of flagella at one or both poles.

lumen The cavity within a tubular organ.

lymphadenitis inflammation of one or more lymph nodes. Also called lymphadenopathy.

lymphatic system A system of vessels and organs that serve as sites for development of immune cells and immune reactions. It includes the spleen, thymus, lymph nodes, and GALT.

lymphokine A soluble substance secreted by sensitized T lymphocytes upon contact with specific antigen. About 50 types exist, and they stimulate inflammatory cells: macrophages, granulocytes and lymphocytes. Examples: migration inhibitory factor, macrophage activating factor, chemotactic factor.

lyophilization Freeze-drying; the separation of a dissolved solid from the solvent by freezing the solution and evacuating the solvent under vacuum. A means of preserving the viability of cultures.

lysin A complement-fixing antibody that destroys specific targeted cells. Examples: hemolysin and bacteriolysin.

lysis The physical rupture or deterioration of a cell.

lysogeny The indefinite persistence of bacteriophage DNA in a host without bringing about the production of virions. A lysogenic cell can revert to a lytic cycle, the process that ends in lysis.

lysosome A cytoplasmic organelle containing lysozyme and other hydrolytic enzymes.

lysozyme An enzyme that attacks the bonds on bacterial peptidoglycan. It is a natural defense found in tears and saliva.

M

macrophage A white blood cell derived from a monocyte that leaves the circulation and enters tissues. These cells are important in nonspecific phagocytosis and in regulating, stimulating, and cleaning up after immune responses.

macule A small, discolored spot or patch of skin that is neither raised above nor depressed below the surrounding surface.

malt The grain, usually barley, that is sprouted to obtain digestive enzymes and dried for making beer.

Mantoux test An intradermal screening test for tuberculin hypersensitivity. A red, firm patch of skin at the injection site greater than 10 mm in diameter after 48 hours is a positive result that indicates current or prior exposure to the TB bacillus.

marker Any trait or factor of a cell, virus, or molecule that makes it distinct and recognizable. Example: a genetic marker.

mash In making beer, this malt grain is steeped in warm water, ground up, and fortified with carbohydrates.

mast cell A nonmotile connective tissue cell implanted along capillaries, especially in the lungs, skin, gastrointestinal tract, and genitourinary tract. Like a basophil, its granules store mediators of allergy.

mastitis Inflammation of the breast or udder.

matrix The dense ground substance between the cristae of a mitochondrion that serves as a site for metabolic reactions.

meiosis The type of cell division necessary for producing gametes in diploid organisms. Two nuclear divisions in rapid succession produce four gametocytes, each containing a haploid number of chromosomes.

memory cell The long-lived progeny of a sensitized lymphocyte that remains in circulation and is genetically programmed to react rapidly with its antigen.

meningitis An inflammation of the membranes (meninges) that surround and protect the brain. It is often caused by bacteria such as *Neisseria meningitidis* (the meningococcus) and *Haemophilus influenzae*.

merozoite The motile, infective stage of an apicomplexan parasite that comes from a liver or red blood cell undergoing multiple fission.

mesosome The irregular invagination of a bacterial cell membrane that is more prominent in gram-positive than in gram-negative bacteria. Although its function is not definitely known, it appears to participate in DNA replication and cell division; in certain cells, it appears to play a role in secretion.

messenger RNA A single-stranded transcript that is a copy of the DNA template that corresponds to a gene.

metabolites Small organic molecules that are intermediates in the stepwise biosynthethsis or breakdown of macromolecules.

metachromatic Exhibiting a color other than that of the dye used to stain it.

metastasis In cancer, the dissemination of tumor cells so that neoplastic growths appear in other sites away from the primary tumor.

MHC Major histocompatibility complex. *See* HLA.

MIC Abbreviation for **m**inimum **i**nhibitory **c**oncentration. The lowest concentration of antibiotic needed to inhibit bacterial growth in a test system.

microaerophile An aerobic bacterium that requires oxygen at a concentration less than that in the atmosphere.

miliary tuberculosis Rapidly fatal tuberculosis due to dissemination of mycobacteria in the blood and formation of tiny granules in various organs and tissues. The term *miliary* means resembling a millet seed.

mineralization The process by which decomposers (bacteria and fungi) convert organic debris into inorganic and elemental form. It is part of the recycling process.

miracidium The ciliated first-stage larva of a trematode. This form is infective for a corresponding intermediate host snail.

mitochondrion A double-membrane organelle of eucaryotes that is the main site for aerobic respiration.

mitosis Somatic cell division that preserves the somatic chromosome number..

mixed culture A container growing two or more different, known species of microbes.

monoclonal antibody An antibody produced by a clone of lymphocytes that respond to a particular antigenic determinant and generate identical antibodies only to that determinant. *See* hybridoma.

monocyte A large mononuclear leukocyte normally found in the lymph nodes, spleen, bone marrow, and loose connective tissue. This type of cell makes up 3% to 7% of circulating leukocytes.

monomer A simple molecule that can be linked by chemical bonds to form larger molecules.

monosaccharide A simple sugar such as glucose that is a basic building block for more complex carbohydrates.

monotrichous Describing a microorganism that bears a single flagellum.

morbidity A diseased condition; the relative incidence of disease in a community.

mordant A chemical that fixes a dye in or on cells by forming an insoluble compound and thereby promoting retention of that dye. Example: Gram's iodine in the Gram stain.

must Juices expressed from crushed fruits that are used in fermentation for wine.

mutagen Any agent that induces genetic mutation. Examples: certain chemical substances, ultraviolet light, radioactivity.

mutation A permanent inheritable alteration in the DNA sequence or content of a cell.

mycelium The filamentous mass that makes up a mold. Composed of hyphae.

mycetoma A chronic fungal infection usually afflicting the feet, typified by swelling and multiple draining lesions. Example: maduromycosis or Madura foot.

mycorrhizae Various species of fungi adapted in an intimate, mutualistic relationship to plant roots.

mycosis Any disease caused by a fungus.

N

narrow spectrum Denotes drugs that are selective and limited in their effects. For example, they inhibit either gram-negative or gram-positive bacteria, but not both.

necrosis A pathologic process in which cells and tissues die and disintegrate.

negative feedback Enzyme regulation of metabolism by the end product of a multienzyme system that blocks the action of a "pacemaker" enzyme at or near the beginning of the pathway.

negative stain A staining technique that renders the background opaque or colored and leaves the object unstained so that it is outlined as a colorless area.

nematode A common name for helminths called roundworms.

neoplasm A synonym for tumor.

neutron An electrically neutral particle in the nuclei of all atoms except hydrogen.

neutrophil A mature granulocyte present in peripheral circulation, exhibiting a multilobular nucleus and numerous cytoplasmic granules that retain a neutral stain. The neutrophil is an active phagocytic cell in bacterial infection.

niche In ecology, an organism's biological role in or contribution to its community.

night soil An archaic euphemism for human excrement collected for fertilizing crops.

nitrogen base A ringed compound of which pyrimidines and purines are types.

nitrogen fixation A process occurring in certain bacteria in which atmospheric N_2 gas is converted to a form (NH_4) usable by plants.

nomenclature A set system for scientifically naming organisms, enzymes, anatomical structures, etc.

nonsense codon A triplet of mRNA bases that does not specify an amino acid but signals the end of a polypeptide chain.

normal flora The native microbial forms that an individual harbors.

nosocomial infection An infection not present upon admission to a hospital but incurred while being treated there.

nucleocapsid In viruses, the close physical combination of the nucleic acid with its protective covering.

nucleoid The basophilic nuclear region or nuclear body that contains the bacterial chromosome.

nucleotide A composite of a nitrogen base (purine or pyrimidine), a 5-carbon sugar (ribose or deoxyribose), and a phosphate group; a unit of nucleic acids.

numerical aperture In microscopy, the amount of light passing from the object and into the object in order to maximize optical clarity and resolution.

nutrient Any chemical substance that must be provided to a cell for normal metabolism and growth. Macronutrients are required in large amounts, and micronutrients in small amounts.

O

obligate Without alternative; restricted to a particular characteristic. Example: An obligate parasite survives and grows only in a host; an obligate aerobe must have oxygen to grow; an obligate anaerobe is destroyed by oxygen.

Okazaki fragment In replication of DNA, a segment formed on the lagging strand in which biosynthesis is conducted in a discontinuous manner dictated by the $5' \rightarrow 3'$ DNA polymerase orientation.

oligodynamic action A chemical having antimicrobial activity in minuscule amounts. Example: Certain heavy metals are effective in a few parts per billion.

oncogene A naturally occurring type of gene that when activated can transform a normal cell into a cancer cell.

oncology The study of neoplasms, their cause, disease characteristics, and treatment.

oocyst The encysted form of a fertilized macrogamete or zygote; typical in the life cycles of apicomplexan parasites.

operon A genetic operational unit that regulates metabolism by controlling mRNA production. In sequence, the unit consists of a regulatory gene, inducer or repressor control sites, and structural genes.

opportunistic In infection, ordinarily nonpathogenic or weakly pathogenic microbes that cause disease primarily in an immunologically compromised host.

opsonization The process of stimulating phagocytosis by affixing molecules (opsonins such as antibodies and complement) to the surfaces of foreign cells or particles.

organelle A small component of eucaryotic cells that is bound by a membrane and that is specialized in function.

osmophile A microorganism that thrives in a medium having high osmotic pressure.

osmosis The diffusion of water across a selectively permeable membrane in the direction of lower water concentration.

oxidation In chemical reactions, the loss of electrons by one reactant.

oxidation-reduction Redox reactions, in which paired sets of molecules participate in electron transfers.

oxidative phosphorylation The synthesis of ATP using energy given off during the electron transport phase of respiration.

P

palindrome A word, verse, number, or sentence that reads the same forward or backward. Palindromes of nitrogen bases in DNA have genetic significance as transposable elements, as regulatory protein targets, and in DNA splicing.

palisades The characteristic arrangement of *Corynebacterium* cells resembling a row of fence posts and created by snapping.

pandemic A disease afflicting an increased proportion of the population over a wide geographic area (often worldwide).

pannus The granular membrane occurring on the cornea in trachoma.

papule An elevation of skin that is small, demarcated, firm, and usually conical.

parasite An organism that lives on or within another organism (the host), from which it obtains nutrients and enjoys protection. The parasite produces some degree of harm in the host.

parenteral Administering a substance into a body compartment other than through the gastrointrointestinal tract, such as via intravenous, subcutaneous, intramuscular, or intramedullary injection.

paroxysmal Events characterized by sharp spasms or convulsions; sudden onset of a symptom such as fever and chills.

passive immunity Specific resistance that is acquired indirectly by donation of preformed immune substances (antibodies) produced in the body of another individual.

pasteurization Heat treatment of perishable fluids such as milk, fruit juices, or wine to destroy heat-sensitive vegetative cells, followed by rapid chilling to inhibit growth of survivors and germination of spores. It prevents infection and spoilage.

pathogen Any agent, usually a virus, bacterium, fungus, protozoan, or helminth, that causes disease.

pathogenicity The capacity of microbes to cause disease.

pathology The structural and physiological effects of disease on the body.

pellicle A membranous cover; a thin skin, film, or scum on a liquid surface; a thin film of salivary glycoproteins that forms over newly cleaned tooth enamel when exposed to saliva.

pelvic-inflammatory disease (PID) An infection of the uterus and fallopian tubes that has ascended from the lower reproductive tract. Caused by gonococci and chlamydias.

penicillinase An enzyme that hydrolyzes penicillin; found in penicillin-resistant strains of bacteria.

penicillins A large group of naturally occurring and synthetic antibiotics produced by *Penicillium* mold and active against the cell wall of bacteria.

pentose A monosaccharide with five carbon atoms per molecule. Examples: arabinose, ribose, xylose.

peptidase An enzyme that can hydrolyze the peptide bonds of a peptide chain.

peptide bond The covalent union between two amino acids that forms between the amine group of one and the carboxyl group of the other. The basic bond of proteins.

peptidoglycan A network of polysaccharide chains cross-linked by short peptides that forms the rigid part of bacterial cell walls. Gram-negative bacteria have a smaller amount of this rigid structure than do gram-positive bacteria.

perinatal In childbirth, occurring before, during, or after delivery.

periodontal Involving the structures that surround the tooth.

periplasmic space The region between the cell wall and cell membrane of the cell envelopes of gram-negative bacteria.

peritrichous In bacterial morphology, having flagella distributed over the entire cell.

petechiae Minute hemorrhagic spots in the skin that range from pinpoint- to pinhead-sized.

Peyer's patches Oblong lymphoid aggregates of the gut located chiefly in the wall of the terminal and small intestine. Along with the tonsils and appendix, Peyer's patches make up the gut-associated-lymphoid tissue that responds to local invasion by infectious agents.

pH The symbol for the negative logarithm of the H ion concentration; p (power) or $[H^+]_{10}$. A system for rating acidity and alkalinity.

phage A bacteriophage; a virus that specifically parasitizes bacteria.

phagocytosis A type of endocytosis in which the cell membrane actively engulfs large particles or cells into vesicles.

phagolysosome A body formed in a phagocyte, consisting of a union between a vesicle containing the ingested particle (the phagosome) and a vacuole of hydrolytic enzymes (the lysosome).

phenotype The observable characteristics of an organism produced by the interaction between its genetic potential (genotype) and the environment.

phlebotomine Pertains to a genus of very small midges or blood-sucking (phlebotomous) sand flies and to diseases associated with those vectors such as kalaazar, Oroya fever, and cutaneous leishmaniasis.

photic zone The aquatic stratum from the surface to the limits of solar light penetration.

photoautotroph An organism that utilizes light for its energy and carbon dioxide chiefly for its carbon needs.

photolysis Literally, splitting water with light. In photosynthesis, this step frees electrons and gives off O_2.

photosynthesis A process occurring in plants, algae, and some bacteria that traps the sun's energy and converts it to the cell. This energy is used to fix CO_2 into organic compounds.

pili Small, stiff filamentous appendages in gram-negative bacteria that function in DNA exchange during bacterial conjugation.

pinocytosis The engulfment or endocytosis of liquids by extensions of the cell membrane.

plankton Minute animals (zooplankton) or plants (phytoplankton) that float and drift in the limnetic zone of bodies of water.

plaque In virus propagation methods, the clear zone of lysed cells in tissue culture or chick embryo membrane that corresponds to the area containing viruses. In dental application, the filamentous mass of microbes that adheres tenaciously to the tooth and predisposes to caries, calculus, or inflammation.

plasma cell A progeny of an activated B cell that actively produces and secretes antibodies.

plasmids Extrachromosomal genetic units characterized by several features. A plasmid is a circular, double-stranded DNA that is smaller than and replicates independently of the cell chromosome; it bears genes that are not essential for cell growth; it can bear genes that code for adaptive traits; and it is transmissible to other bacteria.

pleomorphism Normal variability of cell shapes in a single species.

pluripotential Stem cells having the developmental plasticity to give rise to more than one type. Example: undifferentiated blood cells in the bone marrow.

pneumonia An inflammation of the lung leading to accumulation of fluid and respiratory compromise.

pneumococcus Common name for *Streptococcus pneumoniae,* the major cause of bacterial pneumonia.

pneumonic plague The acute, frequently fatal form of pneumonia caused by *Yersinia pestis.*

point mutation A change that involves the loss, substitution, or addition of one or a few nucleotides.

polyclonal In reference to a collection of antibodies with mixed specificities that arose from more than one clone of B cells.

polymer A macromolecule made up of a chain of repeating units. Examples: starch, protein, DNA.

polymerase An enzyme that produces polymers through catalyzing bond formation between building blocks (polymerization).

polymerase chain reaction (PCR) A technique that amplifies segments of DNA for testing. Using denaturation, primers, and heat-resistant DNA polymerase, the number can be increased several million-fold.

polymorphonuclear leukocytes (PMNLs) White blood cells with variously shaped nuclei. Although this term commonly denotes all granulocytes, it is used especially for the neutrophils.

polymyxin A mixture of antibiotic polypeptides from *Bacillus polymyxa* that are particularly effective against gram-negative bacteria.

polypeptide A relatively large chain of amino acids linked by peptide bonds.

polysaccharide A carbohydrate that can be hydrolyzed into a number of monosaccharides. Examples: cellulose, starch, glycogen.

porin Transmembrane proteins of the outer membrane of gram-negative cells that permit transport of small molecules into the periplasmic space but bar the penetration of larger molecules.

potable Describing water that is relatively clear, odor-free, and safe to drink.

pox The thick, elevated pustular eruptions of various viral infections. Also called pocks.

precipitin An antibody that combines with and aggregates a soluble antigen to bring about it precipitation from solution.

prevalence The total cumulative number of cases of a disease in a certain area and time period.

prion A concocted word to denote "proteinaceous infectious agent"; a cytopathic protein associated with the slow-virus spongiform encephalopathies of humans and animals.

procaryotic cell A small, simple cell lacking a true nucleus, a nuclear envelope, and membrane-enclosed organelles.

prodromium A short period of mild symptoms occurring at the end of the period of incubation. It indicates the onset of an infection.

proglottid The egg-generating segment of a tapeworm that contains both male and female organs.

promastigote A morphological variation of the trypanosome parasite responsible for leishmaniasis.

promoter region The site composed of a short signaling DNA sequence that RNA polymerase recognizes and binds to commence transcription.

properdin A normal serum protein involved in the alternate complement pathway that leads to nonspecific lysis of bacterial cells and viruses.

prophage A lysogenized bacteriophage; a phage that is latently incorporated into the host chromosome instead of undergoing viral replication and lysis.

prophylactic Any device, method, or substance used to prevent disease.

prostaglandin A hormonelike substance that regulates many body functions. Prostaglandin comes from a family of organic acids containing 5-carbon rings that are essential to the human diet.

proton An elementary particle that carries the positive change. It is identical to the nucleus of the hydrogen atom.

protoplast A bacterial cell whose cell wall is completely lacking and that is vulnerable to osmotic lysis.

pseudohypha A chain of easily separated, spherical to sausage-shaped yeast cells partitioned by constrictions rather than by septa.

pseudomembrane A tenacious, noncellular mucous exudate containing cellular debris that tightly blankets the mucosal surface in infections such as diphtheria and pseudomembranous enterocolitis.

pseudopodium A temporary extension of the protoplasm of an ameboid cell. It serves both in ameboid motion and for food gathering (phagocytosis).

psychrophile A microorganism that thrives at low temperature (0°–20°C), with a temperature optimum of 0°–15°C.

pure culture A container of microbial cells whose identity is known.

purine A nitrogen base that is an important encoding component of DNA and RNA. The two most common purines are adenine and guanine.

purpura A condition characterized by bleeding into skin or mucous membrane, giving rise to patches of red that darken and turn purple.

pus The viscous, opaque, usually yellowish matter formed by an inflammatory infection. It consists of serum exudate, tissue debris, leukocytes, and microorganisms.

pyogenic Pertains to pus formers, especially the pyogenic cocci: pneumococci, streptococci, staphylococci, and neisseriae.

pyrimidine Nitrogenous bases that help form the genetic code on DNA and RNA. Uracil, thymine, and cytosine are the most important pyrimidines.

pyrimidine dimer The union of two adjacent pyrimidines on the same DNA strand, brought about by exposure to ultraviolet light. It is a form of mutation.

pyrogen A substance that causes a rise in body temperature. It can come from pyrogenic microorganisms or from polymorphonuclear leukocytes (endogenous pyrogens).

Q

Q fever A disease first described in Queensland, Australia, initially dubbed Q for "query" to denote a fever of unknown origin. Q fever is now known to be caused by a rickettsial infection.

quats A byword that pertains to a family of surfactants called quaternary ammonium compounds. These detergents are only weakly microbicidal and are used as sanitizers and preservatives.

Quellung test The capsular swelling phenomenon; a serological test for the presence of pneumococci whose capsules enlarge and become opaque and visible when exposed to specific anticapsular antibodies.

quinolone A class of synthetic antimicrobic drugs with broad-spectrum effects.

R

rad A unit of measure for absorbed dose of ionizing radiation.

radiation Electromagnetic waves or rays, such as those of light given off from an energy source.

radioactive isotopes Unstable isotopes whose nuclei emit particles of radiation. This emission is called radioactivity or radioactive decay. Three naturally occurring emissions are alpha, beta, and gamma radiation.

radioimmunoassay RIA; a highly sensitive laboratory procedure that employs radioisotope-labeled substances to measure the levels of antibodies or antigens in the serum.

real image An image formed at the focal plane of a convex lens. In the compound light microscope, it is the image created by the objective lens.

receptor In intercellular communication, cell surface molecules involved in recognition, binding, and intracellular signaling.

recombinant DNA A technology, also known as genetic engineering, that deliberately modifies the genetic structure of an organism to create novel products, microbes, animals, plants, and viruses.

redox Denoting an oxidation-reduction reaction.

reduction In chemistry, the gain of electrons.

refraction In optics, the bending of light as it passes from one medium to another with a different index of refraction.

replication In DNA synthesis, the semiconservative mechanisms that ensure precise duplication of the parent DNA strands.

replication fork The Y-shaped point on a replicating DNA molecule where the DNA polymerase is synthesizing new strands of DNA.

repressor The protein product of a repressor gene that combines with the operator and arrests the transcription and translation of structural genes.

reservoir In disease communication, the natural host or habitat of a pathogen.

resident flora The deeper, more stable microflora that inhabit the skin and exposed mucous membranes, as opposed to the superficial, variable, transient population.

resolving power The capacity of a microscope lens system to accurately distinguish between two separate entities that lie close to each other. Also called resolution.

respiratory chain In cellular respiration, a series of electron-carrying molecules that transfers energy-rich electrons and protons to molecular oxygen. In transit, energy is extracted and conserved in the form of ATP.

restriction endonuclease An enzyme present naturally in cells that cleaves specific locations on DNA. it is an important means of inactivating viral genomes, and it is also used to splice genes in genetic engineering.

reticuloendothelial system Also known as the mononuclear phagocyte system, it pertains to a network of fibers and phagocytic cells (macrophages) that permeates the tissues of all organs. Examples: Kupffer cells in liver sinusoids, alveolar phagocytes in the lung, microglia in nervous tissue.

retrovirus A group of RNA viruses (including HIV) that have the mechanisms for converting their genome into a double strand of DNA that can be inserted on a host's chromosome.

reverse transcriptase The enzyme possessed by retroviruses that carries out the reversion of RNA to DNA—a form of reverse transcription.

Reye's syndrome A sudden, usually fatal neurological condition that occurs in children after a viral infection. Autopsy shows cerebral edema and marked fatty change in the liver and renal tubules.

rhizosphere The zone of soil, complete with microbial inhabitants, in the immediate vicinity of plant roots.

ribose A 5-carbon monosaccharide found in RNA.

ribosome A bilobed macromolecular complex of ribonucleoprotein that coordinates the codons of mRNA with tRNA anticodons and, in so doing, constitutes the peptide assembly site.

ringworm A superficial mycosis caused by various dermatophytic fungi. This common name is actually a misnomer.

rolling circle An intermediate stage in viral replication of circular DNA into linear DNA.

rosette formation A technique for distinguishing surface receptors on T cells by reacting them with sensitized indicator sheep red blood cells. The cluster of red cells around the central white blood cell resembles a little rose blossom and is indicative of the type of receptor.

S

salpingitis Inflammation of the fallopian tubes.

sanitize To clean inanimate objects using soap and degerming agents so that they are safe and free of high levels of microorganisms.

saprobe A microbe that decomposes organic remains from dead organisms. Also known as a saprophyte or saprotroph.

sarcina A cubical packet of 8, 16, or more cells; the cellular arrangement of the genus *Sarcina* in the family Micrococcaceae.

sarcoma A fleshy neoplasm of connective tissue. Growth, usually highly malignant, is of mesodermal origin.

satellite phenomenon A type of commensal relationship in which one microbe produces growth factors that favor the growth of its dependent partner. When plated on solid media, the dependent organism appears as surrounding colonies. Example: *Staphylococcus* surrounded by its *Haemophilus* satellite.

schizogony A process of multiple fission whereby first the nucleus divides several times, and subsequently the cytoplasm is subdivided for each new nucleus during cell division.

scolex The anterior end of a tapeworm characterized by hooks and/or suckers for attachment to the host.

SCP Abbreviation for single-cell protein, a euphemistic expression for microbial protein intended for human and animal consumption.

sebaceous glands The sebum- (oily, fatty) secreting glands of the skin.

secondary immune response The more rapid and heightened response to antigen in a sensitized subject due to memory lymphocytes.

secondary infection An infection that compounds a preexisting one.

secretion The process of actively releasing cellular substances into the extracellular environment.

secretory antibody The immunoglobulin (IgA) that is found in secretions of mucous membranes and serves as a local immediate protection against infection.

selective media Nutrient media designed to favor the growth of certain microbes and to inhibit undesirable competitors.

self-limited Applies to an infection that runs its course without disease or residual effects.

semiconservative replication In DNA replication, the synthesis of paired daughter strands, each retaining a parent strand template.

semisolid media Nutrient media with a firmness midway between that of a broth (a liquid medium) and an ordinary solid medium; motility media.

sensitizing dose The initial effective exposure to an antigen or an allergen that stimulates an immune response. Often applies to allergies.

sepsis The state of putrefaction; the presence of pathogenic organisms or their toxins in tissue or blood.

septicemia Systematic infection associated with microorganisms multiplying in circulating blood.

septum A partition or cellular cross wall, as in certain fungal hyphae.

sequela A morbid complication that follows a disease.

seropositive Showing the presence of specific antibody in a serological test. Indicates ongoing infection.

serotonin A vascoconstrictor that inhibits gastric secretion and stimulates smooth muscle.

serotyping The subdivision of a species or subspecies into an immunologic type, based upon antigenic characteristics.

serum The clear fluid expressed from clotted blood that contains dissolved nutrients, antibodies, and hormones but not cells or clotting factors.

sex pilus A conjugative pilus.

sign Any abnormality uncovered upon physical diagnosis that indicates the presence of disease. A sign is an objective assessment of disease, as opposed to a symptom, which is the subjective assessment perceived by the patient.

sodoku A Japanese term pertaining to rat bite fever.

Southern blot A technique that separates fragments of DNA using electrophoresis and identifies them by hybridization.

spheroplast A gram-negative cell whose peptidoglycan, when digested by lysozyme, remains intact but is osmotically vulnerable.

spike A receptor on the surface of certain enveloped viruses that facilitates specific attachment to the host cell.

spirillum A type of bacterial cell with a rigid spiral shape and external flagella.

spirochete A coiled, spiral-shaped bacterium that has endoflagella and flexes as it moves.

sporangium A fungal cell in which asexual spores are formed by multiple cell cleavage.

sporozoite One of many minute elongated bodies generated by multiple division of the oocyst. It is the infectious form of the malarial parasite that is harbored in the salivary gland of the mosquito and is inoculated into the victim during feeding.

start codon The nucleotide triplet AUG that codes for the first amino acid in protein sequences.

starter culture The sizeable inoculation of pure bacterial, mold, or yeast sample for bulk processing, as in the preparation of fermented foods, beverages, and pharmaceuticals.

stasis A state of rest or inactivity; applied to nongrowing microbial cultures. Also called microbistasis.

sterilization Any process that completely removes or destroys all viable microorganisms, including viruses, from an object or habitat. Material so treated is sterile.

strain In microbiology, a set of descendants cloned from a common ancestor that retain the original characteristics. Any deviation from the original is a different strain.

streptolysin A hemolysin produced by streptococci.

stroma The matrix of the chloroplast that is the site of the dark reactions.

structural gene A gene that codes for the amino acid sequence (peptide structure) of a protein.

subacute Indicates an intermediate status between acute and chronic disease.

subcellular vaccine A vaccine against isolated microbial antigens rather than against the entire organism.

subclinical A period of inapparent manifestations that occurs before symptoms and signs of disease appear.

substrate The specific molecule upon which an enzyme acts.

superantigens Bacterial toxins that are potent stimuli for T cells and can be a factor in diseases such as toxic shock.

superinfection An infection occurring during antimicrobic therapy that is caused by an overgrowth of drug-resistant microorganisms.

superoxide ion A toxic radical form oxygen metabolism (O_2^-).

suppressor T cell A class of T cells that inhibits the actions of B cells and other T cells.

surfactant A surface-active agent that form a water-soluble interface. Examples: detergents, wetting agents, dispersing agents, and surface tension depressants.

sylvatic Denotes the natural presence of disease among wild animal populations. Examples: sylvatic (sylvan) plague, rabies.

symbiosis An intimate association between individuals from two species; used as a synonym for mutualism.

symptom The subjective evidence of infection and disease as perceived by the patient.

syncytium A multinucleated protoplasmic mass formed by consolidation of individual cells.

syndrome The collection of signs and symptoms that, taken together, paint a portrait of the disease.

synergism The coordinated or correlated action by two or more drugs or microbes that results in a heightened response or greater activity.

syngamy Conjugation of the gametes in fertilization.

systemic Occurring throughout the body; said of infections that invade many compartments and organs via the circulation.

T

temperate phage A bacteriophage that enters into a less virulent state by becoming incorporated into the host genome as a prophage instead of in the vegetative or lytic form that eventually destroys the cell.

teratogenic Causing abnormal fetal development.

tetanospasmin The neurotoxin of *Clostridium tetani,* the agent of tetanus. Its chief action is directed upon the inhibitory synapses of the anterior horn motor neurons.

tetracyclines A group of broad-spectrum antibiotics with a complex 4-ring structure.

therapeutic index The ratio of the toxic dose to the effective therapeutic dose that is used to assess the safety and reliability of the drug.

thermal death point The lowest temperature that achieves sterilization in a given quantity of broth culture upon a 10-minute exposure. Examples: 55°C for *Escherichia coli,* 60°C for *Mycobacterium tuberculosis,* and 120°C for spores.

thermal death time The least time required to kill all cells of a culture at a specified temperature.

thermocline A temperature buffer zone in a large body of water that separates the warmer water (the epilimnion) from the colder water (the hypolimnion).

thermoduric Resistant to the harmful effects of high temperature.

thermophile A microorganism that thrives at a temperature of 50°C or higher.

thrush *Candida albicans* infection of the oral cavity.

thylakoid Vesicles of a chloroplast formed by elaborate folding of the inner membrane to form "discs." Solar energy trapped in the thylakoids is used in photosynthesis.

tincture A medicinal substance dissolved in an alcoholic solvent.

tinea Ringworm; a fungal infection of the hair, skin, or nails.

titer In immunochemistry, a measure of antibody level in a patient, determined by agglutination methods.

T lymphocyte (T cell) A white blood cell that is processed in the thymus gland and is involved in cell-mediated immunity.

topoisomerases Enzymes that can add or remove DNA twists and thus regulate the degree of supercoiling.

toxemia An abnormality associated with certain infectious diseases. Toxemia is caused by toxins or other noxious substances released by microorganisms circulating in the blood.

toxigenicity The tendency for a pathogen to produce toxins. It is an important factor in bacterial virulence.

toxoid A toxin that has been rendered nontoxic but is still capable of eliciting the formation of protective antitoxin antibodies; used in vaccines.

tracheostomy A surgically created emergency airway opening into the trachea.

transcription mRNA synthesis; the process by which a strand of RNA is produced against a DNA template.

transduction The transfer of genetic material from one bacterium to another by means of a bacteriophage vector.

transformation In microbial genetics, the transfer of genetic material contained in "naked" DNA fragments from a donor cell to a competent recipient cell.

transgenic technology Introduction of foreign DNA into cells or organisms. Used in genetic engineering to create recombinant plants, animals, and microbes.

transients In normal flora, the assortment of superficial microbes whose numbers and types vary depending upon recent exposure. The deeper-lying residents constitute a more stable population.

translation Protein synthesis; the process of decoding the messenger RNA code into a polypeptide.

transposon A DNA segment with an insertion sequence at each end, enabling it to migrate to another plasmid, to the bacterial chromosome, or to a bacteriophage.

trematode A fluke or flatworm parasite of vertebrates.

triglyceride A type of lipid composed of a glycerol molecule bound to three fatty acids.

trophozoite A vegetative protozoan (feeding form) as opposed to a resting (cyst) form.

trypomastigote The infective morphological stage transmitted by the tsetse fly or the reduviid bug in African trypanosomiasis and Chagas' disease.

tubercle In tuberculosis, the granulomatous well-defined lung lesion that can serve as a focus for latent infection.

tuberculin A glycerinated broth culture of *Mycobacterium tuberculosis* that is evaporated and filtered. Formerly used to treat tuberculosis, tuberculin is now used chiefly for diagnostic tests.

tyndallization Fractional (discontinous, intermittent) sterilization designed to destroy spores indirectly. A preparation is exposed to flowing steam for an hour, and then the mineral is allowed to incubate to permit spore germination. The resultant vegetative cells are destroyed by repeated steaming and incubation.

U

uncoating The process of removal of the viral coat and release of the viral genome by its newly invaded host cell.

universal donor In blood grouping and transfusion, a group O individual whose erythrocytes bear neither agglutinogen A nor B.

V

vaccine Originally used in reference to inoculation with the cowpox or vaccinia virus to protect against smallpox. In general, the term now pertains to injection of whole microbes (killed or attenuated), toxoids, or parts of microbes as a prevention or cure for disease.

valence The combining power of an atom based upon the number of electrons it can either take on or give up.

variable region The antigen binding fragment of an immunoglobin molecule, consisting of a combination of heavy and light chains whose molecular conformation is specific for the antigen.

variolation A hazardous, outmoded process of deliberately introducing smallpox material scraped from a victim into the nonimmune subject in the hope of inducing resistance.

VDRL A flocculation test that detects syphilis antibodies. An important screening test. The abbreviation stands for **V**enereal **D**isease **R**esearch **L**aboratories.

vector An animal that transmits infectious agents from one host to another, usually a biting or piercing arthropod like the tick, mosquito, or fly. Infectious agents can be conveyed mechanically by simple contact or biologically whereby the parasite develops in the vector.

vector$_2$ A genetic element such as a plasmid or a bacteriophage used to introduce genetic material into a cloning host during recombinant DNA experiments.

vegetative In describing microbial developmental stages, a metabolically active feeding and dividing form, as opposed to a dormant, seemingly inert, nondividing form. Examples: a bacterial cell versus its spore; a protozoan trophozoite versus its cyst.

vehicle An inanimate material (solid object, liquid, or air) that serves as a transmission agent for pathogens.

verruca A flesh-colored wart. This self-limited tumor arises from accumulations of growing epithelial cells. Example: papilloma warts.

vesicle A blister characterized by a thin-skinned, elevated, superficial pocket inflated with serum.

vibrio A curved, rod-shaped bacterial cell.

viremia The presence of a virus in the bloodstream.

virion An elementary virus particle in its complete morphological and thus infectious form. A virion consists of the nucleic acid core surrounded by a capsid, which can be enclosed in an envelope.

viroid An infectious agent that, unlike a virion, lacks a capsid and consists of a closed circular RNA molecule. Although known viroids are all plant pathogens, it is conceivable that animal versions exist.

virtual image In optics, an image formed by diverging light rays; in the light compound microscope, the second, magnified visual impression formed by the ocular from the real image formed by the objective.

virulence In infection, the relative capacity of a pathogen to invade and harm host cells.

W

wart An epidermal tumor caused by papillomaviruses. Also called a verruca.

Western blot text A procedure for separating and identifying antigen or antibody mixtures by two-dimensional electrophoresis in polyacrylamide gel, followed by immune labeling.

wheal A welt; a marked, slightly red, usually itchy area of the skin that changes in size and shape as it extends to adjacent area. The reaction is triggered by cutaneous contact or intradermal injection of allergens in sensitive individuals.

whey The residual fluid from milk coagulation that separates from the solidified curd.

white piedra A fungus disease of hair, especially of the scalp, face, and genitals, caused by *Trichosporon beigelii*. The infection is associated with soft, mucilaginous, white-to-light-brown nodules that form within and on the hair shafts.

whitlow A deep inflammation of the finger or toe, especially near the tip or around the nail. Whitlow is a painful herpes simplex virus infection that can last several weeks and is most common among health care personnel who come in contact with the virus in patients.

Widal test An agglutination test for diagnosing typhoid.

wort The clear fluid derived from soaked mash that is fermented for beer.

X

xenograft The transfer of a tissue or an organ from an animal of one species to a recipient of another species.

Y

yaws A tropical disease caused by *Treponema pertenue* that produces granulomatous ulcers on the extremities and occasionally on bone, but does not produce central nervous system or cardiovascular complications.

Z

zoonosis An infectious disease indigenous to animals that humans can acquire through direct or indirect contact with infected animals.

zygospore A thick-walled sexual spore produced by the zygomycete fungi. It develops from the union of two hyphae, each bearing nuclei of opposite mating types.

CREDITS

PHOTOGRAPHS

Photo Research by Connie Mueller

Chapter 1

1.1a: Courtesy of General Electric Research and Development Center; **1.1b:** © Mark Joseph/Tony Stone Images; **1.5 c-2:** © Sinclair Stammers/SPL/Photo Researchers, Inc.; **1.5d:** © T.E. Adams/Visuals Unlimited; **1.5e:** © Carolina Biological Supply/Phototake; **1.5f:** © Cabisco/Visuals Unlimited; **1.7:** © Corbis-Bettmann; **1.8 a,b:** © Kathy Talaro/Visuals Unlimited; **1.8c:** © Science/VU/Visuals Unlimited; **1.11:** © Corbis-Bettmann; **1.12:** © Corbis-Bettmann.

Chapter 2

Opener: Courtesy Amersham Pharmacia Biotech, photo by David Andre; **2.10 a,b,c:** © John W. Hole, Jr.; **2.22d:** From A.S. Moffat "Nitrogenase Structure Revealed," *Science* 250:1513, 12/14/90. Copyright 1990 by the AAAS. Photo by M.M. Georgiadis and D.C. Rees, Caltech; **2.25 b:** © Courtesy of Austin S. Grindle, Ealing Corporation; **Page 53a:** © Don Fawcett/Visuals Unlimited.

Chapter 3

Opener: Courtesy of Biolog, Inc.; **3.2 a-d, 3.4 b,d:** © Kathy Talaro/Visuals Unlimited; **3.5:** Kathy Talaro; **3.6b:** © Kathy Talaro/Visuals Unlimited; **3.7b:** Kathy Talaro; **3.8a:** © A. M. Siegelman/Visuals Unlimited; **3.8b:** © Photograph courtesy of Becton Dickinson Microbiology Systems; **3.10 a,b, 3.11a,b:** Kathy Talaro; **3.12:** © Courtesy of Harold J. Benson; **3.13a:** © Kathy Talaro/Visuals Unlimited; **3.14:** © Courtesy Nikon, Inc.; **3.20a:** © Carolina Biological Supply/Phototake, NYC; **3.20 b,c:** © M. Abbey/Visuals Unlimted; **3.21:** © E.C.S. Chan/Visuals Unlimited; **3.22a:** © George J. Wilder/Visuals Unlimited; **3.22b:** © Abbey/Visuals Unlimited; **3.23:** © Molecular Probes, Inc.; **3.24c:** © William Ormerod/Visuals Unlimited; **3.25a:** © Courtesy Cynthia Goldsmith, Charles D. Humphrey and Luanne Elliott, Centers for Disease Control and Prevention; **3.25b:** © W.L. Dentler, University of Kansas/Biological Photo Service; **3.26:** © Dennis Kunkel/CNRI/Phototake, NYC; **Page 61:** © Courtesy of Ken Ho; **Page 66:** © From *ASM News* 47(7): 392, 1981, American Society for Microbiology; **3.27a (top):** Kathy Talaro; **3.27a (bottom):** Courtesy Harold Benson; **3.27b:** Kathy Talaro; **3.27c (gram), 3.27c (acid-fast):** © Jack Bostrack/Visuals Unlimited; **3.27c (spore):** © Manfred Kage/Peter Arnold, Inc.; **3.27d (bottom):** Kathy Talaro; **3.27d (top):** © A. M. Siegelman/Visuals Unlimited; **Page 81:** © Randal Feenstra, IBM Thomas J. Watson Research Center, Yorktown Heights, NY.

Chapter 4

Opener: Courtesy Cynthia F. Norton; **4.1b:** © S.C. Holt/Biological Photo Service; **4.2a:** © Courtesy Julius Adler; **4.3a:** © Courtesy of Dr. Jeffrey C. Burnham; **4.3b:** From Reichelt and Baumann, *Arch. Microbiology*, 94:283–330, © Springer-Verlag, 1973; **4.3c:** © Moel R. Krieg in *Bacteriological Reviews,* March 1976, Vol. 40(1):877; **4.3d:** From Preer et al., *Bacteriological Reviews,* June 1974, 38(2): 121,7. © ASM; **4.7b:** Courtesy Stanley F. Hayes, Rocky Mountain Labs, Microscopy Branch, DHHS, NIAID, NIH, Hamilton, MT; **4.8 a,b:** Courtesy of Dr. S. Knutton from D.R. Lloyd and S. Knutton, *Infection and Immunity*, January 1987, p. 86–92. © ASM; **4.9:** © L. Caro/SPL/Photo Researchers, Inc.; **4.12 a-1, a-2:** © Photographs by Harriet & Ephrussi-Taylor, **4.12b:** © John D. Cunningham/Visuals Unlimited; **4.15 a-2:** © S.C. Holt/ Biological Photo Service; **4.13:** © Science VU-Charles W. Stratton/Visuals Unlimited; **4.15 b-2:** © T. J. Beveridge/ Biological Photo Service; **4.18:** © E.S. Anderson/Photo Researchers, Inc.; **4.20:** © Paul W. Johnson/John Sieburth/ Biological Photo Service; **4.21b:** © Cabisco/Visuals Unlimited; **4.21c:** © Lee D. Simon/Photo Researchers, Inc.; **4.23 a,b:** © David M. Phillips/Visuals Unlimited; **4.23c:** © From *Microbiological Reviews*, 55(1):25, Fig. 2b, March 1991. Courtesy of Jorge Benach; **4.23d:** © R. G. Kessel-C.Y. Shih/Visuals Unlimited; **4.24a:** © A. M. Siegelman/Visuals Unlimited; **4.24b:** From Braude, *Infectious Diseases and Microbiology*, 2E, Fig. 3, Page 257, © Saunders College Publishing Company; **4.27a,b:** © Courtesy of Analytab Products, a division of Sherwood Medical; **4.28d:** Courtesy Steven J. Norris; **4.30:** © Science VU-M.D. Maser/Visuals Unlimited; **Page 116 (top):** © Courtesy Esther R. Angert and Norman R. Pace, Indiana University; **4.32:** From Baca and Paretsky, *Microbiological Reviews*, 47(2):133, Fig. 16, June 1983, © ASM; **4.33:** © David M. Phillips/Visuals Unlimited; **4.34a:** Courtesy John Waterbury, WHOI; **4.34 b,c:** © T. E. Adams/Visuals Unlimited; **4.35a:** © From *ASM News*, 53(2), Feb. 1987, American Society for Microbiology. Photo by H. Kaltwasser; **4.35b:** © Paul W. Johnson/Biological Photo Service; **4.36b:** Courtesy Dr. David Graham, Univ. of Illinois, Urbana; **4.38a:** Kathy Talaro; **4.38b:** © From Kessel and Cohen, "Ultrastructure of Square Bacteria From a Brine Pool in Southern Sinai" in *Journal of Bacteriology* 150(2):851–860, 1982, American Society for Microbiology.

Chapter 5

Opener: Courtesy Howard Glasgow, Dept. Botany, North Carolina State University; **5.1 a,b:** © Andrew Knoll; **5.3a:** © Michael Webb/Visuals Unlimited; **5.4:** © Courtesy of R.G. Garrison, Ph. D. From *Fungal Dimorphism with Emphasis on Fungi Pathogenic for Human,* P. J. Szaniszlo (ed). Reprinted with permission of Plenum Publishing Corp.; **5.5:** © Don W. Fawcett/Visuals Unlimited; **5.6b:** © Science VU/Visuals Unlimited; **5.12b:** © Don W. Fawcett/Visuals Unlimited; **5.15a:** © David M. Phillips/Visuals Unlimited; **5.16 a,b:** © Dr. Judy A. Murphy, San Joaquin Delta College, Dept. of Microscopy, Stockton, CA; **5.17a:** © John D. Cunningham/ Visuals Unlimited; **5.17b:** © Dept. of Plant Pathology, Cornell University, Ithaca, NY, Courtesy of William Fry; **5.17c:** © R. Calentine/Visuals Unlimited; **5.17d:** © Everett S. Beneke/ Visuals Unlimited; **5.23:** © Kathy Talaro/Visuals Unlimited; **5.24a:** © A. M. Siegelman/Visuals Unlimited; **5.24b:** © John D. Cunningham/Visuals Unlimited; **5.25:** From Robert Simmons, *ASM News*, 1991, 57(8):400, © ASM; **Page 146:** © Barbara A. Roy, reprinted by permission from *Nature*, Vol. 362, pg. 57; **5.26b:** © T. E. Adams/Visuals Unlimited; **5.26c, 5.29b:** © David M. Phillips/Visuals Unlimited; **5.31c:** © Biophoto Associates/Photo Researchers, Inc.; **5.32b:** Michael Riggs et al., *Infection and Immunity*, 62(5):1931, May 1994, © ASM.

Chapter 6

Opener: Courtesy Kit Lee; **6.1a:** © K. G. Murti/Visuals Unlimited; **6.1b:** © CDC/Phototake, NYC; **6.1c:** © A. B. Dowsette/SPL/Photo Researchers, Inc.; **6.2a:** © Reprinted from Schaffer et al., *Proceedings of the National Academy of Science*, 41:1020, 1955; **6.2b:** © Omikron/Photo Researchers, Inc.; **6.5b:** © Dennis Kunkel/Phototake, NYC; **6.5d:** © K. G. Murti/Visuals Unlimited; **6.6c:** © Science VU-NIH, R. Feldman/Visuals Unlimited; **6.7a:** © Boehringer Ingelheim International GMBH; **6.7b:** © Veronika Burmeister/Visuals Unlimited; **6.8b:** © Cabisco/Visuals Unlimited; **6.11a:** © Fred Hossler/Visuals Unlimited; **6.13:** © Lee D. Simon/Photo Researchers, Inc.; **6.18:** © K. G. Murti/Visuals Unlimited; **6.19b:** © Science VU/CDC/Visuals Unlimited; **6.20a:** © Patricia Barber/Custom Medical Stock; **6.20b:** © Science VU-Charles W. Stratton/Visuals Unlimited; **6.21a:** © Photo by Ted Heald, State of Iowa Hygenic Laboratory; **6.22:** © Science VU-Fred Marsik/Visuals Unlimited; **6.23a:** © E. C. S. Chan/Visuals Unlimited; **6.23 b,c:** U.S. Department of Health, Education, and Welfare Public Health Service, Centers for Disease Control, Atlanta; **6.24a:** © Carroll N. Weiss/ Camera M. D. Studios; **6.24b:** © A. M. Siegelman/Visuals Unlimited; **6.24 c-1:** Courtesy Derek Garbellini; **6.24 c-2:** © Fred P. Williams, U.S. Environmental Protection Agency; **6.24 d-1:** © Courtesy of Diamedix; **6.24 d-2:** © Courtesy of Reference Laboratory, A PCC Laboratory.

Chapter 7

Opener: © Michael Milstein; **Page 191a:** © Dieter Blum/Peter Arnold, Inc.; **Page 191b:** © From *ASM News*, 53(10): cover, 1987, American Society for Microbiology. Photo by C. W. Sullivan, Marine Biology Research Section, Univ. of Southern California, Los Angeles; **7.2a:** © Ralph Robinson/ Visuals Unlimited; **7.2b:** Courtesy Jack Jones , US EPA; **7.11a:** © Pat Armstrong/Visuals Unlimited; **7.11b:** © Philip Sze/Visuals Unlimited; **7.12a:** © Courtesy of Sheldon Manufacturing, Inc.; **7.8.1 a-a:** © Courtesy of Dr. Hans-Dieter Gortz, Zoologisches Institute, Munster, Germany; **7.8.1a-b:** © Mike Abbey/Visuals Unlimited, **7.1.8 a-c:**

Bauchop et al., *Applied Microbiology*, 30(4): 673, October 1975, © ASM; **7.14:** © Science VU-Fred Marsik/Visuals Unlimited; **7.18a:** © Kathy Talaro/Visuals Unlimited; **7.20b:** © Courtesy of Howard Shapiro.

Chapter 8

Opener: © Courtesy of the Wine Institute; **8.4.1a:** © John D. Cunningham/Visuals Unlimited, **8.4.1b:** © Robert F. Myers/Visuals Unlimited; **8.24b:** Courtesy Donald F. Parsons.

Chapter 9

Opener: © Steven McKnight and D. L. Miller, Department of Biology, University of Virginia; **9.1.1:** © Lawrence Livermore Laboratory/SPL/Custom Medical Stock Photo; **9.3:** © K. G. Murti/Visuals Unlimited.

Chapter 10

Opener (both): © March of Dimes; **10.4b:** Kathy Talaro; **10.5f:** Courtesy NEN Life Technologies; **10.11:** Courtesy Philip Bryan, CARB, Univ. Maryland; **10.12:** © Photo by William Jacobs and Barry Bloom of Albert Einstein College of Medicine; **10.14a:** © Courtesy R. Stalker; **10.14b:** © Richard Shade, Purdue University; **10.14c:** © Courtesy of Ikuo Kimura, Faculty of Agriculture, Hokkaido University, Japan; **10.15a:** © Courtesy R. L. Brinster; **10.15b:** © Kyodo AP/Wide World Photo; **10.15c:** © Courtesy R. L. Brinster; **Page 321c:** © Courtesy of Dr. Michael Baird, Lifecodes Corporation.

Chapter 11

Opener: © Nina Barnett, 1994; **Page 328:** © Corbis-Bettmann; **11.5a:** © Science VU/Visuals Unlimited; **11.6:** © Raymond B. Otero/Visuals Unlimited; **11.9a:** © Science VU/ Nordion International/Visuals Unlimited; **11.9b:** © Leonard Lessin/Peter Arnold, Inc.; **11.11:** Courtesy Trojan Technologies; **11.12a:** © Linda Graziadei and Tom Malone, Cold Spring Harbor Laboratory and Heat Systems, Inc.; **11.12b:** © Linda Graziadei and Tom Malone, Cold Spring Harbor Laboratory and Heat Systems, Inc.; **11.13b:** © Fred Hossler/Visuals Unlimited; **Page 352a:** © David M. Phillips/Visuals Unlimited; **Page 352b:** © Kathy Talaro/Visuals Unlimited; **11.17:** © Kathy Talaro/Visuals Unlimited; **11.19a:** © Raymond B. Otero/Visuals Unlimited.

Chapter 12

Opener: © Will & Deni McIntyre/Photo Researchers, Inc.; **Page 363:** © Kathy Talaro/Visuals Unlimited; **12.3 e,f:** Reprinted with permission of Robert L. Perkins from A. S. Klainer & R. L. Perkins, *Journal of Infectious Diseases*, 122:4, 1970, © University of Chicago Press; **12.15:** © Cabisco/ Visuals Unlimited; **12.19:** © Kenneth E. Greer/Visuals Unlimited; **12.20 b-1, b-2:** © Kathy Talaro; **12.22b:** Courtesy of Dr. Josephine Morello; **13.5:** © David M. Phillips/Visuals Unlimited; **13.8:** © Science Vu/Visuals Unlimited.

Chapter 13

Opener: © AP Photo/Robyn Beck, Pool; **Page 403:** © Centers for Disease Control; **13.22:** © Marshall W. Jennison; **Page 426:** © Corbis-Bettmann.

Chapter 14

Opener: © Manfred Kage/Peter Arnold, Inc.; **14.3b:** © Ellen R. Dirksen/Visuals Unlimited; **14.11:** © Reprinted from Cohen and Warren, *Immunology of Parasitic Infections*, 2/E, Blackwell Scientific Publications. Courtesy of D. L. Molaren; **14.13:** Courtesy Steve Kunkel; **14.21a:** © David M. Phillips/ Visuals Unlimited; **14.24d:** © Reprinted from Wendell F. Rosse et al., "Immune Lysis of Normal Human and Parozysmal Nocturnal Hemogloginuria (PNH) Red Blood Cells," "*Journal of Experimental Medicine*," 123:969, 1966. Rockefeller University Press.

Chapter 15

Opener: © Lennart Nilsson, *The Body Victorious,* Bonnier Fakta; **15.14b:** © R. Feldmann/Rainbow; **15.22:** © Boehringer Ingelheim International GMBH.

Chapter 16

Opener: James Gillroy, British, 1757–1815, *The Cow-Pock,* engraving, 1802, William McCallin McKee Memorial Collection, 1928. 1407. © 1991, The Art Institute of Chicago. All Rights Reserved; **16.7:** © From Gillies and Dodds, *Bacteriology Illustrated*, 5/E, Fig. 13, P. 24. Reprinted by permission of the publisher Churchill Livingstone; **16.9:** © Courtesy of Philip Markham, Advanced BioScience Laboratories, Inc.; **16.13a:** © Miguel Pedraza et al., *Laboratory Medicine*, 14(1), January 1983; **16.13b:** © From F. Latil et al., *Journal of Clinical Microbiology,* 23(6):1018, American Society of Microbiology.

Chapter 17

Opener: © Jeffrey Trent; **17.2b:** © SPL/Photo Researchers, Inc.; **17.2c:** © David M. Phillips/Visuals Unlimited; **17.5:** © Kenneth E. Greer/Visuals Unlimited; **17.6:** © STU/Custom Medical Stock Photo; **17.9 a,b:** © Stuart I. Fox; **17.15a:** © Kenneth E. Greer/Visuals Unlimited; **17.15b:** © SIU/Visuals Unlimited; **17.18a:** Kathy Talaro; **17.18b:** © Kenneth E. Greer/Visuals Unlimited; **Page 537 (left):** © Runk/ Schoenberger/Grant Heilman Photography; **Page 537 (middle):** © Renee Lynn/Photo Researchers, Inc.; **Page 537 (right):** © Norman O. Tomalin/Bruce Coleman, Inc.; **17.22:** © Reprinted from R. Kretschmer, *New England Journal of Medicine* 279:1295, 1968. Courtesy of Fred S. Rosen; **Page 543:** © Courtesy of Baylor College of Medicine, Public Affairs; **Page 559:** © Fred Marsik/Visuals Unlimited; **Page 561:** Courtesy Wadsworth Center, NYS Department of Health.

Chapter 18

Opener: From *Infection and Immunity*, 63(2):442–447, February 1995. Courtesy of C.F. J. Rayner, A.D. Jackson, A. Rutman, A. Dewar, T. J. Mitchell, P.W. Andrew, P. J. Cole, and R. Wilson. © American Society of Microbiology; **18.1:** © David M. Phillips/Visuals Unlimited; **18.2:** © Kathy Talaro/ Visuals Unlimited; **18.3 b,c:** © Carroll H. Weiss/Camera M.D. Studios; **18.4b:** © Science VU-Charles W. Stratton/ Visuals Unlimited; **18.5a:** © Charles Stoer/Camera M. D. Studios; **18.5b:** From Braude, *Infectious Diseases and Medical Microbiology*, 2/E, Fig. 3, p. 1320, © Saunders College Publishing; **18.6:** © Raymond B. Otero/Visuals Unlimited; **18.7:** © Analytab Products; **18.8:** Kathy Talaro; **18.9 a,b:** © G. W. Willis, MD/Biological Photo Service; **18.11 a,b:** © Kenneth E. Greer/Visuals Unlimited; **18.14a:** © Courtesy Dr. David Schlaes; **18.14 b-1, b-2:** © Diagnostic Products Corporation; **18.15a:** © Courtesy of Dr. David Schlaes; **18.15b:** © Courtesy Harold Benson; **18.16:** © Carroll H. Weiss/Camera M.D. Studios; **18.17:** © A. M. Siegelman/ Visuals Unlimited; **18.19:** © Custom Medical Stock Photo; **18.20:** © Raymond B. Otero/Visuals Unlimited; **18.21:** © G. Musil/Visuals Unlimited; **18.25:** From James Bingham, *Pocket Guilde for Clinical Medicine*, © 1984 Williams and Wilkins Co., Baltimore; **18.26:** © George J. Wilder/Visuals Unlimited; **18.28:** © Kenneth E. Greer/Visuals Unlimited; **18.29a:** Kathy Talaro; **18.29b:** © Kathy Talaro/Visuals Unlimited.

Chapter 19

Opener: © Arthur C. Aufderheide, MD; **19.1:** © A. M. Siegelman/Visuals Unlimited; **19.1 b-1:** © George J. Wilder/Visuals Unlimited; **19.1 c-1:** © John D. Cunningham/ Visuals Unlimited; **19.2 a,b:** © Science VU-Charles W. Stratton/Visuals Unlimited; **19.3:** From M.A. Boyd et al., *Journal of Medical Microbiology*, 5:459, 1972. Reprinted by permission of Longarm Group, Ltd. © Pathological Society of Great Britain and Ireland; **19.4:** © Science VU-Charles W. Stratton/Visuals Unlimited; **19.5:** © Los Angeles Times Photo/Rolando Otero; **19.6a:** © Courtesy of Dr. Fred E. Pittman; **19.6b:** Reprinted from Farrar and Lambert, *Pocket Guide for Nurses: Infectious Diseases,* © 1984, Williams and Wilkins, Baltimore; **19.8:** © Courtesy of Dr. T. F. Sellers, Jr.; **19.11:** © Kenneth E. Greer/Visuals Unlimited; **19.12:** © George J. Wilder/Visuals Unlimited; **19.14:** © Centers for Disease Control; MB F: Courtesy Wadsworth Center, NYS Department of Health; **19.15:** © A. M. Siegelman/Visuals Unlimited; **19.16a:** © From Gillies and Dodds, *Bacteriology Illustrated*, 5/e, Fig. 45, p. 58. Reprinted by permission of Churchill Livingstone; **19.16b:** Courtesy of Pascal Meylan © American Society for Microbiology; **19.17a:** © John D. Cunningham/Visuals Unlimited; **19.18:** © Science VU-Charles W. Stratton/Visuals Unlimited; **19.20:** © CNRI/SPL/ Photo Researchers, Inc.; **19.21:** © Elmer Koneman/Visuals Unlimited; **19.22a:** © Kenneth E. Greer/Visuals Unlimited; **19.22b:** © Science VU/Visuals Unlimited; **19.23:** © Kenneth E. Greer/Visuals Unlimited; **19.24:** © NMSB/Custom Medical Stock Photo; **19.25:** Courtesy World Health Organization; **19.26, 19.27a:** © Kenneth E. Greer/Visuals Unlimited; **19.27b:** © Science Vu-Charles W. Stratton/Visuals Unlimited; **19.28:** From J. Walter Wilson, *Fungous Diseases of Man*, Plate 26, © 1965, The Regents of University of California.

Chapter 20

Opener: © Jerry L. Shenep, St. Jude's Childrens Hospital; **20.1:** © Science VU/Visuals Unlimited; **20.2:** © E.J. Bottone and A.A. Perez, II, The Mount Sinai Hospital New York; **20.3:** © Kathy Talaro/Visuals Unlimited; **Page 635:** Courtesy Dr. Barry Pyle, Montana State University; **20.5:** © Jack M. Bostrack/Visuals Unlimited; **20.6:** Centers for Disease Control; **20.9a:** © Kathy Talaro/Visuals Unlimited; **20.9b:** Kathy Talaro; **20.12:** © John D. Cunningham/Visuals Unlimited; **20.13:** © Elmer Koneman/Visuals Unlimited; **20.14:** © A. M. Siegelman/Visuals Unlimited; **20.18, 20.20:** © Science VU-Charles W. Stratton/Visuals Unlimited; **20.21:** © From Gillies and Dodds, *Bacteriology Illustrated*, 5/e, Fig. 78 (right), p. 104. Reprinted by permission of Churchill Livingstone; **20.23:** Reprinted from Farrar and Lambert, *Pocket Guide for Nurses: Infectious Diseases*, © 1984, Williams and Wilkins, Baltimore. Photo by Dr. M. Tapert.

Chapter 21

Opener: Courtesy Didier Raoult; **21.1 b,c,d:** © Science VU/Charles W. Stratton/Visuals Unlimited; **21.2:** © Custom Medical Stock Photo; **21.3:** © Carroll H. Weiss/Camera M.D. Studios; **21.4:** © Science VU/Visuals Unlimited; **21.5:** © Kenneth E. Greer/Visuals Unlimited; **21.6 a,b, 21.7:** © Science VU/CDC/Visuals Unlimited; **21.8a:** From Strickland, *Hunter's Tropical Medicine*, 6/E, © Saunders College Publishing; **21.8b:** © Science VU/Visuals Unlimited; **21.9:** © Science VU-Fred Marsik/Visuals Unlimited; **21.12:** © Centers for Disease Control/Peter Arnold, Inc.; **21.13 a,b:** © From Jacob S. Teppema, "In Vivo Adherence and Colonization of Vibrio Cholerae Strains That Differ in Hemagglutinating Activity and Motility," *Journal of Infection and Immunity*, 55(9):2093–2102, Sept. 1987. Reprinted by permission of American Society of Microbiology; **Page 662:** © AP/Wide World Photos; **21.15:** © From R.R. Colwell and D.M. Rollins, "Viable but Nonculturable Stage of Campylobacter jejuni and Its Role in Survival in the Natural Aquatic Environment," *Applied and Environmental Microbiology*, 52(3):531–538, 1986. Reprinted by permission of American Society for Microbiology; **Page 664:** Courtesy of James G. Fox from *Infection and Immunity*, June 1994, pg. 2367–2374. © American Society of Microbiology; **21.16a:** © Reprinted from R.L. Anaker et al., "Details of the Ultrastructure of Rickettsia Prowazekii Grown in the Chick Yolk Sac" *Journal of Bacteriology*, 94(1): 260, 1967. American Society of Microbiology; **21.16b, Page 667b:** © Science VU/Visuals Unlimited; **Page 667c:** © Richard F. Ashley/Visuals Unlimited; **Page 667d:** © A.M. Siegelman/Visuals

Unlimited; **21.19:** © Dept. of Health and Human Resources, Courtesy of Dr. W. Burgdorfer; **21.20:** © From McCaul and Williams, "Development Cycle of C. Burnetii," *Journal of Bacteriology* 147:1063, 1981. Reprinted by permission of American Society for Microbiology; **21.21:** © Kenneth E. Greer/Visuals Unlimited; **21.23a:** © Armed Forces Institute of Pathology; **21.23b:** © Science VU-Bascom Palmer Institute/Visuals Unlimited; **21.24:** © Harvey Blank/Camera M.D. Studios; **21.25a:** © Science VU/CDC/Visuals Unlimited; **21.25b:** © Science VU-AFIP/Visuals Unlimited; **21.26a:** From Duncan C. Krause, *Trends in Microbiology*, Vol. 6, No. 1, Jan. 1998, Fig. 1, p. 16, © Elsevier Sciences L.T.D.; Photograph courtesy of Duncan C. Krause.; **21.30a:** © R. Gottsegen/Peter Arnold, Inc.; **21.30b:** © Stanley Flegler/Visuals Unlimited; **21.31a:** © Science VU-Max A. Listgarten/Visuals Unlimited; **21.31b:** © Science VU-Max A. Listgarten/Visuals Unlimited; **21.32d:** © Science VU-Max A. Listgarten/Visuals Unlimited.

Chapter 22

Opener: © William R. Harrison; **22.4:** From L.K. Green and D.G. Moore, *Laboratory Medicine*, 18(7): cover, July 1987; **22.6a:** © Elmer W. Koneman/Visuals Unlimited; **22.6b:** © Everett S. Beneke/Visuals Unlimited; **22.7b:** © A.M. Siegelman/Visuals Unlimited; **22.8:** © Science VU-Charles Sutton/Visuals Unlimited; **22.9:** Reprinted from J. Walter Wilson, *Fungous Diseases of Man*, Plate 2, © 1965 The Regents of the University of California; **22.10:** From Rippon, *Medical Mycology*, 2/E, Fig. 17.18, page 419, © Saunders College Publishing; **22.11a:** © Elmer W. Koneman/Visuals Unlimited; **22.11b:** © A.M. Siegelman/Visuals Unlimited; **22.12:** © Reprinted by permission of Upjohn Company from E.S. Beneke, et al., 1984, *Human Mycoses*; **22.13:** From Koneman and Roberts, *Practical Laboratory Mycology*, © Williams and Wilkins Co., Baltimore; **22.14:** © Harkisan Raj/Visuals Unlimited; **22.15:** © Everett S. Beneke/Visuals Unlimited; **22.16:** © Science Vu-Charles W. Stratton/Visuals Unlimited; **22.17:** © Reprinted by permission of Upjohn Company from E.S. Beneke, et al., 1984, *Human Mycoses;* **22.18 a,c:** From Koneman and Roberts, *Practical Laboratory Mycology,* © Williams and Wilkins Co., Baltimore; **22.18b:** © A.M. Siegelman/Visuals Unlimited; **22.19 a,b:** © Everett S. Beneke/Visuals Unlimited; **22.19:** © Kenneth E. Greer/Visuals Unlimited; **Page 705 (top):** Courtesy Toshio Kanbe; **22.20a:** © Kenneth E. Greer/Visuals Unlimited; **22.20b:** Reprinted from J. Walter Wilson, *Fungous Diseases of Man*, Plate 42 (middle right), © 1965, The Regents of the University of California; **22.21:** © Carroll H. Weiss/Camera M.D. Studios; **22.22 a,b, 22.23a:** © Everett S. Beneke/Visuals Unlimited; **22.23b:** © Science VU/CDC/Visuals Unlimited; **22.24a:** © Raymond B. Otero/Visuals Unlimited; **22.24c:** © Elmer Koneman/Visuals Unlimited; **22.24b:** Courtesy Glenn S. Bulmer; **22.25:** Courtesy Gordon Love; **22.26:** Reprinted from J. Walter Wilson, *Fungous Diseases of Man,* Plate 21, © 1965, The Regents of the University of California; **22.27:** © A.M. Siegelman/Visuals Unlimited; **22.28a:** © Everett S. Beneke/Visuals Unlimited; **22.28b:** © Science VU-AFIP/Visuals Unlimited; **22.29:** © Daniel Snyder/Visuals Unlimited; **22.30:** From Rippon, *Medical Mycology*, 2/E, Fig. 25.4b, page 620, © Saunders College Publishing. Photo by L. Calkins; **22.31 a,b:** © M.F. Brown/Visuals Unlimited.

Chapter 23

Opener: Ellen Li et al., *Infection and Immunity,* Nov. 1994, Vol. 62, No. 11, Fig. 5c, p. 5116, © American Society of Microbiology; **23.2:** © Science VU-Charles W. Stratton/Visuals Unlimited; **23.4:** © Science VU-David John/Visuals Unlimited; **23.7b:** Courtesy Stanley Erlandsen; **23.9:** © Reprinted from Katz, Despommier, and Gwadz, *Parasitic Diseases*, Springer-Verlag. Photo by T. Jones; **23.10b:** Courtesy Armed Forces Institute of Pathology; **23.12:** Courtesy Centers for Disease Control; **23.14 a,b:** © Courtesy of Dr. Henry W. Muray. From *Harrison's Principles of Internal Medicine*, 12/e., McGraw-Hill, Inc. Reproduced with permission of the publisher; **23.15:** © Cecil H. Fox/Photo Researchers, Inc.; **23.16:** Courtesy Ynes R. Ortega; **23.18a:** © Lauritz Jensen/Visuals Unlimited; **23.18b:** © A.M. Siegelman/Visuals Unlimited; **23.18c:** © From Rubin and Farbo, *Pathology*. Reprinted by permission of J.B. Lippincott Company; **23.19 b-1:** © R. Calentine/Visuals Unlimited; **23.19 b-2:** © Science VU-Fred Marsik/Visuals Unlimited; **Page 742:** © Lauritz Jensen/Visuals Unlimited; **23.20:** © Carroll H. Weiss/Camera M.D. Studios; **23.22a:** © Science VU-Fred Marsik/Visuals Unlimited; **23.22b:** © Science VU-Fred Marsik/Visuals Unlimited; **23.23:** © Science VU-AFIP/Visuals Unlimited; **23.24a:** © Cabisco/Visuals Unlimited; **23.24b:** © Harvey Blankespoor; **23.24c:** © Science VU/Visuals Unlimited; **23.25:** © A.M. Siegelman/Visuals Unlimited; **23.26a:** © Stanley Flegler/Visuals Unlimited; **23.26b:** © From Katz et al., *Parasitic Diseases*, Springer-Verlag; **23.27:** © Science VU-Charles W. Stratton/Visuals Unlimited.

Chapter 24

Opener: Courtesy NEN Life Science Products; **Page 760 (top):** © World Health Organization; **24.2a:** Courtesy Derek Garbellini; **24.2b:** © Charles Stoer/Camera M.D. Studios; **24.3b:** © Boehringer Ingeheim International GMBH; **24.5a:** © Carroll H. Weiss/Camera M.D. Studios; **24.5b:** © Kenneth E. Greer/Visuals Unlimited; **24.6:** © Carroll H. Weiss/Camera M.D. Studios; **24.7:** © Kenneth E. Greer/Visuals Unlimited; **24.8:** Kathy Talaro; **24.9:** © Science VU-Charles W. Stratton/Visuals Unlimited; **24.10a:** © John D. Cunningham/Visuals Unlimited; **24.10c:** © Carroll H. Weiss/Camera M.D. Studios; **24.11:** From J.S. Nelson and J.P. Wyatt, *Medicine*, 38:22, © 1959 Williams & Wilkins; **24.12:** © Science VU-Charles W. Stratton/Visuals Unlimited; **24.13:** Centers for Disease Control; **24.14:** © Courtesy Barbara O'Connor; **24.17:** © Science VU-NIH/Visuals Unlimited; **24.19a:** © David M. Phillips/Visuals Unlimited; **24.19b:** © K.G. Murti/Visuals Unlimited; **24.20 a,b:** © Kenneth E. Greer/Visuals Unlimited; **24.21:** From Koneman et al., *Diagnostic Microbiology*, 4/e, Fig. 20.20, p. 1037, © J.B. Lippincott Company; **24.22:** © Science VU-Charles W. Stratton/Visuals Unlimited; **24.23:** © Dr. P. Marazzi/SPL/Photo Researchers, Inc.

Chapter 25

Opener: © Robin Thomas; **25.2:** © From C.R. Howard, *Perspectives in Medical Virology*, Vol. 2, 1986, Elsevier Science Publishing Company. Photo by J. L. Maiztegui; **25.3a:** © From Exeen M. Morgand and Fred Rapp, "Measles Virus and Its Assocaites Disease," *Bacteriological Reviews*, 41(3):636–666, 1977. Reprinted by permission of American Society for Microbiology; **25.4:** © Biophoto Associates/Photo Researchers, Inc.; **25.5a:** Centers for Disease Control; **25.5b:** © Kenneth E. Greer/Visuals Unlimited; **25.6:** © Boehringer Ingelheim International GMBH; **25.7d:** © AP/World Wide Photos; **25.8:** © Science VU-AFIP/Visuals Unlimited; **25.9a:** Reprinted by permission of Dr. Kenneth Schiffer from Cooper et al., *Am. J. Dis. Child*, 110:419, © 1965, American Medical Association; **25.9b:** © *Infectious Diseases Magazine*, January 1971; **25.17:** © Hans R. Galderblom/Visuals Unlimited; **25.19:** © Science VU/Visuals Unlimited; **25.22:** © From I. Katayarma, C.Y. Li, and L.T. Yam, "Ultrastructure Characteristics of the Hairy Cells of Leukemic Reticuloen-dotheliosis," *American Journal of Pathology,* 67:361, 1972. Reprinted by permission of J.B. Lippincott Company; **25.23b:** Science VU-CDC/Visuals Unlimited; **25.24:** © Corbis-Bettmann; **25.26 b-1:** © Cabiso/Visuals Unlimited; **25.26 b-2:** © Science VU-Charles W. Stratton/Visuals Unlimited; **25.27:** © Carroll H. Weiss/Camera M. D. Studios; **25.29a:** © K.G. Murti/Visuals Unlimited; **25.29b:** © From Farrar and Lambert, *Pocket Guide for Nurses: Infectious Diseases*, 1984, Williams and Wilkins, Baltimore. Reprinted by permission of Dr. William Edmund Farrar, Jr.; **25.30a:** Reprinted from Fields and Knipe, *Fundamentals of Virology*, 1986, Raven Press; **25.30b:** From Fields and Knipe, *Fundamentals of Virology*, 1986, Raven Press. Photo by Dr. Patricia Merz.

Chapter 26

Opener: From cover of *ASM News*, Vol. 52, #10, October 1986. Photo by Gary C. du Moulin; **26.11b:** © John D. Cunningham/Visuals Unlimited; **26.12 a,b:** Sylvan Wittwer/Visuals Unlimited; **26.14, 26.16:** © John D. Cunningham/Visuals Unlimited; **26.20b:** © Carleton Ray/Photo Researchers, Inc.; **26.21:** © John Cunningham/Visuals Unlimited; **26.23b:** © Kathy Talaro/Visuals Unlimited; **26.23c:** Courtesy IDEXX Laboratories, Inc., Westbrook, Maine, U.S.A.; **26.26:** Courtesy Donald Klein; **BF 26.4a:** © Frank Hanna/Visuals Unlimited; **BF 26.4b:** © John D. Cunningham/Visuals Unlimited; **BF 26.5 a,b:** Kathy Talaro; **26.29:** © Kathy Talaro/Visuals Unlimited; **26.31b:** © Vance Henry/Nelson Henry; **26.32:** © John D. Cunningham/Visuals Unlimited; **26.33b:** © Kevin Schafer/Peter Arnold, Inc.; **26.35a:** © Kathy Talaro/Visuals Unlimited; **26.36b:** © Joe Munroe/Photo Researchers, Inc.; **26.39:** © J.T. MacMillan.

LINE ART

Chapter 1

Opener: From Hubert Lechevalier in *Bacteriological Reviews*, March 1976, Vol. 40, Number 1. American Society of Microbiology, Washington, D.C. Reprinted by permission.

Chapter 3

Figure 3.24: From William A. Jensen and Roderic B. Park. *Cell Ultrastructure*. © 1967 Wadsworth Publishing Company.

Chapter 4

Figure 4.37: Reprinted with permission from *Nature*, Vol. 196, pp. 1189–1192. Copyright © 1962 Macmillan Magazines Ltd.

Chapter 6

Figure 6.8a: From Westwood, et al., *Journal of Microbiology*, 34:67, 1964. Reprinted by permission of The Society for General Microbiology, United Kingdom.

Chapter 11

Figure 11.16b: From Nolte, et al., *Oral Microbiology*, 4/e. Copyright © 1982 Mosby. Reprinted by permission of Gloria-Mae Nolte.

Chapter 15

Figure 15.18: From Joseph A. Bellanti. *Immunology III.* Copyright © 1985 W. B. Saunders and Co. Philadelphia, Pa. Reprinted by permission.

Chapter 20

Figure 20.10: Reprinted courtesy of Betton Dickinson and Company.

ILLUSTRATOR

Page Two Inc.

1.2, 1.3, 1.4, 1.5b, 1.15, 2.22c, 4.25, 8.21, 8.23, 8.24, 9.7, 9.14, 10.13, BX10.2,fig 1, BX11.3,fig 1, 13.16, 15.24, 16.2a,b, BX16.2, 25.20, 26.8c

INDEX

A

B

C

D

E

F

G

H

I

N

O

P

V

W

LE 1

NOTIFIABLE DISEASES — SUMMARY OF REPORTED CASES, UNITED STATES, 1989–1996

Disease	1989	1990	1991	1992	1993	1994	1995	1996
AIDS	33,722	41,595	43,627	45,472	103,691	78,279	71,547	66,885*
Amebiasis	3,217	3,328	2,989	2,942	2,970	2,983	†	†
Anthrax	–	–	–	1	–	–	–	–
Aseptic meningitis	10,274	11,852	14,526	12,223	12,848	8,932	†	†
Botulism, total (including wound and unsp.)	89	92	114	91	97	143	97	119
Foodborne	23	23	27	21	27	50	24	25
Infant	60	65	81	66	65	85	54	80
Brucellosis	95	85	104	105	120	119	98	112
Chancroid	4,692	4,212	3,476	1,886	1,399	773	606	386§
Chlamydia¶	**	**	**	**	**	**	477,638	498,884§
Cholera	–	6	26	103	18	39	23	4
Diphtheria	3	4	5	4	–	2	–	2
Encephalitis, primary	981	1,341	1,021	774	919	717	†	†
Post – infectious	88	105	82	129	170	143	†	†
Escherichia coli O157:H7	**	**	**	**	**	1,420	2,139	2,741
Gonorrhea	733,151	690,169	620,478	501,409	439,673	418,068	392,848	325,883§
Granuloma inguinale	7	97	29	6	19	3	†	†
Haemophilus influenzae, invasive	**	**	2,764	1,412	1,419	1,174	1,180	1,170
Hansen disease (leprosy)	163	198	154	172	187	136	144	112
Hemolytic uremic syndrome, post – diarrheal	**	**	**	**	**	**	**	**
Hepatitis A	35,821	31,441	24,378	23,112	24,238	26,796	31,582	31,032
Hepatitis B	23,419	21,102	18,003	16,126	13,361	12,517	10,805	10,637
Hepatitis, C/non-A, non-B††	2,529	2,553	3,582	6,010	4,786	4,470	4,576	3,716
Hepatitis, unspecified	2,306	1,671	1,260	884	627	444	†	†
Legionellosis	1,190	1,370	1,317	1,339	1,280	1,615	1,241	1,198
Leptospirosis	93	77	58	54	51	38	†	†
Lyme disease	**	**	9,465	9,895	8,257	13,043	11,700	16,455
Lymphogranuloma venereum	189	277	471	302	285	235	†	†
Malaria	1,277	1,292	1,278	1,087	1,411	1,229	1,419	1,800
Measles (rubeola)	18,193	27,786	9,643	2,237	312	963	309	508
Meningococcal disease	2,727	2,451	2,130	2,134	2,637	2,886	3,243	3,437
Mumps	5,712	5,292	4,264	2,572	1,692	1,537	906	751
Murine typhus fever	41	50	43	28	25	†	†	†
Pertussis (whooping cough)	4,157	4,570	2,719	4,083	6,586	4,617	5,137	7,796
Plague	4	2	11	13	10	17	9	5
Poliomyelitis, paralytic§§	11	6	10	6	4	8	6	5
Psittacosis	116	113	94	92	60	38	64	42
Rabies, animal	4,724	4,826	6,910	8,589	9,377	8,147	7,811	6,982
Rabies, human	1	1	3	1	3	6	5	3
Rheumatic fever, acute	144	108	127	75	112	112	†	†
Rocky Mountain spotted fever	623	651	628	502	456	465	590	831
Rubella (German measles)	396	1,125	1,401	160	192	227	128	238
Rubella, congenital syndrome	3	11	47	11	5	7	6	4
Salmonellosis, excluding typhoid fever	47,812	48,603	48,154	40,912	41,641	43,323	45,970	45,471
Shigellosis	25,010	27,077	23,548	23,931	32,198	29,769	32,080	25,978
Syphilis, primary and secondary	44,540	50,223	42,935	33,973	26,498	20,627	16,500	11,387§
Total, all stages	110,797	134,255	128,569	112,581	101,259	81,696	68,953	52,976§
Tetanus	53	64	57	45	48	51	41	36
Toxic-shock syndrome	400	322	280	244	212	196	191	145
Trichinosis	30	129	62	41	16	32	29	11
Tuberculosis	23,495	25,701	26,283	26,673	25,313	24,361	22,860	21,337¶¶
Tularemia	152	152	193	159	132	96	†	†
Typhoid fever	460	552	501	414	440	441	369	396
Varicella (chickenpox)***	185,441	173,099	147,076	158,364	134,722	151,219	120,624	83,511
Yellow fever	†††	†††	†††	†††	†††	†††	†††	1

Source: Data from *Morbidity and Mortality Weekly Report*, Vol. 45, no. 53 (October 31, 1997).

*The total number of acquired immunodeficiency syndrome (AIDS) cases includes all cases reported to the Division of HIV/AIDS Prevention, Surveillance, and Epidemiology, National Center for HIV, STD, and TB Prevention (NCHSTP) through December 31, 1996.

†No longer nationally notifiable.

§Cases were updated through the Division of Sexually Transmitted Diseases Prevention, NCHSTP, as of June 13, 1997.

¶Chlamydia refers to genital infections caused by *C. trachomatis*.

**Not previously nationally notifiable.

††Anti-HCV antibody test was available as of May 1990.

§§Numbers may not reflect changes based on retrospective case evaluations or late reports (see *MMWR* 1986;35:180 – 2).

¶¶Cases were updated through the Division of Tuberculosis Elimination, NCHSTP, as of May 28, 1997.

***Varicella was taken off the nationally notifiable disease list in 1991. Many states continue to report these cases to CDC.

†††Last indigenous case of yellow fever was reported in 1911; before 1996, the last imported case was reported in 1924.

Square-Law Equations

For a triode-operating long-channel NMOS device

$$I_D = KP_n \cdot \frac{W}{L} \cdot \left[(V_{GS} - V_{THN})V_{DS} - \frac{V_{DS}^2}{2} \right]$$

for $V_{GS} \geq V_{THN}$ and $V_{DS} \leq V_{GS} - V_{THN}$

For a long-channel NMOS device operating in the saturation region:

$$I_D = \frac{KP_n}{2} \cdot \frac{W}{L}(V_{GS} - V_{THN})^2[1 + \lambda(V_{DS} - V_{DS,sat})]$$

for $V_{GS} > V_{THN}$ and $V_{DS} \geq V_{GS} - V_{THN}$

On the border between saturation and triode:

$V_{DS,sat} = V_{GS} - V_{THN}$ and the drain current is called $I_{D,sat}$, see Fig. 6.11

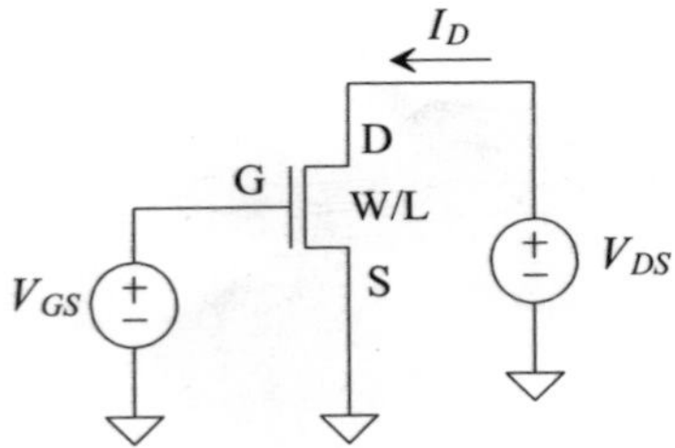

For the PMOS device equations make the following substitutions in the equations listed above

$$V_{DS} \rightarrow V_{SD},\ V_{GS} \rightarrow V_{SG},\ \text{and } V_{THN} \rightarrow V_{THP}.$$

All of the voltages and currents in the PMOS and NMOS equations are **positive**. For example, for the PMOS device to conduct a drain current requires $V_{SG} > V_{THP}$. For the NMOS to conduct a drain current requires $V_{GS} > V_{THN}$.

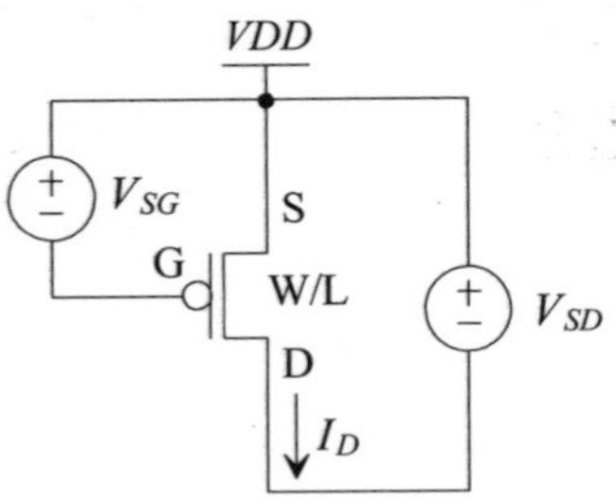

CMOS

Circuit Design, Layout, and Simulation

CMOS

Circuit Design, Layout, and Simulation

Second Edition

R. Jacob Baker

IEEE Press Series on Microelectronic Systems
Stuart K. Tewksbury and Joe. E. Brewer, *Series Editor*

IEEE Solid-State Circuits Society, *Sponsor*

A JOHN WILEY & SONS, INC., PUBLICATION

Published simultaneously in Canada.

For general information on our other products and services please contact our Customer Care Department within the U.S. at 877-762-2974, outside the U.S. at 317-572-3993 or fax 317-572-4002.

Wiley also publishes its books in a variety of electronic formats. Some content that appears in print, however, may not be available in electronic format.

Library of Congress Cataloging-in-Publication Data is available.

ISBN 0-471-70055-X

Printed in the United States of America.

10 9 8 7 6 5 4 3

IEEE Press
445 Hoes Lane
Piscataway, NJ 08855

Books of Related Interest from IEEE Press

CMOS: Mixed Signal Circuit Design
R. Jacob Baker
2002 Hardcover 532pp 0-471-22754-4

DRAM Circuit Design: A Tutorial
Brent Keeth and R. Jacob Baker
2001 Hardcover 200pp 0-780-36014-1

CMOS Electronics: How It Works, How it Fails
Jaume Segura and Charles F. Hawkins
2004 Hardcover 348pp 0-471-47669-2

Silicon Germanium: Technology, Modeling, and Design
Raminderpal Singh, David L. Harame, and Modest M. Oprysko
2004 Hardcover 340pp 0-471-44653-X

Design of High-Performance Microprocessor Circuits
Anantha Chandrakasan, William J. Bowhill, and Frank Fox
2001 Hardcover 557pp 0-780-36001-X

Low-Power CMOS Design
Anantha Chandrakasan and Robert Brodersen
1998 Hardcover 644pp 0-780-36001-X

To Julie

Brief Contents

Contents

Preface

For the last 25 years, CMOS (complementary metal oxide semiconductor) technology has been the dominant technology for fabricating integrated circuits (ICs or chips). For the next 25 years, CMOS will, no doubt, continue to be the dominant integrated circuit technology. Why? CMOS technology is reliable, manufacturable, low power, low cost, and, perhaps most importantly, scalable. The fact that silicon integrated circuit technology is scalable was observed and described in 1965 by Intel founder Gordon Moore. His observations are now referred to as *Moore's law* and state that the number of transistors on a chip will double every 18 months. While originally not specific to CMOS, Moore's law has been fulfilled over the years by scaling down the feature size in CMOS technology (it's getting cut in half every 18 months). Whereas the gate lengths of early CMOS transistors were in the micrometer range (long-channel devices) the feature sizes of current CMOS devices are in the nanometer range (short-channel devices).

To encompass both the long- and short-channel CMOS technologies in this book, I take a two-path approach to custom CMOS integrated circuit design. Design techniques are developed for both and then compared. This comparison gives readers deep insight into the circuit design process.

As an academic text, the book is filled with practical design examples, discussions, and problems. The assumed background of the reader is a knowledge of linear circuits, microelectronics, and digital logic design. The solutions to the end-of-chapter problems (for self-study) and the netlists used when simulating the circuits are found at **http://cmosedu.com**. Those interested in gaining an in-depth knowledge of CMOS analog and digital design will be greatly aided by downloading, modifying, and simulating the netlists of the book's circuits.

Several courses can be taught using this book including VLSI or physical IC design (chapters 1–7 and 10–15), analog IC design (chapters 8–9 and 20–24), memory circuit design (16–19), and advanced analog IC design (25–29). A graduate course in mixed-signal design following the advanced analog IC design course can be taught using the second volume of this book entitled *CMOS Mixed-Signal Circuit Design*.

In this edition, LASI CAD software is used once again to introduce circuit layout and WinSPICE is used for the circuit simulations. The material covering the LASI layout tool can be skipped if other layout software packages are used. I like using LASI in a first course on IC design because it takes very little class time to learn. Further, other SPICE programs can be used to simulate the circuit netlists with little or no modifications. Both the LASI CAD software and the WinSPICE circuit simulator run under Windows and can be downloaded at http://cmosedu.com.

How will this book be useful to the student, researcher, or practicing engineer?

I've put a great deal of effort into making this book useful to an eclectic audience. For the student, the book is filled with hundreds of examples, problems, and practical discussions (according to one of my students there can never be too many examples in a textbook). The layout discussions build a knowledge foundation important for troubleshooting and precision or high-speed design. Layout expertise is gained in a step-by-step fashion by including circuit design details, process steps, and simulation concerns (parasitics). Covering layout in a single chapter and decoupling the discussions from design and simulation was avoided. The digital design chapters emphasize real-world process parameters (e.g., I_{off} , I_{on} , t_{ox} , VDD). The analog chapters provide coherent discussions about selecting device sizes and design considerations. Likewise "cookbook" design procedures for selecting the widths/lengths of MOSFETs and design using long-channel equations in a short-channel process are not present. The focus is on preparing the student to "hit the ground running" when they become a custom CMOS IC designer or product engineer.

For the researcher, topics in circuit design such as noise considerations and sensing using delta-sigma modulation (DSM) have been added to this edition. I've also tried to anticipate future design paradigms. For example, the addition of DSM has been applied to CMOS image sensors, Flash memory, and memory using thin oxides (direct tunneling). Sensing using DSM was included because it makes use of the fact that as CMOS clock speeds are going up, the gain and matching of transistors is deteriorating. Further, I've tried to anticipate noise-limited design issues such as, "Why can't I improve the signal-to-noise ratio in my imaging chip?" or "Why is integrating thermal or flicker noise harmful?"

For the working engineer, I've tried to provide design and layout examples that will be immediately useful in products. At the risk of stating the obvious, matching, power, speed, process shifts, power supply voltage variations, and temperature behavior are extremely important in practical design. I've focused the discussions and examples found in this book on these topics. Phase-locked loops, charge pumps, low-voltage references, single and fully-differential op-amp designs, continuous-time and clocked comparators, memory circuits, etc., are covered in detail with numerous examples. To ensure the most practical computer validation of the designs, the simulations for the nanometer designs (a 50 nm process) use the BSIM4 SPICE models.

Acknowledgments

I would like to acknowledge the support of Micron Technology, Inc., and, in particular, Mary Miller for her assistance with the technical writing review. Further, I would like to thank the reviewers, students, colleagues, and friends that have helped to make this book a possibility: Rupa Balan, David M Binkley, Bill Black, Dave Boyce, Elizabeth Brauer, J. W. Bruce, Kloy Debban, Ahmad Dowlatabadi, Kevin Duesman, Krishna Duvvada, Surendranath Eruvuru, Cathy Faduska, Paul Furth, Neil Goldsman, Tyler Gomm, Kory Hall, Wes Hansford, David Harris, Jeff Jessing, Brent Keeth, Howard Kirsch, Bill Knowlton, Bhavana Kollimarla, Harry W. Li, Matthew Leslie, Song Liu, Amy Moll, Sugato Mukherjee, Ward Parkinson, Terry Sculley, Brian Shirley, Harish Singidi, Mike Smith, Mark Tuttle, Vance Tyree, Gary VanAckern, Indira Vemula, Tony VenGraitis, and Joseph J. Walsh.

R. Jacob (Jake) Baker

Chapter

1

Introduction to CMOS Design

This chapter provides a brief introduction to the CMOS (complementary metal oxide semiconductor) integrated circuit (IC) design process (the design of "chips"). CMOS is used in most very large scale integrated (VLSI) or ultra-large scale integrated (ULSI) circuit chips. The term "VLSI" is generally associated with chips containing thousands or millions of metal oxide semiconductor field effect transistors (MOSFETs). The term "ULSI" is generally associated with chips containing billions, or more, MOSFETs. We'll avoid the use of these descriptive terms in this book and focus simply on "digital and analog CMOS circuit design."

We'll also discuss setting up the LASI (layout system for individuals) layout software for Windows and provide a very brief introduction to SPICE (simulation program with integrated circuit emphasis).

1.1 The CMOS IC Design Process

The CMOS circuit design process consists of defining circuit inputs and outputs, hand calculations, circuit simulations, circuit layout, simulations including parasitics, reevaluation of circuit inputs and outputs, fabrication, and testing. A flowchart of this process is shown in Fig. 1.1. The circuit specifications are rarely set in concrete; that is, they can change as the project matures. This can be the result of trade-offs made between cost and performance, changes in the marketability of the chip, or simply changes in the customer's needs. In almost all cases, major changes after the chip has gone into production are not possible.

This text concentrates on custom IC design. Other (noncustom) methods of designing chips, including field-programmable-gate-arrays (FPGAs) and standard cell libraries, are used when low volume and quick design turnaround are important. Most chips that are mass produced, including microprocessors and memory, are examples of chips that are custom designed.

The task of laying out the IC is often given to a layout designer. However, it is extremely important that the engineer can lay out a chip (and can provide direction to the layout designer on how to layout a chip) and understand the parasitics involved in the layout. Parasitics are the stray capacitances, inductances, pn junctions, and bipolar transistors, with the associated problems (breakdown, stored charge, latch-up, etc.). A fundamental understanding of these problems is important in precision/high-speed design.

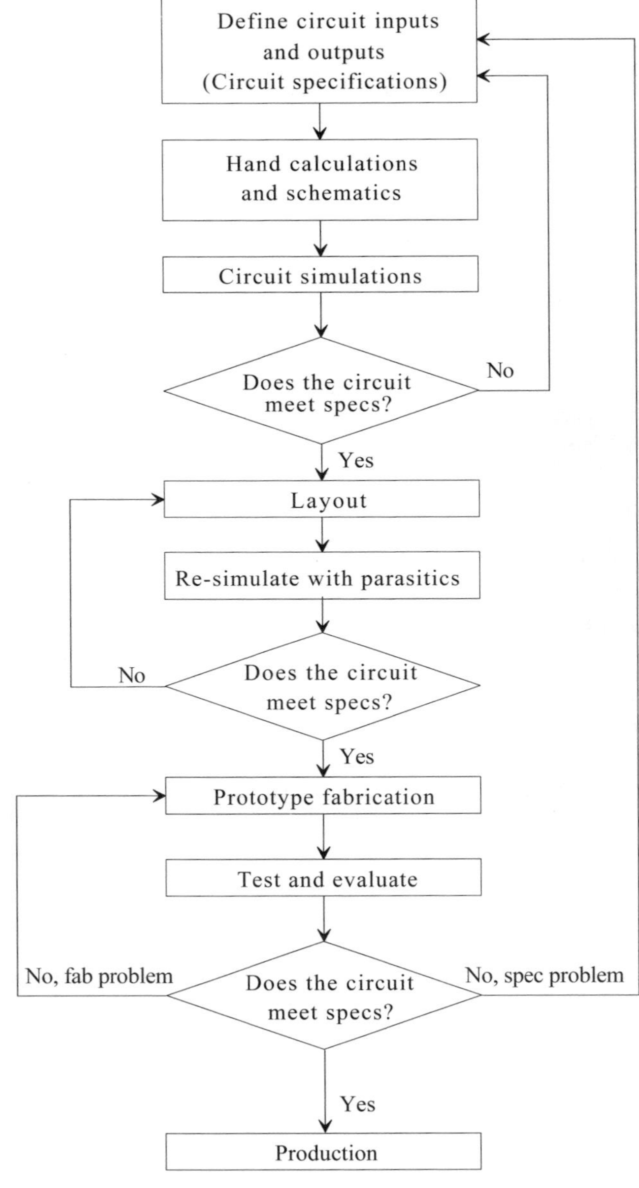

Figure 1.1 Flowchart for the CMOS IC design process.

1.1.1 Fabrication

CMOS integrated circuits are fabricated on thin circular slices of silicon called wafers. Each wafer contains several (perhaps hundreds or even thousands) of individual **chips or "die"** (Fig. 1.2). For production purposes, each die on a wafer is usually identical, as seen in the photograph in Fig. 1.2. In Fig. 1.2, an individual die has been outlined with a black marker and appears next to a dime. Added to the wafer are test structures and process monitor plugs (sections of the wafer used to monitor process parameters).

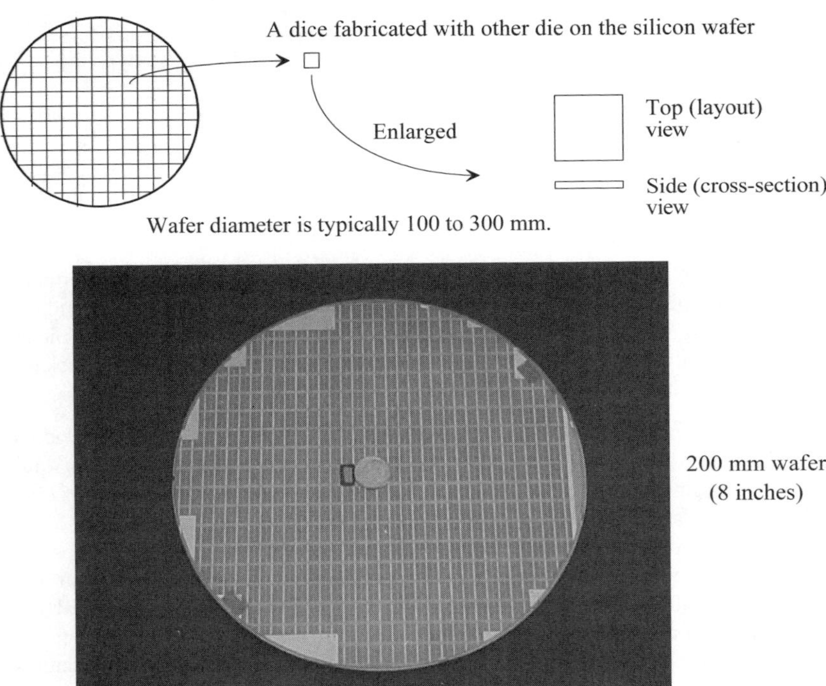

Figure 1.2 CMOS integrated circuits are fabricated on and in a silicon wafer.

The ICs we design and lay out using a layout program can be fabricated through MOSIS (http://mosis.org) on what is called a **multiproject wafer**; that is, a wafer that is comprised of chip designs of varying sizes from different sources (educational, private, government, etc.). MOSIS combines multiple chips on a wafer to split the fab cost among several designs to keep the cost low. MOSIS subcontracts the fabrication of the chip designs (multiproject wafer) out to one of many commercial manufacturers (vendors). MOSIS takes the wafers it receives from the vendors, after fabrication, and cuts them up to isolate the individual chip designs. The chips are then packaged and sent to the originator. A sample package (40-pin ceramic) from a MOSIS-submitted student design is seen in Fig. 1.3. Normally a cover (not shown) keeps the chip from being exposed to light or accidental damage.

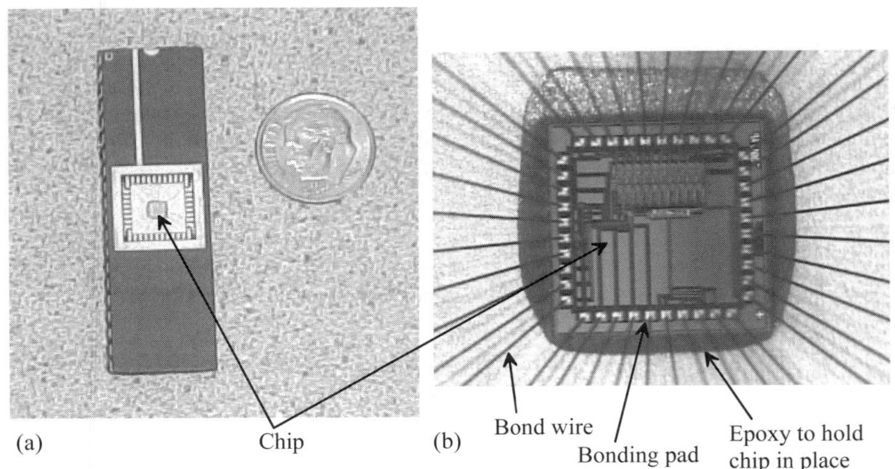

Figure 1.3 How a chip is packaged (a) and (b) a closer view.

Note, in Fig. 1.3, that the chip's electrical signals are transmitted to the pins of the package through wires. These wires (called "bond wires") electrically bond the chip to the package so that a pin of the chip is electrically connected (shorted) to a piece of metal on the chip (called a bonding pad). The chip is held in the cavity of the package with an epoxy resin ("glue") as seen in Fig. 1.3b.

The ceramic package used in Fig. 1.3 isn't used for most mass-produced chips. Most chips that are mass produced use plastic packages. Exceptions to this statement are chips that dissipate a lot of heat or chips that are placed directly on a printed circuit board (where they are simply "packaged" using a glob of resin). Plastic packaged (encapsulated) chips place the die on a lead frame (Fig. 1.4) and then encapsulate the die and lead frame in plastic. The plastic is melted around the chip. After the chip is encapsulated, its leads are bent to the correct position. This is followed by printing information on the chip (the manufacturer, the chip type, and the lot number) and finally placing the chip in a tube or reel for shipping to a company that makes products that use the chips. Example products might include chips that are used in cell phones, computers, microwave ovens, printers.

Layout and Cross Sectional Views

The view that we see when laying out a chip is the top, or layout, view of the die. However, to understand the parasitics and how the circuits are connected together, it's important to understand the chip's cross-sectional view. Since we will often show a layout view followed by a cross-sectional view, let's make sure we understand the difference and how to draw a cross-section from a layout. Figure 1.5a shows the layout (top) view of a pie. In (b) we show the cross-section of the pie (without the pie tin) at the line indicated in (a). To "lay-out" a pie we might have layers called: crust, filling, caramel, whipped-cream, nuts, etc. We draw these layers to indicate how to assemble the pie (e.g., where to place nuts on the top). Note that *the order we draw the layers doesn't matter*. We could draw the nuts (on the top of the pie) first and then the crust. When we fabricate the pie, the order does matter (the crust is baked before the nuts are added).

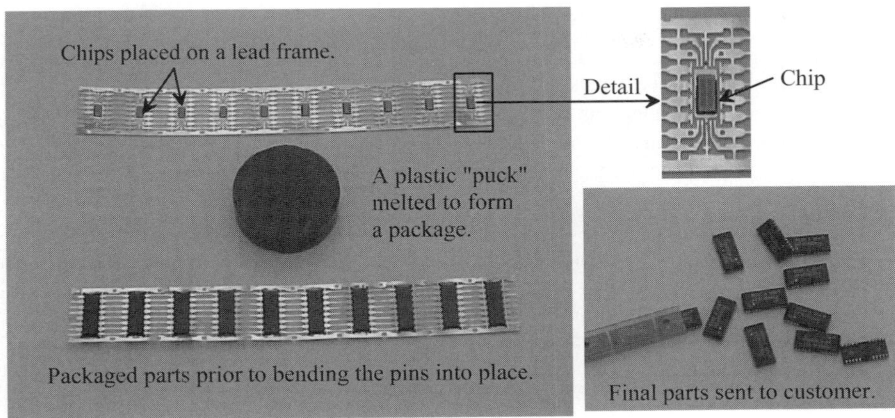

Figure 1.4 Plastic packages are used (generally) when the chip is mass produced.

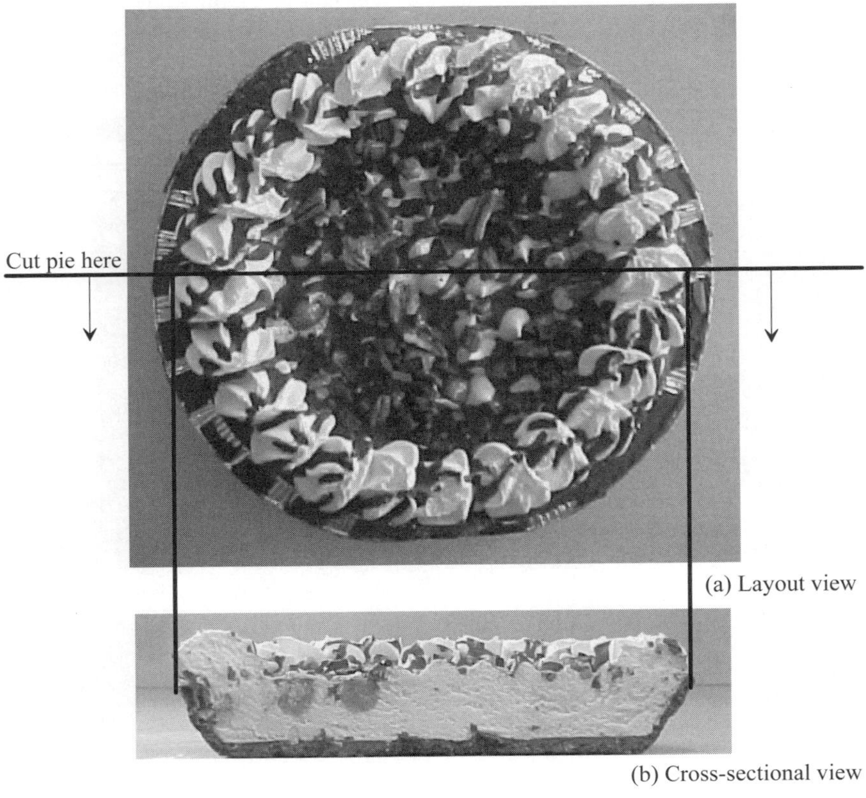

Cut pie here

(a) Layout view

(b) Cross-sectional view

Figure 1.5 Layout and cross sectional view of a pie (minus pie tin).

1.2 Using the LASI Program

The LASI program discussed in this section can be used for the layout and design of CMOS integrated circuits. This section provides a brief tutorial covering the use of LASI. We assume that the reader has downloaded and installed LASI and the MOSIS setups following the instructions at cmosedu.com. We use the MOSIS setups (with a design directory of, for example, C:\Lasi7\Mosis) for our examples in this section.

1.2.1 The Basics of LASI

A chip design should reside in a design directory. The LASI system executables are located in C:\Lasi7 (which *cannot* be used as a design directory). Information on setting up a design directory can be found at cmosedu.com.

Starting LASI

Going to the Window's start button, then Lasi 7, and then MOSIS results in LASI starting, Fig. 1.6. Notice how, at the top of the window in Fig. 1.6, the design directory (Folder) is shown. The menu items (**Help**, **Run**, **Print**, etc.) are executable by either clicking on the word (e.g., **Help**) or on the menu button (e.g., the picture of a question mark). The buttons on the right of the display can be toggled by either pressing the **Menu1** or **Menu2** buttons or by clicking the right mouse button while in the drawing area. We'll talk about the items on the bottom of the screen in a moment.

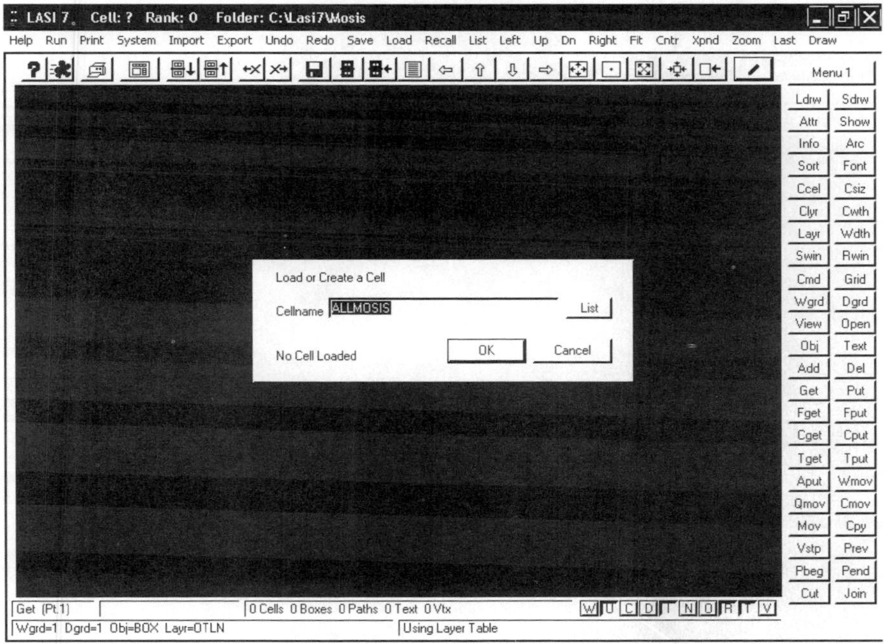

Figure 1.6 Starting LASI using the MOSIS setups.

Getting Help

The LASI layout system comes with a complete on-line manual. This manual is accessible, while LASI is running, by pressing F1 on the keyboard and clicking the command button (simultaneously) with which the user needs help or by pressing the **Help** button. Alternatively, the user can open the help file in the directory C:\Lasi7\help without LASI running.

Cells in LASI I

Complex IC designs are made from simpler objects called cells. A cell might be a logic gate or an op-amp. When LASI starts-up, as seen in Fig. 1.6, the last cell loaded (prior to exiting LASI) appears in the load cell window. In Fig. 1.6, this cell's name is "*ALLMOSIS*." Let's create a new cell called "*test*" with a rank of 1. Figure 1.7 shows the resulting LASI screen. Notice that the current cell's name and rank are displayed at the top of the window. If we want to load or create a different cell, we press either the **Load** menu item or the corresponding picture, as indicated in Fig. 1.7. The drawing area in Fig. 1.7 shows a reference indicator (the origin of the drawing). To toggle displaying this indicator, we can either press **r** on the keyboard or the **R** button in the lower right corner of the display followed by redrawing the cell. To redraw the cell, we press **Enter** on the keyboard or **Draw** or the picture of the pencil at the top of the window.

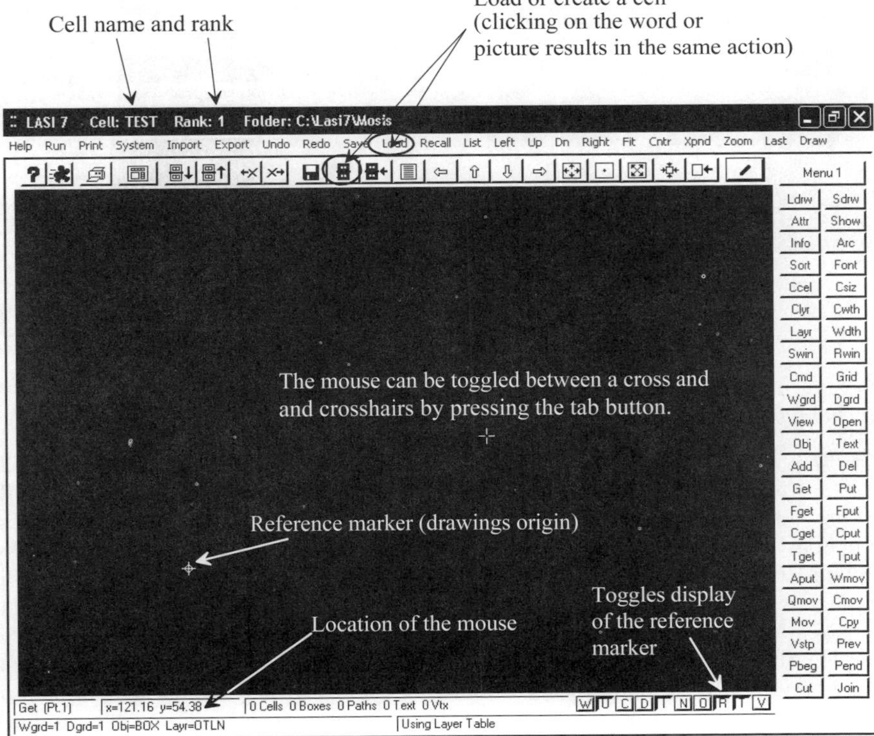

Figure 1.7 Screen after making a cell called "test" with a rank of 1.

Navigating

Pressing the **Grid** button on the right of the screen followed by zooming in around the reference indictor by pressing **Zoom** (and two clicks of the mouse to indicate the zoom area) on the top menu may result in a screen like the one seen in Fig. 1.8. The reader should experiment with the **Fit** (alt+f), **Xpnd** (alt+x or expand), and the arrow (**Left**, **Up**, **Dn**, and **Right**) commands. The center command, **Cntr**, centers the screen around a point set by the mouse. Pressing **Last** on the top menu brings up the previous (or last) view.

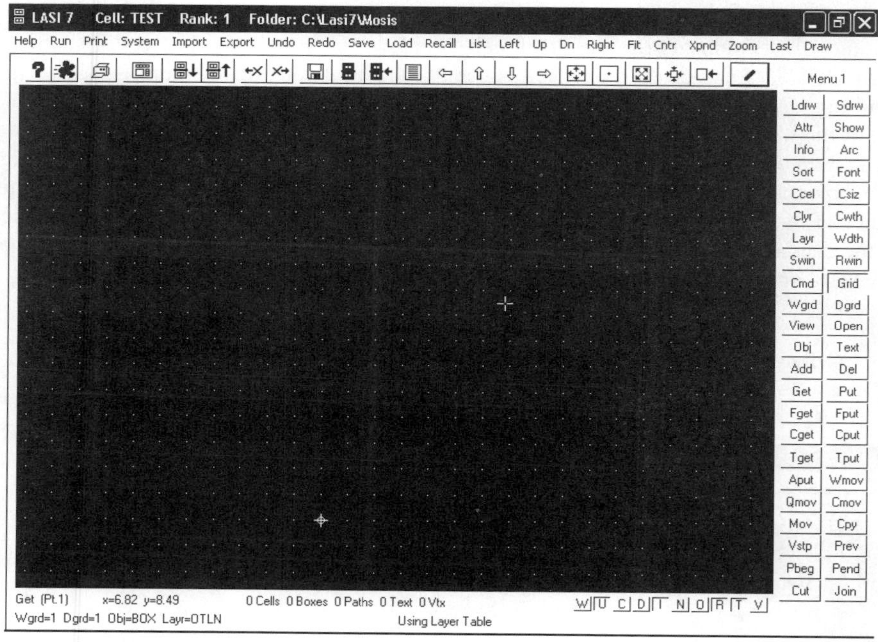

Figure 1.8 Navigating in LASI (see text).

Drawing a Box

Let's draw a box. To start, select the **Layr** command and select "NWEL" or layer 42. This is the n-well layer discussed in the next chapter. Remember that if a command isn't showing, toggle the menus by right clicking the mouse in the drawing area or pressing the buttons **Menu 1** or **Menu 2**. Next select the **Obj** (object) command followed by selecting BOX (double click on the word BOX). The lower left of the display should indicate a working grid (Wgrd) of 1, a dot grid (Dgrd) of 1, a box object, and the n-well layer. Next click on the **Add** button (we want to add a box on the n-well layer). The lower left of the display should indicate that LASI is in the "Add Box" mode. By clicking the left mouse once in the drawing area, moving the mouse, and clicking the left mouse button again, we draw a box on the n-well layer. Figure 1.9 shows a possible resulting display.

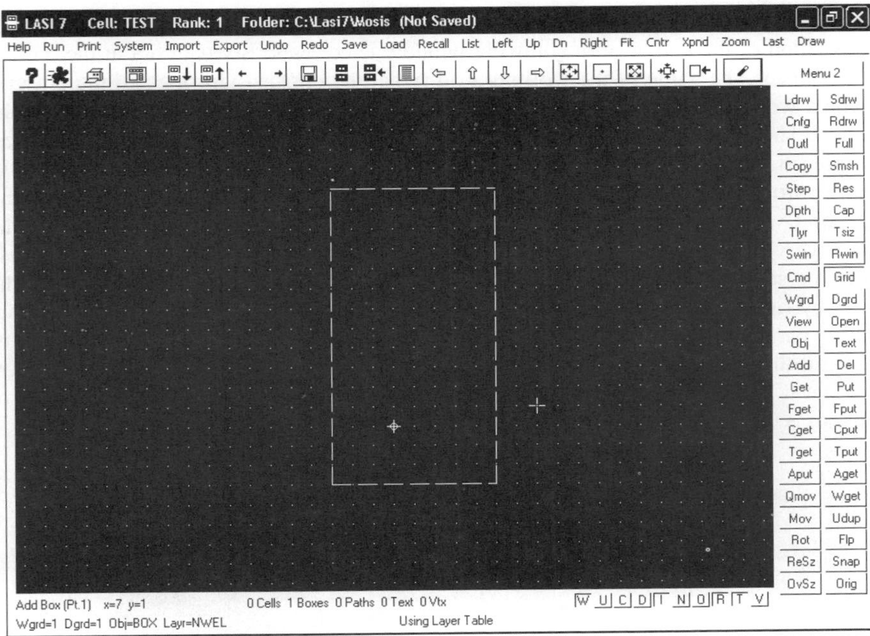

Figure 1.9 Drawing a box using LASI.

Moving and Resizing

Moving or resizing an object consists of *getting* the object using, among others, the **Get** command, *moving* the object using the **Mov** command, and *putting* (or deselecting) the object using the **Put** command. For example, say we want to move the right edge of the n-well box in Fig. 1.9. To begin, select the **Get** command. This is followed by clicking the left button of the mouse on one side of the edge followed by clicking (the mouse) on the other side of the edge (the selected or "Got" edge then becomes highlighted). Using the **Mov** command, we can then use the mouse in the same manner to move the edge. To deselect the edge, we use the **Put** command (again, in the same manner with the two clicks of the left mouse button) or we simply press the all put (**Aput**) command. Note that movement is relative to the clicking of the mouse (we don't have to click on the selected edge). If, for example, the mouse is left-clicked somewhere in the drawing window and then, in a horizontal distance of 5, left-clicked again, the selected edge will move horizontally a distance of 5. (The user should experiment with these commands until they feel comfortable.) To move the entire box, we get all four sides of the box or any part of the box using the **Fget** (full get) command.

To reduce the amount of mouse clicking it is helpful to use the **Qmov** (quick move) command. This command automatically performs the get, move, and put commands in sequence so the user doesn't have to keep clicking menu items. The reader should try out the **Qmov** command in LASI. **Qmov** can save considerable time when making edits to layouts.

The Drawing Grids

The position of the cursor in the drawing window is continuously read out at the bottom of the screen. The value of the working grid (what the cursor snaps to) and the dot grid (the dots we see when the **Grid** button is depressed) are also seen at the bottom of the window. The coordinates of the cursor are either in *working grid* units or in the smallest possible grid unit, the *unit grid* (so that the cursor appears to move smoothly instead of jumping around). For the MOSIS setups, there are 100 grid units between each working grid point when the working grid is 1. By pressing the **Wgrd** and **Dgrd** buttons, the respective working or dot grids are changed between one of ten possible values. These values are set using the **Cnfg** command. (Again, if a command isn't showing, toggle the menus by right clicking in the drawing window.) Figure 1.10 shows the **Cnfg** (configure drawing parameters) window. In the MOSIS setups, we've limited the grids to either 1 or 0.5. Until the user gets familiar with layout, it's not a good idea to change these values. Note that it's possible to have a dot grid of 1 while the working grid (the grid the cursor snaps to) is 0.5.

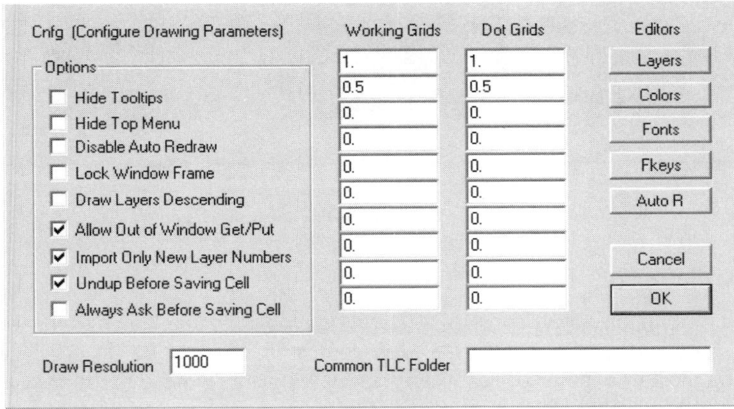

Figure 1.10 Showing how the working and dot grids are set in LASI using the **Cnfg** command.

Keyboard button **a** is a toggle between the working and unit grids. Pressing keyboard button **w** causes the cursor to snap to the working grid while pressing **u** causes the cursor to snap to the unit grid (i.e., appear to move in a continuous movement).

Making Measurements

LASI makes measurements using the keyboard buttons, **z** and **spacebar**. Pressing the **z** button sets a zero reference point. Pressing the **spacebar** displays the distance between the cursor and the zero reference point in the middle bottom of the window. It's useful to remember at this point is that pressing **w** forces the cursor to snap to the working grid so that measurements can be made based on the grid and independent of the state (the current selected command) of LASI. Note that if the **spacebar** is held down, a continuous measurement is displayed on the bottom of the window.

Key Assignments

Below is a summary of the keyboard (button) commands.

Enter does a **Draw** command (redraws the display)
Delete does a **Del** command (deletes the selected objects)
Insert does a **Xpnd** command (expands the current view)
Home does a **Cntr** 0 0 command (centers the display on the origin [reference marker])
End does a **Last** command (switches back to the previous displayed view)
pgup pans up by a full screen
pgdn pans down by a full screen
arrows pan the drawing window in the direction of the pressed arrow
Tab toggles the mouse cursor between a small cross and crosshairs
z sets a measurement zero point
spacebar gives a measurement from the zero point

The following keyboard buttons are also executed by pressing on the button in the lower right portion of the display

w forces the cursor to the working grid
u forces the cursor to the unit grid
c toggles the path center line on and off in a path object
d toggles the distance marker on and off
i toggles the cell image on and off (a dotted line around the silhouette of the cell)
n toggles the outline name on and off
o toggles the octagonal cursor mode on and off
r toggles the 0,0 reference mark on and off
t toggles the text reference point on and off
v toggles viewing vertices of a path or polygon object

The **alt** button can be used with the indicated letter in the top menu commands for execution directly from the keyboard. For example, using **alt+u** (pressing **alt** followed by **u** or pressing both at the same time) results in executing an **Undo** command. Other multipurpose buttons include the **Ctrl**, **Esc**, and **Shift** buttons (see the LASI help manual for additional information).

Finally note that pressing **Esc** (the escape key) aborts a command including a **Draw** command. Using this command can be useful when the layout gets complicated and large. Pressing escape stops the redrawing and shows the cells as simple outlines.

Cells in LASI II

Let's take the *test* cell in Fig. 1.9 and add it to a cell called *test2*. To begin, we press the **Load** command button and create a cell called *test2*. LASI will ask if we want to save the current cell, *test*, (select yes). Next LASI will ask what rank we want to assign the *test2* cell. Select a rank of 2. A new cell will be created and the drawing window will become blank.

The rank is used in the cell collection's (seen by pressing **List**) hierarchy. We might rank a *MOSFET* cell with 1, an *inverter* cell with 2, and a *counter* cell with 5. The *MOSFET* cell can be placed in the *inverter* or the *counter* but the *inverter* or *counter* can't be placed in the *MOSFET*. Similarly, we can place the *inverter* (2) in the *counter* (5) but we can't place the *counter* in the *inverter*. Note that to organize the cell collection we go to the system menu (by pressing **System**) and then press **Organize**.

We can add a box, as we did before, or we can add rank 1 cells to the rank 2 cell. Let's add our *test* cell seen in Fig. 1.9 to this *test2* cell. Press the **Add** command button. Next select the **Obj** command button and double click on the cell *test*. Moving the mouse cursor into the drawing display shows something similar to what's seen in Fig. 1.11. Notice the small outline of the *test* cell. Before adding the cell (by clicking the left mouse button in the drawing window), let's use the zoom command to zoom in around the reference indicator. Adding several *test* cells to the *test2* cell may look something like what is seen in Fig. 1.12.

We can go back to the *test* cell by pressing **List** (saving the *test2* cell) and double clicking on the cell *test*. Press **alt+f** (or **Fit**) at the top of the menu to fit the cell's contents to the drawing window. Let's add another box on the n-well layer (select **Add**, then **Obj**, double clicking on BOX, and then **Layr** followed by selecting the layer NWEL). Adding another n-well box to our layout may result in a layout similar to what appears in Fig. 1.13.

Going back to the *test2* cell, Fig. 1.14, we see that the changes in the layout are automatically updated in the higher ranking cell. If, for example, a change was made to the layout of an inverter, then going to the lower ranking inverter cell and making the change (once) will automatically update the layout wherever the inverter cell is used.

Notice, in Fig. 1.14, that there is a dotted line surrounding the shape of the *test* cell. These dotted lines are the image of the cell. In addition, there are small diamonds in each of the added *test* cells. These diamonds indicate where the reference indicator is in

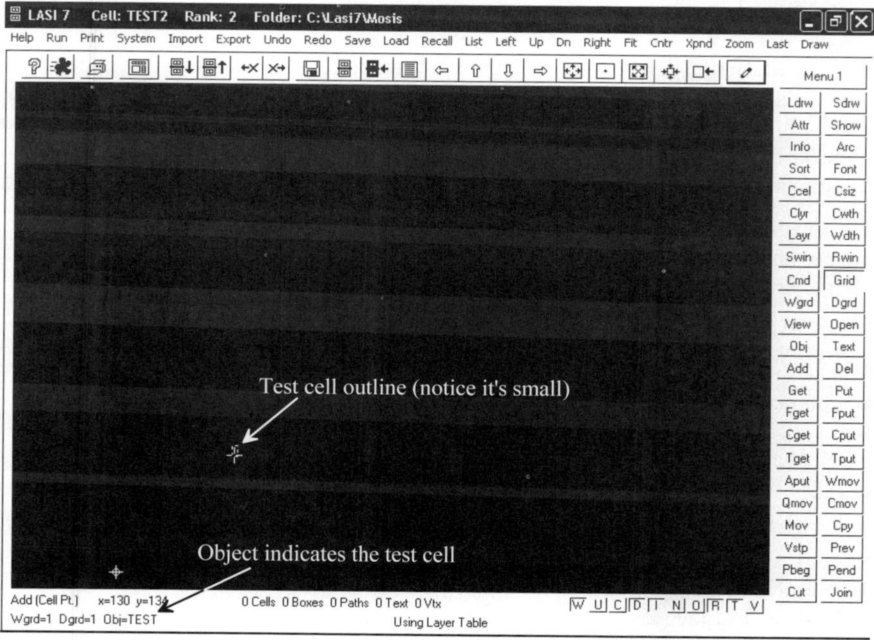

Figure 1.11 Adding the test cell with a rank of 1 to the test2 cell with a rank of 2.

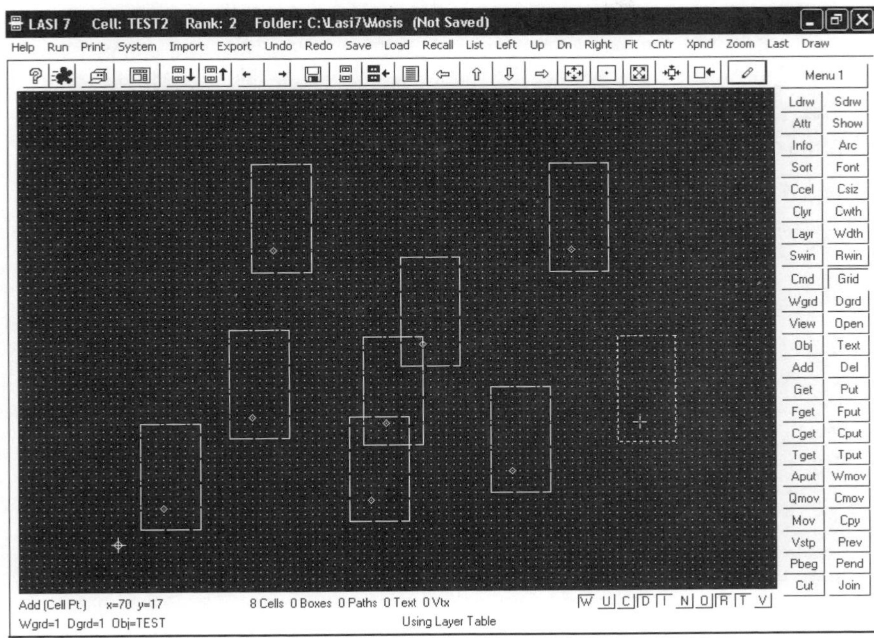

Figure 1.12 Adding several "test" cells with a rank of 1 to the "test2" cell with a rank of 2.

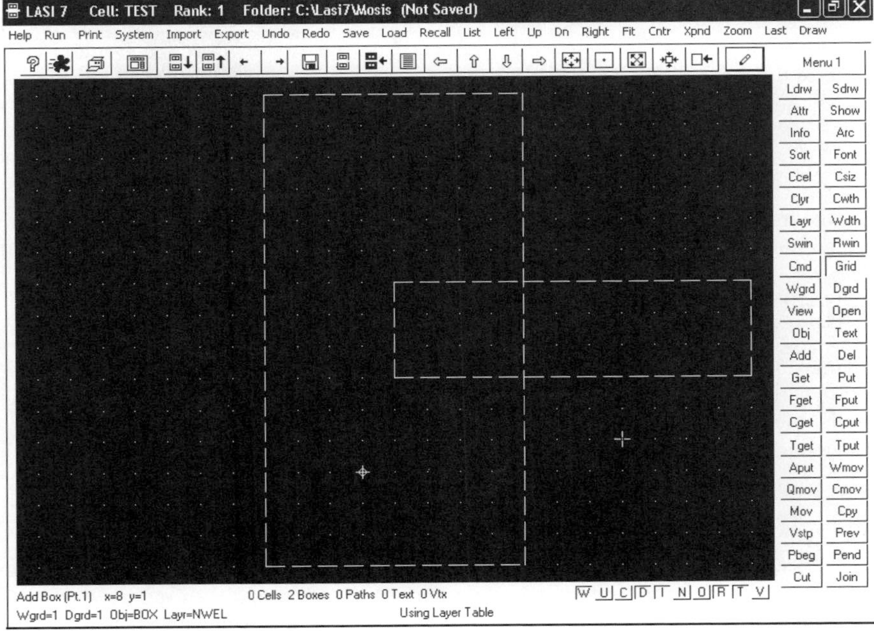

Figure 1.13 Going back to the "test" cell and adding a additional n-well box.

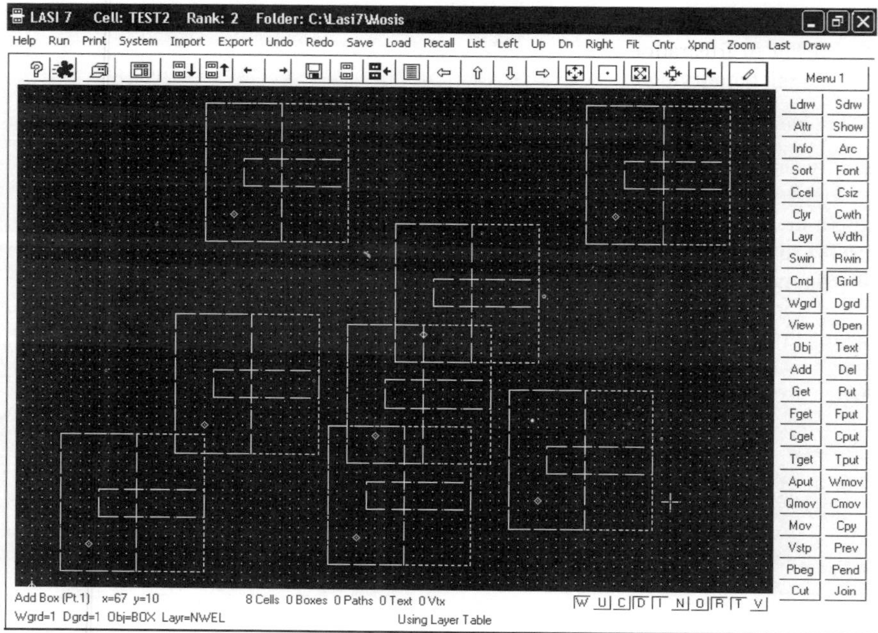

Figure 1.14 How the change to the "test" cell propagates up through the hierarchy.

the *test cell*. Moving or redefining the location of the reference indicator in the lower ranking cell, using the **Orig** command, also shifts the cells' locations in the higher-ranking cells where it is used. Pressing **i** on the keyboard or the **I** button at the bottom of the display toggles the cell's image on and off. (To see the change, the drawing window must be redrawn by pressing on the keyboard or using the **Draw** command.)

Moving a Cell

To move a cell, the **Cget** (cell get) command is used first. Double clicking on the cell or using the mouse to surround a cell (or cells) "gets" the cell. This is followed by using the **Mov** command. Deselecting the cells can be accomplished using the **Cput** (cell put) command or, more often, using the **Aput** (all put) command.

Viewing or Editing Specific Layers

When layouts get complicated, the ability to look at or edit specific layers becomes important. However, someone learning layout can also feel frustrated by not knowing how the viewing and editing of specific layer commands operate. Below we summarize these commands.

Ldrw Draws only one layer of the layout. Useful to quickly view a single layer.

View Selects the layers LASI displays in the drawing window. This command can be frustrating. For example, suppose the NWEL layer is deselected in the view menu. If we were to draw a box on the NWEL layer and then redraw the layout, the box we just drew wouldn't show up!

Open Selects the layers LASI will allow the user to "get." This command, again, can result in frustration. If, for example, the NWEL layer is not selected in the open menu, then trying to "get" the layer will result in a waste of time.

The Polygon and Path Shapes

The only drawing shape we've discussed so far is the box. LASI can also be used to make polygons and path shapes. Let's start out by drawing a polygon (say a triangle). Create a new cell (using the **Load** command) called *test3* with a rank of 1. Select **Add**, then **Obj** (double clicking on PATH/POLY). Let's also select a different layer using the **Layr** command (select POL1 or layer 46). Next click the **Wdth** command and make sure that width is set to 0. When the width is zero, the object is a polygon. When the width is nonzero, the object is a path. Figure 1.15 shows the layout of a polygon (not closed). Notice the location of the polygon's vertices. To move (or change) the shape of a polygon, we must get a vertex. Using the **Get** command on a side of a polygon, and not encompassing a vertex, can result in frustration. To show the polygon's (or path's) vertices in the drawing display, press **v** on the keyboard (or the **V** in the lower right corner of the display) followed by **Enter**. (Try this now.) To close the polygon object in Fig. 1.15, we place the last (fourth) vertex on the first vertex as seen in Fig. 1.16. Note that this last vertex is still active and so it appears that we are still drawing the same polygon (even though our triangle is closed). To draw another shape, click the **Aput** command.

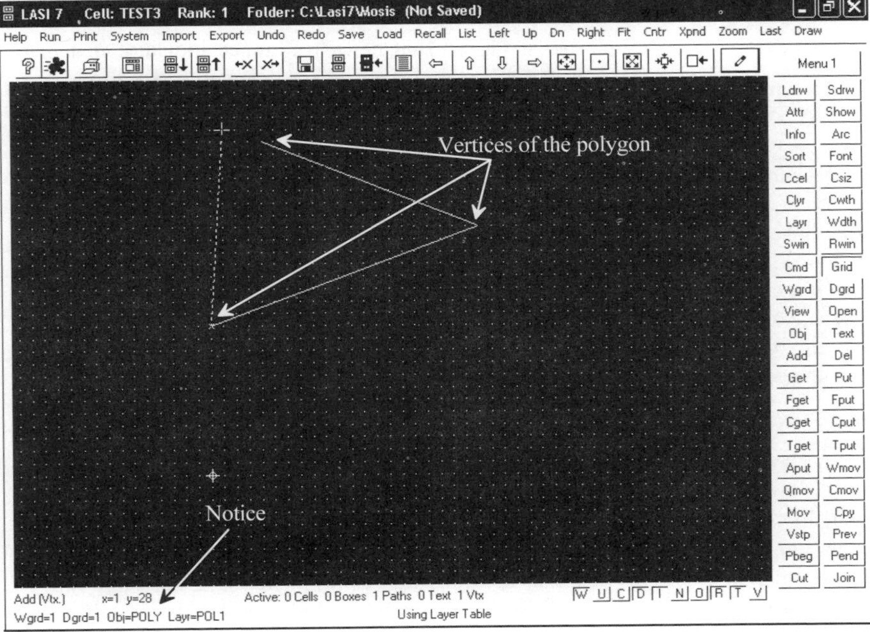

Figure 1.15 Drawing a polygon in LASI on the POL1 (polysilicon) layer.

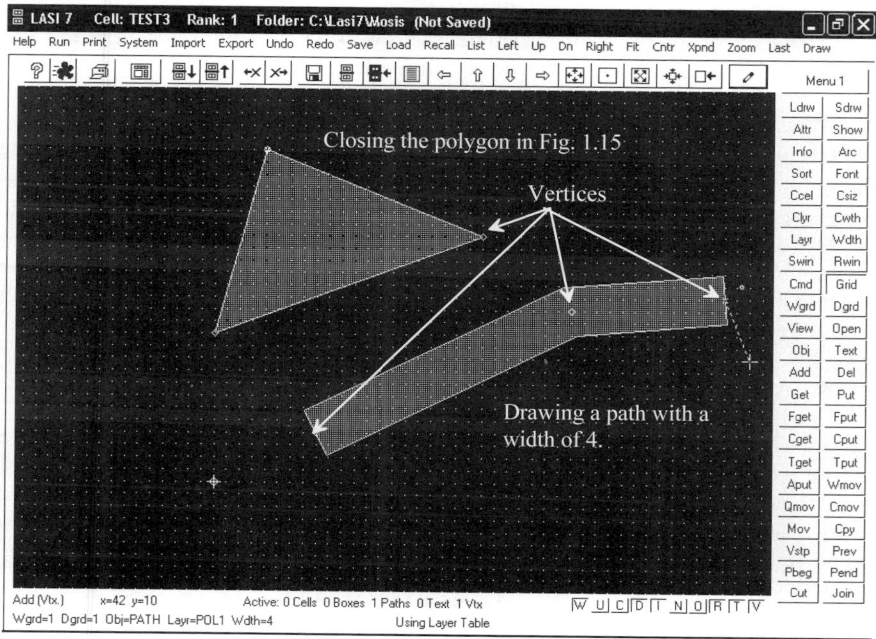

Figure 1.16 Closing the polygon in Fig. 1.15 and drawing a path object with a width of 4.

Also seen in Fig. 1.16 is a path object. To draw a path, after drawing the polygon shape, all we have to do is use the **Wdth** command to set the width of the path to a nonzero number (zero width indicates that we're drawing a polygon). For the path in Fig. 1.16, we used a width of 4. To terminate drawing a path, we can use the **Aput** command. Notice how the vertices of a path are located in the center of the path. Again, to toggle between displaying or not displaying vertices, we can press **v** on the keyboard. Also, again, if we don't encompass a vertex when we are trying to get the object, nothing will get selected. (Trying to get the side of a path or polygon can result in frustration.) To move an entire path or polygon, we can use the **Fget** command and encompass at least one of the path's or polygon's vertices.

Using Text in LASI

Labeling layout with text is important for documenting how the IC is assembled. To add text to a layout, we select the **Text** command. This is followed by selecting the text's layer using the **Tlyr** command (select POL1 for the example to follow). Next the size of the text can be set using **Tsiz**. The command **Font** (remembering if the command isn't showing to right-click the mouse in the drawing display) selects one of three font files. The first font (C:\LASI7\TFF1.DBD) file uses zero-width letters (so they don't have any fabrication significance). The other files (TFF2.DBD and TFF3.DBD) do have widths, and so they can be seen in the fabricated chips. Clicking the left mouse button in the drawing display may result in the text seen in Fig. 1.17 (where we've changed the size

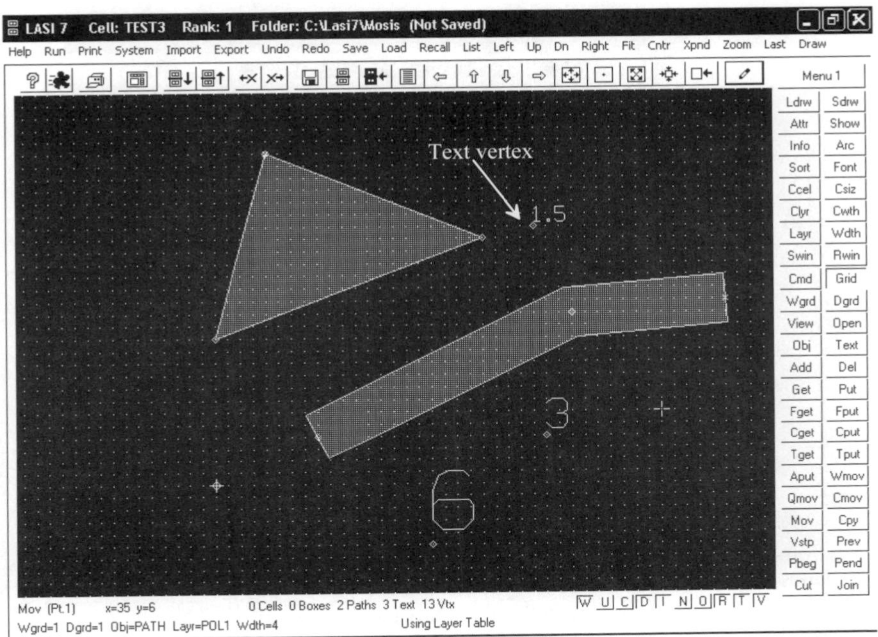

Figure 1.17 Adding text to a layout.

using **Tsiz**). To move the text, we must encompass a text vertex. To show the text vertices, press **t** on the keyboard or **T** in the lower right corner of the display followed by a **Draw** command. Again, not knowing to get a vertex can lead to frustration. To get text in a layout without getting any of the other objects, we can use the **Tget** command. To change the size of the text, we "get" the text by encompassing a vertex using the **Get**, **Fget**, or **Tget** commands, and then using the **Csiz** command. Using the **Text** command and clicking on an existing text vertex results in changing the displayed text but not the size or layer of the existing text. The **Clyr** command can be used to change the layer the text is drawn on or any other shape's layer.

Some Features to Speed Up Layout Design

The reader's right hand should be used for the mouse (assuming the reader is right-handed) while the left hand should be used for pressing keys on the keyboard. To assign more commands to the keyboard, press the **Cnfg** button followed by the **Fkeys** button. Figure 1.18 shows the result. The **F1** key is used for getting help, as discussed earlier. Now, as seen in Fig. 1.18, pressing **F2**, **F3**, and **F4** executes the commands **Aput**, **Qmov**, and **Fget**, respectively. It's important, when adding commands, to ensure that the changes are saved prior to exiting the window. To add a command to the list, click on one of the existing commands. Next type the new command in the command argument field at the bottom of the window. Pressing **Add** places the new command after the selected existing command, while pressing **Insert** places the command before the existing command. To edit an existing command, double click on it. After editing, press **Paste** to update the existing command. Again, it's important to **Save** the changes prior to exiting.

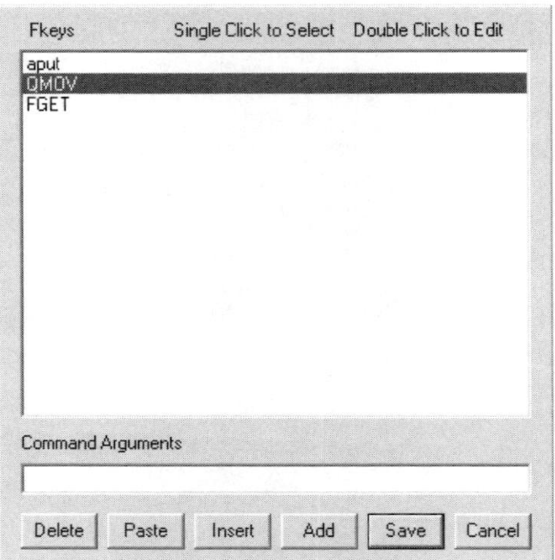

Figure 1.18 Setting up the Fkeys to speed up layout.

Some other useful commands that can speed up layout, when the layout contains a large number of cells, are the **Outl**, **Full**, and **Dpth** commands. To avoid having to redraw the contents of a cell or cells, the **Outl** (outline) command can be used. Either double-clicking on a cell or surrounding the cell's contents with the mouse, when the **Outl** command is selected, causes the cells to be drawn as outlines. Pressing **n** on the keyboard or **N** with the mouse cursor in the lower right portion of the display, toggles the cell's name in its outline on and off. Using the **Full** command shows the contents of the cell.

If we are editing a cell with lower ranking cells nested in it, then we can use the **Dpth** (depth) command to limit the full display of these lower ranking cells. If depth is set to 0, then only the basic objects (boxes, paths, text) in the current-edited cell are shown in detail. The lower level cells are shown as dashed outlines. If the depth is set to 1, then the basic objects in the next level of nested cells are shown ... and so on. Note that cell nesting levels are not the same as cell rank. To show the basic objects of all possible nested cells, the depth should set to 15. To see a listing of nested cells within other cells, use the **Show** command.

Finally, when editing both cells and objects (boxes, polygons, and paths) the window get, **Wget** (window get) and, **Wmov** (window move) commands can be very useful. While in the **Wget** command mode, the mouse is used to select an area of layout containing both cells and drawn objects. The **Wmov** command moves this layout. Note that when "getting" the layout a cell won't be selected unless it is completely surrounded in the mouse selection area.

Understanding the **Cpy** *and* **Copy** *Commands*

The **Cpy** command is used to copy layout within a cell. The layout is selected using, for example, the **Fget**, **Cget**, or **Wget** commands. The **Cpy** command then behaves just like a move command except that the selected layout isn't moved to the new location: it is copied.

It's important to understand that using the **Cpy** command frequently is not a good idea. For example, say a memory cell containing 50 objects (boxes, polygons, or paths) is laid out. If the memory cell is used in a 1 Mbit memory, and the **Cpy** command is used, then the resulting number of objects in the memory cell is 50 million (this is bad). If, on the other hand, the 1 Mbit memory is made with the 1,000,000 memory cells, then the size (the number of objects) is 1,000,050. We can take this a step further. We can make a cell containing a row of memory cells (say 1,000) and use this row cell 1,000 times to make a 1 Mbit memory. The number of objects in the 1 Mbit memory cell is now only 2,050 (1,000 instances of the memory row, 1,000 instances of the memory bit, and 50 objects in the memory bit). We can take this even further to reduce the size of the layout file. **Using cell hierarchy is extremely important.** Do not lay out a large chip unless the material in this paragraph is understood! Else the design file sent to the mask maker will be unreasonably large. Use the **Cpy** command as little as possible.

Using the **Save** command saves the contents of the cell we are working on. It can also be used to save the cell under a different name. However, sometimes we want to take some portion of the layout we are working on and move it (append it) to a different cell (or a new cell). The **Copy** command is useful for this task. To take some layout and append it to a different cell, we first "get" the layout we want to copy to a different cell. Next we select the **Copy** command and the cell we want to move the selected layout into (or the name of a new cell).

Transporting Cells in LASI

To share files between other users or between design directories, *Transportable LASI Cell* files (*.TLC) files or *Transportable LASI Drawing* files (*.TLD) files are used. The files we draw in LASI are saved on the hardisk as text files with the TLC extension.

Copying TLC files into a directory will **not** make them show up in the design directory's cell listing (and so we won't be able to edit the cells). To register a cell in a design directory, we must first **Import** the cell. Note, in the **System** menu, the **Organize** command can be used to organize the cells in a design directory according to rank or name.

The TLC files contain references to other cells but not the actual layouts of the referenced cells themselves. (For example, a reference may indicate "place the inverter cell at the location 12,13.") To get a file that contains the cell's hierarchy, we use TLD files. To generate a TLD file, we go to the **Export** command and select the name of the cell, TLD file, and the location we want to place the TLD file. TLD files can be shared between users or used as backups of work. Note that while LASI imports a TLD file, it still uses TLC files for editing. Every time **Save** is pressed, the TLC file in the design directory (being edited) is updated, leaving the TLD file unchanged (until the **Export** command is used).

Edit-in-Place

Consider the layout seen in Fig. 1.19. In this layout we have two boxes and two cells. Let's say we want to align the cell on the right to the two boxes. The simplest way to do this (keeping in mind that modifying the cell propagates through the hierarchy) is to get the cell using the **Cget** command. Then press **Ctrl** on the keyboard while pressing **Load**. This causes LASI to enter the "Edit-in-Place" (EIP) mode. A lower ranking cell can then be modified while showing the objects from the higher ranking cell. To exit EIP, use the

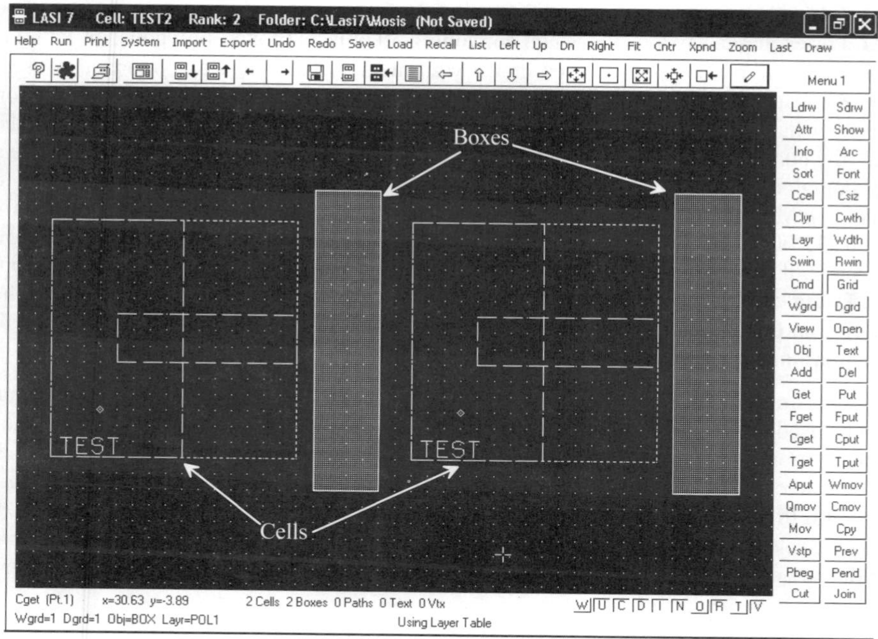

Figure 1.19 Editing a cell in place.

Load or **List** commands. Note that the other cells in the layout will not be shown when in EIP mode (only the cell we are editing and the drawn objects from the higher ranking cell).

Backing Up Your Work

It's very important to constantly back up your work in a location other than the hard disk you are running LASI out of. Backups can be as simple as exporting TLD files to a DVD or as complete as copying the design directory (or zipping it up) to a network drive. The point is layout that is laborious so not backing up the work is a mistake.

1.2.2 Common Problems

After adding an object, the object cannot be seen

Check the **View** layers in the drawing display to ensure that the layer is not in hidden mode. The **Draw** command must be used after using **View**.

Cannot **Get** *an object*

(1) Check, using the **Open** command, that the layer can be opened (or moved). (2) Verify that the object is not part of another cell. (3) When trying to get an object made using the path or polygon object make sure that the cursor encompasses a vertex. (4) If the object is text, then text vertex must be encompassed.

Cells are drawn as outlines, or the perimeter of a cell has a dashed line

(1) Use the **Full** command to show the contents of a cell. (2) Use the **Dpth** command to limit the depth of the cells shown. Increasing the depth to the rank of the cell shows all cells. (3) Press **i** on the keyboard to force an outline to be drawn around a cell. This is indicated by the **I** in the bottom right corner of the drawing display. The letter **n** (or **N** in the lower right of the display) toggles the display of the cell's name.

Fit *command causes the drawing window to expand much larger than the current cell*

There is an unknown object someplace in the cell. Use the **Fget** command to get any objects outside the main cell area. Use the **Del** command to delete the unknown object.

Cursor movement is not smooth

(1) The cursor may be in the octagonal mode. Press **o** on the keyboard or the button **O** in the lower right portion of the display to toggle this mode on and off. (2) Make sure that the display isn't zoomed in too much. (3) Pressing **a** can be used to toggle the mouse cursor between the working and unit grids.

1.2.3 Sending the Layout to the Mask Maker

Once we have a completed layout we can convert the resulting TLC files into a format the mask maker will accept. These formats are either CIF (CalTech intermediate format) or GDS (graphical design system, which is a derivative of the older Calma stream format, CSF). We'll focus on using GDS (at the time of this writing the format is actually a second-generation standard called GDSII) since it is what is used in industry. Generating a *.GDS file in LASI (or any other layout program) is often called *streaming the layout out*. Because GDS files can be large, and they are often stored on magnetic storage tapes, generating and storing the GDS file is often called "taping out" or simply "tape-out."

Assuming the layout is saved, we enter the **System** mode in LASI, Fig. 1.20. Next, we select the **Tlc2Gds** command button, Fig. 1.21. The specific details concerning using this utility or the **Gds2Tlc** utility can be found in the on-line manual. One of the important parameters in the setups is the scale factor Lambda. This takes our layout drawn on a "1" grid and scales it to the appropriate final size.

Checking to Make Sure the Layout Scaled Correctly

To ensure that the layout scales correctly when making the GDS file, copy the design directory into a temporary design directory (TDD). LASI is then set up to run from this TDD. This ensures that all of the original layout isn't corrupted by mistakes in the checking process we're about to discuss (**important**). It's also a good idea to back up the directory at this point as well.

In this TDD we can generate a GDS file, using the **Tlc2Gds** utility. Unless, the user has some atypical needs (e.g., eliminating layers in the final GDS file), this is a simple matter of specifying the name of the TLC file to convert (ensuring the TLC file in the dummy directory is used), specifying the name of the GDS file to create from the TLC file, specifying the scale factor, and starting the conversion (pressing **Go**) after exiting the setups. Both GDS and layer data map (*.ldm) files are generated. (The user will have to select the layers to add to the ldm file while the conversion is in progress unless an ldm file already exists.) The *.ldm file is used when converting the GDS back into a TLC file discussed below.

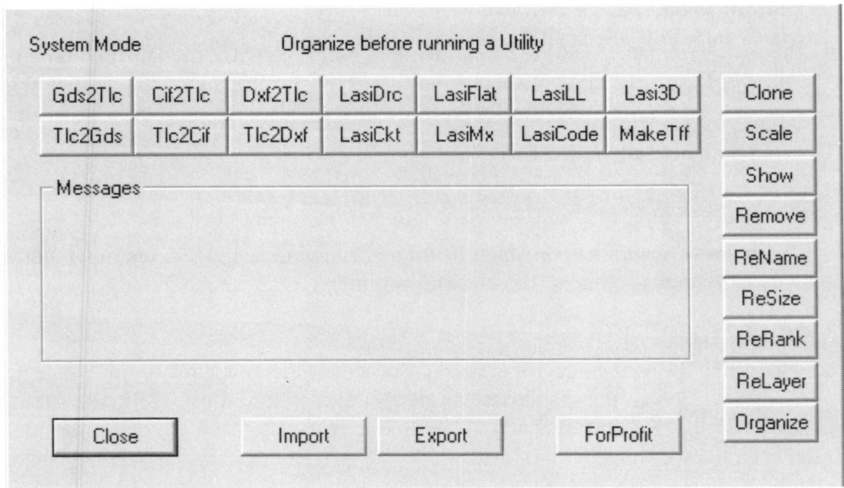

Figure 1.20 The LASI System Screen.

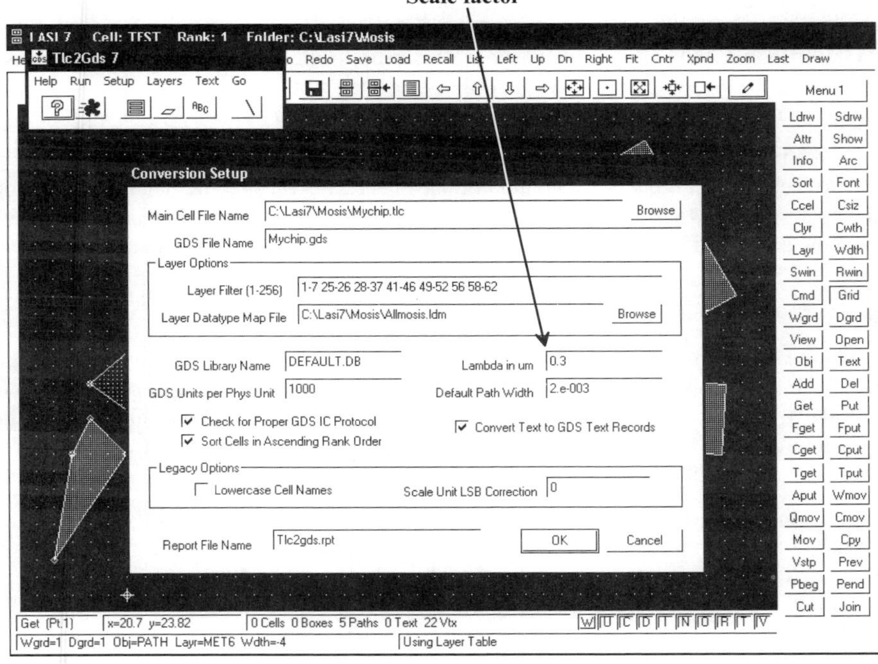

Figure 1.21 Generating a GDS file from a TLC file.

After the GDS file is generated, we can use the **Gds2Tlc** program to convert the GDS file back into TLC files. In the setups we must specify a directory where the TLC files will be written, for example, C:\temp. We can't use a drawing directory because then the existing TLC files would be overwritten. After we've converted the GDS back into TLC files, we can **Import** the (scaled) TLC files into the dummy directory to make sure the generated GDS file is scaled correctly.

Note, for the sake of feeling comfortable with this process, that it's easy to take a simple cell, like the *test* cell in Fig. 1.9 and convert it back and forth between a GDS file and a TLC file (with scale factor). Also note that we can use the **Resize** command on the system menu to change the size of the layout if needed.

1.3 An Introduction to WinSPICE

The simulation program with an integrated circuit emphasis (SPICE) is a ubiquitous software tool for the simulation of circuits. In this book we'll use WinSPICE. See the links at cmosedu.com for download and installation information. WinSPICE, like all SPICE engines, uses a text file netlist for simulation input.

Generating a Netlist File

We can use, among others, the Window's notepad or wordpad programs. WinSPICE likes to see files with a "*.cir" extension. To save a file with this extension, place the file name and extension in quotes as seen in Fig. 1.22. If quotes are not used, then Windows will tack on ".txt" to the filename. This can make finding the file difficult when we open the netlist with WinSPICE (see Fig. 1.23).

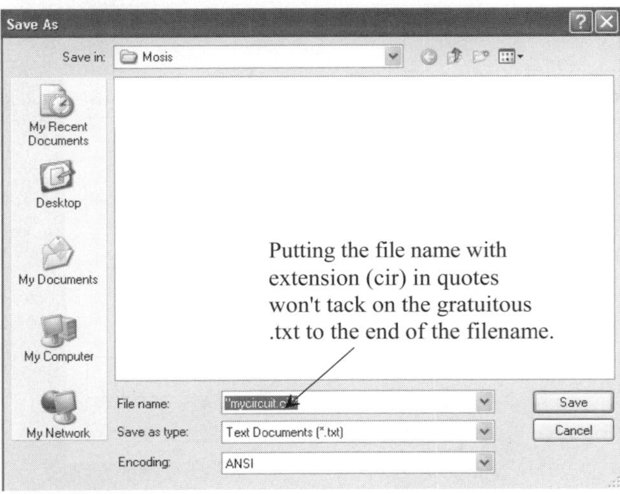

Figure 1.22 Saving a text file with a ".cir" extension.

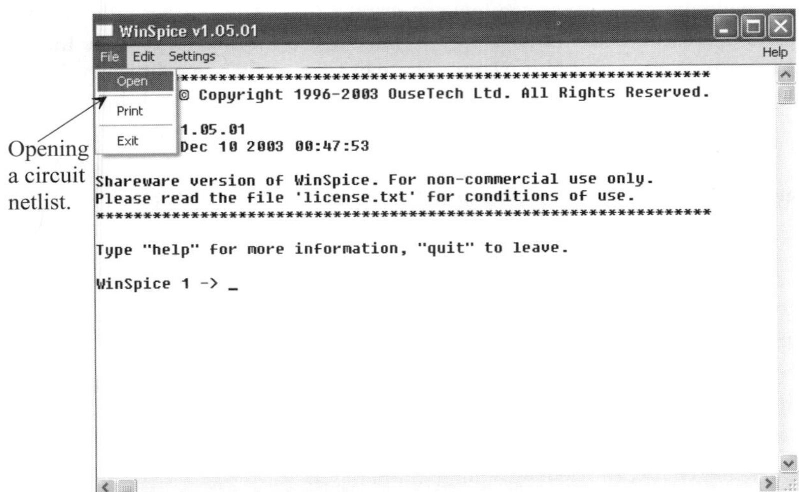

Figure 1.23 Opening a file with WinSPICE.

Transient Analysis

A SPICE transient analysis simulates circuits in the time domain (like an oscilloscope the x-axis is time). Let's simulate the simple circuit seen in Fig. 1.24. A simulation netlist (the text file) may look like:

```
*** Figure 1.25 CMOS: Circuit Design, Layout, and Simulation ***

.control
destroy all
run
plot vin vout
.endc

.tran 100p 100n

Vin     Vin     0       DC      1
R1      Vin     Vout    1k
R2      Vout    0       2k

.end
```

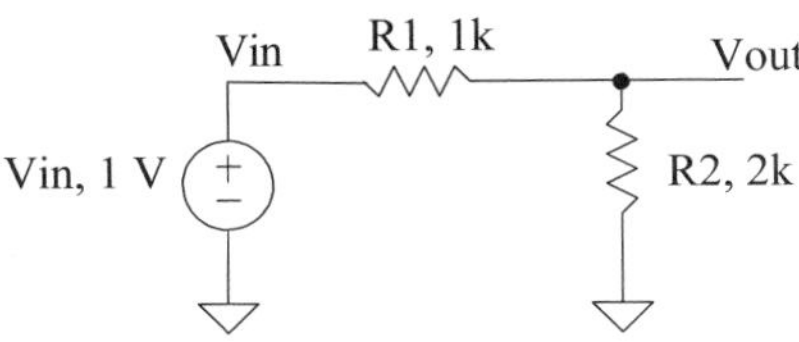

Figure 1.24 Simulating the operation of a resistive divider.

The simulation results using this netlist are seen in Fig. 1.25. The first line in a netlist is a title line. This line is ignored by SPICE (important). The next five lines are control commands. Notice how the end of the control statement is terminated with an ".endc" not an ".end" like at the end of the netlist. Placing ".end" at the end of the control statement causes SPICE to ignore all of the lines containing the circuit information. The statements in the control block can be run directly from the command line in the WinSPICE command window seen in Fig. 1.23. The *destroy all* command destroys all of the previous simulation results (so we don't display old data). The *run* command runs the simulation. The *plot* command plots the voltages on the nodes Vin and Vout.

Note that the DC voltage source Vin is connected to the node Vin. We could have labeled the Vin node with a number like "1." However, it is nice to have node names that correspond with signals.

The connection of the resistors and how they are specified should be easy to determine. A line starting with an "R" indicates a resistor specification. A line beginning with an "*" indicates a comment. Node 0 (zero) is always reserved for ground.

The form of the transient statement (this is the type of analysis) is

```
.TRAN TSTEP TSTOP <TSTART> <TMAX> <UIC>
```

where the terms in < > are optional. The TSTEP term indicates the (suggested) time step to be used in the simulation. The parameter TSTOP indicates the simulation's stop time. The starting time of a simulation is always time equals zero. However, for very large (data) simulations, we can specify a time to start saving data, TSTART (again this term is optional). The TMAX parameter is used to specify the maximum step size. If the plots start to look jagged (like a sinewave that isn't smooth), then TMAX should be reduced.

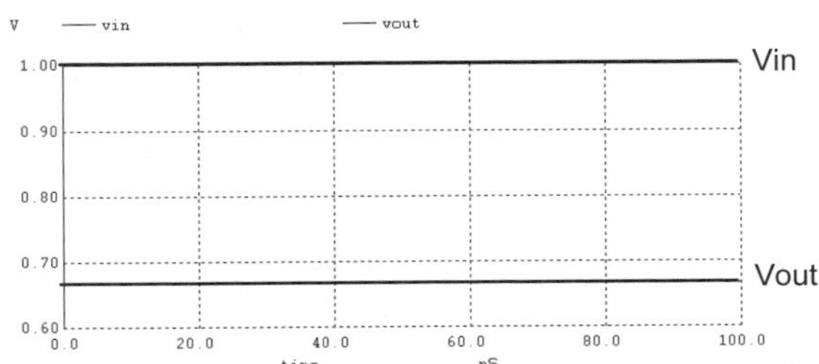

Figure 1.25 Simulating the circuit in Fig. 1.24.

To illustrate a simulation using a sinewave, examine the schematic in Fig. 1.26. The statement for a sinewave in SPICE is

```
SIN VO VA FREQ <TD> <THETA>
```

The parameter VO is the sinusoid's offset (the DC voltage in series with the sinewave). The parameter VA is the peak amplitude of the sinewave. FREQ is the frequency of the sinewave, while TD is the delay before the sinewave starts in the simulation. Finally, THETA is used if the amplitude of the sinusoid has a damped nature. To simulate the circuit in Fig. 1.26, we use a netlist of

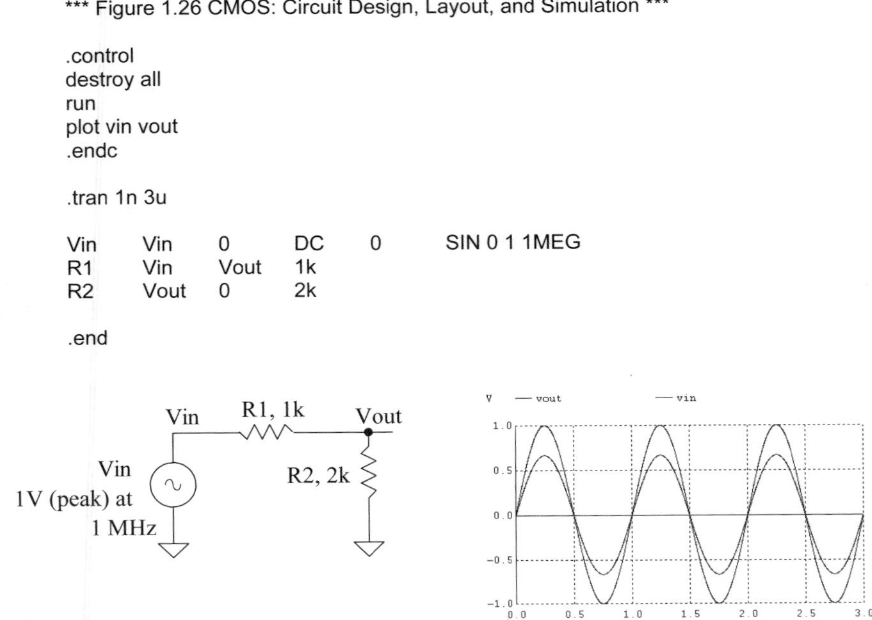

```
*** Figure 1.26 CMOS: Circuit Design, Layout, and Simulation ***

.control
destroy all
run
plot vin vout
.endc

.tran 1n 3u

Vin   Vin    0     DC   0     SIN 0 1 1MEG
R1    Vin    Vout  1k
R2    Vout   0     2k

.end
```


Figure 1.26 Simulating the operation of a resistive divider with a sinewave input.

Some key things to note in this simulation: (1) MEG is used to specify 10^6. Using "m" or "M" indicates milli or 10^{-3}. The parameter 1MHz indicates 1 milliHertz. (Also, f indicates femto or 10^{-15}. A capacitor value of 1f doesn't indicate one farad but rather 1 femto Farad.) (2) Note how we increased the simulation time to 3 μs. If we had a simulation time of 100 ns (as in the previous simulation), we wouldn't see much of the sinewave (one-tenth of the sinewave's period). (3) The "SIN" statement is used in a transient simulation analysis. It is **not** used in an AC analysis.

Before leaving this introduction to transient analysis, let's introduce the SPICE pulse statement. This statement has a format given by

```
PULSE VINIT VFINAL TD TR TF PW PER
```

VINIT is the pulse's initial voltage, VFINAL is the pulse's final (or pulsed) value, TD is the delay before the pulse starts, TR and TF are the rise and fall times, respectively, of the pulse (noting that when these are set to zero the step size used in the transient simulation is used); PW is the pulse's width; and PER is the period of the pulse. Figure 1.27 provides an example of a simulation that uses the pulse statement. A section of the netlist used to generate the waveforms in this figure is seen below.

```
.tran 100p 30n

Vin   Vin    0     DC   0     pulse 0 1 6n 0 0 3n 10n
R1    Vin    Vout  1k
C1    Vout   0     1p
```

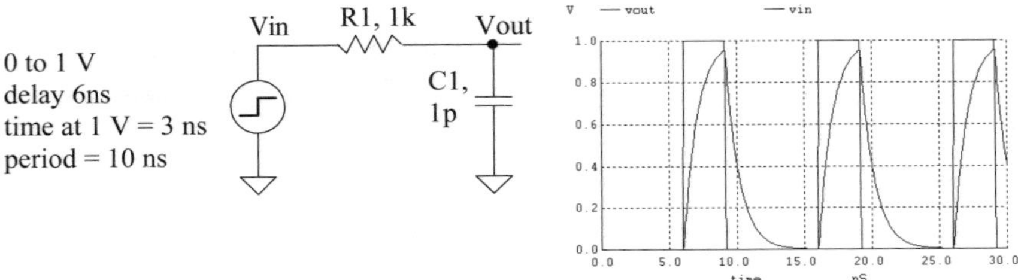

Figure 1.27 Simulating the step response of an RC circuit using a pulsed source voltage.

Other Analysis

Besides the transient analysis presented in this section, we frequently use the SPICE DC and AC analyses. The AC analysis has an x-axis of frequency. This type of analysis is the common "small-signal" analysis used in a basic introductory microelectronics course. The DC analysis has a DC voltage source for the x-axis. The value of the DC source is swept while either a current of voltage is plotted on the y-axis. We don't go into these analyses here but provide numerous examples later in the book.

Convergence

A netlist that doesn't simulate isn't converging numerically. *Assuming* the circuit contains no connection errors, there are basically three parameters that can be adjusted to help convergence: ABSTOL, VNTOL, and RELTOL.

ABSTOL is the absolute current tolerance. Its default value is 1 pA. This means that when a simulated circuit gets within 1 pA of its "actual" value, SPICE assumes that the current has converged and moves onto the next time step or AC/DC value. VNTOL is the node voltage tolerance, default value of 1 μV. RELTOL is the relative tolerance parameter, default value of 0.001 (0.1 percent). RELTOL is used to avoid problems with simulating large and small electrical values in the same circuit. For example, suppose the default value of RELTOL and VNTOL were used in a simulation where the actual node voltage is 1 V. The RELTOL parameter would signify an end to the simulation when the node voltage was within 1 mV of 1 V (1V·RELTOL), while the VNTOL parameter signifies an end when the node voltage is within 1 μV of 1 V. SPICE uses the larger of the two, in this case the RELTOL parameter results, to signify that the node has converged.

Increasing the value of these three parameters helps speed up the simulation and assists with convergence problems at the price of reduced accuracy. To help with convergence, the following statement can be added to a SPICE netlist:

```
.OPTIONS ABSTOL=1uA VNTOL=1mV RELTOL=0.01
```

To (hopefully) force convergence, these values can be increased to

```
.OPTIONS ABSTOL=1mA VNTOL=100mV RELTOL=0.1
```

Note that in some high-gain circuits with feedback (like the op-amp's designed later in the book) decreasing these values can actually help convergence.

Some Common Mistakes and Helpful Techniques

The following is a list helpful techniques for simulating circuits using SPICE.

1. The first line in a SPICE netlist must be a comment line. SPICE ignores the first line in a netlist file.
2. One megaohm is specified using 1MEG, not 1M, 1m, or 1 MEG.
3. One farad is specified by 1, not 1f or 1F. 1F means one femto-farad or 10^{-15} farads.
4. Voltage source names should always be specified with a first letter of V. Current source names should always start with an I.
5. Transient simulations display time data; that is, the x-axis is time. A jagged plot such as a sinewave that looks like a triangle wave or is simply not smooth is the result of not specifying a maximum print step size.
6. Convergence with a transient simulation can usually be helped by adding a UIC (use initial conditions) to the end of a .tran statement.
7. A simulation using MOSFETs must include the scale factor in a .options statement unless the widths and lengths are specified with the actual (final) sizes.
8. In general, the body connection of a PMOS device is connected to *VDD*, and the body connection of an n-channel MOSFET is connected to ground. This is easily checked in the SPICE netlist.
9. Convergence in a DC sweep can often be helped by avoiding the power supply boundaries. For example, sweeping a circuit from 0 to 1 V may not converge, but sweeping from 0.05 to 0.95 will.
10. In any simulation adding .OPTIONS RSHUNT=1E8 (or some other value of resistor) can be used to help convergence. This statement adds a resistor in parallel with every node in the circuit (see the WinSPICE manual for information concerning the GMIN parameter). Using a value too small affects the simulation results.

ADDITIONAL READING

[1] D. E. Boyce, *LASI User's Manual*, available while LASI is running by pressing F1 on the keyboard and the button the user needs help with (at the same time). Alternatively, the user can open the help file in the directory C:\Lasi7\help.

[2] M. Smith, *WinSPICE User's Manual*, Available for download (with the WinSPICE simulation program) at http://www.winspice.co.uk/

PROBLEMS

In the following solutions, it can be very helpful to use the **Prt Sc** (print screen) button on the keyboard to copy contents displayed on the computer's display to the clip board. The image of the display can then be pasted into a document. For ease of viewing (the resulting pasted image in the document), it may also be useful reduce the display resolution prior to using the **Prt Sc** button (right click on the desktop, select properties, then settings).

In the following problems, the solutions should include pictures of the LASI layout screen and discussions indicating that you know what you are doing.

1.1 Create a cell called *mycell* with a rank of 1 using LASI. In this cell, draw a 10 by 10 box using the POL1 layer. Place the lower left corner of the box at the origin. Show how to measure the diagonal distance of the box.

1.2 Explain how the **Qmov** command can be used to edit the box in Problem 1 so that it measures 5 by 8. How would this be accomplished using **Get**, **Mov**, and **Put**?

1.3 What functions do the **Swin** and **Rwin** commands perform? Give an example of where these would be useful.

1.4 What functions do the **Cget** and **Cput** commands perform? Can any other commands be used in place of **Cget**?

1.5 Write the word "test" on the MET1 layer with sizes of 3, 9, and 24 in the cell of Problem 1 (*mycell*). Discuss the possible ways of "getting" the text and "moving " text.

1.6 Using a path and a polygon in the cell (*mycell*) of Problem 1, generate two objects with widths of 10 and lengths of 10 directly next to the box from Problem 1. Discuss the differences in editing boxes, polygons, and paths.

1.7 Explain, in your own words, the difference between the **Cpy** and **Copy** commands. Give examples to demonstrate the differences.

1.8 Demonstrate how the **Wget** and **Wmov** commands function.

1.9 Using the POL1 layer, draw a triangle that measures nominally 10 on each side (make sure the layout is snapped to the working grid). How many vertices does the triangle have? Using LASI's measurement tools, measure the length of the triangle's sides.

1.10 Circles can be drawn using a polygon (path with zero width) and the **Arc** command. Show how to draw a circle using LASI. Make sure the steps are clear.

1.11 What does the **Clone** system command do? Give an example of how to use this command.

1.12 Discuss the different ways cells can be drawn as outlines.

1.13 What do the keyboard buttons **w**, **u**, **a**, **z**, and **space** do in LASI? Give examples to illustrate your understanding of each key's function.

1.14 Describe how to add text in LASI and how to set the text size and layer.

1.15 Explain, in your own words and with examples, how to edit a cell in place.

1.16 Simulate the operation of the circuit in Fig. 1.28. What happens if you use a pulse frequency that's too high? What happens if your simulation time is too short? Give examples to illustrate your understanding.

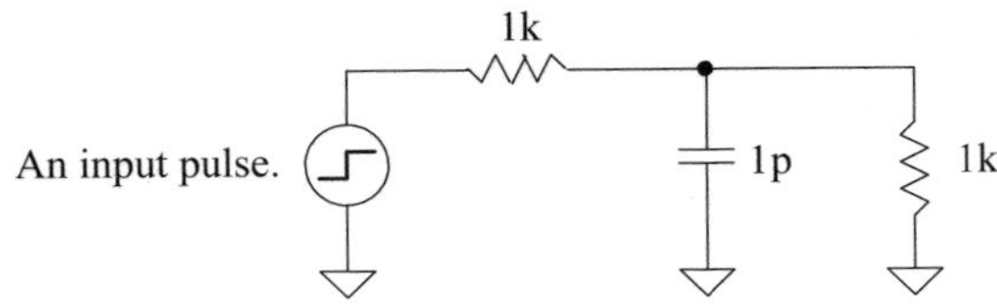

Figure 1.28 Circuit used in Problem 1.16

Chapter

2

The Well

To develop a fundamental understanding of CMOS integrated circuit layout and design, we begin with a study of the well. The well is the first layer fabricated when making a CMOS IC. The approach of studying the details of each fabrication (layout) layer will build a solid foundation for understanding the performance limitations and parasitics (the pn junctions, capacitances, and resistances inherent in a CMOS circuit) of the CMOS process.

The Substrate (The Unprocessed Wafer)

CMOS circuits are fabricated on and in a silicon wafer, as discussed in Ch. 1. This wafer is doped with donor atoms, such as phosphorus for an n-type wafer, or acceptor atoms, such as boron for a p-type wafer. Our discussion centers around a p-type wafer (the most common substrate used in CMOS IC processing). When designing CMOS integrated circuits with a p-type wafer, n-channel MOSFETs (NMOS for short) are fabricated directly in the p-type wafer, while p-channel transistors, PMOS, are fabricated in an "n-well." The substrate or well are sometimes referred to as the bulk or body of a MOSFET. CMOS processes that fabricate MOSFETs in the bulk are known as "bulk CMOS processes." The well and the substrate are illustrated in Fig. 2.1, though not to scale.

Often an epitaxial layer is grown on the wafer. In this book we will not make a distinction between this layer and the substrate. Some processes use a p-well or both n- and p-wells (sometimes called twin tub processes). A process that uses a p-type (n-type) substrate with an n-well (p-well) is called an "n-well process" ("p-well process"). *We will assume, throughout this book, that an n-well process is used for the layout and design discussions.*

A Parasitic Diode

Notice, in Fig. 2.1, that the n-well and the p-substrate form a diode. In CMOS circuits, the substrate is usually tied to the lowest voltage in the circuit (generally, the substrate is grounded) to keep this diode from forward biasing. Ideally, zero current flows in the substrate. We won't concern ourselves with how the substrate is connected to ground at this point (see Ch. 4).

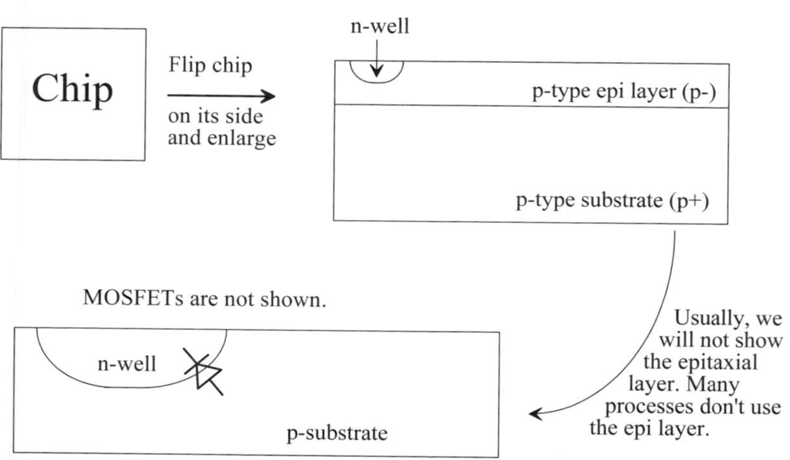

Figure 2.1 The top (layout) and side (cross-sectional) view of a die.

Using the N-well as a Resistor

In addition to being used as the body for p-channel transistors, the n-well can be used as a resistor, Fig. 2.2. The voltage on either side of the resistor must be large enough to keep the substrate/well diode from forward biasing.

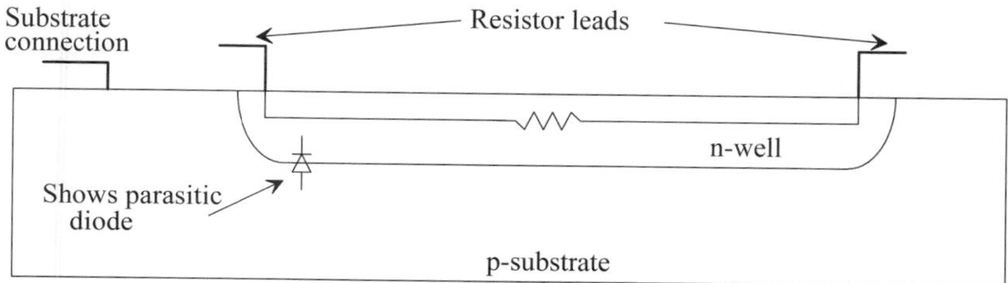

Figure 2.2 The n-well can be used as a resistor.

2.1 Patterning

CMOS integrated circuits are formed by patterning different layers on and in the silicon wafer. Consider the following sequence of events that apply, in a fundamental way, to any layer that we need to pattern. We start out with a clean, bare wafer, as shown in Fig. 2.3a. The distance given by the line A to B will be used as a reference in Figs. 2.3b–j. Figures 2.3b–j are cross-sectional views of the dashed line shown in (a). The small box in Fig. 2.3a is drawn with a layout program (and used for mask generation) to indicate where to put the patterned layer.

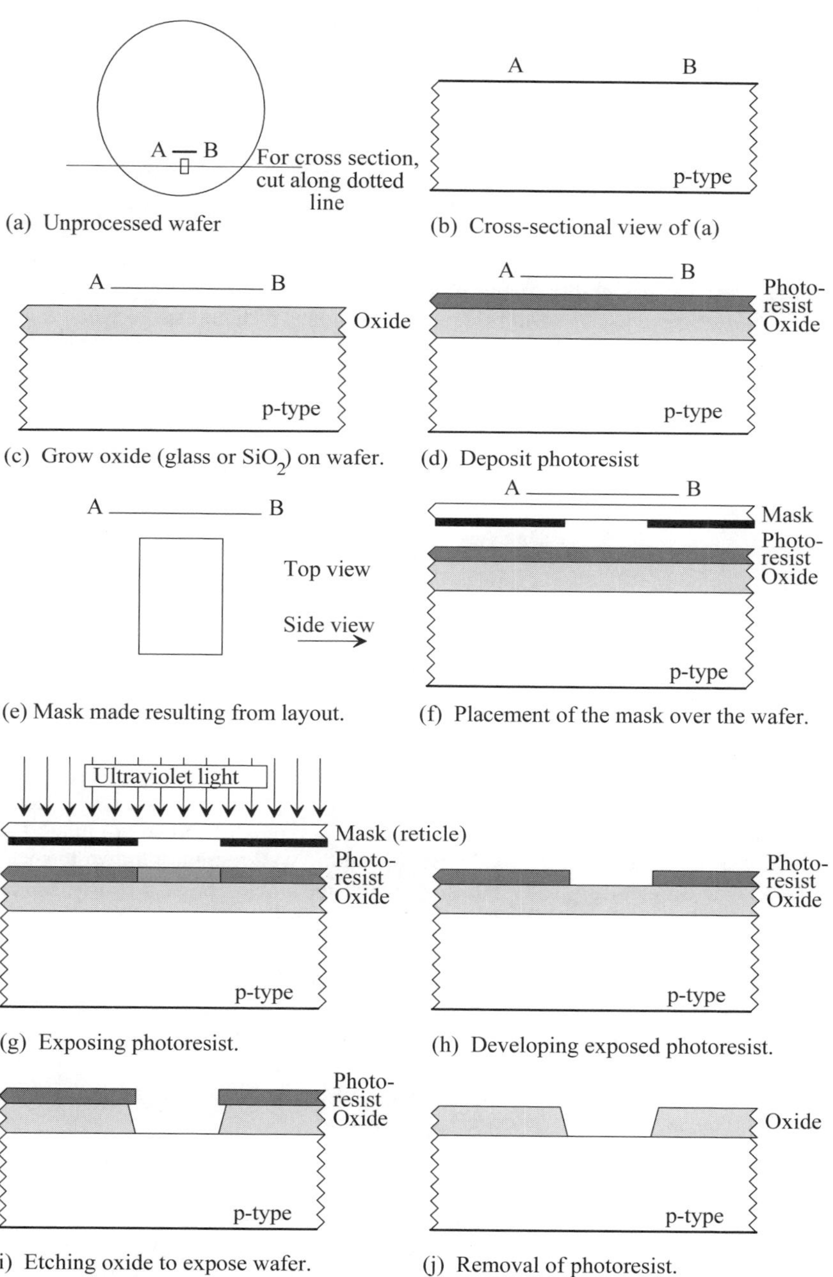

Figure 2.3 Generic sequence of events used in photo patterning.

The first step in our generic patterning discussion is to grow an oxide, SiO_2 or glass, a very good insulator, on the wafer. Simply exposing the wafer to air yields the reaction $Si + O_2 \rightarrow SiO_2$. However, semiconductor processes must have tightly controlled conditions to precisely set the thickness and purity of the oxide. We can grow the oxide using a reaction with steam, H_2O, or with O_2 alone. The oxide resulting from the reaction with steam is called a wet oxide, while the reaction with O_2 is a dry oxide. Both oxides are called thermal oxides due to the increased temperature used during oxide growth. The growth rate increases with temperature. The main benefit of the wet oxide is fast growing time. The main drawback of the wet oxide is the hydrogen byproduct. In general terms, the oxide grown using the wet techniques is not as pure as the dry oxide. The dry oxide generally takes a considerably longer time to grow. Both methods of growing oxide are found in CMOS processes. An important observation we should make when looking at Fig. 2.3c is that the oxide growth actually consumes silicon. This is illustrated in Fig. 2.4. The overall thickness of the oxide is related to thickness of the consumed silicon by

$$x_{Si} = 0.45 \cdot x_{ox} \tag{2.1}$$

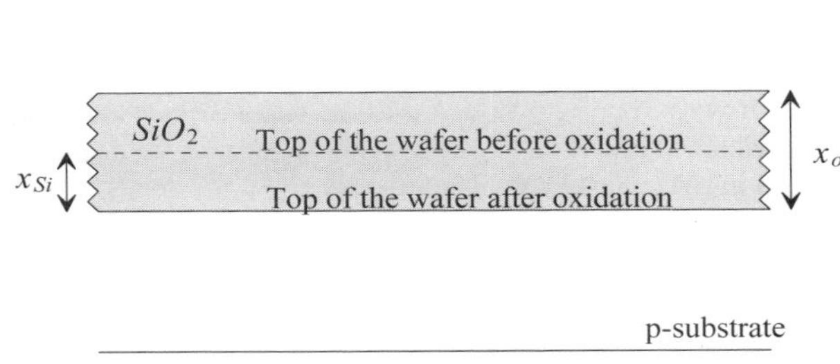

Figure 2.4 How growing oxide consumes silicon.

The next step of the generic CMOS patterning process is to deposit a photosensitive resist layer across the wafer (see Fig. 2.3d). Keep in mind that the dimensions of the layers, that is, oxide, resist, and the wafer, are not drawn to scale. The thickness of a wafer is typically 500 μm, while the thickness of a grown oxide or a deposited resist may be only a μm (10^{-6} m) or even less. After the resist is baked, the mask derived from the layout program, Figs. 2.3e and f, is used to selectively illuminate areas of the wafer, Fig. 2.3g. In practice, a single mask called a reticle, with openings several times larger than the final illuminated area on the wafer, is used to project the pattern and is stepped across the wafer with a machine called a stepper to generate the patterns needed to create multiple copies of a single chip. The light passing through the opening in the reticle is photographically reduced to illuminate the correct size area on the wafer.

The photoresist is developed (Fig. 2.3h), removing the areas that were illuminated. This process is called a positive resist process because the area that was illuminated was removed. A negative resist process removes the areas of resist that were not exposed to the light. Using both types of resist allows the process designer to cut down on the number of masks needed to define a CMOS process. Because creating the masks is expensive, lowering the number of masks is equated with lowering the cost of a process. This is also important in large manufacturing plants where fewer steps equal lower cost.

The next step in the patterning process is to remove the exposed oxide areas (Fig. 2.3i). Notice that the etchant etches under the resist, causing the opening in the oxide to be larger than what was specified by the mask. Some manufacturers intentionally bloat (make larger) or shrink (make smaller) the masks as specified by the layout program. Figure 2.3j shows the cross-sectional view of the opening after the resist has been removed.

2.1.1 Patterning the N-well

At this point we can make an n-well by diffusing donor atoms, those with five valence electrons, as compared to the 4 four found in silicon, into the wafer. Referring to our generic patterning discussion given in Fig. 2.3, we begin by depositing a layer of resist directly on the wafer, Fig. 2.3d (without oxide). This is followed by exposing the resist to light through a mask (Figs. 2.3f and g) and developing or removing the resist (Fig. 2.3h). The mask used is generated with a layout program. The next step in fabricating the n-well is to expose the wafer to donor atoms. The resist blocks the diffusion of the atoms, while the openings allow the donor atoms to penetrate into the wafer. This is shown in Fig. 2.5a. After a certain amount of time, depending on the depth of the n-well desired, the diffusion source is removed (Fig. 2.5b). Notice that the n-well "outdiffuses" under the resist; that is, the final n-well size is not the same as the mask size. Again, the foundry where the chips are fabricated may bloat or shrink the mask to compensate for this lateral diffusion. The final step in making the n-well is the removal of the resist (Fig. 2.5c).

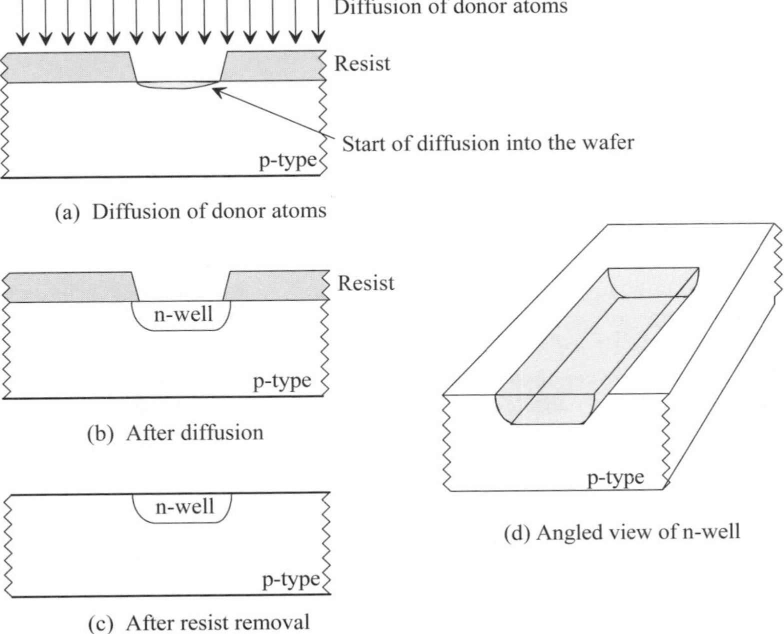

Figure 2.5 Formation of the n-well.

2.2 Laying Out the N-well

When we lay out the n-well, we are viewing the chip from the top. One of the key points in this discussion, as well as the discussions to follow, is that we do layout to a generic **scale factor**. If, for example, the minimum device dimensions are 50 nm (= 0.05 μm = 50×10^{-9} m), then an n-well box **drawn** 10 by 10, see Fig. 2.6, has an actual size after it is fabricated of $10 \cdot 50\, nm$ or half a micron (0.5 μm = 500 nm), neglecting lateral diffusion or other process imperfections. We scale the layout when we generate the GDS (calma stream format) or CIF (Caltech-intermediate-format) file from a layout program. (A GDS or CIF file is what the mask maker uses to make reticles.) Using integers to do layout simplifies things. We'll see in a moment that many electrical parameters are ratios (such as resistance) and so the scale factor cancels out of the ratio.

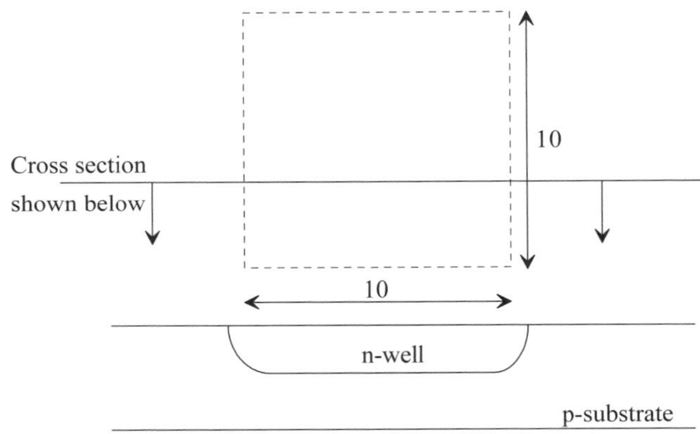

Figure 2.6 Layout and cross-sectional view of a 10 by 10 (drawn) n-well.

2.2.1 Design Rules for the N-well

Now that we've laid out the n-well (drawn a box in a layout program), we might ask the question, "Are there any limitations or constraints on the size and spacing of the n-wells?" That is to say, "Can we make the n-well 2 by 2?" Can we make the distance between the n-wells 1? As we might expect, there are minimum spacing and size requirements for all layers in a CMOS process. Process engineers, who design the integrated circuit process, specify the design rules. The design rules vary from one process technology (say a process with a scale factor of 1 μm) to another (say a process with a scale factor of 50 nm).

Figure 2.7 shows sample design rules for the n-well. The minimum size (width or length) of any n-well is 6, while the minimum spacing between different n-wells is 9. As the layout becomes complicated, the need for a program that ensures that the design rules are not violated is needed. This program is called a *design rule checker* program (DRC program). Note that the minimum size may be set by the quality of patterning the resist (as seen in Fig. 2.5), while the spacing is set by the parasitic npn transistor seen in Fig. 2.7. (We don't want the n-wells interacting to prevent the parasitic npn from turning on.)

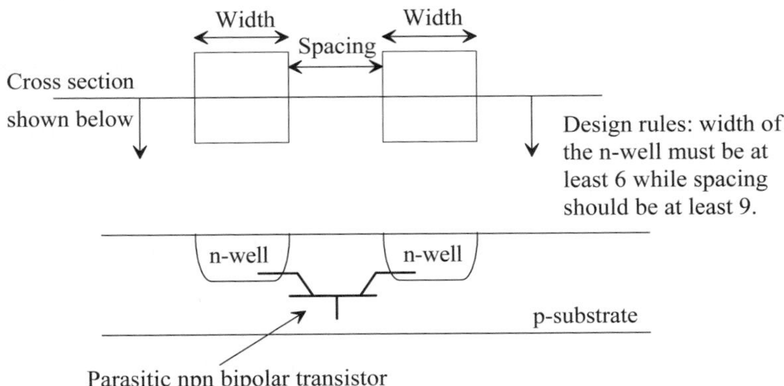

Figure 2.7 Sample design rules for the n-well.

2.3 Resistance Calculation

In addition to serving as a region in which to build PMOS transistors (called the body or bulk of the PMOS devices), n-wells are often used to create resistors. The resistance of a material is a function of the material's resistivity, ρ, and the material's dimensions. For example, the slab of material in Fig. 2.8 between the two leads has a resistance given by

$$R = \frac{\rho}{t} \cdot \frac{L \cdot scale}{W \cdot scale} = \frac{\rho}{t} \cdot \frac{L}{W} \tag{2.2}$$

In semiconductor processing, all of the fabricated thicknesses, t, seen in a cross-sectional view, such as the n-well's, are fixed in depth (**this is important**). When doing layout, we only have control over W (width) and L (length) of the material. The W and L are what we see from the top view, that is, the layout view. We can rewrite Eq. (2.2) as

$$R = R_{square} \cdot \frac{L}{W} \rightarrow R_{square} = \frac{\rho}{t} \tag{2.3}$$

R_{square} is the *sheet resistance* of the material in Ω/square (noting that when $L = W$ the layout is square and $R = R_{square}$).

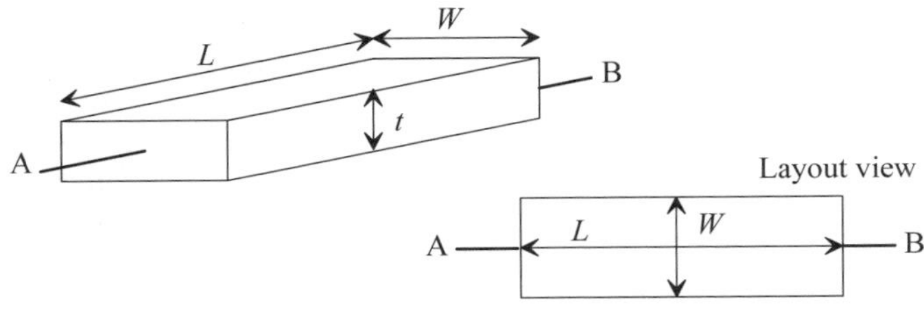

Figure 2.8 Calculation of the resistance of a rectangular block of material.

Example 2.1

Calculate the resistance of an n-well that is 10 wide and 100 long. Assume that the n-well's sheet resistance is typically 2 kΩ/square; however, it can vary with process shifts from 1.6 to 2.4 kΩ/square.

The typical resistance, using Eq. (2.3), between the ends of the n-well is

$$R = 2{,}000 \cdot \frac{100}{10} = 20\ k\Omega$$

The maximum value of the resistor is 24k, while the minimum value is 16k. ■

Layout of Corners

Often, to minimize space, resistors are laid out in a serpentine pattern. The corners, that is, where the layer bends, are not rectangular. This is shown in Fig. 2.9a. All sections in Fig 2.9a are square, so the resistance of sections 1 and 3 is R_{square}. The equivalent resistance of section 2 between the adjacent sides, however, is approximately 0.6 R_{square}. The overall resistance between points A and B is therefore $2.6 \cdot R_{square}$. As seen in Ex. 2.1 the actual resistance value varies with process shifts. The layout shown in Fig. 2.9b uses wires to connect separate sections of unit resistors to avoid corners. Avoiding corners in a resistor is the (generally) preferred method of layout in analog circuit design where the ratio of two resistors is important. For example, the gain of an op-amp circuit may be R_F/R_I.

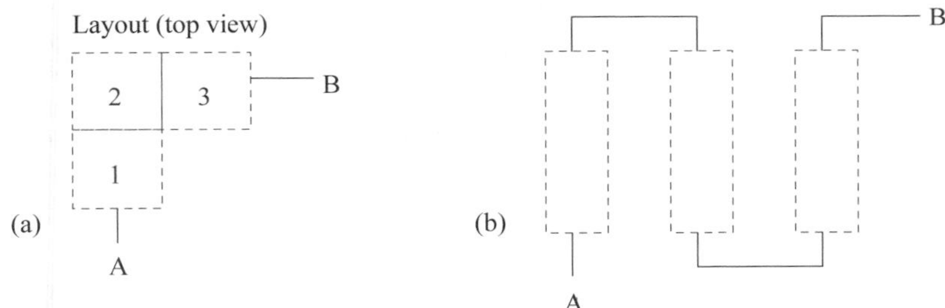

Figure 2.9 (a) Calculating the resistance of a corner section and (b) layout to avoid corners.

2.3.1 The N-well Resistor

At this point, it is appropriate to show the actual cross-sectional view of the n-well after all processing steps are completed (Fig. 2.10). The n+ and p+ implants are used to increase the threshold voltage of the field devices; more will be said on this in Ch. 7. In all practical situations, the sheet resistance of the n-well is measured with the field implant in place, that is, with the n+ implant between the two metal connections in Fig. 2.10. Not shown in Fig. 2.10 is the connection to substrate. The field oxide (FOX; also known as ROX or recessed oxide) are discussed in Chs. 4 and 7 when we discuss the active and poly layers. The reader shouldn't, at this point, feel they should understand any of the cross-sectional layers in Fig. 2.10 except the n-well. Note that the field implants *aren't drawn* in the layout and so their existence is transparent to the designer.

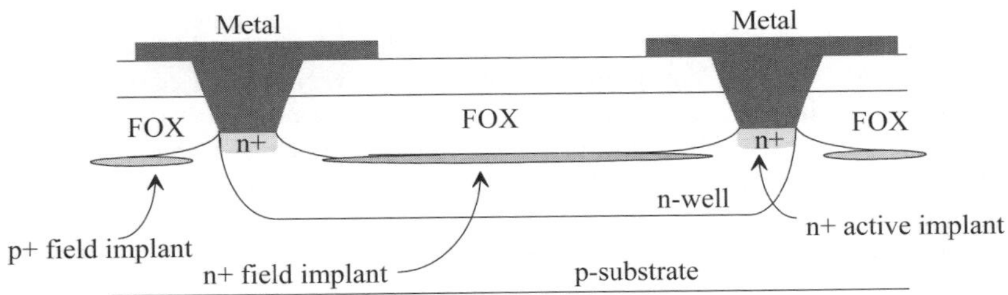

Figure 2.10 Cross-sectional view of n-well showing field implant. The field implantation is sometimes called the "channel stop implant."

2.4 The N-well/Substrate Diode

As seen in Fig. 2.1, placing an n-well in the p-substrate forms a diode. It is important to understand how to model a diode for hand calculations and in SPICE simulations. In particular, let's discuss general diodes using the n-well/substrate pn junction as an example. The DC characteristics of the diode are given by the Shockley diode equation, or

$$I_D = I_S\left(e^{\frac{Vd}{nV_T}} - 1\right) \tag{2.4}$$

The current I_D is the diode current; I_S is the scale (saturation) current; V_d is the voltage across the diode where the anode, A, (p-type material) is assumed positive with respect to the cathode, K, (n-type); and V_T is the *thermal voltage*, which is given by $\frac{kT}{q}$ where k = Boltzmann's constant (1.3806×10^{-23} joules per degree Kelvin), T is temperature in Kelvin, n is the emission coefficient (a term that is related to the doping profile and affects both the exponential behavior of the diode and the diode's turn-on voltage), and q is the electron charge of 1.6022×10^{-19} coulombs. The scale current and thus the overall diode current are related in SPICE by an area factor (not associated with or to be confused with the scale term we use in layouts, Eq. (2.2)). The SPICE (Simulation Program with Integrated Circuit Emphasis) circuit simulation program assumes that the value of I_s supplied in the model statement was measured for a device with a reference area of 1. If an area factor of 2 is supplied for a diode, then I_S is doubled in Eq. (2.4).

2.4.1 A Brief Introduction to PN Junction Physics

A conducting material is made up of atoms that have easily shared orbiting electrons. As a simple example, copper is a better conductor than aluminum because the copper atom's electrons aren't as tightly coupled to its nucleus allowing its electrons to move around more easily. An insulator has, for example, eight valence electrons tightly coupled to the atom's nucleus. A significant electric field is required to break these electrons away from their nucleus (and thus for current conduction). A semiconductor, like silicon, has four valence electrons. Silicon's conductivity falls between an insulator and a conductor (and thus the name "semiconductor"). As silicon atoms are brought together, they form both a periodic crystal structure and bands of energy that restrict the allowable energies an electron can occupy. At absolute zero temperature, ($T = 0$ K), all of the valence electrons in the semiconductor crystal reside in the valence energy band, E_v. As temperature increases, the electrons gain energy (heat is absorbed by the silicon crystal), which causes some of the valence electrons to break free and move to a conducting energy level, E_c. Figure 2.11 shows the movement of an electron from the valence band to the conduction band. Note that there aren't any allowable energies between E_v and E_c in the silicon crystal structure (if the atom were by itself, that is, not in a crystal structure this exact limitation isn't present). Further note that when the **electron** moves from the valence energy band to the conduction energy band, a **hole** is left in the valence band. Having an electron in the conduction band increases the material's conductivity (the electron can move around easily in the semiconductor material because it's not tightly coupled to an atom's nucleus). At the same time a hole in the valence band increases the material's conductivity (electrons in the valence band can move around more easily by simply falling into the open hole). The *key point* is that increasing the number of electrons or holes increases the materials conductivity. Since the hole is more tightly coupled to the atom's nucleus (actually the electrons in the valence band), its mobility (ability to move around) is lower than the electron's mobility in the conduction band. This point is **fundamentally important**. The fact that the mobility of a hole is lower than the mobility of an electron (in silicon) results in, among other things, the size of PMOS devices being larger than the size of NMOS devices (when designing circuits) in order for each device to have the same drive strength.

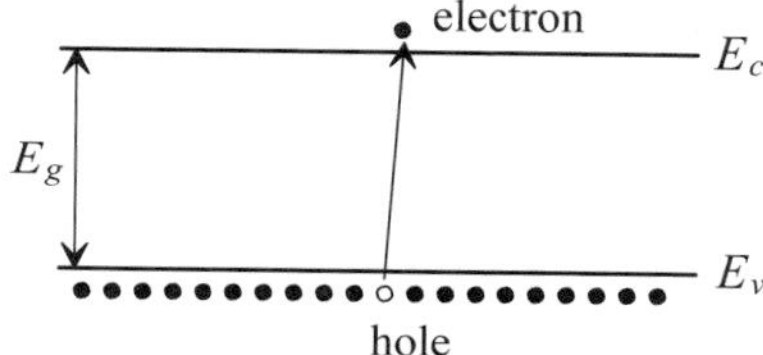

Figure 2.11 An electron moving to the conduction band, leaving behind a hole in the valence band.

Carrier Concentrations

Pure silicon is often called *intrinsic* silicon. As the temperature of the silicon crystal is increased it absorbs heat. Some of the electrons in the valence band gain enough energy to jump the bandgap energy of silicon, E_g (see also Eq. (23.21)), as seen in Fig. 2.11. This movement of an electron from the valence band to the conduction band is called *generation*. When the electron loses energy and falls back into the valence band, it is called *recombination*. The time the electron spends in the conduction band, before it recombines (drops back to the valence energy band), is random and often characterized by the carrier lifetime, τ_T (a root-mean-square, RMS, value of the random times the electrons spend in the conduction band of the silicon crystal). While discussing the actual processes involved with generation-recombination (GR) is outside the scope of this book, the carrier lifetime is a practical important parameter for circuit design. Another important parameter is the number of electrons in the conduction band (and thus the number of holes in the valence band) at a given time (again a random number). These carriers are called intrinsic carriers, n_i. At room temperature

$$n_i \approx 14.5 \times 10^9 \; carriers/cm^3 \qquad (2.5)$$

noting that cm^3 indicates a volume. If we call the number of free electrons (meaning electrons excited up in the conduction band of silicon) n and the number of holes p, then for *intrinsic silicon*,

$$n = p = n_i \approx 14.5 \times 10^9 \; carriers/cm^3 \tag{2.6}$$

This may seem like a lot of carriers. However, the number of silicon atoms, N_{Si}, in a given volume of crystalline silicon is

$$N_{Si} = 50 \times 10^{21} \; atoms/cm^3 \tag{2.7}$$

so there is only one excited electron/hole pair for (roughly) every 10^{12} silicon atoms.

Next let's add different materials to intrinsic silicon (called *doping* the silicon) to change silicon's electrical properties. If we add a small amount of a material containing atoms with five valence electrons like phosphorous (silicon has four), then the added atom would bond with the silicon atoms and the *donated* electron would be free to move around (and easily excited to the conduction band). If we call the density of this added donor material N_D with units of atoms/cm^3 and we assume the number of atoms added to the silicon is much larger than the intrinsic carrier concentration, then we can write the number of free electrons (the electron concentration n) in the material as

$$n \approx N_D \text{ when } N_{Si} >> N_D >> n_i \tag{2.8}$$

A material with added donor atoms is said to be an "n-type" material. Similarly, if we were to add a small material to silicon with atoms having three valence electrons (like boron), the added material would bond with the silicon resulting in a hole in the valence band. Again, this increases the conductivity of silicon because, now, the electrons in the valence band can move into the hole (having the effect of making it look like the hole is moving). The added material in this situation is said to be an *acceptor* material. The added material accepts an electron from the silicon crystal. If the density of the added acceptor material is labeled N_A, then the hole concentration, p, in the material is

$$p \approx N_A \text{ when } N_{Si} >> N_A >> n_i \tag{2.9}$$

A material with added acceptor atoms is said to be a "p-type" material.

If we dope a material with donor atoms, the number of free electrons in the material, n, goes up, as indicated by Eq. (2.8). We would expect, then, the number of free holes in the material to go down (some of those free electrons fall easily into the available holes reducing the number of holes in the material). The relationship between the number of holes, electrons, and intrinsic carrier concentration, is governed by the *mass-action law*

$$pn = n_i^2 \tag{2.10}$$

Consider the following example.

Example 2.2
Suppose silicon is doped with phosphorous having a density, N_D, of 10^{18} atoms/cm^3. Estimate the doped silicon's hole and electron concentration.

The electron concentration, from Eq. (2.8) is, $n = 10^{18}$ electrons/cm^3 (one electron for each donor atom). The hole concentration is found using the mass-action law as

$$p = \frac{n_i^2}{n} = \frac{(14.5 \times 10^9)^2}{10^{18}} = 210 \text{ holes/cm}^3$$

Basically, all of the holes are filled. Note that with a doping density of 10^{18} there is one dopant atom for every 50,000 silicon atoms. If we continue to increase the doping concentration, our assumption that $N_{Si} >> N_D$ isn't valid and the material is said to be *degenerate* (no longer mainly silicon). A degenerate semiconductor doesn't follow the mass-action law (or any of the equations for silicon we present). ■

Fermi Energy Level

To describe the carrier concentration in a semiconductor, the Fermi energy level is often used. The Fermi energy level is useful when determing the contact potentials in materials. For example, the potential that you have to apply across a diode before it turns on is set by the p-type and n-type material contact potential difference. Also, the threshold voltage is determined, in part, by contact potentials.

The Fermi energy level simply indicates the energy level where the probability of occupation by a free electron is 50%. Figure 2.12a shows that for intrinsic silicon the (intrinsic) Fermi level, E_i is close to the middle of the bandgap. In p-type silicon, the Fermi level, E_f, moves towards the valence band, Fig. 2.12b, since the number of free electrons, n, is reduced with the abundance of holes. Figure 2.12c shows the location of the Fermi energy level in n-type silicon. The Fermi level moves towards E_c with the abundance of electrons in the conduction band.

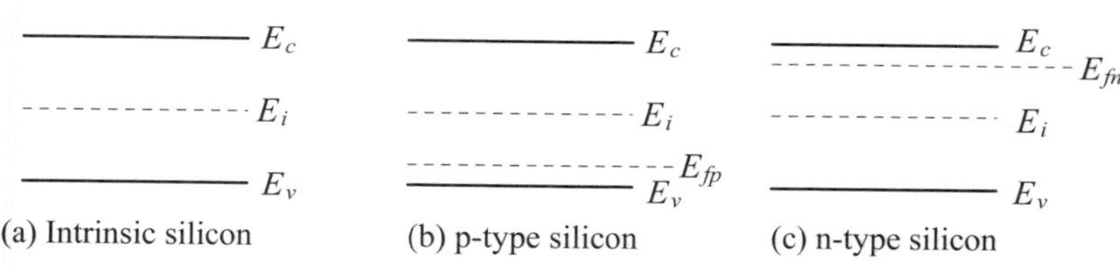

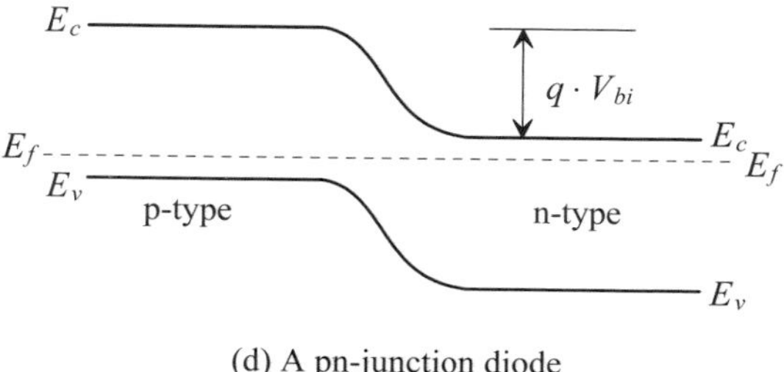

(d) A pn-junction diode

Figure 2.12 The Fermi energy levels in various structures.

The energy difference between the E_i and E_f is given, for a p-type semiconductor, by

$$E_i - E_{fp} = kT \cdot \ln \frac{N_A}{n_i} \tag{2.11}$$

and for an n-type semiconductor by

$$E_{fn} - E_i = kT \cdot \ln \frac{N_D}{n_i} \tag{2.12}$$

The band diagram of a pn junction (a diode) is seen in Fig. 2.12d. Note how the Fermi energy level is constant throughout the diode. A variation in E_f would indicate a nonequilibrium situation (the diode has an external voltage applied across it). To get current to flow in a diode, we must apply an external potential that approaches the diode's contact potential (its built-in potential, V_{bi}). By applying a potential to forward bias the diode, the conduction energy levels in each side of the diode move closer to the same level. The voltage applied to the diode when the conduction energy levels are exactly at the same level is given by

$$V_{bi} = \frac{E_{fn} - E_{fp}}{q} = \frac{kT}{q} \cdot \ln \frac{N_A N_D}{n_i^2} \tag{2.13}$$

noting that $kT/q = V_T$ is the thermal voltage.

2.4.2 Depletion Layer Capacitance

We know that n-type silicon has a number of mobile electrons, while p-type silicon has a number of mobile holes (a vacancy of electrons in the valence band). Formation of a pn junction results in a depleted region at the p-n interface (Fig. 2.13). A depletion region is an area depleted of mobile holes or electrons. The mobile electrons move across the junction, leaving behind fixed donor atoms and thus a positive charge. The movement of holes across the junction, to the right in Fig. 2.13, occurs for the p-type semiconductor as well with a resulting negative charge. The fixed atoms on each side of the junction within the depleted region exert a force on the electrons or holes that have crossed the junction. This equalizes the charge distribution in the diode, preventing further charges from crossing the diode junction and also gives rise to a parasitic capacitance. This parasitic capacitance is called a *depletion* or *junction* capacitance.

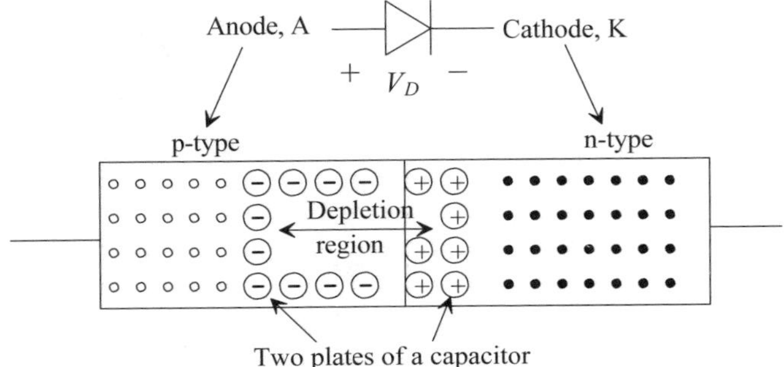

Figure 2.13 Depletion region formation in a pn junction.

The depletion capacitance, C_j, of a pn junction is modeled using

$$C_j = \frac{C_{j0}}{\left[1 - \left(\frac{V_D}{V_{bi}}\right)\right]^m} \tag{2.14}$$

C_{j0} is the zero-bias capacitance of the pn junction, that is, the capacitance when the voltage across the diode is zero. V_D is the voltage across the diode, m is the grading coefficient (showing how the silicon changes from n- to p-type), and V_{bi} is the built-in potential given by Eq. (2.13).

Example 2.3

Schematically sketch the depletion capacitance of an n-well/p-substrate diode 100 × 100 square (with a scale factor of 1 μm), given that the substrate doping is 10^{16} atoms/cm^3 and the well doping is 10^{17} atoms/cm^3. The measured zero-bias depletion capacitance of the junction is 100 aF/μm^2 ($= 100 \times 10^{-18}\ F/\mu m^2$), and the grading coefficient is 0.333. Assume the depth of the n-well is 3 μm.

The n-well doping (n-type side of the diode) is $N_D = 10^{17}$, while the substrate doping is N_A is 10^{16}. We can calculate the built-in potential using Eq. (2.13)

$$V_{bi} = 26\,mV \cdot \ln \frac{10^{16} \cdot 10^{17}}{(14.5 \times 10^9)^2} = 759\ mV$$

The depletion capacitance is made up of a *bottom* component and a *sidewall* component, as shown in Fig. 2.14 (see Eq. (5.17) for the more general form of Eq. (2.14)).

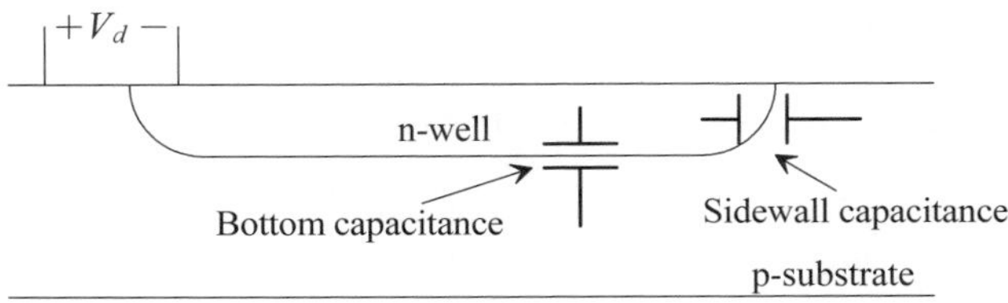

Figure 2.14 A pn junction on the bottom and sides of the junction.

The bottom zero-bias depletion capacitance, C_{j0b}, is given by

$$C_{j0b} = (capacitance/area) \cdot (scale)^2 \cdot (bottom\ area) \tag{2.15}$$

which, for this example, is

$$C_{j0b} = (100\ aF/\mu m^2) \cdot (1\ \mu m)^2 \cdot (100)^2 = 1\ pF$$

The sidewall zero-bias depletion capacitance, C_{j0s}, is given by

$$C_{j0s} = (capacitance/area) \cdot (depth\ of\ the\ well) \cdot (perimeter\ of\ the\ well) \cdot (scale)^2$$

or (2.16)

$$C_{j0s} = (100\ aF/\mu m^2) \cdot (3) \cdot (400) \cdot (1\ \mu m)^2 = 120\ fF$$

The total diode depletion capacitance between the n-well and the p-substrate is the parallel combination of the bottom and sidewall capacitances, or

$$C_j = \frac{C_{j0b}}{\left[1 - \left(\frac{V_D}{V_{bi}}\right)\right]^m} + \frac{C_{j0s}}{\left[1 - \left(\frac{V_D}{V_{bi}}\right)\right]^m} = \frac{C_{j0b} + C_{j0s}}{\left[1 - \left(\frac{V_D}{V_{bi}}\right)\right]^m} \tag{2.17}$$

Substituting in the numbers, we get

$$C_j = \frac{1\,pF + 0.120\,pF}{\left(1 - \left(\frac{V_D}{0.759}\right)\right)^{0.33}} = \frac{1.120\,pF}{\left(1 - \left(\frac{V_D}{0.759}\right)\right)^{0.33}}$$

A sketch of how this capacitance changes with reverse potential is given in Fig. 2.15. Notice that when we discuss the depletion capacitance of a diode, it is usually with regard to a reverse bias (V_D is negative). When the diode becomes forward-biased minority carriers, electrons in the p material and holes in the n material, injected across the junction, form a *stored* or *diffusion* charge in and around the junction and give rise to a storage or diffusion capacitance. This capacitance is usually much larger than the depletion capacitance. Furthermore, the time it takes to remove this stored charge can be significant. ■

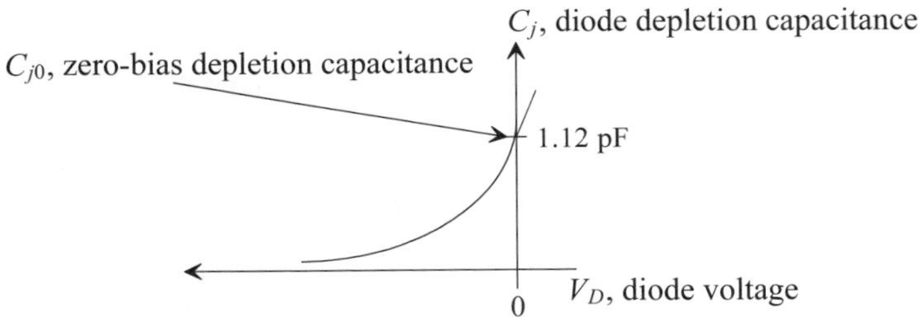

Figure 2.15 Diode depletion capacitance against diode reverse voltage.

2.4.3 Storage or Diffusion Capacitance

Consider the charge distribution of the forward-biased diode shown in Fig. 2.16. When the diode becomes forward biased, electrons from the n-type side of the junction are attracted to the p-type side (and vice versa for the holes). After an electron drifts across the junction, it starts to diffuse toward the metal contact. If the electron recombines, that is, falls into a hole, before it hits the metal contact, the diode is called a *long base diode.* The time it takes an electron to diffuse from the junction to the point where it recombines is called the carrier lifetime, τ_T (see Sec. 2.4.1). For silicon this lifetime is on the order of 10 μs. If the distance between the junction and the metal contact is short, such that the electrons make it to the metal contact before recombining, the diode is said to be a *short base diode.* In either case, the time between crossing the junction and recombining will be labeled τ_T (transit time). A capacitance is formed between the electrons diffusing into the p-side and the holes diffusing into the n-side, that is, formed between the minority carriers. (Electrons are the minority carriers in the p-type semiconductor.) This capacitance is called a **diffusion** capacitance or **storage** capacitance due to the presence of the stored, or diffusing, minority carriers around the forward-biased pn junction.

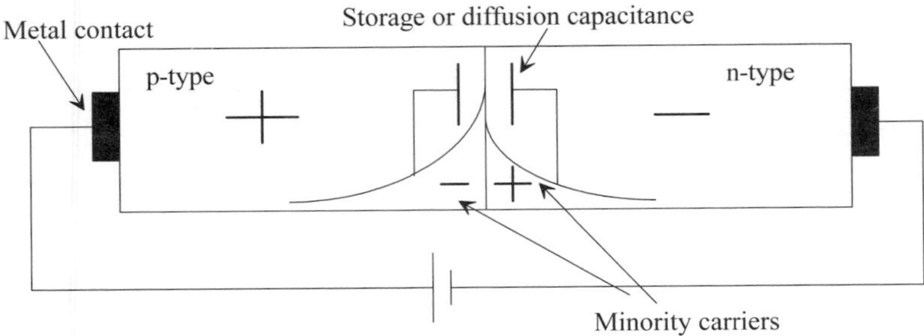

Figure 2.16 Charge distribution in a forward-biased diode.

We can characterize the storage capacitance, C_S, in terms of the minority carrier lifetime. Under DC operating conditions, the storage capacitance is given by

$$C_S = \frac{I_D}{nV_T} \cdot \tau_T \tag{2.18}$$

I_D is the DC current flowing through the forward-biased junction given by Eq. (2.4). Looking at the diode capacitance in this way is very useful for analog AC small-signal analysis. However, for digital applications, we are more interested in the large-signal switching behavior of the diode. It should be pointed out that, in general, for a CMOS process, it is undesirable to have a forward-biased pn junction. If we do have a forward-biased junction, it usually means that there is a problem. For example, electrostatic protection diodes are turning on (Fig. 4.17), capacitive feedthrough is possibly causing latch-up, or such. These topics appear in more detail later in the book.

Consider Fig. 2.17. In the following diode switching analysis, we assume that $V_F >> 0.7$, $V_R < 0$ and that the voltage source has been at V_F long enough to reach steady-state condition; that is, the minority carriers have diffused out to an equilibrium condition. At the time t_1, the input voltage source makes an abrupt transition from a forward voltage of V_F to a reverse voltage of V_R, causing the current to change from $\frac{V_F - 0.7}{R}$ to $\frac{V_R - 0.7}{R}$. The diode voltage remains at 0.7 V, because the diode contains a stored charge that must be removed. At time t_2, the stored charge is removed. At this point, the diode basically looks like a voltage-dependent capacitor that follows Eq. (2.14). In other

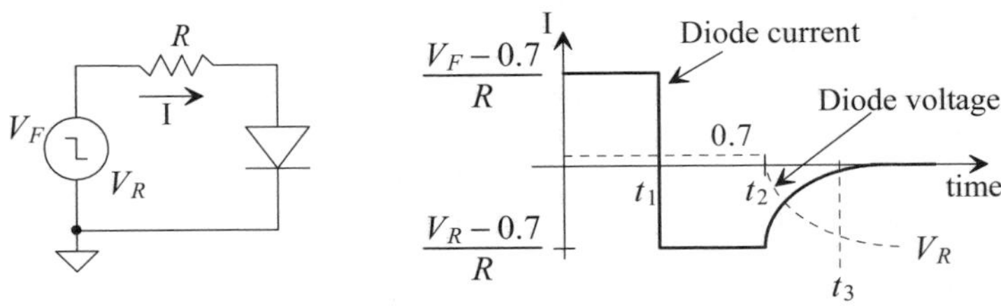

Figure 2.17 Diode reverse recovery test circuit.

words for $t > t_2$, the diode-depletion capacitance is charged through R until the current in the circuit goes to zero and the voltage across the diode is V_R. This accounts for the exponential decay of the current and voltage shown in Fig. 2.17.

The diode storage time, the time it takes to remove the stored charge, t_s, is simply the difference in t_2 and t_1, or

$$t_s = t_2 - t_1 \tag{2.19}$$

This time is also given by

$$t_s = \tau_T \cdot \ln \frac{i_F - i_R}{-i_R} \tag{2.20}$$

where $\frac{V_F - 0.7}{R} = i_F$ and $\frac{V_R - 0.7}{R} = i_R$ = a negative number in this discussion. Note that it is quite easy to determine the minority carrier lifetime using this test setup.

Defining a time t_3, where $t_3 > t_2$, when the current in the diode becomes 10% of $\frac{V_R - 0.7}{R}$, we can define the diode reverse recovery time, or

$$t_{rr} = t_3 - t_1 \tag{2.21}$$

Note that the reverse recovery time (the time it takes to shut off a forward-biased diode) is one of the big reasons that digital circuits made using silicon bipolar transistors don't perform as well, in general, as their CMOS counterparts.

Table 2.1 SPICE parameters related to diode.

Name	SPICE	
I_S	IS	Saturation current
R_S	RS	Series resistance
n	N	Emission coefficient
V_{bd}	BV	Breakdown voltage
I_{bd}	IBV	Current which flows during V_{bd}
C_{j0}	CJ0	Zero-bias pn junction capacitance
V_{bi}	VJ	Built-in potential
m	M	Grading coefficient
τ_T	TT	Carrier transit time

2.4.4 SPICE Modeling

The SPICE diode model parameters are listed in Table 2.1. The series resistance, R_s, results from the finite resistance of the semiconductor used in making the diode and the contact resistance, the resistance resulting from a metal contact to the semiconductor. At this point, we are only concerned with the resistance of the semiconductor. For a reverse-biased diode, the depletion layer width changes, increasing for larger reverse voltages (decreasing both the capacitance and series resistance, of the diode). However, when we model the series resistance, we use a constant value. In other words, SPICE will not show us the effects of a varying R_S.

Example 2.4
Using SPICE, explain what happens when a diode with a carrier lifetime of 10 ns is taken from the forward-biased region to the reverse-biased region. Use the circuit shown in Fig. 2.18 to illustrate your understanding.

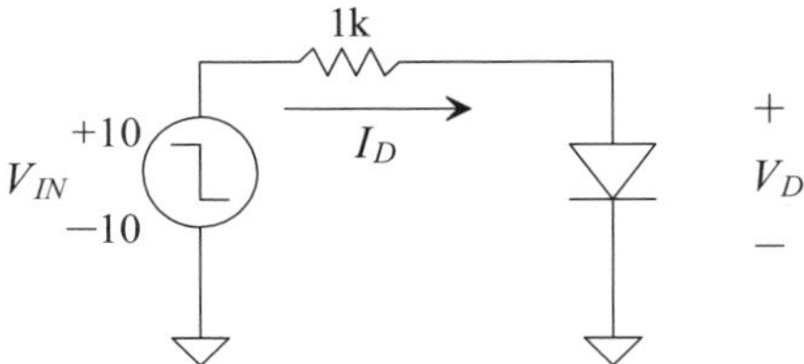

Figure 2.18 Circuit used in Ex. 2.4 to demonstrate simulation of a diode's reverse recovery time.

We assume a zero-bias depletion capacitance of 1 pF. The SPICE netlist used to simulate the circuit in Fig. 2.18 is shown below.

```
*** Figure 2.19 CMOS: Circuit Design, Layout, and Simulation ***

.control
destroy all
run
let id=-i(vin)*1k
plot vd vin id
.endc

D1      vd      0       Dtrr
R1      vin     vd      1k
Vin     vin     0       DC      0       pulse 10 -10 10n .1n .1n 20n 40n

.Model Dtrr D   is=1.0E-15 tt=10E-9 cj0=1E-12 vj=.7 m=0.33
.tran 100p 25n
.end
```

Figure 2.19 shows the current through the diode (I_D), the input voltage step (V_{IN}), and the voltage across the diode (V_D).

What's interesting to notice about this circuit is that current actually flows through the diode in the negative direction, even though the diode is forward biased (has a forward voltage drop of 0.7 V). During this time, the stored minority carrier charge (the diffusion charge) is removed from the junction. The storage time is estimated using Eq. (2.20)

$$t_s = 10ns \cdot \ln \frac{9.3 + 10.7}{10.7} = 6.25\ ns$$

which is close to the simulation results. Note that the input pulse doesn't change until 10 ns after the simulation starts. This ensures a steady-state condition when the input changes from 10 to –10 V. ■

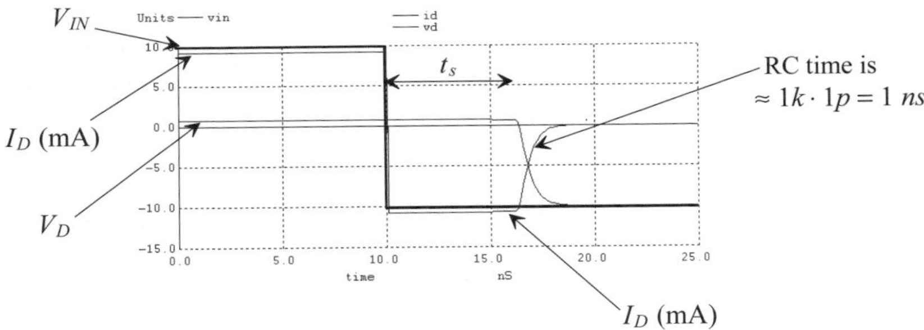

Figure 2.19 The simulation results for Ex. 2.4.

2.5 The RC Delay through an N-well

At this point, we know that the n-well can function as a resistor and as a diode when used with the substrate. Figure 2.20a shows the parasitic capacitance and resistance associated with the n-well. Since there is a depletion capacitance from the n-well to the substrate, we could sketch the equivalent symbol for the n-well resistor, as shown in Fig. 2.20b. This is the basic form of an RC transmission line. If we put a voltage pulse into one side of the n-well resistor, then a finite time later, called the delay time and measured at the 50% points of the pulses, the pulse will appear.

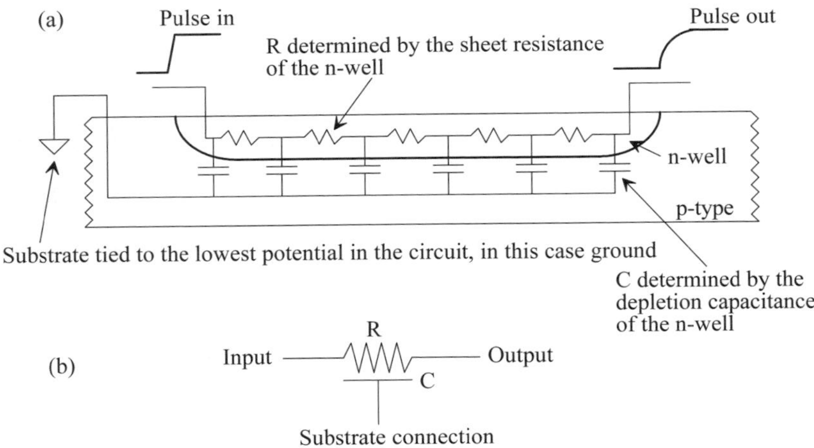

Figure 2.20 (a) Parasitic resistance and capacitance of the n-well and (b) schematic symbol.

RC Circuit Review

Figure 2.21 shows a simple RC circuit driven from a voltage pulse. If the input pulse transitions from 0 to V_{pulse} at a time which we'll call zero, then the voltage across the capacitor (the output voltage) is given by

$$V_{out}(t) = V_{pulse}(1 - e^{-t/RC}) \tag{2.22}$$

The time it takes the output of the RC circuit to reach 50% of V_{pulse} (defined as the circuit's delay time) is determined using

$$\frac{V_{pulse}}{2} = V_{pulse}(1 - e^{-t_d/RC}) \rightarrow t_d \approx 0.7RC \tag{2.23}$$

The rise time of the output pulse is defined as the time it takes the output to go from 10% of the final voltage to 90% of the final voltage (V_{pulse}). To determine the rise time in terms of the RC time constant, we can write

$$0.1V_{pulse} = V_{pulse}(1 - e^{-t_{10\%}/RC}) \tag{2.24}$$

and

$$0.9V_{pulse} = V_{pulse}(1 - e^{-t_{90\%}/RC}) \tag{2.25}$$

Solving these two equations for the rise time gives

$$t_r = t_{90\%} - t_{10\%} \approx 2.2RC \tag{2.26}$$

We will use these results *often* when designing digital circuits. **It's important** that any electrical engineer be able to derive Eqs. (2.23) and (2.26).

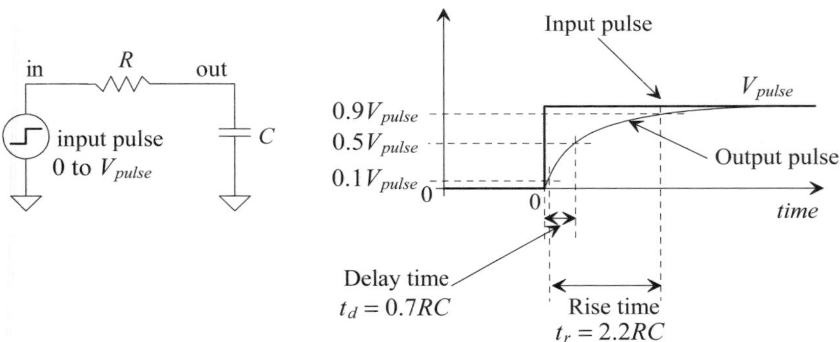

Figure 2.21 Rise and delay times in an RC circuit.

Distributed RC Delay

The n-well resistor seen in Fig. 2.20 is an example of a distributed RC circuit (not a single, RC like the one seen in Fig. 2.21). In order to estimate the delay through a distributed RC, consider the circuit seen in Fig. 2.22. The delay to node A is estimated using

$$t_{dA} = 0.7R_{square}C_{square} \tag{2.27}$$

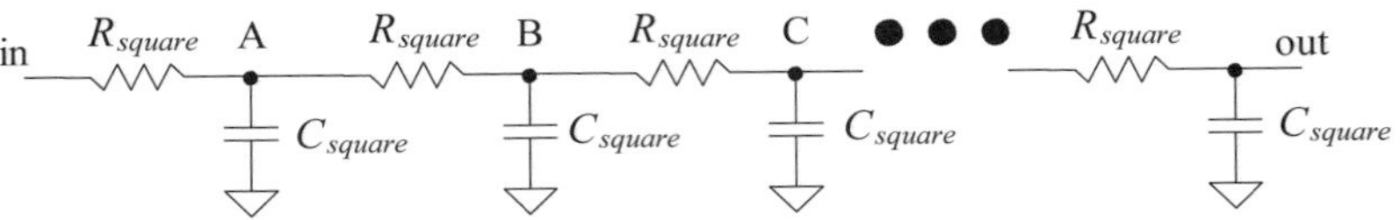

Figure 2.22 Calculating the delay through a distributed RC delay.

The delay to node B is the sum of the delay to point A plus the delay associated with charging the capacitance at node B through $2R_{square}$ or

$$t_{dB} = 0.7(R_{square}C_{square} + 2R_{square}C_{square}) \tag{2.28}$$

Similarly, the delay to node C is

$$t_{dC} = 0.7(R_{square}C_{square} + 2R_{square}C_{square} + 3R_{square}C_{square}) \tag{2.29}$$

For a large number of sections, l, we can write the overall delay through the distributed RC delay as

$$t_d = 0.7R_{square}C_{square} \cdot (1 + 2 + 3 + 4 + ... + l) \tag{2.30}$$

The term in parentheses can be written as

$$(1 + 2 + 3 + 4 + ... + l) = \frac{l(l+1)}{2} \tag{2.31}$$

and so for a large number of sections l

$$t_d \approx 0.35 \cdot R_{square}C_{square} \cdot l^2 \tag{2.32}$$

where the R_{square} and C_{square} are the resistance and capacitance of each square of the distributed RC line.

Example 2.5

Estimate the delay through a 250 kΩ resistor made using an n-well with a width of 10 and a length of 500. Assume that the capacitance of a 10 by 10 square of n-well to substrate is 5 fF. Verify your answer with SPICE.

We can divide the n-well up into 50 squares each having a size of 10 by 10 and a resistance of 5 kΩ. The delay through the resistor is then (remembering 1 femto = 10^{-15})

$$t_d = 0.35 \cdot (5k) \cdot (5f) \cdot (50)^2 \approx 22\ ns$$

Note that the total resistance value is $R_{square} \cdot l$ while the total capacitance of the resistor to substrate is $C_{square} \cdot l$. We can use this result to quickly estimate the delay of a distributed RC line (sometimes called an RC transmission line) as

$$t_d = 0.35 \cdot (\text{total resistance}) \cdot (\text{total capacitance}) \tag{2.33}$$

The simulation results are seen in Fig. 2.23. A SPICE lossy transmission line was used to model the distributed effects of the resistor. ■

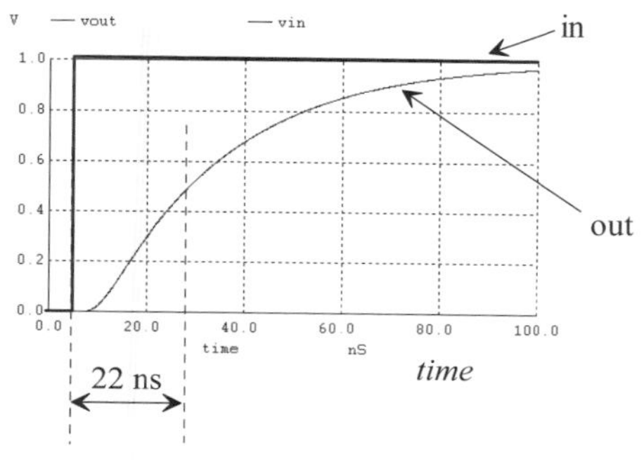


```
*** Figure 2.23 CMOS: Circuit Design, Layout, and Simulation ***
.control
destroy all
run
plot vin vout
.endc
.tran 100p 100n

O1 Vin 0 Vout 0 TRC
Rload Vout 0 1G
Vin vin 0 DC 0 pulse 0 1 5n 0
.model TRC ltra R=5k C=5f len=50
.end
```

Figure 2.23 SPICE simulations showing the delay through an n-well resistor.

Distributed RC Rise Time

A similar analysis to what was used to arrive at Eq. (2.32) can be used to determine the rise time through a distributed RC line. The result is

$$t_r = 1.1 \cdot R_{square} C_{square} \cdot l^2 \qquad (2.34)$$

Using this equation in Ex. 2.5 results in an output rise time of approximately 69 ns. Comparing this estimate to the simulation results in Fig. 2.23, we see good agreement.

2.6 Using the LasiDrc Program

Before presenting the use of the LasiDrc program, let's summarize some of the layout discussions presented in this chapter and discuss some concerns. Examine the cross-sectional views seen in Fig. 2.24. We know that the body of an NMOS transistor is p-type, while the body of a PMOS device is n-type. In the n-well process, Fig. 2.24a, the NMOS are fabricated directly in the p-type substrate and the PMOS are made in the n-well. For a p-well process, Fig. 2.24b, the NMOS are made in the p-well while the PMOS are fabricated in the n-type substrate. Note that sometimes the term **tub** is used in place of well (e.g., an n-tub process) because the resulting semiconductor area has a cross-sectional view like a bathtub's.

When implanting the n-well, in (a) for example, the substrate must be *counter-doped.* This means that the p-substrate must have n-type dopants (such as phosphorous or arsenic) added until its concentration changes from p- to n-type. The problem with counter-doping the p-substrate to make an n-well is that the quality of the resulting semiconductor isn't as good as it would be by simply taking intrinsic silicon and adding donor atoms. The acceptor atoms in the p-substrate become ionized (e.g., electrons fall into the holes) increasing scattering and reducing mobility. Sometimes the effects of counter doping are called *excessive doping effects*. In an n-well process, the PMOS devices suffer from excessive doping and so the quality of the device isn't as good as the quality of a PMOS device in a p-well process. For example, a PMOS device fabricated in an n-well is slower than a PMOS device fabricated in an n-substrate.

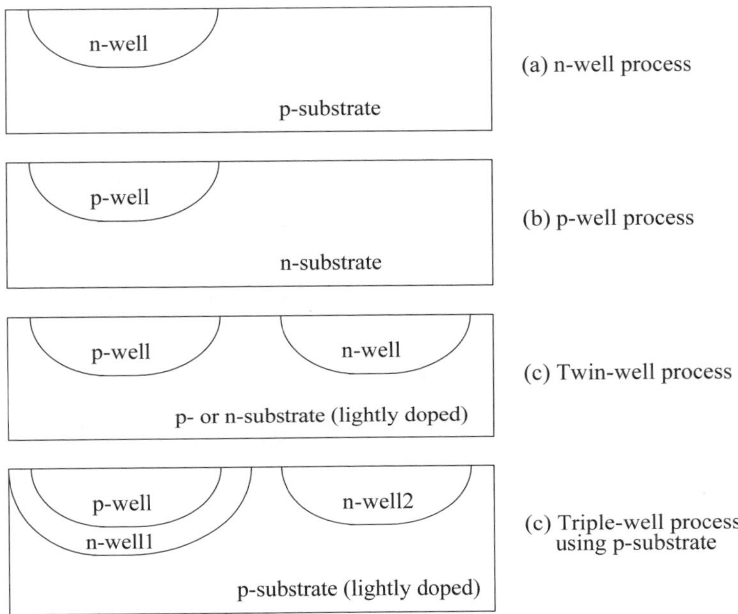

Figure 2.24 The different possible wells used in a bulk CMOS process.

In an attempt to reduce excessive doping effects, a twin-well process, Fig. 2.24c, can be used. When using a lighter doped substrate, the amount of counter doping isn't as significant. We don't use an intrinsic silicon substrate because it is difficult to control the doping at such low levels. If a p-substrate is used, then the p-well is electrically connected to the substrate. The bodies of the NMOS are then all tied to the same potential, usually ground. To allow the bodies of the MOSFETs to be at different potentials a triple-well process can be used, Fig. 2.24c. The added n-well isolates the p-well from the substrate (the n-well1 and p-substrate form a diode that electrically isolates the p-well from the substrate). The p-well can exist directly in the substrate too.

Design Rules for the Well

Figure 2.25 shows the design rules, from MOSIS, for the well. Notice that there are four different sets of rules that the layout designer can follow. In this book **we will use the CMOSEDU rules** to illustrate design examples. Before saying why, let's provide a little history and background. We know from Ch. 1 that MOSIS collects chip designs from various sources (education, private, not-for-profit, etc). These designs are put together to form the masks used for making the chips (on multiproject wafers). The actual vendors used by MOSIS to fabricate chips has changed throughout the years. To make the layouts transferable as well as scalable between different CMOS processes, MOSIS came up with the so-called SCMOS rules (scalable CMOS design rules). A parameter, λ, is used in the rules. All of the layouts are drawn on a λ grid. When making the GDS file (or CIF file), the layout is scaled by this factor. For example, if an n-well box is drawn 10λ by 10λ

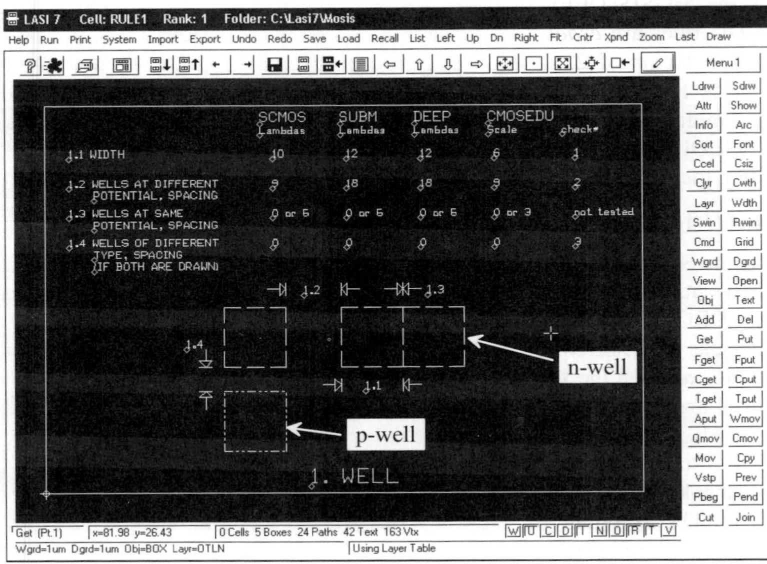

Figure 2.25 LASI layouts showing the MOSIS design rules for the n-well.

with a lambda of 0.3 μm, then when the GDS file is exported (called streaming the layout out) the actual size of the layout is 3 μm square. If the layout were used in a different process, only the value of λ would need to be changed. In other words, the exact same layout can be used in a different technology. Being able use the same layout and simply *scale it* is a significant benefit of CMOS.

When the SCMOS rules were first introduced, the minimum size of a dimension in CMOS was approximately 1 μm. The "fabs" or vendors (the factory where the wafers are actually processed) also have a set of design rules. In general, the fab's design rules are tighter than the SCMOS rules. For example, one fab may specify a minimum n-well width of 3 μm, while another fab, in the same process technology, may specify 4 μm. The SCMOS rules may specify 5 μm to cover all possible situations (the price of using the SCMOS rules over the vendor's rules is larger layout areas). Unfortunately, as the process dimensions have shrunk over time, the MOSIS SCMOS rules weren't relaxed enough, making modifications necessary. This has led to the MOSIS submicron rules (SUBM) and the MOSIS deep-submicron (DEEP) rules seen in Fig. 2.25 (there are three sets of MOSIS scalable design rules, SCMOS, DEEP, and SUBM). Older processes still use the SCMOS rules, while the smaller technologies use the modified rules. Note that if a layout passes the DEEP rules, it will also pass the SCMOS rules (except for the exact via size).

Why use the CMOSEDU rules in this book? Why not use one of the three sets of design rules from MOSIS? The answer comes from how we lay out MOSFETs. The **minimum length of a MOSFET using the MOSIS rules is 2** (2λ, keeping in mind that some scale factor is used when generating the GDS or CIF file). In the CMOSEDU rules we took the MOSIS DEEP rules and divided by two. This means that layouts in the CMOSEDU rules are *exactly the same* as the MOSIS DEEP rules except that they are scaled by a factor of 2. **Using the CMOSEDU rules, the minimum length of a MOSFET is 1**. If MOSIS specifices a scale factor, λ, of 90 nm using the DEEP rules, where the minimum length is 2, then we would use a scale factor of 180 nm when using the CMOSEDU rules with a minimum length of 1. In SPICE we use ".options scale=90n" when using the DEEP rules and ".options scale=180n" when using the CMOSEDU rules.

Running a Design Rule Check

Going to the **System** menu in LASI and then pressing on the **LasiDrc** button starts the LASI DRC utility program. Figure 2.26 shows the setup screen in this utility. When starting a DRC, it's important to make sure that the correct cell name is specified, the correct checks are selected, and the correct DRC file name is used. Also, if the entire cell is to be DRCed, then the **Fit** command button should be pressed (to fit the layout to the DRC check area). If this button isn't pressed, then the last view seen in LASI when the cell is saved, prior to calling the DRC program, will be used (this can be helpful to reduce DRC time when the layout is large). Further, if the DRC is aborted (by pressing **ESC** on the keyboard), then the LasiDRC remembers where the checking left off, allowing the DRC program to continue if errors are found and fixed in the middle of the DRC run. If the DRC program is run on a cell after making edits and the cell is not *first saved*, the edits won't be seen by the DRC program **(this is a common user mistake)**. Finally, errors are both reported in an output file (lasidrc.rpt) and as pictures. Images of the layouts and errors can be viewed with the button indicated in Fig. 2.26.

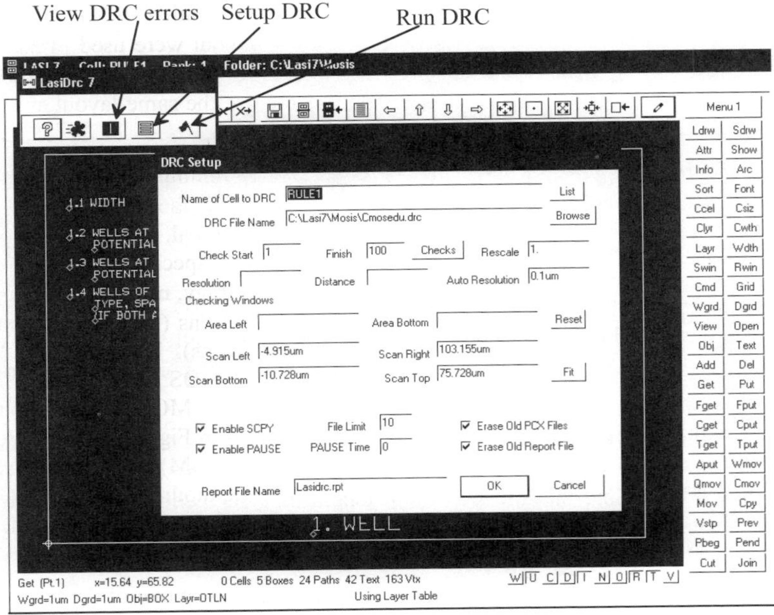

Figure 2.26 Setting up a DRC run.

ADDITIONAL READING

[1] J. P. Uyemura, *Introduction to VLSI Circuits and Systems*, John Wiley and Sons Publishers, 2002. ISBN 0-471-12704-3

[2] Y. Taur and T. H. Ning, *Fundamentals of Modern VLSI Devices*, Cambridge University Press, 1998. ISBN 0-521-55056-4 (hardback) or ISBN 0-521-55959-6 (paperback)

[3] M. Smith, *WinSPICE User's Manual*, Available for download (with the WinSPICE simulation program) at http://www.winspice.co.uk/

[4] J. Jessing, *Introduction to CMOS Process Integration: Front-End of the Line (FEOL)*, course notes at Boise State University.

[5] D. E. Boyce, *LASI User's Manual*, available while LASI is running by pressing F1 on the keyboard and the button the user needs help with (at the same time). Alternatively, the user can open the help file in the directory C:\Lasi7\help.

PROBLEMS

2.1 For the layout seen in Fig. 2.27, sketch the cross-sectional views at the places indicated. Is there a parasitic pn junction in the layout? If so, where? Is there a parasitic bipolar transistor? If so, where?

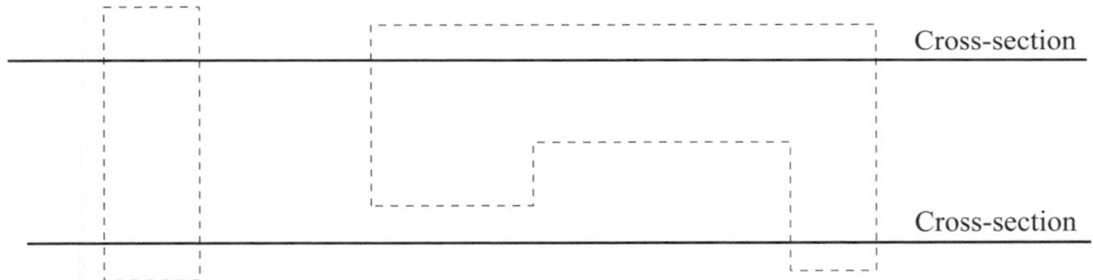

Figure 2.27 Layout used in problem 2.1.

2.2 Sketch (or use LASI) the layout of an n-well box that measures 100 by 10. If the scale factor is 50 nm, what is the actual size of the box after fabrication? What is the area before and after scaling? Neglect lateral diffusion or any other fabrication imperfections.

2.3 Lay out (using a path object in LASI or any other layout program) a nominally 250 kΩ resistor using the n-well in a serpentine pattern similar to what's seen in Fig. 2.28. Assume that the maximum length of a segment is 100 and the sheet resistance is 2 kΩ/square. Design rule check the finished resistor. If the scale factor in the layout is 50 nm, estimate the fabricated size of the resistor.

2.4 If the fabricated n-well depth, t, is 1 μm, then what are the minimum, typical, and maximum values of the n-well resistivity, ρ? Assume that the measured sheet resistances (minimum, typical, and maximum) are 1.6, 2.0, and 2.2 kΩ/square?

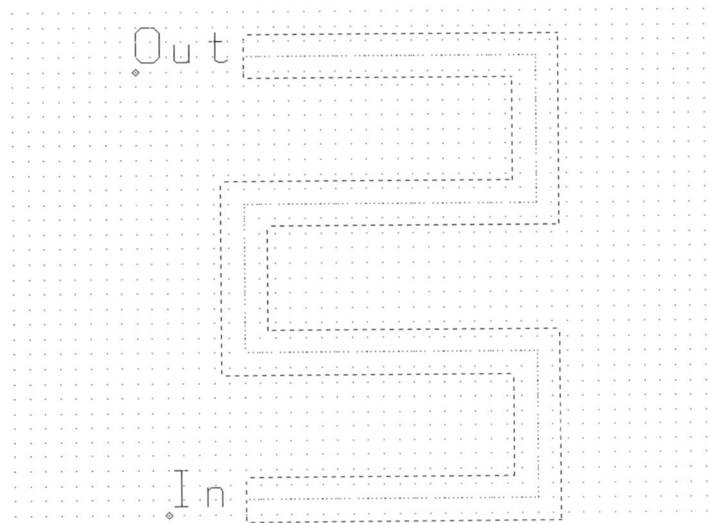

Figure 2.28 Layout of an n-well resistor using a serpentine pattern.

2.5 Normally, the scale current of a pn junction is specified in terms of a scale current density, J_s (A/m^2), and the width and length of a junction (i.e., $I_s = J_s \cdot L \cdot W \cdot scale^2$ neglecting the sidewall component). Estimate the scale current for the diode of Ex. 2.3 if $J_s = 10^{-8}$ A/m^2.

2.6 Repeat problem 2.5, including the sidewall component ($I_s = J_s \cdot L \cdot W \cdot scale^2 + J_s \cdot (2L+2W) \cdot scale \cdot$depth).

2.7 Using the diode of Ex. 2.3 in the circuit of Fig. 2.29, estimate the frequency of the input signal when the AC component of v_{out} is 707 μV (i.e., estimate the 3 dB frequency of the $|v_{out}/v_{in}|$).

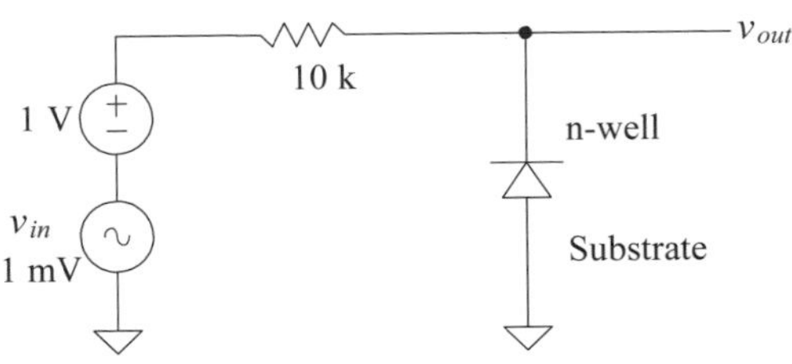

Figure 2.29 Treating the diode as a capacitor. See Problem 2.7.

2.8 Verify the answer given in problem 2.7 with SPICE.

2.9 Using SPICE, show that a diode can conduct significant current from its cathode to its anode when the diode is forward biased.

2.10 Estimate the delay through a 1 MΩ resistor (10 by 2,000) using the values given in Ex. 2.5. Verify the estimate with SPICE.

2.11 If one end of the resistor in problem 2.10 is tied to +1 V and the other end is tied to the substrate that is tied to ground, estimate the depletion capacitance (F/m^2) between the n-well and the substrate at the beginning, the middle, and the end of the resistor. Assume that the resistance does not vary with position along the resistor and that the scale factor is 50 nm, C_{j0} = 25 aF for a 10 by 10 square, m = 0.5, and $V_{bi} = 1$.

2.12 The diode reverse breakdown current, that is, the current that flows when $|V_D| < BV$ (breakdown voltage), is modeled in SPICE by

$$I_D = IBV \cdot e^{-(V_D + BV)/V_T}$$

Assuming that 10 μA of current flows when the junction starts to break down at 10 V, simulate, using a SPICE DC sweep, the reverse breakdown characteristics of the diode. (The breakdown voltage, BV, is a positive number. When the diode starts to break down $-BV = V_D$. For this diode, breakdown occurs when $V_D = -10$ V.)

2.13 Repeat Ex. 2.3 if the n-well/p-substrate diode is 50 square and the acceptor doping concentration is changed to 10^{15} atoms/cm^3.

2.14 Estimate the storage time, that is, the time it takes to remove the stored charge in a diode, when $\tau_T = 5$ ns, $V_F = 5$ V, $V_R = -5$ V, $C_{j0} = 0.5$ pF, and R = 1k. Verify the estimate using SPICE.

Chapter

3

The Metal Layers

The metal layers in a CMOS integrated circuit connect circuit elements (MOSFETs, capacitors, and resistors). In the following discussion we'll discuss a generic CMOS process with two layers of metal. These levels of metal are named metal1 and metal2. The metal in a CMOS process is either aluminum or copper. In this chapter we look at the layout of the bonding pad, capacitances associated with the metal layers, crosstalk, sheet resistance, and electromigration.

3.1 The Bonding Pad

The bonding pad is at the interface between the die and the package or the outside world. One side of a wire is soldered to the pad, while the other side of the wire is connected to a lead frame, as was seen in Fig. 1.3. Figure 3.1 shows a close up of a bonding pad and wire. In this chapter we will not concern ourselves with electrostatic discharge (ESD) protection, which is an important design consideration when designing the pad.

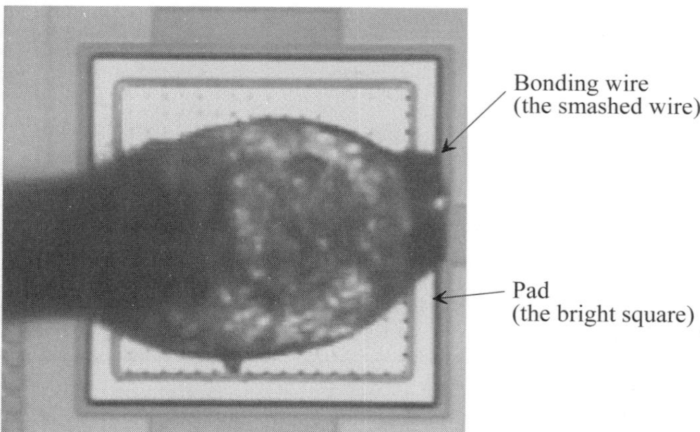

Figure 3.1 The bonding wire connection to a pad.

3.1.1 Laying Out the Pad I

The basic size of the bonding pad specified by MOSIS is a square 100 μm x 100 μm (actual size). For a probe pad, used to probe the circuit with a microprobe station, the size should be greater than 6 μm x 6 μm. In production chips the pads may vary in size (e.g., 75 x 100, or 50 x 75, etc.) depending on the manufacturer's design rules. *The final size of the pads are the only part of a layout that doesn't scale as process dimensions shrink.* The layout of a pad that uses metal2 (MET2 in LASI) is shown in Fig. 3.2. Notice, in the cross-sectional view, the layers of insulator (SiO_2 in most cases) under and above the metal2. These layers are used for isolation between the other layers in the CMOS process.

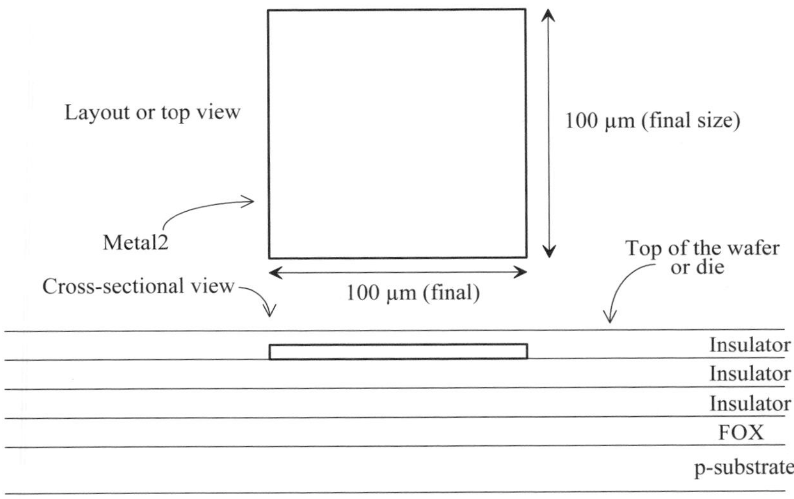

Figure 3.2 Layout of metal2 used for bonding pad with associated cross-sectional view.

Capacitance of Metal-to-Substrate

Before proceeding any further, we might ask the question, "What is the capacitance from the metal2 box (pad) in Fig. 3.2 to the substrate?" The substrate is at ground potential and so, for all intents and purposes, it can be thought of as an equipotential plane. This is important because we have to drive this capacitance to get a signal off the chip. Table 3.1 gives typical values of parasitic capacitances for a CMOS process. Consider the following example.

Example 3.1
Estimate the parasitic capacitance associated with the pad in Fig. 3.2.

The capacitance associated with this pad is the sum of the plate (or bottom) capacitance and the fringe (or edge) capacitance. We can write

$$C_{pad,m2 \to sub} = area \cdot C_{plate} + perimeter \cdot C_{fringe} \qquad (3.1)$$

The area of the pad is 100 μm² square (100 μm by 100 μm), while the perimeter of the pad is 400 μm. Using the typical values of capacitance for metal1 to substrate in Table 3.1 gives

$$C_{pad,m2\rightarrow sub} = 10,000 \cdot 14\ aF + 400 \cdot 81\ aF = 172,400\ aF = 172.4\ fF = 0.172\ pF$$

A significant on-chip capacitance. ■

Table 3.1 Typical parasitic capacitances in a CMOS process. Note that while the physical distance between the layers decreases, as process technology scales downwards, the dielectric constant used in between the layers can be decreased to keep the parasitic capacitances from becoming too significant. The values are representative of the parasitics in both long- and short-channel CMOS processes.

	Plate Cap. aF/μm²			Fringe Cap. aF/μm		
	min	**typ**	**max**	**min**	**typ**	**max**
Poly1 to subs. (FOX)	53	58	63	85	88	92
Metal1 to poly1	35	38	43	84	88	93
Metal1 to substrate	21	23	26	75	79	82
Metal1 to diffusion	35	38	43	84	88	93
Metal2 to poly1	16	18	20	83	87	91
Metal2 to substrate	13	14	15	78	81	85
Metal2 to diffusion	16	18	20	83	87	91
Metal2 to metal1	31	35	38	95	100	104

Example 3.2

The pad layout in Fig. 3.2 is the actual size. However, when we lay out the pad with the other circuit components, it must also be scaled when the layout is streamed out (see Sec. 1.2.3). If the scale factor in a design is 50 nm, what is the size of the box used for a pad that we draw with the layout program? Does the capacitance calculated in Ex. 3.1 change?

Because we want a final pad size of 100 μm by 100 μm, the drawn layout size of the box with a scale factor of 50 nm is

$$\frac{100\ \mu m}{0.05\ \mu m} = 2,000\ \text{(drawn size)}$$

Each side of the pad, in Fig. 3.2, is drawn with a size of 2,000 for a final (actual) size of 100 μm by 100 μm.

The capacitance calculated in Ex. 3.1 doesn't change. We can rewrite Eq. (3.1) as

$$C_{pad,m2\rightarrow sub} = area_{drawn} \cdot (scale)^2 \cdot C_{plate} + perimeter_{drawn} \cdot (scale) \cdot C_{fringe} \quad (3.2)$$

to use the drawn layout size. At this point there should be no confusion between the terms "drawn layout size" and "actual or final layout size." ■

Passivation

Because an insulator is covering the pad (the piece of metal2) in Fig. 3.2, we can't bond (connect a wire) to it. The top layer insulator on the chip is also called passivation. The passivation helps protect the chip from contamination. Openings for bonding pads are called cuts in the passivation. To specify an opening or cut in the glass (insulator) covering the metal2, we use the overglass layer (OVGL). The MOSIS rules specify 6 μm distance between the edge of the metal2 and the overglass box, as seen in Fig. 3.3. The drawn distance between the smaller overglass box and the larger metal2 box, with a scale factor of 50 nm, is 6/0.05 or 120.

There may be another layer in the MOSIS setups for the layout tool called the PAD layer. This layer has no fabrication significance but rather is used by the machine that bonds the chip to the lead frame to indicate the location of the pads. Since MOSIS takes designs of varying sizes and shapes, the locations of the pads change from one project to the next. This layer isn't really necessary since we can use the overglass layer (ensuring the overglass box has a drawn size of 1,760 square or a final size of 88 μm square) to indicate the location of the pads. We won't use the PAD layer in our layouts here.

An Important Note

Here we are using a CMOS process with (only) two layers of metal. In most modern CMOS processes, more than two layers of metal are used. If the process has five layers of metal, then the top layer (just like the top floor in a five-story building) is metal5. Therefore, metal5 is the layer the bonding wire is connected to.

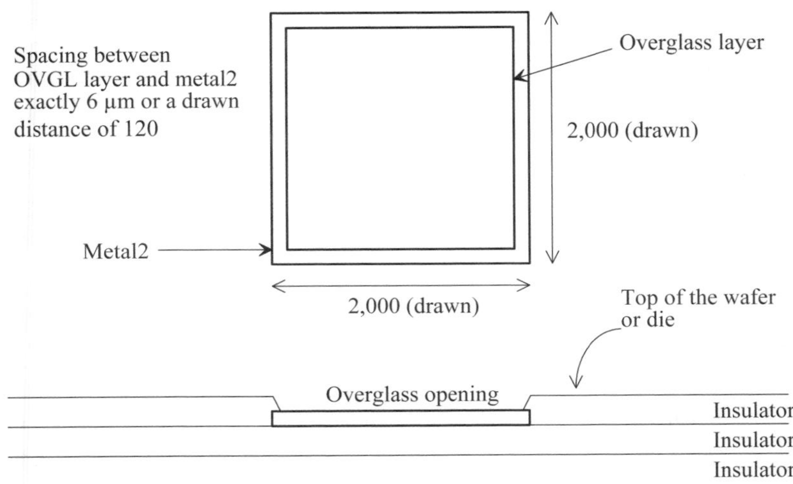

Figure 3.3 Layout of a metal2 pad with pad opening for bonding connection in a 50 nm (scale factor) CMOS process.

3.2 Design and Layout Using the Metal Layers

As mentioned earlier, the metal layers connect the resistors, capacitors, and MOSFETs in a CMOS integrated circuit. So far, in this book, we've learned about the layout layers n-well (NWEL), metal2 (MET2), overglass (OVGL), and pad (PAD). In this section we'll also learn about the metal1 (MET1) and via1 (VIA1) layers and the associated parasitic resistances and capacitances of these layers.

3.2.1 Metal1 and Via1

Metal1 is a layer of metal found directly below metal2. Figure 3.4 shows an example layout and cross-sectional view. The via1 layer connects metal1 and metal2. The via layer specifies that the insulator be removed in the location indicated. Then, for example, a tungsten "plug" is fabricated in the insulator's opening. When the metal2 is laid down, the plug provides a connection between the two metals. Note that if we were to use more than two layers of metal, then via2 would connect metal2 to metal3, via3 would connect metal3 to metal4, etc.

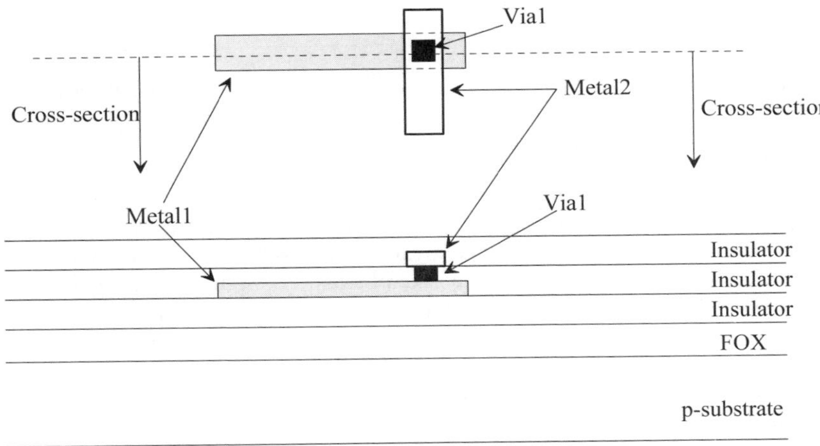

Figure 3.4 Layout and cross-sectional views.

An Example Layout

Figure 3.5 shows an example layout using the n-well, metal1, via1, and metal2 layers. It's important that, before proceeding, this layout and the associated cross-sectional view are understood. For example, how would our cross-sectional view change if we moved the cross-sectional line used in Fig. 3.5 down slightly so that it only intersects the n-well and the metal2 layers? Answer: the cross-sectional view would be the same as seen in Fig. 3.5 except that the metal1 and via1 layers wouldn't be present.

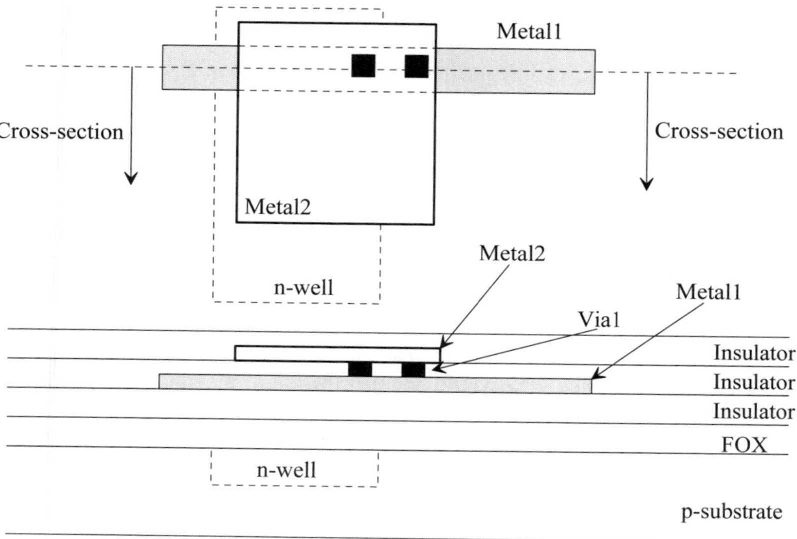

Figure 3.5 An example layout and cross-sectional view using including the n-well.

3.2.2 Parasitics Associated with the Metal Layers

Associated with the metal layers are parasitic capacitances (see Table 3.1) and resistance. Like the n-well in the last chapter, the metal layers are characterized by a sheet resistance. However, the sheet resistance of the metal layers is considerably lower than the sheet resistance of the n-well. For the sake of examples in this book, we'll use metal sheet resistances of **0.1 Ω/square**. Also, there is a finite contact resistance of the via. The following examples illustrate some of the unwanted parasitics associated with these layers.

Example 3.3
Estimate the resistance of a piece of metal1 1 mm long and 200 nm wide. What is the drawn size of this metal line if the scale factor is 50 nm? Also estimate the delay through this piece of metal, treating the metal line as an RC transmission line. Verify your answer with a SPICE simulation.

The drawn size of the metal line is 1 mm/50 nm (= 20,000) by 200/50 (= 4). Figure 3.6 shows the layout of the metal wire (not to scale). The line consists of 1,000/0.2 = 20,000/4 = 5,000 squares of metal1.

To calculate the resistance of the metal line, we use Eq. (2.3)

$$R = \frac{0.1\ \Omega}{\text{square}} \cdot \frac{20,000}{4} = 0.1 \cdot 5,000 = 500\ \Omega$$

To calculate the capacitance, we use the information in Table 3.1 and either Eq. (3.1) or (3.2)

$$C = (1,000 \cdot 0.2) \cdot 23\ aF + (2,000.4) \cdot 79\ aF = 162\ fF$$

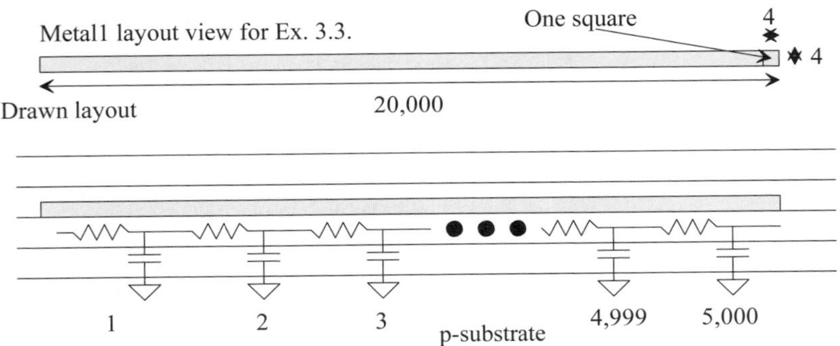

Figure 3.6 Layout and cross-sectional view with parasitics for the metal line in Ex. 3.3.

or the capacitance for each 200 nm by 200 nm square (4 by 4) of metal1 is

$$C_{square} = \frac{162\, fF}{5{,}000} = 32\ aF/\text{square}$$

The delay through the metal line is, using Eqs. (2.32) or (2.33)

$$t_d = 0.35 \cdot R_{square} C_{square} \cdot l^2 = 0.35(0.1)(32\ aF)(5{,}000)^2 = 28\ ps$$

or

$$t_d = 0.35RC = 0.35 \cdot 500 \cdot 162\, fF = 28\ ps$$

The delay of a metal1 line (with nothing connected to it) is 28 ps/mm when the parasitic capacitance and resistance are the limiting factors. The SPICE simulation results are seen in Fig. 3.7. ■

Intrinsic Propagation Delay

The result of this example (a metal delay of 28 ps/mm) should be compared to the intrinsic delay of a signal propagating in a material with a relative dielectric constant, ε_r (no parasitic resistance). The velocity, v, of the signal in this situation is related to the speed of light, c, by

$$v = \frac{c}{\sqrt{\varepsilon_r}}\ \text{(meters/second)} \tag{3.3}$$

If we assume the signal is propagating in silicon dioxide (SiO_2) with a relative dielectric constant of roughly 4, then we can estimate the delay of the metal line as

$$\frac{t_d}{\text{meter}} = \frac{1}{v} = \frac{\sqrt{\varepsilon_r}}{c} = \frac{2}{3 \times 10^8\ m/s} = \frac{6.7\ ns}{\text{meter}} \tag{3.4}$$

or a delay of 6.7 ps/mm. For any practical integrated circuit wire in bulk CMOS, the parasitics (RC delay) dominate the propagation delays.

Note that increasing the width of the wire decreases its resistance and increases its capacitance (resulting in the delay staying relatively constant).

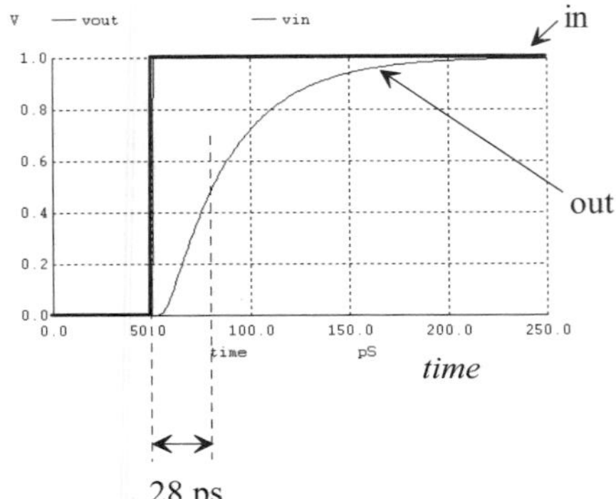


```
*** Figure 3.7 CMOS: Circuit Design, Layout, and Simulation ***
.control
destroy all
run
plot vin vout
.endc
.tran 1p 250p

O1 Vin 0 Vout 0 TRC
Rload Vout 0 1G
Vin vin 0 DC 0 pulse 0 1 5n 0
.model TRC ltra R=0.1 C=32e-18 len=5k
.end
```

Figure 3.7 Simulating the delay through a 1 mm wire made using metal1.

Example 3.4
Estimate the capacitance between a 10 by 10 square piece of metal1 and an equal-size piece of metal2 placed exactly above the metal1 piece. Assume a scale factor of 50 nm. Sketch the layout and the cross-sectional views. Also sketch the symbol of a capacitor on the cross-sectional view.

The plate capacitance, from Table 3.1, between metal1 and metal2 is typically 35 aF/μm², while the fringe capacitance is typically 100 aF/μm. The two layers form a parallel plate capacitor, Fig. 3.8. The capacitance between the plates is given by the sum of the plate capacitance and the fringe capacitance, or

$$C_{12} = 100 \cdot (0.05)^2 \cdot 35\,aF + 40 \cdot (0.05) \cdot 100\,aF = 209\,aF$$

■

Layout view of 10 square metal1 and metal2

Metal2 is the top plate of the capacitor and metal 1 is bottom.

Insulator

Insulator

Figure 3.8 Capacitance between metal1 and metal2.

Example 3.5
In the previous example, estimate the voltage change on metal1 when metal2 changes potential from 0 to 1 V. Verify the result with SPICE.

The capacitance from metal2 to metal1 was calculated as 209 aF. The capacitance from metal1 to substrate is given by

$$C_{1sub} = 100 \cdot (0.05)^2 \cdot (23) + 40 \cdot (0.05) \cdot 79 = 164\ aF$$

The equivalent schematic is shown in Fig. 3.9. The voltage on C_{1sub} is given by

$$\Delta V_{metal1} = \Delta V_{metal2} \cdot \frac{\frac{1}{j\omega C_{1sub}}}{\frac{1}{j\omega C_{1sub}} + \frac{1}{j\omega C_{12}}} = 1 \cdot \frac{C_{12}}{C_{12} + C_{1sub}} = \frac{209}{209 + 164} = 560\ mV$$

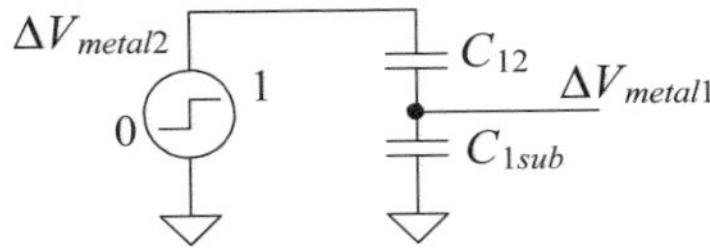

Figure 3.9 Equivalent circuit used to calculate the change in metal1 voltage, see Ex. 3.5.

A displacement current flows through the capacitors, causing the potential on metal1 to change by 560 mV. This may seem significant at first glance. However, one must remember that most metal lines in a CMOS circuit are being driven from a low-impedance source; that is, the metal is not floating but is being held at some potential. This is not the case in some dynamic circuits or in circuits with high-impedance nodes or long metal runs. Figure 3.10 shows the SPICE simulation results and netlist. Notice how we used the "use initial conditions" (UIC) in the transient statement. This sets all nodes that aren't driven by a source to, initially (at the beginning of the simulation), zero volts. Note that SPICE doesn't recognize "a" as atto so we used "e-18." ■

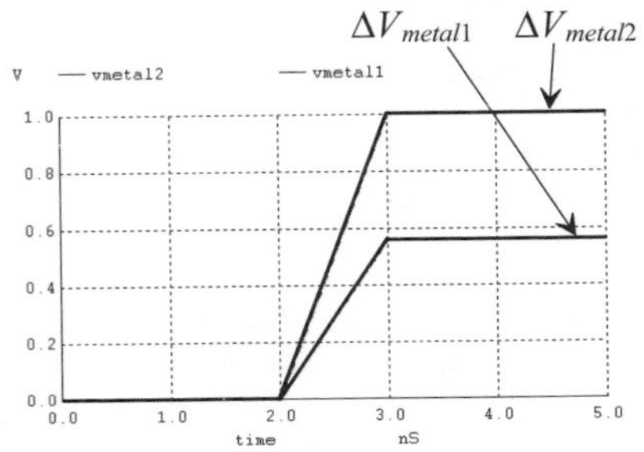


```
*** Figure 3.10 CMOS: Circuit Design, Layout, and Simulation ***

.control
destroy all
run
plot vmetal2 vmetal1
.endc

.tran 10p 5n UIC

vmetal2 vmetal2 0 DC 0 pulse 0 1 2n 1n

C12 vmetal2 vmetal1 209e-18
C1sub vmetal1 0 164e-18

.end
```

Figure 3.10 Simulating the operation of the circuit in Fig. 3.9.

3.2.3 Current-Carrying Limitations

Now that we have some familiarity with the metal layers, we need to answer the question, "How much current can we carry on a given width or length of metal?" The factors that limit the amount of current on a metal wire or bus are metal electromigration and the maximum voltage drop across the wire or bus due to the resistance of the metal layer.

A conductor carrying too much current causes metal electromigration. This effect is similar to the erosion that occurs when a river carries too much water. The result is a change in the conductor dimensions, causing spots of higher resistance and eventually failure. If the current density is kept below the metal migration threshold current density, J_{Al}, metal electromigration will not occur. Typically, for aluminum, the current threshold for migration J_{Al} is $1 \rightarrow 2\ \frac{\text{mA}}{\mu\text{m}}$.

Example 3.6
Assuming a scale factor of 50 nm, estimate the maximum current a piece of metal1 with a drawn width of 3 can carry. Also estimate the maximum current a 100 by 100 μm^2 bonding pad can receive from a bonding wire. Assume that the metal wires are fabricated in aluminum.

The actual width of the metal1 wire in this example is 150 nm. Assuming that $J_{AL} = 1\ \frac{\text{mA}}{\mu\text{m}}$, the maximum current on a 0.15 μm wide aluminum conductor is given by

$$I_{max} = J_{Al} \cdot W = 10^{-3} \cdot 0.15 = 150\ \mu A$$

The maximum current through a bonding pad is then 100 mA. ■

Example 3.7
Estimate the voltage drop across the conductor discussed in the previous example when the length of the conductor is 1 cm and the current flowing in the conductor is 150 μA (I_{max}).

The sheet resistance of metal1 is 0.1 Ω/square. The voltage drop across a metal1 wire that is 3 (0.15 μm) wide and 10,000 μm (1 cm) long carrying 150 μA is

$$V_{drop} = (0.1\ \Omega/\text{square}) \cdot \frac{10,000}{0.15} \cdot 150\ \mu A = 1\ V$$

or a significant voltage drop. If this conductor were used for power, we would want to increase the width significantly; however, if the conductor is used to route data, the size may be fine. ■

In general, the higher levels of metal (metal2, metal3, etc.) should be used for power routing. Metal2 is approximately twice as thick as metal1 and, therefore, has a lower sheet resistance. Metal3 is thicker than metal2, etc. When routing power, the more metal that is used, the fewer problems, in general, that will be encountered. If possible, a ground or power plane should be used across the entire die (entire levels of metal are used for *VDD* and ground). The more capacitance between the power and ground buses, the harder it is to induce a voltage change on the power plane; that is, the DC voltages will not vary.

3.2.4 Design Rules for the Metal Layers

The design rules for the metal1, via1, and metal2 layers are seen in Fig. 3.11. Note that the via1 size must be exactly 1.5 by 1.5. Also note that the minimum allowable spacing between two wires using metal1 is 1.5, while the spacing between wires using metal2 is 2. There isn't a spacing rule between metal1 and metal2 because wires made with metal1 and metal2 are isolated by an insulator (sometimes called an interlayer dielectric, ILD).

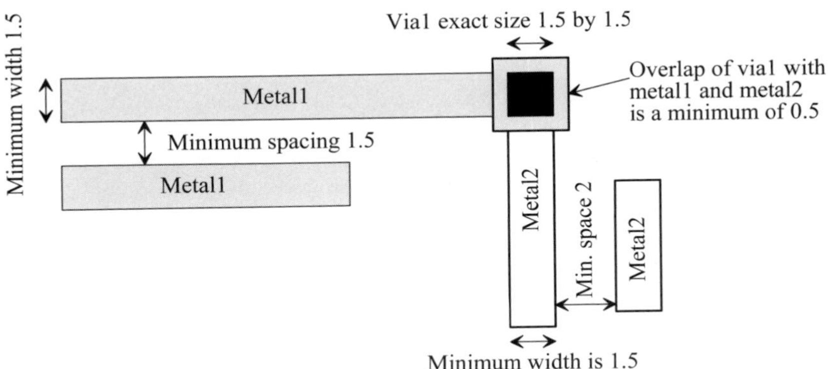

Figure 3.11 Design rules for the metal layers using the CMOSEDU rules.

Layout of Two Shapes or a Single Shape

When learning to do layout, one may wonder about the equivalence of the two layouts seen in Fig. 3.12. In (a) two boxes are used while in (b) a single box is used. When the masks are made the layouts are equivalent.

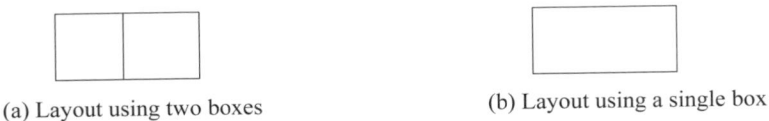

(a) Layout using two boxes (b) Layout using a single box

Figure 3.12 Equivalence of layouts drawn with a different number of shapes.

A Layout Trick for the Metal Layers

Notice that the size of the via is exactly 1.5 by 1.5 and that the minimum metal surrounding the via is 0.5. In order to save time when doing layout, a cell can be made called "via1" with a rank of 1. Instead of drawing boxes on the via1, metal1, and metal2 layers each time a connection between metal1 and metal2 is needed, we simply place the "via1" cell into the layout, Fig. 3.13.

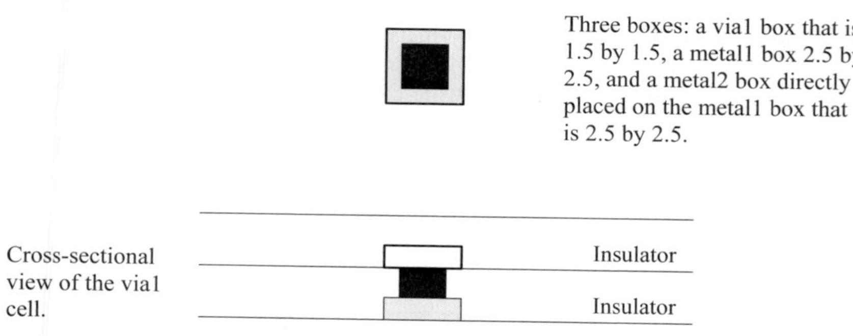
Three boxes: a via1 box that is 1.5 by 1.5, a metal1 box 2.5 by 2.5, and a metal2 box directly placed on the metal1 box that is 2.5 by 2.5.
Cross-sectional view of the via1 cell.
Insulator
Insulator

Figure 3.13 Via1 cell with a rank of 1.

3.2.5 Contact Resistance

Associated with any contact to metal (or any other layer in a CMOS process for that matter) is an associated contact resistance. For the examples using metal layers in this book, we'll use a contact resistance of **10 Ω/contact**. Consider the following example.

Example 3.8
Sketch the equivalent electrical schematic for the layout depicted in Fig. 3.14a showing the via contact resistance. Estimate the voltage drop across the contact resistance of the via when 1 mA flows through the via. Repeat for the layout shown in Fig. 3.14b.

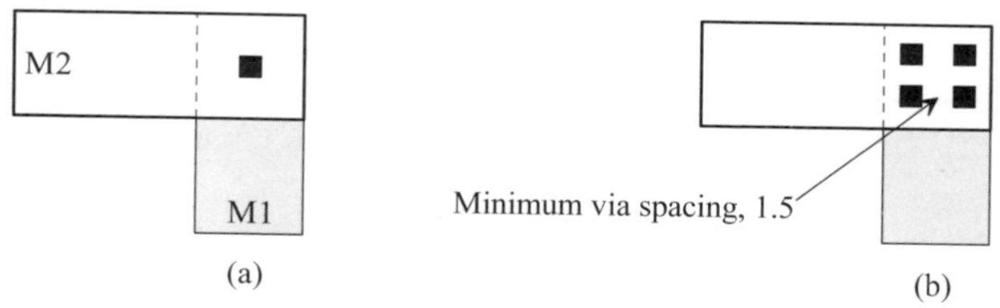
M2
M1
(a)
Minimum via spacing, 1.5
(b)

Figure 3.14 Layouts used in Ex. 3.8.

The equivalent schematics are shown in Fig. 3.15a and (b) for the layouts in Figs. 3.14a and (b) respectively. If the via contact resistance is 10 Ω, and 1 mA flows through the via in (a), then a voltage drop of 10 mV results. Further, the reliability of the single via will be poor with 1 mA flowing through it due to electromigration effects. A "rule-of-thumb" is to allow no more than 100 μA of current flow per via. The four vias shown in Fig. 3.14(b) give an effective contact resistance of 10/4 or 2.5 Ω because the contact resistances of each of the vias are in parallel. The voltage drop across the vias decreases to 2.5 mV with 1 mA flowing in the wires. Increasing the metal overlap and the number of vias will further decrease the voltage drop (and electromigration effects). ■

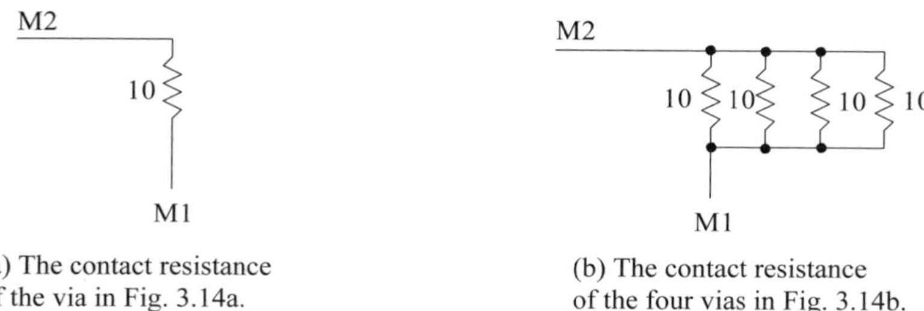

(a) The contact resistance of the via in Fig. 3.14a.

(b) The contact resistance of the four vias in Fig. 3.14b.

Figure 3.15 The schematics of the contact resistances for the layouts in Fig. 3.14.

3.3 Crosstalk and Ground Bounce

Crosstalk is a term used to describe an unwanted interference from one conductor to another. Between two conductors there exists mutual capacitance and inductance, which give rise to signal feedthrough. Ground bounce (and *VDD* droop) are terms describing local variations in the power and ground supplies at a circuit. While crosstalk is only a problem for time-varying signals in a circuit, ground bounce can be problematic for both time varying and DC signals.

3.3.1 Crosstalk

Consider the two metal wires shown in Fig. 3.16. A signal voltage propagating on one of the conductors couples current onto the conductor. This current can be estimated using

$$I_m = C_m \frac{dV_A}{dt} \tag{3.5}$$

where C_m is the mutual capacitance, I_m is the coupled current, and V_A is the signal voltage on the source conductor. Treating the capacitance between the two conductors in this

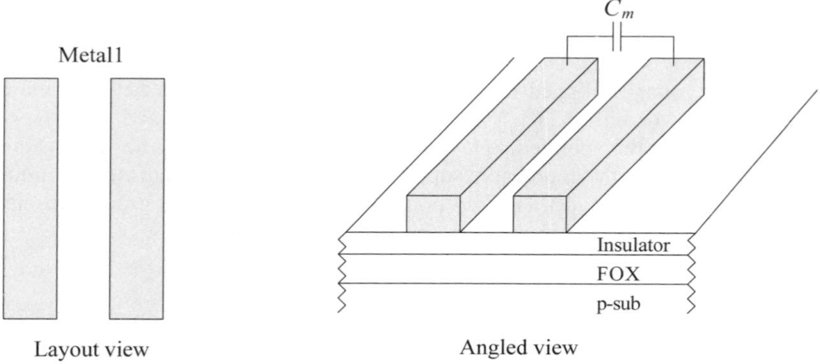

Figure 3.16 Conductors used to illustrate crosstalk.

simple manner is useful in most cases. Determining C_m experimentally proceeds by applying a step voltage to one conductor while measuring the coupled voltage on the adjacent conductor. Since we know the capacitance of any conductor to substrate (see Table 3.1), we can write

$$\Delta V = V_A \cdot \frac{C_m}{C_m + C_{1sub}} \tag{3.6}$$

where ΔV is the coupled noise voltage to the adjacent conductor and C_{1sub} is the capacitance of the adjacent conductor (in this case metal1) to ground (the substrate).

The adjacent metal lines shown in Fig. 3.16 also exhibit a mutual inductance. The effect can be thought of as connecting a miniature transformer between the two conductors. A current flowing on one of the conductors induces a voltage on the other conductor. Measuring the mutual inductance begins by injecting a current into one of the conductors. The voltage on the other conductor is measured. The mutual inductance is determined using

$$V_m = L_m \frac{dI_A}{dt} \tag{3.7}$$

where I_A is the injected (time-varying) current (the input signal), V_m is the induced voltage (the output signal), and L_m is the mutual inductance.

Crosstalk can be reduced by increasing the distance between adjacent conductors. In many applications (e.g., DRAM), the design engineer has no control over the spacing (pitch) between conductors. The circuit designer then attempts to balance the signals on adjacent conductors (see, for example, the open and folded architectures in Ch. 16 concerning DRAM design).

3.3.2 Ground Bounce

DC Problems

Consider the schematic seen in Fig. 3.17a. In this schematic a circuit is connected to *VDD* and ground through two wires measuring 10,000 μm (10 mm) by 150 nm (with a resistance of 6.67 kΩ). Next consider, in (b), what happens if the circuit starts to pull a DC current of 50 μA. Instead of the circuit being connected to a *VDD* of 1 V the actual *VDD* drops to 667 mV. Further, the actual "ground" connected to the circuit increases to 333 mV. The voltage dropped across the circuit is the difference between the applied *VDD* and ground or only 333 mV (considerably less than the ideal 1 V). The obvious solution to making the supplied *VDD* and ground move closer to the ideal values is to increase the widths of the conductors supplying and returning currents to the circuit. This reduces the series resistance. The key point here is that *VDD* and ground are not fixed values; rather, they can vary depending how the circuit is laid out.

AC Problems

It is common, in CMOS circuit design, for a CMOS circuit to draw practically zero current in a static state (not doing anything). This is why, for example, it's possible to use solar power in a CMOS-based calculator. In this situation, conductors with small widths, as those in Fig. 3.17, may be fine. However, consider what happens if the circuit, for a short time, pulls 50 μA. As discussed above, the ground bounces up and *VDD* droops down during this short time. The average current supplied by *VDD* may be well under a

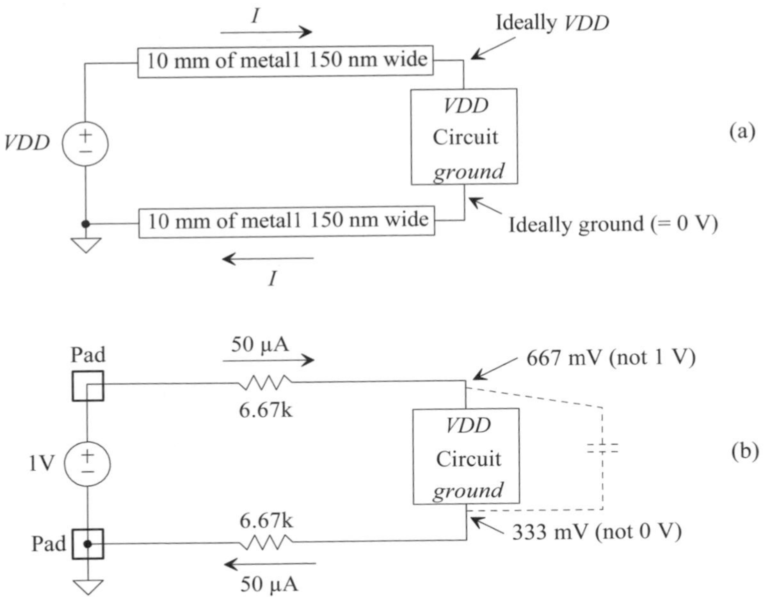

Figure 3.17 Illustrating problems with incorrectly sized conductors.

microamp; however, the occasional need for 50 μA still creates or causes problems. To circumvent these, consider adding an on-chip *decoupling* capacitor physically at the circuit between *VDD* and ground (see dotted lines in Fig. 3.17b). The added capacitor supplies the needed charge during the transient times and keeps the voltage applied across the circuit at *VDD*. Note that a decoupling capacitor should be used external to the chip as well. The capacitor is placed across the *VDD* and ground pins of the chip.

Example 3.9

Suppose that the circuit in Fig. 3.17b needs 50 μA of current for 10 ns. Estimate the size of the decoupling capacitor required if the voltage across the circuit should change by no more than 10 mV during this time.

We can write the charge supplied by the capacitor as

$$Q = I \cdot \Delta t = (50\ \mu A) \cdot 10\ ns = 500 \times 10^{-15}\ Coulumbs$$

The decoupling capacitor must supply this charge

$$\Delta V \cdot C = Q \rightarrow C \geq \frac{Q}{\Delta V} = \frac{I \cdot \Delta t}{\Delta V} = \frac{500 \times 10^{-15}}{10\ mV} \rightarrow C \geq 50\ pF \tag{3.8}$$

A reasonably large capacitor. ■

Example 3.10

To drive off-chip loads, an output buffer (see Ch. 11) is usually placed in between the on-chip logic and the large off-chip load, Fig. 3.18. If *VDD* is 1 V and it is desirable to drive the 30 pF off-chip load in Fig. 3.18 to 900 mV in 1 ns, estimate the size of the decoupling capacitor required. Assume that ground variations are not a concern.

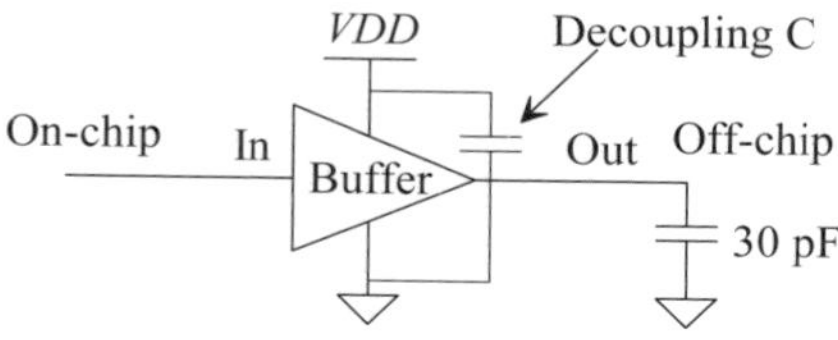

Figure 3.18 Estimating the decoupling capacitance needed in an output buffer.

The charge supplied to the 30 pF capacitor (the load capacitance) by the output buffer is

$$Q = (900\ mV) \cdot (30\ pF) = 27\ pC$$

This charge is supplied by the decoupling capacitor (assuming that the conductors powering the buffer are narrow). Initially, the decoupling capacitor is charged to *VDD* (1 V). If *VDD* (actually the voltage across the decoupling capacitor) drops to 900 mV, then using Eq. (3.8) we can calculate the size of decoupling capacitor as

$$C \geq \frac{27\ pC}{100\ mV} = 270\ pF\ !!!$$

Not a practical value for an on-chip capacitor in most situations. The solution to this problem is to supply *VDD* and ground to the output buffers through wide conductors. Often separate power and ground pads (and very wide wires) are used to power the output buffers separately from the other on-chip circuitry. Using separate pads reduces the size of the decoupling capacitor required and eliminates the noise (ground bounce and *VDD* droop) from interfering with the operation of the other circuitry in the chip. Off-chip decoupling capacitors should still be used across the power and ground pins for the output buffers.

Note that if this buffer is running at 500 MHz (a clock period of 2 ns), the average current supplied to the load is

$$I_{avg} = \frac{27\ pC}{2\ ns} = 13.5\ mA$$

A significant value for a single chip output. ■

A Final Comment

It should be clear that some thought needs to go into the sizing of the metal layers and the number of vias used when transitioning from one metal layer to the next. Ignoring the parasitics associated with the wires used in an IC is an invitation for disaster.

3.4 LASI Layout Examples

In this section we provide some additional LASI layout examples. In the first section we discuss laying out a pad and a padframe. In the next section we introduce the use of LasiCkt.

3.4.1 Laying out the Pad II

Let's say we want to lay out a chip in a 50 nm process. Further let's say that the final die size (chip size) must be approximately 1 mm on a side with a pad size of 100 μm square (again, the pads can be smaller depending on the process). From the MOSIS design rules (the CMOSEDU rules), the distance between pads must be at least 30 μm. Further let's assume a two-metal process (so metal2 is the top layer of metal the bonding wire drops down on). Table 3.2 summarizes the final and scaled sizes for our pads.

Table 3.2 Sizes for an example 1 mm square chip with a scale factor of 50 nm.

	Final size	Scaled size
Pad size	100 μm by 100 μm	2,000 by 2,000
Pad spacing (center to center)	130 μm	2,600
Number of pads on a side (corners empty)	6	6
Total number of pads	24	24
Overglass opening	88 μm by 88 μm	1,760 by 1,760

Let's start out by making a rank 1 cell called "via1" like the one seen in Fig. 3.13. The resulting cell is seen in Fig. 3.19. We'll use this cell in our pad to connect metal1 to metal2. The bond wire will touch the top metal2. However, we'll place metal1 directly beneath the metal2 so that we can connect to the pad using either metal1 or metal2. The layout of the pad is seen in Fig. 3.20. The spacing between the pads is a minimum of 30 μm. We use the outline layer (no fabrication significance) to help when we place the pads together to form a padframe. We've assumed the distance from the pad metal to the edge of the chip is 15 μm. Figure 3.21 shows how the overglass layer is placed in the pad metal area. Also seen in Fig. 3.21 is the placement of the "via1" cell in Fig. 3.19 around the perimeter of the pad. This ensures metal1 is solidly shorted to metal2, Fig. 3.22.

Next let's calculate, assuming we want a chip size of approximately 1 mm on a side, the number of pads we can fit on the chip. The size of the pad in Fig. 3.20 is 130 μm square. Dividing 1/0.13 we get, after rounding, eight pads on a side. However, the corners don't contain a pad so the actual number of pads on a side is six, see Fig. 3.23. Note, in this figure, that we made sure that the pad cell's image was showing (by pressing **i** on the keyboard). The small diamond in the image indicates the location of the reference marker in the cell. As seen in the figure, all of the reference markers in the pads are located close to the edge of the chip (use the **Rot** or LASI rotate command). The CMOS circuits are placed in the area inside the pads, while the part that gets cut up when the wafer is sawed into chips is the area outside the pads (the scribe), see Fig. 1.2.

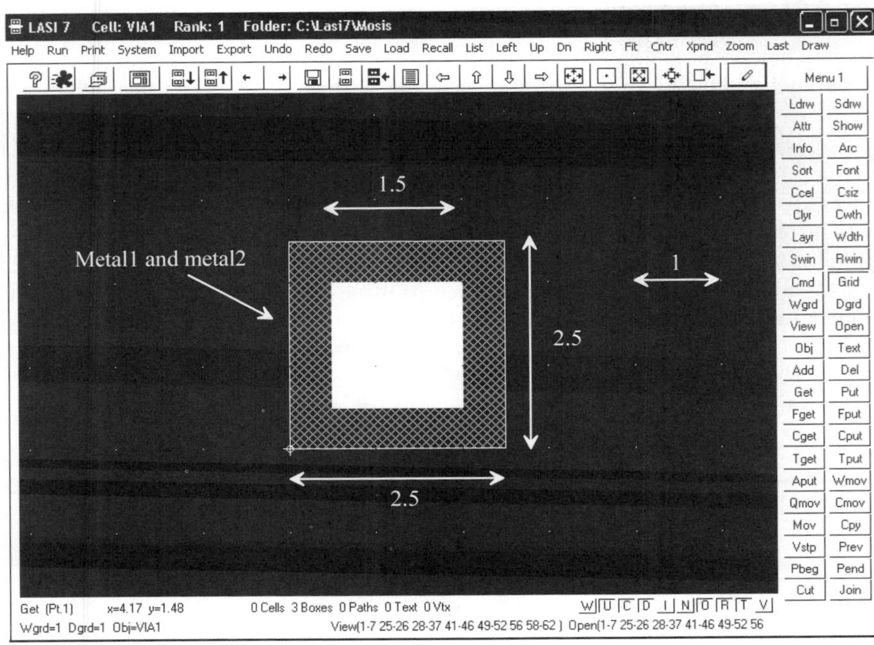

Figure 3.19 Layout of a Via1 cell.

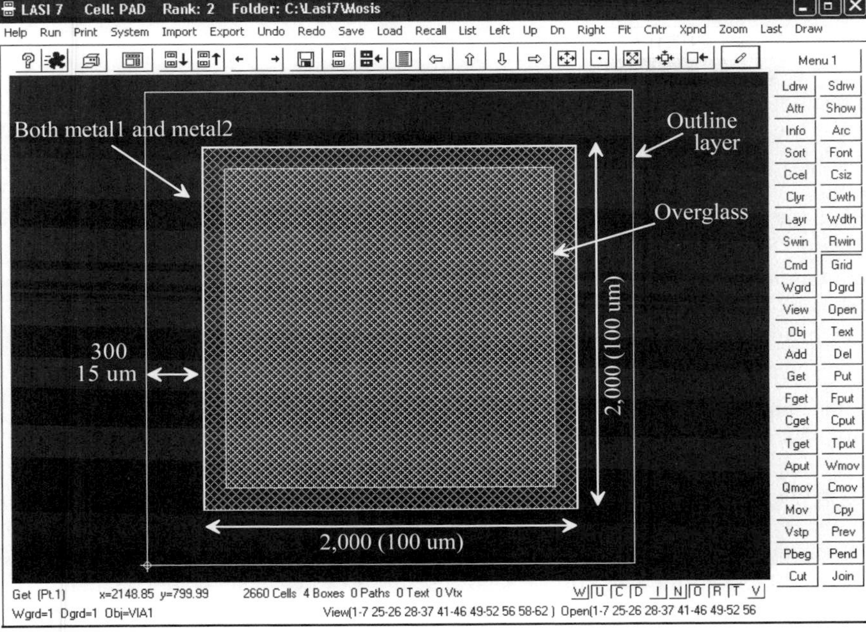

Figure 3.20 Layout of the bonding pad.

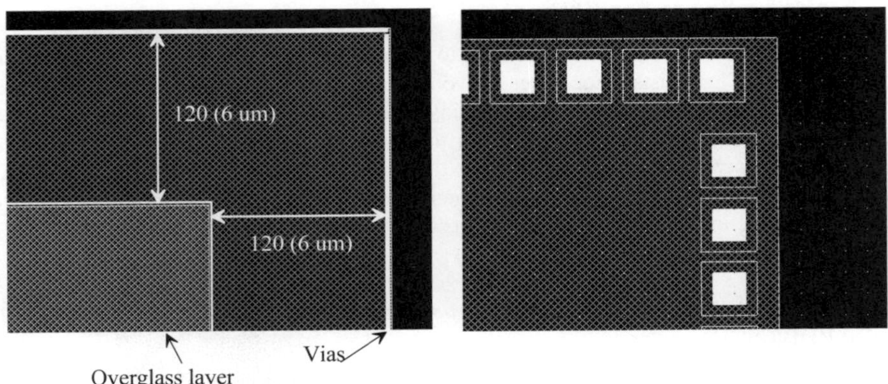

Figure 3.21 Corner detail for the pad in Fig. 3.20.

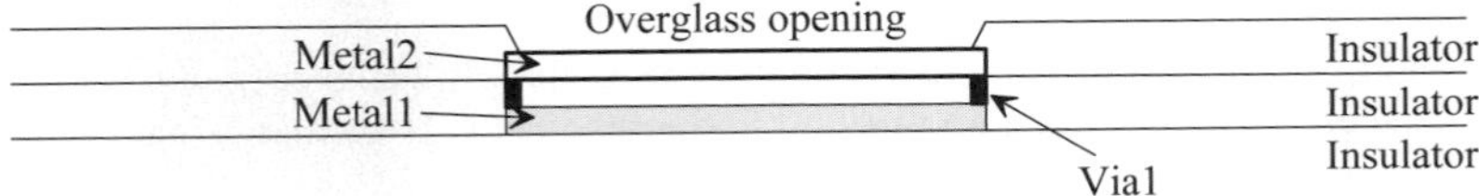

Figure 3.22 Simplified cross-sectional view of the bonding pad discussed in this section.

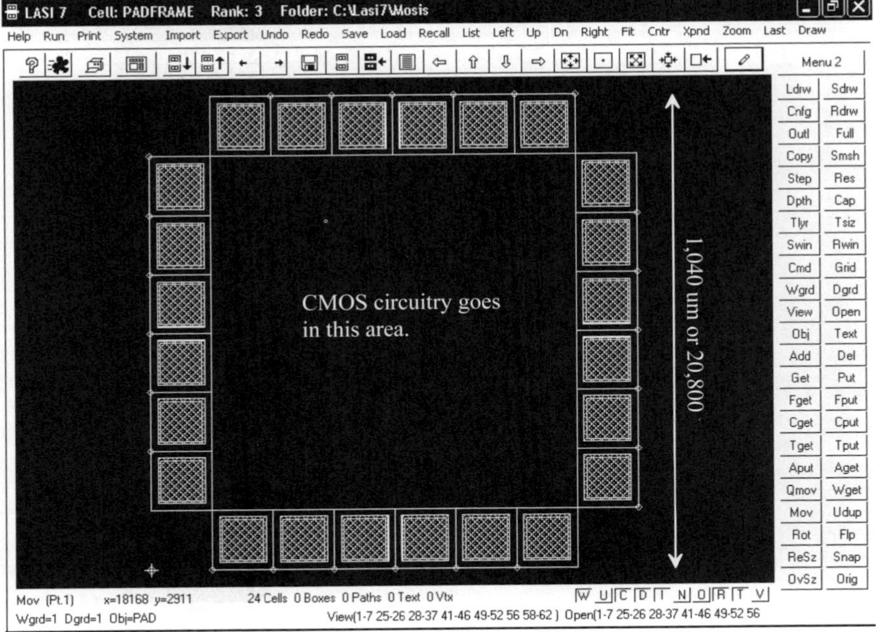

Figure 3.23 The layout of a padframe.

3.4.2 Introduction to LasiCkt

More than a layout tool, LASI can be used to draw schematics. LasiCkt can then be used to generate SPICE netlists from either a layout or a schematic. LASI uses its own format (TLC) for schematics (the schematics are drawn using the same tools and objects used to draw the layouts). Unlike some schematic tools that automatically label MOSFETs or resistors, the user needs to add certain text labels to a LASI drawing using the **Text** command. This text is used by LasiCkt to generate SPICE netlist files. In this section we give an introduction to LasiCkt along with a simple example.

Drawing a Schematic

Let's start by drawing the resistive divider schematic seen in Fig. 3.24 (a rank 2 cell called RDIV_SCH).

1. We can start by adding a rank 1 cell called SCH_RES to RDIV_SCH. This cell (SCH_RES) is a drawing of a resistor using the layer SCHM (schematic layer used for drawing symbols) and two connectors (using the connector text layer, CTXT).

2. Next we use a zero-width path (a polygon) on the MET1 layer (metal1) to connect the cells. We also add a ground cell to the schematic. Note that using a zero width box will **not** connect the symbols.

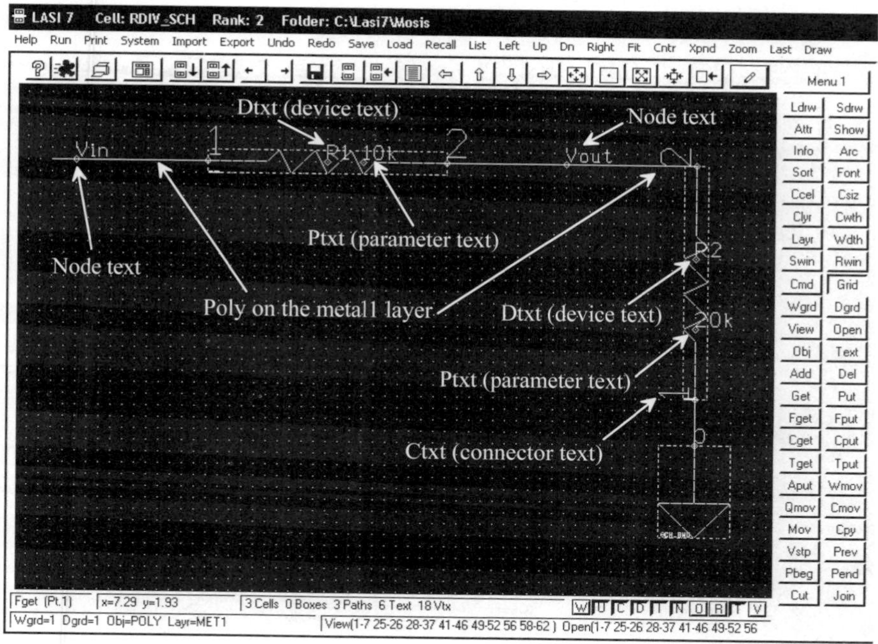

Figure 3.24 Drawing a schematic in LASI.

3. Next we label the nodes (the metal1 polygons) with names using text on the NTXT layer (node text layer). Here we use Vin and Vout for node names. If we don't label the wires, LasiCkt will select a virtual node name (like vn1, vn2, etc.)

4. Next we ensure that each cell's image is visible (the dotted lines around the cells). We place DTXT (device text) and PTXT (parameter text) to indicate the resistor's name and value. The name of a resistor must start with an "R."

We are now ready to run LasiCkt on the schematic. LasiCkt is started from the system menu. Figure 3.25 shows LasiCkt's setup menu. For the moment we'll simply enter the cell's name into the "Name of Cell" field and select "Schematic". Pressing OK and selecting "Go" on the LasiCkt window generates a text file called "rdiv_sch.cir." The contents of the file are

```
*** SPICE Circuit File of RDIV_SCH

* MAIN RDIV_SCH
R1 Vin Vout 10k
R2 0 Vout 20k
.END
```

While this example is simple, we'll see that as our circuits get complicated the ability to generate a netlist from a schematic or a layout allows us to verify that the layout matches the schematic. Sometimes the comparison is called "layout versus schematic" or LVS.

Note that the header file is a text file indicating the type of analysis and the simulation parameters (see the .control line to the .tran line in the netlist on page 24). The footer file is used for models (most often MOSFET models). For more information, see LasiCkt's online help manual.

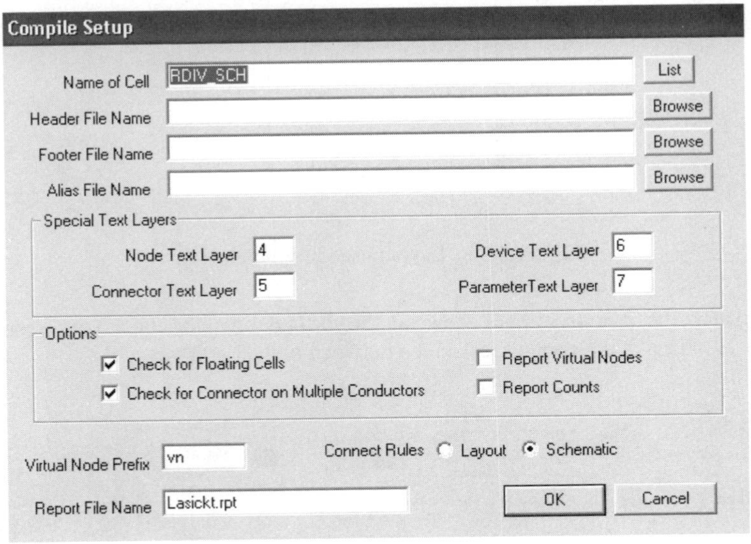

Figure 3.25 LasiCkt setup menu.

ADDITIONAL READING

[1] R. S. Muller, T. I. Kamins, and M. Chan, *Device Electronics for Integrated Circuits*, John Wiley and Sons Publishers, 2002. ISBN 0-471-59398-2

[2] D. E. Boyce, *LASI User's Manual*, available while LASI is running by pressing F1 on the keyboard and the button the user needs help with (at the same time). Alternatively, the user can open the help file in the directory C:\Lasi7\help.

PROBLEMS

Unless otherwise indicated, use the data from Table 3.1, a metal sheet resistance of 0.1 Ω/square, and a metal contact resistance of 10 Ω.

3.1 Redraw the layout and cross-sectional views of a pad, similar to Fig. 3.2, if the final pad size is 50 μm by 75 μm with a scale factor of 100 nm.

3.2 Estimate the capacitance to ground of the pad in Fig. 3.20 made with both metal1 and metal2.

3.3 Suppose a parallel plate capacitor was made by placing a 100 μm square piece of metal1 directly below the metal2 in Fig. 3.2. Estimate the capacitance between the two plates of the capacitor (metal1 and metal2). Estimate the capacitance from metal1 to substrate. The unwanted parasitic capacitance from metal1 to substrate is often called the **bottom plate** parasitic.

3.4 Sketch the cross-sectional view for the layout seen in Fig. 3.26.

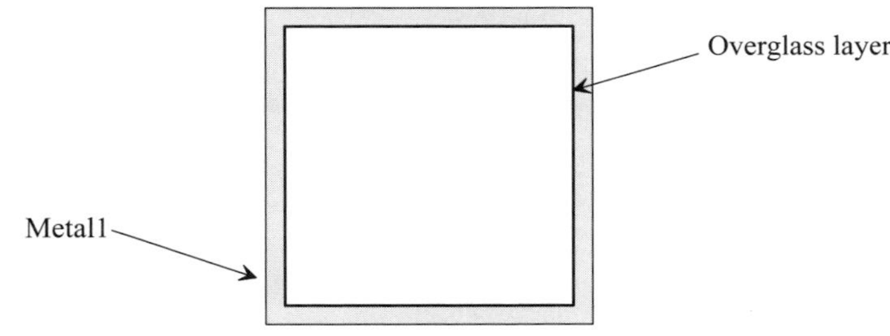

Figure 3.26 Layout used in Problem 3.4.

3.5 Sketch the cross-sectional view, at the dashed line, for the layout seen in Fig. 3.27. What is the contact resistance between metal3 and metal2?

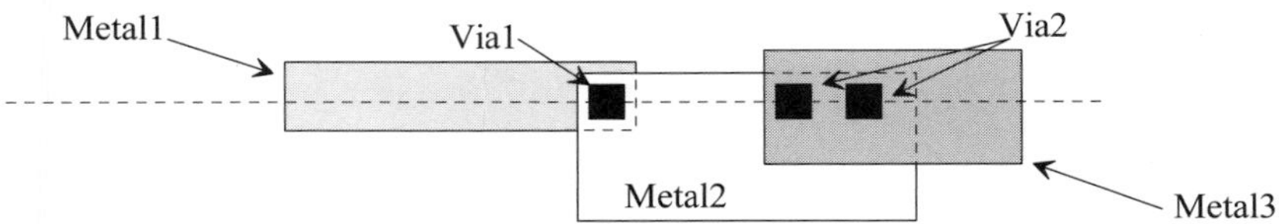

Figure 3.27 Layout for Problem 3.5.

3.6 The insulator used between the metal layers (the interlayer dielectric, ILD) can have a relative dielectric constant well under the relative dielectric constant of SiO_2 (= 4). Estimate the intrinsic propagation delay through a metal line encapsulated in an ILD with a relative dielectric constant of 1.5. What value of metal sheet resistance, using the values from Ex. 3.3, would be required if the RC delay through the metal line is equal to the intrinsic delay?

3.7 Using $CV = Q$, rederive the results in Ex. 3.5.

3.8 For the layout seen in Fig. 3.28, sketch the cross-sectional view (along the dotted line) and estimate the resistance between points A and B. Remember that a via is sized 1.5 by 1.5.

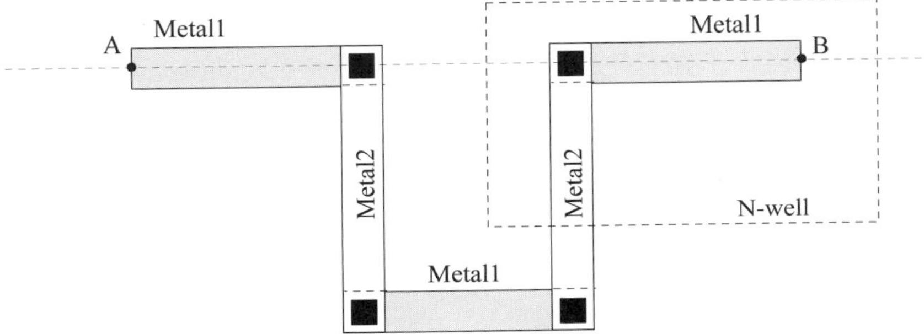

Figure 3.28 Layout for Problem 3.8.

3.9 Laying out two metal wires directly next to each other, and with minimum spacing, for a long distance increases the capacitance between the two conductors, C_m. If the two conductors are *VDD* and ground, is this a good idea? Why or why not?

3.10 Consider the schematic seen in Fig. 3.29. This circuit can be used to model ground bounce and *VDD* droop. Show, using SPICE, that a decoupling capacitor can be used to reduce these effects for various amplitude and duration current pulses.

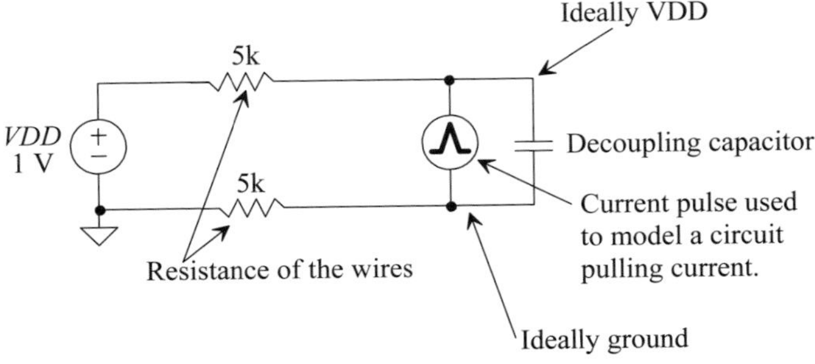

Figure 3.29 Circuit used to show the benefits of a decoupling capacitor.

3.11 Use LasiCkt to draw the schematic seen in Fig. 1.27. The input pulse source should be added into the schematic. Generate a header file that can be used to generate the waveforms seen in the figure. Simulate the output (from LasiCkt) netlist file.

Chapter

4

The Active and Poly Layers

The active, n-select, p-select, and poly layers are used to form n-channel and p-channel MOSFETs (NMOS and PMOS respectively) and so metal1 can make an ohmic contact to the substrate or well. The active layer, in a layout program, defines openings in the silicon dioxide covering the substrate (see Figs. 2.3 and 2.4). The n-select and p-select layers indicate where to implant n-type or p-type atoms, respectively. The active and select layers are always used together. The active defines an opening in the oxide and the select then dopes the semiconductor in the opening either n-type or p-type.

The poly layer forms the gate of the MOSFETs. Poly is a short name for polysilicon (not to be confused with the poly, or polygon, object in a layout program). Polysilicon is made up of small crystalline regions of silicon. Therefore, in the strictest sense, poly is not amorphous silicon (randomly organized atoms), and it is not crystalline silicon (an orderly arrangement of atoms in the material) such as the wafer.

4.1 Layout using the Active and Poly Layers

We've covered the following fabrication layers in Chs. 2 and 3: n-well (NWEL), metal1 (MET1), via1 (VIA1), metal2 (MET2), and overglass (OVGL). In this section we cover the following additional fabrication layers: active (ACTV), n-select (NSEL), p-select (PSEL), poly1 (POL1), silicide block (SILI), and contact (CONT).

The Active Layer

Examine the layout of a box and the corresponding angled view (the fabrication results) seen in Fig. 4.1. The box is drawn on the active layer and indicates where to open a hole in the field oxide (FOX). These openings are called active areas. The field area (the area that isn't the active area, which is the area where the FOX is grown) is used for routing wires (connecting the circuit together). The MOSFETs are fabricated in the bulk (the p-substrate or the n-well) in these active openings. The FOX is used to isolate the devices from one another (the active areas are isolated by the FOX). Note that there will be some resistive connection between active areas (either through the substrate or the n-well). However, the FOX is grown thick enough to keep the interactions between adjacent active areas to a minimum.

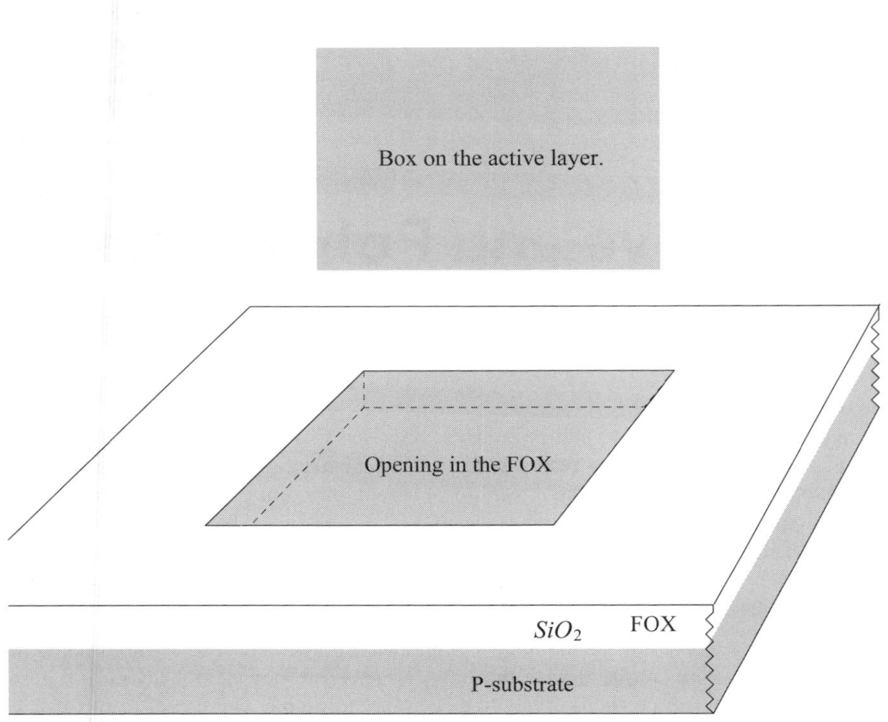

Figure 4.1 How the active layer specifies where to open holes in the field oxide (FOX).

The P- and N-Select Layers

Surrounding the active layer with either the n-select or p-select layers dopes the semiconductor n- or p-type. Figure 4.2 shows several combinations of selects, n-well, and active layers. In (a) and (b) for example, the opening in the FOX is implanted p-type (in the location determined by the p-select mask). When learning to do design and layout, it's important to be able to see a layout and then visualize the corresponding cross-sectional views.

Also seen in this figure (see 4.2i and j) is how a single layer (called the n+ layer) can be used instead of two layers (the active and n-select layers). The n+ layer in (j) is used directly for the active mask (openings in the FOX) in (i). The n-select in (i) is a *derived* mask. It is derived by bloating the size of the n+ layer. The n-select mask must be larger than the active mask due to misalignment. If the implant (select) isn't aligned directly over the active opening in the FOX, then the semiconductor exposed in the active opening won't get doped. If the select and active masks could be aligned perfectly, the select layers wouldn't need to be larger than the active layer. Also note that using a select without active causes the implant to bombard the FOX. Since the FOX is thick, it keeps the implanted atoms from reaching the substrate.

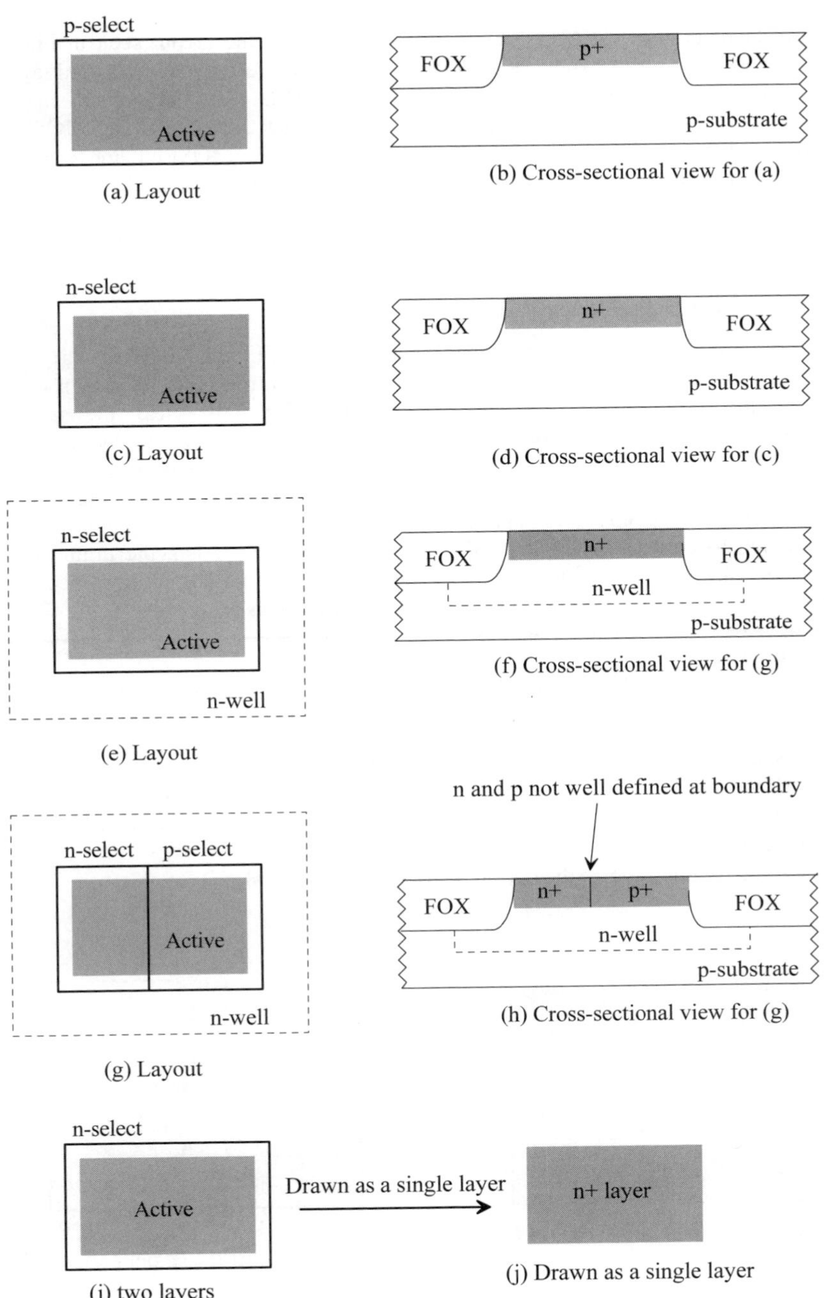

Figure 4.2 Combinations of active, selects, and n-wells.

The Poly Layer

The poly layer is used for MOSFET formation. Consider the layout seen in Fig. 4.3a. Drawing poly over active produces a MOSFET layout. If we see a complicated layout, it is straightforward to determine the number of MOSFETs in the layout simply by counting how many times poly crosses active. Note that the gate of the MOSFET is formed with the polysilicon, and the source and drain of the MOSFET are formed with the n+ implant. Further note that the source and drain of an integrated MOSFET (in a general CMOS process) are interchangeable. We are not showing the (required) contact to the substrate (the body of the MOSFET). The body connection will be covered in a moment (the MOSFET is a four-terminal device).

Self-Aligned Gate

Notice how, Fig. 4.3b or c, the area under the poly gate isn't doped n+. After the opening in the FOX is formed with the active mask, a thin insulating oxide is grown over the opening. This is the MOSFET's gate oxide (GOX), Fig. 4.3b. Next, the poly mask specifies where to deposit the polysilicon gate material. This is followed by applying the implant in the areas specified by the n-select mask. The implant easily penetrates through the thin GOX into the source and drain areas. However, the polysilicon gate acts like a mask to keep the n+ atoms from penetrating under the MOSFET's gate (the poly is made thick enough to ensure the implant doesn't reach the GOX). Also, the drain and gate become self-aligned to the source/drain of the MOSFET. This is important because we know we can't perfectly align the poly mask to the active masks.

Example 4.1
Comment on the problems with the MOSFET layout seen in Fig. Ex4.1.

In Fig. Ex4.1a the active layer defines an opening in the field oxide. The select masks are placed exactly where the desired n+ implants will occur, as seen in Fig. 4.3. However, due to shifts in the select mask, relative to the poly mask, the area directly next to the gate will not get implanted. (Redraw Fig. Ex. 4.1b with the poly layer shifted left or right.) Notice how the incorrect layout in Fig. Ex4.1b looks exactly the same as the correct layout in Fig. 4.3a. ■

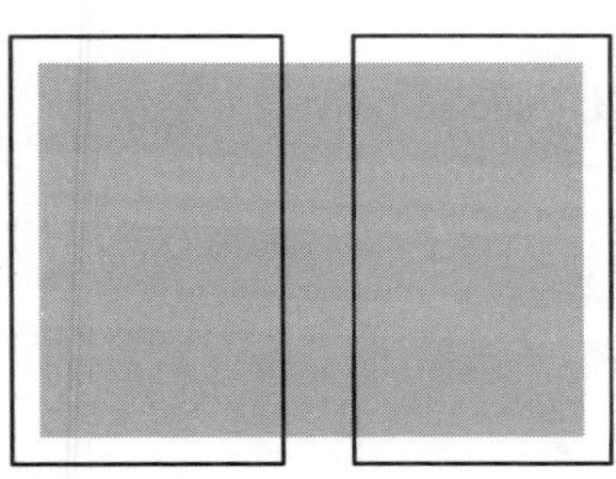

(a) Drawing active and select for a MOSFET (bad).

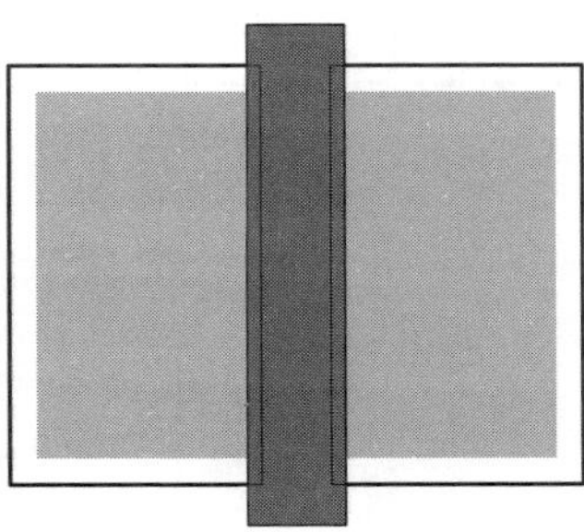

(b) Placing poly over the layout in (a) (bad).

Figure Ex4.1 Bad layout examples (what NOT to do).

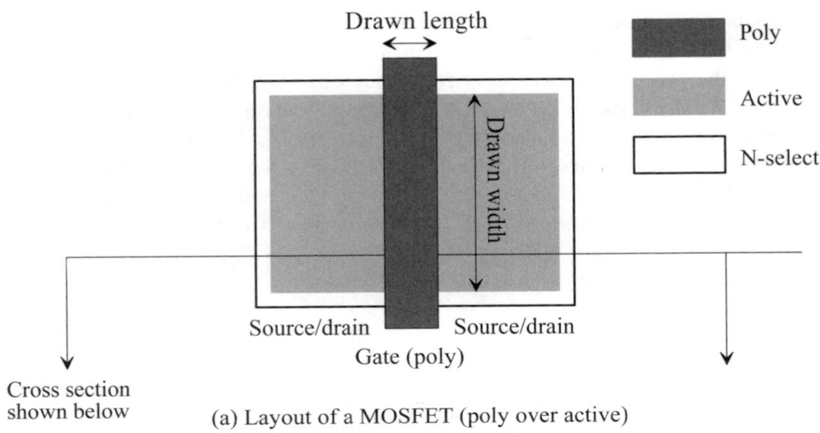

(a) Layout of a MOSFET (poly over active)

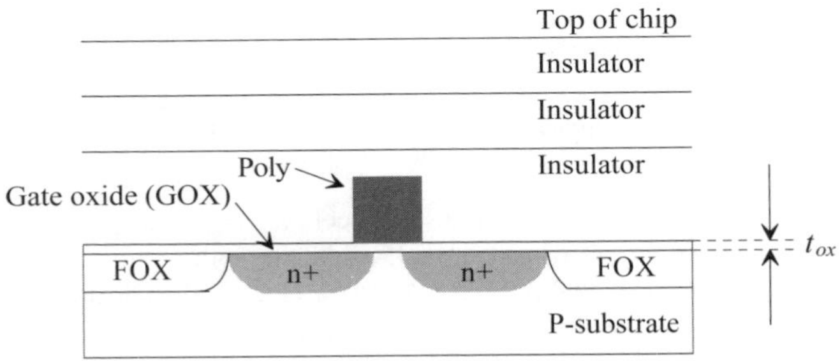

(b) Cross-sectional view of the layout in (a).

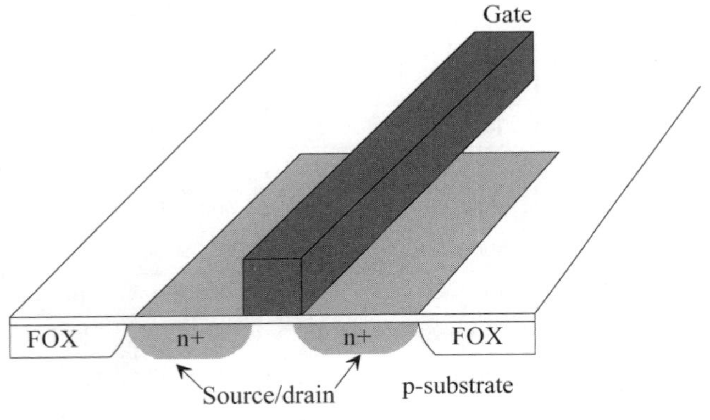

(c) Angled view of the layout in (a)

Figure 4.3 Layout and cross-sectional views of a MOSFET.

The Poly Wire

The poly layer can also be used, like metal1, as a wire. Poly is routed on top of the FOX. The main limitation when using the poly layer for interconnection is its sheet resistance. As we saw in the last chapter, the sheet resistance of the metal layers is approximately 0.1 Ω/square. The sheet resistance of the doped poly can be on the order of 200 Ω/square. The capacitance to substrate is also larger for poly simply because it is closer to the substrate (see Table 3.1). Therefore, the delay through a poly line can be considerably longer than the delay through a metal line. To reduce the sheet resistance of poly (and of the implanted active regions), a silicide (a material that is a mixture of silicon and a refractory metal like tungsten) is deposited over the MOSFET and field region, Fig. 4.4. The silicide and poly gate sandwich is called a polycide.

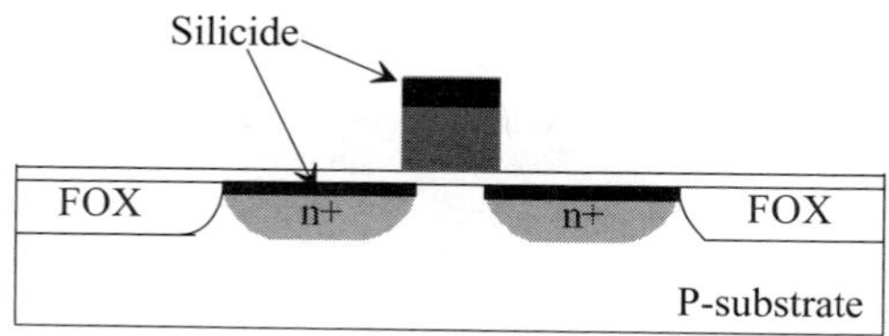

Figure 4.4 How the gate and drain/source of a MOSFET are silicided to reduce sheet resistance.

Table 4.1 gives some typical values of sheet resistance for poly, n+, and p+. Note that when we use the poly as a mask to self-align the source and drain regions of the MOSFET to the gate we dope the poly either n-type (for an NMOS device) or p-type (for a PMOS device). The other specifications (temperature coefficient, TCR, or voltage coefficients, VCR) seen in the table will be used in later chapters.

Table 4.1 Example properties of resistive materials in a nm CMOS process.

Sili-cide	Resistor type	Rs (ohms/sq) AVG.	TCR1 (ppm/C) AVG.	TCR2 (ppm/C²) AVG.	VCR1 (ppm/V) AVG.	VCR2 (ppm/V²) AVG.	Mis-match % ΔR/R
	n-well	500 ± 10	2400 ± 50	7 ± 0.5	8000 ± 200	500 ± 50	< 0.1
	n+ poly	200 ± 1	20 ± 10	0.6 ± 0.03	700 ± 50	150 ± 15	< 0.5
	p+ poly	400 ± 5	160 ± 10	0.8 ± 0.03	600 ± 50	150 ± 15	< 0.2
	n+	100 ± 2	1500 ± 10	0.04 ± 0.1	2500 ± 50	350 ± 20	< 0.4
	p+	125 ± 3	1400 ± 20	0.4 ± 0.1	80 ± 80	100 ± 25	< 0.6
*	n+ poly	5 ± 0.3	3300 ± 90	1.0 ± 0.2	2500 ± 125	3800 ± 400	< 0.4
*	p+ poly	7 ± 0.1	3600 ± 50	1.0 ± 0.2	2500 ± 400	5500 ± 250	< 0.7
*	n+	10 ± 0.1	3700 ± 50	1.0 ± 0.2	350 ± 150	600 ± 60	< 1.0
*	p+	20 ± 0.1	3800 ± 40	1.0 ± 0.2	150 ± 50	800 ± 40	< 1.0

Silicide Block

In some situations (as in making a resistor), it is desirable to keep from depositing the silicide on the gate poly or source/drain regions. A layer called the silicide block (SILI in the LASI setups) can be used for this purpose. Consider the following example.

Example 4.2
Estimate the delay through the poly wire in Fig. 4.5 with and without silicide. The width of the wire is 1 and the length is 1,000. Use a scale factor of 50 nm and the values for capacitance in Table 3.1. Simulate the delay using SPICE.

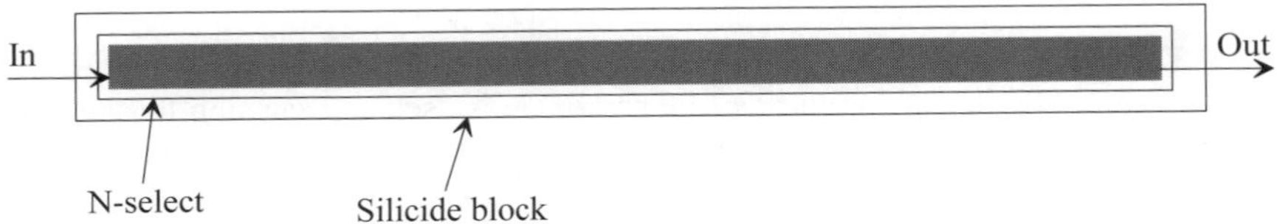

Figure 4.5 Estimating the delay through a polysilicon line with and without a silicide.

The capacitance of the poly wire to substrate doesn't depend on the presence or absence of the silicide. Using the data from Table 3.1, we can estimate the capacitance of the poly wire to substrate as

$$C_{poly} = C_{plate} \cdot area_{drawn} \cdot (scale)^2 + C_{fringe} \cdot perimeter_{drawn} \cdot scale \qquad (4.1)$$

or

$$C_{poly} = (58\ aF) \cdot (1{,}000) \cdot (0.05)^2 + (88\ aF) \cdot (2{,}002) \cdot (0.05) \approx 9\ fF$$

The resistance of the poly wire is calculated using

$$R = R_{square} \cdot \frac{L}{W} \qquad (4.2)$$

From the data in Table 4.1, the resistance of the wire is either 200k (no silicide) or 5k (with silicide). The delays are then calculated as

$$t_d = 0.35 \cdot 9\ fF \cdot 200k = 630\ ps \text{ and } t_d = 0.35 \cdot 9\ fF \cdot 5k = 16\ ps$$

The simulation results are seen in Fig. 4.6. In the simulation netlist we divided the line up into 1,000 squares. The capacitance/square is 9 aF (remembering SPICE doesn't recognize "a" so we use e-18). The resistance/square is either 200 Ω (no silicide) or 5 Ω (with a silicide). ■

4.1.1 Process Flow

A generic CMOS process flow is seen in Fig. 4.7. The fabrication of both PMOS and NMOS devices is detailed in this figure. This figure doesn't show the initial steps taken to fabricate the n-well (and/or p-well) but rather starts with the wells already fabricated.

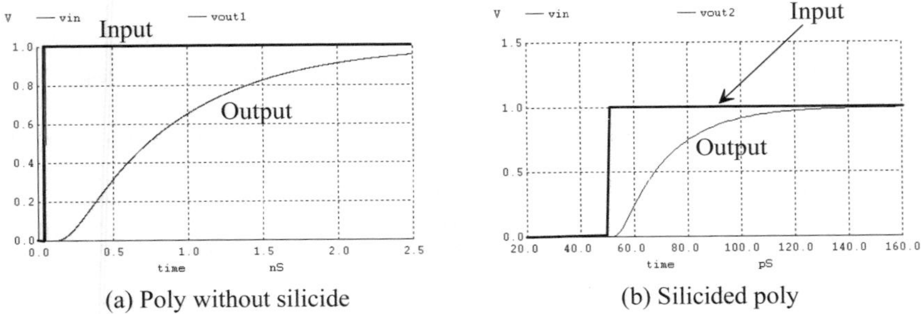

(a) Poly without silicide (b) Silicided poly

Figure 4.6 Simulated delay through poly wires.

The first step, Fig. 4.7a, is to grow a thin pad oxide on top of the entire wafer. This is followed by depositing nitride (the pad oxide is used as a cushion for the nitride) and photoresist layers. The photoresist is then patterned using the active mask. The remaining photoresist, seen in Fig. 4.7a, ultimately defines the openings in the FOX.

In Fig. 4.7b the areas not covered by the photoresist are etched. The etching extends down into the wafer so that *shallow trenches* are formed. In (c) the shallow trenches are filled with SiO_2. These trenches isolate the active areas and form the field regions (FOX). This type of device isolation is called shallow trench isolation (STI).

In (d) two separate implants are performed to adjust the threshold voltages of the devices. A photoresist is patterned (twice) to select the areas for threshold voltage adjust.

Figure 4.7e shows the results after the deposition and patterning of polysilicon (for the MOSFET gate material). This is followed by several implants. In (f) we see a shallow implant to form the MOSFET's lightly doped drains (LDD). The LDD implants prevent the electric field directly next to the source/drain regions from becoming too high (this is discussed further in Ch. 6). Note that the poly gate is used as a mask during this step.

The next step is to grow a spacer oxide on the sides of the gate poly, Fig. 4.7g. After the spacer is grown, the n+/p+ implants are performed. This implant dopes the areas used for the source/drain of the MOSFETs as well as the gate poly. The last step is to silicide the source and drain regions of the MOSFET. This is important for reducing the sheet resistances of the polysilicon and n+/p+ materials, as indicated in Table 4.1.

Finally, note that the process sequence seen in Fig. 4.7 is often called, in the manufacturing process, the front-end of the line (FEOL). The fabrication of the metal layers and associated contacts/vias is called the back-end of the line (BEOL).

Damascene Process Steps

The process of: 1) making a trench, 2) (over) filling the trench with a material, and 3) grinding the material down until the top of the wafer is flat is called a Damascene process (a technique used, and invented, by craftsman in the city of Damascus to inlay gold or silver in swords). The STI process just described is a Damascene process. More often, though, the Damascene process is associated with the metal layers in a CMOS process. Trenches are formed in the insulators. Copper, for example, is then deposited in the trenches. The top of the wafer is then ground down until it is flat.

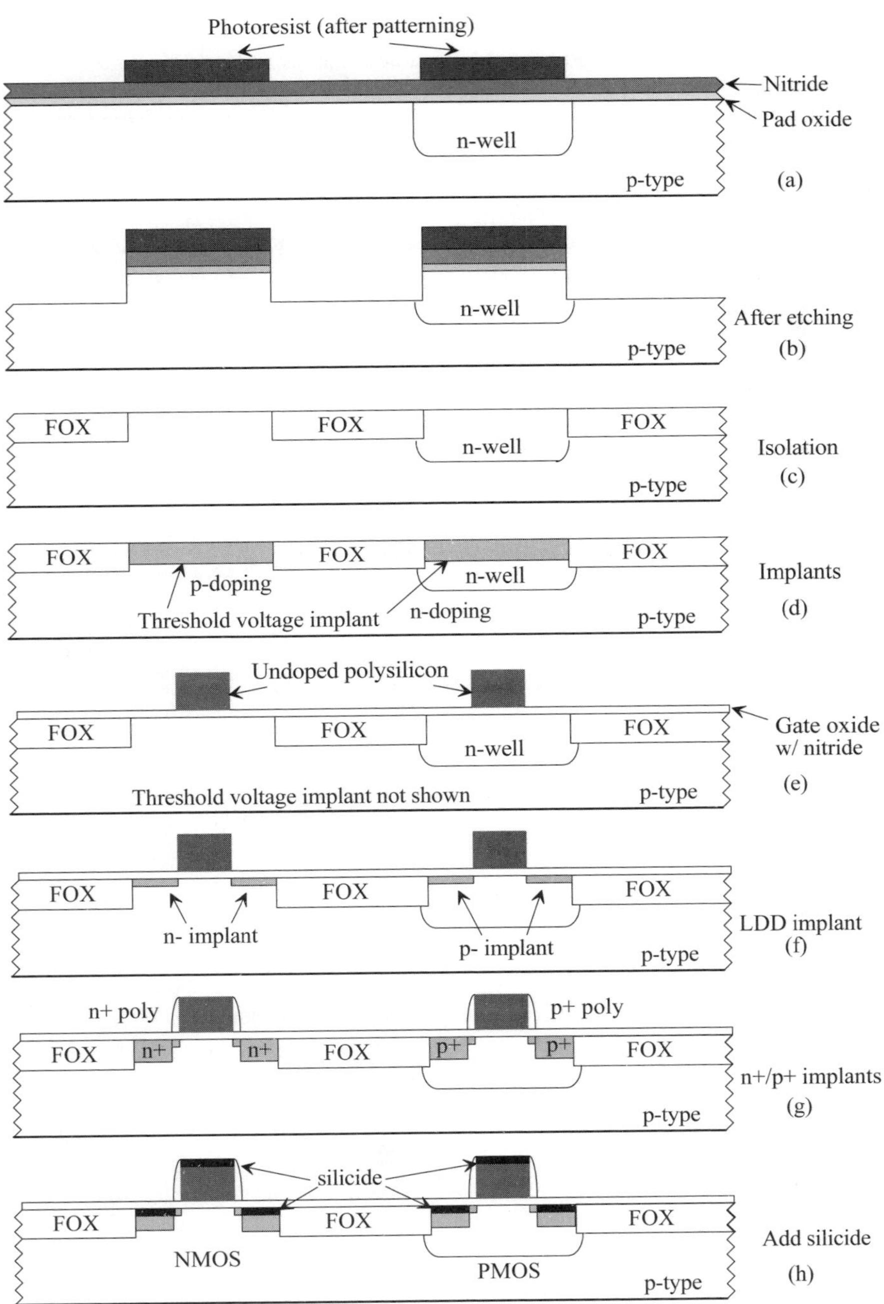

Figure 4.7 General CMOS process flow.

4.2 Connecting Wires to Poly and Active

In the last section we discussed how to lay out active areas and polysilicon. Here, in this section, let's discuss how to connect metal wires to poly and active. The contact layer connects metal1 to either active (n+/p+) or poly. Unless we want to form a rectifying contact (a Schottky diode), *we never connect metal directly to the substrate or well.* Further, we won't connect metal to poly without having the silicide in place. Never put a silicide block around a contact to poly.

Figure 4.8a shows a layout and corresponding cross-sectional view of the layers metal1, contact (CONT), and poly (a contact to poly). Figure 4.8b shows a connection to n+ and p+. Note that, like we did with the via cell in Fig. 3.13, we can layout contact cells to poly, n+, and p+. Further note that metal1 is connected to either metal2 (through via1) or poly/active. Metal2 can't be connected to active/poly without first connecting to metal1 and a contact.

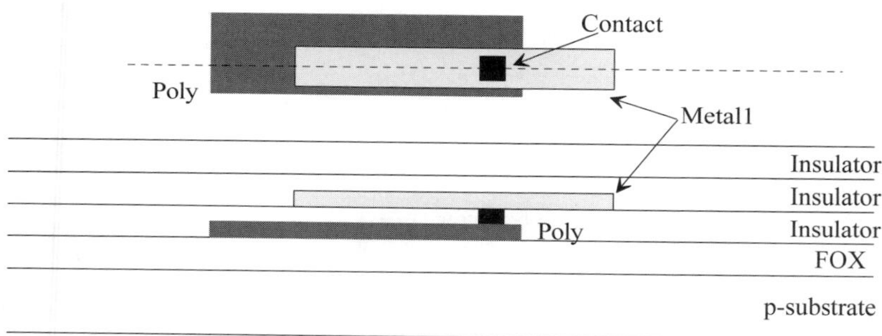

(a) Metal1 connecting to poly through a contact.

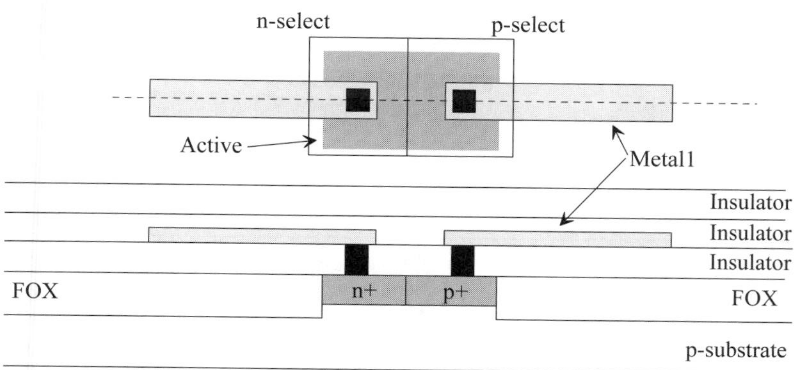

(b) Contacts to active.

Figure 4.8 How metal1 is connected to poly and active.

When etching an opening for a contact to poly or active, an etchant stop layer is used. The etchant stop is put down directly on top of the FOX prior to depositing the insulator.

Connecting the P-Substrate to Ground

So far, throughout the book, we've said that the p-substrate is at ground potential. However, we haven't actually said how we connect the substrate to ground (the substrate must be connected to the ground pad through a wire). The substrate, as seen in Fig. 4.4, is the body of the NMOS devices and is common to all NMOS devices fabricated on the chip (assuming an n-well process, see Fig. 2.24). Towards connecting the substrate to ground, consider the layout seen in Fig. 4.9. Again note that we only connect metal1 to p+ (or n+/poly) and not directly to the substrate. Further notice that poly sits on the FOX while metal1 sits on the insulator above FOX.

An important consideration when "tying down the substrate" is the number of places, around the chip, the substrate is tied to ground. We don't just connect the substrate to ground with one connection, like the one seen in Fig. 4.9, and assume that the entire chip's substrate is grounded. The reason for this is that the substrate is a resistive material. The circuitry fabricated in the substrate (in the bulk) has leakage currents (DC and AC) that flow in the p-type semiconductor of the substrate. The result is an increase in the substrate's potential above ground in localized regions of the chip. Ideally, the current flowing in the substrate is zero. In reality it won't be zero but will have some value that depends on the location and the activity of the on-chip circuitry. A substrate connection provides a place to remove this substrate current (a point of exit), keeping the substrate potential at ground. In practice, substrate connections are used wherever possible (more on this topic when we cover standard cell frames later in the chapter).

The body for the PMOS devices is the n-well. The n-well must also be tied to a known voltage through n+ and metal1. For the PMOS's body connection, an n+ region is placed in the n-well and connected with a metal1 wire to (for digital design) *VDD*. If an n-well is laid out and used to make a resistor or for the body of a PMOS device, then there must be n+ in the n-well.

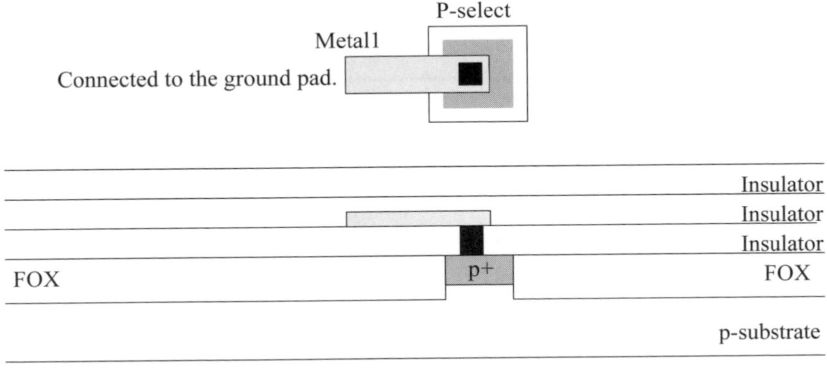

Figure 4.9 Connecting the substrate to ground.

Layout of an N-Well Resistor

The layout of an n-well resistor is seen in Fig. 4.10. On each side of the resistor, n+ regions are implanted so that we can drop metal1 down and make a connection. We aren't showing the silicide in the cross-sectional view (Fig. 4.4). Further, after reviewing Table 4.1, we see that the sheet resistance of the n+ regions is small compared to the sheet resistance of the n-well. When we calculate the number of squares, we measure between edges of active as seen in the figure. This results in a small error in the measured resistance compared to the actual resistance. The variation in the sheet resistance with process shifts (say 20%) makes this error insignificant. The next chapter will discuss the layout of resistors in more detail.

Notice that if the substrate is at ground, we can't apply a potential on either side of the wire less than, say, –0.5 V for fear of turning on the n-well to substrate parasitic diode. These parasitics, as discussed in Ch. 2, are an important concern when laying out resistors.

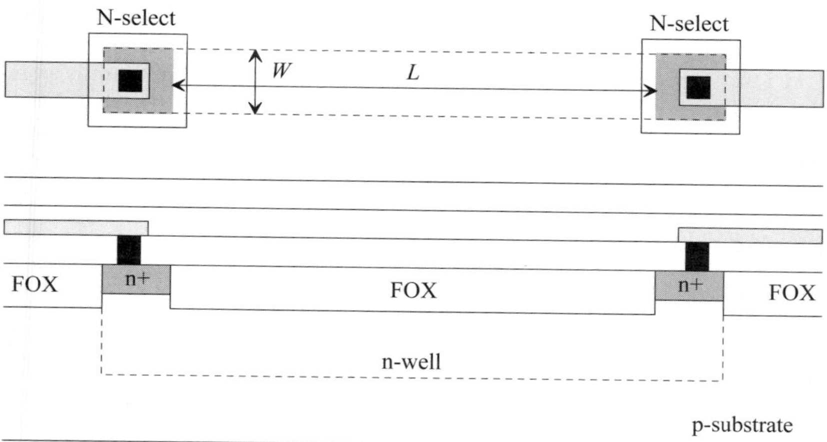

Figure 4.10 Layout of an n-well resistor and the corresponding cross-sectional view.

Example 4.3
Consider the layout for a metal1 connection to an n-well resistor seen in Fig. 4.11. Will the extension of the n+ beyond the n-well affect the resistor's operation? Why or why not?

The cross-sectional view along the dotted line in the layout is also seen in the figure. The n+ active forms a diode with the p-substrate. As long as this diode doesn't forward bias, the connection to the n-well works just like the connection in Fig. 4.10. There is additional junction (depletion) capacitance with this connection (between the n+ and substrate). When the resistor in Fig. 4.10 is fabricated, the active mask won't be perfectly aligned with the n-well. As this example illustrates, it doesn't have to be perfectly aligned for a proper connection to the resistor. ■

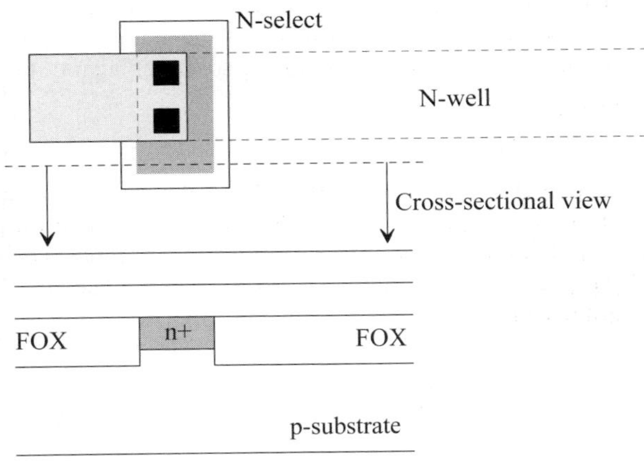

Figure 4.11 Active extending beyond the edges of the n-well.

Layout of an NMOS Device

Figure 4.12 shows the layout, schematic representation, and cross-sectional views of an NMOS device. As seen in Fig. 4.3 and the associated discussion, **poly over active** forms a MOSFET. It's important to remember that the MOSFET is a four-terminal device. In

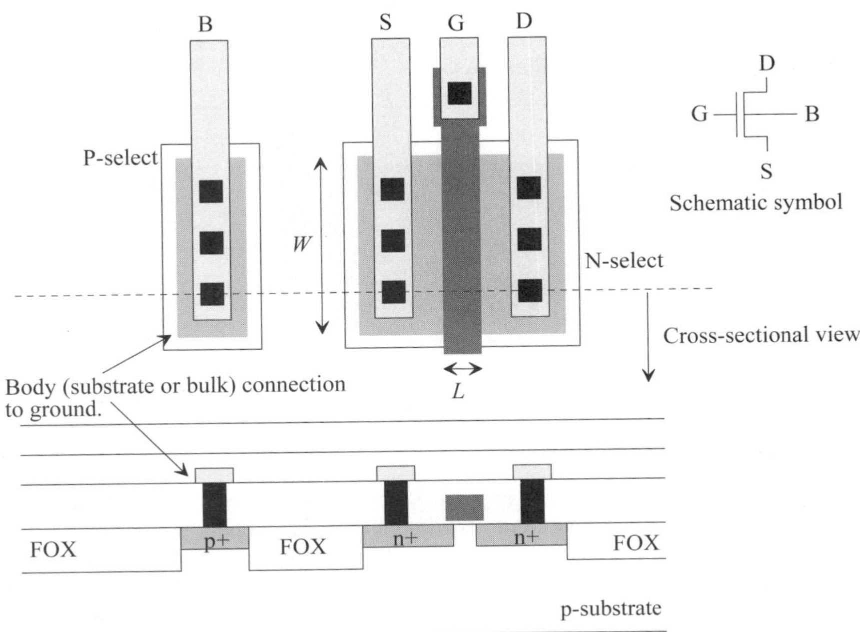

Figure 4.12 Layout and cross-sectional views of an NMOS device.

Fig. 4.12 we show the bulk connection in the layout and in the schematic. In an n-well process, the bulk is tied to ground so the bulk (or body) connection is normally not shown in the schematic symbol. Also note, again, that the source and drain are interchangeable.

Layout of a PMOS Device

Figure 4.13 shows the layout of a PMOS device. Note how we lay the device out in an n-well. Also seen in the figure is the schematic symbol for the PMOS device. Again, the source and drain of the MOSFET are interchangeable. The n-well is normally tied to the highest potential, *VDD*, in the circuit to keep the parasitic n-well/p-substrate diode from forward biasing.

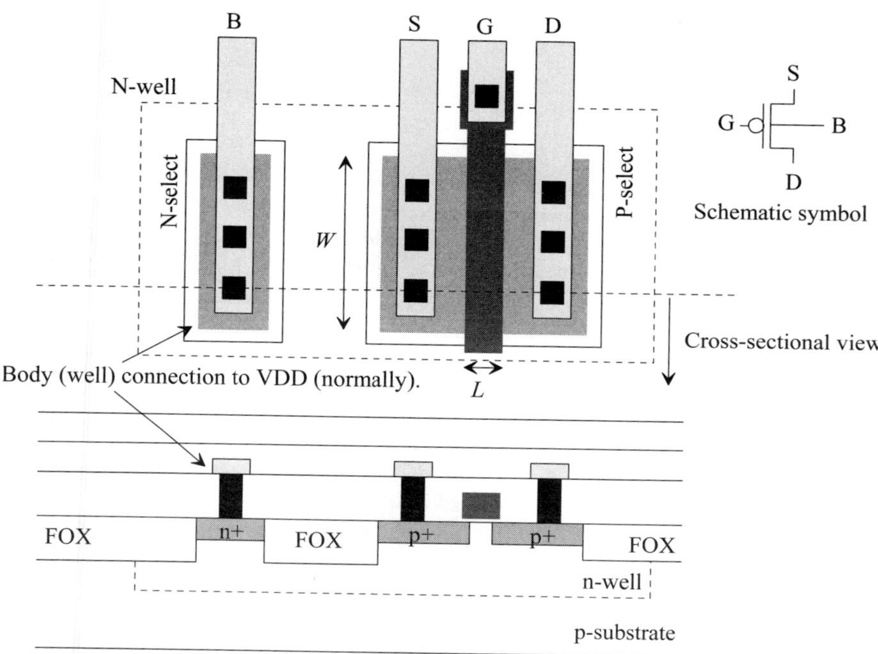

Figure 4.13 Layout and cross-sectional views of an PMOS device.

A Comment Concerning MOSFET Symbols

We'll use the symbols shown in Figs. 4.14a and b to represent NMOS and PMOS devices. Because the bulk terminals aren't drawn in these symbols, it's assumed they are connected to ground (NMOS device) or *VDD* (PMOS device). Notice that the drain and source in this symbol, like the actual layout, are interchangeable. The symbols seen in (c) and (d) are the bipolar derived symbols where the arrow on the MOSFET's source represents the direction of drain current flow (derived from a bipolar junction transistor symbol). Since, for an integrated circuit MOSFET, the current can flow in either direction through the MOSFET, we'll avoid using the bipolar-derived symbols. Finally,

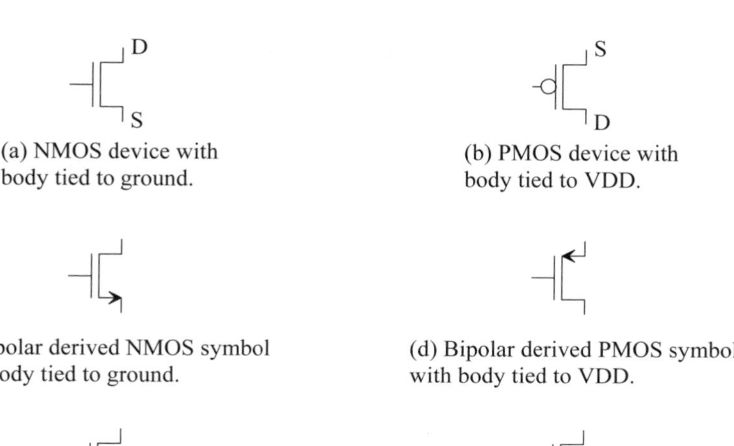

(a) NMOS device with body tied to ground.

(b) PMOS device with body tied to VDD.

(c) Bipolar derived NMOS symbol with body tied to ground.

(d) Bipolar derived PMOS symbol with body tied to VDD.

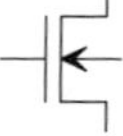

(e) NMOS symbol where the arrow indicates the p-substrate to n-channel diode.

(f) PMOS symbol where the arrow indicates the n-well to p-channel diode.

Figure 4.14 Symbols commonly found in schematics to represent a MOSFET.

the symbols seen in (e) and (f) are also used to draw schematics. The arrow in bulk terminal represents the p-substrate to n-channel diode (NMOS) or the p-channel to n-well diode (PMOS).

Standard Cell Frame

When doing layout, it can be convenient to route power and ground as well as substrate and well connections in a fashion that makes connecting the MOSFETs together simpler. We don't have to worry about the substrate and well connections or the routing of power. Towards this goal, consider the layout seen in Fig. 4.15a. The top half of the layout is an n-well. This n-well is where PMOS devices are placed, as indicated in the figure. The top of the n-well is tied to *VDD* through an n+ implant. In other words, the metal on the top of the layout has n+ underneath it and is routed to the *VDD* pad. Below this layout of the n+ well tie-down, and still in the n-well, is a p-select region. We can draw one or more PMOS devices with active, poly, contacts, and metal1 and place them in this area. The NMOS devices are laid out in the are below this, that is, the area of the n-select layer. Below the NMOS area is the substrate connection, p+, to ground. The metal, at the bottom of the layout, connects the p+ substrate contacts to the ground pad.

Figure 4.15b shows how the standard cell frames can be laid out end-to-end to increase the area available for the layout of the MOSFETs. Notice that the layout area is increased horizontally and not vertically. Standard cell frames have a standard height. Also notice how power, ground, well, and substrate connections are routed through the cell. Finally, note that overlapping layout layers, like the n-well, is fine. It simply indicates a larger layout area (of course the design rule check must be passed).

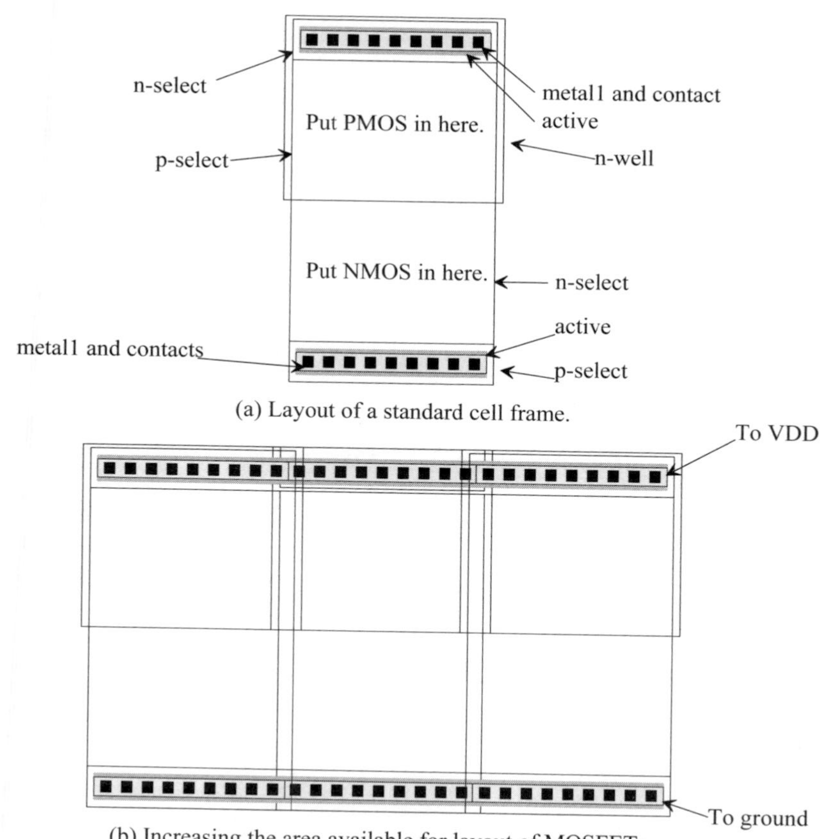

(a) Layout of a standard cell frame.

(b) Increasing the area available for layout of MOSFETs.

Figure 4.15 Layout of standard cell frames.

Design Rules

Figure 4.16 shows the design rules for the active, contact, poly, and select layers. More detailed information on the design rules can be found at MOSIS (http://mosis.org) or with the LASI setups. Seen at the bottom of Fig. 4.16 is an alternative method of laying out an NMOS device. The same active area is used for both the MOSFET formation and the contact to the MOSFET's body (either p-substrate for NMOS or n-well for PMOS). In Fig. 4.12, for example, we laid out the substrate connection a distance away from the source and drain. (The source and drain of the MOSFET are interchangeable.) In the layout seen in Fig. 4.16, we *abut* the n-select directly next to the p-select. This minimizes the layout area. However, as seen in Figs. 4.2g and 4.2h, the n+ and p+ aren't well defined at the boundary between the two materials (the select masks can shift). The p+/n+ can, for all intents and purposes, be thought of as being shorted together. Since the substrate is tied to ground, the layout of the MOSFET in this way (to minimize area) **isn't symmetrical** and the source must also be tied to ground (the source and drain are not interchangeable).

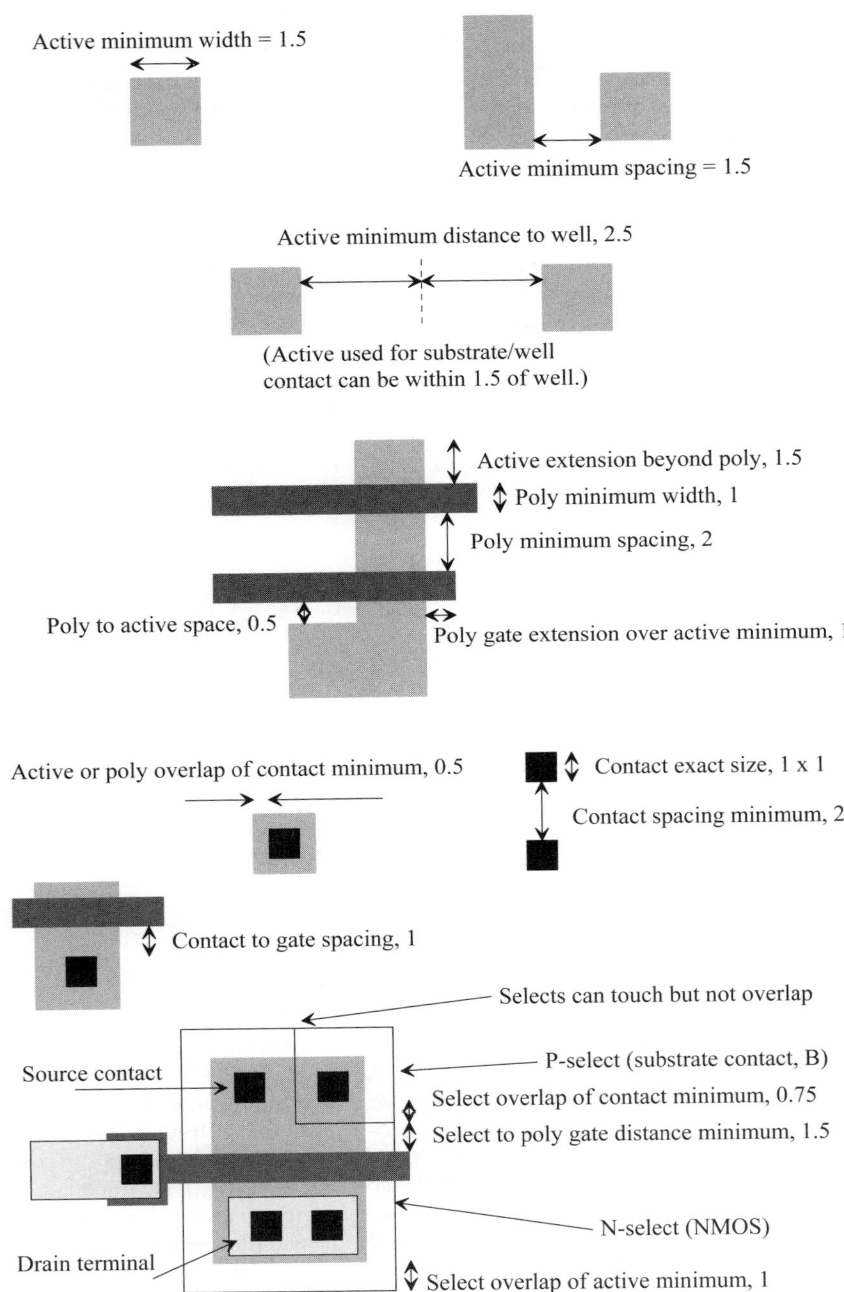

Figure 4.16 Design rules for active, selects, poly, and contacts.

4.3 Electrostatic Discharge (ESD) Protection

A major concern in CMOS technology is the protection of the thin gate oxides, Fig. 4.3b, from electrostatic discharge (ESD). While an in-depth presentation of techniques for preventing ESD damage are outside the scope of this book, here we briefly show how the active layers can be used to form a simple diode-protection structure.

Where does ESD come from? Walking across the floor, for example, causes the buildup of charge on the human body. Touching a conducting object can result in a transfer of charge or a static "shock." If the transfer of this charge (the discharge of the electrostatic charge buildup on the human body) is through the thin gate oxide (GOX) of a MOSFET, it is likely that the GOX will be damaged. While we use the human body as the location of the buildup of electrostatic charge, just about any object can build up charge. To keep from damaging the GOX, consider the circuit seen in Fig. 4.17. Here we connect two diodes to the input pad. If the signals applied to the pad are within ground and *VDD*, neither diode turns on. However, if the voltage on the pad starts to increase above *VDD* + 0.5 V or decrease below – 0.5 V, one of the diodes turns on and provides a low impedance "clamp" to keep the voltage on the GOX from becoming excessive.

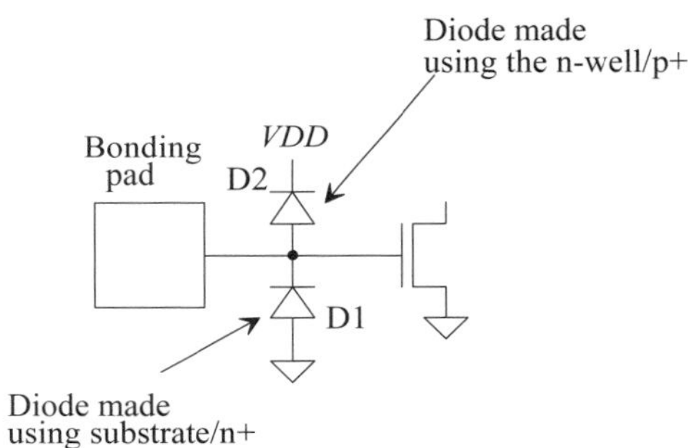

Figure 4.17 Adding diodes to the pads to protect the MOSFET gates from ESD damage.

Layout of the Diodes

Examine the layout seen in Fig. 4.18. This layout shows a conceptual implementation of the diode protection circuitry seen in Fig. 4.17. Diode, D1, is implemented using p-substrate (for the anode) and n+ (for the cathode). The substrate is connected to ground through a metal1 ground bus as seen in the figure. It is desirable to place the substrate connection and n+ as close together as possible to minimize the resistance in series with the diode. Further, it is desirable to increase the size of this n+/p+ diode to both reduce the resistance of the diode and increase its current handling capability. The drawback of increasing the size is the larger capacitive loading on the pad (the depletion capacitances of the diodes). Diode D2 is implemented using p+ in the n-well (for the anode) and the n-well (for the cathode). An n+ implant is needed in the n-well to provide a connection to a wire that is connected to *VDD*. Again, the *VDD* is run through the pad on a bus (the horizontal metal line connected to *VDD* at the bottom of the layout). Again, it's desirable

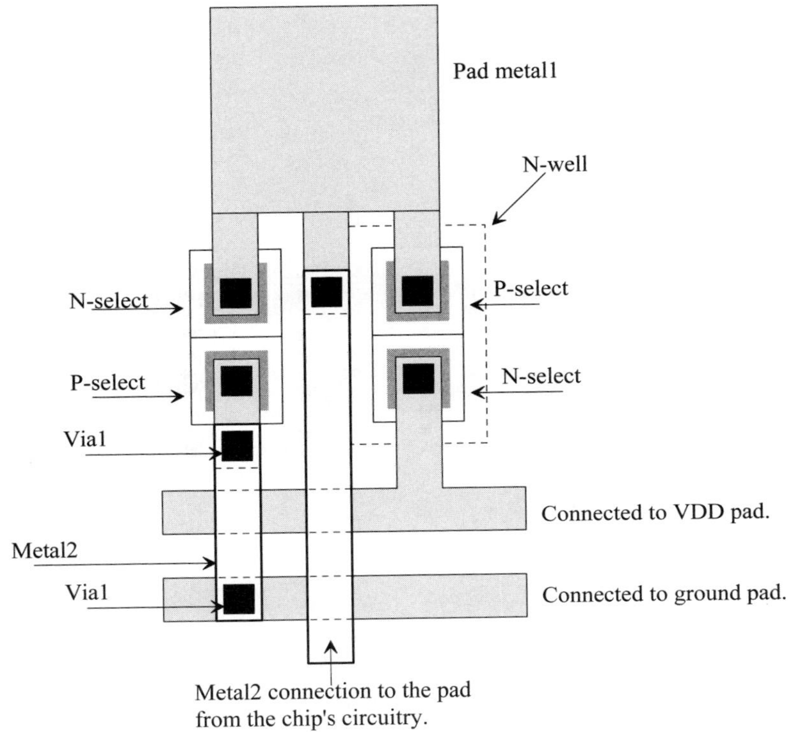

Figure 4.18 Conceptual layout of a pad ESD protection circuit.

to minimize the series resistance between the p+ and n+/n-well. Normally, the ground bus at the bottom of the pad ties the substrate to ground (p+ is implanted under the bus and tied, with contacts, to the metal) and the *VDD* bus is connected to n+/n-well directly beneath it.

Figure 4.19 shows a more realistic view of an I/O (input/output) pad with ESD protection diodes. We've laid the implants directly next to each other to minimize parasitic resistance. We've also staggered the diodes, that is, D1, D2, D1, and D2 to increase their areas. In a real pad layout, the top layer of metal must be used (not metal1 as we've used in the figure, see Figs. 3.20 to 3.23). An example padframe layout is seen in Fig. 4.20. Two of the pads must be connected (separately) to the *VDD* and ground buses.

An important addition to most pads is a buffer to drive the large off-chip capacitances, which are relatively large compared to the on-chip capacitances. We briefly talked about output buffers in Sec. 3.3.2. The circuit design for I/O buffers is covered in Ch. 11 where we talk about inverters. While it's possible to design custom pads for a project, it's generally a better idea to get pads directly from the CMOS vendor where the chips will be fabricated (pads can be downloaded from MOSIS too).

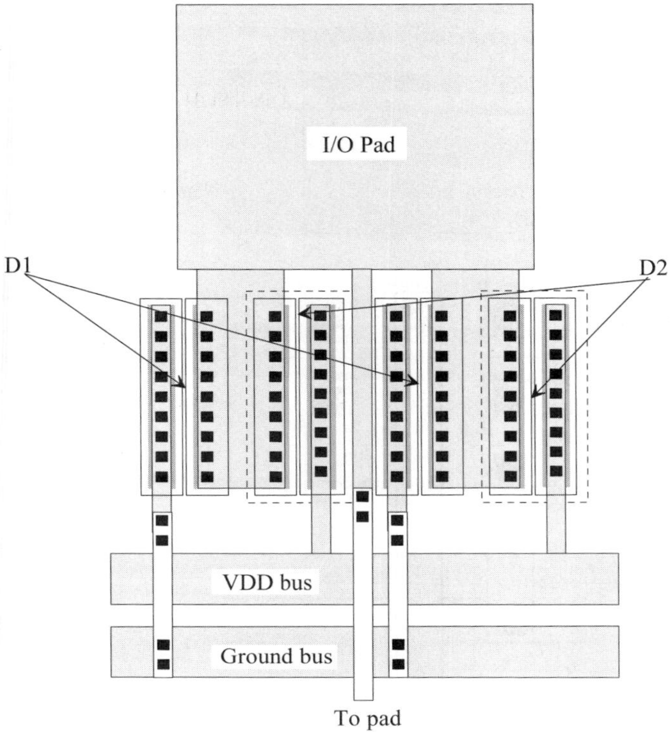

Figure 4.19 More detailed implementation of a pad with ESD protection.

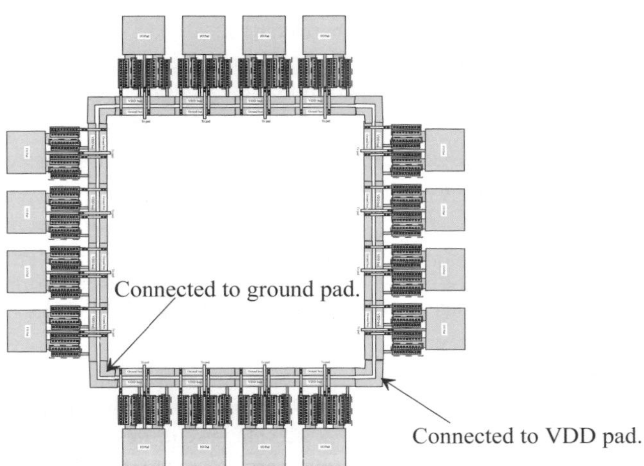

Figure 4.20 Layout of a padframe using pads with ESD diodes.

ADDITIONAL READING

[1] A. Amerasekera and C. Duvvury, *ESD in Silicon Integrated Circuits*, John Wiley and Sons Publishers, 2002. ISBN 0-471-49871-8

[2] D. Clein, *CMOS IC Layout: Concepts, Methodologies, and Tools*, Newnes Publishers, 2000. ISBN 0-750-67194-7

[3] S. Dabral and T. J. Maloney, *Basic ESD and I/O Design*, John Wiley and Sons Publishers, 1999. ISBN 0-471-25359-6

[4] Y. Taur and T. H. Ning, *Fundamentals of Modern VLSI Devices*, Cambridge University Press, 1998. ISBN 0-521-55959-6

PROBLEMS

4.1 Lay out a nominally 200 kΩ resistor with metal1 wire connections. DRC your layout. What would happen if the layout did not include n+ under the contacts? Use the sheet resistances from Table 4.1.

4.2 Sketch the cross-sectional view along the line indicated in Fig. 4.21.

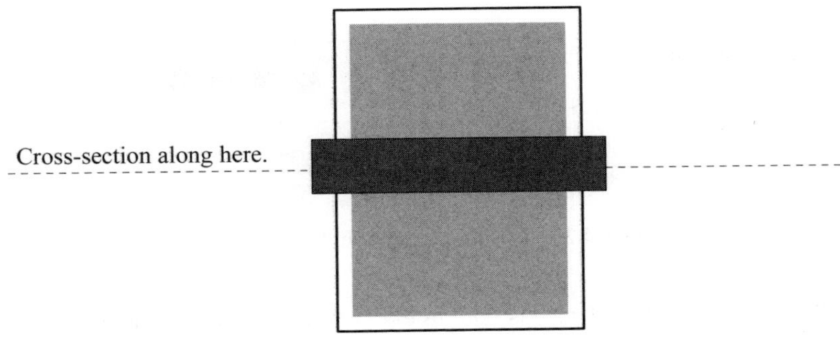

Figure 4.21 Layout used in Problem 4.2.

4.3 Sketch the cross-sectional views across the *VDD* and ground power buses in the standard cell frame of Fig. 4.15.

4.4 Suppose the "bad" layout seen in Fig. Ex4.1 is used to fabricate an NMOS device. Will the poly be doped? Why or why not?

4.5 Why is polysilicon's parasitic capacitance larger than metal1's?

4.6 Lay out an NMOS device with an length of 1 and width of 10. Label all four of the MOSFET terminals.

4.7 Lay out an PMOS device with a length of 1 and width of 20. Label all four of the MOSFET terminals.

4.8 Using the standard cell frame, lay out 10 (length) by 10 (width) NMOS and PMOS transistors. Sketch several cross-sectional views of the resulting layouts showing all four terminals of the MOSFETs.

4.9 Repeat Ex. 4.2 for a poly wire that is 5 wide.

4.10 Sketch the cross-sectional view at the line indicated in Fig. 4.22.

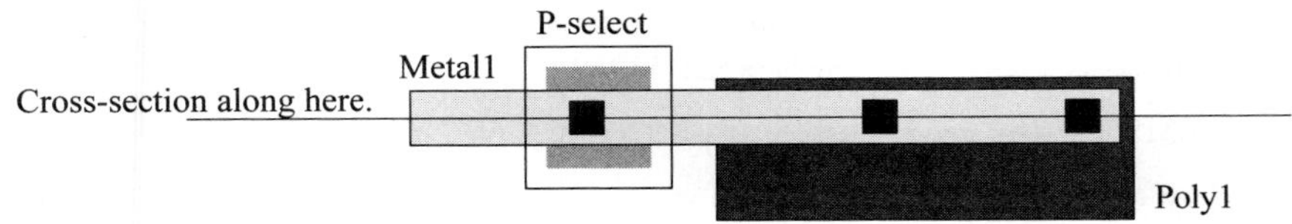

Figure 4.22 Layout used in Problem 4.10.

4.11 Sketch the cross-sectional views across the lines shown in Fig. 4.23.

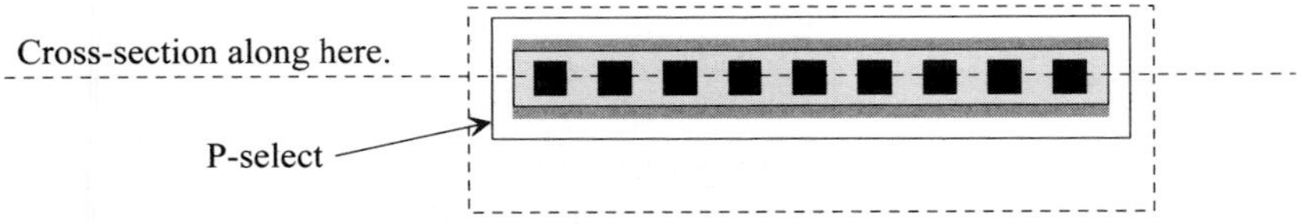

Figure 4.23 Layout used in Problem 4.11.

4.12 The layout of a PMOS device is seen in Fig. 4.24 is incorrect. What is the (fatal) problem?

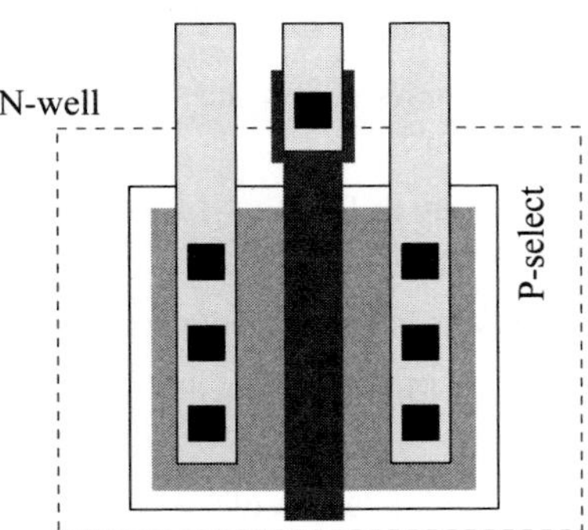

Figure 4.24 Flawed layout of a PMOS device. What is the error?

Chapter

5

Resistors, Capacitors, MOSFETs

This chapter provides more information and examples related to the layout of resistors, capacitors, and MOSFETs. Layout using the poly2 layer and how poly2 is used to make poly-poly capacitors will be covered. We'll also introduce some fundamental layout techniques including using unit cells, layout for matching, and the layout of long length and wide MOSFETs. The temperature and voltage dependence of resistors and capacitors will also be covered.

5.1 Resistors

The resistors and capacitors in a CMOS process have values that change with voltage and temperature. The change is usually listed as ppm/°C (parts per million per degree C). The ppm/°C is equivalent to a multiplier of 10^{-6}/°C.

Temperature Coefficient (Temp Co)

As temperature goes up, so does the value of a resistor, Fig. 5.1 (in general). Generally, the value of a resistor, $R(T_0)$, is specified at room temperature, T_0. To characterize the resistor we use

$$R(T) = R(T_0) \cdot (1 + TCR1 \cdot (T - T_0)) \tag{5.1}$$

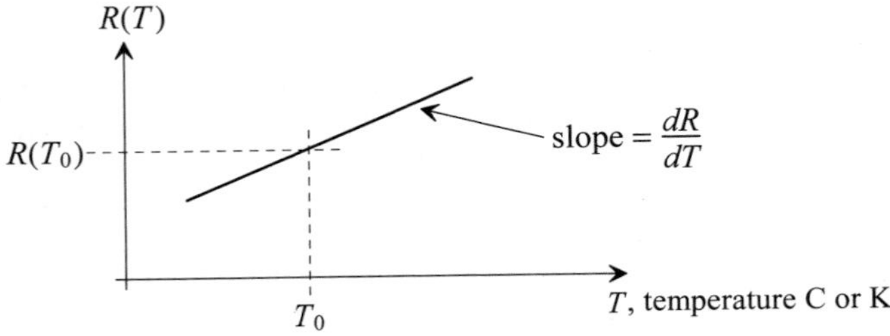

Figure 5.1 Resistor's value change with temperature.

where T is the actual temperature of the resistor. Because of the difference in the equation, it doesn't matter if Celsius or Kelvin are used for the units of T as long as the units match the units used for T_0. The first-order temperature coefficient of a resistor, $TCR1$, is given by

$$TCR1 = \frac{1}{R} \cdot \frac{dR}{dT} \tag{5.2}$$

Notice that the $TCR1$ changes with temperature. For most practical applications, the "temp co" is assumed to be linear. Typical values of $TCR1$ for different layers in the CMOS process are given in Table 4.1. SPICE uses a quadratic, in addition to this first order term, to model the temperature dependence of a resistor

$$R(T) = R(T_0) \cdot [1 + TCR1 \cdot (T - T_0) + TCR2 \cdot (T - T_0)^2] \tag{5.3}$$

For hand calculations, we assume $TCR2$ is 0.

Example 5.1

Using the values in Table 4.1, estimate the minimum and maximum resistance of an n-well resistor with a length of 100 and a width of 5 over a temperature range of 0 to 100 °C. Verify the hand calculations using SPICE.

The n-well resistor sheet resistance is approximately 500 Ω/square (at 27 °C). The value of the resistor in this example (20 squares) is 10 k at 27 °C. The temp co is 2400 ppm/°C (= 0.0024) The minimum resistance is then determined, using Eq. (5.1) by

$$R_{\min} = 10,000 \cdot [1 + 0.0024 \cdot (0 - 27)] = 9.35\ k\Omega$$

and the maximum resistance is determined by

$$R_{\max} = 10,000 \cdot [1 + 0.0024 \cdot (100 - 27)] = 11.75\ k\Omega$$

The SPICE simulations and netlist are shown in Fig. 5.2. ■

Polarity of the Temp Co

We made the comment that in general the temp co of a resistor is positive, which means that the resistor's value increases with increasing temperature. In other words, the resistivity of the silicon material used to fabricate the resistor increases with increasing temperature. If we were to look at the electron concentration in an n-well as temperature is increased, for example, we would see an exponential increase in the number of thermally generated carriers. More carriers, less resistivity, right? Looking at the carrier concentration alone, we would expect silicon's resistivity to decrease with increasing temperature. With the increase in the carrier concentration, however, we get a reduction in the carrier's mobility. The carrier-to-carrier interactions increase the scattering (the average distance a carrier can travel before colliding with some other carrier decreases with a larger number of free carriers). The material parameter "mobility" characterizes how easily carriers can move through a material. The mobility of an electron or hole, μ_n or μ_p respectively, is a material parameter that relates the applied electric field across the material to the velocity the carriers can drift through the material. High velocity indicates that the carriers have high-mobility and can move through the material quickly. Mobility is a measured parameter and is given by

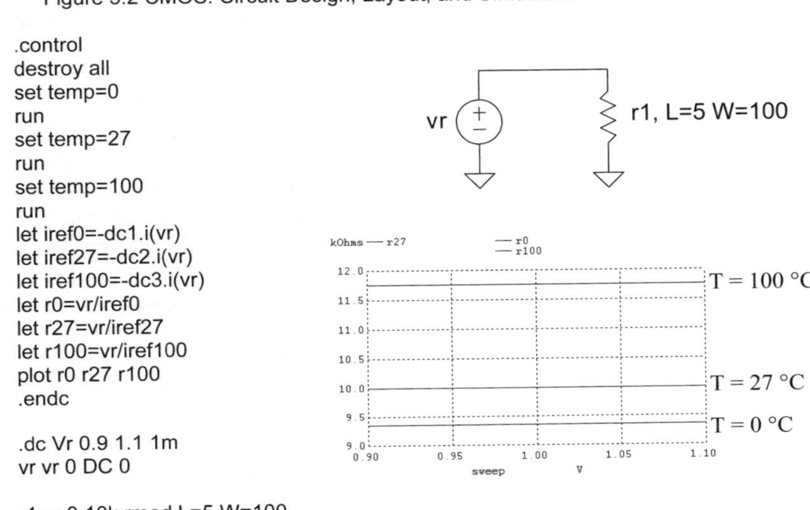

*** Figure 5.2 CMOS: Circuit Design, Layout, and Simulation ***
.control
destroy all
set temp=0
run
set temp=27
run
set temp=100
run
let iref0=-dc1.i(vr)
let iref27=-dc2.i(vr)
let iref100=-dc3.i(vr)
let r0=vr/iref0
let r27=vr/iref27
let r100=vr/iref100
plot r0 r27 r100
.endc
.dc Vr 0.9 1.1 1m
vr vr 0 DC 0
r1 vr 0 10k rmod L=5 W=100
.model rmod R RSH=500 TNOM=27 TC1=0.0024
.end
vr
r1, L=5 W=100
kOhms
r27
r0
r100
T = 100 °C
T = 27 °C
T = 0 °C
sweep
V

Figure 5.2 Simulating the temperature dependence of an n-well resistor.

$$\mu_{n,p} = \frac{\text{Average velocity of carriers, } cm/s}{\text{Applied electric field, } V/cm} \tag{5.4}$$

noting mobility's units are $cm^2/V \cdot s$. The resistivity of the material depends on the number of free carriers (electrons/cm^3, n, and holes/cm^3, p) or

$$\rho = \frac{1}{q(\mu_n n + \mu_p p)} \tag{5.5}$$

where q is the electron charge (see Eqs. [2.2] and [2.3]). This equation is important and shows why we can have a negative or positive resistor temperature coefficient. If the mobilities decrease faster than the carrier concentrations increase, we get a positive temp co. The resistor's value goes up with increasing temperature because the mobility is dropping faster than the carrier concentration is increasing. However, if the increase in carrier concentration with temperature is faster, the resistivity goes down with increasing temperature and we get a negative temp co (the resistor's value goes down with increasing temperature). At room temperature, a positive temperature coefficient is normally observed.

Voltage Coefficient

Another important contributor to a changing resistance is the voltage coefficient of the resistor given by

$$VCR = \frac{1}{R} \cdot \frac{dR}{dV} \tag{5.6}$$

where V is the average voltage applied to the resistor, that is, the sum of the voltages on each end of the resistor divided by two. The resistance as a function of voltage is then given by

$$R(V) = R(V_0) \cdot (1 + VCR1 \cdot (V - V_0) + VCR2 \cdot (V - V_0)^2) \quad (5.7)$$

The value $R(V_0)$ is the value of the resistor at the voltage V_0. A typical value of $VCR1$ is 8000 ppm/V for an n-well resistor. The main contributor to the voltage coefficient is the depletion layer width between the n-well and the p-substrate. The depletion layer extends into the n-well, resulting in an effective change in the sheet resistance. The thickness of the n-well available to conduct current decreases with increasing potential (reverse bias) between the n-well and the substrate.

Example 5.2
Estimate the average resistance of an n-well resistor with a typical value of 10k at 27 °C, for an average voltage across the resistor of 0, 5, and 10 V.

$$R(0) = 10{,}000 \cdot (1 + 0.008 \cdot 0) = 10\ k\Omega$$

$$R(5) = 10{,}000 \cdot (1 + 0.008 \cdot 5) = 10.4\ k\Omega$$

$$R(10) = 10{,}000 \cdot (1 + 0.008 \cdot 10) = 10.8\ k\Omega$$

This is a small change compared to the change due to temperature. However, as the next example will show, the change due to the applied voltage can have a greater influence on the circuit performance than the temperature. ■

Example 5.3
Compare the change in V_{out} in the following circuit due to the VCR with the change due to the TCR.

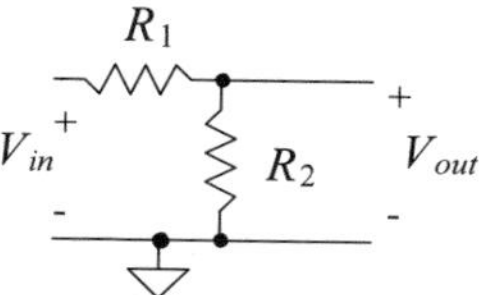

Figure 5.3 Comparing a resistive divider's temperature performance to its voltage performance.

We start by writing the voltage divider equation and then substitute the temperature or voltage dependence. For the temperature dependence

$$V_{out} = V_{in} \cdot \frac{R2(T)}{R1(T) + R2(T)} = V_{in} \cdot \frac{R2 \cdot [1 + TCR(T - T_0)]}{(R1 + R2) \cdot [1 + TCR(T - T_0)]} = V_{in} \cdot \left[\frac{R2}{R1 + R2}\right]$$

showing that the divider is independent of temperature. For the voltage dependence

$$V_{out} = V_{in} \cdot \frac{R2(V)}{R1(V) + R2(V)} = V_{in} \cdot \frac{R2 \cdot (1 + VCR \cdot V_2)}{R1 \cdot (1 + VCR \cdot V_1) + R2 \cdot (1 + VCR \cdot V_2)}$$

where $V_1 = \frac{Vin + Vout}{2}$ and $V_2 = \frac{Vout}{2}$. These results show that the voltage divider has no temperature dependence but that it does have a voltage dependence. ■

Using Unit Elements

The preceding example shows the advantage of ratioing components. For precision design over large temperature ranges, this fact is very important. Usually, a *unit resistor* is laid out with some nominal resistance. Figure 5.4a shows a unit cell resistor layout, say 5 kΩ, using the n-well. Figure 5.4b shows how the nominal 5 kΩ unit cell resistor implements a 4/5 voltage divider. Using the divider in Fig. 5.3 as a reference, the resistor R_1 is 5k and the resistor R_2 is 20k (notice the layout of the 20k resistor is 4 unit cells). The errors due to corners and differing perimeters using a serpentine pattern, see Fig. 2.28, are eliminated when using unit cell layout techniques. Also, the exact calculation of the resistor's value isn't important. Changes in the nominal resistance value because of process shifts, temperature, or how the number of squares is calculated only affect the current flowing in the divider (the power dissipation) and not the output voltage. Amplifier circuits, such as those using op-amps with feedback, are examples of circuits where ratioing is important.

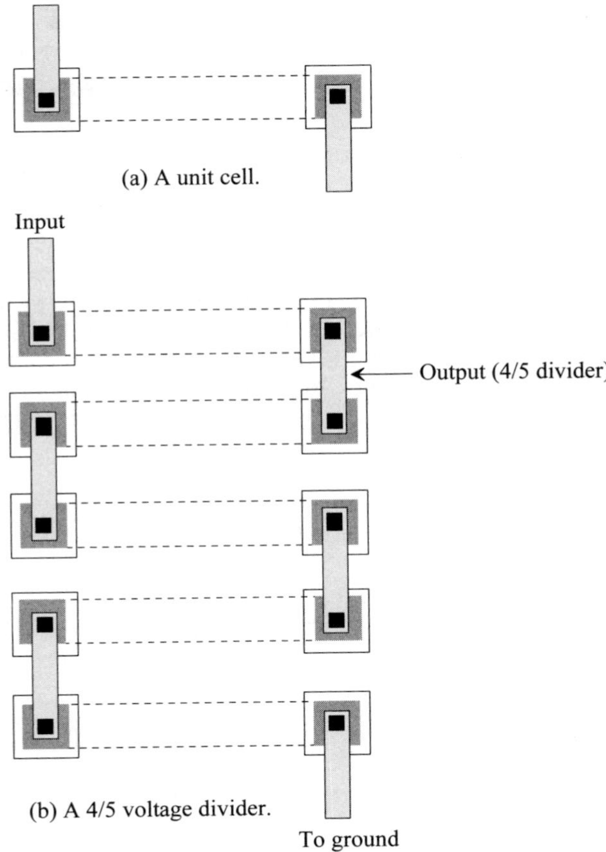

Figure 5.4 (a) Layout of a unit resistor cell, and (b) layout of a divider.

Guard Rings

Any precision circuit is susceptible to substrate noise. Substrate noise results from adjacent circuits injecting current into one another. The simplest method of reducing substrate noise between adjacent circuits is to place a p+ implant (a substrate contact for a p-substrate wafer) between the two circuits. The substrate contact removes the injected carriers and holds the substrate, ideally, at a fixed potential (ground). Figure 5.5 shows the basic idea for a resistor. The p+ implant guards the circuit against carrier injection. Because the implants are laid out in a ring, they are often called *guard rings*.

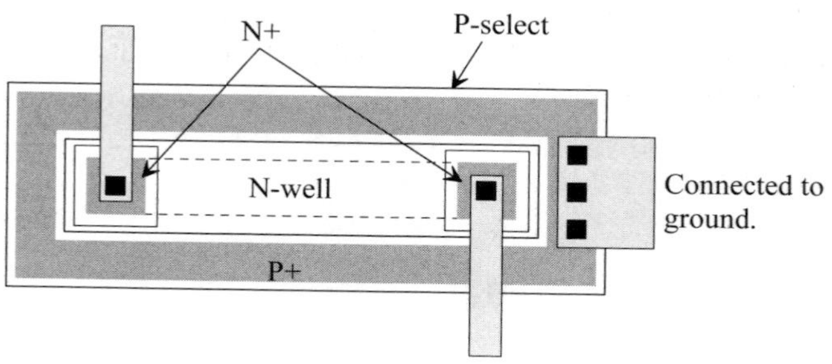

Figure 5.5 Guard ringing an n-well resistor.

Interdigitated Layout

Matching between two different resistors can be improved by using the layout shown in Fig. 5.6. These resistors are said to be interdigitated. Process gradients, in this case changes in the n-well, n+, or p+ doping at different places on the die, are spread between the two resistors more evenly. Notice that the orientation of the resistors is consistent between unit cells; that is, all cells are laid out (here) vertically. Also, each resistor has essentially the same parasitics.

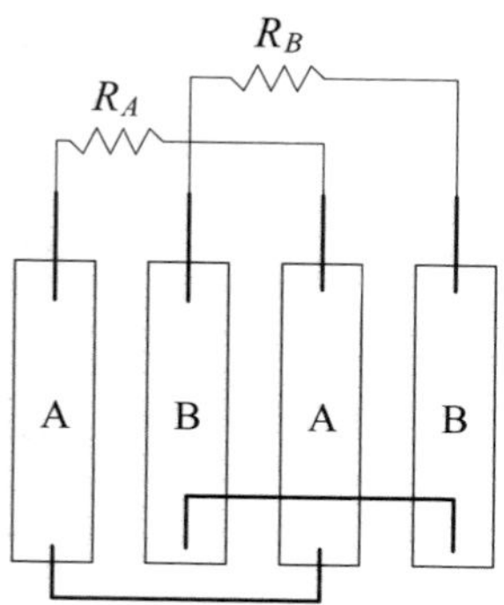

Figure 5.6 Layout of interdigitated resistors.

Common-Centroid Layout

Common-centroid (common center) layout helps improve the matching between two resistors (at the cost of uneven parasitics between the two elements). Consider the arrangement of unit resistors shown in Fig. 5.7a. This interdigitated resistor is the same layout style as just discussed in Fig. 5.6. Consider the effects of a linearly varying sheet resistance on the overall value of each resistor. If we assign a normalized value to each unit resistor, as shown, then resistor A has a value of 16 and resistor B has a value of 20. Ideally, the resistor values are equal.

Next consider the common-centroid layout shown in Fig. 5.7b, noting that resistors A and B share a common center. The value of either resistor in this figure is 18. In other words, the use of a common center (ABBAABBA) will give better matching than the interdigitized layout (ABABABAB). Figure 5.8 shows the common-centroid layout (two different possibilities) of four matched resistors. Note that common-centroid layout can also improve matching in MOSFETs or capacitors.

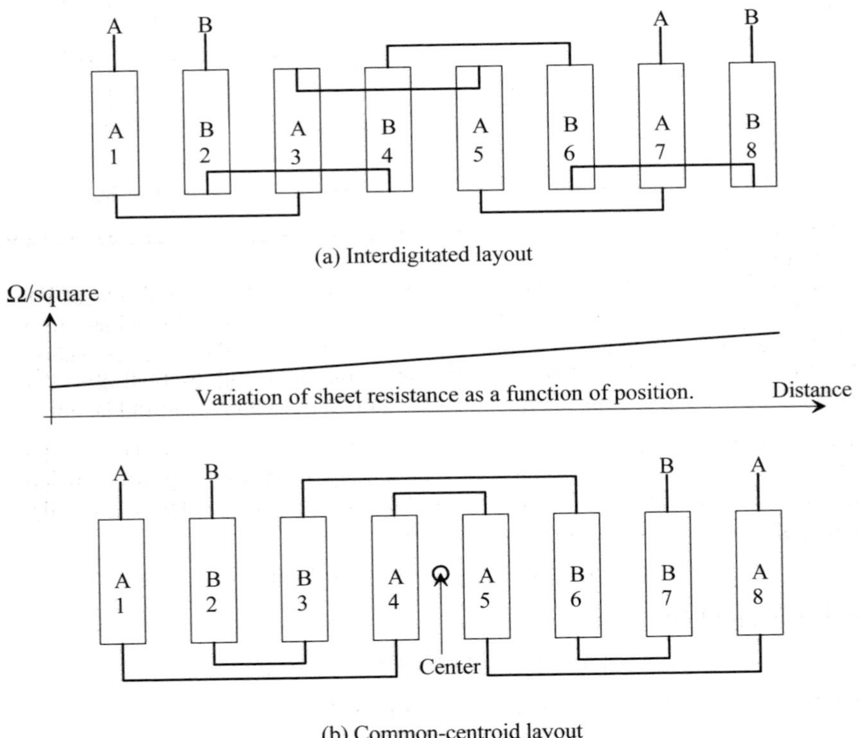

Figure 5.7 (a) Interdigitated layout and (b) common-centroid layout.

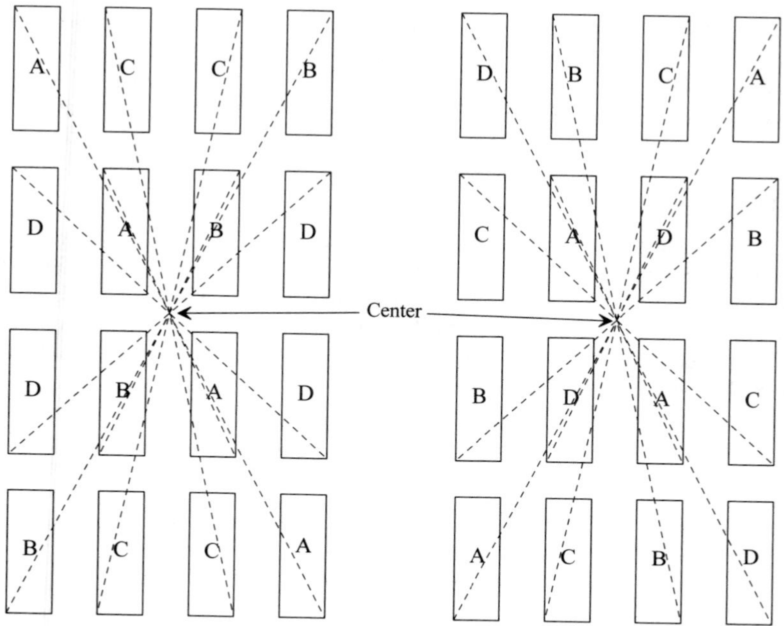

Figure 5.8 Common-centroid layout of four matched resistors (or elements).

Example 5.4

Suppose that the nominal value of the unit resistor, A, on the left side of the layouts in Fig. 5.7 is 5k (a nominal value because we know the actual sheet resistance varies with process shifts and temperature). If the sheet resistance linearly varies across the layout, from the left to the right, and the final resistor's value is 5.07k, compare the interdigitated layout to the common-centroid layout.

The farthest left resistor in the layout has a value of 5k (given). The second resistor's value, because of the linear variation in the sheet resistance from the left to the right in the layout, is 5.01k. The third resistor's value is 5.02k, etc. For the interdigitated layout, the value of resistor A is

$$R_A = 5.0 + 5.02 + 5.04 + 5.06 = 20.12\ k\Omega$$

and the value of resistor B is

$$R_B = 5.01 + 5.03 + 5.05 + 5.07 = 20.16\ k\Omega$$

For the common-centroid layout, resistor A's value is

$$R_A = 5.0 + 5.03 + 5.04 + 5.07 = 20.14\ k\Omega$$

and resistor B's value is

$$R_B = 5.01 + 5.02 + 5.05 + 5.06 = 20.14\ k\Omega$$

exactly the same result. If we had laid out all of the unit elements for resistor A on the left and the all of the unit elements for resistor B on the right, we would get $R_A = 20.06\ k\Omega$ and $R_B = 20.22\ k\Omega$. This shows that the interdigitated layout does help with matching. ■

Dummy Elements

Another technique that improves the matching between two or more elements is the use of dummy elements. Consider the cross-sectional views of the three n-wells shown in Fig. 5.9a. The final amount of diffusion under the resist, on the edges, is different between the outer and the inner unit cells. This is the result of differing dopant concentrations at differing points on the surface (during the diffusion process). This difference results in a mismatch between unit resistors. To compensate for this effect, dummy elements can be added (see Fig. 5.9b) to an interdigitated or common-centroid layout. The dummy element does nothing electrically. It simply ensures that the unit resistors of matched resistors see the same adjacent structures. Normally, these dummy elements are tied off to either ground or *VDD* rather than left floating.

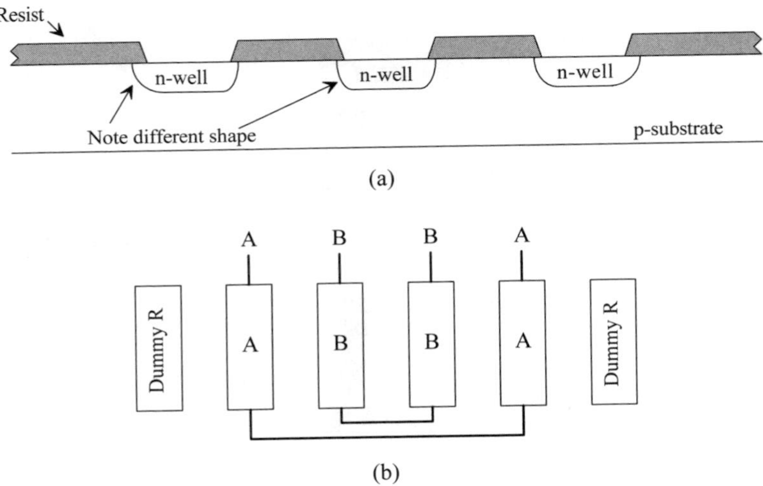

Figure 5.9 (a) Edge effects and (b) a common-centroid layout with dummy R.

5.2 Capacitors

An added layer of polysilicon, called poly2 (POL2), can be present in a CMOS process for capacitor formation (called a poly-poly capacitor), for MOSFET formation (we can use poly2 instead of poly1 for the gate of a MOSFET) and for creating a floating gate device (see Ch. 16's discussion of Flash memory technology for example). In this section we discuss the layout of capacitors using poly2, the parasitics present, and the temperature behavior of the poly-poly capacitor. Many of the layout techniques discussed in the last section can be used when laying out capacitors.

Layout of the Poly-Poly Capacitor

Layout and cross-sectional views of a capacitor using the poly1 and poly2 layers are shown in Fig. 5.10. The silicon dioxide dielectric between the two layers of poly is roughly the same thickness as the gate oxide (GOX), t_{ox} in a MOSFET (see Fig. 4.3b). Table 5.1 shows **typical values for the t_{ox} that we'll use in the 1 μm and 50 nm processes, referred to as long- and short-channel CMOS processes, in this book.** Also seen in the table is the oxide capacitance per area calculated using

$$C'_{ox} = \frac{\varepsilon_r \cdot \varepsilon_0}{t_{ox}} = \frac{\varepsilon_{ox}}{t_{ox}} \tag{5.8}$$

where $\varepsilon_0 = 8.85 \times 10^{-18}\ F/\mu m = 8.85\ aF/\mu m$ and the relative dielectric constant of SiO_2 is 3.97 ($=\varepsilon_r$). To calculate the value of a capacitor, we look at the area where poly1 and poly2 intersect, A, or

$$C_{ox} = C'_{ox} \cdot A \tag{5.9}$$

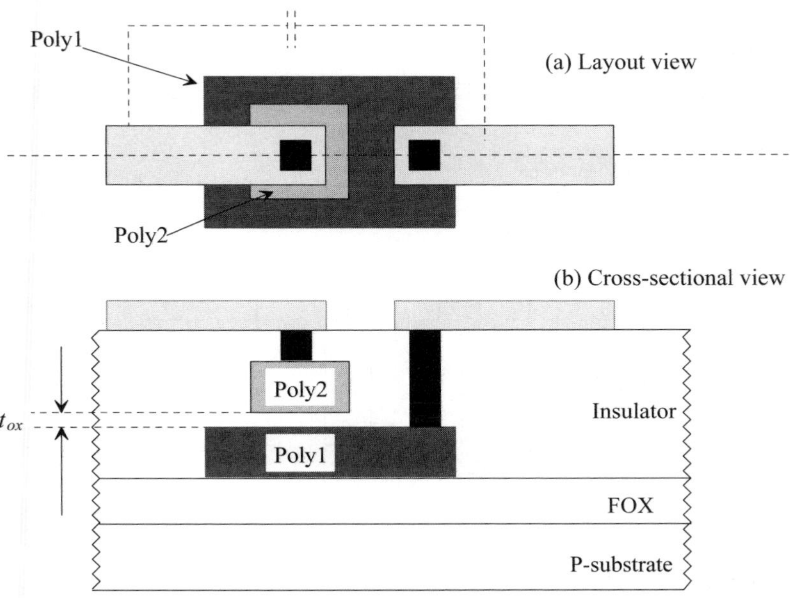

Figure 5.10 Layout and cross-sectional view of a poly-poly capacitor.

If the scale factor is included, then we can write this equation as

$$C_{ox} = C'_{ox} \cdot A \cdot (scale)^2 \tag{5.10}$$

If, in the 50 nm process, a poly-poly capacitor is formed with an intersection of the poly1 and poly2 that measures 10 by 20 then the capacitor's value is

$$C = C_{ox} = 25\ fF/\mu m^2 \cdot 200 \cdot (0.05\ \mu m)^2 = 12.5\ fF \tag{5.11}$$

Table 5.1 Oxide thicknesses and oxide capacitances for the long- and short-channel CMOS processes used in this book.

CMOS technology	**Oxide thickness, t_{ox}**	C'_{ox}
1 µm (long channel)	200 Å	1.75 $fF/\mu m^2$
50 nm (short channel)	14 Å	25 $fF/\mu m^2$

Note that when the poly2 is used to form a floating gate device (see Ch. 16) poly2 can be called an electrode (see the MOSIS design rules). Also note that the practical minimum capacitor value one should try using (laying out) in the long- and short-channel processes described in Table 5.1 is approximately 100 fF and 10 fF, respectively.

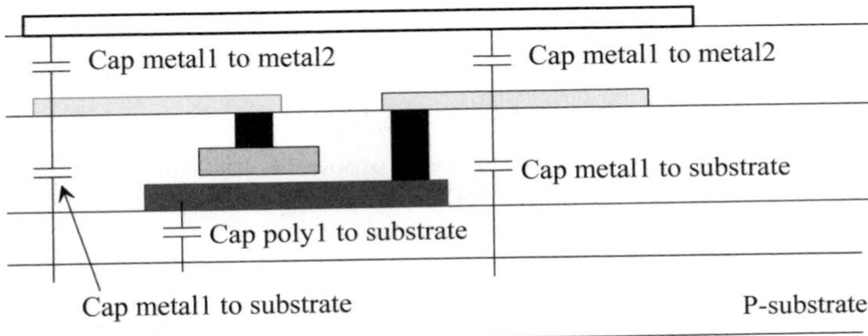

Figure 5.11 The parasitic capacitances of the poly-poly capacitor.

Parasitics

As with any layout structure, we must be concerned with the parasitics. Figure 5.11 shows the parasitics associated with the poly-poly capacitor. The most important (largest) parasitic is the capacitance from poly1 to substrate. This capacitance is called the **bottom plate parasitic** capacitance. Reviewing Table 3.1, this parasitic (plate) capacitance is 58 $aF/\mu m^2$ (there are fringe capacitances as well that can be considered). Keeping in mind that the poly1 area is larger than the poly2 layout area (and the intersection of poly1 and poly2 is the desired capacitance), the bottom plate capacitance can be 20% (or more) of the desired capacitance value. In analog circuits the bottom plate of a capacitor is indicated, as seen in Fig. 5.12, see Ch. 25.

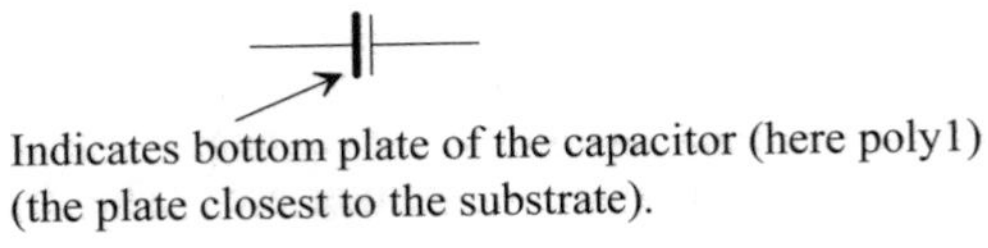

Indicates bottom plate of the capacitor (here poly1)
(the plate closest to the substrate).

Figure 5.12 The bottom plate parasitic of a capacitor.

Temperature Coefficient (Temp Co)

The first-order temperature coefficient of a capacitor, *TCC*, is given by

$$TCC = \frac{1}{C} \cdot \frac{dC}{dT} \tag{5.12}$$

A typical value of *TCC* for a poly-poly capacitor is 20 ppm/°C. Matching of large area poly-poly capacitors on a die is typically better than 0.1% with good layout techniques. The capacitance as a function of temperature is given by

$$C(T) = C(T_0) \cdot [1 + TCC \cdot (T - T_0)] \tag{5.13}$$

where $C(T_0)$ is the capacitance at T_0.

Voltage Coefficient

The voltage coefficient of a capacitor is given by

$$VCC = \frac{1}{C} \cdot \frac{dC}{dV} \tag{5.14}$$

The voltage coefficient of the poly-poly capacitor is in the neighborhood of 10 ppm/V (for a long-channel process). The capacitance as a function of voltage is given by

$$C(V) = C(V_0) \cdot (1 + VCC \cdot V) \tag{5.15}$$

where $C(V_0)$ is the capacitance between the two poly layers with zero applied voltage, and V is the voltage between the two plates.

5.3 MOSFETs

We introduced the layout of MOSFETs in the last chapter. In this section we discuss some electrical parameters of interest and how to lay out MOSFETs to minimize parasitics or with a long length or width.

Lateral Diffusion

When the drain and source regions are implanted, some of the implant dose laterally diffuses out underneath the gate poly, as depicted in Fig. 5.13. If the drawn length is called L_{drawn}, we can write the effective length as

$$L_{effective} = L_{drawn} - 2L_{diff} \tag{5.16}$$

where L_{diff} is the length of the lateral diffusion. To minimize the lateral diffusion, a spacer is normally deposited adjacent to the poly after a light implant (see Figs. 4.7f and g) and before the heavy source/drain implants. This type of structure is called a lightly doped drain (LDD).

Oxide Encroachment

There are also imperfections associated with the width of the MOSFET, Fig. 5.13. When the active area is defined, Fig. 4.3, the FOX won't be precisely patterned as specified by the active mask layer. The oxide may encroach on the active area (called oxide encroachment) and reduce the active opening area. The drawn width, W_{drawn}, of the MOSFET will be different from the effective width, $W_{effective}$ by $2W_{enc}$. To compensate for oxide encroachment, the layout may be bloated before making the active mask.

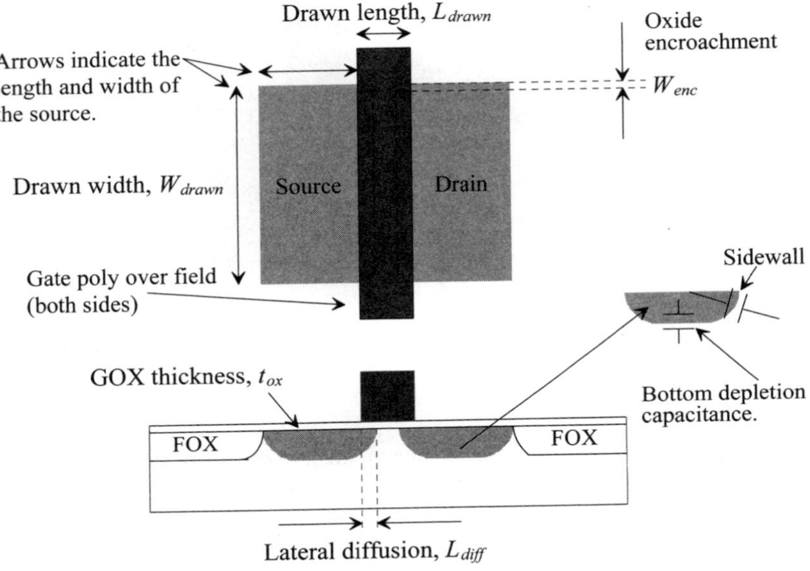

Figure 5.13 Lateral diffusion in a MOSFET.

Source/Drain Depletion Capacitance

As we saw in Ex. 2.3, a pn junction fabricated in the bulk has a depletion capacitance consisting of a bottom component and a sidewall component. For the source and drain junction (depletion) capacitances, this can be written in terms of SPICE parameters as

$$C_{js,d} = \frac{cj \cdot A_{S,D} \cdot (scale)^2}{\left(1 + \frac{V_{S,DB}}{pb}\right)^{mj}} + \frac{cjsw \cdot P_{S,D} \cdot (scale)}{\left(1 + \frac{V_{S,DB}}{pbsw}\right)^{mjsw}} \tag{5.17}$$

where cj is the zero-bias bottom depletion capacitance, $A_{S,D}$ is the area of the source or drain implant, *scale* is the scale factor (1 μm for the long-channel process and 50 nm for the short-channel process), *cjsw* is the zero-bias depletion capacitance for the sidewalls, *pb* and *pbsw* are the built-in potentials for the bottom and sidewall components, respectively, *mj* and *mjsw* are the grading coefficients for the bottom and sidewall components, respectively, and finally $V_{S,DB}$ is the potential from the source or drain to the MOSFET's body (the substrate for the NMOS and the well for the PMOS).

Note that some authors call the source/drain depletion capacitance (incorrectly) diffusion capacitance. As discussed in Sec. 2.4.3, a pn junction only exhibits diffusion capacitance when it becomes forward biased. When CMOS technology was first introduced, the source/drain regions were formed using a diffusion process step (perhaps the reason for the incorrect labeling). CMOS technology developed from, approximately, the mid-80s has used implantation to form the source/drain regions. In any case, this diode junction capacitance should be called either a "junction capacitance" or "depletion capacitance," unless the diode is forward biased.

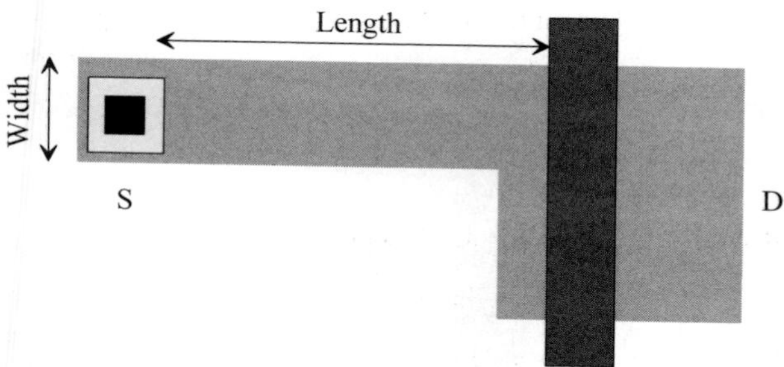

Figure 5.14 Determining a resistance in series with the drain of the MOSFET.

Source/Drain Parasitic Resistance

Examine the layout seen in Fig. 5.14. The active area, in this layout, is used to move the contact to the MOSFET's source or drain away from the gate poly. This can be useful if metal needs to be run over the area directly next to the gate poly. The length of active adds a parasitic resistance in series with the MOSFET. This resistance is determined by specifying the number of squares in the source/drain (NRD/NRS). An estimate for the number of squares, assuming that the extension is on what we label the source of the MOSFET, for the layout in Fig. 5.14, is

$$NRS = \frac{Length}{Width} \quad (5.18)$$

For the other side of the MOSFET (here we call this side the drain), the length is less than the width so we can set *NRD* to zero (or not specify it). To calculate the resistance in series with a MOSFET's source or drain, Fig. 5.15, we use

$$R_S = NRS \cdot R_{sh} \text{ or } R_D = NRD \cdot R_{sh} \quad (5.19)$$

In a SPICE model the paramter rsh is used to specify sheet resistance of the n+ (for the NMOS model) or p+ (for the PMOS model).

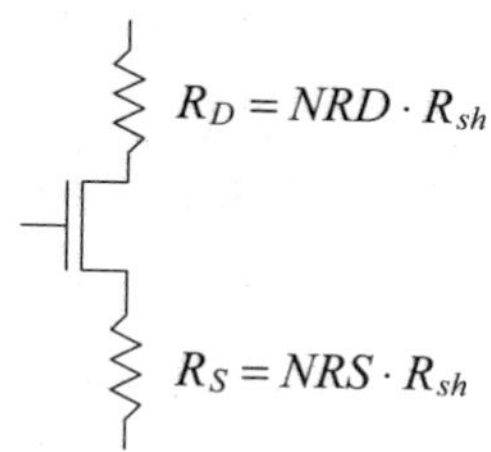

Figure 5.15 How source/drain series parasitic resistance is modeled in SPICE.

Example 5.5
For the MOSFET layout seen in Fig. 5.16, write the SPICE statement including the areas and number of squares in the source/drain regions with or without a scale factor of 50 nm.

Estimates for the areas of the drain/source are

$$A_D = 40 \text{ and } A_S = 45$$

with units of area squared. Note how we didn't include the small area directly adjacent to the gate on the source side. Again, we are estimating (or, more appropriately, approximating) the area. The actual values of capacitance (or resistance) vary with process shifts.

The perimeters of the drain/source are estimated as

$$P_D = 28 \text{ and } P_S = 36$$

The length of the MOSFET is 1 and the width is 10, or,

$$L = 1 \text{ and } W = 10$$

The number of squares of n+ in series with the drain/source is

$$NRD = \frac{4}{10} \to 0 \text{ and } NRS = \frac{11}{3} \approx 4$$

The SPICE statement for the MOSFET is

```
M1  D G S B NMOS L=1 W=10 AD=40 AS=45 PD=28 PS=36 NRD=0 NRS=4
```

The MOSFET device name always starts with an M (a voltage source device name always starts with a V, a resistor R, etc.) The drain, gate, source, and bulk nodes (in this statement) are labeled D, G, S, B, respectively. Note that in an n-well process the bulk is always tied to ground (which is universally zero, 0, in SPICE). The MOSFET's model name is NMOS (same as the technology to make

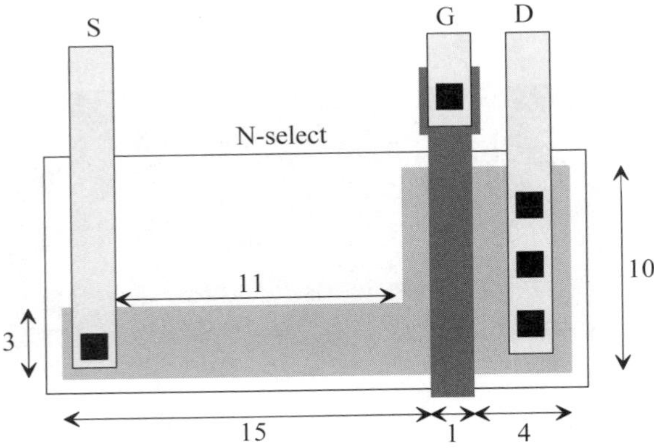

Figure 5.16 Layout of the MOSFET used in Ex. 5.5.

it easier to remember). The statement above is used if we employ a scale factor in the layouts and simulations. Using this MOSFET specification, we would also have to include, in the SPICE netlist,

```
.options scale=50n
```

If we didn't include this statement, then the MOSFET specification would be

```
M1  D G S B NMOS L=50n W=500n AD=100f AS=112.5f PD=1.4u
+ PS=1.8u NRD=0 NRS=4
```

noting that the "+" symbol in the first column of a line indicates that the previous line is continued on the line. **We use drawn sizes in the layout and circuits in this book.** If a SPICE netlist using MOSFETs doesn't include the .options statement with the scale parameter, then the simulation output is likely flawed. ■

Layout of Long-Length MOSFETs

Figure 5.17 shows the layout of a MOSFET with a long length. The active layer is "snaked" back-and-forth under the poly. Each side of this active is contacted to metal1 for the source and drain connections. The width of the MOSFET is equal to the width of the active under the poly. The length of the MOSFET is estimated by looking at the length of the active underneath the poly between the drain and source. In other words, we know that the drain current flows from the drain to the source. Further we know that the drain current flows in the active area. The poly controls the drain current but is isolated from it by the gate oxide. The length is determined by the intersection of the poly and active between the source and drain contacts (the length of the dotted arrows in the figure). Long-length MOSFETs have, as we'll see, a higher effective switching resistance.

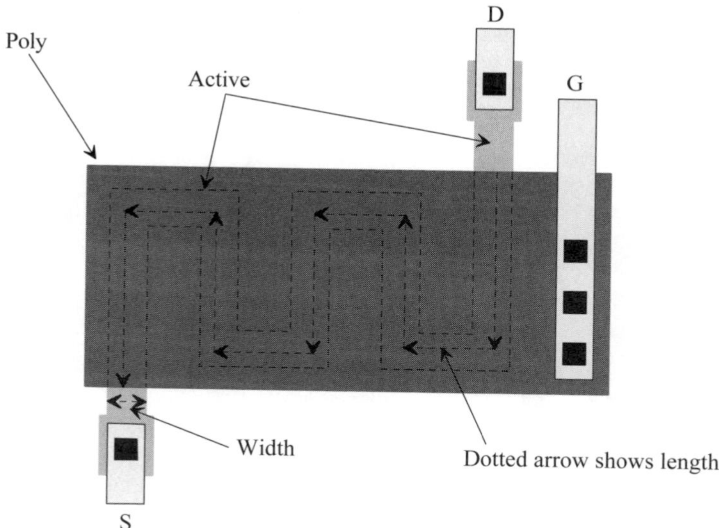

Figure 5.17 Layout of a long-length MOSFET.

Layout of Large-Width MOSFETs

Figure 5.18 shows the layout of a large MOSFET. The length of the MOSFET, L, is still set by the length of poly, as seen in the figure. The MOSFET's overall width is set by the width of poly over active, W, times the number of poly "fingers"

$$\text{Width of MOSFET} = (\text{number of fingers}) \cdot W \tag{5.20}$$

This layout minimizes area by sharing drain and source connections between MOSFETs. Notice how the widths of MOSFET's laid out in parallel (with the gates tied together) add to form an equivalent (to the sum) width MOSFET. This concept can also be used for MOSFET's laid out in series, Fig. 5.19. MOSFETs laid out in series (with their gates tied together) form a MOSFET with an effective length equal to the sum of the individual MOSFET's lengths. We use these concepts often when doing design. For example, we can lay out a MOSFET with a

$$\text{Width-to-length ratio} = W/L = 10/2 \tag{5.21}$$

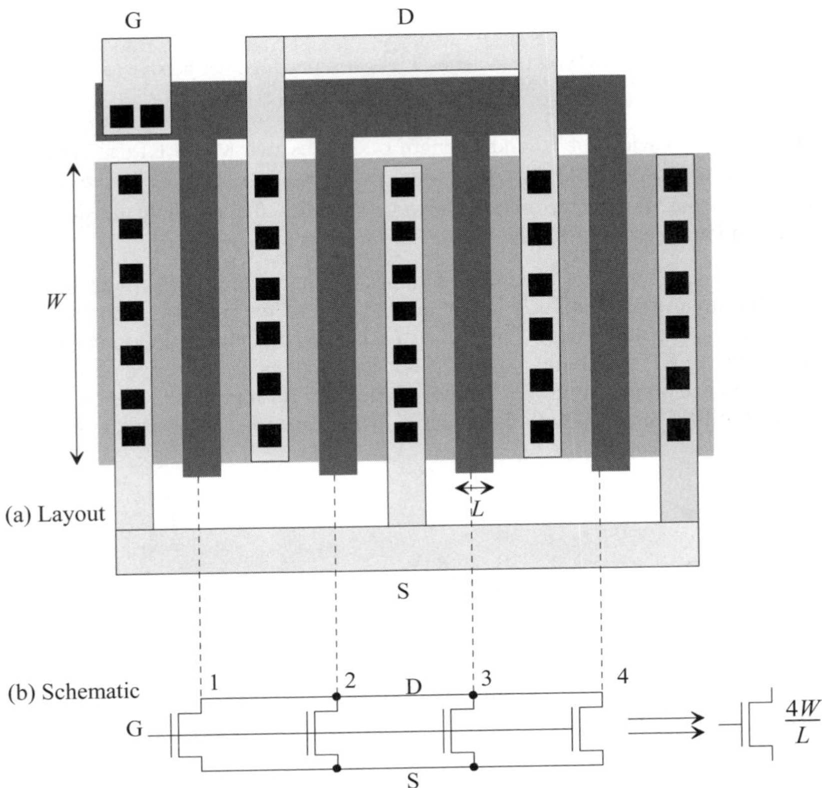

Figure 5.18 Layout and equivalent schematic of a large-width MOSFET.

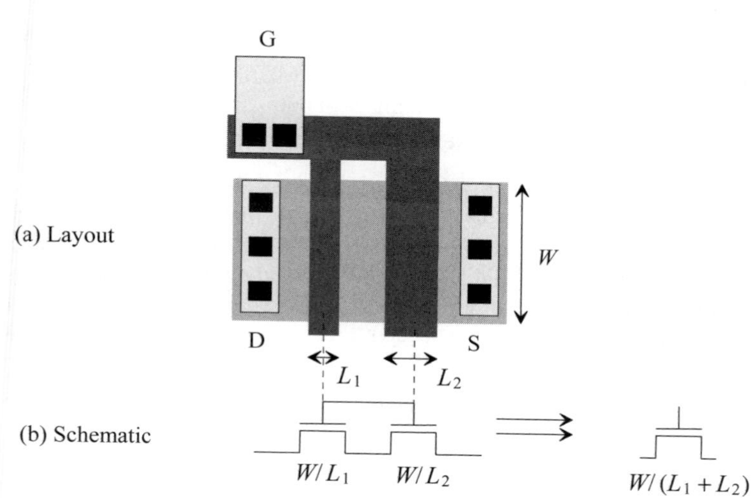

Figure 5.19 MOSFETs in series with gates tied together behave as a single MOSFET with the sum of the lengths.

as a MOSFET with a width of 10 and a length of 2 or as two MOSFETs in series with lengths of 1 and widths of 10. **Notice** that when we write, for a MOSFET size, 10/2, $\frac{15}{3}$, 25/10, or 10/100 the first or top number always specifies the width of the MOSFET, while the second number specifies the MOSFET's length.

Notice how, when there is an even number of fingers in a large width MOSFET (Fig. 5.18), the area of the active is larger on one side of the poly (on one side of the MOSFET) than on the other side. This leads to a larger parasitic depletion capacitance. Consider the MOSFET schematics in Fig. 5.20. In (a) of this figure, we've drawn the parasitic capacitance on the MOSFET symbol. Generally, for high-speed design, we want to place the MOSFET terminal with the larger parasitic capacitance closest to ground (for

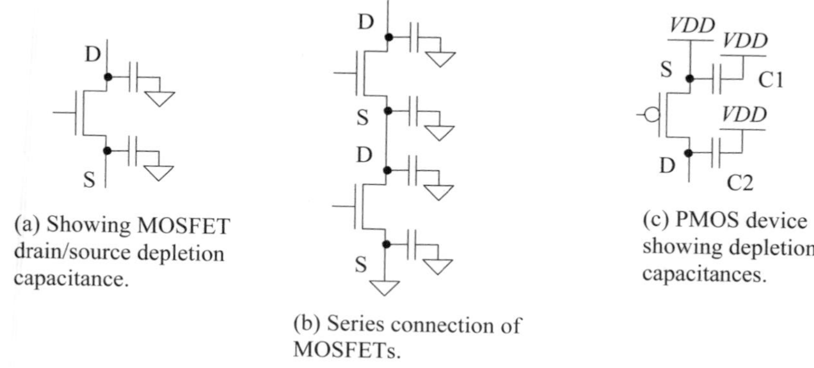

Figure 5.20 Showing depletion capacitance on the MOSFET symbols.

the NMOS) or *VDD* for the PMOS. To understand this, consider the MOSFETs seen in Fig. 5.20b. Thinking of the two NMOS devices as switches, we see that when both switches turn on, the top parasitic capacitance is discharged through both switches. The middle two parasitic capacitors discharge through one switch and the bottom parasitic doesn't charge or discharge (both sides are always tied to ground). The smallest parasitic capacitance should be at the top of the switches because it has the highest resistance discharge path (in this example through two MOSFETs). For the PMOS device in Fig. 5.20c, we would want the larger of C1 or C2 to be called the source terminal and connected to *VDD*.

A Qualitative Description of MOSFET Capacitances

Consider the NMOS device in Fig. 5.21a. Here we apply a voltage to the gate of the NMOS device (the most positive voltage in the circuit, the power supply voltage, *VDD*) while holding the source and drain at the same potential as the substrate (ground). The application of this voltage attracts electrons under the gate oxide. This creates a continuous channel of electrons, effectively shorting the drain/source implants. The capacitance from the gate terminal to ground is calculated as

$$C_{ox} = \frac{\varepsilon_{ox}}{t_{ox}} \cdot W \cdot L = C'_{ox} \cdot W \cdot L \qquad (5.22)$$

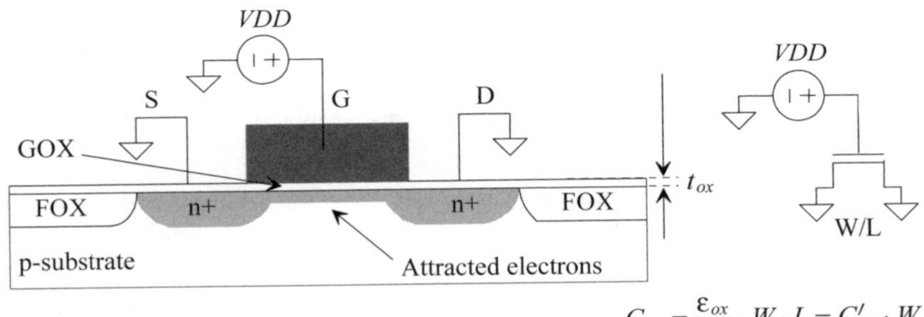

(a) NMOS device in strong inversion (channel formed under the oxide).

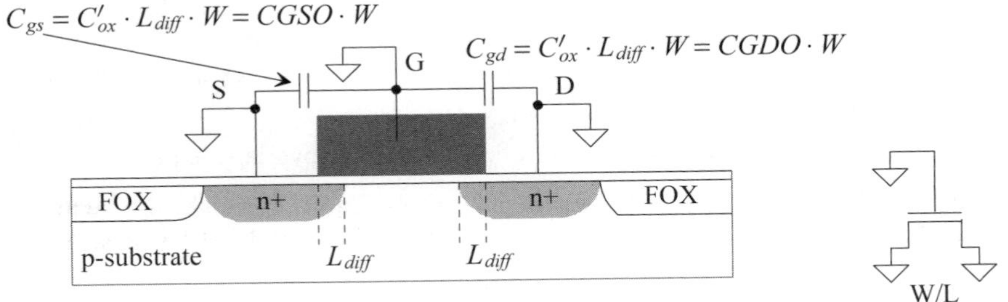

(b) NMOS device in depletion (no channel formed under the oxide).

Figure 5.21 Qualitative description of MOSFET capacitances.

The device is said to be operating in the strong inversion region. The surface under the GOX is p-substrate. When we apply a large positive potential to the gate, we change this material from p-type to n-type (we invert the surface). Note that the oxide capacitance, C_{ox} , does not depend on the extent of the lateral diffusion.

Next examine the configuration in Fig. 5.21b. In this figure all of the MOSFET terminals are grounded. No channel is formed under the GOX. Because of contact potentials, discussed in the next chapter, the area under the GOX is depleted of free carriers. Under these conditions, the MOSFET is operating in the depletion region. The source and drain are not connected as they were in Fig. 5.21a. The capacitance from the gate to the source (or drain) depends on the lateral diffusion and is given by

$$C_{gs} = C'_{ox} \cdot L_{diff} \cdot W = C_{gd} = CGDO \cdot W = CGSO \cdot W \qquad (5.23)$$

The parameter *CGDO* (or *CGSO)* is SPICE parameter called the gate-drain (gate-source) overlap capacitance and is given by

$$CGDO = CGSO = C'_{ox} \cdot L_{diff} \qquad (5.24)$$

When the MOSFET is off, as it is in Fig. 5.21b, the overlap of the gate over the source/drain implant region (the overlap capacitance) is an important component of the capacitance at the MOSFET's gate terminal.

5.4 LASI Design Examples

In this section we present some additional design examples using LASI. We'll show how to draw both schematics and layouts. Then we'll compare the two cells to make sure that the layout matches the schematic (layout versus schematic, LVS). Some things to remember and know when using LasiCkt:

1. Basic SPICE device cells like resistors, capacitors, and MOSFETs are always rank 1 cells. We can't lay out a MOSFET or resistor in a rank 2 cell. This means we can't use a layout like the via1 cell seen in Fig. 3.19 in a MOSFET layout.

2. The device text (e.g., for a resistor name R1) and parameter text (e.g., specifying a resistor's value, 2k) must be placed within the perimeter of the cell they are labeling. It is useful to show the cell's image by pressing **i** on the keyboard.

3. Connectors (connector text) are used on a cell to connect the cell to wires (e.g., metal1) in higher ranking cells. For example, connector text is placed in rank 1 cell, like a resistor, to show where to connect the wires in a rank 2 (or higher) cell.

4. Care must be exercised if cells are placed in a layout where they overlap. The parameter text or device text shouldn't be placed in the overlap area.

5. When drawing schematics, zero-width paths (poly object) should be used with the metal1 layer to connect the cells. Boxes (zero area) should not be present in a schematic. To verify that there aren't any boxes in a schematic, use the **Wget** command on the entire schematic cell followed by the **Info** command.

6. Parameter text (ptxt) labels on rank 2 and higher cells are always the cell's name. Rank one cells have user inputs. For example, a resistor may have a ptxt value of "2k" (the resistor's value) and a MOSFET may use "NMOS L=1 W=10".

A Resistor Divider

An example schematic, similar to the one we implemented back in Fig. 3.24, is seen in Fig. 5.22. Here, we use 2 kΩ unit cells to implement a divider (a 2/3 divider). The corresponding layout is seen in Fig. 5.23. For every text label (node text, parameter text, connector text, and device text) in the schematic there is an identical label in the layout. Once the schematic is drawn, we can use the **Save** command to save the schematic. We can then use the **Save** command again to save the schematic as a layout (followed by deleting all contents of the cell but the text). We can then place the layout of the unit resistor into the cell and connect the cells together with metal1 boxes (and then move the text into the desired locations). **Note** the metal1 boxes must overlap the connectors on the cells. Figure 5.24 shows the rank 2 layout in Fig. 5.23 redrawn with the cells drawn as outlines. The metal1 in the rank 2 cell must overlap the connector in the lower ranking cell (the resistor layout).

One of the important additions to the schematic and layout in Figs. 5.22–5.24 is node text. We use node text with a header file to specify the inputs and outputs of the circuit. A header file may look like

```
.control
destroy all
run
plot vin vout
.endc
.dc vin 0 1 1m
vin vin 0 DC 0
```

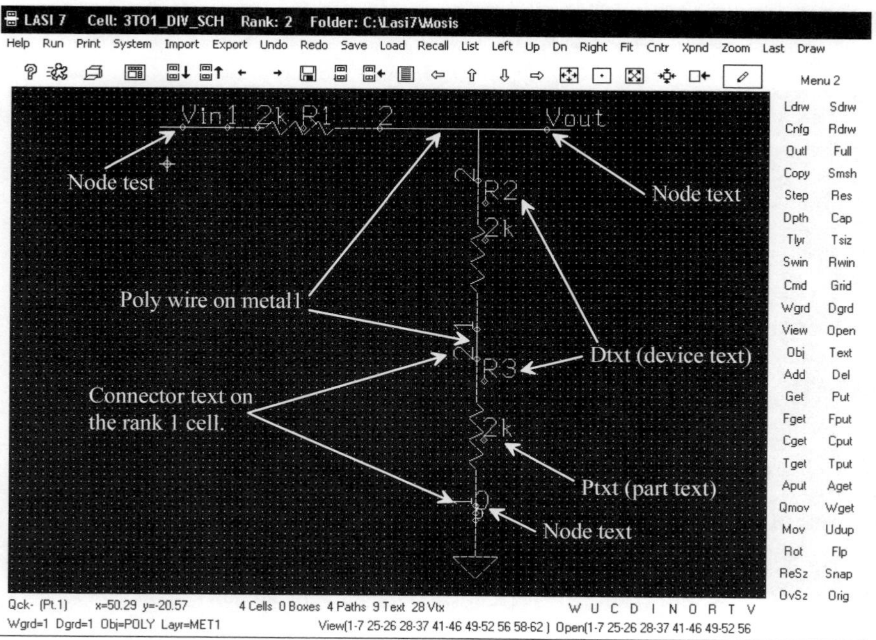

Figure 5.22 Schematic of a resistive divider.

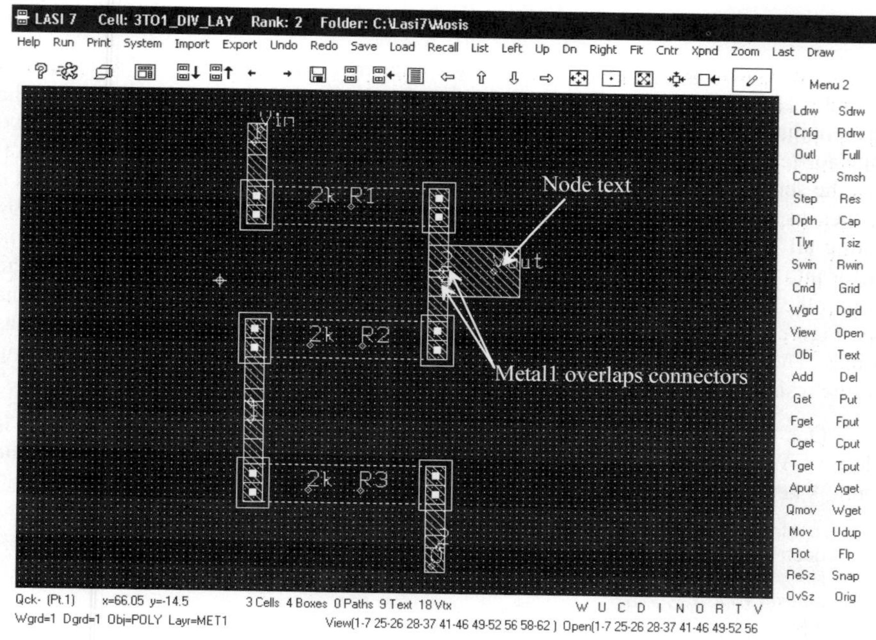

Figure 5.23 Layout corresponding to the schematic in Fig. 5.22.

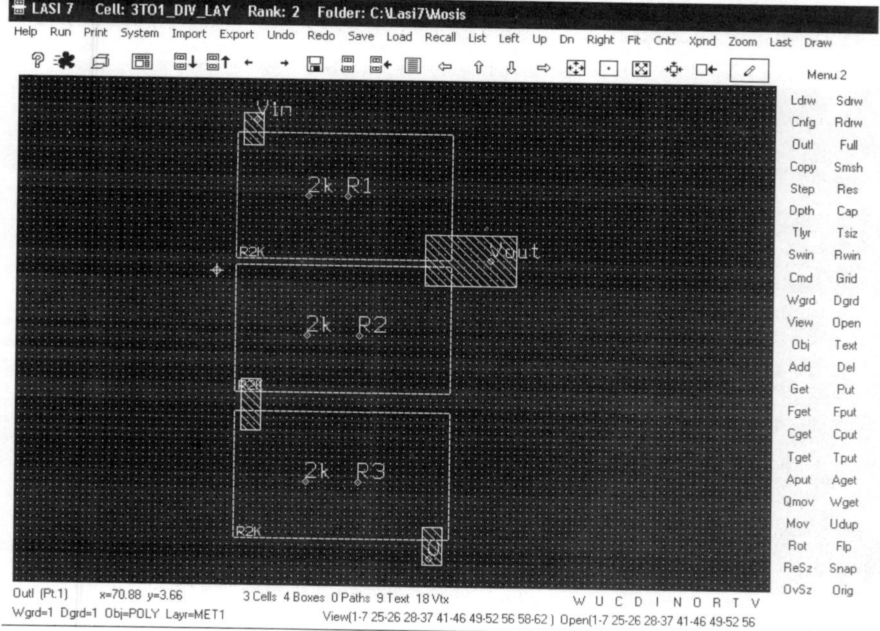

Figure 5.24 Drawing the cells in Fig. 5.23 as outlines.

Layout of a MOSFET

Figure 5.25 shows the schematic of an NMOS transistor with wire connections. This schematic representation is useful for a MOSFET test structure (used to measure the current-voltage characteristics of the MOSFET). The corresponding layout is seen in Fig. 5.26. These cells, as well as the others seen in this book, are available in the file cmosedu_lasi.zip available at cmosedu.com.

An **important point** to note in Fig. 5.25 or any other schematic or layout implemented in LASI is that placing node text directly on (the vertices at the same location) a connector plugs the connector. In other words, if we plug the connector, then running a wire over it has no effect. This can lead to frustration when drawing schematics or layouts and then extracting a netlist.

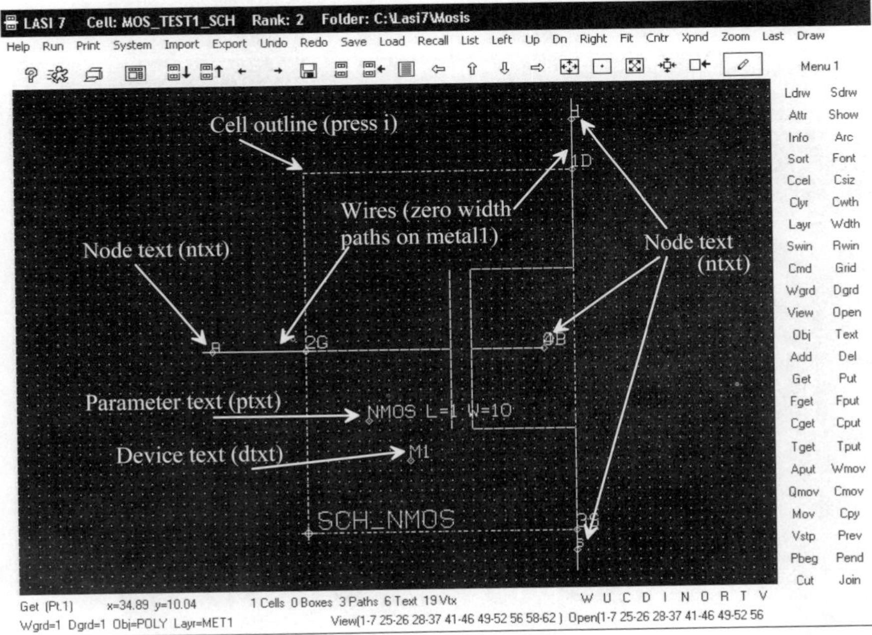

Figure 5.25 Schematic of an NMOS transistor and wire connections.

To simulate the circuit in Fig. 5.25, we can use the following header file (in the file Mos_test1.txt)

```
.control
destroy all
run
let id=-i(vds)
plot id
.endc
```

```
vds d s DC 0
vgs g s DC 0

.dc vds 0 1 1m vgs 0 1 0.25

*Important
.options scale=50nm

*ground the source
Vs  s 0 DC 0
```

We specify the voltage sources in the schematic and the type of analysis. Here, in this header, we sweep the drain-source voltage from 0 to 1 V in 1 mV steps, while the gate-source voltage is stepped in 250 mV increments. Notice the statement showing the scale factor (allowing us to draw the layout with an integer grid). Not remembering to include this specification in a header file can lead to wasted time when running simulations.

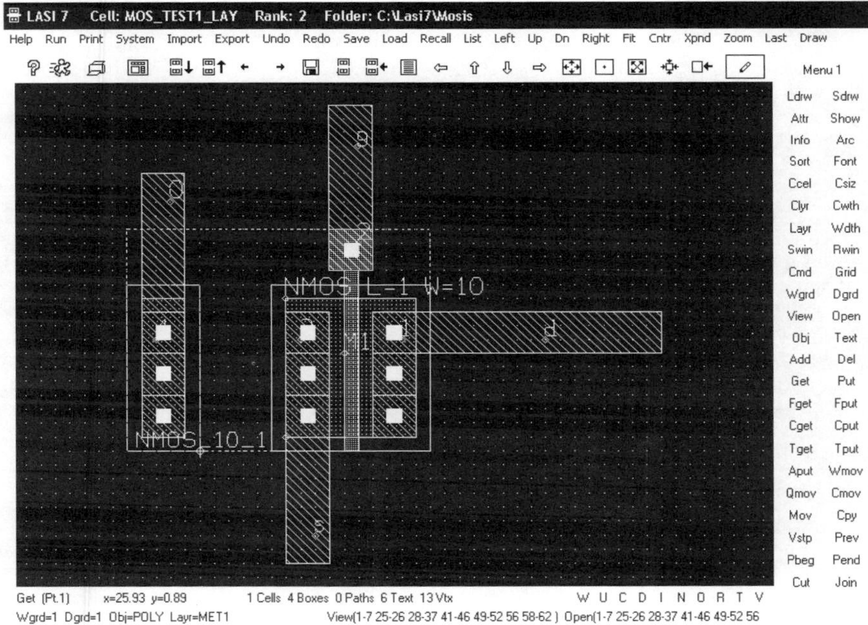

Figure 5.26 Layout of the MOSFET in Fig. 5.25.

Figure 5.27 shows the LasiCkt setup screen for extracting a SPICE netlist from the layout in Fig. 5.26. Note how we specified SPICE models for the footer file. These SPICE models can be downloaded from cmosedu.com. Table 5.2 shows the SPICE file names for the long- and short-channel CMOS processes that **we use in this book**. For the layout in Fig. 5.26, we use the 50 nm CMOS process.

We can extract SPICE netlists from either the layout or the schematic. The SPICE netlists can be compared and the simulation results can be used to perform a heuristic LVS. However, a more complete LVS can be performed by using the node list files in

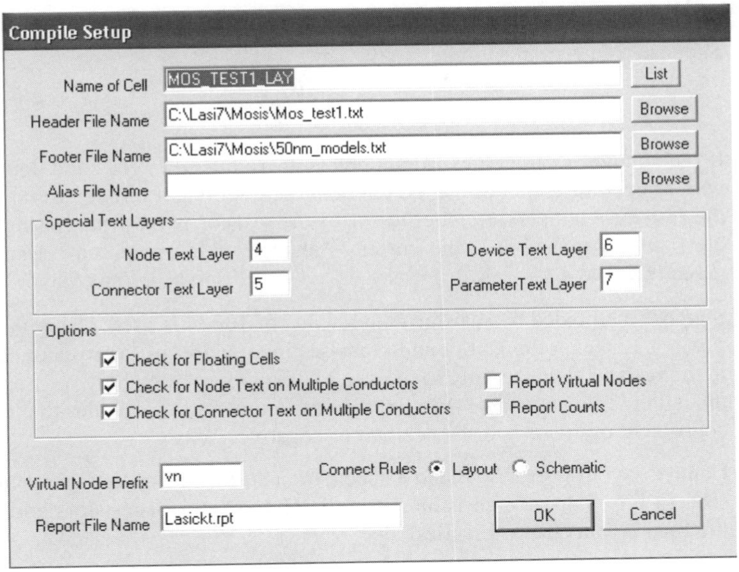

Figure 5.27 LasiCkt setup screen (for extracting the SPICE netlist from a layout).

LASI. By pressing the LVS command in LasiCkt and selecting the file names for the two cells to be compared, LasiCkt can perform an LVS.

Table 5.2 Table listing the SPICE file names and corresponding scale factors (with the .options statement SPICE) used in this book.

CMOS technology	**Scale factor used with .options**	**SPICE file name**	***VDD***
1 μm (long channel)	1um	1u_models.txt	5 V
50 nm (short channel)	50nm	50nm_models.txt	1 V

ADDITIONAL READING

[1] R. A. Pease, J. D. Bruce, H. W. Li, and R. J. Baker, "Comments on Analog Layout Using ALAS!" *IEEE Journal of Solid-State Circuits*, vol. 31, no. 9, September 1996, pp. 1364–1365.

[2] D. J. Allstot and W. C. Black, "Technology Design Considerations for Monolithic MOS Switched-Capacitor Filtering Systems," *Proceedings of the IEEE*, vol. 71, no. 8, August 1983, pp. 967–986.

[3] D. E. Boyce, *LASI User's Manual*, available while LASI is running by the help menu item. This help file (which includes links to help for LasiCkt) can also be accessed by opening the lhh7 file in C:\Lasi7\.

PROBLEMS

5.1 Suppose a current in a circuit is given by

$$I = \frac{V_{REF}}{R}$$

If the voltage, V_{REF}, comes from a precision voltage reference and doesn't change with temperature, determine the temperature coefficient of the current in terms of the resistor's temperature coefficient. If the resistor is fabricated using the n-well plot, similar to Fig. 5.1, the current's change with temperature. Use the TCR1 given in Table 4.1.

5.2 Suppose a silicided n+ resistor with a value of 100 Ω is used. Using the data from Table 4.1, sketch the layout and cross-sectional views of the resistor. The current in the resistor flows mainly in the silicide. Suppose the mobility of free carriers in the silicide is constant with increasing temperature. Would the temperature coefficient of the resistor be positive or negative? Why?

5.3 Using a layout program, make a schematic and layout for the 1/5 voltage divider seen in Fig. 5.4 if the resistor's value is 5k. Use n-well resistors and DRC/LVS the final layout and schematic.

5.4 Using a layout program, make a schematic and layout for an RC lowpass circuit where the resistor's value is 10k and the poly-poly capacitor's value is 100 fF. Use the 50 nm process (see Table 5.1). DRC/LVS the final layout and schematic. Simulate the operation of the circuit with a pulse input (see Fig. 1.27).

5.5 Estimate the areas and perimeters of the source/drain in the layout seen in Fig. 5.18 if the length of the device, L, is 2 and the width of a finger, W, is 20.

5.6 Provide a qualitative discussion for the capacitances of the PMOS device similar to the discussion associated with Fig. 5.21 for the NMOS device. Make sure the descriptions of operation in the strong inversion and depletion regions are clear. Draw the equivalent (to Fig. 5.21) figure for the PMOS devices.

Chapter

6

MOSFET Operation

In this chapter we discuss MOSFET operation. Figure 6.1 shows how we define the voltages, currents, and terminal designations for a MOSFET. When the substrate is connected to ground and the well is tied to *VDD*, we use the simplified models shown at the bottom of the figure. It is important to keep in mind that the MOSFET is a four-terminal device and that the source and drain of the MOSFET are interchangeable. Note that **all voltages and currents are positive** using the naming convention seen in the figure. If we say "the V_{SG} of the MOSFET is...," we know that we are talking about a PMOS device. Also note, the drain current flows from the top of the symbol to the bottom. For the NMOS, the drain is at the top of the symbol, while for the PMOS the source is at the top of the symbol. The devices are complementary.

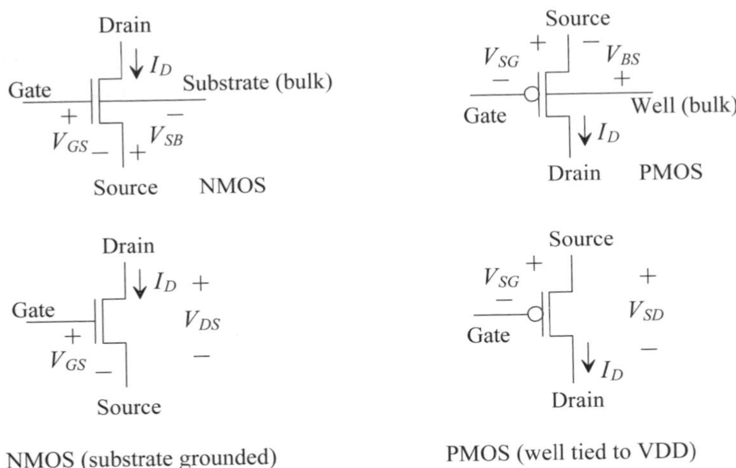

Figure 6.1 Voltage and current designations for MOSFETs in this chapter.

6.1 MOSFET Capacitance Overview/Review

In this section we'll discuss and review the capacitances of a MOSFET operating in the accumulation, depletion (weak inversion), and strong inversion regions.

Case I: Accumulation

Examine the cross-sectional view seen in Fig. 6.2. When $V_{GS} < 0$, mobile holes from the substrate are attracted (or *accumulated*) under the gate oxide. Remembering from Eq. (5.8) that

$$C'_{ox} = \frac{\varepsilon_{ox}}{t_{ox}} \tag{6.1}$$

the capacitance between the *gate* electrode and the *substrate* electrode is given by

$$C_{gb} = \frac{\varepsilon_{ox}}{t_{ox}} \cdot (L_{drawn} - 2 \cdot L_{diff}) \cdot W_{drawn} \cdot (scale)^2 = C'_{ox} \cdot L_{eff} \cdot W_{drawn} \cdot (scale)^2 \tag{6.2}$$

where ε_{ox} (= 3.97 · 8.85 aF/μm) is the dielectric constant of the gate oxide, W_{drawn} is the drawn width (neglecting oxide encroachment), and $L_{drawn} - 2 \cdot L_{diff}$ is the effective channel length, L_{eff}. The capacitance between the gate and drain/source (the overlap capacitances, see Eq. [5.23]) is given by

$$C_{gs} = C'_{ox} \cdot L_{diff} \cdot W_{drawn} \cdot (scale)^2 = C_{gd} \tag{6.3}$$

neglecting oxide encroachment on the width of the MOSFET. The gate-drain overlap capacitance is present in a MOSFET regardless of the biasing conditions.

The total capacitance between the gate and ground in the circuit of Fig. 6.2 is the sum of C_{gd}, C_{gs}, and C_{gb} and is given by

$$C_{gs} + C_{gb} + C_{gd} = C_{ox} = C'_{ox} \cdot L_{drawn} \cdot W_{drawn} \cdot (scale)^2 \tag{6.4}$$

There is a significant resistance in series with C_{gb}. The resistance comes from the physical distance between the substrate connection and the area under the gate oxide. The resistivity of the n+ source and drain regions in series with C_{gs} and C_{gd} tends to be small enough to neglect in most circuit design applications.

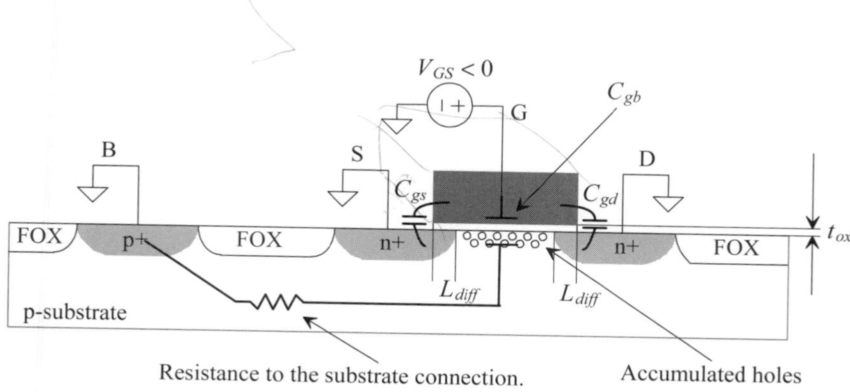

Figure 6.2 Cross-sectional view of a MOSFET operating in accumulation.

Case II: Depletion

Referring again to Fig. 6.2, let's consider the case when V_{GS} is not negative enough to attract a large number of holes under the oxide and not positive enough to attract a large number of electrons. Under these conditions, the surface under the gate is said to be nearly depleted (depeleted of free electrons and holes). Consider the cross-sectional view seen in Fig. 6.3. As V_{GS} is increased from some negative voltage, holes will be displaced under the gate, leaving immobile acceptor ions that contribute a negative charge. We see that as we increase V_{GS} a capacitance between the gate and the induced (n) channel under the oxide exists. Also, a depletion capacitance between the depleted channel and the substrate is formed. The capacitance between the gate and the source/drain is simply the overlap capacitance, while the capacitance between the gate and the substrate is the oxide capacitance *in series* with the depletion capacitance. The depletion layer shown in Fig. 6.3 is formed between the substrate and the induced channel. The MOSFET operated in this region is said to be in *weak inversion* or the *subthreshold region* because the surface under the oxide is not heavily n+.

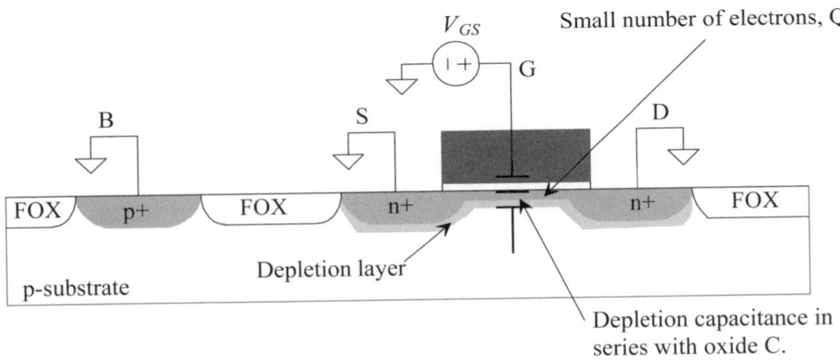

Figure 6.3 Cross-sectional view of a MOSFET operating in depletion.

Case III: Strong Inversion

When V_{GS} is sufficiently large (>> V_{THN}, the threshold voltage of the NMOS device) so that a large number of electrons are attracted under the gate, the surface is said to be inverted, that is, no longer p-type. Figure 6.4 shows how the capacitance at the gate changes as V_{GS} is varied, for an NMOS device, when the source, drain, and bulk are grounded. This figure can be misleading. It may appear that we can operate the MOSFET in accumulation if we need a good capacitor. Remembering that when the MOSFET is in the accumulation region the majority of the capacitance to ground, C_{gb}, runs through the large parasitic resistance of the substrate, we see that to operate the MOSFET in this region we need plentiful substrate connections around the gate oxide (to reduce this parasitic substrate resistance). It's preferable to operate the MOSFET in strong inversion when we need a capacitor. The attracted electrons under the gate oxide short the drain and source together forming a low-resistance bottom plate for the capacitor. We will make a capacitor in this fashion many times when designing circuits.

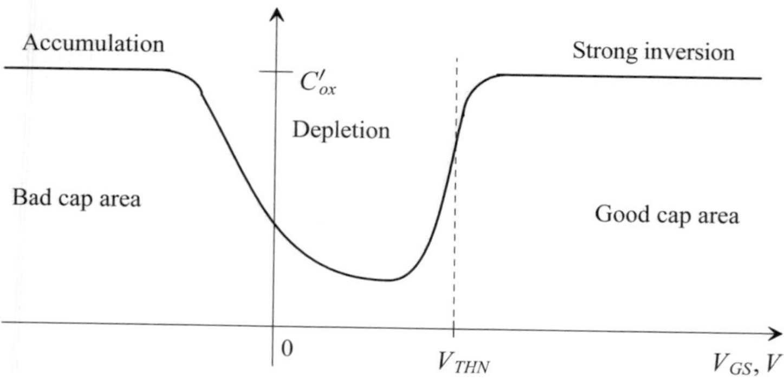

Figure 6.4 The variation of the gate capacitance with DC gate-source voltage.

Example 6.1
Suppose the MOSFET configuration seen in Fig. 6.5 is to be used as a capacitor. If the width and length of the MOSFET are both 100, estimate the capacitance between the gate and the source/drain terminals. Use the long-channel process oxide capacitance listed in Table 5.1. Are there any restrictions on the voltages we can use across the capacitor?

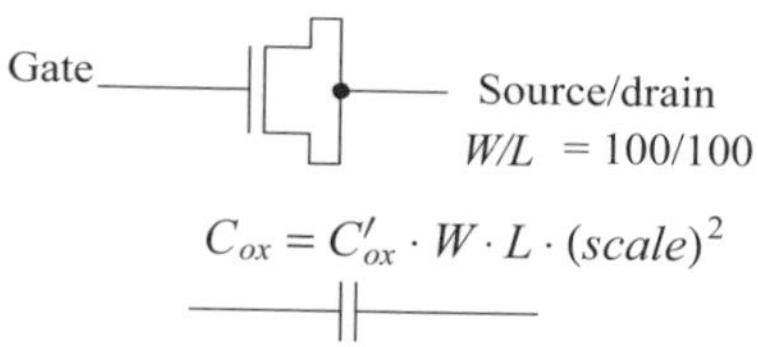

Figure 6.5 Using the MOSFET as a capacitor.

Since the MOSFET is to be used as a capacitor, we require operation in the strong inversion region, that is, $V_{GS} >> V_{THN}$ (the gate potential at least a threshold voltage plus 5% of *VDD* above the source/drain potentials). The capacitance between the gate and the source/drain is then $C_{tot} = C_{ox} = C'_{ox} \cdot W \cdot L$ or from Table 5.1

$$C_{tot} = (1.75\,fF/\mu m^2)(100\,\mu m)(100\,\mu m) = 17.5\,pF$$

Note that we did not concern ourselves with the substrate connection. Since we are assuming strong inversion, the bulk (substrate) connection only affects the capacitances from the drain/source to substrate (those of the source/drain implant regions). We see, however, that the connection of the substrate significantly affects the threshold voltage of the devices and thus the point we label strong inversion. ■

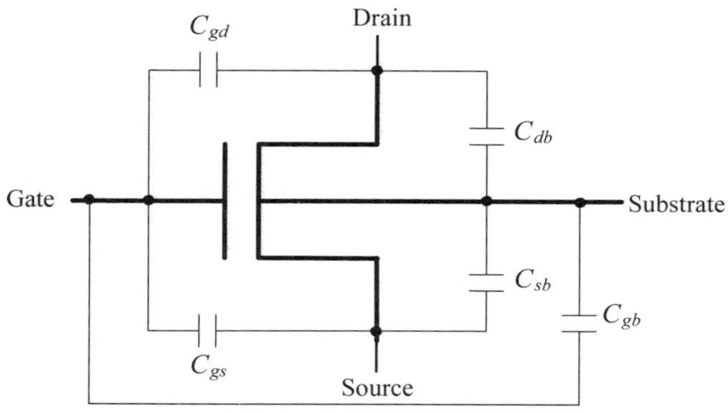

Figure 6.6 MOSFET capacitances.

Summary

Figure 6.6 shows our MOSFET symbol with capacitances. Table 6.1 lists the capacitances based on the region of operation (without a scale factor). The capacitance *CGBO* is the capacitance associated with the gate poly extension over the field region (Fig. 5.13). The gate-drain capacitance, C_{gd} , and the gate-source capacitance, C_{gs} , are determined by the region of operation. For example, as we'll see later, when the MOSFET operates in the triode region, the inverted channel extends between the source and drain implants (the channel resistively connects the source and drain together). The capacitance between the gate and this channel is the oxide capacitance, C_{ox}. We assume that half of this capacitance is between the drain and the other half is between the source.

Table 6.1 MOSFET capacitances.

Name	Off	Triode	Saturation
C_{gd}	$CGDO \cdot W$	$\frac{1}{2} \cdot W \cdot L \cdot C'_{ox}$	$CGDO \cdot W$
C_{db}	C_{jd}	C_{jd}	C_{jd}
C_{gb}	$C'_{ox} WL_{eff} + CGBO \cdot L$	$CGBO \cdot L$	$CGBO \cdot L$
C_{gs}	$CGSO \cdot W$	$\frac{1}{2} \cdot W \cdot L \cdot C'_{ox}$	$\frac{2}{3} \cdot W \cdot L \cdot C'_{ox}$
C_{sb}	C_{js}	C_{js}	C_{js}

6.2 The Threshold Voltage

In the last section we said that the semiconductor/oxide surface is inverted when V_{GS} is greater than the threshold voltage V_{THN}. Under these conditions a channel of electrons is formed under the gate oxide. Below this channel, electrons fill the holes in the substrate giving rise to a depletion region (depleted of free carriers). The thickness of the depletion region (Fig. 6.7) is given from pn junction theory by

$$X_d = \sqrt{\frac{2\varepsilon_{si}|V_s - V_{fp}|}{qN_A}} \tag{6.5}$$

where N_A is the number of acceptor atoms in the substrate, V_s is the electrostatic potential at the oxide-silicon interface (the channel), and the electrostatic potential of the p-type substrate is given by (see Eq. [2.11])

$$V_{fp} = -\frac{E_i - E_{fp}}{q} = -\frac{kT}{q}\ln\frac{N_A}{n_i} \tag{6.6}$$

noting that this is a negative number. As seen in Fig. 6.7, one edge of the depletion region is the MOSFET's gate oxide, while the other edge is the p-substrate (holes). The positive potential on the gate attracts electrons under the gate oxide. This charge is equal and opposite to the charge in the polysilicon gate material. The charge/unit area is given by

$$Q_b' = qN_AX_d = \sqrt{2\varepsilon_{si}qN_A|V_s - V_{fp}|} \tag{6.7}$$

If the surface electrostatic potential at the oxide interface, V_s, is the same as the bulk electrostatic potential V_{fp} (i.e., $V_s = V_{fp}$ and then $Q_b' = 0$), the MOSFET is operating in the accumulation mode, or the MOSFET is OFF in circuit terms. At this point the number of holes at the oxide-semiconductor surface is N_A, the same concentration as the bulk.

As V_{GS} is increased, the surface potential becomes more positive. When $V_s = 0$, the surface under the oxide has become depleted (the carrier concentration is n_i). When $V_s = -V_{fp}$ (a positive number), the channel is inverted (electrons are pulled under the oxide forming a channel), and the electron concentration at the semiconductor-oxide interface is equal to the substrate doping concentration. The value of V_{GS} when $V_s = -V_{fp}$ is arbitrarily defined as the threshold voltage, V_{THN}. Note that the surface potential changed a total of $2|V_{fp}|$ between the strong inversion and depletion cases.

For $V_{GS} = V_{THN}$ ($V_s = -V_{fp}$), the negative charge under the gate oxide is given by

$$Q_{bo}' = \sqrt{2qN_A\varepsilon_{si}|-2V_{fp}|} \tag{6.8}$$

with units of Coulombs/m^2. Up to this point we have assumed that the substrate and source were tied together to ground. If the source of the NMOS device is at a higher potential than the substrate, the potential difference is given by V_{SB}; the negative charge under the gate oxide becomes

$$Q_b' = \sqrt{2qN_A\varepsilon_{si}|-2V_{fp} + V_{SB}|} \tag{6.9}$$

Example 6.2

For a substrate doping of 10^{15} atoms/cm^3, $V_{GS} = V_{THN}$ and $V_{SB} = 0$, estimate the electrostatic potential in the substrate region, V_{fp}, and at the oxide-semiconductor interface, V_s, the depletion layer width, X_d, and the charge contained in the depletion region, Q_b', and thus the inverted region under the gate.

The electrostatic potential of the substrate is

$$V_{fp} = -\frac{kT}{q} \ln \frac{N_A}{n_i} = -26 \text{ mV} \cdot \ln \frac{10^{15}}{14.5 \times 10^9} = -290 \text{ mV}$$

and therefore the electrostatic potential at the oxide semiconductor interface ($V_{GS} = V_{THN}$), V_s, is 290 mV. The depletion layer thickness is given by

$$X_d = \sqrt{\frac{2 \cdot 11.7 \cdot (8.85 \times 10^{-18} F/\mu m)(2 \cdot 0.29 V)}{(1.6 \times 10^{-19} \frac{C}{atom})(10^{15} \frac{atoms}{cm^3})(\frac{cm^3}{10^{12} \mu m^3})}} = 0.866\ \mu m$$

and the charge contained in this region, from Eq. (6.7) or (6.8) with $V_s = -V_{fp}$, by

$$Q'_{bo} = qN_A X_d = \left(1.6 \times 10^{-19} \frac{C}{atom}\right)\left(10^{15} \frac{atoms}{cm^3}\right)\left(\frac{cm^3}{10^{12} \mu m^3}\right)(0.866\ \mu m)$$

$$= 139\ \frac{aC}{\mu m^2}$$

Note that this is true only when $V_{GS} = V_{THN}$. ■

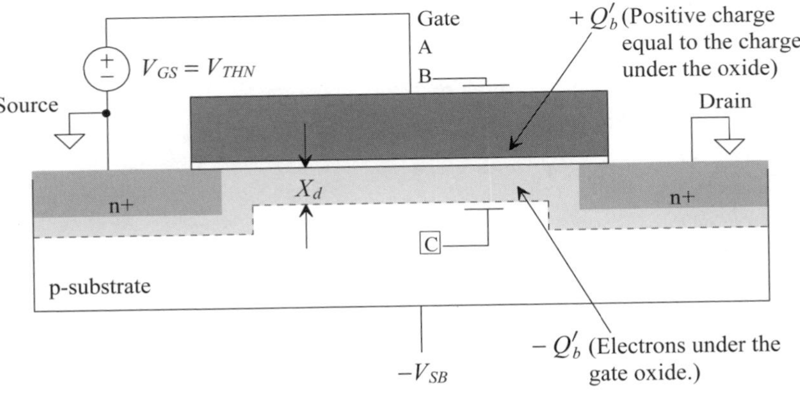

Figure 6.7 Calculation of the threshold voltage.

Contact Potentials

Again, consider the MOSFET shown in Fig. 6.7. We assume that the applied $V_{GS} = V_{THN}$ so that the preceding discussions and assumptions hold. The potential across the gate-oxide capacitance, C'_{ox}, is simply

$$V_{BC} = \frac{Q'_b}{C'_{ox}} \tag{6.10}$$

The surface potential *change*, V_C (= ΔV_s), from the equilibrium case is $|2V_{fp}|$. (The absolute voltage of the channel is 0 V; that is, the source, drain, and channel are at ground.) The potential needed to change the surface potential and fill the fixed holes in the substrate is

$$V_B = \frac{Q'_b}{C'_{ox}} - 2V_{fp} \tag{6.11}$$

An additional charge, Q'_{ss} (columbs/area), can be used to model surface states (aka interface trapped charges, Q'_{it}) that may exist because of dangling bonds at the oxide- silicon interface. Here we assume electrons are attracted to the surface of the semiconductor (and trapped directly under the oxide) causing the threshold voltage to decrease. Equation (6.11) may be rewritten to include these surface-state charges as

$$V_B = \frac{Q'_b - Q'_{ss}}{C'_{ox}} - 2V_{fp} \tag{6.12}$$

The final component needed to determine the threshold voltage is the contact potential between point C (the bulk) and point A (the gate material) in Fig. 6.7. The potential difference between the gate and bulk (p-substrate) can be determined by summing the difference between the materials in the MOS system shown in Fig. 6.8. Adding the contact potentials, we get $(V_G - V_{ox}) + (V_{ox} - V_{fp}) = V_G - V_{fp}$. Note that if we were to include the surface electrostatic potential, V_s, the result would be the same, that is, only the two outer materials are of concern when calculating the contact potential difference. The contact potential between the bulk and the gate poly, we will assume n+ poly with doping concentration $N_{D,\,poly}$, is given by

$$V_{ms} = V_G - V_{fp} = \frac{kT}{q}\ln\left(\frac{N_{D,poly}}{n_i}\right) + \frac{kT}{q}\ln\frac{N_A}{n_i} \tag{6.13}$$

The threshold voltage, V_{THN}, is given by

$$V_{THN} = \frac{Q'_b - Q'_{ss}}{C'_{ox}} - 2V_{fp} - V_{ms} \tag{6.14}$$

$$= -V_{ms} - 2V_{fp} + \frac{Q'_{bo} - Q'_{ss}}{C'_{ox}} - \frac{Q'_{bo} - Q'_b}{C'_{ox}} \tag{6.15}$$

$$= -V_{ms} - 2V_{fp} + \frac{Q'_{bo} - Q'_{ss}}{C'_{ox}} + \frac{\sqrt{2q\varepsilon_{si}N_A}}{C'_{ox}}\left[\sqrt{|2V_{fp}| + V_{SB}} - \sqrt{|2V_{fp}|}\right] \tag{6.16}$$

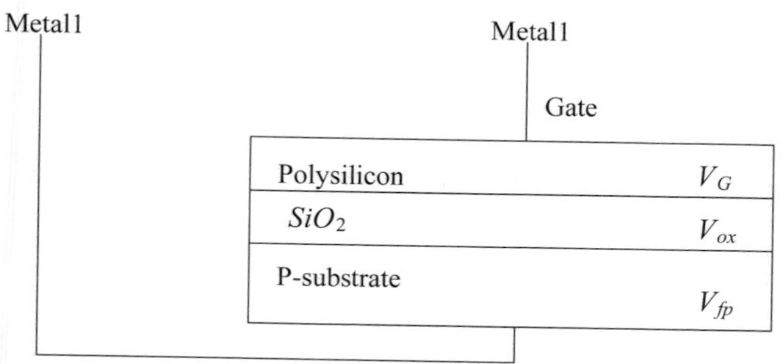

Figure 6.8 Determining the contact potential between poly and substrate.

When the source is shorted to the substrate, $V_{SB} = 0$, we can define the zero-bias threshold voltage as

$$V_{THN0} = -V_{ms} - 2V_{fb} + \frac{Q'_{bo} - Q'_{ss}}{C'_{ox}} \tag{6.17}$$

We can define a body effect coefficient or body factor by

$$\gamma = \frac{\sqrt{2q\varepsilon_{si}N_A}}{C'_{ox}} \tag{6.18}$$

Equation (6.16) can now be written as

$$V_{THN} = V_{THN0} + \gamma\left(\sqrt{|2V_{fp}| + V_{SB}} - \sqrt{|2V_{fp}|}\right) \tag{6.19}$$

It is interesting to note that a voltage, called the flatband voltage V_{FB}, must be applied for the oxide-semiconductor interface surface potential, V_s, to become the same potential as the bulk surface potential, V_{fp}. The flatband voltage is given by

$$V_{FB} = -V_{ms} - \frac{Q'_{ss}}{C'_{ox}} \tag{6.20}$$

The zero-bias threshold voltage may then be written in terms of the flatband voltage as

$$V_{THN0} = V_{FB} - 2V_{fp} + \frac{Q'_{bo}}{C'_{ox}} \tag{6.21}$$

These equations describe how the threshold voltage of the MOSFET is affected by substrate doping, oxide thickness, source/substrate bias, gate material, and surface charge density.

Example 6.3

Assuming $N_A = 10^{16}$ atoms/cm^3 and $C'_{ox} = 1.75$ fF/μm^2, estimate γ (GAMMA, the body effect coefficient).

From Eq. (6.18) the calculated γ is

$$\gamma = \frac{\sqrt{2 \cdot 1.6 \times 10^{-19} \frac{columbs}{atom} \cdot 11.7 \cdot 8.85 \frac{aF}{\mu m} \cdot 10^{16} \frac{atoms}{cm^3} \cdot \frac{cm^3}{10^{12} \mu m^3}}}{1.75 \frac{fF}{\mu m^2}} = 0.330\ V^{1/2} \quad \blacksquare$$

Example 6.4

Estimate the zero-bias threshold voltage for the MOSFET of Ex. 6.2. Assume that the poly doping level is 10^{20} atoms/cm^3. What happens to the threshold voltage if sodium contamination causes an impurity of 40 aC/μm^2 at the oxide-semiconductor interface with $C'_{ox} = 1.75\ fF/\mu m^2$?

The electrostatic potential between the gate and substrate is given by

$$-V_{ms} = V_{fp} - V_G = -290\ mV - 26\ mV \cdot \ln \frac{10^{20}}{14.5 \times 10^9} = -879\ mV$$

$$-2V_{fp} = 580\ mV$$

$$\frac{Q'_{bo}}{C'_{ox}} = \frac{139\ aC/\mu m^2}{1.75\ fF/\mu m^2} = 79\ mV$$

$$\frac{Q'_{ss}}{C'_{ox}} = 23\ mV$$

The threshold voltage, from Eq. (6.17) without the sodium contamination, is –220 mV; with the sodium contamination the threshold voltage is –243 mV. ■

Threshold Voltage Adjust

These threshold voltages would correspond to *depletion devices* (a negative threshold voltage), that is, MOSFETs that conduct when the $V_{GS} = 0$. In CMOS applications, this is highly undesirable. We normally use *enhancement devices* (devices with positive threshold voltages that are off with $V_{GS} = V_{SG} = 0$). To compensate or adjust the value of the threshold voltage (the channel, the area under the gate poly) can be implanted with p+ ions. This effectively increases the value of the threshold voltage by Q'_c/C'_{ox}, where Q'_c is the charge density/unit area due to the implant. If N_I is the ion implant dose in atoms/unit area, then we can write

$$Q'_c = q \cdot N_I \tag{6.22}$$

and the threshold voltage by

$$V_{THN0} = -V_{ms} - 2V_{fp} + \frac{Q'_{bo} - Q'_{ss} + Q'_c}{C'_{ox}} \tag{6.23}$$

Example 6.5

Estimate the ion implant dose required to change the threshold voltage in Ex. 6.4 without sodium contamination, to 1 V.

From Eqs. (6.22) and (6.23) and the results of Ex. 6.4

$$V_{THN0} = -220\ mV + \frac{qN_I}{C'_{ox}} = 1\ V$$

This gives $N_I = 1.3 \times 10^{12}$ atoms/cm^2. ■

These calculations lend some insight into how the threshold voltage is affected by the different process parameters. In practice, the results of these calculations do not exactly match the measured threshold voltage. From a circuit design engineer's point of view, the threshold voltage and the body factor are measured in the laboratory when the SPICE models are extracted.

6.3 IV Characteristics of MOSFETs

Now that we have some familiarity with the factors influencing the threshold voltage of a MOSFET, let's derive the large-signal IV (current/voltage) characteristics of the MOSFET, namely operation in the triode and the saturation regions. The following derivation is sometimes referred to as the *gradual-channel approximation*. The electric field variation in the channel between the source and drain (the y-direction) doesn't vary significantly when compared to the variation in the direction perpendicular to the channel (the x-direction).

6.3.1 MOSFET Operation in the Triode Region

Consider Fig. 6.9, where $V_{GS} > V_{THN}$, so that the surface under the oxide is inverted and $V_{DS} > 0$, causing a drift current to flow from the drain to the source. In our initial analysis, we assume that V_{DS} is sufficiently small so that the threshold voltage and the depletion layer width are approximately constant.

Initially, we must find the charge stored on the oxide capacitance C'_{ox}. The voltage, with respect to the source of the MOSFET, of the channel a distance y away from the source is labeled $V(y)$. The potential difference between the gate electrode and the channel is then $V_{GS} - V(y)$. The charge/unit area in the inversion layer is given by

$$Q'_{ch} = C'_{ox} \cdot [V_{GS} - V(y)] \tag{6.24}$$

However, we know that a charge Q'_b is present in the inversion layer from the application of the threshold voltage, V_{THN}, necessary for conduction between the drain and the source. This charge is given by

$$Q'_b = C'_{ox} \cdot V_{THN} \tag{6.25}$$

The total charge available in the inverted channel, for conduction of a current between the drain and the source, is given by the difference in these two equations, or

$$Q'_I(y) = C'_{ox} \cdot (V_{GS} - V(y) - V_{THN}) \tag{6.26}$$

The differential resistance of the channel region with a length dy and a width W is given by

$$dR = \overbrace{\frac{1}{\mu_n Q'_I(y)}}^{\text{eff. sheet Res.}} \cdot \frac{dy}{W} \tag{6.27}$$

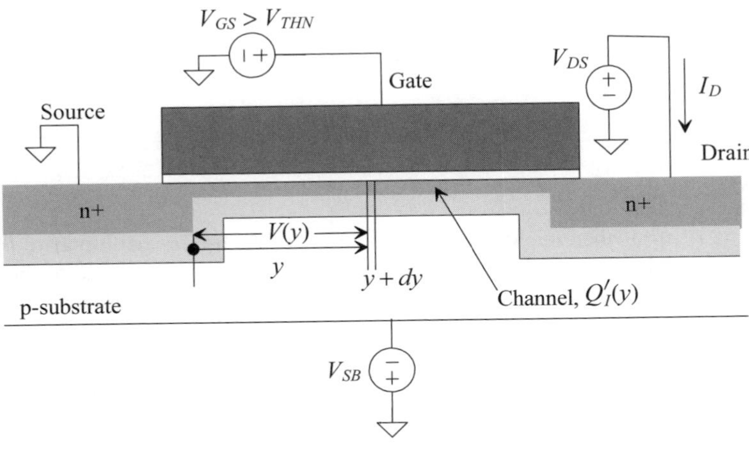

Figure 6.9 Calculation of the large-signal behavior of the MOSFET in the triode (ohmic) region.

where μ_n is the average electron mobility through the channel with units of $cm^2/V\cdot sec$ (see Eq. [5.4]). The mobility is simply a ratio of the electron (or hole) velocity cm/sec to the electric field, V/cm. For short-channel devices, the mobility decreases when the velocity of the carriers starts to saturate. This causes the effective sheet resistance in Eq. (6.27) to increase, resulting in a lowering of the drain current. This (velocity saturation) is discussed in more detail later in the chapter.

The differential voltage drop across this differential resistance is given by

$$dV(y) = I_D \cdot dR = \frac{I_D}{W\mu_n Q_I'(y)} \cdot dy \tag{6.28}$$

or substituting Eq. (6.26) and rearranging

$$I_D \cdot dy = W\mu_n C'_{ox}(V_{GS} - V(y) - V_{THN}) \cdot dV(y) \tag{6.29}$$

At this point, let's define the transconductance parameter, *KP*, for a MOSFET. For an n-channel MOSFET, this parameter is given by

$$KP_n = \mu_n \cdot C'_{ox} = \mu_n \cdot \frac{\varepsilon_{ox}}{t_{ox}} \tag{6.30}$$

and for a p-channel MOSFET, it is given by

$$KP_p = \mu_p \cdot C'_{ox} = \mu_p \cdot \frac{\varepsilon_{ox}}{t_{ox}} \tag{6.31}$$

where μ_p is the mobility of the holes in a PMOS transistor. Typical values of *KP* in the *long-channel* process (with a minimum length of 1 μm) used in this book are 120 $\mu A/V^2$ and 40 $\mu A/V^2$ for n- and p-channel transistors, respectively.

The current can be obtained by integrating the left side of Eq. (6.29) from the source to the drain, that is, from 0 to L and the right side from 0 to V_{DS}. This is shown below:

$$I_D \int_0^L dy = W \cdot KP_n \cdot \int_0^{V_{DS}} (V_{GS} - V(y) - V_{THN}) \cdot dV(y) \tag{6.32}$$

or

$$I_D = KP_n \cdot \frac{W}{L} \cdot \left[(V_{GS} - V_{THN})V_{DS} - \frac{V_{DS}^2}{2} \right] \text{ for } V_{GS} \geq V_{THN} \text{ and } V_{DS} \leq V_{GS} - V_{THN} \tag{6.33}$$

This equation is valid when the MOSFET is operating in the triode (aka **linear** or **ohmic**) region. This is the case when the induced channel extends from the source to the drain. Furthermore, we can rewrite Eq. (6.33) defining the transconductance parameter as

$$\beta = KP_n \cdot \frac{W}{L} \tag{6.34}$$

or

$$I_D = \beta \cdot \left[(V_{GS} - V_{THN})V_{DS} - \frac{V_{DS}^2}{2} \right] \tag{6.35}$$

The equivalent equation for the PMOS device operating in the triode region is

$$I_D = KP_p \cdot \frac{W}{L} \cdot \left[(V_{SG} - V_{THP})V_{SD} - \frac{V_{SD}^2}{2} \right] \text{ for } V_{SG} \geq V_{THP} \text{ and } V_{SD} \leq V_{SG} - V_{THP} \quad (6.36)$$

where the threshold voltage of the p-channel MOSFET is positive (noting, again, that from our sign convention in Fig. 6.1 all voltages and currents are positive.)

6.3.2 The Saturation Region

The voltage at $V(y)$ when $y = L$, that is, $V(L)$, in Eq. (6.26) is simply V_{DS}. In the previous subsection, we said that V_{DS} is always less than $V_{GS} - V_{THN}$ so that at no point along the channel is the inversion charge zero. When $V_{DS} = V_{GS} - V_{THN}$, the inversion charge under the gate at $y = L$ (the drain-channel junction) is zero, Eq. (6.26). This drain-source voltage is called $\boldsymbol{V_{DS,sat}}$ (= $\boldsymbol{V_{GS} - V_{THN}}$), and indicates when the channel charge becomes *pinched off* at the drain-channel interface. Increases in V_{DS} beyond $V_{DS,sat}$ attract the fixed channel charge to the drain terminal depleting the charge in the channel directly adjacent to the drain (again, pinching off the channel). Further increases in V_{DS} do not cause an increase in the drain current[1]. Figure 6.10 shows that the depletion region, with a thickness of X_{dl}, between the drain and substrate increases, causing the channel to pinch off. If V_{DS} is increased until the drain-substrate depletion region extends from the drain to the source, the device is said to be *punched* through. Large currents can flow under these conditions, causing device failure. The maximum voltage, for near minimum-size channel lengths, that can be applied between the drain and source of a MOSFET is set by the "punchthrough" voltage. For long-channel lengths, the maximum voltage is set by the breakdown voltage of the drain (n+) implant/substrate diode.

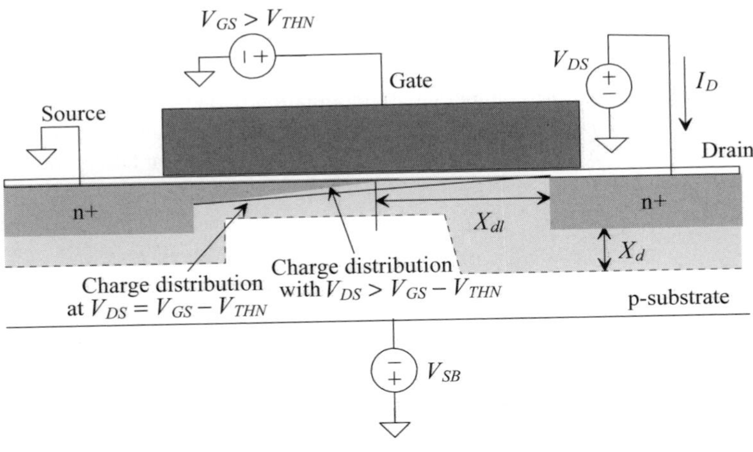

Figure 6.10 The MOSFET in saturation (pinched off).

---

[1] We will see that this is not entirely true. An effect called channel length modulation causes the drain current to increase with increasing drain-source voltage.

When a MOSFET is operated with its channel pinched off, that is, $V_{DS} \geq V_{GS} - V_{THN}$ and $V_{GS} \geq V_{THN}$, it is operating in the saturation region. Substitution of $V_{DS,sat}$ into Eq. (6.33) yields

$$I_D = \frac{KP_n}{2} \cdot \frac{W}{L} \cdot (V_{GS} - V_{THN})^2 = \frac{\beta}{2}(V_{GS} - V_{THN})^2$$

$$\text{for } V_{DS} \geq V_{GS} - V_{THN} \text{ and } V_{GS} \geq V_{THN} \tag{6.37}$$

We can define an electrical channel length of the MOSFET as the difference between the drawn channel length, neglecting lateral diffusion, and the depletion layer width, X_{dl}, between the drain n+ and the channel under the gate oxide by

$$L_{elec} = L_{drawn} - X_{dl} \tag{6.38}$$

Substituting this into Eq. (6.37), we obtain a better representation of the drain current

$$I_D = \frac{KP_n}{2} \cdot \frac{W}{L_{elec}}(V_{GS} - V_{THN})^2 \tag{6.39}$$

Qualitatively, this means that since the depletion layer width increases with increasing V_{DS}, the drain current increases as well. This effect is called *channel length modulation* (CLM). Note that as L_{drawn} is increased the effects of X_{dl} changing (CLM) become negligible.

To determine the change in output current with drain-source voltage, we take the derivative of Eq. (6.39) with respect to V_{DS}, or

$$\frac{\partial I_D}{\partial V_{DS}} = -\frac{KP_n}{2}\frac{W}{L_{elec}^2}(V_{GS} - V_{THN})^2 \cdot \frac{dL_{elec}}{dV_{DS}} = I_D \cdot \left[\frac{1}{L_{elec}}\frac{dX_{dl}}{dV_{DS}}\right] \tag{6.40}$$

where it is common to let

$$\lambda = \frac{1}{L_{elec}} \cdot \frac{dX_{dl}}{dV_{DS}} \tag{6.41}$$

Typical values for λ, called the channel length modulation parameter, range from greater than 0.1 V^{-1} for short-channel devices to 0.01 V^{-1} for long-channel devices. Equation (6.37) can be rewritten for a device operating in the saturation region, taking into account channel length modulation as

$$I_D = \frac{KP_n}{2} \cdot \frac{W}{L}(V_{GS} - V_{THN})^2[1 + \lambda(V_{DS} - V_{DS,sat})]$$

$$\text{for } V_{DS} > V_{DS,sat} = V_{GS} - V_{THN} \text{ and } V_{GS} > V_{THN} \tag{6.42}$$

When the drain-source voltage is $V_{DS,sat}$ (the drain current at the triode/saturation region border), the drain current is sometimes specified as

$$I_{D,sat} = I_D \text{ when } V_{DS} = V_{DS,sat} = V_{GS} - V_{THN} \tag{6.43}$$

In these equations we assumed that the mobility does not vary with V_{DS}. Later in the chapter (in the short-channel MOSFET discussion, Sec. 6.5.2) we'll see that the mobility does indeed vary with V_{DS} making characterizing I_D considerably more challenging. Figure 6.11 shows typical curves for an n-channel MOSFET. Notice how the device *appears to go into saturation earlier* than predicted by $V_{DS,sat} = V_{GS} - V_{THN}$. The bold line in the figure separates the actual triode and saturation regions (and also indicates $I_{D,sat}$).

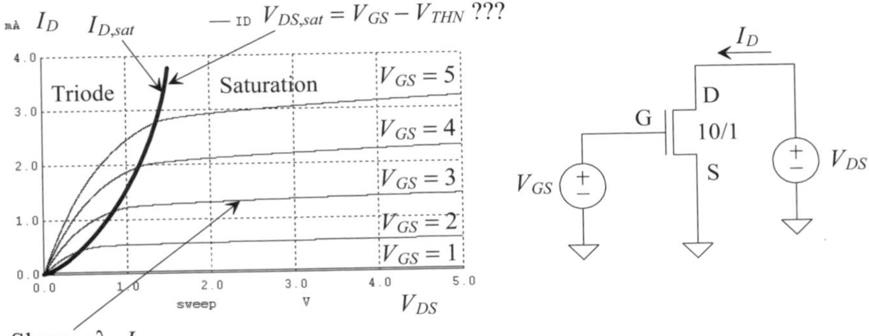

Figure 6.11 Characterisitics of a long-channel NMOS device.

For example, in the simulation results at $V_{GS} = 5$ V (with V_{THN} roughly equal to 1 V), we see that $V_{DS,sat}$ is 1.4 V (not the 4 V we calculate using $V_{GS} - V_{THN}$) The actual charge distribution in the channel is not constant but rather a function of V_{DS}. $Q'_I(y)$ decreases as we move away from the source of the MOSFET, causing $Q'_I(L)$ to become zero earlier, see Fig. 6.10.

C_{gs} Calculation in the Saturation Region

The gate-to-source capacitance of a MOSFET operating in the saturation region can be determined by solving Eq. (6.26) for the total charge in the inverted channel,

$$Q_I = \int_0^L W \cdot Q'_I(y) \cdot dy = WC'_{ox} \int_0^L (V_{GS} - V(y) - V_{THN})\, dy \tag{6.44}$$

or solving Eq. (6.29) for dy and substituting yields

$$Q_I = \frac{(W \cdot C'_{ox})^2 \cdot \mu_n}{I_D} \int_0^{V_{GS}-V_{THN}} (V_{GS} - V(y) - V_{THN})^2 \cdot dV(y) \tag{6.45}$$

where it was used that Q_I goes to zero when $y = L$ occurs when $V_{DS} = V_{GS} - V_{THN}$. Solving this equation using Eqs. (6.30) and (6.37) yields

$$Q_I = \frac{2}{3} \cdot W \cdot L \cdot C'_{ox} \cdot (V_{GS} - V_{THN}) \tag{6.46}$$

We can determine the gate-to-source capacitance while in the saturation region by,

$$C_{gs} = \frac{\partial Q_I}{\partial V_{GS}} = \frac{2}{3} \cdot W \cdot L \cdot C'_{ox} \tag{6.47}$$

See the entry in Table 6.1 (note the discontinuity between saturation and triode).

6.4 SPICE Modeling of the MOSFET

In this section we consider the level 1 SPICE model and how it relates to the equations derived in the last section. The level 1 model is a subset of the level 2 and 3 models.

Level 1 Model Parameters Related to V_{THN}

The following SPICE model parameters are related to the calculation of V_{THN},

Symbol	Name	Description	Default	Typ.	Units
V_{THN0}	VTO	Zero-bias threshold voltage	1.0	0.8	Volts
γ	GAMMA	Body-effect parameter	0	0.4	$V^{1/2}$
$2\lvert V_{fp}\rvert$	PHI	Surface to bulk potential	0.65	0.58	V
N_A	NSUB	Substrate doping	0	1E15	cm^{-3}
Q'_{ss}/q	NSS	Surface state density	0	1E10	cm^{-2}
	TPG	Type of gate material	1	1	

Using Eq. (6.19), we can calculate the threshold voltage, V_{THN}, given the above parameters. If V_{THN0} or γ are not given, then SPICE calculates them using the above information and Eqs. (6.19) – (6.21). TPG specifies the type of gate material: 1 opposite to substrate, –1 same as substrate, and 0 for aluminum gate.

Long-Channel MOSFET Models

The SPICE models used in this book for the "long-channel CMOS process" follow. The scale factor is 1 μm (= minimum drawn channel length).

```
* 1 um Level 3 models
*
* Don't forget the .options scale=1u if using an Lmin of 1
* 1<Ldrawn<200 10<Wdrawn<10000 Vdd=5V

.MODEL NMOS NMOS LEVEL = 3
+ TOX  = 200E-10      NSUB  = 1E17        GAMMA = 0.5
+ PHI  = 0.7          VTO  = 0.8          DELTA = 3.0
+ UO   = 650          ETA  = 3.0E-6       THETA = 0.1
+ KP   = 120E-6       VMAX  = 1E5         KAPPA = 0.3
+ RSH  = 0            NFS  = 1E12         TPG  = 1
+ XJ   = 500E-9       LD   = 100E-9
+ CGDO  = 200E-12     CGSO  = 200E-12     CGBO  = 1E-10
+ CJ   = 400E-6       PB   = 1            MJ   = 0.5
+ CJSW  = 300E-12     MJSW  = 0.5
*
.MODEL PMOS PMOS LEVEL = 3
+ TOX  = 200E-10      NSUB  = 1E17        GAMMA = 0.6
+ PHI  = 0.7          VTO  = -0.9         DELTA = 0.1
+ UO   = 250          ETA  = 0            THETA = 0.1
+ KP   = 40E-6        VMAX  = 5E4         KAPPA = 1
+ RSH  = 0            NFS  = 1E12         TPG  = -1
+ XJ   = 500E-9       LD   = 100E-9
+ CGDO  = 200E-12     CGSO  = 200E-12     CGBO  = 1E-10
+ CJ   = 400E-6       PB   = 1            MJ   = 0.5
+ CJSW  = 300E-12     MJSW  = 0.5
```

Level 1 Model Parameters Related to Transconductance

The following SPICE parameters are related to the calculation of transconductance.

Symbol	Name	Description	Default	Typ.	Units
KP	KP	Transconductance parameter	20E-6	50E-6	A/V^2

t_{ox}	TOX	Gate-oxide thickness	1E-7	40E-10 m
λ	Lambda	Channel-length modulation	0	0.01 V^{-1}
L_{diff}	LD	Lateral diffusion	0	2.5E-7 m
$\mu_{n,p}$	UO	Surface mobility	600	580 cm^2/Vs

SPICE Modeling of the Source and Drain Implants

The following SPICE model parameters are related to calculating the parasitics associated with the source/drain implant regions.

Name	Description	Default	Typical	Units
RD	Drain contact resistance	0	40	Ω
RS	Source contact resistance	0	40	Ω
RSH	Source/drain sheet resistance	0	50	Ω/sq.
CGBO	Gate-bulk overlap capacitance	0	4E-10	F/m
CGDO	Gate-drain overlap capacitance	0	4E-10	F/m
CGSO	Gate-source overlap capacitance	0	4E-10	F/m
PB, PBSW	Bottom, sidewall built-in potential	0.8	0.8	V
MJ, MJSW	Bottom, sidewall grading coefficient	0.6	0.6	
CJ	Bottom zero-bias depletion capacitance	0	3E-4	F/m^2
CJSW	Sidewall zero-bias depletion capacitance	0	2.5E-10	F/m
IS	Bulk-junction saturation current	1E-14	1E-14	A
JS	Bulk-junction saturation current density	0	1E-8	A/m^2
FC	Bulk-junction forward bias coefficient	0.6	0.6	

Summary

Table 6.2 shows a summary of the device characteristics for the long-channel CMOS process.

Table 6.2 Summary of device characteristics for the long-channel CMOS process.

Long-channel MOSFET parameters used in this book. The *VDD* = 5 V and the scale factor is **1 μm** (*scale* = 1e–6)			
Parameter	**NMOS**	**PMOS**	**Comments**
V_{THN} and V_{THP}	800 mV	900 mV	Typical
KP_n and KP_p	120 $\mu A/V^2$	40 $\mu A/V^2$	t_{ox} = 200 Å
$C'_{ox} = \varepsilon_{ox}/t_{ox}$	1.75 $fF/\mu m^2$	1.75 $fF/\mu m^2$	$C_{ox} = C'_{ox} WL \cdot (scale)^2$
λ_n and λ_p	0.01 V^{-1}	0.0125 V^{-1}	at $L = 2$
γ_n and γ_p	0.5 $V^{-1/2}$	0.6 $V^{-1/2}$	Body factor

6.4.1 Some SPICE Simulation Examples

Figure 6.11 shows the I_D-V_{DS} characteristics of an NMOS device in the long-channel process. Figure 6.12 shows the equivalent PMOS device. Notice that the devices are the same size, 10/1, in the two simulations; however, the drain current of the PMOS is less than half the drain current of the NMOS. This is related to the mobility of the holes being two to three times lower than the mobility of the electrons. Electrons in the valence band are more tightly coupled to the nucleus of an atom than are electrons in the conduction band. Because apparent movement of holes is actually the result of electrons moving in the valence band, the hole mobility is lower than electron (in conduction band) mobility.

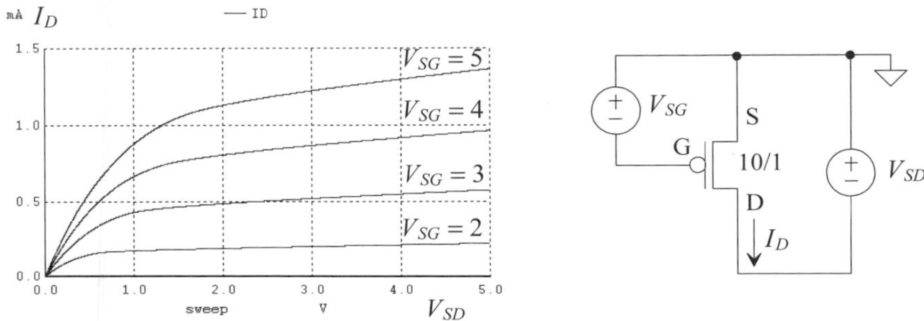

Figure 6.12 Characterisitics of a long-channel PMOS device.

Threshold Voltage and Body Effect

In simple terms, the threshold voltage is the voltage that turns the device on and allows drain current to flow from the drain to the source. Figure 6.13 shows the I_D-V_{GS} curves for an NMOS device. When the source and substrate are at the same potential, $V_{SB} = 0$, the threshold voltage is labeled V_{THN0} (see Eq. [6.17]). When V_{SB} starts to increase, the threshold voltage goes up. This is called the body effect. Figure 6.14 shows two MOSFETs: one with and one without body effect (substrate, or body, is grounded).

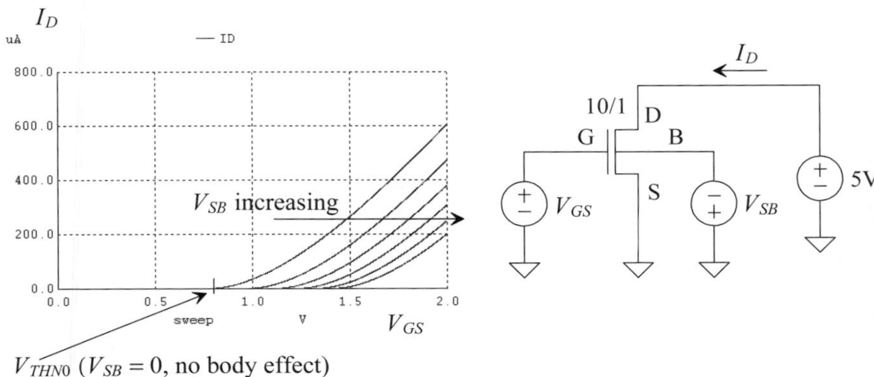

Figure 6.13 Threshold voltage and body effect.

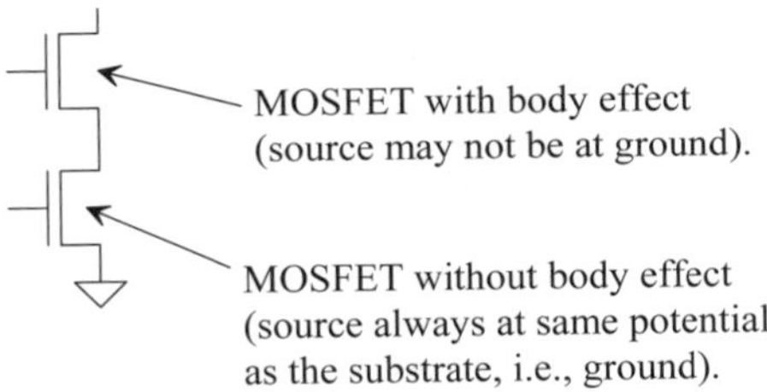

Figure 6.14 How an NMOS can have body effect.

To qualitatively understand the origin of the body effect, consider the MOSFET cross-sectional view seen in Fig. 6.15. As the source potential rises above the bulk (substrate) potential (represented by V_{SB} in the figure), electrons are attracted towards the positive terminal of V_{SB} from the MOSFET's channel. To keep the surface inverted, a larger V_{GS} must be applied to the MOSFET. Thus the effect of the body stealing charge from the channel is an increase in the MOSFET's threshold voltage.

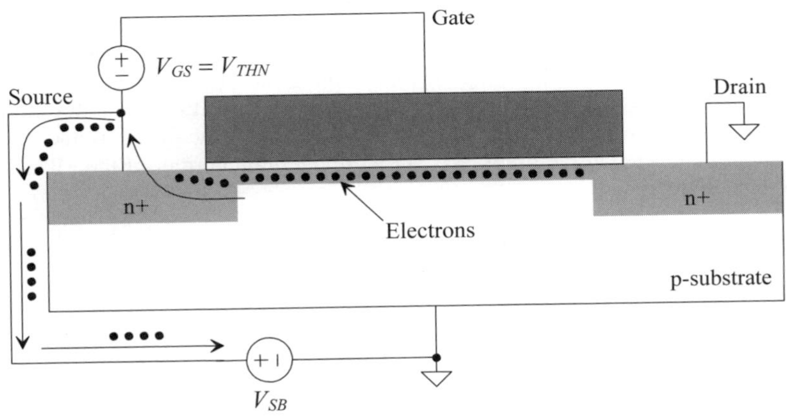

Figure 6.15 Qualitative description of body effect.

6.4.2 The Subthreshold Current

In the last section we said that the MOSFET starts to conduct a current when $V_{GS} = V_{THN}$. In reality there is a drain current, albeit small, when $V_{GS} < V_{THN}$. This current is called subthreshold current. When the MOSFET is operating in the weak inversion region it can also be said to be operating in the subthreshold region. Subthreshold operation can be very useful for low-power operation. Solar-powered calculators, CMOS imagers, or battery- operated watches are examples of devices using CMOS ICs operating in the subthreshold region. The main problems that plague circuits designed to operate in the subthreshold region are matching, noise, and bandwidth. For example, since the drain current is exponentially related to the gate-source voltage (as we'll soon see), any mismatch in these voltages can cause significant differences in the drain current.

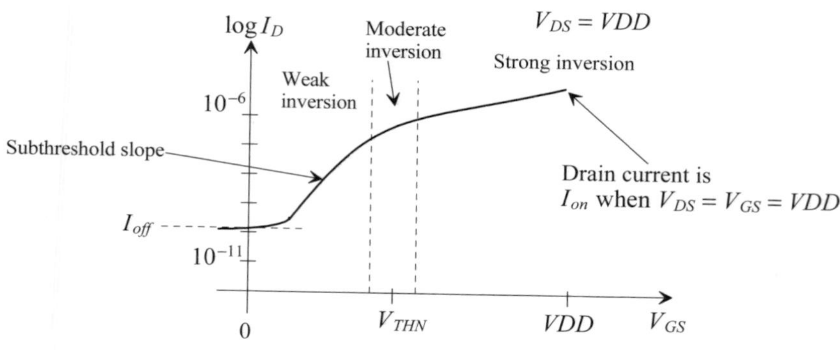

Figure 6.16 Drain current plotted from weak to strong inversion.

The subthreshold region is often characterized using the log I_D plotted against V_{GS} (see Fig. 6.16). The current transport from the drain to source in Sec. 6.3 was via drift. An applied electric field, when the MOSFET is operating in the strong inversion region, causes carriers to drift from the channel to the drain across the depletion region (and hence why we talk about mobility). In the weak inversion, or subthreshold region, the carriers diffuse from the source to the drain just like carrier movement in a bipolar junction transistor, BJT. In a BJT the carriers are emitted from the emitter, diffuse across the base, and are collected at the collector. In a MOSFET operating in subthreshold, the carriers are emitted by the source, diffuse across the body of the device (under the gate oxide) and are collected at the drain. We can write the drain current of the MOSFET in the subthreshold region as

$$I_D = I_{D0} \cdot \frac{W}{L} \cdot e^{q(V_{GS}-V_{THN})/n \cdot kT\ )} \tag{6.48}$$

Taking the log of both sides with $V_T = kT/q$ (the thermal voltage), we get

$$\log I_D = \log\frac{W}{L} + \log I_{D0} + -\frac{V_{THN}}{nV_T}\cdot \log e + \overbrace{\left[\frac{1}{V_T \cdot n}\cdot \log e\right]}^{\text{subthreshold slope}} \cdot V_{GS} \tag{6.49}$$

The reciprocal of the subthreshold slope is given by

$$\text{Subthreshold slope}^{-1} = \frac{V_T \cdot n}{\log e}\ \text{(mV/decade)} \tag{6.50}$$

If $kT/q = 0.026$ V $= V_T$ and n (the slope parameter) = 1, the reciprocal of the subthreshold slope is 60 mV/decade (it can be said the subthreshold slope is 60 mV/decade and it is understood it is actually one over the slope). In bulk CMOS n is around 1.6 and the subthreshold slope is 100 mV/decade at room temperature. For the ideal MOSFET used as a switch when V_{GS} is less than the threshold voltage, the drain current goes to zero. The slope of the curve below V_{THN} in Fig. 6.16 is then infinite (corresponding to zero subthreshold slope^{-1}). The subthreshold slope can be a very important MOSFET parameter in many applications (the design of dynamic circuits). Notice that the drain current that flows with $V_{GS} = 0$ is called I_{off} (with $V_{DS} = VDD$). The drain current that flows when $V_{GS} = V_{DS} = VDD$ (in the strong inversion region) is called I_{on}.

6.5 Short-Channel MOSFETs

The long-channel CMOS process used in the first part of this chapter is useful for illustrating the fundamentals of MOSFET operation. However, modern CMOS transistors have channel lengths that are well below the 1 μm minimum length of this process. The gradual channel approximation used earlier to develop the current-voltage characteristics of the MOSFET falls apart for modern short-channel devices. The electric field under the gate oxide can no longer be treated in a single dimension. In addition, the velocity of the carriers drifting between the channel and the drain of the MOSFET saturates, Fig. 6.17, an effect called *carrier velocity saturation*, v_{sat}. The result is a reduction in electron or hole mobility, μ_n or μ_p, and thus an increase in the channel's effective sheet resistance, Eq. (6.27). Since the mobility of the electron decreases with increasing temperature[2], this effect is sometimes referred to as the hot-carrier effect. Typical values for the low electric field mobilities (electric fields, E, less than the critical electric field, E_{crit}, where the velocity saturates) for electrons and holes are 600 cm²/Vs (μ_n) and 250 cm²/Vs (μ_p). See also Eq. (5.4).

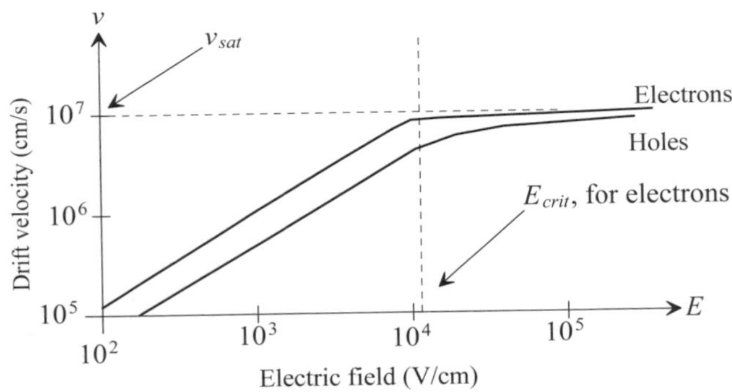

Figure 6.17 Drift velocity plotted against electric field. The slope of these curves is the mobility of the carriers, see Eq. (5.4)

Lightly-Doped Drain (LDD)

A cross-sectional view of an NMOS device using a lightly doped drain (LDD) structure to reduce the effects of hot carriers, is shown in Fig. 6.18. We also show this structure in Fig. 4.7. The LDD is formed by implanting a shallow n-, forming the spacer adjacent to the poly, and implanting the source/drain n+. The addition of the lightly doped n- provides a resistive buffer between the channel and the higher-doped source/drain. The effect of this resistive buffer is to drop more voltage, across a smaller distance, between the channel and the drain. In effect the electric field gets reduced (the drain doesn't see as high of an electric field). The effect, as seen in Fig. 6.17, is to keep (or at least help) the carriers' velocity from saturating.

2 See Ch. 9 for a discussion of the temperature dependence of the mobility.

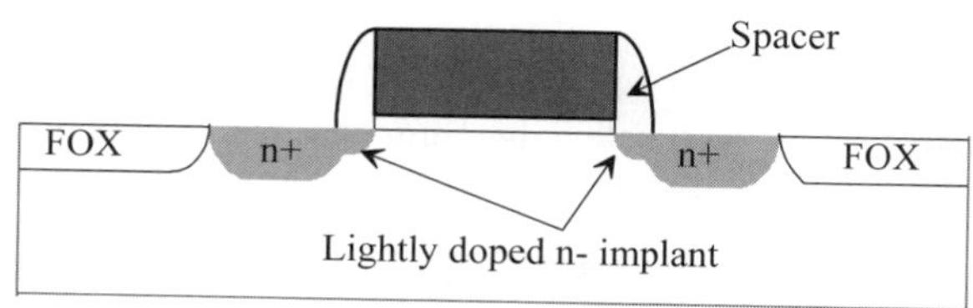

Figure 6.18 Lightly doped drain (LDD) implant.

6.5.1 MOSFET Scaling

Reducing the channel length of a MOSFET can be described in terms of scaling theory. A scaling parameter S ($S < 1$) is used to scale the dimensions of a MOSFET. The value of S is typically in the neighborhood of 0.7 from one CMOS technology generation to the next. For example, if a process uses a VDD of 2 V, a next generation process would use a VDD of 1.4 V. In other words

$$VDD' = VDD \cdot S \tag{6.51}$$

The channel length of the scaled process is reduced to

$$L' = L \cdot S \tag{6.52}$$

while the width is reduced to

$$W' = W \cdot S \tag{6.53}$$

Table 6.3 describes how S affects the MOSFET parameters. The main benefits of scaling are (1) smaller device sizes and thus reduced chip size (increased yield and more parts per wafer), (2) lower gate delays, allowing higher frequency operation, and (3) reduction in power dissipation. Associated with these benefits are some unwanted side effects referred to as short-channel effects. These unwanted effects are discussed in the next section.

Table 6.3 CMOS scaling relationships.

Parameter	Scaling
Supply voltage (VDD)	S
Channel length (L_{min})	S
Channel width (W_{min})	S
Gate-oxide thickness (t_{ox})	S
Substrate doping (N_A)	S^{-1}
On current (I_{on})	S
Gate capacitance (C_{ox})	S
Gate delay	S
Active power	S^3

6.5.2 Short-Channel Effects

The average drift velocity, v, of an electron plotted against electric field, E, is shown in Fig. 6.17. When the electric field reaches a critical value, labeled E_{crit}, the velocity saturates at a value v_{sat}, that is, the velocity ceases to increase with increasing electric field. The ratio of electron drift velocity to applied electric field is the electron mobility (Eq. 5.4), or, again

$$\mu_n = \frac{v}{E} \tag{6.54}$$

Above the critical electric field, the mobility starts to decrease, whereas below E_{crit}, the mobility is essentially constant. Rewriting Eq. (6.29) to determine how the mobility changes with $V(y)$ results in

$$I_D = \mu_n \cdot \frac{dV(y)}{dy} \cdot W \cdot C'_{ox}[V_{GS} - V_{THN} - V(y)] \tag{6.55}$$

We are interested in determining how the drain current of a short-channel MOSFET changes with V_{GS} when operating in the saturation region. (The charge under the gate oxide at the drain channel interface is zero, and the channel is pinched off.) The MOSFET enters the saturation region when $V(L) = V_{DS,sat}$. At high electric fields, the mobility can be approximated by

$$\mu_n = \frac{v_{sat}}{E} = \frac{v_{sat}}{dV(y)/dy} \tag{6.56}$$

so that Eq. (6.55) can be written as

$$I_D = W \cdot v_{sat} \cdot C'_{ox}(V_{GS} - V_{THN} - V_{DS,sat}) \tag{6.57}$$

The drain current of a short-channel MOSFET operating in the saturation region increases linearly with V_{GS}. The long-channel theory, Eq. (6.37), shows the drain current increasing with the square of the gate-source voltage. This result also presents a practical relative figure of merit for the modern CMOS process, the drive current per width of a MOSFET. The on or drive current, I_{on} or I_{drive} (μA/μm), is given by

$$I_{on} = I_{drive} = v_{sat} \cdot C'_{ox}(V_{GS} - V_{THN} - V_{DS,sat}) \tag{6.58}$$

and therefore, see Fig. (6.16),

$$I_D = I_{on} \cdot W = I_{drive} \cdot W \text{ with } V_{GS} = V_{DS} = VDD \tag{6.59}$$

The on (drive) current can be estimated using these equations; however, it is normally measured.

Hot Carriers

Some carriers drifting near the drain can obtain an energy much larger than the thermal energy of carriers under equilibrium conditions. These carriers are termed *hot carriers*. The velocity of these carriers can exceed the saturation velocity indicated in Fig. 6.17. This effect is called *velocity overshoot* and it can enhance the speed of the MOSFETs (which is good). Unfortunately, hot carriers can also tunnel through the gate oxide and cause gate current or become trapped in the gate oxide, having the effect of changing the MOSFET's threshold voltage. Hot carriers can also cause impact ionization (avalanche breakdown).

Oxide Breakdown

For reliable device operation, the maximum electric field across a device gate oxide should be limited to 10 MV/cm. This translates into 1V / 10 Å of gate oxide. A device with t_{ox} of 20 Å should limit the applied gate voltages to 2 V for reliable long-term operation.

Drain-Induced Barrier Lowering

Drain-induced barrier lowering (DIBL, pronounced dibble) causes a threshold voltage reduction with the application of a drain-source voltage. The positive potential at the drain terminal helps to attract electrons under the gate oxide and thus increase the surface potential V_s. In other words V_{DS} helps to invert the channel on the drain side of the device, causing a reduction in the threshold voltage. Since V_{THN} decreases with increasing V_{DS}, the result is an increase in drain current and thus a decrease in the MOSFET's output resistance.

Substrate Current-Induced Body Effect

Substrate current-induced body effect, (SCBE), is the result of hot carriers causing impact ionization and generating a substrate current. This occurs for electric fields greater than 10^5 V/cm. The substrate current flows through the resistance of the substrate, increasing the substrate potential and further decreasing the threshold voltage. The result, again, is an increase in drain current and a decrease in the MOSFET's output resistance.

Gate Tunnel Current

As the oxide thickness scales downwards, the probability of carriers tunneling through the gate oxide increases. For oxide thicknesses less than 15 Å, this gate current can be significant. To reduce the tunnel current, various sandwiches of dielectrics are being explored. Figure 16.67 later in the book presents some results showing values for tunnel currents under various operating conditions.

6.5.3 SPICE Models for Our Short-Channel CMOS Process

Section 6.4 presented some SPICE models for the long-channel CMOS process used in this book. In this section we give the BSIM4[3] models for the 50 nm process we use in the book with *VDD* = 1 V, see also Table 5.2. The model listing is given below.

BSIM4 Model Listing (NMOS)

```
* 50nm BSIM4 models
*
* Don't forget the .options scale=50nm if using an Lmin of 1
* 1<Ldrawn<200   10<Wdrawn<10000 Vdd=1V
* Change to level=54 when using HSPICE

.model       nmos      nmos      level = 14

+binunit = 1        paramchk= 1      mobmod  = 0
+capmod  = 2        igcmod  = 1      igbmod  = 1      geomod  = 1
+diomod  = 1        rdsmod  = 0      rbodymod= 1      rgatemod= 1
```

[3] BSIM4 is a fourth generation MOSFET model developed at the University of California, Berkeley. The acronym stands for Berkeley Short-channel IGFET (insulated gate FET) Model. For more information see: http://www-device.eecs.berkeley.edu

```
+permod  = 1          acnqsmod= 0          trnqsmod= 0

+tnom    = 27         toxe    = 1.4e-009   toxp    = 7e-010     toxm    = 1.4e-009
+epsrox  = 3.9        wint    = 5e-009     lint    = 1.2e-008
+ll      = 0          wl      = 0          lln     = 1          wln     = 1
+lw      = 0          ww      = 0          lwn     = 1          wwn     = 1
+lwl     = 0          wwl     = 0          xpart   = 0          toxref  = 1.4e-009

+vth0    = 0.22       k1      = 0.35       k2      = 0.05       k3      = 0
+k3b     = 0          w0      = 2.5e-006   dvt0    = 2.8        dvt1    = 0.52
+dvt2    = -0.032     dvt0w   = 0          dvt1w   = 0          dvt2w   = 0
+dsub    = 2          minv    = 0.05       voffl   = 0          dvtp0   = 1e-007
+dvtp1   = 0.05       lpe0    = 5.75e-008  lpeb    = 2.3e-010   xj      = 2e-008
+ngate   = 5e+020     ndep    = 2.8e+018   nsd     = 1e+020     phin    = 0
+cdsc    = 0.0002     cdscb   = 0          cdscd   = 0          cit     = 0
+voff    = -0.15      nfactor = 1.2        eta0    = 0.15       etab    = 0
+vfb     = -0.55      u0      = 0.032      ua      = 1.6e-010   ub      = 1.1e-017
+uc      = -3e-011    vsat    = 1.1e+005   a0      = 2          ags     = 1e-020
+a1      = 0          a2      = 1          b0      = -1e-020    b1      = 0
+keta    = 0.04       dwg     = 0          dwb     = 0          pclm    = 0.18
+pdiblc1 = 0.028      pdiblc2 = 0.022      pdiblcb = -0.005     drout   = 0.45
+pvag    = 1e-020     delta   = 0.01       pscbe1  = 8.14e+8    pscbe2  = 1e-007
+fprout  = 0.2        pdits   = 0.2        pditsd  = 0.23       pditsl  = 2.3e+006
+rsh     = 3          rdsw    = 150        rsw     = 150        rdw     = 150
+rdswmin = 0          rdwmin  = 0          rswmin  = 0          prwg    = 0
+prwb    = 6.8e-011   wr      = 1          alpha0  = 0.074      alpha1  = 0.005
+beta0   = 30         agidl   = 0.0002     bgidl   = 2.1e+009   cgidl   = 0.0002
+egidl   = 0.8

+aigbacc = 0.012      bigbacc = 0.0028     cigbacc = 0.002
+nigbacc = 1          aigbinv = 0.014      bigbinv = 0.004      cigbinv = 0.004
+eigbinv = 1.1        nigbinv = 3          aigc    = 0.017      bigc    = 0.0028
+cigc    = 0.002      aigsd   = 0.017      bigsd   = 0.0028     cigsd   = 0.002
+nigc    = 1          poxedge = 1          pigcd   = 1          ntox    = 1

+xrcrg1  = 12         xrcrg2  = 5
+cgso    = 6.238e-010 cgdo    = 6.238e-010 cgbo    = 2.56e-011  cgdl    = 2.495e-10
+cgsl    = 2.495e-10  ckappas = 0.02       ckappad = 0.02       acde    = 1
+moin    = 15         noff    = 0.9        voffcv  = 0.02

+kt1     = -0.21      kt1l    = 0.0        kt2     = -0.042     ute     = -1.5
+ua1     = 1e-009     ub1     = -3.5e-019  uc1     = 0          prt     = 0
+at      = 53000

+fnoimod = 1          tnoimod = 0

+jss     = 0.0001     jsws    = 1e-011     jswgs   = 1e-010     njs     = 1
+ijthsfwd= 0.01       ijthsrev= 0.001      bvs     = 10         xjbvs   = 1
+jsd     = 0.0001     jswd    = 1e-011     jswgd   = 1e-010     njd     = 1
+ijthdfwd= 0.01       ijthdrev= 0.001      bvd     = 10         xjbvd   = 1
+pbs     = 1          cjs     = 0.0005     mjs     = 0.5        pbsws   = 1
+cjsws   = 5e-010     mjsws   = 0.33       pbswgs  = 1          cjswgs  = 3e-010
+mjswgs  = 0.33       pbd     = 1          cjd     = 0.0005     mjd     = 0.5
+pbswd   = 1          cjswd   = 5e-010     mjswd   = 0.33       pbswgd  = 1
+cjswgd  = 5e-010     mjswgd  = 0.33       tpb     = 0.005      tcj     = 0.001
+tpbsw   = 0.005      tcjsw   = 0.001      tpbswg  = 0.005      tcjswg  = 0.001
+xtis    = 3          xtid    = 3

+dmcg    = 0e-006     dmci    = 0e-006     dmdg    = 0e-006     dmcgt   = 0e-007
```

```
+dwj    = 0.0e-008     xgw    = 0e-007       xgl    = 0e-008

+rshg   = 0.4          gbmin  = 1e-010       rbpb   = 5            rbpd   = 15
+rbps   = 15           rbdb   = 15           rbsb   = 15           ngcon  = 1
```

BSIM4 Model Listing (PMOS)

```
.model      pmos       pmos       level = 14

+binunit = 1           paramchk= 1           mobmod  = 0
+capmod  = 2           igcmod  = 1           igbmod  = 1           geomod  = 1
+diomod  = 1           rdsmod  = 0           rbodymod= 1           rgatemod= 1
+permod  = 1           acnqsmod= 0           trnqsmod= 0

+tnom    = 27          toxe   = 1.4e-009     toxp   = 7e-010       toxm   = 1.4e-009
+epsrox  = 3.9         wint   = 5e-009       lint   = 1.2e-008
+ll      = 0           wl     = 0            lln    = 1            wln    = 1
+lw      = 0           ww     = 0            lwn    = 1            wwn    = 1
+lwl     = 0           wwl    = 0            xpart  = 0            toxref = 1.4e-009

+vth0    = -0.22       k1     = 0.39         k2     = 0.05         k3     = 0
+k3b     = 0           w0     = 2.5e-006     dvt0   = 3.9          dvt1   = 0.635
+dvt2    = -0.032      dvt0w  = 0            dvt1w  = 0            dvt2w  = 0
+dsub    = 0.7         minv   = 0.05         voffl  = 0            dvtp0  = 0.5e-008
+dvtp1   = 0.05        lpe0   = 5.75e-008    lpeb   = 2.3e-010     xj     = 2e-008
+ngate   = 5e+020      ndep   = 2.8e+018     nsd    = 1e+020       phin   = 0
+cdsc    = 0.000258    cdscb  = 0            cdscd  = 6.1e-008     cit    = 0
+voff    = -0.15       nfactor = 2           eta0   = 0.15         etab   = 0
+vfb     = 0.55        u0     = 0.0095       ua     = 1.6e-009     ub     = 8e-018
+uc      = 4.6e-013    vsat   = 90000        a0     = 1.2          ags    = 1e-020
+a1      = 0           a2     = 1            b0     = -1e-020      b1     = 0
+keta    = -0.047      dwg    = 0            dwb    = 0            pclm   = 0.55
+pdiblc1 = 0.03        pdiblc2 = 0.0055      pdiblcb = 3.4e-008    drout  = 0.56
+pvag    = 1e-020      delta  = 0.014        pscbe1 = 8.14e+08     pscbe2 = 9.58e-07
+fprout  = 0.2         pdits  = 0.2          pditsd = 0.23         pditsl = 2.3e+006
+rsh     = 3           rdsw   = 250          rsw    = 160          rdw    = 160
+rdswmin = 0           rdwmin = 0            rswmin = 0            prwg   = 3.22e-008
+prwb    = 6.8e-011    wr     = 1            alpha0 = 0.074        alpha1 = 0.005
+beta0   = 30          agidl  = 0.0002       bgidl  = 2.1e+009     cgidl  = 0.0002
+egidl   = 0.8

+aigbacc = 0.012       bigbacc = 0.0028      cigbacc = 0.002
+nigbacc = 1           aigbinv = 0.014       bigbinv = 0.004       cigbinv = 0.004
+eigbinv = 1.1         nigbinv = 3           aigc   = 0.69         bigc   = 0.0012
+cigc    = 0.0008      aigsd  = 0.0087       bigsd  = 0.0012       cigsd  = 0.0008
+nigc    = 1           poxedge = 1           pigcd  = 1            ntox   = 1

+xrcrg1  = 12          xrcrg2 = 5
+cgso    = 7.43e-010   cgdo   = 7.43e-010    cgbo   = 2.56e-011    cgdl   = 1e-014
+cgsl    = 1e-014      ckappas = 0.5         ckappad = 0.5         acde   = 1
+moin    = 15          noff   = 0.9          voffcv = 0.02

+kt1     = -0.19       kt1l   = 0            kt2    = -0.052       ute    = -1.5
+ua1     = -1e-009     ub1    = 2e-018       uc1    = 0            prt    = 0
+at      = 33000

+fnoimod = 1           tnoimod = 0

+jss     = 0.0001      jsws   = 1e-011       jswgs  = 1e-010       njs    = 1
```

```
+ijthsfwd= 0.01      ijthsrev= 0.001     bvs    = 10         xjbvs   = 1
+jsd    = 0.0001     jswd   = 1e-011     jswgd  = 1e-010     njd     = 1
+ijthdfwd= 0.01      ijthdrev= 0.001     bvd    = 10         xjbvd   = 1
+pbs    = 1          cjs    = 0.0005     mjs    = 0.5        pbsws   = 1
+cjsws  = 5e-010     mjsws  = 0.33       pbswgs = 1          cjswgs  = 3e-010
+mjswgs = 0.33       pbd    = 1          cjd    = 0.0005     mjd     = 0.5
+pbswd  = 1          cjswd  = 5e-010     mjswd  = 0.33       pbswgd  = 1
+cjswgd = 5e-010     mjswgd = 0.33       tpb    = 0.005      tcj     = 0.001
+tpbsw  = 0.005      tcjsw  = 0.001      tpbswg = 0.005      tcjswg  = 0.001
+xtis   = 3          xtid   = 3

+dmcg   = 5e-006     dmci   = 5e-006     dmdg   = 5e-006     dmcgt   = 6e-007
+dwj    = 4.5e-008   xgw    = 3e-007     xgl    = 4e-008

+rshg   = 0.4        gbmin  = 1e-010     rbpb   = 5          rbpd    = 15
+rbps   = 15         rbdb   = 15         rbsb   = 15         ngcon   = 1
```

Simulation Results

Figure 6.19 shows 10/1 PMOS and NMOS device simulation results using the topologies seen in Figs. 6.11–6.13. The actual device sizes are 500 nm (width) by 50 nm (length). From the information in this figure and knowing *VDD* is 1 V, we can estimate the on currents for the MOSFETs. For the NMOS device

$$I_{on,n} \approx 300\ \mu A/(W \cdot scale) = 600\ \mu A/\mu m \quad (6.60)$$

For the PMOS device

$$I_{on,p} \approx 150\ \mu A/(W \cdot scale) = 300\ \mu A/\mu m \quad (6.61)$$

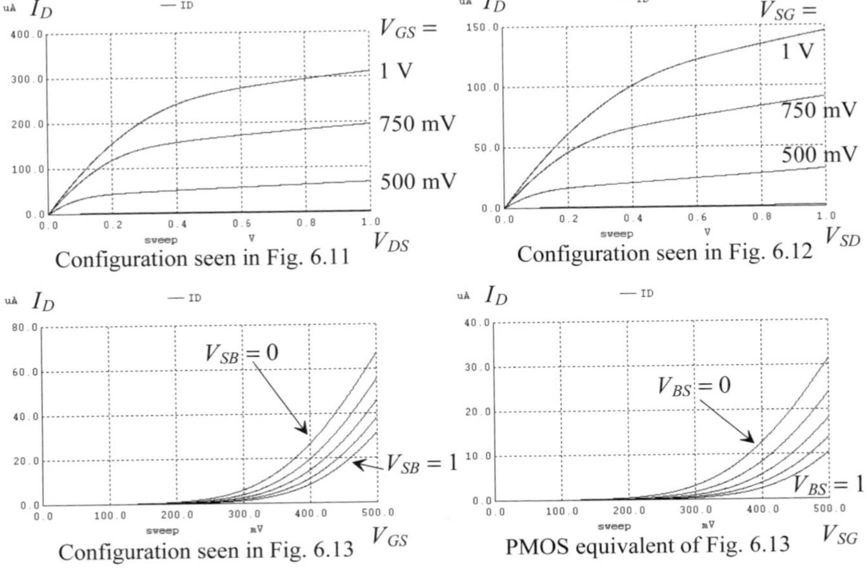

Figure 6.19 Current-voltage characteristics for 50 nm MOSFETs.

The threshold voltages can be estimated as 280 mV. (See Fig. 9.27 and the associated discussion for more information on determining the threshold voltages.) Note that it **doesn't make sense** to try to define a transconductance parameter, *KP*, for a short-channel process (the MOSFETs don't follow the square-law equations, Eqs. [6.33] and [6.37].) Instead we use I_{on}, I_{off}, t_{ox}, *W*, *scale*, $V_{THN,P}$, *VDD*, and plots of measured data. Table 6.4 shows some of the device characteristics for the short-channel CMOS process used in this book. Note that we **don't confuse** I_{on} (Fig. 6.16) with $I_{D,sat}$ (the current at the border between triode and saturation, Figs. 6.11 or 9.4 and Eq. [6.43]).

Table 6.4 Summary of device characteristics for the short-channel CMOS process.

Short-channel MOSFET parameters used in this book. The *VDD* = 1 V and the scale factor is **50 nm** (*scale* = 50e–9)			
Parameter	**NMOS**	**PMOS**	**Comments**
V_{THN} and V_{THP}	280 mV	280 mV	Typical
t_{ox}	14 Å	14 Å	See also Table 5.1
$C'_{ox} = \varepsilon_{ox}/t_{ox}$	25 fF/µm^2	25 fF/µm^2	$C_{ox} = C'_{ox} WL \cdot (scale)^2$
λ_n and λ_p	0.6 V^{-1}	0.3 V^{-1}	At $L = 2$
$I_{on,n}$ and $I_{on,p}$	600 µA/µm	300 µA/µm	On current
$I_{off,n}$ and $I_{off,p}$	7.1 nA/µm	10 nA/µm	Off current, see Fig. 14.2

ADDITIONAL READING

[1] R. S. Muller, T. I. Kamins, and M. Chan, *Device Electronics for Integrated Circuits*, John Wiley and Sons Publishers, 2002. ISBN 0-471-59398-2

[2] W. Liu, "MOSFET Models for Spice Simulation, Including BSIM3v3 and BSIM4," John Wiley and Sons Publishers, 2001. ISBN 0-471-39697-4

[3] Y. Taur and T. H. Ning, *Fundamentals of Modern VLSI Devices*, Cambridge University Press, 1998. ISBN 0-521-55056-4 (hardback) or ISBN 0-521-55959-6 (paperback).

[4] D. Foty, *MOSFET Modeling with SPICE: Principles and Practice,* Prentice-Hall, 1997. ISBN 0-13-227935-5.

[5] M. Bohr, "MOS Transistors: Scaling and Performance Trends," *Semiconductor International*, pp. 75–79, June 1995.

[6] D. A. Neamen, *Semiconductor Physics and Devices-Basic Principles,* Richard D. Irwin, 1992. ISBN 0-256-08405-X.

[7] R. F. Pierret, *Volume IV in the Modular Series on Solid State Devices-Field Effect Devices,* Addison-Wesley, 1990.

[8] R. C. Jaeger, *Modular Series on Solid State Devices-Introduction to Microelectronic Fabrication,* Addison-Wesley, 1989.

[9] D. K. Schroder, *Modular Series on Solid State Devices-Advanced MOS Devices,* Addison-Wesley, 1987.

[10] Y. P. Tsividis, *Operation and Modeling of the MOS Transistor,* McGraw-Hill, 1987. ISBN 0-07-065381-X.

[11] S. M. Sze, *Physics of Semiconductor Devices,* 2nd ed., John-Wiley and Sons, 1981. ISBN 0-471-05661-8.

PROBLEMS

6.1 Plot the magnitude and phase of v_{out} (AC) in the following circuit, Fig. 6.20. Assume that the MOSFET was fabricated using the 50 nm process (see Table 5.1) and is operating in strong inversion. Verify your answer with SPICE.

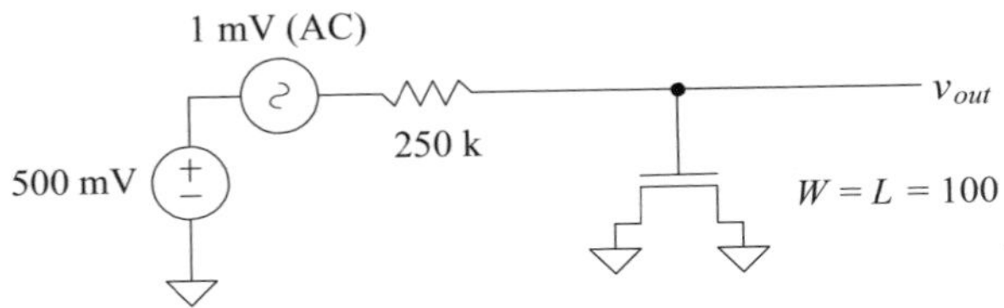

Figure 6.20 Circuit used in Problem 6.1.

6.2 If a MOSFET is used as a capacitor in the strong inversion region where the gate is one electrode and the source/drain is the other electrode, does the gate overlap of the source/drain change the capacitance? Why or not? What is the capacitance?

6.3 Repeat Problem 6.2 when the MOSFET is operating in the accumulation region. Keep in mind that the question is not asking for the capacitance from gate to substrate.

6.4 If the oxide thickness of a MOSFET is 40 Å, what is C'_{ox}?

6.5 Repeat Ex. 6.2 when $V_{SB} = 1$ V.

6.6 Repeat Ex. 6.3 for a p-channel device with a well doping concentration of 10^{16} atoms/cm^3.

6.7 What is the electrostatic potential of the oxide-semiconductor interface when $V_{GS} = V_{THN0}$?

6.8 Repeat Ex. 6.5 to get a threshold voltage of 0.8 V.

6.9 What happens to the threshold voltage in Problem 6.8 if sodium contamination of 100×10^9 sodium ions/cm^2 is present at the oxide-semiconductor interface?

6.10 How much charge (enhanced electrons) is available under the gate for conducting a drain current at the drain-channel interface when $V_{DS} = V_{GS} - V_{THN}$? Why? Assume that the MOSFET is operating in strong inversion, $V_{GS} > V_{THN}$.

6.11 Show the details of the derivation for Eq. (6.33) for the PMOS device.

6.12 Using Eq. (6.35), estimate the small-signal channel resistance (the change in the drain current with changes in the drain-source voltage) of a MOSFET operating in the triode region (the resistance between the drain and source).

6.13 Show, using Eqs. (6.33) and (6.37), that the parallel connection of MOSFETs shown in Fig. 5.18 behave as a single MOSFET with a width equal to the sum of each individual MOSFET width.

6.14 Show that the bottom MOSFET, Fig. 6.21, in a series connection of two MOSFETs cannot operate in the saturation region. Neglect the body effect. *Hint*: Show that M1 is always either in cutoff ($V_{GS1} < V_{THN}$) or triode ($V_{DS1} < V_{GS1} - V_{THN}$).

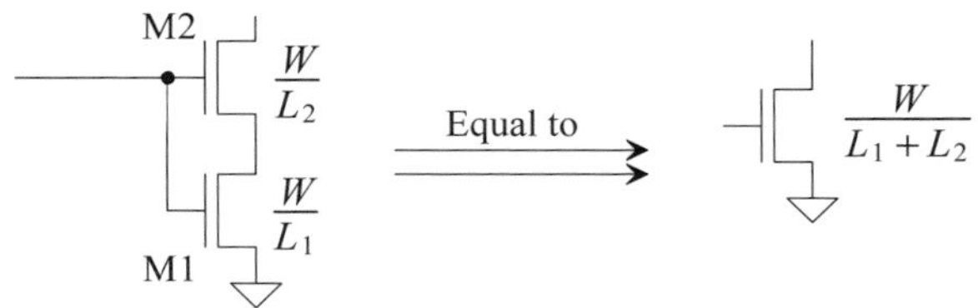

Figure 6.21 MOSFETs operating in series.

6.15 Show that the series connection of MOSFETs shown in Fig. 6.21 behaves as a single MOSFET with twice the length of the individual MOSFETs. Again, neglect the body effect.

This chapter was contributed by

Dr. Jeff Jessing, Boise State University

Chapter
7

CMOS Fabrication

This chapter presents a brief overview of CMOS process integration. Process integration refers to the well-defined collection of semiconductor processes required to fabricate CMOS integrated circuits starting from virgin silicon wafers. This overview is intended to give the reader a fundamental understanding of the processes required in CMOS integrated circuit fabrication. Moreover, there are strong interactions between circuit design and process integration. For instance, the typical design rule set is determined in large part by the limitations in the fabrication processes. Hence, circuit designers, process engineers, and integration engineers are required to communicate effectively. To this end, we first examine the fundamental processes, called unit processes, required for CMOS fabrication. The primary focus is the qualitative understanding of the processes with limited introduction of quantitative expressions. The unit processes are combined in a deliberate sequence to fabricate CMOS. Additionally, the unit processes are typically repeated numerous times in a given process sequence. Here we present a representative modern CMOS process sequence, also called a process flow.

7.1 CMOS Unit Processes

In this section we introduce each of the major processes required in the fabrication of CMOS integrated circuits. We first discuss wafer production. Although wafer production is not a unit process, it is nonetheless important to present the production method which is used by wafer manufacturers. All subsequent discussions are focused on the unit processes incorporated by fabrication facilities to produce integrated circuits. The unit processes are grouped by functionality. Thermal oxidation, doping processes, photolithography, thin-film removal, and thin-film deposition techniques are all presented.

7.1.1 Wafer Manufacture

Silicon is the second most abundant element in the Earth's crust; however, it occurs exclusively in compounds. In fact, elemental silicon is a man-made material that is refined from these various compounds. The most common is silica (impure SiO_2). Modern integrated circuits must be fabricated on ultrapure, defect-free slices of single crystalline silicon called wafers, as discussed in Ch. 1.

Metallurgical Grade Silicon (MGS)

Wafer production requires three general processes: silicon refinement, crystal growth, and wafer formation. Silicon refinement begins with the reduction of silica in an arc furnace at roughly 2000 °C with a carbon source. The carbon effectively "pulls" the oxygen from the SiO_2 molecules, thus chemically reducing the SiO_2 into roughly 98% pure silicon referred to as metallurgical grade silicon (MGS). The overall reduction is governed by the following equation

$$SiO_2 \text{ (solid)} + 2C \text{ (solid)} \rightarrow Si \text{ (liquid)} + 2CO \text{ (gas)} \qquad (7.1)$$

Electronic Grade Silicon (EGS)

MGS is not sufficiently pure for microelectronic device applications. The reason is that the electronic properties of a semiconductor such as silicon are extremely sensitive to impurity concentrations. Impurity levels measured at parts per million or less can have dramatic effects on carrier mobilities, lifetimes, etc. It is therefore necessary to further purify the MGS in what is known as electronic grade silicon (EGS). EGS is produced from the chlorination of grounded MGS as

$$MGS \text{ (solid)} + HCL \text{ (gas)} \rightarrow \text{silane } (SiH_4) \text{ (liquid)} + \text{trichlorosilane } (SiHCL_3) \text{ (liquid} \qquad (7.2)$$

Because the reaction products are liquids at room temperature, ultrapure EGS can be obtained from fractional distillation and chemical reduction processes. The resultant EGS is in the form of polycrystalline chunks.

Czochralski (CZ) Growth and Wafer Formation

To achieve a single crystalline form, the EGS must be subjected to a process called Czochralski (CZ) growth. A schematic representation of the CZ growth process is shown in Fig. 7.1. The polycrystalline EGS is melted in a large quartz crucible where a small seed crystal of known orientation is introduced into the surface of the silicon melt. The seed crystal, rotating in one direction, is slowly pulled from the silicon melt, rotating in the opposite direction. Solidification of the silicon onto the seed forms a growing crystal

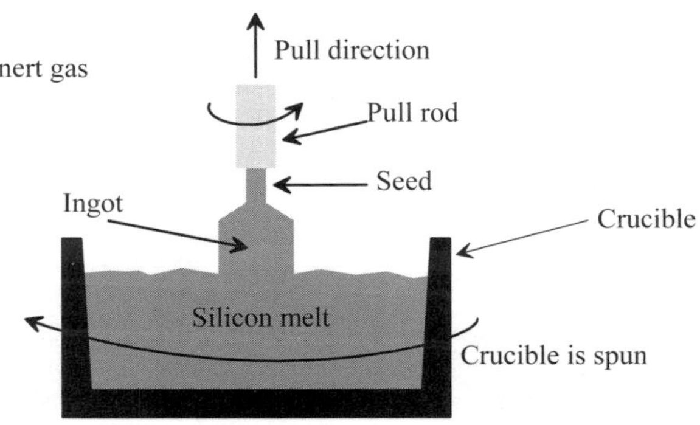

Figure 7.1 Simplified diagram showing Czochralski (CZ) crystal growth.

(called a boule or ingot) that assumes the crystallographic orientation of the seed. In general, the slower the pull-rate (typically mm/hour), the larger the diameter of the silicon crystal. Following CZ growth, the silicon boule is trimmed down to the appropriate diameter. *Flats* or *notches* are ground into the surface of the boule to indicate a precise crystal orientation. Using a special diamond saw, the silicon boule is cut into thin wafers. The wafers are finished by using a chemical mechanical polishing (CMP) process to yield a mirror-like finish on one side of the wafer. Although devices are fabricated entirely within the top couple of micrometers of the wafer, final wafer thicknesses (increasing with wafer diameter) are up to roughly one millimeter for adequate mechanical support.

7.1.2 Thermal Oxidation

Silicon, when exposed to an oxidant at elevated temperatures, readily forms a thin layer of oxide at all exposed surfaces. The native oxide of silicon is in the form of silicon dioxide (SiO_2). With respect to CMOS fabrication, SiO_2 can serve as a high quality dielectric in device structures such as gate oxides. Moreover, during processing, thermally grown oxides can be used as implantation, diffusion, and etch masks. The dominance of silicon as a microelectronic material can be attributed to the existence of this high quality native oxide and the resultant near ideal silicon/oxide interface.

Figure 7.2 depicts the basic thermal oxidation process. The silicon wafer is exposed at high temperatures (typically 900 °C–1200 °C) to a gaseous oxidant such as molecular oxygen (O_2) and/or water vapor (H_2O). For obvious reasons, oxidation in O_2 is called dry oxidation, whereas in H_2O it is called wet oxidation, as discussed in Sec. 2.1. The gas/solid interface forms a stagnant layer through which the oxidant must diffuse to reach the surface of the wafer. Once at the surface, the oxidant again must diffuse through the existing oxide layer that is present. As the oxidant species reaches the silicon/oxide interface, one of two reactions occur

$$Si + O_2 \rightarrow SiO_2 \text{ (dry oxidation)} \quad (7.3)$$

$$Si + 2H_2O \rightarrow SiO_2 + 2H_2 \text{ (wet oxidation)} \quad (7.4)$$

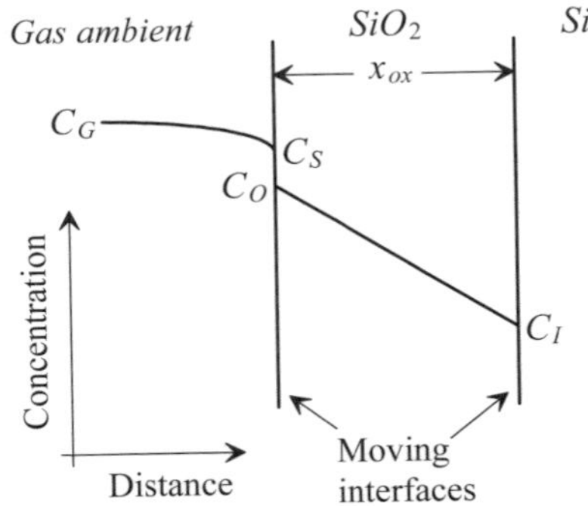

C_G = Oxidant concentration in the bulk gas

C_S = Oxidant concentration at the surface

C_O = Oxidant concentration just inside the oxide surface

C_I = Oxidant concentration at the Si/SiO_2 interface

Figure 7.2 A simple model for thermal oxidation of silicon. Notice the oxidant concentrations (boundary conditions) in the gas, oxide, and silicon.

It should be emphasized that reactions specified by Eqs. (7.3) and (7.4) occur at the silicon/oxide interface where silicon is consumed in the reaction. As Fig. 7.3 illustrates, with respect to the original silicon surface, approximately 45% of the oxide thickness is accounted for by consumption of silicon.

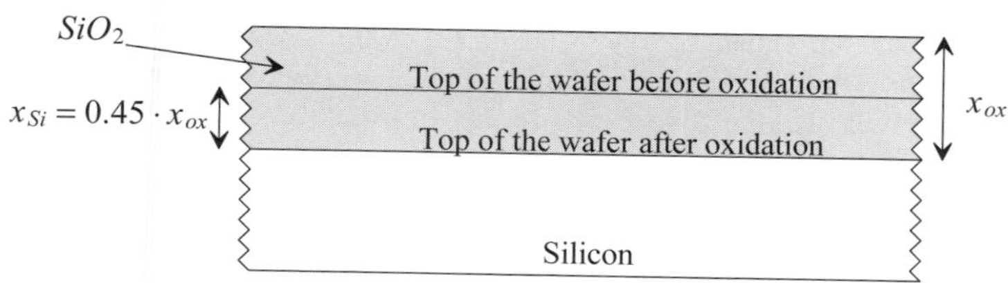

Figure 7.3 Silicon/oxide growth interface. See also Fig. 2.4.

The rate of thermal oxidation is a function of temperature and rate constants. The rate is directly proportional to temperature. The rate constants are, in turn, a function of gas partial pressures, oxidant-type, and silicon wafer characteristics such as doping type, doping concentration, and crystallographic orientation. In general, dry oxidation yields a denser and thus higher quality oxide than does a wet oxidation. However, wet oxidation occurs at a much higher rate compared to dry oxidation. Depending on the temperature and existing thickness of oxide present, the overall oxidation rate can be either diffusion limited (e.g., thick oxides at high temperatures) or reaction rate limited (e.g., thin oxides at low temperatures). Practically, oxide thicknesses are limited to less than a few thousand angstroms and to less than a micron for dry and wet oxidation, respectively.

In a modern fabrication facility, oxidation occurs in either a tube furnace or in a rapid thermal processing (RTP) tool, as schematically shown in Fig. 7.4. The tube furnaces consist of quartz tubes surrounded by heating element coils. The wafers are loaded in the heated tubes where oxidants can be introduced through inlets. The function of the RTP is similar to the tube furnace with the exception that the thermal source is heating lamps.

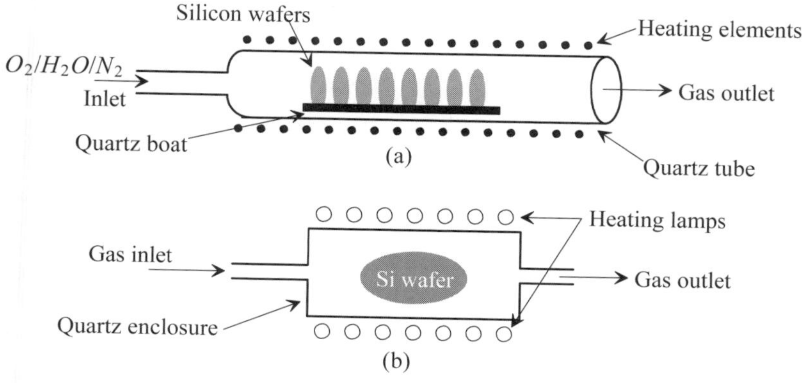

Figure 7.4 (a) Simplified representation of an oxidation tube furnace and (b) simplified diagram for rapid thermal processing.

7.1.3 Doping Processes

The controlled introduction of dopant impurities into silicon is necessary to affect majority carrier type, carrier concentration, carrier mobility, carrier lifetime, and internal electric fields. The two primary methods of dopant introduction are solid state diffusion and ion implantation. Historically, solid state diffusion has been an important doping process; however, ion implantation is the preferred method in modern CMOS fabrication.

Ion Implantation

The workhorse method of introducing dopants into the near-surface region of wafers is a process called ion implantation. In ion implantation, dopant atoms (or molecules) are ionized and then accelerated through a large electric potential (a few kV to MV) towards a wafer. The highly energetic ions bombard and thus implant into the surface. Obviously, this process leads to a high degree of lattice damage, which is generally repaired by annealing at high temperatures. Moreover, the ions do not necessarily come to rest at a lattice site, hence an anneal is required to electrically activate (i.e., thermally agitate the impurities into lattice sites) the dopant impurities.

Figure 7.5 shows a schematic diagram of an ion implanter. The ions are generated by an RF field in the ion source where they are subsequently extracted to a mass spectrometer. The spectrometer only allows ions with a user-selected mass to enter the accelerator, where the ions are passed through a large potential field. The ions are then scanned via electrostatic lens across the surface of the wafer.

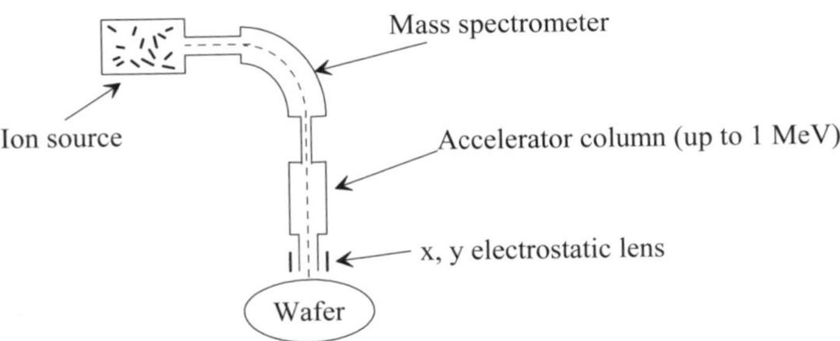

Figure 7.5 Simplified diagram of an ion implanter. The ions are created by an RF field where they are extracted into a mass spectrometer. An electrostatic lens scans the ion beam on the surface of a wafer to achieve the appropriate dose. Electrostatically, the ions can be counted to provide the real-time dose.

A first-order model for an implant doping profile is given by a Gaussian distribution described mathematically as

$$N(x) = N_p \exp\left[-(x - R_p)^2/2\Delta R_p^2\right] \tag{7.5}$$

where N_p is the peak concentration, R_p is the projected range, and ΔR_p is called the straggle. By inspection, R_p should be identified as the mean distance the ions travel into the silicon and ΔR_p as the associated standard deviation. Figure 7.6 illustrates a typical ion implant profile. Obviously, N_p occurs at a depth of R_p. Moreover, the area under the implant curve corresponds to what is referred to as the implant dose Q_{imp} , given mathematically as

$$Q_{imp} = \int_0^{\infty} N(x) \cdot dx \tag{7.6}$$

Localized implantation is achieved by masking off regions of the wafer with an appropriately thick material such as oxide, silicon nitride, polysilicon, or photoresist. Since implantation occurs in the masking layer, the thickness must be of sufficient magnitude to stop the ions prior to reaching the silicon substrate. In comparison to solid state diffusion, ion implantation has the advantage of being a low temperature and highly controlled process.

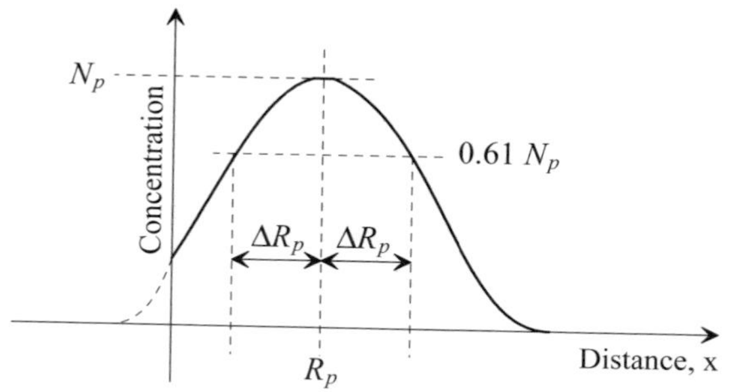

Figure 7.6 Ideal implant profile representing Eq. (7.5). Notice that the peak concentration occurs below the surface and depends on the implant energy.

Solid State Diffusion

Solid state diffusion is a method for introducing and/or redistributing dopants. In this section, we study solid state diffusion primarily to gain insight into "parasitic" dopant redistribution during thermal processes. In typical CMOS process flows, dopants are introduced into localized regions via ion implantation where the subsequent processing often consists of high temperature processing. Solid state diffusion inherently occurs in these high temperature steps, thus spreading out the implant profile in three dimensions. The net effect is to shift the boundary of the implant from its original implant-defined position, both laterally and vertically. This thermal smearing of the implant profiles must be accounted for during CMOS process flow development. If not, the final device characteristics can differ significantly from what was expected.

Solid state diffusion (or simply diffusion) requires two conditions: 1) a dopant concentration gradient, and 2) thermal energy. Diffusion is directly proportional to both. An implanted profile (approximated by a delta function at the surface of the wafer) diffuses to first order as

$$N(x,t) = \frac{Q_{imp}}{\sqrt{\pi \cdot D \cdot t} \cdot \exp(-x^2/4Dt)} \tag{7.7}$$

where Q_{imp} is the implant dose, D is the diffusivity of the dopant, and t is the diffusion time. Figure 7.7 illustrates limited-source diffusion of a one-dimensional implant profile. Notice that the areas under the respective curves for a given time are equal.

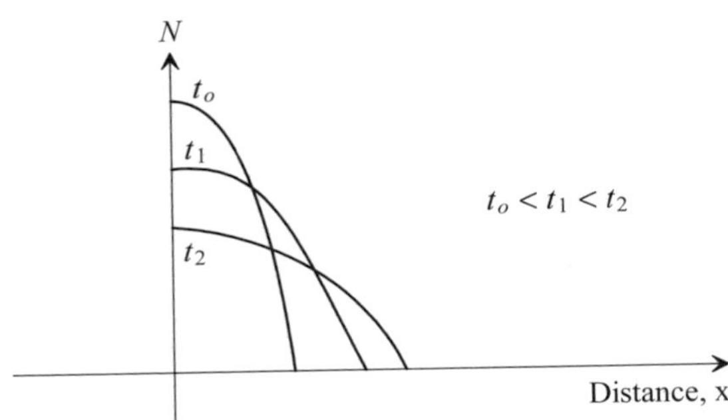

Figure 7.7 Idealized limited-source diffusion profile showing the effects of drive-in time on the profile. Notice that the peak concentration occurs at the surface of the substrate (x = 0) and that the area under the curves is constant.

7.1.4 Photolithography

In the fabrication of CMOS, it is necessary to localize processing effects to form a multitude of features simultaneously on the surface of the wafer. The collection of processes that accomplish this important task using an ultraviolet light, a mask, and a light-sensitive chemical resistant polymer is called *photolithography*. Although there are many different categories of photolithography, they all share the same basic processing steps that result in micron-to-submicron features generated in the light-sensitive polymer called photoresist. The photoresist patterns can then serve as ion implantation masks and etch masks during subsequent processing steps.

Figure 7.8 outlines the major steps required to implement photolithography patterning of a thermally grown oxide. Photoresist, a viscous liquid polymer, is applied to the top surface of the oxidized wafer. The application typically occurs by dropping (or spraying) a small volume of photoresist to a rapidly rotating wafer yielding a uniform thin film on the surface. Following spinning, the coated wafer is softbaked on a hot plate which drives out most of the solvents from the photoresist and improves the adhesion to the underlying substrate. Next, the wafers are exposed to ultraviolet light through a mask (or reticle) that contains the layout patterns for a given drawn layer. Unless the first layer is being printed, the exposure must be preceded by a careful alignment of mask features to existing patterns on the wafer. There are three general methods of exposing (patterning) the photoresist: contact, proximity, and projection photolithography. In both contact and proximity photolithography, the mask and the wafers are in contact and in close proximity, respectively, to the surface of the photoresist. Here the mask features are of the same scale as the features to be exposed on the surface.

In projection photolithography, the dominant type of patterning technology, the mask features are on a larger scale (e.g., 5X or 10X) relative to the features exposed on the surface. This is accomplished with a projection *stepper*. Using reduction with optics, the stepper projects an image through the mask to the photoresist on the surface. For positive-tone photoresist, the ultraviolet light breaks molecular bonds, making the exposed regions more soluble in the developer. In contrast, for negative-tone photoresist, the exposure causes polymerization and thus less solubility. The exposed resist-coated wafer is developed in an alkaline solution. Depending on the formulation of the photoresist, a positive or negative image relative to the mask patterns can be generated.

To harden the photoresist for improved etch-resistance and to improve adhesion, the newly developed wafers are often hardbaked. At this point, the wafer can be etched to transfer the photoresist pattern into the underlying oxide film.

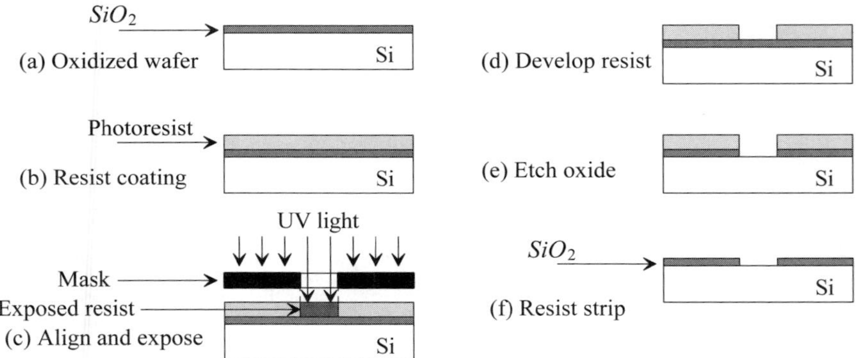

Figure 7.8 Simplified representation of the primary steps required for the implementation of photolighography and pattern transfer.

Resolution

In general, there are three critical parameters associated with a given projection stepper: resolution, depth of focus, and pattern registration. The diffraction of light caused by the various interfaces in its path limits the minimum printable feature size as depicted in Fig. 7.9. Resolution is defined as the minimum feature size, M, that can be printed on the surface of the wafer given by

$$M = \frac{c_1 \cdot \lambda}{NA} \tag{7.8}$$

where lambda, λ, is the wavelength of the ultraviolet light source, NA is the numerical aperture of the projection lens, and c_1 is a constant whose value ranges from 0.5 to 1. The NA of a lens is illustrated in Figure 7.10 and is given mathematically as

$$NA = n \cdot \sin\theta \tag{7.9}$$

where n is the index of refraction of the space between the wafer and the lens and θ is the acute angle between the focal point on the surface of the wafer and the edge of the lens radii. Notice that M is directly proportional to wavelength, hence diffraction effects are the primary limitation in printable feature size. To a limit, the NA of the projection lens can be increased to help combat the diffraction effects because large NA optics have an increased ability to capture diffracted light. At first glance, the minimum feature size cannot be less than the wavelength of the light, however, advance techniques such as optical proximity correction (OPC) and wavefront engineering of the photomasks have been developed to push the resolution limits below the wavelength.

Depth of Focus (DOF)

The depth of focus (DOF) of the projection optics limits one's ability to pattern features at different heights, as illustrated in Fig. 7.11. Mathematically, DOF is given by

$$\text{DOF} = \frac{c_2 \lambda}{NA^2} \tag{7.10}$$

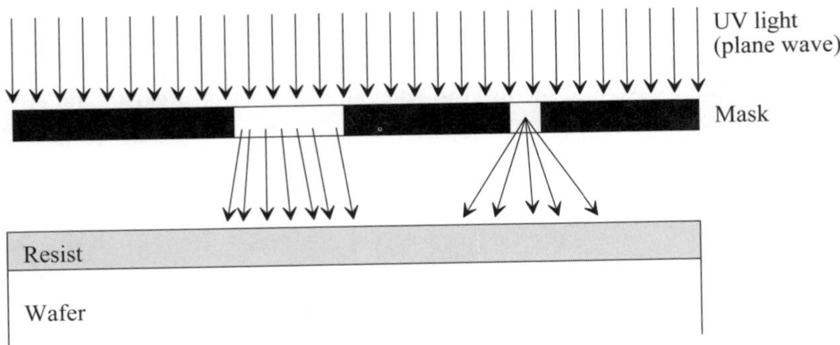

Figure 7.9 The diffraction effects become significant as the mask feature dimensions approach the wavelength of UV light. Notice that the diffraction angle is larger for the smaller opening.

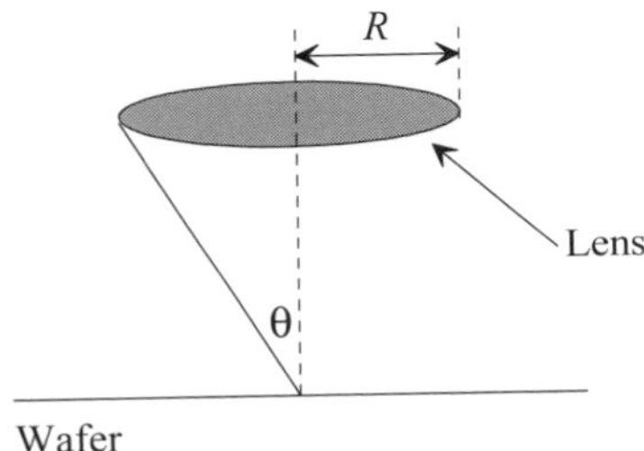

Figure 7.10 The relationship of the lens radii to the angle used to compute NA.

where c_2 is a constant ranging in value from 0.5 to 1. As apparent from Eq. (7.8) and Eq. (7.10), there exists a fundamental trade-off between minimum feature size and DOF. In other words, to print the smallest possible features, the surface topography must be minimized.

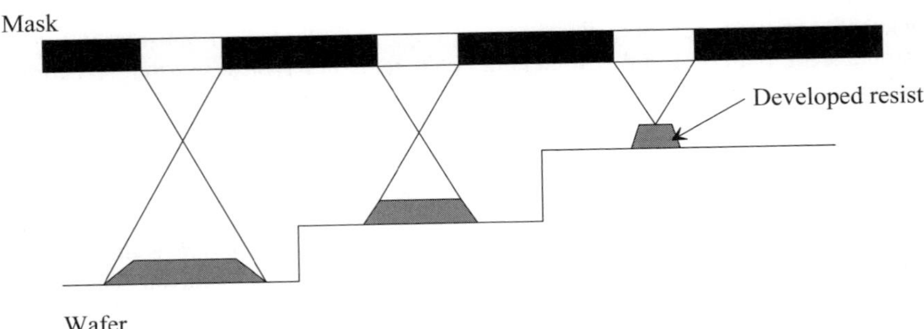

Figure 7.11 Depth of focus diagram illustrating the need to have planar surfaces (minimized topography) during high resolution photopatterning.

Aligning Masks

During CMOS fabrication, numerous mask levels (e.g., active, poly, contacts, etc.) are printed on the wafer. Each of these levels must be accurately aligned to one another. Registration is a measure of the level-to-level alignment error. Registration errors occur in x, y, and z-rotations, as illustrated in Fig. 7.12.

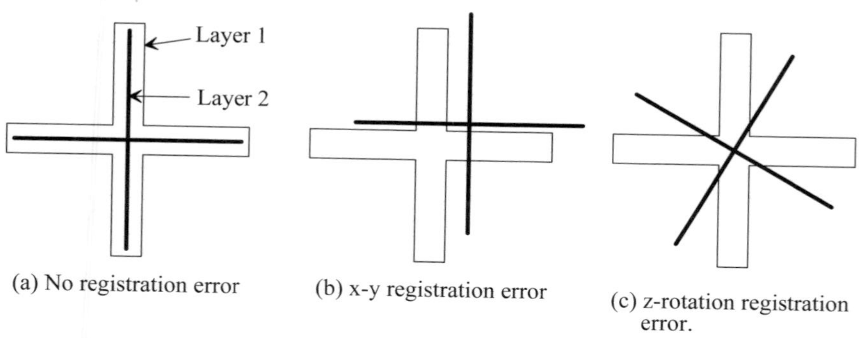

Figure 7.12 Simple registration errors that can occur during wafer-to-mask alignment in photolithography. Other registration errors exist but are not discussed here.

7.1.5 Thin Film Removal

Typically one of two processes are performed following photolithography. One is thin film etching used to transfer the photoresist patterns to the underlying thin film(s). The other is ion implantation using the photoresist patterns to block the dopants from select regions on the surface of the wafer. In this section, we discuss thin-film etching processes based on wet chemical etching and dry etching techniques. Additionally, we discuss a process used to remove unpatterned thin-films called chemical mechanical polishing (CMP).

Thin Film Etching

Once a photoresist pattern is generated there are two commonly employed approaches, wet etching and dry etching, to transfer patterns into underlying films. Etch rate (thickness removed per unit time), selectivity, and degree of anisotropy are key parameters for both wet and dry etching. Etch rates are typically strong functions of solution concentration and temperature. Selectivity, S, is defined as the etch rate ratio of one material to another given by the selectivity equation

$$S = \frac{R_2}{R_1} \tag{7.11}$$

where R_2 is the etch rate of the material intended to be removed and R_1 is the etch rate of the underlying, masking, or adjacent material not intended to be removed. The degree of anisotropy, A_f, is a measure of how rapidly an etchant removes material in different directions, mathematically given by

$$A_f = 1 - \frac{R_l}{R_v} \tag{7.12}$$

where R_l is the lateral etch rate and R_v is the vertical etch rate. Notice that if $A_f = 1$, then the etchant is completely anisotropic. However, if $A_f = 0$, then the etchant is completely isotropic. In conjunction with photolithography, the degree of anisotropy is a major factor in the achievable resolution. Figure 7.13 illustrates the effects of etch bias (i.e., $d_{film} - d_{mask}$) on the final feature size. For the submicron features that are required in CMOS, dry etch techniques are preferred over wet etch processes. This is due to the fact that dry etch techniques can, in general, have a higher degree of anisotropy. Both wet and dry etching are applied to the removal of metals, semiconductors, and insulators.

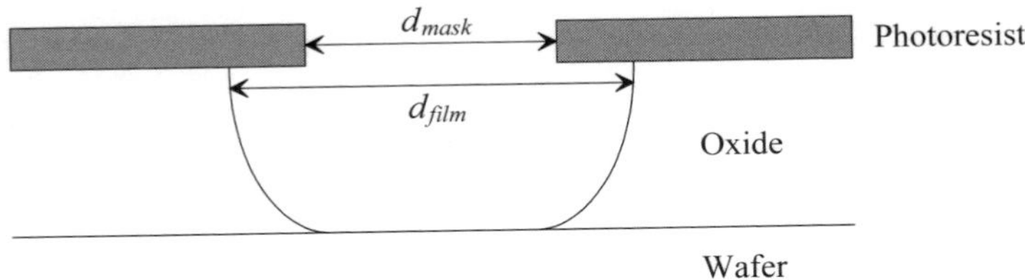

Figure 7.13 Diagram showing a post-etch profile. Notice that because of isotropy in the etch process the mask opening does not match the fabricated opening in the underlying oxide film. The difference between these dimensions is called etch-bias.

Wet Etching

Wet etching consists of using a chemical solution to remove material. In CMOS fabrication, wet processes are used both for cleaning of wafers and for thin-film removal. Wet cleaning processes are repeated numerous times throughout a process flow. Some cleaning processes are targeted to particulate removal, while others are for organic and/or inorganic surface contaminants. Wet etchants can be isotropic (i.e., etch rate is the same in all directions) or anisotropic (i.e., etch rate differs in different directions) although most of the wet etchants used in CMOS fabrication are isotropic. In general, wet etchants tend to be highly selective compared to dry etch processes. A schematic diagram of a wet etch tank is shown in Fig. 7.14. To improve the etch uniformity and to aid particulate removal, it is common to ultrasonically vibrate the etchant, as shown in the figure. Furthermore, microcontrollers accurately control the temperature of the bath. Once the etch is completed, the wafers are rinsed in deionized (DI) water, then spun dried.

Dry Etching

In CMOS fabrication, there are three general categories of dry etch techniques: sputter etching, plasma etching, and reactive ion etching (RIE). Figure 7.15 schematically illustrates a sputter etch process. An inert gas (e.g., argon) is ionized where the ions are accelerated through an electric field established between two conductive electrodes, called the anode and the cathode. A vacuum in the range of millitorr must exist between the plates to allow the appropriate ionization and transfer of ions. Under these conditions, a glow discharge, or plasma, is formed between the electrodes. In simple terms, the plasma consists of positively charged ions and electrons, which respond oppositely to the electric field. The wafer sits on the cathode where it is bombarded by the positively charged ions, causing material to be ejected off of the surface. Essentially, sputter etching

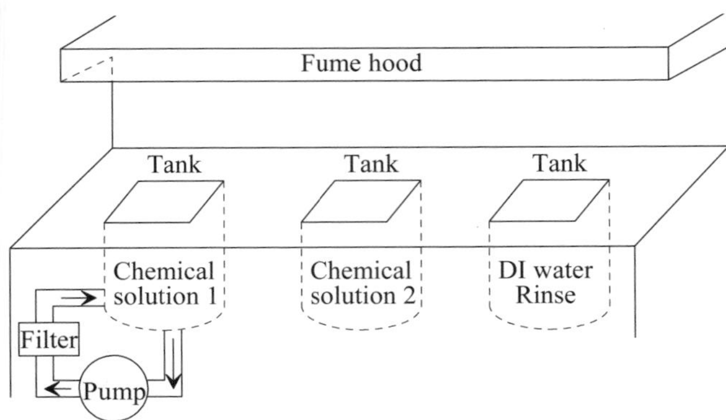

Figure 7.14 Simplfied diagram of a wet bench used for wet chemical cleaning and etching.

is atomic-scale sandblasting. A DC power supply can be used for sputter etching conductive substrates, while an RF supply is required through capacitive coupling for etching non-conductive substrates. Sputter etching tends not to be selective, but it is very anisotropic.

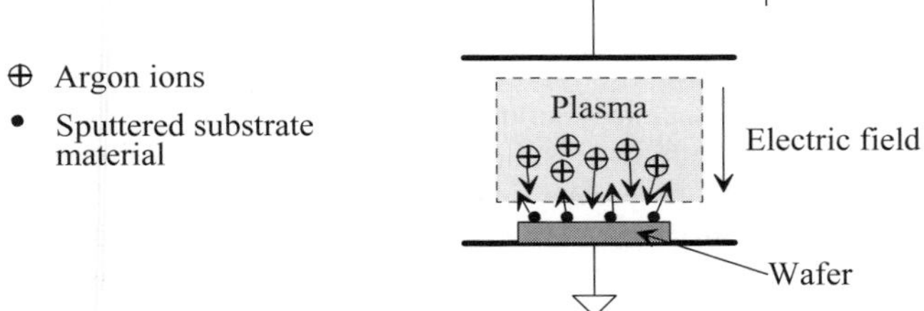

Figure 7.15 Simplified schematic diagram of the sputter etch process. This process is dominated by the physical bombardment of ions on a substrate.

A simplified diagram of a plasma etch system is shown in Figure 7.16. A gas or mixture of gases (e.g., halogens) are ionized, producing reactive species called radicals. A glow discharge or plasma is formed between the electrodes. The radicals chemically react with the surface material forming reaction products in the gas phase which are pumped away through a vacuum system. Plasma etching can be very selective, but is typically highly isotropic.

While sputter etching is a purely mechanical process and plasma etching is purely chemical, RIE is a combination of sputter etching and plasma etching, as schematically shown in Figure 7.17. In RIE, a gas or mixture of gases (e.g., fluorocarbons) are ionized where radicals and ionized species are generated, both of which interact with the surface of the wafer. RIE is the dominant etch process because it can provide the benefits of both sputter etching and plasma etching. In other words, RIE can be highly selective and highly anisotropic.

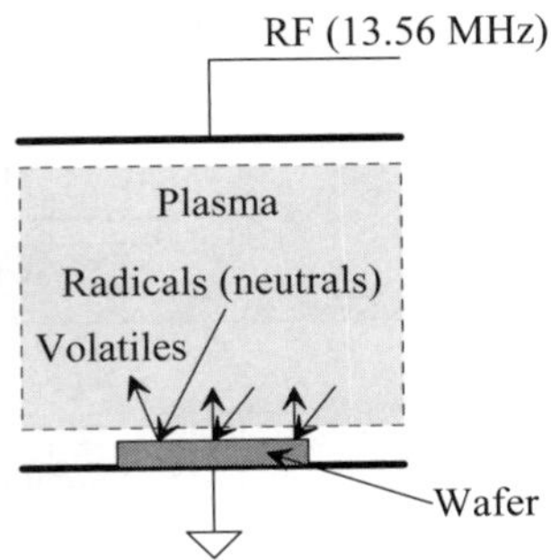

Figure 7.16 Simplified schematic diagram of a plasma etch process. This process is dominated by the chemical reactions of radicals at the surface of the substrate.

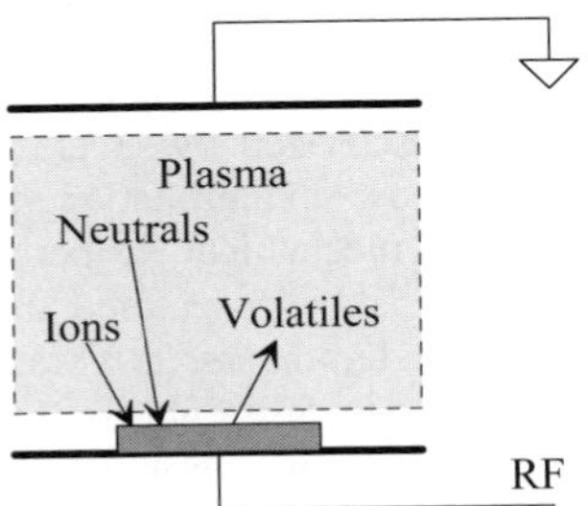

Figure 7.17 Simplified schematic diagram of an RIE etch process. This process has both physical (ion bombardment) and chemical (reaction of radicals) components.

Chemical Mechanical Polishing

Figure 7.18 depicts the key features of chemical mechanical polishing (CMP). In CMP, an abrasive chemical solution, called a slurry, is introduced between a polishing pad and the wafer. Material on the surface of the wafer is removed by both a mechanical polishing component and a chemical reaction component. In modern CMOS fabrication, CMP is a critical process that is used to planarize the surface of the wafer prior to photolithography. The planar surface allows the printed feature size to be decreased. CMP can be used to remove metals, semiconductors, and insulators.

7.1.6 Thin Film Deposition

Insulators, conductors, and semiconductors are all required for CMOS integrated circuits. Semiconductors, such as crystalline silicon for the active areas and polycrystalline silicon for the gate electrodes/local area interconnects, are generally required. Insulators such as Si_3N_4, SiO_2 and doped glasses are used for gate dielectrics, device isolation, metal-to-substrate isolation, metal-to-metal isolation, passivation, etch masks, implantation masks, diffusion barriers, and sidewall spacers. Conductors such as aluminum, copper, cobalt, titanium, tungsten, and titanium nitride are used for local interconnects, contacts, vias, diffusion barriers, global interconnects, and bond pads. In this section, we discuss the various methods to deposit thin films of insulators, conductors, and semiconductors. We

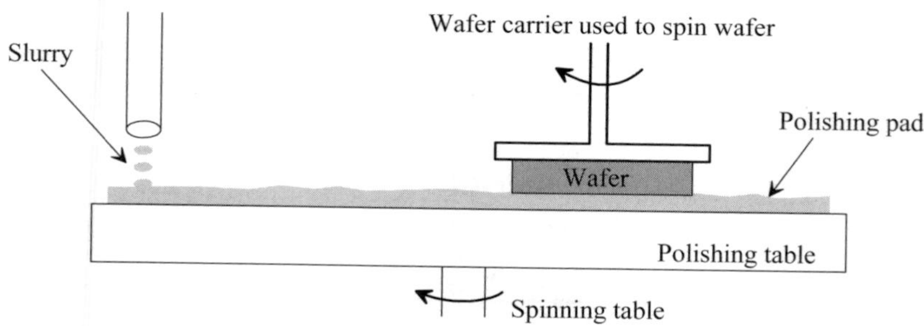

Figure 7.18 Simplified representation of a chemical mechanical polishing process used in the fabrication process.

present two primary categories of thin film deposition: physical vapor deposition and chemical vapor deposition. A third less common category, electrodeposition, for depositing copper for backend interconnects will not be addressed here.

Deposited films are often characterized by several factors. Inherent film quality related to the compositional control, low contamination levels, low defect density, and predictable and stable electrical and mechanical properties are of prime importance. Moreover, film thickness uniformity must be understood and controlled to high levels. To achieve highly uniform CMOS parameters across a wafer, it is common to control the film thickness uniformity to less then +/– 5 nm across the wafer diameter. In addition, film uniformity over topographical features is of critical importance. A measure of this is called step coverage, as depicted in Fig. 7.19. As illustrated, good step coverage results in uniform thickness over all surfaces. In contrast, poor step coverage results in significantly reduced thickness on vertical surfaces relative to surfaces parallel with the surface of the wafer. Related to step coverage is what is referred to as gap fill. Gap fill applies to the deposition of a material into a high aspect ratio opening, such as contacts or gaps between adjacent metal lines. Figure 7.20 illustrates a deposition with good gap fill and a deposition that yields a poor gap fill (also called a keyhole or void).

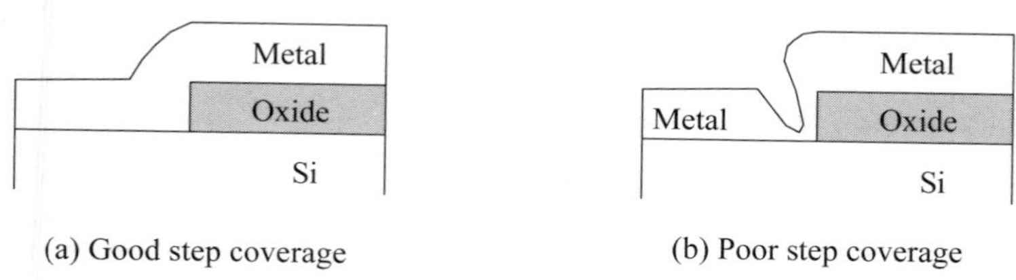

Figure 7.19 Extremes in thin-film deposition coverage over a pre-existing oxide step.

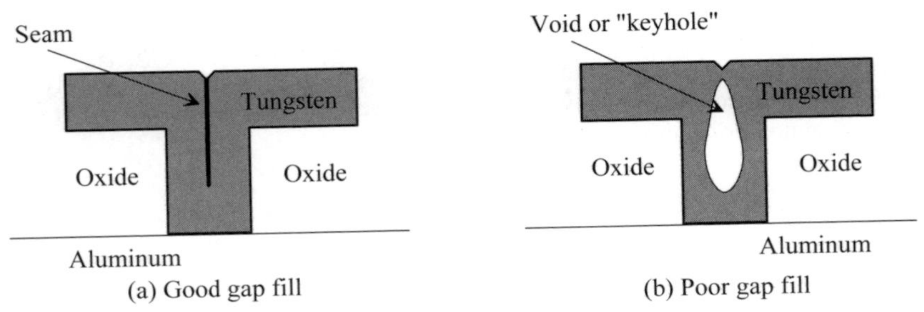

Figure 7.20 Gap-fill profiles (good and bad) of a high aspect ratio opening filled with a deposited film.

Physical Vapor Deposition (PVD)

In physical vapor deposition (PVD), physical processes produce the constituent atoms (or molecules), which pass through a low-pressure gas phase and subsequently condense on the surface of the substrate. The common PVD processes are evaporation and sputter deposition, both of which can be used to deposit a wide range of insulating, conductive, and semiconductive materials. One of the drawbacks of PVD is that the resultant films often have poor step coverage.

Evaporation is one of the oldest methods of depositing thin-films of metals, insulators, and semiconductors. The basic process of evaporation is shown in Fig. 7.21. The material to be deposited is heated past its melting point in a high vacuum chamber where the vapor form of the material coats all exposed surfaces within the mean free path of the evaporant. The heat source can be of one of two types: heating filament or focused electron beam.

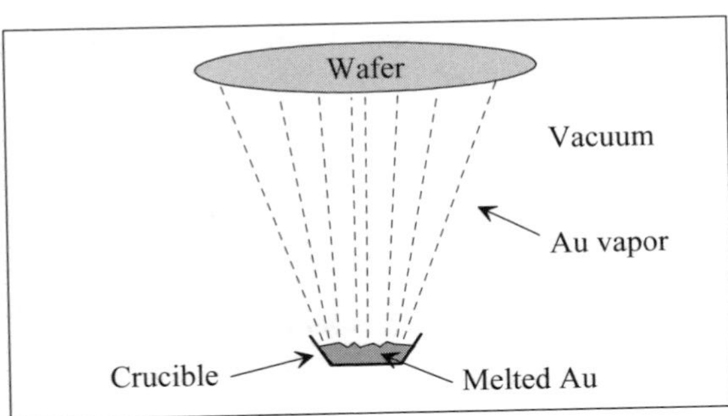

Figure 7.21 Simplified diagram of an evaporation deposition process.

In simple terms, sputter deposition is similar to sputter etching, as discussed in Sec. 7.1.5, with the exception that the wafer serves as the anode, and the cathode is the target material to be deposited. Figure 7.22 outlines a simplified sputter deposition process. An inert gas such as argon is ionized in a low pressure ambient where the positively charged ions are accelerated through the electric field towards the target. The target is an ultra-high purity disk of material to be deposited. The bombardment of ions with the target sputter (or eject) target atoms (or molecules), where they transit to the surface of the wafer forming a thin-film. Similar to sputter etching, a DC supply can be used for sputtering conductors; however, a capacitively coupled RF supply must be used for depositing non-conductive materials.

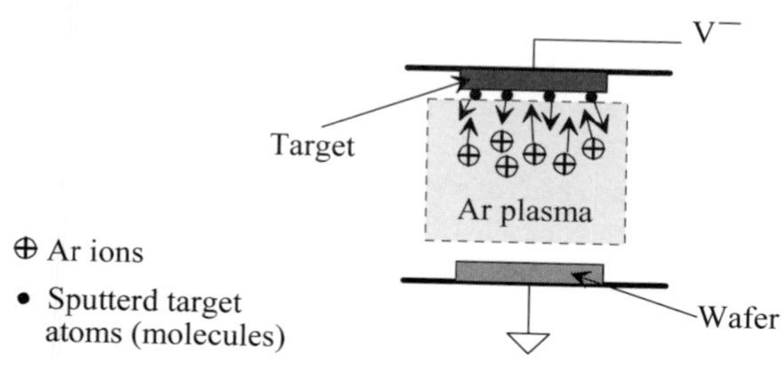

Figure 7.22 Simplified diagram of a sputter deposition process.

Chemical Vapor Deposition (CVD)

In chemical vapor deposition (CVD), reactant gases are introduced into a chamber where chemical reactions between the gases at the surface of the substrate produce a desired film. The common CVD processes are atmospheric pressure (APCVD), low pressure (LPCVD), and plasma enhanced (PECVD). Again, there are a wide variety of insulators, conductors, and semiconductors that can be deposited by CVD. Most importantly, the resultant films tend to have good step coverage compared to PVD processes. APCVD occurs in an apparatus similar to an oxidation tube furnace (see Fig 7.4); however, an appropriate reactive gas is flowed over the wafers. APCVD is performed at relatively low temperatures. As depicted in Fig. 7.23, LPCVD occurs in a reactor in the pressure range of milliTorr to a few Torr. Compared to APCVD, the low pressure process can yield

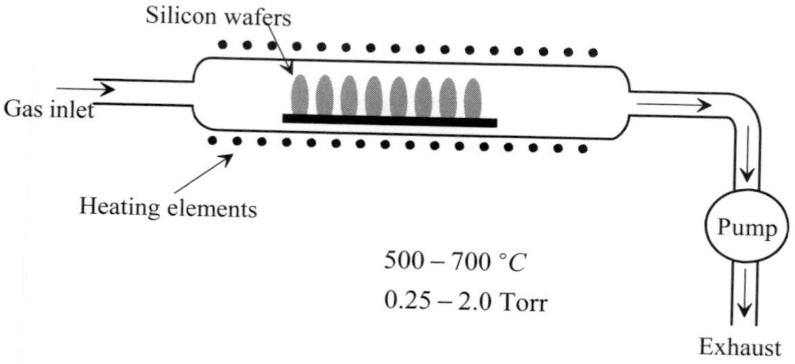

Figure 7.23 Simplified schematic diagram of a LPCVD.

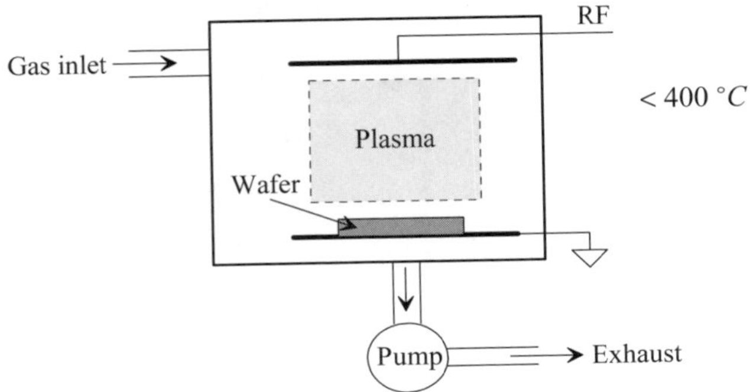

Figure 7.24 Simplified schematic diagram of a PECVD reactor.

highly conformal films, but at the expense of a higher deposition temperature. Fig. 7.24 shows a schematic diagram of a PECVD reactor. In PECVD, a plasma imparts energy for the surface reactions, allowing for lower temperature deposition. By comparison, PECVD has the advantage of being low temperature and highly conformal.

7.2 CMOS Process Integration

Process integration is the task of combining a deliberate sequence of unit processes to fabricate integrated microelectronic circuits (e.g., MOSFETs, resistors, capacitors, etc.) A typical CMOS technology consists of a complex arrangement of unit processes where several hundred steps are required to manufacture integrated circuits on a silicon wafer. Groups of unit processes are combined to form integration modules. For example, the gate module would include a specific sequence of unit processes for yielding a gate electrode on a thin, gate dielectric. The modules could then be combined to yield the overall process flow (or process sequence). The process flow can be divided into frontend-of-the-line (FEOL) and backend-of-the-line (BEOL) processes. A typical process flow consisting of numerous modules is shown in block diagram form in Fig. 7.25.

FEOL

Generally, FEOL refers to all processes preceding salicidation (i.e., silicide formation, see Fig. 4.4). FEOL includes all processes required to fully form isolated CMOS transistors. In Fig. 7.25 we see that the FEOL begins with the selection of the starting material (i.e., type of silicon wafer to be used). Then, the shallow trench isolation (STI) module is implemented to form the regions of dielectric between regions of active area. Next, the wells (or tubs) are formed followed by the gate module, which includes all processes to properly define gate electrodes on a thin oxide. Finally, the FEOL concludes with the source/drain module, which includes the processes required for the formation of the low-doped drain extensions and the source/drain regions themselves.

BEOL

BEOL refers to all processes subsequent to source/drain formation. Hence, BEOL processes are used to "wire" the transistors together using multiple layers of dielectrics

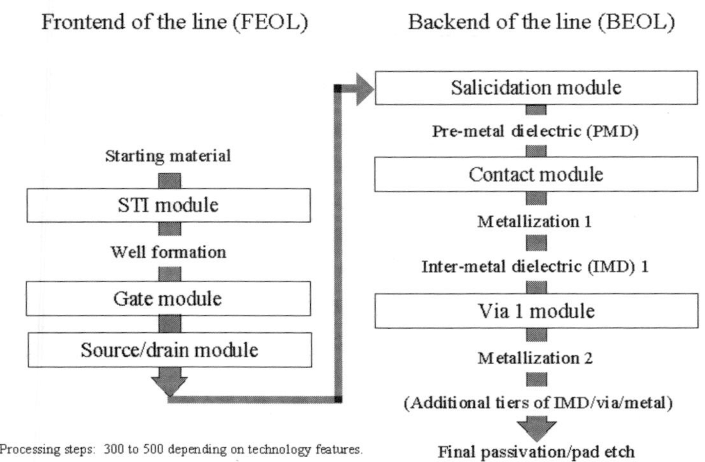

Figure 7.25 A typical CMOS process flow illustrating the difference between FEOL and BEOL processes.

and metals. The BEOL begins with the salicidation of the polysilicon and source/drain regions. The remaining BEOL processes proceed in repetitive sets of modules to yield lateral and vertical interconnects isolated from one another with dielectrics. It is important to understand that there is a high degree of inter-relationship between unit processes within each module and between modules themselves. A seemingly "trivial" change in one unit process in a given module can have dramatic effects on processes in other modules. In other words, **there is no trivial process change**.

CMOS Process Description

Even with a single device type, there are numerous variations in process schemes for achieving similar structures. Hence, it would be virtually impossible to outline all schemes. Therefore, we will describe a generic (but representative) deep-submicron CMOS process flow (deep indicating that a deep, or short wavelength, ultraviolet light source is used when patterning the wafers). Our CMOS technology will have the following features:

1. Frontend of the line (FEOL)
   - (a) Shallow trench isolation (STI)
   - (b) Twin-tubs
   - (c) Single-level polysilicon
   - (d) Low-doped drain extensions
2. Backend of the line (BEOL)
   - (a) Fully planarized dielectrics
   - (b) Planarized tungsten contacts and via plugs
   - (c) Aluminum metallization

Following each major process step, cross sections will be shown. The cross sections were generated with a technology computer-assisted design (TCAD) package called Tsuprem-4 and Taurus-Visual (2D) released by Technology Modeling Associates, Inc. These tools simulate a defined sequence of unit processes, thus allowing one to model a process flow prior to its actual implementation. In the interest of brevity, there are several omissions and consolidations in the process description. These include:

1. Wafer cleaning performed immediately prior to all thermal processes, metal depositions, and photoresist removals.
2. Individual photolithographic process steps such as dehydration bake, wafer priming, photoresist application, softbake, alignment, exposure, photoresist development, hardbake, inspection, registration measurement, and critical dimension (CD) measurement.
3. Backside film removal following select CVD processes.
4. Metrology to measure particle levels, film thickness, and post-etch CDs.

To implement our CMOS technology, we employ a reticle set as outlined in Table 7.1. The masks are labeled as having either a clear or a dark field. Clear field masks are masks with opaque features totaling less than 50% of the mask area. In contrast, dark field refers to a mask that has opaque features that account for greater than 50% of the total area. Using this mask set, we assume the exclusive use of positive tone photoresist processing. When appropriate, representative mask features that yield isolated, complementary transistors adjacent to one another are shown.

Table 7.1 Masks used in our generic CMOS process.

Layer name	Mask	Aligns to level	Times used	Purpose
1 (active)	Clear	aligns to notch	1	Defines active areas
2 (p-well)	Clear	1	2	Defines NMOS sidewall implants and p-well
3 (n-well)	Dark	1	2	Defines PMOS sidewall implants and n-well
4 (poly1)	Clear	1	1	Defines polysilicon
5 (n-select)	Dark	1	2	Defines nLDD and n+
6 (p-select)	Dark	1	2	Defines pLDD and P+
7 (contact)	Dark	4	1	Defines contact to poly and actives areas
8 (metal1)	Clear	7	1	Defines metal1
9 (via1)	Dark	8	1	Defines via1 (connects M1 to M2)
10 (metal2)	Clear	9	1	Defines metal2
passivation	Dark	Top-level metal	1	Defines bond pad opening in passivation

7.2.1 Frontend-of-the-Line Integration

As previously stated, FEOL encompasses all processing required to fabricate the fully-formed, isolated CMOS transistors. In this subsection we discuss the modules and unit processes required for a representative CMOS process flow.

Starting Material

The choice of substrate is strongly influenced by the application and characteristics of the CMOS ICs to be fabricated. Bulk silicon is the least costly but may not be the optimal choice in high performance or harsh environment CMOS applications. Epitaxial (Epi) wafers are heavily doped bulk wafers with a thin, moderately to lightly doped epitaxial silicon layer grown on the surface. The primary advantage of Epi wafers is for immunity to latch-up. Silicon-on-insulator wafers increase performance and eliminate latch-up. However, SOI CMOS is more costly to implement than bulk or epi technologies. The three general types of silicon substrates are shown in Fig. 7.26. In current manufacturing, wafer diameters typically range from 100 to 300 mm. The wafer thickness correspondingly increases with diameter to allow for greater rigidity. The actual CMOS is constructed in the top one micron or less of the wafer, whereas the remaining hundreds of microns are used solely for mechanical support during device fabrication.

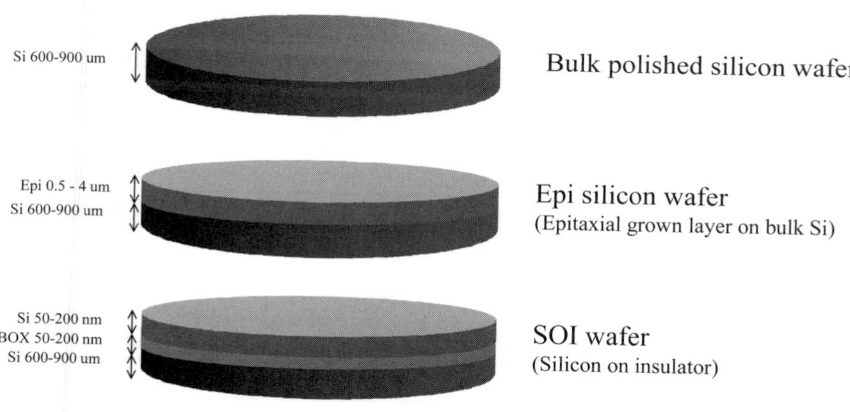

Figure 7.26 The three general types of silicon wafers used for CMOS fabrication.

We use bulk silicon wafers for our CMOS technology in this section. It should be noted that with relatively minor process and integration adjustments our technology could be applicable to epi or SOI CMOS processes. The first of many simulated cross sections are shown in Fig. 7.27. Here only the top two microns of silicon are shown. At the beginning of the fabrication process, the wafer characteristics such as resistivity, sheet resistance, crystallographic orientation, and bow and warp are measured and/or recorded. Furthermore, the wafers are scribed, usually with a laser, with a number which identifies the wafer's lot and number.

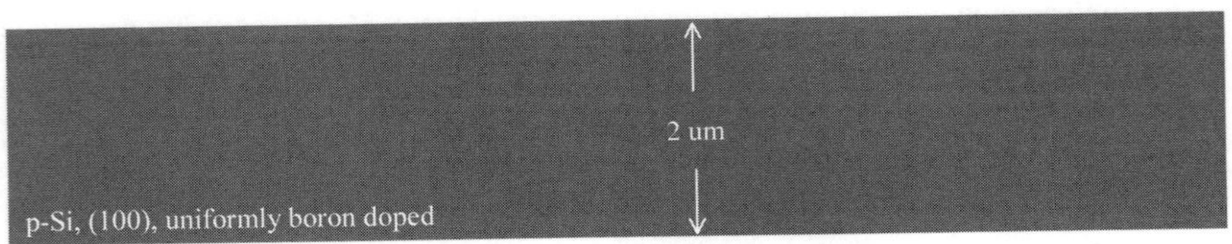

Figure 7.27 Simulated cross-sectional view (the top 2 μm) of the bulk wafer in Fig. 7.26.

Shallow Trench Isolation Module

Devices (e.g., PMOS and NMOS) must be electrically isolated from one another. This isolation is of primary importance for suppressing leakage current between both like and dissimilar devices.

One of the simplest methods of isolation is to fabricate the CMOS such that a reversed-bias pn-junction is formed between the transistors. Oppositely doped regions (e.g., n-well adjacent to a p-well) can be electrically isolated by tying the n-region to the most positive potential in the circuit and the p-region to the most negative. As long as the reverse-bias is maintained and the breakdown voltage is not exceeded for all operating conditions, a small diode reverse saturation current accounts for the leakage current. This junction leakage current is directly proportional to the junction area; hence, for the large p- and n-regions in modern devices, junction isolation alone is not adequate.

The second general method of isolation is related to the formation of thick dielectric regions, called field regions, between transistors. The region without the thick dielectric is where the transistors reside and is known as the active area. The relatively thick oxide that forms between the active areas is called field oxide (FOX). Interconnections of polysilicon are formed over the field regions to provide localized electrical continuity between transistors. This arrangement inherently leads to the formation of parasitic field effect transistors. Effectively, the FOX increases the parasitic transistor's threshold voltage such that the device always remains in the off state. Further, this threshold voltage can be increased by increasing the surface doping concentration, called channel stops, under the FOX. There are two general approaches to forming field oxide regions: LOCOS and STI.

LOCal Oxidation of Silicon (LOCOS) has been used extensively for half-micron or larger minimum linewidth CMOS technologies. In LOCOS, a diffusion barrier of silicon nitride blocks the thermal oxidation of specific regions on the surface of a wafer. Both oxygen and water diffuse slowly through silicon nitride. Hence, nitride can be deposited and patterned to define active and field oxide areas. The primary limitation to LOCOS is bird's beak encroachment, where the lateral diffusion of the oxidant forms an oxide feature that in cross-section resembles a bird's beak. The bird's beak encroaches into the active area, thereby reducing the achievable circuit packing density. Moreover, LOCOS requires a long, high-temperature process, which can result in significant diffusion of previously introduced dopants.

Shallow trench isolation (STI) is the dominant isolation technology for sub-half micron CMOS technologies. As the name implies, a shallow trench is etched into the surface of the wafer and then filled with a dielectric serving as the FOX. A typical STI process sequence follows. From a processing perspective, STI is complex; however, it can be implemented with minimal active-area encroachment. Moreover, it has a relatively low thermal budget.

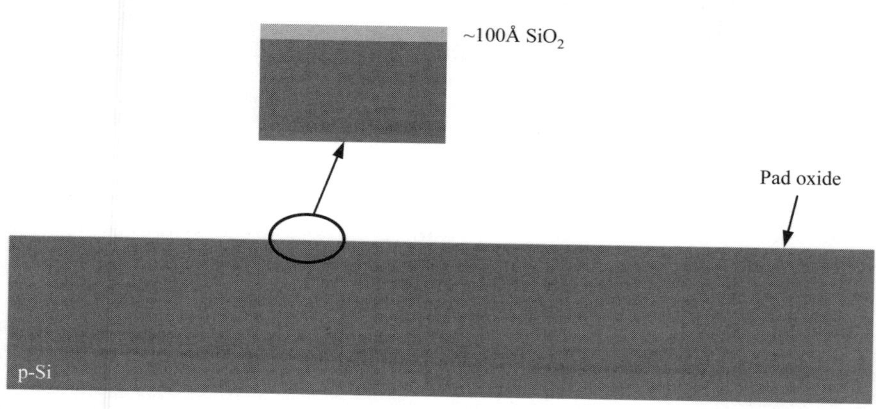

Figure 7.28 STI film stack. The oxide is thermally grown at approximately 900 °C with dry O_2.

The CMOS technology outlined in this chapter uses STI to achieve device isolation. The STI module begins with the thermal oxidation of the wafer surface, as shown in Fig. 7.28. The resultant oxide serves as a film-stress buffer, called a pad oxide, between the silicon and the subsequently deposited Si_3N_4 layer. (In addition, it is also used following the post-CMP nitride strip as an ion implant sacrificial oxide.) Next, silicon nitride is deposited on the oxidized wafer by LPCVD, as shown in Fig. 7.29. Later, this nitride serves as both an implant mask and a CMP stop-layer. As shown in Fig. 7.30, the photolithography (mask layer 1) produces the appropriate patterns in photoresist for defining the active areas. Then, with end-point-detected RIE, the photoresist pattern is transferred into the underlying film stack of nitride and oxide. In Fig. 7.30 notice that the PMOS and NMOS devices will be fabricated on the left side and right side, respectively, under the photoresist. The region cleared of photoresist corresponds to the isolation regions. The 0.4 µm deep silicon trenches are formed by

Figure 7.29 STI film stack. Silicon nitride deposited by LPCVD at approximately 800 °C.

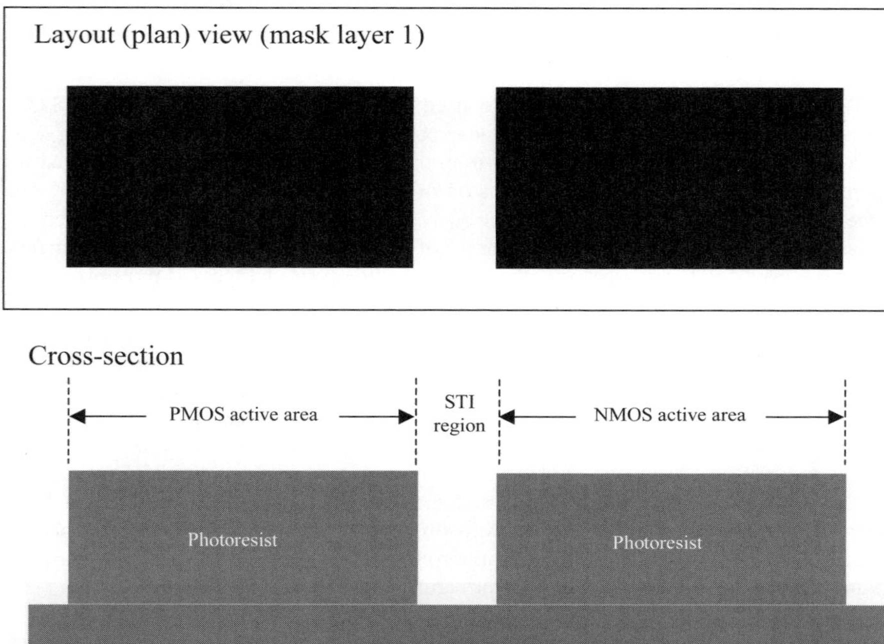

Figure 7.30 STI definition, photolithography and nitride/pad oxide etch with fluorocarbon- based RIE.

timed RIE with the photoresist softmask present, as shown in Fig 7.31. Although the etching can proceed without the resist, the sidewall profile can be tailored to a specific slope with the presence of the polymer during the etch process. At the cost of reduced packing density, the sloped sidewalls aid in the reduction of leakage current from the

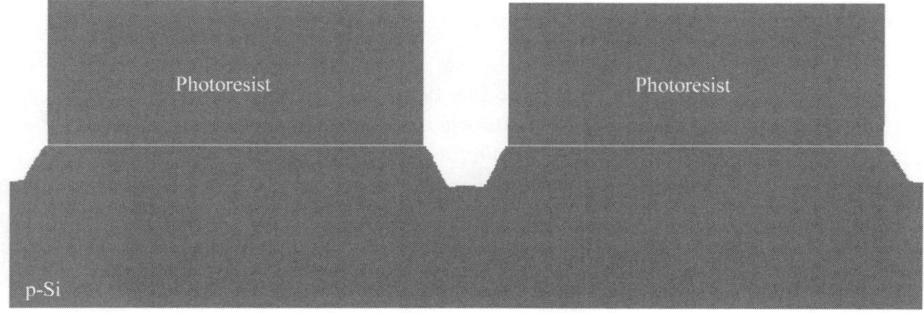

Figure 7.31 Timed silicon trench reactive ion etch.

parasitic corner transistors. Following the silicon etch, O_2 plasma and wet processing strip the photoresist and etch by-products from the surface of the wafer. At this point, the general structural form of the STI is finished.

The next series of processes are used to improve effectiveness of the STI to suppress leakage currents. As shown in the expanded view of the trench seen in Fig. 7.32, a thin, sacrificial oxide is thermally grown on the exposed silicon. It should be noted that the nitride provides a barrier to the diffusion of oxygen, hence the oxidation occurs only in the exposed silicon regions. This oxide serves as a sacrificial oxide for subsequent ion implantation and aids in softening the corner of the trench. In general, implant sacrificial oxides are used to (1) suppress ion channeling in the crystal lattice, (2) minimize lattice damage from the ion bombardment, and (3) protect the silicon surface from contamination. Photolithography (mask layer 2) patterns resist to protect the PMOS sides of the trench during the p-wall implant. A shallow BF_2 implant is performed to dope what will eventually become the p-*well* trench sidewalls (called the p-*walls*), as seen in Fig. 7.33. The p-wall implant increases the threshold voltage of the parasitic corner transistor and minimizes leakage under the trench. The BF_2 implanted resist is stripped using O_2 plasma and wet processing, as shown in Fig. 7.34. Again, photolithography (mask layer 3), Fig. 7.35, is used to produce the complementary pattern for the n-wall implant. For the same reasons, this shallow phosphorous implant is introduced into what will become the n-well trench sidewalls. The phosphorous-implanted resist is stripped using O_2 plasma and wet processing, yielding the structure depicted in Fig. 7.36.

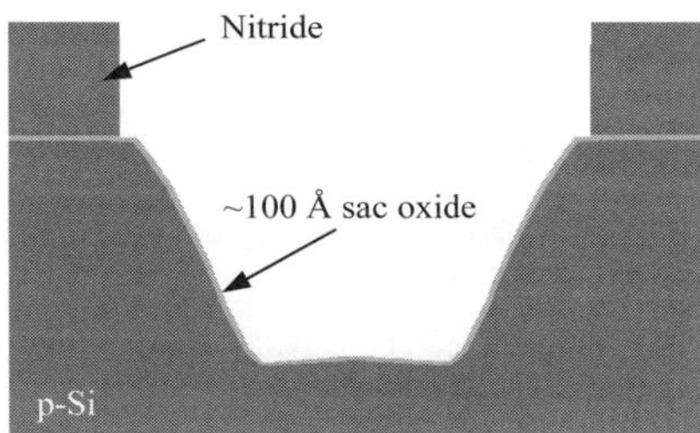

Figure 7.32 Cross-section showing post STI resist strip followed by the dry thermal oxidation (at 900 °C) of a sacrificial (sac) oxide in the trench.

At this point, the sacrificial oxide has been degraded by the implantations and is likewise stripped using a buffered hydrofluoric acid solution. A thin, high-quality thermal oxide is regrown in the trenches to form what is called a *trench liner*. In general, the liner oxide improves the interface quality between the silicon and the subsequent trench fill, thus suppressing the interface leakage current. Specifically, the formation of the trench liner oxide (1) "cleans" the surface prior to trench fill, (2) anneals sidewall implant damage, and (3) passivates interface states to minimize parasitic leakage paths.

Once the liner oxide is grown, CVD is used to overfill the trenches with a dielectric, as shown in Fig. 7.37. The trench fill provides the field isolation required to increase the threshold voltage of the parasitic field transistors. Further, it blocks

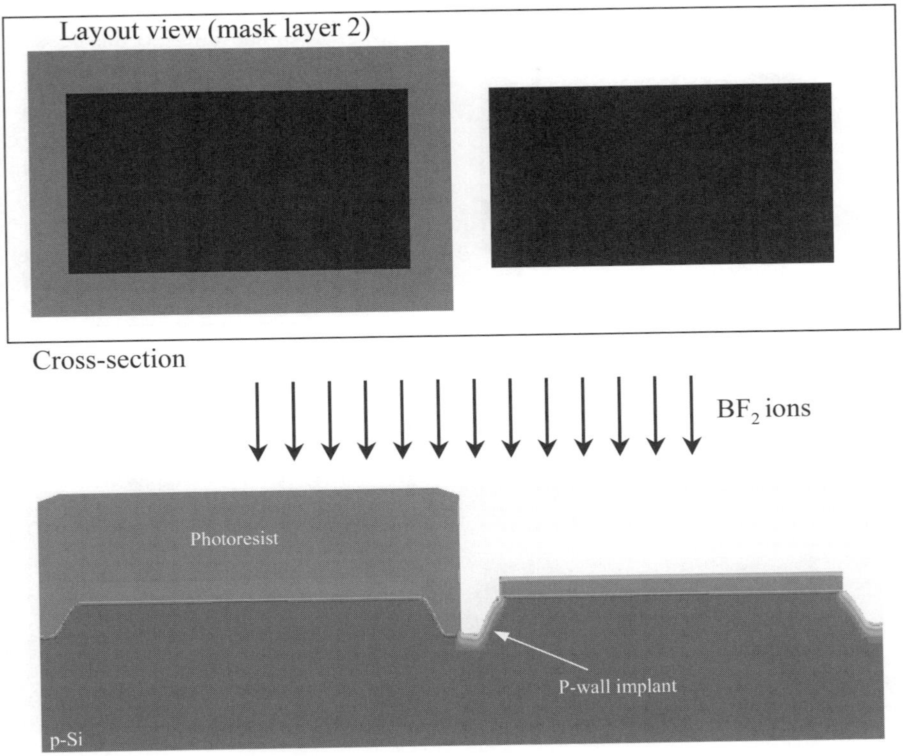

Layout view (mask layer 2)

Cross-section

BF_2 ions

Photoresist

P-wall implant

p-Si

Figure 7.33 P-wall sidewall formation via photolithography and BF_2 implantation.

subsequent ion implants. Although not shown, it is common to use a "block-out" pattern to improve the uniformity of the STI CMP. In Fig. 7.38a, CMP is performed to remove the CVD overfill. The nitride is used as a polish-stop layer. Next, a brief buffered oxide etch removes oxide that may have formed on top of the nitride. Then, the nitride is

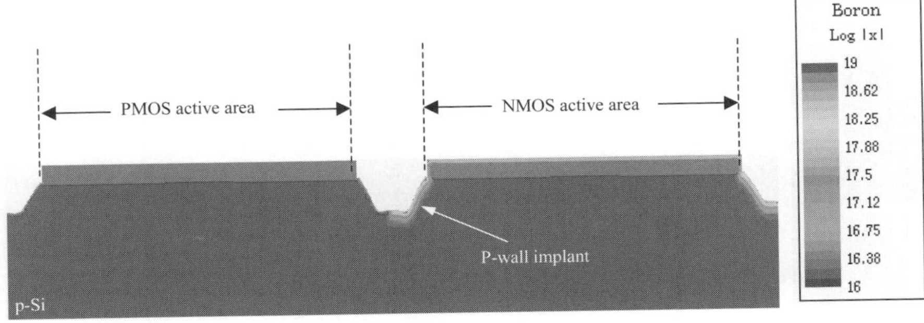

Figure 7.34 Post p-wall photoresist strip using O_2 plasma and wet processing.

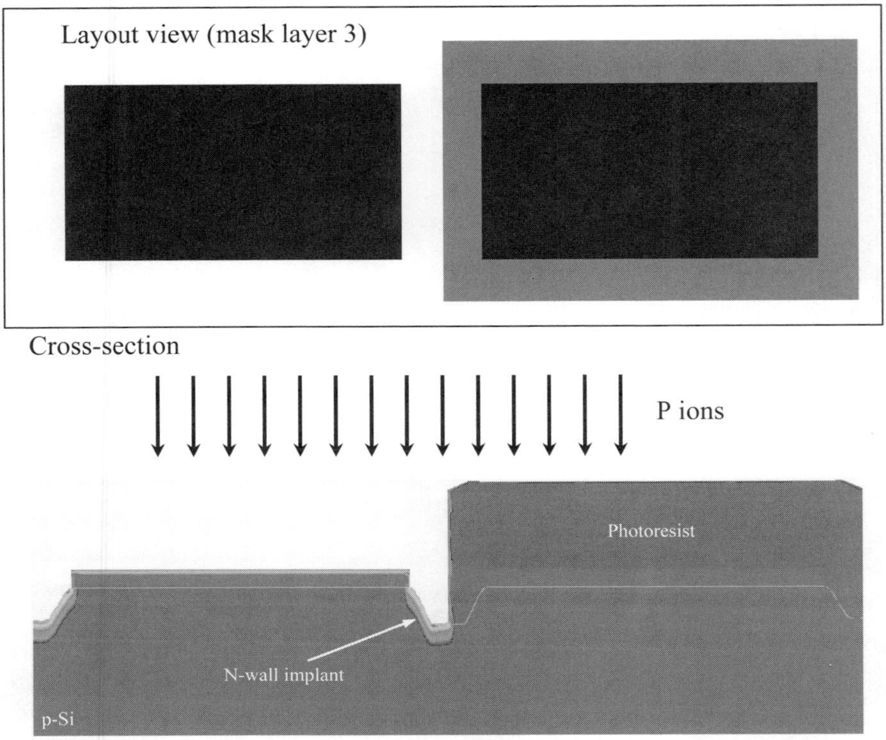

Layout view (mask layer 3)

Cross-section

P ions

Photoresist

N-wall implant

p-Si

Figure 7.35 N-wall sidewall formation via photolithography and P implantation.

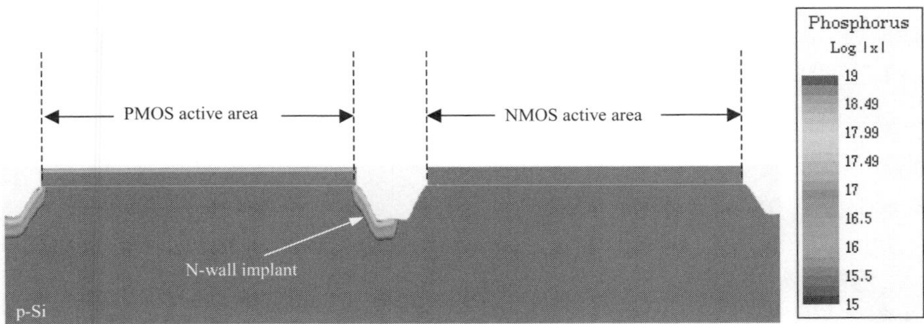

Figure 7.36 Post n-wall photoresist strip using O_2 plasma and wet processing.

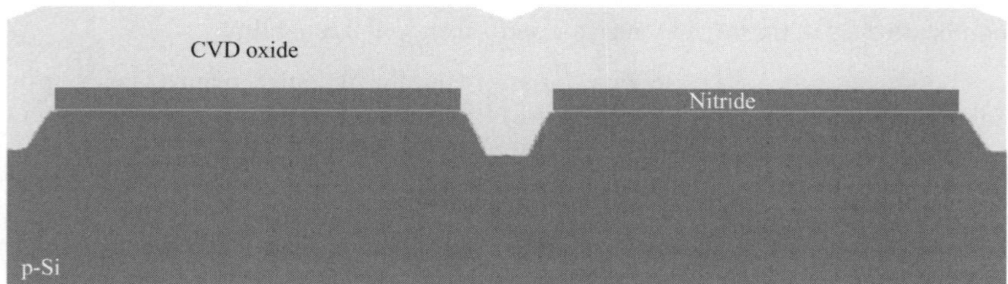

Figure 7.37 High quality, 100 Å thick liner oxide is thermally grown at 900 °C. High density plasma (HDP) CVD trench fill at room temperature. Notice that the trenches are overfilled.

removed from the active areas by using a wet or dry etch process, as illustrated in Fig. 7.38b. Notice that the pad oxide remains after this step. At this point, the STI is fully formed.

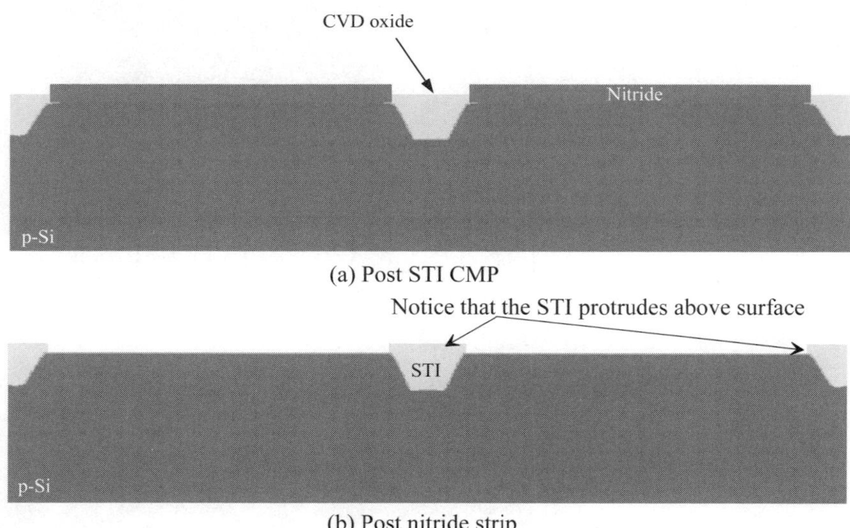

Figure 7.38 STI CMP (a) where the nitride acts like a polish stop and (b) wet nitride etch in hot phosphoric acid and/or dry nitride etch in NF_3/Ar/NO. The remaining oxide will be used as a sacrificial oxide for subsequent implants.

Twin-Tub Module

As explained in Sec. 2.5, CMOS can be implemented in four general forms: n-well, p-well, twin-well (called twin-tub), and triple-well. The CMOS technology discussed in this chapter uses a twin-well approach. The p-well and n-well provide the appropriate dopants for the NMOS and PMOS, respectively. Modern wells are implanted with retrograde profiles to maximize transistor performance and reliability.

Following the STI module, the twin-tub module begins with p-well photolithography (mask layer 2, second use) to generate a resist pattern that covers the PMOS active regions, but exposes the NMOS active areas, as depicted in Fig. 7.39. A relatively high energy boron implant is performed into the NMOS active areas. Here the implant is blocked from the PMOS active area. The pad oxide that remained from the STI module now serves as the implant sacrificial oxide for the well implants. It should be pointed out that the p-well may be formed by a composition of several implants at different doses and energies to achieve the desired retrograde profile. Following the p-well implant, the resist is removed using O_2 plasma and wet processing, resulting in the structure shown in Fig. 7.40.

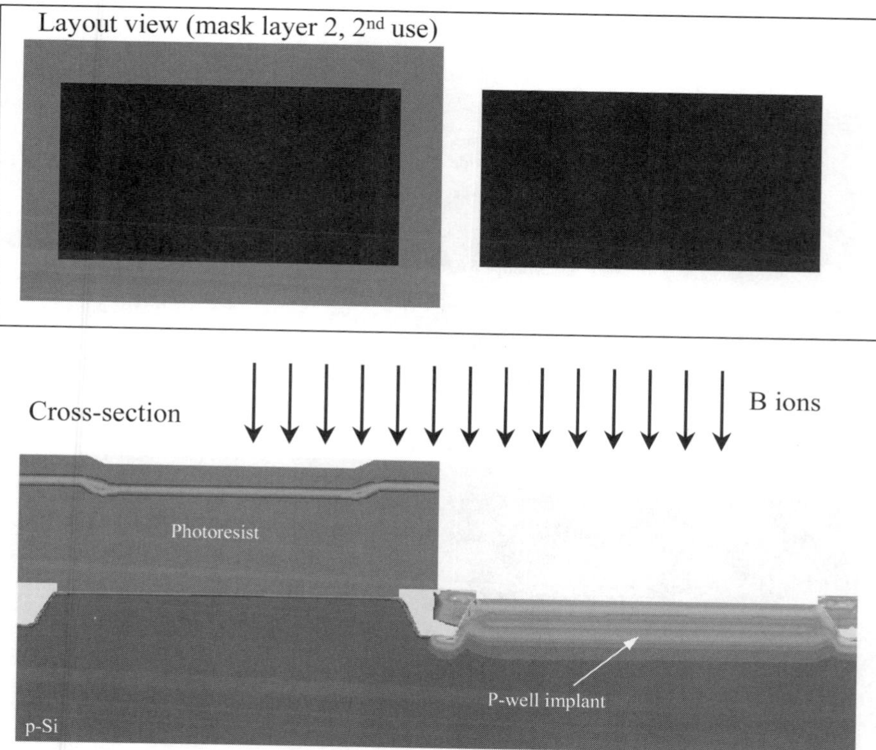

Figure 7.39 P-well formation via photolithography and *B* implantation.

Next, a complementary resist pattern is formed using the n-well mask and photolithography (mask layer 3, second use), as seen in Fig. 7.41. Again, a relatively high energy implant, this time using phosphorus, is performed to generate the n-well. Similar to the p-well, a multitude of implants may be used to achieve the desired retrograde profile. Following the n-well implant, the resist is stripped using O_2 plasma and wet

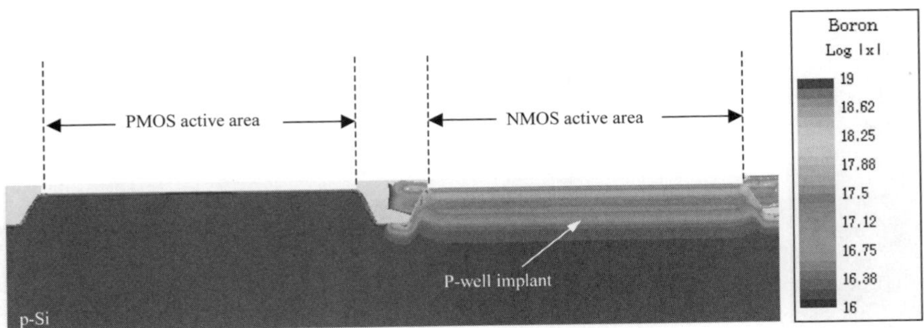

Figure 7.40 Post p-well photoresist strip using O_2 plasma and wet processing.

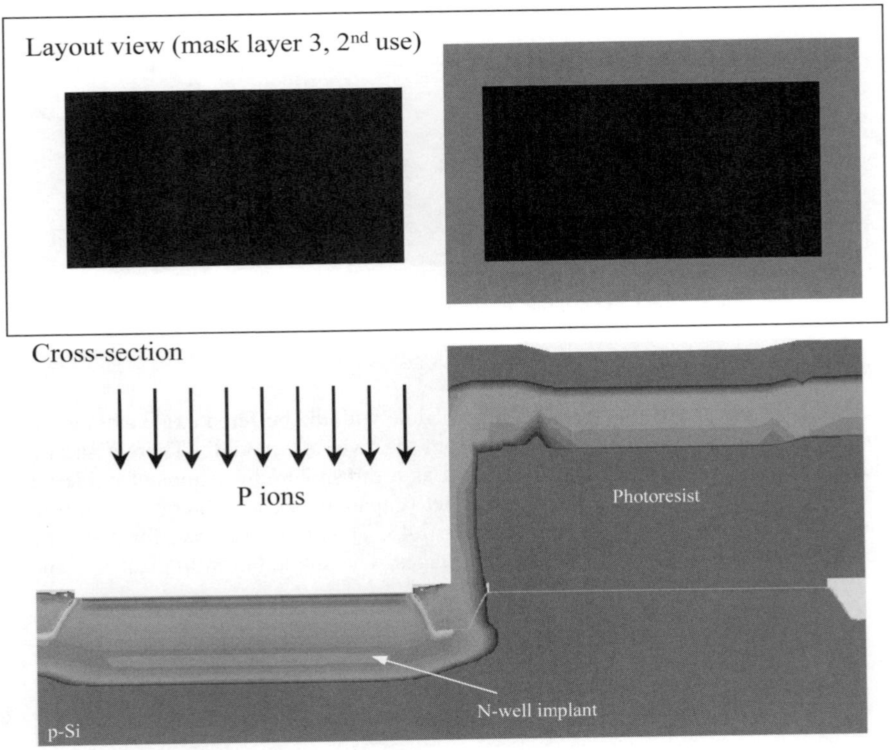

Layout view (mask layer 3, 2nd use)

Cross-section

P ions

Photoresist

N-well implant

p-Si

Figure 7.41 N-well formation via photolithography and *P* implantation.

processing, yielding the structure in Fig. 7.42. At this point, both the isolation and the wells are fully formed. Figure 7.43 shows the cross section of the substrate following the twin-tub module. Notice that the net doping profile is given, thus highlighting both well and wall implants simultaneously. It should be emphasized that the PMOS are fabricated in the n-wells; the NMOS, in the p-wells.

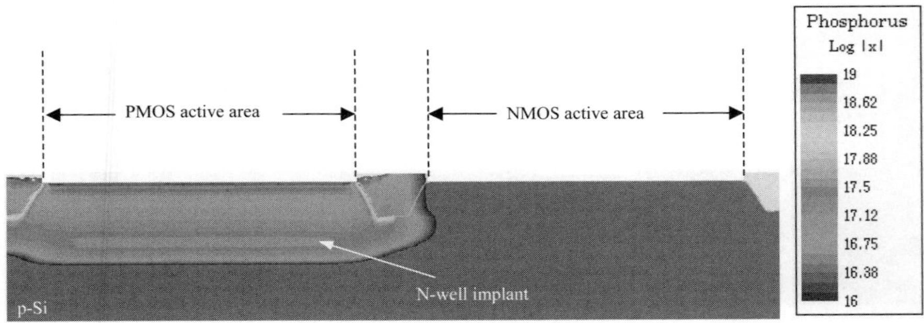

Figure 7.42 Post n-well photoresist strip using O_2 plasma and wet processing.

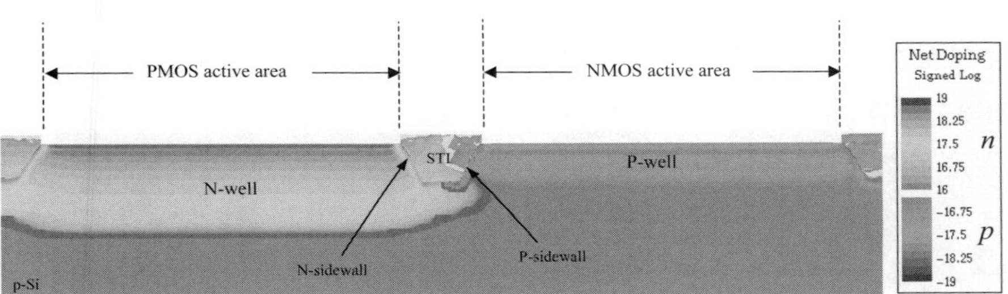

Figure 7.43 Net doping profile of both the n-well and the p-well.

Gate Module

As depicted in Fig. 7.44, we begin the gate module with the buffered oxide etching of the remaining thin oxide in the active areas from the twin-tub module. Then, a sacrificial oxide is thermally grown. This oxide serves as a threshold adjust implant oxide and a pre-gate oxidation "clean-up." Next, a blanket (unpatterned), low energy BF_2 threshold adjust implant is performed, as shown in Fig. 7.45. This implant allows for the "tuning" of both the PMOS and NMOS threshold voltages. The single boron implant is common

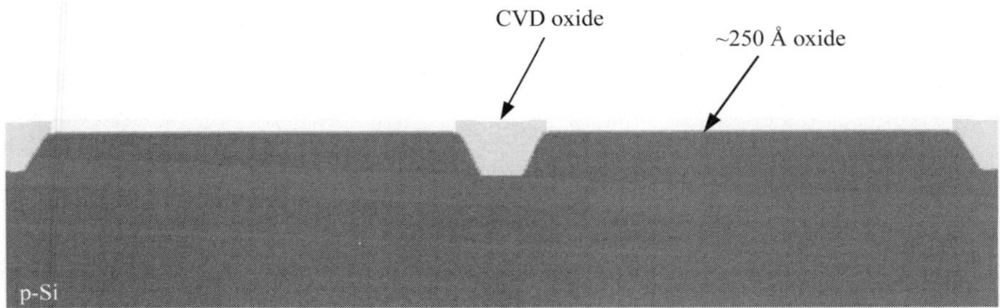

Figure 7.44 Wet etch the remaining trench stack oxide using buffered HF. Sacrificial oxide formation using dry thermal oxidation at approximately 900 °C.

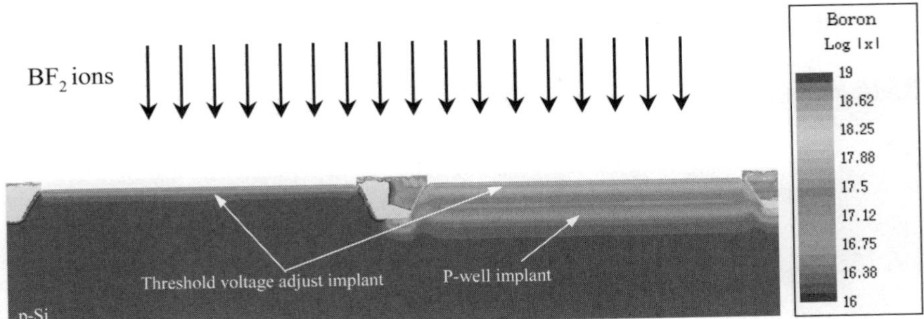

Figure 7.45 Blanket low-energy BF_2 implant for NMOS and PMOS threshold voltage adjust.

for single workfunction gates. However, for dual workfunction gates (common in technologies with minimum gate lengths of 250 nm or less), separate p-type and n-type implants are required for the threshold adjustment in the NMOS and PMOS, respectively.

To form the gate stack (i.e., the gate dielectric and polysilicon gate electrode) the next set of processes are required. Of course, the gate stack provides for the capacitive coupling to the channel. Using wet processing, the sacrificial oxide is stripped from the active areas. As shown in Fig. 7.46, a high quality, thin oxide is thermally grown, which serves as the gate dielectric. In modern CMOS, it is common to use nitrided gate oxide by performing the oxidation in O_2 and NO or N_2O. It can be argued that the gate oxidation is the most critical step in the entire process sequence, as the characteristic of the resultant film greatly determines the behavior of the CMOS transistors. The gate oxidation is immediately followed by an LPCVD polysilicon deposition, as depicted in Fig. 7.47. For single workfunction gates, the polysilicon can be doped with phosphorous during poly deposition or subsequently implanted. For dual workfunction gates, the NMOS and PMOS can be doped during the n+ and p+ source/drain implants, respectively.

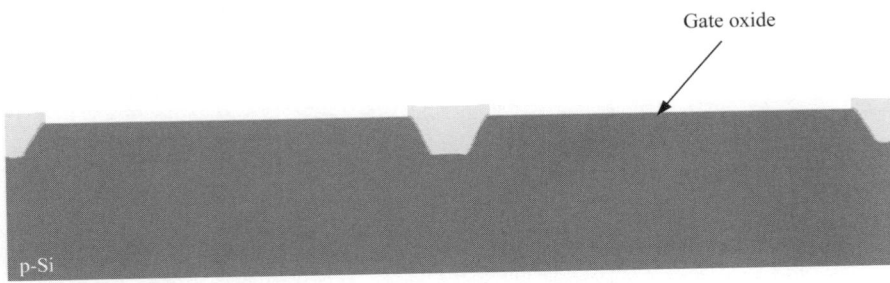

Figure 7.46 Removal of sacrificial oxide using buffered HF followed by gate dielectric formation using dry oxidation in an ambient of O_2 , NO and/or N_2O.

Once the gate stack is formed, the transistor gates and local interconnects are patterned using photolithography (mask layer 4) to generate the appropriate patterns in photoresist, as seen in Fig. 7.48. The gate patterning must be precisely controlled as it

Figure 7.47 Polysilicon deposition via LPCVD at approximately 550 °C. Note that the polysilicon deposition must occur immediately following gate oxidation.

determines the gate lengths. Deviations in the resultant physical gate lengths can cause severe performance issues with the CMOS. Seen in Fig. 7.48 are the ideal gate profiles following the RIE of polysilicon and subsequent resist strip.

The gate module concludes with the poly reoxidation as shown in Fig. 7.49. Here the thermal oxidation of the polysilicon and active silicon is performed to (1) grow a buffer pad oxide for the subsequent nitride spacer deposition and (2) electrically activate the implanted dopants in the polysilicon. Notice that since the polysilicon oxidizes at a faster rate than the crystalline silicon, the resultant oxide thickness is greater on the polysilicon than the active silicon.

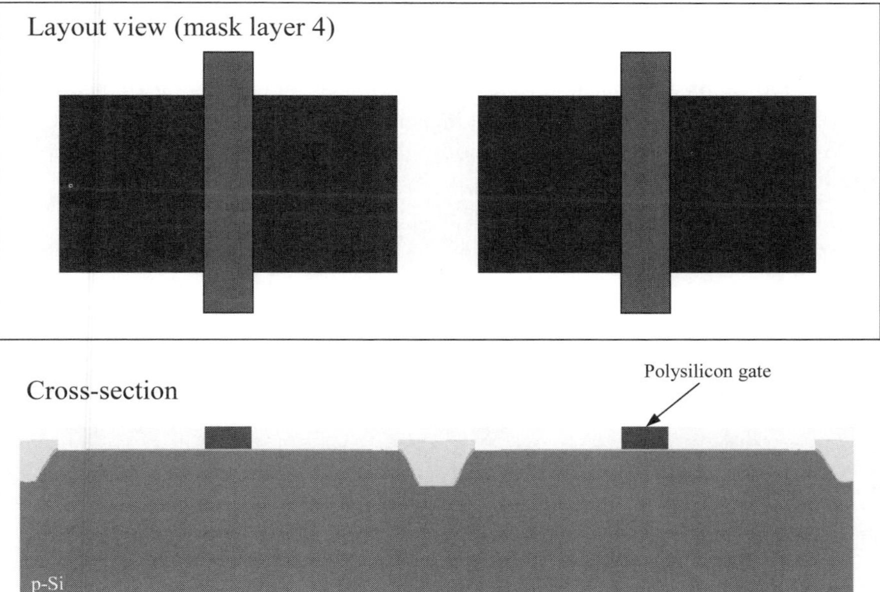

Figure 7.48 Gate electrode and local interconnect photolithography and polysilicon reactive ion etching.

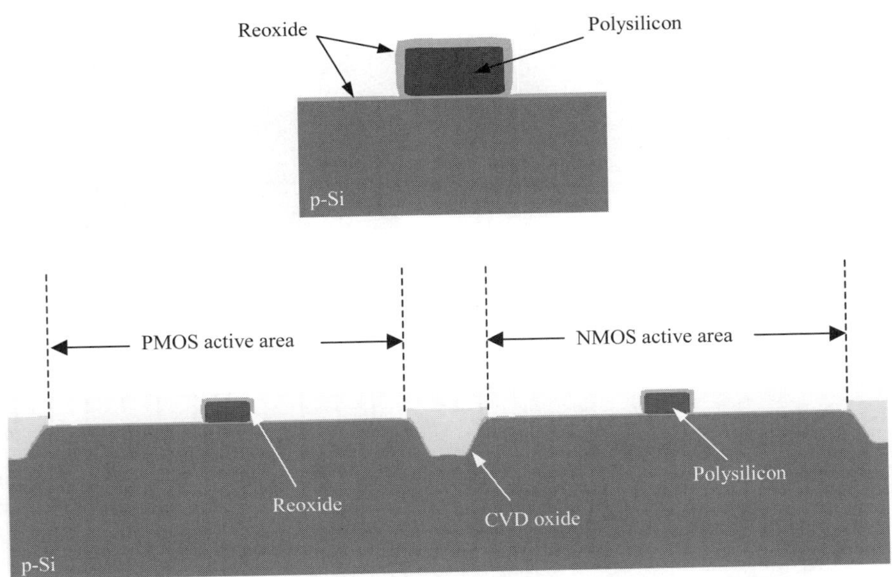

Figure 7.49 Polysilcion reoxidation using dry O_2 at approximately 900 °C. Notice that the resulting oxide is thicker on the polysilicon than on the active silicon.

Source/Drain Module

At the onset of the source/drain module, the source/drain extensions are formed by a series of processes. Photolithography (mask layer 5) is used to pattern resist such that the NMOS devices are exposed, as shown in Fig. 7.50. Then, a low energy phosphorus implant is performed to form the n-channel, low doped drain (nLDD) extensions. Notice that the presence of the polysilicon gate inherently leads to the self-alignment of the extensions with respect to the gate. The nLDD suppresses hot carrier injection into the gate and reduces short-channel effects in the NMOS. At this point in the process sequence, a deep boron pocket implant is often used to prevent source/drain punchthrough in the NMOS. The photoresist is stripped. The resultant structure is shown in Fig. 7.51.

In a similar manner, the p-channel source/drain extensions are formed. Photolithography (mask layer 6) is used to protect NMOS devices with resist, as shown in Fig. 7.52. Boron is implanted at low energy to form the p-channel low doped drain (pLDD) extensions. Again, the polysilicon serves to self-align the implant with respect to the gate electrodes. As was the case with the NMOS, it is common to use a deep phosphorus pocket implant to suppress PMOS punchthrough. Once the photoresist is stripped, the cross-section shown in Fig. 7.53 is achieved.

To complete the source/drain extensions, the gate sidewall spacers must be formed prior to the actual source/drain implants. As shown in Fig. 7.54, conformal silicon nitride is deposited using LPCVD. A CVD oxide may be used in lieu of the nitride. Following the subsequent nitride etch, this nitride will form the LDD sidewall spacers.

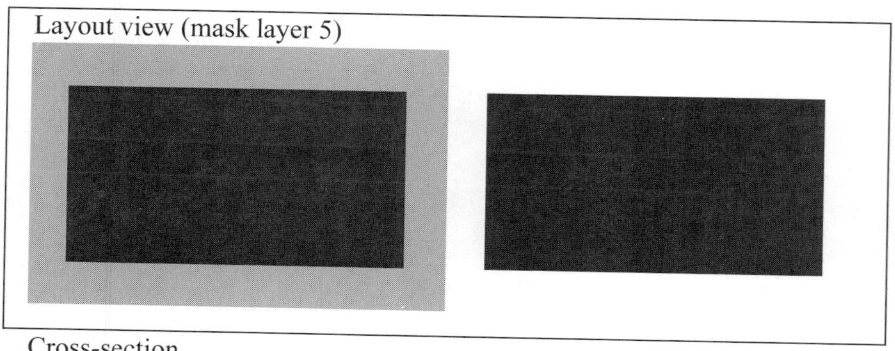

Cross-section

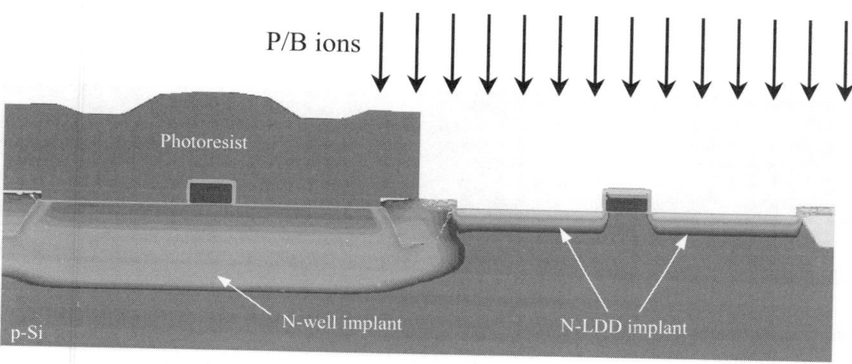

Figure 7.50 N-LDD/n-pocket formation using low energy implantation of P and B, respectively.

The spacers function as (1) a mask to the source/drain implants and (2) a barrier to the subsequent salicide formation. The actual spacers are formed by an unpatterned anisotropic RIE of nitride, as illustrated in Fig. 7.55. The spacer etch is end-pointed on the underlying oxide. Notice that since the nitride is thickest along the polysilicon

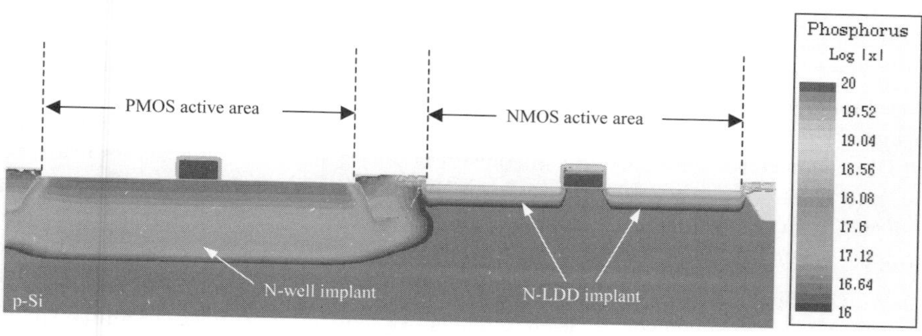

Figure 7.51 Post-n-LDD resist strip using O_2 plasma and wet processing.

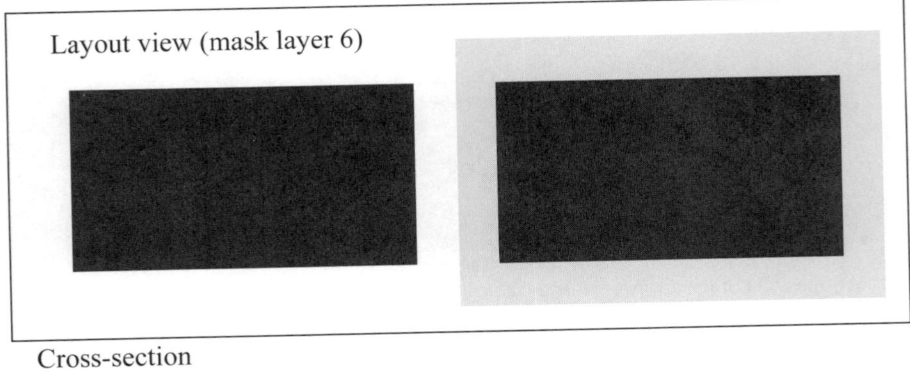

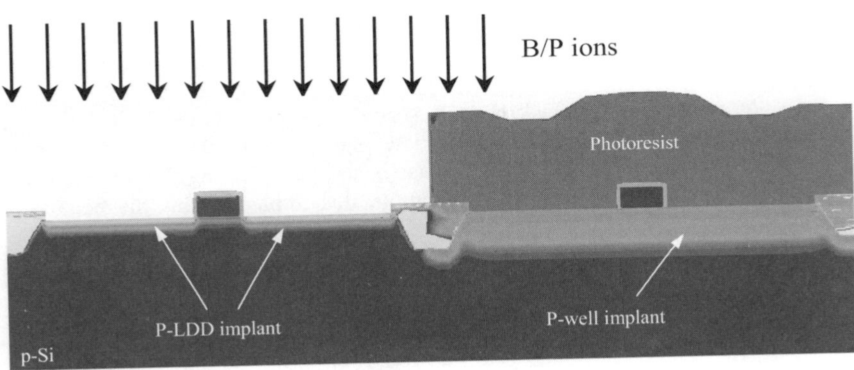

Figure 7.52 P-LDD/p-pocket formation using low energy implantation of B and P, respectively.

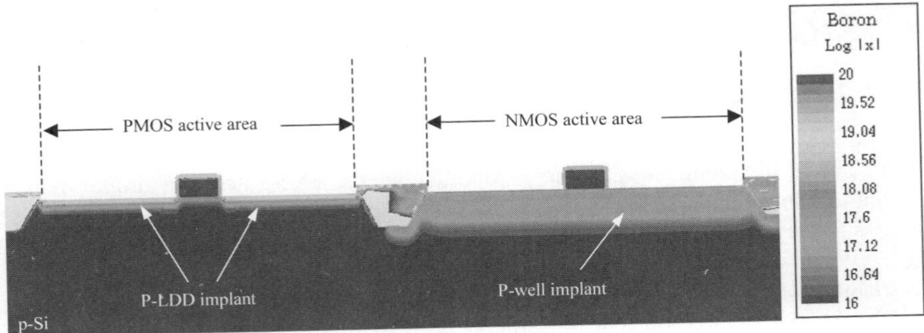

Figure 7.53 Post-p-LDD resist strip using O_2 plasma and wet processing.

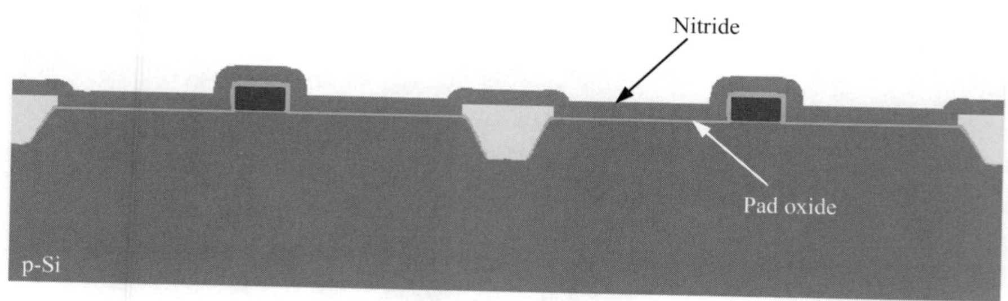

Figure 7.54 Sidewall spacer nitride deposition using LPCVD at approximately 800 °C.

sidewall, a well-formed insulating region remains on both sides of the polysilicon. This structure is called a spacer.

During the source/drain implants, the combination of the polysilicon and spacers block the implantation, thus allowing for self-alignment to not only the gate but also to the LDD extensions. With this stated, the NMOS source/drains are formed, as shown in Fig. 7.56. Photolithography (mask layer 5, second use) protects the PMOS with resist while exposing the NMOS. A relatively low energy, high dose arsenic implant is performed to form the n+ regions. The resist is stripped yielding the structure in Fig. 7.57. In addition to the source/drain formation, this implant forms the necessary n+ ohmic contacts.

Figure 7.55 Dry, anisotropic, end-pointed reactive ion etch of spacer nitride yielding gate sidewall spacers.

The PMOS source/drains and p+ ohmic contacts are formed in a similar manner. Photolithography (mask layer 6, second use) and a low energy, high dose BF_2 implant is used, as illustrated in Fig. 7.58. The resist is stripped resulting in the structure shown in Fig. 7.59. The source/drain module concludes with a high temperature anneal that electrically activates the implants and re-crystallizes the damaged silicon. In modern CMOS, the primary reason that polysilicon is chosen as the gate electrode material as opposed to metal is that the poly can withstand the high temperatures required to activate the source/drain implants.

At this point in our CMOS process sequence we have fully formed the CMOS transistors and their isolation. This marks the completion of the FEOL. Figure 7.60 provides a summary of the main features generated in the FEOL.

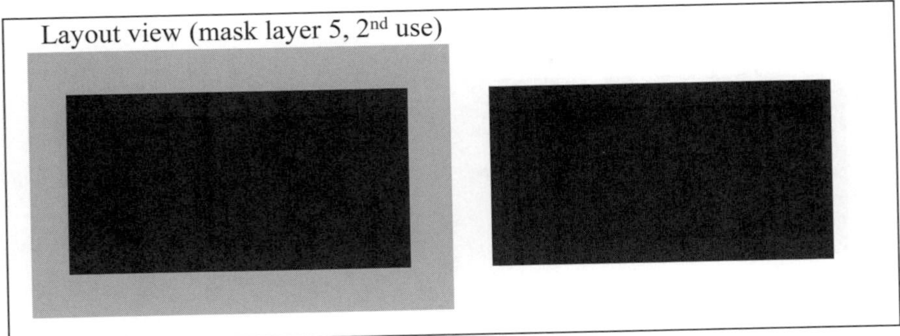

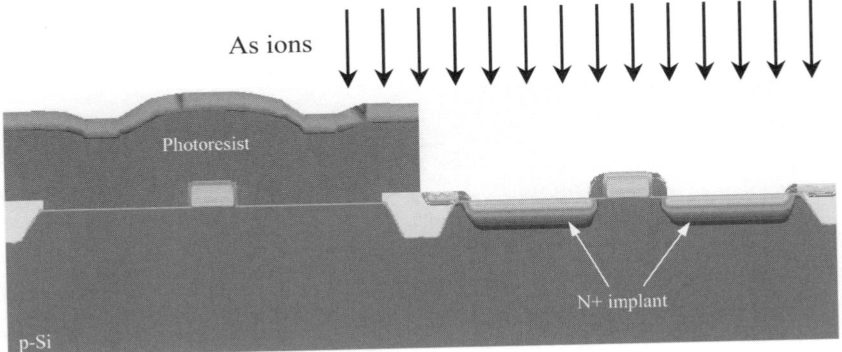

Figure 7.56 N+ source/drain formation using a low energy, high dose implantation of *As*.

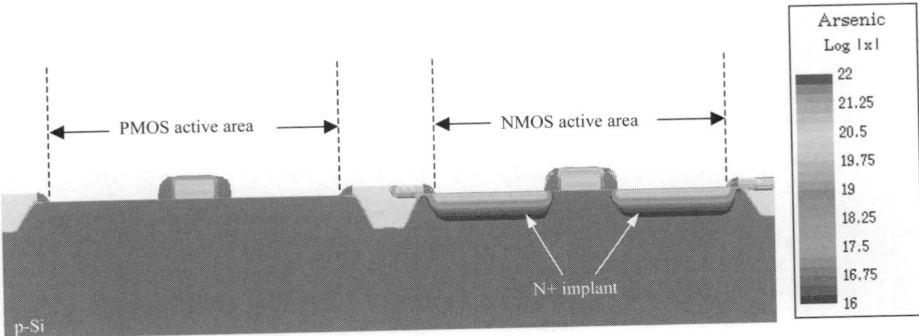

Figure 7.57 Post-n+ resist strip using O_2 plasma and wet processing.

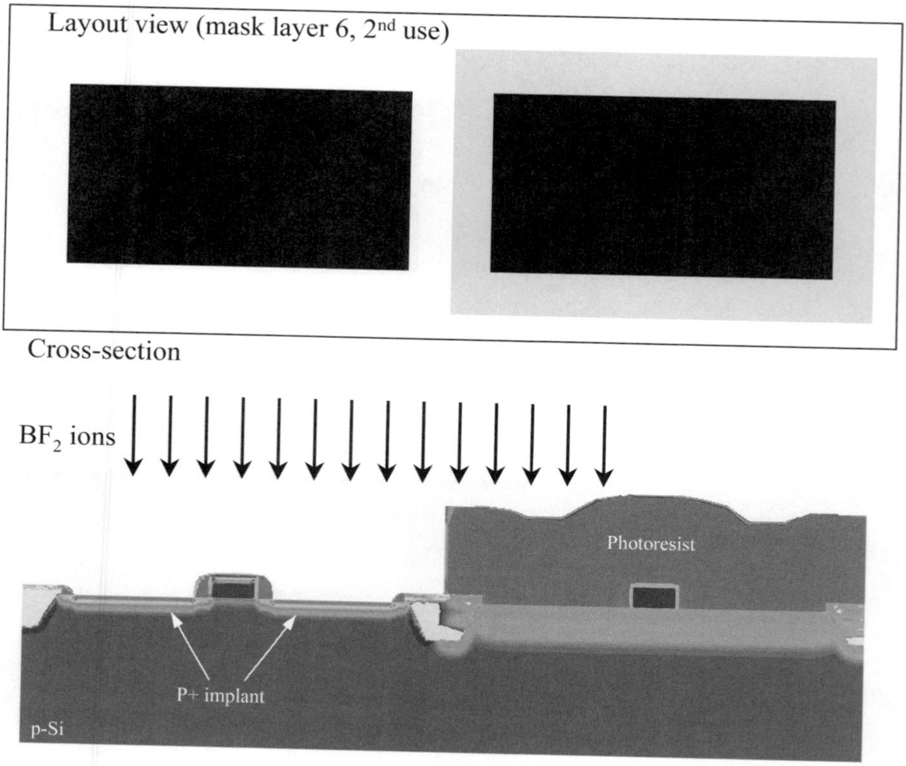

Layout view (mask layer 6, 2nd use)
Cross-section
BF2 ions
Photoresist
P+ implant
p-Si

Figure 7.58 P+ source/drain formation using a low energy, high dose implantation of BF_2.

PMOS active area
NMOS active area
P+ implant
p-Si
Boron
Log |x|
21
20.2
19.6
19
18.4
17.8
17.2
16.6
16

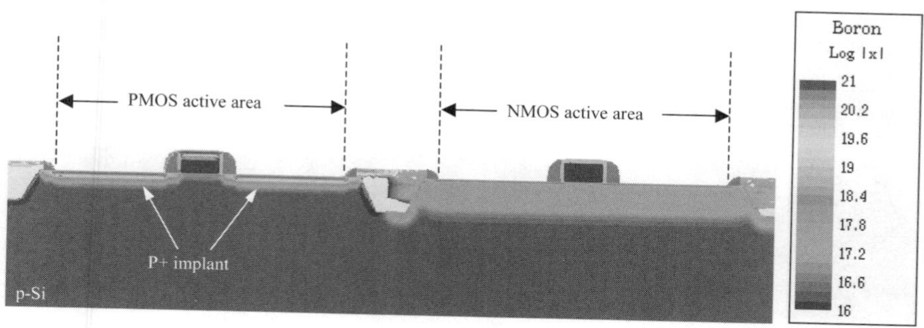

Figure 7.59 Post p+ resist strip using O_2 plasma and wet processing.

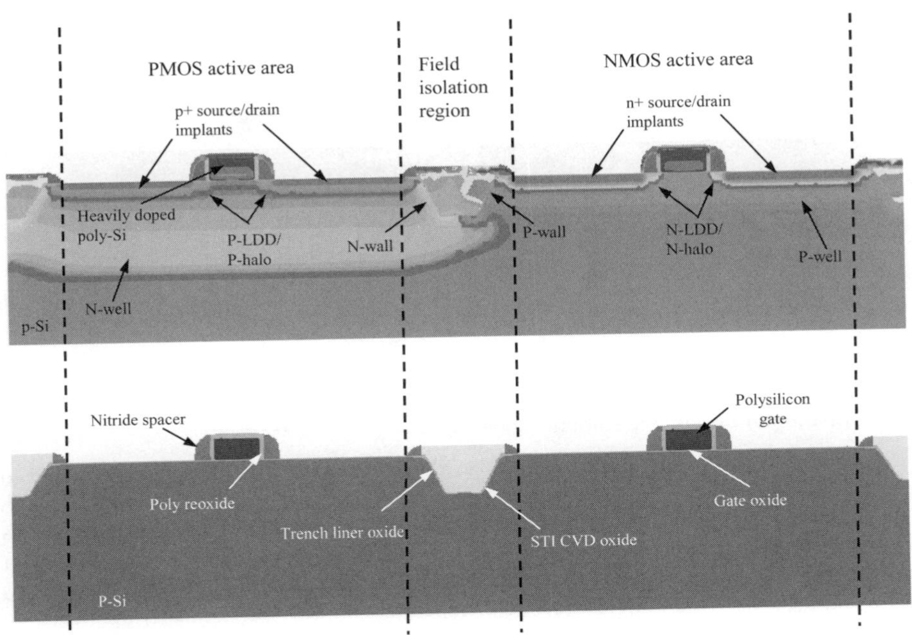

Figure 7.60 Summary of the FEOL features.

7.2.2 Backend-of-the-Line Integration

In this section we continue our CMOS process flow through the BEOL. The BEOL encompasses all processes required to "wire" the transistors to one another and to the bond pads. CMOS requires several metal layers to achieve the interconnects necessary for modern designs. We discuss the processing through the first two metal layers to give an appreciation of the overall BEOL integration.

Self-Aligned Silicide (Salicide) Module

At the boundary of the FEOL and BEOL is the self-aligned silicide, called salicide, formation. Silicide lowers the sheet resistance of the polysilicon and active silicon regions. The self-aligned silicide (salicide) relies on the fact that metal silicide will generally not form over dielectric materials such as silicon nitride. Therefore, a metal such as titanium or cobalt can be deposited over the entire surface of the wafer, then annealed to selectively form silicide over exposed polysilicon and silicon. Because of the presence of the trench fill and sidewall spacers, the silicide become self-aligned without the need for photopatterning.

The salicide module begins with the removal of the thin oxide, present from the FEOL, using buffered HF, as shown in Fig. 7.61. Next, a refractory metal (e.g., titanium or cobalt) is deposited by sputtering, as depicted in Fig. 7.62. To minimize contamination, a thin layer of TiN is deposited as a cap. A relatively low temperature, nitrogen ambient, rapid thermal annealing (RTA) is used to react titanium (or cobalt) with the silicon, forming $TiSi_2$ (C49 phase) or ($CoSi_x$). The resultant silicide (i.e., C49) is

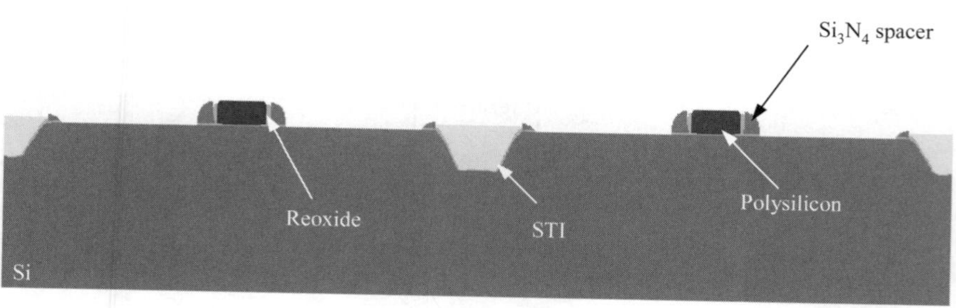

Figure 7.61 Removal of the exposed reoxide (present from the FEOL) using buffered-*HF*.

a high resistivity phase. Also, notice that the underlying nitride and oxide serves to block the formation of the silicide from the sidewalls and trenches, respectively.

To prevent spacer overgrowth of the silicide, the low resistivity phase is achieved by processing with two separate anneals. The first, as described above, forms the high resistivity phase without the risk of silicide formation on the nitride. The second, as described below, occurs following the wet chemical etching of the unreacted titanium (or cobalt) using a higher temperature, which causes a phase change (C49 to C54 for $TiSi_2$) with a much lower resistivity. If one high temperature anneal was originally performed to achieve the low resistivity phase, then significant overgrowth could occur, leading to leakage current from the source and drain to the gate of the transistors.

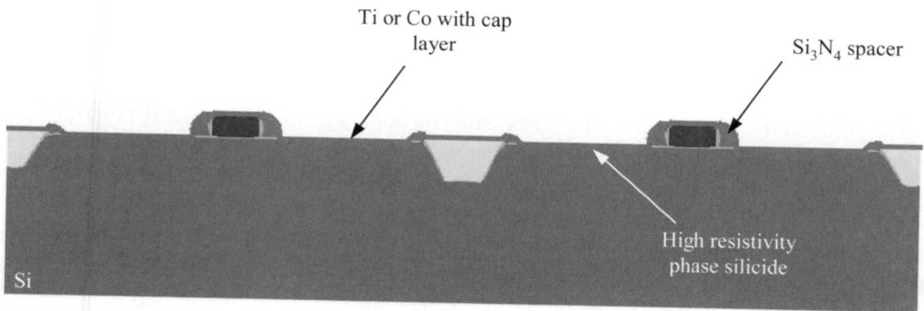

Figure 7.62 Titanium or cobalt deposited by PVD followed by the first salicide rapid thermal anneal.

To continue the salicide module, following the first anneal, the unreacted titanium (or cobalt) is wet chemically etched from the wafer. The second RTA, in argon ambient at a slightly higher temperature, achieves the low resistivity phase, as shown in Fig. 7.63.

Pre-Metal Dielectric

The pre-metal dielectric (PMD) provides electrical isolation between metal1 and polysilicon/silicon. To aid the subsequent contact etch process, a thin layer of silicon nitride is deposited as an etch stop. This is followed by a high density plasma deposition of the PMD oxide, as shown in Fig. 7.64. The resultant surface of the PMD must be

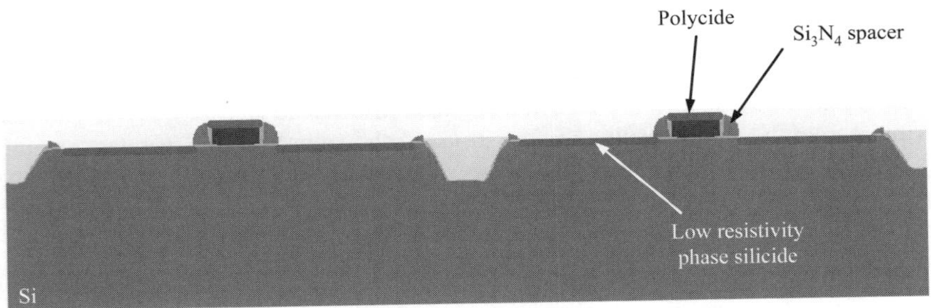

Figure 7.63 Wet chemical etch of the unreacted titanium or cobalt followed by the second salicide rapid thermal anneal.

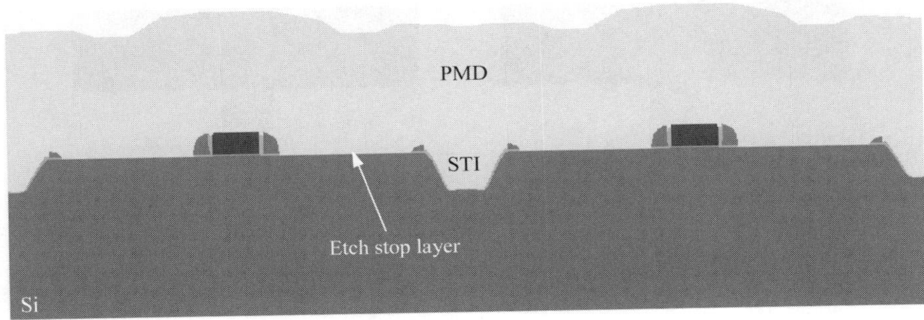

Figure 7.64 Pre-metal dielectric (PMD) deposition using high density plasma

planarized to allow for improved depth-of-focus for the subsequent high resolution photopatterning of metal1. With CMP, the PMD is planarized, producing the cross-section shown in Fig. 7.65.

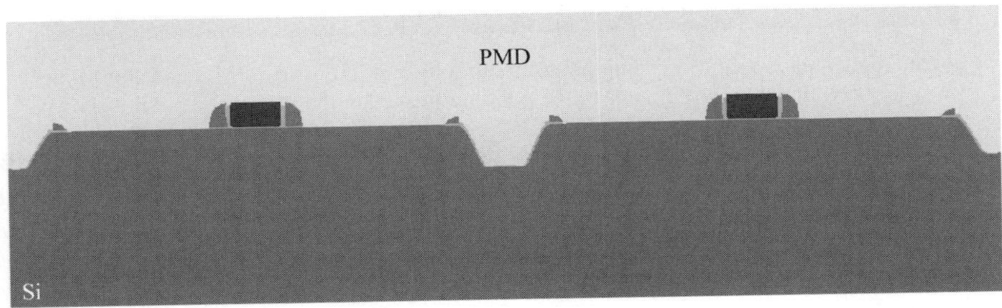

Figure 7.65 Planarizing of the PMD using CMP.

Contact Module

The contacts provide the electrical coupling between metal1 and polysilicon/silicon. The first BEOL photolithography (mask layer 7) step patterns contact openings in the resist. The PMD and nitride is then dry etched using the nitride as an etch-stop layer. The resist is stripped from the wafer, resulting in the structure shown in Fig. 7.66. Contact openings to the source and drains are shown; however, contacts to polysilicon (not shown) over field oxide are simultaneously formed.

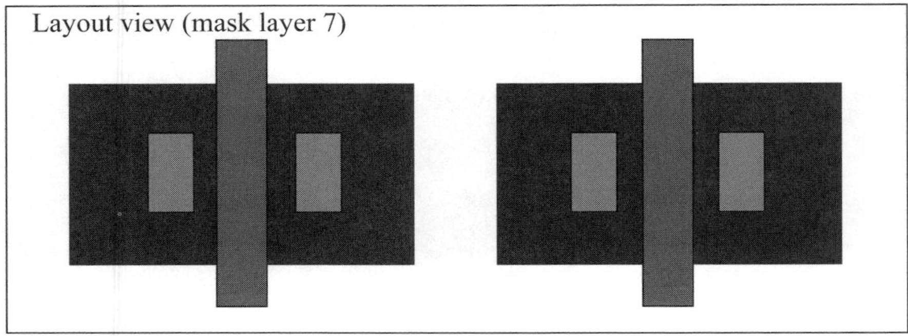

Cross-section

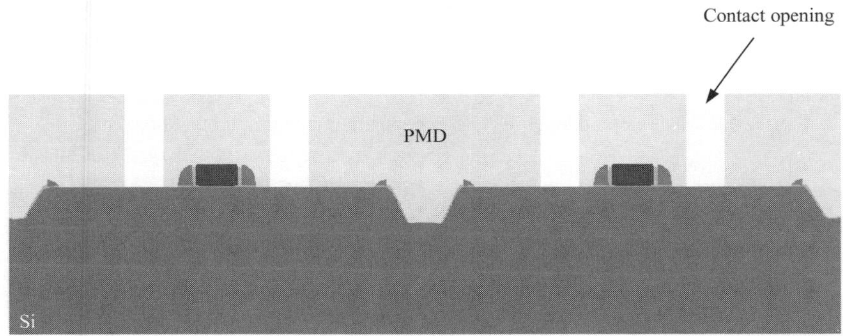

Figure 7.66 Contact definition using photolithography and RIE of the PMD and stop layer. Notice that the contacts to the poly are formed at the same time but not shown.

Next, a thin layer of titanium is deposited by ionized metal plasma (IMP) sputtering, preceded by an in-situ argon sputter etch to clean the bottom of the high aspect ratio contact openings. The titanium is the first component of the contact liner and functions to chemically reduce oxides at the bottom of the contacts. The second component of the liner is a thin CVD *TiN*. The primary purpose of this layer is to act as a diffusion barrier to the fluorine (which readily etches silicon) used in the subsequent tungsten deposition. The contact openings are filled (actually overfilled) with tungsten using a WF_6 CVD process, as shown in Fig. 7.67 As depicted in Fig. 7.68, the overfilled tungsten is polished back to the top of the planarized PMD using CMP. At this point, the surface of the wafer is ultra-smooth and essentially free of all topography. To aid the photolithographic alignment of the metal layer to the contacts, a tungsten recess etch is

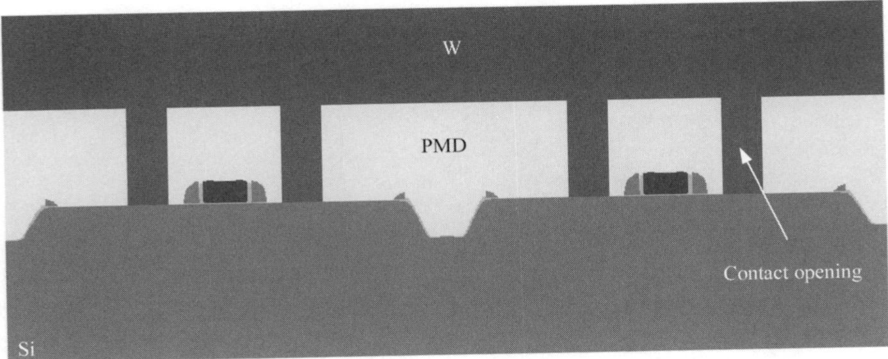

Figure 7.67 *Ti/TiN* liner deposition using IMP and CVD, respectively. *W* contact fill deposition using WF_6 CVD.

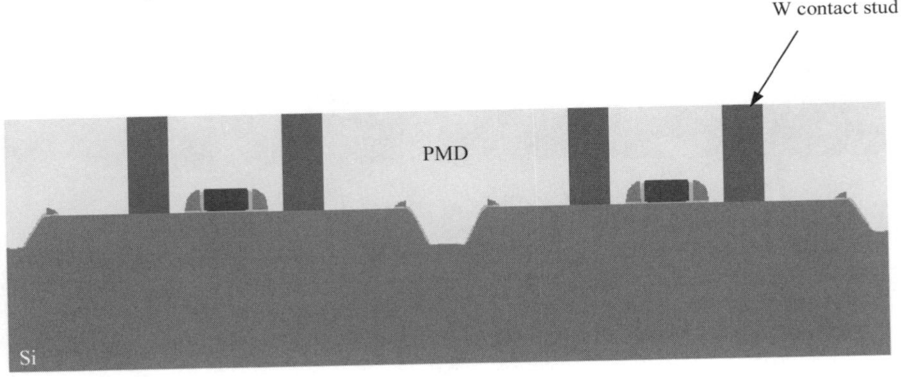

Figure 7.68 *W* CMP to form defined contacts.

performed to provide adequate alignment reference. The recessed contacts are shown in Fig. 7.69.

Metallization 1

To allow for electrical signal transmission from contact-to-contact and from contact-to-via1, defined metallization must be formed. Following the recess etch, sputtering is used to deposit a film stack consisting of *Ti/TiN/Al/TiN*, as seen in Fig. 7.70. The Ti provides adhesion of the TiN and reduction in electromigration problems. The bottom TiN serves primarily as a diffusion barrier to $TiAl_3$ formation. The topmost TiN acts as an anti- reflective coating for the metal photolithography as well as an etch stop for the subsequent via formation.

Using photolithography (mask layer 8), the metal 1 pattern in resist is generated, as depicted in Fig. 7.71. A dry metal etch transfers the pattern into the metal. To prevent metal corrosion, the resist is plasma stripped in an $O_2/N_2/H_2O$ ambient, producing the cross-section shown in Fig. 7.71.

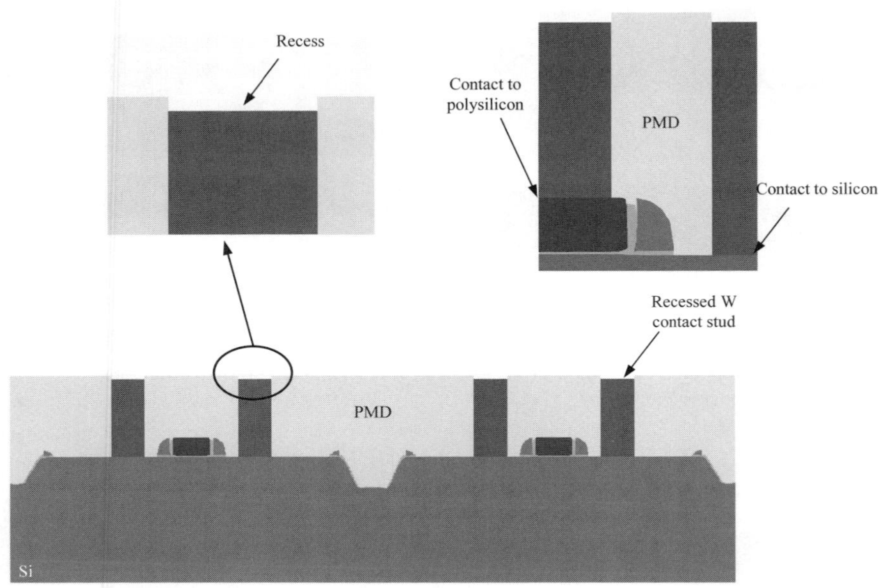

Figure 7.69 *W* recess etch using "buff" polish or dry *W* etch.

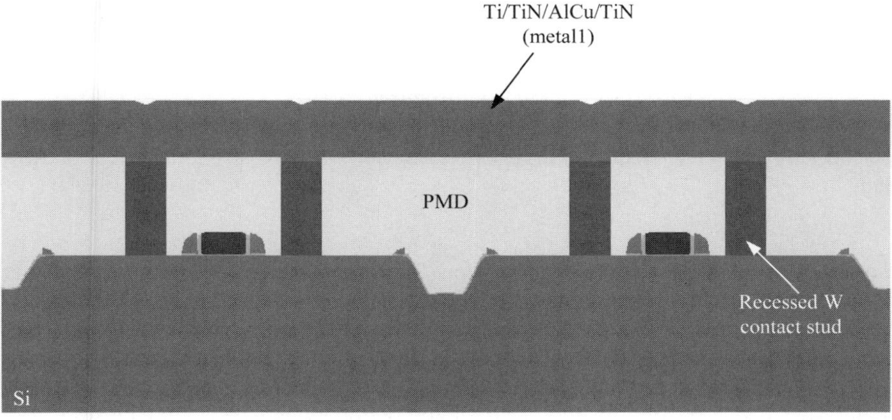

Figure 7.70 Metal1 stack deposition using PVD.

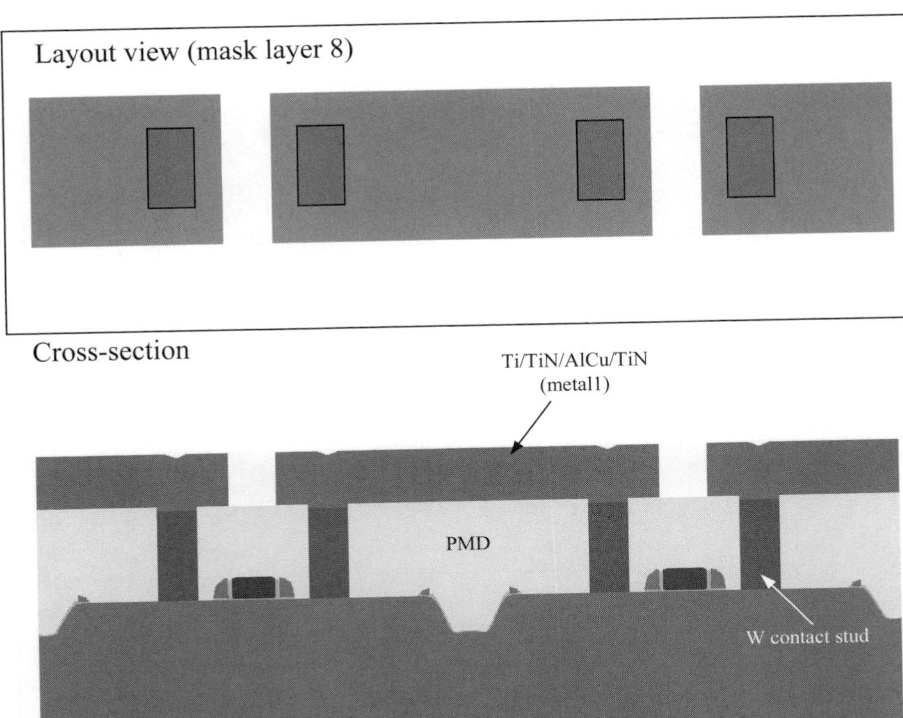

Figure 7.71 Metal1 definition using photolithography and dry metal etch.

Intra-Metal Dielectric 1 Deposition

The intra-metal dielectric 1 (IMD1) provides the electrical isolation between metal1 and metal2. It is common to deposit this film using high density plasma CVD, as shown in Fig. 7.72. As can be seen, the conformal deposition results in a surface topography that must be planarized by CMP. The depth-of-focus is improved for the subsequent photolithographic steps as it is for the PMD planarization.

Via 1 Module

Electrical coupling between metal1 and metal2 is achieved by the via1 module. The planarized IMD is photolithographically defined (mask layer 9), and an RIE opens the vias. The resist is stripped using O_2 plasma and wet processing, producing the cross-section shown in Fig. 7.73. Next, similar to the contact fill, an argon sputter etch is performed followed by the deposition of thin layers of IMP titanium and CVD TiN. Using a WF_6 CVD process, the vias are deposited (overfilled), as seen in Fig. 7.74. The excess tungsten is removed by CMP utilizing the IMD1 as a "polish stop," as depicted in Fig. 7.75. Again, as to provide observable alignment features, a tungsten recess etch is often required (also shown in Fig. 7.75).

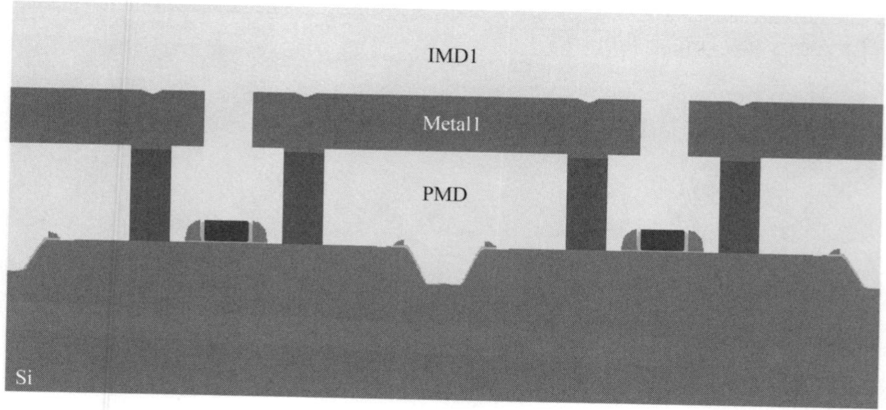

Figure 7.72 Intra-metal dielectric 1 (IMD1) deposition using HDP CVD. This is followed by IMD1 planarization using CMP.

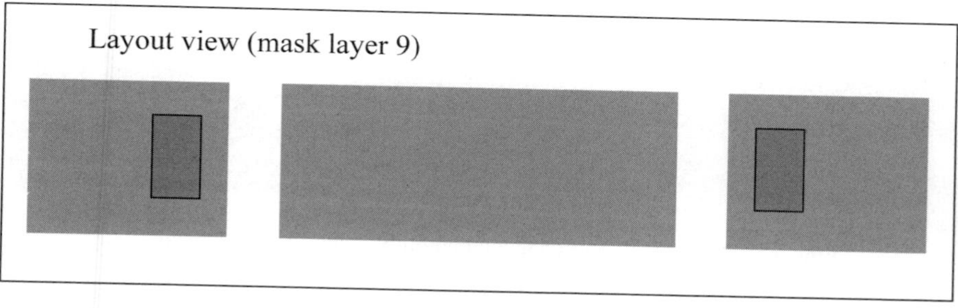

Cross-section

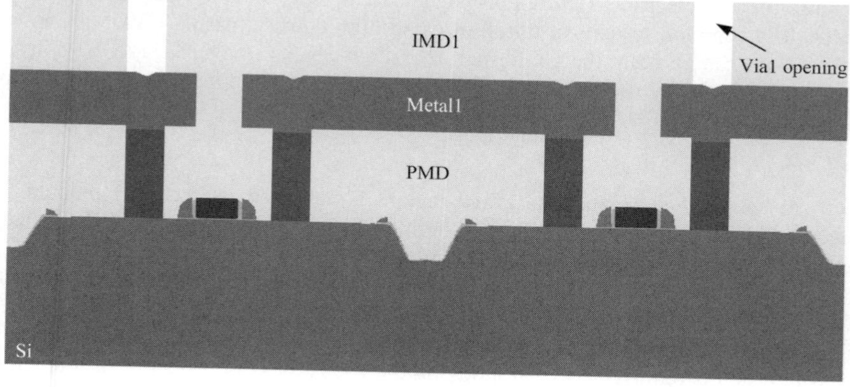

Figure 7.73 Via1 definition using photolithography and dry IMD1 etch.

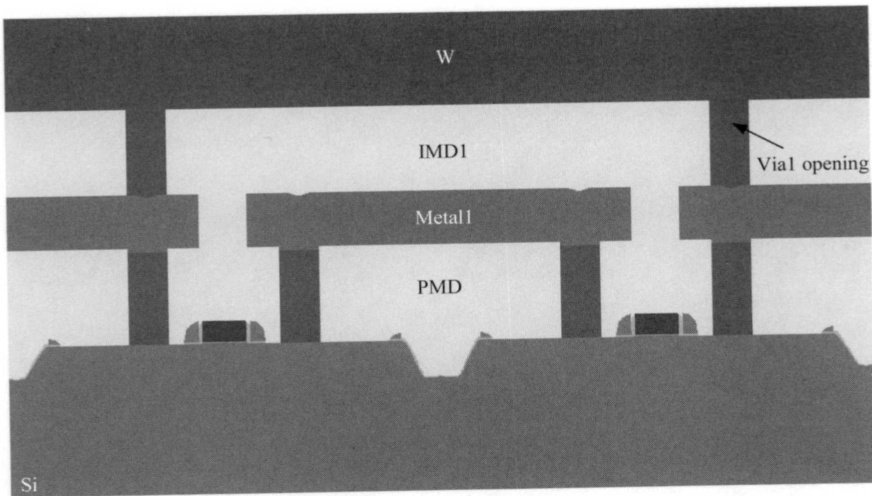

Figure 7.74 *Ti/TiN* liner deposition using IMP and CVD, respectively. *W* via1 fill deposition using WF_6 CVD.

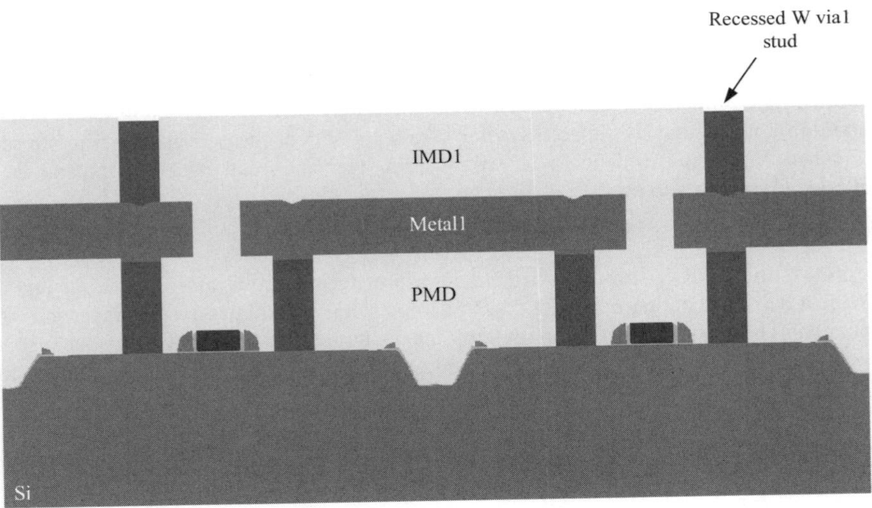

Figure 7.75 *W* CMP to form defined vias. This is followed by *W* recess etch using "buff" polish or dry *W* etch.

Metallization 2

In a similar manner as the metal1 process, the metal 2 stack is deposited (Fig. 7.76) and photolithographically (mask layer 10) defined, as shown in Fig. 7.77. Note that we are not discussing metal implementation using copper. Copper wiring is often implemented with dual-Damascene techniques, see Ch. 4, where both vias and metal layers are simultaneously formed in a series of process steps.

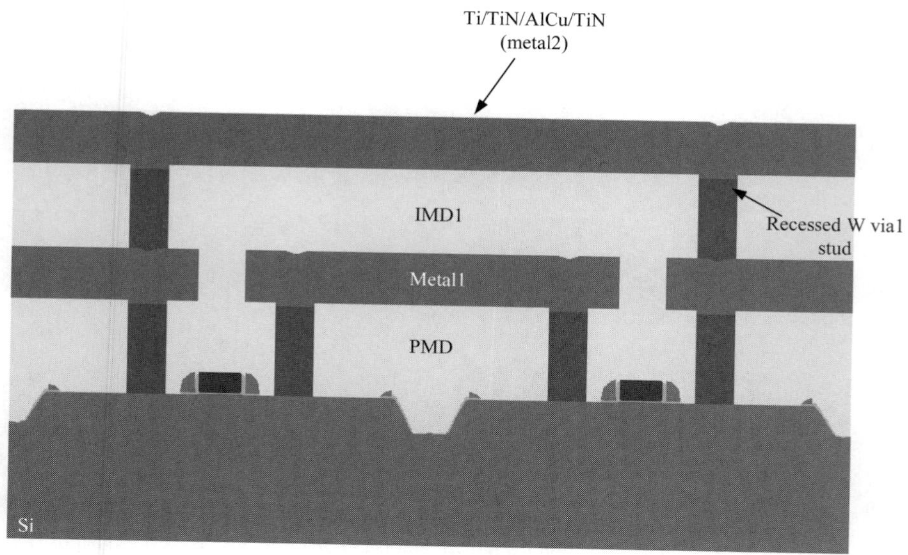

Figure 7.76 Metal2 stack deposition using PVD.

Additional Metal/Dielectric Layers

At this point, additional tiers of dielectric/metal layers can be formed by replicating the aforementioned processes. In modern CMOS there may be more than eight metal layers. It should be noted that as dielectric/metal layers are added, the cumulative film stresses can cause significant bow/warp in the wafers. Hence, great effort is expended to minimize the stresses in the BEOL films.

Final Passivation

To protect the CMOS from mechanical abrasion during probe and packaging and to provide a barrier to contaminants (e.g., H_2O, ionic salts), a final passivation layer must be deposited. The passivation type is determined in large part by the type of package in which the CMOS IC will be placed. Common passivation layers are (1) doped glass and (2) silicon nitride on deposited oxide. Figure 7.78 shows the cross-section of our CMOS process flow following the deposition of the passivation. Finally, the bond pads are opened using photolithography (mask layer n) and by dry etching the passivation. Following photoresist strip, the final CMOS cross-section is shown in Fig. 7.79.

7.3 Backend Processes

Following completion of the final passivation, the wafers are removed from the cleanroom in preparation for a series of backend (i.e., post-fab) processes. These processes include wafer probe, die separation, packaging, and final test/burn-in.

Wafer Probe

Generally, dedicated die with parametric structures and devices are stepped into various positions of the wafer. Alternatively, parametric structures are placed in the dividing

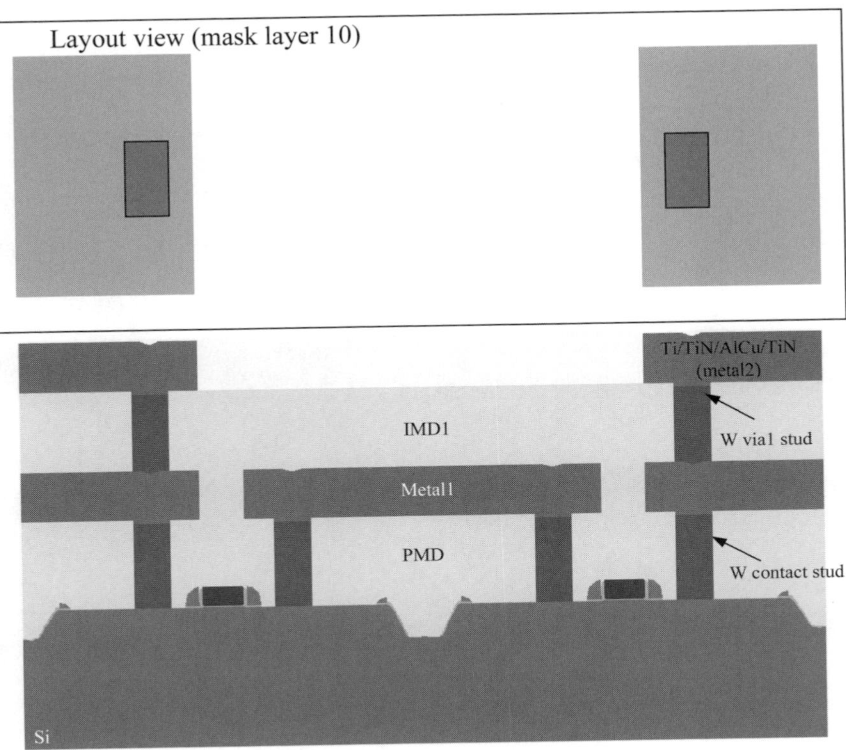

Figure 7.77 Metal2 definition using photolithography and dry metal etch.

regions, called *streets* or *scribe lines*, between die. Electrical characterization of these parametric structures and devices is often performed at select points in the fabrication process flow (such as following metal 1 patterning). Parameters such as contact resistance, sheet resistance, transistor threshold voltage, saturated drain current, off-current, sub-threshold slope, etc., are measured. If problems are observed from the in-line parametric tests, troubleshooting can begin sooner than if the testing were only performed following final passivation. Furthermore, wafers that do not meet the parametric standards can be removed (or "killed") from the fabrication sequence (and thus money is saved).

After the completed wafers are removed from the fab, wafer-level probing is performed to check final device parameters and to check CMOS integrated circuit functionality and performance. Wafer probe is accomplished by using sophisticated testers that can probe individual die (or sets of die) and apply test vectors to determine circuit behavior. Inevitablity, there are a percentage of die that will not pass all the vector sets and thus are considered failed die. The ratio of good die-to-total die represents the wafer *yield* given by

$$Y = \frac{\text{\# of die passing all tests}}{\text{total \# of die}} \tag{7.13}$$

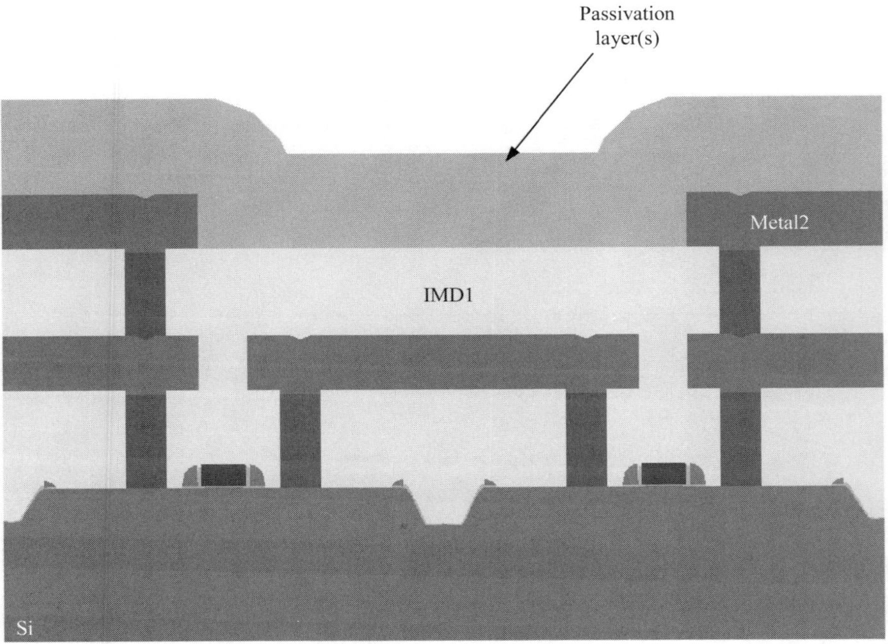

Figure 7.78 Deposition of final passivation.

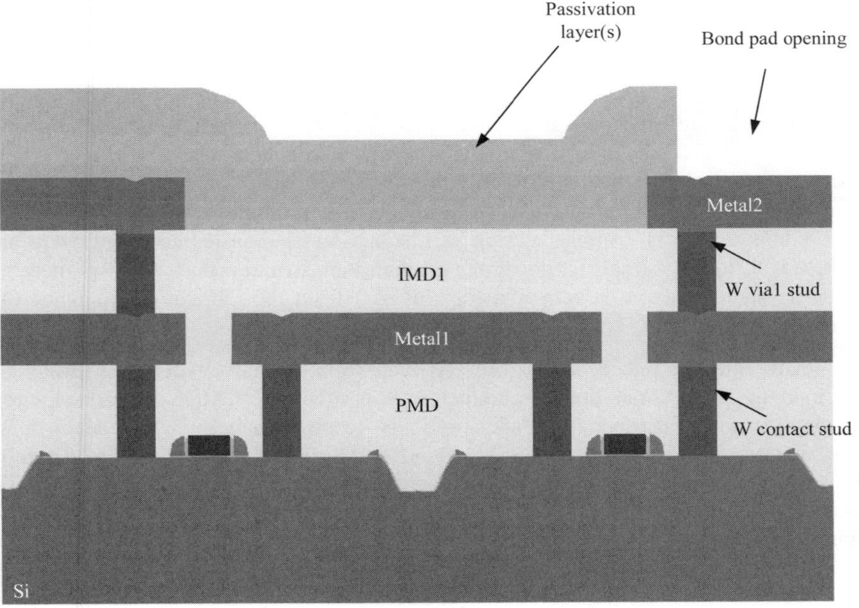

Figure 7.79 Bond pad definition using photolithography and dry etch of passivation.

In these cases, the die are marked, often with an ink dot, to indicate a nonfunctional circuit. Since the processing costs of a wafer are fixed, higher yield equates to higher profit.

Die Separation

Prior to separating the individual die, the backs of the wafers are often thinned using a lapping process similar to the CMP process discussed in Sec. 7.1. This thinning is often required for specific types of packages. Furthermore, thinning can aid in improving the removal of heat from the CMOS circuits.

Next, the die are separated from the wafers using a dicing saw comprised of a diamond-coated blade. The cutting paths are aligned to the *streets* or *scribe lines* on the wafer. Great care is given to minimize damage to the die during this mechanical separation. Along with the inked die, die that are observed to be damaged from the dicing are discarded. Obviously, the separation process can reduce the yield.

Packaging

The good die are now attached to a header in the appropriate package type. The die attach can be accomplished by either eutectic or epoxy attachment. Next, the bond pads are wired to the leads of the package. Common wire bonding techniques include thermocompression, thermosonic, and ultrasonic bonding. At this point, the packaging is completed by a wide range of different processes, greatly dependent on the type of package in which the IC resides. For instance, in plastic dual in-line packages (DIPs), a process similar to injection molding is performed to form a relatively inexpensive package. If ceramic packaging is required, then the attached, wire bonded die will reside in a cavity that is sealed by a metal lid. In general, plastic (or epoxy) packages are inexpensive, but do not provide a hermetic seal. On the contrary, ceramic packages are more costly, but do provide a hermetic seal. For information on other packaging schemes, the reader is referred to the list of additional reading at the end of the chapter. Finally, it should be noted that the packaging process can add to the overall yield loss.

Final Test and Burn-In

Once packaged, the CMOS parts are tested for final functionality and performance. When this is completed, it is common for the parts to go through a *burn-in* step. Here they are operated at extreme temperatures and voltages to weed out infant failures. Additional yield loss can be observed.

7.4 Summary

In this chapter the fundamental unit processes required in the manufacture of CMOS integrated circuits were introduced. These unit processes include thermal oxidation, solid state diffusion, ion implantation, photolithography, wet chemical etching, dry (plasma) etching, chemical mechanical polishing, physical vapor deposition, and chemical vapor deposition. Additionally, a brief overview of substrate preparation was given. With this foundation, a representative, deep-submicron CMOS process flow was provided and the significant issues in both the FEOL and BEOL integration were discussed. Finally, an overview of the backend processes was presented including wafer probe, die separation, packaging, and final test and burn-in.

ADDITIONAL READING

[1] S. A. Campbell, *The Science and Engineering of Micrelectronic Fabrication*, 2nd ed, Oxford University Press, 2001. ISBN 0-19-513605-5

[2] R. C. Jaeger, *Introduction to Microelectronic Fabrication*, 2nd ed, volume 5 of the Modular Series on Solid State Devices, Prentice-Hall Publishers, 2002. ISBN 0-20-144494-1

[3] J. D. Plummer, M. D. Deal, and P. B. Griffin, *Silicon VLSI Technology, Fundamentals, Practice, and Modeling*, Prentice-Hall Publishers, 2000. ISBN 0-13-085037-3

Chapter 8

Electrical Noise: An Overview

The word "noise," in everyday use, is any wanted or unwanted sound that can be heard. Electrical noise has a significantly different meaning. Electrical noise is a current or voltage signal that is *unwanted* in an electrical circuit. Real signals are the sum of this unwanted noise and the desired signals. Types of noise include 1) noise resulting from the discrete and random movement of charge in a wire or device (which we'll call inherent circuit noise, e.g., thermal noise, shot noise, or flicker noise), 2) quantization noise (resulting from the finite digital word size when changing an analog signal into a digital signal), and 3) coupled noise (resulting from the signals of adjacent circuits feeding into each other and interfering). We focus the discussion in this chapter on inherent noise. Quantization and coupled noise are discussed in other parts of the book, e.g., Ch. 28.

Electrical noise is introduced early in our study of circuits in this book because it is often the limiting factor in a system's performance. For example, the distance a radio receiver can pick up a transmitted signal increases as the noise performance of the receiver improves. When studying noise, the units, terminology, and procedures used in a noise analysis can present a barrier to understanding what circuit element limits the noise performance (that is, how the circuit's noise performance can be improved). To obviate this barrier, we'll take an unusual approach: we'll use a few simple signals and measurements to illustrate some of the noise analysis procedures.

8.1 Signals

In this section we briefly review some fundamentals related to electrical signals.

8.1.1 Power and Energy

Figure 8.1 shows a simple sine wave. At the risk of stating the obvious, the average value of this signal is zero. Does this mean that if this signal is applied across a resistor zero average power is dissipated? Of course not (or else many people wouldn't be able to cook food or heat their homes). In order to characterize the *average* amount of power dissipated by a resistor when this sinusoidal signal is applied across it, Fig. 8.2, we can begin by writing the instantaneous power dissipated by the resistor (the power dissipated by the resistor at a specific time t) as

$$P_{inst}(t) = \frac{v^2(t)}{R} = \frac{V_P^2 \sin^2(2\pi \cdot \frac{t}{T})}{R} \tag{8.1}$$

Summing this power over time results in

$$Energy = \int_0^\infty \frac{V_P^2 \sin^2(2\pi \cdot \frac{t}{T})}{R} \cdot dt \tag{8.2}$$

with units of $W \cdot s$ or Joules (energy). Clearly, summing the instantaneous power doesn't result in the average power dissipated by the resistor but rather the energy the sine wave source supplies to the resistor. Because we didn't place an upper bound on the integration in Eq. (8.2), the energy supplied tends to infinity. (A resistor dissipating power over an infinite period of time is being driven by a source with an infinite amount of energy.)

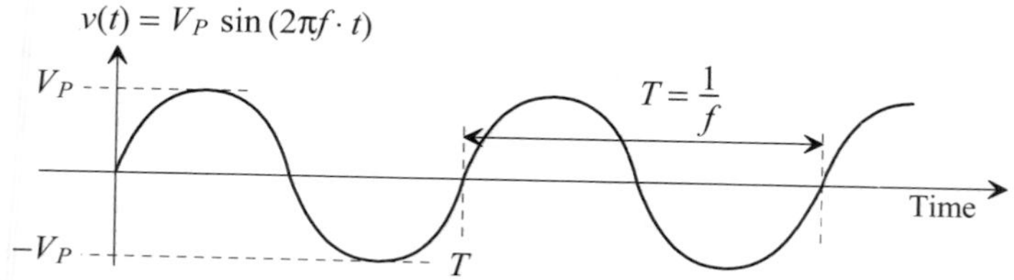

Figure 8.1 A sine wave.

Examining Fig. 8.1 for a moment, we realize that because the sine wave is periodic, we need only estimate the average power over one cycle (say from time zero to T) to estimate the average power supplied to a resistor by the entire waveform. We can do this by summing the instantaneous power for one cycle (which results in the energy supplied by the sine wave source during this cycle) and then dividing by the cycle time T. This can be written as

$$\text{Average power dissipated} = \frac{1}{T}\int_0^T \frac{V_P^2 \sin^2(2\pi \cdot \frac{t}{T})}{R} \cdot dt \tag{8.3}$$

noting that the resistor's value can be moved outside of the integration. This result, with units of watts, is the mean (average) squared voltage of the sine wave signal dropped across the resistor divided by the resistor's value. For the sine wave in Figs. 8.1 and 8.2, we can write

$$\text{Mean-squared voltage} = \overline{v^2} = \frac{1}{T}\int_0^T V_P^2 \sin^2(2\pi \cdot \frac{t}{T}) \cdot dt = \frac{V_P^2}{2} \tag{8.4}$$

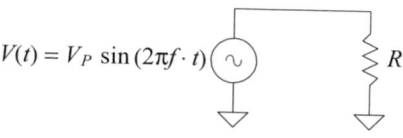

$V(t) = V_P \sin(2\pi f \cdot t)$ R

Figure 8.2 A sine wave driving a resistor.

with units of volts squared. Equation (8.3) simplifies to the mean-squared voltage divided by R, i.e., $\overline{v^2}/R$, or, for a sinusoid $V_P^2/2R$. This should be compared to the power dissipated by a resistor with a DC voltage across it, i.e., V_{DC}^2/R. When $V_P^2/2 = V_{DC}^2$, the power dissipation is the same. If we take the square root of the mean-squared voltage (the result is called the root-mean-squared, RMS, voltage) or

$$\text{RMS voltage of a sine wave, } \sqrt{\overline{v^2}} = V_{RMS} = \sqrt{\frac{1}{T}\int_0^T V_P^2 \sin^2(2\pi f\cdot t)\cdot dt} = \frac{V_P}{\sqrt{2}} \quad (8.5)$$

then we can say that when $V_{RMS} = V_{DC}$, the power dissipated by the resistor with either a sinusoidal or DC source is the same.

Comments

Power is heat. If a circuit is battery-powered, then the less heat it generates the longer the battery life. Batteries are often characterized by their capacity to supply power (the potential energy they hold with units of joules, watthours, or if the battery voltage is known, mA·hours). A battery can also be characterized by its ability to supply a given amount of power in a period of time (kinetic energy).

For example, two 1.5 V batteries may be rated with the same capacity, i.e., potential energies (say 1 W·hr, 3600 J, or 667 mA·hr) but one may only be capable of supplying a maximum current of 1 mA, while the other may supply a maximum of 10 mA. In a one-hour period, and at maximum current draw, the first battery can supply a power of 1.5 mW to a circuit, while the second battery can supply up to 15 mW. The energy (kinetic) supplied at these maximums in one hour is then 5.40 J (the battery supplying 1.5 mW for one hour) and 54 J (the battery supplying 15 mW for one hour). Question: Which battery will discharge first? Answer: The one supplying 10 mA of current will discharge first because both started with the same capacity. It will discharge in 66.7 hours.

When a voltage signal containing many sinusoids (and perhaps DC) is applied across a resistor, we add the power from each sinusoid (and DC) to obtain the total power dissipated by the resistor. In other words, we sum the mean-squared voltages from each sine wave (the mean-squared voltage of a DC source, V_{DC}, is V_{DC}) and divide by the resistor's value to obtain the total power dissipated. Taking the square-root of the sum of the mean-squared voltages, we obtain the RMS (root OF THE mean OF THE square) value of the signal applied to the resistor. This is *important* because we will regularly sum mean-squared voltages when performing noise analysis. (We regularly add the power [mean-squared value] contributed from each noise source to a circuit.) We *never* sum RMS voltages.

8.1.2 Power Spectral Density

Figure 8.3a shows spectral plots of the sine wave in Fig. 8.1. Notice how we use either a single line or a dot to indicate spectral content at a frequency of f. Figure 8.3b shows the spectrum of a signal containing two sinewaves and a DC component (with arbitrary amplitudes) using lines and dots. In Fig. 8.3c we show a signal where the spectrum is occupied with sinusoids from DC to some maximum f_{max}. Even though we show a constant amplitude (a continuous spectral density from DC to f_{max}), it is understood that we are representing a spectrum with spectral components that are, or may be, changing.

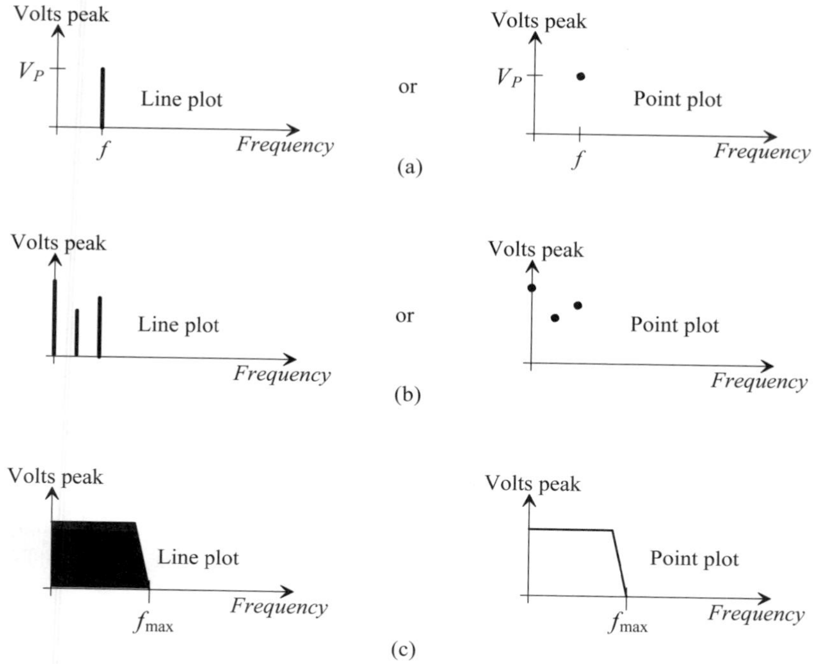

Figure 8.3 Spectrums of various signals (see text).

Spectrum Analyzers

A spectrum analyzer (SA) is an instrument for evaluating the spectral content of a signal. The SA outputs plots similar to those seen in Fig. 8.3 (point-by-point). A block diagram of an SA is seen in Fig. 8.4. The input of the SA is multiplied by a sinusoid to frequency shift the input signal. After the multiplication, the bandpass filter limits the range of frequencies to a bandwidth of f_{res} (the resolution bandwidth of the measurement). The power in the output signal of the bandpass filter is then measured. The counting index, n, varies the measurement from a start frequency, f_{start}, to a stop frequency, f_{stop}. This range of frequencies makes up the x-axis in our spectral plots (as seen in Fig. 8.3). The power measured at each corresponding point sets the y-axis values.

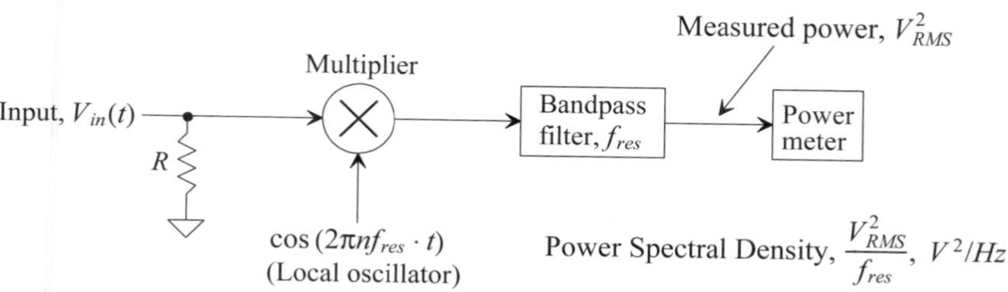

Figure 8.4 Block diagram of a spectrum analyzer.

Let's make some assumptions and give an example to illustrate how the SA operates. Let's assume that we want to look at the spectrum of a signal from DC ($f = 0 = f_{start}$) to 10 kHz ($= f_{stop}$) with a resolution of 100 Hz ($= f_{res}$). Further let's assume that the signal we want to look at is a 1 V peak amplitude sine wave at 4.05 kHz in series with 1 V DC, that is, $V_{in}(t) = 1 + \sin 2\pi \cdot 4.05\,kHz \cdot t$ V. To begin, we calculate the range of the counting index, n,

$$f_{stop} = 10\;kHz = n_{\max} \cdot f_{res} \rightarrow n_{\max} = 100 \tag{8.6}$$

The counting index will vary from 0 to 100. In other words, the frequency of the cosine signal in Fig. 8.4 will vary from 0 to 10 kHz in 100 Hz steps. Our plot will have 101 points.

With our input signal applied, the power at the first point, $n = 0$ (or for the x-axis $f = 0$), is measured. The cosine term applied to the multiplier is 1. The entire input signal is applied to the bandpass filter. In a general SA, a bandpass filter is used. In this example, we use a lowpass filter that passes DC to <100 Hz. The signal applied to the power meter is $(1\text{ V})^2$. We are ready to plot the first point in the spectral analysis. For this first point at $f = 0$ (and $n = 0$), the y-axis units can be peak voltages (=1 V), mean-squared (= 1/2 V^2), RMS (= $1/\sqrt{2}$ V = V_{RMS}), power (= $1/2R = V_{RMS}^2/R$ watts where the R is the input resistance of the SA seen in Fig. 8.4), power spectral density (PSD) (= $1/[2Rf_{res}]$ with units of W/Hz or joules[1]), or voltage spectral density (= $V_{RMS}/\sqrt{f_{res}} = 1/\sqrt{2 \cdot 100}$ with units of $V/\sqrt{Hz}$).

Often we'll use a PSD with units of V^2/Hz. This is the result of eliminating the resistance from the calculation **so that the PSD is** V_{RMS}^2/f_{res} ($= 1/2f_{res} = 5 \times 10^{-3}\;V^2/Hz$ for this first point). We use the PSD in noise analysis (and elsewhere) because, for a continuous spectrum (something that we don't have in this example but that is common for noise signals), increasing f_{res} increases the power we measure (V_{RMS}^2), resulting in a constant number when we take the ratio V_{RMS}^2/f_{res}. (The PSD of a continuous spectrum doesn't change with measurement resolution.) Unfortunately, when a single sine wave is measured, increasing f_{res} decreases the sinewave's amplitude PSD amplitude, as seen in Fig. 8.5.

Continuing on with our measurement when $n = 1$ and the multiplying (local oscillator) frequency is 100 Hz, the output of the multiplier, in Fig. 8.4, is

$$\cos(2\pi \cdot 100 \cdot t) \cdot (1 + \sin[2\pi \cdot 4.05\,kHz \cdot t])\text{ volts} \tag{8.7}$$

or knowing $\cos A \cdot \sin B = \frac{1}{2}(\sin[B-A] + \sin[A+B])$, we get

$$\cos(2\pi \cdot 100t) + \tfrac{1}{2}(\sin[2\pi(3.95k)t] + \sin[2\pi(4.15k)t])\text{ volts} \tag{8.8}$$

This multiplier output signal (remembering $n = 1$) contains three sinusoids with frequencies of 100 Hz, 3.95 kHz, and 4.15 kHz. Passing this signal through the filter, which allows frequencies from DC to <100 Hz through, results in zero measured power. This assumes an ideal filter that doesn't pass 100 Hz (see the comment at the end of this

---

[1] From the units, W/Hz = W·s = J, we can tell that this is the amount of power dissipated by R (for the frequency range of sine waves that pass through the bandpass filter) in one second (the energy used). This shouldn't be confused with energy spectral density (ESD) which has units of $V^2/Hz/s$ (or V^2), for transient signals (for a finite amount of time, i.e., one second).

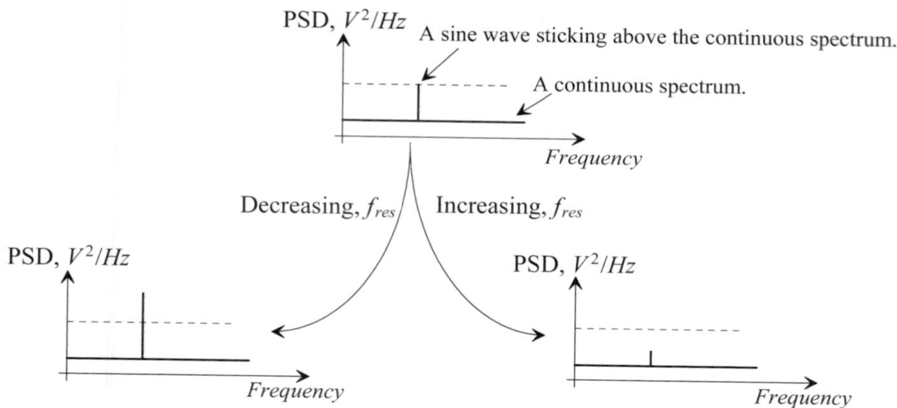

Figure 8.5 Changing the resolution bandwidth doesn't affect the PSD amplitude of the continuous spectrum but does affect the amplitude of the sine wave sticking up above the continuous spectrum.

section). In fact, it's easy to show that we get zero measured signal until $nf_{res} = 4\,kHz$. When the local oscillator frequency is 4 kHz, the output of the multiplier is

$$\cos(2\pi \cdot 4kt) + \tfrac{1}{2}(\sin[2\pi(50)t] + \sin[2\pi(8.05k)t]) \text{ volts} \quad (8.9a)$$

that is, a signal containing power at frequencies of 50 Hz, 4 kHz, and 8.05 kHz. Again, the filter passes the signal content in the range of $0 \le f < 100\ Hz$ to the power meter. What comes out of the filter is a sine wave at 50 Hz with a peak amplitude of 0.5 V. The next measured point occurs when $nf_{res} = 4.1\ kHz$. The signal applied to the filter (the output of the multiplier) is

$$\cos(2\pi \cdot 4.1kt) + \tfrac{1}{2}(\sin[2\pi(-50)t] + \sin[2\pi(8.15k)t]) \text{ volts} \quad (8.9b)$$

The *negative frequency* signal present here can be thought of as a phase-shifted positive frequency signal. Because the sine function is an odd function, $\sin(-2\pi f \cdot t) = -\sin(2\pi f \cdot t) = \sin(2\pi f \cdot t \pm \pi)$, there is still spectral content at 50 Hz. The signal applied to the meter is, again, a 50 Hz sinewave with an amplitude of 0.5 V. The total (measured) signal amplitude from the 4.05 kHz component of the input is the sum of the measured signal amplitudes when nf_{res} is 4 kHz and 4.1 kHz (that is, a sum of 1 V). When plotting points, the amplitudes at measured adjacent points are summed. For example, we can plot a point at 4.05 kHz with an amplitude of 1 V by summing the measured signals when nf_{res} is 4 kHz and 4.1 kHz. We can plot points at 3.95 and 4.15 kHz with amplitudes of 0.5 V because at 3.9 kHz and 4.2 kHz ($= nf_{res}$), we have zero measured signal.

Looking at the spectrums of signals using a discrete Fourier transform (DFT) of analog waveforms is analogous to using an SA, to look at spectrums. For more detailed information on using both a DFT and an SA the interested reader is referred to the second volume of this book entitled *CMOS Mixed-Signal Circuit Design*. In this book additional information and discussions are given.

8.2 Circuit Noise

The electrical noise coming out of a circuit can be measured using a SA, as seen in Fig. 8.6. The circuit under test (CUT) is connected to a spectrum analyzer with no source signal applied. If the noise coming out of a CUT is lower than the noise floor of the SA, a low-noise amplifier (LNA) is inserted between the CUT and SA to help with the noise measurement (the measured noise is then increased by the LNA's gain). As mentioned in the previous section, we generally plot PSD versus frequency to ultimately characterize the power in a signal or the signal's RMS value.

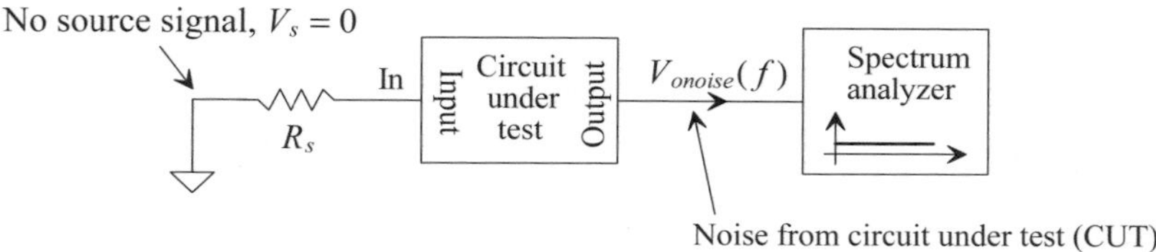

Figure 8.6 Circuit used for noise measurement.

8.2.1 Calculating and Modeling Circuit Noise

Figure 8.7 shows a time domain representation of a noise signal. If this noise signal's PSD is called $V^2_{noise}(f)$ (units, V^2/Hz), we can determine its RMS value using

$$V_{RMS} = \sqrt{\int_{f_L}^{f_H} V^2_{noise}(f) \cdot df} \qquad \text{Volts} \tag{8.10}$$

where f_L and f_H are the lowest and highest frequencies of interest. If the noise PSD is flat (often called a "white noise spectrum" after white light that contains spectral content at all visible wavelengths), then this equation reduces to

$$V_{RMS} = \sqrt{(f_H - f_L) \cdot V^2_{noise}(f)} = \sqrt{B} \cdot \sqrt{V^2_{noise}(f)} \tag{8.11}$$

where the bandwidth of the measurement, B, is $f_H - f_L$.

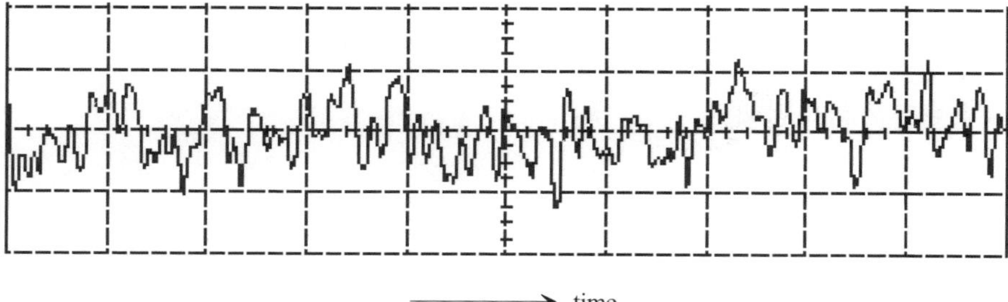

Figure 8.7 Time domain view of circuit noise.

Input-Referred Noise I

Noise is always measured on the output of a circuit. It can, however, be referred back to the input of the circuit for comparison with an input signal. This input-referred noise isn't really present on the input of the CUT (and so it can't be measured directly). This is especially true for CMOS circuits where the input to a CUT may be the polysilicon gate of a MOSFET (which is, of course, surrounded by an insulating dielectric).

Consider the amplifier models seen in Fig. 8.8. Figure 8.8a shows the measured output noise PSD. We can calculate the input-referred PSD, Fig. 8.8b, by simply dividing the output PSD by the gain, A, of the amplifier squared (or divide the voltage spectral density, $= \sqrt{PSD}$, by the gain).

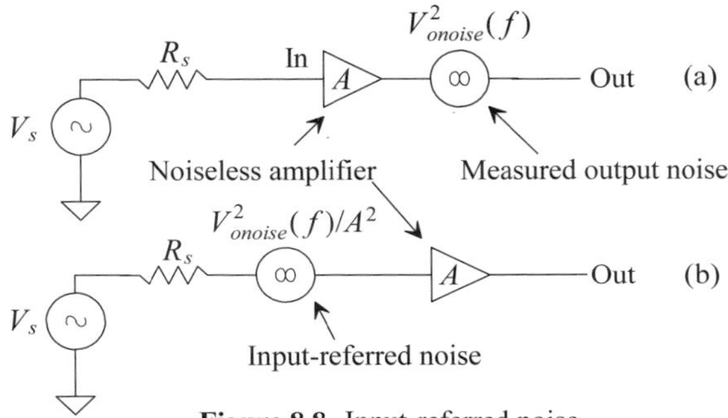

Figure 8.8 Input-referred noise.

Example 8.1

Suppose the measured noise voltage spectral density on the output of an amplifier is a white spectrum from DC to 100 MHz ($B = f_H - f_L = 100\ MHz$) and has a value of 10 $nV/\sqrt{Hz}$. Estimate the amplifier's input-referred noise if its gain is 100.

The amplifier's output noise PSD is

$$V^2_{onoise}(f) = \left(10\ nV/\sqrt{Hz}\right)^2 = 100 \times 10^{-18}\ V^2/Hz$$

The RMS value of this spectral density is

$$V_{onoise,RMS} = \sqrt{\int_0^{100MHz} 100 \times 10^{-18} df} = 100\ \mu V$$

Referring this noise back to the input of the amplifier results in

$$V_{inoise,RMS} = \frac{V_{onoise,RMS}}{A} = 1\ \mu V \qquad (8.12)$$

■

Noise Equivalent Bandwidth

In the previous example, the bandwidth of the noise calculation was given. However, we may wonder, in a real circuit, how we determine f_L and f_H. In the ideal situation, we would use an infinite bandwidth when calculating the noise. Figure 8.9a shows a white noise spectrum. If we were to calculate the RMS value of the noise voltage from this spectrum, over an infinite bandwidth, we would end up with an infinite RMS noise voltage. In real circuits the signals, and noise, are bandlimited (their spectral content goes to zero at some frequencies). This bandlimiting can be the result of intentionally added or parasitic capacitances present in the circuit. Figure 8.9b shows a noise spectrum if the CUT shows a simple single-pole roll-off. In this case, our noise is no longer white at high frequencies but rolls off above the 3-dB frequency of the circuit, f_{3dB}.

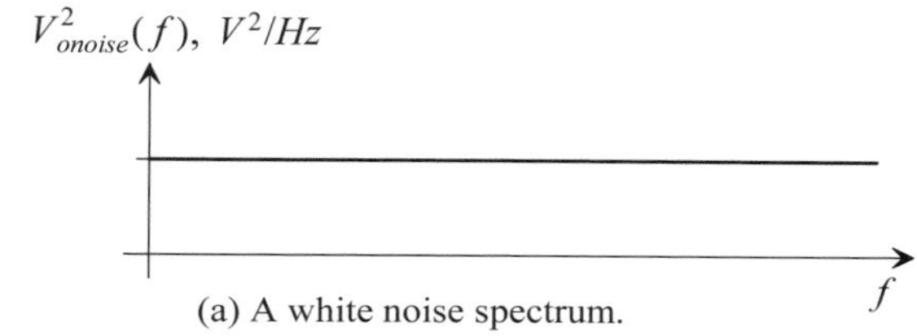

(a) A white noise spectrum.

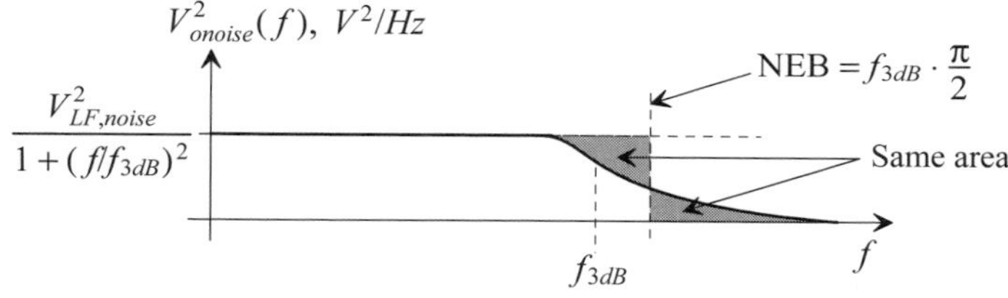

(b) White noise rolling off with amplifier bandwidth.

Figure 8.9 Noise PSDs with (a) a white spectrum and (b) a spectrum that rolls-off with circuit bandwidth (assuming a dominant pole circuit) .

If we assume that a CUT has a single-pole roll-off, as seen in Fig. 8.9b, and the low-frequency noise of the CUT is white (again as seen in Fig. 8.9b), then we can calculate the RMS output noise using

$$V^2_{onoise,RMS} = \int_0^\infty \overbrace{\frac{V^2_{LF,noise}}{\left(\sqrt{1+(f/f_{3dB})^2}\right)^2}}^{V^2_{onoise}(f)} df \tag{8.13}$$

If we use

$$\int \frac{du}{a^2+u^2} = \frac{1}{a}\tan^{-1}\frac{u}{a} + C \tag{8.14}$$

then

$$V^2_{onoise,RMS} = V^2_{LF,noise} \cdot f_{3dB} \cdot [\tan^{-1} 2\pi f/f_{3dB}]_0^\infty$$

$$= V^2_{LF,noise} \cdot \overbrace{f_{3dB} \cdot \frac{\pi}{2}}^{NEB} \tag{8.15}$$

where *NEB* is the noise-equivalent bandwidth. Rewriting Eq. (8.11), we get

$$V_{onoise,RMS} = \sqrt{NEB} \cdot \sqrt{V^2_{LF,noise}} \tag{8.16}$$

Of course this assumes $V^2_{LF,noise}$ is constant until the frequency approaches f_{3dB}.

Example 8.2
Repeat Ex. 8.1 if the amplifier's output noise power spectral density starts to roll off at 1 MHz as seen in Fig. 8.10.

We begin by using Eq. (8.16) directly, that is,

$$V_{onoise,RMS} = \sqrt{f_{3dB} \cdot \frac{\pi}{2}} \cdot \sqrt{V^2_{LF,noise}} \tag{8.17}$$

which evaluates to 12.5 μV. The gain of the amplifier is 100 so $V_{inoise,RMS} = 125$ nV. ■

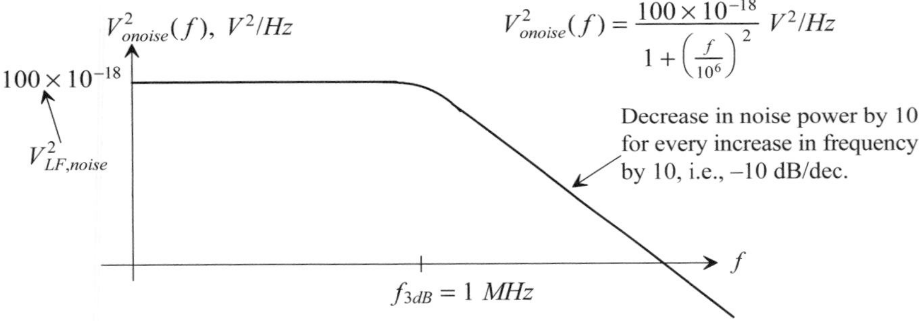

Figure 8.10 Measured output noise spectrum for the CUT discussed in Ex. 8.2.

In this example, we assumed the voltage gain of the amplifier was 100 independent of frequency when we calculated the input-referred RMS noise voltage. In practice, the frequency response of the noise may follow (be correlated with) the amplifier's frequency response. If this is the case, we can write the frequency response of the amplifier in Ex. 8.2 as

$$A(f) = \frac{V_{out}}{V_{in}} = \frac{A_{DC}}{1 + j\frac{f}{f_{3dB}}} = \frac{100}{1 + j\frac{f}{f_{3dB}}} \tag{8.18}$$

where $f_{3dB} = 1$ MHz. We can determine the input-referred noise PSD using

$$V^2_{inoise}(f) = \frac{V^2_{onoise}(f)}{|A(f)|^2} = \frac{V^2_{LF,noise}}{1 + (f/f_{3dB})^2} \cdot \frac{1 + (f/f_{3dB})^2}{A^2_{DC}} = \frac{V^2_{LF,noise}}{A^2_{DC}} \tag{8.19}$$

showing the input-referred PSD of the output noise is a flat spectrum from DC to infinity! The problem with this is that if, to determine the RMS input-referred noise, we integrate this PSD over an infinite bandwidth, we get an infinite RMS voltage. We must remember that the input-referred noise is used to model (with obvious limitations) the circuit's output noise (again, you can't measure input-referred noise because it isn't really there!).

Input-referred noise is only used to get an idea of how a circuit will corrupt an input signal. This latter point is important because two amplifiers with identical bandwidths may have the same output noise PSD but different gains. The amplifier with the larger gain will have a smaller input-referred noise and thus result in an output signal with a better signal-to-noise ratio. Knowing that the output noise is bandlimited and that the input-referred RMS voltage should reflect this, we can write

$$V_{inoise,RMS} = \frac{\sqrt{NEB \cdot V_{LF,noise}^2}}{A_{DC}} = \frac{V_{onoise,RMS}}{A_{DC}} \quad (8.20)$$

which is, of course, what we used in Ex. 8.2.

Input-Referred Noise in Cascaded Amplifiers

Figure 8.11a shows a cascade of noisy amplifiers with corresponding input-referred noise sources. The output noise power in (a) can be written as

$$V_{onoise,RMS}^2 = (A_1A_2A_3)^2V_{inoise,RMS1}^2 + (A_2A_3)^2V_{inoise,RMS2}^2 + A_3^2V_{inoise,RMS3}^2 \quad (8.21)$$

again remembering that we add the noise powers from each amplifier. The input-referred noise in (b) is given by

$$V_{inoise,RMS}^2 = V_{inoise,RMS1}^2 + V_{inoise,RMS2}^2/A_1^2 + V_{inoise,RMS3}^2/(A_1A_2)^2 \quad (8.22)$$

The key point to notice here is that the noise of the first amplifier has the largest effect on the noise performance of the amplifier chain. For good noise performance, the design of the first stage is critical.

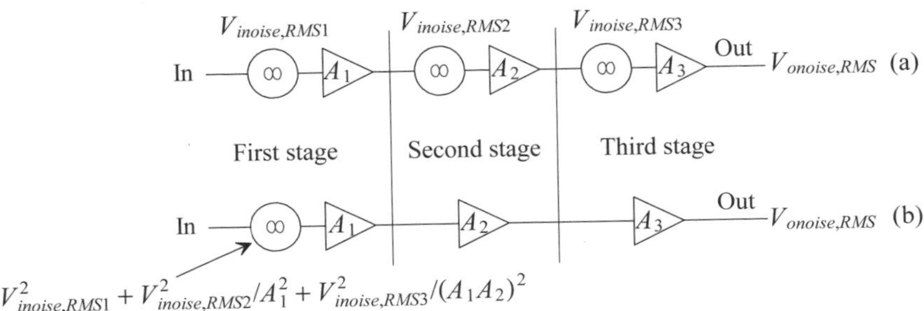

Figure 8.11 Noise performance of a cascade of amplifiers.

Example 8.3
Comment on the limitations of Eqs. (8.21) and (8.22) for calculating noise.

Measured output noise usually includes the thermal noise of the source resistance. If the effective source resistance changes when we cascade the amplifiers, the value calculated for input-referred noise, $V_{inoise,RMS}$, will also change.

A perhaps more important concern is the change in the bandwidth of the noise. Cascading amplifiers results in a reduction in the circuit's bandwidth. The point of Eqs. (8.21) and (8.22) is still valid, that is, that the first stage's output noise and gain (equivalent to saying "input-referred noise") are critical for overall low-noise performance. However, to accurately determine the input-referred noise, it is best to measure the noise on the output of the cascade and then refer it back to the cascade's input. At the risk of stating the obvious, the gain of the cascaded amplifiers is determined by applying a small sine wave signal to the cascade's input at a frequency that falls within the amplifier's passband (not too high or too low). Taking the ratio of the cascade's output sine wave amplitude to the input sine wave amplitude is the gain. The input-referred noise is then the output RMS noise divided by the gain of the overall cascade. ■

Calculating $V_{onoise,RMS}$ from a Spectrum: A Summary

Before leaving this section, let's show, Fig. 8.12, a summary of how the output RMS noise voltage is calculated from a noise spectrum.

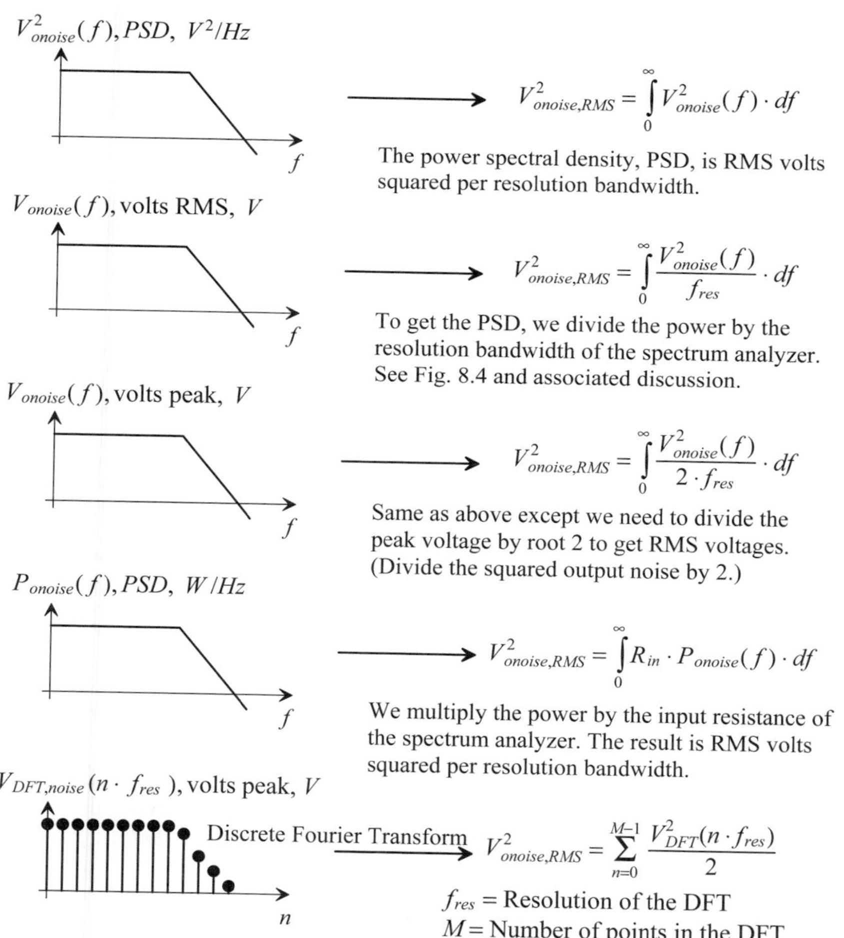

$V^2_{onoise}(f), PSD,\ V^2/Hz$
$V^2_{onoise,RMS} = \int_0^\infty V^2_{onoise}(f) \cdot df$
The power spectral density, PSD, is RMS volts squared per resolution bandwidth.
$V_{onoise}(f)$, volts RMS, V
$V^2_{onoise,RMS} = \int_0^\infty \frac{V^2_{onoise}(f)}{f_{res}} \cdot df$
To get the PSD, we divide the power by the resolution bandwidth of the spectrum analyzer. See Fig. 8.4 and associated discussion.
$V_{onoise}(f)$, volts peak, V
$V^2_{onoise,RMS} = \int_0^\infty \frac{V^2_{onoise}(f)}{2 \cdot f_{res}} \cdot df$
Same as above except we need to divide the peak voltage by root 2 to get RMS voltages. (Divide the squared output noise by 2.)
$P_{onoise}(f), PSD,\ W/Hz$
$V^2_{onoise,RMS} = \int_0^\infty R_{in} \cdot P_{onoise}(f) \cdot df$
We multiply the power by the input resistance of the spectrum analyzer. The result is RMS volts squared per resolution bandwidth.
$V_{DFT,noise}(n \cdot f_{res})$, volts peak, V
Discrete Fourier Transform
$V^2_{onoise,RMS} = \sum_{n=0}^{M-1} \frac{V^2_{DFT}(n \cdot f_{res})}{2}$
f_{res} = Resolution of the DFT
M = Number of points in the DFT
f
n

Figure 8.12 Calculating RMS output noise from a noise spectrum.

8.2.2 Thermal Noise

Noise in a resistor is primarily the result of random motion of electrons due to thermal effects. This type of noise is termed *thermal noise*, or Johnson noise, after J. B. Johnson who is credited for first observing the effect. Thermal noise in a resistor can be characterized by a PSD of

$$V_R^2(f) = 4kTR \text{ with units of V}^2\text{/Hz} \tag{8.23}$$

(noting thermal noise is white and independent of frequency) where

k = Boltzmann's constant = 13.8 x 10^{-24} watt ·sec/° K (or J/° K)

T = temperature in °K

R = resistance in Ω

Example 8.4
Verify that the thermal noise PSD given by Eq. (8.23) does indeed have units of V²/Hz. The term kT has units of joules (or watt·sec) and can be thought of as thermal energy. The units for joules can be written as

$$kT\text{, units} = \text{joules} = \text{watt·sec} = \text{volts·amps·sec} = \text{volts·columbs}$$

knowing amps = columbs/sec. The units of resistance, R, can be written as

$$R\text{, units of } \Omega = \frac{\text{volts}}{\text{amps}} = \frac{\text{volts}}{\text{columbs/sec}} = \frac{\text{volts}}{\text{columbs} \cdot \text{Hz}}$$

The units for thermal noise PSD, $4kTR$, are then V²/Hz. ■

Figure 8.13 shows how the thermal noise from a resistor is modeled and added to a circuit. The output noise PSD in (a) is $(4kT/R) \cdot R^2$ (we multiply by R^2). Heat (power) is absorbed by the resistor when it is not at 0° K. The heat causes lattice vibrations in the resistive material and thus randomizes the motion of electrons moving in or traveling through the resistor. The net current that flows out of the resistor due to heat is zero but,

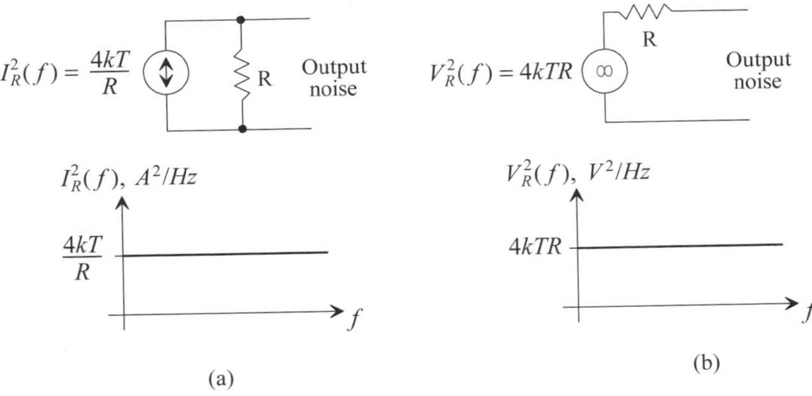

Figure 8.13 Circuit representations and their corresponding PSDs for thermal noise.

over a finite period of time, the current crossing the terminal of a resistor may have a net nonzero value (either into or out of the resistor). In other words, electrons move back and forth from the resistor to the metal wire with an overall net zero charge transfer but, over a short finite period of time, a net nonzero current flow is possible.

The term kT, at 300 °K, is 4.14×10^{-21} joules. For a one-second time frame, again at 300 °K, the resistor dissipates 4.14×10^{-21} watts (absorbs this much heat from a heat source). For one hour, the energy supplied to keep the resistor heated to 300 °K is $(3600\text{ s}) \cdot (4.14 \times 10^{-21}\text{ watts}) = 14.9 \times 10^{-18}\text{ joules}$.

Example 8.5

For the circuit shown in Fig. 8.14, determine the RMS output and input-referred noise over a bandwidth from DC to 1 kHz. Verify your answer with SPICE.

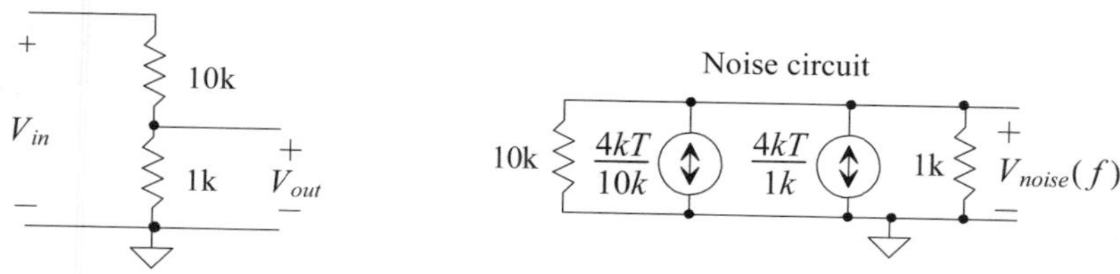

Figure 8.14 Circuit used in Ex. 8.5.

The PSD of the noise currents are given by (again noting, as seen in Fig. 8.6, that we apply zero volts to the input when calculating output noise PSD, that is, we ground the input)

$$I_{10k}^2(f) = \frac{4 \cdot (13.8 \times 10^{-24}) \cdot (300)}{10{,}000} = 1.66 \times 10^{-24}\,\frac{A^2}{Hz} \rightarrow I_{10k}(f) = 1.29 \times 10^{-12}\,\frac{A}{\sqrt{Hz}}$$

$$I_{1k}^2(f) = 16.66 \times 10^{-24}\,\frac{A^2}{Hz} \rightarrow I_{1k}(f) = 4.1 \times 10^{-12}\,\frac{A}{\sqrt{Hz}}$$

The output thermal noise PSD due to the 10k resistor is

$$I_{10k}^2(f) \cdot \left[\frac{1k \cdot 10k}{1k + 10k}\right]^2 \quad \text{units } V^2/Hz$$

In order to avoid analyzing circuits in a different way, (for example, $V^2 = I^2R^2$, $i_d^2 = g_m^2 v_{gs}^2$, $V^2 = I^2(R_1 + R_2)^2$, etc.), we can use the voltage or current spectral densities (square-root of PSD) of the noise when doing circuit noise calculations

$$V_{10k}(f) = I_{10k}(f) \cdot \frac{1k \cdot 10k}{1k + 10k} \quad \text{units } V/\sqrt{Hz}$$

which evaluates to 1.2 $nV/\sqrt{Hz}$. The output noise voltage-spectral density $V_{1k}(f)$ from the 1k resistor is 3.7 $nV/\sqrt{Hz}$. The total output noise PSD, $V_{onoise}^2(f)$, is the sum of $V_{10k}^2(f)$ and $V_{1k}^2(f)$. Again, we sum the PSDs (or power) but **not** the voltage-spectral densities (or the RMS voltages).

The mean-squared output noise voltage over a bandwidth of 1 kHz is

$$V^2_{onoise,RMS} = \int_{f_L}^{f_H} V^2_{onoise}(f) \cdot df = \int_0^{1\text{kHz}} \left(V^2_{10k}(f) + V^2_{1k}(f) \right) \cdot df = 15.1 \times 10^{-15} \ V^2$$

The RMS output noise voltage is

$$V_{onoise,RMS} = 123 \ nV$$

The input-referred noise voltage is

$$V_{inoise,RMS} = 123nV \cdot \frac{10k + 1k}{1k} = 1.35 \ \mu V$$

SPICE simulation gives an output mean squared noise of 1.5053×10^{-14} V^2 (= 123 nV RMS) and an input-referred noise referenced to an input voltage of 1 V of 1.8215×10^{-12} (= 1.35 μV RMS). The SPICE netlist is shown below.

```
*** Example 8.5 CMOS: Circuit Design, Layout, and Simulation ***

.control
destroy all
run
print all
.endc

.noise   v(2,0)  Vin dec  100 1 1k
R1       1 2 10k
R2       2 0 1k
Vin      1 0     dc 0 ac 1
.print noise all
.end
```

The SPICE output is seen below

```
TEMP=27 deg C
Noise analysis ... 100%
inoise_total = 1.821510e-12
onoise_total = 1.505380e-14
```

A bandwidth of 1 to 1,000 Hz was used in this simulation rather than DC to 1 kHz. Also note that the reference supply, Vin in the netlist, was a voltage, so the units of the SPICE output are V^2. The input AC source has a magnitude of 1 V, so the input noise SPICE gives is divided by 1 V squared. If we had used a 1 mV AC supply for Vin, then the inoise_total above would be 1.8215×10^{-6} V^2.

Here, in SPICE, we used a single input-referred voltage source to model the input-referred noise, see Eq. (8.22) and the associated discussion. In SPICE we always refer the input noise to a source (voltage or current) and not to a node (like we do in our analysis). This means that the input-referred noise in SPICE is always a single source in series (voltage input) or parallel (current input) with the input of the circuit. ■

Example 8.6
Estimate $V_{onoise,RMS}$ for the circuit seen in Fig. 8.15. Verify the answer with SPICE.

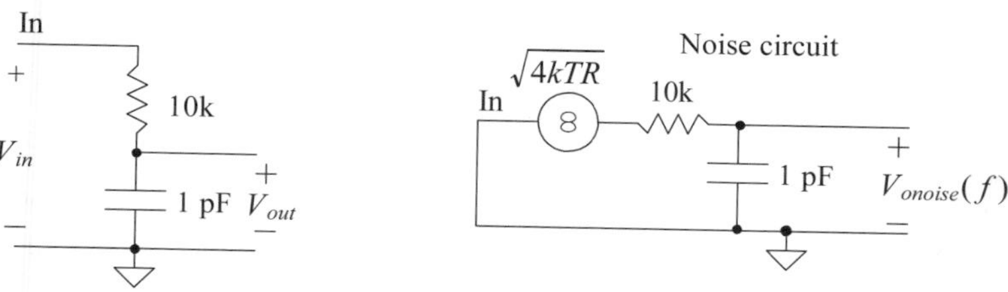

Figure 8.15 Circuit used in Ex. 8.6.

The only element in this circuit that generates noise is the resistor. It generates thermal noise. The resistor's noise voltage spectral density is $\sqrt{4kTR}$. The output noise spectral density is then

$$V_{onoise}(f) = \sqrt{4kTR}\,\frac{1/j\omega C}{1/j\omega C + R} = \frac{\sqrt{4kTR}}{1 + j\frac{f}{f_{3dB}}} \text{ units, } V/\sqrt{Hz}$$

or

$$V^2_{onoise}(f) = \frac{4kTR}{|1 + j\frac{f}{f_{3dB}}|^2} = \frac{4kTR}{\left(\sqrt{1 + (f/f_{3dB})^2}\right)^2} = \frac{4kTR}{1 + (f/f_{3dB})^2} \text{ units, } V^2/Hz$$

where $f_{3dB} = 1/2\pi RC$. This single-pole roll-off was the reason we discussed noise-equivalent bandwidth (NEB) earlier, Ex. 8.2. Using Eq. (8.17), the output RMS noise voltage is

$$V_{onoise,RMS} = \sqrt{\frac{1}{2\pi RC} \cdot \frac{\pi}{2} \cdot 4kTR} = \sqrt{\frac{kT}{C}} \tag{8.24}$$

The RMS value of the thermal noise in this circuit is limited by the size of the capacitor and independent of the size of the resistor. This result is very useful. This "Kay Tee over Cee" noise is frequently used to determine the size of the capacitors used in filtering or sampling circuits.

The SPICE netlist and output are seen below.

```
*** Example 8.6 CMOS: Circuit Design, Layout, and Simulation ***

.noise   v(Vout,0)      Vin    dec    100    1    1G
R1       Vin            Vout   10k
C1       Vout   0       1p
Vin      Vin    0       dc     0      ac     1
.print noise all
.end

TEMP=27 deg C
Noise analysis ... 100%
```

```
inoise_total = 1.695223e-07
onoise_total = 4.101864e-09
```

Using Eq. (8.24) at 300 °K, we get $V^2_{onoise,RMS} = 4.14 \times 10^{-9}\ V^2$. This is close to what SPICE gives above for onoise_total (close but not exact; we stopped the simulation at 1 GHz not infinity). The output RMS noise is 64 μV (keeping in mind that the peak-to-peak value of the thermal noise will be larger than this).

The input-referred noise, inoise_total, in this simulation example is somewhat meaningless. Changing the stop frequency in the simulation from 1 GHz to 10 GHz has little effect on the output RMS noise but does cause the input-referred noise to increase. As indicated in Eq. (8.19) and the associated discussion, the input-referred noise spectral density increases indefinitely. Integrating this spectral density from DC to infinity results in an infinite RMS input-referred voltage. Since the low-frequency gain here is one, we would specify $V_{inoise,RMS} = V_{onoise,RMS}$. ■

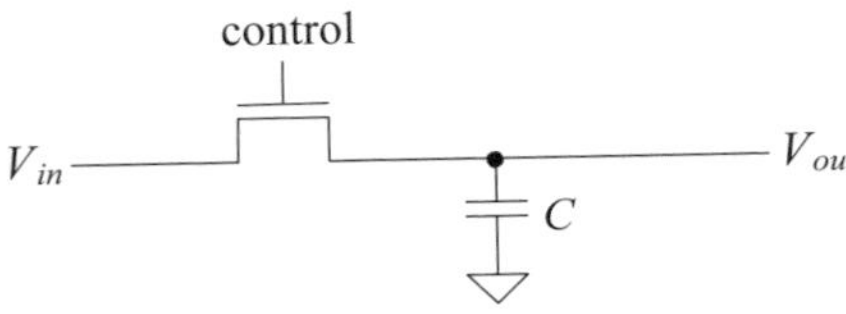

Figure 8.16 A sample-and-hold circuit.

For an example of the usefulness of Eq. (8.24), consider the sample-and-hold circuit seen in Fig. 8.16. When the MOSFET gate is driven to *VDD,* the MOSFET behaves like a resistor and permits V_{in} to charge the capacitor. When the MOSFET is off, the capacitor remains charged and the MOSFET behaves like an open. The voltage on the capacitor doesn't change again until the MOSFET turns back on. We can think of the sampling operation as sampling both the input signal and the *kT*/*C* noise onto the capacitor. Table 8.1 provides a comparison between capacitor sizes and *kT*/*C* noise.

Table 8.1 Capacitor size and corresponding *kT*/*C* noise at 300 °K.

Capacitor size, pF	$\sqrt{kT/C}$, μV
0.01	640
0.1	200
1	64
10	20
100	6.4

8.2.3 Signal-to-Noise Ratio

Signal-to-noise ratio, *SNR*, can be defined in general terms by

$$SNR = \frac{\text{desired signal power, } P_s}{\text{undesired signal power (noise), } P_{noise}} \tag{8.25}$$

SNR can be specified using dB as

$$SNR = 10\log\frac{P_s}{P_{noise}} \tag{8.26}$$

If the power is normalized to a 1-Ω load (e.g., $P_s = V^2_{s,RMS}/(1\ \Omega)$), we can write

$$SNR = 10\log\frac{P_s}{P_{noise}} = 20\log\frac{\sqrt{P_s}}{\sqrt{P_{noise}}} = 20\log\frac{V_{s,RMS}}{V_{noise,RMS}} \tag{8.27}$$

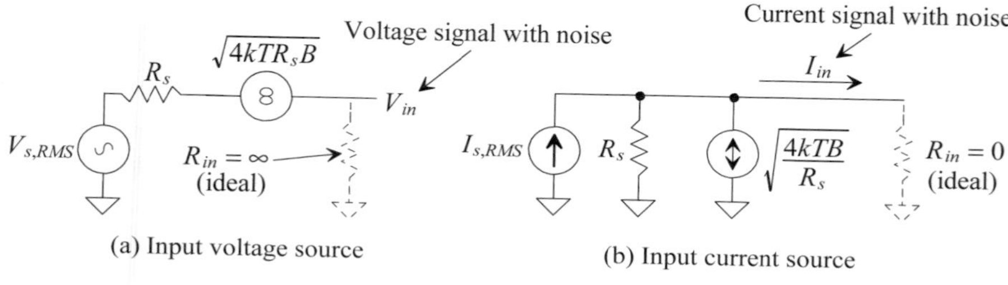

Figure 8.17 Calculating input SNR.

Figure 8.17 shows two equivalent input sources with associated thermal noise models. With the ideal R_{in} the *SNR* associated with these circuits is (noting $V^2_{s,RMS} = I^2_{s,RMS}\cdot R^2_s$) an open (voltage input) or a short (current input)

$$SNR_{in} = \frac{V^2_{s,RMS}}{4kTR_sB} = \frac{I^2_{s,RMS}}{4kTB/R_s} \tag{8.28}$$

In Fig. 8.17a, V_{in} is dropped across an infinite R_{in} (an open). In Fig. 8.17b, I_{in} drives zero R_{in} (a short). In the practical case, R_{in} is finite and nonzero. For Fig. 8.17a, the input signal and noise are attenuated by the voltage divider formed between R_s and R_{in}

$$SNR_{in} = \frac{V^2_{s,RMS}\cdot\left[\frac{R_{in}}{R_{in}+R_s}\right]^2}{4kTR_sB\cdot\left[\frac{R_{in}}{R_{in}+R_s}\right]^2} = \frac{V^2_{s,RMS}}{4kTR_sB} \tag{8.29}$$

which has no effect on the SNR_{in}. We can also show that there is no change in SNR_{in} if R_{in} is nonzero in Fig. 8.17b.

Examine the amplifier model with noise seen in Fig. 8.18. The output noise, $V_{onoise,RMS}$, includes the thermal noise contributions from R_s. We measured the output noise with the source connected to the amplifier (see Figs. 8.6 and 8.8). The *SNR* associated with the output of the amplifier (where the noise is due to the amplifier and the thermal noise from R_s) is then

$$SNR_{out} = \frac{V_{s,RMS}^2 \cdot \left[\frac{R_{in}}{R_{in}+R_s}\right]^2 \cdot A^2}{V_{onoise,RMS}^2} \tag{8.30}$$

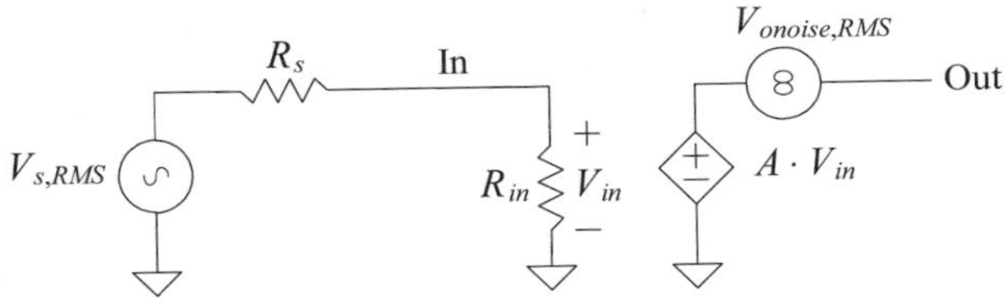

Figure 8.18 Calculating output SNR.

Input-Referred Noise II

When we discussed input-referred noise in Fig. 8.8, we tacitly assumed that the input resistance of the amplifier R_{in} was infinite. The input-referred noise could then be modeled with a single voltage source in series with the input signal source V_s. Keeping in mind that we are measuring $V_{onoise,RMS}$ at a fixed value of R_s, Fig. 8.19a, we can show that if the amplifier's input resistance R_{in} is not infinite, this single input-referred voltage source can still manage to model the amplifier's output noise. Looking at Fig. 8.19b, we can write

$$A \cdot V_{inoise,RMS} \cdot \frac{R_{in}}{R_{in}+R_s} = V_{onoise,RMS} \tag{8.31}$$

If the input resistance R_{in} goes to infinity (common for low-frequency CMOS amplifiers) or R_s is zero (a voltage source is connected to the amplifier's input), a single input-referred noise voltage is ideal for modeling the output noise. (And we'll use this simpler model in most of our noise discussions in this book.)

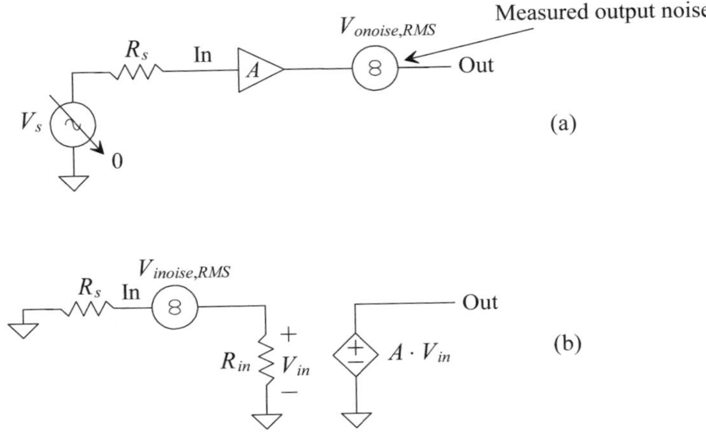

Figure 8.19 Modeling measured output noise with a single input-referred source.

Using a single voltage source to model input-referred noise does have some limitations though. For example, what happens if our input signal is a current (meaning R_s is infinite), as seen in Fig. 8.20a? In this case, the input-referred noise voltage is irrelevant. The input voltage is independent of $V_{inoise,RMS}$ (the output of the circuit is free of noise). To avoid this situation and to make the input-referred noise sources *independent of* R_s, we can use the noise model seen in Fig. 8.20b. The output noise using this model is

$$V^2_{onoise,RMS} = 4kTR_sB \cdot \left(\frac{AR_{in}}{R_s + R_{in}}\right)^2 + I^2_{inoise,RMS} \cdot \left(\frac{AR_sR_{in}}{R_s + R_{in}}\right)^2 + V^2_{inoise,RMS} \cdot \left(\frac{AR_{in}}{R_s + R_{in}}\right)^2 \quad (8.32)$$

noting that we sum the power from each source to get the total output noise power. It's important to note that our input-referred sources no longer depend on the thermal noise contributions from the source resistance (we've added it separately). Looking at Eq. (8.32), we see that if R_s is zero or infinite, the thermal noise contributions from R_s to the output noise are zero. We should also see that if R_{in} is large (or R_s is small), $V_{inoise,RMS}$ alone is sufficient to model the input-referred noise. Similarly, if R_s is large (or R_{in} is small), $I_{inoise,RMS}$ alone can be used to model the input-referred noise.

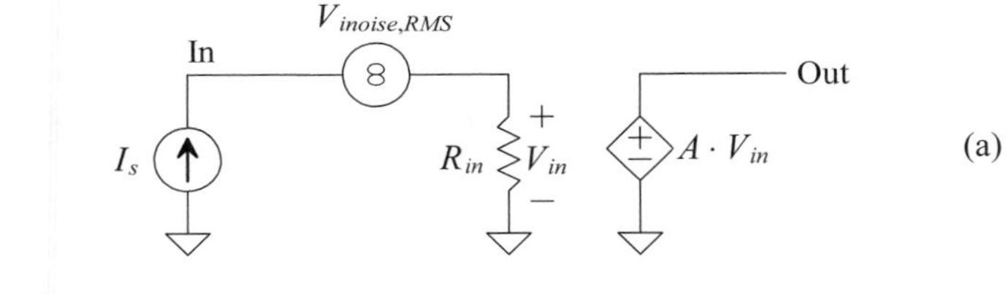

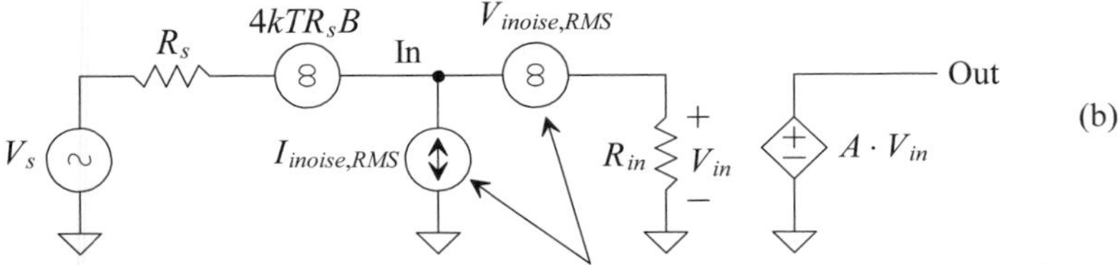

Figure 8.20 Using two input-referred noise sources to model an amplifier's output noise.

Example 8.7
Discuss how to determine $V_{inoise,RMS}$ and $I_{inoise,RMS}$.

Looking at Eq. (8.32), we see that if R_s is zero, then $V_{inoise,RMS}$ is sufficient to model the output noise. We can therefore write

$$V_{inoise,RMS} = \frac{V_{onoise,RMS,Rs=0}}{A} \quad (V_{onoise,RMS} \text{ measured with } R_s \text{ shorted}) \quad (8.33)$$

If R_s is infinite (think of the input as seen in Fig. 8.17b), then $I_{inoise,RMS}$ is sufficient to model the amplifier's output noise

$$I_{inoise,RMS} = \frac{V_{onoise,RMS,Rs=\infty}}{AR_{in}} \quad (V_{onoise,RMS} \text{ measured with } R_s \text{ opened}) \tag{8.34}$$

■

Noise Figure

The noise figure, NF, of an amplifier is given by

$$NF = 10\log\frac{SNR_{in}}{SNR_{out}} = 10\log[\text{noise factor, } F] \tag{8.35}$$

If the amplifier doesn't degrade the SNR, then the input and output SNRs are the same, the noise factor, F, is 1, and the NF is 0 dB. Using Eqs. (8.29) and (8.30), we can write the NF as

$$NF = 10\log\frac{V^2_{onoise,RMS}}{4kTR_sB\cdot\left[\frac{R_{in}}{R_{in}+R_s}\right]^2\cdot A^2} \tag{8.36}$$

When the measuring bandwidth B is small (common for narrowband amplifiers used in communication circuits), we are measuring the spot NF (noting, from Eq. [8.10], the bandwidth used to calculate $V_{onoise,RMS}$ is reduced). If the measuring bandwidth is large (common in general analog CMOS design), we are measuring the average NF.

The numerator in Eq. (8.36) is the total output noise power and the denominator is the output noise power due to the source resistance, that is,

$$F = \frac{\text{total output noise power}}{\text{output noise power due to source resistance}} \tag{8.37}$$

If the amplifier is noise-free ($V_{onoise,RMS} = \sqrt{4kTR_sB}\cdot A\cdot R_{in}/[R_{in}+R_s]$, that is, the output contains only the thermal noise from the source resistance), the NF is, again, 0 dB. For a low-noise amplifier (LNA), the NF may typically range from 0.5 to 5 dB.

Note that if SNR_{in} is infinite, the NF is meaningless. Infinite SNR_{in} could occur if we use the impractical case, see Fig. 8.17, of $R_s = 0$ (an ideal voltage source input where, also, ideally $R_{in} = \infty$) or $R_s = \infty$ (an ideal current source input where, also, ideally $R_{in} = 0$).

An Important Limitation of the Noise Figure

Looking at Eq. (8.36), we see that if R_{in} is large compared to R_s ($R_{in} >> R_s$), F approaches

$$F = \frac{V^2_{onoise,RMS}}{4kTR_sB\cdot A^2} \tag{8.38}$$

This equation shows why CMOS amplifiers work so well for low-noise amplification with large source resistances. At low frequencies, the gate of the MOSFET is an open. If we take $R_s \to \infty$, then F goes to 1 which is a perfect noiseless amplifier right? After reviewing how R_s affects SNR_{in} in Eq. (8.29), we see that the cost for this good amplifier NF is an SNR_{in} that approaches zero. The input voltage signal is so noisy that it makes the amplifier noise irrelevant and thus the NF is 0 dB. Note that if R_{in} is infinite, we must use the model seen in Fig. 8.17a. In Fig. 8.17b I_{in} would be zero if $R_{in} = \infty$.

Example 8.8
Suppose an input signal source has an *SNR* of 60 dB and a source resistance R_s. Further, suppose this signal is amplified with an amplifier having an *NF* of 1 dB when driven from a source with a resistance R_s. Estimate the *SNR* on the output of the amplifier.

We can begin by rewriting Eq. (8.31)

$$NF = 10\log SNR_{in} - 10\log SNR_{out} \tag{8.39}$$

illustrating the beauty of using the *NF*. That is, SNR_{out} is calculated by subtracting the *NF* from SNR_{in}. In this case, the SNR_{out} is 59 dB. ■

Example 8.9
Calculate the input-referred noise, *F*, and *SNR*s for the circuit seen in Fig. 8.21.

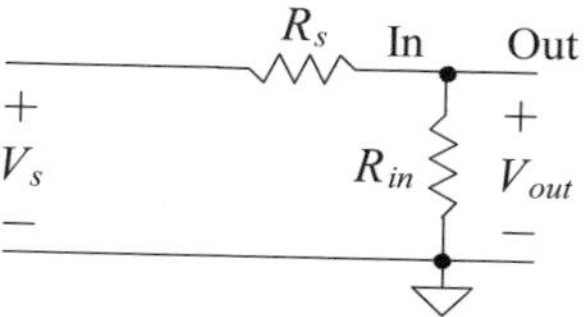

Figure 8.21 Circuit used in Ex. 8.9.

Let's begin by adding the noise voltage spectral density to the circuit, Fig. 8.22a. The output noise PSD is

$$V^2_{onoise}(f) = 4kTR_s\left[\frac{R_{in}}{R_{in}+R_s}\right]^2 + 4kTR_{in}\left[\frac{R_s}{R_{in}+R_s}\right]^2$$

To determine $V_{onoise,RMS}$, we integrate this PSD over the bandwidth of interest *B* or

$$V^2_{onoise,RMS} = \int_{f_L}^{f_H} V^2_{onoise}(f)\cdot df = 4kTBR_s\left[\frac{R_{in}}{R_{in}+R_s}\right]^2 + 4kTBR_{in}\left[\frac{R_s}{R_{in}+R_s}\right]^2$$

Noting our gain A ($= V_{out}/V_{in}$ not V_{out}/V_s) is one, we can use the model shown in Fig. 8.22b. To determine the input-referred noise sources, we can use Eq. (8.32) and the results in Ex. 8.7. To determine $V_{inoise,RMS}$, we short the input to ground ($R_s = 0$ in Fig. 8.21 and the equation above), Fig. 8.22c, and equate the circuit output to $V_{onoise,RMS}$. This gives

$$V_{onoise,RMS,Rs=0} = V_{inoise,RMS} = 0$$

To determine $I_{inoise,RMS}$, we open the input ($R_s = \infty$), Fig. 8.22d, and equate the circuit's output to $V_{onoise,RMS}$ (from the equation above). This gives

$$R^2_{in}\cdot I^2_{inoise,RMS} = V^2_{onoise,RMS,Rs=\infty} = 4kTBR_{in} \rightarrow I_{inoise,RMS} = \sqrt{\frac{4kTB}{R_{in}}}$$

The input SNR is given in Eq. (8.29). The output SNR, Fig. 8.22e, is

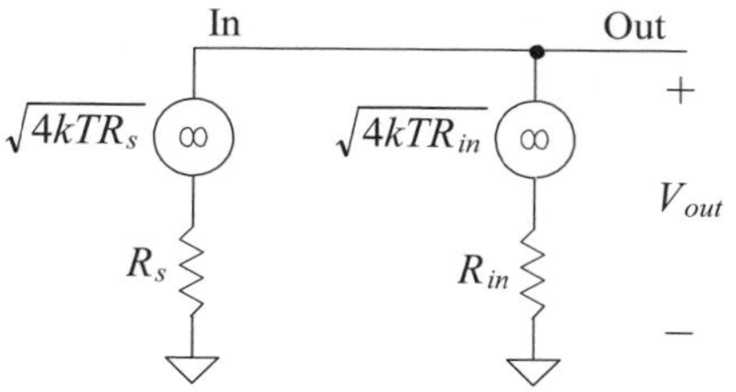

(a) Adding noise sources to the circuit.

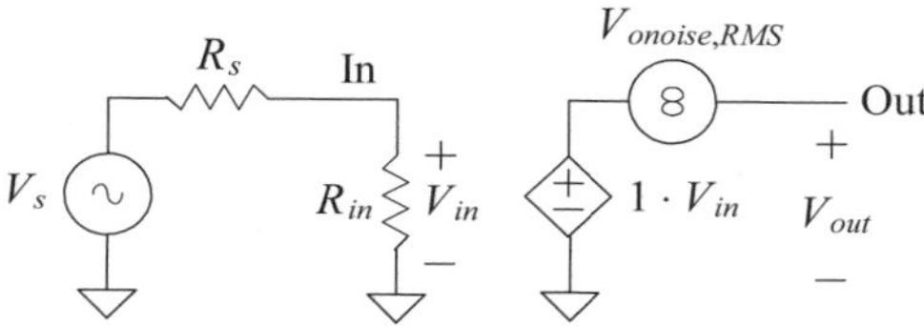

(b) Model showing output noise.

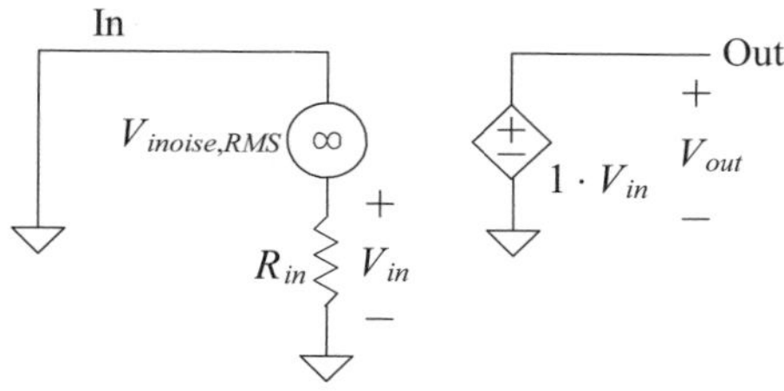

(c) Shorting the input.

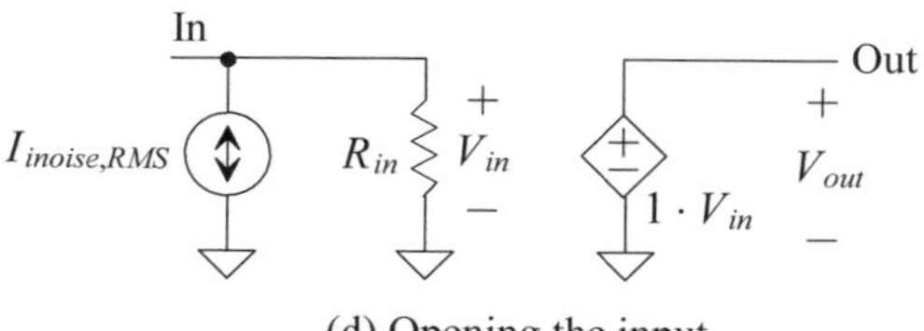

(d) Opening the input.

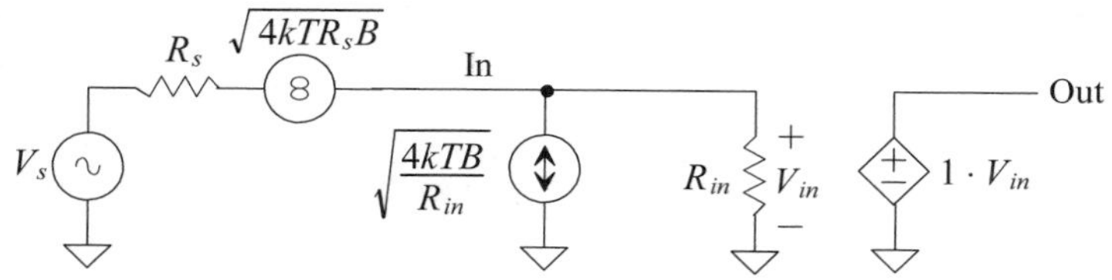

(e) Input-referred noise model.

Figure 8.22 Noise analysis procedure for Ex. 8.9.

$$SNR_{out} = \frac{V^2_{s,RMS} \cdot \left[\frac{R_{in}}{R_s+R_{in}}\right]^2}{V^2_{onoise,RMS}} = \frac{V^2_{s,RMS}}{4kTB \cdot R_s(1+R_s/R_{in})} \tag{8.40}$$

The noise factor is then

$$F = 1 + \frac{R_s}{R_{in}} \tag{8.41}$$

To minimize the NF, we can decrease R_s or increase R_{in}. Decreasing R_s causes SNR_{in} and SNR_{out} to increase, as seen in Eqs. (8.29) and (8.40). At the same time, increasing R_{in} causes SNR_{out} to move towards SNR_{in}, Eq. (8.40), resulting in F moving towards 1. ■

Optimum Source Resistance

Looking at Eq. (8.36), notice that if R_s approaches a short or an open, the *NF* gets really large. To determine the optimum source resistance (to minimize *NF*), let's write F using Eqs. (8.32) and (8.36) as

$$F = \frac{4kTR_sB + I^2_{inoise,RMS} \cdot R_s^2 + V^2_{inoise,RMS}}{4kTR_sB} \tag{8.42}$$

Taking the derivative with respect to R_s and setting the result equal to zero (to determine the minimum), results in

$$I^2_{inoise,RMS} + 2 \cdot I_{inoise,RMS} \cdot R_s \cdot \frac{\partial I_{inoise,RMS}}{\partial R_s} - \frac{V^2_{inoise,RMS}}{R_s^2} + \frac{2V_{inoise,RMS}}{R_s} \cdot \frac{\partial V_{inoise,RMS}}{\partial R_s} = 0 \tag{8.43}$$

Since the input-referred noise sources do not vary with R_s (the derivatives are zero), we can write the optimum source $R_{s,opt}$ as

$$R_{s,opt} = \frac{V_{inoise,RMS}}{I_{inoise,RMS}} \tag{8.44}$$

From Ex. 8.9 we can use this equation to see that $R_{s,opt} = 0$.

To determine the optimum noise factor F_{opt}, we can substitute Eq. (8.44) into Eq. (8.42) to get

$$F_{opt} = 1 + \frac{(V_{inoise,RMS} \cdot I_{inoise,RMS})/2}{kTB} \tag{8.45}$$

The best noise performance is characterized by a ratio of the input-referred noise power of the amplifier, $(V_{inoise,RMS} \cdot I_{inoise,RMS})/2$, to the thermal power (the thermal energy kT in a bandwidth of B). This result isn't too useful because it indicates that for good noise performance we need to use an amplifier with low input-referred noise power (which is as profound as saying "to get good noise performance use a low-noise amplifier").

Simulating Noiseless Resistors

To verify the *NF* of a circuit using SPICE, it may be useful to replace one or more resistors with elements that don't generate noise. To simulate a noiseless resistor, a voltage-controlled current source (VCCS) can be used, Fig. 8.23. Current flows from node1 to node2 in the resistor of (a) while it flows from nplus to nminus in the VCCS

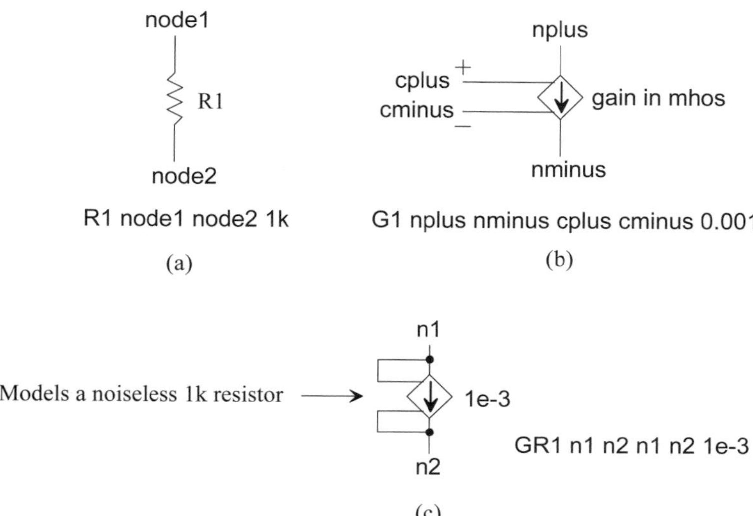

Figure 8.23 Modeling a noiseless resistor in SPICE.

in (b). The current in (b) is controlled by the voltage applied to the controlling nodes cplus and cminus. The input to the VCCS is a voltage, and the output is a current (meaning the gain is current/volts or transconductance). To implement a noiseless resistor, we can use the VCCS where the controlling nodes are tied across the VCCS, as seen in (c). The value of the resistor is one over the transconductance of the VCCS.

Example 8.10
Suppose, in Fig. 8.21, R_s is 10k and R_{in} is 1k. Use SPICE to verify the value of noise factor derived in Ex. 8.9 (that is, Eq. [8.41]).

Using Eq. (8.41), the noise factor is 11 (NF = 10.4 dB). Note that Eq. (8.41) is not dependent on bandwidth. We simulated this circuit already in Ex. 8.5 over a bandwidth of 1 to 1kHz. The result was a $V^2_{onoise,RMS}$ of $1.5053 \times 10^{-14} V^2$. Referring to Eq. (8.37), this is the "total output noise power." To determine the "output noise power due to source resistance," we can replace R_{in} with a noiseless resistor, as seen in Fig. 8.24. The SPICE netlist is given below.

```
*** Example 8.10 CMOS: Circuit Design, Layout, and Simulation ***

.control
destroy all
run
print all
.endc

.noise   V(Vout,0)     Vs     dec    100    1     1k

Rs       Vs     Vout   10k
Gin      Vout   0      Vout   0      1e-3
```

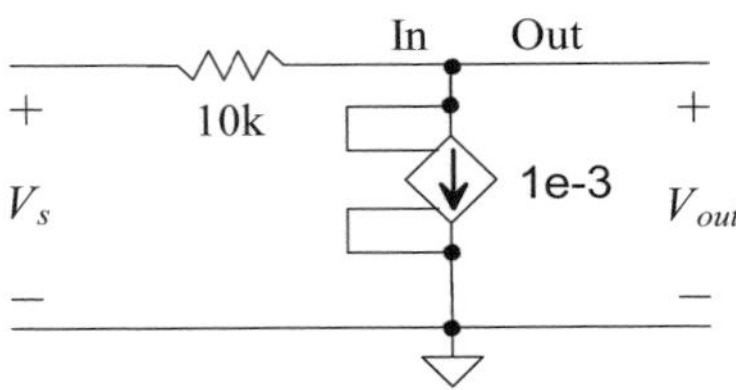

Figure 8.24 Replacing Rin with a noiseless resistor.

```
Vs      Vs    0      dc     0      ac     1

.print noise all
.end

TEMP=27 deg C
Noise analysis ... 100%
inoise_total = 1.655918e-13
onoise_total = 1.368527e-15
```

The noise factor is calculated from the simulation results as

$$F = \frac{1.5053 \times 10^{-14} V^2}{1.3685 \times 10^{-15} V^2} = 11$$

which is what we calculated using Eq. (8.41). ■

In a general SPICE simulation, we can determine F by eliminating the noise contributed by R_s. To do so, we first need to rewrite Eq. (8.37) as

$$F = \frac{\text{total output noise power}}{\text{total output noise power } - \text{total output noise power with } R_s \text{ noiseless}} \tag{8.46}$$

Example 8.11

Using SPICE and Eq. (8.46), calculate the noise factor of the circuit in Ex. 8.10 (verify, once again, that it is 11).

Again, as determined in Ex. 8.5, the total output noise power for this circuit is $1.5053 \times 10^{-14} V^2$. If we simulate the noise performance of the circuit in Fig. 8.21 with R_s being noiseless, we get

```
TEMP=27 deg C
Noise analysis ... 100%
inoise_total = 1.655918e-12
onoise_total = 1.368527e-14
```

Using Eq. (8.46), we get

$$F = \frac{1.5053 \times 10^{-14}}{1.5053 \times 10^{-14} - 1.3685 \times 10^{-14}} = 11$$

Simulating with a noiseless source resistor is the preferred way to simulate NF with SPICE. ■

Noise Temperature

Sometimes the term *noise temperature* is used to characterize the noise performance of an amplifier. In Fig. 8.25a we measure an amplifier's output RMS noise, $V_{onoise,RMS}$, with the input source connected but with $V_s = 0$ (the source contributes thermal noise to the output noise PSD). In Fig. 8.25b we refer this noise back to the input of the amplifier (this input-referred voltage includes both the amplifier and source resistance noise). After referring the noise back to the input, we think of the amplifier as being noiseless. In Fig. 8.23c we remove the input-referred source but, to have the same output noise as in (a) or (b), we change the effective temperature of the source's resistor and include its corresponding thermal noise model. By adjusting the temperature of the source resistor in our calculations (called the *noise temperature*, T_n), we can get the same total output noise power as in (a) or (b). We can relate the noise factor to the noise temperature by re-writing Eq. (8.36) as

$$F = \frac{\overbrace{4kTR_sB \cdot A^2 \cdot \left(\frac{R_{in}}{R_{in}+R_s}\right)^2}^{\text{Output noise from } Rs} + \overbrace{4kT_nR_sB \cdot A^2 \cdot \left(\frac{R_{in}}{R_{in}+R_s}\right)^2}^{\text{Amplifier's output noise alone}}}{\underbrace{4kTR_sB \cdot A^2 \cdot \left(\frac{R_{in}}{R_{in}+R_s}\right)^2}_{}\ \ ^{\text{Output noise from } Rs}} = 1 + \frac{T_n}{T} \quad \text{where } T_n \geq 0 \tag{8.47}$$

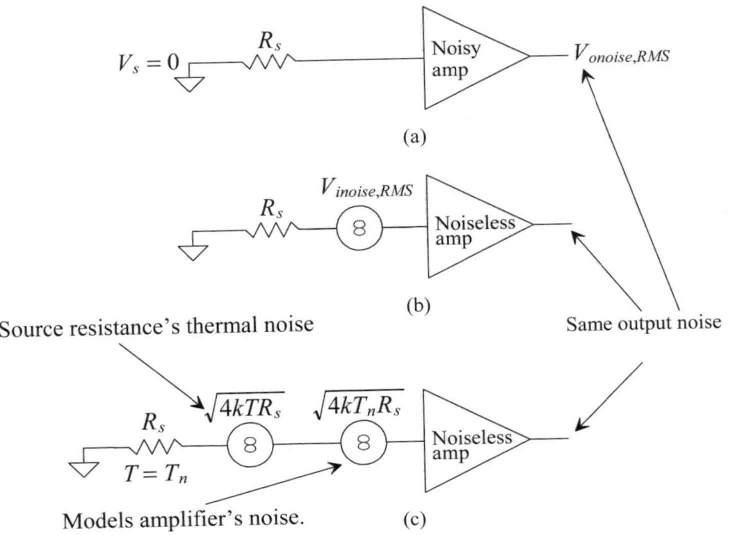

Figure 8.25 Determining the noise temperature of an amplifier.

where T is the temperature, in Kelvin, at which the total output noise was measured in (a). We know that if the amplifier is noiseless, $F = 1$. A noiseless amplifier has $T_n = 0$. Note that we could now relate Eq. (8.47) to Eq. (8.45) to write the optimum noise temperature of an amplifier in terms of $R_{s,opt}$. Again this wouldn't be too useful because it would state that we should use an amplifier with a low input-referred noise power to get a low value of T_n (good amplifier noise performance).

Table 8.2 provides a listing of *NF*s and the corresponding values of F and T_n. Note that an *NF*, less than 3 dB indicates that more than half of the output noise is due to the thermal noise of the source resistance.

Table 8.2 *NF*s and corresponding values of F and T_n (with $T = 300°$ K).

Noise figure, *NF*, dB	Noise factor, *F*	Noise temp, T_n, ° K
0	1	0
0.5	1.12	36.61
0.75	1.19	56.55
1	1.26	77.68
1.5	1.41	123.8
2	1.59	175.5
3	2	298.6
4	2.51	453.6
5	3.16	648.7

Averaging White Noise

Before moving on, let's comment on what happens if we average white noise (like thermal noise or shot noise discussed next). If the number of samples averaged is K, then the spectral density, and RMS value, of the noise is reduced by K. For thermal noise, we would rewrite Eq. (8.23) as

$$V_{RMS}^2 = \frac{4kTRB}{K} \text{ with units of } V^2 \tag{8.48}$$

Averaging can be thought of as lowpass filtering the random signal. Looking at Eq. (8.48), we see that the RMS value of the noise is reduced by the square root of the number of points averaged. Reviewing Ex. 8.6 (kT/C noise), we see the same result. Increasing C reduces the bandwidth of the thermal noise in the output of the RC circuit. This then causes the RMS value of the output noise to decrease by the root of C.

Figure 8.26a shows a noisy voltage signal with no averaging. The peak-to-peak amplitude of this waveform is roughly 5 divisions. In (b) we are averaging 4 of the waveforms in (a). We expect the amplitude to go to $5/\sqrt{4}$ or 2.5 divisions. If 16 averages are used, we should get an amplitude of 1.25 divisions, Fig. 8.26c. Finally, Fig. 8.26d shows how the noise is reduced if 64 waveforms are averaged. We would expect an amplitude of $5/\sqrt{64}$ or 0.625 divisions.

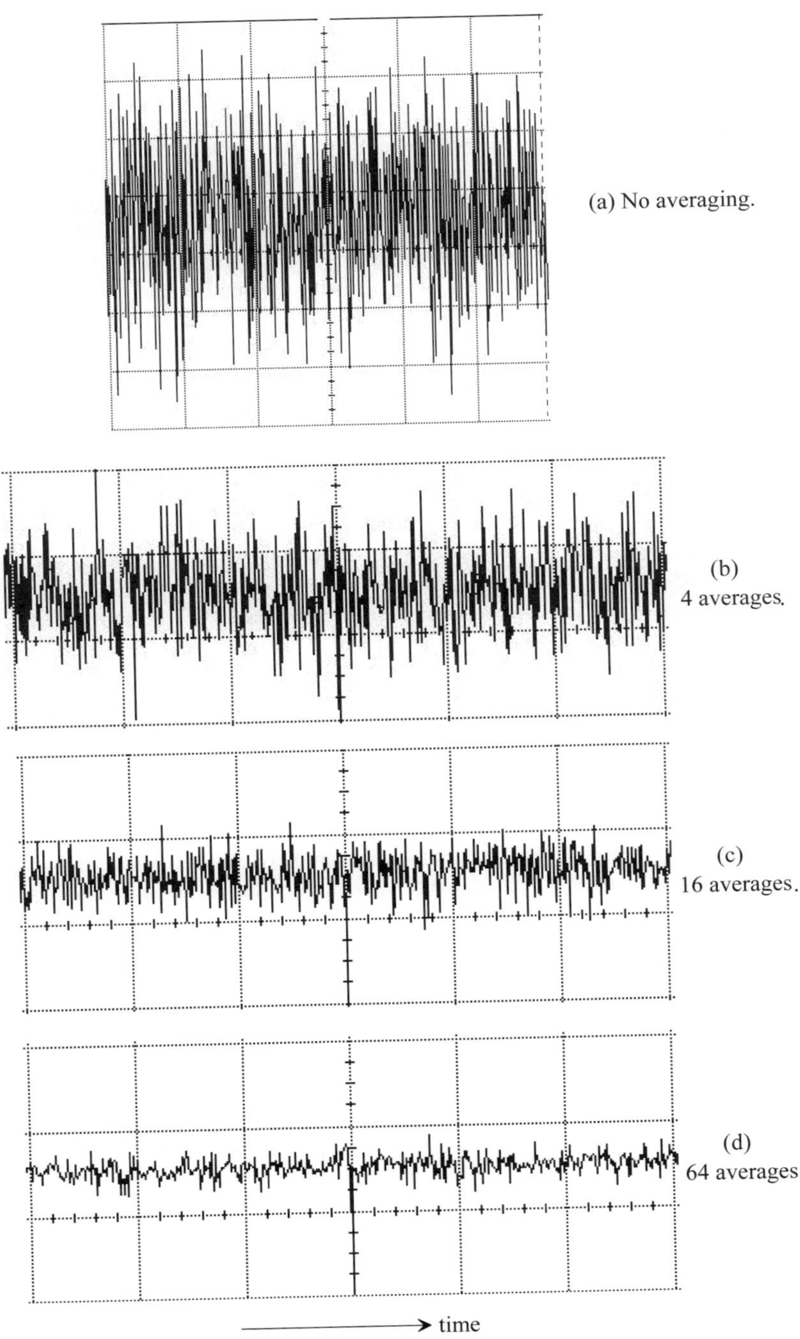

Figure 8.26 Averaging reduces the amplitude of white noise.

8.2.4 Shot Noise

Shot noise results from the discrete movement of charge across a potential barrier (e.g., a diode). To understand the origin of shot noise, imagine dumping marbles onto a table. As the marbles move around on the table, some will fall to the floor. The rate the marbles drop to the floor is random even though the rate we dump them onto the table may be constant. When current flows in a diode (when we dump the marbles onto the table), the movement of charge across the depletion region is random (the rate the marbles fall to the floor is random). For example, when a majority carrier, say an electron (a marble on top of the table), moves into the space-charge region (falls off the table), the electric field sweeps it across the junction (gravity causes the marble to fall to the floor). On the p-type side (the floor), where it is now a minority carrier, it diffuses out until it recombines (the marble moves until it stops). Note the name "shot" has to do with using shot from a shotgun shell (little balls of lead or, nowadays for waterfowl, steel used in shotgun shells) in the example above instead of marbles.

The power spectral density of shot noise can be determined empirically and is given by

$$I_{shot}^2(f) = 2qI_{DC} \text{ with units of } A^2/Hz \tag{8.49}$$

Shot noise is different from thermal noise (although both are modeled with a white noise PSD). Recall that the carriers can randomly move either into or out of the resistive material (with net zero charge transfer) due to thermal noise. Thermal noise is present even without a current flowing in the resistor. For shot noise to be present, we must have both a potential barrier and a current flowing. (There is net charge transfer with an average equal to the current flowing across the barrier.) The movement across the barrier is random and in one direction (the marbles don't hop back up on the table).

Shot noise is not present in long-channel MOSFETs (however, see the discussion in [9]). Shot noise is present in short-channel MOSFETs ($t_{ox} < 20$ nm) because of the gate tunneling current. When a long-channel MOSFET is operating in the saturation region, the entrance of charge into the depletion region, between the channel and the drain, has a discrete nature. However, this discrete movement is the result of thermal variations in the MOSFET's channel resistance and not carriers crossing a potential barrier and being swept up by an electric field.

Example 8.12

Estimate the output noise in the circuit seen in Fig. 8.27a. Verify your answer with SPICE. Assume that the diode's minority carrier lifetime (see Ch. 2) is 10 ns.

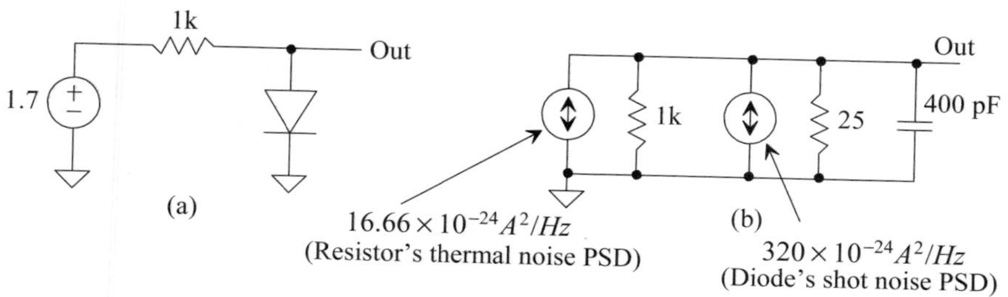

Figure 8.27 Calculating output noise for Ex. 8.12.

Figure 8.27b shows the circuit used for noise analysis. The diode is forward-biased at roughly 0.7 V. One volt is dropped across the 1k resistor and so 1 mA of current flows in the circuit. The diode's small-signal resistance is

$$r_d = \frac{V_T}{I_{DC}} = \frac{kT}{qI_{DC}} \approx 25\ \Omega$$

noting that the small-signal resistance of the diode does not generate thermal noise (it's a model) but that any diode series resistance will. (We set the diode's series resistance to zero in the SPICE model statement for this example.) The diode's storage capacitance is

$$C_S = \frac{I_{DC}}{V_T} \cdot \tau_T = \frac{\tau_T}{r_d} = 400\ pF$$

The diode's shot noise PSD is

$$I^2_{shot}(f) = 2qI_{DC} = 2 \cdot (1.6 \times 10^{-19}) \cdot 1mA = 320 \times 10^{-24} A^2/Hz$$

The resistor's thermal noise PSD was calculated in Ex. 8.5 as $16.66 \times 10^{-24} A^2/Hz$. The circuit output PSD is then the sum of the thermal and the shot noise contributions times the parallel connection of the 1k and 25 ohm resistors or

$$V^2_{onoise}(f) = 336.6\ \times 10^{-24} \cdot (25||1k)^2 = 200 \times 10^{-21}\, V^2/Hz$$

To calculate the output RMS noise voltage, we need to integrate this PSD from DC to infinity as indicated in Fig. 8.12. However, notice that this circuit has a single time constant of $(1k||25) \cdot 400pF$ or approximately 10 ns (the diodes minority carrier lifetime). The noise equivalent bandwidth, *NEB* from Eq. (8.15), is roughly

$$NEB = \frac{1}{2\pi \cdot 10ns} \cdot \frac{\pi}{2} = 25\ MHz$$

The RMS output noise is then

$$V_{onoise,RMS} = \sqrt{25 \times 10^6 \times 200 \times 10^{-21}} = 2.23\,\mu V$$

The SPICE netlist and simulation output are seen below.

```
*** Example 8.12 CMOS: Circuit Design, Layout, and Simulation ***

.control
destroy all
run
print all
.endc

.noise  V(Vout,0)       Vs      dec     100     1       100G

Vs      Vs      0       dc      1.7     ac      1
Rs      Vs      Vout    1k
D1      Vout    0       Diode

.model Diode D TT=10n Rs=0
.print noise all
.end
```

```
TEMP=27 deg C
Noise analysis ... 100%
inoise_total = 3.592225e-05
onoise_total = 5.257612e-12
```

The RMS output noise calculated using SPICE is $\sqrt{5.26 \times 10^{-12}} = 2.28\ \mu V$. Again note that the input-referred noise is meaningless in this example, as seen in Eq. (8.19) and the associated discussion. ■

Example 8.13

Estimate the output noise in the circuit seen in Fig. 8.27a if the diode's anode and cathode are swapped. Assume that the diode's zero bias depletion capacitance CJ0 is 25 fF, its built-in potential VJ is 1 V, and its grading coefficient m is 0.5.

The diode is now reverse-biased. The small reverse leakage current that flows in the diode will result in shot noise but it will be tiny compared to the thermal noise generated by the resistor. To calculate the shot noise PSD, we use the reverse leakage current for I_{DC} in Eq. (8.49). Because the diode is reverse-biased, it can be thought of as a capacitor. The value of the diode's depletion capacitance is calculated as

$$C_j = \frac{C_{j0}}{\left(1 + \frac{V_d}{VJ}\right)^m} = \frac{25\ fF}{\sqrt{1 + \frac{1.7}{1}}} = 15.2\ fF$$

We can use the circuit seen in Fig. 8.28 for the noise analysis. After examining this figure for a moment, we see that the output RMS noise is simply, from Ex. 8.6 and Eq. (8.24), kT/C noise. For this example then, $V_{onoise,RMS} = 521\ \mu V$ (verified with SPICE). The fundamental way to reduce the noise is to decrease the bandwidth of the circuit by increasing the capacitance shunting the output. Changing the resistor size has practically no effect on the RMS output noise. ■

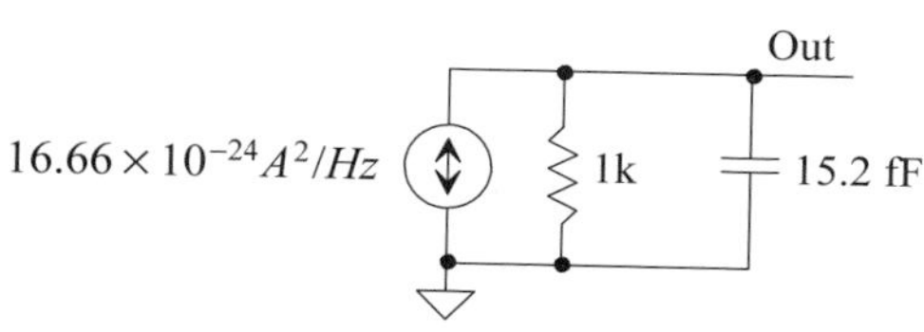

Figure 8.28 Calculating output noise for Ex. 8.12.

8.2.5 Flicker Noise

Flicker noise is a low-frequency noise and is probably the most important, and most misunderstood, noise source in CMOS circuit design. Flicker noise is also known as pink noise (a reddish color present in the lower range of the visible spectrum) or $1/f$ noise (pronouced "one over f" because its PSD, as we shall see, is inversely proportional to frequency). Before going too much further, let's do an example.

Example 8.14

Estimate the output noise PSD and the RMS value for the integrator circuit seen in Fig. 8.29 (which includes the only noise source, that is, the thermal noise from the resistor). Assume that the op-amp is ideal (noiseless, infinite gain, and infinite bandwidth).

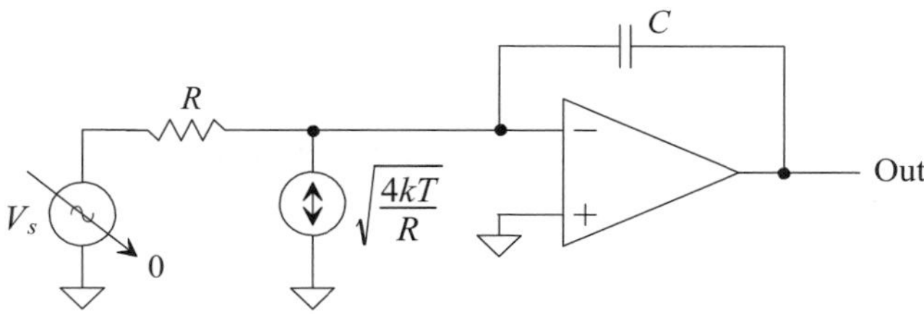

Figure 8.29 Integrating thermal noise.

The op-amp will keep the inverting input at the same potential as the noninverting input, that is, ground. This means that all of the thermal noise from the resistor will flow through the feedback capacitor (both sides of R are at ground). The integrator's output noise PSD is then

$$V_{onoise}^2(f) = \left|\frac{1}{j\omega C}\right|^2 \cdot \frac{4kT}{R} = \frac{4kT}{R} \cdot \frac{1}{(2\pi C)^2} \cdot \frac{1}{f^2} \text{ units of } V^2/Hz$$

Noise with a $1/f^2$ PSD shape is often called red noise, Fig. 8.30. Looking at this spectrum, we see that the noise PSD becomes infinite as we approach DC. Remember, from our discussion at the beginning of the chapter concerning how a spectrum analyzer operates, that a point representing the PSD at a particular frequency is the measured power, V_{RMS}^2, divided by the resolution bandwidth of the spectrum analyzer, f_{res} (that is, V_{RMS}^2/f_{res}). To make the low-frequency measurements, f_{res} must decrease, e.g., go from 1 Hz to 0.01 Hz to 0.00001 Hz, etc. A DC signal (or a sine wave for that matter, see Fig. 8.5) results in an infinite PSD point if we take $f_{res} \to 0$.

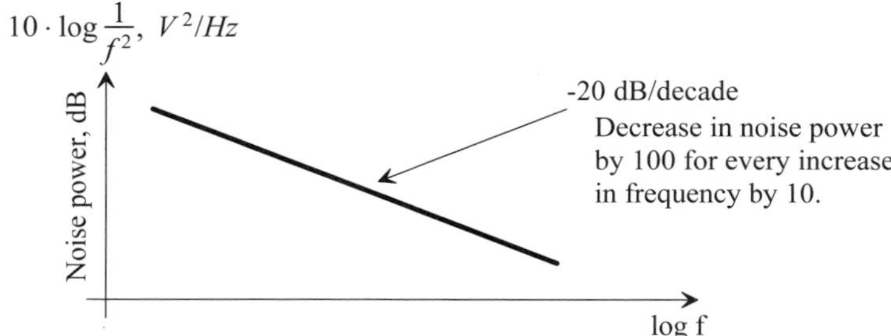

Figure 8.30 PSD of integrated thermal noise (red noise).

Let's use Eq. (8.10) to determine the RMS output noise voltage for this circuit

$$V^2_{onoise,RMS} = \int_{f_L}^{f_H} V^2_{onoise}(f)\cdot df = \frac{4kT}{R}\cdot\frac{1}{(2\pi C)^2}\cdot\left(\frac{1}{f_L}-\frac{1}{f_H}\right) \tag{8.50}$$

If we take $f_H \to \infty$, this equation becomes

$$V_{onoise,RMS} = \sqrt{\frac{4kT}{R}}\cdot\frac{1}{2\pi C}\cdot\frac{1}{\sqrt{f_L}} \tag{8.51}$$

If we call the length of time we integrate (or measure an input signal since integrators are often used for sensing) T_{meas}, then we can get an approximation for lower integration frequency as

$$T_{meas} \approx \frac{1}{f_L} \tag{8.52}$$

or, rewriting Eq. (8.51),

$$V_{onoise,RMS} \approx \sqrt{\frac{4kT}{R}}\cdot\frac{1}{2\pi C}\cdot\sqrt{T_{meas}} \tag{8.53}$$

This result is practically very important. If we average (the same as integrating, which is just summing a variable) a signal with a $1/f^2$ noise spectrum (integrated thermal noise), the RMS output noise voltage actually gets bigger the longer we average (unlike averaging thermal noise, see Eq. [8.48]). A signal containing $1/f^3$ noise (e.g., integrating flicker noise) shows a linear increase in its RMS noise voltage with measurement time. The result is an SNR that doesn't get better, and may get worse, by increasing the measurement time. These results pose a limiting factor when imaging in astronomy. The night sky contains this type of noise (flicker, red, etc.). We can make images brighter by exposing our imaging system for longer periods of time but the images don't get clearer. ■

Flicker noise (a name used because when viewed in a light source flickering is observed) is modeled using

$$I^2_{1/f}(f) \text{ or } V^2_{1/f}(f) = \frac{FNN}{f} \text{ units of } A^2/Hz \text{ or } V^2/Hz \tag{8.54}$$

where *FNN* is the flicker noise numerator. The RMS value of a $1/f$ noise signal is

$$V^2_{noise,RMS} = \int_{f_L}^{f_H} \frac{FNN}{f}\cdot df = FNN\cdot\ln\frac{f_H}{f_L} \text{ units of } V^2 \tag{8.55}$$

If f_H is 100 GHz and f_L is once every 10^{-10} Hz (roughly once every 320 years), this equation becomes

$$V_{noise,RMS} = 7\cdot\sqrt{FNN} \text{ Volts} \tag{8.56}$$

where *FNN* has units of V^2. After reviewing Ex. 8.14, we can write

$$V_{noise,RMS} \propto \sqrt{\ln T_{meas}} \tag{8.57}$$

For all intents and purposes, $V_{noise,RMS}$ stabilizes as measurement time increases.

Example 8.15
The input-referred noise voltage spectral density of the TLC220x (a low-noise CMOS op-amp) is seen in Fig. 8.31. The input-referred noise current is not shown but listed on the data sheet as typically $0.6\ fA/\sqrt{Hz}$ (a tiny number and so it will be neglected in the following example). For the unity follower op-amp seen in Fig. 8.32, estimate the RMS output noise voltage.

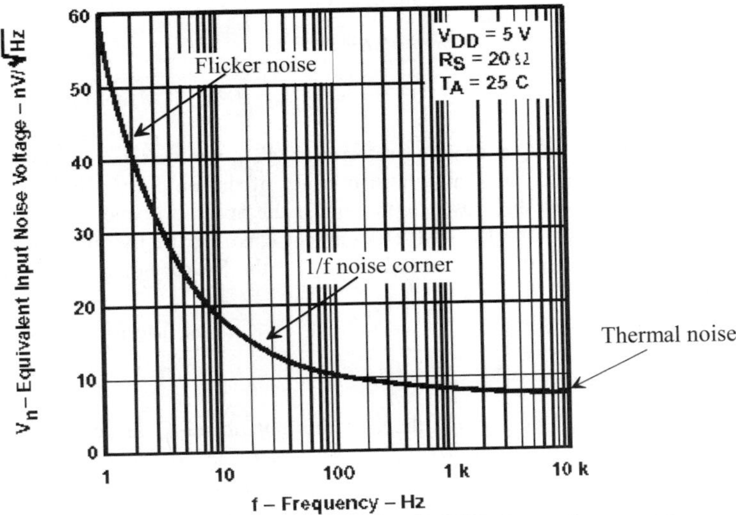

Figure 8.31 Input-referred noise for the TLC220x low-noise op-amp.

The input-referred thermal noise PSD is roughly $\left(8\ nV/\sqrt{Hz}\right)^2$. The square root of the flicker noise input-referred PSD is

$$V_{inoise,1/f}(f) = \sqrt{\frac{FNN}{f}} = \frac{56\,nV}{\sqrt{Hz}} \text{ by looking at } f = 1\text{ Hz}$$

then

$$FNN = (56\ nV)^2 = 3.14 \times 10^{-15}\,V^2$$

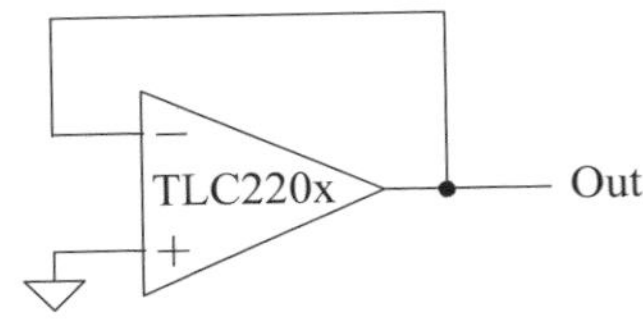

Figure 8.32 Estimating RMS output noise in a voltage follower.

As a quick check we see, in Fig. 8.31, that at 4 Hz the spectral density is roughly 28 nV/$\sqrt{Hz}$ or

$$V_{inoise,1/f}(f) = \sqrt{\frac{FNN}{f}} = \sqrt{\frac{3.14\times10^{-15}V^2}{4\ Hz}} = \frac{56\ nV}{\sqrt{4\ Hz}} = 28\ nV/\sqrt{Hz}$$

Note, in Fig. 8.31, that the point where the thermal noise PSD gets comparable to the 1/*f* noise PSD is sometimes called the 1/*f* noise corner. Because the gain of the op-amp is one, we can write the input or output noise PSD as

$$V_{onoise}^2(f) = V_{inoise}^2(f) = \overbrace{\frac{3.14\times10^{-15}}{f}}^{V_{inoise,1/f}^2(f),\ \text{flicker noise}} + \overbrace{64\times10^{-18}}^{\text{thermal noise}} \text{ with units of } V^2/Hz$$

To determine the output or input-referred RMS noise voltage, we integrate this noise spectrum. The gain-bandwidth product of the op-amp is roughly 2 MHz. This means, because the closed-loop gain of the op-amp is one, its f_{3dB} frequency is 2 MHz. Knowing the f_{3dB} of the circuit, at least for the thermal noise, we can use Eq. (8.15). With the help of Eq. (8.56) for the flicker noise contributions, we can then write

$$V_{onoise,RMS}^2 = V_{inoise,RMS}^2 = \overbrace{49\cdot(3.14\times10^{-15})}^{\text{using Eq. (8.56)}} + \overbrace{(64\times10^{-18})\cdot(2\times10^6)\cdot\frac{\pi}{2}}^{\text{using Eq. (8.15)}}$$

or noting the main contributor here is thermal noise. (The contributions from flicker noise, relative to the thermal noise present, fall off quickly above 10 Hz. Therefore, not using the amplifier's bandwidth when calculating the flicker noise contributions doesn't result in any significant error.) Solving this equation gives

$$V_{onoise,RMS} = V_{inoise,RMS} = 15\ \mu V$$

or roughly 100 μV peak-to-peak if the noise has a Gaussian probability distribution function, Fig. 8.33, (where the standard deviation 1σ is the RMS value of the noise voltage and the peak-to-peak value is roughly 6σ).

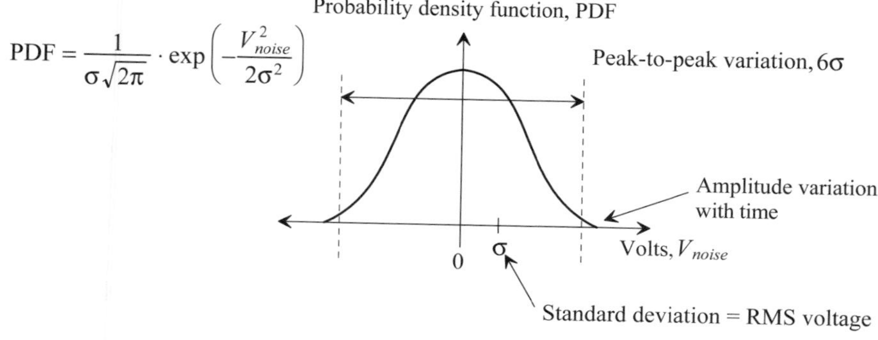

Figure 8.33 Gaussian probability distribution.

Note that this last result is useful for any of the RMS values we've calculated. To get an estimate for the peak-to-peak amplitude of the noise, we simply multiply by 6. Going the other way, we can estimate the RMS value of a noise waveform by looking at its peak-to-peak amplitude in the time domain and dividing by 6. ■

Example 8.16
On the TLC220x data sheet, the equivalent RMS input noise voltage for frequencies between 0.1 Hz and 1 Hz is 0.5 μV peak-to-peak. Verify this result with the data given in Fig. 8.31.

Using the numbers from the previous example, we see, in this 0.9 Hz bandwidth, that the contributions from thermal noise are negligible. The contributions from Flicker noise, using Eq. (8.55), are

$$V^2_{inoise,RMS} = \int_{f_L}^{f_H} \frac{3.14 \times 10^{-15}}{f} \cdot df = 3.14 \times 10^{-15} \cdot \ln \frac{1}{0.1}$$

or

$$V_{inoise,RMS} = 85 \; nV$$

The peak-to-peak input-referred noise voltage is roughly six times this or 510 nV (verifying the specification matches the data we get with the plot). ■

Example 8.17
Estimate the input-referred noise for the circuit seen in Fig. 8.34.

This amplifier is a transimpedance amplifier, that is, current input and voltage output. The amplifier takes the current generated by the photodiode and converts it into an output voltage. The low-frequency gain of the amplifier is R_F or 100k. The 1,000 pF feedback capacitor limits the bandwidth of the circuit to reduce noise. The photodiode is reverse-biased. Its noise contributions are related to the conversion of light to electrons (which we won't add to our noise analysis). The diode's reverse leakage current will be small and so the shot noise contributions to the circuit noise will be insignificant. The op-amp holds the diode's anode at a fixed potential (ground). The diode's junction capacitance won't affect the noise in the circuit (both sides of the capacitance are at AC ground).

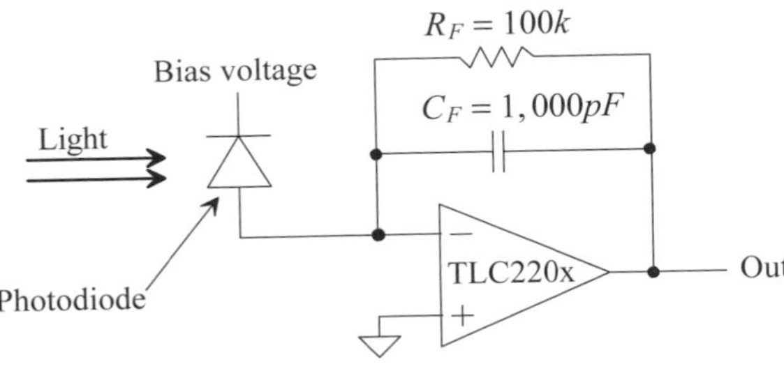

Figure 8.34 Estimating RMS output noise in a transimpedance amplifier.

The basic noise circuit for Fig. 8.34 is seen in Fig. 8.35. All of the output is fed back to the input so that the op-amp's thermal and flicker noise sources appear directly in the output signal (to keep the inverting and noninverting terminals at the same potential). The resistor's thermal noise (keeping in mind that we only look at one noise source at a time so that the inverting and noninverting inputs of the op-amp are at 0V), will create an output voltage that is the product of the noise current and the parallel combination of the 100k and 1,000pF.

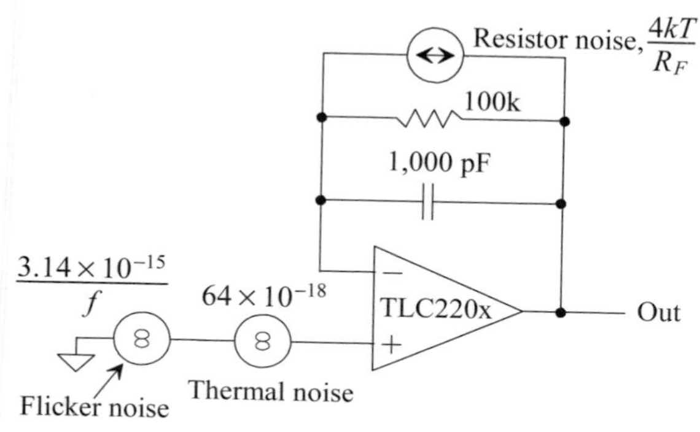

Figure 8.35 Noise model for the transimpedance amplifier in Fig. 8.34.

The output noise PSD can then be written as

$$V_{onoise}^2(f) = \frac{3.14\times10^{-15}}{f} + 64\times10^{18} + \overbrace{\frac{4kT}{R_F}\cdot\left|\frac{R_F\cdot 1/j\omega C_F}{R_F+1/j\omega C_F}\right|^2}^{\frac{4kTR_F}{1+(2\pi f\cdot R_F C_F)^2}}$$

We should recognize the last term in this PSD as simply kT/C noise (Ex. 8.6). The bandwidth of the amplifier can be determined using Fig. 8.36 as

$$A(f) = R_m = \frac{v_{out}}{i_s} = R_F || \frac{1}{j\omega C_F} = \frac{R_F}{1+j\omega R_F C_F}$$

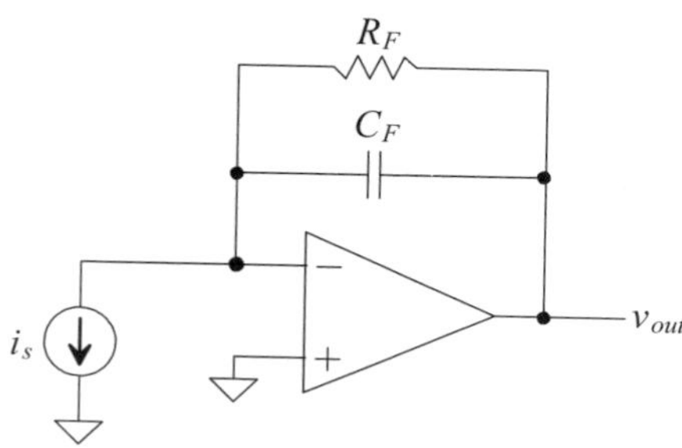

Figure 8.36 Determining the transfer function of a transimpedance amplifier.

or using

$$f_{3dB} = \frac{1}{2\pi R_F C_F}$$

which, for the present example, is 1.59 kHz. We can estimate the output RMS noise as

$$V^2_{onoise,RMS} = \overbrace{49 \cdot 3.14 \times 10^{-15}}^{=154\times10^{-15}} + \overbrace{64 \times 10^{-18} \cdot \frac{\pi}{2} \cdot 1.59\ kHz}^{=160\times10^{-15}} + \overbrace{\frac{kT}{C_F}}^{=4\times10^{-12}}$$

or

$$V_{onoise,RMS} \approx 2.07\ \mu V$$

noting that the thermal noise from R_F dominates the output RMS noise (the kT/C noise). Our estimate for the flicker noise is high. However, it's easy to throw some numbers in for high and low frequencies in Eq. (8.55) and convince ourselves that the result isn't impacted too much. For example, we used $\ln\frac{10^{11}}{10^{-10}} \approx 49$ (yes, it's closer to 48 but we use 49 because the square-root is a whole number, i.e., 7) when we derived Eq. (8.55). If we were to use instead $\ln\frac{10^5}{10^{-10}}$ (a million times smaller upper frequency), we get 34.5 (a square root of roughly 6). The impact on the result is too small to worry about (unless, of course, the bandwidth is very narrow as in Ex. 8.16).

Dropping the feedback capacitance to 1 pF increases the bandwidth to 1.59 MHz and increases the RMS output noise to approximately 64 μV.

Our output signal is determined using $i_s \cdot R_F$. An SNR_{out} of 0 dB, in this example with C_F = 1,000 pF, would correspond to an i_s of roughly 20.7 pA. If we wanted an input signal to produce an output that is larger than the peak-to-peak value of the output noise (here, from Fig. 8.33, $> 6 \times 2.07\,\mu V$), then $i_s > 125\ pA$. ■

Example 8.18
Verify Ex. 8.17 using SPICE. Use a voltage-controlled voltage source for the op-amp (neglect the op-amp noise).

The schematic used for simulations is seen in Fig. 8.37. Instead of connecting the op-amp's noninverting input (the + input) to ground directly, we connect it to

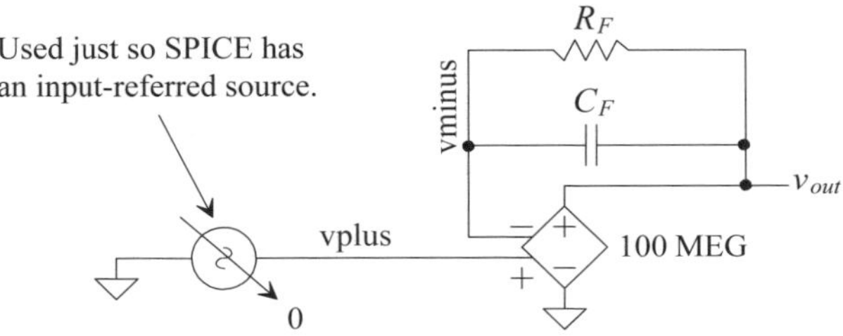

Figure 8.37 Using a voltage-controlled voltage source for an op-amp in SPICE.

ground via a zero volt DC supply so that, in the simulation, we have an "input." (Recall comments about how the AC magnitude of this input source affects the input-referred noise calculation on the bottom of page 227.) The SPICE netlist is

```
Comment line; the first line of a netlist is ignored by SPICE
.noise  V(Vout,0)        Vplus  dec    100    1        1000G
Vplus   Vplus   0        dc     0      ac     1
Rf      Vout    Vminus 100k
Cf      Vout    Vminus          1000pf
Eopamp          Vout     0      Vplus  Vminus 100MEG
.print noise all
.end
```

The SPICE output is

```
TEMP=27 deg C
Noise analysis ... 100%
inoise_total = 4.142190e-12
onoise_total = 4.142190e-12
```

For the 1,000 pF capacitor, this gives 2.03 μV. Resimulating with the 1 pF capacitor, gives 64.3 μV. ■

8.2.6 Other Noise Sources

Thermal and flicker noise are the dominant noise sources in CMOS circuit design; however, there are other noise mechanisms. In this section we briefly discuss other types of noise, including burst (popcorn) and avalanche noise.

Burst (Popcorn) Noise

Burst noise is seen, in the time-domain, in Fig. 8.38. Burst noise is often called popcorn noise. When a signal containing burst noise is played on a speaker, the popcorn noise component sounds like popping corn. The origin of popcorn noise is thought to be related to how generation-recombination, GR, centers affect the diffusion of carriers in a pn junction. If gold, for example, is introduced into a semiconductor, its inclusion results in nonuniformity in the silicon crystal lattice and thus additional GR sites. With a large number of these sites present, the variation in the diffusion rate, that is, the rate that the carriers diffuse away from a depletion region, can be significant. Since a popcorn noise burst, Fig. 8.38, can last hundreds of microseconds, it is likely that at times more sites are filled and thus more carriers can diffuse further into the diode with a smaller number landing in a recombination site. This results in a build up of stored charge in the diode. The popping in the signal occurs when this additional stored charge exits the diode.

Although popcorn noise doesn't have a Gaussian noise distribution, its PSD can be modeled using

$$I^2_{popcorn}(f) = \frac{K_{popcorn} \cdot I_{DC}}{1 + \left(\frac{f}{f_{3dB}}\right)^2} \text{ units of } A^2/Hz \tag{8.58}$$

where $K_{popcorn}$ has units of A/Hz and f_{3dB} (e.g., 100 Hz) is the point where the noise starts to roll off and is related to the number of pops per second, Fig. 8.39. Note that at very low frequencies, popcorn noise is white (like thermal and shot noises). However, at higher frequencies, the noise rolls off as $1/f^2$.

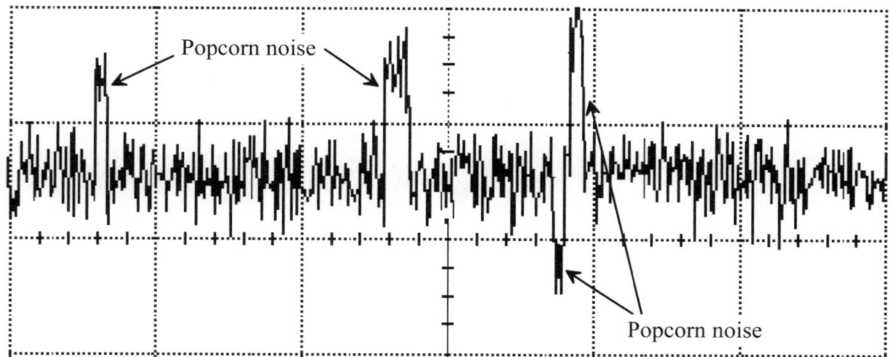

Figure 8.38 Popcorn (burst) noise viewed in the time domain.

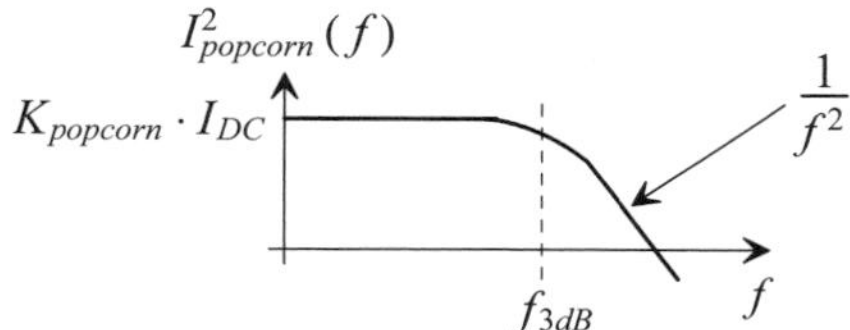

Figure 8.39 PSD of popcorn noise.

Excess Noise (Flicker Noise)

Thermal noise in a resistor is present independent of the amount of current flowing in the resistor, as seen in Eq. (8.23). However, there is also flicker noise present in the resistor, dubbed *excess* noise, see Eq. (8.54). Flicker noise is present whenever a direct current flows in a discontinuous material (e.g., a material's surface or the interface between a MOSFET's gate oxide and the silicon used as the channel of the MOSFET). The electrons "jump" from one location to the next while sometimes being randomly trapped and released. Excess (flicker) noise is not dependent on temperature, unlike thermal noise, and so it can be especially troublesome when making low noise, temperature, and frequency measurements.

Avalanche Noise

When the electric field in a reverse-biased pn junction gets really large, carriers are accelerated to a point where they can strike the lattice and dislodge additional electron-hole pairs. The result is an increase in diode current. This additional generation of current is termed avalanche multiplication and is used in many Zener diodes to break the diode down. Tunneling is used in the lower voltage, < 6 V, Zener diodes which may have a shot noise spectrum because carriers tunnel across a potential barrier. Avalanche breakdown is a very noisy process. A typical PSD (noting a dependence on the current flowing in the reverse-biased junction) is $10^{-16}\, V^2/Hz$. The PSD is generally white, which has led to the use of Zener diodes as white noise generators, Fig. 8.40. In low-noise design it can be challenging to design a low-noise voltage reference using a Zener diode.

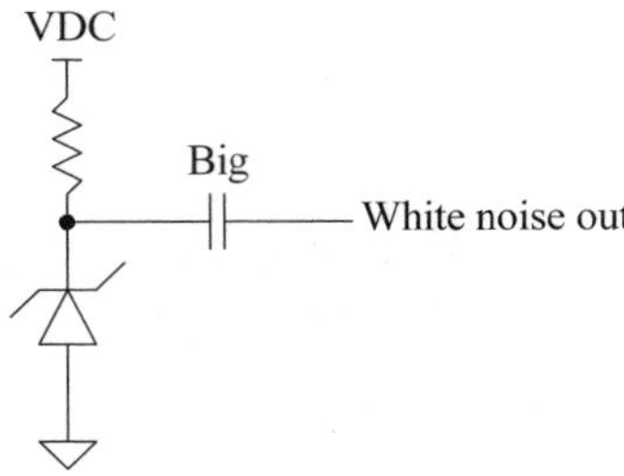

Figure 8.40 A Zener diode-based white noise generator.

Avalanche noise can also be troublesome in small geometry CMOS processes where a large drain-source voltage can lead to avalanche multiplication (and thus make MOSFET thermal noise appear to be very large).

8.3 Discussion

In this section we'll provide additional discussions and examples to help explain electrical noise.

8.3.1 Correlation

Consider the two signals seen in Fig. 8.41. If the RMS values of $v_1(t)$ and $v_2(t)$ are called V_{1RMS} and V_{2RMS}, respectively, then the power dissipated by the resistor is

$$P_{AVG} = \frac{V_{1RMS}^2 + V_{2RMS}^2}{R} \tag{8.59}$$

as discussed in Sec. 8.1.1 and used throughout the chapter. Let's look at this in more detail because it's easy to show that under certain circumstances this equation is misleading (and wrong!). In particular, if the two voltages sources are correlated, we can get a totally different value for the power the resistor dissipates. For example, we know that the voltage dropped across the resistor is $v_1(t) + v_2(t)$. What happens if $v_1(t) = V_P \cdot \sin(2\pi f \cdot t)$ and $v_2(t) = V_P \cdot \sin(2\pi f \cdot t + \pi)$ (which, of course, is the same as $v_2(t) = -V_P \cdot \sin[2\pi f \cdot t]$). The two sinewaves sum to zero. The voltage dropped across the resistor is zero and so is the power dissipated by the resistor. In a noise analysis, if our

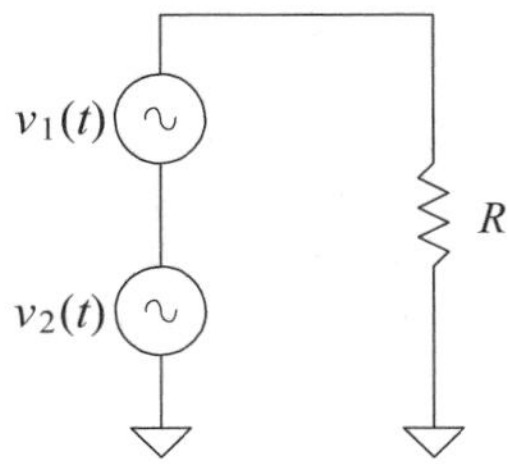

Figure 8.41 Series combination of voltage signals.

noise sources are correlated, then we can't simply add their noise contributions on the output of a network as we've done throughout the chapter. If the noise sources are correlated, the analysis can be much more challenging.

To describe how signals can be correlated in a little more detail, we can write, from Eq. (8.1),

$$P_{inst}(t) = \frac{V^2(t)}{R} = \frac{[v_1(t)+v_2(t)]^2}{R} \tag{8.60}$$

$$= \frac{v_1^2(t)}{R} + \frac{v_2^2(t)}{R} + \frac{2v_1(t)v_2(t)}{R} \tag{8.61}$$

The first two terms in the equation are independent. We can calculate the RMS value of these terms, and thus the average power supplied to R from each source, using

$$P_{AVG} = \frac{V_{RMS}^2}{R} = \frac{1}{T_{meas}} \cdot \int_0^{T_{meas}} \frac{v^2(t)}{R} \cdot dt \tag{8.62}$$

where T_{meas} is the amount of time used to measure the average power. Of course, we use this equation twice (once for each source) and then sum the average powers to get the average power supplied to R (which is Eq. [8.59]). If no correlation exists between $v_1(t)$ and $v_2(t)$, then the third term in Eq. (8.61) is zero. A simple example of two signals with no correlation are sine waves with different frequencies. Because each sine wave will go positive and negative (the signals are no longer squared and thus always positive), the summation of the signal (the integration) results in zero power added or subtracted from the total power. An example of 100% correlation is when

$$v_1(t) = v_2(t) = V_P \cdot \sin 2\pi f \cdot t \tag{8.63}$$

The signal applied to R in Fig. 8.41 has a peak amplitude of $2V_P$, an RMS value of $2V_P/\sqrt{2}$, and supplies a power to R of $2V_P^2/R$. Using Eqs. (8.61) and (8.62), we can write

$$P_{AVG} = \frac{V_P^2}{2R} + \frac{V_P^2}{2R} + \overbrace{\frac{2V_P^2}{2R}}^{\text{100\% correlation}} = \frac{2V_P^2}{R} \tag{8.64}$$

We showed an example of -100% correlation a moment ago when

$$v_1(t) = V_P \cdot \sin 2\pi f \cdot t \text{ and } v_2(t) = -V_P \sin 2\pi f \cdot t \tag{8.65}$$

or

$$P_{AVG} = \frac{V_P^2}{2R} + \frac{V_P^2}{2R} + \overbrace{\frac{2(V_P)(-V_P)}{2R}}^{-100\% \text{ correlation}} = 0 \tag{8.66}$$

In general, the correlation between two waveforms is specified using C as

$$V_{1RMS}^2 + V_{2RMS}^2 + 2CV_{1RMS}V_{2RMS} \tag{8.67}$$

where $-1 \le C \le 1$. If C is zero, no correlation exists (again what we've used throughout the chapter). An example of two signals partially correlated are two sine waves at the same frequency but different amplitudes or phases. Note that this discussion used RMS values for the signals. However, it is also possible to discuss correlation in PSDs.

Correlation of Input-Referred Noise Sources

When we discussed input-referred noise and derived Eq. (8.32), we assumed zero correlation between $I_{inoise,RMS}$ and $V_{inoise,RMS}$. We can rewrite Eq. (8.32) without the thermal noise from the source resistance and with the help of Fig. 8.42 as

$$V^2_{onoise,RMS} = \left(\frac{AR_{in}}{R_s + R_{in}}\right)^2 [V_{inoise,RMS} + I_{inoise,RMS} \cdot R_s]^2 \tag{8.68}$$

If the correlation is zero, Eq. (8.68) is equivalent to Eq. (8.32) (without thermal noise contributions from R_s). With correlation, we can write

$$V^2_{onoise,RMS} = \left(\frac{AR_{in}}{R_s + R_{in}}\right)^2 \left[V^2_{inoise,RMS} + I^2_{inoise,RMS} \cdot R_s^2 + 2C \cdot V_{inoise,RMS} \cdot I_{inoise,RMS}\right] \tag{8.69}$$

To use the input-referred noise models, we generally assume no correlation ($C = 0$) between $I_{inoise,RMS}$ and $V_{inoise,RMS}$. In a practical circuit, the correlation can be significant because both are derived from the same noise mechanisms.

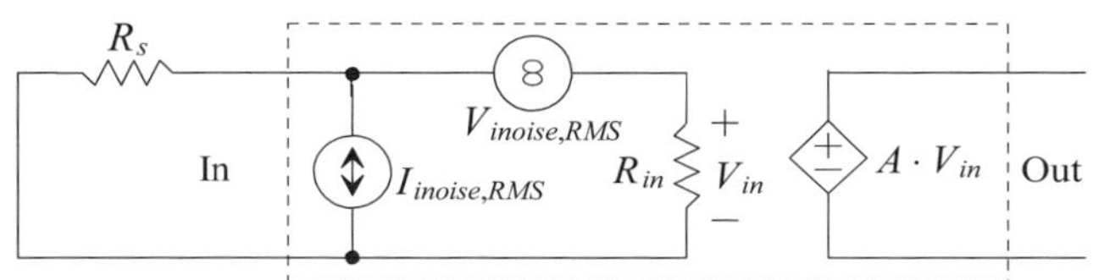

Figure 8.42 Correlation calculation of input-referred noise sources.

Complex Input Impedance

We've used a real input resistance in all of our calculations in this chapter. It's straightforward to extend our discussions and derivation to a general input impedance, Z_{in}. Our input voltage divider changes to

$$\frac{R_{in}}{R_s + R_{in}} \rightarrow \frac{Z_{in}}{R_s + Z_{in}} \tag{8.70}$$

When we multiply this divider by, say, V_s, to get V_{in}, we use

$$V_{in} = V_s \cdot \frac{|Z_{in}|}{|R_s + Z_{in}|} \text{ or } V^2_{in} = V^2_s \cdot \frac{|Z_{in}|^2}{|R_s + Z_{in}|^2} \tag{8.71}$$

It's important to take the magnitude before squaring. For example, if $Z_{in} = 1/j\omega C_{in} = -j/\omega C_{in}$(the input impedance is a capacitor), then the correct way of calculating the denominator in Eq. (8.71) is

$$|R_s + Z_{in}|^2 = \left(\sqrt{R_s^2 + (1/\omega C_{in})^2}\right)^2 = R_s^2 + (1/\omega C_{in})^2 \tag{8.72}$$

The incorrect way is

$$|R_s + Z_{in}|^2 \neq |(R_s + 1/j\omega C_{in})(R_s + 1/j\omega C_{in})| = |\overbrace{R_s^2 + 1/\omega^2 C_{in}^2}^{\text{real term}} + j\cdot\overbrace{(-2/\omega C_{in})}^{\text{imaginary term}}|$$

$$= \sqrt{\left(R_s^2 + 1/\omega^2 C_{in}^2\right)^2 + (2/\omega C_{in})^2} \text{ (again wrong!)}$$

(8.73)

It appears that there may be correlation in the equation above but it is really just bad complex algebra. We used the magnitude squared earlier in Eq. (8.13), that is, $|1 + j(f/f_{3dB})|^2$ or $1 + (f/f_{3dB})^2$. The point is to always take the magnitude of the expressions prior to squaring.

Let's use a complex input impedance, Z_{in}, to derive the optimum source resistance, Eq. (8.44). As we saw in Ex. 8.7 and 8.9, we measure the output noise with R_s shorted (or a big capacitor placed across the input of the circuit, which is essentially a short for all frequencies of interest) or opened to determine the input-referred noise sources. If the measured output noise (excluding thermal noise contributions from R_s) doesn't change with variations in R_s (quite common in amplifiers whose bias is independent of R_s like AC-coupled input amplifiers, but not the case for the circuit in Ex. 8.9), that is,

$$V_{onoise,RMS,Rs=\infty} = V_{onoise,RMS,Rs=0} = V_{onoise,RMS} \quad (8.74)$$

and the input-referred noise sources are uncorrelated then, for a real input resistance,

$$R_{s,opt} = R_{in} \quad (8.75)$$

We may think that we can get both maximum power transfer and the best noise performance by matching the input resistance to the source resistance. However, if we use a complex input impedance, Eq. (8.75) becomes

$$R_{s,opt} = |Z_{in}| = |R_{in} + jX_{in}| = \sqrt{R_{in}^2 + X_{in}^2} \quad (8.76)$$

which can be valid without regard for maximum power transfer. Knowing that the requirement for maximum power transfer is making the source impedance the complex conjugate of the load impedance, we see that if the source impedance is purely real, it will be impossible to achieve maximum power transfer when the input resistance has an imaginary component. In radio-frequency circuits (narrow bandwidths where the impedances are relatively constant), impedance transformation circuits are used to improve both power transfer and noise performance. From the source side, the input impedance is transformed to maximize the transfer of power. From the input side, the source impedance is transformed to provide the best noise performance.

Example 8.19

Suppose an amplifier has an output noise PSD, $V_{onoise}(f)$, of 1 $\mu V/\sqrt{Hz}$ (we get this PSD if R_s is a short or an open), a gain of 100, and an input capacitance of 1 pF (infinite input resistance). Determine the input-referred noise sources for the amplifier.

With the input shorted, Fig. 8.43a, we get an input-referred noise voltage spectral density of

$$A \cdot V_{inoise}(f) = V_{onoise}(f) = 1\ \mu V/\sqrt{Hz}$$

or since $A = 100$

$$V_{inoise}(f) = 10\ nV/\sqrt{Hz}$$

With the input opened, Fig. 8.43b, we get

$$I_{inoise}(f) \cdot |Z_{in}| \cdot A = \frac{I_{inoise}(f)}{2\pi f \cdot 1pF} \cdot 100 = 1\ \mu V/\sqrt{Hz}$$

or

$$I_{inoise}(f) = 62.8 \times 10^{-21} \cdot f\ A/\sqrt{Hz}$$

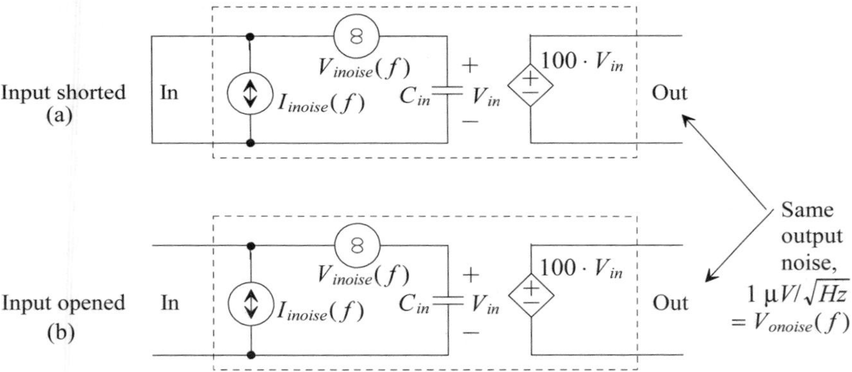

Figure 8.43 Determining input-referred noise in Ex. 8.19.

Note that the spectral density increases with f (Fig. 8.44). The noise current contributions to the output noise is insignificant until the frequency gets comparable to $1/(2\pi R_s C_{in})$ (noting that squaring $I_{inoise}(f)$ and integrating to find the RMS value results in an f^3 term). Unless the source resistance, frequencies of interest, or input capacitance are relatively large, the single input-referred noise voltage source is all that is needed to model the amplifier's output noise. ■

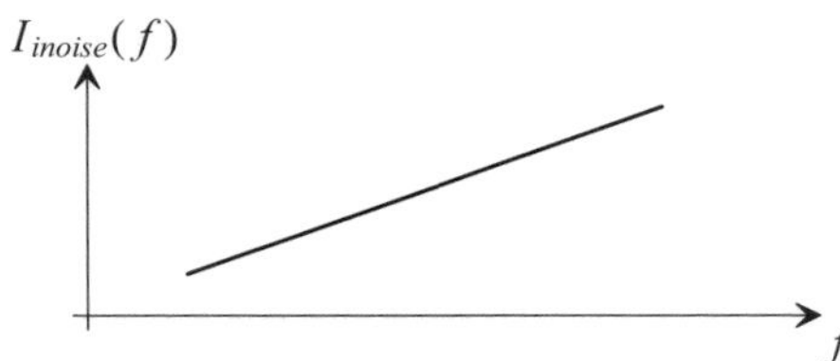

Figure 8.44 Input-referred noise current increasing with frequency when the input impedance is capacitive.

8.3.2 Noise and Feedback

Figure 8.45a shows the basic block diagram of a feedback circuit. In Fig. 8.45b we show the input-referred noise source. The output of the summer in (a) or (b) can be written as $V_{in} - \beta \cdot V_{out}$. The output of the circuit in (b) is

$$V_{out} = [(V_{in} - \beta V_{out}) + V_{inoise,RMS}] \cdot A \tag{8.77}$$

or

$$V_{out} = \frac{A}{1+\beta A} \cdot (V_{in} + V_{inoise,RMS}) \tag{8.78}$$

In other words, feedback doesn't affect the circuit's noise performance. The input-referred noise adds directly to the input signal independent of the feedback. In practical circuits, the addition of resistors to provide feedback degrades the noise performance of the amplifier.

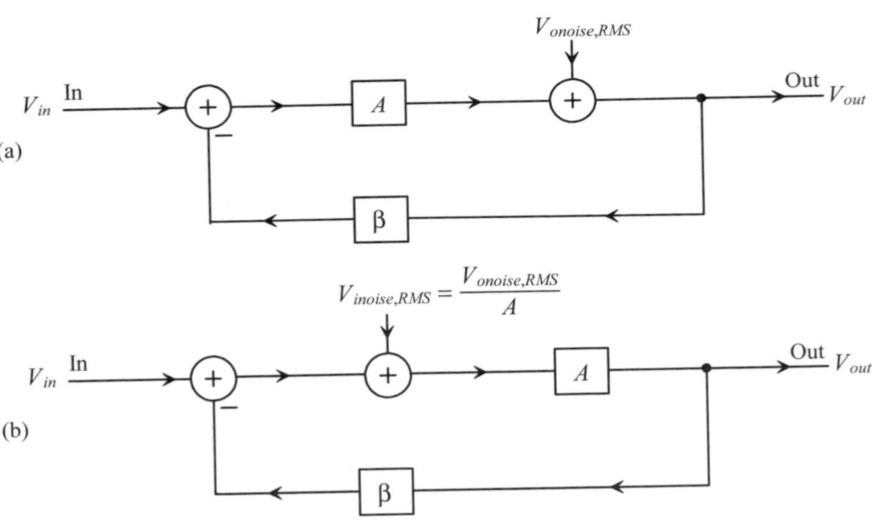

Figure 8.45 Noise in a feedback circuit.

Op-Amp Noise Modeling

Figure 8.46a shows how input-referred noise sources are added to an op-amp to model output noise. The input-referred noise current is connected to the inverting terminal of the op-amp, while the input-referred noise voltage is connected in series with the noninverting terminal. Figure 8.46b shows the general implementation of a feedback amplifier using an op-amp, including noise sources modeling the resistor's thermal noise. If the input to the op-amp circuit is on the left side of R_1, then the gain to the output is inverting, $(-R_2/R_1)$. If the input is connected directly to the noninverting terminal of the op-amp, then the gain is positive, $(1 + R_2/R_1)$. Remember, we look at only one noise source at a time (superposition) when we calculate the total output noise. Also, remember that the op-amp, through the feedback action, holds the inverting terminal at ground.

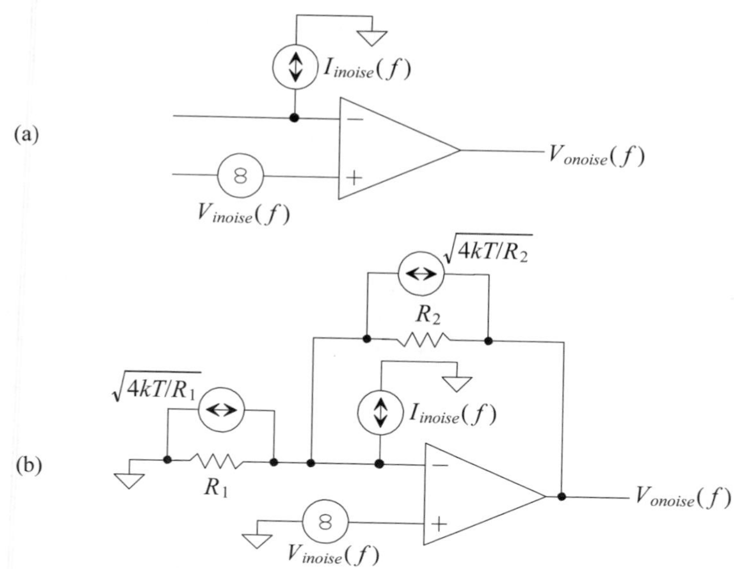

Figure 8.46 Modeling op-amp noise.

Let's write the output noise, for Fig. 8.46b, in terms of the input-referred noise sources and the thermal noise from the resistors. Because both sides of R_1 are at ground (one side is physically tied to ground while the other side is held at ground by the op-amp), we can write

$$V_{onoise}^2(f) = V_{inoise}^2(f) \cdot \left(1 + \frac{R_2}{R_1}\right)^2 + \left[I_{inoise}^2(f) + \frac{4kT}{R_1} + \frac{4kT}{R_2}\right] \cdot R_2^2 \qquad (8.79)$$

The gain of this circuit for the sake of calculating bandwidth is $(1 + R_2/R_1)$. The closed-loop value of bandwidth is written in terms of the op-amp's gain-bandwidth product, f_{un} (= the frequency when the open-loop gain of the op-amp is unity, see pages 792-793) as

$$\text{op-amp's gain·bandwidth} = f_{un} = \left(1 + \frac{R_2}{R_1}\right) \cdot f_{3dB} \qquad (8.80)$$

The NEB, Eq. (8.15), can be written as

$$NEB = \frac{\pi}{2} \cdot f_{un} \cdot \frac{R_1}{R_1 + R_2} \qquad (8.81)$$

Assuming that the input-referred sources are not correlated and the NEB of the output noise limits the RMS values, see Eq. (8.15) and associated discussion, we can write

$$V_{onoise,RMS} = \sqrt{V_{onoise}^2(f) \cdot \frac{\pi}{2} \cdot f_{un} \frac{R_1}{R_1 + R_2}} \qquad (8.82)$$

where $V_{onoise}^2(f)$ is assumed to be white until it roll-offs with the amplifer's bandwidth, Fig. 8.9b.

As an example, Fig. 8.47 shows the input-referred noise sources for the LT1364 op-amp. Neglecting the low-frequency (flicker) noise, we see that $V_{inoise}(f) = 9\ nV/\sqrt{Hz}$ (= e_n in Fig. 8.47) and $I_{inoise}(f) = 8.5\ pA/\sqrt{Hz}$ (= i_n in Fig. 8.47). From the datasheet, the gain-bandwidth product (f_{un}) of this op-amp is nominally 70 MHz (and so an op-amp in a gain of 10 (= $1 + R_2/R_1$) configuration has f_{3dB} = 7 MHz).

Equations (8.79) and (8.82) can be interpreted in many ways. If our op-amp is used as a transimpedance amplifier as in Fig. 8.34 (Ex. 8.17), then we want R_1 to be infinite to minimize the output noise. We think of the input source, i_s , as a current connected to the inverting op-amp input, the amplifier as having a gain of $|R_2|$, and the output voltage as $i_s \cdot R_2$. (The inverting input to the op-amp is a virtual ground, which is the ideal input resistance for an input current signal.) Looking at the SNR_{out} or $(i_s \cdot R_2)/V_{onoise,RMS}$, we see that reducing R_2 lowers both the gain and the noise while at the same time causing the SNR_{out} to drop.

If the op-amp is used as an inverting voltage amplifier with a gain of $-R_2/R_1$ (input signal connected to the left side of R_1), or as a noninverting voltage amplifier with a gain of $1 + R_2/R_1$ (input signal connected to the noninverting op-amp input terminal) then increasing R_1, as seen in Eq. (8.79), reduces the output noise. If R_2 is held constant then increasing R_1 also reduces the amplifier gain causing the SNR_{out} to decrease (where $SNR_{out} = (v_s \cdot R_2/R_1)/V_{onoise,RMS}$ or $(v_s \cdot [1 + R_2/R_1])/V_{onoise,RMS}$. In this situation, for large SNR_{out} , we want both R_1 (for large gain) and R_2 (to minimize the output noise as seen in Eq. [8.79]) to be as small as possible. For example, if we want a gain of 10, it is preferable to use an R_1 and R_2 of 10k and 1k over using 100k and 10k. Of course, using lower values results in more power dissipation in these feedback resistors.

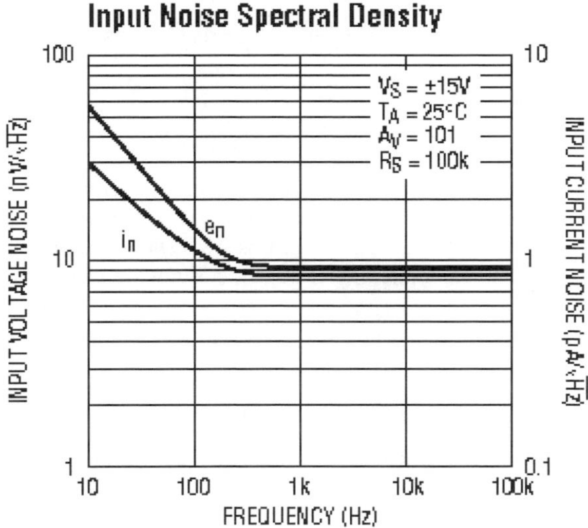

Figure 8.47 Input-referred noise sources for the LT1364 high-speed op-amp.

8.3.3 Some Final Notes Concerning Notation

When we calculate the mean-squared value of a noise voltage signal (which means that the desired signal components are not present in the signal), we used

$$\overline{v^2} = V_{RMS}^2 = \frac{1}{T_{meas}} \int_0^{T_{meas}} v^2(t) \cdot dt \tag{8.83}$$

The square root of the mean-squared voltage is the RMS value of a noise waveform. In Ex. 8.14 we showed that the RMS value of the integrated thermal noise increases with measuring time. Does this mean that if we measure, or look, at the value of this noise signal at a specific time it can't be zero? As seen in Fig. 8.26, a noise voltage can be positive, zero, or negative. What we are saying when we say the RMS value is increasing is that the voltage excursions away from zero are growing (but v(t) is still zero at some times).

The next thing we should note is that the mean-squared value is a number not a spectral density. Strictly speaking, it is incorrect to write, for example, the thermal noise of a resistor as

$$\overline{v^2} = 4kTR \text{ or } \overline{i^2} = \frac{4kT}{R} \tag{8.84}$$

A number (the left side of the equation) is not equal to a spectrum (the right side of the equation is a PSD). When we talked about measuring PSD with a spectrum analyzer in Sec. 8.1.2, we said that we plot a point by dividing the measured power by the resolution bandwidth, that is, V_{RMS}^2/f_{res}, at a particular frequency. Sometimes the notation, Δf, is used for the resolution bandwidth f_{res}. It is correct to write the PSD of thermal noise as

$$V_R^2(f) = \frac{V_{RMS}^2}{f_{res}} = \frac{\overline{v^2}}{\Delta f} = 4kTR \text{ units of } \frac{V^2}{Hz} \tag{8.85}$$

It is also correct to write

$$\overline{v^2} = 4kTR \cdot \Delta f \ \text{ volts} \tag{8.86}$$

where the resolution bandwidth is assumed narrow (for a general noise signal that may or may not be white). What makes this more confusing is that the mean-squared values are used for both narrow band signals, as seen in Eqs. (8.85) and (8.86), and for entire noise spectrums. For example, to calculate the RMS value of a wideband noise signal, we use

$$V_{RMS}^2 = \overline{v^2} = \int_{f_L}^{f_H} V_R{}^2(f) \cdot df = V^2(f) \cdot B = 4kTRB \tag{8.87}$$

where $B = f_H - f_L \neq \Delta f = f_{res}$. Note that Δf has nothing to do with *NEB* or *B*.

As another example, if we were to measure the PSD of flicker noise, we would get a value of

$$V_{1/f}^2(f) = \frac{\overline{v^2}}{\Delta f} = \frac{FNN}{f} \tag{8.88}$$

at each frequency f. The measured power in the narrow bandwidth Δf (= f_{res}), $\overline{v^2}$ (= V_{RMS}^2), decreases as f goes up resulting in the $1/f$ PSD.

ADDITIONAL READING AND/OR INFORMATION

[1] R. J. Baker, *CMOS: Mixed-Signal Circuit Design*, John Wiley and Sons, 2002. ISBN 0-471-22754-4. Volume 2 of this book. Covers averaging and how it affects circuit noise.

[2] W. T. (Tim) Holman, Private communication, Feb. 2002. Information on how the RMS value of a signal containing $1/f$ noise (and $1/f^2$, $1/f^3$, etc.) grows with measuring time.

[3] C. D. Motchenbacher and J. A. Connelly, *Low-Noise Electronic System Design*, John Wiley and Sons, 1993. ISBN 0-471-57742-1. Excellent reference for low-noise design.

[4] D. M. Binkley, J. M. Rochelle, M. J. Paulus, and M. E. Casey, "A Low-Noise, Wideband, Integrated CMOS Transimpedance Preamplifier for Photodiode Applications," in *IEEE Transactions on Nuclear Science*, vol. NS-39, no. 4, August 1992, pp. 747–752. Covers the design of low-noise transimpedance amplifiers.

[5] H. L. Krauss, C. W. Bostian, and F. H. Raab, *Solid State Radio Engineering*, John Wiley and Sons, 1980. ISBN 0-471-03018-X. Reference for the design of radio circuits.

[6] W. R. Bennett, *Electrical Noise*, McGraw-Hill, 1960. Fundamental reference on electrical noise. Written by a pioneer in noise theory. No longer in print but available used through various book sellers.

[7] A. Van der Ziel, *Noise; Sources, Characterization, Measurement*, Prentice-Hall. Also written by a pioneer in noise theory.

[8] H. F. Friis, "Noise Figures of Radio Receivers," in *Proceedings of the IRE* (Institute of Radio Engineers), vol. 32, no. 7, 1944, pp. 419–422. Fundamental paper covering noise figure.

[9] R. Sarpeshkar, T. Delbrück, and C. A. Mead, "White Noise in MOS Transistors and Resistors," *IEEE Circuits and Devices Magazine*, Nov. 1993. Shows that both shot and thermal noise mechanisms in MOSFETs have the same origins.

LIST OF SYMBOLS/ACRONYMS

A - Voltage gain of a circuit.

A_{DC} - DC gain of a circuit.

B - Bandwidth of a noise measurement, i.e., $f_H - f_L$.

C - Correlation term, see Eq. (8.67).

C_{in} - Input capacitance.

C_j - Diodes junction capacitance (also called depletion capacitance).

C_{j0} - Depletion capacitance measured with zero volts across the diode.

CUT - Circuit under test.

Δf - Resolution bandwidth, same as f_{res}.

DFT - Discrete Fourier transform.

ESD - Energy spectral density.

F - Noise factor.

FNN - Flicker noise numerator.

F_{opt} - Optimum noise factor.

f - Frequency of a sine wave.

f_{3dB} - 3 dB frequency of a circuit (the power is half the low frequency value at f_{3dB} or the voltage is 0.707 its low-frequency value at a frequency of f_{3dB}).

f_H - Higher frequency in an RMS noise calculation, see Eq. (8.10).

f_L - Lower frequency in an RMS noise calculation, see Eq. (8.10).

f_{res} - Resolution bandwidth of a spectrum analyzer or a DFT (same as Δf).

f_{start} - Starting frequency in a DFT or a spectrum analyzer measurement.

f_{stop} - Final frequency in a DFT or a spectrum analyzer measurement.

f_{un} - Unity gain frequency of an op-amp (the open-loop gain is 1).

GR - Generation-recombination.

$\overline{i^2}$ - Mean-squared current$\left(= I_{RMS}^2\right)$.

$I(f)$ - Square root of a current PSD, units of $A/\sqrt{Hz}$.

$I^2(f)$ - PSD of a noise current, units of A^2/Hz.

$I_{1/f}^2(f)$ - PSD of flicker noise current, units of A^2/Hz.

$I_{inoise}^2(f)$ - PSD of the input-referred noise of a circuit, units of A^2/Hz.

$I_{inoise,RMS}$ - Input-referred RMS noise current, unit of A.

$I_{popcorn}^2(f)$ PSD of popcorn (burst) noise current, units of A^2/Hz.

I_{RMS} - Root mean-squared value of a current waveform, units of A, also written in this chapter as $\sqrt{\overline{i^2}}$.

$I_R^2(f)$ - PSD of thermal noise, units of A^2/Hz.

I_{RMS}^2 - Mean-squared value of a current waveform, units of A^2, also written as $\overline{i^2}$.

$I_{shot}^2(f)$ - Shot noise current PSD, units of A^2/Hz.

j - $\sqrt{-1}$.

K - Number of averages, see Eq. (8.48).

k - Boltzmann's constant, $13.8 \times 10^{-24} J/K$.

LNA - Low-noise amplifier.

n - A counting index.

NEB - Noise-equivalent bandwidth, see Eq. (8.15).

NF - Noise figure, see Eq. (8.35).

$P_{noise}(f)$ - PSD of a signal with units of W/Hz.

P_{AVG} - Average power dissipated.

PDF - Probability density function, see Fig. 8.33.

$P_{inst}(t)$ - Instantaneous power dissipation (the power dissipated at a specific time). Units of watts (joules/s).

PSD - Power spectral density, units of V^2/Hz.

q - Electron charge of 1.6×10^{-19} coulombs.

R - Resistance in ohms.

R_s - Resistance of an input source signal.

RMS - Root mean square.

SA - Spectrum analyzer.

SNR_{in} - Input signal-to-noise ratio, see Eq. (8.28).

SNR_{out} - Output signal-to-noise ratio, see Eq. (8.40).

t - Time in seconds.

T - Temperature in Kelvin.

T_{meas} - Time a measurement is taken, units of seconds.

$\overline{v^2}$ - Mean-squared voltage $\left(= V_{RMS}^2\right)$.

$V(f)$ - Square root of a PSD, units of $V/\sqrt{Hz}$ (also called a voltage spectral density).

$v(t)$ - Time domain voltage waveform.

V_{in} - Input voltage.

$V_{inoise}^2(f)$ - PSD of the input-referred noise of a circuit.

$V_{inoise,RMS}$ - Input-referred RMS noise voltage, units of V.

$V_{LF,noise}^2(f)$ - Low-frequency white noise, see Fig. 8.9.

$V_{noise}^2(f)$ - PSD of a noise signal.

$V_{onoise}^2(f)$ - PSD of the output noise of a circuit.

$V_{onoise,RMS}$ - Output RMS noise voltage.

V_P - Peak voltage of a sine waves wave.

$V_R^2(f)$ - PSD of thermal noise, units of V^2/Hz.

$V_{1/f}^2(f)$ - PSD of flicker noise voltage, units of V^2/Hz.

V_{RMS} - Root mean-squared value of a voltage waveform, units of V, also written in this chapter as $\sqrt{\overline{v^2}}$.

V_{RMS}^2 - Mean-squared value of a voltage waveform, units of V^2, also written in this chapter as $\overline{v^2}$.

V_s - Source input voltage.

V_T - Thermal voltage, kT/q or 26 mV at room temperature.

ω - $2\pi f$.

Z_{in} - A complex input impedance.

PROBLEMS

8.1 Suppose that the power company is charging $ 0.25 for a kilowatthour (1,000 W of power supplied for one hour) of energy. How much are you paying for a Joule of energy? How much energy, in joules, is supplied to a 100 W lightbulb in one hour? Is it potential energy or kinetic energy the power company sells?

8.2 The power company attempts to hold, in the United States, the RMS value of the voltage supplied to households to 120 V RMS with $\pm 3V$ at a frequency of 60 Hz. What is the peak-to-peak value of this voltage? If we were to look at the PSD of this signal in a 10 Hz bandwidth, what amplitude would we see?

8.3 Suppose a SA measures a noise voltage spectral density of 1 $\mu V/\sqrt{Hz}$. What is the RMS value of this noisy signal over a bandwidth of DC to 1 MHz?

8.4 Suppose the noise signal in problem 3 shows a 3-dB frequency of 5 MHz. What is its RMS value over an infinite bandwidth?

8.5 Estimate the RMS output noise in the following circuit over a bandwidth of 1 to 1kHz. Verify your answer with SPICE. (Hint: simplify the circuit by combining resistors.)

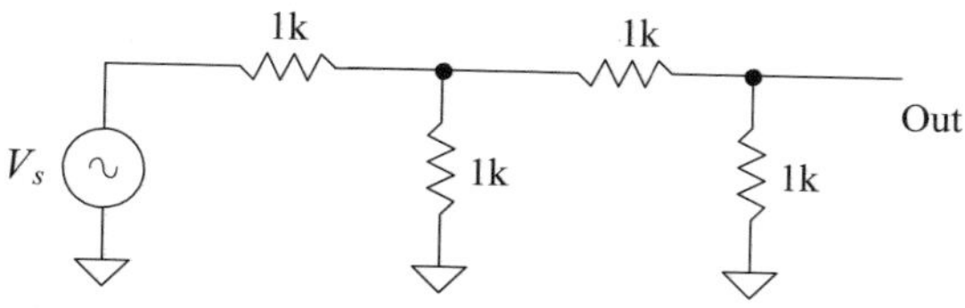

Figure 8.48 Circuit for Problem 8.5.

8.6 Estimate the RMS output noise over an infinite bandwidth for the circuit in Fig. 8.48 if the output is shunted with a 1 pF capacitor.

8.7 Show, Fig. 8.49, that the calculation of SNR_{in} results in the same value independent of treating the input as a voltage or a current.

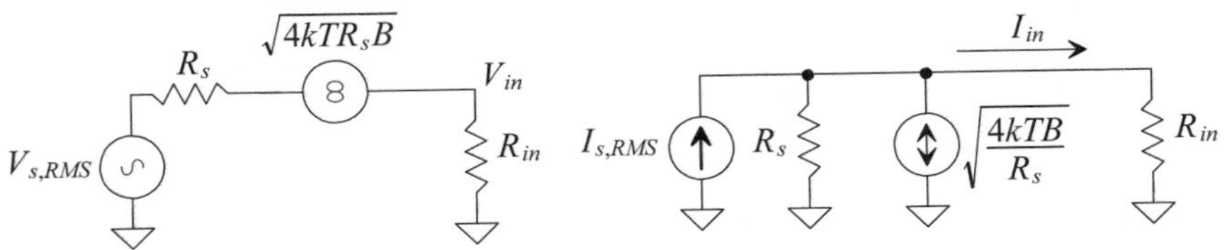

Figure 8.49 Calculating input SNR in Problem 8.7.

8.8 Using the input-referred noise model seen in Fig. 8.20b, verify that if the input resistance becomes infinite, the output noise is adequately modeled using a single input-referred noise voltage.

8.9 If an amplifier has a 0 dB NF, does that indicate the amplifier's output is free of noise? Why or why not?

8.10 Verify, as seen in Eq. (8.45), that the input-referred noise power of an amplifier is $(V_{inoise,RMS} \cdot I_{inoise,RMS})/2$.

8.11 Verify that the units for the shot noise PSD seen in Eq. (8.49) are indeed A^2/Hz.

8.12 Repeat Ex. 8.12 if the input voltage is reduced from 1.7 V to 1 V.

8.13 Verify the comment made in Sec. 8.2.5 that the RMS value of integrated flicker noise signal increases linearly with measurement time.

8.14 If the maximum allowable RMS output noise of a transimpedance amplifier built using the TLC220x is 100 µV in a bandwidth of 1 MHz is needed, what are the maximum values of C_F and R_F? What is the peak-to-peak value of the noise in the time domain?

8.15 Estimate the RMS output noise for the circuit seen in Fig. 8.50. Compare the hand-calculated result to SPICE simulation results using an ideal op-amp (as used in Ex. 8.18).

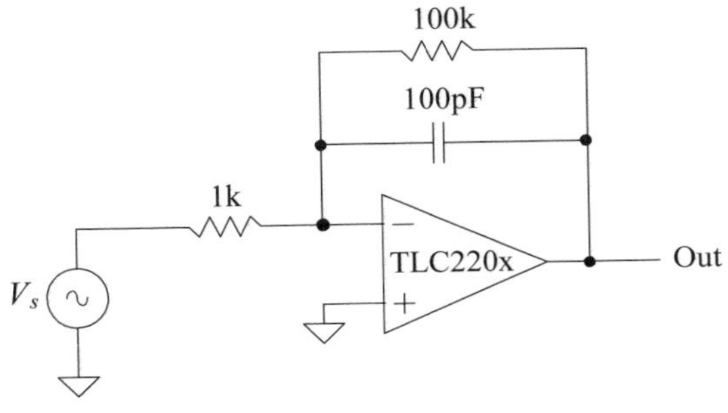

Figure 8.50 Circuit for Problem 8.15.

8.16 Suppose a 4.7 V Zener diode is used in a noise generator, as seen in Fig. 8.40. Estimate the output noise voltage PSD in terms of the diode's noise current and the biasing resistor (use 1k) if the power supply is 9 V. (Note: the noise mechanism is shot noise not avalanche noise.)

8.17 Rewrite Eq. (8.72) if $Z_{in} = R_{in} + 1/j\omega C_{in}$.

8.18 Estimate the RMS output noise for the amplifier seen in Fig. 8.51. What would placing a capacitor across the feedback resistor do to the RMS output noise and to the speed (bandwidth) of the amplifier? Name two ways to lower the RMS output noise for this amplifier. What is the cost for the lower output noise?

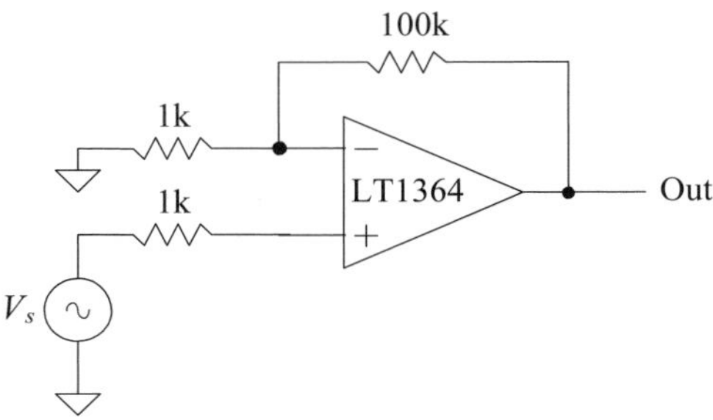

Figure 8.51 Circuit for Problem 8.18.

Chapter 9

Models for Analog Design

In this chapter we develop models for analog design using the MOSFET. We'll break this discussion up into three sections. The first section covers long-channel MOSFET models with the assumption that the MOSFET follows the "square-law" equations derived in Ch. 6. In the second section we discuss models using modern MOSFETs with short-channel lengths (< 1 μm). Models for these short-channel MOSFETs are developed with graphs showing device characteristics, (e.g., output resistance, transconductance, etc.). Finally, at the end of the chapter, we introduce MOSFET noise modeling.

9.1 Long-Channel MOSFETs

When we do analog design we often say things like "the MOSFET looks like a current source when operating in the saturation region" or "it looks like a resistor." Before going too far, let's make sure that we understand these statements. Examine the current-voltage (IV) plot in Fig. 9.1. In this figure we've plotted the (DC) current-voltage characteristics of a resistor, a current source, and a voltage source. Often the controlling parameter in a semiconductor device is a voltage. The controlled parameter is then the device's output current (and this is why current is on the y-axis and voltage is on the x-axis in an IV

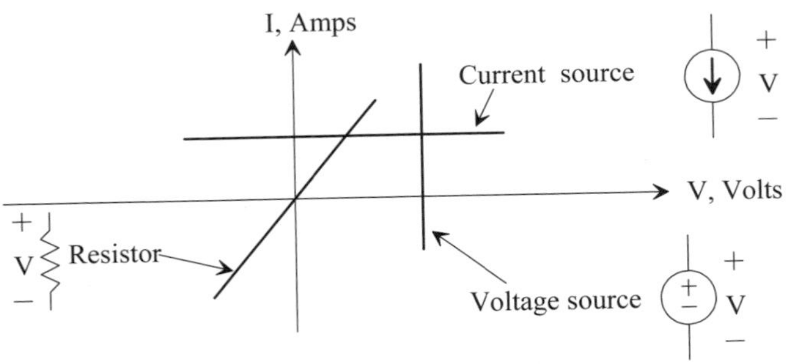

Figure 9.1 Current-voltage (IV) plots for various electrical components.

plot). The voltage across the voltage source, for example, doesn't vary with changes in current running through it. The voltage across the resistor is linearly related to the current flowing through the resistor (Ohm's law). An important thing to note is that resistance can be calculated by taking the reciprocal of the IV plot slope. (So the voltage source in this figure has zero resistance; the current source, infinite resistance.) Also note that the x-axis corresponds to plotting the IV characteristics of an open circuit (no current with changes in voltage). The y-axis corresponds to a short (no changes in the voltage across a wire [a short], with changing current).

Example 9.1
Plot the IV characteristics for the circuit seen in Fig. 9.2.

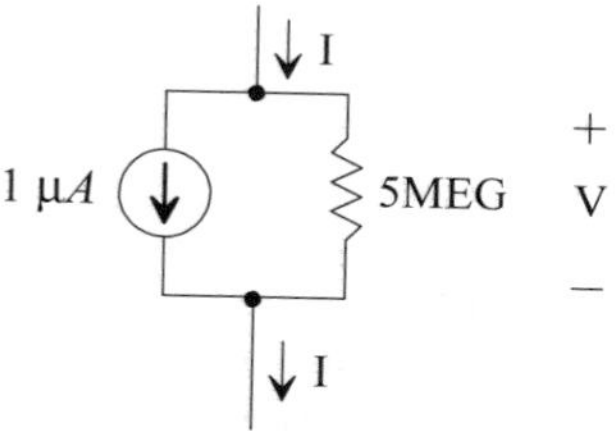

Figure 9.2 Circuit for Ex. 9.1.

Figure 9.3a shows the IV curves for each component of the circuit where we use a single quadrant of the IV plotting plane. The slope of the resistor is 200 $nA/1\ V$. The resistance value is the reciprocal of this slope (5MEG). The combined IV curve is seen in Fig. 9.3b. ■

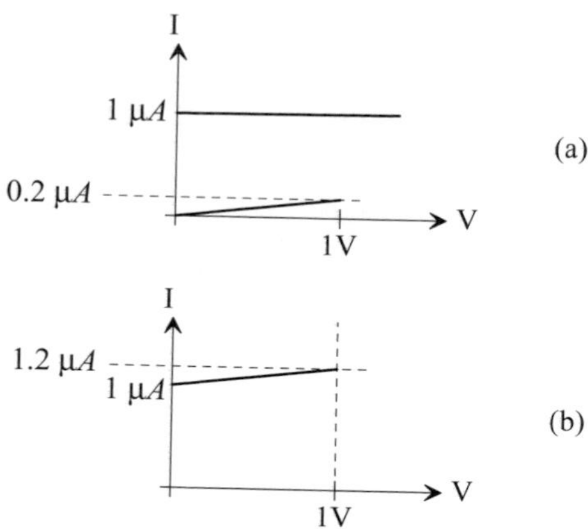

Figure 9.3 IV plots for Ex. 9.1.

Example 9.2
Figure 9.4 shows the IV curves (drain current versus drain-source voltage with constant gate source voltage) for a MOSFET. Comment on what the MOSFET looks like in the triode and saturation regions.

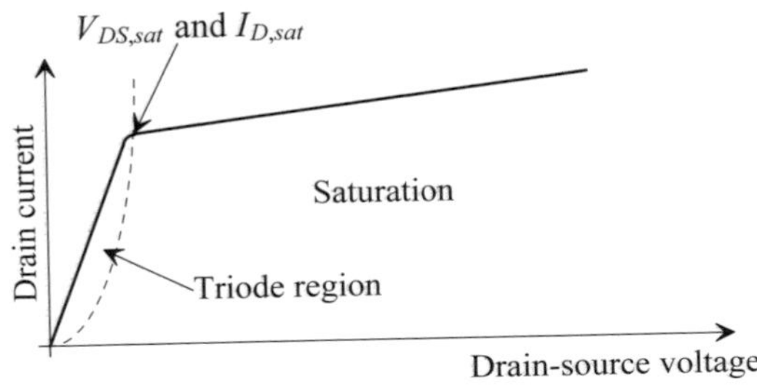

Figure 9.4 IV curves of a MOSFET.

Clearly, in the triode region (also known as the linear or ohmic region), the MOSFET behaves like a resistor. In the saturation region, the MOSFET behaves, as seen in Fig. 9.3b, like a current source in parallel with a resistor. The resistive component, whether in the triode or saturation regions, is often called the MOSFET's output resistance. ■

9.1.1 The Square-Law Equations

The drain current, I_D, is related to the gate-source voltage, V_{GS}, and the drain-source voltage, V_{DS}, using

$$I_D = \frac{KP_n}{2} \cdot \frac{W}{L}(V_{GS} - V_{THN})^2(1 + \lambda(V_{DS} - V_{DS,sat})) \quad (9.1)$$

for $V_{DS} \geq V_{DS,sat}$ and $V_{GS} \geq V_{THN}$. As seen in Fig. 9.4, $V_{DS,sat}$ is the voltage where the MOSFET moves from the triode region to the saturation region. For long-channel MOSFETs, this can be written as

$$V_{DS,sat} = V_{GS} - V_{THN} \quad (9.2)$$

This term is very important and will be used frequently when doing analog design. Notice that $V_{DS,sat}$ represents the amount of gate-source voltage that we have in *excess* or *over* the threshold voltage. For this reason, it is sometimes called

$$V_{DS,sat} = \text{excess gate voltage} = \text{gate overdrive voltage} \quad (9.3)$$

Note that while **Eq. (9.2) is only valid for long-channel MOSFETs**, the voltage, $V_{DS,sat}$, simply indicates, for long- or short-channel MOSFETs, the V_{DS} at the boundary between triode and saturation. When $V_{DS} = V_{DS,sat}$, the drain current is labeled $I_{D,sat}$ or

$$I_{D,sat} = \frac{KP_n}{2} \cdot \frac{W}{L}(V_{GS} - V_{THN})^2 = \frac{KP_n}{2} \cdot \frac{W}{L}(V_{DS,sat})^2 \quad (9.4)$$

Equation (9.1) can be rewritten as

$$I_D = I_{D,sat} + I_{D,sat}\lambda \cdot (V_{DS} - V_{DS,sat}) \tag{9.5}$$

Using the results from Exs. 9.1 and 9.2, we see that the MOSFET behaves, while in the saturation region, like a current source $I_{D,sat}$ in parallel with a resistor of value

$$r_o = \frac{1}{\lambda I_{D,sat}} \tag{9.6}$$

Figure 9.5 shows a gate-drain connected MOSFET, something we'll see often in analog design. Notice that $V_{GS} = V_{DS}$. If $V_{GS} > V_{THN}$ (a current is flowing through the device), then, for the MOSFET to operate in the saturation region, we must have $V_{DS} \geq V_{GS} - V_{THN}$ or $0 \geq -V_{THN}$ (indicating that a gate-drain-connected MOSFET with a current flowing through it is always operating in the saturation region [remember this]). We can also write the requirement for operation in the saturation region as

$$\overbrace{V_D - V_S}^{V_{DS}} \geq \overbrace{V_G - V_S}^{V_{GS}} - V_{THN} \text{ or } V_D \geq V_G - V_{THN} \tag{9.7}$$

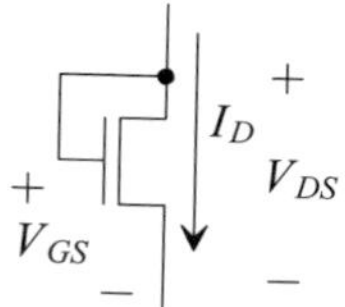

Figure 9.5 Gate-drain connected MOSFET. Also known as a diode-connected MOSFET.

PMOS Square-Law Equations

The PMOS equivalents of Eqs. (9.1) and (9.2) are (for completeness)

$$I_D = \frac{KP_p}{2} \cdot \frac{W}{L}(V_{SG} - V_{THP})^2(1 + \lambda(V_{SD} - V_{SD,sat})) \tag{9.8}$$

and

$$V_{SD,sat} = V_{SG} - V_{THP} \tag{9.9}$$

Note that all we did was swap the subscripts of the symbols used in the NMOS equations. Using this notation, the terminal currents and voltages of the MOSFET (both NMOS and PMOS) **are always positive**.

Qualitative Discussion

To develop some intuitive understanding for MOSFET operation, let's look at the circuits in Fig. 9.6. In the following discussion it is assumed that the MOSFETs are operating in the saturation region and that the terminal voltages (V_{GS} and V_{DS}) of the device do not exceed the power supply rails.

Imagine injecting a current into the drain of the NMOS device in Fig. 9.6a. What happens to the device's drain current? Answer: it goes up. What happens to the device's V_{DS}? Answer: it goes up. If we hold V_{GS} constant then, as seen in Eq. (9.1), we must see an increase in V_{DS} if I_D increases. Stealing current from the NMOS's drain (pulling more

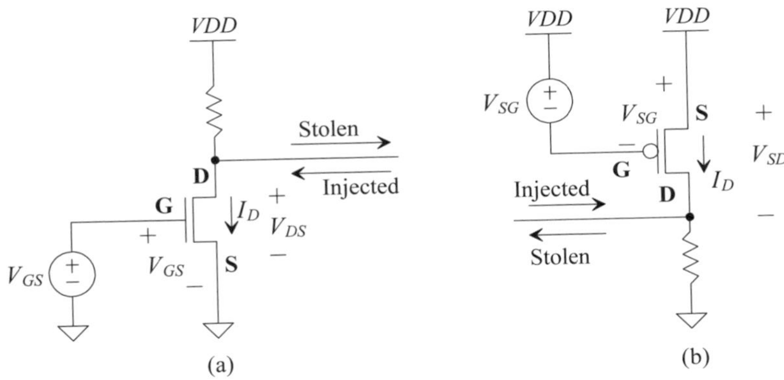

Figure 9.6 Movement of voltages and currents in a MOSFET, see text.

of the current flowing down from the resistor away from the MOSFET's drain terminal) results in both the drain current and V_{DS} going down (until the device enters the triode region and ultimately turns off). Make sure that these statements are clearly understood. We will frequently sum or take the difference of currents using MOSFETs; how the voltages and currents change should be "felt" when looking at an analog design, without referring to the equations.

For the PMOS device in Fig. 9.6b, injecting a current into the drain causes the drain current to go down. In the PMOS device, drain current flows out of the drain terminal of the MOSFET. Injecting a current into the drain results in the drain current going down. When we inject this current into the PMOS drain, the source-drain voltage, V_{SD}, decreases as well (meaning that the drain voltage moves towards the power supply voltage *VDD*). Ultimately the device will shut off and the drain current will go to zero. If we steal current from the drain, the drain voltage will move towards ground (noting that $V_{SD} = VDD - V_D$) and the PMOS drain current will increase.

Question: If the channel length modulation parameter, λ, is zero and the MOSFETs stay in the saturation region, will the drain current change with drain-source voltage? Answer: no, the drain current is then independent of V_{DS}, Eq. (9.1). We won't be able to inject or steal a current from the MOSFET without either pushing the MOSFET into triode or moving the drain terminals beyond *VDD* or ground until the MOSFET breaks down. As seen in Eq. (9.1), with $\lambda = 0$, I_D is only dependent on V_{GS} as long as the MOSFET is in saturation.

Example 9.3
For the circuit in Fig. 9.7, describe qualitatively what will happen if we inject a current at the location seen in the figure of $-1\,\mu A \leq I \leq 1\,\mu A$. Verify your answer with SPICE. (Use the long-channel CMOS models from Ch. 6.)

This circuit, as we will see in Ch. 20, is a cascode current mirror. M1 and M3 are operating in the saturation region with a drain current of 1 μA. Neglecting body effect (the change in threshold voltage when the source and body of the MOSFET aren't at the same potential), we know $V_{GS1} = V_{GS3}$ when M1 and M3 are sized

All MOSFETs are 10/1

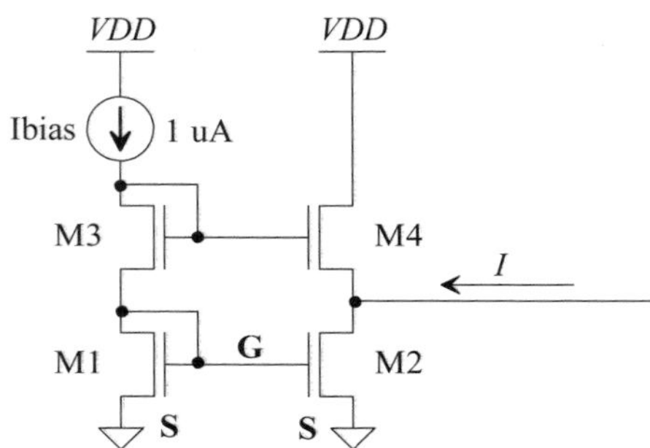

Figure 9.7 Schematic for Ex. 9.3.

the same and have the same drain currents. Because the gates and sources of M2 and M1 are physically tied together, we can write $V_{GS1} = V_{GS2} = V_{GS3} = V_{GS4}$, meaning that 1 μA flows in all MOSFETs.

For the injected current seen in the figure, a positive 1 μA indicates that current flows to the left into the drain of M2 (the source of M4). A positive current also indicates that we are injecting a current into the node. A −1 μA indicates that current flows to the right out of the node (we are stealing current from the node).

If we inject 1 μA into the drain of M2, then M2's drain voltage and M4's source voltage increase causing V_{GS4} to decrease. Since M2 wants to (meaning its V_{GS} is setting) sink a current of 1 μA, our injected current goes entirely through M2 to ground. M4 will shut off ($V_{GS4} \leq V_{THN}$).

If we steal 1 μA from the node ($I = -1$ μA), then the drain voltage of M2 moves downwards towards ground. Then M4 supplies 2 μA of current (V_{GS4} increases).

Figure 9.8 shows the SPICE simulation results. As discussed, M4 starts to shut off (its gate-source voltage falls below V_{THN}) when we inject 1 μA of current. When we steal 1 μA of current (the left side of the plot), V_{GS4} increases to supply 2 μA of current.

To look at the currents flowing in the circuit, zero-volt DC sources can be added. Then, in SPICE, the current through these added sources can be plotted. An example is seen in Fig. 9.9. Simulating the circuit in Fig. 9.7 with the added DC source, we can plot the current through M4, as seen in Fig. 9.10. As mentioned, the drain current shuts off when 1 μA is injected into the node, and it goes to 2 μA when we pull (steal) 1 μA from the node. To monitor the current flowing in M2, we can either add another zero-volt DC source or simply move the injection point to the top of the added DC voltage source in Fig. 9.9.

It's highly recommended that the reader simulate the circuit in Fig. 9.7 under various operating conditions until its operation is fully understood. ■

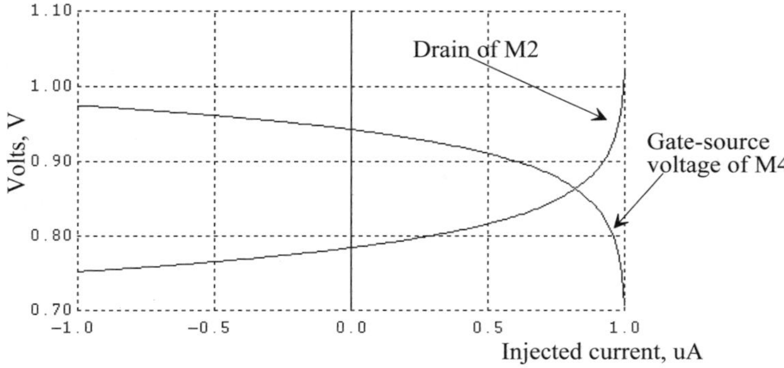

Figure 9.8 SPICE simulations verifying the discussions in Ex. 9.3.

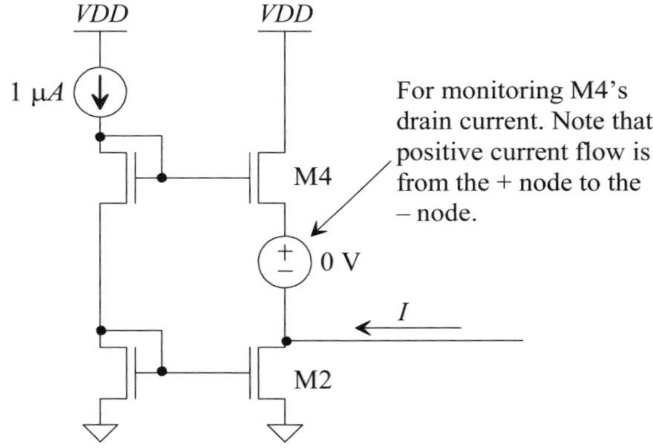

Figure 9.9 Adding zero volt sources to monitor currents in SPICE.

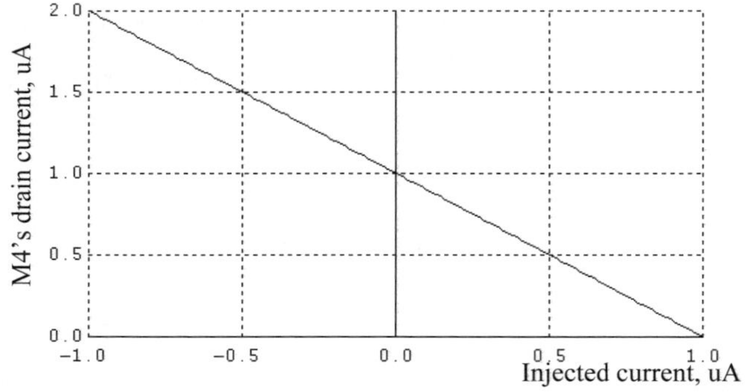

Figure 9.10 How the drain current of M4 changes with injected current.

Threshold Voltage and Body Effect

For an NMOS device, the threshold voltage increases when the source is at a higher potential than the NMOS body (the p-substrate, or ground, in this book). This change in threshold voltage is called the *body effect*. A simple example of a MOSFET operating with body effect (an increased threshold voltage) is seen in Fig. 9.7. Both M3 and M4 have a larger threshold voltage than M1 and M2.

From Ch. 6 the change in threshold voltage can be written as a function of the source to bulk potential, V_{SB}, Fig. 9.11, as

$$V_{THN}(V_{SB}) = V_{THN0} + \gamma_n\left(\sqrt{2|V_{fp}| + V_{SB}} - \sqrt{2|V_{fp}|}\right) \quad (9.10)$$

and for the PMOS (again we simply switch the subscripts so that all voltages and currents are positive)

$$V_{THP}(V_{BS}) = V_{THP0} + \gamma_p\left(\sqrt{2|V_{fn}| + V_{BS}} - \sqrt{2|V_{fn}|}\right) \quad (9.11)$$

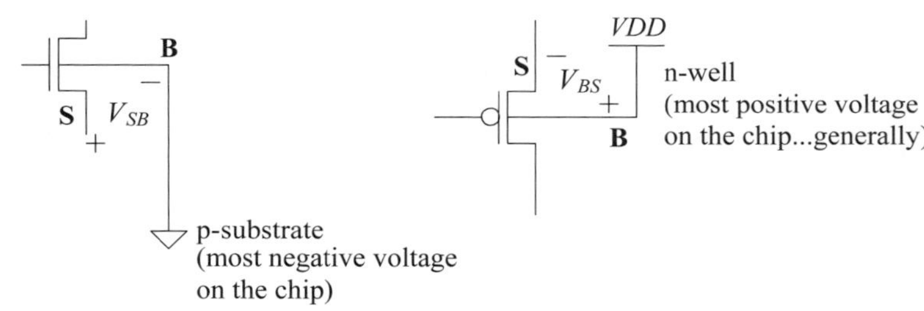

Figure 9.11 The body connection of a MOSFET.

Qualitative Discussion

It's useful to have a feeling for how the threshold voltage changes with increasing source-to-body potential. Figure 9.12 shows a plot of the threshold variation with V_{SB}. For large values of V_{SB}, the threshold voltage change isn't very significant, while for small values it is significant. The circuit seen in Fig. 9.13 is an example where we might want small variations in the threshold voltage. Because the resistor is large, the gate-source voltage will be close to the threshold voltage. If the threshold voltage didn't vary, then the drain current would be linearly related to the input voltage. This circuit is useful for voltage-to-current conversions.

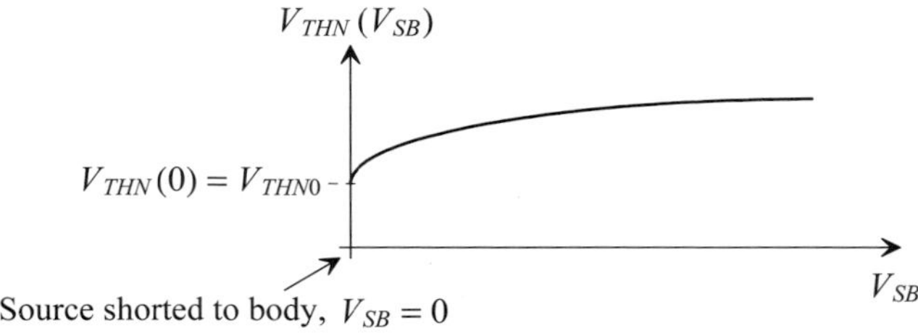

Figure 9.12 Variation in threshold voltage with source to bulk potential.

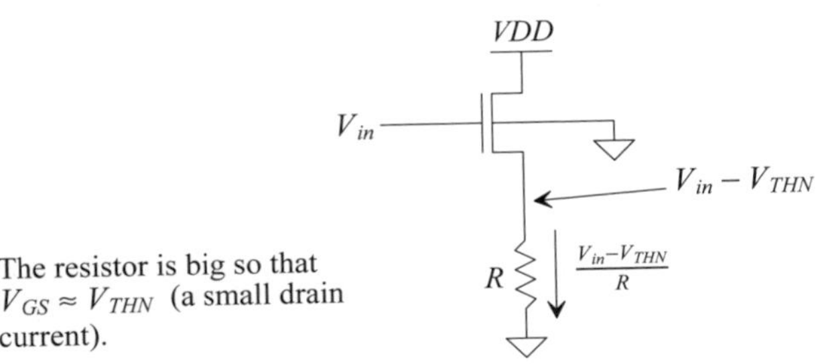

Figure 9.13 An example of a circuit where body effect is important.

Example 9.4
Simulate the operation of the voltage-to-current converter in Fig. 9.13 if the size of the MOSFET is 10/1 and the resistor is 10MEG. Use long-channel devices.

Figure 9.14a shows how the current varies through the MOSFET with V_{in} changing from 1 V to 5 V. We start at 1 V V_{in} because the current will shut off when $V_{in} < V_{THN}$, resulting in a large nonlinearity. Looking at the linearity of the current, it looks really good (straight). However, in Fig. 9.14b we take the derivative of the line in (a) to see the slope. The variation in the slope is approximately 20%. For precision design of voltage-to-current converters, the variation in the linearity of the threshold voltage can be a limiting factor. We should note that channel length modulation also contributes to nonlinearities (so increasing the length, to increase r_o, of the device can help improve the linearity).

Question: what is the ideal slope for Fig. 9.14; that is, what is the ideal value of dI/dV_{in}? Answer: $1/R$ or, in this example, $100 \times 10^{-9}\ A/V$. ■

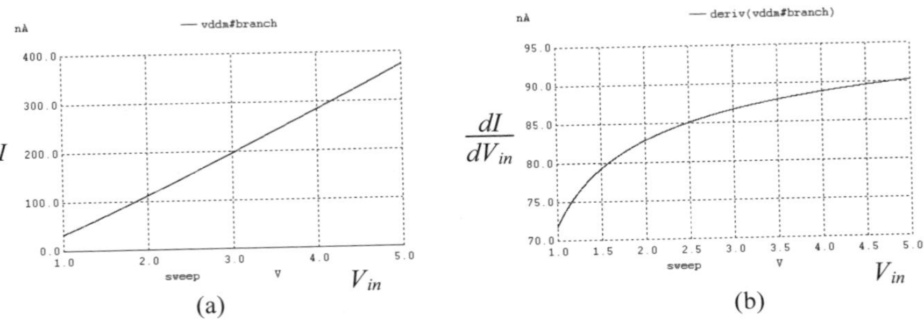

Figure 9.14 (a) Current flowing in the circuit of Fig. 9.13 and (b) its linearity.

There isn't any way to eliminate the body-effect in an n-well process where the NMOS device body is the substrate (though, if a triple well process is used, the NMOS devices can sit in their own wells). However, the PMOS devices can sit in their own wells and have separate bodies (and so we can design without body effect).

The Triode Region

The relationship between I_D, V_{GS}, and V_{DS} for an NMOS device operating in the triode region is

$$I_D = KP_n\frac{W}{L}\cdot\left((V_{GS}-V_{THN})V_{DS}-\frac{V_{DS}^2}{2}\right) \quad (9.12)$$

where $V_{GS} \geq V_{THN}$ and $V_{DS} \leq V_{DS,sat}\ (= V_{GS}-V_{THN})$. The equivalent expression for the PMOS device is

$$I_D = KP_p\frac{W}{L}\cdot\left((V_{SG}-V_{THP})V_{SD}-\frac{V_{SD}^2}{2}\right) \quad (9.13)$$

where $V_{SG} \geq V_{THP}$ and $V_{SD} \leq V_{SD,sat}\ (= V_{SG}-V_{THP})$.

We said earlier, in Fig. 9.4 and the associated discussion, that the MOSFET looks like a resistor when it is operating in the triode region. To estimate the value of the resistance, we can use

$$R_{ch}^{-1} = \frac{\partial I_D}{\partial V_{DS}} = KP_n\cdot\frac{W}{L}\cdot(V_{GS}-V_{THN}) - KP_n\cdot\frac{W}{L}\cdot V_{DS} \quad (9.14)$$

or

$$R_{ch} = \frac{1}{KP_n\cdot\frac{W}{L}\cdot(V_{DS,sat}-V_{DS})} \quad (9.15)$$

If $V_{DS,sat} >> V_{DS}$, this equation can be written as

$$R_{ch} \approx \frac{1}{KP_n\cdot\frac{W}{L}(V_{GS}-V_{THN})} \quad (9.16)$$

The Cutoff and Subthreshold Regions

For general design, we normally assume that the device is off (meaning zero drain current) when $V_{GS} < V_{THN}$ or $V_{SG} < V_{THP}$. However, it would be more correct to say that the device is operating in the subthreshold region (instead of the strong inversion region we've discussed so far in this chapter). The subthreshold current of a MOSFET operating in the active region (the amplifying region, which is similar to the saturation region for a MOSFET operating in strong inversion) is modeled using

$$I_D = I_{D0}\cdot\frac{W}{L}\cdot e^{(V_{GS}-V_{THN})/nV_T} \quad (9.17)$$

$V_{GS} \leq V_{THN}$ and $V_{DS} > 4V_T$ (remembering the thermal voltage, V_T, is kT/q or 26 mV at room temperature). The current I_{D0} is the (scaled) current that flows when $V_{GS}-V_{THN}=0$ (the gate-source voltage is equal to the threshold voltage). Note that Eq. (9.17) shows no dependence on V_{DS}. This indicates that the MOSFET has infinite output resistance when operating in the active region (which isn't the case, and so this equation has limitations).

9.1.2 Small Signal Models

A quick note concerning symbols: throughout the book we'll represent a signal containing both AC and DC components by a lowercase letter with uppercase subscripts. AC signals are represented with both lowercase letters and subscripts, while DC signals are represented with both uppercase letters and subscripts, see Fig. 9.15. (Note that if the maximum value of v_{GS} is VDD in Fig. 9.15, then the MOSFET is either in the saturation region or off [$V_{GS} < V_{THN}$].)

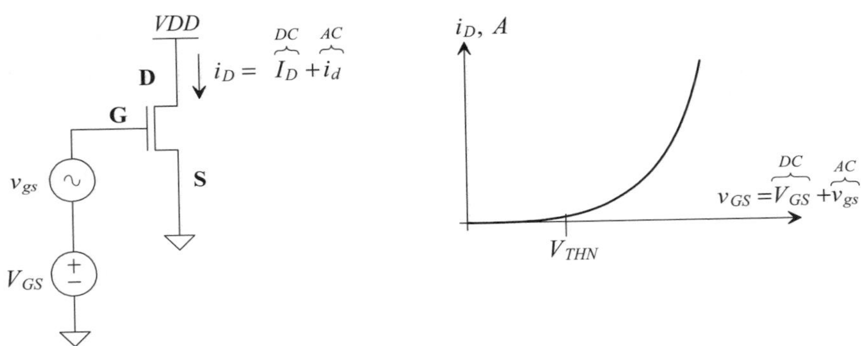

Figure 9.15 IV curves of a MOSFET in saturation.

Small-signal models are used to calculate AC gains. Figure 9.16 shows the basic idea. We adjust the DC gate-source voltage, V_{GS}, to a value that corresponds to a DC drain current I_D. At this bias point, we apply a small AC signal where $|v_{gs}| << V_{GS}$ and $|i_d| << I_D$. Because the signals are small, the change in drain current, i_d, with gate voltage, v_{gs} is essentially linear (as seen in the blown-up view in Fig. 9.16). If our AC signal amplitudes get comparable to the DC operating (or bias) points, we get high nonlinearity (which makes feedback necessary for any highly linear amplifier).

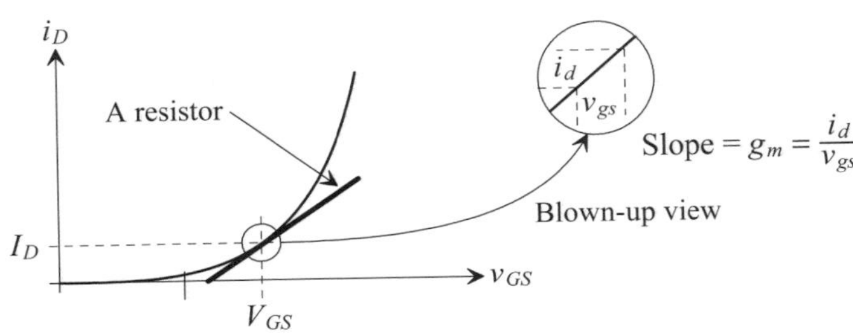

Figure 9.16 How small-signal parameters are calculated.

We often treat DC voltage sources as shorts when doing AC analysis (and current sources as opens). Looking at Fig. 9.15, we see that it's impossible for the AC signal to generate a voltage across the DC source. The voltage across the DC source is fixed with zero AC component so that the DC voltage source is an AC short.

Finally note that performing a small-signal analysis consists of the following steps:

(1) Calculating the bias point of the circuit using the DC equations from Sec. 9.1.1.

(2) Using the DC values from (1) to calculate the small-signal parameters. Small-signal AC parameters are always a function of the DC operating point.

(3) Replacing the active elements (e.g., MOSFETs) with their small-signal models. At the same time, the DC sources are removed (that is, short out all DC voltage sources and open up all DC current sources).

An AC analysis doesn't include any DC voltages or currents. For example, suppose we have an (AC) v_{gs} = 1 mV. It *doesn't make sense* to say that the MOSFET is off because $v_{gs} < V_{THN}$. To perform a small-signal analysis (to calculate small-signal parameters), the MOSFETs are in saturation or triode (meaning $V_{GS} > V_{THN}$).

Transconductance

An extremely important parameter in analog design is a device's transconductance, g_m. The g_m of a device is an AC small-signal parameter that relates the AC gate voltage to the AC drain current, that is,

$$i_d = g_m \cdot v_{gs} \tag{9.18}$$

From Figs. 9.15 and 9.16, g_m is simply the slope of the line at the intersection of the DC operating values V_{GS} and I_D. Using Eq. (9.1), without concerning ourselves with channel-length modulation, we can write

$$i_D = i_d + I_D = \frac{KP_n}{2} \cdot \frac{W}{L} \cdot \left(\overbrace{v_{gs} + V_{GS}}^{v_{GS}} - V_{THN} \right)^2 \tag{9.19}$$

To find the slope (g_m) of the i_D-v_{GS} curve at the fixed bias points V_{GS} and I_D (Fig. 9.16), we take the derivative of this equation with respect to the x-axis variable (v_{GS})

$$g_m = \left[\frac{\delta i_D}{\delta v_{GS}} \right]_{V_{GS} = \text{constant}}^{I_D = \text{constant}} = KP_n \cdot \frac{W}{L} \cdot (v_{gs} + V_{GS} - V_{THN}) \tag{9.20}$$

If we remember

$$\beta_n = KP_n \cdot \frac{W}{L} \text{ and } |v_{gs}| << V_{GS} \tag{9.21}$$

then we can write

$$g_m = \beta_n \left(\overbrace{V_{GS} - V_{THN}}^{V_{DS,sat}} \right) = \sqrt{2\beta_n I_D} \tag{9.22}$$

The key points are that g_m goes up as the root of the drain current and linearly with $V_{DS,sat}$.

Example 9.5
Calculate the DC and AC voltages and currents for the circuit seen in Fig. 9.17. Use the long-channel MOSFET parameters from Ch. 6.

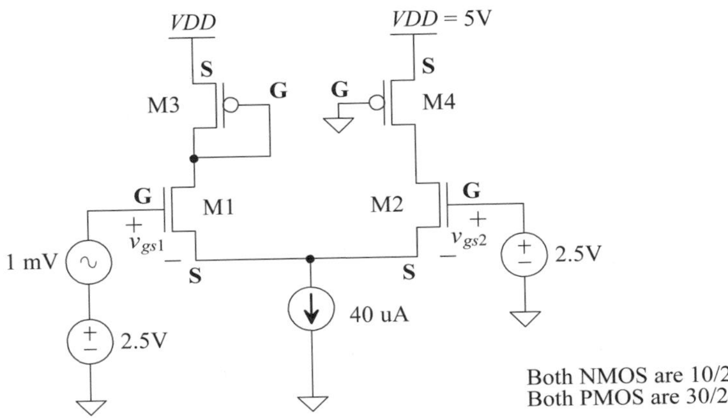

Figure 9.17 Circuit discussed in Ex. 9.5.

We begin by looking at the DC operating point. The gates of both M1 and M2 are at 2.5V. This is necessary to keep them turned on. Since the sources of M1 and M2 are physically tied together, $V_{GS1} = V_{GS2}$ and $I_{D1} = I_{D2} = 20\ \mu A$. Rewriting Eq. (9.1), neglecting channel-length modulation, results in

$$V_{GS} = \overbrace{\sqrt{\frac{2I_D}{KP_n} \cdot \frac{L}{W}}}^{V_{DS,sat}} + V_{THN} \tag{9.23}$$

Assuming both M1 and M2 are operating in the saturation region (we'll verify this in a moment), we get

$$V_{GS1} = V_{GS2} = \sqrt{\frac{40}{120} \cdot \frac{2}{10}} + 0.8 = 1.058\ V \rightarrow V_{DS,sat} \approx 250\ mV$$

The source-to-gate voltage for M3 (knowing its drain current is 20 μA) is

$$V_{SG3} = \sqrt{\frac{2 \cdot 20}{40} \cdot \frac{2}{30}} + 0.9 = 1.158\ V \rightarrow V_{SD,sat} \approx 250\ mV$$

The drain potential of M1 and M3 is

$$V_{D1} = V_{D3} = VDD - V_{SG3} = 3.842\ V$$

From Fig. 9.5 and the associated discussion, we know that M3 is in saturation. To see if M1 is in saturation, we use Eq. (9.7)

$$V_{D1} \overset{?}{\geq} V_{G1} - V_{THN} \rightarrow 3.842 \geq 1.7\ V \text{ (yes, M1 is in saturation)}$$

Next we look at M4. M4's source-to-gate voltage is 5 V. It's very likely that it is in triode. We know, for a PMOS to be operating in the saturation region, we must have

$$\overbrace{V_{SD}}^{V_S - V_D} \geq \overbrace{V_{SG}}^{V_S - V_G} - V_{THP} \rightarrow V_D \leq V_G + V_{THP} \tag{9.24}$$

M4's gate is grounded ($V_G = 0$), so for M4 to be in saturation, $V_D \leq 0.9\ V$. Since the gate of M2 is at 2.5 V and $V_{GS2} = 1.058$, then its source is 1.442 V, which makes it impossible for M4 to be saturated. To estimate the drain-to-source voltage of M4, we use Eq. (9.13), knowing that M4's drain current is 20 μA and its gate-source voltage is 5 V

$$20 = 40 \cdot \frac{30}{2} \cdot \left((5 - 0.9) V_{SD} - \frac{V_{SD}^2}{2} \right)$$

which results in V_{SD} = 8.13 mV. The drains of M4 and M2 are then 4.992 V or essentially at *VDD*. Clearly M2 is operating in the saturation region. M4 can be thought of as a resistor with a value of (Eq. [9.16] noting $V_{SD,sat} >> V_{SD}$, that is, 4.1 V >> 8.13 mV)

$$R_{chM4} = \frac{1}{(40\ \mu A/V) \cdot \frac{30}{2} \cdot 4.1} = 407\ \Omega$$

Note that we could have estimated the resistance using $V_{SD}/I_D = 8.13m/20\mu = 407\ \Omega$. Before moving on, let's do an operating point analysis for this circuit (an .op analysis). The SPICE netlist is seen below.

```
*** Example 9.5 CMOS: Circuit Design, Layout, and Simulation ***

.control
destroy all
run
** for the operating point analysis
print all
*
** for the AC analysis
*plot mag(vd13) mag(vs12) mag(vg1) mag(vd34)
*
** for the transient analysis
*plot vd13
*plot vs12
*plot vg1
.endc

.option scale=1u
.op
**.ac dec 100 1 10k
**.tran 1u 300u

VDD     VDD     0    DC    5
VG1     VG1     0    DC    2.5    AC    1m    SIN 2.5 1m 10k
VG2     VG2     0    Dc    2.5
Ibias   VS12    0    DC    40u
```

```
M1     VD13   VG1    VS12   0      NMOS L=2 W=10
M2     VD24   VG2    VS12   0      NMOS L=2 W=10
M3     VD13   VD13   VDD    VDD    PMOS L=2 W=30
M4     VD24   0      VDD    VDD    PMOS L=2 W=30

.MODEL NMOS NMOS LEVEL  = 3
+ TOX    = 200E-10          NSUB   = 1E17         GAMMA  = 0.5
+ PHI    = 0.7              VTO    = 0.8          DELTA  = 3.0
+ UO     = 650              ETA    = 3.0E-6       THETA  = 0.1
+ KP     = 120E-6           VMAX   = 1E5          KAPPA  = 0.3
+ RSH    = 0                NFS    = 1E12         TPG    = 1
+ XJ     = 500E-9           LD     = 100E-9
+ CGDO   = 200E-12          CGSO   = 200E-12      CGBO   = 1E-10
+ CJ     = 400E-6           PB     = 1            MJ     = 0.5
+ CJSW   = 300E-12          MJSW   = 0.5
*
.MODEL PMOS PMOS LEVEL  = 3
+ TOX    = 200E-10          NSUB   = 1E17         GAMMA  = 0.6
+ PHI    = 0.7              VTO    = -0.9         DELTA  = 0.1
+ UO     = 250              ETA    = 0            THETA  = 0.1
+ KP     = 40E-6            VMAX   = 5E4          KAPPA  = 1
+ RSH    = 0                NFS    = 1E12         TPG    = -1
+ XJ     = 500E-9           LD     = 100E-9
+ CGDO   = 200E-12          CGSO   = 200E-12      CGBO   = 1E-10
+ CJ     = 400E-6           PB     = 1            MJ     = 0.5
+ CJSW   = 300E-12          MJSW   = 0.5

.end
```

A portion of the operating point analysis output is

```
vd13 = 3.854977e+00   (we hand calculated 3.842)
vd24 = 4.989641e+00   (we hand calculated 4.992)
vs12 = 1.171189e+00   (we hand calculated 1.442)
```

The only significant difference between our hand calculations and the simulation results is the potential calculated for the sources of M1/M2. The smaller value in the simulation is because we didn't include the body effect in our calculations. We used $V_{THN0} = 0.8\ V$ when the actual threshold voltage was closer to 1.1 V.

Let's now turn our attention towards the AC analysis. The transconductance of M1 and M2 is

$$g_{m1} = g_{m2} = \sqrt{2 \cdot KP_n \frac{W}{L} \cdot I_D} = \sqrt{2 \cdot 120\mu \cdot \frac{10}{2} \cdot 20\mu} \approx 150\ \mu A/V$$

The transconductance of M3 is

$$g_{m3} = \sqrt{2 \cdot KP_p \frac{W}{L} \cdot I_D} = \sqrt{2 \cdot 40\mu \cdot \frac{30}{2} \cdot 20\mu} \approx 150\ \mu A/V$$

M4 is operating in the triode region and so we think of it as a resistor (407 Ω). Figure 9.18 shows the simplified AC schematic of Fig. 9.17. Notice how we can replace M3 with a resistor of $1/g_{m3}$. This is because the AC voltage across M3 is $v_{sg3} = v_{sd3}$ and the AC current through it is i_d or

$$\frac{1}{g_m} = \frac{v_{sd}}{i_d} = \frac{v_{sg}}{i_d} \tag{9.25}$$

A gate-drain connected MOSFET with a current flowing through it is always in saturation, as discussed earlier, and can be thought of as a small-signal resistance of $1/g_m$ (again, remember this).

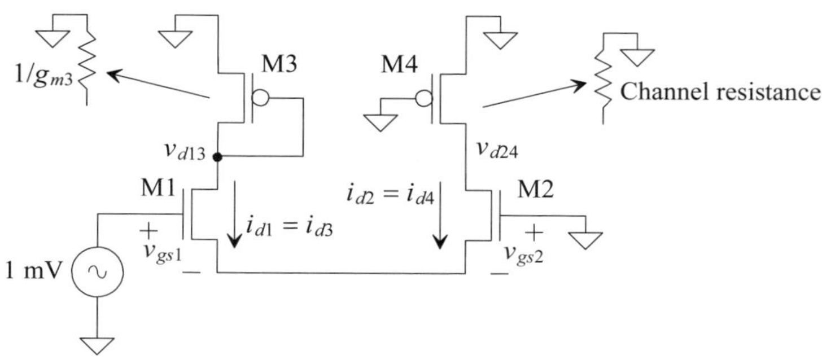

Figure 9.18 AC schematic for the circuit in Fig. 9.17.

When we talk about negative currents or voltages in an AC small-signal analysis, we are indicating that the overall AC + DC signal is decreasing. For example, we'll see in a moment that the AC small-signal drain current in M2, Fig. 9.18, is negative. This means that the current may be going from 20 μA (its DC operating point) to, say, 19.995 μA with an AC input. We would then say that the AC drain current is, i_{d2} , – 5 nA.

To analyze the circuit, we can begin by writing

$$1\ mV = v_{gs1} - v_{gs2} = \frac{i_{d1}}{g_{m1}} - \frac{i_{d2}}{g_{m2}}$$

We know that $g_{m1} = g_{m2}$ and, from Fig. 9.18,

$$i_{d1} = -i_{d2}$$

so

$$v_{gs1} = -v_{gs2} = 0.5\ mV$$

The sources of M1 and M2 are at an AC voltage of 0.5 mV. The AC drain currents are

$$i_{d1} = i_{d3} = -i_{d2} = -i_{d4} = g_{m1} \cdot v_{gs1} = (150\ \mu A/V) \cdot 0.5\ mV = 75\ nA$$

This means the overall (AC + DC) drain current of M1/M3 is

$$i_{D1} = 20\ \mu A + 0.075 \sin 2\pi f t$$

where f is the frequency of our 1 mV input signal and

$$i_{D2} = 20\ \mu A - 0.075 \sin 2\pi f t$$

noting that i_{D1} and i_{D2} must sum to the DC bias of 40 μA.

The (AC) drain voltage of M1 and M3 is

$$v_{d1} = v_{d3} = -i_{d1} \cdot \frac{1}{g_{m3}} = -\frac{75\ nA}{150\ \mu A/V} = -0.5\ mV$$

The voltages on the drains of M2 and M4 are

$$v_{d2} = v_{d4} = -i_{d4} \cdot R_{chM4} = (75\ nA) \cdot 407 = 0.03\ mV$$

We'll verify these voltages with SPICE in a moment. This lengthy example illustrates several key concepts that will be used throughout the analysis and design of analog CMOS circuits in this book. Before moving on, make sure that the concepts are clear and well understood (SPICE can be very helpful for this).

Question: How would we look at the SPICE simulated currents (AC and DC) flowing in the MOSFETs in Fig. 9.17? Answer: add zero-volt voltage sources as seen in Fig. 9.9 and the associated discussion. ■

AC Analysis

We can perform a small-signal analysis using a SPICE ".ac" statement. Just like in our hand calculations, SPICE first calculates the operating point using the DC sources (no AC sources are present in the circuit). Then SPICE replaces the active devices with their small-signal equivalent circuits (no DC sources in the circuit, now just the AC sources). The key point here is that SPICE assumes the user knows that the AC components should be much smaller than the DC components. If we were to simulate, using an AC analysis, the operation of the circuit in Fig. 9.17, but used a 1,000 V AC input instead of a 1 mV AC input, SPICE would simply scale all of the outputs (the simulation won't tell the user that the AC input is too big).

The possible syntax for an .ac analysis statement is

```
.AC DEC ND FSTART FSTOP
.AC OCT NO FSTART FSTOP
.AC LIN NP FSTART FSTOP
```

where DEC stands for decade (our x-axis in an .ac analysis is frequency) and ND is the number of points per decade (OCT stands for octave and LIN stands for linear). FSTART and FSTOP are the starting and ending frequencies for the small-signal AC analysis.

We can do a SPICE AC analysis for the circuit in Fig. 9.17 by changing the .op analysis used in the netlist to an AC analysis control statement

```
.ac dec 100 1 10k
```

where we've picked a frequency range of 1 Hz to 10k Hz with 100 points calculated per decade. (The first decade is 1 to 10 Hz, the second, 10 Hz to 100 Hz, etc. If we had used a start frequency of 2 Hz, then the first decade would be 2 Hz to 20 Hz, the second would be 20 Hz to 200 Hz, noting that increasing by a decade is multiplying by 10 and decreasing by a decade is dividing by 10.) The input signal (the gate of M1) is

```
VG1    VG1    0    DC    2.5    AC    1m    SIN 2.5 1m 10k
```

The operating point analysis ignores the AC voltage of 1 mV and the sinusoid specification (a sine wave with a DC offset of 2.5 V, a peak amplitude of 1 mV, and a frequency of 10 kHz). The AC analysis uses the DC value to calculate the operating point and the AC value to calculate the small-signal AC voltages and currents. We'll discuss the SIN portion in a moment when we talk about transient analysis. The simulation results are seen in Fig. 9.19. The values should be compared to the values calculated in Ex. 9.5. The only (small) difference is the voltage calculated at the sources of M1 and M2. We calculated 0.5 mV, while SPICE gives 0.42 mV. This is due to the body effect transconductance, g_{mb}, which is discussed on the next page.

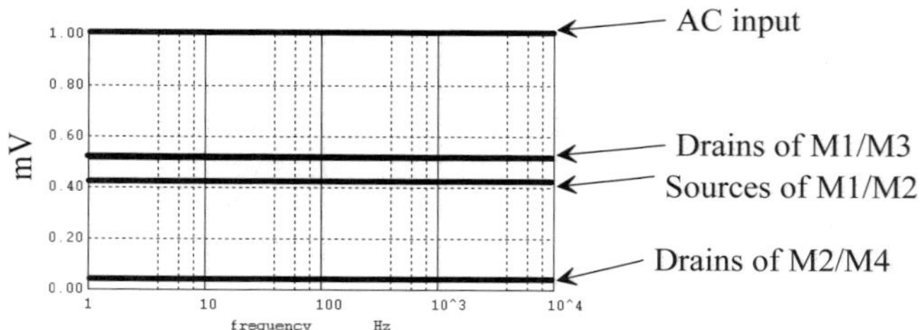

Figure 9.19 AC simulation results for the circuit in Fig. 9.17.

Transient Analysis

The AC analysis is frequently used to look at the frequency response of a circuit. However, as we have just discussed, it will not show large signal nonlinearity. A transient analysis, however, will show large signal nonlinearity (the x-axis is time just like the x-axis on an oscilloscope display). The control statement for a transient analysis is

```
.TRAN TSTEP TSTOP <TSTART <TMAX> <UIC>
```

where TSTEP is the step size (say 1/1,000) of the stop time TSTOP. TSTART is an optional parameter to specify a later starting time. The simulation always starts at 0 seconds. However, in some simulations there are start-up transient signals that we are not interested in. To avoid saving this uninteresting data in our output file (to reduce the output file size), we can specify a start-up time later than 0. In this book we won't use this option. The TMAX parameter specifies a maximum step size for the simulation. If, for example, a sine wave is plotted and it looks *jagged* (meaning that our step size is too coarse), we would use a smaller step size (TMAX) in the simulation to smooth it out. UIC indicates, "Use Initial Conditions." If UIC isn't present in a simulation, SPICE ignores all initial conditions. A sample transient control statement is

```
.TRAN 1n 1000n 0 1n UIC
```

To simulate the circuit in Fig. 9.17 using a transient analysis, we use the input signal source of

```
VG1    VG1    0    DC    2.5    AC    1m    SIN 2.5 1m 10k
```

In a transient analysis, SPICE always ignores the AC component. It ignores the DC component as well if pulse, sinusoid, or some other signal is specified. Note that the SIN specification **has nothing to do with AC analysis**. For the transient simulation of the circuit in Fig. 9.17, we use a sinusoid with a DC offset of 2.5 V, a peak amplitude of 1mV, and a frequency of 10 kHz (picked arbitrarily for this simulation). The simulation results showing the drain potential of M1 and M3 are seen in Fig. 9.20.

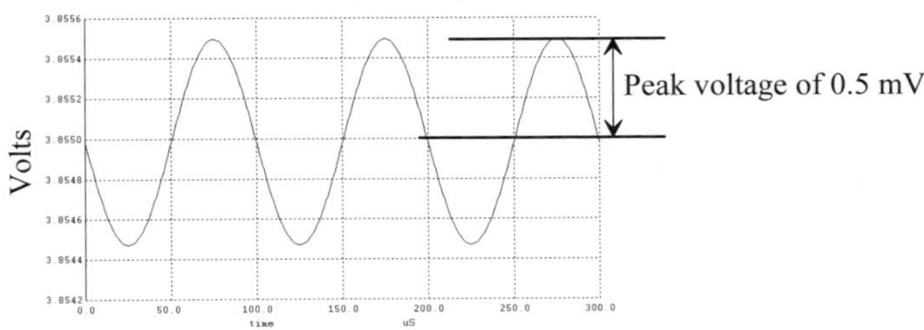

Figure 9.20 Transient simulation showing the drain voltages of M1 and M3 from Fig. 9.17.

Body Effect Transconductance, g_{mb}

Figure 9.21 shows the setup to determine how the drain current varies with source-to-bulk potential V_{SB}. If we raise the potential of the source, we eventually run into the point where the source potential is less than a V_{THN} below the gate potential, and the MOSFET shuts off.

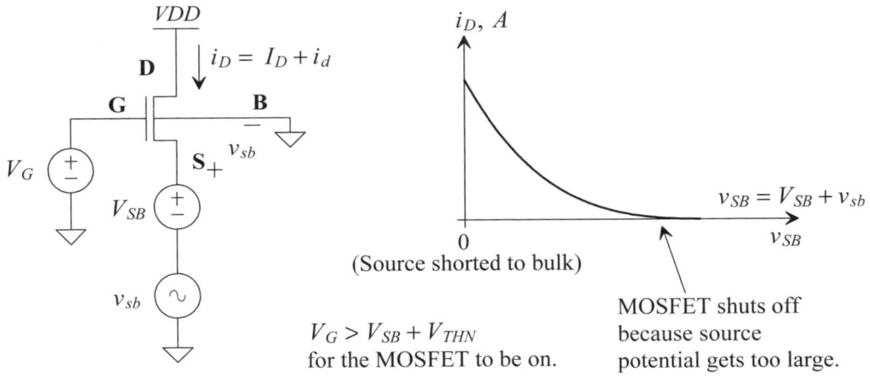

Figure 9.21 How the drain current changes with body-effect.

Remembering that the body-effect is the variation of the threshold voltage with V_{SB}, we can write

$$g_{mb} = \left[\frac{\partial i_D}{\partial v_{SB}}\right]_{V_{SB}=\text{constant}}^{I_D=\text{constant}} = \frac{\partial}{\partial v_{SB}}\left[\frac{KP_n}{2}\cdot\frac{W}{L}(V_{GS}-V_{THN})^2\right]_{V_{SB}=\text{constant}}^{I_D=\text{constant}} \tag{9.26}$$

or

$$g_{mb} = \overbrace{KP_n \cdot \frac{W}{L} \cdot (V_{GS} - V_{THN})}^{g_m} \cdot \left(-\frac{\partial V_{THN}}{\partial v_{SB}}\right) \tag{9.27}$$

or (noting there is a variation in v_{GS} with v_{SB} but this is simply the forward g_m).

$$g_{mb} = g_m \cdot \eta \tag{9.28}$$

The factor η describes how the threshold voltage changes with V_{SB} and generally ranges from 0 (no body effect) to 0.5. The minus sign in Eq. (9.27) simply indicates that the AC drain current contributions from changes in the threshold voltage with V_{SB} (body effect) flow in the opposite direction of the contributions from the forward transconductance, g_m. Figure 9.22 shows the AC small-signal model of the MOSFET with both small-signal transconductances included.

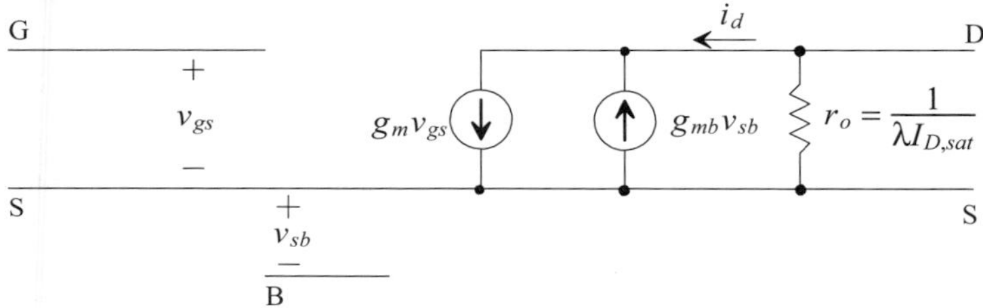

Figure 9.22 Small-signal MOSFET with both transconductances.

Output Resistance

We've already calculated the output resistance of a MOSFET operating in the saturation region in Eq. (9.6) (added to Fig. 9.22). Let's calculate this value again using the circuit seen in Fig. 9.23. We can write

$$r_o^{-1} = \left[\frac{\partial i_D}{\partial v_{DS}}\right]_{V_{DS}=\text{constant}}^{I_D=\text{constant}} = \frac{\partial}{\partial v_{DS}}\left(\frac{KP_n}{2} \cdot \frac{W}{L}(V_{GS} - V_{THN})^2\left(1 + \lambda\left(\overbrace{v_{ds} + V_{DS}}^{v_{DS}} - V_{DS,sat}\right)\right)\right) \tag{9.29}$$

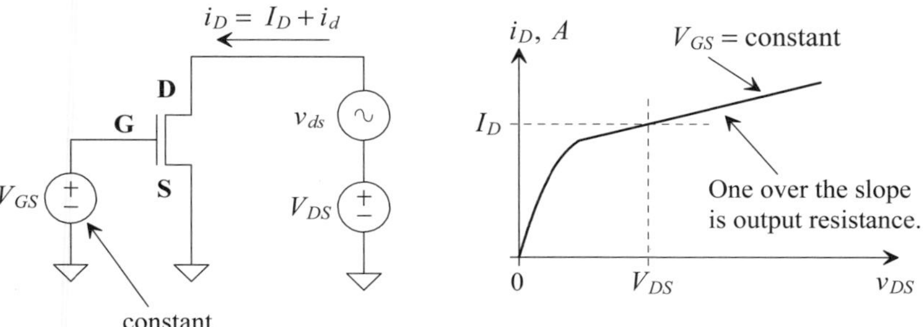

Figure 9.23 How the drain current changes with drain-to-source voltage.

or, once again,

$$r_o = \frac{1}{\lambda I_{D,sat}} \tag{9.30}$$

Example 9.6
Determine λ, using simulations, for the MOSFETs in Ex. 9.5.

The V_{GS} (NMOS) and V_{SG} (PMOS) of the MOSFETs in Ex. 9.5 are 1.05 V and 1.15 V, respectively (roughly for 20 μA of bias current). Figure 9.24 shows the IV plots for the MOSFETs and the reciprocal of the derivative of the drain current (which gives us the output resistance). Because the NMOS threshold voltage is 0.8 V and the PMOS threshold voltage is 0.9 V, both MOSFETs have a $V_{DS,sat}$ of 250 mV and a channel length of 2. The channel-length modulation parameter is calculated using

$$\lambda_n = \frac{1}{I_{D,sat} \cdot r_o} = \frac{1}{20\mu \cdot 5MEG} = 0.01\ V^{-1}$$

and

$$\lambda_p = \frac{1}{20\mu \cdot 4MEG} = 0.0125\ V^{-1}$$

noting that the values are approximations. Also note that the output resistance is very dependent on drain-to-source voltage. ■

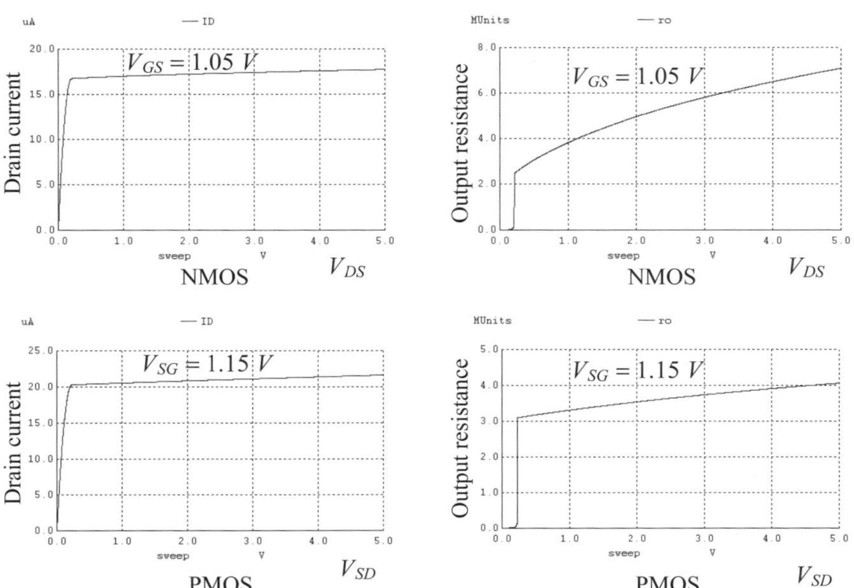

Figure 9.24 Using simulations to determine lambda.

It's of interest to determine how output resistance changes with channel length and $V_{DS,sat}$. Remember

$$I_{D,sat} = \frac{KP_N}{2} \cdot \frac{W}{L} \cdot V_{DS,sat}^2 \tag{9.31}$$

and, from Ch. 6,

$$\lambda \propto \frac{1}{L} \tag{9.32}$$

so we can write

$$r_o \propto \frac{L^2}{V_{DS,sat}^2} \tag{9.33}$$

If the length of the MOSFET is increased, for fixed V_{GS} (equivalent to saying for fixed $V_{DS,sat}$, since $V_{DS,sat} = V_{GS} - V_{THN}$), then the drain current decreases and the output resistance increases. If the length is held constant, then decreasing $V_{DS,sat}$ causes the drain current to decrease and the output resistance to increase. We might think that, to get large output resistance, all we have to do is use a very long-length device. However, as we'll show next, this causes the inherent speed of the MOSFETs to decrease.

MOSFET Transition Frequency, f_T

Examine the circuit in Fig. 9.25. The drain of the MOSFET is at AC ground (shorted through the DC drain-source voltage). This causes, from the gate terminal, C_{gs} and C_{gd}, to appear as though they are in parallel. We can then write

$$v_{gs} = \frac{i_g}{j\omega \cdot (C_{gs} + C_{gd})} \tag{9.34}$$

Knowing $i_d = g_m \cdot v_{gs}$, we can write the current gain of the MOSFET as

$$\left|\frac{i_d}{i_g}\right| = \frac{g_m}{2\pi f \cdot (C_{gs} + C_{gd})} \tag{9.35}$$

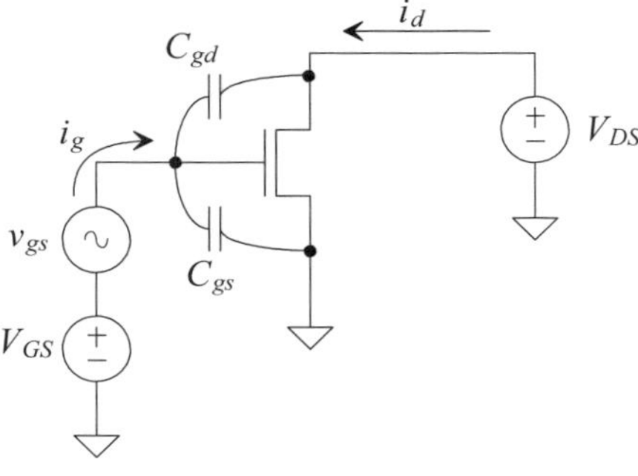

Figure 9.25 Determining the transition frequency of an NMOS transistor.

If we call the frequency where the current gain of the MOSFET is one, Fig. 9.26, the transition frequency, f_T (the transistor transitions from an amplifier to an attenuator) and we remember that $C_{gs}\left(=\frac{2}{3}WLC'_{ox}\right) >> C_{gd}$, then we can write (see Eq. [9.22])

$$f_T \approx \frac{g_m}{2\pi C_{gs}} = \frac{3KP_n \cdot (V_{GS} - V_{THN})}{4\pi \cdot L^2 C'_{ox}} = \frac{3\mu_n}{4\pi} \cdot \frac{V_{DS,sat}}{L^2} \tag{9.36}$$

This equation is *fundamentally important*. To get high speed, we need to use minimum channel lengths and design with a large $V_{DS,sat}$. However, as seen in Eq. (9.33), using minimum lengths results in lower output resistances (and, as we'll see later, lower gain). What this indicates is a constant gain-bandwidth product (higher speed resulting in lower gain). In using a large $V_{DS,sat}$, the MOSFETs enter the triode region earlier (resulting in reduced output swing in amplifiers or mirrors). Note that the f_T for the PMOS device is inherently smaller than that of the NMOS device due to the lower value of hole mobility.

For short-channel devices, the mobility is no longer constant but starts to decrease (velocity saturation as discussed in Ch. 6) with decreasing length (increasing electric field between the drain and channel). Because of this, we treat the term μ_n/L as a relatively constant value and rewrite Eq. (9.36) as

$$f_T \propto \frac{V_{DS,sat}}{L} \text{ (Short-channel devices)} \tag{9.37}$$

Again, for high-speed, we need to use the smallest possible channel length and large $V_{DS,sat}$.

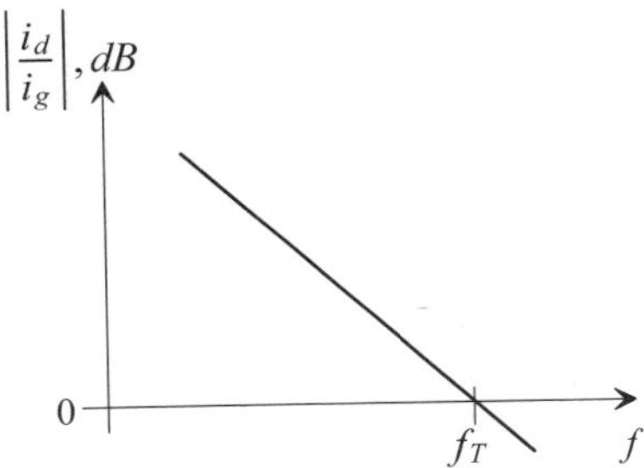

Figure 9.26 The transition frequency of a MOSFET.

General Device Sizes for Analog Design

Based on Eqs. (9.33) and (9.36), we can make some general statements about selecting device L, W, and $V_{DS,sat}$. For general analog design, use an L of 2–5 times minimum (in this book L_{min} = 1, since we scale the MOSFET sizes by either 1 μm [long-channel MOSFETs] or 50 nm [short-channel MOSFETs] in the SPICE netlists or layout). We'll use an L, for analog design, of 2 as a good trade-off between speed and gain.

For general design, use a $V_{DS,sat}$ of 5% of VDD. For our long-channel MOSFETs (VDD = 5 V), we'll use a $V_{DS,sat}$ of 250 mV, while for the short-channel process (VDD = 1V), we'll use 50 mV. Table 9.1 shows the parameters for general analog design using the long-channel process discussed in this chapter.

Table 9.1 Typical parameters for analog design using the *long-channel* CMOS process discussed in this book. Note that the parameters may change with temperature or drain-to-source voltage (e.g., Fig. 9.24).

Long-channel MOSFET parameters for general analog design $VDD = 5$ V and a scale factor of **1 μm** (*scale* = 1e–6)			
Parameter	**NMOS**	**PMOS**	**Comments**
Bias current, I_D	20 μA	20 μA	Approximate
W/L	10/2	30/2	Selected based on I_D and $V_{DS,sat}$
$V_{DS,sat}$ and $V_{SD,sat}$	250 mV	250 mV	For sizes listed
V_{GS} and V_{SG}	1.05 V	1.15 V	No body effect
V_{THN} and V_{THP}	800 mV	900 mV	Typical
$\partial V_{THN,P}/\partial T$	–1 mV/C°	–1.4 mV/C°	Change with temperature
KP_n and KP_p	120 μA/V^2	40 μA/V^2	$t_{ox} = 200$ Å
$C'_{ox} = \varepsilon_{ox}/t_{ox}$	1.75 fF/μm^2	1.75 fF/μm^2	$C_{ox} = C'_{ox}WL \cdot (scale)^2$
C_{oxn} and C_{oxp}	35 fF	105 fF	PMOS is three times wider
C_{gsn} and C_{sgp}	23.3 fF	70 fF	$C_{gs} = \frac{2}{3}C_{ox}$
C_{gdn} and C_{dgp}	2 fF	6 fF	$C_{gd} = CGDO \cdot W \cdot scale$
g_{mn} and g_{mp}	150 μA/V	150 μA/V	At $I_D = 20$ μA
r_{on} and r_{op}	5 MΩ	4 MΩ	Approximate at $I_D = 20$ μA
$g_{mn}r_{on}$ and $g_{mp}r_{op}$	750 V/V	600 V/V	Open circuit gain
λ_n and λ_p	0.01 V^{-1}	0.0125 V^{-1}	At $L = 2$
f_{Tn} and f_{Tp}	900 MHz	300 MHz	For L =2, f_T goes up if $L = 1$

Subthreshold g_m and V_{THN}

Before leaving the topic of small-signal models, let's derive the forward transconductance of the MOSFET operating in the subthreshold region. Using Eqs. (9.17) and (9.20), we can write

$$g_m = \left[\frac{\delta i_D}{\delta v_{GS}}\right]_{V_{GS}\,=\,\text{constant}}^{I_D\,=\,\text{constant}} = I_{D0} \cdot \frac{W}{L} \cdot e^{\left(\overbrace{V_{GS}+v_{gs}}^{v_{GS}} - V_{THN}\right)/nV_T} \cdot \left(\frac{1}{nV_T}\right) \quad (9.38)$$

If, (as always for a small-signal analysis) $v_{gs} << V_{GS}$, then

$$g_m = \frac{I_D}{nV_T} \quad (9.39)$$

The transconductance increases linearly with bias current when operating in the subthreshold region (compare to Eq. (9.22)). Unfortunately, the speed (f_T) of MOSFETs operating in this region is considerably slower. The small currents charging the device's own capacitances limit the speeds to, in general, < MHz. As CMOS technology scales downwards, the inherent speeds increase. This, together with the need for lower power, increases the number of designs operating in, or near, the subthreshold region.

As I_D (V_{GS}) increases, the MOSFET moves from operating in the subthreshold region, $g_m = I_D/nV_T$ (g_m is exponentially dependent on V_{GS}), to moderate inversion, and then to the strong inversion region, $g_m = \sqrt{2I_D\beta_n} = \beta_n(V_{GS} - V_{THN})$ (g_m is linearly dependent on V_{GS}). As seen in Eq. (9.16), using small V_{DS} results in a channel resistance, R_{ch} , of $1/g_m$ (assuming that the MOSFET is operating in the *triode* region). Fig. 9.27a shows a plot of I_D against V_{GS} for a MOSFET in the short-channel process discussed later in the chapter. The threshold voltage in (a) is estimated by linearly extrapolating back to the x-axis (again from Eq. [9.16] the slope is R_{ch}^{-1}). In (b) we take the derivative of (a) to get the g_m of the device. Knowing $g_m = \beta_n(V_{GS} - V_{THN})$ (for a MOSFET operating in the *saturation* region), we can linearly extrapolate back to the x-axis to estimate when g_m goes to zero (to get V_{THN}). The two methods give different results (there are other methods as well, such as taking the derivative of g_m with respect to V_{GS} and looking at the V_{GS} [= V_{THN}] when this second derivative of I_D becomes constant). Because, even for small V_{DS} (> a couple of kT/q), the MOSFET will be operating in the saturation region when $V_{GS} \approx V_{THN}$ we'll use method (b).

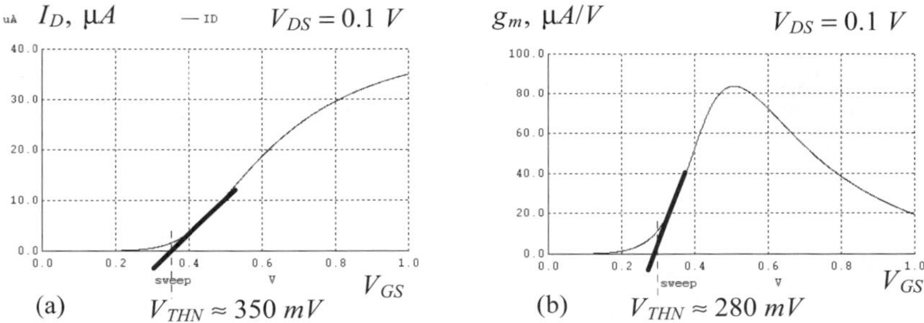

Figure 9.27 (a) Drain current plotted against gate-source voltage with small Vds and (b) transconductance plotted against gate-source voltage.

9.1.3 Temperature Effects

In this section we look at how the drain current of a MOSFET operating in the saturation region changes with temperature. Looking at the long-channel expression for the drain current

$$I_{DS,sat} = \frac{\mu_n \cdot \varepsilon_{ox}}{2t_{ox}} \cdot \frac{W}{L} \cdot (V_{GS} - V_{THN})^2 \tag{9.40}$$

(where we've written KP_n as $\mu_n\varepsilon_{ox}/t_{ox}$), the variables that change with temperature are the mobility and the threshold voltage.

Threshold Variation with Temperature

From Ch. 6, Eq. (6.17), we can write

$$V_{THN} = -V_{ms} - 2V_{fp} + \frac{Q'_{b0} - Q'_{ss}}{C'_{ox}} \tag{9.41}$$

noting Q'_{ss} is a constant and $Q'_{b0} = \sqrt{2qN_A\varepsilon_{si}|-2V_{fp}|}$. Knowing

$$V_{fp} = -\frac{kT}{q}\ln\frac{N_A}{n_i} \text{ and } V_{ms} = V_G - V_{fp} = \frac{kT}{q}\ln\frac{N_{D,poly}}{n_i} - V_{fp} \tag{9.42}$$

the change in the threshold voltage with temperature is

$$\frac{\partial V_{THN}}{\partial T} = -\frac{k}{q}\cdot \ln\frac{N_{D,poly}}{N_A} + \frac{Q'_{b0}}{C'_{ox}\cdot 2T} \approx -\frac{k}{q}\cdot \ln\frac{N_{D,poly}}{N_A} \tag{9.43}$$

The term k/q is the change in the thermal voltage with temperature, that is,

$$\frac{\partial V_T}{\partial T} = \frac{\partial}{\partial T}\left(\frac{kT}{q}\right) = \frac{k}{q} = 0.085\ mV/C° \tag{9.44}$$

(keeping in mind that we don't confuse the temperature behavior of the thermal voltage, V_T, with the behavior of the threshold voltage, V_{THN}). If $N_{D,poly} = 10^{20}$ and $N_A = 10^{15}$ (a ballpark value for a long-channel MOSFET), then

$$\frac{\partial V_{THN}}{\partial T} \approx -1\ mV/C° \tag{9.45}$$

If $N_A = 10^{17}$ (N_A scales upwards as the channel length of the CMOS technology decreases, see Table 6.3 in Ch. 6), then

$$\frac{\partial V_{THN}}{\partial T} \approx -0.6\ mV/C° \tag{9.46}$$

noting that the threshold voltage decreases with increasing temperature, while the thermal voltage increases with increasing temperature. The temperature coefficient of the threshold voltage is defined as

$$TCV_{THN} = \frac{1}{V_{THN}}\cdot\frac{\partial V_{THN}}{\partial T} \tag{9.47}$$

so that the threshold voltage can be written as a function of temperature as

$$V_{THN}(T) = V_{THN}(T_0)\cdot(1 + TCV_{THN}\cdot(T - T_0)) \tag{9.48}$$

where the threshold voltage is measured at the temperature T_0, Fig. 9.28. The units for temperature can be Kelvin or Celsius because of the difference used in this equation. For our long-channel process with $V_{THN} = 0.8$ V, the temperature coefficient is –1,250 ppm/$C°$. For a short-channel process with $V_{THN} = 0.28$ V, we get a temperature coefficient of –2,143 ppm/$C°$. For a 100-degree increase in temperature, our short-channel threshold voltage decreases by 60 mV, resulting in a larger I_{off} (a factor of 10 larger if the subthreshold slope is the ideal 60 mV/decade at room temperature).

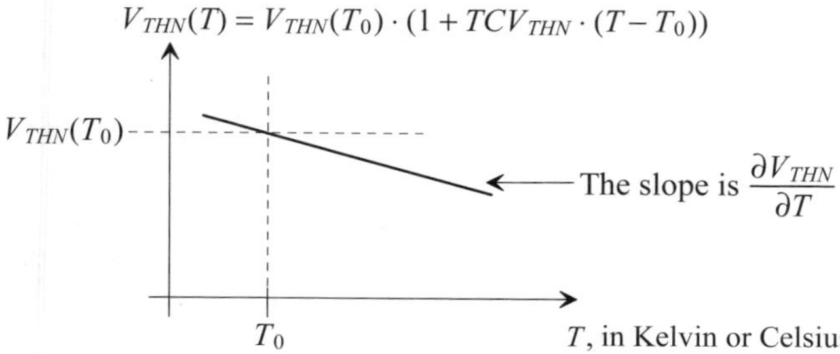

Figure 9.28 Temperature dependence of the threshold voltage.

Mobility Variation with Temperature

The reduction in mobility with increasing temperature is modeled using

$$\mu(T) = \mu(T_0) \cdot \left(\frac{T_0}{T}\right)^{1.5} \tag{9.49}$$

and so it follows that the reduction in the transconductance parameter with increasing temperature is

$$KP(T) = KP(T_0) \cdot \left(\frac{T_0}{T}\right)^{1.5} \tag{9.50}$$

where, once again, the mobility is measured at temperature T_0. Note that both T and T_0 have units of Kelvin. The change in the MOSFET transconductance parameter, KP, with temperature around T_0 is

$$\left[\frac{\partial KP(T)}{\partial T}\right]_{T_0 = \text{constant}} = KP(T_0) \cdot (-1.5)\left(\frac{T_0}{T}\right)^{2.5} \cdot \frac{1}{T_0} \tag{9.51}$$

The temperature coefficient of the transconductance parameter around the temperature T_0 is then

$$\frac{1}{KP(T)} \cdot \frac{\partial KP(T)}{\partial T} = -\frac{1.5}{T} \tag{9.52}$$

The transconductance parameter at a particular temperature close to T_0 is given by

$$KP(T) = KP(T_0) \cdot \left(1 - 1.5 \cdot \frac{T - T_0}{T}\right) \tag{9.53}$$

As temperature increases, the mobility and KP decrease. Note that our derivation used the change in the temperature (the slope of the line) around the measured temperature T_0 (just like we used for small-signal analysis). Equations (9.52) and (9.53) work well for hand calculations. However, for wide temperature changes, we'll need to use simulations (which can include the nonlinear variations in the mobility). Note, again, that unlike Eq. (9.48) where a difference in temperatures is present, and so either Kelvin or Celsius can be used, we must use Kelvin when using Eq. (9.53).

Drain Current Change with Temperature

We now know that as temperature goes up, the threshold voltage and mobility go down. A decrease in mobility, see Eq. (9.40), causes the drain current to go down. At the same time, a decrease in threshold voltage causes the drain current to go up. At low V_{GS}, the changes in V_{THN} dominate and the drain current increases with increasing temperature. At higher V_{GS}, the mobility dominates and the drain current decreases with increasing temperature. When the effects cancel, the drain current doesn't change with temperature, Fig. 9.29. For a long-channel device like the one seen in Fig. 9.29, again, $\partial V_{THN}/\partial T = -1\ mV/C°$.

Figure 9.30a shows how the drain current changes with temperature in a short-channel CMOS technology. Again, we see where the effects of the mobility changing with temperature cancel the effects of the threshold voltage changing with temperature. Looking at Fig. 9.30b, we see that, for a constant current of 10 μA, V_{GS} changes at a rate of $\approx -0.6\ mV/C°$ ($= \partial V_{THN}/\partial T$ because the V_{GS} is small).

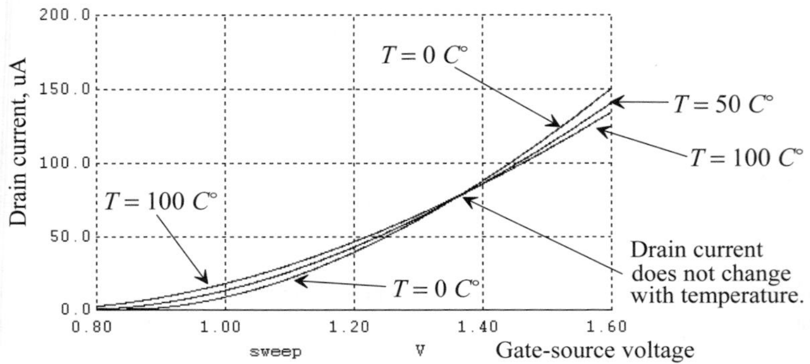

Figure 9.29 Long-channel drain current change with temperature.

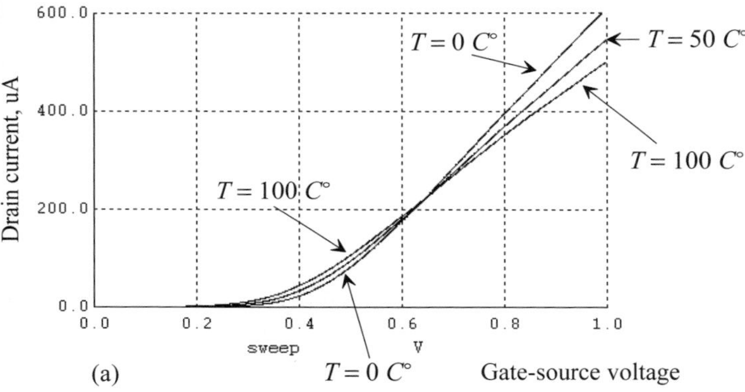

(a)

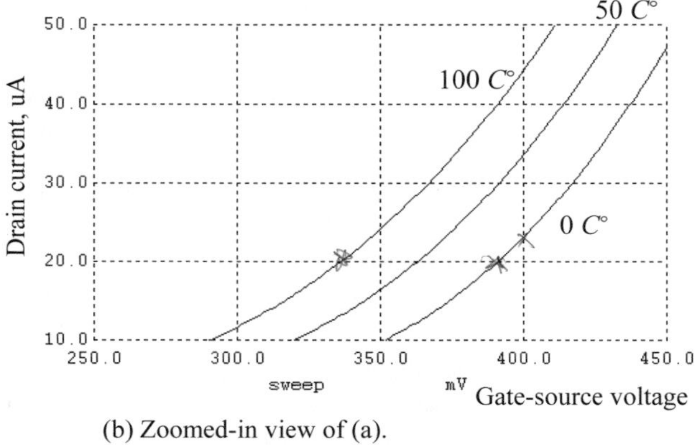

(b) Zoomed-in view of (a).

Figure 9.30 Short-channel drain current change with temperature.

9.2 Short-Channel MOSFETs

The last section covered the fundamental models used for MOSFETs in analog design. In this section we use these results to help with our selection of device sizes and biasing currents to meet certain design requirements: speed, power, output or input swing, etc. Our approach is to develop plots of device parameters based on simulations that can be used when designing with short-channel MOSFETs. The CMOS process technology that we use in this section has a minimum length of 50 nm and a *VDD* of 1 V.

9.2.1 General Design (A Starting Point)

We labeled the V_{DS} where the MOSFET enters the saturation region, for a fixed V_{GS} and for either long- or short-channel MOSFETs, $V_{DS,sat}$ (Fig. 9.4). For long-channel MOSFETs this voltage was determined using Eq. (9.2), that is, $V_{GS} - V_{THN}$. However, for short-channel devices this relationship ($V_{DS,sat} = V_{GS} - V_{THN}$) isn't meaningful. For these devices, we'll call the difference between V_{GS} and V_{THN} the *gate overdrive voltage*

$$V_{ovn} = V_{GS} - V_{THN} \neq V_{DS,sat} \tag{9.54}$$

As we saw in Fig. 9.27 the threshold voltage for this 50 nm process is 280 mV. We said earlier that for general analog design we set the overdrive voltage to roughly 5% of *VDD*. We might use, as a starting point,

$$V_{ovn} = 70\ mV \rightarrow V_{GS} = 350\ mV$$

For higher speed we increase V_{ovn} at the price of reduced output swing (a higher V_{DS} is needed to move the MOSFET into saturation). Using the overdrive voltage, we can rewrite, perhaps more correctly now, Eq. (9.37) as

$$f_T = \frac{g_m}{2\pi C_{gs}} \propto \frac{V_{ovn}}{L} \tag{9.55}$$

Now we must select the biasing current and the width/length of the MOSFETs. As mentioned earlier, we use 2–5 times minimum length for general design (we use minimum length for high-speed design). As we did with the long-channel process, let's use twice minimum length ($L = 2$ or, with the scale factor, $L = 100$ nm) to get started as a good trade-off between speed and gain. To select the bias current and width, let's remember the on current for a short-channel MOSFET, from Ch. 6, is

$$i_D = v_{sat} \cdot C'_{ox} \cdot W \cdot (v_{GS} - V_{THN} - V_{DS,sat}) \tag{9.56}$$

The transconductance, following the same procedure used earlier, for a short-channel MOSFET is then

$$g_m = \left[\frac{\partial i_D}{\partial v_{GS}}\right]_{V_{GS} = \text{constant}}^{I_D = \text{constant}} = v_{sat} \cdot C'_{ox} \cdot W \tag{9.57}$$

The g_m of a short-channel device depends only on the MOSFET's width if the velocity of the carriers saturates (is a constant). Fortunately, effects such as velocity overshoot **cause g_m to increase with increases in V_{GS} or V_{DS}.** The width of the MOSFET is selected to ensure the MOSFET has enough current drive for a particular load (for a more detailed discussion see Sec. 26.1). Here we'll select a bias current of 10 μA and a g_m of 150 μA/V. As a first cut, let's use $V_{ovn} = 70$ mV, $W = 50$, and $L = 2$ (selected based on simulations). Figure 9.31 shows the IV curves for a 50/2 (actual size of 2.5μm/100nm device) with a

V_{GS} of 350 mV. The current shows significant variation as V_{DS} changes. Knowing I_D does change with V_{DS}, we'll still simply say the drain current is 10 μA (as an approximation).

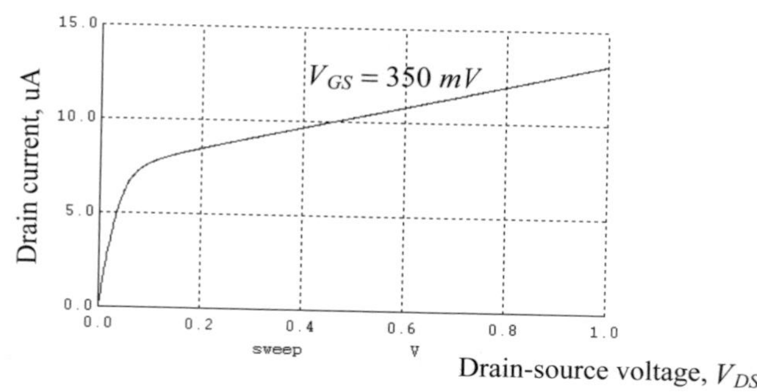

Figure 9.31 IV curves for a 50/2 NMOS with VGS = 350 mV.

Output Resistance

Figure 9.32 shows the output resistance of this 50/2 NMOS device. We get r_o by taking the reciprocal of the drain current's derivative in Fig. 9.31. To determine $V_{DS,sat}$, we can look at the point where the output resistance starts to increase. Here this is approximately 50 mV (= $V_{DS,sat}$). However, notice that if we use larger V_{DS}, we get considerably higher output resistances. This is an **important point** when we design current mirrors in Ch. 20. The lambda can be estimated using Figs. 9.31 and 9.32 as 0.6 V^{-1}.

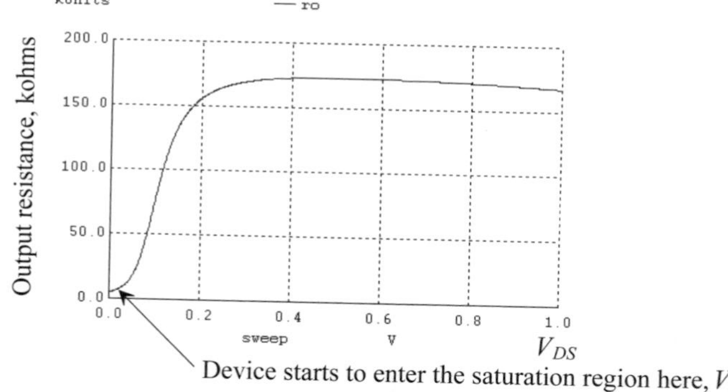

Figure 9.32 Output resistance of the MOSFET in Fig. 9.31 against V_{DS} .

Forward Transconductance

The forward transconductance, g_m , is plotted against V_{GS} in Fig. 9.33. At our V_{GS} of 350 mV (gate overdrive, V_{ovn} , of 70 mV), we get, as designed for, a g_m of 150 μA/*V*. The open circuit gain is $g_m r_o$ and is only 25. This is *considerably lower* than the open circuit gains seen in Table 9.1 for the long-channel devices and results in design "challenges" when designing with such small devices. Also notice that the g_m does change with V_{GS} , unlike

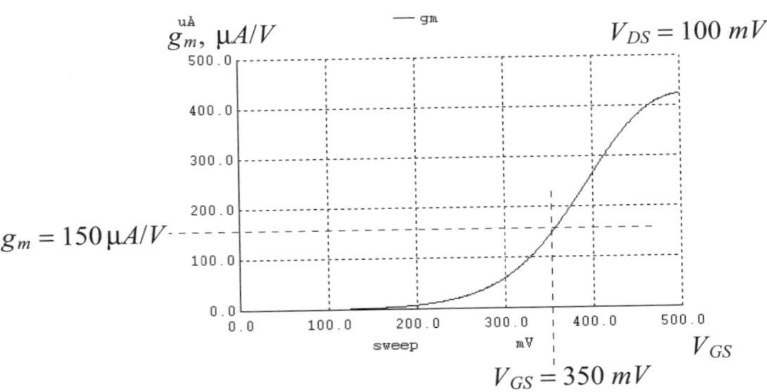

Figure 9.33 How transconductance changes with gate-source bias.

what was indicated in Eq. (9.57). This is because the saturation velocity isn't exactly constant and depends on both V_{GS} and V_{DS}.

Transition Frequency

From Fig. 9.34 we see that the transition frequency of the 50/2 NMOS is approximately 6 GHz. While the equations show that increasing the gate overdrive increases f_T, it can be educational to change the V_{GS} in the netlist and look at how the speed (f_T) changes.

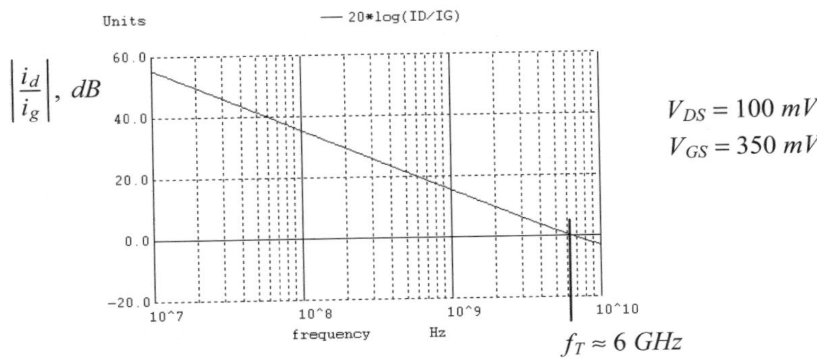

Figure 9.34 MOSFET transition frequency.

Table 9.2 shows the typical parameters for the sizes and biasing current we've used in this section. These values are a good starting point for general design. However, if the design must be either high-speed or low-power, the biasing current, device sizes, and thus overdrive voltages **will be** different from what's listed. They'll also be different with shifts in process, voltage, and temperature (PVT).

Finally, we generated the PMOS data from simulation netlists that are available at cmosedu.com. These simulation netlists are invaluable for quickly regenerating the data in Table 9.2 for different device sizes and overdrive voltages.

Table 9.2 Typical parameters for analog design using the *short-channel* CMOS process discussed in this book. These parameters are valid only for the device sizes and currents listed.

Short-channel MOSFET parameters for general analog design *VDD* = 1 V and a scale factor of **50 nm** (*scale* = 50e–9)			
Parameter	**NMOS**	**PMOS**	**Comments**
Bias current, I_D	10 μA	10 μA	Approximate, see Fig. 9.31
W/L	50/2	100/2	Selected based on I_D and V_{ov}
Actual W/L	2.5μm/100nm	5μm/100nm	L_{min} is 50 nm
$V_{DS,sat}$ and $V_{SD,sat}$ V_{ovn} and V_{ovp}	50 mV 70 mV	50 mV 70 mV	However, see Fig. 9.32 and the associated discussion
V_{GS} and V_{SG}	350 mV	350 mV	No body effect
V_{THN} and V_{THP}	280 mV	280 mV	Typical
$\partial V_{THN,P}/\partial T$	– 0.6 mV/C°	– 0.6 mV/C°	Change with temperature
v_{satn} and v_{satp}	110 x 10^3 m/s	90 x 10^3 m/s	From the BSIM4 model
t_{ox}	14 Å	14 Å	Tunnel gate current, 5 A/cm^2
$C'_{ox} = \varepsilon_{ox}/t_{ox}$	25 *f*F/μm^2	25 *f*F/μm^2	$C_{ox} = C'_{ox}WL \cdot (scale)^2$
C_{oxn} and C_{oxp}	6.25 *f*F	12.5 *f*F	PMOS is two times wider
C_{gsn} and C_{sgp}	4.17 *f*F	8.34 *f*F	$C_{gs} = \frac{2}{3}C_{ox}$
C_{gdn} and C_{dgp}	1.56 *f*F	3.7 *f*F	$C_{gd} = CGDO \cdot W \cdot scale$
g_{mn} and g_{mp}	150 μA/V	150 μA/V	At I_D = 10 μA
r_{on} and r_{op}	167 kΩ	333 kΩ	Approximate at I_D = 10 μA
$g_{mn}r_{on}$ and $g_{mp}r_{op}$	25 V/V	50 V/V	!!Open circuit gain!!
λ_n and λ_p	0.6 V^{-1}	0.3 V^{-1}	$L = 2$
f_{Tn} and f_{Tp}	6000 MHz	3000 MHz	Approximate at $L = 2$

9.2.2 Specific Design (A Discussion)

A figure-of-merit (FOM) that is useful when comparing designs with different *W/Ls* and bias currents is the gain-f_T product (GFT) of an amplifier. The GFT can be written as a product of the open-circuit gain of the MOSFET with its f_T, that is,

$$\text{GFT} = g_m r_o \cdot f_T \tag{9.58}$$

For a *long-channel* process we can write this equation (see Eqs. [9.22], [9.30], and [9.36]) knowing $C_{gs} = \frac{2}{3}C'_{ox}WL$ as

$$\text{GFT} = g_m r_o \cdot f_T = \frac{g_m^2}{2\pi C_{gs}} \cdot \frac{1}{\lambda I_D} = \frac{3\mu_n}{2\pi \cdot L^2 \lambda} \tag{9.59}$$

Notice that the expression is *independent of drain current*. It is based entirely on the channel length of the MOSFET. This result is important because it shows that if the gain goes up, the speed (f_T) goes down. Not understanding this equation will lead to *wasted time* when trying to increase both the gain and bandwidth of an amplifier.

We know from Eq. (9.36) that increasing the drain current (or increasing V_{GS}) results in an increase in speed. But the cost for the higher speed is a reduction in the open-circuit gain. This gain can be written, in strong inversion, as

$$g_m r_o = \frac{\sqrt{2KP_n \frac{W}{L} I_D}}{\lambda I_D} = \frac{\sqrt{2KP_n \frac{W}{L}}}{\lambda} \cdot \frac{1}{\sqrt{I_D}} \tag{9.60}$$

showing that gain decreases with increasing drain current. We can also write

$$g_m r_o = \frac{KP_n \frac{W}{L}(V_{GS} - V_{THN})}{\lambda \cdot \frac{KP_n}{2} \frac{W}{L}(V_{GS} - V_{THN})^2} = \frac{2}{\lambda(V_{GS} - V_{THN})} = \frac{2}{\lambda V_{DS,sat}} \tag{9.61}$$

showing gain decreases with increasing gate-source voltage (but, of course, f_T goes up). When operating in the *subthreshold region,* we can write

$$g_m r_o = \frac{I_D}{nV_T} \cdot \frac{1}{\lambda I_D} = \frac{1}{nV_T \lambda} \tag{9.62}$$

showing that the gain is independent of drain current when the long-channel MOSFET is operating in the subthreshold region, see Fig. 9.35.

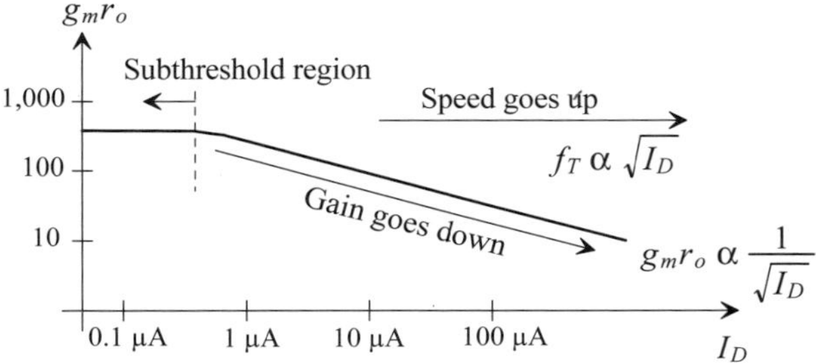

Figure 9.35 Gain falling off with bias current.

While these results were derived using long-channel models, we can use the outcomes for design in a short-channel process. For high speed we want to use higher biasing currents and thus overdrive voltages. As the speed of the design increases, the gain falls (but, as we'll see, not at the same rate). As an example, if we decrease the width of the NMOS device used to generate Table 9.2 from 50 to 20 (using the same gate-source voltage of 350 mV), we would expect I_D to fall, f_T to remain unchanged (the device is narrower so its capacitance decreases but so does the g_m), r_o to increase, and g_m to decrease. The GFT shouldn't change much by simply reducing the width of the device. First, from Table 9.2 the GFT is

$$g_m r_o \cdot f_T = \overbrace{150\tfrac{\mu A}{V} \cdot 167\ k\Omega}^{\text{open circuit gain} = 25} \cdot 6\ GHz = 150\ GHz$$

Next, with the decrease in the device's width to 20, we get (with the help of the simulations that generated Figs. 9.31–9.34): $I_D = 4\ \mu A$ (V_{ovn} is still 70 mV), $r_o = 435\ k\Omega$, $g_m = 55\ \mu A/V$, and $f_T = 6$ GHz. The GFT with the device width of 20 is

$$g_m r_o \cdot f_T = \overbrace{55\frac{\mu A}{V} \cdot 435\ k\Omega}^{\text{open circuit gain} = 24} \cdot 6\ GHz = 144\ GHz$$

practically no difference. If we now increase the gate-source voltage to 450 mV (almost half of *VDD*) so that the overdrive goes from 70 mV to 170 mV while using 50/2 NMOS, we would expect the f_T to increase, g_m to increase, I_D to increase, and r_o to decrease. Again, with the help of the simulations that generated Figs. 9.31–9.34, $I_D = 45\ \mu A$ (V_{ovn} is now 170 mV), $r_o = 55\ k\Omega$, $g_m = 390\ \mu A/V$, and $f_T = 10$ GHz. The GFT with the increased V_{ovn} is

$$g_m r_o \cdot f_T = \overbrace{390\frac{\mu A}{V} \cdot 55\ k\Omega}^{\text{open circuit gain} = 21.5} \cdot 10\ GHz = 215\ GHz$$

Note that the open circuit gain only decreased a modest 3.5 from the value in Table 9.2. Using the long-channel theory we would expect, with a roughly 4 increase in the drain current, the open circuit gain to be cut in half (Eq. [9.60]).

By using higher gate overdrives, in a short-channel process, we can improve the speed of our circuits at the cost of power and overhead (the devices will go into triode at higher V_{DS}) with small degradation to the gain (the GFT increases with increasing gate overdrive). It might appear that, because the GFT in a long-channel CMOS technology is independent of bias conditions (see Eq. [9.59]), it is better to use a short-channel process. However, we must remember that the open circuit gains of designs using the short-channel processes are considerably lower than those of designs using the long-channel technology (see Tables 9.1 and 9.2). The GFT, for the NMOS device in Table 9.1 for example, is 675 GHz (compared to 150 GHz for the one in Table 9.2). The short-channel device is better for high-speed design because its f_T is higher (it will still behave like an amplifier in the GHz frequency region) but worse for lower speed design.

9.3 MOSFET Noise Modeling

In this section we cover modeling MOSFET noise using SPICE. It is assumed that the reader is familiar with the material presented in Ch. 8, that is, the spectral characteristics and origins of thermal and flicker noise.

Drain Current Noise Model

The noise generators in MOSFETs are due to thermal and flicker noise. The thermal noise due to the channel resistance, modeled by a resistor of $\frac{3}{2} \cdot \frac{1}{g_m}$ while in the saturation region and R_{CH} in the triode region, is given, in the saturation region, by a power spectral density, PSD, of

$$I_R^2(f) = \frac{4kT}{\frac{3}{2} \cdot \frac{1}{g_m}} = \frac{8kT}{3} \cdot g_m \tag{9.63}$$

This noise current source is placed across the drain and source of the MOSFET (so the output current is the sum of the desired and undesired [noise] components).

Flicker noise (one-over-f, that is, $1/f$) results from the trapping of charges at the oxide/semiconductor interface. As indicated in Ch. 8, flicker noise is present whenever a direct current flows in a discontinuous material. The electrons jump from one location to the next while sometimes being randomly trapped and released. This trapping gives rise to a flickering in the drain current. Since the carrier lifetime in silicon is on the order of tens of microseconds (relatively long), the resulting current fluctuations are concentrated at lower frequencies. Flicker noise can be modeled in SPICE (with NLEV = 0, the default) by a PSD of

$$I_{1/f}^2(f) = \frac{KF \cdot I_D^{AF}}{f \cdot (C'_{ox} \cdot L)^2} \tag{9.64}$$

where KF is the flicker noise coefficient, a typical value is 10^{-28} $A^{2-AF}(F/m)^2$, I_D is the DC drain current, AF is the flicker noise exponent, a value ranging from 0.5 to 2 (generally very close to 1), and f is the frequency variable we integrate over. Setting NLEV = 1 in SPICE changes the L^2 term in the denominator to LW (the area of the MOSFET), that is,

$$I_{1/f}^2(f) = \frac{KF \cdot I_D^{AF}}{f \cdot (C'_{ox})^2 LW} \tag{9.65}$$

Figure 9.36 shows measured data comparing a straight-line fit to the PSD generated using this noise model. The device used to collect this data is 10 μm wide and 2.8 μm long. Note that the noise PSD was averaged 20 times and still looks jagged. The 1/f noise spectrums we drew in Ch. 8, e.g., Fig. 8.31, show smooth PSDs. We only get smooth spectrums after much averaging (it is, after all, noise).

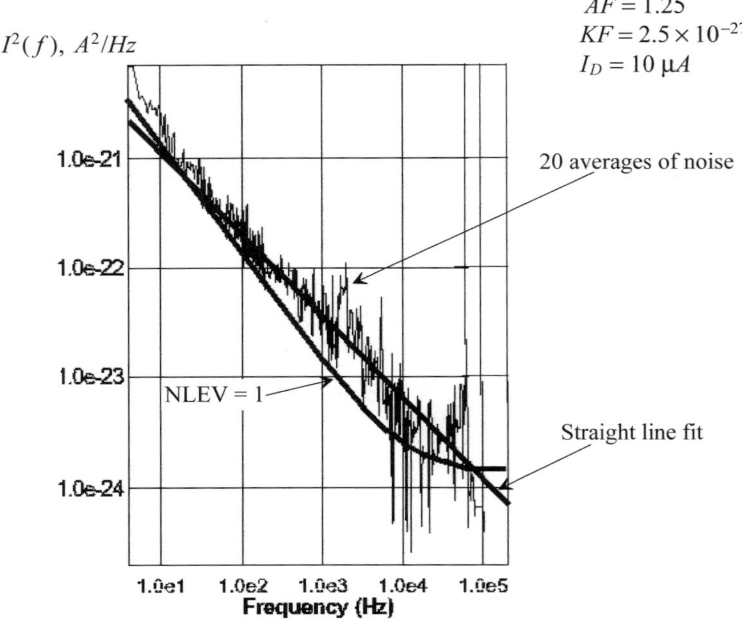

Figure 9.36 Measured MOSFET flicker noise spectrum.

The total PSD of MOSFET drain current noise (the sum of the flicker and thermal noise contributions) is given by

$$I_M^2(f) = I_{1/f}^2(f) + I_R^2(f) = \frac{KF \cdot I_D^{AF}}{f \cdot (C'_{ox})^2 LW} + \frac{8kT}{3} \cdot g_m \quad (9.66)$$

Remembering noise is always measured on the output of a circuit and referred back to the input of a circuit so that it can be compared to the input signal, we can write (knowing $i_d = g_m v_{gs}$) the MOSFET's input-referred noise PSD as

$$V_{inoise}^2(f) = \frac{KF \cdot I_D^{AF}}{f \cdot (C'_{ox})^2 LW \cdot g_m^2} + \frac{8kT}{3g_m} \quad (9.67)$$

Figure 9.37 shows how the noise models are added to the schematic representation of the MOSFET. Notice that increasing the MOSFET's forward transconductance, g_m, reduces the input-referred thermal noise. If the g_m is increased by making the device wider (that is, it is not increased by simply raising the drain current), then the $1/f$ noise decreases as well. Note, as discussed in Ch. 8, that looking at the (size of the) output noise alone gives no indication of the noise performance of the amplifier or circuit. The input-referred noise should be compared to the input signal for an indication of noise performance. (A notable exception to this is a circuit that doesn't have an input signal like a current mirror. In these types of circuits we *do* care about reducing the size of the output noise.)

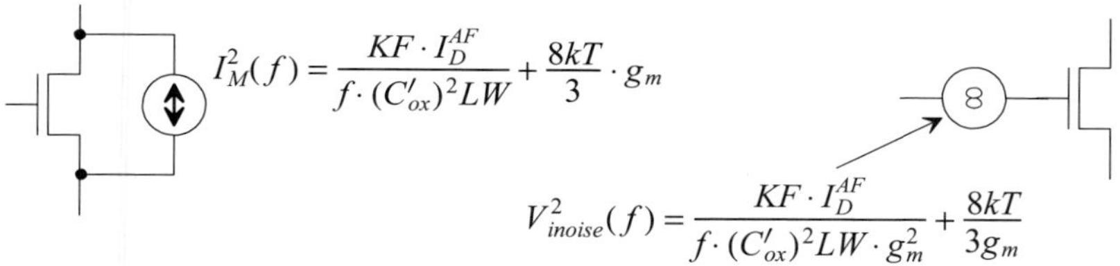

Figure 9.37 (a) Modeling MOSFET noise current and (b) input-referred noise voltage.

ADDITIONAL READING

Textbooks Covering Small-Signal Analysis

[1] R. Jaeger and T. Blalock, *Microelectronic Circuit Design*, McGraw-Hill Publishers, 2nd ed., 2003. ISBN 0-072-50503-6.

[2] R. Spencer and M. Ghausi, *Introduction to Electronic Circuit Design*, Prentice-Hall Publishers, 2003. ISBN 0-201-36183-3.

[3] D. J. Comer and D. T. Comer, *Fundamentals of Electronic Circuit Design*, John Wiley and Sons, 2002. ISBN 0-471-41016-0.

[4] D. A. Neamen, *Electronic Circuit Analysis and Design*, McGraw-Hill Publishers, 2001. ISBN 0-072-45194-7.

[5] A. R. Hambley, *Electronics*, Prentice-Hall Publishers, 2nd ed, 2000. ISBN 0-136-91982-0.

[6] M. H. Rashid, *Microelectronic Circuits: Analysis and Design*, Brookes-Cole Publishing, 1998. ISBN 0-534-95174-0.

[7] A. Sedra and K. Smith, *Microelectronic Circuits*, Oxford University Press, 4th ed., 1998. ISBN 0-195-11663-1.

[8] R. T. Howe and C. G. Sodini, *Microelectronics: An Integrated Approach*, Prentice-Hall Publishers, 1997. ISBN 0-135-88518-3.

[9] N. Malik, *Electronic Circuits: Analysis, Simulation, and Design*, Prentice-Hall Publishers, 1995. ISBN 0-023-74910-5.

[10] M. N. Horenstein, *Microelectronic Circuits and Devices*, Prentice-Hall Publishers, 2nd ed., 1990. ISBN 0-135-83170-9.

[11] M. S. Roden, G. L. Carpenter, and C. J. Savant, *Electronic Design: Circuits and Systems*, Pearson Benjamin Cummings Publishers, 2nd ed, 1990.

[12] J. Millman and A. Grabel, *Microelectronics*, McGraw-Hill Publishers, 2nd ed. 1987. ISBN 0-070-42330-X.

Textbooks Covering Noise Analysis

[13] F. Bonani and G. Ghione, *Noise in Semiconductor Devices: Modeling and Simulation*, Springer-Verlag Publishers, 2002. ISBN 3-540-66583-8.

[14] C. D. Motchenbacher and J. A. Connelly, *Low-Noise Electronic System Design*, John Wiley and Sons, 1993. ISBN 0-471-57742-1.

[15] H. L. Krauss, C. W. Bostian, and F. H. Raab, *Solid State Radio Engineering*, John Wiley and Sons, 1980. ISBN 0-471-03018-X.

A Good Paper on the Origins of Flicker Noise

[16] Reimbold, G., "Modified 1/f Trapping Noise Theory and Experiments in MOS Transistors Biased from Weak to Strong Inversion - Influence of Interface States," *IEEE Transactions on Electron Devices*, vol. ED-31, no. 9, pp. 1190–1198, September, 1984.

PROBLEMS

For the following problems use the long-channel process information given in Table 9.1 for *KP*, V_{TH}, and C'_{ox}, unless otherwise indicated.

9.1 Calculate and simulate the values of I_D and V_{GS} in the following circuit.

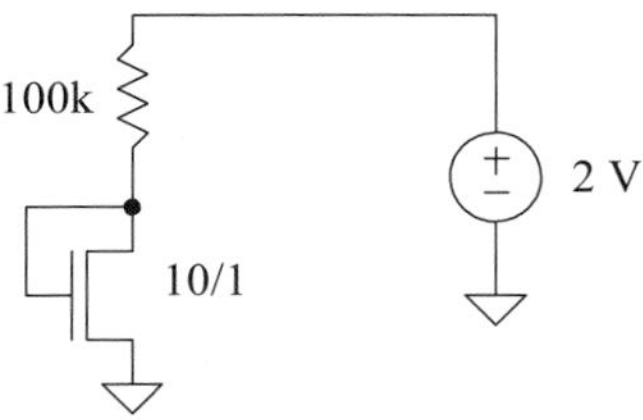

Figure 9.38 Circuit used in Problem 9.1

9.2 Repeat Problem 9.1 if the MOSFET size is changed to 10/10.

9.3 Calculate and simulate the values of I_D and V_{SG} in the following circuit.

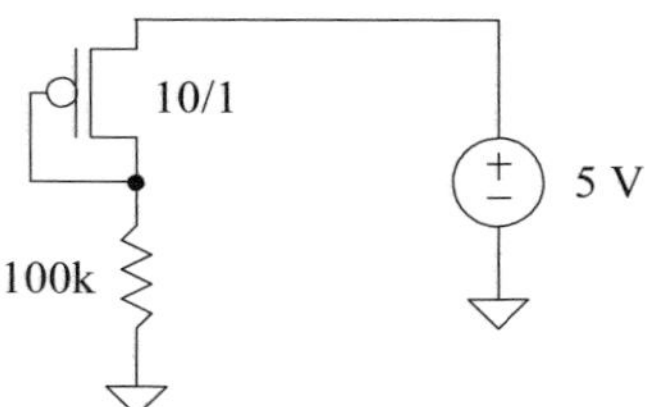

Figure 9.39 Circuit used in Problem 9.3

9.4 Repeat Problem 9.3 if the MOSFET size is changed to 10/10.

9.5 Calculate I_D, V_{DS}, and estimate the small-signal resistance looking into the drain of the MOSFET in the following circuit.

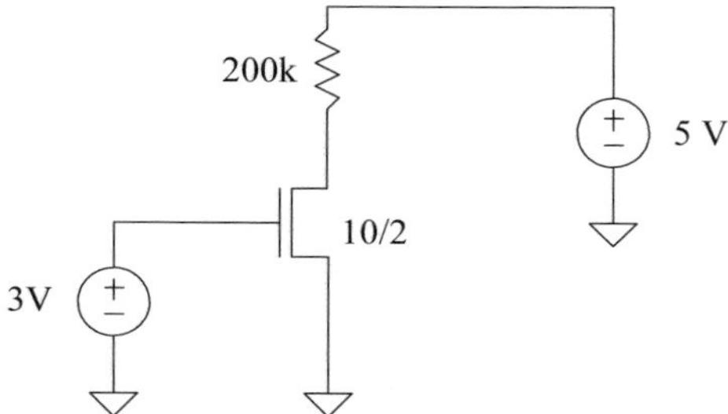

Figure 9.40 Circuit used in Problem 9.5

9.6 To determine the value of a small-signal resistance in a simulation, the circuit seen in Fig. 9.41 is used. The ratio of the test voltage, v_T, to the current that flows in the source, i_T, that is, v_T/i_T, is the small signal resistance. Using this circuit determine, with SPICE, the value of the resistance looking into the drain of the circuit in Fig. 9.40. How does the 200k resistor affect i_T in the simulation?

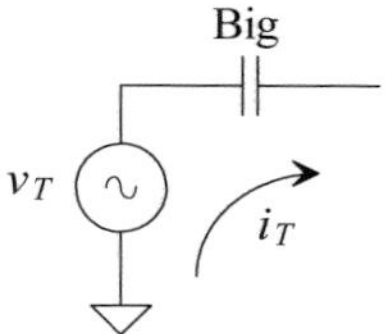

Figure 9.41 Using a test voltage to determine a resistance in a simulation.

9.7 Explain qualitatively what happens to V_{GS4} and V_{DS4} in Fig. 9.42 as the bias current is increased.

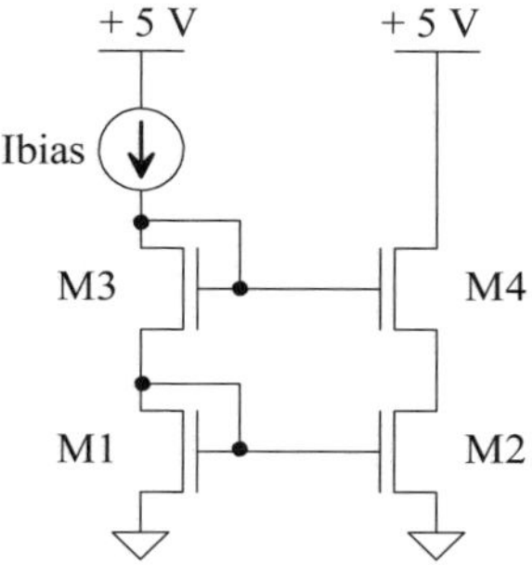

Figure 9.42 Schematic for Problem 9.7.

9.8 Describe qualitatively what happens if we steal or inject a current at the point indicated in Fig. 9.43. How does this affect the operation of M1 and M2? Verify your answer with SPICE.

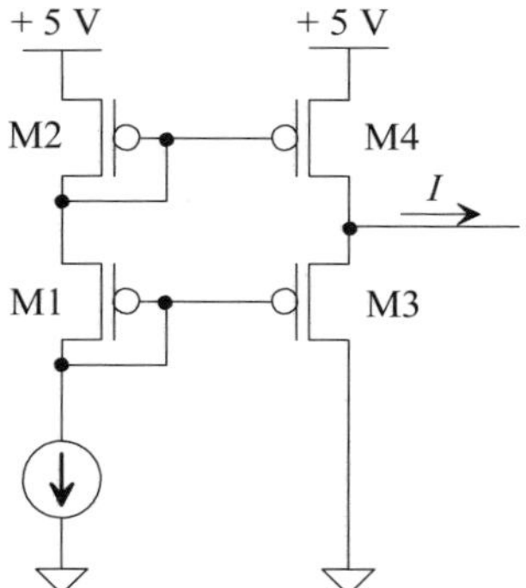

Figure 9.43 Schematic for Problem 9.8.

9.9 Using simulations, generate the plot seen in Fig. 9.12 for both NMOS and PMOS devices.

9.10 Design a circuit that will linearly convert an input voltage that ranges from 0 to 4V into a current that ranges from roughly 50 μA to 0. Simulate the operation of the design showing the linearity of the voltage-to-current conversion. How does the MOSFET's length affect the linearity?

9.11 Using a PMOS device, discuss and show with simulations how it can be used to implement a 10k resistor. Are there any limitations to the voltage across the PMOS resistor? Explain.

9.12 Using SPICE (and ensuring the MOSFET is operating in the saturation region with sufficient V_{DS}), generate the i_D versus v_{GS} curve seen in Fig. 9.15. Using

SPICE take the derivative of i_D (e.g., "plot deriv(ID)") to get the device's g_m (versus V_{GS}). How does the result compare to Eq. (9.22)? Does the level 3 model used in the simulation show a continuous change from subthreshold to strong inversion?

9.13 Estimate the AC, i_d, drain current that flows in the circuit seen in Fig. 9.44. Verify your answer with SPICE in using both transient and AC simulations.

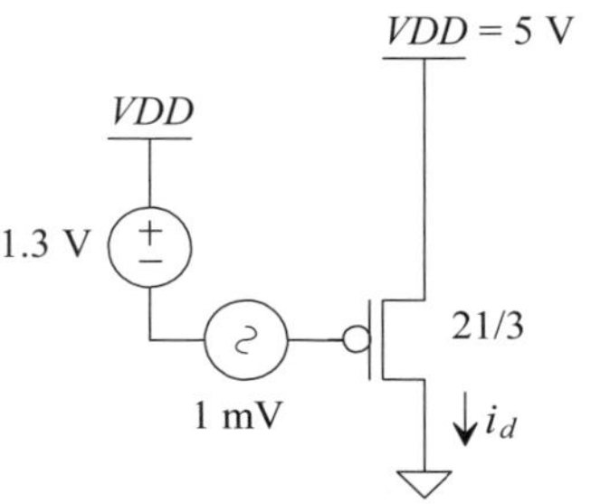

Figure 9.44 Schematic for Problem 9.13.

9.14 Calculate the DC and AC voltages and currents for the circuit seen in Fig. 9.45. Verify your answers with a SPICE AC analysis simulation.

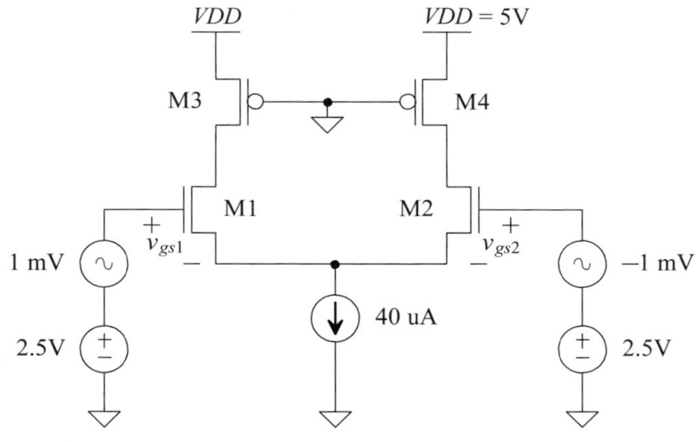

Figure 9.45 Circuit used in Problem 9.14.

9.15 Repeat Problem 9.14 if the bias current is reduced to 20 μA.

9.16 Using SPICE (and zero volt sources), verify the drain currents calculated in Ex. 9.5. Show both your AC analysis and transient analysis simulation results.

9.17 Using SPICE, generate the i_D versus v_{SB} curve seen in Fig. 9.21.

9.18 Repeat Ex. 9.6 if the widths and lengths of the PMOS and NMOS devices are multiplied by 10 (that is the NMOS is now 100/20 and the PMOS is now 300/20). How does the drain current change?

9.19 Compare the f_T of a 10/2 NMOS to a 100/20 NMOS. Verify your hand calculations using simulations.

9.20 Equation (9.39) is used to calculate the forward transconductance of a MOSFET operating in the subthreshold region. Is it possible to have subthreshold operation when I_D is 100 μA? If so, how?

9.21 Using the simulations that generated Fig. 9.27, estimate the threshold voltage for 50/5 PMOS and NMOS devices (both in the short-channel CMOS process).

9.22 Show the details leading to Eq. (9.43). Show, as an example, that the approximation is valid if $Q'_{b0}/C'_{ox} = 30\ mV$. (Remember: temperature is in Kelvin.)

9.23 For the circuits seen in Fig. 9.46, estimate both the output voltage at room temperature and how it will change with temperature. Verify your answer with SPICE. Use the short-channel CMOS process.

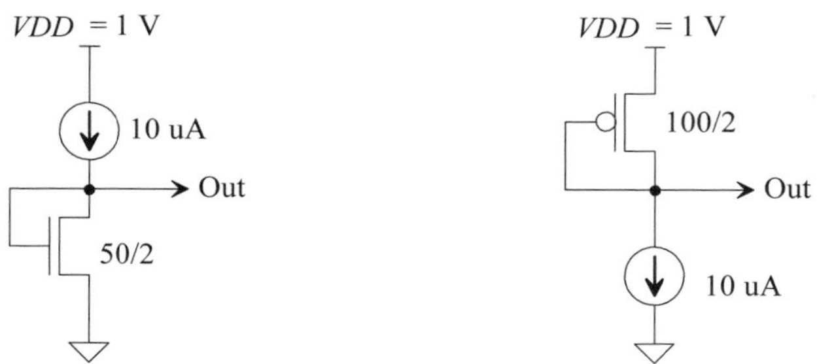

Figure 9.46 Circuits for Problem 9.23.

9.24 Verify, using simulations, the result given in Eq. (9.59), that is, that the GFT of a long-channel process is independent of biasing conditions. Make sure that the MOSFET is biased away from the subthreshold and triode regions.

9.25 When Eq. (9.36) was derived it was assumed that $C_{gs} >> C_{gd}$. Looking at the data in Table 9.2, is this a good assumption? Compare the hand-calculated value of f_T to the simulation results in Table 9.2.

9.26 Using the short-channel parameters in Table 9.2, calculate the drain voltage of M1 and its drain current (both AC and DC components) for the circuit seen in Fig. 9.47.

9.27 What is the PSD of a MOSFET's thermal noise when it is operating in the triode region?

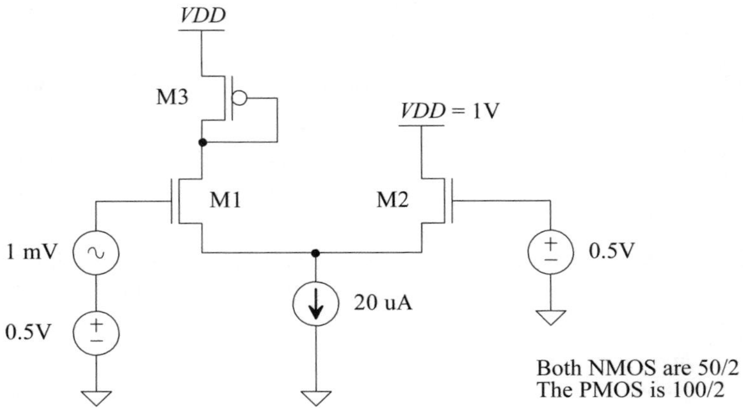

Figure 9.47 Circuit used in Problem 9.26.

9.28 Looking at Eq. (9.66), we see that decreasing the MOSFET's g_m reduces the MOSFET's drain current noise PSD. If the MOSFET is to be used as an amplifier where the input signal is on the MOSFET's gate, is reducing g_m a good idea? Why or why not?

Chapter 10

Models for Digital Design

In this chapter we develop a digital model for the MOSFET. For a simple logical analysis of a digital circuit we think, Fig. 10.1, of the MOSFETs as simple switches. When the gate of an NMOS device is a logic 1 (*VDD*), the NMOS device is on. When the gate is a logic 0 (ground), the NMOS device is off. The PMOS device operates in a complementary way to the NMOS device. When its gate is a logic zero, it is on (hence why we draw a bubble at its gate). It's important, for a quick logical analysis of a complex digital circuit, to think of the MOSFETs as simple logic-controlled switches.

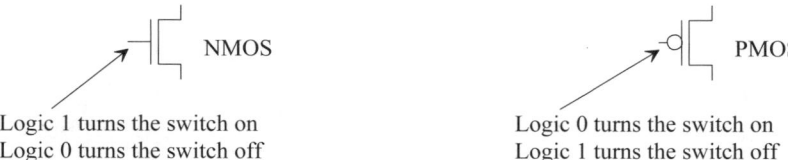

Figure 10.1 Logical operation of the switches.

Miller Capacitance

One of the important concepts we'll use in this chapter has to do with the "Miller effect" discussed in detail in Ch. 21 (see Fig. 21.5 and the associated discussion). To understand this effect here in our development of digital models, consider the simple schematic in Fig. 10.2. If, initially, the input node is at 0 V and the output node is at *VDD*, the charge on the capacitor is

$$Q_{init} = C \cdot (0 - VDD) = -C \cdot VDD \tag{10.1}$$

If the input node changes to *VDD* and the output node transitions to ground, the charge on the capacitor is

$$Q_{final} = C \cdot (VDD - 0) = C \cdot VDD \tag{10.2}$$

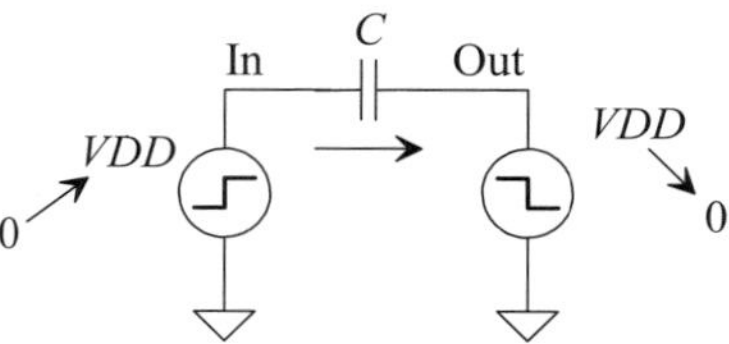

Figure 10.2 Determining the charge through a capacitor.

The total charge supplied by the input or output voltage source to the capacitor after the transition is then

$$Q_{tot} = Q_{final} - Q_{init} = 2C \cdot VDD \tag{10.3}$$

Remembering, in this discussion, that the input source transitions from 0 to *VDD* while, at the same time, the output source transitions from *VDD* to 0, what is the effective capacitance that each source sees? Towards answering this question, consider the models seen in Fig. 10.3. In (a) the input source supplies a charge of

$$Q_{tot} = 2C \cdot VDD \tag{10.4}$$

In (b) the output sinks the same amount of charge. The point here is that the input or output capacitance of the circuits in 10.3a or b is twice the capacitance value connecting the input to the output in Fig. 10.2. We'll use this result in a moment as we develop a digital model for the MOSFET.

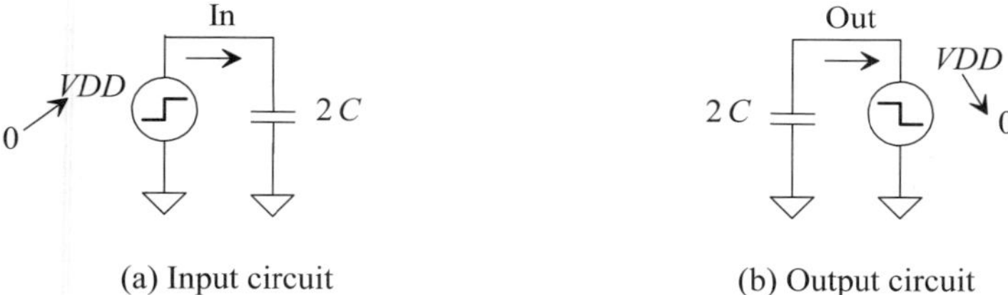

Figure 10.3 Splitting the capacitor in Fig. 10.2 up into two equivalent capacitors for developing a model.

10.1 The Digital MOSFET Model

Effective Switching Resistance

Consider the MOSFET circuit shown in Fig. 10.4. Initially, the MOSFET is off, $V_{GS} = 0$, and the drain of the MOSFET is at *VDD*. If the gate of the MOSFET is taken instantaneously from 0 to *VDD*, a current given by

$$I_D = \frac{KP_n}{2} \cdot \frac{W}{L} \cdot (VDD - V_{THN})^2 = \frac{\beta}{2} \cdot (VDD - V_{THN})^2 \tag{10.5}$$

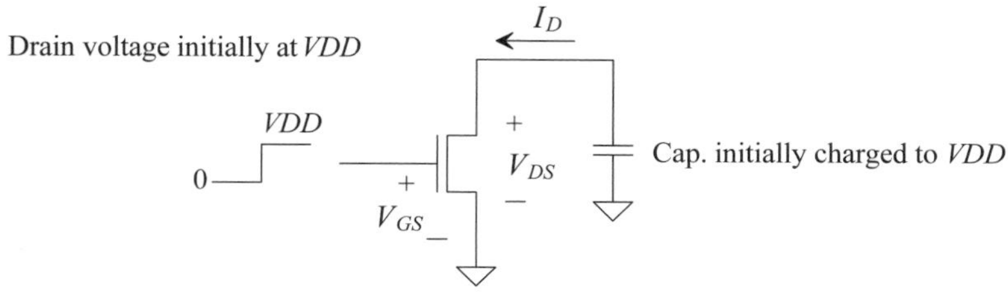

Figure 10.4 MOSFET switching circuit.

initially flows through the MOSFET. Point A in Fig. 10.5 shows the operating point of the MOSFET prior to switching for VDD = 5 V. After switching takes place, the operating point moves to point B and follows the curve $V_{GS} = VDD$ down to $I_D \approx 0$ and $V_{DS} = 0$, point C. At this point, the NMOS switch is on.

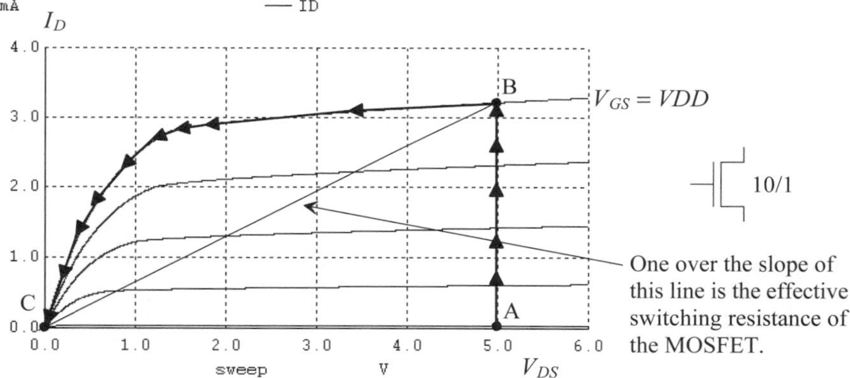

Figure 10.5 IV plot for a 10/1 NMOS device to estimate average switching resistance.

An estimate for the resistance between the drain and source of the MOSFET is given by the reciprocal slope of the line BC in Fig. 10.5, or

$$R_n = \frac{VDD}{\frac{KP_n}{2}\frac{W}{L} \cdot (VDD - V_{THN})^2} = R'_n \cdot \frac{L}{W} \tag{10.6}$$

The MOSFET is modeled by the circuit shown in Fig. 10.6. When $V_{GS} > VDD/2$, the switch is closed; when $V_{GS} < VDD/2$ the switch is open. In the derivation of this model, we assumed that the input step transition occurred in zero time; that is, that the rise time was zero, so that the point at which the switch was opened or closed was well defined. In practice, we will not encounter a zero rise time pulse; therefore, the model has limitations. Nevertheless, the model works remarkably well in designing and analyzing digital circuits by hand, giving results that are usually within a factor of two of simulation or measurement in general applications.

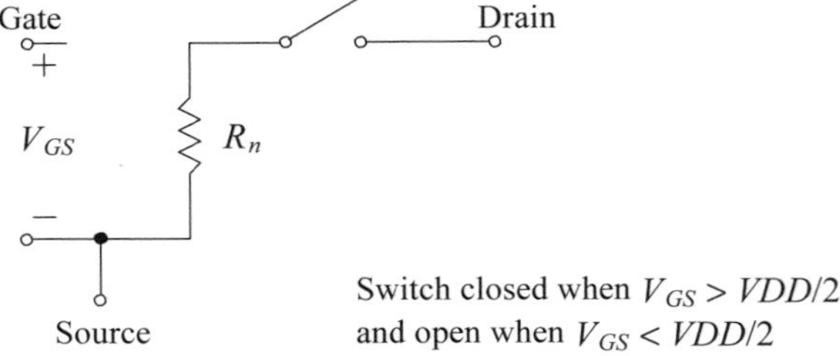

Figure 10.6 Simple digital MOSFET model.

From the data seen in Fig. 10.5 we can estimate the effective digital resistance as

$$R_n = R'_n \cdot \frac{L}{W} = \frac{VDD}{I_{D,sat}} = \frac{5}{3.3\ mA} \rightarrow R'_n \approx 15\ k\Omega \tag{10.7}$$

For an NMOS device with a specific width, W, or length, L, we can estimate the effective switching resistance of the device in the *long-channel CMOS* process in this book (scale factor of 1 μm) as

$$R_n \approx 15k \cdot \frac{L}{W} \tag{10.8}$$

For the PMOS device, the transconductance parameter, KP, is three times smaller than the NMOS's KP (see Table 9.1), so we can write

$$R_p \approx 45k \cdot \frac{L}{W} \tag{10.9}$$

The effective resistance of the PMOS device is three times as large as the NMOS's effective switching resistance due to the mobility of the electrons being larger than the mobility of the holes.

Short-Channel MOSFET Effective Switching Resistance

As discussed in Chs. 6 and 9, the short-channel MOSFET doesn't follow the square-law long-channel MOSFET models. We can't use Eq. (10.6) to estimate the effective switching resistance. However, we can use the on current for the devices (see Sec. 6.5.2). For the *short-channel CMOS* process used in this book (scale factor of 50 nm and a VDD of 1 V), we can write, see Eqs. (6.59), (6.60), and (6.61),

$$R_n = \frac{VDD}{I_{D,n}} = \frac{VDD}{I_{on,n} \cdot W \cdot scale} = \frac{1\ V}{600\ \frac{\mu A}{\mu m} \cdot W \cdot scale} \rightarrow R_n \approx \frac{1.7k \cdot \mu m}{W \cdot scale} \tag{10.10}$$

and

$$R_p = \frac{VDD}{I_{D,p}} = \frac{VDD}{I_{on,p} \cdot W \cdot scale} = \frac{1\ V}{300\ \frac{\mu A}{\mu m} \cdot W \cdot scale} \rightarrow R_p \approx \frac{3.4k \cdot \mu m}{W \cdot scale} \tag{10.11}$$

noting that the effective switching resistance of a short-channel MOSFET is independent of the MOSFET's length. In practice this is true to a point. Increasing the channel length above a certain value causes the MOSFET's resistance to start to increase.

Note that in the long-channel equations, Eqs. (10.8) and (10.9), the scale factor doesn't affect the resistance. However, in the short-channel equations, the scale factor is important and does affect the switching resistance. Knowing, for the short-channel process used in this book, that the scale factor is 50 nm, we can rewrite Eqs. (10.10) and (10.11)

$$R_n \approx \frac{34k}{W} \tag{10.12}$$

and

$$R_p \approx \frac{68k}{W} \tag{10.13}$$

where W is the drawn width of the devices.

10.1.1 Capacitive Effects

At this point, we need to add the capacitances of the switching MOSFET to our model of Fig. 10.6. Consider the MOSFET circuit shown in Fig. 10.7 with capacitance $\frac{C_{ox}}{2}$ between the gate-drain and the gate-source electrodes. This is the capacitance when the MOSFET is in the triode region and is an *overestimate* for the capacitances of the MOSFET. For example, the capacitance between the gate and drain of the MOSFET is the overlap capacitance from the lateral diffusion when the MOSFET is operating in the saturation region (see Table 6.1). Because of this overestimate, we will neglect the depletion capacitances of the source and drain implants to substrate when doing hand calculations. Simulations using SPICE are required for better estimates of switching behavior (e.g., showing the nonlinearities of a depletion capacitance, see Fig. 2.15 or Eq. [5.17]).

Returning to Fig. 10.7, when the input pulse transitions from 0 to VDD, the output transitions from VDD to 0 (review Figs. 10.2 and 10.3). The current through C_{gd} ($= C_{ox}/2$) is given by

$$I = C_{gd} \cdot \frac{dV_{gd}}{dt} = \frac{C_{ox}}{2} \cdot \frac{VDD - (-VDD)}{\Delta t} = C_{ox} \cdot \frac{VDD}{\Delta t} = C_{ox} \cdot \frac{dV_{DS}}{dt} \tag{10.14}$$

The voltage across C_{gd} changes by $2 \cdot VDD$. The current that flows through this capacitance is the drain current of the MOSFET in Fig. 10.7. As seen in Fig. 10.3, we can break C_{gd} into a component from the gate to ground and from the drain to ground of value $2C_{gd}$ or C_{ox}. The complete model of a switching MOSFET is shown in Fig. 10.8.

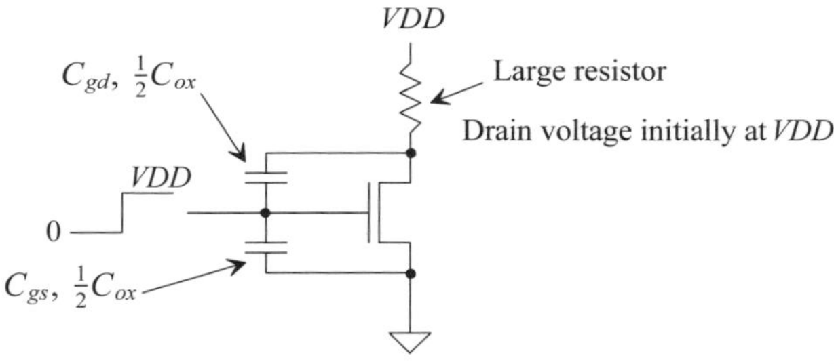

Figure 10.7 MOSFET switching circuit with capacitances.

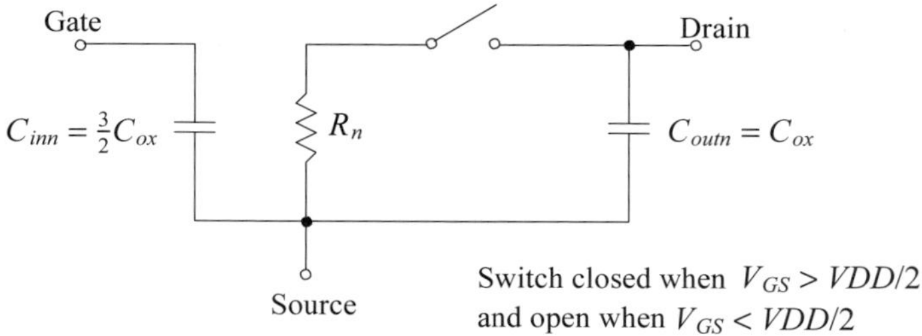

Figure 10.8 Simple digital MOSFET model.

10.1.2 Process Characteristic Time Constant

An important question we can answer at this point is, "What is the intrinsic switching speed of a MOSFET?" Looking at Figs. 10.7 and 10.8, we can see an intrinsic *process characteristic time constant,* τ_n, of R_nC_{ox}. That is, if the drain is charged to *VDD* as in Fig. 10.7 and the input switches from 0 to *VDD*, the output voltage will decay with a time constant of R_nC_{ox}. For an NMOS device in the long-channel process, this is given by

$$\tau_n = R_nC_{ox} = \frac{2L \cdot VDD}{KP_nW(VDD - V_{THN})^2} \cdot C'_{ox}WL \cdot (scale)^2 = \frac{2C'_{ox} \cdot VDD \cdot (L \cdot scale)^2}{KP_n \cdot (VDD - V_{THN})^2} \tag{10.15}$$

Notice that the "speed" of a process increases as the square of the channel length and that the speed is independent of the channel width, *W*. The process characteristic time constant is specified using the process minimum length. Also note that the larger the *VDD,* the faster the process. This is very similar to the unity current gain frequency, f_T, that we discussed in the last chapter. For a short-channel process, Eq. (10.15) can be rewritten as

$$\tau_n = R_nC_{ox} = \frac{VDD}{I_{on,n} \cdot W \cdot scale} \cdot C'_{ox}WL \cdot (scale)^2 = \frac{VDD \cdot C'_{ox} \cdot L \cdot scale}{I_{on,n}} \tag{10.16}$$

For the short-channel process, the time constant decreases linearly with decreasing channel length.

Example 10.1
Estimate the process characteristic time constants for the long- and short-channel CMOS processes used in this book.

For the NMOS device in the long-channel CMOS process, using the data in Table 5.1 and Eq. (10.8),

$$\tau_n = R_nC'_{ox}WL \cdot (scale)^2 = 15k \cdot \frac{L}{W} \cdot 1.75\frac{fF}{\mu m^2} \cdot LW \cdot (scale)^2 \approx 25\ ps$$

Knowing from Eqs. (10.8) and (10.9) that the effective resistance of the PMOS device is three times as large as the resistance of the NMOS device, we can write

$$\tau_p \approx 75\ ps$$

For the short-channel process, using the data in Table 5.1 and Eq. (10.10), we can write

$$\tau_n = R_n C'_{ox} \cdot WL \cdot (scale)^2 = \frac{1.7k \cdot \mu m}{W \cdot scale} \cdot 25\frac{fF}{\mu m^2} \cdot WL \cdot (scale)^2 = 2.1\ ps$$

and for the PMOS device

$$\tau_p = 4.2\ ps$$

Table 10.1 summarizes the effective switching resistances and oxide capacitances for the long- and short-channel CMOS processes used in this book. ■

Table 10.1 Digital model parameters used for hand calculations in the long- and short-channel CMOS processes used in this book. Note that the widths, W, and lengths, L, seen in this table are drawn lengths (minimum length is 1 while minimum width is 10).

Technology	R_n	R_p	Scale factor	$C_{ox} = C'_{ox} WL \cdot (scale)^2$
1 µm (long-channel)	$15k\frac{L}{W}$	$45k\frac{L}{W}$	1 µm	$(1.75\ fF) \cdot WL$
50 nm (short-channel)	$\frac{34k}{W}$	$\frac{68k}{W}$	50 nm	$(62.5\ aF) \cdot WL$

10.1.3 Delay and Transition Times

Before we go any further in the discussion of the digital models, let's define delays and transition times in logic circuits. Consider Fig. 10.9. The top trace represents the input to a logic gate, while the bottom trace represents the output. Note that there is no logic inversion between the input and output; however, the following definitions apply equally well to the case when there is an inversion. The input rise and fall times are labeled t_r and t_f respectively. The output rise and fall times are labeled t_{LH} and t_{HL}, respectively. The

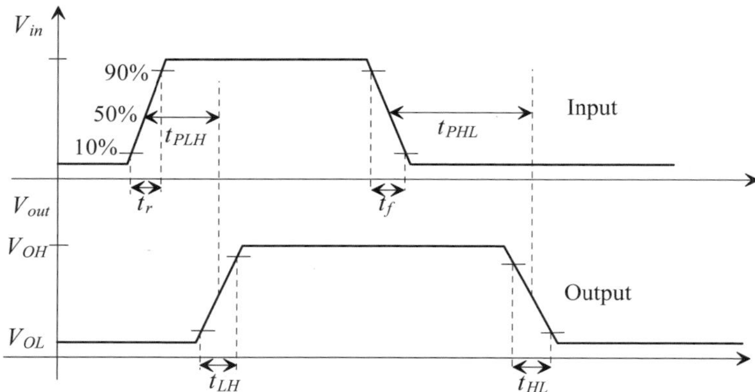

Figure 10.9 Definition of delays and transition times.

delay time between the 50% points of the input and output are labeled t_{PLH} and t_{PHL}, depending on whether the output is changing from a high to a low or from a low to a high. These definitions are extremely important in characterizing the time-domain characteristics of digital circuits.

For the simple *RC* circuit shown back in Fig. 2.21, the delay time is given by

$$t_{delay} = 0.7RC \tag{10.17}$$

and the rise or fall time is given by

$$t_{rise} = 2.2RC \tag{10.18}$$

For our simple digital model of Fig. 10.8, we will assume that the propagation delay time, whether high-to-low or low-to-high, is given by

$$t_{PHL} \approx 0.7 \cdot R_n \cdot C_{tot} \text{ and } t_{PLH} \approx 0.7 \cdot R_p \cdot C_{tot} \tag{10.19}$$

and the output rise and fall times are given by

$$t_{HL} \approx 2.2 \cdot R_n \cdot C_{tot} \text{ and } t_{LH} \approx 2.2 \cdot R_p \cdot C_{tot} \tag{10.20}$$

where C_{tot} is the total capacitance from the drain of the MOSFET to ground. These models for hand calculations **do not give exact results**. The models are useful for determining approximate delay and transition times, usually within a factor of two. They can be used to reveal the location of a speed limitation in a circuit.

Example 10.2

Using hand calculations, estimate the rise, fall, and delay times of the following circuits (Fig. 10.10) in both the long- and short-channel CMOS processes. Compare your results to SPICE simulations.

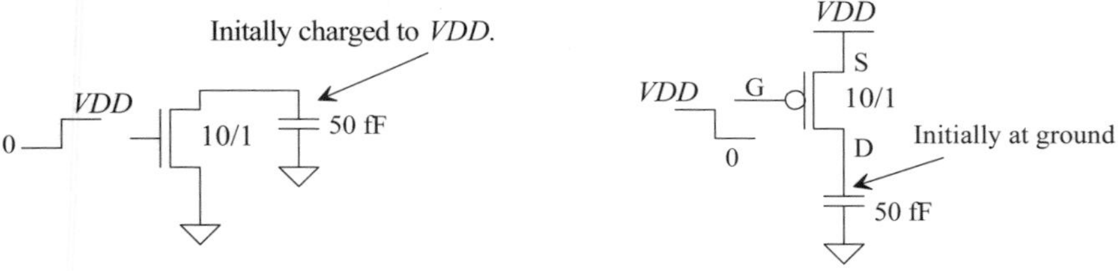

Figure 10.10 Circuits used in Example 10.2.

The models used for calculating delays are seen in Fig. 10.11. The time constant associated with discharging a load capacitor (the NMOS switch) is

$$t_{PHL} = 0.7 \cdot R_n C_{tot} = 0.7 \cdot R_n \cdot (C_{ox} + C_L) \tag{10.21}$$

The time constant associated with charging a load capacitance (the PMOS switch) is

$$t_{PLH} = 0.7 \cdot R_p C_{tot} = 0.7 \cdot R_p \cdot (C_{ox} + C_L) \tag{10.22}$$

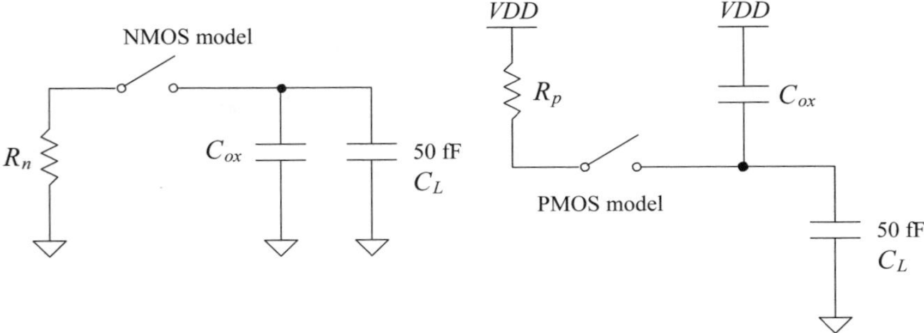

Figure 10.11 Models used to determine switching times in Example 10.2.

In many practical situations the load capacitance is much larger than the output capacitances of the switches. These equations can then be reduced to

$$t_{PHL} = 0.7 \cdot R_n \cdot C_L \text{ and } t_{PLH} = 0.7 \cdot R_p \cdot C_L \text{ for } C_L >> C_{ox} \tag{10.23}$$

Using Eqs. (10.21) and (10.22) with the parameters in Table 10.1, we can estimate the delays using the long-channel process as

$$t_{PHL} = 0.7 \cdot 15k \cdot \frac{1}{10} \cdot (1.75\,fF \cdot 10 + 50\,fF) = 70\,ps$$

and

$$t_{PLH} = 0.7 \cdot 45k \cdot \frac{1}{10} \cdot (1.75\,fF \cdot 10 + 50\,fF) = 210\,ps$$

For the short-channel process, we get

$$t_{PHL} = 0.7 \cdot \frac{34k}{10} \cdot (62.5\,aF \cdot 10 + 50\,fF) = 120\,ps$$

and

$$t_{PLH} = 0.7 \cdot \frac{68k}{10} \cdot (62.5\,aF \cdot 10 + 50\,fF) = 240\,ps$$

The simulation results are seen in Fig. 10.12. We can also estimate the output rise and fall times using Eqs. (10.19) and (10.20). For the long-channel devices

$$t_{HL} = \frac{2.2}{0.7} \cdot t_{PHL} = 220\,ps$$

and

$$t_{LH} = \frac{2.2}{0.7} \cdot t_{PLH} = 660\,ps$$

For the short-channel devices

$$t_{HL} = 377\,ps \text{ and } t_{LH} = 754\,ps \blacksquare$$

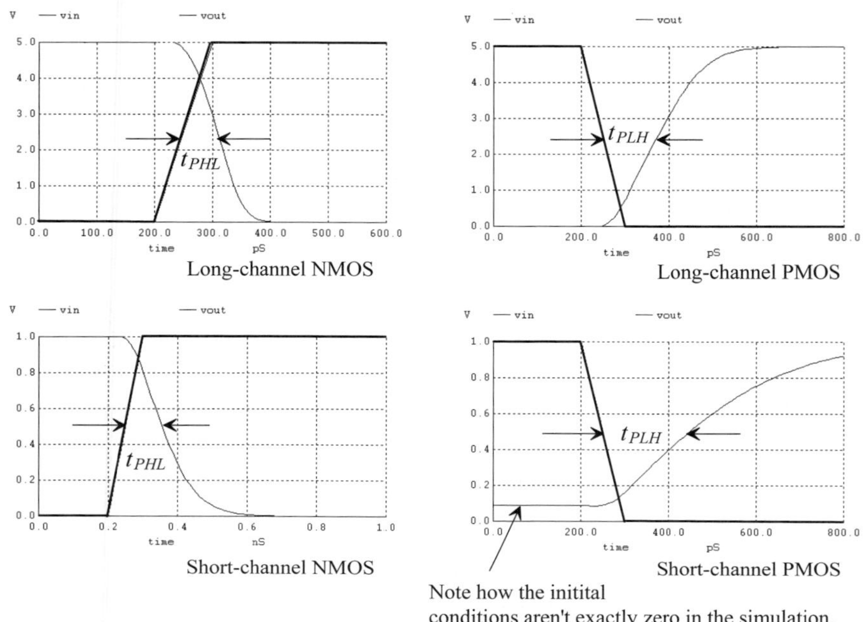

Figure 10.12 Simulating the circuits in Fig. 10.10.

10.1.4 General Digital Design

For general digital design, using the long or short-channel CMOS processes discussed in this book, we can set the drawn size of the NMOS devices to 10/1. To match the effective resistance of the PMOS device to the NMOS device in the long-channel process, Table 10.1, we set the PMOS size to 30/1 (the width of the PMOS is three times the width of the NMOS). For the short-channel process, we use a PMOS device of 20/1 to match the effective switching resistances. Table 10.2 summarizes the parameters for general digital design in the short- and long-channel processes used in this book. For specific digital design solutions we may use a longer device if a higher resistance is needed or a wider device if more drive current through the MOSFET is required.

Table 10.2 Parameters for general digital design using the long-channel (scale factor is 1 μm) or short-channel (scale factor of 50 nm) CMOS process **used in this book**.

Technology	Drawn	Actual size	$R_{n,p}$	$C_{ox,n,p}$
NMOS (long-channel)	10/1	10 μm by 1 μm	1.5k	17.5 *fF*
PMOS (long-channel)	30/1	30 μm by 1 μm	1.5k	52.5 *fF*
NMOS (short-channel)	10/1	0.5 μm by 50 nm	3.4k	625 *aF*
PMOS (short-channel)	20/1	1 μm by 50 nm	3.4k	1.25 *fF*

10.2 The MOSFET Pass Gate

Consider the NMOS device seen in Fig. 10.13a. This is the configuration we are accustomed to looking at, that is, where the NMOS device pulls the output to ground. In (b) we flip the MOSFET in (a) on its side and change the labels from ground to "logic 0" and from *VDD* to "logic 1." Note the locations of the source and drain in these configurations. Also note that in (b) the output is pulled all the way to ground (to a good "0"). It can be said that *an NMOS device is good at passing a* "0."

Next look at the configuration in (c). The configurations in (b) or (c) are sometimes called the *pass gate (PG) configuration*. The MOSFET passes the logic level on its input to its output when its gate is driven to *VDD* (when the PG is enabled). Note that when the PG is disabled (its control gate is driven to ground), the outputs are in a high-impedance state (a *Hi-Z* state). The PG can be useful when sharing a bus or a logic circuit. Also note that the inputs and outputs of the PG can be swapped. Logic flow through the PG can be *bi-directional.*

Returning to (c), we see that if the input to the PG is a "1" we can no longer think of the input as the source of the MOSFET like we did in (a) and (b). If we were to do so, then the V_{GS} of the MOSFET would be zero and the MOSFET would be off. If the NMOS device were off while its gate was driven to *VDD*, then this would contradict our comments at the beginning of the chapter associated with Fig. 10.1. To keep the MOSFET on, we need at least a V_{GS} of V_{THN}. As seen in (d) of the figure, this means that the NMOS PG passes a "1" with a threshold voltage drop. It can be said that *an NMOS device is not good at passing a* "1." Figure 10.14 shows how the output of a PG varies as its input transitions between ground and *VDD* using the 50 nm CMOS process.

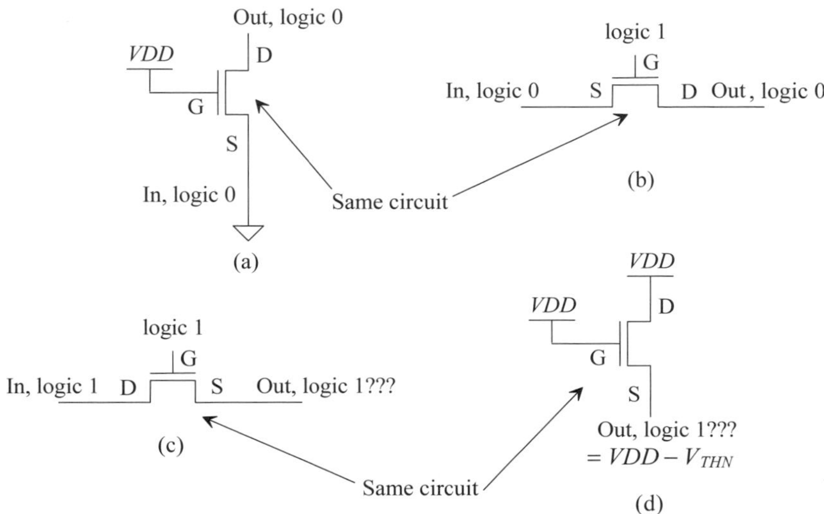

Figure 10.13 Using the NMOS switch as a pass gate.

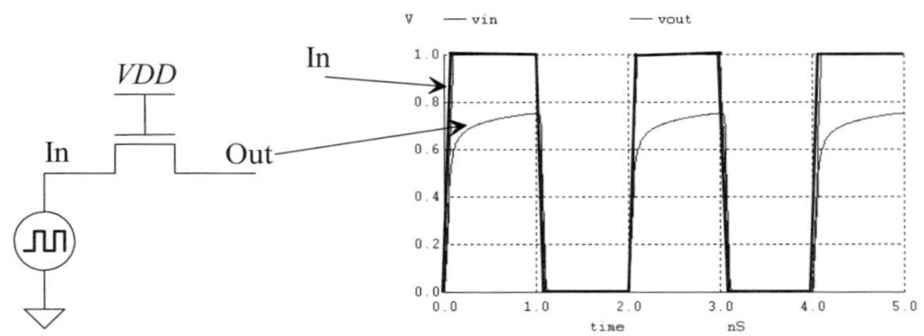

Figure 10.14 The input and output of an NMOS pass gate.

The PMOS Pass Gate

Figure 10.15 shows the operation of the PMOS PG. As expected the operation of the PMOS device is complementary to the NMOS's operation. The PMOS device turns on when its gate is driven to ground. If its gate is pulled to *VDD*, the device is off (and the output is in the Hi-Z state). In Fig. 10.15a the PG is passing a "1" to the output (the V_{SG} is *VDD*). In (b) a "0" is passed to the output. However, noting that the terminals we label drain and source are swapped from (a), the output only gets pulled down to V_{THP}. In (b) the V_{SG} of the MOSFET is V_{THP}. It can be said that a *PMOS PG is good at passing a 1 and bad at passing a 0.*

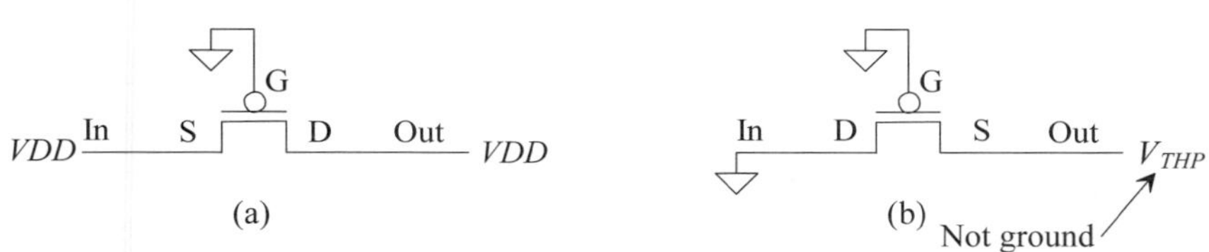

Figure 10.15 How the PMOS device does not pass a logic 0 well.

Example 10.3
Estimate the output voltages in the circuits seen in Fig. 10.16.

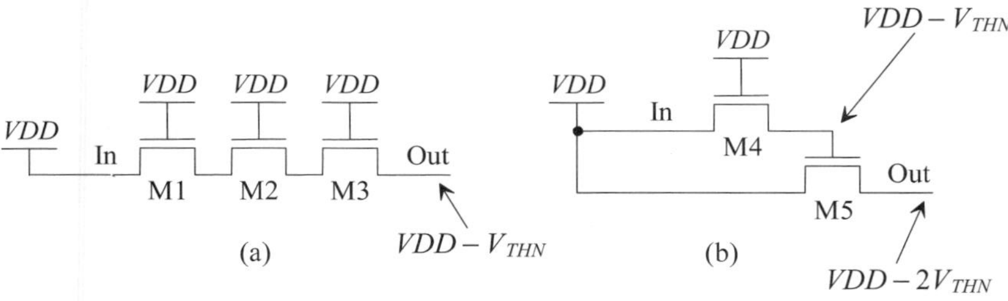

Figure 10.16 Circuits used in Ex. 10.3.

In (a) the output of M1 is $VDD - V_{THN}$. To keep M2 and M3 on, each MOSFET must have a V_{GS} of at least V_{THN}. Because the gates of M2 and M3 are already at VDD, the output of M1 gets passed through M2 and M3 to the final output of the circuit. As seen in the figure this means the overall output is also $VDD - V_{THN}$. We only take one threshold voltage hit.

In (b) the output of the M4 is $VDD - V_{THN}$. This is the gate voltage of M5. For M5 to be on its gate-source voltage must be greater than V_{THN}. The final output is then $VDD - 2V_{THN}$ (again as seen in the figure). ■

10.2.1 Delay through a Pass Gate

Consider the PG configuration seen in Fig. 10.17a. Let's estimate the delay between the input and the output of the PG. Note that if the input to the circuit is a "0" the PG behaves like the configuration seen in Fig. 10.4 and the output gets pulled all the way down to ground. Let's consider the configuration seen in Fig. 10.17b where the input is transitioning from a "0" to a "1" (VDD). The capacitance that the input sees is $C_{ox}/2$. The total load capacitance is

$$C_{tot} = C_L + \frac{C_{ox}}{2} \tag{10.24}$$

The delay through the PG can be estimated as

$$t_{delay} = 0.7 \cdot R_n C_{tot} = 0.7 \cdot R_n \cdot \left(C_L + \frac{C_{ox}}{2}\right) \tag{10.25}$$

Let's use this result in an example.

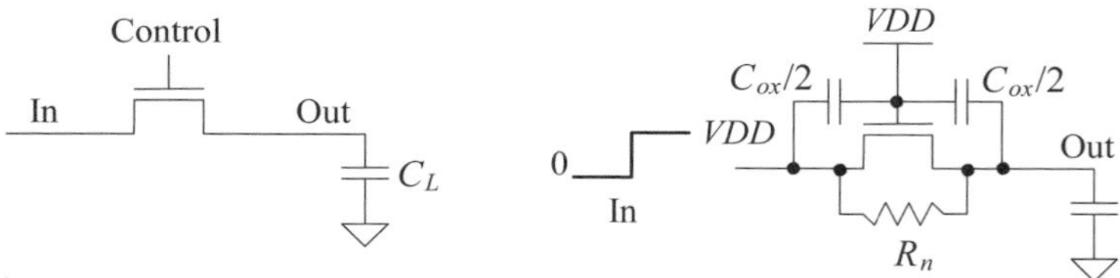

(a) Delay from the input to the output through a pass gate.

(b) The capacitances and effective resistance when turning the MOSFET on with an input 1.

Figure 10.17 Estimating the delay through a pass transistor.

Example 10.4
Estimate the delays through the PGs shown in Fig. 10.18. Verify your estimates with simulations. Use the 50 nm (short-channel) CMOS process.

The MOSFETs used in this example have the effective resistances and oxide capacitances listed in Table 10.2. Because the oxide capacitance (the MOSFET's capacitance) is much less than the load capacitance, we can write

$$t_{delay} \approx 0.7 \cdot R_{n,p} C_L \tag{10.26}$$

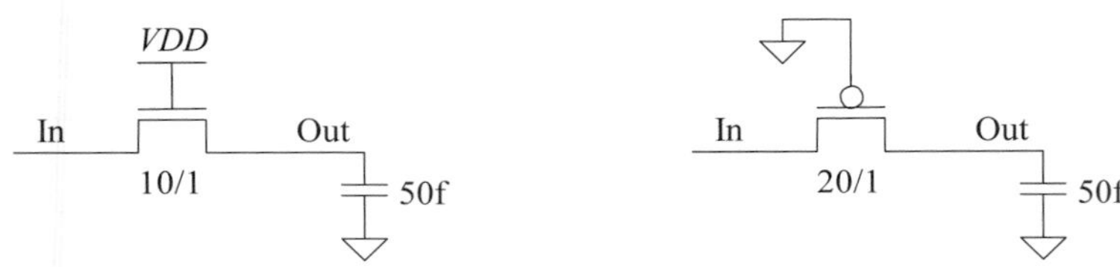

Figure 10.18 Circuits used to calculate delays in Ex. 10.4.

For the NMOS or PMOS PGs in Fig. 10.18 the delays are estimated as

$$t_{delay} \approx 0.7 \cdot 3.4k \cdot 50fF = 120\, ps$$

The simulation results are seen in Fig. 10.19. If we measure the output delays in Fig. 10.19 at 50% of *VDD* (500 mV), as defined in Fig. 10.9, then we get considerably different values from our hand calculations. The fact that the outputs of the PGs don't swing all the way to the power supply rails changes the points where we would measure the delays (to, say, 50% of the output swing). Again, it's important to note that our hand calculations give approximate delays and, more importantly, can indicate the location of a speed limitation in a digital circuit. The hand calculations won't provide exact delay values. Even if they could, the results would be subjective, dependent on the locations (voltage levels) on the input and output waveforms where we measure the delay. ■

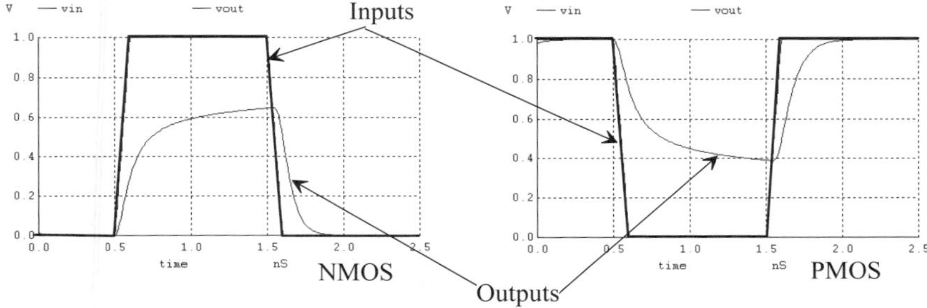

Figure 10.19 The delay through the PGs in Fig. 10.18.

The Transmission Gate (The TG)

The observant reader may be wondering: "If an NMOS PG passes a 0 well and a PMOS PG passes a 1 well, can't we put the two together and pass full logic levels?" The resulting circuit is called a *transmission gate*, *TG*, and is seen in Fig. 10.20. When the select control signal, *S*, is high, the TG is on and the input is passed to the output. The drawbacks of the TG over the PG are increased layout area and the need for two control signals (*S* and its complement). The benefit of using the TG is its rail-to-rail output swing. We'll discuss the TG in more detail in Ch. 13.

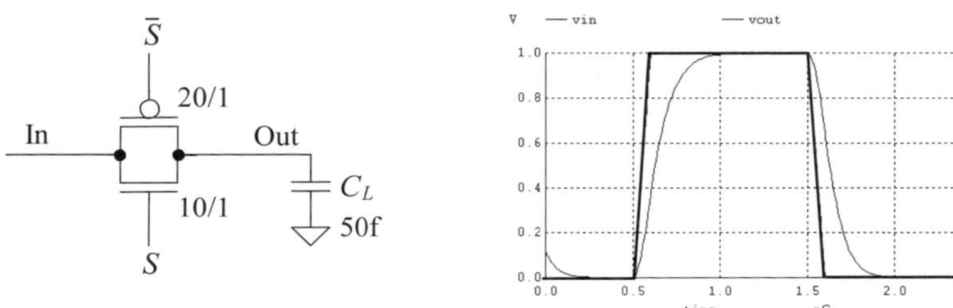

Figure 10.20 Simulating the operation of a transmission gate.

Also seen in Fig. 10.20 are simulation results showing the delay through the TG. For the values seen, the load capacitance is, again, much larger than the oxide capacitances of the MOSFETs, so we can calculate the delay as

$$t_{delay} = 0.7 \cdot (R_n || R_p) \cdot C_L \tag{10.27}$$

Using the results from Ex. 10.4, the delay is estimated as 60 ps (very close to the simulation results).

10.2.2 Delay through Series-Connected PGs

Consider the series connection of (identically sized) NMOS PGs seen in Fig. 10.21. Reviewing Fig. 10.17, we see that capacitance on the internal nodes, in between MOSFETs, is C_{ox} (a contribution of half of C_{ox} from each MOSFET). To approximate the delay through the MOSFETs, we can use Eq. (2.32) or

$$t_{delay} \approx 0.35 \cdot R_n \cdot C_{ox} \cdot l^2 \tag{10.28}$$

Noting $R_n C_{ox}$ is the process characteristic time constant, τ_n , we can quickly estimate delays through series-connected PGs without doing much of a calculation. For example,

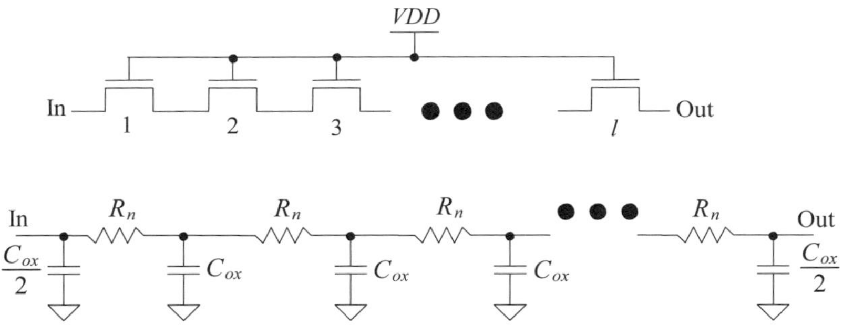

Figure 10.21 How a series connection of MOSFETs behaves like an RC transmission line (see Fig. 2.22).

we know that τ_n , for the 50 nm process, is 2.1 ps from Ex. 10.1. If we have 10 NMOS PGs in a row, the delay through the string is estimated as 73.5 ps.

Example 10.5

Estimate the delay through the circuit in Fig. 10.22. Verify the estimate with a SPICE simulation. Use the 50 nm process with 10/1 NMOS devices.

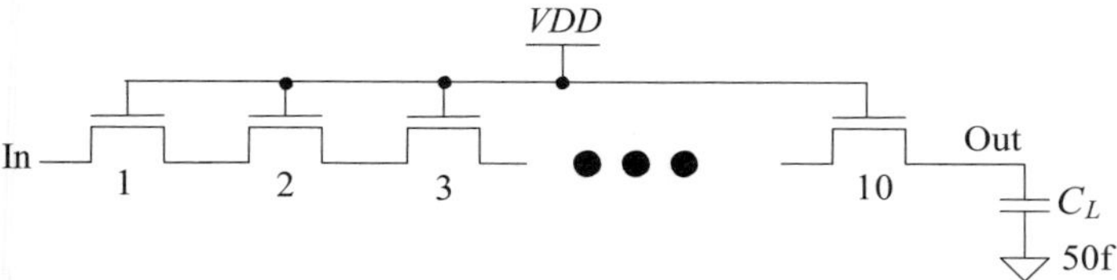

Figure 10.22 Circuit used in Ex. 10.5.

The total delay is the sum of the RC transmission line delay with the delay in charging the load capacitance through the 10 PGs. This delay can be written for the general case as

$$t_{delay} \approx 0.35 \cdot R_n \cdot C_{ox} \cdot l^2 + 0.7 \cdot l \cdot R_n \cdot C_L \tag{10.29}$$

For the present example, the delay is

$$t_{delay} \approx \overbrace{0.35 \cdot 2.1\,ps \cdot (10)^2}^{73.5\,ps} + 0.7 \cdot 10 \cdot 3.4k \cdot 50f \approx 1.2\ ns \tag{10.30}$$

The simulation results are seen in Fig. 10.23. ■

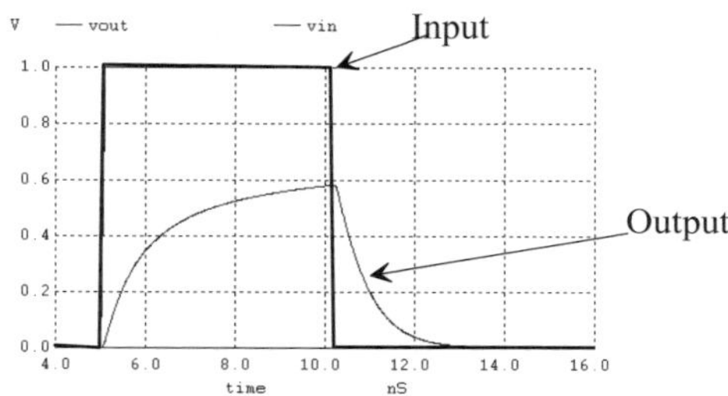

Figure 10.23 Simulating the operation of the circuit in Fig. 10.22.

10.3 A Final Comment Concerning Measurements

Notice that the capacitances we are discussing in this chapter are relatively small. When making measurements it is extremely easy to add a capacitance to the circuit that significantly increases the delays (and may cause circuit failure). Towards understanding this comment in more detail consider the compensated scope probe shown in Fig. 10.24. In (a) we see the input impedance of the oscilloscope (o-scope) is a 1 MΩ resistor in

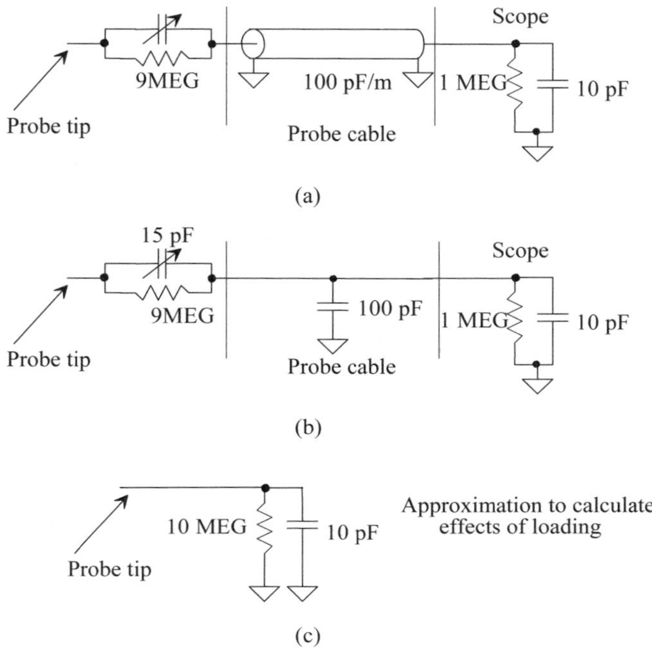

Figure 10.24 Showing how a common scope probe is assembled.

parallel with a 10 pF capacitor. (The actual values are indicated next to the connector on the front of the particular o-scope.) If we connect a piece of cable (co-axial cable or simply coax) from the circuit under test to the input of the scope, we introduce significant capacitance into the circuit. As seen in (a) the cable's capacitance may be as much as 100 pF/meter. Using a piece of coax to probe the circuit (alone) would then add a 110 pF capacitor and a 1 MΩ resistor to ground at each point we probe! **Understanding this is important.** It's common to see new engineering students, in a digital logic lab for example, probing with a cable (that is, without an o-scope probe). They may wonder why their circuits don't work or only work at slow speeds.

To compensate for the cable capacitance, the probe tip has a series resistor and capacitor added in between the cable and the probe tip, (b). This combination of a cable and probe tip RC is called a *compensated scope probe*. The RC in the probe tip is adjusted to have nine times the impedance of the RC from the scope's input to ground (the cable capacitance in parallel with the scope's input impedance) over all frequencies of interest. In other words, a 10:1 voltage divider exists between the probe tip and the input of the scope (and so the minimum signal we can measure increases when using a compensated scope probe). The big benefit, as seen in the approximation for the loading in (c), is that the size of the capacitance introduced into the circuit is reduced. For probing on-chip (or on-wafer), special probes are used with active devices in the probe tips (to reduce the probe's loading on the circuit it is measuring). Active probe tips, called fetoprobes, can be purchased that only have femtofarads amounts of loading.

ADDITIONAL READING

[1] J. P. Uyemura, *Introduction to VLSI Circuits and Systems*, John Wiley and Sons Publishers, 2002. ISBN 0-471-12704-3.

[2] R. L. Geiger, P. E. Allen, and N. R. Strader, *VLSI-Design Techniques for Analog and Digital Circuits,* McGraw-Hill Publishing Company, 1990. ISBN 0-07-023253-9.

PROBLEMS

10.1 Using the parameters in Table 6.2, compare the hand-calculated effective digital switching resistance from Eq. (10.6) to the empircially derived values given in Table 10.1.

10.2 Regenerate Fig. 10.14 for the PMOS device.

10.3 Using SPICE verify the results of Ex. 10.3.

10.4 Replacing the NMOS PGs in Fig. 10.16 with PMOS PGs and changing the *VDD*-connected nodes to ground-connected nodes, show, and verify with simulations, the outputs of the two modified circuits.

10.5 For the following circuits estimate the delay between the input and the output. Use the 50 nm (short-channel CMOS) process. Verify the estimates with SPICE.

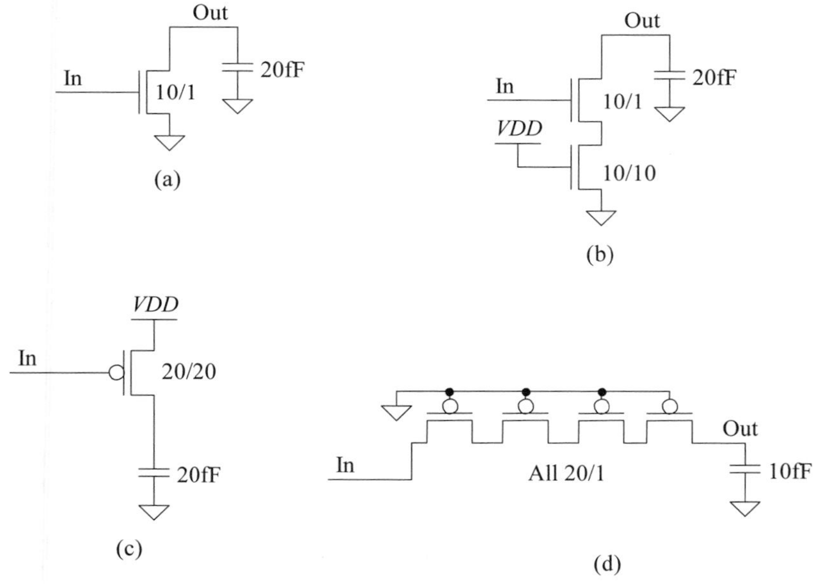

Figure 10.25 Circuits used in Problem 10.5.

Possible Student Projects

This section lists some possible student projects for fabrication through the MOSIS service (see Ch. 1). Generally, two to four student projects should be implemented on one chip. MOSIS will return to the MOSIS liaison (generally, the course instructor) several copies of each chip design submitted.

The design rule-checked designs should be turned in along with (1) one sheet of paper showing the logic level diagrams *and* pin connections so that whoever is evaluating the chip can quickly determine functionality, and (2) final reports that consist of a block diagram, schematic diagram, layout information, hand calculations, SPICE simulations, and clear explanations (and trade-offs) of the operation of the circuit.

1. Quad 2-input MUX
2. Clock-doubling circuit using exclusive OR gate
3. Buffer with tristate outputs
4. SR flipflop with tristate outputs
5. Edge-triggered T flipflop
6. Edge-triggered D flipflop
7. Schmitt trigger
8. 1-of-16 decoder
9. Up counter with asynchrounous reset
10. 4-bit static shift register
11. 4-bit dynamic shift register
12. 2-bit adder with carryout
13. Current-starved VCO with center frequency of 20 MHz
14. 2-bit bidirectional transceiver
15. PE gate to implement $X = \overline{A + BCD + EF}$
16. One-shot whose output pulse width is determined by external RC
17. Buffer for driving a 20 pF load with minimum delay
18. Buffer for driving a 20 pF load with smaller layout area

Advanced projects

19. A 64-bit static RAM including a storage cell, addressing and decoding circuitry, buffers, a write/read enable, and a chip select.
20. Charge pump (voltage generator). The input to the charge pump is *VDD* (= 5 V) and the output is –3 V. The circuit should be fully simulated. The reference, oscillator, and feedback should be fully simulated and discussed in the final report.

21. A 64-bit DRAM, including a storage cell, addressing and decoding circuitry, buffers, a write/read enable, and a chip select.

22. A DPLL which will take a 1 MHz input and generate a 4 MHz output. The output should follow the input for frequency changes from 900 kHz to 1.1 MHz. You should discuss the transient properties of the DPLL, as well as present a detailed design of the phase detector, VCO, and loop filter. The entire design should be monolithic; that is, no external components should be used.

Chapter

11

The Inverter

The CMOS inverter is a basic building block for digital circuit design. As Fig. 11.1 shows, the inverter performs the logic operation of A to $\overline{A}$. When the input to the inverter is connected to ground, the output is pulled to *VDD* through the PMOS device M2 (and M1 shuts off). When the input terminal is connected to *VDD*, the output is pulled to ground through the NMOS device M1 (and M2 shuts off). The CMOS inverter has several important characteristics that are addressed in this chapter: for example, its output voltage swings from *VDD* to ground unlike other logic families that never quite reach the supply levels. Also, the static power dissipation of the CMOS inverter is practically zero, the inverter can be sized to give equal sourcing and sinking capabilities, and the logic switching threshold can be set by changing the size of the device.

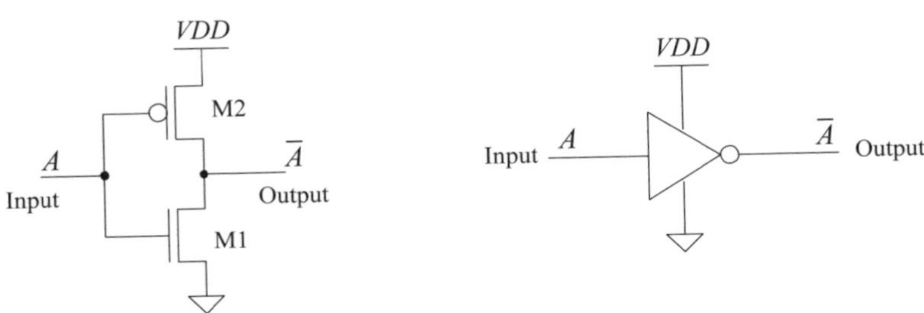

Figure 11.1 The CMOS inverter, schematic, and logic symbol.

11.1 DC Characteristics

Consider the inverter shown in Fig. 11.2 and the associated transfer characteristic plot. In region 1 of the transfer characteristics, the input voltage is sufficiently low (typically less than the threshold voltage of M1), so that M1 is off and M2 is on ($V_{SG} >> V_{THP}$). As V_{in} is increased, both M2 and M1 turn on (region 2). Increasing V_{in} further causes M2 to turn off and M1 to fully turn on, as shown in region 3.

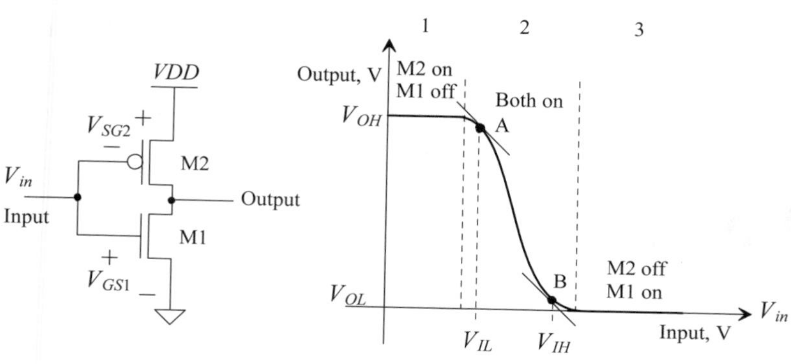

Figure 11.2 The CMOS inverter transfer characteristics.

The maximum output "high" voltage is labeled V_{OH} and the minimum output "low" voltage, V_{OL}. Points A and B on this curve are defined by the slope of the transfer curves equaling −1. Input voltages less than or equal to the voltage V_{IL}, defined by point A, are considered a logic low on the input of the inverter. Input voltages greater than or equal to the voltage V_{IH}, defined by point B, are considered a logic high on the input of the inverter. Input voltages between V_{IL} and V_{IH} do not define a valid logic voltage level. Ideally, the difference in V_{IL} and V_{IH} is zero; however, this is never the case in real logic circuits.

Example 11.1
Using SPICE, plot the transfer characteristics for the inverter seen in Fig. 11.3 in both the long- and short-channel CMOS processes used in this book. From the plot, determine V_{IH}, V_{IL}, V_{OH}, and V_{OL}.

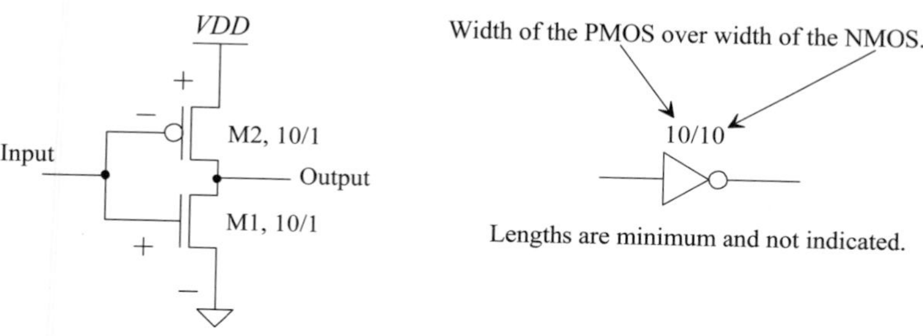

Figure 11.3 Inverter used in Ex. 11.1.

The inverter *voltage transfer curves* (VTCs) are shown in Fig. 11.4. Notice how the *VDD* used for the long-channel process is 5 V, while the *VDD* used in the short-channel, process is 1 V. The output high voltage, V_{OH}, is *VDD* and the output low voltage, V_{OL}, is ground (for both inverters). For the inverter using the

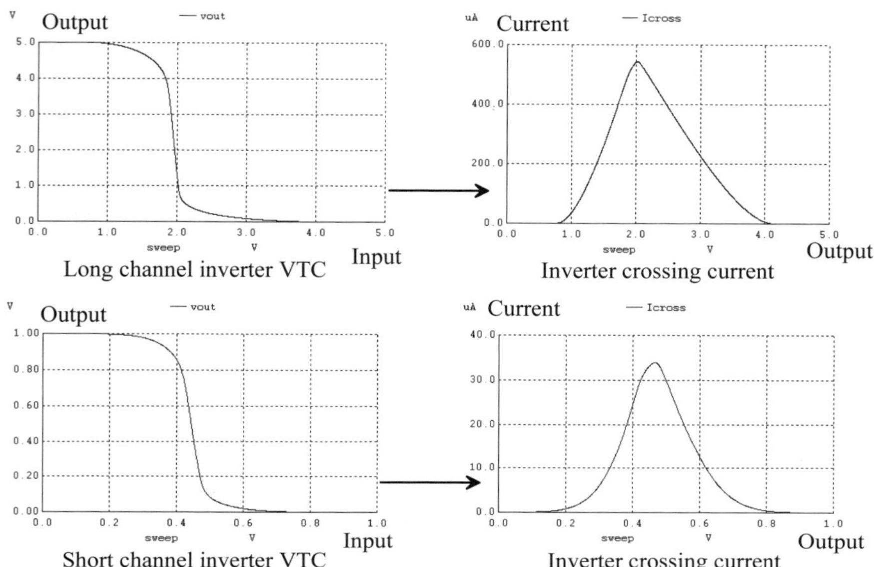

Figure 11.4 Voltage transfer curves (VTCs) for a long and a short channel inverter.

long-channel process, V_{IL} is approximately 1.8 V while, when using the short-channel process, V_{IL} is approximately 400 mV. For V_{IH} we get 2.1 V and 500 mV for the long- and short-channel processes, respectively.

Note that in Fig. 11.4 we also plotted the crossing current (the inverter's output voltage crossing between a logic 1 and a logic 0). This is the current that flows when the inverter is operating in region 2 in Fig. 11.2 (the inverter's input transitioning from a high to a low or from a low to a high). If the inverter's input transitions quickly, the amount of charge pulled from *VDD* (the amount of time the inverter is operating in region 2) is small. However, if the inverter's input logic signal transitions slowly or the logic levels don't swing all the way to the power supply rails (like what we get with the pass gates discussed in the last chapter, see Fig. 10.19), it's possible for a significant current to flow through the inverter (important!) ■

Noise Margins

The noise margins of a digital gate or circuit indicate how well the gate will perform under noisy conditions. The noise margin for the high logic levels is given by

$$NM_H = V_{OH} - V_{IH} \tag{11.1}$$

and the noise margin for the low logic levels is given by

$$NM_L = V_{IL} - V_{OL} \tag{11.2}$$

For $VDD = 1$ V, the ideal noise margins are 500 mV; that is, $NM_L = NM_H = VDD/2$.

Inverter Switching Point

Consider the transfer characteristics of the basic inverter as shown in Fig. 11.5. Point C corresponds to the point on the curve when the input voltage is equal to the output voltage. At this point, the input (or output) voltage is called the *inverter switching point voltage*, V_{SP}, and both MOSFETs in the inverter are in the saturation region. Since the drain current in each MOSFET must be equal, the following is true:

$$\frac{\beta_n}{2}(V_{SP}-V_{THN})^2=\frac{\beta_p}{2}(VDD-V_{SP}-V_{THP})^2 \tag{11.3}$$

Solving for V_{SP} gives

$$V_{SP}=\frac{\sqrt{\frac{\beta_n}{\beta_p}}\cdot V_{THN}+(VDD-V_{THP})}{1+\sqrt{\frac{\beta_n}{\beta_p}}} \tag{11.4}$$

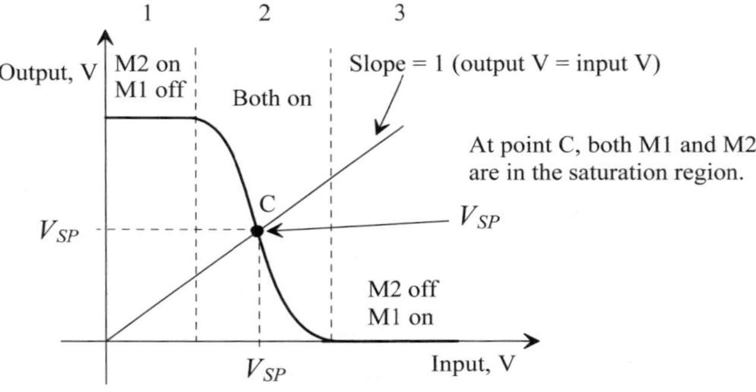

Figure 11.5 Transfer characteristics of the inverter showing the switching point.

Ideal Inverter VTC and Noise Margins

The ideal voltage transfer curves for an inverter are seen in Fig. 11.6. The ideal switching point voltage, V_{SP}, is *VDD*/2. As seen in Eqs. (11.1) and (11.2), this makes the noise margins equal to ensure the best performance (a noise margin, say the logic low level, isn't improved at the cost of the other margin). When looking at Fig. 11.6, notice that, unlike what is seen in Fig. 11.5, the inverter never operates where both MOSFETs are on. The input to the inverter is recognized as either a 1 or a 0.

Example 11.2

Estimate β_n and β_p so that the switching point voltage of a CMOS inverter designed in the long-channel CMOS process is 2.5 V (= *VDD*/2).

Solving Eq. (11.4) with V_{SP} = 2.5 V for the ratio β_n/β_p gives a value of approximately unity. That is,

$$\beta_n = \beta_p = KP_n \frac{W_1}{L_1} = KP_p \frac{W_2}{L_2}$$

Since $KP_n = 3KP_p$, the width of the PMOS device must be three times the width of the NMOS, assuming equal-length MOSFETs. For $V_{SP} = 2.5$ V, this requires

$$W_2 = 3W_1$$

which is also the requirement for making $R_n = R_p$. This is an important practical result. It shows how the electron and hole mobilities are related to both the effective switching resistances and the switching point voltage. As seen in Table 10.2, we often increase the widths of the PMOS devices to try to center V_{SP} and equate the propagation delays (the pull-up resistance, R_p , is the same as the pull-down resistance R_n). ■

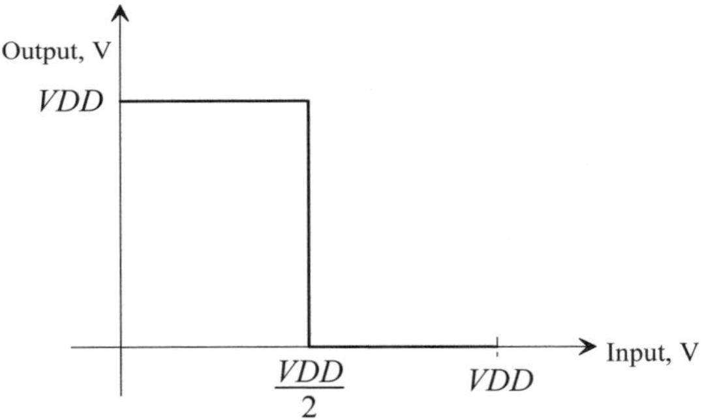

Figure 11.6 Ideal VTCs for the inverter.

Example 11.3
Show, using SPICE, that the inverter seen in Fig. 11.7 and implemented in the short-channel CMOS process has a switching point voltage close to the ideal value of *VDD*/2. Comment on the sizes of the devices.

The simulation results are seen (also) in Fig. 11.7. Because we've sized the width of the PMOS to twice the width of the NMOS (see Table 10.2), the switching point is close to the ideal value of *VDD*/2. We know that short-channel devices don't follow the square-law models and so we can't use Eq. [11.4] to calculate the V_{SP} in our 50 nm process. We can, however, get a good estimate using the effective switching resistances as seen in the figure (a voltage divider). When $R_p = R_n$, the switching point voltage is close to ideal. By changing the widths of the devices, we can adjust the switching point voltage. In other words, for a short-channel CMOS process we can use

$$V_{SP} = VDD \cdot \frac{R_n}{R_n + R_p} \tag{11.5}$$

to estimate V_{SP}. This equation does have limitations. For example, if $R_n >> R_p$, then this equation indicates V_{SP} is *VDD*. However, we know that V_{SP} has to be

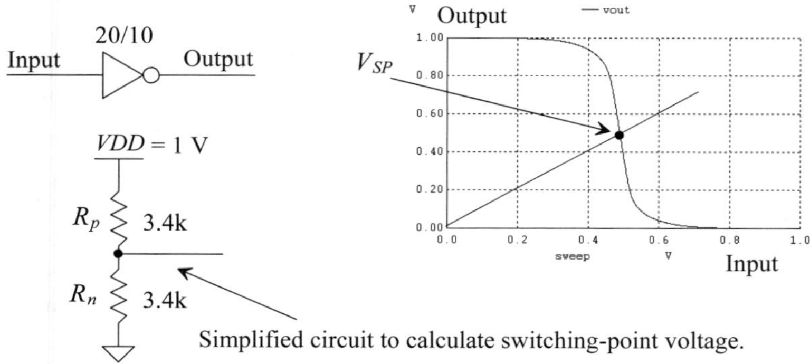

Figure 11.7 Switching point voltage for an inverter in the short-channel process.

greater than V_{THN} and less than $VDD - V_{THP}$ (make sure that this is understood). ■

Example 11.4
Show, using SPICE and the long-channel process, the transfer curves for the CMOS inverter with transconductance ratios β_n/β_p of 3, 1, and 1/3. Explain what changing the inverter ratio does to the transfer characteristics.

The simulation results are seen in Fig. 11.8. For all three DC sweeps the MOSFET's lengths are 1. For the case when $\beta_n/\beta_p = 1$, the $W_n = 10$ and $W_p = 30$. For the case when $\beta_n/\beta_p = 3$, the $W_n = 10$ and $W_p = 10$, etc. Increasing the strength of the NMOS (increasing the NMOS's width, which decreases R_n) causes the switching point voltage to decrease. We also get a decrease in V_{SP} by decreasing the strength of the PMOS device (which increases R_p). ■

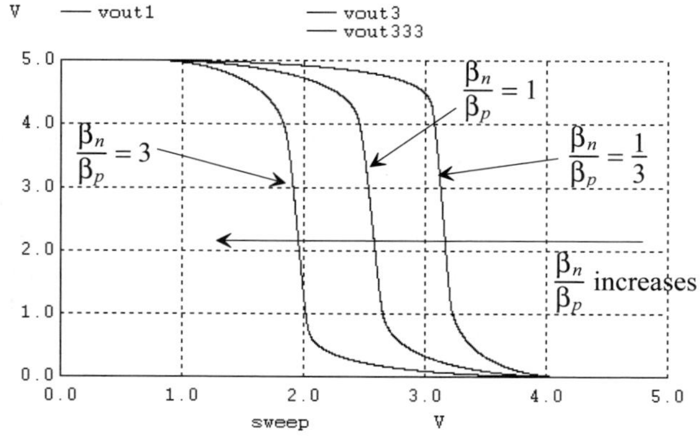

Figure 11.8 Sizing the inverter changes the switching point voltages.

11.2 Switching Characteristics

The switching behavior of the inverter can be generalized by examining the parasitic capacitances and resistances associated with the inverter. Consider the inverter shown in Fig. 11.9 with its equivalent digital model. Although the model is shown with both switches open, in practice one of the switches is closed, keeping the output connected to *VDD* or ground. The effective input capacitance of the inverter is

$$C_{in} = \frac{3}{2}(C_{ox1} + C_{ox2}) = C_{inn} + C_{inp} \tag{11.6}$$

The effective output capacitance of the inverter is simply

$$C_{out} = C_{ox1} + C_{ox2} = C_{outn} + C_{outp} \tag{11.7}$$

The intrinsic propagation delays of the inverter are

$$t_{PLH} = 0.7 \cdot R_{p2} \cdot C_{out} \text{ and } t_{PHL} = 0.7 \cdot R_{n1} \cdot C_{out} \tag{11.8}$$

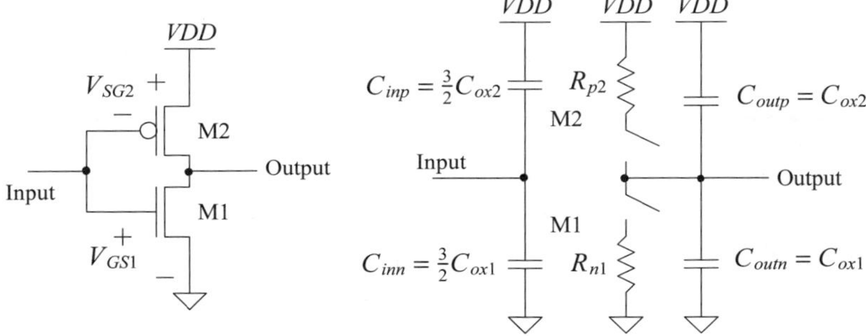

Figure 11.9 The CMOS inverter switching characteristics using the digital model.

Example 11.5

Estimate and simulate the intrinsic propagation delays of the inverter seen in Fig. 11.7. Estimate the inverter's input capacitance.

From the data in Table 10.2 and Eqs. (11.6) to (11.8), we can write

$$t_{PHL} = t_{PLH} = 0.7 \cdot 3.4k \cdot (0.625 + 1.25)\,fF = 4.5\,ps$$

The simulation results are seen in Fig. 11.10. The intrinisic delays are considerably larger than this calculation (around 20 ps).

Using Eq. (11.6), the inverter's input capacitance is

$$C_{in} = \frac{3}{2}(0.625 + 1.25)\,fF = 2.8\,fF$$

Sizing up the width of the PMOS so that its effective resistance is equal to R_n has the unwanted effect of increasing the inverter's input capacitance. ■

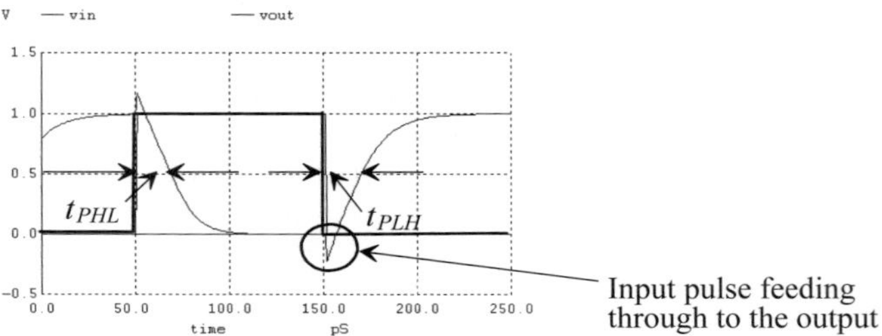

Figure 11.10 The intrinsic propagation delays of an inverter.

The propagation delays for an inverter driving a capacitive load are

$$t_{PLH} = 0.7 \cdot R_{p2} \cdot C_{tot} = 0.7 \cdot R_{p2} \cdot (C_{out} + C_{load}) \tag{11.9}$$

and

$$t_{PHL} = 0.7 \cdot R_{n1} \cdot C_{tot} = 0.7 \cdot R_{n1} \cdot (C_{out} + C_{load}) \tag{11.10}$$

where C_{tot} is the total capacitance on the output of the inverter, that is, the sum of the output capacitance of the inverter, any capacitance of interconnecting lines, and the input capacitance of the following gate(s).

Example 11.6
Estimate and simulate the propagation delays for the circuit seen in Fig. 11.11. Use the 50 nm CMOS process.

Because the load capacitance is much larger than the output capacitance of the inverter, we can rewrite

$$t_{PLH} = 0.7 \cdot R_{p2} \cdot C_{tot} \approx 0.7 \cdot R_{p2} \cdot C_{load} = 120\, ps$$

and

$$t_{PHL} = 0.7 \cdot R_{n1} \cdot C_{tot} \approx 0.7 \cdot R_{n1} \cdot C_{load} = 120\, ps$$

The simulation results are seen in Fig. 11.11. Notice that we are treating the MOSFETs used in the digital circuits as resistors and calculating the delays with a simple product of these resistors with the load capacitance. ■

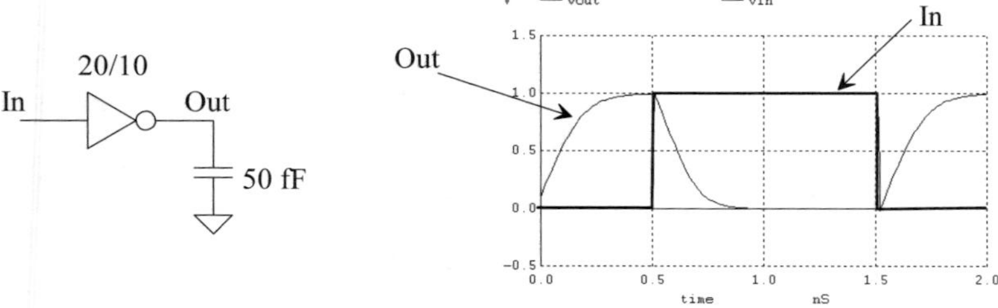

Figure 11.11 The delay associated with an inverter driving a 50 fF load.

The Ring Oscillator

The odd number of inverters in the circuit shown in Fig. 11.12 forms a closed loop with positive feedback and is called a ring oscillator. The oscillation frequency is given by

$$f_{osc} = \frac{1}{n \cdot (t_{PHL} + t_{PLH})} \tag{11.11}$$

assuming that the inverters are identical and n is the number (odd) of inverters in the ring oscillator. Since the ring oscillator is self-starting, it is often added to a test portion of a wafer to indicate the speed of a particular process run. The sum of the high-to-low and low-to-high delays is used to calculate the period of the oscillation because each inverter switches twice during a single oscillation period.

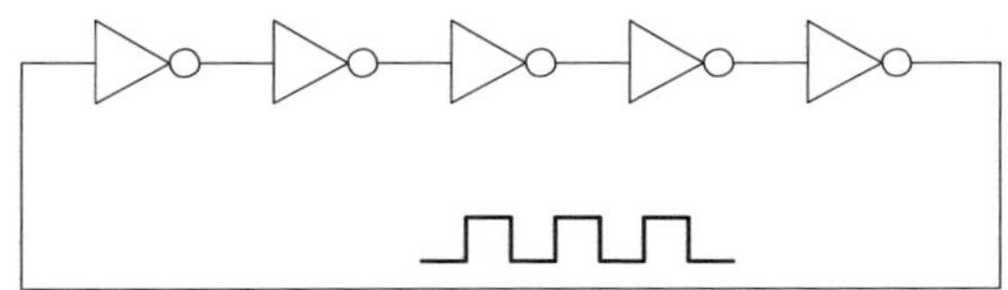

Figure 11.12 A five-stage ring oscillator.

When identical inverters are used the capacitance on the inverter's input/output is the sum of an inverter's input capacitance with the inverter's output capacitance, see Fig. 11.9, or

$$C_{tot} = \overbrace{C_{oxp} + C_{oxn}}^{C_{out}} + \overbrace{\frac{3}{2} \cdot (C_{oxp} + C_{oxn})}^{C_{in}} = \frac{5}{2} \cdot (C_{oxp} + C_{oxn}) \tag{11.12}$$

where, again, $C_{oxp} = C'_{ox} \cdot W_p \cdot L_p \cdot (scale)^2$ and $C_{oxn} = C'_{ox} \cdot W_n \cdot L_n \cdot (scale)^2$. The delay is then calculated using

$$t_{PHL} + t_{PLH} = 0.7 \cdot (R_n + R_p) \cdot C_{tot} \tag{11.13}$$

Dynamic Power Dissipation

Consider the CMOS inverter driving a capacitive load shown in Fig. 11.13. Each time the inverter changes states, it must either supply a charge to C_{tot} or sink the charge stored on C_{tot} to ground. If a square pulse is applied to the input of the inverter with a period T and frequency, f_{clk}, the average amount of current that the inverter must pull from *VDD*, recalling that current is being supplied from *VDD* only when the PMOS device is on, is

$$I_{avg} = \frac{Q_{Ctot}}{T} = \frac{VDD \cdot C_{tot}}{T} \tag{11.14}$$

The average dynamic power dissipated by the inverter is

$$P_{avg} = VDD \cdot I_{avg} = \frac{C_{tot} \cdot VDD^2}{T} = C_{tot} \cdot VDD^2 \cdot f_{clk} \tag{11.15}$$

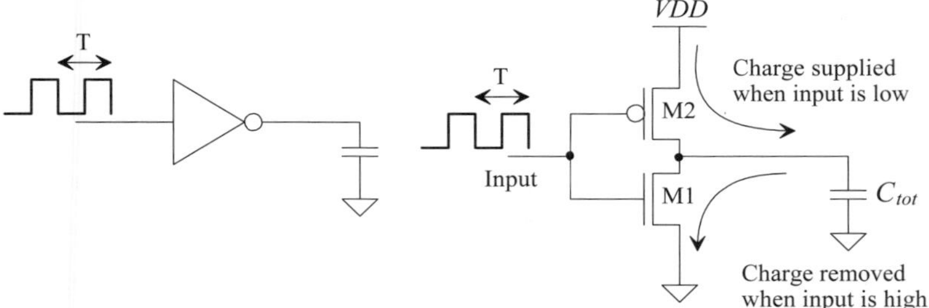

Figure 11.13 Dynamic power dissipation of the CMOS inverter.

Notice that the power dissipation is a function of the clock frequency, power supply voltage, and load capacitance. A great deal of effort is put into reducing the power dissipation in CMOS circuits. One of the major advantages of dynamic logic (Ch. 14) is its lower power dissipation.

To characterize the speed of a digital process, a term called the power delay product (*PDP)* is often used. The *PDP,* measured in joules, is defined by

$$PDP = P_{avg} \cdot (t_{PHL} + t_{PLH}) \tag{11.16}$$

These terms can be determined from a ring oscillator. The *PDP* is frequently used to compare different technologies or device sizes; for example, a GaAs process can be compared with a 50 nm CMOS process. Although the GaAs process may have a lower propagation delay, the power dissipation may be larger and result in a larger *PDP*.

Example 11.7

Estimate the oscillation frequency of an 11-stage ring oscillator in the 50 nm process using the inverter seen in Fig. 11.7 (see Table 10.2). Verify the estimate with simulations.

Using the data in Table 10.2 and Eqs. (11.11) – (11.13), we can write

$$C_{tot} = \frac{5}{2} \cdot (1.25 + 0.625)\, fF = 4.7\, fF$$

and

$$t_{PHL} + t_{PLH} = 0.7 \cdot (3.4k + 3.4k) \cdot 4.7\, fF = 22\, ps$$

For an 11-stage ring oscillator, we get an oscillation frequency of

$$f_{osc} = \frac{1}{11 \cdot (22\, ps)} = 4.1\, GHz$$

The simulation results are seen in Fig. 11.14. The simulated oscillation frequency is close to 1.25 GHz or considerably different from our hand calculations (as will be the case when delays are close to the intrinsic values). The average power, P_{avg}, dissipated by a single inverter in this ring oscillator, using Eq. (11.15), is estimated as 19.6 μW. The PDP for the 50 nm process is then $431 \times 10^{-18}\, J$. ■

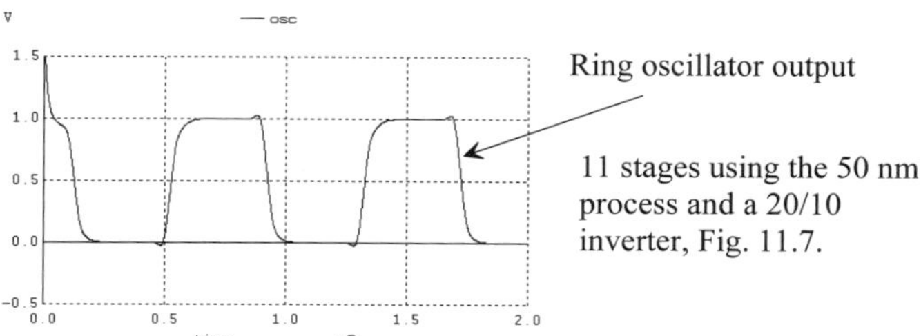

Figure 11.14 Oscillation frequency for the ring oscillator described in Ex. 11.7.

11.3 Layout of the Inverter

If care is not taken when laying out CMOS circuits, the parasitic devices present can cause a condition known as latch-up. Once latch-up occurs, the inverter output will not change with the input; that is, the output may be stuck in a logic state. To correct this problem, the power must be removed. Latch-up is especially troubling in output driver circuits.

Latch-Ip

Figure 11.15 illustrates two methods of laying out an inverter. Notice how the cell's inputs and outputs are on metal2, while the power and ground conductors are routed on metal1 in the standard cell frame (see Fig. 4.15 and the associated discussion). The cross-sectional view in Fig. 11.16 shows both the NMOS and PMOS devices that make up an inverter (and associated parasistics). Notice that in Fig. 11.10, the input pulse feeds through the gate-drain capacitance of the MOSFETs to the output of the inverter. This causes the output to change in the same direction as the input before the inverter starts to switch. This feedthrough and the parasitic bipolar transistors can cause the latch-up.

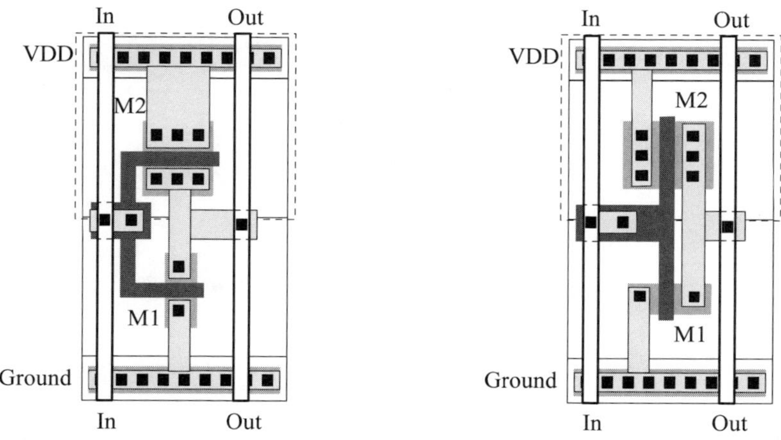

Figure 11.15 Layout styles for inverters.

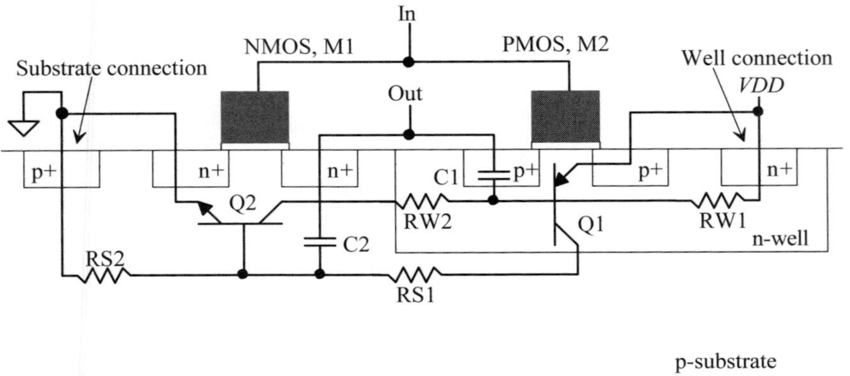

Figure 11.16 Cross-sectional view of an inverter showing parasitic bipolar transistors and resistors.

In Fig. 11.16, the emitter, base, and collector of transistor Q1 are the source of the PMOS, n-well, and substrate, respectively. Transistor Q2's collector, base, and emitter are the n-well, substrate, and source of the NMOS transistor. Resistors RW1 and RW2 represent the effects of the resistance of the n-well, and resistors RS1 and RS2 represent the resistance of the substrate. The capacitors C1 and C2 represent the drain implant depletion capacitance, that is, the capacitance between the drains of the transistors and the n-well (for C1) and substrate (for C2). The parasitic circuit resulting from the inverter layout is shown in Fig. 11.17.

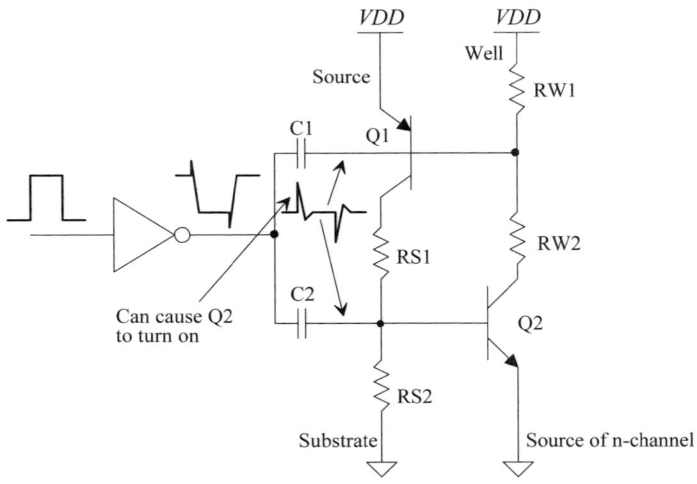

Figure 11.17 Schematic used to describe latch-up.

If the output of the inverter switches fast enough, the pulse fed through C2 (for positive-going inputs) can cause the base-emitter junction of Q2 to become forward biased. This then causes the current through RW2 and RW1 to increase, turning on Q1. When Q1 is turned on, the current through RS1 and RS2 increases, causing Q2 to turn on harder. This positive feedback will eventually cause Q2 and Q1 to turn on fully and remain that way until the power is removed and reapplied. A similar argument can be given for negative-going inputs feeding through C1, *VDD* bouncing upwards, or ground bouncing downwards.

Several techniques reduce the latch-up problem. One technique is to slow the rise and fall times of the logic gates, reducing the amount of signal fed through C1 and C2. Reducing the areas of M1 and M2's drains lowers the size of the depletion capacitance and the amount of signal fed through. Probably the best method of reducing latch-up effects is to reduce the parasitic resistances RW1 and RS2. If these resistances are zero, Q1 and Q2 never turn on. The value of these resistances, as seen in Fig. 11.16, is a strong function of the distance between the well and substrate contacts. Simply put, the closer these contacts are to the MOSFETs used in the inverter, the less likely it is that the inverter will latch up. These contacts should be plentiful as well as close. Placing substrate and well contacts between the PMOS and NMOS devices provides a low-resistance connection to *VDD* and ground, significantly helping to reduce latchup (see Fig. 11.18 for a simple layout example). Placing n+ and p+ areas between or around circuits reduces the amount of signal reaching a given circuit from another circuit. These implants are sometimes called guard rings (see Fig. 5.5). Notice that poly cannot be used to connect the gates of the MOSFETs, since poly over the n+ or p+ will be interpreted as a MOSFET. Therefore, metal2 is used to connect the MOSFETs together. The cost of reducing the possibility of latch-up is a more complicated layout in a larger area.

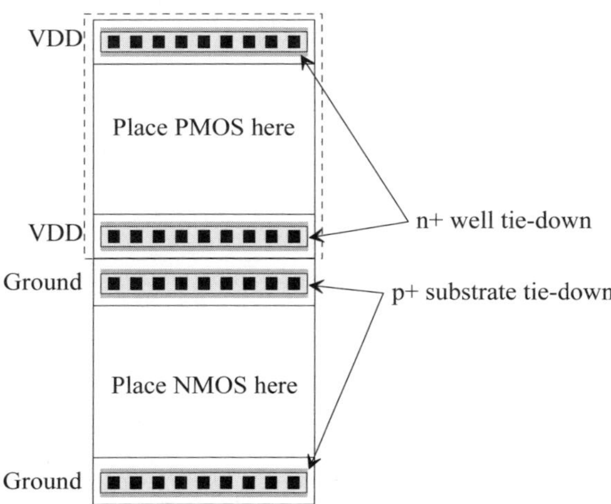

Figure 11.18 Adding an extra implant between NMOS and PMOS to reduce latch-up.

11.4 Sizing for Large Capacitive Loads

Designing a circuit to drive large capacitive loads with minimum delay is important when driving off-chip loads. As we saw in Ex. 11.6, a large load capacitance can drastically affect the delay through an inverter. Remembering our discussion in Sec. 10.3, we see that using a standard scope probe to measure an output signal can result in delays that are microseconds in length. In order to avoid this situation, we add a buffer circuit (a string of inverters) between the on-chip logic and the bonding pads. In this section we discuss how to design (select the widths of the MOSFETs) for low delays.

Buffer Topology

Consider the inverter string (a buffer) driving a load capacitance, labeled C_{load} and shown in Fig. 11.19. Moving towards the load in a cascade of the N inverters, each inverter larger than the previous by a factor A (that is, the width of each MOSFET is multiplied by A), a minimum delay can be obtained as long as A and N are picked correctly. Each inverter's input capacitance is larger than the previous inverter's input capacitance by a factor of A,

$$C_{in2} = A \cdot C_{in1} \text{ and } C_{in3} = A \cdot C_{in2} = A^2 \cdot C_{in1}\text{, etc.} \tag{11.17}$$

The effective switching resistances are also divided by a factor of A, resulting in the same delay for each stage of the buffer,

$$R_{n,p2} = \frac{R_{n,p1}}{A} \text{ and } R_{n,p3} = \frac{R_{n,p2}}{A} = \frac{R_{n,p1}}{A^2} \tag{11.18}$$

In other words, the effective switching resistance of the NMOS in the third inverter is $1/A^2$ smaller than the effective switching resistance of the NMOS in the first inverter.

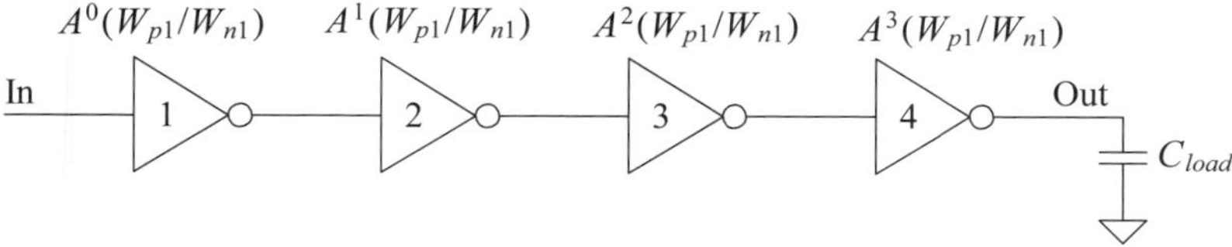

Figure 11.19 Cascade of inverters used to drive a large load capacitance.

If the load capacitance equals the input capacitance of the last inverter multiplied by A (so that the load capacitance has a value equal to the input capacitance of the next inverter if an additional inverter were used), then

$$\text{Input C of the final inverter} = C_{in1} \cdot A^N = C_{load} \tag{11.19}$$

or

$$A = \left[\frac{C_{load}}{C_{in1}}\right]^{\frac{1}{N}} \tag{11.20}$$

Again, the delays of each stage in the buffer are equal. The total delay of the inverter string is given by

$$(t_{PHL}+t_{PLH})_{total}=\underbrace{0.7\cdot(R_{n1}+R_{p1})(C_{out1}+AC_{in1})}_{\text{First-stage delay}}+\underbrace{0.7\cdot\frac{(R_{n1}+R_{p1})}{A}\cdot(AC_{out1}+A^2C_{in1})}_{\text{Second-stage delay}}\ldots \quad (11.21)$$

Because as the inverters are increased in size by *A,* their capacitances, both input and output, increase by *A*, while their resistances decrease by a factor *A*. This equation can be rewritten as

$$(t_{PHL}+t_{PLH})_{total}=0.7\cdot\sum_{k=1}^{N}(R_{n1}+R_{p1})(C_{out1}+AC_{in1})=0.7\cdot N(R_{n1}+R_{p1})(C_{out1}+AC_{in1}) \quad (11.22)$$

or with the help of Eq. (11.20):

$$(t_{PHL}+t_{PLH})_{total}=0.7\cdot N(R_{n1}+R_{p1})\cdot\left(C_{out1}+\left(\frac{C_{load}}{C_{in1}}\right)^{\frac{1}{N}}\cdot C_{in1}\right) \quad (11.23)$$

The minimum delay can be found by taking the derivative of this equation with respect to *N*, setting the result equal to zero, and solving for *N*. Taking the derivative of Eq. (11.23) with respect to *N* gives

$$0.7\cdot\left((R_{n1}+R_{p1})C_{out1}+(R_{n1}+R_{p1})C_{in1}\left(\left(\frac{C_{load}}{C_{in1}}\right)^{\frac{1}{N}}+N\cdot\left(\frac{C_{load}}{C_{in1}}\right)^{\frac{1}{N}}\frac{\ln(C_{load}/C_{in1})}{-N^2}\right)\right)=0 \quad (11.24)$$

The first term in this equation is the intrinsic delay of the first inverter in our cascade of inverters. *This first stage represents, generally, the on-chip logic gate and* ***is not part*** *of the buffer.* The first inverter is included in the calculations to represent the limited drive of the on-chip logic. If we assume that this delay is small, solving for *N* gives

$$N=\ln\frac{C_{load}}{C_{in1}} \quad (11.25)$$

Equations (11.25) and (11.20) are used in the buffer design to drive a large capacitance. Note that the larger the first inverter, the fewer the number of inverters needed to drive a given capacitive load.

Example 11.8
Estimate $t_{PHL}+t_{PLH}$ for the inverter shown in Fig. 11.11 driving a load capacitance of 20 pF. Design a buffer to drive the load capacitance with a minimum delay.

The total propagation delay of the unbuffered inverter, Fig. 11.11, is given by

$$t_{PHL}+t_{PLH}=0.7\cdot(3.4k+3.4k)\cdot(20\,pF)\ =95\,ns\ !$$

Designing a buffer begins with determining C_{in1}. For our 20/10 inverter in the 50 nm process, the C_{in1} (see Table 10.2) is $\frac{3}{2}(1.25+0.625)\,fF=2.81\,fF$ and C_{out1} is $1.875\,fF$. To determine the number of inverters, we use

$$N=\ln\left(\frac{20\,pF}{2.81\,fF}\right)=8.87\rightarrow\ 9\text{ stages}$$

To maintain the same logic, that is, an inversion of the input signal, we use seven inverters. In practice, the difference in delay between eight and nine inverters is negligible. If we did not want a logic inversion, we would use eight stages. The area factor is then

$$A = \left[\frac{20\,pF}{2.81\,fF}\right]^{\frac{1}{8.87}} = 2.718 = e$$

noting that for the *minimum delay* in all cases the widths are increased by e.

The total delay, using Eq. (11.23), is then

$$(t_{PHL} + t_{PLH})_{total} = 0.7 \cdot 9 \cdot (3.4k + 3.4k)(1.875\,fF + 2.718 \cdot 2.81\,fF) = 407\,ps$$

or over 200 times faster. Since the PMOS width is twice the width of the NMOS, the propagation delay times, t_{PHL} and t_{PLH}, are equal, or

$$t_{PHL} = t_{PLH} = \frac{(t_{PHL} + t_{PLH})_{total}}{2N} = 22.5\,ps$$

A schematic of the design is shown in Fig. 11.20. ■

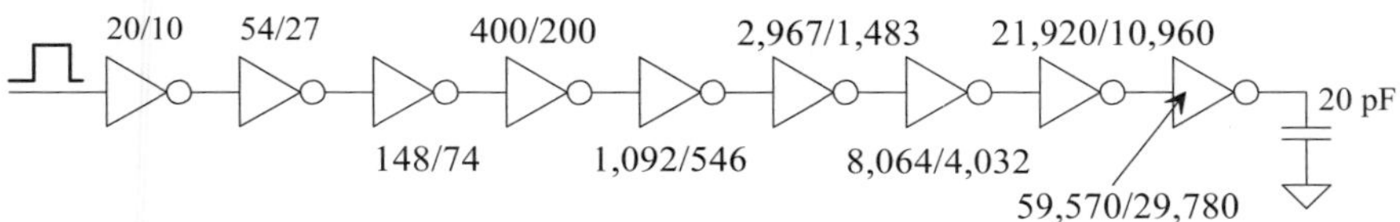

Figure 11.20 Buffer designed in Ex. 11.8. Attains the minimum delay but the buffer is not practical.

It should be clear that, although this technique results in the least delay in driving the 20 pF load, the MOSFETs needed are very large. In many applications, the very minimum delay through a buffer is not required. The value of A (ideally e) can be considerably larger and have little impact on the delay of the buffer (reducing the number of stages and their widths). Consider the following.

Example 11.9

Redesign the buffer of Ex. 11.8 with an A of 8. Compare the delay of the modified (practical) buffer to the delay of the ideal buffer calculated in Ex. 11.8.

We can rewrite Eq. (11.20) as

$$N \cdot \ln A = \ln \frac{C_{load}}{C_{in1}}$$

The natural logarithm of 8 is roughly 2 so we can solve for the number of stages using

$$N = \frac{1}{2} \cdot \ln \frac{C_{load}}{C_{in1}} = 4.43$$

To maintain the logic inversion, we'll use five stages. The delay is (roughly, because we are using 5 instead of 4.43) calculated as

$$(t_{PHL} + t_{PLH})_{total} = 0.7 \cdot 5 \cdot (3.4k + 3.4k)(1.875\, fF + 8 \cdot 2.81\, fF) = 580\, ps$$

(not too much larger than the 407 ps calculated in Ex. 11.7). The resulting buffer is shown in Fig. 11.21. ■

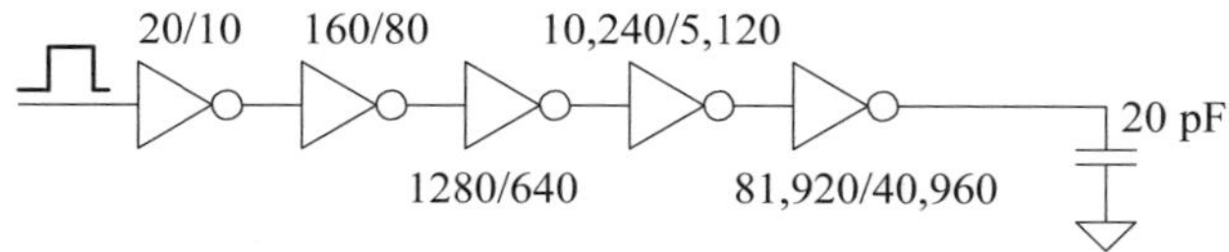

Figure 11.21 Buffer designed in Ex. 11.9.

Distributed Drivers

Consider the driver circuit shown in Fig. 11.22a containing 11 inverters. If all of the inverters shown in the figure are the same size, the delay from the input to the output is

$$t_{PHL} + t_{PLH} = 0.7 \cdot (R_n + R_p)(C_{out} + 10C_{in}) \qquad (11.26)$$

Now consider the circuit shown in Fig. 11.22b with 13 inverters. Again, assuming all of the inverters are the same size, the delay from the input to the output is

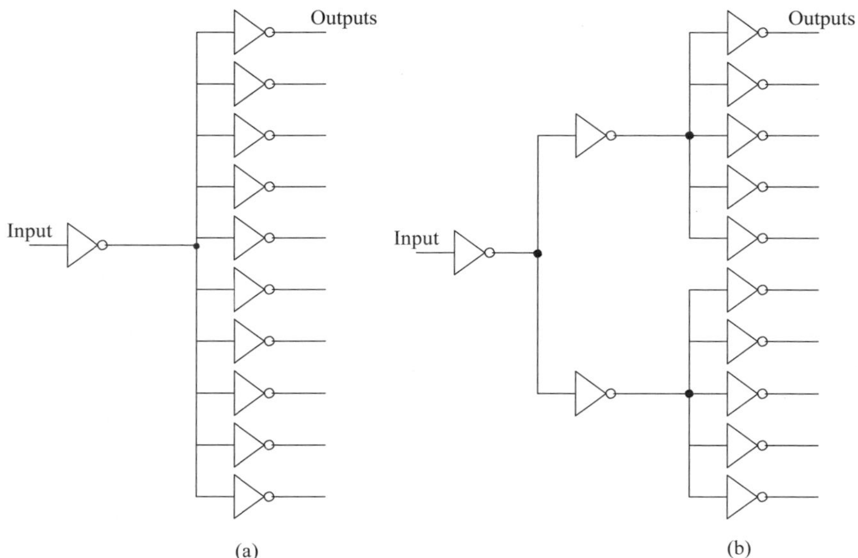

Figure 11.22 Distributed drivers.

$$t_{PHL} + t_{PLH} = 0.7 \cdot (R_n + R_p)[(C_{out} + 2C_{in}) + (C_{out} + 5C_{in})] = 0.7 \cdot (R_n + R_p)[2C_{out} + 7C_{in}] \quad (11.27)$$

which is less delay than the circuit with 11 inverters. Distributing the signal into different paths can reduce the propagation delays. Using the results from the last section, we can get minimum overall delay when the delays through each layer of logic are equal. This occurs when $A = e$ (each inverter drives 2.718 other inverters) and Eq. (11.25) is used to select the (number of) layers of logic. The load capacitance is equal to the number of outputs multiplied by the capacitance on each output. In practice, again as we saw in the last section, the change in delays isn't too significant, as long as one logic gate (one path) isn't loaded too much (compare actual numbers in Eqs. [11.26] and [11.27]).

At this point we can ask the question, "Why not make the first inverter in the circuits of Fig. 11.22 really large (small R_n and R_p) so that it has small effective resistances for driving the ten inverters quickly?" The answer is simply that as we increase the size of an inverter, we also increase its input capacitance. In SPICE simulations, we use ideal voltage sources to drive the first gate in our circuit. In practice, this inverter is driven from another gate somewhere on the chip. Increasing the size will slow the propagation delay-time of the gate driving this inverter.

Driving Long Lines

Often when designing large systems, a signal may need to be driven across the chip. In some cases, for example, in dynamic random-access memory (DRAM), the signal must be transmitted over a line that has a large parasitic resistance and capacitance. We need to develop a method of determining the delay through this line using hand calculations. This will lend insight to the design and help to determine exactly how to design the driver (or drivers).

Consider the driver circuit shown in Fig. 11.23. The inverter is driving an RC transmission line with resistance/unit length, r, capacitance/unit length, c, and unit length, l. We can estimate the delay from the input to voltage across the capacitor by adding the delays. This is given by

$$t_{PHL} + t_{PLH} = 0.7 \cdot [(R_n + R_p)(C_{out} + c \cdot l + C_{load}) + (r \cdot l)(C_{load})] + 2 \cdot 0.35 \cdot rcl^2 \quad (11.28)$$

where the first term in this equation is the delay associated with the inverter driving the total capacitance at its output to ground. The second term is the delay through the RC line, while the last term is an estimate of the delay associated with driving a capacitive load through the line's resistance. The most common method of reducing the delay

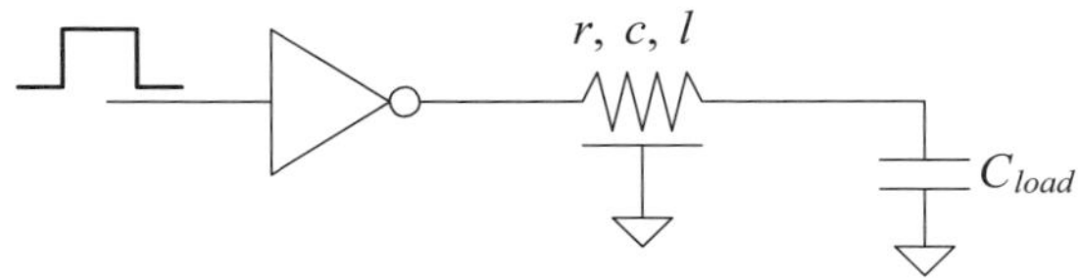

Figure 11.23 Driving an RC transmission line.

through the line is to place buffer stages at different locations along the line. This effectively breaks the line up and can lower the overall delay. If C_{load} is a major contributor to the delay, a buffer can be inserted between the *RC* line and C_{load} to reduce the delay.

11.5 Other Inverter Configurations

Three other inverter configurations are shown in Fig. 11.24. The inverter shown in Fig. 11.24a is an NMOS-only inverter, useful in avoiding latch-up. There's no PMOS device so there's no parasitic bipolar junction transistors. The inverters shown in Fig. 11.24b and c use a PMOS load, which is, in general, most useful in logic gates with a large number of inputs (more on this in the next chapter). In general, the selection of the MOSFET sizes in (a) and (b) follows the 4-to-1 rule; that is, the resistance (R_n or R_p) of the load is made four times larger than the resistance of M1. The resistance of the PMOS in (c) can be made eight times the resistance of the NMOS. Because $R_p > R_n$, the t_{PLH} will always be greater than the t_{PHL}. In other words, the switching times will be asymmetric.

For all inverter configurations in Fig. 11.24, a logic 1 input signal results in a DC current flowing in both MOSFETs. Making sure that this DC current isn't too large is an important design concern when sizing the MOSFETs. The output logic low will never reach 0 V in these inverters (V_{OL} doesn't reach ground), and thus the noise margins are poorer than the basic CMOS inverter of Fig. 11.1. The output high level of the inverter of Fig. 11.24c will reach *VDD*, while the other inverter's output high level will be a threshold voltage drop below *VDD*. It might be concluded that the power dissipation of the inverters shown in Fig. 11.24 is greater than the basic CMOS inverter. However, since the input capacitance of these inverters is less than the basic CMOS inverter and the output voltage swing is reduced, the inverter with the greatest power dissipation is determined by the operating frequency. At high operating frequencies, the basic CMOS inverter dissipates the most power. At DC or low frequencies, the inverter configurations seen in Fig. 11.24 dissipate more power.

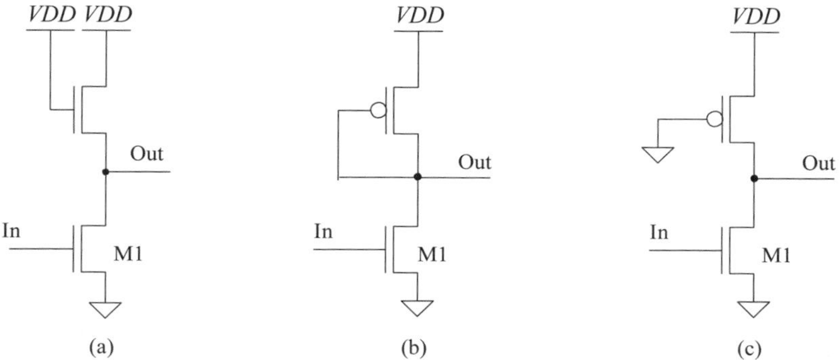

Figure 11.24 Other inverter configurations.

NMOS-Only Output Drivers

Because of the susceptibility of the basic CMOS inverter to latch-up, output drivers consisting of only NMOS devices are used. Figure 11.25 shows the basic "NMOS super buffer." When the input signal is low, M1 and M4 are off while M2 and M3 are on. The output is pulled to ground through M2. A high on the input to the buffer causes M1 and M4 to turn on pulling the output to $VDD - V_{THN}$, assuming that the input high-signal amplitude is *VDD*.

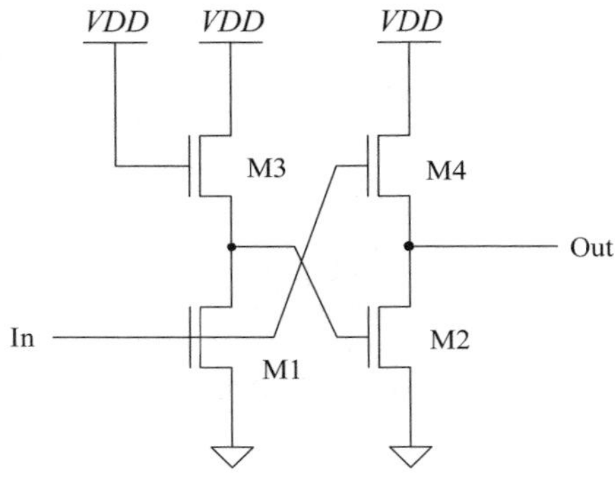

Figure 11.25 NMOS super buffer.

The reduced output voltage of the NMOS-only output buffer can be improved using the circuit of Fig. 11.26. The inverter driving the gate of M2 uses an on-chip generated DC voltage of *VDD* + *Vadd* (where *Vadd* is large enough to ensure that M2 turns fully on and the output is driven to *VDD*). Thus, the output swings from 0 to *VDD* similar to the CMOS output buffer. Note that with the addition of an enabling logic gate, the gates of M1 and M2 can be held at ground, forcing the output into the high-impedance (Hi-Z) state.

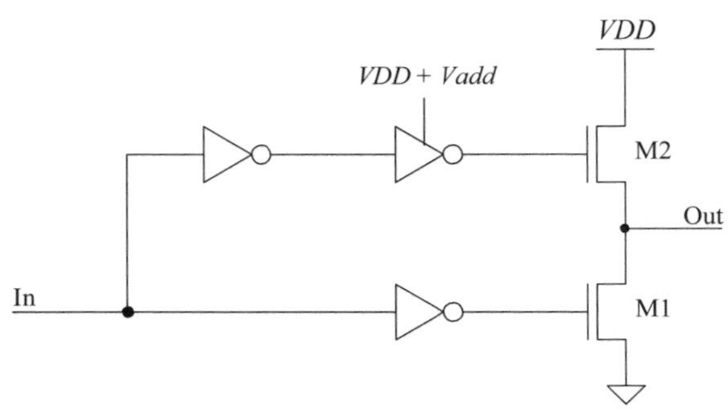

Figure 11.26 Output buffer using a pumped voltage.

Inverters with Tri-State Outputs

Two configurations used in the design of an inverter with tri-state outputs are shown in Fig. 11.27. A high on the *S* input allows the circuit to operate normally, that is, as an inverter. A low on the S input forces the output into the Hi-Z, or high-impedance state. These circuits are useful when data is shared on a communication bus. The logic symbol for the tri-state inverter is shown in Fig. 11.27c. Note that the circuit in (a) dissipates power even when the the select signal, *S* , is a low, while the circuit in (b) does not. However, the circuit in (a) has faster switching times because the effective resistance seen at the output to ground or *VDD* is lower (the NMOS and PMOS devices are in parallel).

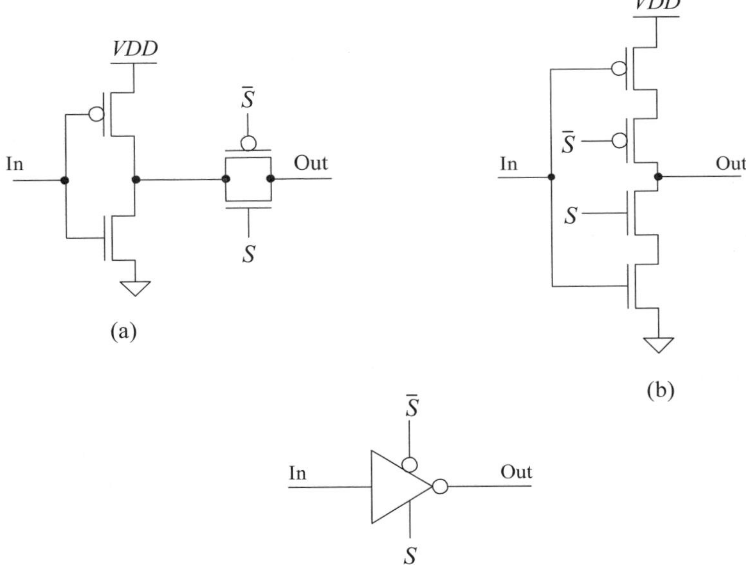

Figure 11.27 Circuits and logic symbol for the tri-state inverter.

Additional Examples

Additional delay calculations using inverters (and other CMOS circuit building blocks) can be found in Sec. 13.4.

ADDITIONAL READING

[1] J. P. Uyemura, *Introduction to VLSI Circuits and Systems*, John Wiley and Sons Publishers, 2002. ISBN 0-471-12704-3.

[2] N. H. E. Weste and D. Harris, *Principles of CMOS VLSI Design*, Addison-Wesley, 3rd ed., 2004. ISBN 0-321-14901-7.

[3] M. I. Elmasry, *Digital MOS Integrated Circuits II,* IEEE Press, 1992. ISBN 0-87942-275-0, IEEE order number: PC0269-1.

[4] R. L. Geiger, P. E. Allen, and N. R. Strader, *VLSI-Design Techniques for Analog and Digital Circuits,* McGraw-Hill Publishing Co., 1990. ISBN 0-07-023253-9.

PROBLEMS

Use the 50 nm CMOS process for the following problems unless otherwise stated.

11.1 Estimate the noise margins for the inverters used to generate Fig. 11.4.

11.2 Design and simulate the DC characteristics of an inverter with V_{SP} approximately equal to V_{THN}. Estimate the resulting noise margins for the design.

11.3 Show that the switching point of three inverters in series is dominated by the V_{SP} of the first inverter.

11.4 Repeat Ex. 11.6 using a PMOS device with a width of 10.

11.5 Repeat Ex. 11.6 using the long-channel process with a 30/10 inverter.

11.6 Estimate the oscillation frequency of a 11-stage ring oscillator using inverters 30/10 inverters in the long-channel CMOS process. Compare your hand calculations to simulation results.

11.7 Using the long-channel process, design a buffer with minimum delay (A = 2.718) to insert between a 30/10 inverter and a 50 pF load capacitance. Simulate the operation of the design.

11.8 Repeat Problem 11.7 using an area factor, A, of 8.

11.9 Derive an equation for the switching point voltage, similar to the derivation of Eq. (11.4), for the NMOS inverter seen in Fig. 11.24a.

11.10 Repeat Problem 11.9 for the inverter in Fig. 11.24c. Note that the PMOS transistor is operating in the triode region when the input/output are at V_{SP}.

Chapter

12

Static Logic Gates

In this chapter we discuss the DC characteristics, dynamic behavior, and layout of CMOS static logic gates. Static logic means that the output of the gate is always a logical function of the inputs and always available on the outputs of the gate regardless of time. We begin with the NAND and NOR gates.

12.1 DC Characteristics of the NAND and NOR Gates

The two basic input NAND and NOR gates are shown in Fig. 12.1. Before we get into the operation, notice that each input into the gate is connected to both a PMOS and an NMOS device similar to the inverter of the last chapter. We will make use of the results of Ch. 11 to explain the operation of these gates.

12.1.1 DC Characteristics of the NAND Gate

The NAND gate of Fig. 12.1a requires both inputs to be high before the output switches low. Let's begin our analysis by determining the voltage transfer curve (VTC) of a NAND gate with PMOS devices that have the same widths, W_p, and lengths, L_p, and NMOS devices with equal widths of W_n and lengths of L_n. If both inputs of the gate are tied together, then the gate behaves like an inverter.

To determine the gate switching point voltage, V_{SP}, we must remember that two MOSFETs in parallel behave like a single MOSFET with a width equal to the sum of the individual widths. For the two parallel PMOS devices in Fig. 12.1a, we can write

$$W_3 + W_4 = 2W_p \tag{12.1}$$

again assuming that all PMOS devices are of the same size. The transconductance parameters can also be combined into the transconductance parameter of a single MOSFET, or

$$\beta_3 + \beta_4 = 2\beta_p \tag{12.2}$$

The two NMOS devices in series (with their gates tied together) behave like a single MOSFET with a channel length equal to the sum of the individual MOSFET lengths. We can write for the NMOS devices

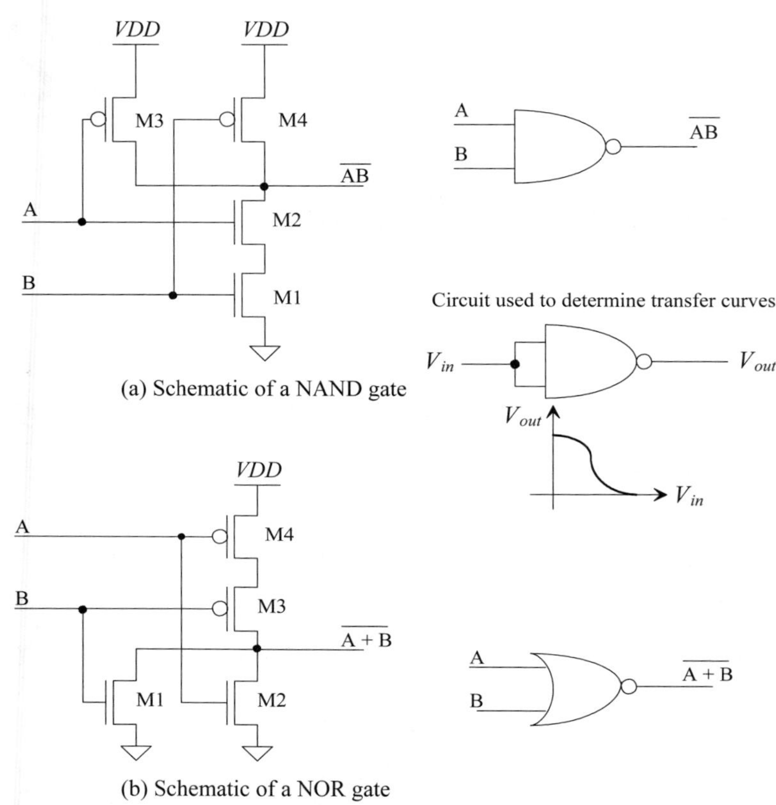

Figure 12.1 NAND and NOR gate circuits and logic symbols.

$$L_1 + L_2 = 2L_n \tag{12.3}$$

and the transconductance of the single MOSFET is given by

$$\beta_1 + \beta_2 = \frac{\beta_n}{2} \tag{12.4}$$

If we model the NAND gate with both inputs tied together as an inverter with an NMOS device having a width of W_n and length $2L_n$ and a PMOS device with a width of $2W_p$ and length L_p, then we can write the transconductance ratio as

$$\text{Transconductance ratio of NAND gate} = \frac{\beta_n}{4\beta_p} \tag{12.5}$$

The V_{SP}, with the help of Eq. (11.4), of the two-input NAND gate is then given by

$$V_{SP} = \frac{\sqrt{\frac{\beta_n}{4\beta_p}} \cdot V_{THN} + (VDD - V_{THP})}{1 + \sqrt{\frac{\beta_n}{4\beta_p}}} \tag{12.6}$$

or in general for an n-input NAND gate (see Fig. 12.2), we get

$$V_{SP} = \frac{\sqrt{\frac{\beta_n}{N^2 \cdot \beta_p}} \cdot V_{THN} + (VDD - V_{THP})}{1 + \sqrt{\frac{\beta_n}{N^2 \cdot \beta_p}}} \tag{12.7}$$

These equations are derived under the assumption that all inputs are tied together. If, for example, only one input is switching, the V_{SP} will vary from what is calculated using Eq. (12.7) (assuming the single input switching does indeed cause the gate's output to switch). This equation is used to show why NAND gates are preferred in CMOS design. If equal- sized NMOS and PMOS devices are used, then, since the mobility of the hole is less than the mobility of the electron, $\beta_n > \beta_p$. Using NMOS devices in series and PMOS in parallel (as in the NAND gate) makes it easier to design a logic gate with the ideal switching point voltage of *VDD*/2.

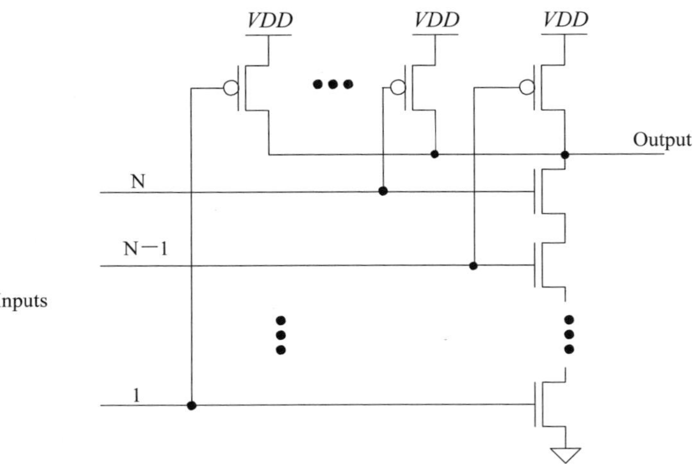

Figure 12.2 Schematic of an n-input NAND gate.

Example 12.1
Determine V_{SP} by hand calculations and compare to a SPICE simulation for a three-input NAND gate using 10/1 devices in the long-channel CMOS process used in this book, see Table 6.2. Compare the hand calculations to simulation results.

The switching point voltage is determined by calculating the transconductance ratio of the gate, or

$$\sqrt{\frac{\beta_n}{N^2\beta_p}} = \sqrt{\frac{\frac{120\ \mu A/V^2 \cdot 10}{1}}{9 \cdot \frac{40\ \mu A/V^2 \cdot 10}{1}}} = 0.58$$

and then using Eq. (12.7),

$$V_{SP} = \frac{0.58 \cdot (0.8) + (5 - 0.9)}{1 + 0.58} = 2.9\ V$$

The SPICE simulation results are shown in Fig. 12.3. The simulation also gives a V_{SP} of approximately 2.9 V. ■

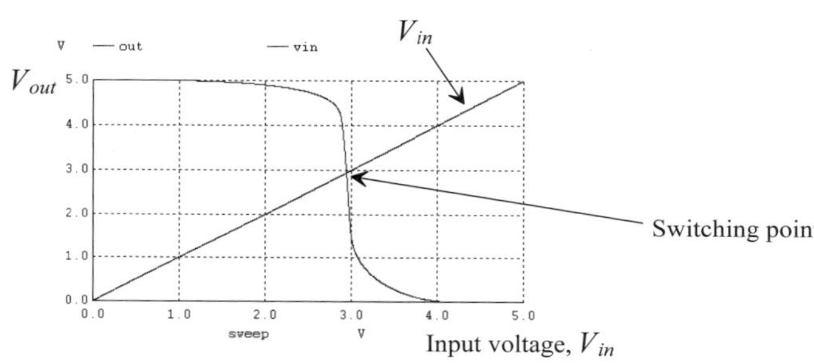

Figure 12.3 VTCs of the three-input minimum-size (using 10/1 MOSFETs) NAND gate.

12.1.2 DC Characteristics of the NOR gate

Following a similar analysis for the n-input NOR gate (see Fig. 12.4) gives a switching point voltage of

$$V_{SP} = \frac{\sqrt{\frac{N^2 \cdot \beta_n}{\beta_p}} \cdot V_{THN} + (VDD - V_{THP})}{1 + \sqrt{\frac{N^2 \cdot \beta_n}{\beta_p}}} \tag{12.8}$$

Example 12.2

Compare the switching point voltage of a three-input NOR gate made from minimum-size MOSFETs to that of the three-input NAND gate of Ex. 12.1. Comment on which gate's V_{SP} is closer to ideal, that is, $V_{SP} = VDD/2$.

The transconductance ratio is calculated as

$$\sqrt{N^2 \cdot \frac{\beta_n}{\beta_p}} = \sqrt{9 \cdot \frac{\frac{120\ \mu A/V^2 \cdot 10}{1}}{\frac{40\ \mu A/V^2 \cdot 10}{1}}} = 5.2$$

The V_{SP} of the minimum-size three-input NOR gate is 1.33 V, while the V_{SP} of the minimum-size three-input NAND gate was calculated to be 2.9 V. For an ideal

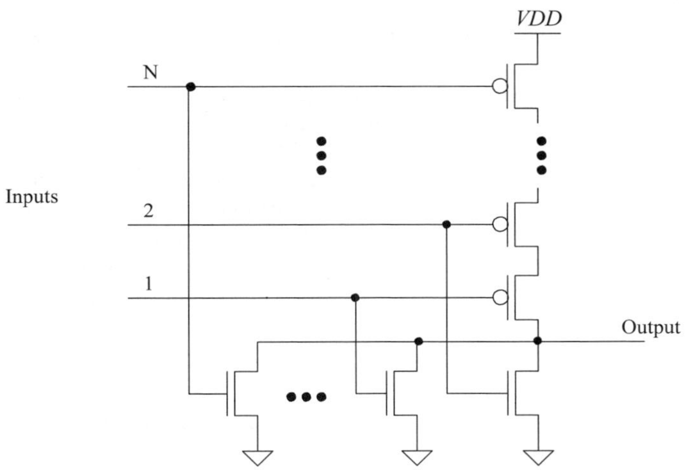

Figure 12.4 Schematic of an n-input NOR gate.

gate, $V_{SP} = 2.5$ V, so that the NAND gate is closer to ideal than the NOR gate. In CMOS digital design, the NAND gate is used most often. This is due to the DC characteristics, better noise margins, and the dynamic characteristics. We will also see shortly that the NAND gate has better transient characteristics than the NOR gate. ■

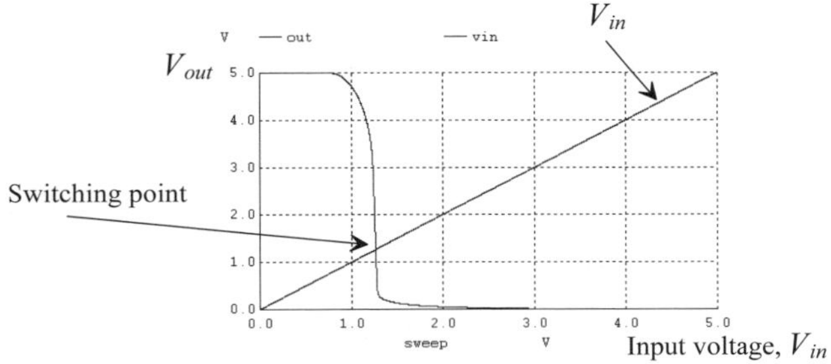

Figure 12.5 VTCs of the three-input minimum-size (using 10/1 MOSFETs) NOR gate.

A Practical Note Concerning V_{SP} and Pass Gates

Reviewing Fig. 10.19, we see that passing a logic signal through a pass gate (PG) can result in a reduction in the logic signal's amplitude. Using an NMOS PG, for example, results in an output signal swing from ground to $VDD - V_{THN}$. If this logic signal is connected to an inverter or a logic gate, we will want to set the V_{SP} to maximize the noise margins. Using the NMOS PG, our inverter/gate would have a $V_{SP} = (VDD - V_{THN})/2$.

12.2 Layout of the NAND and NOR Gates

Layout of the three-input minimum-size NOR and NAND gates is shown in Fig. 12.6, using the standard-cell frame. MOSFETs in series, for example, the NMOS devices in the NAND gate, are laid out using a single-drain and a single-source implant area. The active area between the gate poly is shared between two devices. This has the effect of reducing the parasitic drain/source implant capacitances. MOSFETs in parallel, for example, the NMOS devices in the NOR gate, can share a drain area or a source area. The inputs of the gates are shown on the poly layer while the outputs of the two logic gates are on metal2. To make the inputs easy to connect to we would route them up to metal2 like the outputs. Note how metal1 is used inside the cell and horizontally to connect power and ground to the cells, while metal2 is used for vertical running wires (inputs and outputs).

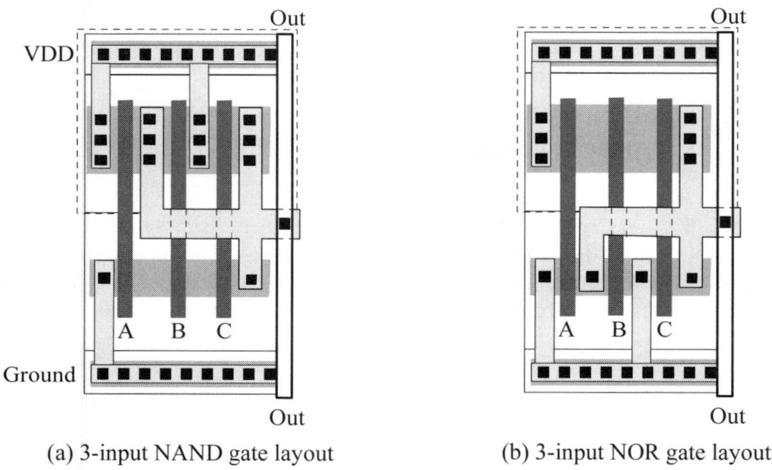

Figure 12.6 Layouts of NAND (a) and NOR (b) gates.

12.3 Switching Characteristics

In this section we discuss the switching characteristics of static logic gates.

Parallel Connection of MOSFETs

Consider the parallel connection of identical MOSFETs shown in Fig. 12.7 *with their* gates tied together. From the equivalent digital models, also shown, we can determine the propagation delay associated with this parallel connection of N MOSFETs as

$$t_{PLH} = 0.7 \cdot \frac{R_p}{N} \cdot (N \cdot C_{oxp}) = 0.7 \cdot R_p C_{oxp} \tag{12.9}$$

where $C_{oxp} = C'_{ox} \cdot W \cdot L \cdot (scale)^2$. With an external load capacitance, the low-to-high delay-time becomes

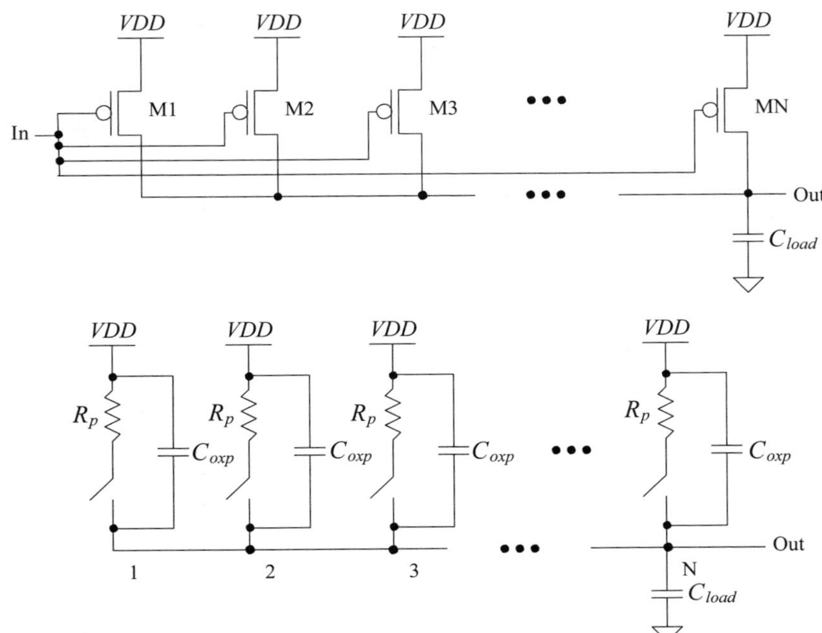

Figure 12.7 Parallel connection of MOSFETs and equivalent digital model.

$$t_{PLH} = 0.7 \cdot \frac{R_p}{N} \cdot (N \cdot C_{oxp} + C_{load}) \tag{12.10}$$

This again assumes that the MOSFET's gates are tied together (all are switching at the same time). For NMOS devices in parallel, a similar analysis yields

$$t_{PHL} = 0.7 \cdot \frac{R_n}{N} \cdot (N \cdot C_{oxn} + C_{load}) \tag{12.11}$$

The load capacitance, C_{load}, consists of all capacitances on the output node except the output capacitances of the MOSFETs in parallel.

Series Connection of MOSFETs

Consider the series connection of identical NMOS devices shown in Fig. 12.8. We can *estimate* the intrinsic switching time of series-connected MOSFETs by

$$t_{PHL} = 0.35 \cdot R_n C_{oxn} \cdot N^2 \tag{12.12}$$

as discussed back in Sec. 10.2.2. With an external load capacitance, the high-to-low delay-time becomes

$$t_{PHL} = 0.35 \cdot R_n C_{oxn} \cdot N^2 + 0.7 \cdot N \cdot R_n \cdot C_{load} \tag{12.13}$$

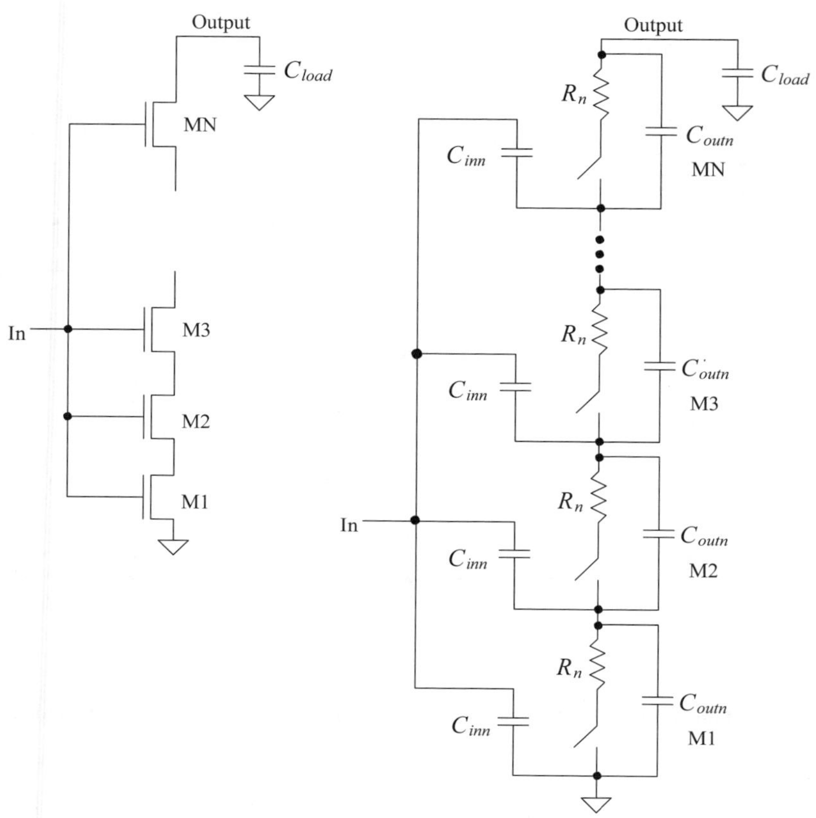

Figure 12.8 Series connection of MOSFETs and equivalent digital model.

For PMOS devices in series, a similar analysis yields

$$t_{PLH} = 0.35 \cdot R_p C_{oxp} \cdot N^2 + 0.7 \cdot N \cdot R_p \cdot C_{load} \tag{12.14}$$

These equations are approximations for the propagation delays which give results usually to within a factor of two of the measurements.

12.3.1 NAND Gate

Consider the n-input NAND gate of Fig. 12.9 driving a capacitive load C_{load}. The low-to-high propagation time, using Eq. (12.10), is

$$t_{PLH} = 0.7 \cdot \frac{R_p}{N}\left(N \cdot C_{outp} + \frac{C_{outn}}{N} + C_{load}\right) \tag{12.15}$$

where here C_{load} represents the capacitance external to the gate, whereas in Eq. (12.10) C_{load} represented the capacitance external to the parallel PMOS devices. If the load

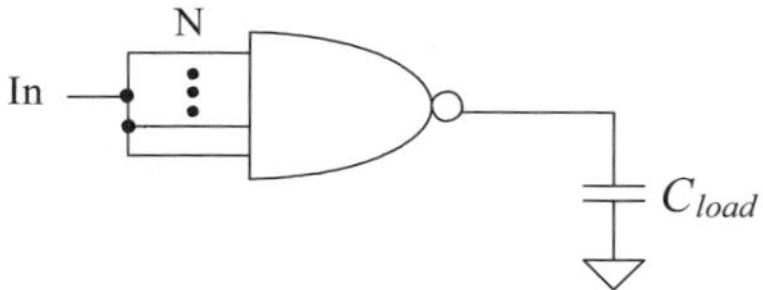

Figure 12.9 An n-input NAND gate driving a load capacitance.

capacitance is much greater than the output capacitance of the NAND gate, the low-to-high propagation time can be estimated by

$$t_{PLH} \approx 0.7 \cdot \frac{R_p}{N} \cdot C_{load} \tag{12.16}$$

The high-to-low propagation time, using Eq. (12.13), is given by

$$t_{PHL} = 0.7 \cdot N \cdot R_n \left[N \cdot C_{outp} + \frac{C_{outn}}{N} + C_{load} \right] + 0.35 \cdot R_n C_{oxn} \cdot N^2 \tag{12.17}$$

If C_{load} is much larger than the output capacitance of the NAND gate, then

$$t_{PHL} \approx 0.7 \cdot N \cdot R_n \cdot C_{load} \tag{12.18}$$

Example 12.3

Estimate the intrinsic propagation delays, $t_{PHL} + t_{PLH}$, of a three-input NAND gate made using 10/1 NMOS and 20/1 PMOS in the short-channel process. Estimate and simulate the delay when the gate is driving a load capacitance of 50 fF. Assume that the inputs are tied together.

Using the data in Table 10.2 and Eqs. (12.15) and (12.17),

$$t_{PLH} = 0.7 \cdot \frac{3.4k}{3} \cdot \left(3 \cdot 1.25\,fF + \frac{0.625\,fF}{3} + 50\,fF \right) = 43\,ps$$

and

$$t_{PHL} = 0.7 \cdot 3 \cdot 3.4k \cdot \left[3 \cdot 1.25\,fF + \frac{0.625\,fF}{3} + 50\,fF \right] + 0.35 \cdot 3.4k \cdot 0.625\,fF \cdot 9 = 393\,ps$$

The simulation results are seen in Fig. 12.10. ■

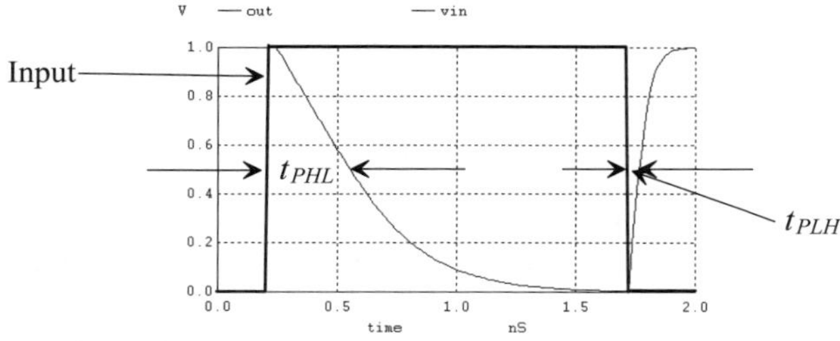

Figure 12.10 Simulating the operation of a 3-input NAND gate in 50 nm CMOS driving a 50 fF load capacitance.

Quick Estimate of Delays

The delay equations derived in this section are useful in understanding the limitations on the number of MOSFETs used in a NAND gate for high-speed design. Notice, in Ex. 12.3, how the load capacitance term dominates the gate's delay. A more useful, though not as precise, method of determining delays can be found by considering the fact that whenever the output changes from *VDD* to ground the discharge path is through *N* resistors of value R_n. This is true if all or only one of the inputs to the NAND gate changes, causing the output to change. Under these circumstances, Eq. (12.18) predicts the high-to-low delay-time, or for *series* connection of *N* NMOS devices as,

$$t_{PHL} \approx 0.7 \cdot N \cdot R_n \cdot C_{load} \tag{12.19}$$

The case when the output of the NAND gate changes from a low to a high is somewhat different then the high-to-low case. Referring to Fig. 12.7, we see that if one of the MOSFETs turns on, it can pull the output to *VDD* independent of the number of MOSFETs in parallel. Under these circumstances Eq. (12.16) can be used with $N = 1$ to predict the low-to-high delay-time

$$t_{PLH} \approx 0.7 \cdot R_p \cdot C_{load} \tag{12.20}$$

We will try to use Eqs. (12.19) and (12.20) as much as possible because of their simplicity. The further simplified digital models of MOSFETs are shown in Fig. 12.11. (Input capacitance is not shown.)

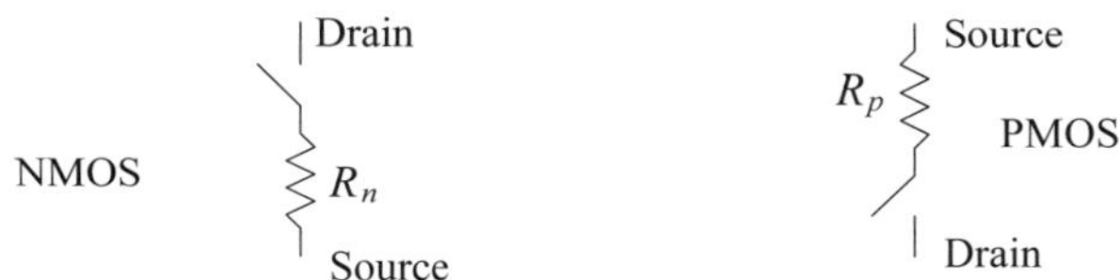

Figure 12.11 Further simplification of digital models not showing input capacitance.

Example 12.4
Repeat Ex. 12.3 using Eqs. (12.19) and (12.20) with only one input switching. Compare the results to simulations.

Using Eq. (12.19), we get

$$t_{PHL} = 0.7 \cdot 3 \cdot 3.4k \cdot 50\,fF = 357\,ps$$

and

$$t_{PLH} = 0.7 \cdot 3.4k \cdot 50\,fF = 119\,ps$$

Figure 12.12 shows the simulation results. The delay through the NMOS devices doesn't change much from what was calculated in Ex. 12.3 since the load capacitance dominates the delay. However, for the PMOS turning on, the delay is considerably longer since only a single PMOS device is switching pulling the output high. This situation (one input changing) is the more practical case. ■

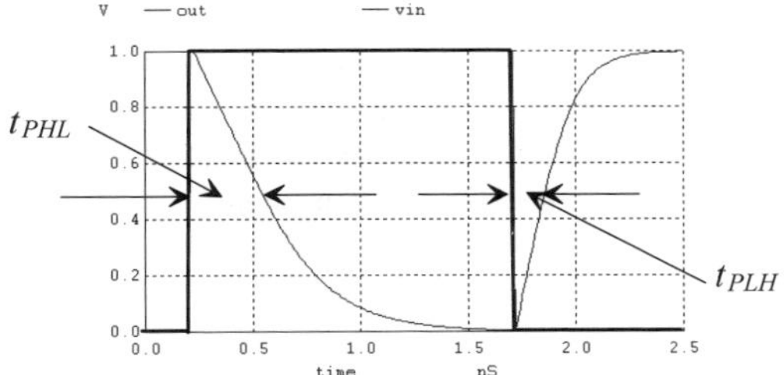

Figure 12.12 Switching delays in a 3-input NAND gate with only one changing states and driving a 50 fF load capacitance.

12.3.2 Number of Inputs

As the number of inputs, *N*, to a static NAND (or NOR) gate increases, the scheme shown in Fig. 12.2 (Fig. 12.4) becomes difficult to realize. Consider a NOR gate with 100 inputs. This gate requires PMOS devices in series and a total of 200 MOSFETs (2*N* MOSFETs). The delay associated with the series PMOS devices charging a load capacitance is too long for most practical situations.

Now consider the schematic of an *N* input NOR gate shown in Fig. 12.13, which uses $N + 1$ MOSFETs. If any input to the NOR gate is high, the output is pulled low through the corresponding NMOS device to a voltage, when designed properly, well below V_{THN}. If all inputs are low, then all NMOS are off and the PMOS pulls the output high (to *VDD*). A simple (long-channel) analysis of the output low voltage, V_{OL}, with *one* input at *VDD* yields

$$\frac{\beta_p}{2}(VDD - V_{THP})^2 = \beta_n\left[(VDD - V_{THN})V_{OL} - \frac{V_{OL}^2}{2}\right] \qquad (12.21)$$

The drawback of using this topology is the long pull-up time (t_{PLH}) resulting from the large output capacitance of the parallel NMOS (and the load capacitance) together with the large resistance of the long length pull-up device.

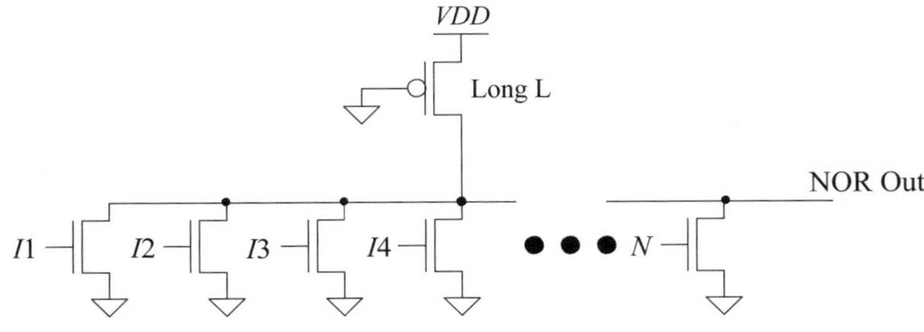

Figure 12.13 NOR configuration used for a large number of inputs.

12.4 Complex CMOS Logic Gates

Implementing complex logic functions in CMOS requires the basic building blocks shown in Fig. 12.14. We have already used the circuits to implement NAND and NOR gates. In general, any And-Or-Invert (AOI) logic function can be implemented using these techniques. A major benefit of AOI logic is that for a relatively complex logic function the delay can be significantly lower than a logic gate implementation. Consider the following example.

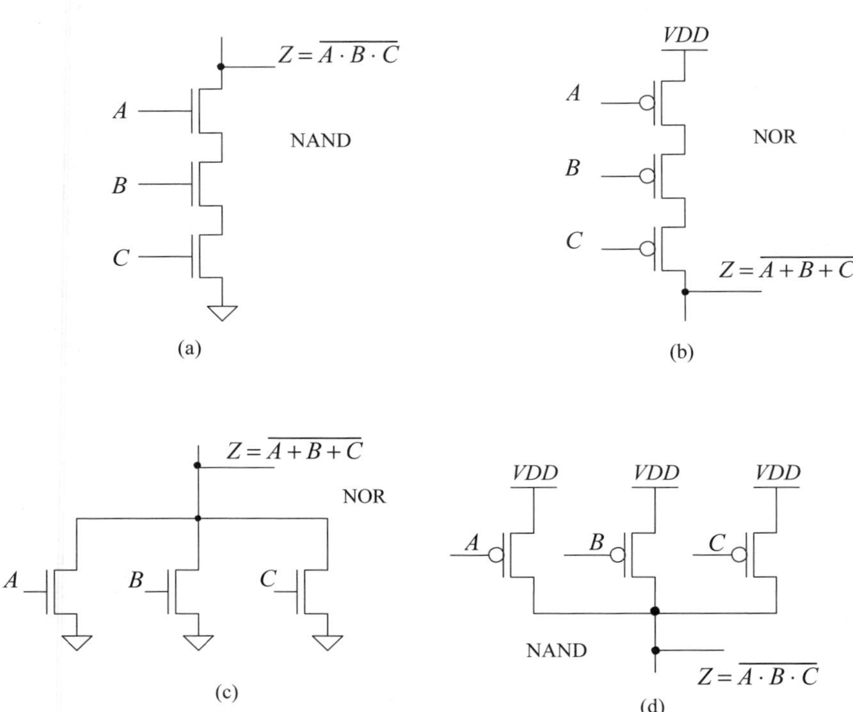

Figure 12.14 Logic implementation in CMOS.

Example 12.5
Using AOI logic, implement the following logic functions:

$$Z = \overline{A} + BC \qquad \text{and} \qquad Z = A + \overline{B}C + CD$$

The implementation of the first function is shown in Fig. 12.15a. Notice that the PMOS configuration is complementary with the NMOS configuration. The function we obtain is the complement of the desired function, and, therefore, an inverter is used to obtain *Z*. Using an inverter is, in general, undesirable if both

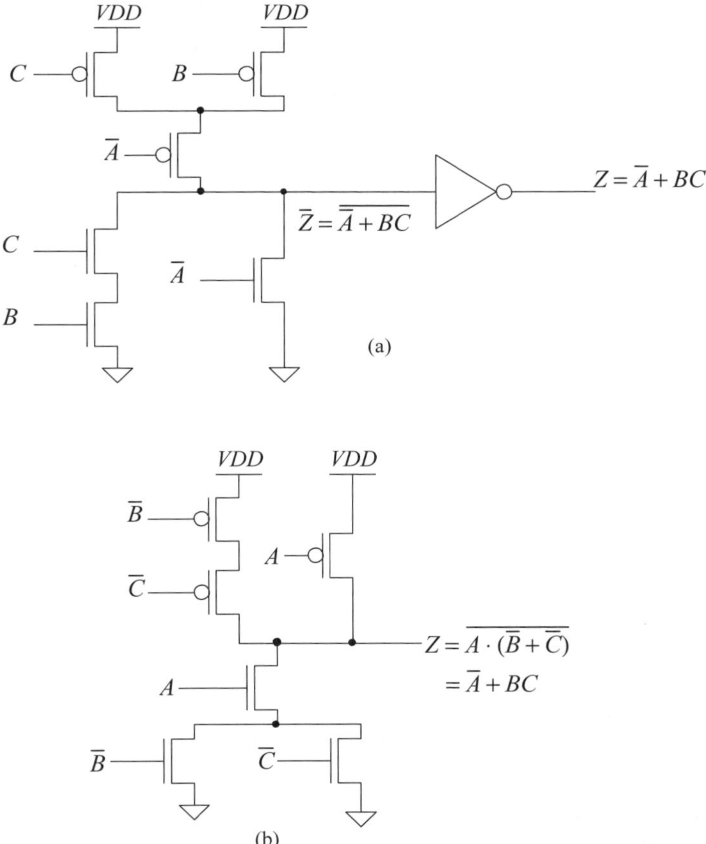

Figure 12.15 First logic gate of Ex. 12.5.

true and complements of the input variables are available. Applying Boolean algebra to the logic function, we obtain

$$Z = \overline{A} + BC \Rightarrow \overline{Z} = \overline{\overline{A} + BC} = A \cdot (\overline{B} + \overline{C}) \Rightarrow Z = \overline{A \cdot (\overline{B} + \overline{C})}$$

The AOI implementation of the result is shown in Fig. 12.15b. Logically, the circuits of Figs. 12.15a and b are equivalent. However, the circuit of Fig. 12.15b is simpler and thus more desirable. **Note** that to reduce the output capacitance and thus decrease the switching times, the parallel combination of NMOS devices is placed at the bottom of the logic block.

The second logic function is given by

$$Z = A + \overline{B}C + CD = A + C(\overline{B} + D) \Rightarrow \overline{Z} = \overline{A + C(\overline{B} + D)} = \overline{A} \cdot (\overline{C} + B\overline{D})$$

or

$$Z = \overline{\overline{A} \cdot (\overline{C} + B\overline{D})}$$

The logic implementation is given in Fig. 12.16. ■

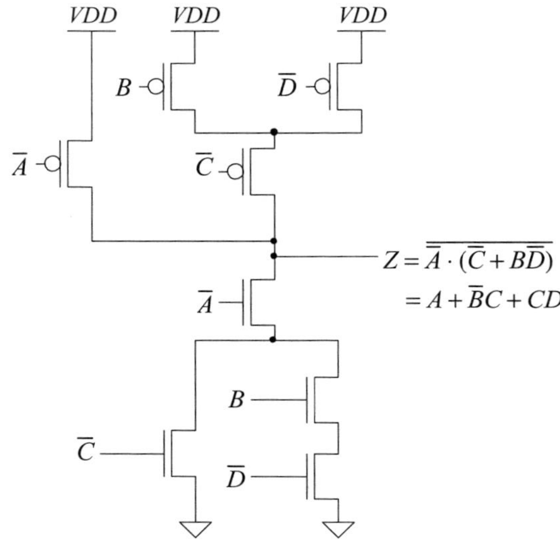

Figure 12.16 Second logic gate of Ex. 12.5.

Example 12.6
Using AOI logic, implement an exclusive OR gate (XOR).

The logic symbol and truth table for an XOR gate are shown in Fig. 12.17. From the truth table, the logic function for the XOR gate is given by

$$Z = A \oplus B = (A + B) \cdot (\overline{A} + \overline{B}) \tag{12.22}$$

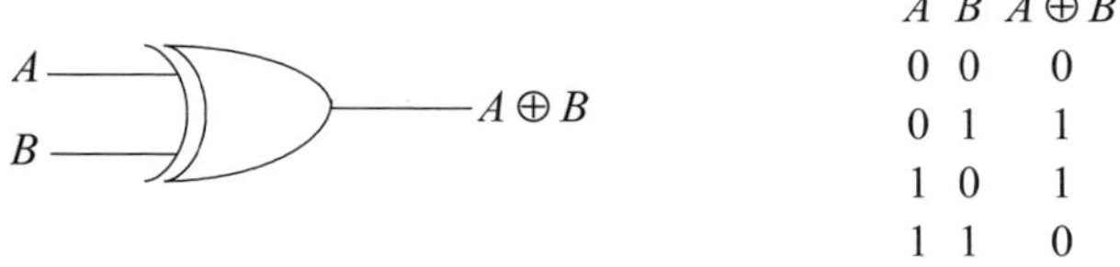

A	B	$A \oplus B$
0	0	0
0	1	1
1	0	1
1	1	0

Figure 12.17 Exclusive OR gate.

or

$$\overline{Z} = \overline{A \oplus B} = \overline{(A+B) \cdot (\overline{A}+\overline{B})} = \overline{A} \cdot \overline{B} + A \cdot B$$

and finally

$$Z = \overline{\overline{A} \cdot \overline{B} + A \cdot B} = A \oplus B \qquad (12.23)$$

The CMOS AOI implementation of an XOR gate is shown in Fig. 12.18. ■

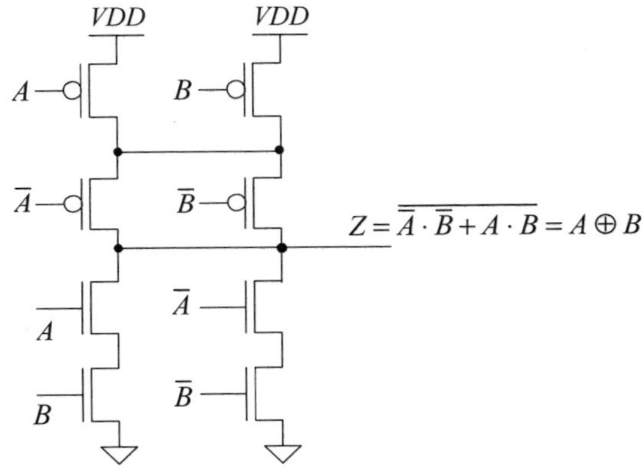

Figure 12.18 CMOS AOI XOR gate.

Example 12.7
Design a CMOS full adder using CMOS AOI logic.

The logic symbol and truth table for a full adder circuit are shown in Fig. 12.19. The logic functions for the sum and carry outputs can be written as

$$S_n = A_n \oplus B_n \oplus C_n$$

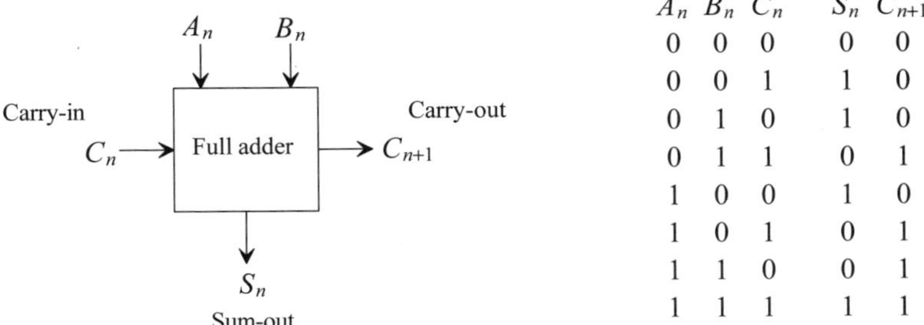

A_n	B_n	C_n	S_n	C_{n+1}
0	0	0	0	0
0	0	1	1	0
0	1	0	1	0
0	1	1	0	1
1	0	0	1	0
1	0	1	0	1
1	1	0	0	1
1	1	1	1	1

Figure 12.19 Full adder.

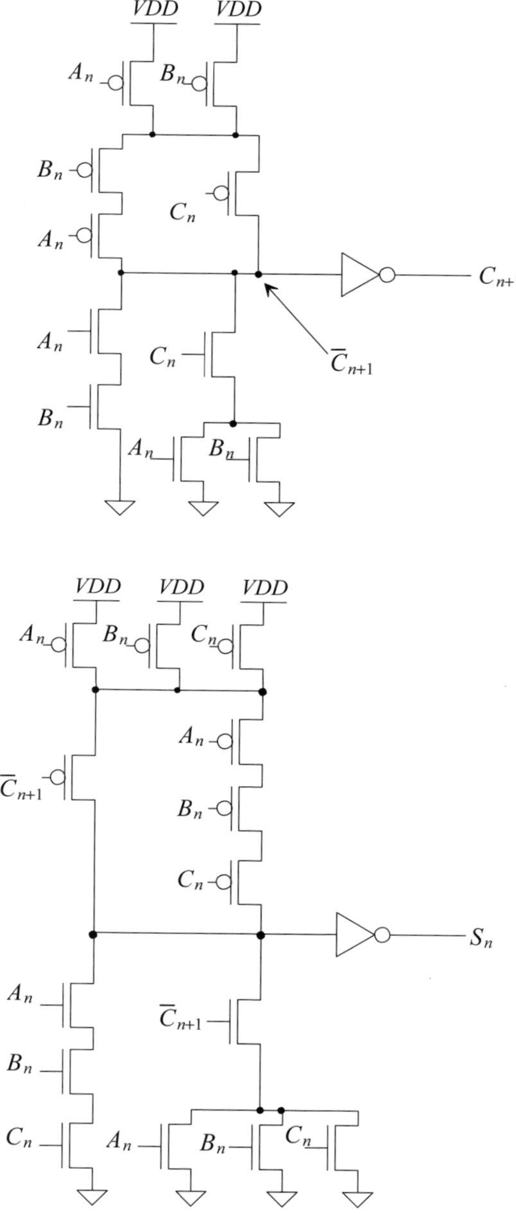

Figure 12.20 AOI implementation of a full adder.

and

$$C_{n+1} = A_n \cdot B_n + C_n(A_n + B_n)$$

The logic expression for the sum can be rewritten as a sum of products

$$S_n = \overline{A}_n\overline{B}_nC_n + \overline{A}_nB_n\overline{C}_n + A_n\overline{B}_n\overline{C}_n + A_nB_nC_n$$

or since

$$\overline{C}_{n+1} = \left(\overline{A}_n + \overline{B}_n\right) \cdot \left(\overline{C}_n + \overline{A}_n \cdot \overline{B}_n\right)$$

the sum of products can be rewritten as

$$S_n = (A_n + B_n + C_n)\overline{C}_{n+1} + A_nB_nC_n$$

The AOI implementation of the full adder is shown in Fig. 12.20. ■

Cascode Voltage Switch Logic

Cascode voltage switch logic (CVSL) or differential cascode voltage switch logic (DVSL) is a differential output logic that uses positive feedback in the load of the logic gate to speed up the switching times (in some cases). Figure 12.21 shows the basic idea. A PMOS gate cross-connected load is used instead of PMOS switches, as in the AOI logic, to pull the output high. Consider the implementation of $Z = \overline{A} + BC$. (This logic function was implemented in AOI in Fig. 12.15.) Figure 12.22 shows how NMOS devices can be used to implement Z and $\overline{Z}$. The concern with this implementation is the contention current. When one branch of the NMOS starts to turn on, a significant current can flow through the "on" PMOS device. The amount of current flowing in the conducting PMOS load device(s) can be reduced by increasing their lengths. However, this has the unwanted side effects of lengthening the delay times.

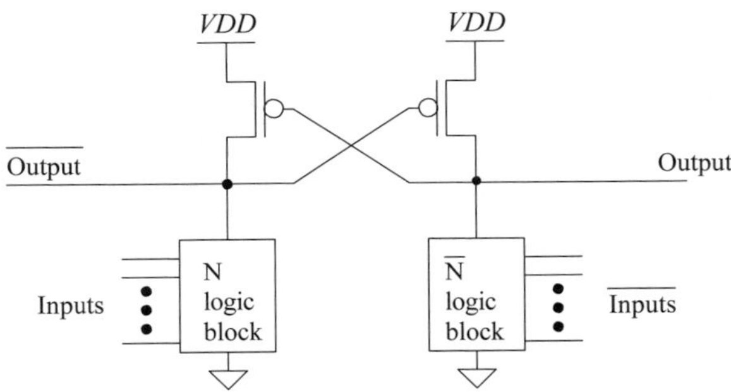

Figure 12.21 CVSL block diagram.

As another example, Fig. 12.23a shows the implementation of a CVSL two-input XOR/XNOR gate, while Fig. 12.23b shows a CVSL three-input XOR/XNOR gate useful in adder design.

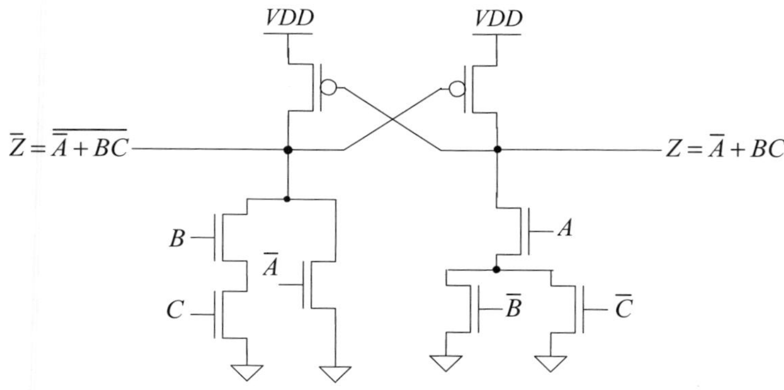

Figure 12.22 CVSL logic gate.

Differential Split-Level Logic

Differential split-level logic (DSL logic) is a scheme wherein the load is used to reduce output voltage swing and thus lower gate delays (at the cost of smaller noise margins). The basic idea is shown in Fig. 12.24. The reference voltage V_{ref} on the gates of M1 and M2 is set to $VDD/2 + V_{THN}$. The sources of M1 and M2 are then at a maximum voltage of *VDD*/2. This has the effect of limiting the output voltage swing to a maximum of *VDD* and a minimum of *VDD*/2. The main drawback of this logic implementation is the increased power dissipation resulting from the continuous power draw through the output leg at a voltage of *VDD*/2. The output leg at *VDD* draws no DC power.

Tri-State Outputs

A final example of a static logic gate, a tri-state buffer, is shown in Fig. 12.25. When the *Enable* input is high, the NAND and NOR gates invert and pass *A* (*VDD* or ground) to the gates of M1 and M2. Under these circumstances, M1 and M2 behave as an inverter. The combination of M1 and M2 with the inversion NAND/NOR gate causes the output to be the same polarity as *A*. When *Enable* is low, the gate of M1 is held at ground and the gate of M2 is held at *VDD*. This turns both M1 and M2 off. Under these circumstances, the output is said to be in the high-impedance or Hi-Z state. This circuit is preferable to the inverter circuits of Fig. 11.27 because only one switch is in series with the output to *VDD* or ground. An inverting buffer configuration is shown in Fig. 12.26.

Additional Examples

Additional delay calculations using static logic gates (and other CMOS circuit building blocks) can be found in Sec. 13.4.

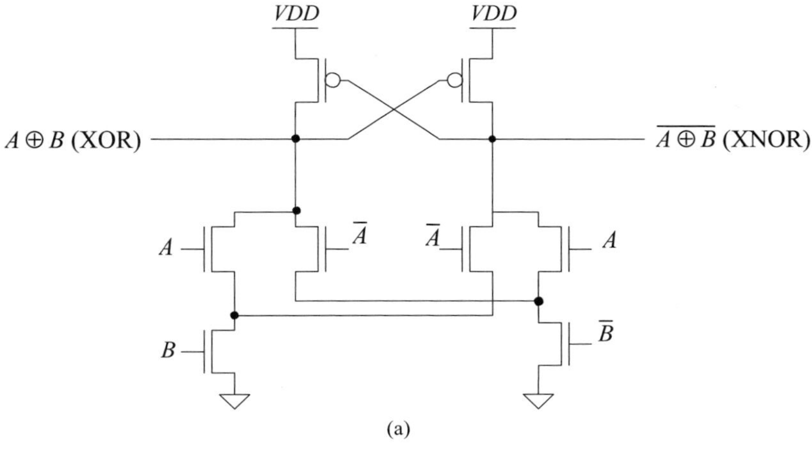

(a)

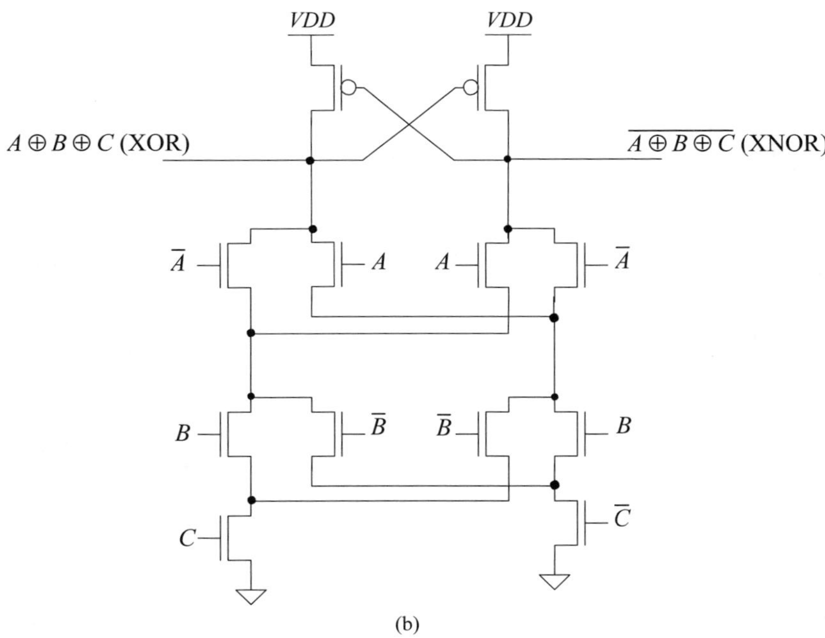

(b)

Figure 12.23 (a) Two-input and (b) three-input XOR/XNOR gates.

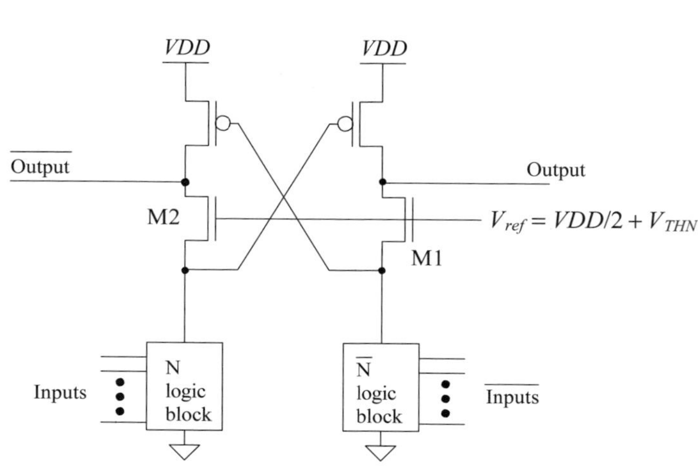

Figure 12.24 DSL block diagram.

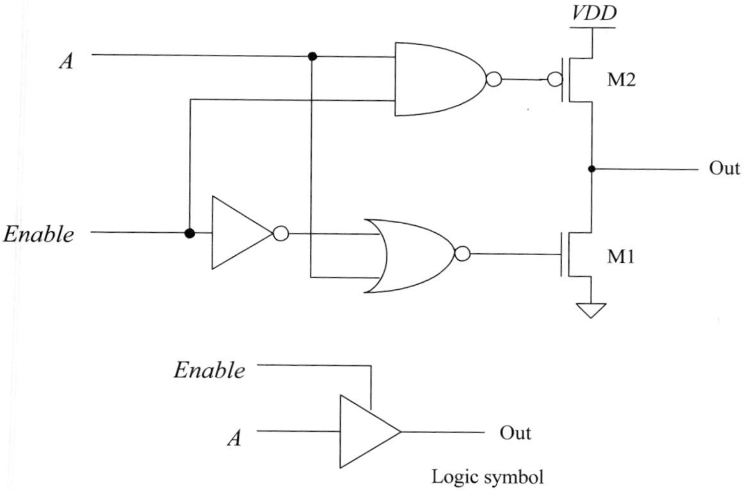

Figure 12.25 Tri-state buffer.

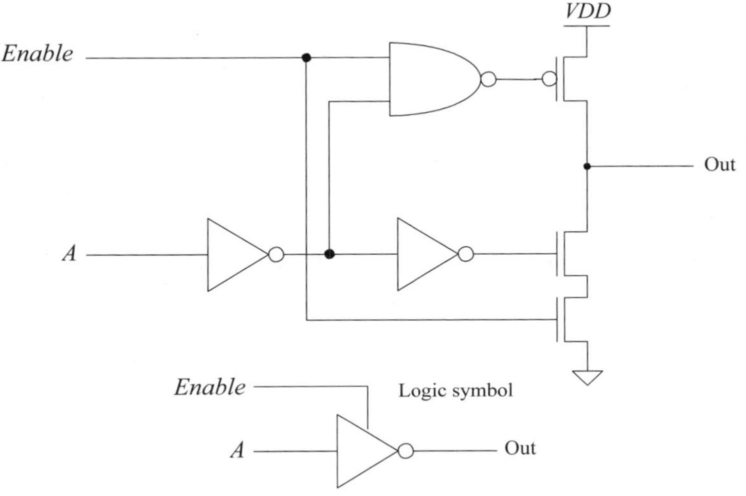

Figure 12.26 Tri-state inverting buffer.

ADDITIONAL READING

[1] M. I. Elmasry, *Digital MOS Integrated Circuits II,* IEEE Press, 1992. ISBN 0-87942-275-0, IEEE order number: PC0269-1.

[2] J. P. Uyemura, *Circuit Design for Digital CMOS VLSI,* Kluwer Academic Publishers, 1992.

[3] M. Shoji, *CMOS Digital Circuit Technology,* Prentice-Hall, 1988. ISBN 0-13-138850-9.

PROBLEMS

Use the 50 nm, short-channel process unless otherwise indicated.

12.1 Design, lay out, and simulate the operation of a CMOS AND gate with a V_{SP} of approximately 500 mV. Use the standard-cell frame discussed in Ch. 4 for the layout.

12.2 Design and simulate the operation of a CMOS AOI half adder circuit using static logic gates.

12.3 Repeat Ex. 12.3 for a three-input NOR gate. (Use the effective resistances to estimate the V_{SP}.)

12.4 Repeat Ex. 12.4 for a three-input NOR gate. (Use the effective resistances to estimate the V_{SP}.)

12.5 Sketch the schematic of an OR gate with 20 inputs. Comment on your design.

12.6 Sketch the schematic of a static logic gate that implements $\left(A+B\cdot\overline{C}\right)\cdot D$. Estimate the worst-case delay through the gate when driving a 50 fF load capacitance.

12.7 Design and simulate the operation of a CSVL OR gate made with minimum-size devices.

12.8 Design and simulate the operation of a tri-state buffer that has propagation delays under 5 ns when driving a 1 pF load. Assume that the maximum input capacitance of the buffer is 100 fF.

12.9 Sketch the schematic of a three-input XOR gate implemented in AOI logic.

12.10 The circuit shown in Fig. 12.27 is an edge detector. Discuss, and simulate, the operation of the circuit.

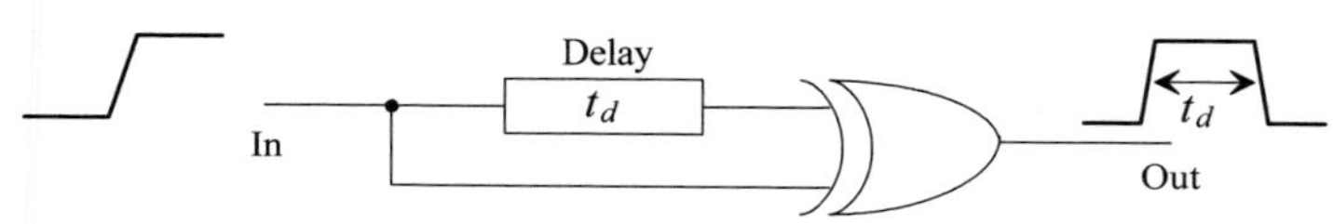

Figure 12.27 An edge detector circuit.

Chapter

13

Clocked Circuits

The transmission gate (TG) is used in digital CMOS circuit design to pass or not pass a signal, see Sec. 10.2.1. The schematic and logic symbol of the transmission gate (TG) are shown in Fig. 13.1. The gate is made up of the parallel connection of an NMOS and a PMOS device. Referring to the figure when S (for select) is high, we observe that the transmission gate passes the signal on the input to the output (noting the nodes we define as the input or output are interchangeable). The resistance between the input and the output can be estimated as $R_n \| R_p$. We begin this chapter with a description of the CMOS TG.

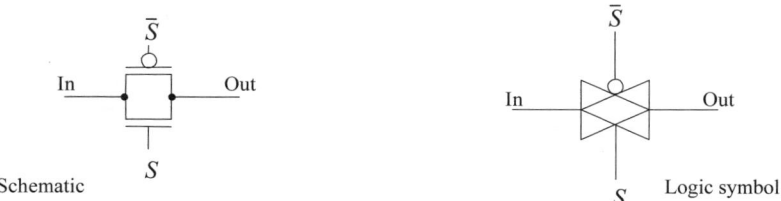

Figure 13.1 The transmission gate.

13.1 The CMOS TG

Since the NMOS pass gate (PG) passes logic lows well and the PMOS PG passes logic highs well, putting the two complementary MOSFETs in parallel, as seen in Fig. 13.1, produces a TG that passes both logic levels well. The propagation delay-times of the CMOS TG in the configuration seen in Fig. 13.2 (with a large load capacitance) are estimated as

$$t_{PHL} = t_{PLH} = 0.7 \cdot (R_n || R_p) \cdot C_{load} \tag{13.1}$$

The capacitance on the S input of the TG is the input capacitance of the NMOS device, or C_{inn} (= $1.5C_{oxn}$). The capacitance on the $\overline{S}$ input of the TG is the input capacitance of the PMOS device, or C_{inp}.

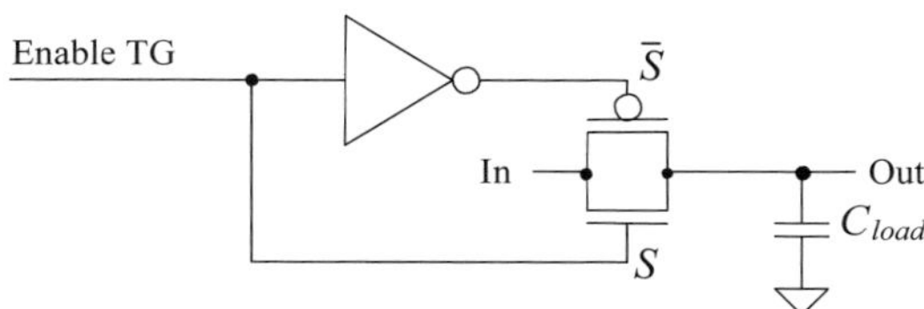

Figure 13.2 The transmission gate with control signals shown.

Increasing the widths of the MOSFETs used in the TG reduces the propagation delay-times from the input to the output of the TG when driving a specific load capacitance. However, the delay-times in turning the TG on, the select lines going high, increase because of the increase in input capacitance. This should be remembered when simulating. Using a voltage source in SPICE for the select lines, which can supply infinite current to charge the input capacitance of the TG, gives the designer a false sense that the delay through the TG is limited by R_n and R_p. Often, when simulating logic of any kind, the SPICE-generated control signals are sent through a chain of inverters so that the control signals more closely match what will actually control the logic on die (and the control signals have a finite source driving resistance).

Example 13.1

Estimate and simulate the delays through the TG circuit shown in Fig. 13.3 using the short-channel CMOS process and a load capacitance of 50 fF.

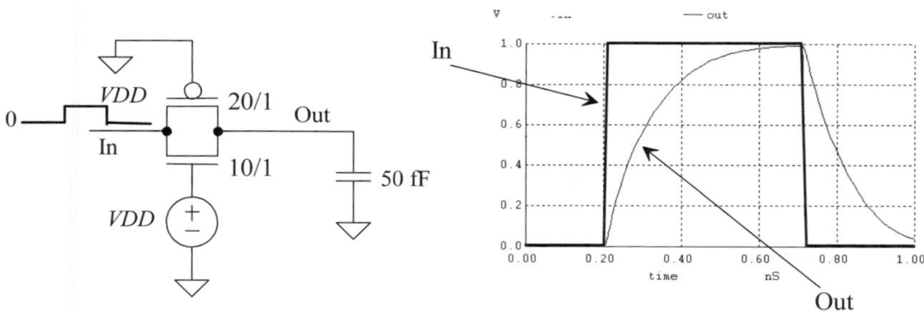

Figure 13.3 TG circuit discussed in Ex. 13.1.

From Table 10.2 $R_n||R_p$ = 1.7 kΩ so that using Eq. (13.1) we get $t_{PLH} = t_{PHL} = 60$ ps. The SPICE simulation results are also seen in Fig. 13.3. ■

Example 13.2

Repeat Ex. 13.1 for the circuit seen in Fig. 13.4

The output load is initially at *VDD* and then discharged to ground when the TG turns on. The delay time calculation is exactly the same as the one used in Ex. 13.1. Also seen in the figure are the simulation results. ■

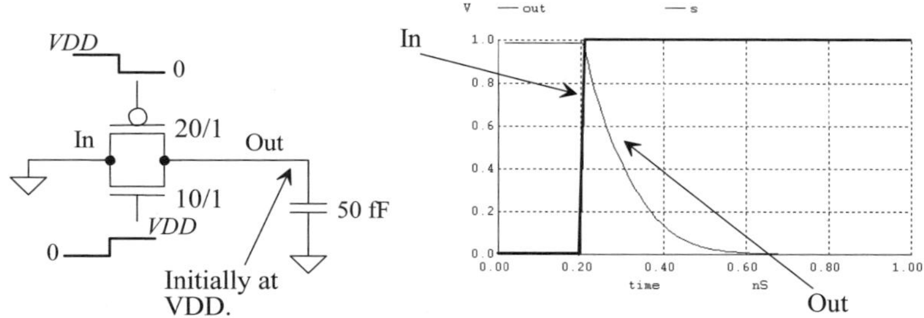

Figure 13.4 TG circuit discussed in Ex. 13.2.

Series Connection of TGs

Consider the series connection of the CMOS transmission gates shown in Fig. 13.5. The equivalent digital model is also depicted in this figure. As seen in Figs. 10.17, 10.21, and the associated discussion, the capacitance on each MOSFET's source/drain (assuming triode operation) is $C_{ox}/2$. The total capacitance on each internal node in a series connection of TGs is then the sum of the oxide capacitances from each adjacent PG, that is, $C_{oxn} + C_{oxp}$. The delay through the series connection of TGs can be estimated using

$$t_{PHL} = t_{PLH} = 0.7 \cdot N \cdot (R_n || R_p)(C_{load}) + 0.35 \cdot (R_n || R_p)(C_{oxn} + C_{oxp})(N)^2 \tag{13.2}$$

The first term in this equation is simply the time needed to charge C_{load} through the sum of the TG effective resistances, while the second term in the equation describes the RC transmission line effects.

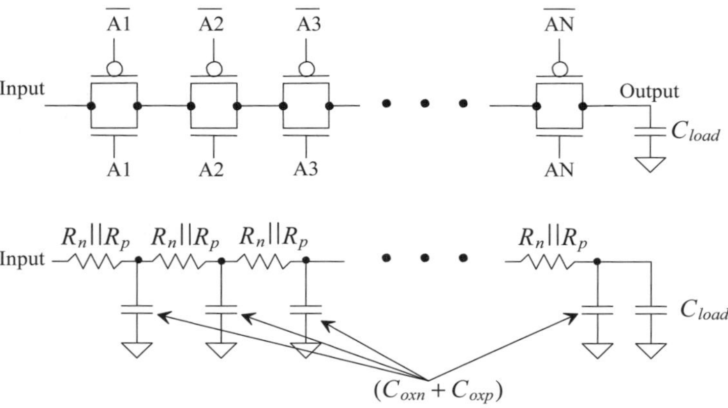

Figure 13.5 Series connection of TGs with digital model.

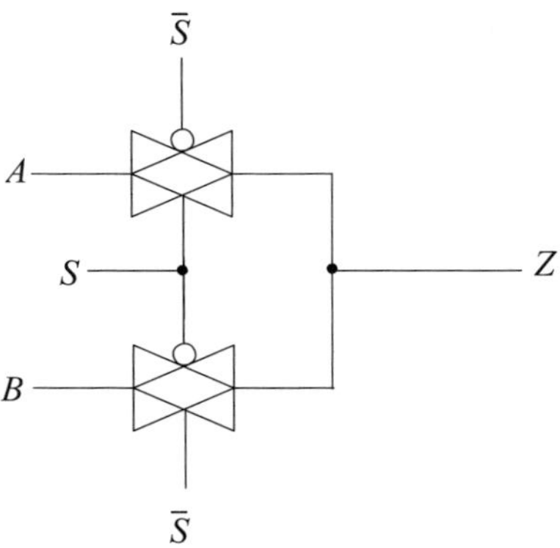

Figure 13.6 Path selector.

13.2 Applications of the Transmission Gate

In this section we present some of the applications of the TG.

Path Selector

The circuit shown in Fig. 13.6 is a two-input path selector. Logically, the output of the circuit can be written as

$$Z = AS + B\overline{S} \tag{13.3}$$

When the selector signal S is high, A is passed to the output while a low on S passes B to the output.

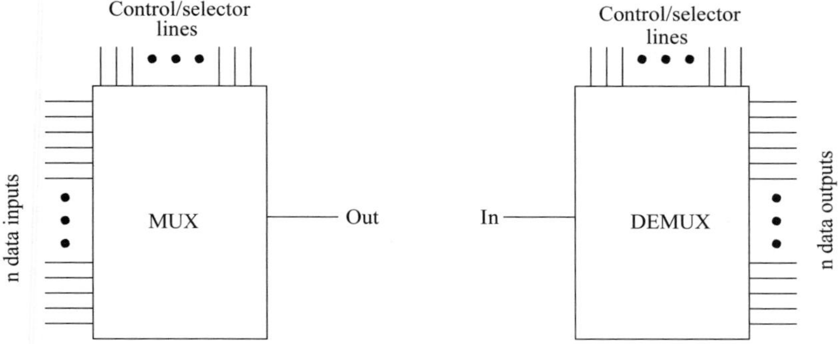

Figure 13.7 Block diagram of MUX/DEMUX.

This same idea can be used to implement multiplexers/demultiplexers (MUX/DEMUX). Consider the block diagrams of a MUX and DEMUX shown in Fig. 13.7. The number of control lines is related to the number of input lines by

$$2^m = n \tag{13.4}$$

where n is the number of inputs (outputs) to the MUX (DEMUX) and m is the number of control lines. A 4-to-1 MUX/DEMUX is shown in Fig. 13.8. Note that the MUX is bi-directional; that is, it can be used as a MUX or a DEMUX. The logic equation describing the operation of the MUX is given by

$$Z = A(S1 \cdot S2) + B(S1 \cdot \overline{S2}) + C(\overline{S1} \cdot S2) + D(\overline{S1} \cdot \overline{S2}) \tag{13.5}$$

Figure 13.9 shows the transistor-level implementation of the circuit in Fig. 13.8. Notice how the PMOS are grouped together (and laid out in the same n-well) while the NMOS are grouped together.

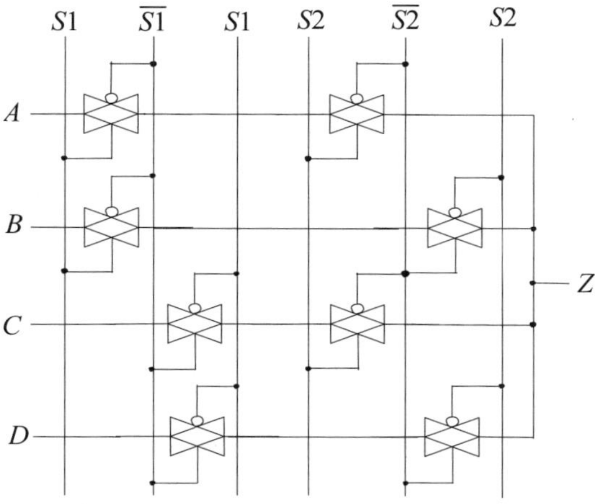

Figure 13.8 Circuit implementation of a 4-to-1 MUX.

Figure 13.10a shows an NMOS PG implementation of the 4-to-1 MUX. The pass transistor implementation is simpler, using fewer transistors, at the price of a threshold voltage drop from input to output when the input is a high (*VDD*). A simplified version of the circuit of Fig. 13.10a is shown in Fig. 13.10b. Here the MOSFETs connected to *S2* and $\overline{S2}$ are combined to reduce the total number of MOSFETs used. The reduction of the total number of MOSFETs used can be extended to an n-input (output) MUX (DEMUX) (see Fig. 16.43). Again, it should be remembered that a DEMUX can be formed using the circuits of Figs. 13.8 or 13.10 by switching the inputs with the outputs.

Static Gates

The TG can be used to form static logic gates. Consider the OR gate shown in Fig. 13.11. To understand the operation of the gate, consider the case when both *A* and *B* are low. Under these circumstances the pass transistor, M1 is off, and the TG is on. Since the input *B* is low, a low is passed to the output. If *A* is high, M1 is on and *A* is passed to the output. If *B* is high and *A* is low, *B* is passed to the output through the TG. If both *A* and *B* are high, the TG is off and M1 is on passing *A*, a high, to the output.

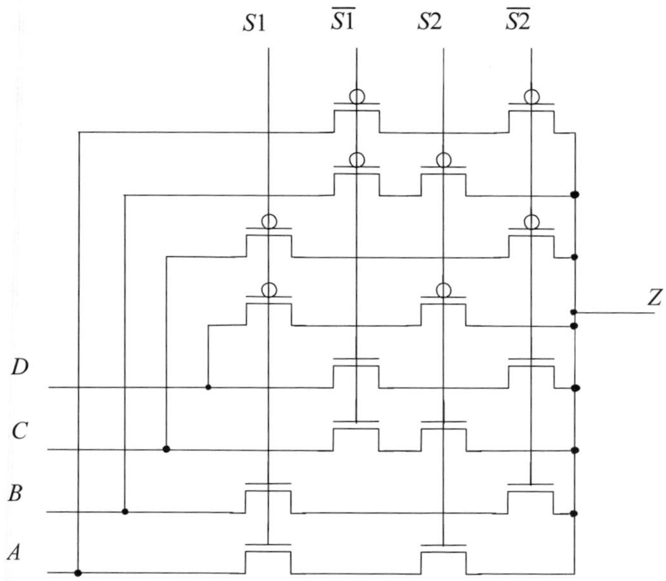

Figuer 13.9 Transistor implementation of Fig. 13.8.

Figure 13.12 shows an XOR and an XNOR gate made using TGs. Consider the XOR gate with both A and B low. Under these circumstances, the top TG is on and its output is connected to A, a low. If both inputs are high, the bottom TG connects the output to $\overline{A}$, again a low. If A is high and B is low, the top TG is on and the output is connected to A, a high. Similarly, if A is low and B is high, the bottom TG is on and connects the output to $\overline{A}$, a high.

13.3 Latches and Flip-Flops

Basic Latches

Consider the set-reset latch (SR latch) shown in Fig. 13.13 made using NAND gates. The logic symbol and truth table are also shown in this figure. Consider the case when S is high and R is low. Forcing R low causes Q to go high. Since S is high and $Q$is high, the $\overline{Q}$output is low. Now consider the case when both S and R are low. Under these circumstances, the latch's outputs are both high. This latch can easily be designed and laid out with the techniques of Ch. 12 (see also Fig. 15.6).

An alternative implementation of the SR latch is shown in Fig. 13.14 using NOR gates. Consider the case when S is high and R is low. For the NOR gate, a high input forces the output of the gate low. Therefore, the $\overline{Q}$ output is low whenever the S input is high. Similarly, whenever the R input is high, the Q output must be low. The case of both inputs being high causes both Q and $\overline{Q}$ to go low, or in other words the outputs of the latch are no longer complements.

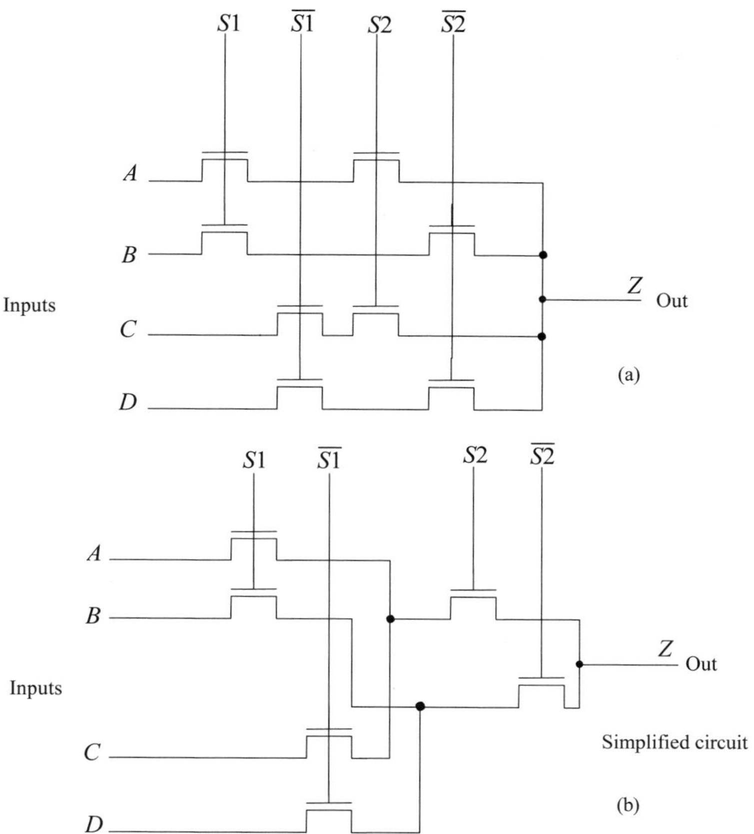

Figure 13.10 MUX/DEMUX using pass transistors.

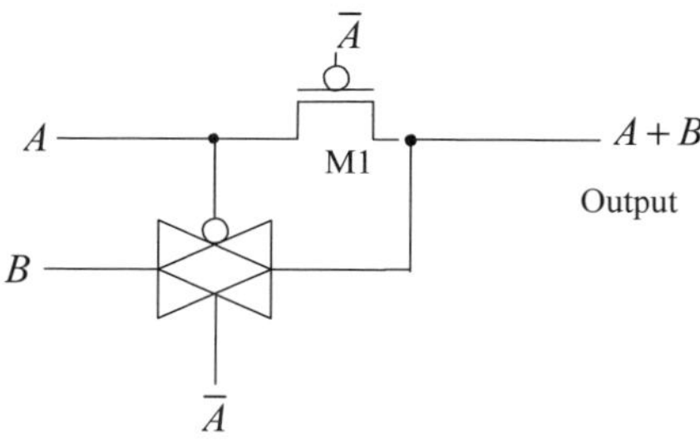

Figure 13.11 TG-based OR gate.

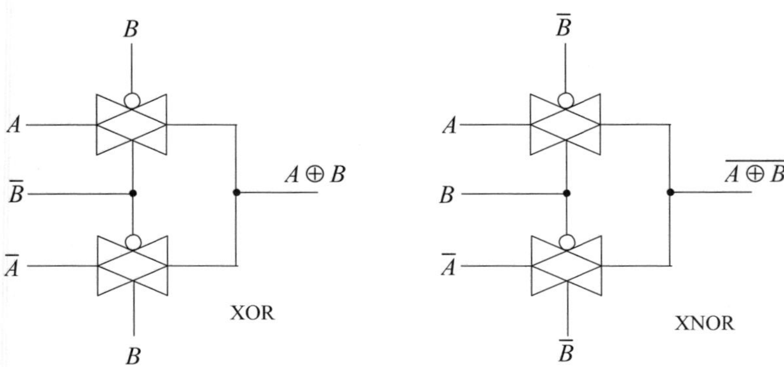

Figure 13.12 TG implementation of XOR/XNOR gate.

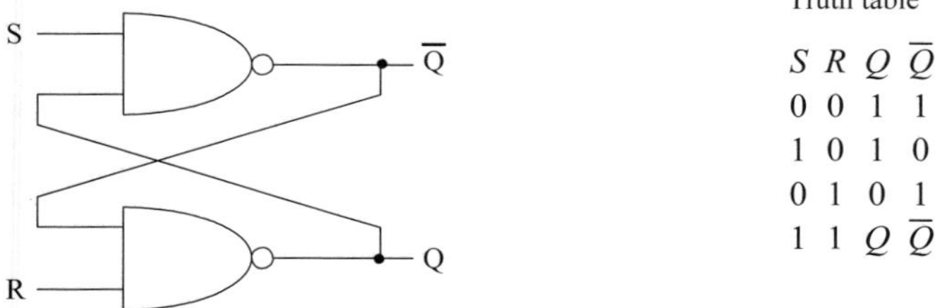

Truth table

S	R	Q	$\overline{Q}$
0	0	1	1
1	0	1	0
0	1	0	1
1	1	Q	$\overline{Q}$

Figure 13.13 Set-reset latch made using NAND gates.

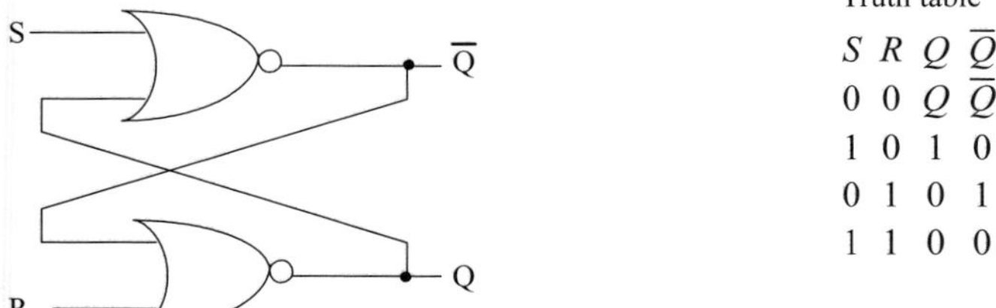

Truth table

S	R	Q	$\overline{Q}$
0	0	Q	$\overline{Q}$
1	0	1	0
0	1	0	1
1	1	0	0

Figure 13.14 Set-reset latch made using NOR gates.

An Arbiter

One of the uses of the NAND latch is in an arbitration circuit (called an arbiter, Fig. 13.15). This circuit can be very useful in asynchronous circuit design or in clock synchronization circuits (see Ch. 19). To understand the operation of the arbiter in Fig. 13.15, let's consider the case when both In1 and In2 are low. Both NAND gates' outputs are high and the two inverters' outputs, that is Out1 and Out2, are low. When In1 goes high, the output of X1 goes low, while the output of X2 remains high. This causes Out1 to go high and Out2 to remain low. Notice that with the output of X1 going low the power supplied to the PMOS in the inverter connected to Out2 is removed (making it impossible for Out2 to go high). When In2 goes high, while In1 is already high, the fact that the output of X1 is a low keeps the output of X2 high and Out2 low. When In1 goes low, with In2 high, Out1 goes low and Out2 goes high. The usefulness of this circuit occurs when In1 and In2 transition high at nearly the same times. The way the inverters are powered by the NAND gates makes it impossible for both Out1 and Out2 to go high at the same time. *Thus an arbiter is useful for determining which input arrives first.*

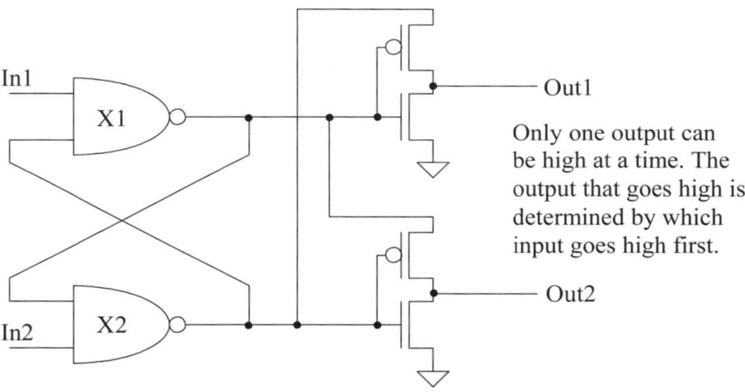

Figure 13.15 An arbiter made using NAND gates.

To ensure that the arbiter makes the decision quickly (there isn't a long delay or *metastability* when the two inputs are arriving at the same time), two inverters can be added in series with the NAND gate outputs. This addition of inverters simply increases the gain of the NAND gates (their VTCs get steeper at the switching point, see Fig. 12.3 in the last chapter).

Flip-Flops and Flow-through Latches

A flow-through latch is a storage circuit whose output changes with a clock signal level (so a flow-through latch is often called a level-sensitive latch). A flip-flop (FF) is a storage circuit that changes states on the rising or falling edge of a clock signal.

Most FFs in CMOS IC design are based on the cross-coupled inverters seen in Fig. 13.16. Also seen in this figure are the VTCs for the two inverters. I1's characteristics are drawn with its input on the x-axis and its output on the y-axis (the normal curves as

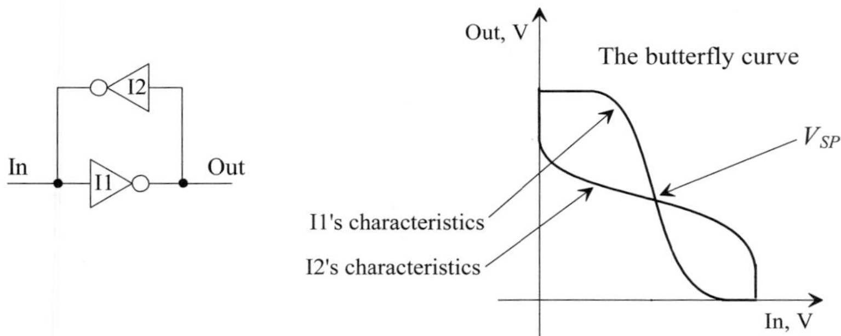

Figure 13.16 A cross-coupled inverter (a latch) and the input output characteristics.

seen in Fig. 11.2 back in Ch. 11). I2's input is labeled Out in the figure while its output is labeled In (and so its input is drawn as the y-axis and its output is drawn as the x-axis). The big concern when designing a FF with cross-coupled inverters is *metastability* (meaning changing stability). When the input signal moves to the switching-point of the inverters, both In and Out will be at V_{SP} (a stable state). This isn't where we want the circuit to operate because V_{SP} isn't a valid logic level voltage. Driving the input higher or lower enables the positive feedback present in the cross-coupled inverters to drive the In/Out nodes to valid logic levels (the other stable states). To avoid metastability, we can 1) use longer length inverters (which increases the gains of the inverters at the cost of longer delays), 2) use two inverters in series with the outputs of each inverter in Fig. 13.16 (again, to boost the gain) so that three inverters are used between In and Out, 3) ensure the In signal is driven with good logic levels (and a low impedance source to overdrive the output of I2), or 4) add a switch on the output of I2 to disconnect it when the In signal is changing to ensure that the input signal doesn't have to "fight" with the output of I2 to drive the signal to a valid logic level (the input signal range must still swing to valid logic levels). Figure 13.17 shows how metastability can result in a long delay before the outputs move to valid logic states. In this simulation the two capacitors are charged to 500 mV. It takes approximately 2.5 ns for the In/Out signals to change states.

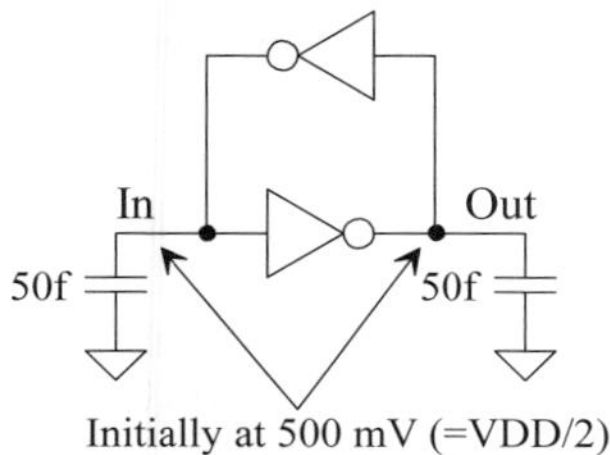

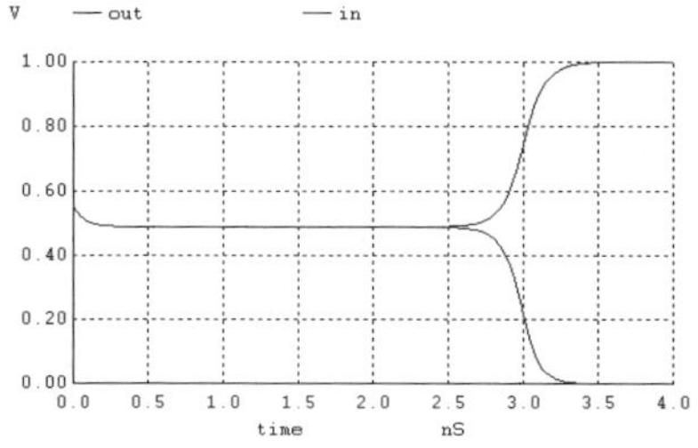

Figure 13.17 The delay associated with metastability.

Note that trying to move the switching point voltages of each inverter to differing levels makes the problem of metastability worse. When each inverter's V_{SP} is the same the gain around the loop is the highest (and so it's possible noise alone can cause the inverters to snap to a valid logic state).

Figure 13.18 shows a level-sensitive latch. When the clock signal is high, the data or D input passes through the NMOS PG and drives the cross-coupled inverters. The feedback inverter's lengths are made long to ensure that the source driving the D input can "overpower" the feedback inverter and allow the output to switch states. This circuit, while simple, has several fundamental issues. To begin, we know that an NMOS pass gate's output will swing from 0 to $VDD - V_{THN}$. To optimize the noise margins, we lower the V_{SP} of I1. Here we reduce the width of the PMOS from 20 to 10. Another concern is the contention current that flows when the D input is fighting with the output of I2 for control of I1's input. By increasing the lengths in I2 to 10, we lower this wasted current.

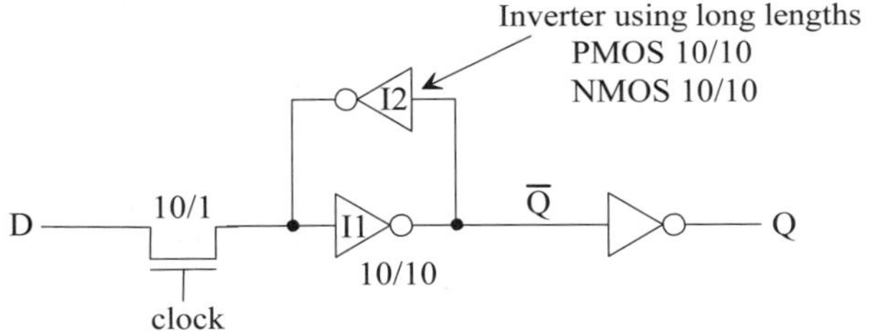

Figure 13.18 A level-sensitive latch.

Figure 13.19 shows the input and output simulation results for the latch in Fig. 13.18. When the clock signal is high, the input flows to the output (with a delay). When the clock signal goes low, the value on the D input is captured and remains on the Q output until the clock signal goes back high again. Notice that at 8 ns, when the clock signal goes high, the delay between the D input going high and the Q output going high is considerable.

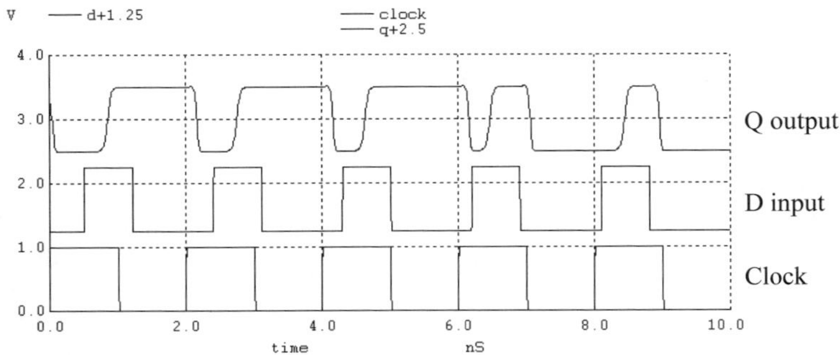

Figure 13.19 Simulating the level-sensitive latch in Fig. 13.18.

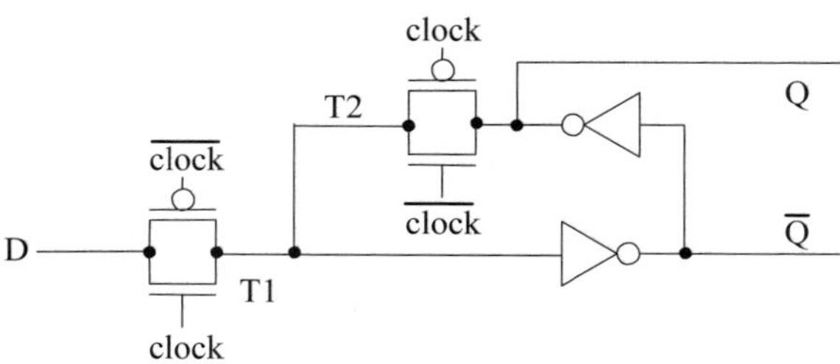

Figure 13.20 A higher performance level-sensitive latch.

Figure 13.20 shows a modification of the latch in Fig. 13.18. A TG is used on the input of the latch instead of a PG. This improves the noise performance of the circuit (at the cost of an additional clock signal, that is, the complement of clock). We've also added a TG in series with the feedback inverter. When T1 is on, the Q output follows the D input. When T1 is on, T2 is off so that the input doesn't have to fight with the feedback inverter. When T1 turns off, T2 turns on and the value of D at that instance is stored in the inverters. Figure 13.21 shows the simulation results using this latch with the input signals used to generate Fig. 13.19. Notice, for example, the improvement in performance over what's seen in Fig. 13.19 when the clock transitions high at 8 ns.

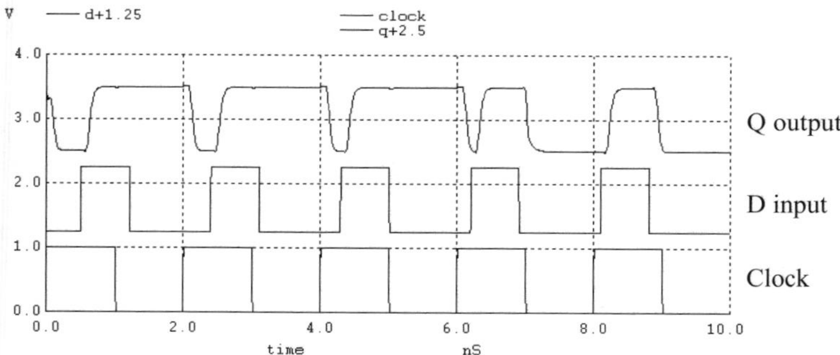

Figure 13.21 Simulating the level-sensitive latch in Fig. 13.20.

An Edge-Triggered D-FF

Notice that the value of the D input in the circuit seen in Fig. 13.20 is stored or captured when clock goes low. To implement an edge-triggered FF, we can use two of these level-sensitive latches in cascade. When the clock is low, the first stage tracks the D input and the second stage holds the previous output. When clock goes high, the first stage captures the input and transfers it to the second stage. The first stage is often called the "master" latch, while the second stage is the "slave" latch. Figure 13.22 shows an implementation of an edge-triggered D-FF. When clock is low, T1 and T4 are on. T2 and T3 are off. The D input flows through to point A and its complement to point B. When clock goes high, T1 and T4 shut off while T2 and T3 turn on. This causes the value of the D input, when

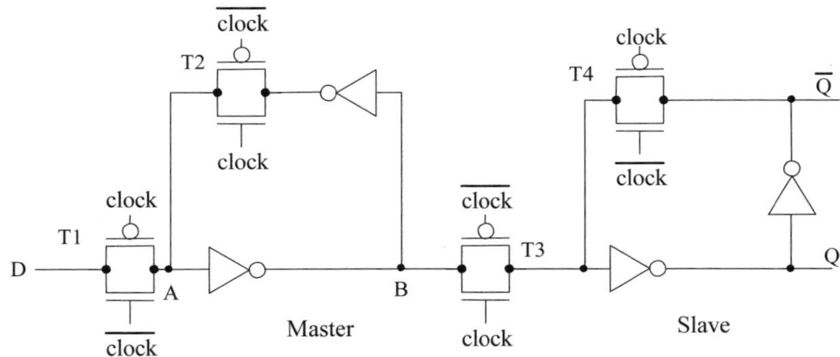

Figure 13.22 An edge-triggered D-FF.

clock transitioned high, to be captured and passed to the Q output. When clock goes back low, T1 and T4 turn on. The value of D on the Q output is then circulated around the two inverters. The Q output can change states again when clock goes back high. Figure 13.23 shows results of simulating this FF with SPICE. Note the output only changes on the rising edge of the clock signal.

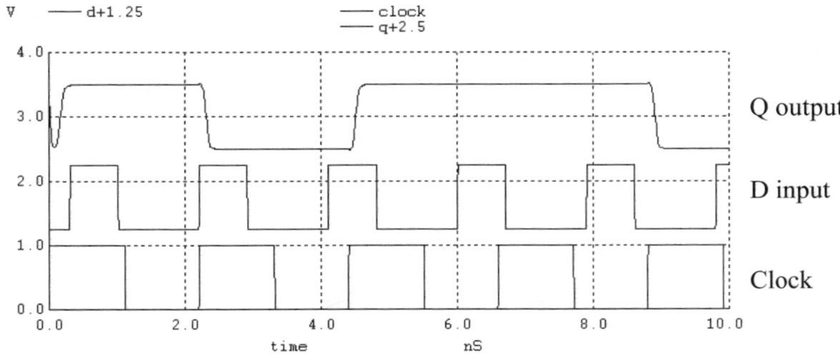

Figure 13.23 Simulating the operation of the edge-triggered FF in Fig. 13.22.

Figure 13.24 shows the addition of NAND gates to the circuit of Fig. 13.22 for clear and set inputs. When clear goes low, the outputs of N1 and N4 go high. If T3 is on or T4 is on, the outputs of the FF are forced to Q = 0 (and thus its complement is set to a 1). If the set input goes low, then Q is forced to a 1. If both clear and set are pulled low, then both the Q output and its complement are driven high. This feature can useful in some situations.

Note that it is **very important** for the clock signals we use to have quick rise times. If they don't, then the FF may not function correctly. Often the clock signals are applied to a string of inverters both to sharpen their transition times and to generate the complement clock required for the TGs.

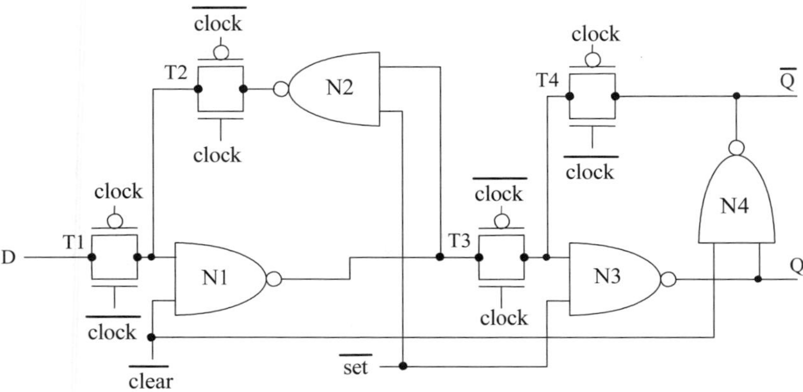

Figure 13.24 An edge-triggered FF with asynchrounous set and clear.

Flip-Flop Timing

The data must be *set up* or present on the D input of the FF (see Fig. 13.22) a certain time before we apply the clock signal. This time is defined as the setup time of the FF. To understand the origin of this time, consider the time it takes the signal at D to propagate through T1 and the inverter to node B. Before the clock pulse can be applied, the logic level D must be settled on node B. Consider the waveforms of Fig. 13.25. The time between D going high (or low) and the clock rising edge is termed the setup time of the FF and is labeled, t_s.

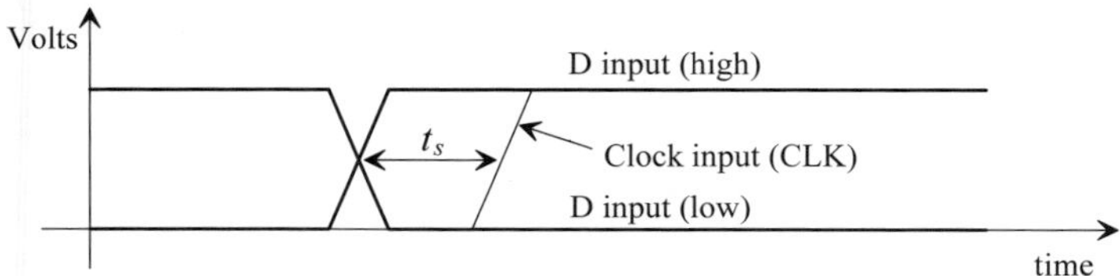

Figure 13.25 Illustrating D FF setup time.

The wanted D input must be applied t_s before the clock pulse is applied. Now the question becomes "How long does the wanted D input have to remain applied to, or *held* on, the input of the FF after the clock pulse is applied?" This time is called the hold time, t_h, and is illustrated in Fig. 13.26. Shown in this figure, t_h is a positive number. However, inspection of Fig. 13.22 shows that if the D input is removed slightly before the clock pulse is applied, node B will remain unchanged because of the propagation delay from the D input to node B. Analysis of this FF would yield a negative hold time. In other words, for the point B to charge to D, a time labeled t_s is needed. Once point B is charged, the D input can be changed as long as the clock signal occurs within t_h.

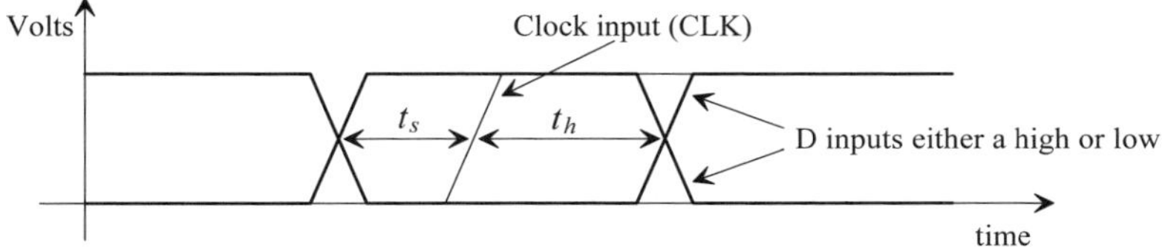

Figure 13.26 Illustrating D FF hold time.

One final important comment regarding the clock input of an FF is (again) in order. If the clock input risetime is slow, the FF will not function properly. There will not be an abrupt transition between the sets of transmission gates turning on and off. The result will be logic levels at indeterminate states. What is usually done to eliminate this problem is to buffer the clock input through several inverters. This has the effect of speeding up the leading and trailing edges of a slow input pulse and presenting a lower input capacitance on the clock input to whatever is driving the FF. The main disadvantage is the increase in delay times, t_{PHL} and t_{PLH} (defined by clock to output), of the FF. In general, the FF of Fig 13.22 should not be laid out without buffering the clock inputs.

The minimum pulse width of the clock, set, or clear inputs in Figs. 13.22 and 13.24 is labeled t_w. The minimum width is determined by the delay through (referring to Fig. 13.20) two NAND gates and a TG. The last timing definition we will comment on here is the recovery time, that is, the time between removing the set or clear inputs and a valid clock input. This variable is labeled t_{rec}.

13.4 Examples

In this section we give examples of delay calculations and comments on how to decrease the delays and the associated performance costs. Throughout this section we use the 50 nm short-channel process.

Example 13.3
Estimate the input capacitance and the delay through the circuit seen in Fig. 13.27. Compare the hand calculation estimates to simulation results.

Using the information seen in Table 10.2, we can calculate the first inverter's input capacitance, and thus the input capacitance of the circuit,

$$C_{in1} = \frac{3}{2} \cdot (C_{oxn1} + C_{oxp1}) = 0.938\,fF + 1.875\,fF = 2.81\,fF$$

Since the widths of the second inverter are 10 times larger than the widths of the first inverter, the input capacitance of the second inverter is 28.1 fF.

The delay through the first inverter, that is, the delay from the node labeled "In" to the node labeled "N1" in Fig. 13.27 is calculated as

$$t_{PHL1} + t_{PLH1} = 0.7 \cdot (3.4k + 3.4k) \cdot (0.625 + 1.25 + 9.38 + 18.75)\,fF = 143\,ps$$

noting, because $t_{PHL} = t_{PLH}$, that $t_{PHL1} = t_{PLH1} = 71.5\,ps$.

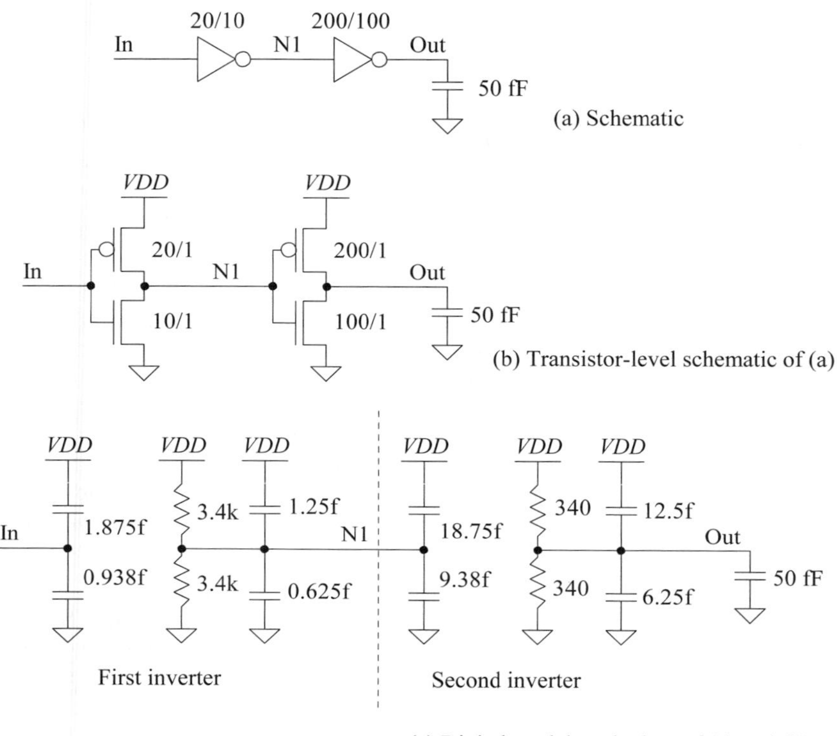

Figure 13.27 Circuit used in Ex. 13.3.

The delay through the second stage, that is, from N1 to the output is calculated as

$$t_{PHL2} + t_{PLH2} = 0.7 \cdot (340 + 340) \cdot 68.75\, fF = 33\ ps$$

Again as seen in the first inverter, the effective switching resistance of the PMOS is equal to the NMOS device's effective resistance so $t_{PHL2} = t_{PLH2} = 16.5\ ps$.

The overall delay through both inverters is

$$t_{PHL} = t_{PLH} = 88\ ps$$

Figure 13.28 shows the SPICE simulation results. The delays, from the simulations, are approximately 120 ps.

To decrease the delay through the circuit, the simplest solution is to increase the widths of the first inverter. Of course, this increases the circuit's input capacitance. Note that the *analysis* of this circuit has nothing to do with the *design* equations used for implementing a buffer, Sec. 11.4. ■

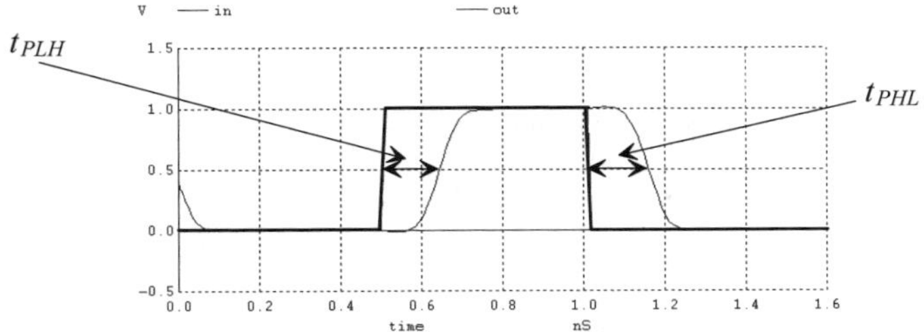

Figure 13.28 Simulating the circuit in Ex. 13.3 and Fig. 13.27.

Example 13.4

In Ex. 13.3 the devices were sized to give (ideally) equal propagation delays. Repeat Ex. 13.3 for the circuit seen in Fig. 13.29.

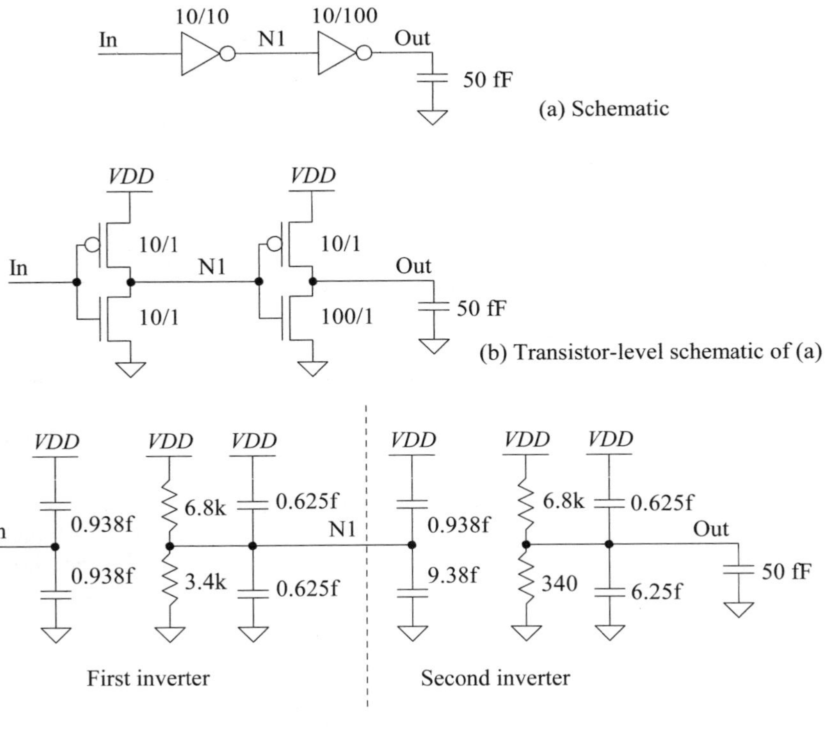

Figure 13.29 Circuit used in Ex. 13.4.

The PMOS device has a width of 10 which is half of the width used for the data in Table 10.2. We can write the effective switching resistance and oxide capacitance for this device as

$$R_p = 6.8k \text{ and } C_{oxp} = 0.625\, fF$$

As seen in Fig. 13.29, the input capacitance of the circuit is 1.876 fF.

Calculating t_{PLH} begins by noting that when the input goes high the NMOS device in the first inverter turns on. This turns the PMOS device in the second inverter on. The overall delay is calculated as

$$t_{PLH} = \overbrace{0.7 \cdot 3.4k \cdot 11.57\, fF}^{\text{First inverter's delay}} + \overbrace{0.7 \cdot 6.8k \cdot 56.88\, fF}^{\text{Second inverter's delay}} \approx 300\, ps$$

Similarly, when the input goes low, the PMOS in the first inverter turns on and the NMOS in the second inverter turns on, pulling the output low

$$t_{PHL} = \overbrace{0.7 \cdot 6.8k \cdot 11.57\, fF}^{\text{First inverter's delay}} + \overbrace{0.7 \cdot 340 \cdot 56.88\, fF}^{\text{Second inverter's delay}} \approx 70\, ps$$

The simulation results are seen in Fig. 13.30. ■

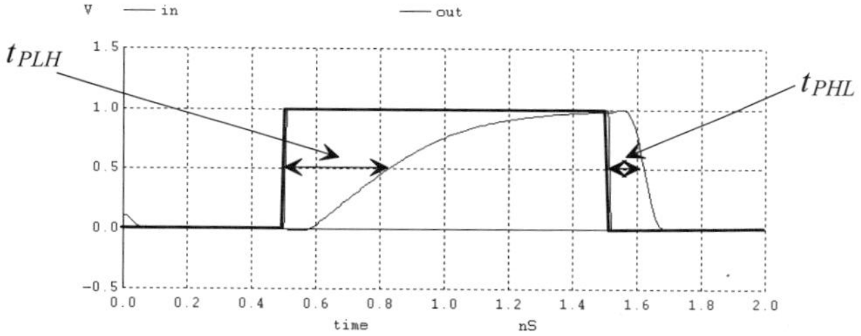

Figure 13.30 Simulating the operation of the circuit in Fig. 13.29.

Example 13.5
Estimate the delay through the circuit seen in Fig. 13.31 using 20/1 PMOS and 10/1 NMOS devices.

Since the load capacitance, 50 fF, is much larger than the output capacitance of the NAND gate, we don't include the NAND gate's capacitance contributions in the digital model seen in Fig. 13.31c. When the input goes high, the NMOS device in the inverter turns on and the PMOS device in the NAND gate turns on (causing the output to go high). The low-to-high delay time is calculated as

$$t_{PLH} = 0.7 \cdot 3.4k \cdot 4.7f + 0.7 \cdot 3.4k \cdot 50f = 130\, ps$$

The high-to-low delay time is calculated, noting the two NMOS in series when pulling the output low, as

$$t_{PHL} = 0.7 \cdot 3.4k \cdot 4.7f + 0.7 \cdot 6.8k \cdot 50f = 250\, ps$$

Figure 13.32 shows the simulation results. ■

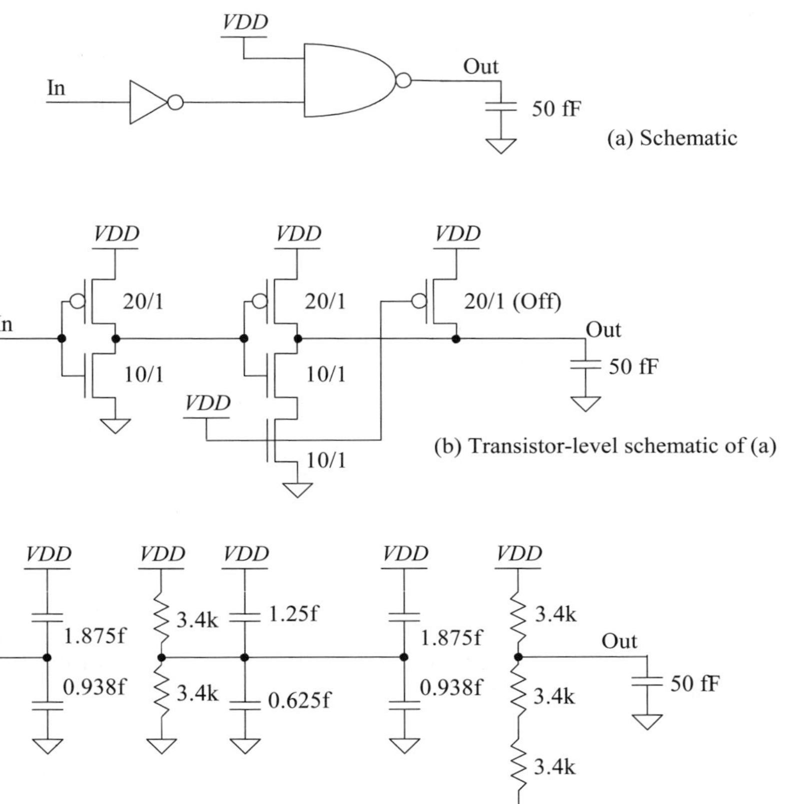

Figure 13.31 Circuit used in Ex. 13.5.

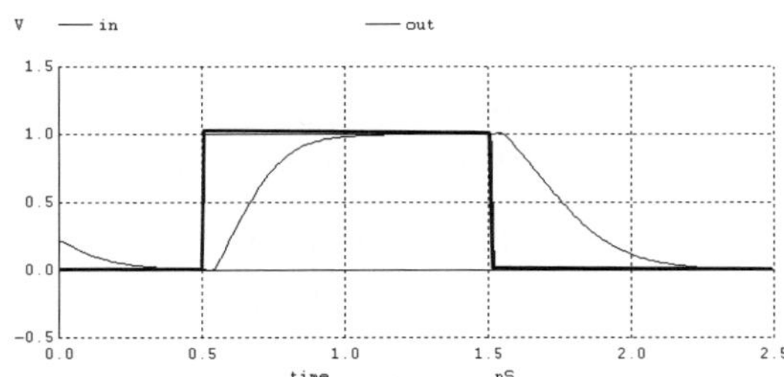

Figure 13.32 Simulating the circuit in Fig. 13.31.

Example 13.6
The circuit seen in Fig. 13.33 is the input portion of the edge triggered D-FF seen in Fig. 13.22. Estimate the delay from the D input to points A and B. Compare the estimate to SPICE simulations. Note that this delay is important because it directly determines the setup time of the FF.

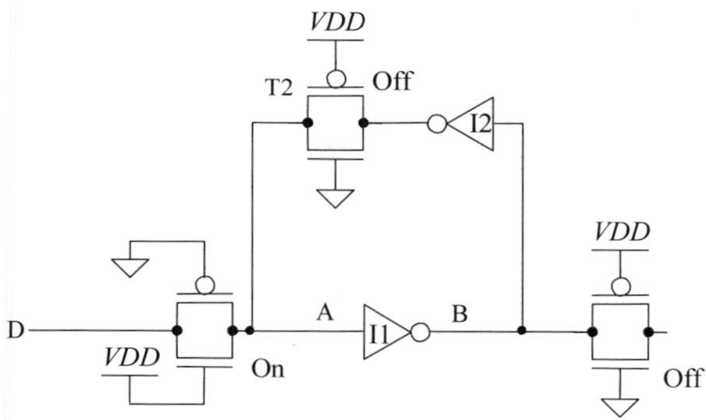

Figure 13.33 Section of a D-FF used in Ex. 13.5.

Node A in Fig. 13.33 is driven through the on TG. The resistance of the TG is $R_n||R_p$. The capacitance on node A is

$$C_A = \overbrace{\frac{C_{oxn}}{2}+\frac{C_{oxp}}{2}}^{\text{Capacitance from the on TG}} + \overbrace{\frac{3}{2}\cdot(C_{oxn}+C_{oxp})}^{\text{Input capacitance of the inverter}} = 3.75\,fF$$

The first term is the capacitance on node A due to the on TG (see Fig. 10.7). The second term is the input capacitance of the inverter. Note that we have not included the capacitive loading from the off TG. This causes the actual delays to be longer than what we calculate here. (A good exercise at this point is to estimate the capacitive loading of the TGs and include them in the calculations here.) The capacitance on node B is made up of an inverter input capacitance and an inverter output capacitance or

$$C_B = \overbrace{\frac{3}{2}\cdot(C_{oxn}+C_{oxp})}^{\text{Inverter input capacitance}} + \overbrace{(C_{oxn}+C_{oxp})}^{\text{Inverter output capacitance}} = 4.7\,fF$$

The delay from the D input to point A is estimated as

$$t_{PLHA} = t_{PHLA} = 0.7\cdot R_n||R_p\cdot C_A = 4.5\,ps$$

The delay from point A to point B through the inverter is

$$t_{PLHB} + t_{PHLB} = 0.7\cdot(R_n+R_p)\cdot C_B = 22.5\,ps$$

The propagation delay from D to B is then

$$t_{PLH} = t_{PHL} = 15.75\ ps$$

The simulation results are seen in Fig. 13.34. The delay to point A is approximately 30 ps, while the delay to point B is about 90 ps. It's important to reiterate the importance of simulations when determining delays. Hand calculations, as this example shows, have limitations. Hand calculations are still important, however, because they reveal the location of the dominant delays in a digital circuit. ■

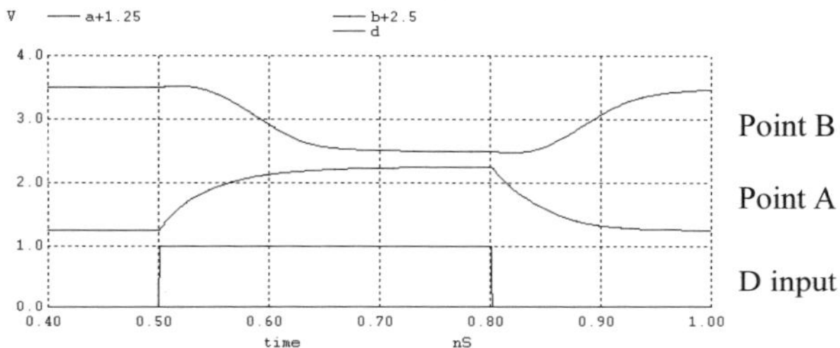

Figure 13.34 Simulating the delays through the latch seen in Fig. 13.33.

ADDITIONAL READING

[1] M. I. Elmasry, *Digital MOS Integrated Circuits II,* IEEE Press, 1992. ISBN 0-87942-275-0, IEEE order number: PC0269-1.

[2] J. P. Uyemura, *Circuit Design for Digital CMOS VLSI,* Kluwer Academic Publishers, 1992.

[3] M. Shoji, *CMOS Digital Circuit Technology,* Prentice-Hall, 1988. ISBN 0-13-138850-9.

PROBLEMS

Unless otherwise stated, use the 50 nm, short-channel CMOS process with the parameters seen in Table 10.2.

13.1 Estimate and simulate the delay through 10 TGs connected a 50 fF load capacitance.

13.2 Design and simulate the operation of a half adder circuit using TGs.

13.3 Sketch the schematic of an 8-to-1 DEMUX using NMOS PGs. Estimate the delay through the DEMUX when the output is connected to a 50 fF load capacitance.

13.4 Verify, using SPICE, that the circuit seen in Fig. 13.12 operates as an XOR gate.

13.5 Simulate the operation of an SR latch made with NAND gates. Show all four possible input logic combinations.

13.6 Simulate the operation of the arbiter seen in Fig. 13.15. Show how two inputs arriving at nearly the same time results in only one output going high.

13.7 Show, using simulations, how making the feedback inverter, I2, in Fig. 13.18 stronger (decrease the lengths) can result in the output having either a long delay or not fully switching.

13.8 Redesign the FF in Fig. 13.22 without T2 and T4 present. Simulate the operation of your design. Show, by using a limited amplitude on the D input, how point B can have a metastability problems.

13.9 In your own words describe setup and hold times. Use the D-FF in Fig. 13.22 and simulations to help support your clear descriptions.

Chapter

14

Dynamic Logic Gates

Dynamic or clocked logic gates are used to decrease complexity, increase speed, and lower power dissipation. The basic idea behind dynamic logic is to use the capacitive input of the MOSFET to store a charge and thus remember a logic level for use later. Before we start looking into the design of dynamic logic gates, let's discuss leakage current and the design of clock circuits.

14.1 Fundamentals of Dynamic Logic

Consider the NMOS pass gate (PG) driving an inverter, as shown in Fig. 14.1. If we clock the gate of the PG high, the logic level on the input, point A, will be passed to the input of the inverter, point B. If this logic level is a "0," the input of the inverter will be forced to ground, while a logic "1" will force the input of the inverter to $VDD - V_{THN}$. When the clock signal goes low, the PG shuts off and the input to the inverter "remembers" the logic level. In other words, when the PG turns on, the input capacitance of the inverter is either charged to $VDD - V_{THN}$ or discharged to ground through the PG. As long as this charge is present on the parasitic input capacitance of the inverter, the logic value is remembered. What we are concerned with at this point are the leakage mechanisms which can leak the stored charge off the node. A node, such as the one labeled B in Fig. 14.1, is called a dynamic node or a storage node. Note that this node is a high-impedance node and is easily susceptible to noise (see Ex. 3.5).

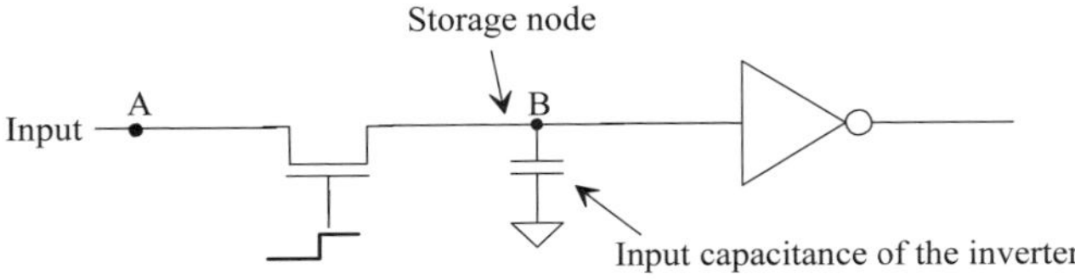

Figure 14.1 Example of a dynamic circuit and associated storage capacitance.

14.1.1 Charge Leakage

One of the important concerns with designing dynamic circuits is the MOSFET's off current, see Sec. 6.4.2. In Fig. 14.2 we show the simulated drain current of a 10/1 MOSFET in the 50 nm process plotted against gate-source voltage. The off current, with $V_{GS} = 0$, is taken from the plot as

$$\log I_D = -8.45 \rightarrow I_D = 3.55\ nA = I_{off} \cdot W \cdot scale \tag{14.1}$$

noting that the worst case off current occurs when the minimum channel length is used ($L = 1$). The off current can be estimated as

$$I_{off} = 7.1\ nA/\mu m \tag{14.2}$$

The off current is made up of leakage through the source/drain implant to substrate (the formed diode and the associated saturation currents), leakage around the edges of the poly at the edge of the field oxide, and by the source-to-drain leakage currents.

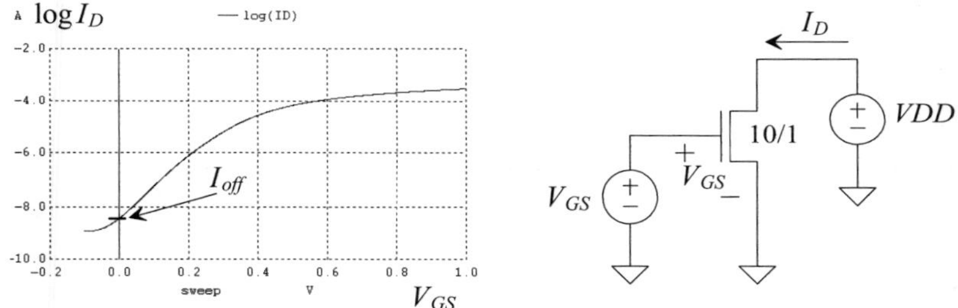

Figure 14.2 Drain current of an NMOS device plotted from weak to strong inversion. See Fig. 6.16. PMOS netlist is found at cmosedu.com.

The rate at which the storage node seen in Fig. 14.1 discharges is given by

$$\frac{dV}{dt} = \frac{I_{off} \cdot W \cdot scale}{C_{node}} \tag{14.3}$$

The node capacitance, C_{node}, is the sum of the input capacitance of the inverter, the capacitance to ground of the metal or poly line connecting the inverter to the pass transistor, and the capacitance of the drain implant to substrate (the depletion capacitance). For practical applications, we assume that

$$C_{node} \approx C_{in} \text{ of the inverter} \tag{14.4}$$

Example 14.1
Estimate the discharge rate of the 50 fF capacitor shown Fig. 14.3.

Using Eqs. (14.1) and (14.3), we can write the capacitor's discharge rate as

$$\frac{dV}{dt} = \frac{3.55\ nA}{50\ fF} = 71\ mV/\mu s$$

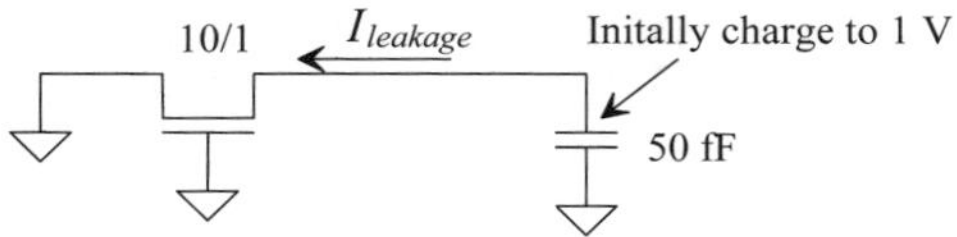

Figure 14.3 Circuit used in Ex. 14.1.

For the capacitor to discharge to ground (from an initial voltage of 1 V) takes, roughly, $(1\ V)/(71\ mV/\mu s) = 14\ \mu s$. Figure 14.4 shows the simulation results. Note how the time it takes the output voltage to discharge to ground is twice as long as what we calculated or roughly 30 μs. This is due to the fact that as the voltage between the drain and source of the MOSFET decreases so does the off current. The reduction in the off current causes the capacitor to discharge slower. To keep the voltage across the capacitor from discharging too much we might try holding one side of the MOSFET switch at *VDD*/2. Consider the following. ■

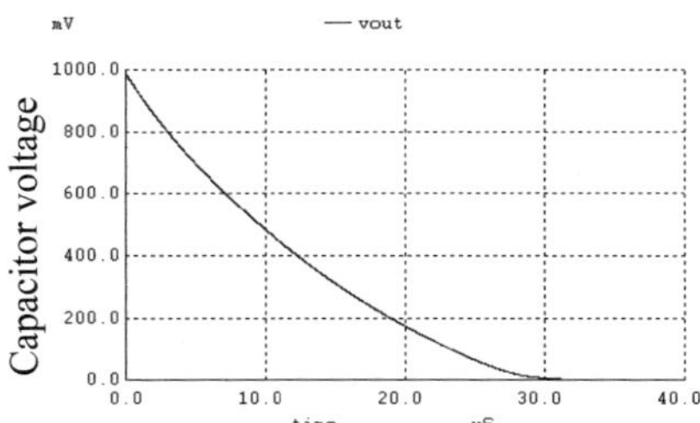

Figure 14.4 Time it takes the capacitor to discharge due to the off current of the MOSFET in Fig. 14.3.

Example 14.2

Repeat Ex. 14.1 with the input tied to *VDD*/2 instead of ground.

The schematic and simulation results are seen in Fig. 14.5. The leakage current is minimized by lowering the V_{DS} of the MOSFET. This is a common technique to

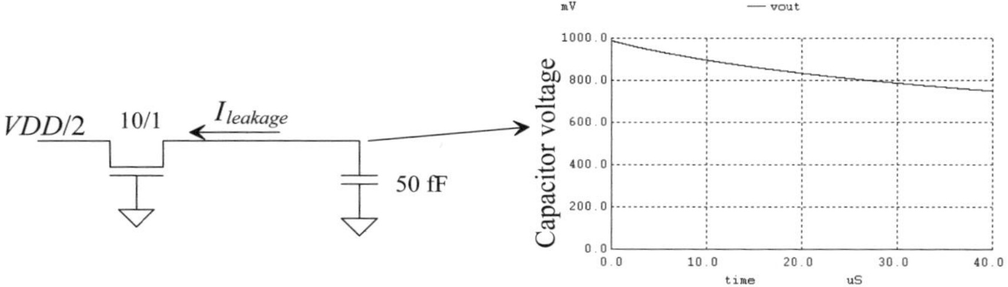

Figure 14.5 Circuit used in Ex. 14.2.

reduce leakage currents in DRAM (see Fig. 16.7 where the bitlines are equilibrated to *VDD*/2). Note that when the voltage across the capacitor (the storage node) is zero and the PG is off the leakage through the MOSFET causes the storage node to charge upwards (see Problem 14.2). ■

Example 14.3
Figure 14.6 shows a dynamic level-sensitive latch. Estimate the maximum time PG can be off before data is lost on the charge storage node. Compare the estimate to SPICE. Use the 50 nm process with 20/1 PMOS devices and 10/1 NMOS devices.

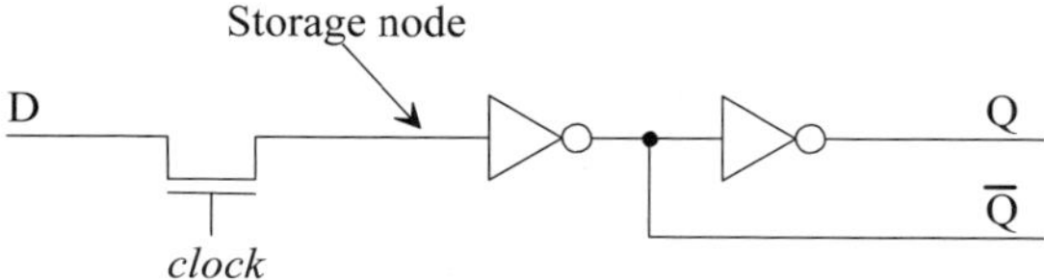

Figure 14.6 A dynamic level-sensitive latch.

The input capacitance of the inverter, using Table 10.2, is estimated as

$$C_{in} = \frac{3}{2} \cdot (1.25 + 0.625)\, fF = 2.8\, fF$$

Using Eq. (14.3), we can estimate the rate the storage node decays as

$$\frac{dV}{dt} = \frac{3.55\, nA}{2.8\, fF} = 1.27\ V/\mu s$$

If we want, at most, 100 mV of droop in the storage node's voltage, then we would want to ensure that the PG is clocked, at the minimum, every 100 ns. Figure 14.7 shows the simulation results. Compared to our hand calculations, the discharge rate is considerably faster. This is due to our calculation of the input capacitance. The values used for C_{in} were the switching input capacitance (see Figs. 10.2, 10.7, and 10.8), not the static input capacitance (see Fig. 6.4). ■

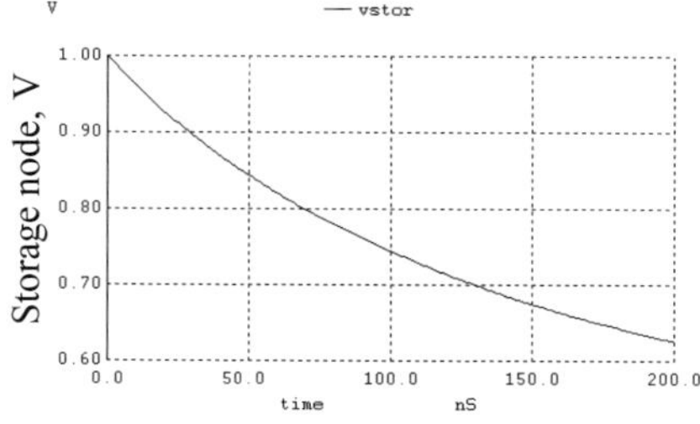

Figure 14.7 How the voltage on the storage node varies in the dynamic latch circuit seen in Fig. 14.6.

14.1.2 Simulating Dynamic Circuits

Because of the extremely small leakage currents involved, simulating dynamic circuits can be challenging. To begin, when SPICE simulates any circuit, it puts a resistor with a conductance value given by the parameter GMIN across every pn junction and MOSFET drain-to-source. The default value of GMIN is 10^{-12} mhos or a 1 TΩ resistor. A charge storage node at a potential of 1 V has a leakage current, due to GMIN, of 1 pA. The value of GMIN can be set, and increased, using the .OPTIONS command, at the cost of a longer or more difficult convergence time, to a smaller value, say 10^{-15}. Some SPICE simulators (like WinSPICE) are also capable of adding a resistor at every node to ground called RSHUNT (to help with convergence, see page 28). If RSHUNT is 10^9 and a node voltage is 1 V, then a current of 1 nA flows through the added RSHUNT resistor.

In practice, we can use the default values of SPICE when simulating dynamic circuits. The SPICE leakage paths (using the default values of GMIN and RSHUNT) have little effect on the estimates for the discharge times of storage nodes when simulating dynamic circuits. For example, the leakage current, for $VDD = 1$ V, due to the default value of GMIN is given by

$$I_{leakage} \approx 1\ pA = VDD \cdot GMIN \tag{14.5}$$

and

$$\frac{dV}{dt} = \frac{1\ pA}{C_{node}} = \frac{VDD \cdot GMIN}{C_{node}} \tag{14.6}$$

If $C_{node} = 5$ fF, then $dV/dt = 200 = 200\ \mu V/\mu s$. It takes approximately 5 ms for the voltage on the charge storage node to drop to ground. In all practical simulations, the leakage through a modern nanometer MOSFET is much more significant than the leakage through GMIN.

14.1.3 Nonoverlapping Clock Generation

Consider the string of PGs/inverters shown in Fig. 14.8. This circuit is called a dynamic shift register. When ϕ_1 goes high, the first and third stages of the register are enabled. Data is passed from the input to point A0 and from point A1 to A2. If ϕ_2 is low while ϕ_1 is high, the data cannot pass from A0 to A1 and from A2 to A3. If ϕ_1 goes low and ϕ_2 goes high, data is passed from A0 to A1 and from A2 to A3. If both ϕ_1 and ϕ_2 are high at the same time, the input of the shift register and the output are connected together, which is not desirable in a shift register application. The purpose of the inverter between PGs is to restore logic levels, since the NMOS PG passes a high with a threshold voltage drop. Two inverters would be used to eliminate the logic inversion between stages. *The clocks used in this dynamic circuit must be nonoverlapping, or logically*

$$\phi_1 \cdot \phi_2 = 0 \tag{14.7}$$

There should be a period of dead time between transitions of the clock signals, labeled Δ in Fig. 14.8. The rise and fall times of the clock signals should not occur at the same time.

Since the design and layout of the dynamic shift register is straightforward, let's concentrate on the generation of clock signals, ϕ_1 and ϕ_2. Note that a simple logic inversion will not generate nonoverlapping clock signals. Consider the schematic of the

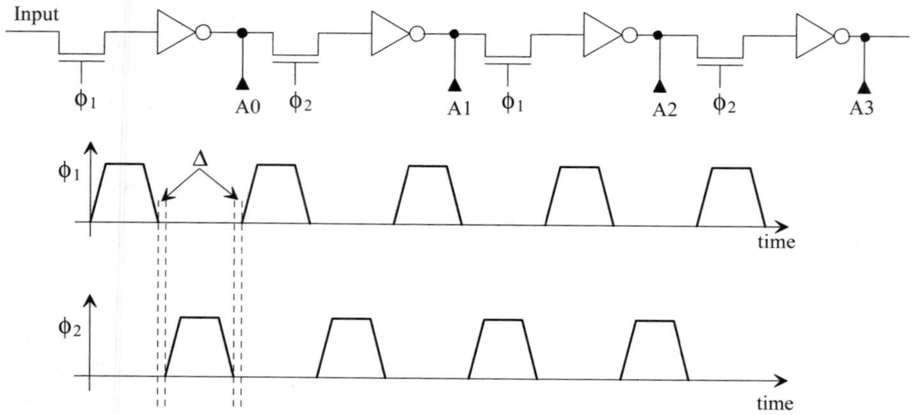

Figure 14.8 Dynamic shift register with associated nonoverlapping clock signals.

nonoverlapping clock generator shown in Fig. 14.9. This circuit takes a clock signal and generates a two-phase nonoverlapping clock. The amount of separation is set by the delay through the NAND gate and the two inverters on the NAND gate output. Consider the input clock going high. This forces ϕ_1 high and ϕ_2 low. When the input clock goes low, ϕ_1 goes low. After ϕ_1 goes low, ϕ_2 can go high. When driving long transmission lines, like a wire implemented using polysilicon, where the rise time of the signals can be significant, a large number of inverters may be needed. Line drivers, a string of inverters used to drive a large capacitance, can be used as part of the delay in the nonoverlapping clock generation circuit.

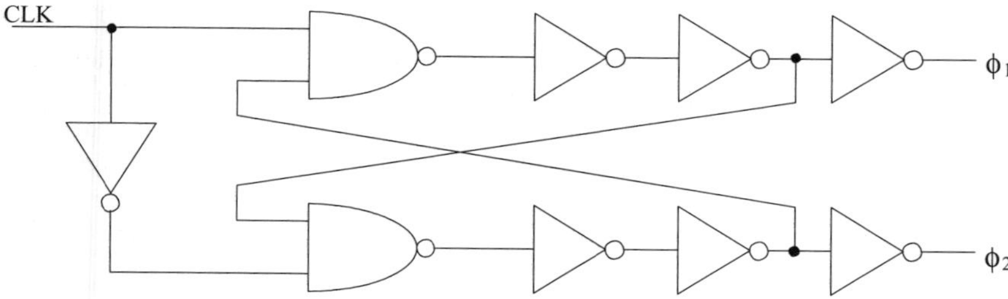

Figure 14.9 Nonoverlapping clock generation circuit.

14.1.4 CMOS TG in Dynamic Circuits

The CMOS TG used as a switch to charge or discharge the node capacitance of the charge storage node is shown in Fig. 14.10. Because understanding the charging and discharging of the input capacitance of the inverter follows many of the same analysis and discussions of Ch. 13, we concentrate here on the charge leakage from the TG.

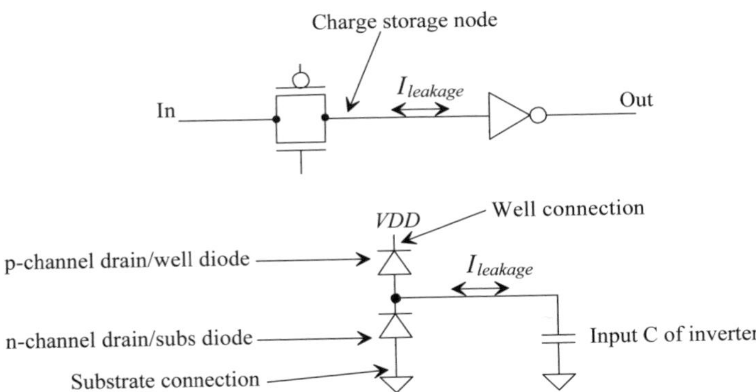

Figure 14.10 CMOS TG used in dynamic logic.

The leakage of charge off of or onto the input capacitance of the inverter in Fig. 14.10 can be attributed to the drain-well diode of the PMOS device and the drain-substrate diode of the NMOS device used in the TG. If these leakage currents were equal, then the leakage of charge off of the storage node would be zero. Notice that unlike the NMOS PG, using a TG can result in the charge storage node leaking to *VDD* or ground, depending on the size of the drain areas and the leakage currents in each device.

14.2 Clocked CMOS Logic

In this section we provide a brief overview of dynamic, or clocked, CMOS logic design.

Clocked CMOS Latch

Figure 14.11 shows the schematic of a clocked CMOS latch. When ϕ_1 is a low, M2 and M3 are on. The master stage simply behaves like an inverter. The D input drives, through the enabled master stage, the node N1. During this time, both M6 and M7 are off. The latch's output, Q, is a charge storage node. When ϕ_1 goes high, M2 and M3 shut off and N1 is the charge storage node. M6 and M7 are on, and the Q output is actively driven either high or low.

An Important Note

Notice, in Fig. 14.11, that if the node N1 starts to move away from either *VDD* or ground, when ϕ_1 is a high, that the slave stage can move towards its switching point. This can cause a significant current to flow from *VDD* to ground in the slave stage. Another example of where this problem can occur is seen in Fig. 14.6. If the storage node starts to wander, the inverter's input can move towards its switching point voltage. As seen in Fig. 11.4, again, a significant current can flow.

Note that the storage node's voltage can wander because of the MOSFET's off current (Fig. 14.2), a gate tunnel current leaking into the node (Fig. 16.67), or because of capacitive coupling from a noisy node (as discussed in Ex. 3.5). The issues of gate tunneling current and the reduction in parasitic capacitances present challenges when implementing dynamic logic in nanometer CMOS technologies.

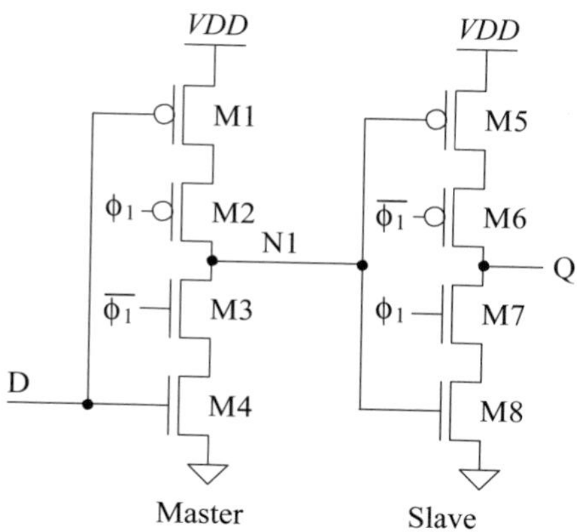

Figure 14.11 A clocked CMOS latch. The clock signals can be generated with an RS latch so that the edges occur essentially at the same moment in time.

PE Logic

This section presents precharge-evaluate logic, or PE logic. Consider the three-input NAND gate shown in Fig. 14.12. The operation of this gate relies on a single clock input. When ϕ_1 is low, the output node capacitance is charged to *VDD* through M5. During the evaluate phase, ϕ_1 is high, M1 is on, and if A0, A1, and A2 are high, the output is pulled low. The logic output is available only when ϕ_1 is high. The output is a logic one when ϕ_1 is low. One disadvantage of PE logic is that the gate logic output is available part of the time and not all of the time as in the static gates.

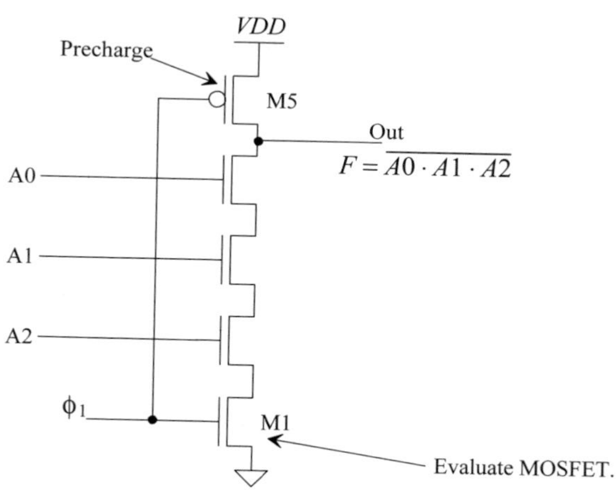

Figure 14.12 Precharge-evaluate three-input NAND gate.

Several important characteristics of the PE gate should be pointed out. The input capacitance of the PE gate is less than that of the static gate. Each input is connected to a single MOSFET where the static gate inputs are tied to two MOSFETs. Potentially the PE gate operates faster and dissipates less power.

The sizes of the MOSFETs used in a PE gate does not need adjusting for symmetrical switching point voltage. The absence of complementary devices and the fact that the output is pulled high during each half cycle makes the gate V_{SP} meaningless. However, we may need to size the devices to attain a certain speed for a given load capacitance. If the sizes of all NMOS transistors used in Fig. 14.12 are equal, then the t_{PHL} is approximately $4R_nC_{node}$ and the t_{PLH} is R_pC_{node}, where C_{node} is the total capacitance on the output node. This may include the interconnecting capacitance and the input capacitance of the next stage. Here we have neglected both the transmission line effects through a series connection of MOSFETs and the intrinsic switching speeds. A more complex logic function, $F = \overline{A0 + A1 \cdot A2 + A3 \cdot A4}$, implemented in PE logic is shown in Fig. 14.13.

Domino Logic

Consider the cascade of PE gates shown in Fig. 14.14. During the precharge phase of the clock, the output of each PE gate is a logic high. This high-level output is connected to the input of the next PE gate. Suppose that the logic out of the first PE gate during the evaluate phase is a low. This output will turn off any MOSFETs in the second PE gate. However, during the precharge phase, those same MOSFETs in the second PE gate will be turned on. The delay between the clock pulse going high and the valid output of the first gate will cause the second gates, output to glitch or show an invalid logic output. If we can hold the output voltage of the PE gate low instead of high, we can eliminate this race condition. Upon adding an inverter to the PE gate (Fig. 14.15), the condition for glitch-free operation is met. The PE gate with the addition of an inverter is called Domino logic. The name *Domino* comes from the fact that a gate in a series of Domino logic gates cannot change output states until the previous gate changes states. The change in output of the gates occurs similar to a series of falling dominoes. The inverter used in the Domino gate has the added advantage that it can be sized to drive large capacitive loads.

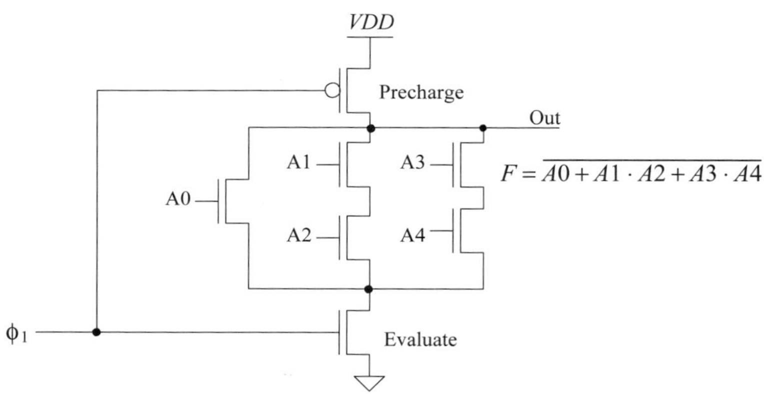

Figure 14.13 A complex PE gate.

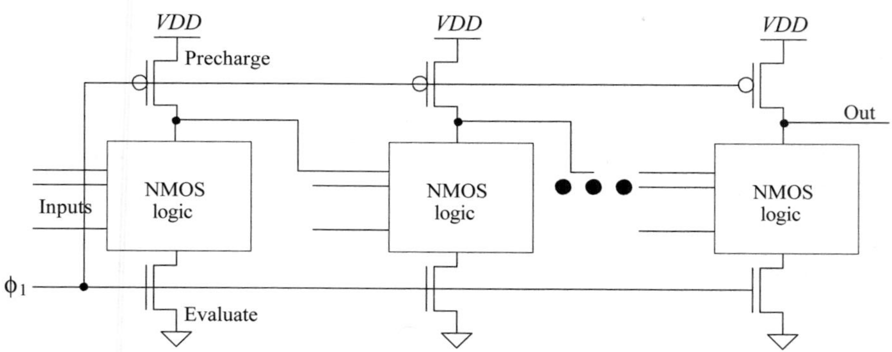

Figure 14.14 Problems with a cascade of PE gates.

One problem does exist with this scheme, however. Referring to Fig. 14.15, note that during the precharge phase, node A is charged to *VDD*. If the NMOS logic results in a logic high on node A during the evaluate phase, then that node is at a high impedance with no direct path to *VDD* or ground. The result is charge leakage off of node A when the PE output is a logic high. The circuit of Fig. 14.16a eliminates this problem. A "keeper" PMOS device is added to help keep node A at *VDD* when the NMOS logic is off. The *W*/*L* of this MOSFET is small (long length and minimum width), so that it provides enough current to compensate for the leakage but not so much that the NMOS logic can't drive node A down to ground. The long-length keeper MOSFET is said to be weak. Sometimes, to implement the keeper MOSFET, two MOSFETs are used in series, as seen in Fig. 14.16b. Connecting a long-length MOSFET, and thus large area MOSFET, to the inverter's output can add unneeded load capacitance. By using a regular switch in series with a long-length PMOS device, the loading on the output of the inverter can be eliminated.

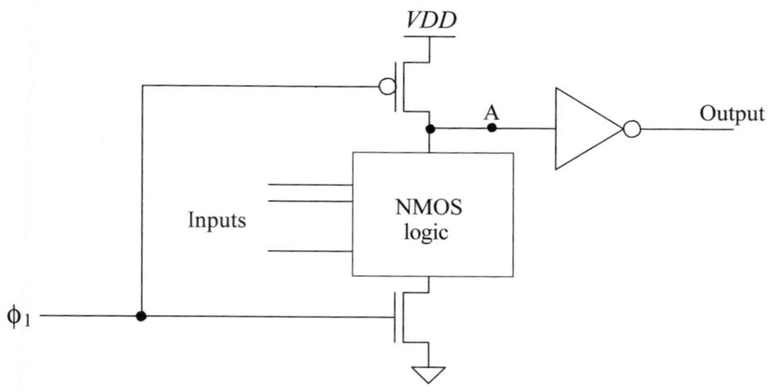

Figure 14.15 Domino logic gate.

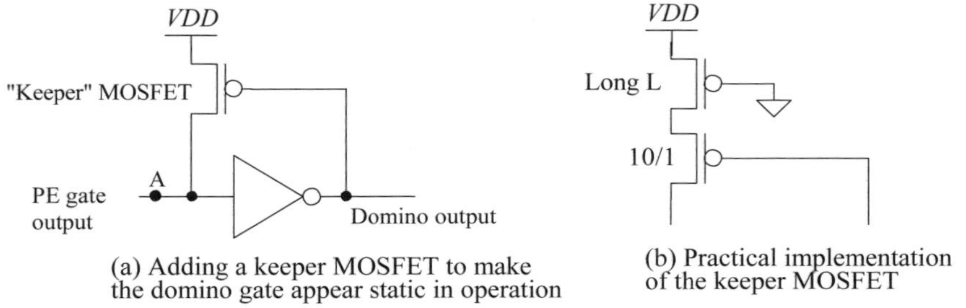

Figure 14.16 Keeper MOSFET used to hold node A in Fig. 14.15 at VDD when PE gate output is high.

NP Logic (Zipper Logic)

The idea behind implementing a logic function using NP logic is shown in Fig. 14.17. Staggering NMOS and PMOS stages eliminates the need for, and delay associated with, the inverter used in Domino logic, making higher speed operation possible. A circuit that can easily be implemented in NP logic is the full adder circuit of Fig. 12.20. The NMOS section of the carry circuit is implemented in the first section of the NP logic, while the PMOS section of the sum circuit is implemented in the PMOS section of the NP logic gate.

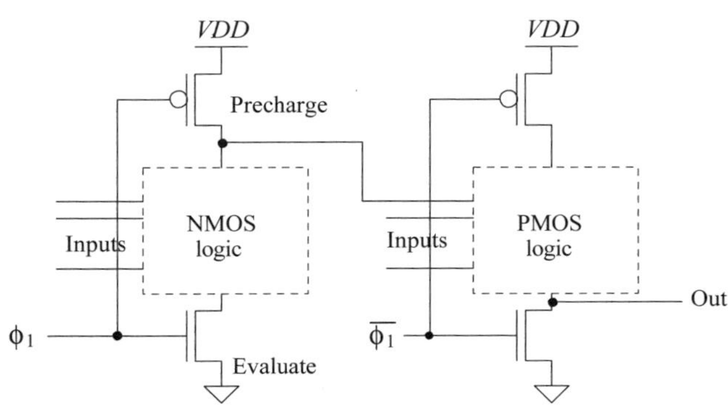

Figure 14.17 NP logic.

Pipelining

The NP logic adder just described adds two 1-bit words with carry during each clock cycle. Adding two 4-bit words can use pipelining, see Fig. 14.18. The bits of the word are delayed, both on the input and output of the adder, so that all bits of the sum reach the output of the adder at the same time. Note, however, that two new 4-bit words can be input to the adder at the beginning of each clock cycle and that it takes four clock cycles

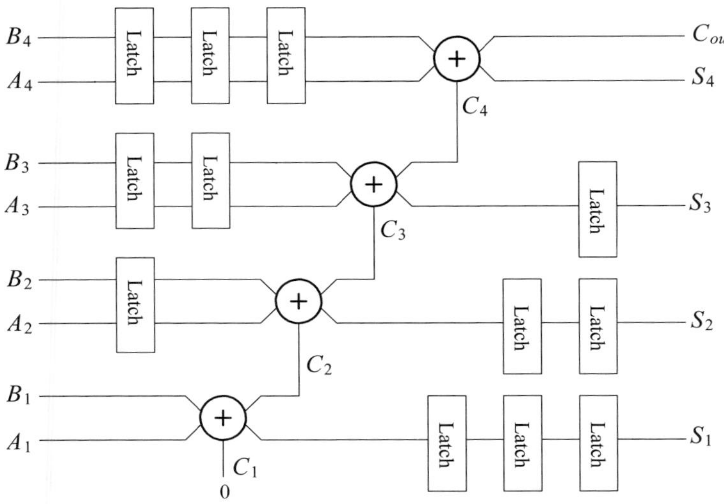

Figure 14.18 A pipelined adder. The latches (clocked) behave as delay elements.

to finish the addition of the two words. If this circuit were dedicated to continually performing the addition of two words, we could input the words at a very fast rate. However, since performing a single addition requires four clock cycles, applications of pipelining where two numbers are not added continuously can result in longer delay-times.

ADDITIONAL READING

[1] J. Yuen and C. Svensson, "New Single-Clock CMOS Latches and Flipflops with Improved Speed and Power Savings," *IEEE Journal of Solid-State Circuits,* vol. 32, no. 1, pp. 62–69, 1997.

[2] N. H. E. Weste and D. Harris, *Principles of CMOS VLSI Design,* Addison-Wesley, 3rd ed., 2004. ISBN 0-321-14901-7.

[3] M. I. Elmasry, *Digital MOS Integrated Circuits II,* IEEE Press, 1992. ISBN 0-87942-275-0, IEEE order number: PC0269-1.

[4] J. P. Uyemura, *Circuit Design for Digital CMOS VLSI,* Kluwer Academic Publishers, 1992.

[5] R. L. Geiger, P. E. Allen, and N. R. Strader, *VLSI-Design Techniques for Analog and Digital Circuits,* McGraw-Hill Publishing Co., 1990. ISBN 0-07-023253-9.

PROBLEMS

Unless otherwise stated, use the 50 nm, short-channel process.

14.1 Regenerate Fig. 14.2 for the PMOS device. From the results, determine the PMOS's I_{off}.

14.2 Repeat Ex. 14.2 if the storage node is a logic 0 (ground). Explain why the charge storage node charges up. What would happen if the PG's input were held at *VDD* instead of *VDD*/2.

14.3 Comment on the usefulness of dynamic logic in our 50 nm CMOS process-based on the results given in Ex. 14.3 with a clock frequency of 10 MHz.

14.4 Using the circuit in Fig. 14.6 and the SPICE simulation, show how the current drawn from *VDD*, by the inverters, changes with time. Do you see any concerns? If so, what?

14.5 Simulate the operation of the nonoverlapping clock generator circuit in Fig. 14.9. Assume that the input clock signal is running at 100 MHz. Show how both ϕ_1 and ϕ_2 are nonoverlapping.

14.6 Simulate the operation of the clocked CMOS latch shown in Fig. 14.11.

14.7 Design and simulate the operation of a PE gate that will implement the logical function $F = \overline{ABCD + E}$.

14.8 If the PE gate shown in Fig. 14.13 drives a 50 fF capacitor, estimate the worst-case t_{PHL}. Use a 20/1 PMOS and a 10/1 NMOS.

14.9 Implement an XOR gate using Domino logic. Simulate the operation of the resulting implementation.

14.10 The circuit shown in Fig. 14.19 results from the implementation of a high-speed adder cell (1-bit). What type of logic was used to implement this circuit? Using timing diagrams, describe the operation of the circuit.

14.11 Discuss the design of a 2-bit adder using the adder cell of Fig. 14.19 If a clock, running at 200 MHz, is used with the 2-bit adder, how long will it take to add two words? How long will it take if the word size is increased to 32 bits?

14.12 Sketch the implementation of an NP logic half adder cell.

14.13 Design (sketch the schematic of) a full adder circuit using PE logic.

14.14 Simulate the operation of the circuit designed in Problem 14.10.

14.15 Show that the dynamic circuit shown in Fig. 14.20 is an edge-triggered flip-flop [1]. Note that a single-phase clock signal is used.

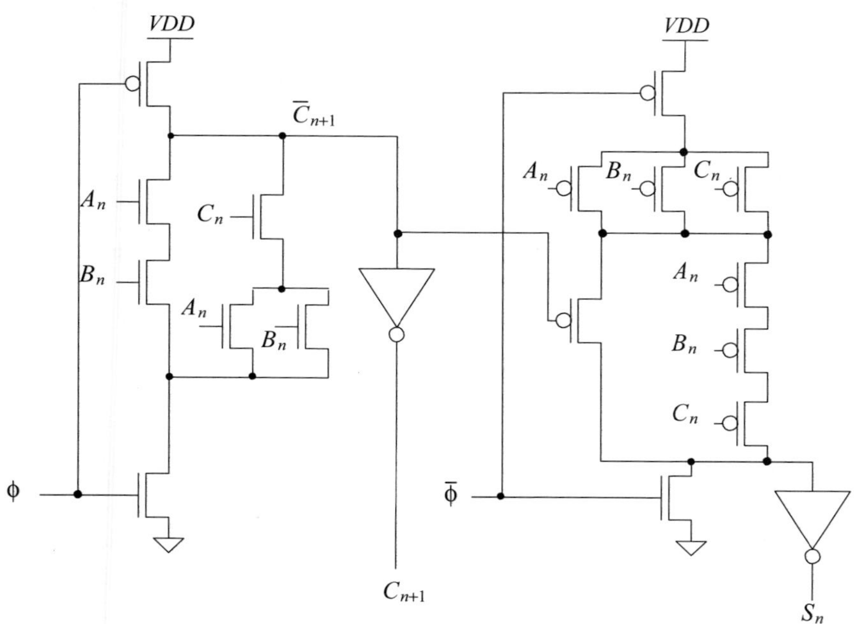

Figure 14.19 A high speed adder cell. See Problem 14.10.

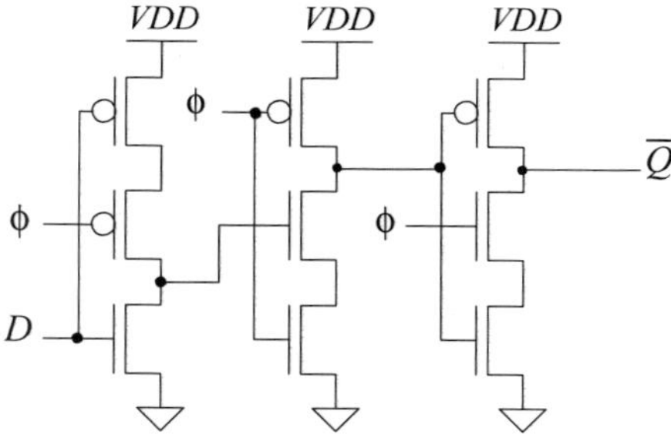

Figure 14.20 A true-single phase clocked FF, see Problem 14.15.

Chapter

15

VLSI Layout Examples

In the past chapters we have concentrated on basic logic-gate design and layout. In this chapter we discuss the implementation of logic functions on a chip where the size and organization of the layouts are important. The number of MOSFETs on a chip, depending on the application, can range from tens (an op-amp) to more than hundreds of millions (a 256 Mbit DRAM). Designs where thousands of MOSFETs or more are integrated on a single die are termed *very-large-scale-integration* (VLSI) designs.

To help us understand why chip size is important, examine Fig. 15.1. The dark dots indicate and, thus, a failing chip. Figure 15.1a shows a wafer with nine full die. The partial die around the edge of the wafer are wasted. Five of the nine die do not contain a defect and thus can be packaged and sold. Next consider a reduction in the die size (Fig. 15.1b). We are assuming each die, whether discussing the die of Fig. 15.1a or b, performs the same function. This reduction can be the result of having better layout (resulting in a smaller layout area) or fabricating the chips in a process with smaller device dimensions

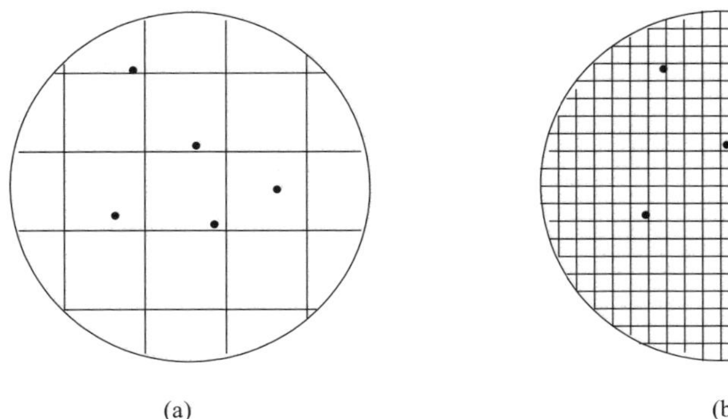

Figure 15.1 Defect density effects on yield.

(e.g., going from a 130 nm process to a 50 nm process). The total number of die lost (see Fig. 15.1b), due to defects is five; however, the number of good die is significantly larger than the five good die of Fig. 15.1a. The yield (number of good die/total number of die on the wafer) is increased with smaller die size. The result is more die/wafer available for sale. Another benefit of reducing die size comes from the realization that processing costs per wafer are relatively constant. Increasing the number of die on a wafer decreases the cost per die.

15.1 Chip Layout

VLSI designs can be implemented using many different techniques including gate-arrays, standard-cells, and full-custom design. Because designs based on gate-arrays are used, in general, where low volume and fast turnaround time are required and the chip designers need know little to nothing[1] about the actual implementation of the CMOS circuits, we will concentrate on full-custom design and design using standard cells

Regularity

An important consideration when implementing a VLSI chip design is regularity. The layout should be an orderly arrangement of cells. Toward this goal, the first step in designing a chip is drawing up a chip (or section of the chip) floor plan. Figure 15.2 shows a simple floor plan for an adder data-path. This floor plan can be added to the floor plan of an overall chip, which includes output buffers, control logic, and memory. At this point, we may ask the question, "How do we determine the size of the blocks in Fig. 15.2?" The answer to this question leads us into the design and layout of the cells used to implement each of the logic blocks in Fig. 15.2.

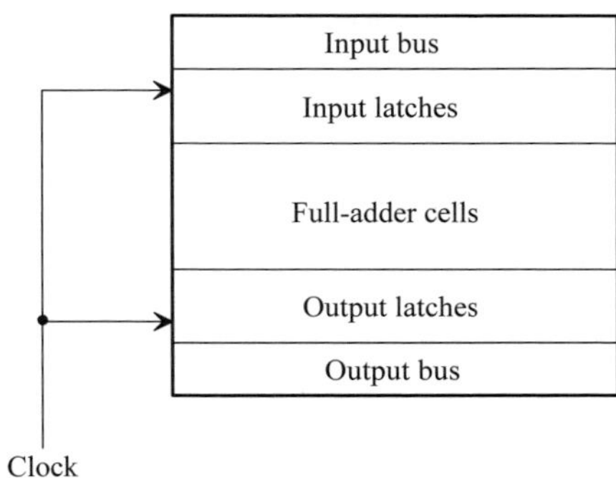

Figure 15.2 Floor plan for an adder.

[1] At many universities, design using a hardware description language (HDL) with field-programmable-gate arrays (FPGAs) is discussed in the first (or perhaps second) course on digital systems design.

Standard-Cell Examples

Standard cells are layouts of logic elements including gates, flip-flops, and ALU functions that are available in a cell library for use in the design of a chip. *Custom design* refers to the design of cells or standard cells using MOSFETs at the lowest level. *Standard-cell design* refers to design using standard cells; that is, the designer connects wires between standard cells to create a circuit or system. The difference between the two types of design can be illustrated using a printed circuit board-level analogy. A standard-cell design is analogous to designing with packaged parts. The design is accomplished by connecting wires between the pins of the packaged parts. Custom design is analogous to designing the "insides" of the packaged parts themselves

Figure 15.3 shows an example of an inverter. In addition to keeping the layout size as small as possible, an important consideration when laying out a standard cell is the routing of signals. Keeping this in mind, we can state the following general guidelines for standard-cell design:

1. Cell inputs and outputs should be available, at the same relative horizontal distance, on the top and bottom of the cell.

2. Horizontal runs of metal are used to supply power and ground to the cell, a.k.a., power and ground buses. Also, well and substrate tie downs should be under these buses.

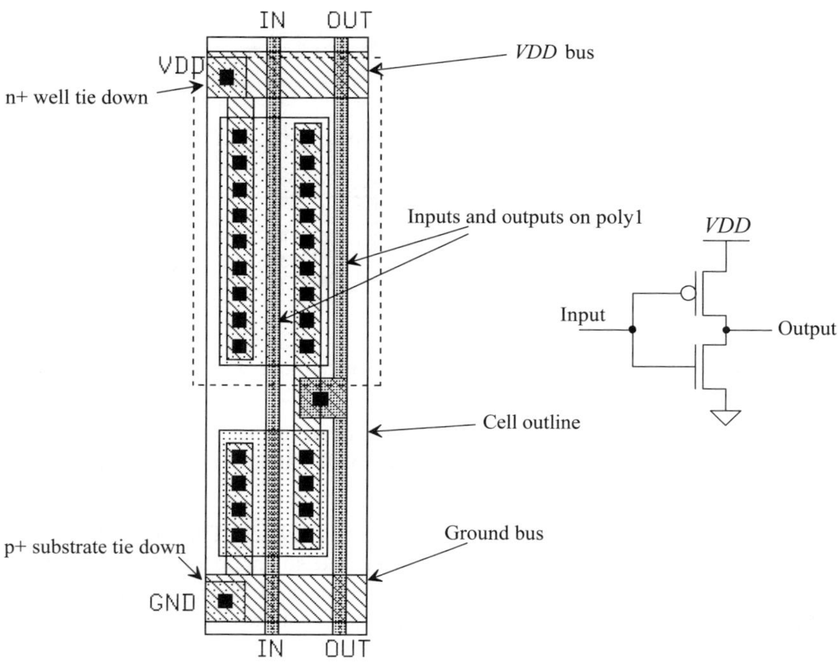

Figure 15.3 Standard cell layout of an inverter.

3. The height of the cells should be a constant, so that when the standard cells are placed end to end the power and ground buses line up. The width of the cell should be as narrow as the layout will allow. However, the absolute width is not important and can be increased as needed.

4. The layout should be labeled to indicate power, ground, and input and output connections. Also, an outline of the cell, useful in alignment, should be added to the cell layout.

Figure 15.4 illustrates the connection of standard cells to a bus. Note that poly, which runs vertically, can cross the metal1 lines, which run horizontally without making contact. This fact is used to route signals and interconnect standard cells in a VLSI design. Also, in this figure, note how the two inverter standard cells are placed end to end. The result is that power and ground are automatically routed to each cell.

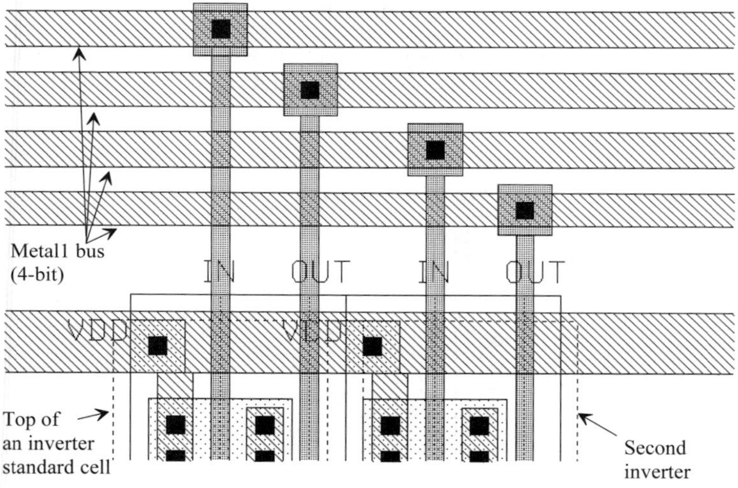

Figure 15.4 Connection of two inverter standard cells to a bus.

Other examples of static standard cells are shown in Fig. 15.5. A double inverter standard cell is shown in Fig. 15.5a, while NAND, NOR, and transmission gate standard cells are shown in Figs. 15.5b, c, and d.

Figure 15.6 shows the layout of a NAND-based SR latch. This layout differs from the others we have discussed. In all layouts discussed so far metal1 and contacts are adjacent to the gate poly. Also, the gate poly has been laid down without bends. The expanded view of a PMOS device used in the SR latch is shown in Fig. 15.7. Keeping in mind that whenever poly crosses active (n+ or p+), a MOSFET is formed, we see that the source of the MOSFET is connected to metal through two contacts, while the p+ implant forms a resistive connection to metal1 along the remainder of the device. The layout size,

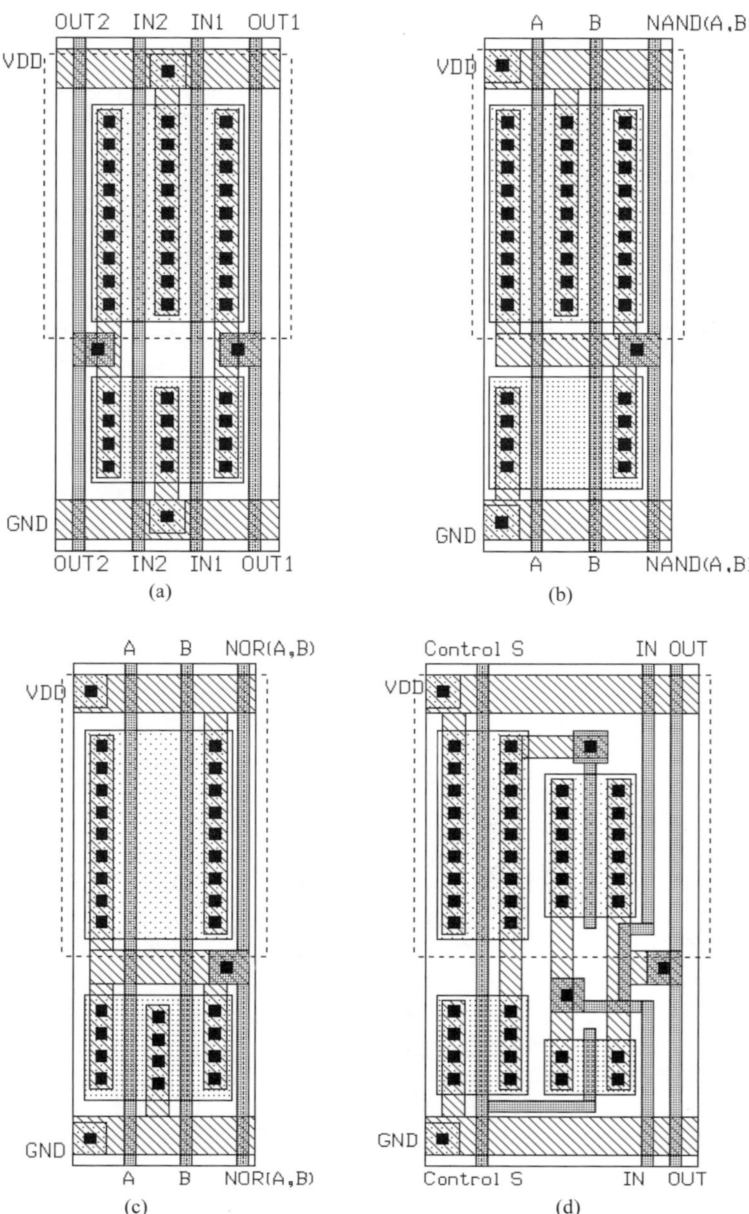

Figure 15.5 (a) Double inverter, (b) two-input NAND, (c) two-input NOR, and (d) transmission gate.

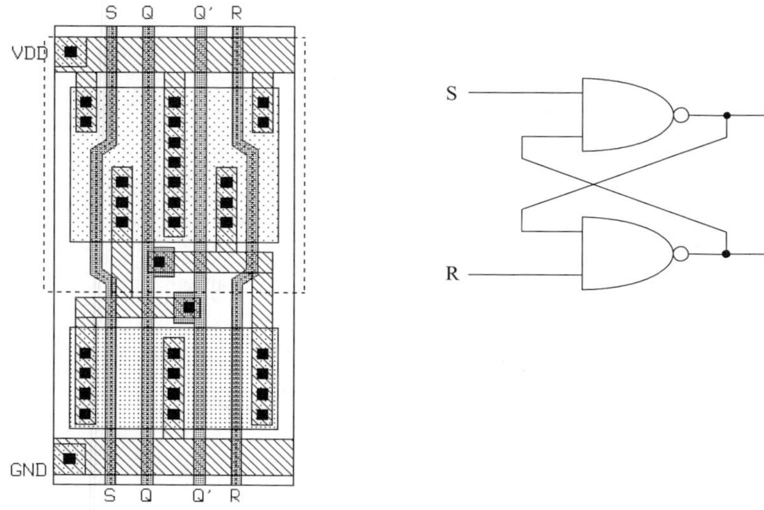

Figure 15.6 SR latch using NAND gates.

in this case the width of the standard cell, can be reduced using this technique. Because of the bend in the gate, the width of this MOSFET is longer than the adjacent MOSFET. This additional width is of little importance and has little effect on the DC and transient properties of the gate. Figure 15.8 shows the NOR implementation of an SR latch.

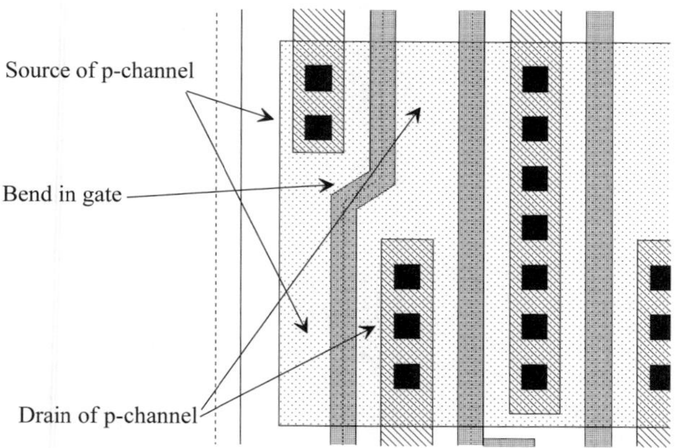

Figure 15.7 Section of the layout shown in Fig. 15.6.

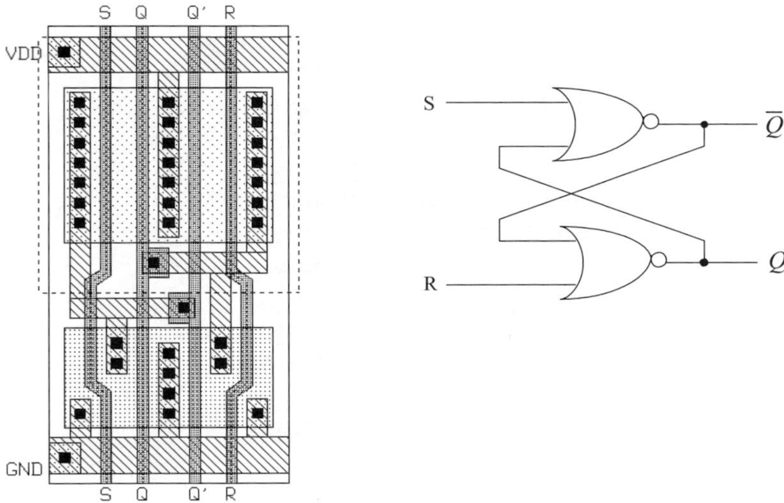

Figure 15.8 SR latch using NOR gates.

Power and Ground Connections

Many of the problems encountered when designing a chip can be related to the distribution of power and ground. When power and ground are not distributed properly, noise can be coupled from one circuit onto the power and ground conductors and injected into some other circuit.

Consider the placement of standard cells in a padframe shown in Fig. 15.9a, without connections to power and ground. Approximately 600 standard cells are shown in this figure. The space between the rows of standard cells is used for the routing of signals. A line drawing of a possible power and ground busing architecture is shown in Fig. 15.9b. Consider the section of bus shown in Fig. 15.9c. Wire A connects the standard cells in the top row to *VDD*, while wire B is used for the connection to ground. Ideally, the current supplied on A (*VDD*) is returned on B (ground). In practice, there is coupling between conductors B and C, which gives rise to an unwanted signal (noise) on either conductor. This coupling can be reduced by increasing the space between B and C, which reduces the inductive and capacitive coupling between the conductors. Another solution is to increase the capacitance between A and B. A standard cell decoupling capacitor (Fig. 15.10) can be used toward this goal. The capacitor is placed in the middle of a standard-cell row. Also, the AC resistive drop effects discussed in Ch. 3 (see Fig. 3.17 and the associated discussion) are greatly reduced by including this capacitor.

Coupling is a problem on signal buses as well. Figure 15.11 shows a simple scheme to reduce coupling. The length of a section, where two wires are adjacent, is reduced by routing the wire to other locations at varying distances along the bus. The inductive or capacitive coupling between two conductors is directly related to the length of the wire runs.

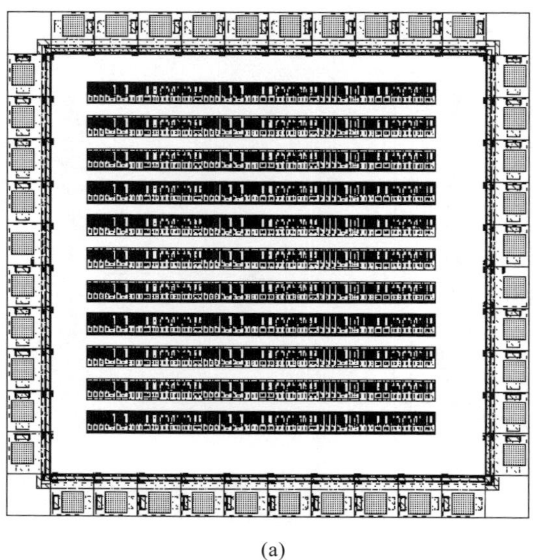

(a)

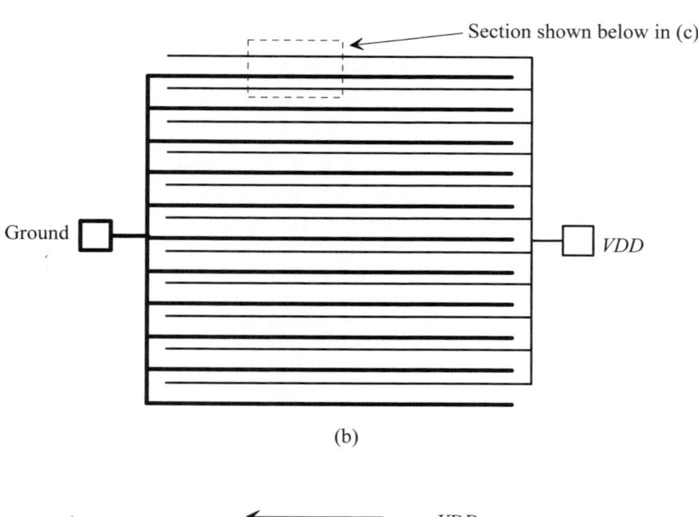

(b)

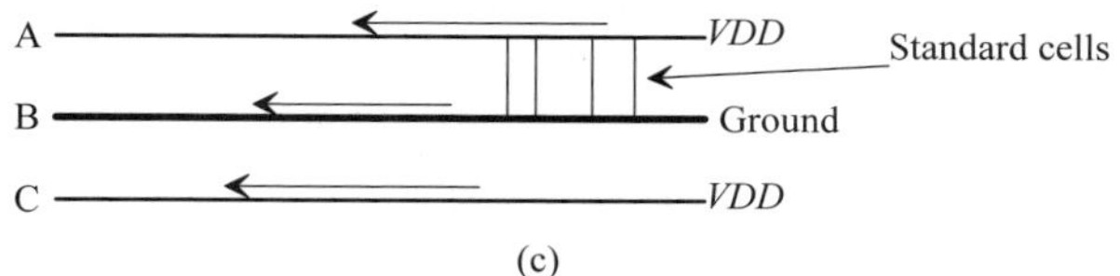

(c)

Figure 15.9 Connection of power and ground to standard cells.

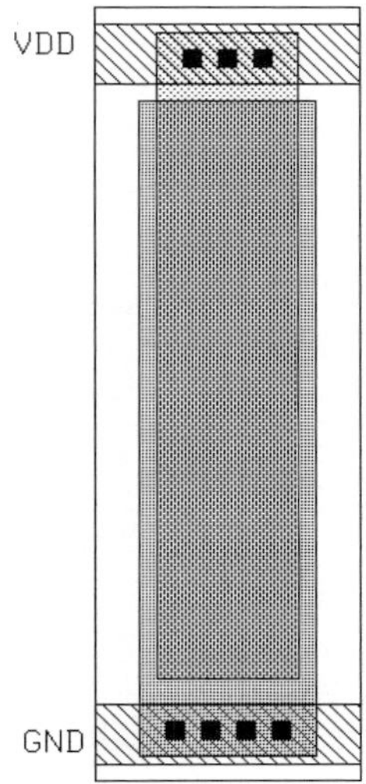

Figure 15.10 Decoupling capacitor.

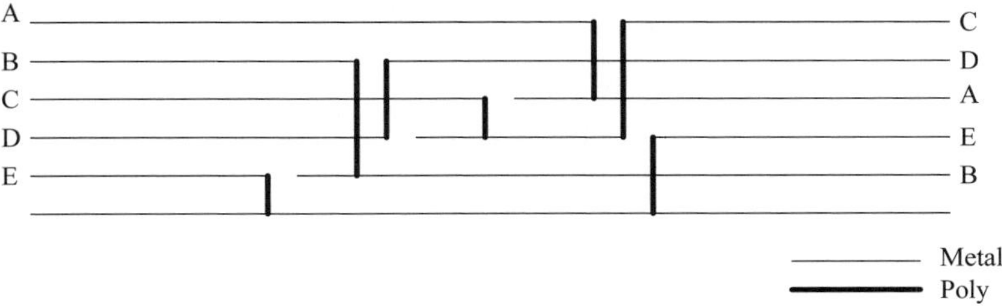

Figure 15.11 Busing structure used to decrease signal coupling.

An Adder Example

As another example, let's consider the implementation of a 4-bit adder. (The floorplan for this adder was shown in Fig. 15.2.) The first components that must be designed are the input and output latches. Figure 15.12a shows the schematic of the latches. This latch is the level-sensitive type discussed in Ch. 13. When CLK is high, the output, Q, changes states with the input, D. The inverter, I4, provides positive feedback and is sized with a small W/L ratio so that I1 does not need to supply a large amount of DC current to force the latch to change states. The layout of the latch is shown in Fig. 15.12b. The layout size and the size of the MOSFETs in these examples may be larger than normal to make it easier to understand and view the layouts.

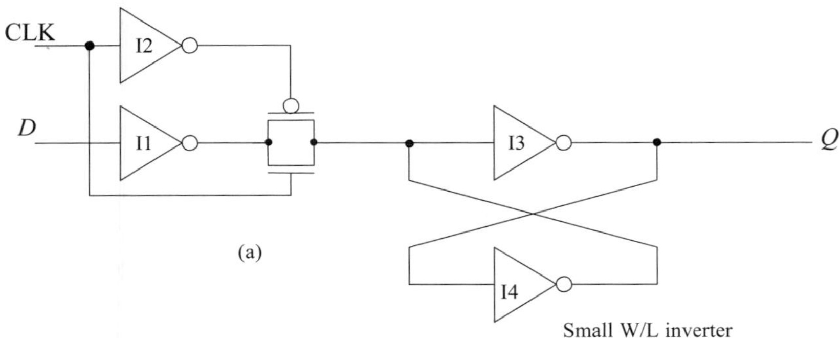

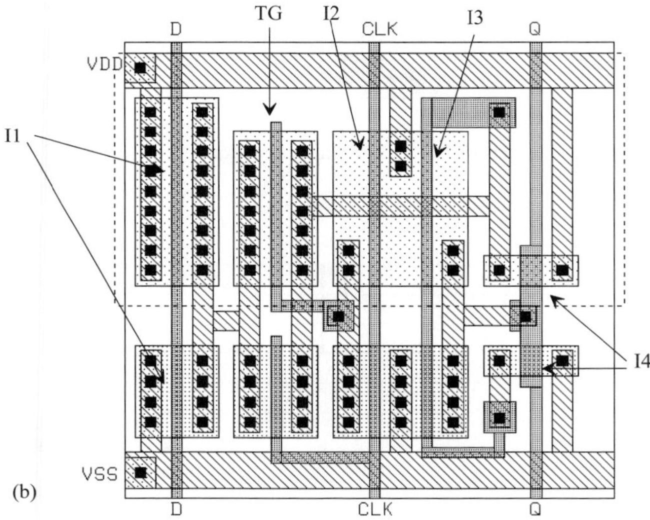

Figure 15.12 Schematic and layout of a latch.

The layout of the static adder is shown in Fig. 15.13. This is the implementation, using near minimum-size MOSFETs, of an AOI (and-or-inverter) static adder. Both the carry-out and sum-out logic functions are implemented in this cell.

The complete layout of the adder is shown in Fig. 15.14. The two 4-bit words, Word-A and Word-B, are input to the adder on the input bus. These data are clocked into the input latch when CLK is high, while the results of the addition are clocked into the output latch when CLK is low. The inverter standard cell of Fig. 15.3 is placed at the end of the output latches and generates $\overline{\text{CLK}}$ for use in the output latches. The inputs and outputs of the adder cells are run on poly because of the short distances involved. The carry-in of the adders is connected to ground, as shown in the figure.

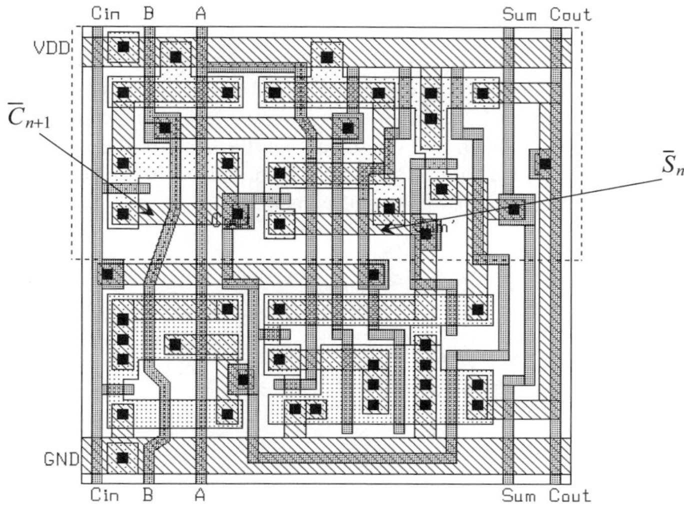

Figure 15.13 Layout of an AOI static adder.

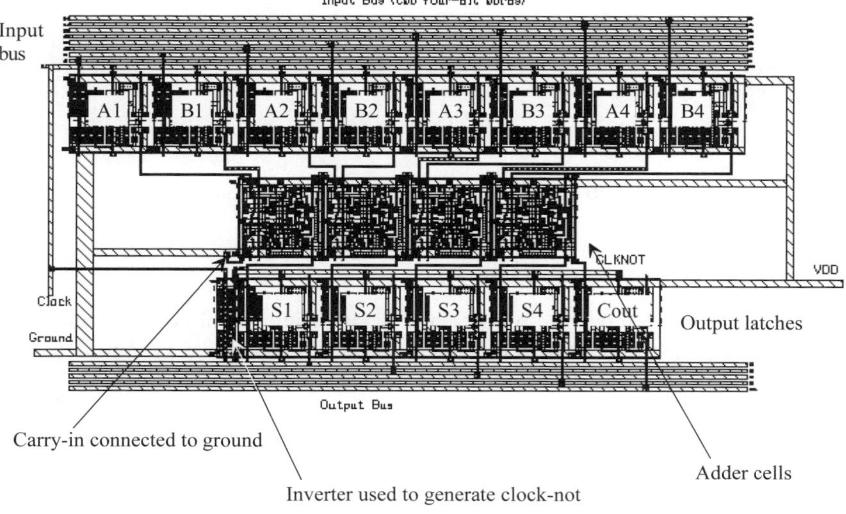

Figure 15.14 Layout of the complete adder.

A 4-to-1 MUX/DEMUX

The layout of a 4-to-1 MUX/DEMUX is shown in Fig. 15.15 (based on the use of NMOS pass gates). Notice that the (required) p+ substrate connections are not shown. This layout is different from the layouts discussed so far as the circuit does not require power and ground connections and the input/output signals are connected on n+. The select signals are supplied to the circuit on metal1 at the top of the layout. For A to be connected to the output, the signals S1 and S2 should be high. For a large MUX, the propagation delay through the n+ should be considered.

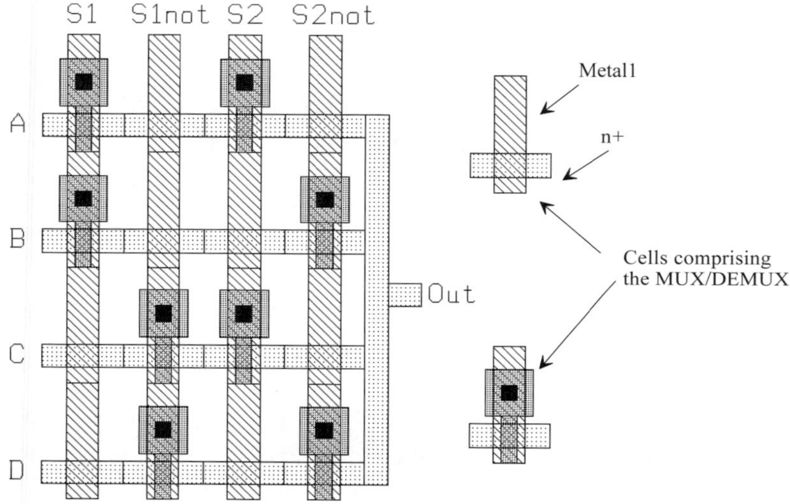

Figure 15.15 Layout of a 4-to-1 MUX/DEMUX.

15.2 Layout Steps *by Dean Moriarty*

The steps involved in rendering a schematic diagram into its physical layout are plan, place, connect, polish, and verify. Let's illustrate each of these steps in some detail.

Planning and Stick Diagrams

The planning steps start with paper and pencil. Colored pencils are useful for distinguishing one object from another. You can use graph paper to help achieve a sense of proportion in the cell plan but don't get too bogged down in the details of design rules or line widths at this point; we just want to come up with a general plan. A "stick diagram" is a paper and pencil tool that you can use to plan the layout of a cell. The stick diagram resembles the actual layout but uses "sticks" or lines to represent the devices and conductors. When used thoughtfully, it can reveal any special hook-up problems early in the layout, and you can then resolve them without wasting any time.

Figure 15.16a shows the schematic of an inverter. To realize the layout of this circuit, it is first necessary to define the direction and metallization of the power supply, ground, input, and output. Since the standard-cell template was discussed earlier (see also Fig. 4.15 and the associated discussion), we'll use it. Power and ground run horizontally and provide the substrate and well connections. The input and output are accessible from the top or bottom of the cell and will be in metal2 running vertically. Figure 15.16b shows the completed stick diagram. Note the use of "X" and "O" to denote contacts and vias, respectively. The stick diagram should be compared to the resulting layout of Fig. 15.17.

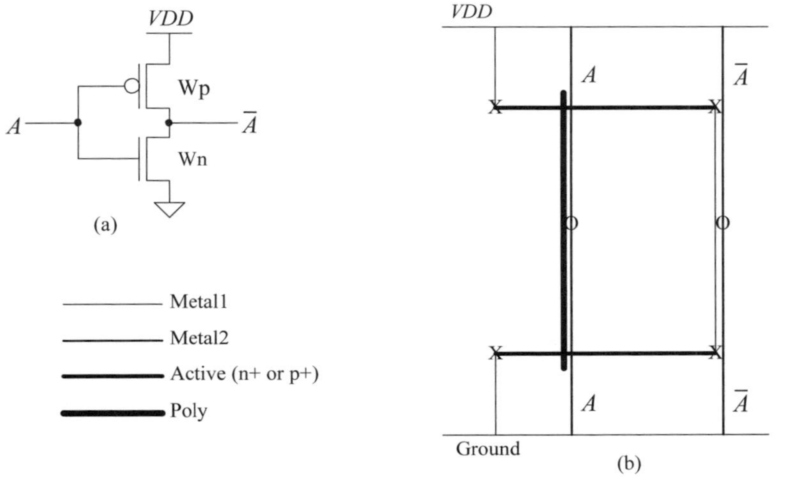

Figure 15.16 (a) Inverter and (b) stick diagram used for layout.

Suppose the device widths of the inverter circuit in Figure 15.16a were quadrupled. Furthermore, let's assume that the maximum recommended poly1 gate width is 20 (due to the sheet resistance of the poly) and that exceeding that maximum could introduce significant unwanted RC delays. Let's also suppose that we are to optimize the layout for size and speed (as most digital circuits are). To meet these criteria, it will be necessary to split transistors M1 and M2 in half and lay them out as two parallel "stripes." Figures 15.18a–d show the schematics, stick diagram, and layout for this scenario. The output node (drain of M1 and M2) is shared between the stripes so as to minimize the output capacitance. Taking the output in metal2 also helps in this regard. Notice that the stick diagram for this circuit looks like the previous inverter plus its mirror image along the output node. Also observe that the layout of this inverter is mirrored, as shown in the stick diagram. This is a common layout technique.

Figure 15.19 shows stick diagrams and layouts for two more common circuits: the two-input NAND and the two-input NOR. Compare the stick diagrams of Figs. 15.19a and c to the layouts of Figs. 15.19b and d. Observe that the output nodes share the active area just as in the previous example. Also note that the spacing between the gates of the series-connected devices is minimal.

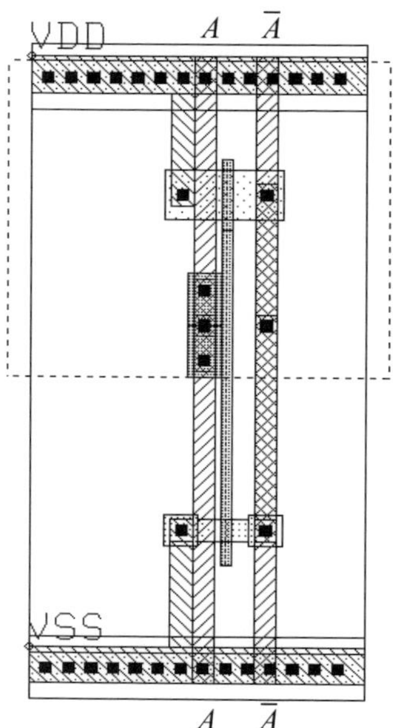

Figure 15.17 Layout of the inverter shown in Fig. 15.16.

Take another look at the two circuits from a geometrical rather than an electrical viewpoint. Compare the NAND gate layout to the NOR gate layout. Each can be created from the other by simply "flipping" the metal and poly connections about the x-axis.

Device Placement

Figure 15.20 shows the schematic of a dynamic register cell, while Figs. 15.21 a–c show the stick diagrams and layout for a dynamic register. Compare the schematic of Fig. 15.20 to the stick diagram of Fig. 15.21a. We have labeled this stick diagram "preliminary" for reasons that will soon become apparent. Notice that there is a break or gap in the active area, which will form our NMOS devices. Also note that the clock signals CLK and $\overline{\text{CLK}}$must be "cross-connected" from one side of the layout to the other. We don't have to think this through very far to notice that, with this placement of devices, hooking up the clock signals is going to be very difficult. Now look at the stick diagram shown in Fig 15.21b. Notice that we have rearranged the devices so that the active area is a continuous unbroken line. Normally, this "unbroken line" approach to device placement is preferred. It usually results in the most workable device placement. We say "usually" because at times your layout has to fit in an area defined by other blocks around it and you have no control over it. Also observe from Fig. 15.21b that the clock signal hook-up is more straightforward. Compare this stick diagram to the layout of Fig. 15.22c. Obviously, the device sizes used for this circuit are not practical; its purpose is merely to illustrate a layout concept. We can also see that the stick diagram is a useful tool throughout the layout process.

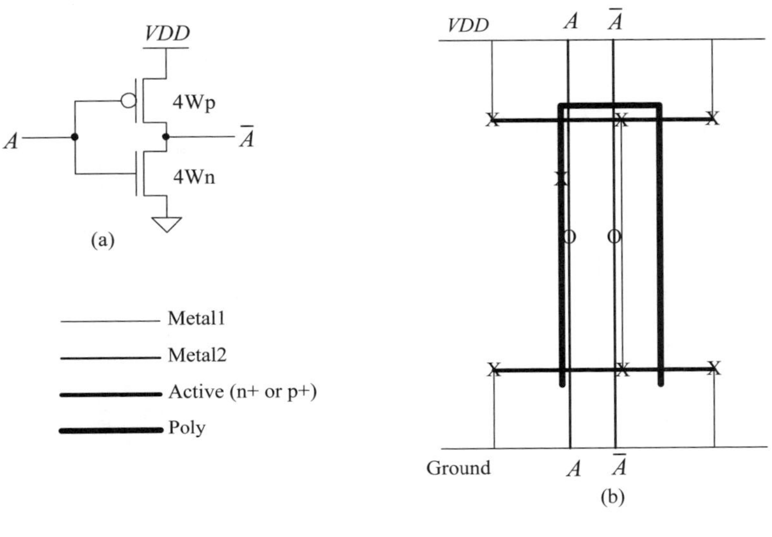

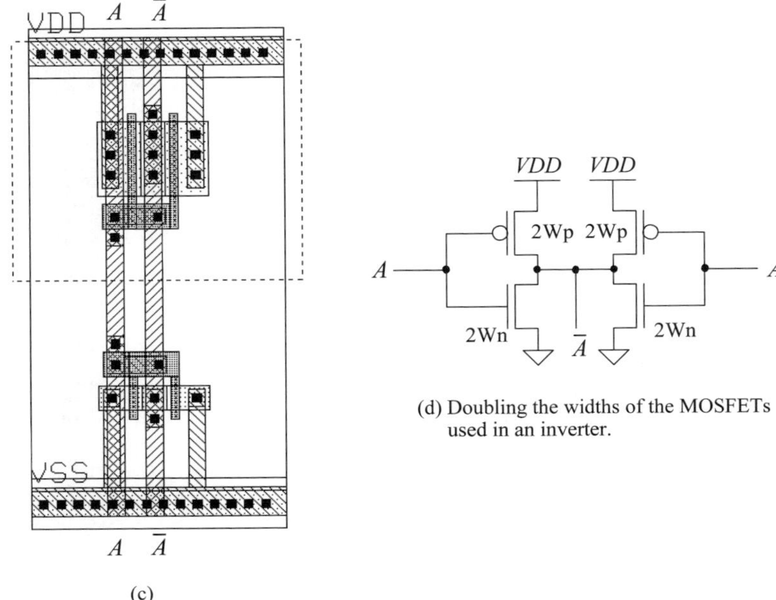

(d) Doubling the widths of the MOSFETs used in an inverter.

Figure 15.18 (a) Inverter, (b) stick diagram used for layout, (c) layout, and (d) equivalent schematic.

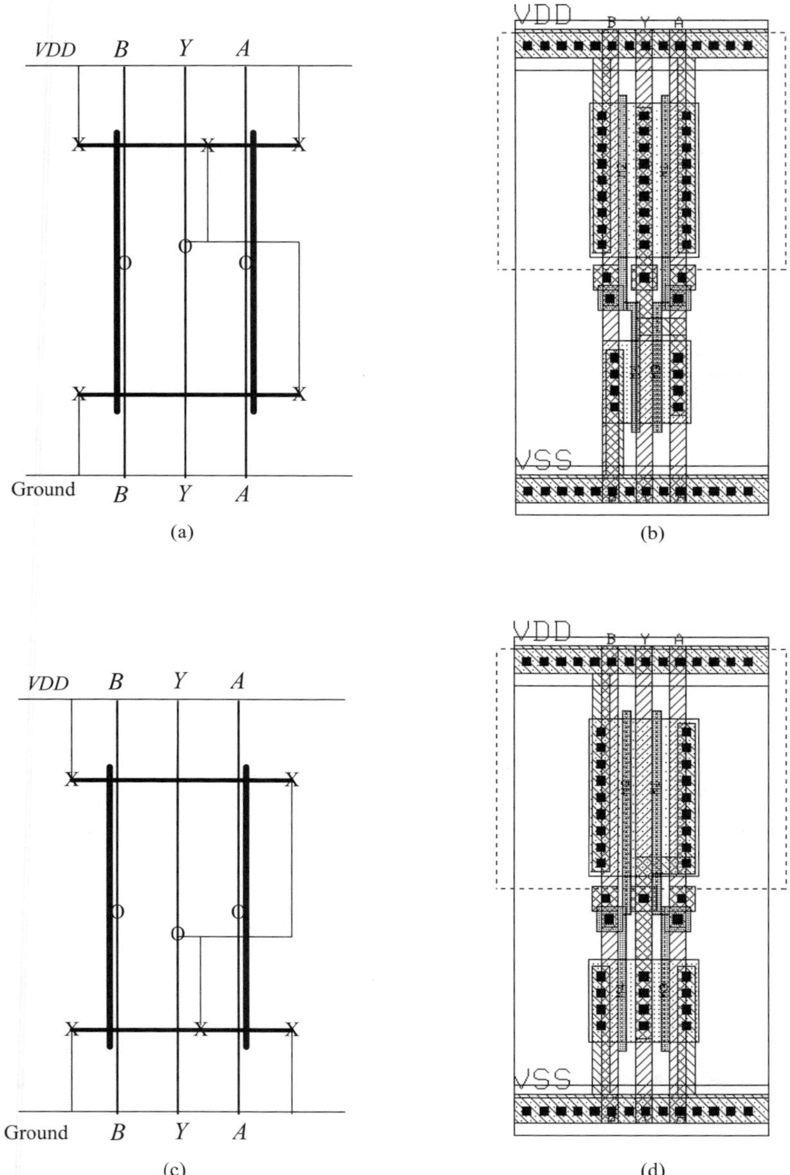

Figure 15.19 (a) NAND stick diagram, (b) layout, (c) NOR stick diagram, and (d) layout.

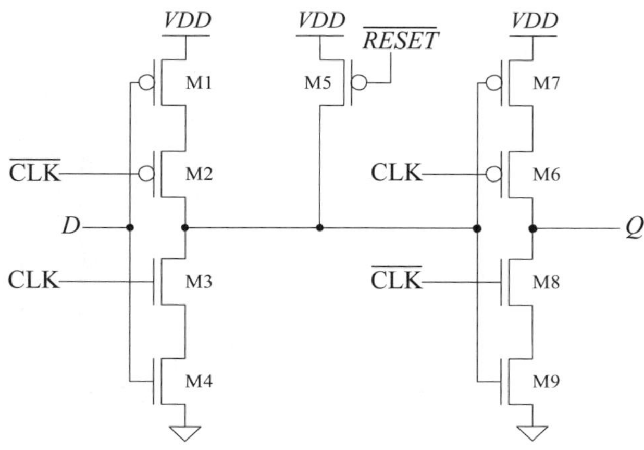

Figure 15.20 Schematic of a dynamic register cell.

Polish

After your layout is basically finished, it is time to step back and take a look at it from a purely aesthetic point of view. Is it pleasing to the eye? Is the hook-up as straightforward as possible, or is it "busy" and hard to follow? Are the spaces between poly gates and contacts minimum? What about the space between diffusions? Are there enough contacts? Did you share all of the source and drain implants that can be shared? Are there sufficient well and substrate ties? If you have planned well and followed the plan described here, you shouldn't run into too many problems.

Standard Cells Versus Full-Custom Layout

The standard cell approach to physical design usually dictates a fixed cell height and variable width when implementing the circuit. Furthermore, standard cells are designed to abut on two sides, usually left and right, and that abutment scheme must be quite regular so that any cell can reside next to any other cell without creating a design rule violation. The standard cell approach to layout is very useful and is always an excellent place to start. However, in the real world, area on a wafer translates directly into profit and loss (money). Wafer costs are relatively fixed whether they're blank or as tightly packed with circuitry as possible. Therefore, it follows that we want a layout that is as small as possible so that there can be as many die per wafer as possible. These are the economics of the situation. There are also technical advantages to be gained from having as small a layout as possible: interconnecting wires can be as short as possible, thereby reducing parasitic loading and crosstalk effects.

Figure 15.22 shows a typical standard-cell block that has been placed and routed by an automatic tool. Most of the individual cells have been omitted for clarity. Notice the interconnect channels between the rows of standard cells. Power, ground, and clock signal trunks run vertically to both sides of the block by means of a special cell called an "end cap." Cell rows are connected to power and ground through horizontal buses that are part of the standard cells themselves. All remaining connections are made via the

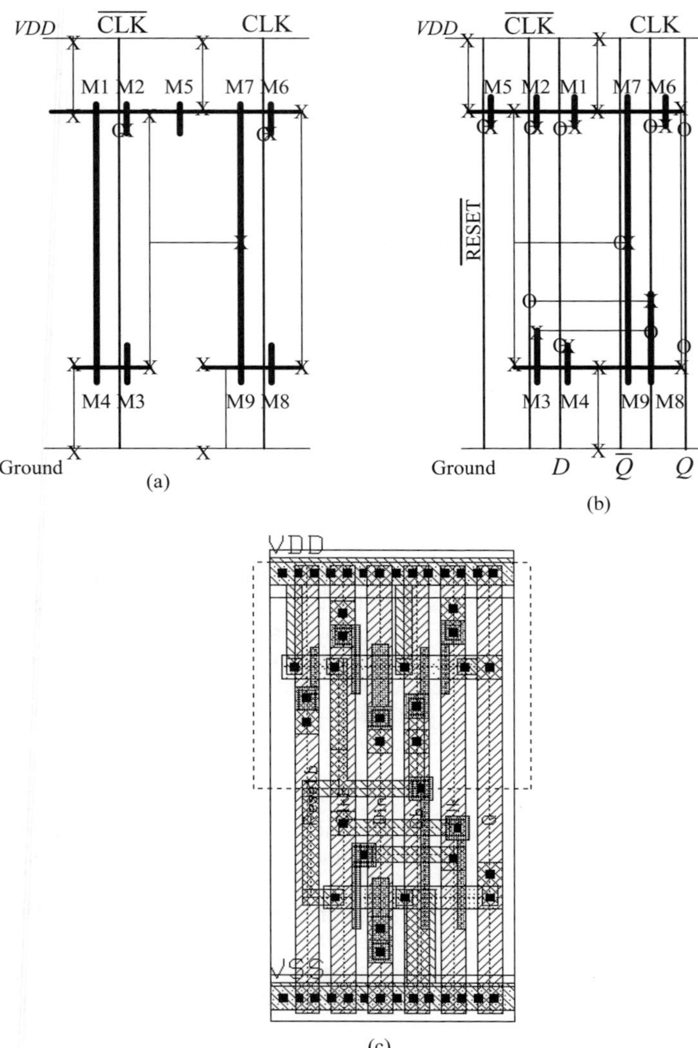

Figure 15.21 Layout of a dynamic register cell.

routing channels. The standard-cell layouts are designed to accommodate metal2 feedthroughs that run vertically through each cell. The autorouter makes use of this space and adds the feedthroughs as needed to connect or pass signals from one routing channel to another. The routing channels and their associated interconnecting wires are the limiting factors for both the density and circuit performance of this type of layout.

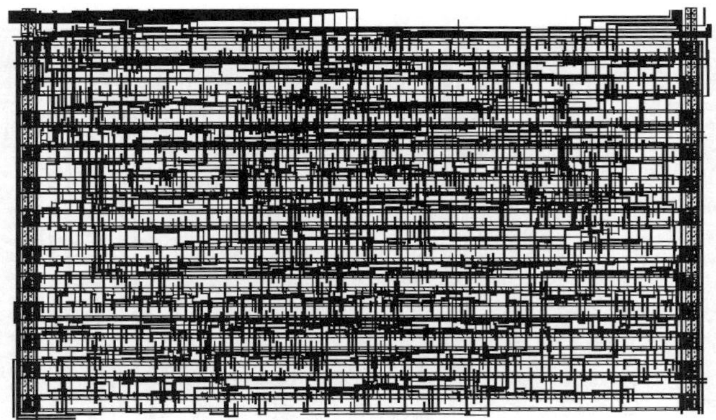

Figure 15.22 Layout based on standard cells.

Before we continue our discussion of relative layout density, we need to define a metric with which to quantify the matter. It is customary to use the number of transistors per square millimeter of area for this purpose. Because it is a raw number and the common denominator of all circuit layouts, we can use it even when comparing different types of circuitry or even unlike processes.

The density of the standard-cell route shown in Fig. 15.22 is approximately 5,000 transistors per square millimeter. This is fairly representative of the possible density for the channel-based routing approach and the process used (0.8 μm). Figure 15.23 shows a full-custom layout for a digital filter. The circuit area is approximately 2.1 square millimeters. The density is approximately 17,500 transistors per square millimeter, representing a 3.5-fold increase. This circuit, too, is fairly representative of the attainable density of full-custom layout using this particular 0.8 μm process. Both of these circuits were laid out using the same process, and in fact they are from the same die. The device sizes within each block would probably average out to minimum or close to minimum. The main difference affecting density is the interconnect wiring. This overhead associated with interconnect wiring is commonly referred to as the "interconnect burden." The designer must bear this burden in terms of both physical (wasted area) and electrical parameters (parasitic loading). Let us examine one method of creating a high-density custom layout that will minimize interconnect burden and circuit area.

Figures 15.24a–c show a small section of the interpolation filter from Fig. 15.23. In Fig. 15.24a we see an exploded view of four cells that form part of a data-path: an input data register, a t-gate, a full adder, and an output data register. These are instantiated (placed as a cell) twice, creating a view of eight cells. The two adder cells are slightly different: the carry inputs and outputs are on opposite sides, so that the carry-out can cascade to the carry-in of the next adder by abutting (placing next to one another) the cells. Unlike standard cells, the height and width constraints placed on custom layouts are contextual. In other words, a cell's aspect ratio depends on that of its neighbors. In this case, the width of each cell depended on the maximum allowable width of the widest cell

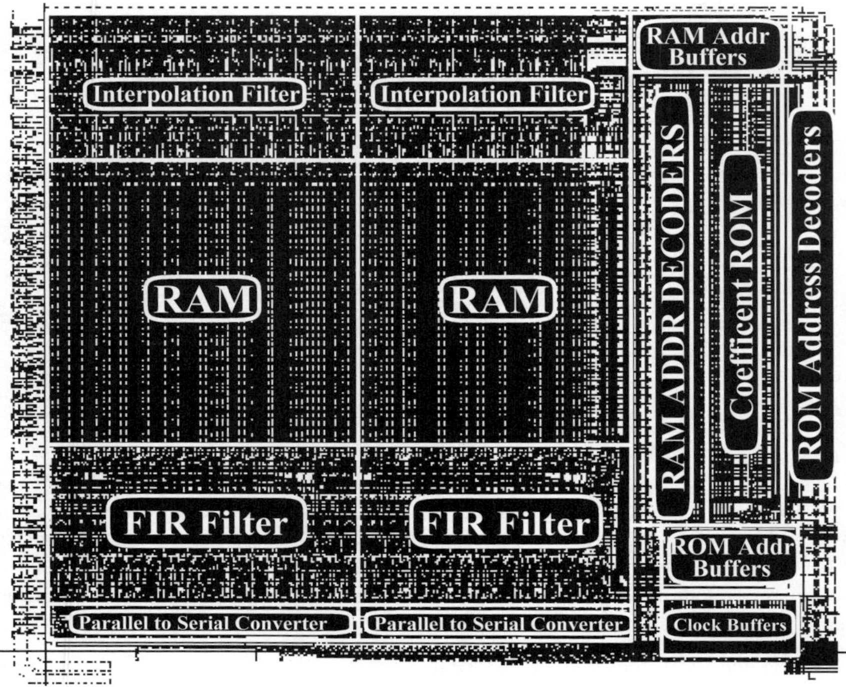

Figure 15.23 Full custom layout of a digital filter.

in the group: the data register. Notice the top, bottom, left, and right boundaries of each cell in Fig. 15.24a. Data enter the register cell from the top and are output at the bottom. Clocks, power, ground, and control signals route across all the cells. The adder receives its A and B inputs from the top and outputs their SUM at the bottom. As already mentioned, carry-out and carry-in are available on the left and right edges of the adder, respectively. Figure 15.24b shows a 2-bit slice of this data-path with all of the connections made by cell abutment. Figure 15.24c illustrates how all four edges of each cell join together to complete the hook-up.

We have seen how circuits can be implemented by means of standard cells or custom layout. The time needed to produce a standard-cell route is far less than that of a full-custom implementation. The trade-offs are area and performance. Automatic place and route tools based on routing area rather than routing channels are now coming into use. These promise a compromise solution between the two extremes. The density of their results rivals that of full custom layout. Perhaps the hand-rendered, full custom layout will someday become a thing of the past. Nevertheless, process technology continues to advance, circuit designers continue to design circuits that test the outermost limits of this technology, and the marketplace will still be there demanding ever cheaper, more powerful, and faster products. It is likely then that we will all still have the

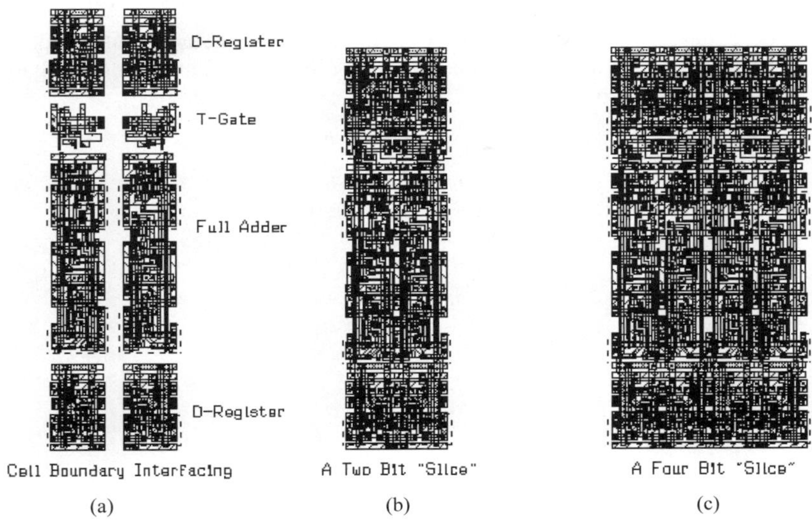

Figure 15.24 Sections of the digital interpolation filter.

opportunity to "push a polygon" or two for the foreseeable future. There remains no doubt that the future will bring us ever more powerful software tools that will take over the tedious aspects of placing and connecting layouts, leaving to us the more creative aspects of planning and polishing them.

ADDITIONAL READING

[1] N. H. E. Weste and D. Harris, *Principles of CMOS VLSI Design*, Addison-Wesley, 3rd ed., 2004. ISBN 0-321-14901-7.

[2] J. P. Uyemura, *Introduction to VLSI Circuits and Systems*, John Wiley and Sons Publishers, 2002. ISBN 0-471-12704-3.

[3] Kerth, Donald A. *"Floorplanning-Lecture Notes,"* Crystal Semiconductor, Inc.

[4] Kerth, Donald A. *"Analog Tricks of the Trade-Lecture Notes."* Crystal Semiconductor, Inc.

[5] J. Uyemura, *Physical Design of CMOS Integrated Circuits Using L-EDIT*, PWS Publishing Co., 1995. ISBN 0-534-94326-8.

[6] D. V. Heinbuch, *CMOS3 Cell Library*, Addison-Wesley, 1988. ISBN 0-201-11257-4.

[7] D. Clein, *CMOS IC Layout: Concepts, Methodologies, and Tools*, Newnes Publishers, 2000. ISBN 0-750-67194-7. Excellent introductory book for CMOS IC layout. Covers the entire layout process.

[8] C. Saint and J. Saint, *IC Layout Basics: A Practical Guide*, McGraw-Hill, 2001. ISBN 0-071-38625-4. Very good introductory book on IC Layout.

[9] C. Saint and J. Saint, *IC Mask Design: Essential Layout Techniques*, McGraw-Hill, 2002. ISBN 0-071-38996-2. Companion book for reference [8].

[10] G. Petley, *The Art of Standard Cell Library Design*, to be published. See the website: http://www.vlsitechnology.org/ for more information (including example layouts and schematics).

Chapter 16

Memory Circuits

In this chapter we turn our attention towards the design of semiconductor memory circuits. This includes the array design, sensing, and, finally, the operation of the memory cells themselves. The memories we look at in this chapter are termed random access memories or RAM because any bit of data can be accessed at any time. A block diagram of a RAM is shown in Fig. 16.1. At the intersection of a row line (a.k.a., word line) and a column line (a.k.a., a digit or bit line) is a memory cell. External to the memory array are

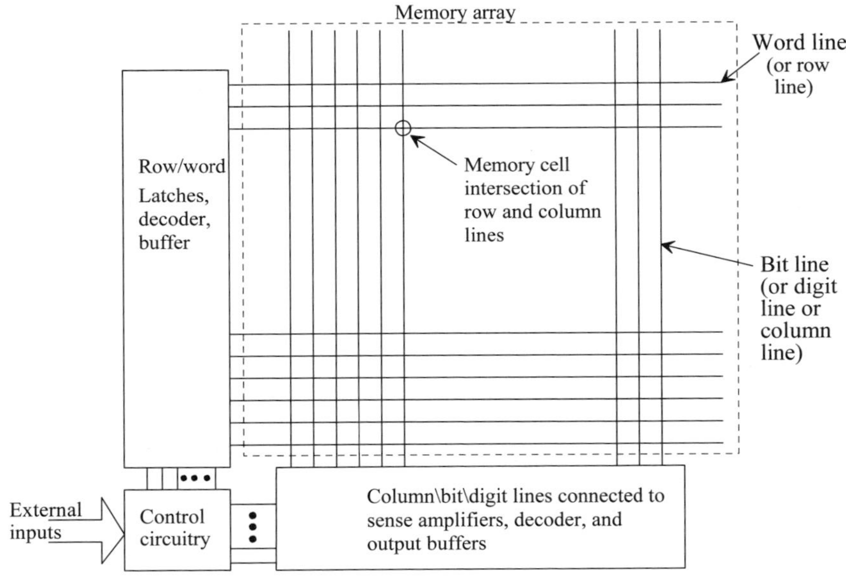

Figure 16.1 Block diagram of random access memory.

the row and column logic. Referring to the row lines, the row address is latched, decoded, and then buffered. A particular row line will, when selected, go high. This selects the entire row of the array. Since the row line may be long and loaded periodically with the capacitive memory cells, a buffer is needed to drive the line. The address is latched with signals from the control logic. After a particular row line is selected, the column address is used to decode which of the bits from the row are the addressed information. At this point, data can be read into or out of the array through the column decoder. The majority of this chapter presents the circuits used to implement a RAM.

16.1 Array Architectures

Examine the long length of metal shown in Fig. 16.2. Let's treat this metal line as one of the bit lines seen in Fig. 16.1. The parasitic capacitance of this bit line to ground (substrate) can be calculated using

$$C_{col1sub} = Area \cdot C_{1sub} \tag{16.1}$$

If the capacitance from the metal lines to substrate is 100 aF/μm^2, then

$$C_{col1sub} = (0.1)(100)(100\ aF) = 1\ fF \tag{16.2}$$

Not that significant of a capacitance. However, at the intersection of the bit line with every word line we have a memory cell. Let's say that we have a memory cell every 400 nm (250 total cells or word lines) and that each memory cell is connected through an NMOS device to the bit line. As seen in Fig. 16.3, this results in a periodic (depletion or junction) capacitance on the bit line from each MOSFET's source or drain implant. If this capacitance is 0.4 fF, then the total capacitance hanging on the bit line is

$$C_{col} = (\text{number of word lines}) \cdot (\text{capacitance of the MOSFET's source/drain}) + C_{col1sub} \tag{16.3}$$

or $C_{col} = 101\,fF \approx 100\,fF$ for this discussion. For the majority of the discussions in this chapter, we'll treat the column conductor as a capacitor, C_{col}. Note that increasing the number of word lines (the number of memory cells connected to a bit line) increases the bit line capacitance.

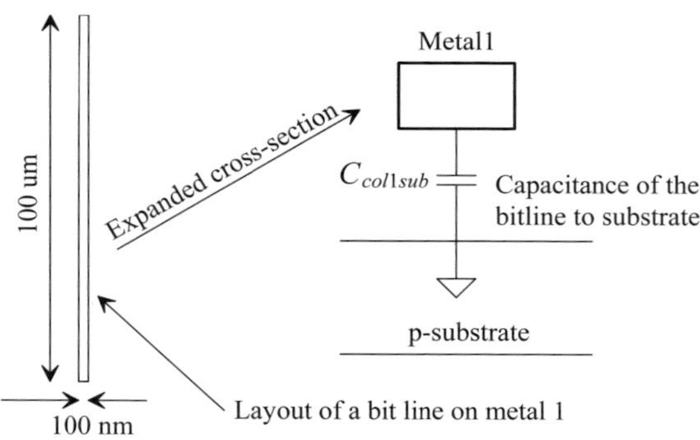

Figure 16.2 The parasitic capacitance of a bitline to ground (substrate).

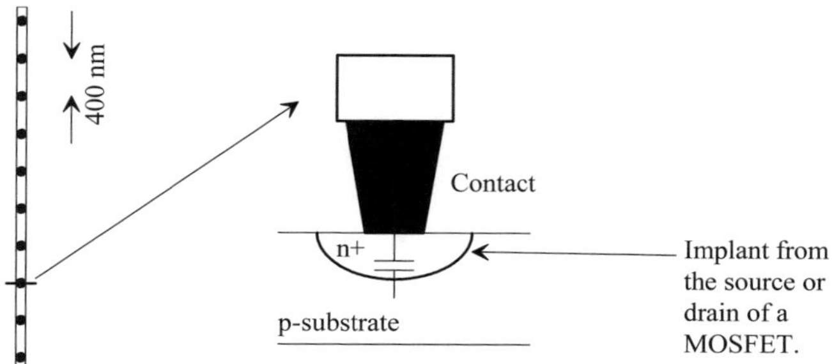

Figure 16.3 How a bit line is loaded with implants (depeletion capacitance).

16.1.1 Sensing Basics

In this section we discuss sensing the datum from a memory cell. Examine Fig. 16.4. When a memory cell is accessed (the word [row] line goes high), the datum from the cell (a charge) is placed on the bit line. The bit line voltage changes. At this point we know the bit line looks like a capacitor and that only one word line can go high at a time in a memory array (so that we don't have two memory cells trying to dump their data onto the same bit line at the same time). The voltage movement on the bit line, ΔV_{bit}, may be very small, e.g., 50 mV or less, and so determining if the voltage is moving upwards or downwards can be challenging. In addition, we would like our sense amplifier to drive the bit line to full, valid, logic levels (for speed reasons in some memories and to refresh the cell in a dynamic RAM, DRAM). Let's consider some sense amplifier topologies.

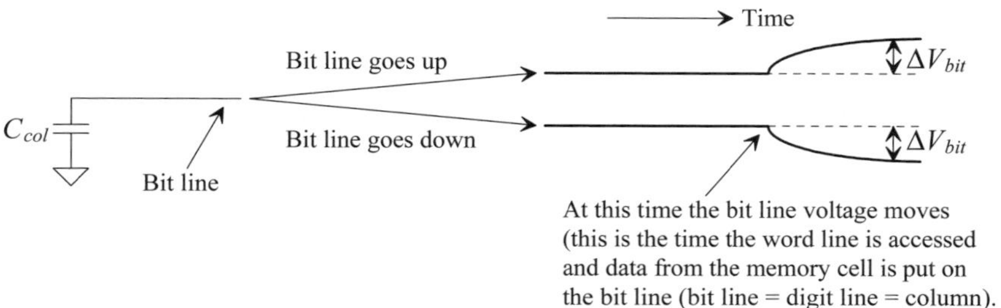

Figure 16.4 Sensing a change on the bit line in an array.

NMOS Sense Amplifier (NSA)

Consider the schematic of an NMOS sense amplifier (NSA) seen in Fig. 16.5. M1 and M2 form the NMOS portion of an inverter-based latch. The idea is to develop an imbalance on the gates of M1/M2 so that one MOSFET turns on harder than the other. Before we start sensing, we set the drains of M1/M2 to the same potential (an equilibrium potential). When *sense_N* goes high (indicating that we are starting the sense operation), the signal *NLAT* (N-latch) goes to ground (the sources of M1/M2 move to ground). While this circuit clearly can't pull the bit line high (we'll add a PMOS sense amplifier later to do this), we should see a problem. How do we get good sensing if the transistor loads are not balanced?

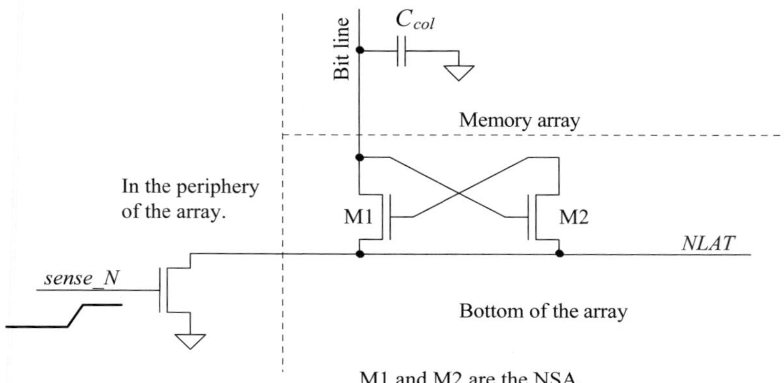

Figure 16.5 Development of an NMOS sense amplifier (NSA).

The Open Array Architecture

To provide the same load capacitance for each input of the NSA, two arrays can be used, as seen in Fig. 16.6. The array architecture in Fig. 16.6, from the side, looks like an open book and so it is called an "open array architecture." Let's use some numbers and the partial schematic seen in Fig. 16.7 to illustrate the sense amplifier's operation.

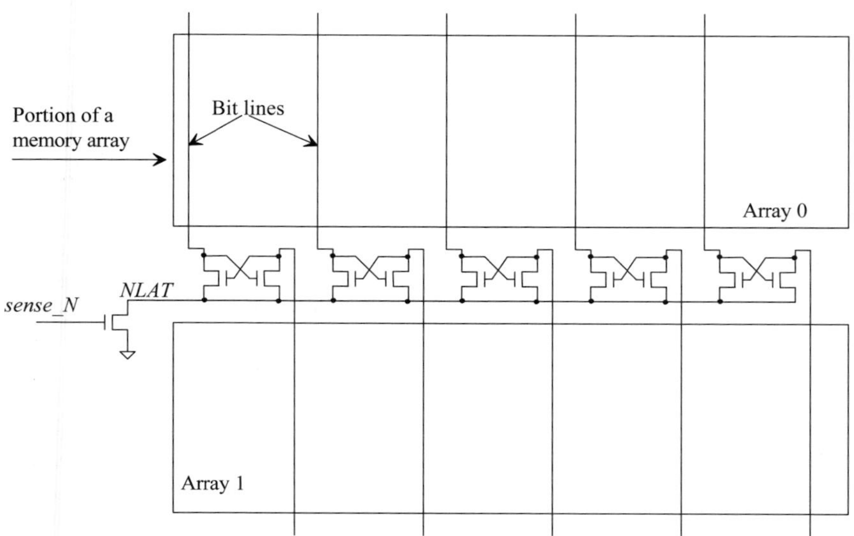

Figure 16.6 How the NSA is placed between two memory arrays in the so-called open memory array architecture.

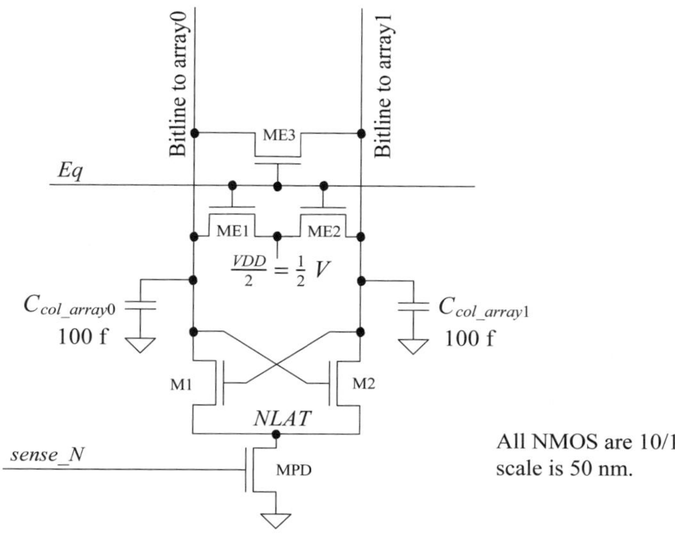

Figure 16.7 An NSA with equilibration circuitry and connections to two bit lines.

In the following we assume $VDD = 1$ V and the column capacitance is 100 fF. We start any sense operation by equilibrating the bit lines (the inputs to the sense amplifier). In Fig. 16.7 ME1–ME3 short the bit lines together and to $VDD/2$ (= 0.5 V). When the equilibrate signal, *Eq*, goes high, all of the word lines are low. We are not accessing any data in the array but, rather, getting ready for the sense (read) operation. Figure 16.8 shows how the *Eq* signal is asserted for a short period of time and how, during this time, the bit lines are equilibrated together and to 0.5 V.

During a sense operation, one of the bit lines will be pulled from $VDD/2$ to VDD. The other line will be pulled from $VDD/2$ to ground. The amount of power used during a sense depends on frequency and given by

$$P_{avg} = (\text{number of sense amplifiers}) \cdot C_{col} \cdot (VDD/2)^2 \cdot f \qquad (16.4)$$

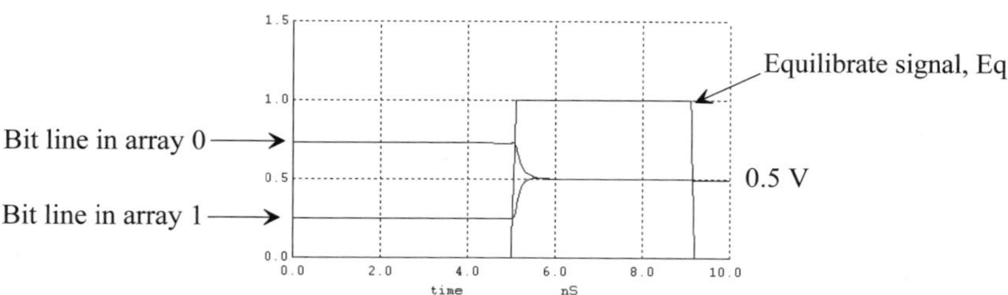

Figure 16.8 How the equilibrate circuitry operates.

Note, however, that when we equilibrate, no power is consumed. The two bit line capacitances simply share their charge (one going from ground to *VDD*/2 while the other goes from *VDD* to *VDD*/2).

To show a sensing operation, let's use the basic one-transistor, one-capacitor (1T1C) DRAM memory cell seen in Fig. 16.9. To fully turn on the access MOSFET, we need to drive the word line to $VDD + V_{THN}$ (with body effect). A typical value for the memory bit capacitance, C_{mbit}, is 20 fF. If the voltage on this capacitance is called V_{mbit} and the bit line is precharged to *VDD*/2, then we can write the total charge on both capacitors *before* the access MOSFET turns on as

$$Q_{tot} = C_{mbit}V_{mbit} + (VDD/2) \cdot C_{col\_array} \tag{16.5}$$

After the access MOSFET turns on, the voltage across each capacitor will be the same. We'll call this voltage, V_{final}. Since charge must be conserved

$$V_{final}(C_{mbit} + C_{col\_array}) = C_{mbit}V_{mbit} + (VDD/2) \cdot C_{col\_array} \tag{16.6}$$

or

$$V_{final} = \frac{C_{mbit}V_{mbit} + (VDD/2) \cdot C_{col\_array}}{C_{mbit} + C_{col\_array}} \tag{16.7}$$

If a logic one is stored on C_{mbit} (which means V_{mbit} is *VDD*, or here, 1 V) and C_{mbit} = 20 fF and C_{col_array} = 100 fF, then V_{final} = 0.583 V. The change in the bit line voltage is

$$\Delta V_{bit} = V_{final} - VDD/2 \tag{16.8}$$

or here $\Delta V_{bit} = 83\ mV$.

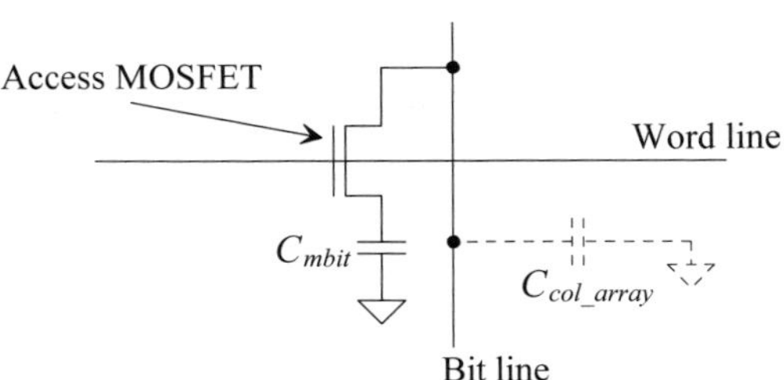

Figure 16.9 The one-transistor, one-capacitor (1T1C) DRAM memory cell.

Figure 16.11 shows the signals resulting from the operation of the sense amplifier in Fig. 16.10. The 1T1C DRAM (mbit) cell in array1 (the DRAM cell at the bottom of the schematic) has its word line held at ground when we sense the data in the cell in array0. The bit line from array1 is simply used as a reference. When we sense the data in array1, the bit line in array0 is used as the reference (and so the word line in array0 is held at ground).

For the simulation results in Fig. 16.11, we start out by equilibrating the bit lines (as in Fig. 16.8). Next our word line, in array0, goes to a voltage greater than $VDD + V_{THN}$. This causes charge sharing between C_{mbit} and the digit line capacitance. For the simulation results seen in Fig. 16.11, we set the voltage on C_{mbit} to zero so we are reading

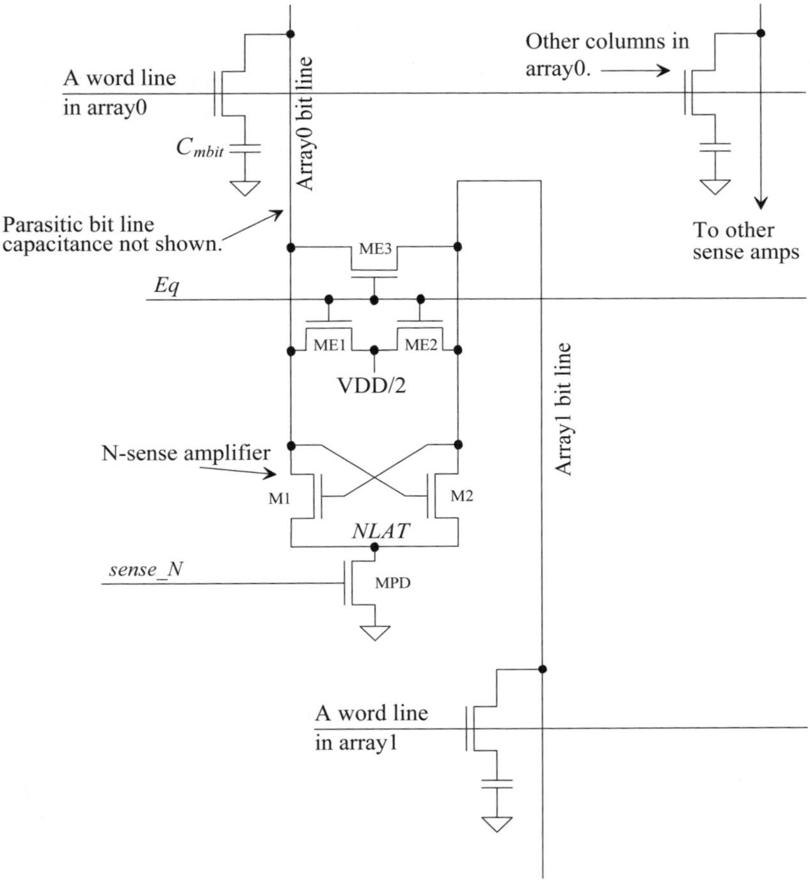

Figure 16.10 The connection of the NSA to the memory arrays.

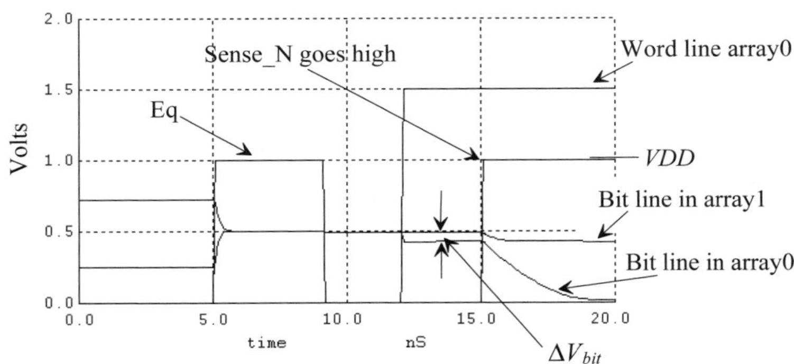

Figure 16.11 N-sense amp's operation.

out zero. When *sense_N* goes high, the NSA "fires" causing the bit line in array0 to move to ground. Note that our reference bit line (from array1) droops a little. Figure 16.12 shows the signals in a sensing operation if the mbit cell contains a "1." The bit line voltage in array0 now increases. The reference bit line in array1 is pulled to ground (this doesn't harm the data in array1). To pull the bit line in array0 high, we'll now add a PMOS sense amplifier (PSA).

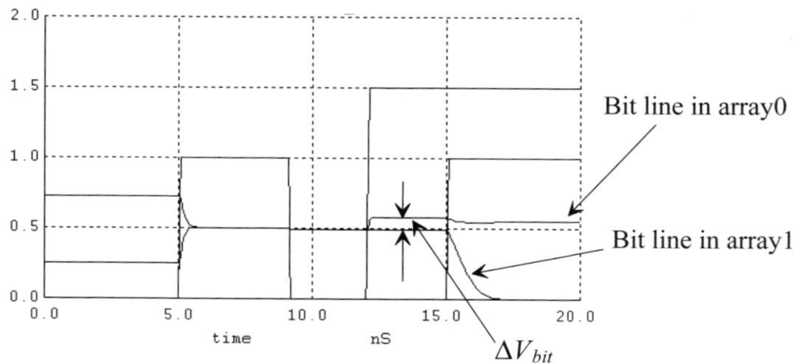

Figure 16.12 Reading out a "1" from the cell in array0.

PMOS Sense Amplifier (PSA)

To pull the bit lines up to *VDD*, we can add a PSA to the periphery of the array. Figure 16.13 shows the schematic of the PSA. The signal *ACT* (active pullup) is common to all of the PSAs on the periphery of the array (as is *NLAT* for the NSAs seen in Fig. 16.6). The signal $\overline{sense\_P}$ is active low and indicates that the PSA is firing. The PSA is usually fired after the NSA because the matching of the NMOS devices is, generally, better than the matching of the PMOS. It's not common to fire the sense amplifiers at the same time

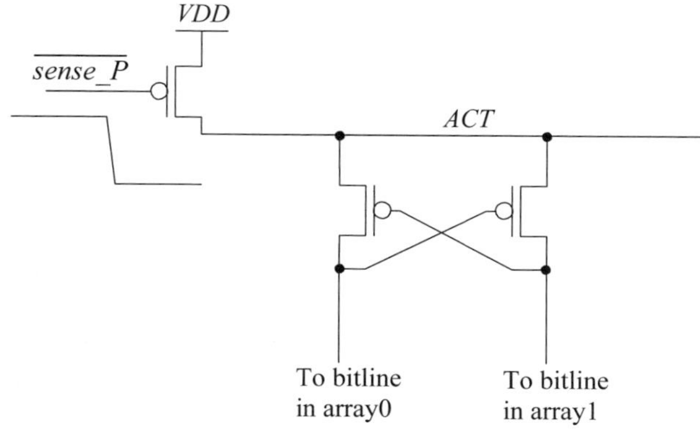

Figure 16.13 Schematic diagram of a PMOS sense amplifier.

because of the potentially significant contention current that can flow. Figure 16.14 shows the full operation of a sense when the cell we are reading out has a one stored in it.

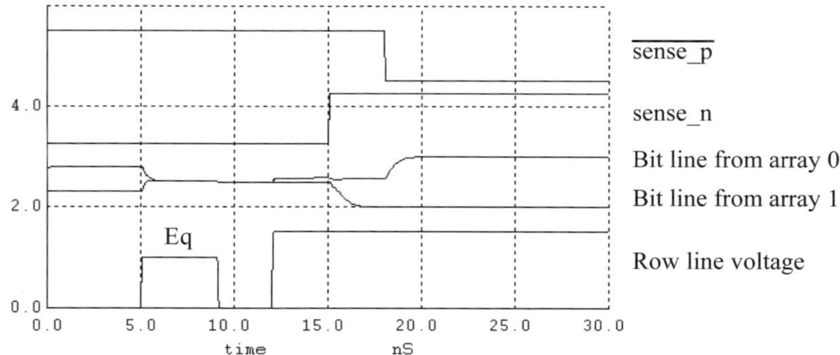

Figure 16.14 How the PSA pulls the bit line from array 0 high.

Refresh Operation

Our sensing operation determined whether a 1 or a 0 was stored in the memory bit. For this particular sense amplifier topology, the inputs and outputs are the same terminals. This is fundamentally important for DRAM operation where the charge can leak off the capacitor because of the finite subthreshold slope of the access transistors or leakage through the MOSFET's source/drain implant to substrate (hence the name "dynamic"). To ensure long-lasting data retention in a DRAM, the cells must be periodically refreshed. When we fire the sense amplifiers, with the access device still conducting, the mbit capacitor is refreshed through the access MOSFET (the bit line is driven to ground or *VDD* by the sense amp).

16.1.2 The Folded Array

The open array architecture seen in Fig. 16.6 has a memory cell at the location of every intersection of a word line and a bit line. The open array architecture results in the most dense array topology. Unfortunately, because of the physical distance between the bit lines used by the sense amplifier, this architecture is sometimes not used. Figure 16.15 shows the basic problem. Coupled noise (e.g., from the substrate) feeds unequal amounts

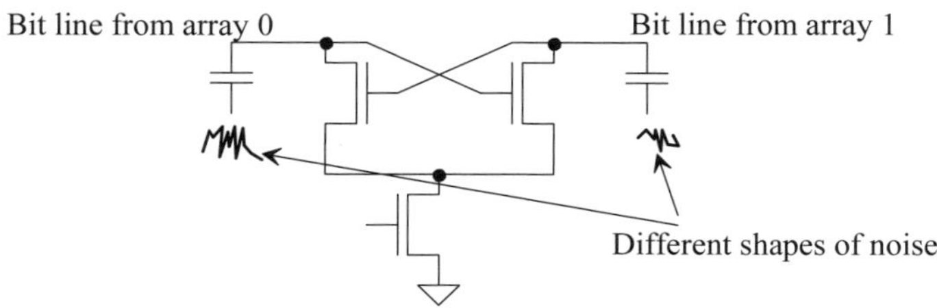

Figure 16.15 Different amplitudes of coupled noise into physically separated bit line.

of charge into the bit lines. This can cause the sense amplifier to make wrong decisions. To attempt to make each bit line see the same sources of noise, we can lay the bit lines out next to each other by folding array 1 on top of array 0 (see Fig. 16.6). The result, called a *folded array architecture*, is seen in Fig. 16.16. The bold lines in this figure are the bit lines from array 1.

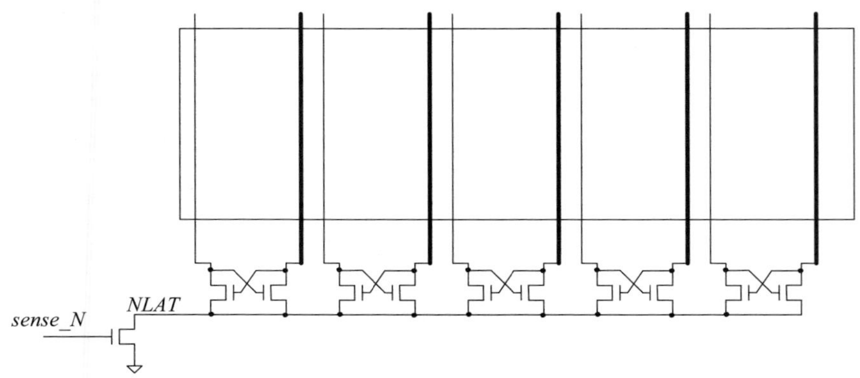

Figure 16.16 The folded array is formed by taking the open array architecture (open book) topology seen in Fig. 16.6 and "closing the book," that is, folding array 1 on top of array 0. Note that the bold lines indicate the bitlines from array 1 in the newly formed array.

What is the cost for this improved noise performance? We know that when a sense amplifier is used, one input is varied by the memory cell we are sensing, while the other input is simply used as a reference. What this means is that instead of having a memory cell at the intersection of every row and column, as in the open array, now we have a memory cell at the location of every other row and column, Fig. 16.17.

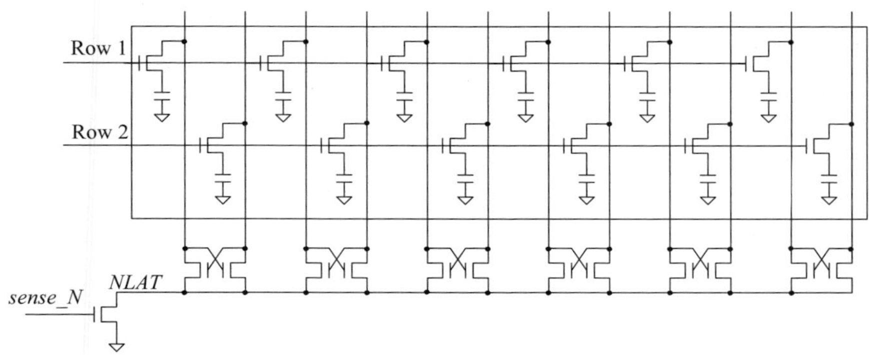

Figure 16.17 How a memory cell is located at every other intersection of a row line and a column line in a folded-area architecture.

Layout of the DRAM Memory Bit (Mbit)

Layout area, especially in a periodic array like a memory, is very important. Reducing the chip size increases the number of die on a wafer. Since the processing cost for a wafer is fixed, having more chips to sell per wafer increases the manufacturer's profit. To reduce the size of an mbit, often (always in DRAM) the contact to the bit line is shared between two mbit cells, Fig. 16.18. The word lines (row lines) are made using silicided polysilicon. Using polysilicon for the word lines can lead to significant delays. As seen in Fig. 16.17, for example, a signal applied to row 1 must propagate down a distributed R (the resistance of the polysilicon) C (the gate oxide capacitance of the MOSFET) line. The propagation delay through a word line can be estimated using

$$t_d = \overbrace{(\text{number of columns}) \cdot \left(WL\frac{\varepsilon_{ox}}{t_{ox}} + C_{parasitic}\right)}^{\text{total capacitance on the word line}} \cdot \overbrace{(\text{number of columns}) \cdot R_{gat}}^{\text{total resistance of the word line}} \quad (16.9)$$

The number of columns is simply the number of MOSFETs in the row. The term $WL\frac{\varepsilon_{ox}}{t_{ox}}$ is C_{ox}, while R_{gate} is the resistance from one end of the polysilicon gate to the other end in the memory cell layout seen in Fig. 16.18. The term $C_{parasitic}$ is the parasitic capacitance associated with the cell (such as the capacitance from the word line to the bit line). If C_{ox} is 400 aF, $C_{parasitic}$ = 100 aF, R_{gate}= 4 Ω, and there are 512 bit lines in the array then the delay time through the word line is estimated as 500 ps. To fully turn the word line on (not just to the 50% point where delay is measured), we would probably want 3 ns.

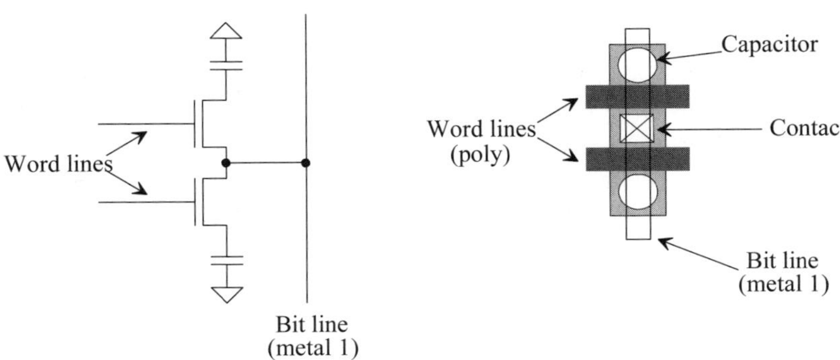

Figure 16.18 Two mbits sharing a contact to the bit line.

A section of the area layout for an open architecture DRAM is seen in Fig. 16.19. Notice that at the intersection of every bit and word line is a memory cell. A common term used to describe the density of, or distance in, a periodic array is *pitch*. As seen in Fig. 16.19, the pitch is defined as the distance between like points in the array. We used the distance between the right side of the digit lines to show pitch in this figure. A figure of merit for memory cell layout is its area. The area of the mbit DRAM cell seen in this figure is $6F^2$ where F = pitch/2. Question: How many MOSFETs are laid out in Fig. 16.19? Answer: Because polysilicon over active forms a MOSFET, we simply count the number of times poly crosses active (12 MOSFETs).

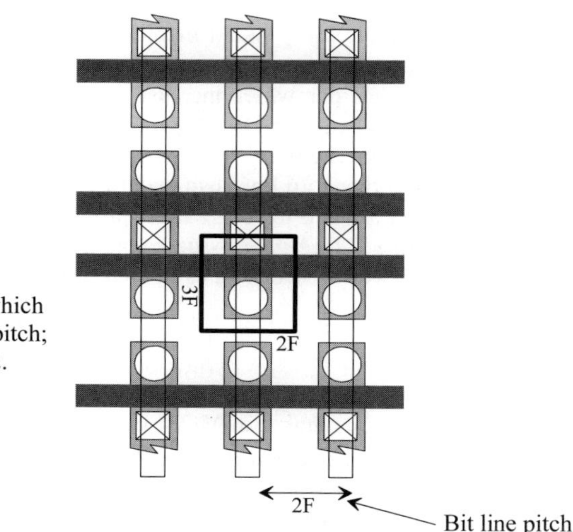
F is feature size, which is half the bit line pitch; that is, F = pitch/2.
3F
2F
2F
Bit line pitch

Figure 16.19 Layout of mbit used in an open bit line configuration.

Figure 16.20 shows the layout of the cell used in the folded area architecture. The schematic seen in this figure is different from the one seen in Fig. 16.17. In Fig. 16.20 our mbits share the contact to the bit line. This means that memory cells are located at the intersection of every other mbit pair with the bit lines. Figure 16.21 shows a section of the array layout in the folded architecture. The cell size is $8F^2$. The increase in the cell size is due to the needed poly interconnect between adjacent cells (the poly that runs over the field oxide, FOX).

Figures 16.22 and 16.23 show the process cross sectional views for mbits using trench capacitors and buried capacitors, respectively. The trench capacitor is formed by etching a hole in the substrate. For high-density memory, the aspect ratio (the ratio of the

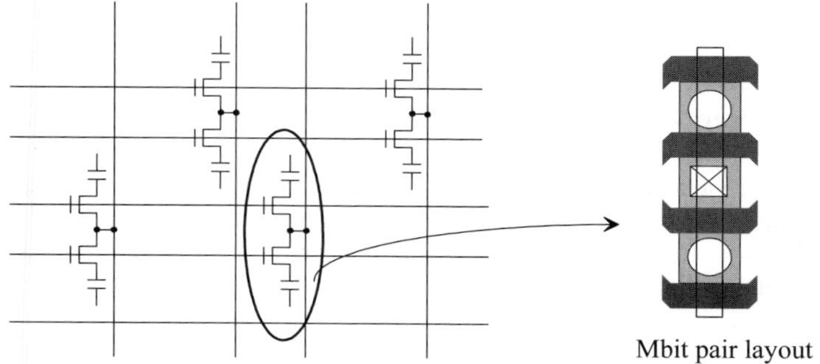
Mbit pair layout

Figure 16.20 The mbit pair used for a folded architecture.

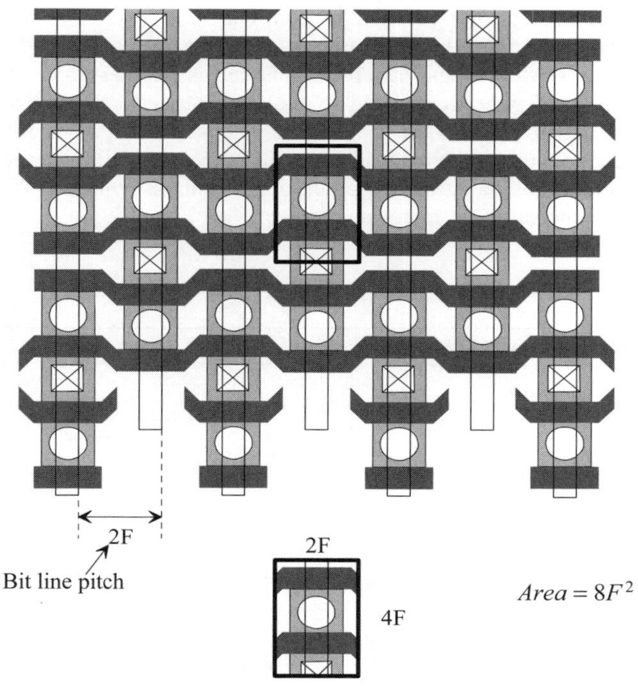

Figure 16.21 Folded architecture layout and cell size.

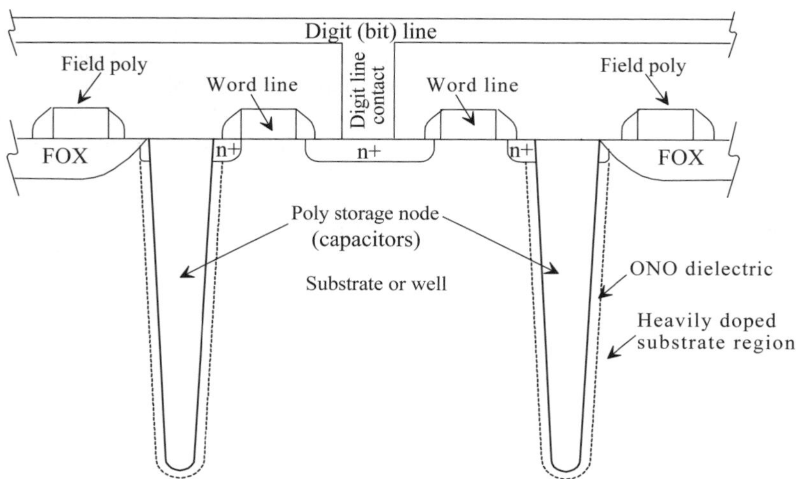

Figure 16.22 Cross-sectional view of a trench capacitor cell.

hole's depth to its diameter) can be quite high, which leads to processing concerns. In Fig. 16.23 the buried capacitor cell pair is seen. Unlike the trench capacitor-based cell that places the cell directly in the substrate, the capacitor is "buried" under the digit line but still above the substrate. The benefit of this cell is simpler processing steps. One problem with this cell, when compared to the trench-based cell, is that the parasitic capacitances are higher (such as the capacitance loading the (digit) bit line). Also, for a given bit capacitance, C_{mbit} , the area of the buried capacitor cell may need to increase, while the area of the cell using the trench capacitor remains constant. The depth of the trench capacitor is increased for more capacitance.

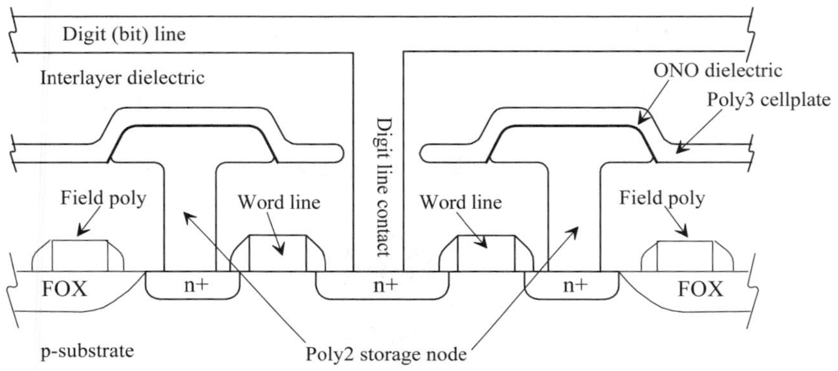

Figure 16.23 Cross-sectional view of a buried capacitor cell.

As we saw in Eq. (16.7), the value of C_{mbit} is very important. Thin dielectrics with high dielectric constants are used to implement C_{mbit}. Tricks like roughing up the dielectric to increase the surface area are often employed. To minimize the stress on the oxide, the common node of the capacitor is usually tied to $VDD/2$, as seen in Fig. 16.24. When a logic 1 (VDD) is written to the cell, the charge on the capacitor is $(VDD/2) \cdot C_{mbit}$. When a logic 0 (ground) is written, the charge on the capacitor is $-(VDD/2) \cdot C_{mbit}$. The difference in these charges is $VDD \cdot C_{mbit}$, which is the same difference we would have if the common node were connected to ground.

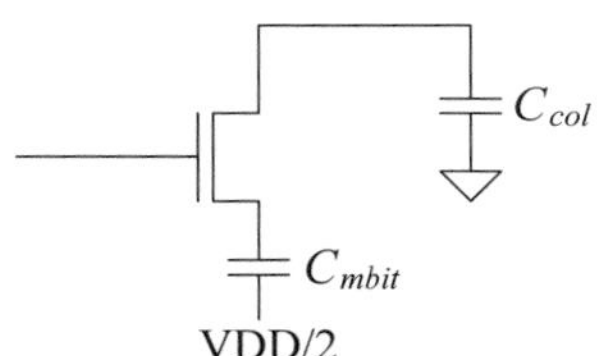

Figure 16.24 Holding the capacitor's common plate at VDD/2 to minimize oxide stress.

16.1.3 Chip Organization

The number of rows in a DRAM is limited because each additional row adds capacitance to the bit line. The power goes up, as seen in Eq. (16.4), with increasing C_{col}. The number of bit lines is limited by the delay through the row line (the length of the row line), as indicated in Eq. (16.9). For these reasons, the size of a DRAM array is limited. A typical array size is 512 word line pairs and 512 bit lines for a total memory size of 256 kbits. We use the term "pairs" to indicate that, as seen in Fig. 16.20, there are actually 1,024 row lines in a 256k array. We can arrange these 256k arrays to form a larger memory array. Figure 16.25 shows the basic idea for an 8-Mbit array block. A 1-Gbit DRAM uses 8,192 256k arrays. These arrays may be further subdivided into "banks" of memory. We might, for a 1-Gbit memory, have 8 banks of 128-Mbits. We subdivide the memory so that an operation (say a read) can be done in one bank while, at the same time, some other operation (say a refresh) can be performed in a different bank.

The *array efficiency* is defined as the ratio of the area of the memory blocks to the total chip area. A typical value ranges from 50 to 60%. The peripheral circuitry seen in Fig. 16.25 (the sense amplifier and row decoder blocks) take up a significant amount of chip real estate. Because of their importance, we'll devote the entire next section to the design of peripheral circuits including sense amplifiers, row drivers, and decoder circuits.

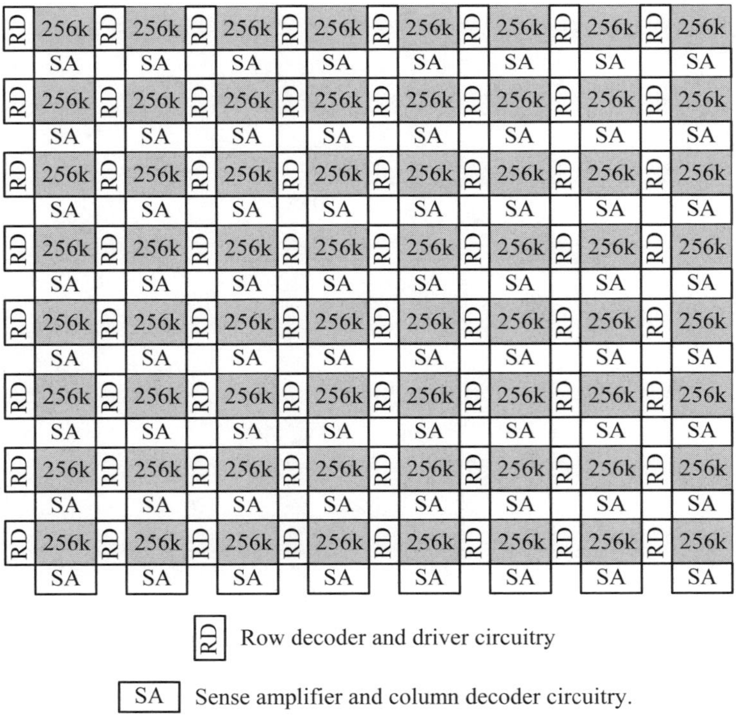

RD	256k	RD	256k	RD	256k	RD	256k	RD	256k	RD	256k	RD	256k	RD	256k
	SA		SA		SA		SA		SA		SA		SA		SA
RD	256k	RD	256k	RD	256k	RD	256k	RD	256k	RD	256k	RD	256k	RD	256k
	SA		SA		SA		SA		SA		SA		SA		SA
RD	256k	RD	256k	RD	256k	RD	256k	RD	256k	RD	256k	RD	256k	RD	256k
	SA		SA		SA		SA		SA		SA		SA		SA
RD	256k	RD	256k	RD	256k	RD	256k	RD	256k	RD	256k	RD	256k	RD	256k
	SA		SA		SA		SA		SA		SA		SA		SA
RD	256k	RD	256k	RD	256k	RD	256k	RD	256k	RD	256k	RD	256k	RD	256k
	SA		SA		SA		SA		SA		SA		SA		SA
RD	256k	RD	256k	RD	256k	RD	256k	RD	256k	RD	256k	RD	256k	RD	256k
	SA		SA		SA		SA		SA		SA		SA		SA
RD	256k	RD	256k	RD	256k	RD	256k	RD	256k	RD	256k	RD	256k	RD	256k
	SA		SA		SA		SA		SA		SA		SA		SA
RD	256k	RD	256k	RD	256k	RD	256k	RD	256k	RD	256k	RD	256k	RD	256k
	SA		SA		SA		SA		SA		SA		SA		SA

RD Row decoder and driver circuitry

SA Sense amplifier and column decoder circuitry.

Figure 16.25 A 16-Mbit array block.

16.2 Peripheral Circuits

In this section we discuss the design of general sense amplifiers (for general use in memory design), row drivers (remembering, for a DRAM, the row voltage needs to be driven above *VDD*), and column and row decoder circuits.

16.2.1 Sense Amplifier Design

Examine the clocked sense amplifier seen in Fig. 16.26. When *clock* is low, MS3 is off while MS1 and MS2 are on. Our input signals can't go below V_{THP} (with body effect) without shutting MS1 and MS2 off. Assuming that the inputs stay above V_{THP}, the drains of M1/M3 and M2/M4 are charged to In+ or In– (creating an imbalance). When *clock* goes high, the imbalance causes the circuit to latch high or low depending on the state of the inputs. Figure 16.27 is the simulation output showing the typical operation of the circuit. When *clock* is low, the circuit outputs are not valid logic signals but rather, ideally, track the input signals. This circuit is plagued with problems including: kickback noise, memory, and significant potential contention current.

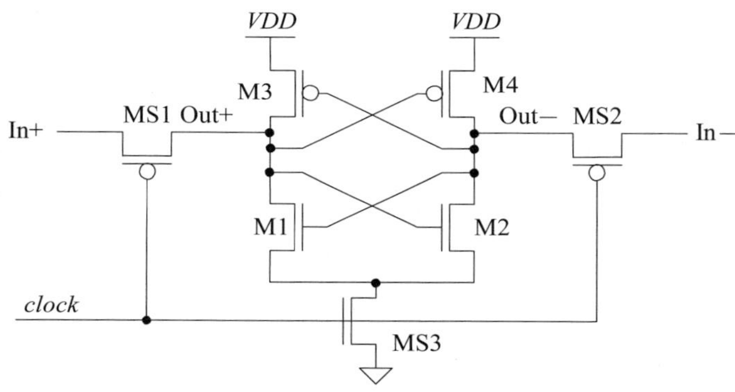

Figure 16.26 Clocked sense amplifier.

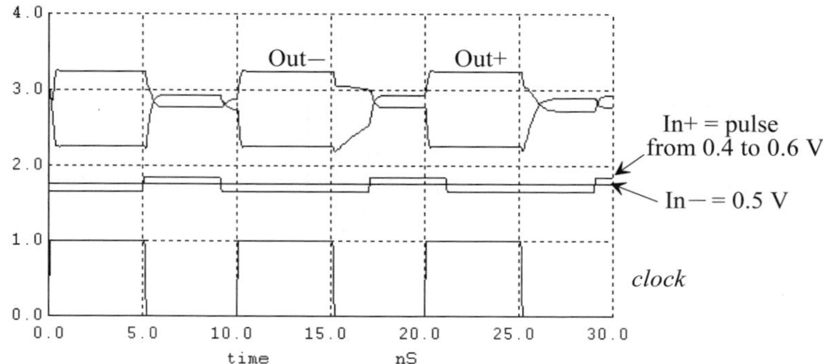

Figure 16.27 Simulating the operation of the circuit in Fig. 16.26.

Kickback Noise and Clock Feedthrough

In the simulation that generated Fig. 16.27 we used, for our inputs, voltage sources. In a real application, the inputs would come from some other logic or a bit line (a source with a finite driving impedance). Consider including logic for driving the sense amplifier seen in Fig. 16.28. We use long-length inverters to simulate a weak driver circuit (such as a decoder made with a series of transmission gates). Seen in the figure are simulation results at the time when the clock goes high (we get similar glitches when the clock goes low). The large glitch seen in these figures is called *clock feedthrough* noise. It is present when a clock signal has a capacitive path directly to the input of the sensing or comparator circuit. The clock feedthrough noise in Fig. 16.28 is close to 50 mV on both inputs.

Another type of noise, called *kickback noise*, is present and injected into the inputs of the circuit when the latch switches states. *It's important to simulate the operation of the sense amplifier with nonideal sources (with finite source resistance) to determine the significance of clock feedthrough and kickback noise.* Using voltage sources with 10k resistors is a reasonable source for general circuit simulations. This noise is an important specification for a sense amplifier. It can often be the *limiting factor* when making sensitive measurements. If the kickback noise, for example, is too large, it can interfere with sensing. A simple example of a potential problem occurs when 256 sense amps are all being clocked at the same time (e.g., on the bottom of a memory array) and they are directly adjacent to each other. We don't want one of the amplifiers interfering with any of the others by feeding noise into the others' inputs.

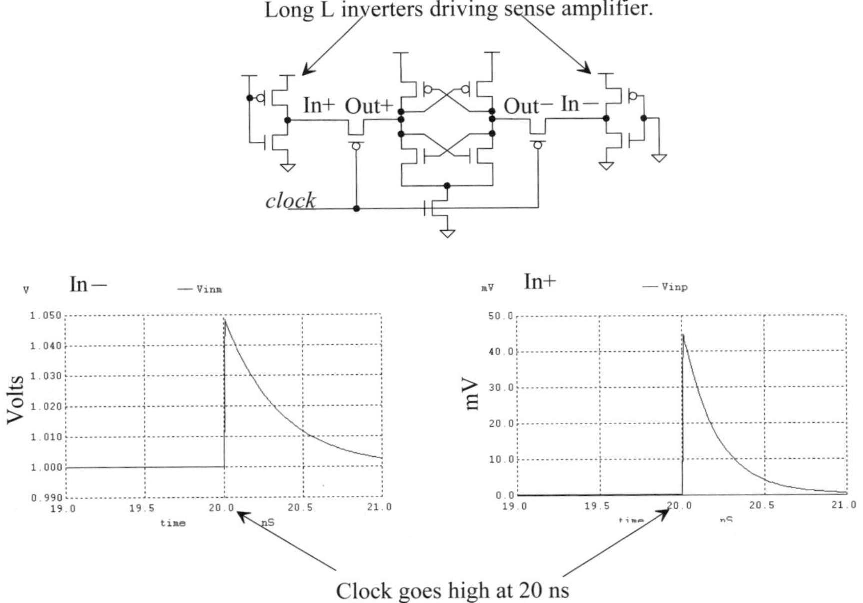

Figure 16.28 Clock feedthrough noise.

Memory

For a good comparison, the sense amplifier shouldn't have a memory of previous sensing operations. For the sense amplifier in Fig. 16.26, the outputs are actively driven by the input signals prior to clocking. However, the drain of MS3 (sources of M1/M2) is floating. This node is a dynamic node that floats to a voltage dependent on the previous decision and the input signals. *For a precision sense operation, all nodes in the sense amplifier must be driven or equilibrated, prior to clocking, to a known voltage.*

Current Draw

Because there may be thousands of sense amplifiers operating on a chip at the same time, it is extremely important to minimize the power used by these circuits. Let's take a look at the current supplied by *VDD* to the circuit in Fig. 16.26 (with the signals seen in Fig. 16.27). Because the inputs actively drive the gates of M3 and M4, it is possible M3/M4 source a significant current into the inputs. In this case, Fig. 16.27, the V_{SG} voltages are roughly 0.5 V (and so both M3 and M4 are conducting a drain current). The current supplied by *VDD* is seen in Fig. 16.29. *Clock* is high during 0 to 5ns, 10ns to 15ns, etc. (when the current supplied by *VDD* is small). If the average current is 50 μA and there are 1,000 sense amplifiers operating on the chip, then *VDD* supplies 50 mA of current (just to the sense amplifiers). *For minimum power, it's important that there are no DC paths from VDD to ground except during switching times.*

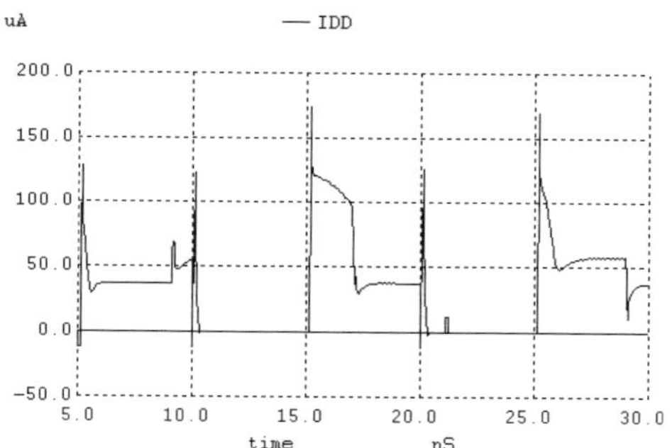

Figure 16.29 The current that flows in the sense amplifier in Fig. 16.26.

Contention Current (Switching Current)

When switching takes place in the sense amplifier of Fig. 16.26, both M3 and M4 are conducting. When *clock* goes high, both M1 and M2 turn on and conduct current as well. In this situation there is a direct path from *VDD* to ground. If the inputs are at relatively the same voltages, the time that M1–M4 remain conducting can be very long (resulting in significant current being pulled from *VDD*). To eliminate this *metastable* condition and force the comparator to make a decision (switch to valid logic levels), the gain in series with the input signals can be increased. A simple example of increasing input signal gain is to place an amplifier on the inputs of the latch.

Note that if our input signals amplitudes are within a V_{THP} of *VDD*, the PMOS devices are off and we don't have this problem. The inputs can generate an imbalance on the drains/gates of the NMOS transistors when clock goes high. Neither PMOS device will turn on until either M1 or M2's drain moves below $VDD - V_{THP}$. *For low power (low contention current), try to keep one side of the latch shut off until a significant imbalance is present in the circuit.*

Removing Sense Amplifier Memory

Figure 16.30 shows how the sense amplifier's memory can be erased. M1–M4 form a latch (Fig. 13.16). To remove the sense amplifier's memory, *all* nodes in the sense amplifier must be actively driven to a known voltage (no floating or dynamically charged nodes). When *clock* is low, the sense amplifier's outputs are pulled to *VDD* through MS3 and MS4. The MOSFETs MS1 and MS2 are off, breaking the connection between *VDD* and ground (so no current flows in the latch). The gates of M1 and M2 are at *VDD* so their drains are actively driven to ground. In other words, when *clock* is low, all nodes in the circuit are either pulled high to *VDD* or low to ground.

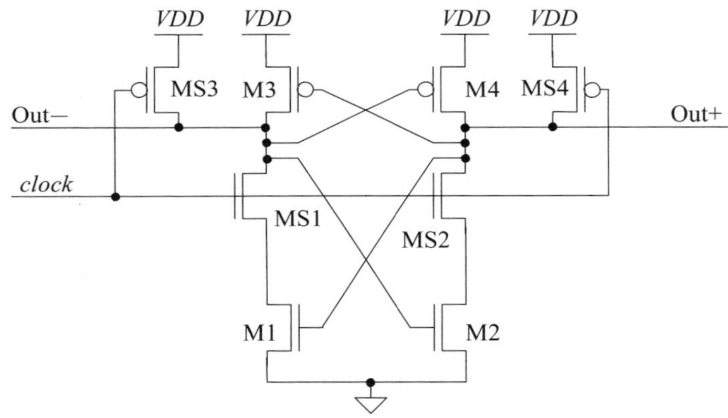

Figure 16.30 Removing memory in a sense amplifier.

Creating an Imbalance and Reducing Kickback Noise

When *clock* goes high, the latch action can take place. For the comparison to function as desired, we need to generate an imbalance. This, alone, isn't too difficult. What is difficult is creating an imbalance for a wide range of input signals while not causing an excessive amount of current to flow in the circuit. Consider the additions to our sensing circuit seen in Fig. 16.31. MB1/MB2 can be connected to either the drain or the source of MS1/MS2, as seen in the figure. Let's first consider connecting them to the drains. When *clock* is low, the drains of MB1 and MB2 are pulled to *VDD*. If the input voltages are significant, a *large* current can be drawn from *VDD* (when *clock* is low). If the input voltages are relatively small, then the current pulled from *VDD* may be tolerable (keeping in mind that the sense amplifier doesn't function correctly when the input voltages are less than V_{THN}). The benefit of connecting to the drains is high gain. Very small voltage differences can cause quick sensing (the outputs swing quickly to valid logic voltages). Note that the possibility for large current flowing through either MB1/MB2 is present independent of the state of *clock*.

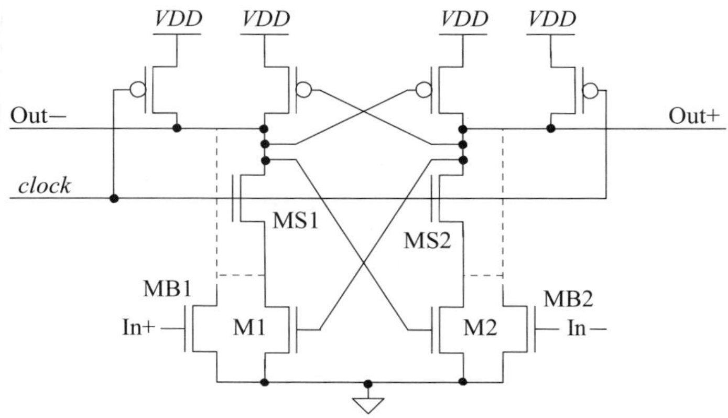

Figure 16.31 Creating an imbalance in a sense amplifier.

Next let's consider connecting MB1/MB2 to the sources of MS1/MS2 (drains of M1/M2). When *clock* goes low, the drains of MB1/MB2 are pulled low through M1/M2. No current flows in the comparator when *clock* is low (unlike when the drains of MB1/MB2 are connected to the drains of MS1/MS2). The gain is much lower because now MB1/MB2 are operating in the triode region when *clock* goes high. The potential for large current flow is still present when *clock* is high but, now, the maximum value on the drain of MB1/MB2 is $VDD - V_{THN}$. And so the potential for MB1/MB2 to move into the triode region can result in a smaller output current.

We seem to have a problem. To reduce the possibility of metastability, we need to increase the gain in series with the signal path (increase the W/L ratio of MB1 and MB2; that is, reduce their switching resistance). However, this results in larger power dissipation. How can we keep power down while maximizing the sensitivity of our sense amplifier? While we can use long L devices for MB1 and MB2 so they never pull significant current (at the cost of sensitivity and speed) or have a separate pre-amplifier to generate the input signals, let's look at some other ideas (noting that no design is perfect but a trade-off between power, speed, and sensitivity).

Figure 16.32 shows one idea for creating an imbalance without the possibility of significant current flow. MB1 and MB2 are used to create an imbalance in the gate-source voltages of M1/M2. Since, prior to switching, the voltage on the gates of M1/M2 is *VDD*, we can still have good sensitivity even though MB1 and MB2 are operating in the triode region (look like resistors). The large voltage dropped across $V_{GS,M1}$ and $V_{DS,MB1}$ ensure that good gain. It is important, however, that the inputs signals are $> V_{THN}$ to ensure that the circuit operates properly. The sensing fails if this isn't the case because neither MB1 or MB2 can turn on and provide the needed path to ground for proper logic level signals. As long as the inputs are above the threshold voltage, there aren't any dynamic nodes (*no memory in the sense amplifier*). The sources of MS1 and MS2 are still pulled to ground when *clock* is low. *Kickback noise* is reduced using this topology because the inputs are isolated from the latch by MB1/MB2. Kickback noise is still present and must be considered when designing and simulating this sense amplifier.

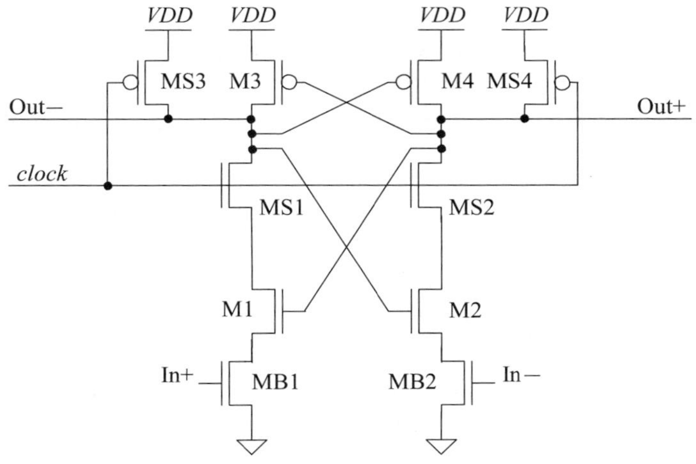

Figure 16.32 Reducing power in a sense amplifier.

Note that we might try to design our comparator to operate over a wide range of input voltages by adding PMOS devices in series with the sources of M3/M4, as seen in Fig. 16.33. We know that MB1–MB4 must be capable of turning on when *clock* goes high. If either pair can't turn on, then the latch won't be able to generate full logic levels. For the design in Fig. 16.33, the input signals would have to fall within V_{THN} and $VDD - V_{THP}$ (more restricted than the circuit in Fig. 16.32). Also, and perhaps more importantly, MB3/MB4 can't generate an imbalance in the latch. If the sources of M1/M2 are connected to ground so that MB1/MB2 don't generate the imbalance, then when *clock* goes high, MB3 and MB4 are off. They won't turn on until the NMOS pair M1/M2 turns on and drops the gates of M3/M4 below $VDD - V_{THP}$. For great differences in the input signals, the comparator may still function correctly. However, for the majority of the input signal differences, the resulting output logic levels is determined by the matching between M1/M2, e.g., the one with the lower "actual" V_{THN} will turn on faster.

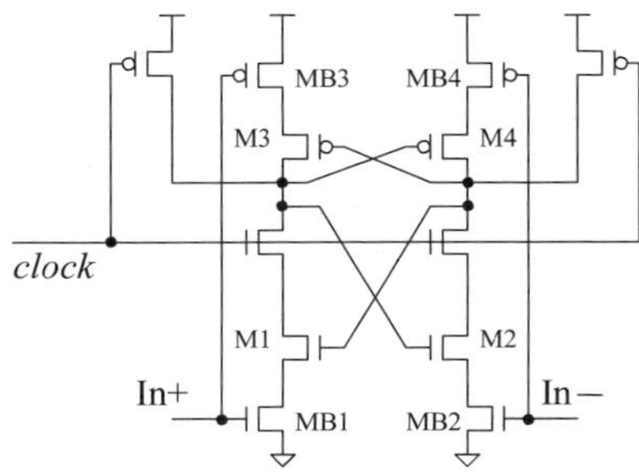

Figure 16.33 A bad design: how not to get wide-swing operation.

Example 16.1
Suppose that it is suggested that the gates of M3/M4 be tied to the drains of M1/M2 or MB1/MB2, respectively, in the sense amplifier (clocked comparator) seen in Fig. 16.32. Comment on the concerns.

If the gates of M3/M4 are tied to the drains of M2/M1, then, when *clock* goes high, a significant contention current flows in the latch. As discussed earlier, it is desirable that one side of the latch (either the NMOS or PMOS pairs) is off when the latch is clocked in order to reduce power. The same concerns are present when these gates are connected to the drains of MB1/MB2.

We might try to eliminate MS3/MS4 when we tie the gates of M3/M4 to the drains of MB1/MB2. Because the drains of MB1/MB2 always move to ground when *clock* is low (again assuming the inputs are > V_{THN}), M3/M4 will turn on and pull the outputs high. When *clock* goes high, though, it's impossible to shut M3/M4 off. Because the maximum voltage on the sources of MS1/MS2 is $VDD - V_{THN}$, the gates of M3/M4 can't go to *VDD* to shut one of the devices off. ■

Increasing the Input Range

Figure 16.34 shows adding, to our basic sense amplifier in Fig. 16.32, MOSFETs MB3–MB8 so that the circuit will function with input signals ranging from 0 to *VDD* (actually it will operate correctly with input signals beyond the power supply rails). We know current can only flow out of the sources of M1/M2, so our additional circuitry, for creating the imbalance, must be capable of sinking current from these sources (and so we'll add MB3 and MB4 for this reason). Next, we know that MB1/MB2 function fine for creating the imbalance if the input signals are above V_{THN}. However, if the input signals are below V_{THN}, they turn off. To level-shift the input, let's use the MB5–MB8. If the input signals are < V_{THN}, then MB7 and MB8 are on. A difference in the input voltages causes different currents to flow in MB5 and MB6. The different currents flowing in these transistors results in different voltages across each one. This voltage difference is then used to create an imbalance in MB3 and MB4. Unfortunately, the addition of MB5–MB6 contradicts our earlier statement that, "*For minimum power it's important that there are no DC paths from VDD to ground except during switching times.*" There is a DC path through MB8/MB6 and MB7/MB5. To lower the power, we can increase the length of MB5 and MB6 so that the continuous current flowing through the DC path is lessened.

Simulation Examples

Let's simulate the operation of the circuit in Fig. 16.32. We know that the outputs of the sense-amp go high every time *clock* goes low. To make the outputs change only on the rising edge of *clock*, we can use the SR latch discussed in Ch. 13 on the output of the circuit, Fig. 16.35. Now, with the addition of the NAND gates, when the sense amplifier's outputs are high, the outputs of the latch don't change from the previous decision (made on the rising edge of clock).

Figure 16.36 shows the current supplied by *VDD* for the circuit in Fig. 16.35. While the current seen in this figure includes the current supplied to the NAND gates, it still should be compared to Fig. 16.29. Notice in Fig. 16.36 that current flows only during the times the clock changes states (unlike the current flow in Fig. 16.29).

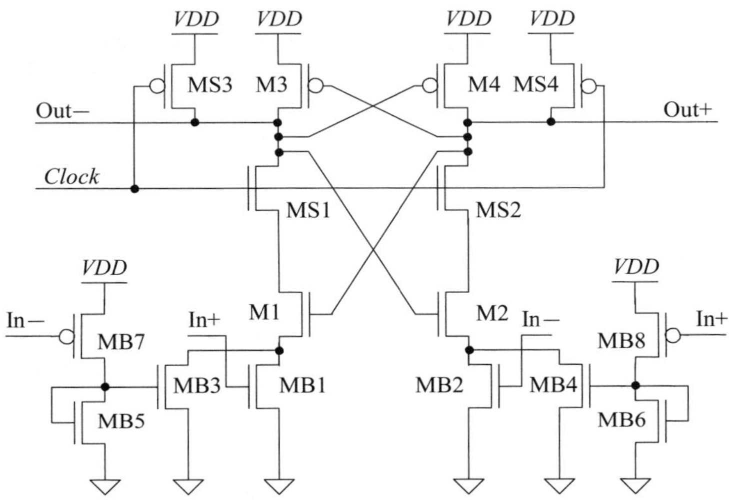

Figure 16.34 Rail-to-rail input signal range. See also Fig. 27.16.

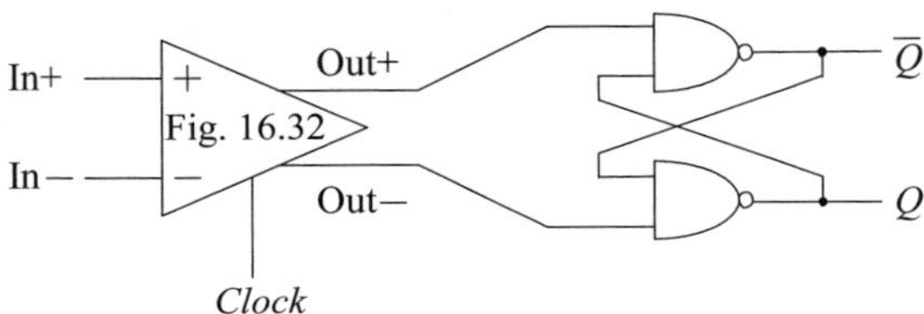

Figure 16.35 Adding an SR latch so output only changes on rising edge of the clock.

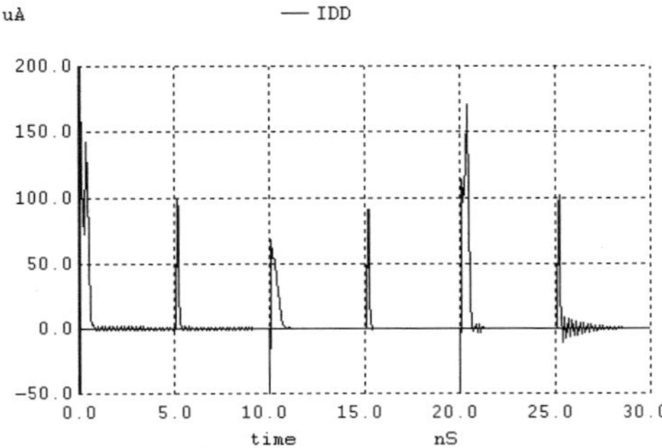

Figure 16.36 Regenerating the plot seen in Fig. 16.29 with the sense amplifier from Fig. 16.35. The current includes the switching current from the NAND gates.

Next let's look at the kickback noise. The clock feedthrough noise is small for this topology because the *clock* is isolated from the input by three transistors. Let's put the sense amplifier in the topology seen in Fig. 16.28 where the inputs are driven to the rails by long-length inverters. The simulation results are seen in Fig. 16.37. The kickback noise is approximately 5 mV.

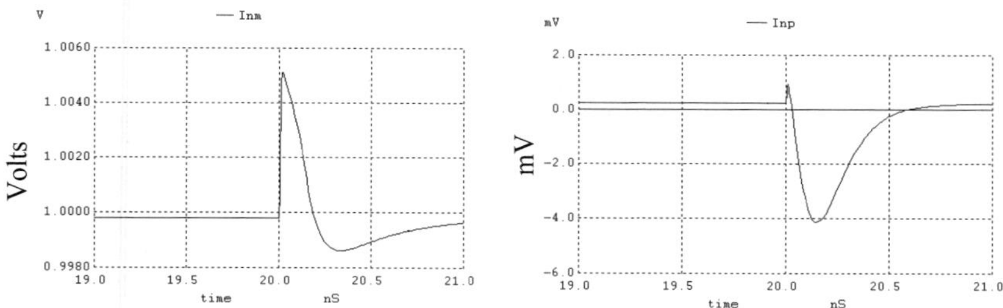

Figure 16.37 Kickback noise.

Figure 16.38 shows the operation of the circuit in Fig. 16.35 where the outputs only change on the rising edge of *clock*. The input signals are the same ones that were used in generating Fig. 16.27 (although the simulation was 20 ns longer in Fig. 16.38). The sensitivity of this sense amplifier is better than 10 mV. The sensitivity can be improved by increasing the lengths of MB1/MB2 (so that they have a large resistance). The drawback of the improved sensitivity is the increased time it takes to pull one of the outputs to ground (the output transition time). The sensitivity can be increased without this penalty by adding a preamplifier (e.g., a differential amplifier) between the sense amplifier and the inputs, see Fig. 27.16.

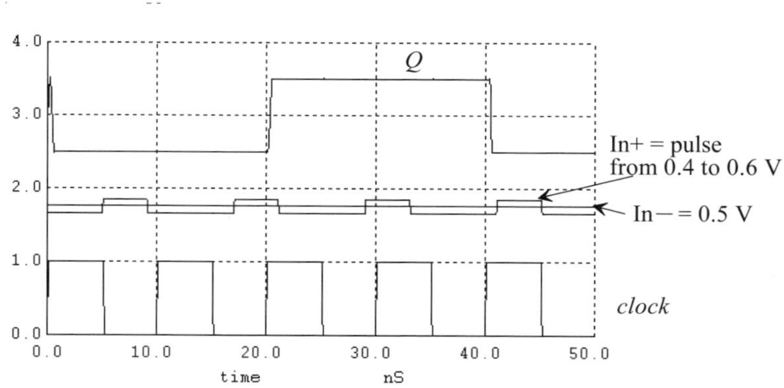

Figure 16.38 The operation of the circuit in Fig. 16.35.

16.2.2 Row/Column Decoders

Reviewing the memory block organization in Fig. 16.25, we may wonder how to address one or more of the 256kbit subarrays. When we say "address," for the row lines, what we mean is that we are driving one, and only one, of the row lines in one or more of the 256k subarrays to a voltage greater than $VDD + V_{THN}$ (to fully turn the access MOSFET on), while holding all other row (word) lines at ground. To accomplish this selection, a *row address decoder* is used. Once the word line has transitioned high and the sensing is complete, the data from the memory cells is present on the bit lines (as discussed in detail in Sec. 16.1). To select the addressed data from one or more of these bit lines (in one or more memory arrays), a *column decoder* is used.

In a large (capacity) memory the chip's address pins are multiplexed, that is, the same set of pins are used for both the row and column addresses. A separate clock signal is used to strobe in either the row or the column addresses, at different times, into some hold registers. In older DRAMs, for example, the falling edge of $\overline{RAS}$ (row address strobe) was used to clock in the row address and the falling edge of $\overline{CAS}$ was used to clock in the column address. The outputs of the row address hold register (column address hold register) are decoded and used to select a row (column) line, Fig. 16.39.

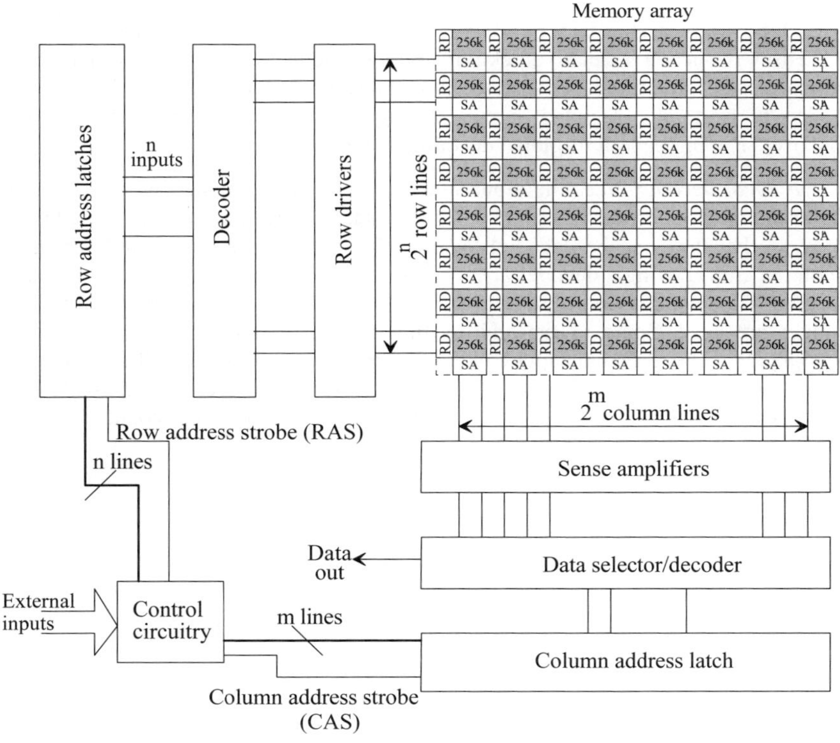

Figure 16.39 Detailed block diagram of a RAM.

The length of the column (m-bits in the Fig. 16.39) and row (n-bits in Fig. 16.39) address words depends on the size of the memory and the word size. For example, a 1-Gbit memory can be organized in x1 (by one), x2 (by two), x4 (by four), etc., configuration. The "by four" for example, simply signifies the word size used in the memory's input/output data path is 4-bits. A 1-Gbit memory in a x4 configuration simply indicates that 256-Mwords can be addressed where each word is 4-bits. In a x1 configuration 1-Gbits of data can be accessed (2^{30} = 1-G = 1,073,741,824). We would then have 15 address pins for 2^{15} different row and column addresses (remembering that the address pins are shared). The next question then becomes how to decode the addresses and select the appropriate row(s) and column(s).

Global and Local Decoders

Looking at Fig. 16.39, we see that we decode the outputs of the row address latches and the column address latches and then feed the outputs to the subarrays. This is called a *global decode*. If the memory size is, once again, 1-Gbit (in a x1 configuration), then 2^{15} (= 32,768) row line wires and 2^{15} column line wires are fed to the memory arrays. If each memory array is 256-kbits (512 word lines and 512 column lines), then we need 4,096 arrays to get a 1-Gbit memory (64 by 64 memory arrays in Fig. 16.39). When the output of the row decoder goes high, it turns on a word line in the 64 memory arrays of the addressed row. A total of 32k columns then have data sitting on them. Because our memory is going to take only 1 bit of data and feed it to the output, we have, perhaps, wasted a lot of power. (In many topologies this large amount of data is called a "page." Once the row is opened it can be quickly read out of the memory by simply changing the column address.) The other issue with the global decoder is that a separate layer of metal is needed to route the decoded signals throughout the entire chip (adding process complexity). The big benefit of global decoding is reduced chip size.

A *local decoder* takes the input addresses and, as the name says, decodes them locally at the memory array. The 15 bits, for the example given above, are routed to each subarray so that they can be decoded locally. This takes up a considerable amount of space on the chip (having a decoder at each 256k array, for example) but doesn't call for increased process complexity. In most practical designs a combination of local and global techniques are used. Some of the address bits are decoded globally while others are decoded locally. The global decoder enables an array or (arrays), while the local decoder selects the array's row line and column line. A static decoder is seen in Fig. 16.40. Figure 16.41 shows how the 10-bit address (for 1,024 word or row lines) is connected to each decoder element.

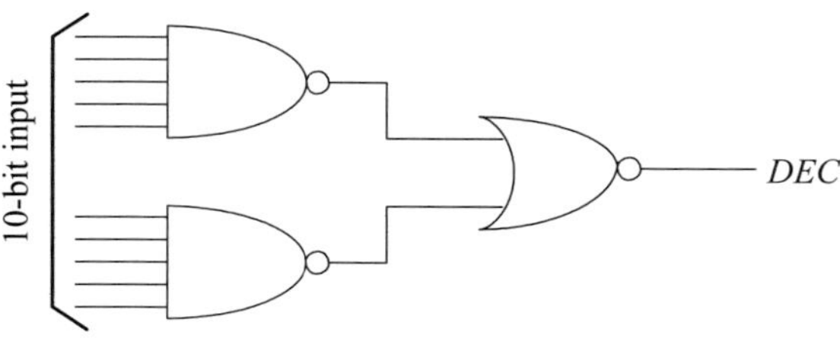

Figure 16.40 A static decoder.

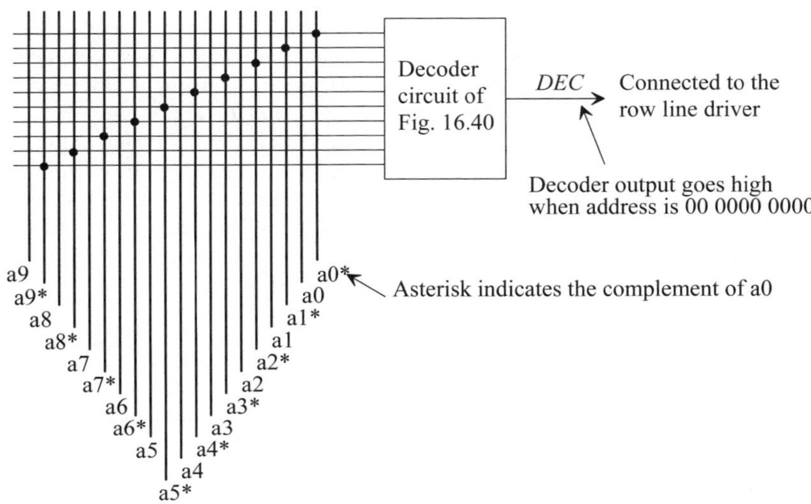

Figure 16.41 How the address lines are connected to a decoder element.

For the column decoder, we know that we have to be able to write or read data to the memory cells (not just select a column). For this reason pass transistors are used on the output of the decoder, as seen in Fig. 16.42. The pass transistor doesn't pass a logic one to full levels. We lose a V_{THN} with body effect when writing or reading a one. The sense amplifier may be used to pull the bit line up to a full logic one. Similarly, on the output of the column decoder a sense amplifier may be used to pull the output line to full logic levels. Finally, note that we drew the column decoder and its outputs going horizontally. In practice, the outputs of the column decoder run in the same direction as the columns themselves in order to save space.

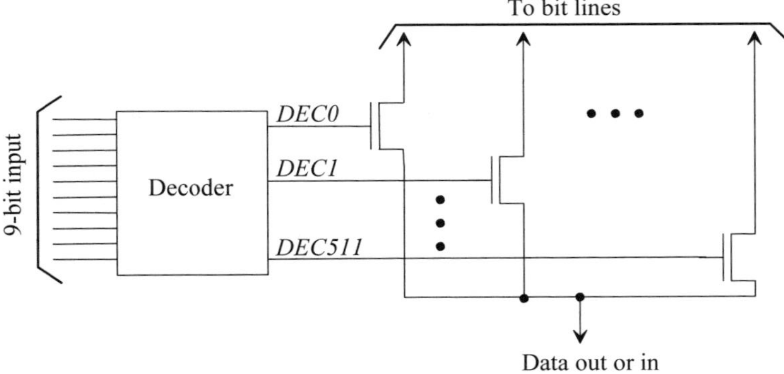

Figure 16.42 Addition of pass transistors to implement a column decoder.

Reducing Decoder Layout Area

To reduce the layout area of address decoders, a pass-transistor based decoder, see Fig. 16.43a, can be used. The concerns with using this decoder are that the unselected outputs are not actively driven high or low (they are floating) and, once again, as NMOS devices they don't pass a logic one to a full logic level (that is, *VDD*). This means that on the output of the decoder, as part of our row driver, we need a circuit that, when not selected, pulls the output of the decoder low. At the same time, to maximize the noise margins, the switching point of the driver needs to be reduced. For example, with $VDD = 1$ V and $V_{THN} = 0.35$ V (high because of body effect), the selected output of the decoder swings from 0 to 0.65 V. The switching point of the row driver should lie in the middle of this swing at 0.3125 (ideally), to maximize the noise margins, Fig. 16.43b.

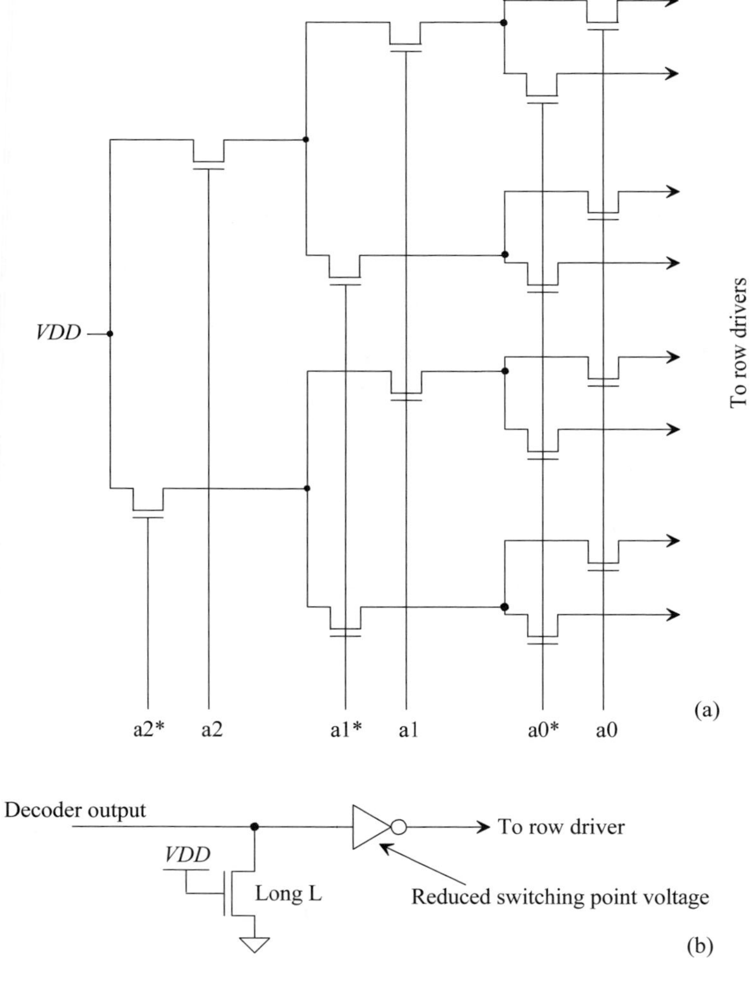

Figure 16.43 A tree decoder used to reduce layout area.

Another technique to reduce decoder layout size uses the precharge-evaluate (PE) logic, Fig. 16.44, discussed in Ch. 14. This figure shows a 3-bit input decoder. The output of the circuit goes high when the input address is 000. Using the long-length keeper MOSFET makes the dynamic circuit operate as a static circuit. Variations of this circuit are common in commercially available memories.

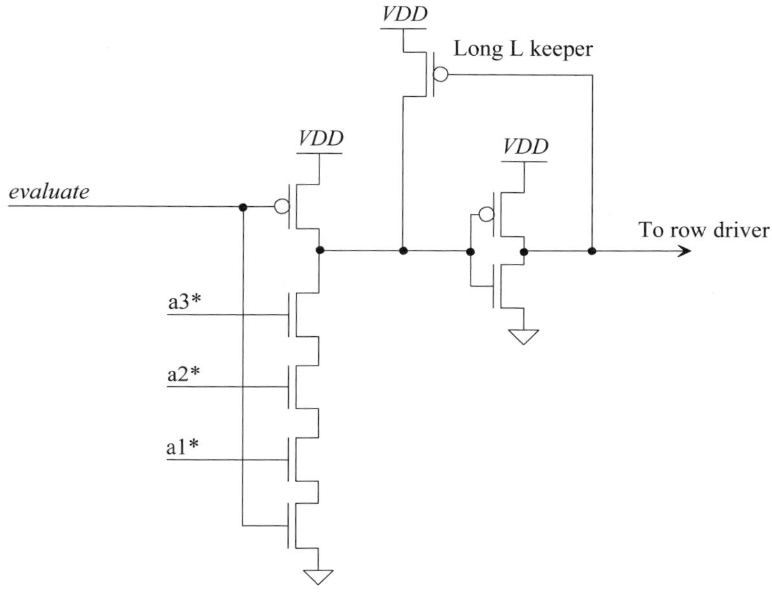

Figure 16.44 3-bit PE decoder.

16.2.3 Row Drivers

In most memories a single NMOS pass transistor switches data into or out of the memory cell. To fully turn this pass transistor on, the row line is driven to a pumped voltage (a voltage outside of the power supply rails that is generated using the on-chip charge pump circuit discussed in Ch. 18). We'll call this pumped voltage *VDDP*. For our purposes this voltage must be greater than $VDD + V_{THN}$ (with body effect). In the following discussion we'll assume a *VDDP* of 1.5 V (since our *VDD* is 1 V and V_{THN} is 280 mV without body effect).

Examine the inverter circuit seen in Fig. 16.45. The input to the inverter connected to the word line swings from 0 to *VDD* (1 V). Note that the PMOS is sitting in its own well. Both its source and body are tied to *VDDP*. When the input to this inverter is low (0 V), the PMOS is on and the NMOS is off. The word line is driven to *VDDP* (which is what we want). Unfortunately when this inverter input is high (= *VDD*), the NMOS turns on but the PMOS can't shut off. The source gate voltage of the PMOS is 0.5 V (above the PMOS's threshold voltage). In this section we discuss word-line drivers, knowing that a simple inverter can't be used.

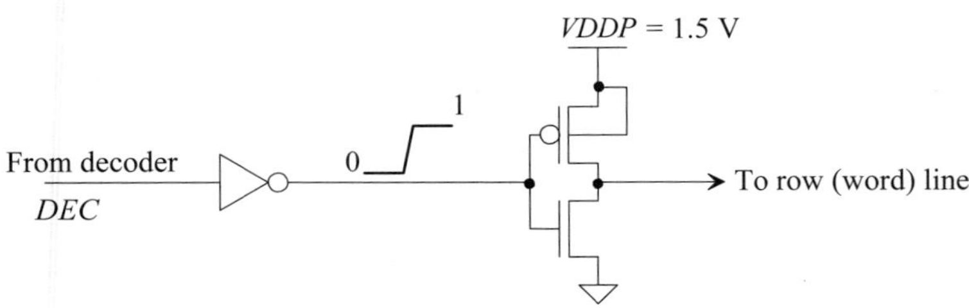

Figure 16.45 Problems with using an inverter for a row driver.

A row driver based on the inverter is seen in Fig. 16.46. When the decoder output is low, M1 is off and M2 is on. This causes M3 to turn on, driving the gate of M4 to *VDDP* so that it shuts off. When the decoder output goes high, M1 turns on and M2 shuts off. The gate of M4 is pulled to ground turning it on. This causes the word line to move to *VDDP* and M3 to shut off. This circuit works well but there can be a significant contention current when M3/M4 are turning on. To avoid this, the pumped voltage can be a clocked signal that goes high *after* the output of the decoder has transitioned. The idea is seen in Fig. 16.47.

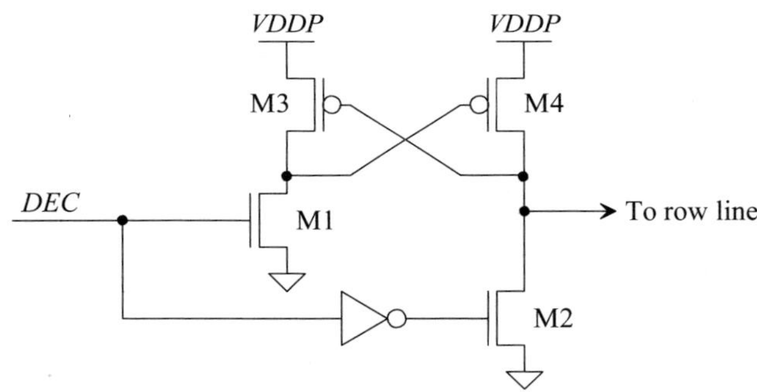

Figure 16.46 A CMOS word line driver.

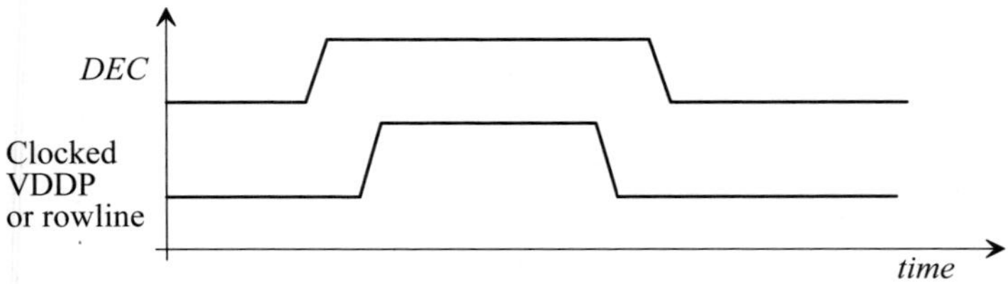

Figure 16.47 Using a clocked VDDP to reduce contention current in a row line driver.

16.3 Memory Cells

We've already looked at the DRAM memory cell in detail. In this section we'll take a look at the static RAM cell (SRAM), the erasable programmable read-only memory (EPROM) cell, the electrically erasable programmable read-only memory (EEPROM) cell, and the Flash memory cell.

16.3.1 The SRAM Cell

The schematic and layout of a six-transistor SRAM memory cell is seen in Fig. 16.48. This is, as its name implies, static, meaning that as long as power is applied to the cell it will remember its contents (unlike the DRAM cell, which loses its memory after a short time). The basic cross-coupled inverter latch should be recognized in the topology. To access the cell, the word line goes high and turns on the access MOSFETs. The bit lines are driven in complementary directions with a "strong" driver for writing. When the word line goes low, the datum is latched in the cell.

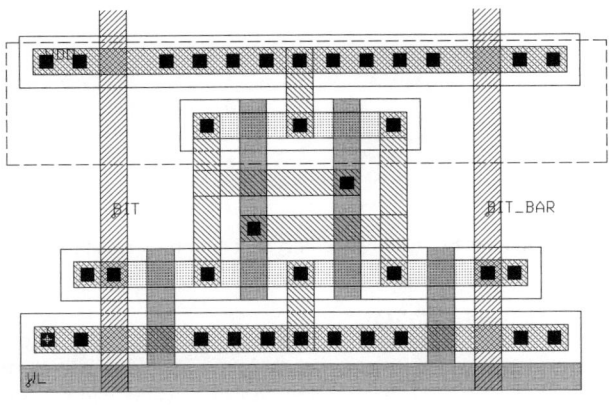

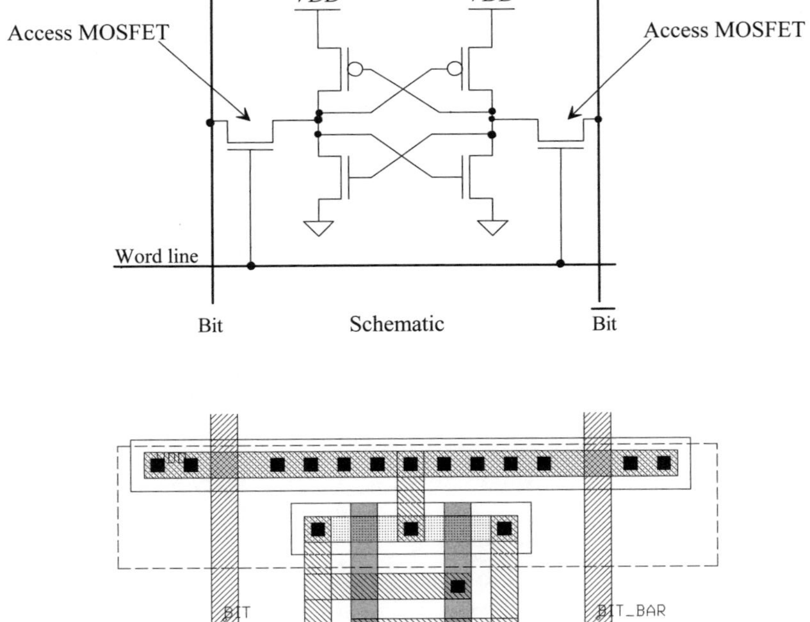

Figure 16.48 The six-transistor SRAM memory cell.

The sizing of the access MOSFETs is important. If they are too weak, the bit lines won't be able to flip the state of the cell when writing. If they are too strong, the layout area can be large (noting that bit lines are often precharged high in an SRAM so that the access devices are initially off during sensing). To reduce the layout area, a cell that doesn't use PMOS devices can be used, Fig. 16.49. The cell size is decreased by eliminating the n-well and using high resistance poly n+/p+ resistors. The layout of the polysilicon resistor is also seen in Fig. 16.49. The resistor can be thought of as a leaky bipolar transistor. Typical resistance values approximately 10 MΩ (or more). The CMOS SRAM cell dissipates little static power, while the resistor/n-channel SRAM cell dissipates VDD^2/R_{poly}.

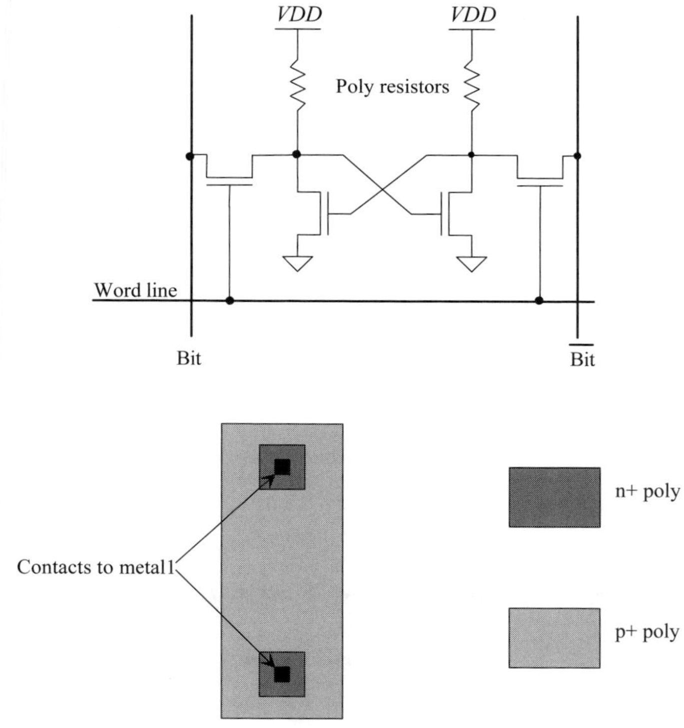

Figure 16.49 SRAM memory cell with poly resistors.

16.3.2 Read-Only Memory (ROM)

ROM is the simplest semiconductor memory. It is used primarily to store instructions or constants in a digital system. The basic operation of a ROM can be explained with the ROM memory schematic shown in Fig. 16.50. Remembering that only one word line (row line) can be high at a time, we see that R_1 going high causes the column lines C_1, C_2, and C_4 to be pulled low. Column lines C_3 and C_5 are pulled high through the long L

MOSFET loads at the top of the array. If the information that is to be stored in the ROM memory is not known prior to fabrication, the memory array is fabricated with an n-channel MOSFET at every intersection of a row and column line (Fig. 16.51a). The ROM is programmed (PROM) by cutting (or never fabricating) the connection between the drain of the MOSFET and the column line Fig. (16.51b). Because it is not easy to program ROM, it is limited to applications where it is mass produced.

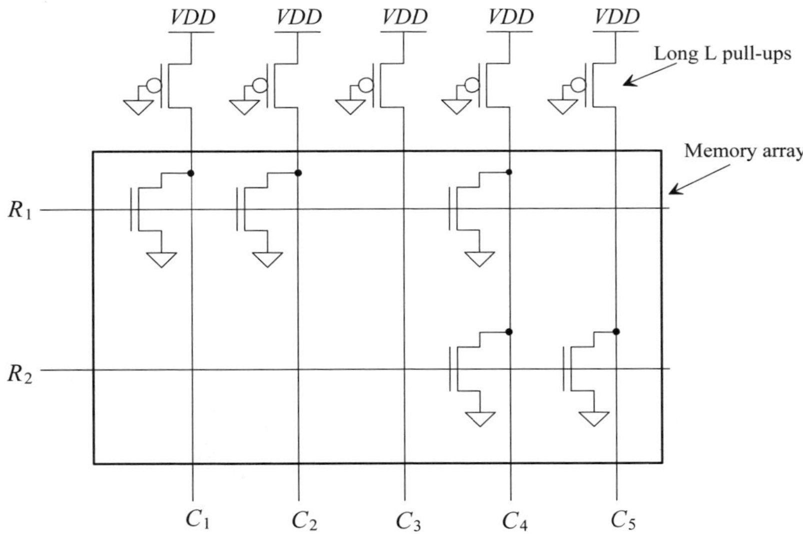

Figure 16.50 A ROM memory array.

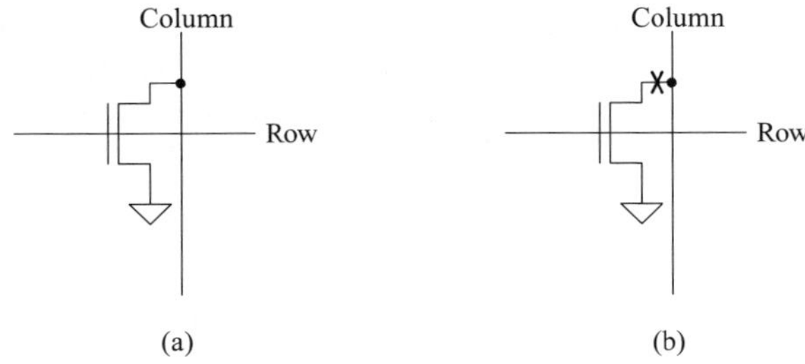

Figure 16.51 (a) n-channel MOSFET at the intersection of every column and row line and (b) eliminating the connection between the drain and column line to program the ROM.

16.3.3 Floating Gate Memory

A MOSFET made with two layers of poly is seen in Fig. 16.52 (along with its schematic symbol and a typical layout). As seen in the figure, the poly1 is floating, that is, not electrically connected to anything. A dielectric surrounds this floating island of poly1. With the gate oxide on the bottom, a thin oxide insulates the MOSFET from the poly2 (word/row line) above. Poly2 is the *controlling gate* of the transistor (the terminal we drive to turn the MOSFET on). We are going to make a memory element out of this cell by changing, adding or removing, the charge stored on the floating gate.

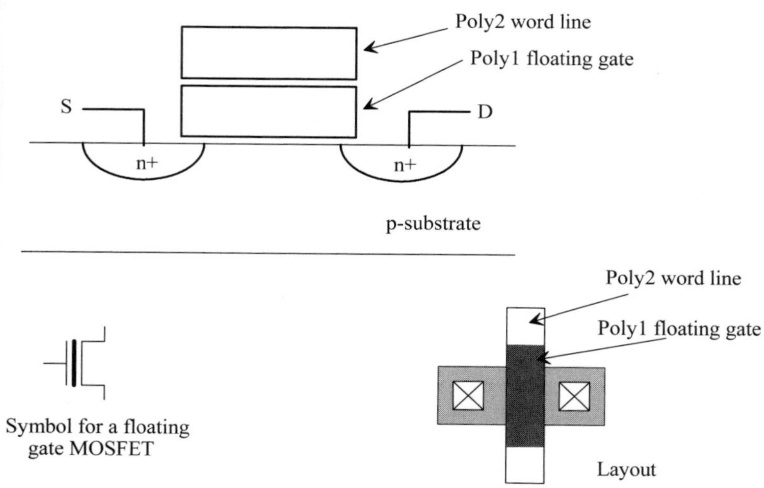

Figure 16.52 A floating gate MOSFET, its symbol and layout.

Figure 16.53 shows the difference between the *erased* state and the *programmed* state in a floating gate memory. The erased state, the state of the memory when it is fabricated (that is, before we force charge onto the floating gate), shows normal MOSFET behavior. Above the threshold voltage, the device turns on and conducts a current. When the cell is programmed, we force a negative charge (electrons) onto the floating gate (how we force this charge will be discussed shortly). This negative charge attracts a positive charge beneath the gate oxide. The result is that a larger controlling gate voltage must be applied to turn the device on (the threshold voltage increases).

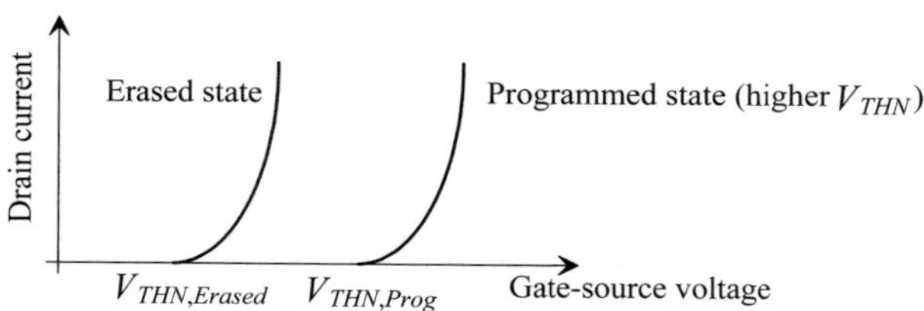

Figure 16.53 Programmed and erased states of a floating gate memory.

The Threshold Voltage

From Ch. 6 we can write the threshold voltage of a single-poly gate MOSFET as

$$V_{THN} = -V_{ms} - 2V_{fp} + \frac{Q'_{b0}}{C'_{ox}} \tag{16.10}$$

where we haven't included the shifts due to the threshold voltage implant, Q'_c, or any unwanted surface state charge, Q'_{ss} (to keep the equations shorter). Looking at Fig. 16.54, we see that the effective oxide capacitance from the controlling gate to the channel has decreased from C'_{ox} for a single-poly gate MOSFET to $C'_{ox}/2$ for a dual poly gate MOSFET. Our threshold voltage, for a floating gate MOSFET, can then be written as

$$V_{THN,Erased} = -V_{ms} - 2V_{fp} + 2 \cdot \frac{Q'_{b0}}{C'_{ox}} \tag{16.11}$$

If the term Q'_{b0}/C'_{ox} is approximately 50 mV (a typical oxide thickness used in a floating gate memory is 100 Å), then the erased threshold voltage, $V_{THN,Erased}$ (see Fig. 16.53) is only 50 mV larger than the V_{THN} of a normal (single poly gate) MOSFET.

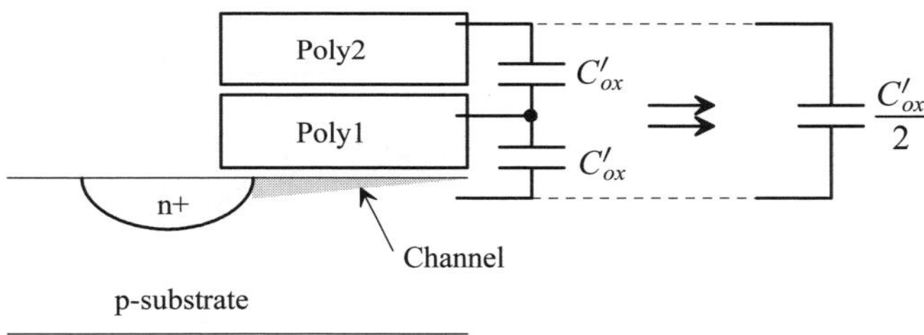

Figure 16.54 Oxide capacitance estimation for calculating threshold voltage..

Next consider what happens if we trap a negative charge on the floating poly1 gate, Fig. 16.55. The threshold voltage with this trapped charge, Q'_{poly1}, is shifted to

$$V_{THN,Prog} = -V_{ms} - 2V_{fp} + 2 \cdot \left(\frac{Q'_{b0}}{C'_{ox}} + \frac{|Q'_{poly1}|}{C'_{ox}} \right) \tag{16.12}$$

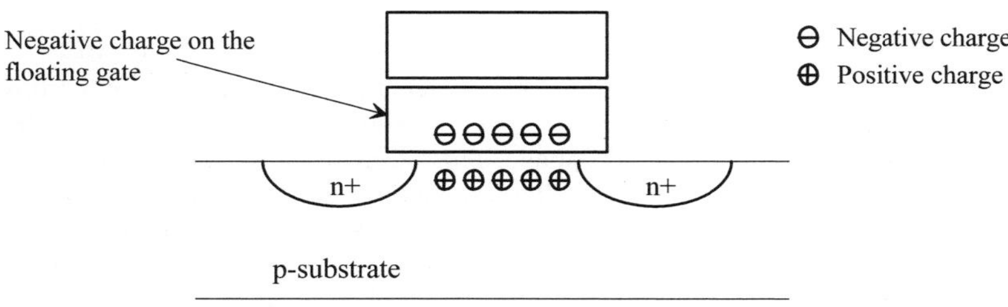

Figure 16.55 Trapped negative charge on the floating gate.

Erasable Programmable Read-Only Memory

Erasable programmable ROM (EPROM) was the first floating gate memory that could be programmed electrically. To erase (return the cell to its fabricated state, that is, leave no trapped charge on the floating gate), the cell is exposed to ultra-violet light through a quartz window in the top of the chip's package. The ultra-violet light increases the conductivity of the silicon-dioxide surrounding the floating gate and allows the trapped charge to leak off. The inability to erase the EPROM electrically has resulted in its being replaced by Flash memory (discussed later).

Programming the EPROM relies on *channel hot-electron* (CHE) injection. CHE is accomplished by driving the gate and the drain of the MOSFET to high voltages (say a pumped voltage of 25 V), Fig. 16.56. The high voltage on the drain of the device causes hot electrons (those with significant kinetic energy) to flow in the channel. A large positive potential applied to the gate attracts some of these electrons to the floating gate. The electrons can penetrate the potential barrier between the floating gate and channel because of their large energies.

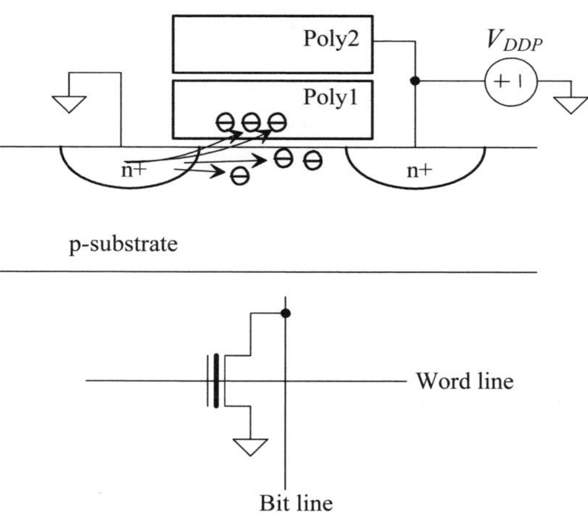

Figure 16.56 How charge is trapped on the floating gate using channel hot electron (CHE) injection.

Two Important Notes

When we are programming the floating gate device, the accumulation of electrons on the floating gate causes an increase in the device's threshold voltage. The more electrons that are trapped the higher the threshold voltage. This increase in threshold voltage causes the drain current to decrease. The decrease in drain current then reduces the rate at which the electrons are trapped on the floating gate oxide. If we apply the programming voltages for a long period of time, the drain current drops to zero (or practically a small value). Because of this feedback mechanism, the programming is said to be *self-limiting*. We simply apply the high voltages for a long enough time to ensure that the selected devices are programmed.

Next examine Fig. 16.57. When we are programming a row of cells, we drive the word line to a high voltage. If the cell is to remain erased, we simply leave the corresponding bit line at ground. If the cell is to be programmed, we drive the bit line to the high voltage. For CHE, both the drain and gate terminals of the floating device must be at a high voltage. This is important because it removes the need for a select transistor (something we'll have to use in other programming methods to keep from programming an unselected cell).

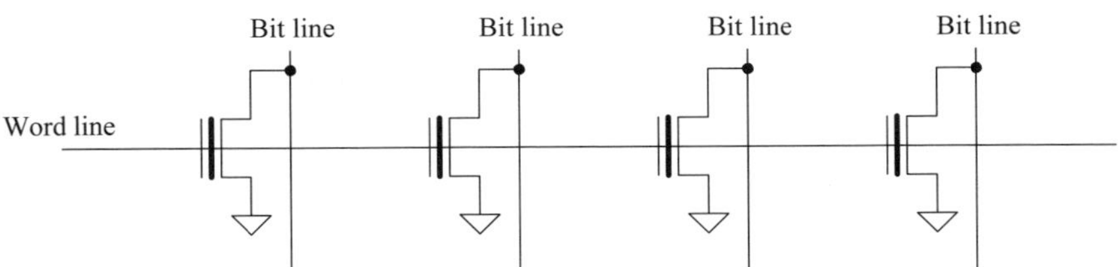

Figure 16.57 Row of floating gate devices. When programming the word line is driven to a high voltage. To program a specific cell, its bit line is also driven to a high voltage. To leave the cell erased, the bit line is held at ground.

Flash Memory

By reducing the thickness of the oxide from, say, 300 Å (a typical value used in an EPROM) to 100 Å, Fowler-Nordheim tunneling (FNT) can be used to program or erase the memory cell. Floating gate memory that can be both electrically erased and programmed is called electrically erasable programmable ROM (EEPROM). Note that this name is an oxymoron. If we can electrically *write* to the memory, then it isn't a *read*-only memory. For this reason, and because the rows of memory are generally erased in a *flash* (that is, large amounts of memory, say a memory array, are erased simultaneously and, when compared to the EPROM method of removing the chips from the system and exposing to ultra-violet light to erase, very quickly), we call floating gate memory that can be electrically programmed and erased *Flash* memory.

While CHE and FNT can be used together (CHE for the programming and FNT for erasing) to implement a memory technology, we assume FNT is used for both programming and erasing in the remaining discussion in this chapter.

Figure 16.58 shows the basic idea of using FNT to *program* a device (recall that this means to trap electrons on the floating gate so that the devices threshold voltage increases). The control gate (poly2) is driven to a large positive voltage. For a 100 Å gate oxide this voltage is somewhere between 15 and 20 V. Note that we are assuming that our NMOS devices are sitting in a p-well that is sitting in an n-well (so we can adjust the p-well [body of the NMOS] potential). The electrons tunnel through the thin oxide via FNT and accumulate on the floating gate. Like programming in an EPROM device, this mechanism is self-limiting. As electrons accumulate on the floating gate, the amount of tunneling current falls. Note that if we didn't want to program the device when poly2 is at 20 V we could hold the drain at a higher voltage (not ground) to reduce the potential across the thin gate oxide (aka tunnel oxide).

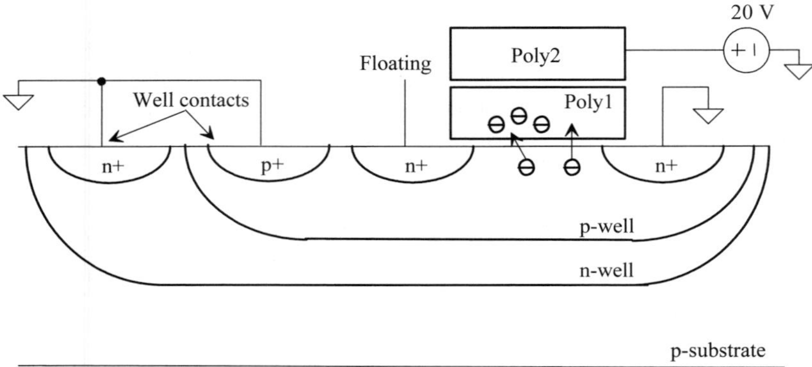

Figure 16.58 FNT of electrons from the p-well to a floating gate to increase threshold voltage (showing programming).

To *erase* the device using FNT, examine Fig. 16.59. Both the p-well and the n-well are driven to 20 V while the control gate (poly2) is grounded. Electrons tunnel via FNT off of the floating gate (poly1) to the p-well. The source and drain contacts to the device are floating (to accomplish this we will float the bit line and the source n+ outside of the array). Again, the movement of charge is self-limiting (however, there are device issues that can result in over erasing). The tunnel current drops as positive charge accumulates on the floating gate. If the erasing time is long, a significant amount of positive charge can accumulate on the floating gate. This will *decrease* the threshold voltage of the MOSFET. Figure 16.60 shows the programmed and erase states for a Flash memory where a positive threshold voltage indicates that the device is programmed and a negative threshold voltage indicates that the device is erased (we show $\pm 3\ V$ as typical values).

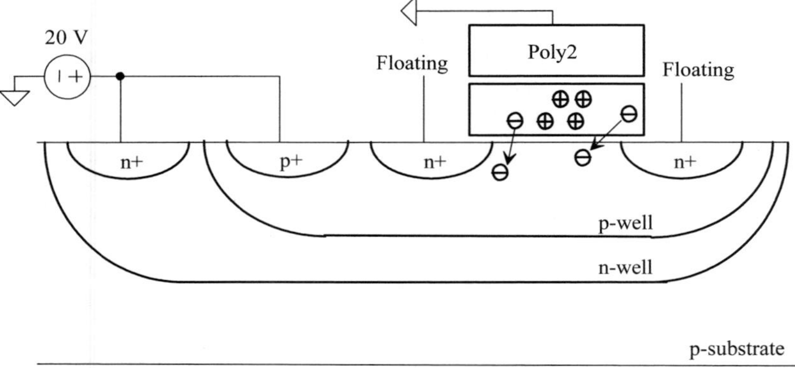

Figure 16.59 FNT of electrons from the floating gate to p-well to decrease threshold voltage (showing erasing).

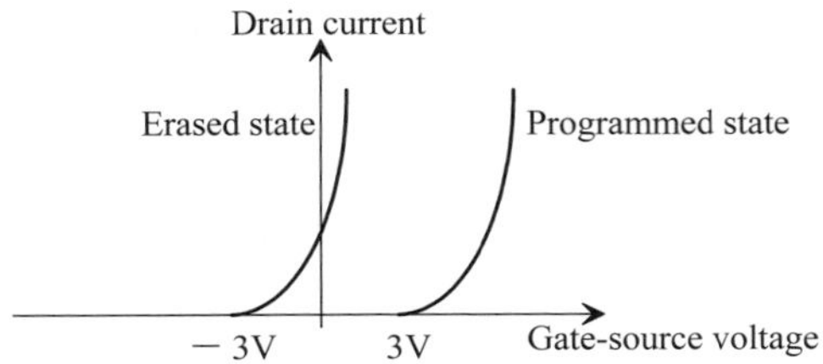

Figure 16.60 Programmed and erased states of a flash memory.

The schematic and layout of a 4-bit NAND Flash memory cell is seen in Fig. 16.61. The select transistors are made using single poly (normal) MOSFETs. When the cell (all four bits) is erased, the p-well and the n-well are driven to 20 V external to the memory array via the p+ implant. The bit lines and the n+ source connection at the bottom of the layout are floated. The four control gates and the two select MOSFET gates are pulled to ground (so all six poly gates, aka, word lines, are at ground for an erase).

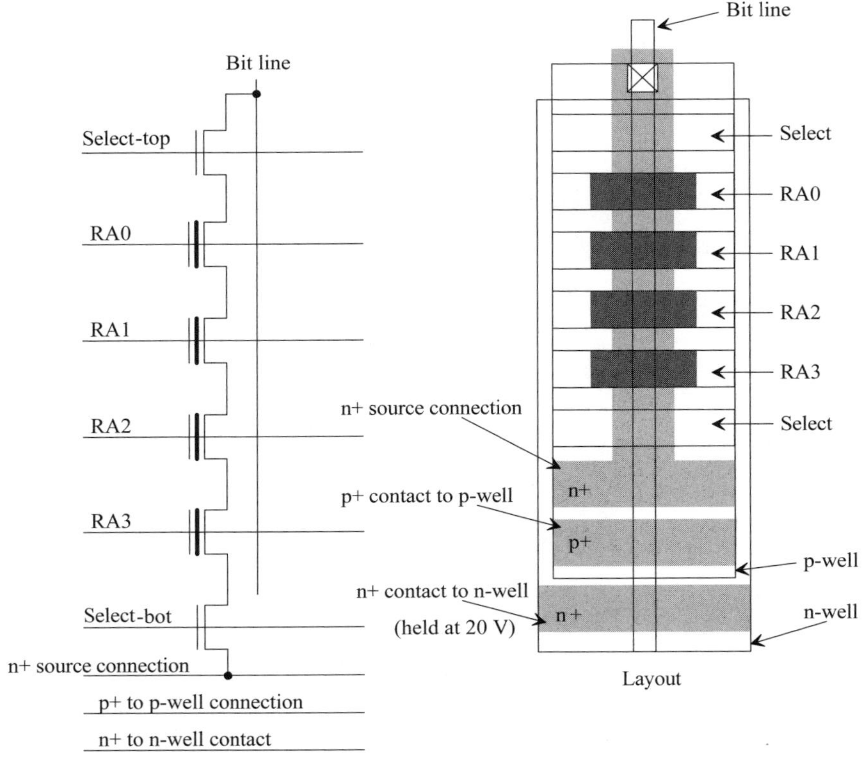

Figure 16.61 A 4-bit NAND Flash memory cell.

To illustrate programming the floating gate MOSFET connected to RA0 (row address 0), examine the connections to the NAND cell seen in Fig. 16.62. The bit line is driven to ground. A voltage, say 20 V, is applied to the gate of the top select gate. The gates of the floating gate MOSFETs connected to RA1–RA3 are driven to 5V. This 5 V signal turns on these devices but isn't so big that FNT will occur in them. The bottom select MOSFET remains off so that there is no DC path from the bit line to ground, Fig. 16.58. The p-well (which is common to all of the cells in the memory array, that is, not just the four in the memory cell) is pulled to ground external to the memory array via the p+ implant. Because the gate, RA0, is pulled to 20 V, as seen in Fig. 16.58, and the drain implant is pulled to ground through the bit line, the device will be programmed (electrons will tunnel through the oxide and accumulate on the floating gate).

The next thing we need to look at before talking about reading the cell is how we keep from programming the adjacent floating gate devices, those also connected to *RA*0, if they are to remain erased. What we need to do is ensure that no FNT occurs in these unselected devices. Figure 16.63 shows how we keep from programming a device. The bit line of the cell that is to remain erased is driven to a voltage that is large enough to keep FNT from occurring. The bottom select MOSFET is off, so there won't be a DC path from the bit line to ground (this is important because all other MOSFETs in the memory cell will be on).

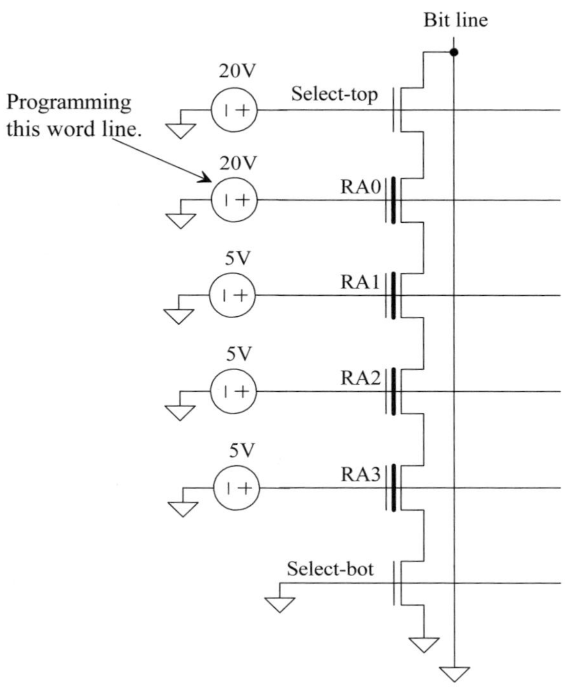

Figure 16.62 Programming in a Flash NAND cell.

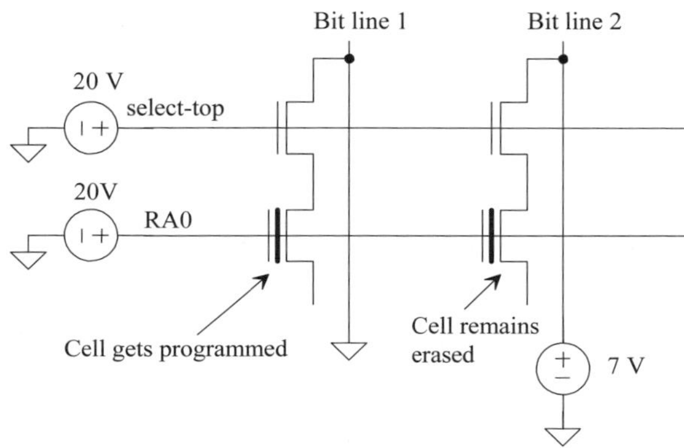

Figure 16.63 How cells are programmed (or not) in a NAND Flash memory array.

To understand reading a NAND Flash cell, consider the expanded view of Fig. 16.60 shown in Fig. 16.64. If both select transistors are turned on (say 5 V on their gates), the unselected row lines are driven high (say, again, to 5 V), and the selected row line is held at zero volts, then the current difference between I_{erased} and I_{prog} can be used to determine if the (selected) floating gate MOSFET is erased or programmed. If an average current, that is, $(I_{erased} + I_{prog})/2$ is driven into the bit line, an erased cell will keep the bit line at a low voltage. The selected (erased) MOSFET will want to sink a current of I_{erased} when its gate is zero volts. A programmed cell won't be able to sink this current (it will want to sink a current of I_{prog}) and so the bit line will go high.

Table 16.1 shows a summary of erasing, programming, and reading a NAND Flash memory cell. Notice that in Figs. 16.58 and 16.59 (during either erasing or programming the cell) the source of the MOSFET is floating. In Fig. 16.58 we can float the source by shutting off the bottom select MOSFET in the NAND stack. However, in Fig. 16.59 if we try to isolate the cell by shutting off the select transistors we see an issue, that is, when 20 V is applied to the p-well, the n+ source/drain implants can forward bias. This is why we must float both the bit line and the n+ source external to the array. Also, note that the top select MOSFET is used to provide better isolation between the memory cell and the bitline (it lowers the capacitive loading on the bit line). Going through the

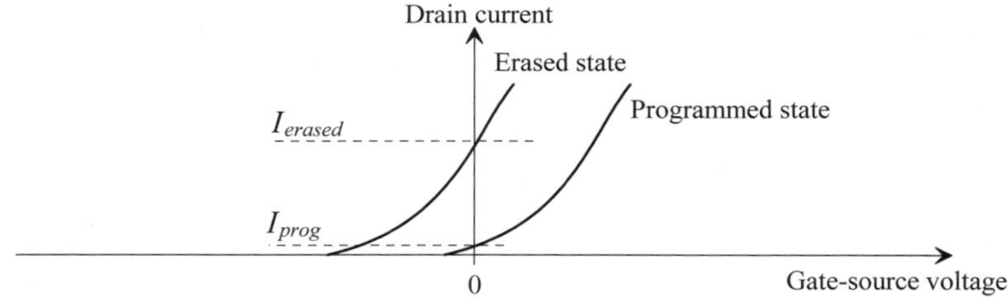

Figure 16.64 Expanded view showing erased and programmed IV curves.

operation of the NAND memory cells, we see that the lower select transistor, from a logical point of view, is all that is needed for basic cell operation.

Table 16.1 Summarizing NAND Flash cell operation

Inputs	Erase	Program	Read
Bit line	Floating	0 V	High or low
select_top	0 V	20 V	5 V
RA0	0 V	20 V	0 V
RA1	0 V	5 V	5 V
RA2	0 V	5 V	5 V
RA3	0 V	5 V	5 V
select_bot	0 V	0 V (so the source of the cell floats)	5 V
n+ source	Floating	0 V	0 V
p+ well tie-down	20 V	0 V	0 V
Comments	Erases entire array since well and word lines are common (Flash). The p-well at 20V will forward bias the n+ source and drain regions. This requires the bit line and n+ source float external to the array.	Programming the RA0 cell. Bit lines of the cells not to be programmed, but on RA0, are driven to 7 V to avoid FNT. Unused word lines driven to 5V.	Reading the contents of RA0. An average current is put into the bit line.

Before leaving this topic, let's use our short-channel process, which, of course, has only a single poly layer to show how direct tunneling gate current varies with terminal voltages. For our 50 nm, short-channel process used in this book (t_{ox} = 14 Å), the gate current density is (roughly) 5 A/cm^2. A 500 nm by 50 nm device will have a gate current of (typically) 50 pA. However, if the potentials of either the drain or source terminals of the MOSFET move further away from the gate's potential, the gate tunnel current will increase. Figure 16.65 shows how the gate current changes in a 10/1 NMOS device (actual width of 500 nm and a length of 50 nm) with the drain and bulk held at ground while the gate voltage is swept from 0 to 2 V. Notice that at a V_{GS} above *VDD* (= 1 V), the gate current becomes much more significant than the 50 pA we calculated for a typical value a moment ago. Figure 16.66 shows that if we hold the drain at 1 V, the gate current is significantly less than the values seen in Fig. 16.65. This is important because we can keep from programming an erased cell (if we were using a floating gate in this technology with oxide thickness of 14 Å) simply by driving its drain (the bit line) to *VDD*. Finally, Fig. 16.67 shows the current when erasing the device. The problem with using direct tunneling for Flash memory is retention. While it is easy to program/erase the floating gate device, it is equally easy for trapped charge to tunnel back off of a floating gate over time.

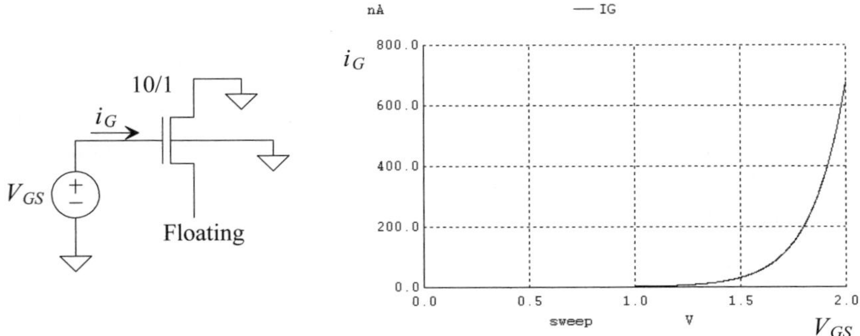

Figure 16.65 How gate current changes with gate voltage with either the gate or source grounded.

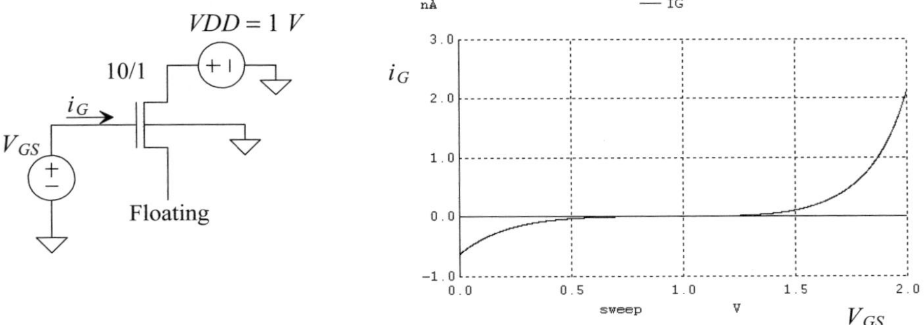

Figure 16.66 How gate current changes with gate voltage when drain is floating and drain is at VDD.

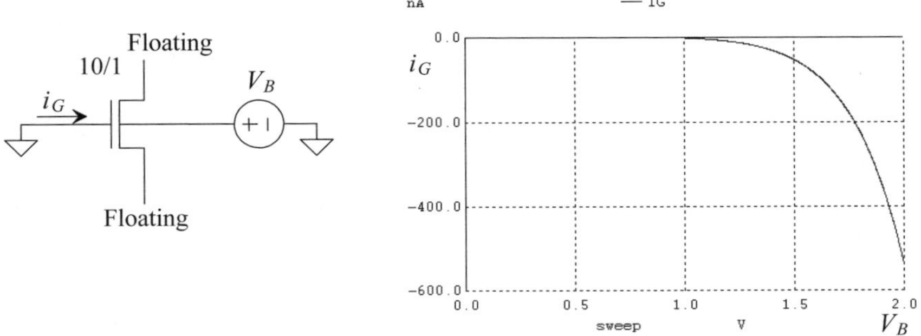

Figure 16.67 Erasing. Since electrons are being removed from the gate (they are tunneling through the gate oxide), we know the gate current will flow out of the device.

An interesting use for floating gate MOSFETs is in the design of analog circuits. Being able to adjust the threshold voltage of a MOSFET can be very useful for low-power or low-voltage design. For example, the input common-mode range of a differential amplifier can be widened by reducing the threshold voltage of the input diff-pair. Also, trimming (removing offsets) can be accomplished when floating gate MOSFETs are used. The interested reader is referred to the references on the following page for additional information.

ADDITIONAL READING

[1] B. Keeth and R. J. Baker, *DRAM Circuit Design: A Tutorial*, John Wiley and Sons Publishers, 2001. ISBN 0-7803-6014-1

[2] W. D. Brown and J. E. Brewer (eds.), *Nonvolatile Semiconductor Memory Technology: A Comprehensive Guide to Understanding and Using NVSM Devices*, John Wiley and Sons Publishers, 1998. ISBN 0-7803-1173-6

[3] B. Prince, *Semiconductor Memories: A Handbook of Design Manufacture, and Application,* (second edition), John Wiley and Sons Publishers, 1991. ISBN 0-471-92465-2

[4] A. K. Sharma, *Semiconductor Memories: Technology, Testing and Reliability*, John Wiley and Sons Publishers, 1997. ISBN 0-7803-1000-4

[5] K. Itoh, *VLSI Memory Chip Design*, Springer-Verlag Publishers, 2001. ISBN 3-5406-7820-4

[6] T. Hamada, "A Split-Level Diagonal Bit-Line (SLDB) Stacked Capacitor Cell for 256Mb DRAMs," *1992 IEDM Technical Digest*, pp. 799–802.

[7] T. Mohihara, et al., "Disk-Shaped Capacitor Cell for 256Mb Dynamic Random-Acess Memory," *Japan Journal of Applied Physics*, vol. 33, part 1, no. 8, pp. 4570–4575, August 1994.

[8] J. H. Ahn et al., "Micro Villus Patterning (MVP) Technology for 256Mb DRAM Stack Cell," *1992 Symposium on VLSI Technical Digest of Technical Papers*, pp. 12–13.

[9] K. Sagara, et al., "Recessed Memory Array Technology for a Double Cylindrical Stacked Capacitor Cell of 256M DRAM," *IEICE Trans. Electron.*, vol. E75-C, No. 11, pp. 1313–1322, November 1992.

[10] M. I. Elmasry, *Digital MOS Integrated Circuits II,* IEEE Press, 1992. ISBN 0-87942-275-0

[11] R. D. Pashley and S. K. Lai, "Flash Memories: The Best of Two Worlds," *IEEE Spectrum,* 1989, pp. 30–33.

[12] D. Frohman-Bentchkowsky, "FAMOS-A New Semiconductor Charge Storage Device," *Solid-State Electronics,* vol. 17, pp. 517–529, 1974.

[13] E. H. Snow, "Fowler-Nordheim Tunneling in SiO_2 Films," *Solid-State Communications,* vol. 5, pp. 813–815, 1967.

Additional Readings Covering Analog Applications of Floating Gate Devices

[14] L. R. Carley, "Trimming Analog Circuits Using Floating-Gate Analog MOS Memory," *IEEE Journal of Solid-State Circuits*, vol. SC-24, pp. 1569–1575, 1989.

[15] J. Sweeney and R. L. Geiger, "Very High Precision Analog Trimming Using Floating Gate MOSFETs," *Proceedings of the European Converence on Circuit Theory and Design*, Brighton, United Kingdom, September 1989, pp. 652–655.

[16] C. G. Yu and R. L. Geiger, "Very Low Voltage Operational Amplifiers Using Floating Gate MOS Transistors," *IEEE Symp. Circuits Sys.*, vol. 2, pp. 1152–1155, 1993.

[17] J. Ramírez-Angulo, S.C. Choi, and G. Gonzalez-Altamirano, "Low-Voltage OTA Architectures Using Multiple Input Floating gate Transistors," *IEEE Transactions on Circuits and Systems*, vol. 42, no. 12, pp. 971–974, November 1995.

[18] J. Ramírez-Angulo, "Ultracompact Low-voltage Analog CMOS Multiplier Using Multiple Input Floating Gate Transistors," 1996 European Solid State Circuits Conference, pp. 99–103.

[19] P. M. Furth and H. A. Ommani, "A 500-nW floating-gate amplifier with programmable gain," 41st Midwest Symp. Circuits and Systems, South Bend, IN, August 1998.

[20] E. Sánchez-Sinencio and A. G. Andreou, "Low-Voltage/Low-Power Integrated Circuits and Systems: Low-Voltage Mixed-Signal Circuits," *IEEE Press*, 1999. ISBN 0-7803-3446-9

[21] R. R. Harrison, J. A. Bragg, P. Hasler, B. A. Minch, and S. P. Deweerth, "A CMOS programmable analog memory-cell array using floating-gate circuits," *IEEE Transactions on Circuits and Systems-II,* vol. 48, no. 1, pp. 4–11, 2001.

[22] F. Munoz, A. Torralba, R. G. Carvajal, J. Tombs, and J. Ramírez Angulo, "Floating-Gate based tunable CMOS low-voltage linear transconductor and its application to HF g_m-*C* filter design," *IEEE Transactions on Circuits and Systems*, vol. 48, no. 1, January 2001, pp. 106–110.

[23] C. T. Charles and R. R. Harrison, "A Floating Gate Common Mode Feedback Circuit for Low Noise Amplifiers," *Proceedings of the Southwest Symposium on Mixed-Signal Design*, Las Vegas, NV, pp. 180–185, February 23–25, 2003.

[24] F. Adil, G. Serrano, and P. Hasler, "Offset Removal Using Floating-Gate Circuits for Mixed-Signal Systems," *Proceedings of the Southwest Symposium on Mixed-Signal Design*, Las Vegas, NV, pp. 190–195, February 23-25, 2003.

PROBLEMS

Unless otherwise indicated use the short-channel CMOS BSIM4 models for all simulations.

16.1 Estimate the bit line capacitance if there are 256 word lines and we include the gate-drain overlap capacitance from each MOSFET, as seen in Fig. 16.68.

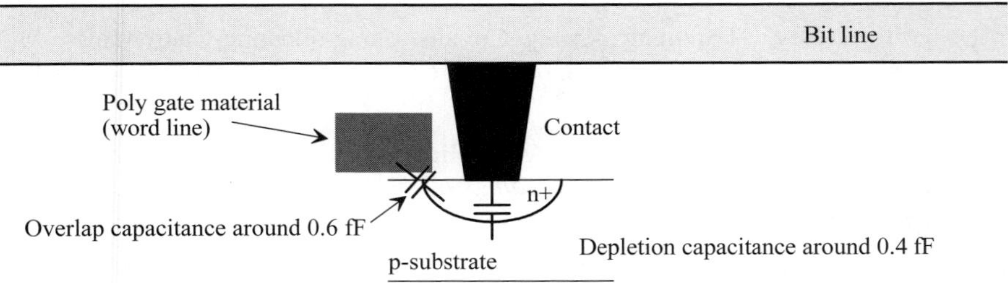

Figure 16.68 How the gate-drain overlap capacitance loads the bit line. Note that this cross-sectional view is rotated 90 degrees from the one seen in Fig. 16.3.

16.2 Consider the NSA seen in Fig. 16.69. Suppose that the load capacitance is mismatched by 20%, as seen in the figure. Assuming that both caps are equilibrated to 0.5 V prior to sensing, which capacitor will fully discharge to ground (which MOSFET will fully turn on)? What voltage difference on the capacitors is needed to cause metastability? Verify your answers with SPICE.

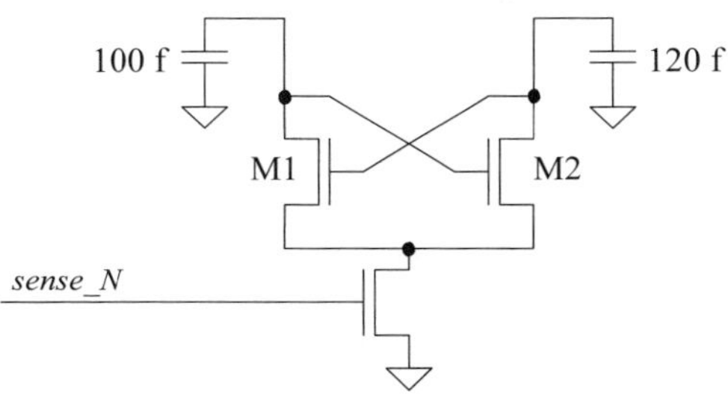

Figure 16.69 Problem 16.2 showing how a mismatch in the load capacitance of an NSA can result in in sensing errors.

16.3 Repeat Problem 16.2 with the circuit in Fig. 16.70. In this figure M1 and M2 experience a threshold voltage mismatch (modeled by the DC voltage source seen in the figure).

16.4 Examining Fig. 16.17 we see that there will be a voltage drop along the metal line labeled *NLAT* when the sense amplifiers fire. Re-sketch this metal line as resistors between each NSA. If the voltage drop along the line is significant (the length of the line is very long), errors can result. Would we make things better by using individual MOSFETs on the bottom of each NSA connected to *sense_N*? Why or why not?

16.5 Suppose a memory array has 1024 columns. If the word line resistance of one cell is 2 Ω and the capacitance per cell (to ground) is 500 aF, estimate the delay to open a row line. Sketch the equivalent RC circuit of the word line.

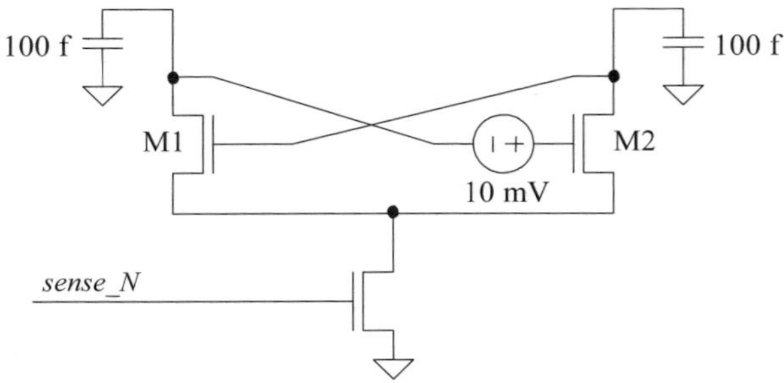

Figure 16.70 Circuit for Problem 16.3.

16.6 In Fig. 16.71, if the top plate of capacitor C_1 is initially charged to V_1 and the top plate of capacitor C_2 is initially charged to V_2, estimate the final voltage, V_{final}, on the top plates of the capacitors after the switch closes.

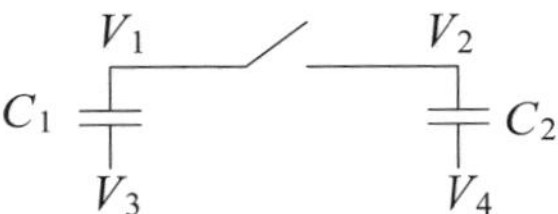

Figure 16.71 Circuit for Problem 16.6.

16.7 Figure 16.72 shows a clocked comparator topology based on the topology seen in Fig. 16.32. The location of the imbalance MOSFETs has been moved in this figure. Comment on the merits and disadvantages of this topology compared to the one in Fig. 16.32. Simulate the operation of this circuit.

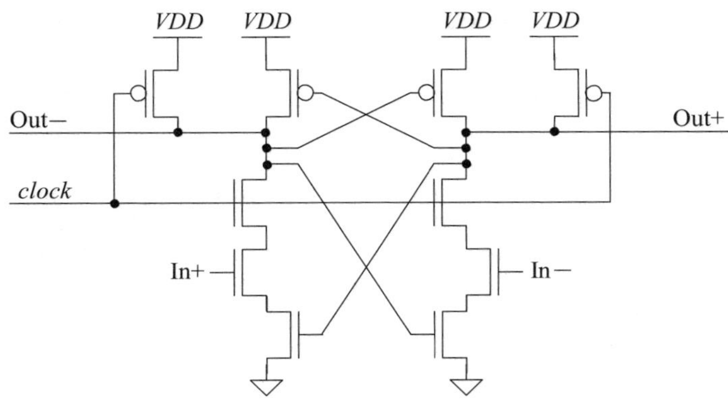

Figure 16.72 An alternative clocked sense amplifier.

16.8 Figure 16.73 shows the addition of I/O (input/output) transistors to the memory array and NSA seen in Fig. 16.17. Sketch a column decoder design using static logic. Show how the 3-bit input address is connected to each stage.

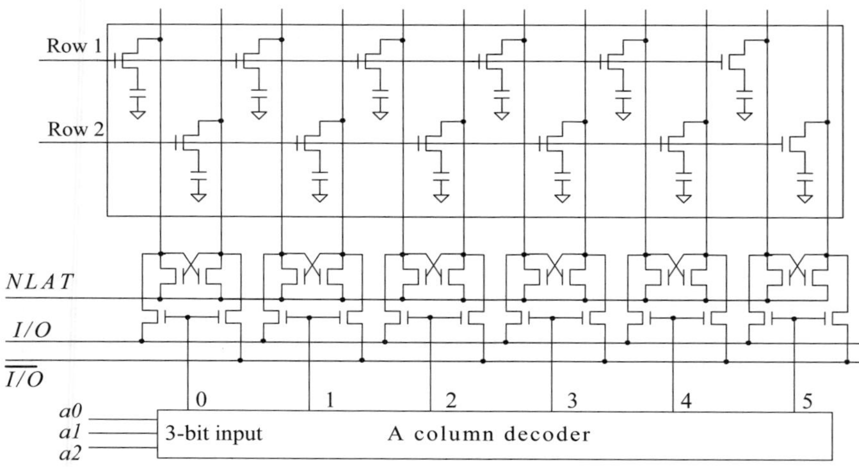

Figure 16.73 Design of a column decoder for problem 16.8.

16.9 The I/O lines in the previous problem won't swing to full logic levels. To restore a full *VDD* level on these lines and to speed up the signals, a *helper flip-flop* is used. Sketch a possible implementation of the helper flip-flop. Why can it be used to speed up the signals?

16.10 Suppose a 2Mbit memory is to be designed as a x2 part (2-bit input/output words). Further suppose that 10 address pins are available to access the memory. Sketch a block diagram of how the row and column addresses are multiplexed together and stored in separate registers. Use $\overline{RAS}$ and $\overline{CAS}$, as discussed in the chapter to clock in the addresses. Comment on the validity of using 20-bits to access 2Mbits of data.

16.11 Suppose that the memory in Problem 16.10 is made up of 8-256k memory arrays (assume 1024 row lines, a folded array, and 512 column lines). How many I/O lines are needed for each array? If three bits of the address are used to select one of the memory arrays (so that only a single row is open in a single memory array at a time), how big is a page of data? Sketch a block diagram of a possible decoding scheme for the chip. Assume that only the three bits of the address are globally decoded and that the remaining 17 bits are locally decoded. How many of these bits must be routed to each array assuming an enable (one for each of the eight memory arrays) signal from the global decoder is routed to each memory array.

16.12 Suppose that it is suggested that the word line driver in Fig. 16.46 be modified to simplify it as seen in Fig. 16.74. What is the problem with this design? Use SPICE to illustrate the driver not functioning properly.

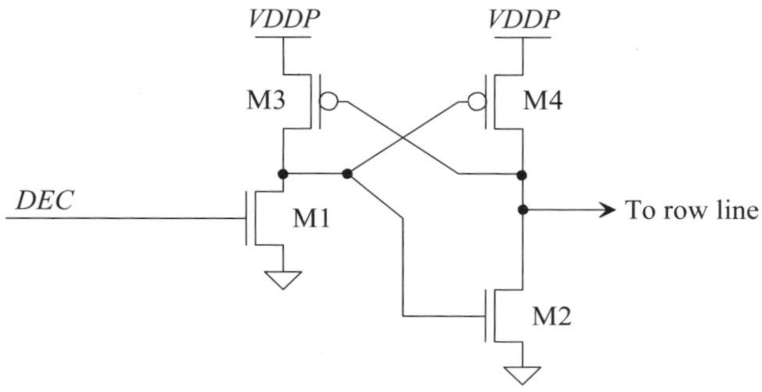

Figure 16.74 A (bad) CMOS word line driver, Problem 16.12.

16.13 Simulate the operation of the SRAM cell seen in Fig. 16.48. Use the 50 nm process with NMOS of 10/1 and PMOS of 20/1. Is it wise for the access MOSFETs to be the same size as the latch MOSFETs? Why or why not? Use simulations to verify your answers.

16.14 Suppose an array of EPROM cells, see Fig. 16.57, consisting of two row lines and four bit lines is designed. Further assume $V_{THN,Erased} = 1\ V$ and $V_{THN,Prog} = 4\ V$. Will there be any problems reading the memory out of the array if an unused row line is grounded while the accessed row line is driven to 5 V ? Explain why or why not.

16.15 Suppose that it is suggested that the n-well in a NAND Flash memory cell should be grounded at all times except when the array is being erased. Is this OK? What would be a potential benefit?

16.16 Explain in your own words and with the help of pictures why a top select MOSFET is needed in a NAND Flash memory cell.

16.17 Suppose that the transistor connected to RA3 in Fig. 16.61 is to be programmed. Further suppose that the n+ source implant seen in the layout is to remain grounded as indicated in Table 16.1. How do we float the source of the transistor connected to RA3 so that it can be programmed as seen in Fig. 16.58?

16.18 Reviewing Fig. 16.62, is it necessary that the gates of the floating gate MOSFETs connected to RA1 – RA3 be driven to 5 V? Can the gates of these MOSFETs remain grounded? Why or why not? If the RA1 MOSFET is to be programmed, in this figure, must the gate of RA0 be driven to 5 V?

16.19 If the I_{erased} of a NAND Flash memory cell is 20 μA and the I_{prog} is 2 μA, explain what the bit line voltage will do when reading out the cell in the following configuration, Fig. 16.75. Will the bit line go all the way to *VDD*? All the way to ground? Explain. If the bit line capacitance is 200 fF, estimate the length of time it will take the data on the bit line to settle before it can be read out. Assume that the *VDD* is 5 V and the bit line is equilibrated to 2.5 V prior to sensing.

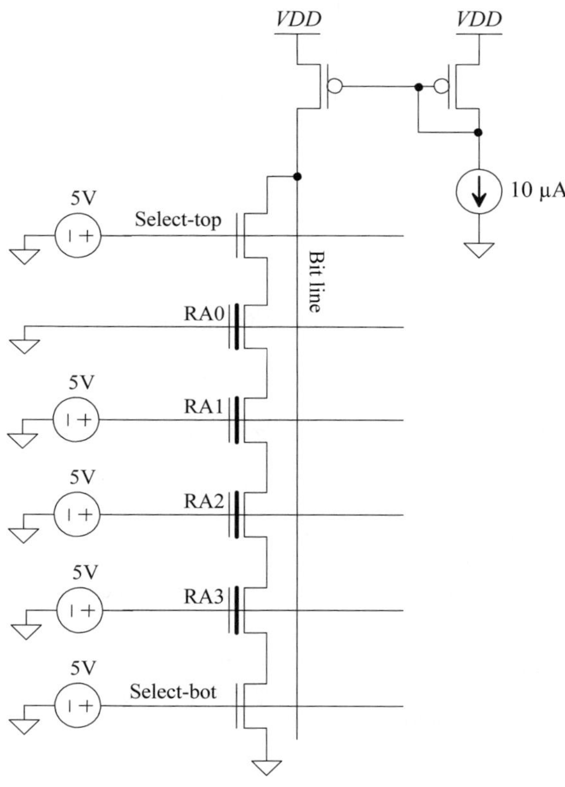

Figure 16.75 Reading the contents of RA0 in a NAND Flash cell.

16.20 Show, using a configuration similar to the one seen in Fig. 16.66, if it is possible to get negative gate tunnel gate current (used for erasing) by grounding the gate and raising the potential on the drain of the MOSFET. Explain why it works or why it doesn't work.

Chapter

17

Sensing Using ΔΣ Modulation

In the last chapter we performed sensing by clocking a sense amplifier that compared two inputs and determined which one had the greater value. To illustrate some concerns with this approach, as signals get smaller or noisier, consider the water analogy seen in Fig. 17.1. In this figure we show two buckets filled with water. Our sensing circuit should determine if the water is above or below the line indicated on the bucket. In (a) our sense-amp can clearly determine that the water is below the line. In (b), however, the water is sloshing around, so it's difficult to determine if the level is above or below the line. To make a more accurate determination in (b), we might try averaging, at different times, several sense amplifier outputs. For example, we might get, after strobing the sense-amp four times, outputs of: yes (it's above the line), no (it's not), yes, yes. Averaging these responses results in the answer "yes, it is above the line." The averaging can be thought of as reducing the noise in the signal (the variations in the water level because of sloshing).

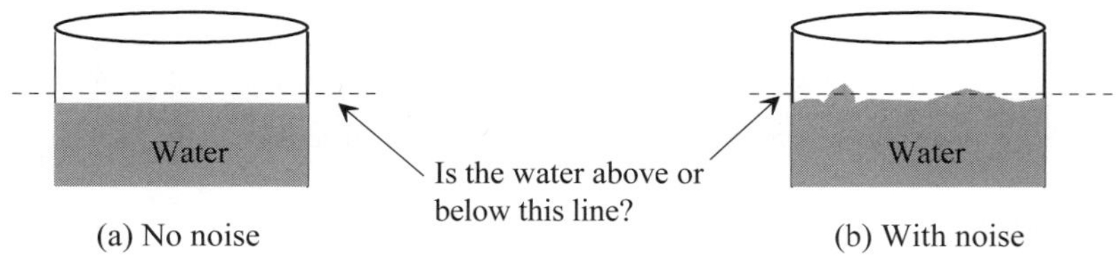

Figure 17.1 Using a water analogy to illustrate the problems with sensing.

Let's take this a step further. The output in the previous analogy resulted in answering the question, "Is the water above or below the line?" In many applications (multi-level memory cells, imaging sensors, analog-to-digital conversion, etc.) this isn't enough information. We need to determine the actual height of the water in the bucket. This means we need more than just a "1" (yes) or "0" (no) output. The purpose of this chapter is to present a powerful and practical circuit technique called *delta-sigma modulation* (also known as *sigma-delta modulation*) for sensing applications (where the desired signal we are sensing is a constant but may be corrupted with noise).

17.1 Qualitative Discussion

Figure 17.2 shows how delta-sigma modulation (DSM) works. The height of the water in the bucket that is "sensed" is changed into a rate of water flow. If the height of the water level increases, the float moves upwards causing the valve to open up and more water to fall into the "sigma" bucket. If the water level gets too high (above an arbitrary line in the sigma bucket), we remove a cup of water. The delta is the difference in how many times we remove a cup of water per time with the rate the water flows into the sigma bucket. By averaging the number of times we remove a cup of water over time, we can determine the rate the water flows out of the valve and thus the relative height of the water in the bucket we are sensing. The best way to understand the fundamentals of DSM is with examples.

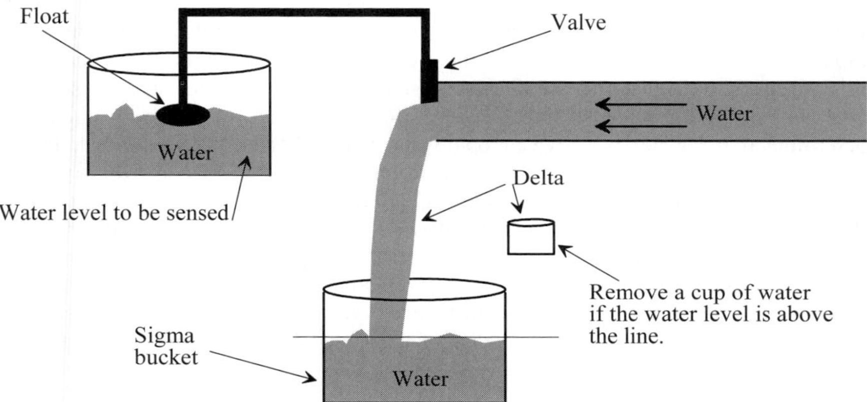

Figure 17.2 Changing the water height (pressure or voltage) into a water flow (current) using the float and valve. The delta comes from the difference in the number of cups we remove from the bucket with the water we add to the bucket. The sigma (sum) is the storage of the difference in the bucket.

17.1.1 Examples of DSM

Let's say the rate the water is flowing into the sigma bucket is one cup every 40 seconds (0.25 cups per 10 seconds). Further, let's say we check the height of the water in this bucket every 10 seconds. If we start the sense off with the water height in the sigma bucket at (arbitrarily) 5 cups (our reference line), we can generate the data in Table 17.1. At a time of 10 seconds 0.25 cups have fallen into the sigma bucket and so the water level is 5.25 cups. Since this water level is > 5 cups, we remove a cup of water from the bucket (at 10 seconds) leaving 4.25 cups in the bucket. At 20 seconds, after another 0.25 cups of water have fallen into the bucket, we have 4.5 cups. Notice that the longer we average, the closer our output moves to 0.25 cups/10 seconds.

Another key point to notice is that if we make a mistake when determining if the water level in the sigma bucket is > 5 cups, it doesn't really matter. The error averages out over time. What does matter though is how carefully we fill up the cup when removing water. If we don't fill it up all the way or if the water spills out of the cup, the level of the water in the sigma bucket changes. The result limits the precision of the sense. One other limiting factor is the sigma bucket. If it is "leaky" and the water it holds leaks out, the quality of the sense will be affected.

Table 17.1 An example of DSM.

Time, seconds	**Water level in the sigma bucket (cups)**	**Remove cup? (water level > 5?)**	**Running average**
0	5		0
10	5.25	Yes	1
20	4.5	No	0.5
30	4.75	No	0.33
40	5	No	0.25
50	5.25	Yes	0.4
60	4.5	No	0.33
70	4.75	No	0.29
80	5	No	0.25
90	5.25	Yes	0.33
100	4.5	No	0.3
110	4.75	No	0.27
120	5	No	0.25
130	5.25	Yes	0.31
140	4.5	No	0.29
150	4.75	No	0.26
160	5	No	0.25
170	5.25	Yes	0.29

The Counter

We might wonder how, in a practical sensing circuit, we can get decimal numbers like the ones seen in Table 17.1? The answer to this question is seen in Fig. 17.3. If a "yes, remove a cup of water" is indicated by the output of the DSM sensing circuit going high, a counter can be used to generate the number. If N is the total number of times we clock the DSM, then

$$\text{Output number} = \frac{\text{number of Yes outputs}}{N} \tag{17.1}$$

In Table 17.1 N is 17. The number of "yes" outputs is 5. The final output number is then 5/17 or 0.29. If a 10-bit counter is used (assuming it was reset to zeroes at the beginning of the sense), then its digital output is 00 0000 0101. To find the absolute value of the input, we multiply this number by the size of the feedback signal (here one cup) to get

$$\text{Input signal} = \text{Ouput number} \times \text{fed back signal size} \quad (17.2)$$

or here 0.29 cups (per 10 seconds). Note that to converge on the actual rate of 0.25 cups/10 seconds we would need to average more points.

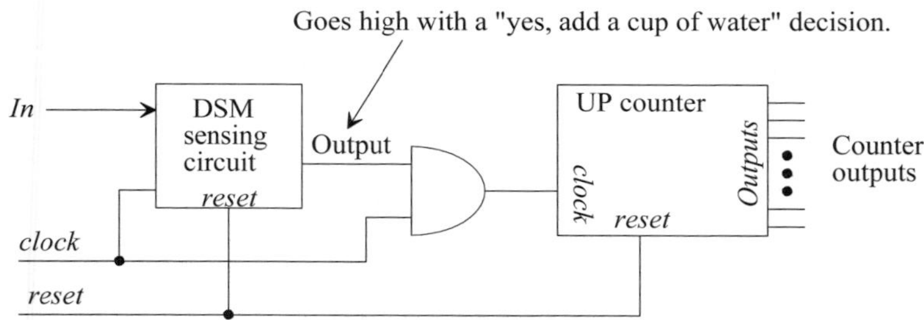

Figure 17.3 How a counter is used to average the outputs of a DSM sensing circuit.

Cup Size

The size of our fed back signal, here the cup size, is very important. If we use a small cup, we can converge on the correct digital representation (the counter output) of our analog signal (the flow of water into the sigma bucket) quicker. However, if we use too small of a cup, then we can't remove the water fast enough from the bucket and it will overfill. For a large range, we need a big cup. Using a big cup means that we have to average longer (the sensing lasts for a longer period of time).

Another Example

Figure 17.4 shows another water analogy where DSM can be used for measuring an analog quantity (in this case the water flowing out of the sigma bucket). When the water level gets too low, we add a cup of water to the bucket. Again, averaging how often we add the cup of water to the bucket gives a digital representation of the rate at which water leaves the bucket.

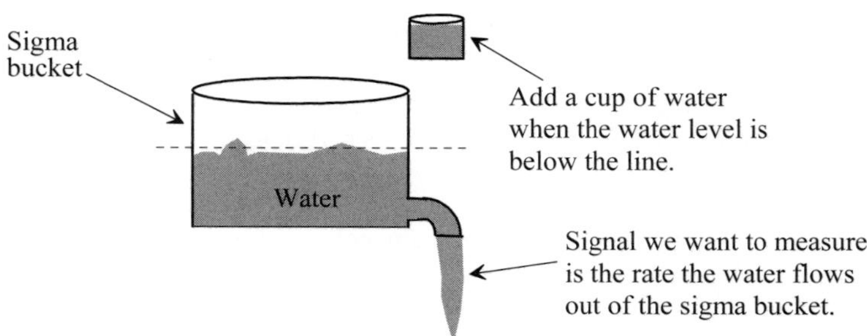

Figure 17.4 Using DSM to measure the water leaving a bucket.

17.1.2 Using DSM for Sensing in Flash Memory

Let's discuss sensing using DSM in a floating gate memory cell technology (Flash). Figure 17.5a shows the characteristics of the Flash memory cell discussed in the last chapter (see Fig. 16.64). When we are sensing the state of the Flash cell, that is, erased or programmed, we hold the row line at ground so that the cell's V_{GS} is 0. If the bit line is held at a potential above ground, say 1 V, then a current of either I_{erased} or I_{prog} (ideally) flows in the cell. In reality variations in the production of the cell and the consistency of the erase from cell to cell affect these values. Using DSM we can more precisely determine the drain current in the cell. This allows us to make a more intelligent decision about the state of the cell. Further, DSM can also be used to program the cell to precisely set the programmed current flow. This allows us to make a memory cell out of a single transistor, which can be used to store several logic levels (values of programmed current). Because of the ability to precisely control the programming operation in a floating gate technology, programming using Fowler-Nordheim Tunneling (FNT) can be abandoned (gate oxides of around 80 Å) for direct tunneling (gate oxides < 20 Å). Of course, the issues with data retention (the charge leaking off the floating gate) are still a concern. In this section we qualitatively discuss how DSM can be used for sensing in Flash technology.

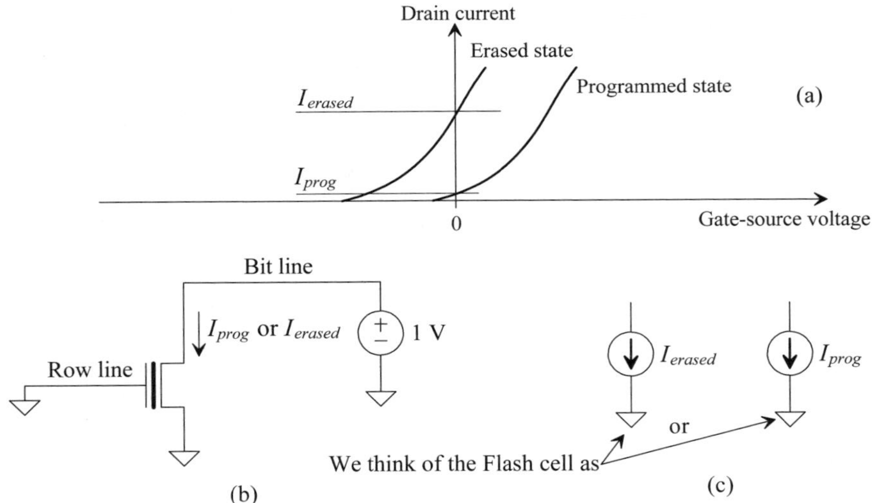

Figure 17.5 The IV characteristics of a NAND memory cell and how we think of the cell when sensing.

The Basic Idea

Figure 17.6 shows the basic idea. The sigma bucket in this figure is the bit line capacitance, C_{bit}. As seen in Fig. 17.4, the signal we are measuring is the rate current flows out of this bucket, that is I_{erased} or I_{prog}. The comparator is used to determine if the voltage on the bit line is below 1 V (again an arbitrary reference). The memory cell is continuously removing charge from the bit line. When the bit line voltage is below 1 V,

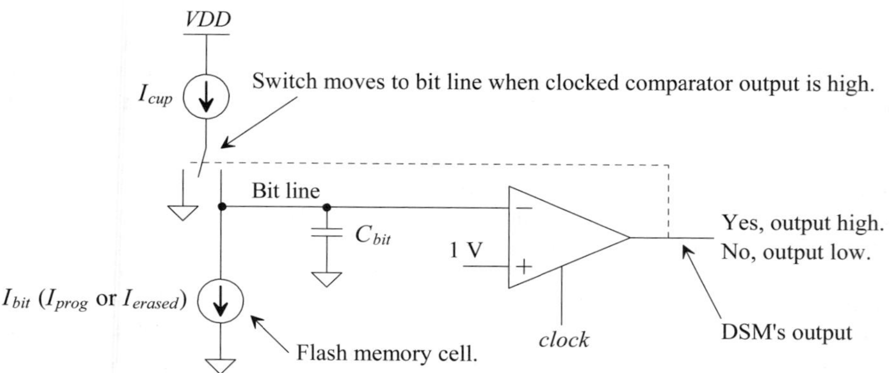

Figure 17.6 Sensing a Flash memory cell using DSM.

the current source, I_{cup} , is connected to C_{bit} to provide the water to the bucket. Note that *we cannot connect the bit line to a voltage source when the output of the comparator goes high.* A voltage source, in our water analogy, is a pressure with an infinite supply of water. The connection of the voltage source would fill the bucket up (the bit line capacitance) to the pressure of the source (the voltage of the source). Again, by looking at the number of times the output of the DSM sensing circuit goes high, we can determine, very precisely, the value of I_{bit}.

If we are clocking the comparator at a rate of f at a period of T (= $1/f$), then the rate that charge is removed from the bit line capacitance (the sigma bucket) is

$$\frac{I_{bit}}{C_{bit}} = \frac{\Delta V_{bit}}{T} \tag{17.3}$$

where ΔV_{bit} is the change in the bit line voltage. The amount of charge removed from the bit line in one clock cycle T is

$$Q_{bit} = I_{bit} T = C_{bit} \cdot \Delta V_{bit} \tag{17.4}$$

These quantities are constant. However, the rate at which we add charge to the bit line isn't a constant (unless the output of the comparator is always the same). If N is the total number of clock cycles (the total number of times we clock the comparator) and M is the number of times the output of the comparator goes high (the number of "Yes, add charge to the capacitor" signals), then the rate we add charge to the bit line from I_{cup} is

$$Q_{cup} = I_{cup} \cdot \frac{M}{N} \cdot T \tag{17.5}$$

In order for the voltage on the bit line (the water level in the bucket) to remain, on average, constant ,we require that the amount of charge leaving the bucket (Q_{bit}) equal the amount of charge entering the bucket (Q_{cup}) or

$$Q_{bit} = I_{bit} T = Q_{cup} = I_{cup} \cdot \frac{M}{N} \cdot T \tag{17.6}$$

or

$$\frac{I_{bit}}{I_{cup}} = \frac{M}{N} \tag{17.7}$$

This result is important. It relates the counter output code, M, and the total number of times the DSM sensing circuit is clocked, N, to the ratio of the Flash memory cells current, I_{bit} and the fed-back signal I_{cup}. Note again that the charge leaving the bit line capacitance can't be greater than the charge entering the bit line capacitance, that is, we require $I_{cup} \geq I_{bit}$. As mentioned earlier, using a large cup (large I_{cup}) increases the sensing time for a required resolution (we're averaging a larger variable).

Notice that Eq. (17.7) doesn't include the clock frequency or period. We might think that these quantities aren't important. While, if the DSM is designed correctly, it doesn't directly affect the output of the sensing operation, we can have the situation where the bucket (bit line capacitance) empties or overflows (goes to ground or *VDD*). To avoid this situation, we may require that the maximum deviation on the bit line, $\Delta V_{bit,\max}$ be less than some value over a time T. We can write, assuming $I_{cup} \geq I_{bit}$,

$$\Delta V_{bit,\max} = \frac{I_{cup}T}{C_{bit}} = \frac{I_{cup}}{C_{bit} \cdot f_{clk}} = I_{cup} \cdot R_{sc} \tag{17.8}$$

For example, if the bit line capacitance is 500 *fF*, the clock frequency is 100 MHz ($T = 10$ ns), and I_{cup} is 10 μA, then $\Delta V_{bit,\max} = 0.2\ V$. If we clock the DSM slower, we must increase our bucket size (add capacitance in parallel with C_{bit} on the input of our DSM).

Example 17.1

Suppose that the DSM sensing circuit in Fig. 17.6 is used to determine the current flowing in a programmed Flash memory cell. If the single transistor Flash memory cell is programmed to conduct 1, 3, 5, or 7 μA of current, estimate the counter output codes if the DSM is clocked 15 times (a 4-bit counter is used) and I_{cup} is 10 μA. Estimate the maximum bit line voltage change if the DSM is clocked at 100 MHz and the bit line capacitance is 500 *fF*.

For the 1 μA program current, using Eq. (17.7), we get

$$\frac{1}{10} = \frac{M}{15} \rightarrow M = 1.5 \text{ so the counter output would be 2 (0010)}$$

For the 3 μA program current through the Flash memory cell,

$$\frac{3}{10} = \frac{M}{15} \rightarrow M = 4.5 \text{ so the counter output would be 5 (0101)}$$

For the 5 μA program current,

$$\frac{5}{10} = \frac{M}{15} \rightarrow M = 7.5 \text{ so the counter output would be 8 (1000)}$$

Finally, for the 7 μA program current,

$$\frac{7}{10} = \frac{M}{15} \rightarrow M = 10.5 \text{ so the counter output would be 11 (1011)}$$

The maximum deviation of the voltage on the bit line would be when the program current is 1 μA (leaving the bit line) and I_{cup} is connected (10 μA charging the bit line). This would give a net current of 9 μA into the bit line so

$$\Delta V_{bit,\max} = \frac{9\mu A \cdot 10\ ns}{500\ fF} = 180\ mV$$

Let's show the output decisions for the case when I_{prog} is 1 μA, Table 17.2. Note that if 1 μA is removed from the 500 *fF* bit line capacitor for 10 ns, the bit line voltage drops 20 mV. If (net) 9 μA of current is put into the bit line for 10 ns, the bit line voltage increases by 180 mV. Note that as calculated, the maximum change in the bit line voltage is 180 mV. Also note that if the comparator had output a Yes at 100 ns instead of at 110 ns, it wouldn't have affected the final output. ■

Table 17.2 See Ex. 17.1.

Time, nanoseconds	Bit line voltage	Add current? ($V_{bit} < 1$ V?)	Bit line current	Counter output
0	1		−1 μA	0 (0000)
10	0.98	Yes (1)	9 μA	1 (0001)
20	1.16	No (0)	−1 μA	1 (0001)
30	1.14	No (0)	−1 μA	1 (0001)
40	1.12	No (0)	−1 μA	1 (0001)
50	1.1	No (0)	−1 μA	1 (0001)
60	1.08	No (0)	−1 μA	1 (0001)
70	1.06	No (0)	−1 μA	1 (0001)
80	1.04	No (0)	−1 μA	1 (0001)
90	1.02	No (0)	−1 μA	1 (0001)
100	1	No (0)	−1 μA	1 (0001)
110	0.98	Yes (1)	9 μA	2 (0010)
120	1.16	No (0)	−1 μA	2 (0010)
130	1.14	No (0)	−1 μA	2 (0010)
140	1.12	No (0)	−1 μA	2 (0010)
150	1.1	No (0)	−1 μA	2 (0010)

Notice, in the previous example, that we can clock the DSM indefinitely. The only limitation is the size of the counter (after a time the counter will roll over). This is important because, to get better resolution, all we have to do is sense for a longer period of time. We can define the *dynamic range* of the sense operation, in dB, as the maximum counter output code (N) to the smallest output code (1)

$$\text{dynamic range (DR)} = 20 \cdot \log \frac{N}{1} = 20 \cdot \log N \tag{17.9}$$

noting that no signal would give a counter output code of zero. Clocking the DSM 1,000 times results in, ideally, a DR of 60 dB. Clocking the DSM 15 times results in a DR of 23.5 dB.

The maximum input signal, $I_{bit,\max}$, equals the fed-back signal I_{cup} ($I_{cup} \geq I_{bit}$). The minimum resolvable signal is then

$$\text{minimum resolvable signal} = \frac{\text{fed-back signal}}{N} \tag{17.10}$$

Again, this illustrates that the larger the fed-back signal, the more averages, N, are needed for a given resolution.

Example 17.2
Determine the minimum resolvable programmed current, I_{bit}, in Ex. 17.1. Verify the answer with a table similar to Table 17.2.

Using Eq. (17.10), the minimum resolvable current is 10 μA/15 or 0.666 μA. If this current is removed from the bit line capacitance for 10 ns, the voltage drop is 13.33 mV. If the bit line capacitance is charged with 9.333 μA for 10 ns, its voltage increases to 186.66 mV. Table 17.3 tabulates the operation of the DSM sensing circuit when I_{prog} is 666.6 nA. Note that in all situations where I_{prog} is not zero the first decision is a Yes. However, after neglecting this decision, N clock cycles later the counter increments to 2 when the cell is sinking the minimum resolvable signal (here 666.6 nA). ■

Table 17.3 See Ex. 17.2.

Time, nanoseconds	Bit line voltage (millivolts)	Add current? (V_{bit} < 1 V?)	Bit line current	Counter output
0	1,000		–0.666 μA	0 (0000)
10	986.67	Yes (1)	9.333 μA	1 (0001)
20	1,173.33	No (0)	–0.666 μA	1 (0001)
30	1,160	No (0)	–0.666 μA	1 (0001)
40	1,146.67	No (0)	–0.666 μA	1 (0001)
50	1,133.34	No (0)	–0.666 μA	1 (0001)
60	1,120	No (0)	–0.666 μA	1 (0001)
70	1,106.67	No (0)	–0.666 μA	1 (0001)
80	1,093.33	No (0)	–0.666 μA	1 (0001)
90	1,080	No (0)	–0.666 μA	1 (0001)
100	1,066.67	No (0)	–0.666 μA	1 (0001)
110	1,053.33	No (0)	–0.666 μA	1 (0001)
120	1,040	No (0)	–0.666 μA	1 (0001)
130	1,026.67	No (0)	–0.666 μA	1 (0001)
140	1,013.33	No (0)	–0.666 μA	1 (0001)
150	1,000	No (0)	–0.666 μA	1 (0001)
160	986.67	Yes (1)	9.333 μA	2 (0010)

The Feedback Signal

The precision of a sense operation in a DSM is limited by how precisely current can be guided into or out of the bit line capacitance. The bit line capacitance can be thought of as integrating (an integrator, *sigma*) the difference (*delta*) between the fed-back signal (I_{cup} in Fig. 17.6) and the signal we are measuring (I_{bit}). If the amount of charge (current) steered into the bit line isn't consistent from one clock period to the next, errors will occur. To illustrate the possibility of errors, examine the circuit seen in Fig. 17.7. When the switch is connected to ground $C_{parasitic}$ (the parasitic capacitance on the output of the current source), is discharged. Now when the switch is connected to the bit line I_{cup} must charge both $C_{parasitic}$ and C_{bit}. Because the bit line voltage will be moving around and the switch may stay connected to the bitline for two or more consecutive cycles, errors in the sense will occur that limit the resolution. For example, the charge supplied to $C_{parasitic}$ may be $1.05 \cdot C_{parasitic}$ (a bit line voltage of 1.05V) in one clock cycle, while in a different clock cycle it may be $1.02 \cdot C_{parasitic}$. In our water analogy in Fig. 17.4 the amount of water in the cup effectively changes due to this parasitic. If the size of the bit line capacitance is large (or if capacitance is added to the input of the DSM to purposely increase this capacitance) or only a course sense is used (say 128 or less clock cycles), then $C_{parasitic}$ will probably not affect the sense in any significant way. Note that if the switch were connected to 1V in Fig. 17.7 instead of ground, the errors from $C_{parasitic}$ would be reduced.

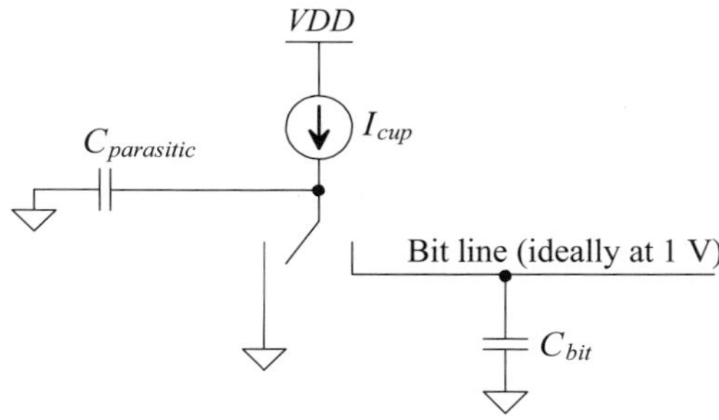

Figure 17.7 How the parasitic capacitance on the output of the current source causes errors.

In an attempt to minimize the power (power is burned unnecessarily when I_{cup} is connected to ground in Figs. 17.6 or 17.7) and the effects of parasitics consider the switched-capacitor (SC) circuit seen in Fig. 17.8. In the following we assume the bit line reference voltage (here 1 V) is greater than the threshold voltage of the PMOS transistors so that the PMOS switches used in the SC can turn fully on. The two clock signals, ϕ_1 and ϕ_2 form nonoverlapping clock signals (they are never low at the same time). This is important because we never want to connect the bit line directly to *VDD* (which occurs if the clock signals are all low at the same time). When ϕ_1 is low, C_{cup} charges to *VDD*. Next, ϕ_1 goes high. Between ϕ_1 going high and ϕ_2 going low, the comparator is clocked (on the rising edge of *clock*). If charge needs to be added to the bit line, the comparator output goes high and the gate of M3 is driven low. Note that the inverter can be removed

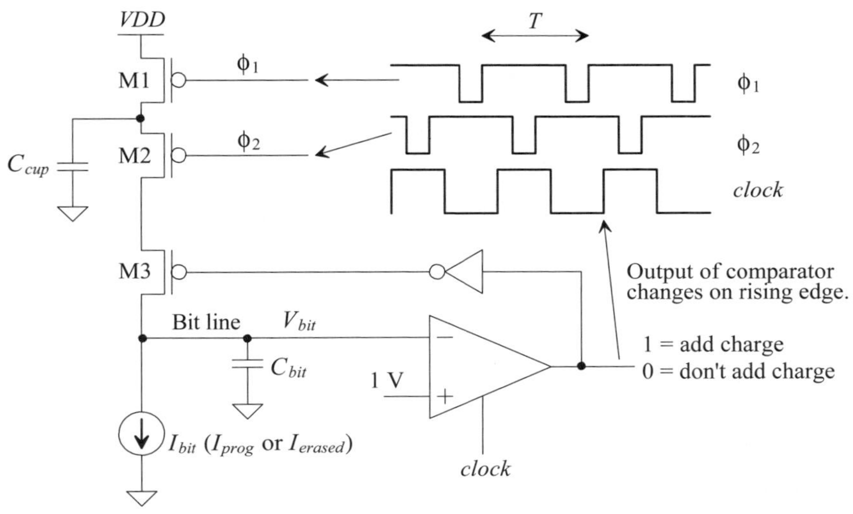

Figure 17.8 Using a switched-capacitor to add charge to the bit line.

if we swap the inputs of the clocked comparator so that a 1 output indicates "don't add charge" and a 0 output indicates "add charge." When ϕ_2 goes low, the charge on C_{cup} is dumped into the bit line. The charge leaving the bit line is still given by Eq. (17.4). The charge dumped into the bit line from C_{cup} is

$$Q_{cup} = C_{cup} \cdot (VDD - V_{bit}) \tag{17.11}$$

When we compare this amount of charge to Eq. (17.5), we see a benefit when Q_{cup} is not a function of the clock period. However, the big drawback is that the charge we add to the bit line is a function of the bit line voltage. This limits the accuracy of the sense. The water analogy to this problem is that the cup is being filled up to a level dependent on the sigma bucket water level. The amount of water dumped into the bucket from the cup should be independent of the water level in the bucket.

To make the charge we add to the bit line independent of the bit line voltage, consider the addition of another PMOS device in Fig. 17.9. When both the signal from the comparator and ϕ_2 are low, M2 and M3 are on. The charge from C_{cup} is dumped into the source of M4. The result is an initial increase in M4's source potential (M4 turns on). However, after the charge is dumped into the bit line, M4 shuts off. Its V_{SG} goes to V_{THP} (M4's source potential goes to $V_{REF} + V_{THP}$). This keeps the potential across C_{cup} constant. Equation (17.11) can be rewritten as

$$Q_{cup} = C_{cup} \cdot (VDD - V_{REF} - V_{THP}) \tag{17.12}$$

knowing, to keep M4 from being fully on, that is, $V_{SD,sat}$ not zero

$$V_{REF} + V_{THP} > V_{bit,\max} \tag{17.13}$$

where $V_{bit,\max}$ is the maximum voltage on the bit line (and, again, $V_{bit,\min} > V_{THP}$).

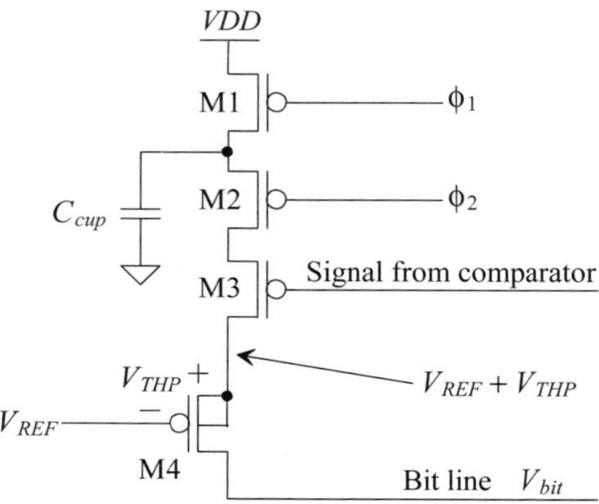

Figure 17.9 Adding a MOSFET to set the swing across the capacitor.

Example 17.3
Using SPICE demonstrate the validity of Eqs. (17.11) and (17.12).

To demonstrate the validity of Eq. (17.11), examine the circuit in Fig. 17.10. We know that a PMOS device passes a voltage from *VDD* to V_{THP}, so we make sure that the bit line is always at a potential greater than V_{THP} (= 280 mV for the short channel process used in this book). If it's not, the PMOS devices don't behave like switches. In the simulation we'll sweep the bit line voltage from 0.3 to 1V and look at the current through the bit line voltage source, V_{bit}. M3 is always on so we can look at the SC's operation alone (without the effects of the comparator). If Eq. (17.11) is valid, we should see a linear decrease in the current pulses (the charge) as V_{bit} is increased. Figure 17.11a shows the nonoverlapping clocks used in the simulation. Note how neither clock (ϕ_1 or ϕ_2) is low at the same time (again, this is important). Figure 17.11b shows the current through V_{bit}. As expected it decreases as V_{bit} increases. As Eq. (17.11) indicates. the current goes to zero when

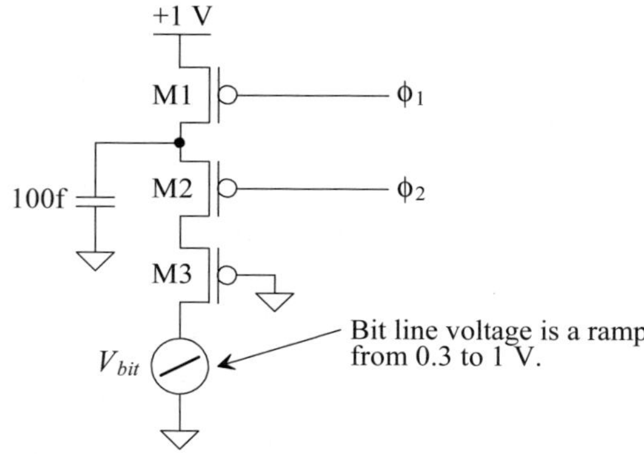

Figure 17.10 Verifying Eq. (17.11), see Ex. 17.3.

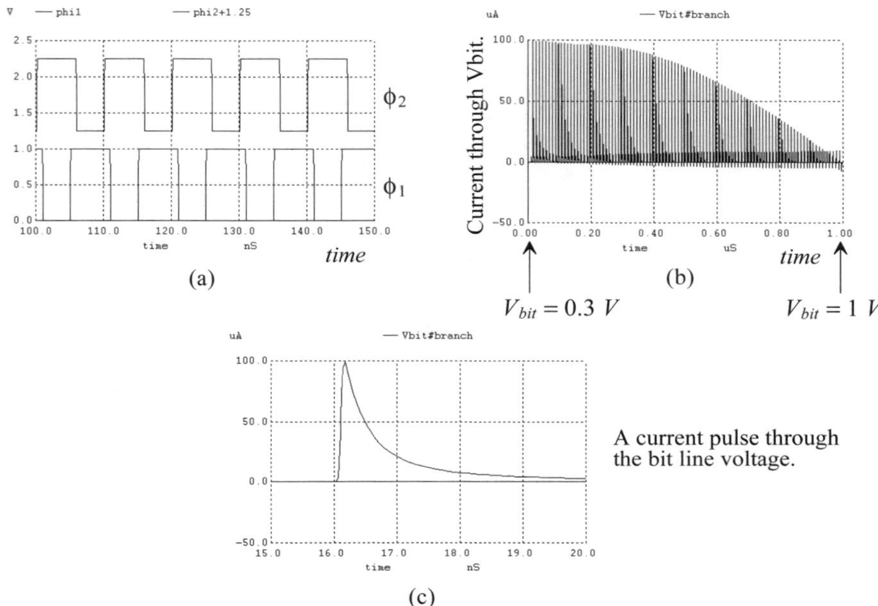

Figure 17.11 The operation of the circuit in Fig. 17.10.

V_{bit} is 1 V (= VDD). Let's use Eq. (17.11) to calculate the charge supplied by C_{cup} when V_{bit} is 300 mV

$$Q_{cup} = 100\,fF \cdot (1 - 0.3) = 70\,fC \tag{17.14}$$

Figure 17.11c shows the pulse of current that is dumped into V_{bit} at the beginning of the simulation. To estimate the amount of charge in this pulse, we take the amplitude and multiply it by an estimate of the pulse's width. That is, $100\ \mu A \cdot 0.7\ ns = 70\,fC$ or the same result as seen in Eq. (17.14).

To validate Eq. (17.12), we can add M4 from Fig. 17.9 to the circuit seen in Fig. 17.10. The only parameter we need to calculate before simulating is the value of V_{REF}. Using Eq. (17.13) and knowing V_{THP} = 280 mV, we see that if we set V_{REF} to a large voltage, Q_{cup} gets small. If, for example, $V_{REF} = VDD - V_{THP}$ (= 750 mV in this example), Q_{cup} goes to zero. We can write

$$V_{REF} < VDD - V_{THP} \tag{17.15}$$

Let's use a V_{REF} of $VDD/2$ or 500 mV. Figure 17.12 shows the simulation results. Note, in Fig. 17.9, that we place M4 in its own well. Since we care about how the threshold voltage varies, we've eliminated the body effect in this device. The simulation results are seen in Fig. 17.12. In this simulation we've swept the bit line voltage from 0.3 to 0.6 V. Comparing the plot to Fig. 17.11b, we see much better linearity (the charge added to the bit line doesn't change with bit line voltage). The cost for this improvement is more limited voltage swing on the bit line. ■

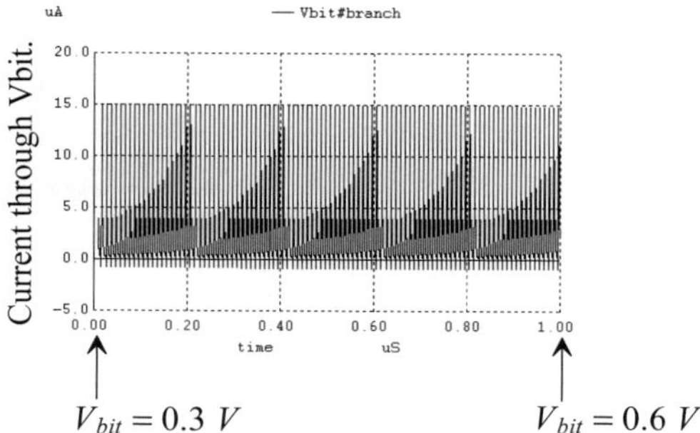

Figure 17.12 How the charge supplied to Vbit becomes linear if we add M4, from Fig. 17.9, to the simulation in Fig. 17.10. Note the reduced bit line voltage swing and the reduction in the current pulse amplitudes.

Incomplete Settling

If we zoom in on the current pulses in Fig. 17.12, we get a view like the one seen in Fig. 17.13. The glitches at 30 and 40 ns are the result of the ϕ_2 clock going high and M2 shutting off. Notice how the current through M4 (and V_{bit}) isn't zero when M2 shuts off. Using our water analogy, this is equivalent to stopping the pouring of the water out of the cup before the cup is empty. This incomplete emptying of the cup or capacitor is termed *incomplete settling*. The currents in the circuit haven't gone to zero before the clock transitions. The parasitic capacitances on the sources of M3/M4 continue to discharge through V_{bit} after M2 shuts off, resulting in nonzero current. Incomplete settling has the effect of making the capacitor, C_{cup} , appear smaller. It won't affect the linearity of the sense but it can affect the maximum value of the sensed current. To eliminate the incomplete settling behavior we can increase the width of M4. This results in the source of M4 being held closer to $V_{REF} + V_{THP}$ when M2 and M3 turns on. We could also slow the clock frequency down to ensure that the circuit settles. Because the amount of charge transferred from C_{cup} is constant from one clock cycle to the next, the effects of incomplete settling on DSM are simply a limit on the maximum current that can be removed from the bit line and a modification to the digital output for the absolute value of the input signal, that is, Eq. (17.2) (a smaller value of fed-back signal size is used).

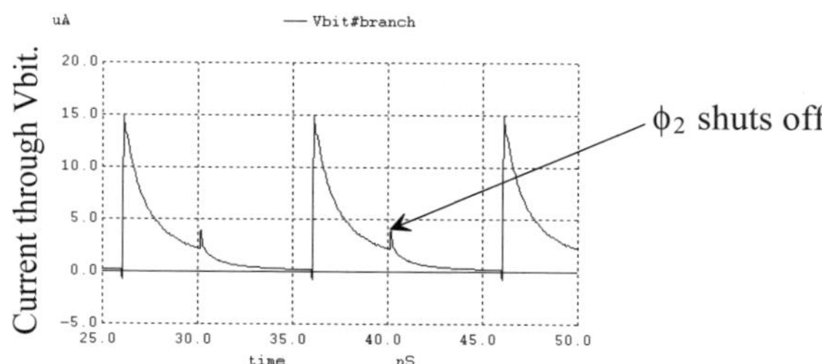

Figure 17.13 The incomplete settling in Fig. 17.12.

17.2 Sensing Resistive Memory

Let's apply the techniques we've just discussed to sensing a resistive memory cell. Figure 17.14a shows a schematic of one-transistor, one-resistor (1T1R) memory cell. Notice the similarity to the 1T1C DRAM memory cell seen in Fig. 16.9. *Ideally* the resistor is either zero ohms (which we'll call *programmed*) or infinite (which we'll call *erased*), Fig. 17.14b. To sense in this ideal case, we simply precharge the bit line to *VDD*. When the word line is driven high, the access device turns on, resulting in the bit line either moving to *VDD*/2 (if the cell is programmed) or not moving at all (if the cell is erased). In a practical circuit, however, the cell's resistance will be nonzero and not infinite (say 10kΩ to 100kΩ as seen in Fig. 17.14c).

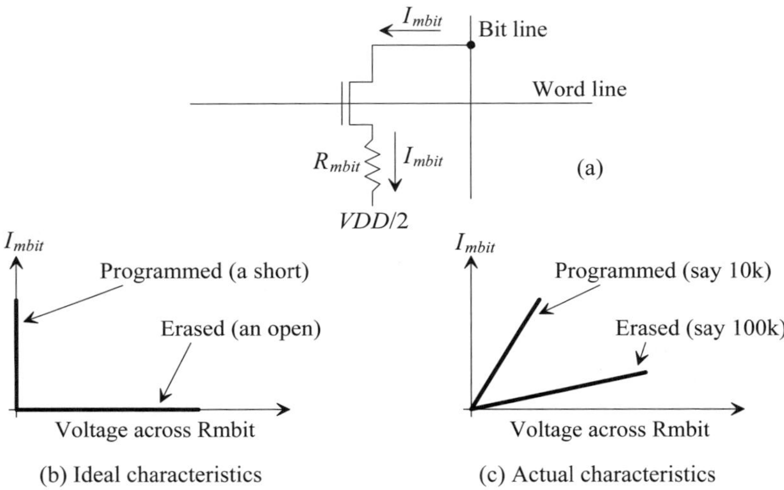

Figure 17.14 The one-transistor, one-resistor (1T1R) RAM memory cell.

The Bit Line Voltage

Consider the block diagram of a DSM seen in Fig. 17.15 with a resistive memory cell. To keep the schematic simpler, we won't include the access MOSFET. We know from the previous discussions that the DSM tries to hold the bit line at precisely the reference voltage used by the comparator. In Fig. 17.15 if the reference voltage used by the comparator is *VDD*/2, then the current that flows through the memory resistor is zero. To avoid this, we might use a reference that is offset from *VDD*/2 by V_{os}. The current that flows in the resistor is then

$$I_{mbit} = \frac{V_{os}}{R_{mbit}} \tag{17.16}$$

In the ideal condition, the erased resistor results in zero I_{mbit}. The output of the DSM is then simply a string of zeroes (we never have to add current to the bit line because none is leaving). If the resistor is programmed, the bitline gets pulled down towards *VDD*/2. The current source can't supply enough charge to pull the bit line up and the DSM output is a constant string of ones (yes, add current).

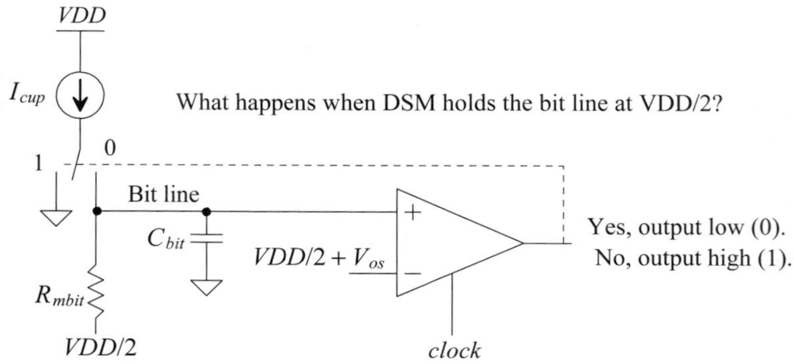

Figure 17.15 Sensing a resitive memory cell using DSM. Notice that we've eliminated the inverter from Fig. 17.8 by switching the input terminals of the comparator.

Adding an Offset to the Comparator

It's important to minimize the number of reference voltages used in a memory sense scheme. We'll use *VDD* and *VDD*/2 and then design a comparator with a built-in offset voltage. The comparator design is seen in Fig. 17.16 (see Figs. 16.32 and 16.35 from the last chapter). Simulation results showing the offset are seen in Fig. 17.17. When In+ gets 50 mV above In–, the output of the comparator changes states. It's possible to design the comparator to simplify the sense circuitry, e.g., clock the comparator on the falling edge of ϕ_2, remove the NAND gates on the output, and combine M2 and M3 in Fig. 17.9. In a

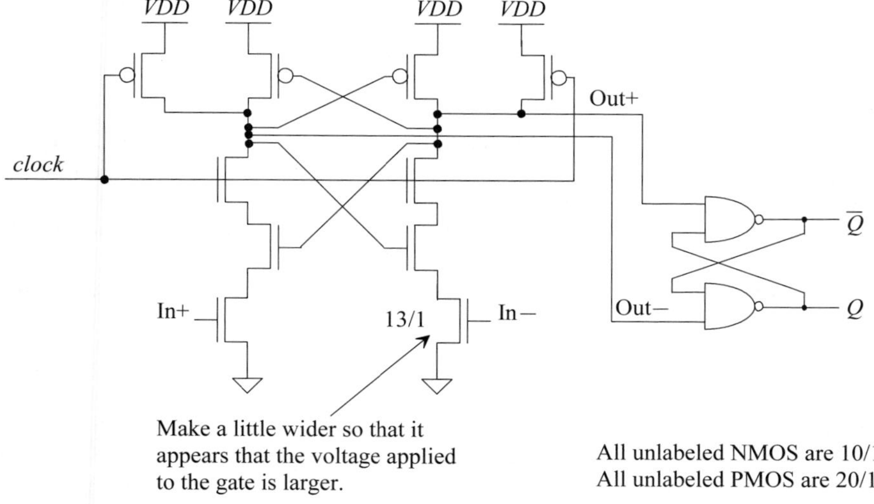

Figure 17.16 Designing a clocked comparator with a built-in offset.

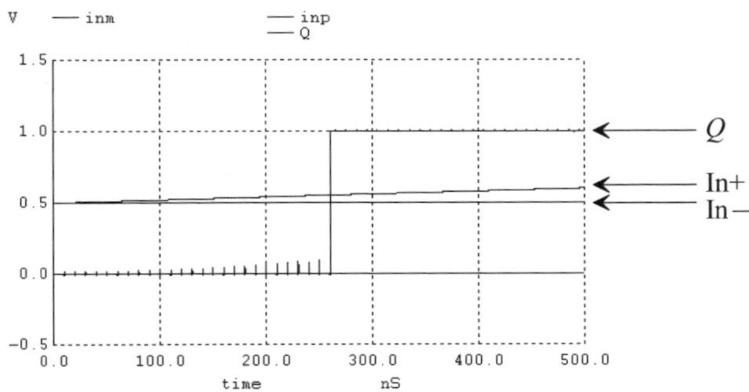

Figure 17.17 The output of the comparator switching states when the + input is 50 mV above the - input.

production part we might want to do this. However, here, where we want to minimize the glitches that may turn some switches inadvertently partially on, we don't try to simplify the circuitry (see problems at the end of the chapter).

Schematic and Design Values

The schematic for a DSM sensing circuit for resistive memories is seen in Fig. 17.18. The capacitance associated with the bit line is shown explicitly (even though it's a parasitic associated with the line). When we are sensing the value of a resistive cell, a word line goes high and connects the resistance, R_{mbit} , to the digit line. The comparator, with its built-in offset, tries to hold the bit line, through the feedback loop, at $VDD/2 + V_{OS}$. The

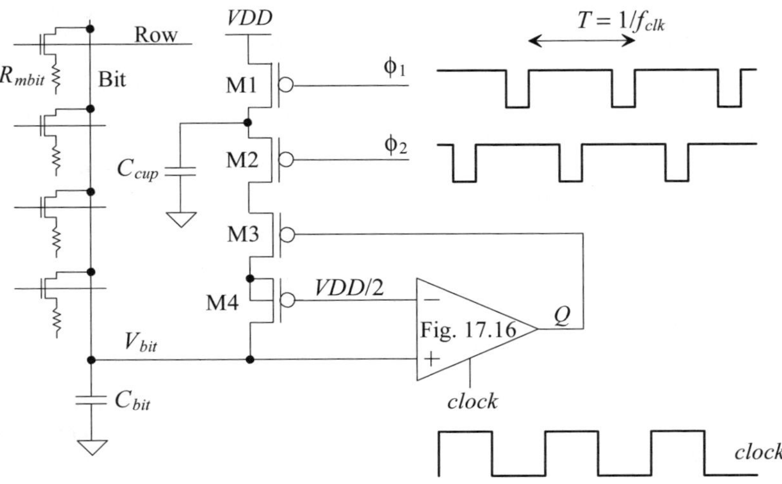

Figure 17.18 Sensing circuit for resistive memory.

current through the cell is then given by Eq. (17.16). Again, if N is the total number of times the DSM is clocked and M is the number of times M3 is enabled so that C_{cup} can dump its charge into the bit line, then we can write (reviewing Eqs. [17.3] to [17.7], [17.12], and [17.16])

$$\frac{V_{OS}}{R_{mbit}} = Q_{cup} \cdot \frac{M}{N} \cdot \frac{1}{T} = (VDD - VDD/2 - V_{THP}) \cdot C_{cup} \cdot \frac{M}{N} \cdot \frac{1}{T} \tag{17.17}$$

or

$$R_{mbit} = \frac{V_{OS} \cdot T}{(VDD/2 - V_{THP}) \cdot C_{cup} \cdot \frac{M}{N}} \tag{17.18}$$

Notice that if R_{mbit} goes to zero (programmed), the output of the DSM is always high (M is always a 1). If R_{mbit} is infinite, then the output of the DSM is always low (M is always a 0). To design the DSM, we need to pick an R_{mbit} value that separates what we define as a low resistance (a programmed state) from what we define as a high resistance (an erased state). Let's use 50k. When the resistance is 50k, then $M = N/2$ (half of the clock cycles the output of the DSM goes high). If $M > N/2$, then the cell is programmed. If $M < N/2$, the cell is erased. If V_{OS} is 50 mV, $VDD = 1\ V$, $V_{THP} = 280$ mV, and $f_{clk} = 100$ MHz then

$$C_{cup} = \frac{0.05}{0.22 \cdot 50k \cdot \frac{1}{2}} \cdot 10\ ns = 90\ fF \tag{17.19}$$

Since offset won't be precisely 50 mV, we'll round this up to 100 fF (and it will vary with process runs). To estimate the value of the resistor given both N and M, we use Eq. (17.18)

$$R_{mbit} \approx 25k \cdot \frac{N}{M} \tag{17.20}$$

Again, this is an approximation since our value for the offset won't be exactly 50 mV. Note that the minimum value of resistor we can sense occurs when $N = M$ ($R_{mbit} = 25k$).

One more thing that we must calculate before simulating the design is the variation of the bit line voltage (see Eq. [17.8]). The minimum voltage on the bit line is $VDD/2$ (when R_{mbit} is small). The maximum voltage will be $VDD/2 + V_{os} + \Delta V_{bit}$ (where the last term is the maximum change in the bit line voltage). To determine ΔV_{bit}, we can write (knowing charge must be conserved)

$$C_{cup} \cdot (VDD - VDD/2 - V_{THP}) = \Delta V_{bit} \cdot (C_{bit} + C_{cup}) \tag{17.21}$$

or

$$\Delta V_{bit} = \frac{C_{cup}}{C_{cup} + C_{bit}} \cdot (VDD/2 - V_{THP}) \tag{17.22}$$

Using the numbers above with a C_{bit} of 500 fF gives a $\Delta V_{bit} = 37\ mV$. Note that we were very concerned about how the bit line voltage change affected the amount of charge supplied by C_{cup} (see Figs. 17.8, 17.9, and the associated discussion). Here (and Eq. [17.16]) we aren't concerned with how changes in the bit line voltage affect the current through R_{mbit}. The *average* current through R_{mbit} is V_{OS}/R_{mbit}. The DSM holds the bit line, on average, at $VDD/2 + V_{OS}$. The charge supplied by C_{cup}, however, is not constant on average (as discussed earlier), and that is why M4 was added to the DSM.

Figure 17.19 shows the DSM sensing circuit's output for various values of R_{mbit}. Let's compare the theoretical estimate, that is, Eq. (17.20) to the simulated results. In all of the simulations seen in Fig. 17.19 we clock the DSM 50 times (a 100 MHz clock for 500 ns). In (a), with $R_{mbit} = 25\text{k}$, we get an output of 14 zeroes and 36 ones (= M). We can then write

$$R_{mbit} = 25k \cdot \frac{50}{36} = 35k \text{ (actual value 25k)}$$

For (b), we see $M = 17$ so

$$R_{mbit} = 25k \cdot \frac{50}{17} = 73k \text{ (actual value 50k)}$$

and for (c) and (d),

$$R_{mbit} = 25k \cdot \frac{50}{10} = 125k \text{ (actual value 100k)}$$

$$R_{mbit} = 25k \cdot \frac{50}{6} = 208k \text{ (actual value 200k)}$$

Notice that as we increase the value of R_{mbit}, we start to get nonlinear. The maximum finite value we can sense with 50 clock pulses is

$$R_{mbit,\max} = \frac{V_{OS} \cdot T \cdot N}{(VDD/2 - V_{THP}) \cdot C_{cup}} \text{ (= 1.25M here)} \tag{17.23}$$

Again, the minimum resistance is

$$R_{mbit,\min} = \frac{V_{OS} \cdot T}{(VDD/2 - V_{THP}) \cdot C_{cup}} \text{ (= 25k here)} \tag{17.24}$$

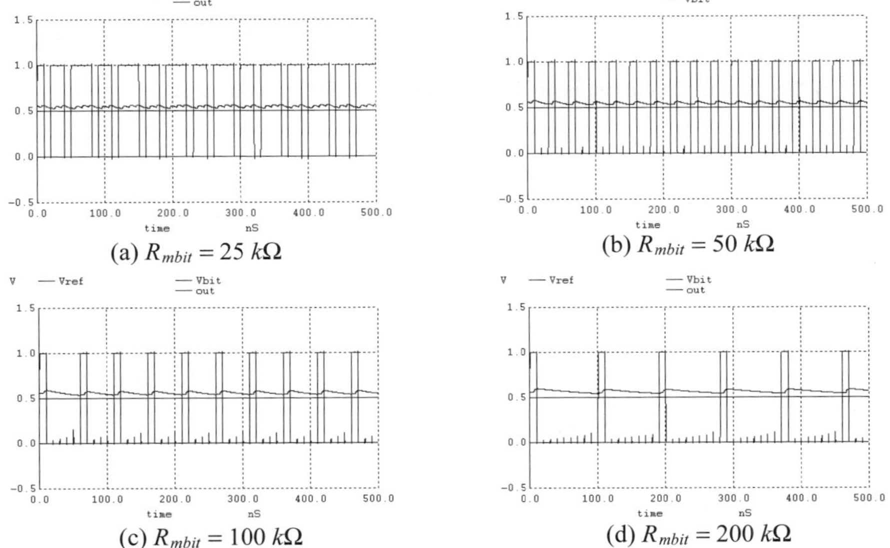

Figure 17.19 Outputs of the DSM for various Rmbit values.

Notice that we get good linearity with (actual) resistance values ranging from 25k to 75k. To sense larger values of resistances with good linearity, we must clock the DSM sensing circuit more times (increase N). However, if we are simply trying to separate a "big" resistor from a "small" resistor, we see that the scheme works very well.

A Couple of Comments

We might wonder why the simulated values are different from the actual values. The answers come from incomplete settling (making C_{cup} appear smaller than it actually is) and from the fact that the offset voltage is not precisely 50 mV. For the incomplete settling problem we can clock the circuit slower or try using a larger width for M4. For the offset voltage problem we might try to generate a precise reference voltage ($VDD/2 + V_{OS}$) for the comparator. Practically, this could lead to problems. In a real memory system, the voltages are noisy. Reviewing the designs in Figs. 17.15 and 17.18, we see that if there is noise on $VDD/2$ it feeds evenly into the comparator circuit and doesn't affect the operation (the – input of the comparator is connected to $VDD/2$ and so is R_{mbit}). Further, even if we could generate noise-free reference voltages, the comparator will exhibit an offset because the MOSFETs won't be perfectly matched. In most sensing applications, the relative values are usually more important than the absolute values.

Example 17.4

Show that M1, M2, M3 and C_{cup} in Fig. 17.18 can be thought of as a resistor. What is the resistor's value?

Figure 17.20 shows the circuit portion from Fig. 17.18. The current that flows in the resistor is

$$I_{avg} = \frac{VDD - VDD/2 - V_{THP}}{R_{SC}} \tag{17.25}$$

The amount of charge that flows during one clock cycle, if M3 is on, is

$$Q_{cup} = (VDD - VDD/2 - V_{THP}) \cdot C_{cup} \tag{17.26}$$

or, if this amount of charge is only allowed to flow M times out of N

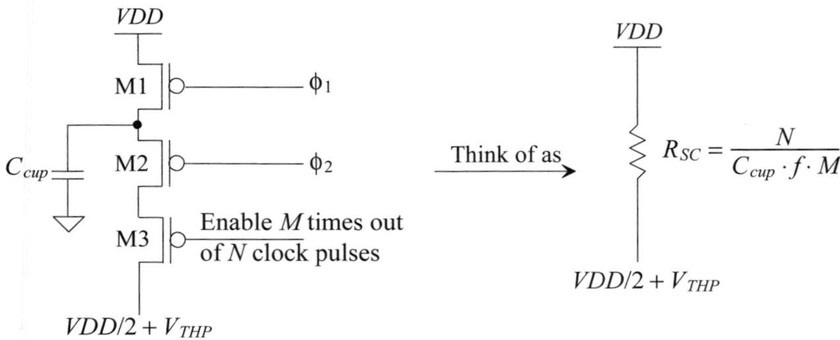

Figure 17.20 How a switched-capacitor resistor is modeled.

$$I_{avg} = \frac{Q_{cup}}{T} \cdot \frac{M}{N} = Q_{cup} \cdot f \cdot \frac{M}{N} \tag{17.27}$$

or

$$R_{SC} = \frac{N}{C_{cup} \cdot f \cdot M} \tag{17.28}$$

Noting that if M3 is always enabled (the circuit with only M1 and M2) we get a switched-capacitor resistance of

$$R_{SC} = \frac{1}{C_{cup} \cdot f} \tag{17.29}$$

which, for Fig. 17.18 and the associated discussion, is 25k. ■

Example 17.5
Show that M1, M2, and C_{cup} can be replaced with a resistor in the DSM sensing circuit of Fig. 17.18. Show the DSM's output when R_{mbit} is 50k and 200k.

The schematic for the DSM modulator is seen in Fig. 17.21. We might think that the errors due to incomplete settling would be eliminated using this scheme. However, the source of M4 will not be precisely at $VDD + V_{THP}$, so an error will still be present. Again, increasing the width of M4 will lessen the error's effects.

Using the same values as used to generate Fig. 17.19b and d, we get the simulation results seen in Fig. 17.22. The big benefit of this topology over the one in Fig. 17.18 using the switched capacitor resistor is simpler design (no nonoverlapping clocks are needed) and lower power (there are not as many parasitic capacitances to charge and discharge). The current pulled from VDD in Fig. 17.18 is 26 μA when clocked at 100 MHz, while the circuit in Fig. 17.21 uses 20 μA. The big drawback of using the simple resistor is the inability to adjust the resistance by changing the clock frequency. In a practical circuit the fabrication

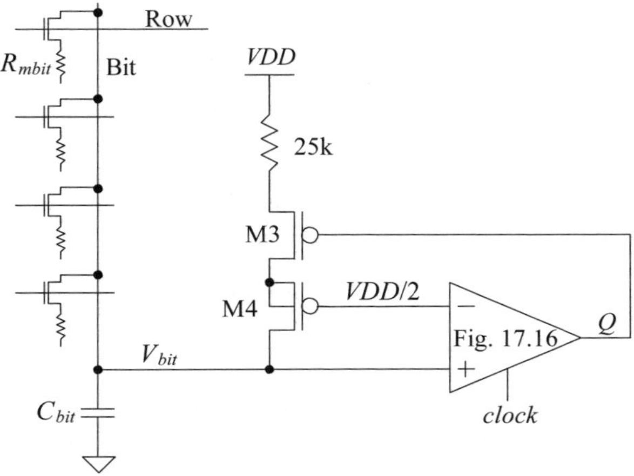

Figure 17.21 Simpler DSM for sensing resistive memory.

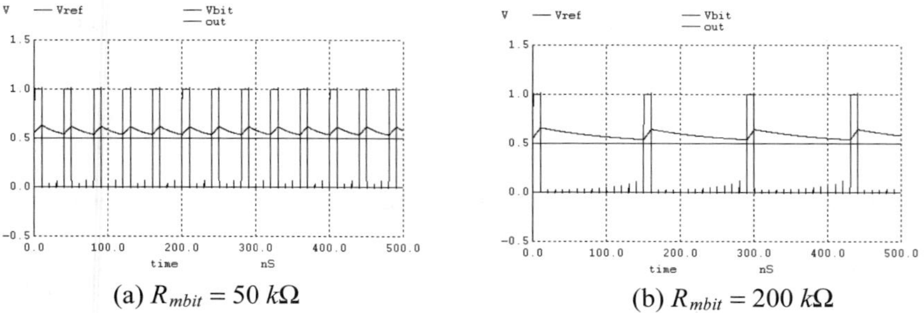

(a) $R_{mbit} = 50\ k\Omega$ (b) $R_{mbit} = 200\ k\Omega$

Figure 17.22 Simulation results for the circuit in Fig. 17.21.

process characteristics and temperature vary. To center the value of sense to half way between a programmed and an erased resistor value, the clock frequency can be adjusted. Adjusting the clock frequency in Fig. 17.21 simply adjusts the variation in the bit line voltage. ■

17.3 Sensing in CMOS Imagers

Another area where DSM (delta-sigma modulation) can be used for sensing is in CMOS imaging chips that acquire images in cameras or video recording. A schematic of a CMOS active pixel sensor (APS) is seen in Fig. 17.23. Light is applied through a filter (so that only red, green, or blue light passes) to the photodiode. The photodiode changes the light intensity into a charge. The charge is converted into a voltage and passed to the column line. The brighter the light, the bigger voltage change we get on the column line.

Resetting the Pixel

Prior to acquiring an image each pixel in the imaging chip is reset. This is accomplished by driving the reset row line (*ResetN*) to a voltage, *VDDP*, greater than *VDD* + V_{THN} (with body effect). This turns M1 on and sets the voltage across the photodiode to *VDD*. This condition is called the *dark or reference level* of the pixel. We can then turn M3 on by driving *RowN* to *VDDP* and, with M2 behaving as a source follower, driving the column line to the reference voltage level, V_R. This is important because each pixel in an imaging array will have slightly different characteristics. The differences in the pixels can result in speckles in the image. We can subtract this reference level from the actual measured signal level to get an accurate idea of the light intensity applied to the pixel (to eliminate the pixel gain differences).

The point here is that our DSM circuit will have to sense, and store, a reference level, V_R, at the beginning of the sense.

The Intensity Level

After we have stored the reference level voltage V_R (on a capacitor), the *ResetN* signal goes low. This allows the light striking the reverse-biased photodiode (through the color filter) to generate electron-hole pairs that cause the voltage across the diode to decrease. After a time, the information stored on the gate of M2 (the decrease in the voltage across the photodiode) is sampled on the column line (assuming M3 is on). The time between

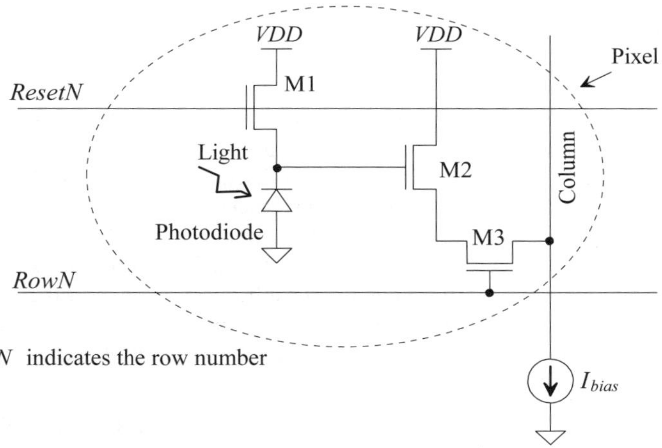

Figure 17.23 A CMOS active pixel sensor (APS).

the *ResetN* going low and the data (the column voltage, V_I, corresponding to the intensity of the light striking the photodiode) being sampled on the column line is called the *aperture* time. Note that a dark signal corresponds to a large voltage on the column line ($VDD - V_{THN}$), while a bright signal corresponds to a lower voltage (less than $VDD - V_{THN}$).

Sampling the Reference and Intensity Signals

When *ResetN* and *RowN* are high, the pixel's reference (or dark) voltage, V_R, is placed on the column line, Fig. 17.24. At this time the sample and hold reference signal, *SHR*, goes high and V_R is sampled onto a hold capacitor. Next, *ResetN* goes low and the photodiode changes light into charge. After the aperture time, the information from the photodiode (the intensity of the light) is on the column line, V_I. This voltage is then sampled and held on a hold capacitor when the signal *SHI* (sample and hold the intensity

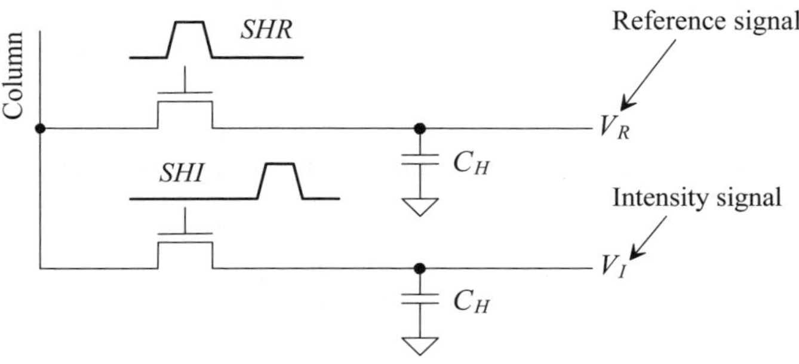

Figure 17.24 Sampling the reference and intensity signals.

of the light) goes high. What we want is to design a circuit that takes the difference in V_R and V_I and generates a digital number.

Noise Issues

Circuit noise limits the dynamic range of the sense, resulting in the blurring of the images or the inability to detect low light levels or distinguish between bright or high-intensity light levels. The major noise sources in CMOS, as discussed in Chs. 8 and 9, are flicker and thermal noises. The first design value we must select is the size of the hold capacitor in Fig. 17.24. The size of this capacitor limits the thermal noise in the sample. As seen in Table 8.1, using a hold capacitor of 1 pF results in an RMS noise voltage, just due to thermal noise, of 64 μV. This corresponds to a peak-to-peak voltage in the time domain of roughly 400 μV (six times the RMS value as seen in Fig. 8.33). Using a larger capacitor takes up more layout area and takes longer to charge but lowers the amount of thermal noise sampled onto C_H.

The output current of the pixel (when M3 is on and M2 is a source follower) also contains flicker noise. When this pixel is connected to the large hold capacitance and the capacitance of the bit line, the flicker noise current will be integrated. The result is a noise power spectral density with a $1/f^3$ spectral shape. As discussed in Ex. 8.14, the RMS value of the resulting noise signal will grow linearly with measurement time. What this means is that to *achieve a low noise* sample onto C_H we want to minimize the amount of time *SHR* and *SHI* are high *and* the time difference between the two signals. *SHR* and *SHI* should only go high long enough to charge C_H. The amount of time between M1 turning on and *SHI* shutting off should be as small as possible.

The amount of noise in V_R and V_I cannot be reduced after they are captured on the hold capacitors. For example, ideally V_I may be 0.5 V. However, because of noise V_I may be 0.501 V or 0.495 V, etc. We *may* be able to reduce the noise by averaging successful samples though (see Eq. [8.48] and Ex. 8.14).

Ideally, the sensing circuitry (the circuitry used to change the analog voltages, V_R and V_I, into a digital number) doesn't introduce any additional noise. When using the DSM, the counter can be thought of as a lowpass digital filter, Fig. 17.25. (See also the second volume of this book entitled *CMOS Mixed-Signal Circuit Design* for much more detailed description of the frequency response of a counter.) As seen in this figure, if the sense time is the total number of times the DSM is clocked, *N*, multiplied by the clock's period *T*, that is $T_{meas} = NT$, then increasing the sense times lowers the bandwidth of the digital filter. This has the effect of lowering the noise bandwidth (so that the DSM sensing circuit does not contribute any further noise to the measured signal).

An example where a counter is used that doesn't result in filtering the measured signal is seen in Fig. 17.26. Here the column voltage corresponding to the intensity of the light is used as the – input to a comparator. For the + input, a constant current is used to charge a capacitor to generate a voltage ramp. When the ramp's voltage gets above V_I, the output of the comparator goes high and stops the counter. In this scenario the counter is simply used to give a digital representation of a time or ramp voltage. The counter doesn't provide any filtering (or more precisely averaging as used in DSM, see Tables 17.1 to 17.3). Notice also that this sensing topology very likely adds noise to the measured signal. For example, we know that using a constant current to charge a capacitor results in a voltage signal containing $1/f^3$ noise. The ramp, with noise, voltage

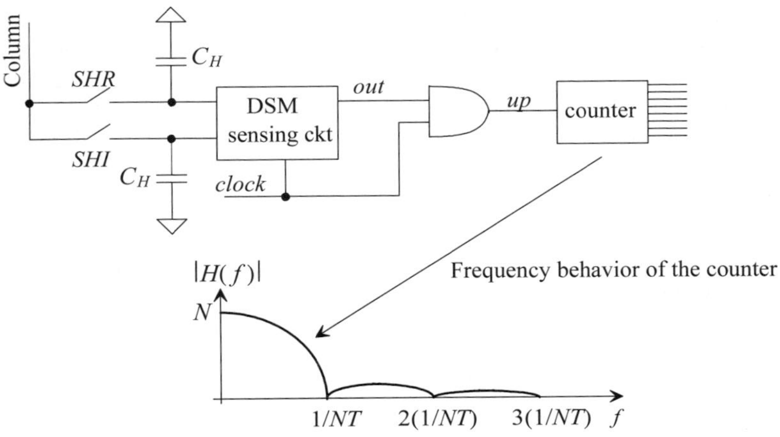

Figure 17.25 Thinking of the counter as a digital low pass filter

is connected directly to the comparator and causes noise to be added to the measured signal. Further, if the comparator makes an error and switches states too early or too late because of an offset or noise coupling from the adjacent column sensing circuits, again, the sense adds noise to the measured signal (the digital output code isn't constant but rather moves around even though the inputs to the sense amplifier may be constant).

Also note that using the DSM, we can run the sense operation indefinitely, while the scheme in Fig. 17.26 is limited by the ramp rate. Further increasing the clock frequency in Fig. 17.26 won't increase the resolution if the sense is noise limited. In the DSM sensing circuit, increasing the clock frequency, as seen in Fig. 17.25, lowers the bandwidth of the digital filter and increases the resolution of the sense.

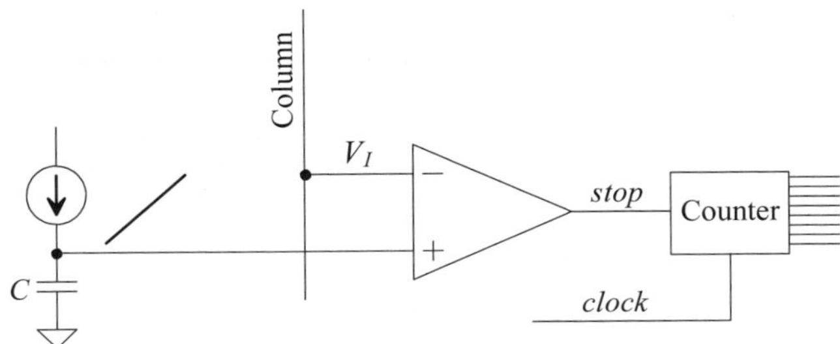

Figure 17.26 A circuit where the counter doesn't behave like a lowpass filter.

Subtracting V_R from V_I

As we mentioned earlier, each pixel has slightly different characterisitics. For example, the threshold voltage of M2 in one pixel may be 10 mV different from the threshold voltage of M2 in a different pixel. To remove this error, we subtract the measured reference voltage, V_R , from the measured signal intensity, V_I. It's important, during the sense, not to change these voltages with our sensing circuit. Let's convert these voltages to currents and then subtract the currents to get the difference (and not try to subtract the voltages directly).

Examine the voltage-to-current converter seen in Fig. 17.27. This circuit is simply a source follower that is made wide so that its V_{SG} is always approximately the threshold voltage. The relationship between the drain current and the column voltage V_{col} (V_I or V_R) is

$$I = \frac{VDD - V_{THP} - V_{col}}{R} \tag{17.30}$$

for

$$V_{col} < VDD - V_{THP} \tag{17.31}$$

Reviewing Fig. 17.23, we see that if the threshold voltage drop of M2 (with body effect) is more than V_{THP} (without body effect because we placed the PMOS device in Fig. 17.27 in its own well), Eq. (17.31) will always be satisfied.

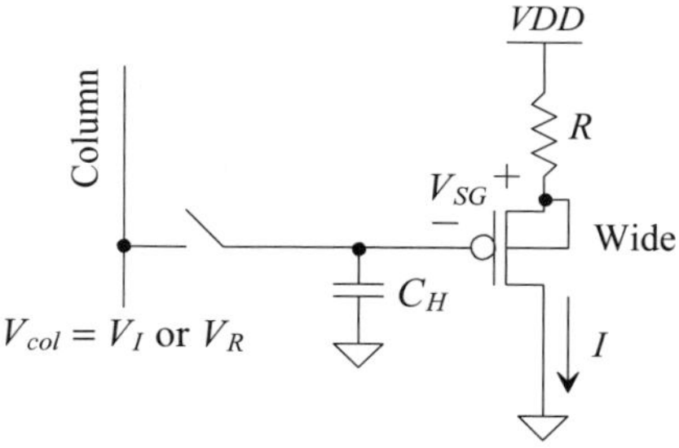

Figure 17.27 Linear voltage-to-current conversion.

Example 17.6
Using SPICE determine the linearity of the current in the circuit seen in Fig. 17.28. Use the short-channel CMOS process and compare the simulation results to hand caculations.

If *VDD* is 1 V, $V_{THP} = 250$ mV, and *R* is 10MEG, then

$$I = 75 - 100 \cdot V_{col}\ nA$$

Noting that the ideal slope of the line is $-1/R$ ($= -100 \times 10^{-9}$). Figure 17.29 shows the simulation results. In (a) we see from the transfer curve how the output, I , changes with the input, V_{col}. In (b) we take the derivative of *I* to see if the slope is perfectly flat. The linearity is pretty good (1%) for $V_I < 400$ mV. ■

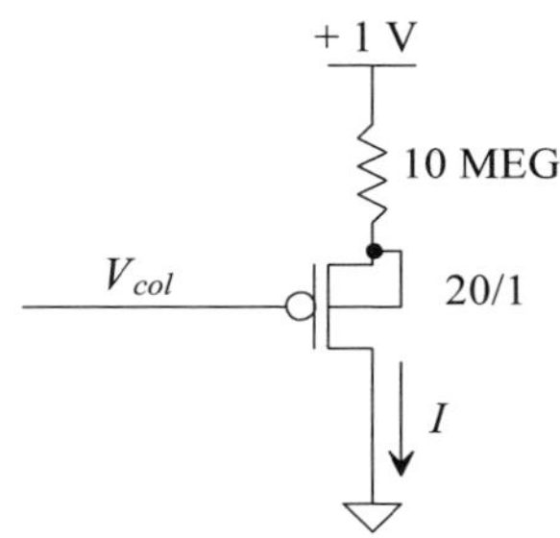

Figure 17.28 Circuit used in Ex. 17.6.

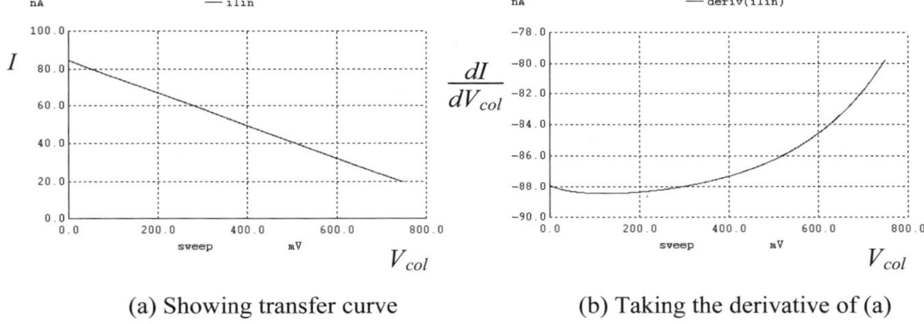

(a) Showing transfer curve (b) Taking the derivative of (a)

Figure 17.29 Simulating the operation and linearity of the voltage-to-current converter in Fig. 17.28.

We might wonder, from the last example, if it would be a good idea to try to make the linearity of our voltage-to-current converter even better. Before doing this, let's look at the linearity of the source follower in the pixel itself. Figure 17.30 shows a simplified schematic of the source follower (M2) used in the APS. We've modeled the finite output resistance of the current source with a 10MEG resistor. The simulation results in Fig. 17.31 show the transfer relation of this circuit and its linearity. For column voltages between 100 mV and 600 mV, the linearity is 0.5% (which is comparable so we won't concern ourselves any further with trying to better the linearity at this point).

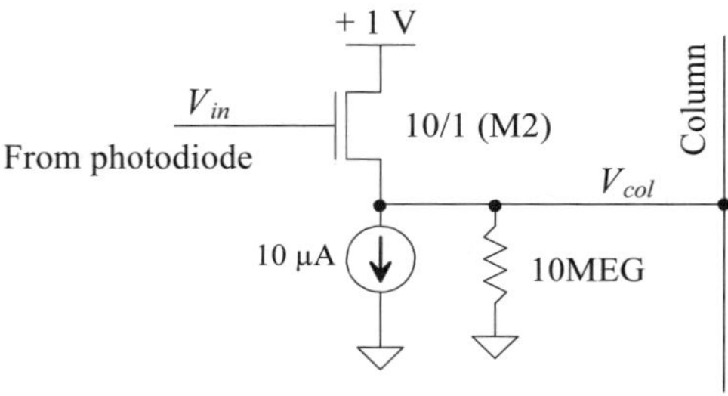

Figure 17.30 Looking at the linearity of the source follower in the pixel.

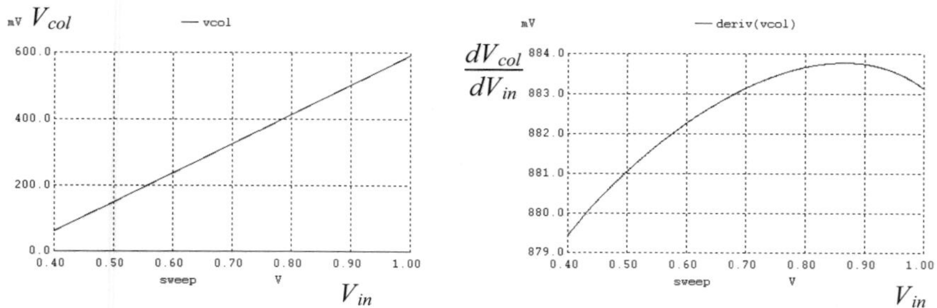

Figure 17.31 Showing the transfer curve and linearity of the circuit in Fig. 17.30.

To take the difference in the currents, let's use a current mirror as seen in Fig. 17.32. The current corresponding to the reference voltage is

$$I_R = \frac{VDD - V_{THP} - V_R}{R_R} = \frac{V_{R,shift}}{R_R} \tag{17.32}$$

and the current corresponding to the intensity of light is

$$I_I = \frac{VDD - V_{THP} - V_I}{R_I} = \frac{V_{I,shift}}{R_I} \tag{17.33}$$

The difference in these currents is summed (sigma) in the bucket capacitor, as seen in Fig. 17.32. When we add our comparator and feedback to form a DSM, the charge on the capacitor, averaged over time, is a constant. This occurs when

$$I_R = I_I = \frac{V_{R,shift}}{R_R} = \frac{V_{I,shift}}{R_I} \tag{17.34}$$

or

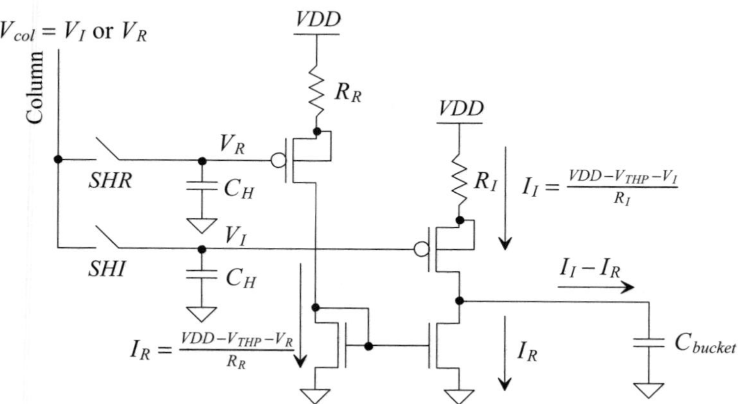

Figure 17.32 Subtracting the currents.

$$V_{I,shift} = \frac{R_I}{R_R} \cdot V_{R,shift} \tag{17.35}$$

The ratio of the resistances gives us the information we need to determine the relative (to the reference level, V_R) intensity of light on the pixel. To implement the resistors, let's use the switched-capacitor resistor as seen in Fig. 17.20. For the reference voltage we know

$$V_{I,shift} \geq V_{R,shift} \tag{17.36}$$

and so $(R_I \geq R_R)$. Let's use (M3 always on in Fig. 17.20)

$$R_R = \frac{1}{f \cdot C_{cup}} \tag{17.37}$$

and for R_I (which *will* be enabled via M3 when its gate goes low)

$$R_I = \frac{1}{f \cdot C_{cup} \cdot \frac{\overline{M}}{N}} \tag{17.38}$$

Rewriting Eq. (17.35), knowing $\overline{M} = N - M$, gives

$$V_{I,shift} = V_{R,shift} \cdot \frac{N}{N-M} \tag{17.39}$$

If the DSM is clocked 1,000 times (*N*), then *M* can range from 0 (the pixel is not illuminated, that is, the dark or reference level) to 1,000 (very bright). Figure 17.33 shows a schematic of the sensing circuit with some typical values.

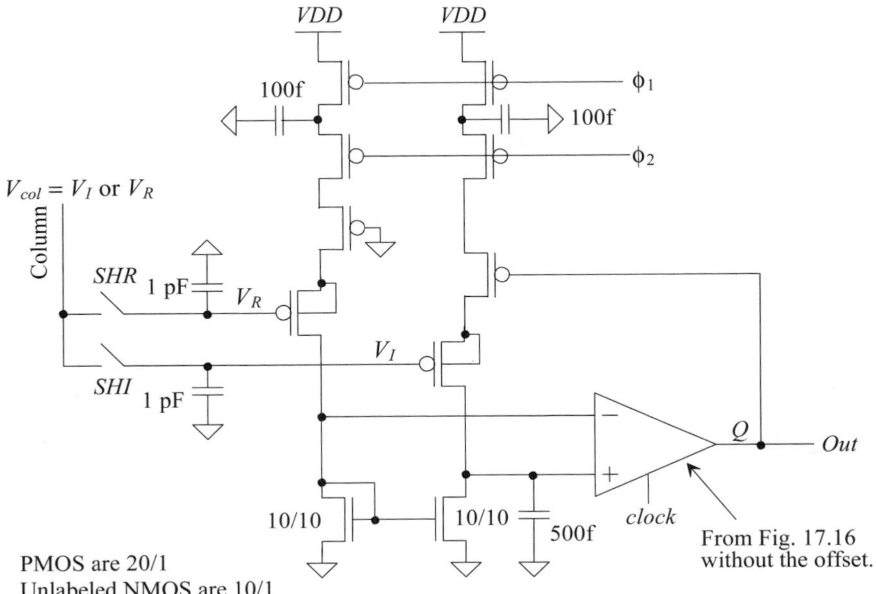

Figure 17.33 Schematic of a DSM for sensing in CMOS imaging chips.

The circuit in Fig. 17.33 was made as symmetrical as possible so that power supply or ground noise affected each signal path equally. The MOSFETs in the current mirror are made long (10 drawn) so that the voltages on the input of the comparator are greater than the NMOS threshold voltage. The size of the capacitors isn't that critical. We want the C_{cup} capacitors to be less than the C_{bucket} capacitor. We don't have to worry about overcharging C_{bucket} in this scheme because the input signal contributions are limited by the switched capacitor resistors. (We only get one C_{cup} every clock cycle.)

Figure 17.34 shows the simulation results for the DSM circuit of Fig. 17.33. We assume that the counter, Fig. 17.25, is enabled after 500 ns. Prior to this time, the reference and intensity signals are sampled onto the hold capacitors. If $V_R = 650$ mV, then

$$V_{R,shift} \approx 1 - 0.25 - 0.65 = 100\ mV$$

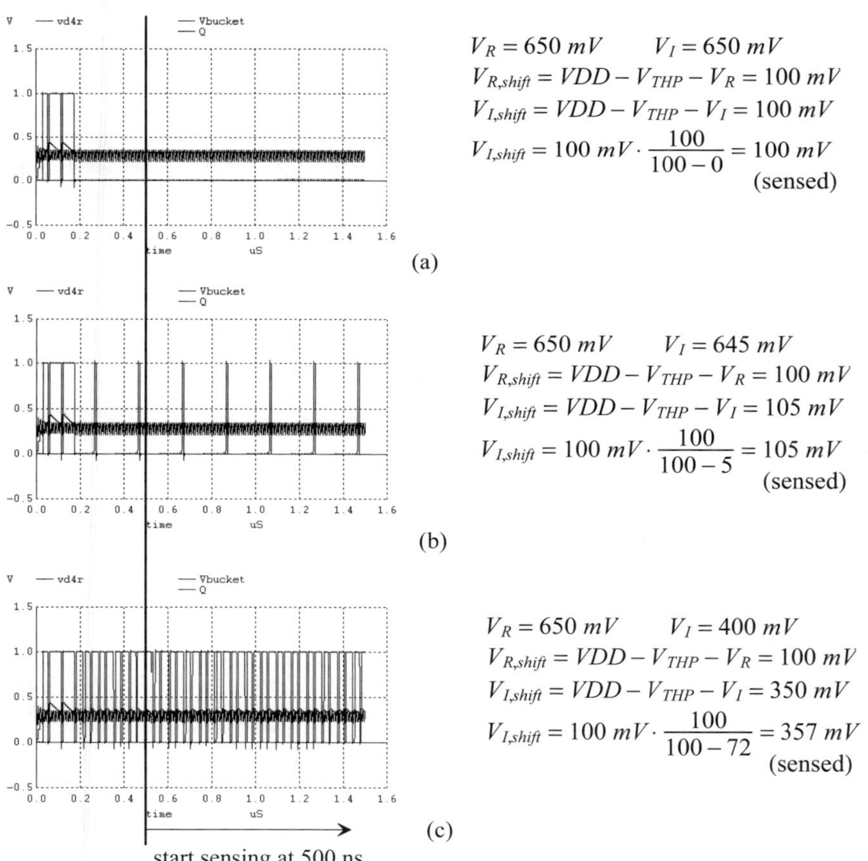

Figure 17.34 How the DSM sensing circuit in Fig. 17.33 operates.

In Fig. 17.34a we apply the same signal to the DSM sensing circuit, that is 650 mV, for V_I. As expected, the output stays low for all times. In (b) we drop V_I to 645 mV and see, during the 500 ns to 1,500 ns sensing time, 5 output ones. As seen in the figure, the sensed value indicates the intensity is 5 mV below the reference. In (c) we drop the intensity signal to 400 mV resulting in $V_{I,shift} = 350$ mV. The sensed value with 72 of the 100 clock cycles being high is 357 mV. Let's look at what would happen if we sensed 71 ones

$$100\ mV \cdot \frac{100}{100-71} = 345\ mV$$

In either case (71 or 72 ones) the resolution was so coarse that we couldn't resolve the actual signal. If we think about this for a moment, we see that if the counter output code is small then the resolution of the measurement is better. For example, counter outputs of 1 and 2 result in

$$100\ mV \cdot \frac{100}{100-1} = 101\ mV \text{ and } 100\ mV \cdot \frac{100}{100-2} = 102\ mV$$

Looking at Eq. (17.39), we see that the dependence on M (the number of times the output of the DSM goes high) is *not linearly related* to the light intensity, V_I. What we want is an equation like (17.7), that is,

$$V_{I,shift} = V_{R,shift} \cdot \frac{M}{N} \tag{17.40}$$

To get a relationship like this, we might try to control the value of R_R too as seen in Fig. 17.35. The complementary output of the comparator is fed back so that

$$R_R = \frac{1}{f \cdot C \cdot \frac{M}{N}} \tag{17.41}$$

and thus

$$V_{I,shift} = V_{R,shift} \cdot \frac{M}{N-M} \tag{17.42}$$

Again, we do not have a linear relationship. Further, half or more of our outputs must be used to enable the switched-capacitor resistor in series with the reference signal. If, for example, $V_{R,shift} = V_{I,shift}$, then $M = N/2$. Since $V_R \geq V_I$, then $0 \leq M \leq N/2$ (not good design).

If we review the derivations leading up to Eqs. (17.7) and (17.20), the common theme is that the feedback signal controlled by the comparator is a constant addition to the capacitive bucket (or bit line). In Fig. 17.18, for example, we used M4 to ensure that the charge from C_{cup} was a constant added to C_{bit}. In our current sense amplifier, Fig. 17.33, the signal we feedback is not a constant but rather a function of V_I. When sensing in a CMOS imager, the signal fed back should be a function of the reference level, V_R. The comparator's output should control a fed-back signal that is derived from V_R.

Note that it's a bad idea to ground the gate of M3I in Fig. 17.35 and only have a single feedback path. Since $V_R > V_I$, we won't be able to supply enough current through the reference signal path. The charge supplied by the intensity path will always be greater than the charge supplied from the reference path (and so C_{bucket} will overfill). A new topology is needed.

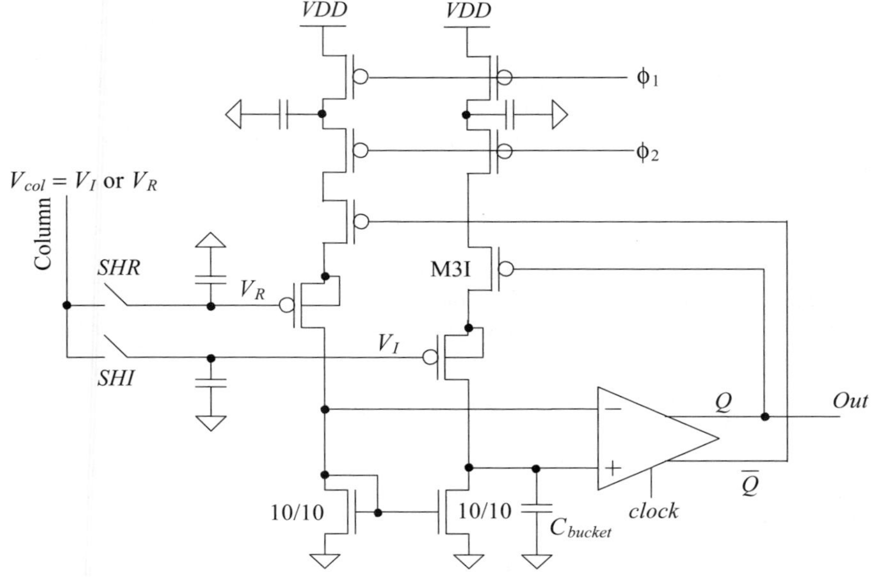

Figure 17.35 Using both comparator outputs for a DSM sensing circuit (bad if $V_R \geq V_I$).

Figure 17.36 shows an NMOS version of Fig. 17.33. We've replaced the capacitors with MOSFETs to show that the DSM sensor can be implemented using a single-poly CMOS process. The PMOS devices are used for the "cup" capacitors. We use PMOS instead of NMOS because, for the topology seen, the PMOS devices always remain in strong inversion. The NMOS devices, for example, move towards accumulation mode when ϕ_2 turns on and discharges the capacitors. The result is a nonlinear capacitance (the size of the cup varies). Similarly, we use NMOS for the bucket capacitors because, for the topology used, they will always remain in strong inversion (the capacitance won't vary with the changes in the voltage on their gates). Note that a second "bucket" capacitor was added across the 10/10 diode connected (gate and drain tied together) PMOS device. This addition serves two purposes. The first is to ensure that ground noise affects the comparator inputs equally (noise on ground will feed evenly through the bucket capacitors to the input of the comparator). The second reason is that it smoothes out the summation of the currents. Note that if the added capacitor is too big, stability can be an issue (the added capacitor adds a delay in series with the feedback path).

In Fig. 17.36 we show a comparator design using PMOS imbalance MOSFETs (as seen in Fig. 16.32 for the NMOS flavor). When ϕ_2 goes high, the outputs of the comparator, *Out* and $\overline{Out}$ are driven low. When ϕ_2 goes low, on the falling edge, the comparator makes a decision concerning which of the bucket capacitors has the higher potential across it. Based on this decision, the outputs of the comparator are latched with the NOR-based SR latch. The *Q* output is fed back to enable or disable the summation of charge via the reference path. We've used the *Qi* output as the DSM's output. When the

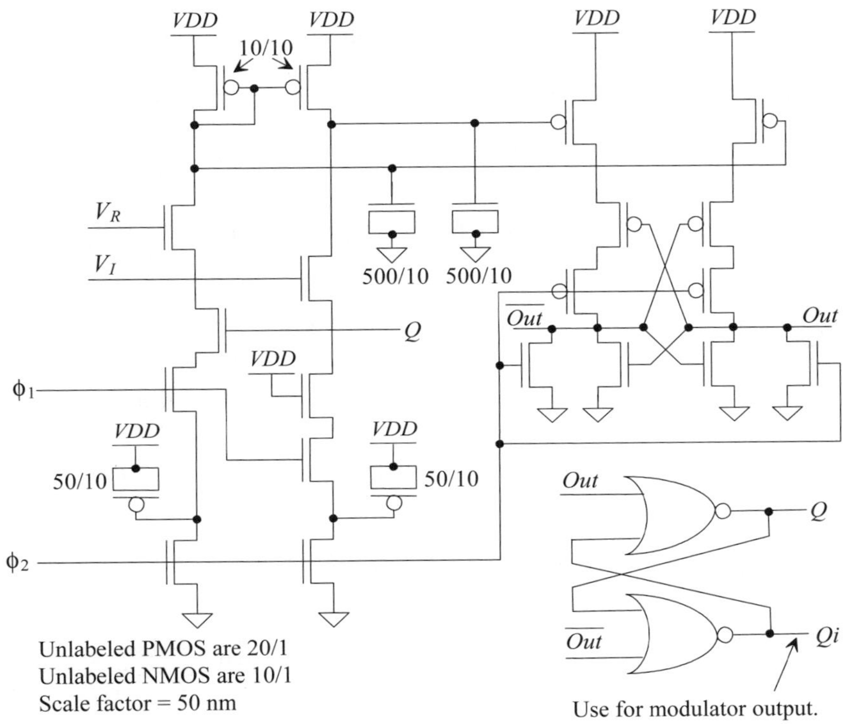

Figure 17.36 DSM circuit for sensing in a CMOS imaging chip.

intensity level is the same as the reference level, the output stays low for all times. It may be possible to eliminate this SR latch in a production part. However, we include it here because it makes the simulation results easier to look at (the glitches in the comparator's output are reduced).

Figure 17.37 shows some simulation results for the DSM circuit seen in Fig. 17.36. Notice, in this figure, that we've *assumed* the threshold voltage of the NMOS device is 250 mV. Since our reference level (the black level for the pixel) is 650 mV, our shifted reference level, $V_{R,shift}$, is 400 mV. For 100 samples then we can estimate the resolution as

$$V_{res} = \frac{V_{R,shift}}{N} \rightarrow 4\ mV \tag{17.43}$$

Looking at the figure, we may think that there is a large nonlinearity in the sense because, for lower input voltages, $V_{I,shift}$, the sensed value doesn't exactly match the shifted value. However, notice that the input changes from, say, 650 mV to 600 mV we get a code difference of 12 (24/100 mV), or from 600 mV to 500 mV (23/100 mV), from 400 mV to 300 mV (20/100 mV). To get a better estimate for the resolution, let's use an average change of 23 counts per 100 mV to estimate the resolution, that is,

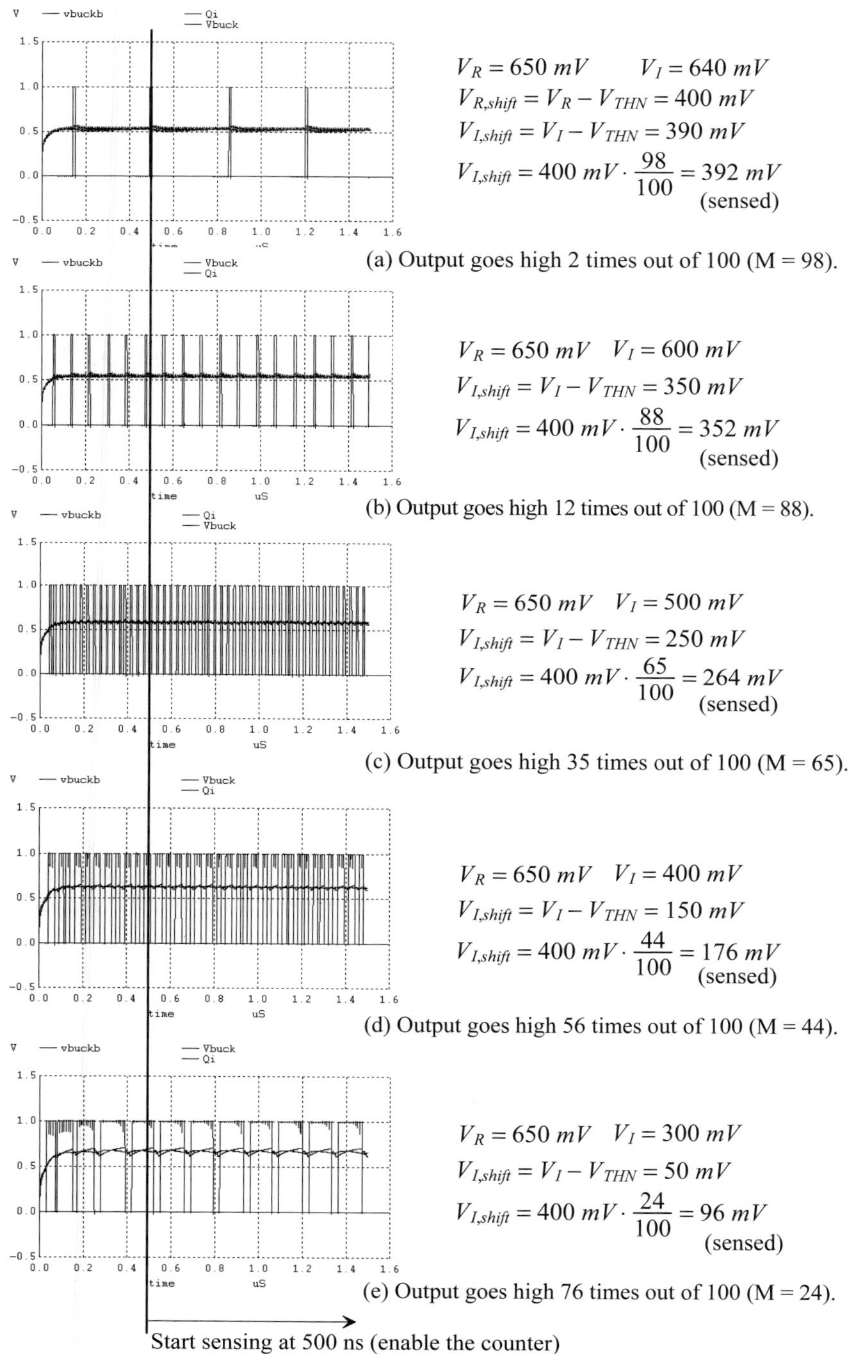

vbuckb
Qi
Vbuck
time
uS
$V_R = 650\ mV \qquad V_I = 640\ mV$
$V_{R,shift} = V_R - V_{THN} = 400\ mV$
$V_{I,shift} = V_I - V_{THN} = 390\ mV$
$V_{I,shift} = 400\ mV \cdot \frac{98}{100} = 392\ mV$ (sensed)
(a) Output goes high 2 times out of 100 (M = 98).
$V_R = 650\ mV \quad V_I = 600\ mV$
$V_{I,shift} = V_I - V_{THN} = 350\ mV$
$V_{I,shift} = 400\ mV \cdot \frac{88}{100} = 352\ mV$ (sensed)
(b) Output goes high 12 times out of 100 (M = 88).
$V_R = 650\ mV \quad V_I = 500\ mV$
$V_{I,shift} = V_I - V_{THN} = 250\ mV$
$V_{I,shift} = 400\ mV \cdot \frac{65}{100} = 264\ mV$ (sensed)
(c) Output goes high 35 times out of 100 (M = 65).
$V_R = 650\ mV \quad V_I = 400\ mV$
$V_{I,shift} = V_I - V_{THN} = 150\ mV$
$V_{I,shift} = 400\ mV \cdot \frac{44}{100} = 176\ mV$ (sensed)
(d) Output goes high 56 times out of 100 (M = 44).
$V_R = 650\ mV \quad V_I = 300\ mV$
$V_{I,shift} = V_I - V_{THN} = 50\ mV$
$V_{I,shift} = 400\ mV \cdot \frac{24}{100} = 96\ mV$ (sensed)
(e) Output goes high 76 times out of 100 (M = 24).
Start sensing at 500 ns (enable the counter)

Figure 17.37 The operation of the DSM sensing circuit in Fig. 17.36.

$$V_{res} = \frac{100\ mV}{23} = 4.35\ mV \tag{17.44}$$

and so the shift in the reference voltage can be more accurately predicted as

$$V_{R,shift} = 435\ mV \text{ because } (= N \cdot V_{res}) \text{ so } V_{THN} = 215\ mV \tag{17.45}$$

Using this, the sensed outputs in Fig. 17.37 can be rewritten as:

(a), $V_{I,shift} = 640\ mV - 215\ mV = 425\ mV$ and the sensed value $435\ mV \cdot \dfrac{98}{100} = 426\ mV$

(b), $V_{I,shift} = 600\ mV - 215\ mV = 385\ mV$ and the sensed value $435\ mV \cdot \dfrac{88}{100} = 382\ mV$

(c), $V_{I,shift} = 500\ mV - 215\ mV = 285\ mV$ and the sensed value $435\ mV \cdot \dfrac{65}{100} = 283\ mV$

(d), $V_{I,shift} = 400\ mV - 215\ mV = 185\ mV$ and the sensed value $435\ mV \cdot \dfrac{44}{100} = 191\ mV$

(e), $V_{I,shift} = 300\ mV - 215\ mV = 85\ mV$ and the sensed value $435\ mV \cdot \dfrac{24}{100} = 104\ mV$

Indicating a linear sense that becomes nonlinear at the edges of operation (when V_I becomes comparable to the V_{THN}).

It's important to understand the robustness of this sensing scheme. If the comparator makes a mistake, it is averaged out (comparator gain and offset aren't important). If noise is coupled into the sense amplifier, it will be averaged out. The sensing operation can be indefinite (noting that the hold capacitor voltages changing because of charge leaking off of the capacitors will ultimately limit the length of the sense). To increase the resolution of the sense, the clock frequency can be increased (noting the size of the counter used must be increased too). Finally, the topology requires little power. For the topology in Fig. 17.36 the current supplied by *VDD* is approximately 25 μA. If 1,000 of these sense amplifiers are used at the same time (say on the bottom of an imaging array with 1,000 columns), then the current required from *VDD* is 25 mA. The current can be further reduced by designing the DSM without the NOR latch seen in Fig. 17.36 or by using smaller capacitors.

The thermal noise from the capacitors is characterized using *kT/C* (see Table 8.1) and the associated discussion. We might think that using smaller capacitors results in an increase in the thermal noise. However this noise is averaged by *N*, the number of clock cycles (the counter), so we can rewrite Eq. (8.24) to show the decrease in the thermal noise with averaging as

$$V_{onoise,RMS} = \sqrt{\frac{kT}{NC}} \tag{17.46}$$

Sensing Circuit Mismatches

The point of sampling the reference or dark level and then subtracting the desired signal (the intensity of light on the pixel) was to subtract out mismatches in the pixel. For example, M2 in one pixel, may have a threshold voltage of 250 mV, while in a different pixel the threshold voltage may be 230 mV. The result, without the subtraction, would be two pixels with different output voltages even though the light applied to each is exactly the same. After thinking about this for a moment, we might realize that the DSM sensing circuit, having two separate paths for the reference and intensity signal paths, will also be subject to a mismatch. If, for example, one sensing circuit on the bottom of a column in an imaging chip has different characteristics than the sensing circuits directly adjacent to it, then the image will show vertical streaks. We might try, using layout techniques, to reduce the mismatch. However, the human eye is very perceptive and will likely detect any differences in the sense (especially if the image is single color). This is why the majority of imaging chips used a single pipeline ADC (see Ch. 29) operating at a high conversion rate at the time of this writing. Each pixel sees the exact same sensing circuit.

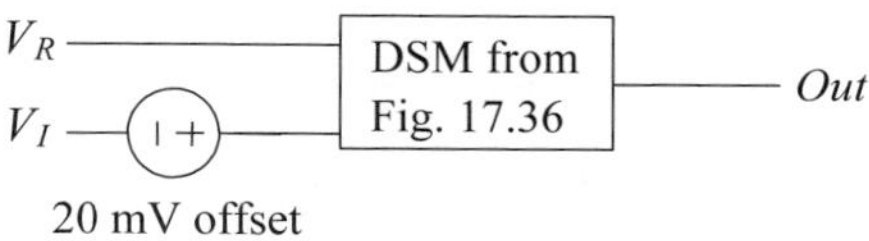

Figure 17.38 Modeling the differences in the signal paths in the DSM of Fig. 17.36 with a voltage in series with the intensity signal path.

To illustrate this problem, let's resimulate the DSM in Fig. 17.36 with an offset, as seen in Fig. 17.38. This offset voltage, which is an unknown that may be positive or negative, simply models a random difference in the signal paths. Using the input values seen in Fig. 17.37a and a 20 mV offset, we get the simulation results seen in Fig. 17.39. Instead of getting two outputs going high, we now get seven.

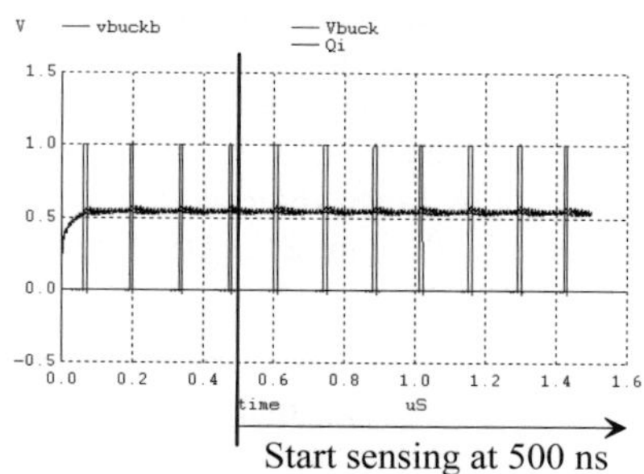

Figure 17.39 How a 20 mV offset changes the outputs in Fig. 17.37a.

To reduce the effects of mismatch on the sense, consider, halfway through the sense time, switching the inputs to the DSM, as seen in Fig. 17.40. By switching the inputs halfway through the sense, the effects of path mismatch average, ideally, to zero. Note that by switching the inputs we also have to change the feedback internal to the DSM. As seen in Fig. 17.36, we must switch the gate connections of Q and VDD in series with the switched capacitor resistors, Fig. 17.41, (the outputs of the NOR latch are switched).

Note at signal intensities *close to the reference*, V_R , that the averaging will not work out with just an up counter (as seen in Fig. 17.3). The signal, V_I, must move away from the reference by more than V_{OS}. For example, if we swap the inputs to the DSM at

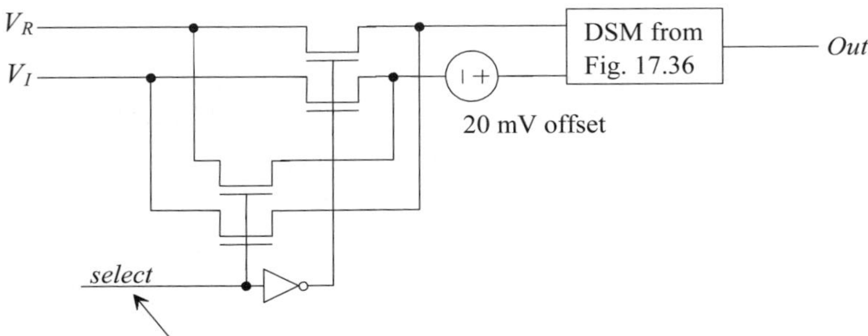

Figure 17.40 Switching the inputs of a DSM to eliminate path mismatch.

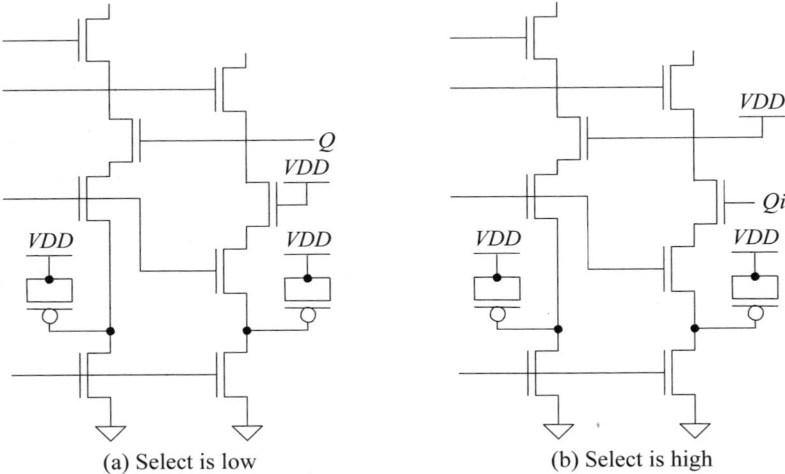

Figure 17.41 The change in the feedback with change in select.

one μs in the simulation seen in Fig. 17.39, then the output of the DSM after 1 μs is always low and the output code is 3 (the ideal output code from Fig. 17.37a is 2). This (the averaging won't work unless the $|V_R - V_I| < V_{OS}$) shouldn't be a problem since the output codes at these levels correspond to dark signals. Again note that feeding back both Q and Qi, Fig. 17.35, results in a nonlinearity, as seen in Eq. (17.42).

Finally note that whenever starting or switching the inputs during a sense operation there will be a start-up transient (see Fig. 17.34 for example). What this means is that the counter should be disabled at the beginning of the sense or in the middle (if the inputs to the DSM are swapped, as seen in Fig. 17.40, halfway through the sense operation). Not disabling the counter during these times can result in sensing errors.

ADDITIONAL READING

See Volume 2 of this book entitled *CMOS Mixed-Signal Circuit Design*, John Wiley and Sons, 2002. ISBN 0-471-22754-4

PROBLEMS

17.1 Regenerate Table 17.1 in which the water level where a cup of water is removed is 4.7 instead of 5. How are the results affected? If the sensing time is increased, how are the final results affected (compare a water level of 5 against 4.7).

17.2 Generate a table, similar to Table 17.1, for the situation seen in Fig. 17.4 if the amount of water leaving the bucket is 0.3 cups per 10 seconds.

17.3 Rederive Eqs. (17.3) – (17.8) if I_{cup} in Fig. 17.6 is replaced with a resistor. Assume that the clock frequency is large (why?) to simplify the equations. What is the requirement for the current through the resistor when it is connected to the bit line in terms of the maximum bit current, I_{bit}.

17.4 Using SPICE simulations demonstrate that the error because of parasitics, as seen in Fig. 17.7, is reduced by connecting the switch to a 1 V source instead of ground. Illustrate, with drawings, what is happening.

17.5 Show, using simulations, that if the output of the comparator swings from *VDD* to *VDD*/2 we can eliminate M4 in Fig. 17.9 and still have $Q_{cup} = C_{cup} \cdot (VDD - V_{REF} - V_{THP})$. Why? Does the amount of current supplied by *VDD*/2 increase? Could this be a problem?

17.6 Show how the incomplete settling seen in Fig. 17.13 can be made more complete by reducing the clock frequency or increasing the width of M4.

17.7 Demonstrate that the operation of the circuit in Fig. 17.42 may be used in place of the comparator and M3 in Fig. 17.18. How do output glitches affect the sensing circuit's operation?

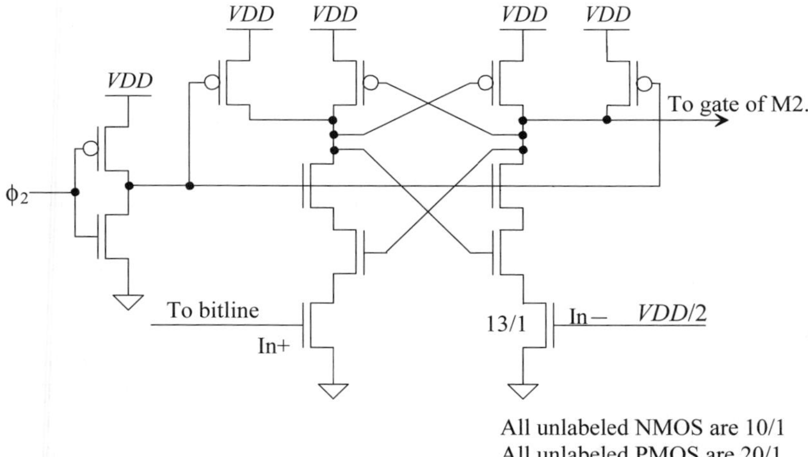

Figure 17.42 Simplifying the comparator in Figs. 17.16 and 17.18.

17.8 Design a DSM sensing circuit that will determine the value of a resistor that may range from 100k to 10 MΩ. Simulate your design with SPICE and comment on the design trade-offs concerning operating frequency and resistor (both the sensed and, if used in the DSM, the feedback resistance) changes with process variations or temperature.

17.9 Suppose that it is desired to have a noise floor of 100 μV RMS in a CMOS imager. Further suppose that the sensing circuit doesn't contribute any noise to the sense (the transformation from the analog column voltage to a digital word). Estimate the size of the hold capacitors used to sample both the reference and the intensity signals.

17.10 Using simulations, determine if the linearity of the voltage-to-current converter can be made better by adjusting the length and width of the PMOS device seen in Fig. 17.28. Why does, or doesn't, the performance get better?

17.11 Using the short-channel CMOS devices determine the average current that flows in the following circuit, Fig. 17.43. The clocks are nonoverlapping (never low at the same time) as used throughout the chapter and have a frequency of 100 MHz. Verify your answer using SPICE (the WinSPICE command "print mean(vdd# branch)" can be used to get the average current supplied by *VDD*).

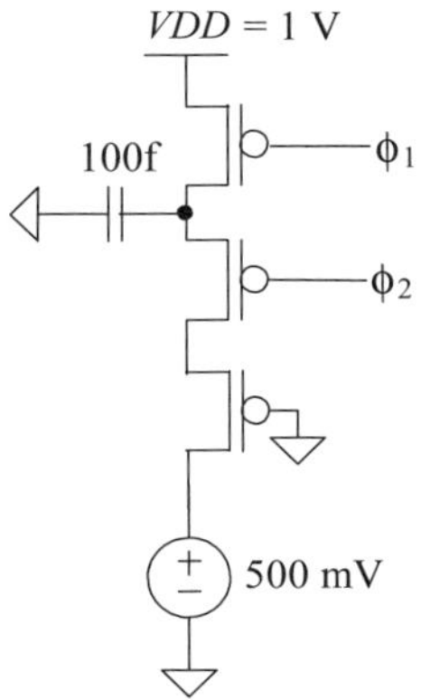

Figure 17.43 Determining the average current that flows in a switched-capacitor resistor. See Problem 17.11.

17.12 What happens, to the simulation results seen in Fig. 17.34, if the time step used in the transient simulation is increased to 1 ns? How do the nonoverlapping clocks look with this time step?

17.13 Simulate the operation of the circuit seen in Fig. 17.44. Do the comparator outputs make full logic transitions? Are glitches a concern? Why? How do the simulation results compare to the results seen in Fig. 17.37? Note that the outputs of the comparator go low each time ϕ_2 goes high so that $\overline{Out}$ can be used to clock a counter directly.

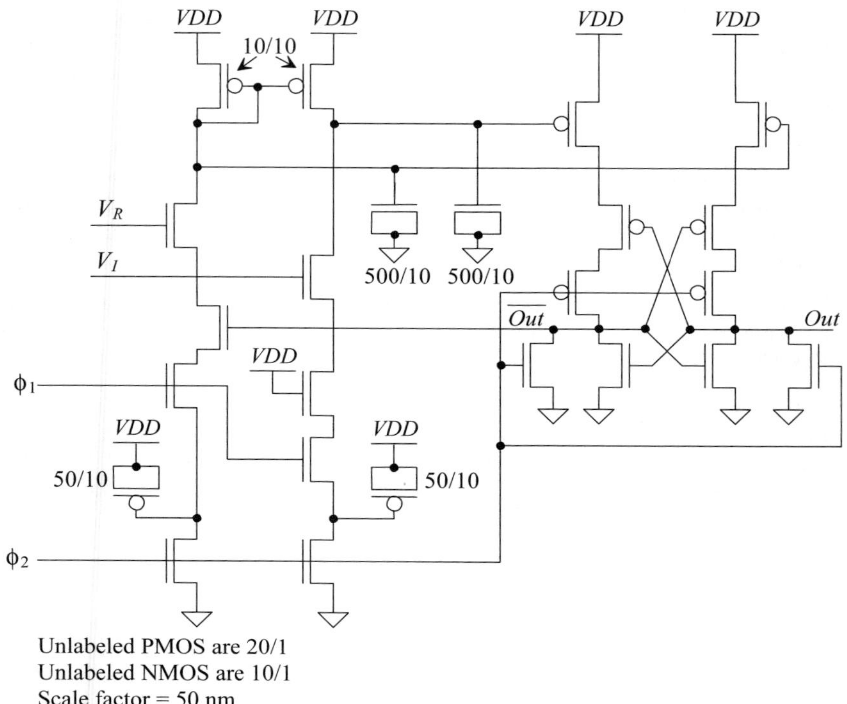

Figure 17.44 Simplifying the DSM sensing circuit, see Problem 17.13.

Chapter

18

Special Purpose CMOS Circuits

In this chapter we discuss some special-purpose CMOS circuits. We begin with the Schmitt trigger, a circuit useful in generating clean pulses from a noisy input signal or in the design of oscillator circuits. Next, we discuss multivibrator circuits, both astable and monostable types. This is followed by a discussion concerning input buffer design. Good receiver circuits (input buffers) in CMOS chips are required in any high-speed, board-level design to change the distorted signals transmitted between chips (because of the imperfections in the interconnecting signal paths) into well-defined digital signals with the correct pulse widths and amplitudes. Finally, we end this chapter with a discussion of on-chip voltage generators.

18.1 The Schmitt Trigger

The schematic symbol of the Schmitt trigger is shown in Fig. 18.1 along with typical transfer curves. We should note the similarity to the inverter transfer characteristics with the exception of a steeper transition region (and hysteresis). Curve A in Fig. 18.1 corresponds to the output of the Schmitt trigger changing from a low to a high, while curve B corresponds to the output changing from a high to a low. The hysteresis present in the transfer curves is what sets the Schmitt trigger apart from the basic inverter.

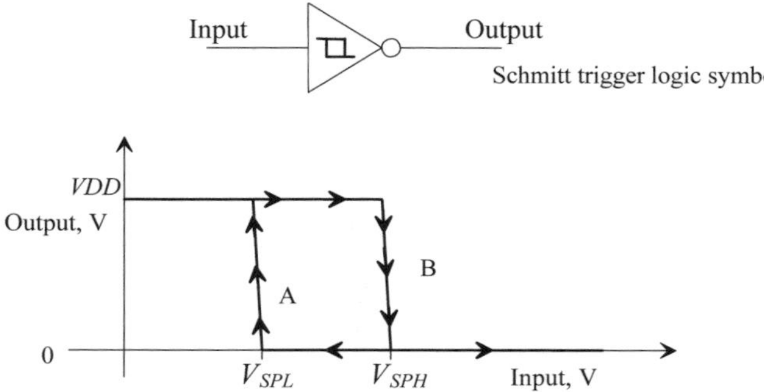

Figure 18.1 Transfer characteristics of a Schmitt trigger.

Figure 18.2 shows a possible input to a Schmitt trigger and the resulting output. When the output is high and the input exceeds V_{SPH}, the output switches low. However, the input voltage must go below V_{SPL} before the output can switch high again. Note that we get normal inverter operation when $V_{SPH} = V_{SPL}$. The hysteresis of the Schmitt trigger is defined by

$$V_H = V_{SPH} - V_{SPL} \tag{18.1}$$

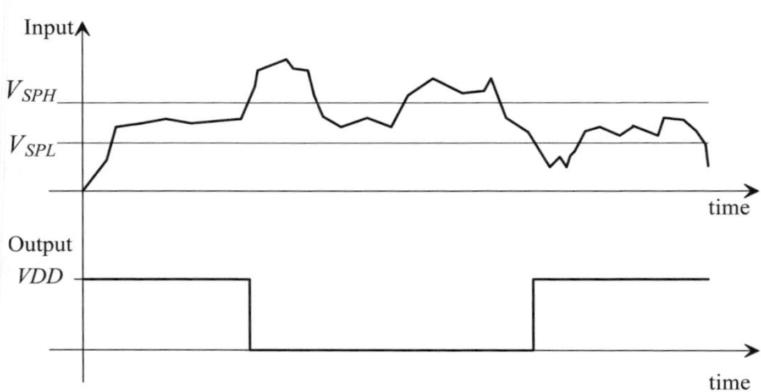

Figure 18.2 Input, top trace, and output of a Schmitt trigger.

18.1.1 Design of the Schmitt Trigger

The basic schematic of the Schmitt trigger is shown in Fig. 18.3. We can divide the circuit into two parts, depending on whether the output is high or low. If the output is low, then M6 is on and M3 is off and we are concerned with the p-channel portion when calculating the switching point voltages, while if the output is high, M3 is on and M6 is off and we are concerned with the n-channel portion. Also, if the output is high, M4 and M5 are on, providing a DC path to *VDD*.

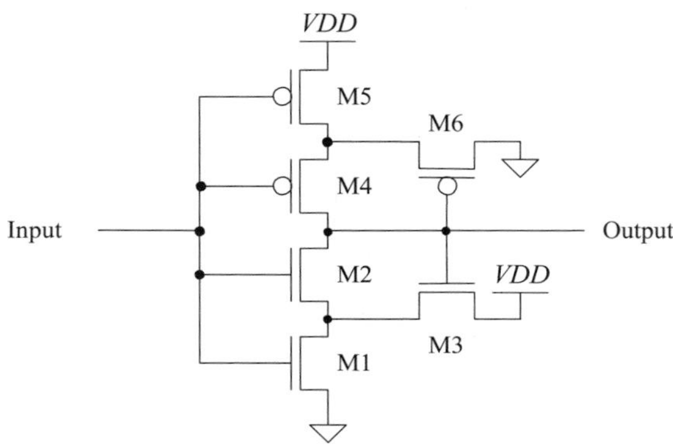

Figure 18.3 Schematic of the Schmitt trigger.

Let's begin our analysis of this circuit, assuming that the output is high (= *VDD*) and the input is low (= 0 V). Figure 18.4 shows the bottom portion of the Schmitt trigger used in calculating the upper switching point voltage, V_{SPH}. MOSFETs M1 and M2 are off, with V_{in} = 0 V while M3 is on. The source of M3 floats to $VDD - V_{THN}$, or approximately 4 V for *VDD* = 5 V. We can label this potential V_x, as shown in the figure.

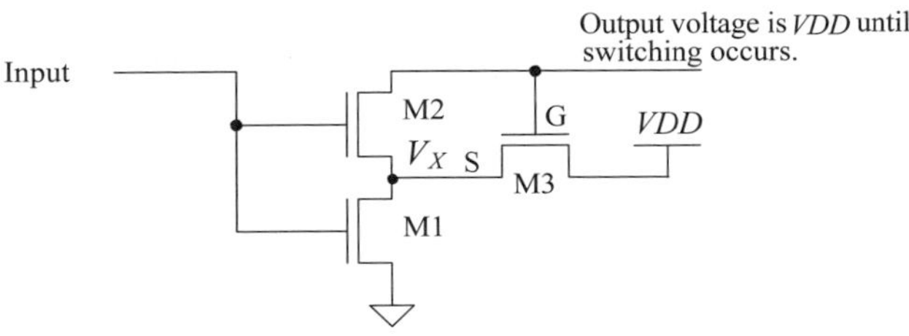

Figure 18.4 Portion of the Schmitt trigger schematic used to calculate upper switching point voltage.

With V_{in} less than the threshold voltage of M1, V_X remains at $VDD - V_{THN3}$. As V_{in} is increased further, M1 begins to turn on and the voltage, V_X, starts to fall toward ground. The high switching point voltage is defined when

$$V_{in} = V_{SPH} = V_{THN2} + V_X \tag{18.2}$$

or when M2 starts to turn on. As M2 starts to turn on, the output starts to move toward ground, causing M3 to start turning off. This in turn causes V_X to fall further, turning M2 on even more. This continues until M3 is totally off and M2 and M1 are on. This positive feedback causes the switching point voltage to be very well defined.

When Eq. (18.2) is valid, the currents flowing in M1 and M3 are essentially the same. Equating these currents gives

$$\frac{\beta_1}{2}(V_{SPH} - V_{THN})^2 = \frac{\beta_3}{2}(VDD - V_X - V_{THN3})^2 \tag{18.3}$$

Since the sources of M2 and M3 are tied together, $V_{THN2} = V_{THN3}$, the increase in the threshold voltages from the body effect is the same for each MOSFET. The combination of Eqs. (18.2) and (18.3) yields

$$\frac{\beta_1}{\beta_3} = \frac{W_1 L_3}{L_1 W_3} = \left[\frac{VDD - V_{SPH}}{V_{SPH} - V_{THN}}\right]^2 \tag{18.4}$$

The threshold voltage of M1, given by V_{THN} in this equation, is the zero body bias threshold voltage (= 0.8 V in our long-channel CMOS process and 0.25 V in the short-channel process). Given a specific upper switching point voltage, the ratio of the MOSFET transconductors is determined by solving this equation. A general design rule for selecting the size of M2, that is, β_2, is to require that

$$\beta_2 \geq \beta_1 \text{ or } \beta_3 \tag{18.5}$$

since M2 is used as a switch.

A similar analysis can be used to determine the lower switching point voltage, V_{SPL}, resulting in the following design equation:

$$\frac{\beta_5}{\beta_6} = \frac{W_5 L_6}{L_5 W_6} = \left[\frac{V_{SPL}}{VDD - V_{SPL} - V_{THP}}\right]^2 \tag{18.6}$$

The following example illustrates the design procedure for a Schmitt trigger.

Example 18.1
Design and simulate a Schmitt trigger using the short-channel CMOS process with $V_{SPL} = 400$ mV and $V_{SPH} = 700$ mV.

We begin by solving Eqs. (18.4) and (18.6) for the transconductance ratios. For the upper switching point voltage,

$$\frac{W_1 L_3}{W_3 L_1} = \left[\frac{1-0.7}{0.7-0.25}\right]^2 = 0.444 \rightarrow L_1 = L_3 = 1 \text{ and } W_1 = 10, W_3 = 22.5$$

and for the lower switching point voltage,

$$\frac{W_5 L_6}{W_6 L_5} = \left[\frac{0.4}{1-0.4-0.25}\right]^2 = 1.3 \rightarrow L_5 = L_6 = 1 \text{ and } W_6 = 20, W_5 = 26$$

M2 is set to 10/1 and M4 is set to 20/1. Simulation results are seen in Fig. 18.5. This figure reveals the benefit of using a Schmitt trigger, namely, it allows slow moving inputs to be made into good solid logic high and low values. In a practical circuit, connecting the ramp-shaped input seen in Fig. 18.5 to an inverter would produce oscillations in the inverter's output (because of noise on the ramp). ■

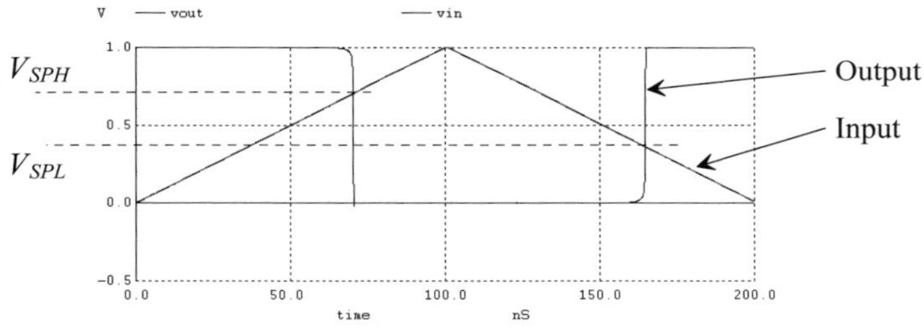

Figure 18.5 The input and output of the Schmitt trigger designed in Ex. 18.1.

Switching Characteristics

The propagation delays of the Schmitt trigger can be calculated in much the same way as the inverter of Ch. 11. Defining equivalent digital resistances for M1, M2, M4, and M5 as R_{n1}, R_{n2}, R_{p4}, and R_{p5}, respectively, gives a high-to-low propagation delay-time, neglecting the Schmitt trigger output capacitance, of

$$t_{PHL} = 0.7 \cdot (R_{n1} + R_{n2}) \cdot C_{load} \tag{18.7}$$

and

$$t_{PLH} = 0.7 \cdot (R_{p4} + R_{p5}) \cdot C_{load} \tag{18.8}$$

18.1.2 Applications of the Schmitt Trigger

Consider the waveform shown in Fig. 18.6. A pulse with ringing is a common voltage waveform encountered in buses or lines interconnecting systems. If this voltage is applied directly to a logic gate or inverter input with a V_{SP} of 0.5 V, the output of the gate will vary with the period of the ringing on top of the pulse. Using a Schmitt trigger with properly designed switching points can eliminate this problem.

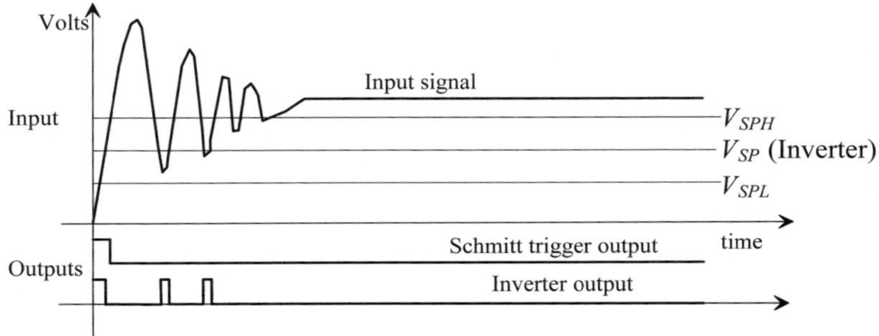

Figure 18.6 Applying a Schmitt trigger to clean up an interconnecting signal.

The Schmitt trigger can also be used as an oscillator (Fig. 18.7). The delay-time in charging and discharging the capacitor sets the oscillation frequency. At the moment in time when the output of the Schmitt trigger switches low, the voltage across the capacitor is V_{SPH}. The capacitor will start to discharge toward ground. The voltage across the capacitor is given by

$$V_c(t) = V_{SPH} \cdot e^{-t/RC} \tag{18.9}$$

At the time when $V_c(t) = V_{SPL}$, the output of the Schmitt trigger changes state. This time is given by solving Eq. (18.9) by

$$t_1 = RC \cdot \ln \frac{V_{SPH}}{V_{SPL}} \tag{18.10}$$

A similar analysis for the case when the capacitor is charged from V_{SPL} to V_{SPH} gives

$$V_c(t) = V_{SPL} + (VDD - V_{SPL})\left(1 - e^{\frac{-t}{RC}}\right) \tag{18.11}$$

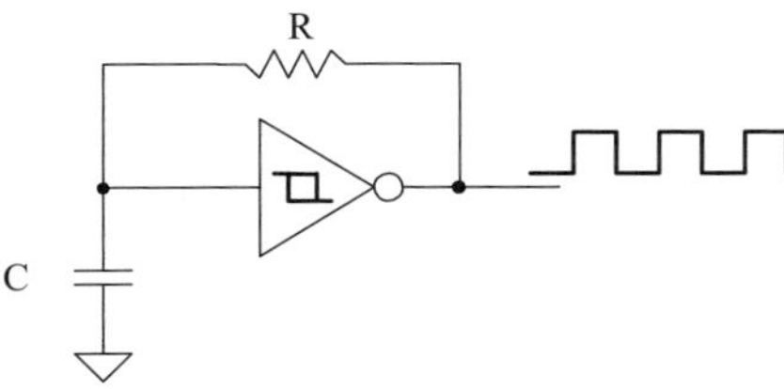

Figure 18.7 Oscillator design using a Schmitt trigger.

and

$$t_2 = RC \cdot \ln \frac{VDD - V_{SPL}}{VDD - V_{SPH}} \tag{18.12}$$

The oscillation frequency, neglecting the intrinsic delay of the Schmitt trigger, is given by

$$f_{osc} = \frac{1}{t_1 + t_2} \tag{18.13}$$

The capacitance used in these equations is the sum of the input capacitance of the Schmitt trigger and any external capacitance.

An alternative oscillator using the Schmitt trigger is shown in Fig. 18.8. Here the MOSFETs M1 and M4 behave as current sources (see Ch. 20) mirroring the current in M5 and M6. When the output of the oscillator is low, M3 is on and M2 is off. This allows the constant current from M4 to charge C. When the voltage across C reaches V_{SPH}, the output of the Schmitt trigger swings low. This causes the output of the oscillator to go high and allows the constant current from M1 to discharge C. When C is discharged down to V_{SPL}, the Schmitt trigger changes states. This series of events continues, generating the square wave output.

If we label the drain currents of M1 and M4 as I_{D1} and I_{D4}, we can estimate the time it takes the capacitor to charge from V_{SPL} to V_{SPH} as

$$t_1 = C \cdot \frac{V_{SPH} - V_{SPL}}{I_{D4}} \tag{18.14}$$

and the time it takes to charge from V_{SPH} to V_{SPL} is

$$t_2 = C \cdot \frac{V_{SPH} - V_{SPL}}{I_{D1}} \tag{18.15}$$

The period of the oscillation frequency is, as before, the sum of t_1 and t_2.

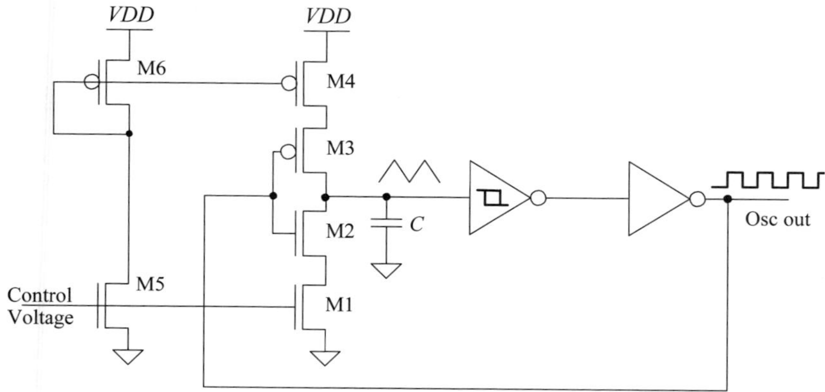

Figure 18.8 Voltage-controlled oscillator using Schmitt trigger and current sources. MOSFETs M2 and M3 are used as switches.

This type of oscillator is termed a voltage-controlled oscillator (VCO) since the output frequency can be controlled by an external voltage. The currents I_{D1} and I_{D4}, (Fig. 18.8) are directly controlled by the control voltage. As we will see in Ch. 20, the current in M5 is mirrored in M1, M4, and M6, with an appropriate scaling factor dependent on the size of the transistors.

18.2 Multivibrator Circuits

Multivibrator circuits (Fig. 18.9) are circuits that employ positive feedback. The name "multivibrator" is a vestige from early-time electronics development (prior to the ubiquitous term "digital") where the circuits' outputs vibrate between two states. There are three types of multivibrators: astable, bistable, and monostable. Astable multivibrator circuits are unstable in either output (high or low) state. The oscillators that we have discussed are examples of astable multivibrators. The bistable multivibrator is stable in either the high or low state. Flip-flops and latches are examples of the bistable multivibrator. Monostable multivibrators are stable in a single state. Monostable multivibrators are also called one-shots. In this section we discuss the monostable and astable multivibrators. We distinguish the material in this section from the other material in the book by using resistor-capacitor time constants to set time intervals.

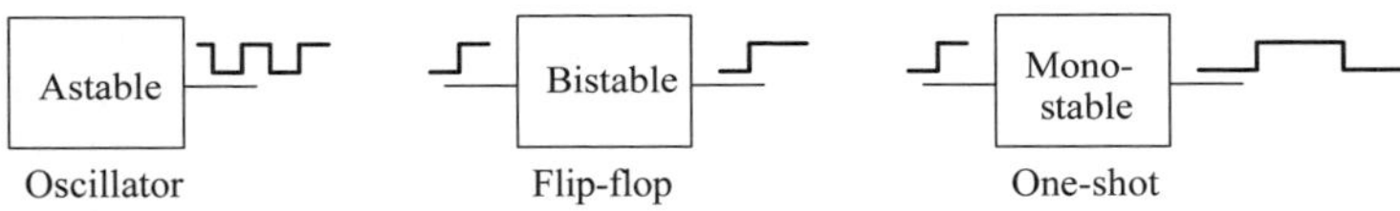

Figure 18.9 Multivibrator circuits.

18.2.1 The Monostable Multivibrator

A CMOS implementation of the monostable multivibrator is shown in Fig. 18.10. Under normal conditions, V_{in} is low and the output of the NOR gate, V_1, is high. The voltage V_2 is pulled high through the resistor, and the output of the inverter, V_3, is a low. Upon application of a trigger pulse, that is, V_{in} going high, both V_1 and V_2 drop to zero volts and the output of the inverter, which is also the output of the monostable, goes high. This output is fed back to the input of the NOR gate holding V_1 at ground potential.

After triggering, the potential V_2 will start to increase because C is charged through R. The potential across the capacitor after triggering takes place is given by

$$V_c(t) = V_2(t) - \overbrace{V_1(t)}^{=0} = VDD \cdot (1 - e^{\frac{-t}{RC}}) \tag{18.16}$$

If we assume that the V_{SP} of the inverter is $VDD/2$, then the time it takes for the capacitor to charge to V_{SP} is given by

$$t = RC \cdot \ln \frac{VDD}{VDD - V_{SP}} = RC \cdot \ln(2) \approx 0.7RC \tag{18.17}$$

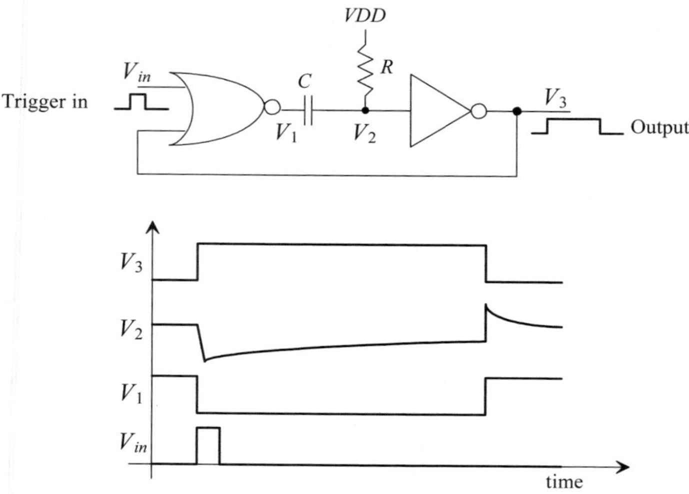

Figure 18.10 Operation of the monostable multivibrator.

This time also defines the output pulse width, neglecting gate delays, since the inverter switches low. The inverter output, V_3, going low causes V_1 to go back to *VDD* and V_2 to go to *VDD* + *VDD*/2. If the resistor or capacitor is bonded out (connected to the output pads), the ESD diodes may keep V_2 from going much above *VDD* + 0.7. The time it takes V_2 to decay back down to *VDD* limits the rate at which the one-shot can be retriggered. Also note that the trigger input can be longer than the output pulse width. Longer output pulse widths may cause V_2 to go as high as 10 V as well as limit the maximum trigger rate.

18.2.2 The Astable Multivibrator

An example of an astable multivibrator is shown in Fig. 18.11. This circuit has no stable state and thus oscillates. To analyze the behavior of this multivibrator, let's begin by assuming that the output, V_3, has just switched high. The output going high causes V_1 to go high (to $VDD + V_{SP1}$), forcing V_2 low. The voltage across the capacitor after this switching takes place is given by

$$V_c(t) = V_1(t) - \overbrace{V_3(t)}^{=VDD} = (VDD + V_{SP1}) \cdot e^{\frac{-t}{RC}} - VDD \tag{18.18}$$

The output of the astable will go low, $V_3 = 0$, when $V_1 = V_{SP1}$. Substituting this condition into the previous equation gives the time the output is high (or low) and is given by

$$t_1 = RC \cdot \ln \frac{VDD + V_{SP1}}{V_{SP1}} = t_2 \tag{18.19}$$

If $V_{SP1} = VDD/2$, then

$$t_1 = t_2 = 1.1RC \tag{18.20}$$

and the frequency of oscillation is

$$f_{osc} = \frac{1}{t_1 + t_2} = \frac{1}{2.2RC} \quad (18.21)$$

Again, if the resistor and capacitor are bonded out, the ESD diodes on the pads will limit the voltage swing of V_1.

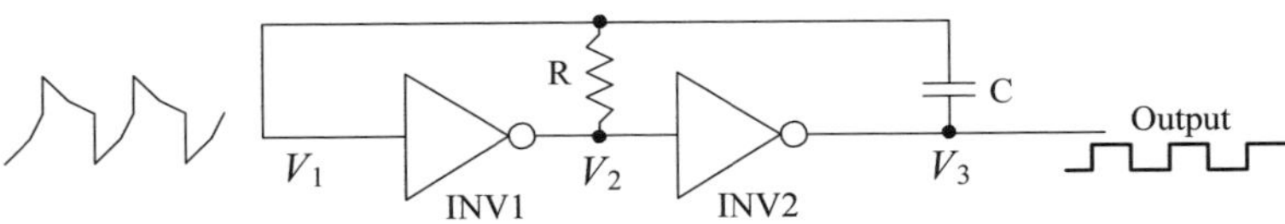

Figure 18.11 An astable multivibrator.

18.3 Input Buffers

Input buffers are circuits that take a chip's input signal, with imperfections such as slow rise and fall times, and convert it into a clean digital signal for use on-chip. If the buffer doesn't "slice" the data in the correct position, timing errors can occur. For example, consider the waveforms seen in Fig. 18.12. If the input signal is sliced too high or too low, the output signal's width is incorrect. In high-speed systems this reduces the timing budget in the system and can result in errors.

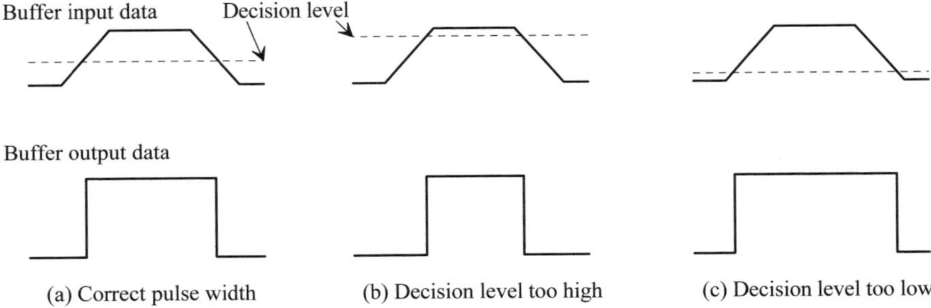

Figure 18.12 Timing errors in regenerating digital data.

18.3.1 Basic Circuits

If not careful, simple circuits can introduce unwanted pulse width variations because of differences in rise and fall times. Consider the inverter circuit seen in Fig. 18.13. The switching point voltage of the inverter, because of process, *VDD* (voltage), or temperature variations (PVT) will move around. Further, differences in the PMOS and the NMOS resistance will affect the rise and fall times of the inverter's output signal. The output of the first inverter, in Fig. 18.13, sees a much larger load capacitance than the second inverter. The delay time from the input going high to the output going high is roughly 400 ps. However, the delay time between the input and the output going low is

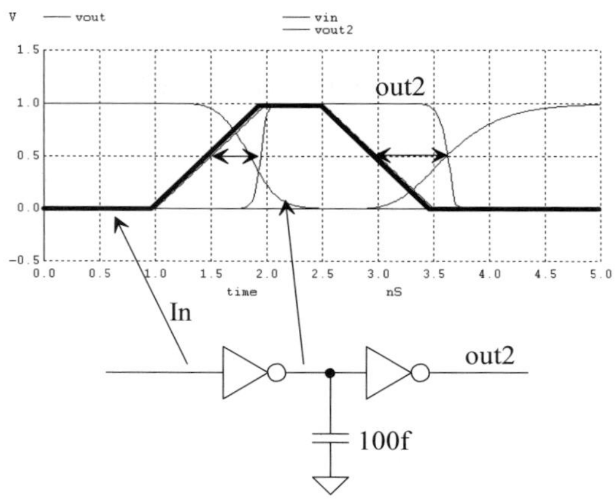

Figure 18.13 How skew is introduced into a high-speed signal.

600 ps. As the input signal travels through the digital system, the time skew added to the input signal can add up and result in timing errors (the data is received too early or too late at different points in the system). Note that had we matched the loading of each inverter, the propagation delays would be more similar, Fig. 18.14. When the input signal goes high, the output of the first inverter goes low (t_{PHL}) and the output of the second inverter goes high (t_{PLH}). The total delay is the sum of $t_{PHL} + t_{PLH}$. However, when the input signal goes low, the output of the first inverter goes high (t_{PLH}) and the output of the second inverter goes low (t_{PHL}), resulting in the same sum ($t_{PHL} + t_{PLH}$). *If possible, make the number of t_{PHL} delays in series with a high-speed signal even and equal to the number of t_{PLH} delays.*

Of course, to better equalize the delays, the second inverter in Fig. 18.14 would need an additional bit of loading (an inverter connected to its output). Also, the input signal should transition faster in an attempt to match the transition times of the inverters.

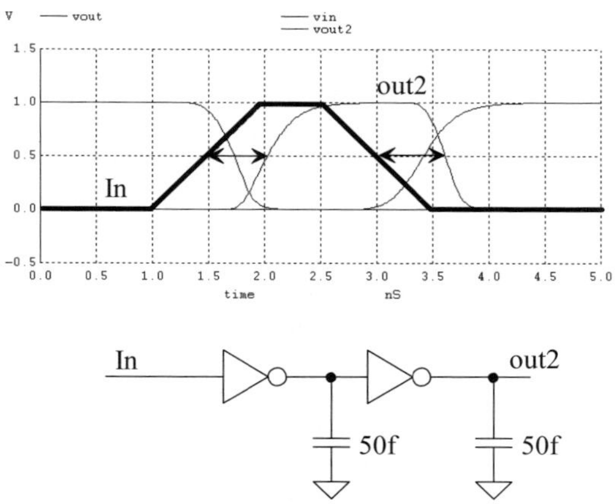

Figure 18.14 How equalizing delays can be used to reduce skew.

Skew in Logic Gates

Consider the NAND gate in Fig. 18.15. A different delay will be added to an input signal depending on which input of the NAND gate is used. As seen in the figure, the *A* input, with the *B* input high, propagates to the output slightly quicker than the *B* input (with the *A* input high). For the NAND gate, this difference in propagation delays is only a factor for the NMOS devices (because they are in series with the output of the gate).

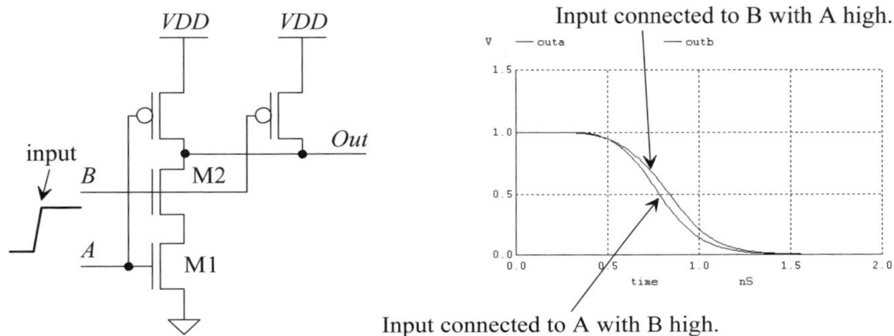

Figure 18.15 The skew introduced by using different inputs.

Figure 18.16 shows how two NAND gates can be used in parallel to eliminate the differences in the propagation delay between inputs. The series NMOS are arranged so that no matter which input is toggled the output changes at the same rate. Note that two of the PMOS devices can be eliminated to simplify the circuit without affecting the propagation delays.

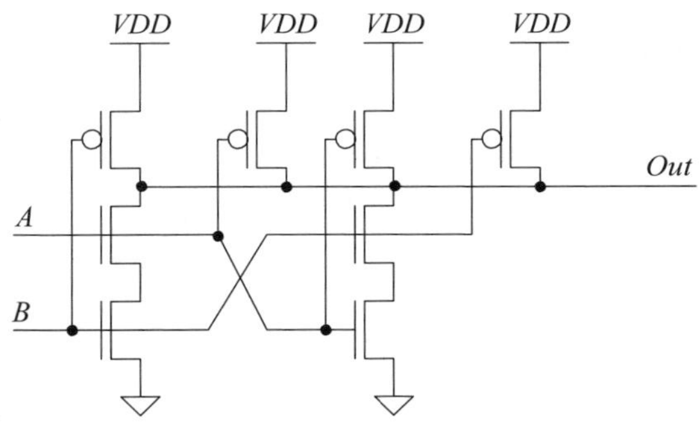

Figure 18.16 Using two NAND gates in parallel to reduce input-dependent skew.

The previous discussion neglected the effects of the gate's switching point voltage. If the input signals are not transitioning to full logic levels (*VDD* and ground) or the rise and fall times are not fast (compared to the gate delay time), then the point where the input circuit switches is critical. This leads us to our next topic.

18.3.2 Differential Circuits

In order to precisely slice input data, as seen in Fig. 18.12, a reference voltage may be transmitted, on a different signal path, along with the data. Alternatively, the data may be transmitted differentially (an input and its complement). In either case a differential amplifier is needed, Fig. 18.17. A differential amplifier input buffer (which we'll simply call an input buffer from this point on) amplifies the difference between the two inputs. In the simplest case one input to the input buffer is a DC voltage, say 0.5 V (*Vinm* in Fig. 18.17). When the other input (*Vinp*) goes above 0.5 V, the output of the buffer changes states (goes from a low to a high) or

$$Vinp > Vinm \rightarrow \text{out} = \text{"1"} \tag{18.22}$$

and

$$Vinp < Vinm \rightarrow \text{out} = \text{"0"} \tag{18.23}$$

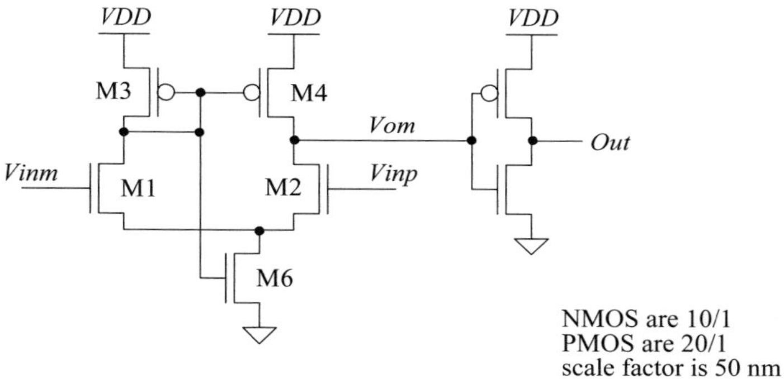

Figure 18.17 An (n-flavor) input buffer for high-speed digital design.

The diff-amp in Fig. 18.17 is based on the topologies seen in Ch. 22. This circuit is *self-biased* because no external references are used to set the current in the circuit (the gate of M6 is tied up to the gate of M3). When *Vinp* is larger than *Vinm*, the current in M2 is larger than the current in M1 ($V_{GS2} > V_{GS1}$). The current in M1 flows through M3 and is mirrored by M4 (and so M4's current is less than M2's current). This causes the diff-amp's output, *Vom*, to go towards ground (until the current in M2 equals the current in M4) and the output of the inverter, *Out*, to go high. Note that the gain from the *Vinp* input to the output of the circuit is larger than the gain from *Vinm* to the output (M3, being diode-connected, is a lower resistance than M4). It is generally a good idea to connect the reference voltage to the *Vinm* input.

The DC simulation results for the buffer in Fig. 18.17 are shown in Fig. 18.18. The x-axis is the *Vinp* input swept from 0 to 1 V. The *Vinm* input is held from 0 to 1 V in 200 mV steps. When *Vinm* is held at 400 mV and *Vinp* goes above 400 mV, the output changes from 0 to 1 (although there is a small offset which necessitates that *Vinp* go to approximately 415 mV before the output transitions).

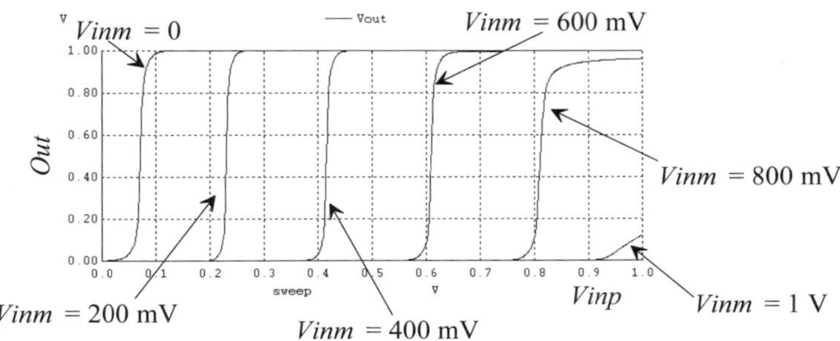

Figure 18.18 Simulating the DC behavior of the buffer in Fig. 18.17.

Transient Response

An example transient response for the buffer in Fig. 18.17 is seen in Fig. 18.19. A very small increase in *Vinp* above *Vinm* is required to make the output of the buffer switch states. We have several practical questions that should be answered with variations of this simulation. For example, what are the minimum and maximum values of *Vinm* allowed. Next, why are the delays between *Vinp* going high and going low different? Is it because of the offset seen in Fig. 18.18 (the output doesn't precisely switch at a *Vinm* 400 mV)? Let's provide some discussion concerning these concerns.

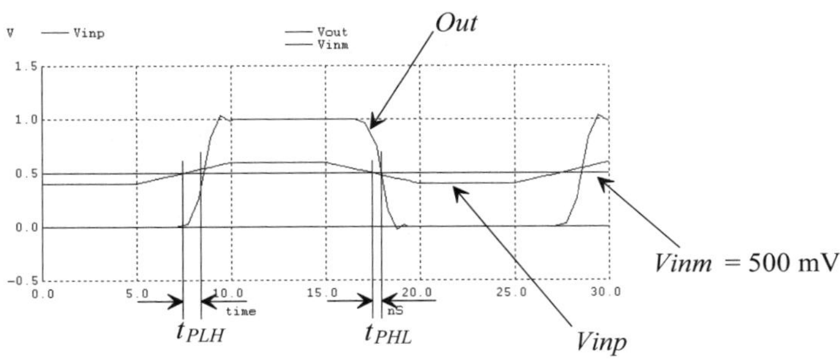

Figure 18.19 Transient response of the input buffer in Fig. 18.17 driving a 50 fF load.

Looking at Fig. 18.17, we can see that if the inputs fall below V_{THN} (= 250 mV here), then the circuit won't work very quickly (the MOSFETs move into the subthreshold region). So we would expect the propagation delays to increase. Indeed, as seen in Fig. 18.20, the delays go up. Ideally, the delay of the buffer is independent of power supply voltage, temperature, or input signal amplitudes (or pulse shape). To get better performance for lower input level signals, we might use the PMOS version of the buffer in Fig. 18.17, as seen in Fig. 18.21. Resimulating this buffer with the signals seen in Fig. 18.20 gives the results seen in Fig. 18.22. The delays are considerably better, however, there is an offset that appears rather large (because the output changes at the same time as *Vinp* going past *Vinm*, indicating that an offset is present). To avoid this

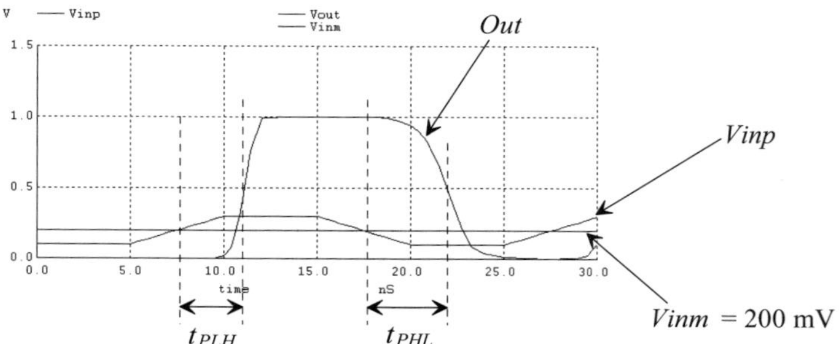

Figure 18.20 Resimulating with lower input signal voltages.

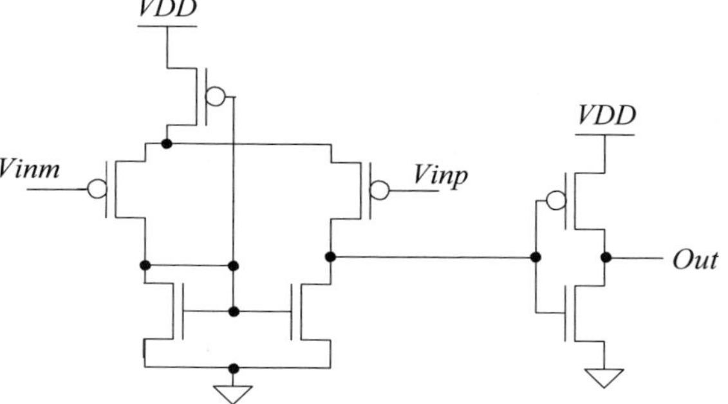

Figure 18.21 A PMOS input buffer for high-speed digital design.

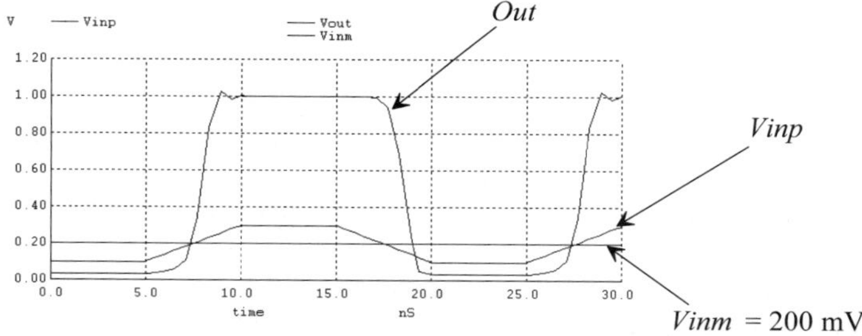

Figure 18.22 Repeating the simulation in Fig. 18.20 with the buffer in Fig. 18.21.

offset, we might use the NMOS buffer in Fig. 18.17 with the PMOS buffer in Fig. 18.21 to form a buffer that operates well with input signals approaching ground or *VDD*. The result is seen in Fig. 18.23. By using the buffers in parallel, the complementary nature results in a buffer that is robust and works over a wide range of operating voltages. Figure 18.24 shows a DC sweep simulation for the buffer with the *Vinp* input swept and the *Vinm* changed in increments of 100 mV. Note the smaller offset.

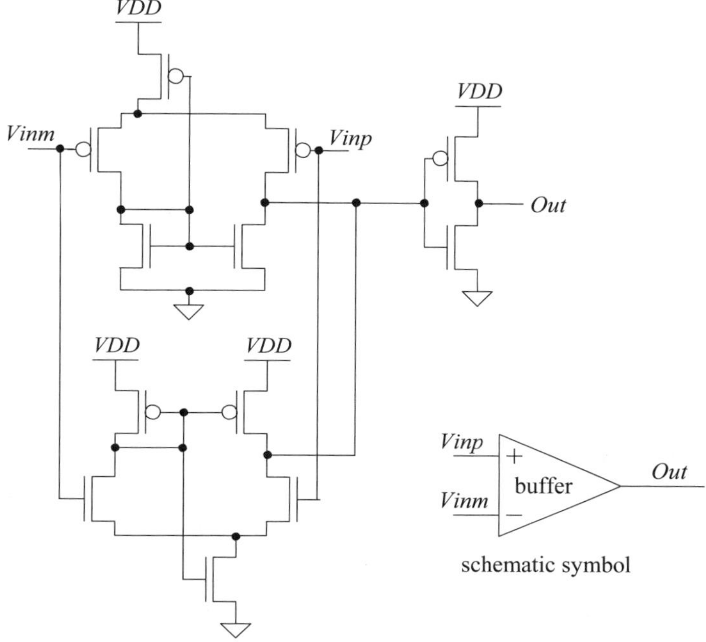

Figure 18.23 A rail-to-rail input buffer based on the topologies in Figs. 18.17 and 18.21.

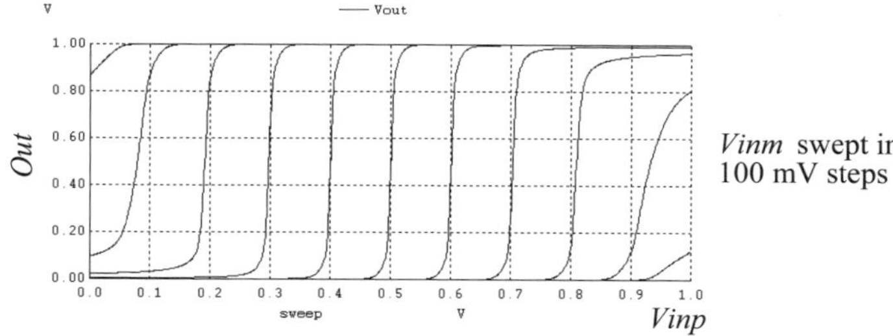

Figure 18.24 DC characterisitics of the buffer in Fig. 18.23.

Example 18.2
Design a high-speed input buffer based on the topology in Fig. 18.17 with a small amount of hysteresis.

In any circuit that has hysteresis we have some positive feedback. We can introduce positive feedback into the buffer in Fig. 18.17 by adding a long L MOSFET across the output inverter, Fig. 18.25. When the output is low, this turns the long L MOSFET on and make the self-biased diff-amp work harder to pull the input of the inverter to ground. When the output is high, the long L MOSFET is off and doesn't affect the circuit. ■

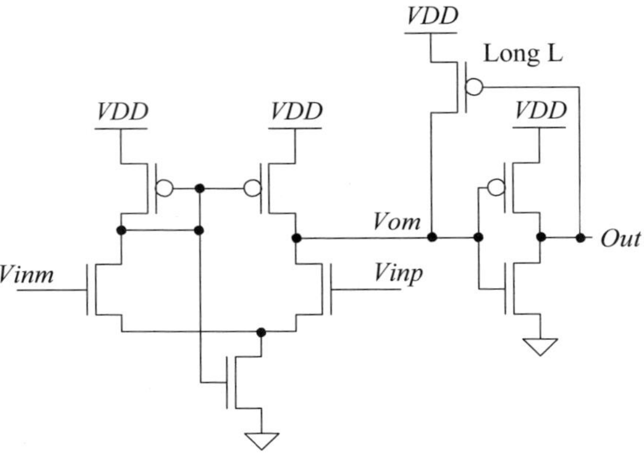

Figure 18.25 An input buffer for high-speed digital design with hysteresis.

18.3.3 DC Reference

In the previous section we assumed we had a DC voltage, *Vinm* , at precisely the correct voltage to slice the data (in the middle). In a real system the data varies and the communication channel is bandlimited (behaves like a lowpass filter). The result is that the amplitudes of the data vary depending on the data and the channel frequency response. Consider the simple RC lowpass filter and data used to model a transmission system as seen in Fig. 18.26. The ideal point where we slice the output data changes with the input data. What we need is a circuit that determines the maximum and minimum of an input waveform and outputs the average of the two to slice the buffer's input data in the center.

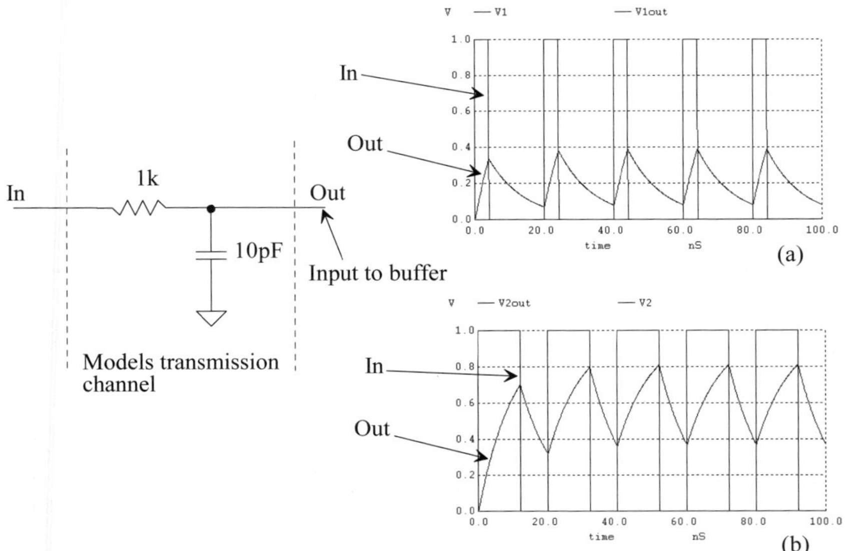

Figure 18.26 How the shape of the data changes through a communication channel.

Towards this goal consider the peak detector circuit seen in Fig. 18.27. The buffer from Fig. 18.23 can be used in this circuit. When the input, v_{in} , goes above the voltage stored on the capacitor, v_{peak} , the output of the buffer goes low, turning on the long-length MOSFET and pulling the output towards *VDD*. As v_{peak} approaches v_{in} , the MOSFET starts to shut off. As a result the output voltage across the capacitor, v_{peak} , corresponds to the peak voltage of the input signal. This peak detector can be used to generate a reference voltage that falls within the middle of the input data, Fig. 18.28. The peak and valley detectors are used to find the minimum and maximum of the input signal. The two resistors average the minimum voltages and feed the result (the DC average of the input) to the bottom buffer circuit. The resistors also are used to leak charge off of the capacitors so that the averaging circuit can follow changes in the input data (the averaging is actually a *running average*). Figure 18.29 shows some simulation results. In the top plots the output signal from Fig. 18.26a is applied to the input of the DC generation circuit. As expected, the output of the DC generation circuit, after a start-up delay, goes right to the middle of the input data.

The speed (response time) of this generation circuit can be increased by reducing the resistors and capacitors in the circuit. However, if the values are reduced too much, a long string of zeroes or ones will cause the DC generation circuit's output (labeled "average" in Fig. 18.28) to go to ground or *VDD*. The bottom traces in Fig. 18.29 show the results of applying the output signal from Fig. 18.26b to the DC generation circuit. Again, the output moves quickly to the center of the data. What's more interesting in these figures is a comparison between the original input data and the regenerated output data (labeled *In* and *Out* in Fig. 18.29). The output data pulse widths are considerably different from the input pulse widths (remembering that there is a delay through both the channel RC seen in Fig. 18.26 and the buffer in Fig. 18.28). Ideally, the pulse widths (the data) of the input and output are the same. Looking at the responses in Fig. 18.26a, we see that the effect of the finite channel bandwidth is to distort the channel's output data. *The only solution to this problem is to make the channel transmission bandwidth effectively wider.* An equalizer (discussed in the next chapter) can be used for this purpose at the cost of reduced signal swing. We can also reduce the input impedance of the input buffer to lower the time constant associated with the channel. Let's discuss this second approach.

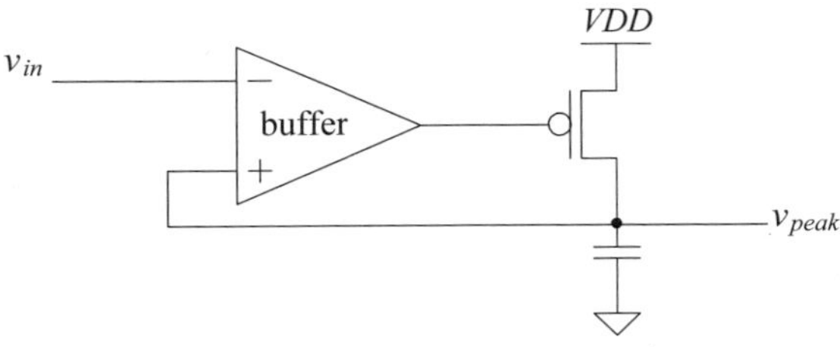

Figure 18.27 CMOS peak detector.

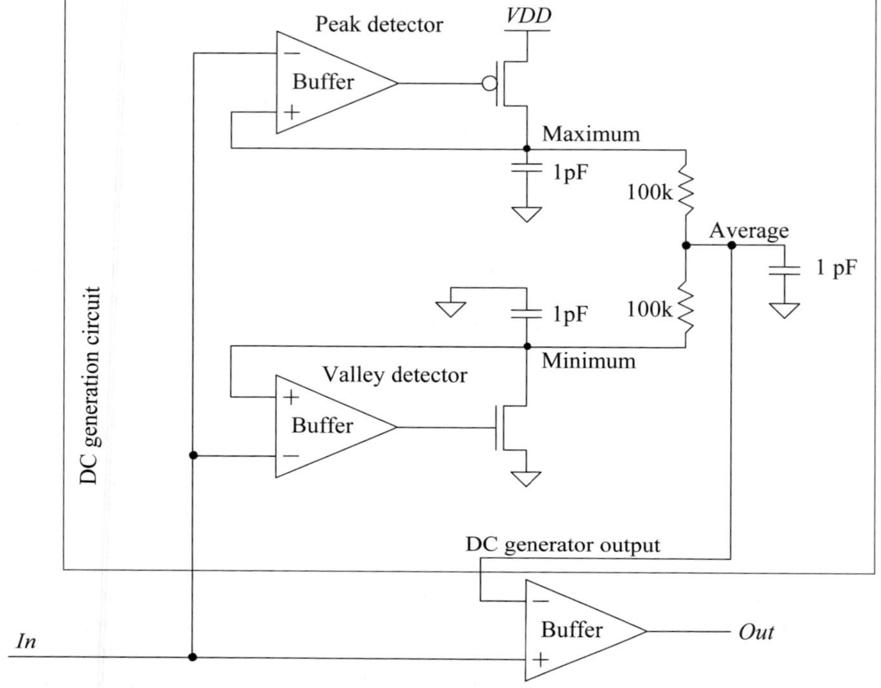

Figure 18.28 A DC generation circuit.

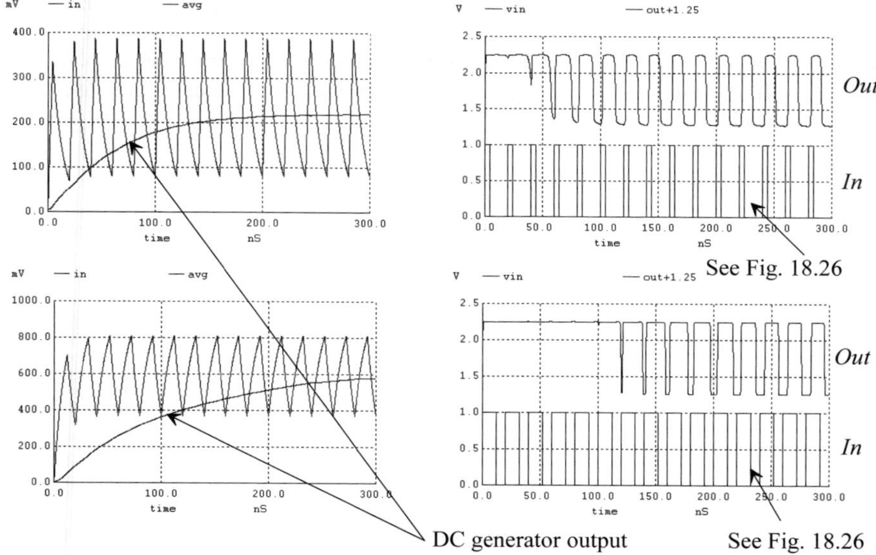

Figure 18.29 The output of the DC restore circuit in Fig. 18.28.

18.3.4 Reducing Buffer Input Resistance

A simple solution to increasing the effective bandwidth of the transmission channel is to reduce the input resistance of the input buffer. The time constant associated with the channel in Fig. 18.30 is 10 ns. When we connect the buffer to the transmission channel, the time constant drops to 3.33 ns (the three 1k resistors in parallel multiplied by the 10 pF capacitance). The drawbacks of this approach are increased power dissipation and reduced input signal amplitudes. Figure 18.31 shows how the buffer in Fig. 18.30 behaves with the input signals seen in Fig. 18.26a. The distortion (difference in the pulse widths of ones and zeroes when comparing the communication channel input, *In* , and the buffer output, *Out*) is much better.

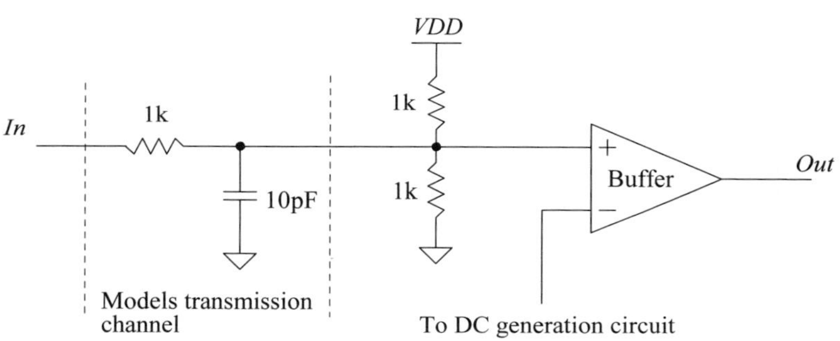

Figure 18.30 Reducing the input resistance of the input buffer.

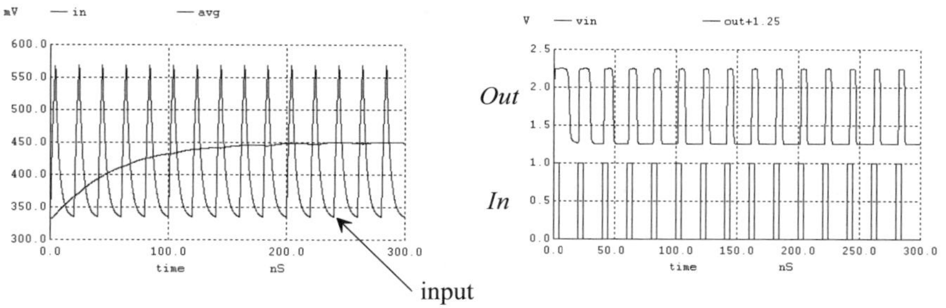

Figure 18.31 How reducing the input resistance of the buffer reduces the input signal amplitude (bad) and distortion (good).

Figure 18.32 shows an alternative low input resistance input buffer. When the input goes high, M1 turns on and M4 shuts off. This causes M2 to turn on and M5 to shut off. The output then goes high. When the input goes low, M1 turns off and M4 turns on (with M2 shutting off and M5 turning on). The result: the output goes low. Again, the power burned by M1 and M4 can be high if the transistors aren't sized properly. Further, the input swing is limited to a threshold voltage away from the supply voltages. This reduces the input voltage swing and further enhances the speed.

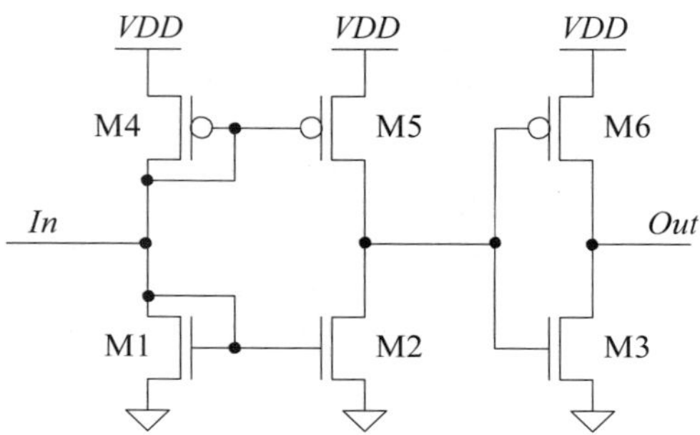

Figure 18.32 A low-input resistance input buffer.

18.4 Charge Pumps (Voltage Generators)

Often, when designing CMOS circuits, positive and negative DC voltages are needed that do not lie between ground and *VDD*. An example of an application where a larger DC voltage source is needed was seen in Ch. 16 when we discussed Flash memory (see Fig. 16.59). A simple circuit, sometimes called a voltage pump (or more often, a *charge pump*), useful in generating a voltage greater than *VDD,* is seen in Fig. 18.33.

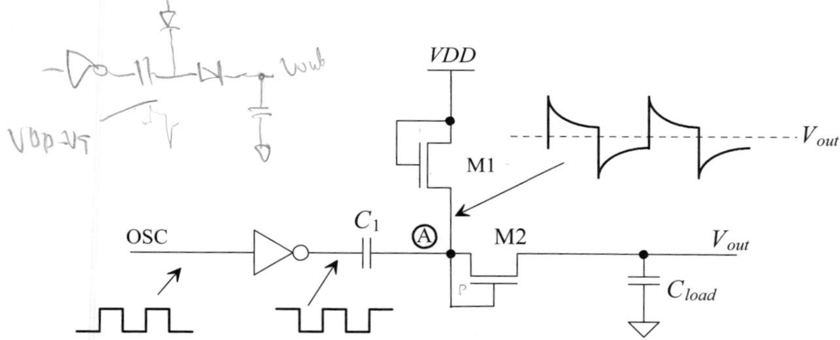

Figure 18.33 Pump used to generate a voltage greater than *VDD.*

The operation of the voltage pump of Fig. 18.33 can be explained by first realizing that both M1 and M2 operate like a diode. M1 is simply used to pull point A to a voltage of $VDD - V_{THN}$. M2 allows the charge from C_1 to charge C_{load} but not vice-versa. Let's begin the description of the circuit operation by assuming that the output of the inverter is low and point A is at a potential of $VDD - V_{THN}$. When the output of the inverter goes high, the potential at point A increases to $VDD + (VDD - V_{THN}) = 2 \cdot VDD - V_{THN}$. This turns on M2 and charges C_{load} to $2 \cdot (VDD - V_{THN})$, [an extra V_{THN} because of M2's gate-source voltage drop], provided $C_1 >> C_{load}$ and the oscillator frequency allows the capacitors to fully charge or discharge before changing states. In most practical situations, C_{load} and C_1 are comparable in size, and the output of the pump is loaded with a DC load. The result is an output voltage with a startup time; that is, V_{out}

does not immediately rise to $2 \cdot (VDD - V_{THN})$ but requires several oscillator cycles to reach steady state. Also, the output has a ripple dependent on the DC load. Figure 18.34 shows the simulation results for the circuit of Fig. 18.13 using 10/1 NMOS (scale factor of 50 nm) with $C_{load} = C_1 = 1$ pF and an oscillator frequency of 10 MHz. Using this simple pump with a DC load, such as a resistor, may require employing larger MOSFETs and capacitors to avoid excessive voltage droop.

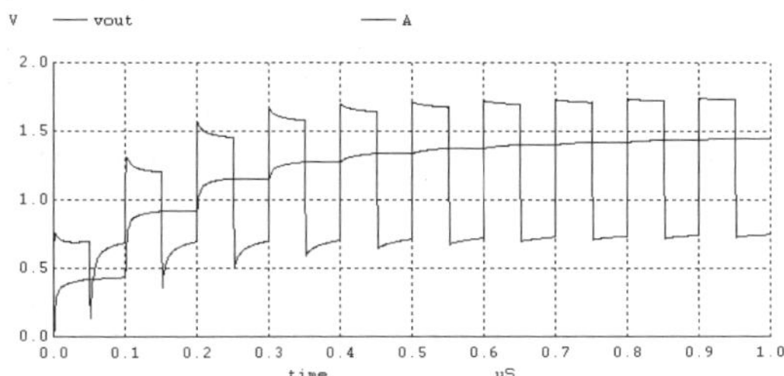

Figure 18.34 Simulating the operation of the voltage pump in Fig. 18.33.

Negative Voltages

The positive voltage pump uses n-channel MOSFETs, while the negative voltage pump, Fig. 18.35, uses p-channel MOSFETs. The reason for this comes from the requirement that the diode formed with the n+ (p+) implant used in the drain/source of the MOSFET combined with the p-substrate (n-well) does not become forward-biased. Forward biasing this parasitic diode is an *important concern* and is often the reason why some more exotic pump topologies can't be implemented. For example, in Fig. 18.35, we connect the well of the PMOS devices to ground instead of to their respective sources. Consider what would have happened had we connected the well to the source of the MOSFET. When the output voltage goes negative, the (n-type) well goes negative too. If the substrate (p-type) is at ground, this forward biases the n-well to substrate diode acting to clamp the output of the pump at a negative diode drop. While we could have left the bodies of the PMOS (the n-wells) tied to *VDD*, here we chose to connect them to ground so that the body effect wasn't so severe (a lower threshold voltage).

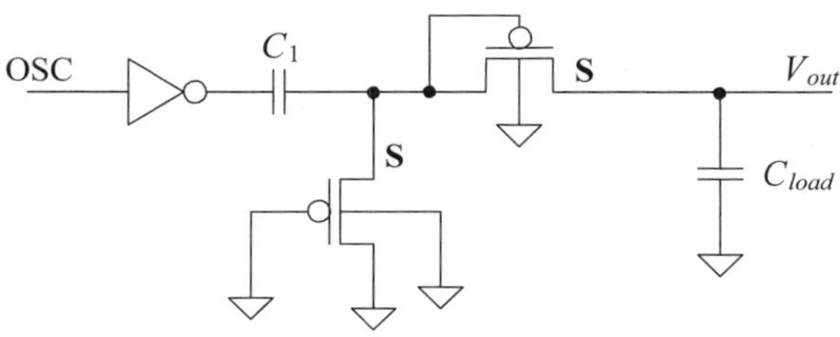

Figure 18.35 Negative voltage pump.

Using MOSFETs for the Capacitors

The capacitors used in Figs. 18.33 and 18.35 can be replaced with MOSFETs. An example of using n-channel MOSFETs in place of the capacitors in Fig. 18.33 is shown in Fig. 18.36. The main requirement on a MOSFET used as a capacitor is that its V_{GS} remain greater than V_{THN} at all possible operating conditions. In other words, the MOSFET must remain in the strong inversion region so that its capacitance is a constant $C'_{ox} \cdot W \cdot L$. The capacitor, C_{load}, in Fig. 18.36 remains in strong inversion because $V_{out} >> V_{THN}$. The capacitor C_1 remains in strong inversion because, when the inverter output is low, the voltage on the gate of C_1 is $VDD - V_{THN}$ (= V_{GS}). When the output of the inverter is high (VDD), the voltage on the gate of C_1 is $2 \cdot VDD - V_{THN}$. In both cases (the output of the inverter high and low), $V_{GS} = VDD - V_{THN}$, and the MOSFET is in the strong inversion region.

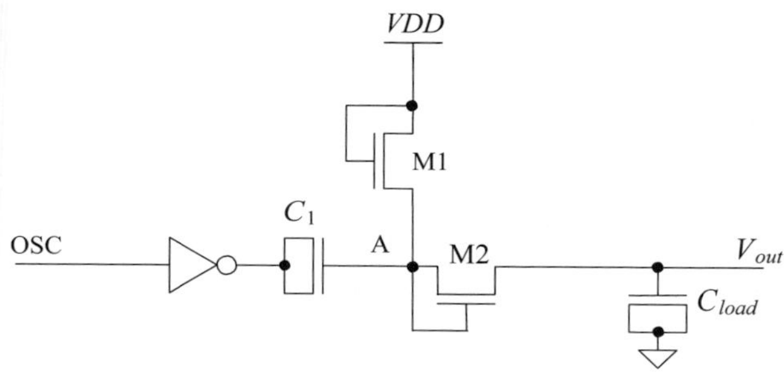

Figure 18.36 Using MOSFETs as capacitors.

18.4.1 Increasing the Output Voltage

Figure 18.37 shows a voltage pump with a higher output voltage. The increase in the output voltage comes from eliminating the threshold voltage drop at the gate and drain of M7. This allows the output to swing up to $2 \cdot VDD - V_{THN}$.

To understand the operation of the circuit, let's assume that the voltage at point A is low and the voltage at point C is $VDD - V_{THN}$. When the output of INV1 goes high, point A is VDD and the voltage at point C swings up to $2 \cdot VDD - V_{THN}$. This causes M4, M5, and M6 to turn on and pull points D and E to VDD. Now when point B goes high (point A goes back to zero), points D and E swing up to $2 \cdot VDD$ and the output goes to $2 \cdot VDD - V_{THN}$. Note that MOSFETs M2 and M3 are not needed; unless the pump drives a DC load , they never turn on. Also, separating points D and E is unnecessary unless the pump supplies a DC current.

18.4.2 Generating Higher Voltages: The Dickson Charge Pump

The voltage pumps of the last section are limited to voltages less than $2 \cdot VDD$ and greater than $-2 \cdot VDD$. Figure 18.38 shows a scheme for generating arbitrarily high voltages (limited by the breakdown voltage of the capacitors or the oxide breakdown of the MOSFETs). Again, the MOSFETs are used as diodes in this configuration.

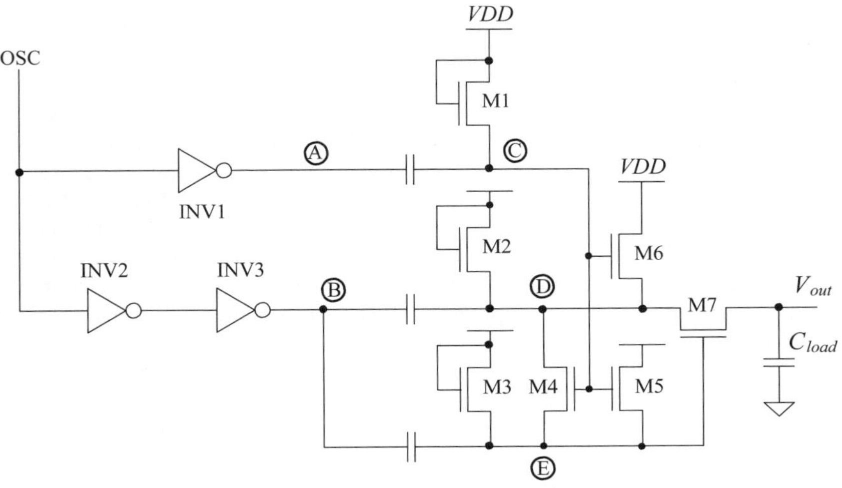

Figure 18.37 Increased output voltage pump.

In the following description of circuit operation, we assume steady-state operation with no DC load present. When CLK is low, point A is pulled, by M1, to $VDD - V_{THN}$. When CLK goes high, point A swings up to $2 \cdot VDD - V_{THN}$. Diode M2 turns on, and point B charges to $2 \cdot VDD - 2 \cdot V_{THN}$. When CLK goes low, $\overline{CLK}$ goes high and point B swings up to $3 \cdot VDD - 2 \cdot V_{THN}$. When $\overline{CLK}$ goes low, point B swings back down to $2 \cdot VDD - 2 \cdot V_{THN}$. This operation proceeds through the circuit to the last stage of the multiplier. The output of the multiplier swings from $N \cdot VDD - N \cdot V_{THN}$ down to $N \cdot VDD - (N+1) \cdot V_{THN}$. C_N can be made larger than the other capacitors to reduce this ripple.

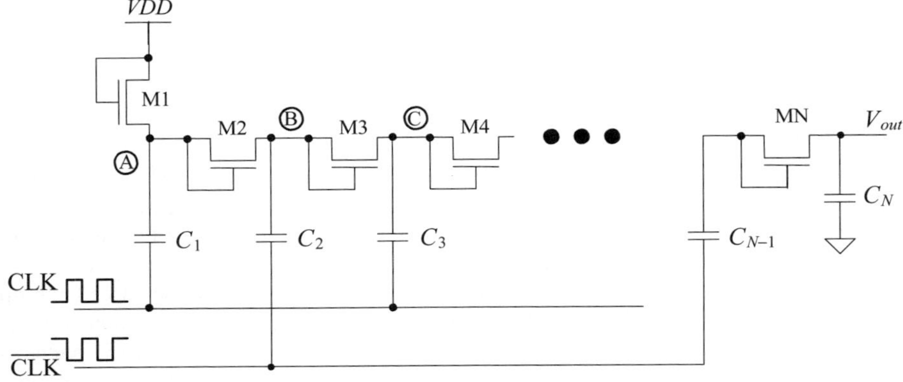

Figure 18.38 N-stage voltage multiplier (the Dickson charge pump).

Clock Driver with a Pumped Output Voltage

To fully turn on NMOS switches, a driver configuration is needed with a pulsed output voltage greater than $VDD + V_{THN}$ (with body effect). One such configuration is seen in Fig. 18.39. When the input to the circuit is a logic low, the output of INV1 is a high and the output of INV2 is a low. Node B in the figure is at roughly *VDD*, and node A is at roughly 2*VDD*. M1 is off and M2 is on. The output of the driver is a logic low (ground). When the input of the circuit goes high, the output of INV1 goes low. This causes node A to go to *VDD*. The output of INV2 goes high, and node B boots up to 2*VDD* (M1 on M2 off). The output of the driver circuit then goes to 2*VDD* as well. Notice that one side of the cross-coupled NMOS devices is sized smaller than the other side. This is to minimize layout area and power (the charging and discharging of the parasitic capacitors doesn't waste power). Node A doesn't supply any power to the output of the circuit. It is simply used to turn M2 on so that node B is precharged to *VDD* when the input to the circuit is low. A larger capacitor is used on node A because the charge that goes to load must be supplied by this capacitor. If the output load capacitance gets large, then the size of this capacitor must be increased. When designing the driver, it's a good idea to characterize the performance as a function of the load capacitance.

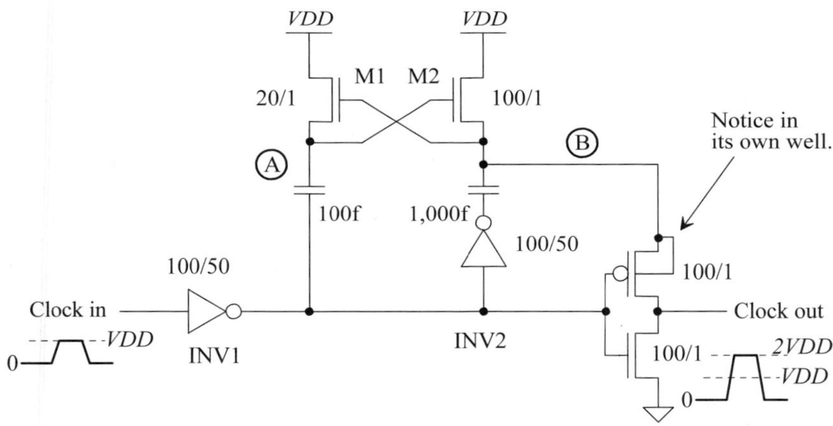

Figure 18.39 Charge-pump clock driver.

18.4.3 Example

A common application of a voltage generator in digital circuits is generating a negative substrate bias; that is, instead of tying the substrate to ground, the substrate is held at some negative voltage. Typically, this negative voltage is between –0.5 and –1 V. "Pumping" the substrate negative is found in some DRAMs. A negative substrate bias has several benefits. It (1) stabilizes n-channel threshold voltages, (2) increases latch-up immunity after power up, (3) prevents forward biasing n+ to p-substrate pn junction, (4) allows chip inputs to go negative without forward biasing a pn junction, (5) prevents substrate from going locally above ground, (6) reduces depletion capacitances associated with the n+ to p-substrate junction, and (7) reduces subthreshold leakage current.

A simple substrate pump is shown in Fig. 18.40. In this circuit we can use n-channel MOSFETs to generate a negative potential, unlike Fig. 18.35, since the negative voltage is connected to the substrate. In this situation, the drain/source implants of the n-channel MOSFETs cannot become forward biased. Note the absence of the load capacitance. It turns out that the capacitance of the substrate to everything else in the circuit presents a very large capacitance to the pump. In other words, the substrate itself is a very large capacitor.

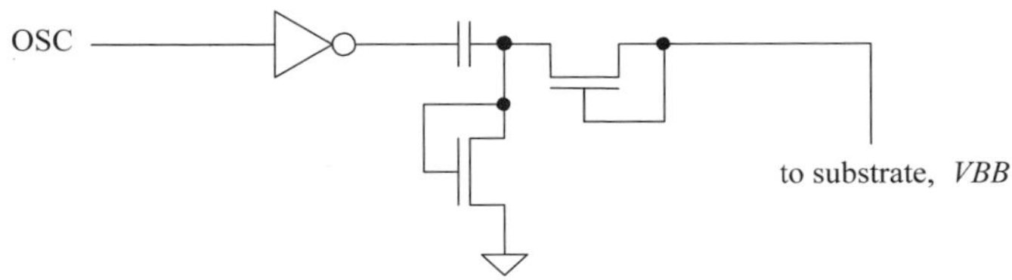

Figure 18.40 Simple substrate pump.

An example oscillator circuit that is used to drive the substrate pump is shown in Fig. 18.41. This is a standard ring oscillator with a NAND gate to enable/disable the oscillator. Capacitors are added at a few points in the middle of the oscillator to increase the delay and lower the frequency of the oscillator so that the pump's capacitors fully charge.

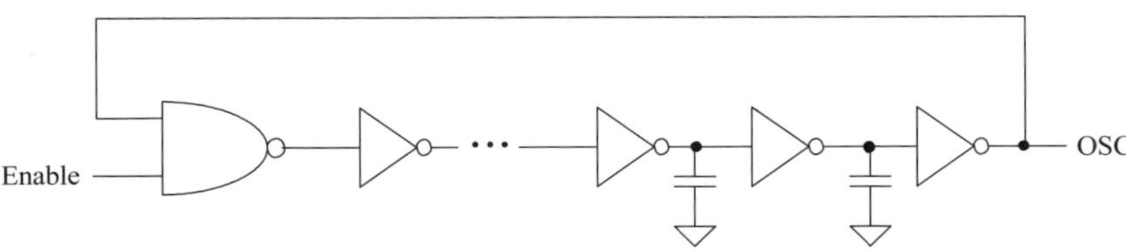

Figure 18.41 Ring oscillator with enable.

The final component in a substrate pump is the regulator, a circuit that senses the voltage on the substrate and enables or disables the substrate pump (the oscillator). Using a comparator with hysteresis, a precision voltage reference, and a level shifting circuit, the enable signal can be generated with the circuit of Fig. 18.42. The hysteresis of the comparator determines the amount of ripple on the substrate voltage. Forcing a constant current through the two MOSFETs, M1 and M2, compels their source-gate voltages to remain constant (note how the body effect is eliminated by using p-channel MOSFETs), causing the MOSFETs to behave as if they were batteries. The battery action of M1 and M2 shifts the substrate voltage up so that it lies in the input range of the comparator. The number of MOSFETs (in this case two) and the magnitude of the current I determine the substrate voltage.

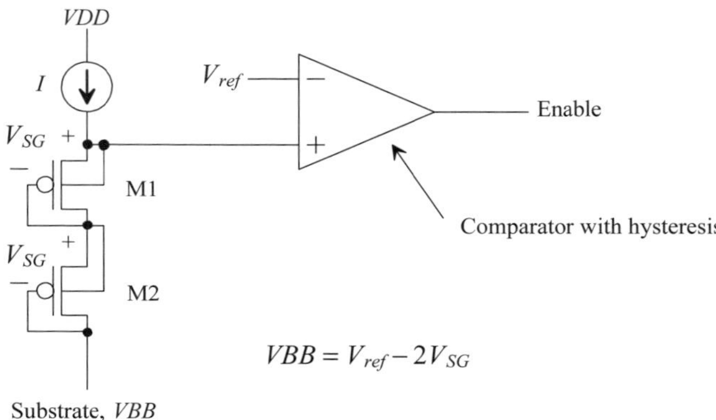

Figure 18.42 Regulator circuit used in substrate pump.

A simpler and less accurate implementation of the regulator is shown in Fig. 18.43. The comparator and voltage reference are implemented with an inverter and M1–M3. MOSFET M3 causes the inverter to have hysteresis, that is, behave like a Schmitt trigger. MOSFETs M1 and M2 form an inverter with a switching point voltage of approximately V_{THN}. The current source of Fig. 18.42 is implemented with the long L MOSFET M4 in Fig. 18.43. The level shifting is performed with the n-channel MOSFETs M5 and M6. When point A gets above a potential of V_{THN}, the Enable output goes high, causing the substrate pump to turn on and drive the substrate voltage negative. When point A gets pulled, via the substrate voltage through M5 and M6, below V_{THN}, the Enable output goes low and the pump shuts off. The substrate voltage generated with this circuit is approximately $-V_{THN}$ with body effect.

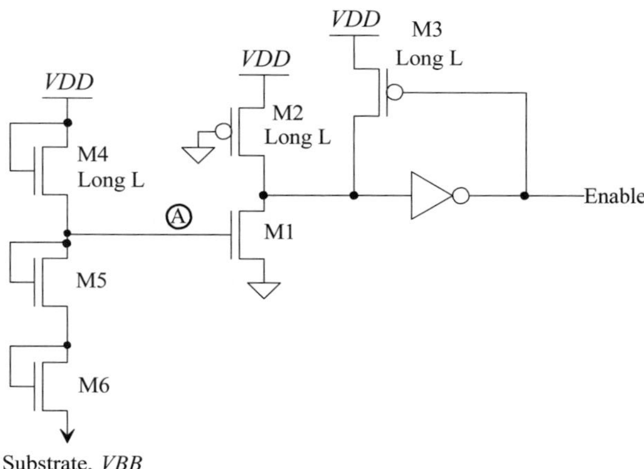

Figure 18.43 Simpler substrate regulator.

ADDITIONAL READING

[1] O. H. Schmitt, "A Thermionic Trigger," *Journal of Scientific Instruments*, Vol. XV, pp. 24–26, no. 1, Jan. 1938. Information on the "Schmitt trigger."

[2] J. Millman and H. Taub, *Pulse, Digital, and Switching Waveforms*, McGraw-Hill Publishers, 1965. ISBN 07-042386-5. Information on multivibrators and oscillators.

[3] J. F. Dickson, "On-Chip High-Voltage Generation in MNOS (sic) Integrated Circuits Using an Improved Voltage Multiplier Technique," *IEEE Journal of Solid State Circuits*, vol. SC-11, no. 3, June 1976.

[4] K. Sawada, Y. Sugawara, and S. Masui, "An on-chip high-voltage generator circuit for EEPROMs with a power supply voltage below 2 V," *Symp. VLSI Circuits Dig.* 9, pp. 75–76, June 1995.

[5] N. Otsuka and M. A. Horowitz, "Circuit techniques for 1.5-V power supply flash memory," *IEEE Journal of Solid-State Circuits,* vol. 32, pp. 1217–1230, August 1997.

PROBLEMS

For the following problems use the short-channel CMOS process with a scale factor of 50 nm and a *VDD* of 1 V.

18.1 Design a Schmitt trigger with $V_{SPL} = 0.35$ and $V_{SPH} = 0.55$ V. Use SPICE to verify your design.

18.2 Estimate t_{PHL} and t_{PLH} for the Schmitt trigger of Ex. 18.1, driving a 100 fF load capacitance. Compare your hand calculations to simulation results.

18.3 Design and simulate the operation of a Schmitt trigger-based oscillator with an output frequency of 10 MHz. Simulate the design using SPICE.

18.4 Estimate the total input capacitance on the control voltage input for the VCO shown in Fig. 18.8.

18.5 Design an astable multivibrator with an output oscillation frequency of 20 MHz.

18.6 Design and simulate the operation of a one-shot that has an output pulse width of 100 ns. Comment on the maximum rate of retrigger and how ESD diodes connected to bonding pads will affect the circuit operation if the resistor and capacitor are bonded out.

18.7 Design and simulate a 100 MHz oscillator using the astable multivibrator in Fig. 18.11. Comment how process shifts in the resistor and capacitor affects the oscillation frequency. If the resistor and capacitor are bonded out, will the ESD diodes affect the circuit's operation?

18.8 Using the buffer seen in Fig. 18.23 to drive a load capacitance of 100 *fF*, plot the propagation delays against the *Vinp* input signal amplitude when *Vinm* is 250, 500, and 700 mV. (*Vinp* is centered around *Vinm.*)

18.9 The DC restore circuit seen in Fig. 18.28 has limitations, such as it stops working if a long string of 1s or 0s is applied to the input buffer or the input data moves too quickly and the long RC time constants keep the DC restore output from following the center of the data. Comment, with the help of SPICE simulation results, on these limitations. Show how changing the values of the resistors and capacitors can compensate for these limitations. (For example, using longer time constants allows for longer strings of ones are zeroes before the DC restore signal is invalid.)

18.10 Using the model for the transmission channel seen in Figs. 18.26 and 18.30, show that the circuit in Fig. 18.32 will work as an input buffer. Using SPICE show the limitations of the buffer (signal amplitude and speed). Also, comment on the power pulled by the buffer.

18.11 Design a nominally 2 V voltage generator that can supply at most 1 μA of DC current (2 MEG resistor). Simulate the operation of your design with SPICE.

18.12 Design and simulate the operation of a nominally –1 V substrate pump. Comment on the design trade-offs.

Chapter

19

Digital Phase-Locked Loops

The digital phase-locked loop, DPLL, is a circuit that is used frequently in modern integrated circuit design. Consider the waveform and block diagram of a communication system shown in Fig. 19.1. Digital data[1] is loaded into the shift register at the transmitting end. The data is shifted out sequentially to the transmitter output driver. At the receiving end, where the data may be analog (and, thus, without well-defined amplitudes) after passing through the communication channel, the receiver amplifies and changes the data back into digital logic levels. The next logical step in this sequence is to shift the data back into a shift register at the receiver and process the received data. However, the absence of a clock signal makes this difficult. The DPLL performs the function of generating a clock signal, which is locked or synchronized with the incoming signal. The generated clock signal of the receiver clocks the shift register and thus recovers the data. This application of a DPLL is often termed *a clock-recovery circuit* or *bit synchronization circuit.*

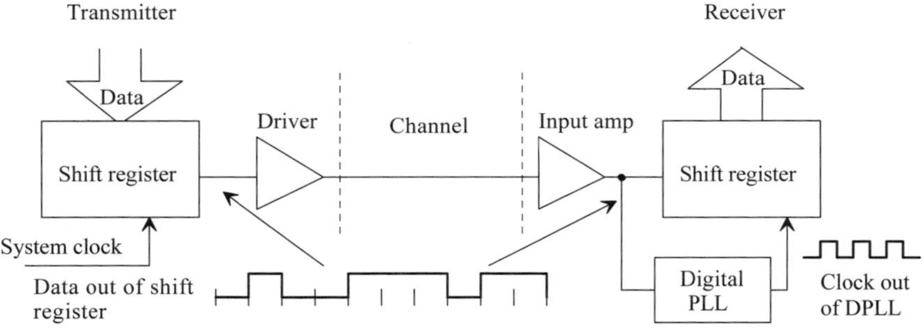

Figure 19.1 Block diagram of a communication system using a DPLL for the generation of a clock signal.

---

[1] While "data" is plural and "datum" is singular, we will not use, in this chapter, the grammatically correct "data are" in favor of the colloquial "data is".

A more detailed picture of the incoming data and possible clock signals out of the DPLL are shown in Fig. 19.2. The possible clock signals are labeled XOR *clock* and PFD (phase frequency detector) *clock*, corresponding to the type of phase[2] detector (PD) used. For the XOR PD, the rising edge of the clock occurs in the center of the data, while for the PFD, the rising edge occurs at the beginning of the data. The phase of the clock signal is determined by the PD used[3].

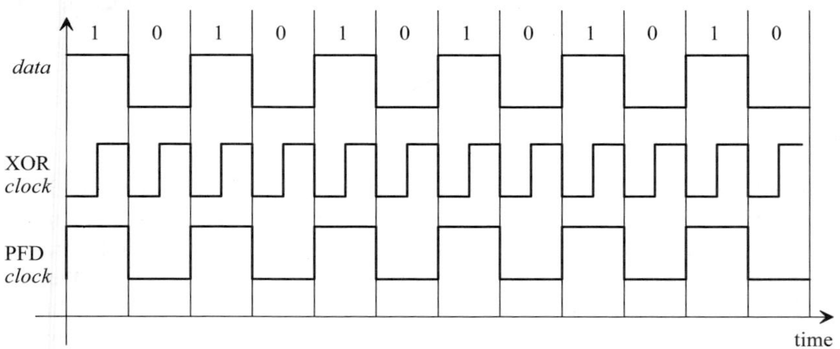

Figure 19.2 Data input to DPLL in lock and possible clock outputs using the XOR phase detector and PFD.

A block diagram of a DPLL is shown in Fig. 19.3. The PD generates an output signal proportional to the time difference between the *data in* and the divided down clock, *dclock*. This signal is filtered by a loop filter. The filtered signal, V_{inVCO}, is connected to the input of a voltage-controlled oscillator (VCO). Each one of these blocks is discussed in detail in the following sections. Once each block is understood, we will put them together and discuss the operation of the DPLL.

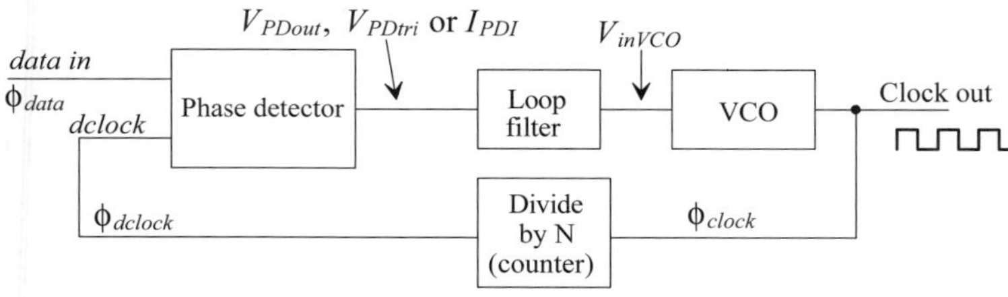

Figure 19.3 Block diagram of a digital phase-locked loop.

---

2 A more correct name for digital applications is time difference detector (TDD).

3 The PFD is rarely used in clock recovery applications. Figure 19.2 simply illustrates the different phase relations resulting from an XOR PD or PFD in a locked DPLL.

19.1 The Phase Detector

The first component in our DPLL is the phase detector. The two types of phase detectors, an XOR gate and a phase frequency detector (PFD), have significantly different characteristics. It is therefore very important to understand their performance capabilities and limitations. Selection of the type of phase detector is the first step in a DPLL design.

19.1.1 The XOR Phase Detector

The XOR PD is simply an exclusive OR gate. When the output of the XOR is a pulse train with a 50% duty cycle (square wave), the DPLL is said to be in lock; or in other words, the clock signal out of the DPLL is synchronized to the incoming data, provided the following conditions are met. Consider the XOR PD shown in Fig. 19.4. Let's begin by assuming that the incoming data is a string of zeros and that a divide by two is used in the feedback loop. The output of the phase detector is simply a replica of the *dclock* signal. Since the *dclock* signal has a 50% duty cycle, it would appear that the DPLL is in lock. If a logic "1" is suddenly applied, there is no way to know if the clock signal is synchronized (the clock rising edge coincides with the center of the data bit) to the data. This leads to the first characteristic of an XOR PD;

1. The incoming data must have a minimum number of transitions over a given time interval.

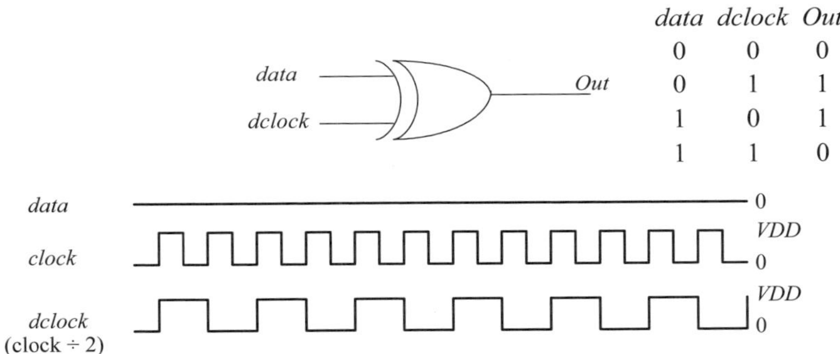

Figure 19.4 Operation of the XOR phase detector.

Now consider the situation when the output of the phase detector, with a string of zeros as the *data* input, is applied to a simple RC low-pass filter (Fig. 19.5). If RC >> period of the clock signal, the output of the filter is simply *VDD*/2. This leads to the second characteristic of the XOR phase detector.

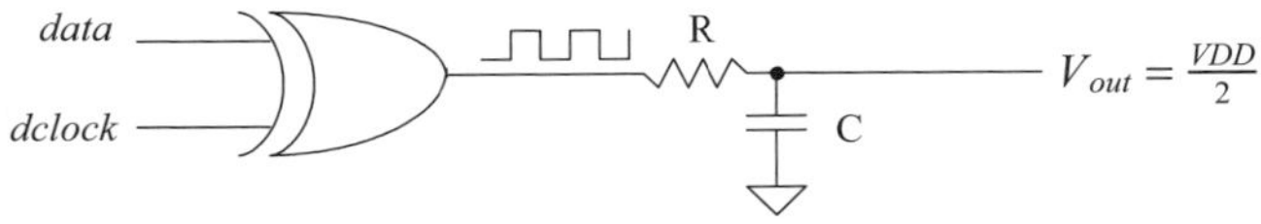

Figure 19.5 How the filtered output of the phase detector becomes VDD/2.

2. With no input data, the filtered output of the phase detector is *VDD*/2.

The voltage out of the loop filter is connected to the input of the VCO, as seen in Fig. 19.3. Consider the typical characteristics of a VCO shown in Fig. 19.6. The frequency of the square wave output of the VCO is f_{center} when V_{in} ($= V_{center}$) is *VDD*/2 (typically). The other two frequencies of interest are the minimum and maximum oscillator frequencies, $f_{\min}$ and $f_{\max}$ possible, with input voltage $V_{\min}$ and $V_{\max}$, respectively. It is important that the VCO continues to oscillate with no input data. Normally, the VCO is designed so that the nominal data input rate and the VCO center frequency are the same. This minimizes the time it takes the DPLL to lock (and is critical for proper operation of the XOR PD).

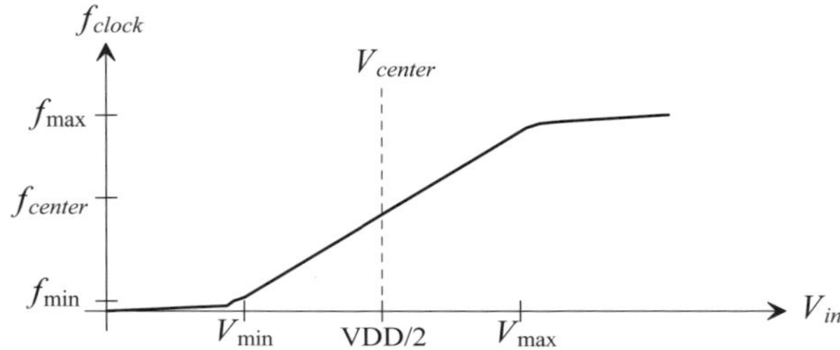

Figure 19.6 Output frequency of VCO versus input control voltage.

Now let's consider an example input to the phase-detector and corresponding output. The data shown in Fig. 19.7 is leading the *dclock* signal. The corresponding output of the phase detector is also shown. If this output is applied to a low-pass filter, the result is an average voltage less than *VDD*/2. This causes the VCO frequency to decrease until the edge of the *dclock* is centered on the data.

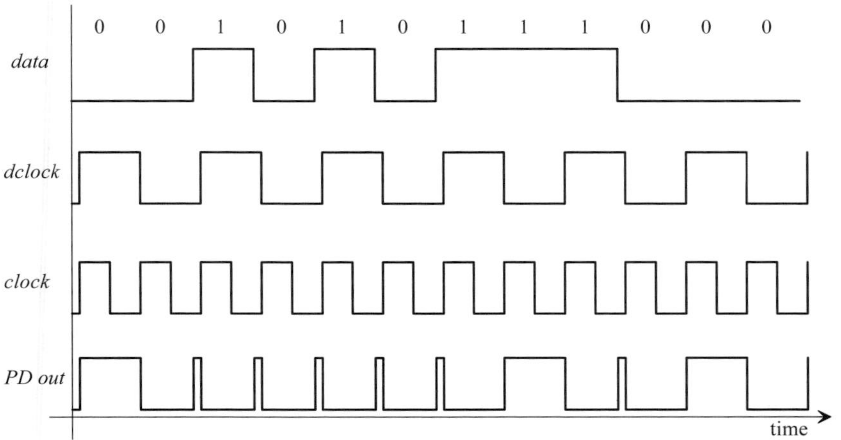

Figure 19.7 Possible XOR phase-detector inputs and the resulting output.

3. The time it takes the loop to lock depends on the data pattern input to the DPLL and the loop-filter characteristics.

Since the output of the PD is averaged, or more correctly integrated, noise injected into the data stream (a false bit) can be rejected. A fourth characteristic is:

4. The XOR DPLL has good noise rejection.

Another important characteristic of a DPLL is whether or not it will lock on a harmonic of the input data. The XOR DPLL will lock on harmonics of the data. To prove that this is indeed the case, consider replacing any of the clock signals of Figs. 19.2, 19.4, or 19.7 with a clock at twice or half the frequency. The average of the waveforms will remain essentially the same. A fifth characteristic of a DPLL using an XOR gate is:

5. The VCO operating frequency range should be limited to frequencies much less than $2f_{clock}$ and much greater than $0.5f_{clock}$, where f_{clock} is the nominal clock frequency for proper lock with a XOR PD.

The loop filter used with this type of PD is a simple RC low-pass filter, as shown in Fig. 19.5. Since the output of the PD is oscillating, the output of the filter shows a ripple as well, even when the loop is locked. This modulates the clock frequency, an unwanted characteristic of a DPLL using the XOR PD. This characteristic can be added to our list:

6. A ripple on the output of the loop filter with a frequency equal to the clock frequency modulates the control voltage of the VCO.

To characterize the phase detector (see Fig. 19.8), we can define the time difference between the rising edge of the *dclock* and the beginning of the *data* as Δt. The phase difference between the *dclock* and *data*, $\phi_{data} - \phi_{dclock}$, is given by

$$\Delta\phi = \phi_{data} - \phi_{dclock} = \frac{\Delta t}{T_{dclock}} \cdot 2\pi \text{ (radians)} \tag{19.1}$$

or, in terms of the DPLL output clock frequency,

$$\Delta\phi = \frac{\Delta t}{2T_{clock}} \cdot 2\pi \tag{19.2}$$

$$f_{clock} = \frac{1}{T_{clock}} = 2f_{dclock} = \frac{2}{T_{dclock}} \tag{19.3}$$

When the loop is locked, the *clock* rising edge is centered on the data; the time difference, Δt, between the *dclock* rising edge and the beginning of the data is simply $T_{clock}/2$ or $T_{dclock}/4$ (see Fig. 19.8c). Therefore, the phase difference between *dclock* and the *data*, under locked conditions, may be written as

$$\Delta\phi = \frac{\pi}{2} \tag{19.4}$$

Note that the phase difference between *clock* and *data* when in lock is π. The average voltage out of the phase detector (Fig. 19.8) may be expressed by

$$V_{PDout} = VDD \cdot \frac{\Delta\phi}{\pi} = K_{PD} \cdot \overbrace{\Delta\phi}^{input} \tag{19.5}$$

where the gain of the PD may be written as

$$K_{PD} = \frac{VDD}{\pi} \text{ (V/radians)} \tag{19.6}$$

As an aid in the understanding of these equations, consider the diagrams shown in Fig. 19.8. If the edges of the clock and data are coincident in time (Fig. 19.8a), the XOR output, V_{PDout}, is 0 V and the phase difference is 0. The loop filter averages the output of the PD and causes the VCO to lower its output frequency. This causes $\Delta\phi$ to increase, thus increasing V_{PDout}. Depending on the selection of the loop filter, this increase could cause the rising edges to increase beyond, or *overshoot*, the desirable center point, as shown in Fig. 19.8b. In this case, the phase difference is $-\frac{3}{4}\pi$, and V_{PDout} is $\frac{3}{4}VDD$. Figure 19.8c shows the condition when the loop is in lock and the phase difference is $\pi/2$.

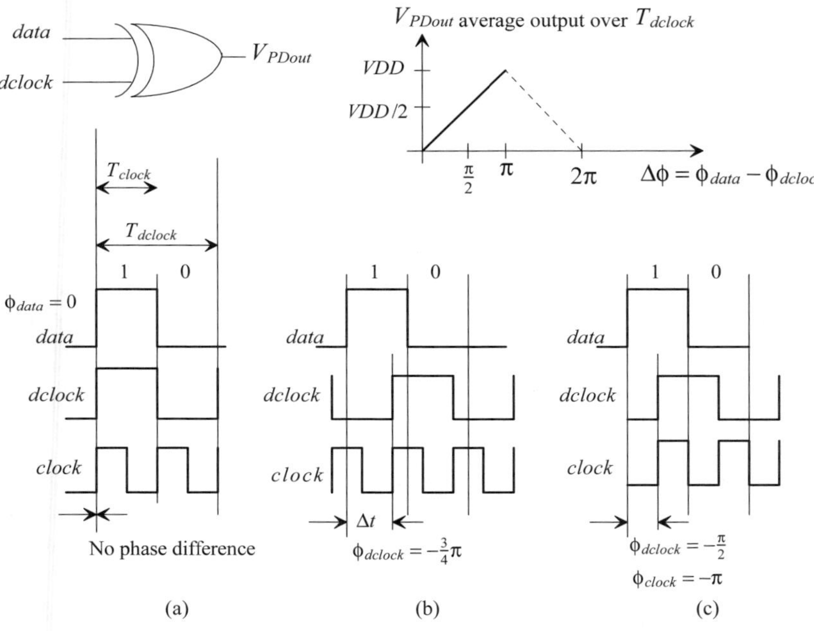

Figure 19.8 XOR PD output for various inputs (assuming input data are a string of alternating ones and zeros).

Here we should point out the importance of the VCO center frequency, f_{center}, being equal to the desired clock frequency. If $f_{center} = f_{clock}$, the VCO control voltage, depending on the loop filter, will look similar to Fig. 19.9a during acquisition (the loop trying to lock). If these frequencies are not equal, the control voltage will oscillate, causing the clock to move, in time, around some other point than the center of the data bit (Fig. 19.9b). The actual control voltages will look different from those portrayed in Fig. 19.9 because the VCO control voltage depends on the input data pattern, as discussed earlier.

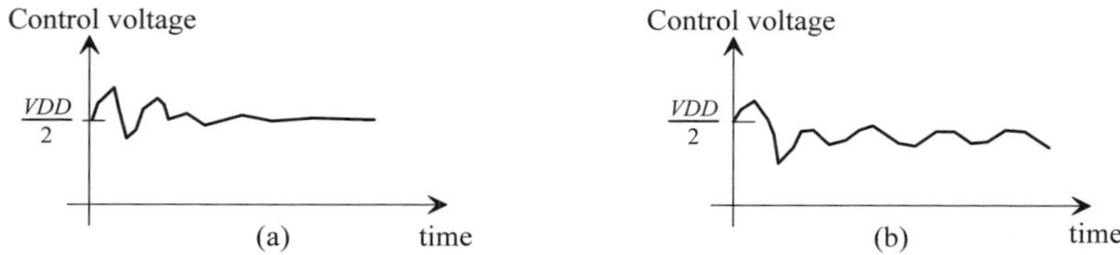

Figure 19.9 Average output voltage of phase detector during acquisition.

To summarize the design criteria of the VCO with the XOR PD, we desire that:

1. The center frequency, f_{center}, should equal the clock frequency when the VCO control voltage is *VDD*/2.

2. The maximum and minimum oscillation frequencies, f_{max} and f_{min}, of the VCO should be selected to avoid locking on harmonics of the input data.

3. The VCO duty cycle is 50%. If this is not the case, the DPLL will have problems locking, or once locked the clock will jitter (move around in time).

19.1.2 The Phase Frequency Detector

A schematic diagram of the phase frequency detector is shown in Fig. 19.10. The output of the PFD depends on both the phase and frequency of the inputs. This type of phase detector is also termed a sequential phase detector. It compares the leading edges of the *data* and *dclock*. A *dclock* rising edge cannot be present without a data rising edge. To aid in understanding the PFD, consider the examples shown in Fig. 19.11. The first thing we notice is that the *data* pulse width and the *dclock* pulse width do not matter. If the rising edge of the *data* leads the *dclock* rising edge (Fig. 19.11a), the "up" output of the phase detector goes high, while the *Down* output remains low. This causes the *dclock* frequency to increase, having the effect of moving the edges closer together. When the *dclock* signal leads the data (Fig. 19.11b), *Up* remains low, while the *Down* goes high a time equal to the phase difference between *dclock* and *data*. Figure 19.11c shows the condition of the locked loop. Notice that, unlike the XOR PD, the outputs remain low when the loop is locked. Again, several characteristics of the PFD can be described:

1. A rising edge from the *dclock* and *data* must be present when making a phase comparison.

2. The widths of the *dclock* and the *data* are irrelevant.

3. The PFD will not lock on a harmonic of the data.

4. The outputs (*Up* and *Down*) of the PFD are both logic low when the loop is in lock, eliminating ripple on the output of the loop filter.

5. This PFD has poor noise rejection; a false edge on either the *data* or the *dclock* inputs drastically affects the output of the PFD.

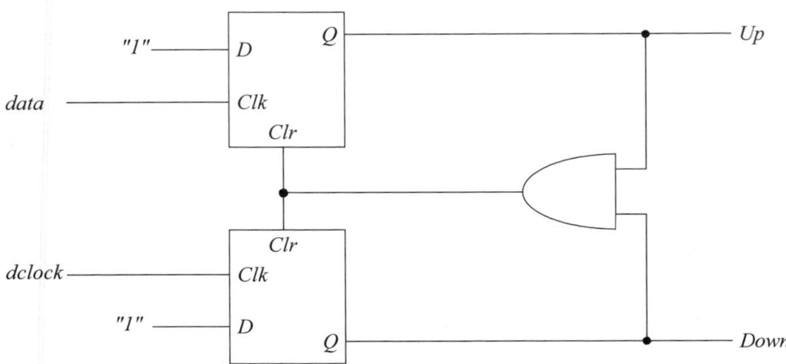

Figure 19.10 Phase frequency detector (PFD).

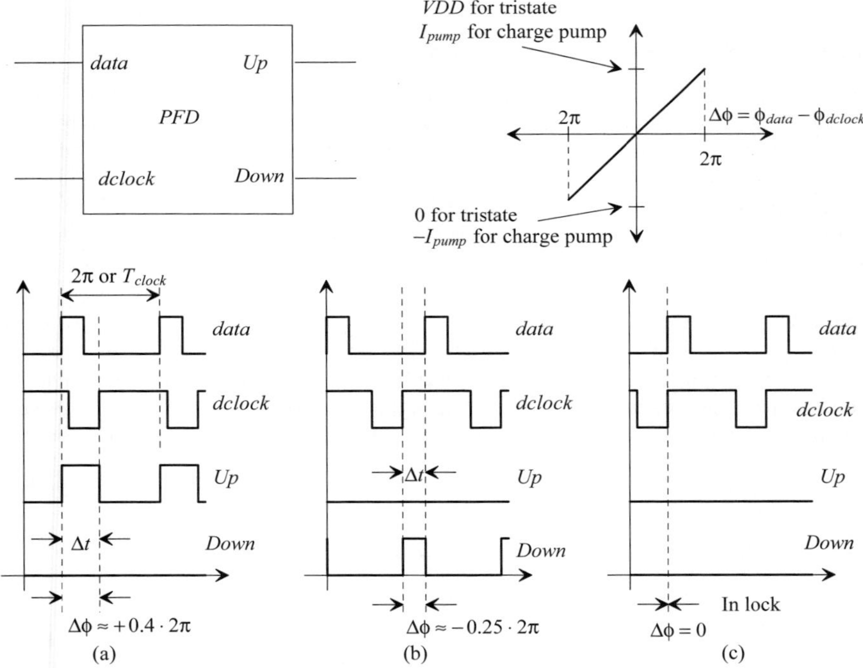

Figure 19.11 PFD phase-detector inputs and outputs.

The output of the PFD should be combined into a single output for driving the loop filter. There are two methods of doing this, both of which are shown in Fig. 19.12. The first method is called a *tri-state* output. When both signals, *Up* and *Down*, are low, both MOSFETs are off and the output is in a high-impedance state. If the *Up* signal goes high, M2 turns on and pulls the output up to *VDD*, while if the *Down* signal is high, the output is pulled low through M1. The main problem that exists with this configuration is that power supply variations can significantly affect the output voltage when M2 is on. The effect is to modulate the VCO control voltage. This wasn't as big a problem when the XOR PFD was used due to the averaging taking place.

The second configuration shown in this figure is the so-called *charge pump*. MOS current sources are placed in series with M1 and M2. When the PFD *Up* signal goes high, M2 turns on, connecting the current source to the loop filter. Because the current source can be made insensitive to supply variations, modulation of the VCO control voltage is absent.

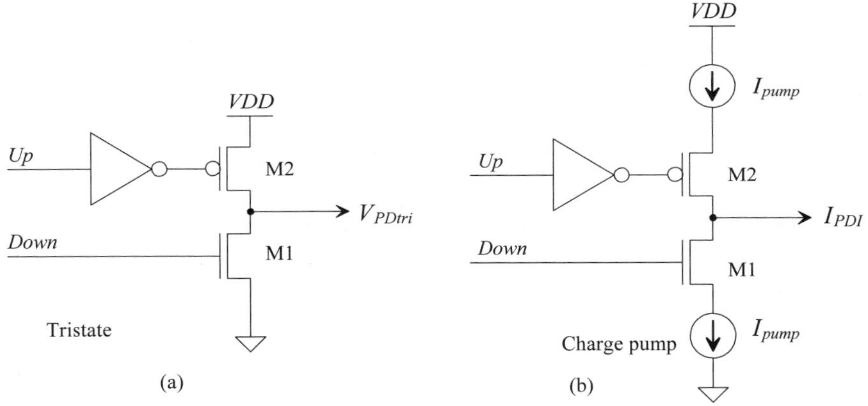

Figure 19.12 (a) Tri-state and (b) charge pump outputs of the PFD.

We can characterize the PFD in much the same manner as we did the XOR PD. We will assume that f_{clock} and f_{dclock} are equal in this analysis. Therefore, the feedback loop, Fig. 19.3, divides by one ($N = 1$). If we again assume that the time difference between the rising edges of the *data* and *dclock* is labeled Δt and the time between leading edges of the *clock* (or the time differences between the leading edges of the *data* since we must have both leading edges present to do a phase comparison) is labeled T_{clock}, then we can write the phase ($T_{clock} = T_{dclock}$) as

$$\Delta\phi = \frac{\Delta t}{T_{clock}} \cdot 2\pi \text{ (radians)} \tag{19.7}$$

The phase difference, $\Delta\phi$, is zero when the loop is in lock. The output voltage of the PFD using the tri-state output configuration (see Fig. 19.11) is

$$V_{PDtri} = \frac{VDD - 0}{4\pi} \cdot \Delta\phi = K_{PDtri} \cdot \Delta\phi \tag{19.8}$$

where the gain is,

$$K_{PDtri} = \frac{VDD}{4\pi} \text{ (volts/radian)} \tag{19.9}$$

If the output of the PFD uses the charge-pump configuration, the output current can be written (again see Fig. 19.11) as

$$I_{PDI} = \frac{I_{pump} - (-I_{pump})}{4\pi} \cdot \Delta\phi = K_{PDI} \cdot \Delta\phi \tag{19.10}$$

where

$$K_{PDI} = \frac{I_{pump}}{2\pi} \text{ (amps/radian)} \tag{19.11}$$

The loop filters used with tri-state and the charge-pump outputs are shown in Fig. 19.13. The first loop filter has a transfer function given by

$$V_{inVCO} = \frac{1 + j\omega R_2 C}{1 + j\omega(R_1 + R_2)C} \cdot V_{PDtri} = K_F \cdot V_{PDtri} \tag{19.12}$$

The charge-pump loop-filter transfer function is given (noting the input variable is a current while the output is a voltage) by

$$V_{inVCO} = I_{PDI} \cdot \frac{1 + j\omega RC_1}{j\omega(C_1 + C_2) \cdot \left[1 + j\omega R\frac{C_1 C_2}{C_1 + C_2}\right]} = K_F \cdot I_{PDI} \tag{19.13}$$

To qualitatively understand how these loop filters work, let's begin by considering the loop filter for use with the tri-state output. For slow variations in the phase difference, the filter acts like an integrator averaging the output of the PD. For fast variations, however, the filter looks like a resistive divider without any integration. This allows the loop filter to track fast variations in the time difference between the rising edges. A similar discussion can be made for the loop filter used with the charge pump. For slow variations in the phase, the current, I_{pump}, linearly charges C_1 and C_2. This gives an averaging effect. For fast variations, the charge pump simply drives the resistor R (assuming C_2 is small), eliminating the averaging and allowing the VCO to track quickly moving variations in the input data.

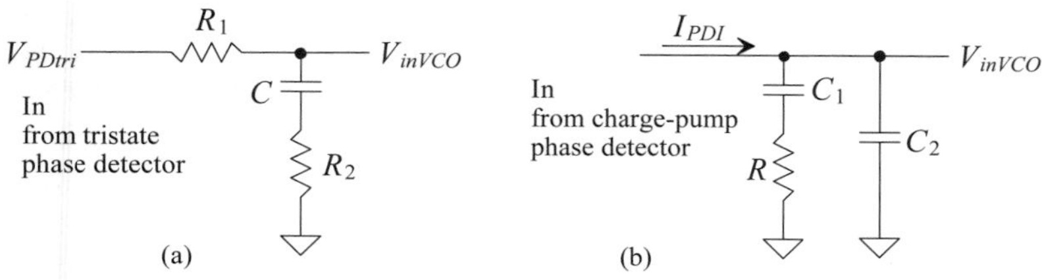

Figure 19.13 Loop filters for (a) tristate output and (b) charge-pump output.

The design requirements of the VCO used with the PFD can be much more relaxed than those for the XOR PD; therefore, it is preferred over the XOR PD. Because the output of the PD can be a voltage from 0 to *VDD,* the requirement that $f_{center} = f_{clock}$ is not present, although it is still a good idea to design the VCO so that this is true. The oscillation range, $f_{max} - f_{min}$, is not limited by the harmonics of f_{clock} because the PFD will not lock on a harmonic of the clock frequency. The VCO clock duty cycle is irrelevant with the PFD because the PD looks only at rising edges. In high-speed or data communications applications, however, one may be forced to use the XOR PD.

19.2 The Voltage-Controlled Oscillator

The gain of the voltage-controlled oscillator is simply the slope of the curves given in Fig. 19.6. This gain can be written as

$$K_{VCO} = 2\pi \cdot \frac{f_{\max} - f_{\min}}{V_{\max} - V_{\min}} \text{ (radians/s} \cdot \text{V)} \tag{19.14}$$

The VCO output frequency, f_{clock}, is related to the VCO input voltage (see Fig. 19.6) by

$$\omega_{clock} = 2\pi \cdot f_{clock} = K_{VCO} \cdot V_{inVCO} + \omega_o \text{ (radians/s)} \tag{19.15}$$

where ω_o is a constant. However, the variable we are feeding back is not frequency but phase (hence the name of the circuit). The phase of the VCO clock output is related to f_{clock} by

$$\phi_{clock} = \int \omega_{clock} \cdot dt = \frac{K_{VCO}}{j\omega} \cdot V_{inVCO} \text{ (radians)} \tag{19.16}$$

where this signal can be related to the ϕ_{dclock} by

$$\phi_{dclock} = \frac{1}{N} \cdot \phi_{clock} = \beta \cdot \phi_{clock} \tag{19.17}$$

where N is the divide by count and β is the feedback factor.

19.2.1 The Current-Starved VCO

The current-starved VCO is shown schematically in Fig. 19.14. Its operation is similar to the ring oscillator discussed earlier. MOSFETs M2 and M3 operate as an inverter, while MOSFETs M1 and M4 operate as current sources. The current sources, M1 and M4, limit the current available to the inverter, M2 and M3; in other words, the inverter is starved for current. The drain currents of MOSFETs M5 and M6 are the same and are set by the input control voltage. The currents in M5 and M6 are mirrored in each inverter/current source stage. An important property of the VCO used in any of the CMOS DPLLs discussed in this chapter is the input impedance. The filter configurations we have discussed rely on the fact that the input resistance of the VCO is practically infinite and the input capacitance is small compared to the capacitances present in the loop filter. Attaining infinite input resistance is usually an easy part of the design. For the charge-pump configuration, the input capacitance of the VCO can be added to C_2.

To determine the design equations for use with the current-starved VCO, consider the simplified schematic of one stage of the VCO shown in Fig. 19.15. The total capacitance on the drains of M2 and M3 is given by

$$C_{tot} = C_{out} + C_{in} = \overbrace{C'_{ox}(W_pL_p + W_nL_n)}^{C_{out}} + \overbrace{\frac{3}{2}C'_{ox}(W_pL_p + W_nL_n)}^{C_{in}} \quad (19.18)$$

which is simply the output and input capacitances of the inverter. This equation can be written in a more useful form as

$$C_{tot} = \frac{5}{2}C'_{ox}(W_pL_p + W_nL_n) \quad (19.19)$$

The time it takes to charge C_{tot} from zero to V_{SP} with the constant-current I_{D4} is given by

$$t_1 = C_{tot} \cdot \frac{V_{SP}}{I_{D4}} \quad (19.20)$$

while the time it takes to discharge C_{tot} from VDD to V_{SP} is given by

$$t_2 = C_{tot} \cdot \frac{VDD - V_{SP}}{I_{D1}} \quad (19.21)$$

If we set $I_{D4} = I_{D1} = I_D$ (which we will label $I_{Dcenter}$ when $V_{inVCO} = VDD/2$), then the sum of t_1 and t_2 is simply

$$t_1 + t_2 = \frac{C_{tot} \cdot VDD}{I_D} \quad (19.22)$$

The oscillation frequency of the current-starved VCO for N (an odd number ≥ 5) of stages is

$$f_{osc} = \frac{1}{N(t_1 + t_2)} = \frac{I_D}{N \cdot C_{tot} \cdot VDD} \quad (19.23)$$

which is $= f_{center}(@V_{inVCO} = VDD/2 \text{ and } I_D = I_{Dcenter})$

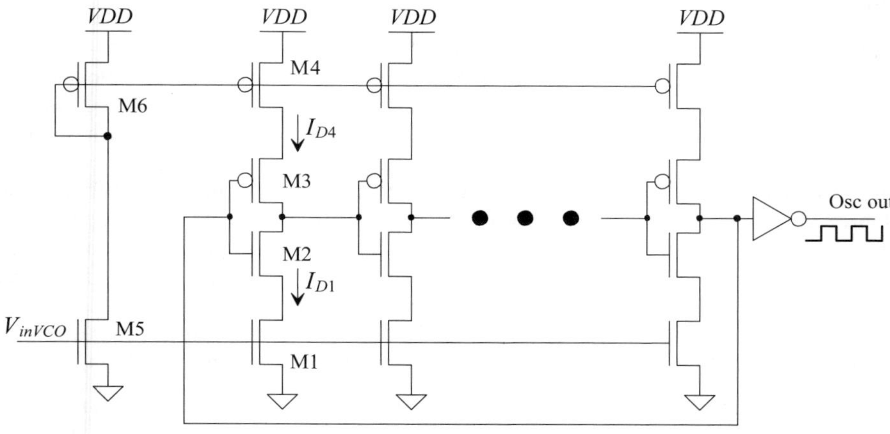

Figure 19.14 Current-starved VCO.

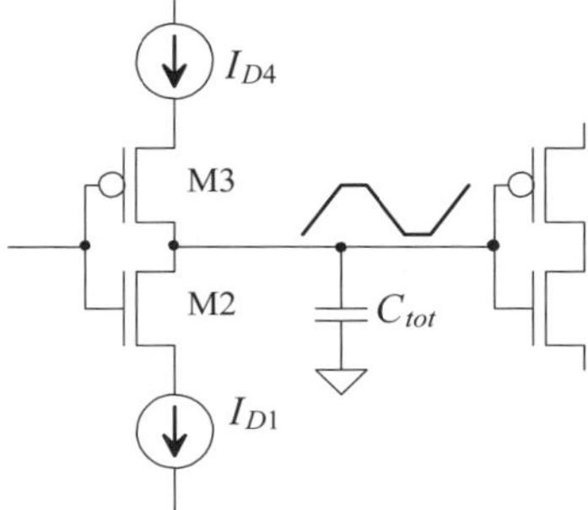

Figure 19.15 Simplified view of a single stage of the current-starved VCO.

Equation (19.23) gives the center frequency of the VCO when $I_D = I_{Dcenter}$. The VCO stops oscillating, neglecting subthreshold currents, when $V_{inVCO} < V_{THN}$. Therefore, we can define

$$V_{\min} = V_{THN} \text{ and } f_{\min} = 0 \tag{19.24}$$

The maximum VCO oscillation frequency, f_{max}, is determined by finding I_D when $V_{inVCO} = VDD$. At the maximum frequency then, $V_{max} = VDD$.

The output of the current-starved VCO shown in Fig. 19.14 is normally buffered through one or two inverters. Attaching a large load capacitance on the output of the VCO can significantly affect the oscillation frequency or lower the gain of the oscillator enough to kill oscillations altogether.

The average current drawn by the VCO is

$$I_{avg} = N \cdot \frac{VDD \cdot C_{tot}}{T} = N \cdot VDD \cdot C_{tot} \cdot f_{osc} \tag{19.25}$$

or

$$I_{avg} = I_D \tag{19.26}$$

The average power dissipated by the VCO is

$$P_{avg} = VDD \cdot I_{avg} = VDD \cdot I_D \tag{19.27}$$

If we include the power dissipated by the mirror MOSFETs, M5 and M6, the power is doubled from that given by Eq. (19.27), assuming that $I_D = I_{D5} = I_{D6}$. For low-power dissipation we must keep I_D low, which is equivalent to stating that for low-power dissipation we must use a low-oscillation frequency.

Example 19.1
Design a current-starved VCO with f_{center} = 100 MHz in the short-channel process. Simulate the design using SPICE.

We begin by calculating the total capacitance, C_{tot}. Using Eq. (19.19) and assuming the inverters, M2 and M3, are sized for equal drive, that is, $L_n = L_p = 1$, $W_n = 10$ and $W_p = 20$, the capacitance is

$$C_{tot} = \frac{5}{2} \cdot 25 \frac{fF}{\mu m^2} \cdot (10 \cdot 1 + 20 \cdot 1) \cdot \overbrace{(0.050\ \mu m)^2}^{\text{scale factor}} = 4.7\ fF$$

Let's use a center drain current of 10 μA based on the I_D - V_{GS} characteristics of the MOSFETs. The selection of this current is important. We want, when V_{inVCO} is *VDD*/2, the oscillation frequency to be 100 MHz. Here, just to show the design procedure, we will simply adjust the V_{inVCO} to set the 100 MHz output frequency.

The number of stages, using Eq. (19.23), is given by

$$N = \frac{I_D}{f_{osc} \cdot C_{tot} \cdot VDD} = \frac{10\ \mu A}{100\ MHz \cdot 4.7\ fF \cdot 1\ V} \approx 21$$

The simulation results are shown in Fig. 19.16. For an output frequency of 100 MHz, the input voltage, V_{inVCO} , was set to 450 mV. The big problem with this design is that the output oscillation frequency is not linearly related to the control voltage (easy to verify using the simulation that generated Fig. 19.16). Having a nonlinear VCO gain can greatly reduce the quality of the performance of the DPLL (*this is important*). The output of the phase-locked loop can jitter (move around) or may not lock at all when the VCO gain is nonlinear. ■

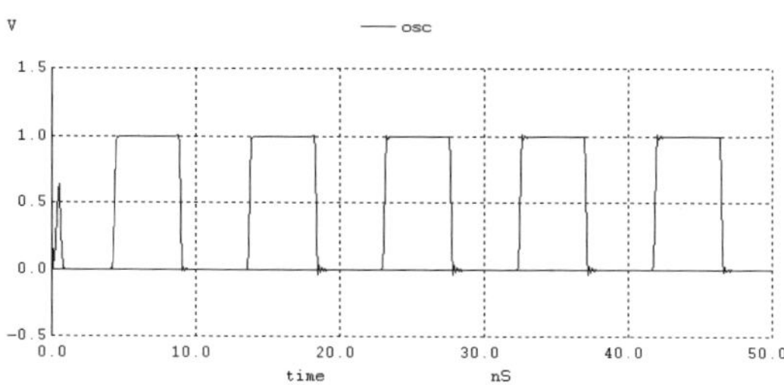

Figure 19.16 Output of the oscillator designed in Ex. 19.1.

Linearizing the VCO's Gain

After studying Fig. 19.14, we see that applying V_{inVCO} directly to the gate of M5 causes the current in M5 (and M6) to be nonlinear. In the long-channel case the drain current of a MOSFET is related to the square of the MOSFET's V_{GS}. To make the current in a MOSFET linearly related to the VCO's input voltage, consider the circuit seen in Fig. 19.17. The width of M5R is made wide so that its V_{GS} is always (independent of V_{inVCO}) approximately V_{THN}. Note that the current in M6R is mirrored over to M6 and M5 to control the current used in the current-starved VCO. Figure 19.18 shows the output of the oscillator (and its transfer curves) in Fig. 19.16 if the linearizing scheme in Fig. 19.17 is used with *R* of 10k and a wide device (M5R) with a size of 100/1. The gain of the VCO is $K_{VCO} = 2\pi \cdot 25 MHz/100 mV = 1.57 \times 10^9\ radians/V \cdot s$. We'll use this VCO in examples later in this chapter. Note that the gain of this *VCO* is rather high. A 10 mV voltage variation in V_{inVCO} gives a 2.5 MHz variation in the output frequency (e.g., the output frequency varies from 100 to 102.5 MHz). In the time domain, the variation in the period (jitter) is then $1/100\ MHz - 1/102.5\ MHz = 244\ ps$ (roughly 2.5% of the ideal 10 ns period). *For any low jitter PLL design, a VCO with low gain must be used.*

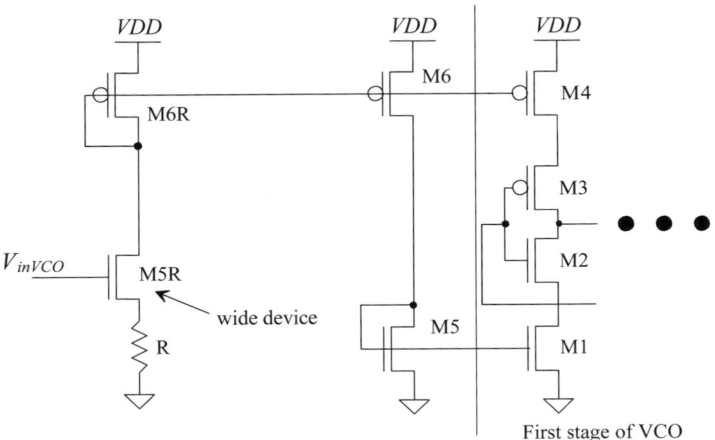

Figure 19.17 Linearizing the current in a current-starved VCO.

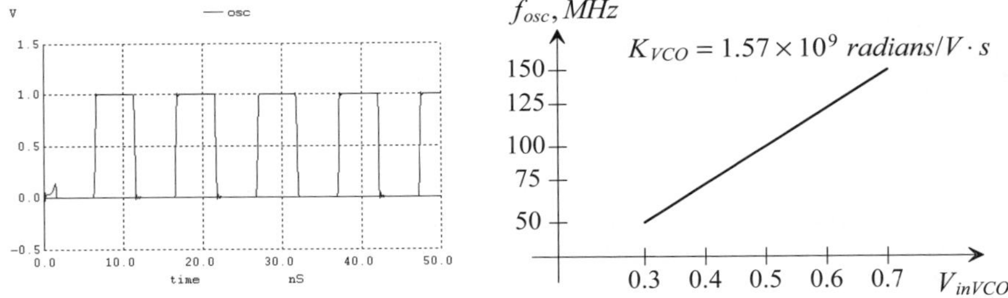

Figure 19.18 Gain of the VCO in Fig. 19.16 after linearizing with Fig. 19.17.

19.2.2 Source-Coupled VCOs

Another variety of VCO, source-coupled VCOs, is shown in Fig. 19.19. These VCOs can be designed to dissipate less power than the current-starved VCO of the last section for a given frequency. The major disadvantage of these configurations is the need for a capacitor, something that may not be available in a single-poly pure digital process without using parasitics for example, a metal1 to metal2 capacitor, and a reduced output voltage swing. However, this configuration is useful when the VCO center frequency is set by an external capacitor; that is, the capacitor shown in the figure is bonded out.

To understand the operation, let's consider the NMOS source-coupled VCO of Fig. 19.19a. The operation of the CMOS source-coupled VCO of Fig. 19.19b is identical to (a) except for the fact that the load MOSFETs M3 and M4 pull the outputs to $VDD - V_{THN}$ (for the NMOS-load VCO) and VDD (for the PMOS-load VCO). The buffer in Fig. 18.17 of the last chapter can be used to restore full logic levels.

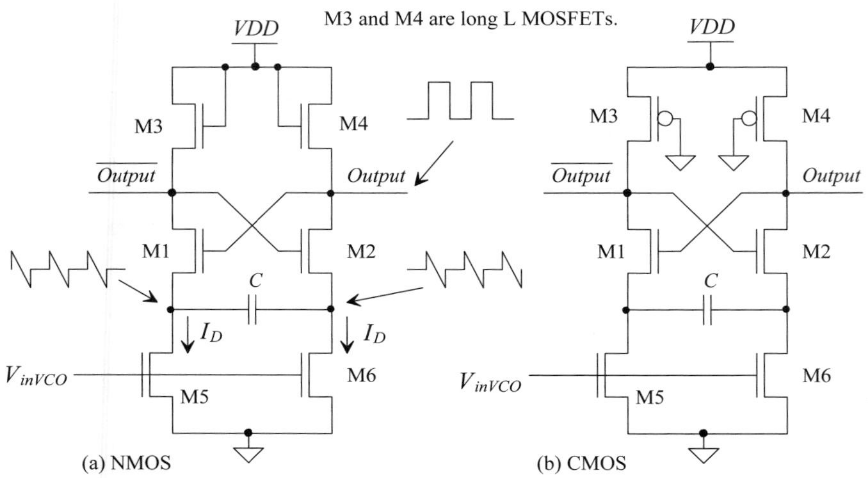

Figure 19.19 Source coupled voltage-controlled oscillators (also known as source coupled multivibrators).

For the circuit in Fig. 19.19a, MOSFETs M5 and M6 behave as constant-current sources sinking a current I_D. MOSFETs M1 and M2 operate as switches. If M1 is off and M2 is on, the drain of M1 is pulled to $VDD - V_{THN}$ by M3. Since the gate of M2 is at $VDD - V_{THN}$, the source and drain (the *Output*) of M2 are approximately $VDD - 2 \cdot V_{THN}$. This is the minimum output voltage. The output voltage swing is limited to V_{THN}. A simplified schematic shown in Fig. 19.20 with M1 off and M2 on is helpful in determining the oscillator frequency. The *Output* gate of M1 is approximately $VDD - 2 \cdot V_{THN}$ and is held at this voltage through M2 until M1 turns on and M2 turns off. Initially, at the moment when M1 turns off and M2 turns on, point X is $VDD - V_{THN}$. The current through C, I_D, causes point X to discharge down toward ground. When point X gets down to $VDD - 3 \cdot V_{THN}$, M1 turns on and M2 turns off. In other words, the voltage at point X changed a

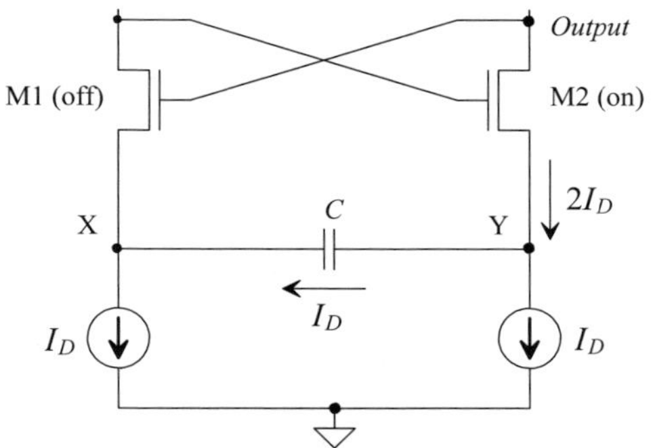

Figure 19.20 Simplified schematic of source coupled oscillator, M1 is on and M2 is off.

total of $2 \cdot V_{THN}$ before switching took place. The time it takes point X to change $2 \cdot V_{THN}$ is given by

$$\Delta t = C \cdot \frac{2 \cdot V_{THN}}{I_D} \tag{19.28}$$

Since the circuit is symmetrical, two of these discharge times are needed for each cycle of the oscillation. The frequency of oscillation is given by

$$f_{osc} = \frac{1}{2\Delta t} = \frac{I_D}{4 \cdot C \cdot V_{THN}} \tag{19.29}$$

The waveforms at the points X, Y, and *Output* are shown in Fig. 19.21 for continuous time operation.

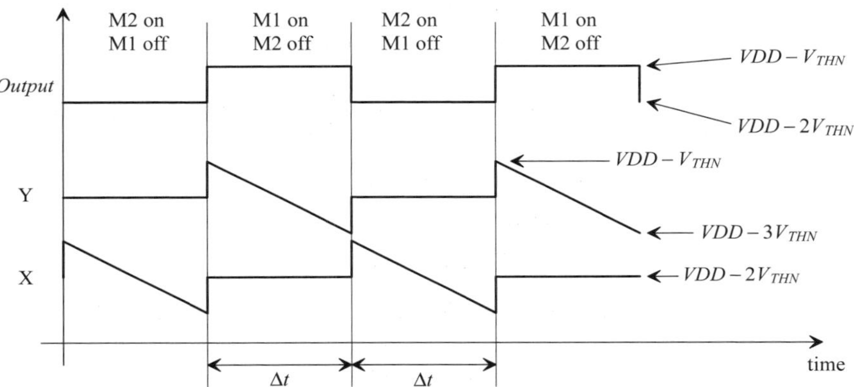

Figure 19.21 Voltage waveforms for the NMOS source-coupled VCO.

19.3 The Loop Filter

The loop filter is the brain of the DPLL. In this section, we discuss how to select the loop-filter values in order to keep the DPLL from oscillating (i.e., keep the V_{inVCO} voltage from oscillating, causing the frequency out of the VCO to wander). If the loop-filter values are not selected correctly, it may take the loop too long to lock, or once locked, small variations in the input data may cause the loop to unlock.

In the following discussion, we will be concerned with the *pull-in range* and the *lock range*. The pull-in range, $\pm\Delta\omega_P$, is defined as the range of input frequencies that the DPLL will lock to. The time it takes the loop to lock is labeled T_P and can be a very long time. If the center frequency of the DPLL is 10 MHz and the pull-in range is 1 MHz, the DPLL will lock on an input frequency from 9 to 11 MHz in a time T_P (assuming $N = 1$). The lock range, $\pm\omega_L$, is the range of frequencies in which the DPLL locks within one single beat note between the divided down output (*dclock*) and input (*data*) of the DPLL. *The operating frequency of the DPLL should be limited to the lock range for normal operation.* Once the DPLL is locked, it will remain locked as long as abrupt frequency changes, $\Delta\omega$, in the input frequency (input frequency steps) over a time interval t are much smaller than the natural frequency of the system squared, that is, $\Delta\omega/t < \omega_n^2$.

19.3.1 XOR DPLL

Consider the block diagram of the DPLL using the XOR DPLL shown in Fig. 19.22. The phase transfer function (neglecting the static or DC behavior) is given by

$$H(s) = \frac{\phi_{clock}}{\phi_{data}} = \frac{K_{PD}K_F K_{VCO}}{s + \beta \cdot K_{PD}K_F K_{VCO}} \tag{19.30}$$

with $s = j\omega$ and the feedback factor, β, is

$$\beta = \frac{1}{N} \tag{19.31}$$

The transfer function of the loop filter is given by

$$K_F = \frac{1}{1 + j\omega RC} = \frac{1}{1 + sRC} \tag{19.32}$$

Substituting Eqs. (19.31) and (19.32) into Eq. 19.30 yields

$$H(s) = \frac{\phi_{clock}}{\phi_{data}} = \frac{K_{PD}K_{VCO} \cdot \frac{1}{1+sRC}}{s + \frac{1}{N} \cdot K_{PD}K_{VCO} \cdot \frac{1}{1+sRC}} \tag{19.33}$$

This is a second-order system. $H(s)$ can be rewritten as

$$H(s) = \frac{\frac{K_{PD}K_{VCO}}{RC}}{s^2 + \frac{s}{RC} + \frac{1}{N} \cdot \frac{K_{PD}K_{VCO}}{RC}} = \frac{f_{clock}}{f_{data}} = \frac{(K_{VCO}V_{inVCO})/2\pi + f_o}{f_{data}} \tag{19.34}$$

since $j\omega \cdot \phi = f$. The natural frequency, ω_n, of this system is given by

$$\omega_n = \sqrt{\frac{K_{PD}K_{VCO}}{N \cdot RC}} \tag{19.35}$$

and the damping ratio, ζ, is given by

$$\zeta = \frac{1}{2RC\omega_n} = \frac{1}{2} \cdot \sqrt{\frac{N}{K_{PD}K_{VCO} \cdot RC}} \tag{19.36}$$

The pull-in range is given by

$$\Delta\omega_P = \frac{\pi}{2} \cdot \sqrt{2\zeta\omega_n K_{VCO}K_{PD} - \omega_n^2} \tag{19.37}$$

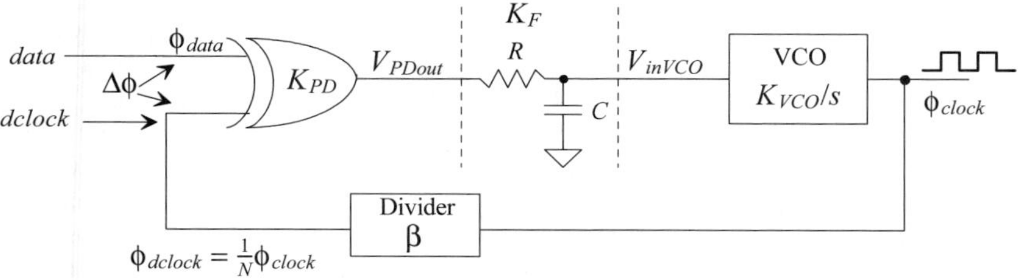

Figure 19.22 Block diagram of a DPLL using a XOR phase detector

while the pull-in time is given by

$$T_P = \frac{4}{\pi^2} \cdot \frac{\Delta\omega_{center}^2}{\zeta\omega_n^3} \tag{19.38}$$

The lock range of the loop is given by

$$\Delta\omega_L = \pi\zeta\omega_n = \frac{\pi}{2} \cdot \frac{1}{RC} \tag{19.39}$$

while the lock time is

$$T_L = \frac{2\pi}{\omega_n} \tag{19.40}$$

Probably the best way to understand how these equations affect the performance of a DPLL is through an example.

Example 19.2

Design and simulate the operation of a DPLL with the topology seen in Fig. 19.22 using the VCO from Fig. 19.18. Assume that the input data rate is 100 Mbits/s and has a format as in Fig. 19.7 (so the DPLL's output is a 100 MHz square wave).

Let's start off by drawing the schematic of the XOR phase detector, Fig. 19.23, and the divide by two circuit, Fig. 19.24. Notice that the XOR gate won't have zero output resistance, so the value of R used in the loop filter may need to be modified accordingly. The divide by two circuit is implemented using true single phase logic.

To calculate the values for the loop filter, let's write the gain of the VCO in Fig. 19.18 as

$$K_{VCO} = 1.57 \times 10^9 \; radians/V \cdot s$$

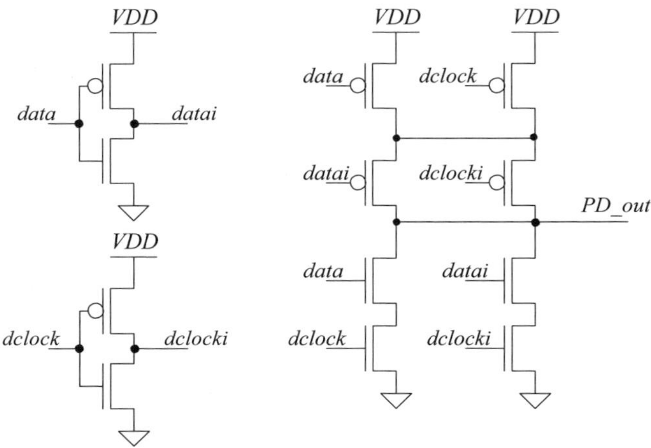

Figure 19.23 XOR phase detector.

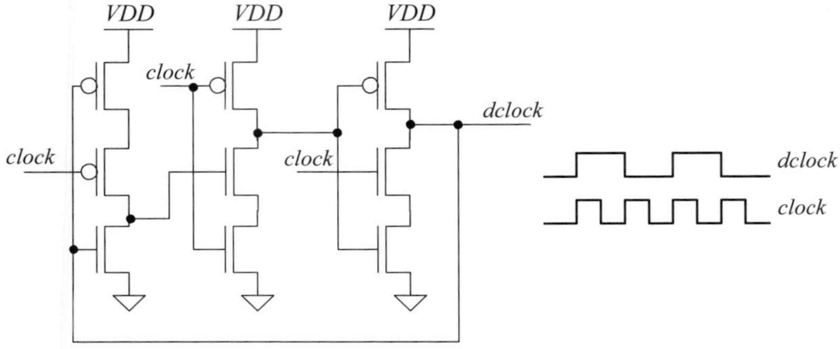

Figure 19.24 A divide by two circuit using true single phase clocking (TSPC) logic.

and the gain of the XOR phase detector

$$K_{PD} = \frac{VDD}{\pi} = \frac{1}{\pi} \text{ (V/radians)}$$

Let's begin by setting the damping ratio ζ to 1. Using Eq. (19.36) results in

$$1 = \frac{1}{2} \cdot \sqrt{\frac{2}{\frac{1}{\pi} \cdot (1.57 \times 10^9) RC}} \rightarrow RC = 500\ ps$$

clearly not practical! We must reduce the gain of the VCO. This means the VCO output frequency range must be reduced. The problem with reducing the output frequency range is that it becomes very difficult to fabricate the VCO in CMOS at the precise center oscillation frequency. Nonetheless, let's modify the VCO from Fig. 19.18 so that the output frequency range is limited.

Consider the schematic seen in Fig. 19.25 of the current generator portion of the VCO (see Fig. 19.17). Here, we've added a resistor to set the lower frequency range of the VCO (which makes the VCO very dependent on the power supply voltage). The range is still set by M5R. However, now we've increased the value of the resistor to limit the VCO's output frequency range. The resulting VCO gain, with the values seen in Fig. 19.25, appears in Fig. 19.26. Using this modified VCO gain in the above equation gives

$$RC = 5 \text{ ns}$$

The natural frequency of the second-order system is, from Eq. (19.36),

$$\omega_n = \frac{1}{2 \cdot RC \cdot \zeta} = 100 \times 10^6 \text{ radians/s}$$

The lock range is

$$\Delta\omega_L = \pi\zeta\omega_n = 314 \times 10^6 \text{ radians/s or } \Delta f_L = 50\ MHz$$

which is outside the VCO operating range. Therefore, this DPLL will lock over the entire VCO operating range, that is, from 95 to 105 Mbits/s. We do not need

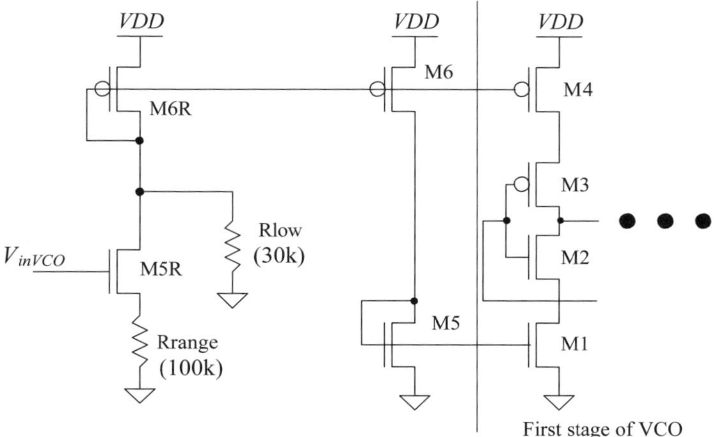

Figure 19.25 Limiting the current in a current-starved VCO.

to calculate the pull-in range since it is greater than the lock-in range. Again, the design of the oscillator is important in this DPLL. The lock time, T_L, of the DPLL is 40 ns. SPICE simulation results of this DPLL are seen in Fig. 19.27. The top waveform is the input control voltage of the VCO. The time that it takes the loop to pull-in the frequency and lock, from this figure, is roughly 250 ns. The lock time would be the time it takes the loop if the input frequency jumped after the loop was locked. Note how, in the bottom traces, the output clock rising edge is centered on the data. Notice that the VCO's control voltage varies by 200 mV. From Fig. 19.26, we can see a 5 MHz variation, Δf_{VCO}, in the VCO's output frequency. To estimate the jitter, Δt_{jitter}, in the output, we can use

$$\Delta t_{jitter} = \frac{1}{f_{center}} - \frac{1}{f_{center} + \Delta f_{VCO}} = \frac{1}{100\ MHz} - \frac{1}{105\ MHz} = 480\ ps \qquad (19.41)$$

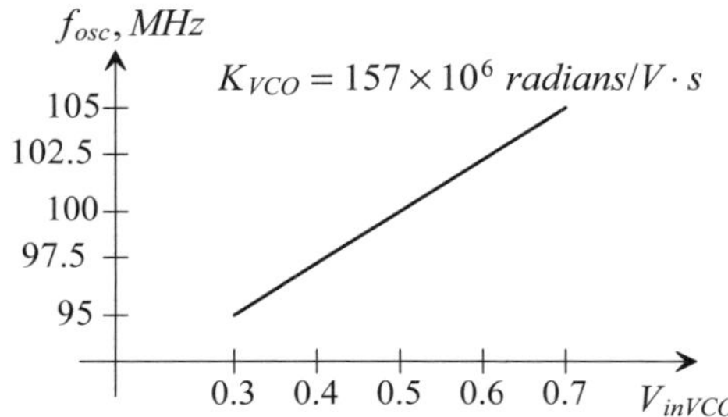

Figure 19.26 Gain of the VCO in Fig. 19.16 after linearizing and limiting with Fig. 19.25.

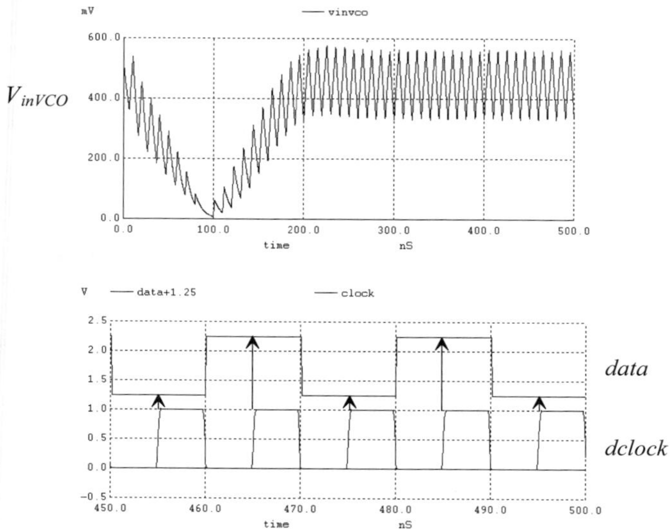

Figure 19.27 Simulating the DPLL in Ex. 19.2.

In this simulation (Fig. 19.27), we used the ideal input data, that is, a string of alternating ones and zeroes. Figure 19.28 shows what happens when the data is a one followed by seven zeroes. The rising edge of the clock is occurring a little offset from the center of the data. This error is sometimes called a *static phase error*. As mentioned earlier, when the input data is not changing, the VCO control voltage starts to wander back towards *VDD*/2.

Finally, *it's very important* to realize that if we were to increase the *RC* time constant of the filter, as seen above, the damping factor, ζ , would decrease and the loop would never lock. This is counter-intuitive and the reason we should *always use the design equations when designing DPLLs*. ■

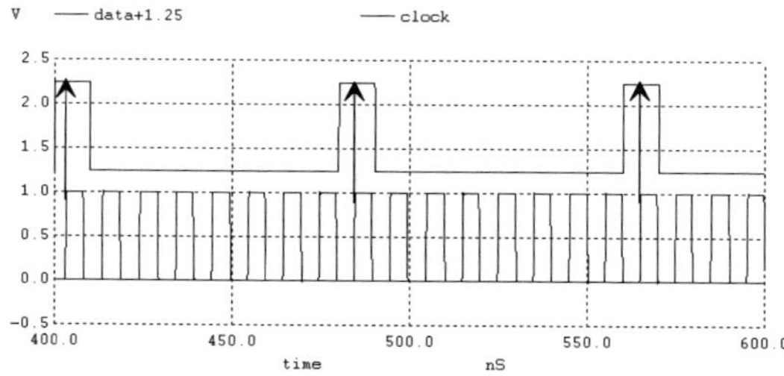

Figure 19.28 Showing how the DPLL doesn't lock up to the center of the data when the data isn't alternating ones and zeroes.

Adding a zero to the simple passive RC loop filter, Fig. 19.29 (called a passive lag loop filter), the loop-filter pole can be made small (and thus the gain of the VCO can be made larger) while at the same time a reasonable damping factor can be achieved. The result is an increase in the lock range of a DPLL using the XOR PD and a shorter lock time see Eqs. (19.37) – (19.40). The passive lag loop filter is, in most situations preferred over the simple RC. Again, as Ex. 19.2 showed, if the center frequency of the VCO doesn't match the input frequency, the clock will not align at π/2 (in the center of the data).

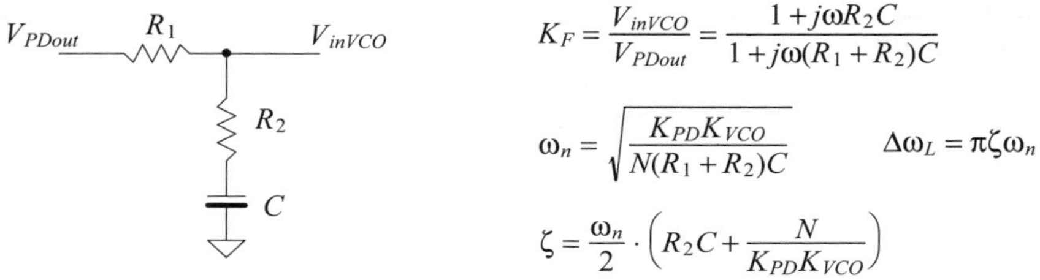

V_{PDout}
R_1
V_{inVCO}
R_2
C
$K_F = \frac{V_{inVCO}}{V_{PDout}} = \frac{1 + j\omega R_2 C}{1 + j\omega (R_1 + R_2) C}$
$\omega_n = \sqrt{\frac{K_{PD} K_{VCO}}{N(R_1 + R_2) C}}$
$\Delta\omega_L = \pi \zeta \omega_n$
$\zeta = \frac{\omega_n}{2} \cdot \left(R_2 C + \frac{N}{K_{PD} K_{VCO}} \right)$

Figure 19.29 Passive lag loop filter used to increase DPLL lock range.

Active PI Loop Filter

The clock misalignment encountered in a DPLL using an XOR PD and passive loop filter can be minimized by using the active proportional + integral (active PI) loop filter shown in Fig. 19.30. The integrator allows the V_{inVCO} voltage to move, and stay, away from *VDD*/2. This then eliminates the static phase error (the clock will align to the middle of the data). The transfer function of this filter is given by

$$K_F = \frac{1 + sR_2 C}{sR_1 C} \tag{19.42}$$

The natural frequency of the resulting second-order system is given by

$$\omega_n = \sqrt{\frac{K_{PD} K_{VCO}}{NR_1 C}} \tag{19.43}$$

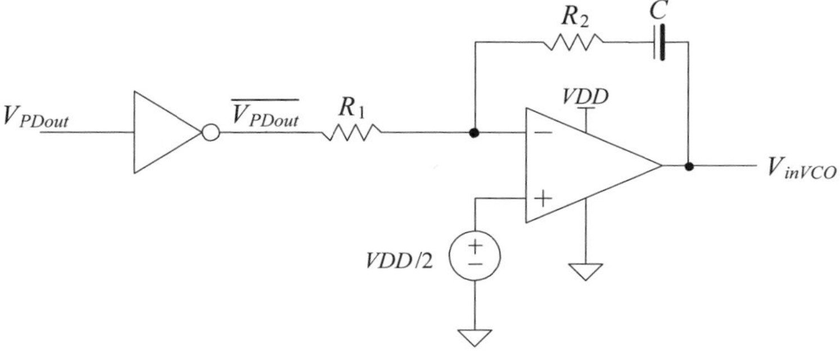

Figure 19.30 Active PI loop filter.

and the damping ratio is given by

$$\zeta = \frac{\omega_n R_2 C}{2} \tag{19.44}$$

The lock range is

$$\Delta\omega_L = 4\pi\zeta\omega_n \tag{19.45}$$

while the lock time remains $2\pi/\omega_n$. The pull-in range, using the active PI loop filter, is limited by the VCO oscillator frequency. Consider the following example.

Example 19.3

Repeat Ex. 19.2 using the active PI loop filter. The SPICE model for an op-amp (a voltage-controlled voltage source) seen in Fig. 20.19 can be used for the op-amp.

Using Eq. (19.43) in (19.44) gives

$$\zeta = \frac{R_2 C}{2}\sqrt{\frac{K_{PD}K_{VCO}}{NR_1C}}$$

If we set the damping factor again to 1 we can write

$$4 = \frac{157 \times 10^6 R_2^2 C}{2\pi R_1}$$

Setting C to 10 pF and R_2 to 25k, then R_1 is 39k. The simulation results are seen in Fig. 19.31. The output clock is aligned to the center of the data. It's important to make sure that the step size used in the SPICE transient simulation isn't too big. For example, if a 100 MHz clock is used with a step size of 1 ns, it is likely that the loop either won't lock or if it does lock, both the VCO control voltage and thus the output data will oscillate. A maximum step size for 100 MHz signals may be 100 ps.

Finally, it is important to remember that the VCO tuning scheme in Fig. 19.25 is not practical unless the components can be set manually (either off-chip or using fuses or switches on chip). Also, something very important that we are not discussing (yet) is power supply noise. The current-starved VCO's output frequency is not very stable with changes in *VDD*. ■

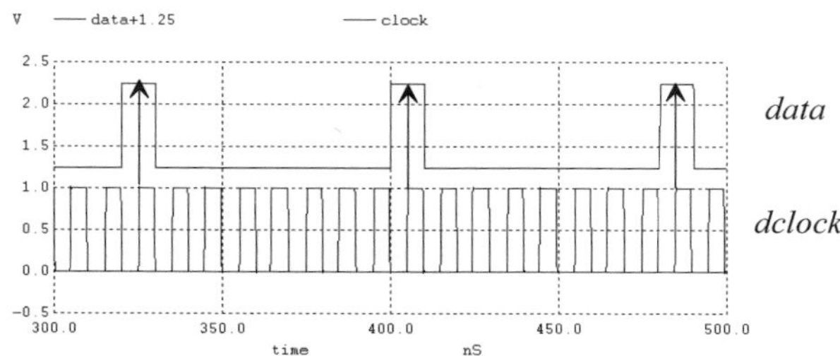

Figure 19.31 How the loop locks up in the center of the data.

19.3.2 PFD DPLL

Tri-State Output

A block diagram of a DPLL using the PFD with tri-state output is shown in Fig. 19.32. The phase transfer function is the same as Eq. (19.30)

$$H(s) = \frac{\phi_{clock}}{\phi_{data}} = \frac{K_{PDtri}K_F K_{VCO}}{s + \beta \cdot K_{PDtri}K_F K_{VCO}} \tag{19.46}$$

where

$$K_F = \frac{1 + sR_2C}{1 + s(R_1 + R_2)C} \tag{19.47}$$

When this filter is driven with the tri-state output, no current flows in R_1 or R_2 with the output in the high impedance state. The voltage across the capacitor remains unchanged. We can think of the filter, tri-state output as an ideal integrator with a transfer function

$$K_F' = \frac{1 + sR_2C}{s(R_1 + R_2)C} \tag{19.48}$$

Substituting this equation into Eq. 19.46 and rearranging results in

$$H(s) = \frac{K_{PDtri}K_{VCO}\frac{1+sR_2C}{(R_1+R_2)C}}{s^2 + s\frac{K_{PDtri}K_{VCO}R_2C}{N(R_1+R_2)C} + \frac{K_{PDtri}K_{VCO}}{N(R_1+R_2)C}} = \frac{\phi_{clock}}{\phi_{data}} = \frac{f_{clock}}{f_{data}} \tag{19.49}$$

From this equation, the natural frequency is

$$\omega_n = \sqrt{\frac{K_{PDtri}K_{VCO}}{N(R_1 + R_2)C}} \tag{19.50}$$

and the damping factor is determined by solving

$$2\zeta\omega_n = \frac{K_{PDtri}K_{VCO}R_2C}{N(R_1 + R_2)C} \tag{19.51}$$

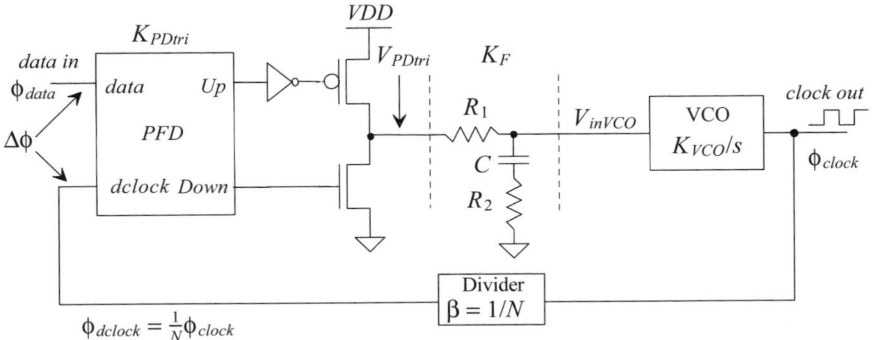

Figure 19.32 Block diagram of a DPLL using a sequential phase detector (PFD).

which results in a damping factor given by

$$\zeta = \frac{\omega_n}{2} \cdot R_2 C \tag{19.52}$$

The lock range is given by

$$\Delta\omega_L = 4\pi\zeta\omega_n \tag{19.53}$$

while the lock time, T_L , remains $2\pi/\omega_n$. The pull-in range is limited by the VCO operating frequency. The pull-in time is given by

$$T_P = 2R_1 C \cdot \ln \frac{(K_{VCO}/N) \cdot (VDD/2)}{(K_{VCO}/N)(VDD/2) - \Delta\omega} \tag{19.54}$$

where $\Delta\omega$ is the magnitude of the *input* frequency step.

Implementing the PFD in CMOS

The PFD seen in Fig. 19.10 can be implemented using inverters and nand gates as seen in Fig. 19.33. Simulating this circuit gives the results seen in Fig. 19.34. When the *dclock* is lagging, the *data* the output of the PFD is the *up* pulse (indicating that the edge of *dclock* needs to speed up or occur earlier in time, that is, the VCO's control voltage should increase). When *data* is lagging *dclock* the *down* pulse goes high indicating *dclock* should slow down.

We have some fine points that need to be discussed here. For example, what happens to *up* and *down* as the rising edges of *dclock* and *data* move close together? What we may get is some small glitches in the *up* and *down* signals. Or we may see both *up* and *down* staying low as the two pulses move closer together. In the PFD of Fig. 19.33, the delay through the two inverters can be used to set how *up* and *down* behave as

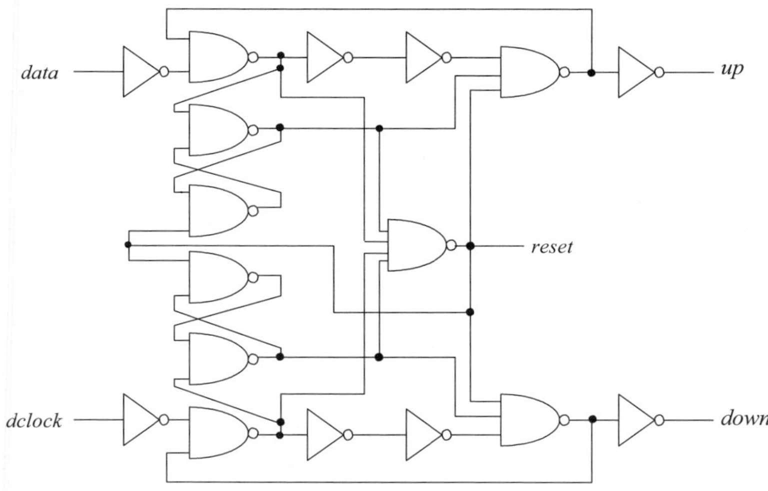

Figure 19.33 CMOS implementation of a PFD.

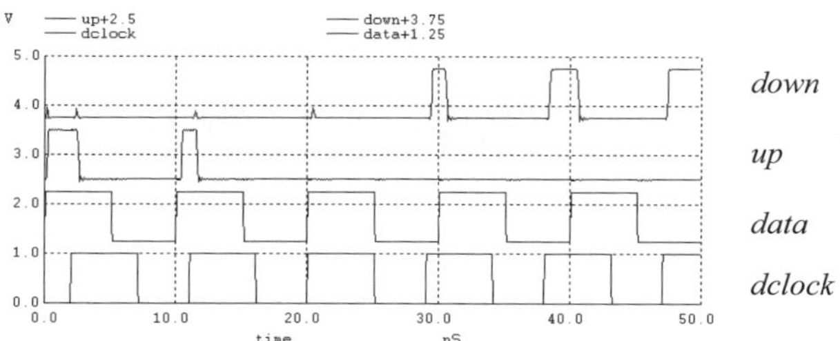

Figure 19.34 Simulation results for the PFD in Fig. 19.33.

the PFD's inputs move close together. If, for example, both stay low when the PFD's inputs get within 100 ps of each other, we get a *static phase error*. The loop won't lock any tighter than within 100 ps of its ideal position. Another way of looking at this is that as the phase difference, $\Delta\phi$, moves towards zero (see Fig. 19.11), the gain (the slope of the curve in Fig. 19.11) decreases. The phase detector is then said to have a *dead zone* where the gain of the phase detector (the slope of the curves) decreases.

Example 19.4

Design a DPLL using the tri-state topology seen in Fig. 19.32 that generates a clock signal at a frequency of 100 MHz from a 50 MHz square wave input. This application of the DPLL is called *frequency synthesis*.

The feedback path contains a divide by 2 circuit ($N = 2$). Let's use the VCO with the characteristics seen in Fig. 19.18

$$K_{VCO} = 1.57 \times 10^9 \ radians/V \cdot s$$

The gain of the phase detector (knowing *VDD* is 1 V) is

$$K_{PDtri} = \frac{1}{4\pi}$$

The lock range, Δf_L will be set to 20 MHz. Again let's set $\zeta = 1$. Using Eq. (19.53)

$$\Delta\omega_L = 4\pi\zeta\omega_n \rightarrow \omega_n = 10 \times 10^6 \ radians/V \cdot s$$

Using Eq. (19.52) gives

$$R_2 C = 200 \ ns$$

Let's set the capacitor to 10 pF and R_2 to 20k. Solving Eq. (19.50) for R_1 gives 42.5k ($= R_1$).

The simulation results are seen in Fig. 19.35. We should first notice that the VCO's control voltage doesn't have the excessive ripple like we had in the DPLL using the XOR gate phase detector. The response of the loop shows a nice $\zeta = 1$ shape. Again note that a DPLL loop's pull-in range is limited by the VCO's operating frequency (which, in this example uses the VCO from Fig. 19.18, which ranges from 50 to 150 MHz). A good exercise to perform at this point is to change the divider in the feedback path (to divide by 1, 2, 4, etc.) and the input frequency (a signal we've called *data*) and look at the robustness of the loop. ■

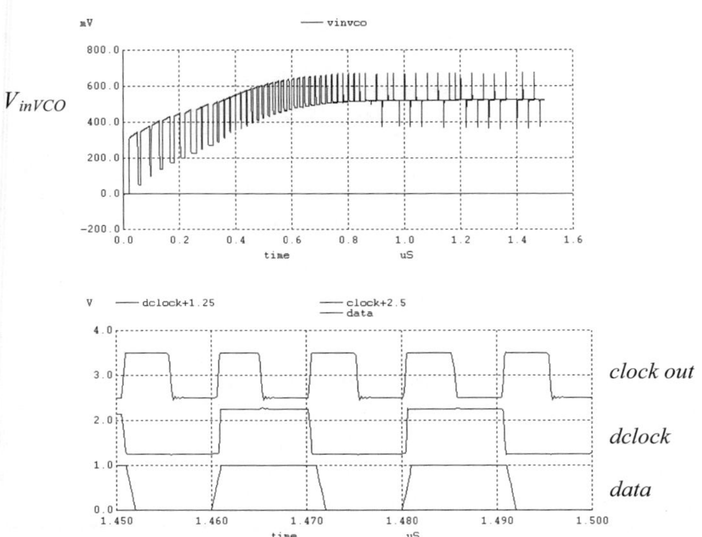

Figure 19.35 Simulation results for Ex. 19.4.

PFD with a Charge Pump Output

The PFD with a charge-pump output is seen in Fig. 19.36. A CMOS implementation of a DPLL using this configuration is, in general, preferred over the tri-state output because of the better immunity to power supply variations. In the tristate configuration seen in Fig. 19.32 note that when either the NMOS or PMOS switches are on, either *VDD* or ground are connected directly to the loop filter. Power supply or ground noise can thus feed directly into the loop filter and then into the VCO's control voltage.

The loop filter integrates the charge supplied by the charge pump. The capacitor C_2 prevents $I_{pump} \cdot R$ from causing voltage jumps on the input of the VCO and thus frequency jumps in the DPLL output. In general, C_2 is set at about one-tenth (or less) of C_1. The loop-filter transfer function neglecting C_2 is given by

$$K_F = \frac{1 + sRC_1}{sC_1} \tag{19.55}$$

The feedback loop transfer function is given by

$$H(s) = \frac{\phi_{clock}}{\phi_{data}} = \frac{K_{PDI}K_{VCO}(1 + sRC_1)}{s^2 + s(\frac{K_{PDI}K_{VCO} \cdot R}{N}) + \frac{K_{PDI}K_{VCO}}{NC_1}} \tag{19.56}$$

From the transfer function the natural frequency is given by

$$\omega_n = \sqrt{\frac{K_{PDI}K_{VCO}}{NC_1}} \tag{19.57}$$

and the damping factor is

$$\zeta = \frac{\omega_n}{2} \cdot RC_1 \tag{19.58}$$

The lock range and lock time remain the same (using the different values for the natural frequency and damping ratio) as the PFD with the tri-state output. Again, the pull-in range is set by the VCO oscillator frequency range. The pull-in time is given by

$$T_P = 2RC_1 \ln\left[\frac{(K_{VCO}/N)\cdot(I_{pump})}{(K_{VCO}/N)\cdot(I_{pump})-\Delta\omega}\right] \tag{19.59}$$

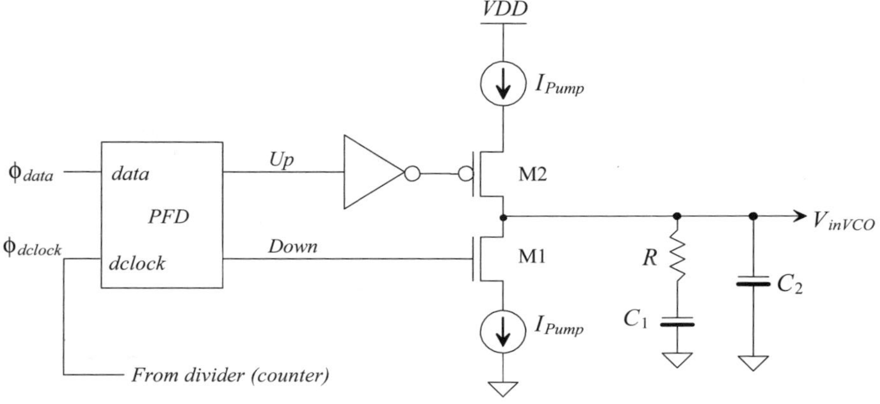

Figure 19.36 PFD using the charge pump.

Practical Implementation of the Charge Pump

The circuit seen in Fig. 19.36 is useful for illustrating the concept of the PFD with charge pump. However, in a practical circuit the fact that the sources of M1 and M2 charge to ground and *VDD* respectively when M1 and M2 are off creates some design issues. For example, suppose that the source of M1 is discharged to ground when the *down* signal goes high. When M1 turns on, it doesn't supply the charge set by I_{pump} to the loop filter but rather, until the voltage across the current sink increases, behaves like a switch simply connecting the filter's input to ground (meaning that we are not controlling the signal that is applied to the loop filter). To get around this problem, consider the circuit seen in Fig. 19.37. The bias voltages come from diode-connected MOSFETs (to form current mirrors, see Ch. 20). When *up* and *down* are low, M1L and M2L are on. The pump-up current source, MPup, is driving the pump-down current source, MNdwn. When either *up* or *down* goes high, the output is connected to one of the current sources. The MOSFETs M1L,R and M2L,R simply steer the current to the loop filter. The only other concern we have with this topology is the fact that when M1L and M2L are both on, the voltage on their drains won't precisely match the voltage across the loop filter (V_{inVCO}). In other words, the voltage across each of the current sources (MPup and MNdwn) won't be the same as it is when they are connected to the output node. The result is that charge sharing between the parasitic capacitance on the drains of MPup and MNdwn and the capacitance used in the loop filter cause a static phase error or jitter. To eliminate this problem, an amplifier (see dashed lines in the figure) can be inserted to set the drain voltage of M1L and M2L to the same value V_{inVCO}.

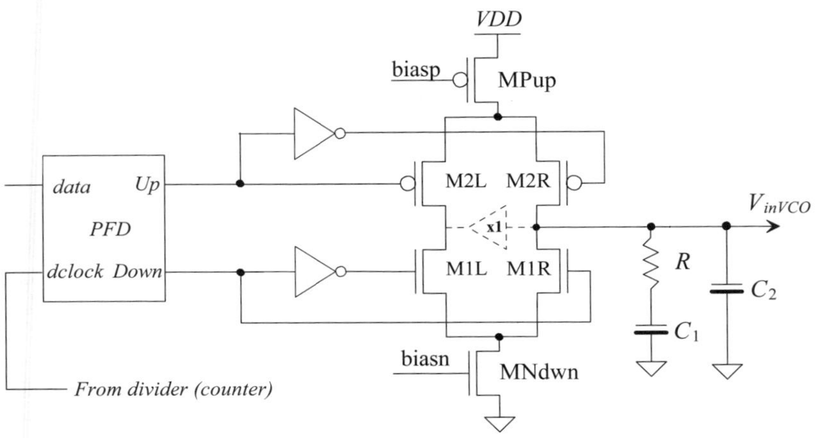

Figure 19.37 Practical implementation of the charge pump.

Example 19.5
Repeat Ex. 19.4 using the PFD and charge pump seen in Fig. 19.37.

One of the other benefits of using the charge pump is the fact that we can select I_{pump}, that is,

$$K_{PDI} = \frac{I_{pump}}{2\pi}$$

Again, let's set the lock range to 20 MHz. Using Eq. (19.53) with $\zeta = 1$ gives, again, $\omega_n = 10 \times 10^6$ *radians/V · s*. From Eq. (19.58) we get

$RC_1 = 200$ *ns* so let's use $R = 20\text{k}$, $C_1 = 10$ pF, and $C_2 = 1$ pF

remembering that we generally set C_2 to one-tenth of C_1. Using Eq. (19.57), we can now find the value of I_{pump}

$$10 \times 10^6 = \sqrt{\frac{I_{pump} \cdot 1.57 \times 10^9}{2\pi \cdot 2 \cdot 10 \times 10^{-12}}} \rightarrow I_{pump} = 8\ \mu A$$

Because this value isn't that critical, we'll round it up to 10 µA. Figure 19.38 shows the simulation results. Notice the $\zeta = 1$ (or maybe a little less because the voltage does overshoot its final value by a little bit) shape of the VCO's control voltage. Let's modify the loop filter to show what V_{inVCO} would look like if $\zeta = 0.1$. All we have to do, from Eq. (19.58), is drop R by a factor of 10. The result is seen in Fig. 19.39. The control voltage is oscillating and the loop isn't locking. Of course, we can also increase the damping factor. The loop now behaves sluggishly and does not respond to fast changes in the input signal. For general design, where the process and temperature vary, it's better to center the design on a larger damping factor to avoid instability. ■

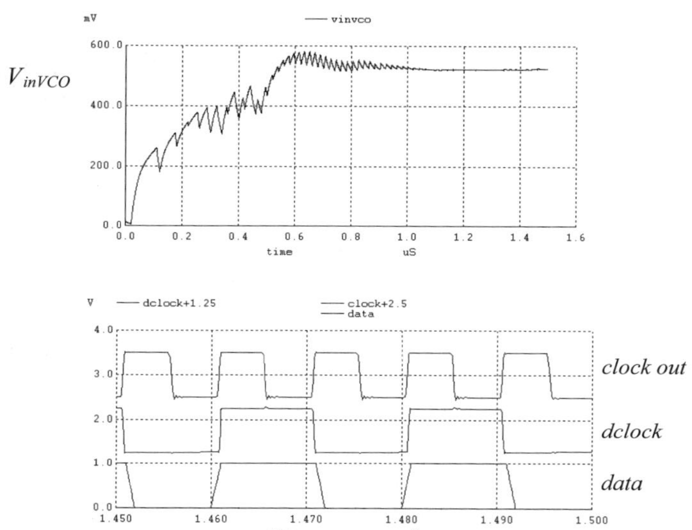

Figure 19.38 Simulation results for Ex. 19.5.

Discussion

When selecting values for the loop filter, we assumed that the output resistance of the phase detector was small (for the XOR and tri-state PD) compared to the impedances used in the loop filter. We also assumed that the input resistance of the VCO was infinite and the input capacitance of the VCO was small compared to the capacitance used in the loop filter. Considering the parasitics present in the DPLL is an important part of the design.

The center frequency of the VCO is critical for good DPLL performance when using the XOR gate with *RC* loop filter. If the center frequency, f_{center}, of the VCO (i.e., $V_{inVCO} = VDD/2$) does not match twice the input data rate, the DPLL will lock up at a phase different from $\pi/2$ (the input frequencies to any phase detector must be equal). The need for a precision center frequency is eliminated by using the XOR PD with an active PI loop filter or by using the PFD. The other big benefit of using the PFD is that the VCO's gain can be larger.

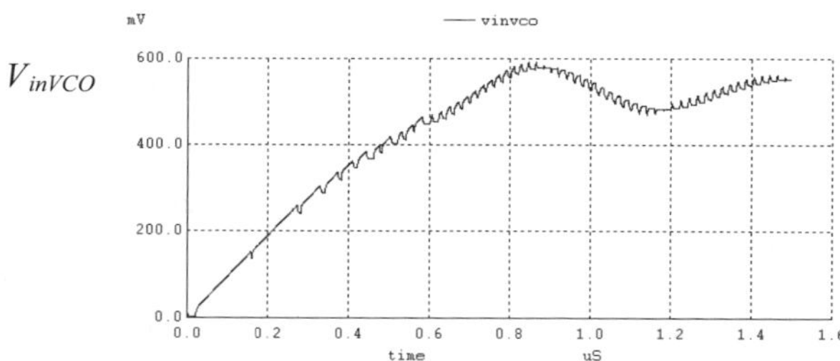

Figure 19.39 Showing what happens when the damping factor is reduced to 0.1.

Finally, selecting of the loop's damping factor, ζ , is very important. If the value of ζ is too small, the loop will have trouble locking or the output will jitter excessively when the loop is locked. This last problem is sometimes (gratuitously) called *jitter peaking* since a step in the DPLL's input frequency causes excessive overshoot in V_{inVCO} with small ζ.

19.4 System Considerations

System concerns are often the driving force behind the design of a DPLL. Referring to Fig. 19.1, we observe that the data transmitted through the channel should ideally arrive at the receiver with the same shape with which it was transmitted. In reality, the data becomes distorted. Distortion arises from nonlinearities in the receiver input amplifier and the finite bandwidth of the channel. To understand the conditions for distortionless transmission, consider the block diagram shown in Fig. 19.40. The system has a transfer function in the frequency domain of $H(f)$ and in the time domain $h(t)$. For distortionless transmission, we can relate the input and output of the system by

$$y(t) = K \cdot x(t - t_o) \tag{19.60}$$

where t_o is the time delay through the system and K is a constant. This equation shows that for distortionless transmission through a system the output is simply a scaled, time-delayed version of the input. An interesting observation can be made by taking the Fourier Transform of both sides of this equation,

$$Y(f) = K \cdot X(f) e^{-j2\pi f t_o} \tag{19.61}$$

The transfer function of a distortionless system can then be written as

$$H(f) = \frac{Y(f)}{X(f)} = K e^{-j2\pi f t_o} \tag{19.62}$$

Figure 19.41 shows the magnitude and phase responses of a distortionless system. A system is distortionless when its amplitude response, $|H(f)|$, is a constant, K, and its phase response, $\angle H(f)$, is linear with a slope of $-2\pi t_o$ over all frequencies of interest.

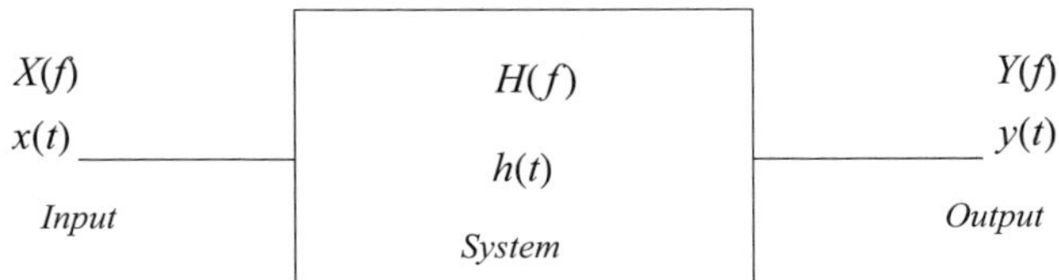

Figure 19.40 Representation of a system with input and output.

The responses shown in Fig. 19.41 are ideal. In practice, the magnitude response of a system may look similar to Fig. 19.42a. At higher frequencies, the magnitude rolls off. To compensate for this roll-off, or other imperfections, a circuit called an equalizer is added in series with the system (Fig. 19.42b). The equalizer has a transfer function in which its magnitude response increases with increasing frequency beyond a point (Fig. 19.42c). If the low-frequency gain of the equalizer is A/K and the low-frequency gain of the system is K, then the resulting gain of the system/equalizer combination is A.

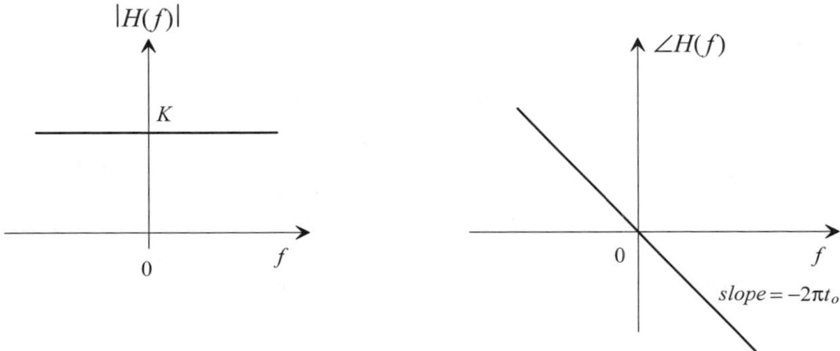

Figure 19.41 Magnitude and phase response of a distortionless system.

Another source of potential distortion occurs when the receiver input data is regenerated into digital levels. This was discussed back in Sec. 18.3. Timing errors occur when the input data is not precisely sliced through its middle (see Fig. 18.12). What makes slicing the data correctly even more difficult is the fact that the amplitude response of the channel can change with time and the data pattern can affect the average level of the data. There are two solutions to this problem. The first uses a circuit (see Fig. 18.29) that determines the peak positive and negative input analog amplitudes, averages the values, and feeds back the result to the comparator in the decision-making circuit. The second method encodes the digital data so that the duty cycle of the resulting encoded data is 50%. The encoding increases the channel bandwidth for a constant data rate.

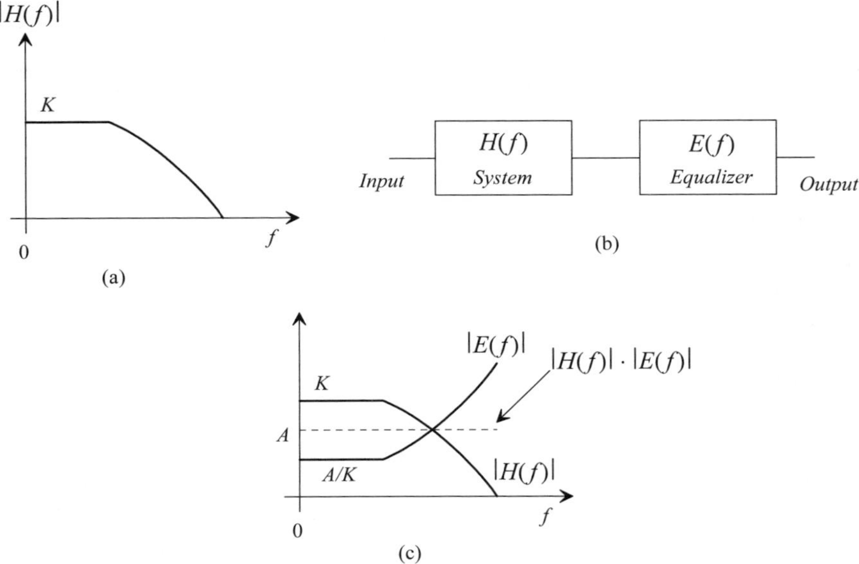

Figure 19.42 Using an equalizer to lower distortion in a system.

Encoding the data can eliminate the need for a decision circuit. If the resulting encoded data has a 50% duty cycle, it can be passed through a capacitor in the receiver, resulting in an analog signal centered around ground. The noninverting input of the comparator will then be connected to ground. The comparator will then be able to slice the data at the correct moments in time (in the middle of the data bit). An example of an encoding scheme is shown in Fig. 19.43. Encoding occurs in the transmitter prior to transmission over the channel. This particular encoding scheme is referred to as the bi-phase format, or more precisely, the bi-phase-level (sometimes called bi-phase-L or Manchester NRZ) format. The cost of using this scheme over the NRZ data format is increased channel bandwidth. Other encoding schemes are shown in Fig. 19.44.

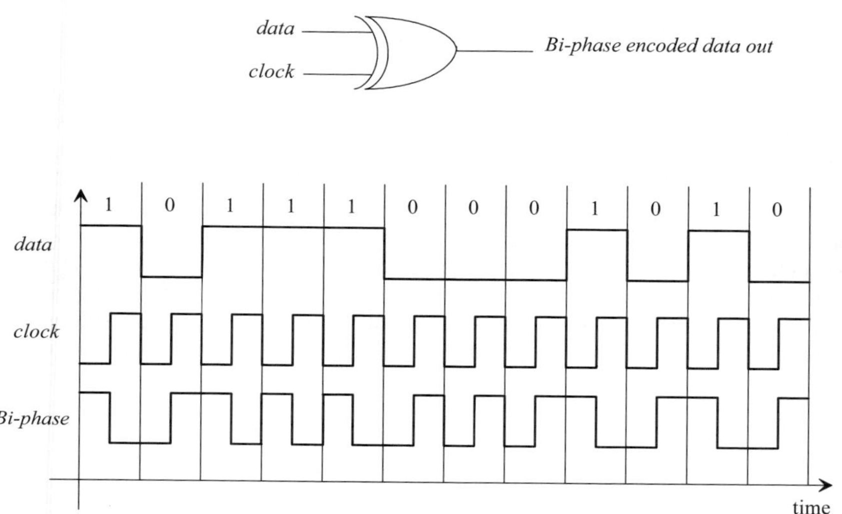

Figure 19.43 Bi-phase data encoding.

19.4.1 Clock Recovery from NRZ Data

One of the most important steps in the design of a communication system is the selection of the transmission format, that is, NRZ, bi-phase, or some other format, together with use of parity, cyclic redundancy code, or some other encoding format. In this section we discuss some of the considerations that go into the design of a clock-recovery DPLL in a system that uses NRZ.

Let's begin this discussion by considering the NRZ *data* and *clock* shown in Fig. 19.45. Let's further assume that these signals are the inputs to an XOR PD in a DPLL, which is not in lock since the clock is not aligned properly to the data. The resulting output of the XOR PD is shown in this figure as well. If we were to average this output using a loop filter, we would get *VDD*/2. In fact, it is easy to show that shifting the *clock* signal in time has no effect on the average output of the PD. Why? To answer this question, let's use some numbers. Assume that a bit width of *data* is 10 ns (which is also the period of the *clock*). The frequency of the square wave resulting from the alternating

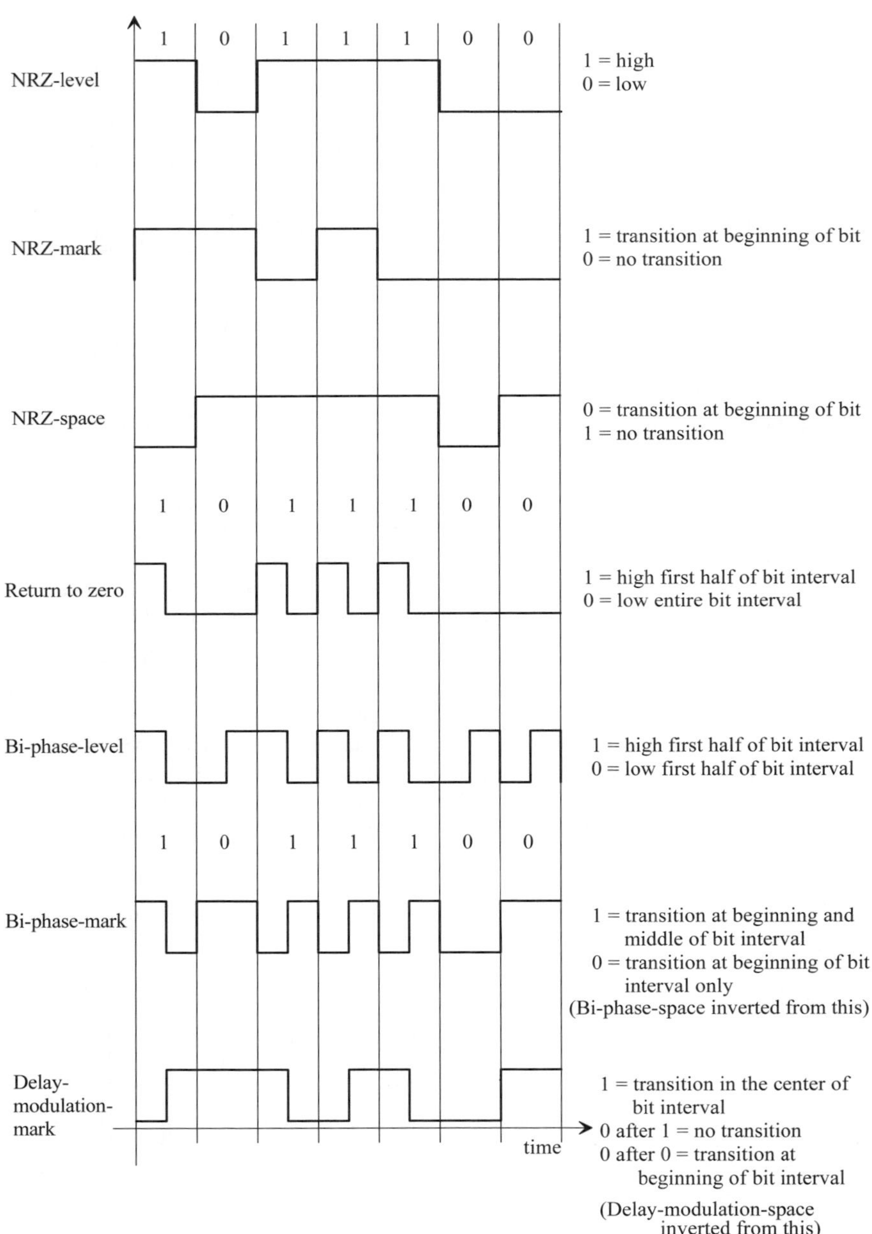

Figure 19.44 Data transmission formats.

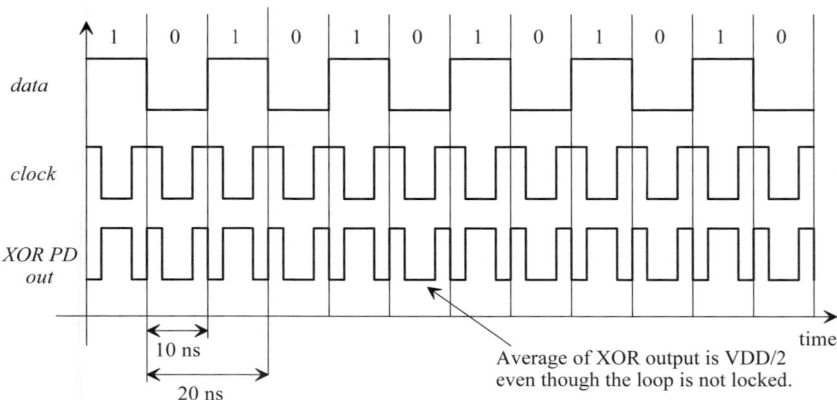

Figure 19.45 The problems of using clock without the divide by 2 to lock on data.

strings of ones and zeros is 50 MHz. We know that if we take the Fourier Transform of a square-wave, only the odd harmonics (i.e., 50, 150, and 250 MHz) are present. Since the *clock* signal is at 100 MHz, there is no energy or information common between the *clock* and *data* signals. To remedy this, we divide the clock down in frequency, *dclock*, so that it is at the same frequency as the alternating ones and zeros of the data (divide by two).

The next problem we encounter, if we use the divide by two in the feedback loop, occurs when we get *data* that is a repeating string of 2 ones followed by 2 zeroes, Fig. 19.46 (the *dclock* is not locked to the data). Again, there is no common information between the two inputs, and the resulting XOR PD output will always have an average of *VDD*/2. In this case, however, the *dclock* is running at 50 MHz, and the *data* is a square-wave of 25 MHz. Should we divide the *clock* down further to avoid this situation? The answer is no. However many times we divide the clock down, we can still come up

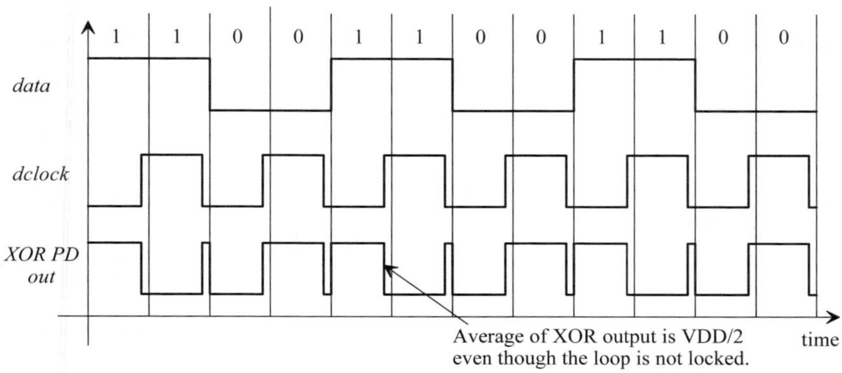

Figure 19.46 Problems trying to lock on a data stream that is one-half the dclock frequency.

with a data string that will not allow the loop to lock. Also, it is the actual edge transitions (the frequency of the *data* and *dclock*) that is used when the inputs are not pure squarewaves. Increasing the width of *dclock* has the effect of removing information and making it more difficult to lock up to the data. One solution to this problem is to use odd-parity with an 8-bit word (9 bits total) and eliminate, at the transmitting end, the possibility of all 8 bits being high, that is, 11111111 or 255. It is then impossible to generate a square-wave.

The restrictions on the data pattern in a communications system using NRZ data can be reduced by detecting the edges of the input data with an edge detector circuit (Fig. 19.47). The delay through the inverters sets the width of the output pulses, labeled "*Edge out*" in this circuit. The frequency content of the output pulses will always contain energy at the *clock* frequency and thus the loop can lock up on the data.

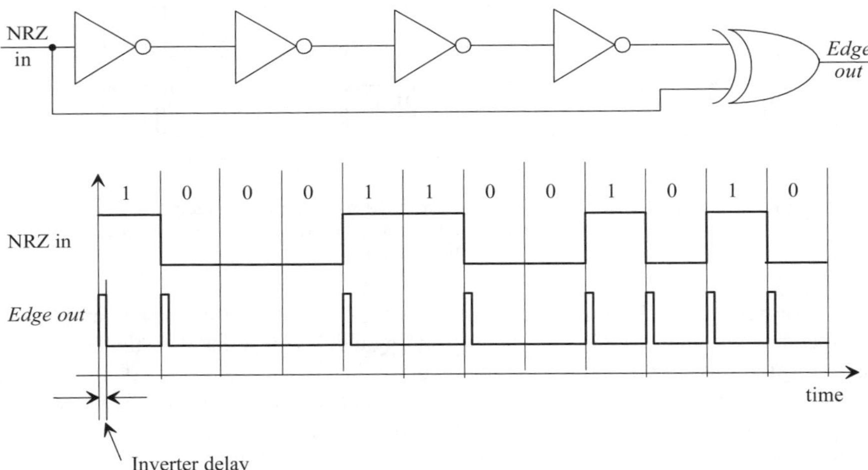

Figure 19.47 Detecting the edges in NRZ data.

As an example, consider the block diagram and data shown in Fig. 19.48. The *Edge output* is connected as the *input* of an XOR-based DPLL. The output of the VCO, *clock*, will lock up on the center of the *Edge output*, that is, the rising edge of the *clock* signal will become aligned with the center of *Edge out*. Averaging *PD out*, in this figure, results in *VDD*/2. If the clock is shifted to the left or right in time, the average value of *PD out* will shift down or up, causing the VCO frequency to change and keep the *clock* aligned to the center of *Edge out*. Several practical problems exist with this configuration. The delay through the inverters should be constant whether a high-to-low or a low-to-high transition is propagating through the inverters. Also, for best performance the delay of the inverters, or whatever element is used for the delay (one common element for high-speed applications is a microstrip line) should be close to one-half of the bit-interval time. This delay is important as it directly affects the gain of the phase detector and therefore the transient properties of the DPLL.

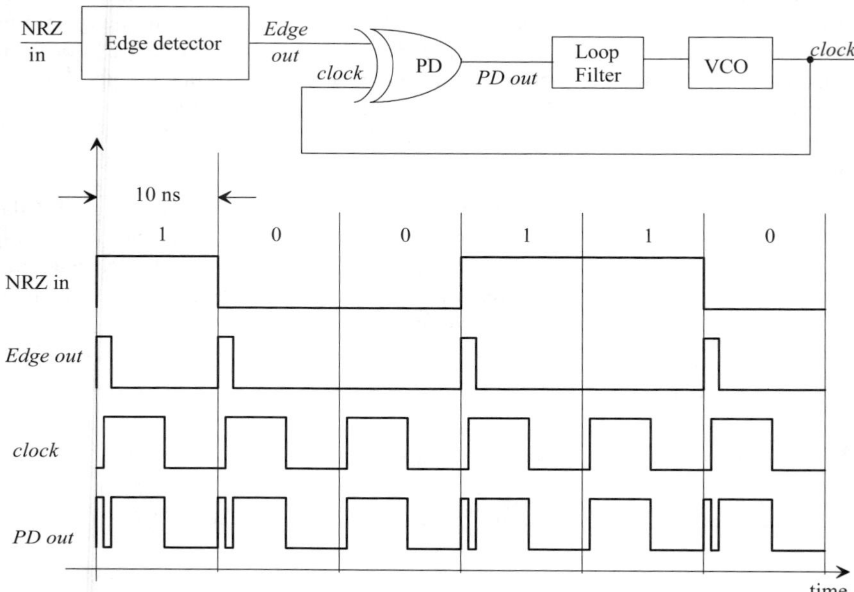

Figure 19.48 Clock-recovery circuit for NRZ using an edge detector. Note that the DPLL is in lock, when the rising edge of clock is centered on the edge output pulse.

Not having the *clock* aligned to the center of the data bit can cause problems in high-speed, clock-recovery circuits. Simply adding a delay in series with the *clock* signal does not solve this problem since the temperature dependence and process variations of the associated circuit do not guarantee proper alignment. What is needed is a circuit that is *self-correcting*, causing the clock signal to align to the center of the data bit independent of the data-rate, the temperature, or process variations.

The Hogge Phase Detector

The PD portion of a self-correcting, clock-recovery circuit is shown in Fig. 19.49 along with associated waveforms for a locked DPLL. Nodes A and B are simply the input NRZ data shifted in time by one-half bit-interval and one bit-interval, respectively. The outputs of the phase detector are labeled *Increase* and *Decrease*. If *Increase* is low more often than *Decrease*, the average voltage out of the loop filter and thus the frequency out of the VCO will decrease. A loop filter that can be used in a self-correcting DPLL is shown in Fig. 19.50a. This filter is the active-PI loop filter discussed in Sec. 19.3.1, with an added input to accommodate both outputs of the PD. Figure 19.50b shows the resulting waveforms in a DPLL where the clock is leading the center of the NRZ data, and thus *Increase* is high less often than *Decrease*. If the NRZ data was lagging the center of the data bit, *Decrease* would be high less often than *Increase,* resulting in an increase in the loop-filter output voltage. Note that in this discussion we have neglected the propagation delays present in the circuit. For a high-speed, self-correcting PD design, we would have to analyze each delay in the PD to determine their effect on the performance of the DPLL.

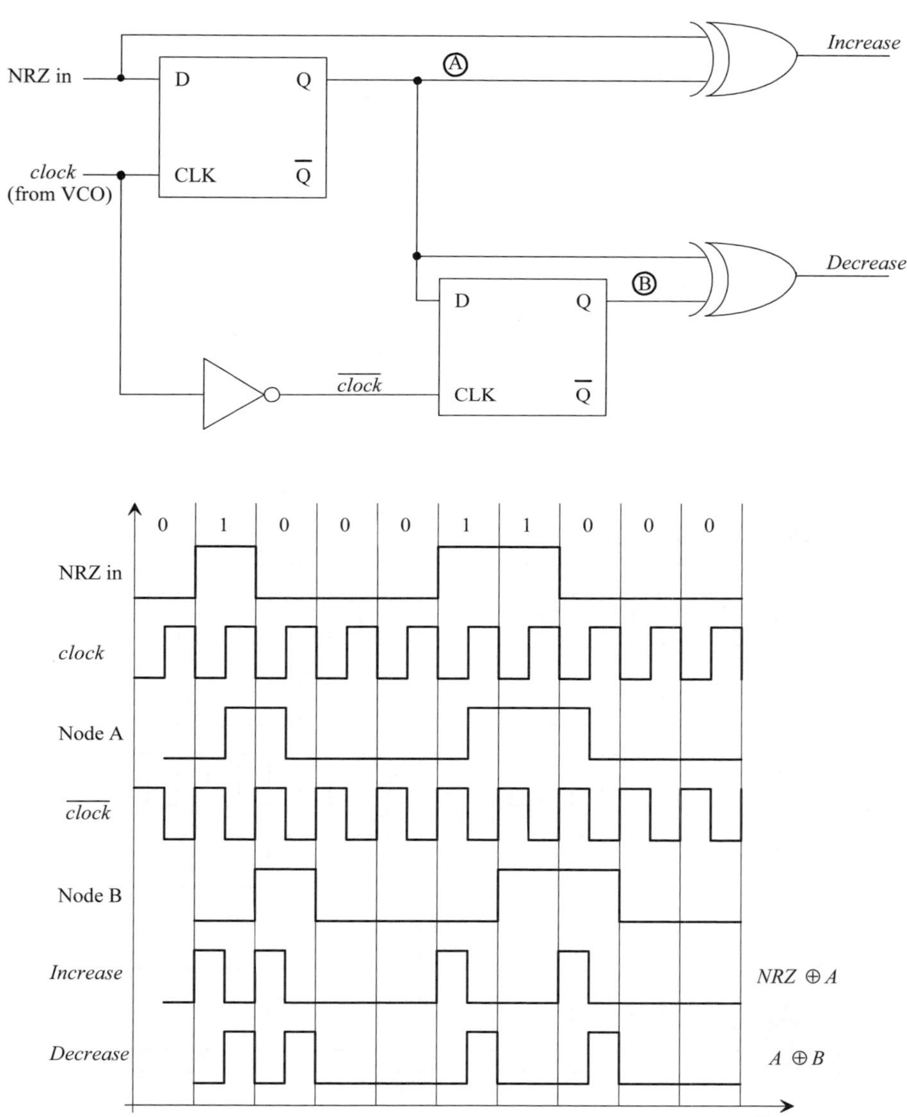

Figure 19.49 The PD (Hogge) portion of a self-correcting, clock-recovery circuit in lock.

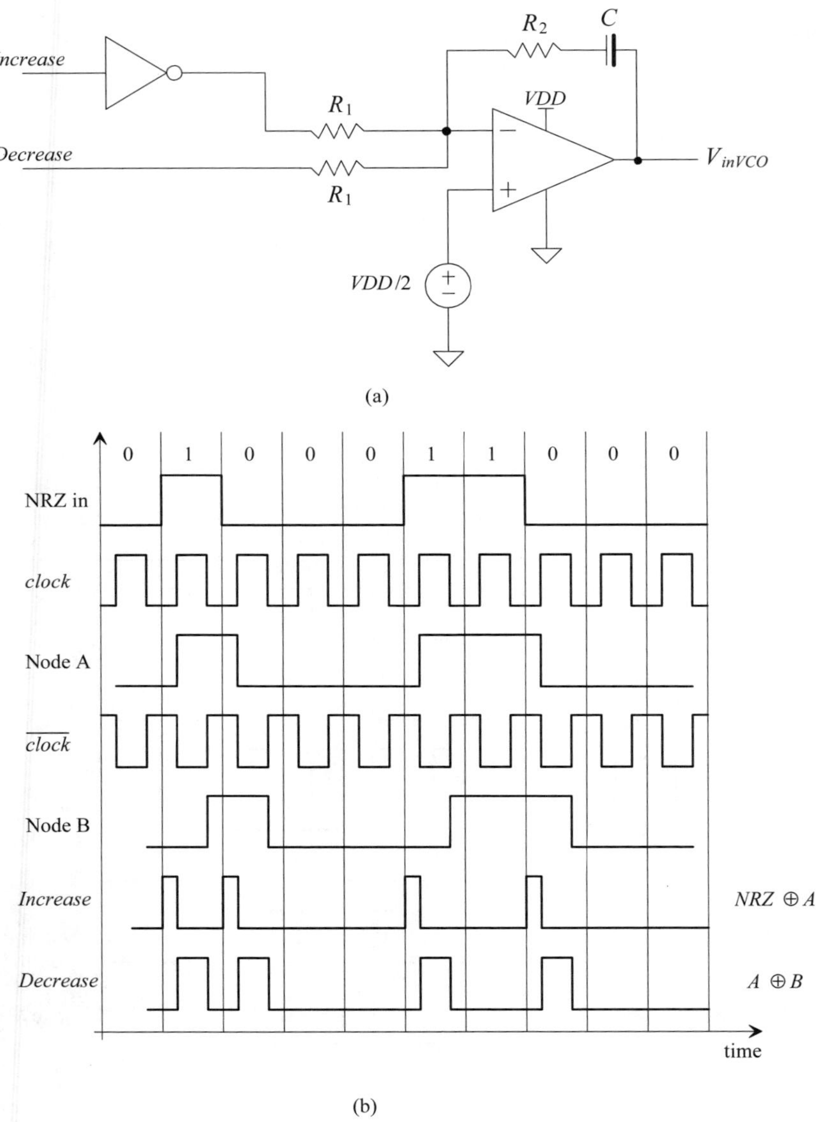

Figure 19.50 (a) Possible loop filter used in a self-correcting (Hogge) DPLL and (b) waveforms when the loop is not in lock and the clock leads the center of the data.

Jitter

Jitter, in the most general sense, for clock-recovery and synchronization circuits, can be defined as the amount of time the regenerated clock varies once the loop is locked. Figure 19.51a shows the idealized case when the *clock* doesn't jitter, while Fig. 19.51b shows the actual situation where the clock-rising edge moves in time (jitters). In these figures, the oscilloscope is triggered by the rising edge of the *data*. In the following discussion, we neglect power supply and oscillator noise, that is, we assume that the oscillator frequency is an exact number that is directly related to the VCO input voltage. In the section following this one, we cover delay-locked loops and further discuss the limitations of the VCO.

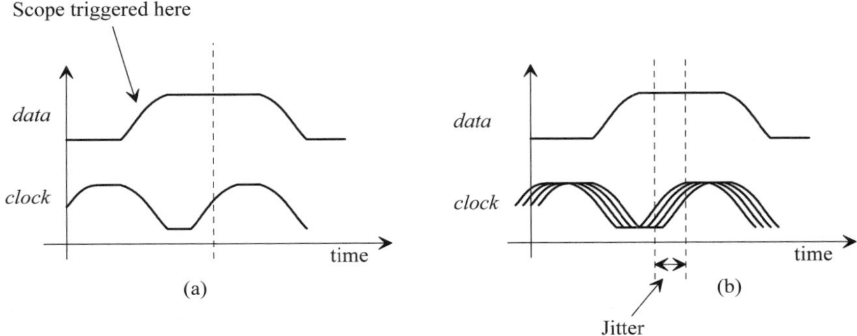

Figure 19.51 (a) Idealized view of clock and data without jitter and (b) with jitter.

Consider using the charge pump with the self-correcting PD shown in Fig. 19.52. When the loop is locked (from Fig. 19.49), both increase and decrease occur (with the same width) for every transition in the incoming data. Note that unlike a PFD/charge pump combination where the output of the charge pump remains unchanged when the loop is locked, the self-correcting PD/charge-pump combination generates a voltage ripple on the input of the VCO (similar to the XOR PD with RC or Active PI filter)[4]. Let's assume that this ripple is 10 mV and use the values for VCO gain and frequencies given in Ex. 19.5 to illustrate the resulting jitter introduced into the output clock. The change in output frequency resulting from this ripple is $10\text{mV}\cdot(1.57 \times 10^9 \text{ radians/V}\cdot\text{s})\cdot(1/2\pi)$ or 2.5 MHz. This means that the output of the DPLL will vary from, say, 100 MHz to 102.5 MHz. In terms of a jitter specification (see Eq. (19.41)), the clock jitter is (roughly) 250 ps (2.5% of the output clock's period).

From this example, *data dependent jitter*, can be reduced by

1. *Reducing the gain of the VCO.* Ripple on the input of the VCO has less of an effect on the output frequency. The main disadvantage, in the most general sense, of reducing the gain of the VCO is that the range of frequencies the DPLL can

[4] Of course, the self-correcting PD should not be used in most frequency synthesis applications. Similarly, the PFD should not be used in most clock-recovery applications.

lock up on is reduced. Also, the VCO gain strongly affects the ability to fabricate the VCO without postproduction tuning.

2. *Reducing the bandwidth of the loop filter.* This has the effect of reducing the actual ripple on the input of the VCO. The main drawback here is the increase in chip real estate needed to realize the larger components used in the filter.

3. *Reducing the gain of the PD.* This also has the effect of reducing the ripple on the input of the VCO. This method is, in general, the easiest when using the charge pump since reducing the gain is a simple matter of reducing I_{pump}. Substrate noise can become more of a factor in the design.

Another way of stating the above methods of reducing jitter is simply to say that the forward loop gain of the DPLL, that is, $K_{PD}K_FK_{VCO}$, should be made small. The main problems with using small forward loop gain are the reductions in lock range and pull-in range coupled with the associated increases in pull-in and lock times.

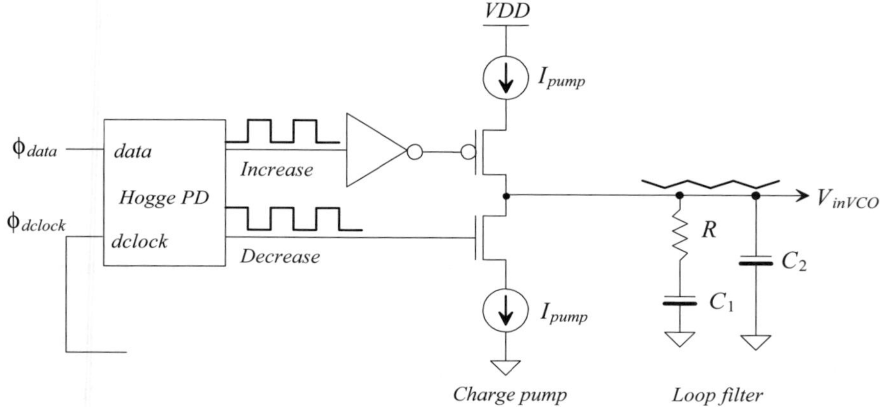

Figure 19.52 Self-correcting PD with charge pump output.

19.5 Delay-Locked Loops

Problems with PLL output jitter resulting from the VCO output frequency changing (often called oscillator or phase noise) with a constant input voltage (V_{inVCO} = constant) has led to the concept of a delay-locked loop (DLL). Figure 19.53 shows the basic block diagram of a DLL. *Assuming* that a reference clock is available at exactly the correct frequency, the input data is delayed through a voltage-controlled delay line (VCDL) a time t_o until it is synchronized with the reference clock. Jitter is reduced by using an element, the VCDL, that does not generate a signal (like the VCO did). The transfer function ϕ_{clock}/ϕ_{out} is zero (the phase of the reference clock is taken as the reference for the other signals in the DLL, i.e., $\phi_{clock} = 0$), so that oscillator noise and the resulting jitter are not factors in DLL design. The jitter considerations discussed in the last section, however, are still a concern since any ripple on the output of the loop filter will cause jitter.

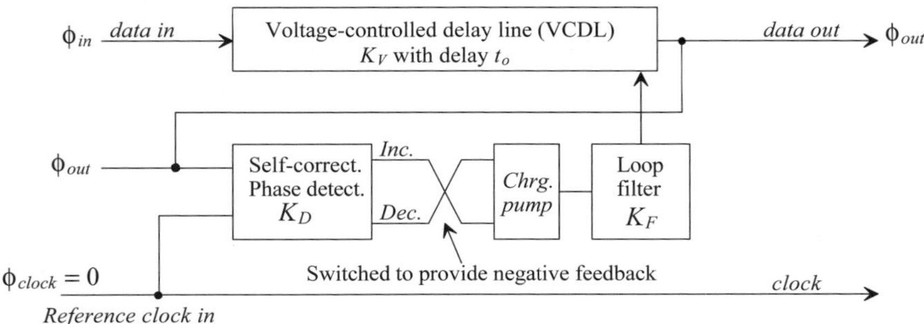

Figure 19.53 Block diagram of a delay-locked loop.

The phase (in radians) of the input data is related to the phase of the output data by

$$\phi_{out} = \phi_{in} + t_o \cdot \frac{2\pi}{T_{clock}} \tag{19.63}$$

where T_{clock} is the period of the reference clock (or half of the period of the *data in* for a string of alternating ones and zeros). The gain of the VCDL can be written in terms of the delay, t_o, by

$$t_o = K_V \cdot V_{indel} \tag{19.64}$$

where K_V has units of seconds/V and V_{indel} is the voltage input to the VCDL from the loop filter. The minimum and maximum delays of the VCDL should, in general, lie between $T_{clock}/2$ and $1.5T_{clock}$ for proper operation. The output of the loop filter (input to the VCDL) can be written as

$$V_{indel} = \phi_{out} \cdot K_D \cdot K_F \tag{19.65}$$

The overall transfer function may now be written as

$$\frac{\phi_{out}}{\phi_{in}} = \frac{1}{1 - K_D K_F K_V \cdot \omega_{clk}} \tag{19.66}$$

where $\omega_{clk} = 2\pi/T_{clock}$. The gain of the self-correcting PD with charge-pump output, with the help of Fig. 19.54 and noting that *Increase* and *Decrease* can occur at the same time, is

$$K_D = -\frac{I_{pump}}{\pi} \text{ (amps/radian)} \tag{19.67}$$

The negative sign is the result of switching the *Increase* and *Decrease* outputs of the self-correcting PD when connecting to the charge pump. This switch is required to provide negative feedback around the loop. Another benefit of the DLL is that the loop filter can be a simple capacitor, which results in a first-order feedback loop, that is,

$$K_F = \frac{1}{sC_1} \tag{19.68}$$

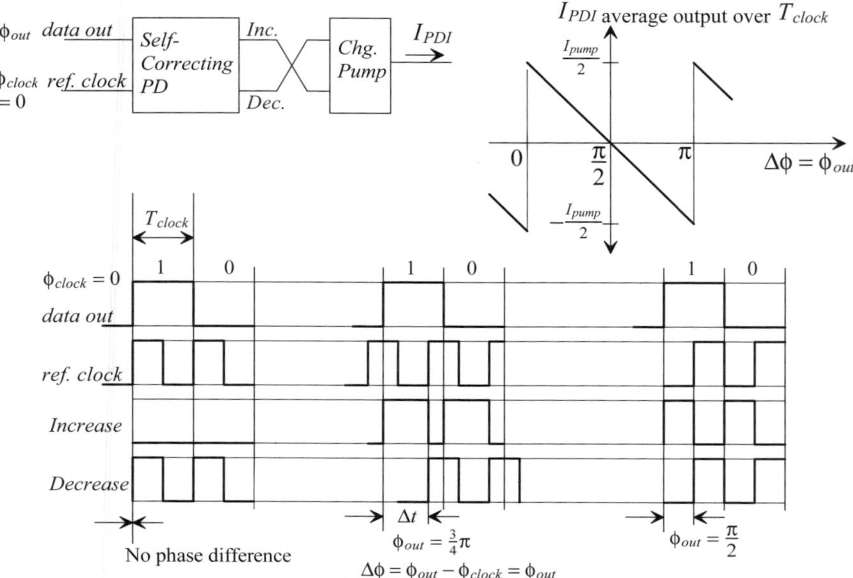

Figure 19.54 Self-correcting PD output for various inputs (assuming input data is a string of alternating ones and zeros).

The transfer function of the DLL relating the input data to the time-shifted output is

$$\frac{\phi_{out}}{\phi_{in}} = \frac{1}{1 + \frac{I_{pump}}{\pi} \cdot \frac{1}{sC_1} \cdot K_V \cdot \omega_{clk}} = \frac{s}{s + K_V \cdot \frac{2I_{pump}}{C_1 T_{clock}}} \tag{19.69}$$

We know that the frequency of the reference clock must be exactly related to the frequency of the input data. However, there will exist instantaneous changes in the phase of the input data which the output of the DLL should follow. Modeling instantaneous changes in ϕ_{in} by $\Delta\phi_{in}/s$ (a step function with an amplitude of $\Delta\phi_{in}$), we get a change in output phase given by

$$\Delta\phi_{out} = \frac{\Delta\phi_{in}}{s + K_V \cdot \frac{2I_{pump}}{C_1 T_{clock}}} \tag{19.70}$$

The time it takes the DLL to respond to an input step in phase is simply

$$T_r = 2.2 \cdot \frac{C_1 T_{clock}}{K_V \cdot 2I_{pump}} = \text{number of clock cycles} \cdot T_{clock} \tag{19.71}$$

This time can be decreased by making C_1/I_{pump} small, which from our discussion in the last section, has the result of increasing the output pulse jitter (i.e., jitter dependent on the input data pattern). Decreasing C_1/I_{pump}, increases the ripple on the control voltage of the VCDL. Similarly, increasing K_V (the time/volt delay of the VCDL) increases jitter since a given ripple on the control voltage of the VCDL will have a larger effect on the delay. Again, trade-offs must be made between responses to input variations and output jitter.

Delay Elements

The VCDL is an important component of the DLL. Figure 19.55a shows the basic implementation of a VCDL using adjustable delay inverters. The last two inverters in the VCDL ensure that clean digital signals are output from the line. Figure 19.55b shows the circuit schematics for possible delay elements. The first delay element should be recognized as a current-starved inverter discussed earlier in the chapter. The second delay element is nothing more than an inverter with a variable load. In practice, these delay elements are rarely used because of the susceptibility to noise and power supply variations. Rather, fully-differential delay elements are used.

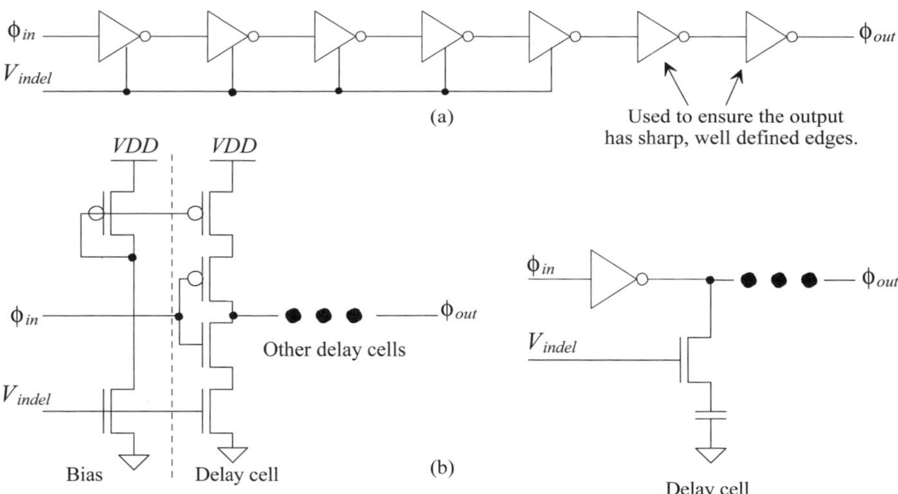

Figure 19.55 (a) VCDL made using inverter delay cells and (b) possible delay cells.

Figure 19.56 shows the connection of a fully differential VCDL. Using any number of stages, a fully differential VCO can be implemented with the delay elements (and the proper feedback), while using an even number with the inverting (noninverting) output fed back and connected to the noninverting (inverting) input in-phase (I) and quadrature (Q) signals can be generated.

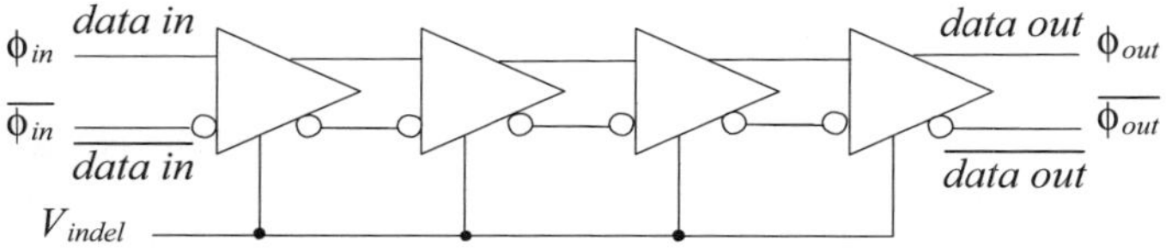

Figure 19.56 Implementation of a fully differential VCDL.

Practical VCO and VCDL Design

An example of a practical delay stage is shown in Fig. 19.57. The control voltage (that controls the delay if used in a VCDL or the oscillation frequency if the element is used in a VCO) can be generated from some linear voltage to current converter (like the one seen in Fig. 19.17 using a resistor, M5R, and M6R). The reason that this circuit is labeled "practical" can be understood by first considering what happens when there is noise on *VDD*. The noise causes a voltage variation across the PMOS current source, MPC. Ideally, this doesn't change the current through MPC. The output voltages of the delay cell swing between V_{REF} and ground. If there is noise on ground, it feeds equally into each output. This common-mode noise is then, ideally, rejected by the differential amplification action of the next stage. The bias circuit is a half-replica of the delay stage and is used to bias the delay elements so that the swing is up to V_{REF} when the gate of one of the PMOS switches is at ground.

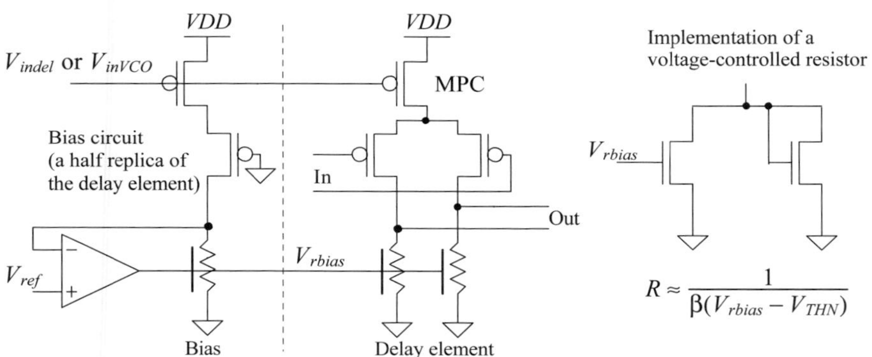

Figure 19.57 A differential delay element based on a voltage-controlled resistor. The bias circuit adjusts the value of the resistors used in the delay elements to sink the current sourced by the p-channel MOSFETs.

19.6 Some Examples

To finish up this chapter, let's present some examples with discussions and simulations.

19.6.1 A 2 GHz DLL

To begin, let's design a 500 ps delay line for a DLL that is used for *clock synchronization*. Let's design the delay line so that when it is used in a DLL, with the input and output of the delay line synchronized, we have eight phases of the input clock (we need eight stages in the delay line).

The first circuits we need to design are the bias circuit and the voltage-to-current converter used to generate $V_{outfilter}$ and V_{rbias} (see Fig. 19.57). We'll use the basic schemes of Fig. 19.17 and 19.57, as seen in Fig. 19.58. The important issues concerning this circuit are: 1) the linearity of the V_{indel} voltage against the generated I_{REF} , 2) the sensitivity of I_{REF} to changes in *VDD*, and 3) how well the amplifier regulates node n2 to V_{REF}.

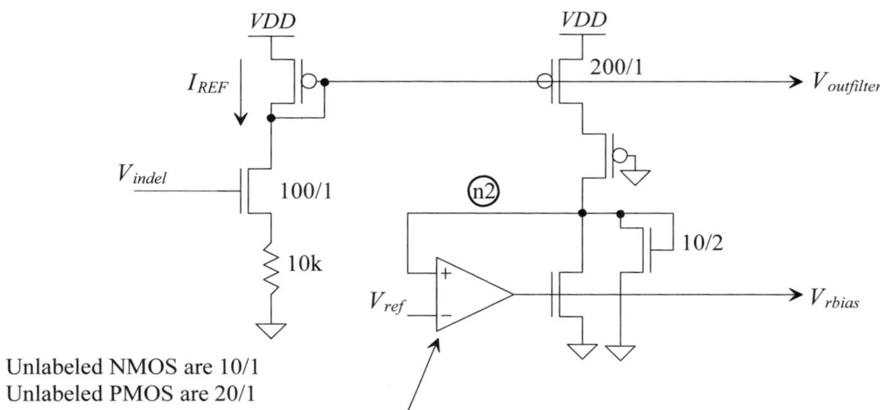

Figure 19.58 Bias circuit for a 500 ps delay line.

Figure 19.59 shows the simulation results for the bias circuit in Fig. 19.58. In (a) the x-axis is the delay line's control voltage, V_{indel}, while the y-axis is the generated reference current. Also seen in the figure, although hard to discern, is the change in *VDD* from 900 mV to 1.1 V (*VDD* is changed from 900 mV to 1.1 V in 50 mV steps, while V_{indel} is swept from 300 mV to 800 mV). Figure 19.59b shows how well node n2 is regulated to V_{REF}. V_{REF} can be generated with the beta-multiplier discussed in the next chapter (see Eq. (20.38) and Fig. 20.22).

Note that we aren't setting a lower value of current in this circuit like we did in Fig. 19.25. If we were to do this (which may be useful to reduce jitter since the resulting delay line or VCO would have lower gain), we would use a topology like the one seen in Fig. 19.60. The reference voltage is used to set the minimum current (when V_{indel} is small) through Rlow. The problem with setting the lower current with the method seen in Fig. 19.25 is that any changes in *VDD* feed directly across Rlow and change the bias current. Using the method in Fig. 19.60 with wide NMOS devices, the lower current is (roughly) $(V_{REF} - V_{THN})$/Rlow.

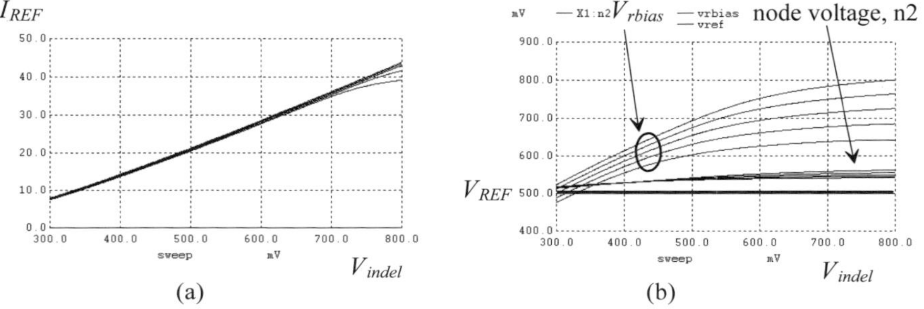

Figure 19.59 The performance of the bias circuit in Fig. 19.58.

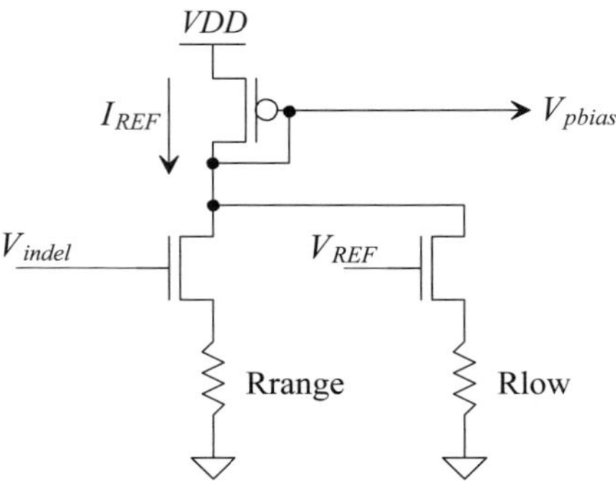

Figure 19.60 Lowering the current range in the bias circuit.

A schematic of the delay line is seen in Fig. 19.61. Notice that the current in the each delay stage was sized up by a factor of ten (the current sources in the delay cell are sized 200/1). This VCDL was simulated using the topology seen in Fig. 19.62. The control voltage was set to *VDD*/2 while the differential inputs were generated using an inverter and a transmission gate (to attempt to equalize the delays that the input signal sees to the VCDL). The outputs of the even stages are shown in this figure. Notice how they are evenly spaced. Notice, also, that the outputs only swing up to V_{REF} (= 500 mV here). As discussed earlier, this was done to minimize the effects of power supply and ground noise on the delay of the circuit.

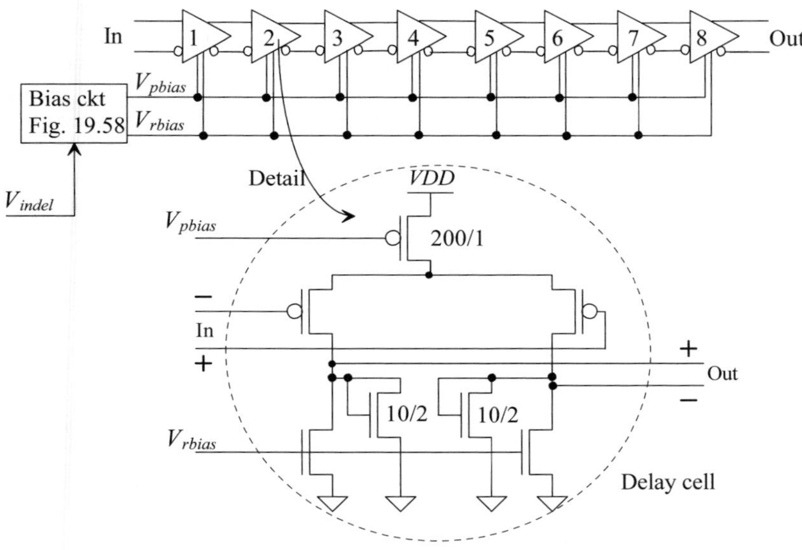

Figure 19.61 An eight-stage VCDL.

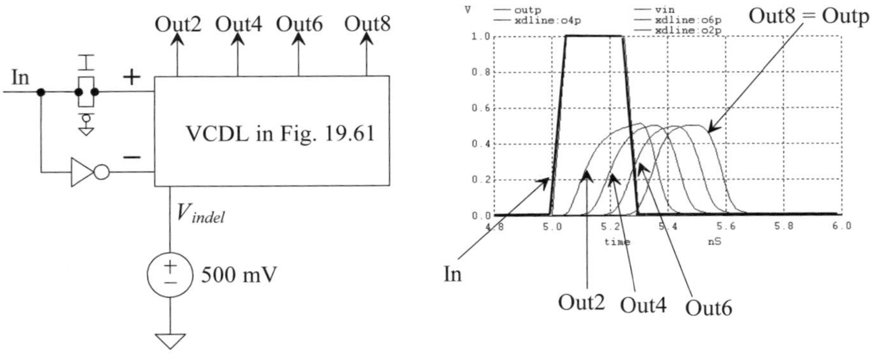

Figure 19.62 Simulating the VCDL in Fig. 19.61.

An important concern is how we regenerate full logic levels without introducing skew into our signals (see Sec. 18.3). Consider the modified VCDL seen in Fig. 19.63. Here we've changed the last stage to two diff-amps with swapped input signals (so we can generate the positive and minus outputs). Figure 19.64 shows the simulation results with this change. The shift in the eighth stage's output is small, in (a), and not much different than what is seen in Fig. 19.62. However, as needed, the output amplitude is larger. To get full logic levels, we pass these signals through inverters (that add more delay as seen Fig. 19.64b).

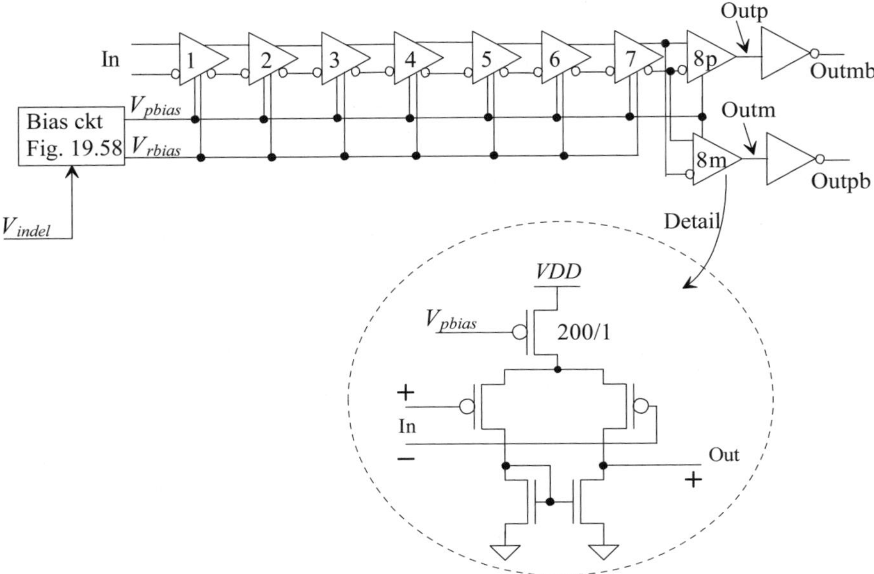

Figure 19.63 Modifying the VCDL to generate full output logic levels.

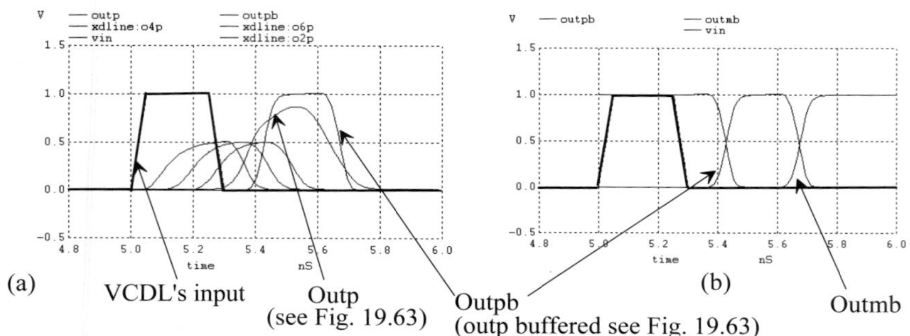

Figure 19.64 The operation of the VCDL in Fig. 19.63.

Using the simulations that generated Fig. 19.64, we can get a reasonable estimate for the VCDL's gain as

$$K_V = \frac{75\ ps}{100\ mV} = 750\,\frac{ps}{V}$$

The minimum delay through the VCDL is roughly 300 ps, while the maximum delay is infinite when V_{indel} moves towards V_{THN}. As seen in Fig. 19.60, we can set the minimum current that flows, I_{REF} , and thus the maximum delay. While this is an important concern we won't address it any further here.

For the PD in our DLL, let's use the PFD detector and charge pump from Ex. 19.5. The PFD's gain can be written as

$$K_D = \frac{I_{pump}}{2\pi} = \frac{10\ \mu A}{2\pi} = 1.59\times10^{-6}\ amps/radian$$

Using Eq. (19.71) with a lock time of 50 cycles gives

$$C_1 = \frac{(750\times10^{-12})\cdot 2\cdot(10\times10^{-6})\cdot 50}{2.2} = 340\,fF$$

As indicated in the discussion following Eq. (19.71), however, this capacitor's selection is based, in some designs, not on lock time but rather the allowable jitter in the output signal once the loop is locked. A block diagram of our DLL is seen in Fig. 19.65. Simulations show that, indeed, the loop locks within 50 cycles with this value of loop filter capacitor (340 *fF*). However, the ripple on the control voltage is 20 mV. With the VCDL gain given above, we can estimate the output jitter as 15 ps. Since the period is 500 ps, this is 3% of the period. This much jitter, in a general application, may not be too bad. When we derived Eq. (19.71), we assumed that the gain of the VCDL was linear. If it is not linear the DLL can exhibit some second-order locking effects (meaning the response isn't truly first-order), resulting in a static phase error or a dead zone (the loop doesn't lock tightly to the input signal). To reduce the jitter and to linearize the loop's response, let's use a large capacitor for the loop filter (a 5 pF) in this example.

The simulated input and output signals of the DLL in Fig. 19.65 using a 5 pF capacitor are seen in Fig. 19.66. Notice that the static phase error. The center of the rising edge of the input signal isn't exactly occurring at the same time as the center of the rising edge of the output signal. Further, notice the asymmetry in the output signal, which stays

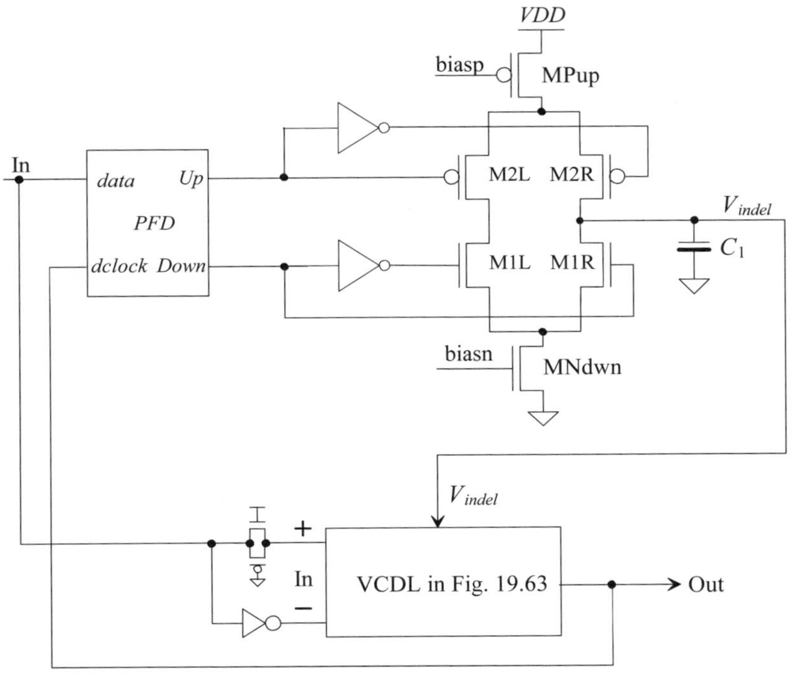

Figure 19.65 Block diagram of a DLL.

high longer than it stays low. This skew, as discussed in Sec. 18.3, is caused by different drive strengths of NMOS and PMOS devices or by differing slopes of input signals. We used the half-replica bias circuit in Fig. 19.58 to make the delay line more tolerant of power supply and ground noise. The drawback is that, at the end of the line, we do have to generate full logic levels. This causes skew between the final output of the VCDL (stage 8 in Fig. 19.63) and the outputs of the other stages in the line. The simple solution

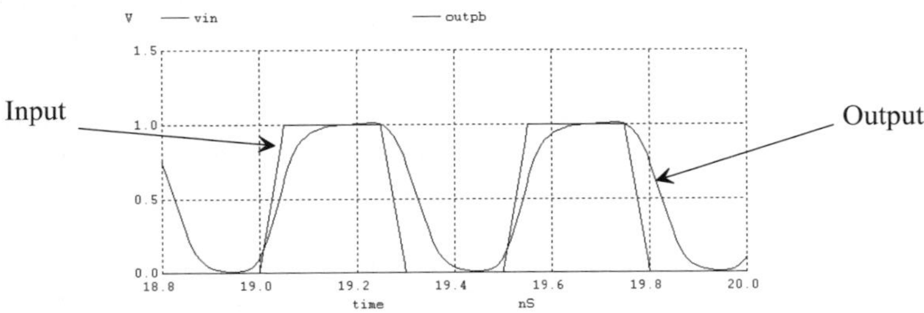

Figure 19.66 Input and output of the DLL in Fig. 19.65.

to this problem is to use a delay stage where the outputs swing to full logic levels. To reduce the DLL's susceptability to power supply and ground noise, we can power it with its own on-chip regulated power supply, supply power and ground to the DLL from separate *VDD* and ground pads, or use some simple filtering to ensure the noise on *VDD* is slow enough that the DLL can respond and correct for variations in the VCDL's delay circuits. For example, we might, in Fig. 19.61, reduce the number of stages to four and add, on the output of each of the delay cells seen in this figure, two of the PMOS buffers seen in Fig. 18.21. We need two buffers because we swap the inputs of each buffer to generate both true and compelement output logic levels. An example of this delay cell is seen in Fig. 19.67. We only use four stages now (for the 500 ps delay line in this section) because of the extra capacitive loading on the output of each stage. The buffered delay cell outputs drive an external load while the basic delay cell outputs are used to drive the input of the next delay cell in the VCDL (only).

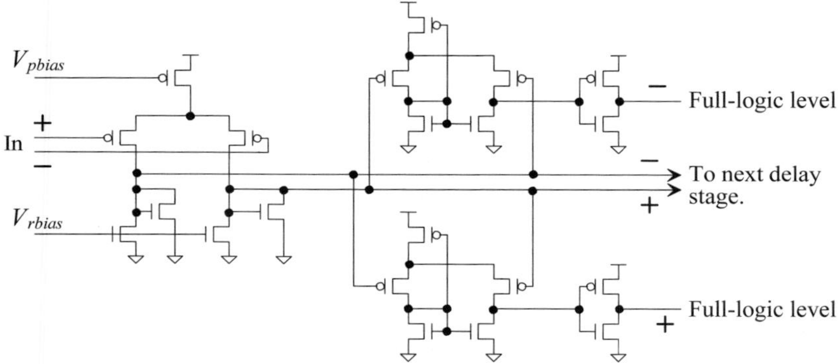

Figure 19.67 Generating full-logic levels in each delay stage.

19.6.2 A 1 Gbit/s Clock-Recovery Circuit

As another example, let's discuss and simulate the design of a 1 Gbit/s clock recovery circuit using the NRZ data format. Let's use the VCDL we've already developed in the VCO (see Fig. 19.68). The output of the VCDL is fed back to its input (with an inversion) to get the positive feedback needed for oscillations. Note that the VCO oscillates at 1 GHz when V_{inVCO} is (roughly) 350 mV (which is not *VDD*/2). While, in a production part, centering the VCO's operating frequency is important, we don't modify the VCDL in Fig. 19.63 for the PLL design here.

We will be using the Hogge phase detector in this design. Since the Hogge PD doesn't perform frequency detection (like the PFD), we must be concerned with locking on harmonics. For example, we might have an input NRZ data stream of 1000 0000. Our loop locks with the clock centered on the one and then six other edges centered on the seven zeroes. The loop is locked just not at the right clock frequency (it's locked at 7/8 of the correct frequency). While this is an **important practical** concern, we won't discuss it further until the end of the section. We will use initial conditions on the loop filter to set the initial clock frequency close to the correct value to avoid locking on a harmonic.

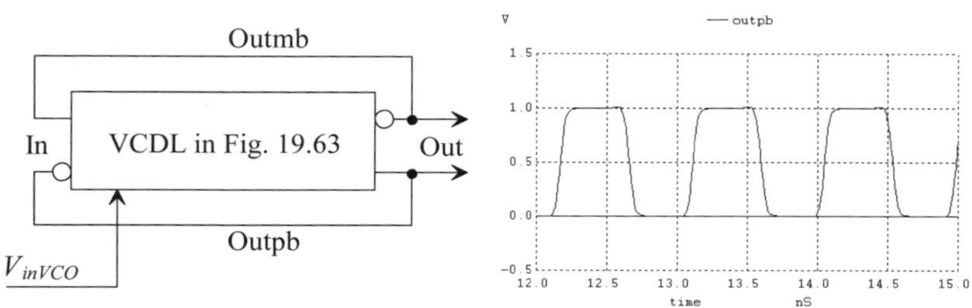

Figure 19.68 Making a VCO with the delay line in Fig. 19.63 and its output when Vinvco = 350 mV.

We know that the gain of the VCDL is 750 ps/V. Since, for one complete oscillation of the VCO, the signal must travel through the VCDL twice, we can estimate the gain of the VCO as

$$K_{VCO} = 2\pi \cdot \frac{1.175\,GHz - 1\,GHz}{100\,mV} = 11 \times 10^9\ radians/V \cdot s$$

where we assume at 350 mV the output frequency is 1 GHz (period is 1 ns). If we increase V_{inVCO} by 100 mV, we get a decrease in the period by 150 ps ($2 \times 100\,mV \cdot 750\,ps/V$) resulting in an output frequency of 1.175 GHz (period is 850 ps).

To implement the Hogge PD seen in Fig. 19.49, we'll use the same TSPC edge-triggered latch that we used for the divide-by-2 seen in Fig. 19.24. The result is seen in Fig. 19.69. We've been very careful to buffer the inputs and outputs of the PD to ensure high-speed output edges and to square up the input signals. To verify the operation, we've simulated the PD with the NRZ and clock signals seen in Fig. 19.49, Fig. 19.70. Looking at the *increase* and *decrease* signals, we see that the width of the *increase* signal is too large. On careful inspection, the cause of this error is the delay through the DFF the NRZ data sees to node A. To compensate for this delay, we'll add a delay in parallel with this path to the XOR gate, as seen in Fig. 19.71. Figure 19.72 shows the resulting simulation output. Again note that unlike the PFD, which only looks at the edges, the Hogge PD, like the XOR PD, uses the widths of its inputs to drive the loop filter.

When the Hogge PD is used with a charge pump the gain is, see Eq. (19.67),

$$K_D = \frac{I_{pump}}{\pi}$$

If we use the charge pump from the DLL in Fig. 19.65 with $I_{pump} = 10\ \mu A$, the gain of the PD is 3.2 μA/radian. If we set the natural frequency, ω_n, to, 100×10^6 radians/s and $\zeta = 1$, then using Eq. (19.58)

$$RC_1 = 20\ ns$$

and using Eq. (19.57) we can solve for C_1 (knowing that there isn't a feedback divider so N is 1) as

$$C_1 = \frac{K_D K_{VCO}}{\omega_n^2} = \frac{(3.2 \times 10^{-6})(11 \times 10^9)}{(100 \times 10^6)^2} = 3.5\ pF$$

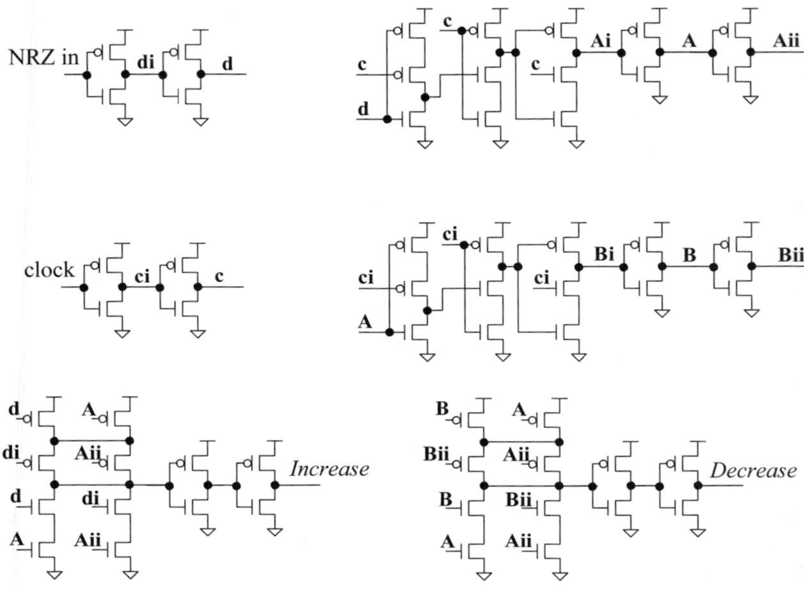

Figure 19.69 CMOS implementation of the Hogge PD.

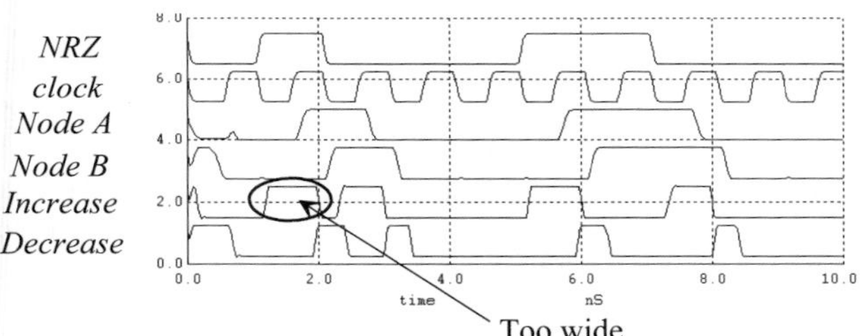

Figure 19.70 Simulating the Hogge PD in Fig. 19.69. These results should be compared to Fig. 19.49. Notice how the increased pulse widths are too wide. The result is that the Vinvco control voltage will increase above the desired value (and cause a static phase error or false locking).

Let's set C_1 to 3.5 pF, R to 5k, and C_2 to 0.35 pF. The schematic of the complete DPLL clock-recovery circuit is seen in Fig. 19.73.

In the first simulation, Fig. 19.74, we apply an alternating string of ones and zeroes to the PLL clock-recovery circuit. Since the data rate is 1 Gbit/s, the width of a one or a zero is 1 ns. In this simulation we didn't apply any initial conditions to the loop filter to start the simulation out. However, because the NRZ data is full of transitions, V_{inVCO} quickly moves to the correct voltage (around 350 mV). The important thing to note is that, again, the lock time depends on the input data.

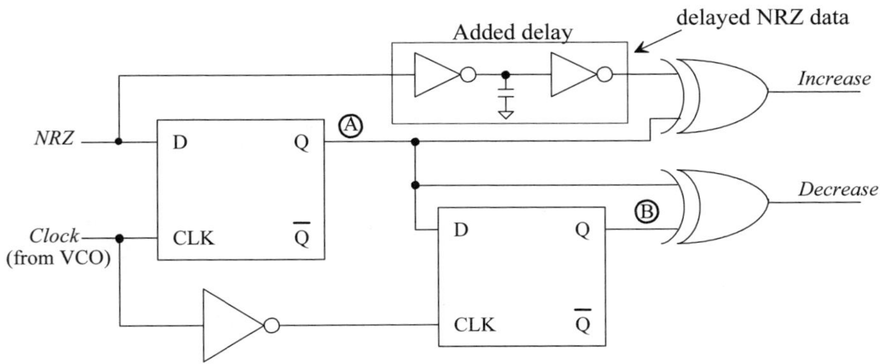

Figure 19.71 Adding a delay to the Hogge PD to compensate for the delay the NRZ data sees to node A.

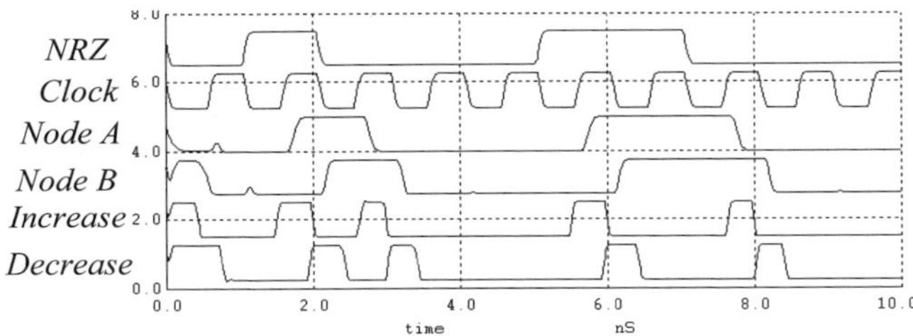

Figure 19.72 Resimulating the operation of the Hogge PD with the delay seen in Fig. 19.71 included.

For the second simulation, Fig. 19.75, we start the voltage across the loop filter, V_{inVCO}, out at 340 mV (slightly below the final value of around 350 mV). After approximately 600 ns, the loop locks. Had we not used this initial condition, the lock time (and the simulation time required to attain lock) would be quite long. Because of the large gain of the VCO, not using an initial voltage for V_{inVCO} would result in the VCO oscillating at the wrong frequency and the loop not locking correctly (locking on the wrong frequency) when the data is, as seen, a single one followed by seven zeroes (or some other sparse pattern). To avoid this problem, we can require the PLL sees a long pattern of alternating ones and zeroes to ensure lock prior to sending data (called sending a preamble) or we can reduce the gain of the VCO. The problem with sending a preamble is that it reduces the data rate through the channel. Further, large VCO gain makes it easier for the PLL to lose lock with steps in the phase shift through a communication channel. The practical solution is to reduce the gain of the VCO. In this example if we were to limit the VCO's oscillation range to, say, 950 MHz to 1050 MHz, then (with proper design of the loop filter) the PLL could acquire lock quickly and stay locked with reasonable variations in the phase shifts of the input NRZ data (these phase shifts are common in channels whose path length is not fixed, as in wireless communications). The practical problem with low VCO gain, again, is manufacturability. It's impossible to

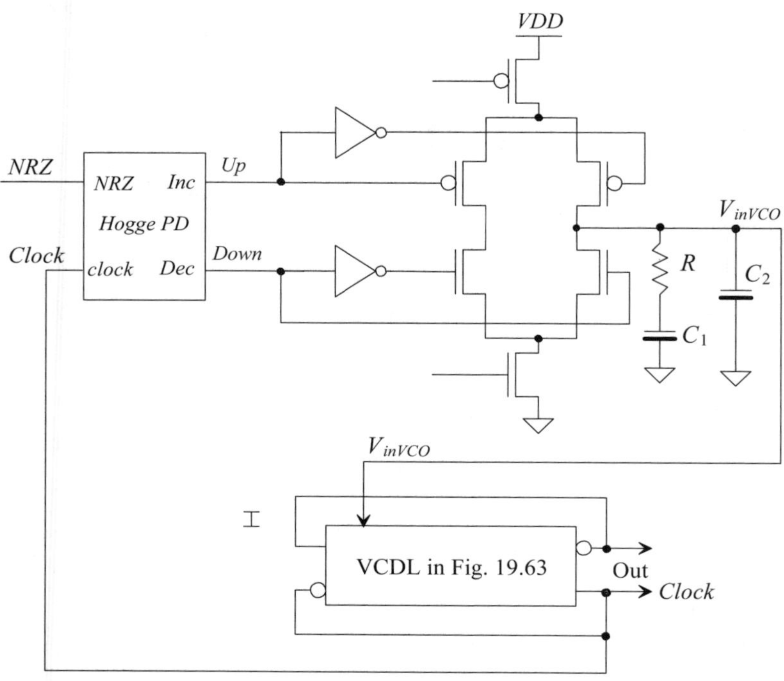

Figure 19.73 Block diagram of the DPLL used for clock-recovery discussed in this section.

design a purely monolithic CMOS oscillator (meaning no off-chip components) without some sort of hand tuning (meaning using fuses or some other sort of digital adjustment to center the oscillation frequency and limit the gain after the chips are made). Further, even if we hand-tailor the performance of each VCO in a production line, the temperature and power supply sensitivity will limit the minimum gain we can attain. Tuned circuits

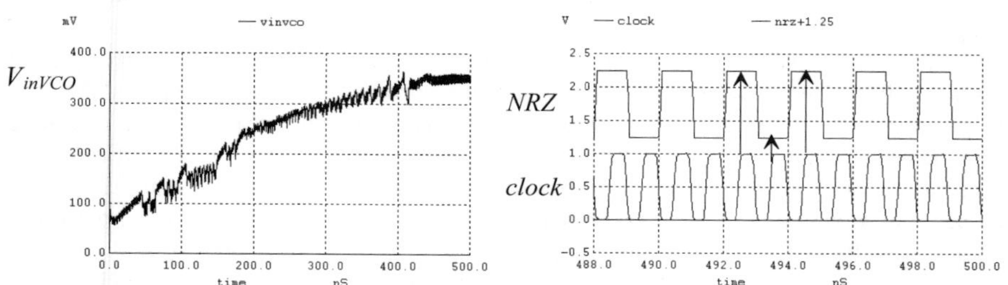

Figure 19.74 Simulating the PLL in Fig. 19.73 when the input NRZ data is an alternating string of ones and zeroes.

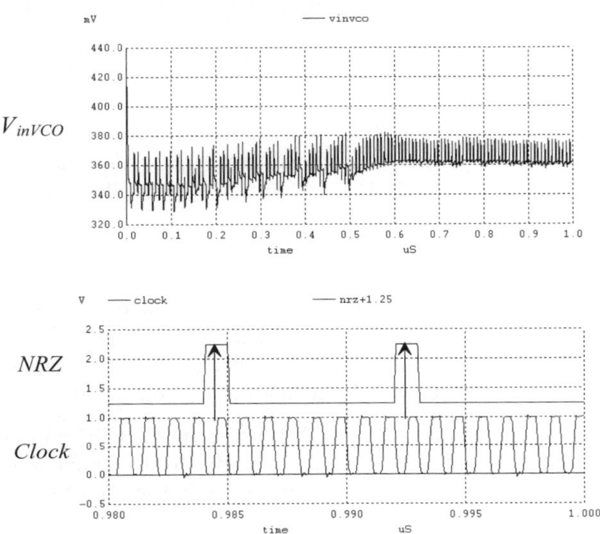

Figure 19.75 Simulating the PLL in Fig. 19.73 when the input data is a string of seven zeroes followed by a single one.

(parallel inductor-capacitor "tanks") are commonly used to help make purely monolithic clock-recovery circuits manufacturable. However, hand-tuning during Probe is still common to set the oscillation frequency and oscillation range.

ADDITIONAL READING

[1] R. E. Best, *Phase-Locked Loops, Theory, Design and Applications,* McGraw-Hill, 2nd ed., 1993. ISBN 0-07-911386-9. Excellent book covering PLL design.

[2] D. H. Wolaver, *Phase-Locked Loop Circuit Design*, Prentice-Hall, 1991. ISBN 0-1366-2743-9. Another excellent book covering PLL design.

[3] W. F. Egan, *Phase-Lock Basics*, John Wiley and Sons, 1998. ISBN 0-4712-4261-6 Good introduction to PLLs. See also the same author's *Frequency Synthesis by Phase Lock.*

[4] F. M. Gardner, "Charge-Pump Phase-Lock Loops," *IEEE Transactions on Communications,* COM-28, no. 11, pp. 1849–1858, November 1980. The paper first discussing charge pump PLLs.

[5] B. Razavi, *Monolithic Phase-Locked-Loops and Clock Recovery Circuits,* IEEE Press, 1996. ISBN 0-7803-1149-3. Good tutorial and collection of papers.

[6] S. Haykin, *An Introduction to Analog and Digital Communications,* John Wiley and Sons, 1989. ISBN 0-471-85978-8. Covers distortionless transmission.

[7] C. R. Hogge, Jr., "A Self Correcting Clock Recovery Circuit," *IEEE Journal of Lightwave Technology*, vol. LT-3, pp. 1312–1314, December 1985. Paper presenting the "Hogge" phase detector.

[8] M. G. Johnson and E. L. Hudson, "A Variable Delay Line PLL for CPU - Coprocessor Synchronization," *IEEE Journal of Solid-State Circuits,* vol. SC-23, pp. 1218–1223, October, 1988. Classic paper presenting the concept of a delay-locked loop.

[9] I. A. Young, J. K. Greason, and K. L. Wong, "A PLL Clock Generator with 5 to 110 MHz of Lock Range for Microprocessors," *IEEE Journal of Solid-State Circuits,* vol. SC-27, pp. 1599–1607, November 1992. Presents the practical design of a CMOS delay element.

PROBLEMS

19.1 The XOR gate PD seen in Fig. 19.4 can exhibit input-dependent skew. In other words, the delay from one of the inputs changing to the output of the PD will not be precisely the same as the delay from the other input to the output. Design an XOR PD that doesn't exhibit input-dependent skew. Hint: see Fig. 18.16.

19.2 Verify, using simulations, that a locked PLL using an XOR PD will exhibit, after RC filtering, an average value of *VDD*/2. Show, using simulations and hand calculations, the filter's average output if the XOR PD sees a phase difference in its inputs of $-\pi/4$.

19.3 Why, in your own words, is it so important for the VCO's center frequency to match the input NRZ data rate when a passive loop filter is used. Use simulations to show the problems. Why does using the active loop filter eliminate this requirement?

19.4 Suppose, for a robust high-speed PLL using an XOR PD, that it desirable to adjust the PD's gain. Show a charge pump can be used towards this goal. Should the loop filter have two outputs? Discuss the PD's gain and how it would be adjusted. What topology would be used for a passive loop filter? for an active loop filter?

19.5 Demonstrate, using simulations, how the outputs of the XOR PD can be equivalent when the loop is true or false locking (locking on a harmonic).

19.6 Describe, in your own words and using simulations, what the dotted line in Fig. 19.8 indicates.

19.7 We know that a PFD isn't generally used in clock recovery applications because both edges (*data* and *dclock*) must present when doing a phase comparison. Suggest a scheme where the edge of the input *data* enables a phase comparison. When an edge of data is not present your circuit will "swallow" the pulse from *dclock* resulting in no edges being applied to the PFD (no phase- frequency comparison).

19.8 Demonstrate, using simulations, charge sharing between the charge pump and loop filter using the topology seen in Fig. 19.12b (and Fig. 19.13b). Show how the topology in Fig. 19.37 helps with charge sharing (simulate with and without the x1 amplifier). For the x1 amplifier, use the n-type diff-amp from Fig. 18.17 without the inverter on its output. Connect the loop filter to the gate of M1 (in the diff-amp). To make the gain "1," tie the output (which is connected to drains of M1L/M2L in Fig. 19.37) of the diff-amp (the drains of M2/M4) to the gate of M2.

19.9 Using the VCO that generated the simulation data in Fig. 19.18, plot the VCO's center frequency against changes in *VDD*. Is this VCO insensitive to changes in *VDD*? Where does the sensitivity come from?

19.10 Discuss, and demonstrate with simulations, how to reduce the gain of the VCO used to generate the data in Fig. 19.18 (see Fig. 19.25 and the associated discussion). Your discussion should include some insight into the manufacturability of low-gain VCOs.

19.11 Design and simulate the operation of a 100 MHz VCO using the topology seen in Fig. 19.19a. The output of your VCO should be full logic levels. Your design should show a linearly frequency against V_{inVCO} curve. What is the gain of your design?

19.12 Using the step response of an RLC circuit, Fig. 19.76, demonstrate how selection of the resistor, inductor, and capacitor affect the output voltage's damping factor and natural frequency. From this plot show how natural frequency and lock time are related.

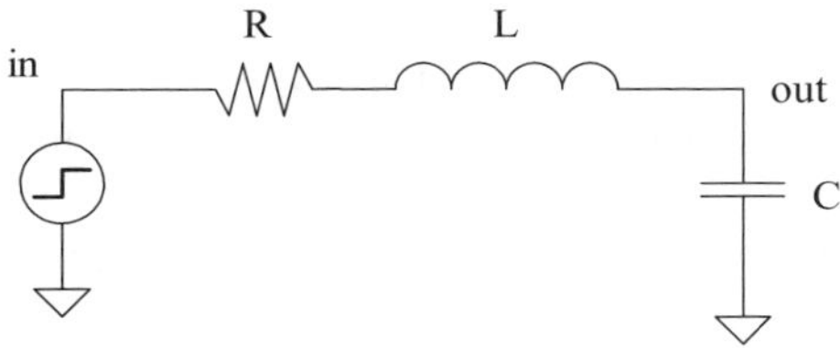

Figure 19.76 Using a second-order circuit to demonstrate how lock time and natural frequency are related.

19.13 Replace, in the DPLL that generated the waveforms in Figs. 19.27 and 19.28, the simple RC loop filter with the passive lag loop filter in Fig. 19.29. Show how the filter's component values are selected. Comment on, using simulations to support your conclusions, how the performance of the DPLL is affected.

19.14 Using the PFD in Fig. 19.33 and the charge pump in Fig. 19.36, demonstrate, with simulations, how the gain of PFD/charge-pump configuration can have a dead zone (the gain, K_{PDI}, decreases) as seen in Fig. 19.77.

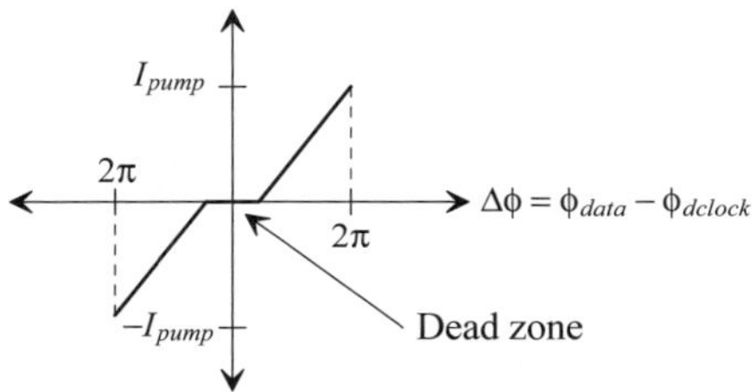

Figure 19.77 Showing the dead zone in a PFD/charge-pump.

19.15 Suppose that the pump currents used in the DPLL in Ex. 19.5 are mismatched by 50%. Will this cause a static phase error? Why or why not? Use simulations to support your answer.

19.16 Redesign the DPLL in Ex. 19.5 so that the output remains a 100 MHz square wave signal when the input signal is changed to 25 MHz.

19.17 We used pF values for the capacitors in the loop filters in this chapter. Why not reduce the filter's layout area by using fF size capacitors? Demonstrate the problems with using such small loop filter capacitors.

19.18 What are we sacrificing by using an equalizer? What does the minus sign indicate in the slope of the phase in Fig. 19.41?

19.19 Using the DPLL from Ex. 19.2, demonstrate the false locking as seen in Fig. 19.46.

19.20 Design an edge detector, like the one seen in Fig. 19.47, for use in the DPLL in Ex. 19.2. Regenerate the simulation data seen in Figs. 19.27 and 19.28. Remember to eliminate the divide by two in the feedback path.

19.21 Derive Eq. (19.71).

19.22 Design a nominal delay line of 1 ns (when V_{indel} = 500 mV) using the current starved delay element seen in Fig. 19.55. Determine the delay's sensitivity to variations in *VDD*.

19.23 Repeat Problem 19.22 for the inverter delay cell in Fig. 19.55.

19.24 Suppose, as seen in Fig. 19.78, that instead of using a half-replica of the delay cell in Fig. 19.58, the full delay cell is used to generate V_{rbias}. Electrically is there any difference in V_{rbias} when comparing the full- and half-replica circuits? What may be the benefit of using a full-replica of the delay cell?

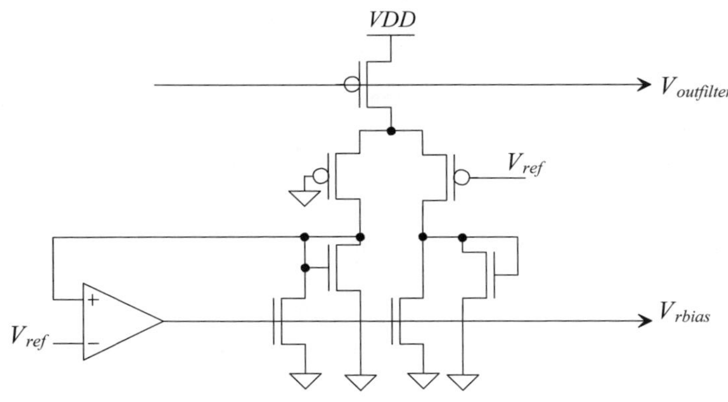

Figure 19.78 Using a full-replica of the delay element for generating Vrbias.

19.25 For the delay line that generated the simulation results in Fig. 19.62 determine, using simulations, the delay's sensitivity to changes *VDD*. Plot the VCDL's delay as a function of *VDD* with V_{indel} held at 500 mV.

19.26 The delay element seen in Fig. 19.79 doesn't use a reference voltage. Show how this element can be used in the VCDL of Fig. 19.61. Note that the input capacitance of this delay stage is twice as large as the input capacitance of the element in Fig. 19.61 (and so you may have to use fewer stages to attain the same overall delay). How does the NMOS gate potential change with the inputs/outputs switching? Why? Is the output amplitude a function of *VDD*?

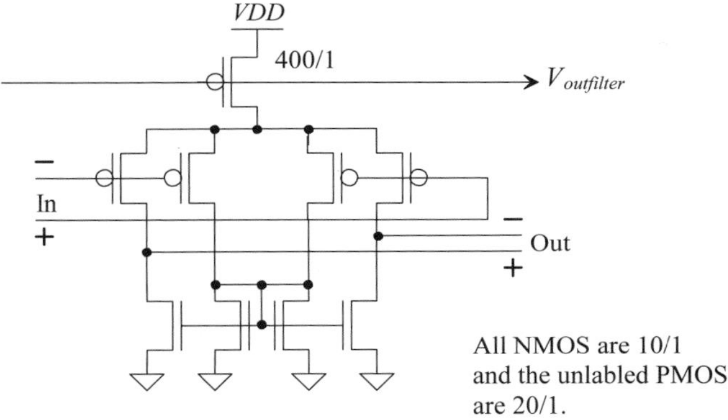

Figure 19.79 A delay element that doesn't use a reference voltage.

19.27 Use the delay element in Fig. 19.67 to implement a VCDL. Use the VCDL in the DLL seen Fig. 19.65 to generate the waveforms in Fig. 19.66. Show the 90, 180, 270, and 360 degree outputs of the VCDL swinging to full-logic levels.

19.28 The VCO output seen in Fig. 19.68 doesn't have an exactly 50% duty cycle. Will this affect a clock-recovery circuit's operation? Why or why not? Use simulations to support you answer.

19.29 Suggest another method, other than the one seen in Fig. 19.72, to equalize the widths of the *Increase* and *Decrease* outputs of the Hogge PD. Verify your design with simulations.

19.30 Must the currents in the charge pump in Fig. 19.73 be equal for proper DPLL clock-recovery operation? Why or why not? What happens if the currents aren't equal? What happens if the NMOS switches turn on at different speeds than the PMOS switches? Use SPICE to support your answers.

19.31 Modify the simulation inputs in Fig. 19.74 to show false locking. Comment on what can be done in a practical clock-recovery circuit to eliminate false locking.

19.32 Replace the charge pump used in the DPLL in Fig. 19.73 with the active-PI loop filter seen in Fig. 19.50. Calculate the loop filter component values. Using the new loop filter, regenerate Figs. 19.74 and 19.75.

Chapter 20

Current Mirrors

In this chapter we turn our attention towards the design, layout, and simulation of current mirrors (a circuit that sources [or sinks] a constant current). As we observed back in Fig. 9.1, and the associated discussion, the ideal output resistance, r_o, of a current source is infinite. Achieving high output resistance (meaning that the output current doesn't vary much with the voltage across the current source) will be the main focus of this chapter.

It's very important that the reader first understand the material in Ch. 9 concerning the selection of biasing currents and device sizes and how they affect the gain/speed of the analog circuits. We'll use the parameters found in Tables 9.1 and 9.2 in many of the examples in this chapter.

20.1 The Basic Current Mirror

The basic NMOS current mirror, made using M1 and M2, is seen in Fig. 20.1. Let's assume that M1 and M2 have the same width and length and note that $V_{GS1} = V_{DS1} = V_{GS2}$. Because the MOSFETs have the same gate-source voltages, we expect (neglecting channel-length modulation) them to have the same drain current. If the two resistors in the drains of M1/M2 are equal, the drain of M2 will be at the same potential as the drain of M1 (this is important). By matching the size, V_{GS}, and I_D of two transistors, we are assured that the two MOSFETs have the same drain-source voltage, $(V_{GS1} = V_{DS1} = V_{GS2} = V_{DS2})$.

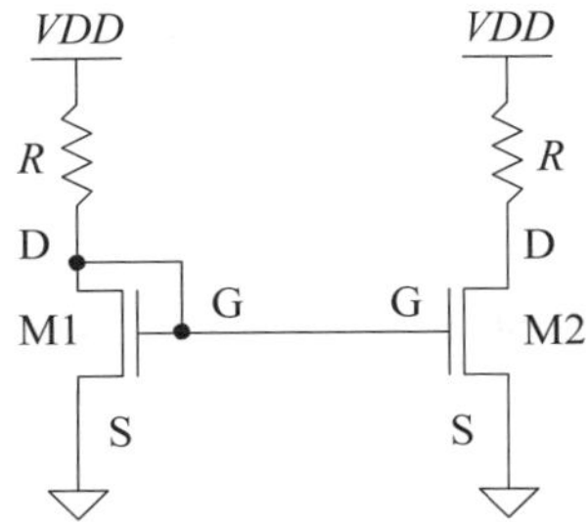

Figure 20.1 A basic current mirror.

20.1.1 Long-Channel Design

Examine Fig. 20.2. In this figure we show a current mirror and the equivalent circuit representation of a current source. Looking at M1, we can write

$$I_{REF} = I_{D1} = \frac{KP_n}{2}\frac{W_1}{L_1}(V_{GS1} - V_{THN})^2(1 + \lambda(V_{DS1} - V_{DS1,sat})) \tag{20.1}$$

knowing $V_{DS1} = V_{GS1}$ and $V_{DS1,sat} = V_{GS1} - V_{THN}$. For M2, we write

$$I_O = I_{D2} = \frac{KP_n}{2}\frac{W_2}{L_2}(V_{GS1} - V_{THN})^2(1 + \lambda(V_O - V_{DS1,sat})) \tag{20.2}$$

noting $V_{GS1} = V_{GS2}$, $V_{DS1,sat} = V_{DS2,sat}$, and V_o is the voltage across the current source. Looking at the ratio of the drain currents, we get

$$\frac{I_O}{I_{REF}} = \frac{W_2/L_2}{W_1/L_1} \cdot \frac{1 + \lambda(V_O - V_{DS1,sat})}{1 + \lambda(V_{DS1} - V_{DS1,sat})} \tag{20.3}$$

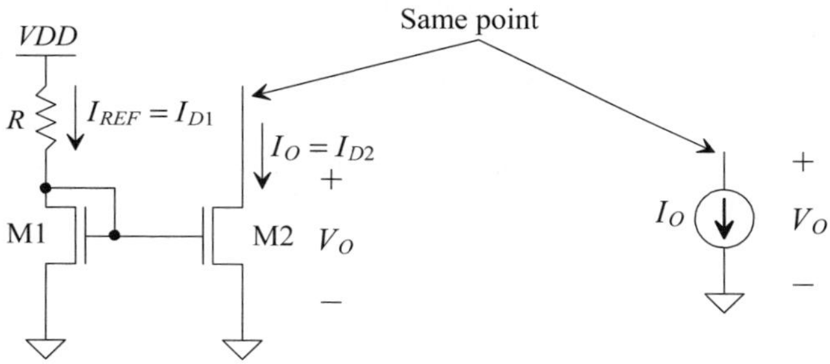

Figure 20.2 The current mirror and how we think about it.

Generally, the lengths of the devices in the current mirror are equal (let's assume they are for the moment). If, also at this time, we don't concern ourselves with channel-length modulation ($\lambda = 0$), we get a very useful result, that is,

$$\frac{I_O}{I_{REF}} = \frac{W_2}{W_1} \tag{20.4}$$

By simply scaling the width of M2, we can adjust the size of our output current. Figure 20.3 shows an example of this scaling (using PMOS devices).

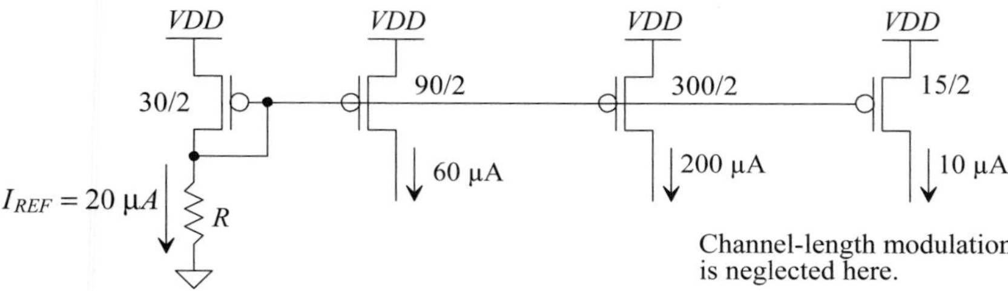

Figure 20.3 How current mirrors are ratioed.

Example 20.1
Determine the value of the resistor, R, needed in Figs. 20.2 and 20.3 so that the reference drain currents are 20 μA. Use the long-channel parameters seen in Table 9.1. Simulate the operation of the NMOS mirror with the calculated resistor value.

In Fig. 20.2 (the NMOS current mirror), we can write

$$I_{REF} = 20\ \mu A = \frac{VDD - V_{GS1}}{R} \approx \frac{KP_n}{2} \cdot \frac{10}{2} \cdot \left(\overbrace{V_{GS1}}^{1.05\ V} - \overbrace{V_{THN}}^{0.8} \right)^2 = \frac{KP_n}{2} \cdot \frac{10}{2} \cdot (0.25)^2$$

(notice how we indicate "approximately" because channel-length modulation is not included in the equations) or

$$R = \frac{5 - 1.05}{20\ \mu A} \approx 200\ k\Omega$$

For Fig. 20.3 (the PMOS current mirror), we can write

$$I_{REF} = 20\ \mu A = \frac{VDD - V_{SG}}{R} \approx \frac{KP_p}{2} \cdot \frac{30}{2} \left(\overbrace{V_{SG}}^{1.15\ V} - \overbrace{V_{THP}}^{0.9} \right)^2 = \frac{KP_p}{2} \cdot \frac{30}{2} \cdot (0.25)^2$$

or

$$R = \frac{5 - 1.15}{20\ \mu A} \approx 200\ k\Omega$$

Simulation of the operation of the NMOS current mirror is seen in Fig. 20.4. The reference current isn't exactly 20 μA (and we shouldn't expect it to be). The x-axis is a sweep of the voltage across the current source, V_O. Note that below $V_{DS,sat}$ (= 250 mV here) M2 triodes and the output current, I_O, goes to zero. The output *compliance* range for this current source (the range of output voltages where the current source behaves like a current source, that is, not an open or a resistor) is between VDD and $V_{DS,sat}$. The point where $V_O = V_{DS1} = V_{GS1}$ is where $I_O = I_{REF}$ (again this is important for matching two currents). Finally, note that I_{REF} and V_{GS1} are not dependent on V_O. ■

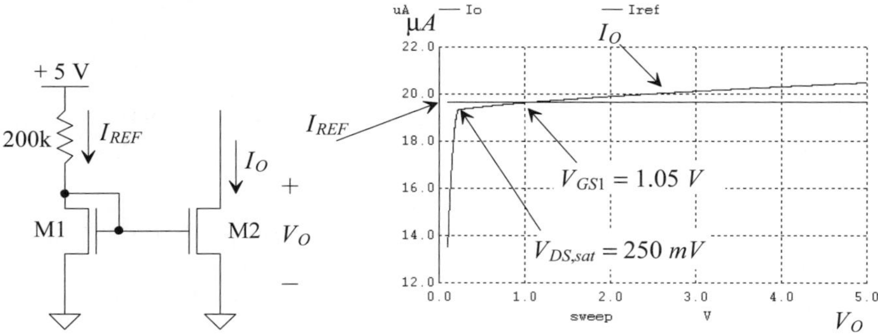

Figure 20.4 The operation of an NMOS current mirror.

20.1.2 Matching Currents in the Mirror

Many analog applications are susceptible to errors due to layout. In circuits in which devices need to be matched, layout becomes a critical factor. For example, in the basic current mirror shown in Fig. 20.2, first-order process errors can cause the output current, I_O, to be significantly different from the reference current. Process parameters such as gate-oxide thickness, lateral diffusion, oxide encroachment, and oxide charge density can drastically affect the performance of a device. Layout methods can be used to minimize the first-order effects of these parameter variations.

Threshold Voltage Mismatch

In a given current mirror application, the values for the threshold voltages are critical in determining the overall accuracy of the mirror. Again, examine the basic current mirror shown in Fig. 20.2. Since both devices have the same value for V_{GS} and assuming the sizes and transconductance parameters for both devices are equal, let's examine the effect of a mismatch in threshold voltages between the two devices. If it is assumed that the threshold mismatch is distributed across both devices such that

$$V_{THN1} = V_{THN} - \frac{\Delta V_{THN}}{2} \tag{20.5}$$

$$V_{THN2} = V_{THN} + \frac{\Delta V_{THN}}{2} \tag{20.6}$$

where V_{THN} is the average value of V_{THN1} and V_{THN2} and ΔV_{THN} is the mismatch, then

$$\frac{I_O}{I_{REF}} = \frac{\frac{KP_n}{2}\frac{W}{L}(V_{GS} - V_{THN} - \frac{\Delta V_{THN}}{2})^2}{\frac{KP_n}{2}\frac{W}{L}(V_{GS} - V_{THN} + \frac{\Delta V_{THN}}{2})^2} = \frac{\left[1 - \frac{\Delta V_{THN}}{2(V_{GS}-V_{THN})}\right]^2}{\left[1 + \frac{\Delta V_{THN}}{2(V_{GS}-V_{THN})}\right]^2} \tag{20.7}$$

If both expressions are squared and the higher order terms are ignored, then the first-order expression for the ratio of currents becomes

$$\frac{I_O}{I_{REF}} \approx 1 - \frac{2\Delta V_{THN}}{V_{GS} - V_{THN}} = 1 - \frac{2\Delta V_{THN}}{V_{DS,sat}} \tag{20.8}$$

Equation (20.8) is quite revealing because it shows that as V_{GS} decreases, the difference in the mirrored currents increases due to threshold voltage mismatch. This is particularly critical for devices that are separated by relatively long distances because the threshold voltage is susceptible to process gradients. *To attain high speed and to reduce the effects of threshold voltage mismatch, a large gate overdrive voltage should be used* (remembering for a long-channel process that $V_{ovn} = V_{DS,sat} = V_{GS} - V_{THN}$). Of course the drawback, for a current mirror, is a reduced range of compliance (the MOSFET enters the triode region earlier).

Transconductance Parameter Mismatch

The same analysis can be performed on the transconductance parameter, KP_n. If $KP_{n1} = KP_n - \Delta KP_n/2$ and $KP_{n2} = KP_n + \Delta KP_n/2$, where KP_n is the average of KP_{n1} and KP_{n2}, then assuming perfect matching on all other parameters, the difference in the currents becomes

$$\frac{I_O}{I_{REF}} = \frac{KP_n + 0.5\Delta KP_n}{KP_n - 0.5\Delta KP_n} \approx 1 + \frac{\Delta KP_n}{KP_n} \tag{20.9}$$

Since KP_n is a process parameter, we might think that as CMOS scales downwards (C'_{ox} goes up) the mismatch due to oxide variations and mobility differences might get better. However, there is less averaging of these variations with the smaller layout sizes. Practically, as we're about to see, differences in V_{DS} dominate the matching behavior.

Drain-to-Source Voltage and Lambda

One aspect of current mirror design that is *critical* for generating accurate currents is the drain-to-source voltage. As seen in Fig. 20.4, the only point where the currents of the devices are actually the same is when their V_{DS} values are equal. In Eq. (20.3) the ratio of the output current to the reference current is affected by both the matching in the drain-to-source voltages (V_O and V_{DS1}) and the device λs. If, for example in the short-channel process (see Table 9.2), V_{DS1} = .35 V, $V_{DS2} = V_O$ = .75 V, and $\lambda_1 = \lambda_2 = 0.6$ V^{-1}, then

$$\frac{I_O}{I_{REF}} = \frac{1+\lambda_2 \cdot V_O}{1+\lambda_1 \cdot V_{DS1}} = \frac{1+0.6 \cdot 0.75}{1+0.6 \cdot 0.35} = 1.20 \qquad (20.10)$$

resulting in 20% error! It is extremely important, for good matching, that the V_{DS} values of the MOSFETs in the current mirror are equal.

Layout Techniques to Improve Matching

Most general analog applications require the length of the gate to be longer than minimum since the channel-length modulation, λ, has less effect on longer devices than on shorter ones (as discussed in Ch. 9). As a result, the minimum-sized devices found in digital circuits are not used as often in general analog design (however, see Ch. 26). However, the larger devices can result in larger parasitics if some layout issues are not considered. Figure 20.5a illustrates a basic MOSFET device with a large *W/L*. The implant resistance of the source and drain can be modeled as shown in Fig. 20.5b. The implant resistance can easily be reduced by simply adding as many contacts as possible along the width of both the source and the drain as seen in Fig. 20.5c. The increase in the number of contacts results in lower resistance, more current capability, and a more distributed current load throughout the device. However, as the device width increases, another technique is used which distributes the parasitics (both resistive and capacitive) into smaller contributions.

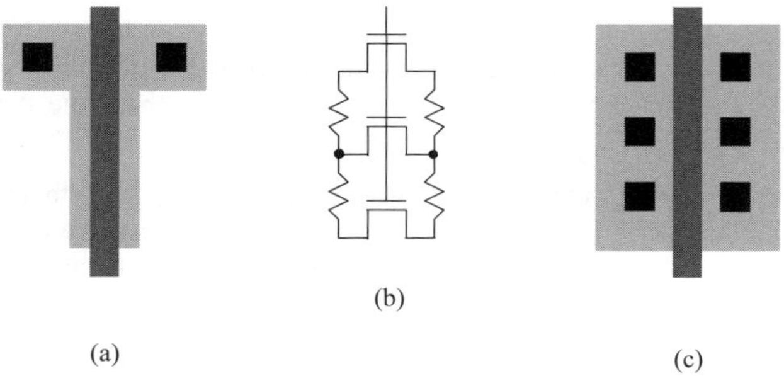

Figure 20.5 (a) Large device with a single contact and (b) its equivalent circuit. (c) Adding more contacts to reduce parasitic resistance.

Examine Fig. 20.6. In this figure, a single device with a large *W*/*L* is split into several parallel devices, each with a width one-fourth of the original *W*. One result of splitting the device into several parts is smaller overall parasitic capacitance associated with the reverse-biased implant substrate diode (the drain or source depletion capacitance to substrate). Since values of C_{db} and C_{sb} are proportional to *W*, the split device reduces these parasitics by a factor of $(n + 1)/2n$ where *n* is the number of parallel devices and is odd. If *n* is even, then the C_{sb} is reduced by one-half and C_{db} is reduced by $(n + 2)/2n$. As seen in Fig. 20.6b, putting the devices in parallel also reduces the parasitic resistance in series with both the source and the drain (it gets cut roughly in half too).

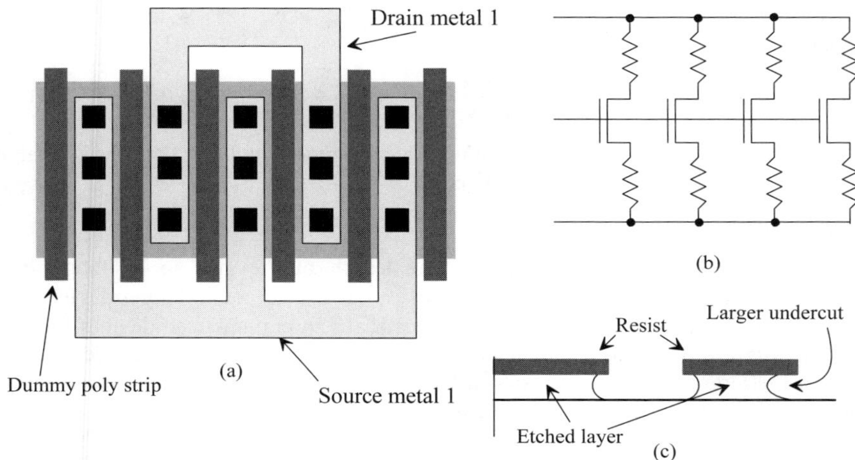

Figure 20.6 (a) A parallel device with dummy strips, (b) the equivalent circuit, and (c) undercutting.

Notice also, that Fig. 20.6a has dummy poly strips on both sides of the device. These strips are used to help minimize the effects of undercutting the poly on the outer edges after patterning, Fig. 20.6c. If the dummy strips had not been used, the poly would have been etched out more under the outermost gates, resulting in a mismatch between the four parallel devices.

When matching two devices, it is imperative that the two devices be as symmetrical as possible. Always orient the two devices in the same direction, unlike those illustrated in Fig. 20.7.

Splitting the devices into parallel devices and interdigitizing them can distribute process gradients across both devices and thus improve matching. An example of this is seen in Fig. 20.8. In (a) the current mirror seen in (b) is laid out. Each MOSFET in (b) is split up into four MOSFETs. If the W/L of each MOSFET in (b) is 80/2, then the size of each MOSFET (finger) in (a) is 20/2. Note the use of dummy poly strips on this layout. A good exercise at this point is to lay out the mirror of Fig. 20.8 in a common-centroid arrangement (see Ch. 5). Using a large layout area (long lengths and wide widths) and common-centroid layouts can result in significant improvements in matching.

Figure 20.7 Devices with differing orientation (bad).

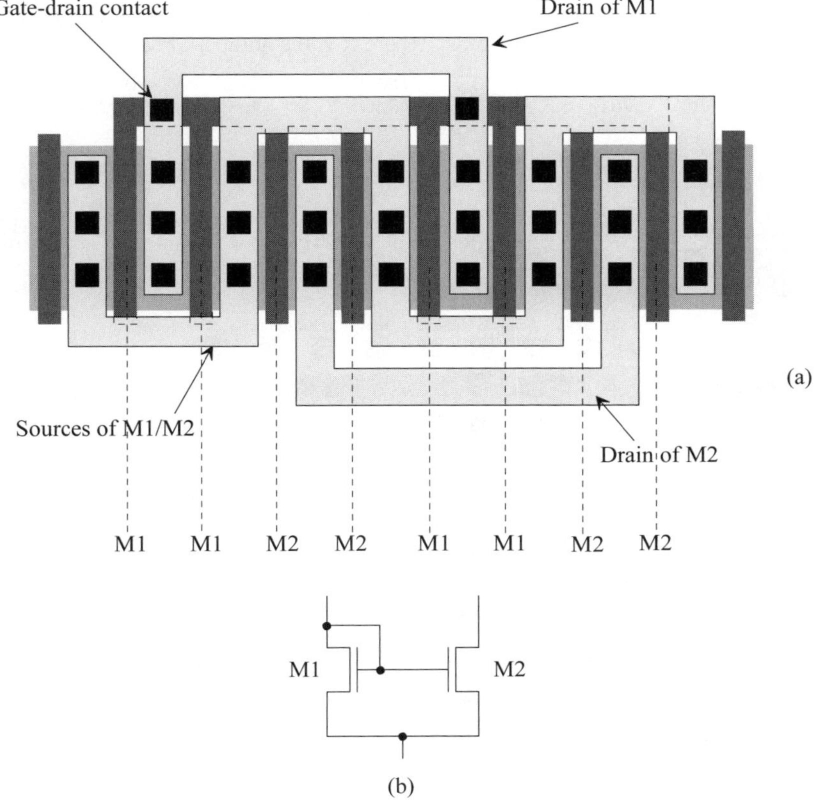

Figure 20.8 (a) Layout of a simple current mirror using interdigitation and (b) equivalent circuit.

Layout of the Mirror with Different Widths

When laying out the current mirror, the lateral diffusion, L_{diff}, under the gate oxide (Fig. 5.13) and the oxide encroachment, W_{enc}, can affect the actual length and width, respectively, of the MOSFET. If not careful with the layout, this will affect the mirror's ratio. Equation 20.3 can be rewritten, without including the differences due to drain-source voltage and lambda, as

$$\frac{I_O}{I_{REF}} = \frac{(W_{2drawn} - 2W_{enc}) \cdot (L_{1drawn} - 2L_{diff})}{(W_{1drawn} - 2W_{enc}) \cdot (L_{2drawn} - 2L_{diff})} \tag{20.11}$$

If the requirement that $L_{1drawn} = L_{2drawn}$ is imposed, then the widths of the devices determine the relative currents in the mirror. Figure 20.9a shows a current mirror layout without width compensation. If W_{enc} is 0.1, then for this layout,

$$\frac{I_O}{I_{REF}} = \frac{40 - .2}{20 - .2} = 2.01 \text{ (a 1\% error due to poor layout)}$$

Figure 20.9b shows how to lay out a current mirror to avoid these problems. The layout of M2 is two MOSFETs in parallel. This can be specified in SPICE by adding M = X after the MOSFET statement in the netlist, where X is the number of MOSFETs,

```
M1    Vd1    Vd1    0    0    NMOS    L=2    W=20
M2    Vd2    Vd1    0    0    NMOS    L=2    W=20   M=2
```

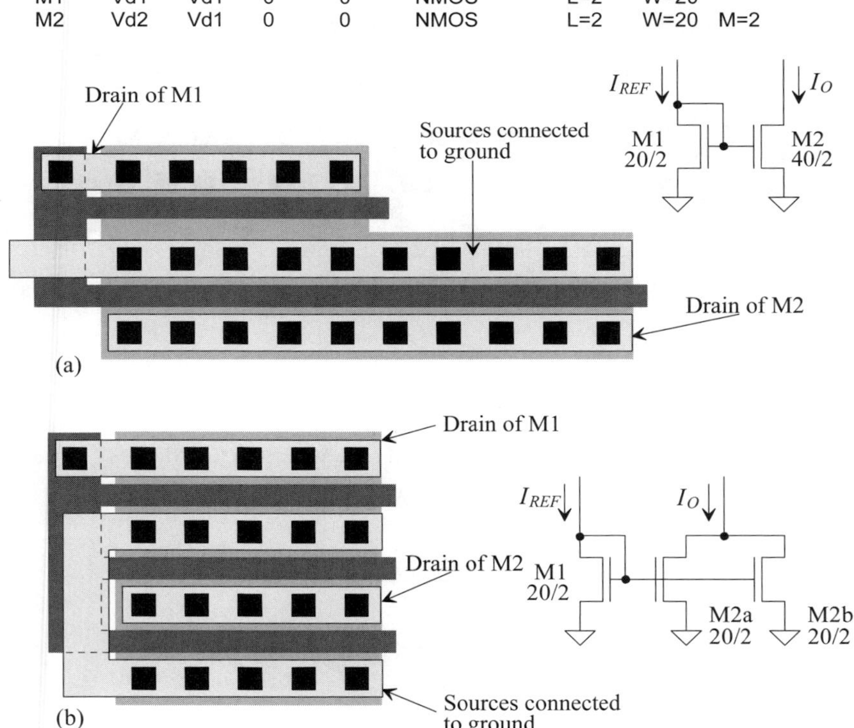

Figure 20.9 Layout of a current mirror (a) without width correction and (b) with width correction.

20.1.3 Biasing the Current Mirror

Using a resistor to set the bias current, as seen in Figs. 20.2–20.4, can result in currents that are too dependent on the power supply value and temperature. Consider the current mirror seen in Fig. 20.10a. In this design we've used the sizes and bias current given in Table 9.2 (the short-channel CMOS process) to select the resistor. In particular, $V_{SG} = 0.35\ V$ so the gate potential is 0.65 V. Figure 20.10b shows how the reference and output currents vary if VDD is swept from 900 mV to 1 V. The reference current is linearly dependent on VDD (as seen in Ex. 20.1). The output current is dependent on both the reference current and the V_{DS} (λ) of M2. In general, we want a reference current that isn't dependent on power supply or ground variations (noise). This is an important point. Consider the following example.

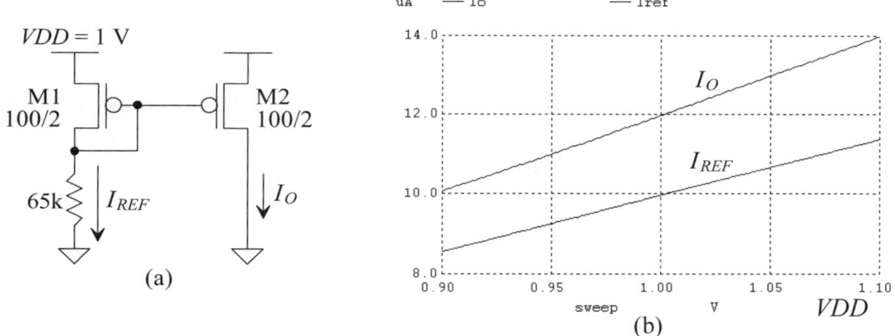

Figure 20.10 How reference and output current vary with VDD.

Example 20.2

Regenerate Fig. 20.10 if the 65k resistor is replaced with an ideal 10 μA current source, Fig. 20.11a. Explain the results.

The simulation results are seen in Fig. 20.11b. Note how the output current variation with VDD is much better than what is seen in Fig. 20.10. The reference current through M1 is, of course, a constant. The output current is larger than the

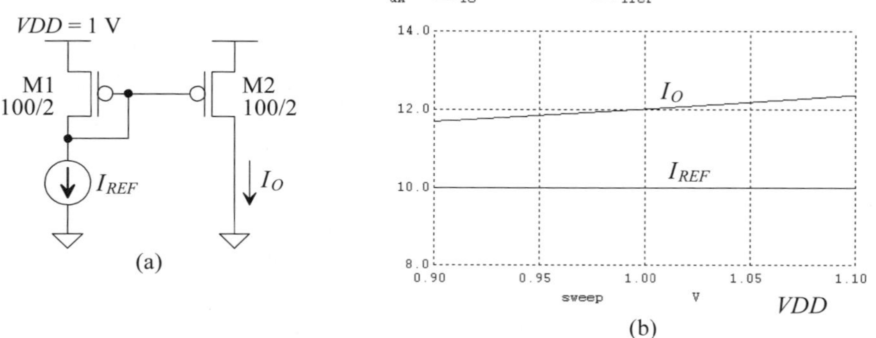

Figure 20.11 Showing that VDD variations don't affect the reference current.

reference current because $V_{SD1} < V_{SD2}$ (notice how often this point is coming up in the design of current mirrors). The reason I_O varies is due to the finite output resistance ($\lambda \neq 0$) of M2, Eq. (20.2).

The point of this example is that it is desirable to have a bias current that doesn't vary with changes in *VDD* or ground. ■

Using a MOSFET-Only Reference Circuit

Let's try replacing the resistor bias with a MOSFET, Fig. 20.12. If we assume long-channel behavior, then

$$VDD = V_{SG3} + V_{GS1} \tag{20.12}$$

or

$$VDD = \sqrt{\frac{2I_{REF}}{KP_p \frac{W_3}{L_3}}} + V_{THP} + \sqrt{\frac{2I_{REF}}{KP_n \frac{W_1}{L_1}}} + V_{THN} \tag{20.13}$$

If we are going to think of M3 as a resistor, as in Fig. 20.2, so that we mirror the current in M1, then we solve for the size of M3. Using the data from Table 9.1, we get

$$5 = \sqrt{\frac{2 \cdot 20}{40 \cdot \frac{W_3}{L_3}}} + 0.9 + \sqrt{\frac{2 \cdot 20}{120 \cdot \frac{10}{2}}} + 0.8 \rightarrow \frac{W_3}{L_3} = 0.11 \approx \frac{10}{90} \tag{20.14}$$

If we are going to think of M1 as a resistor, as in Fig. 20.10a, so that we mirror the current in M3, then we solve for the size of M1 as we did in Eq. (20.14) and get

$$\frac{W_1}{L_1} \approx \frac{10}{270} \tag{20.15}$$

Figure 20.13 shows the operation of these MOSFET bias circuits with *VDD* swept from 4.5 to 5.5 V. Note that, at 5 V, we should have a bias current of 20 μA. What we get is 14 μA. The reason that we have such a large difference is due to neglecting the output resistance of the MOSFETs (and the effects of mobility degradation). Note that the sensitivities, how the output current changes with *VDD*, are roughly 8 μA/V. (Let's compare this to hand-calculated values of sensitivity next.)

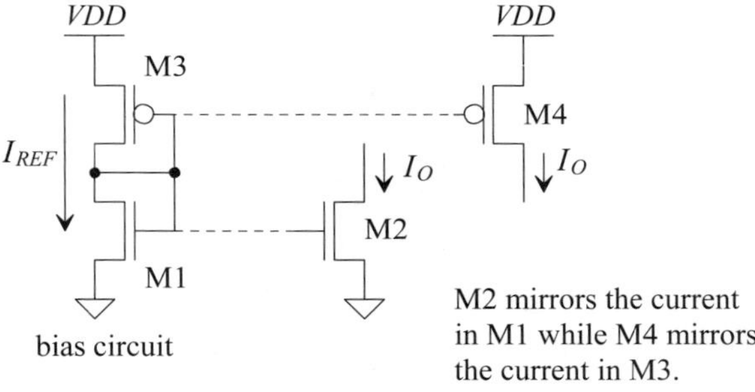

Figure 20.12 A MOSFET-only bias circuit.

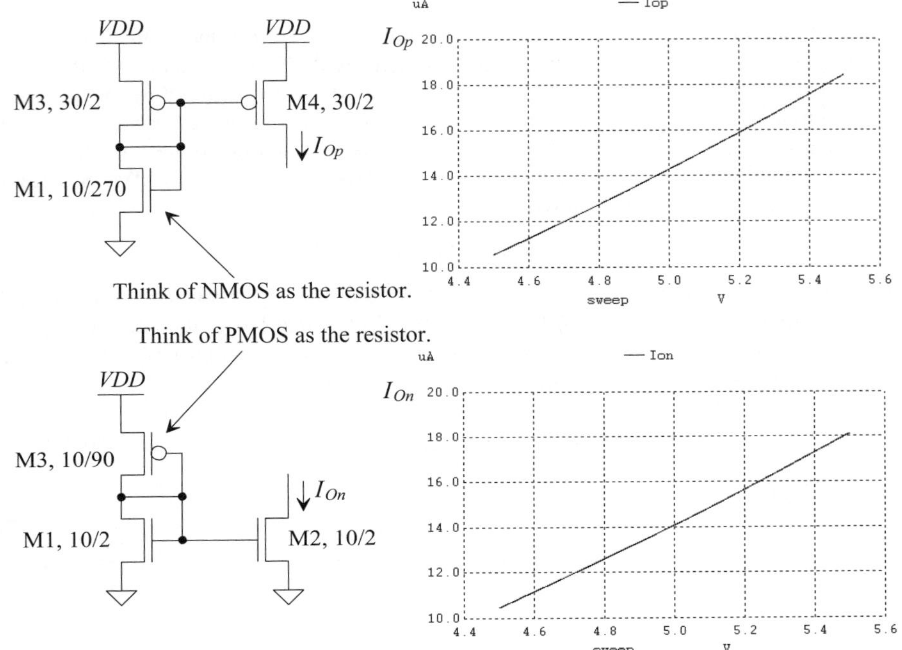

Figure 20.13 Behavior of MOSFET-only bias circuits with changes in VDD.

To determine I_{REF}'s sensitivity to *VDD*, let's take the derivative of the I_{REF} in Eq. (20.13) with respect to *VDD*. First let's write

$$(VDD - V_{THN} - V_{THP})^2 = I_{REF} \cdot \overbrace{\left[\sqrt{\frac{2L_3}{KP_p \cdot W_3}} + \sqrt{\frac{2L_1}{KP_n \cdot W_1}} \right]^2}^{K} \qquad (20.16)$$

and next

$$\frac{\partial I_{REF}}{\partial VDD} = \frac{2 \cdot VDD}{K} - \frac{2 \cdot (V_{THN} + V_{THP})}{K} \qquad (20.17)$$

Using the values in Table 9.1 and Eqs. (20.14) or (20.15), K is approximately 547×10^3 V^2/A so

$$\frac{\partial I_{REF}}{\partial VDD} = 12\ \mu A/V \qquad (20.18)$$

For every millivolt change in *VDD* (around $VDD = 5$ V), we get 12 nA change in I_{REF}. As a comparison from Ex. 20.1, we can write (assuming the change in V_{GS} with *VDD* small)

$$\frac{\partial I_{REF}}{\partial VDD} \approx \frac{1}{R} = \frac{1}{200k} = 5\ \mu A/V \qquad (20.19)$$

For every millivolt change in *VDD* we get 5 nA change in bias current.

Supply Independent Biasing

Instead of putting the resistor in the drain side of the current mirror, let's consider placing it in the source side, Fig. 20.14a. One of the problems with this approach is that if we try to mirror the current in M2 using M5 we won't know the value of the mirrored current. It's not obvious how V_{GS2} and V_{GS5} are related. In (b) we add a diode connected M1 so that its current can be mirrored (by M5). The next question is how do we force the same current through both M1 and M2? Figure 20.14c shows the addition of a PMOS current mirror to do this. From (c) we can write

$$V_{GS1} = V_{GS2} + I_{REF} \cdot R \tag{20.20}$$

which can only be valid if $V_{GS1} > V_{GS2}$. To ensure that this is the case, we use a larger value of β in M2, that is, we *multiply-up* M1's β in M2 so that less gate-source voltage is needed to conduct I_{REF}. Generally, this is done by simply using a larger width in M2. The resulting circuit is called a *Beta-multiplier* reference circuit. Knowing

$$V_{GS} = \sqrt{\frac{2I_D}{\beta}} + V_{THN} \tag{20.21}$$

$\left(\beta = KP_n \cdot \frac{W}{L}\right)$ and

$$\beta_2 = K \cdot \beta_1 \text{ (which is satisfied by } W_2 = K \cdot W_1\text{)} \tag{20.22}$$

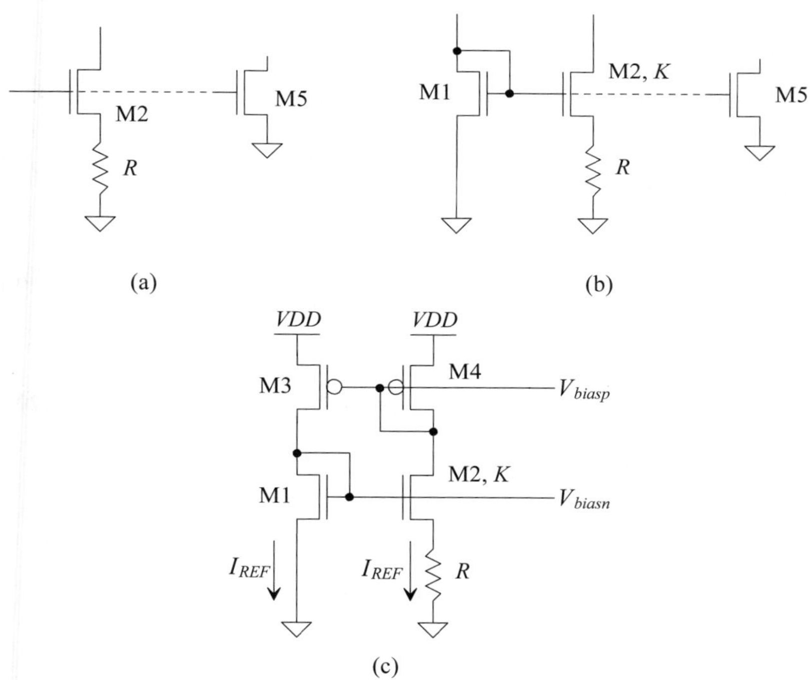

Figure 20.14 Developing the Beta-multiplier reference.

we can write

$$I_{REF} = \frac{2}{R^2 KP_n \cdot \frac{W_1}{L_1}}\left(1 - \frac{1}{\sqrt{K}}\right)^2 \text{ or } V_{DS,sat} = V_{GS} - V_{THN} = \frac{2}{R \cdot KP_n \cdot \frac{W_1}{L_1}}\left(1 - \frac{1}{\sqrt{K}}\right) \quad (20.23)$$

It's important to note I_{REF} and V_{ovn} ($= V_{GS} - V_{THN}$), neglecting the finite output resistance of the MOSFETs which wasn't included in the derivation of Eq. (20.23), are independent of *VDD*. Solving for R with the values given in Table 9.1 and a K of 4 gives $R = 6.5\ k\Omega$. Note that this circuit, when $K = 4$, is sometimes called a *constant-*g_m bias circuit because

$$g_m = \sqrt{2KP_n \frac{W}{L} \cdot I_{REF}} = \frac{1}{R} \quad (20.24)$$

a constant independent of MOSFET process shifts. Figure 20.15 shows the schematic of a bias circuit based on the sizes, bias currents, and $V_{DS,sat}$ given in Table 9.1.

Note that a "start-up circuit" was included in Fig. 20.15. In any self-biased circuit there are two possible operating points: the one we just described and the unwanted one where zero current flows in the circuit. This unwanted state occurs when the gates of M1/M2 are at ground while the gates of M3/M4 are at *VDD*. When in this state, the gate of MSU1 is at ground and so it is off. The gate of MSU2 is somewhere between *VDD* and *VDD* – V_{THP}. MSU3, which behaves like an NMOS switch, turns on and leaks current into the gates of M1/M2 from the gates of M3/M4. This causes the current to snap to the desired state and MSU3 to turn off. *Note that during normal operation the start-up circuit should not affect the Beta-multiplier's operation.* The current through MSU3 should be zero (or very small).

The Beta-multiplier is an example of a circuit that uses positive feedback. The addition of the resistor kills the closed loop gain (a positive feedback system can be stable if its closed loop gain is less than one). However, if we decrease the size of the resistor, we increase the gain of the loop and push the feedback system closer to instability. An example of when this could occur is if the parasitic capacitance on the source of M2 to ground is large (effectively shorting M2's source to ground). If the resistor, for example, is bonded out off-chip to set the current, it is likely that this bias circuit will oscillate.

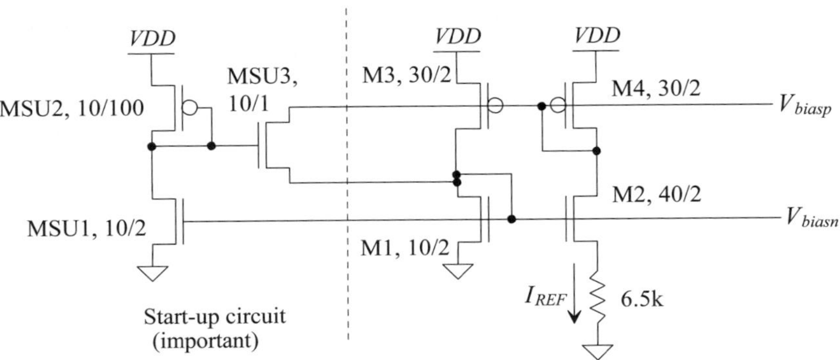

Figure 20.15 Beta-multiplier reference for biasing in the long-channel process described in Table 9.1.

Figure 20.16 shows how the reference currents through M1 and M2 vary with *VDD*. The minimum value of *VDD* can be estimated by looking at the minimum value of voltage across the drain-source of M3 and M1. For M3 this is $V_{SD3,sat}$ or 250 mV (because we are using the parameters from Table 9.1 in the design of this current reference). For M1 this is $V_{DS1} = V_{GS1} = 1.05$ V. We can then write

$$VDD_{\min} = V_{SD3,sat} + V_{GS1} = 1.3\ V \tag{20.25}$$

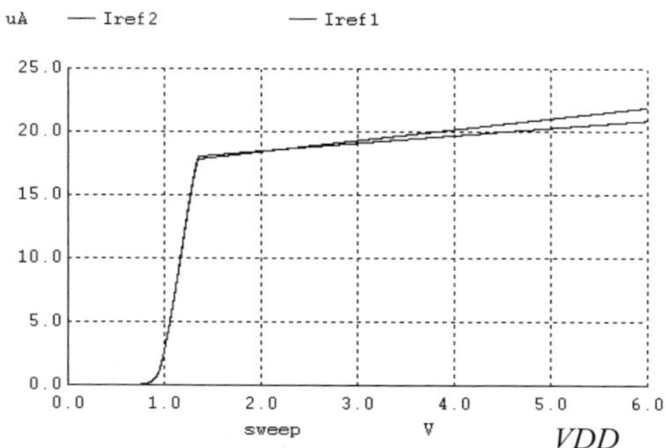

Figure 20.16 The reference currents through M1 and M2 in the Beta-multiplier.

Finally, the sensitivity of I_{REF} is directly dependent on the output resistance of the MOSFETs. From the simulations

$$\frac{\partial I_{REF}}{\partial VDD} \approx 800\ nA/V$$

or almost an order of magnitude better than the previous reference circuits

---

Example 20.3
Estimate the voltage on the gate of M6 and its drain current in Fig. 20.17.

M3 and M4 are biased to source 20 μA of current. Since $V_{GS1} = V_{GS2} = V_{DS1}$ (and M1/M2 have the same drain current, see Fig. 20.1 and the associated discussion), it follows that $V_{GS6} = V_{DS2} = V_{GS1} = 1.05$ V. *We treat M6, for biasing purposes, as if its gate were tied to the gates of M1 or M2.* It follows then that $I_{D6} = 20$ μA. ■

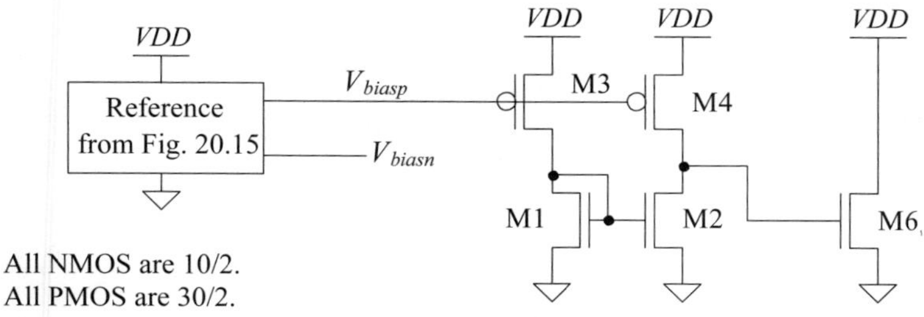

Figure 20.17 Circuit used in Ex. 20.3

20.1.4 Short-Channel Design

Figure 20.18 shows a Beta-multiplier biasing circuit based on the values given in Table 9.2 as well as simulation results showing how the reference currents vary with *VDD*. From Eq. (20.25) and Table 9.2 we would expect VDD_{min} to be 400 mV. At *VDD* = 1 V the reference currents are indeed 10 μA. However, what we see is a horrible sensitivity to changes in *VDD*. Reviewing Figs. 9.31 and 20.10 we see the low output resistance of the short-channel devices causes the drain current to change significantly with changes in drain-to-source voltage. In some analog applications this variation isn't that harmful. However, if the Beta-multiplier circuit is to behave like a true current reference, the reference currents shouldn't vary with changes in *VDD*.

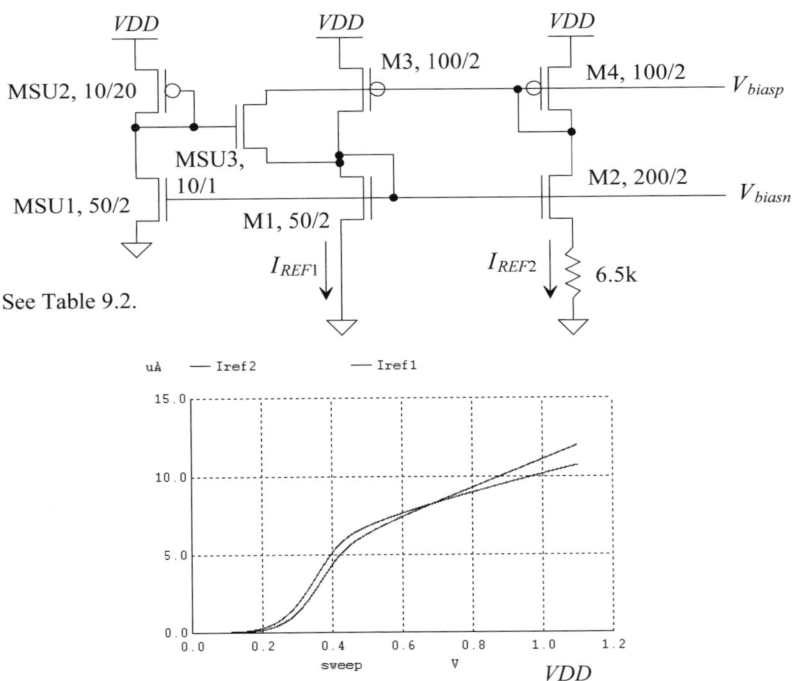

Figure 20.18 Beta-multiplier reference for short-channel design (see Table 9.2).

To reduce the sensitivity, we need to reduce the variations in the drain-to-source voltages of the NMOS devices with changes in *VDD*. Consider adding a differential amplifier (diff-amp) to the basic Beta-multiplier seen in Fig. 20.19. Note that M4 is no longer diode-connected so its drain can move to the same potential as M2's drain, that is, V_{biasn}. The start-up circuit (required) is not shown. The idea is to use the amplifier to compare the drain voltage of M1 (V_{biasn}) with the drain voltage of M2 (V_{reg}) and regulate them to be equal. The result is an effective increase in M2's output resistance. For example, if V_{reg} is above V_{biasn} , the amplifier's output voltage increases. This drives the gate of M4 upwards, lowering the current it supplies and causing V_{reg} to drop back down.

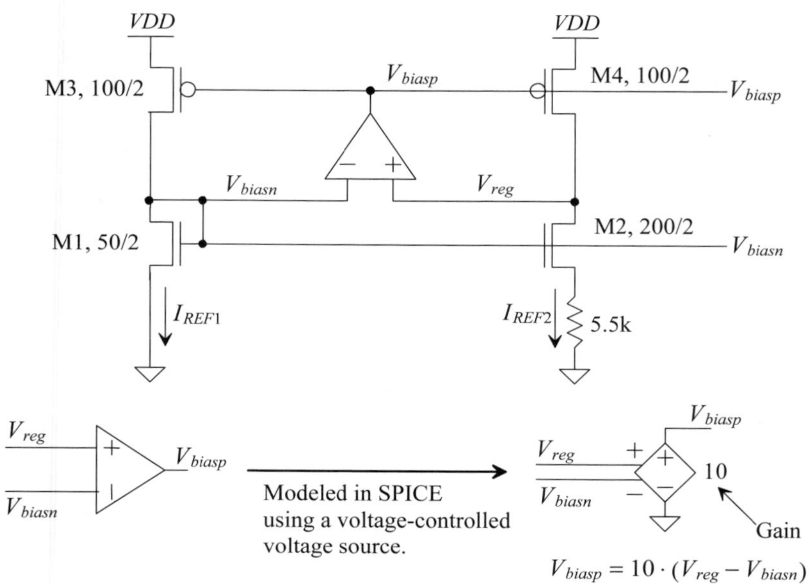

Figure 20.19 Increasing the output resistance of short-channel MOSFETs using feedback. The result, for the Beta-multiplier circuit, is better power supply sensitivity.

At the same time the gate of M3 is also increased, causing it to source less current. This causes a drop in V_{biasn} (the same as V_{reg} because of the symmetry as discussed in Fig. 20.1 or Ex. 20.3). Figure 20.20 shows how the reference current, in Fig. 20.19, changes with *VDD* when the added amplifier has a gain of 10. We've lowered *R* to 5.5k to more accurately set the current and used, in the simulation, a voltage-controlled voltage source for the amplifier.

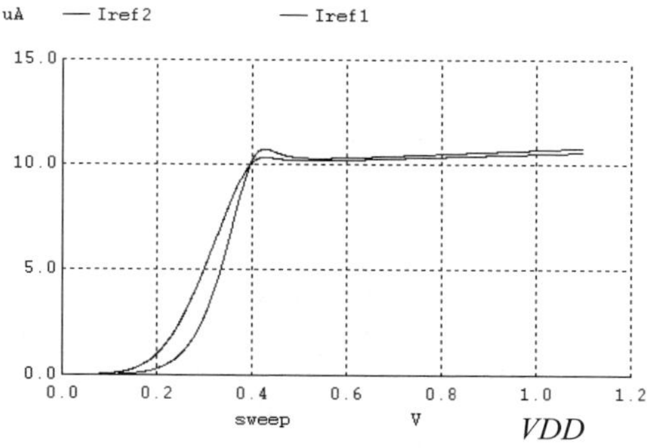

Figure 20.20 Improvement with the added amplifier.

Figure 20.21 shows a possible differential amplifier configuration. The PMOS devices form a current mirror while the NMOS devices will, when $V_{biasn} = V_{reg}$, simply mirror the current in M1. If $V_{biasn} \neq V_{reg}$ then the imbalance causes the amplifier output to swing up or down providing the desired action.

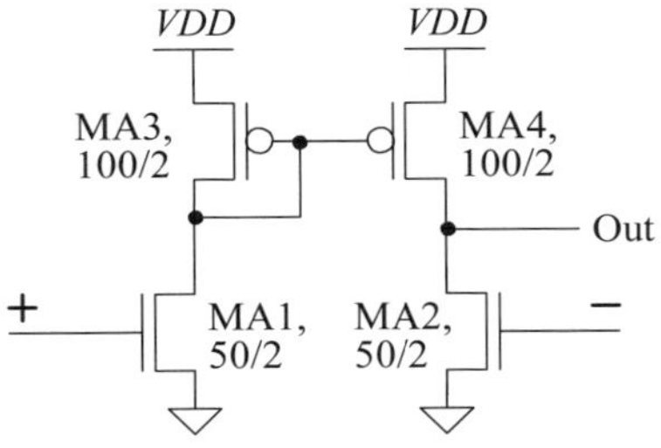

Figure 20.21 A possible implementation of the diff-amp in Fig. 20.19.

Figure 20.22 shows the improved current reference for use in a short-channel CMOS process (where devices have very low output resistances). The amplifier in Fig. 20.21 is placed in the reference, as indicated in Fig. 20.19. As with any feedback scheme, stability is of critical importance. As we'll see in the coming chapters, a feedback amplifier with only a single high-impedance node (meaning no diode-connected MOSFETs or MOSFET-source terminals are connected to the node) is straightforward to compensate (make stable). So that the reference has a single high-impedance node, that is V_{biasp}, we connected M2's gate to its drain (instead of to V_{biasn}). We get the same effect, as

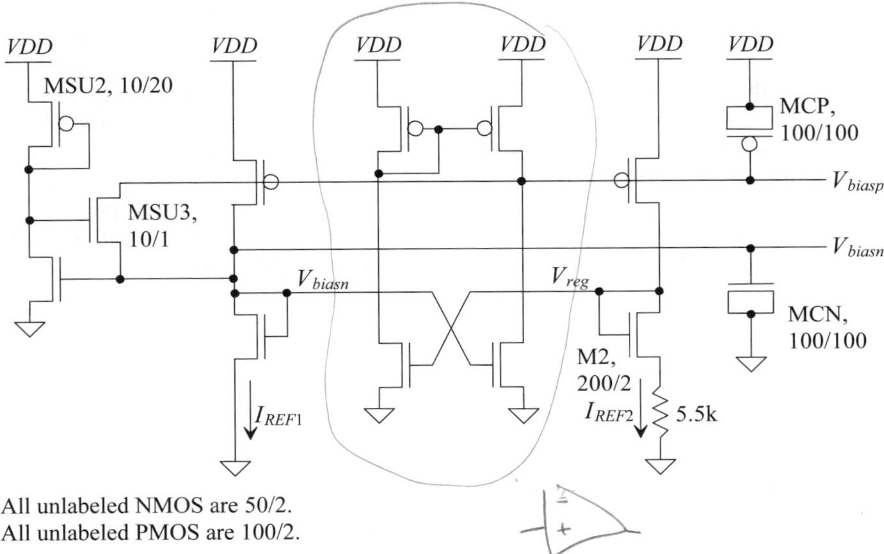

Figure 20.22 Improved current reference for short-channel devices. See also Sec. 23.1.3 for more information.

seen in Fig. 20.19 but the gain around the loop is reduced. To make the reference stable, we add capacitors (MCP and MCN) to the circuit. The high-impedance node is the critical point for adding the compensation capacitor (that is MCP is the critical capacitance). If the reference drives a significant number of MOSFETs (so these driven MOSFETs provide the load capacitance), then MCP and MCN may be excluded from the design. Figure 20.23 shows how the reference currents change with *VDD*.

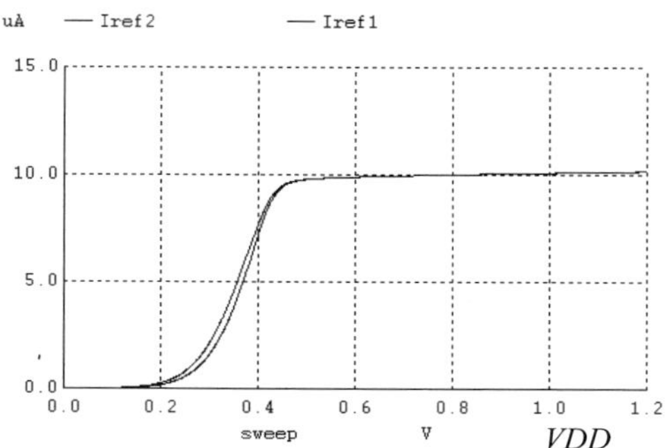

Figure 20.23 Variation of reference currents with VDD for the circuit in Fig. 20.22.

An Important Note

It's extremely important to understand that together with the benefits of better power supply sensitivity are the undesired traits of feedback, that is, the potential for an unstable circuit. For example, Fig. 20.24a shows what happens to the reference currents if we remove MCP and MCN in Fig. 20.22 and then apply a step voltage to *VDD* (that is, *VDD* steps from 0 to 1 V at 50 ns in the simulation). Clearly, the reference is not stable and the currents oscillate. In (b) we do the same thing but with MCP and MCN present. The response shows first-order behavior, and oscillations are not present. We might think we are done with the reference and move on to the other analog circuit designs. However, further characterization of the current reference using simulations is warranted. For example, what happens if *VDD* has a 50 mV squarewave signal coupled to it at 100 MHz? How does the reference behave with this signal (a squarewave *VDD* that oscillates between 1 and 1.05 V at 100 MHz)? In general, no analog circuit has good power supply noise rejection at high frequencies. To remedy this, the power supply is decoupled (a large capacitor is placed from *VDD* to ground to remove high-frequency noise).

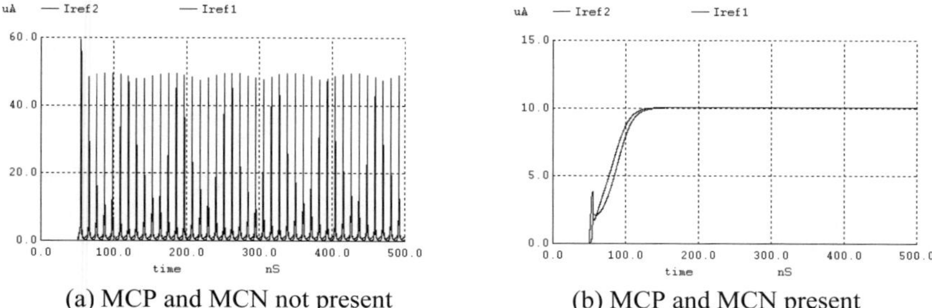

(a) MCP and MCN not present (b) MCP and MCN present

Figure 20.24 What happens when VDD is pulsed from 0 to 1 at 50 ns.

20.1.5 Temperature Behavior

While we've been focusing on how the reference current changes with *VDD,* it's also important to know how the current changes with temperature. Looking at the basic current mirror in Fig. 20.11a, note that if I_{REF} is a constant, independent of temperature, then the output current will also be independent of temperature. That's not to say that the MOSFET characteristics aren't changing because they are; they change at the same rate. Because they change together, the current mirror relationship is still valid and M2 mirrors the current in M1. Of course, if M1 is located at a different physical location than M2 and each is heated differently, a mismatch in the currents will result. *The point is that the temperature behavior of the reference current determines the temperature behavior of all mirrored currents.*

Using the short-channel CMOS parameters we know, from Ch. 9 (Fig. 9.30b), that the change in V_{THN} with temperature is –0.6 mV/C° (for an NMOS device). For the PMOS device we can vary temperature and look at the change in V_{THP}. Using the mirror in Fig. 20.11 and looking at the V_{SG} at different temperatures results in the plots seen in Fig. 20.25. We see that it also varies at a rate of –0.6 mV/C°.

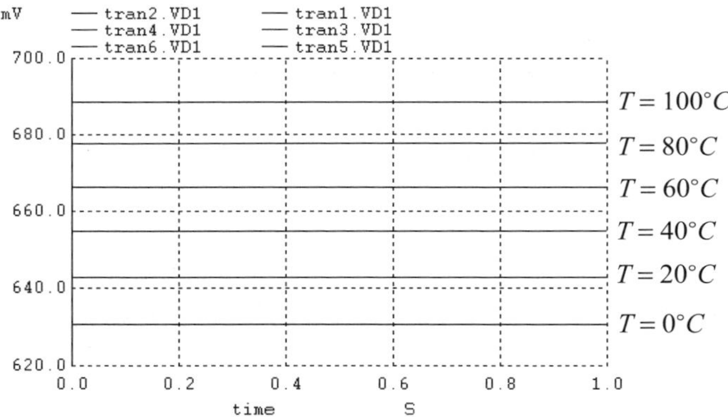

Figure 20.25 Variation in the reference gate voltage in the PMOS mirror seen in Fig. 20.11.

Resistor-MOSFET Reference Circuit

For a resistor-MOSFET reference circuit (Fig. 20.2), we can write (neglecting the MOSFET's finite output resistance)

$$I_{REF} = \frac{VDD - V_{GS}}{R} \tag{20.26}$$

The change in I_{REF} with temperature is

$$\frac{\partial I_{REF}}{\partial T} = -\frac{VDD - V_{GS}}{R^2} \cdot \frac{\partial R}{\partial T} - \frac{1}{R} \cdot \frac{\partial V_{GS}}{\partial T} \tag{20.27}$$

or

$$\frac{\partial I_{REF}}{\partial T} = -I_{REF} \cdot \frac{1}{R}\frac{\partial R}{\partial T} - \frac{1}{R} \cdot \frac{\partial V_{GS}}{\partial T} \tag{20.28}$$

We know that to write the reference current as a function of temperature we use

$$I_{REF}(T) = I_{REF}(T_0) \cdot (1 + TCI_{REF} \cdot (T - T_0)) \tag{20.29}$$

where the temperature coefficient of the reference current is

$$TCI_{REF} = \frac{1}{I_{REF}} \cdot \frac{\partial I_{REF}}{\partial T} \tag{20.30}$$

Rewriting Eq. (20.28) using Eq. (20.26) yields

$$TCI_{REF} = \frac{1}{I_{REF}} \cdot \frac{\partial I_{REF}}{\partial T} = -\frac{1}{R}\frac{\partial R}{\partial T} - \frac{1}{VDD - V_{GS}} \frac{\partial V_{GS}}{\partial T} \tag{20.31}$$

Noting, again, that the temperature coefficient is not a constant but rather changes with temperature (even though this isn't indicated in Eq. (20.29)) just like a small-signal parameter, e.g., g_m, changes with the DC operating point.

Example 20.4
Determine the temperature behavior of the reference current in the mirror seen in Fig. 20.10. Verify the answer with SPICE.

If the temperature coefficient of the resistor, $TCR = \frac{1}{R}\frac{\partial R}{\partial T} = 2000\ ppm/C^\circ\ (= 0.002)$ and we assume after looking at Fig. 9.30 that $\frac{\partial V_{GS}}{\partial T} \approx \frac{\partial V_{THN}}{\partial T} \approx -0.6\ mV/C^\circ$ then, knowing the parameters used in Table 9.2 were the basis for the design of the current mirror in Fig. 20.10,

$$TCI_{REF} = -0.002/C^\circ - \frac{1}{1 - 0.35} \cdot (-0.6\ mV/C^\circ) \approx -1{,}000\ ppm/C^\circ$$

and so

$$I_{REF}(T) = 10 \cdot (1 - 0.001 \cdot (T - 27))\ \mu A$$

At 100 C° the reference current is 9.27 μA and at 0 C° it's 10.27 μA (of course, the reference current is approximately 10 μA at 27 C°). Figure 20.26 shows the SPICE simulation results. ■

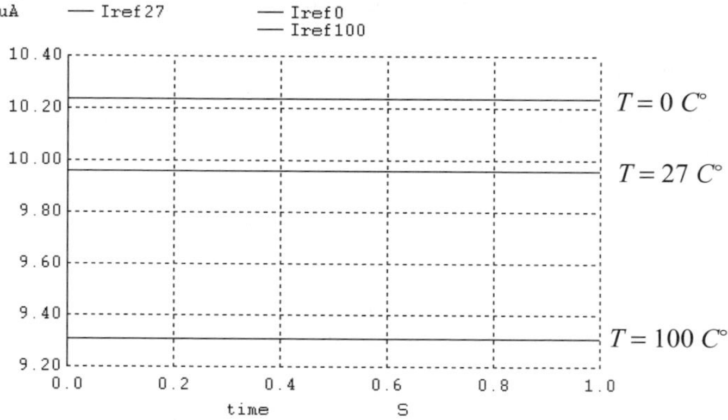

Figure 20.26 Example 20.4 showing the temperature behavior of the reference current in Fig. 20.10.

MOSFET-Only Reference Circuit

For the MOSFET-only bias circuit, Fig. 20.12, we can write the reference current as (see Eq. [20.16])

$$I_{REF} = \frac{VDD^2 - 2VDD \cdot (V_{THN} + V_{THP}) + (V_{THN} + V_{THP})^2}{\left(\sqrt{\frac{2L_3}{KP_p \cdot W_3}} + \sqrt{\frac{2L_1}{KP_n \cdot W_1}}\right)^2} \tag{20.32}$$

Knowing $KP_n = M \cdot KP_p$, where M is 3 for our long-channel process and 2 for our short-channel process (see Tables 9.1 and 9.2), we can write

$$I_{REF} = KP_n \cdot \frac{VDD^2 - 2VDD \cdot (V_{THN} + V_{THP}) + (V_{THN} + V_{THP})^2}{\left(\sqrt{\frac{M \cdot 2L_3}{W_3}} + \sqrt{\frac{2L_1}{W_1}}\right)^2} \tag{20.33}$$

or

$$\frac{\partial I_{REF}}{\partial T} = \frac{\partial KP_n}{\partial T} \cdot \frac{VDD^2 - 2VDD \cdot (V_{THN} + V_{THP}) + (V_{THN} + V_{THP})^2}{\left(\sqrt{\frac{M \cdot 2L_3}{W_3}} + \sqrt{\frac{2L_1}{W_1}}\right)^2} +$$

$$\frac{KP_n}{\left(\sqrt{\frac{M \cdot 2L_3}{W_3}} + \sqrt{\frac{2L_1}{W_1}}\right)^2} \cdot \left(-2VDD \cdot \left(\frac{\partial V_{THN}}{\partial T} + \frac{\partial V_{THP}}{\partial T}\right) + 2(V_{THN} + V_{THP})\left(\frac{\partial V_{THN}}{\partial T} + \frac{\partial V_{THP}}{\partial T}\right)\right) \tag{20.34}$$

Dividing both sides by Eq. (20.33) gives

$$\frac{1}{I_{REF}} \frac{\partial I_{REF}}{\partial T} = \frac{1}{KP_n} \frac{\partial KP_n}{\partial T} + \frac{2\left(\frac{\partial V_{THN}}{\partial T} + \frac{\partial V_{THP}}{\partial T}\right)(V_{THN} + V_{THP} - VDD)}{VDD^2 - 2VDD \cdot (V_{THN} + V_{THP}) + (V_{THN} + V_{THP})^2} \tag{20.35}$$

noting, from Eq. (9.52), that the first term is simply –1.5/T.

Example 20.5

Determine the temperature behavior of the reference current in the MOSFET-only bias circuit seen in Fig. 20.13.

We note that both references in Fig. 20.13 have the same temperature behavior since Eq. (20.35) doesn't show a width or length dependence (there is a dependence in the reference current though). Figure 20.13 used the long-channel MOSFET parameters from Table 9.1 where

$$V_{THN} = 0.8,\ V_{THP} = 0.9,\ \frac{\partial V_{THN}}{\partial T} = -1\ mV/C^\circ,\ \frac{\partial V_{THP}}{\partial T} = -1.4\ mV/C^\circ$$

and VDD = 5 V with I_{REF} = 13.5 µA (measured in the simulation that generated Fig. 20.13) at T = 300 K°. Plugging the numbers into Eq. (20.35) gives

$$TCI_{REF} = \frac{1}{I_{REF}} \frac{\partial I_{REF}}{\partial T} = -\frac{1.5}{300} + \frac{2(-0.0024)(-3.3)}{25 - 17 + 2.89} = -0.005 + 0.00146 \approx -3{,}500\ ppm/C^\circ$$

and so $I_{REF}(T) = 13.5(1 - 0.0035(T - T_0))\ \mu A$. ■

Temperature Behavior of the Beta-Multiplier

To determine the temperature behavior of the Beta-multiplier, we take the derivative of Eq. (20.23) with respect to temperature

$$\frac{\partial I_{REF}}{\partial T} = \frac{-4}{R^3 KP_n \cdot \frac{W_1}{L_1}}\left(1-\frac{1}{\sqrt{K}}\right)^2 \cdot \frac{\partial R}{\partial T} - \frac{2}{R^2 KP_n^2 \cdot \frac{W_1}{L_1}}\left(1-\frac{1}{\sqrt{K}}\right)^2 \frac{\partial KP_n}{\partial T} \quad (20.36)$$

Dividing both sides by I_{REF} (Eq. [20.23]) gives

$$TCI_{REF} = \frac{1}{I_{REF}} \cdot \frac{\partial I_{REF}}{\partial T} = -2\left(\frac{1}{R}\frac{\partial R}{\partial T}\right) - \frac{1}{KP_n}\frac{\partial KP_n}{\partial T} \quad (20.37)$$

Example 20.6

Determine the temperature behavior of the Beta-multiplier seen in Fig. 20.22.

If the temperature coefficient of the resistor, $TCR = \frac{1}{R}\frac{\partial R}{\partial T} = 2000\ ppm/C°$ then

$$TCI_{REF} = -0.004 + \frac{1.5}{300} = 1000\ ppm/C°$$

and so

$$I_{REF}(T) = 10 \cdot (1 + 0.001 \cdot (T-27))\ \mu A$$

The positive TC of the resistor subtracts from the negative TC of the mobility to stabilize the current reference. If the mobility change with temperature is complicated by velocity saturation effects, then simulations become invaluable to determine the actual temperature performance of the circuit. ■

Voltage Reference Using the Beta-Multiplier

Let's try designing the Beta-multiplier where the gate voltage of M1 (see Figs. 20.15 or 20.22) is a constant independent of temperature. In this type of design we aren't using this voltage directly to bias anything so we'll change the label from V_{biasn} to V_{REF}. Using Eqs. (20.22) and (20.23) gives

$$V_{REF} = V_{GS1} = \frac{2}{R \cdot KP_n \cdot \frac{W}{L}}\left(1-\frac{1}{\sqrt{K}}\right) + V_{THN} \quad (20.38)$$

Taking the derivative with respect to temperature gives

$$\frac{\partial V_{REF}}{\partial T} = \frac{\partial V_{THN}}{\partial T} - \frac{2}{R \cdot KP_n \cdot \frac{W}{L}}\left(1-\frac{1}{\sqrt{K}}\right) \cdot \left(\frac{1}{R}\frac{\partial R}{\partial T} + \frac{1}{KP_n} \cdot \frac{\partial KP_n}{\partial T}\right) \quad (20.39)$$

If we want the change with temperature to be zero, $\partial V_{REF}/\partial T = 0$, then we can select the resistor based on

$$R = \frac{2}{\frac{\partial V_{THN}}{\partial T} \cdot KP_n \cdot \frac{W}{L}}\left(1-\frac{1}{\sqrt{K}}\right) \cdot \left(\frac{1}{R}\frac{\partial R}{\partial T} + \frac{1}{KP_n} \cdot \frac{\partial KP_n}{\partial T}\right) \quad (20.40)$$

Using the parameters from Table 9.1 with $K = 4$ and a resistor temperature coefficient of 0.002 gives a resistor value of 5 kΩ. Figure 20.27 shows the SPICE simulation results. The variation is approximately 60 mV/100 C° or 600 μV/C°.

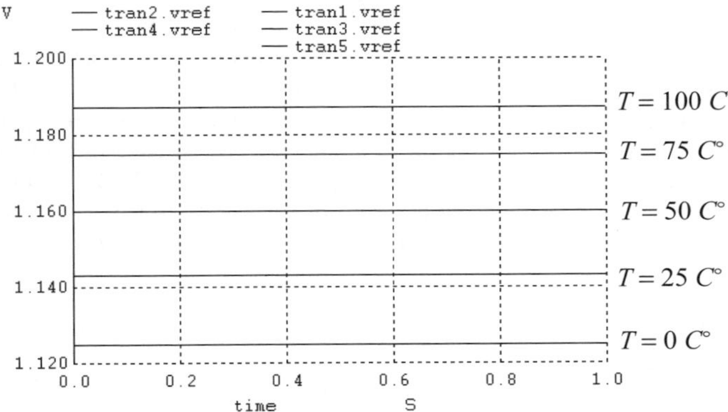

Figure 20.27 Temperature performance of a voltage reference using the Beta-multiplier.

20.1.6 Biasing in the Subthreshold Region

In some CMOS designs (for example, low-power designs) the MOSFETs may be operated in the weak or subthreshold regions. If a resistor is used to bias a current mirror, as seen in Fig. 20.10, its value can be very large. For example, if we use the short-channel process with $VDD = 1$ V and a V_{SG} of approximately V_{THP} (= 280 mV) with an I_D of 10 nA, we need a resistor with a value of

$$I_{REF} = \frac{VDD - V_{SG}}{R} \rightarrow R = 72\ M\Omega\ ! \tag{20.41}$$

Clearly, this isn't practical. To bias circuits for subthreshold operation, let's use the Beta-multiplier. From Eq. (9.17) we can write

$$V_{GS1} = nV_T \cdot \ln\left(\frac{I_{REF} \cdot L_1}{I_{D0} \cdot W_1}\right) + V_{THN} \tag{20.42}$$

and

$$V_{GS2} = nV_T \cdot \ln\left(\frac{I_{REF} \cdot L_1}{I_{D0} \cdot K \cdot W_1}\right) + V_{THN} \tag{20.43}$$

Knowing

$$I_{REF} = \frac{V_{GS1} - V_{GS2}}{R} \tag{20.44}$$

we get

$$I_{REF} = \frac{n \cdot V_T}{R} \cdot \ln K \tag{20.45}$$

or

$$R = \frac{n \cdot V_T}{I_{REF}} \cdot \ln K \tag{20.46}$$

Using the values above, that is, $K = 4$ and an I_{REF} of 10 nA, gives an R (assuming $n = 1$) of approximately 3.5 MΩ. Still large but much better than 72 MΩ.

20.2 Cascoding the Current Mirror

In this section we discuss increasing the output resistance of a current mirror so that it behaves more ideally. Figure 20.28 shows the problem with using a single MOSFET for a current source. Instead of being a constant, the output current, I_O , increases as the voltage across the current source, V_O , increases. We model this increase using the MOSFET's output resistance, r_o (see Table 9.2). If we can hold the drain-source voltage of the MOSFET constant, then the current doesn't vary. However, this requires a fixed V_O. To avoid this, we'll add circuitry in between the current-mirrored MOSFET and V_O.

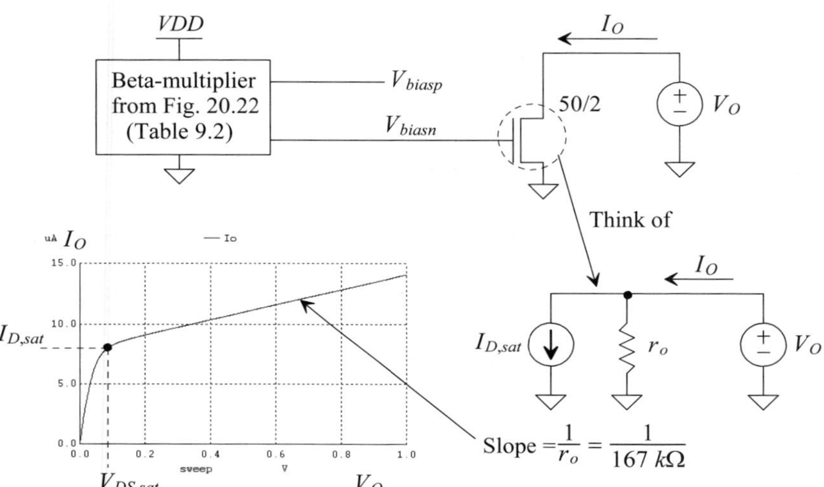

Figure 20.28 How the finite output resistance of the MOSFET affects the output current.

20.2.1 The Simple Cascode

The simple cascode current mirror made with M1–M4 is seen in Fig. 20.29. The name "cascode" is a vestige from the days of vacuum tubes when a common-cathode amplifier was cascaded (in series with) a common-grid amplifier. Drawing an analogy with MOSFETs, the cathode of a tube is equivalent to the source of a MOSFET. The grid of a tube corresponds to the gate of a MOSFET and the anode to the drain of the MOSFET.

It's important to realize that I_O is still determined by the gate-source voltages of M1 and M2. Changing the sizes of M3 and M4 simply changes the drain-source voltages of M1/M2, affecting the matching of their drain currents, as indicated by Eq. (20.10). Again, this (the gate-source voltages of M1/M2 determine the currents) is important to understand since, in the next several pages, we present different modifications to this basic cascode circuit. These modifications attempt to hold the drain source voltages of M1/M2 more constant to increase the current mirror's output resistance (make I_O change less with changes in V_O).

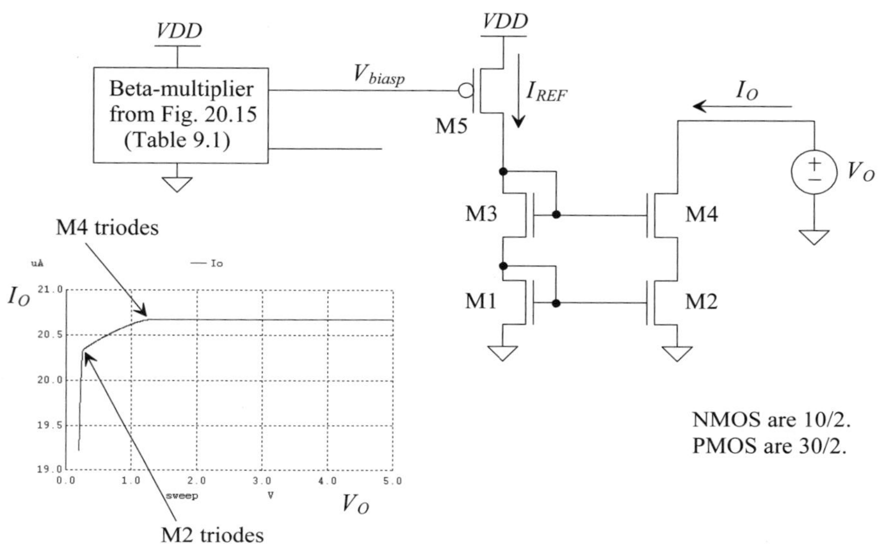

Figure 20.29 Biasing of the cascode current source and its operation.

DC Operation

The voltage on the gate of M4 in Fig. 20.29 is (remembering the actual values for these specific biasing conditions is given in Table 9.1)

$$2V_{GS} = 2(V_{DS,sat} + V_{THN}) \tag{20.47}$$

The voltage on the drain of M2, assuming that it is operating in the saturation region, is V_{GS}. The minimum voltage across the current source is then

$$V_{O,\min} = V_{DS,sat4} + V_{GS} = 2V_{DS,sat} + V_{THN} \tag{20.48}$$

Plugging in the numbers from Table 9.1, we get $V_{O,\min} = 1.3\ V$. Reviewing Fig. 20.29, we see that M4 starts to move into the triode region at this voltage.

To keep both M2 and M4 operating in the saturation region, we need only a $V_{DS,sat}$ across each one. In the next section we'll use this fact to design a *low-voltage cascode* current mirror.

Cascode Output Resistance

To estimate the output resistance of the cascode current mirror, let's apply a test voltage to the simplified cascode schematic seen in Fig. 20.30. We treat M1 and M3 as DC bias sources (AC grounds) and assume that the DC voltage on the drain of M4 is large enough to ensure that M2 and M4 are operating in the saturation. Further we draw the output resistances external to the devices. Note that M2's AC gate source voltage is zero and its drain voltage is $-v_{gs4}$. The resistance seen looking into the drain of M4 is

$$R_o = \frac{v_T}{i_T} \tag{20.49}$$

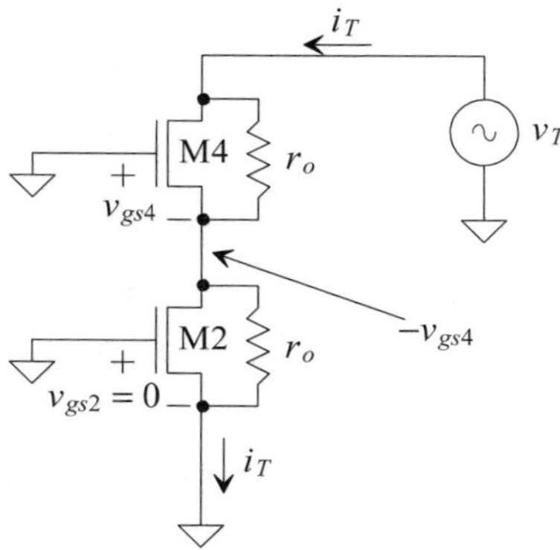

Figure 20.30 AC circuit used to determine the output resistance of a cascode current source.

Looking at Fig. 20.30, we can write

$$-v_{gs4} = i_T \cdot r_o \tag{20.50}$$

The current through M4 is then

$$i_T = g_m v_{gs4} + \frac{v_T - (-v_{gs4})}{r_o} \tag{20.51}$$

Substituting (20.50) into (20.51) gives

$$i_T = g_m(-i_T \cdot r_o) + \frac{v_T}{r_o} - i_T \tag{20.52}$$

Solving for the output resistance

$$R_o = (2 + g_m r_o) r_o \approx g_m r_o^2 \tag{20.53}$$

This result should be remembered because it will be referred to often. Using the numbers from Tables 9.1 and 9.2, we can tabulate the output resistance of cascode structures as seen in Table 20.1.

Table 20.1 Output resistances for cascode structures.

Cascode type	Long-channel process, Table 9.1, R_o	Short-channel process, Table 9.2, R_o
PMOS	2.4 GΩ	16.6 MΩ
NMOS	3.75 GΩ	4.2 MΩ

Notice the large difference (almost three orders of magnitude) between the output resistance of the short- and the long-channel devices. Again, this is the reason we need to use special circuit techniques when designing in short-channel processes.

Finally, note that the output resistance can be determined from a plot like the one seen in Fig. 20.29 using, in WinSPICE, `plot 1/deriv(Io)`. As seen in Fig. 20.28, the output resistance is the reciprocal of the derivative of the output current.

20.2.2 Low-Voltage (Wide-Swing) Cascode

Figure 20.31a shows the regularly biased cascode structure. The key point of this figure is that the drain of M2 is not at the minimum voltage required to keep it in saturation, $V_{DS,sat}$, but rather it is held at a threshold voltage above this minimum. In (b) the cascode is biased for lowest voltage operation. What this means is that the drain of M4, V_o, can go to the minimum possible voltage that keeps M2/M4 in saturation, that is, $2V_{DS,sat}$. Again, it's important to notice that the gate-source voltages of M1/M2 are what set the currents in the mirror.

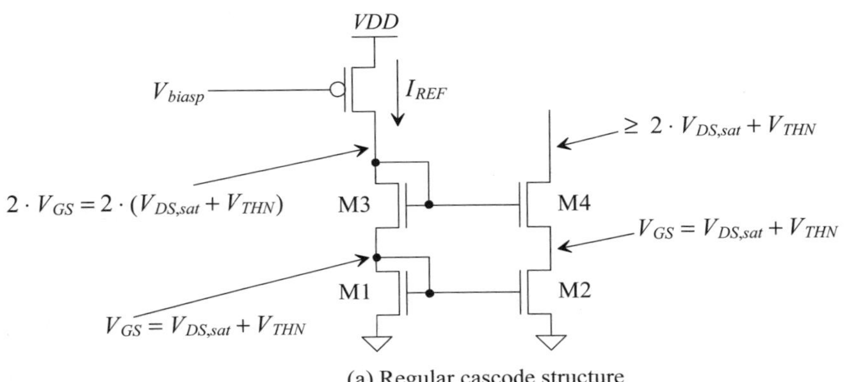

(a) Regular cascode structure

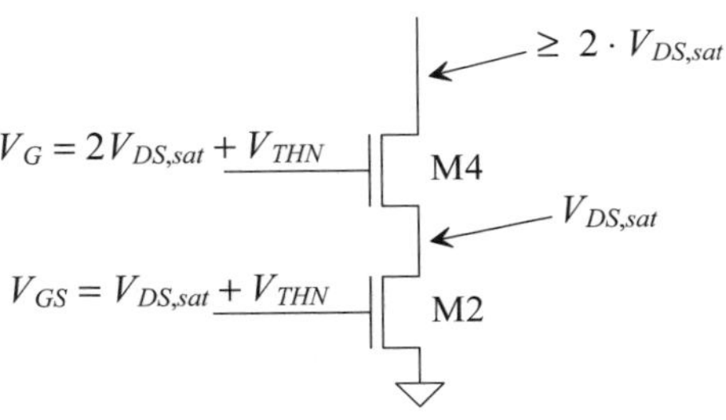

(b) Low-voltage (aka wide-swing) structure

Figure 20.31 DC voltages in (a) a cascode current mirror and in (b) a low-voltage cascode.

To generate the gate voltage of M2, we simply use a diode-connected MOSFET as seen in almost every schematic in this chapter. From Eq. (20.1), for example, we can write

$$I_{REF} = \frac{KP_n}{2} \cdot \frac{W}{L}(V_{GS} - V_{THN})^2 \tag{20.54}$$

neglecting channel-length modulation. To generate the bias voltage for M4, we can adjust the width and length of a MOSFET, MWS, as seen in Fig. 20.32. We can write, knowing $V_{DS,sat} = V_{GS} - V_{THN}$

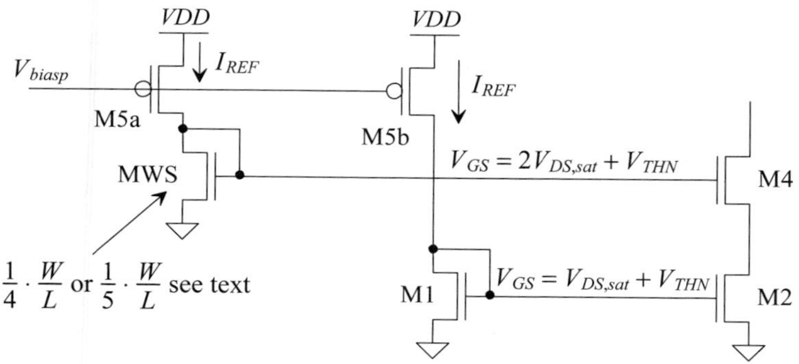

Figure 20.32 Generating a bias voltage for M4.

$$I_{REF} = \frac{KP_n}{2} \cdot \frac{W_{MWS}}{L_{MWS}}(2(V_{GS} - V_{THN}) + V_{THN} - V_{THN})^2 \tag{20.55}$$

or

$$I_{REF} = \frac{KP_n}{2} \cdot \frac{W_{MWS}}{L_{MWS}} \cdot 4(V_{GS} - V_{THN})^2 \tag{20.56}$$

Using our device sizes from either Tables 9.1 or 9.2, we can make the length of MWS four times the length of the other MOSFETs or

$$\frac{W}{L} = \frac{W_{MWS}}{L_{MWS}} \cdot 4 \tag{20.57}$$

If we use the same widths, then

$$L_{MWS} = 4 \cdot L \tag{20.58}$$

If Eq. (20.58) is valid, then Eq. (20.56) is equal to Eq. (20.54).

Note that, as seen in Fig. 20.31b, we are biasing M2 right on the edge of the triode region when we use Eq. (20.58) to size MWS. In many situations we may want to move M2 further into the saturation region. Intuitively, we would think that by increasing the length of MWS we increase its effective resistance and so the voltage drop across it goes up. If we increase MWS's length further, the gate voltage of M4 goes up and thus so does the drain voltage of M2. M2 is moved further away from the edge of the triode region into the saturation region. The minimum voltage across the current source increases. It's important to understand that it's OK for M2 to be biased away from the triode region by a small amount.

Example 20.7
Regenerate Fig. 20.29 using a wide-swing cascode structure.

The simulation results are seen in Fig. 20.33. It appears that the output resistance is really large until $V_O < V_{DS,sat}$. However, if we zoom in around $2V_{DS,sat}$ (= 500 mV here), we would see that the slope increases and M4 triodes. ■

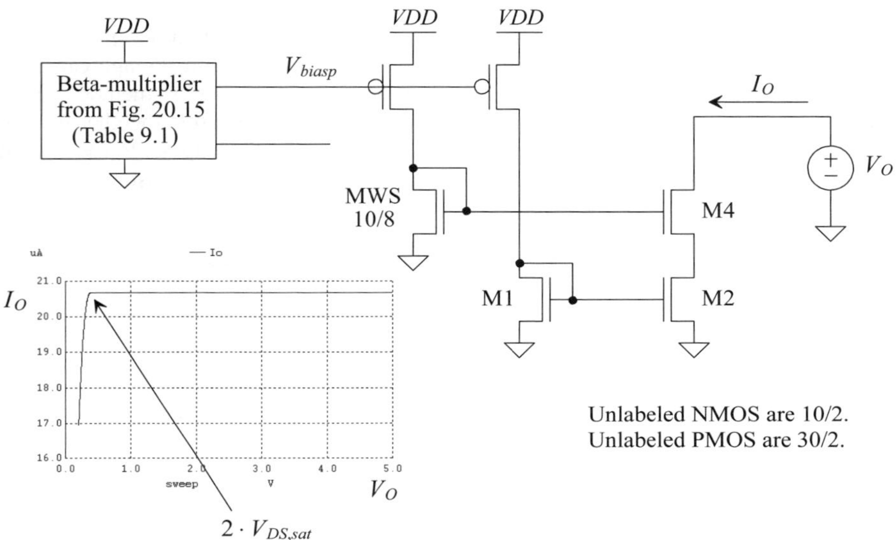

Figure 20.33 Wide-swing cascode current source in the long-channel process.

An Important Practical Note

In general we won't design a wide-swing current sink by setting the length to four times the length used in the rest of the design. As just mentioned, this biases M2 right on the edge of the triode region (point A) in Fig. 20.34. Stealing a current from the drain of M2 will move M2 into the triode region, point B. If we bias M2 further into the saturation region, point C, then more current must be removed before M2 triodes and the output resistance of the current source decreases. When we design folded-cascode amplifiers later, we will steal or add current in this way. For general *long-channel design,* we'll use

$$\frac{W_{MWS}}{L_{MWS}} = \frac{1}{5} \cdot \frac{W}{L} \tag{20.59}$$

or using the same widths then

$$L_{MWS} = 5 \cdot L \tag{20.60}$$

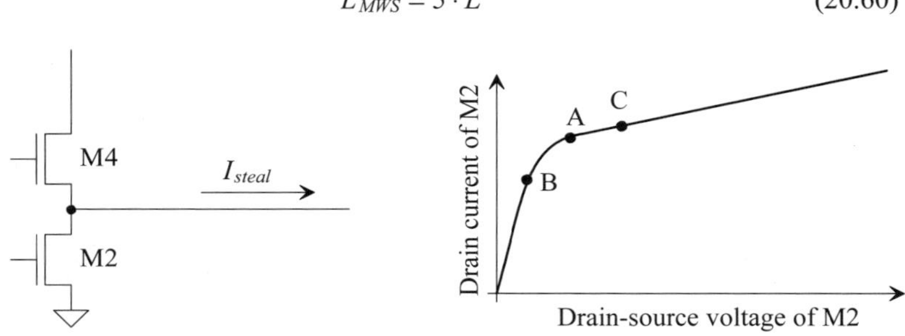

Figure 20.34 Showing how stealing current from M2 moves it into the triode region.

Layout Concerns

When we go to layout the long L device seen in Fig. 20.32, we might simply layout a single MOSFET with the appropriate length. However, the threshold voltage can vary significantly with the length of the device. Figure 20.35 shows a method where MOSFETs connected in series with the same widths and their gates tied together behave like a single MOSFET with the sum of the individual MOSFET's lengths. Because each device is identical, changes in the threshold voltage shouldn't affect the biasing circuit.

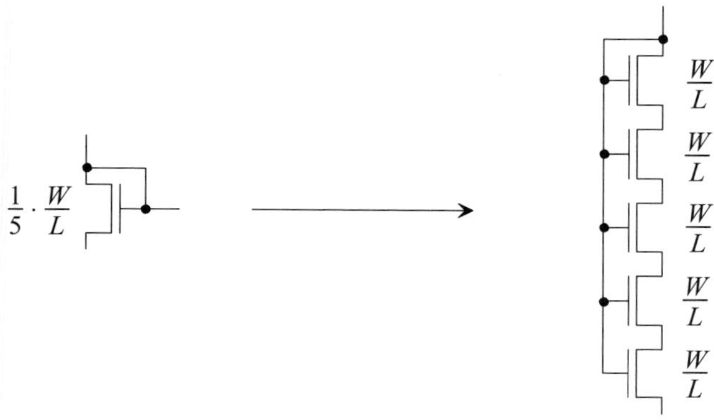

Figure 20.35 Using equal length devices to implement a long L MOSFET.

20.2.3 Wide-Swing, Short-Channel Design

When designing a wide-swing current mirror in a short-channel CMOS process, we know

$$V_{DS,sat} \neq V_{GS} - V_{THN} = V_{ovn} \tag{20.61}$$

Further, as seen in Fig. 9.32 in Ch. 9, the output resistance of a short-channel MOSFET depends on the drain-to-source voltage. Figure 20.36 shows a wide-swing current mirror designed in the short-channel CMOS process (see Table 9.2). Note that the current is approximately 4 µA when it should be, as seen in Fig. 20.23, 10 µA. Further, if we were to measure the output resistance, we would get approximately 600k, which is considerably less than the 4 MΩ seen in Table 20.1. As seen in Fig. 9.32, the output resistance at $V_{DS,sat}$ isn't 166k as indicated in Table 9.2 but much less. To increase the output resistance, let's attempt to bias M2 deeper into the saturation region. Let's do this by using a larger value of voltage on the gate of M4. For our short-channel design, let's try

$$\frac{W_{MWS}}{L_{MWS}} = \frac{1}{25} \cdot \frac{W}{L} \tag{20.62}$$

or a five times smaller ratio than what we used for the long-channel devices (Eq. [20.59]). The simulation results for the circuit in Fig. 20.36 are seen in Fig. 20.37 when MWS is sized 10/10. Notice that our output current is closer to 10 µA. Studying Fig. 20.36 and remembering our constant theme in this chapter that good matching requires both equal V_{GS} and V_{DS}, we see a concern. M1's drain-to-source voltage isn't the same as

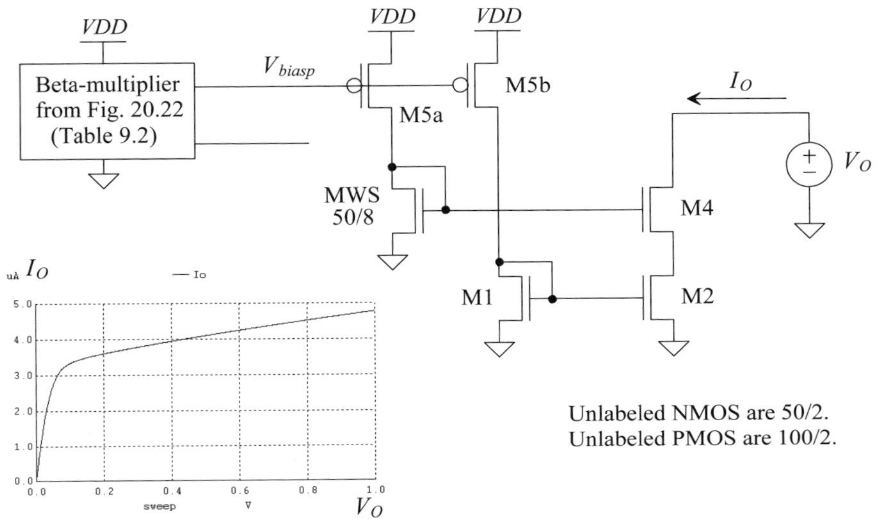

Figure 20.36 Wide-swing cascode current source in the short-channel process (bad).

M2's. By making the gate voltage of M4 larger, we act to equalize the two drain-to-source voltages (we increase the drain voltage of M2) causing the output current to approach the current that flows in M1 (10 μA). Let's attempt to make the current mirrors more symmetrical in an effort to equilibrate the drain-to-source voltages of M1 and M2.

Figure 20.38 shows the addition of a MOSFET, M3, to lower M1's V_{DS} so that it matches M2's V_{DS}. The 10 μA current through M5b can now be mirrored accurately. M1's gate voltage increases until M1 (and thus M2) can sink the current supplied by M5b with the smaller V_{DS}. *For all wide-swing current mirrors the topology seen in Fig. 20.38 using M3 should be used.* Note also, in Fig. 20.38, the larger voltage needed across the current source to keep M2/M4 saturated. The drains of M1 and M2 are at 150 mV or considerably above $V_{DS,sat}$ (which is 50 mV from Fig. 9.32). M3 is moving close to the triode region. Equation (20.62) can be modified to move M3 away from triode by using a larger W/L device (say 1/10); however, the output resistance will decrease (again as seen in Fig. 9.32, the decrease in V_{DS} results in smaller output resistance, r_o).

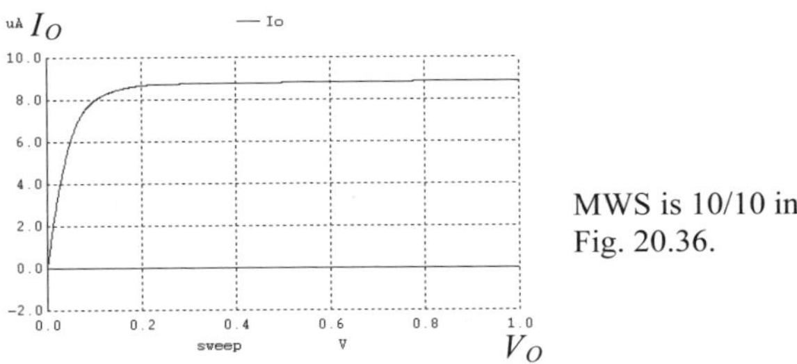

Figure 20.37 Increasing the W/L of MWS to 1/25 the other W/Ls.

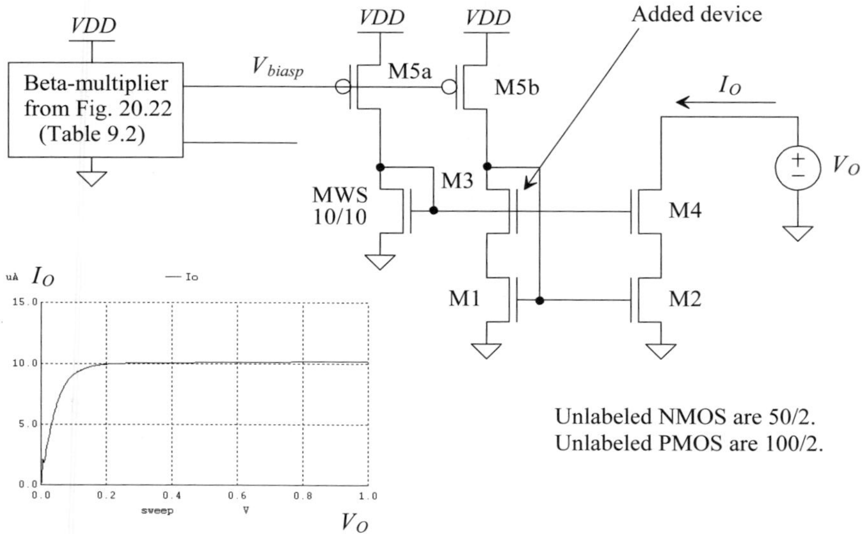

Figure 20.38 Wide-swing cascode current source in the short-channel process (good). Notice the drain-to-source voltages of M1 and M2 are the same.

Example 20.8

Regenerate the plot seen in Fig. 20.38 if MWS is sized using

$$\frac{W_{MWS}}{L_{MWS}} = \frac{1}{10} \cdot \frac{W}{L} \tag{20.63}$$

The simulation results are seen in Fig. 20.39 where MWS is sized 10/4. The output resistance is roughly four times smaller than what we got in Fig. 20.38 using Eq. (20.62). M3 is biased further away from the triode region. The point here is that there isn't any single equation that governs the selection of MWS size for all processes. Design trade-offs must be made.

Note that an interesting simulation to run at this point is to verify that the current mirror still functions correctly with small *VDD* (as seen in Fig. 20.23 with *VDD* down to 0.5 V). The output voltage, V_O, can still go to 1 V. ■

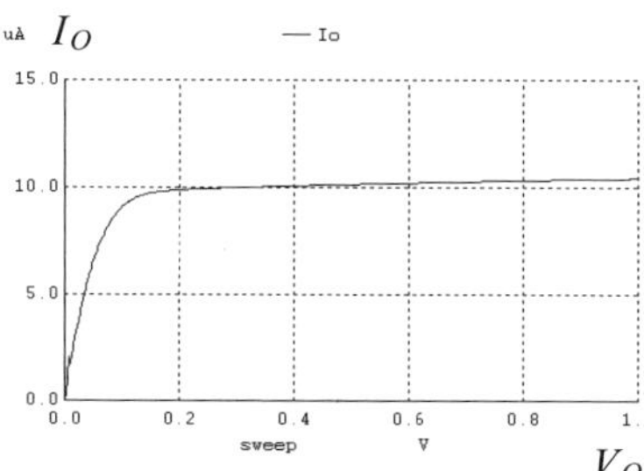

Figure 20.39 Resimulating the circuit in Fig. 20.39 using a 10/4 device for MWS.

20.2.4 Regulated Drain Current Mirror

Consider adding an amplifier to the basic wide-swing current mirror as seen in Fig. 20.40. The purpose of the amplifier is to regulate or hold the drain of M2 at a fixed potential (in this case, the drain potential of M1). If we can hold the drain potential of M2 perfectly fixed, then M2's drain current won't vary and the output resistance of the current mirror will be infinite. In the next section, we'll discuss general biasing circuits. We'll use the drain voltage of M1 as a bias voltage simply for regulating the drain in current mirrors.

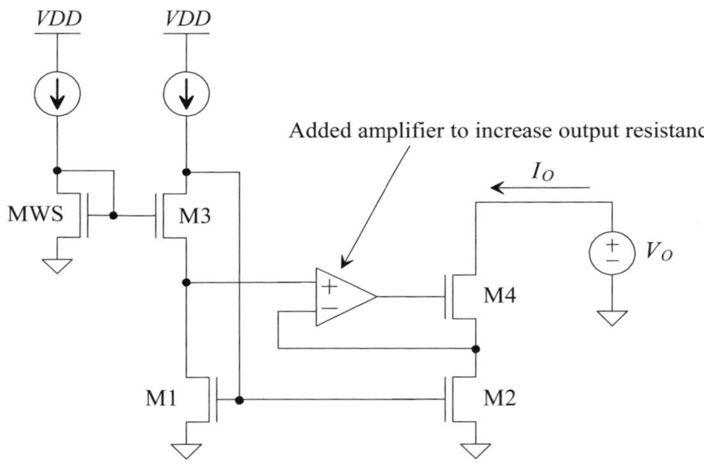

Figure 20.40 Regulating the drain of M2 using an amplifier.

We can estimate the enhancement in the output resistance of the current mirror by assuming the output of the added amplifier is related to its inputs using

$$v_{out} = A \cdot (v_+ - v_-) \tag{20.64}$$

where A is the gain of the added amplifier. Using Fig. 20.41, we can write, knowing the drain and gate of M1 are DC bias voltages (AC grounds),

$$v_{gs4} = -i_T r_o (A+1) \tag{20.65}$$

and

$$i_T = g_m(-i_T r_o(A+1)) + \frac{v_T - i_T r_o}{r_o} \tag{20.66}$$

and finally,

$$R_o = \frac{v_T}{i_T} = 2r_o + g_m r_o^2 (A+1) \approx g_m r_o^2 \cdot A \tag{20.67}$$

When compared to Eq. (20.53), the output resistance is increased by the gain of the added amplifier. This result can be very useful (and we will use it often) in practical circuit design.

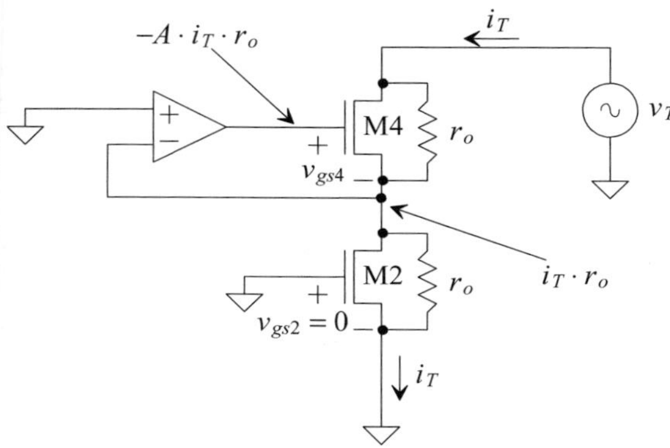

Figure 20.41 Determining the output impedance of a regulated drain current mirror.

A practical example of a current mirror that uses this technique is seen in Fig. 20.42. The drains of M1 and M2 are held at a V_{GS} by MA1 and MA2. If the drain potential of M2, for example, starts to decrease, MA2 starts shutting off, causing the gate voltage of M4 to increase, and pulling the drain potential of M2 back up. MA1 and MA2 are used to ensure that the circuit is symmetrical (M1 and M2 have the same V_{DS}). The drawback of this topology, as seen in the simulation results in the figure, is that the minimum voltage across the current mirror is $V_{DS,sat} + V_{GS}$ (just like the basic cascode, see Eq. [20.48]). The gain of the added amplifier (A) is, as we'll see in the next chapter, $g_m r_o/2$. We can estimate the output resistance as $g_m^2 r_o^3/2$ or 52 MΩ. Note that M3 is operating in the triode region.

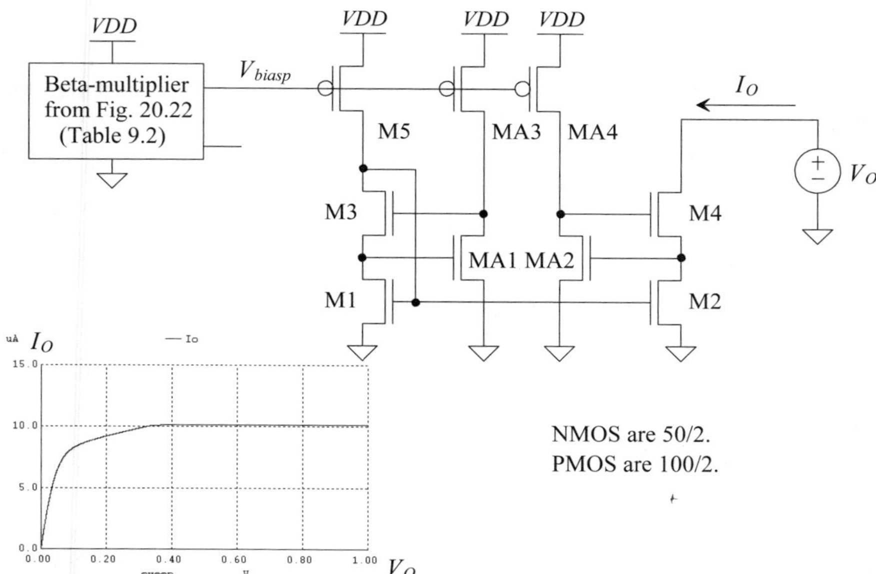

Figure 20.42 Using amplifiers to regulate the drain potentials in a current mirror.

20.3 Biasing Circuits

In this section we provide biasing circuit examples for the chapters that follow based on the data in Tables 9.1 and 9.2.

20.3.1 Long-Channel Biasing Circuits

Figure 20.43 shows a general biasing circuit for the long-channel CMOS process based on the sizes and currents given in Table 9.1. The Beta-multiplier, self-referenced circuit from Fig. 20.15 is used for biasing wide-swing current mirrors (V_{bias1} to V_{bias4}). V_{high} and V_{low} will be used later with amplifiers for regulating the drain of a MOSFET, as seen in Fig. 20.40. V_{ncas} and V_{pcas} are used for a circuit called a "floating current source" (which is discussed later). Perhaps the best way to see how this circuit can be used is to give a couple of examples.

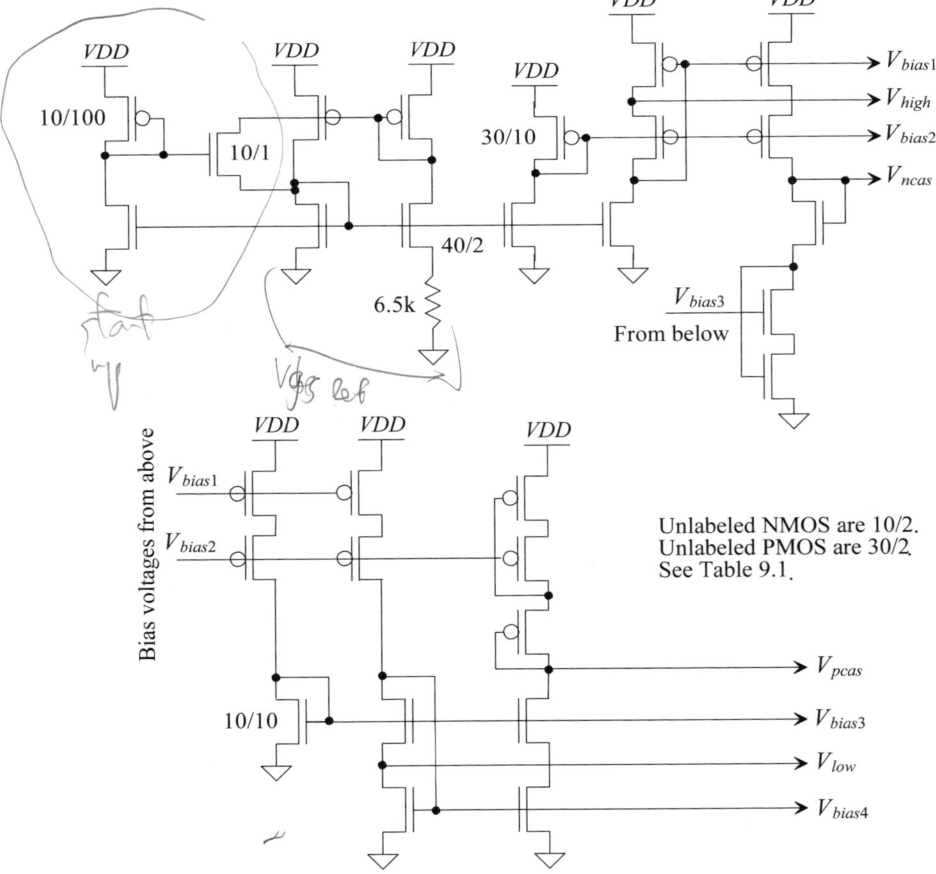

Figure 20.43 General biasing circuit for long-channel CMOS design using the data in Table 9.1

Basic Cascode Biasing

Figure 20.44 shows how the bias voltages generated in Fig. 20.43 can be used to bias both PMOS and NMOS cascode current sources. As indicated in Table 9.1 and Fig. 20.31b, the minimum voltage across the current sources is $2V_{DS,sat}$ or 500 mV. The current that flows in the current sources is nominally 20 μA.

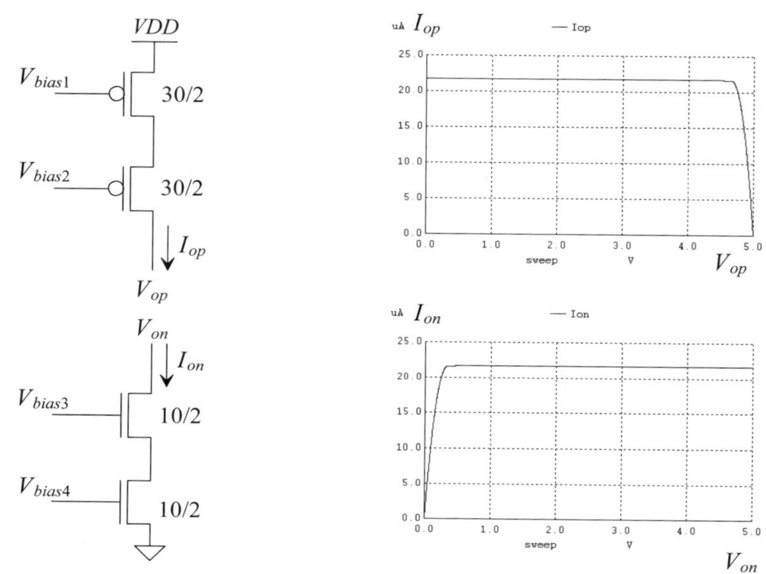

Bias voltages come from Fig. 20.43 (long-channel parameters in Table 9.1).

Figure 20.44 How cascode currents are biased and how they operate.

The Folded-Cascode Structure

Figure 20.45a shows a structure where an NMOS current source is connected to a PMOS current source. In this topology there is no way we can guarantee that the current flowing in the PMOS transistors is precisely equal to the current flowing in the NMOS transistors. This will cause the output of the circuit to move towards either *VDD* or ground (depending on which cascode structure is sourcing the most current). In (b) we fold the cascode structure over and diode-connect the folded NMOS devices (we could, just as well, have diode-connected the PMOS devices instead). Now, the gate potentials of MB1 and MB2 are set by the current that is sourced from MB5 and MB7. If the transistors are perfectly matched, the output voltage will equal the gate voltage of MB1 (the drain voltage of MB3) because of the circuit's symmetry. (We also saw this in Fig. 20.1.)

In Figure 20.46 we add some MOSFETs to steal 10 μA from MB7 and MB8. To make sure that the currents are equal, we size up MB7 and MB8 so that they source 30 μA. Note how we doubled the length of M1 and M2 so that they sink 10 μA from the same bias voltages.

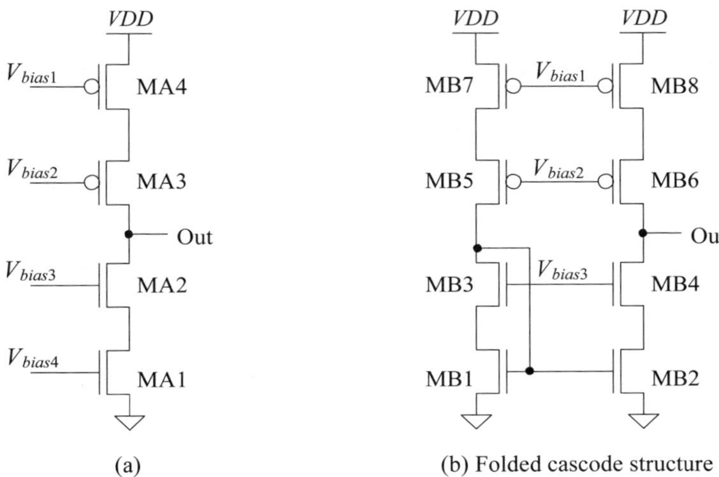

Bias voltages come from Fig. 20.43 (long-channel parameters in Table 9.1).
All NMOS are 10/2, while all PMOS are 30/2.

Figure 20.45 Using a folded cascode structure to make sure that the current sourced by the PMOS equals the current sourced by the NMOS.

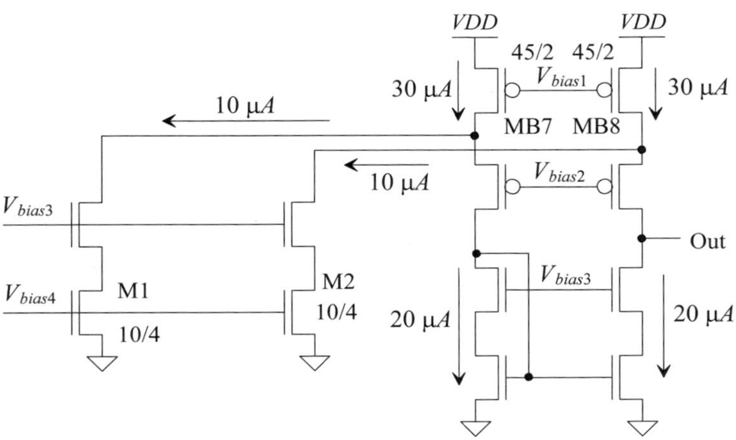

Bias voltages come from Fig. 20.43 (long-channel parameters in Table 9.1).
All unlabeled NMOS are 10/2, while all unlabeled PMOS are 30/2.

Figure 20.46 Stealing current from the folded cascode structure.

20.3.2 Short-Channel Biasing Circuits

Figure 20.47 shows a biasing circuit for a general analog design short-channel process. It uses the Beta-multiplier, self-referenced bias circuit from Fig. 20.22. As indicated in Sec. 20.1.4, the critical capacitance for stability is MCP. Notice that we are using the PMOS devices in the Beta-multiplier to bias the current mirrors, whereas in Fig. 20.43 we used the NMOS devices. We picked the PMOS here simply to increase the capacitance on V_{biasp} and so to further stabilize the circuit. Figure 20.48 shows the operation of the cascode current mirrors biased with the circuit in Fig. 20.47. It's important to realize that the minimum voltage across the current source is considerably above $2V_{DS,sat}$. Again, this was a design choice (see Eqs. [20.62] and [20.63]) to increase output resistance.

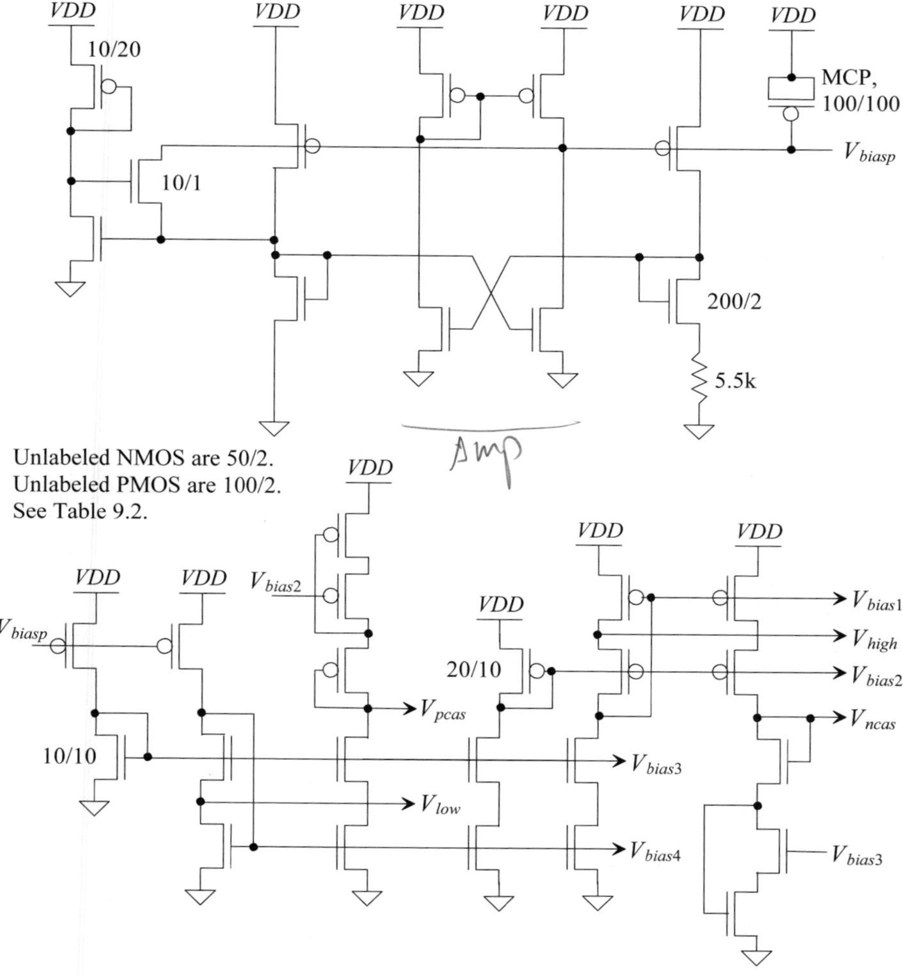

Figure 20.47 General biasing circuit for short-channel design using the data in Table 9.2.

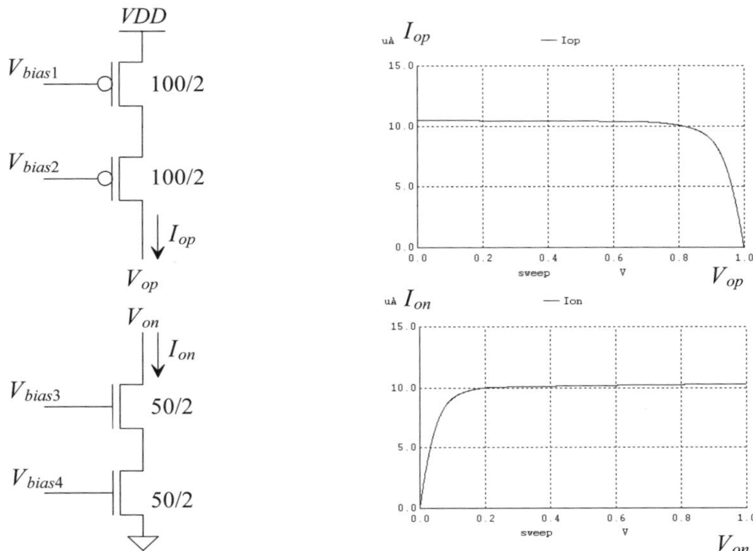

Bias voltages come from Fig. 20.47 (short-channel parameters in Table 9.2).

Figure 20.48 Cascode current sources operating in a short-channel process.

Floating Current Sources

In Fig. 20.49 we add two MOSFETs, MFCP and MFCN, to the cascode structure to form a *floating current source*. The addition of MFCN, for example, allows the voltage across the PMOS cascode structure to become V_{bias1} (assuming that matching between the bias circuit and this circuit is perfect). The voltage across the NMOS cascode becomes V_{bias4}. This can be very useful for biasing the next stage in an amplifier. For example, if we connect an NMOS (MON in the figure) device to the source of MFCN, then, because this potential is ideally V_{bias4}, we treat the MOSFET as if it were biased with V_{bias4}. Since the width of MON in the figure is ten times the widths of the other NMOS, the current in the output stage will be 100 μA. Figure 20.50 shows the simulation results with currents that flow in stage 1 and stage 2 of the circuit.

Note that the circuit in Fig. 20.49 uses all of the bias voltages except for V_{low} and V_{high}. These voltages are used with circuits that employ drain regulation, as discussed in Sec. 20.2.4.

20.3.3 A Final Comment

The bias circuits in Figs. 20.43 and 20.47 will be used in the next several chapters to provide biasing for amplifier design examples. It's important to remember our constant theme when we design current mirrors, that is, if we can set both the gate-source and the drain-source voltages of a MOSFET, we set the drain current. Similarly if we can set the drain current and gate-source voltage, then we set the drain-source voltage.

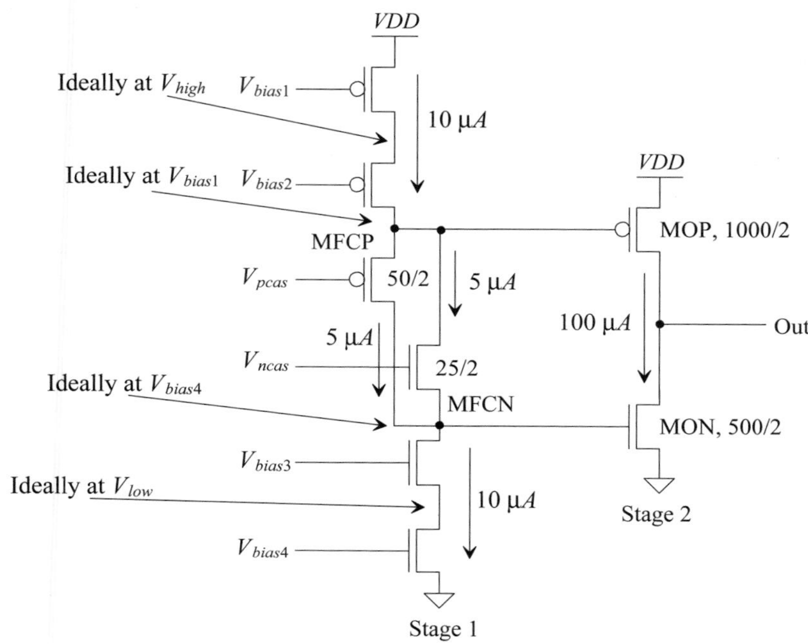

Bias voltages come from Fig. 20.47 (short-channel parameters in Table 9.2).
Unlabeled NMOS are 50/2, while unlabeled PMOS are 100/2.

Figure 20.49 Biasing with a floating current source.

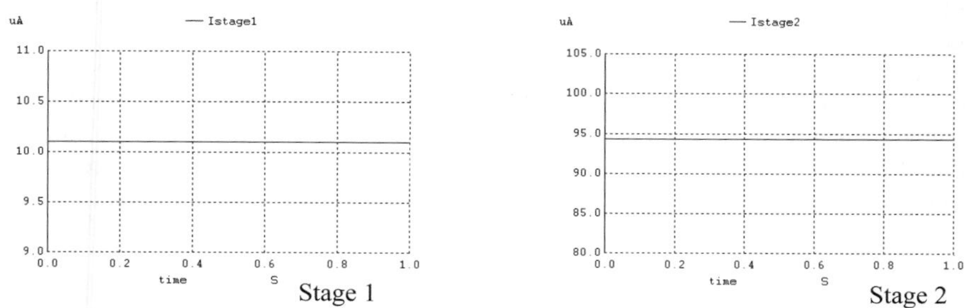

Figure 20.50 The simulated currents that flow in stages 1 and 2 of the circuit in Fig. 20.49.

ADDITIONAL READING

[1] S. J. Lovett, M. Welten, A. Mathewson, and B. Mason, "Optimizing MOS transistor mismatch," *IEEE Journal of Solid-State Circuits,* vol. 33, pp. 147–150, January 1998.

[2] K. de Langen and J. H. Huijsing, "Compact low-voltage power-efficient operational amplifier cells for VLSI," *IEEE Journal of Solid-State Circuits,* vol. 33, pp. 1482–1496, October 1998.

[3] P. Larsson, "Parasitic resistance in an MOS transistor used as on-chip decoupling capacitance," *IEEE Journal of Solid-State Circuits,* vol. 32, pp. 574–576, April 1997.

[4] C. Yu and R. L. Geiger, "An automatic offset compensation scheme with ping-pong control for CMOS operational amplifiers," *IEEE Journal of Solid-State Circuits,* vol. 29, pp. 601–610, May 1994.

[5] R. G. Eschauzier, R. Hogervorst, and J. H. Huijsing, "A programmable 1.5 V CMOS class-AB operational amplifier with hybrid nested Miller compensation for 120 dB gain and 6 MHz UGF," *IEEE Journal of Solid-State Circuits,* vol. 29, pp. 1497–1504, December 1994.

[6] R. Hogervorst, J. P. Tero, R. G. H. Eschauzier, and J. H. Huijsing, "A Compact Power-Efficient 3 V CMOS Rail-to-Rail Input/Output Operational Amplifier for VLSI Cell Libraries," *IEEE Journal of Solid State Circuits,* vol. 29, pp. 1505–1513, December 1994.

[7] E. Säckinger and W. Guggenbühl, "A High-Swing, High-Impedance MOS Cascode Circuit," *IEEE Journal of Solid-State Circuits,* vol. 25, pp. 289–298, February 1990.

[8] M. J. M. Pelgrom, A. C. J. Duinmaijer, and A. P. G. Welbers, "Matching properties of MOS transistors," *IEEE Journal of Solid-State Circuits*, vol. 24, pp. 1433–1439, October 1989.

[9] K. Lakshmikumar, R. A. Hadaway, and M. A. Copeland, "Characterisation and Modeling of Mismatch in MOS Transistors for Precision Analog Design," *IEEE Journal of Solid-State Circuits*, vol. 21, pp. 1057–1066, December 1986.

[10] J. Shyu, F. Krummenacher, and G. C. Temes, "Random error effects in matched MOS capacitors and current sources," *IEEE Journal of Solid-State Circuits,* vol. 19, pp. 948–956, December 1984.

[11] P. R. Gray and R. G. Meyer, "MOS operational amplifier design - A tutorial overview," *IEEE Journal of Solid-State Circuits,* vol. 17, pp. 969–982, December 1982.

[12] J. L. McCreary, "Matching properties, and voltage and temperature dependence of MOS capacitors," *IEEE Journal of Solid-State Circuits,* vol. 16, pp. 608–616, December 1981.

[13] W. Steinhagen and W. L. Engl, "Design of integrated analog CMOS circuits - A multichannel telemetry transmitter," *IEEE Journal of Solid-State Circuits,* vol. 13, pp. 799–805, December 1978.

[14] E. Vittoz and J. Fellrath, "CMOS analog integrated circuits based on weak inversion operations," *IEEE Journal of Solid-State Circuits,* vol. 12, pp. 224–231, June 1977.

PROBLEMS

20.1 Using the CMOS long-channel process (1 μm), determine the current flowing in the circuit seen in Fig. 20.51. Verify your answer with SPICE.

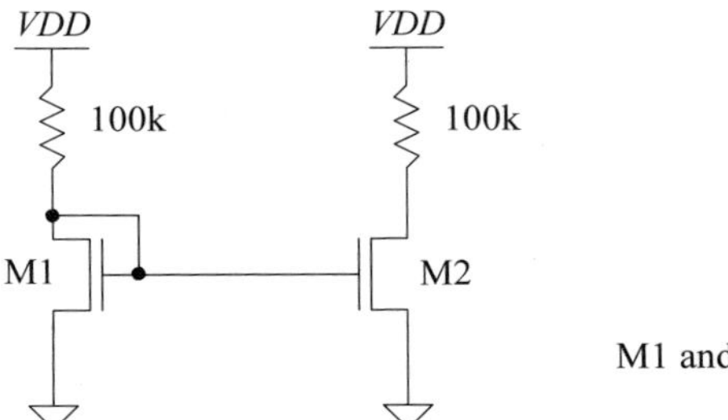

Figure 20.51 Current mirror used in Problem 20.1.

20.2 Repeat problem 20.1 for the circuit in Fig. 20.52. Can M1 and M2 be replaced with a single MOSFET? If so how and what size? If not why?

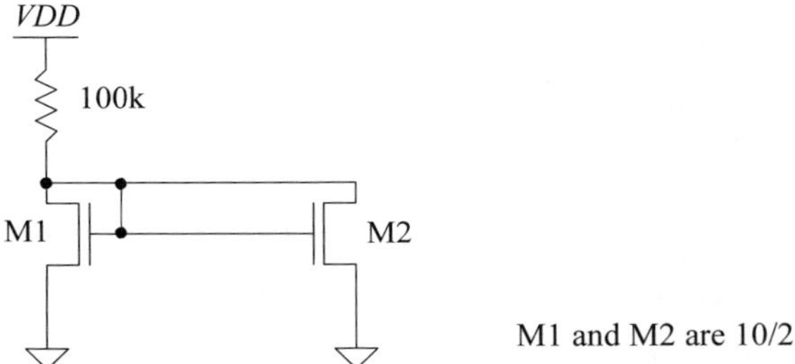

Figure 20.52 Circuit used in Problem 20.2.

20.3 Show the SPICE simulations for the PMOS devices in Ex. 20.1.

20.4 A threshold voltage mismatch in SPICE can be modeled by adding a small voltage source in series with the gate of one of the MOSFETs, as seen in Fig. 20.53. Show, using SPICE, that Eq. (20.8) is valid.

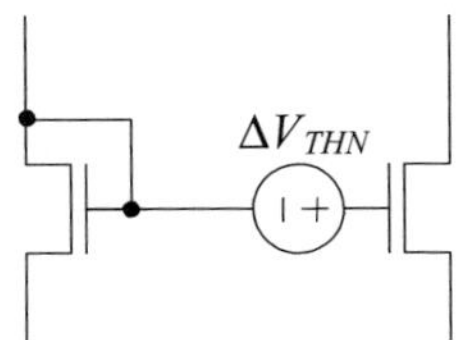

Figure 20.53 Modeling an offset voltage in SPICE (or for hand calculations).

20.5 Show, using simulations and hand-calculations, that by using a larger value of $V_{DS,sat}$ when designing bias circuits, the MOSFETs enter the triode region earlier.

20.6 In Ex. 20.2, how does the gate voltage of M1/M2 change as *VDD* is decreased? How does the V_{SG} change? Use SPICE to verify your answers.

20.7 For the same bias current used in Table 9.1 (nominally 20 μA), regenerate Fig. 20.13 if minimum-length MOSFETs are used in the current mirror. Show the hand calculations leading to the selection of the MOSFET sizes.

20.8 Suppose that the bias circuit used to generate Fig. 20.16 shows a reference current change with *VDD,* as seen in Fig. 20.54. Knowing that the reference current should be constant with changes in *VDD,* what is wrong with the reference? (What important portion of the reference is causing the problem?)

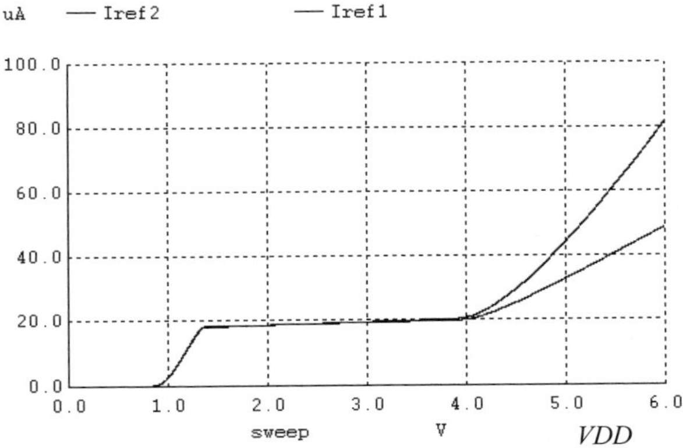

Figure 20.54 Problems with reference current increasing with VDD.

20.9 Using hand calculations, estimate the value of the voltage across *R* for the current reference in Fig. 20.15.

20.10 Using SPICE, determine if the reference circuit seen in Fig. 20.22 can become unstable with changes in *VDD*. Show that reducing the size of MCP and MCN affects the stability of the circuit.

20.11 Estimate the temperature behavior of the gate voltage of M1 in Fig. 20.10.

20.12 Design a voltage reference using the Beta-multiplier in the short-channel process and the data from Table 9.2. Generate a figure similar to Fig. 20.27 to show the temperature performance of your design.

20.13 *K* can be used to minimize *R* in Eq. (20.46) for a fixed-reference current. If M1/M2 experience a threshold voltage mismatch (see problem 20.4), discuss and show with simulations how much ΔV_{THN} can be tolerated and the error in the reference current that results.

20.14 Using an AC test voltage, determine the output resistance of the MOSFET current mirror seen in Fig. 20.28.

20.15 Based on the data in Table 9.1, what are the voltages on the drain, gate, and source terminals of M1–M4 in Fig. 20.29, assuming that the devices are operating in the saturation region? Compare your hand calculations to simulations.

20.16 If the MOSFET in Fig. 20.55 is operating in the saturation region, determine the small-signal resistance looking into its drain.

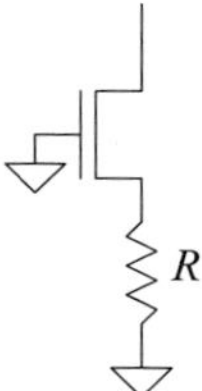

Figure 20.55 Resistance looking into the drain of a MOSFET with a source resistance. See Problem 20.16.

20.17 Suppose that the wide-swing current mirror seen in Fig. 20.38 is implemented in the long-channel CMOS process (Table 9.1). Estimate the size of MWS when M3 triodes. Verify your answer with simulations.

20.18 Label the voltages, based on the values in Table 9.1, in the schematic of the general biasing circuits seen in Fig. 20.43. Compare the tabulated values of V_{GS}, V_{DS}, etc., to the simulated values.

20.19 How would the current in Fig. 20.44 change if the width of the device connected to V_{bias2} were doubled? Were halved? Why? Explain what is going on in the circuit.

20.20 Should the voltage labeled "Out" in Fig. 20.49 be at a specific value? Why or why not?

Chapter

21

Amplifiers

In this chapter we turn our attention towards amplifiers. Single-stage amplifiers are used in virtually every op-amp design. By replacing a passive load resistor with a MOSFET transistor (called an *active load*), significant amounts of chip area can be saved. An active load can also produce higher values of resistance when compared with a passive resistor, resulting in higher gains.

Several types of active loads are studied in this chapter. The *gate-drain load* consists of a MOSFET with the gate and drain shorted and provides large bandwidths and low-output impedance at the cost of reduced gain. The *current source load* amplifier has a higher gain and output impedance at the expense of lower bandwidth. Current source load amplifiers are preferred when external feedback is applied to the amplifier to set the gain. This chapter explores basic single-stage amplifiers with active loads, as well as the trade-offs associated with each. The cascode amplifier is examined in detail along with several configurations of output stages, including the push-pull amplifier.

21.1 Gate-Drain-Connected Loads

A gate-drain-connected MOSFET (load) is half of a current mirror as seen in Fig. 20.1 in the last chapter. Here we move the discussion from the DC operating conditions and biasing presented in the last chapter to AC small-signal analysis.

21.1.1 Common-Source (CS) Amplifiers

A gate-drain-connected MOSFET (operating with a nonzero drain current) can be thought of as a resistor of value $1/g_m$ (see Fig. 9.18 and Eq. [9.25]). The four possible common-source (CS) amplifiers using gate-drain-connected loads are seen in Fig. 21.1. In each configuration, M1 and M2 are assumed to be biased in the saturation region. Notice how the source of the amplifying MOSFET (i.e., not the load or gate-drain MOSFET) is common to both the input and the output in each configuration.

Examine the circuit shown in Fig. 21.1a. Since many MOSFET amplifiers are analyzed in this chapter and the next several chapters using small-signal analysis, an intuitive analysis approach will be used. This approach allows the designer to quickly analyze the gain of a circuit. To analyze the gain of the circuit in Fig. 21.1, begin by

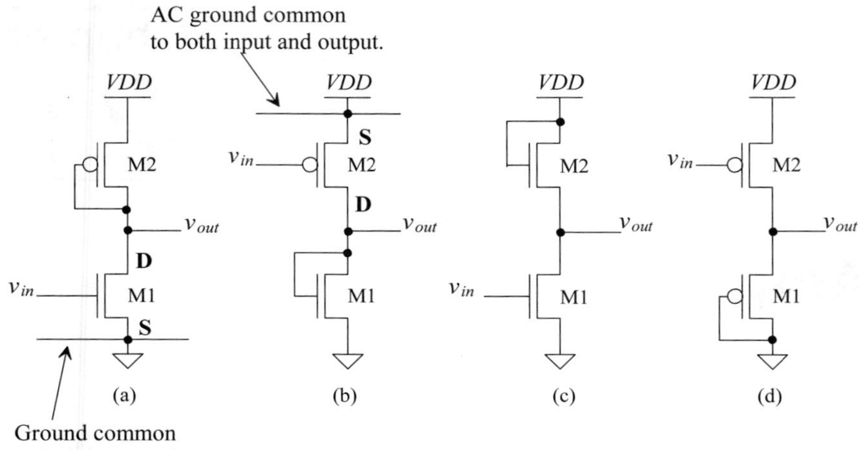

Figure 21.1 Four possible configurations of common-source amplifiers with gate-drain loads.

replacing M2 with a resistor of value $\frac{1}{g_{m2}}$ and replacing M1 with a current source of value $g_{m1} \cdot v_{in}$, provided $\frac{1}{g_{m2}} << r_{o1} || r_{o2}$. The equivalent circuit is shown in Fig. 21.2. Note that there is no body effect in either MOSFET and that only the low-frequency model is shown. Again, as seen in Ch. 9, a gate-drain-connected MOSFET has a small-signal resistance of value $\frac{1}{g_m}$. The small-signal gain of the **common-source amplifier** is given by

$$\frac{v_{out}}{v_{in}} = \frac{-i_d \cdot \frac{1}{g_{m2}}}{i_d \cdot \frac{1}{g_{m1}}} = -\frac{\frac{1}{g_{m2}}}{\frac{1}{g_{m1}}} = -\frac{\text{resistance in the drain}}{\text{resistance in the source}} = -\frac{g_{m1}}{g_{m2}} \qquad (21.1)$$

This result is very important in the intuitive analysis. It states that the small-signal gain of a common-source amplifier is simply the resistance in the drain of M1 divided by the resistance in the source (the resistance looking into the source of M1 added to any resistance from the source of M1 to ground). The actual value of the resistance in the

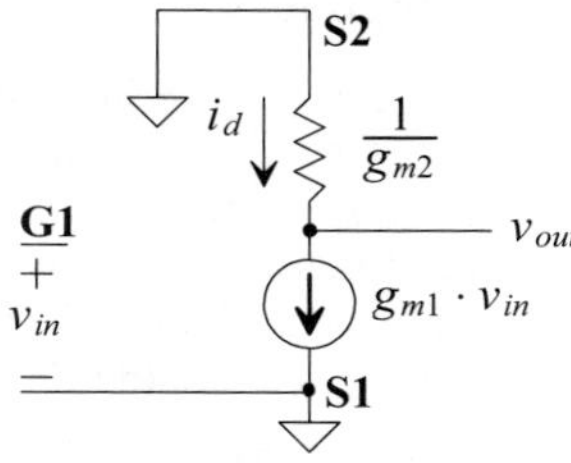

Figure 21.2 Simplified circuit of Fig. 21.1a.

drain, as will be soon seen, is $\frac{1}{g_{m2}}||r_{o1}||r_{o2}$. In most cases, however, r_o will be much greater than $\frac{1}{g_m}$, and the approximation is an accurate one.

Example 21.1
Determine the small-signal AC gain of the circuit shown in Fig. 21.3 using the intuitive method presented in Eq. (21.1).

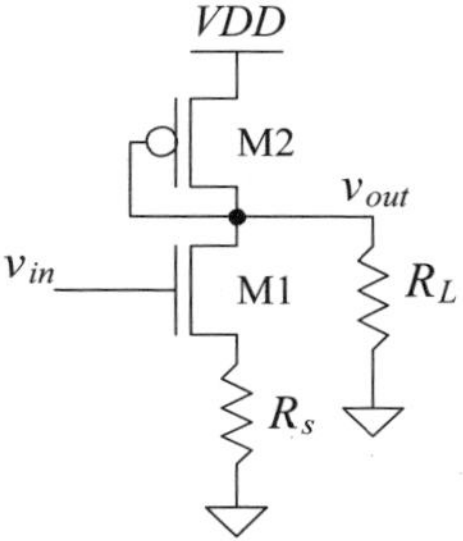

Figure 21.3 Amplifier analyzed in Ex. 21.1.

The small-signal gain of this circuit is given by

$$A_v = \frac{v_{out}}{v_{in}} = -\frac{\text{resistance in the drain}}{\text{resistance in source}}$$

The resistance in the drain is the parallel combination of all resistances connected to the drain of M1, given by

$$\text{Resistance in the drain} = \frac{1}{g_{m2}}||R_L \text{ for } r_{o1}||r_{o2} >> \frac{1}{g_{m2}}$$

The resistance in the source is the sum of the resistance looking into the source of M1 ($1/g_{m1}$) and the resistance connected from the source of M1 to ground. For the present problem, this resistance is given by

$$R_{\text{in the source}} = \frac{1}{g_{m1}} + R_s$$

The voltage gain is then given by

$$A_v = -\frac{\frac{1}{g_{m2}}||R_L}{\frac{1}{g_{m1}} + R_s} \text{ which for } R_L \to \infty \text{ and } R_s \to 0 \text{ reduces to } -\frac{g_{m1}}{g_{m2}} \blacksquare$$

We can determine the exact gain of the amplifier of Fig. 21.1a using the full small-signal model shown in Fig. 21.4. Using KCL at the output of the amplifier gives

$$g_{m1}v_{in} + \frac{v_{out}}{r_{o2}||r_{o1}} = -g_{m2}v_{out} \tag{21.2}$$

or

$$A_v = \frac{v_{out}}{v_{in}} = -\frac{g_{m1}}{g_{m2} + \frac{1}{r_{o1}||r_{o2}}} = -\frac{\frac{1}{g_{m2}}||r_{o1}||r_{o2}}{\frac{1}{g_{m1}}} = -\frac{\text{resistance in the drain}}{\text{resistance in the source}} \tag{21.3}$$

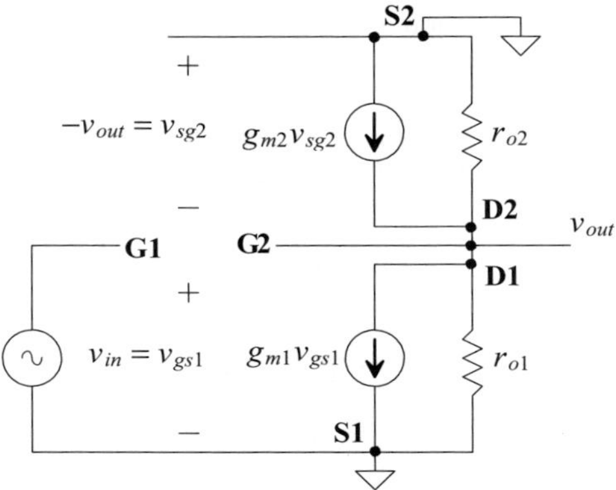

Figure 21.4 Small-signal model of the amplifier shown in Fig. 21.1(a).

When $\frac{1}{g_{m2}} << r_{o1}||r_{o2}$, this reduces to

$$A_v = -\frac{\frac{1}{g_{m2}}}{\frac{1}{gm1}} = -\frac{g_{m1}}{g_{m2}} \tag{21.4}$$

which is the same form as Eq. (21.1).

Miller's Theorem

Consider the inverting amplifier seen in Fig. 21.5a with a gain of $-A_v$ and a feedback capacitance between the output and the input of the amplifier. The current supplied by v_{in} or v_{out} to the capacitor is

$$i = \frac{v_{in} - v_{out}}{1/j\omega C_F} = j\omega C_F \cdot (1 + |A_v|) \cdot v_{in} = j\omega C_F \cdot \left(1 + \frac{1}{|A_v|}\right) \cdot v_{out} \tag{21.5}$$

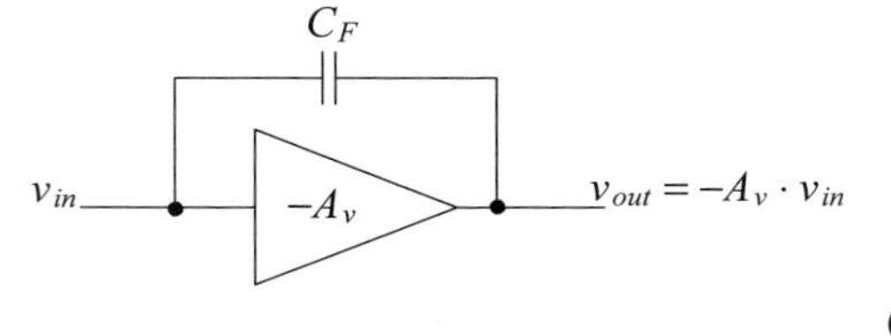

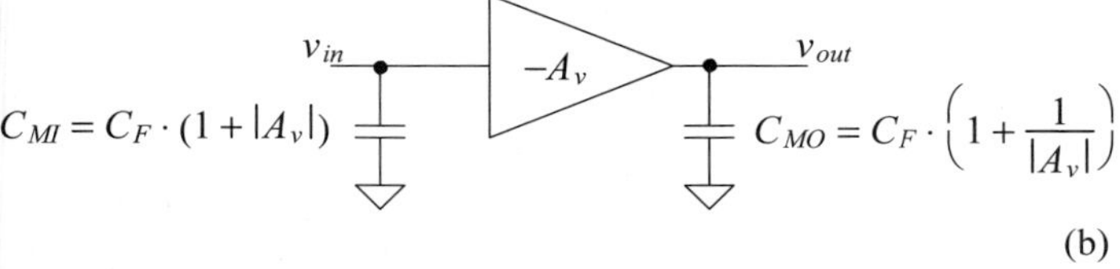

Figure 21.5 The derivation of Miller's theorem.

In Fig. 21.5b we replace the feedback capacitor with capacitors on the input and the output of the amplifier to ground. These (adjusted size) capacitors require the same currents from the input source and the output of the amplifier. Note that this substitution, Eq. (21.5), is often called *Miller's theorem,* and the feedback capacitor, C_F, is called the *Miller capacitance*.

Notice that the effective capacitance on the input of the amplifier is multiplied by the gain of the amplifier. For big gains, $|-A_v|$, this means that the input capacitance of the amplifier can be large and can result in slow circuits. Intuitively, this can be understood by realizing that the effective voltage across C_F is $A_v + 1$ times larger than the input signal itself. This large voltage drop across C_F (one side going up by v_{in} while the other side goes down by $A_v \cdot v_{in}$) results in a large displacement current, which makes the capacitor appear larger than it really is.

Frequency Response

Now consider the frequency response of the CS amplifier of Fig. 21.1a. The high-frequency (meaning that the device capacitances are included in the circuit) equivalent circuit is shown in Fig. 21.6a. The gate-drain capacitance of M2 isn't drawn because the gate and drain of M2 are shorted together. A source resistance is added to the circuit to model the effects of the driving source impedance. Miller's theorem is used to break C_{gd1} into two parts: a capacitance at the gate of M1 to ground and a capacitance from the drain of M1 to ground, as seen in Fig. 21.6b (note that the low-frequency gain of the amplifier in Fig. 21.1a is $-g_{m1}/g_{m2}$). Two *RC* time constants exist in this circuit: one on the input of the circuit and one on the output of the circuit. The input time constant is given by

$$\tau_{in} = R_s(C_{MI} + C_{gs1}) \tag{21.6}$$

where the Miller capacitance at the input, C_{MI}, is

$$C_{MI} = C_{gd1}\left(1 + \frac{g_{m1}}{g_{m2}}\right) \tag{21.7}$$

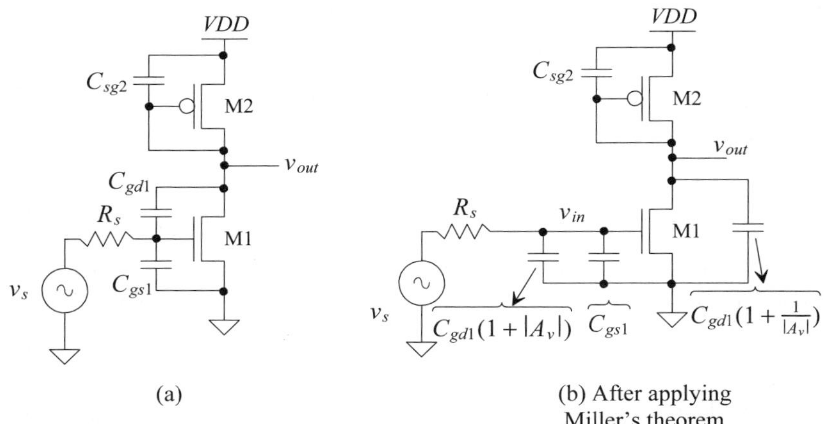

Figure 21.6 Frequency response of the CS amplifier with gate-drain-connected load.

The output time constant is found in a similar way and is given by

$$\tau_{out} = \frac{1}{g_{m2}} \cdot (C_{sg2} + C_{MO}) \tag{21.8}$$

The Miller capacitance which appears at the output, C_{MO}, now becomes

$$C_{MO} = C_{gd1}(1 + \frac{g_{m2}}{g_{m1}}) \tag{21.9}$$

The frequency response of the amplifier in Fig. 21.1a (Fig. 21.6a) is given by

$$A_v(f) = \frac{-\frac{g_{m1}}{g_{m2}}}{\left(1 + j\frac{f}{f_{in}}\right)\left(1 + j\frac{f}{f_{out}}\right)} \tag{21.10}$$

where the pole associated with the amplifier's input is located at

$$f_{in} = \frac{1}{2\pi\tau_{in}} \tag{21.11}$$

and the pole associated with the amplifier's output is located at

$$f_{out} = \frac{1}{2\pi\tau_{out}} \tag{21.12}$$

The Right-Half Plane Zero

It should be noted that when using Miller's theorem a zero is neglected. If we take a look at Fig. 21.5a, we see that at high frequencies C_F shorts the amplifier's output to its input. Our model in Fig. 21.5b doesn't show this shorting. What this means for the amplifier in Fig. 21.1a or its model in Fig. 21.6a is that C_{gd1} shorts the output to the input at high frequencies.

To characterize this high-frequency effect, where we can't use Miller's theorem, consider the circuit seen in Fig. 21.7. In this figure we model the amplifier in Fig. 21.6 (or Fig. 21.1) with a small-signal resistance on the output of the circuit called $R_o = \frac{1}{g_{m2}}||r_{o2}||r_{o1} \approx \frac{1}{g_{m2}}$ and a capacitance on the output, without including C_{gd1}, of $C_o = C_{sg2}$. We don't concern ourselves with the input time constant but rather just look at the output of the circuit. Summing the currents on the output of the model results in

$$\frac{v_{out} - v_{in}}{1/j\omega C_{gd1}} + \frac{v_{out}}{R_o||1/j\omega C_o} + g_{m1} \cdot v_{in} = 0 \tag{21.13}$$

Solving for v_{out}/v_{in} gives

$$\frac{v_{out}}{v_{in}} = -g_{m1}R_o \cdot \frac{1 - j\omega\frac{C_{gd1}}{g_{m1}}}{1 + j\omega(C_{gd1} + C_o) \cdot R_o} \tag{21.14}$$

The pole should be recognized as what we wrote in Eq. (21.12) if the gain of the amplifier is small ($A_v = -g_{m1}/g_{m2} \approx 0$). In the numerator we see a right-half plane zero at

$$f_z = \frac{g_{m1}}{2\pi C_{gd1}} \tag{21.15}$$

A zero in the right-half plane (RHP) results in the same magnitude response as a zero in the left-half plane (LHP). However, the phase response is different. A zero in the RHP has the same influence on the phase of a system as a pole in the LHP. Understanding this

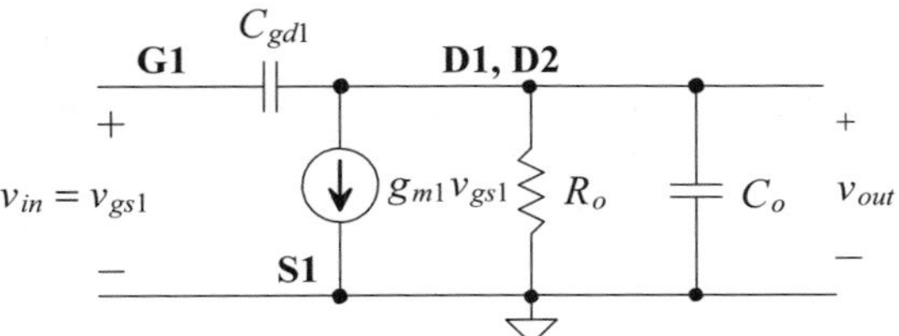

Figure 21.7 Small-signal model used to calculate the RHP zero for amplifiers in Fig. 22.1a–d.

point is important. For the inverting amplifiers (which mean that at low frequencies their phase shift is 180°) in Fig. 21.1, for example, the input feeding directly to the output because of the RHP zero, without the inversion, can result in instability when the amplifier is used with feedback. The output can feed back and add to the input (positive feedback). The result is that the output of the amplifier will grow without limit (and, ultimately, the output of the amplifier will oscillate back and forth between some voltages). The magnitude response of the RHP zero is usually ignored because its value typically places it well beyond the pole frequency. However, in CMOS op-amp design, the phase shift of this RHP zero can, and will, result in stability and/or settling time issues.

Example 21.2

Determine the gain, magnitude, and phase shift of the amplifier shown in Fig. 21.8. Using an AC SPICE simulation, verify your hand calculations.

The bias circuit is used to bias the amplifier at the operating point seen in Table 9.1, that is, at a drain current of 20 μA. The big resistor and capacitor set the DC bias point but don't affect the AC operation of the circuit.

$$A_v = -\frac{\frac{1}{g_{m2}}}{\frac{1}{g_{m1}}} = -\frac{g_{m1}}{g_{m2}} = -\frac{150\ \mu A/V}{150\ \mu A/V} = -1$$

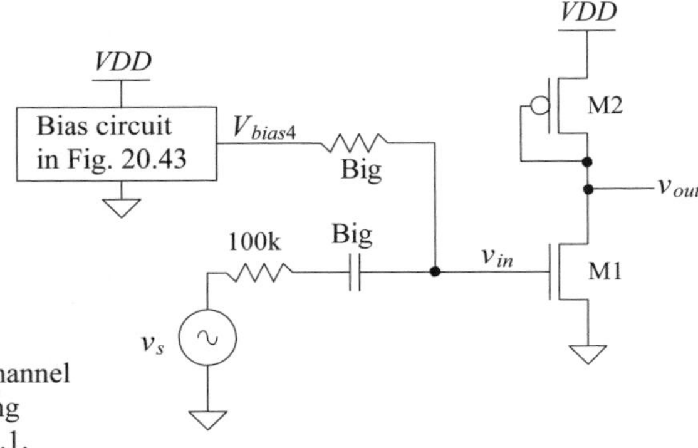

Use the long-channel sizes and biasing seen in Table 9.1.

Figure 21.8 Amplifier used in Ex. 21.2.

From Table 9.1, $C_{gs1} = 23.3\ fF$ and $C_{gd1} = 2\ fF$. The Miller input capacitance, because the gain of the amplifier is –1, is then 4 fF. The input pole is located at

$$f_{in} = \frac{1}{2\pi \cdot R_s(C_{MI} + C_{gs1})} = \frac{1}{2\pi \cdot 100k \cdot (27.3\ fF)} = 58\ MHz \qquad (21.16)$$

From Table 9.1, $C_{sg2} = 70\ fF$. The Miller output capacitance is, again, 4 fF. The output pole is located at

$$f_{out} = \frac{1}{2\pi \cdot \frac{1}{g_{m2}} \cdot (C_{MO} + C_{sg2})} = \frac{1}{2\pi \cdot 6.5k \cdot (74\ fF)} = 331\ MHz \qquad (21.17)$$

The zero is located at

$$f_z = \frac{g_{m1}}{2\pi C_{gd1}} = \frac{150\ \mu A/V}{2\pi \cdot 2\ fF} = 11.9\ GHz$$

The transfer function of the amplifier is then

$$A_v(f) = \frac{v_{out}(f)}{v_s(f)} = -\frac{\left(1 - j\frac{f}{11.9\ GHz}\right)}{\left(1 + j\frac{f}{58\ MHz}\right)\left(1 + j\frac{f}{331\ MHz}\right)}$$

Figure 21.9 shows the AC analysis simulation results for the amplifier of Fig. 21.8. What would happen if we used v_{in} instead of v_s? We wouldn't see the input

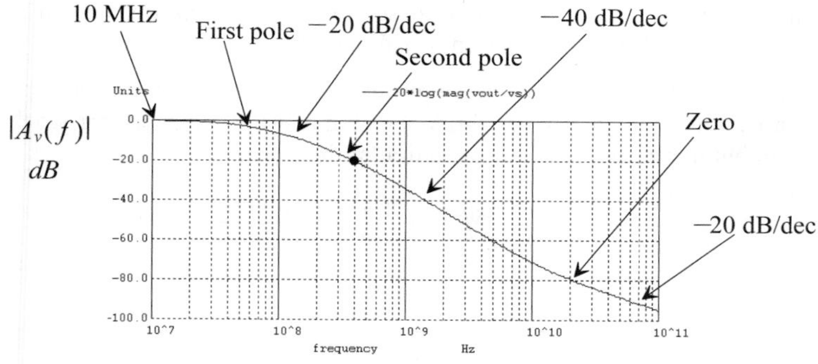

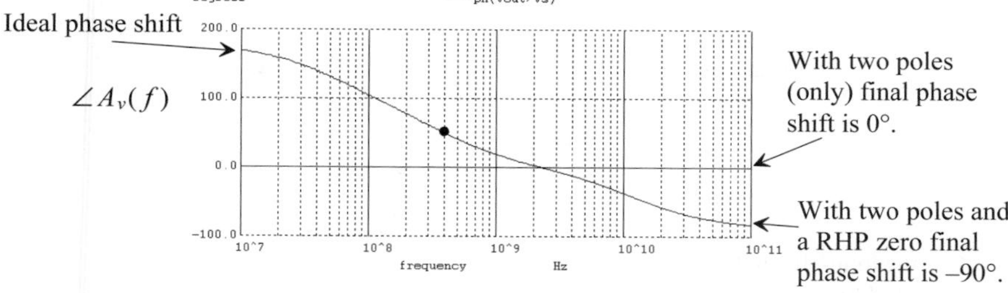

Figure 21.9 Frequency response of the amplifier in Fig. 21.8.

pole's effects on the frequency response. In the magnitude response we see that right after the first pole the gain decreases by a factor of 10 for every increase in frequency by 10 (equivalent to saying –20 dB/decade). We could also say that the gain decreases by 2 for every increase in frequency by 2 (equivalent to saying –6 dB/octave). After the second pole, the gain decreases by 100 for every increase in frequency by 10. The zero causes the frequency response to go back to –20 dB/decade. (The output and input of M1 are shorted together so that all we see are the effects of R_s charging the input capacitance of the amplifier.) It is useful to look at, and understand, the behavior of the amplifier from v_{in} to v_{out} in the simulation.

The effects of the RHP zero occur one decade before f_z. This *excess phase* shift is the problem we discussed earlier. It makes the amplifier appear not like an inverting amplifier (phase shift of –180 degrees) but rather like a noninverting amplifier sooner (at lower frequencies).

Note that magnitude of voltage gains in the neighborhood of 1 are common when using gate-drain-connected loads. At this point it may appear that this amplifier is almost worthless; however, we will see that when this amplifier is used with additional circuitry (e.g., a current source load for a differential pair), it can be very useful. ■

Example 21.3

Determine the magnitude and phase of the time-domain signal v_{out}, at 400 MHz, for the amplifier in Ex. 21.2 based on the frequency plots seen in Fig. 21.9. Verify the results using SPICE.

The magnitude response at 400 MHz is

$$\left|\frac{v_{out}}{v_s}\right| = -20\ dB = 0.1$$

and the phase response is

$$\angle\frac{v_{out}}{v_{in}} = 50° = -310°$$

The positive value means that the output is leading, or occurring earlier in time, than the input. Remembering that phase shift is simply an indication of a time delay at a particular frequency, we can write

$$50° = 360 \cdot \frac{t_d}{T} = 360 \cdot t_d \cdot 400\ MHz \rightarrow t_d \approx 350\ ps$$

for a period of 2.5 ns. We'll use an input voltage of 1 mV to ensure small-signal operation. In the simulation that generated Fig. 21.9 our input AC signal was 1 V. Again, remember that SPICE replaces the active devices with linear models in an AC simulation so any value of amplitude can be used for the input signals. Using 1 V in an AC simulation is convenient because of the ratios often used when calculating gains. Figure 21.10 shows the SPICE simulation results. Notice how the peak amplitude of the input is 1 mV while the peak amplitude of the output (remembering the DC output voltage is $VDD - V_{SG}$ or, from Table 9.1, approximately 3.85 V) is 0.1 mV (–20 db down from the low-frequency value). In a transient simulation we have to wait some time before the circuit settles to a

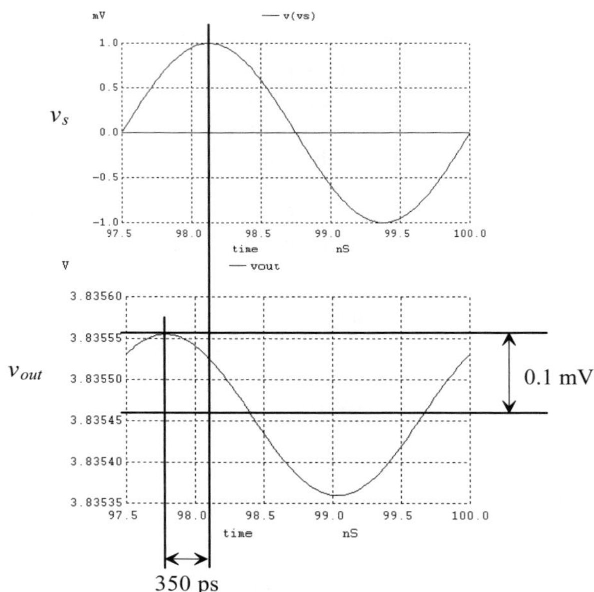

Figure 21.10 Time-domain behavior of the amplifier in Fig. 21.8 at 400 MHz.

stable operating point. In Fig. 21.10 we've waited 97.5 ns before looking at the output voltage. This gives the bias circuit, for example, time to turn on and generate the bias voltage needed in the amplifier (the value of the big capacitor was also reduced in this transient simulation so that its charging wouldn't increase the simulation time.) ■

A Common-Source Current Amplifier

The previous discussion was centered around CS voltage amplifiers. Next consider the circuit shown in Fig. 21.11. This amplifier can be thought of as a transimpedance amplifier, that is, voltage out and current in. The ideal input impedance of an amplifier with a current input is zero (all of the input current flows into the amplifier with this ideal input impedance). Note that the biasing of the amplifier is controlled by the current mirror formed between M3 and M1 (drawn in a different way to ensure that the reader recognizes the topology when it is seen).

The input impedance of the amplifier in Fig. 21.11 is given by

$$R_{in} = \frac{1}{g_{m3}} || r_{o3} || r_{o4} \approx \frac{1}{g_{m3}} \tag{21.18}$$

The input current, i_{in}, is determined by

$$i_{in} = g_{m3} v_{in} = g_{m3} v_{gs3} = g_{m3} v_{gs1} \tag{21.19}$$

The current through M1, and thus M2, is given by

$$i_d = g_{m1} v_{in} = \frac{g_{m1}}{g_{m3}} \cdot i_{in} \tag{21.20}$$

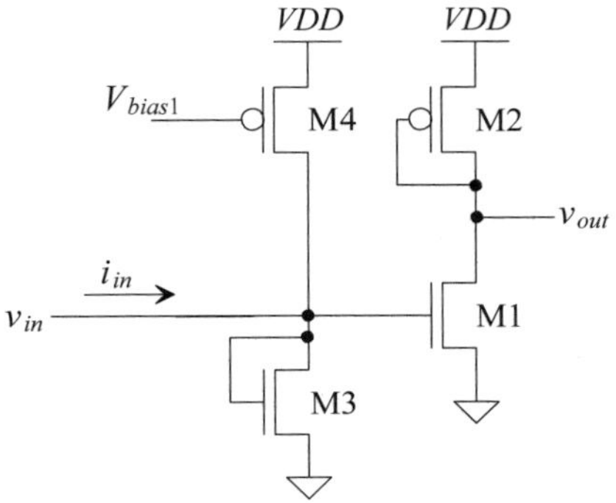

Figure 21.11 A transimpedance amplifier.

This equation can be rewritten, $g_{m3} = \beta_3(V_{GS3} - V_{THN})$ and $g_{m1} = \beta_1(V_{GS3} - V_{THN})$, as

$$\frac{i_d}{i_{in}} = \frac{\beta_1}{\beta_3} = \frac{W_1 L_3}{W_3 L_1} \tag{21.21}$$

That is, we can get a current gain simply by adjusting the size of M1 and M3. The current in M2 can then be mirrored out to a load. The transresistance gain of this configuration is given by

$$A_R = \frac{v_{out}}{i_{in}} = -\frac{i_d \frac{1}{g_{m2}}}{i_d \cdot \frac{g_{m3}}{g_{m1}}} = -\frac{g_{m1}}{g_{m2} g_{m3}} \tag{21.22}$$

Note that adding M3 and thus the resistor of value $\frac{1}{g_{m3}}$ to ground lowers the input time constant and increases the bandwidth of the amplifier.

Common-Source Amplifier with Source Degeneration

The amplifier in Fig. 21.3 is an example of a CS amplifier with source degeneration. The gain of the amplifier is reduced because the effective transconductance of the amplifier becomes

$$\frac{1}{g_{m,eff}} = \frac{1}{g_m} + R_s \tag{21.23}$$

as seen in Ex. 21.1. Adding the source resistance is useful, especially if $R_s >> 1/g_m$, to precisely set the transconductance of an amplifier. Note that we didn't, in Ex. 21.1, include the resistance looking into the drain of M1 when we wrote the "resistance in the drain." *Let's review, or again verify, how to determine the small-signal, low-frequency resistances looking into the drain and source of a MOSFET.* Using this information will greatly aid in our intuitive analysis of CMOS circuits.

Consider the test circuit shown in Fig. 21.12. Here the DC voltages ensure that the MOSFET is operating in the saturation region. We are concerned with the AC current that flows through v_t. This test voltage divided by i_d gives us the small-signal AC resistance looking into the drain of the MOSFET. In other words, this is the resistance connected to ground at the drain node.

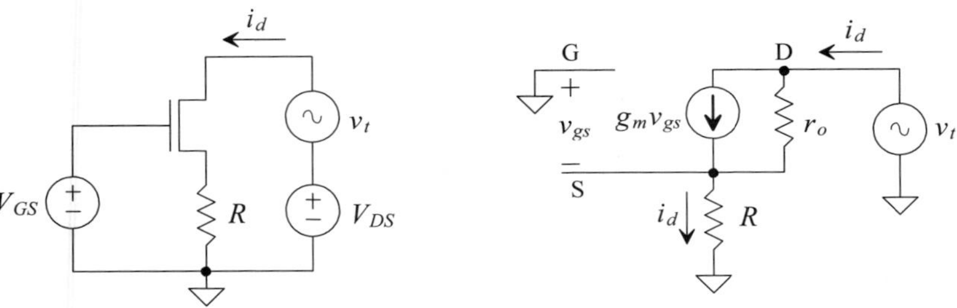

Figure 21.12 Circuit used to determine the resistance looking into the drain of a MOSFET.

Neglecting the body effect, we can write the test voltage as

$$v_t = (i_d - g_m v_{gs}) r_o + i_d R \tag{21.24}$$

where (noting $v_g = 0$ and $v_s = i_d R$)

$$v_{gs} = -\, i_d R \tag{21.25}$$

The resistance looking into the drain of the MOSFET is then given by

$$R_o = r_d = \frac{v_t}{i_d} = (1 + g_m R) r_o + R \approx (1 + g_m R) r_o \tag{21.26}$$

keeping in mind that $r_o \approx \frac{1}{\lambda I_D}$. For a cascoded MOSFET topology ($R = r_o$), this is approximately, as seen in Eq. (20.53), $g_m r_o^2$. For a *triple cascode*, three MOSFETs in series, ($R = g_m r_o^2$), the output resistance is

$$R_o \approx g_m^2 r_o^3 \tag{21.27}$$

noting that the output resistance goes up with each added MOSFET by the open-circuit gain, $g_m r_o$.

Example 21.4
Repeat Ex. 21.1 without neglecting the loading effects of the output resistances of M1 and M2.

The resistance looking into the drain of M2 is $\frac{1}{g_{m2}} || r_{o2}$, while the resistance looking into the drain of M1 is $r_{o1}(1 + g_{m1} R_s)$. The exact gain of the circuit is

$$A_v = \frac{v_{out}}{v_{in}} = -\frac{\frac{1}{g_{m2}} || r_{o2} || [r_{o1}(1 + g_{m1} R_s)] || R_L}{R_s + \frac{1}{g_{m1}}}$$

or when $r_{o1} || r_{o2} >> \dfrac{1}{g_{m2}}$, this reduces to

$$A_V = \frac{\frac{1}{g_{m2}} || R_L}{R_s + \frac{1}{g_{m1}}} \quad \blacksquare$$

The resistance in the source of the MOSFET can be found using the circuit shown in Fig. 21.13. Neglecting the body effect and output resistance of the MOSFET gives

$$v_{in} = v_{gs} + i_d R \tag{21.28}$$

Since $g_m v_{gs} = i_d$, this equation can be written as

$$v_{in} = \frac{i_d}{g_m} + i_d R = i_d \left[\frac{1}{g_m} + R \right] \tag{21.29}$$

We can look at this in a simplified manner; that is, $\frac{1}{g_m}$ is the resistance looking into the source of a MOSFET, while R is the resistance connected from the source of the MOSFET to ground.

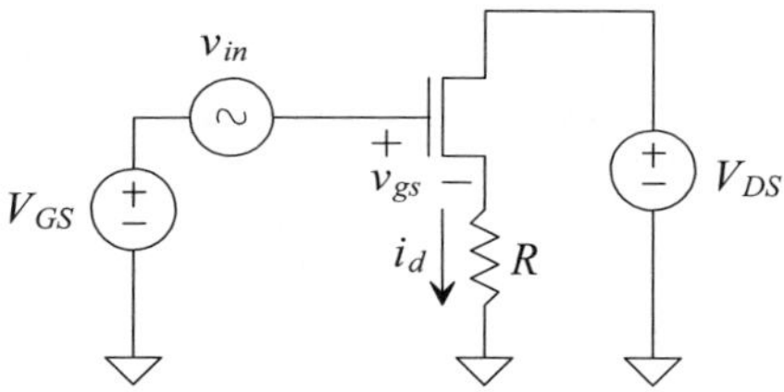

Figure 21.13 Circuit used to determine the resistance looking into the source of a MOSFET.

Noise Performance of the CS Amplifier with Gate-Drain Load

Figure 21.14 shows the CS amplifier of Fig. 21.1a with each MOSFET's model for noise power spectral density included (see Eq. [9.66]). Remembering that we only add the power from each noise source, we can write the output noise power spectral density as

$$V_{onoise}^2(f) = \left(\frac{1}{g_{m2}} \right)^2 \cdot \left(I_{M1}^2 + I_{M2}^2 \right) \tag{21.30}$$

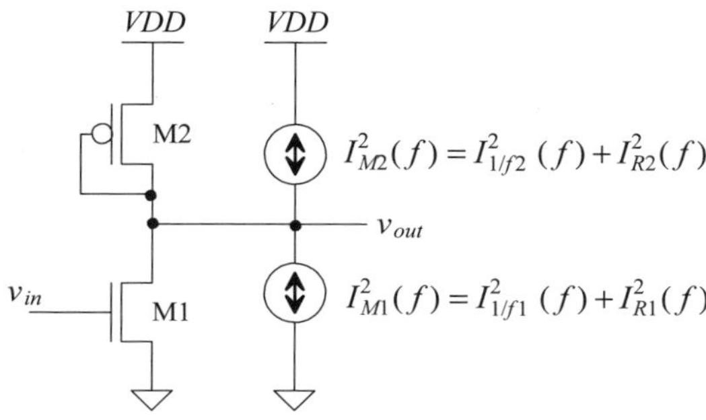

Figure 21.14 Noise modeling in a CS amplifier.

The input-referred noise power spectral density is then

$$V_{inoise}^2(f) = \frac{V_{onoise}^2(f)}{A_v^2} = \left(\frac{g_{m2}}{g_{m1}}\right)^2 \cdot \left(\frac{1}{g_{m2}}\right)^2 \cdot \left(I_{M1}^2 + I_{M2}^2\right) \tag{21.31}$$

Minimizing the input-referred noise is accomplished by making g_{m1} large. Using the expressions in Eq. (9.66), we also see that minimizing g_{m2} reduces the input-referred noise. This is equivalent to saying large A_v results in minimum input-referred noise.

21.1.2 The Source Follower (Common-Drain Amplifier)

Source-follower configurations using a gate-drain-connected load are shown in Fig. 21.15. The drain of the amplifying device, that is, the MOSFET *not* gate-drain-connected (the load), is common to both the input and the output, so this topology is often called a common-drain amplifier. Setting the bias current in this topology can be challenging. In the following we don't concern ourselves with this issue.

A source follower implemented in CMOS has an asymmetric drive capability; that is, the ability of the follower to source current is not equal to the ability to sink current for a given bias condition and AC input signal. The small-signal gain of the NMOS source follower shown in Fig. 21.15 is simply determined by a voltage divider between the resistance looking into the source of M2 ($1/g_{m2}$) with the resistance of the gate-drain-connected load, M1 ($1/g_{m1}$). The gain of the **source follower amplifier** is

$$A_v = \frac{v_{out}}{v_{in}} = \frac{\text{resistance connected to the source}}{\text{resistance connected to the source} + \text{resistance looking into the source}} \tag{21.32}$$

or for the NMOS source follower in Fig. 21.15

$$v_{in} = \overbrace{v_{gs2}}^{i_d/g_{m2}} + \overbrace{v_{out}}^{i_d/g_{m1}} \text{ where } (v_{gs1} = v_{out}) \text{ and so } v_{out} = v_{in} \cdot \frac{\frac{1}{g_{m1}}}{\frac{1}{g_{m1}} + \frac{1}{g_{m2}}} \tag{21.33}$$

The output resistance of the source follower shown in this figure is

$$R_{out} = \frac{1}{g_{m1}} || \frac{1}{g_{m2}} \tag{21.34}$$

Note that M2 can only source current, while M1 can only sink current, and the gain of the source follower is always less than one.

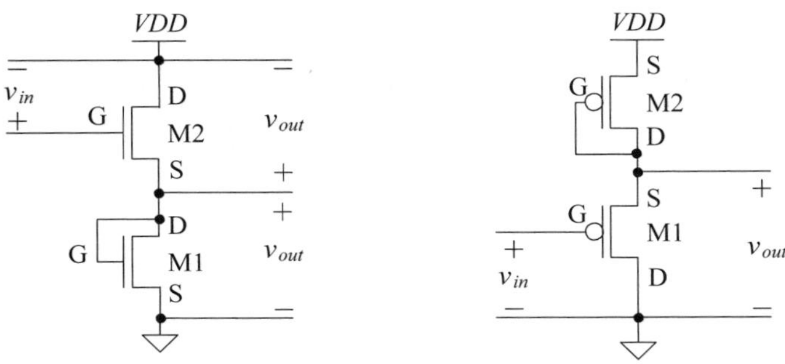

Figure 21.15 Source followers (common drain) using gate-drain loads.

21.1.3 Common Gate Amplifier

The common gate amplifier with gate-drain load is shown in Fig. 21.16. The input resistance of this amplifier is simply the resistance looking into the source of M1, or

$$R_{in} = \frac{1}{g_{m1}} \tag{21.35}$$

The gain of this **common-gate amplifier** is

$$A_v = \frac{v_{out}}{v_{in}} = \frac{-i_d \cdot \frac{1}{g_{m2}}}{-v_{gs1}} = \frac{-i_d \cdot \frac{1}{g_{m2}}}{-i_d \cdot \frac{1}{g_{m1}}} = \frac{\text{resistance in the drain}}{\text{resistance in the source}} = \frac{\frac{1}{g_{m2}}}{\frac{1}{g_{m1}}} \tag{21.36}$$

or noting the positive gain

$$A_v = \frac{g_{m1}}{g_{m2}} = \sqrt{\frac{KP_n}{KP_p} \cdot \frac{W_1 L_2}{W_2 L_1}} \tag{21.37}$$

which is the same form as the gain for the CS amplifier with gate-drain load.

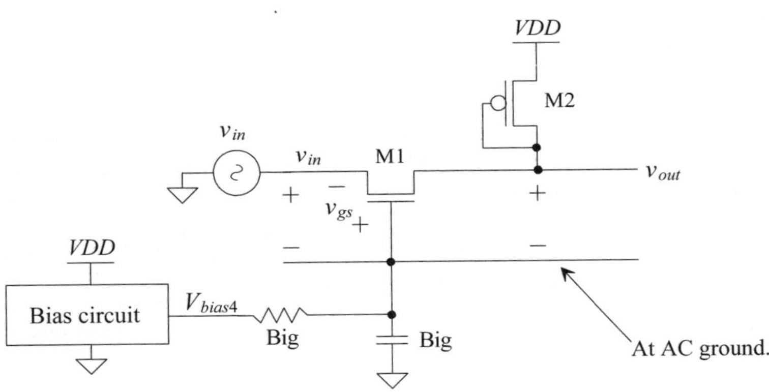

Figure 21.16 The common gate amplifier.

21.2 Current Source Loads

In the last section we talked about the gate-drain MOSFET part of a current mirror. In this section we talk about the current-source side of the mirror. The current source load provides an amplifier with the largest possible load resistance and thus the largest gain.

21.2.1 Common-Source Amplifier

Consider the CS amplifier with current source load shown in Fig. 21.17. The MOSFET M1 is the common-source component of the amplifier, while the MOSFET M2 is the current source load. The DC transfer characteristics are also seen in this figure. The amplifying portion of the curve occurs when the output is not close to *VDD* or ground (both M1 and M2 are saturated). The slope of the line when both transistors are saturated corresponds to the gain of the amplifier.

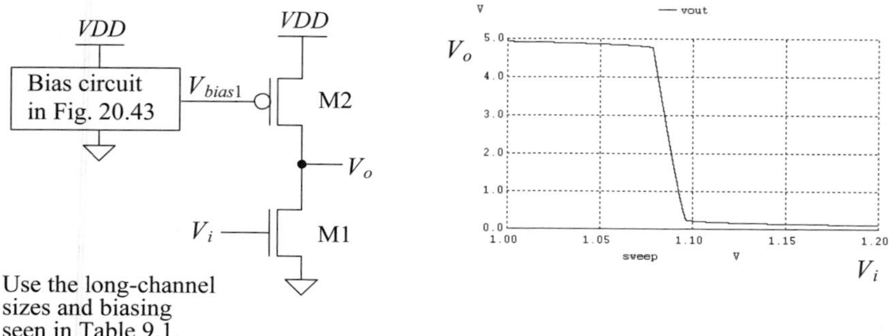

Figure 21.17 Common-source amplifier with current source load.

Class A Operation

We know from Table 9.1, and the bias circuit in Fig. 20.43, that M2 in Fig. 21.17 is biased to behave like a 20 μA current source. This means that if M1 shuts off, the maximum current we can supply to the output is 20 μA: this is important to understand. The CS amplifier in Fig. 21.17 is called a *class A* amplifier because, for proper operation, both MOSFETs are always conducting a current. In a *class B* amplifier, only one MOSFET is conducting a current at a given time. In *class AB* amplifiers (which we'll talk about later), both MOSFETs or a single MOSFET conduct a current at a given time.

Example 21.5

Estimate the maximum rate the load capacitor in Fig. 21.18 can be charged (estimate the slew rate across the load capacitor). Verify the estimate with SPICE.

The maximum current that the amplifier can source is 20 μA. The slew rate across the load is calculated using

$$I = C_L \cdot \frac{dV_{out}}{dt} \text{ or slew rate} = \frac{dV_{out}}{dt} = \frac{I}{C_L} \tag{21.38}$$

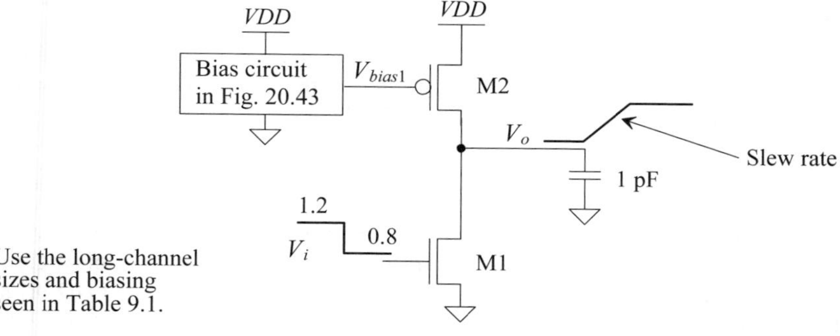

Figure 21.18 Slew rate limitations in a class A amplifier.

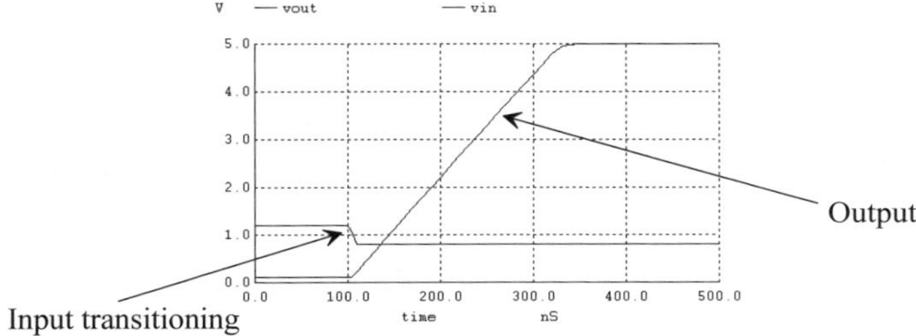

Figure 21.19 Verifying the results in Ex. 21.5

or

$$\frac{dV_{out}}{dt} = \frac{20\ \mu A}{1\ pF} = 20\ V/\mu s$$

For the output to transition 5 V requires 250 ns. ■

Small-Signal Gain

If we follow the intuitive method discussed in the last section, Eq. (21.1), the voltage gain of a common-source amplifier is the total parallel resistance at the drain of M1 divided by the resistance in the source of M1, or

$$A_v = \frac{v_o}{v_i} = -\frac{r_{o1}||r_{o2}}{\frac{1}{g_{m1}}} = -g_{m1}(r_{o1}||r_{o2}) = -\frac{g_{m1}}{g_{o1}+g_{o2}} \tag{21.39}$$

where $r_{o1} = 1/g_{o1}$.

Open Circuit Gain

The gain of the amplifier can be increased by using a cascode current load in place of M2. The resistance looking into the drain of the cascode current source/load is much larger than the output resistance of M1. This situation is sometimes referred to as the *open circuit gain* of a common-source amplifier (see the specification in Tables 9.1 and 9.2). This specification indicates the maximum possible gain attainable with a single MOSFET (the MOSFET's load is its own output resistance r_o). The open circuit gain of M1 can be written as

$$\text{Open circuit gain} = -\frac{r_{o1}}{\frac{1}{g_{m1}}} = -g_{m1}r_{o1} = -\frac{\sqrt{2\beta_1 I_D}}{I_D\lambda_1} \tag{21.40}$$

High-Impedance and Low-Impedance Nodes

In Fig. 21.17, the output node at the drains of M1 and M2 is termed a *high-impedance node*, that is, a node with only drain connections. The effective resistance at this node to ground is $r_{o1}||r_{o2}$. A node connected to the source of a MOSFET or a MOSFET with its drain and gate connected is termed a *low-impedance node*. The small-signal resistance looking into the source of a MOSFET and the small-signal resistance of the gate-drain (diode)-connected MOSFET are both $1/g_m$. High-impedance nodes usually have a low-frequency pole associated with them that limits the speed of the amplifier.

Frequency Response

Figure 21.20 shows the AC frequency response circuit for the CS amplifier with current source load in Fig. 21.17. *VDD* in this figure is drawn as an AC ground. The source-gate capacitance of M2 isn't shown because both sides of it are at AC ground (DC voltages), so it doesn't affect the frequency behavior of the circuit. The pole associated with the input node of the circuit is

$$f_{in} = \frac{1}{2\pi(C_{gs1} + C_{gd1}(1 + |A_v|)) \cdot R_s} \tag{21.41}$$

while the pole associated with the output node of the circuit is located at

$$f_{out} = \frac{1}{2\pi\left(C_{dg2} + C_{gd1}\left(1 + \frac{1}{|A_v|}\right)\right) \cdot r_{o1}||r_{o2}} \tag{21.42}$$

where $A_v = -g_{m1} \cdot r_{o1}||r_{o2}$ (see Eq. [21.39]). Using Eq. (21.42) to calculate the output pole, results in a *wrong estimate* when A_v is large. As we'll see in a moment, the fact that A_v decreases above f_{in} causes v_{in} to "flatten out" because the Miller capacitance, $C_{gd1}(1+|A_v|)$, gets smaller. The effect of v_{in} ceasing to decrease causes the output pole to appear much larger than what is predicted by Eq. (21.42). This important phenomenon is called *pole splitting*.

The zero in the transfer function (see Eq. (21.15)) is located at

$$f_z = \frac{g_{m1}}{2\pi C_{gd1}} \tag{21.43}$$

The transfer function of the amplifier is then estimated as

$$A_v(f) = -g_{m1} \cdot (r_{o1}||r_{o2}) \cdot \frac{\left(1 - j\frac{f}{f_z}\right)}{\left(1 + j\frac{f}{f_{in}}\right)\left(1 + j\frac{f}{f_{out}}\right)} \tag{21.44}$$

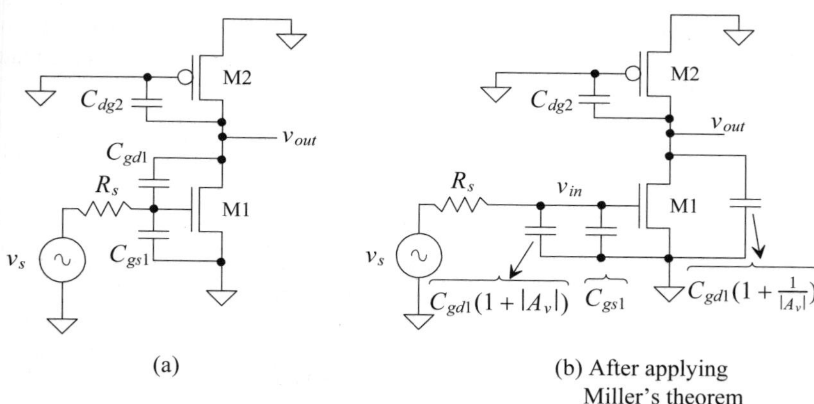

Figure 21.20 Frequency response of the CS amplifier with current source load.

We can then write the magnitude as

$$|A_v(f)| = \frac{g_{m1} \cdot (r_{o1}||r_{o2}) \cdot \sqrt{1+\left(\frac{f}{f_z}\right)^2}}{\left(\sqrt{1+\left(\frac{f}{f_{in}}\right)^2}\right) \cdot \left(\sqrt{1+\left(\frac{f}{f_{out}}\right)^2}\right)} \tag{21.45}$$

and the phase shift through the amplifier as

$$\angle A_v = 180 - \tan^{-1}\left(\frac{f}{f_z}\right) - \tan^{-1}\left(\frac{f}{f_{in}}\right) - \tan^{-1}\left(\frac{f}{f_{out}}\right) \tag{21.46}$$

Example 21.6

Determine the gain, magnitude, and phase shift of the amplifier shown in Fig. 21.21. Using an AC SPICE simulation, verify your hand calculations.

For the AC small-signal analysis, the big resistor appears like an open and the big capacitor a short. These components are added to ensure that M1 is biased to sink the current supplied by M2.

The low-frequency gain of the circuit is, from Eq. (21.39) and Table 9.1,

$$|A_v| = (150\ \mu A/V) \cdot (5M\Omega||4M\Omega) = 333\ V/V \rightarrow 50\ dB$$

The input capacitance is

$$C_{in} = (C_{gs1} + C_{gd1}(1 + |A_v|)) = (23.3 + 2 \cdot 333)\, fF = 691\, fF\ ! \tag{21.47}$$

The output capacitance is

$$C_{out} = \left(C_{dg2} + C_{gd1}\left(1 + \frac{1}{|A_v|}\right)\right) \approx 6 + 2\, fF = 8\, fF \tag{21.48}$$

Using Eqs. (21.41)–(21.43), we get f_{in} = 2.3 MHz, f_{out} = 9 MHz, and f_z = 11.9 GHz. Figure 21.22 shows the simulation results. As the discussion following Eq. (21.42) indicated, the estimate for f_{out} is wrong. To determine the correct value, let's discuss pole splitting (see Eq. [21.62]). ■

Use the long-channel sizes and biasing seen in Table 9.1.

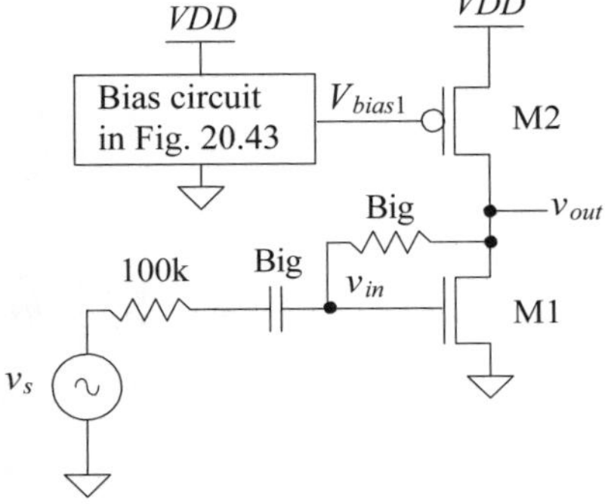

Figure 21.21 Amplifier used in Ex. 21.6.

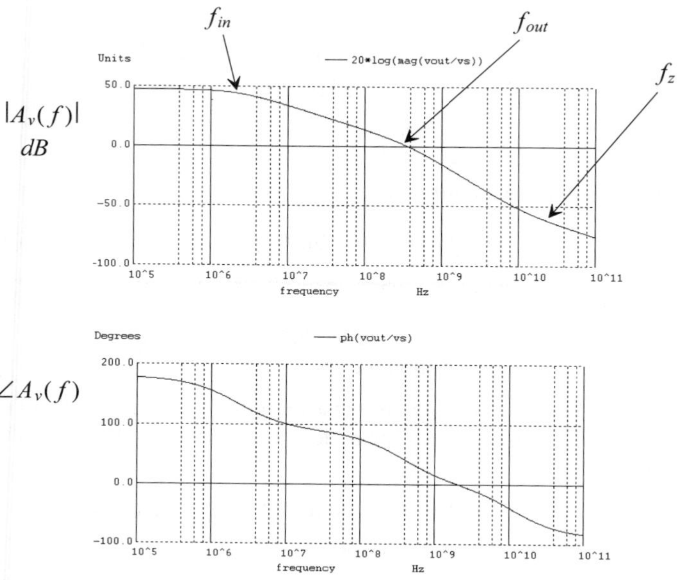

Figure 21.22 Frequency response of the amplifier in Fig. 21.21.

Pole Splitting

As discussed after Eq. (21.42), using Miller's theorem to calculate the location of the output pole when the gain of the amplifier is large results in error. Clearly, as just demonstrated in Ex. 21.6, the calculated value of f_{out} is 9 MHz while the SPICE simulation in Fig. 21.22 shows that f_{out} is approximately 200 MHz. The source of this discrepancy can be traced to the fact that above f_{in} the gain of the amplifier, A_v, decreases causing the effective input capacitance of the amplifier to decrease. As seen in Fig. 21.23, this causes the input voltage to "flatten out" instead of continuing to fall at –20 dB/decade. The result is the second, or output pole, appears to split from (leave) the first,

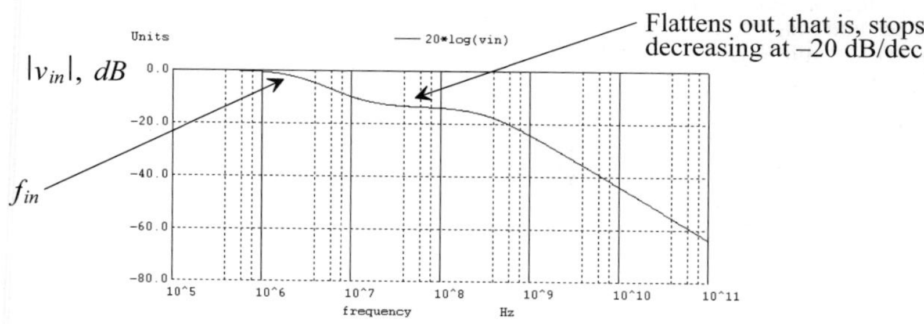

Figure 21.23 Showing how the amplifier's, in Fig. 21.21, input voltage flattens out as the gain decreases.

or input pole, and move to a higher frequency. Note that if we resimulate the amplifier in Fig. 21.21 with the source resistance set to zero ohms (so there isn't an input pole), pole splitting isn't present and Eq. (21.42) can be used to calculate the location of the output pole.

To characterize this effect, let's use the circuit in Fig. 21.7 with the addition of a source resistance, R_s , as seen in Fig. 21.24. In this figure, $R_o = r_{o1} \| r_{o2}$ and $C_o = C_{dg2}$. From Eq. (21.14), we can write

$$\frac{v_{out}}{v_{in}} = -g_{m1}R_o \cdot \frac{1 - j\omega\frac{C_{gd1}}{g_{m1}}}{1 + j\omega(C_{gd1} + C_o) \cdot R_o} \tag{21.49}$$

At the gate of M1, we can write

$$\frac{v_{in} - v_s}{R_s} + \frac{v_{in}}{1/j\omega C_{gs1}} + \frac{v_{in} - v_{out}}{1/j\omega C_{gd1}} = 0 \tag{21.50}$$

or

$$v_{in} = \frac{\frac{v_s}{R_s} + v_{out} \cdot j\omega C_{gd1}}{\frac{1}{R_s} + j\omega C_{gs1} + j\omega C_{gd1}} \tag{21.51}$$

Substituting Eq. (21.51) into Eq. (21.49) gives

$$v_{out} = -g_{m1}R_o \cdot \frac{1 - j\omega\frac{C_{gd1}}{g_{m1}}}{1 + j\omega(C_{gd1} + C_o) \cdot R_o} \cdot \frac{\frac{v_s}{R_s} + v_{out} \cdot j\omega C_{gd1}}{\frac{1}{R_s} + j\omega C_{gs1} + j\omega C_{gd1}} \tag{21.52}$$

or, with $s = j\omega$,

$$\frac{v_{out}}{v_s} = \frac{-g_{m1}R_o \cdot \left(1 - s\frac{C_{gd1}}{g_{m1}}\right)}{s^2[R_oR_s(C_{gd1}C_{gs1} + C_oC_{gs1} + C_oC_{gd1})] + s[(C_{gd1} + C_o)R_o + (C_{gs1} + C_{gd1})R_s + C_{gd1}g_{m1}R_oR_s] + 1} \tag{21.53}$$

The zero is still, as indicated in Eqs. (21.15) and (21.43), located at

$$f_z = \frac{g_{m1}}{2\pi C_{gd1}} \tag{21.54}$$

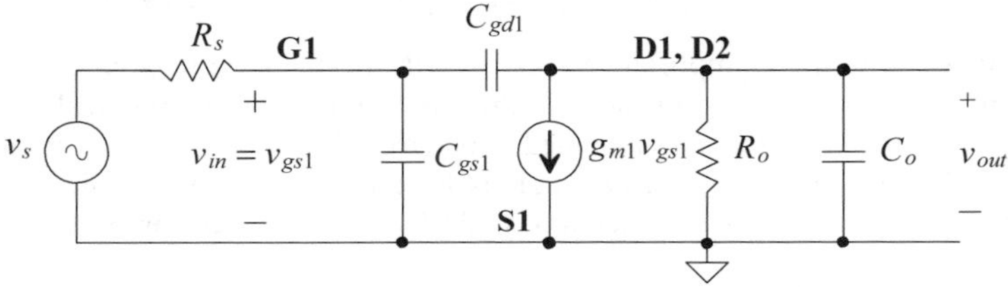

Figure 21.24 Circuit used to calculate the transfer function of the amplifier in Fig. 21.21.

At low frequencies the denominator is approximately (the coefficient of the s^2 term is small)

$$1 + j\omega \cdot [(C_{gd1} + C_o)R_o + (C_{gs1} + C_{gd1})R_s + C_{gd1}g_{m1}R_oR_s] \qquad (21.55)$$

and so the low-frequency pole, using $|A_v| = g_{m1}(r_{o1}||r_{o2}) = g_{m1}R_o$, $C_o = C_{dg2}$, is located at

$$f_1 \approx \frac{1}{2\pi[(C_{gd1} + C_{dg2}) \cdot r_{o1}||r_{o2} + (C_{gs1} + C_{gd1}(1 + |A_v|)) \cdot R_s]} \qquad (21.56)$$

If $C_{gd1}(1 + |A_v|)$ is much larger than the other capacitances,

$$f_1 \approx \frac{1}{2\pi C_{gd1}(1 + |A_v|) \cdot R_s} \qquad (21.57)$$

Notice the similarity to Eq. (21.41), where the pole is associated with the input of the amplifier. Further notice that if we had used this result in Ex. 21.6, the input pole would have changed very little. Also notice that if $R_s = 0$, then

$$f_1 \approx \frac{1}{2\pi(C_{gd1} + C_{dg2}) \cdot r_{o1}||r_{o2}} \qquad (21.58)$$

Notice the similarity to Eq. (21.42), where the pole is associated with the output of the amplifier. To determine the location of the second root of the denominator (the second pole location, f_2), let's factor Eq. (21.55) from the denominator of Eq. (21.53) or

$$[1 + s[(C_{gd1} + C_o)R_o + (C_{gs1} + C_{gd1})R_s + C_{gd1}g_{m1}R_oR_s]] \cdot \left(1 + \frac{s^2[R_oR_s(C_{gd1}C_{gs1} + C_oC_{gs1} + C_oC_{gd1})]}{s[(C_{gd1} + C_o)R_o + (C_{gs1} + C_{gd1})R_s + C_{gd1}g_{m1}R_oR_s] + 1}\right) \qquad (21.59)$$

where this equation is in the form

$$\left(1 + j \cdot \frac{f}{f_1}\right) \cdot \left(1 + j \cdot \frac{f}{f_2}\right) \qquad (21.60)$$

Note again that if $R_s = 0$, the second term in Eq. (21.59) goes to unity. Looking at the second term of Eq. (21.59), let's divide the numerator and denominator by sR_oR_s

$$1 + j \cdot \frac{2\pi f \cdot (C_{gd1}C_{gs1} + C_oC_{gs1} + C_oC_{gd1})}{(C_{gd1} + C_o)/R_s + (C_{gs1} + C_{gd1})/R_o + C_{gd1}g_{m1} + 1/sR_oR_s} \qquad (21.61)$$

In any practical MOSFET amplifier, $g_m >> 1/r_o$ (if not, then the magnitude of the open circuit gain, Eq. (21.40), is too small for the MOSFET to be of practical value as an amplifying device). Also, the R_s is assumed to be large (if not, then f_2 is not used), that is, the amplifier's transfer function has a single-pole response, where the pole is associated with the output of the amplifier as seen in Eq. (21.58). We can therefore write

$$f_2 \approx \frac{g_{m1}C_{gd1}}{2\pi \cdot (C_{gd1}C_{gs1} + C_oC_{gs1} + C_oC_{gd1})} \qquad (21.62)$$

Calculating the location of the second pole in Ex. 21.6 with this equation results in f_2 = 240 MHz (which is very close to the simulated location).

Why the name pole splitting? If we increase the effective size of C_{gd1} by placing a capacitor, C_c , from the amplifier's input to its output (so that C_c is in parallel with C_{gd1}, that is, the effective value is $C_c + C_{gd1}$), then, as Eq. (21.56) shows, the low-frequency location of the pole, f_1 , decreases. At the same time, increasing the effective size of C_{gd1} causes the location of the high-frequency pole f_2, as seen in Eq. (21.62), to increase. Thus the name *pole splitting.*

Pole Splitting Summary

Because of the ubiquity of pole splitting in CMOS amplifier design, let's summarize our discussion using a generic model, Fig. 21.25. In terms of the parameters seen in this figure the location of the zero is now

$$f_z = \frac{g_{m2}}{2\pi \cdot C_c} \tag{21.63}$$

The first pole is located at

$$f_1 \approx \frac{1}{2\pi[(C_c + C_2) \cdot R_2 + (C_1 + C_c(1 + g_{m2}R_2)) \cdot R_1]} \tag{21.64}$$

For large gain, $g_{m2}R_2$, we can simplify this to

$$f_1 \approx \frac{1}{2\pi g_{m2}R_2R_1C_c} \tag{21.65}$$

The second pole is located at

$$f_2 \approx \frac{g_{m2}C_c}{2\pi \cdot (C_cC_1 + C_1C_2 + C_cC_2)} \tag{21.66}$$

The transfer function is then written as

$$A_v(f) = \frac{v_{out}(f)}{v_s(f)} = g_{m1}R_1g_{m2}R_2 \cdot \frac{\left(1 - j \cdot \frac{f}{f_z}\right)}{\left(1 + j \cdot \frac{f}{f_1}\right) \cdot \left(1 + j \cdot \frac{f}{f_2}\right)} \tag{21.67}$$

where $A_v = g_{m1}R_1g_{m2}R_2$ (the gain is now positive because of the defined direction of the current source in Fig. 21.25).

Again note that if C_c is increased, f_1 moves downwards and f_2 moves upwards (pole splitting). Also note that if the capacitor, C_c , is made large (quite common), then f_1 is much lower than the location of the other pole or the zero (the pole associated with f_1 is

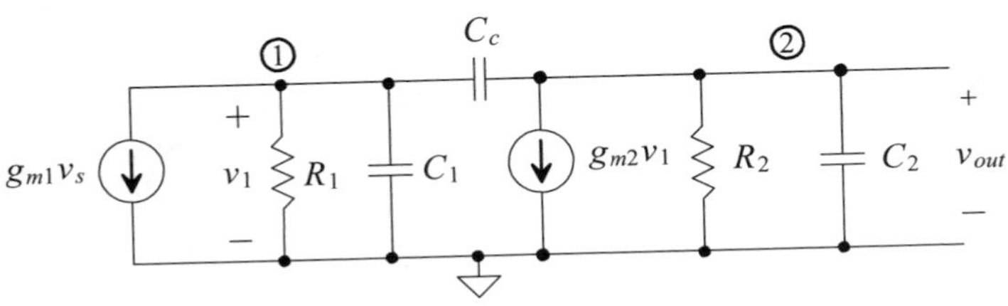

Figure 21.25 Generic model used to estimate bandwidth in a CMOS amplifier.

said to be the *dominant pole*). The amplifier's transfer function, with the help of Eq. (21.65), can be written as

$$A_v(f) = \frac{v_{out}(f)}{v_s(f)} \approx \frac{g_{m1}R_1g_{m2}R_2}{\left(1+j\cdot\frac{f}{f_1}\right)} = \frac{g_{m1}R_1g_{m2}R_2}{(1+j\cdot 2\pi f\cdot g_{m2}R_2R_1C_c)} \tag{21.68}$$

This equation is in the form

$$A_v(f) = \frac{v_{out}(f)}{v_s(f)} = \frac{A_{DC}}{1+j\cdot\frac{f}{f_{3dB}}} \tag{21.69}$$

where

$$A_{DC} = g_{m1}R_1g_{m2}R_2 \text{ and } f_{3dB} = \frac{1}{2\pi g_{m2}R_2R_1C_c} \tag{21.70}$$

The magnitude and phase responses of a generic CMOS amplifier are seen in Fig. 21.26. Note that at frequencies much larger than the amplifier's 3 dB frequency, that is, $f/f_{3dB} \gg 1$, Eq. (21.68) can be written as

$$A_v(f) = \left|\frac{v_{out}(f)}{v_s(f)}\right| \approx \frac{g_{m1}}{2\pi f\cdot C_c} \tag{21.71}$$

To determine the frequency, f_{un}, where the transfer function is unity, we set Eq. (21.71) to one and solve to yield

$$f_{un} = \frac{g_{m1}}{2\pi C_c} \tag{21.72}$$

We've derived some very useful equations, so let's apply them in some examples.

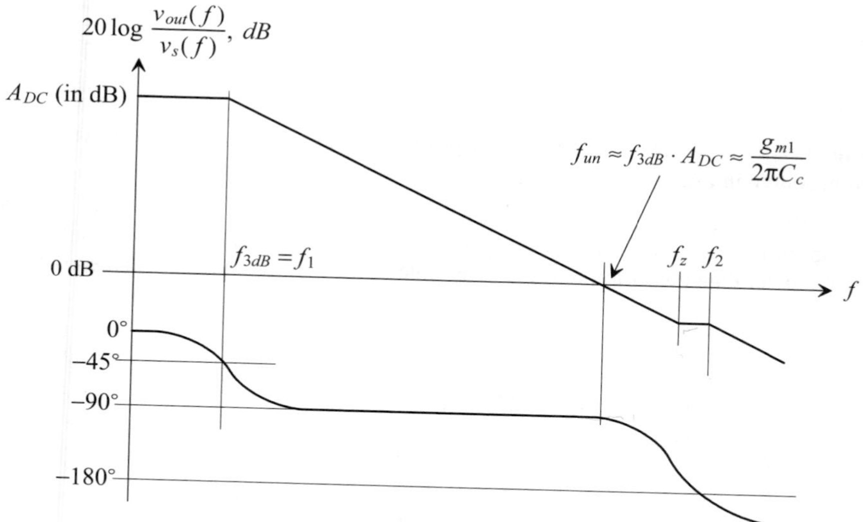

Figure 21.26 Magnitude and phase responses of a generic CMOS amplifier.

Example 21.7
Repeat Ex. 21.6 using the model in Fig. 21.25.

Using the model parameters in Fig. 21.25 and looking at Figs. 21.20a and 21.21, we can write (with the help of Table 9.1)

$$R_1 = R_s = 100\text{k},\ C_1 = C_{gs1} = 23.3\,fF,\ C_c = C_{gd1} = 2\,fF,$$

$$C_2 = C_{dg2} = 6\,fF,\ R_2 = r_{o1} \,||\, r_{o2} = 2.22\ \text{M}\Omega$$

$$g_{m2}\ (\text{in Fig. 21.25}) = g_{m1}\ (\text{in Fig. 21.21}) = 150\ \mu A/V$$

$$g_{m1}\ (\text{in Fig. 21.25}) = 1/R_s\ (\text{in Fig. 21.21}) = 10\ \mu A/V$$

Using Eq. (21.67) or (21.70), we have $|A_v| = 333\ V/V \rightarrow 50\ dB$. Using Eq. (21.63), we get $f_z = 11.9\ GHz$; Eq. (21.65), we get $f_1 = 2.3\ MHz$; and finally using Eq. (21.66), we get $f_2 = 240\ MHz$. These calculations should be compared to the simulation results seen in Fig. 21.22. ■

Example 21.8
Repeat Ex. 21.7 if R_s is zero. Verify the hand calculations using SPICE.

Again we use the model seen in Fig. 21.25. From Ex. 21.7, $g_{m1}R_1$ is unity (and so R_s doesn't affect the gain). Further, the location of the zero doesn't change. Since R_s is zero ($R_1 = 0$ so the current source, $g_{m1}v_s$, and the resistor, R_1 in Fig. 21.25 are replaced with a voltage source, v_s), the location of f_2 is moved to an infinite frequency (as seen in Eq. [21.59]). Using Eq. (21.64), we get

$$f_1 \approx \frac{1}{2\pi(C_{gd1} + C_{dg2}) \cdot r_{o1} || r_{o2}} = \frac{1}{2\pi(8\,fF) \cdot 2.22M} = 9\ MHz$$

The simulation results are seen in Fig. 21.27. Note the small perturbation in the frequency response above f_1. The response is not perfectly decreasing at –20 dB/decade. This imperfection can be traced to the bias voltage, V_{bias1}, used on the gate of M2. It is not a perfect AC ground as was assumed in our hand calculations. By placing a capacitor on the gate of M2 to ground, the variation in V_{bias1} is reduced and the amplifier behaves as expected. This is an *important point.* It is often a good idea to "bypass" the bias voltages to AC ground to reduce their impedance in practical design. This also keeps circuits that are biased from the same bias circuit from interacting. ■

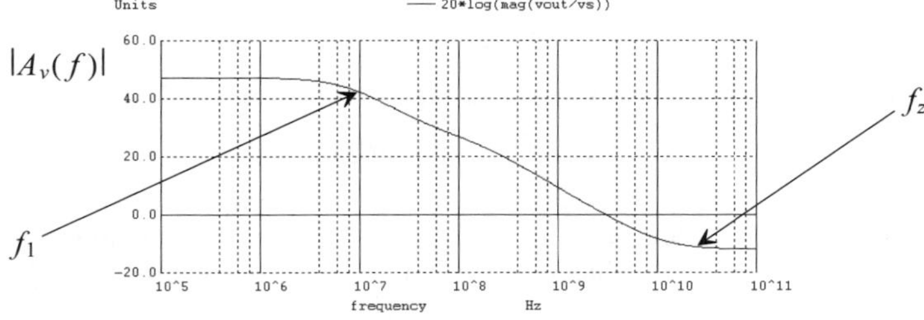

Figure 21.27 Simulation results for Ex. 21.8.

Example 21.9
Determine the frequency response of the amplifier seen in Fig. 21.28. Verify the hand calculations using simulations.

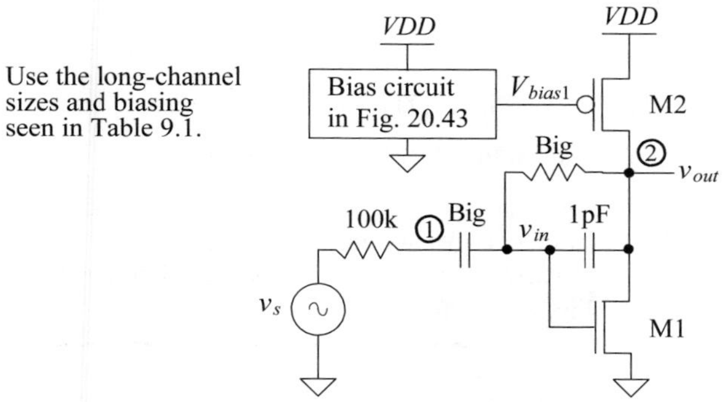

Use the long-channel sizes and biasing seen in Table 9.1.

Figure 21.28 Amplifier used in Ex. 21.9.

The only thing that is different between this amplifier and the one in Ex. 21.7 is the value of C_c. In Fig. 21.28

$$C_c = 1\ pF + C_{gd1} \approx 1\ pF$$

We can then write, using Eq. (21.65)

$$f_1 \approx \frac{1}{2\pi \cdot (150\ \mu A/V) \cdot 100k \cdot 2.2MEG \cdot 1pF} = 4.8\ kHz$$

using Eq. (21.66)

$$f_2 \approx \frac{(150\ \mu A/V) \cdot 1p}{2\pi \cdot (1p \cdot 23.3f + 23.3f \cdot 6f + 1p \cdot 6f)} = 811\ MHz$$

At the risk of stating the obvious, these two poles are split apart by a great distance with the addition of the 1 pF capacitor. The zero, from Eq. (21.63), is located at

$$f_z = \frac{150\ \mu A/V}{2\pi \cdot 1p} = 23\ MHz$$

The unity-gain frequency is found using Eq. (21.72) with $R_s = 1/g_{m1}$.

$$f_{un} = \frac{1}{2\pi R_s C_c} = 1.59\ MHz$$

The simulation results are seen in Fig. 21.29. The value of the gain is 47 dB (we calculated 50 dB). The value of the first pole is approximately 6 kHz (we calculated 4.8 kHz). The small discrepancies are most likely the result of differences in the actual circuit parameters compared to the values we used in the formulas. ■

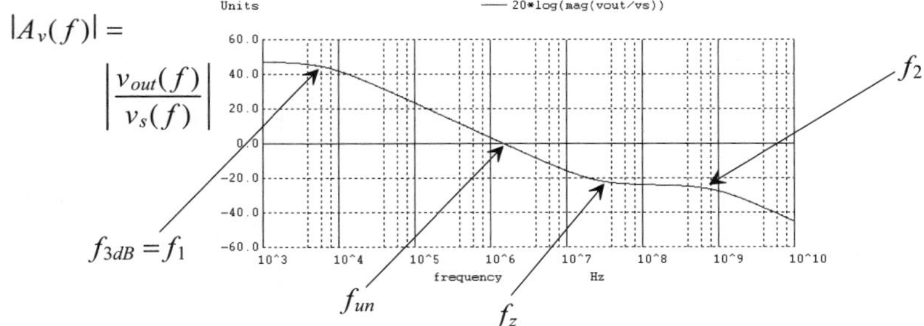

Figure 21.29 Frequency response of the amplifier in Fig. 21.28.

Example 21.10
Estimate the frequency response of the amplifier seen in Fig. 21.30. Verify the estimates with SPICE.

MA's gate and drain, at DC, are at the same potential. This biases MB to sink the right amount of current from MD.

Using the model in Fig. 21.25, we see that the source voltage, v_s, modulates the drain current in MA. We can therefore write

$$g_{mA}v_s = g_{m1}v_s \rightarrow g_{m1} \text{ (in Fig. 21.25)} = g_{mA} \text{ (in Fig. 21.30)} = 150\ \mu A/V$$

$$g_{m2} \text{ (in Fig. 21.25)} = g_{mB} \text{ (in Fig. 21.30)} = 150\ \mu A/V$$

$$R_1 = r_{oA}||r_{oC} = R_2 = r_{oB}||r_{oD} = 2.22\ M\Omega$$

To determine the capacitance on nodes 1 and 2, we can use Fig. 21.31. At node 1

$$C_1 = C_{gdA} + C_{dgC} + C_{gsB} = 31.3\ fF$$

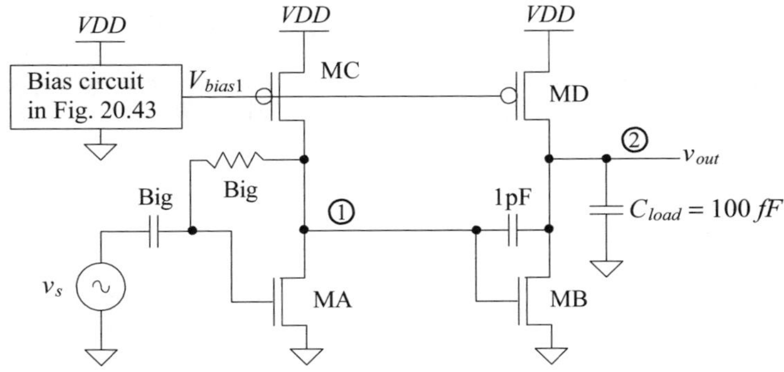

Use the long-channel sizes and biasing seen in Table 9.1.

Figure 21.30 Amplifier used in Ex. 21.10.

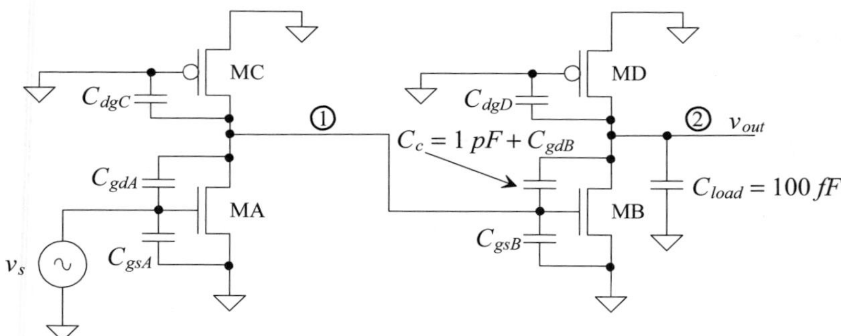

Figure 21.31 The capacitances for the amplifier in Fig. 21.30.

where C_{gdA} is the output Miller capacitance of MA ($C_{gdA} \approx C_{gdA}\left(1+\frac{1}{|A|}\right)$ when the gain of MA is large). At node 2

$$C_2 = C_{load} + C_{dgD} = 106\ fF$$

The value of C_c is

$$C_c = 1\ pF + C_{gdB} \approx 1\ pF \text{ (just the added capacitor)}$$

The gain of the topology is estimated as

$$A_{DC} = g_{m1}R_1g_{m2}R_2 = (150\ \mu A/V)^2(2.22\ M\Omega)^2 = 110{,}889\ V/V \rightarrow 101\ dB$$

The poles and zero can be calculated using Eqs. (21.63), (21.65), and (21.66)

$$f_z = \frac{150\ \mu A/V}{2\pi \cdot 1.002\ pF} = 23\ MHz$$

$$f_1 \approx \frac{1}{2\pi \cdot (150\ \mu A/V)(2.22\ M\Omega)^2(1.002\ pF)} = 214\ Hz$$

$$f_2 \approx \frac{(150\ \mu A/V) \cdot 1.002}{2\pi \cdot (1.002 \cdot 0.0313 + 0.0313 \cdot 0.106 + 1.002 \cdot 0.106)\ pF} = 170\ MHz$$

The location of the unity gain frequency is calculated using Eq. (21.72) as 23 MHz. The simulation results are seen in Fig. 21.32. ■

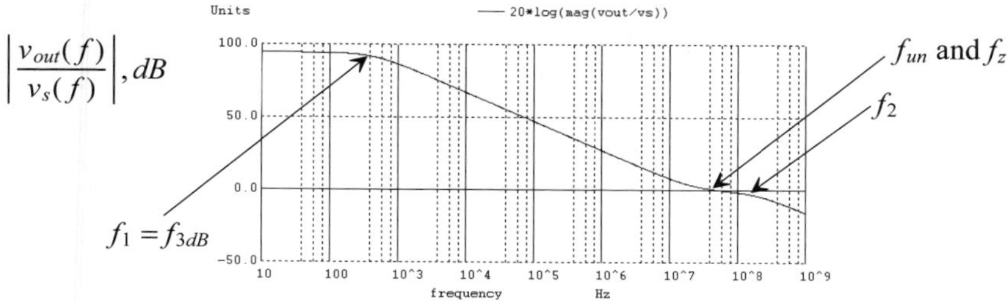

Figure 21.32 Frequency response of the amplifier in Fig. 21.30.

Canceling the RHP Zero

Notice in Ex. 21.10 that the unity-gain frequency (the frequency where the gain is one) is equal to the frequency of the zero. As mentioned earlier, the input of the amplifier feeds directly through C_c to the amplifier's output without the phase inversion. If the gain of the amplifier is larger than one (or even slightly less than one), then the lack of inversion, when the amplifier uses feedback, can result in an unstable amplifier. The amplifier's output signal feeds back and adds to its input signal without the inversion through the amplifier. To avoid this situation, we might add an amplifier in series with C_c that allows the output to feed back to the input through C_c but not vice versa, Fig. 21.33a. A simpler solution is to add a resistor in series with the capacitor, Fig. 21.33b, to attenuate the higher frequency signals (where the zero occurs) and push the zero out to a higher frequency. Adding the resistor moves the zero to a frequency (see Eq. [21.63])

$$f_z = \frac{1}{2\pi \cdot C_c \cdot \frac{1}{g_{m2}}} \overset{\text{With a resistor}}{\rightarrow} f_z = \frac{1}{2\pi \cdot C_c \cdot \left(\frac{1}{g_{m2}} - R_z\right)} \tag{21.73}$$

If $R_z = 1/g_{m2}$, the zero disappears (is pushed to an infinite frequency). If $R_z > 1/g_{m2}$, the zero is pushed back into the LHP (phase shift is opposite from the poles). *Any practical design where pole splitting is used should include the zero-nulling resistor R_z.* Consider the following example.

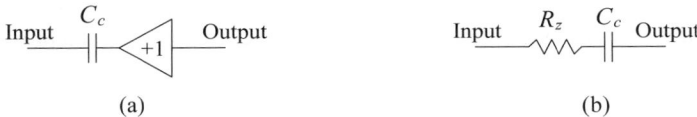

Figure 21.33 Removing the zero in the amplifier's transfer function.

Example 21.11
Using a zero-nulling resistor, show that the location of the zero in the frequency response of the amplifier in Fig. 21.30 can be eliminated.

In this amplifier, $R_z = 1/g_{m2} = 6.5\ k\Omega$. By adding this resistor in series with the 1 pF capacitor, we get the frequency response seen in Fig. 21.34. Note that the zero is still seen but at a considerably higher frequency. ■

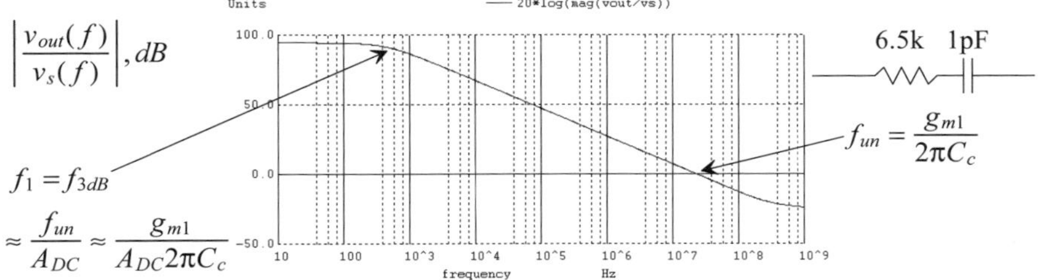

Figure 21.34 Pushing the zero in Fig. 21.32 to a higher frequency.

Noise Performance of the CS Amplifier with Current Source Load

To determine the noise performance of the CS amplifier with current source load, we can use the schematic in Fig. 21.14 with the gate of M2 at AC ground. The output noise power spectral density is then

$$V^2_{onoise}(f) = (r_{o1}||r_{o2})^2 \cdot \left(I^2_{M1} + I^2_{M2}\right) \tag{21.74}$$

The input-referred noise is then

$$V^2_{onoise}(f) = \frac{V^2_{onoise}(f)}{A^2_v} = \frac{1}{g^2_{m1}} \cdot \left(I^2_{M1} + I^2_{M2}\right) \tag{21.75}$$

showing once again that to minimize the input-referred noise we need to make the transconductance of M1 large (or, in other words, make A_v large).

21.2.2 The Cascode Amplifier

Consider a CS amplifier with current source load as seen Fig. 21.17 but implemented using the short-channel CMOS process. Using the parameters in Table 9.2, the gain of the topology, Eq. (21.39), is only 16.7 (it was 333 using the long-channel process). To boost the gain of the single-stage amplifier and to eliminate the Miller effect, consider the cascode amplifier seen in Fig. 21.35. The resistance looking into the drain of M3 is $g_{mp} \cdot r^2_{op}$ (see Table 20.1) with a value of 16.6 MΩ. The resistance looking into the drain of M2 is 4.2 MΩ (again from Table 20.1). The gain of the cascode amplifier is the resistance in the drain divided by the resistance in the source of the amplifying device (M1) or

$$\frac{v_{out}}{v_{in}} = -\frac{g_{mn}r^2_{on}||g_{mp}r^2_{op}}{1/g_{mn}} = -g_{mn} \cdot R_{ocas} \tag{21.76}$$

Using the bias circuit from Fig. 20.47 and the sizes in Table 9.2 (and thus the small-signal parameters in this table).

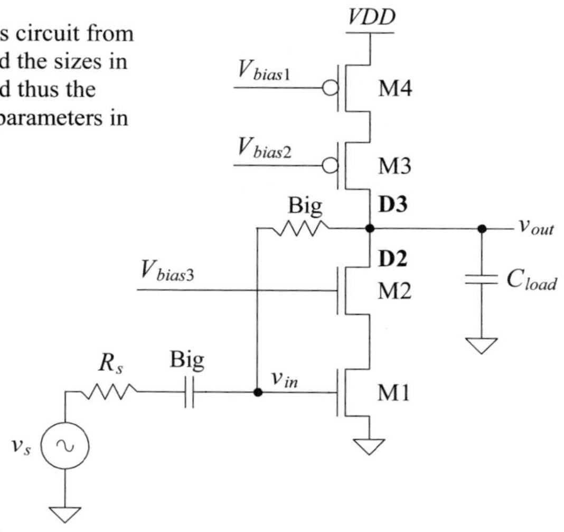

Figure 21.35 A cascode amplifier.

where

$$R_{ocas} = g_{mn} r_{on}^2 || g_{mp} r_{op}^2 \quad (21.77)$$

Using the parameters from Table 9.2, the gain of the cascode amplifier in Fig. 21.35 is, roughly, 500.

Frequency Response

The resistance looking into the source of M2 is $1/g_{m2}$ ($= 1/g_{mn}$). The resistance in the drain of M1 is then approximately $1/g_{m2}$. The gain from the gate of M1 (the amplifier's input) to the drain of M1 is then

$$A_{v1} = \frac{v_{d1}}{v_{in}} = \frac{-v_{gs2}}{v_{gs1}} = \frac{-i_d/g_{m2}}{i_d/g_{m1}} = -\frac{1/g_{m2}}{1/g_{m1}} = -\frac{g_{m1}}{g_{m2}} = -1 \quad (21.78)$$

If we calculate the Miller capacitance, Eq. (21.5), we see that the low gain of M1 eliminates the large loading (from the Miller capacitance) on the input of the amplifier. For this reason, the cascode connection is often used in high-frequency design. Consider the following.

Example 21.12

Estimate the frequency response of the cascode amplifier in Fig. 21.35 if R_s is 100k and $C_L = 100\,fF$. Verify the hand calculations with simulations.

The input time constant is calculated as

$$\tau_{in} = R_s \cdot (C_{gs1} + (1 + |A_{v1}|) \cdot C_{gd1})$$

Using the values from Table 9.2

$$\tau_{in} = 100k \cdot (4.17\,fF + 2 \cdot 1.56\,fF) = 729\ ps$$

The pole associated with this input time constant is then

$$f_{in} = \frac{1}{2\pi\tau_{in}} = 218\ MHz$$

Since the load capacitance is large, $C_{load} >> C_{gd2} + C_{dg3}$, we can write

$$\tau_{out} = R_{ocas} \cdot (C_{load} + C_{gd2} + C_{dg3}) \approx g_{mn} r_{on}^2 || g_{mp} r_{op}^2 \cdot C_{load} = 335\ ns$$

and so the pole associated with the output node is located at

$$f_{out} = \frac{1}{2\pi\tau_{out}} = 475\ kHz$$

Again, the gain of the topology is 500 (54 dB). The simulation results are seen in Fig. 21.36. The magnitude response is

$$\left|\frac{v_{out}}{v_s}\right| = \frac{500}{\sqrt{1 + \left(\frac{f}{475kHz}\right)^2} \cdot \sqrt{1 + \left(\frac{f}{218\ MHz}\right)^2}}$$

and the phase response (in degrees) is

$$\angle\frac{v_{out}}{v_{in}} = 180 - \tan^{-1}\frac{f}{475\ kHz} - \tan^{-1}\frac{f}{218\ MHz}$$ ■

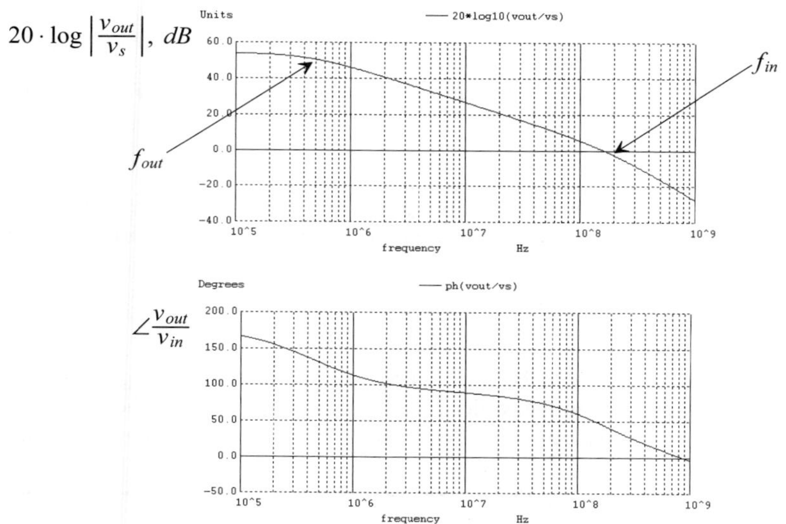

Figure 21.36 The simulation results for Ex. 21.12.

Class A Operation

The cascode amplifier in Fig. 21.35 is another example of a class A amplifier. When the input of the amplifier goes sufficiently negative, M1 shuts off. M3 and M4 are a current source and so the maximum rate that the load capacitance can be charged is given by Eq. (21.38). For the amplifier in Fig. 21.35, this is 10 μA/100 *fF* or 100 mV/ns.

Noise Performance of the Cascode Amplifier

The cascode's output noise power spectral density is given by

$$V_{onoise}^2(f) = \left(g_{mn}r_{on}^2 || g_{mp}r_{op}^2\right)^2 \left(I_{M1}^2 + I_{M2}^2 + I_{M3}^2 + I_{M4}^2\right) \quad (21.79)$$

The input-referred noise power is then ($g_{m1} = g_{mn}$)

$$V_{inoise}^2(f) = \frac{V_{onoise}^2(f)}{A_v^2} = \frac{\left(I_{M1}^2 + I_{M2}^2 + I_{M3}^2 + I_{M4}^2\right)}{(g_{m1})^2} \quad (21.80)$$

Again, maximizing the transconductance of the amplifying device (equivalent to saying maximizing the gain of the amplifier) reduces the input-referred noise.

Operation as a Transimpedance Amplifier

Figure 21.37 shows a transimpedance amplifier (current input and voltage output). Note that the AC voltages between the gates and sources of M1 and M4 are zero (both the gates and the sources are at AC ground). The gate-source voltage of M2 is

$$i_{in} = g_{m2}v_{gs2} = g_{m2}(-v_{ds1}) \quad (21.81)$$

See Table 9.2 and Fig. 20.47

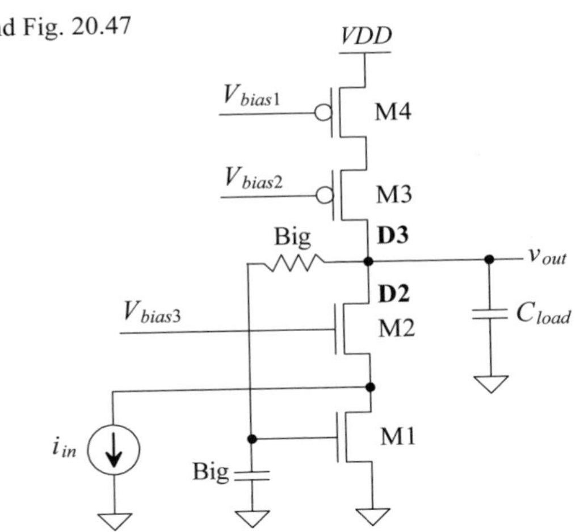

Figure 21.37 A transimpedance amplifier.

since the gate of M2 is at AC ground. The drain current that flows through M2 is i_{in}. The output voltage is

$$v_{out} = \left(g_{mn}r_{on}^2 || g_{mp}r_{op}^2\right) \cdot (-i_{in}) \tag{21.82}$$

or the gain is

$$\frac{v_{out}}{i_{in}} = -g_{mn}r_{on}^2 || g_{mp}r_{op}^2 = -R_{ocas} \tag{21.83}$$

The input resistance of the transimpedance amplifier is

$$R_{in} = \frac{1}{g_{m2}} = \frac{1}{g_{mn}} \tag{21.84}$$

The ideal input resistance of a transimpedance amplifier (or any current input amplifier) is zero ohms.

21.2.3 The Common-Gate Amplifier

M2 in Fig. 21.37 is an example of a common-gate (CG) amplifier. We can redraw this circuit, as seen in Fig. 21.38 with a voltage source input, to show how the gate of the amplifying device, M2, is common to both the input and the output of the amplifier. Though the gate of M2 is at a DC voltage of V_{bias3}, we think of it as being at AC ground. M1 is simply an ordinary current source while M3 and M4 are a cascode current source load. The input resistance of this amplifier is given by Eq. (21.84), and the output resistance is given by Eq. (21.77). The voltage gain can be calculated by first writing

$$v_{in} = -v_{gs2} = -i_d/g_{m2} = i_{in}/g_{mn} \tag{21.85}$$

See Table 9.2 and Fig. 20.47

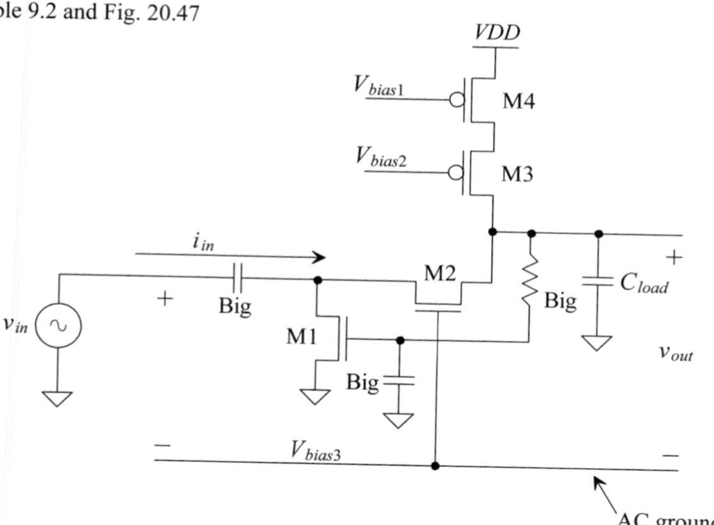

Figure 21.38 A common-gate amplifier.

where the drain current of M2 is $-i_{in}$. The output voltage is given by Eq. (21.82) and so the gain is

$$A_v = \frac{v_{out}}{v_{in}} = \frac{-i_d\left(g_{mn}r_{on}^2 || g_{mp}r_{op}^2\right)}{-i_d/g_{m2}} = g_{mn} \cdot R_{ocas} \tag{21.86}$$

21.2.4 The Source Follower (Common-Drain Amplifier)

The source follower (SF) with current source load is seen in Fig. 21.39. Looking at the NMOS SF, we can write, for AC small signals,

$$v_{in} = v_{gs2} + v_{out} \tag{21.87}$$

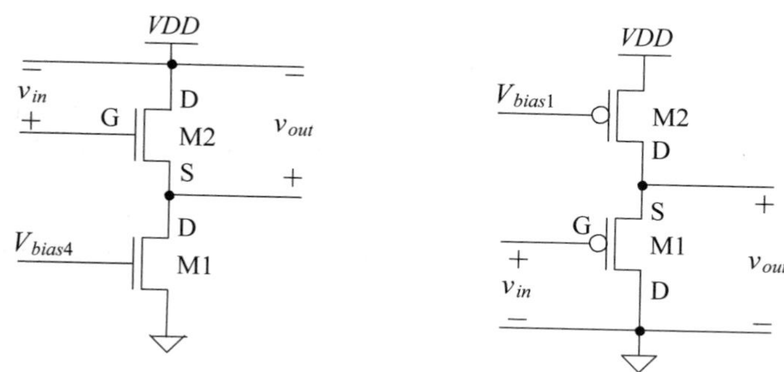

Figure 21.39 Source followers (common drain) using current source loads.

Knowing the resistance looking into the drain of M1 is r_o and the current that flows in M1 is i_d, we can write

$$v_{out} = i_d \cdot r_o \text{ and } v_{gs2} = \frac{i_d}{g_{m2}} \tag{21.88}$$

Solving for the gain, we get

$$A_v = \frac{v_{out}}{v_{in}} = \frac{r_o}{r_o + 1/g_{m2}} = \frac{g_{m2} r_o}{g_{m2} r_o + 1} \tag{21.89}$$

If we use a current source for the load with a very large small-signal output resistance, the gain goes to one. (Limitations concerning the body effect will be discussed in a moment.) Looking at Eq. (21.89), we see that the gain is a voltage divider between the resistance looking into the source of a MOSFET ($1/g_m$) and the output resistance of the current source (here using the simple, single MOSFET r_o) as seen in Eq. (21.32).

Note that the maximum current the NMOS SF can sink is limited by the size of the current flowing in the current source load M1. The SF is a class A amplifier where, when discharging a load capacitance, the current through M1 limits the rate at which the capacitance can discharge (as indicated in Eq. [21.38]). M2 only sources current and M1 only sinks current. When the SF is sourcing a current, M2 provides the current to both the amplifier's output and to M1. When the SF is sinking a current, the current in M2 decreases while the current in M1 is constant (resulting in a net sinking of current on the amplifier's output). Finally, note that the maximum output voltage of an NMOS SF occurs when the gate of M2 goes to *VDD*. The maximum output voltage then goes to $VDD - V_{GS2}$. The minimum output voltage is set by the minimum voltage across the current source load (to keep the MOSFETs in the saturation region).

Body Effect and Gain

We might think, after looking at Eq. (21.89), that by using a cascode current source load for the source follower we can get an AC small-signal gain very close to unity. However, the body effect ultimately limits the gain of the amplifier. Figure 21.40 shows an SF with an ideal current source load. If we use the AC small-signal model seen in the figure where the output resistance, r_o, is infinite, we see that the AC output voltage, v_{out}, equals the AC source to bulk potential v_{sb}. Further, we can write

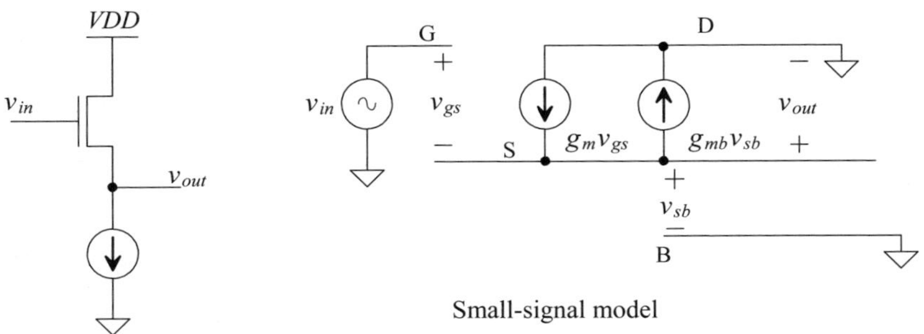

Figure 21.40 SF gain with body effect

$$v_{gs} = v_{in} - v_{out} \tag{21.90}$$

and, summing the currents at the output node,

$$g_m v_{gs} = g_{mb} v_{sb} = g_{mb} v_{out} \tag{21.91}$$

Solving for the gain of the SF with body effect and knowing $g_{mb} = \eta \cdot g_m$ (Eq. (9.28)), we get

$$\frac{v_{out}}{v_{in}} = \frac{g_m}{g_m + g_{mb}} = \frac{1}{1+\eta} \tag{21.92}$$

If $\eta = 0.25$, the gain is 0.8. The obvious way of eliminating the body effect (and thus move the gain closer to one) is to put the common-drain device in its own well, Fig. 21.41.

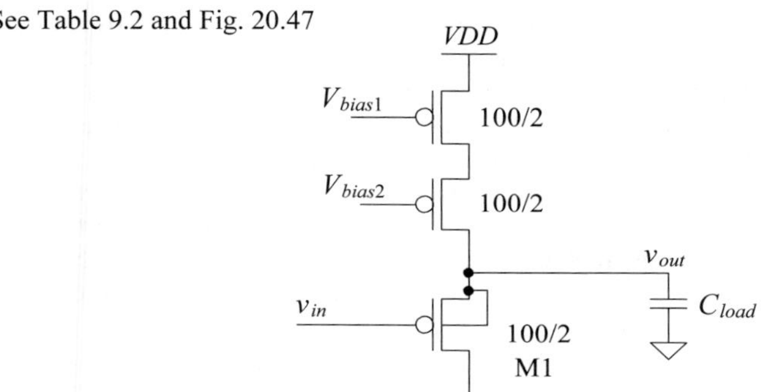

Figure 21.41 PMOS SF without body effect.

Level Shifting

One of the important uses of an SF is to provide a DC level shift to an input voltage. For example, M1 in Fig. 21.41 will remain in saturation as long as

$$v_{SD} = v_{OUT} \geq v_{SG} - V_{THP} = \overbrace{v_{OUT} - v_{IN}}^{v_{SG}} - V_{THP} \tag{21.93}$$

or

$$v_{IN} \geq -V_{THP} \tag{21.94}$$

As long as the input voltage is greater than $-V_{THP}$, M1 operates in the saturation region. This result is practically important. Note the output and input of the SF in Fig. 21.41 are related by

$$v_{OUT} = v_{IN} + V_{SG} \tag{21.95}$$

Using the parameters in Table 9.2 where VDD is 1 V, $V_{SG} = 350$ mV, and $V_{THP} = 250$ mV, the input voltage is shifted upwards, on the output of the amplifier, by 350 mV. The input signal can go down to –250 mV before M1 triodes. The minimum voltage across the current source, as seen in Fig. 20.48, is approximately 150 mV. This means that the range of the input signal is from VDD – 150 mV – 350 mV (= 600 mV = $V_{in,max}$) down to – 250 mV. Simulation results are seen in Fig. 21.42.

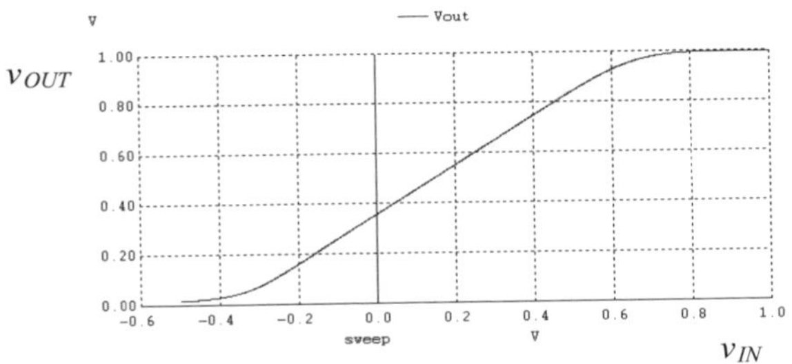

Figure 21.42 How the SF in Fig. 21.41 can shift negative input voltages upwards. This circuit is very useful when input signals are centered around ground.

Input Capacitance

Consider the partial schematic of an SF seen in Fig. 21.43. The capacitance on the input can be determined by looking at how much charge is supplied by the input source for an input voltage change Δv_{IN}

$$Q_{IN} = \Delta v_{IN} \cdot C_{dg} + (\Delta v_{IN} - A_v \cdot \Delta v_{IN}) \cdot C_{sg} \tag{21.96}$$

where $A_v \leq 1$. The input capacitance of an SF is then

$$C_{IN} = \frac{Q_{IN}}{\Delta v_{IN}} = C_{dg} + C_{sg}(1 - A_v) \approx C_{dg} \tag{21.97}$$

As the gain, A_v , of the SF approaches one, the input capacitance approaches the drain-gate capacitance of the MOSFET. In other words the source-gate capacitance doesn't affect the input capacitance (unlike the CS amplifier, which can have a very large input capacitance due to the Miller effect; see Ex. 21.6). Intuitively we can understand why C_{sg} doesn't affect the input capacitance by realizing that the displacement current through it is zero when the gate and source potentials move at the same rate. When the input voltage change and the output voltage changes are equal, the current through C_{sg} is zero. The SF is often used on the input of amplifiers that must have low input capacitance (like in charge-amplification applications). Unfortunately, the noise performance of the SF, because it doesn't have gain, is poorer than the CS or CG amplifiers.

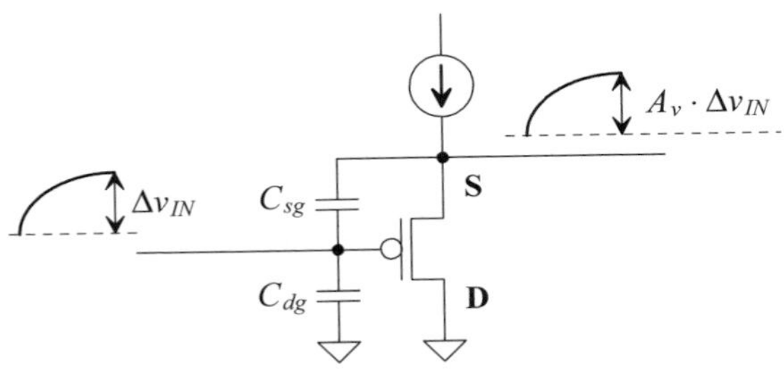

Figure 21.43 Input capacitance of a SF.

Noise Performance of the SF Amplifier

The SF amplifier with MOSFET noise sources is seen in Fig. 21.44. The noise current from M3 flows through M2 to the output node. The resistance on the output node (the output resistance of the SF) is the resistance looking into the source of M1 in parallel with the resistance looking into the drain of M2 or

$$R_{oSF} = \frac{1}{g_{m1}} || R_o \approx \frac{1}{g_{m1}} \tag{21.98}$$

The output noise power spectral density is

$$V^2_{onoise}(f) = \frac{I^2_{M1}(f) + I^2_{M2}(f) + I^2_{M3}(f)}{g^2_{m1}} \tag{21.99}$$

Noting that the SF's gain is close to one, the input-referred noise is nearly equal to the output noise PSD. Again, by using a large value of g_{m1}, we can minimize both the input-referred and output noise.

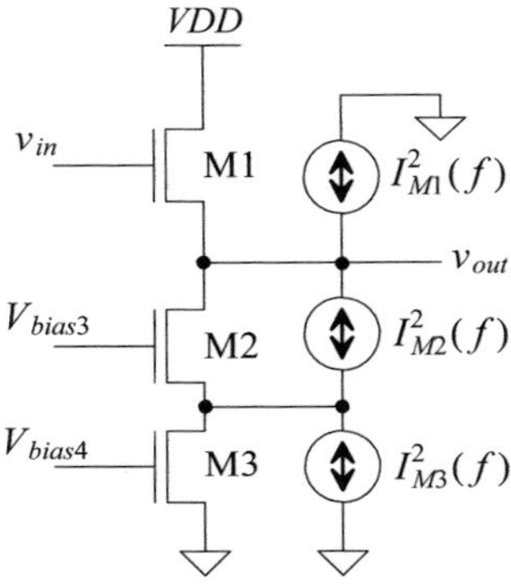

Figure 21.44 Noise model of the source follower amplifier.

Frequency Behavior

Using Eq. (21.97), we see that if we drove the SF with an input whose source resistance was R_s, then we would have a pole associated with the input at

$$f_{in} = \frac{1}{2\pi \cdot R_s C_{dg}} \tag{21.100}$$

a very high frequency. On the output of the SF driving a capacitive load, we can write, using Eq. (21.98),

$$f_{out} = \frac{1}{2\pi \cdot R_{oSF} C_{load}} = \frac{g_{m1}}{2\pi C_{load}} \tag{21.101}$$

However, both Eqs. (21.100) and (21.101) assume that the transconductance doesn't vary with frequency. At very high frequencies, the input and output of the SF are shorted together through the gate-source capacitance. To calculate the variation of the transconductance with frequency, consider the circuit seen in Fig. 21.45. We can write

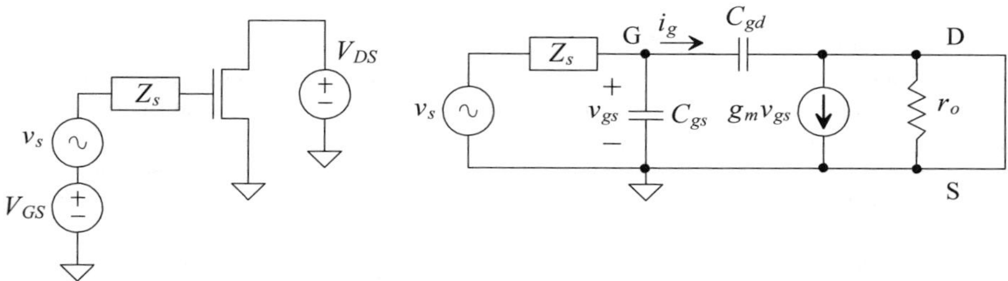

Figure 21.45 Determining the variation of the transconductance with frequency.

$$v_{gs}(f) = v_s \cdot \frac{\frac{1}{j\omega\cdot(C_{gs}+C_{gd})}}{\frac{1}{j\omega\cdot(C_{gs}+C_{gd})}+Z_s} = \frac{v_s}{1+Z_s\cdot j\omega\cdot(C_{gs}+C_{gd})} \tag{21.102}$$

and

$$i_d(f) = g_m \cdot v_{gs}(f) = \frac{g_m v_s}{1+Z_s\cdot j\omega\cdot(C_{gs}+C_{gd})} \tag{21.103}$$

The effective transconductance as a function of frequency is then

$$g_m(f) = \frac{g_m}{1+j\omega\cdot Z_s(C_{gs}+C_{gd})} \tag{21.104}$$

The impedance looking into the source of a MOSFET as a function of frequency is then

$$R_{\text{into source}} = \frac{1}{g_m(f)} = \frac{1}{g_m}\cdot(1+j\omega\cdot Z_s(C_{gs}+C_{gd})) \tag{21.105}$$

We note that at low frequencies the impedance is resistive and has a value of $1/g_m$. However, at higher frequencies and with a resistive source impedance, that is, $Z_s = R_s$, the impedance looking into the source appears to be the series connection of a resistor with a value of $1/g_m$ and an inductor of value

$$L_s = \frac{R_s(C_{gs}+C_{gd})}{g_m} \tag{21.106}$$

An SF driving a capacitive load can exhibit ringing because of the effective RLC circuit formed by its output impedance driving a load capacitance. If the impedance driving an SF is inductive (as we would have in the cascade of two SFs), the output impedance can become negative. For example, if $Z_s = j\omega L$, then substituting into Eq. (21.105), we get

$$R_{\text{into source}} = \frac{1}{g_m} - \omega^2\cdot L\cdot(C_{gs}+C_{gd}) \tag{21.107}$$

A battery or voltage source is an example of a circuit with a negative resistance (any source of power has a negative resistance). At higher frequencies the resistance looking into the source of the MOSFET becomes negative. This means that the circuit will oscillate or simply have a poor step response (the voltage across the capacitive load will ring). Often, to implement a microwave frequency oscillator using a MOSFET, an inductor is added from the gate of the MOSFET to AC ground to create the negative resistance.

SF as an Output Buffer

One of the common uses of an SF is as an output buffer. Connecting a resistive load to a high-impedance node kills the gain of a single-stage CMOS amplifier. A buffer amplifier is often inserted between the high-impedance node and the load resistance to keep the gain high. Consider the following example.

Example 21.13
Suppose that the cascode amplifier in Fig. 21.35, with R_s = 100k, must drive a 1 pF capacitor in parallel with a 10 kΩ resistor. Design an SF buffer to ensure that the 10 kΩ resistor doesn't kill the gain of the amplifier. Estimate the frequency response of the amplifier. Verify your design with SPICE simulations.

Looking at Eq. (21.76) and the associated discussion, we see that the output resistance of the cascode amplifier is in the megaohms region. Connecting a 10 kΩ load resistor on the amplifier's output would cause the cascode gain to drop from 500 to less than 1. The DC output voltage of the cascode amplifier in Fig. 21.35 is the V_{GS} of M1 (which, from Table 9.2 is 350 mV). If we use an NMOS SF, this DC voltage isn't enough to ensure that all of the MOSFETs in the SF operate in the saturation region. Looking at the NMOS SF in Fig. 21.39, a voltage of 350 mV on the gate of M2 would be just enough to turn it on but leave little voltage to drop across the current source M1. We'll use a PMOS source follower in this design. The schematic of the design is seen in Fig. 21.46. The current source (M5 and M6) in the SF must drive 1 V across the resistor of 10 kΩ (100 μA), so we've bumped up their size. Knowing, from Table 9.2, that when using the bias circuit in Fig. 20.47 a 100/2 PMOS conducts 10 μA of current, we increased the widths of M5–M7 so that they conduct 125 μA of current with the same bias voltages (the current source, M5/M6, sources 125 μA of current).

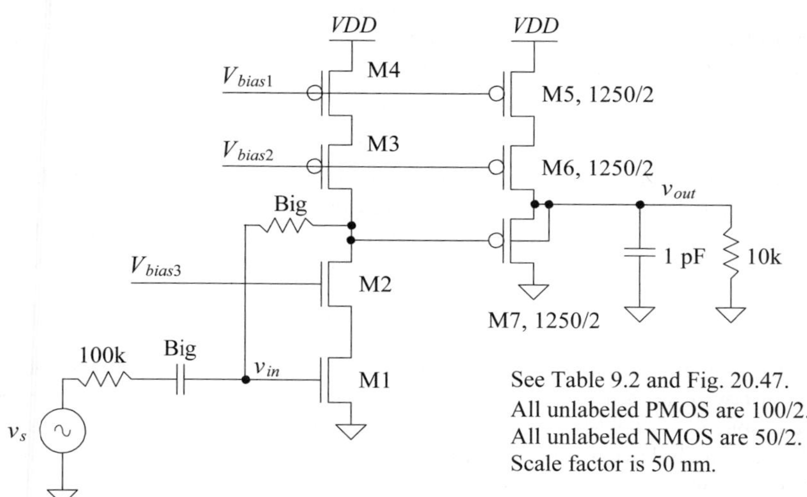

Figure 21.46 A cascode amplifier with SF output buffer.

The location of the input pole, f_{in} , is still (from Ex. 21.12) located at 218 MHz. The SF won't have much affect on the overall frequency response (because of its low input capacitance and small output resistance). However, now the cascode amplifier's load capacitance changes. Using the result in Eq. (21.97) and the data from Table 9.2, we can estimate the capacitance on the output of the cascode amplifier as the gate-drain capacitance of M7 ($3.7\,fF \cdot 12.5 \approx 50\,fF$). The factor of 12.5 comes from the increase in the PMOS's width by 12.5. Again, using the results from Ex. 21.12, the output pole, f_{out} , (output of the cascode amplifier) is at approximately 1 MHz (double what it was in Ex. 21.12).

The gain of the cascode amplifier is approximately 500. The transconductance of the wider PMOS devices is estimated using (see Eq. [9.22])

$$g_{m,wide} = W \cdot g_{m,Table\,9.2} = 1.875\ mA/V \qquad (21.108)$$

The gain of the SF driving a resistive load, see Eq. (21.89), is

$$A_{v,SF} = \frac{R_L}{R_L + 1/g_{m,wide}} = \frac{10k}{10k + 533} = 0.95 \qquad (21.109)$$

The overall gain from the input of the cascode amplifier to the output of the SF is then estimated as $500 \cdot 0.95 = 475 \rightarrow 53.5\ dB$. The simulation results are seen in Fig. 21.47. Note that the bandwidth of this amplifier can be increased by reducing the width of M7. This reduces the input capacitance of the SF output amplifier and thus pushes the pole location out to a higher frequency. ■

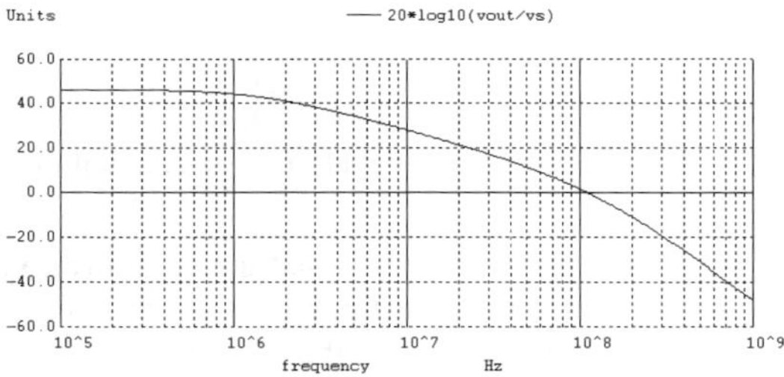

Figure 21.47 Simulating the operation of the amplifier in Fig. 21.46.

A Class AB Output Buffer Using SFs

In order to drive the 10k resistor in the previous example, we had to increase the size of the current source used in the SF. The practical problem with this approach is that as the load resistor gets small, the current source must source a significant current. The current source is sourcing this current all of the time (class A operation) either to the load or, if the load capacitance is being discharged, to the common-drain MOSFET (M7 in Fig. 21.46). What we need, for better drive capability, is for one side of the output buffer to shut off if the other side is supplying a large amount of current. For example, if, in Fig. 21.46, M7 starts to pull a large amount of current from the load, we would want M5 and M6 to turn off so that power wasn't wasted (class AB operation).

A class AB output buffer using SFs is seen in Fig. 21.48. The output buffer is comprised of M1–M4. M5 and M6 form a common-source amplifier with a current source load. When the circuit is in steady state, the AC input, v_{in} , is zero and the gate of M6 is at a DC potential of V_{bias1}. The current through M3 and M4 is mirrored by M1 and M2. This sets the DC current in M1/M2 to a known value (important). As v_{in} goes up, M6 shuts off. However, the current in M5 is constant, so the gates of M3 and M4 move towards ground. This turns M1 on and shuts M2 off (thus the class AB action). Similarly if v_{in} decreases, M6 turns on and pulls the gates of M3 and M4 up. The result is that M2 turns on and M1 shuts off.

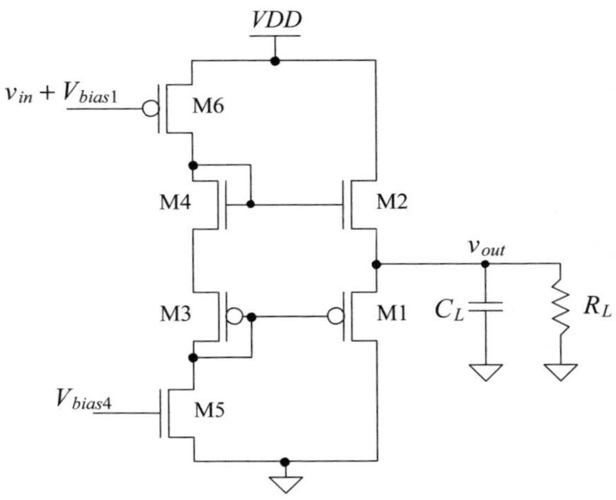

Figure 21.48 Class AB output buffer (M1–M4) using source followers M1 and M2.

The *practical problem* with this buffer is that the output can't swing very close to the power supply voltages (rails). The lowest potential we can get on the gate of M1 is ground, and the highest potential we can get on the gate of M2 is *VDD*. Knowing that M1 has to have a source-to-gate voltage greater than V_{THP} and M2 has to have a gate-source voltage greater than V_{THN} , we can write

$$VDD - V_{THN} \geq v_{out} \geq V_{THP} \tag{21.110}$$

For our short-channel CMOS process (neglecting body effect which will make things worse), the range of output voltages is half the power supply voltage. For example, with $VDD = 1$ V, our output voltage may swing up to (roughly) 750 mV and down to 250 mV. Losing half of the supply voltage often makes this output buffer impractical.

21.3 The Push-Pull Amplifier

What is needed for an output buffer is a topology like an inverter where the output can swing from rail (*VDD*) to rail (ground). Figure 21.49 shows a possible topology with biasing circuitry. The current source, I_{bias} , is a floating current source as seen in Fig. 20.49 of the last chapter. In Fig. 21.49, with zero AC input current, i_{in}, M1 and M2 mirror

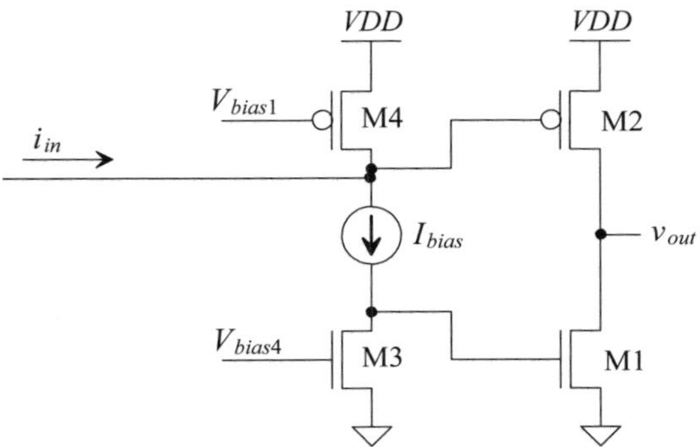

Figure 21.49 Class AB amplifier using an inverter output structure (push-pull).

the current in M3 and M4 (I_{bias}). This precisely sets the current in the output stage (again, important). A positive AC input current causes both the gates of M1 and M2 to go up (shutting M2 off and turning M1 on). M1 and M2 are pushing or pulling a current to/from the output (and so this topology is often called a *push-pull amplifier*). Because the output can swing very close to ground and *VDD* before M1 and M2 triode or shut off, this topology is very useful in modern CMOS output buffer design.

Note that the AC input current can be connected to the gate of M2, as shown in Fig. 21.49, or the gate of M1 (or both).

21.3.1 DC Operation and Biasing

The biasing of the push-pull amplifier can be accomplished, as seen in Fig. 21.50 (see also Fig. 20.49). MFCP and MFCN form the floating current source. M3 (A and B) with M4 (A and B) are cascode current sources. If the input current is positive, the gate of MOP is charged up and thus MOP shuts off. At the same time, MFCP turns on (more), causing the gate of MON to go up, turning it on further. If the input current is negative, MOP turns on and MON shuts off.

Figure 21.51 shows the amplifier in Fig. 21.50 with a voltage input (V_{bias4} replaced with an input voltage, V_{in}). M3 and M4 now form a common-source amplifier with a current source load. Shown in Fig. 21.51 is a plot of V_{out} for varying V_{in} with no load. The slope of the curve in this figure is the gain. To show the slope (the gain) as a function of V_{in} , we can take the derivative of the output voltage (also shown in the figure). The gain of this topology (without a load) is roughly 5,000 V/V. If we place a 1k resistor on the output of the amplifier, the gain drops to 1,500 V/V. Using a 100-ohm resistor results in a gain of only 150. The push-pull amplifier, MON/MOP, has a gain less than one when the load resistor is only 100 ohms. If the amplifier did need to drive such a heavy load (small resistance), it would be a good idea to increase the widths of MON and MOP.

Power Conversion Efficiency

The power conversion efficiency (PCE) is defined as the ratio of the power supplied to the load to the power delivered to the amplifier and load by a power supply. In other words, a perfectly efficient amplifier doesn't dissipate any power; rather, all of the power

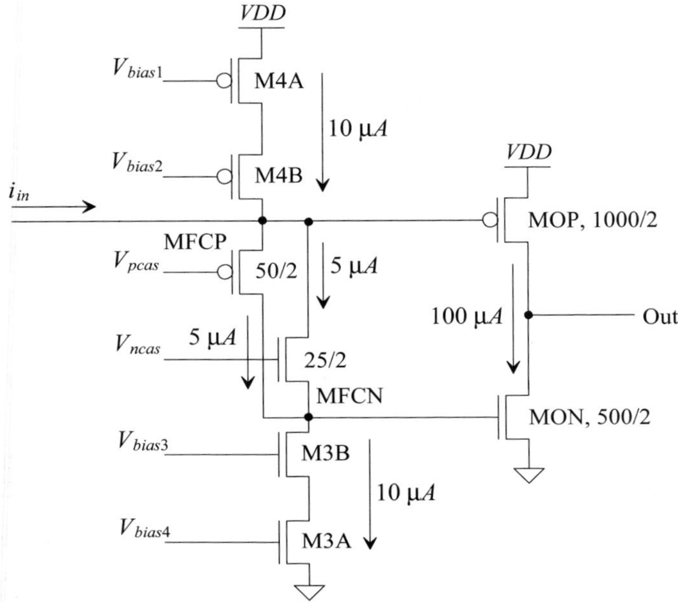

Bias voltages come from Fig. 20.47 (short-channel parameters in Table 9.2).
Unlabeled NMOS are 50/2, while unlabeled PMOS are 100/2.

Figure 21.50 Biasing the push-pull amplifier.

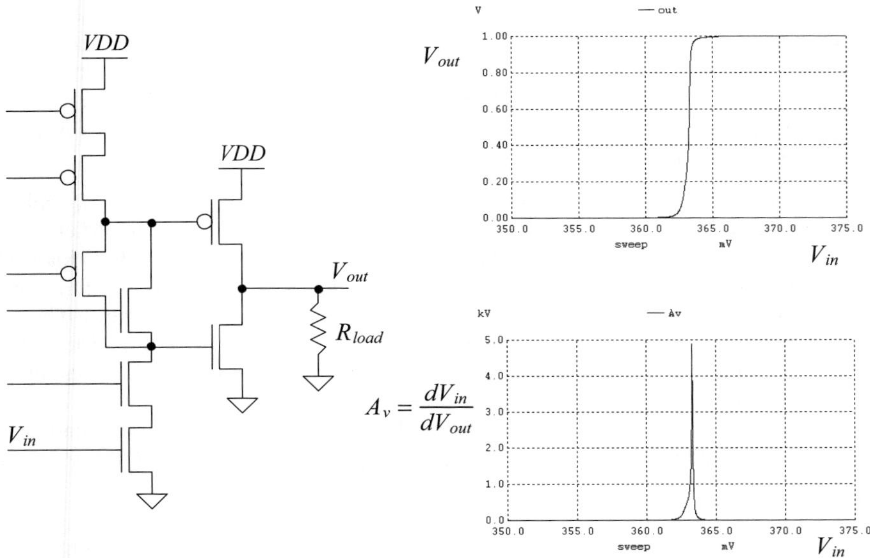

Figure 21.51 Simulating the amplifier in Fig. 21.50 with a voltage input.

from the power supply (*VDD*) is supplied to the load. The PCE is usually written as a percentage

$$\% \text{ PCE} = \frac{\text{Load power, } P_{load}}{\text{Supply power, } P_{supply}} \times 100 \ \% \qquad (21.111)$$

For example, the source follower seen in Fig. 21.41 (a class A amplifier) pulls a fixed current of I_{bias} (= 10 μA for the sizes seen in this figure and Table 9.2). The power supplied (pulled or delivered) from the power supply is

$$P_{supply} = VDD \cdot I_{bias} \qquad (21.112)$$

If the SF is driving a resistive load, R_{load}, the peak value (under ideal conditions) of a sinewave voltage applied to this resistor is *VDD*/2 (the sinewave can go up or down by *VDD*/2). The RMS value of this sinewave is then $VDD/\left(2\sqrt{2}\right)$ and the power supplied to the load is then

$$P_{load} = \frac{\left(VDD/\left(2\sqrt{2}\right)\right)^2}{R_L} \qquad (21.113)$$

The current source, I_{bias}, when the output voltage goes to *VDD*, drives the load resistor directly (M1 in Fig. 21.41 shuts off), that is

$$VDD = I_{bias} \cdot R_L \qquad (21.114)$$

This is the best efficiency because no power is wasted in the amplifier (M1 is off and all of the supply power is delivered to the load). Calculating the PCE of the SF gives

$$\% \text{ PCE} = \frac{1}{VDD \cdot I_{bias}} \cdot \frac{\left(VDD/\left(2\sqrt{2}\right)\right)^2}{VDD/I_{bias}} \times 100 \ \% \qquad (21.115)$$

So the PCE % is, at best, 12.5%. This is the ideal efficiency of a class A amplifier. If the output voltage swing is reduced, the PCE goes down as well.

For the class AB amplifier, the PCE can approach 100%. (Typical values for general designs with low distortion are approximately 75%.) If the current supplied to the load is much larger than the current burned in the amplifier, the PCE is large (much larger than the ideal 12.5% of the class A amplifier). Consider the following example.

Example 21.14

Suppose the amplifier in Figs. 21.50 and 21.51 drives a 1 kΩ load. Using SPICE plot the current supplied to the amplifier and load, the current supplied to only the load, and the current supplied to only the amplifier as a function of the input voltage, V_{in}. Using the results estimate the PCE.

The simulation results are seen in Fig. 21.52. The current supplied to the amplifier includes the bias circuit current. Note how, when the output is 1 V (= *VDD*), the current supplied to the load is 1 mA. The total current supplied by *VDD* is approximately 1.2 mA. At the risk of stating the obvious, the push-pull amplifier is much more power-efficient than the SF (and has a wider output swing).

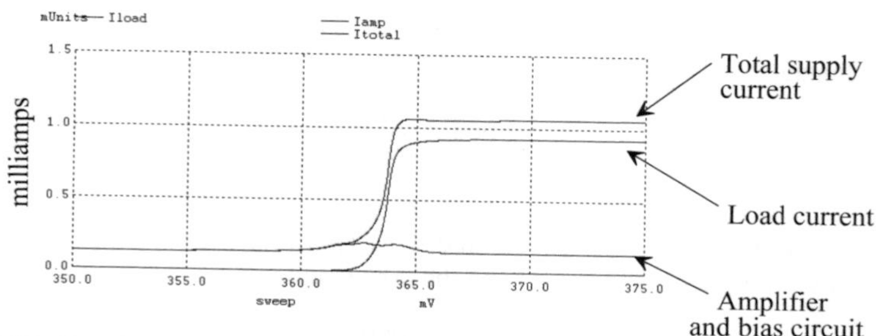

Figure 21.52 Supply, load, and total currents when the amplifier in Fig. 21.51 drives a 1k load resistor.

The PCE efficiency can be estimated by writing the power dissipated by the load

$$P_{load} = \left(\frac{VDD}{2\sqrt{2}\,R_{load}}\right)^2 = \left(\frac{1}{2\sqrt{2}\,1k}\right)^2 = 353\ \mu W$$

The power supplied by VDD (assuming the amplifier and bias circuit pull a constant 200 μA of current) is then

$$P_{supply} = P_{load} + VDD \cdot 200\ \mu A = 553\ \mu W$$

The PCE is then

$$\%\ \mathrm{PCE} = \frac{353}{553} \times 100\% = 64\%$$

Note that the bias circuit pulls 140 μA. If we recalculate the PCE without the bias circuit current included, we get

$$\%\ \mathrm{PCE} = \frac{353}{413} \times 100\% = 85\% \quad \blacksquare$$

21.3.2 Small-Signal Analysis

The simplified schematic of the push-pull amplifier is seen in Fig. 21.53. The resistance on the output of the amplifier (the drains of MOP and MON) is $r_{op}||r_{on}||R_{load} \approx R_{load}$. The drain current of MON is $g_{mon}v_{in}$, and the drain current of MOP is $g_{mop}v_{in}$. The output voltage is then the sum of these two currents or

$$v_{out} = -(g_{mon} + g_{mop}) \cdot v_{in} \cdot R_L \tag{21.116}$$

Note the "resistance in the source" in this amplifier is the parallel combination of the resistance looking into the sources of MON and MOP (both sources are connected to AC ground). We can rewrite Eq. (21.116) as

$$A_{v,push-pull} = \frac{v_{out}}{v_{in}} = -\frac{R_L}{\frac{1}{g_{mon}}||\frac{1}{g_{mop}}} \tag{21.117}$$

If the load resistance is 1k and the sum of the transconductances (using wider devices than what is indicated in Tables 9.1 or 9.2, as seen in Fig. 21.50) is 1 mA/V, then the gain of the push-pull amplifier is –1 V/V. Note that if the load resistance changes, the gain of the output stage push-pull amplifier changes as well.

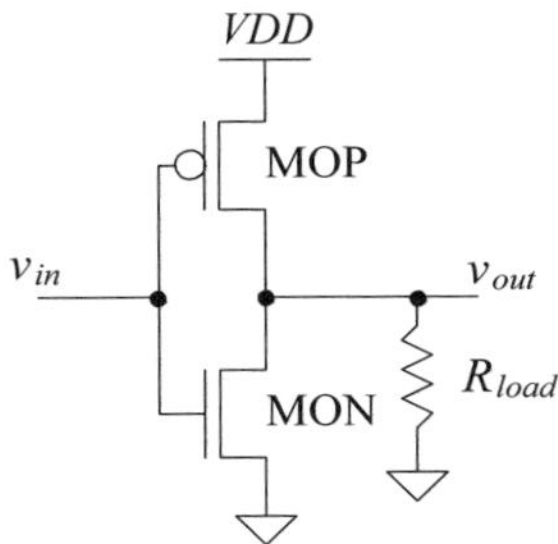

Figure 21.53 Small-signal analysis of the push-pull amplifier.

The gain of the CS amplifier with the cascode load portion of the amplifier in Figs. 21.50 and 21.51 (M3 and M4) is given by Eq. (21.76). The overall small-signal gain of the topology in Fig. 21.51 is given by the product of the two gains or

$$A_v = g_{mn} \cdot \left(g_{mn}r_{on}^2 || g_{mp}r_{op}^2\right) \cdot (g_{mon} + g_{mop}) \cdot R_L \tag{21.118}$$

We might wonder how the floating current source, MFCP and MFCN, affects the gain. If we look at Fig. 21.50, we see that one or the other has a source connection to the drains of either M3 (gate of MON) or M4 (gate of MOP). Further we might think that this source connection would load the gates with a $1/g_m$ small-signal resistance. However, because MFCP and MFCN form a feedback loop, their addition doesn't reduce the output resistance of the CS amplifier (M3 and M4). Consider the following.

Figure 21.54 shows a test circuit used for determining the small-signal resistance that the PMOS current source (M4) sees on its output. To determine this resistance, we apply a test voltage and look at the current that flows. Looking at the figure and noting the current that flows in $r_{on}||r_{op}$ is $(i_t - i_{dn} - i_{dp})$, we can write

$$v_t = (i_t - i_{dn} - i_{dp}) \cdot r_{on}||r_{op} + i_t \cdot R_{ncas} \tag{21.119}$$

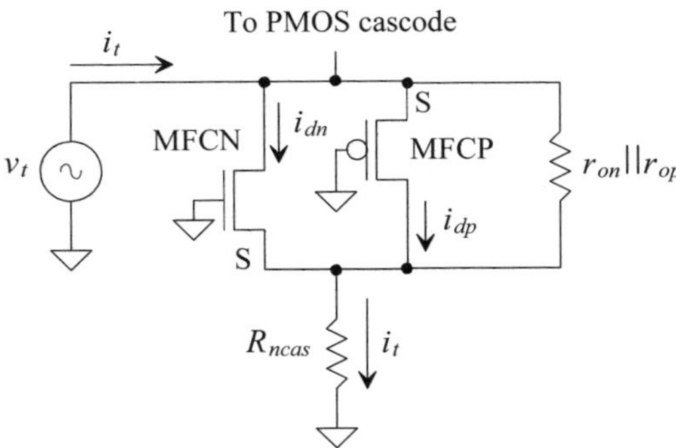

Figure 21.54 Determining the loading affects of the floating current source..

where R_{ncas} is the output resistance of the cascode stack. The gate-source AC voltage of MFCN is $-i_t R_{ncas}$ and so

$$i_{dn} = g_{mn} \cdot (-i_t R_{ncas}) \tag{21.120}$$

Similarly, the drain current through MFCP is

$$i_{dp} = g_{mp} \cdot v_t \tag{21.121}$$

Substituting these equations into Eq. (21.119) gives

$$v_t = (i_t + g_{mn} \cdot i_t R_{ncas} - g_{mp} \cdot v_t) \cdot r_{on} || r_{op} + i_t \cdot R_{ncas} \tag{21.122}$$

or

$$v_t \cdot \overbrace{(1 + g_{mp} \cdot r_{on} || r_{op})}^{\approx g_{mp} \cdot r_{on} || r_{op}} = i_t \cdot \overbrace{(1 + g_{mn} R_{ncas} \cdot r_{on} || r_{op} + R_{ncas})}^{\approx g_{mn} \cdot r_{on} || r_{op} \cdot R_{ncas}} \tag{21.123}$$

The resistance the PMOS cascode sees is then

$$\frac{v_t}{i_t} \approx R_{ncas} \tag{21.124}$$

In other words, the floating current source doesn't load the cascode structure.

21.3.3 Distortion

Small-signal analysis works remarkably well for all of the internal nodes in an op-amp. However, the output buffer must, ideally, swing from rail-to-rail. To illustrate the problem with this, consider the basic gain of a CS amplifier with current source load, Eq. (21.39),

$$|A_v| = g_{m1} \cdot (r_{o1} || r_{o2}) = \frac{\sqrt{2\beta_1 (I_D + i_d)}}{2(\lambda_n + \lambda_p)(I_D + i_d)} = \frac{\sqrt{2\beta_1}}{2(\lambda_n + \lambda_p)\sqrt{(I_D + i_d)}} \tag{21.125}$$

Normally, the AC component of the drain current, i_d, is assumed to be much less than the DC component of the drain current, I_D, and the amplifier gain is essentially constant (small-signal approximation). If the AC component is comparable to the DC component, noticeable distortion results. The voltage gain for large inputs depends on the input signal amplitude.

Characterizing an amplifier begins by applying a pure single-tone sinusoid of the form

$$V_{in}(t) = V_p \sin 2\pi f \cdot t \tag{21.126}$$

to the input of the amplifier. The output of the amplifier is a series of tones at an integer multiple of the input tone given by

$$V_{out}(t) = a_1 V_p \sin(\omega t) + a_2 V_p \sin(2\omega t) + \cdots + a_n V_p \sin(n\omega t) \tag{21.127}$$

The magnitude of the fundamental or wanted signal is $a_1 V_p$. Ideally, a_2 through a_n are zero, and the amplifier is free of distortion. The nth term harmonic distortion is given by

$$HD_n = \frac{a_n}{a_1}, \text{ for } n > 1 \tag{21.128}$$

The *total harmonic distortion* (THD) is given by

$$THD = \sqrt{\frac{a_2^2 + a_3^2 + a_4^2 + \cdot \cdot \cdot + a_n^2}{a_1^2}} \tag{21.129}$$

Again, output buffers, amplifiers used to drive a large load capacitance or low resistance, are examples of amplifiers where low THD is important. If the output amplifier is biased so that the DC component of the drain current is large compared to the transient or time-varying current, the buffer will dissipate too much power for most applications. Therefore, in almost all situations, the biasing current in an output buffer is comparable, or even smaller for class B operation, than the time-varying current. We reduce the distortion by employing feedback around the amplifier. If the open-loop gain of an op-amp varies from 1,000 to 10,000 depending on the amplitude of the input signal, then the feedback around the amplifier reduces the gain sensitivity. In fact, it is nearly impossible to design a linear output amplifier with low distortion without using feedback.

Modeling Distortion with SPICE

SPICE can be used to simulate distortion using a transient analysis and the .FOUR (Fourier) statement. The general form of this statement is .FOUR FREQ OVl <OV2 OVI . . . where FREQ is the frequency of the fundamental and OV1 . . . are the outputs of the circuit (the voltage or current outputs for which SPICE will calculate distortion). As a simple example, consider the circuit and netlist shown in Fig. 21.55. The input sine wave must have at least one full period for the .FOUR statement to calculate distortion. In the case where there's more than one period, SPICE uses the output over the last full period. Also, the maximum transient step size should be less than the period of the input divided by 100. For the example of Fig. 21.55 with a 1 kHz input (a period of 1 ms), the maximum print size should be 10 μs.

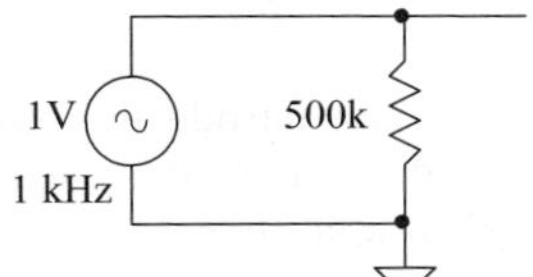


```
*** SPICE Fourier Analysis Example ***
.four 1k Vinout
R1 Vinout 0 500k
Vinout Vinout 0 DC 0 AC 0 0 SIN(0 1 1khz 0 0)
.tran 10u 2m 0 10u
.end
```

Figure 21.55 Simple circuit to demonstrate the use of the .FOUR statement.

The resulting simulation output follows.

Fourier analysis for vinout:
No. Harmonics: 10, THD: 3.17012e-06 %, Gridsize: 200, Interpolation Degree: 1

Harmonic	Frequency	Magnitude	Phase	Norm. Mag	Norm. Phase
0	0.000000e+00	-5.60197e-09	0.000000e+00	0.000000e+00	0.000000e+00
1	1.000000e+03	9.996317e-01	-9.39744e-06	1.000000e+00	0.000000e+00
2	2.000000e+03	1.120393e-08	-8.64000e+01	1.120806e-08	-8.64000e+01
3	3.000000e+03	1.120393e-08	-8.46000e+01	1.120806e-08	-8.46000e+01
4	4.000000e+03	1.120394e-08	-8.28000e+01	1.120806e-08	-8.28000e+01
5	5.000000e+03	1.120394e-08	-8.10000e+01	1.120806e-08	-8.10000e+01
6	6.000000e+03	1.120394e-08	-7.92000e+01	1.120806e-08	-7.92000e+01

```
7    7.000000e+03   1.120393e-08  -7.74000e+01  1.120806e-08  -7.74000e+01
8    8.000000e+03   1.120393e-08  -7.56000e+01  1.120806e-08  -7.56000e+01
9    9.000000e+03   1.120393e-08  -7.38000e+01  1.120806e-08  -7.38000e+01
```

Notice that, as we would expect, the output of this circuit doesn't show any harmonic distortion. SPICE calculates the magnitude and phase of the first nine harmonics and DC. Also note that SPICE automatically calculates the THD for a circuit.

As a more practical example, consider the push-pull amplifier shown in Fig. 21.56. The SPICE netlist for simulating the distortion performance of this output amplifier is shown below. Note that to avoid a long start-up transient we started the big capacitors at the DC bias voltages (V_{bias1} for the PMOS and V_{bias4} for the NMOS). When the input sinewave has an amplitude of 2 mV, the THD is 0.6%. When the input drops to 100 μV (small-signal approximation is more valid), the THD drops to 0.046%. If the input amplitude increases to 10 mV (so the output of the push-pull amplifier swings close to the rails), the THD increases to 9.2%.

```
*** Figure 21.56 CMOS: Circuit Design, Layout, and Simulation ***

.option scale=50n
.tran 10n 2u UIC
.four 1MEG Vout
VDD     VDD     0       DC      1
Vin     Vin     0       DC      0       sin 0 2m 1MEG
RL      out     0       1k
Rbigp   Vbias1  vgp     1G
Cbigp   vgp     vin     1 IC=0.643
Rbign   Vout    vgn     1G
Cbign   Vgn     vin     1 IC=0.362
Xbias   VDD Vbias1 Vbias2 Vbias3 Vbias4 Vhigh Vlow Vncas Vpcas bias
MON     Vout    vgn     0       0       NMOS L=2 W=500
MOP     Vout    vgp     VDD     VDD     PMOS L=2 W=1000
(subcircuit and SPICE models not shown)
```

Note that another useful SPICE tool to evaluate the distortion introduced into a signal by a circuit is a Discrete Fourier Transform (DFT). The spec command is used in SPICE (spectral analysis) to determine the output spectrum of an amplifier. Using this command is covered in depth in the second volume of this book entitled, *CMOS Mixed Signal Circuit Design*.

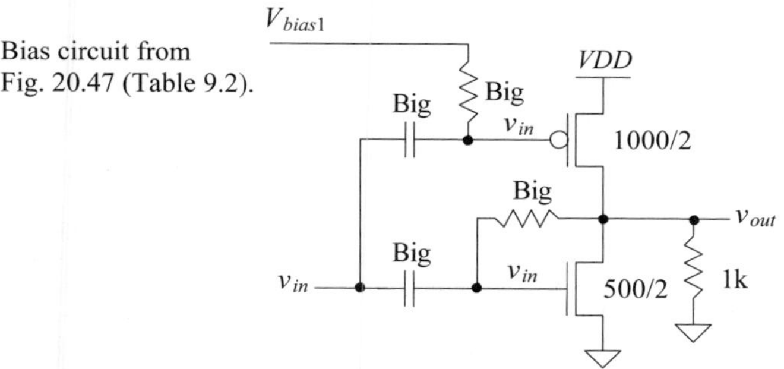

Figure 21.56 Simulating the distortion in a push-pull amplifier.

ADDITIONAL READING

[1] K. N. Leung, P. K. T. Mok, W. Ki, and J. K. O. Sin, "Three-stage large capacitive load amplifier with damping-factor-control frequency compensation," *IEEE Journal of Solid-State Circuits,* vol. 35, pp. 221–230, February 2000.

[2] A. B. Dowlatabadi, "A robust, load-insensitive pad driver," *IEEE Journal of Solid-State Circuits,* vol. 35, pp. 660–665, April 2000.

[3] B. Razavi, "CMOS technology characterization for analog and RF design," *IEEE Journal of Solid-State Circuits,* vol. 34, pp. 268–276, March 1999.

[4] K. de Langen and J. H. Huijsing, "Compact low-voltage power-efficient operational amplifier cells for VLSI," *IEEE Journal of Solid-State Circuits,* vol. 33, pp. 1482–1496, October 1998.

[5] P. E. Allen, B. J. Blalock, and G. A. Rincon, "A 1V CMOS Opamp using Bulk-Driven MOSFETs," *IEEE International Solid-State Circuits Conference,* vol. 38, pp. 192–193, February 1995.

[6] R. G. Eschauzier, R. Hogervorst, and J. H. Huijsing, "A programmable 1.5 V CMOS class-AB operational amplifier with hybrid nested Miller compensation for 120 dB gain and 6 MHz UGF," *IEEE Journal of Solid-State Circuits,* vol. 29, pp. 1497–1504, December 1994.

[7] R. Hogervorst, J. P. Tero, R. G. H. Eschauzier, and J. H. Huijsing, "A Compact Power-Efficient 3 V CMOS Rail-to-Rail Input/Output Operational Amplifier for VLSI Cell Libraries," *IEEE Journal of Solid State Circuits,* vol. 29, pp. 1505–1513, December 1994.

[8] A. A. Abidi, "On the operation of cascode gain stages," *IEEE Journal of Solid-State Circuits,* vol. 23, pp. 1434–1437, December 1988.

[9] S. L. Wong and C. A. T. Salama, "An efficient CMOS buffer for driving large capacitive loads," *IEEE Journal of Solid-State Circuits,* vol. 21, pp. 464–469, June 1986.

[10] D. B. Ribner and M. A. Copeland, "Design techniques for cascoded CMOS op amps with improved PSRR and common-mode input range," *IEEE Journal of Solid-State Circuits,* vol. 19, pp. 919–925, December 1984.

[11] V. R. Saari, "Low-power high-drive CMOS operational amplifiers," *IEEE Journal of Solid-State Circuits,* vol. 18, pp. 121–127, February 1983.

[12] B. K. Ahuja, "An Improved Frequency Compensation Technique for CMOS Operational Amplifiers," *IEEE Journal of Solid-State Circuits*, vol. 18, pp. 629–633, December 1983.

[13] P. R. Gray and R. G. Meyer, "MOS operational amplifier design - A tutorial overview," *IEEE Journal of Solid-State Circuits,* vol. 17, pp. 969–982, December 1982.

[14] B. J. Hosticka, "Dynamic CMOS amplifiers," *IEEE Journal of Solid-State Circuits,* vol. 15, pp. 887–894, October 1980.

PROBLEMS

21.1 Using simulations, show that the small-signal resistance of a gate-drain-connected PMOS device, Fig. 21.57, behaves like a resistor with a value of $1/g_m$.

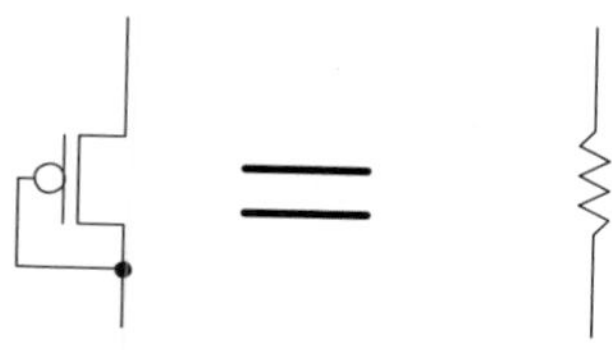

Figure 21.57 The small-signal behavior of a diode-connected MOSFET.

21.2 Estimate the frequency response of the circuit seen in Fig. 21.58. Verify your hand calculations using SPICE.

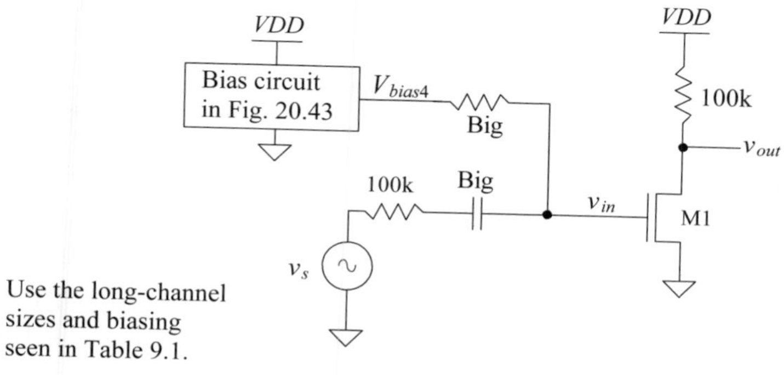

Figure 21.58 Amplifier used in Problem 21.2.

21.3 Repeat problem 21.2 if the amplifier drives a 100 *fF* load capacitance.

21.4 For the circuit in Fig. 21.12, estimate the effective transconductance of the circuit, g_{meff}, that relates the MOSFET drain current to the AC input voltage, v_{in}. Verify your solution with SPICE.

21.5 Simulate the operation of the common-gate amplifier in Fig. 21.16 using the bias circuit in Fig. 20.43 and the device sizes seen in Table 9.1. Compare the simulation results to hand calculations.

21.6 Estimate the transfer function of the amplifier in Ex. 21.6 if it drives a 100 fF load. Verify your hand calculations using simulations.

21.7 Determine the frequency response of the amplifier seen in Fig. 21.59. Verify your hand calculations using simulations.

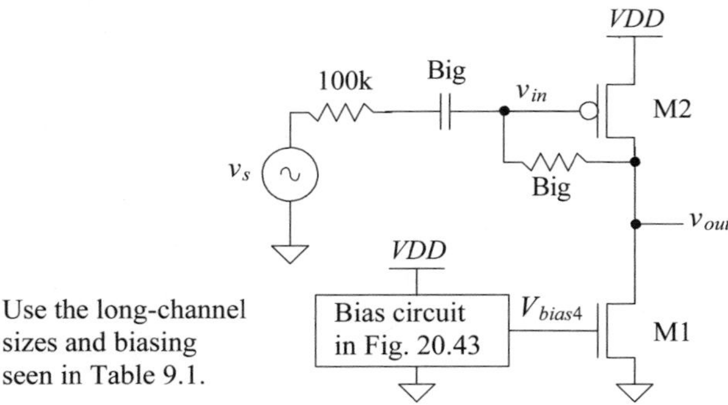

Figure 21.59 Amplifier for Problem 21.7.

21.8 Repeat Ex. 21.9 using the biasing circuit from Fig. 20.47 and the sizes in Table 9.2.

21.9 Repeat Ex. 21.10 using the biasing circuit from Fig. 20.47 and the sizes in Table 9.2.

21.10 Repeat Ex. 21.12 using the biasing circuit from Fig. 20.43 and the sizes in Table 9.1.

21.11 Show, using SPICE simulations and an ideal current source of 10 μA (device size from Table 9.2), the difference in the voltage gains with and without body effect for the circuit seen in Fig. 21.60.

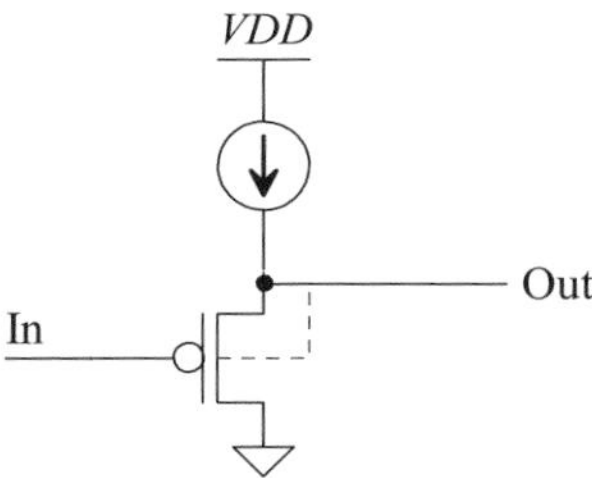

Figure 21.60 Gain of an SF with and without body effect for Problem 21.11.

21.12 Does the SF in Fig. 21.46 exhibit slew-rate limitations if it needs to discharge the 1-pF load capacitance quickly? Why or why not? Verify your answer using simulations.

21.13 Simulate the operation of the class AB output buffer in Fig. 21.48 using the bias circuit from Fig. 20.47 and the sizes in Table 9.2. If the buffer is driving a load resistor of 10k, plot v_{in} against v_{out}. What is the linear output range?

21.14 In the class AB output buffer in Fig. 21.50 or 21.51, a load resistor connected to ground causes the PMOS device to conduct more current then the NMOS. Why? Resimulate the buffer in Fig. 21.51 if it drives a 1k resistor and MON is reduced in size to 50/2. Is this a better design? Why or why not?

21.15 Using simulations, show the problem of using an inverter output buffer without a floating current source as seen in Fig. 21.53. (The quiescent current that flows in the MOSFETs is huge and not accurately controlled as it is in Fig. 21.50.)

21.16 Is the distortion the output buffer in Fig. 21.56 introduces into a signal a function of the load resistance? Verify your answer with simulations and show some time-domain waveforms with and without distortion.

Chapter

22

Differential Amplifiers

In the last chapter big resistors and capacitors were used to bias the circuits to the correct operating point, as seen in Fig. 21.21. The DC operating voltage on the gate of M1 (in Figs. 21.17 or 21.21) is extremely important when biasing the amplifier. If it's not at precisely the correct value, then the current sourced by M2 won't equal the current in M1 when both MOSFETs are operating in the saturation region.

The differential amplifier (*diff-amp*) is used on the input of an amplifier to allow input voltages to move around so that biasing of the gain stages isn't affected (that is, so it isn't a function of the input voltage). The diff-amp is a fundamental building block in CMOS analog integrated circuit design, and an understanding of its operation and design is extremely important. In this chapter we discuss three basic types of differential amplifiers: the source-coupled pair, the source cross-coupled pair, and the current differential amplifier.

22.1 The Source-Coupled Pair

The source-coupled pair comprised of M1 and M2 is shown in Fig. 22.1. When M1 and M2 are used in this configuration they are sometimes called a *diff-pair*.

22.1.1 DC Operation

The diff-pair in Fig. 22.1 is biased with a current source so that

$$I_{SS} = i_{D1} + i_{D2} \tag{22.1}$$

If we label the input voltages at the gates of M1 and M2 as v_{I1} and v_{I2}, we can write the input difference as

$$v_{DI} = v_{I1} - v_{I2} = v_{GS1} - v_{GS2} \tag{22.2}$$

or in terms of the AC and DC components of the differential input voltage, v_{DI},

$$v_{DI} = V_{GS1} + v_{gs1} - V_{GS2} - v_{gs2} \tag{22.3}$$

When the gate potentials of M1 and M2 are equal, then (assuming both are operating in the saturation region)

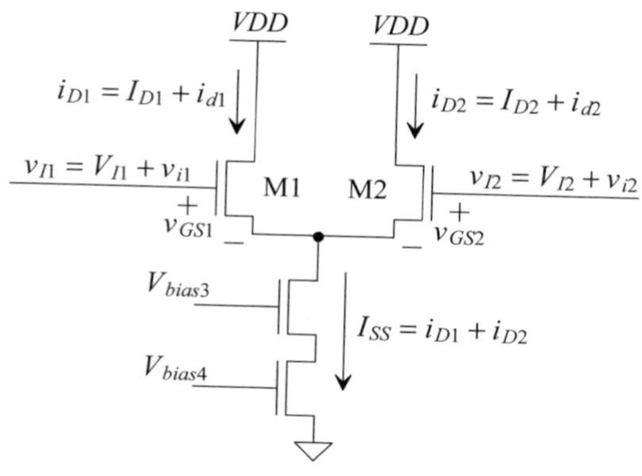

Figure 22.1 A basic NMOS source-connected pair made using the diff-pair M1 and M2.

$$I_{D1} = I_{D2} = \frac{I_{SS}}{2} \tag{22.4}$$

Maximum and Minimum Differential Input Voltage

Since we know that a saturated MOSFET follows the relation

$$i_D = \frac{\beta_n}{2}(v_{GS} - V_{THN})^2 \tag{22.5}$$

the difference in the input voltages may be written as

$$v_{DI} = \sqrt{\frac{2}{\beta_n}}\left(\sqrt{i_{D1}} - \sqrt{i_{D2}}\right) \tag{22.6}$$

The maximum difference in the input voltages, v_{DIMAX} (maximum differential input voltage), is found by setting i_{D1} to I_{SS} (M1 conducting all of the tail bias current) and i_{D2} to 0 (M2 off)

$$v_{DIMAX} < v_{I1} - v_{I2} = \sqrt{\frac{2 \cdot L \cdot I_{SS}}{KP_n \cdot W}} \tag{22.7}$$

The minimum differential input voltage, v_{DIMIN}, is found by setting i_{D2} to I_{SS} and i_{D1} to 0

$$v_{DIMIN} = -v_{DIMAX} = -(v_{I1} - v_{I2}) = -\sqrt{\frac{2 \cdot L \cdot I_{SS}}{KP_n \cdot W}} \tag{22.8}$$

Example 22.1

Estimate the maximum and minimum voltage on the gate of M1 in Fig. 22.2 that ensures that neither M1 or M2 shut off. Verify your hand calculations with SPICE simulations.

Note that the diff-amp tail current, I_{SS}, is 40 μA (the widths of the MOSFETs were doubled from the sizes indicated in Table 9.1). When $v_{I1} = v_{I2} = 2.5$ V, the differential input voltage, v_{DI}, is 0 and the current flowing in M1 and M2 is 20 μA ($= i_{D1} = i_{D2}$).

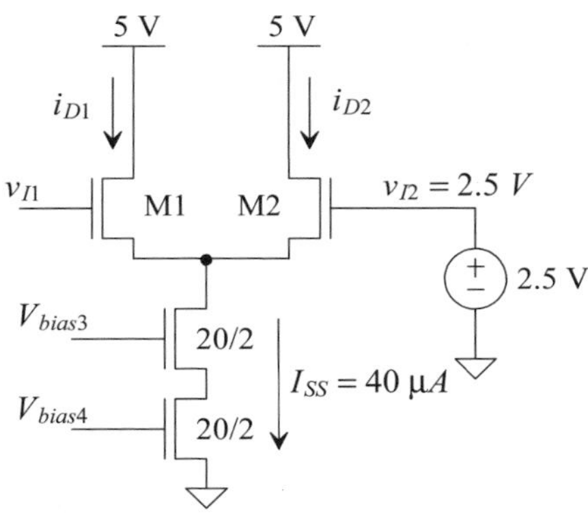

Figure 22.2 Diff-amp used in Ex. 22.1.

Using Eq. (22.7), we can estimate the maximum voltage on the gate of M1 as

$$v_{I1MAX} = \sqrt{\frac{2 \cdot L \cdot I_{SS}}{KP_n \cdot W}} + v_{I2} = \sqrt{\frac{2 \cdot 2 \cdot 40}{120 \cdot 10}} + 2.5 = 2.865\ V$$

or v_{I1} is 365 mV above v_{I2}. The minimum voltage allowed on the gate of M1 (to keep some current flowing in M1) is then 365 mV below v_{I2} or 2.135 V. The simulation results are seen in Fig. 22.3. Notice how, when the inputs are equal (both are 2.5 V), the currents in M1 and M2 are equal and approximately 20 μA (= $I_{SS}/2$). ■

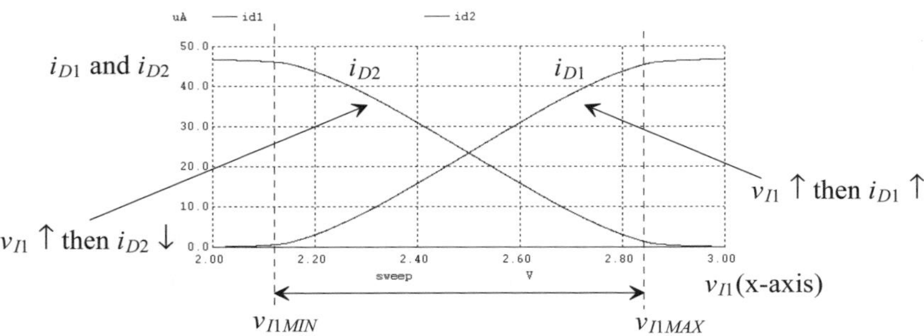

Figure 22.3 Simulating the operation of the diff-amp in Fig. 22.2.

Maximum and Minimum Common-Mode Input Voltage

When the diff-amp is used on the input of an op-amp, the inputs are forced, via feedback around the op-amp, to the same values (or very nearly the same values). This value (or more precisely the average of the two inputs) is called the *common-mode voltage*. Figure 22.4 shows the diff-amp with both inputs tied together. We're interested in the maximum and minimum voltage that will keep both M1 and M2 operating in the saturation region (that is, not off or in the triode region). We'll call the maximum common-mode voltage, v_{CMMAX} and the minimum common-mode voltage, v_{CMMIN}. Note that when the common-mode voltage (the potential on the gates of M1 and M2) is too high, both M1 and M2 triode (behave like resistors). When the common-mode voltage is too low, M1 and M2 shut off.

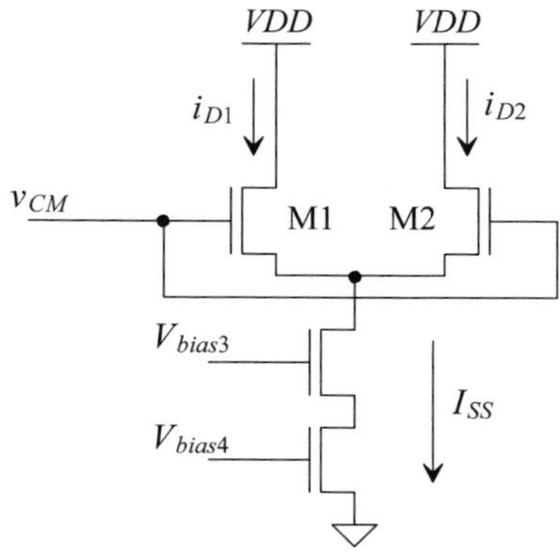

Figure 22.4 Diff-amp used to calculate the minimum and maximum diff-amp input voltages.

To determine the maximum input voltage, we know that for M1 and M2 to remain in the saturation region

$$V_{DS} \geq V_{GS} - V_{THN} \rightarrow V_D \geq V_G - V_{THN} \tag{22.9}$$

Because $V_D = VDD$, we can write

$$V_{CMMAX} = VDD + V_{THN} \tag{22.10}$$

The input common-mode voltage can actually be higher than VDD before M1/M2 move into the triode region.

For the minimum input voltage, we can write

$$V_{CMMIN} = V_{GS1,2} + 2 \cdot V_{DS,sat} \tag{22.11}$$

where the minimum voltage across the current source is assumed to be $2V_{DS,sat}$. For the long-channel process, from Table 9.1, the minimum input voltage for an NMOS diff-amp is 1.55 V.

Example 22.2

Figure 22.5 shows a folded cascode amplifier (see Fig. 20.45). Assuming that all MOSFETs are operating in the saturation region, estimate the minimum and maximum input voltage of the amplifier.

Note how the widths of M5–M6 are doubled to sink the additional current from the diff-amp. It's important to understand how the sizes of the MOSFETs are adjusted in the current mirrors so that the currents sum correctly. The currents, as seen in Fig. 22.5, are labeled assuming $Vplus = Vminus$.

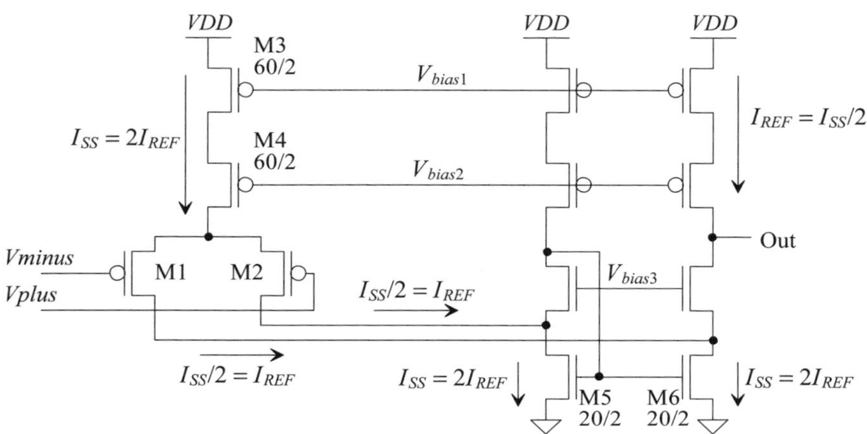

Figure 22.5 Folded cascode amplifier.

The maximum allowable input common-mode voltage is determined by

$$V_{CMMAX} = VDD - V_{SG} - 2V_{SD,sat} = 5 - 1.15 - 0.5 = 3.35\ V$$

The minimum input common-mode voltage is determined by noting the drain voltage of M5 and M6 (or M1 and M2) is a $V_{DS,sat}$

$$V_{SD} \geq V_{SG} - V_{THP} \rightarrow V_D \leq V_G + V_{THP}$$

and so

$$V_{CMMIN} = V_D - V_{THP} = 0.25 - 0.9 = -0.65$$

In other words, the input common-mode voltage can actually be below ground and the amplifier will still function correctly. This result is *very useful*. The folded- cascode amplifier with PMOS diff-amp can be used when the input voltage swings around ground. Note, however, that the output voltage of the amplifier will not swing below ground but is, rather, limited to $2V_{DS,sat}$ above ground and $2V_{SD,sat}$ below *VDD* if the MOSFETs are to remain in saturation (the high-gain region of operation). ■

Current Mirror Load

The previous uses of the diff-amp have exploited the fact that the output of the diff-amp is a current controlled by a differential input voltage, v_{DI}. In some applications, it is desirable to have a diff-amp with a voltage output. Towards this goal, consider the diff-amp with a current mirror load, as seen in Fig. 22.6. An imbalance in the drain currents of M1 and M2 causes the output of the diff-amp to swing either towards *VDD* or

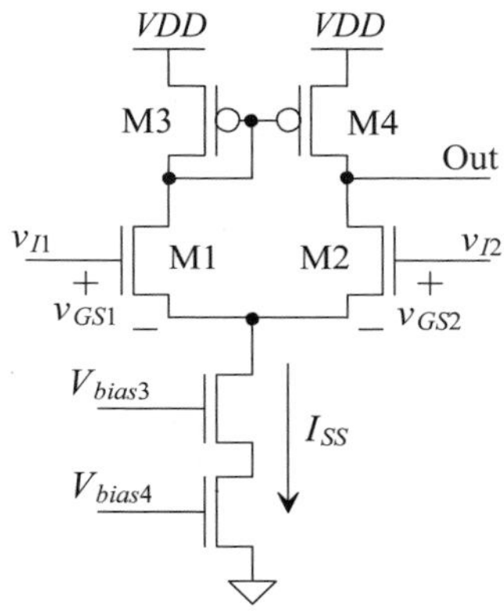

Figure 22.6 Diff-amp with a current mirror load.

ground. The minimum input common mode voltage is, once again, given by Eq. 22.11. The maximum input common mode voltage is determined knowing that the drain voltage of M2 is the same as the drain voltage of M1 (when both diff-amp inputs are the same potential), that is, $VDD - V_{SG}$ of the PMOS. We can therefore write

$$V_{DS} \geq V_{GS} - V_{THN} \rightarrow V_D \geq V_G - V_{THN} \rightarrow V_{CMMAX} = VDD - V_{SG} + V_{THN} \quad (22.12)$$

For the parameters in Table 9.1, $V_{CMMAX} = 5 - 1.15 + 0.8 = 4.65\ V$.

The voltage output swing can be determined by noting that the maximum voltage output is limited by keeping M4 in saturation. Therefore,

$$V_{OUTMAX} = VDD - V_{SD,sat} \quad (22.13)$$

The minimum output voltage is determined by the voltage on the gate of M2 (M2 must remain in saturation).

$$V_D \geq V_G - V_{THN} \rightarrow V_{OUTMIN} = V_{I2} - V_{THN} \quad (22.14)$$

Example 22.3

Suppose the diff-amp in Fig. 22.6 is implemented with the sizes and bias currents seen in Table 9.1 (I_{SS} = 40 μA). If the gate of M2 is held at 4 V, estimate the diff-amp's output swing. Verify the answers with SPICE.

From Eq. (22.13), the output can swing up to 4.75 V before M4 triodes. From Eq. (22.14), the output can swing down to 4–0.8 or 3.2 V before M2 triodes. The simulation results are seen in Fig. 22.7. Looking at the results, we see the output goes down to approximately 2.8 V or 400 mV less than what we predicted. This can be attributed to the body effect in M1 and M2. With body effect, the threshold voltage is 1.2 V (instead of 0.8 V). ■

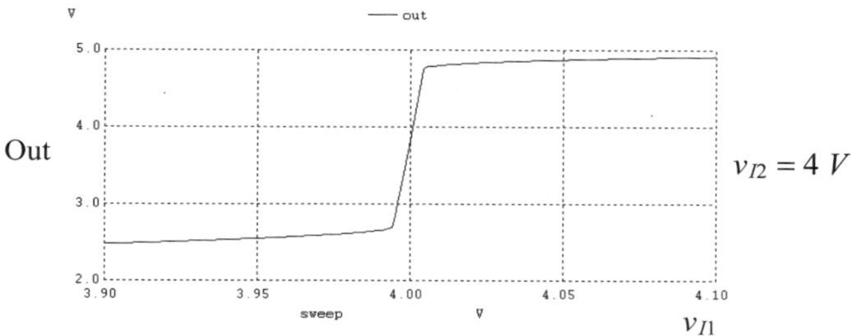

Figure 22.7 Simulation results for Ex. 22.3

Biasing from the Current Mirror Load

Consider the connection of the common-source amplifier, M7, to the output of the diff-amp in Fig. 22.8. When the inputs to the diff-amp are at the same potential, the currents that flow in M3 and M4 are equal ($= I_{SS}/2$). We know from Ch. 20 that the drain of M4 is then at the same potential as its gate. This means, for biasing purposes, that the gate of M7 can be treated as if it were tied to the gate of M3 (M4).

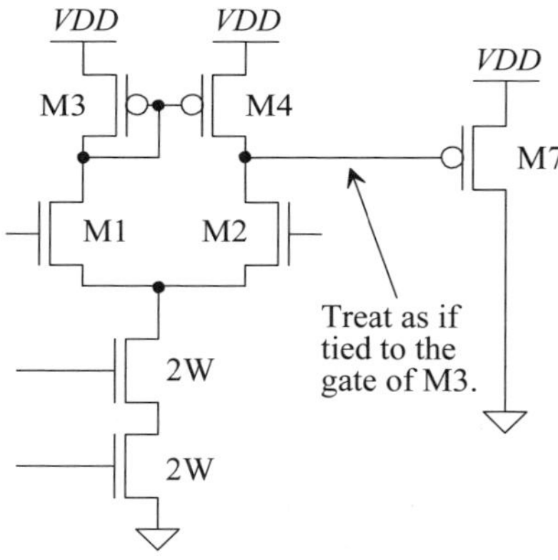

Figure 22.8 Using a diff-amp to bias the next stage amplifier.

Minimum Power Supply Voltage

Looking at the diff-amp in Fig. 22.6, we can estimate the minimum allowable power supply voltage, VDD_{min}, as

$$VDD_{min} = V_{SG3} + \overbrace{V_{DS,sat1}}^{\text{minimum voltage across M1}} + \overbrace{2V_{DS,sat}}^{\text{minimum voltage across Iss}} \quad (22.15)$$

Looking at this equation, we see that the V_{SG} of M3 is the "weak link" in the minimum power supply voltage. Using the parameters from Table 9.2 (the short-channel devices with a VDD of 1 V), we get a VDD_{min} of 500 mV. We can lower this value by using a single MOSFET to bias the diff-amp (instead of a cascode circuit).

22.1.2 AC Operation

To describe the AC small-signal operation of the diff-amp in Fig. 22.1, consider the AC schematic seen in Fig. 22.9. We've replaced the current source with an open and the DC supply (*VDD*) with a short (to ground). In the following discussion, it's important to remember that a negative AC current simply means that the overall current (AC + DC) is decreasing (see Ex. 9.5 in Ch. 9). We can write

$$v_{di} = v_{i1} - v_{i2} = v_{gs1} - v_{gs2} = \frac{i_{d1}}{g_m} - \frac{i_{d2}}{g_m} \tag{22.16}$$

Reviewing Fig. 22.9, we see that

$$i_{d1} = -i_{d2} = i_d \tag{22.17}$$

and thus

$$v_{gs1} = -v_{gs2} = \frac{v_{di}}{2} \tag{22.18}$$

so finally

$$g_m v_{gs1} = g_m(-v_{gs2}) = i_d = g_m \cdot \frac{v_{di}}{2} \tag{22.19}$$

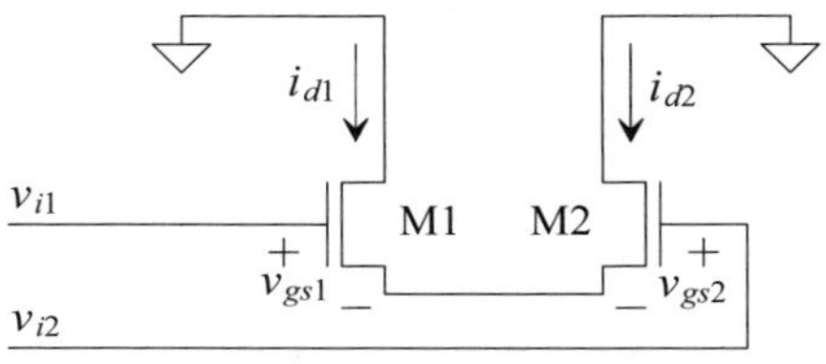

Figure 22.9 AC circuit for a diff-amp.

Example 22.4
Estimate the AC components of the drain currents of M1 and M2 for the circuit in Fig. 22.2 if $v_{I1} = 2.5 + 1mV \cdot \sin(2\pi \cdot 1kHz \cdot t)$. Verify your hand calculations using SPICE.

From Table 9.1 we know the g_m of the NMOS devices used in the diff-pair is 150 $\mu A/V$. The AC component of v_{I2} is zero so we know

$$v_{gs1} = -v_{gs2} = 0.5\ mV$$

and thus

$$i_d = g_m v_{gs} = (150 \times 10^{-6})(500 \times 10^{-6}) = 75\ nA$$

This is the AC component of the drain currents. When i_{d1} = 75 nA, then i_{d2} = −75 nA. The overall drain currents are written as

$$i_{D1} = 20\ \mu A + (75\ nA) \cdot \sin(2\pi 1k \cdot t) \text{ and } i_{D2} = 20\ \mu A - (75\ nA) \cdot \sin(2\pi 1k \cdot t)$$

The simulation results are seen in Fig. 22.10. ■

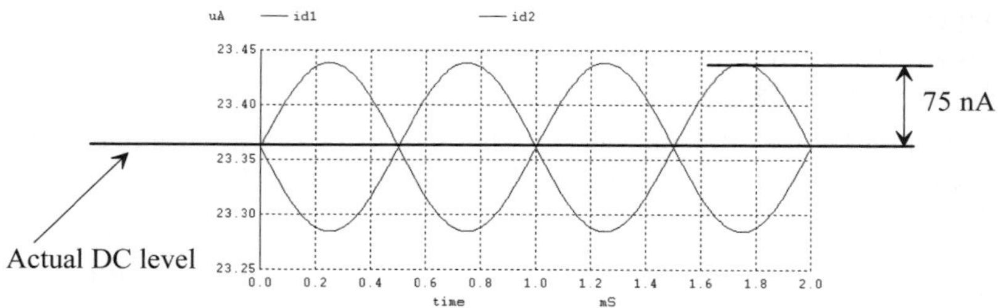

Figure 22.10 Simulation results for Ex. 22.4.

AC Gain with a Current Mirror Load

To determine the AC gain of the diff-amp with a current mirror load, Fig. 22.6, consider the small signal model seen in Fig. 22.11. We've replaced the diode-connected MOSFET, M3, with a $1/g_{m3}$ resistor. Also the resistance looking into the drain of M4 (r_{o4}) and the resistance looking into the drain of M2 (r_{o2}) are drawn explicitly. We assumed that the small-signal resistance looking into the drain of M2 is simply r_{o2} (for all practical design cases this is a good assumption). Note how, because of the current mirror action, the AC current flowing in r_{o4} is i_{d1}. The current in M3 is mirrored by M4. The output voltage can be written as

$$v_{out} = (i_{d1} - i_{d2}) \cdot (r_{o2} || r_{o4}) \tag{22.20}$$

Knowing $i_{d1} = -i_{d2} = i_d$

$$v_{out} = 2i_d \cdot (r_{o2} || r_{o4}) \tag{22.21}$$

The differential mode gain is then written with the help of Eq. (22.19) as

$$A_d = \frac{v_{out}}{v_{di}} = \frac{v_{out}}{v_{i1} - v_{i2}} = g_m \cdot (r_{o2} || r_{o4}) \tag{22.22}$$

Notice that as the voltage on the gate of M1 increases, the current in M1 (and M3/M4) increases. This causes the output voltage to increase. When the gate potential of M2 increases, so does its drain current, causing the output voltage to decrease. Sometimes the gate potential of M1 is called the *noninverting* input and the gate potential of M2 is called the *inverting* input.

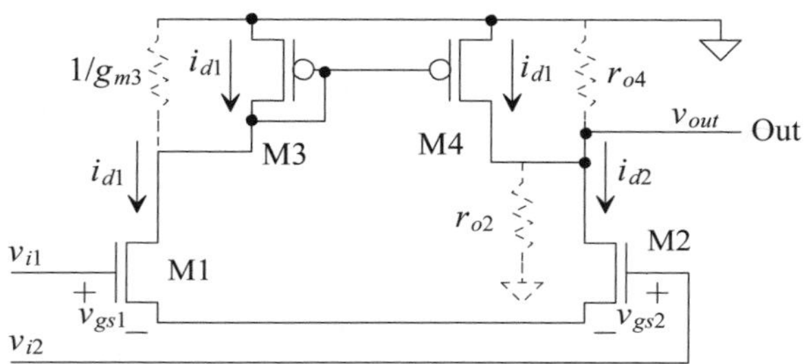

Figure 22.11 AC circuit of the diff-amp with a current-source load.

Example 22.5
Determine the output voltage for the diff-amp seen in Fig. 22.12. Verify your hand calculations using SPICE.

The tail current of the diff-amp conducts 20 μA so M1 and M2, when the inputs are equal, each conduct 10 μA. Using the parameters from Table 9.2 and Eq. (22.22), the gain of the diff-amp is

$$A_d = g_m \cdot (r_{on} || r_{op}) = (150 \times 10^{-6}) \cdot \left(\frac{167 \cdot 333}{167 + 333} \times 10^3\right) = 16.7\ V/V$$

This means that the 1 mV AC input will appear as a 16.7 mV AC output. The DC level on the output is $VDD - V_{SG} = 650\ mV$. The output voltage is then

$$v_{out}(t) = 0.65 + (16.7\ mV) \cdot \sin(2\pi 10 MHz \cdot t)$$

The simulation results are seen in Fig. 22.13. ■

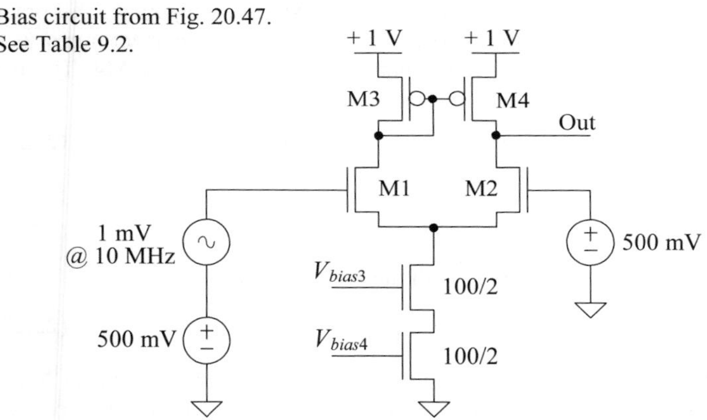

Figure 22.12 Circuit used in Ex. 22.5.

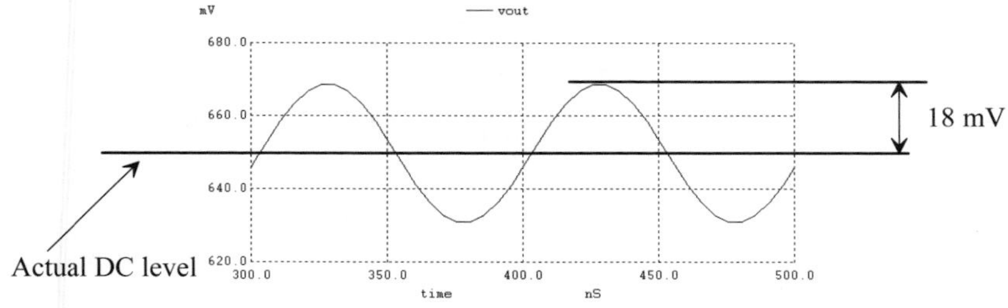

Figure 22.13 The output of the circuit in Fig. 22.12. Note that the transient analysis doesn't show the beginning of the simulation where the reference circuit is starting up.

Example 22.6
Estimate the f_{3dB} of the diff-amp in Ex. 22.5 if it drives a load capacitance of 1 pF. Verify your hand calculations with SPICE.

The resistance on the output node is $r_{o2}||r_{o4} = 111\ k\Omega$; therefore, since this is the *high-impedance node*, we can estimate the circuit's 3-dB frequency in the circuit as

$$f_{3dB} = \frac{1}{2\pi \cdot 111k \cdot 1\ pF} = 1.4\ MHz$$

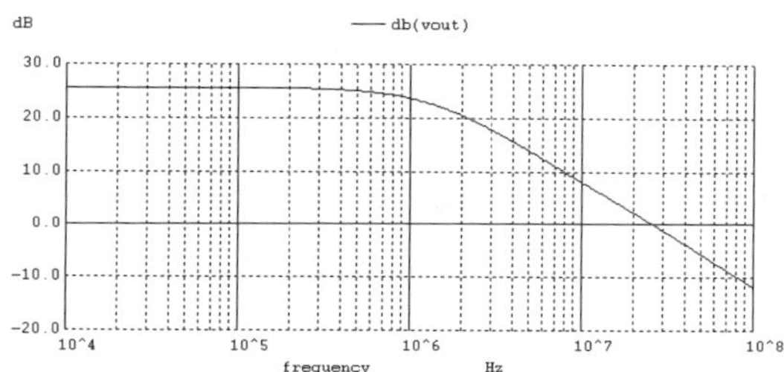

Figure 22.14 The AC response of the diff-amp in Fig. 22.12 when driving a 1 pF load capacitance.

The simulation results are seen in Fig. 22.14. ■

22.1.3 Common-Mode Rejection Ratio

An important aspect of the differential amplifier is its ability to reject a common signal applied to both inputs. Often, in analog systems, signals are transmitted differentially, and the ability of an amplifier to reject coupled noise into each line is very desirable. Consider the amplifier shown in Fig. 22.15. The bias current at the sources of M1/M2 has been replaced with its small-signal output resistance. The common-mode signal (the "noise" labeled in the figure) on both inputs is, ideally, rejected by the diff-amp. The diff-amp's output voltage doesn't vary. The noise causes the source potentials of M1/M2 to vary (the voltage across the tail current source).

If we apply an AC signal, v_c, to the gates of M1/M2, equivalent to saying that a common signal is applied to the input of the differential amplifier, we can calculate the common-mode gain. We begin by writing the AC small-signal, common-mode input voltage, v_c, as

$$v_c = v_{gs1,2} + 2i_d R_o \qquad (22.23)$$

The MOSFETs M1 and M2 each source a current, i_d, through the output resistance of the current source, R_o (hence the factor of two in this equation). This equation may be rewritten as

$$v_c = i_d\left(\frac{1}{g_m} + 2R_o\right) \approx i_d 2R_o \qquad (22.24)$$

The output voltage, because of the symmetry of the circuit, is given by

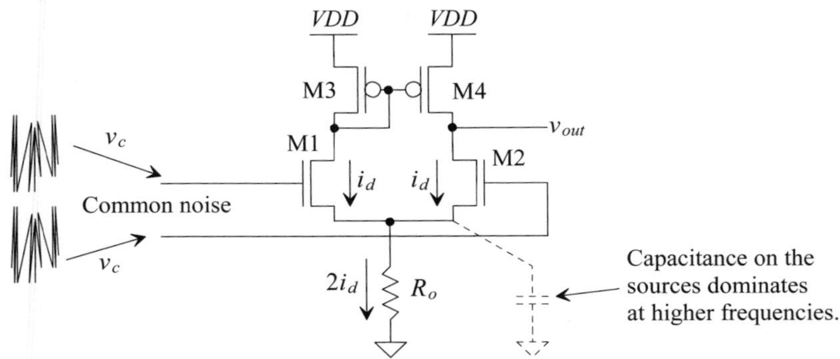

Figure 22.15 Calculating the CMRR of a diff-amp.

$$v_{out} = -i_d \cdot \frac{1}{g_{m3}} = -i_d \cdot \frac{1}{g_{m4}} \tag{22.25}$$

The common-mode gain is then

$$A_c = \frac{v_{out}}{v_c} = \frac{-1/g_{m3,4}}{2R_o} = -\frac{1}{2g_{m3,4}R_o} \tag{22.26}$$

Notice that by increasing the output resistance of the biasing current source, the common-mode gain drops. (This is good because we don't want any common signal to be amplified by the diff-amp.) The common-mode rejection ratio (*CMRR*), in dB, of a diff-amp with a current mirror load is given by

$$CMRR = 20 \cdot \log\left|\frac{A_d}{A_c}\right| = 20 \cdot \log\left[g_{m1,2}(r_{o2}||r_{o4}) \cdot 2g_{m3,4}R_o\right] \tag{22.27}$$

The larger the *CMRR,* the better the performance of the diff-amp. For the diff-amp in Fig. 22.12, Eq. (22.27) evaluates to 86 dB. As the frequency of the input common-mode signal, v_c , increases, the *CMRR* goes down. This is the result of the impedance of the capacitance to ground on the sources of the diff-pair becoming much smaller than R_o.

Example 22.7

Simulate the *CMRR* of the diff-amp in Fig. 22.12. Show that the *CMRR* falls with increasing frequency.

As mentioned above, the (low-frequency) calculated *CMRR* is 86 dB. To simulate the *CMRR,* we perform AC simulations on two of the diff-amps in Fig. 22.12 so we can simultaneously determine A_d and A_c. The ratio of these gains is the *CMRR.* Figure 22.16 shows the simulation results. The common-mode gain, A_c , depends on the DC common-mode voltage. The results seen in Fig. 22.16 are with a common-mode (DC) voltage of 700 mV. Dropping the common-mode voltage down to 500 mV causes the low-frequency *CMRR* to drop down to 50 dB. The reason for this is that, with higher DC common-mode voltages, the voltage across the tail current source is higher, so its output resistance is larger. ■

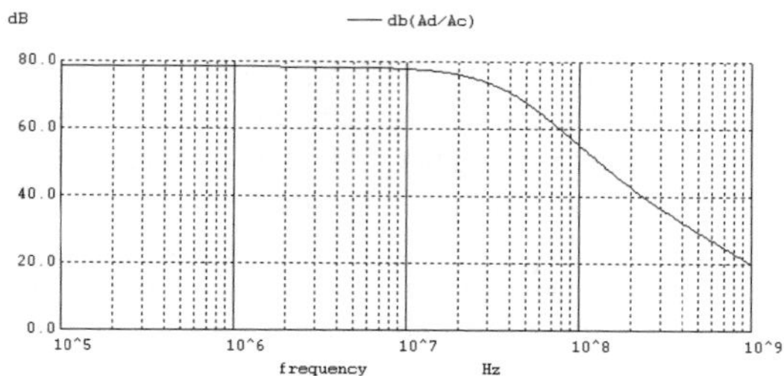

Figure 22.16 The simulated CMRR for the diff-amp in Fig. 22.12.

Input-Referred Offset from Finite CMRR

It's important we understand, in Eq. (22.26), why large R_o reduces A_c. Looking at Fig. 22.15, notice that as v_c varies so too will the voltage across the biasing current source (represented in this figure as R_o). As seen in Ch. 20, Fig. 20.4, variations in the voltage across a current source changes the output of the current source (unless its output resistance is infinite). This variation in current causes the current in M1 and M2 to change with variations in v_c. Changes in the drain current of M1/M2 (and thus M3/M4) causes the output voltage to change. Figure 22.17 shows how the output voltage changes as a function of the DC common-mode voltage. Understanding this circuit is *very important*. If we wanted to hold the diff-amp output voltage constant while varying the common mode voltage, we would have to apply a small voltage difference between the gates of M1 and M2 (*an offset voltage*). We'll revisit this important topic when we discuss op-amp *CMRR*.

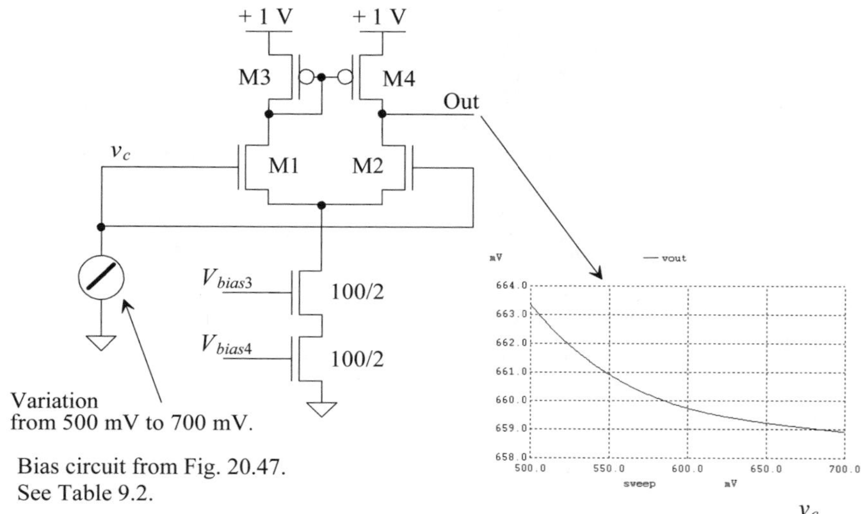

Figure 22.17 How variation in common-mode voltage affects the output voltage.

22.1.4 Matching Considerations

The fact that performance depends on the matching of devices is another important aspect of differential amplifiers. Layout techniques can be used to minimize the first-order effects of mismatch due to oxide gradients and other process variations. Figure 22.18 shows how the diff-pair would be laid out in a *common-centroid* configuration (see Ch. 5). Common-centroid layout, as the name implies, constructs two devices symmetrically about a common center in the layout. This allows the two devices to cancel process gradients in both the x and y directions and exposes both devices to heat sources in an identical manner. Note how we've attempted to match the parasitic capacitances of the pair (on the gates and drains) and how we've minimized the capacitance at the sources (to keep the *CMRR* high at higher frequencies).

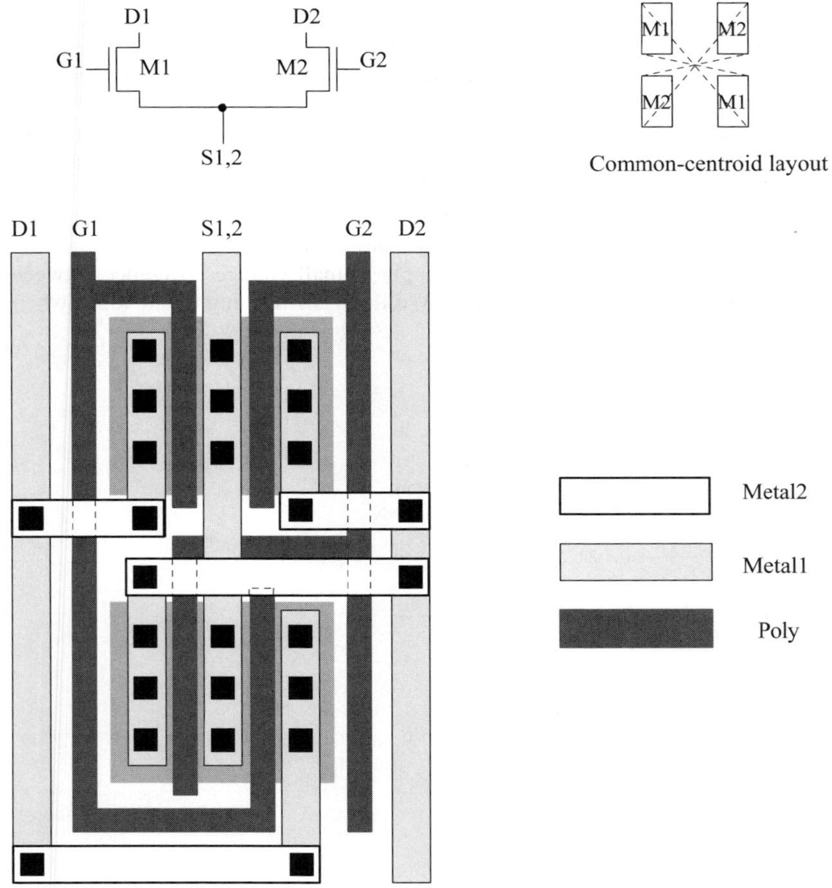

Figure 22.18 Common-centroid layout of a diff-pair.

The input-referred offset voltage of a diff-pair, resulting from mismatches in threshold voltage, geometries, and load resistances, can be determined with the help of Fig. 22.19. When the diff-pair is offset-free, connecting the diff-amp's inputs together causes the drain currents and drain voltages of M1 and M2 to be equal ($V_o = 0$). When the load resistors or M1 and M2 are mismatched, V_o is not zero. (We have mismatch and thus V_o is offset from its ideal value.). To drive V_o to zero, we apply a voltage difference between the gates of M1 and M2 (an input-referred offset).

$$V_{OS} = V_{GS1} - V_{GS2} = V_{THN1} + \sqrt{\frac{2I_{D1}}{\beta_1}} - V_{THN2} - \sqrt{\frac{2I_{D2}}{\beta_2}} \qquad (22.28)$$

We can define the differences and averages of the threshold voltage, load resistance, and geometry as ΔV_{THN} ($V_{THN1}-V_{THN2}$), V_{THN} (average of V_{THN1} and V_{THN2}), ΔR_L ($R_{L1}-R_{L2}$), R_L (average of R_{L1} and R_{L2}), $\Delta(W/L)$ [$(W_1/L_1)-(W_2/L_2)$] and (W/L) [average of (W_1/L_1) and (W_2/L_2)]. Thus, we can make the appropriate substitutions into Eq. (22.28), requiring $I_{D1}R_{L1} = I_{D2}R_{L2}$, with the result:

$$V_{OS} = \Delta V_{THN} + \frac{V_{GS} - V_{THN}}{2} \cdot \left[\frac{-\Delta R_L}{R_L} - \frac{\Delta(W/L)}{(W/L)}\right] \qquad (22.29)$$

The threshold voltage mismatch must be reduced using layout techniques. Mismatches resulting from unequal geometry and differences in the load resistance can be reduced by designing with a small V_{GS} (V_{GS} close to V_{THN}). This should be compared with Eq. (20.8), which shows that using a small V_{GS} in a current mirror results in a large mirrored current difference.

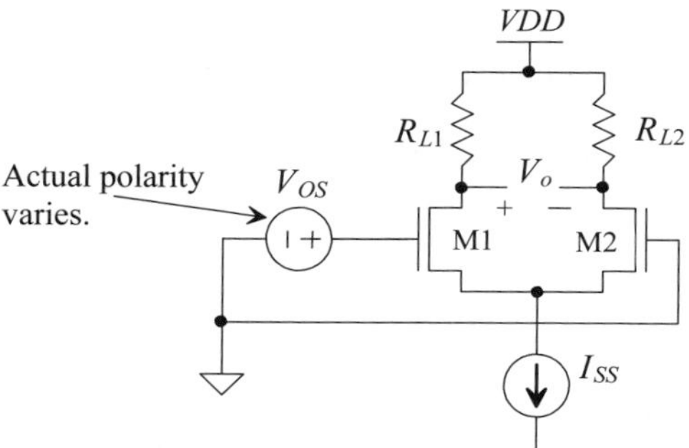

Figure 22.19 Determining diff-amp input offset voltage.

Input-Referred Offset with a Current Mirror Load

Suppose the diff-amp with current mirror load, Fig. 22.12, has an offset (the devices aren't perfectly matched). What this means is that the drain potential of M4 (M2) is not equal to the drain potential of M3 (M1 and most likely the drain currents in M1 and M2 aren't equal as well). The input-referred offset would then be the voltage difference on the gates of M1 and M2 needed to drive the output voltage to the correct value. Using Eq. (22.22), we can write the input-referred offset in terms of the difference-mode gain as

$$|V_{OS}| = \frac{V_{o,ideal}}{|A_d|} = \frac{V_{o,ideal}}{g_m \cdot (r_{o2}||r_{o4})} \tag{22.30}$$

which indicates that to reduce the input-referred offset voltage we simply need to design the diff-amp with large gain. This is the same result as that of Eq. (22.29), if we pause and think about it. To reduce the offset, we said we design with a small V_{GS}. If we hold the diff-amp's bias current constant and reduce V_{GS} (by increasing the widths of M1 and M2), we are increasing g_m.

Finally, although not clearly indicated by Eq. (22.30) as it is in Eq. (22.29), a mismatch in the diff-pair's threshold voltage can only be reduced by layout techniques (e.g., using common-centroid layout). Using large devices (larger-width MOSFETs), reduces variation in device characteristics and leads to better matching.

22.1.5 Noise Performance

The diff-amp with noise sources is seen in Fig. 22.20. The noise from M5 and M6 feeds equally into M1 and M2 and thus, because of the current mirror action in M3 and M4, ideally doesn't affect the output voltage. Therefore, we will ignore M5 and M6 in the following discussion.

The diff-amp's output noise power spectral density (PSD) is given by

$$V^2_{onoise}(f) = \left(I^2_{M1}(f) + I^2_{M2}(f) + I^2_{M3}(f) + I^2_{M4}(f)\right) \cdot (r_{o2}||r_{o4})^2 \tag{22.31}$$

The input-referred noise PSD is then

$$V^2_{inoise}(f) = \frac{V^2_{onoise}(f)}{A^2_d} \tag{22.32}$$

Substituting in the difference-mode gain of the diff-amp, Eq. (22.22), we get

$$V^2_{inoise}(f) = \frac{I^2_{M1}(f) + I^2_{M2}(f) + I^2_{M3}(f) + I^2_{M4}(f)}{g^2_m} \tag{22.33}$$

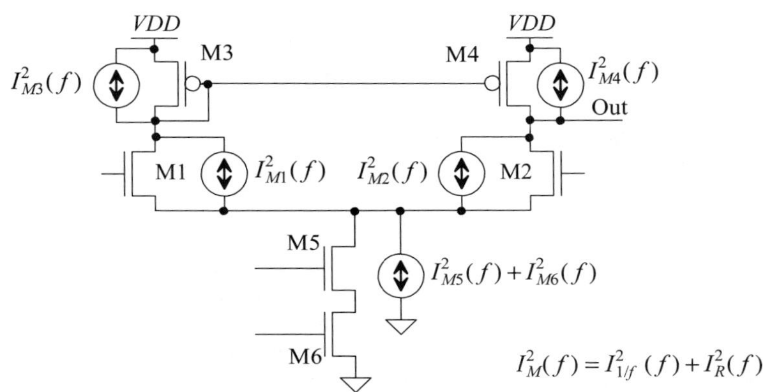

Figure 22.20 Noise model for the diff-amp.

By maximizing the transconductance, g_m, of the diff-pair, we can reduce the input-referred noise. This statement alone is a little misleading. As we saw in Eqs. (9.63)–(9.65), increasing the drain current increases the noise PSD. So to reduce the input-referred noise, we would want to maximize the transconductance by increasing the diff-pair's width (make M1 and M2 wide), not by increasing I_{SS}.

22.1.6 Slew-Rate Limitations

Like the class A amplifiers studied in the last chapter, the basic diff-pair also exhibits slew-rate limitations. As seen in Fig. 22.21, when either M1 or M2 turns off, the total current available to charge a load capacitance is I_{SS}. The slew-rate is then (see also Eq. (21.38))

$$\text{slew rate} = \frac{dV_{out}}{dt} = \frac{I_{SS}}{C_L} \tag{22.34}$$

For high-speed, we need a large bias current. However, this causes excess power dissipation. Once again, to get high-speed (large current drive capability) with low quiescent power dissipation, let's look at a class AB configuration.

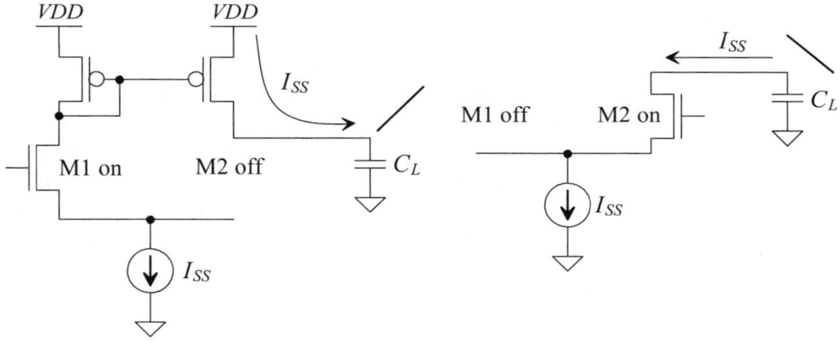

Figure 22.21 Slew-rate limitations in diff-amps.

22.2 The Source Cross-Coupled Pair

The source cross-coupled pair is shown in Fig. 22.22. This configuration is very useful in practical circuit design because it can eliminate slew-rate limitations. A diff-amp built with the source cross-coupled pair can operate in the class AB mode with significant output drive current. Assuming that both inputs to the pair are connected to a (the same) voltage within the pair's common-mode input range and that all NMOS are sized the same and that all PMOS are sized the same, a current I_{SS} flows in all the MOSFETs in the circuit. Note that M11, M21, M31, and M41 are simply behaving as biasing batteries. Their gate-source voltages are mirrored by M1–M4 to set the DC operating current. Using the parameters in Table 9.1 with the gates of M1 and M2 (v_{I1} and v_{I2}) at 3.5 V and I_{SS} of 20 μA, the gate of M3 is at a constant potential that we will label $v_{I2}-V_{bias}$ (or $V_{bias} = V_{GS21} + V_{SG31} = 2.2$ V neglecting body effect).

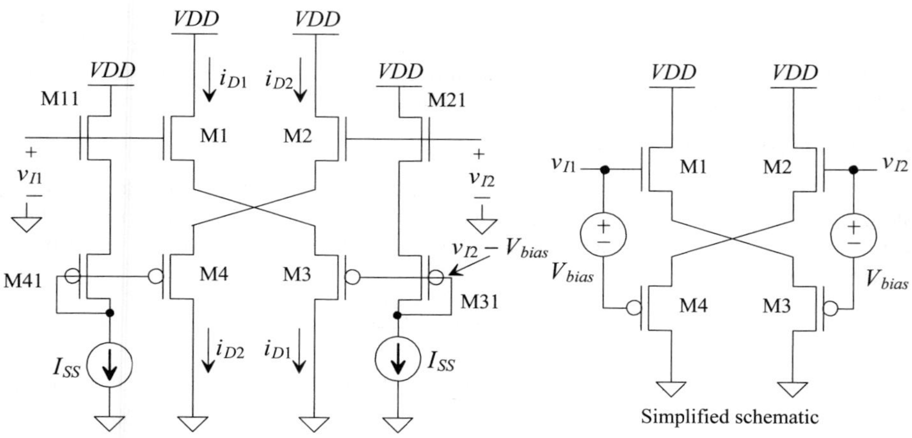

Figure 22.22 Source cross-coupled differential amplifier with NMOS input devices.

Operation of the Diff-Amp

An important characteristic of the source cross-coupled differential amplifier is that as v_{DI} is increased, i_{D1} continues to increase and i_{D2} shuts off, while the opposite is true when v_{DI} decreases. In other words, the amplifier is operating class AB where neither of the output currents is zero as long as their magnitudes remain less than I_{SS}. Looking at the simplified schematic in Fig. 22.22, if v_{I1} increases, then M1 turns on. Because the gate potential of M3 is constant, it turns on as well. In other words, an increase in v_{I1} causes the gate-source voltages of both M1 and M3 to increase (i_{D1} increases). At the same time, the gate potential of M4 increases. This cause both M4 and M2 to shut off (i_{D2} shuts off).

Example 22.8

Simulate the operation of the diff-amp in Fig. 22.22 using the parameters from Table 9.1.

The simulation results are seen in Fig. 22.23. Note that when both inputs are at 3.5 V, the output currents are equal. This figure should be compared to Fig. 22.3.

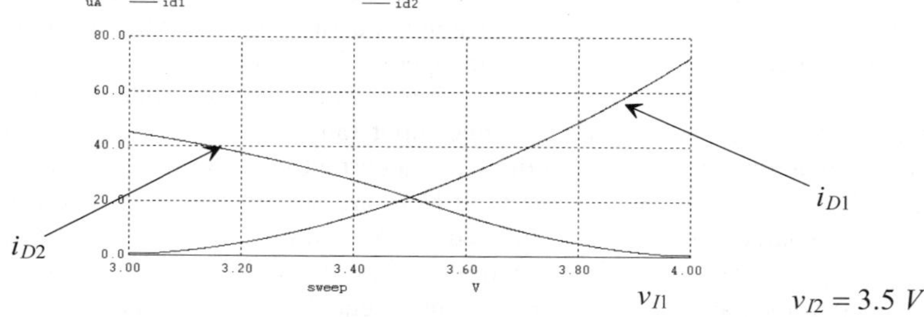

Figure 22.23 The output currents of the diff-amp in Fig. 22.22.

Note how in Fig. 22.3, the output currents flatten out above a maximum input voltage. Here the current continues to increase. ■

After looking at the results of this example, we might wonder: why use such a complicated diff-amp if the output current is only three times more than what we get with the basic diff-amp discussed in the first section? We can always just scale up the current in a basic diff-amp to get more drive. The benefits of the source cross-coupled diff-amp are seen in special-purpose, *low-power* design. In low-power design, the gate source voltages are close to the threshold voltages (and so low power equates to low speed, see Eq. [9.36]). Consider the amplifier and simulation results seen in Fig. 22.24. Here we've increased the widths of the amplifying devices and reduced their bias current (by sizing up the lengths of the biasing current sources as seen in the figure). When the two inputs are the same voltage, the quiescent current that flows in the diff-amp is a few microamps. However, if one input gets a little larger than the other input, the drain-current increases dramatically. In a low-power design, this topology would be useful to quickly charge a capacitor while burning little quiescent power.

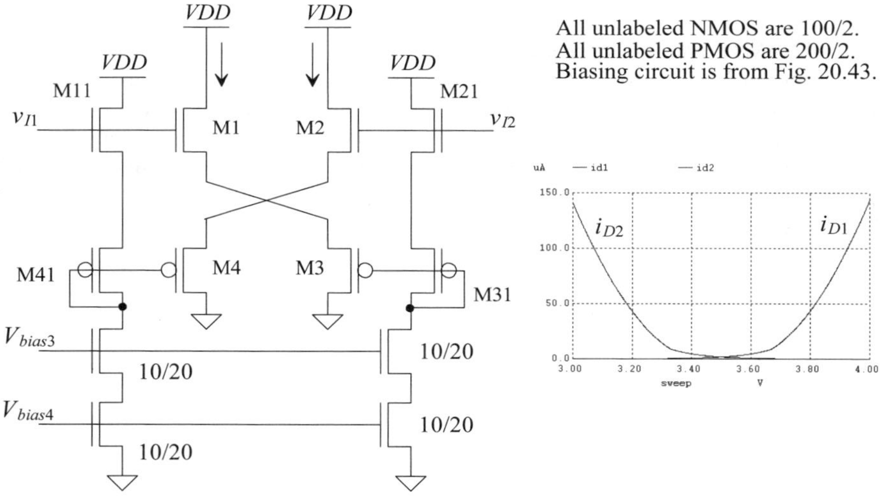

Figure 22.24 Lowering the power in the source cross-coupled diff-amp.

Input Signal Range

The drawback of the source cross-coupled diff-amp is the limited input range. The PMOS version of the source cross-coupled differential amplifier is shown in Fig. 22.25. Normally the version (PMOS or NMOS) of this differential amplifier depends on the input signal range. The NMOS version has the best positive input signal range, while the PMOS version has the best negative input signal range. For centering the input signal range the topology seen in Fig. 22.26 can be used. Again, both an NMOS V_{GS} and a PMOS V_{SG} are used to bias the cross-coupled diff-amp. Note that the biasing devices

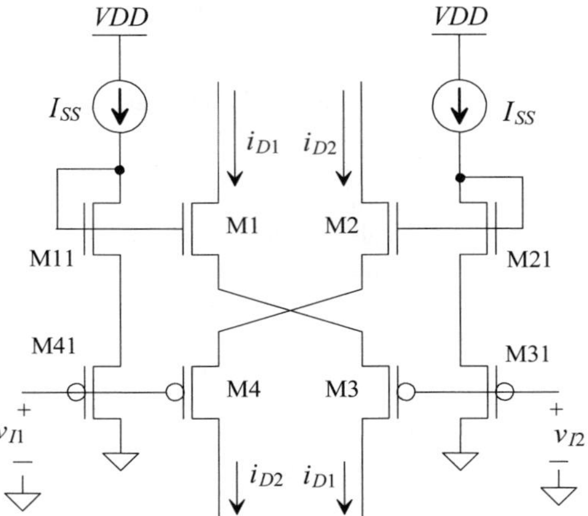

Figure 22.25 Source cross-coupled differential amplifier, PMOS.

(M11, M21, M31, and M41) experience body effect differently than M1–M4. This causes the currents in M1–M4 to be less than the current in the biasing devices. If possible, place M1–M4 and the biasing devices in their own wells to eliminate the body effect. Note that when very low currents are used, the constant current sources driving the gates of M1–M4 can cause slew-rate limitations. To eliminate this problem, capacitors can be placed between the gates of M1/M4 and M2/M3 to make the biasing appear more battery-like.

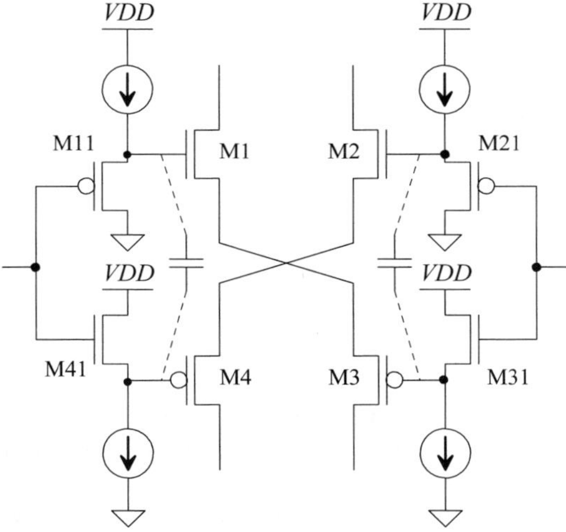

Figure 22.26 Biasing scheme used to shift input common-mode range. Note that the body effect affects the biasing, causing the current in M1–M4 to decrease.

22.2.1 Current Source Load

The source cross-coupled differential amplifier with active loads is shown in Fig. 22.27. Note that we could have also added a pair of active loads in series with the drains of M1 and M4 to obtain a differential output amplifier (two complementary amplifier outputs). Superposition can be used to find the AC small-signal gain from the output back to each input. First, determine the small-signal gain, $\frac{v_o}{v_{i1}}$, with v_{i2} at AC ground. Neglecting the bulk effect, we can see that the voltage gain from the gate of M1 to the source of M1 is simply a common drain amplifier of the form

$$\frac{v_{s1}}{v_{i1}} = \frac{g_{m1}R_{Leq}}{1+g_{m1}R_{Leq}} \text{ V/V} \tag{22.35}$$

where R_{Leq} is simply the load seen by the source of M1. This can be found by using a test source (as seen in Fig. 22.28) to determine the effective impedance seen looking into the source of M3. This results in

$$R_{Leq} = \frac{v_t}{i_t} = \frac{1+\frac{1}{g_{m5}r_{o3}}}{g_{m3}+\frac{1}{r_{o3}}} \approx \frac{1}{g_{m3}} \tag{22.36}$$

Therefore, the voltage gain from v_{i1} to the source of M1 becomes

$$\frac{v_{s1}}{v_{i1}} = \frac{g_{m1}\frac{1}{g_{m3}}}{1+g_{m1}\frac{1}{g_{m3}}} \tag{22.37}$$

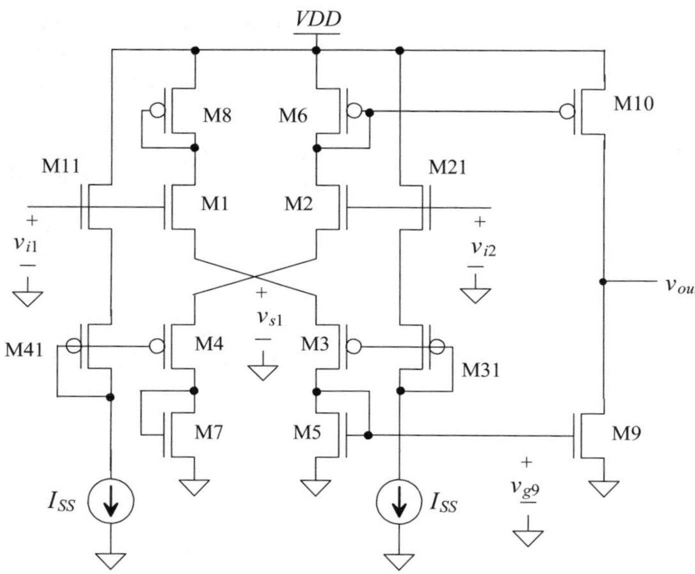

Figure 22.27 Source cross-coupled differential amplifier, with current source loads.

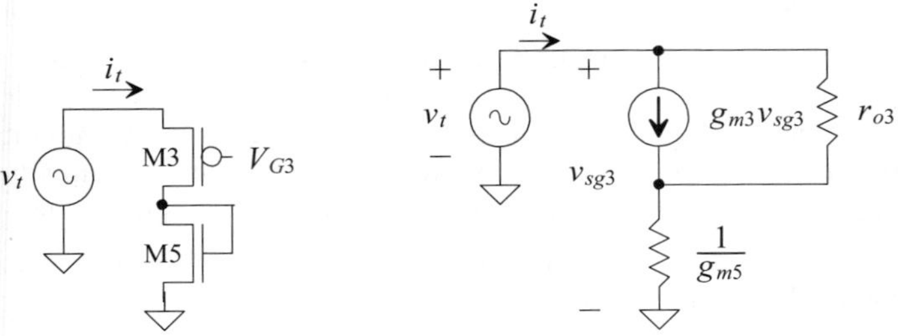

Figure 22.28 (a) Simplified schematic and (b) small-signal model.

The voltage gain from v_{s1} to v_{g9} is the same form as a common gate amplifier. This gain is

$$\frac{v_{g9}}{v_{s1}} = g_{m3}\frac{1}{g_{m5}} \tag{22.38}$$

And finally, the gain from the gate of M9 to the output is simply $-g_{m9}(r_{o9}||r_{o10})$, and

$$\frac{v_o}{v_{i1}} = -\frac{g_{m1}\frac{1}{g_{m5}}}{1+g_{m1}\frac{1}{gm3}}g_{m9}(r_{o9}||r_{o10}) \text{ for } v_{i2}=0 \tag{22.39}$$

Using the same type of analysis, we find that the expression from v_{i2} to the output is

$$\frac{v_o}{v_{i2}} = \frac{g_{m2}\frac{1}{g_{m6}}}{1+g_{m2}\frac{1}{gm4}}g_{m10}(r_{o9}||r_{o10}) \text{ for } v_{i1}=0 \tag{22.40}$$

Superposition can now be used to determine the total effect of each input on the output. Therefore, if M1 and M2 are matched, as are M3 and M4, and if M5 and M6 are designed so that their g_ms are equal as are M7 and M8 (respectively), then the differential gain can be expressed as

$$v_o = (v_{i2}-v_{i1})\frac{2\cdot g_{m1,2}\frac{1}{g_{m5,6}}}{1+g_{m1,2}\frac{1}{gm3,4}}g_{m9,10}(r_{o9}||r_{o10}) \Rightarrow \frac{v_o}{v_{di}} = \frac{2\cdot g_{m1,2}\frac{1}{g_{m5,6}}}{1+g_{m1,2}\frac{1}{gm3,4}}g_{m9,10}(r_{o9}||r_{o10}) \tag{22.41}$$

Input Signal Range

The positive-input, common-mode range (CMR) is determined in exactly the same manner as in the source coupled diff-amp of the first section. The positive CMR is given by

$$v_{CMMAX} \approx VDD - \sqrt{\frac{2I_{SS}}{\beta_6}} = VDD - V_{DS,sat} \tag{22.42}$$

while the negative CMR is given by

$$v_{CMMIN} = \overbrace{\sqrt{\frac{2I_{SS}}{\beta_1}} + V_{THN}}^{V_{GS1}} + \overbrace{\sqrt{\frac{2I_{SS}}{\beta_3}}}^{V_{DS3}} + \overbrace{\sqrt{\frac{2I_{SS}}{\beta_5}} + V_{THN}}^{V_{GS5}} \tag{22.43}$$

An alternative current source (current mirror) load configuration is shown in Fig 22.29. The gain of this configuration is given by

$$A_d = \frac{v_{out}}{v_{i1} - v_{i2}} = 2 \cdot \frac{r_{o2}||r_{o6}}{\frac{1}{g_{m1}} + \frac{1}{g_{m3}}} \tag{22.44}$$

assuming $g_{m1} = g_{m2}$ and $g_{m3} = g_{m4}$.

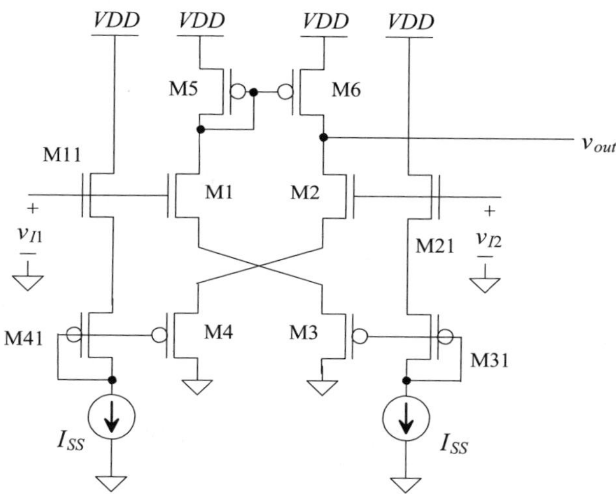

Figure 22.29 Alternative current source load configuration.

22.3 Cascode Loads (The Telescopic Diff-Amp)

The differential amplifier with current source load of Fig. 22.6 gave a voltage gain of $g_m(r_{o2}||r_{o4})$. For many applications, this gain may be too low. Using a cascode current source load in place of M3/M4, results in a gain of approximately $g_m r_{o2}$ (the output resistance of the load becomes much larger than the resistance looking into the drain of M2). This modest improvement in gain can be increased by also cascoding M1 and M2. Figure 22.30 shows the resulting circuit configuration (sometimes called a *telescopic diff-amp*). The MOSFET M6 is selected so that its V_{GS} keeps M1/M2 and MC1/MC2 in the saturation region. The gain of this configuration is given by

$$A_d = g_{m1} \cdot (R_{into\,DC2}||R_{into\,DC4}) \tag{22.45}$$

The resistance looking into the drain of MC2, assuming M2 and MC2 are the same size, is given by

$$R_{into\,DC2} \approx g_{m2} \cdot r_{o2}^2 \tag{22.46}$$

and the resistance looking into the drain of MC4, again assuming MC4 and M4 are the same size, is given by

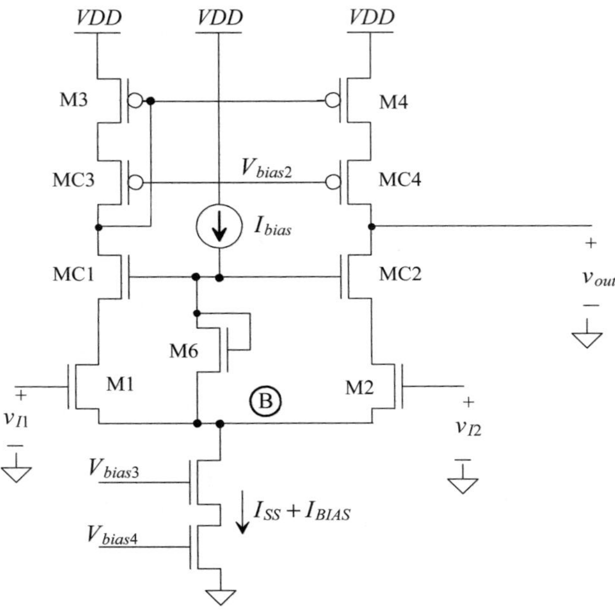

Figure 22.30 Cascode differential amplifier.

$$R_{into\ DC4} \approx g_{m4} \cdot r_{o4}^2 \tag{22.47}$$

The gain of the cascode differential amplifier may be written as

$$A_d = g_{m1}\left(g_{m2}r_{o2}^2 || g_{m4}r_{o4}^2\right) \tag{22.48}$$

The main drawback of using the cascode differential amplifier is the reduction in positive common-mode range. In the negative direction, V_{CMMIN} is identical to the source-coupled diff-amp discussed in Sec. 22.1. The positive common-mode signal is limited by the additional circuitry required by the cascode loads. Notice that V_{CMMAX} in this case is limited by the amount of voltage necessary to keep M1, MC1, MC3, and M3 in their respective saturation regions.

An interesting issue concerning this circuit is the DC feedback through M6. Because the current through M6 is constant, V_{GS6} will be constant. Normally, an increase in the common-mode voltage on the gates of M1 and M2 causes their drain voltages to decrease. However, because the currents through M1 and M2 are constant, the common source node, marked as node B in Fig. 22.30, increases with the increasing input voltage. This causes the gate potential of MC1 and MC2 to increase. The result is that the drain voltages of M1 and M2 go up (since the gate-source voltages of MC1 and MC2 are constant). This feedback action attempts to keep M1 and M2 from going into nonsaturation. The maximum common-mode voltage, V_{CMMAX}, will be limited by the output voltage and by MC1/MC2 going into nonsaturation.

Example 22.9
Determine the minimum and maximum input common-mode voltage for the diff-amp seen in Fig. 22.31. Note that the voltage across M6 is $V_{GS} + V_{DS,sat}$, where V_{GS} is the gate-source voltage of the other NMOS devices in this figure (see Fig. 20.32 and the associated discussion).

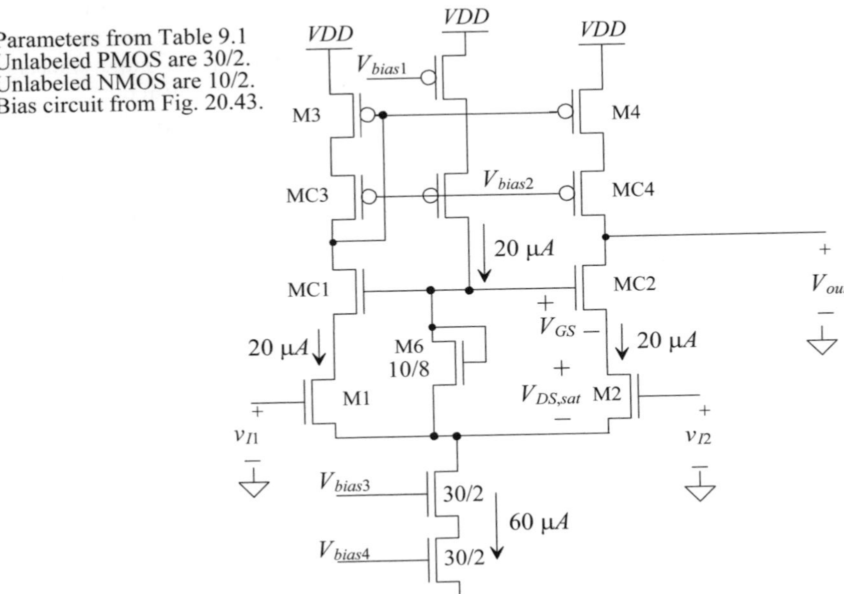

Figure 22.31 Cascode differential amplifier.

The minimum input common-mode voltage is given by

$$V_{CMMIN} = V_{GS1,2} + 2V_{DS,sat} = 1.05 + 0.5 = 1.55\ V$$

As mentioned a moment ago, MC1 and MC2 must be kept in the saturation region. If we assume that the drains of MC1 and MC2 are at $VDD - V_{SG}$ (= 3.85 V), then when MC1 and MC2 are on the verge of trioding (their drain-source voltages are $V_{DS,sat}$), the drains of M1 and M2 are at

$$V_{D1,2} = VDD - V_{SG} - V_{DS,sat} = 3.6\ V$$

We know for an NMOS device (M1 or M2) to be in saturation

$$V_D \geq V_G - V_{THN} \text{ and so } V_{D1,2} \geq V_{CMMAX} - V_{THN}$$

giving a $V_{CMMAX} = 4.4\ V$. In terms of an arbitrary output voltage, we can write

$$V_{CMMAX} = V_{out} - V_{DS,sat} + V_{THN}$$

so if V_{out} is 2.45 V, then $V_{CMMAX} = 3\ V$. ■

22.4 Wide-Swing Differential Amplifiers

An increasingly important concern with the advent of low-voltage design is the need for improved common-mode range. A technique for extending the allowable input swing of a diff-amp is to use two complementary diff-amp stages in parallel, as shown in Fig. 22.32. To understand the operation of this circuit, consider the case when the DC component of the input signal, v_{in}, is such that both diff-amps are on (and the AC component is small compared to the DC). The current through the diff-pairs M1/M2 and M9/M10 is I, while the current through the summing MOSFETs M4 and M12 is $2I$. If M5 is the same size as M4 and if M7 is the same size as M12, then a current $2I$ flows in the output transistors. The small-signal voltage gain, assuming $g_{m1} = g_{m2}$ and $g_{m9} = g_{m10}$, is given by

$$A_v = (g_{m1} + g_{m9})[r_{o7}(2I) || r_{o5}(2I)] = \frac{(g_{m1} + g_{m9})}{\lambda_7 2I + \lambda_5 2I} = \frac{\sqrt{2\beta_1 I} + \sqrt{2\beta_9 I}}{2I(\lambda_7 + \lambda_5)} \tag{22.49}$$

If the input is such that the p-channel diff-amp is on and the n-channel diff-amp is off, then a current I flows in the summing MOSFETs M4 and M12 and zero current flows in M1, M2, M3, and M6. The current that flows in M5 and M7 is now I. The small-signal voltage gain is

$$A_v = g_{m9}[r_{o7}(I) || r_{o5}(I)] = \frac{\sqrt{2\beta_9 I}}{I(\lambda_7 + \lambda_5)} \tag{22.50}$$

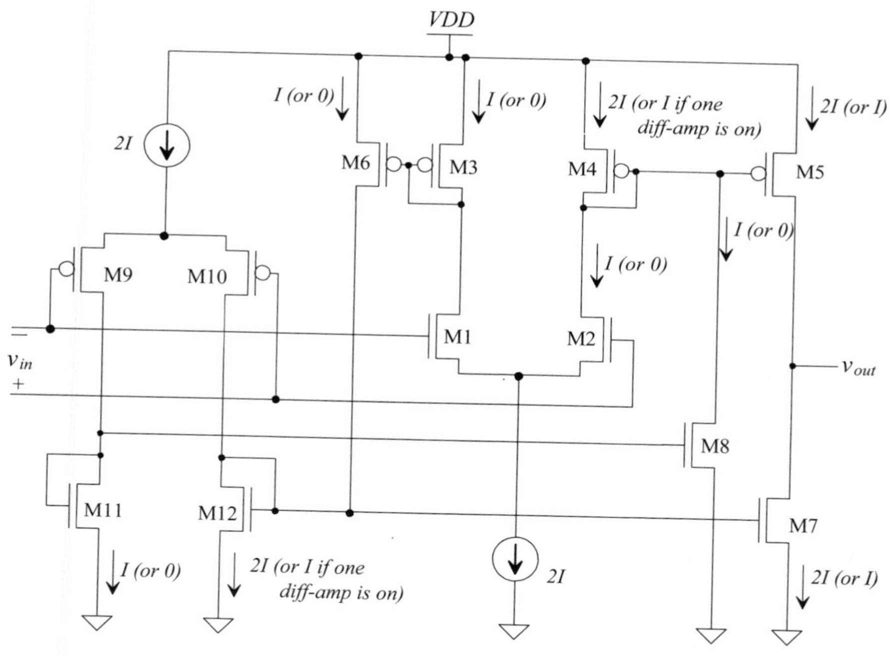

Figure 22.32 Two parallel differential amplifiers used to increase input swing.

The small-signal gain when the p-channel diff-amp is off and the n-channel diff-amp on is given by

$$A_v = g_{m1}[r_{o7}(I)||r_{o5}(I)] = \frac{\sqrt{2\beta_1 I}}{I(\lambda_7 + \lambda_5)} \tag{22.51}$$

It is desirable to transition smoothly from having a single diff-amp on to having both diff-amps on. If we require

$$\beta_1 = \beta_9 = \beta \Rightarrow G_m = \sqrt{2\beta I} \tag{22.52}$$

then Eqs. (22.49) through (22.51) can be rewritten as

$$A_v = G_m \cdot [r_{o7}(I)||r_{o5}(I)] \tag{22.53}$$

For low distortion, a constant gain is important. Also, for a stable amplifier, it is important that the amplifier can be compensated to ensure stability. An amplifier whose gain depends on the input signal amplitude may be more challenging to compensate for.

22.4.1 Current Differential Amplifier

Another wide-swing, differential amplifier is the current differencing amplifier. The current diff-amp is shown schematically in Fig. 22.33. In the following discussion, let's assume that M1 through M4 are the same size. If both i_1 and i_2 are zero, then a current I_{SS} flows in all MOSFETs in the circuit. Now assume that i_1 is increased above zero, but less than I_{SS}. This causes the current in both M1 and M2 to increase. As a result, the drain current in M3 decreases to keep $i_{D2} + i_{D3} = 2I_{SS}$. The decrease in M3 drain current is mirrored in M4, forcing the current i_1 out of the diff-amp. A similar argument can be made for increasing i_2; that is, it causes the output of the differential amplifier to sink a current equal to i_2. The sizes of M1–M4 can be ratioed to give the diff-amp a gain or to scale the input currents.

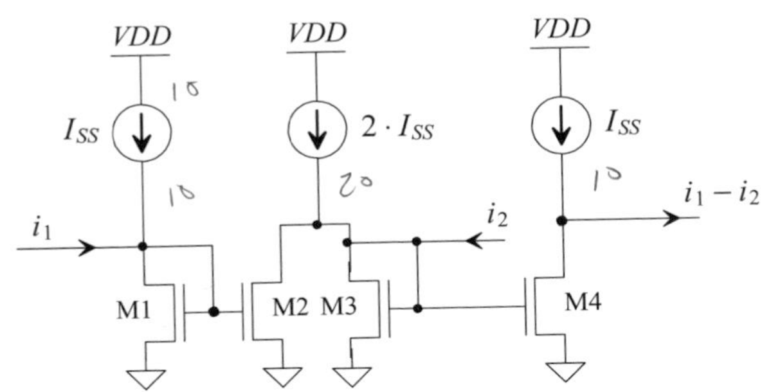

Figure 22.33 Current differential amplifier.

The input impedance of the current differential amplifier is simply the small-signal resistance of a diode-connected MOSFET, or

$$R_{in} = \frac{1}{g_m} \tag{22.54}$$

This configuration finds applications in both low-power and high-speed circuit design.

22.4.2 Constant Transconductance Diff-Amp

A rail-to-rail differential amplifier made using n- and p-channel diff-amps is shown in Fig. 22.34. It is desirable, for good distortion and proper compensation, that the overall transconductance of the diff-amp remain constant independent of the region of operation, that is, operation with both n- and p-channel diff-amps on or only a single diff-amp on. A constant g_m is guaranteed over the input range if

$$g_m = g_{mn} + g_{mp} = \sqrt{2\beta_n I_n} + \sqrt{2\beta_p I_p} = \text{constant} \tag{22.55}$$

where g_{mn} and g_{mp} are the transconductances of the n- and p-channel diff-amps and g_m is the overall transconductance of the input stage. Since β_n and β_p are constant and can be made equal, Eq. (22.55) can be rewritten as (assuming both diff-amps are operating),

$$\sqrt{I_n} + \sqrt{I_p} = \text{constant} \tag{22.56}$$

This equation always holds if both differential amplifiers are on. The problem with nonconstant transconductance occurs if only one diff-amp is on. If, for example, the common-mode input (common input voltage on the + and – inputs of the differential amplifier) is large enough to shut the p-channel diff-amp off, then $I_p = 0$ and the transconductance of the overall input diff-amp changes.

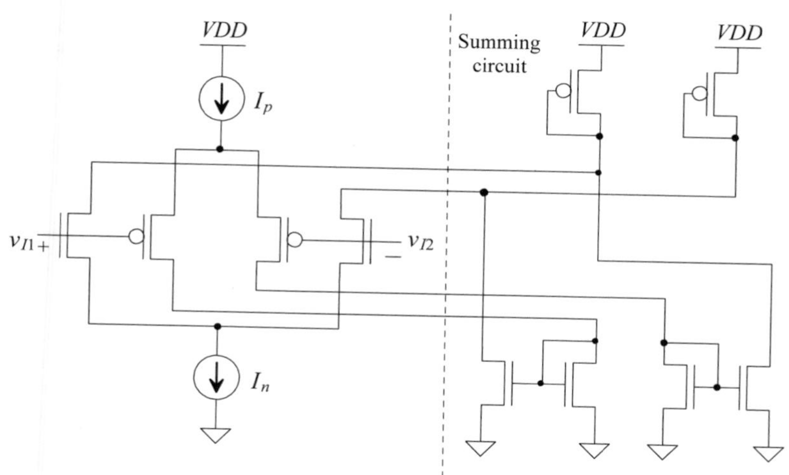

Figure 22.34 Rail-to-rail differential amplifier.

The solution to the problem of non-constant g_m begins by making

$$I_n = I_p = I_o \tag{22.57}$$

When both diff-amps are on (keeping in mind that we have already set the requirement that $\beta_n = \beta_p$), Eq. (22.56) reduces to

$$2\sqrt{I_o} = \text{constant} \tag{22.58}$$

If we add a current of $3I_o$ to I_n (or to I_p) when the p-channel diff-pair (n-channel diff-pair) is off, the transconductance of the pair is constant, or

$$\overbrace{2\sqrt{I_o}}^{\text{both on}} = \overbrace{\sqrt{3I_o+I_n}}^{\text{n diff-amp on}} = \overbrace{\sqrt{3I_o+I_p}}^{\text{p diff-amp on}} \text{ if } I_o = I_p = I_n \tag{22.59}$$

An example of a constant-g_m, rail-to-rail input stage is shown in Fig. 22.35. The summing circuit of Fig. 22.34 is not shown in this figure. MOSFETs M1–M4 make up the p- and n-channel diff-pairs, while MP1 and MN1 source the constant current I_o when both diff-pairs are on. (MP1 and MN1 represent the constant current sources shown in Fig. 22.28.) When both diff-pairs are on, the current out of each current diff-amp is approximately 0. (The drain current of M6 and M7 is approximately 0.) If the common-mode input voltage becomes large enough to shut the p-channel diff-pair off, then MOSFETs MS1 and MS2 are off as well. This causes the current in M5 to become I_o. The current in M6 mirrors the current in M5. Since M6 is three times larger than M5, the current in M6 becomes $3I_o$.

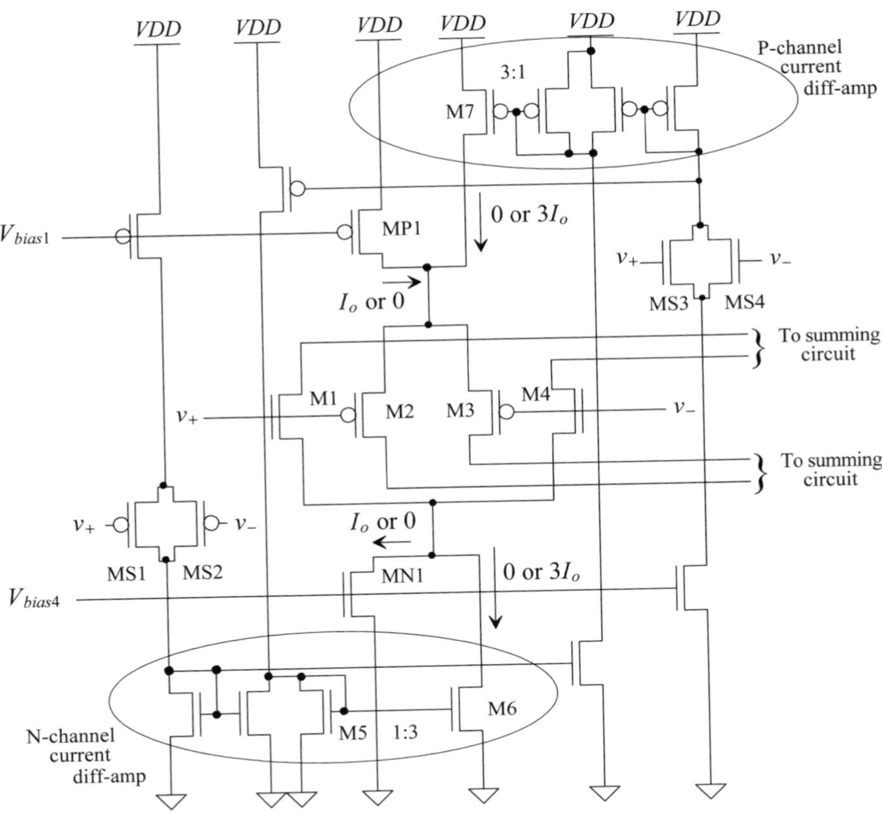

Figure 22.35 A wide-swing diff-amp with constant gm.

Discussion

In general, the constant-g_m diff-amp is used in an op-amp to prevent overcompensation of the op-amp and to avoid distortion when the op-amp is used with large signals where the diff-amps, on the input of the op-amp, are turning on and off with variations in the input signal. Several practical problems exist with these uses of the constant-g_m diff-amp. Because the value of g_m may vary with shifts in the process, we may still overcompensate an op-amp design. Using the constant-g_m stage to avoid overcompensation is useful, but since we may overcompensate (or undercompensate) the op-amp with process variations, the added complexity and power draw may not justify the additional circuitry. Using the constant-g_m stage to avoid distortion is based on keeping the unity gain frequency of an op-amp, f_{un}, a constant independent of the common-mode input voltage. In practice, mismatches (threshold voltage and geometry) in the input diff-pair and changes in DC biasing conditions, resulting from the diff-amps turning off and on, can cause distortion. In some cases, this distortion can be worse than the distortion resulting from the nonconstant f_{un} (which we get with the nonconstant g_m diff-amp). The distortion resulting from mismatches can be modeled as an offset-voltage that is dependent on the input common-mode voltage. Since the DC currents sourced or sunk from a constant-g_m diff-amp are not constant, the DC operating point of the circuit summing the currents changes with the input signal's amplitude. The result is a change in the low-frequency gain and added distortion.

ADDITIONAL READING

[1] P. E. Allen and D. R. Holberg, *CMOS Analog Circuit Design*, 2nd ed., Oxford University Press, 2002. ISBN 0-19-511644-5.

[2] P. R. Gray, P. J. Hurst, and S. H. Lewis, and R. G. Meyer, *Analysis and Design of Analog Integrated Circuits,* 4th ed., Wiley, 2001. ISBN 0-47-132168-0.

[3] R. Castello and P. R. Gray, "A High-Performance Micropower Switched-Capacitor Filter," *IEEE Journal of Solid State Circuits,* vol. SC-20, no. 6, pp. 1122–1132, December, 1985.

[4] E. Seevinck and R. F. Wassenaar, "A Versatile CMOS Linear Transconductor/Square-Law Function Circuit," *IEEE Journal of Solid State Circuits,* vol. SC-22, no. 3, pp. 366–377, June 1987.

[5] M. Steyaert and W. Sansen, "A High-Dynamic-Range CMOS Op Amp with Low-Distortion Output Structure," *IEEE Journal of Solid State Circuits,* vol. SC-22, no. 6, pp. 1204–1207, December 1987.

[6] T. S. Fiez, H. C. Yang, J. J. Yang, C. Yu, and D. J. Allstot, "A Family of High-Swing CMOS Operational Amplifiers," *IEEE Journal of Solid State Circuits,* vol. 22, no. 6, pp. 1683–1687, December 1989.

[7] J. H. Botma, R. F. Wassenaar, and R. J. Wiegerink, "A Low-Voltage CMOS Op Amp with a Rail-to-Rail Constant-g_m Input Stage and a Class AB Rail-to-Rail Output Stage," *Proceedings of the 1993 IEEE ISCAS,* p. 1314.

[8] A. L. Coban, P. E. Allen, and X. Shi, "Low-Voltage Analog IC Design in CMOS Technology," *IEEE Transactions on Circuits and Systems,* vol. 42, no. 11, November 1995.

PROBLEMS

22.1 Determine the drain current of M1 as a function of the input voltage, $v_{I1} - v_{I2}$, for the diff-amp shown in Fig. 22.36. Neglect body effect. What does your derivation give for i_{D1} when v_{I1} is much larger than v_{I2}?

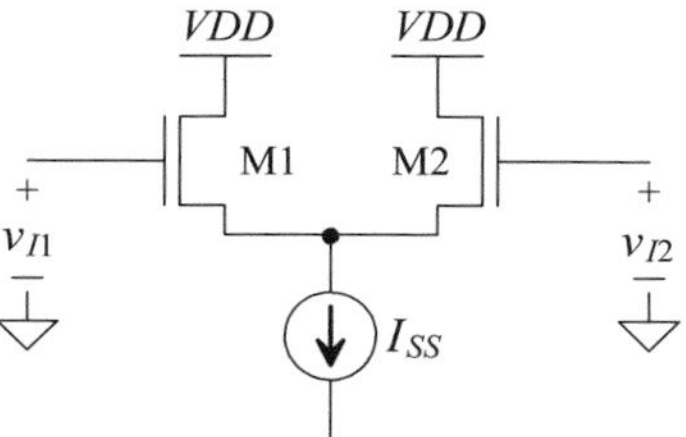

Figure 22.36 Diff-amp used in Problem 22.1

22.2 Repeat Ex. 22.1 if the widths of M1 and M2 are increased from 10 to 100. Determine the transconductance of the diff-amp. Write i_{d2} as a product of g_m (= $g_{m1} = g_{m2}$) and v_{I1} (with v_{I2} = AC ground), v_{I2} (with v_{I1} = AC ground), and $v_{I1} - v_{I2}$.

22.3 Determine the maximum and minimum common mode voltages for the PMOS version of the diff-amp seen in Fig. 22.4.

22.4 Determine the AC currents flowing in the circuit of Fig. 22.5 if the gate of M1 is grounded and the gate of M2 is an AC signal of 1 mV. Verify your answers with an AC SPICE simulation.

22.5 Determine the small-signal gain and the input common-mode-range (CMR) for the diff-amps shown in Fig. 22.37. Verify your answers with SPICE.

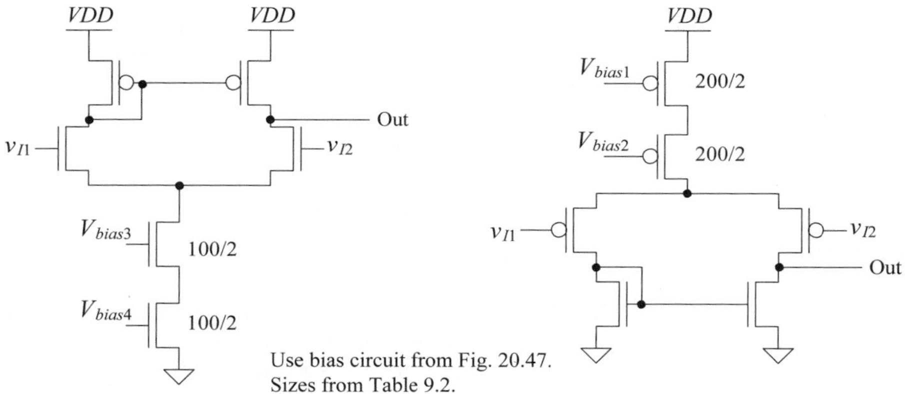

Figure 22.37 Diff-amps for Problem 22.5.

22.6 Show that the capacitance on the sources of M1/M2 in Ex. 22.6 causes the *CMRR* to roll off quicker with increasing frequency. (Add a capacitance to the sources of M1/M2 in a simulation and show that *CMRR* decreases at a lower frequency.)

22.7 Estimate the slew-rate limitations in charging and discharging a 1 pF capacitor tied to the outputs of the diff-amps shown in Fig. 22.37. Verify with SPICE.

22.8 For the n-channel diff-pair shown in Fig. 22.38, show that the following relationships are valid if the body effect is included in the analysis of the transconductance.

$$i_{d1} = \frac{g_m}{2}\left[v_{i1}\left(2 - \frac{g_m}{g_m + g_{mb}}\right) - v_{i2} \cdot \frac{g_m}{g_m + g_{mb}}\right] - \frac{g_m \cdot g_{mb}}{g_m + g_{mb}} \cdot \frac{[v_{i1} + v_{i2}]}{2}$$

and

$$i_{d2} = \frac{g_m}{2}\left[v_{i2}\left(2 - \frac{g_m}{g_m + g_{mb}}\right) - v_{i1} \cdot \frac{g_m}{g_m + g_{mb}}\right] - \frac{g_m \cdot g_{mb}}{g_m + g_{mb}} \cdot \frac{[v_{i1} + v_{i2}]}{2}$$

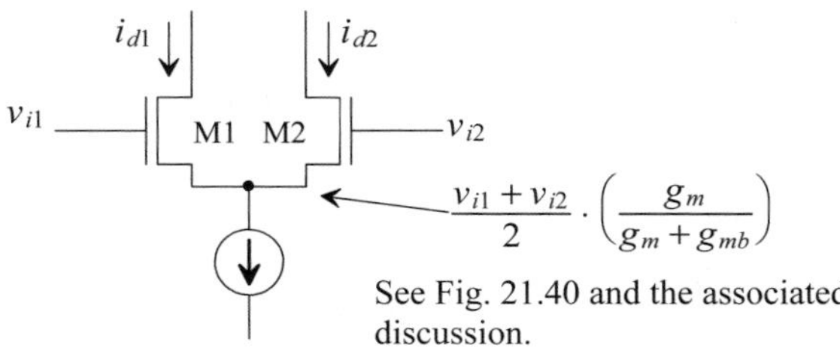

Figure 22.38 How body effect changes the AC behavior of the diff-amp.

22.9 The diff-amp configuration shown in Fig. 22.39 is useful in situations where a truly differential output signal is needed. Determine the following: (a) the transconductance of the diff-amp, (b) the AC small-signal drain currents of all MOSFETs in terms of the input voltages and g_{mn} (the transconductance of an n-channel MOSFET), and (c) the small-signal voltage gain, $(v_{O+} - v_{O-})/(v_{I+} - v_{I-})$. Verify your answers with SPICE.

22.10 Using the diff-amp topology in Fig. 22.24 and a current mirror load, show how quickly (or slowly) the diff-amp can drive a 1 pF load. Does this diff-amp exhibit slew-rate limitations? Verify your answers with SPICE.

22.11 Using the topology seen in Fig. 22.26 with the long-channel process (Table 9.1) without body effect (all MOSFET's bodies tied to their respective sources), show that the currents in M1–M4 match the biasing currents used in the source followers. Show how the currents become mismatched when the bodies of the NMOS are tied to ground and the bodies of PMOS are tied to *VDD*.

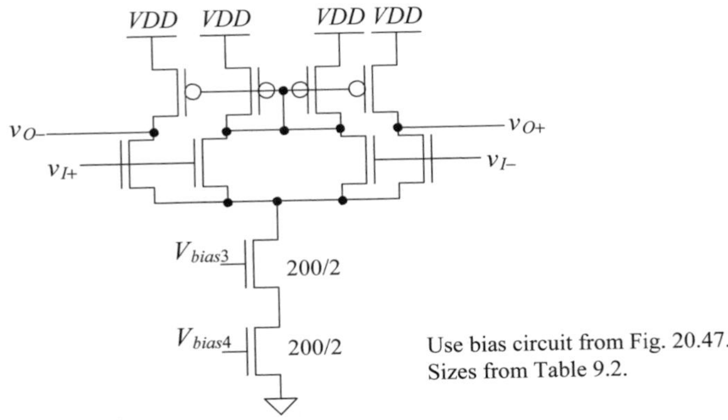

Figure 22.39 Fully differential diff-amp used in Problem 22.9.

22.12 Simulate the operation of the diff-amp seen in Fig. 22.31. Use a common-mode voltage of 2.5 V and an AC input voltage of 100 μV at 1 kHz. Compare the simulation results to your hand calculations.

22.13 Using SPICE with current sources for inputs, show the operation of the diff-amp in Fig. 22.33.

22.14 The circuit seen in Fig. 22.40 is an another example of a circuit that sums the currents from NMOS and PMOS diff-amps (so the input common-mode range can extend from beyond the power supply rails). Describe and simulate the operation of this circuit using the short-channel parameters in Table 9.2 and the bias circuit from Fig. 20.47.

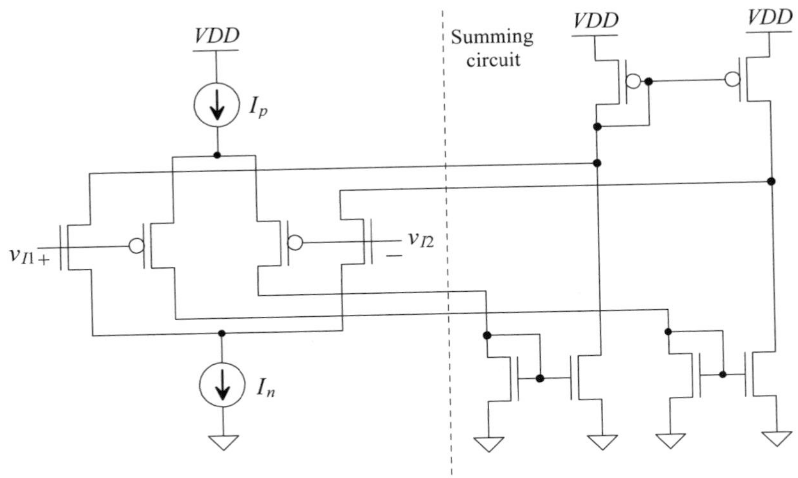

Figure 22.40 Summing circuit for Problem 22.14.

22.15 Using symbolic analysis, determine the small-signal gain of the self-biased diff-amp seen in Fig. 22.41 (see also Fig. 18.17).

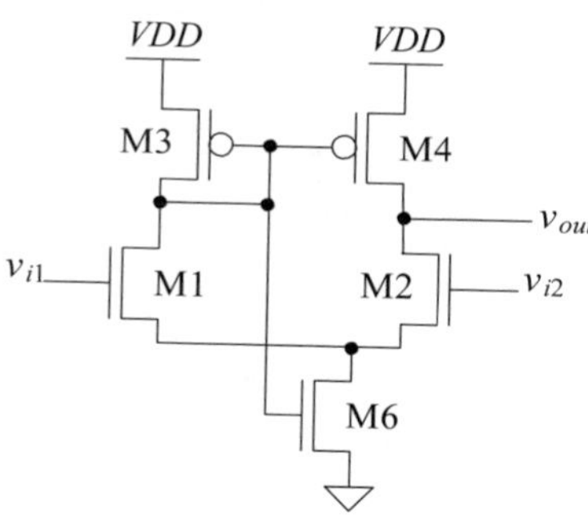

Figure 22.41 A self-biased NMOS diff-amp.

Chapter

23

Voltage References

A general-use (ideal) voltage reference is a circuit used to generate a fixed voltage, V_{REF}, that is independent of the power supply voltage *VDD* (where $V_{REF} < VDD$), temperature, and process variations. In other words, the ideal reference voltage is independent of PVT. In some cases, we want to design a reference that varies with temperature. For example, if V_{REF} increases with temperature, Fig. 23.1a, we say that the reference voltage is *proportional to absolute temperature* or PTAT. If the reference voltage decreases with increasing temperature, Fig. 23.1b, the reference is said to be *complementary to absolute temperature* or CTAT. The PTAT and CTAT references can be used to design a voltage reference that changes very little with temperature called a *bandgap* reference. Unfortunately, the generation of PTAT and CTAT reference voltages requires using parasitic diodes. In a CMOS process, the electrical characteristics of the parasitic pn junctions are not monitored and controlled (like, say, the threshold voltage) during manufacturing. Therefore, if possible, the implementation of a reference voltage using MOSFET-resistor circuits is desirable.

This chapter is split into two sections. The first section covers the design of voltage references using MOSFETs and resistors, while the second section covers the design of voltage references using parasitic diodes. The first section also offers some overview material common to all voltage references.

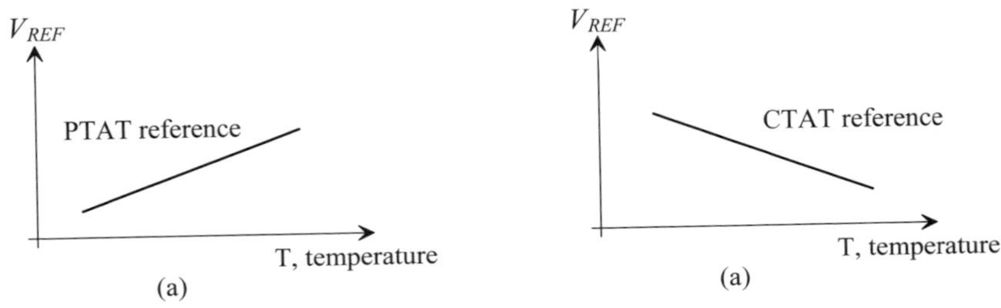

Figure 23.1 (a) PTAT and (b) CTAT voltage references.

23.1 MOSFET-Resistor Voltage References

We can derive reference voltages from the power supplies using the resistor and the MOSFET, as seen in Fig. 23.2. The voltage divider formed with two resistors has the advantage of simplicity, temperature insensitivity (as was shown in Ch. 5), and process insensitivity; that is, changes in the sheet resistance have no effect on the voltage division. The main problem with this circuit is that in order to reduce the power dissipation (i.e., the current through the resistors), the resistors must be made large. Since large resistors require a large area on the die, this voltage divider may not be practical in many cases. One situation where we will use this simple voltage divider is in generating a voltage halfway between *VDD* and ground, *VDD*/2, (sometimes called the *common-mode* voltage of an analog circuit or system).

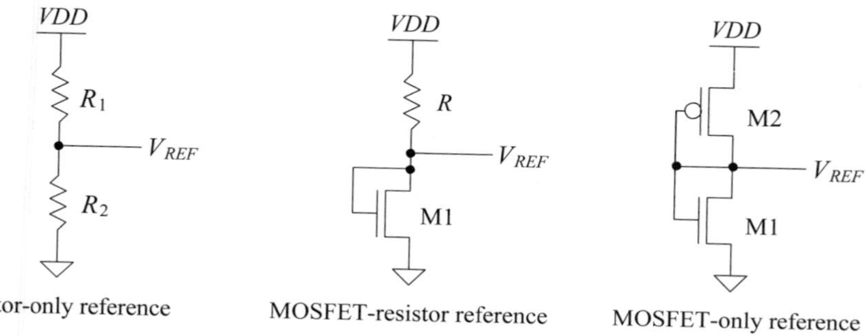

Figure 23.2 Voltage dividers implemented in CMOS.

The voltage divider formed between the resistor and the MOSFET can be recognized as the same circuit we used for a bias in the current mirror back in Ch. 20 (see Fig. 20.2 and Eqs. (20.26)–(20.31)). The final reference, a voltage divider between NMOS and PMOS devices, has the advantage that the layout can be small (see Fig. 20.13 and Eqs. (20.12)–(20.19)). In the following two subsections, we analyze the behavior of these last two voltage dividers.

23.1.1 The Resistor-MOSFET Divider

The reference voltage used in the resistor-MOSFET divider is equal to the V_{GS} of the MOSFET. We can write for this circuit

$$I_D = \frac{VDD - V_{REF}}{R} = \frac{\beta_1}{2}(V_{REF} - V_{THN})^2 \tag{23.1}$$

or

$$V_{REF} = V_{THN} + \sqrt{\frac{2I_D}{\beta_1}} = V_{THN} + \sqrt{\frac{2(VDD - V_{REF})}{R \cdot \beta_1}} \tag{23.2}$$

If V_{REF} is designed so that it is close to V_{THN}, then the reference voltage will be insensitive to changes in *VDD* and its temperature behavior will follow the threshold voltage (see

Eqs. (9.43)–(9.48)). However, for the general case, the temperature coefficient of the resistor-MOSFET voltage divider is determined using

$$TCV_{REF} = \frac{1}{V_{REF}} \cdot \frac{\partial V_{REF}}{\partial T} \tag{23.3}$$

and if we assume $VDD >> V_{REF}$, then

$$TCV_{REF} = \frac{1}{V_{REF}}\left[V_{THN} \cdot TCV_{THN} - \frac{1}{2}\sqrt{\frac{2L_1}{W_1} \cdot \frac{VDD}{R \cdot KP(T)}} \cdot \left[\frac{1}{R}\frac{\partial R}{\partial T} - \frac{1.5}{T}\right]\right] \tag{23.4}$$

Example 23.1
Estimate the temperature performance of the resistor-MOSFET voltage references seen in Fig. 23.3. Assume that the temperature coefficient of the resistor is 2,000 *ppm/C*, the long-channel process is used where $KP_n = 120\ \mu A/V^2$, and the nominal *VDD* is 5 V. Use simulations to verify the answers. Also show how the reference voltages change with *VDD*.

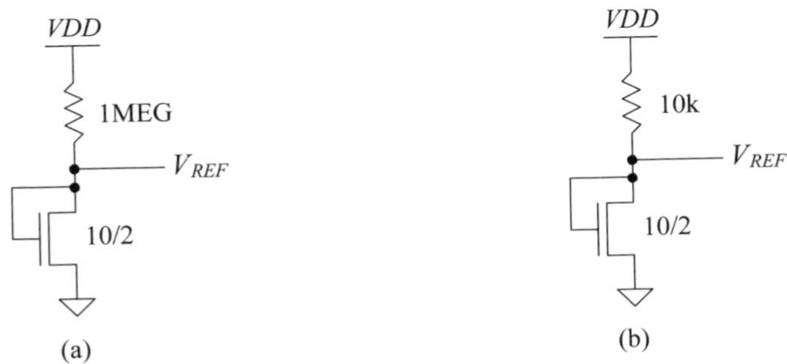

Figure 23.3 Resistor-MOSFET references used in Ex. 23.1

For the reference made using the 1MEG resistor, Fig. 23.3a, the current that flows in the circuit is

$$I = \frac{VDD - V_{REF}}{10^6} = \frac{KP_n}{2} \cdot \frac{W}{L}(V_{REF} - V_{THN})^2$$

which can be solved to determine I is around 4 μA and

$$V_{REF} = V_{GS} \approx 900\ mV$$

From Sec. 9.1.3 the rate the threshold voltage changes with temperature is

$$\frac{\partial V_{THN}}{\partial T} \approx -1\ mV/C° \approx \frac{\partial V_{REF}}{\partial T}$$

The temperature coefficient of the reference voltage is

$$TCV_{REF} = \frac{1}{V_{REF}}\frac{\partial V_{REF}}{\partial T} = \frac{-0.001}{0.9} = -1,111\ ppm/C° \tag{23.5}$$

and thus

$$V_{REF}(T) = V_{REF} \cdot (1 + TCV_{REF}(T - T_0)) = 0.9 \cdot (1 - 0.00111(T - 25))) \tag{23.6}$$

where we assume that the threshold voltage was measured at 25 °C. The simulation results are seen in Fig. 23.4. Note that at higher temperatures the threshold voltage decreases and thus so does V_{REF}. Further note that a change in temperature of 25 °C corresponds to a change in the reference voltage of 31.25 mV $(= 25 \cdot 0.00125)$.

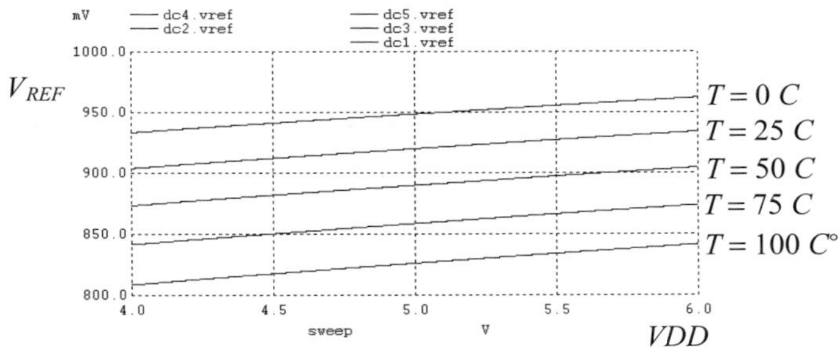

Figure 23.4 How the reference voltage changes with VDD and temperature for the reference in Fig. 23.3a.

To estimate the temperature behavior of the circuit in Fig. 23.3b, we must first determine the value of V_{REF}. Using Eq. (23.2), we can write

$$V_{REF} = V_{THN} + \sqrt{\frac{2(VDD - V_{REF})}{R \cdot \beta_1}} = 0.8 + \sqrt{\frac{2(5 - V_{REF})}{10k \cdot \frac{10}{2} \cdot 120\mu A/V^2}}$$

After a few iterations, we can determine $V_{REF} \approx 1.85\ V$. The temperature coefficient is calculated using Eq. (23.4)

$$TCV_{REF} = \frac{1}{1.85} \cdot \left[-1\ mV/C^\circ - \frac{1}{2}\sqrt{\frac{2 \cdot 2}{10} \cdot \frac{5}{10k \cdot 120\ \mu A/V^2}} \cdot \left[0.002 - \frac{1.5}{300} \right] \right]$$

which evaluates to $TCV_{REF} = 500\ ppm/C$. The simulation results are seen in Fig. 23.5. Note how, when comparing Figs. 23.4 and 23.5, one reference circuit performs well with regard to temperature (23.3b), while the other reference circuit (23.3a) performs well with regard to VDD variations. ■

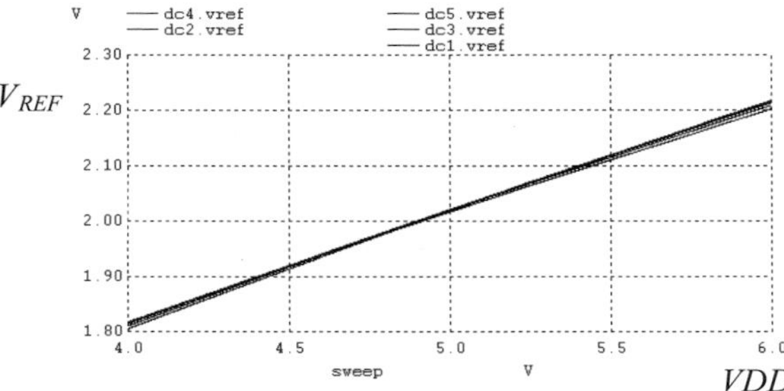

Figure 23.5 How the reference voltage changes with VDD and temperature for the reference in Fig. 23.3b.

A modification to the basic resistor-MOSFET divider is shown in Fig. 23.6. The reference voltage in this circuit is given by

$$V_{REF} = V_{GS}\left(\frac{R_1}{R_2} + 1\right) \tag{23.7}$$

noting the ratio of the resistors (and so their temperature behavior or sheet resistance shift with process variations) doesn't affect the reference voltage. When V_{GS} is designed to be approximately V_{THN}, the resulting reference circuit is said to be a threshold-voltage multiplier reference circuit.

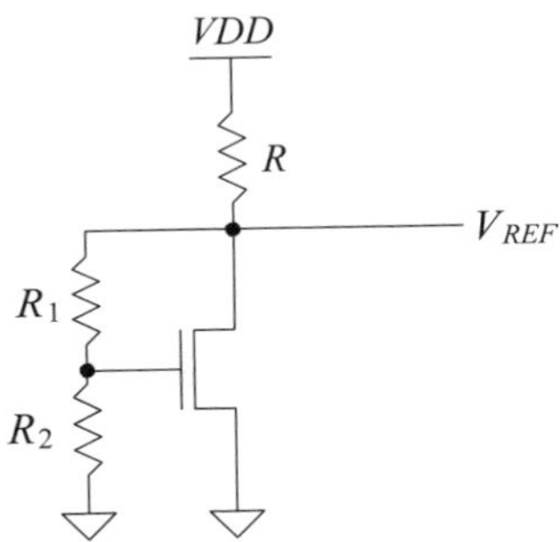

Figure 23.6 Modification of the resistor-MOSFET voltage divider.

23.1.2 The MOSFET-Only Voltage Divider

The MOSFET-only voltage divider shown in Fig. 23.2 generates a reference voltage that is equal to the voltage on the gates of the MOSFETs with respect to ground. Since $I_{D1} = I_{D2}$, we can write

$$\frac{\beta_1}{2}(V_{REF} - V_{THN})^2 = \frac{\beta_2}{2}(VDD - V_{REF} - V_{THP})^2 \tag{23.8}$$

or the reference voltage is given by

$$V_{REF} = \frac{VDD - V_{THP} + \sqrt{\frac{\beta_1}{\beta_2}} \cdot V_{THN}}{\sqrt{\frac{\beta_1}{\beta_2}} + 1} \tag{23.9}$$

or knowing the desired reference voltage and the power supply voltage

$$\frac{\beta_1}{\beta_2} = \left[\frac{VDD - V_{REF} - V_{THP}}{V_{REF} - V_{THN}}\right]^2 \tag{23.10}$$

The temperature dependence of the MOSFET-only voltage divider, assuming the temperature dependence of the ratio of the transconductance parameters, $\frac{\beta_1}{\beta_2}$, is negligible, is given by

$$TCV_{REF} = \frac{1}{V_{REF}} \cdot \frac{\partial V_{REF}}{\partial T} = \frac{1}{V_{REF}} \cdot \frac{1}{\sqrt{\frac{\beta_1}{\beta_2}} + 1} \cdot \left[\frac{\partial(-V_{THP})}{\partial T} + \sqrt{\frac{\beta_1}{\beta_2}}\frac{\partial V_{THN}}{\partial T}\right] \tag{23.11}$$

From Ch. 9, we know for the long-channel devices

$$\frac{\partial V_{THN}}{\partial T} = -1\ mV/C^{\circ} \tag{23.12}$$

and

$$-\frac{\partial V_{THP}}{\partial T} = 1.4\ mV/C^{\circ} \tag{23.13}$$

To achieve $TCV_{REF} = 0$, requires

$$-\frac{\partial V_{THP}}{\partial T} = -\sqrt{\frac{\beta_1}{\beta_2}} \cdot \frac{\partial V_{THN}}{\partial T} \Rightarrow 1.4\ mV/C^{\circ} = \sqrt{\frac{\beta_1}{\beta_2}} \cdot 1\ mV/C^{\circ} \tag{23.14}$$

or

$$\sqrt{\frac{\beta_1}{\beta_2}} = 1.4 \tag{23.15}$$

Zero temperature coefficient, to a first order, can be met by satisfying this equation. However, this ratio is most often set by the desired V_{REF}.

23.1.3 Self-Biased Voltage References

We've already presented a voltage reference back in Ch. 20 using the beta-multiplier reference (BMR) (see Figs. 20.15, 20.19, and 20.22). Consider the BMR seen in Fig. 23.7. The added amplifier forces the drain/gates of M1 and M2 to the same potential. Because the gates and sources of M3/M4 are at the same potential, the same current is forced through each side of the reference. *This is an important common theme for all of the self-biased references discussed in this chapter.* We'll discuss this in greater detail in a moment. Before moving on to this topic, let's discuss why we connect the inverting and noninverting inputs to the added amplifier in the way seen in Fig. 23.7. (Why is M2 connected to the + amplifier input instead of the – input?) For the amplifier to be stable, we want the amount of signal fed back and subtracted from the input to be larger than the

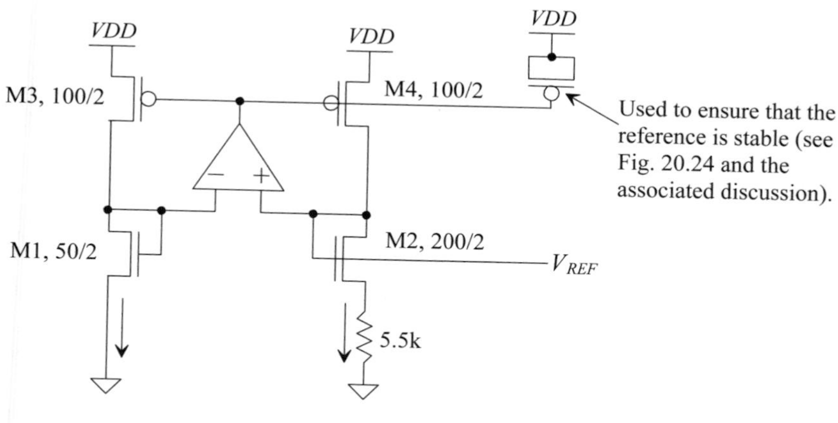

Figure 23.7 BMR from Figs. 20.19 and 20.22.

amount of signal fed back and added to the input. Because M3 and M4 are *inverting* common-source amplifiers, we want the signal fed back to the + input of the amplifier to be larger than the signal fed back to the – input of the amplifier. Since the (AC) currents flowing in M3 and M4 are equal and the small-signal resistance of M2 + R is larger than the small-signal resistance of M1, we connect M2 to the + amplifier input.

Forcing the Same Current through Each Side of the Reference

The simplest method to force the same current through each side of a reference is to use a simple current mirror, Fig. 23.8a. As we saw in Fig. 20.15, the low-output resistance found in short-channel devices makes the resulting reference very sensitive to changes in *VDD*. By cascoding the current mirror, Fig. 23.8b, the currents can be made more equal and the sensitivity to *VDD* can be reduced. The problem with using a cascode current mirror is that the minimum allowable *VDD* increases. Consider the following.

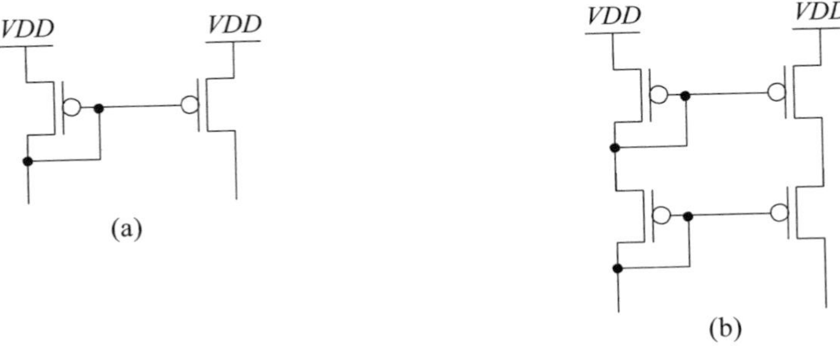

Figure 23.8 Using a current mirror to force the same current through each side of a reference.

Example 23.2
Resimulate the BMR in Fig. 20.18 using a cascode current mirror for the PMOS devices as seen in Fig. 23.9.

Using the cascode current mirror, we expect the currents through each branch of the reference to be very nearly equal. As seen in the simulation results in Fig. 23.10, the currents *are* equal. However, they are not constant and independent of changes in *VDD*. Similarly, if we were to use V_{biasn} as a reference voltage (see Sec. 20.1.5), then V_{biasn} would show a large sensitivity to changes in *VDD*. The problem with cascoding only the PMOS devices is that while the currents through each branch are equal (but not constant), the variation with *VDD* is still present. The voltages at the drains of M1 and M2 change with *VDD*. To reduce this voltage variation, we might consider cascoding the NMOS devices as well (see Fig. 23.11). The PMOS devices are still used to force the same current through each side of the reference, while now the NMOS cascode stack is used to keep the voltages across M1 and M2 constant with changes in *VDD*. As the simulation results, Fig. 23.12, show, the currents do stabilize but at a voltage 20% higher than *VDD*. Clearly, cascoding devices won't be useful in a short-channel CMOS process (and so *we'll stick with using the added amplifier to both set the currents*

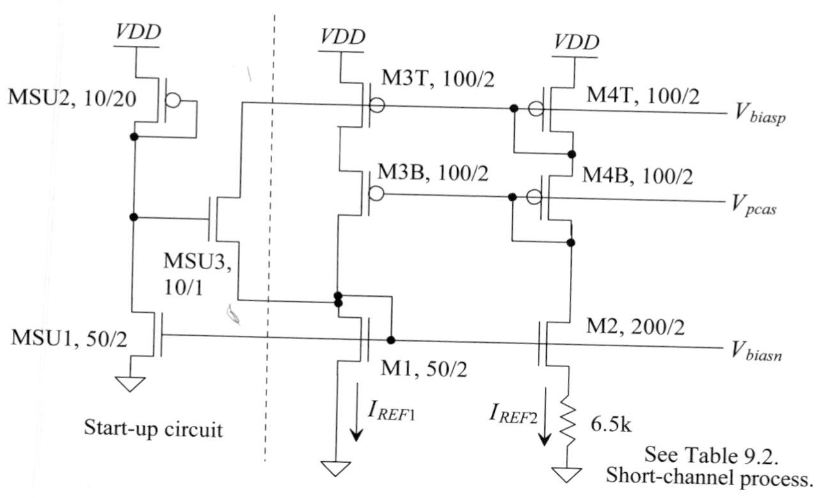

Figure 23.9 Cascoding the PMOS devices in the BMR circuit.

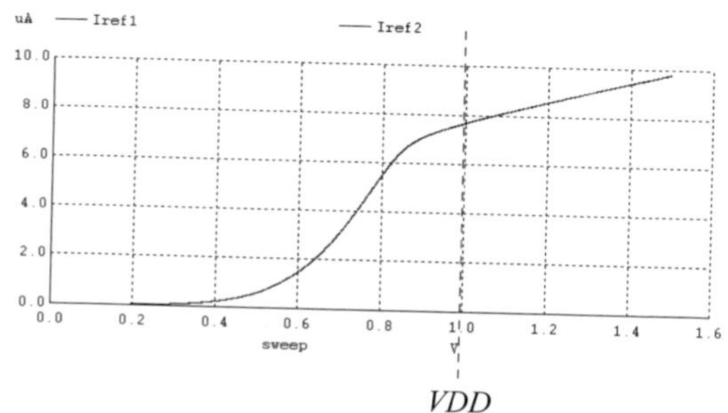

Figure 23.10 Simulating the behavior of the currents in the BMR of Fig. 23.9.

and hold the voltage across M1 and M2 constant in the remainder of the ***short-channel CMOS*** *designs presented in this chapter*). Note that cascoding devices can be useful when designing in long-channel CMOS processes. In these processes, the threshold voltage is a smaller percentage of the power supply voltage. For example, in our short-channel process, Table 9.2, this percentage is 28%. In the long-channel process, Table 9.1, it is 16%. Also note that using NMOS cascodes alone won't make the reference more tolerant to changes in *VDD*. The NMOS cascodes will keep the voltages across M1 and M2 constant (but not equal). Since the currents through each branch aren't equal (because the PMOS devices aren't cascoded), we still get significant sensitivity to *VDD*.

Finally, we need to, again (see Fig. 20.15 and the associated discussion), mention the importance of the start-up circuit. The start-up circuit is often overlooked and is often a cause of problems in practical designs. We include a start-up circuit in every self-biased reference to avoid the situation where zero

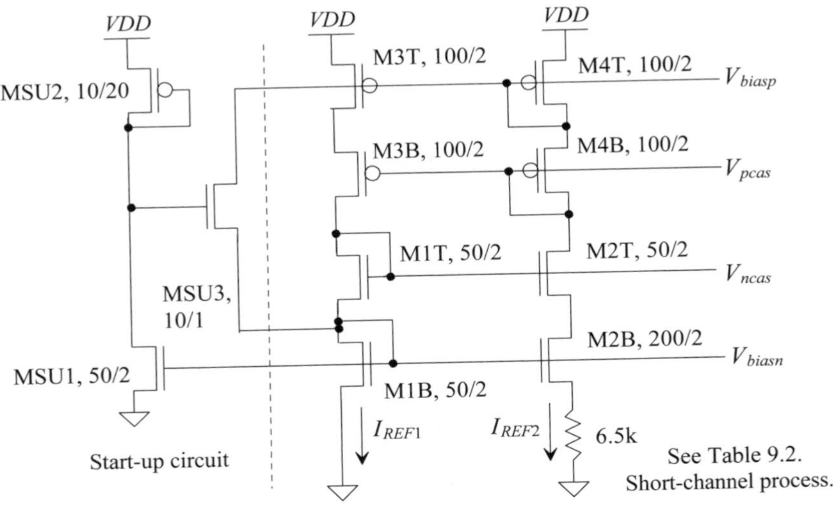

Figure 23.11 Cascoding both NMOS and PMOS devices in the BMR circuit.

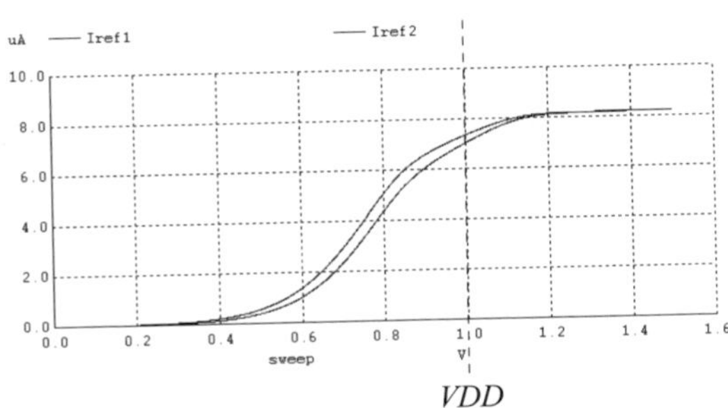

Figure 23.12 Simulating the behavior of the currents in the BMR of Fig. 23.11.

current flows in the reference. It's important to ensure that the start-up circuit doesn't affect normal operation or draw too much current from *VDD*. For the start-up circuit seen in Figs. 23.9 and 23.11, we use the V_{biasn} to set the current drawn by the bias circuit. In normal operation, MSU3 should be off (it should have a negative V_{GS}). ■

Example 23.3
Using the topology seen in Fig. 20.22, design a voltage reference with, ideally, zero temperature coefficient. Simulate the design to determine the reference's sensitivity to changes in temperature and *VDD*. Comment on the repeatability of the V_{REF} (with process variations from one process run to the next) and methods (and concerns) with trimming V_{REF}. Assume $TCR = 0.002$ (2,000 ppm/C).

We begin by writing the long-channel equation, Eq. (20.40), for the resistance needed for zero temperature coefficient (ZTC)

$$R = \frac{2}{\frac{\partial V_{THN}}{\partial T} \cdot KP_n \cdot \frac{W}{L}} \left(1 - \frac{1}{\sqrt{K}}\right) \cdot \left(\frac{1}{R}\frac{\partial R}{\partial T} + \frac{1}{KP_n} \cdot \frac{\partial KP_n}{\partial T}\right) \quad (23.16)$$

While this equation won't directly apply for the design in a short-channel process, we can use it, with simulations, to help point us to the factors that influence the temperature behavior of the reference. Substituting in the appropriate numbers with $K = 4$, we get

$$R = \frac{1}{(-0.0006) \cdot KP_n \cdot \frac{W}{L}} \cdot \left((0.002) - \frac{1.5}{300}\right) \quad (23.17)$$

or

$$R = \frac{5}{KP_n \cdot \frac{W}{L}} \quad (23.18)$$

To move the reference towards a ZTC, let's set the resistor to nominally 5.5k (the same value we used before, see Fig. 20.22 and the associated discussion) and adjust the widths and lengths of the NMOS devices until the simulations show good temperature behavior (keeping in mind that we must keep the W/L of M2 K times the W/L of M1). The resulting circuit is seen in Fig. 23.13. Simulation results are seen in Fig. 23.14. (Note that the currents and gate-source voltages have nothing to do with the values listed in Table 9.2.) The reference voltage is nominally 500 mV (= $VDD/2$). The reference, according to simulations, moves 40 mV over a 100 °C temperature change (–400 μV/C or a TC of –800 ppm/C, not that great when compared to the references described later). Further, the reference is fairly insensitive to variations in VDD once VDD gets above 600 mV.

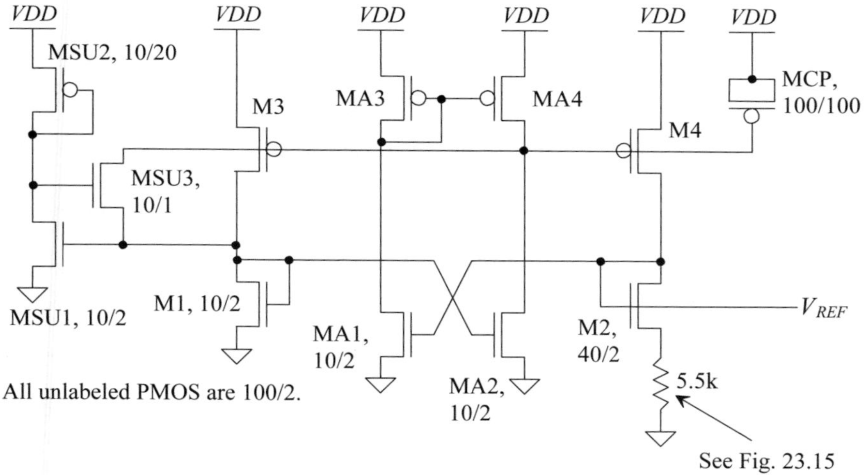

Figure 23.13 Voltage reference using the beta-multiplier.

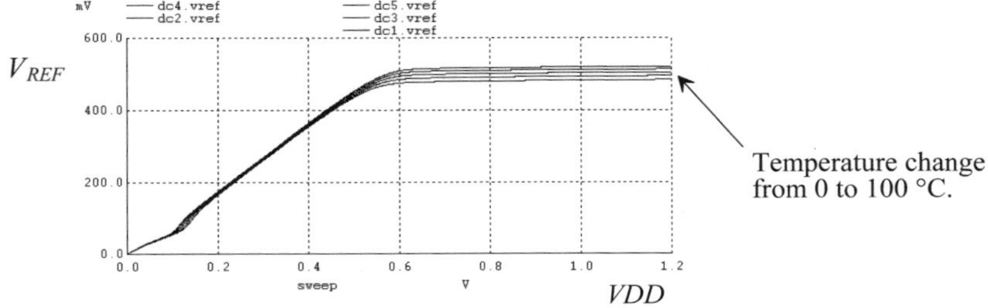

Figure 23.14 Simulating the operation of the reference in Fig. 23.13.

A couple of practical notes are needed at this point. While we've used simulations to zero in on a design with decent temperature behavior and a reference voltage at *VDD*/2, actually fabricating this reference would result in a V_{REF} different from the one seen in Fig. 23.14 (in most fabrication runs). The MOSFET characteristics and the sheet resistance vary with each process run.

To adjust the reference voltage to *VDD*/2, the resistor must be trimmed using fuses, Fig. 23.15. It can be shown that the temperature behavior doesn't vary significantly with small changes in V_{REF} when trimming. As seen in Eq. (20.38), lowering the resistor value causes V_{REF} to increase, while increasing the resistor value causes V_{REF} to decrease. The fuses in Fig. 23.15 short across the resistor until they are blown. To trim the resistor, we start blowing the resistors (electrically or with a laser). With each blown fuse we add (nominally) 200 Ω in series with the nominally 4k resistor. If the processes' sheet resistance increases by 20% (so that the resistors are now 4.8k and 240 Ω), then we only need to blow three fuses (neglecting the change in the MOSFET characteristics) to trim the resistor to 5.5k. If the sheet resistance decreases by 20%, then the resistors are 3.2k and 160 Ω. Fifteen fuses would need to be blown to trim the resistor to 5.5k.

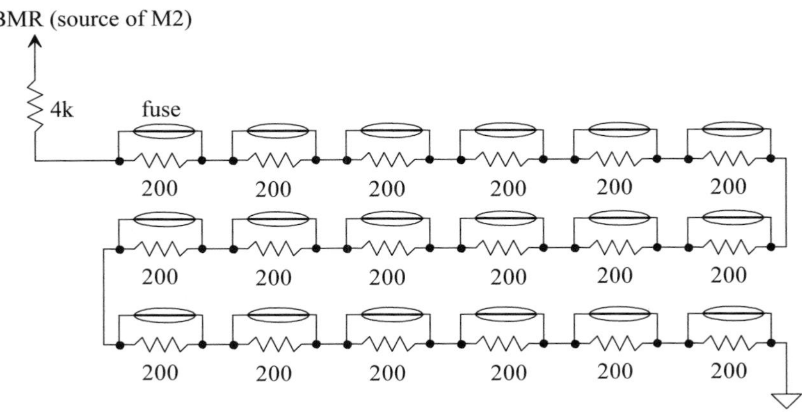

Figure 23.15 Trimming a resistor with fuses.

Finally, it's important to remember that stability is of great importance when using feedback. As discussed in Sec. 20.1.4, the feedback loop is made stable by adding a capacitance on the output of the added amplifier. The bigger this capacitance, the more stable the reference. We stabilize the reference in Fig. 23.13 with the addition of MCP. ■

An Alternate Topology

Instead of using PMOS devices in the top of each branch of the reference, we might try holding the voltage across resistors constant to force the current through each branch of the reference to be the same. Figure 23.16 shows the idea drawn two different ways. In (a) the amplifier tries to hold both of its inputs at the same potential. This causes the voltage across each R to be the same and thus each branch of the reference has the same current. Note now we've tied M2 and the resistor (a resistance larger than M1 alone) to the – amplifier input. This is because the inverting PMOS common-source amplifiers (M3 and M4) are no longer in the feedback signal path; we need to ensure that the largest signal is fed back to the inverting amplifier terminal. Note also, in (b), that the series order of the resistor and MOSFET is not important. In fact, having both sources connected to ground eliminates the body-effect mismatch and may result in better temperature performance. We won't use these topologies in any of the MOSFET-resistor topologies in this chapter because the amplifier has to drive a resistive load (which, as we saw in Chs. 22 and 23, kills the amplifier's gain). A two-stage op-amp, discussed in the next chapter, can be used. However, then we have to be concerned with compensating the op-amp. Finally, note that a start-up circuit is still needed with these topologies. The start-up circuit would leak current into the node connected to the + amplifier input to start the reference circuit up (and then shut off after the reference turns on). Connecting the start-up circuit to the – amplifier input would move the amplifier output in the wrong direction and the reference would never start up (noting in Figs. 23.7 and 23.13 that we do connect the start-up circuit to the – input because M3 and M4 are inverting amplifiers).

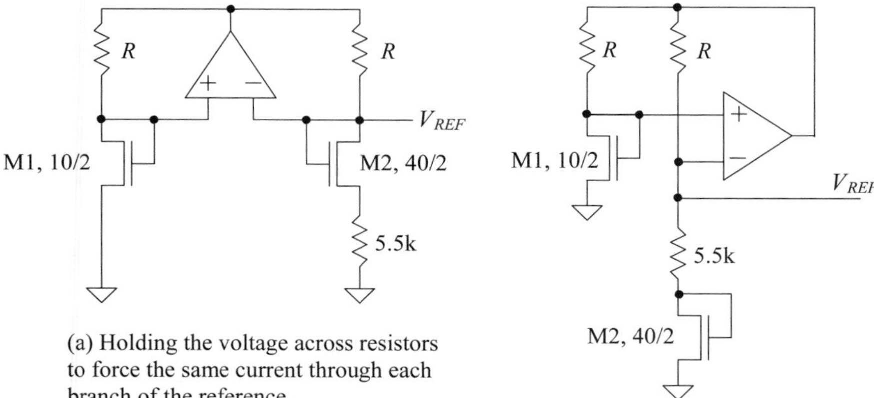

(a) Holding the voltage across resistors to force the same current through each branch of the reference.

Note start-up circuit not shown (see text).

(b) Redrawing (a) and switching the places of the resistor and M2.

Figure 23.16 Other references useful in MOSFET-resistor circuits.

23.2 Parasitic Diode-Based References

By using a junction diode, the variability encountered in the references of the last section that used a MOSFET's threshold voltage can be reduced. One practical implementation of a parasitic diode available in CMOS is seen in Fig. 23.17. A diode is formed between the p+ implant and the n-well. However, in a practical circuit the parasitic vertical PNP bipolar device formed by the p+, n-well, and p-substrate causes current to be injected into the substrate. Good guard rings are used to surround the p+ in both the n-well and the p-substrate to collect this current. This ensures that the injected carriers are collected and not causing substrate current to flow in other portions of the chip. Because the substrate is connected to ground, the diode's cathode (K) must be tied to ground. We might try to use an n+ implant directly in the substrate for diode formation. However, assuming the substrate (which forms the anode of the resulting diode) is grounded, we would have to apply a negative voltage to the n+ (K) to forward-bias the diode. One other practical parasitic diode available in CMOS is made using the lateral PNP device seen in Fig. 23.18. In this device, the p+ forms the emitter/collector of the resulting device (the n-well is the base). However, the vertical PNP (and the resulting substrate current) is still present. A significant portion of the lateral PNP's emitter current will flow into the substrate. In the following example, we'll assume vertical PNP (diodes) are used.

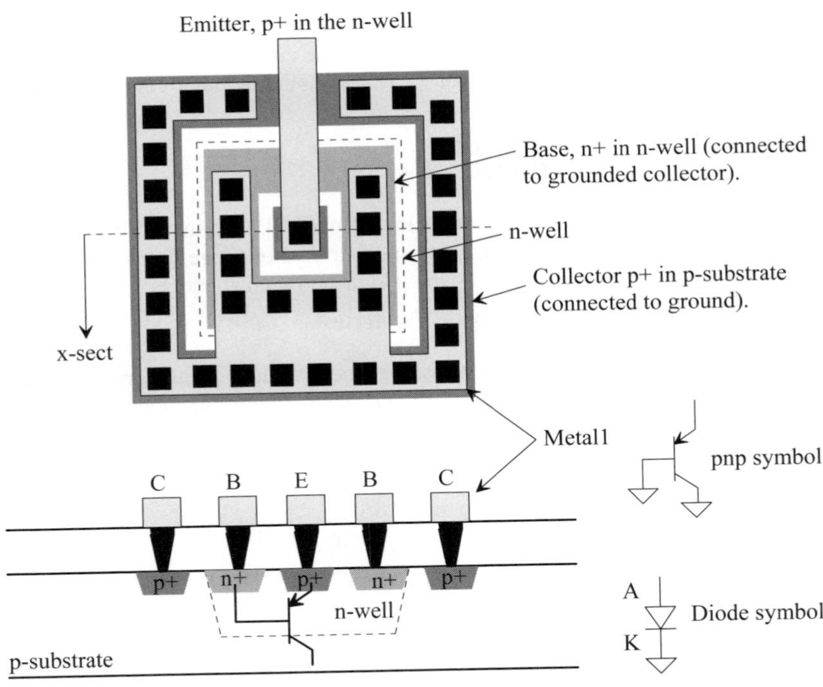

Figure 23.17 A vertical parasitic pnp bipolar junction transistor used as a diode.

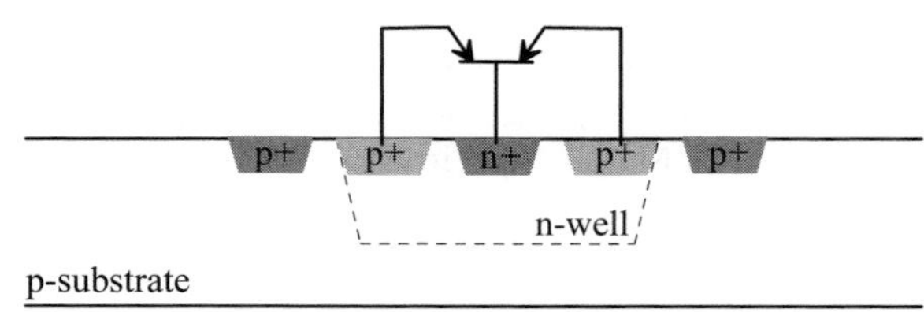

Figure 23.18 A lateral parasitic bipolar junction transistor.

Diode Behavior

It's important to realize that the parasitic diode's behavior must be thoroughly characterized in a particular CMOS process to determine its voltage-current (and temperature) characteristics. For the sake of illustrating design procedures, considerations, and examples in this chapter, we can develop a general diode model. The current through a forward-biased diode is given by

$$I_D = I_S \cdot e^{V_D/n \cdot V_T} \tag{23.19}$$

where V_D is the voltage across the diode, n is the emission coefficient (used to shape the diode's current-voltage curve to fit experimental data), V_T is the thermal voltage (kT/q or 26 mV at room temperature), and I_S is the diode's scale current. Here, we'll use a value of 10^{-18} A for the scale current and an n of 1. The SPICE statement corresponding to the schematic seen in Fig. 23.19 is

```
D1    1    0    PNPDIODE
D2    2    0    PNPDIODE    4
```

where the 4 indicates 4 diodes are connected in parallel. We may use, in schematics drawn in later in the chapter, the term K to indicate K diodes are connected in parallel. The model statement for the diode, using the parameters seen above, is

```
.MODEL      PNPDIODE    D     is=1e-18      n=1
```

Figure 23.20 shows how the voltage across the diode changes if the current through the diode is held constant and the temperature is changed. Using our model (n =1 and I_S =

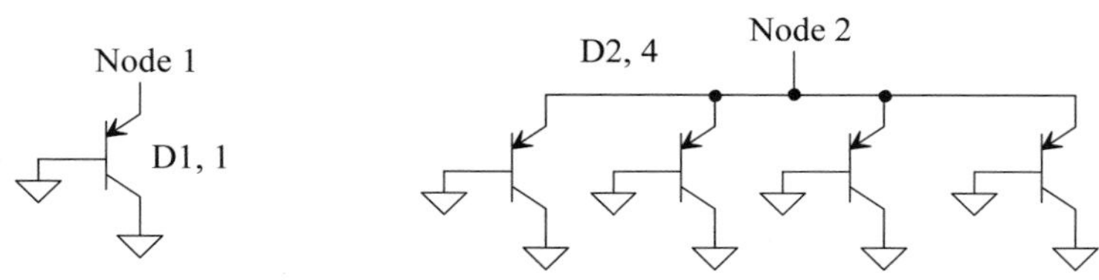

Figure 23.19 A schematic of diodes.

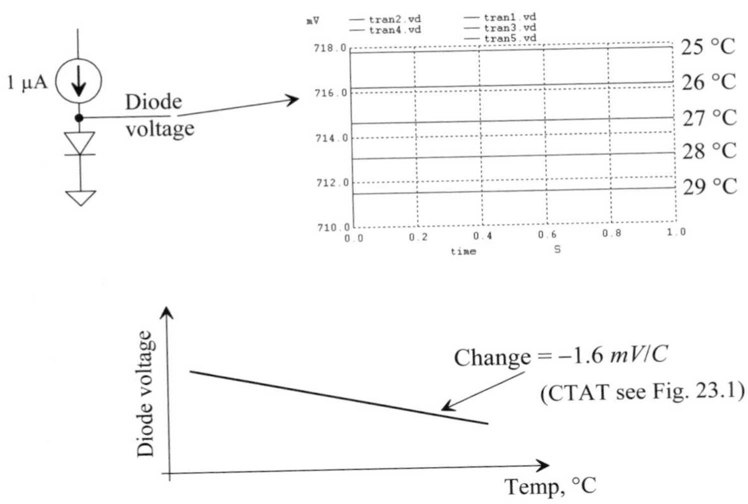

Figure 23.20 Change in diode voltage with temperature.

10^{-18}), we can estimate the change in the diode's voltage with temperature, from the simulation data, as

$$\frac{dV_D}{dT} = -1.6\ mV/C \tag{23.20}$$

noting that this number is very dependent on the DC current through the diode. (Also note this voltage change is complementary to absolute temperature, CTAT.) Increasing the bias current in Fig. 23.20 to 10 μA, causes the diode's voltage drop to increase (of course) and its voltage change with temperature to become $-1.4\ mV/C$.

The Bandgap Energy of Silicon

The silicon bandgap energy as a function of temperature is given by

$$E_g(T) = 1.16 - (702 \times 10^{-6}) \cdot \frac{T^2}{T + 1108}\ \text{(eV)} \tag{23.21}$$

At room temperature, the bandgap of silicon is approximately 1.1 eV (where 1 electron-volt, eV, is 1.6×10^{-19} J). Looking at Eq. (23.21), notice that the bandgap energy decreases with increasing temperature. If we put a constant current through a diode and increase the temperature, the barrier height between the n and p sides of the diode decreases and thus so does the diode's voltage drop (and that is why the diode's voltage change with temperature in Fig. 23.20 is negative).

In the next two sections we'll talk about *BandGap References* (BGR). BGR's combine the CTAT behavior of the bandgap of silicon (the diode's forward voltage drop) with the PTAT behavior of the thermal voltage (the thermal voltage, kT/q, increases with increasing temperature see Fig. 23.1) to form a voltage reference (a BGR) that doesn't vary (much) with temperature.

Lower Voltage Reference Design

While, at the time of this writing, power supply voltages are well above the forward turn-on voltage of the pn junction diode (parasitic PNP device), it is clear that the near future will bring *VDD* voltages below 0.7 V. To implement a reference in these lower supply voltage processes, consider the Schottky diode layout of Fig. 23.21. In this layout, we connect the metal (the anode) directly to the n-well (that is, without the p+) to form the Schottky diode. To connect the metal directly to the substrate, we need an opening in the oxide. To form this opening, we use the "active" layer without the select layer (without the implant). It's important that no select layer surrounds this Schottky contact.

The turn-on voltage of a Schottky contact is considerably lower than a standard pn junction. In Fig. 23.21 we draw the turn on voltage as 300 mV. In real silicon this voltage will depend highly on the doping of the n-well (which, as we saw in Ch. 6, scales up as process dimensions shrink). The design discussions and procedures developed in the next two sections can be applied to Schottky diode-based references. Again, though, thorough characterization of the parasitic diode is required prior to designing the reference. One concern, among others, is the repeatability of the diode's current-voltage characteristics from one process run to the next. We don't discuss Schottky diode-based references any further in this chapter, but, rather, leave it to the refereed literature.

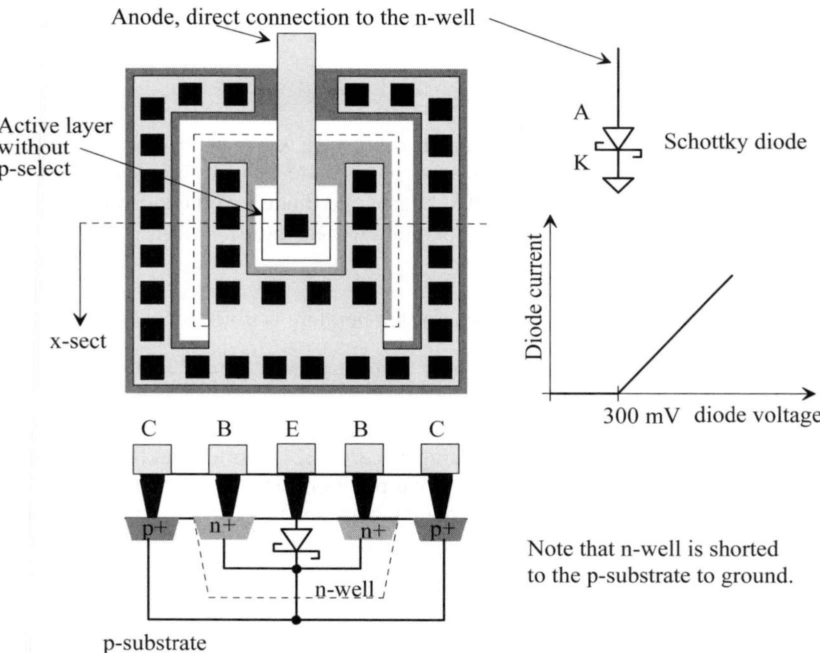

Figure 23.21 Connecting metal directly to n-well to make a Schottky diode.

23.2.1 Long-Channel BGR Design

In this section we present some long-channel designs for bandgap references (BGR) using the cascode structure seen in Fig. 23.11. We will not show the (**required**) start-up circuit in the following schematics.

Diode-Referenced Self-Biasing (CTAT)

A diode-referenced, self-biasing circuit is seen in Fig. 23.22. The cascode structures force the same current through each branch of the reference. The voltage across the resistor now equals the forward voltage drop of the diode. The current in the circuit (the current flowing in each branch of the reference) is then

$$I_{REF} = \frac{V_D}{R} = I_S e^{V_D/nV_T} \tag{23.22}$$

Solving for a resistor to set the reference current level gives

$$R = \frac{nV_T}{I} \ln \frac{I}{I_S} \tag{23.23}$$

To determine the reference current's variation with temperature, we can write

$$\frac{\partial I_{REF}}{\partial T} = \frac{\partial}{\partial T}\left(\frac{V_D}{R}\right) = \frac{1}{R} \cdot \overbrace{\frac{\partial V_D}{\partial T}}^{\text{change in } V_D \text{ with } T} - \overbrace{\frac{V_D}{R^2}\frac{\partial R}{\partial T}}^{\text{TC of the resistor}} \tag{23.24}$$

Remembering that the overall current change with temperature is written as

$$I_{REF}(T) = I_{REF}(T_0) \cdot (1 + TCI_{REF}(T - T_0)) \tag{23.25}$$

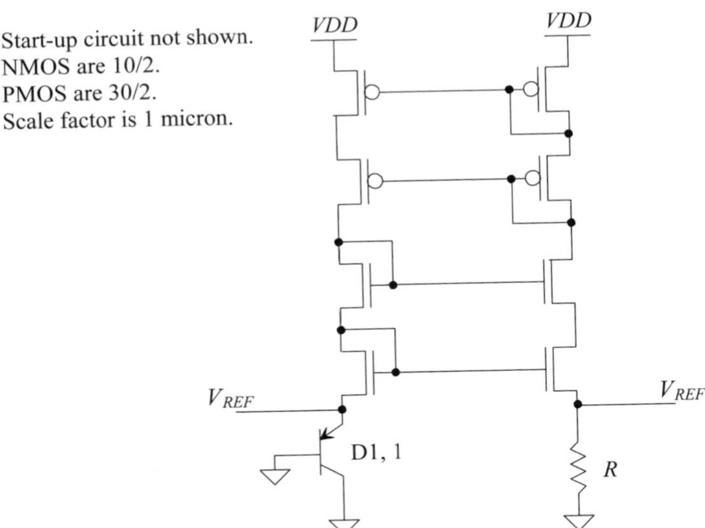
Start-up circuit not shown.
NMOS are 10/2.
PMOS are 30/2.
Scale factor is 1 micron.
VDD
VDD
VREF
VREF
D1, 1
R

Figure 23.22 Diode-referenced, self-biasing circuit.

where

$$TCI_{REF} = \frac{1}{I_{REF}} \cdot \frac{\partial I_{REF}}{\partial T} \tag{23.26}$$

we can, with the help of Eq. (23.24), write

$$\frac{1}{I_{REF}} \frac{\partial I_{REF}}{\partial T} = \frac{1}{V_D} \frac{\partial V_D}{\partial T} - \frac{1}{R} \frac{\partial R}{\partial T} \tag{23.27}$$

The behavior of the diode-referenced, self-biased circuit shows a large negative coefficient. If the TC of the resistor is 0.002 and the change in the diode voltage with T (Eq. (23.20)) is –0.0016 (or if the diode voltage, V_D , is nominally 0.7, the first term in Eq. (23.27) is –0.0023), then the overall TC of the current is, roughly, –0.004 or –4,000 ppm/C. If we were to use the voltage across the diode (or resistor) as a voltage reference, the fact that the current through the diode is decreasing with increasing temperature (causing the diode's voltage to drop) and the bandgap of silicon is decreasing with increasing temperature, causes V_D to drop (relatively quickly) with increasing temperature (CTAT). Consider the following example.

Example 23.4
Simulate the operation of the reference in Fig. 23.22 where the voltage across the diode is used as a reference voltage. Use a nominal reference current of 1 μA.

Using Eq. (23.23), we can set R to 700k. The simulation results are seen in Fig. 23.23. The reference voltage change is essentially set by the decrease in the diode voltage with increasing temperature, that is, –1.6 mV/C. At the risk of stating the obvious, this isn't a very good voltage reference. ■

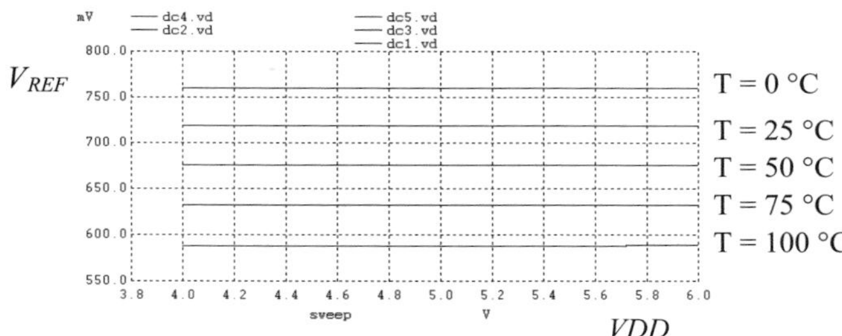

Figure 23.23 Temperature behavior of the diode-referenced circuit of Fig. 23.22. See Ex. 23.4.

Thermal Voltage-Referenced Self-Biasing (PTAT)

A thermal voltage, V_T , referenced, self-biasing circuit is seen in Fig. 23.24. In this configuration the voltage across D1 must equal the voltage across D2 and the resistor

$$V_{D1} = V_{D2} + I_{D2}R \tag{23.28}$$

Notice that D2 must be larger than D1 if the current flowing in the reference is to be nonzero. The larger D2 will drop a smaller voltage than D1 for the same current through each branch. As indicated in Eq. (23.28), the difference in the diode voltages is dropped

Start-up circuit not shown.
NMOS are 10/2.
PMOS are 30/2.
Scale factor is 1 micron.
VDD
VDD
$I = \frac{nk \cdot \ln K}{qR} \cdot T$
R
D1, 1
D2, K (K diodes in parallel)

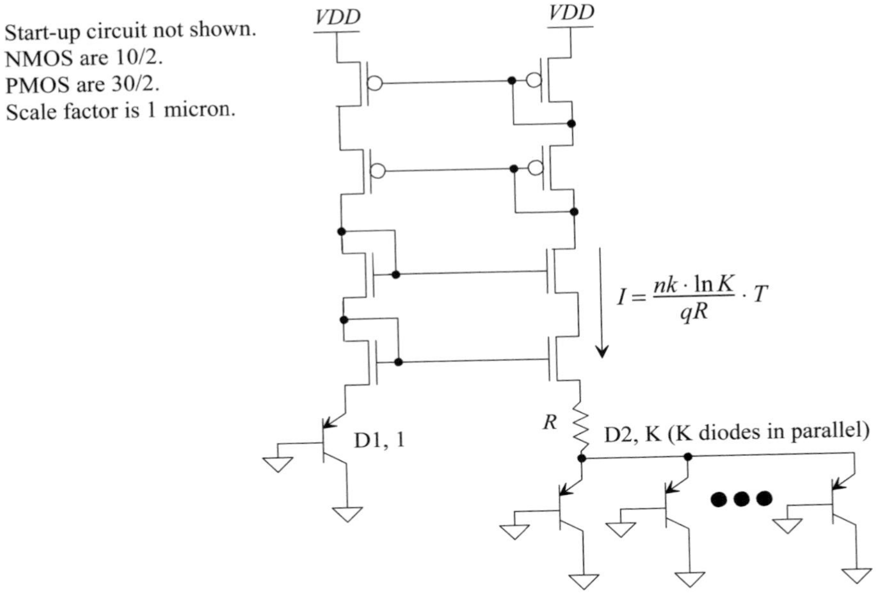

Figure 23.24 Thermal voltage-referenced, self-biasing circuit.

across R. Alternatively, we could size the diodes the same and increase the current flowing in the left branch of the reference by making the MOSFETs K times wider. The current in the left branch would be K times more than the current in the right branch of the reference.

We know

$$I_{D1} = I_S e^{V_{D1}/nV_T} \rightarrow V_{D1} = nV_T \cdot \ln \frac{I_{D1}}{I_S} \tag{23.29}$$

and

$$I_{D2} = K \cdot I_S e^{V_{D2}/nV_T} \rightarrow V_{D2} = nV_T \cdot \ln \frac{I_{D2}}{K \cdot I_S} \tag{23.30}$$

Knowing, because of the cascode structures, $I_{D1} = I_{D2} = I$, we can write

$$R = \frac{nV_T \cdot \ln K}{I} \text{ or } I = \frac{nk \cdot \ln K}{qR} \cdot T \tag{23.31}$$

Notice that the current is proportional to absolute temperature (PTAT). To get a voltage reference that is also PTAT, consider the reference seen in Fig. 23.25. Here the reference voltage is given by

$$V_{REF} = I \cdot L \cdot R = \frac{nk \cdot L \cdot \ln K}{q} \cdot T \tag{23.32}$$

Notice how the temperature behavior of the resistor falls out of the equation.

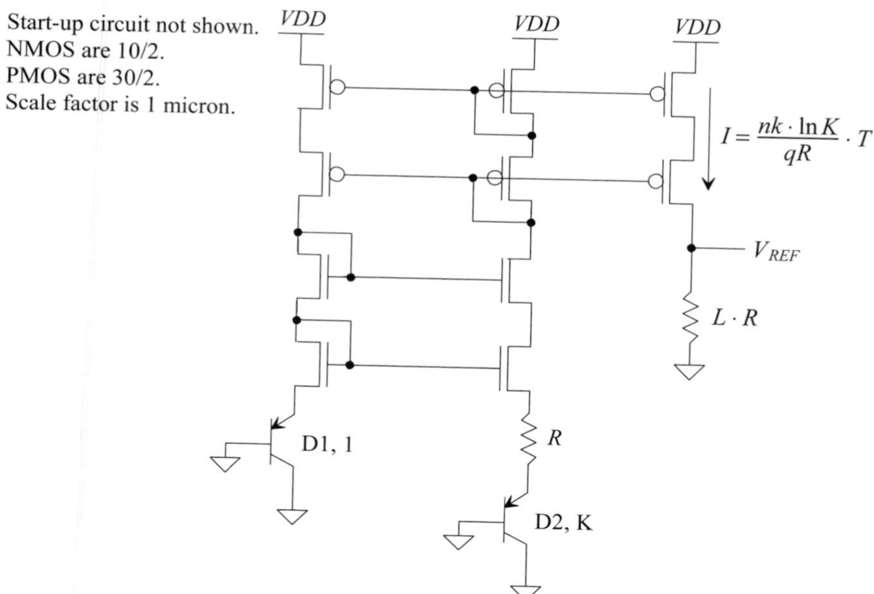

Figure 23.25 PTAT voltage reference based on the thermal voltage-referenced, self-biased circuit.

Example 23.5

Design a nominally 2.500 V voltage reference (at room temperature of 27 °C or 300 °K) PTAT voltage reference. Use the topology seen in Fig. 23.25 with a bias current of 1 μA. Simulate the operation of the design.

Let's set K (the number of diodes connected in parallel) to 8. We do this since it is easy to remember ln 8 = 2. Solving for R using Eq. (23.31) gives

$$R = \frac{1 \cdot 0.026 \cdot 2}{10^{-6}} = 52k$$

The voltage drop across the resistor is $1\ \mu A \cdot 52k = 52\ mV$. Using Eq. (23.32), we can solve for L as 48 (so $L \cdot R = 2.5\ M\Omega$). The simulation results are seen in Fig. 23.26. At room temperature, the reference voltage is roughly 2.5 V. Note how, as temperature increases, the reference voltage (PTAT) increases. Using Eq. (23.32), we can estimate the reference voltage's change with temperature as

$$\frac{\partial V_{REF}}{\partial T} = \frac{nk \cdot L \cdot \ln K}{q} \tag{23.33}$$

Using the numbers from this example with $k = 1.38 \times 10^{-23} J/K$ and $q = 1.6 \times 10^{-19} C$ gives

$$\frac{\partial V_{REF}}{\partial T} = \frac{1 \cdot 1.38 \cdot 10^{-23} \cdot 48 \cdot 2}{1.6 \times 10^{-19}} = 8.26\ mV/C$$

For every 25 °C increase in temperature, we expect V_{REF} to go up by 206 mV. ■

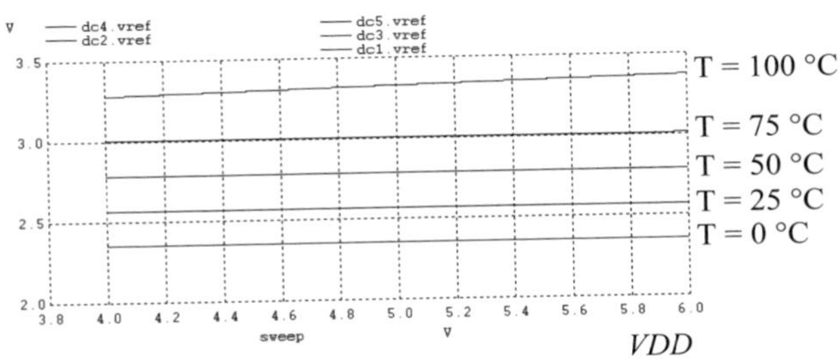

Figure 23.26 A PTAT voltage reference. See Ex. 23.5.

Bandgap Reference Design

The bandgap reference (BGR) is formed using CTAT and PTAT references. When designed properly, the TC of the BGR can be very small. Figure 23.27 shows a schematic of a BGR. The PTAT current generated from the circuit in Fig. 23.24 is driven into a resistive load (to give a PTAT voltage drop, see Eq. (23.32)) and a diode (to give a CTAT voltage drop, see Eq. (23.20)). The reference voltage is then the sum of the PTAT and CTAT voltage drops or

$$V_{REF} = V_{D3} + I \cdot L \cdot R = V_{D3} + L \cdot n \cdot \ln K \cdot V_T \qquad (23.34)$$

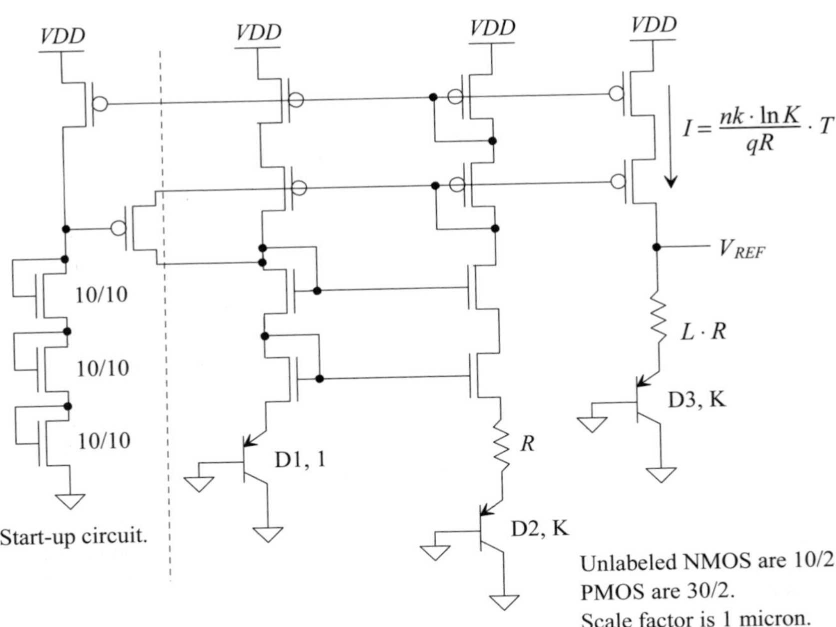

Figure 23.27 A bandgap reference circuit.

The change in the reference voltage with temperature is

$$\frac{\partial V_{REF}}{\partial T} = \overbrace{\frac{\partial V_{D3}}{\partial T}}^{-1.6\, mV/C} + L \cdot n \cdot \ln K \cdot \overbrace{\frac{\partial V_T}{\partial T}}^{0.085\, mV/C} \tag{23.35}$$

where the change in the thermal voltage with temperature was taken from Eq. (9.44). To determine the value of L that causes the change in V_{REF} with temperature to go to zero, we set Eq. (23.35) to zero and solve

$$L = \frac{1.6}{n \cdot \ln K \cdot 0.085} \tag{23.36}$$

If $K = 8$ and $n = 1$, then $L = 9.41$. Using Eq. (23.34) with these values, gives a V_{REF} of 1.2V. Figure 23.28 shows the simulation results using these values and the schematic seen in Fig. 23.27. The start-up circuit causes the reference to turn on at a *VDD* of approximately 3V. The change in the reference voltage with temperature is roughly 1.5 mV per 75 °C or 20 μV/C. The TC of the reference, Eq. (23.3), is then 16.7 ppm/C. At higher temperatures the variation in V_{REF} (with *VDD*) is slightly poorer likely due to not keeping the currents in each branch equal. Note that if a precise voltage is needed, trimming the $L \cdot R$ resistor (see Fig. 23.15 and the associated discussion) is required.

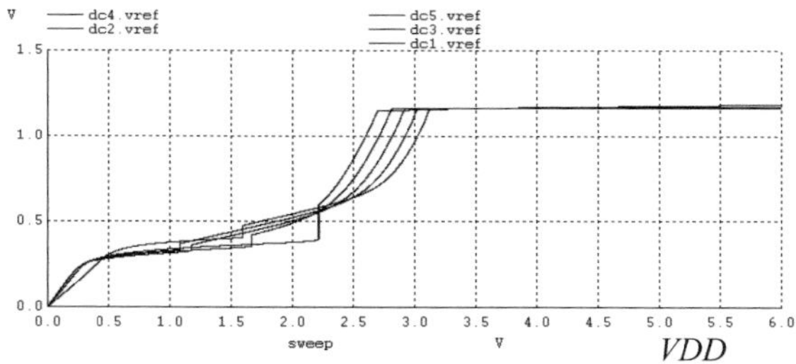

Figure 23.28 Simulating the BGR in Fig. 23.27 from 0 to 100 C.

Alternative BGR Topologies

We might wonder if there is a way to simplify the BGR of Fig. 23.27 to reduce power and layout area. Figure 23.29a shows one possibility. In this circuit we've combined the three branches in Fig. 23.27 into two branches. The same voltage, as in Fig. 23.25, is dropped across the single resistor *R*. The $L \cdot R$ resistors are a common element to both branches. The current is still PTAT, while the reference voltage at the top of the $L \cdot R$ resistor is still a bandgap voltage. The drawback of this topology is the fact that the minimum value of *VDD* for proper reference operation increases.

Figure 23.29b shows a topology based on the configuration, using resistors, seen in Fig. 23.16. As with all of the topologies in this chapter (again), the point of the added amplifier is to force the same current through each side of the reference. The benefit of this topology over the topology in (a) is that *VDD* can move lower before it affects V_{REF}. The drawback of this topology is that the amplifier must be capable of driving a resistive load (which we'll see in the next chapter requires two gain stages for the amplifier's overall gain to be reasonably high). In (c) PMOS devices are added to isolate the

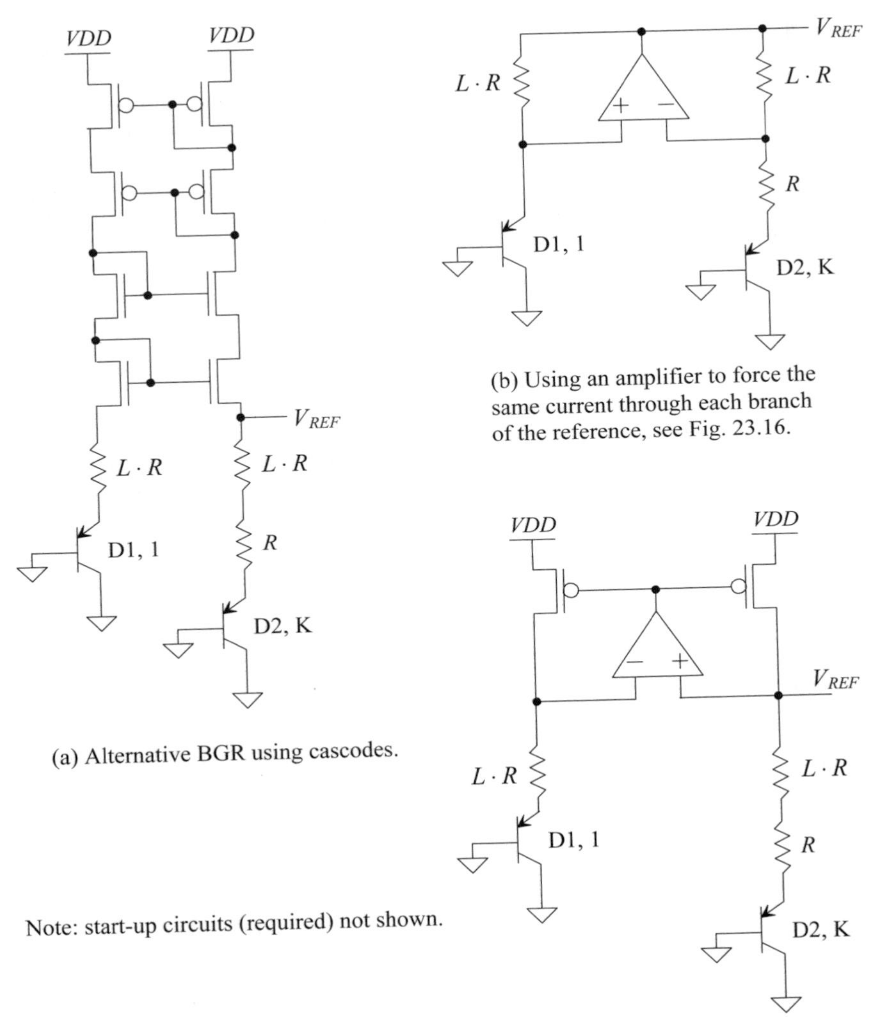
VDD
VDD
V_{REF}
$L \cdot R$
$L \cdot R$
R
D1, 1
D2, K
(a) Alternative BGR using cascodes.
V_{REF}
$L \cdot R$
$L \cdot R$
R
D1, 1
D2, K
(b) Using an amplifier to force the same current through each branch of the reference, see Fig. 23.16.
VDD
VDD
V_{REF}
$L \cdot R$
$L \cdot R$
R
D1, 1
D2, K
(c) Using PMOS devices to isolate the amplifier from the resistive loading.
Note: start-up circuits (required) not shown.

Figure 23.29 Alternative BGR topologies.

amplifier's output from the resistors. This makes using a diff-amp for the added amplifier possible. The currents are equal in (c) because the source-gate voltages of the PMOS devices are equal. Note how the branch containing D2 is a higher resistance (than the branch containing D1) and so it (the D2 branch) is always connected to an inverting point in the feedback loop. As seen in (c) the addition of the PMOS devices (which are inverting) means that we need to switch the inverting and noninverting amplifier inputs from (b). Finally, note that a start-up circuit is required for all three of these references.

23.2.2 Short-Channel BGR Design

The bandgap reference voltage of 1.2 V developed in the last section is greater than *VDD* (= 1 V) in our short-channel process. To develop a BGR for short-channel processes, consider the schematic seen in Fig. 23.30. The diodes D1 and D2 together with the resistor, *R*, form a PTAT current generator, as seen in Fig. 23.24. To provide a CTAT current to sum with the PTAT current, consider the addition of the $L \cdot R$ resistors, as seen in the schematic. As temperature increases, the diode voltage decreases, causing the current through the $L \cdot R$ resistors to decrease (CTAT). We know that the current due to the PTAT portion of the circuit, see Eq. (23.31), is

$$I_{PTAT} = \frac{nV_T \cdot \ln K}{R} \tag{23.37}$$

The current through the added resistors (the CTAT portion of the circuit) is

$$I_{CTAT} = \frac{V_{D1}}{L \cdot R} \tag{23.38}$$

The total current is driven through the $N \cdot R$ to generate the reference voltage

$$V_{REF} = nV_T \cdot N \cdot \ln K + \frac{N}{L} \cdot V_{D1} \tag{23.39}$$

The temperature behavior of the BGR is

$$\frac{\partial V_{REF}}{\partial T} = n \cdot N \cdot \ln K \cdot \overbrace{\frac{\partial V_T}{\partial T}}^{0.085\, mV/C} + \frac{N}{L} \cdot \overbrace{\frac{\partial V_{D1}}{\partial T}}^{-1.6\, mV/C} \tag{23.40}$$

For zero TC, we get an *L* of

$$L = \frac{1.6}{n \cdot \ln K \cdot 0.085} \tag{23.41}$$

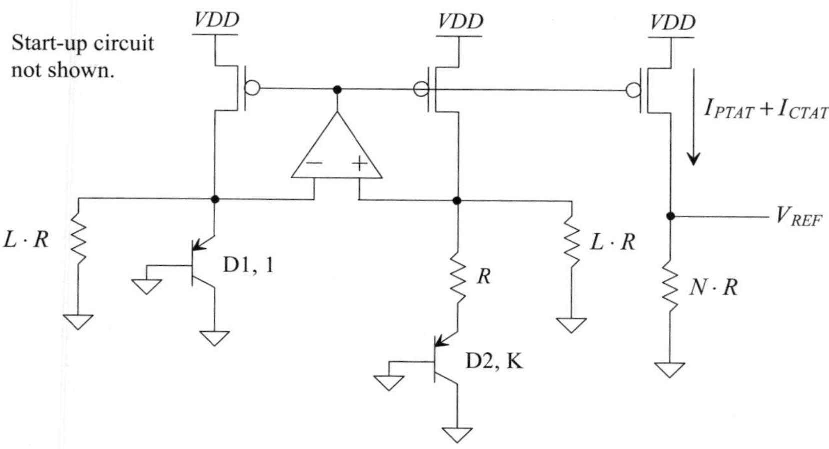

Figure 23.30 Lower voltage BGR.

Using a K of 8 results in, again, an L of 9.41. To get a particular reference voltage, we use Eq. (23.39) to determine N. For example, if we want a reference voltage of 500 mV (half of VDD), we get an N of (using a K of 8)

$$N = \frac{V_{REF}}{nV_T \cdot \ln K + \frac{V_{D1}}{L}} = \frac{0.5}{0.052 + \frac{0.7}{9.41}} = 3.91 \qquad (23.42)$$

Figure 23.31 shows some simulation results using these numbers (the schematic of the full design is seen in Fig. 23.32). In (a) the reference turns on at a VDD of approximately 900 mV. The temperature behavior of the reference is seen in (b). Notice that the reference voltage is higher than what we designed for. In all practical situations, the $N \cdot R$ resistor (the 205k resistor in Fig. 23.32) would need trimming (see Fig. 23.15) to set the reference voltage to a precise value.

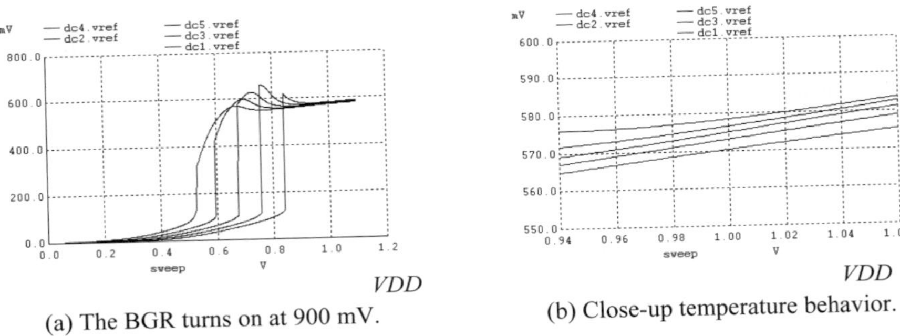

(a) The BGR turns on at 900 mV.

(b) Close-up temperature behavior.

Figure 23.31 Simulating the behavior of the reference in Fig. 23.30.

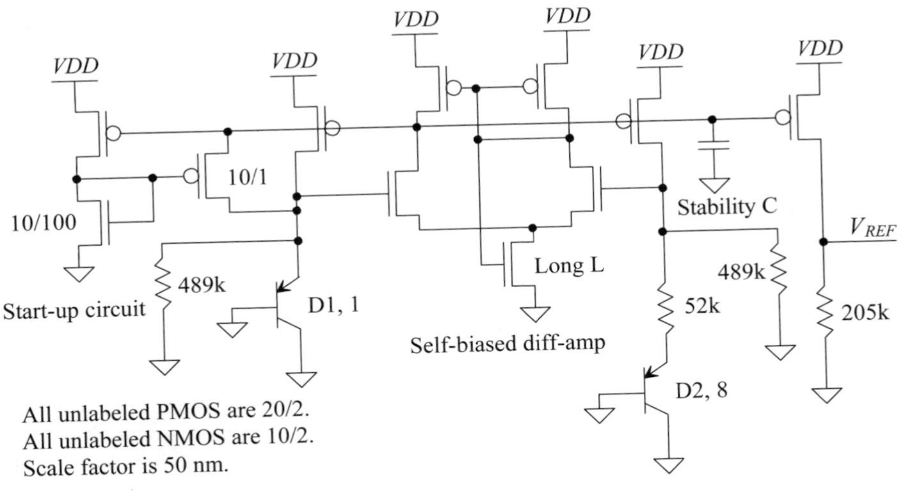

Figure 23.32 Lower voltage BGR used to generate the simulation data seen in Fig. 23.31.

The Added Amplifier

In Fig. 23.32 we used a self-biased amplifier (see Fig. 18.17) to hold the voltage across D1 to the same value as the voltage across the 52k resistor and D2. This amplifier is critical for good BGR performance. For example, notice how the reference voltage in Fig. 23.31b varies with changes in *VDD*. By using a higher-gain amplifier, we can reduce this variation. While we'll discuss op-amp design in detail in the next chapter, one concern is the added amplifier's input common-mode voltage. Because the forward voltage drop of the diodes is approximately 0.7 V (a significant percentage of *VDD*), the performance (gain) of the amplifier may drop if any of the MOSFETs in the diff-amp triode or shut off. To lower the input common-mode voltage, consider the topology of Fig. 23.33. Here we set

$$R_1 + R_2 = L \cdot R \tag{23.43}$$

The amplifier sets the voltages across each R_2 to the same value and so the current flowing in R_1 and R_2 is the same as in Fig. 23.30, provided Eq. (23.43) is satisfied. The big benefit of using the topology is that the DC input common-mode voltage of the added amplifier can be reduced.

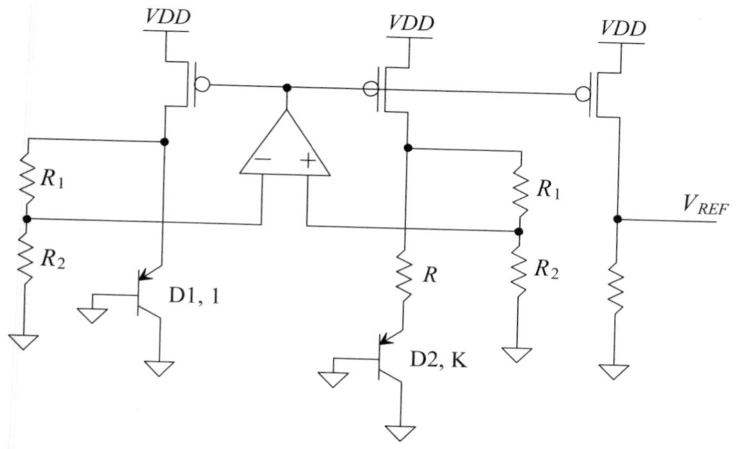

Figure 23.33 Lowering the input common-mode range of the added amplifier.

Lower Voltage Operation

Fundamentally, the limitation on how low *VDD* can be in the BGR reference of Fig. 23.30 (or 23.33) is the forward voltage drop of the diodes. As the dimensions of CMOS technology continue to scale downwards, so does *VDD*. While the MOSFET-resistor references, such as the beta-multiplier reference (BMR), discussed in the first section of this chapter may provide a solution to many reference circuit needs at lower *VDD*s, it is unlikely that they will replace the BGR in all applications. Even though both the BMR and BGR require trimming to set V_{REF}, the temperature performance of the BGR is, in general, better than that of the BMR. What may happen is that the pn junction used in the BGR is replaced by a junction with a lower built-in potential (and thus lower forward turn-on voltage) like the Schottky diode (see Fig. 23.21).

ADDITIONAL READING

[1] E. Vittoz and J. Fellrath, "CMOS Analog Integrated Circuits Based on Weak Inversion Operation," *IEEE Journal of Solid-State Circuits*, vol. SC-12, no. 3, June 1977, pp. 224–231. Covers the BMR design operating in the weak inversion region.

[2] K. E. Kuijk, "A Precision Reference Voltage Source," *IEEE Journal of Solid-State Circuits*, vol. SC-8, no. 3, June 1973, pp. 222–226. Development of the BGR seen in Fig. 23.29b.

[3] G. Tzanateas, C. A. T. Salama, and Y. P. Tsividis, "A CMOS Bandgap Voltage Reference," *IEEE Journal of Solid-State Circuits*, vol. SC-13, no. 3, June 1979, pp. 655–657. Good paper discussing BGR reference design.

[4] H. Banba, H. Shiga, A. Umezawa, T. Miyaba, T. Tanzawa, S. Atsumi, and K. Sakui, "A CMOS Bandgap Reference Circuit with Sub-1-V Operation," *IEEE Journal of Solid-State Circuits*, vol. 34, no. 5, May 1999, pp. 670–674. Presents the design of the lower voltage BGR seen in Fig. 23.30.

[5] K. N. Leung and P. K. T. Mok, "A Sub-1-V 15-ppm/C CMOS Bandgap Voltage Reference Without Requiring Low Threshold Voltage Device," *IEEE Journal of Solid-State Circuits*, vol. 37, no. 4, April 2002, pp. 526–530. Develops the scheme seen in Fig. 23.33 for lowering the input common-mode range of the added amplifier.

PROBLEMS

23.1 Using the MOSFET-only reference seen in Fig. 23.2, design a nominally 500 mV reference in the short-channel CMOS process. Using simulations, characterize the sensitivity of the reference voltage to changes in *VDD* and temperature.

23.2 Use the long-channel CMOS process and the topology seen in Fig. 23.6 to design a voltage reference of $3V_{THN}$. Simulate your design and show the *VDD* sensitivity and temperature behavior of the reference. Do TCs of R_1 and R_2 affect the performance of the reference? Why or why not?

23.3 Suppose it was desired, in Fig. 23.7 (see also Fig. 20.22), to make M1 and M2 the same size. However, to increase the gate source voltage of M1, relative to M2, the width of M3 is increased by *K*. How do the equations governing the operation of the BMR change? How does the current flowing in the BMR change?

23.4 Verify that if the PMOS devices in Fig. 23.11 are not cascoded (that is, they have only NMOS cascodes), the currents in each branch will not be equal and there will be significant sensitivity to *VDD* (a sensitivity similar to what is seen in Fig. 23.10).

23.5 In a CMOS process, several of the layers including poly, n+, p+, and n-well can be used for resistor formation. Each of these layers has a different temperature coefficient (TC). For the BMR that generated Fig. 23.14, use simulations to determine the optimum resistor TC.

23.6 Derive the equations that govern the operation of the reference in Fig. 23.16b.

23.7 Show why n+ directly in the p-substrate cannot be used as a diode in a CMOS process.

23.8 Generate a diode model that produces a forward voltage drop of 700 mV when driven with 1 μA and has a change with temperature, dV_D/dT, near room temperature, of −2 mV/C. Use simulations to verify your model meets the requirements.

23.9 Generate a SPICE model for the Schottky diode seen in Fig. 23.21. Assume that the series resistance of the diode is 1 kΩ.

23.10 Estimate, using hand calculations, the minimum allowable *VDD* for the reference of Ex. 23.4. What are the PMOS and NMOS gate-source and drain-source voltages when the reference current is 1 μA? Note that the parameters in Table 9.1 have nothing to do with the operating conditions in this question. Verify your answers using SPICE.

23.11 Show that *K* forward-biased diodes in parallel behave like a single diode with a scale current of $K \cdot I_S$, as assumed in Eq. (23.30).

23.12 Suggest a reference design that would output a voltage of $n \cdot V_T$.

23.13 Determine whether the performance of the BGR of Fig. 23.32 can be enhanced by using the topology of Fig. 23.33. Use simulations to verify your answer.

Chapter

24

Operational Amplifiers I

The operational amplifier (op-amp) is a fundamental building block in analog integrated circuit design. A block diagram of the two-stage op-amp with output buffer is shown in Fig. 24.1. The first stage of an op-amp is a differential amplifier. This is followed by another gain stage, such as a common source stage, and finally by an output buffer. If the op-amp is intended to drive a small purely capacitive load, which is the case in many switched capacitor or data conversion applications, the output buffer is not used. If the op-amp is used to drive a resistive load or a large capacitive load (or a combination of both), the output buffer is used.

Design of the op-amp consists of determining the specifications, selecting device sizes and biasing conditions, compensating the op-amp for stability, simulating and characterizing the op-amp A_{OL} (open-loop gain), CMR (common-mode range on the input), CMRR (common-mode rejection ratio), PSRR (power supply rejection ratio), output voltage range, current sourcing/sinking capability, and power dissipation.

We'll start this chapter off with a very simple two-stage op-amp (without an output buffer). By pointing out the weaknesses with this two-stage topology, we'll set the stage for developing practical op-amps in the rest of the chapter.

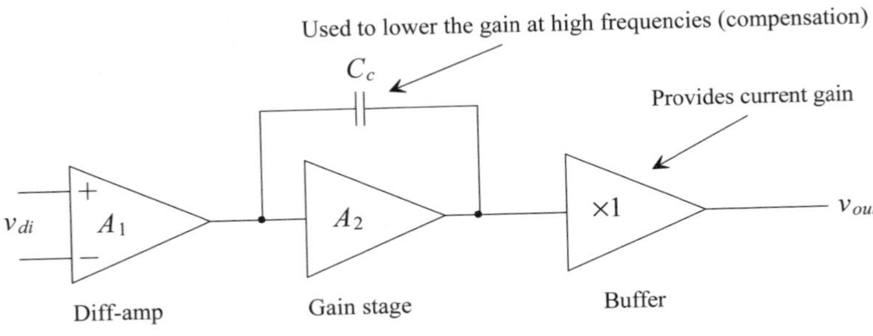

Figure 24.1 Block diagram of two-stage op-amp with output buffer.

24.1 The Two-Stage Op-Amp

Figure 24.2 shows the basic two stage op-amp made using an NMOS diff-amp and a PMOS common-source amplifier (M7). As seen in Fig. 22.8 M7 is biased to have the same current as M3 and M4 (10 μA from Table 9.2). Note also the addition of the compensating network consisting of a compensation capacitor, C_c, (*Miller compensation*) and a zero-nulling resistor R_z (see Figs. 21.25 and 21.33 along with the associated discussion). Because the op-amp doesn't have an output buffer, it is limited to driving capacitive loads and very large resistances (comparable to the output resistance of a MOSFET, that is, megaohms).

Low-Frequency, Open Loop Gain, A_{OLDC}

The low-frequency, open loop gain of the op-amp is calculated as the product of each stage gain, that is,

$$A_{OLDC} = A_1 \cdot A_2 = \overbrace{g_{mn} \cdot (r_{on} || r_{op})}^{A_1 = \text{diff-amp's gain}} \cdot \overbrace{g_{mp} \cdot r_{op}}^{A_2,\ \text{M7's gain}} \tag{24.1}$$

where the output resistance of the cascode current source load, M8, is assumed to be much larger than the M7 output resistance, r_{op}. Using the values from Table 9.2, we get an A_{OLDC} of 832 *V/V*.

Input Common-Mode Range

The minimum input common-mode voltage is given by Eq. (22.11) or 450 mV. The maximum input common-mode voltage is given by Eq. (22.12) or 900 mV. This means, for proper operation of our two-stage op-amp, the input voltages (v_p and v_m) should fall within the range of 450 to 900 mV. If they go outside this range, the op-amp gain drops, and it is likely that the circuit employing the op-amp will not function properly. Figure 24.3 shows a SPICE DC sweep where the *inverting* input (v_m) is held at 500 mV and the *noninverting* input (v_p) is swept from 495 to 505 mV. The slope of this transfer curve is the DC open-loop gain of the op-amp, A_{OLDC}.

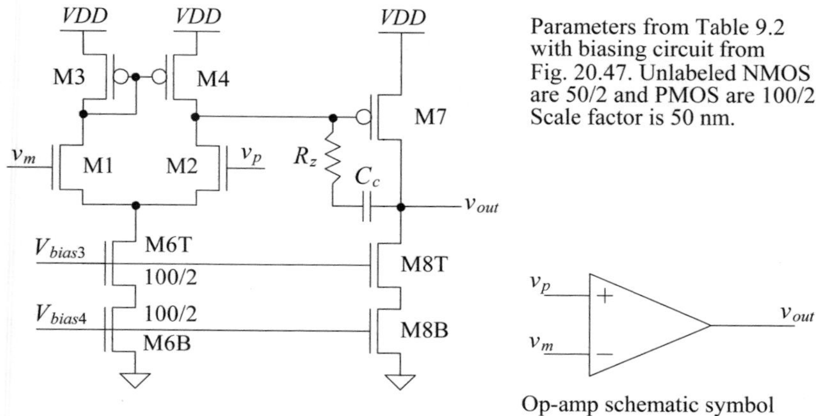

Figure 24.2 Basic two-stage op-amp.

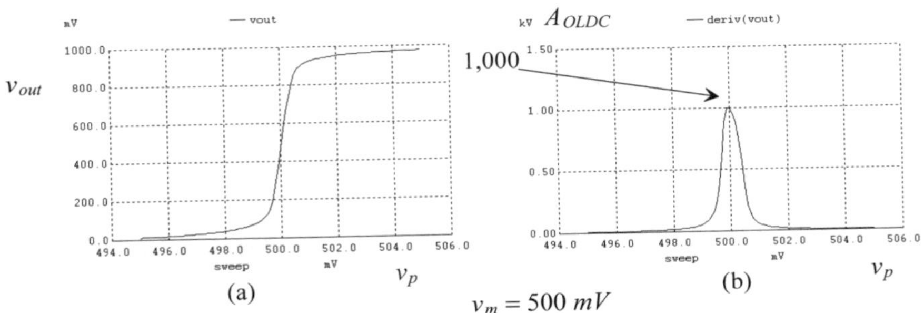

Figure 24.3 (a) DC transfer curves for the op-amp in Fig. 24.2 and (b) its gain (the derivation of (a)).

Power Dissipation

To determine the power dissipated by the op-amp, we sum the currents supplied by the constant current sources and multiply the result by *VDD*. For the op-amp in Fig. 24.2, the current through M6 is 20 µA and the current through M8 is 10 µA. The total power dissipation is then 30 µW ($VDD = 1\ V$).

Output Swing and Current Sourcing/Sinking Capability

The maximum output swing (for the op-amp in Fig. 24.2) is limited by M7 going into the triode region. If we must keep at least 100 mV across M7, then the maximum output voltage is 900 mV. When M8 goes into triode or roughly 100 mV (see Fig. 20.48), the minimum output voltage is set. As seen in Fig. 24.3a, the high-gain (or large slope) region falls between v_{out} of 100 and 900 mV. Note that the maximum amount of current that this op-amp can sink is limited by the constant current sink M8 or 10 µA. The op-amp can source considerably more than 10 µA by pulling the gate of M7 downwards. Because this topology can source a considerable amount of current, it is useful as a voltage regulator (because the op-amp is always only sourcing current in a voltage regulator application).

Offsets

We've talked in great length (earlier in the book) about random offsets and how to design and layout circuits to minimize their effects. Another type of offset (that is not random) is termed a *systematic offset*. When we sized the MOSFETs in Fig. 24.2, for example, we made sure that M7 was sized to source 10 µA and M8 was sized to sink 10 µA. What would happen if we sized M7 to source 100 µA of current instead (changed its size from 100/2 to 1000/2)? Because M8 is a constant bias of 10 µA, M7 would move into the triode region until it was sourcing the 10 µA of current that M8 wants to sink. The output voltage would be very close to *VDD* with M7 in the triode region. Effectively we would get a shift or offset in the transfer curves seen in Fig. 24.3a (see Fig. 24.4a). To model this output voltage shift, we can refer it back to the op-amp input as an input-referred offset voltage, Fig. 24.4b. Note that unlike a random offset, which may be positive or negative in value, a systematic offset will always be a known polarity. Finally, note that while we chose to model the offset in series with the noninverting input terminal, in Fig. 24.4, we could just as easily have placed it in series with the inverting op-amp terminal (with a change in polarity).

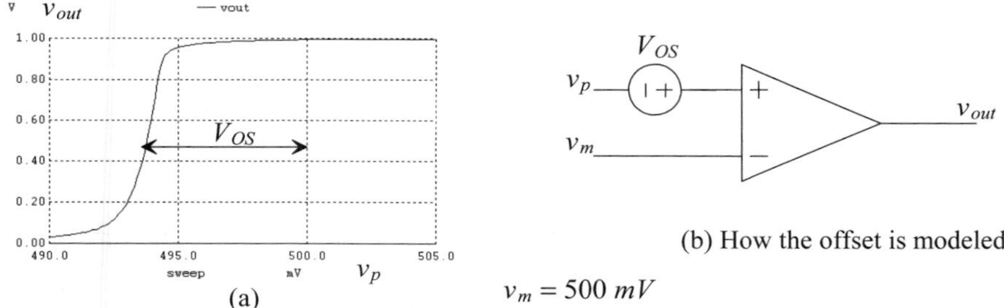

Figure 24.4 Showing how increasing M7's width to 1000 in Fig. 24.2 causes an input-referred (systematic) offset voltage.

Compensating the Op-Amp

An important step in op-amp design is the design of the compensation network. The op-amp takes the difference between the inverting and noninverting input terminal voltages and, ideally, multiplies the result by a very large number (the gain). A block diagram representation of an op-amp is seen in Fig. 24.5. The open-loop gain as a function of frequency is labeled $A_{OL}(f)$.

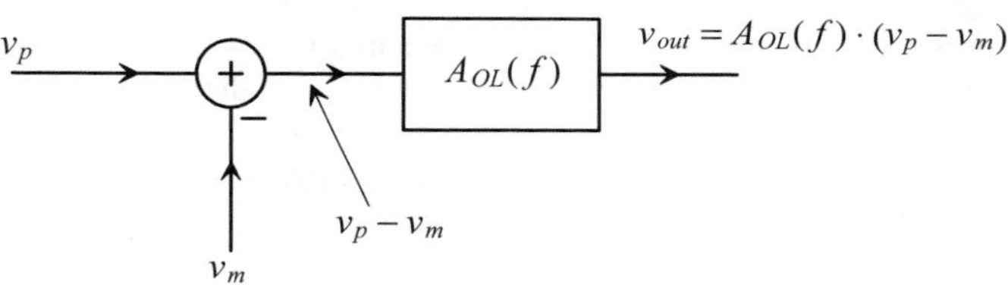

Figure 24.5 Block level diagram of an op-amp.

In all practical situations, the op-amp is used with feedback, Fig. 24.6. While our op-amp in Fig. 24.2 cannot drive a resistive load (unless it is megaohms, as a resistor on the op-amp output will kill the gain of the second stage), we'll still use this figure (Fig. 24.6) to illustrate the concept of feedback and compensation. We can write

$$v_{out} = A_{OL}(f) \cdot (v_{in} - v_f) \tag{24.2}$$

and

$$v_f = v_{out} \cdot \frac{R_2}{R_1 + R_2} \tag{24.3}$$

The amount of the output that is fed back is often called the feedback factor β or

$$\beta = \frac{R_2}{R_1 + R_2} \tag{24.4}$$

Substituting Eqs. (24.3) and (24.4) into Eq. (24.2) and solving for the closed loop gain gives

$$A_{CL}(f) = \frac{v_{out}}{v_{in}} = \frac{A_{OL}(f)}{1 + \beta \cdot A_{OL}(f)} \tag{24.5}$$

If we take $A_{OL}(f) \to \infty$, then the closed loop gain becomes

$$A_{CL}(f) \to \frac{1}{\beta} = 1 + \frac{R_1}{R_2} \tag{24.6}$$

We have several important points that we need to discuss. To begin, notice that in Eq. (24.5) if

$$\beta \cdot A_{OL}(f) = -1 \tag{24.7}$$

or more precisely

$$|\beta \cdot A_{OL}(f)| = 1 \text{ and } \angle \beta \cdot A_{OL}(f) = \pm\, 180° \tag{24.8}$$

the closed loop gain blows up (the feedback amplifier becomes unstable). The worst case situation (the largest value of β) occurs when all of the output is fed back to the op-amp input. The voltage follower, Fig. 24.7, is an example of this situation. *To determine the stability of an op-amp, we'll look at the open loop gain when the feedback factor is one, that is,*

$$|A_{OL}(f)| = 1 \text{ and } \angle A_{OL}(f) = 180° \tag{24.9}$$

We'll discuss this in greater detail in a moment. Notice that the larger the closed loop gain, the smaller the value of β (the less output signal we feed back) and the more likely the op-amp circuit, with feedback, will be stable. *This is important*. While feedback helps to desensitize an amplifier's gain to variations in an op-amp's A_{OL}, the drawback is stability. In a high-performance op-amp that will never operate in a unity-follower configuration, we can get an enhancement in speed by reducing the amount of compensation (the amount of the reduction guided by Eq. (24.8) with the actual value of β used).

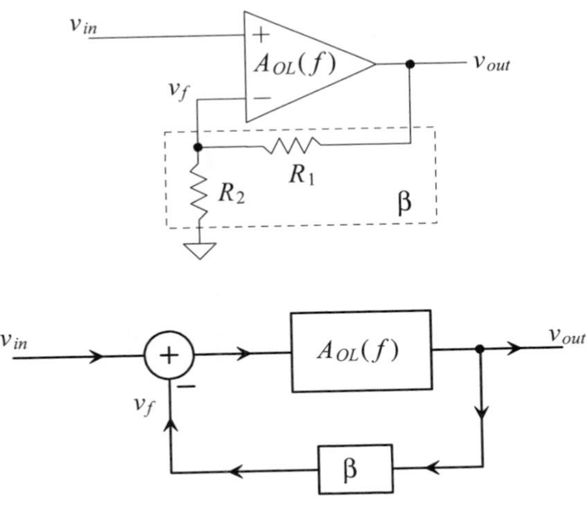

Figure 24.6 An example of feedback in an op-amp.

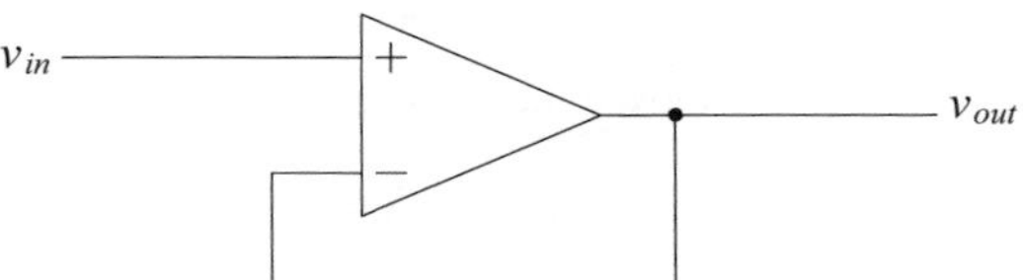

Figure 24.7 Voltage follower configuration, an example of a closed-loop amplifier with unity feedback factor.

To estimate the open-loop frequency response of the op-amp, let's use the generic model seen in Fig. 21.25. Figure 24.8 shows the location of nodes 1 and 2 on our two-stage op-amp. We've included a load capacitance in this figure. With the help of Table 9.2 we can write

$$R_1 = r_{on}||r_{op} = 111\ k\Omega$$

$$R_2 = r_{op}||R_{ocasn} \approx r_{op} = 333\ k\Omega$$

$$g_{m1} = g_{mn} = 150\ \mu A/V \text{ (diff-amp)}$$

$$g_{m2} = g_{mp} = 150\ \mu A/V \text{ (common-source)}$$

$$C_1 = C_{dg4} + C_{gd2} + C_{gs7} = 13.6\ fF$$

$$C_2 = C_L + C_{gd8} \approx C_L + 1.56\ fF$$

Let's calculate the open-loop response with a load and compensation capacitance of 100 fF, that is, $C_L = C_c = 100\ fF$. The pole associated with node 1, from Eq. (21.65), is

$$f_1 \approx \frac{1}{2\pi g_{m2} R_1 R_2 C_c} = 287\ kHz$$

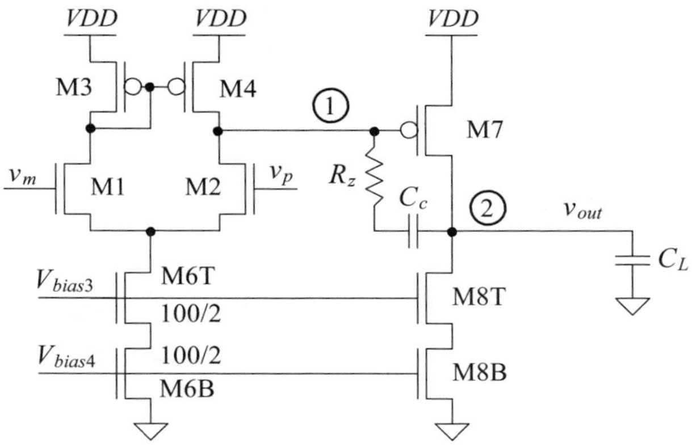

Figure 24.8 Calculating the frequency response of the op-amp.

The location of the pole associated with the output node (node 2) is, from Eq. (21.66),

$$f_2 = \frac{g_{m2}C_c}{2\pi(C_cC_1 + C_1C_2 + C_cC_2)} = 210\ MHz$$

The location of the zero, f_z in Eq. (21.63), is at 240 MHz (very near the second pole).

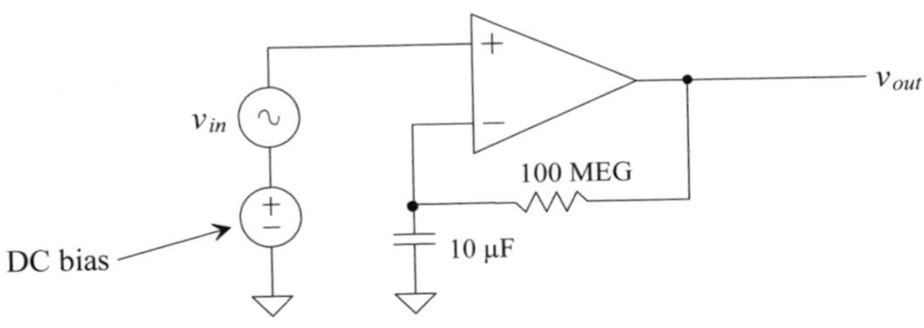

Figure 24.9 Circuit configuration used to simulate open-loop frequency response.

To simulate the open-loop response of the op-amp, we can use the configuration seen in Fig. 24.9. The feedback resistor and capacitor form a time constant so large that for all intents and purposes none of the AC output voltage is fed back to the inverting input. However, the DC bias level is fed back so that the op-amp biases up correctly (all MOSFETs are operating in the saturation region). With a DC bias voltage of 500 mV, Fig. 24.10 shows the open loop responses of the op-amp in Fig. 24.8 with 100 fF compensation capacitance and load capacitance. The simulated values are close to the

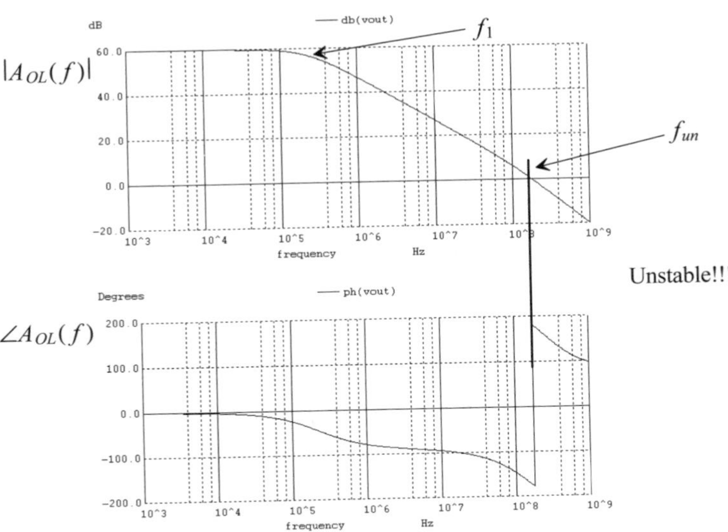

Figure 24.10 The open-loop frequency response of the op-amp in Fig. 24.8 with a 100 fF compensation capacitance and load capacitance.

hand-calculated values given above. However, from the criteria given in Eq. (24.9), the op-amp will clearly be unstable. *We want the open loop gain of the op-amp to be much less than one when the phase shift is 180°*. Looking at Fig. 24.10, we effectively want to shift the lower frequency pole (f_1) downwards in frequency. At the same time, we want to move the higher frequency pole, f_2, higher in frequency (we want to *split the poles*). Remember this (pole-splitting) was the point of adding the compensation capacitor to an amplifier back in Ch. 21. Let's use Eq. (21.72) to select C_c. Looking at Fig. 24.10, we see that the phase shift is –100° at 10 MHz, so let's set the unity gain frequency to this value (in an attempt to make sure that the open loop gain is well under one when the phase shift is 180°).

$$f_{un} = \frac{g_{m1}}{2\pi C_c} = \frac{150\ \mu A/V}{2\pi \cdot C_c} = 10\ MHz \rightarrow C_c = 2.4\ pF \tag{24.10}$$

The simulation results are seen Fig. 24.11. The unity-gain frequency, f_{un}, is close to 10 MHz. However, while the lower frequency moved downwards and the higher frequency pole moved upwards, the zero moved down to unity-gain frequency, see Eq. (21.63). This causes the open-loop gain to hover around unity instead of continuing to decrease (this is bad). The stability and step response of the op-amp will be degraded because of the zero. To illustrate this, consider the simulation results seen in Fig. 24.12. Notice that, in a transient simulation, we have to wait some time to let the bias circuit start up. Here, in this simulation, we've waited 500 ns before applying the input signal. Next, notice that our step input signal has a small amplitude change (5 mV). If we apply a larger step, the small-signal analysis used to derive pole-splitting is no longer valid (and we'll see the slew-rate limitations). Clearly the output of the op-amp doesn't behave well with the zero present. To eliminate the zero, we add the zero-nulling resistor, as seen in Fig. 21.33b. Figure 24.11 is regenerated in Fig. 24.13 with the addition of a zero-nulling resistor ($1/g_{mn}$ = 6.5k). Figure 24.14 shows the well-behaved step response.

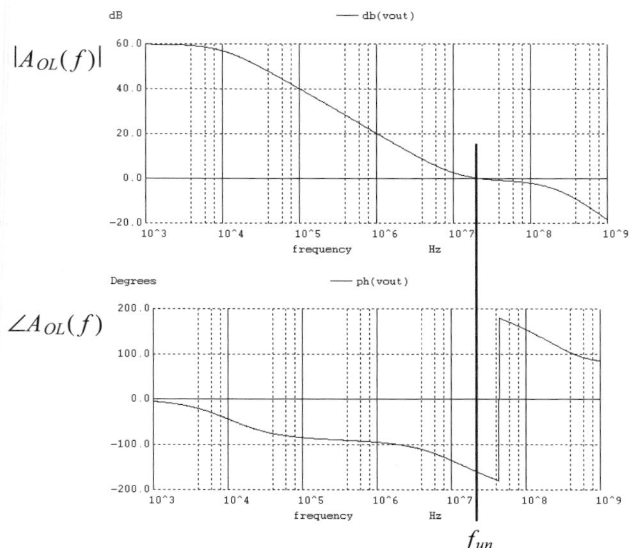

Figure 24.11 Increasing the compensation capacitor's value to 2.4 pF.

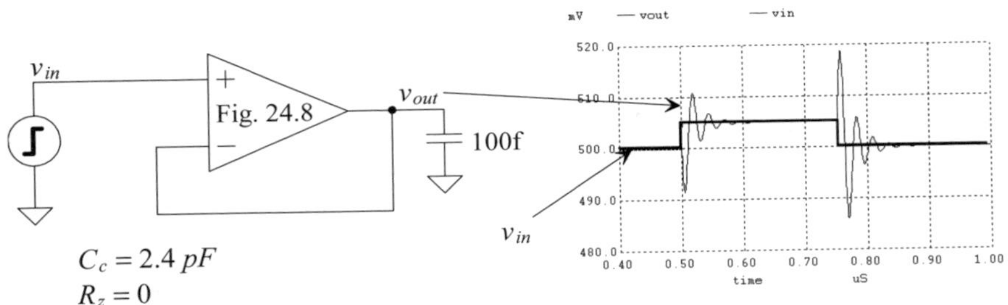

Figure 24.12 The poor step response of the op-amp with the zero present.

Gain and Phase Margins

In the previous discussions we assumed a load of 100 fF. If the load capacitance varies, the stability can (will) be affected. Further, with temperature, process, and power supply variations, the stability of the op-amp can change. In order to specify "how stable" the op-amp is at a given set of operating conditions, the parameters gain margin (GM) and phase margin (PM) are used. To determine an op-amp's PM, we look at the phase shift when the open-loop gain is unity. The amount of phase shift away from 180° is the PM of the op-amp. As seen in Fig. 24.13, the phase shift when the gain is unity is –90°. Taking the difference between this value and 180° gives a PM of 90°. To calculate the GM, we look at the difference between the open-loop gain and unity when the phase of the op-amp is ±180°. For the op-amp response in Fig. 24.13, the GM is approximately 25 dB. Note that as seen in Fig. 24.14, it is nice to have a PM of 90° because the step response of the op-amp has a first-order response (like an RC circuit).

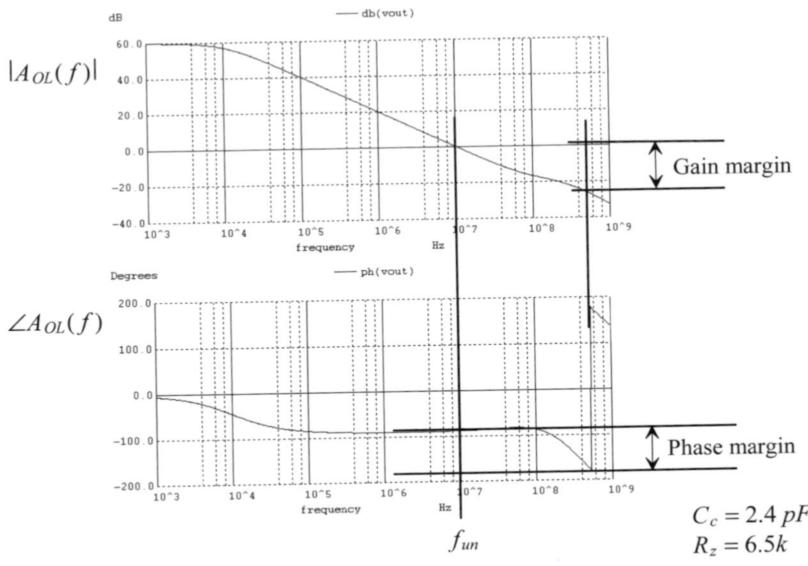

Figure 24.13 Adding a zero nulling resistor to the op-amp in Fig. 24.8.

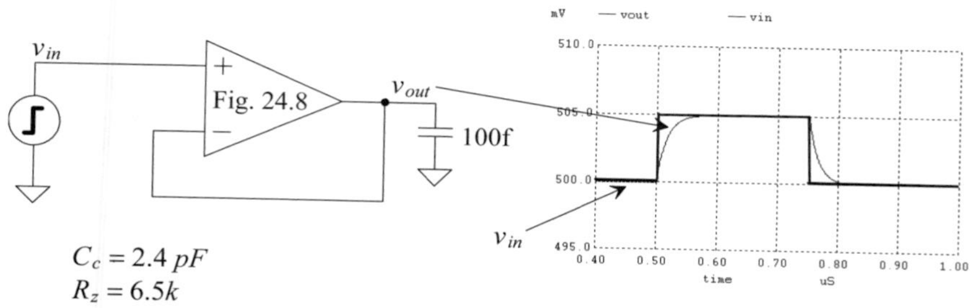

Figure 24.14 Good step response of the op-amp with the zero absent.

Removing the Zero

As the previous discussion illustrated, the RHP zero can raise issues concerning stability and settling time (this is important). As seen in Eq. (21.73) and the associated discussion, R_z can be added to eliminate ($R_z = 1/g_{m1}$) or move the zero into the LHP ($R_z > 1/g_{m1}$). When the zero is moved to the LHP, the phase response of the zero adds to the overall phase response, increasing the PM (this is called *lead compensation*). The practical problem with using R_z is that setting its value to a precise number (say $1/g_{m1}$) is challenging with shifts in process, temperature, or voltage. One solution to this problem is to replace the resistor with a MOSFET operating in the triode region, Fig. 24.15. Mz behaves like a resistor with a value of $1/g_{m1}$. Ideally, the source-gate voltage of M7 is the same as the V_{SG} of MP1. It then follows that the source-gate potential of MP2 equals the source-gate potential of Mz. The channel resistance of Mz is then, from Eq. (9.16), set to $1/g_{m1}$. The practical issues with this method are the wasted power dissipated by the additional circuitry and the fact that if the output swing becomes large (especially at higher frequencies where C_c has a small impedance), Mz can move out of the triode region, which can affect the large-signal behavior of the op-amp.

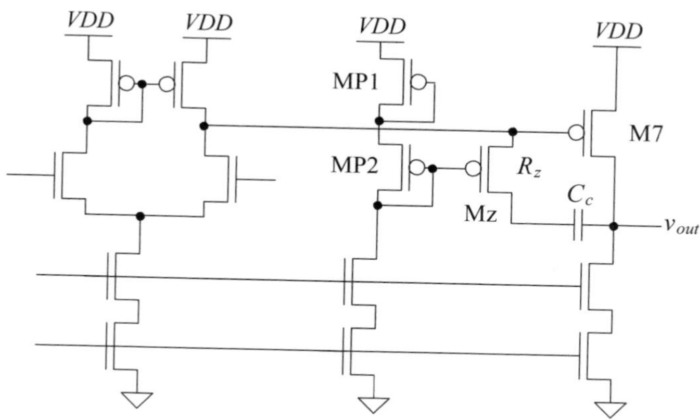

Figure 24.15 Making the zero-nulling resistor process independent.

The other method to remove the RHP zero seen in Fig. 21.33a is to add an amplifier with a gain of +1 in series with the compensation capacitor. Figure 24.16 shows the idea. A source follower allows the output signal to feed back through the compensation capacitor (so that the pole-splitting effect is still present). However, the root cause of the RHP zero, namely, C_c shorting the input of the second stage (the output of the diff-amp) to the output of the second stage (the output of the op-amp) at higher frequencies is removed. The concerns with this topology are, again, power dissipation and, perhaps more importantly, large signals. If the overall output voltage swings too low, it causes the source-follower to shut off.

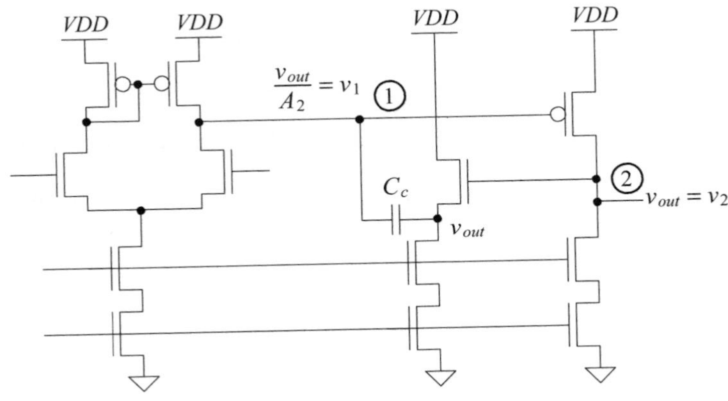

Figure 24.16 Using an amplifier to eliminate forward signal feedthrough via the compensation capacitor.

Compensation for High-Speed Operation

Notice, in Fig. 24.16 (or in any other of the topologies we've discussed up to this point), that the current fed back through C_c is

$$i_{Cc} = \frac{v_{out} - \frac{v_{out}}{A_2}}{1/j\omega C_c} \tag{24.11}$$

If the second-stage gain, A_2 ($g_{m2}R_2 = g_{m7}r_{o7}$) is reasonably large, then this equation can be approximated using

$$i_{Cc} \approx \frac{v_{out}}{1/j\omega C_c} \tag{24.12}$$

If we can feed back this current *indirectly* to the output of the diff-amp, we can still compensate the op-amp (and have pole-splitting). Further, if we do it correctly, we avoid connecting the compensation capacitor directly to the output of the diff-amp and thus avoid the RHP zero. Towards this goal, consider the modified op-amp schematic seen in Fig. 24.17. The added MOSFETs form a common-gate amplifier. (Note that here we are assuming $v_p \ll v_{out}$, so the source-follower action of MCG is negligible. MCG is connected to v_p to set its gate at the DC voltage of the input.) The current i_{Cc} is fed back through the common-gate MOSFET, MCG, to node 1 (the output of the diff-amp).

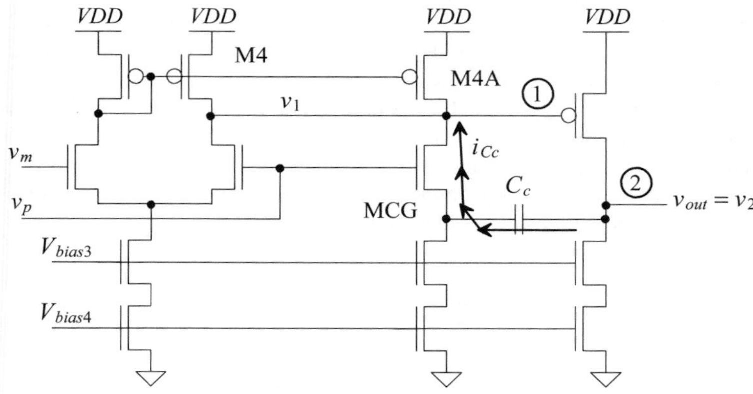

Figure 24.17 Feeding back a current indirectly to avoid the RHP zero.

To determine the frequency response of this amplifier, consider the model seen in Fig. 24.18 (modified from Fig. 21.25). Summing the currents at node 1 gives

$$-g_{m1}v_s + \frac{v_1}{R_1 || \frac{1}{j\omega C_1}} - \overbrace{\frac{v_{out}}{1/j\omega C_c + 1/g_{mcg}}}^{i_{Cc}} = 0 \tag{24.13}$$

where the resistance looking into the source of MCG is $1/g_{mcg}$. Solving for v_1 gives

$$v_1 = \frac{g_{m1}R_1}{1 + j\omega R_1 C_1} \cdot \left(\frac{j\omega \frac{C_c}{g_{m1}}}{1 + j\omega \frac{C_c}{g_{mcg}}} \cdot v_{out} + v_s \right) \tag{24.14}$$

For the output node (node 2), we can write (noting that now C_2 includes both C_c and C_L)

$$v_{out} = -g_{m2}v_1 \cdot \left(\frac{R_2}{1 + j\omega R_2 C_2} \right) \tag{24.15}$$

Plugging Eq. (24.14) into this equation gives

$$v_{out} = -g_{m1}R_1 g_{m2}R_2 \cdot \frac{\frac{j\omega \frac{C_c}{g_{m1}}}{1 + j\omega \frac{C_c}{g_{mcg}}} \cdot v_{out} + v_s}{(1 + j\omega R_1 C_1)(1 + j\omega R_2 C_2)} \tag{24.16}$$

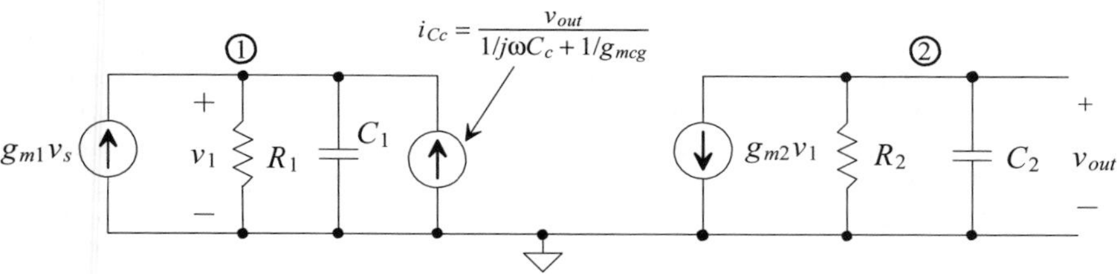

Figure 24.18 Model used to estimate bandwidth when indirect feedback current is used.

Letting

$$K = \frac{-g_{m1}R_1 g_{m2}R_2}{(1+j\omega R_1C_1)(1+j\omega R_2C_2)} \tag{24.17}$$

we can rewrite Eq. (24.16) as

$$v_{out} = \frac{j\omega\frac{C_c}{g_{m1}}}{1+j\omega\frac{C_c}{g_{mcg}}}\cdot K\cdot v_{out} + K\cdot v_s \tag{24.18}$$

and thus

$$\frac{v_{out}}{v_s} = \frac{K\cdot\left(1+j\omega\frac{C_c}{g_{mcg}}\right)}{1-j\omega\frac{C_cK}{g_{m1}}} \tag{24.19}$$

Resubstituting in K gives

$$\frac{v_{out}}{v_s} = \frac{-g_{m1}R_1g_{m2}R_2\left(1+j\omega\frac{C_c}{g_{mcg}}\right)}{(1+j\omega R_1C_1)(1+j\omega R_2C_2)+j\omega\frac{C_c}{g_{m1}}\cdot g_{m1}R_1g_{m2}R_2} \tag{24.20}$$

or, with $s = j\omega$, we can write

$$\frac{v_{out}}{v_s} = \frac{-g_{m1}R_1g_{m2}R_2\left(1+s\frac{C_c}{g_{mcg}}\right)}{s^2\cdot(R_1C_1R_2C_2)+s\cdot(R_1C_1+R_2C_2+R_1g_{m2}R_2C_c)+1} \tag{24.21}$$

Notice there is a LHP zero at

$$f_z = \frac{g_{mcg}}{2\pi C_c} \tag{24.22}$$

Since this zero is in the LHP, it will add to the phase response and enhance the speed of the op-amp. Intuitively, we can think that at high speeds the phase shift through C_c will cause the output signal to feed back and add to the signal at node 1. This *positive feedback* enhances the speed of the op-amp. To determine the location of the second pole, let's assume $R_1g_{m2}R_2C_c >> R_1C_1$ or R_2C_2 or $R_1C_1R_2C_2$ so

$$s_{1,2} \approx \frac{-R_1g_{m2}R_2C_c \pm R_1g_{m2}R_2C_c}{2(R_1C_1R_2C_2)} \tag{24.23}$$

Our approximation won't tell us the location of the lower frequency pole (it's given by Eq. (21.65)). The location of the second pole is

$$f_2 \approx \frac{g_{m2}C_c}{2\pi\cdot C_1C_2} \approx \frac{g_{m2}C_c}{2\pi\cdot C_1(C_L+C_c)} \tag{24.24}$$

This result should be compared to Eq. (21.66). The location of the second pole is at a considerably higher frequency using this technique. *The result is that we can set the unity gain frequency to a higher value and still have a stable op-amp.* Further, the load capacitance (which is included in C_2) can be considerably larger for a given PM or GM. Again, Eq. (21.72) can be used to set the unity-gain frequency, f_{un}, assuming $f_2 \approx f_z$

$$f_{un} = \frac{g_{m1}}{2\pi C_c}\ (\approx f_z \text{ if } g_{m1} \approx g_{mcg}) \tag{24.25}$$

Using the *indirect compensation* method seen in Fig. 24.17 in the op-amp of Fig. 24.8 with C_c = 240 *fF* (f_{un} = 100 MHz) gives the results seen in Fig. 24.19. The small-signal step response is seen in Fig. 24.20. This response should be compared to Figs. (24.12) and (24.14).

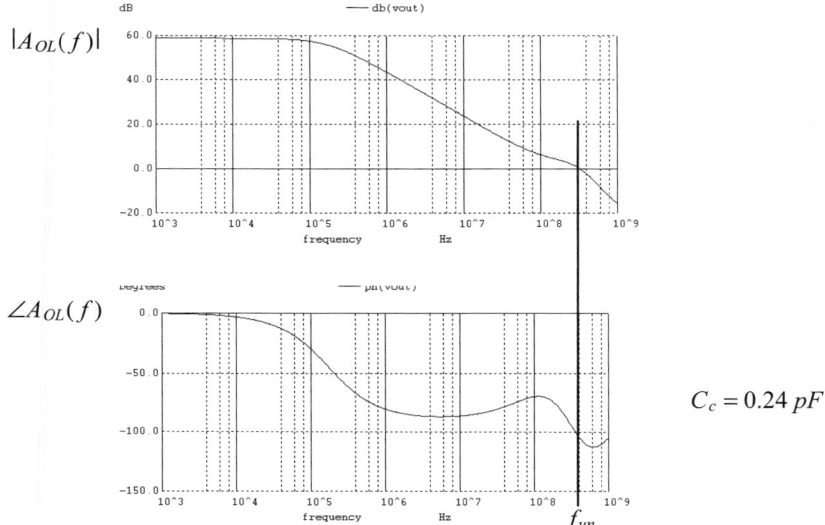

Figure 24.19 Simulating the op-amp of Fig. 24.8 using the indirect compensation scheme seen in Fig. 24.17 with a compensation capacitor of 240 fF.

The indirect feedback of the current through the compensation capacitor results in faster op-amp circuits and less layout area (the compensation capacitor generally dominates the layout area of an op-amp). However, the added circuit in Fig. 24.17 dissipates more power. To eliminate the additional power dissipation, consider the op-amp topology seen in Fig. 24.21. Here we've used the fact that a 100/2 PMOS device can be laid out as two 100/1 PMOS in series (see also Fig. 20.35). The current through the compensation capacitor is fed back through M4B to the output of the diff-amp. Note how we've also split M7 into two devices. It's important, for small offsets, to try to match both the drain-source and the gate-source voltages of the MOSFETs used in the op-amp. The step-response of the op-amp is seen in Fig. 24.22.

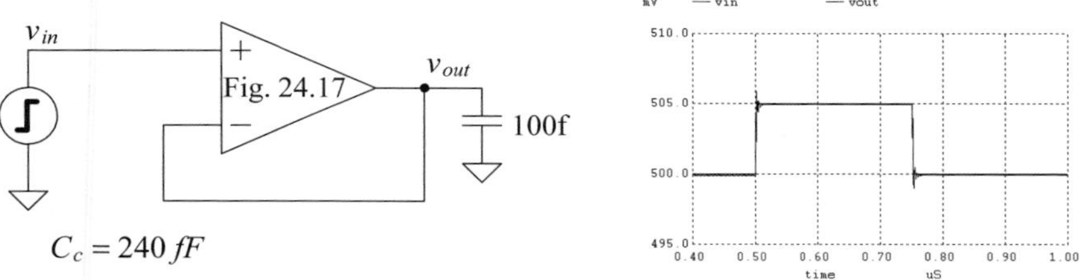

Figure 24.20 The step response of the op-amp in Fig. 24.17 with frequency response seen in Fig. 24.19.

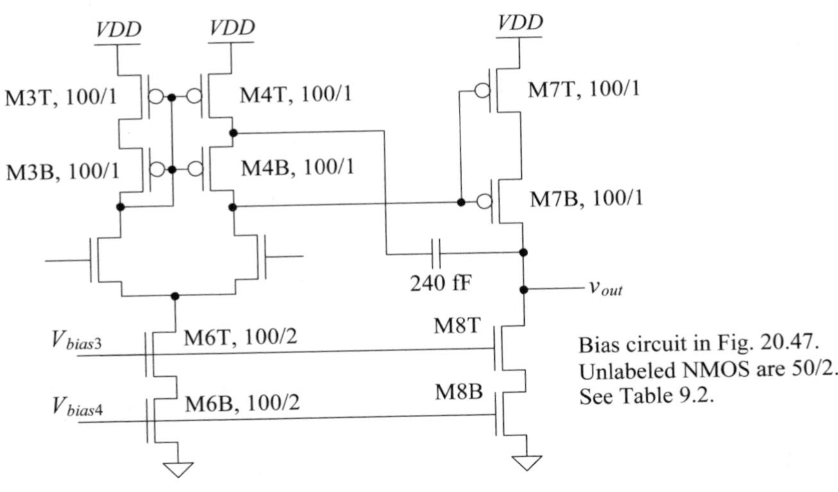

Figure 24.21 Implementing indirect feedback compensation without additional power dissipation in a two-stage op-amp.

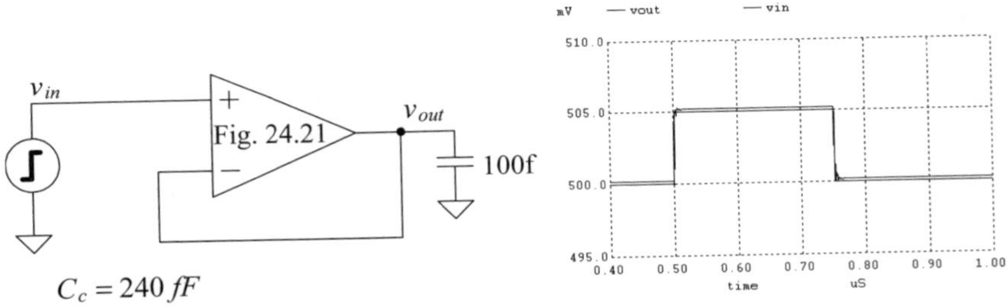

Figure 24.22 The step response of the op-amp in Fig. 24.21 driving 100 fF.

Slew-Rate Limitations

When we talked about step response, we limited the amplitude of the signals to small steps (5 mV) to avoid slew-rate limitations. Let's take the op-amp from Fig. 24.8 and put it in the configuration seen in Fig. 24.14, driving a 1 pF load capacitance. Further, let's increase the input pulse amplitude to (close to) the maximum allowable range, that is, from 500 mV to 900 mV (limited by the op-amp's input common-mode range). The simulated results are seen in Fig. 24.23. Referring to Fig. 24.8, notice that when the op-amp's noninverting input terminal (v_p) is driven high, M2 turns on and pulls the gate of M7 down. Since there isn't a current source in series with M7, it can quickly charge the 1 pF load capacitance. The only large-signal limitation is the bias current of the diff-amp, I_{SS} (= 20 μA here), charging C_c (= 2.4 pF here). This limits the output rate of change since any variation in v_{out} causes a displacement current through C_c, which must be sourced/sunk by the diff-amp. This slew-rate limitation is calculated as 20 μA/2.4 pF or 8.3 mV/ns. The change in the output voltage is 400 mV, so we would require roughly 50 ns to change the voltage across C_c.

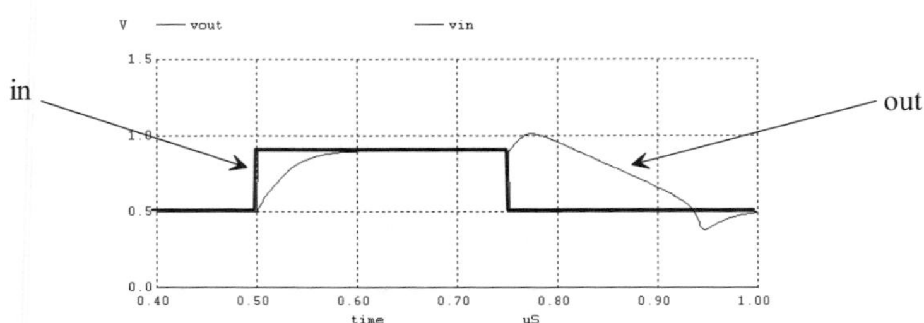

Figure 24.23 The large-signal (500 mV to 900 mV) performance of the op-amp in Fig. 24.8 with C_c = 2.4 pF and R_z = 6.5 k driving a 1pF load. The low-to-high settling time is roughly 100 ns, while the high-to-low settling time, which is slew-rate limited, is roughly 500 ns.

The more interesting behavior occurs when the input signal drops from 900 mV to 500 mV. This causes M2 to shut off, allowing all of the current from M4 to charge the gate of M7. The result is a positive pulse on the gate of M7 that feeds directly to the output through the compensation capacitor (this is bad, and it is another example of why it is desirable to use the indirect compensation method). This pulse accounts for the (unwanted) positive movement in the output voltage seen in Fig. 24.23 just after the input pulse transitions negative. Now the load capacitor and the compensation capacitor must both be discharged through the constant current source M8. The output slew-rate is estimated by how fast this constant current (here 10 μA) can discharge 3.4 pF (the sum of the compensation capacitor and the load capacitor). This rate is calculated as (roughly) 3 mV/ns. For the output to transition 400 mV requires 133 ns (this time will be longer because of the positive movement in the output voltage just after the input switches). Note that we might think that simply by sizing up the current conducting in M8 we can enhance the op-amp's slew-rate. This is true to the point where the diff-amp's current charging the compensation capacitor becomes the limiting factor.

Finally, note how the output voltage shoots past the final (desired) voltage level of 500 mV. This is because the gate of M7 is pulled to *VDD* to shut it off when slewing is taking place (both M4 and M7 are shut off because of this). When the inputs of the op-amp move to the same value (500 mV here), the gate of M7 must be pulled downwards to the quiescent value. The time it takes for this to happen results in the undershoot in the op-amp's output voltage seen in Fig. 24.23.

The same test setup used to generate the data in Fig. 24.23 is also used to generate the data in Fig. 24.24 except that here the op-amp in Fig. 24.21 is used. Note the compressed time scale. The settling time for the low-to-high transition is now 10 ns, while the high-to-low transition time is 60 ns. We would expect much faster settling because the compensation capacitor is now ten times smaller than the value used in the op-amp of Fig. 24.8. Again, the settling time is limited by how fast M8 can discharge the load capacitance and the compensation capacitance. This can be estimated as 10 μA/1.24 pF or 8 mV/ns. It takes 50 ns for the output to transition 400 mV. The size of M8 can be increased to meet a specific load capacitance and settling time requirement.

Because of the improved speed performance, smaller layout area, and (as we'll see) better power supply noise rejection, we'll use indirect compensation in place of Miller compensation in the remaining two-stage op-amps that we discuss in this book.

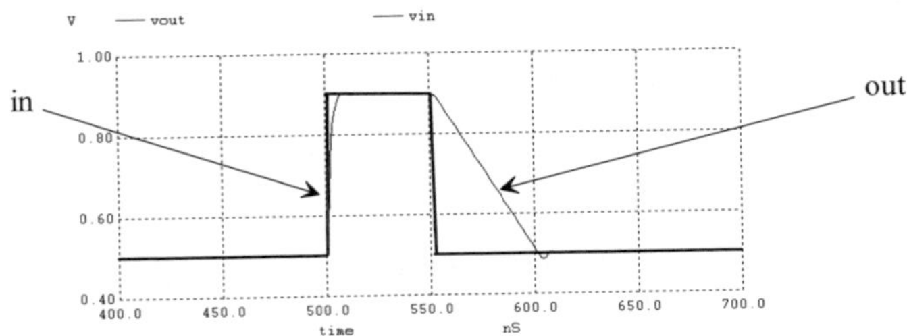

Figure 24.24 The large-signal (500 mV to 900 mV) performance of the op-amp in Fig. 24.21 with $C_c = 0.24$ pF driving a 1 pF load. The low-to-high settling time is roughly 10 ns, while the high-to-low settling time, which is slew-rate limited, is roughly 60 ns. Note the different time scale when compared to Fig. 24.23.

Common-Mode Rejection Ratio (CMRR)

The *CMRR* of an op-amp is calculated in the same way as the diff-amp in Sec. 22.1.3. The common-mode gain of the diff-amp is A_c. The common-mode gain of the op-amp is $A_c A_2$. The differential gain of the op-amp is $A_{OL}(f) = A_d \cdot A_2$ (where $A_d = A_1$). The *CMRR* of an op-amp in dB is given by

$$CMRR = 20 \cdot \log \left| \frac{A_{OL}(f)}{A_c \cdot A_2} \right| = 20 \cdot \log \left| \frac{A_d}{A_c} \right| \tag{24.26}$$

which shows that the op-amp *CMRR* is determined by the differential stage. Simulating the *CMRR* of an op-amp can be accomplished with the circuits seen in Fig. 24.25. For the op-amps discussed so far in this chapter, the *CMRR* is approximately 50 dB (see Ex. 22.7, where the common-mode voltage is 500 mV). We can think of A_c as the open-loop gain of the op-amp for common-mode signals. If the applied differential voltage is zero, and we change the common-mode voltage by ΔV_c, then the output voltage will change by ΔV_o

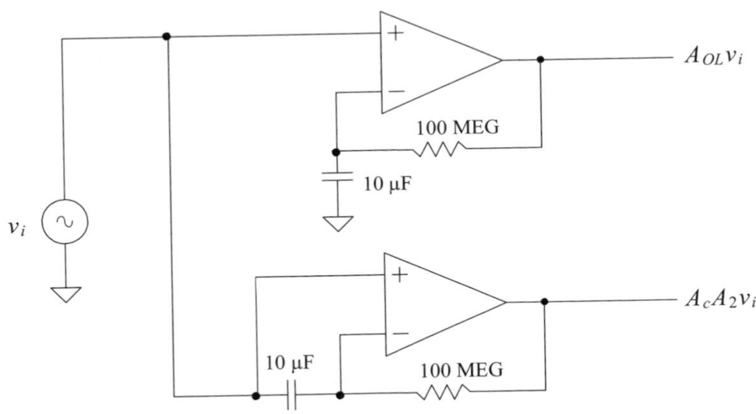

Figure 24.25 Circuit configuration used to simulate CMRR.

$= A_{cm} \cdot \Delta V_c$ (see Fig. 22.17). To compensate for the change in the output voltage, a nonzero input differential voltage develops on the input of the op-amp (an offset voltage that is a function of the common-mode voltage). This offset voltage can be estimated by

$$\Delta V_{OS} = \frac{\Delta V_o}{A_{OL}} = \frac{\Delta V_c \cdot A_{cm}}{A_{OL}} = \frac{\Delta V_c}{CMRR} \tag{24.27}$$

Knowing that 50 dB = 316, a change in the common-mode level on the inputs of the op-amp by 500 mV results in an input-referred offset voltage change of 500 mV/316 or 1.6 mV (if the gain of the diff-amp is 10, then its output voltage will change by 16 mV). This may not seem like such a big deal. However, notice that at higher frequencies, in Fig. 22.16, the *CMRR* falls off, indicating that the common-mode gain is increasing. This leads to distortion and forces the use of op-amp topologies in which common-mode voltage doesn't vary (the inverting op-amp topology, see Fig. 24.26).

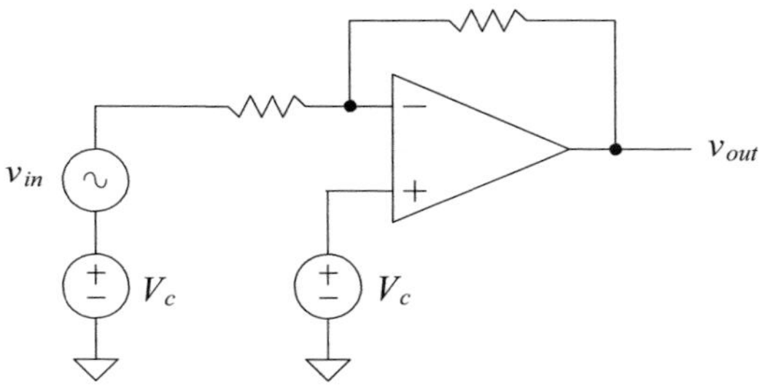

Figure 24.26 An inverting op-amp topology. The common mode voltage is held constant.

Power Supply Rejection Ratio (PSRR)

The power supply rejection ratio (PSRR) is a term used to describe how well an amplifier rejects noise or changes on the *VDD* and or ground power buses. This parameter can be extremely important in precision analog design. Consider the test setup shown in Fig. 24.27. The positive PSRR is defined by

$$PSRR^{+} = \frac{A_{OL}(f)}{v_{out}/v^{+}} \tag{24.28}$$

while the negative PSRR is defined by

$$PSRR^{-} = \frac{A_{OL}(f)}{v_{out}/v^{-}} \tag{24.29}$$

Ideally, v_{out} doesn't vary with changes in *VDD* and ground (and so the *PSRR* is infinite).

To understand how variations in *VDD* or ground can feed to the output of the amplifier, consider the op-amp in Fig. 24.8. Variations in *VDD* feed through to the gate of M3. Because of the circuit's symmetry, this causes the drain of M4, and thus the gate of M7, to move around. However, the source of M7 is moving the same amount, causing the v_{sg} of M7 to remain constant. Again, because of the symmetry, the output voltage moves at the same rate. Thus, noise on *VDD* feeds directly to the output of the op-amp. This is indicated in Fig. 24.27c, where v_{out} / v^{+} is one. At higher frequencies (in the kHz range), the gate and drain of M7 get shorted together through the compensation capacitor. This, again, causes all of the noise from *VDD* to feed directly to the output of the amplifier (see netlists at cmosedu.com). The indirect compensation scheme used in the op-amp of Fig. 24.21 can be designed for larger f_{un} and, thus, has better *PSRR* at higher frequencies. Using the common-gate stage, Fig. 24.17, can result in an even higher *PSRR* by isolating the compensation capacitor from the power supplies. For this reason op-amp's that use a cascode-load diff-amp, Fig. 24.29, are very useful in practical design.

Noise on ground ideally won't affect the operation of the op-amp (in either of the topologies of Fig. 24.8 or 24.21). The noise appears as a voltage variation across the current sources. In reality, at low frequencies, some of the ground noise, v^{-}, appears at the output (this is especially true in nanometer processes that have low r_o). However, at higher frequencies, because of the compensation capacitance connected to the output of the op-amp, ground noise contributions to the output signal decrease.

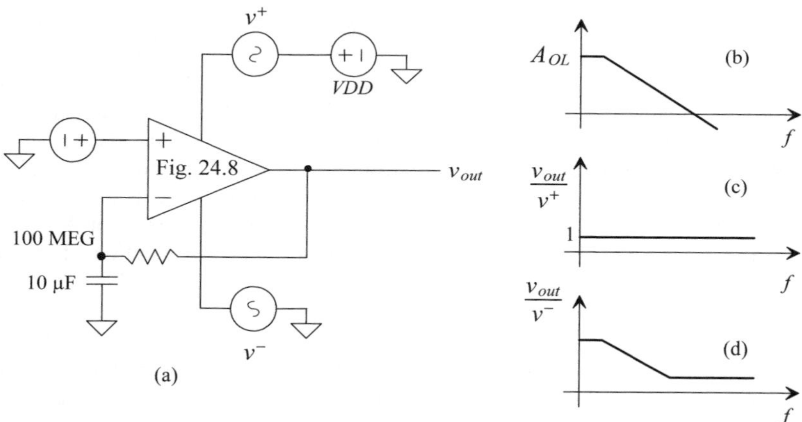

Figure 24.27 (a) Test setup to determine PSRR, (b) open-loop gain, (c) gain from AC signal on VDD to output, and (d) gain from AC signal on ground to output.

Increasing the Input Common-Mode Voltage Range

It may be useful in some situations to have an input common-mode voltage range that extends close to the power supply rails. For our op-amp in Fig. 24.21, the allowable input common-mode voltage range is from (roughly) 450 mV to 900 mV (basically half of *VDD*). Towards the goal of extending the input common-mode voltage range, consider the addition of a PMOS diff-amp to Fig. 24.21, as seen in Fig. 24.28. The circuitry on the right side of the schematic is the op-amp in Fig. 24.21 drawn a little differently. Here, we've drawn the current source of the diff-amp as two parallel current sources. In Fig. 24.21 the current source biasing the NMOS diff-amp used MOSFETs with 100/2 sizes. Here we supply the same current but use two 50/2-sized current sources. Also, we've bumped up the size of the PMOS devices (M3, M4, and M7) by two. This is to accommodate the extra current supplied by the added PMOS diff-amp (the left side of the schematic). We could keep the PMOS devices the same size as seen in Fig. 24.21 and the

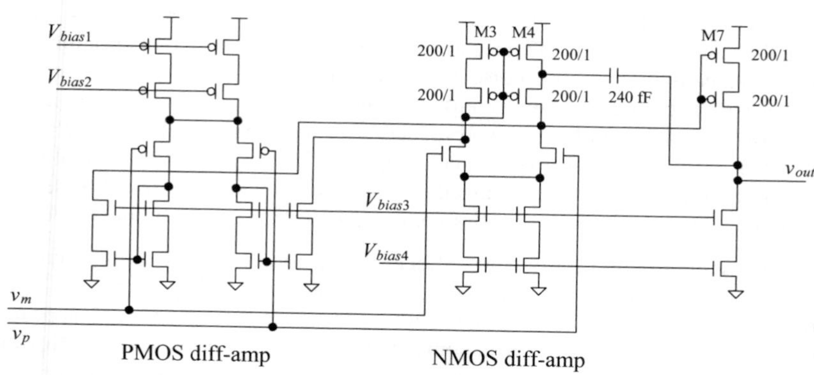

Figure 24.28 Two-stage op-amp of Fig. 24.21 with rail-to-rail input range.

circuit would still work fine (the V_{SG} voltages of the PMOS would simply be a little larger when both diff-amps are on). Two other notes: notice that the DC currents flowing in M3, M4, and M7 change depending on the input common-mode voltage. When the input common-mode voltage is large, the PMOS diff-pair is off; when it's small, the NMOS diff-pair is off. In the middle, both are conducting current. This means that the transconductance of the diff-amp will vary from g_{mn} to g_{mn} + g_{mp} to g_{mp}. When both diff-amps are on (assuming $g_{mn} = g_{mp}$), the transconductance is twice as high as when a single diff-amp is on. This may require doubling the compensation capacitor. Finally, the change in the DC biasing conditions result in an offset that is a function of the input common-mode voltage.

Estimating Bandwidth in Op-Amp Circuits

When we finish designing an op-amp, its frequency response can be written using a single dominant-pole response (see Fig. 21.26 on page 680 with $A_{DC} = A_{OLDC}$) as

$$A_{OL}(f) = \frac{A_{OLDC}}{1 + j \cdot \frac{f}{f_{3dB}}} \tag{24.30}$$

For the op-amp response in Fig. 24.19, we can write

$$A_{OL}(f) = \frac{1,000}{1 + j\frac{f}{10\,kHz}} \tag{24.31}$$

To estimate the bandwidth of a closed-loop op-amp circuit where $f >> f_{3dB}$, we can write Eq. (24.30) as

$$|A_{OL}(f)| \approx \frac{A_{OLDC}}{\frac{f}{f_{3dB}}} = \frac{A_{OLDC} \cdot f_{3dB}}{f} = \frac{f_{un}}{f} \tag{24.32}$$

noting the unity-gain frequency is calculated using

$$f_{un} = A_{OLDC} \cdot f_{3dB} \tag{24.33}$$

From this equation, we should see why the unity-gain frequency, f_{un}, is called the *gain-bandwidth product*. Note that for every increase in frequency by 10 (decade) above f_{3dB}, we get a decrease in A_{OL} by 10 (–20 dB). Alternatively, for every increase in frequency by 2 (octave) above f_{3dB}, we get a decrease in A_{OL} by 2 (– 6 dB).

The closed-loop bandwidth (which we'll call f_{3dBCL}) of the op-amp circuit cannot be larger than the bandwidth of the op-amp. Therefore, to estimate the bandwidth of a closed-loop op-amp circuit, assuming the op-amp is the limiting factor, we can write

$$A_{CL} \cdot f_{3dBCL} = f_{un} = \text{gain-bandwidth product} \qquad (24.34)$$

If the op-amp circuit has a closed-loop gain of one (a follower configuration as seen in Fig. (24.7)), the bandwidth is f_{un}. The op-amp in Fig. 24.21 used in a follower configuration would have a bandwidth of (roughly) 100 MHz. If this op-amp were used in a closed-loop configuration with a gain of 10, then the bandwidth of the op-amp circuit would be estimated as 10 MHz.

24.2 An Op-Amp with Output Buffer

In the previous section, the low-frequency open loop gain of the op-amps in Figs. 24.8 or 24.21 was determined by a product of the diff-amp's gain, A_1, and the common-source's gain, A_2 or

$$A_{OLDC} = A_1 \cdot A_2 = \overbrace{g_{mn}(r_{o2}||r_{o4})}^{A_1} \cdot \overbrace{g_{mp}r_{o7}}^{A_2} \qquad (24.35)$$

which, from the simulations (see Fig. 24.19 for example) was 1,000. The small-signal resistance on the output of the op-amp, r_{o7}, is (from Table 9.2) 333 kΩ. If we were to connect a 10 kΩ resistor to the op-amp output, then A_{OLDC} would drop to 33. As seen in Fig. 24.1, we can add a buffer to the output of second stage to isolate it from a load resistance (or a large capacitance). We could use a source-follower (that has a voltage gain of 1), as seen in Figs. 21.46 or 21.48. However, this addition limits the output swing of the op-amp (which may be OK in some situations). Next we could try using a push-pull topology, as seen in Fig. 21.49. However, the voltage gain of the push-pull amplifier is >1 (see Eq. (21.116)) so if we were to include it in our basic op-amp of Fig. 24.21, we would have a three-stage op-amp (meaning all three stages have voltage gains greater than one). A three-stage op-amp can be challenging to compensate over production corners (and temperature). Also, as seen in Eq. (21.116), a small load resistor (kΩ) would still kill the push-pull amplifier's gain. To keep the gain high, let's try to use a first-stage topology with a large gain (increase A_1) while using a push-pull amplifier for the second stage. The gain of the second stage can still drop but, since the gain of the first stage is large, we still end up with a reasonable overall gain.

Towards the goal of increasing the first-stage gain, consider the op-amp schematic seen in Fig. 24.29. For the first stage, we've used the cascode diff-amp originally seen in Fig. 22.30. The gain of this diff-amp (now A_1) is 500 (from Tables 9.2, and 20.1 and Eq. (22.48)). The second stage of the op-amp is the topology seen in Fig. 21.50. Its gain depends on the load resistance it drives. If no load is connected to its output, then

$$A_2 = -10 \cdot (g_{mn} + g_{mp}) \cdot \frac{(r_{op}||r_{on})}{10} = -31.6\ V/V \qquad (24.36)$$

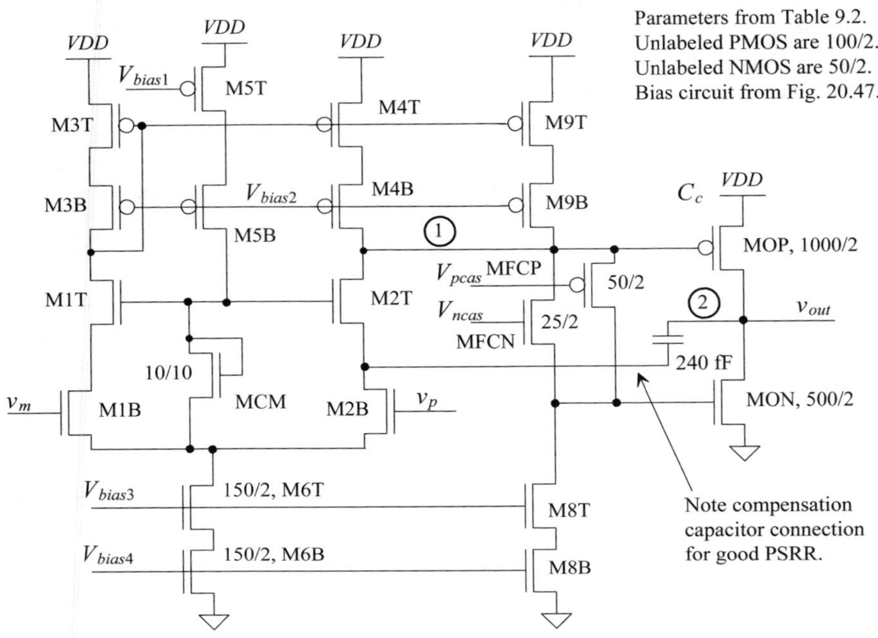

Figure 24.29 A CMOS op-amp with output buffer.

giving an overall low-frequency gain, A_{OLDC}, of 15,800 (= 84 dB). Note that the widths of MON and MOP are ten times wider than the values specified in Table 9.2 and so, as seen in Fig. 21.50, they conduct 100 μA of current. In Eq. (24.36) we multiplied the g_ms in Table 9.2 by 10 ($g_{mon} = 10 \cdot \beta_n(V_{GS} - V_{THN}) = 10 \cdot g_{mn,Table\,9.2}$) and divided the output resistances given in Table 9.2 by 10 $\left(r_{omon} = \frac{1}{\lambda 10 \cdot 10 \mu A} = \frac{r_{on}}{10} \right)$.

Compensating the Op-Amp

We increased the gain of the first stage by increasing R_1 (the resistance at node 1 is now set by cascoded current sources). As seen in Eq. (21.65), this pushes the pole lower in frequency. As just described in Eq. (24.33), this decrease in f_1 (= f_{3dB}) and increase in low-frequency gain *doesn't have any effect* on the gain-bandwidth product (= f_{un}) of the op-amp. What this indicates, to a certain extent, is that we can still compensate the op-amp with a 240 fF capacitor as we did in the last section. To feed the indirect compensation current to node 1 we connect the compensation capacitor to the source of M2T, as seen in Fig. 24.29. This ensures good PSRR.

The open-loop response of the op-amp in Fig. 24.29 (without a load) is seen in Fig. 24.30. The phase-margin is roughly 70 degrees. The step response of the op-amp driving a 1k resistor and a 10 pF capacitor is seen in Fig. 24.31. The input signal is a pulse from 100 mV to 900 mV. Since the gain of the amplifier is −1 and the common-mode voltage is 500 mV, the output swings from 900 mV to 100 mV.

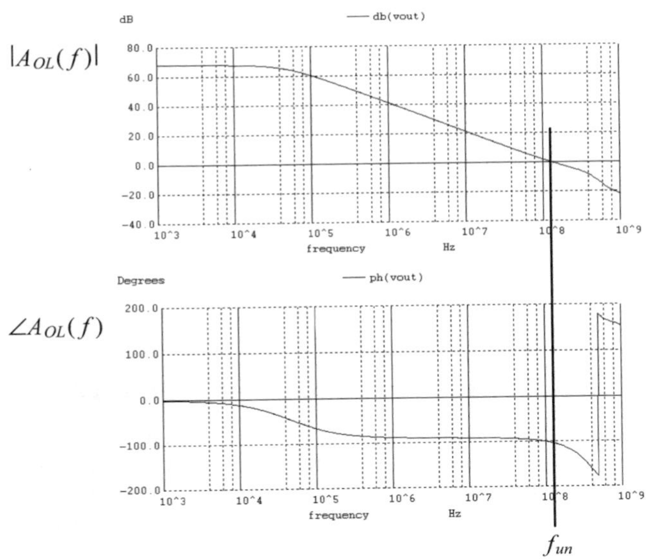

Figure 24.30 Open-loop response of the op-amp in Fig. 24.29.

This is a good point to remember one of the fundamentals we presented in Ch. 9: that is, for high-speed design, we must use minimum length devices (see Eq. (9.59)). If we reduce all of the length of 2 (100 nm) devices in Fig. 24.29 to length of 1 (50 nm) devices, we can increase the speed of the op-amp. Figure 24.32 shows how the step response in Fig. 24.31 changes when we make this reduction in length. The big problem with using minimum channel lengths is the reduction in gain. Close inspection of Fig. 24.32 shows that the output is never quite pulled up to the correct voltage (it is either slightly below 900 mV or 100 mV). Simulating the AC open-loop response results in a gain of 44 dB (= 159). An output voltage of 900 mV would result in a difference on the inputs of the op-amp of 5.66 mV ($0.9/159 = v_p - v_m$).

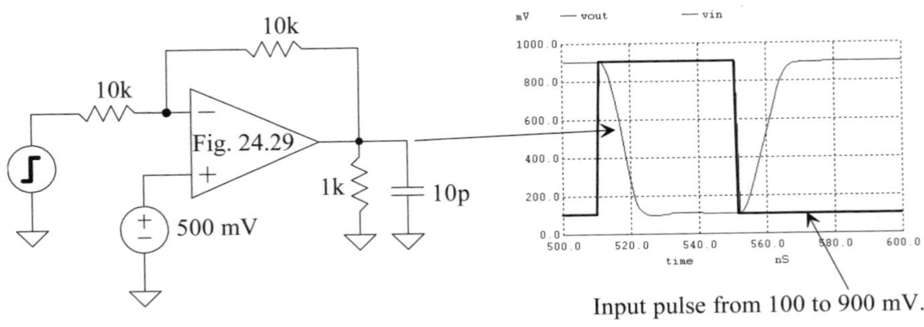

Figure 24.31 Step response of the op-amp in Fig. 24.29 driving 1k and 10 pF.

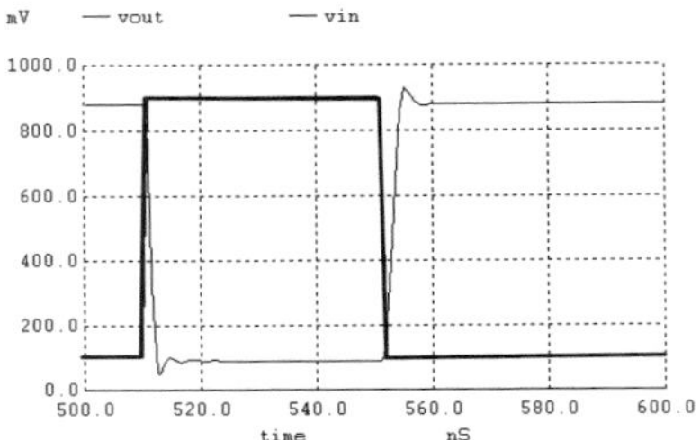

Figure 24.32 Step response of the amplifier in Fig. 24.29 when the lengths of the devices are reduced from 2 to 1 (from 100 nm to 50 nm). The load is still, as seen in Fig. 24.31, 1k and 10 pF.

24.3 The Operational Transconductance Amplifier (OTA)

The operational transconductance amplifier (OTA) can be defined as an amplifier where all nodes are low impedance except the input and output nodes. A simple example of an OTA is the diff-amp with current mirror load seen in Fig. 22.6. *An OTA without buffer can only drive capacitive loads.* A resistive load (unless the resistor is very large) will kill the gain of the OTA.

A sample OTA is shown in Fig. 24.33. Note that the basic op-amp in Fig. 24.2 is not considered an OTA because the drain of M4 is a high-impedance node and not the input or output of the amplifier. As seen in Fig. 24.33, except for the input and output nodes, all nodes have either a gate-drain-connected device or a source connected to them. The terminology "1:K" indicates that M4 and M5 can be sized K times wider (K > 1) than the other MOSFETs in the circuit. Assuming that $\beta_1 = \beta_2$, $\beta_{31} = \beta_{41}$, we observe that the current i_{d31} or i_{d41} is given by

$$-i_{d31} = i_{d41} = \frac{g_{mn}}{2}(v_p - v_m) = i_d \tag{24.37}$$

Furthermore, if $\beta_4 = K \cdot \beta_{41} = K \cdot \beta_{31} = K \cdot \beta_3$ and $K \cdot \beta_{51} = \beta_5$, then $i_{d4} = -i_{d5} = K \cdot i_{d41} = -K \cdot i_{d31}$. If the impedance of the capacitor is large compared to $r_{o4} \| r_{o5}$, then the output voltage of the OTA is given by

$$v_{out} = 2Ki_d(r_{o4} \| r_{o5}) \tag{24.38}$$

and the voltage gain is given by

$$A_v = \frac{v_{out}}{v_p - v_m} = K \cdot g_m \cdot (r_{o4} \| r_{o5}) \tag{24.39}$$

noting that the noninverting input of the OTA is the gate of M2. While the derivation is interesting, as the name implies, we are interested in the transconductance of OTA. If the impedance of the capacitor is small compared to the OTA output resistance (at higher frequencies), we can write

$$i_{out} = i_{d4} - i_{d5} = 2Ki_d \tag{24.40}$$

The transconductance of the OTA is given by

$$g_{mOTA} = \frac{i_{out}}{v_p - v_m} = K \cdot g_m \tag{24.41}$$

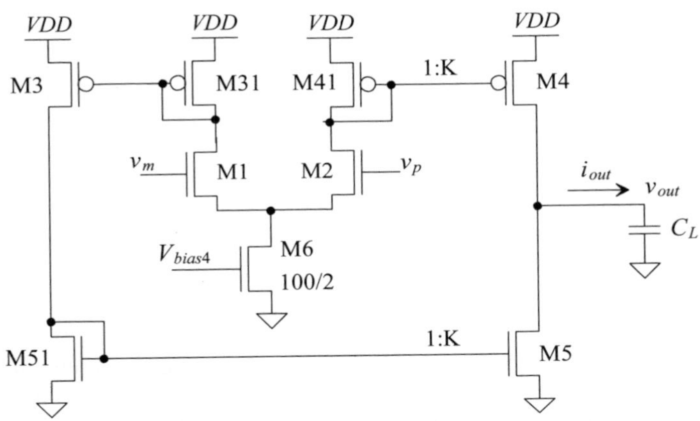

Unless otherwise indicated, parameters from Table 9.2 with biasing circuit in Fig. 20.47.

Figure 24.33 Example of an operational transconductance amplifier (OTA).

Unity-Gain Frequency, f_{un}

To simulate the unity-gain frequency, f_{un}, (where the open loop gain is one) for the OTA, we can use the DC stability scheme seen in Fig. 24.34. We can write (for higher frequencies) the output voltage as

$$v_{out} = i_{out} \cdot \frac{1}{j\omega \cdot C_L} = g_{mn} v_{in} \cdot \frac{1}{j\omega \cdot C_L} \tag{24.42}$$

or

$$\frac{v_{out}}{v_{in}} = \frac{g_{mn}}{j\omega C_L} \rightarrow \left|\frac{v_{out}}{v_{in}}\right| = \frac{g_{mn}}{2\pi f \cdot C_L} \tag{24.43}$$

As before, we define the unity-gain frequency as the point where the open-loop gain is unity

$$\frac{g_{mn}}{2\pi f_{un} \cdot C_L} = 1 \rightarrow f_{un} = \frac{g_{mn}}{2\pi C_L} \tag{24.44}$$

For the OTA in Fig. 24.33 driving a 1 pF load capacitance, we can estimate the unity-gain frequency (with $K = 1$) as

$$f_{un} = \frac{150\ \mu A/V}{2\pi \cdot 1\,pF} = 24\ MHz$$

The simulation results are seen in Fig. 24.34. The location of the pole is estimated using

$$f_{3dB} = \frac{1}{2\pi(r_{o4}||r_{o5})C_L} \tag{24.45}$$

For the OTA in Fig. 24.33, f_{3dB} = 1.4 MHz using the values in Table 9.2. The low-frequency gain is $g_m(r_{o4}||r_{o5})$ = 16.65 V/V (= 24.4 dB).

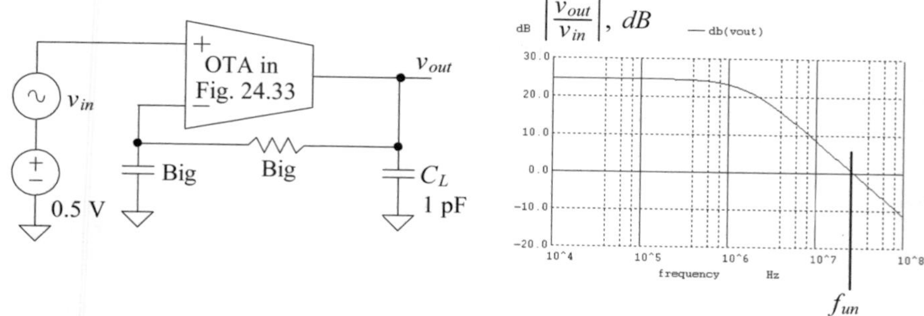

Figure 24.34 Simulating the open-loop response of the OTA in Fig. 24.33.

Note that the maximum value of i_{out} is KI_{SS} (the drain current of M6 multiplied by the increase in widths, K, in the current mirrors M41/M4 and M51/M5). This is important to understand because using an OTA for the first stage of an op-amp results in the same slew-rate limitations of I_{SS} driving C_c that we had when using a diff-amp.

Increasing the OTA Output Resistance

The ideal OTA has infinite output resistance. All of i_{out} flows in the external capacitive load and none flows in the OTA's own output resistance (which is $r_{o4}||r_{o5}$ for the OTA in Fig. 24.33). Towards increasing the OTA output resistance, consider the configuration seen in Fig. 24.35. This topology is based on the one seen in Fig. 24.33 except that now we've cascoded the current mirrors. Note how we've drawn the diff-amp's tail current source as four MOSFETs instead of two with twice the width (this is the way we would lay it out too). The low-frequency gain of the OTA is now increased to

$$A_v = g_{mn} \cdot (R_{ocasn}||R_{ocasp}) \tag{24.46}$$

The 3-dB frequency is now

$$f_{3dB} = \frac{1}{2\pi(R_{ocasn}||R_{ocasp}) \cdot C_L} \tag{24.47}$$

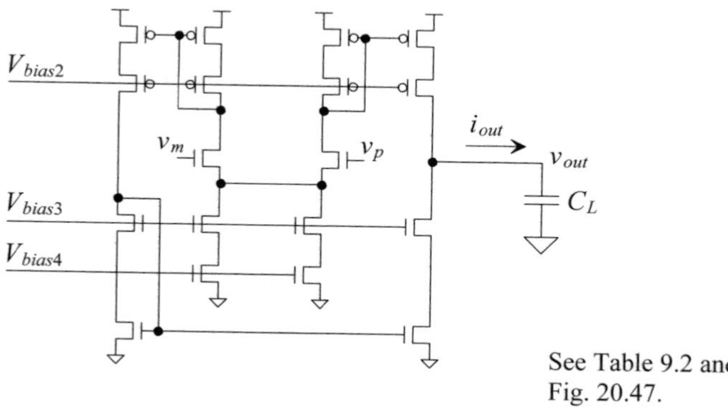

Figure 24.35 A cascode OTA circuit (higher output resistance).

As discussed before (Sec. 24.2 in the discussion concerning compensating the op-amp), when we get a decrease in the 3-dB frequency and an increase in the low-frequency gain, the effects cancel and the gain-bandwidth product (f_{un}) remains constant. The unity-gain frequency is still given by Eq. (24.44). For the OTA in Fig. 24.35, we can estimate the low-frequency gain, using Tables 9.2 and 20.1, as

$$A_v = (150) \cdot (16.6||4.2) = 500\ (54\ dB)$$

and the 3-dB frequency, driving a 1 pF load, is

$$f_{3dB} = \frac{1}{2\pi(16.6||4.2 \times 10^6)(1\ pF)} = 47\ kHz$$

Simulation results are seen in Fig. 24.36. The PM is roughly 85°.

An Important Note

Note, in Eq. (24.44), that increasing the load capacitance, decreases the unity-gain frequency, making the OTA more stable. Unlike the two-stage op-amps discussed in the first two sections of this chapter, where the op-amp can become unstable with a large load capacitance, the OTA only becomes more stable with large load capacitance. Also, the PM approaches 90° as the load capacitance increases. The step response of the OTA has a first-order shape (just as in an RC circuit). In many on-chip applications, the load is purely capacitive and so the OTA works great as an "op-amp" in a closed-loop configuration (thus we'll often call OTAs op-amps). Note that adding a second stage to the basic OTA forms a (two-stage) op-amp like those discussed earlier. We can use the methods already presented to compensate the resulting structure (but, again, driving a load capacitance that's too large with these two-stage structures can result in instability).

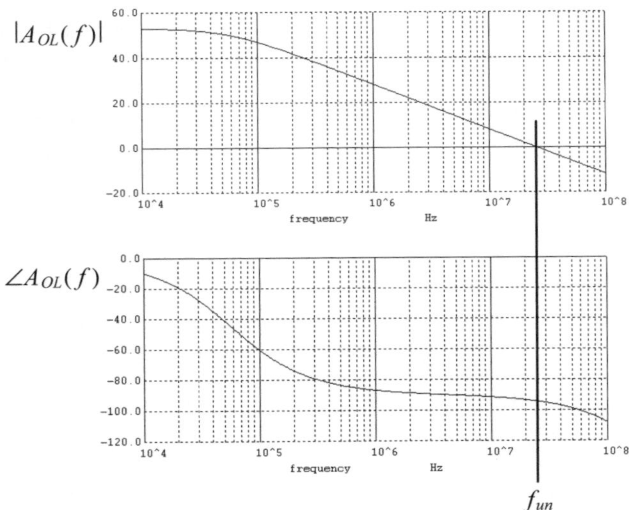

Figure 24.36 The open-loop gain and phase response for the OTA in Fig. 24.35.

OTA with an Output Buffer (An Op-Amp)

If we use the OTA in Fig. 24.35 (or Fig. 24.33) to drive a 1 pF load, we're limited to charging/discharging the capacitor at a rate of

$$\frac{I_{SS}}{C_L} = \frac{dv_{out}}{dt} = \frac{20\ \mu A}{1\ pF} = \frac{20\ mV}{ns} \tag{24.48}$$

For the OTA's output to change from ground to *VDD* (1 V) requires (just due to slewing) 50 ns. This length of time (the slew-rate limited portion of the settling time) may be fine for many applications. However, for high-speed design, we want faster settling times. Looking at Eq. (24.48), we see that we can decrease the settling time by increasing I_{SS}. However, as discussed in Ch. 9, this causes the gain of the circuit to decrease. To avoid the reduction in gain, we can make a corresponding increase in the size of the devices to keep the overdrive voltage constant. But this is doing nothing more than effectively reducing the size of the load capacitor (the load capacitance becomes comparable in size to the parasitic transistor capacitances).

To speed up the circuit, we may add an output stage, Fig. 24.37. The added stage is a common-source amplifier (with a negative gain indicating **we swap the inverting and noninverting terminals** of the op-amp or else we waste a lot of time). While M7 can turn on and charge a load capacitor quickly, the current source, M8, still limits the rate of load capacitance discharge. For this reason, we've bumped up the current in the output stage to 100 μA. The open-loop simulations driving a 1-pF load are seen in Fig. 24.38. The PM is 45° (too low if the op-amp will be used in a unity follower configuration). Figure 24.39 shows the step response of the op-amp.

We should compare this figure to the response we saw in Fig. 24.31. The op-amp in Fig. 24.31 is driving a heavier load but still has cleaner settling behavior. Further, a comparison of the current pulled from *VDD* reveals, under the same loading and operating conditions, that the op-amp in Fig. 24.37 pulls more current (see Fig. 24.39). The main difference in these two op-amps is the output buffer used. Let's consider adding a class AB buffer to the OTA in Fig. 24.35 (like the one used in the op-amp of Fig. 24.29). Since we are only driving a 1 pF load, we'll leave the output devices, MON

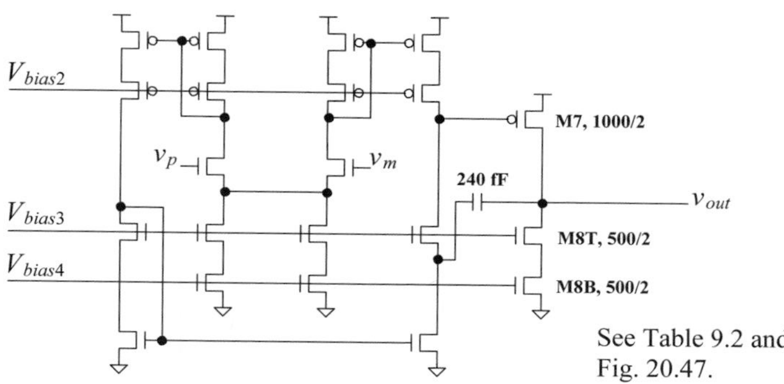

Figure 24.37 A cascode OTA circuit with common-source output buffer (an op-amp). Note the inverting and noninverting terminals are swapped from Fig. 24.35.

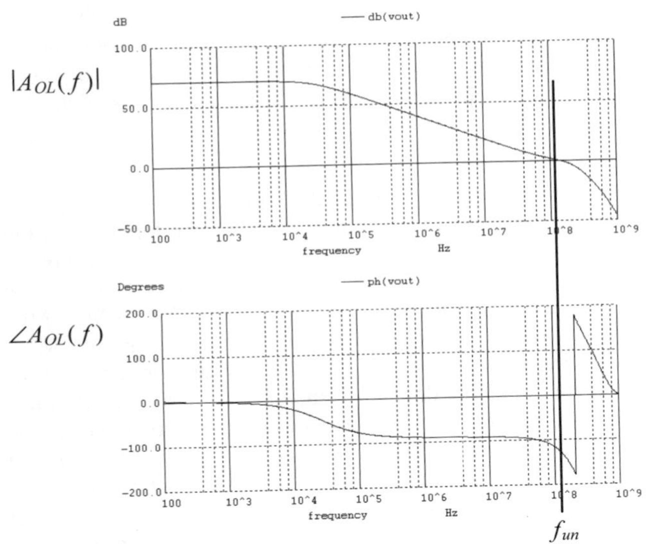

Figure 24.38 Open-loop response of the op-amp in Fig. 24.37 driving a 1-pF load.

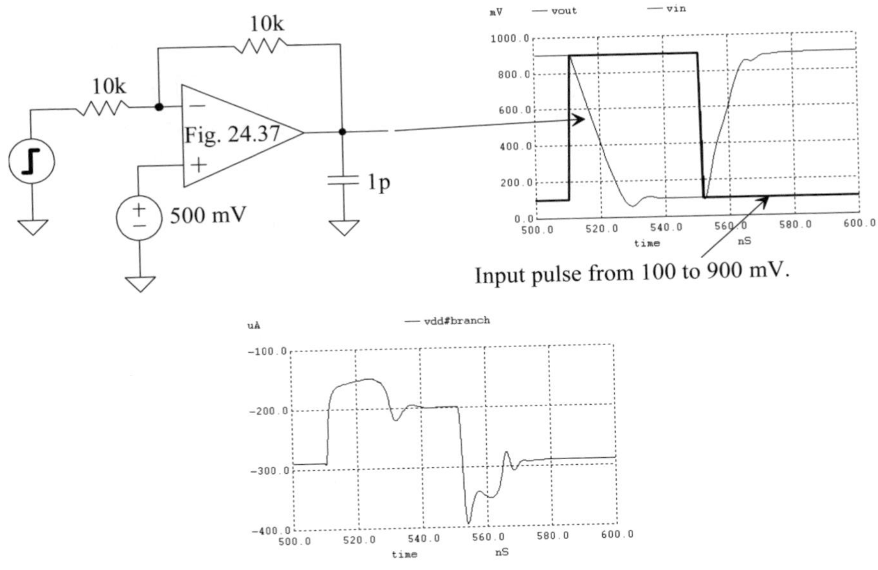

Figure 24.39 Step response of the op-amp in Fig. 24.37 driving a 1 pF.

and MOP, at the same size as the other MOSFETs in the circuit (PMOS are 100/2 and NMOS are 50/2). The resulting op-amp is seen in Fig. 24.40. The step response (in the same configuration used to generate Fig. 24.39) is seen in Fig. 24.41. Notice that the current pulled from *VDD* by the op-amp in Fig. 24.40 is 100 μA less than the current pulled by the op-amp in Fig. 24.37. This lower current is the result of using a class AB output stage.

The performance of the op-amps in Figs. 24.37 and 24.40, with regard to high-speed operation and settling time, won't be quite as good as the topology in Fig. 24.29 (assuming the same biasing conditions, device sizes, and overdrive voltages). To understand why, look at the path, in Fig. 24.40, that the noninverting op-amp input, v_p, takes to get to the op-amp output, v_{out} (that is, M1 - M31 - M3 - M51 - M5 - MON). The delay through this path is longer than the delay, in Fig. 24.29, through M2 - MOP. Further, for high speed, M5 of the op-amp in Fig. 24.40 must turn on (above its quiescent current level) to pull the gate of MON down. For the op-amp in Fig. 24.29, the fact that both the gate of M9T and MOP are driven (in opposite directions) by the diff-amp allows us to use a simple pull-down current (M8) without losing speed. The issues with the cascode diff-amp-based, op-amp topology of Fig. 24.29 are the more limited input common-mode range and the limited voltage swing of node 1. Towards getting better input common-mode voltage range and output swing as well as improved high-speed operation, let's discuss the folded-cascode OTA (see Fig. 22.5).

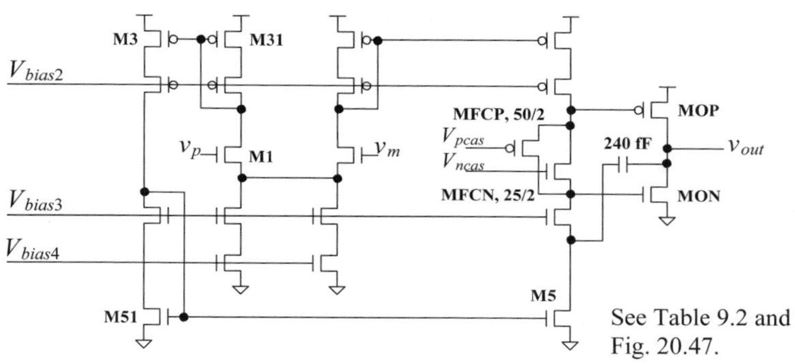

Figure 24.40 Using a class AB output stage with the OTA.

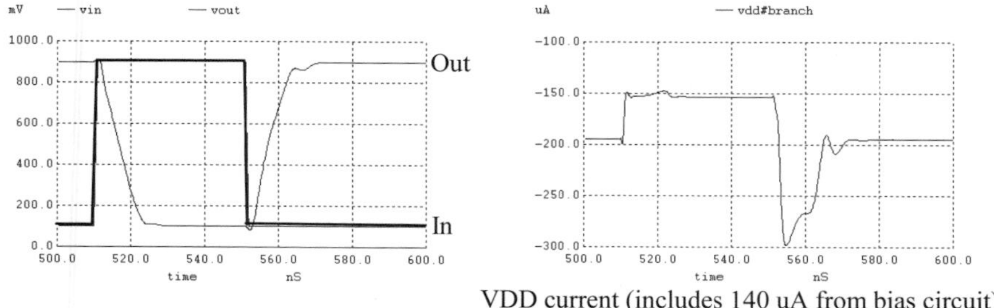

Figure 24.41 The step response and current in the topologies seen in Fig. 24.39 using the op-amp in Fig. 24.40.

The Folded-Cascode OTA and Op-Amp

An example of an NMOS diff-amp-based folded cascode OTA is seen in Fig. 24.42. It is assumed that the reader understands the biasing and operation of the circuits seen in Figs. 20.45, 20.46, and 22.5. To avoid labeling MOSFETs with twice the width, we've drawn the schematic in Fig. 24.42 with MOSFETs in parallel in the branches that supply or sink more current (for example, the diff-amp's tail current). All MOSFETs in Fig. 24.42 conduct 10 μA of current under equilibrium conditions. Before moving on, it may be a good idea to pause and make sure that the reader understands how the currents sum in this OTA.

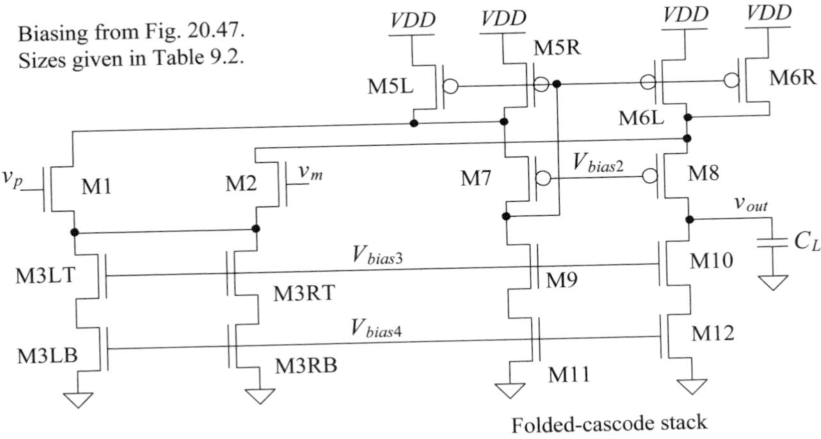

Figure 24.42 A folded-cascode OTA.

To qualitatively describe the operation of the OTA, let's look at what happens when v_p goes high (above v_m). This causes M1 to turn on and M2 to shut off. When M1 turns on, it pulls the drain of M5 down, shutting M7 off. As M7 shuts off, M9 and M11 pull the gate of M5 down. With the gate of M5 being pulled down, the current in M6 increases. At the same time, because the current in M2 is decreasing, the current in M8 is increasing and the output voltage increases. The maximum current the OTA output can supply (out of the OTA to the load) is in the neighborhood of 20 μA, while the maximum current the OTA can sink (into the OTA) is 10 μA (set by the current sink M10/M11).

The same AC equations presented at the beginning of this section also apply to this OTA. For example, the AC drain currents of M1 and M2 can be written as

$$i_{d1} = -i_{d2} = g_{mn} v_{gs1} = -g_{mn} v_{gs2} \tag{24.49}$$

which, again, indicates $v_{gs1} = -v_{gs2}$. However, we know

$$v_p - v_m = v_{gs1} + (-v_{gs2}) \tag{24.50}$$

so

$$i_{d1} = i_{d5} = i_{d6} = (v_p - v_m) \cdot \frac{g_m}{2} \tag{24.51}$$

noting that the AC current through M9/M11 (and thus M7) is ideally zero. The current through M8 is then

$$i_{d8} = i_{d6} - i_{d2} = 2 \cdot i_{d1} \tag{24.52}$$

and thus the output voltage (again, the AC current in M10/M11 is ideally zero) is

$$v_{out} = i_{d8} \cdot (R_{ocasn} || R_{ocasp}) \tag{24.53}$$

and thus the gain is

$$A_v = \frac{v_{out}}{v_p - v_m} = g_{mn} \cdot (R_{ocasn} || R_{ocasp}) \tag{24.54}$$

This result should be compared to Eq. (24.46). Equation (24.47) gives the 3-dB frequency of the OTA's open-loop response, while Eq. (24.44) gives the unity-gain frequency. Figure 24.43 shows the open-loop response of the folded-cascode OTA driving a 1 pF load capacitance.

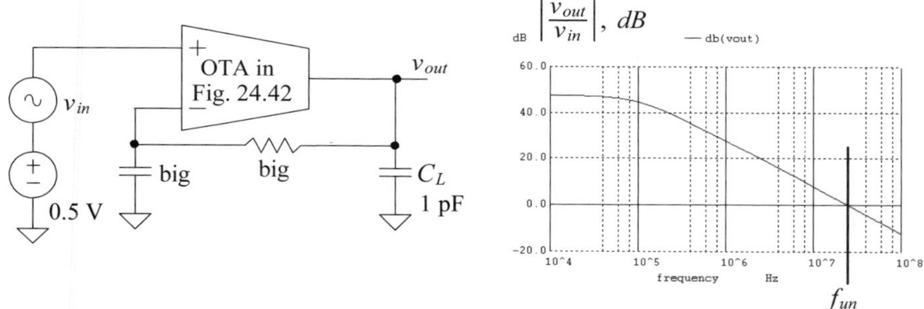

Figure 24.43 Simulating the open-loop response of the OTA in Fig. 24.42.

Figure 24.44 shows the schematic of a folded-cascode op-amp using a class AB output buffer. We've added floating current sources to equalize the voltages across M9/M10 (to minimize the input-referred offset). In the open-loop response of the op-amp (see Fig. 24.45), the load of the op-amp is a 1 pF capacitor. The step response is seen in Fig. 24.46, using the topology of Fig. 24.39 (gain of –1 driving 1 pF and the 10k feedback resistor). Reviewing the previous op-amp responses, we don't see much difference in performance. We can't seem to push the gain-bandwidth product, f_{un}, much above 100 MHz. Again, the only way to increase the speed is to reduce the channel length and/or increase the overdrive voltage.

Figure 24.47 shows what happens if we reduce the lengths of the MOSFETs used in the op-amp in Fig. 24.44 from 2 (100 nm) to 1 (50 nm). The gain-bandwidth product jumps to 400 MHz. However, the low-frequency gain drops to approximately 50 dB. The step response is clean, and the settling time is approximately 5 ns. While we can increase the speed further by redesigning the bias circuit and increasing the currents and overdrive voltage, let's, instead, concentrate on increasing the low-frequency gain of the op-amp. This won't increase the gain-bandwidth product of the op-amp, but it's necessary for precision amplification. To *enhance the gain,* we'll try to increase the OTA output resistance (R_1 or the resistance on node 1) by regulating the drain in the cascode current sources. We've already discussed this technique (see Fig. 20.40), but we haven't applied it to amplifiers. Before leaving this section, let's present one more example.

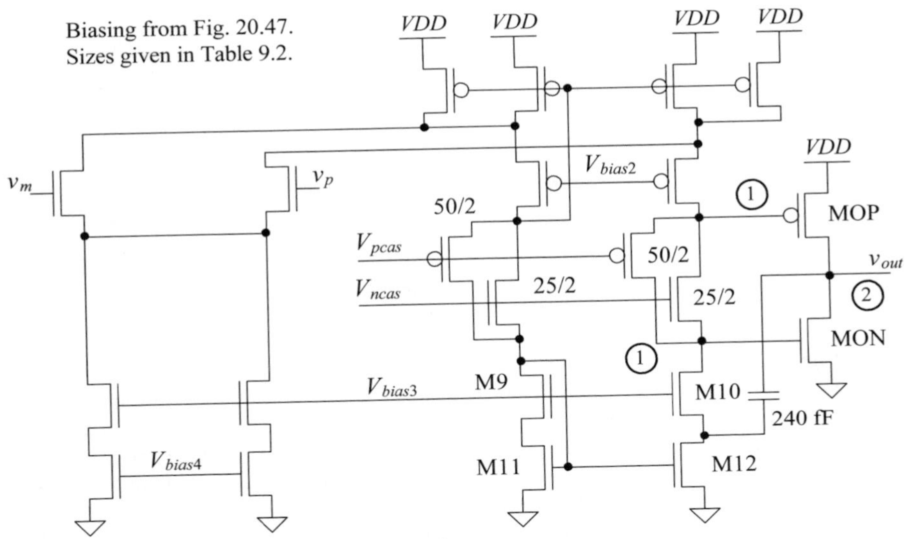

Figure 24.44 Folded-cascode op-amp with class AB output buffer.

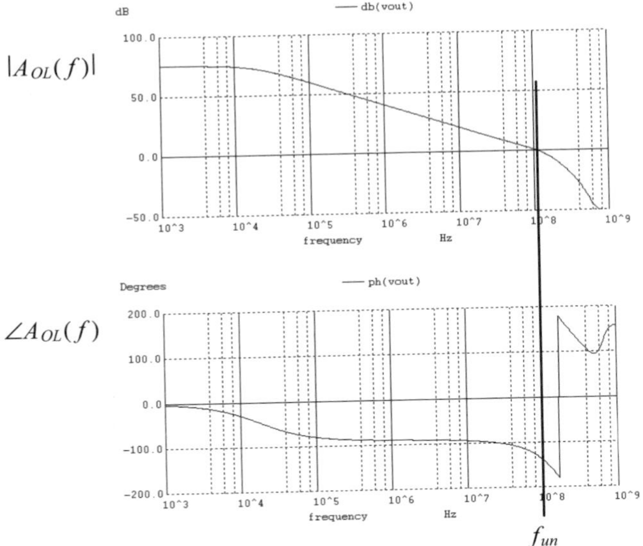

Figure 24.45 Open-loop response of the op-amp in Fig. 24.44 driving a 1-pF load.

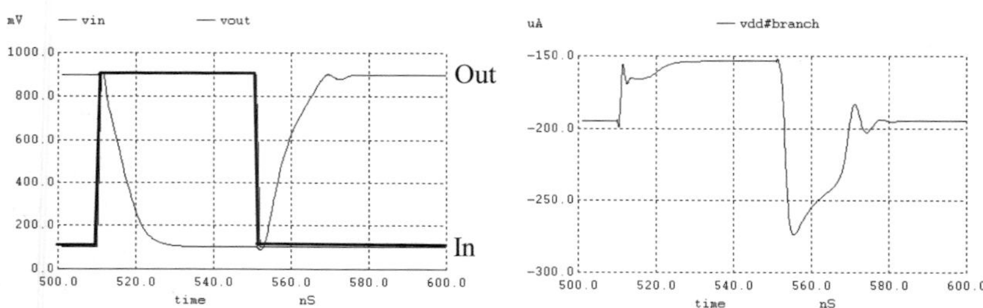

VDD current (includes 140 uA from bias circuit).

Figure 24.46 The step response and current in the topologies seen in Fig. 24.39 using the op-amp in Fig. 24.44.

Figure 24.48 shows a wide-swing op-amp based on the topology seen in Fig. 24.44. To get wide-swing operation (which means that the input common-mode voltage extends beyond the power supply rails and the op-amp still functions), we added a PMOS diff-amp stage. Further, to sink the additional current supplied by the PMOS diff-amp when it's on, we add two NMOS devices at the bottom of the folded-cascode stack. When, for example, the PMOS diff-amp shuts off with the input common-mode voltage moving towards *VDD,* the added two devices don't really affect the circuit's operation. The only result is that the V_{GS} of the bottom NMOS devices is slightly smaller when the PMOS diff-amp shuts off. A similar conclusion can be drawn when the NMOS diff-amp shuts off with the common-mode voltage moving towards ground. Note that when both diff-amps are on, the first-stage transconductance becomes, $g_{mn} + g_{mp}$. This means the gain-bandwidth product, f_{un} , moves to $(g_{mn} + g_{mp})/2\pi C_c$. The value of the compensation

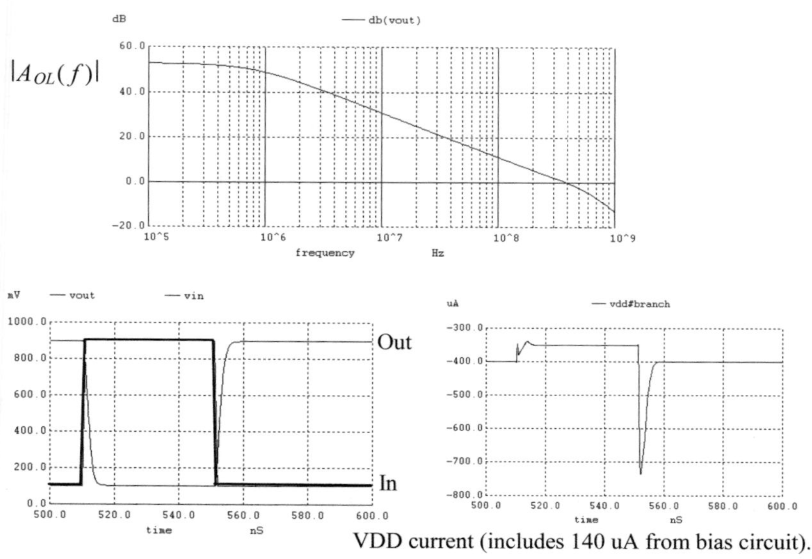

VDD current (includes 140 uA from bias circuit).

Figure 24.47 Reducing the length of the MOSFETs used in the op-amp in Fig. 24.44 from 2 to 1 and resimulating open-loop and step responses.

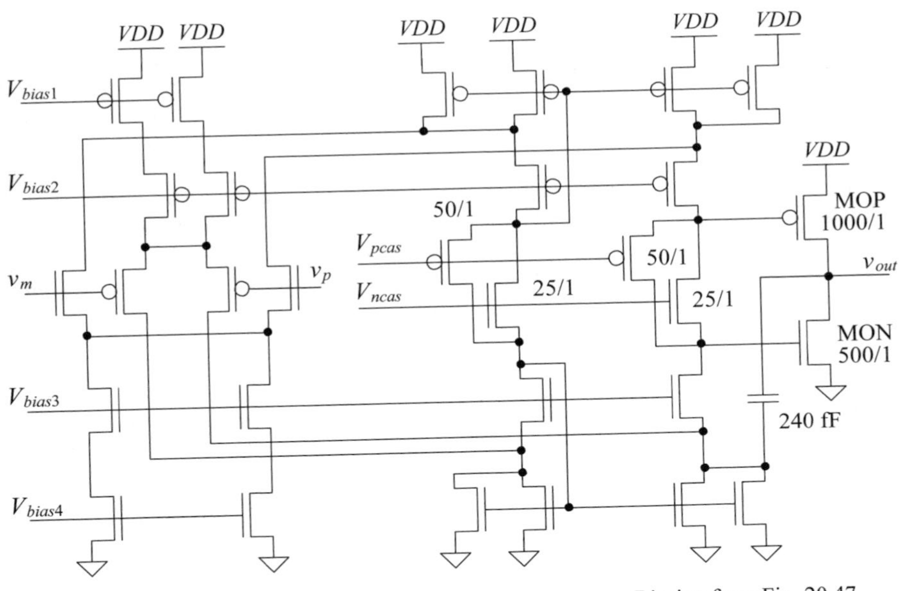

Biasing from Fig. 20.47.
Unlabeled NMOS are 50/1.
Unlabeled PMOS are 100/1.

Figure 24.48 An op-amp with an input common-mode range that extends beyond the power supply rails and that can drive heavy loads.

capacitor may need to be increased. When only one diff-amp is on, the transconductance of the other diff-amp goes to zero, resulting in a smaller gain-bandwidth product. The issue here is that the gain of the op-amp is changing with variations in the input common-mode voltage. This, as mentioned before, can lead to distortion and is the reason that the inverting topology, as seen in Fig. 24.39, is preferred (noting that the input common-mode voltage is fixed when using the inverting topology).

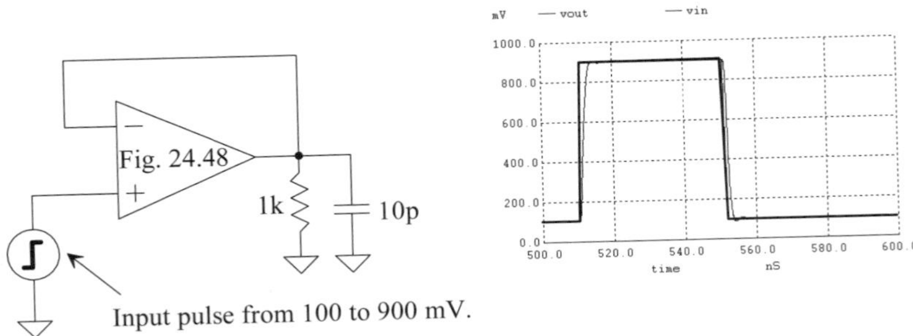

Figure 24.49 Step response of the op-amp in Fig. 24.48 driving 1k and 10 pF. Note that the op-amp is in the follower configuration and that the input common-mode voltage is changing.

Finally, in some applications, an op-amp that can supply significant amounts of current, say 100 mA, may be needed. To get such a large current, we may need to increase the widths of the output MOSFETs in the push-pull amplifier to, for example, 10,000 and 5,000 (for the PMOS and NMOS devices, respectively). The problem with this is that the quiescent current that flows in these devices, using the op-amp configuration in Fig. 24.44 as an example, is 1 mA. To reduce this current, we can increase the lengths of the floating current sources used in the folded-cascode section. This increases their gate-source voltage drops and moves the gates of the push-pull MOSFETs towards the power supply rails (shutting them off). Care must be exercised when doing this because shutting off the output MOSFETs can result in moving g_{m2} towards zero, causing, as indicated in Eq. (24.24), f_2 to move down in frequency, thus making the op-amp unstable.

24.4 Gain-Enhancement

While using minimum channel lengths ($L = 1$) results in the fastest op-amps, the open-loop gain, A_{OLDC}, is considerably reduced. Using gain-enhancement (GE) techniques (regulating the drain node in a cascode structure, see Fig. 20.40), we boost the low-frequency gain by increasing the cascode output resistance. However, the gain-bandwidth product, f_{un}, is still $g_m/2\pi C_c$. (We get the speed by using a smaller compensation capacitor and/or a larger diff-amp transconductance.) Note, in Fig. 24.47 for example, that our gain-bandwidth product went up by the increase in g_m with the reduction in the channel length. Reviewing Eqs. (24.30)–(24.34), we see that if A_{OLDC} goes up, f_{3dB} goes down (and f_{un} is a constant).

In the following discussion, we only concern ourselves with two-stage op-amps in which we are using GE to compensate for the low gain caused by our minimum channel lengths. While GE can be used in an OTA, the fact that the speed of the OTA is limited by the capacitance it is driving keeps them from being as fast as the two-stage op-amps we've discussed (unless, of course, the load capacitance is small).

As seen in Fig. 20.40, we need an additional amplifier to help regulate the drain of the cascode device. Consider the two differential amplifiers (OTAs) seen in Fig. 24.50. The source-followers on the inputs of the diff-amps are used to allow for the amplification of signals near ground (for the P amplifier) or *VDD* (for the N amplifier). The gains of the diff-amps are

$$A_P = g_{mp} \cdot (r_{on} || r_{op}) \text{ and } A_N = g_{mn} \cdot (r_{on} || r_{op}) \tag{24.55}$$

To simplify the equations, we'll assume

$$A_{GE} = A_P = A_N \tag{24.56}$$

The low-frequency gain of the op-amp in Fig. 24.44 is

$$A_{OLDC} = \overbrace{g_{mn} \cdot (R_{ocasn} || R_{ocasp})}^{\text{diff-amp gain}} \cdot \overbrace{(g_{mp} + g_{mn}) \cdot (r_{on} || r_{op})}^{\text{push-pull output stage}} \tag{24.57}$$

Using Eq. (20.67), we can write the open-loop gain *with gain-enhancement* as

$$A_{OLDC,GE} = A_{OLDC} \cdot A_{GE} \tag{24.58}$$

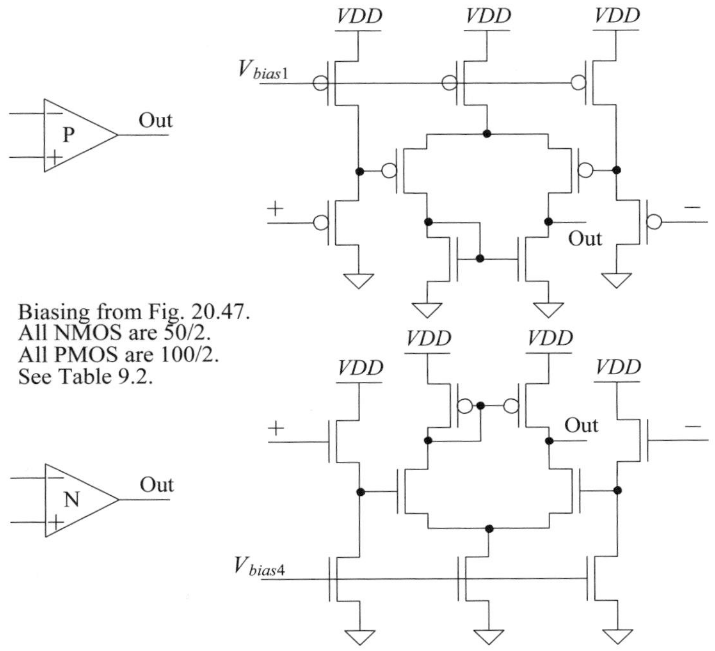

Figure 24.50 Diff-amps with source-follower level shifters for use in GE.

The open-loop gain is increased by the gain of the added amplifier. Figure 24.51 shows the op-amp of Fig. 24.44 with GE.

The simulation results in Fig. 24.47 were generated using the op-amp of Fig. 24.44 with the lengths reduced from 2 to 1. The low-frequency, open-loop gain was 53 dB. In Fig. 24.51 we added GE and used the resulting op-amp to generate the simulation results seen in Fig. 24.52 (with the GE compensation capacitors, C_{cGE}, set to zero). The low-frequency, open-loop gain is now 74 dB. The step response is seen as well. The current pulled from *VDD* is now 60 μA higher than the op-amp without GE. By replacing the diff-amps in the N and P amplifiers in Fig. 24.50 with folded-cascode OTAs (like the one seen in Fig. 24.42), we can get a greater increase in open loop gain. However, the stability of the regulated-drain feedback loop becomes even more important.

Bandwidth of the Added GE Amplifiers

Let's model the frequency response of the added amplifiers in Fig. 24.50 by

$$A_{GE}(f) = \frac{A_{GEDC}}{1 + j\frac{f}{f_{3dBGE}}} \tag{24.59}$$

Using this equation with Eqs. (24.30) and (24.58), we can write

$$A_{OLGE}(f) = \frac{A_{OLDC}}{1 + j \cdot \frac{f}{f_{3dB}}} \cdot \overbrace{\frac{A_{GEDC}}{1 + j \cdot \frac{f}{f_{3dBGE}}}}^{A_{GE}(f)} \tag{24.60}$$

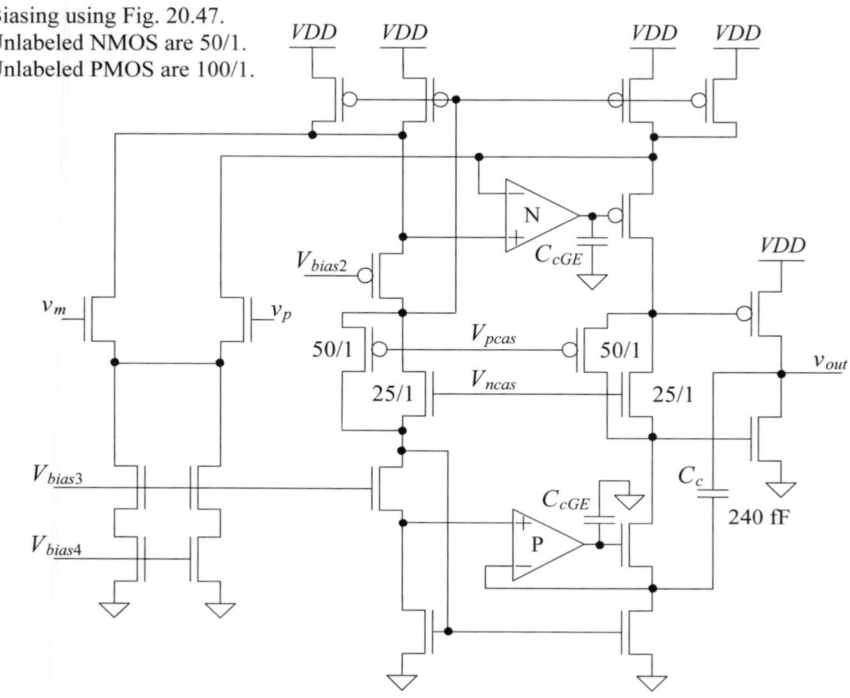

Figure 24.51 Folded-cascode op-amp with class AB output buffer and gain-enhancement.

However, the pole ($f_1 = f_{3dB}$) on the output of the folded-cascode OTA can be written, see Eqs. (20.67) and (24.47), with GE as

$$f_{3dB} = \frac{1}{2\pi(R_{ocasn}||R_{ocasp}) \cdot A_{GE}(f) \cdot C_c} \tag{24.61}$$

assuming the current through C_c is larger than the other currents, the output of the OTA may have to charge. If we substitute Eq. (24.61) into Eq. (24.60) and look at frequencies much larger than f_{3dB} (which from Fig. 24.52 is only a 100 kHz or so), we see that the frequency response of the added GE amplifiers cancels out of $A_{OLGE}(f)$. What this means is that the bandwidth of the added amplifiers doesn't need to be wide. As long as the bandwidths are larger than the f_{3dB} of the op-amp, the GE works as desired. Also, notice that when we substitute Eq. (24.61) into Eq. (24.60), the resulting transfer function shows a *doublet* (a pole and zero at the same frequency). If the pole and zero aren't at exactly the same frequency, then the settling time of the amplifier can be affected by the change in phase shift (noting that in Eq. (20.67) we are approximating the output resistance of the cascode structure). The output resistance of the cascode structures is decreasing with increasing frequency, while, at the same time, the decrease in the output resistance results in an increase in the circuit bandwidth (causing the gain-bandwidth product to be a constant).

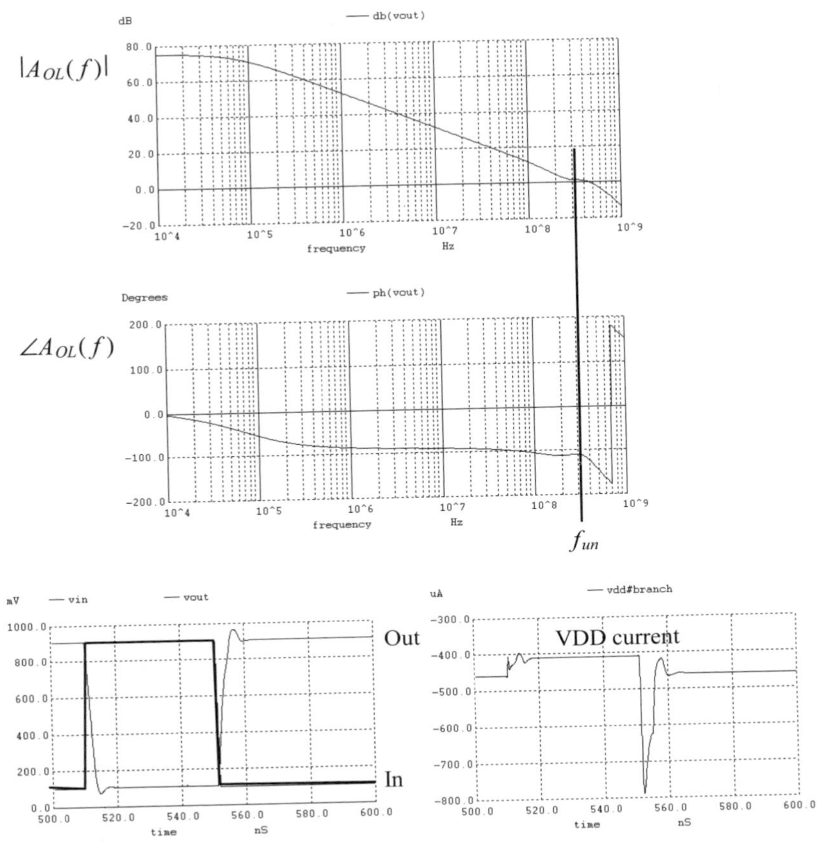

Figure 24.52 Operation of the op-amp in Fig. 24.51 driving a 1-pF load.

Compensating the Added GE Amplifiers

When we generated the simulations in Fig. 24.52, we didn't have compensation capacitors on the output of the added GE OTAs (N and P); that is, C_{cGE} were zero. The problem with this is that the resulting feedback loops aren't necessarily stable. As seen in Fig. 24.52, the step response shows some overshoot (indicating a low phase margin). Also, the frequency response isn't as monotonic as we would like (–20 dB/dec) around f_{un}. As with any feedback structure, compensation is important. *So we need to add capacitors to the outputs of the GE OTAs.* Figure 24.53 shows the response of the op-amp in Fig. 24.51 when C_{cGE}s are set to 240 fF (the GE OTAs then have f_{un} of 100 MHz, as calculated using Eq. (24.44)). The frequency response is more monotonic and the step response doesn't show overshoot and ringing. The gain-bandwidth product of the op-amp is 400 MHz. Note, it may be useful to regenerate Fig. 24.53, using larger values of C_{cGE}, to prove to oneself that indeed the bandwidth of the added GE amplifiers isn't critical. Also, it may be useful to replace the basic diff-amps with higher-gain OTAs to show that the low-frequency gain of the op-amp does indeed increase as Eq. (24.58) shows.

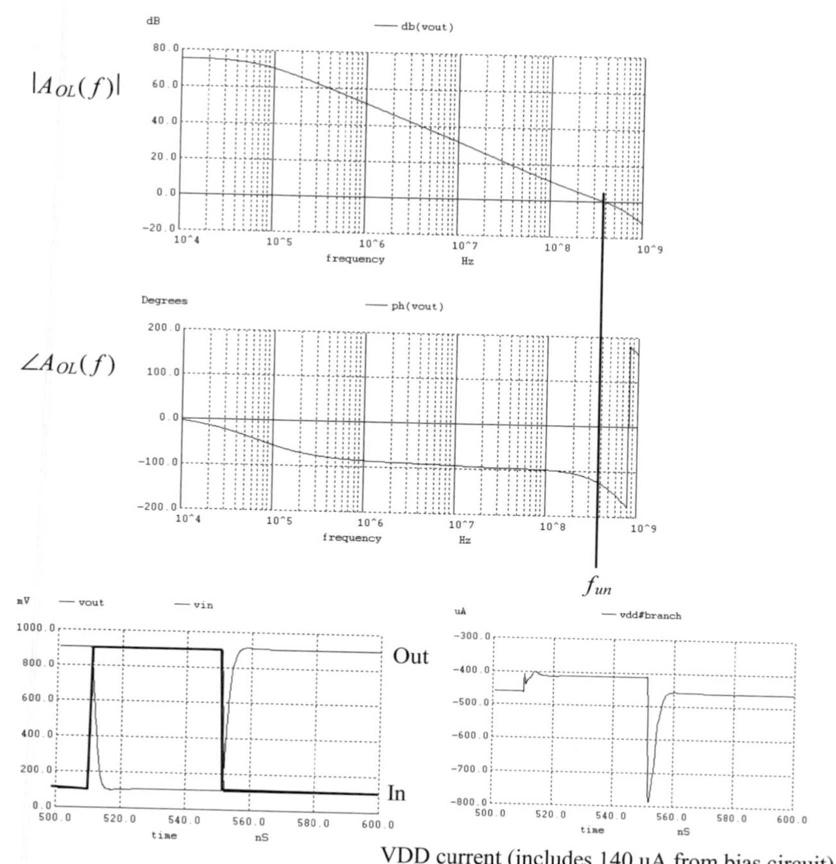

Figure 24.53 Operation of the op-amp in Fig. 24.51 driving a 1-pF load with GE compensation capacitors of 240 fF.

24.5 Some Examples and Discussions

A Voltage Regulator

One of the applications of an op-amp is in regulating a voltage on-chip. Figure 24.54 shows the basic topology. A voltage reference is used with the op-amp to generate a regulated voltage, V_{REG}. If the voltage reference is stable with temperature, the fact that the V_{REG} is a function of a ratio of resistors (so process or temperature changes in the resistance value don't affect the ratio) and the variation in the op-amp's open loop gain is desensitized using feedback makes the regulated voltage stable with process and temperature changes. The ideal (meaning that the op-amp has infinite open-loop gain) regulated voltage is

$$V_{REG} = V_{REF} \cdot \left(1 + \frac{R_A}{R_B}\right) \tag{24.62}$$

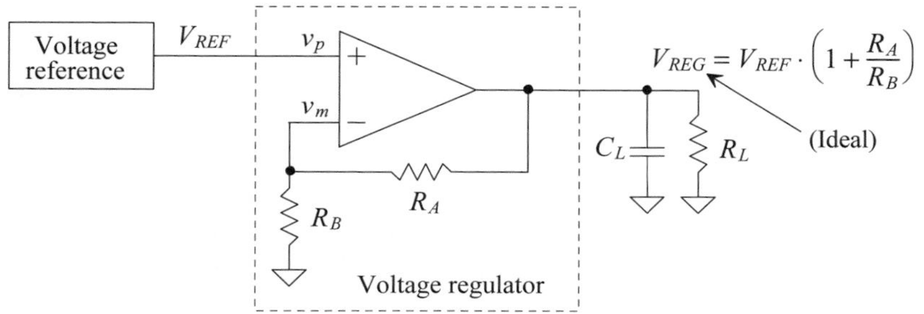

Figure 24.54 Schematic of a voltage regulator.

If the op-amp's open-loop gain is finite, then we can write

$$V_{REG} = A_{OL} \cdot (v_p - v_m) \quad (24.63)$$

and

$$v_m = V_{REG} \cdot \frac{R_B}{R_A + R_B} \text{ and } v_p = V_{REF} \quad (24.64)$$

Solving for the actual regulated voltage, gives

$$V_{REG} = V_{REF} \cdot \frac{1}{\frac{1}{A_{OL}} + \frac{R_B}{R_A + R_B}} \quad (24.65)$$

noting that when A_{OL} is infinite (or very large) this equation simplifies to Eq. (24.62). Equation (24.65) can be very revealing. Let's say that V_{REF} is 500 mV, R_B is an open, and R_A is a short (the op-amp is a voltage follower). Ideally, V_{REG} is also 500 mV. If we use a diff-amp with an open-loop gain of 20, then the actual regulated voltage is

$$V_{REG} = 0.5 \cdot \frac{1}{1.05} = 476\ mV \quad (24.66)$$

This simple example illustrates *why open-loop gain is important.* When op-amps are used in data converters (discussed in Chs. 28 and 29), they must amplify signals to within mV, which requires op-amp open-loop gains in the thousands.

Figure 24.55 shows a basic regulator topology using the two-stage op-amp. The regulator only sources current and so we've made the PMOS device, on the output, very large. One concern with using the large PMOS device is the op-amp offset (which can be considerable, see Fig. 24.4, for example). Another concern occurs when simulating. In some simulators both NRD and NRS (number of squares of implant area in the source or drain of a MOSFET) defaults to one. If the sheet resistance, rsh, is 50 ohms, then the resistance in series with the MOSFET is 100 ohms (limiting the current the MOSFET can supply). For large MOSFETs, it is generally a good idea to set NRD and NRS to zero. An example of a (large) SPICE MOSFET statement is

```
Mbig   vout   vn1   VDD   VDD   PMOS  L=1   W=10000   NRD=0   NRS=0
```

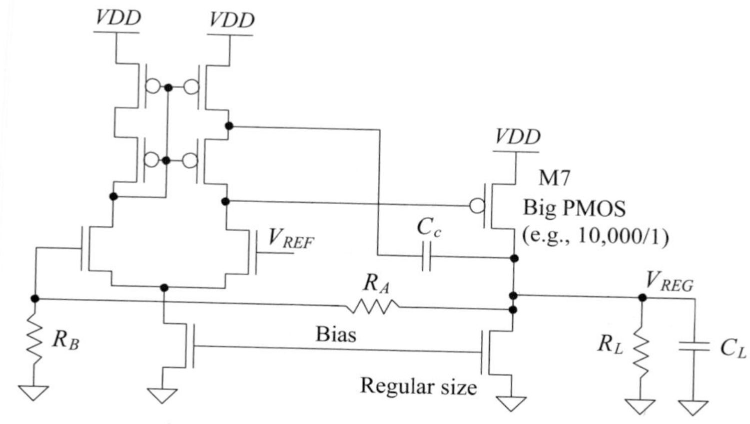

Figure 24.55 Schematic of a voltage regulator.

The point of making M7 big was to ensure that the regulator could supply a significant amount of current. However, notice that the gate of M7 cannot swing any lower (assuming that all MOSFETs remain in saturation) than

$$V_{G7,\min} = V_{REF} - V_{THN} \tag{24.67}$$

For example, if V_{REF} is 500 mV and V_{THN} is 250 mV, then the minimum voltage on the gate of M7 is 250 mV (25% of *VDD*). Ideally, we would like the gate of M7 to be able to swing to ground, fully turning it on. Towards this goal, consider the modified voltage regulator seen in Fig. 24.56. In this figure, we've used the OTA from Fig. 24.33 for the first stage of the op-amp. Now, when M5 turns on and M4 shuts off, the gate of M7 can get yanked down towards ground, allowing M7 to turn fully on.

In any practical regulator, a bypass capacitor (a big capacitor placed from V_{REG} to ground) is used to supply charge to the load for fast current transients. The required bandwidth of the regulator is reduced since the bypass capacitor (part of C_L in Figs. 24.55 and 24.56) takes care of the fast transients. It is desirable to have a large C_L for "filtering" the load's fast current needs. However, from our discussions earlier, this (large load capacitance) makes compensating the op-amp more challenging. In all of our compensation schemes, the pole associated with the output of the first stage was at a frequency much lower than the pole associated with the output stage (this was one of the reasons we used pole-splitting). To isolate the load from the second-stage gain, we might insert an NMOS source-follower (SF) between the load and the output of the second stage. The problem with this is that we won't be able to turn the SF on enough to supply significant amounts of current. Next, we might try replacing the common-source amplifier (M7) with a class AB stage (like the topology seen in Fig. 24.29). Here, again, the output won't be able to supply as much current as what we get with a full *VDD* across M7's V_{SG}. (Note that one of the reasons for using a voltage regulator is to hold V_{REG} constant even if *VDD* approaches V_{REG}.) If we must use a large load capacitor to supply charge for the fast variations in the load current and if the topology in Fig. 24.56 is used, then *we must try to compensate the op-amp with the load capacitance.*

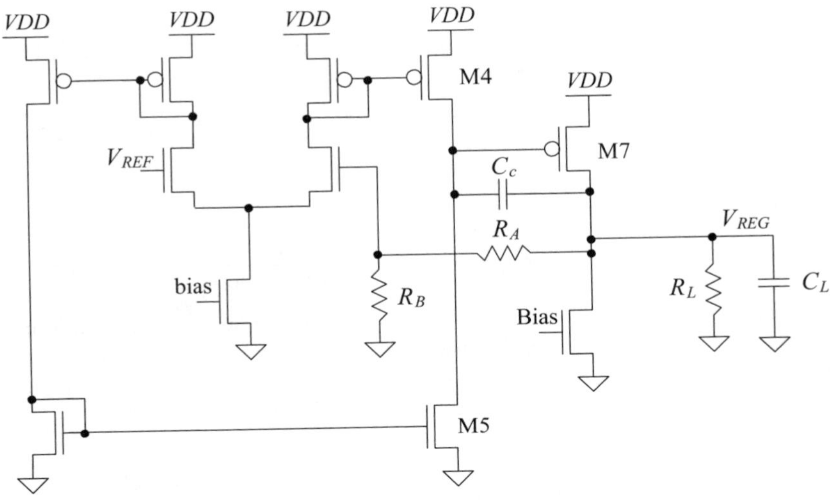

Figure 24.56 Schematic of a voltage regulator where the gate of M7 can be pulled all the way to ground.

If we are compensating the op-amp (the voltage regulator) using the load capacitance, then we've got to make sure that there is a minimum value of C_L. Further, increasing the load capacitance should make the regulator more stable, slowing the response of the op-amp. The reduction in the speed of the op-amp is offset by the fact that a larger load capacitance can supply more charge to a fast current transient before the op-amp must respond.

Why use a compensation capacitor, C_c , if we are compensating with the load capacitance? The compensation capacitor pushes the pole associated with the output node to a higher frequency (exactly what we don't want to do). We still use the capacitor (but with a smaller value) for large-signal reasons. If V_{REG} drops suddenly, the decrease in voltage is fed back directly to the gate of M7 through C_c. This turns M7 (quickly) on and allows it to pull V_{REG} back up (bypassing the slower feedback action of the op-amp). Note then that C_c is *not used* for compensating the op-amp (and actually makes the regulator more unstable). Further note that increasing the gain of the op-amp by making the first-stage gain larger (by, say, using a cascode diff-amp to increase the output resistance of the diff-amp, R_1) also hurts the regulator's stability. (By using a larger R_1, the pole associated with the output of the first stage is moved lower in frequency.)

To estimate the location of the output pole, assuming that it is the dominant pole in the system, let's calculate the closed-loop output resistance of the op-amp $R_{out,CL}$ (the open- loop output resistance is R_2). We know that the low-frequency, open-loop gain of the op-amp is

$$A_{OLDC} = \frac{v_{out}}{v_p - v_m} = g_{m1} R_1 g_{m2} R_2 \qquad (24.68)$$

If we replace the load in Fig. 24.54 with a test voltage, v_t , and note that V_{REF} $(= v_p = 0)$ is AC ground, then with the help of Fig. 24.57 we can write

$$i_t = \frac{v_t}{R_2} + \frac{v_t}{R_A + R_B} + g_{m1}R_1 g_{m2} \cdot \frac{v_t}{A_{CL}} \tag{24.69}$$

If we assume that the current through the feedback path (R_A + R_B) is small, then the output resistance of the regulator is

$$R_{out,CL} \approx \frac{v_t}{i_t} = \frac{1}{\left(\frac{1}{R_2} + \frac{g_{m1}R_1 g_{m2}}{A_{CL}}\right)} \approx \frac{A_{CL}}{A_{OLDC}} \cdot R_2 \tag{24.70}$$

Our output pole is located at

$$f_2 = \frac{1}{2\pi R_{out,CL} \cdot C_L} = \frac{A_{OLDC}}{2\pi R_2 C_L \cdot A_{CL}} \tag{24.71}$$

For this pole to compensate the op-amp, we want to use a low value of open-loop gain, A_{OLDC}. Again, this means that we don't want a large value of resistance on node 1 (the output of the diff-amp). The problem with this is that the lower the open-loop gain, the more difficult it is to regulate the output voltage, as indicated in Eq. (24.66).

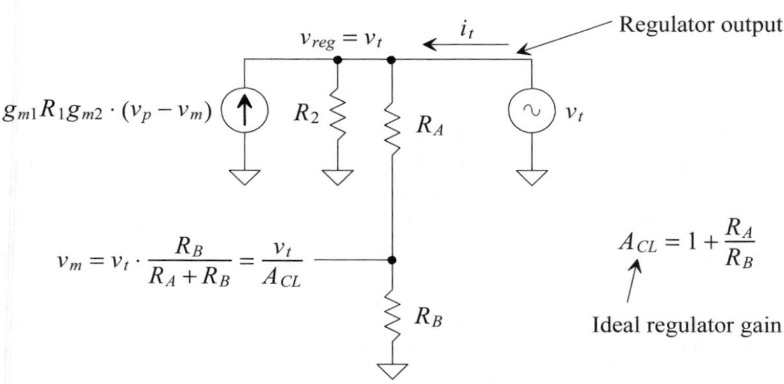

Figure 24.57 Determining the closed-loop output resistance of a two-stage op-amp.

To get an idea for the minimum size load capacitance required for a stable op-amp, let's use the parameters from Table 9.2. To begin, note that the more current the regulator supplies to a load (the more current flowing through M7), the smaller the value of R_2. This (decreasing R_2) means that the regulator becomes more likely to be unstable (f_2 increases) as it supplies more current. However, at the same time, the open-loop gain, A_{OLDC}, drops and the ability of the op-amp to regulate the output declines.

Let's assume M7's r_o and g_m are the values given in Table 9.2. Because M7 is so wide, this means (using these values for r_o and g_m) that it is conducting very little current. The value of f_1 can be estimated (assuming the $C_{gs7} >> C_c$ and the width of M7 is 10,000 or 100 times larger than the widths of the PMOS in Table 9.2) as

$$f_1 \approx \frac{1}{2\pi R_1 C_{gs7}} = \frac{1}{2\pi \cdot (167k || 333k) \cdot 100 \cdot 8.34f} = 1.7\ MHz \tag{24.72}$$

The DC open-loop gain of the op-amp is calculated using Eq. (24.68) as 277 (no load). If we use the op-amp in the follower configuration where R_B is infinite and R_A is zero (A_{CL} is 1), then (knowing $R_2 = 167k||333k = 111k$)

$$f_2 = \frac{A_{OLDC}}{2\pi R_2 C_L \cdot A_{CL}} = \frac{277}{2\pi(111k)\cdot C_L} \tag{24.73}$$

Knowing we want $f_2 << f_1$, A_{OLDC} = 277, and f_1 = 1.7 MHz, we can set the unity-gain frequency to 1 MHz (making sure that it is less than f_1). If we want a –20 dB/decade roll-off above f_2 (now the low-frequency pole) then, as seen in Eq. (24.33), we can write

$$f_{un} = A_{OLDC}\cdot f_2 = \frac{A_{OLDC}^2}{2\pi R_2 C_{L\min}\cdot A_{CL}} = 1\ MHz \tag{24.74}$$

Solving for the minimum value of $C_{L\min}$ using these equations gives 110 nF. Clearly, this value is too large if we want the regulator to be completely on-chip. Rewriting Eq. (24.74)

$$C_{L\min} = \frac{(g_{m1}R_1 g_{m2})^2 R_2}{2\pi\cdot f_{un}\cdot A_{CL}} \tag{24.75}$$

Note that linearly increasing the current through M7 causes R_2 to decrease linearly (assuming that M7's output resistance dominates R_2) and g_{m2}^2 to increase linearly (see Eqs. (9.6) and (9.22)). The result is that $C_{L\min}$ doesn't change. However, notice what happens if we include R_L in the calculations. If $R_L << r_{o7}$, then we can write

$$\frac{C_{L\min}}{R_{L\max}} = \frac{(g_{m1}R_1 g_{m2})^2}{2\pi\cdot f_{un}\cdot A_{CL}} \tag{24.76}$$

If we select $C_{L\min} = 1000\,pF$, then $R_{L\max} = 1\,k\Omega$. We use the variable $R_{L\max}$ to indicate that this is the maximum value of load resistance possible for a stable regulator (we want the regulator to always supply at least $V_{REG}/R_{L\max}$ current). If the regulator's output voltage is 500 mV, it must be supplying at least 500 μA of current. This resistor keeps the open-loop gain low when the regulator is supplying small amounts of current. We might think, after looking at Eq. (24.73), that by supplying more current (R_L or R_2 decreasing further), the location of f_2 increases and the regulator moves towards instability. However, for further decreases in the load resistance, the current supplied by M7 increases and thus so does its g_m. What happens is this: we get the canceling of effects. As the current supplied to the output increases linearly with a linear decrease in R_L , the value of g_m^2 increases linearly and the effects cancel.

A final comment: because this design is different from the previous op-amp designs, thorough characterization using simulations is needed. Steps in the load current as well as a wide range of load capacitances and resistances should be simulated to ensure that the regulator remains stable under all possible *VDDs*, temperatures, and process shifts.

Bad Output Stage Design

In this section we'll discuss some problems with biasing in Class AB output stages. To begin, consider the op-amp design seen in Fig. 24.58. This topology is a diff-amp driving an inverter. The problem with this design is that the current flowing in the push-pull output stage (an inverter) isn't set by a bias circuit. Because of this, the current may be

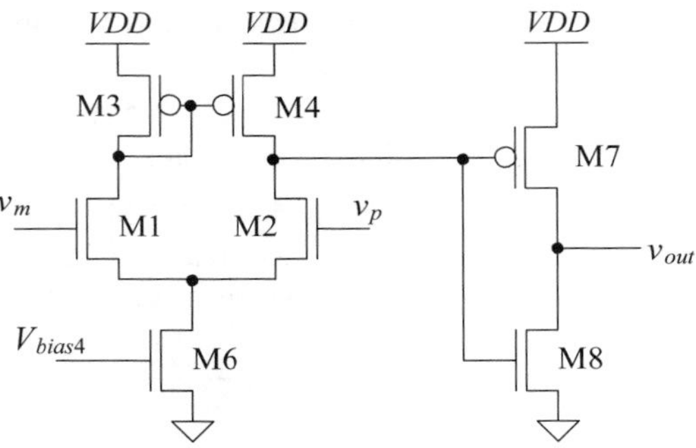

Figure 24.58 Bad op-amp design.

significant and vary greatly with both temperature and process shifts. Further, notice that when the inputs to the op-amp are equal, the drain of M4 is at the same potential as the drain of M3 (the currents in M3 and M4 are equal). This, as we've already discussed, can be used to set the bias current in M7 (we treat M7, for biasing purposes, as if its gate were tied to the gates of M3 and M4). If we call the gate potential of M7 (and thus the gate potential of M8) $VDD - V_{SG}$, then for M8 to be in saturation

$$V_{out} \geq VDD - V_{SG} - V_{THN} \tag{24.77}$$

If VDD is 1 V, $V_{SG} = 350$ mV, and $V_{THN} = 250$ mV, then V_{out} must be greater than 400 mV for M8 to be saturated. What this means is that for wide output swings, M8 will move into triode and the push-pull output amplifier gain will drop (killing the overall gain of the op-amp). What we need to do is drop the gate potential of M8 down so that its drain can swing to a lower voltage without it trioding. Towards this goal, consider adding a source-follower level shifter to the output of the op-amp, as seen in Fig. 24.59. Now the output voltage in the amplifier can swing lower without M8 trioding. However, we still aren't controlling the current in the output stage. By not controlling this current, as we've already discussed, we produce an op-amp with poor systematic, input-referred offset voltage. If, for example, we have an input-referred offset of 10 mV and the diff-amp gain is 20, then the change in the output voltage on the gate of M7 is 200 mV. This can cause

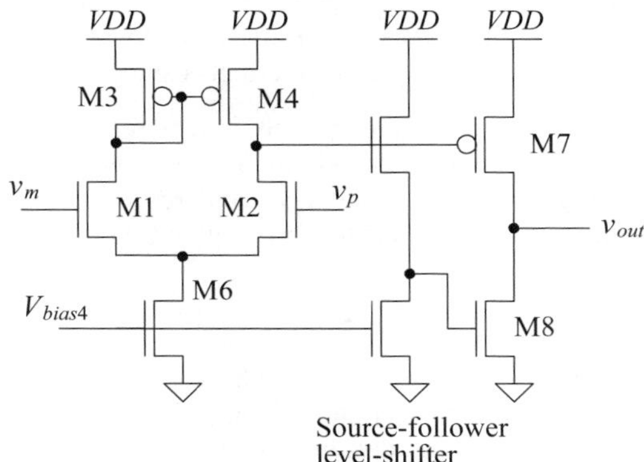

Figure 24.59 Another bad op-amp design.

the output current flowing in M7/M8 to be hundreds of μA above the current in the diff-amp.

We might try connecting the gate of M8 to the gates of M3/M4. When the noninverting input voltage, v_p , goes high, M2 turns on and causes the gate potential of M7 to drop (turning M7 further on). At the same time, the gate potential of M3 increases, causing M8 to also turn further on. The result is that M7 and M8 both turn on and fight each other for control of v_{out}. For proper operation, the gates of the output push-pull stage should move in the same direction.

To precisely control the current in the output stage, consider the topology seen in Fig. 24.60. Since, for biasing purposes, we treat the drain of M4 as if it's at the same potential as the gate of M3, the currents in M7, M9, and M10 are equal when the op-amp inputs are at the same potential. We also treat M8 as if its gate were tied to the gate of M11. The result is that the quiescent currents in M7 and M8 are equal and set by the biasing of M6. We seem to have solved the problem with biasing the class AB output stage.

However, a different problem arises. We now have three high-impedance nodes: the output of the diff-amp (a pole we've called f_1), the output of the op-amp (a pole we've called f_2), and, now, the gate of M8. If we try to make the node at the gate of M7 the dominant node, then the path that the op-amp inputs sees through M3 to M9 to the gate of M8 (the other high-impedance node) isn't affected. If we add compensation capacitors from the op-amp output to both the gates of M7 and M8, the current fed back isn't necessarily balanced. This imbalance may only be noticed when the op-amp's large signal step response is simulated. (It is generally not seen in an AC simulation.) The result is that the op-amp's settling time and stability become marginal in some situations. Having said this, the topology can still be useful in some situations (e.g., low power or low *VDD*).

In an attempt to improve or simplify the op-amp in Fig. 24.60, we might try to make M12 diode connected and remove M10/M11 from the circuit. With this change, however, we don't get class AB action (M9 mirrors the current in M3, which flows in the diode-connected M12 and is mirrored by M8). Finally, again, why not eliminate M9 and M12 followed by connecting the gate of M8 to the diode-connected M11? The gate potentials of M8 and M7 would then move in opposite directions, causing M7 to fight with M8 for control of the output voltage.

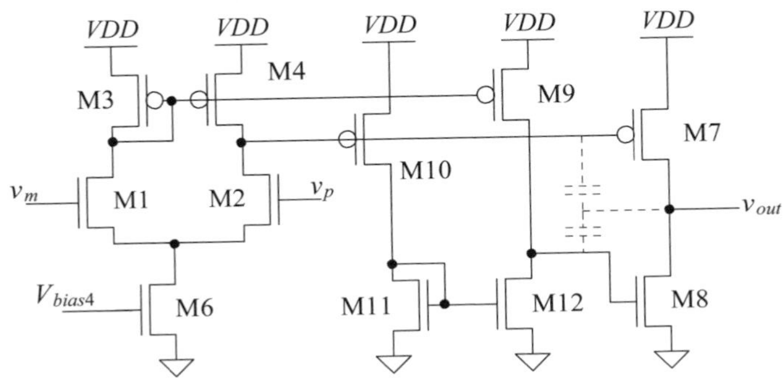

Figure 24.60 Controlling the current in a class AB output stage.

Three-Stage Op-Amp Design

In some applications we need an op-amp that can operate with a low power-supply voltage or with minimal power from *VDD*. The bias circuit we developed in Fig. 20.47 can pull more current from *VDD* than the op-amp it biases. (So, if possible, use the same bias circuit for multiple op-amps, if power and layout area are of concern.) If we look at the heart of this bias circuit (Fig. 20.47) and its operation, the Beta-multiplier in Figs. 20.22 and 20.23, we see that *VDD* can be decreased down to 500 mV with the bias currents relatively unaffected. The problem in using a *VDD* which is half of the normal *VDD* (= 1 V) is that our class AB output stage (Fig. 24.44 for example) fails to operate. For proper operation of the class AB output stage using the floating current sources (biased with V_{ncas} and V_{pcas}), we need a *VDD* > 800 mV (perhaps higher over process corners and temperature). Also, as already discussed, if we connect a resistive load to the output of the two-stage op-amp in Fig. 24.2, the gain is reduced.

Towards keeping a large gain and lowering the power supply voltage, consider the three-stage op-amp in Fig. 24.61. This design is a cascade of two diff-amps followed by a common-source amplifier. If we connect a resistive load to the output of the common-source stage, the overall op-amp gain remains relatively high due to the cascaded gain of the diff-amps. The concern with this topology (an op-amp with more than two stages), as alluded to earlier, is compensation. We still compensate the op-amp so that the pole associated with node 1 is dominant. The nesting of the compensation capacitors within the op-amp in Fig. 24.61 is sometimes called *nested Miller compensation*. We'll avoid the use of this label here since we aren't strictly using Miller compensation (Fig. 24.8 among others) but rather the indirect compensation method discussed earlier (see Fig. 24.18 and the associated discussion). Note how the current through C_{c1} feeds back to node 1 through the diode-connected load of the input diff-amp. This is necessary to ensure that the signal through C_{c1} adds with the signal fed back directly to node 1 through C_{c2}.

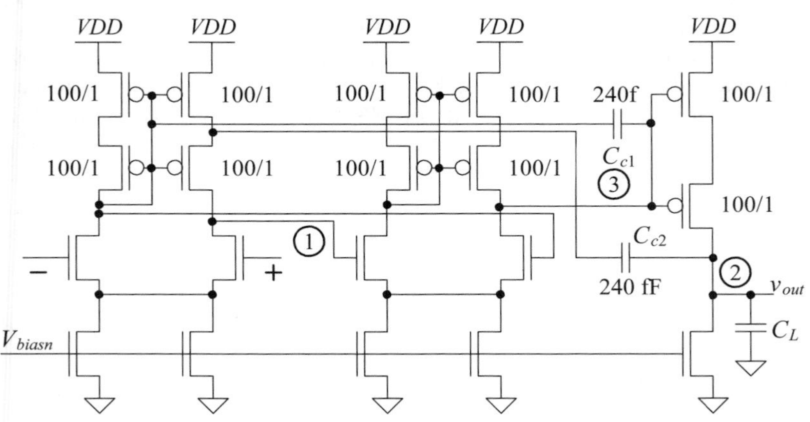

Figure 24.61 A three-stage op-amp.

Because of pole-splitting, the location of the pole at node 3 (the output of the second diff-amp) is pushed out to, see Eq. (24.24),

$$f_3 = \frac{g_{mn} \cdot C_{c1}}{2\pi \cdot C_1(C_{L2} + C_{c1})} \approx \frac{g_{mn}}{2\pi C_1} \quad (24.78)$$

a large frequency. Here we are assuming that the transconductance of the second diff-amp is g_{mn} and that C_{c1} is much larger than the input capacitance of common-source amplifier (which is the second diff-amp's C_{L2}). The location of the pole on the output of the amplifier is still given by

$$f_2 = \frac{g_{mp} \cdot C_{c2}}{2\pi \cdot C_1(C_L + C_{c2})} \quad (24.79)$$

where g_{mp} is the transconductance of the common-source amplifier. Note that we aren't discussing the LHP zeroes whose locations are specified using Eq. (24.22). We'll discuss these zeroes in a moment. We'll compensate, as usual, the op-amp so that the unity-gain frequency is less than the frequencies of the poles at f_2 and f_3.

The low-frequency, open-loop gain of the op-amp is given by

$$A_{OLDC} = \overbrace{g_{mn}(r_{on}||r_{op})}^{\text{first diff-amp's gain, } A_1} \cdot \overbrace{g_{mn}(r_{on}||r_{op})}^{\text{second diff-amp's gain, } A_2} \cdot \overbrace{g_{mp}(r_{on}||r_{op})}^{\text{common-source gain, } A_3} \quad (24.80)$$

At frequencies above the op-amp's f_{3dB} and below the gain-bandwidth product (the unity-gain frequency, f_{un}), the gain of the first-stage is rolling off. For all intents and purposes, the AC signal on the output of the first diff-amp is considerably smaller than the AC outputs of the second or third stages. In this situation the current fed back to node 1, knowing that the output voltage of the second diff-amp (at node 3) is v_{out}/A_3 , is

$$i_{Cctot} \approx \frac{v_{out}}{1/j\omega C_{c2}} + \frac{v_{out}/A_3}{1/j\omega C_{c1}} \quad (24.81)$$

The output current of the diff-amp is

$$i_{out1} = g_{mn}(v_p - v_m) \quad (24.82)$$

Because this current must equal the current fed back to node 1, we can equate Eqs. (24.81) and (24.82) to write

$$\frac{v_{out}}{v_p - v_m} = \frac{g_{mn}}{j\omega(C_{c2} + C_{c1}/A_3)} \quad (24.83)$$

The unity-gain frequency is then

$$f_{un} = \frac{g_{mn}}{2\pi(C_{c2} + C_{c1}/A_3)} \approx \frac{g_{mn}}{2\pi C_{c2}} \quad (24.84)$$

noting that the unity-gain frequency is a function of the third-stage gain. Here, we'll set C_{c2} to 240 fF to get a unity-gain frequency of 100 MHz. We still need C_{c1} to push node 3's pole to a higher frequency (pole splitting). We'll also set C_{c1} to 240 fF.

Figure 24.62 shows the frequency response of the op-amp in Fig. 24.61 without driving a load capacitance. As designed for using Eq. (24.84), the unity-gain frequency is 100 MHz. Also, as specified in Eqs. (24.22) and (24.25), we have zeroes around the unity-gain frequency. It appears from the frequency response in Fig. 24.62 that the zeroes occur above the unity-gain frequency and cause the open-loop response to flatten out.

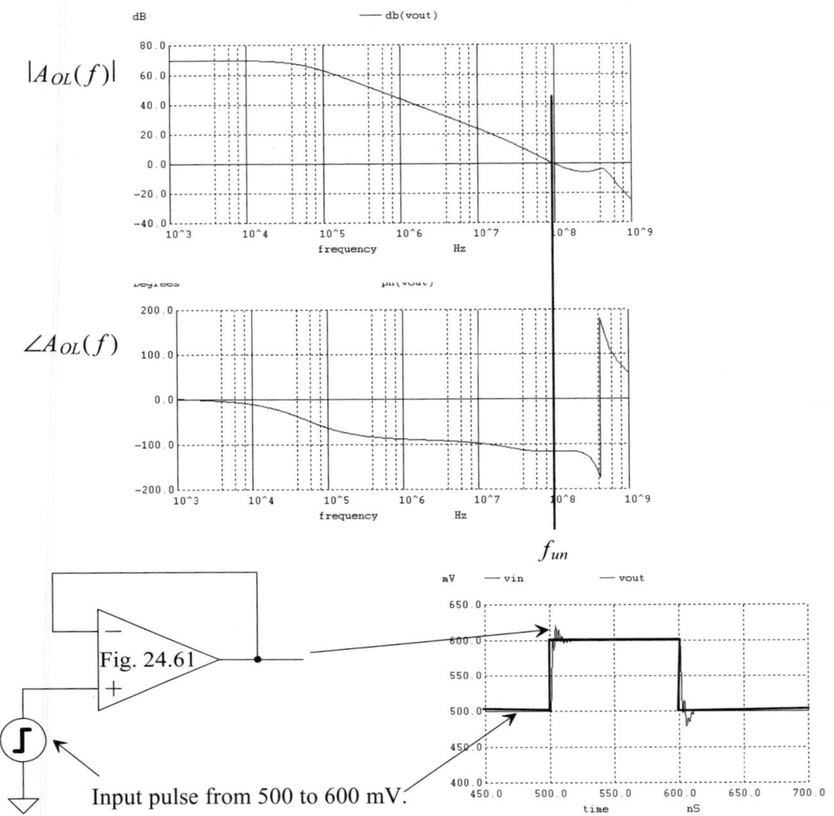

Figure 24.62 Unloaded AC and step responses for the op-amp in Fig. 24.61.

What this indicates is that one of the zeroes is canceling one of the higher-frequency poles. Note that the PM is 45°. The step response seen in Fig. 24.62 shows the characteristic ringing associated with a PM of 45°.

If we drive a load capacitance, then we expect the third-stage gain, A_3 , to start to decrease at a lower frequency (it starts to decrease sooner with increasing frequency). Looking at Eq. (24.84), we then expect f_{un} to drop in value. Figure 24.63 shows the simulation where the op-amp in Fig. 24.61 is driving a 1 pF load. As expected, the addition of the 1 pF load causes f_{un} to drop in value. The PM is approximately 25°. With such a low PM, we expect the output to show significant ringing (and, as seen in Fig. 24.63, it does). To increase the PM, we can increase the value of the compensation capacitor. We can increase C_{c2} to, say, 2400 fF, for a unity gain frequency of (roughly) 10 MHz. We don't necessarily need to increase C_{c1} to decrease f_{un} , as seen in Eq. (24.84).

Finally, why did we choose the topology seen in Fig. 24.61 for our three-stage op-amp example? The main reason for using this topology comes from biasing the

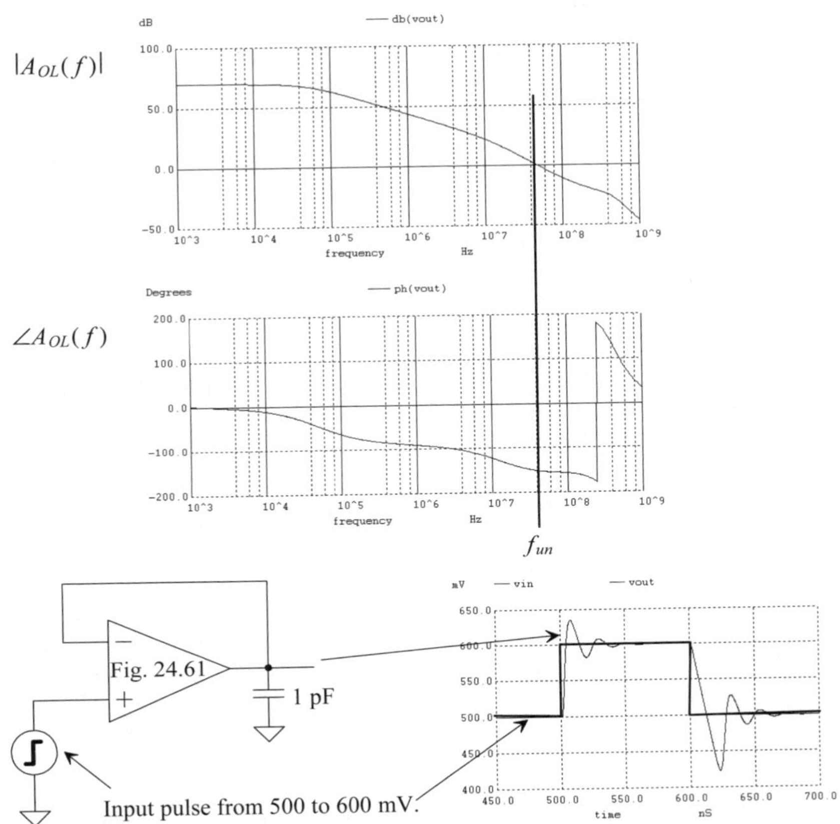

Figure 24.63 AC and step responses for the op-amp in Fig. 24.61 driving 1 pF.

common-source stage. If we were to replace the second diff-amp with a common-source stage (so our last two stages in the op-amp are common-source amplifiers), we wouldn't have a known second stage output voltage to bias the final common-source stage. The exact output voltage of a common-source amplifier, without feedback to set it, is unknown (the two drains of the NMOS and PMOS will either float up or float down, depending on which device is sourcing or sinking the most current). By using the diff-amp stages, as seen in Fig. 22.8, we ensure that each stage is biased with a known current. We don't use a diff-amp for the final stage of the op-amp because of the diff-amp's limited output swing.

Another benefit of using two diff-amps on the input of the op-amp is the increase in the *CMRR*. The overall common-mode gain is the product of each diff-amp's common-mode gain. If a single diff-amp has a *CMRR* of 80 dB, then the cascade of two diff-amps results in a *CMRR* of 160 dB.

ADDITIONAL READING

[1] R. Harjani, R. Heineke, and F. Wang, "An integrated low-voltage class AB CMOS OTA," *IEEE Journal of Solid-State Circuits,* vol. 34, pp. 134–142, February 1999.

[2] L. Moldovan and H. H. Li, "A rail-to-rail, constant gain, buffered op-amp for real time video applications," *IEEE Journal of Solid-State Circuits,* vol. 32, pp. 169–176, February 1997.

[3] F. You, S. H. K. Embabi, and E. Sánchez-Sinencio, "A multistage amplifier topology with nested Gm-C compensation for low-voltage application," *IEEE International Solid-State Circuits Conference,* vol. XL, pp. 348–349, February 1997.

[4] W. S. Wu, W. J. Helms, J. A. Kuhn, and B. E. Byrkett, "Digital-compatible high-performance operational amplifier with rail-to-rail input and output ranges," *IEEE Journal of Solid-State Circuits,* vol. 29, pp. 63–66, January 1994.

[5] R. Hogervorst, J. P. Tero, R. G. H. Eschauzier, and J. H. Huijsing, "A Compact Power-Efficient 3 V CMOS Rail-to-Rail Input/Output Operational Amplifier for VLSI Cell Libraries," *IEEE Journal of Solid State Circuits,* vol. 29, pp. 1505–1513, December 1994.

[6] J. Ramírez-Angulo and E. Sánchez-Sinencio, "Active compensation of operational transconductance amplifier filters using partial positive feedback," *IEEE Journal of Solid-State Circuits,* vol. 25, pp. 1024–1028, August 1990.

[7] K. Bult and G. J. G. M. Geelen, "A Fast-Settling CMOS Op-Amp for SC Circuits with 90-dB DC Gain," *IEEE Journal of Solid State Circuits,* vol. 25, pp. 1379–1384, December 1990.

[8] S. M. Mallya and J. H. Nevin, "Design procedures for a fully differential folded-cascode CMOS operational amplifier," *IEEE Journal of Solid-State Circuits,* vol. 24, pp. 1737–1740, December 1989.

[9] M. Banu, J. M. Khoury, and Y. Tsividis, "Fully Differential Operational Amplifiers with Accurate Output Balancing," *IEEE Journal of Solid State Circuits,* vol. 23, No. 6, pp. 1410–1414, December 1988.

[10] B. K. Ahuja, "An Improved Frequency Compensation Technique for CMOS Operational Amplifiers," *IEEE Journal of Solid-State Circuits*, vol. 18, pp. 629–633, December 1983.

[11] P. R. Gray and R. G. Meyer, "MOS Operational Amplifier Design: A Tutorial Overview," *IEEE Journal of Solid-State Circuits,* vol. 17, pp. 969–982, December 1982.

[12] B. Y. Kamath, R. G. Meyer, and P. R. Gray, "Relationship Between Frequency Response and Settling Time of Operational Amplifiers," *IEEE Journal of Solid State Circuits,* vol. SC-9, pp. 347–352, December 1974.

PROBLEMS

24.1 Suggest, and verify with simulations, a method for reducing the minimum input common-mode voltage of the op-amp in Fig. 24.2.

24.2 Redesign the bias circuit for the op-amp in Fig. 24.2 for minimum power. Compare the power dissipation of your new op-amp design (actually the bias circuit) to the design in Fig. 24.2. Using your redesign, generate the plots seen in Fig. 24.3.

24.3 Show, using simulations, how a 1% mismatch in the widths of M1 and M2 in the op-amp of Fig. 24.2 affect the op-amp's input-referred offset voltage. Compare this offset to the offset caused by a 1% mismatch in the widths of M3 and M4. Quantitatively explain why one is worse than the other.

24.4 Simulate the use of the "zero-nulling" circuit in Fig. 24.15 in the op-amp of Fig. 24.8. Show AC, operating-point, and transient (step) operation of the resulting op-amp. Verify with the .op analysis that the gate of MP1 is at the same potential as the gate of M7 in quiescent conditions.

24.5 When we derived Eq. (24.24), we lumped C_c into C_2 and neglected the effects of MCG. A more accurate model for the indirect compensation, assuming $C_L >>$ the output capacitance of the amplifier, is seen in Fig. 24.64. Using this model, estimate the location of the output pole.

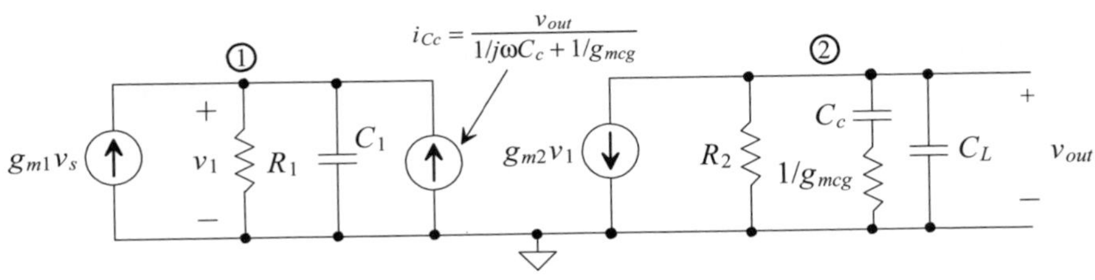

Figure 24.64 Model used to estimate bandwidth when indirect feedback current. See Problem 24.5.

24.6 Regenerate Fig. 24.19 using a 2.4 pF compensation capacitor. On the resulting simulation output, label the location f_1 , f_2 , f_z , and f_{un}. How do the simulated results compare to the hand calculated values?

24.7 For the op-amp in Fig. 24.21, determine the *CMRR* using hand calculations. Verify your hand calculations using simulations. How does the *CMRR* change based on the DC common-mode voltage?

24.8 Simulate the *PSRR*s for the op-amp in Fig. 24.8 (with an R_z of 6.5k and a C_c of 2.4 pF) and compare the results to the op-amp in Fig. 24.21 when C_c is set to (also) 2.4 pF (so each op-amp has the same gain-bandwidth).

24.9 Simulate the operation of the op-amp in Fig. 24.28. Show the open-loop frequency response of the op-amp. What is the op-amp's PM? Show the op-amp's step response when it is put into a follower configuration driving a 100 fF load with an input step in voltage from 100 mV to 900 mV.

24.10 The op-amp seen in Fig. 24.29 has a gain-bandwidth product (f_{un}) of about 100 MHz. Suppose that this op-amp is used in the amplifier seen in Fig. 24.65 (gain of –5). Estimate the amplifier's closed loop bandwidth (where the output of the amplifier is –3 dB down from its low-frequency value) using Eq. (24.34). Verify your results using simulations (transient analysis). What are the maximum and minimum voltages allowable for an input sinewave if the output voltage of the amplifier must lie between 100 and 900 mV?

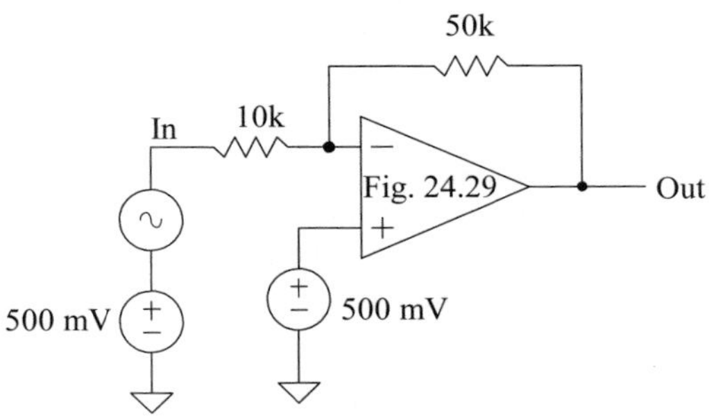

Figure 24.65 Amplifier used for Problem 24.10.

24.11 Suppose it is decided to eliminate the 500 mV common-mode voltage in the amplifier seen in Fig. 24.65 and use ground, as seen in Fig. 24.66. Knowing that the input voltage can only fall between ground and *VDD*, what is the problem one will encounter?

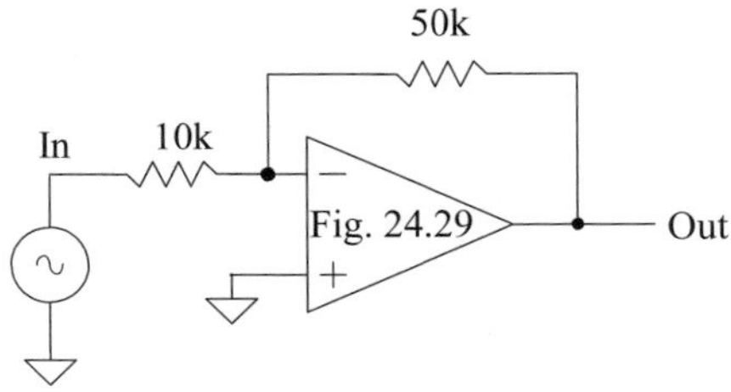

Figure 24.66 Amplifier used for Problem 24.11. What's wrong with this topology?

24.12 To limit the current flowing in MOP or MON in the op-amp of Fig. 24.29 (to protect the op-amp from destruction if its output is shorted to ground, for example), we may add 100 Ω resistors, as seen in Fig. 24.67. What is the maximum amount of current this modified op-amp can source/sink? How is the closed-loop output resistance of the op-amp affected. (Hint: see Eq. (24.70).) Simulate the operation of the op-amp using the topology seen in Fig. 24.31. Is there any noticeable difference between the simulation output in Fig. 24.31 and the output with the 100 Ω resistors present?

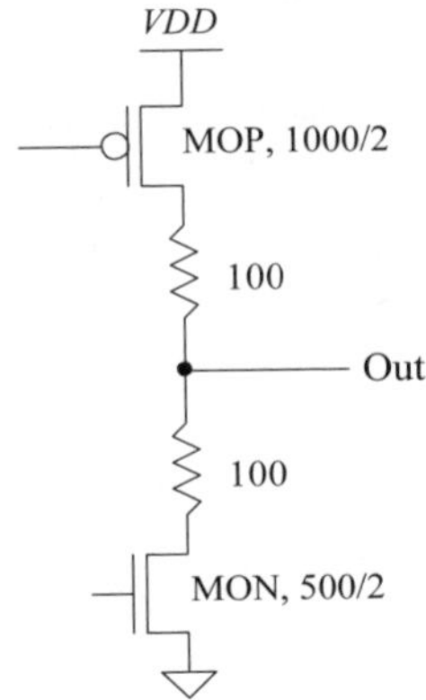

Figure 24.67 Adding resistors to the output of the op-amp in Fig. 24.29 for short circuit protection. See Problem 24.12.

24.13 Resimulate the OTA in Fig. 24.33 driving a 1 pF load (to determine f_{un}) if K = 10. How do the simulation results compare to the hand calculations using Eq. (24.41)? Estimate the parasitic poles associated with the gates of M4 and M5 (for example, the pole associated with the gate of M4 is $\approx 1/g_{m41} \cdot C_{sg4}$). Are these poles comparable to f_{un}?

24.14 Using the OTA in Fig. 24.35, design a lowpass filter with a 3 dB frequency of 1 MHz.

24.15 Suppose M8T in the op-amp of Fig. 24.37 is removed and replaced with a short from the output to the drain of M8B. How will the gain be affected? Verify your answer with SPICE.

24.16 Suppose, to simulate the open-loop gain of an OTA, the big resistor and capacitor used in Fig. 24.43 are removed and the inverting input is connected to 500 mV. Will this work? Why or why not? What happens if the OTA doesn't have an offset voltage? Will it work then?

24.17 Why is the noninverting topology (Fig. 24.49) inherently faster than the inverting topology (Fig. 24.39). What are the feedback factors, β, for each topology. Use the op-amp in Fig. 24.48 to compare the settling times for a +1 and a −1 amplifier driving 10 pF.

24.18 Suppose, to reduce power, the lengths of the current sources used in the amplifiers seen in Fig. 24.50 are increased from 2 to 10. Will the AC performance of the op-amp used to generate Fig. 24.53 change with this modification? Why or why not? Verify your answer using a SPICE simulation. Compare the currents used in the modified op-amp (with the lower power GE diff-amps) to the current seen in Fig. 24.53.

24.19 To increase the gain of the op-amp in Fig. 24.51, we may replace the GE diff-amps with folded-cascode OTAs. Will we need source-follower level-shifters in the new design? Regenerate the simulation data seen in Fig. 24.53 using the folded-cascode OTAs.

24.20 Suppose an op-amp is to be used to amplify 250 mV to 500 mV with an error less than 1 mV. Estimate the minimum required op-amp open loop gain.

24.21 Design a voltage regulator to supply at least 50 mA of current at 500 mV with a *VDD* as low as 600 mV. Assume that a 500 mV voltage reference is available and that the load capacitance is, minimum, 1,000 pF. How does the design respond to a load current pulse from 0 to 50 mA? Use SPICE to verify your design.

24.22 Using the nominal sizes from Table 9.2 and the bias circuit in Fig. 20.47, simulate, using an .op analysis, the operation of the op-amp in Fig. 24.58 in the configuration seen in Fig. 24.9. What is the current flowing in M7 and M8 when *VDD* is 1 V? is 1.2 V?

24.23 Repeat Problem 24.22 for the op-amp in Fig. 24.59.

24.24 Using the information from Table 9.2 and the bias circuit in Fig. 20.47, demonstrate the operation (AC and transient) of the op-amp in Fig. 24.60 driving a 100 fF load. Show that the bias current pulled from *VDD* is relatively constant for a *VDD* of 1 or 1.2 V (put the op-amp in the configuration seen in Fig. 24.9).

24.25 Replace the common-source output stage in the op-amp of Fig. 24.61 with a class AB output stage like the one seen in Fig. 24.60. Simulate the operation of the amplifier (AC and transient).

Chapter

25

Dynamic Analog Circuits

In Chapter 14 we discussed dynamic logic gates. Dynamic logic is useful for reducing power dissipation and the number of MOSFETs used to perform a given circuit operation (layout area). Dynamic analog circuits exploit the fact that information can be stored on a capacitor or gate capacitance of a MOSFET for a period of time. In this chapter, we discuss analog circuits such as sample and holds, current mirrors, amplifiers, and filters using dynamic techniques.

25.1 The MOSFET Switch

A fundamental component of any dynamic circuit (analog or digital) is the switch (Fig. 25.1). An important attribute of the switch, in CMOS, is that under DC conditions the gate of the MOSFET does not draw a current. Therefore, neglecting capacitances from the gate to the drain/source, we find that the gate control signal does not interfere with information being passed through the switch. Figure 25.2 shows the small-signal resistance of the switches of Fig. 25.1 plotted against input voltage. The benefits of using the CMOS transmission gate are seen from this figure, namely, lower overall resistance. Another benefit of using the CMOS TG is that it can pass a logic high or a logic low without a threshold voltage drop. The largest voltage that an NMOS switch can pass is $VDD - V_{THN}$, while the lowest voltage a PMOS switch can pass is V_{THP}.

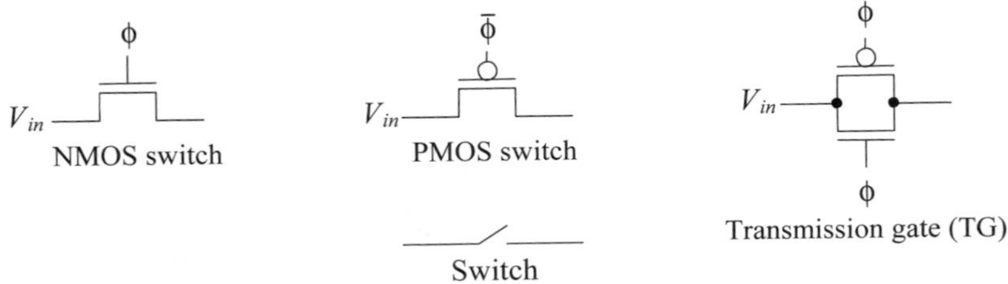

Figure 25.1 MOSFETs used as switches.

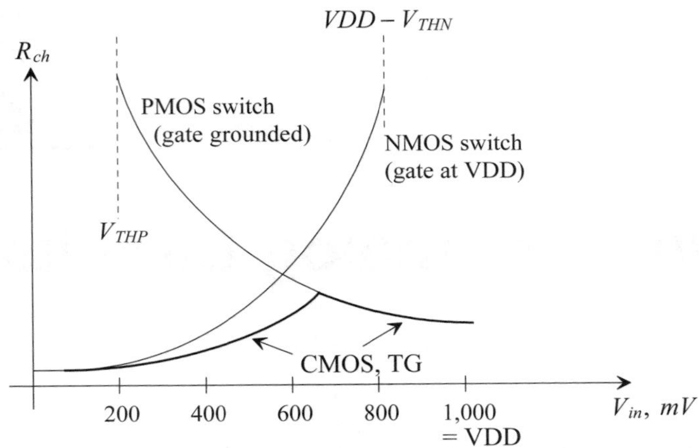

Figure 25.2 Small-signal on-resistance of MOSFET switches .

While MOS switches may offer substantial benefits, they are not without some detraction. Two nonideal effects typically associated with these switches may ultimately limit the use of MOS switches in some applications (particularly sampled-data circuits such as data converters). These two effects are known as *charge injection* and *clock feedthrough*.

Charge Injection

Charge injection can be understood with the help of Fig. 25.3. When the MOSFET switch is on and V_{DS} is small, the charge under the gate oxide resulting from the inverted channel is (from Ch. 6) Q'_{ch}. When the MOSFET turns off, this charge is injected onto the capacitor and into V_{in}. Because V_{in} is assumed to be a low-impedance, source-driven node, the injected charge has no effect on this node. However, the charge injected onto C_{load} results in a change in voltage across it. Note that we also have charge injection (in the opposite direction) when we turn the switch on. However, the fact that the input voltage is connected to C_{load} through the channel resistance makes this error unimportant (the voltage across C_{load} charges to V_{in} through the MOSFET's channel resistance).

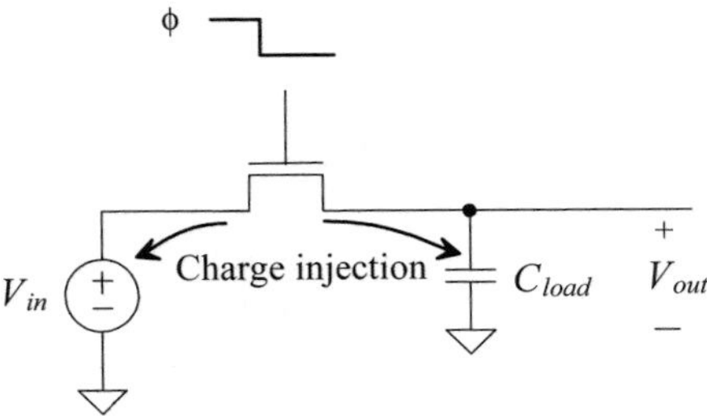

Figure 25.3 Simple configuration using an NMOS switch to show charge injection.

Although the charge injection mechanism is itself a complex one, many studies have sought to characterize and minimize its effects. It has been shown that if the clock signal turns off fast, the channel charge distributes fairly equally between the adjacent nodes. Thus, half of the channel charge is distributed onto C_{load}. From Ch. 6, the charge/unit area of an inverted channel can be approximated as

$$Q_I'(y) = C_{ox}' \cdot (V_{GS} - V_{THN}) \tag{25.1}$$

The total charge in the channel must then be multiplied by the area of the channel resulting in

$$Q_I(y) = C_{ox}' \cdot W \cdot L \cdot (V_{GS} - V_{THN}) \tag{25.2}$$

Therefore, the change in voltage across C_{load} (if an NMOS switch is used) is

$$\Delta V_{load} = -\frac{C_{ox}' \cdot W \cdot L \cdot (V_{GS} - V_{THN})}{2C_{load}} \tag{25.3}$$

which can be written as

$$\Delta V_{load} = -\frac{C_{ox}' \cdot W \cdot L \cdot (VDD - V_{in} - V_{THN})}{2C_{load}} \tag{25.4}$$

if it is assumed that the clock swings between *VDD* and ground. The threshold voltage, Eq. (6.19), can also be substituted into Eq. (25.4) to form

$$\Delta V_{load} = -\frac{C_{ox}' \cdot W \cdot L \cdot (VDD - V_{in} - [V_{THN0} + \gamma(\sqrt{|2V_{fp}| + V_{in}} - \sqrt{|2V_{fp}|})])}{2C_{load}} \tag{25.5}$$

Note that Eq. (25.5) illustrates the problem associated with charge injection. The change in voltage across C_{load} is nonlinear with respect to V_{in} due to the threshold voltage. Thus, it can be said that if the charge injection is signal-dependent, harmonic distortion results. In sampled-data systems, charge injection results in nonlinearity errors. In the case where the charge injection is signal-independent, a simple offset occurs, which is much easier to manage than harmonic distortion. These will be discussed in more detail in Chs. 28 and 29, but it should be obvious here that charge injection effects should be minimized as much as possible.

Capacitive Feedthrough

Consider the schematic of the NMOS switch shown in Fig. 25.4. Here the capacitances between the gate/drain and gate/source of the MOSFET are modeled with the assumption that the MOSFET is operating in the triode region. When the gate clock signal, ϕ, goes high, the clock signal feeds through the gate/drain and gate/source capacitances. However, as the switch turns on, the input signal, V_{in}, is connected to the load capacitor through the NMOS switch. The result is that C_{load} is charged to V_{in} and the capacitive feedthrough has no effect on the final value of V_{out}. However, now consider what happens when the clock signal makes the transition low, that is, the n-channel MOSFET turns off. A capacitive voltage divider exists between the gate-drain (source) capacitance and the load capacitance. As a result, a portion of the clock signal, ϕ, appears across C_{load} as

$$\Delta V_{load} = \frac{C_{overlap} \cdot VDD}{C_{overlap} + C_{load}} \tag{25.6}$$

where $C_{overlap}$ is the overlap capacitance value,

$$C_{overlap} = C'_{ox} \cdot W \cdot LD \tag{25.7}$$

and LD is the length of the gate that overlaps the drain/source.

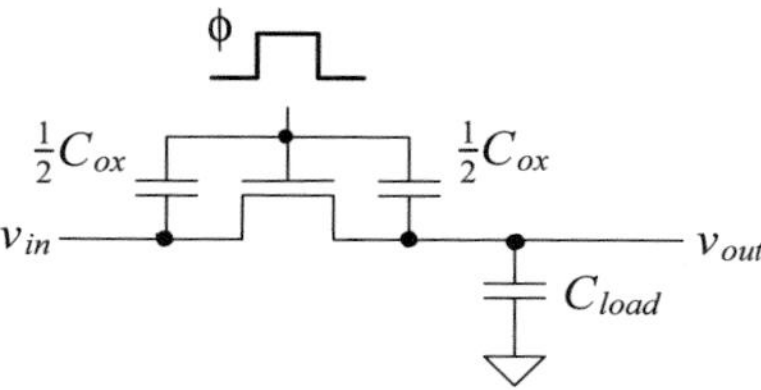

Figure 25.4 Illustration of capacitive feedthrough.

Reduction of Charge Injection and Clock Feedthrough

Many methods have been reported that reduce the effects of charge injection and capacitive feedthrough. One of the most widely used is the dummy switch, as seen in Fig. 25.5. Here, a switch, M2, with its drain and source shorted is placed in series with the desired switch M1. Notice that the clock signal controlling the dummy switch is the complement of the signal controlling M1, and in addition, should also be slightly delayed.

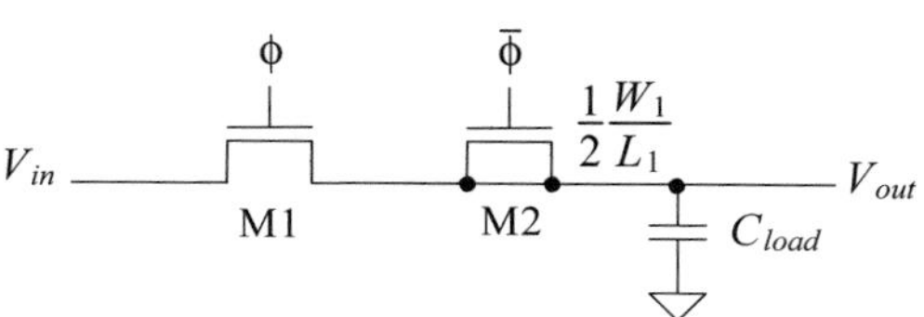

Figure 25.5 Dummy switch circuit used to minimize charge injection.

When M1 turns off, half of the channel charge is injected toward the dummy switch, thus explaining why the size of M2 is one-half that of M1. Although M2 is effectively shorted, a channel can still be induced by applying a voltage on the gate. Therefore, the charge injected by M1 is essentially matched by the charge induced by M2, and the overall charge injection is canceled. Note what happens when M2 turns off. It will inject half of its charge in both directions. However, because the drain and source are shorted and M1 is on, all of the charge from M2 will be injected into the low-impedance, voltage-driven source, which is also charging C_{load}. Therefore, M2's charge injection will not affect the value of voltage on C_{load}.

Another method for counteracting charge injection and clock feedthrough is to replace the switch with a CMOS transmission gate (TG). This results in lower changes in V_{out} because the complementary signals that are used will act to cancel each other.

However, this approach requires precise control on the complementary clocks (the clocks must be switched at exactly the same time) and assumes that the input signal, V_{in}, is small, since the symmetry of the turn-on and turn-off waveforms depend on the input signal.

Fully-differential circuit topologies are used to cancel these effects to a first order, as seen in Fig. 25.6. Since the nonideal charge injection and clock feedthrough effects appear as a common-mode signal to the amplifier, they will be reduced by the CMRR of the amplifier. However, the second-order effects resulting from the input signal amplitude dependence will ultimately limit the dynamic range of operation, neglecting coupled and inherent noise in the dynamic circuits. This subject will be discussed in more detail in the following sections and in the next chapter.

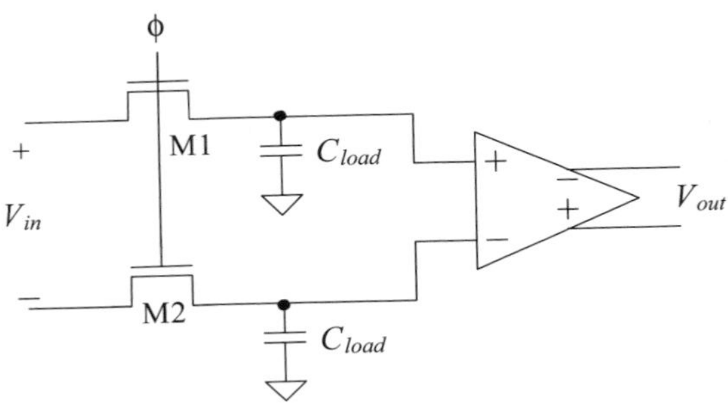

Figure 25.6 Using a fully-differential circuit to minimize charge injection and clock feedthrough.

kT/C Noise

In Ch. 8 we saw that the maximum RMS output noise generated from a simple *RC* circuit was $\sqrt{kT/C}$ (see Table 8.1). If we think of the MOSFET in Fig. 25.7 as a resistor (when the MOSFET is on), then we can add this RMS noise source in series with the output of the capacitor. The noise can be regarded as a sampled (random) voltage onto the capacitor each time the switch is turned on. The RMS noise generated, at room temperature, when using a 1 pF capacitor is 64 μV, while a 100 fF capacitor results in a noise voltage of 200 μV. In other words, the larger the capacitor, the smaller the noise voltage sampled on to the storage capacitor. For high-speed systems, it is desirable to use small capacitors since they take less time to charge. When designing a high-speed and low-noise circuit, trade-offs must be made when selecting the capacitor size.

Note, as mentioned above, the MOSFET is thought of as a resistor (and so *kT/C* noise in the circuit in Fig. 25.7 is from MOSFET thermal noise). If we review the noise mechanisms found in a MOSFET in Ch. 9, we also see that Flicker noise may be present. However, for a MOSFET's drain current to contain Flicker noise, the MOSFET must be conducting a DC current. Because the MOSFET in Fig. 25.7 doesn't conduct any DC current after C_H is charged, Flicker noise doesn't affect the sampled voltage.

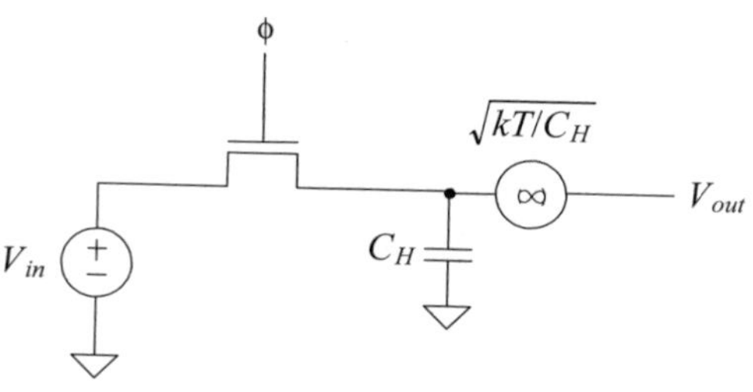

Figure 25.7 How kT/C noise adds to a sampled signal.

25.1.1 Sample-and-Hold Circuits

An important application of the switch is in the sample-and-hold circuit. The sample-and-hold circuit finds extensive use in data converter applications as a sampling gate. A variety of topologies exist, each with their own benefits. The simplest is shown in Fig. 25.8. A narrow pulse is applied to the gate of the MOSFET, enabling v_{in} to charge the hold capacitor, C_H. The width of the strobing gate pulse should allow the capacitor to fully charge before being removed. The op-amp simply acts as a unity gain buffer, isolating the hold capacitor from any external load. This circuit suffers from the clock feedthrough and charge injection problems mentioned in the previous discussion.

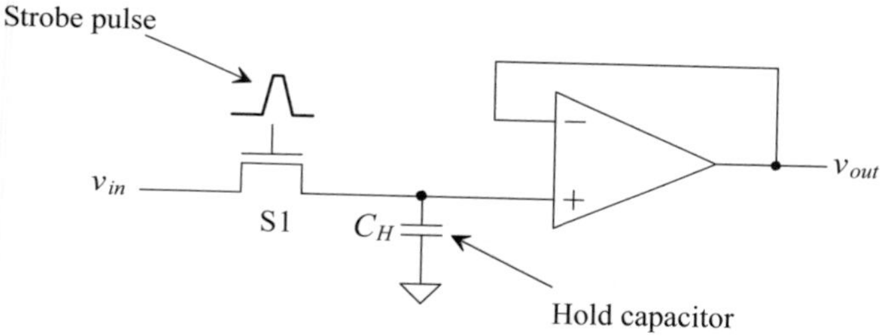

Figure 25.8 A basic sample-and-hold circuit (more correctly called a track-and-hold, see Fig. 28.5b).

A possible improvement in the basic S/H circuit is seen in Fig. 25.9. Here, two amplifiers buffer the input and the output. Notice that switch S_2 ensures that amplifier A1 is stable while in hold mode. If the switch were not present, A1 would be open loop during hold mode and would swing to one of the rails. During the next sample mode, it would then be slew-limited while going from the supply to the value of v_{in}. However, with the addition of S_2, the output of A1 tracks v_{in} even while in hold mode. The switch S_3 also disconnects A1 from the output during hold mode. This S/H has its disadvantages, however. The capacitor is still subjected to charge injection and clock-feedthrough problems. In addition, during sample mode, the circuit may become unstable since there are now two amplifiers in the single-loop feedback structure. Although compensation

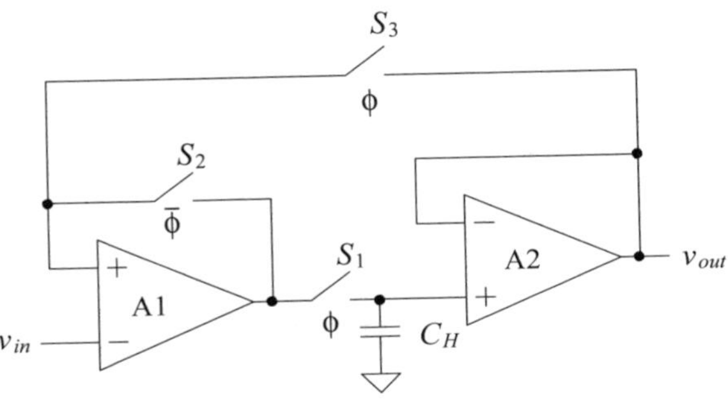

Figure 25.9 A closed-loop S/H circuit.

capacitors can be added to stabilize its performance, the size and placement of the capacitors depend solely on the type and characteristics of the op-amps.

Another S/H circuit is seen in Fig. 25.10. Here, a transconductance amplifier is used to charge the hold capacitor. A control signal turns the amplifier, A1, on or off digitally, thus eliminating the need for the switches S_2 and S_3 (from Fig. 25.9). Since CMOS op-amps are well suited for high-output impedance applications, this configuration would seem to be a popular one. However, the speed of this topology is dictated by the maximum current output of the transconductance amplifier and the size of the hold capacitor.

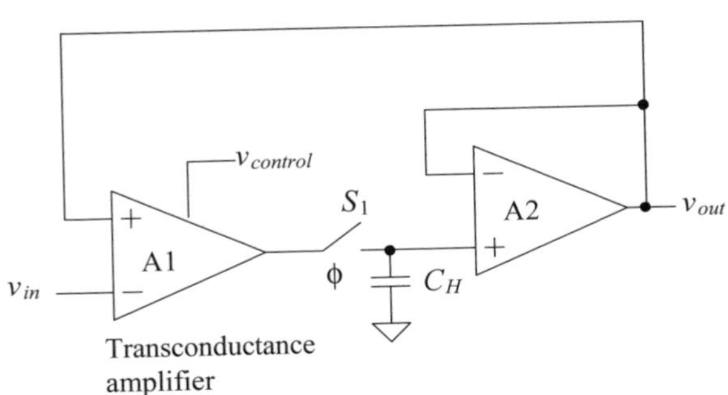

Figure 25.10 A closed-loop S/H circuit using a transconductance amplifier.

A third S/H circuit can be seen in Fig. 25.11. The advantage of this circuit may not be completely obvious and warrants further explanation. First, notice that the hold capacitor is actually in the feedback path of the amplifier, A2, with one side connected to the output of the amplifier and the other connected to a virtual ground. When switch S_1 turns off, any charge injected onto the hold capacitor results in a slight change in the output voltage. However, now that one side of the switch is at virtual ground, the change in voltage is no longer dependent on the threshold voltage of the switch itself. Therefore,

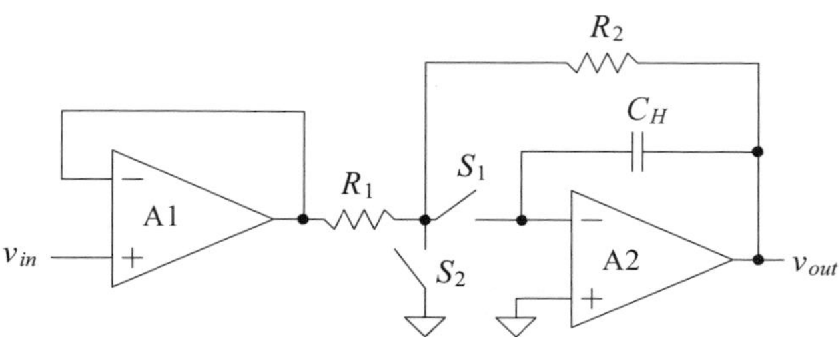

Figure 25.11 A closed-loop S/H circuit using a transconductance amplifier.

the charge injection will be independent of the input signal and will result as a simple offset at the output. An offset error is much easier to tolerate than a nonlinearity error, as will be seen in Chs. 28 and 29.

When sampling, S_1 is closed and S_2 is open, and the equivalent circuit is simply a low-pass filter with a buffered input. The overall transfer function becomes

$$\frac{v_{out}}{v_{in}} = -\frac{R_2}{R_1} \cdot \frac{1}{(sR_2C_H + 1)} \tag{25.8}$$

Therefore, this circuit performs a low-pass filter function while sampling. The buffer A1 can be eliminated when we desire a low-input impedance. Once hold mode commences, the output will stay constant at a value equal to v_{in}, while the switch S_2 isolates the input from the hold capacitor. One important issue to note here is that A2 will need to be a buffered CMOS amplifier because of the resistive load attached at v_{out} during hold mode. Notice also that during both sample mode and hold mode, there is only one op-amp in each feedback loop, so this S/H topology is much more stable than the closed-loop structure introduced in Fig. 25.10.

25.2 Fully-Differential Circuits

As we saw in Fig. 25.6 and the associated discussion, using a fully-differential topology (an op-amp with both differential inputs and outputs) can reduce the effects from imperfect switches. However, using a fully-differential topology requires the use of a *common-mode-feedback* (CMFB) circuit. In the following we present an overview of fully-differential op-amps and their application in a sample-and-hold circuit.

Gain

The differential output op-amp symbol is shown in Fig. 25.12. The open-loop gain of the op-amp is related to its inputs and outputs by

$$v_{out} = v_{op} - v_{om} = A_{OL} \cdot (v_p - v_m) \tag{25.9}$$

This should be compared to the single-ended output op-amp (the op-amp we have been discussing up to this point), which has a gain given by

$$v_{out} = v_{op} = A_{OL} \cdot (v_p - v_m) \tag{25.10}$$

If we ignore v_{om}, then the differential output op-amp behaves just like a single-ended op-amp. For linear applications, the op-amp is used with a feedback network, the inputs are related (assuming the open-loop gain, A_{OL} is large) by

$$v_p \approx v_m \tag{25.11}$$

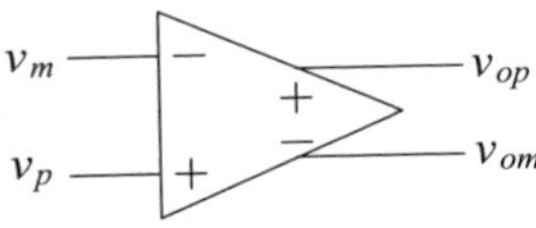

Figure 25.12 Differential output op-amp.

Common-Mode Feedback (CMFB)

Normally, the average of the op-amp outputs is called the common-mode output voltage. The two op-amp outputs swing around the common-mode voltage. If the largest op-amp output voltage is *VDD* and the minimum output voltage is ground, then the common-mode voltage is

$$V_{CM} = \frac{VDD}{2} \tag{25.12}$$

Figure 25.13a shows a simple differential output op-amp gain configuration (inverting or noninverting, depending on which output is used as the positive output). Because the input voltages to the circuit are equal, the output voltages of the op-amp should remain at V_{CM}. Due to the op-amp's high gain and the negative feedback, we always have $v_p \approx v_m$ (Eq. [25.11]). However, if $v_{op} = v_{om} = VDD$ or $v_{op} = v_{om} =$ ground or $v_{op} = v_{om} =$ anything, then $v_p = v_m$ and Eq. (25.11) is satisfied. *This is a problem* and the reason why we need a CMFB circuit. In Fig. 25.13b, we add a CMFB circuit to monitor the outputs of the op-amp and make adjustments in the op-amp (how these adjustments are made is discussed in the next chapter) to keep the two outputs balanced around V_{CM} ($VDD/2$).

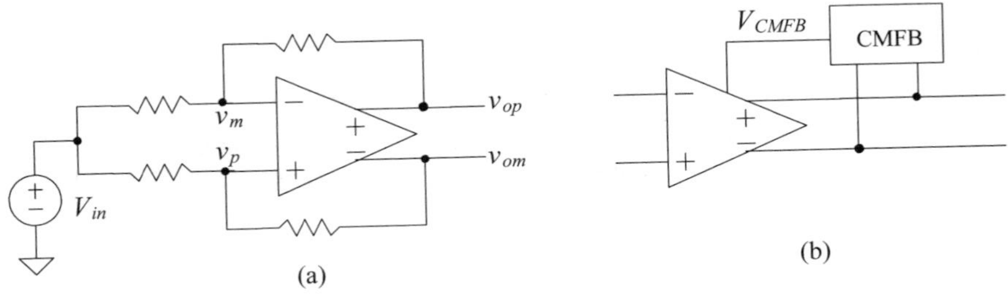

Figure 25.13 (a) Simple gain configuration using differential output op-amp and (b) the use of a common-mode feedback circuit to adjust the common-mode output voltage.

Coupled Noise Rejection

Differential topologies are required to minimize the effects of charge injection and capacitive feedthrough from switches. They also are used to help reduce the effects of coupled noise. Consider the cascade of differential output op-amps, with no feedback network shown, seen in Fig. 25.14. The stray capacitance between the interconnecting metal lines (the metal lines that carry the signals between op-amps) and the substrate or any other noise source are shown. If the metal lines are run close to one another, then the noise voltage will couple even amounts (ideally) of noise into each signal wire. Since the diff-amp, on the input of the op-amp, rejects common signals (signals that are present on both inputs), the coupled noise is not passed to the next op-amp in the string. Variations on the power supply rails are rejected as well. If the differential op-amp is symmetric, then changes on the power supply couple evenly into both outputs, having little effect on the difference, or desired signal, coming out of the op-amp. For these reasons, that is, good coupled noise rejection and PSRR, the differential op-amp is a necessity in any dynamic analog integrated circuit.

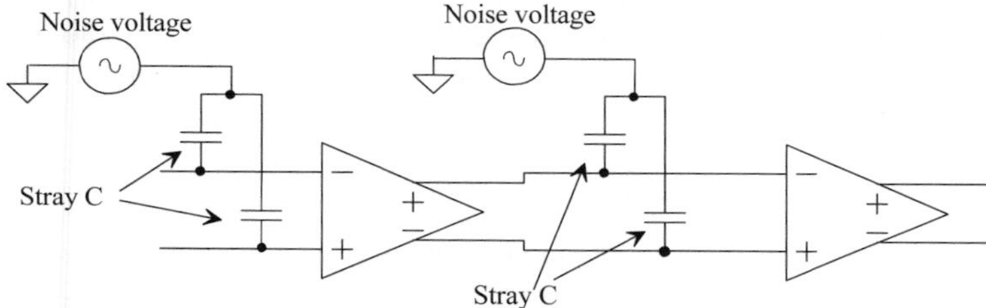

Figure 25.14 Differential output op-amps showing parasitic capacitance and noise.

Other Benefits of Fully-Differential Op-Amps

Another benefit of fully-differential circuits is a doubling in output voltage swing. For example, if *VDD* is 1 V, then the output of a single-ended op-amp can swing from 1 V to ground. In a fully-differential topology, both v_{op} and v_{om} can swing from 0 to 1. Using Eq. (25.9), we can write

$$v_{out,\max} = v_{op} - v_{om} = 1 - 0 = 1\ V \text{ and } v_{out,\min} = v_{op} - v_{om} = 0 - 1 = -1\ V \quad (25.13)$$

The output swing is $2VDD$ when using fully-differential topologies.

A further benefit of fully-differential topologies is the fact that the input common-mode voltage of the op-amp remains at V_{CM} (and so the op-amp's first stage diff-amp's common-mode voltage range requirements are easy to meet).

25.2.1 A Fully-Differential Sample-and-Hold

Figure 25.15 shows a fully-differential sample-and-hold circuit and the associated clock waveforms that eliminate clock feedthrough and charge injection to a first order. The switches in this figure are closed when their controlling clock signals are high. The basic operation can be understood by considering the state of the circuit at t_0. At this time, the input signals charge the sampling capacitors. The bottom plates of the capacitors (poly1) are tied directly to the input signals, for reasons that will be explained below. The op-amp is operating in a unity-follower configuration in which both inputs of the op-amp are held at V_{CM}. At this particular instance in time, prior to t_1, the amplifier is said to be operating in the sample mode of operation.

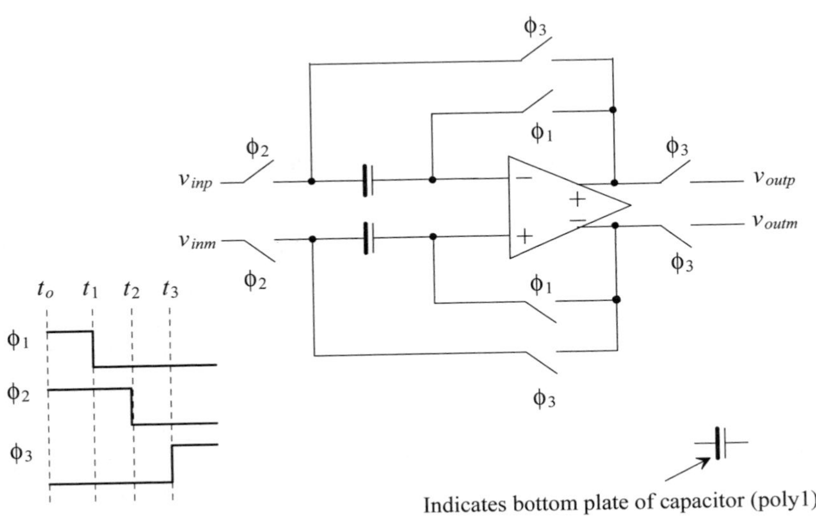

Figure 25.15 Sample-and-hold using differential topology.

At t_1, the ϕ_1 switches turn off. The resulting charge injection and clock feedthrough appear as a common-mode signal on the inputs of the op-amp and are ideally rejected. Since the top plates of the hold capacitors (the inputs to the op-amp) are always at V_{CM}, at this point in time the charge injection and clock feedthrough are independent of the input signals. This produces an increase in the dynamic range of the sample-and-hold (the minimum measurable input signal decreases). The voltage on the inputs of the op-amp (the top plate of the capacitor) between t_1 and t_2 is $V_{OFF1} + V_{CM}$, a constant voltage. Note that the op-amp is operating open loop at this time so the time between t_1 and t_3 should be short.

At t_2 the ϕ_2 switches turn off. At this point in time, the voltages on the bottom plates of the sampling (or hold) capacitors (poly1) are v_{inp} and v_{inm}. The voltages on the top plates of the capacitors (connected to the op-amp) are $V_{OFF1} + V_{OFF2} + V_{CM}$ (assuming that the storage capacitors are much larger than the input capacitance of the op-amp). The term V_{OFF2} is ideally a constant that results from the charge injection and capacitive feedthrough from the ϕ_2 switches turning off. The time between t_1 and t_2 should be short compared to variations in the input signals.

At time t_3 the ϕ_3 switches turn on and the op-amp behaves like a voltage follower; the circuit is said to be in the hold mode of operation. The charge injection and clock feedthrough resulting from the ϕ_3 switches turning on causes the top plate of the capacitor to become $V_{OFF1} + V_{OFF2} + V_{OFF3} + V_{CM}$, again assuming that the storage capacitors are much larger than the input capacitance of the op-amp. The outputs of the sample-and-hold are v_{inp} and v_{inm}, assuming infinite op-amp gain since these offsets appear as a common-mode voltage on the input of the op-amp. Note that the terms V_{OFF2} and V_{OFF3} are dependent on the input signals.

Connecting the Inputs to the Bottom (Poly1) Plate

The reason for connecting the input signals to the bottom plate of the capacitor can be explained with the help of Fig. 25.16. This figure is a simplified, single-ended version of Fig. 25.15 where the capacitance, C_p, is the parasitic capacitance from the bottom plate to the substrate. With regard to Fig. 25.16a, coupled noise from the substrate sees either the input voltage from the op-amp driving the sample-and-hold (in the sample mode) or the output voltage of the op-amp used in the sample-and-hold itself (in the hold mode). Since the op-amps in either mode directly set this voltage, the substrate noise has, ideally, little effect on the circuit's operation.

In Fig. 25.16b, coupled substrate noise feeds directly into the input of the op-amp and can thus drastically affect the output of the sample-and-hold. Another more subtle problem occurs in the circuit of Fig. 25.16b. When the circuit makes the transition to the hold mode at t_3, the output of the op-amp should quickly change to the voltage sampled on the input capacitors. The time it takes the output of the op-amp to change and settle to this final voltage is called the *settling time*. The parasitic capacitance on the input of the op-amp in (b) reduces the feedback factor from unity to $C_H/(C_p + C_H)$. This slows the settling time and causes a gain error in the circuit's transfer function. For these reasons, the parasitic capacitance on the top plate of the capacitor should be small. Nothing should be laid out over or in near proximity of poly2.

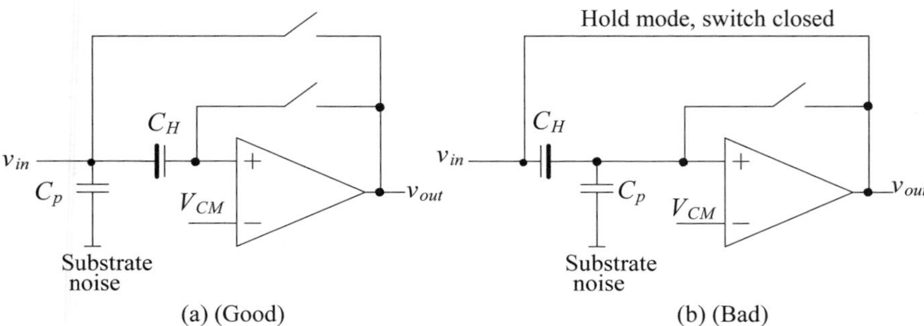

Figure 25.16 Explanation for connecting the bottom plate of the capacitor to the input.

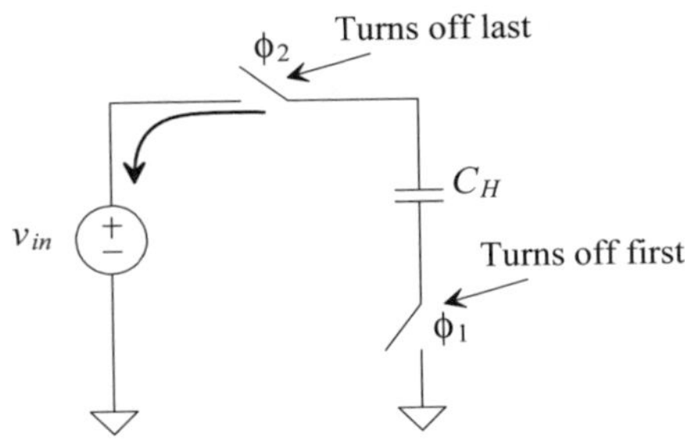

Figure 25.17 Bottom plate sampling.

Bottom Plate Sampling

Turning the ϕ_1 controlled switches off in Fig. 25.15 slightly before the ϕ_2 controlled switches is sometimes called *bottom plate sampling*. Figure 25.17 illustrates why. When the ϕ_1 switches turn off, the charge injected into C_H is a constant independent of the input signal amplitude. The resulting offset voltage across C_H is then constant. When the ϕ_2 switches turn off, the resulting charge injection then takes the path of least resistance, that is, into the input source v_{in}. The voltage across C_H is then, ideally, independent of the input signal voltage.

SPICE Simulation

Let's simulate the operation of the sample-and-hold in Fig. 25.15. To begin, let's use voltage-controlled voltage sources and a DC voltage to model the op-amp (again, we'll discuss differential output op-amps in the next chapter). The ideal fully-differential output op-amp SPICE model is seen in Fig. 25.18. The SPICE netlist statements used to model this circuit (we'll assume an open-loop gain of 10^6 and *VDD* of 1 *V*) are

```
E1     vop    vcm    vp     vm      1e6
E2     vcm    vom    vp     vm      1e6
Vcm    vcm    0      DC     500m
```

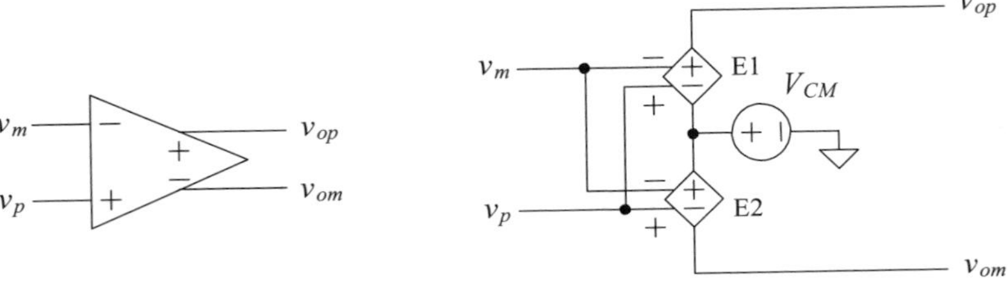

Figure 25.18 SPICE modeling a differential input/output op-amp with common-mode voltage.

For the switches we'll use NMOS devices as seen in the simulation schematic of Fig. 25.19. For the input signals, we've used a 5 MHz sinewave signal with a peak amplitude of 200 mV. Note that each of the inputs is referenced to the common-mode voltage, V_{CM} (which is 500 mV here).

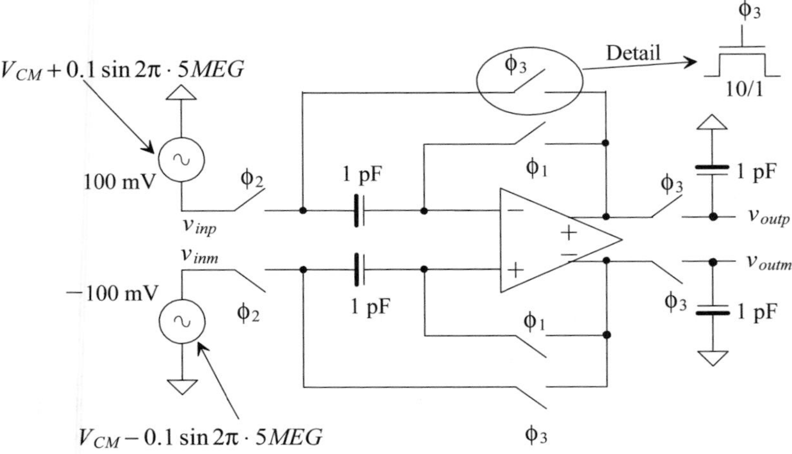

Figure 25.19 Simulating the operation of the fully differential sample-and-hold.

The simulation results are seen in Fig. 25.20. Let's first focus on the clock signals. Notice how ϕ_3 isn't high at the same time as ϕ_1 or ϕ_2. If all switches were "on" at the same time, the charge stored (or held) on the 1 pF capacitors at the output of the circuit would change (remember when the ϕ_1 switches close the op-amp is in the follower configuration with the inputs and outputs pulled to V_{CM}). We make sure to disconnect these output capacitors before putting the circuit into the sample mode. The output of the sample-and-hold seen in Fig. 25.20 doesn't exactly match the input signal. There is a one-clock cycle delay and we can see the effects of the finite clock frequency (finite sample points). Note, these topics are discussed in detail in Volume II of this book entitled *CMOS Mixed-Signal Circuit Design*.

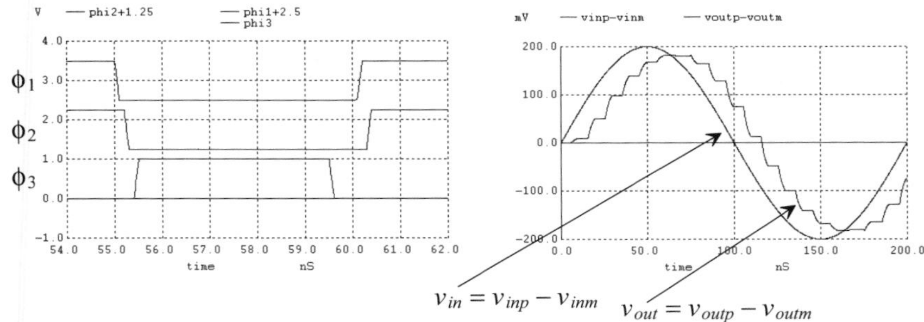

Figure 25.20 Simulating the operation of the sample-and-hold in Fig. 25.19.

25.3 Switched-Capacitor Circuits

Consider the circuit shown in Fig. 25.21a. This dynamic circuit, named a *switched-capacitor resistor*, is useful in simulating a large value resistor, generally >1 MΩ. The clock signals ϕ_1 and ϕ_2 form two phases of a non-overlapping clock signal with frequency f_{clk} and period T as seen in the figure. Let's begin by considering the case when S1 is closed. When ϕ_1 is high, the capacitor C is charged to v_1. The charge, q_1, stored on the capacitor during this interval, Fig. 25.21c, is

$$q_1 = Cv_1 \tag{25.14}$$

while if S2 is closed, the charge stored on the capacitor is

$$q_2 = Cv_2 \tag{25.15}$$

If v_1 and v_2 are not equal, keeping in mind that S1 and S2 cannot be closed at the same time due to the nonoverlapping clock signals, then a charge equal to the difference between q_1 and q_2 is transferred between v_1 and v_2 during each interval T. The difference in the charge is given by

$$q_1 - q_2 = C(v_1 - v_2) \tag{25.16}$$

If v_1 and v_2 vary slowly compared to f_{clk}, then the average current transferred in an interval T is given by

$$I_{avg} = \frac{C(v_1 - v_2)}{T} = \frac{v_1 - v_2}{R_{sc}} \tag{25.17}$$

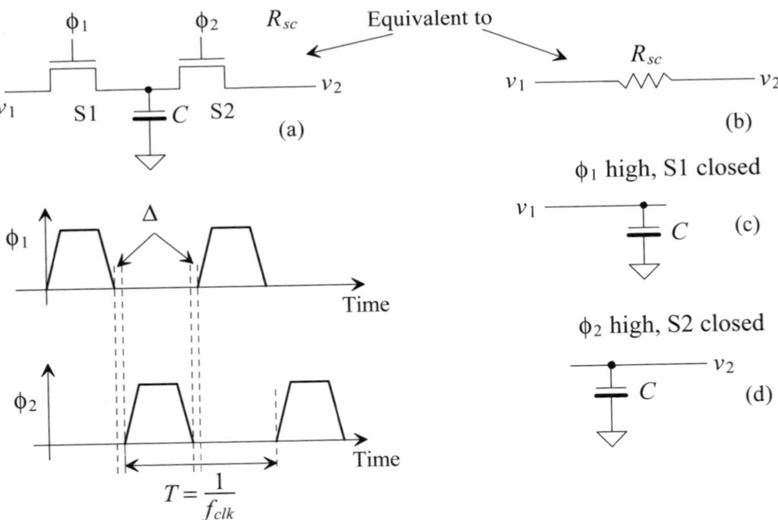

Figure 25.21 Switched-capacitor resistor (a) and associated waveforms and (b, c, d) the equivalent circuits.

The resistance of the switched-capacitor circuit is given by

$$R_{sc} = \frac{T}{C} = \frac{1}{C \cdot f_{clk}} \tag{25.18}$$

In general, the signals v_1 and v_2 should be bandlimited to a frequency at least ten times less than f_{clk} (more on this later). Note that we derived a similar result back in Ch. 17, see Eq. (17.29).

Example 25.1

Using switched-capacitor techniques, implement the circuit shown in Fig. 25.22a so that the product of RC is 1 ms, that is, the 3-dB frequency of $|v_{out}/v_{in}|$ is 159 Hz.

The switched-capacitor implementation of this circuit is shown in Fig. 25.22b. The product of RC may now be written in terms of Eq. (25.18) as

$$RC_2 = \frac{C_2}{C_1} \cdot \frac{1}{f_{clk}} \tag{25.19}$$

This result is important! The product RC_2 is determined by f_{clk}, which may be an accurate frequency derived from a crystal oscillator and the ratio of C_2 to C_1, which will be within 1% on a chip. This means that even if the values of the capacitors change by 20% from wafer to wafer, the ratio of the capacitors relative to one another, on the same wafer, will remain constant within 1%.

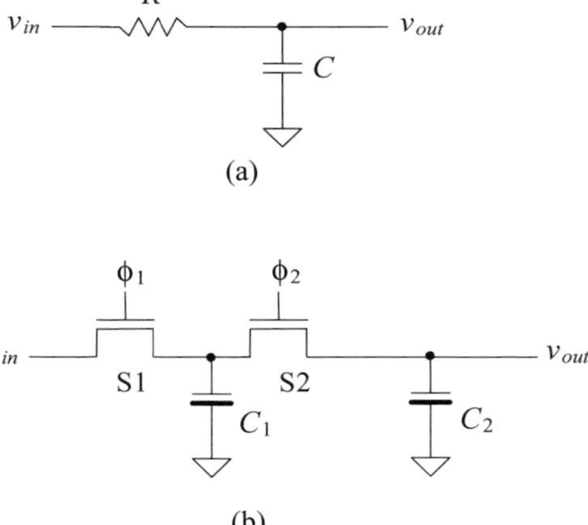

Figure 25.22 (a) Circuit used in Ex. 27.1 and (b) its implementation using a SC resistor.

It is also desirable to keep C_1 larger than the associated parasitics present in the circuit (e.g., depletion capacitances of the source/drain implants and the stray capacitances to substrate). For the present example, we will set C_1 to 1 pF. The selection of f_{clk} is usually determined by what is available. For the present design, a value of 100 kHz will be used. This selection assumes that the energy present in v_{in} at frequencies above 10 kHz is negligible. The value of C_2 is determined solving Eq. (25.19) and is 100 pF. Note that the value of the switched-capacitor resistor is 10 MΩ. Implementing this resistor in a CMOS process using n-well with a sheet resistance of 1,000 ohms/square would require 10,000 squares! The resulting delay through the n-well resistor, because of the capacitance to substrate, may cause a significant phase error in the transfer function. ■

25.3.1 Switched-Capacitor Integrator

Because the switched-capacitor resistor of Fig. 25.21a is sensitive to parasitic capacitances, it finds little use, by itself, in analog switched-capacitor circuits. Consider the circuit of Fig. 25.23a. This circuit, a switched-capacitor integrator, is the heart of the circuits we will be discussing in the remainder of the section. The portion of the circuit consisting of switches S1 through S4 and C_I forms a switched-capacitor resistor with a value given by

$$R_{sc} = \frac{1}{C_I f_{clk}} \tag{25.20}$$

The equivalent continuous time circuit for the switched-capacitor integrator is shown in Fig. 25.23b. Notice that v_{in} is now negative. It may be helpful in the following discussion to remember that the combination of switches and C_I of the switched-capacitor integrator can be thought of as a simple resistor. The transfer function of the switched-capacitor integrator is given by

$$\frac{v_{out}}{v_{in}} = \frac{1/j\omega C_F}{R_{sc}} = \frac{1}{j\omega\left(\frac{C_F}{C_I}\cdot\frac{1}{f_{clk}}\right)} \tag{25.21}$$

Again, the ratio of capacitors is present, allowing the designer to precisely set the gain of the amplifier and the integration time constant.

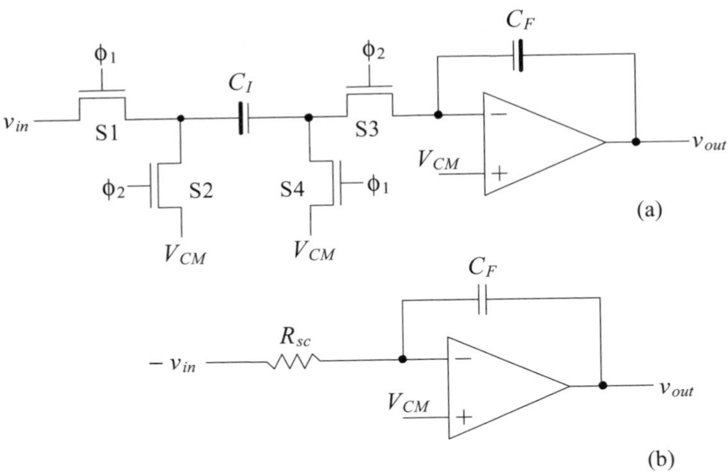

Figure 25.23 (a) A stray insensitive switched-capacitor integrator (noninverting) and (b) the equivalent continuous time circuit.

Parasitic Insensitive

The switched-capacitor integrator in Fig. 25.23 is not sensitive to parasitic or stray capacitances. This can be understood with the use of Fig. 25.24. To begin, if we realize that C_{p2} (the parasitic capacitance on the right side of C_I) is always connected to V_{CM} either through S4 or through the connection to the inverting input of the op-amp, then C_{p2} doesn't see a change in the charge stored on it. Next, the capacitance C_{p1} is charged to v_{in} when S1 is closed and then charged to V_{CM} when S2 closes. Since none of the charge stored on C_{p1} when S1 is closed is transferred to C_I, it does not affect the integrating function. A practical minimum for C_I is 100 fF set by kT/C noise (see Table 8.1).

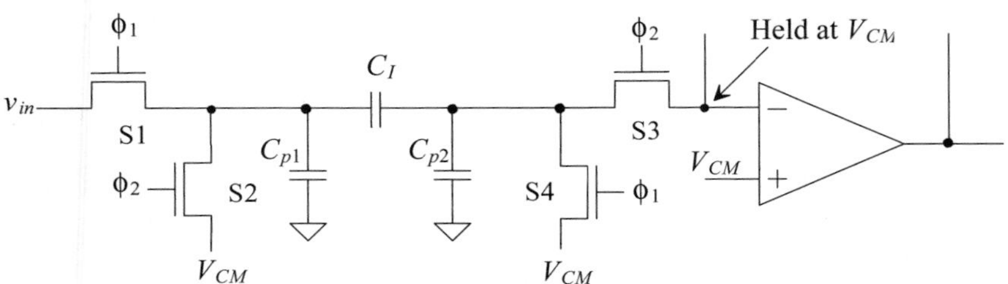

Figure 25.24 Parasitic capacitances associated with a switched-capacitor resistor.

Other Integrator Configurations

An inverting integrator configuration can be formed by simply swapping the clock signals used with S1 and S2 in Fig. 25.23 (the noninverting configuration). The gate of S1 is connected, for the inverting configuration, to ϕ_2 while the gate of S2 is connected to ϕ_1. The gain of this configuration is given by

$$\frac{v_{out}}{v_{in}} = -\frac{1}{j\omega\left(\frac{C_F}{C_I}\cdot\frac{1}{f_{clk}}\right)} \tag{25.22}$$

An example of a switched-capacitor integrator circuit that combines input signals is shown in Fig. 25.25a. Remembering that each switched-capacitor section can be thought of as a resistor, we note that the relationship between the inputs and output is

$$v_{out} = \frac{v_1}{j\omega\left(\frac{C_F}{C_1}\frac{1}{f_{clk}}\right)} + \frac{v_2}{j\omega\left(\frac{C_F}{C_2}\frac{1}{f_{clk}}\right)} - \frac{v_3}{j\omega\left(\frac{C_F}{C_3}\frac{1}{f_{clk}}\right)} \tag{25.23}$$

Figure 25.25b shows how redundant switches can be combined to reduce the number of devices used. Used alone, the basic integrator has the practical problem of integrating not only the input signal but also the offset voltage of the op-amp. In many applications, a reset switch or resistor is placed across the feedback capacitor (Fig. 25.25b). An example of a lossy integrator circuit useful in first-order filter design is shown in Fig. 25.26. The transfer function of this circuit is given by

$$\frac{v_{out}}{v_{in}} = \frac{R_4}{R_3}\left(\frac{1+j\omega R_3 C_1}{1+j\omega R_4 C_2}\right) = \frac{C_3}{C_4}\left(\frac{1+j\omega\left(\frac{C_1}{C_3}\cdot\frac{1}{f_{clk}}\right)}{1+j\omega\left(\frac{C_2}{C_4}\cdot\frac{1}{f_{clk}}\right)}\right) \tag{25.24}$$

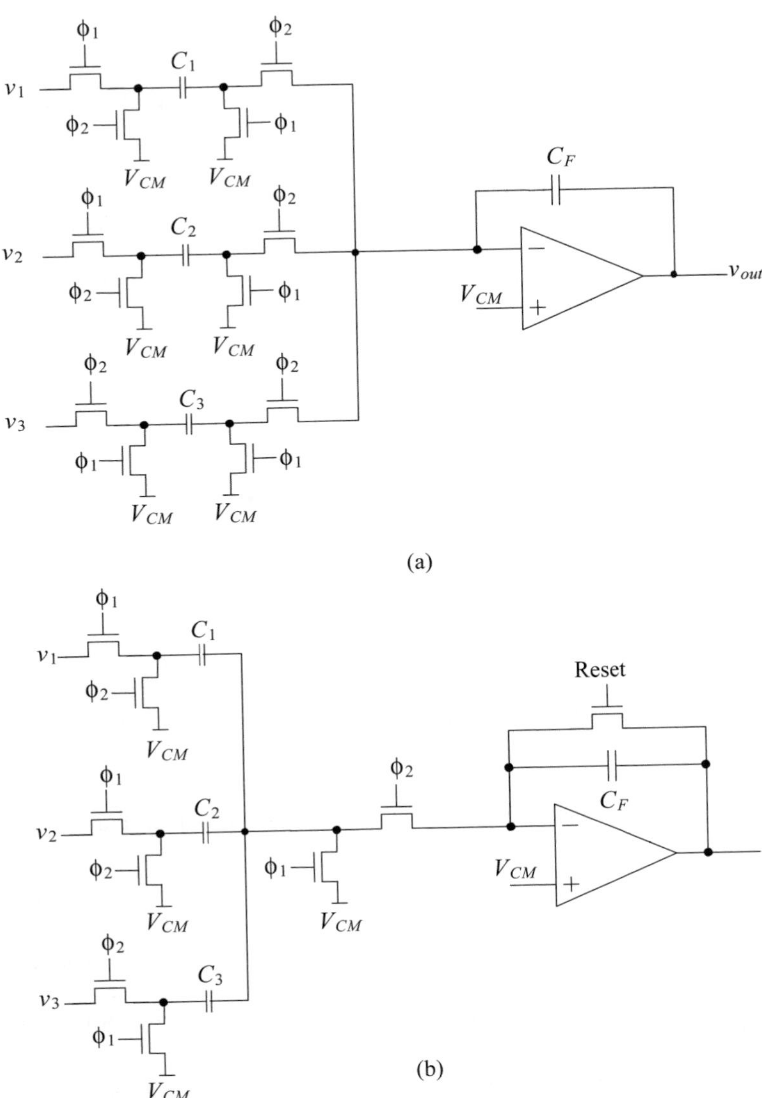

Figure 25.25 (a) Switched-capacitor implementation of a summing integrator and (b) practical implementation of the circuit combining switches and adding reset.

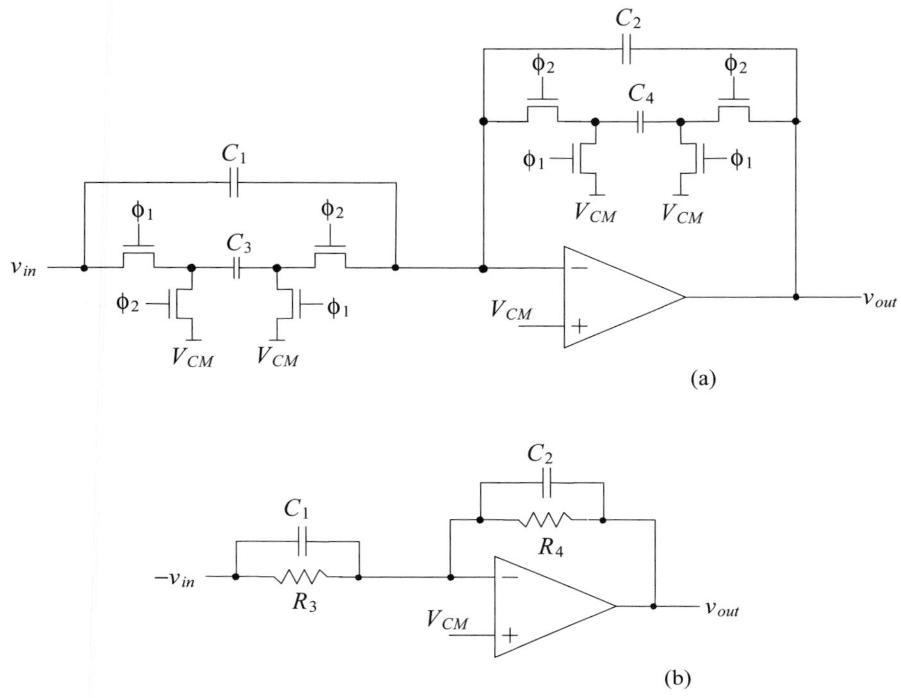

Figure 25.26 Lossy integrator (a) switched-capacitor implementation and (b) continuous time circuit.

For low-frequency input signals (low frequencies compared to the pole and zero given in Eq. (25.24), the gain of the lossy integrator is simply

$$\frac{v_{out}}{v_{in}} = \frac{C_3}{C_4} \tag{25.25}$$

which is again a precise number due to the ratio of the capacitors. Also note that the switched-capacitor resistor in the feedback loop is stray insensitive. The left side of C_4 is always connected to V_{CM}, while the right side is either connected to V_{CM} or to the output of the op-amp.

Note that we cannot eliminate the capacitor across the switched-capacitor resistor in the feedback path. If we were to do so, then the op-amp would be operating open-loop when ϕ_2 goes low. The outputs of the op-amp would then rail up at *VDD* or down at ground.

Example 25.2

Design a switched-capacitor filter with the transfer characteristics shown in Fig. 25.27.

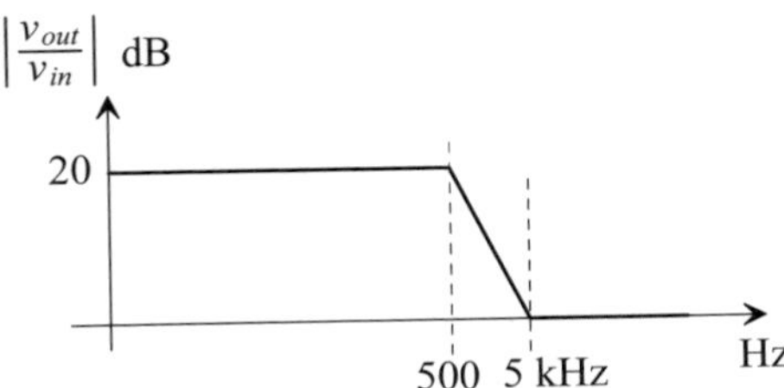

Figure 25.27 Filter characteristics for Ex. 25.2.

We can see that this transfer function has a pole at 500 Hz and a zero at 5 kHz. The lossy integrator of Fig. 25.26 will be used to realize this filter. The low-frequency gain of this circuit is 10 (20 dB). Using Eq. (25.25), we have

$$\frac{C_3}{C_4} = 10$$

while the pole and zero locations are given by

$$f_p = \frac{1}{2\pi\left(\frac{C_2}{C_4}\cdot\frac{1}{f_{clk}}\right)} = 500 \text{ and } f_z = \frac{1}{2\pi\left(\frac{C_1}{C_3}\cdot\frac{1}{f_{clk}}\right)} = 5\text{ kHz}$$

If we set f_{clk} to 100 kHz and C_4 to 100 fF, then C_3 = 1.0 pF, C_2 = 3.2 pF, and C_1 = 3.2 pF. ■

Exact Frequency Response of a Switched-Capacitor Integrator

We will now develop an exact relationship between the switching frequency, f_{clk}, and the signal frequency ω. Referring to Fig. 25.28, we can write the output of the integrator as the sum of the previous output voltage, $v_{out(n)}$, at a time nT and the contribution from the current sample as

$$v_{out(n+1)} = v_{out(n)} + \frac{C_I}{C_F}\cdot v_{in(n)} \tag{25.26}$$

Since a delay in the time domain of T corresponds to a phase shift of ωT in the frequency domain, we can take the Fourier transform of this equation and get

$$e^{j\omega T}v_{out}(j\omega) = v_{out}(j\omega) + \frac{C_I}{C_F}\cdot v_{in}(j\omega) \tag{25.27}$$

Solving this equation for v_{out}/v_{in} gives

$$\frac{v_{out}}{v_{in}}(j\omega) = \frac{C_I}{C_F}\left(\frac{1}{e^{j\omega T}-1}\right) = \frac{C_I}{C_F}\left(\frac{e^{-j\omega T/2}}{e^{j\omega T/2}-e^{-j\omega T/2}}\right) = \frac{C_I}{C_F}\left[\frac{1}{z-1}\right] \tag{25.28}$$

where $z = e^{j\omega T}$. Remembering $f_{clk} = 1/T$ and $\omega = 2\pi f$, we get

$$\frac{v_{out}}{v_{in}}(j\omega) = \frac{1}{j\omega\left(\frac{C_F}{C_I}\cdot\frac{1}{f_{clk}}\right)}\left(\frac{\frac{\pi\cdot f}{f_{clk}}}{\sin\frac{\pi\cdot f}{f_{clk}}}\cdot e^{-j\pi\cdot f/f_{clk}}\right) \tag{25.29}$$

Ideally, the term on the right in parentheses is unity. This occurs when f is much less than f_{clk}. This equation describes how the magnitude and phase of the integrator are affected by finite f_{clk}.

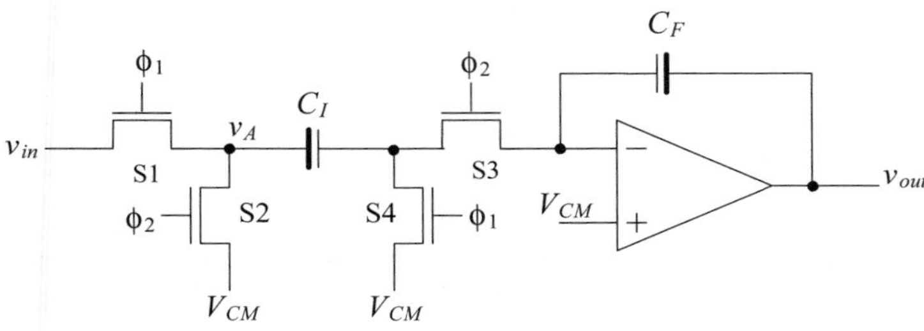

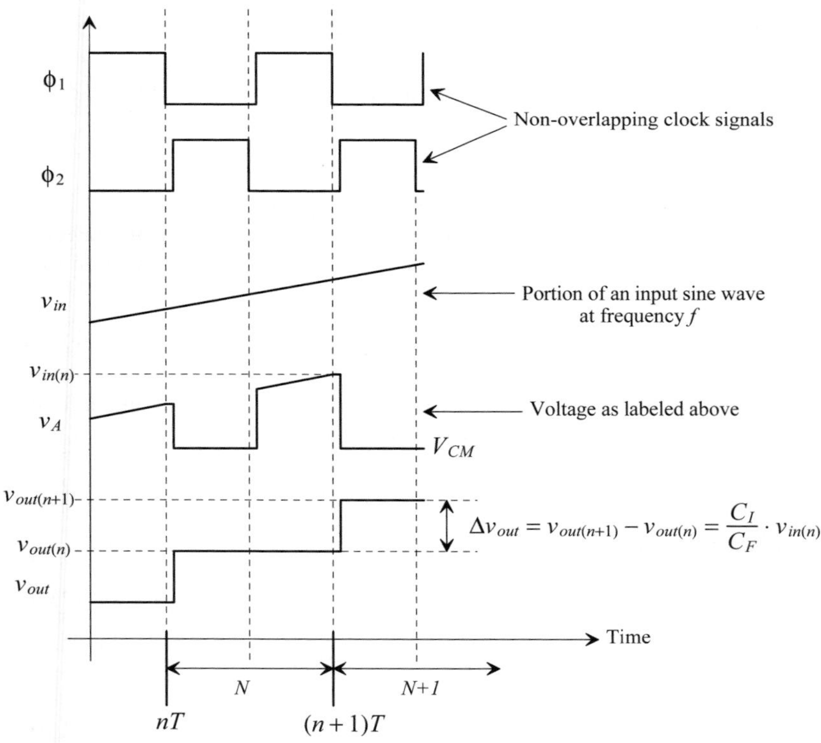

Figure 25.28 Switched-capacitor integrator used to determine the relationship between input frequency and switch clock frequency.

Capacitor Layout

An important step in the implementation of any switched-capacitor design is the layout of the capacitors. Normally, a unit-size capacitor is laid out and then replicated to the desired capacitance (as discussed in Ch. 6). For Ex. 25.2, the unit size capacitance would nominally be 100-fF (using the 1 μm process) or 10 *fF* (in the 50 nm process), as in Fig. 25.29. Note that here, in Fig. 25.29a, we are assuming that a circle can be accurately reproduced on the reticle and patterned on the wafer. In practice, effects such as the finite e-beam size (and the granularity of the grid) used to make the reticle can make this assumption questionable. Figure 25.29b shows a layout where we've tried to minimize the number of 90° corners in an effort to avoid pattern errors when etching poly2.

Again, the absolute value of the capacitors isn't important; rather, the important value is the ratio. A total of 32 of these unit-size capacitors (using a unit-size cell of 100 *fF*) would be used to achieve the larger nominally 3.2 pF capacitors in Ex. 25.2 (see Fig. 25.30). This approach (using unit elements to make the large capacitors) eliminates errors due to uneven patterning of poly to a first order. A p+ guard ring can be placed around the capacitor to help reduce coupled substrate noise. Substrate noise can also be reduced by laying the capacitor out over an n-well that is tied to *VDD*. Injected minority carriers are collected either by the p+ or the n-well (or a combination of both). If matching of the capacitors is critical, schemes that use a common-centroid layout can be used. Also, as was discussed in Ch. 20, dummy poly strips or capacitors can be placed around the array of capacitors so that the edge differences from underetching poly are eliminated. In both cases, what we are doing is trying to ensure that all capacitors see the same adjacent structures.

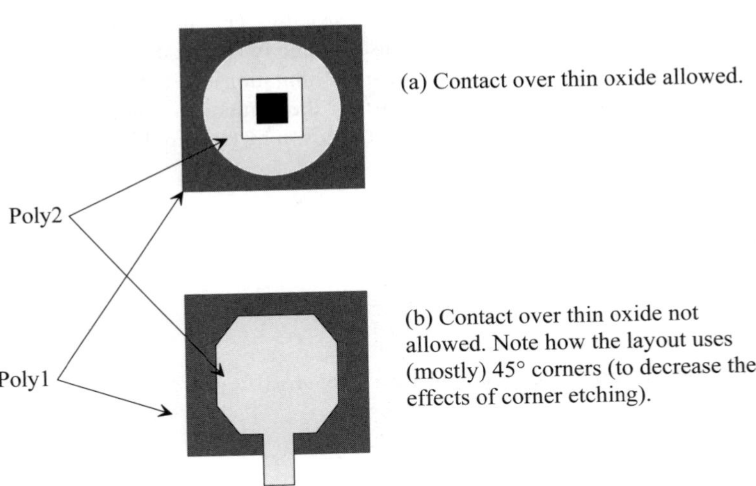

(a) Contact over thin oxide allowed.
Poly2
Poly1
(b) Contact over thin oxide not allowed. Note how the layout uses (mostly) 45° corners (to decrease the effects of corner etching).

Figure 25.29 Layout of a unit cell capacitor.

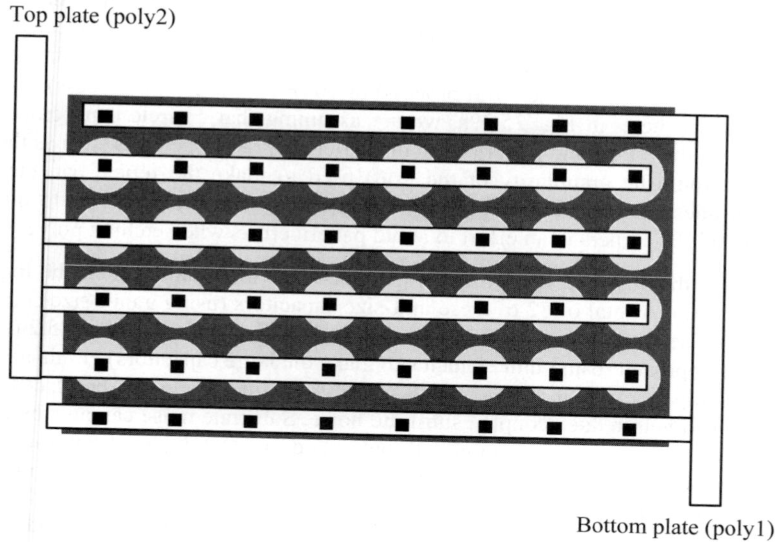

Figure 25.30 Layout of a 3.2 pF capacitor using a 100 fF unit cell.

Op-Amp Settling Time

Figure 25.31 shows an op-amp configuration in which the op-amp can source or sink current to a switched capacitor and a feedback capacitor. The time it takes to charge and discharge these capacitors is important because it directly affects the maximum switched capacitor clocking frequency, f_{clk}. The slew-rate limitations of the op-amp have been discussed in detail already. Let's now consider the limitations due to op-amp finite bandwidth. The closed-loop gain of the op-amp is given by (see Eq. [24.5])

$$A_{CL} = \frac{A_{OL}}{1 + A_{OL} \cdot \beta} \tag{25.30}$$

while the open-loop gain of an op-amp is given, with units of A/A, V/V, V/A, or A/V, by

$$A_{OL} = \frac{A_{OL}(0)}{1 + j\frac{f}{f_{3dB}}} \tag{25.31}$$

Combining these equations and assuming $1 >> 1/[\beta \cdot A_{OL}(0)]$, we get

$$A_{CL} = \frac{\frac{1}{\beta}}{1 + j\frac{f}{f_{un} \cdot \beta}} \tag{25.32}$$

where $f_{3dB} \cdot A_{OL}(0) = f_{un}$, where f_{un} may have units of Hz, Hz/Ω, or Hz·Ω ($f_u\beta$ is in Hz). The closed-loop gain reduces to a simple single-pole transfer function (see Eqs. [24.30]–[24.34]). The low-frequency gain of the circuit is $1/\beta$, while the product of f_{un} and β gives the circuit time constant of

$$\tau = \frac{1}{2\pi f_{un} \cdot \beta} \tag{25.33}$$

keeping in mind the unity-gain frequency of the op-amp, f_{un}, is a strong function of the load capacitance. For a step-input to the op-amp, a common occurrence in switched-capacitor circuits (Fig. 25.28), the output voltage of the op-amp, again neglecting slew-rate limitations, is given by

$$v_{out} = V_{outfinal}(1 - e^{-t/\tau}) \tag{25.34}$$

For the output voltage of the op-amp to settle to less than 1% of its final value requires 5τ. A "rule-of-thumb" estimate for the settling time is simply $1/(f_u \cdot \beta)$.

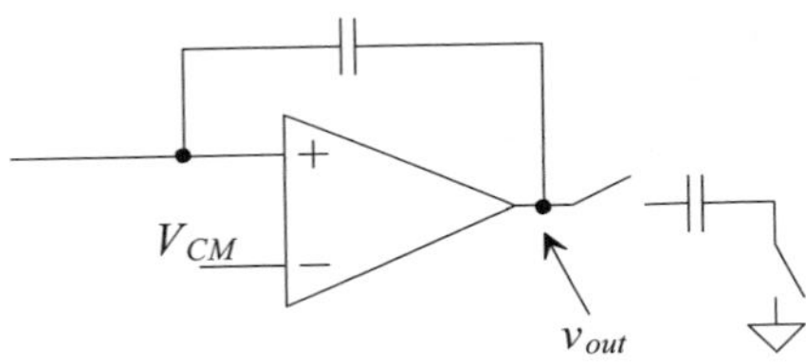

Figure 25.31 Charging and discharging a switched capacitor.

25.4 Circuits

This section presents several examples of dynamic analog circuits.

Reducing Offset Voltage of an Op-Amp

As seen in Fig. 24.4, the op-amp's offset voltage can be modeled by adding a DC voltage in series with the noninverting input of the op-amp, Fig. 25.32a. The basic idea behind eliminating the offset voltage is shown in Fig. 25.32b. A capacitor is charged to a voltage equal and opposite to the comparator offset voltage. The voltage across the capacitor is then added in series with the noninverting op-amp input to subtract away the op-amp's offset.

The dynamic analog circuit shown in Fig. 25.33 is used to implement this subtraction. For this method to be effective, *the op-amp must be stable* in the unity gain configuration. The clock signals ϕ_1 and ϕ_2 are the non-overlapping clock signals discussed earlier (see Fig. 25.28). The nonoverlapping clocks keep switches S1, S2, and S3 from being on at the same time as switches S4 and S5. Let's consider the case shown in Fig. 25.33b where ϕ_1 is high and ϕ_2 is low. The op-amp, via the negative feedback, tries to force its inverting input to V_{CM}. However, because of the offset, the inverting input is actually charged to $V_{OS} + V_{CM}$. Note that under these conditions the op-amp is removed from the inputs. The voltage across the capacitor is then V_{OS}. When ϕ_2 is high and ϕ_1 is low, Fig. 25.33c, the op-amp functions normally, assuming the storage capacitance C is much larger than the input capacitance of the op-amp.

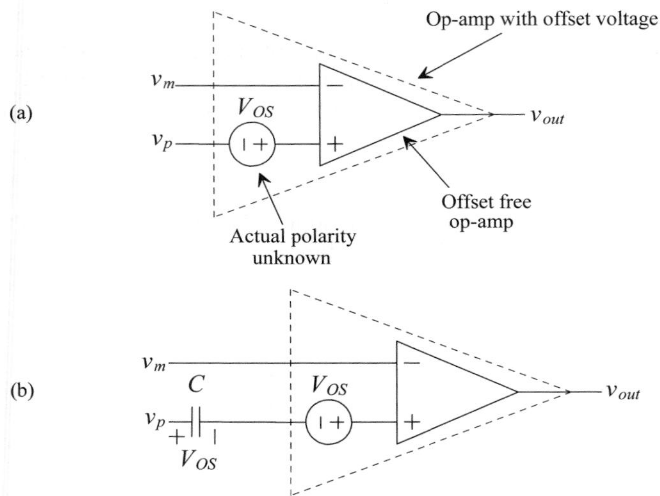

Figure 25.32 (a) Offset voltage of an op-amp modeled by a DC voltage source in series with the noninverting input of the op-amp and (b) using a capacitor to cancel the offset voltage.

Dynamic Comparator

Before we present a dynamic comparator, let's consider the *RC* switch circuit of Fig. 25.34a. When node A is connected to +1 V, node B is connected to ground. Consider what happens when the switches change positions, that is, when node A is connected to ground and the switch at node B is connected to an open. At the moment just after switching takes place, the potential at node B becomes –1 V. In other words, the voltage across the capacitor does not change instantaneously. If node A is connected back to +1 V and node B is connected back to ground a short time compared to the product of *R* and *C*, then the voltage across the capacitor remains +1 V.

A more useful circuit for CMOS is shown in Fig. 25.34b. If the switch across C_B is connected to ground when node A is connected to $V1$, then V_B, when the switches change positions, is given by

$$V_B = (V2 - V1) \cdot \frac{C_A}{C_A + C_B} \tag{25.35}$$

Figure 25.35 shows a dynamic comparator based on the inverter. When ϕ_1 is high, the voltage on the v_m input is connected to node A, while the voltage on node B is set via S3 so that the input and output voltages of the inverter are equal. (The inverter is operating as a linear amplifier where both M1 and M2 are in the saturation regions.) When ϕ_2 goes high (ϕ_1 is low since the clocks are nonoverlapping), the v_p input is connected to node A. If C_A is much larger than the input capacitance of the inverter (C_B), then the voltage change on the input of the inverter (V_B) is

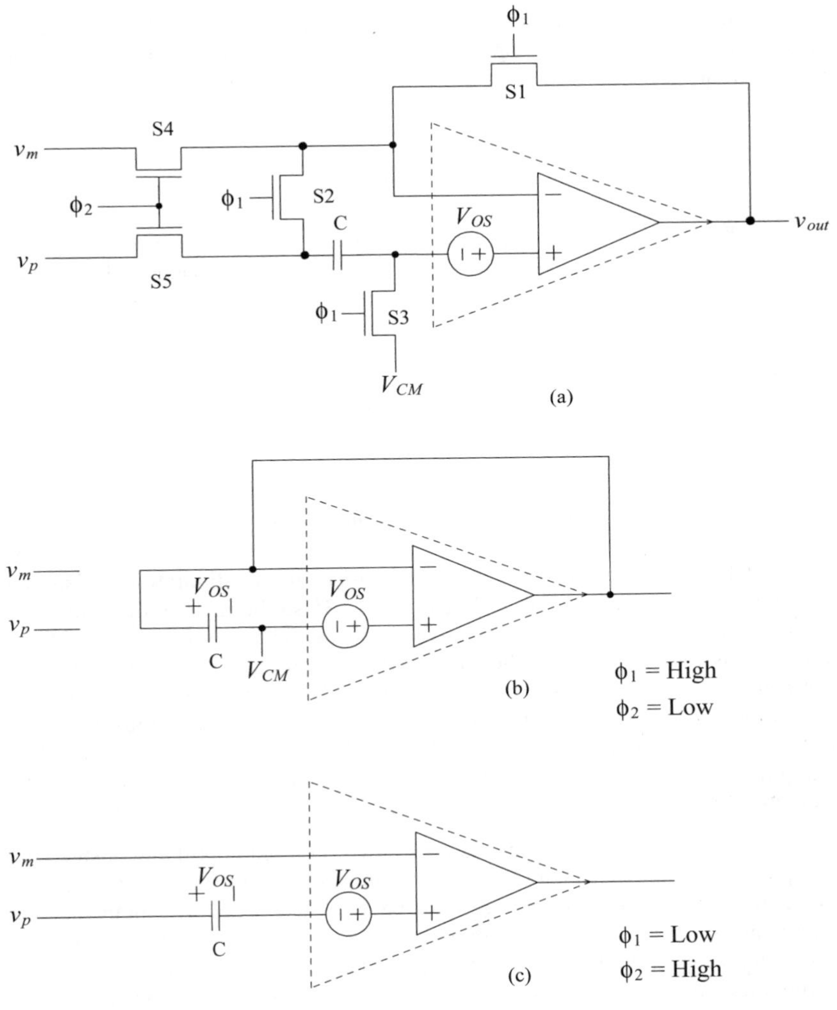

Figure 25.33 Dynamic reduction of the offset voltage.

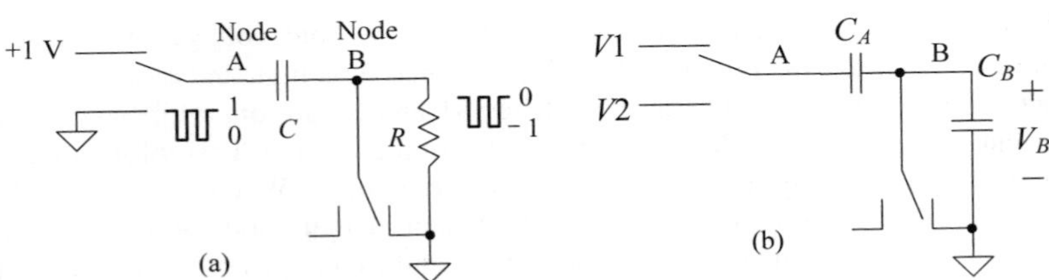

Figure 25.34 Circuits used to illustrate switching in dynamic circuits.

$$v_{in} = v_p - v_m \tag{25.36}$$

Provided the gain of the inverter is large, this change causes the inverter output to rail, that is, go to either *VDD* or ground. The output is then latched and available during ϕ_1. The gain of the comparator can be increased by using additional inverter stages.

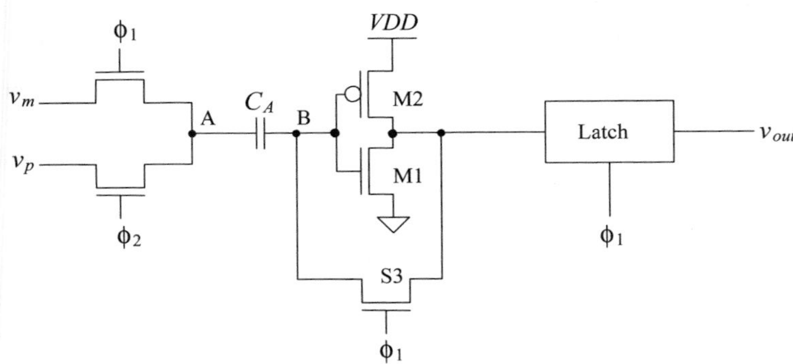

Figure 25.35 A dynamic comparator.

Another high-performance dynamic comparator configuration is based on the sense amplifier in Figs. 16.26 or 16.32 (or Fig. 16.35 so the final outputs only change on the clock's rising edge) that use positive feedback. The offset voltage of the overall comparator is reduced, using either input offset storage (IOS) or output offset storage (OOS) around the comparator preamp. Figure 25.36 shows the two types of offset cancellation techniques. In the IOS configuration in (a), the preamp must be stable in the unity feedback configuration. In the OOS configuration in (b) the MOSFETs in the preamp must remain in saturation when the offset voltage is stored on the capacitors. If a differential amplifier is used as the preamp, this condition is usually easily met.

The size of the storage capacitors is based on three important considerations: (1) preamp or latch input capacitance, (2) charge injection, and (3) kT/C noise. For the IOS scheme, the input storage capacitance must be much larger than the input capacitance of the preamp, so that the storage capacitors don't attenuate the input signals. For example, if the storage capacitors have the same capacitance value as the input capacitance of the preamp, then one-half of the input signals reaches the preamp. For the OOS scheme, the storage capacitors should be much larger than the input capacitance of the dynamic latch.

Dynamic Current Mirrors

Using dynamic techniques can reduce the effects of threshold voltage mismatches in current mirrors. Consider the circuit of Fig. 25.37. When ϕ_1 is high and ϕ_2 is low (again, these clock signals are nonoverlapping), switches S1 and S3 are on, while switch S2 is off. A current I_{ref} flows through M1, setting its gate-source voltage. This information [the gate-source voltage (actually the charge) of M1] is stored on *C*. When S2 closes with S1 and S3 off, a current I_{out} equal to I_{ref}, neglecting channel length modulation, flows. This circuit behaves like a current source when ϕ_2 is high and as an open when ϕ_2 is low. The circuit shown in Fig. 25.38 shows a dynamic current mirror that operates continuously.

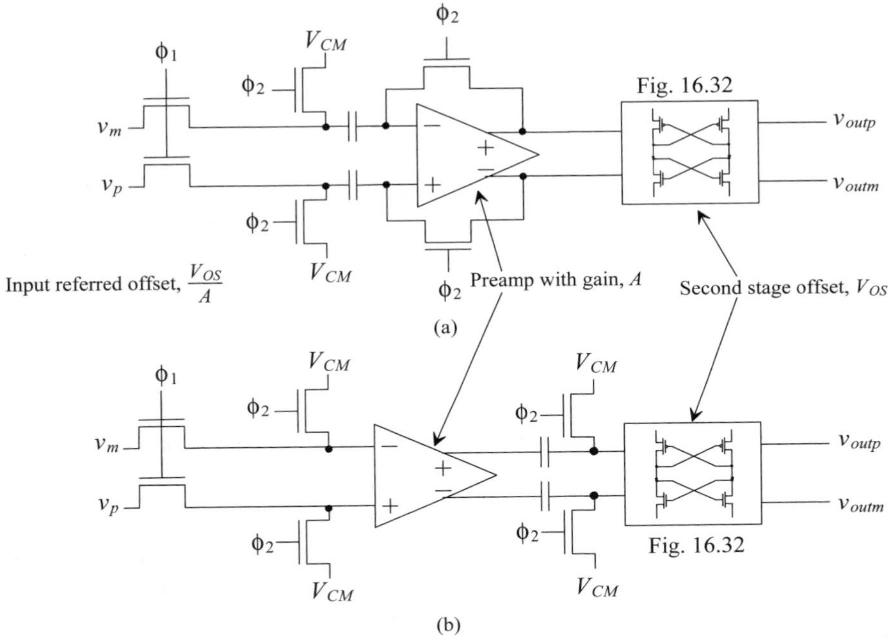

Figure 25.36 (a) Input offset storage (IOS) and (b) output offset storage (OOS).

When ϕ_1 is high, M2 sinks current, and when ϕ_2 is high, M1 sinks current. These circuits are useful in eliminating the mismatch effects and, thus, differences in the output currents, resulting from threshold voltage and transconductance parameter differences between devices. Since a single-reference current can be used to program the current in a string of current mirrors, only the finite output resistance of the mirrors causes current differences.

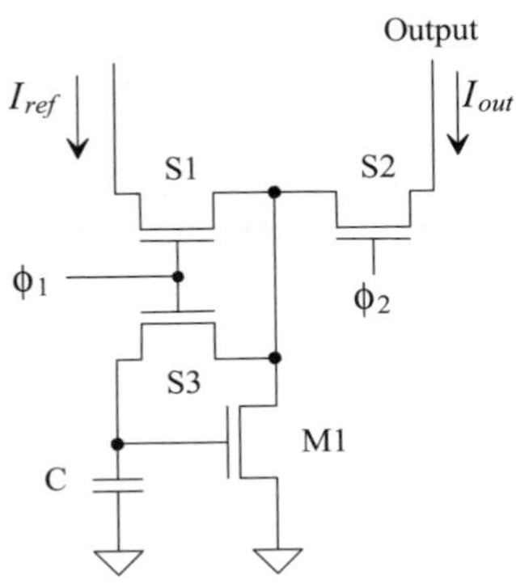

Figure 25.37 Dynamic biasing of a current mirror.

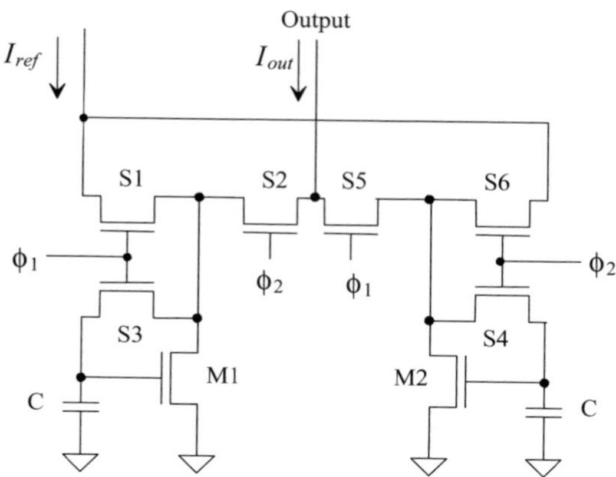

Figure 25.38 Dynamic current mirror that operates during both clock phases.

Dynamic Amplifiers

Figure 25.39 shows a dynamic amplifier. The circuit amplifies when ϕ is low and dynamically biases M1 and M2, and therefore does not amplify, when ϕ is high. If C1 and C2 are large compared to the input capacitance of M1 and M2, then the input AC signal, v_{in}, is applied to both gates. This biasing scheme makes the amplifier less sensitive to threshold and power supply variations. Other dynamic amplifier configurations exist, which have differential inputs and operate over both clock cycles.

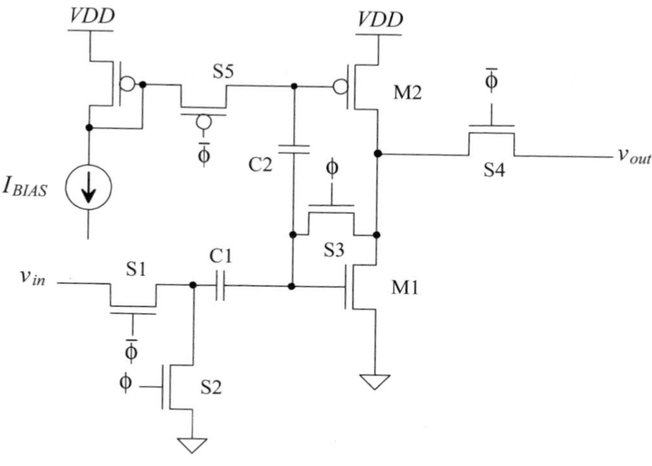

Figure 25.39 Dynamic amplifier used to reduce biasing sensitivity to power supply and threshold voltage.

ADDITIONAL READING

[1] D. J. Allstot and W. C. Black, "Technology Design Considerations for Monolithic MOS Switched-Capacitor Filtering Systems," *Proceedings of the IEEE*, vol. 71, no. 8, pp. 967–986, August 1983.

[2] J. Shieh, M. Patil, and B. Sheu, "Measurement and Analysis of Charge Injection in MOS Analog Switches," *IEEE Journal of Solid State Circuits*, vol. 22, no. 2, pp. 277–281, April 1987.

[3] G. Wegmann, E. Vittoz, and F. Rahali, "Charge Injection in Analog MOS Switches," *IEEE Journal of Solid State Circuits*, vol. 22, no. 6, pp. 1091–1097, December 1987.

[4] C. Eichenberger and W. Guggenbuhl, "On Charge Injection in Analog MOS Switches and Dummy Switch Compensation Techniques," *IEEE Transactions on Circuits and Systems*, vol. 37, no. 2, pp. 256–264, February 1990.

[5] J. McCreary and P. R. Gray, "All MOS Charge Redistribution Analog-to-Digital Conversion Techniques - Part 1," *IEEE Journal of Solid State Circuits*, vol. 10, pp. 371–379, December 1975.

[6] P. W. Li, M. J. Chin, P. R. Gray, and R. Castello, "A Ratio-Independent Algorithmic Analog-to-Digital Conversion Technique," *IEEE Journal of Solid-State Circuits,* vol. SC-19, no. 6, pp. 828–836, December 1984.

[7] E. J. Kennedy, *Operational Amplifier Circuits: Theory and Applications*, Holt, Rinehart and Winston, New York, 1988.

[8] D. Johns and K. Martin, *Analog Integrated Circuit Design*, John Wiley and Sons, New York, 1997.

[9] A. B. Grebene, *Bipolar and MOS Integrated Circuit Design*, John Wiley and Sons, New York, 1984.

[10] R. W. Broderson, P. R. Gray, and D. A. Hodges, "MOS Switched-Capacitor Filters," *Proceedings of the IEEE,* vol. 67, no. 1, January 1979.

[11] K. Martin, "Improved Circuits for the Realization of Switched-Capacitor Filters," *IEEE Transactions on Circuits and Systems,* vol. CAS-25, no. 4, pp. 237–244, April 1980.

[12] P. R. Gray and R. G. Meyer, *Analysis and Design of Analog Integrated Circuits,* 2nd ed., John Wiley and Sons, 1984. ISBN 0-471-87493-0.

[13] R. Gregorian, K. W. Martin, and G. Temes, "Switched-Capacitor Circuit Design," *Proceedings of the IEEE,* vol. 71, no. 8, pp. 941–966, August 1983.

[14] D. J. Allstot, R. W. Broderson, and P. R. Gray, "MOS Switched-Capacitor Ladder Filters," *IEEE Journal of Solid-State Circuits,* vol. SC-13, no. 6, pp. 806–814, December 1978.

[15] R. Castello and P. R. Gray, "A High-Performance Micropower Switched -Capacitor Filter," *IEEE Journal of Solid-State Circuits,* vol. SC-20, no. 6, pp. 1122–1132, Dec. 1987.

[16] P. E. Allen and D. R. Holberg, *CMOS Analog Circuit Design*, 2nd ed., Oxford University Press, 2002. ISBN 0-19-511644-5.

[17] A. G. Dingwall and V. Zazzu, "An 8-MHz Subranging 8-bit A/D Converter," *IEEE Journal of Solid-State Circuits,* vol. SC-20, no. 6, pp. 1138–1143, December 1985.

[18] B. Razavi and B. A. Wooley, "Design Techniques for High-Speed, High-Resolution Comparators," *IEEE Journal of Solid-State Circuits,* vol. 25, no. 12, pp. 1916–1926, December 1992.

[19] S. Masuda, Y. Kitamura, S. Ohya, and M. Kikuchi, "CMOS Sampled Differential Push-Pull Cascode Operational Amplifier," *IEEE International Symposium on Circuits and Systems,* vol. 3, pp. 1211–1214, 1983.

PROBLEMS

In the following problems, where appropriate, use the short-channel CMOS process with a scale factor of 50 nm and a *VDD* of 1 V.

25.1 Using SPICE simulations, show the effects of clock feedthrough on the voltage across the load capacitor for the switch circuits shown in Fig. 25.40. How does this voltage change if the capacitor value is increased to 100fF?

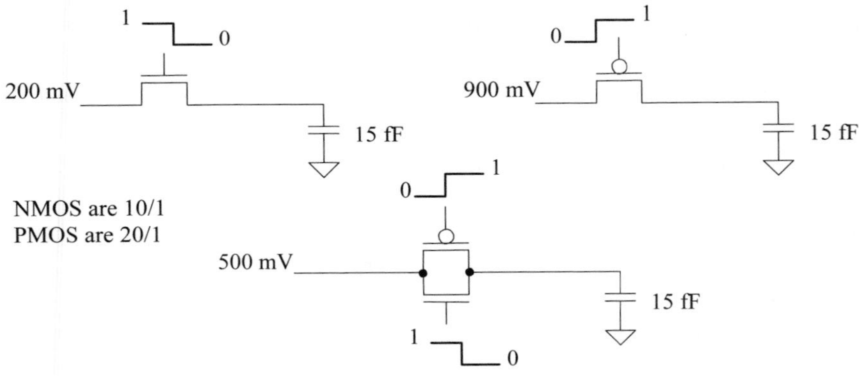

Figure 25.40 Circuits used in Problem 25.1 to show clock feedthrough.

25.2 Repeat problem 25.1 if dummy switches are used. Show schematics of how the dummy switches are added to the schematics.

25.3 Using a voltage-controlled voltage source for the op-amp (see Fig. 20.19 for example) with an open-loop gain of 10^6, use SPICE to show how the track-and-hold seen Fig. 25.8 operates with a sinewave input. What happens if the input sinewave's amplitude is above $VDD{-}V_{THN}$? Use a 100 MHz clock (strobe) pulse with a 50% duty cycle and an input sinewave frequency of 5 MHz. Note that the input sinewave should be centered around V_{CM} (= 500 mV).

25.4 Using the topology seen in Fig. 25.13a and the SPICE op-amp model in Fig. 25.18, show how both V_{in} and V_{CM} can be varied while the op-amp's inputs are equal, that is, $v_p \approx v_m$. Is the output common-mode voltage always at V_{CM} using the simple SPICE model for the fully-differential output op-amp in Fig. 25.18? Why or why not? Give an example supporting your answer.

25.5 Suppose, in Fig. 25.19, that instead of the two input sine waves being connected to ground they are tied (together) to a common mode signal (say a noise voltage). Show that a common mode signal (like a sine wave) won't change the circuits' output signals. The amplitude of the common-mode signal shouldn't be so large that the NMOS switches shut off.

25.6 Show that the switched-capacitor circuits shown in Fig. 25.41 behave like resistors, for $f \ll f_{clk}$, with the resistor values shown.

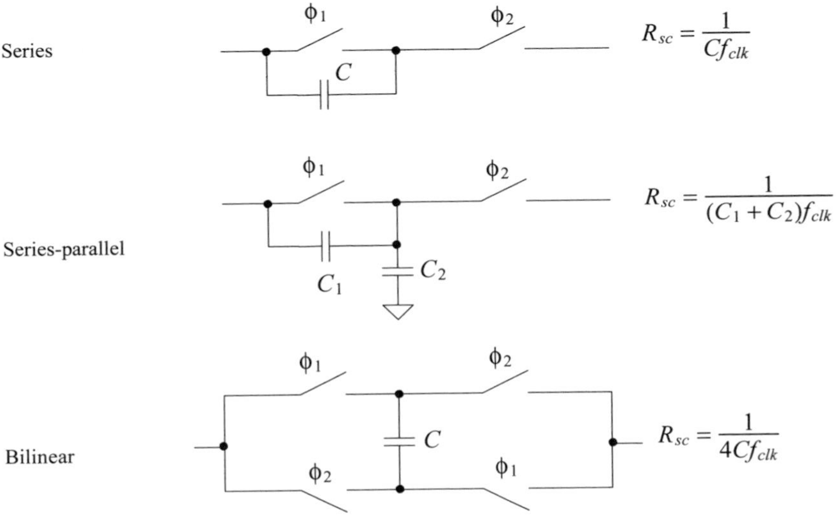

Figure 25.41 Alternative forms of switched-capacitor resistors.

25.7 Simulate the operation of the switched-capacitor resistor seen in Fig. 25.21. Plot the mean of the current flowing in the voltage sources v_1 or v_2 to show that the circuit actually behaves like a resistor. Comment on the selection of the bottom plate of the capacitor shown in Fig. 25.21a.

25.8 Comment on the selection of the bottom plate of C_F shown in Fig. 25.23.

25.9 Sketch the schematic, similar in form to Fig. 25.23, of the fully-differential switched-capacitor integrator made using a differential input/output op-amp. What is the transfer function of this topology?

25.10 Repeat Ex. 25.2 if the low-frequency gain is 40 dB and the zero is located at 50 kHz.

25.11 Using the results given in Eq. (25.29), plot the magnitude of v_{out}/v_{in} against f/f_{clk}. Comment on the resulting plot.

25.12 An important consideration in SC circuits is the slew-rate requirements of the op-amps used. In the derivation in Fig. 25.28, we assumed that a voltage source was connected to the input of the circuit. In reality, the input of the circuit is provided by an op-amp. When ϕ_1 goes high, in this figure, the capacitor C_I is charged to the input voltage v_{in} $(= v_A)$. If C_I is 5 pF and f_{clk} is 100 kHz, estimate the minimum slew-rate requirements for the op-amp providing v_{in}.

25.13 Suppose that the op-amp in problem 25.12 is used with a feedback factor of 0.5. Estimate the minimum unity gain frequency, f_{un}, that the op-amp must possess.

25.14 Simulate the operation of the dynamic comparator shown in Fig. 25.35.

Chapter 26

Operational Amplifiers II

In the last chapter we saw that MOSFET switches cause charge injection and clock feedthrough in the circuits where they are used. To reduce the effects of these problems (and others), fully-differential op-amp topologies are used. As discussed in Sec. 25.2, the fully-differential output op-amp requires the design of a common-mode feedback (CMFB). The CMFB circuit keeps the op-amp's outputs balanced around a known voltage (generally the common-mode voltage of $V_{CM} = VDD/2$).

In this chapter we discuss the design of fully-differential output op-amps and CMFB circuits. Our discussion is centered around practical design where power, speed, offsets, and gain are (as usual) of importance. Throughout the chapter, the 50 nm process (with a *VDD* of 1 V) is used to illustrate the design techniques.

26.1 Biasing for Power and Speed

The biasing circuits we developed earlier were used for general analog design. In this chapter we want to design circuits that are used for very high speed with the least amount of power dissipation possible.

For high-speed design, we must use the minimum channel length (L = 1). However, using minimum channel lengths results in large mismatches between devices and low MOSFET output resistance (hence, why we used $L = 2$ for the designs presented earlier). The results are low gain and large input-referred offset voltages. Further, for low power design we want to use the lowest biasing currents possible. This (low biasing currents) is in direct conflict with high-speed design. As discussed in Ch. 9, the device speed figure-of-merit, FOM, was the transition frequency, f_T. Low biasing current is the same as low overdrive voltage. As seen in Eq. (9.55), using a low value of overdrive voltage results in slower circuits. Of course, if the overdrive voltage is too high, the MOSFET enters the triode region too soon. For general analog design, we set the overdrive voltage to 5% of *VDD*. For high-speed design, we might set the overdrive voltage to 10% of *VDD* or larger. *To minimize power and maximize speed, we will use minimum size devices.* For the NMOS, we'll use a 10/1. To match the drive, we'll use 20/1 for the PMOS devices. The question we need to answer having made these selections is: "Will these devices have the strength to drive a load capacitance quickly?"

26.1.1 Device Characteristics

Figure 26.1 shows the IV characteristics of our selected devices (an NMOS of 10/1 and a PMOS of 20/1). If we set the overdrive voltages to roughly 10% of *VDD* (here 100 mV), then knowing, from Table 9.2, the threshold voltages are 280 mV (typical), we can use gate-source voltages of, nominally, 400 mV with a corresponding drain current of 20 μA. If we want to increase the speed, then we must use a higher overdrive voltage (say bias the devices up at 50 μA).

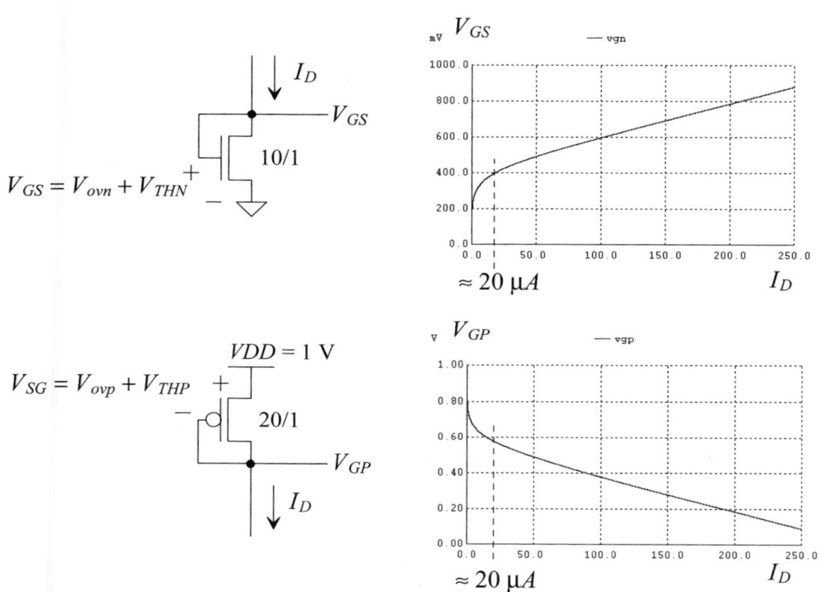

Figure 26.1 Gate-source voltages plotted against drain currents.

Next we need to determine if these devices will have the drive current needed to charge a specific size capacitor in the time required. Looking at Table 8.1 as a guide for selecting capacitor sizes, we see that using a 100 fF capacitor results in an RMS noise (because of the MOSFET switch thermal noise) of 200 μV. From Fig. 8.33 (the assumed PDF for thermal noise), the peak-to-peak value of this noise is (roughly) 1.2 mV. In this chapter *we'll use a load capacitance of 250 fF* for a peak-to-peak noise of 750 μV. Looking at the simulation results in Fig. 26.1, we see that a reasonable pulsed drain current estimate for the devices is 100 μA. The rate we can charge the load capacitor is then

$$\frac{dV_{out}}{dt} = \frac{I}{C_L} = \frac{100\ \mu A}{250\ fF} = 400\ mV/ns \tag{26.1}$$

If our clock frequency, f_{clk}, is 100 MHz, then half a clock cycle is 5 ns and the device sizes and bias conditions will likely be adequate (remembering that *VDD* is 1 V and V_{CM} is 500 mV). However, if our clock frequency approaches 1 GHz (or the load capacitance increases), then we must increase the widths of the devices (to get more drive) and the overdrive voltages (to get more speed).

26.1.2 Biasing Circuit

When designing the bias circuits earlier, Fig. 20.47 for example, our goals were to provide biasing for general analog design. Here, in this chapter, our goal is to design a bias circuit for our op-amp designs with minimum power dissipation. When designing biasing circuits fully-differential topologies have some advantages over the single output op-amps presented in Ch. 24. For example, our earlier bias circuit provided biasing, V_{pcas} and V_{ncas} , for the floating current sources used in the class AB output stages (see Fig. 24.29 for example). These bias voltages won't be needed here.

Towards understanding this last statement, consider the two-stage op-amp seen in Fig. 26.2. Consider what happens if the noninverting op-amp input, v_p , increases relative to the inverting input, v_m. The drain voltage of M1R drops while the drain voltage of M1L rises. The decrease in the gate voltage of M6R causes it to turn on and the output to go high. At the same time, the increase in the drain voltage of M1L causes M4R to shut off and thus so does M7R (giving class AB operation). In simple terms, we can yank the gates of M4 or M6 down independent of the diff-amp's tail current.

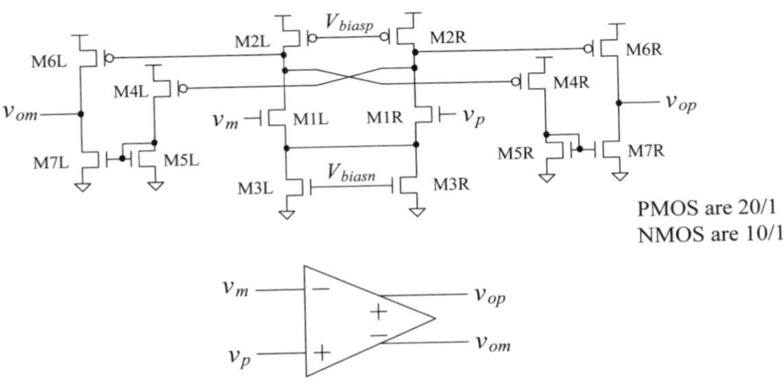

Figure 26.2 A two-stage fully-differential op-amp. Compensation and CMFB are not shown. Output stage operates class AB. See discussion in the next section concerning the output voltage of the diff-amp.

Layout of Differential Op-Amps

One of the common things we do in this chapter is draw schematics that are symmetrical. If we were to draw a line down the middle of the op-amp in Fig. 26.2, separating the left and right sides of the schematic, we could fold the left side of the schematic directly over onto the right side of the schematic and see a perfect match. For example, instead of drawing the diff-amp's tail current source (M3) as a single MOSFET with twice the width, we've drawn it as two MOSFETs in parallel, each having the same width. *Drawing schematics in this fashion is useful when doing layout.* We can fold the schematic in half and lay out like devices, e.g., M6L and M6R, directly next to each other. Further, then the outputs (and inputs) of the op-amp are laid out right next to each other.

Self-Biased Reference

Figure 26.3 shows the bias circuit we'll use in this chapter (see also Fig. 20.22 and the associated discussion). We used the length of 2 MOSFETs for the added amplifier to minimize power dissipation and boost the amplifier's gain. The current pulled from *VDD* is approximately 50 μA. Figure 26.4 shows the simulation results of current variations vary with changes in *VDD*. Notice how V_{biasn} is close to the 400 mV, and V_{biasp} is close to *VDD* – 400 mV (of course the absolute value of V_{biasp} varies with *VDD* but the ideal value of V_{SG} for the PMOS devices will be 400 mV). Note that, as discussed in Sec. 20.1.4, it is important to ensure that the reference is stable over all possible operating conditions and loads (connected to V_{biasn} and V_{biasp}). Notice that we aren't discussing how the reference currents vary with process shifts in the MOSFETs and the resistor. This is an important practical concern.

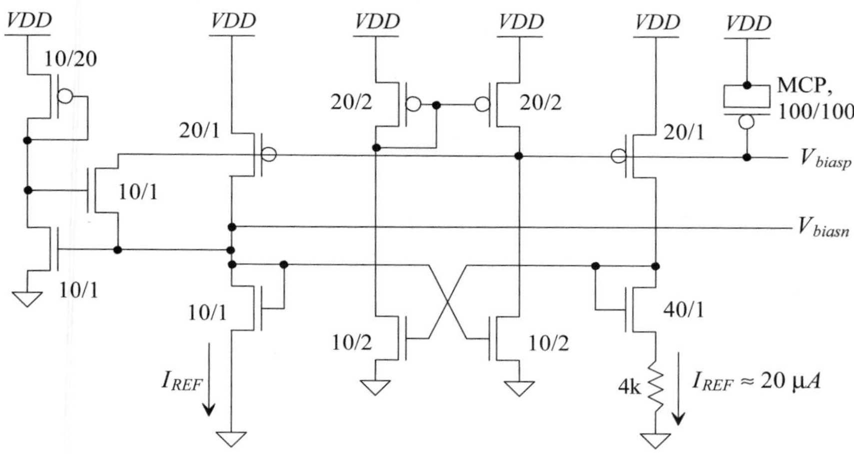

Figure 26.3 Biasing circuit used in this chapter. This bias circuit pulls approximately 50 microamps.

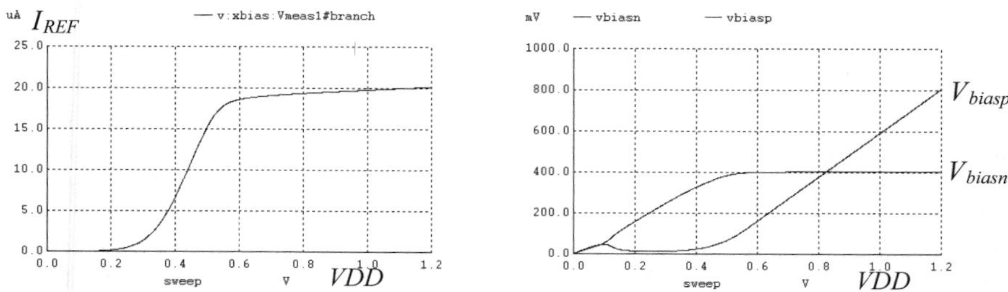

Figure 26.4 Simulating how the reference current changes with VDD.

26.2 Basic Concepts

Before we discuss the design of op-amps, let's look at some basic concepts that will be useful when evaluating a specific op-amp topology.

Modeling Offset

In a real circuit, especially an analog circuit designed with minimum length devices, mismatches can be a significant factor in the selection of an op-amp topology. In a SPICE simulation, all of the MOSFETs are perfectly matched. We have to come up with a method, in SPICE, to determine an op-amp's sensitivity to offsets.

Consider the circuit in Fig. 26.5. In (a) the measured currents should be equal. Both MOSFETs V_{SG}s and V_{SD}s are equal (and they are the same size). However, for whatever reason (e.g., threshold voltage mismatch), there exists a mismatch between the two devices that causes a difference in their drain currents. We can model this mismatch in a SPICE simulation, as seen in (b), by adding a DC voltage source in series with the gate of M2. For a general design, we can insert these offset voltages at various points in the circuit and verify that the circuit still functions correctly. The next question that needs answering is: "What value of V_{OS} should be used?" While no absolute answer can be given here, **we'll use a V_{OS} of 50 mV** (5% of *VDD* or roughly 20% of the threshold voltage). The reader might feel that this value is way too high. However, if we can design circuits that function properly over the process, voltage, and temperature (PVT) variations with an offset this large, it is likely the op-amp will be difficult to "break."

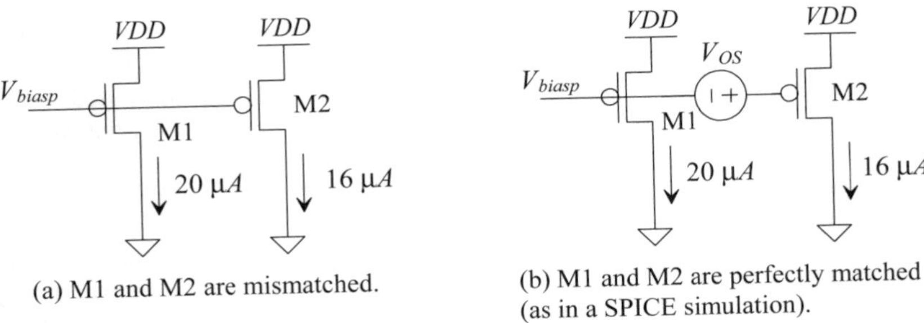

(a) M1 and M2 are mismatched.

(b) M1 and M2 are perfectly matched (as in a SPICE simulation).

Figure 26.5 How we add an offset into the circuit to model mismatch.

A Diff-Amp

Figure 26.6a shows a basic diff-amp without an offset. This offset-free diff-amp has simulated output voltages of 700 mV. The diff-amp in (b) is simulated with an input-referred offset voltage of 50 mV. The polarity of the offset doesn't matter because it simply swaps the two output voltage values. In other words, we could have put the same polarity offset on the other input of the diff-amp and swapped the voltages on the output of the op-amp. We put the offset voltage on the op-amp's input because the input has the greatest effect on the diff-amp's output voltages. Note: we can think of the 50 mV as an input signal causing an output signal difference, from the ideal 700 mV, of 900 mV –550 mV or 350 mV (assuming linear operation). This indicates the gain of the diff-amp is only 7!

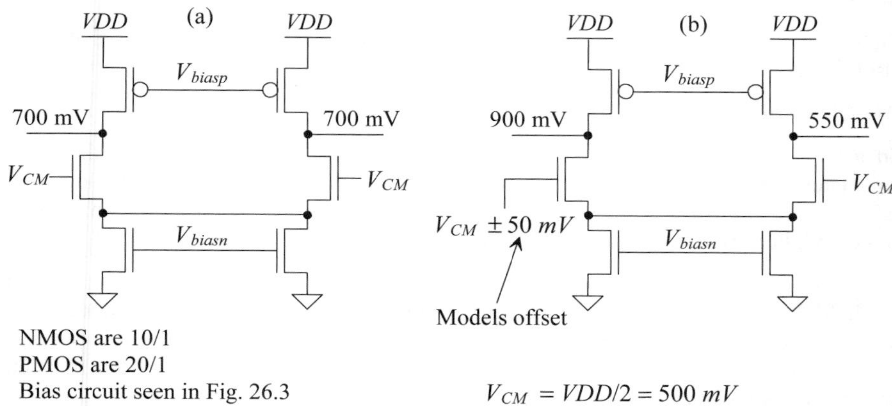

Figure 26.6 Comparing the diff-amp's output voltages with and without an offset.

The simulation results seen in Fig. 26.6 give some practical information that we need to consider. To begin, remember from Ch. 24 that we frequently use the diff-amp's output to bias the next stage (see Figs. 22.8 and 24.2). With a *VDD* of 1 V and a V_{SG} of nominally 400 mV, V_{biasp} is roughly 600 mV, as seen in Fig. 26.4. However, as seen in Fig. 26.6a, the diff-amp's output is 700 mV. The result, when the diff-amp is used with a second-stage and feedback, is an additional input-referred offset. It would be nice to know that the outputs of the diff-amp, in the ideal case, go to a known value (preferably a voltage that can be used to bias the next stage). The diff-amp in Fig. 26.6 is an example of two current sources, the PMOS biased with V_{biasp} and the NMOS biased with V_{biasn}, fighting each other for control of the output voltage. In general, we want to avoid this situation.

A Single Bias Input Diff-Amp

Figure 26.7 shows a diff-amp that generates its own bias reference for the PMOS devices. Notice that the two gate-drain-connected PMOS devices behave simply like a MOSFET with twice the width of the other two PMOS devices. Similarly, the four NMOS devices that are biased from V_{biasn} behave like a MOSFET with four times the width of the other NMOS devices. When the diff-amp's inputs are equal, the same current, *I*, flows in all of the MOSFETs. If the + diff-amp input is raised significantly above the − input, all of the bias tail current flows in the left two NMOS devices of the diff-amp (each will conduct 2*I*). The current flowing in the gate-drain-connected PMOS device, in either case, is the same, 2*I*, keeping the PMOS's gate voltage constant. The outputs of this diff-amp can be used to bias the next stage. Offsets, however, cause the diff-amp's outputs to vary from their ideal values. One drawback to using this diff-amp is that it dissipates twice the power of the diff-amp in Fig. 26.6. Another drawback is the larger input capacitance.

The Diff-Amp's Tail Current Source

Notice that we are not using a cascode tail current source to bias the diff-amps in this chapter, as we did in earlier chapters. We're avoiding the cascode structure because we'd need an additional bias voltage (resulting in more power dissipation). In a

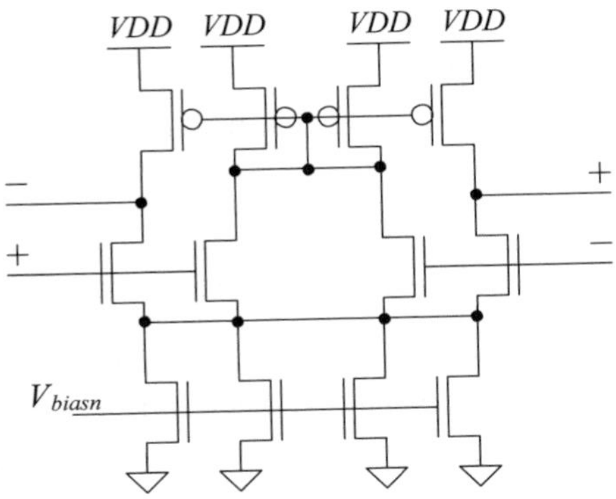

Figure 26.7 A fully-differential diff-amp that generates its own bias for the PMOS.

fully-differential op-amp topology (both the inputs and the outputs of the op-amp, with feedback, are double-ended signals swinging around V_{CM}), the input common-mode voltage of the op-amp is constant (= V_{CM}). The common-mode rejection ratio isn't as important when the input common-mode voltage of the op-amp doesn't vary.

Using a CMFB Amplifier

Another possible way to set the diff-amp's output voltages to a known value is seen in Fig. 26.8. An amplifier (called a common-mode feedback amplifier or CMFB amplifier) is used to amplify the difference between the average of the diff-amp's outputs and V_{biasp}. If the gain of the CMFB amplifier is large, then the average of the two outputs will be very close to V_{biasp}. Note that the CMFB amplifier's output signal, V_{CMFB} , is common, through M1L and M1R, to both outputs. Any variation in V_{CMFB} affects each output by the same amount. This is important because all we want the CMFB amplifier to do is make

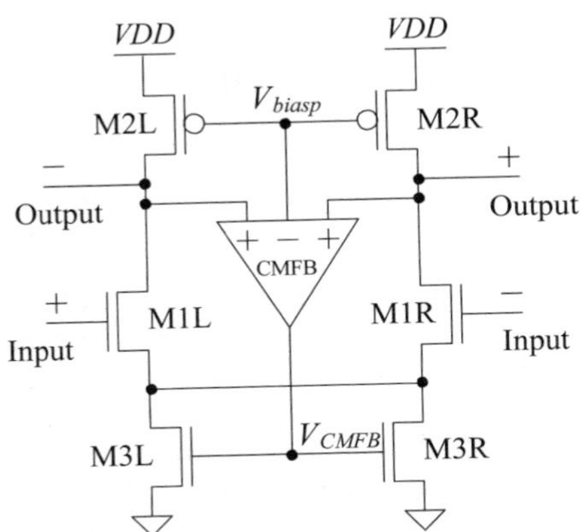

Figure 26.8 Using a common-mode feedback (CMFB) amplifier to set the output voltages.

sure that the diff-amp's outputs vary around V_{biasp}. The CMFB amplifier shouldn't affect the differential amplification in the diff-amp. When the diff-amp's outputs are equal (the inputs to the diff-amp are equal and neglecting offsets), they should be V_{biasp}.

Figure 26.9 shows one possible implementation of a CMFB amplifier. When the + inputs (noticing the PMOS connected to these inputs have half the width of the other PMOS) are equal to V_{biasp}, the currents that flow in M1 and M2 are 20 μA. This provides the proper mirroring action to the diff-amp (M3L, R) in Fig. 26.8. If the average of the + inputs moves above the – input, the current flowing in M2 decreases and the current in M1 increases. This causes V_{CMFB} to increase. The result is that M3L and M3R turn on more and pull both of the outputs in Fig. 26.8 down (until their average is equal to V_{biasp}). Several concerns exist with this topology. To begin, whenever we employ feedback, we must be concerned with stability. To stabilize the CMFB loop, we can add capacitors to the outputs of the diff-amp (inputs to the CMFB amplifier). We may be able to stabilize the CMFB loop by adding capacitors to the CMFB amplifier's outputs. However, since we will need compensation capacitors for compensating the differential action of the op-amp, we might as well use these capacitors for both loops. The other concern with the CMFB amplifier in Fig. 26.9 is the input common-mode range. When the CMFB amplifier's + inputs move significantly away from the – input, we don't get the correct balancing. This keeps large signal swings on the diff-amp's outputs in Fig. 26.8 from being balanced around V_{biasp}.

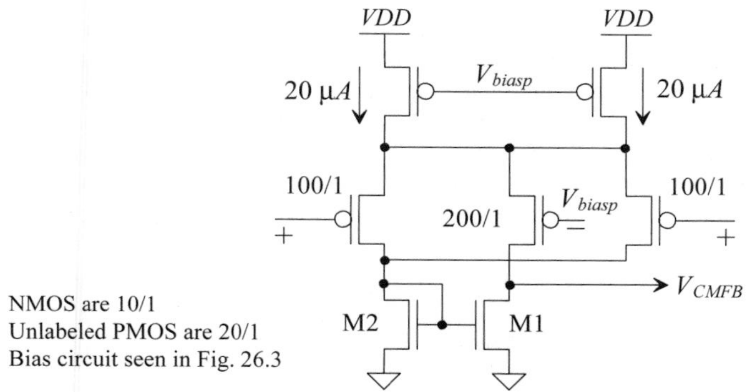

Figure 26.9 Implementation of the CMFB amplifier in Fig. 26.8.

Figure 26.10 shows simulation results using the CMFB amplifier in Fig. 26.9 with the amplifier in Fig. 26.8. The addition of the capacitors stabilizes the CMFB loop *(we need to discuss this further)*. In (a) we see the correct balancing. The CMFB loop adjusts the diff-amp's output voltages to V_{biasp}. In (b) we re-simulate with an offset and see that the CMFB circuit isn't balancing the outputs. If it were, we would see one output above V_{biasp} by some amount and the other output below V_{biasp} by the same amount. With these output voltages (the inputs to the CMFB amplifier), one of the PMOS devices on the input of the CMFB is off and we don't get proper amplifier action. *We must use a CMFB circuit that can balance the outputs over the entire range of diff-amp output voltages.*

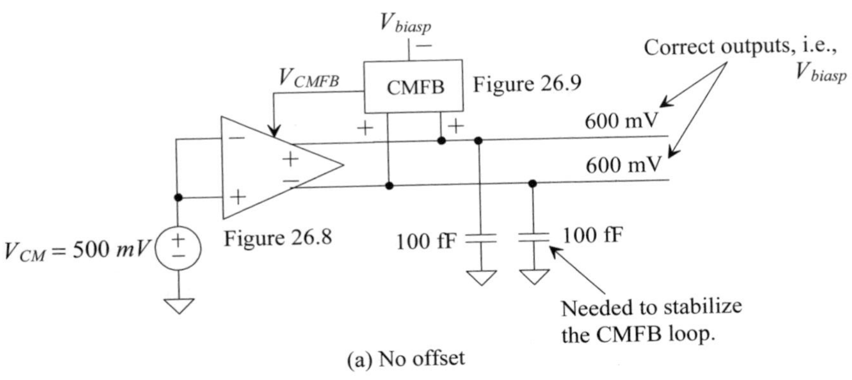

(a) No offset

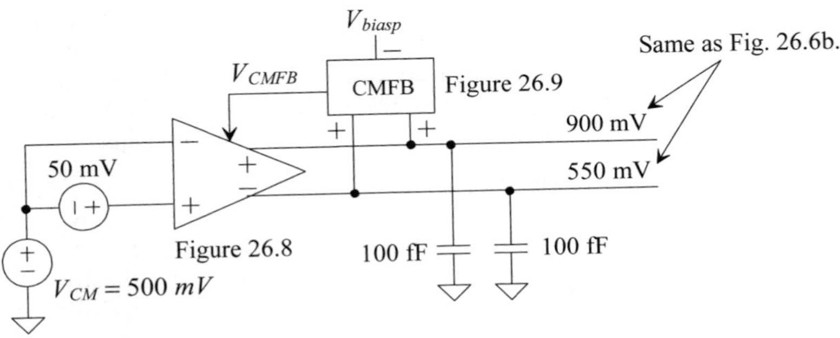

(b) With a 50 mV offset. Note how the CMFB isn't doing anything.

Figure 26.10 Simulating the operation of the CMFB circuit in Fig. 26.9.

Compensating the CMFB Loop

Consider the schematic seen in Fig. 26.11. The diff-amp behaves like an operational transconductance that can be compensated, as discussed in Sec. 24.3 (see Eq. [24.44]). The CMFB loop can be compensated in a similar fashion. The AC common-mode signal is represented in this schematic as v_c. If the gain of the CMFB amplifier is A_c, then following the procedure leading to Eq. (24.44), we can write the unity-gain frequency of the CMFB loop as

$$f_{un,cm} = \frac{A_{cm} \cdot g_{mn}}{2\pi C_L} \tag{26.2}$$

If we want to compensate the CMFB loop with the same load capacitance used to compensate the differential forward signal path, then we must ensure that the gain of the CMFB amplifier is less than or equal to unity,

$$A_{cm} \leq 1 \tag{26.3}$$

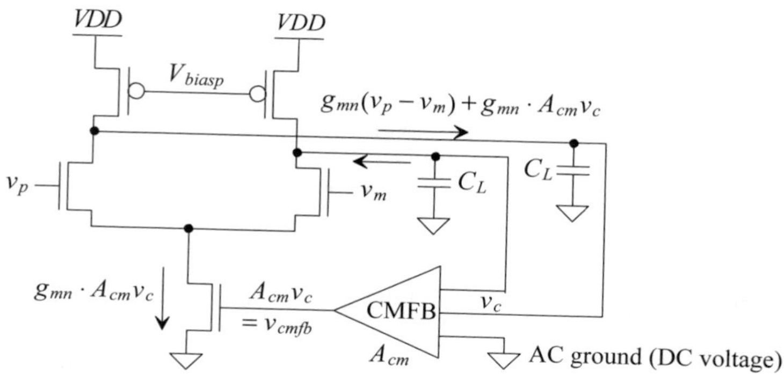

Figure 26.11 Schematic view of differential and CM feedback.

Reviewing the CMFB amplifier in Fig. 26.9 that has a current mirror active load, we see that A_{cm} is greater than 1. Removing the capacitors in Fig. 26.10 and resimulating will show that the CMFB loop is unstable. The added capacitors are relatively large and will overcompensate the differential signal path of the diff-amp. What we need to do is reduce the gain of the CMFB amplifier or reduce the CMFB loop's forward gain.

Towards reducing the gain, examine the CMFB amplifier seen in Fig. 26.12. This is the same amplifier topology seen in Fig. 26.9 except that here we've used a diode-connected load instead of a current mirror load (to reduce the gain). The schematic is drawn symmetrical around its center for ease of layout, as discussed earlier. Note that we used the common-mode voltage in this schematic, V_{CM}, as the voltage that the outputs of the amplifier will swing around (the more general case) rather than V_{biasp} as used in Fig. 26.10. Using this CMFB amplifier in the circuits of Fig. 26.10 (where the diff-amp of Fig. 26.8 has a low gain) won't precisely balance the outputs. The loop gain around the CMFB loop isn't large enough for proper operation. As we saw in Ch. 24, low first-stage gain can be remedied by using a telescopic input (cascode-load diff-amp) or a folded-cascode OTA. The other problem with this CMFB amplifier, again, is the limited allowable swing on the + inputs. As we've already seen, the input common-mode range limits the range of output voltages this CMFB can balance properly.

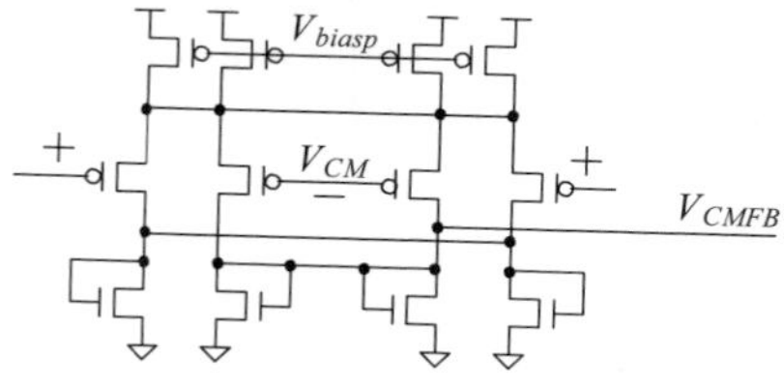

Figure 26.12 A CMFB amplifier with a gain of nominally unity.

Towards reducing the CMFB loop's forward gain, consider breaking the diff-amp's tail current up into parts, as seen in Fig. 26.13. The CMFB signal is applied to only one gate of the tail current. When compared to the topology in Fig. 26.8, the forward gain of the CMFB loop is halved. Further reduction can be implemented by adjusting the sizes of the transistors to further reduce the strength of the V_{CMFB} signal. *This is a common practical way of making CMFB loops stable.*

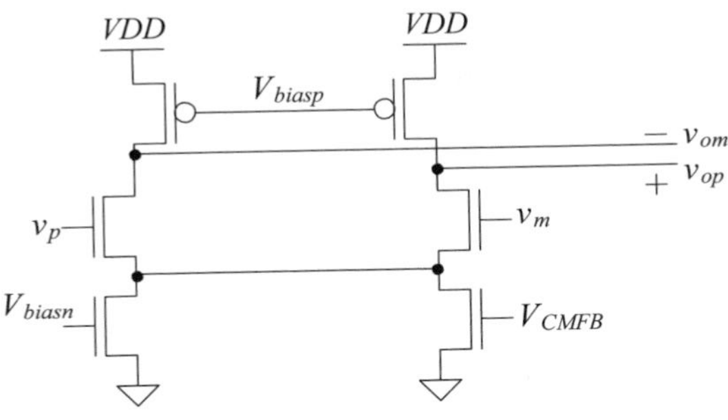

Figure 26.13 Reducing the forward gain of the CMFB loop.

Extending the CMFB Amplifier Input Range

The problem with the previous CMFB amplifier topologies based on a diff-amp is the diff-amp's limited input range. It's desirable to have a CMFB amplifier that functions over the entire range of possible amplifier output voltages. Towards this goal, consider the conceptual schematic in Fig. 26.14a. The resistors average the two outputs. This average is compared with the common-mode voltage (or a bias voltage as used in Fig. 26.8). Figure 26.14b shows the practical implementation of the amplifier for symmetry (at the cost of extra power dissipation). The practical problem with this topology is the

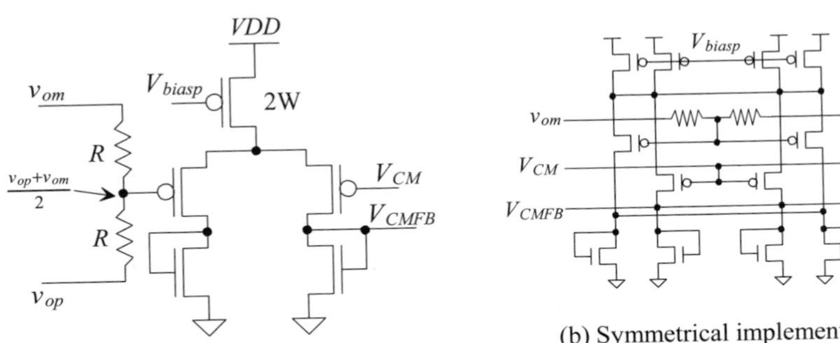

(a) Using resistors to average differential output signals.

(b) Symmetrical implementation of the CMFB circuit in (a).

Figure 26.14 Increasing CMFB amplifier input range.

loading by the resistors. If we were to connect this CMFB amplifier in the circuit configuration of Fig. 26.8, the resistors, unless they are huge (>100k) would load the differential amplifier and lower its gain. This topology is used, most often, on the output of an op-amp that has output buffers (and can thus drive resistive loads).

When using resistors for averaging in high-speed applications, we may have some parasitic effects that should be considered. The output signals have to charge, through the averaging resistors, the input capacitance of the MOSFET, as seen in Fig. 26.15. To ensure that the balancing action works at high speeds, capacitors (shown dashed in the figure) can be added, shunting the resistors. These capacitors can be very important if the size of the resistors is increased to reduce their loading on the output of the amplifier.

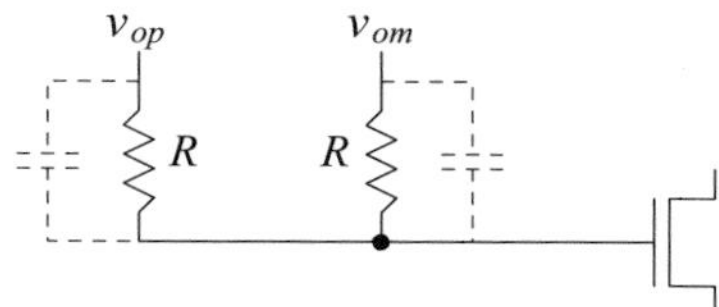

Figure 26.15 Adding parasitic capacitances across the resistors to compensate for the input capacitance of the MOSFET.

Dynamic CMFB

Figure 26.16 shows a switched-capacitor (SC) implementation of a CMFB circuit. The clocks, as in all SC circuits, are nonoverlapping (never high at the same time) clock signals, as seen in Fig. 25.28. The SC resistors are formed with the C_1 capacitors. The C_2 capacitors are used for the high-speed averaging just discussed (the dashed capacitors in Fig. 26.15). The SC resistors, as we'll see in a moment, perform both the averaging and the differencing needed in a CMFB amplifier. If the ϕ_2 controlled switches connected to amplifier's outputs, v_{op} and v_{om}, are transmission gates, the circuit can provide balancing from *VDD* to ground.

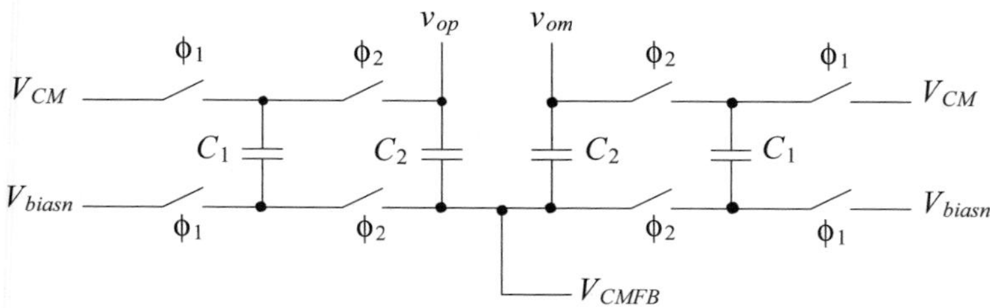

Figure 26.16 A switched-capacitor CMFB circuit.

To describe the operation of this circuit, consider the case when ϕ_1 is high. During this time, the total charge stored on each C_1 capacitor is

$$q_1 = 2 \cdot (V_{biasn} - V_{CM}) \cdot C_1 \tag{26.4}$$

When the ϕ_1 switches shut off, the ϕ_2 switches turn on. The total charge on both C_1 capacitors is then

$$q_2 = (V_{CMFB} - v_{op}) \cdot C_1 + (V_{CMFB} - v_{om}) \cdot C_1 \tag{26.5}$$

The change in V_{CMFB} is proportional to the difference in q_1 and q_2 or

$$\Delta V_{CMFB} \cdot 2(C_1 + C_2) \propto (q_1 - q_2) \tag{26.6}$$

This equation is important because it shows the CMFB voltage will continue to change until the two charges, q_1 and q_2 , are equal. Note that if v_{op} and v_{om} are balanced around V_{CM} , their net contributions, when ϕ_2 goes high, to V_{CMFB} are zero. Looking at the difference in the charges, we get

$$q_1 - q_2 = 2C_1\left(V_{biasn} - V_{CMFB} + \frac{v_{op} + v_{om}}{2} - V_{CM}\right) \tag{26.7}$$

This equation is quite interesting. Ideally, V_{CMFB} is equal to V_{biasn}, and the average of the outputs is equal to the common-mode voltage. If the actual value of V_{CMFB}, for balanced outputs, is 10 mV offset from V_{biasn} , then the average of the outputs will be 10 mV offset from V_{CM}. The offsets can occur because of improper device sizing (under ideal conditions the currents don't sum correctly) or mismatches.

As an example of where this offset can come from, consider the amplifier seen in Fig. 26.17 (also in Figs. 26.6 and 26.13). As seen in Fig. 26.6a, when V_{CMFB} is V_{biasn} or roughly 400 mV, the outputs of the diff-amp are 700 mV. As seen in Fig. 26.10 and the associated discussions, it can be useful, for next stage biasing, if the outputs are set to V_{biasp} (600 mV). From Fig. 26.17, the value of V_{CMFB} at this output voltage is roughly 425 mV or a 25 mV offset.

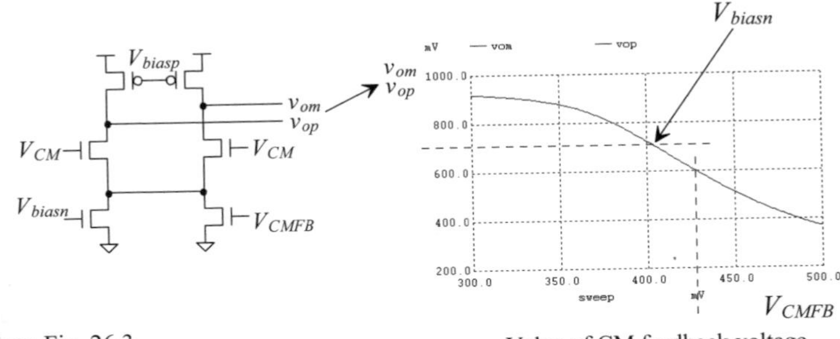

Figure 26.17 Plotting the output voltages as a function of the CM feedback voltage.

26.3 Basic Op-Amp Design

Reviewing the data in Table 9.2, we see that the open circuit gains are 25 (NMOS) and 50 (PMOS). In this chapter we both increased the biasing current and reduced the channel length from the values used in Table 9.2. Each of these changes has the effect of reducing the MOSFET's open circuit gain. If, for example, the open circuit gains are both now 10, then a common-source amplifier with current source load will have a gain of 5. A cascode amplifier will have a gain of 25, and a two-stage op-amp using a cascoded first-stage a gain of only 125. In other words, we adjusted our biasing for high-speed operation but we are going to face some issues with getting large open-loop gain.

This is a good time to remember the useful simulation netlists that we've developed. Figure 26.18 shows the IV curves, output resistance, and transconductance for a 10/1 NMOS and a 20/1 PMOS based on Figs. 9.31 to 9.33. Notice in (a) that the NMOS's drain current at a V_{GS} of 400 mV and a V_{DS} of 100 mV is approximately 13 μA. The PMOS's drain current under the same conditions, (b), is closer to 9 μA. In (c) and

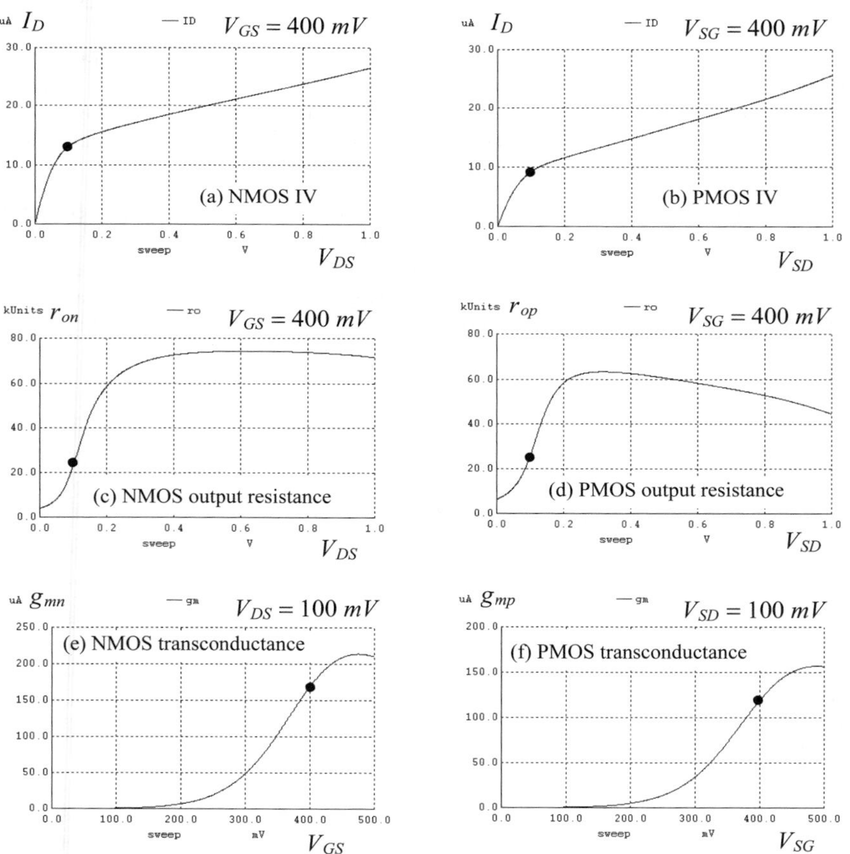

Figure 26.18 Characteristics of NMOS (10/1) and PMOS (20/1) devices.

(d), at a V_{DS} of 100 mV, the output resistance is only 25k. Using the transconductances in (e) and (f), the open circuit gains are, roughly, 4.375 (NMOS) and 3.125 (PMOS). In the actual circuits, the biasing points will vary (but in any case the gain of single stages will be low).

The Differential Amplifier

Figure 26.19 shows a cascode load diff-amp based on the topology seen in Fig. 26.7. Seen in the figure are typical values for the voltages in the circuit assuming gate-source voltages of 400 mV and, for the bottom two rows of NMOS devices, drain-source voltages of 100 mV. Notice how we used V_{biasn} to bias the second row of PMOS devices, which puts 200 mV across the drain-source voltages of the PMOS devices. This results in larger gain and reduced output swing. Diff-amp output swing is not an issue if this amplifier is used as the first-stage in a two-stage op-amp.

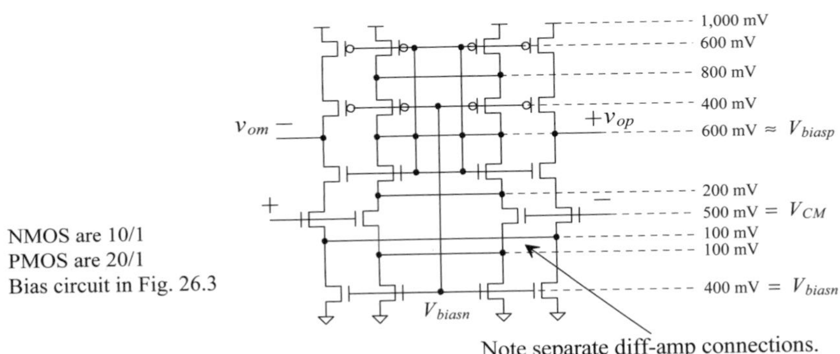

Figure 26.19 Fully-differential cascode diff-amp.

Figure 26.20 shows how the output voltages of the diff-amp in Fig. 26.19 vary with changes in V_{CM}. A 200 mV change in the common-mode voltage results in a, roughly, 50 mV change in the diff-amp's common-mode output voltages. Looking at Fig. 26.1, we see that the change in the drain current will be, again roughly, ± 5 μA around the quiescent value set with V_{CM} equal to 500 mV.

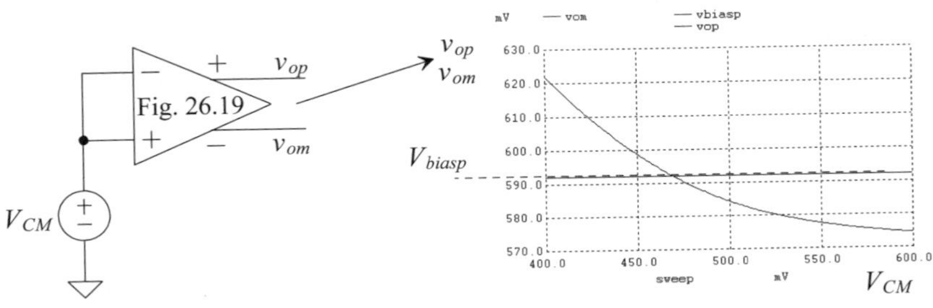

Figure 26.20 Varying the common-mode voltage and looking at the output.

Figure 26.21 shows the DC characteristics of the diff-amp. The gain is approximately 40. An important concern is how the mismatches in the MOSFETs used in the diff-amp affect the operation and biasing of the amplifier. Before discussing this issue, let's add the second stage and CMFB circuitry to form an op-amp. Note that, with a diff-amp gain of 40, and a second stage gain of 10, our op-amp's open loop gain will only have a value in the hundreds.

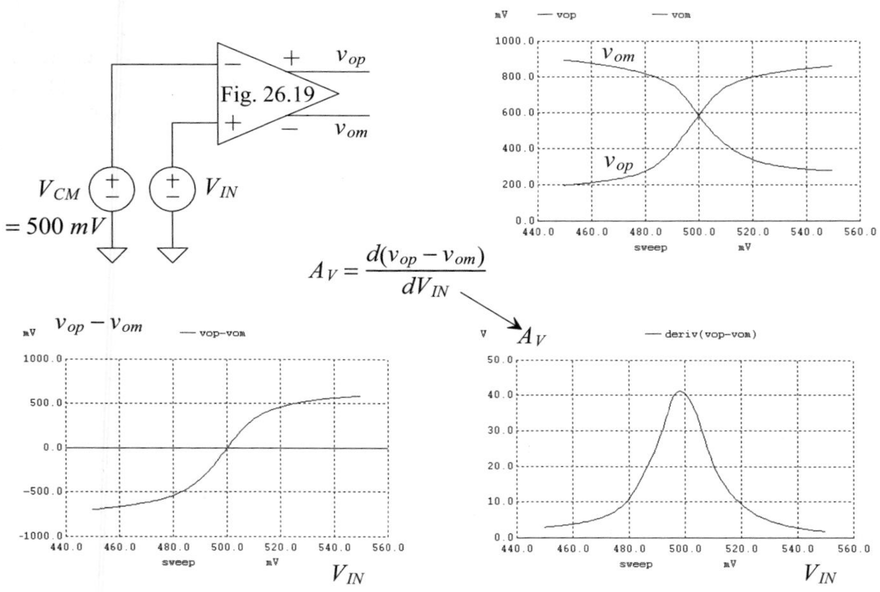

Figure 26.21 DC behavior and gain of the diff-amp in Fig. 26.19.

Adding a Second Stage (Making an Op-Amp)

Figure 26.22 shows a two-stage op-amp without CMFB circuit. The second stage of the op-amp operates class AB, as seen in Fig. 26.2, and the associated discussion. We've spent a considerable amount of time discussing how the output voltages of the diff-amp are approximately V_{biasp}. It should be clear after studying the op-amp in Fig. 26.22 why this is important. This voltage sets the quiescent current flowing in the output stages. There are eight vertical branches in this op-amp, so we can estimate the current pulled from *VDD* under quiescent conditions as 160 μA.

We used 50 fF capacitors for compensation in this op-amp. We can estimate the slew-rate limitations caused by the diff-amp driving the compensation capacitor, Eq. (22.34), as $20\,\mu A/50\,fF = 400\,mV/ns$. Using a class AB output stage, we don't have slew-rate limitations associated with driving a load capacitance from a constant current source. As discussed at the beginning of the chapter, the output MOSFET's drain currents can be pulsed to a value greater than 100 μA. When the op-amp is driving a 250 fF capacitive load, the speed limitations associated with charging the load capacitance are similar to the limitations we get when the diff-amp drives the compensation capacitor.

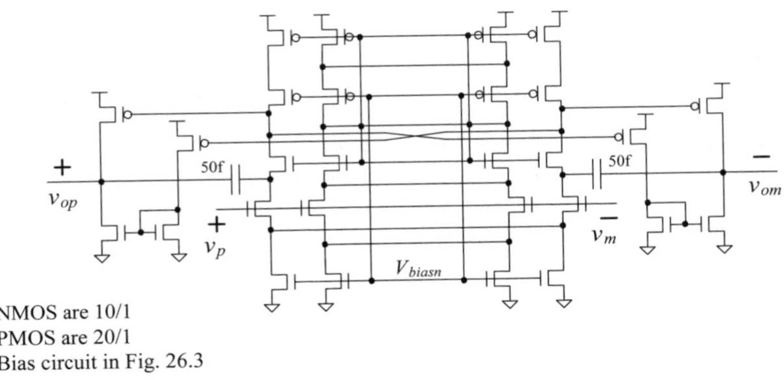

Figure 26.22 Basic two-stage op-amp without CMFB.

Figure 26.23 shows the DC characteristics of the op-amp in Fig. 26.22 where, once again, we've held the inverting op-amp input at the common-mode voltage and swept the voltage on the noninverting input. Notice how the outputs swing all the way from ground to *VDD* (= 1V). Further notice how the differential output voltage swings from −1 to +1 V (a doubling in the output swing as discussed in Sec. 25.2.) The op-amp's DC gain is approximately 500 without a DC load. Notice how the two op-amp outputs cross at 400 mV (not at V_{CM} where they should). We'll discuss the CMFB in a moment.

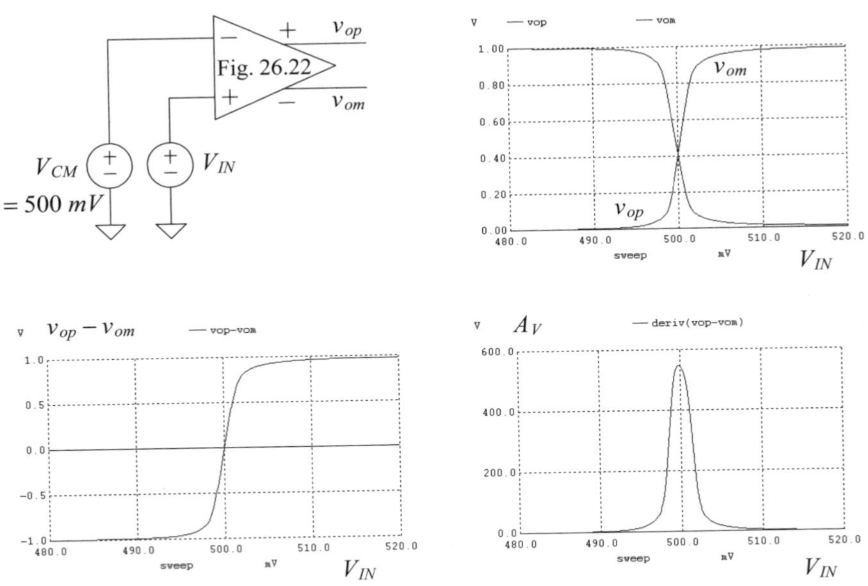

Figure 26.23 DC behavior and gain of the op-amp in Fig. 26.22.

Step Response

To determine both the stability of the op-amp in Fig. 26.22 and the settling time under certain loading conditions, consider the configuration seen in Fig. 26.24. We used relatively small resistors, 20k, in this circuit to reduce the RC time constant associated with the outputs of the op-amp charging the input capacitance of the op-amp. If, for example, the input capacitance of the op-amp is 25 fF (from parasitics and the MOSFETs used on the op-amp's input), then the RC time associated with charging this capacitance through a 20k resistor is 0.5 ns. As seen in Fig. 26.24, the settling time is approximately 2.5 ns so this RC time can have a significant effect on the settling time and stability of the circuit (important).

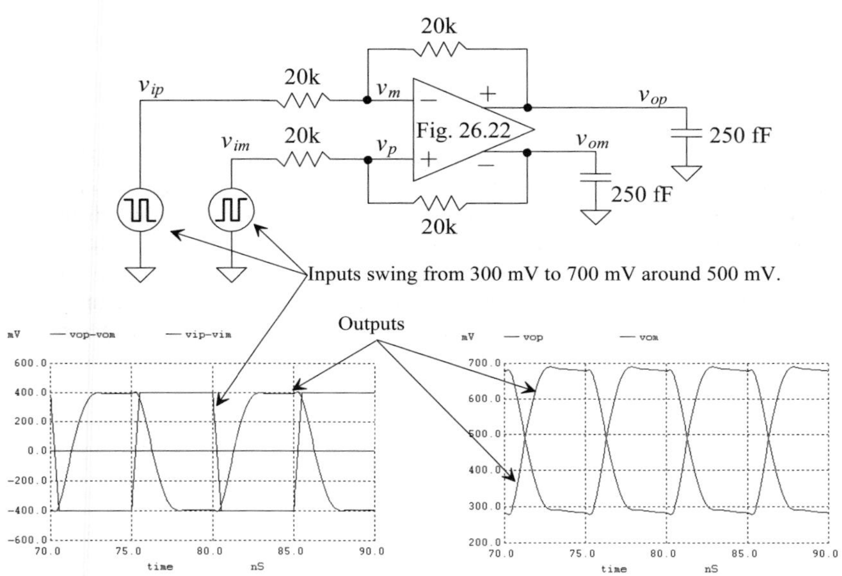

Figure 26.24 Step response of the op-amp in Fig. 26.22 driving 250 fF load capacitors and 20k feedback resistors.

Notice how, in Fig. 26.24, we used input signals that don't swing rail-to-rail. Since we don't have a CMFB circuit in the op-amp, the exact common-mode output voltage is an unknown. It may be 400 mV or it may be 600 mV. We get some help in setting the circuit's output common-mode level by using input signals with common-mode voltages of 500 mV and DC feedback. Looking at the simulation results in Fig. 26.24, we might get a false sense of not needing a CMFB circuit. To illustrate this is indeed a false sense, consider the sample-and-hold seen in Fig. 26.25 (see also Fig. 25.19). When the ϕ_1 switches are closed, the op-amp's inputs and outputs should be held to $V_{CM} \pm V_{OS}$ (the op-amp is placed in the follower configuration). As seen in the figure, the outputs during this time are driven close to 400 mV. On closer inspection, we see the output common-mode level is wandering downwards until eventually the op-amp shuts off.

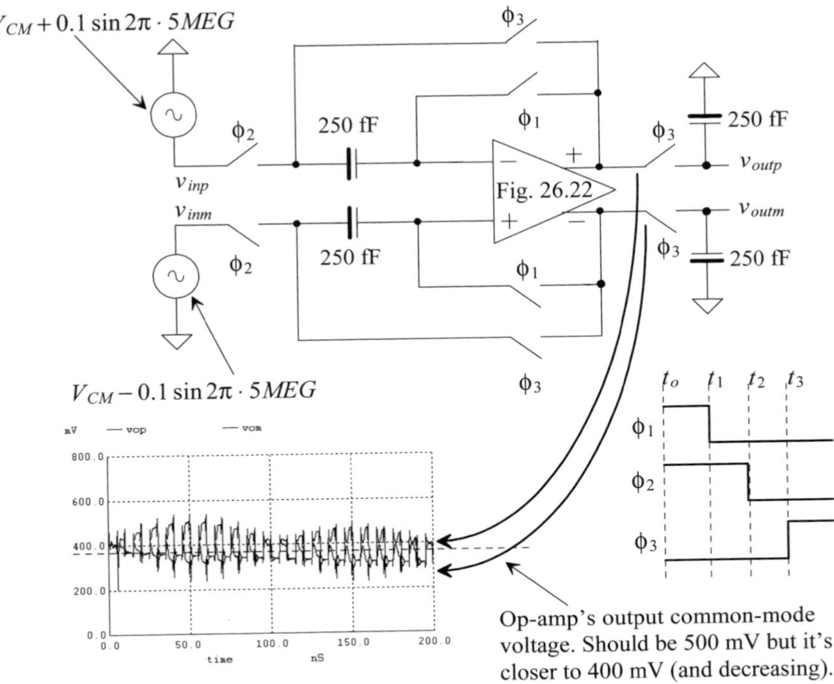

Figure 26.25 A sample-and-hold circuit. Notice how the output common-mode voltage is wandering.

Adding CMFB

Figure 26.26 shows how we can modify the basic op-amp to allow for a CMFB signal input. If the outputs of the op-amp (their average or common-mode voltage) are too high,

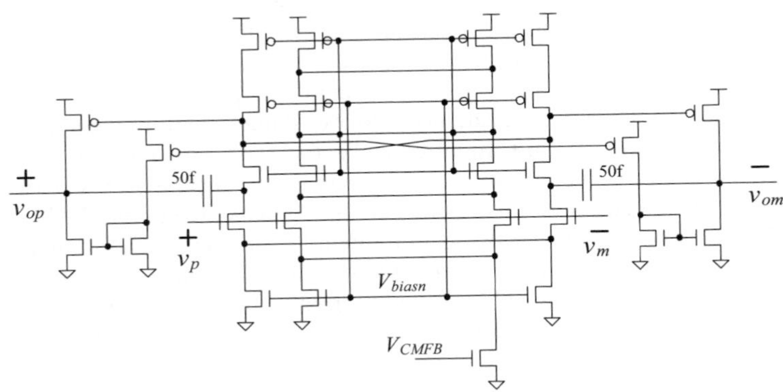

Figure 26.26 Modifying the op-amp for a CMFB input signal.

V_{CMFB} goes up. This causes the output voltage of the diff-amp to go up (increasing V_{CMFB} causes the bias current in the center MOSFETs of the diff-pair to increase). An increase in the diff-amp's output voltage lowers the quiescent current flowing in the output buffers, causing the output voltage to move downwards. This is a good time to remember one of the fundamentals from Ch. 20, namely, if two MOSFET gate-source voltages are equal and they have the same drain current, then their drain-source voltages must be equal. For the output buffer in Fig. 26.26, this means that as we reduce the drain currents flowing in the output buffer (by driving V_{CMFB} high), the gate-source voltages of the NMOS devices decrease. Because of the fundamental concept just mentioned, this causes the output voltages of the op-amp to decrease. Figure 26.27 shows the op-amp's output voltage change with V_{CMFB}. For stability concerns, it's of interest to determine the gain from the CMFB input to the outputs. From the simulation, the gain is approximately 25 (about 10 times less than the differential gain). The lower forward CMFB gain allows us to use a diff-amp with a current mirror load for the CMFB amplifier (discussed next).

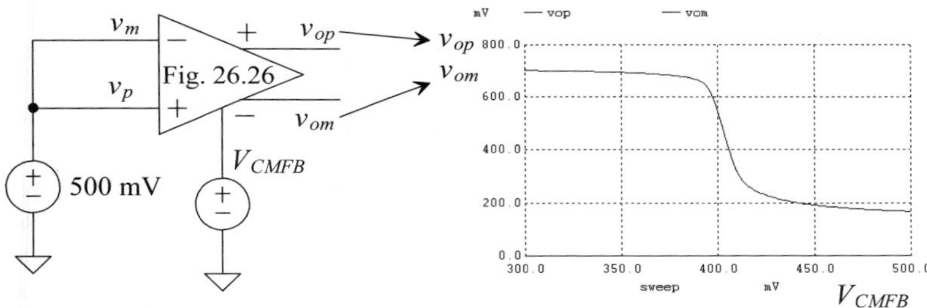

Figure 26.27 The CMFB input to output relationship. The gain is approximately 25 (considerably less than the forward differential gain).

CMFB Amplifier

Now that we know how our CMFB signal, V_{CMFB}, in the op-amp of Fig. 26.26 affects the output voltages and that the gain this signal sees is approximately 25 ($v_{op,n}/V_{CMFB}$), we need to discuss the CMFB amplifier. Consider the two diff-amps seen in Fig. 26.28. In the top diff-amp, gate-drain-connected loads are used. As seen in the simulations, the gain from the averaged input, V_{CMA}, to the diff-amp's output, V_{CMFB} is considerably less than 1 (and so the CMFB is guaranteed to be stable). Unfortunately, the CMFB amplifier's output voltage isn't high enough. (As seen in Fig. 26.27, we need approximately 400 mV.) The diff-amp's tail current can be increased in size (use at least two PMOS for the tail current) to increase the output voltage. However, the low gain, say around 0.3, combined with the CMFB gain through the op-amp, again around 25, means that the CMFB loop's overall gain is only around 7. This isn't large enough to precisely balance the outputs. Using the diff-amp with current mirror load, Fig. 26.28b, gives a gain of approximately 8. This combined with the op-amp's CMFB gain gives an overall CMFB loop gain of 200. This is less than the op-amp's differential gain (so we can use the same compensation capacitors to stabilize the CMFB loop), Fig. 26.23, and large enough to precisely balance the op-amp's outputs.

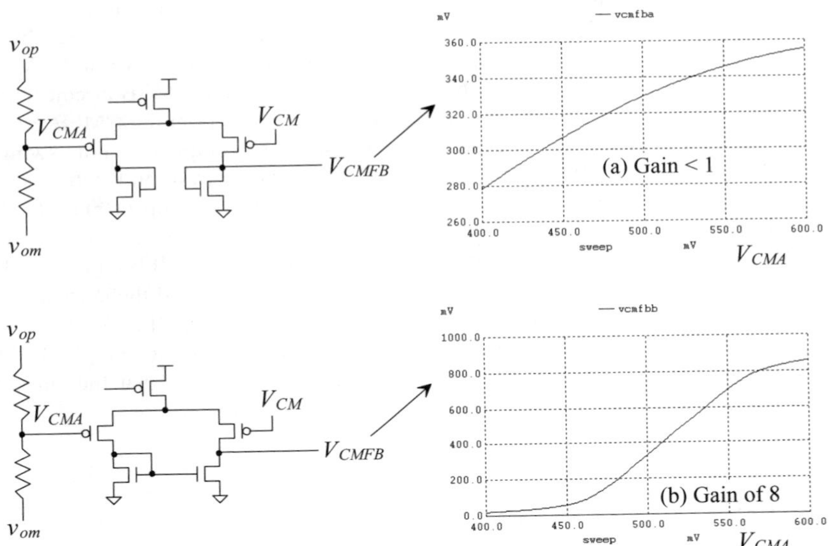

Figure 26.28 Gains of CMFB amplifiers.

The Two-Stage Op-Amp with CMFB

Figure 26.29 shows the complete schematic of the op-amp. Notice that we've doubled the width of the MOSFETs in the output buffer. The added loading from the CMFB averaging resistors will lower the open-loop gain. To compensate for this reduction, the

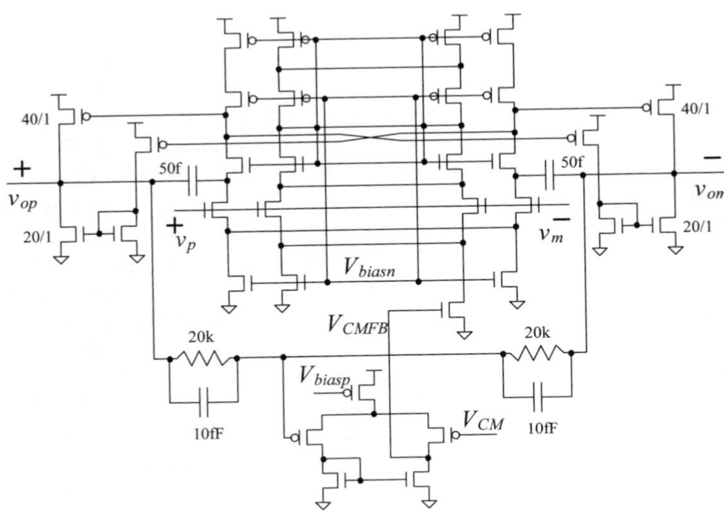

Figure 26.29 Complete schematic of op-amp with CMFB.

drive strength of the MOSFETs on the output of the op-amp was increased. Figure 26.30 shows the DC behavior of the op-amp when placed in the configuration seen in Fig. 26.23. Notice that the outputs cross at the ideal common-mode voltage of 500 mV. The gain is reduced because of the loading by the 20k resistors in the CMFB circuit (and so the gain can be increased by increasing the values of the resistors in the CMFB circuit). However, **we see a problem** with these simulation results. The outputs are only swinging from 200 to 800 mV (not from 0 to *VDD* as possible with a push-pull output stage). Further, we can estimate the current pulled from *VDD* for the op-amp in Fig. 26.29 by counting the number of branches in the op-amp (noting that the output stage counts twice because we've doubled the widths of these devices). Including the CMFB amplifier, there are 11 branches. If the bias current through each branch is 20 μA and the current pulled by the bias circuit is 50 μA, then we would expect the op-amp to pull 270 μA (perhaps even a little less because many of the MOSFETs are biased near the triode region). When we look at the simulation results that generated Fig. 26.30, we see that the current is closer to 500 μA (way off indicating, again, that we've got a problem).

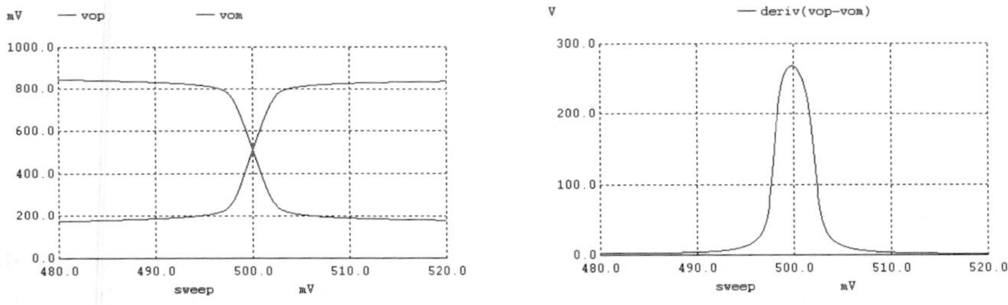

Figure 26.30 Simulating the operation of the op-amp in Fig. 26.29.

Origin of the Problem

The origin of the problem (that is, the op-amp drawing too much current and the output swing not reaching the rails) can be traced to the output buffer. We spent a considerable amount of time discussing how we want to set the outputs of the diff-amp to V_{biasp}. When we look at the simulations, we see the diff-amp's outputs are considerably less than the ideal 600 mV of V_{biasp}. To understand why, let's look at the output buffer in Fig. 26.31. Since the problems appear with the addition of the CMFB circuit, we assume that the two gates of the PMOS devices are moving at the same potential (tied together), that is, that the inputs to the buffer are moving with the diff-amp's output common-mode level. When the output of the diff-amp is V_{biasp} (600 mV), 20 μA flows in all of the MOSFETs. (To keep things simpler, we don't include the doubling in the widths used in the op-amp output devices.) The gate potentials of the NMOS are, roughly, V_{biasn} (400 mV). Because of the symmetry of the circuit, this means that the output is also at 400 mV. To increase the output voltage to 500 mV, we must drop the potential on the gates of the PMOS until the gate-drain-connected NMOS has a V_{GS} of 500 mV (again because of the symmetry). This increases the current (significantly) flowing in the output buffer, lowers its gain, and reduces the linear output swing. The overdrive voltages essentially change from the desired 100 mV when 20 μA of current flows to 200 mV when the output is driven to 500 mV.

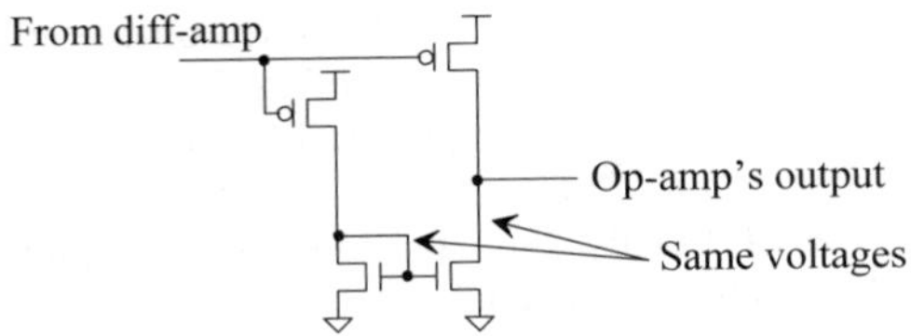

Figure 26.31 Output buffer used in the op-amp of Fig. 26.29.

Figure 26.32 shows one solution to this problem. We've added a device to cascode the output buffer's NMOS (the one connected to the output terminal). The speed shouldn't be affected by the addition of the device (which operates near or in the triode region). The added device allows the op-amp's output to swing more freely (but the current still won't be precisely set). The added device won't affect the output swing range of the op-amp. Figure 26.33 shows the op-amp with modified output buffer.

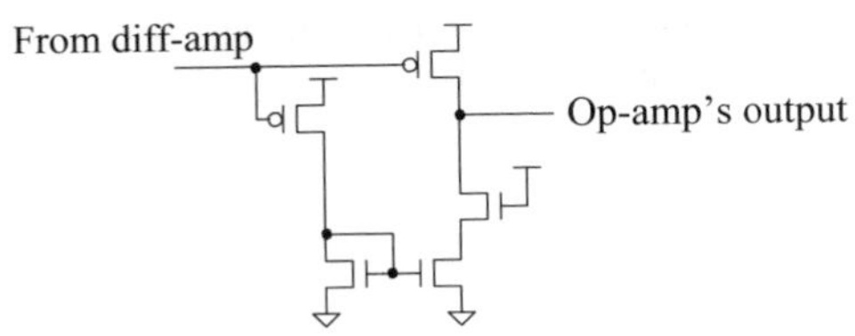

Figure 26.32 Adding a device to allow the output voltage to swing.

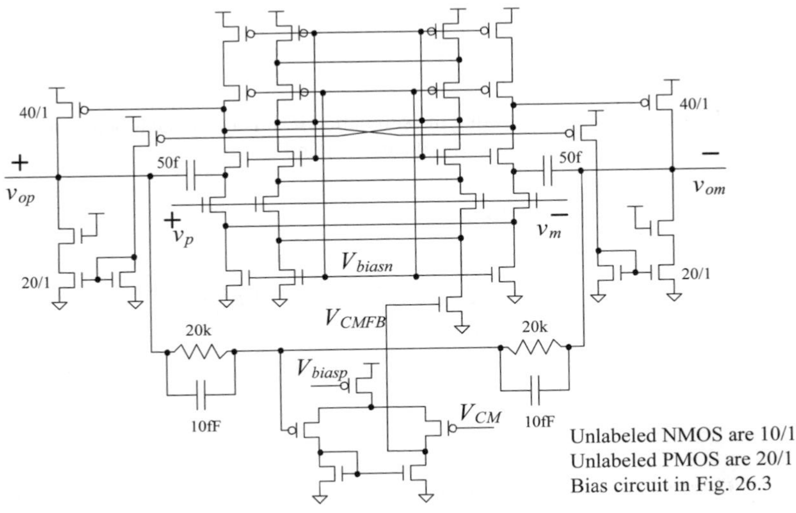

Figure 26.33 Op-amp with modified output buffer.

Simulation Results

Figures 26.34 to 26.36 show simulation results based on the circuit topologies in Figs. 26.23 to 26.25, respectively, using the op-amp in Fig. 26.33. In Fig. 26.34 we see that the outputs now swing close to the power supply rails (unlike what we saw in Fig. 26.30). Also, the gain is higher. Figure 26.35 shows the step response of the amplifier in the topology seen in Fig. 26.24. Finally, Fig. 26.36 shows the outputs of the sample-and-hold in Fig. 26.25 using the op-amp in Fig. 26.33.

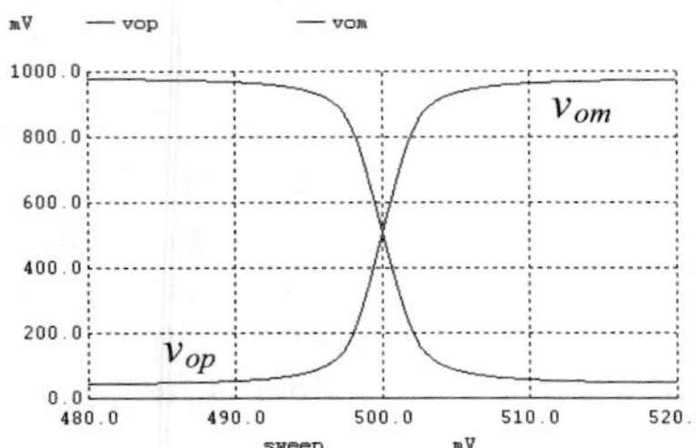

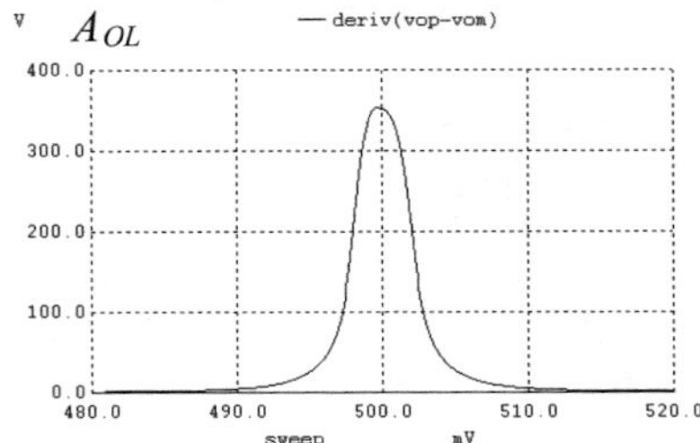

Figure 26.34 Resimulating the op-amp in Fig. 26.33 in the configuration seen in Fig. 26.23.

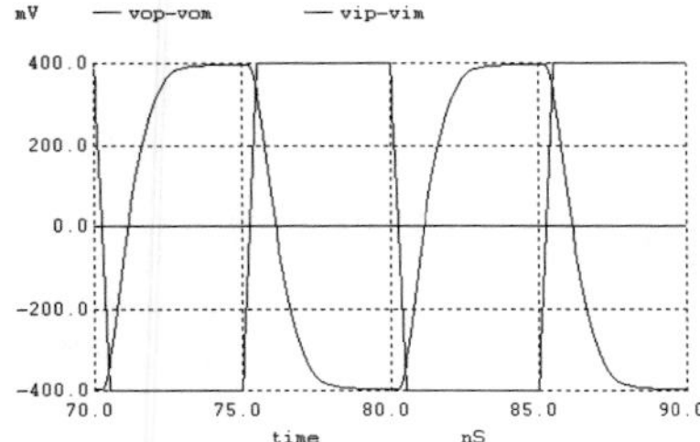

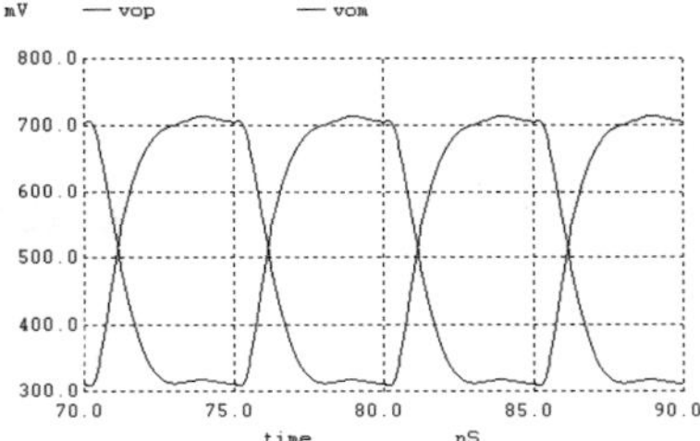

Figure 26.35 Regenerating the simulation results using the topology in Fig. 26.24 with the op-amp in Fig. 26.33.

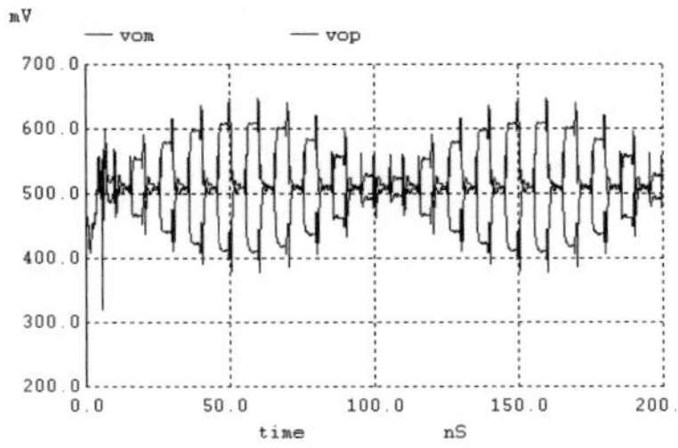

Figure 26.36 Using the op-amp in Fig. 26.33 in the sample-and-hold circuit of Fig. 26.25. Figure shows the outputs of the op-amp.

Using MOSFETs Operating in the Triode Region

For the output buffer in Fig. 26.32, we added a transistor in the drain portion of the circuit. We might wonder if we can accomplish better control of the current flowing in the output buffer by adding circuitry to the source side of the buffer. Towards answering this question, consider the portion of the output buffer seen in Fig. 26.37. If both outputs are at V_{CM}, the gate-source voltages of M2 and M3 are the same and so are the gate-source voltages of M1 and M4. If one output goes high and the other output goes low (and the outputs are centered around V_{CM}), then the net drain current of each M3 is constant. If one output goes below the threshold voltage of an NMOS device, then the balancing stops working. The practical problem of using triode-operating MOSFETs (in a CMFB circuit or any type of amplifying configuration) is that the gain through a MOSFET operating in the triode region is low (so it's difficult to provide control).

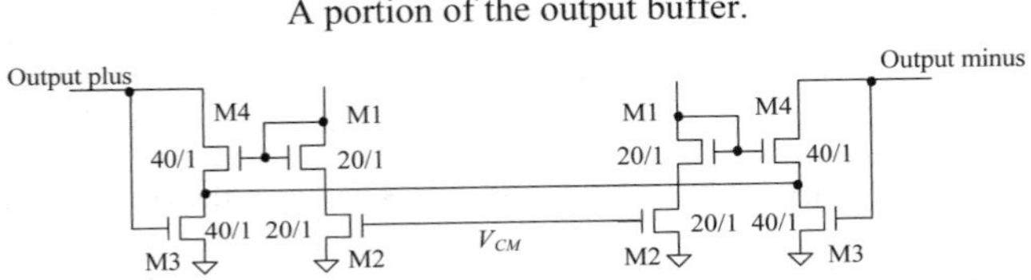

Figure 26.37 Using triode-operating MOSFETs to balance the outputs (bad).

Start-up Problems

Examine the circuit in Fig. 26.38 using the op-amp in Fig. 26.33. Because the op-amp's inputs are at 0 V, the diff-amps on the input of the op-amp are off. This causes the gates of the PMOS devices in the output buffer to be pulled to *VDD*. The circuit remains in this state and doesn't move the outputs or the inputs up to V_{CM}. To avoid this start-up problem, we need to ensure that there is some DC path to *VDD* or V_{CM} to "start-up" the op-amp.

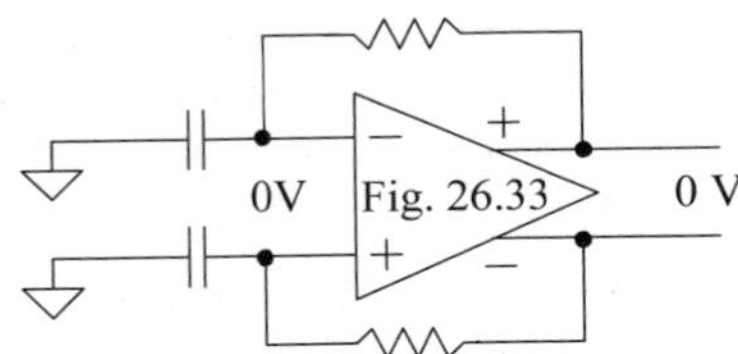

Figure 26.38 The op-amp in Fig. 26.33 won't turn on in this topology.

Lowering Input Capacitance

Consider connecting the gates of the NMOS diff-amp used for biasing to V_{CM} (see the bold line in Fig. 26.39). This reduces the input capacitance of the op-amp and shouldn't affect the normal operation as long as the common-mode voltage of the op-amp (the gates

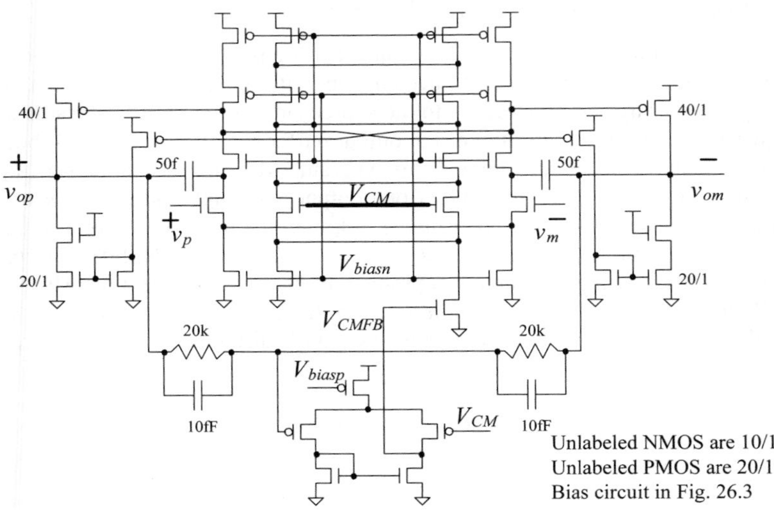

Figure 26.39 Connecting the bias circuit diff-amp's inputs to the common-mode voltage.

of the other diff-amp) is V_{CM}. Simulating the operation of the op-amp in Fig. 26.39 to generate the data in Figs. 26.34 to 26.36 (in the same topologies), we see no change in the simulation data (the netlists used to verify this statement can be found at cmosedu.com).

Looking at Fig. 26.39, we might now wonder why we need two biasing branches down the middle of the op-amp. Further, looking at the voltages in Fig. 26.19, we might wonder if we can make the op-amp more tolerant to changes in *VDD*. The way the op-amp is biased now the open-loop gain drops significantly if *VDD* drops to 900 mV.

Making the Op-Amp More Practical

As just mentioned, having two identical branches down the middle of our op-amp wastes power (although it is useful for a symmetrical layout). We can cut one of the branches out of the design and reduce the power dissipated by the op-amp. We might further wonder if using V_{biasn} for biasing the cascode PMOS current sources is such a good idea. Reviewing Fig. 26.4, we see that V_{biasn} essentially stays at 400 mV after the reference turns on (*VDD* gets above a certain value). In a practical op-amp, we want this voltage to decrease as *VDD* drops in an effort to keep the PMOS devices operating in the saturation region.

Examine the op-amp in Fig. 26.40. We've added a wide-swing bias circuit, see Fig. 20.38, and cut out one of the biasing branches used in the op-amp seen in Fig. 26.39. The power dissipation remains essentially the same as in the previous op-amp topologies. Note how we apply the CMFB to one side of the bias circuit (not to the 10/3 wide-swing bias branch). Further notice how we've reduced the gain of the CMFB loop by using two 10/2 MOSFETs. It may be a good idea to reduce the strength of the CMFB loop even further by increasing the lengths of the NMOS devices. For example, we might change these devices from 10/2 to 10/4. The ability to drive V_{CMFB} considerably above V_{biasn} eliminates the concern that the NMOS device can sink the current needed to bias the circuit at the correct point (and why the CMFB loop can so easily become unstable).

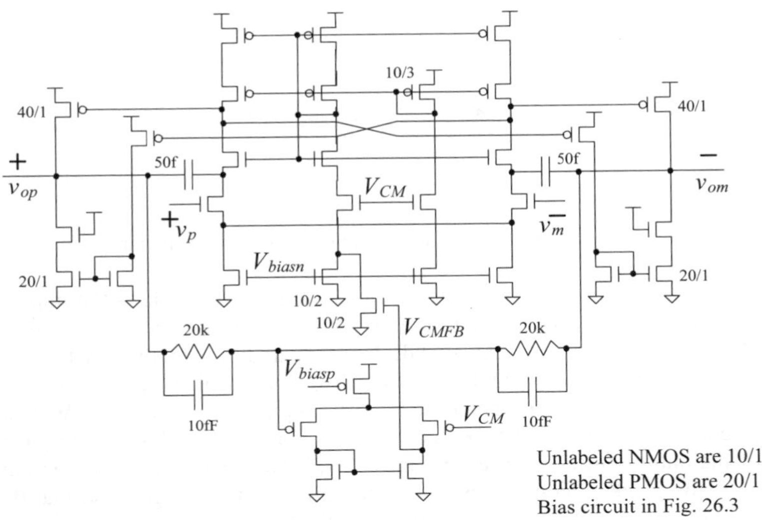

Figure 26.40 Making the op-amp more practical.

A more practical problem is the change in the MOSFET's drain currents with drain-source voltages. For example, looking at Figs. 26.18a and b, we see that it's possible for the NMOS's drain current to be twice the PMOS's drain current for the same gate-source voltages. This can cause op-amp failure in the CMFB circuit. If, for example, the current flowing in the 10/2 MOSFET connected to V_{biasn} is larger than the current sourced by the PMOS, the voltage V_{CMFB} goes to zero and the CMFB loop doesn't work properly. A good "rule-of-thumb" is for the fixed current flowing in the CMFB-controlled bias circuit to be 25–50% of the total expected current.

Increasing the Op-Amp's Open-Loop Gain

The op-amp that we've developed in this section has an open-loop gain in the hundreds. As we'll see in Ch. 29 (Eq. [29.59]), the open-loop gain of the op-amps used in a data converter has a direct effect on the maximum attainable resolution. Towards increasing the gain of the op-amp, let's use the gain-enhancement (GE) techniques presented in Sec. 24.4.

Figure 26.41 shows how we should *not* implement GE. An amplifier with fully-differential outputs, like the amplifier in Fig. 26.7, regulates the drains of the top PMOS devices. We know that this is bad because we would need a CMFB circuit to balance the outputs of the added amplifier.

Figure 26.42 shows how we can add GE to the op-amp developed in this section. Notice that GE is implemented with amplifiers having single-ended outputs. When designing the added amplifiers, as discussed in Sec. 24.4, the bandwidth isn't important for high-speed operation. The lengths of the MOSFET in the added amplifiers can be increased to reduce power and boost gain. An important concern is the added amplifier's input common-mode range.

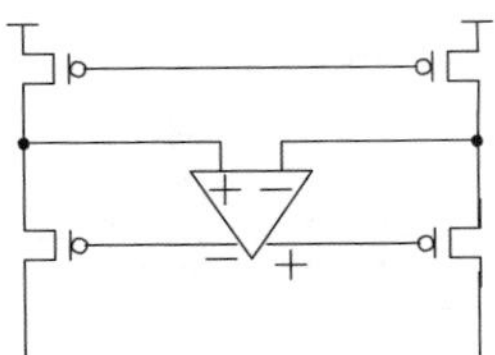

Figure 26.41 How not to implement GE in an op-amp.

The practical problem with the topology seen in Fig. 26.42 is the implementation of the CMFB. With the GE added to the op-amp we now have four additional feedback loops. Variations in V_{CMFB} affect all GE loops in addition to the CMFB through the op-amp. Making these loops stable becomes extremely challenging. What we need is to implement the CMFB without including the diff-amp and GE amplifiers. We can only do this by controlling the output buffer common-mode level, Fig. 26.43. This circuit includes the output buffer we used in Fig. 26.37. Now, however, we add an amplifier in series with the triode-operating MOSFETs to boost the CMFB gain (so that the outputs can be balanced around V_{CM}). The problem with this approach is that the CMFB isn't compensated using the same capacitors as the differential forward signal path (i.e., no Miller effect). We must be concerned with CMFB stability.

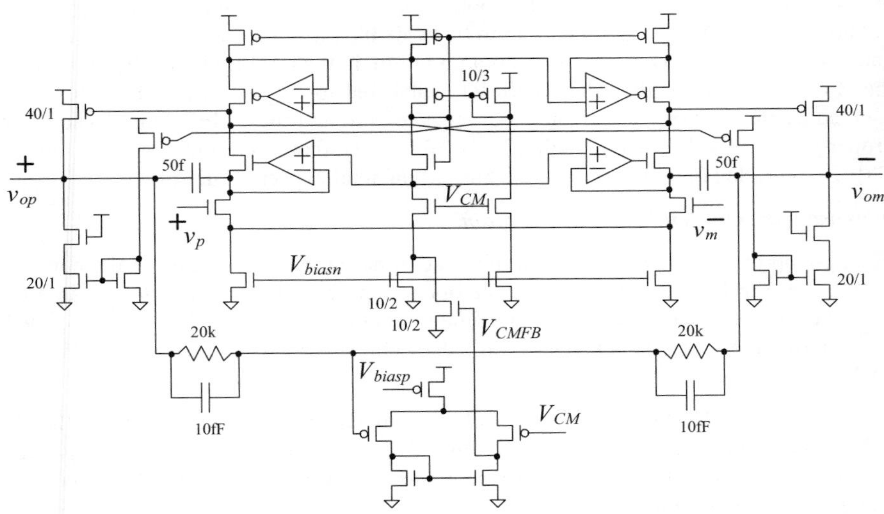

Figure 26.42 Adding gain-enhancement to the op-amp.

Let's discuss the op-amp design in Fig. 26.43. We begin by showing the DC behavior of the op-amp (to show that the outputs are indeed balanced). Figures 26.44a and (b) show the DC behavior and gain of the op-amp in Fig. 26.43 in the topology seen in Fig. 26.24. Again note that by increasing the values of the 20k resistors used to

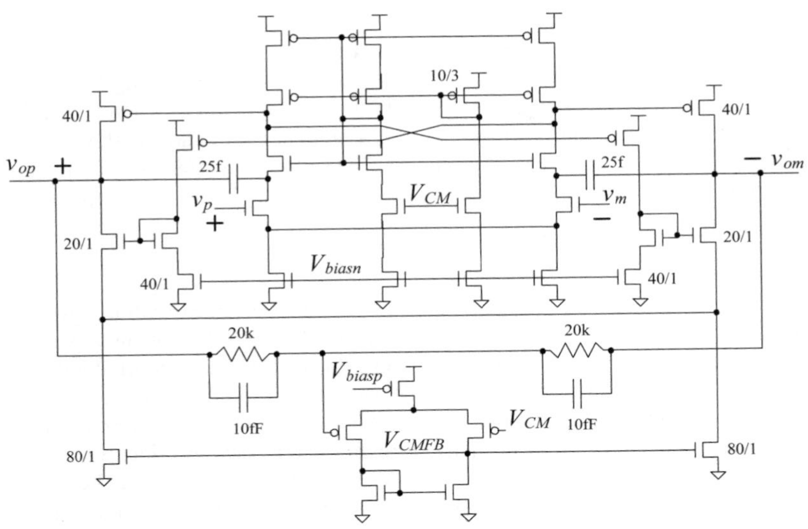

Figure 26.43 Providing CMFB through just the output buffer. Using an amplifier with triode-operating MOSFETs for CMFB (good).

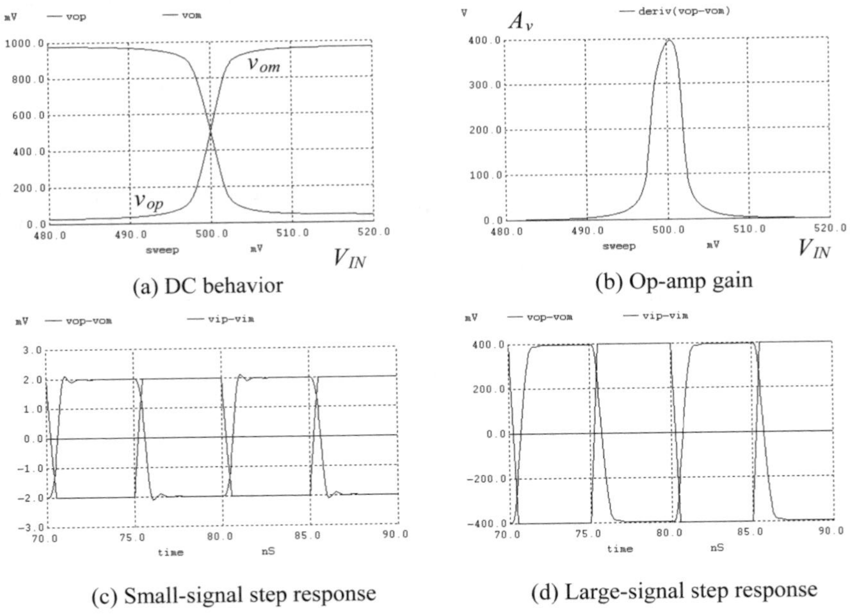

(a) DC behavior
(b) Op-amp gain
(c) Small-signal step response
(d) Large-signal step response

Figure 26.44 Behavior of the op-amp in Fig. 26.43 (see text).

average the outputs we can increase the DC gain of the op-amp. Because the forward differential path compensation (now) doesn't include the CMFB path, we've reduced the compensation capacitors from 50 fF to 25 fF (there are practical issues with this change that we'll discuss next). Note that the widths of the added triode-operating MOSFETs are increased to ensure that they operate well within the triode region and don't significantly affect the output drive capability of the op-amp. Figures 26.44c and (d) show the small- and large-signal step responses (see Fig. 26.24). The settling times are approximately 1–2 ns. The current drawn from *VDD* (including the the bias circuit current) is approximately 250 μA (200 μA for the op-amp alone).

Offsets

We said in Sec. 26.2 that we want to add offsets of 50 mV in series with the gates of our MOSFETs to see if the op-amp "breaks." Consider the addition of an input-referred offset voltage in the schematic seen in Fig. 26.45. This offset is used to model the overall offset voltage of the op-amp (see Fig. 24.4). If we look at the effects of the offset on the circuitry in Fig. 26.43, we see that one of the NMOS devices in the diff-pair will have a higher overdrive voltage (larger g_m) than the other NMOS device. Since the unity-gain frequency of an op-amp is given by $g_m/2\pi C_c$, the effect is a shift in the unity-gain frequency (and potential instability). As seen in Fig. 26.45, using the 25 fF compensation capacitors destabilizes the op-amp (as indicated by the ringing). Increasing the compensation capacitor's value (back) to 50 fF makes the step response cleaner. We could argue that 50 mV is an unrealistically high offset voltage, so it's better to leave the capacitors at 25 fF (and this may be the case). However, in a practical CMOS process, the op-amp's characteristics shift with the process shifts (and temperature). It's better to overcompensate than to have an unstable op-amp.

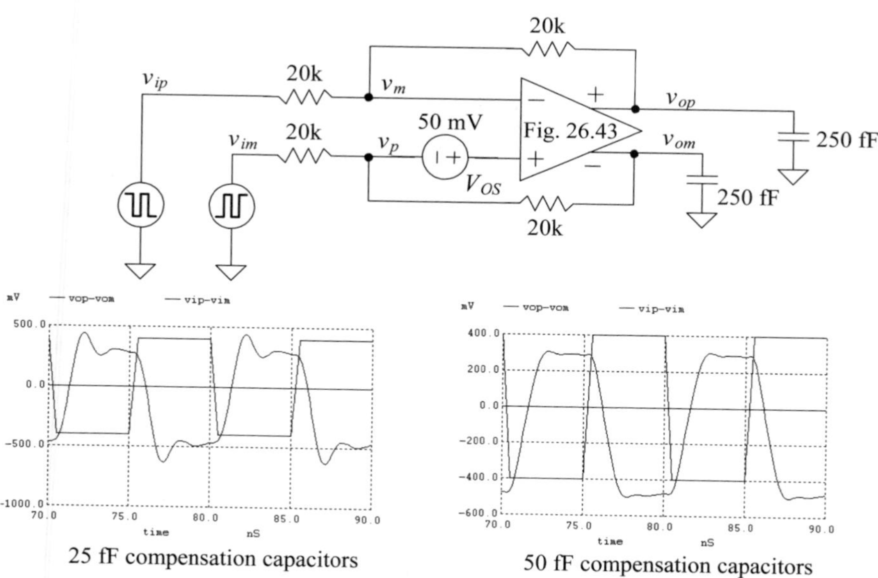

Figure 26.45 How an offset can affect the step response (compensation).

Op-Amp Offset Effects on Outputs

Notice, in Fig. 26.45, that an offset shifts the differential output signal by twice the offset voltage. The individual op-amp output voltages remain centered around V_{CM}. To describe this in more detail, consider the test setup seen in Fig. 26.46. We know the CMFB forces

$$\frac{v_{op}+v_{om}}{2}=V_{CM} \text{ or } v_{op}=2V_{CM}-v_{om} \tag{26.8}$$

Further, assuming large op-amp open-loop gain,

$$v_p+V_{OS}=v_{pos}\approx v_m \tag{26.9}$$

Equating currents, we can write

$$\frac{V_{CMI}-v_p}{R_{in}}=\frac{v_p-v_{om}}{R_f} \text{ or } \frac{V_{CMI}-v_m+V_{OS}}{R_{in}}=\frac{v_m-V_{OS}-v_{om}}{R_f} \tag{26.10}$$

and

$$\frac{V_{CMI}-v_m}{R_{in}}=\frac{v_m-v_{op}}{R_f} \tag{26.11}$$

Subtracting Eq. (26.11) from Eq. (26.10), we get

$$\frac{V_{OS}}{R_{in}}=\frac{-V_{OS}+v_{op}-v_{om}}{R_f} \tag{26.12}$$

or

$$v_{op}-v_{om}=V_{OS}\cdot\left(\frac{R_f}{R_{in}}+1\right) \tag{26.13}$$

The voltages on the outputs of the op-amp are then

$$v_{op}=\frac{V_{OS}}{2}\cdot\left(1+\frac{R_f}{R_{in}}\right)+V_{CM} \text{ and } v_{om}=V_{CM}-\frac{V_{OS}}{2}\cdot\left(1+\frac{R_f}{R_{in}}\right) \tag{26.14}$$

The op-amp's input voltages are

$$v_m\approx v_p+V_{OS}=\frac{V_{OS}}{2}+V_{CM}\cdot\left(\frac{R_{in}}{R_{in}+R_f}\right)+V_{CMI}\cdot\left(\frac{R_f}{R_{in}+R_f}\right) \tag{26.15}$$

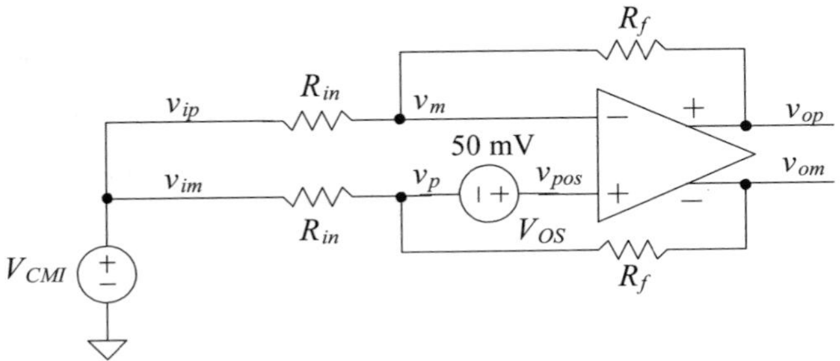

Figure 26.46 How an offset voltage causes an imbalance in the outputs.

If the input signal's common-mode voltage, V_{CMI} , is the same as the op-amp's common-voltage, V_{CM} , and V_{OS} is zero, then both op-amp inputs are held at V_{CM}. If $V_{CM} = V_{CMI}$ but the offset isn't zero, then $v_m = V_{OS}/2 + V_{CM}$ and $v_p = V_{CM} - V_{OS}/2$. Neglecting the offset voltage, if the input signal's common mode voltage isn't V_{CM} , then the op-amp's input common-mode voltage ($v_p = v_m$) will be different from the ideal value of V_{CM}. This can shut the op-amp off and cause undesirable behavior.

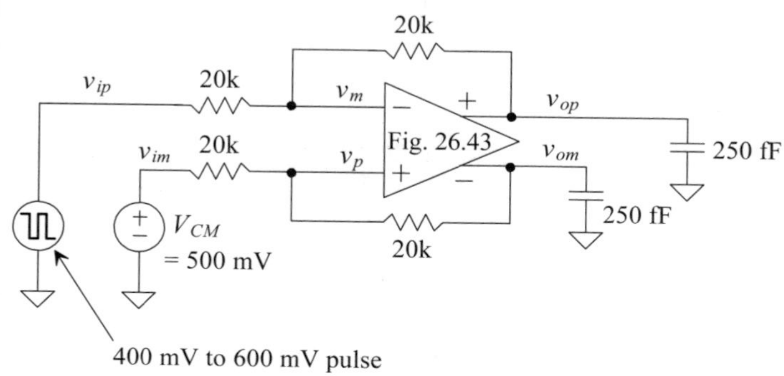

Figure 26.47 Problems with single-ended input signals.

Single-Ended to Differential Conversion

Even if the input signals are referenced around V_{CM} and the op-amp is offset free, we can still have problems. Consider the circuit in Fig. 26.47. In this circuit the input signal is single-ended. The other input to the op-amp is tied to V_{CM}. Even though the input is balanced around 500 mV, it is not a truly differential signal. When the input signal is up at 600 mV, the effective input common-mode voltage, V_{CMI} , is 550 mV (the average of the two inputs). When the input signal is at 400 mV, V_{CMI} is 450 mV. Figure 26.48 shows the simulation results using the circuit in Fig. 26.47. These results are very interesting. Consider what's happening between 70 and 75 ns. During this time, v_{ip} is 600 mV and v_{im} is 500 mV (and so $v_{ip} - v_{im} = 100\ mV$). The op-amp's inputs move above the ideal 500 mV, as seen in the figure. This has the effect of increasing the transconductance of the diff-amp (the diff-amp's tail current is operating near/in the triode region so the tail

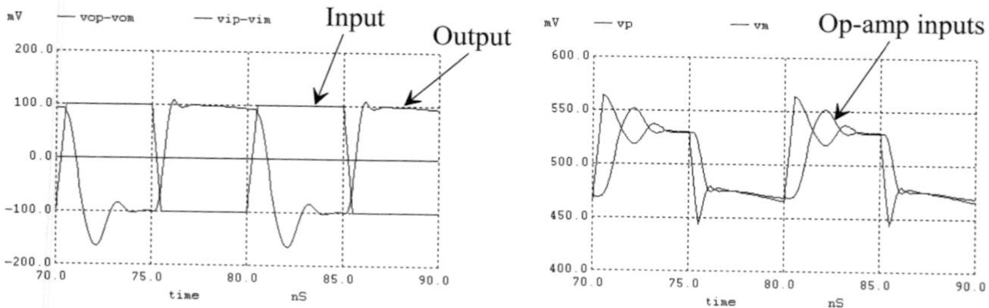

Figure 26.48 Simulating the operation of the circuit in Fig. 26.47.

current increases as the input common-mode voltage goes up). The increase in the diff-amp's g_m causes the unity gain frequency to increase and, with a fixed compensation capacitor, the phase margin to decrease (causing overshoot and ringing in the circuit's step response). Next consider what happens between 75 and 80 ns. The input signal switches to $400\ mV - 500\ mV = -100\ mV$ and the op-amp's inputs drop down to around 475 mV. This causes the diff-amp's g_m to decrease and the op-amp to slow down (become more stable). However, at such a low input common-mode voltage the gain of the op-amp will drop and the outputs may wander.

CMFB Settling Time

If we look at the voltage, V_{CMFB}, in the op-amps in Fig. 26.40 or 26.43, we may see that they aren't settling as quickly as the differential mode signals. Is this bad? As seen in Eq. (26.13), variations in V_{CMI} don't (ideally) affect the differential output signal. However, if the variations in V_{CMFB} and thus V_{CMI} are large, one of the output signals can saturate at close to *VDD* or ground (the op-amp's gain drops), affecting the differential signal path.

CMFB in the Output Buffer (Fig. 26.43) or the Diff-Amp (Fig. 26.40)?

In this section we've presented the design of op-amps using continuous-time CMFB. Various design trade-offs and topologies were presented. At this point a good question is: "Which CMFB scheme is better?" Controlling the output common-mode level through the output buffer (Fig. 26.43) is simple, easy to ensure stability, and fairly robust. However, consider what happens if the op-amp in Fig. 26.43 is used in the topology in Fig. 26.46 (without an offset). If V_{CMI} is ground, then, according to Eq. (26.15), the input voltages to the op-amp will move to 250 mV while the outputs of the op-amp remain at 500 mV. Thinking about this for a moment and neglecting the fact that an op-amp input common-mode voltage of 250 mV will shut the op-amp off, we see that the outputs of the op-amp must source a current back through the feedback resistors to the inputs. Further, anytime $V_{CMI} < V_{CM}$, the PMOS in the output buffer must source a DC current back to the inputs. Similarly, if $V_{CMI} > V_{CM}$, the NMOS in the output buffer must sink a current from the inputs of the op-amp. When using CMFB in the output buffers, we have no way of increasing the quiescent or DC current flowing in the output buffer to source/sink current from the input source (or to a DC load connected to ground). The output voltage of the diff-amp in the op-amp of Fig. 26.43 biases, or sets, the quiescent current in the output buffer. The CMFB scheme used in the output buffer simply adjusts the drive strength of the NMOS (in the output buffer) to set the op-amp's output common-mode level.

Using the CMFB scheme in Fig. 26.40, we can adjust the output voltage of the diff-amp (and ultimately the op-amp's output voltage) and thus the bias current in the output buffer (both the NMOS and the PMOS connected to the output buffer). However, if the output buffer must source/sink a significant amount of current, we can have problems with this topology too. For example, if the diff-amp's output voltage increases (turning off the PMOS in the output buffer and attempting to pull the op-amp's outputs down) while the NMOS devices connected to the outputs are sinking significant current, then it's possible that the outputs will get pulled upwards and the CMFB will fail (likely causing the gain of the op-amp to drop and the outputs to have limited swing).

To ensure the most robust op-amp design (biasing tolerant to offsets and easy to compensate the CMFB loops), CMFB circuits can be placed around both op-amp stages. This method is used in our last op-amp design example discussed next.

26.4 Op-Amp Design Using Switched-Capacitor CMFB

In this section we turn our attention towards op-amp designs for switched-capacitor (SC) circuits. We'll develop an op-amp design based on the topologies discussed in the last section. In this section, however, we'll use SC CMFB instead of continuous-time CMFB. Again, we'll use the nanometer CMOS process with minimum lengths and (roughly) 100 mV overdrive voltages (for drain currents of 20 μA, see Fig. 26.18). We'll stick with two-stage designs because of the low open-circuit gains present when doing high-speed design.

Clock Signals

The SPICE listing for the clock signals used in the simulations in this section is seen below. Note that the 100 MHz phi1 and phi2 clocks are used with the NMOS transistors (phi1 and phi2 are not high at the same times), while the complements of these clock signals are used with the PMOS switches (not low at the same times).

```
*Clock Signals
Vphi1  phi1   0 DC 0 Pulse  0 1 0  200p 200p 4n 10n
Vphi1b phi1b  0 DC 0 Pulse  1 0 0  200p 200p 4n 10n
Vphi2  phi2   0 DC 0 Pulse  0 1 5n 200p 200p 4n 10n
Vphi2b phi2b  0 DC 0 Pulse  1 0 5n 200p 200p 4n 10n
R1     phi1   0        1MEG
R2     phi1b  0        1MEG
R3     phi2   0        1MEG
R4     phi2b  0        1MEG
```

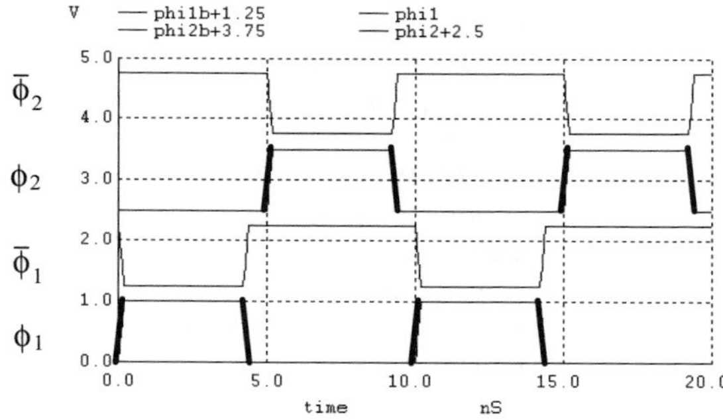

Figure 26.49 Generating nonoverlapping clocks for SC circuits.

Switched-Capacitor CMFB

Figure 26.50 shows the SC CMFB circuit from Fig. 26.16 along with a symbolic representation. We've selected 25 fF capacitors for the high-speed averaging capacitors. The kT/C noise associated with these capacitors is a common-mode signal and so it shouldn't affect the differential-mode signal. We don't want these capacitors to be too large because they will load the output of the amplifier. We made the switched-capacitors associated with the differencing and averaging 10 fF so that the changes in V_{CMFB} during one clock cycle won't be too large due to differences between the ideal V_{CMFB} and the actual V_{CMFB}. When the average of the outputs, v_{op} and v_{om} , is above their ideal value, the

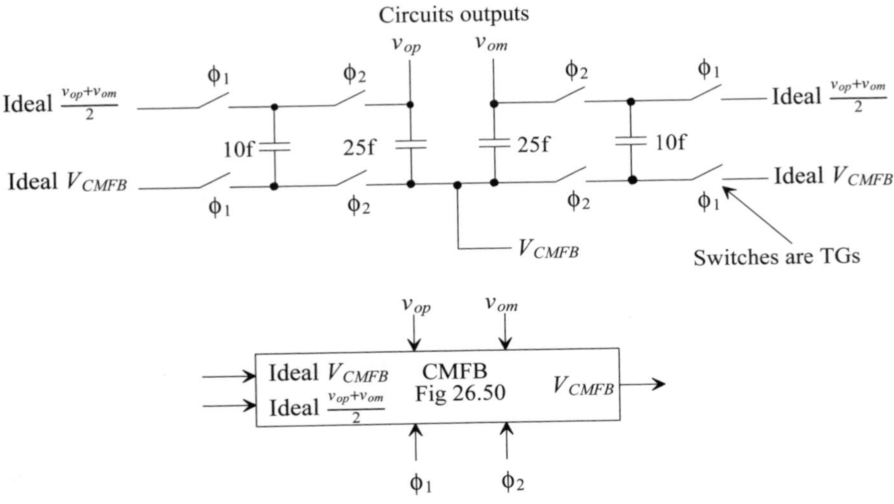

Figure 26.50 A switched-capacitor CMFB circuit (see Fig. 26.16). Switches implemented with transmission gates.

CMFB circuit's outputs moves upwards. It's important that this causes the average value of the outputs to move downwards. In other words, we've got to ensure that negative feedback is used in the CMFB loop.

Figure 26.51 shows some simulation results using the SC CMFB in Fig. 26.50. The ideal V_{CMFB} is 400 mV. The ideal average output, $\frac{v_{op}+v_{om}}{2}$, is 500 mV (V_{CM}). Prior to 150 ns, both v_{op} and v_{om} are 500 mV. At 150 ns, these two voltages jump (in the simulation) to 700 mV (they jump 200 mV away from the ideal average output of 500 mV). The voltage, V_{CMFB}, also jumps at this time. The key point to notice is that V_{CMFB} moves in the same direction as do changes in the output common-mode voltage. Knowing this is important when we are ensuring that our CMFB loop employs negative feedback.

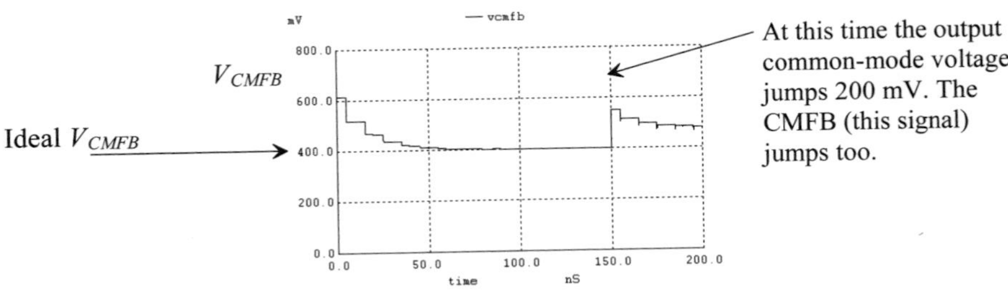

Figure 26.51 How the common-mode feedback signal increases to pull the outputs down.

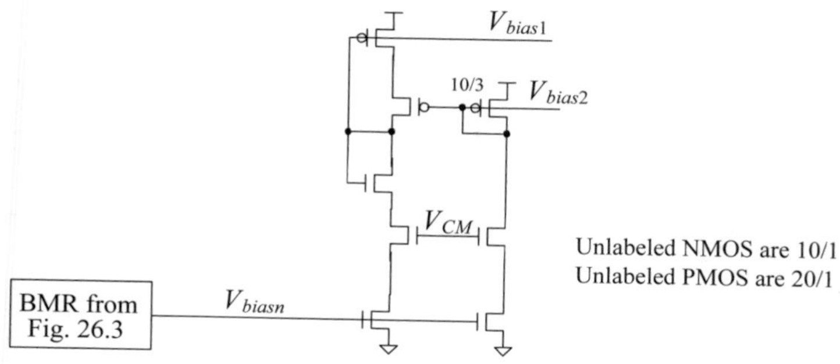

Figure 26.52 Biasing circuit for the op-amp developed in this section.

The Op-Amp's First Stage

As just mentioned, the output of the SC CMFB circuit, V_{CMFB} , moves in the same direction as movement in the average of the amplifier's outputs, v_{op} and v_{om} (the output common-mode level). Keeping this in mind, consider the bias circuit and diff-amp schematics seen in Figs. 26.52 and 26.53. We've separated the bias circuit out from the diff-amp, see Fig. 26.43, for a couple of reasons. To begin, an increase in the CMFB signal in Fig. 26.43 caused the outputs of the diff-amp to increase. As just mentioned, we want the diff-amp's outputs to move in the opposite direction of V_{CMFB}. Moving the CMFB-controlled MOSFET into the tail current of the diff-amp provides this control. Next, our bias circuit may now be shared with several op-amps. The benefit of sharing the bias circuit is a reduction in power dissipation.

Before adding the CMFB circuit, let's look at a SC circuit using an op-amp. Reviewing the sample and hold in Fig. 26.25, we see that when the ϕ_1 switches are closed, the op-amp's inputs and outputs are at (ideally) V_{CM}. The op-amp, during this

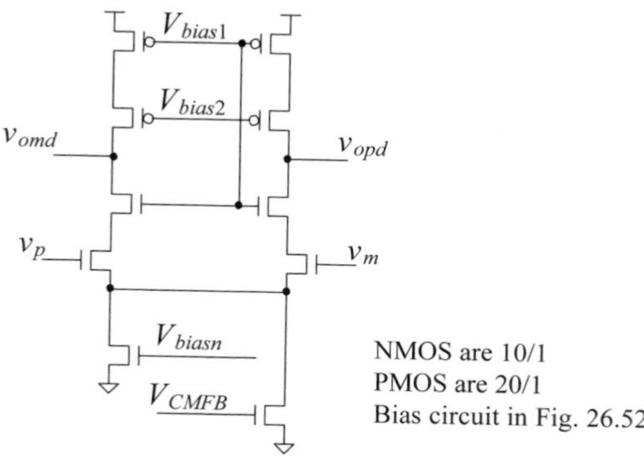

Figure 26.53 Diff-amp used with the bias circuit of Fig. 26.52.

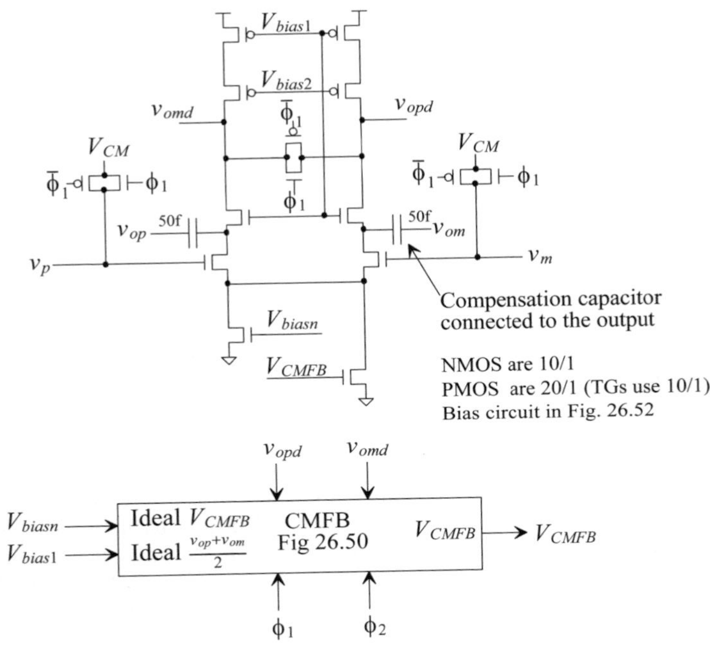

Figure 26.54 First-stage diff-amp with SC CMFB.

time, is in the unity-follower configuration. Instead of placing the op-amp in the follower configuration during this time, let's: 1) short the inputs to op-amp (Fig. 26.54) to V_{CM}, 2) short the outputs of the diff-amp (the inputs to the output buffer) together, 3) use a SC CMFB circuit around the diff-amp to ensure a balance condition (and to set the diff-amp's output voltage to V_{bias1}), 4) short the outputs of the op-amp together (so now both the output buffers inputs are shorted and outputs are shorted), and 5) use a SC CMFB around the output buffer to ensure balanced outputs. In other words, during ϕ_1, the differential inputs are shorted together. At the same time, the outputs of the op-amp are shorted together.

Figure 26.55 shows the simulated output of the diff-amp in Fig. 26.54. The SC CMFB circuit sets the outputs of the diff-amp at V_{bias1} or roughly 600 mV. By shorting the two diff-amp outputs together and connecting the inputs to V_{CM}, we are ensuring that even with horrible mismatch, the outputs of the diff-amp will be equal and can be used to bias the output buffer. If, for example, the diff-pair shows a 50 mV mismatch (V_{OS}) then, when ϕ_1 is high, the drain currents in each side of diff-amp are set differently (making this scheme very tolerant to offsets). When ϕ_1 goes low and the op-amp is placed in a feedback configuration, each of the diff-amp's inputs will move (in opposite directions) by $V_{OS}/2$ (and so this topology doesn't store the op-amp's offset voltage).

Note that an important concern, as discussed in the last section, is the input common-mode voltage of the diff-amp. If, for example, the input common-mode voltage

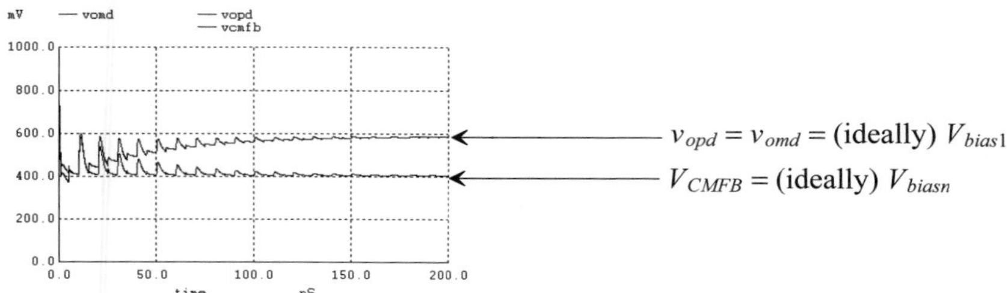

Figure 26.55 Simulating the operation of the circuit in Fig. 26.54.

drops from 500 mV, in Fig. 26.54, to 400 mV the NMOS devices will start shutting off. The op-amp will not settle or behave properly.

The Output Buffer

Figure 26.56 shows a schematic of the output buffer and SC CMFB stage. The nodes v_{omd} and v_{opd} are connected to the diff-amp outputs in Fig. 26.54. When ϕ_1 is high, the outputs of the buffer are shorted together. The SC CMFB is used to set the outputs to the common-mode voltage during this time. Figure 26.57 shows the simulation results for both stages of the op-amp (the op-amp is made with Figs. 26.54 and 26.56) with the inputs and outputs floating. When ϕ_1 is high, the inputs to the op-amp are connected to V_{CM}.

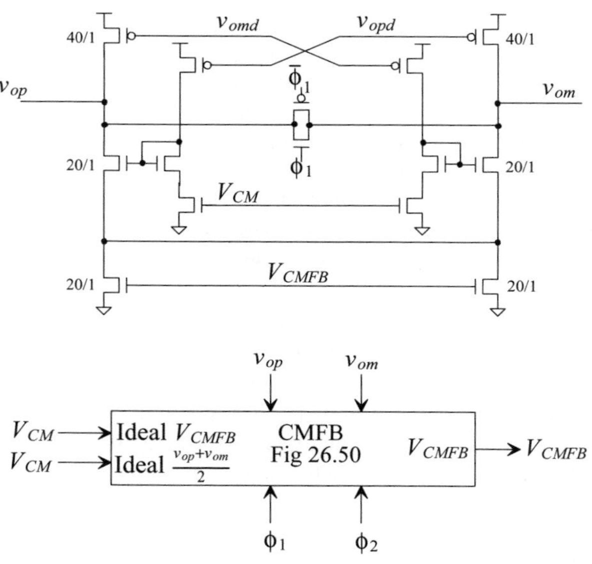

Figure 26.56 Output buffer and CMFB.

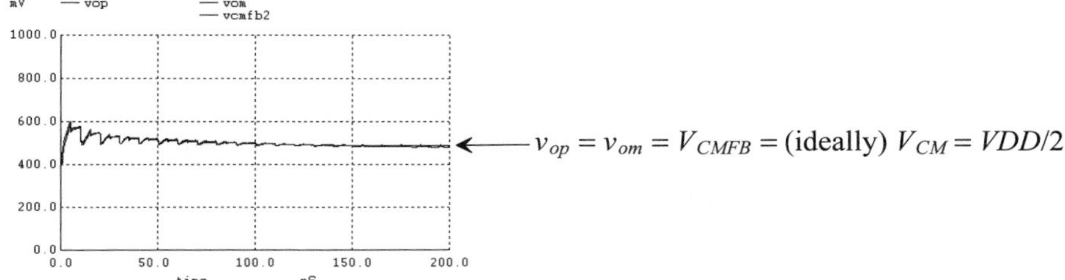

Figure 26.57 Simulating the operation of the op-amp (Figs. 26.54 and 26.56).

Is shorting the outputs of the buffer (or diff-amp) necessary? If the op-amp is operating open-loop (when ϕ_1 goes high), we don't want the outputs to float. By shorting the diff-amp and output buffer output terminals together, we ensure that they are set to a known value. If we were to use this op-amp in a SC integrator, we wouldn't short the inputs of the diff-amp to V_{cm}, the outputs of the diff-amp together, or the output buffer outputs together since there is always feedback around the op-amp. In other words, the same op-amp topology can be used in a SC integrator but without the three TGs seen in Figs. 26.54 and 26.56.

An Application of the Op-Amp

An application of the op-amp we've just developed is seen in Fig. 26.58. This is the sample and hold developed earlier (Figs. 25.15 and 26.25) except that now, during ϕ_1 (high) the inputs to the op-amp are shorted to V_{CM} (Fig. 26.54) and the outputs are shorted together (Fig. 26.56). When ϕ_2 goes high (ϕ_1 is low since the two clocks are nonoverlapping), the op-amp moves into the follower configuration and holds its outputs at the value of the inputs when the ϕ_1 switches shut off.

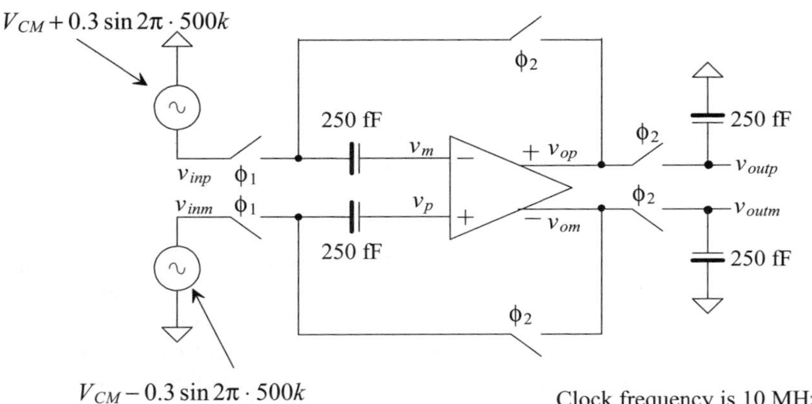

Figure 26.58 Simulating the operation of the op-amp formed with the diff-amp in Fig. 26.54 and buffer in Fig. 26.56.

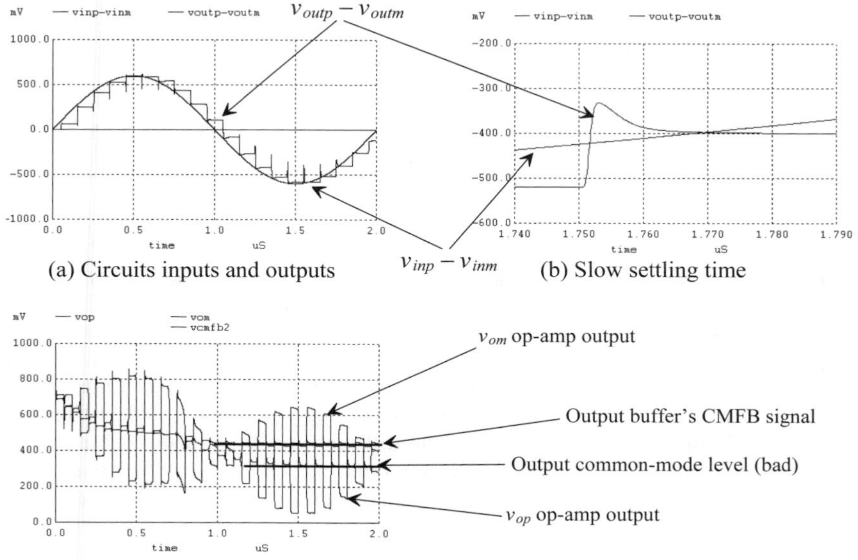

(a) Circuits inputs and outputs

(b) Slow settling time

(c) Op-amp's output and buffer's CMFB signal

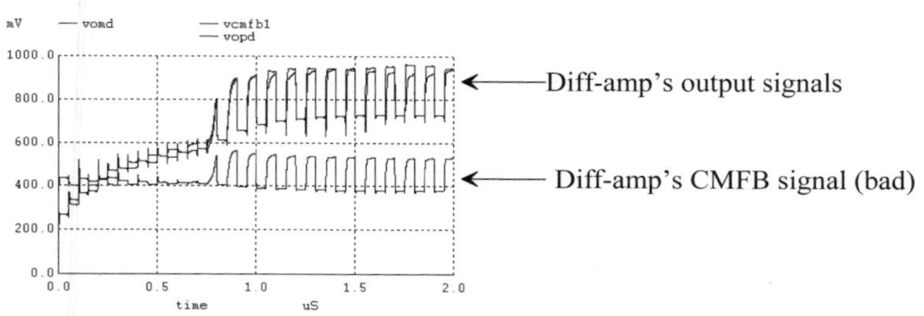

(d) Diff-amp's outputs and CMFB signal

Figure 26.59 Simulating the operation of the circuit in Fig. 26.58 with the op-amp made with the diff-amp in Fig. 26.54 and the output buffer in Fig. 26.56.

Simulation Results

The simulation results are seen in Fig. 26.59. To ensure that we see the full settling behavior of the op-amp and any other issues or concerns, we use a slow clock (10 MHz) for the initial simulations. As we gain confidence that our op-amp is settling correctly and stable, we can increase the clock frequency and the input signal frequency to see how the sample and hold performs. The main thing that's different with this design compared to previous designs is that the op-amp isn't operating with feedback all of the time. When the op-amp does get put into a feedback configuration (ϕ_2 goes high), there will be a start-up time (e.g., the inputs of the op-amp will move away from V_{CM} because of the op-amp's offset).

Returning to the simulation results in Fig. 26.59, we see, in (b), that the settling time is quite long (approximately 10 ns). Letting the simulation run for a longer period of time, we see that the settling time increases and, ultimately, that the op-amp stops functioning correctly. In (c) we see the problem: the outputs of the op-amp aren't balanced around V_{CM}. Further, movement downwards causes the common-mode voltage on the input of the diff-amp to drop. The result is that the input diff-amp starts to shut off. The voltage across the diff-amp's tail current source drops and makes it difficult to control the diff-amp's output common-mode level. As seen in (d), the diff-amp's CMFB signal is increasing in an attempt to increase the current flowing in the diff-amp and pull the diff-amp's output voltages downwards. However, because the common-mode voltage on the input of the diff-amp has decreased, the voltage across the tail currents is small (tens of mV). For example, if the input common-mode voltage drops from 500 mV (V_{CM}) to 450 mV and the V_{GS} of the input diff-amp is 400 mV, then only 50 mV is left to drop across the tail current sources (one biased with V_{biasn} and the other with V_{CMFB}). We've discussed this problem earlier.

We can do one of two things to make this decrease in input common-mode voltage less of a problem: 1) reduce the diff-amp's bias current or 2) increase the widths of the input diff-pair. In both cases we need to increase the voltage dropped across the tail current sources by decreasing the V_{GS} of the NMOS diff-pair (the inputs to the diff-amp are, ideally, at V_{CM}). If we decrease the bias current, the g_m of the diff-pair decreases and so does the speed of the op-amp (remembering $f_{un} = g_m/2\pi C_c$). By increasing the width of the diff-pair, we increase g_m and thus the unity-gain frequency of the op-amp. The issue with increasing the width of the diff-pair is that the overdrive voltage for these MOSFETs decreases and thus so does their f_T. If we keep the increase in the width to a modest level, the parasitic pole associated with these diff-pair MOSFETs shouldn't, in any significant way, affect the stability of the op-amp.

Increasing the Widths of the Diff-Pair

Figure 26.60 shows the operation of the op-amp made with the diff-amp in Fig. 26.54 and the output buffer in Fig. 26.56 if we increase the widths of the diff-pair from 10 to 30. We also change the MOSFETs in the biasing circuit Fig. 26.52 with gates to V_{CM} to widths of 30 to maintain symmetry (important for reducing the offset voltage). The settling time, as seen in (b), is approximately 4 ns. In (c) we see the outputs are swinging around the correct value (V_{CM}). In (d) the outputs of the diff-amp are swinging around V_{bias1} and the common-mode feedback voltage is relatively stable.

Reviewing the op-amp designs in the last section, we see that their performance may also be improved by a small increase in the widths of the diff-pair (with an increase in the compensation capacitors as needed). However, as just mentioned, we have to be careful to maintain symmetry. For example, if we take the op-amp in Fig. 26.43, increase the widths of the first stage diff-pair from 10 to 30, and simulate, we see that the outputs don't swing all the way up (or down) to the correct values. Further investigation reveals that the diff-pair moved into the triode region. To move the MOSFETs back into the saturation region, we must also increase the widths of the MOSFETs in the bias circuit (with gates at V_{CM}) from 10 to 30. This adjusts the bias voltages for the NMOS/PMOS diff-amp loads. By providing the CMFB back through the diff-amp (instead of through the output buffer), we can adjust the diff-amp's output voltage. Thus the op-amp is more tolerant to issues in the bias circuit.

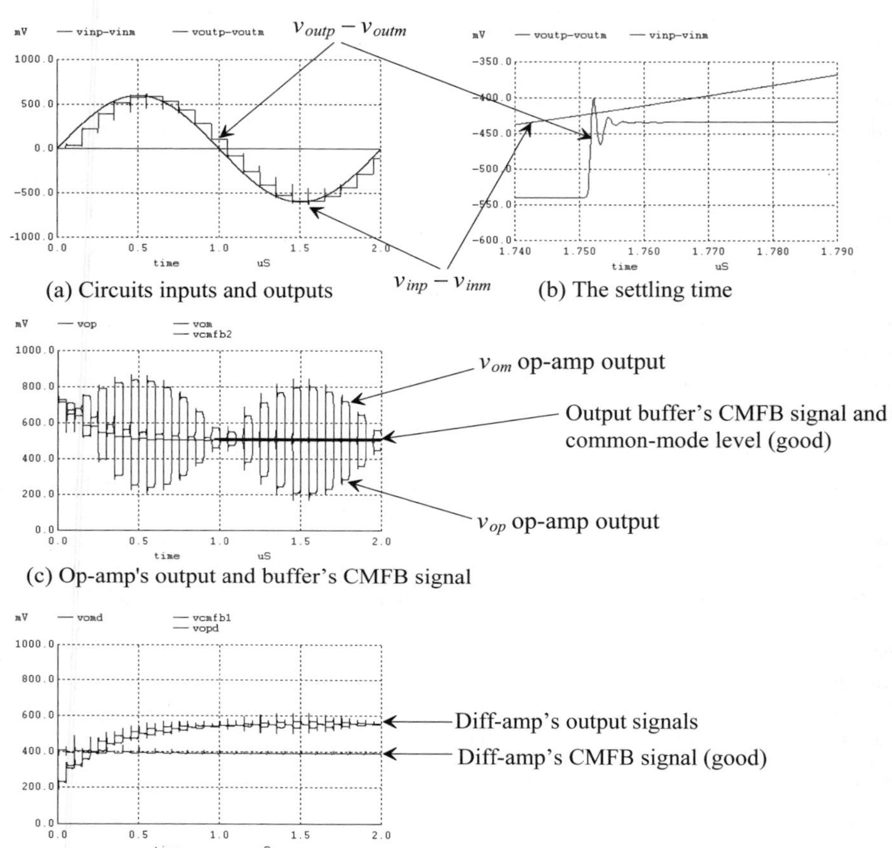

Figure 26.60 Regenerating the data in Fig. 26.59 after increasing the width of the input diff-pair to 30 (from 10).

A Final Note Concerning Biasing

In (most) of the designs presented in this chapter, we biased the gates of the NMOS cascode devices in the diff-amp's load at the same (ideally) voltage as the diff-amp's output. For example, in Fig. 26.53, the gate of the NMOS device is tied to V_{bias1}, while its drain (the output of the diff-amp) is also (ideally) at V_{bias1}. This selection is fine for general design. It minimizes power by avoiding an extra bias reference circuit. However, if the power supply voltage, *VDD*, starts to drop, the drain-source voltages across the diff-pair and tail current sources can drop to the point where the op-amp shuts off. Again, increasing the widths of the diff-pair helps with this concern. A more general solution is to bias the gates of the NMOS with a gate-drain-connected NMOS, as seen in Fig. 22.30. This allows the gate voltage of the NMOS to move to a higher voltage than its drain (allowing the op-amp to function with lower *VDD*). For example, the diff-amp in Fig. 26.53 has a DC output voltage of roughly 600 mV (= V_{bias1}). A more appropriate voltage for the gate of the NMOS device is 800 mV. Remembering that the gate-source voltages of the NMOS diff-pair (Fig. 26.18) are roughly 400 mV puts the drains of the diff-pair at 400 mV. The result: *VDD* is more evenly divided across the devices in the diff-amp.

ADDITIONAL READING

[1] V. S. L. Cheung, H. C. Luong, M. Chan, and W. H. Ki, "A 1-V 3.5mW CMOS Switched-Opamp Quadrature IF Circuitry for Bluetooth Receivers," *Symp. VLSI Circuits Dig.* 16, pp. 140–143, June 2002.

[2] M. Keskin, U. Moon, and G. C. Temes, "A 1-V 10-MHz clock-rate 13-bit CMOS DS modulator using unity-gain-reset opamps," *IEEE Journal of Solid-State Circuits,* vol. 37, pp. 817–824, July 2002.

[3] V. S. L. Cheung, H. C. Luong, and W. H. Ki, "A 1V CMOS switched-opamp switched-capacitor pseudo-2-path filter," *IEEE International Solid-State Circuits Conference,* vol. 35, pp. 154–155, February 2000.

[4] I. E. Opris, L. D. Lewicki, and B. C. Wong, "A single-ended 12-bit 20 Msample/s self-calibrating pipeline A/D converter," *IEEE Journal of Solid-State Circuits,* vol. 33, pp. 1898–1903, December 1998.

[5] D. C. Thelen Jr. and D. D. Chu, "A low noise readout detector circuit for nanoampere sensor applications," *IEEE Journal of Solid-State Circuits,* vol. 32, pp. 337–348, March 1997.

[6] P. D. Walker and M. M. Green, "A tuneable pulse-shaping filter for use in a nuclear spectrometer system," *IEEE Journal of Solid-State Circuits,* vol. 31, pp. 850–855, June 1996.

[7] G. Caiulo, F. Maloberti, G. Palmisano, and S. Portaluri, "Video CMOS power buffer with extended linearity," *IEEE Journal of Solid-State Circuits,* vol. 28, pp. 845–848, July 1993.

[8] K. Nakamura and L. R. Carley, "An enhanced fully differential folded-cascode op amp," *IEEE Journal of Solid-State Circuits,* vol. 27, pp. 563–568, April 1992.

[9] R. Castello, G. Nicollini, and P. Monguzzi, "A high-linearity 50-Ω CMOS differential driver for ISDN applications," *IEEE Journal of Solid-State Circuits,* vol. 26, pp. 1809–1816, December 1991.

[10] S. M. Mallya and J. H. Nevin, "Design procedures for a fully differential folded-cascode CMOS operational amplifier," *IEEE Journal of Solid-State Circuits,* vol. 24, pp. 1737–1740, December 1989.

[11] M. Banu, J. M. Khoury, and Y. Tsividis, "Fully Differential Operational Amplifiers with Accurate Output Balancing," *IEEE Journal of Solid State Circuits,* vol. 23, No. 6, pp. 1410–1414, December 1988.

[12] R. Castello and P. R. Gray, "A high-performance micropower switched-capacitor filter," *IEEE Journal of Solid-State Circuits,* vol. 20, pp. 1122–1132, Dec. 1985.

[13] T. C. Choi, R. T. Kaneshiro, R. Broderson, and P. R. Gray, "High-Frequency CMOS Switched Capacitor Filters for Communication Applications," *IEEE Journal of Solid State Circuits,* vol. SC-18, pp. 652–664, December 1983.

[14] D. Senderowicz, S. F. Dreyer, J. H. Huggins, C. F. Rahim, and C. A. Laber, "A family of differential NMOS analog circuits for a PCM codec filter chip," *IEEE Journal of Solid-State Circuits,* vol. 17, pp. 1014–1023, December 1982.

PROBLEMS

Unless otherwise indicated, use the 50 nm CMOS process from the examples in this chapter, the biasing circuit in Fig. 26.3, and 10/1 NMOS and 20/1 PMOS.

26.1 Using simulations, determine the transition frequencies, f_T , for the NMOS and PMOS devices seen in Fig. 26.1 at the nominal operating conditions indicated in the figure. Show that by increasing the MOSFET's overdrive voltage, the f_T s of the MOSFETs increases.

26.2 Simulate the operation of the two-stage op-amp in Fig. 26.2. Show that the quiescent current in the output buffers is considerably below the desired 20 μA. (Note that the inputs of the op-amp should be held at V_{CM} in the simulation to keep the diff-amp conducting current.) Does this affect the speed of the output buffer? Why or why not?

26.3 Plot reference current against resistor value for the BMR seen in Fig. 26.3. Use simulations to determine the I_{REF} for each value of resistance.

26.4 Comment on the benefits and/or concerns with the CMFB input connections seen in Fig. 26.61. Use simulations to support your answers. Note the similarity to the way CMFB was implemented in Fig. 26.17.

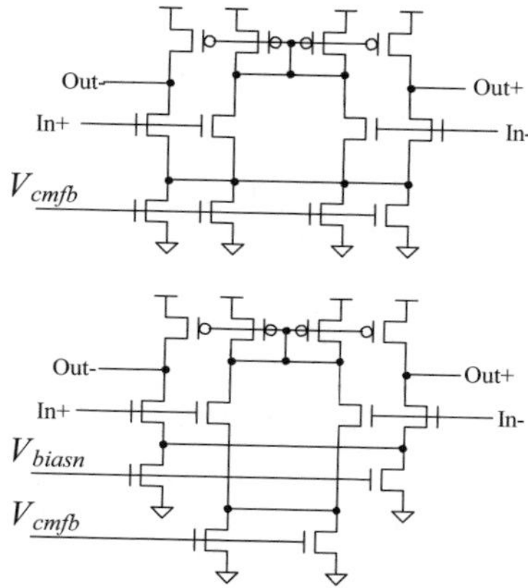

Figure 26.61 Circuits for Problem 26.4.

26.5 Verify, using simulations, that the circuit in Fig. 26.9 does indeed amplify the difference between V_{biasp} and the average on the + inputs of the amplifier.

Comment on the operation of the circuit, making sure it is clear that the limitations, uses, and operation of the amplifier are understood.

26.6 Simulate the operation of the CMFB circuit in Fig. 26.16.

26.7 Is *VDD* divided evenly amongst the drain-source voltages of the MOSFETs in Fig. 26.19? Suggest a method to better divide *VDD* amongst the MOSFETs used in the diff-amp.

26.8 Suggest a simple method to speed up the step response of the op-amp in Fig. 26.22 in the circuit of Fig. 26.24. Verify the validity of your suggestion using SPICE simulations.

26.9 Suggest an alternative method to the one seen in Fig. 26.32 for controlling the output buffer's current. Using the modification, regenerate the results seen in Fig. 26.36.

26.10 Is the CMFB loop stable in the op-amp of Fig. 26.40? Use the large-signal test circuit seen in Fig. 26.24 (at a slower frequency) and simulations to look at the stability of this loop. Suggest, and verify with simulations, methods to improve the stability of the CMFB loop.

26.11 Repeat Problem 26.10 for the op-amp in Fig. 26.43.

26.12 As discussed at the end of the chapter and in Fig. 26.59 and the associated discussion, the input common-mode voltage which drops below V_{CM}, can shut off the op-amp's input diff-amp and cause problems. The diff-amp's NMOS devices have a nominal V_{GS} of 400 mV, leaving only 100 mV across the diff-amp's tail current. Show, using the op-amp in Fig. 26.39 in the configuration seen in Fig. 26.24 with an input common-mode voltage less than 500 mV (the input pulse waveforms average to a voltage less than V_{CM}), the resulting problems. Show that increasing the widths of the NMOS diff-pair (and the mirrored NMOS in the bias circuit with gates tied to V_{CM}) from 10 to 30 helps to increase the operating range (four MOSFET widths are increased from 10 to 30). What happens if only the diff-pair widths (two MOSFETs) are increased?

26.13 Repeat Problem 26.12 for the op-amp in Fig. 26.40. What happens if the input common-mode voltage becomes greater than 500 mV?

26.14 Repeat Problem 26.12 for the op-amp in Fig. 26.43. What happens if the input common-mode voltage becomes greater than 500 mV? Why does the op-amp in Fig. 26.40 perform better with variations in the input common-mode voltage than the op-amp in Fig. 26.43?

26.15 Design and simulate the operation of an op-amp using gain-enhancement (Fig. 26.42) and with an open-loop DC gain greater than 2,000, based on the topologies seen in Figs. 26.40 or 26.43. Simulate the operation of your design and generate outputs like those seen in Fig. 26.44.

26.16 Figure 26.62 shows an op-amp based on the topology seen in Fig. 26.40 but biased for lower *VDD* operation. Select the size of the added gate-drain connected MOSFET to allow for proper operation. Simulate the operation of the design showing the DC gain and large signal step responses (as in Fig. 26.34 and 26.35).

26.17 Repeat Problem 26.16 for the op-amp in Fig. 26.43.

26.18 Using the diff-amp in Fig. 26.63 in the configuration seen in Fig. 26.54 (with SC CMFB) and the output buffer in Fig. 26.56 (again with SC CMFB), regenerate the waveforms seen in Fig. 26.60.

26.19 A switched-capacitor integrator is an example of a circuit that uses an op-amp that can't have the outputs of its diff-amp or output buffer shorted when ϕ_1 goes high. Using the diff-amp in Fig. 26.63 (with SC CMFB) and the output buffer in Fig. 26.56 without the switches (again with SC CMFB), demonstrate the operation of a SC integrator. For examples of fully-differential SC circuits, see the second volume of this book entitled, *CMOS Mixed-Signal Circuit Design*.

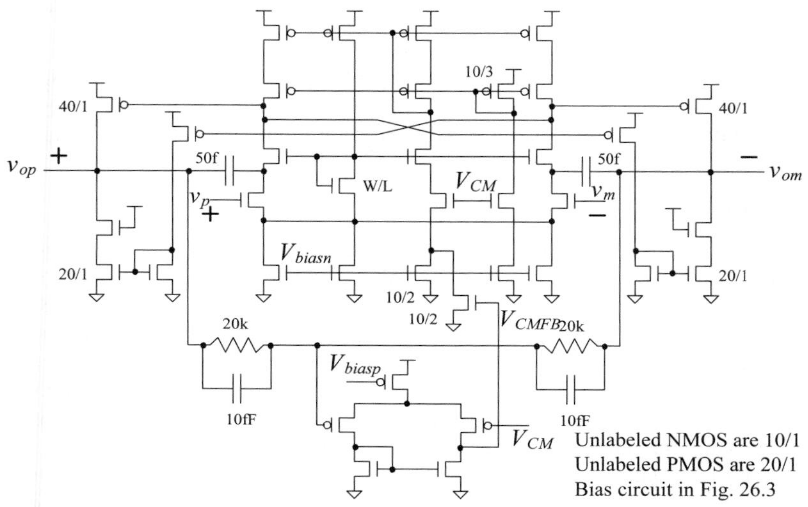

Figure 26.62 An op-amp with better biasing (but more power) for lower VDD operation.

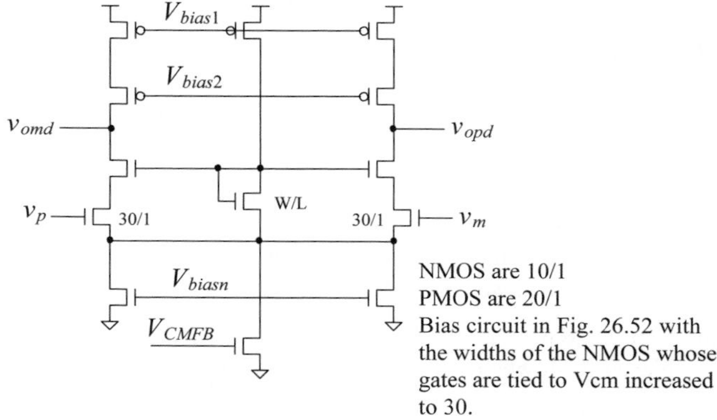

Figure 26.63 Modifying the diff-amp seen in Fig. 26.53 for wider swing operation.

Chapter

27

Nonlinear Analog Circuits

In this book we've studied digital circuits (two discrete amplitude levels) and linear analog circuits (circuits whose input signals are linearly related to the circuits' output signals). In this chapter, we discuss several circuits that are not purely analog or digital. We term these circuits *nonlinear analog circuits* (the inputs are not linearly related to the outputs). In particular, we discuss voltage comparator analysis and design, adaptive biasing, and analog multiplier design.

27.1 Basic CMOS Comparator Design

The schematic symbol and basic operation of a voltage comparator are shown in Fig. 27.1. The comparator can be thought of as a decision-making circuit. If the +, v_p, input of the comparator is at a greater potential than the –, v_m, input, the output of the comparator is a logic 1, whereas if the + input is at a potential less than the – input, the output of the comparator is at a logic 0. Although the basic op-amps developed in Ch. 24 can be used as a voltage comparator, in some less demanding low-frequency or speed applications, we do not consider the op-amp as a comparator. Instead, in this chapter, we discuss practical comparator design and analysis where propagation delay and sensitivity are important.

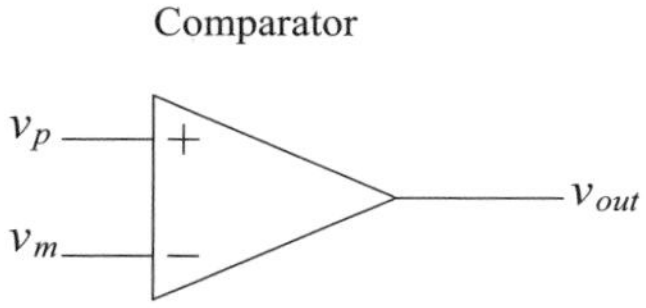

$v_p > v_m$ then $v_{out} = VDD =$ logic 1

$v_p < v_m$ then $v_{out} = 0 =$ logic 0

Figure 27.1 Comparator operation.

A block diagram of a high-performance comparator is shown in Fig. 27.2. The comparator consists of three stages: the input preamplifier, a positive feedback or decision stage, and an output buffer. The pre-amp stage (or stages) amplifies the input signal to improve the comparator sensitivity (i.e., increases the minimum input signal with which the comparator can make a decision) and isolates the input of the comparator from switching noise (often called kickback noise) coming from the positive feedback stage (*this is important*, see the discussion in Sec. 16.2.1). The positive feedback stage determines which of the input signals is larger. The output buffer amplifies this information and outputs a digital signal. Designing a comparator can begin with considering input common-mode range, power dissipation, propagation delay, and comparator gain.

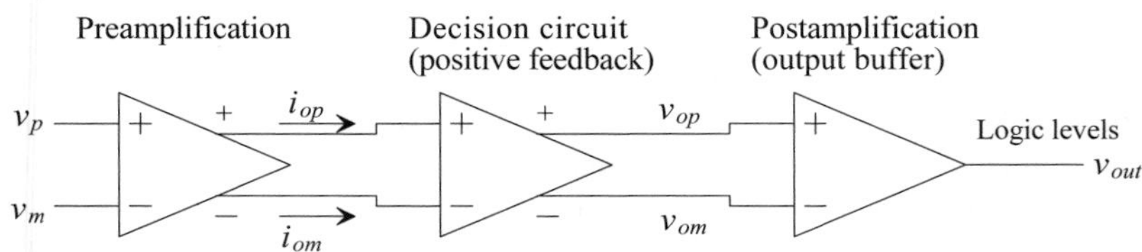

Figure 27.2 Block diagram of a voltage comparator.

Preamplification

For the preamplification (pre-amp) stage, we chose the circuit of Fig. 27.3. For this first section we'll use the long-channel CMOS process to illustrate the design procedures (since long-channel MOSFETs follow the square-law equations). This circuit is a differential amplifier with active loads. The sizes of M1 and M2 are set by considering the diff-amp transconductance, g_m, and the input capacitance. The transconductance sets the gain of the stage, while the input capacitance of the comparator is determined by the sizes of M1 and M2. Notice that there are no high-impedance nodes in this circuit, other than the input and output nodes. This is important to ensure high speed. Using the sizes given in the schematic, we can relate the input voltages to the output currents (noting i_{op} and i_{om} are the small-signal AC currents in the circuit) by

$$i_{op} = \frac{g_m}{2}(v_p - v_m) + \frac{I_{SS}}{2} = I_{SS} - i_{om} \quad (27.1)$$

Noting that if $v_p > v_m$, then i_{op} is positive i_{om} is negative ($i_{op} = -i_{om}$).

Decision Circuit

The decision circuit is the heart of the comparator and should be capable of discriminating mV-level signals. We should also be able to design the circuit with some hysteresis (see Ch. 18) for use in rejecting noise on a signal. The circuit that we use in the comparator under development is shown in Fig. 27.4. The circuit uses positive feedback from the cross-gate connection of M6 and M7 to increase the gain of the decision element.

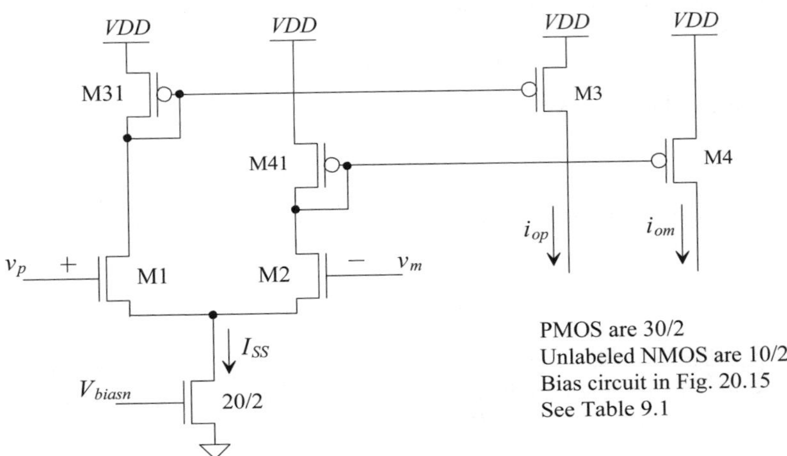

Figure 27.3 Preamplification stage of comparator.

Let's begin by assuming that i_{op} is much larger than i_{om} so that M5 and M7 are on and M6 and M8 are off. We also assume that $\beta_5 = \beta_8 = \beta_A$ and $\beta_6 = \beta_7 = \beta_B$. Under these circumstances, v_{om} is approximately 0 V and v_{op} is

$$v_{op} = \sqrt{\frac{2i_{op}}{\beta_A}} + V_{THN} \tag{27.2}$$

If we start to increase i_{om} and decrease i_{op}, switching starts to take place when the gate-source voltage of M8 is equal to V_{THN}. As we increase M8's V_{GS} beyond V_{THN} (by further increasing i_{om} with the corresponding decrease in i_{op}), M6 starts to take current away from M5. This decreases the drain-source voltage of M5/M6 and thus turns M7 off.

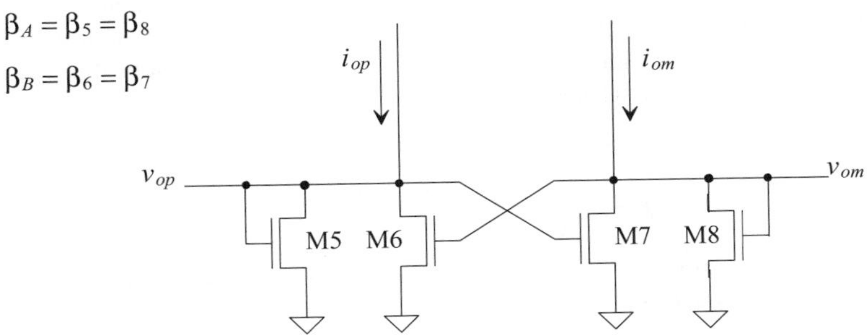

Figure 27.4 Positive feedback decision circuit.

When M8 is just about to turn on (M8's V_{GS} is approaching V_{THN} but the drain currents of M8 and M6 are still zero), the current flowing in M7 is

$$i_{om} = \frac{\beta_B}{2}(v_{op} - V_{THN})^2 \tag{27.3}$$

and the current flowing in M5 is

$$i_{op} = \frac{\beta_A}{2}(v_{op} - V_{THN})^2 \tag{27.4}$$

Noting that the current in M7 (at the switching point) mirrors the current in M5, we can write

$$i_{op} = \frac{\beta_A}{\beta_B} \cdot i_{om} \tag{27.5}$$

If $\beta_A = \beta_B$, then switching takes place when the currents, i_{op} and i_{om}, are equal. Unequal βs cause the comparator to exhibit hysteresis. Relating these equations to Eq. (27.1) yields the switching point voltages (review Ch. 18), or

$$V_{SPH} = v_p - v_m = \frac{I_{SS}}{g_m} \cdot \frac{\frac{\beta_B}{\beta_A} - 1}{\frac{\beta_B}{\beta_A} + 1} \text{ for } \beta_B \geq \beta_A \tag{27.6}$$

and

$$V_{SPL} = -V_{SPH} \tag{27.7}$$

Consider the following example.

Example 27.1

For the circuit shown in Fig. 27.5, estimate and simulate the switching point voltages for two designs: (1) $W_5 = W_6 = W_7 = W_8 = 10$ with $L = 1$ and (2) $W_5 = W_8 = 10$ and $W_6 = W_7 = 12$ with $L = 1$.

For the first case (using the data in Table 9.1)

$$\beta_A = \beta_B = 120\,\frac{\mu A}{V^2} \cdot \frac{10}{1} = 1.2\ mA/V^2$$

so that, as seen in Eq. (27.6) $V_{SPH} = V_{SPL} = 0$. In other words, the comparator does not exhibit hysteresis. Simulation results are shown in Fig. 27.6. The input v_m is set to 2.5 V, while the v_p input is swept from 2.48 to 2.52 V. Note that the amplitudes of v_{op} and v_{om} are limited.

For the second case

$$\beta_A = 120\,\frac{\mu A}{V^2} \cdot \frac{10}{1} \text{ and } \beta_B = 120\,\frac{\mu A}{V^2} \cdot \frac{12}{1}$$

or $\beta_B = 1.2\beta_A$. Using Eqs. (27.6) and (27.7) with $I_{SS} = 40\ \mu A$ and $g_m = 150\ \mu A/V$ (again, see Table 9.1), we get

$$V_{SPH} = -V_{SPL} = \frac{I_{SS}}{g_m} \cdot \frac{\frac{\beta_B}{\beta_A} - 1}{\frac{\beta_B}{\beta_A} + 1} = \frac{40}{150} \cdot \frac{1.2 - 1}{1.2 + 1} = 24\ mV$$

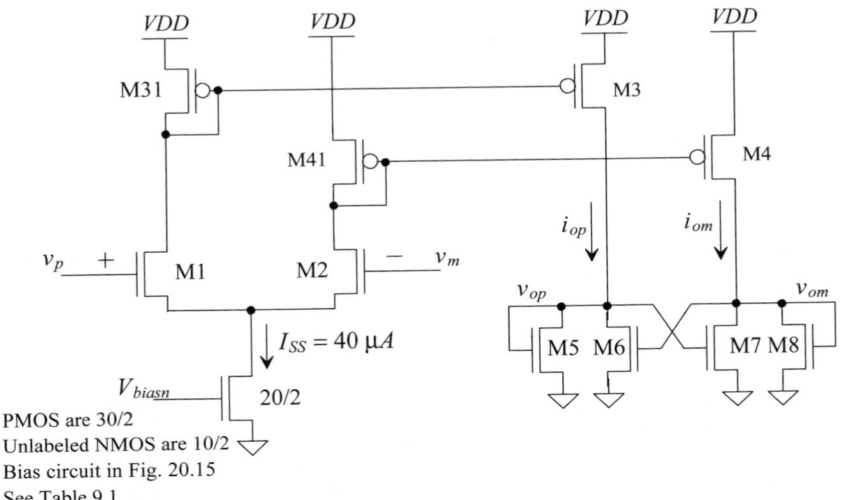

Figure 27.5 Schematic of the pre-amp and decision circuit used in Ex. 27.1.

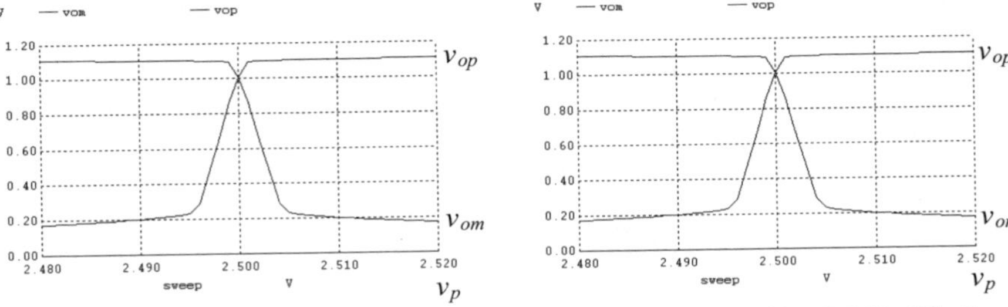

(a) Minus input held at 2.5 V while the positive input is swept from 2.4 to 2.6 V.

(b) Minus input held at 2.5 V while the positive input is swept from 2.52 to 2.4.

Figure 27.6 The outputs of the decision circuit in Fig. 27.5 without hysteresis.

The simulation results are shown in Fig. 27.7. Figure 27.7a shows a sweep of v_p from 2.4 to 2.6 V, with v_m held at 2.5 V. Since V_{SPH} is 24 mV, the decision circuit switches states when v_p is 24 mV above v_m (i.e., when $v_p = 2.524$ V). The case of v_p being swept from 2.6 to 2.4 V is shown in part (b) of the figure. Switching occurs in this situation when v_p is approximately 2.476 V, or 24 mV, less than 2.5 V. ■

Output Buffer

The final component in our comparator design is the output buffer or post-amplifier. The main purpose of the output buffer is to convert the output of the decision circuit into a logic signal (i.e., 0 or *VDD*). The output buffer should accept a differential input signal

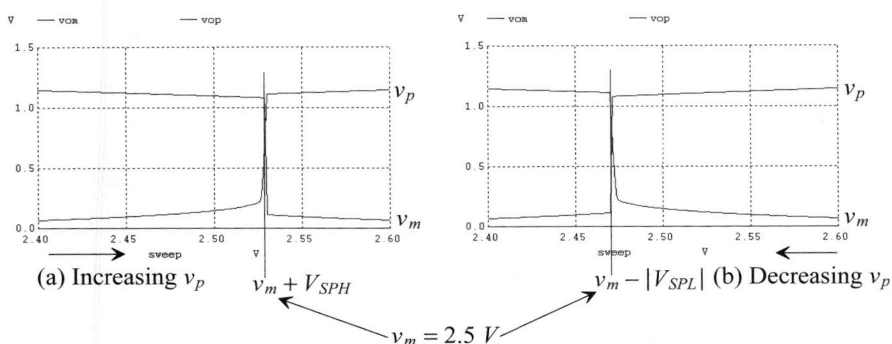

Figure 27.7 Simulating the pre-amp and decision circuit with hysteresis.

and not have slew-rate limitations. We might try to use a NAND SR latch as seen back in Fig. 16.35. However, with process shifts we might run into the situation where the switching points of the gates aren't centered in the middle of the limited swing of the decision circuit's outputs. For a general comparator design, we'll use a diff-amp to help regenerate the digital output signals. For a simple design for the output buffer, we can use the self-biased diff-amp seen back in Fig. 18.17. To move the decision circuit's output swing into the common-mode range of the diff-amp, we can add a gate-drain connected device as seen in Fig. 27.8. An inverter was added on the output of the amplifier as an additional gain stage and to isolate any load capacitance from the self-biasing differential amplifier. This comparator works very well. However, the current in the self-biased stage can be (relatively) huge. This can be a problem if power is a concern (as it is in the Flash ADC discussed in Ch. 29).

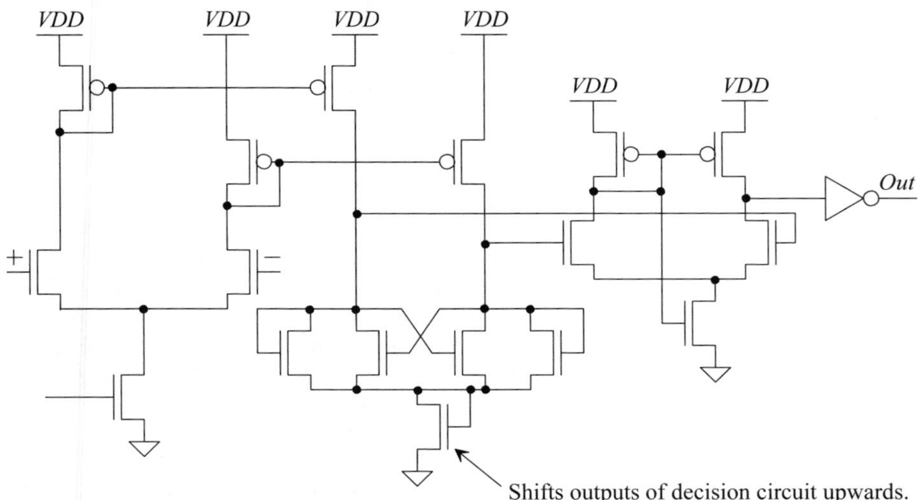

Figure 27.8 Using a self-biased diff-amp for the output buffer.

Figure 27.9 shows a comparator design with both a PMOS and NMOS diff-amp feeding the decision circuit (for input common-mode range beyond the power supply rails). The output buffer is a PMOS diff-amp driving an inverter. While this op-amp won't be as fast as the op-amp seen in Fig. 27.8, it is a good general purpose design. The current drawn from *VDD* will be considerably less than the design in Fig. 27.8. Note that the output currents of the PMOS diff-amp, on the input of the comparator, are added to the NMOS (current) outputs to ensure that current is always sourced to the decision circuit. We might be tempted to connect the PMOS diff-amp directly to the decision circuit. However, this would result in a larger minimum input common-mode voltage.

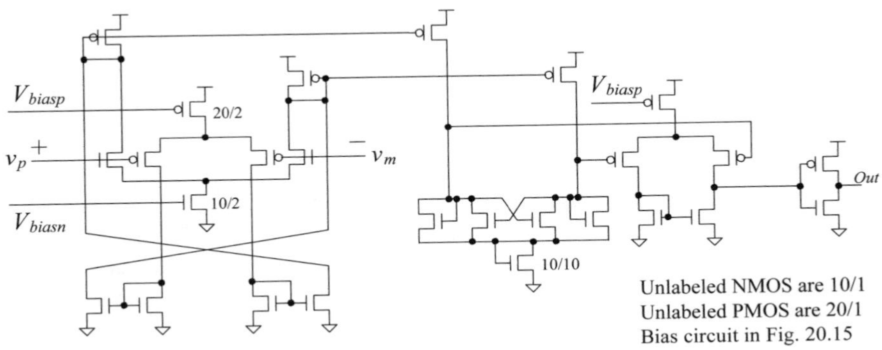

Figure 27.9 General-purpose comparator with rail-to-rail input common-mode range.

27.1.1 Characterizing the Comparator

Comparator DC Performance

Figure 27.10 shows the DC simulated performance of the comparator in Fig. 27.9. The positive comparator input is swept from 0 to *VDD* (= 5 V) while the negative input is stepped for each simulation from 0 to 5 in 500 mV increments (to show wide input common-mode range). Also seen in this figure is the comparator's current draw from *VDD*. Included in this current is the current supplied to the bias circuit.

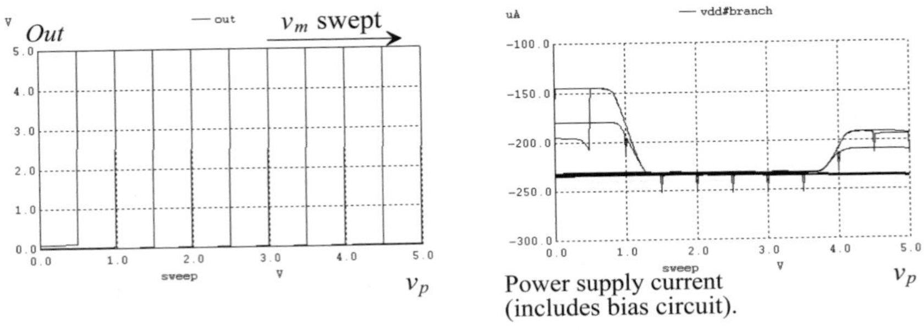

Figure 27.10 DC performance of the comparator in Fig. 27.9

If we take the derivative of these transfer curves, the gain of the comparator and thus the smallest difference that can be discriminated between v_p and v_m becomes known. We have to be careful here with the step size used in the DC simulation. If, for example, we were to use a step size of 1 mV, then the maximum gain we get from the simulation is 1,000. Figure 27.11 shows the expanded view of the comparator with the minus input held at 2.5 V while the positive input is swept from 2.499 to 2.501 V. The offset seen in the figure is a systematic offset (as discussed earlier). The gain of the comparator is, roughly, 175,000. Notice that the comparator was designed without hysteresis. However, in a practical comparator mismatch in the decision circuit results in hysteresis (and so if hysteresis isn't desired, the layout of the decision circuit is critical and may require common-centroid techniques).

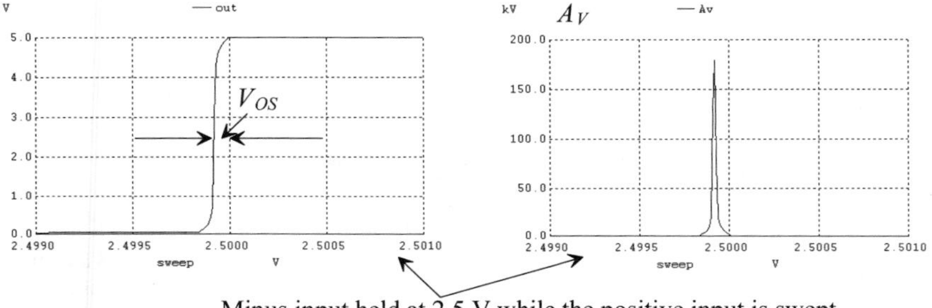

Figure 27.11 The gain of the comparator in Fig. 27.9

Transient Response

The transient response of a comparator can be significantly more difficult to characterize than the DC characteristics. Let's begin by considering v_m = 2.5 V DC, with the v_p input to the comparator a 10 ns wide pulse and an amplitude varying from 2.45 to 2.55 V. This is termed a narrow pulse with a 50 mV overdrive; we are driving the + input of the comparator 50 mV over the negative input. The simulation results for these inputs are shown in Fig. 27.12. If the pulse amplitude or width is reduced much beyond this, the comparator does not make a full transition. Notice how we only show simulation data after 120 ns. This was done to ensure the bias circuit had time to start-up and stabilize before changing the inputs of the comparator.

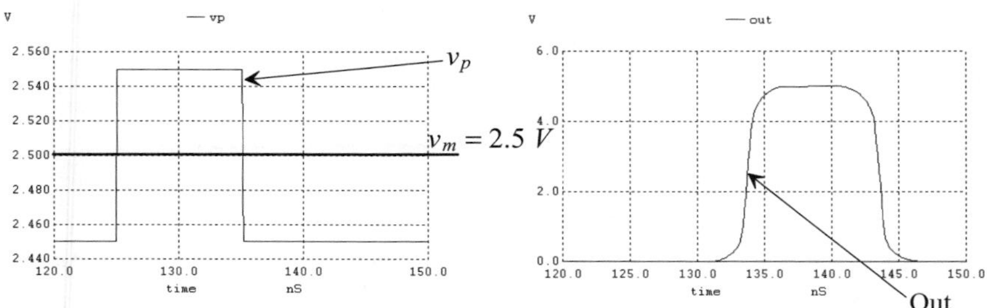

Figure 27.12 Transient response of the comparator in Fig. 27.9.

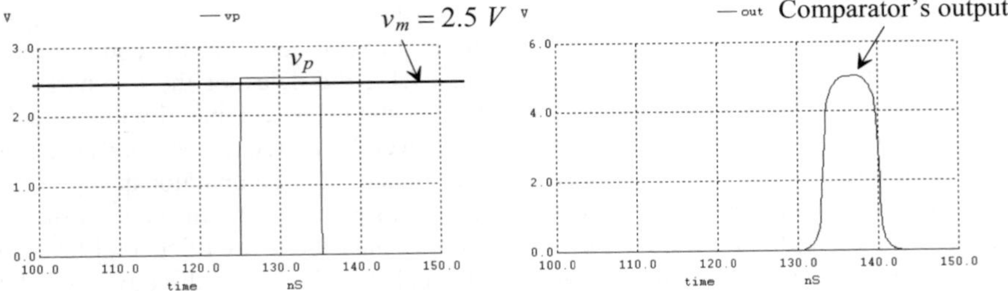

Figure 27.13 Transient response of the comparator in Fig. 27.9 with one side of the pre-amp initially shut off.

Although these results are interesting, they are not practically useful in some situations. In few cases will the comparator discriminate between signals similar to what is seen in Fig. 27.12. A better indication of comparator performance is to apply a signal that starts at a voltage where one side of the differential amplifier is cut off and to finish at a voltage slightly larger than the reference voltage (2.5 V in the above example), Fig. 27.13. In applying a 0 to 2.6 V pulse to the v_p input, the left side of the diff-amp is initially off and then on. The internal comparator nodes must be charged over a wider voltage range with a 0–2.6 V input when compared to the 2.49–2.51 V example of Fig. 27.12. In order to keep one side of the pre-amp from turning off, the circuit of Fig. 27.14 can be used. The added MOSFET ensures that the voltage between the drains of M31 and M41 don't deviate too much. This configuration is sometimes referred to as a clamped input stage. Note that the size of the added MOSFET should be as small as possible to avoid unneeded loading in the diff-amp (which slows down the comparator's performance).

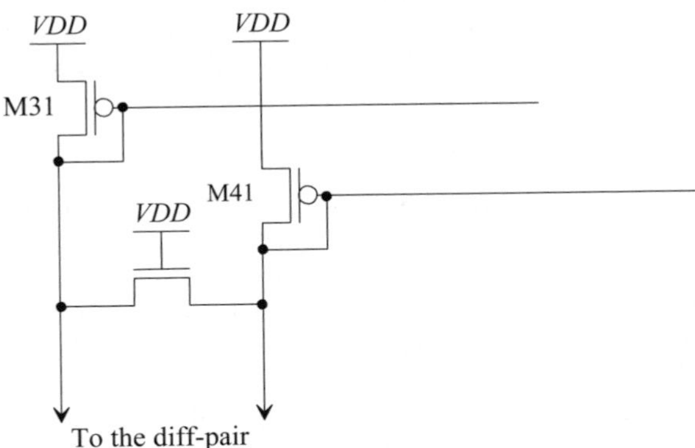

Figure 27.14 Adding a balancing resistor to the input stage of a comparator. See Fig. 27.5

Propagation Delay

Ideally, the propagation delay (the time difference between the input, v_p, crossing the reference voltage, v_m, and the output changing logic states) is zero. For the comparator of Fig. 27.9 with simulation results shown in Fig. 27.12, the delay of the comparator is approximately 9 ns (or less with an overdrive). This should be compared with the delay of an op-amp used as a comparator, which may be several hundred nanoseconds (because of the compensation capacitor). Another interesting fact about comparator design is that the delay of a comparator can be reduced by cascading gain stages. In other words, the delay of a single high-gain stage is in general longer than the delay of several low-gain stages. Improvements in comparator sensitivity and speed can be directly related to improvements in the preamplifier. A derivation (similar to that given in Ch. 11 for minimum delay through an inverter string driving a load capacitance, C_L) of the number of stages, N, needed in a preamplifier to reach minimum delay for a given load capacitance and input capacitance results in

$$N = \ln\frac{C_L}{C_{in}} \tag{27.8}$$

In the derivation of this equation, it was assumed that the driving resistance of any stage is $1/g_{mn}$ (the transconductance of the nth differential amplifier stage). In practice, Eq. (27.8) is of little use because one side of the differential amplifier can be off. For this reason (and others such as size and power draw), comparators with more than two or three pre-amp stages are not common.

Minimum Input Slew Rate

The last characteristic we discuss is the minimum input slew rate of a comparator. If the input signals to the comparator vary at a slow rate (e.g., a sine wave generated from the AC line), the output of the comparator may very well oscillate resulting in an output with a metastable state. If a comparator is to be used with slowly varying signals, or in a noisy environment, the decision circuit should have hysteresis. The minimum input slew rate is difficult to simulate with SPICE because of the slow and fast varying signals present at the same instant of time in the circuit. This same situation (metastability) can also occur if the input overdrive is small. The result, for this case, however, is an increase in the delay time of the comparator.

27.1.2 Clocked Comparators

We developed "sense amplifiers" back in Sec. 16.2.1. The sense amplifier can be used as a clocked comparator, that is a comparator whose outputs change on the rising (or falling) edge of a clock signal. Consider the clocked comparator in Fig. 27.15. This circuit is directly taken from Figs. 16.32 and 16.35. It works very well for signal differences that are tens of millivolts. When *clock* is low, the inputs to the NAND SR latch are pulled high. The outputs of the comparator don't change. When *clock* goes high, the two inputs are compared causing the output of the circuit to register which one is higher. As discussed in detail back in Ch. 16, the minimum input voltages for this comparator are above the threshold voltage of the NMOS devices. Further if the input signals get too large, MB1 and MB2 move deep into the triode region. The small differences in their channel resistances adversely affect the quality of the comparator's decisions. Let's modify this topology for better sensitivity, wider input signal swing, and better immunity to kickback.

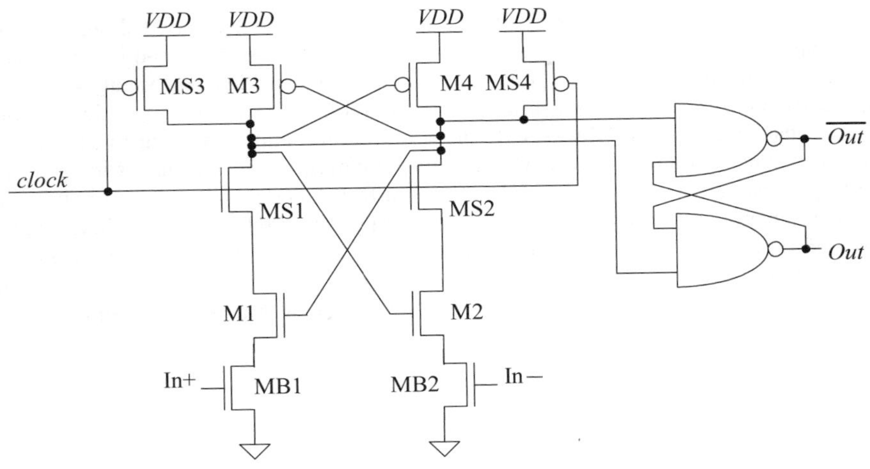

Figure 27.15 A clocked comparator based on Figs. 16.32 and 16.35.

Our modification is seen in Fig. 27.16. Instead of biasing the PMOS and NMOS diff-amps with a bias circuit, here, to be different, we simply use long-length MOSFETs. The bias currents aren't too critical in this application (like they are in an op-amp design). We can still use a bias circuit if controlling the bias current is, for some reason, important. We've removed the triode-operating MOSFETs MB1 and MB2 from the basic cross-coupled latch section. Now we are steering currents from one side of the latch to the other. The differences in the currents causes the latch to switch dependent on the input signals. Note that if either of the diff-amps isn't present in the comparator or if one is off because the comparator's input common-mode voltage is too high or too low, the comparator still works as desired. Only one diff-amp is needed to create the imbalance. Also note, again, that the SR latch is used to make the outputs of the circuit change on the rising edge of the clock signal. If the outputs of the comparator can go high, for a particular application, when the clock signal is low then the NAND gates can be removed.

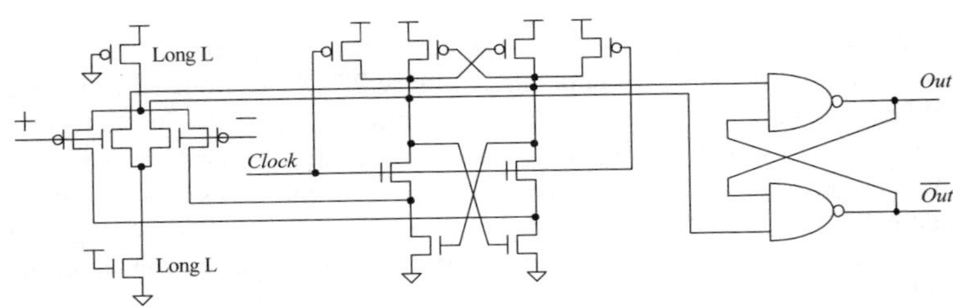

Figure 27.16 Wide-swing clocked comparator. Outputs change on the rising edge of the clock signal.

27.1.3 Input Buffers Revisited

When we discussed input buffers back in Ch. 18 we used a self-biased design for the highest speed (see Fig. 18.17 and the associated discussions). As discussed in Ch. 18, it's desirable to have symmetrical propagation delays independent of input slew-rate, amplitudes, or direction (high-to-low or low-to-high). Towards the more ideal input buffer, consider using two NMOS self-biased buffers in parallel, as seen in Fig. 27.17. Here, to maintain good symmetry, we've split the current sources in half and used one side to generate a common-mode feedback signal to balance the outputs. To reduce the power dissipation in the buffer, the triode-operating MOSFETs can have their lengths increased. Further, for rail-to-rail input common-mode range, we can place the PMOS version of this buffer in parallel with the NMOS version seen in Fig. 27.17, as we did in Fig. 18.23. This input buffer can be very useful for low-skew, high-speed design.

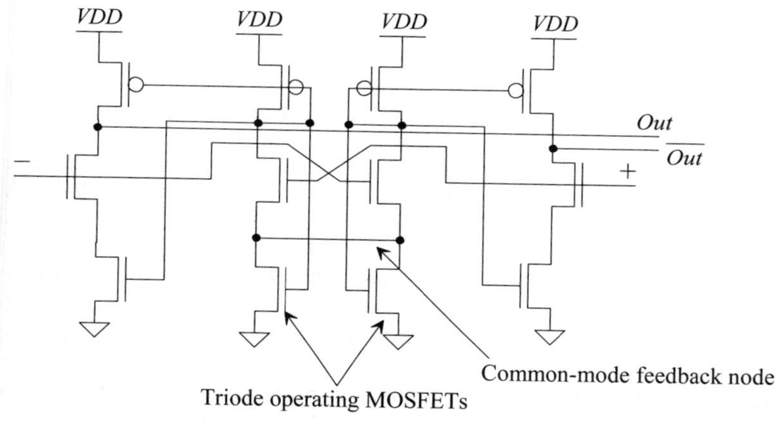

Figure 27.17 A fully-differential input buffer based on the topology discussed in Sec. 18.17.

27.2 Adaptive Biasing

Adaptive biasing can reduce power dissipation in an amplifier while at the same time increasing output current drive capability. Figure 27.18 can be used to help illustrate the idea. When v_{I1} and v_{I2} are equal, the current sources I_{SS1} and I_{SS2} are zero (an open). The diff-amp DC tail current is simply I_{SS}, the same as an ordinarily biased diff-amp. If v_{I1} becomes larger than v_{I2}, the current source I_{SS1} increases above zero, effectively increasing the diff-amp DC bias current. Similarly, if v_{I2} becomes larger than v_{I1}, the current source I_{SS2} increases above zero. The diff-amp output current is normally limited to I_{SS} when one side of the diff-amp shuts off. However, now that the maximum output current is limited to either $I_{SS} + I_{SS1}$ or $I_{SS} + I_{SS2}$. Power dissipation can be reduced using an adaptive bias, and slew-rate problems can be eliminated.

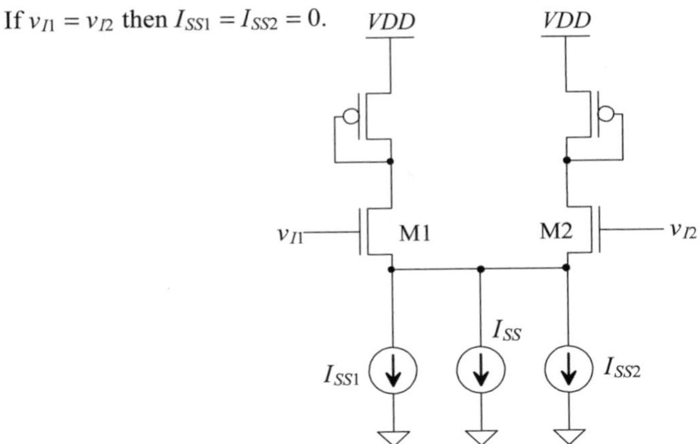

Figure 27.18 Adaptively biased diff-amp.

The current diff-amp of Fig. 27.19 can be used to implement the current source I_{SS1} or I_{SS2}. If the currents I_1 and I_2 are equal, then zero current flows in M3 and M4. Also, if I_2 is greater than I_1, zero current flows in M3 and M4. If I_1 is larger than I_2, the difference between these two currents $(I_1 - I_2)$ flows in M3. Since M4 is K times wider than M3, a current of $K(I_1 - I_2)$ flows in M4 (normally $K < 1$). Two of these diff-amps are needed to implement the adaptive biasing of the diff-amp of Fig. 27.18.

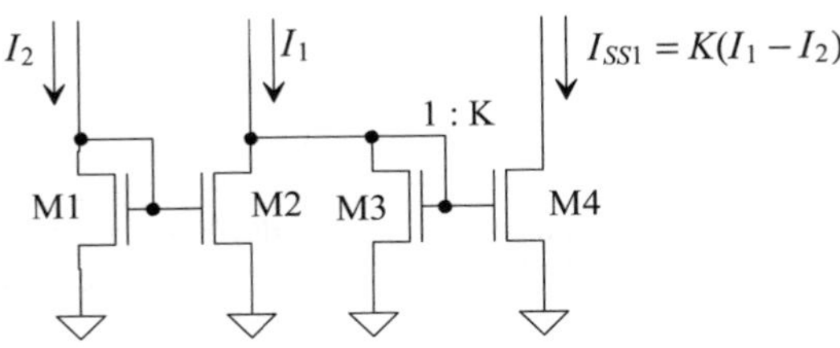

Figure 27.19 Current diff-amp used in adaptive biasing.

Figure 27.20 shows the implementation of adaptive biasing into the diff-amp of Fig. 27.18. P-channel MOSFETs are added adjacent to M3 and M4 to mirror the currents through M1 and M2 (I_1 and I_2). The maximum total current available through M1 occurs when M2 is off. Positive feedback exists through the loop M1, M3, M5–M7. Initially, when M2 shuts off, the current in M1 and M3 is I_{SS}. This is mirrored in M5 and M6, and thus I_{SS1} becomes $K \cdot I_{SS}$. At this particular instance in time, the tail current, which flows through M1, is now $I_{SS} + K \cdot I_{SS}$. However, provided the MOSFETs M1, M3–M7 remain in

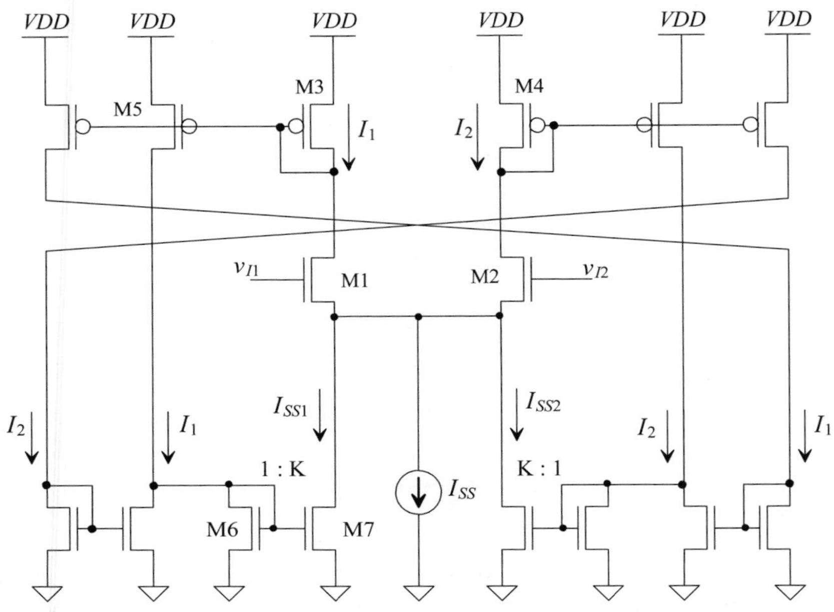

Figure 27.20 Adaptively biased diff-amp.

saturation, this current circles back around the positive feedback loop and increases by K. This continues, resulting in a final or total tail current of

$$I_{tot} = I_{SS} \cdot (1 + K + K^2 + K^3 + ...) \tag{27.9}$$

If $K < 1$, this geometric series can be written as

$$I_{tot} = \frac{I_{ss}}{1 - K} \tag{27.10}$$

Setting $K = 0$ (MOSFET M7 doesn't exist) results in no adaptive biasing and a total tail current of I_{SS}. Setting $K = 1/2$ (M7 half the size of M6) results in a total available tail current of $2 \cdot I_{SS}$. Because the tail current limits the slew rate, when the diff-amp is driving a capacitive load, making K equal to one eliminates slew-rate limitations while at the same time not increasing static power dissipation. In practice, shutting off one side of the diff-pair, M1/M2, is difficult since the adaptive biasing has the effect of lowering the source potentials of the diff-pair, keeping both MOSFETs on. Adaptive biasing can be used in a comparator where M1 or M2 can shut off. However, the static power dissipation will be large; therefore, there is no benefit over the comparators discussed earlier in the chapter. When applying adaptive biasing to an OTA design, the value of K should be unity or less. [Using a K of 1 or 2 can still result in a finite I_{tot} since the MOSFETs have a finite output resistance and the MOSFETs in the diff-pair will not shut off, as was assumed in the derivation of Eq. (27.10)].

A final example of an adaptive voltage-follower amplifier is shown in Fig. 27.21. This amplifier can only source current to a load. If v_{in} and v_{out} are equal, the current that flows in M1 and M2 is $I_{SS} + I_{D6}$. If v_{in} is increased, the current in M1 and M3 increases. This causes the currents in M4–M6 to increase, effectively increasing the tail current of the diff-pair. The result is a large current available to drive the load. Note that M7 can be sized larger than the other MOSFETs to increase maximum output current.

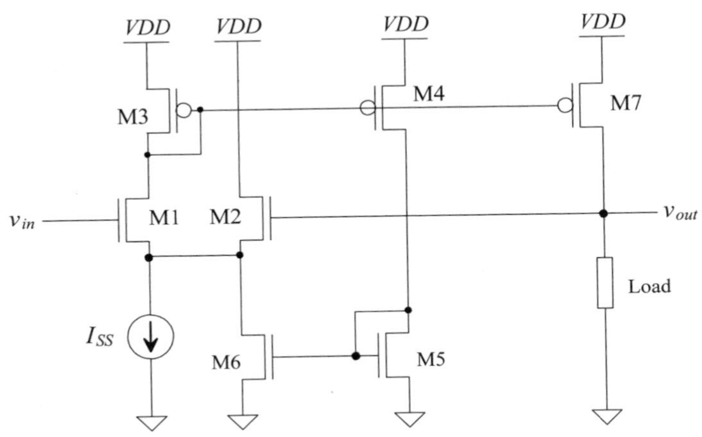

Figure 27.21 Adaptive voltage follower.

27.3 Analog Multipliers

Analog multipliers find extensive use in communication systems. Figure 27.22 shows the voltage characteristics of a four-quadrant multiplier. This multiplier is termed a four-quadrant multiplier because both inputs can be either positive or negative around a common-mode voltage, V_{CM}. The ideal output of the multiplier is related to the inputs by

$$v_{out} = K_m \cdot v_x v_y \tag{27.11}$$

where K_m is the multiplier gain with units of V^{-1}. In reality, imperfections exist in the multiplier gain, resulting in offsets and nonlinearities. The output of the multiplier can be written as

$$v_{out} = K_m(v_x + V_{OSx})(v_y + V_{OSy}) + V_{OSout} + v_x^n + v_y^m \tag{27.12}$$

where V_{OSx}, V_{OSy}, and V_{OSout} are the offset voltages associated with the x-, y-inputs, and the output, respectively. The terms v_x^n and v_y^m represent nonlinearities in the multiplier. Normally, these nonlinearities are specified in terms of the total harmonic-distortion or by specifying the maximum deviation in percentages between a straight line and the actual characteristic curves shown in Fig. 27.22 over some range of input voltages. Although many different techniques exist for implementing analog multipliers in CMOS, we concentrate on a technique useful in high- and low-frequency multiplication.

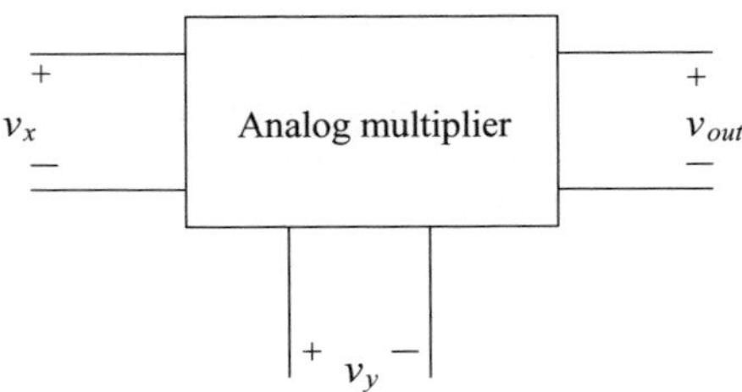

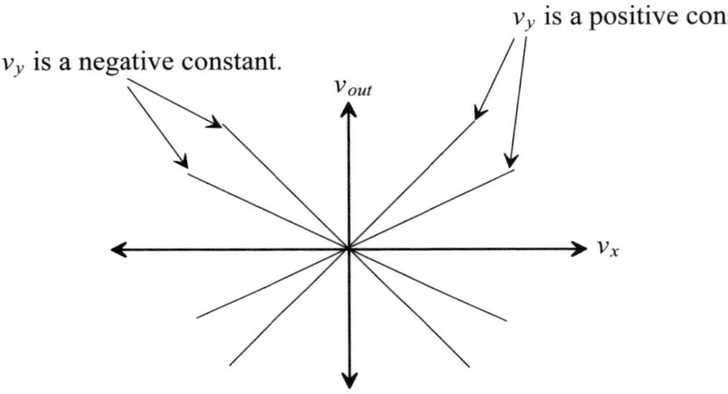

Figure 27.22 Operation of a four-quadrant analog multiplier.

27.3.1 The Multiplying Quad

A CMOS multiplier employing a multiplying quad (M1–M4) is shown in Fig. 27.23. The multiplying quad operates in the triode region, and thus MOSFETs M1–M4 can be thought of as resistors. For the moment we will not consider the biasing of the quad. The negative output voltage of the multiplier is given by

$$v_{om} = -R \cdot (i_{D1} + i_{D2}) \tag{27.13}$$

while the positive output voltage is

$$v_{op} = -R \cdot (i_{D3} + i_{D4}) \tag{27.14}$$

The output voltage of the multiplier is

$$v_{out} = v_{op} - v_{om} = R \cdot (i_{D1} + i_{D2} - i_{D3} - i_{D4}) \tag{27.15}$$

A simplified schematic of the multiplying quad with biasing is shown in Fig. 27.24. The op-amp inputs are at an AC virtual ground and at a DC voltage of V_{CM} (the op-amp output common-mode voltage). In order to minimize the DC input current on the x-axis inputs, the common-mode DC voltage on this input is set to V_{CM}. The DC biasing voltage on the y-input is set to a value large enough to keep the quad in triode. The input signals have been broken into two parts (e.g., $v_x/2$ and $-v_x/2$) to maintain generality. In practice, the minus inputs can be connected directly to the bias voltages at the cost of

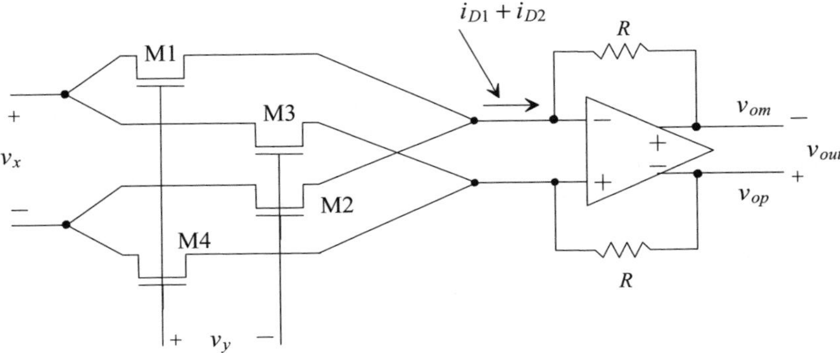

Figure 27.23 CMOS analog multiplier.

large-signal linearity. (The input is not truly differential in this situation.) However, as discussed earlier, a fully differential system has much better coupled noise immunity.

Using Eq. (9.12) and noticing that the DC gate-source voltage of all MOSFETs is the same, the drain currents can be written as

$$i_{D1} = \beta_1\left[\left(V_{GS} + \frac{v_y}{2} - V_{THN1}\right)\left(\frac{v_x}{2}\right) - \frac{1}{2}\left(\frac{v_x}{2}\right)^2\right] \tag{27.16}$$

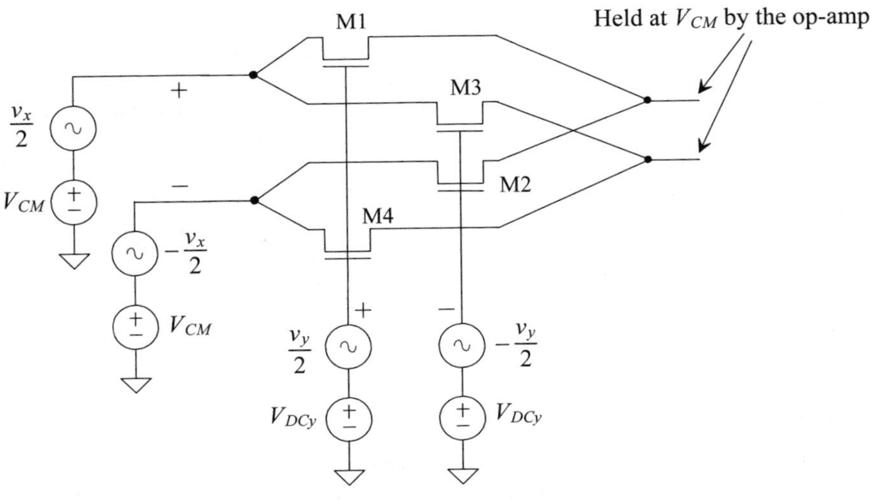

Figure 27.24 Biasing of the multiplying quad.

$$i_{D2} = \beta_2\left[\left(V_{GS} - \frac{v_y}{2} - V_{THN2}\right)\left(-\frac{v_x}{2}\right) - \frac{1}{2}\left(-\frac{v_x}{2}\right)^2\right] \tag{27.17}$$

$$i_{D3} = \beta_3\left[\left(V_{GS} - \frac{v_y}{2} - V_{THN3}\right)\left(\frac{v_x}{2}\right) - \frac{1}{2}\left(\frac{v_x}{2}\right)^2\right] \tag{27.18}$$

$$i_{D4} = \beta_4\left[\left(V_{GS} + \frac{v_y}{2} - V_{THN4}\right)\left(-\frac{v_x}{2}\right) - \frac{1}{2}\left(-\frac{v_x}{2}\right)^2\right] \tag{27.19}$$

We can design so that $\beta = \beta_1 = \beta_2 = \beta_3 = \beta_4$. We can use Eq. (27.15) together with Eqs. (27.16)–(27.19) to rewrite the output voltage of the multiplier as

$$v_{out} = R\beta \cdot \left(\frac{v_x}{2}\right)\left[\frac{v_y}{2} - V_{THN1} + \frac{v_y}{2} + V_{THN2} + \frac{v_y}{2} + V_{THN3} + \frac{v_y}{2} - V_{THN4}\right] \tag{27.20}$$

We can see that if $V_{THN1} = (V_{THN2}$ or $V_{THN3})$ and $V_{THN4} = (V_{THN3}$ or $V_{THN2})$, this equation can be rewritten as

$$v_{out} = R\beta \cdot v_x v_y \tag{27.21}$$

The source of a MOSFET (the terminal we label "source" depends on which way current flows in the MOSFET) in the multiplying quad is connected either to the op-amp or to the x inputs. When the sources of the MOSFETs are connected to the op-amp, all of the MOSFETs in the multiplying quad have the same threshold voltage. (Since the source of each MOSFET is tied to the same potential, the body effect changes each MOSFET's threshold voltage by the same amount.) If the positive x-input is sinking a current, then the sources of M1 and M3 are the "+" x-input and thus $V_{THN1} = V_{THN3}$. In any case, the threshold voltages of the MOSFETs cancel and Eq. (27.21) holds. Comparing Eqs. (27.21) and (27.11) results in defining the gain of this multiplier as

$$K_m = R \cdot \beta \tag{27.22}$$

Simulating the Operation of the Multiplier

Simulating the performance and understanding the analog multiplier operation is an important step in the design process. The design of a multiplier consists of designing the op-amp, selecting the sizes of the multiplying quad, and designing the biasing network. Because we covered the design of differential input/output op-amps in the last chapter, it is not covered here. In order to simulate the performance of a multiplier in SPICE without including the limitations of the op-amp, the simple model shown in Fig. 27.25 is used. The sum of the multiplying factors associated with the voltage-controlled voltage sources, E1 and E2, is the open-loop gain of the op-amp. A typical SPICE statement for these VCVS (voltage-controlled voltage source) where the op-amp open loop gain is 20,000 is

```
E1    Voplus 8 4 3    1E4
E2    8 Vominus 4 3  1E4
```

where the nodes correspond to those labeled in Fig. 27.25.

The next problem we encounter in simulating the operation of the multiplier is implementing the differential voltages (e.g., $\pm\ v_x/2$), in addition to the DC biasing voltages. The setup shown in Fig. 27.26 is used to implement the biasing and the differential voltage sources. The op-amp common-mode output voltage, V_{CM}, and the x

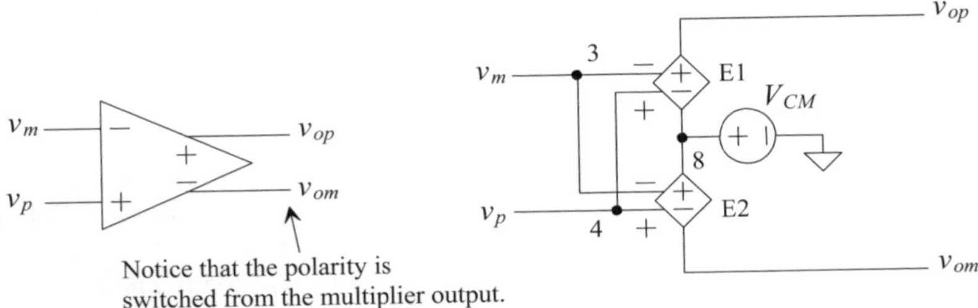

Figure 27.25 SPICE modeling a differential input/output op-amp with common-mode voltage.

input voltage are set to 1.5 V. The lower this voltage, the easier it is to bias the multiplying quad into the triode region. On the other hand, a reduction in the value of V_{CM} limits the op-amp output voltage swing and thus the multiplier output range. The size of the multiplying quad was set to 10/2. The larger the W/L ratio of the MOSFETs used in this quad, the easier it is to keep the quad in the triode region. On the other hand, using a large W/L increases the required input current. The channel length can be increased to ensure the device operates as a long-channel device and follows Eq. (9.12). Since the quad is part of the feedback around the op-amp, long-channel devices do not affect the

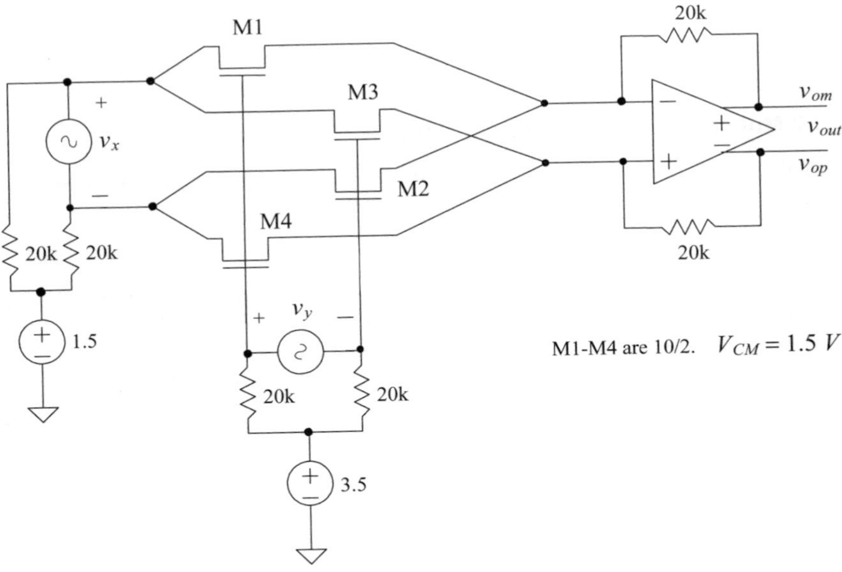

Figure 27.26 SPICE simulation schematic.

speed. The DC voltage at the y-inputs was set to 3.5 V as a compromise between keeping the multiplying quad in triode and the y-input voltage range. The gain of the multiplier in Fig. 27.26 is, from Eq. (27.22) and Table 6.2,

$$K_m = 20k \cdot 120\frac{\mu A}{V^2} \cdot \frac{10}{2} = 12\ V^{-1}$$

A DC sweep showing the operation of the multiplier is shown in Fig. 27.27. The x-input, v_x, was swept from –1 to +1, while at the same time the y-input was stepped from –1 to 1 V in 0.5 V increments. Keeping in mind that the output of the multiplier is $v_{op} - v_{om}$, we can understand the data presented in Fig. 27.27 by considering points A and B. At point A, the y-input is 1 V while the x-input is 0.5 V. The output voltage of the multiplier is the product of the multiplier gain and these two voltages (i.e., 12·1·0.25 = 3 V). The output voltage at point B is 12·0.5·(–0.3) = 1.8 V. Note that this figure was generated using an almost ideal op-amp. The characteristics do not show the limitations of the op-amp. In particular, the limited output swing.

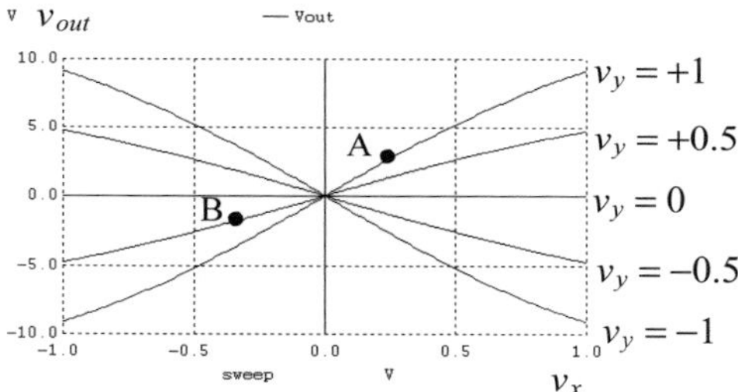

Figure 27.27 DC characteristics of the multiplier of Fig. 26.26.

27.3.2 Multiplier Design Using Squaring Circuits

An analog multiplier can be designed based on the difference between the sum of two voltages squared and the difference of two voltages squared, or

$$V_o = (V_1 + V_2)^2 - (V_1 - V_2)^2 = 4V_1V_2 \tag{27.23}$$

The basic sum-squaring and difference-squaring circuits are shown in Fig. 27.28. MOSFETs M1 and M4 are source-followers, while MOSFETs M2 and M4 are called squaring MOSFETs. This circuit is designed so that $\beta_1 = \beta_4 = \beta_{14}$, $\beta_2 = \beta_3 = \beta_{23}$, and $\beta_{14} >> \beta_{23}$. This makes almost all of the DC bias currents, I_{S12} and I_{S34}, flow in the MOSFETs M1 and M4, respectively. The squaring current, I_{SQ}, assuming that zero current flows through the resistor when both inputs are zero volts (or whatever the common-mode voltage when a single supply is used), is given by

$$I_{SQ(a)} = \frac{\beta_{23}}{4}(V_1 + V_2)^2 \tag{27.24}$$

Similarly, the squaring current in the difference-square circuit of Fig. 27.28b is given by

$$I_{SQ(b)} = \frac{\beta_{23}}{4}(V_1 - V_2)^2 \tag{27.25}$$

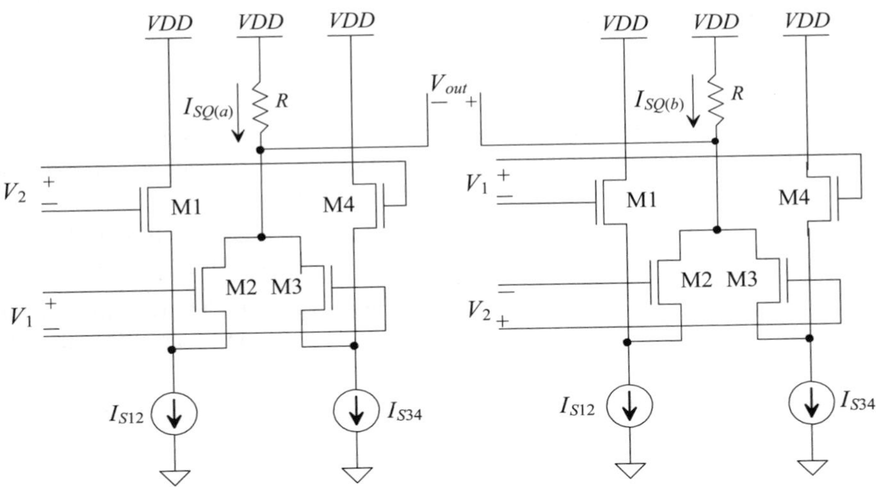

Figure 27.28 (a) Sum-squaring circuit and (b) difference squaring circuit.

The output voltage of the sum-square circuit is given by

$$V_{o-} = VDD - I_{SQ(a)}R \tag{27.26}$$

while the output voltage of the difference-square circuit is given by

$$V_{o+} = VDD - I_{SQ(b)}R \tag{27.27}$$

A multiplier is formed by taking the difference between these voltages. The output voltage of the multiplier of Fig. 27.27 is given by

$$V_{out} = V_{o+} - V_{o-} = R\frac{\beta_{23}}{4}\left[(V_1 + V_2)^2 - (V_1 - V_2)^2\right] \tag{27.28}$$

or, using Eq. (27.23)

$$V_{out} = R\beta_{23} \cdot V_1 V_2 \tag{27.29}$$

The fundamental concern with using this type of multiplier is the fact that modern short-channel CMOS devices don't follow the square-law equations (Eq. [27.24]) we used to derive this result.

ADDITIONAL READING

[1] D. J. Allstot, "A Precision Variable-Supply CMOS Comparator," *IEEE Journal of Solid-State Circuits,* vol. SC-17, no. 6, pp. 1080–1087, December 1982.

[2] M. G. Degrauwe, J. Rijmenants, E. A. Vittoz, and H. J. DeMan, "Adaptive Biasing CMOS Amplifiers," *IEEE Journal of Solid-State Circuits*, vol. SC-17, no. 3, pp. 522–528, June 1982.

[3] E. A. Vittoz, "Micropower Techniques," Chapter 3 in J. E. Franca and Y. Tsividis (eds.) *Design of Analog-Digital VLSI Circuits for Telecommunications and Signal Processing,* 2nd ed., Prentice Hall, 1994. ISBN 0-13-203639-8.

[4] S. Soclof, *Applications of Analog Integrated Circuits*, Prentice Hall, 1985. ISBN 0-13-039173-5.

[5] M. Ismail, S-C. Huang, and S. Sakurai, "Continuous-Time Signal Processing," Chapter 3 in M. Ismail and T. Fiez (eds.), *Analog VLSI: Signal and Information Processing*, McGraw Hill, 1994. ISBN 0-07-032386-0.

[6] B-S. Song, "CMOS RF Circuits for Data Communications Applications," *IEEE Journal of Solid-State Circuits*, vol. SC-21, no. 2, pp. 310–317, April 1986.

[7] J. Crols and M. S. J. Steyaert, "A 1.5 GHz Highly Linear CMOS Downconversion Mixer," *IEEE Journal of Solid-State Circuits*, vol. 30, no. 7, pp. 736–742, July 1995.

[8] H-J. Song and C-K. Kim, "A MOS Four-Quadrant Analog Multiplier Using Simple Two-Input Squaring Circuits with Source Followers," *IEEE Journal of Solid-State Circuits*, vol. 25, no. 3, pp. 841–848, June 1990.

PROBLEMS

27.1 Using the long-channel CMOS process, compare the performance (using simulations) of the comparator in Fig. 27.8 with the comparator in Fig. 27.9. Your comparison should include DC gain, systematic offset, delay, sensitivity, and power consumption.

27.2 Show, using simulations, how the addition of a balancing resistor in Fig. 27.14 can be used to improve the response seen in Fig. 27.13.

27.3 Simulate the operation of the comparator in Fig. 27.15 in the short-channel CMOS process. Determine the comparators sensitivity and the kickback noise.

27.4 Repeat problem 27.3 for the comparator in Fig. 27.16. Show that the input common-mode range of the comparator in Fig. 27.16 extends beyond the power supply rails.

27.5 Simulate the operation of the input buffer in Fig. 27.17 in the short-channel CMOS process. How sensitive is the buffer to input slew-rate? How symmetrical are the output rise and fall times? Suggest, and verify with simulations, a method to reduce the power consumed by the input buffer.

27.6 Design a low power clocked comparator for use with a Flash ADC (discussed in Ch. 29). Use the short-channel CMOS process and a clocking frequency of 250 MHz estimate the power dissipated by 256 of these comparators.

Chapter

28

Data Converter Fundamentals

Data converters (a circuit that changes analog signals to digital representations or vice-versa) play an important role in an ever-increasing digital world. As more products perform calculations in the digital or discrete time domain, more sophisticated data converters must translate the digital data to and from our inherently analog world. This chapter introduces concepts of data conversion and sampling which surround this useful circuit.

28.1 Analog Versus Discrete Time Signals

Analog-to-digital converters, also known as A/Ds or ADCs, convert analog signals to discrete time or digital signals. Digital-to-analog converters (D/As or DACs) perform the reverse operation. Figure 28.1 illustrates these two operations.To understand the functionality of these data converters, it would be wise first to compare the characteristics of analog versus digital signals.

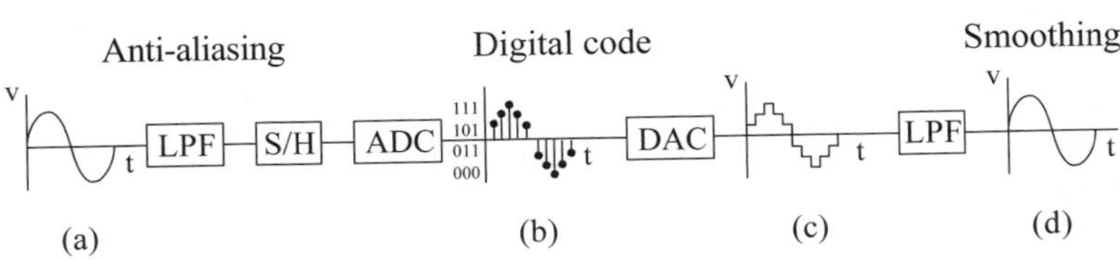

Figure 28.1 Signal characteristics caused by A/D and D/A conversion.

In Fig. 28.1 the original analog signal (a) is filtered by an anti-aliasing filter to remove any high-frequency components that may cause an effect known as aliasing (see Sec. 28.5). The signal is sampled and held and then converted into a digital signal (b). Next the DAC converts the digital signal back into an analog signal (c). Note that the output of the DAC is not as "smooth" as the original signal. A low-pass filter returns the analog signal back to its original form (plus phase shift introduced from the conversions) after eliminating the higher order signal components caused by the conversion. This example illustrates the main differences between analog and digital signals. Whereas the analog signal in Fig. 28.1a is *continuous* and *infinite* valued, the digital signal in (b) is *discrete* with respect to time and *quantized*. The term *continuous-time signal* refers to a signal whose response with respect to time is uninterrupted. Simply stated, the signal has a continuous value for the entire segment of time for which the signal exists. By referring to the analog signal as infinite valued, we mean that the signal can possess any value between the parameters of the system. For example, in Fig. 28.1a, if the peak amplitude of the sine wave was +1V, then the analog signal can be any value between –1 and 1 V (such as 0.4758393848 V). Of course, measuring all the values between –1 and 1 V would require a piece of laboratory equipment with infinite precision.

The digital signal, on the other hand, is discrete with respect to time. This means that the signal is defined for only certain or discrete periods of time. A signal that is quantized can only have certain values (as opposed to an infinitely valued analog signal) for each discrete period. The signal illustrated in Fig. 28.1b illustrates these qualities.

28.2 Converting Analog Signals to Digital Signals

We have already established the differences between analog and digital signals. How is it possible to convert from an analog signal to a digital signal? An example will illustrate the process.

Where you live the temperature in the winter stays between 0° F and 50° F (Fig. 28.2a). Suppose you had a thermometer with only two readings, hot and cold, and you wanted to record the weather patterns and plot the results. The two quantization levels can be correlated with the actual temperature as follows:

If $0° F \leq T < 25° F$ Temperature is recorded as cold

If $25° F \leq T < 50° F$ Temperature is recorded as hot

You take a measurement every day at noon and plot the results after one week. From Fig. 28.2b, it is apparent that your discretized version of the weather is not an accurate representation of the actual weather.

Now suppose that you find another thermometer with four possible temperatures (hot, warm, cool, and cold) and you increase the number of readings to two per day. The result of this reading is seen in Fig. 28.3a. The quantization levels represent four equal bands of temperature as seen below:

If $0° F \leq T < 12.5° F$ Temperature is recorded as cold

If $12.5° F \leq T < 25° F$ Temperature is recorded as cool

If $25° F \leq T < 37.5° F$ Temperature is recorded as warm

If $37.5° F \leq T < 50° F$ Temperature is recorded as hot

Here, the digital version of the weather still looks nothing like the actual weather pattern, but the critical issues in digitizing an analog signal should be apparent. The actual weather pattern is the analog signal. It is continuous with respect to time, and its value can be between 0° F and 50° F (even 33.9638483920398439° F!). The accuracy of the digitized signal is dependent on two things: the number of samples taken and the

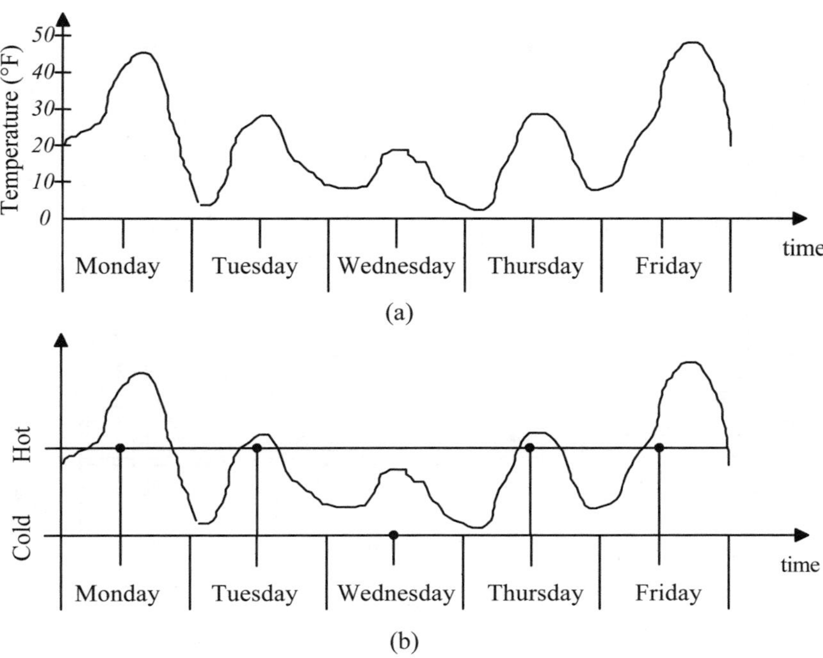

Figure 28.2 (a) An analog signal representing the temperature where you live and (b) a digital representation of the analog signal taking one sample per day with two quantization levels.

resolution, or number of quantization levels, of the converter. In our example, we need to increase both the number of samples and the resolution of thermometer.

Suppose that finally we obtain a thermometer with 25 temperature readings and that we take a reading eight times per day. Each of the 25 quantization levels now represents a 2° F band of temperature. From Fig. 28.3b, we can see that the digital version of the weather is approaching that of the actual analog signal. If we kept increasing both sampling time and resolution, the difference between the analog and the digital signals would become negligible. This brings up another critical issue: how many samples should one take in order to accurately represent the analog signal?

Suppose a sudden rainstorm swept through your town and caused a sharp decrease in temperature before returning to normal. If that storm had occurred between our sampling times, our experiment would not have shown the effects of the storm. Our sampling time was too slow to catch the change in the weather. If we had increased the number of samples, we would have recognized that something happened which caused the temperature to drop dramatically during that period.

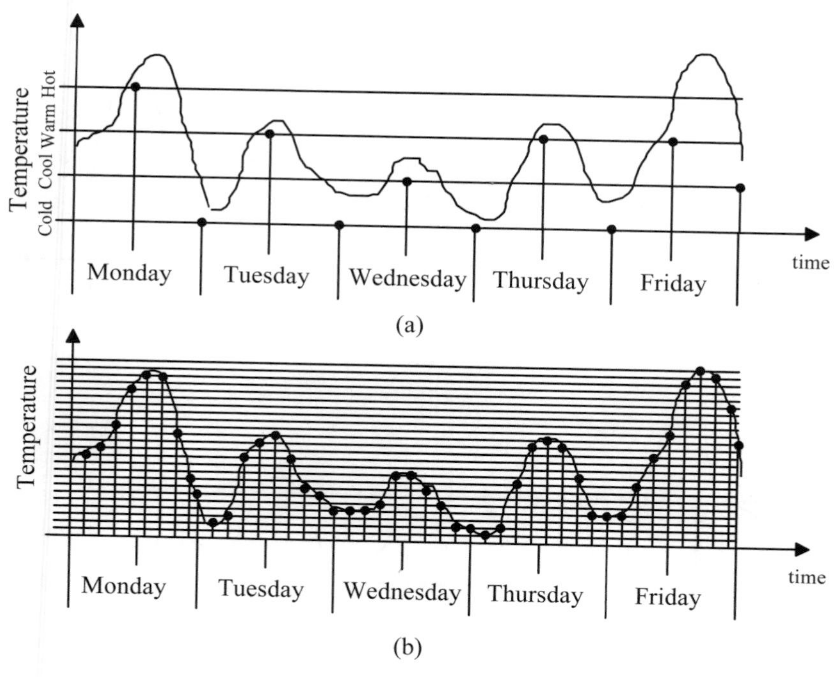

Figure 28.3 Digital representation of the temperature taking (a) two samples per day with four quantization levels and (b) nine samples per day with 25 quantization levels.

As it turns out, the *Nyquist Criterion* defines how fast the sampling rate needs to be to represent an analog signal accurately. This criterion requires that the sampling rate is at least two times the highest frequency contained in the analog signal. In our example, we need to know how quickly the weather can change and then take samples twice as fast as that value. The Nyquist Criterion can be described as

$$F_{sampling} = 2\,F_{MAX} \tag{28.1}$$

where $F_{sampling}$ is the sampling frequency required to accurately represent the analog signal and F_{MAX} is the highest frequency of the sampled signal.

How much resolution should we use to represent the analog signal accurately? There is no absolute criterion for this specification. Each application will have its own requirements. In our weather example, if we were only interested in following general trends, then the 25 quantization levels would more than suffice. However, if we were interested in keeping an accurate record of the temperature to within ±0.5° F, we would need to double the resolution to 50 quantization levels so that each quantization level would correspond to each degree ±0.5° F (Fig. 28.4).

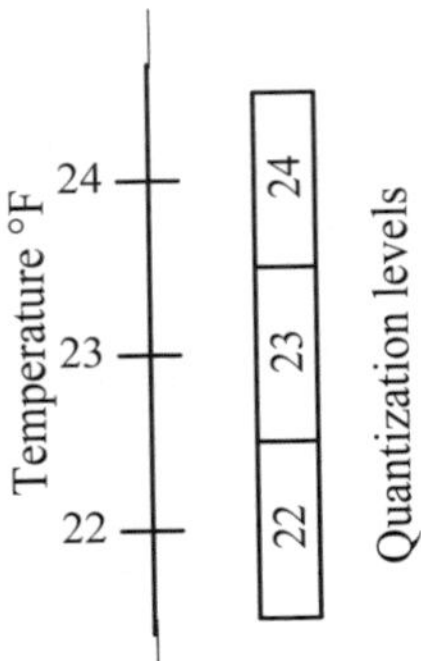

Figure 28.4 Quantization levels overlap actual temperature by ±½° F.

28.3 Sample-and-Hold (S/H) Characteristics

Sample-and-hold (S/H) circuits are critical in converting analog signals to digital signals. The behavior of the S/H is analogous to that of a camera. Its main function is to "take a picture" of the analog signal and hold its value until the ADC can process the information. It is important to characterize the S/H circuit when performing data conversion. Ignoring this component can result in serious error, for both speed and accuracy can be limited by the S/H. Ideally, the S/H circuit should have an output similar to that shown in Fig. 28.5a. Here, the analog signal is instantly captured and held until the next sampling period. However, a finite period of time is required for the sampling to occur. During the sampling period, the analog signal may continue to vary; thus, another type of circuit is called a track-and-hold, or T/H. Here, the analog signal is "tracked" during the time required to sample the signal, as seen in Fig. 28.5b. It can be seen that S/H circuits operate in both static (hold mode) and dynamic (sample mode) circumstances. Thus, characterization of the S/H will be discussed in the context of these two categories. Figure 28.6 presents a summary of the major errors associated with a S/H [1–5]. A discussion of each error follows.

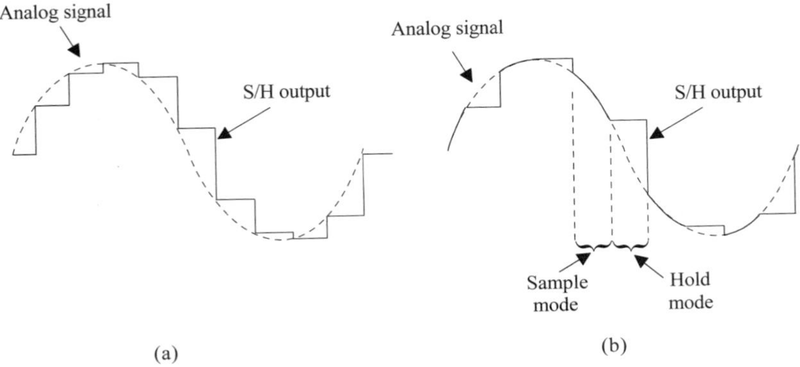

Figure 28.5 The output of (a) an ideal S/H circuit and (b) a track-and-hold (T/H).

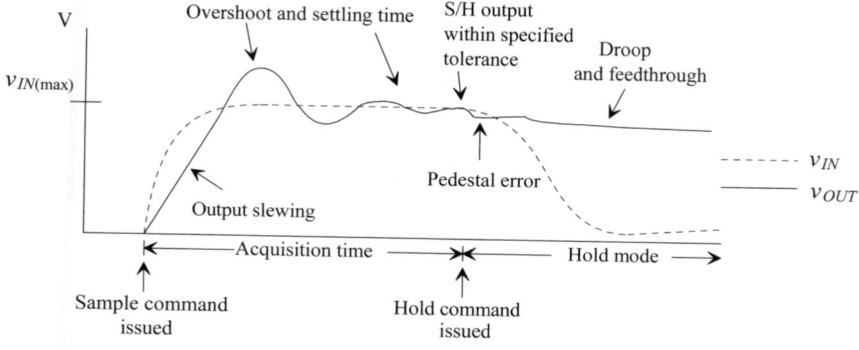

Figure 28.6 Typical errors associated wth an S/H.

Sample Mode

Once the sampling command has been issued, the time required for the S/H to track the analog signal to within a specified tolerance is known as the *acquisition time*. In the worst-case scenario, the analog signal would vary from zero volts to its maximum value, $v_{IN(max)}$. And the worst-case acquisition time would correspond to the time required for the output to transition from zero to $v_{IN(max)}$. Since most S/H circuits use amplifiers as buffers (as seen in Fig. 28.7), it should be obvious that the acquisition will be a function of the amplifier's own specifications. For example, notice that if the input changes very quickly, then the output of the T/H could be limited by the amplifier's slew rate. The amplifier's stability is also extremely critical. If the amplifier is not compensated correctly, and the phase margin is too small, then a large *overshoot* will occur. A large overshoot requires a longer *settling time* for the S/H to settle within the specified tolerance. The error tolerance at the output of the S/H also depends on the amplifier's *offset*, *gain error* (ideally, the S/H should have a gain of 1) and *linearity* (the gain of the S/H should not vary over the input voltage range).

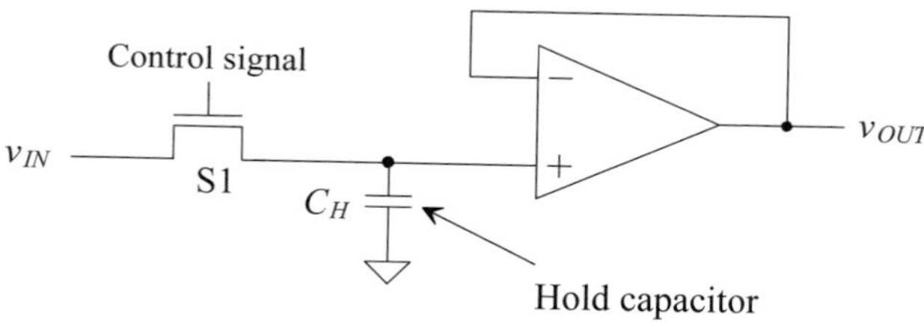

Figure 28.7 Track-and-hold circuit using an output buffer.

Hold Mode

Once the hold command is issued, the S/H faces other errors. Pedestal error occurs as a result of charge injection and clock feedthrough. Part of the charge built up in the channel of the switch is distributed onto the capacitor, thus slightly changing its voltage. Also, the clock couples onto the capacitor via overlap capacitance between the gate and the source or drain. Another error that occurs during the hold mode is called *droop*. This error is related to the leakage of current from the capacitor due to parasitic impedances and to the leakage through the reverse-biased diode formed by the drain of the switch. This diode leakage can be minimized by making the drain area as small as can be tolerated. Although the input impedance of the buffer amplifier is very large, the switch has a finite OFF impedance through which leakage can occur. Current can also leak through the substrate. The key to minimizing droop is increasing the value of the sampling capacitor. The trade-off, however, is increased time that's required to charge the capacitor to the value of the input signal.

Aperture Error

A transient effect that introduces error occurs between the sample and the hold modes. A finite amount of time, referred to as aperture time, is required to disconnect the capacitor from the analog input source. The aperture time actually varies slightly as a result of noise on the hold-control signal and the value of the input signal, since the switch will not turn off until the gate voltage becomes less than the value of the input voltage less one threshold voltage drop. This effect is called *aperture uncertainty* or *aperture jitter*. As a result, if a periodic signal were being sampled repeatedly at the same points, slight variations in the hold value would result, thus creating *sampling error*. Figure 28.8 illustrates this effect. Note that the amount of aperture error is directly related to the frequency of the signal and that the worst-case aperture error occurs at the zero crossing, where dV/dt is the greatest. This assumes that the S/H circuit is capable of sampling both positive and negative voltages (bipolar). The amount of error that can be tolerated is directly related to the resolution of the conversion. Aperture error will be discussed again in Sec. 28.5 as it relates to the error in an ADC.

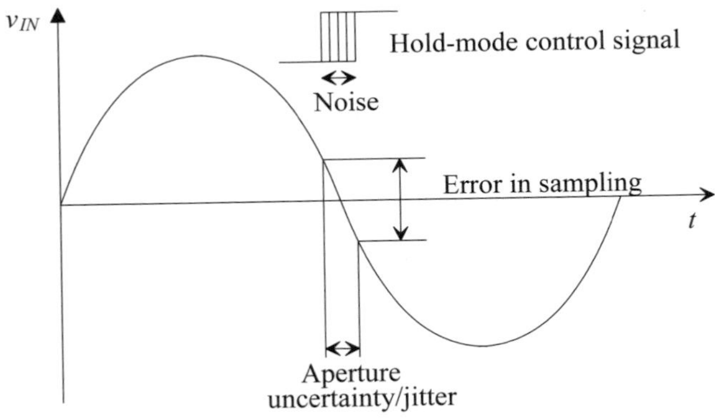

Figure 28.8 Aperture error.

Example 28.1

Find the maximum sampling error for a S/H circuit that is sampling a sinusoidal input signal that could be described as

$$v_{IN} = A\sin 2\pi ft$$

where A is 2 V and f = 100 kHz. Assume that the aperture uncertainty is equal to 0.5 ns.

The sampling error due to the aperture uncertainty can be thought of as a slew rate such that

$$\frac{dV}{dt} = \frac{d}{dt}A\sin 2\pi ft = 2\pi fA\cos 2\pi ft$$

with maximum slewing occurring when the cosine term is equal to 1. Therefore,

$$\frac{dV}{dt}(max) = 2\pi fA = (2\pi \cdot 100\text{ kHz})(2\text{ V})$$

and the maximum sampling error is

$$Maximum\ Sampling\ Error = dV(max) \quad \text{or}$$

$$(0.5\times 10^{-9}\text{ s})(2\pi \cdot 100\text{ kHz})(2\text{ V}) = 0.628\text{ mV} \ \blacksquare$$

28.4 Digital-to-Analog Converter (DAC) Specifications

Probably the most popular digital-to-analog converter application is the digital audio compact disc player. Here digital information stored on the CD is converted into music via a high-precision DAC. Many characteristics define a DAC's performance. Each characteristic will be discussed before we look at the basic architectures in Ch. 29. This "top-down" approach allows a smoother transition from the data converter characteristics to the actual architectures, since most data converters have similar performance limitations. A discussion of some of the basic definitions associated with DACs follows. It should be noted that DACs and ADCs can use either voltage or current as their analog signal. For purposes of describing specifications, it will be assumed that the analog signal is a voltage.

A block diagram of a DAC can be seen in Fig. 28.9. Here an N-bit digital word is mapped into a single analog voltage. Typically, the output of the DAC is a voltage that is some fraction of a reference voltage (or current), such that

$$v_{OUT} = FV_{REF} \tag{28.2}$$

where v_{OUT} is the analog voltage output, V_{REF} is the reference voltage, and F is the fraction defined by the input word, D, that is N bits wide. The number of input combinations represented by the input word D is related to the number of bits in the word by

$$\text{Number of input combinations} = 2^N \tag{28.3}$$

A 4-bit DAC has a total of 2^4 or 16 total input values. A converter with 4-bit resolution must be able to map a change in the analog output, which is equal to 1 part in 16. The maximum analog output voltage for any DAC is limited by the value of some reference voltage, V_{REF}. If the input is an N-bit word, then the value of the fraction, F, can be determined by,

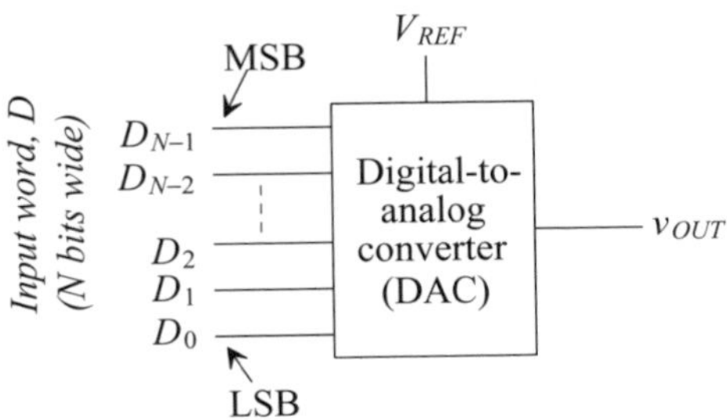

Figure 28.9 Block diagram of the digital-to-analog converter.

$$F = \frac{D}{2^N} \tag{28.4}$$

Therefore, if a 3-bit DAC is being used, the input, D, is $100 = 4_{10}$, and V_{REF} is 5 V, then the value of F is

$$F = \frac{100}{2^3} = \frac{4}{8} \tag{28.5}$$

and the analog voltage that appears at the output becomes,

$$v_{OUT} = \frac{4}{8}(5) = 2.5 \text{ V} \tag{28.6}$$

By plotting the input word, D, versus v_{OUT} as D is incremented from 000 to 111, the *transfer curve* seen in Fig. 28.10 would be generated. The y-axis has been normalized to V_{REF} ; therefore, the graduated marks also represent F by Eq. (28.2). Some important characteristics need to be discussed here. First, notice that the transfer curve is not continuous. Since the input is a digital signal, which is inherently discrete, the input signal can only have eight values that must correspondingly produce eight output voltages. If a straight line connected each of the output values, the slope of the line would ideally be one increment/input code value. Also note that the maximum value of the output is 7/8. Since the case where $D = 000$ has to result in an analog voltage of 0 V, and a 3-bit DAC has eight possible analog output voltages, then the analog output will increase from 0 V to only 7/8 V_{REF}.

Again, using Eq. (28.2), this means that the maximum analog output that can be generated by the 3-bit DAC is

$$v_{OUT(\max)} = \frac{7}{8} \cdot V_{REF} \tag{28.7}$$

This maximum analog output voltage that can be generated is known as *full-scale voltage,* V_{FS}, and can be generalized to any N-bit DAC as

$$V_{FS} = \frac{2^N - 1}{2^N} \cdot V_{REF} \tag{28.8}$$

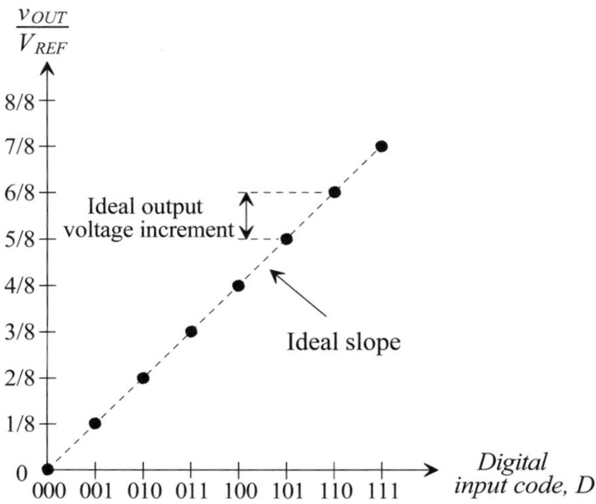

Figure 28.10 Ideal transfer curve for a 3-bit DAC.

The *least significant bit* (*LSB*) refers to the rightmost bit in the digital input word. The LSB defines the smallest possible change in the analog output voltage. The LSB will always be denoted as D_0. One LSB can be defined as

$$1\ LSB = \frac{V_{REF}}{2^N} \tag{28.9}$$

In the previous case of the 3-bit DAC, 1 LSB = 5/8 V, or 0.625 V. Generating an output in multiples of 0.625 V may not seem difficult, but as the number of bits increases, the voltage value of one LSB decreases for a fixed value of V_{REF}.

The *most significant bit* (*MSB*) refers to the leftmost bit of the digital word, D. In the previous example, $D = 100$ or $D_2D_1D_0$, with D_2 being the MSB. Generalizing to the N-bit DAC, the MSB would be denoted as $D_{N\text{-}1}$. (Since the LSB is denoted as bit 0, the MSB is denoted as N-1.) Note that when discussing DACs, the MSB causes the output to change by 1/2 V_{REF}.

When discussing data converters, the term *resolution* describes the smallest change in the analog output with respect to the value of the reference voltage, V_{REF}. This is slightly different from the definition of LSB in that resolution is typically given in terms of bits and represents the *number of unique output voltage levels*, i.e., 2^N.

Example 28.2
Find the resolution for a DAC if the output voltage is desired to change in 1 mV increments while using a reference voltage of 5 V.

The DAC must resolve

$$\frac{1\ \text{mV}}{5\ \text{V}} = 0.0002 \text{ or } 0.02\% \text{ adjustability}$$

Therefore, the *accuracy* required for 1 LSB change over a range of V_{REF} is

$$\frac{1\,LSB}{V_{REF}} = \frac{1}{2^N} = 0.0002 \tag{28.10}$$

and solving N for the resolution yields

$$N = \text{Log}_2\left(\frac{5\text{ V}}{1\text{ mV}}\right) = 12.29 \text{ bits}$$

which means that a 13-bit DAC will be needed to produce the accuracy capable of generating 1 mV changes in the output using a 5 V reference. ■

Example 28.3
Find the number of input combinations, values for 1 LSB, the percentage accuracy, and the full-scale voltage generated for a 3-bit, 8-bit, and 16-bit DAC, assuming that $V_{REF} = 5$ V.

Using Eqs. (28.3), (28.8), (28.9), and (28.10), we can generate the following information:

Resolution	Input combinations	1 LSB	% accuracy	V_{FS}
3	8	0.625 V	12.5	4.375 V
8	256	19.5 mV	0.391	4.985 V
16	65,536	76.29 μV	0.00153	4.9999 V

The value of 1 LSB for an 8-bit converter is 19.5 mV, while 1 LSB for a 16-bit converter is 76.3 μV (a factor of 256)! Increasing the resolution by 1 bit increases the accuracy by a factor of 2. The precision required to map the analog signal at high resolutions is very difficult to achieve. We will examine some of these issues as we examine the limitations of the data converter in Ch. 29.

Note that a data converter may have a resolution of 8 bits, where an LSB is 19.5 mV as above, while having a much higher accuracy. For example, we could require the 8-bit data converter above to have an accuracy of 0.1%. The higher accuracy results in a more ideal (linear) DAC. A typical specification for DAC accuracy is ±½ LSB for reasons discussed below. ■

Differential Nonlinearity

As seen in the ideal DAC in Fig. 28.10, each adjacent output increment should be exactly one-eighth. Since the y-axis is normalized, the values for the increment heights will be unitless. However, the increment heights can be easily converted to volts by multiplying the height by V_{REF}. This corresponds to the ideal increment corresponding to 0.625 V = 1 LSB (assuming $V_{REF} = 5$ V).

Nonideal components cause the analog increments to differ from their ideal values. The difference between the ideal and nonideal values is known as *differential nonlinearity*, or *DNL* and is defined as

$$DNL_n = \text{Actual increment height of transition } n - \text{Ideal increment height} \tag{28.11}$$

where n is the number corresponding to the digital input transition. The DNL specification measures how well a DAC can generate uniform analog LSB multiples at its output.

Example 28.4
Determine the DNL for the 3-bit nonideal DAC whose transfer curve is shown in Fig. 28.11. Assume that $V_{REF} = 5$ V.

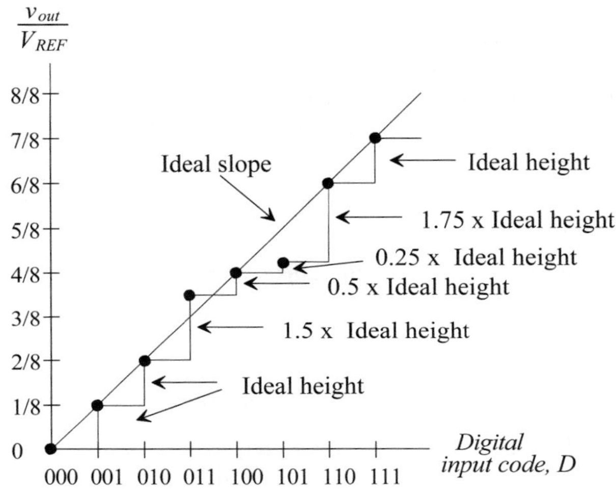

Figure 28.11 Example of differential nonlinearity for a 3-bit DAC.

The actual increment heights are labeled with respect to the ideal increment height, which is 1 LSB, or 1/8 of $\frac{v_{OUT}}{V_{REF}}$. Notice that there is no increment corresponding to 000, since it is desirable to have zero output voltage with a digital input code of 000. The increment height corresponding to 001, however, is equal to the corresponding height of the ideal case seen in Fig. 28.10; therefore, $DNL_1 = 0$. Similarly, DNL_2 is also zero since the increment associated with the transition at 010 is equal to the ideal height. Notice that the 011 increment, however, is not equal to the ideal curve but is 3/16, or 1.5 times the ideal height.

$$DNL_3 = 1.5 \text{ LSB} - 1 \text{ LSB} = 0.5 \text{ LSB}$$

Since we have already determined in Eq. (28.9) that for a 3-bit DAC, 1 LSB = 0.625 V, we can convert the DNL_3 to volts as well. Therefore, $DNL_3 = 0.5$ LSB = 0.3125 V. However, it is popular to refer to DNL in terms of LSBs. The remainder of the digital output codes can be characterized as follows:

$$DNL_4 = 0.5 \text{ LSB} - 1 \text{ LSB} = -0.5 \text{ LSB}$$

$$DNL_5 = 0.25 \text{ LSB} - 1 \text{ LSB} = -0.75 \text{ LSB}$$

$$DNL_6 = 1.75 \text{ LSB} - 1 \text{ LSB} = 0.75 \text{ LSB}$$

$$DNL_7 = 1 \text{ LSB} - 1 \text{ LSB} = 0$$

If we were to plot the value of DNL (in LSBs) versus the input digital code, Fig. 28.12 would result. The DNL for the entire converter used in this illustration is ±0.75 LSB since the overall error of the DAC is defined by its worst-case DNL. ■

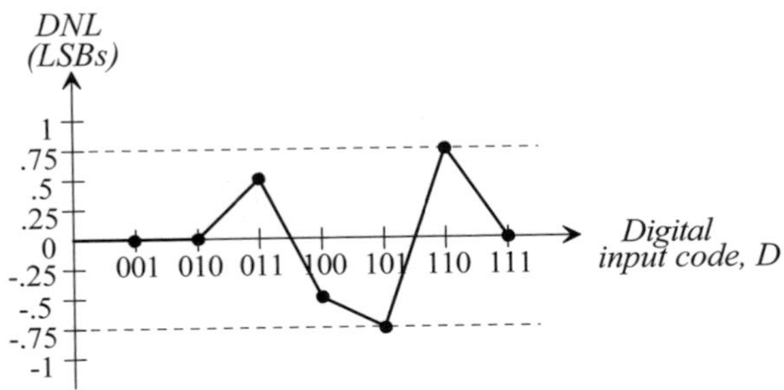

Figure 28.12 DNL curve for the nonideal 3-bit DAC.

Generally, a DAC will have less than ±½ LSB of DNL if it is to be N-bit accurate. A 5-bit DAC with 0.75 LSBs of DNL actually has the resolution of a 4-bit DAC. If the DNL for a DAC is less than –1 LSBs, then the DAC is said to be *nonmonotonic*, which means that the analog output voltage does not always increase as the digital input code is incremented. A DAC should always exhibit *monotonicity* if it is to function without error.

Integral Nonlinearity

Another important static characteristic of DACs is called *integral nonlinearity* (*INL*). Defined as the difference between the data converter output values and a reference straight line drawn through the first and last output values, INL defines the linearity of the overall transfer curve and can be described as

$$\text{INL}_n = \text{Output value for input code } n - \text{Output value of the reference line at that point} \tag{28.12}$$

An illustration of this measurement is presented in Fig. 28.13. It is assumed that all other errors due to offset and gain (these will be discussed shortly) are zero. An example follows shortly.

It is common practice to assume that a converter with N-bit resolution will have less than ±½ LSB of DNL and INL. The term, ½ *LSB*, is a common term that typically denotes the maximum error of a data converter (both DACs and ADCs). For example, a 13-bit DAC having greater than ±½ LSB of DNL or INL actually has the resolution of a 12-bit DAC. The value of ½ LSB in volts is simply

$$0.5\ LSB = \frac{V_{REF}}{2^{N+1}} \tag{28.13}$$

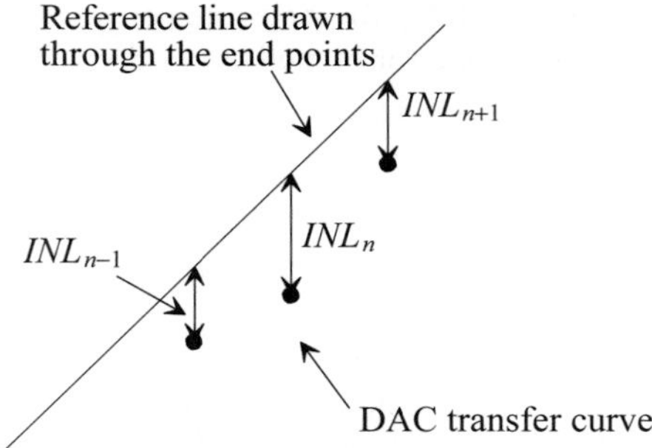

Figure 28.13 Measuring the INL for a DAC transfer curve.

Example 28.5
Determine the INL for the nonideal 3-bit DAC shown in Fig. 28.14. Assume that $V_{REF} = 5$ V.

First, a reference line is drawn through the first and last output values. The INL is zero for every code in which the output value lies on the reference line; therefore, $INL_2 = INL_4 = INL_6 = INL_7 = 0$. Only outputs corresponding to 001, 011, and 101 do not lie on the reference. Both the 001 and the 011 transitions occur ½ LSB higher than the straight-line values; therefore, $INL_1 = INL_3 = 0.5$ LSB. By the same reasoning, $INL_5 = -0.75$ LSB. Therefore, the INL for the DAC is considered to be its worst-case INL of +0.5 LSB and –0.75 LSB. The INL plot for the nonideal 3-bit DAC can be seen in Fig. 28.15. ■

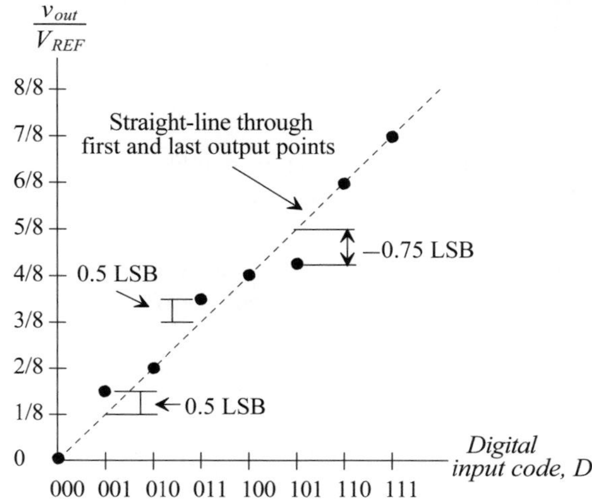

Figure 28.14 Example of integral nonlinearity for a DAC.

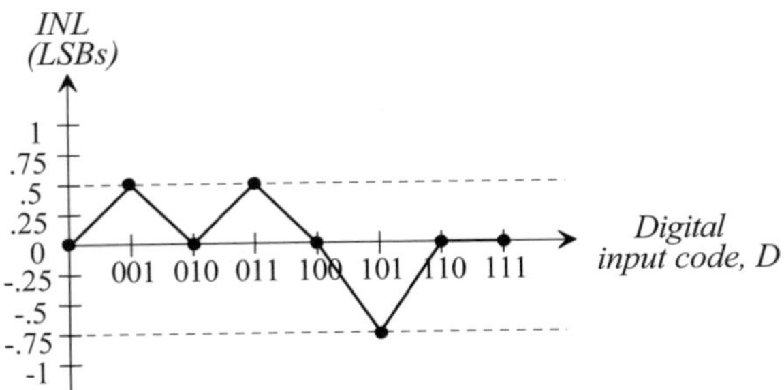

Figure 28.15 INL curve for the nonideal 3-bit DAC.

It should be noted that other methods are used to determine INL. One method compares the output values to the ideal reference line, regardless of the positions of the first and last output values. If the DAC has an offset voltage or gain error, this will be included in the INL determination. Usually, the offset and gain errors are determined as separate specifications.

Another method, described as the "best-fit" method, attempts to minimize the INL by constructing the reference line so that it passes as closely as possible to a majority of the output values. Although this method does minimize the INL error, it is a rather subjective method that is not as widely used as drawing the reference line through the first and last output values.

Offset

The analog output should be 0 V for $D = 0$. However, an offset exists if the analog output voltage is not equal to zero. This can be seen as a shift in the transfer curve as illustrated in Fig. 28.16. This specification is similar to the offset voltage for an operational amplifier except that it is not referred to the input.

Gain Error

A gain error exists if the slope of the best-fit line through the transfer curve is different from the slope of the best-fit line for the ideal case. For the DAC illustrated in Fig. 28.17, the gain error becomes

$$\text{Gain error} = \text{Ideal slope} - \text{Actual slope} \tag{28.14}$$

Latency

This specification defines the total time from the moment that the input digital word changes to the time the analog output value has settled to within a specified tolerance. Latency should not be confused with settling time, since latency includes the delay required to map the digital word to an analog value plus the settling time. It should be noted that settling time considerations are just as important for a DAC as they are for a S/H or an operational amplifier.

Signal-to-Noise Ratio (*SNR*)

Signal-to-noise (SNR) is defined as the ratio of the signal power to the noise at the analog output. In amplifier applications, this specification is typically measured using a

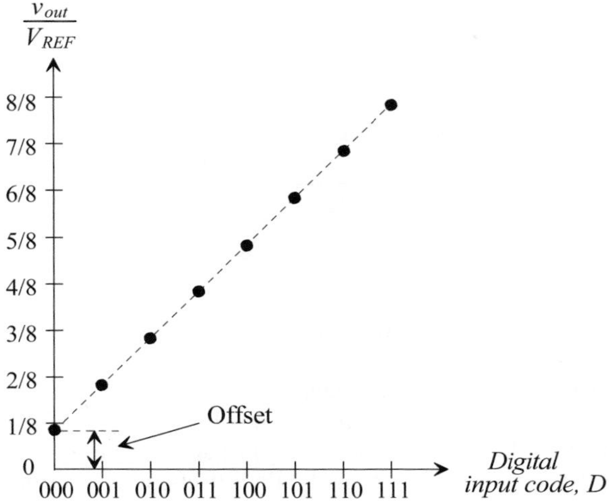

Figure 28.16 Illustration of offset error for a 3-bit DAC.

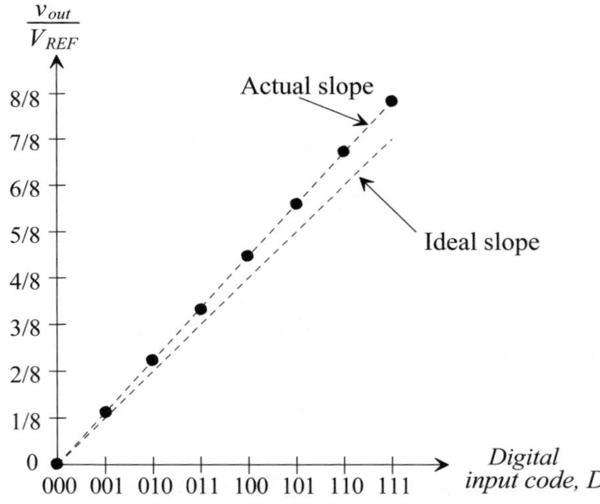

Figure 28.17 Illustration of gain error for a 3-bit DAC.

sinewave input. For the DAC, a "digital" sinewave is generated through instrumentation or through an A/D. The SNR can reveal the true resolution of a data converter as the effective number of bits can be quantified mathematically. A detailed derivation of the SNR is presented in Sec. 28.6, on the discussion of ADC specifications.

Dynamic Range

Dynamic range is defined as the ratio of the largest output signal over the smallest output signal. For both DACs and ADCs, the dynamic range is related to the resolution of the converter. For example, an N-bit DAC can produce a maximum output of $2^N - 1$ multiples of LSBs and a minimum value of 1 LSB. Therefore, the dynamic range in decibels is simply

$$DR = 20\text{Log}\left(\frac{2^N - 1}{1}\right) \approx 6.02 \cdot N \text{ dB} \qquad (28.15)$$

A 16-bit data converter has a dynamic range of 96.33 dB.

28.5 Analog-to-Digital Converter (ADC) Specifications

Many of the specifications that describe the ADC are similar to those that describe the DAC. However, there are subtle differences. Since the DAC is converting a discrete signal into an analog representation that is also limited by the resolution of the converter, a fixed number of inputs and outputs are generated. However, with the ADC, the input is an analog signal with an infinite number of values, which then has to be quantized into an N-bit digital word (Fig. 28.18). This process is much more difficult than the digital-to-analog process. In fact, many ADC architectures use a DAC as a critical component.

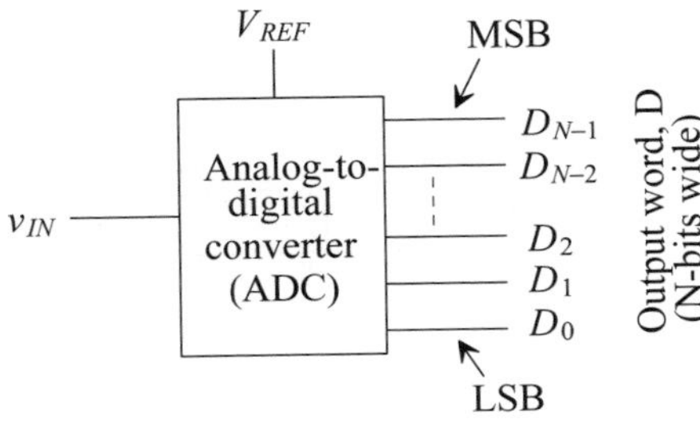

Figure 28.18 Block diagram of the analog-to-digital converter.

For example, in the previous discussion of DACs, it was determined that for a 16-bit DAC, the converter would need to generate output voltages in multiples of 76 μV. However, for the ADC, the converter needs to resolve differences in the analog signal of 76 μV. This means that the ADC must be able to detect changes in the input signal on the order of 1 part in 65,536! In contrast, the DAC had a finite number of input combinations (2^N). The ADC, however, has to "quantize" the infinite-valued analog signal into many segments so that

$$\text{Number of quantization levels} = 2^N \qquad (28.16)$$

This distinction is subtle but must be recognized to understand the differences between the two types of conversion.

Examine Fig. 28.19a. The digital output, D, of an ideal, 3-bit ADC is plotted versus the analog input, v_{IN}. Note the difference in the transfer curve for the ADC versus the DAC (Fig. 28.10). The y-axis is now the digital output, and the x-axis has been normalized to V_{REF}. Since the input signal is a continuous signal and the output is discrete, the transfer curve of the ADC resembles that of a staircase. Another fact to observe is that the 2^N quantization levels correspond to the digital output codes 0 to 7. Thus, the maximum output of the ADC will be 111 ($2^N - 1$), corresponding to the value for which $\frac{v_{IN}}{V_{REF}} \geq \frac{7}{8}$. Figure 28.19b corresponds to the error caused by the quantization.

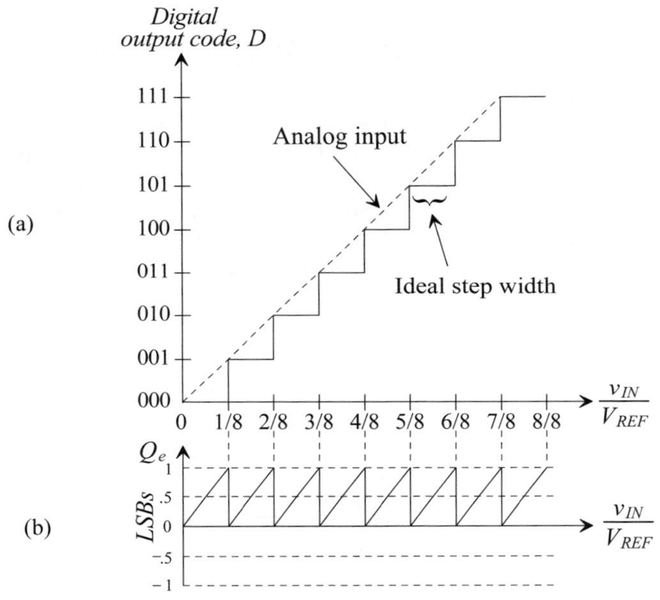

Figure 28.19 (a) Transfer curve for an ideal ADC and (b) its corresponding quantization error.

The value of 1 LSB for this ADC can be calculated using Eq. (28.9) and is the ideal step width (1/8) in Fig. 28.19 (versus the height for the DAC) multiplied by V_{REF}. Therefore, assuming that $V_{REF} = 5$ V,

$$1 \text{ LSB} = 0.625 \text{ V} \tag{28.17}$$

Quantization Error

Since the analog input is an infinite valued quantity and the output is a discrete value, an error will be produced as a result of the quantization. This error, known as *quantization error*, Q_e, is defined as the difference between the actual analog input and the value of the output (staircase) given in voltage. It is calculated as

$$Q_e = v_{IN} - V_{staircase} \tag{28.18}$$

where the value of the staircase output, $V_{staircase}$, can be calculated by

$$V_{staircase} = D \cdot \frac{V_{REF}}{2^N} = D \cdot V_{LSB} \tag{28.19}$$

where D is the value of the digital output code and V_{LSB} is the value of 1 LSB in volts, in this case 0.625 V. We can also easily convert the value of Q_e in units of LSBs. In Fig. 28.19a, Q_e can be generated by subtracting the value of the staircase from the dashed line. The result can be seen in Fig. 28.19b. A sawtooth waveform is formed centered about ½ LSBs. Ideally, the magnitude of Q_e will be no greater than one LSB and no less than 0. It would be advantageous if the quantization error were centered about zero so that the error would be at most ±½ LSBs (as opposed to +1LSB). This is easily achieved as seen in Fig. 28.20a and b. Here, the entire transfer curve is shifted to the left by ½ LSB, thus making the codes centered around the LSB increments on the x-axis. This drawing illustrates that at best, an ideal ADC will have quantization error of ±½ LSB.

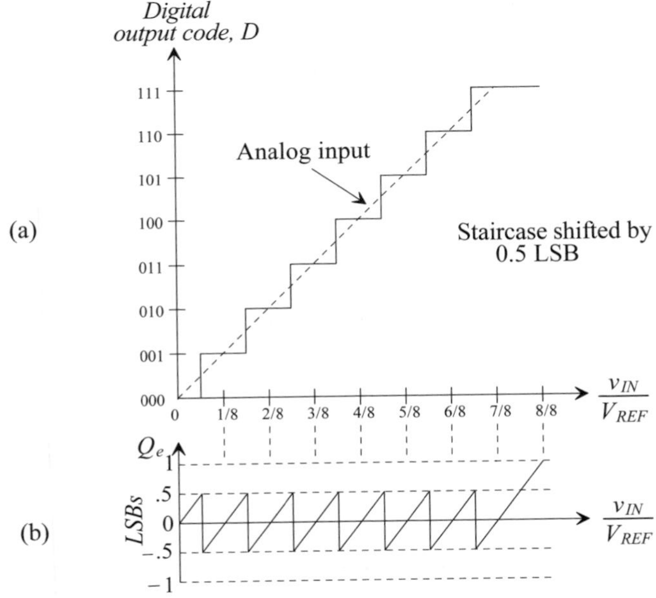

Figure 28.20 (a) Transfer curve for an ideal 3-bit ADC with (b) quantization error centered about zero.

In shifting this curve to the left, notice that the first code transition occurs when $\frac{v_{IN}}{V_{REF}} \geq \frac{1}{16}$. Therefore, the range of $\frac{v_{IN}}{V_{REF}}$ for the digital output corresponding to 000 is half as wide as the ideal step width. The last code transition occurs when $\frac{v_{IN}}{V_{REF}} \geq \frac{13}{16}$ (between 6/8 and 7/8). Note that the step width corresponding to this last code transition is 1.5 times larger than the ideal width and that the quantization error extends up to 1 LSB when $\frac{v_{IN}}{V_{REF}} = 1$. However, the converter would be considered to be out of range once $\frac{v_{IN}}{V_{REF}} \geq \frac{15}{16}$ (halfway between 7/8 and 8/8), so the problem is moot.

Differential Nonlinearity

Differential nonlinearity for an ADC is similar to that defined for a DAC. However, for the ADC, DNL is the difference between the actual code *width* of a nonideal converter and the ideal case. Figure 28.21 illustrates the transfer curve for a nonideal 3-bit ADC. The values for the DNL can be solved as follows:

$$\text{DNL} = \text{Actual step width} - \text{Ideal step width} \tag{28.20}$$

Since the step widths can be converted to either volts for LSBs, DNL can be defined using either units. The value of the ideal step is 1/8. Converting to volts, this becomes

$$V_{idealstepwidth} = \frac{1}{8} \cdot V_{REF} = 0.625\ V = 1\ \text{LSB} \tag{28.21}$$

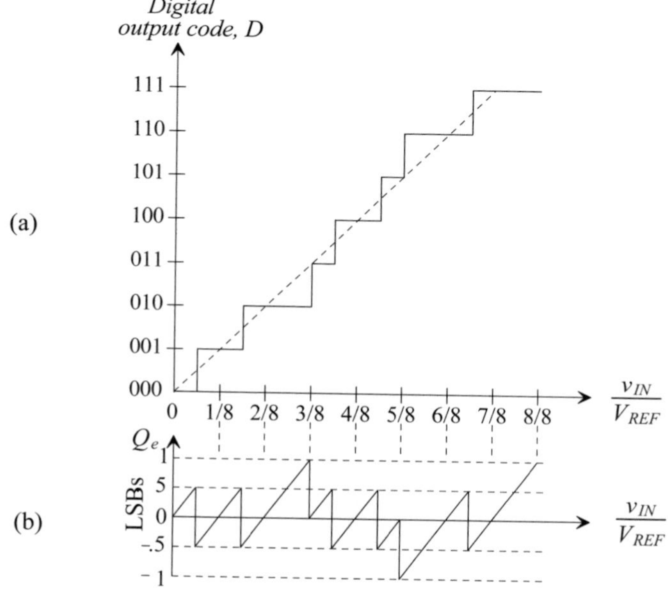

Figure 28.21 (a) Transfer curve for a nonideal 3-bit ADC used in Ex. 28.4 with (b) quantization error illustrating differential nonlinearity.

Example 28.6

Using Fig. 28.21a, calculate the differential nonlinearity of the 3-bit ADC. Assume that $V_{REF} = 5$ V. Draw the quantization error, Q_e, in units of LSBs.

The DNL of the converter can be calculated by examining the step width of each digital output code. Since the ideal step width of the 000 transition is ½ LSB, then $\text{DNL}_0 = 0$. Also note that the step widths associated with 001 and 100 are equal to 1 LSB; therefore, both DNL_1 and DNL_4 are zero. However, the remaining values code widths are not equal to the ideal value but can be calculated as

$$DNL_2 = 1.5\text{ LSB} - 1\text{ LSB} = 0.5\text{ LSB}$$

$$DNL_3 = 0.5\text{ LSB} - 1\text{ LSB} = -0.5\text{ LSB}$$

$$DNL_5 = -0.5\text{ LSB}$$

$$DNL_6 = 0.5\text{ LSB}$$

$$DNL_7 = 0\text{ LSB (since the ideal step width is 1.5 LSB wide at this code transition)}$$

The overall DNL for the converter used in this illustration is ±0.5 LSB. Note that the quantization error illustrated in Fig. 28.21b is directly related to the DNL. As DNL increases in either direction, the quantization error worsens. Each "tooth" in the quantization error waveform should ideally be the same size. ■

Missing Codes

It is of interest to note the consequences of having a DNL that is equal to –1 LSB. Figure 28.22 illustrates an ADC for which this is true. The total width of the step corresponding to 101 is completely missing; thus, the value of DNL_5 is –1 LSB. Any ADC possessing a DNL that is equal to –1 LSB is guaranteed to have a missing code. Notice that the step width corresponding to 010 is 2 LSBs and that the value for DNL_2 is +1 LSB. However, there is not a missing code corresponding to 011, since the step width of code 011 depends on the 100 transition. Therefore, an ADC having a DNL greater than +1 LSB is not guaranteed to have a missing code, though in all probability a missing code will occur.

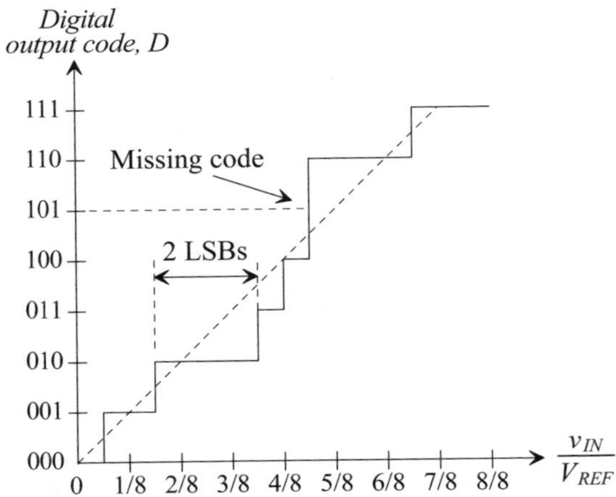

Figure 28.22 Transfer curve for a nonideal 3-bit ADC with a missing code.

Integral Nonlinearity

Integral nonlinearity (INL) is defined similarly to that for a DAC. Again, a "best-fit" straight line is drawn through the end points of the first and last code transition, with INL being defined as the difference between the data converter code transition points and the straight line with all other errors set to zero.

Example 28.7
Determine the INL for the ADC whose transfer curve is illustrated in Fig. 28.23a. Assume that V_{REF} = 5 V. Draw the quantization error, Q_e, in units of LSBs.

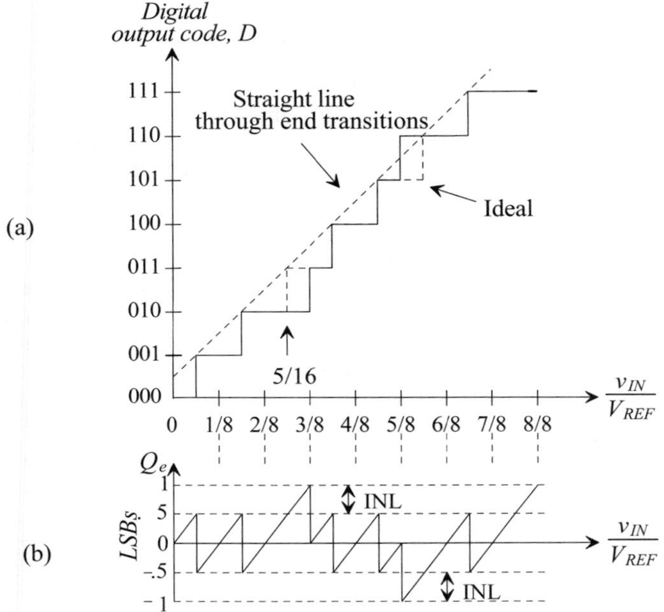

Figure 28.23 (a) Transfer curve of a nonideal 3-bit ADC and (b) its quantization error illustrating INL.

By inspection, it can be seen that all of the transition points occur on the best-fit line except for the transitions associated with code 011 and 110. Therefore,

$$INL_0 = INL_1 = INL_2 = INL_4 = INL_5 = INL_7 = 0$$

The INL corresponding to the remaining codes can be calculated as

$$INL_3 = 3/8 - 5/16 = 1/16 \text{ or } 0.5 \text{ LSB}$$

Similarly, INL_6 can be calculated in the same manner and is found to be –0.5 LSB. Thus, the overall INL for the converter is the maximum value of INL corresponding to ±0.5 LSB.

The INL can also be determined by inspecting the quantization error in Fig. 28.23b. Here, the INL will be the magnitude of the quantization error which lies outside the ±½ LSB band of Q_e. It can be seen that Q_e = 1 LSB, corresponding to the point at which INL = 0.5 LSB for digital output code 011, and that Q_e = –1 LSB at the output code corresponding to INL = –0.5 LSB for digital output code 110. ■

Offset and Gain Error

Offset and gain error are identical to the DAC case. *Offset error* occurs when there is a difference between the value of the first code transition and the ideal value of ½ LSBs. As seen in Fig. 28.24a, the offset error is a constant value. Note that the quantization error becomes ideal after the initial offset voltage is overcome. *Gain error* or *scale factor error*, seen in Fig. 28.24b, is the difference in the slope of a straight line drawn through the transfer characteristic and the slope of 1 of an ideal ADC. Causes of offset and gain error are discussed in Ch. 29, but it is important here to understand their overall effects on ADC transfer curves.

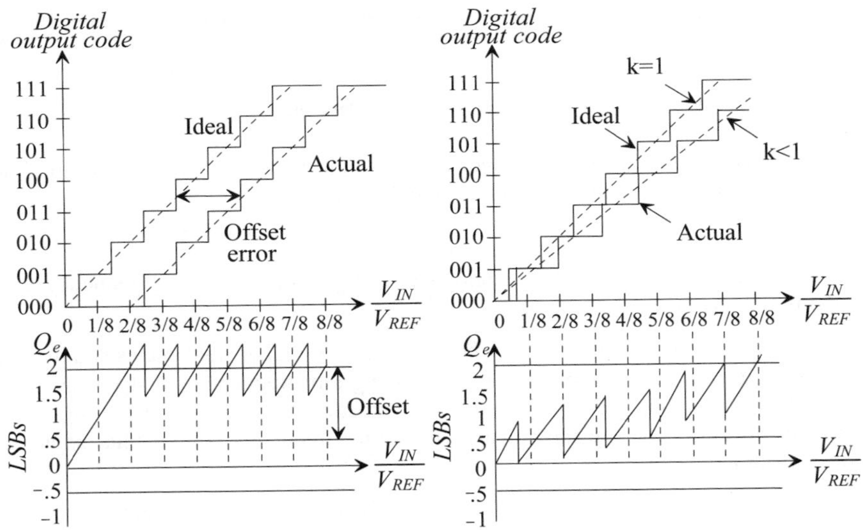

Figure 28.24 Transfer curve illustrating (a) offset error and (b) gain error.

So far, we have examined only the DC characteristics of an ADC. However, examining the dynamic aspects of the converter will lead to a whole new set of errors. Sampling is inherently a dynamic process since the accuracy of the sample is dependent on the speed of the analog signal. Many effects that occur during sampling limit the overall performance of the converter.

Aliasing

As mentioned earlier in the chapter, the Nyquist Criterion requires that a signal be sampled at least two times the highest frequency contained in the signal. What would happen if this criterion were ignored and the sampling rate was actually less than that amount? A phenomenon known as aliasing would occur.

Examine Fig. 28.25. Here, an analog signal is being sampled at a rate slower than the Nyquist Criterion requires. As a result, it appears that a totally different signal (see

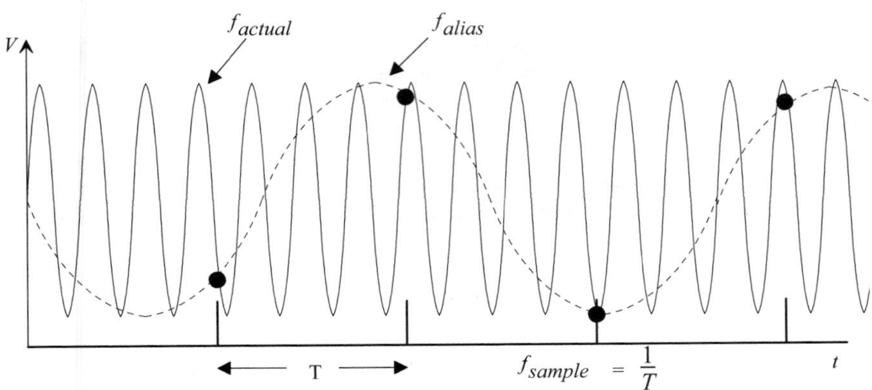

Figure 28.25 Aliasing caused by undersampling.

example dashed line) is being sampled. The different frequency signal is an "alias" of the original signal, and its frequency can be calculated using

$$f_{alias} = f_{actual} + k f_{sample} \quad (k = \ldots -2, -1, 0, 1, 2, 3 \ldots) \tag{28.22}$$

where f_{actual} is the frequency of the analog signal, f_{sample} is the sampling frequency, and f_{alias} is the frequency of the alias signal.

Aliasing can be eliminated by both sampling at higher frequencies and by filtering the analog signal before sampling and removing any frequencies that are greater than one-half the sampling frequency. It is good practice to filter the analog signal before sampling to eliminate any unknown higher order frequency components or noise that could result in aliasing.

A frequency domain analysis may further illustrate the concepts of aliasing. Figure 28.26 shows the analog signal, the *sampling function* (represented by a unit impulse train) and the resulting sampled signal in both the time and frequency domains. The analog signal in Fig. 28.26a is represented as a simple band-limited signal with center frequency, f_o. This simply means that the signal is contained within the frequency range shown. In Fig. 28.26b, the sampling function is shown in both the time and frequency domain. The sampling function simply represents the action of sampling at discrete points in time. The frequency domain version of the sampling function is similar to its time domain counterpart, except that the x-axis is now represented as $f = 1/T$. Since each of the impulses has a value of 1, the resulting sampled signal shown in Fig. 28.26c is the impulse function multiplied by the amplitude of the analog signal at each discrete point in time. Remembering that multiplication in the time domain is equivalent to convolution in the frequency domain, we note that the frequency domain representation of the sampled signal reveals that the overall signal consists of multiple versions of the band-limited signal at multiples of the sampling frequency.

Note in Fig. 28.26b that as the sampling time increases, the sampling frequency decreases and the impulses in the frequency domain become more closely spaced. This results in 28.26d, which illustrates the aliasing as the multiple versions of the

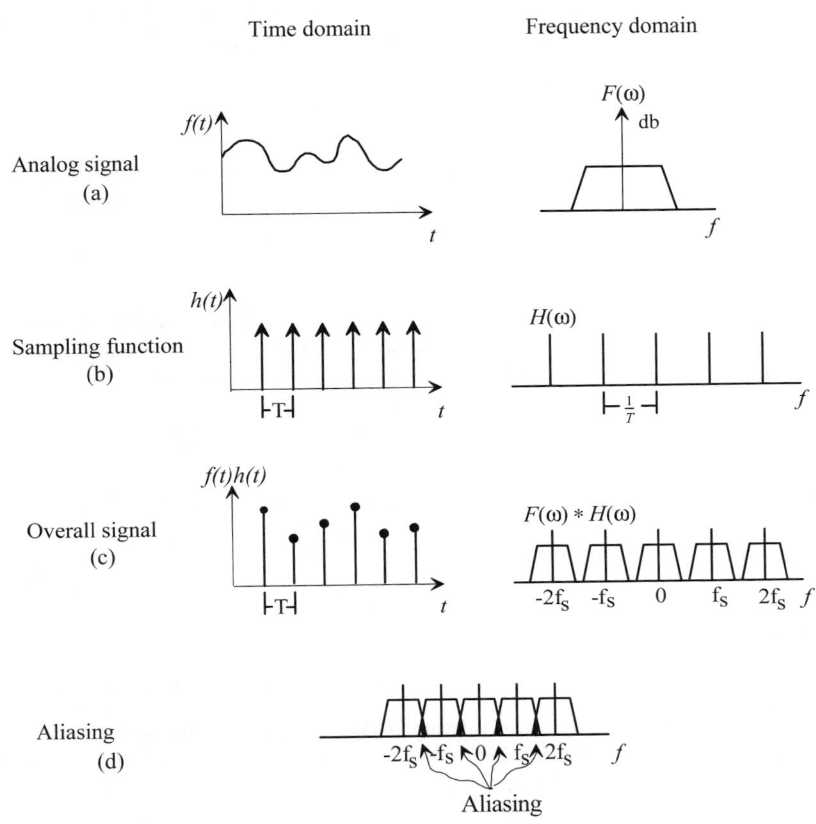

Figure 28.26 Illustration of aliasing in the time and frequency domain. (a) The analog signal; (b) the sampling function; (c) the overall signal; and (d) aliasing in the frequency domain.

band-limited signal begin to overlap. One could also filter the signal "post-sampling" and eliminate the frequencies for which overlap occurs. The point at which the spectra overlap is called the *folding frequency*.

As mentioned earlier, the solutions to aliasing are higher sampling frequency and filtering. Focusing on just one of the solutions may worsen the situation for several reasons. Some noise signals are wide band, which means that they have a large bandwidth. Attempting to increase only the sampling frequency to eliminate the aliasing effects of the noise would be a practical impossibility, not to mention a costly one. However, simply filtering the input signal and the sampled signal adds delays to the overall conversion and increases the expense of the circuit. It is best to use a combination of the two to minimize the problem most efficiently.

Signal-to-Noise Ratio

Signal-to-noise (SNR) ratios of ADCs represent the value of the largest RMS input signal into the converter over the RMS value of the noise. Typically given in dB, the expression for SNR is

$$SNR = 20\text{Log}\left(\frac{v_{in(max)}}{v_{noise}}\right) \tag{28.23}$$

If it is assumed that the input signal is a sinewave with a peak-to-peak value equal to the full-scale reference voltage of the converter, then the RMS value for $v_{in(max)}$ becomes

$$v_{in(max)} = \frac{V_{REF}}{2\sqrt{2}} = \frac{2^N(V_{LSB})}{2\sqrt{2}} \tag{28.24}$$

where V_{LSB} is the voltage value of 1 LSB. The value of the noise (if the data converter is considered to be ideal) will be equivalent to the RMS value of the error signal, Q_e (in volts), shown in Fig. 28.20b. The RMS value of Q_e can be calculated to be

$$Q_{e,RMS} = \left[\frac{1}{V_{LSB}}\int_{-0.5VLSB}^{0.5VLSB}(V_{LSB})^2 dV_{LSB}\right]^{0.5} = \frac{V_{LSB}}{\sqrt{12}} \tag{28.25}$$

Therefore, the SNR for the ideal ADC will be the ratio of these two RMS values,

$$SNR = 20\cdot\text{Log}\frac{\frac{2^N(V_{LSB})}{2\sqrt{2}}}{Q_{e,RMS}} \tag{28.26}$$

which can be written in terms of N as simply

$$SNR = 20N\text{Log}(2) + 20\text{Log}\sqrt{12} - 20\text{Log}(2\sqrt{2}) = 6.02N + 1.76 \tag{28.27}$$

Equation (28.27) is an important one relating SNR to the resolution of the ADC. For 16-bit data conversion, one must design a circuit that will have an SNR of (6.02)(16) + 1.76 = 98.08 dB! Equation (28.27) can also be used in calculating the *signal-to-noise plus distortion ratio*, also known as *SNDR*. Since the output data is digital, we cannot use a spectrum analyzer to calculate this ratio but must instead use a *Discrete Fourier Transform* (*DFT*) and examine the data in the digital domain.

Another useful application of Eq. (28.27) is the determination of effective number of bits given a system with a known SNR or SNDR. For example, if a 16-bit ADC yielded an SNDR of 88 dB, then the effective resolution of the converter would be

$$N = \frac{88 - 1.76}{6.02} = 14.32 \text{ bits} \tag{28.28}$$

and the ADC would be producing the resolution equivalent to that of a 14-bit converter.

Aperture Error

The aperture error described in Sec. 28.3 (S/H) should be related to the errors associated with the ADC. In the previous discussion, the aperture error resulted in sampling error (Fig. 28.8). However, now that ADC characteristics have been discussed, we can relate the sampling error to the ADC. Since we know that the maximum errors associated with an ADC are related to ½ LSB, we can assume that the maximum sampling error associated with the aperture uncertainty can be no larger than ½ LSB.

Example 28.8
Find the maximum resolution of an ADC which can use the S/H described in Ex. 28.1 while maintaining a sampling error less than ½ LSB.

Since it was determined that the maximum sampling error produced by the given aperture uncertainty was 0.628 mV, we can relate this value to the highest resolution of an ADC by assuming that 0.628 mV will be less than or equal ½ LSB. Therefore,

$$0.628 \text{ mV} \leq .5\ LSB = \frac{V_{REF}}{2^{N+1}} = \frac{5}{2^{N+1}}$$

or

$$2^{N+1} \leq 7961.8$$

which, solving for N (limited to an integer), yields a maximum resolution of 11 bits. ■

28.6 Mixed-Signal Layout Issues

Naturally, analog ICs are more sensitive to noise than digital ICs. For any analog design to be successful, careful attention must be paid to layout issues, particularly in a digital environment. Sensitive analog nodes must be protected and shielded from any potential noise sources. Grounding and power supply routing must also be considered when using digital and analog circuitry on the same substrate. Since a majority of ADCs use switches controlled by digital signals, separate routing channels must be provided for each type of signal.

Techniques used to increase the success of mixed-signal designs vary in complexity and priority. Strategies regarding the systemwide minimization of noise should always be considered foremost. A mixed-signal layout strategy can be modeled as seen in Fig. 28.27. The lowest issues are foundational and must be considered before each succeeding step. The successful mixed-signal design will always minimize the effect of the digital switching on the analog circuits.

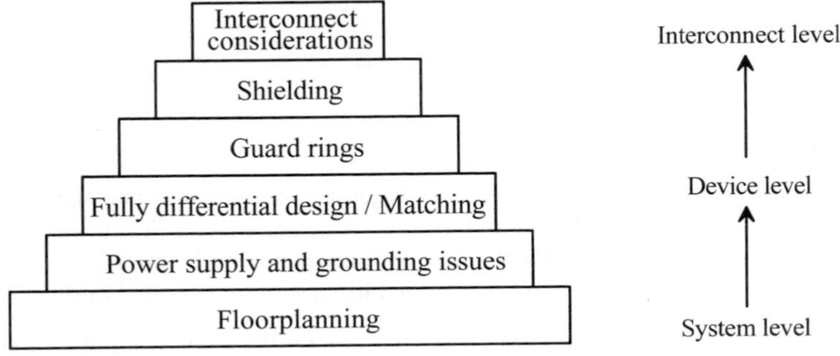

Figure 28.27 Mixed-signal layout strategy.

Floorplanning

The placement of sensitive analog components can greatly affect a circuit's performance. Many issues must be considered. In designing a mixed-signal system, strategies regarding the "floorplan" of the circuitry should be thoroughly analyzed well before the layout is to begin.

The analog circuitry should be categorized by the sensitivity of the analog signal to noise. For example, low-level signals or high-impedance nodes typically associated with input signals are considered to be sensitive nodes. These signals should be closely guarded and shielded, especially from digital output buffers. High-swing analog circuits such as comparators and output buffer amplifiers should be placed between the sensitive analog and the digital circuitry.

The digital circuitry should also be categorized by speed and function. Obviously, since digital output buffers are usually designed to drive capacitive loads at very high rates, they should be kept farthest from the sensitive analog signals. Next, the high and lower speed digital should be placed between the insensitive analog and the output buffers. An example of this type of strategy can be seen in Fig. 28.28 [6]. Notice that *the sensitive analog is as far away as possible from the digital output buffers* and that the least sensitive analog circuitry is next to the least offensive digital circuitry.

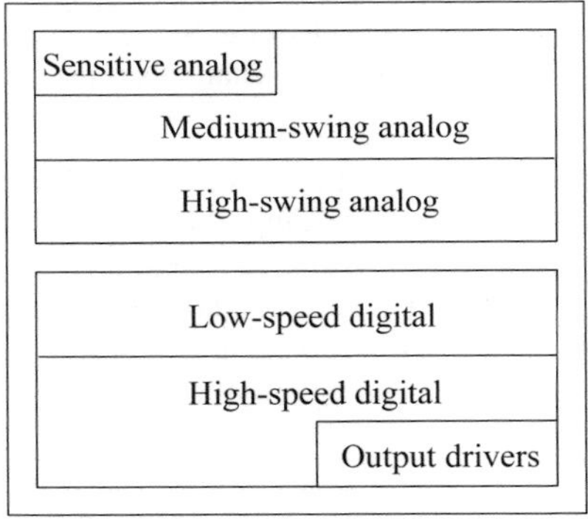

Figure 28.28 Example of a mixed-signal floorplan [6].

Power Supply and Grounding Issues

Whenever analog and digital circuit reside together on the same die, danger exists of injecting noise from the digital system to the sensitive analog circuitry through the power supply and ground connections. Much of the intercoupling can be minimized by carefully considering how power and ground are supplied to both the analog and digital circuits.

In Fig. 28.28a analog and digital circuitry share the same routing to a single pad for power and ground. The resistors, R_{I1} and R_{I2}, represent the small, nonnegligible resistance of the interconnect to the pad. The inductors, L_{B1} and L_{B2}, represent the inductance of the bonding wire which connects the pads to the pin on the lead frame.

Since digital circuitry is typified by high amounts of transient currents due to switching, a small amount of resistance associated with the interconnect can result in significant voltage spikes. Low-level analog signals are very sensitive to such interference, thus resulting in a contaminated analog system. Another significant voltage spike can occur due to the inductance of the bonding wire. Since the voltage across the inductor is proportional to the change in current through it, voltage spikes equating to hundreds of millivolts can result! Both of these voltage effects are true for both the power and ground connection.

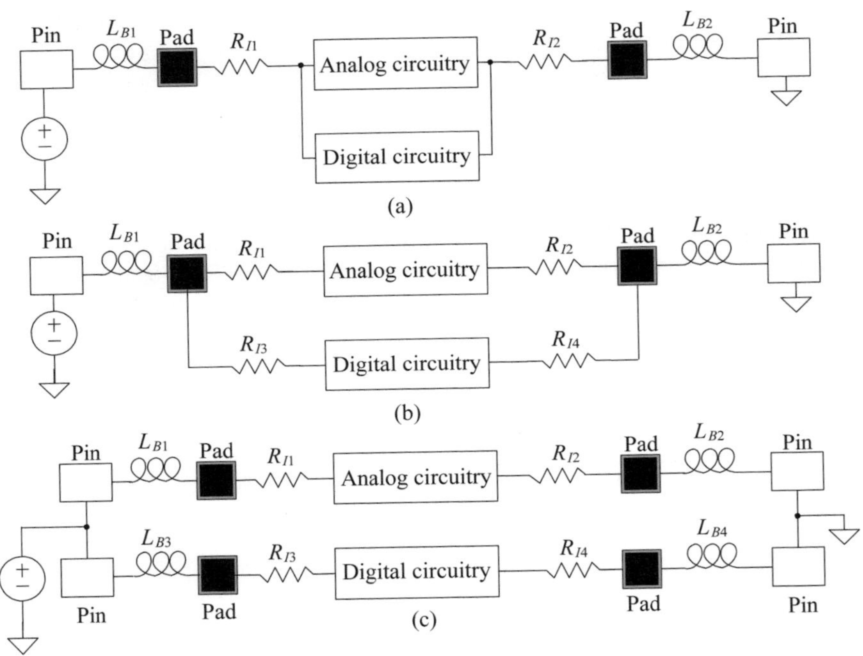

Figure 28.29 Power and ground connection examples, (a) poor noise immunity and (b) better noise immunity, and (c) using separate power and ground pins to achieve even better immunity.

One way to reduce the interference, seen in Fig. 28.29b, is to prohibit the analog and digital circuit from sharing the same interconnect. The routing for the supply and ground for both the analog and digital sections are provided separately. Although this eliminates the parasitic resistance due to the common interconnect, there is still a common inductance due to the bonding wire which causes interference.

Another method that minimizes interference even more than the previous case is seen in Fig. 28.29c. *By using separate pads and pins, the analog and the digital circuits are completely decoupled.* The current through the analog interconnect is much less abrupt than the digital; thus, the analog circuitry now has a "quiet" power and ground. However, this technique depends on whether extra pins and pads are available for this use. The separate power supply and ground pins are then connected externally. *It is not wise to use two separate power supplies because if both types of circuits are not powered up simultaneously, latch-up could easily result.*

In cases presented in Fig. 28.29b and c, *the resistance associated with the analog connection to ground or supply can be reduced by making the power supply and ground bus as wide as feasible.* This reduces the overall resistance of the metal run, thus decreasing the voltage spikes that occur across the resistor. The inductor itself is impossible to eliminate, though it can be minimized with careful planning. Since the length of the bonding wire depends on the distance from the pad to the lead frame, *one could reduce the effect of the wire inductance by reserving pins closest to the die for sensitive connections such as analog supply and ground.* This, again, illustrates the importance of floorplanning.

Fully Differential Design

Fully differential operational amplifiers were discussed earlier, Fig. 28.30. The noise sources represent the noise from digital circuitry coupled through the parasitic stray capacitors. If equal amounts of noise are injected into the differential amplifiers, then the common-mode rejection inherent in the amplifiers will eliminate most or all of the noise. This, of course, depends on the symmetry of the amplifiers, meaning that matching the transistors in the amplifier becomes crucial. Therefore, *in a mixed-signal environment, layout techniques should be used to improve matching such as common-centroid and interdigitated techniques*.

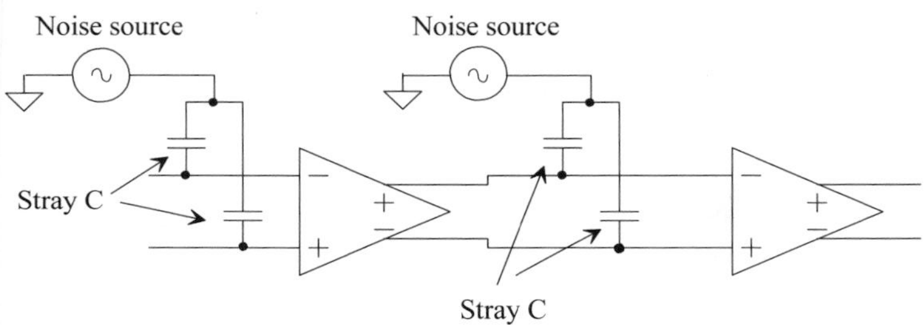

Figure 28.30 Differential output op-amps showing parasitic coupling to noise sources.

Guard Rings

Guard rings should be used wisely throughout a mixed-signal environment. Circuits that process sensitive signals should be placed in a separate well (if possible) with guard rings attached to the analog *VDD* supply. In the case of an n-well (only) process, the n-type devices outside the well should have guard rings attached to analog ground placed around them. Digital circuits should be placed in their own well with guard rings attached to digital *VDD*. Guard rings placed around the n-channel digital devices also help minimize the amount of noise transmitted from the digital devices.

Shielding

A number of techniques exist which can shield sensitive, low-level analog signals from noise resulting from digital switching. A shield can take the form of a layer tied to analog ground placed between two other layers, or it can be a barrier between two signals running in parallel.

If at all possible, one should avoid crossing sensitive analog signals, such as low-level analog input signals, with any digital signals. The parasitic capacitance coupling the two signal lines can be as much as a couple of fF, depending on the process. If it cannot be avoided, then attempt to carry the digital signal using the top layer of metal (such as metal2). If the analog signal is an input signal, then it will most likely be carried by the poly layer (or a lower level of metal). A strip of metal1 can be placed between the two layers and connected to analog ground (see Fig. 28.31).

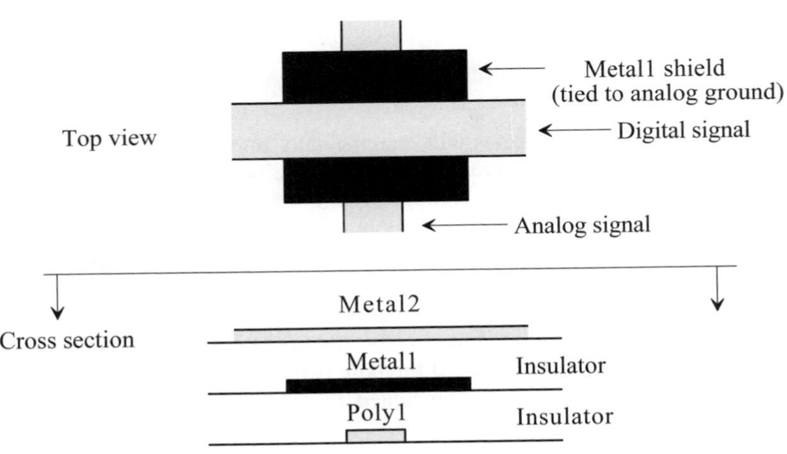

Figure 28.31 Shielding a sensitive analog signal from a digital signal crossover using a metal 1 shield layer.

Another situation that should be avoided is running interconnect containing sensitive analog signals parallel and adjacent to any interconnect carrying digital signals. Coupling occurs due to the parasitic capacitance between the lines. If this situation cannot be avoided, then an additional line connected to analog ground should be placed between the two signals, as seen in Fig. 28.32. This method can also be used to partition the analog and digital sections of the chip.

In addition, the n-well can be used as a bottom-plate shield to protect analog signals from substrate noise. Poly resistors (or capacitors) used for sensitive analog signals can be shielded by placing an n-well beneath the components and connecting the well to analog *VDD*.

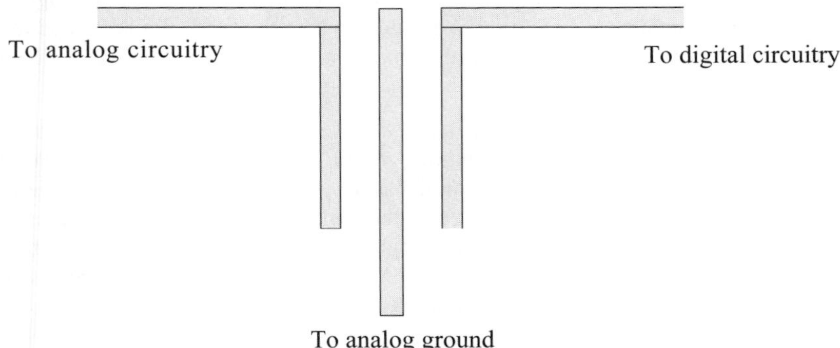

Figure 28.32 Using a dummy metal strip to provide shielding to two parallel signals.

Other Interconnect Considerations

Finally, some other layout strategies will incrementally improve the performance of the analog circuitry. However, if the previous strategies are not followed, these suggestions will be useless. *When routing the analog circuitry, minimize the lengths of current carrying paths.* This will simply reduce the amount of voltage drop across the path due to the metal1 or metal2 resistance. *Contacts should also be used very liberally whenever changing layers.* Not only does this minimize resistance in the path, but it also improves fabrication reliablity. *Avoid using poly to route current carrying signal paths.* Not only is the poly higher in resistance value, but also the additional contact resistance required to change layers will not be insignificant. If the poly is made wider to lower the resistance, additional parasitic capacitance will be added to the node. *Use poly to route only high-impedance gate nodes that carry virtually no current.*

REFERENCES

[1] M. J. Demler, *High-Speed Analog-to-Digital Conversion*, Academic Press, 1991.

[2] B. Razavi, *Principles of Data Conversion System Design*, IEEE Press, 1995.

[3] R. L. Geiger, P. E. Allen, and N. R. Strader, *VLSI Design Techniques for Analog and Digital Circuits*, McGraw-Hill Publishing Co., 1990.

[4] D. H. Sheingold, *Analog-Digital Conversion Handbook*, Prentice-Hall Publishing, 1986.

[5] S. K. Tewksbury, et al., "Terminology Related to the Performance of S/H, A/D and D/A Circuits," *IEEE Transactions on Circuits and Systems*, CAS-25, vol. CAS-25, pp. 419–426, July 1978.

[6] Y. Tsividis, *Mixed Analog-Digital VLSI Devices and Technology: An Introduction*, McGraw-Hill Publishing Co., 1996.

PROBLEMS

28.1 Determine the number of quantization levels needed if one wanted to make a digital thermometer that was capable of measuring temperatures to within 0.1 °C accuracy over a range from –50 °C to 150 °C. What resolution of ADC would be required?

28.2 Using the same thermometer as above, what sampling rate, in samples per second, would be required if the temperature displayed a frequency of $15° \cdot \sin(0.01 \cdot 2\pi t)$?

28.3 Determine the maximum droop allowed in an S/H used in a 16-bit ADC assuming that all other aspects of both the S/H and ADC are ideal. Assume $V_{ref} = 5$ V.

28.4 An S/H circuit settles to within 1 percent of its final value at 5 μs. What is the maximum resolution and speed with which an ADC can use this data assuming that the ADC is ideal?

28.5 A digitally programmable signal generator uses a 14-bit DAC with a 10-volt reference to generate a DC output voltage. What is the smallest incremental change at the output that can occur? What is the DAC's full-scale value? What is its accuracy?

28.6 Determine the maximum DNL (in LSBs) for a 3-bit DAC, which has the following characteristics. Does the DAC have 3-bit accuracy? If not, what is the resolution of the DAC having this characteristic?

Digital Input	**Voltage Output**
000	0 V
001	0.625 V
010	1.5625 V
011	2.0 V
100	2.5 V
101	3.125 V
110	3.4375 V
111	4.375 V

28.7 Repeat Problem 28.6 calculating the INL (in LSBs).

28.8 A DAC has a reference voltage of 1,000 V, and its maximum INL measures 2.5 mV. What is the maximum resolution of the converter assuming that all the other characteristics of the converter are ideal?

28.9 Determine the INL and DNL for a DAC that has a transfer curve shown in Fig. 28.33.

28.10 A DAC has a full-scale voltage of 4.97 V using a 5 V reference, and its minimum output voltage is limited by the value of one LSB. Determine the resolution and dynamic range of the converter.

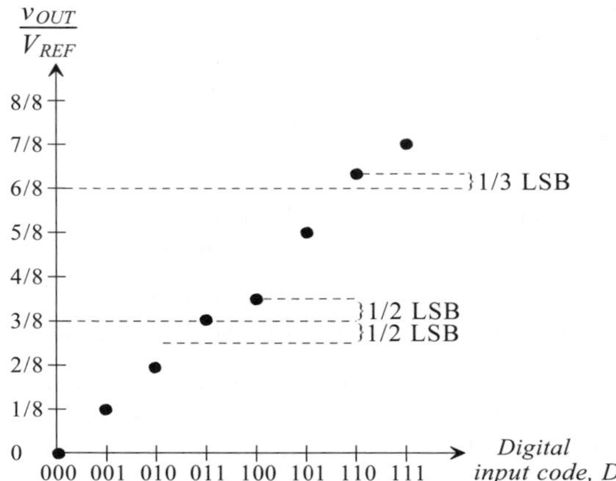

Figure 28.33 Transfer curves for Problem 28.9.

28.11 Prove that the RMS value of the quantization noise shown in Fig. 28.20b is as stated in Eq. (28.26).

28.12 An ADC has a stated SNR of 94 dB. Determine the effective number of bits of resolution of the converter.

28.13 Discuss the methods used to prevent aliasing and the advantages and disadvantages of each.

Chapter 29

Data Converter Architectures

Applications such as wireless communications and digital audio and video have created the need for cost-effective data converters that will achieve higher speed and resolution. The needs required by digital signal processors continually challenge analog designers to improve and develop new ADC and DAC architectures. There are many different types of architectures, each with unique characteristics and different limitations. This chapter presents a basic overview of the more popular data converter architectures and discusses the advantages and disadvantages of each along with their limitations.

Now that we have defined the operating characteristics of ADCs in Ch. 28, a more detailed examination of the basic architectures will be discussed using a top-down approach. Because many of the converters use op-amps, comparators, and resistor and capacitor arrays, the top-down approach will allow a broader discussion of the key component limitations in later sections.

29.1 DAC Architectures

A wide variety of DAC architectures exist, ranging from very simple to complex. Each, of course, has its own merits. Some use voltage division, whereas others employ current steering and even charge scaling to map the digital value into an analog quantity.

29.1.1 Digital Input Code

In many cases, the digital signal is not provided in binary code but is any one of a number of codes: binary, BCD, thermometer code, Gray code, sign-magnitude, two's complement, offset binary, and so on. (See Fig. 29.1 for a comparison of some of the more commonly used digital input codes.) For example, it may be desirable to allow only one bit to change value when changing from one code to the next. If that is the case, a Gray code will suffice. The thermometer code is used quite frequently and can also be seen in Fig. 29.1. Notice that it requires $2^N - 1$ bits to represent an N-bit word. The choice of code depends on the application, and the reader should be aware that many types of codes are available.

Decimal	Binary	Thermometer	Gray	Two's Complement
0	000	0000000	000	000
1	001	0000001	001	111
2	010	0000011	011	110
3	011	0000111	010	101
4	100	0001111	110	100
5	101	0011111	111	011
6	110	0111111	101	010
7	111	1111111	100	001

Figure 29.1 Comparison of digital input codes.

29.1.2 Resistor String

The most basic DAC is seen in Fig. 29.2a. Comprised of a simple resistor string of 2^N identical resistors and switches, the analog output is simply the voltage division of the resistors at the selected tap. Note that a $N:2^N$ decoder is required to provide the 2^N signals controlling the switches. This architecture typically results in good accuracy, provided that no output current is required and that the values of the resistors are within the specified error tolerance of the converter. One big advantage of a resistor string is that the output is always guaranteed to be monotonic.

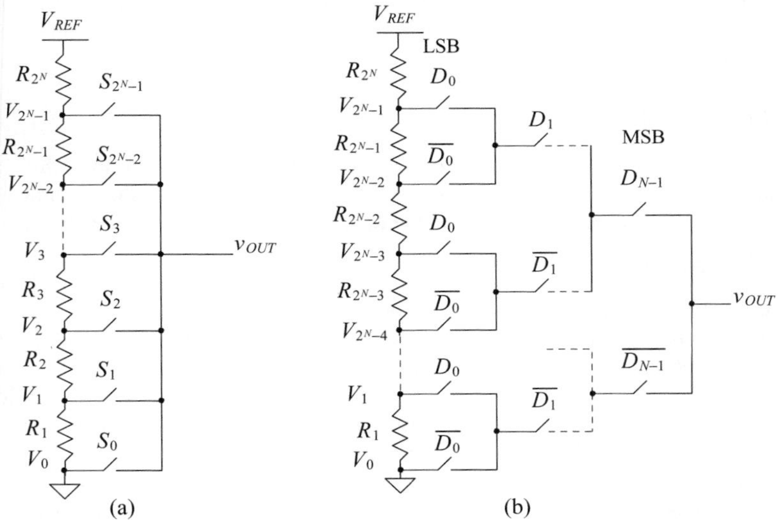

Figure 29.2 (a) A simple resistor-string DAC and (b) the use of a binary switch array to lower the output capacitance.

One problem with this converter is that the converter output is always connected to $2^N - 1$ switches that are off and one switch that is on. For larger resolutions, a large parasitic capacitance appears at the output node, resulting in slower conversion speeds. A better alternative for the resistor-string DAC is seen in Fig. 29.2b. Here, a binary switch array ensures that the output is connected to at most N switches that are on and N switches that are off, thus increasing conversion speed. The input to this switch array is a binary word since the decoding is inherent in the binary tree arrangement of the switches.

Another problem with the resistor-string DAC is the balance between area and power dissipation. An integrated version of this converter leads to a large chip area for higher bit resolutions because of the large number of passive components needed. Active resistors such as the n-well resistor can be used for low-resolution applications. However, as the resolution increases, the relative accuracy of the resistors becomes an important factor. Although the value of R could always be made small to minimize the chip area required, power dissipation would then become the critical issue as current flows through the resistor string at all times.

Example 29.1

Design a 3-bit resistor-string ladder using a binary switch array. Assume that V_{REF} = 5 V and that the maximum power dissipation of the converter is to be 5 mW (not including the power required by the digital logic). Determine the value of the analog voltage for each of the possible digital input codes.

The power dissipation will determine the current flowing through the resistor string by

$$I_{MAX} = \frac{5 \times 10^{-3}\ \text{W}}{5\ \text{V}} = 1\ \text{mA}$$

Since a 3-bit converter will have eight resistors, the value of R is

$$R = \frac{1}{8} \cdot \frac{5\ \text{V}}{1\ \text{mA}} = 625\ \Omega$$

The converter can be seen in Fig. 29.3. Examine the switch array if the input code is $D_2D_1D_0$ = 100 or 4_{10}. Since D_2 is high, the top switch will be closed and the lower switch, $\overline{D_2}$, will be open. In the row corresponding to D_1, since $D_1 = 0$, both of the switches marked $\overline{D_1}$ will be closed and the other two will be open. The LSB controls the largest number of switches; therefore, since D_0 is low, all of the $\overline{D_0}$ switches will be closed and all of the D_0 switches will be open. There should be only one path connecting a single tap on the resistor string to the output. This is bolded, with the resistor string tapped in the middle of the string. Therefore, $v_{OUT} = ½\, V_{REF} = 2.5$ V. The remaining outputs can be seen in Fig. 29.4. ■

Mismatch Errors Related to the Resistor-String DAC

The accuracy of the resistor string is obviously related to matching between the resistors, which ultimately determines the INL and DNL for the entire DAC. Suppose that the *i*-th resistor, R_i, has a mismatch error associated with it so that

$$R_i = R + \Delta R_i \tag{29.1}$$

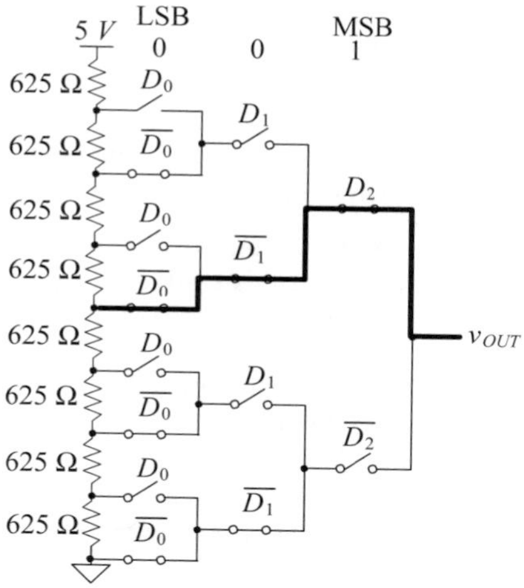

Figure 29.3 A 3-bit resistor-string DAC used in Ex. 29.1

where R is the ideal value of the resistor and ΔR_i is the mismatch error. Also suppose that the mismatches were symmetrical about the string so that the sum of all the mismatch terms were zero, or

$$\sum_{i=1}^{2^N} \Delta R_i = 0 \tag{29.2}$$

The value of the voltage at the tap associated with the i-th resistor should ideally be

$$V_{i,ideal} = \frac{(i)V_{REF}}{2^N}, \text{ for } i = 0, 1, 2, \ldots.., 2^N - 1 \tag{29.3}$$

$D_2D_1D_0$	v_{OUT}
000	0
001	0.625
010	1.25
011	1.875
100	2.5
101	3.125
110	3.75
111	4.375

Figure 29.4 Output voltages generated from the 3-bit DAC in Ex. 29.1.

However, including the mismatch, the actual value of the i-th voltage is the sum of all of the resistances up to and including resistor i, divided by the sum of all of the resistances in the string. This can be represented by

$$V_i = V_{REF} \cdot \frac{\sum_{k=1}^{i} R_k}{\sum_{k=1}^{2^N} R_k} = V_{REF} \cdot \frac{\sum_{k=1}^{i} R + \Delta R_k}{2^N R} \qquad (29.4)$$

The denominator does not include any mismatch error since it was assumed that the mismatches sum to zero as defined in Eq. (29.2). Notice that there is no resistor, R_0, corresponding to V_0 (see Fig. 29.2), and it is assumed that V_0 is ground. Equation (29.4) can be rewritten as

$$V_i = \frac{V_{REF}}{2^N R}\left[(i)R + \sum_{k=1}^{i} \Delta R_k\right] = \frac{(i)V_{REF}}{2^N} + \frac{V_{REF}}{2^N R}\sum_{k=1}^{i} \Delta R_k \qquad (29.5)$$

or finally, the value of the voltage at the i-th tap is

$$V_i = V_{i,ideal} + \frac{V_{REF}}{2^N} \cdot \sum_{k=1}^{i} \frac{\Delta R_k}{R} \qquad (29.6)$$

Equation (29.6) is not of much importance by itself, but it can be used to help determine the nonlinearity errors.

Integral Nonlinearity of the Resistor-String DAC

Integral nonlinearity (INL) is defined as the difference between the actual and ideal switching points, or

$$INL = V_i - V_{i,ideal} \qquad (29.7)$$

and plugging in Eqs. (29.6) and (29.3) into (29.7) yields,

$$INL = \frac{V_{REF}}{2^N} \cdot \sum_{k=1}^{i} \frac{\Delta R_k}{R} \qquad (29.8)$$

Equation (29.8) is a general expression for the INL for a given resistor, R_i, and requires that the mismatch of all the resistances used in the summation are known. However, this equation does not illuminate how to determine the *worst-case* or maximum INL for a resistor string.

Intuitively, one would think that the worst-case INL would occur at the top of the resistor string (i=2^N) with all of the ΔR_k's at their maximum values. However, the previous derivation was performed with the assumption that the mismatches summed to zero. With this restriction, the maximum INL occurs at the midpoint of the string where $i = 2^{N-1}$, corresponding to the case where the MSB was a one and all other bits were zero. Another condition that will ensure a worst-case scenario is to consider the lower half resistors at their maximum positive mismatch value and the upper half resistors at their maximum negative mismatch value, or vice versa.

If the resistors on a string were known to have 2% matching, then ΔR_k would be constrained to the bounds of

$$-0.02R \le \Delta R_k \le 0.02R \qquad (29.9)$$

and the worst-case INL (again, % matching = 0.02) using Eq. (29.8) would be,

$$|INL|_{max} = \frac{V_{REF}}{2^N} \cdot \sum_{k=1}^{2^{N-1}} \frac{\Delta R_k}{R} = \frac{V_{REF}}{2^N} \cdot \frac{2^{N-1} \cdot \Delta R_k}{R} = \frac{1}{2} LSB \cdot 2^N \cdot (\% \text{ matching}) = 0.01 V_{REF} \tag{29.10}$$

which for $INL < 0.5\ LSB$ requires $1/2^N >$ (% matching). For 2% matching, the maximum number of bits, N, is then 5! For better than 0.2% matching $N = 9$ bits.

Because the worst-case analysis was performed, the maximum INL occurs at the middle of the string. We can improve this specification on paper by using the "best-fit" approach to measuring INL. In this case, the reference line is simply shifted up slightly (refer to Ch. 28) so that it no longer passes through the end points, but instead minimizes the INL.

Example 29.2

Determine the effective number of bits for a resistor-string DAC, which is assumed to be limited by the INL. The resistors are passive poly resistors with a known relative matching of 1%, and $V_{REF} = 5$ V.

Using Eq. (29.10), the maximum INL will be

$$|INL|_{max} = 0.005 \cdot V_{REF} = 0.025 \text{ V}$$

Since we know that this maximum INL should be equal to ½ LSB in the worst case,

$$\frac{1}{2} LSB = \frac{5}{2^{N+1}} = 0.025 \text{ V}$$

and solving for N yields

$$N = \log_2\left(\frac{5}{0.025}\right) - 1 = 6.64 \text{ bits}$$

This means that the resolution for a DAC containing a resistor string matched to within 1% will be, at most 6 bits. ■

Differential Nonlinearity of the Worst-Case Resistor-String DAC

Resistor-string matching is not as critical when determining the DNL. Remembering that the definition of DNL is simply the actual height of the stair-step in the DAC transfer curve minus the ideal step height, we can write this in terms of the voltages at the taps of adjacent resistors on the string. Using Eq. (29.5), we can express this as,

$$|V_i - V_{i-1}| = \left| \left[\frac{(i) V_{REF}}{2^N} + \frac{V_{REF}}{2^N} \cdot \sum_{k=1}^{i} \frac{\Delta R_k}{R} \right] - \left[\frac{(i-1) V_{REF}}{2^N} + \frac{V_{REF}}{2^N} \cdot \sum_{k=1}^{i-1} \frac{\Delta R_k}{R} \right] \right|$$

which can be simplified to

$$|V_i - V_{i-1}| = \left| \frac{V_{REF}}{2^N}\left(1 + \frac{\Delta R_i}{R}\right) \right| \tag{29.11}$$

The DNL can then be determined by subtracting the ideal step height from Eq. (29.11),

$$DNL_i = \left| \frac{V_{REF}}{2^N}\left(1 + \frac{\Delta R_i}{R}\right) - \frac{V_{REF}}{2^N} \right| = \left| \frac{V_{REF}}{2^N} \cdot \frac{\Delta R_i}{R} \right| \tag{29.12}$$

and the maximum DNL will occur at the value of i for which ΔR is at its maximum value. If it is assumed once again that the resistors are matched to within 2 percent, the worst-case DNL will be

$$DNL_{max} = \left| 0.02 \cdot \frac{V_{REF}}{2^N} \right| = 0.02\ LSB \tag{29.13}$$

which is well below the ½ LSB limit. The INL is obviously the limiting factor in determining the resolution of a resistor-string DAC as its maximum value is 2^N times larger than the DNL.

29.1.3 *R*-2*R* Ladder Networks

Another DAC architecture that incorporates fewer resistors is called the *R*-2*R* ladder network [1]. This configuration consists of a network of resistors alternating in value of *R* and 2*R*. Figure 29.5 illustrates an *N*-bit *R*-2*R* ladder. Starting at the right end of the network, notice that the resistance looking to the right of any node to ground is 2*R*. The digital input determines whether each resistor is switched to ground (noninverting input) or to the inverting input of the op-amp. Each node voltage is related to V_{REF}, by a binary-weighted relationship caused by the voltage division of the ladder network. The total current flowing from V_{REF} is constant, since the potential at the bottom of each switched resistor is always zero volts (either ground or virtual ground). Therefore, the node voltages remain constant for any value of the digital input.

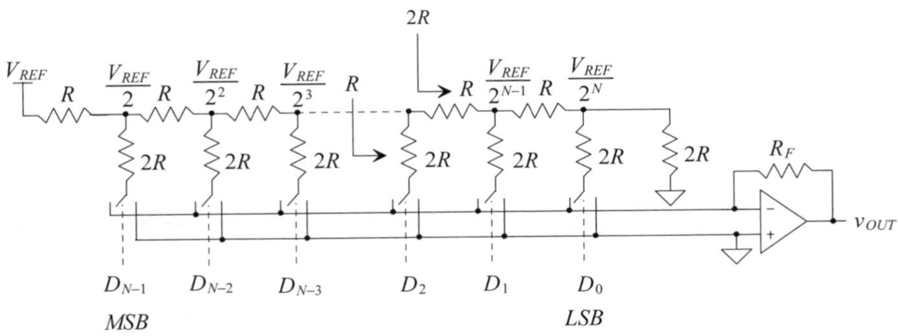

Figure 29.5 An *R*-2*R* digital-to-analog converter.

The output voltage, v_{OUT}, depends on currents flowing through the feedback resistor, R_F, such that

$$v_{OUT} = -i_{TOT} \cdot R_F \tag{29.14}$$

where i_{TOT} is the sum of the currents selected by the digital input by

$$i_{TOT} = \sum_{k=0}^{N-1} D_k \cdot \frac{V_{REF}}{2^{N-k}} \cdot \frac{1}{2R} \tag{29.15}$$

where D_k is the *k*-th bit of the input word with a value that is either a 1 or a 0.

This architecture, like the resistor-string architecture, requires matching to within the resolution of the converter. Therefore, the switch resistance must be negligible, or a small voltage drop will occur across each switch, resulting in an error. One way to eliminate this problem is to add dummy switches. Assume that the resistance of each switch connected to the $2R$ resistors is ΔR, as seen in Fig. 29.6. Dummy switches with one-half the resistance of the real switches are "hard-wired" so that they are always on and placed in series with each of the horizontal resistors. The total resistance of any horizontal branch, R', is

$$R' = R + \frac{\Delta R}{2} \tag{29.16}$$

The resistance of any vertical branch is $2R + \Delta R$, which is twice the value of the horizontal branch. Therefore, a $R' - 2R'$ relationship is maintained. Of course, a dummy switch equal to the switch size of a $2R$ switch will have to be placed in series with the terminating resistor as well.

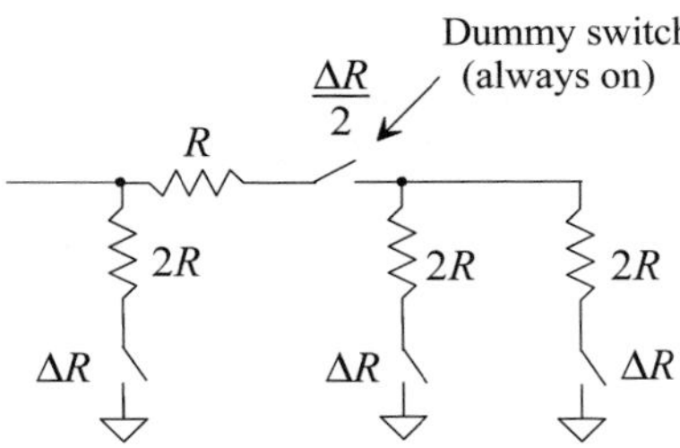

Figure 29.6 Use of dummy switches to offset switch resistance.

Example 29.3

Design a 3-bit DAC using an R-$2R$ architecture with R = 1 kΩ, R_F = 2 kΩ, and V_{REF} = 5 V. Assume that the resistances of the switches are negligible. Determine the value of i_{TOT} for each digital input and the corresponding output voltage, v_{OUT}.

Figure 29.7 shows the 3-bit DAC for a digital input of 001. The voltages at each node in the resistor network are labeled. For each switch, if the digital input bit is a zero, then the resistor is attached to the ground. If the bit is a one, then the resistor is attached to the virtual ground of the inverting input and current flows to the output of the op-amp. Therefore, for $D_2D_1D_0$ = 000, all of the switches are connected to ground, no current flows through the feedback resistor, and the output voltage, v_{OUT}, is zero.

When $D_2D_1D_0$ = 001, the rightmost resistor is switched to the op-amp inverting input and the other two resistors remain attached to ground. Therefore, the total current flowing through the feedback resistor is simply the current through the rightmost resistor, which is defined by Eq. (29.15) as

$$\frac{V_{REF}}{8} \cdot \frac{1}{2000} = 0.3126 \text{ mA}$$

and the output voltage, by Eq. (29.14) becomes,

$$v_{OUT} = -(0.3126 \text{ mA})(2000 \ \Omega) = -0.625 \text{ V}$$

which is to be expected. The other values for the output voltage can be calculated using Eqs. (29.14) and (29.15) and are seen in Fig. 29.8. ■

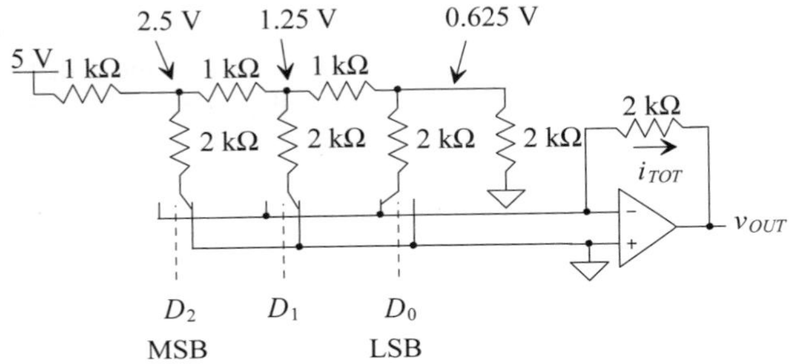

Figure 29.7 A 3-bit R-$2R$ digital-to-analog converter used in Ex. 29.3.

$D_2D_1D_0$	i_{TOT} (mA)	v_{OUT} (V)
000	0	0
001	0.3125	– 0.625
010	0.625	– 1.25
011	0.625 + 0.3125 = 0.9375	– 1.875
100	1.25	– 2.5
101	1.25 + 0.3125 = 1.5625	– 3.125
110	1.25 + 0.625 = 1.875	– 3.75
111	1.25 + 0.625 + 0.3125 = 2.1875	– 4.375

Figure 29.8 Output voltages generated from the 3-bit DAC in Example 29.3.

29.1.4 Current Steering

In the previous section, a voltage was converted into a current, which then generated a voltage at the output. Another DAC method uses current throughout the conversion. Known as *current steering*, this type of DAC requires precision current sources that are summed in various fashions.

Figure 29.9 illustrates a generic current-steering DAC. This configuration requires a set of current sources, each having a unit value of current, I. Since there are no current sources generating i_{OUT} when all the digital inputs are zero, the MSB, D_{2^N-2}, is offset by two index positions instead of one. For example, for a 3-bit converter, seven current sources will be needed, labeled from D_0 to D_6. The binary signal controls whether or not the current sources are connected to either i_{OUT} or some other summing node (in this case ground). The output current, i_{OUT}, has the range of

$$0 \leq i_{OUT} \leq (2^N - 1) \cdot I \tag{29.17}$$

and can be any integer multiple of I in between. An interesting issue to note is the format of the digital code required to drive the switches. Since there are $2^N - 1$ current sources, the digital input will be in the form of a *thermometer code*. This code will be all ones from the LSB up to the value of the k-th bit, D_k, and all zeros above it. The point at which the input code changes from all ones to all zeros "floats" up or down and resembles the action of a thermometer, hence the name. Typically, a thermometer encoder is used to convert binary input data into a thermometer code.

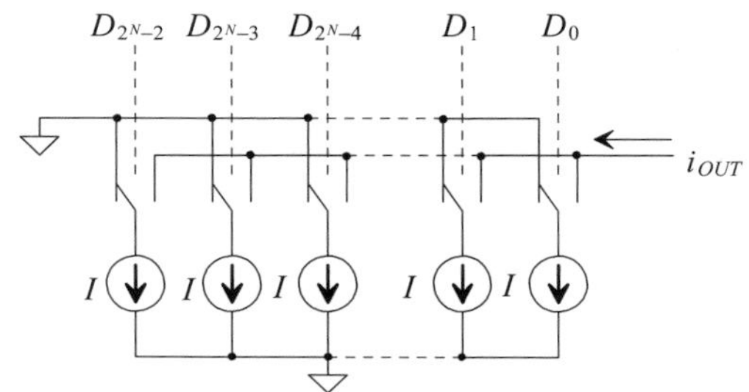

Figure 29.9 A generic current-steering DAC.

Another current-steering architecture is seen in Fig. 29.10. This architecture uses binary-weighted current sources, thus requiring only N current sources of various sizes versus $2^N - 1$ sources in the previous example. Since the current sources are binary weighted, the input code can be a simple binary number with no thermometer encoder needed.

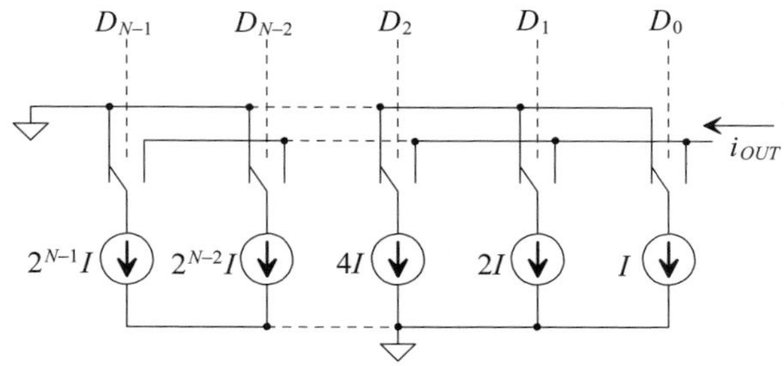

Figure 29.10 A current-steering DAC using binary-weighted current sources.

One advantage of the current-steering DACs is the high-current drive inherent in the system. Since no output buffers are necessary to drive resistive loads, these DACs are typically used in high-speed applications. Traditionally, high-speed current-steering DACs have been fabricated using bipolar technology. However, the ability to generate matched current mirrors makes CMOS an enticing alternative. Of course, the precision needed to generate high resolutions depends on how well the current sources can be matched or the degree to which they can be made binary weighted. For example, if a 13-bit DAC was designed using these architectures, there would have to be 8,191 current sources resident on the chip, not an insignificant amount. For the binary-weighted sources, only 13 current sources would be needed. Yet the size of the largest current source would have to be 4,096 or 2^{N-1} times larger than the smallest. Even if the unit current, I, was chosen to be 5 μA, the largest current source would be 20.48 mA!

Another problem associated with this architecture is the error due to the switching. Since the current sources are in parallel, if one of the current sources is switched off and another is switched on, a "glitch" could occur in the output if the timing was such that both of them were on or both were off for an instant. While this may not seem significant, if the converter is switching from 0111111 to 10000000, the output will spike toward ground and then back to the correct value if all the switches turn off for an instant. If the DAC is driving a resistive load and the output current is converted to a voltage, a substantial voltage spike will occur at the output.

Example 29.4

Construct a table showing the thermometer code necessary to generate the output shown in Fig. 29.11a for a 3-bit current-steering DAC using unit current sources.

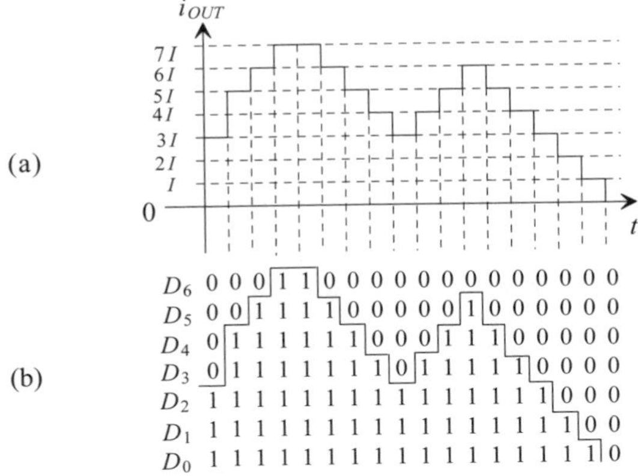

Figure 29.11 (a) Output of a 3-bit current-steering DAC and (b) the thermometer code input.

The thermometer code can be seen in Fig. 29.11b. When the code is all zeros, the output is 0 volts. Therefore, only 7 bits are needed to represent the 2^N or 8 states of a 3-bit DAC. Note how the interface between all ones and all zeros actually resembles the output signal itself. ■

Mismatch Errors Related to Current-Steering DACs

Analysis of the mismatch associated with the current sources is similar to the resistor string analysis. It is assumed that each current source in Fig. 29.9 is

$$I_k = I + \Delta I_k \text{ for } k = 1, 2, 3, \ldots.., 2^N - 1 \tag{29.18}$$

where I is the ideal value of the current and ΔI_k is the error due to mismatch. If it is again assumed that the ΔI_K terms sum to zero and that one-half of the current sources contain the maximum positive mismatch, ΔI_{max}, and the other half contains the maximum negative mismatch, $-\Delta I_{max}$, (or vice versa), then the worst-case condition will occur at midscale with the actual output current being

$$I_{out} = \sum_{k=1}^{2^{N-1}} (I + \Delta I_k) = 2^{N-1} \cdot I + 2^{N-1} \cdot |\Delta I|_{max} = I_{out,ideal} + 2^{N-1} \cdot |\Delta I|_{max} \tag{29.19}$$

Since the INL is simply the actual output current minus the ideal, the worst-case INL is

$$|INL|_{max} = 2^{N-1} \cdot |\Delta I|_{max,INL} \tag{29.20}$$

The term, $|\Delta I|_{max,INL}$ represents the maximum current source mismatch error that will keep the INL less than ½ LSB. Each current source represents the value of 1 LSB; therefore, ½ LSB is equal to 0.5 I. Because the maximum INL should correspond to the ½ LSB, equating Eq. (29.20) to ½ I results in the value for $|\Delta I|_{max,INL}$,

$$|\Delta I|_{max,INL} = \frac{0.5I}{2^{N-1}} = \frac{I}{2^N} \tag{29.21}$$

Equation (29.21) illustrates the difficulty of using this architecture at high resolutions. If the value of I is set to be 5 μA, and the N is desired to be 12 bits, then

$$|\Delta I|_{max,INL} = \frac{5 \times 10^{-6}}{2^{12}} = 1.221 \text{ nA!} \tag{29.22}$$

which means that each of the 5 μA current sources must lie between the bounds of,

$$4.99878\ \mu A \le I_k \le 5.001221\ \mu\text{A} \tag{29.23}$$

to achieve a worst-case INL, which is within ½ LSB error.

The DNL is easily obtained since the step height in the transfer curve is equivalent to the value of the ideal current source, I. The maximum difference between any two adjacent values of output current will simply be the value of the single source, I_k, which contains the largest mismatch error for which the DNL will be less than ½ LSB, $|\Delta I|_{max,DNL}$:

$$I_{out(x)} - I_{out(x-1)} = I_k + |\Delta I|_{max,DNL} \tag{29.24}$$

Therefore, the DNL is simply

$$|DNL|_{max} = I_k + |\Delta I|_{max,DNL} - I_k = |\Delta I|_{max,DNL} \tag{29.25}$$

Equating the maximum DNL to the value of ½ LSB,

$$|\Delta I|_{max,DNL} = \tfrac{1}{2} LSB = \tfrac{1}{2} I \tag{29.26}$$

which is much easier to attain than the requirement for the INL.

For the binary-weighted current sources seen in Fig. 29.10, a slightly different analysis is needed to determine the requirements for INL and DNL. In this case, it will be assumed that the current source corresponding to the MSB (D_{N-1}) has a maximum positive mismatch error value and that the remainder of the bits (D_0 to D_{N-2}) contain a maximum negative mismatch error, so that the sum of all the errors equals zero. The INL is

$$|INL|_{max} = 2^{N-1}(I + |\Delta I|_{max,INL}) - 2^{N-1} \cdot I = 2^{N-1} \cdot |\Delta I|_{max,INL} \tag{29.27}$$

which is equivalent to the value of the current steering array in Fig. 29.9.

The DNL is slightly different because of the binary weighting of the current sources. One cannot add a single current source with each incremental increase in the digital input code. However, the worst-case condition for binary-weighted arrays tends to occur at midscale when the code transitions from 011111....111 to 100000....000. The worst-case DNL at this point is

$$DNL_{max} = \left[2^{N-1} \cdot (I + |\Delta I|_{max,DNL}) - \sum_{k=1}^{N-1} 2^{k-1} \cdot (I - |\Delta I|_{max,DNL}) \right] - I \tag{29.28}$$

which can be written as

$$DNL_{max} = 2^{N-1} \cdot (I + |\Delta I|_{max,DNL}) - (2^{N-1} - 1) \cdot (I - |\Delta I|_{max,DNL}) - I = (2^N - 1) \cdot |\Delta I|_{max,DNL} \tag{29.29}$$

and setting this value equal to ½ LSB and solving for ΔI_{max},

$$|\Delta I|_{max,DNL} = \frac{0.5I}{2^N - 1} = \frac{I}{2^{N+1} - 2} \tag{29.30}$$

Therefore, the DNL requirements for the binary-weighted current source array is more stringent than the INL requirements.

One interesting issue regarding the previous derivation is that the challenging accuracy requirements in Eq. (29.30) are placed only on the MSB current source. For each of the remaining binary-weighted sources, the DNL requirements become more relaxed. This is simply because the size of the MSB source is equivalent to all of the other sources combined, and so its value plays the most important role in the DAC's accuracy.

Example 29.5
Determine the tolerance of the MSB current source on a 10-bit binary-weighted current source array with a unit current source of 1 μA, which will result in a worst-case DNL that is less than ½ LSB.

Since Eq. (29.30) defines the maximum $|\Delta I|$ needed to keep the DNL less than ½ LSB, we must first use this equation,

$$|\Delta I|_{max,DNL} = \frac{1 \times 10^{-6}}{2^{11} - 2} = 0.4888 \text{ nA}$$

For a 10-bit DAC, the MSB current source will have a value that is 2^9 times larger than the unit current source, or 0.512 mA. Therefore, the range of values for which this array will have a DNL that is less than ½ LSB is

$$0.51199995 \text{ mA} \le I_{MSB} \le 0.5120004888 \text{ mA} \quad \blacksquare$$

29.1.5 Charge-Scaling DACs

A very popular DAC architecture used in CMOS technology is the charge-scaling DAC. Shown in Fig. 29.12a, a parallel array of binary-weighted capacitors, totaling $2^N C$, is connected to an op-amp. The value, C, is a unit capacitance of any value. After initially being discharged, the digital signal switches each capacitor to either V_{REF} or ground, causing the output voltage, v_{OUT}, to be a function of the voltage division between the capacitors.

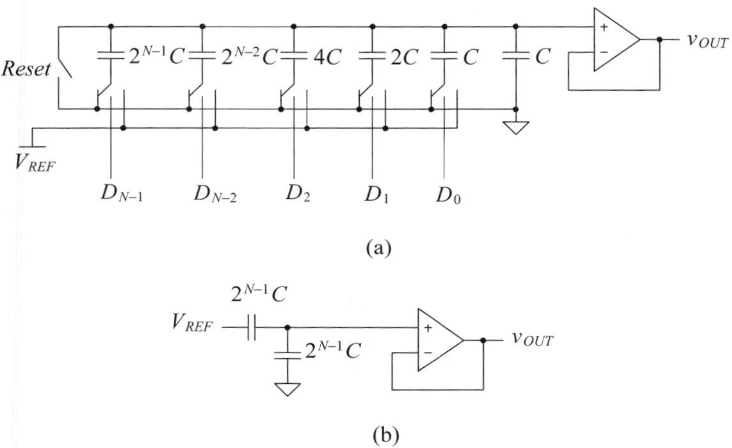

Figure 29.12 (a) A charge-scaling DAC, (b) the equivalent circuit with the MSB = 1, and all other bits set to zero.

The capacitor array totals $2^N C$. Therefore, if the MSB is high and the remaining bits are low, then a voltage divider occurs between the MSB capacitor and the rest of the array. The analog output voltage, v_{OUT}, becomes

$$v_{OUT} = V_{REF} \cdot \frac{2^{N-1}C}{(2^{N-1}+2^{N-2}+2^{N-3}+\ldots+4+2+1+1)C} = V_{REF} \cdot \frac{2^{N-1}C}{2^N C} = \frac{V_{REF}}{2} \tag{29.31}$$

which confirms the fact that the MSB changes the output of a DAC by ½ V_{REF}. Figure 29.12b shows the equivalent circuit under this condition. The ratio between v_{OUT} and V_{REF} due to each capacitor can be generalized to

$$v_{OUT} = \frac{2^k C}{2^N C} \cdot V_{REF} = 2^{k-N} \cdot V_{REF} \tag{29.32}$$

where it is assumed that the k-th bit, D_k, is one and all other bits are zero. Superposition can then be used to find the value of v_{OUT} for any digital input word by

$$v_{OUT} = \sum_{k=0}^{N-1} D_k 2^{k-N} \cdot V_{REF} \tag{29.33}$$

One limitation of this architecture as shown in Fig. 29.12a is the existence of a parasitic capacitance at the top plate of the capacitor array due to the op-amp. This will prohibit its use as a high-resolution data converter. A better implementation would include the use of a parasitic insensitive, switched-capacitor integrator (see Ch. 25) as the driving circuit. However, the capacitor array itself is the critical component of this data converter and is used in charge redistribution ADCs (Sec. 29.2.5).

The INL and DNL calculations for the binary-weighted capacitor array are identical to those for the binary-weighted current source array, except that the unit current source, I, and its corresponding error term, ΔI, are replaced by C and ΔC in Eqs. (29.27) – (29.30).

Example 29.6
Design a 3-bit charge-scaling DAC and find the value of the output voltage for $D_2D_1D_0 = 010$ and 101. Assume that $V_{REF} = 5$ V and $C = 0.5$ pF.

The 3-bit DAC can be seen in Fig. 29.13a. The equivalent circuits for the capacitor array can be seen in Fig. 29.13b and c. The value of the output voltage can be calculated by either using Eq. (29.32) or the equivalent circuits and performing the voltage division. For $D = 010$, the equivalent circuit in Fig. 29.13b yields

$$v_{OUT} = V_{REF} \cdot \left(\frac{1}{4}\right) = 1.25 \text{ V}$$

Using Eq. (29.33) to calculate v_{OUT} for $D = 101$ yields

$$v_{OUT} = \sum_{k=0}^{N-1} D_k 2^{k-N} \cdot V_{REF} = [1 \cdot (2^{-3}) + 0 \cdot (2^{-2}) + 1 \cdot (2^{-1})] \cdot 5 = \left(\frac{1}{8} + \frac{1}{2}\right) \cdot 5 = 3.125 \text{ V}$$

which is the result expected. ■

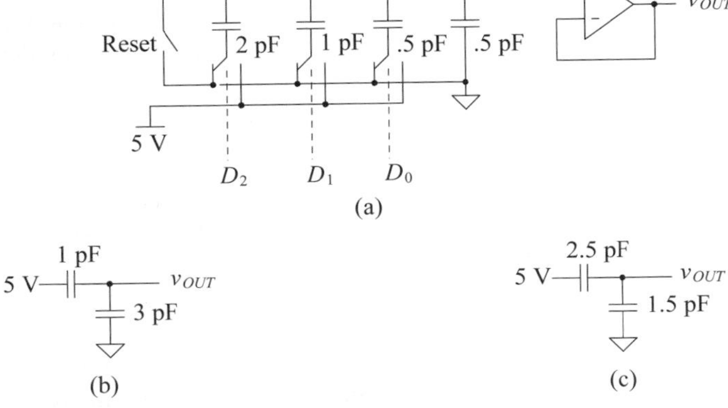

Figure 29.13 (a) A 3-bit charge-scaling DAC used in Ex. 29.6 and the equivalent circuits inputs equal to (b) 010 (c) 101.

Layout Considerations for a Binary-Weighted Capacitor Array

One problem with this converter is the need for precisely ratioed capacitors. As the number of bits increase, the ratio of the MSB capacitor to the LSB capacitor becomes more difficult to control. For example, Fig. 29.14a shows a 3-bit binary capacitor array using three capacitors. When the capacitor is fabricated, *undercutting* of the mask [2, 3] causes an error in the ratio of the capacitors, creating potentially large DNL and INL errors as N increases.

One solution to this problem is seen in Fig. 20.14b. Here, each capacitor in the array is constructed out of a unit capacitance. Undercutting then affects all of the capacitors in the same way, and the ratio between capacitors is maintained. Another problem that affects even this layout strategy is a nonuniform oxide growth. Gradients result in errors in the ratios of the capacitors. Figure 29.14c illustrates another layout strategy that overcomes this issue. The capacitors are laid out in a common-centroid scheme so that the first-order oxide errors average out to be the same for each capacitor.

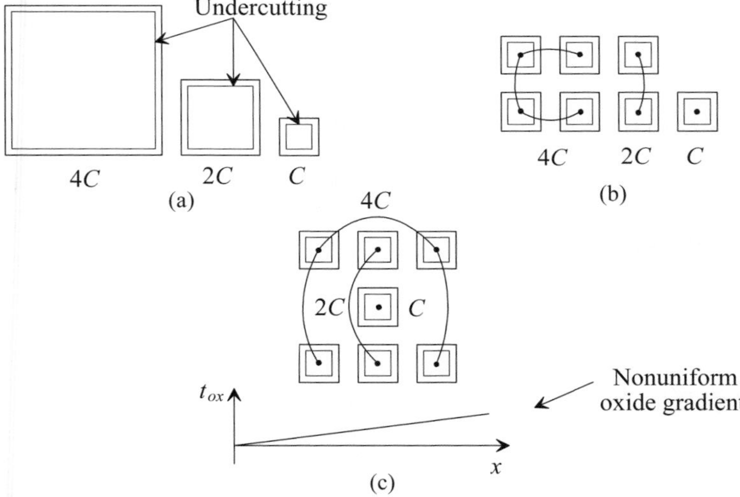

Figure 29.14 Layout of a binary-weighted capacitor array using (a) single capacitors (b) unit capacitors to minimize undercutting effect, and (c) common-centroid to minimize oxide gradients.

The Split Array

The charge-scaling architecture is very popular among CMOS designers because of its simplicity and relatively good accuracy. Although a linear capacitor is required using poly2, high resolutions in the 10- to 12-bit range can be achieved. Passive, double-poly capacitors have good matching accuracy as well. However, as the resolution increases, the size of the MSB capacitor becomes a major concern. For example if the unit capacitor, C, were 0.5 pF, and a 16-bit DAC were to be designed, the MSB capacitor would need to be

$$C_{MSB} = 2^{N-1} \cdot 0.5 \text{ pF} = 16.384 \text{ nF} \tag{29.34}$$

If the capacitance between poly1 and poly2 is nominally 25 fF/μm^2, then the area required for this one capacitor is (roughly) 800 by 800 μm^2.

One method of reducing the size of the capacitors is to use a split array. A 6-bit example of the array is pictured in Fig. 29.15. This architecture [1] is slightly different from the charge-scaling DAC pictured in Fig. 29.13 in that the output is taken off a different node and an additional attenuation capacitor is used to separate the array into a LSB array and a MSB array. Note that the LSB, D_0, now corresponds to the leftmost switch and that the MSB, D_5, corresponds to the rightmost switch. The value of the attenuation capacitor can be found by

$$C_{atten} = \frac{\textit{sum of the LSB array capacitors}}{\textit{sum of the MSB array capacitors}} \cdot C \tag{29.35}$$

where the sum of the MSB array equals the sum of LSB capacitor array minus C. The value of the attenuation capacitor should be such that the series combination of the attenuation capacitor and the LSB array, assuming all bits are zero, equals C.

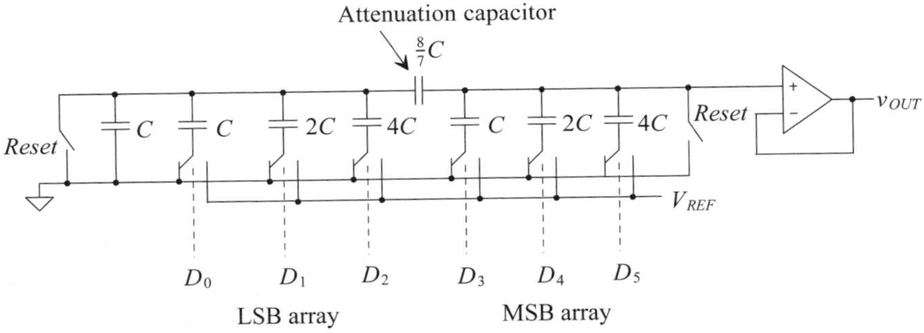

Figure 29.15 A charge-scaling DAC using a split array.

Example 29.7

Using the 6-bit charge-scaling DAC shown in Fig. 29.15, (a) show that the output voltage will be $\frac{1}{2} \cdot V_{REF}$ if (a) $D_5 D_4 D_3 D_2 D_1 D_0 = 100000$ and (b) the output will be $\frac{1}{64} \cdot V_{REF}$ if $D_5 D_4 D_3 D_2 D_1 D_0 = 000001$.

(a) If $D_5 = 1$ and the remaining bits are all zero, then the equivalent circuit for the DAC can be represented by Fig. 29.16a. The expression for the output voltage then becomes

$$v_{OUT} = \frac{4}{\left(\frac{8}{7} \text{ in series with } 8\right) + 3 + 4} \cdot V_{REF} = \frac{1}{2} \cdot V_{REF}$$

(b) For the second case, the equivalent circuit can be seen in Fig. 29.16b. The intermediate node voltage, V_A, is simply the voltage division between the C associated with D_0 and the remainder of the circuit, or

$$V_A = V_{REF} \cdot \frac{1}{\left(7 + \frac{\frac{8}{7} \cdot 7}{\frac{8}{7} + 7}\right) + 1} = \frac{1}{8 + \frac{56}{57}} \cdot V_{REF} \tag{29.36}$$

The output voltage can be written as

$$v_{OUT} = V_A \cdot \frac{\frac{8}{7}}{\frac{8}{7} + 7} = \frac{8}{57} \cdot V_A \tag{29.37}$$

Plugging Eq. (29.36) into Eq. (29.37) yields

$$v_{OUT} = V_{REF} \cdot \frac{8}{(8 \cdot 57) + 56} = \frac{V_{REF}}{64} \tag{29.38}$$

which is the desired result. ■

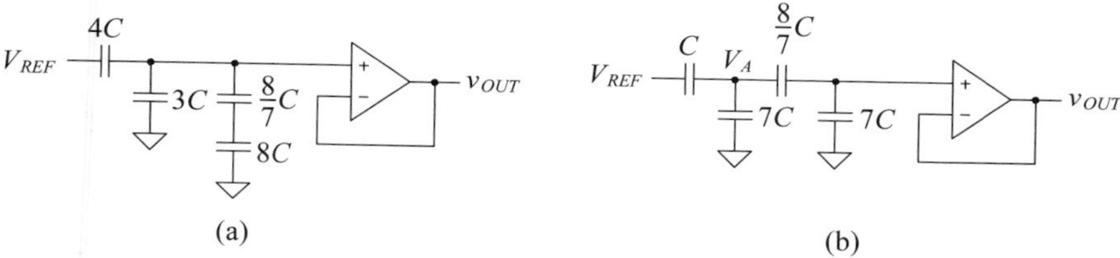

Figure 29.16 Equivalent circuits for Example 29.7.

29.1.6 Cyclic DAC

The cyclic DAC uses only a couple of simple components to perform the conversion. As seen in Fig. 29.17, a summer adds V_{REF} or ground to the feedback signal depending on the input bits. An amplifier with a gain of 0.5 feeds the output voltage back to the summer such that the output at the end of each cycle depends on the value of the output during the cycle before. Notice that the input bits must be read in a serial fashion. Therefore, the conversion is performed one bit at a time, resulting in N cycles required for each conversion. The voltage output at the end of the n-th cycle of the conversion can be written as

$$v_{OUT}(n) = \left(D_{n-1} \cdot V_{REF} + \frac{1}{2} \cdot v_A(n-1)\right) \cdot \frac{1}{2} \tag{29.39}$$

with a condition such that the output of the S/H is initially zero [$v_A(0) = 0$ V].

The accuracy of this converter is dependent on several factors. The gain of the 0.5 amplifier needs to be highly accurate (to within the accuracy of the DAC) and is usually generated with passive capacitors. Similarly, the summer and the sample-and-hold also need to be N-bit accurate. Limitations of the converter due to these fundamental building blocks will be discussed in more detail in Sec. 29.3. Since this converter uses a pseudo-"sampled-data" approach, implementing this architecture using switched capacitors is relatively easy.

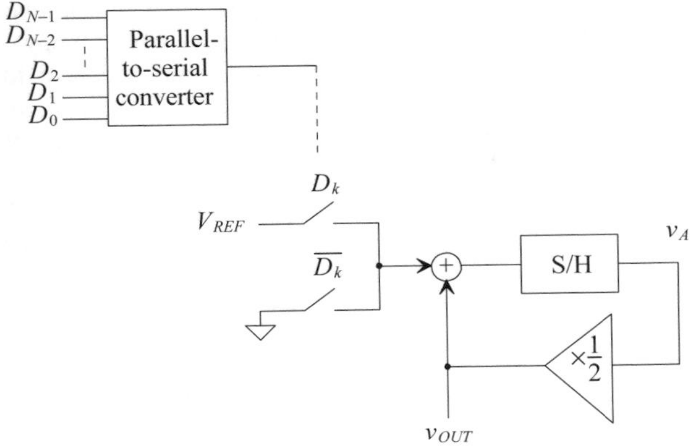

Figure 29.17 A cyclic digital-to-analog converter.

Example 29.8

Show the value of the output voltage at the end of each cycle for a 6-bit cyclic DAC with an input value of $D_5 D_4 D_3 D_2 D_1 D_0 = 110101$. Assume that $V_{REF} = 5$ V.

We can predict the value of the output based on our previous experience with DACs. The digital input 110101 corresponds to 53_{10}. Therefore, the output voltage due to this input should be

$$v_{OUT} = \frac{53}{64} \cdot V_{REF} = 4.140625 \text{ V}$$

Now examine the cyclic converter in Fig. 29.17. By performing a 6-bit conversion and using Eq. (29.39), the outputs occurring at the end of each cycle can be seen in Fig. 29.18.

The output voltage at the end of the sixth cycle is precisely what was predicted. Note that had this been a 3-bit conversion, the output voltage at the end of cycle 3 would correspond to the value of the 3-bit DACs studied previously with an input of 101. ■

Cycle Number, n	D_{n-1}	$v_A(n-1)$	$v_{OUT}(n)$
1	1	0	½ (5 + 0) = 2.5 V
2	0	5	½ (0 + 2.5) = 1.25 V
3	1	2.5	½ (5 +1.25) = 3.125 V
4	0	6.25	½ (0 + 3.125) = 1.5625 V
5	1	3.125	½ (5 + 1.5625) = 3.28125 V
6	1	6.5625	½ (5 + 3.28125) = 4.140625 V

Figure 29.18 Output from the 6-bit cyclic DAC used in Ex. 29.8.

29.1.7 Pipeline DAC

The cyclic converter presented in the last section takes N clock cycles per N-bit conversion. Instead of recycling the output back to the input each time, we could extend the cyclic converter to N stages, where each stage performs one bit of the conversion. This extension of the cyclic converter is called a *pipeline* DAC and is seen in Fig. 29.19. Here, the signal is passed down the "pipeline," and as each stage works on one conversion, the previous stage can begin processing another. Therefore, an initial N clock cycle delay is experienced as the signal makes its way down the pipeline the very first time. After the N clock cycle delay, a conversion takes place at every clock cycle.

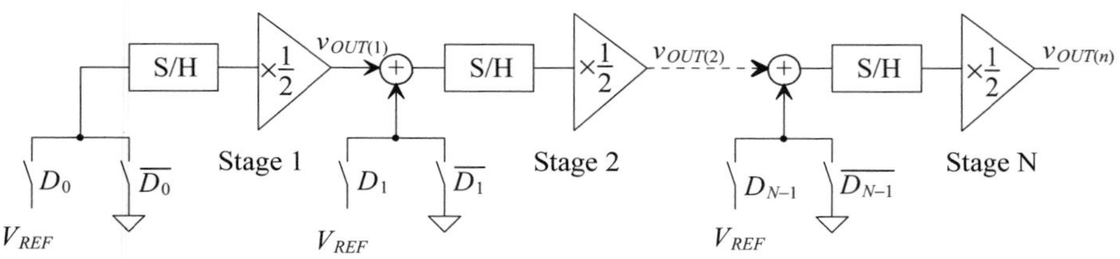

Figure 29.19 A pipeline digital-to-analog converter.

Besides the N clock cycle delay, this architecture can be very fast. However, the amplifier gains must be very accurate to produce high resolutions. Also, this architecture uses N times more circuitry than that of the cyclic, so there is a trade-off between speed and chip area. The output voltage of the n-th stage in the converter can be written as

$$v_{OUT(n)} = [D_{n-1} \cdot V_{REF} + v_{OUT(n-1)}] \cdot \frac{1}{2} \tag{29.40}$$

The operation of each stage in the pipeline can be summarized as follows: if the input bit is a 1, add V_{REF} to the output of the previous stage, divide by two, and pass the value to the next stage. If the input bit is a 0, simply divide the output of the previous stage by two and pass along the resulting value.

Example 29.9

Find the output voltage for a 3-bit pipeline DAC for three cases: $D_A = 001$, $D_B = 110$, and $D_C = 101$. Show that the conversion time to perform all three conversions is five clock cycles using the pipeline approach. Assume that $V_{REF} = 5$ V.

The first stage operates on the LSBs of each word; the second stage operates on the middle bits; and the last stage, the MSBs. Based on the pipeline strategy, once the LSB of the first input word is performed and passed on, the LSB of the second word, D_B, can begin its conversion. Similarly, once the LSB of the second stage is completed and passed on, the LSB of the third word, D_C, can begin. The conversion cycle for all three input words produces the output shown in Fig. 29.20. The items that are in bold are associated with the first input word, D_A, whereas the italicized numbers represent the values associated with D_B and the underlined items, D_C.

Clock Cycle	$v_{OUT(1)}$	$v_{OUT(2)}$	$v_{OUT(3)}$	D_0	D_1	D_2
1	**2.5**	0	0	**1**	0	0
2	*0*	**1.25**	0	*0*	**0**	0
3	2.5	*2.5*	**0.625**	1	*1*	**0**
4		1.25	*3.75*		0	*1*
5			3.125			1

Figure 29.20 Output from the 3-bit pipeline DAC used in Example 29.9.

The first output of the DAC is not valid until the end of the third clock cycle and should look familiar as the 3-bit DAC output for an input word of $D_2 D_1 D_0 = 001$. The following two clock cycles that produce outputs for $D_2 D_1 D_0$ equal 110 and 101, respectively. ■

29.2 ADC Architectures

A survey of the field of current A/D converter research reveals that a majority of effort has been directed to four different types of architectures: pipeline, flash-type, successive approximation, and oversampled ADCs. Each has benefits that are unique to that architecture and span the spectrum of high speed and resolution.

Since the ADC has a continuous, infinite-valued signal as its input, the important analog points on the transfer curve x-axis for an ADC are the ones that correspond to changes in the digital output word. These input transitions determine the amount of INL and DNL associated with the converter.

29.2.1 Flash

Flash or parallel converters have the highest speed of any type of ADC. As seen in Fig. 29.21, they use one comparator per quantization level ($2^N - 1$) and 2^N resistors (a resistor-string DAC). The reference voltage is divided into 2^N values, each of which is fed into a comparator. The input voltage is compared with each reference value and results in a thermometer code at the output of the comparators. A thermometer code exhibits all zeros for each resistor level if the value of v_{IN} is less than the value on the resistor string, and ones if v_{IN} is greater than or equal to voltage on the resistor string. A simple $2^N - 1{:}N$ digital thermometer decoder circuit converts the compared data into an N-bit digital word. The obvious advantage of this converter is the speed with which one conversion can take place. Each clock pulse generates an output digital word. The advantage of having high speed, however, is counterbalanced by the doubling of area with each bit of increased resolution. For example, an 8-bit converter requires 255 comparators, but a 9-bit ADC requires 511! Flash converters have traditionally been limited to 8-bit resolution with conversion speeds of 10–40 Ms/s using CMOS technology [4–6]. The disadvantages of the Flash ADC are the area and power requirements of the $2^N - 1$ comparators. The speed is limited by the switching of the comparators and the digital logic.

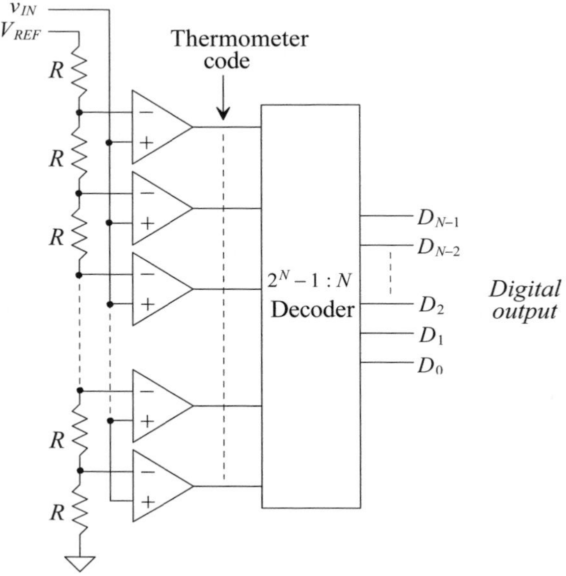

Figure 29.21 Block diagram of a Flash ADC.

Example 29.10

Design a 3-bit Flash converter, listing the values of the voltages at each resistor tap, and draw the transfer curve for v_{IN} = 0 to 5 V. Assume V_{REF} = 5 V. Construct a table listing the values of the thermometer code and the output of the decoder for v_{IN} = 1.5, 3.0, and 4.5 V.

The 3-bit converter can be seen in Fig. 29.22. As the values of all the resistors are equal, the voltage of each resistor tap, V_i, will be $V_i = V_{REF}\left(\frac{i}{8}\right)$ where i is the number of the resistor in the string for i = 1 to 7. Obviously, V_1 = 0.625 V, V_2 = 1.25 V, V_3 = 1.875 V, V_4 = 2.5 V, V_5 = 3.125, V_6 = 3.75 V, V_7 = 4.375 V. Therefore, when v_{IN} first becomes equal or greater than each of these values, a transition will occur in the transfer curve. The transfer curve can be seen in Fig. 29.23 and should look similar to those seen in Ch. 28. The quantization levels and their corresponding thermometer codes are summarized in Fig. 29.24.

The transfer curve of this ADC corresponds to the ADC with quantization error centered about +½ LSB, as discussed in Ch. 28 (Fig. 28.20). To shift the curve by ½ LSB so that the code transitions occur around the LSB values and the quantization error is centered around 0 LSB, the value of the last resistor in the string would have to be adjusted to $\frac{R}{2}$ and the value of the MSB resistor, closest to the reference voltage, would have to be made 1.5R. Then the first code transition would occur at v_{IN} = 0.3125 V, and the last code transition would occur at v_{IN} = 4.0625, and so the transfer curve would exactly match that of Fig. 28.20.

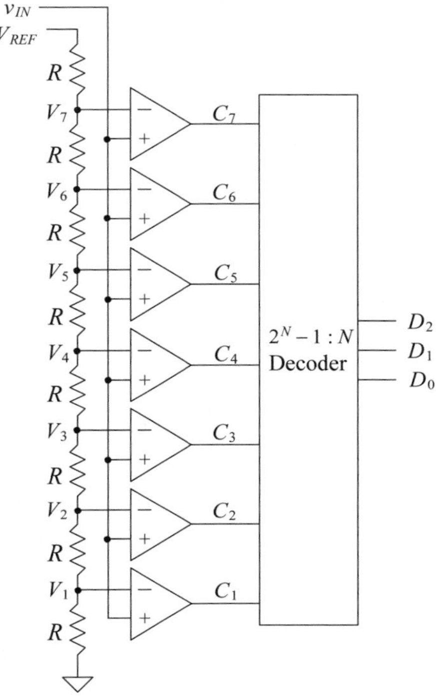

Figure 29.22 Three-bit Flash A/D converter to be used in Ex. 29.10.

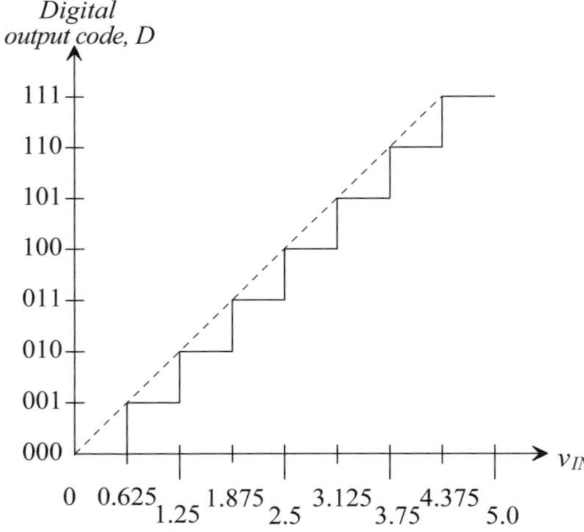

Figure 29.23 Transfer curve for the 3-bit Flash converter in Example 29.10.

v_{IN}	$C_7C_6C_5C_4C_3C_2C_1$	$D_2D_1D_0$
$0 \le v_{IN} < 0.625$ V	0000000	000
0.625 V $\le v_{IN} <$ 1.25 V	0000001	001
1.25 V $\le v_{IN} <$ 1.875 V	0000011	010
1.875 V $\le v_{IN} <$ 2.5 V	0000111	011
2.5 V $\le v_{IN} <$ 3.125 V	0001111	100
3.125 V $\le v_{IN} <$ 3.75 V	0011111	101
3.75 V $\le v_{IN} <$ 4.375 V	0111111	110
$4.375 \le v_{IN}$	1111111	111

Figure 29.24 Code transitions for the Flash ADC used in Ex. 29.10.

Based on Fig. 29.24, when v_{IN} = 1.5 V, only comparators C_1 and C_2 will have outputs of 1, since both V_1 and V_2 are less than 1.5 V. The remaining comparator outputs will be 0 since V_3 through V_8 will be greater than 1.5 V, thus generating the thermometer code, 0000011. The encoder must then convert this into a 3-bit digital word, resulting in 010. The same reasoning can be used to construct the data shown in Fig. 29.25. It should be obvious that if the polarity of the comparators were reversed, the thermometer code would be inverted. ■

v_{IN}	$C_7C_6C_5C_4C_3C_2C_1$	$D_2D_1D_0$
1.5	0000011	010
3.0	0001111	100
4.5	1111111	111

Figure 29.25 Output for the Flash ADC used in Ex. 29.10.

Accuracy Issues for the Flash ADC

Accuracy depends on the matching of the resistor string and the input offset voltage of the comparators. From our discussions earlier, we know that an ideal comparator should switch at the point at which the two inputs, v_+ and v_-, are the same potential. However, the offset voltage, V_{os}, prohibits this from occurring as the comparator output switches states as follows:

$$v_o = 1 \qquad \text{when } v_+ \ge v_- + V_{os} \tag{29.41}$$

$$v_o = 0 \qquad \text{when } v_+ < v_- + V_{os} \tag{29.42}$$

The resistor-string DAC was analyzed and presented in Sec. 29.1.2; the voltage on the i-th tap of the resistor string was found to be

$$V_i = V_{i,ideal} + \frac{V_{REF}}{2^N} \cdot \sum_{k=1}^{i} \frac{\Delta R_k}{R} \tag{29.43}$$

where $V_{i,ideal}$ is the voltage at the i-th tap if all the resistors had an ideal value of R. The term, ΔR_k, is the value of the resistance error (difference from ideal) due to the mismatch. Note that for the resistor-string DAC, the sum of the mismatch terms plays an important factor in the overall voltage at each tap.

The switching point for the i-th comparator, $V_{sw,i}$, then becomes

$$V_{sw,i} = V_i + V_{os,i} \tag{29.44}$$

where $V_{os,i}$ is the input referred offset voltage of the i-th comparator. The INL for the converter can then be described as

$$INL = V_{sw,i} - V_{sw,ideal} = V_{sw,i} - V_{i,ideal} \tag{29.45}$$

which becomes

$$INL = \frac{V_{REF}}{2^N} \cdot \sum_{k=1}^{i} \frac{\Delta R_k}{R} + V_{os,i} \tag{29.46}$$

The worst-case INL will occur at the middle of the string ($i = 2^{N-1}$), as described in Sec. 29.1.2 and Eq. (29.10). Including the offset voltage, the maximum INL will be

$$|INL|_{max} = \frac{V_{REF}}{2^N} \cdot \sum_{k=1}^{2^{N-1}} \frac{\Delta R_k}{R} + |V_{os,i}|_{max} = V_{REF} \cdot \frac{2^{N-1}}{2^N R} \cdot |\Delta R_k|_{max} + |V_{os,i}|_{max} \tag{29.47}$$

which can be rewritten as

$$|INL|_{max} = \frac{V_{REF}}{2} \cdot \left|\frac{\Delta R_k}{R}\right|_{max} + |V_{os,i}|_{max} \tag{29.48}$$

where it is assumed that the maximum positive mismatch occurs in all the resistors in the lower half of the string and the maximum negative mismatch occurs in the upper half (or vice versa) and that the comparator at the i-th tap contains the maximum offset voltage, $|V_{os,i}|_{max}$. Notice that the offset contributes directly to the maximum value for the INL. This explains another limitation to using Flash converters at high resolutions. The offset voltage alone can make the INL greater than ½ LSB.

Example 29.11

If a 10-bit Flash converter is designed, determine the maximum offset voltage of the comparators which will make the INL less than ½ LSB. Assume that the resistor string is perfectly matched and $V_{REF} = 5$ V.

Equation (29.48) requires that the offset voltage equal ½ LSB. Therefore,

$$|V_{os}|_{max} = \frac{5}{2^{11}} = 2.44 \text{ mV} \quad \blacksquare$$

The DNL calculation for the Flash converter is also attained using the analysis first presented in Sec. 29.1.2. Using the definition of DNL,

$$DNL = V_{sw,i} - V_{sw,i-1} - 1 \text{ LSB (in volts)} \tag{29.49}$$

Plugging in Eq. (29.44),

$$DNL = V_i + V_{os,i} - V_{i-1} - V_{os,i-1} - 1 \text{ LSB} \tag{29.50}$$

which can be written by using Eq. (29.6) as

$$DNL = V_{i,ideal} - V_{i-1,ideal} + \frac{V_{REF}}{2^N} \cdot \frac{\Delta R_i}{R} + V_{os,i} - V_{os,i-1} - 1\ LSB \tag{29.51}$$

which becomes

$$DNL = \frac{V_{REF}}{2^N} \cdot \frac{\Delta R_i}{R} + V_{os,i} - V_{os,i-1} \tag{29.52}$$

The maximum DNL will occur, assuming ΔR_i is at its maximum, $V_{os,i}$ is at its maximum positive value, and $V_{os,i-1}$ is at its maximum negative voltage. Thus,

$$|DNL|_{max} = \frac{V_{REF}}{2^N} \cdot \left|\frac{\Delta R_i}{R}\right|_{max} + 2|V_{os}|_{max} \tag{29.53}$$

which assumes that the maximum offset voltage in the positive and negative directions are symmetrical. Therefore, both resistor-string matching and offset voltage affect the DNL of the converter.

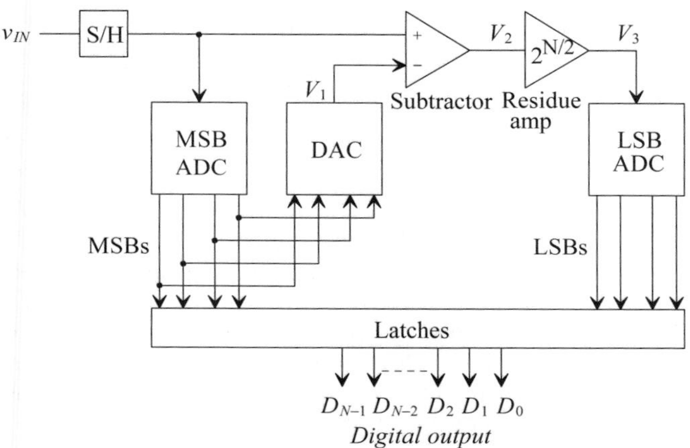

Figure 29.26 Block diagram of a two-step Flash ADC.

29.2.2 The Two-Step Flash ADC

Another type of Flash converter is called the two-step Flash converter or the parallel, feed-forward ADC [7–10]. The basic block diagram of a two-step converter is seen in Fig. 29.26. The converter is separated into two complete Flash ADCs with feed-forward circuitry. The first converter generates a rough estimate of the value of the input, and the second converter performs a fine conversion. The advantages of this architecture are that the number of comparators is greatly reduced from that of the Flash converter—from $2^N - 1$ comparators to $2(2^{N/2} - 1)$ comparators. For example, an 8-bit Flash converter requires 255 comparators, while the two-step Flash requires only 30. The trade-off is that the conversion process takes two steps instead of one, with the speed limited by the bandwidth and settling time required by the residue amplifier and the summer. The conversion process is as follows:

1. After the input is sampled, the most significant bits (MSBs) are converted by the first Flash ADC.

2. The result is then converted back to an analog voltage with the DAC and subtracted with the original input.

3. The result of the subtraction, known as the *residue*, is then multiplied by $2^{N/2}$ and input into the second ADC. The multiplication not only allows the two ADCs to be identical, but also increases the quantum level of the signal input into the second ADC.

4. The second ADC produces the least significant bits through a Flash conversion.

Some architectures use the same set of comparators in order to perform both steps. The multiplication mentioned in step 3 can be eliminated if the second converter is designed to handle very small input signals. The accuracy of the two-step ADC depends primarily on the linearity of the first ADC.

Figure 29.27 illustrates the two-step nature of the converter. A more intuitive approach can be explained with this picture. The first conversion identifies the segment in which the analog voltage resides. This is also known as a *coarse conversion* of the MSBs. The results of the coarse conversion are then multiplied by $2^{N/2}$ so that the segment within which V_{IN} resides will be scaled to the same reference as the first conversion. The second conversion is known as the *fine conversion* and will generate the final LSBs using the same Flash approach. One can see why the accuracy of the first converter is so important. If the input value is close to the boundary between two coarse segments and the first ADC is unable to choose the correct coarse segment, then the second conversion will be completely erroneous. The following example further illustrates the two-step algorithm.

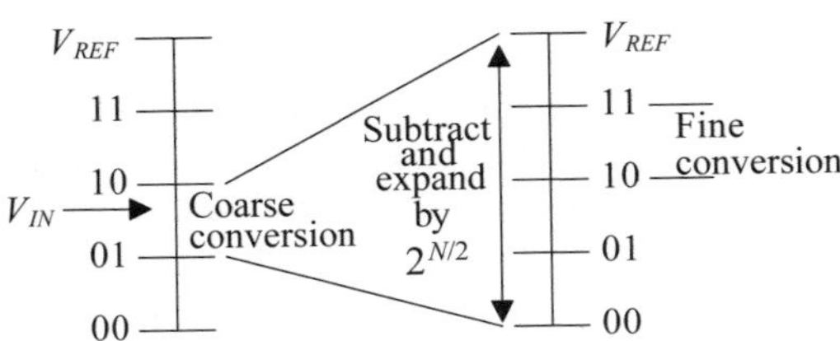

Figure 29.27 Coarse and fine conversions using a two-step ADC.

Example 29.12

Assume that the two-step ADC shown in Fig. 29.26 has four bits of resolution. Make a table listing the MSBs, V_1, V_2, V_3, and the LSBs for V_{IN} = 2, 4, 9, and 15 V assuming that V_{REF} = 16 V.

Since V_{REF} was conveniently made 16 V, each LSB will be 1 V. If V_{IN} = 2 V, the output of the first 2-bit Flash converter will be 00 since V_{REF} = 16 V and each resistor drops 4 V. The output of the 2-bit DAC, V_1, will therefore be 0, resulting in V_2 = 2 V. The multiplication of V_2 by the 4 results in V_3 = 8 V. Remember that each 2-bit Flash converter resembles that of Fig. 29.21. The thermometer code from the second Flash converter will be 0011, which results in 10 as the LSBs. The other values can be calculated as seen in Fig. 29.28. ■

V_{IN}	D_3D_2 (MSBs)	V_1	V_2	V_3	D_1D_0 (LSBs)
2	00	0	2	8	10
4	01	4	0	0	00
9	10	8	1	4	01
15	11	12	3	12	11

Figure 29.28 Output for the Flash ADC used in Ex. 29.12.

Accuracy Issues Related to the Two-Step Flash Converters

As stated previously, the overall accuracy of the converter depends on the first ADC. The second Flash must have only the accuracy of a stand-alone Flash converter. This means that if an 8-bit, two-step Flash converter contains two 4-bit Flash converters, the second Flash needs only to have the resolution of a 4-bit Flash, which is not difficult to achieve. However, the first 4-bit Flash must have the accuracy of an 8-bit Flash, meaning that the worst-case INL and DNL for the first bit Flash must be less than ±½ LSB for an 8-bit ADC. Thus, the resistor matching and comparators contained in the first ADC must possess the accuracy of the overall converter. Refer to Sec. 29.2.1 for derivations on INL and DNL for a Flash. The DAC must also be accurate to within the resolution of the ADC.

Accuracy Issues Related to the Operational Amplifiers

With the addition of the summer and the amplifier, other sources of accuracy errors are present in this converter. The summer and the amplifier must add and amplify the signal to within ±½ LSB of the ideal value. It is difficult to implement standard operational amplifiers within high-resolution data converters because of these accuracy requirements. The nonideal characteristics of the op-amp are well known and in many cases alone limit the accuracy of the data converter. In this case, the amplifier is required to multiply the residue signal by some factor of two. Although this may not seem difficult at first glance, a closer examination will reveal a dependency on the open-loop gain.

Suppose that the amplifier were being used in a 12-bit, two-step data converter. Remember that in order for a data converter to be N-bit accurate, the INL and DNL need to be kept below ±½ LSB and one-half of an LSB can be defined as

$$0.5\text{ LSB} = \frac{V_{REF}}{2^{N+1}} \tag{29.54}$$

Since the output of the amplifier gets quantized to 6 bits, the amplifier would need to be 6-bit accurate to within ±½ LSB, resulting in an accuracy of

$$\text{Accuracy} = \frac{0.5\ LSB}{\text{Full scale range }(V_{REF})} = \frac{1}{2^{6+1}} = \frac{1}{128} = 0.0078 = 0.78\% \tag{29.55}$$

And suppose that a feedback amplifier with a gain of 64, or $2^{N/2}$, is used as the residue amplifier. The gain would need to be within the following range:

$$63.5 \text{ V/V} < A_{CL} < 64.5 \text{ V/V} \tag{29.56}$$

where A_{CL} is the closed-loop gain of the amplifier. Already, one can see the limitations of using operational amplifiers with feedback in high-accuracy applications. Designing an op-amp based amplifier with a high degree of gain accuracy can be difficult.

Generalizing this concept for an N-bit application requires knowledge of feedback theory discussed in Ch. 24. The closed-loop gain of the amplifier is expressed as

$$A_{CL} = \frac{v_o}{v_i} = \frac{A_{OL}}{1 + A_{OL}\beta} \tag{29.57}$$

where A_{OL} is the open-loop gain of the amplifier and β is the feedback factor. Also, from Ch. 24, it is known that as A_{OL} increases in value, the closed-loop gain, A_{CL}, approaches the value of $1/\beta$. Therefore, if it is assumed the closed-loop gain of the amplifier equals the ideal value of $1/\beta$ minus some maximum deviation from the ideal, ΔA, then,

$$A_{CL} = \frac{v_o}{v_i} = \frac{A_{OL}}{1 + A_{OL}\beta} = \frac{1}{\beta} - \Delta A \tag{29.58}$$

where $1/\beta$ is the desired value of the closed-loop gain (usually some factor of 2^N) and ΔA is the required accuracy (±½ LSB) of the gain (i.e., $(1/\beta)\cdot(1/2^{N+1})$). The right two terms of Eq. (29.58) can be solved for the open-loop gain of the amplifier,

$$|A_{OL}| = \frac{1}{\beta}(2^{N+1} - 1) \approx \frac{2^{N+1}}{\beta} \tag{29.59}$$

If the op-amp is used as a gain of 64 ($1/\beta$) and is required to amplify signals with 6-bit accuracy, then the open-loop gain of the amplifier must be at least $|A_{OL}| \geq 128 \cdot 64 = 8,192$ V/V. This is certainly an achievable specification. However, notice that for every bit increase in resolution, the open-loop gain requirement doubles. This is one reason two-step Flash converters are limited in resolution to approximately 12 bits [9–12].

The unity-gain frequency, f_{un}, required of an op-amp used in or with a data converter for a specific settling time t, (where $t < 1/f_{clk}$) can be estimated assuming linear settling, and requiring the output of the op-amp be ½ LSB accurate, by

$$v_{out} = V_{outfinal}(1 - \frac{1}{2^{N+1}}) = V_{outfinal}(1 - e^{-t/\tau}) \text{ or } f_{un} \geq \frac{f_{clk} \cdot \ln 2^{N+1}}{2\pi \cdot \beta} \tag{29.60}$$

This equation can be used to determine the minimum op-amp gain-bandwidth product (= f_{un}) needed to achieve a specific settling time provided the op-amp slew-rate doesn't come into play and the op-amp can be modeled as a first-order system.

Linearity of the amplifier is another aspect of amplifier performance that must be considered when designing ADCs. The amplifier must be able to linearly amplify the input signal over an input voltage range to within ½ LSB of the number bits that its output is quantized. If the amplifier is not designed correctly, nonlinearity is introduced as devices in the amplifier go into nonsaturation. Harmonic distortion occurs, resulting in an error within the ADC. Linearity is typically measured in terms of total harmonic distortion, or THD, (refer to Sec. 21.3.3). However, the transfer curve illustrates the

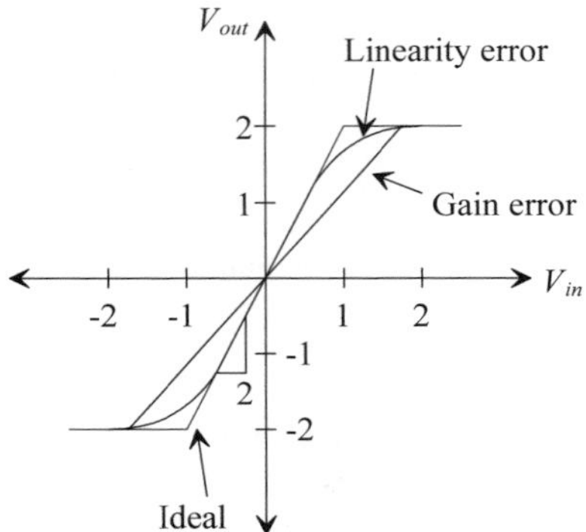

Figure 29.29 An op-amp transfer curve that distinguishes between gain error and linearity error.

limitation more effectively. Figure 29.29 shows a transfer curve of an op-amp with a gain of two. The ideal transfer curve is shown if the input range is known to be between –1 and 1 V. The actual transfer curve shows nonlinearity introduced at both ends of the input range. In order for the amplifier to be *N*-bit accurate, the slope of the actual transfer curve may not vary from the ideal by more than the accuracy required at the output of the amplifier. Note also in Fig. 29.29 the subtle difference between a gain error and nonlinearity. However, a gain error is much less harmful to an ADC's performance than harmonic distortion.

29.2.3 The Pipeline ADC

After examining the two-step ADC, one might wonder whether there is such a converter as a three-step or four-step ADC. In actuality, one could divide the number of conversions into many steps. The pipeline ADC is an *N*-step converter, with 1 bit being converted per stage. Able to achieve high resolution (10–13 bits) at relatively fast speeds [11–15], the pipeline ADC consists of *N* stages connected in series (Fig. 29.30). Each stage contains a 1-bit ADC (a comparator), a sample-and-hold, a summer, and a gain of two amplifier. Each stage of the converter performs the following operation:

1. After the input signal has been sampled, compare it to $\frac{V_{REF}}{2}$. The output of each comparator is the bit conversion for that stage.

2. If $v_{IN} > \frac{V_{REF}}{2}$ (comparator output is 1), $\frac{V_{REF}}{2}$ is subtracted from the held signal and pass the result to the amplifier. If $v_{IN} < \frac{V_{REF}}{2}$ (comparator output is 0), then pass the original input signal to the amplifier. The output of each stage in the converter is referred to as the *residue*.

3. Multiply the result of the summation by 2 and pass the result to the sample-and-hold of the next stage.

A main advantage of the pipeline converter is its high throughput. After an initial *latency* of *N* clock cycles, one conversion will be completed per clock cycle. While the residue of the first stage is being operated on by the second stage, the first stage is free to operate on the next samples. Each stage operates on the residue passed down from the

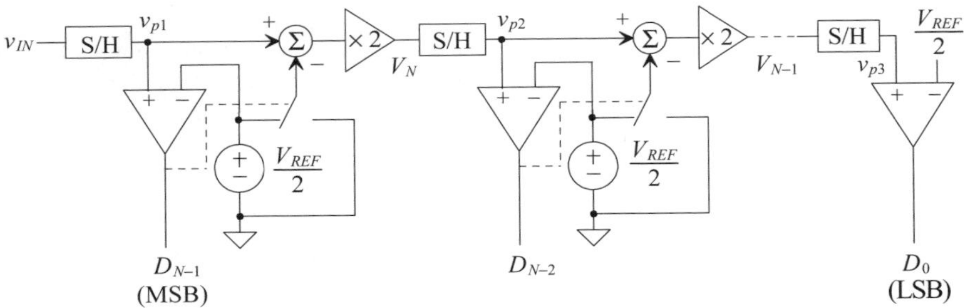

Figure 29.30 Block diagram of a pipeline ADC.

previous stage, thereby allowing for fast conversions. The disadvantage is having the initial N clock cycle delay before the first digital output appears. The severity of this disadvantage depends, of course, on the application.

One interesting aspect of this converter is its dependency on the most significant stages for accuracy. A slight error in the first stage propagates through the converter and results in a much larger error at the end of the conversion. Each succeeding stage requires less accuracy than the one before, so special care must be taken when considering the first several stages.

Example 29.13
Assume that the pipeline converter shown in Fig. 29.30 is a 3-bit converter. Analyze the conversion process by making a table of the following variables: D_2, D_1, D_0, V_2, V_1, for $v_{IN} = 2$, 3, and 4.5 V. Assume that $V_{REF} = 5$ V, V_3 is the residue voltage out of the first stage, and V_2 is the residue voltage out of the second stage.

The output of the first comparator, $D_2 = 0$, since $v_{IN} < 2.5$ V. Since $D_2 = 0$, $V_3 = 2(2) = 4$ V. Passing this voltage down the pipeline, since $V_3 > 2.5$ V, $D_1 = 1$ and V_2 becomes

$$V_1 = \left(V_2 - \frac{V_{REF}}{2}\right) \times 2 = 3 \text{ V}$$

The LSB, $D_0 = 1$, since $V_2 > 2.5$ V, and the digital output corresponding to $V_{IN} = 2$ V, is $D_2D_1D_0 = 011$. The actual digital outputs are simply the comparator outputs, and the data can be completed as seen in Fig. 29.31. ■

v_{IN}	V_3 (V)	V_2 (V)	**Digital Out ($D_2D_1D_0$)**
2.0	4.0	3.0	011
3.0	1.0	2.0	100
4.5	4.0	3.0	111

Figure 29.31 Output for the pipeline ADC used in Ex. 29.13.

Accuracy Issues Related to the Pipeline Converter

The 1-bit per stage ADC can be analyzed by examining the switching point of each comparator for the ideal and nonideal case. Using Fig. 29.30, and assuming that all of the components are ideal, let $v_{IN,1}$ represent the value of the input voltage when the first comparator switches. This occurs when

$$v_{IN,1} = \frac{1}{2} V_{REF} \tag{29.61}$$

The positive input voltage on the second comparator, v_{p2}, can be written in terms of the previous stage, or

$$v_{p2} = [v_{IN} - \tfrac{1}{2} \cdot D_{N-1} \cdot V_{REF}] \cdot 2 \tag{29.62}$$

where D_{N-1} is the MSB output from the first comparator and is either a 1 or a 0. The second comparator switches when $v_{p2} = \frac{1}{2} V_{REF}$. The value of v_{IN} at this point, denoted as $v_{IN,2}$, is

$$v_{IN,2} = \tfrac{1}{2} \cdot D_{N-1} \cdot V_{REF} + \tfrac{1}{4} V_{REF} \tag{29.63}$$

Continuing on in a similar manner, we can write the value of the voltage on the positive input of the third comparator in terms of the previous two stages as

$$v_{p3} = \left[[v_{IN} - \frac{1}{2} \cdot D_{N-1} \cdot V_{REF}] \cdot 2 - [\frac{1}{2} \cdot D_{N-2} \cdot V_{REF}] \right] \cdot 2 \tag{29.64}$$

and the third comparator will switch when $v_{p3} = \frac{1}{2} V_{REF}$, which corresponds to the point at which v_{IN} becomes

$$v_{IN,3} = \tfrac{1}{2} \cdot D_{N-1} \cdot V_{REF} + \tfrac{1}{4} \cdot D_{N-2} \cdot V_{REF} + \tfrac{1}{8} V_{REF} \tag{29.65}$$

By now, a general trend can be recognized and the value of v_{IN} can be derived for the point at which the comparator of the N-th stage switches. This expression can be written as

$$v_{IN,N} = \tfrac{1}{2} \cdot D_{N-1} \cdot V_{REF} + \tfrac{1}{4} \cdot D_{N-2} \cdot V_{REF} + \tfrac{1}{8} \cdot D_{N-3} \cdot V_{REF} + \ldots\ldots + \tfrac{1}{2^{N-1}} \cdot D_1 \cdot V_{REF} + \tfrac{1}{2^N} \cdot V_{REF} \tag{29.66}$$

Notice that the preceding equation does not include D_0. This is because D_0 is the output of the N-th stage comparator.

Now that we have derived the switching points for the ideal case, the nonideal case can be considered. Only the major sources of error will be included in the analysis so as not to overwhelm the reader. These include the comparator offset voltage, $V_{COS,x}$, and the sample-and-hold offset voltage, $V_{SOS,x}$. The variable, x, represents the number of the stage for which each of the errors is associated, and the "prime" notation will be used to distinguish between the ideal and nonideal case. The reader should also be aware that the offset voltages can be of either polarity. It will be assumed that all of the residue amplifiers have the same gain, denoted as A.

The positive input to the first nonideal comparator, v'_{p1}, will include the offset from the first sample-and-hold, such that

$$v'_{p1} = v_{IN} + V_{SOS,1} \tag{29.67}$$

Now the first comparator will not switch until the voltage on the positive input overcomes the comparator offset as well. This occurs when

$$v_{p1}' = \tfrac{1}{2}V_{REF} + V_{COS,1} \tag{29.68}$$

Thus, equating Eqs. (29.67) and (29.68) and solving for the value of the input voltage when the switching occurs for the first comparator yields

$$v_{IN,1}' = \tfrac{1}{2}V_{REF} + V_{COS,1} - V_{SOS,1} \tag{29.69}$$

The input to the second comparator, v_{p2}', can be written as

$$v_{p2}' = [v_{IN} + V_{SOS,1} - \tfrac{1}{2}\cdot D_{N-1}\cdot V_{REF}]\cdot A + V_{SOS,2} \tag{29.70}$$

and the value of input voltage at the point which the second comparator switches occurs when

$$v_{IN,2}' = \frac{1}{2}\cdot D_{N-1}\cdot V_{REF} + \frac{1}{2}\frac{V_{REF}}{A} - V_{SOS,1} - \frac{1}{A}(V_{SOS,2} - V_{COS,2}) \tag{29.71}$$

Continuing in the same manner, we can write the value of the input voltage that causes the third comparator to switch as

$$v_{IN,3}' = \tfrac{1}{2}\cdot D_{N-1}\cdot V_{REF} + \tfrac{1}{2}\cdot D_{N-2}\cdot \tfrac{V_{REF}}{A} - V_{SOS,1} - \tfrac{1}{A}V_{SOS,2} - \tfrac{1}{A^2}V_{SOS,3} - \tfrac{1}{A^2}\left[V_{COS,3} - \tfrac{1}{2}V_{REF}\right] \tag{29.72}$$

which can be generalized to the N-th switching point as

$$v_{IN,N}' = \tfrac{1}{2}\cdot D_{N-1}\cdot V_{REF} + \tfrac{1}{2}\cdot D_{N-2}\cdot \tfrac{V_{REF}}{A} + \ldots + \tfrac{1}{2}\cdot D_1\cdot \tfrac{V_{REF}}{A^{N-2}} + \tfrac{1}{2}\cdot \tfrac{V_{REF}}{A^{N-1}} + \tfrac{V_{COS,N}}{A^{N-1}} - \sum_{k=1}^{N}\tfrac{V_{SOS,k}}{A^{k-1}} \tag{29.73}$$

The INL can be calculated by subtracting switching point between the nonideal and ideal case. Therefore, the INL of the first stage is found by subtracting Eqs. (29.69) and (29.61).

$$INL_1 = v_{IN,1}' - v_{IN,1} = V_{COS,1} - V_{SOS,1} \tag{29.74}$$

The second stage INL is

$$INL_2 = v_{IN,2}' - v_{IN,2} = \frac{V_{REF}}{2}\left(\frac{1}{A} - \frac{1}{2}\right) - V_{SOS,1} - \frac{V_{SOS,2}}{A} + \frac{V_{COS,2}}{A} \tag{29.75}$$

and the INL for the N-th stage is

$$INL_N = \frac{1}{2}\cdot D_{N-2}\cdot V_{REF}\cdot\left(\frac{1}{A} - \frac{1}{2}\right) + \frac{1}{2}\cdot D_{N-3}\cdot V_{REF}\cdot\left(\frac{1}{A^2} - \frac{1}{4}\right) + \ldots$$
$$+\frac{1}{2}\cdot D_1\cdot V_{REF}\cdot\left(\frac{1}{A^{N-2}} - \frac{1}{2^{N-2}}\right) + \frac{1}{2}\cdot V_{REF}\cdot\left(\frac{1}{A^{N-1}} - \frac{1}{2^{N-1}}\right) + \frac{V_{COS,N}}{A^{N-1}} - \sum_{k=1}^{N}\frac{V_{SOS,K}}{A^{k-1}} \tag{29.76}$$

Equations (29.74)–(29.76) are very important to understanding the limitations of the pipeline ADC. Notice the importance of the comparator and summer offsets in Eq. (29.74). The worst-case addition of the offsets must be less than ½ LSB to keep the ADC

N-bit accurate. The second stage is more dependent on the gain of the residue amplifier as seen in Eq. (29.75). The gain error discussed in the previous section plays an important role in determining the overall accuracy of the converter. Now examine the effects of the offsets on the INL of the N-th stage. In Eq. (29.76), both the comparator and summer offsets of the N-th stage (when $k = N$) are divided by a large gain. Therefore, the latter stages in a pipeline ADC are not as critical to the accuracy as the first stages, and die area and power can be reduced by using less accurate designs for the least significant stages. The summation term in Eq. (29.76) also reveals that the summer offset of the first stage ($k = 1$) has a large effect on the N-th stage. However, this point is inconsequential since $V_{SOS,1}$ must be minimized to achieve N-bit accuracy for the first stage anyway. Typically, if the INL and DNL specifications can be made N-bit accurate in the first few stages, the latter stages will not adversely affect overall accuracy.

The DNL can be found by calculating the difference between the worst-case switching points and subtracting the ideal value for an LSB. As defined earlier, the worst case will occur at midscale when the output switches from 0111...111 to 1000...000 as v_{IN} increases. Thus, the DNL is

$$DNL_{max} = v'_{IN,1} - v'_{IN,N} - \frac{V_{REF}}{2^N} \tag{29.77}$$

where $v'_{IN,N}$ is calculated using Eq. (29.73) and assuming that D_{N-1} is a zero and that all of the other bits are ones. Plugging in Eqs. (29.69) and (29.73) into Eq. (29.77) yields

$$DNL_{max} = \frac{1}{2}V_{REF}\left(1 - \sum_{k=1}^{N-1}\frac{1}{A^k}\right) + V_{COS,1} - \frac{V_{COS,N}}{A^{N-1}} + \sum_{k=2}^{N}\frac{V_{SOS,K}}{A^{k-1}} - \frac{V_{REF}}{2^N} \tag{29.78}$$

Again, the term that dominates this expression is the comparator offset associated with the first stage and the summer offset of the second stage. The entire expression in Eq. (29.78) must be less than ½ LSB for the ADC to have N-bit resolution.

29.2.4 Integrating ADCs

Another type of ADC performs the conversion by integrating the input signal and correlating the integration time with a digital counter. Known as single- and dual-slope ADCs, these types of converters are used in high-resolution applications but have relatively slow conversions. However, they are very inexpensive to produce and are commonly found in slow-speed, cost-conscious applications.

Single-Slope Architecture

Figure 29.32 illustrates the single-slope converter in block level form. A counter determines the number of clock pulses that are required before the integrated value of a reference voltage is equal to the sampled input signal. The number of clock pulses is proportional to the actual value of the input, and the output of the counter is the actual digital representation of the analog voltage.

Since the reference is a DC voltage, the output of the integrator should start at zero and linearly increase with a slope that depends on the gain of the integrator. Notice that the reference voltage is defined as negative so that the output of the inverting integrator is positive. At the time when the output of the integrator surpasses the value of the S/H output, the comparator switches states, thus triggering the control logic to latch

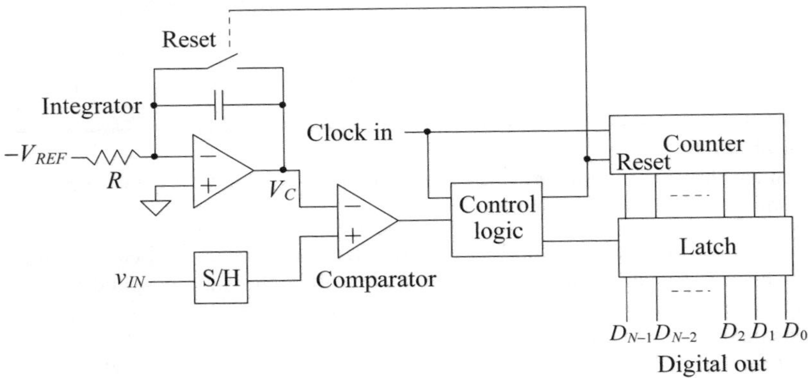

Figure 29.32 Block diagram of a single-slope ADC.

the value of the counter. The control logic also resets the system for the next sample. Figure 29.33 illustrates the behavior of the integrator output and the clock.

Note that if the input voltage is very small, the conversion time is very short, as the counter has to increment only a few times before the comparator latches the data. However, if the input voltage is at its full-scale value, the counter must increment to its maximum value of 2^N clock cycles. Thus, the clock frequency must be many times faster than the bandwidth of the input signal. The conversion time, t_c, depends on the value of the input signal and can be described as

$$t_c = \frac{v_{IN}}{V_{REF}} \cdot 2^N \cdot T_{CLK} \tag{29.79}$$

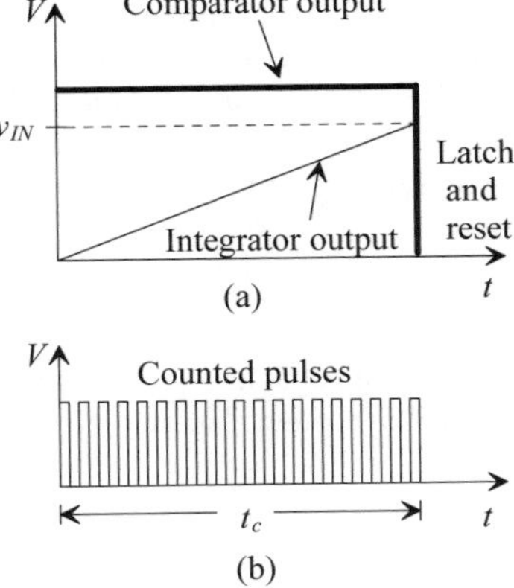

Figure 29.33 Single-slope ADC timing diagrams for (a) the comparator inputs and outputs and (b) the resulting counted pulses.

where T_{CLK} is the period of the clock. The sampling rate is inversely proportional to the conversion time and can be written as

$$f_{Sample} = \frac{V_{REF}}{V_{IN} \cdot 2^N} \cdot f_{CLK} \tag{29.80}$$

Example 29.14

Determine the clock frequency needed to form an 8-bit, single-slope converter, if the analog signal bandwidth is 20 kHz.

Since the sampling rate required is 40 kHz, then the worst-case situation would occur for a full-scale input, in which event the integrator output would have to climb to its maximum value and the counter would increment 2^N times during the corresponding 25 μs period between samples. Therefore, the clock frequency would need to be 2^N times faster than the sampling rate or 10.24 MHz. ■

Accuracy Issues Related to the Single-Slope ADC

Obviously, many potential error sources abound in this architecture. At the end of the conversion, the voltage across the integrating capacitor, V_C, assuming no initial condition, will be

$$V_C = \frac{1}{C}\int_0^{t_c} \frac{V_{REF}}{R} dt = \frac{V_{REF} \cdot t_c}{RC} \tag{29.81}$$

where t_c is the conversion time. Plugging Eq. (29.79) into Eq. (29.81) yields

$$V_C = \frac{2^N \cdot T_{CLK} \cdot v_{IN}}{RC} = \frac{2^N \cdot v_{IN}}{f_{CLK} \cdot RC} \tag{29.82}$$

Equation (29.82) is a revealing one in that the final voltage on the integrator output depends not only on the value of the input voltage, which is to be expected, but also on the value of *R*, *C*, and f_{CLK}. Therefore, any nonideal effects affecting these values will have an influence on the accuracy of the integrator output from sample to sample. For example, if an integrated diffused-resistor is used, then the voltage coefficient of the resistor could limit the accuracy, since the resistor will be effectively nonlinear. Similarly, the capacitor may have charge leakage or aging effects associated with it. Also, any jitter in the clock will affect the overall accuracy. The integrator must have a linear slope to within the accuracy of the converter, which depends on the specifications of the op-amp (open- loop gain, settling time, offset, etc.) and must be considered accordingly.

Offset voltages on the comparator, the S/H, or the integrator result in additional or fewer clock pulses, depending on the polarity of the offset. A delay also exists from the time that the inputs to the comparator are equal and the time that the output of the counter is actually latched. The reference voltage must also stay constant to within the accuracy of the converter.

Dual-Slope Architecture

A slightly more sophisticated design known as the dual-slope, integrating ADC (Fig. 29.34) eliminates most of the problems encountered when using the single-slope converter. Here, two integrations are performed, one on the input signal and one on V_{REF}. The input voltage in this case is assumed to be negative, so that the output of the

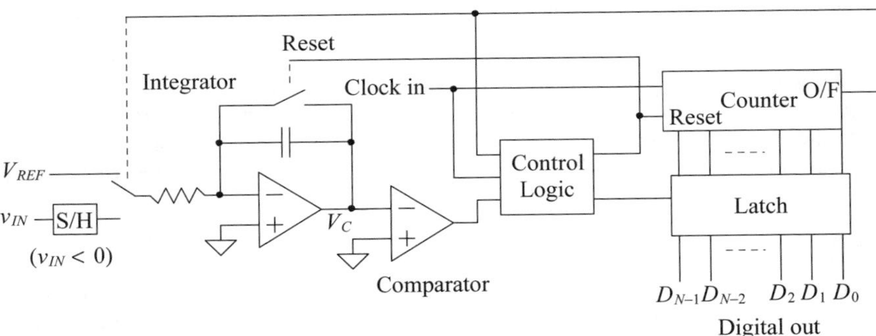

Figure 29.34 Block diagram of a dual-slope ADC.

inverting integrator results in a positive slope during the first integration. Figure 29.35 illustrates the behavior for two separate samples. The first integration is of fixed length, dictated by the counter, in which the sample-and-held signal is integrated, resulting in the first slope. After the counter overflows and is reset, the reference voltage is connected to the input of the integrator. Since v_{IN} was negative and the reference voltage is positive, the inverting integrator output begins discharging back down to zero at a constant slope. A counter again measures the amount of time for the integrator to discharge, thus generating the digital output.

For Fig. 29.35, a 3-bit ADC is being used. Thus, the first integration period continues until the beginning of the eighth (2^3) clock pulse, which corresponds to the counter's overflow bit. Note that the integrator's output corresponding to V_B is twice the value of the output corresponding to V_A. Thus, it requires twice as many clock pulses for

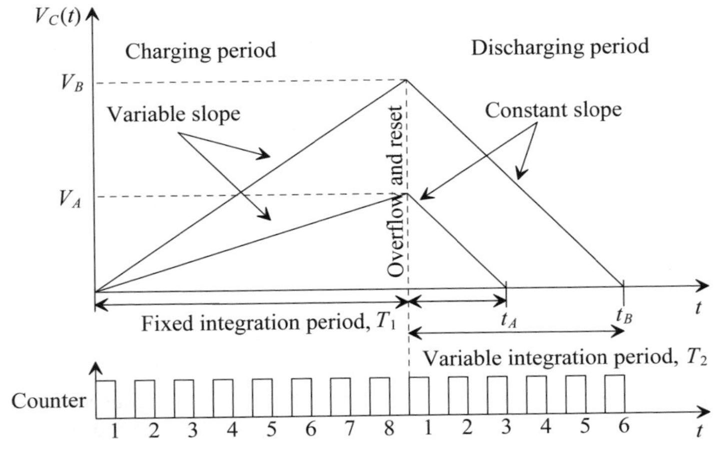

Figure 29.35 Integration periods and counter output for two separate samples of a 3-bit dual-slope ADC.

the integrator to discharge back to zero from V_B than from V_A. The output of the counter at t_A is three or 011, while the counter output at t_B is twice that value or six (110) and the quantization is complete.

Notice that the first slope varies according to the value of the input signal, while the second slope, dependent only on V_{REF}, is constant. Similarly, the time required to generate the first slope is constant, since it is limited by the size of the counter. However, the discharging period is variable and results in the digital representation of the input voltage.

Accuracy Issues Related to the Dual-Slope ADC

One may wonder how the dual-slope converter is an improvement over the single-slope architecture, since a significantly longer conversion time is required. The first integration period requires a full 2^N clock cycle and cannot be decreased, because the second integration might require the full 2^N clock cycles to discharge if the maximum value of v_{IN} is being converted. However, the dual slope is the preferred architecture because the same integrator and clock are used to produce both slopes. Therefore, any nonidealities will essentially be canceled. For example, assuming that the S/H is ideal, the gain of the integrator at the end of the first integration period, T_1, becomes

$$V_C = -\frac{1}{C}\int_0^{T_1} \frac{v_{IN}}{R}dt = \frac{|v_{IN}| \cdot T_1}{RC} \tag{29.83}$$

The output at the end of T_1 is positive since the input voltage is considered to be negative and the integrator is inverting. After the clock has been reset, the discharging commences, with the initial condition defined by the value of the integrator output at the end of the charging period, or

$$V_C = \frac{|v_{IN}| \cdot T_1}{RC} - \frac{1}{C}\int_0^{T_2} \frac{V_{REF}}{R}dt \tag{29.84}$$

Once the value of the integrator output, V_C, reaches zero volts, Eq. (29.84) becomes

$$V_C = \frac{|v_{IN}| \cdot T_1}{RC} - \frac{V_{REF} \cdot T_2}{RC} = 0 \tag{29.85}$$

or,

$$|v_{IN}| \cdot T_1 = V_{REF} \cdot T_2 \tag{29.86}$$

At the end of the conversion, the dependencies on R and C have canceled out. Since we also know that the counter increments 2^N times at time, T_1, and the counter increments D times at time, T_2, Eq. (29.86) can be rewritten as

$$\frac{D}{2^N} = \frac{|v_{IN}|}{V_{REF}} \tag{29.87}$$

where D is the counter output that is actually the digital representation of the input voltage. Thus, it can be written that the ratio of the input voltage and the reference voltage is proportional to the ratio of the binary value of the digital word, D, and 2^N. Therefore, since the same clock pulse is responsible for the charging and discharging times, any irregularities will also cancel out.

29.2.5 The Successive Approximation ADC

The successive approximation converter performs basically a binary search through all possible quantization levels before converging on the final digital answer. The block diagram is seen in Fig. 29.36. An N-bit register controls the timing of the conversion where N is the resolution of the ADC. V_{IN} is sampled and compared to the output of the DAC. The comparator output controls the direction of the binary search, and the output of the successive approximation register (SAR) is the actual digital conversion. The successive approximation algorithm is as follows.

1. A 1 is applied to the input to the shift register. For each bit converted, the 1 is shifted to the right 1-bit position. $B_{N-1} = 1$ and B_{N-2} through $B_0 = 0$.

2. The MSB of the SAR, D_{N-1}, is initially set to 1, while the remaining bits, D_{N-2} through D_0, are set to 0.

3. Since the SAR output controls the DAC and the SAR output is 100...0, the DAC output will be set to $\frac{V_{REF}}{2}$.

4. Next, v_{IN} is compared to $\frac{V_{REF}}{2}$. If $\frac{V_{REF}}{2}$ is greater than v_{IN}, then the comparator output is a 1 and the comparator resets D_{N-1} to 0. If $\frac{V_{REF}}{2}$ is less than v_{IN}, then the comparator output is a 0 and the D_{N-1} remains a 1. D_{N-1} is the actual MSB of the final digital output code.

5. The 1 applied to the shift register is then shifted by one position so that $B_{N-2} = 1$, while the remaining bits are all 0.

6. D_{N-2} is set to a 1, D_{N-3} through D_0 remain 0, while D_{N-1} remains the value from the MSB conversion. The output of the DAC will now either equal $\frac{V_{REF}}{4}$ (if $D_{N-1} = 0$) or $\frac{3V_{REF}}{4}$ (if $D_{N-1} = 1$).

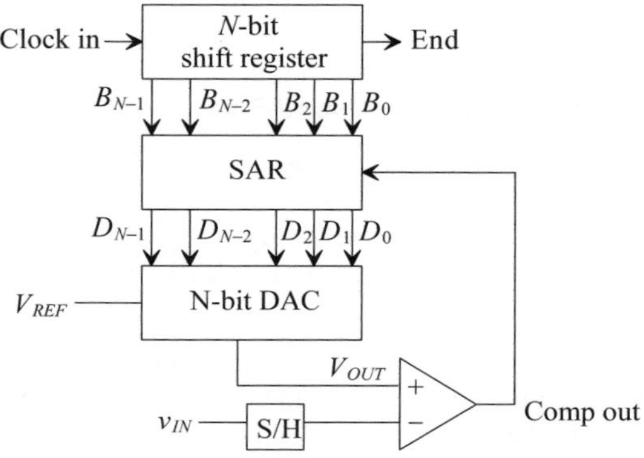

Figure 29.36 Block diagram of the successive approximation ADC.

7. Next, v_{IN} is compared to the output of the DAC. If the DAC output is greater than v_{IN}, the comparator the D_{N-2} is reset to 0. If v_{IN} is less than the DAC output, D_{N-2} remains a 1.

8. The process repeats until the output of the DAC converges to the value of v_{IN} within the resolution of the converter.

Figure 29.37 shows an example of the binary search nature of the converter. The bolded line shows the path of the conversion for 101, corresponding to $\frac{5}{8}V_{REF}$. All possible quantization levels are represented in the binary tree. With each bit decided, the search space decreases by one-half until the correct answer is converged upon.

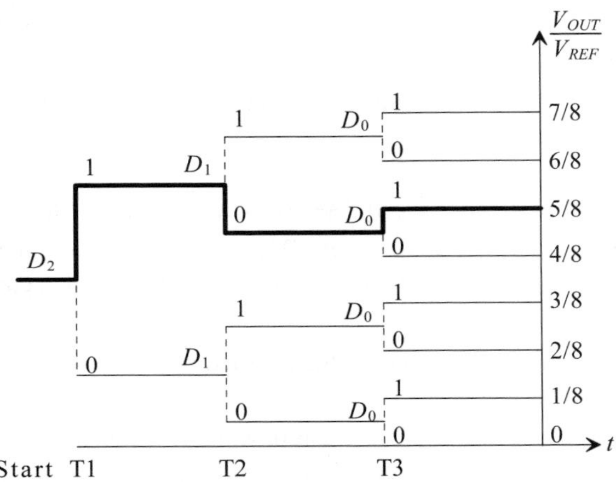

Figure 29.37 Binary search performed by a 3-bit successive approximation ADC for D=101.

Example 29.15

Perform the operation of a 3-bit successive approximation ADC similar to Fig. 29.36 with V_{REF} = 8. Make a table that consists of $D_2D_1D_0$, $B_2B_1B_0$, V_{OUT} (the output from the DAC) and the comparator output, which shows the binary search algorithm of the converter for v_{IN} = 5.5 V and 2.5 V.

We will designate $D_{2'}D_{1'}D_{0'}$ as the initial output of the SAR before the comparator makes its decision. The final value is designated as $D_2D_1D_0$. Notice that if the comparator is a 1, $D_{2'}D_{1'}D_{0'}$ differs from $D_2D_1D_0$, but if the comparator outputs a 0, then $D_{2'}D_{1'}D_{0'}$ = $D_2D_1D_0$. The output of the shift register is designated as $B_2B_1B_0$.

Following the algorithm discussed previously, initially v_{IN} = 5.5 V and is compared with 4 V. Since the comparator output is 0, the MSB remains a 1. The next bit is examined, and the output of the DAC is now 6 V. Since $V_{OUT} > v_{IN}$, the comparator output is 1, which resets the current SAR bit, D_1, to a 0 at the end of period T2. Lastly, the LSB is examined, and v_{IN} is compared with 5 V. Since $v_{IN} > V_{OUT}$, the comparator output is a 0, and the current SAR bit, D_0, remains a 1. The results can be examined in Fig. 29.38a. The final value for $D_2D_1D_0$ is 101, which is what is expected considering that 101 in binary is equivalent to 5_{10}. Figure 29.38b shows the data for the ADC using v_{IN} = 2.5 V. The final value for v_{IN} = 2.5 is 010, which again is what is expected for 3-bit resolution. ■

Step	v_{IN}	$B_2B_1B_0$	$D_2{}'D_1{}'D_0{}'$	V_{OUT}	Comp Out	$D_2D_1D_0$
T1	5.5	100	100	1/2 V_{REF} = 4 V	0	100
T2	5.5	010	110	(1/2+1/4)V_{REF} = 6 V	1	100
T3	5.5	001	101	(1/2+1/8)V_{REF} = 5 V	0	101

(a)

Step	v_{IN}	$B_2B_1B_0$	$D'_2D'_1D'_0$	V_{OUT}	Comp Out	$D_2D_1D_0$
T1	2.5	100	100	1/2 V_{REF} = 4 V	1	000
T2	2.5	010	010	1/4 V_{REF} = 2 V	0	010
T3	2.5	001	011	(1/4+1/8)V_{REF} = 3 V	1	010

(b)

Figure 29.38 Results from the 3-bit successive approximation ADC using (a) v_{IN} = 5.5 and (b) 2.5 V.

The successive approximation ADC is one of the most popular architectures used today. The simplicity of the design allows for both high speed and high resolution while maintaining relatively small area. The limit to the ADC's accuracy depends mainly on the accuracy of the DAC. If the DAC does not produce the correct analog voltage with which to compare the input voltage, the entire converter output will contain an error. Referring again to Fig. 29.37, we can see that if a wrong decision is made early, a massive error will result as the converter attempts to search for the correct quantization level in the wrong half of the binary tree.

The Charge-Redistribution Successive Approximation ADC

One of the most popular types of successive approximation architectures uses the binary-weighted capacitor array (analyzed in Sec. 29.1.5) as its DAC. Called a charge-redistribution, successive-approximation ADC [2,16,17], this converter samples the input signal and then performs the binary search based on the amount of charge on each of the DAC capacitors. Figure 29.39 shows an *N*-bit architecture. A comparator has replaced the unity gain buffer used in the DAC architecture. The binary-weighted capacitor array also samples the input voltage, so no external sample-and-hold is needed.

The conversion process begins by discharging the capacitor array, via the reset switch. Although this may appear to be an insignificant action, the converter is also performing automatic offset cancellation. Once the reset switch is closed, the comparator acts as a unity gain buffer. Thus, the capacitor array charges to the offset voltage of the comparator. This requires that the comparator is designed to be unity-gain stable, which means that internal compensation may have to be switched in during the reset period. Next, the input voltage, v_{IN}, is sampled onto the capacitor array. The reset switch is still closed, for the top plate of the capacitor array needs to be connected to virtual ground of the unity gain buffer. The equivalent circuit is seen in Fig. 29.40a. The reset switch is

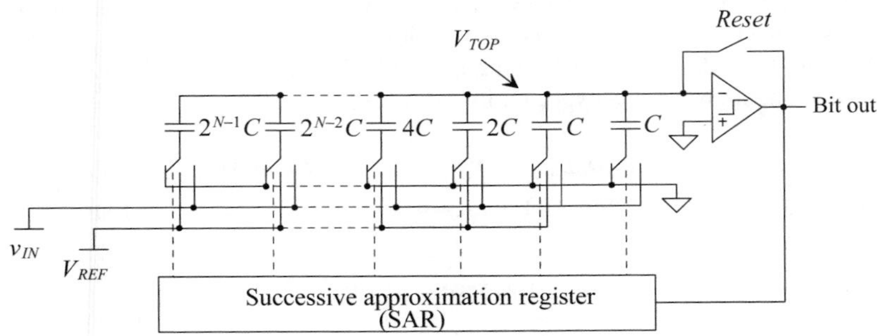

Figure 29.39 A charge redistribution ADC using a binary-weighted capacitor array DAC.

then opened, and the bottom plates of each capacitor in the array are switched to ground, so that the voltage appearing at the top plate of the array is now $V_{OS} - v_{IN}$ (Fig. 29.40b). The conversion process begins by switching the bottom plate of the MSB capacitor to V_{REF} (Fig. 29.40c). If the output of the comparator is high, the bottom plate of the MSB capacitor remains connected to V_{REF}. If the comparator output is low, the bottom plate of the MSB is connected back to ground. The output of the comparator is D_{N-1}. The voltage at the top of the capacitor array, V_{TOP}, is now

$$V_{TOP} = -v_{IN} + V_{OS} + D_{N-1} \cdot \frac{V_{REF}}{2} \tag{29.88}$$

The next largest capacitor is tested in the same manner as seen in Fig. 29.40d. The voltage at the top plate of the capacitor after the second capacitor is tested becomes

$$V_{TOP} = -v_{IN} + V_{OS} + D_{N-1} \cdot \frac{V_{REF}}{2} + D_{N-2} \cdot \frac{V_{REF}}{4} \tag{29.89}$$

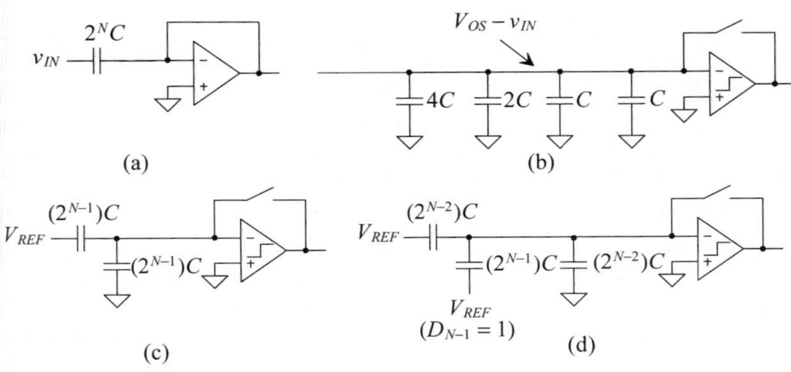

Figure 29.40 The charge redistribution process: (a) Sampling the input while autozeroing the offset, (b) the voltage at the top plate after sampling, (c) the equivalent circuit while converting the MSB, and (d) the equivalent circuit while converting the next largest capacitor with the MSB result equal to one.

The conversion process continues on with the remaining capacitors so that the voltage on the top plate of the array, V_{TOP}, converges to the value of the offset voltage, V_{OS} (within the resolution of the converter), or

$$V_{TOP} = -v_{IN} + V_{OS} + D_{N-1} \cdot \frac{V_{REF}}{2} + D_{N-2} \cdot \frac{V_{REF}}{4} + ... + D_1 \cdot \frac{V_{REF}}{2^{N-2}} + D_0 \cdot \frac{V_{REF}}{2^{N-1}} \approx V_{OS} \tag{29.90}$$

Note that the initial charge stored on the capacitor array is now redistributed onto only those capacitors that have their bottom plates connected to V_{REF}.

Accuracy Issues Related to the Charge-Redistribution, Successive-Approximation ADC

Obviously, the limitation of this architecture is the capacitor matching. The mismatch is analyzed in the same manner as the binary-weighted current source array of Sec. 29.1.4. Thus, substituting the value of the unit capacitance, C, for the value of the unit current source, I, and using Eqs. (29.27)–(29.30),

$$|INL|_{max} = 2^{N-1}(C + |\Delta C|_{max,INL}) - 2^{N-1} \cdot C = 2^{N-1} \cdot |\Delta C|_{max,INL} \tag{29.91}$$

where the maximum ΔC that will result in an INL that is less than ½ LSB is

$$|\Delta C|_{max,INL} = \frac{0.5C}{2^{N-1}} = \frac{C}{2^N} \tag{29.92}$$

and DNL is defined by

$$DNL_{max} = (2^N - 1) \cdot |\Delta C|_{max,DNL} \tag{29.93}$$

with the maximum ΔC, which results in a DNL less than ½ LSB:

$$|\Delta C|_{max,DNL} = \frac{0.5C}{2^N - 1} = \frac{C}{2^{N+1} - 2} \tag{29.94}$$

29.2.6 The Oversampling ADC

ADCs can be separated into two categories depending on the rate of sampling. The first category samples the input at the Nyquist rate, or $f_N = 2F$ where F is the bandwidth of the signal and f_N is the sampling rate. The second type samples the signal at a rate much higher than the signal bandwidth. This type of converter is called an oversampling converter. Traditionally, successive approximation or dual-slope converters are used when high resolution is desired. However, trimming is required when attempting to achieve higher accuracy. Dual-slope converters require high-speed, high-accuracy integrators that are only available using a high f_T bipolar process. Having to design a high-precision sample-and-hold is another factor that limits the realization of a high-resolution ADC using these architectures.

The oversampling ADC [18–20] is able to achieve much higher resolution than the Nyquist rate converters. This is because digital signal processing techniques are used in place of complex and precise analog components. The accuracy of the converter does not depend on the component matching, precise sample-and-hold circuitry, or trimming, and only a small amount of analog circuitry is required. Switched-capacitor implementations are easily achieved, and, as a result of the high sampling rate, only simplistic anti-aliasing circuitry needs to be used. However, because of the amount of time required to sample the input signal, the throughput is considerably less than the Nyquist rate ADCs.

Differences in Nyquist Rate and Oversampling ADCs

The typical process used in analog-to-digital conversion is seen in Fig. 29.41a, while the block diagram for the oversampling ADC is seen in Fig. 29.41b. After filtering the signal to help minimize aliasing effects, the signal is sampled, quantized, and encoded or decoded using simple digital logic to provide the digital data in the proper format. When using oversampling ADCs, little if any, anti-alias filtering is needed, no dedicated S/H is required, the quantization is performed with a modulator, and the encoding usually takes the form of a digital filter.

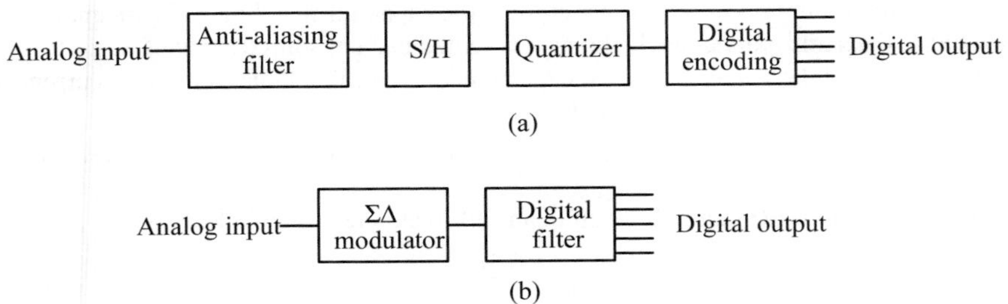

Figure 29.41 Typical block diagram for (a) Nyquist rate converters and (b) oversampling ADCs.

Since the oversampling converter samples the signal bandwidth at many times, aliasing is not a serious problem. A discussion of the frequency characteristics of aliasing was presented in Ch. 28. Figure 29.42a shows that when using Nyquist rate converters, a sampled signal in the frequency domain appears as a series of band-limited signals at multiples of the sampling frequency (see Fig. 28.26 for more details). As the sampling frequency decreases, the frequency spectra begin to overlap, and aliasing (Fig. 29.42b) occurs. Complex, "brickwall" filters are needed to correct the problem.

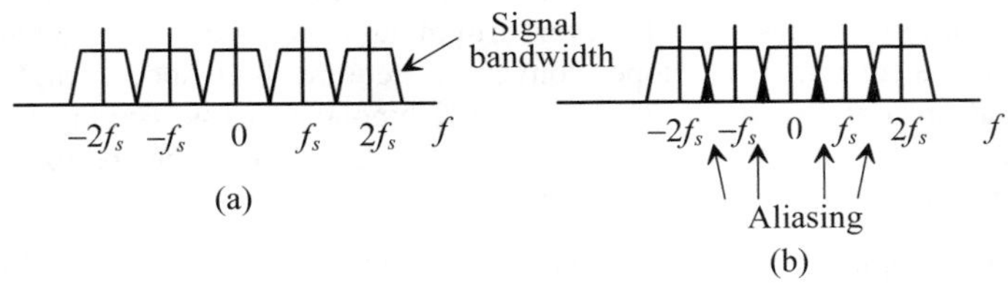

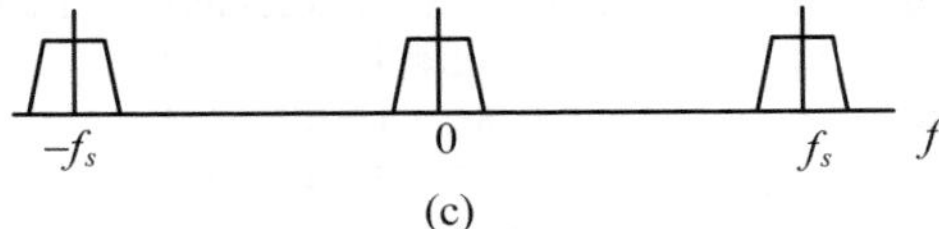

Figure 29.42 Frequency domain for (a) Nyquist rate converters, (b) the aliasing that occurs, and (c) an oversampling converter.

For oversampled ADCs, aliasing becomes much less of a factor. Since the sampling rate is much greater than the bandwidth of the signal, the frequency domain representation shows that the spectra are widely spaced, as seen in Fig. 29.42c. Therefore, overlapping of the spectra, and thus aliasing, will not occur, and only simple, first-order filters are required.

Oversampling converters typically employ switched-capacitor circuits and therefore do not need sample-and-hold circuits. The output of the modulator is a pulse-density modulated signal that represents the average of the input signal. The modulator constructs these pulses in real time, and so it is not necessary to hold the input value and perform the conversion.

As stated previously, the modulator actually provides the quantization in the form of a pulse-density modulated signal. Referred to as sigma-delta ($\Sigma\Delta$) or delta-sigma ($\Delta\Sigma$) modulation, the density of the pulses represents the average value of the signal over a specific period. Figure 29.43 illustrates the output of the modulator for the positive half of a sine wave input. Note that for the peak of the sine wave, most of the pulses are high. As the sine wave decreases in value, the pulses become distributed between high and low according to the sine wave value.

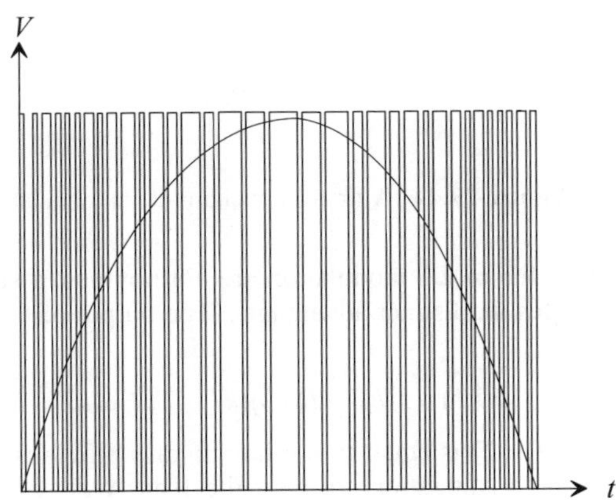

Figure 29.43 Pulse-density output from a sigma-delta modulator for a sine wave input.

If the frequency of the sine wave represented the highest frequency component of the input signal, a Nyquist rate converter would take only two samples. The oversampling converter, however, may take hundreds of samples over the same period to produce this pulse-density signal.

Digital signal processing is then used, which has two purposes: to filter any out-of-band quantization noise and to attenuate any spurious out-of-band signals. The output of the filter is then downsampled to the Nyquist rate so that the resulting output of the ADC is the digital data. This data represents the average value of the analog voltage over the oversampling period. The effective resolution of oversampling converters is determined by the values of signal-to-noise ratio and dynamic range obtained.

The First-Order ΣΔ Modulator

Now that the basic function of the ΣΔ modulator has been described, it would be useful to examine its inner workings and determine why ΣΔ modulation is so beneficial for generating high-resolution data. A basic first-order ΣΔ modulator can be seen in Fig. 29.44. Here, an integrator and a 1-bit ADC are in the forward path, and a 1-bit DAC is in the feedback path of a single-feedback loop system. The variables labeled are in terms of time, T, which is the inverse of the sampling frequency and k, which is an integer. The 1-bit ADC is simply a comparator that converts an analog signal into either a high or a low. The 1-bit DAC uses the comparator output to determine if $+V_{REF}$ or $-V_{REF}$ is summed with the input.

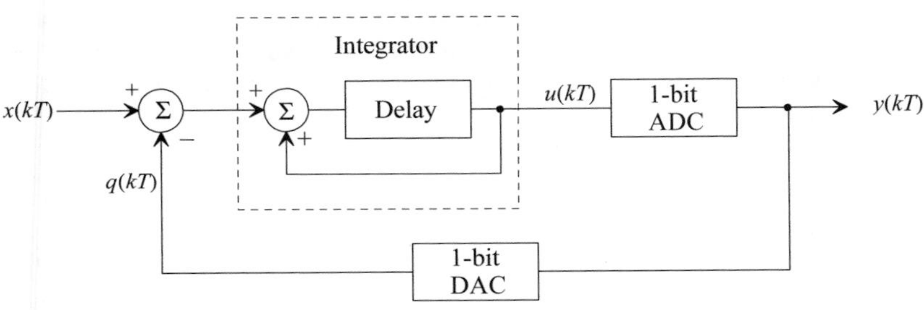

Figure 29.44 A first-order sigma-delta modulator.

While the benefits of ΣΔ modulation are not obvious, a simple derivation of the output, $y(kT)$, illuminates its distinct advantages. The output of the integrator, $u(kT)$, can be described as

$$u(kT) = x(kT - T) - q(kT - T) + u(kT - T) \tag{29.95}$$

where, $x(kT - T) - q(kT - T)$ is equal to the integrator's previous input, and $u(kT - T)$ is its previous output. The quantization error for the 1-bit ADC, as discussed in Ch. 28, is again defined as the difference between its output and input such that

$$Q_e(kT) = y(kT) - u(kT) \tag{29.96}$$

Plugging Eq. (29.95) into Eq. (29.96), the output response, $y(kT)$ is

$$y(kT) = Q_e(kT) + x(kT - T) - q(kT - T) + u(kT - T) \tag{29.97}$$

An ideal 1-bit DAC has the following characteristic: if the input, $y(kT) = 0$, the output, $q(kT) = -V_{REF}$, and if $y(kT) = 1$, then $q(kT) = V_{REF}$. In reality, a 1-bit DAC consists of a couple of switches connecting V_{REF} or $-V_{REF}$ to a common node, so it is not difficult to assume that the DAC is ideal. Therefore,

$$y(kT) = q(kT) \tag{29.98}$$

Utilizing Eq. (29.96) and Eq. (29.97), we find that Eq. (29.98) becomes

$$y(kT) = x(kT - T) + Q_e(kT) - Q_e(kT - T) \tag{29.99}$$

Therefore, the output of the modulator consists of a quantized value of the input signal delayed by one sample period, plus a differencing of the quantization error between the present and previous values. Thus, the real power of ΣΔ modulation is that the quantization noise, Q_e, cancels itself out to the first order.

A frequency domain example further illuminates this important fact. Suppose that the first-order modulator can be modeled in the s domain, as seen in Fig. 29.45, with an ideal integrator represented with transfer function of $\frac{1}{s}$, the 1-bit ADC modeled as a simple error source, $Q_e(s)$, and again the DAC considered to be ideal, such that $y(s)$ is equal to $q(s)$. It is also assumed that the bandwidth of the input signal is much less than the bandwidth of the modulator. Therefore, using simple feedback theory, $v_{OUT}(s)$ becomes

$$v_{OUT}(s) = Q_e(s) + \frac{1}{s} \cdot [v_{IN}(s) - v_{OUT}(s)] \tag{29.100}$$

and solving for v_{OUT} yields,

$$v_{OUT}(s) = Q_e(s) \cdot \frac{s}{s+1} + v_{IN}(s)\frac{1}{s+1} \tag{29.101}$$

Note that the transfer function from v_{IN} to v_{OUT} follows that of a low-pass filter and that the transfer function of the quantization noise follows that of a high-pass filter. Plotted together in Fig. 29.46, it is seen that in the region where the signal is of interest, the noise has a small value while the signal has a high gain, and that at higher frequencies, beyond the bandwidth of the signal, the noise increases. The modulator has essentially pushed the power of the noise out of the bandwidth of the signal. This high-pass characteristic is known as *noise shaping* and is a powerful concept used within oversampling ADCs. Low-pass filtering is then performed by the digital filter in order to remove all of the out-of-band quantization noise, which then permits the signal to be downsampled to yield the final high-resolution output.

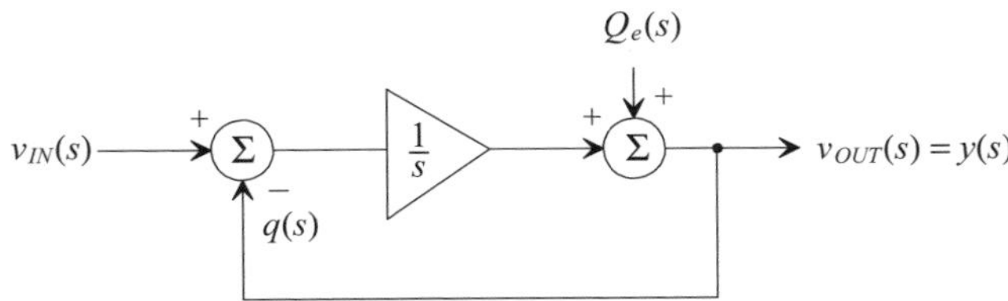

Figure 29.45 A frequency domain model for the first-order sigma-delta modulator.

As the ΣΔ modulator is generating the pulse-density modulated output, it is interesting to examine the mechanics occurring in the loop, which result in an average of the input. An actual ΣΔ modulator might resemble Fig. 29.47. A switched-capacitor integrator provides the summing as well as the delay needed. The 1-bit ADC is a simple comparator, and 1-bit DAC is simply two voltage-controlled switches that select either V_{REF} or $-V_{REF}$ to be summed with the input. A latched comparator provides the necessary loop delay. Notice that the variables are voltage representations of the variables used in

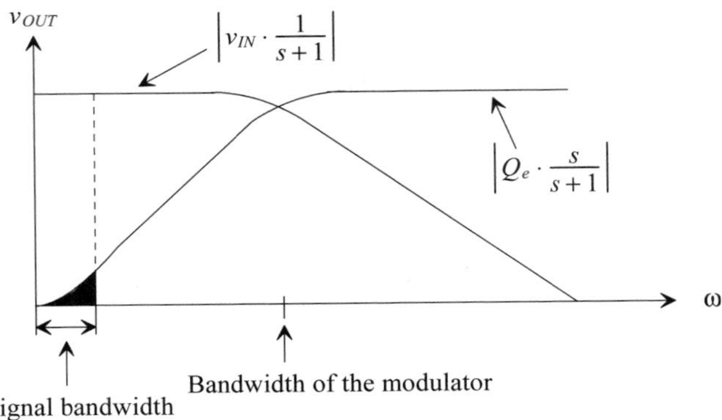

Figure 29.46 Frequency response of the first-order sigma-delta modulator.

Fig. 29.44. Remember that the function of the integrator is to accumulate differences between the input signal and the output of the DAC. If it is assumed that the input, $v_x(kT)$, is a positive DC voltage, then the output of the integrator should increase. However, the feedback mechanism is such that the 1-bit ADC (the comparator) has a low output if the integrator output, $v_u(kT)$, is positive. Thus, V_{REF} appears at the output of the DAC and is subtracted from the input, and the integrator output is driven back toward zero. The opposite occurs when $v_u(kT)$ is negative such that the integrator output is always driven toward zero by the feedback mechanism. An example will illustrate the operation of the modulator in more detail.

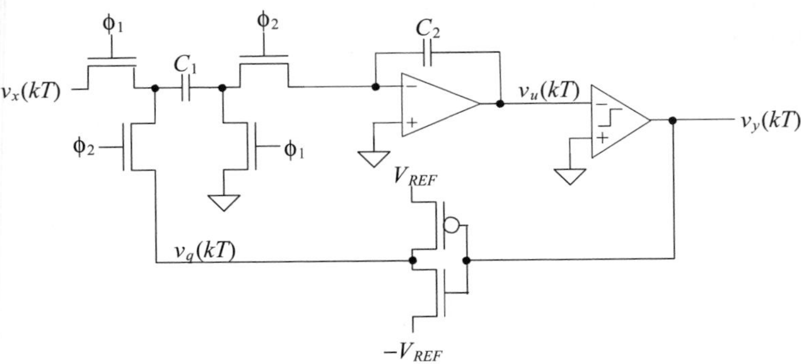

Figure 29.47 Implementation of a first-order sigma-delta modulator using a switched capacitor integrator.

Example 29.16

Using a general first-order ΣΔ modulator, assume that the input to the modulator, $v_x(kT)$ is a positive DC voltage of 0.4 V. Show the values of each variable around the ΣΔ modulator loop and prove that the overall average output of the DAC approaches 0.4 V after 10 cycles. Assume that the DAC output is ±1 V, and that the integrator output has a unity gain with an initial output voltage of 0.1 V, and that the comparator output is either ±1 V.

The present integrator output will be equal to the sum of the previous integrator output and the previous integrator input. Therefore, Eq. (29.95) becomes

$$v_u(kT) = v_u(kT - T) + v_a(kT - T) \qquad (29.102)$$

where

$$v_a(kT) = v_x(kT) - v_q(kT) \qquad (29.103)$$

and the quantizing error, $Q_e(kT)$, is defined by Eqs. (29.96) and (29.98) as

$$Q_e(kT) = v_q(kT) - v_u(kT) \qquad (29.104)$$

The initial conditions define the values of the variable for $k = 0$. The output of the integrator is given to be 0.1 V. Thus, the ADC output is low, the DAC output is V_{REF}, and the output of the summer, $v_a(0)$, is $0.4 - V_{REF} = -0.6$ V.

The output for $k = 1$ begins again with the integrator output. Using Eq. (29.102), $v_u(kT)$ becomes

$$v_u(T) = 0.1 + (-0.6) = -0.5 \text{ V}$$

Since the output of the integrator is negative, the output of the comparator is positive and $-V_{REF}$ is subtracted from 0.4 to arrive at the value for $v_a(T)$.

Continuing in the same manner and using the previous equations, we note the voltages for each cycle in Fig. 29.48. After 10 cycles through the modulator, the average value of $v_q(kT)$ becomes,

$$\overline{v_q(kT)} = \frac{7-3}{10} = 0.4\ V$$

k	$v_a(kT)$	$v_u(kT)$	$v_q(kT) = v_y(kT)$	$Q_e(kT)$	$\overline{v_q(kT)}$
0	–0.6	0.1	1.0	0.9	1.0
1	1.4	–0.5	–1.0	–0.5	0
2	–0.6	0.9	1.0	0.1	0.333
3	–0.6	0.3	1.0	0.7	0.50
4	1.4	–0.3	–1.0	–0.7	0.20
5	–0.6	1.1	1.0	–0.1	0.333
6	–0.6	0.5	1.0	0.5	0.429
7	1.4	–0.1	–1.0	–0.9	0.25
8	–0.6	1.3	1.0	–0.3	0.333
9	–0.6	0.7	1.0	0.3	0.40

Figure 29.48 Data from the first-order ΣΔ modulator.

Notice that the behavior of $\overline{v_q(kT)}$ swings around the desired value 0.4 V. If we were to continue computing values, as k increases, the amount that $\overline{v_q(kT)}$ differs from 0.4 V would decrease. Ideally, we could make the deviation of $\overline{v_q(kT)}$ as small as desired by allowing the modulator to take as many samples as necessary to meet that accuracy. ■

It is interesting to examine the effects of using a nonideal comparator. Suppose the integrator's output was smaller than the offset voltage of the comparator. A wrong decision would be made, causing $v_y(kT)$ to be the opposite of the desired value. However, as k increases, this error is averaged out, and the modulator still converges on the correct answer. Therefore, the comparator does not have to be very accurate in its ability to distinguish between two voltages, in contrast to Nyquist rate comparators.

The Higher Order ΣΔ Modulators

Higher order ΣΔ modulators exist which provide a greater amount of noise shaping. A second-order ΣΔ modulator can be seen in Fig. 29.49. A derivation of the second-order transfer function would reveal that the output contained a delayed version of the input plus a second-order differencing of the quantization noise, Q_e (see Problem 29.38). A third-order modulator would contain third-order differencing of the quantization and can be constructed by adding another integrator similar to integrator A into the system.

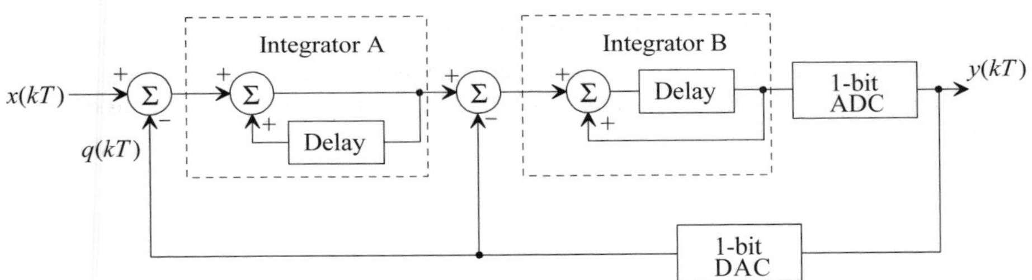

Figure 29.49 A second-order, sigma-delta modulator.

Figure 29.50 shows the noise-shaping functions of a first-, second-, and third-order modulator. The cross-hatched area under each of the curves represents the noise that remains in the signal bandwidth and is a magnified version of the blackened area of Fig. 29.46. As the order increases, notice that more of the noise is pushed out into the higher frequencies, thus decreasing the noise in the signal bandwidth. It should be reiterated that ΣΔ modulators do not attenuate noise at all. In fact, they add quantization noise that is very large at high frequencies. But because almost all of the noise is out of the signal bandwidth, it can easily be filtered, leaving only a small portion within the signal bandwidth. This point is important because the ΣΔ modulator should not be construed as a filtering circuit.

The resolution also increases as the order of the ΣΔ modulator and the oversampling ratio increases, as seen in Fig. 29.51 [20]. Using a first-order modulator, one can expect an increase in dynamic range of 9 dB with every doubling of the

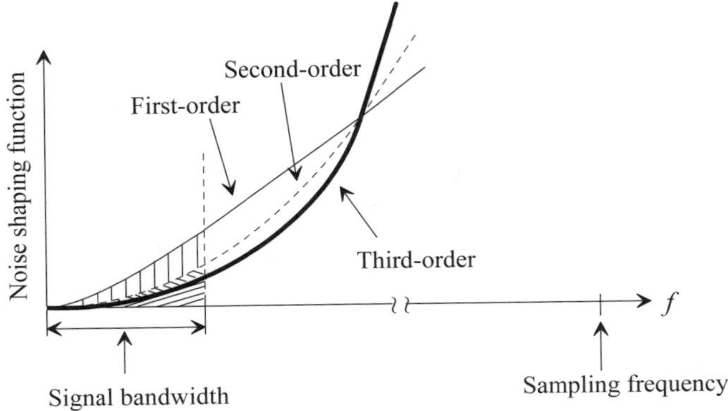

Figure 29.50 Noise shaping comparison of a first-, second- and third-order modulator.

oversampling ratio. This correlates to an approximate increase in resolution of 1.5 bits according to Eq. (28.28). The higher-order modulators have even greater gains in resolution as a 2.5-bit increase is attained with each doubling of the oversampling ratio using a second-order modulator, while the third-order modulator increases 3.5 bits.

One could essentially construct a high-order ΣΔ modulator with many integrators. However, as with any system employing feedback, stability becomes a critical issue. The same holds true for the high-order ΣΔ modulators. Several other topologies have been developed which can implement modulators in a cascaded fashion and are guaranteed to be stable [21, 22]. However, considerable matching requirements need to be overcome.

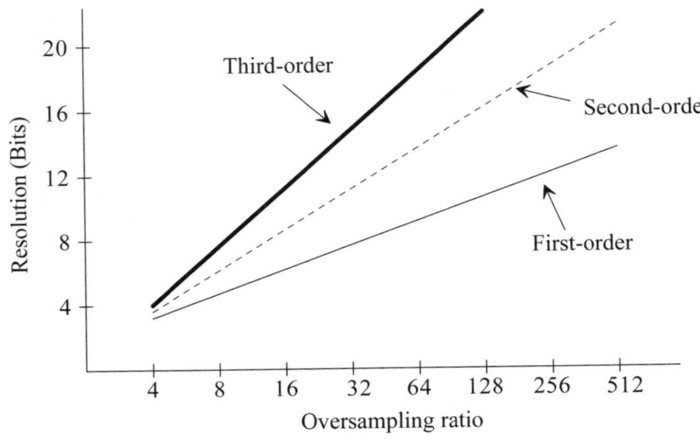

Figure 29.51 Comparison of first-, second-, and third-order modulators versus oversampling ratio and resolution.

REFERENCES

[1] R. L. Geiger, P. E. Allen, and N. R. Strader, *VLSI - Design Techniques for Analog and Digital Circuits*, McGraw-Hill Publishing Co., 1990.

[2] R. E. Suarez, P. R. Gray, and D. A. Hodges, "All-MOS Charge Redistribution Analog-to-Digital Conversion Techniques - Part II," *IEEE Journal of Solid State Circuits*, vol. 10, no. 6, pp. 379–385, December 1975.

[3] J. Shyu, G. C. Temes, and F. Krummenacher, "Random Errors in MOS Capacitors and Current Sources," *IEEE Journal of Solid State Circuits*, vol. 16, no. 6, pp. 948–955, December 1984.

[4] M. J. M. Pelgrom, et. al, "25-Ms/s 8-bit CMOS A/D Converter for Embedded Application," *IEEE Journal of Solid-State Circuits*, vol. 29, no. 8, pp. 879–886, August 1994.

[5] D. Choi, et. al, "Analog Front-End Signal Processor for a 64 Mbits/s PRML Hard-Disk Drive Channel," *IEEE Journal of Solid-State Circuits*, vol. 29, no. 12, pp. 1596–1605, December 1994.

[6] N. Shiwaku, "A Rail-to-Rail Video-band Full Nyquist 8-bit A/D Converter," *Proceedings of the 1991 Custom Integrated Circuits Conference.*

[7] B. Razavi and B. A. Wooley, "A 12-b, 5-MSample/s Two-Step CMOS A/D Converter," *IEEE Journal of Solid State Circuits*, vol. 27, no. 12, pp. 1667–1678, December 1992.

[8] J. Dornberg, P. R. Gray, and D. A. Hodges, "A 10-bit, 5-Msample/s CMOS Two-Step Flash ADC," *IEEE Journal of Solid State Circuits*, vol. 24, no. 2, pp. 241–249, April 1989.

[9] T. Shimizu, et al., "A 10-bit, 20 MHz Two-Step Parallel A/D Converter with Internal S/H," *IEEE Journal of Solid State Circuits*, vol. 24, no. 1, pp. 13–20, February 1989.

[10] B. S. Song, S. H. Lee, and M. F. Tompsett, "A 10-bit 15 MHz CMOS Recycling Two-Step A/D Converter," *IEEE Journal of Solid State Circuits*, vol. 25, no. 12, pp. 1328–1338, December 1990.

[11] B. S. Song, M. F. Tompsett, and K. R. Lakshmikumar, "A 12-bit, 1-MSample/s Capacitor Error-Averaging Pipelined A/D Converter," *IEEE Journal of Solid State Circuits*, vol. 23, no. 6, pp. 1324–1333, December 1988.

[12] S. H. Lewis and P. R. Gray, "A Pipelined 5-Msample/s 9 bit Analog-to-Digital Converter," *IEEE Journal of Solid State Circuits*, vol. 22, no. 6, pp. 954–961, December 1987.

[13] S. Sutarja and P. R. Gray, "A Pipelined 13-bit, 250-ks/s, 5-V Analog-to-Digital Converter," *IEEE Journal of Solid State Circuits*, vol. 23, no. 6, pp. 1316–1323, December 1988.

[14] P. Vorenkamp and J. P. M. Verdaasdonk, "A 10 b 50 Ms/s Pipelined ADC," *IEEE ISSCC Digest of Technical Papers*, pp. 34–35, February 1992.

[15] M. Yotsuyanagi, T. Etoh, and K. Hirata, "A 10 Bit 50 MHz Pipelined CMOS A/D Converter with S/H," *IEEE Journal of Solid State Circuits*, vol. 28, no. 3, pp. 292–300, March 1993.

[16] J. L. McCreary and P. R. Gray, "All-MOS Charge Redistribution Analog-to-Digital Conversion Techniques - Part I," *IEEE Journal of Solid State Circuits*, vol. 10, no. 6, pp. 371–379, December 1975.

[17] K. Bacrania, "A 12 Bit Successive-Approximation ADC with Digital Error Correction," *IEEE Journal of Solid State Circuits*, vol. 21, no. 6, pp. 1016–1025, December 1986.

[18] M. Ismail and T. Fiez, *Analog VLSI Signal and Information Processing*, McGraw-Hill, 1994.

[19] B. E. Boser, "Design and Implementation of Oversampled Analog-to-Digital Converters," Ph.D. Dissertation, Stanford University, 1988.

[20] B. P. Brandt, *Oversampled Analog-to-Digital Conversion,* Integrated Circuits Laboratory, Technical Report No. ICL91-009, Stanford University, 1991.

[21] Y. Matsuya, K. Uchimura, et al, "A 16-bit Oversampling A/D Conversion Technology Using Triple Integration Noise Shaping," *IEEE Journal of Solid State Circuits*, vol. 22, no. 6, pp. 921–929, December 1987.

[22] K. Uchimura et al, "Oversampling A-to-D and D-to-A Converters with Multistage Noise Shaping Modulators," *IEEE Transactions on Acoustics, Speech and Signal Processing*, pp. 1899–1905, December 1988.

[23] J. W. Bruce, "Meeting the analog world challenge: Nyquist rate analog to digital converter architectures," *IEEE Potentials,* vol. 17, no. 5, pp. 36–39, January 1999. Very good introduction to ADCs.

[24] J. W. Bruce, "Nyquist-rate digital to analog converter architectures," *IEEE Potentials*, vol. 20, no. 3, pp. 24–28, August 2001. Good overview of digital to analog converters.

[25] R. J. Baker, *CMOS Mixed-Signal Circuit Design*, John Wiley and Sons, 2002. ISBN 0-471-22754-4. Volume II of this book. Provides much more detail concerning the implementation of data converters discussed in this chapter and the characterizations presented in Ch. 28.

Special Thanks and Credit

The material presented in this chapter and Ch. 28 was written by Dr. Harry Li, a former Associate Professor of Electrical Engineering at the University of Idaho. In 2002, he left academics to become a pastor of the Mosaic Church of Central Arkansas in Little Rock, a church that specializes in cross-cultural ministry. Special thanks and credit goes to Harry for authoring this material. For more information consult www.mosaicchurch.net.

The author would also like to extend special thanks to Dr. Terry Sculley. Dr. Sculley derived the INL and DNL calculations used in Ch. 29 (not a trivial task) while he was at Washington State University. Thanks and credit go to him for allowing us (again) to include this material in the second edition of the book.

PROBLEMS

29.1 A 3-bit, resistor-string DAC similar to the one shown in Fig. 29.2a was designed with a desired resistor of 500 Ω. After fabrication, mismatch caused the actual value of the resistors to be

$R_1 = 500, R_2 = 480, R_3 = 470, R_4 = 520, R_5 = 510, R_6 = 490, R_7 = 530, R_8 = 500$

Determine the maximum INL and DNL for the DAC assuming $V_{REF} = 5$ V.

29.2 An 8-bit resistor string DAC similar to the one shown in Fig. 29.2b was fabricated with a nominal resistor value of 1 kΩ. If the process was able to provide matching of resistors to within 1%, find the effective resolution of the converter. What is the maximum INL and DNL of the converter? Assume that $V_{REF} = 5$ V.

29.3 Compare the digital input codes necessary to generate all eight output values for a 3-bit resistor string DAC similar to those shown in Fig. 29.2a and b. Design a digital circuit that will allow a 3-bit binary digital input code to be used for the DAC in Fig. 29.2a. Discuss the advantages and disadvantages of both architectures.

29.4 Plot the transfer curve of a 3-bit *R*-2*R* DAC if all *R*s = 1.1 kΩ and 2*R*s = 2 kΩ. What is the maximum INL and DNL for the converter? Assume all of the switches to be ideal and $V_{REF} = 5$ V.

29.5 Suppose that a 3-bit *R*-2*R* DAC contained resistors that were perfectly matched and that $R = 1$ kΩ and $V_{REF} = 5$ V. Determine the maximum switch resistance that can be tolerated for which the converter will still have 3-bit resolution. What are the values of INL and DNL?

29.6 The circuit illustrated in Fig. 29.5 is known as a current-mode *R*-2*R* DAC, since the output voltage is defined by the current through R_F. Shown in Fig. P29.6 is an *N*-bit voltage-mode *R*-2*R* DAC. Design a 3-bit voltage mode DAC and determine the output voltage for each of the eight input codes. Label each node voltage for each input. Assume that $R = 1$ kΩ and that $R_2 = R_1 = 10$ kΩ and $V_{REF} = 5$ V.

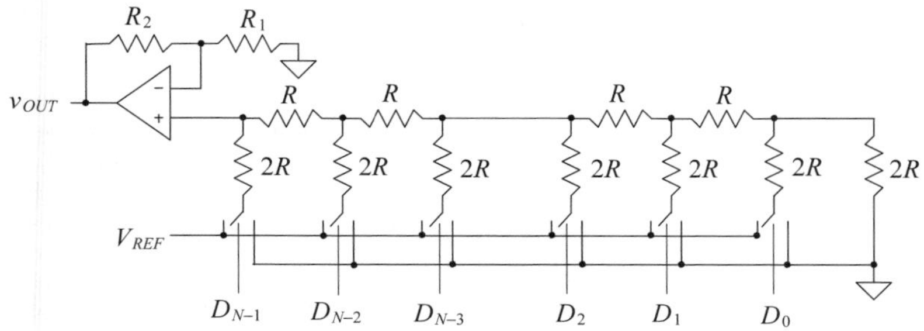

Figure 29.52 DAC used in Problem 29.6.

29.7 Design a 3-bit, current-steering DAC using the generic current-steering DAC shown in Fig. 29.9. Assume that each current source, I, is 5 mA, and find the total output current for each input code.

29.8 A certain process is able to fabricate matched current sources to within 0.05%. Determine the maximum resolution that a current-steering (nonbinary-weighted) DAC can attain using this process.

29.9 Design an 8-bit current-steering DAC using binary-weighted current sources. Assume that the smallest current source will have a value of 1 μA. What is the range of values that the current source corresponding to the MSB can have while maintaining an INL of ½ LSB? Repeat for a DNL less than or equal to ½ LSB.

29.10 Prove that the 3-bit charge-scaling DAC used in Ex. 29.6 has the same output voltage increments as the R-$2R$ DAC in Ex. 29.3 for $V_{REF} = 5$ V and $C = 0.5$ pF.

29.11 Determine the output of the 6-bit, charge-scaling DAC used in Ex. 29.7 for each of the following inputs: $D = 000010$, 000100, 001000, and 010000.

29.12 Design a 4-bit, charge-scaling DAC using a split array. Assume that $V_{REF} = 5$ V and that $C = 0.5$ pF. Draw the equivalent circuit for each of the following input words and determine the value of the output voltage: $D = 0001$, 0010, 0100, 1000. Assuming the capacitor associated with the MSB had a mismatch of 4 percent, calculate the INL and DNL.

29.13 For the cyclic converter shown in Fig. 29.17, determine the gain error for a 3-bit conversion if the feedback amplifier had a gain of 0.45 V/V. Assume that $V_{REF} = 5$ V.

29.14 Repeat Problem 29.13 assuming that the output of the summer was always 0.2 V greater than the ideal and that the amplifier in the feedback path had a perfect gain of 0.5 V/V.

29.15 Repeat Problem 29.13 assuming that the output of the summer was always 0.2 V greater than ideal and that the amplifier in the feedback path had a gain of 0.45 V/V.

29.16 Design a 3-bit pipeline DAC using $V_{REF} = 5$ V. (a) Determine the maximum and minimum gain values for the first-stage amplifier for the DAC to have less than ±½ LSBs of DNL assuming that the rest of the circuit is ideal. (b) Repeat for the second-stage amplifier. (c) Repeat for the last-stage amplifier.

29.17 Using the same DAC designed in Problem 29.16, (a) determine the overall error (offset, DNL, and INL) for the DAC if the S/H amplifier in the first stage produces an offset at its output of 0.25 V. Assume that all of the remaining components are ideal. (b) Repeat for the second-stage S/H. (c) Repeat for the last-stage S/H.

29.18 Design a 3-bit Flash ADC with its quantization error centered about zero LSBs. Determine the worst-case DNL and INL if resistor matching is known to be 5%. Assume that $V_{REF} = 5$ V.

29.19 Using the ADC designed in Problem 29.18, determine the maximum offset that can be tolerated if all of the comparators have the same magnitude of offset, but with different polarities, to attain a DNL of less than or equal to ±½ LSB.

29.20 A 4-bit Flash ADC converter has a resistor string with mismatch as shown in Table 29.1. Determine the DNL and INL of the converter. How many bits of resolution does this converter possess? $V_{REF} = 5$ V.

Resistor	Mismatch (%)
1	2
2	1.5
3	0
4	−1
5	−0.5
6	1
7	1.5
8	2
9	2.5
10	1
11	−0.5
12	−1.5
13	−2
14	0
15	1
16	1

Table 29.1 Mismatch in resistors used in Problem 29.20

29.21 Determine the open-loop gain required for the residue amplifier of a two-step ADC necessary to keep the converter to within ½ LSB of accuracy with resolutions of (a) 4 bits, (b) 8 bits, and (c) 10 bits.

29.22 Assume that a 4-bit, two-step Flash ADC uses two separate Flash converters for the MSB and LSB ADCs. Assuming that all other components are ideal, show that the first Flash converter needs to be more accurate than the second converter. Assume that $V_{REF} = 5$ V.

29.23 Repeat Ex. 29.12 for $V_{IN} = 3, 5, 7.5, 14.75$ V.

29.24 Repeat Ex. 29.13 for $V_{IN} = 1, 4, 6, 7$ V and $V_{REF} = 8$ V.

29.25 Assume that an 8-bit pipeline ADC was fabricated and that all the amplifiers had a gain of 2.1 V/V instead of 2 V/V. If $V_{IN} = 3$ V and $V_{REF} = 5$ V, what would be the resulting digital output if the remaining components were considered to be ideal? What are the DNL and INL for this converter?

29.26 Show that the first-stage accuracy is the most critical for a 3-bit, 1-bit per stage pipeline ADC by generating a transfer curve and determining DNL and INL for the ADC for three cases: (1) The gain of the first-stage residue amplifier set equal to 2.2 V/V, (2) the second-stage residue amplifier set equal to 2.2 V/V, and (3) the third-stage residue amplifier set equal to 2.2 V/V. For each case, assume that the remaining components are ideal. Assume that $V_{REF} = 5$ V.

29.27 An 8-bit single-slope ADC with a 5 V reference is used to convert a slow-moving analog signal. What is the maximum conversion time assuming that the clock frequency is 1 MHz? What is the maximum frequency of the analog signal? What is the maximum value of the analog signal which can be converted?

29.28 An 8-bit single slope ADC with a 5 V reference uses a clock frequency of 1 MHz. Assuming that all of the other components are ideal, what is the limitation on the value of RC? What is the tolerance of the clock frequency which will ensure less than 0.5 LSB of INL?

29.29 An 8-bit dual slope ADC with a 5 V reference is used to convert the same analog signal in Problem 29.27. What is the maximum conversion time assuming that the clock frequency is 1 MHz? What is the minimum conversion time that can be attained? If the analog signal is 2.5 V, what will be the total conversion time?

29.30 Discuss the advantages and disadvantages of using a dual-slope versus a single slope ADC architecture.

29.31 Repeat Ex. 29.15 for a 4-bit successive approximation ADC using $V_{REF} = 5$ V for $v_{IN} = 1$, 3, and full-scale.

29.32 Assume that $v_{IN} = 2.49$ V for the ADC used in Problem 29.31 and that the comparator, because of its offset, makes the wrong decision for the MSB conversion. What will be the final digital output? Repeat for $v_{IN} = 0.3025$, assuming that the comparator makes the wrong decision on the LSB.

29.33 Design a 3-bit, charge-redistribution ADC similar to that shown in Fig. 29.39 and determine the voltage on the top plate of the capacitor array throughout the conversion process for $v_{IN} = 2$, 3, and 4 V, assuming that $V_{REF} = 5$ V. Assume that all components are ideal. Draw the equivalent circuit for each bit decision.

29.34 Determine the maximum INL and maximum DNL of the ADC designed in Problem 29.33 assuming that the capacitor array matching is 1%. Assume that the remaining components are ideal and that the unit capacitance, C, is 1 pF.

29.35 Show that the charge redistribution ADC used in Problems 29.32 and 29.33 is immune to comparator offset by assuming an initial offset voltage of 0.3 and determining the conversion for $v_{IN} = 2$ V.

29.36 Discuss the differences between Nyquist rate ADCs and oversampling ADCs.

29.37 Write a simple computer program or use a math program to perform the analysis shown in Ex. 29.16. Run the program for $k = 200$ clock cycles and show that the average value of $v_q(kT)$ converges to the correct answer. How many clock cycles will it take to obtain an average value if $v_q(kT)$ stays within 8-bit accuracy of the ideal value of 0.4 V? 12-bit accuracy? 16-bit accuracy?

29.38 Prove that the output of the second-order ΣΔ modulator shown in Fig. 29.49 is,

$$y(kT) = x(kT - T) + Q_e(kT) - 2Q_e(kT - T) + Q_e(kT - 2T)$$

29.39 Assume that a first order ΣΔ ADC used on a satellite in a low earth orbit experiences radiation in which an energetic particle causes a noise spike resulting in the comparator making the wrong decision on the 10th clock period. Using the program written in Problem 29.37, determine the number of clock cycles required before the average value of $v_q(kT)$ is within 12-bit accuracy of the ideal value of 0.4 V. How many extra clock cycles were required for this case versus the ideal conversion used in Prob. 36?

Index

A

B

C

D

E

F

G

M

N

O

P

R

S

T

U

V

W

X

Y

Z

About the Author

Russel Jacob (Jake) Baker was born in Ogden, Utah, on October 5, 1964. He received the B.S. and M.S. degrees in electrical engineering from the University of Nevada, Las Vegas, and the Ph.D. degree in electrical engineering from the University of Nevada, Reno.

From 1981 to 1987, he was in the United States Marine Corps Reserves. From 1985 to 1993, he worked for E. G. & G. Energy Measurements and the Lawrence Livermore National Laboratory designing nuclear diagnostic instrumentation for underground nuclear weapons tests at the Nevada test site. During this time, he designed over 30 electronic and electro-optic instruments including high-speed (750 Mb/s) fiber-optic receiver/transmitters, PLLs, frame- and bit-syncs, data converters, streak-camera sweep circuits, micro-channel plate gating circuits, and analog oscilloscope electronics. From 1993 to 2000, he was a faculty member in the department of electrical engineering at the University of Idaho. In 2000, he joined a new electrical and computer engineering program at the Boise State University where he is currently a professor and chair of the department. Also, since 1993, he has consulted for various companies and laboratories including Micron Technology, Amkor Wafer Fabrication Services, Tower Semiconductor, Rendition, Lawrence Berkeley Laboratory, and the Tower ASIC Design Center.

Professor Baker holds over 75 granted or pending patents in integrated circuit design. He is a member of the electrical engineering honor society Eta Kappa Nu and an author/coauthor of the books *CMOS: Circuit Design, Layout, and Simulation*, *DRAM Circuit Design: A Tutorial*, and *CMOS: Mixed-Signal Circuit Design*. His research interests are in the areas of CMOS mixed-signal integrated circuit design and the design of memory in new and emerging fabrication technologies. Professor Baker was a co-recipient of the 2000 Prize Paper Award of the IEEE Power Electronics Society.

About the Author